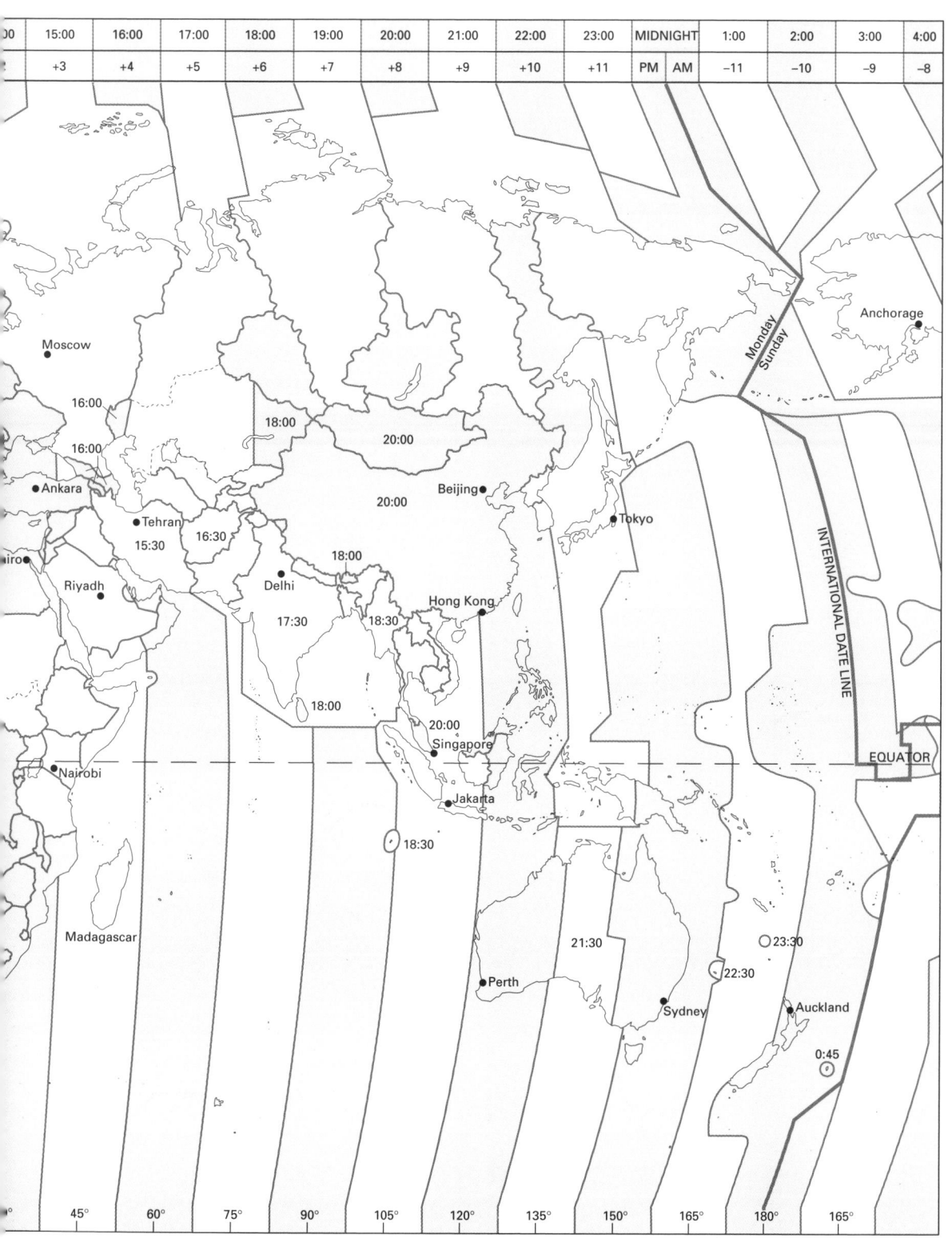

15:00	16:00	17:00	18:00	19:00	20:00	21:00	22:00	23:00	MIDNIGHT		1:00	2:00	3:00	4:00
+3	+4	+5	+6	+7	+8	+9	+10	+11	PM	AM	−11	−10	−9	−8

Moscow

16:00

16:00

Ankara

Tehran 16:30

15:30

Riyadh

Delhi

17:30

18:00

18:30

18:00

18:00

20:00

Singapore

Jakarta

18:30

Madagascar

Nairobi

18:00

20:00

20:00

Beijing

Tokyo

Hong Kong

Monday Sunday

Anchorage

INTERNATIONAL DATE LINE

EQUATOR

21:30

23:30

22:30

Perth

Sydney

Auckland

0:45

45° 60° 75° 90° 105° 120° 135° 150° 165° 180° 165°

THE
STATESMAN'S
YEARBOOK
2014

'It is often easier to fight for principles than to live up to them.'

Adlai Stevenson II (1940—65)

Editors

Frederick Martin	1864—1883
Sir John Scott-Keltie	1883—1926
Mortimer Epstein	1927—1946
S. H. Steinberg	1946—1969
John Paxton	1969—1990
Brian Hunter	1990—1997
Barry Turner	1997—

Credits

Publisher	David Bull (London)
	Farideh Koohi-Kamali (New York)
Editor	Barry Turner
Senior Research Editor	Nicholas Heath-Brown
Assistant Research Editor	Marta Kask
Researchers	Kavita Malhotra
	Daniel Smith
	Richard German
	Saif Ullah
	Ingeborg Farstad
	Robert McGowan
	Liane Jones
	Justine Foong
	Ben Eastham
	James Wilson
	Dominic Frost
	William Goodfellow
	Jill Fenner
	Maximilian Kressner
	Sharita Oomeer
	Martha Nyman
Assistant Editor for Reference	Susan Povey
Index	Richard German
Print production	Phillipa Davidson-Blake
Online production	Semantico
Technical support	Sheeja Sanoj
Marketing	Beverley Millar (London)
	Denise De La Rosa (New York)

email: sybcomments@palgrave.com

THE STATESMAN'S YEARBOOK

THE POLITICS, CULTURES AND ECONOMIES OF THE WORLD

2014

Edited by

BARRY TURNER

palgrave
macmillan

© Macmillan Publishers Ltd 2013

Published annually since 1864

This edition published 2013 by
PALGRAVE MACMILLAN

Palgrave Macmillan in the UK is an imprint of Macmillan Publishers Limited, registered in England, company number 785998, of Houndmills, Basingstoke, Hampshire RG21 6XS.

Palgrave Macmillan in the US is a division of St Martin's Press LLC, 175 Fifth Avenue, New York, NY 10010.

Palgrave Macmillan is the global academic imprint of the above companies and has companies and representatives throughout the world.

Palgrave® and Macmillan® are registered trademarks in the United States, the United Kingdom, Europe and other countries.

ISBN 978-0-230-37769-1
ISSN 0081-4601

This book is printed on paper suitable for recycling and made from fully managed and sustained forest sources. Logging, pulping and manufacturing processes are expected to conform to the environmental regulations of the country of origin.

A catalogue record for this book is available from the British Library.

A catalog record for this book is available from the Library of Congress.

PREFACE

It is with some pride that I welcome you to the 150th edition of *The Statesman's Yearbook*. Published every year since 1864, *The Statesman's Yearbook* is a unique and assured reference work. It was in the middle of the nineteenth century that the then British Prime Minister, Sir Robert Peel, suggested to Alexander Macmillan that he should publish 'a handbook presenting in a compact shape a picture of the actual condition, political and social, of the various states in the civilised world'. So it was that the first edition of *The Statesman's Yearbook* appeared in print for the first time in 1864.

The longevity and success of *The Statesman's Yearbook* owes much to two contrasting features. In the first instance, continuity and tradition: we have had the services of only seven editors in our long and rich history, while our adherence to accuracy and strict levels of editorial quality remain paramount. This is as true today as when the inaugural editor, Frederick Martin, wrote in the preface to the first edition, 'The great aim has been to ensure an absolute correctness of the multiplicity of fact and figures in *The Statesman's Yearbook*.' Yet, this consistency of mission and approach contrasts with an ability to adapt, and prosper, in changing times. The structure and coverage of *The Statesman's Yearbook* has been successfully amended and enhanced over time to reflect changes in the world order, a need to provide varied and new data and a shifting focus upon key international actors and institutions.

This 150th edition of *The Statesman's Yearbook* features an essay by our current editor, Barry Turner, tracing the origins and history of the publication. We also include some fascinating maps that appeared in earlier editions and interesting advertisements from our archive. I also encourage you to visit our online service at www.statesmansyearbook.com where you will find not only a wealth of information and data, regularly updated, but also complementary initiatives including chronologies and timely and incisive essays and features.

Do please get in touch if you have any comments or suggestions concerning *The Statesman's Yearbook*. Either email us at sybcomments@palgrave.com or write to *The Statesman's Yearbook*, Palgrave Macmillan, Houndmills, Basingstoke RG21 6XS, UK.

David Bull
Publisher, *The Statesman's Yearbook*

CONTENTS

Part II: Countries of the World A−Z

Explore the world at
www.statesmansyearbook.com

A PROUD HERITAGE

Barry Turner traces the 150-year history of The Statesman's Yearbook

THE

STATESMAN'S YEAR-BOOK

A STATISTICAL, GENEALOGICAL, AND HISTORICAL
ACCOUNT OF THE STATES AND SOVEREIGNS
OF THE CIVILISED WORLD

FOR THE YEAR

1864

BY FREDERICK MARTIN

London and Cambridge

MACMILLAN. AND CO.
1864

The Right of Translation and Reproduction is reserved

The Victorians had a passion for collecting knowledge in handy-sized packs, otherwise known as reference books. The reasons for this are not hard to find. In an age of rapid change when, as Matthew Arnold observed, 'points of moral and intellectual matters settled for centuries' were thrown open in an 'atmosphere of unrest and paradox', the desire to keep up with the latest intelligence was met by writers and editors capable of reducing essential information to manageable proportions.

The *Encyclopaedia Britannica*, first published in three volumes between 1768 and 1771, came into its own in the nineteenth century with contributions from recognized experts on a wide range of topical subjects. By 1810, it had expanded to 20 volumes. For those in need of a rounded view of political developments at home and abroad, *The Annual Register* offered an authoritative record.

But, as Sir Robert Peel observed in the 1840s, this left a gap in the market. There was no user-friendly compilation of international political, economic and social statistics. Such a project was dear to the heart of William Gladstone too. The future British Prime Minister was, in the mid-1860s, in his second term as Chancellor of the Exchequer. He was acutely aware that his powers of persuasion rested on having all the relevant facts and figures at his command. His great rival, Benjamin Disraeli, begged to differ. As he remarked famously, 'There are three kinds of lies: lies, damned lies and statistics.' But the weight of intellectual opinion was with Gladstone. It was surely possible to create in 'compact shape a picture of the actual condition, political and social, of the various states of the civilised world'.

One who certainly believed this to be a worthy objective was the historian and cultural sage, Thomas Carlyle. Moreover, he knew just the man to edit such a volume. Born in Berlin, Frederick Martin was a German national, a linguist and academic who had come over to London to assist Carlyle in writing his biography of Frederick the Great. Having proved his worth in reducing a mass of information to digestible form, he was ready for a bigger challenge. It only remained to find a publisher.

Macmillan was a likely prospect. As head of the firm, Alexander Macmillan was keen on books of lasting appeal that brought in a regular annual income. Among his successes were Palgrave's *Golden Treasury* series and the *Globe Shakespeare*. He calculated that an annually updated world gazetteer would have an equally strong appeal.

The question then was what to call this new publication. Settling on *The Statesman's Yearbook* (or 'SYB') was an inspiration. On the face of it, the title had limited pulling power. After all, when the first edition appeared 150 years ago, there were not many potential customers who qualified as statesmen. But like all successful publishers, Alexander Macmillan knew a thing or two about aspirational buyers. Though few had reached the political or diplomatic heights, many more were eager to get there. It was an incentive to go out and buy. Around 700 did so, paying 8s 4d (around a third of the working class average weekly wage) for a volume of 684 pages. It made its first appearance in January 1864. Two months later Frederick Martin was granted British citizenship.

The format of the early SYB tells us much about the state of world politics. Half the book was devoted to Europe with details of the constitutions and governments of 15 states plus the 33 members of the German confederation, still some way short of unification. The second half covered America in the throes of civil war (the Confederate States had separate entries) and other 'principal States not in Europe' including the five colonies contained within Australia.

As far as we can tell, Martin was left to his own devices. He worked alone, collecting information from official agencies. He quickly learned to detect lies or downright deceit. His first tussle was with the Russian embassy which provided a highly suspect estimate of expenses at the Tsarist court given the 'boundless pomp and splendour displayed on all occasions'. Credit goes to Martin for setting standards of independent editorial judgement, the enduring hallmark of SYB.

Martin remained editor for 20 years. By then the sole responsibility of turning out an annual publication of increasing complexity was taking its toll. Though a master at juggling world affairs, he was less skilled at organizing his own life. In an effort to solve his money problems, for several years he took on other editorial work, adding to his mental strain. Fears that he was no longer up to the job were confirmed when he left the revised proofs of the 1883 edition on a train. This in the days before duplicates were made. Martin was eased into what turned out to be a short-lived retirement. He died less than a year later.

For Macmillan the challenge was to find someone to finish the editorial work and to publish SYB on time, a matter of pride and practical business to the Macmillans who knew that the weakness of their rival, *The Annual Register*, was its appearance at irregular intervals. The solution was to appoint as Martin's successor a 43-year-old Scottish journalist, John Scott-Keltie.

Macmillan Best Cellars

Vintage 1864

THE LIBRARY ASSOCIATION · THE McCOLVIN MEDAL

The Statesman's Year-Book 1973-1974
Edited by John Paxton

A classic blend for discerning palates
£4.30   333 13202 5

The Music Yearbook 1973-4
Edited by Arthur Jacobs

A prize-winning young vintage
£6.50   333 13356 0

Tasting begins September

* Awarded the Library Association's McColvin Medal for the outstanding reference book published in 1972

The new editor was a graduate of St Andrews and Edinburgh universities. Highly intelligent and with an ability to soak up information, Scott-Keltie was originally destined to be a United Presbyterian minister but his love of science conflicted with the fundamentalist beliefs of his church. With his family on hard times and unable to afford training for the law or medicine, he took to a literary career. After contributing to early editions of *Chambers Encyclopaedia*, he was employed as an assistant editor on *Nature*, another Macmillan enterprise, founded five years after SYB. So successful was he that soon half the weekly journal was directly attributable to him. Scott-Keltie was the obvious choice to take on SYB.

During his tenure, SYB was expanded to include every country 'that can be regarded as a country, however rudimentary'. Maps were introduced to illustrate, as Scott-Keltie put it, 'subjects of great moment'. SYB grew to more than double its original size and with annual sales of over 3,000 copies was regarded as one of the jewels in the Macmillan crown.

Since compiling so many facts and figures was too much for one man, Scott-Keltie put together an editorial team of freelance specialists. At the same time, his scholarly reputation enabled him to branch out into other activities. He was already established on *The Times*. The first of his many articles appeared in 1875. The subject was Socotra, an archipelago of four islands in the Indian Ocean. Recently annexed by Britain, Socotra was a haven for exotic wildlife. It was said that a third of the plants found on the islands were unknown anywhere else on earth. Scott-Keltie never went there; indeed, though increasingly preoccupied with geography, he was not much of a traveller. But *Times* readers were thrilled to learn about the latest addition to the British Empire and demanded more of the same. Scott-Keltie obliged with a series of articles about the exploration and colonization of Africa.

The Royal Geographical Society proved to be another time-consuming interest for Scott-Keltie. In 1885, as Inspector of

Geographical Education, he launched a lengthy and successful campaign for his subject to be a regular part of the school curriculum. Within ten years he was virtually running the Society, first as an assistant secretary and then, from 1896, as secretary.

It was fortunate that Scott-Keltie was a workaholic. In his time, SYB continued to expand and with the setting up of the Macmillan Company of New York in 1896, plans were advanced for the USA to have its own section. In contrast to Martin, who had argued that the American market was too small to worry about, Scott-Keltie was enthusiastic. 'Several US correspondents have written to me suggesting that larger space might be given to so great a country.' It was a measure of the status of SYB that among those who offered discreet encouragement to expand the US coverage was President Theodore Roosevelt. And so, starting in 1906, each of the then 46 states of the Union was allotted its own section.

By now, Scott-Keltie had so many strings to his bow that editing SYB was essentially a part-time activity. But such was his distinction (he was knighted in 1918) Macmillan was keen to retain his services. The solution was to appoint a deputy editor.

The job went to Mortimer Epstein, a Lithuanian who had emigrated to England and settled in Manchester. As Scott-Keltie's health began to decline, his young assistant proved more than capable of gathering and collating a miscellany of abstruse facts. When he eventually became editor, on Scott-Keltie's death in 1927, he also had other roles in life, notably as editor of *The Annual Register* which he ran in tandem with SYB from 1922 to 1945.

SYB continued to be published throughout the war, albeit in slimmer format. The Macmillan archive contains several letters of apology from hard-pressed embassies and government departments. Accurate, up-to-date statistics were hard to come by. Needless to say, no attempt was made to persuade Germany and Japan to revise their entries. The exiled Norwegian government in London advised 'showing the position when the Germans invaded the country', adding optimistically that 'great changes in the administration could soon be expected', presumably after the occupiers had been expelled.

Letters to the editor were often homely and chatty, an indication of the close relationship Scott-Keltie and Epstein had built up with their sources. From the state librarian in Nevada, along with regrets for a 'tardy return of corrections' came a request for 'foreign postage stamps if you have any on hand. I have two boys who collect and they would be glad to exchange.'

Relations with the official censor, a hangover from the days when Britain was threatened with invasion, were less congenial. A disagreement on figures for submarines under the command of the Royal Navy led to a lengthy correspondence involving not only the Censor and the Admiralty but also the editor of *Jane's Fighting Ships* who backed Epstein. Even so, the pressure built up until Epstein agreed to changes 'under protest'.

War-time and post-war restrictions on the supply of paper limited the print run of SYB, giving the book a scarcity value. Epstein had to urge the British Council to hurry along an order for overseas branches and libraries warning that 'last year's edition was exhausted within a couple of days after publication' and that 'literally hundreds of later applications had to be given cold comfort'.

When Epstein died in 1946 the world order was in disarray. SYB needed a thorough revision to take account of new countries, old countries under new regimes and a rash of international agencies. This was achieved by the fourth editor, Henry Steinberg, an academic refugee from Nazi Germany, whose literary credits included a history of printing and an account of the Thirty Years War.

Archive correspondence shows Steinberg to have been an assiduous seeker after information which was just as well since SYB depended increasingly on easy access to those best equipped to supply reliable, if unattributable, facts and figures denied to

IS YOUR HEAD SPINNING?

WORLD LEGISLATURES
JOHN PAXTON
£4.95 333 14549 5

THE STATESMAN'S YEAR-BOOK 1974/1975
JOHN PAXTON
£4.95 333 14105 5

THE STATESMAN'S YEAR-BOOK WORLD GAZETTEER
JOHN PAXTON
£9.95 333 13359 5

YOU SHOULD ORDER THE STATESMAN'S YEAR-BOOK & ITS NEW COMPANIONS

TO BE PUBLISHED IN SEPTEMBER

MACMILLAN

with official handbooks, correspondence with civil servants of all nationalities and collections of news cuttings.

By now SYB had expanded to 1,700 pages. Even so, much had to be excluded such as details of major cities. Readers were quick to react when they spotted what they thought was an error. A long-running correspondence related to the height of Everest. SYB opted for 29,040 feet, a figure supported by *The Geographical Journal*. Those who favoured a lower estimate, if only by a mere 12 feet, were said to be relying on a lower than average snowfall. The editor temporized. 'The most noteworthy result of the latest researches is the fact that the height of Everest varies over different years and seasons.'

John's successor was Brian Hunter, the retired chief librarian of the London School of Economics and a long-time SYB contributor on Eastern Europe. It fell to Brian to administer drastic surgery to SYB when the collapse of the Soviet Union transformed Europe's territorial and political landscape.

To add to his problems, the news, welcome on all other counts, was of democracy on the march. The number of countries that were more or less democratic burgeoned from 66 in 1987 to 121 at the millennium and the one certainty of democracy in whatever form it takes is the passion for sharing knowledge. Statistics and interpretation poured out from a thousand ministries. The rush to inform was further stimulated by globalization and a booming economy which enhanced interest in China, India and other developing countries.

As seventh in the editorial line starting with Frederick Martin, I took on the mantle in 1997. It was a critical time for SYB as, indeed, for every major reference work. The potential for the internet to spark off a knowledge explosion was about to be realized. To meet this formidable opposition, the process of collecting and sifting material for SYB had to be raised to a new level. A good start was made by creating a database which made regular updating of entries easier and more systematic. But then the size of the book had to be increased to accommodate more than a million words.

Each country was given an extended historical introduction and an economic overview along with profiles of political leaders. In total, all this and more amounted to a 20 per cent increase in statistics and commentary. The in-house and freelance staff of researchers was strengthened by the arrival of Nicholas Heath-Brown, a senior editor with an unfailing eye for detail who came to us from *The Guinness Book of Records*.

But the biggest change came with the launch of the website containing all that is in the print edition plus city profiles, profiles of past leaders, extracts from relevant Palgrave Macmillan books and journals, and spotlight articles on significant anniversaries. The recent addition of the entire SYB archive has proved to be a boon for comparative studies.

One thing does not change and that is SYB's dedication to accuracy and objectivity, a welcome contrast to the thinly disguised bias and propaganda encountered across the internet.

Barry Turner

other reference works. If Steinberg had a fault it was his tendency to be easily offended. As a contributor and copy editor for SYB in New York, Karl Brown had to work hard at pacifying Steinberg when he felt slighted. On one occasion a box of Havana cigars, sent over in a package of books, did the trick.

In 1963 he was joined by John Paxton who, six years later, became the first home-grown editor in almost half a century. It was John (I can be forgiven reverting to first names as I have known John for nearly 40 years) who oversaw the change from hot metal printing to computer setting. This allowed for a rearrangement of entries into alphabetical order. Gone were the days when the book was divided into four parts – International Organizations, the Commonwealth, the USA and Other Countries. With its new format, to quote John, 'SYB lost its imperialistic feel and became truly international'.

The opportunity was also taken to reorder the sections within each country into a more logical sequence. But the tools of the trade remained essentially the same. John's office was piled high

'The Statesman's Yearbook in the course of time has become an English institution.'

Scientific American, 1913

FACT SHEET: 150 YEARS OF THE STATESMAN'S YEARBOOK

- The world population at the time of the first edition in 1864 was 1·25bn., as compared to over 7bn. today.

- In 1861 the United States reported revenue and expenditure of US$83·2m. and US$84·6m. respectively (so even then the USA had a budget deficit). In 2011 the comparable figures were US$2·3trn. and US$3·6trn.

- The two most populous states in the USA today, California and Texas, ranked only 26th and 23rd respectively in 1860.

- The 'Constitution and Government' section for the Confederate States noted: 'The "institution of negro slavery" as it now exists shall be recognised and protected by Congress and by the Territorial Governments in the territory.'

- In 1860 Virginia was the US state with most slaves (490,865). However, the states with the highest proportion of slaves were South Carolina (57·2% of the population) and Mississippi (55·2%).

- Japan was a land of considerable mystery in 1864, its entry beginning: 'The system of government of the Japanese empire is as yet but imperfectly known.' It was later noted that: 'The port of Hakodadi, in the north of Japan, was deserted, after a lengthened trial, by all the foreign merchants settled there, it having been found impossible to establish any satisfactory intercourse with the natives. Centuries of isolation appear to have made the people of Japan entirely independent of the outer world, and of the immense trade and commerce of western civilisation.'

- Nor was France's education system well regarded: 'Popular education, although having made great progress of late, stands, as yet, rather low in France. From the returns of marriages in... 1861... 83,905 bridegrooms and 136,447 brides, out of a total of 270,896 marriages, were unable to write their names. Consequently, about one-third of all the men, and considerably more than one-half of the women, married without the fundamental elements of instruction.'

- As for France's criminal statistics, of the 7,960 convicts sentenced to hard labour in 1861 there were among them 'five ecclesiastics, three comedians, six notaries, and one professor of literature'.

- A 'curious official return' from Spain for 1859 recorded that the 70,811 citizens of the city of Cádiz consumed 80,646 stones of soap, while the 94,293 citizens of Málaga consumed a mere 98 stones.

- The 1864 entry for Queensland (an entry for the Commonwealth of Australia not appearing until 1903) recorded that: 'Great gold fields have hitherto not been discovered, though the metal is believed to be extant in large quantities.' The colony would enjoy a gold rush later in the decade and in 1871 the Charters Towers goldfield was discovered, which proved to be among the richest in Australian history.

- The Indians of Mexico, accounting for around half the population at the time of the first edition, were described as 'reduced to a state of abject misery and servitude' in the country's 1864 entry.

- Switzerland, constitutionally barred from having a standing army, provided military instruction for the 'greater number of pupils at the upper and middle class schools. They not only go through the infantry exercises, but practise gunnery, the necessary rifles and cannon—the latter 2- and 4-pounders—being furnished by the Federal government.'

- The financial outlook for Greece in 1864 was, as today, stormy: 'At the time of King Otho's departure from Greece [1862], the exchequer was not only empty, but exhibited a deficit of 6,000,000 drachmas.'

- Coverage of African states did not begin until the Cape of Good Hope, Liberia and Natal were included in the second edition of 1865. Of the 194 nation states covered in the book today, African countries make up the largest share with 55.

- Of the 18 European nations featured in the 1864 edition, Switzerland was alone in not having a monarch. Of the 50 states in Europe today, only 12 are monarchies.

- The first mention of the telephone came in the 1883 edition.

- The 25th anniversary edition of 1888 reported a world population of 1,483m., divided as follows: Africa, 197m., America, 112m.; Asia, 789m.; Europe, 347m.; Oceania, 38m.

- The same year, the British Empire was recorded as the world's largest in terms of area (9,339,000 sq. miles). However, in terms of population, the Chinese Empire was bigger, with 404m. inhabitants (against 307m. in the British Empire).

- The population of New York City in 1890 was put at 1,515,301. Its 2010 census population was 8,175,133.

- The first mention of electricity came in 1891.

- In New Zealand's 1894 entry it was recorded that females had been 'admitted to the franchise' the previous year, making New Zealand the first country to grant all women of appropriate age the vote in national elections.

- France's asylums for imbeciles were reported as having 64,639 inmates at the end of 1897.

- In 1898, 4·1% of births in England and Wales were classified as 'illegitimate'. In the UK in 2008, 45·4% of births were to unmarried women.

- The salary of the President of the USA is US$400,000 today. In 1900 it was US$50,000—equivalent to about US$1·4m. at current prices.

- The mortality rate among children under one year of age was estimated at over 26% in Russia in 1900. In 2005 the rate was 1·1%.

- In 1900 Russia had the world's largest army, numbering 896,000 in peace time but growing to 3·5m. when on a war footing. China, whose standing army of 1·6m. is the world's largest today, had around 300,000 troops in 1900 but its forces were described as 'having no unity or cohesion; there is no proper discipline, the drill is mere physical exercise, the weapons are long since obsolete, and there is no transport, commissariat, or medical service'.

- In England and Wales in 1901, the average number of persons per inhabited building was 5·2. By 2010, the average British household consisted of 2·4 people.

- The first mention of the motor car came in 1905.

- The UK had by far the largest navy at the outbreak of the First World War, numbering 681 vessels either built or being built. The total was almost double the size of its nearest rival, Germany. Today, the USA has the largest navy, bigger in terms of tonnage than those of the next six largest combined.

- In 1919, provisional casualties of the First World War were put at 6,886,411 killed, 12,616,017 wounded and 6,477,761 missing or imprisoned.

- The first reference to Hitler came in relation to the election of '2 Hitlerites' to the Diet of Saxony at the election of 31 Oct. 1926.

- The first mention of the atomic bomb came in 1946.

- In 1960 details of funding for the National Aeronautics and Space Administration (NASA) were included, the first indicator of the Cold War space race.

- Internet usage statistics were first included in 2000, two years before *The Statesman's Yearbook*'s own website went live.

- Seven states have disappeared since the 1978 edition (the first in which all the countries of the world were included in a simple A-Z format): Czechoslovakia; German Democratic Republic; Germany, Federal Republic of; the Soviet Union; Yemen Arab Republic; Yemen, People's Democratic Republic of; and Yugoslavia.

- Thirty-nine new states have been added since the 1978 edition: Antigua and Barbuda; Armenia; Azerbaijan; Belarus; Bosnia and Herzegovina; Croatia; Czech Republic; Dominica; Eritrea; Estonia; Georgia; Germany; Kazakhstan; Kiribati; Kyrgyzstan; Latvia; Lithuania; Macedonia; Marshall Islands; Micronesia; Moldova; Montenegro; Namibia; Palau; Russia; St Kitts and Nevis; St Lucia; St Vincent and the Grenadines; Serbia; Slovakia; Slovenia; South Sudan; Tajikistan; Timor-Leste; Turkmenistan; Ukraine; Uzbekistan; Vanuatu; and Yemen.

- Nine states have been officially renamed since the 1978 edition: Burma (now Myanmar); Central African Empire (now Central African Republic); Fiji (now Fiji Islands); Ivory Coast (now Côte d'Ivoire); Kampuchea (now Cambodia); Rhodesia (now Zimbabwe); Upper Volta (now Burkina Faso); Western Samoa (now Samoa); and Zaïre (now the Democratic Republic of the Congo).

'It is not too much to say this book is one of the most valuable and important publications which is issued anywhere in the world. Libraries, newspapers, etc., absolutely require every volume of this splendid compendium …'

Scientific American, 1898

The world in focus at
www.statesmansyearbook.com

WORLD POPULATION DEVELOPMENTS

<table>
<tr><td colspan="2">1864 STATESMAN'S YEARBOOK FIGURES</td></tr>
<tr><td>1. China</td><td>367,632,907</td></tr>
<tr><td>2. India</td><td>135,634,244</td></tr>
<tr><td>3. Russian Empire</td><td>73,992,373</td></tr>
<tr><td>4. German Confederation</td><td>45,013,034</td></tr>
<tr><td>5. France</td><td>37,382,225</td></tr>
<tr><td>6. Turkey</td><td>35,350,000</td></tr>
<tr><td>7. Japan[1]</td><td>35,000,000</td></tr>
<tr><td>8. USA</td><td>31,445,080</td></tr>
<tr><td>9. UK</td><td>28,887,519</td></tr>
<tr><td>10. Italy</td><td>21,894,925</td></tr>
</table>

[1] Population for Japan first included in 1867 edition.

2010	
1. China	1,341,335,000
2. India	1,224,614,000
3. USA	310,384,000
4. Indonesia	239,871,000
5. Brazil	194,946,000
6. Pakistan	173,953,000
7. Nigeria	158,423,000
8. Bangladesh	148,692,000
9. Russia	142,958,000
10. Japan	126,536,000

2050 PROJECTIONS	
1. India	1,692,008,000
2. China	1,295,604,000
3. USA	403,101,000
4. Nigeria	389,615,000
5. Indonesia	293,456,000
6. Pakistan	274,875,000
7. Brazil	222,843,000
8. Bangladesh	194,353,000
9. Philippines	154,939,000
10. Congo (Dem. Republic of)	148,523,000

Sources: The Statesman's Yearbook (1864 edition); United Nations World Population Prospects (2010 Revision)

LARGEST URBAN AGGLOMERATIONS

1870	
1. London, United Kingdom	3,890,000
2. Paris, France	1,852,000
3. New York, USA	1,478,000
4. Vienna, Austria	834,000
5. Berlin, Germany	826,000
6. Calcutta (Kolkata), India	742,000
7. Philadelphia, USA	674,000
8. St Petersburg, Russia	667,000
9. Bombay (Mumbai), India	644,000
10. Moscow, Russia	612,000

2010	
1. Tokyo, Japan	36,933,000
2. Delhi, India	21,935,000
3. Mexico City, Mexico	20,142,000
4. New York-Newark, USA	20,104,000
5. São Paulo, Brazil	19,649,000
6. Shanghai, China	19,554,000
7. Mumbai (Bombay), India	19,422,000
8. Beijing, China	15,000,000
9. Dhaka, Bangladesh	14,930,000
10. Kolkata (Calcutta), India	14,283,000

2025 PROJECTIONS	
1. Tokyo, Japan	38,661,000
2. Delhi, India	32,935,000
3. Shanghai, China	28,404,000
4. Mumbai (Bombay), India	26,557,000
5. Mexico City, Mexico	24,581,000
6. New York-Newark, USA	23,572,000
7. São Paulo, Brazil	23,175,000
8. Dhaka, Bangladesh	22,906,000
9. Beijing, China	22,633,000
10. Karachi, Pakistan	20,190,000

Sources: International Historical Statistics (Palgrave Macmillan); United Nations Department of Economic and Social Affairs/Population Division, World Urbanization Prospects (2011 Revision)

North America 1837

The Edinburgh Geographical Institute

Reproduced from the 34th edition of The Statesman's Yearbook (1897)

English Miles

0 100 200 300 400 500   1000

J.G.Bartholomew,

Reproduced from the 34th edition of The Statesman's Yearbook (1897)

Europe 1837

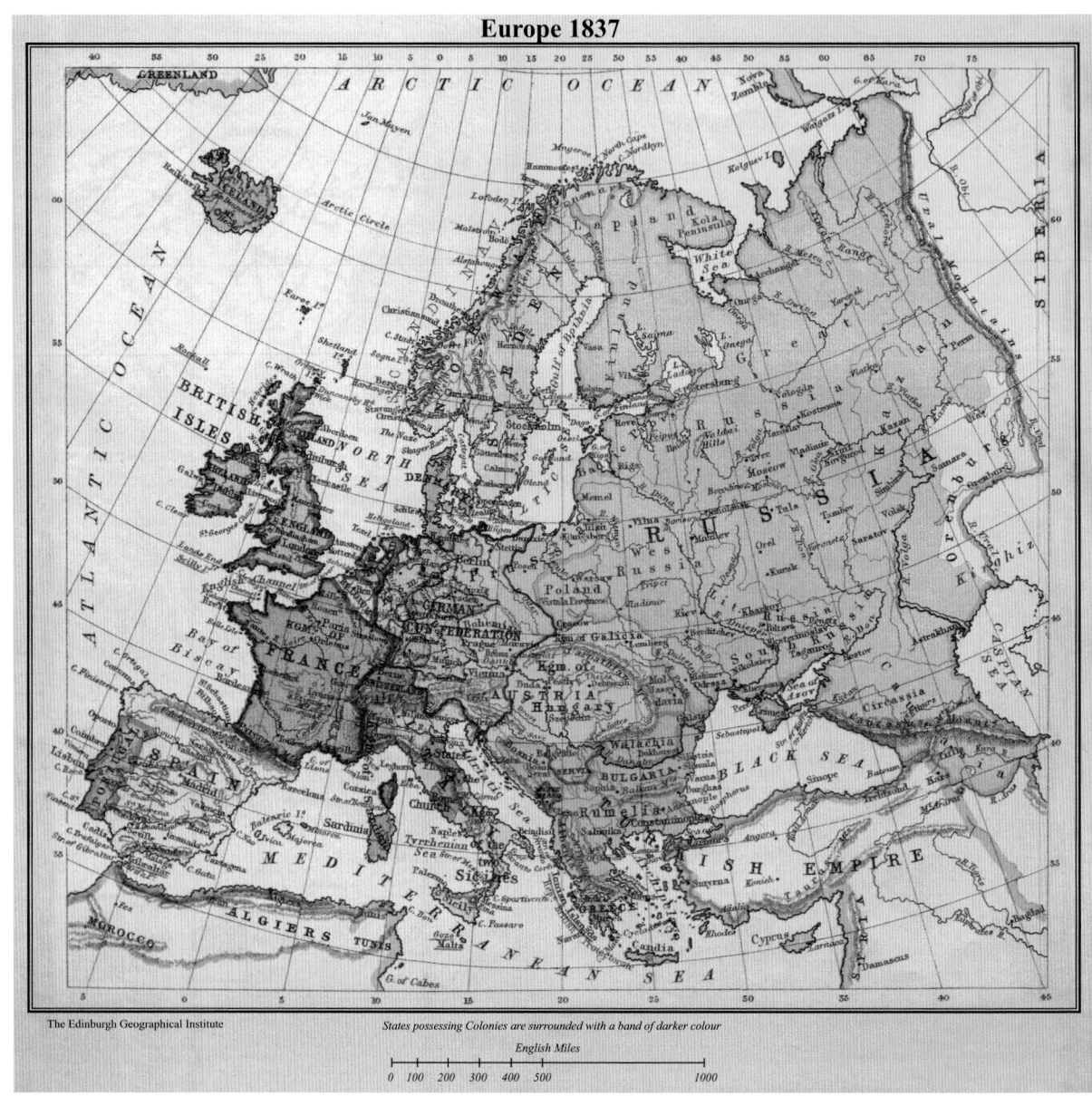

Reproduced from the 34th edition of The Statesman's Yearbook (1897)

Europe 1897

States possessing Colonies are surrounded with a band of darker colour
English Miles

0 100 200 300 400 500   1000

J.G.Bartholomew

Reproduced from the 34th edition of The Statesman's Yearbook (1897)

COMPARATIVE GROWTH OF POPULATION OF COUNTRIES DURING XIXth CENTURY

All the Diagrams are drawn to the same Scale

UNITED KINGDOM

1801	1811	1821	1831	1841	1851	1861	1871	1881	1891	1901
16·345·646	17·370·000	21·272·187	24·392·485	27·036·450	27·724·056	29·321·288	31·845·379	35·241·482	38·104·975	41·605·220

UNITED STATES

The Area of the United States for each of the Census Periods is indicated in Square Miles in italic figures.

1800	1810	1820	1830	1840	1850	1860	1870	1880	1890	1900
602,224 sq. m.	*755,435 sq. m.*	*958,705 sq. m.*	*1,017,973 sq. m.*	*1,126,942 sq. m.*	*1,943,581 sq. m.*	*2,170,461 sq. m.*	*3,580,242 sq. m. Including Alaska.*	*3,602,990 sq. m.*	*3,602,990 sq. m.*	*3,622,933 sq. m. Including Hawaii.*
5·308·483	7·239·881	9·633·822	12·866·020	17·069·453	23·191·876	31·443·321	38·558·371	50·155·783	62·622·250	75·620·859

GERMANY

	1810	1820	1830	1840	1850	1860	1871	1880	1890	1900
NO RELIABLE ESTIMATE	24·833·000	26·294·000	29·520·000	32·828·000	35·397·000	37·747·000	41·058·804	45·236·061	49·428·135	56·356·246

FRANCE

1801		1821		1841	1851	1861	1872	1881	1891	1901
26·930·756		29·871·176		35·400·486	35·783·206	35·844·902	36·102·921	37·672·048	38·343·192	38·641·333

Reproduced from the 39th edition of The Statesman's Yearbook (1902)

COMPARATIVE GROWTH OF POPULATION OF CITIES DURING XIXth CENTURY

All the Diagrams are drawn to the same Scale

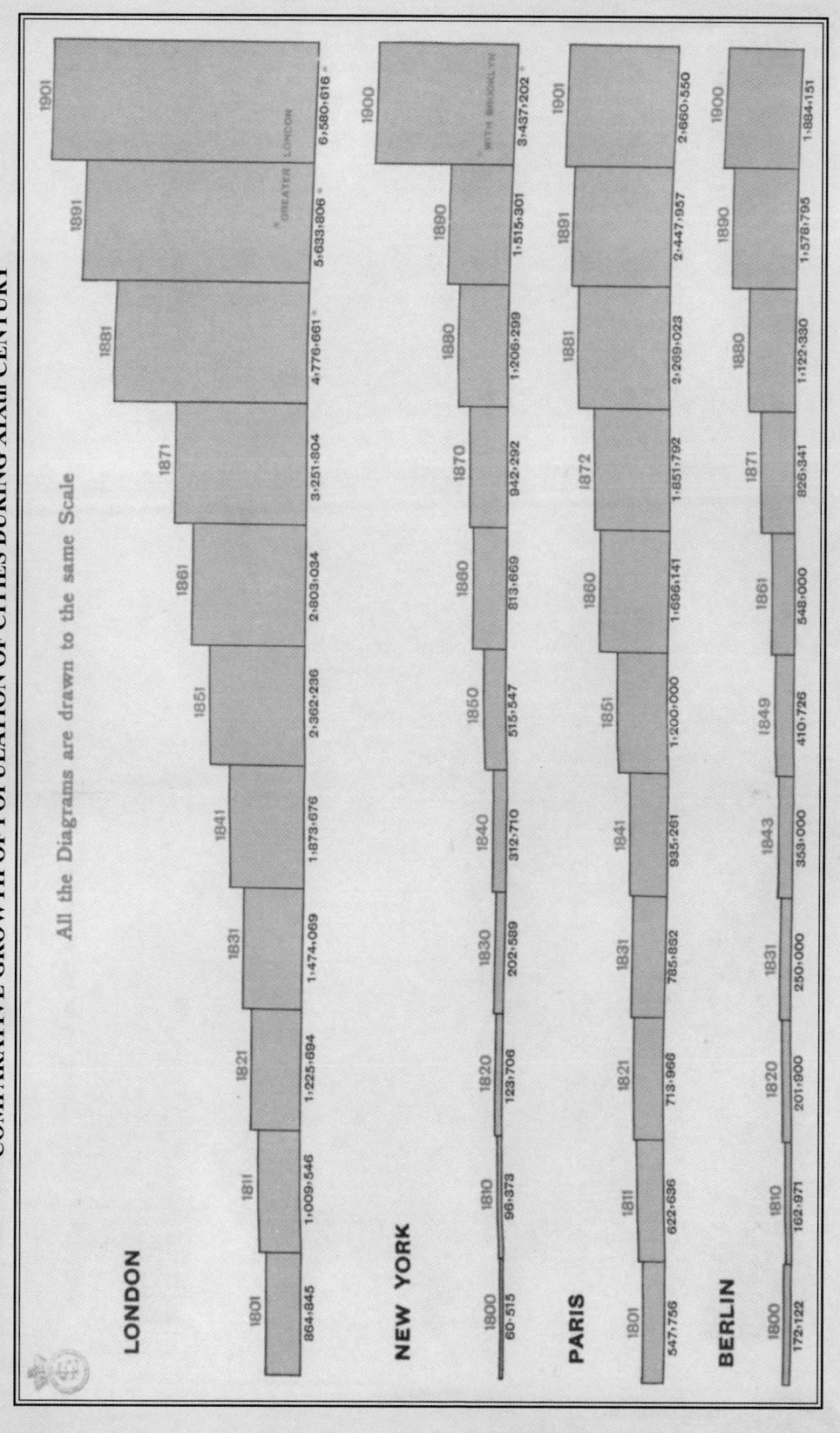

LONDON

1801	1811	1821	1831	1841	1851	1861	1871	1881	1891	1901
864·845	1·009·546	1·225·694	1·474·069	1·873·676	2·362·236	2·803·034	3·251·804	4·776·661 *	5·633·806 *	6·580·616 =

* GREATER LONDON

NEW YORK

1800	1810	1820	1830	1840	1850	1860	1870	1880	1890	1900
60·515	96·373	123·706	202·589	312·710	515·547	813·669	942·292	1·206·299	1·515·301	3·437·202 *

* WITH BROOKLYN

PARIS

1801	1811	1821	1831	1841	1851	1860	1872	1881	1891	1901
547·756	622·636	713·966	785·832	935·261	1·200·000	1·696·141	1·851·792	2·269·023	2·447·957	2·660·550

BERLIN

1800	1810	1820	1831	1843	1849	1861	1871	1880	1890	1900
172·122	162·971	201·900	250·000	353·000	410·726	548·000	826·341	1·122·330	1·578·795	1·884·151

Reproduced from the 39th edition of The Statesman's Yearbook (1902)

PART I

INTERNATIONAL ORGANIZATIONS

United Nations (UN)

Origin and Aims. The United Nations is an association of states, or intergovernmental organizations, pledged to maintain international peace and security and to co-operate in solving international political, economic, social, cultural and humanitarian problems. The name 'United Nations' was devised by US President Franklin D. Roosevelt and was first used in the Declaration by United Nations of 1 Jan. 1942, during the Second World War, when 26 nations pledged to continue fighting the Axis Powers.

The United Nations Charter was drawn up by the representatives of 50 countries at the United Nations Conference on International Organization, which met in San Francisco from 25 April to 26 June 1945. Delegates started with proposals worked out by the representatives of China, the Soviet Union, the United Kingdom and the United States at Dumbarton Oaks (Washington, D.C.) from 21 Aug. to 28 Sept. 1944. The Charter was signed on 26 June 1945 by the representatives of the 50 countries. Poland, which was not represented at the Conference, signed later and became one of the original 51 member states. The United Nations came into existence officially on 24 Oct. 1945, with the deposit of the requisite number of ratifications of the Charter with the US Department of State. United Nations Day is celebrated on 24 Oct.

In recent years, most of the UN's work has been devoted to helping developing countries. Major goals include the protection of human rights; saving children from starvation and disease; providing relief assistance to refugees and disaster victims; countering global crime, drugs and disease; and assisting countries devastated by war and the long-term threat of landmines.

Members. New member states are admitted by the General Assembly on the recommendation of the Security Council. The Charter provides for the suspension or expulsion of a member for violation of its principles, but no such action has ever been taken. The UN has 193 member states, comprising every internationally recognized sovereign state, with the exception of the Holy See. (For a list of these, *see* below.)

Finance. Contributions from member states constitute the main source of funds. These are in accordance with a scale specified by the Assembly, and determined primarily by the country's share of the world economy and ability to pay, in the range 22%–0·001%. The Organization is prohibited by law from borrowing from commercial institutions.

A Working Group on the Financial Situation of the United Nations was established in 1994 to address the long-standing financial crisis caused by non-payment of assessed dues by many member states. As of 31 May 2012 member states owed the UN a total of US$23,458m., of which the USA owed US$1,135m. (46%). Total regular budget debts as of 31 May 2012 were US$1,006m., of which the USA's share was US$744m. (74%).

Official languages: Arabic, Chinese, English, French, Russian and Spanish.

Structure. The UN has five principal organs established by the founding Charter (formerly six). All have their headquarters in New York except the International Court of Justice, which has its seat in The Hague. These core bodies work through dozens of related agencies, operational programmes and funds, and through special agreements with separate, autonomous, intergovernmental agencies, known as Specialized Agencies, to provide a programme of action in the fields of peace and security, justice and human rights, humanitarian assistance, and social and economic development. The five principal UN organs are:

1. **The General Assembly**, composed of all members, with each member having one vote. Meeting once a year, proceedings begin on the Tuesday of the third week of Sept. The 67th Session opened on 18 Sept. 2012.

At least three months before the start of each session, the Assembly elects a new President, 21 Vice-Presidents and the chairs of its six main committees, listed below. To ensure equitable geographical representation, the presidency of the Assembly rotates each year among the five geographical groups of states: Africa, Asia, Eastern Europe, Latin America and the Caribbean, and Western Europe and other States. Special sessions may be convoked by the Secretary-General if requested by the Security Council, by a majority of members, or by one member if the majority of members concur. Emergency sessions may be called within 24 hours at the request of the Security Council on the vote of any nine Council members, or a majority of United Nations members, or one member if the majority of members concur. Decisions on important questions, such as peace and security, new membership and budgetary matters, require a two-thirds majority; other questions require a simple majority of members present and voting.

The work of the General Assembly is divided between six Main Committees, on which every member state is represented: the Disarmament and International Security Committee (First Committee); the Economic and Financial Committee (Second Committee); the Social, Humanitarian and Cultural Committee (Third Committee); the Special Political and Decolonization Committee (Fourth Committee); the Administrative and Budgetary Committee (Fifth Committee); and the Legal Committee (Sixth Committee).

There is also a General Committee charged with the task of co-ordinating the proceedings of the Assembly and its Committees, and a Credentials Committee, which examines the credentials of representatives of Member States. The General Committee consists of 28 members: the President and 21 Vice-Presidents of the General Assembly and the chairs of the six main committees. The Credentials Committee consists of nine members appointed by the Assembly on the proposal of the President at each session. In addition, the Assembly has two standing committees—an Advisory Committee on Administrative and Budgetary Questions and a Committee on Contributions—and may establish subsidiary and *ad hoc* bodies when necessary to deal with specific matters. These include the Special Committee on Peacekeeping Operations (144 members), the Human Rights Council (47 members), the Committee on the Peaceful Uses of Outer Space (74 members), the Committee on the Exercise of the Inalienable Rights of the Palestinian People (24 members), the Conference on Disarmament (65 members), the International Law Commission (34 independent members), the Scientific Committee on the Effects of Atomic Radiation (21 members), the Special Committee on the Situation with Regard to the Implementation of the Declaration on the Granting of Independence to Colonial Countries and Peoples (known as the Special Committee of 24 on Decolonization; 29 members), and the Commission on International Trade Law (60 members).

The General Assembly has the right to discuss any matters within the scope of the Charter and, with the exception of any situation or dispute on the agenda of the Security Council, may make recommendations accordingly. Occupying a central position in the UN, the Assembly receives reports from other organs, admits new members, directs activities for development, sets policies and determines programmes for the Secretariat and approves the UN budget. The Assembly appoints the Secretary-General, who reports annually to it on the work of the Organization.

Under the 'Uniting For Peace' resolution (377) adopted by the General Assembly in Nov. 1950, the Assembly is also empowered to take action if the Security Council, because of a lack of unanimity of its permanent members, fails to exercise its primary responsibility for the maintenance of international peace and security in any case where there appears to be a threat to the peace, breach of the peace or act of aggression. In this event, the General Assembly may consider the matter immediately with a view to making appropriate recommendations to members for collective measures, including, in the case of a breach of the peace or act of aggression, the use of armed force to maintain or restore international peace and security.

The first Emergency Special Session of the Assembly was called in 1956 during the Suez Crisis by Yugoslavia, which cited Resolution 377; demands were made for the withdrawal of British, French and Israeli troops from Egypt. On the Assembly's recommendations, the United Nations Emergency Force (UNEF I) was formed as the UN's first armed peacekeeping force.

Over the years a number of programmes and funds have been established to address particular humanitarian and development concerns. These bodies usually report to the General Assembly. They include: the United Nations Children's Fund (UNICEF); Office of the United Nations High Commissioner for Refugees (UNHCR); United Nations Conference on Trade and Development (UNCTAD); United Nations Development Programme (UNDP) and Population Fund (UNFPA); United Nations Environment Programme (UNEP); United Nations World Food Programme (WFP); United Nations Office on Drugs and Crime (UNODC).

Website: http://www.un.org/ga
President: Vuk Jeremić (Serbia) was elected President for the Sixty-Seventh Session in 2012.

2. **The Security Council** has primary responsibility for the maintenance of international peace and security. Under the Charter, the Security Council alone has the power to take decisions that member states are obligated to carry out. A representative of each of its members must be present at all times at UN Headquarters, but it may meet elsewhere as best facilitates its work.

The Presidency of the Council rotates monthly, according to the English alphabetical order of members' names. The Council consists of 15 members: five permanent and ten non-permanent elected for a two-year term by a two-thirds majority of the General Assembly. Each member has one vote. Retiring members are not eligible for immediate re-election. Any other member of the United Nations may participate without a vote in the discussion of questions specially affecting its interests.

Decisions on procedural questions are made by an affirmative vote of at least nine members. On all other matters, the affirmative vote of nine members must include the concurring votes of all permanent members (subject to the provision that when the Council is considering methods for the peaceful settlement of a dispute, parties to the dispute abstain from voting). Consequently, a negative vote from a permanent member has the power of veto. If a permanent member does not support a decision but does not wish to veto it, it may abstain. From 1945–91 the USSR employed its veto 119 times, the USA 69 times, the UK 32 times, France 18 times and China three times (including once before the People's Republic of China took over Taiwan's seat at the UN in 1971). From 1992–Feb. 2013 the USA vetoed 14 resolutions, Russian Federation eight and China six; France and the UK did not veto any resolutions.

The Council has three standing committees—the Committee of Experts, the Committee on Admission of New Members and the Committee on Council Meetings away from Headquarters. It may establish *ad hoc* committees and commissions, which include all Council members—there were three in Feb. 2013: the Governing Council of the United Nations Compensation Commission, established in 1991 by Security Council Resolution 692 to compensate for losses related to the Iraqi invasion of Kuwait; the Counter-Terrorism Committee, established pursuant to Resolution 1373 (2001); and the 1540 Committee, established pursuant to Resolution 1540 (2004).

When a threat to peace is brought before the Council, it may undertake mediation, setting out principles for a settlement, and may take measures to enforce its decisions by ceasefire directives, economic sanctions, peacekeeping missions or, in some cases, by collective military action. For the maintenance of international peace and security, the Council can, subject to special agreements, call on the armed forces, assistance and facilities of the member states. It is assisted by a Military Staff Committee consisting of the Chiefs of Staff of the permanent members of the Council or their representatives.

The Council also makes recommendations to the Assembly on the appointment of the Secretary-General and, with the Assembly, elects the judges of the International Court of Justice.

Peacekeeping. The Charter contains no explicit provisions for peacekeeping operations (PKOs), yet they have the highest profile of all the UN's operations. PKOs are associated with humanitarian intervention though their emergence was primarily a result of the failure of the Charter's collective security system during the Cold War and the absence of a UN Force. The end of the Cold War and the rise of intra-state conflict led to a proliferation of PKOs from the late 1980s and a greater proportion of armed missions. However, notable failures in the early and mid-1990s, such as the missions to Somalia in 1993 and to Rwanda in 1994, account for a drop in PKOs and shorter mandates. In 1992 then Secretary-General Boutros Boutros-Ghali presented the 'Agenda for Peace', which laid out four phases to prevent or end conflict: preventative diplomacy; peacemaking with civilian and military means; peacekeeping, in its traditional sense of operations in the field; and post-conflict peace-building, an area seen as comparatively neglected in previous missions. Then Secretary-General Kofi Annan presented a report aimed at conflict prevention in July 2001 emphasizing inter-agency co-operation and long-term strategies to prevent regional instability. There have been 67 peacekeeping operations in total since 1948. The first mission—consisting of unarmed observers—was to monitor the ceasefire during the First Arab–Israeli War.

Recent History. In Nov. 2002 the Security Council held Iraq in 'material breach' of disarmament obligations. Weapons inspectors, led by Hans Blix (Sweden), returned to Iraq four years after their last inspections. Amid suspicion that Iraq was failing to comply, the USA, the UK and Spain reserved the right to disarm Iraq without a further Security Council resolution. Other Council members, notably China, France, Germany and Russia, opposed such action. In April 2003 US forces, supported by the UK, brought an end to Saddam Hussein's rule. In June 2004 the UN recognized the transfer of sovereignty to the interim government of Iraq. In 2005 the Council referred cases to the International Criminal Court (ICC) for the first time, asking the court to investigate the situation in Darfur, Sudan. Instability in Sudan saw the establishment of peacekeeping missions in Darfur (2007), the disputed Abyei region and newly independent South Sudan (both 2011). Other peacekeeping missions established in recent years include Liberia (2003), Côte d'Ivoire (2004), Haiti (2004) and Timor-Leste (2006). In March 2011 the UN approved a no-fly zone in Libya for the protection of civilians, an action pivotal to the success of the rebel movement, and later in the year established a mission to assist in nation rebuilding in the post-Gaddafi era. In Jan. 2013 sanctions were tightened against North Korea in response to rocket tests carried out by Pyongyang the previous month.

Reform. The composition of the Security Council, with its five permanent members having qualified as the principal Second World War victors, has been subject to intense debate in recent years. The lack of permanent representation from Latin America and the Caribbean or from Africa and the Islamic World is frequently cited to demonstrate that the Council is unrepresentative.

However, reform is in the hands of the permanent members and a unanimous agreement has proved elusive. In Sept. 2004 Brazil, Germany, India and Japan (the G4) launched a joint bid for permanent membership, along with a seat for an African state. In March 2005 then Secretary-General Annan proposed either six new permanent members and three new non-permanent members or the election of a new type of member, eight of which would be elected for a four-year period. The World Summit in Sept. 2005 failed to agree on Security Council reform but pledged to continue negotiations.

Permanent Members. China, France, Russian Federation, UK, USA (Russian Federation took over the seat of the former USSR in Dec. 1991).

Non-Permanent Members. Azerbaijan, Guatemala, Morocco, Pakistan and Togo (until 31 Dec. 2013); Argentina, Australia, Luxembourg, South Korea and Rwanda (until 31 Dec. 2014).

Finance. The budget for UN peacekeeping operations in 2012–13 was US$7·3bn. In Dec. 2012 outstanding contributions to peacekeeping totalled US$1·3bn.

3. **The Economic and Social Council (ECOSOC)** is responsible under the General Assembly for co-ordinating international economic, social, cultural, educational, health and related matters.

The Council consists of 54 member states elected by a two-thirds majority of the General Assembly for a three-year term. Members are elected according to the following geographic distribution: Africa, 14 members; Asia, 11; Eastern Europe, 6; Latin America and Caribbean, 10; Western Europe and other States, 13. A third of the members retire each year. Retiring members are eligible for immediate re-election. Each member has one vote. Decisions are made by a majority of the members present and voting.

The Council holds one five-week substantive session a year, alternating between New York and Geneva, and one organizational session in New York. The substantive session includes a high-level meeting attended by Ministers, to discuss economic and social issues. Special sessions may be held if required. The President is elected for one year and is eligible for immediate re-election.

The subsidiary machinery of ECOSOC includes:
Nine Functional Commissions. Statistical Commission; Commission on Population and Development; Commission for Social Development; Commission on the Status of Women; Commission on Narcotic Drugs (and Subcommission on Illicit Drug Traffic and Related Matters in the Near and Middle East); Commission on Science and Technology for Development; Commission on Crime Prevention and Criminal Justice; Commission on Sustainable Development; United Nations Forum on Forests.

Five Regional Economic Commissions. ECA (Economic Commission for Africa, Addis Ababa, Ethiopia); ESCAP (Economic and Social Commission for Asia and the Pacific, Bangkok, Thailand); ECE (Economic Commission for Europe, Geneva, Switzerland); ECLAC (Economic Commission for Latin America and the Caribbean, Santiago, Chile); ESCWA (Economic Commission for Western Asia, Beirut, Lebanon).

Three Standing Committees. Committee for Programme and Co-ordination; Commission on Non-Governmental Organizations; Committee on Negotiations with Intergovernmental Agencies.

In addition, the Council may consult international non-governmental organizations (NGOs) and, after consultation with the member concerned, with national organizations. Over 3,000 organizations have consultative status. NGOs may send observers to ECOSOC's public meetings and those of its subsidiary bodies, and may also submit written statements relevant to its work. They may also consult with the UN Secretariat on matters of mutual concern. The term of office of the members listed below expires on 31 Dec. of each year.

Members. Albania (2015), Austria (2014), Belarus (2014), Benin (2015), Bolivia (2015), Brazil (2014), Bulgaria (2013), Burkina Faso (2014), Cameroon (2013), Canada (2015), China (2013), Colombia (2015), Croatia (2015), Cuba (2014), Denmark (2013), Dominican Republic (2014), Ecuador (2013), El Salvador (2014), Ethiopia (2014), France (2014), Gabon (2013), Haiti (2015), India (2014), Indonesia (2014), Ireland (2014), Japan (2014), South Korea (2013), Kuwait (2015), Kyrgyzstan (2015), Latvia (2013), Lesotho (2014), Libya (2014), Malawi (2013), Mauritius (2015), Mexico (2013), Nepal (2015), Netherlands (2015), New Zealand (2013), Nicaragua (2013), Nigeria (2014), Pakistan (2013), Qatar (2013), Russia (2013), San Marino (2015), Senegal (2013), South Africa (2015), Spain (2014), Sudan (2015), Sweden (2013), Tunisia (2015), Turkey (2014), Turkmenistan (2015), United Kingdom (2013), United States of America (2015).

Finance. In 2009, US$22,001m. in socio-economic development assistance grants was provided through the organizations of the UN system.

4. **The International Court of Justice** is the principal judicial organ of the UN. It has a dual role: to settle in accordance with international law the legal disputes submitted to it by States; and to give opinions on legal questions referred to it by authorized international organs and agencies.

The Court operates under a Statute of the United Nations Charter. Only States may apply to and appear before the court. The Court is composed of 15 judges, each of a different nationality, elected by an absolute majority by the General Assembly and the Security Council to nine-year terms of office. The composition of the Court must reflect the main forms of civilization and principal legal systems of the world. Elections are held every three years for one-third of the seats; retiring judges may be re-elected. Judges do not represent their respective governments but sit as independent magistrates. They must have the qualifications required in their respective countries for appointment to the highest judicial offices, or be jurists of recognized competence in international law. Candidates are nominated by the national panels of jurists in the Permanent Court of Arbitration established by The Hague Conventions of 1899 and 1907. The Court elects its own President and Vice-President for a three-year term, and is permanently in session.

Decisions are taken by a majority of judges present, subject to a quorum of nine members, with the President having a casting vote. Judgment is final and without appeal, but a revision may be applied for within ten years from the date of the judgment on the ground of new decisive evidence. When the Court does not include a judge of the nationality of a State party to a case, that State has the right to appoint a judge *ad hoc* for that case. While the Court normally sits in plenary session, it can form chambers of three or more judges to deal with specific matters. Judgments by chambers are considered as rendered by the full Court. In 1993, in view of the global expansion of environmental law and protection, the Court formed a seven-member Chamber for Environmental Matters.

Judges. The nine-year terms of office of the judges currently serving end on 5 Feb. of each year indicated: Peter Tomka, President (Slovakia; 2021), Bernardo Sepúlveda Amor, Vice-President (Mexico; 2015), Mohamed Bennouna (Morocco; 2015), Kenneth Keith (New Zealand; 2015), Leonid Skotnikov (Russian Federation; 2015), Joan E. Donoghue (USA; 2015), Antônio Augusto Cançado Trindade (Brazil; 2018), Ronny Abraham (France; 2018), Dalveer Bhandari (India; 2018), Abdulqawi A. Yusuf (Somalia; 2018), Sir Christopher Greenwood (UK; 2018), Xue Hanqin (China; 2021), Giorgio Gaja (Italy; 2021), Hishashi Owada (Japan; 2021), Julia Sebutinde (Uganda; 2021).

Competence and Jurisdiction. In contentious cases, only States may apply to or appear before the Court. The conditions under which the Court will be open to non-member states are laid down by the Security Council. The jurisdiction of the Court covers all matters that parties refer to it and all matters provided for in the Charter or in treaties and conventions in force. Disputes concerning the jurisdiction of the Court are settled by the Court's own decision. The Court may apply in its decision:

(a) international conventions;

(b) international custom;

(c) the general principles of law recognized by civilized nations;

(d) as subsidiary means for the determination of the rules of law, judicial decisions and the teachings of highly qualified publicists. If the parties agree, the Court may decide a case *ex aequo et bono*.

Since 1946 the Court has delivered 112 judgments on disputes concerning *inter alia* land frontiers and maritime boundaries, territorial sovereignty, the use of force, interference in the internal affairs of States, diplomatic relations, hostage-taking, the right of asylum, nationality, guardianship, rights of passage and economic rights.

The Court may also give advisory opinions on legal questions to the General Assembly, the Security Council, certain other organs of the UN and 16 agencies of the UN family.

Since 1946 the Court has given 27 advisory opinions, concerning *inter alia* admission to United Nations membership, reparation for injuries suffered in the service of the United Nations, the territorial status of South-West Africa (Namibia) and Western Sahara, expenses of certain United Nations operations, the status of human rights informers, the threat or use of nuclear weapons and legal consequences of the construction of a wall in the Occupied Palestinian Territory.

Finance. The expenses of the Court are borne by the UN. No court fees are paid by parties to the Statute.

Official languages: English, French.
Headquarters: Peace Palace, Carnegieplein 2, 2517 KJ The Hague, Netherlands.
Website: http://www.icj-cij.org
Registrar: Philippe Couvreur (Belgium).

5. **The Secretariat** services the other four organs of the UN, carrying out their programmes, providing administrative support and information. It has a staff of 8,900 at the UN Headquarters in New York and around the world. At its head is the Secretary-General, appointed by the General Assembly on the recommendation of the Security Council for a five-year, renewable term. The Secretary-General acts as chief administrative officer in all meetings of the General Assembly, Security Council and Economic and Social Council. An Office of Internal Oversight, established in 1994 under the tenure of former Secretary-General Boutros Boutros-Ghali (Egypt), pursues a cost-saving mandate to investigate and eliminate waste, fraud and mismanagement within the system. The Secretary-General is assisted by Under-Secretaries-General and Assistant Secretaries-General. A new position of Deputy Secretary-General was agreed by the General Assembly in Dec. 1997 to assist in the running of the Secretariat and to raise the economic, social and development profile of the UN. Peacekeeping operations (PKOs) are chiefly run by Secretariat officials, who present a report to, and are authorized by, the Security Council.

Finance. The financial year coincides with the calendar year. The budget for the two-year period 2012–13 is US$5·15bn., compared to expenditure of US$5·4bn for 2010–11.

Headquarters: United Nations Plaza, New York, NY 10017, USA.
Website: http://www.un.org
Secretary-General: Ban Ki-moon (sworn in 1 Jan. 2007 and re-elected unanimously 21 June 2011, South Korea). *Deputy Secretary-General:* Asha-Rose Migiro (appointed 5 Jan. 2007, Tanzania).

Secretaries-General since 1945

1945–46	UK	Gladwyn Jebb (acting)
1946–52	Norway	Trygve Halvdan Lie
1953–61	Sweden	Dag Hammarskjöld
1961–71	Burma	Sithu U Thant
1972–81	Austria	Kurt Waldheim
1982–91	Peru	Javier Pérez de Cuéllar
1992–96	Egypt	Boutros Boutros-Ghali
1997–2006	Ghana	Kofi Atta Annan
2007–	South Korea	Ban Ki-moon

The Trusteeship Council was one of the principal organs, but has been inactive since 1994. It was established to ensure that governments responsible for administering Trust Territories take adequate steps to prepare them for self-government or independence. It consisted of the five permanent members of the Security Council. The task of decolonization was completed in 1994, when the Security Council terminated the Trusteeship Agreement for the last of the original UN Trusteeships (Palau), administered by the USA. All Trust Territories attained self-government or independence either as separate States or by joining neighbouring independent countries. The Council formally suspended operations on 1 Nov. 1994 following Palau's independence. By a resolution adopted on 25 May 1994 the Council amended its rules of procedure to drop the obligation to meet annually and agreed to meet as occasion required.

The proposal from then Secretary-General Kofi Annan, in the second part of his reform programme, in July 1997, was that it should be used as a forum to exercise their 'trusteeship' for the global commons, environment and resource systems. However, in his 2005 report, *In Larger Freedom*, Annan called for the deletion of the Council from the UN Charter.

Current Leaders

Ban Ki-moon

Position
Secretary-General

Introduction
A career diplomat and politician, Ban Ki-moon served as South Korea's foreign minister before becoming the eighth UN Secretary-General in Jan. 2007 and the first Asian to head the UN since 1971. He has a reputation as a mediator and administrator, and has prioritized UN structural reform to deliver operations more effectively. In 2011 he ran unopposed for a second five-year term and was unanimously re-elected in June to serve until the end of 2016.

Early Life
Ban Ki-moon was born on 13 June 1944 in Chungju, Chungcheongnam province, South Korea. After graduating in international relations from Seoul University in 1970, he joined the ministry of foreign affairs and in 1972 was posted to New Delhi. In the early 1980s he studied at the Kennedy School of Government at Harvard University, gaining a masters degree in public affairs in 1984. He joined the South Korean permanent observer mission at the UN in New York and rose to become first secretary.

From July 1987–May 1990 Ban was posted to the Korean embassy in Washington, becoming director-general of American affairs in 1990. In 1992, after South and North Korea adopted the joint declaration on the denuclearization of the Korean peninsula, Ban was vice chairman of the South–North joint nuclear control commission. He was made deputy minister for policy planning and international organizations in 1995 and in 1996 became national security adviser to the president.

Ban was appointed ambassador to Vienna in May 1998, where he resumed his involvement in nuclear control issues as chairman of the preparatory commission for the comprehensive nuclear-test-ban treaty in 1999. From Jan. 2000–March 2001 he served as South Korea's vice minister of foreign affairs. When South Korea took the presidency of the 56th UN General Assembly in 2001, he worked as *chef de cabinet* to the president of the assembly.

In Feb. 2003 Ban became foreign policy adviser to the then South Korean president, taking over the foreign affairs portfolio the following year and leading negotiations on the North Korean nuclear issue.

Ban declared his candidacy for the post of UN Secretary-General in Feb. 2006. He was considered an experienced diplomat with a non-confrontational style, although some critics suggested that his low-key manner might prove a weakness. He was formally appointed on 13 Oct. 2006, taking office on 1 Jan. 2007.

Career in Office

With UN peacekeeping activities at an unprecedented high, Ban has prioritized more effective operational delivery, although some of his plans have met with resistance. He has also supported the principle of reforming the UN structure. Ongoing challenges during his tenure have included the long-running division of Cyprus, Iran's and North Korea's defiance of the UN over their nuclear development programmes, the rehabilitation of war-torn Afghanistan, instability in the eastern Democratic Republic of the Congo, conflict in Sudan and the new state of South Sudan, civil war in Syria and stalemate in the Israeli–Palestinian peace process. He has overseen some progress in the global approach to climate change at a series of international conferences.

In his annual review in Dec. 2012, Ban said it had been a 'tumultuous year' that saw 'tension from Syria to the Sahel, and from eastern Congo to the Middle East', and he urged the international community to back UN efforts to pursue peace through dialogue and negotiation.

Member States of the UN

The 193 member states, with percentage scale of contributions to the Regular Budget in 2012 and year of admission:

	% contribution	Year of admission		% contribution	Year of admission		% contribution	Year of admission
Afghanistan	0·004	1946	Ecuador[1]	0·040	1945	Maldives	0·001	1965
Albania	0·010	1955	Egypt[1, 2]	0·094	1945	Mali	0·003	1960
Algeria	0·128	1962	El Salvador[1]	0·019	1945	Malta	0·017	1964
Andorra	0·007	1993	Equatorial Guinea	0·008	1968	Marshall Islands	0·001	1991
Angola	0·010	1976	Eritrea	0·001	1993	Mauritania	0·001	1961
Antigua and			Estonia	0·040	1991	Mauritius	0·011	1968
Barbuda	0·002	1981	Ethiopia[1]	0·008	1945	Mexico[1]	2·356	1945
Argentina[1]	0·287	1945	Fiji Islands	0·004	1970	Micronesia	0·001	1991
Armenia	0·005	1992	Finland	0·566	1955	Moldova	0·002	1992
Australia[1]	1·933	1945	France[1]	6·123	1945	Monaco	0·003	1993
Austria	0·851	1955	Gabon	0·014	1960	Mongolia	0·002	1961
Azerbaijan	0·015	1992	Gambia	0·001	1965	Montenegro[3]	0·004	2006
Bahamas	0·018	1973	Georgia	0·006	1992	Morocco	0·058	1956
Bahrain	0·039	1971	Germany[4]	8·018	1973	Mozambique	0·003	1975
Bangladesh	0·010	1974	Ghana	0·006	1957	Myanmar[5]	0·006	1948
Barbados	0·008	1966	Greece[1]	0·691	1945	Namibia	0·008	1990
Belarus[1, 6]	0·042	1945	Grenada	0·001	1974	Nauru	0·001	1999
Belgium[1]	1·075	1945	Guatemala[1]	0·028	1945	Nepal	0·006	1955
Belize	0·001	1981	Guinea	0·002	1958	Netherlands[1]	1·855	1945
Benin	0·003	1960	Guinea-Bissau	0·001	1974	New Zealand[1]	0·273	1945
Bhutan	0·001	1971	Guyana	0·001	1966	Nicaragua[1]	0·003	1945
Bolivia[1]	0·007	1945	Haiti[1]	0·003	1945	Niger	0·002	1960
Bosnia and			Honduras[1]	0·008	1945	Nigeria	0·078	1960
Herzegovina[7]	0·014	1992	Hungary	0·291	1955	Norway[1]	0·871	1945
Botswana	0·018	1966	Iceland	0·042	1946	Oman	0·086	1971
Brazil[1]	1·611	1945	India[1]	0·534	1945	Pakistan	0·082	1947
Brunei	0·028	1984	Indonesia[8]	0·238	1950	Palau	0·001	1994
Bulgaria	0·038	1955	Iran[1]	0·233	1945	Panama[1]	0·022	1945
Burkina Faso	0·003	1960	Iraq[1]	0·020	1945	Papua New Guinea	0·002	1975
Burundi	0·001	1962	Ireland, Rep. of	0·498	1955	Paraguay[1]	0·007	1945
Cambodia	0·003	1955	Israel	0·384	1949	Peru[1]	0·090	1945
Cameroon	0·011	1960	Italy	4·999	1955	Philippines[1]	0·090	1945
Canada[1]	3·207	1945	Jamaica	0·014	1962	Poland[1]	0·828	1945
Cape Verde	0·001	1975	Japan	12·530	1956	Portugal	0·511	1955
Central African			Jordan	0·014	1955	Qatar	0·135	1971
Rep.	0·001	1960	Kazakhstan	0·076	1992	Romania	0·177	1955
Chad	0·002	1960	Kenya	0·012	1963	Russia[1, 9]	1·602	1945
Chile[1]	0·236	1945	Kiribati	0·001	1999	Rwanda	0·001	1962
China[1]	3·189	1945	Korea, North	0·007	1991	St Kitts and Nevis	0·001	1983
Colombia[1]	0·144	1945	Korea, South	2·260	1991	St Lucia	0·001	1979
Comoros	0·001	1975	Kuwait	0·263	1963	St Vincent and the		
Congo, Dem. Rep.			Kyrgyzstan	0·001	1992	Grenadines	0·001	1980
of the[10]	0·003	1960	Laos	0·001	1955	Samoa	0·001	1976
Congo, Rep. of the	0·003	1960	Latvia	0·038	1991	San Marino	0·003	1992
Costa Rica[1]	0·034	1945	Lebanon[1]	0·033	1945	São Tomé e		
Côte d'Ivoire	0·010	1960	Lesotho	0·001	1966	Príncipe	0·001	1975
Croatia[11]	0·097	1992	Liberia[1]	0·001	1945	Saudi Arabia[1]	0·830	1945
Cuba[1]	0·071	1945	Libya	0·129	1955	Senegal	0·006	1960
Cyprus	0·046	1960	Liechtenstein	0·009	1990	Serbia[1, 12, 13]	0·037	1945
Czech Republic[14]	0·349	1993	Lithuania	0·065	1991	Seychelles	0·002	1976
Denmark[1]	0·736	1945	Luxembourg[1]	0·090	1945	Sierra Leone	0·001	1961
Djibouti	0·001	1977	Macedonia[1]	0·007	1993	Singapore[15]	0·335	1965
Dominica	0·001	1978	Madagascar	0·003	1960	Slovakia[14]	0·142	1993
Dominican			Malawi	0·001	1964	Slovenia[11]	0·103	1992
Republic[1]	0·042	1945	Malaysia[16]	0·253	1957	Solomon Islands	0·001	1978

	% contribution	Year of admission		% contribution	Year of admission		% contribution	Year of admission
Somalia	0·001	1960	Tanzania[17]	0·008	1961	United Arab Emirates	0·391	1971
South Africa[1]	0·385	1945	Thailand	0·209	1946	UK[1]	6·604	1945
South Sudan	—	2011	Timor-Leste	0·001	2002	USA[1]	22·000	1945
Spain	3·177	1955	Togo	0·001	1960	Uruguay[1]	0·027	1945
Sri Lanka	0·019	1955	Tonga	0·001	1999	Uzbekistan	0·010	1992
Sudan	0·010	1956	Trinidad and Tobago	0·044	1962	Vanuatu	0·001	1981
Suriname	0·003	1975	Tunisia	0·030	1956	Venezuela[1]	0·314	1945
Swaziland	0·003	1968	Turkey[1]	0·617	1945	Vietnam	0·033	1977
Sweden	1·064	1946	Turkmenistan	0·026	1992	Yemen[18]	0·010	1947
Switzerland	1·130	2002	Tuvalu	0·001	2000	Zambia	0·004	1964
Syria[1, 19]	0·025	1945	Uganda	0·006	1962	Zimbabwe	0·003	1980
Tajikistan	0·002	1992	Ukraine[1]	0·087	1945			

[1]Original member. [2]As United Arab Republic, 1958–71, following union with Syria (1958–61). [3]Pre-independence (2006) as part of Yugoslavia, which was an original member, from 1945–2003 and Serbia and Montenegro from 2003–06. [4]Pre-unification (1990) as two states: the Federal Republic of Germany and the German Democratic Republic. [5]As Burma, 1948–89. [6]As Byelorussia, 1945–91. [7]Pre-independence (1992) as part of Yugoslavia, which was an original member. [8]Withdrew temporarily, 1965–66. [9]As USSR, 1945–91. [10]As Zaïre, 1960–97. [11]Pre-independence (1991) as part of Yugoslavia, which was an original member. [12]As Yugoslavia, 1945–2003, and Serbia and Montenegro, 2003–06. [13]Excluded from the General Assembly in 1992; readmitted in Nov. 2000. [14]Pre-partition Czechoslovakia (1945–92) was an original member. [15]As part of Malaysia, 1963–65. [16]As the Federation of Malaya till 1963, when the new federation of Malaysia (including Singapore, Sarawak and Sabah) was formed. [17]As two states: Tanganyika, 1961–64, and Zanzibar, 1963–64, prior to union as one republic under new name. [18]As Yemen, 1947–90, and Democratic Yemen, 1967–90, prior to merger of the two. [19]As United Arab Republic, by union with Egypt, 1958–61.

The USA is the leading contributor to the Peacekeeping Operations Budget, with 27·1415% of the total in 2012, followed by Japan (12·5300%), UK (8·1474%), Germany (8·0180%), France (7·5540%), Italy (4·9990%), China (3·9343%), Canada (3·2070%), Spain (3·1770%) and South Korea (2·2600%). All other countries contribute less than 2%.

Publications. Yearbook of the United Nations. New York, 1947 ff.—United Nations Chronicle. Quarterly.—Monthly Bulletin of Statistics.—General Assembly: Official Records: Resolutions.—Reports of the Secretary-General of the United Nations on the Work of the Organization. 1946 ff.—Charter of the United Nations and Statute of the International Court of Justice.—Official Records of the Security Council, the Economic and Social Council, Trusteeship Council and the Disarmament Commission.—Demographic Yearbook. New York.—The United Nations Today. New York, 2008.—Statistical Yearbook. New York, 1947 ff.—Yearbook of International Statistics. New York, 1947 ff.—Economic Survey of Asia and the Far East. New York, 1946 ff.—Economic Survey of Latin America. New York, 1948 ff.—Economic Survey of Europe. New York, 1948 ff.—Economic Survey of Africa. New York, 1960 ff.—United Nations Reference Guide in the Field of Human Rights. UN Centre for Human Rights, 1993.

Further Reading

Baehr, Peter R. and Gordenker, Leon, The United Nations: Reality and Ideal. 2005

Beigbeder, Y., The Internal Management of United Nations Organizations: the Long Quest for Reform. 1996

Carnegie Commission on Preventing Deadly Conflict, Preventing Deadly Conflict: Final Report with Executive Summary. 1997

Cortright, D. and Lopez, G. A., The Sanctions Decade: Assessing UN Strategies in the 1990s. 2000

Durch, W. J., The Evolution of UN Peacekeeping: Case Studies and Comparative Analysis. 1993

Gareis, S. B., The United Nations: An Introduction. 2nd ed. 2012

Hoopes, T. and Brinkley, D., FDR and the Creation of the UN. 1998

Kennedy, Paul, The Parliament of Man: The Past, Present, and Future of the United Nations. 2006

Knight, W. Andy, Adapting the United Nations to a Postmodern Era. 2nd ed. 2005

Meisler, S., United Nations: The First Fifty Years. 1998

New Zealand Ministry of Foreign Affairs and Trade, United Nations Handbook 2012/13. 2012

Parsons, A., From Cold War to Hot Peace: UN Interventions, 1947–94. 1995

Price, Richard and Zacher, Mark W., United Nations and Global Security. 2004

Pugh, M., The UN, Peace and Force. 1997

Simma, B. (ed.) The Charter of the United Nations: a Commentary. 1995

Universal Declaration of Human Rights

On 10 Dec. 1948 the General Assembly of the United Nations adopted and proclaimed the Universal Declaration of Human Rights.

Preamble

Whereas recognition of the inherent dignity and of the equal and inalienable rights of all members of the human family is the foundation of freedom, justice and peace in the world,

Whereas disregard and contempt for human rights have resulted in barbarous acts which have outraged the conscience of mankind, and the advent of a world in which human beings shall enjoy freedom of speech and belief and freedom from fear and want has been proclaimed as the highest aspiration of the common people,

Whereas it is essential, if man is not to be compelled to have recourse, as a last resort, to rebellion against tyranny and oppression, that human rights should be protected by the rule of law,

Whereas it is essential to promote the development of friendly relations between nations,

Whereas the peoples of the United Nations have in the Charter reaffirmed their faith in fundamental human rights, in the dignity and worth of the human person and in the equal rights of men and women and have determined to promote social progress and better standards of life in larger freedom,

Whereas Member States have pledged themselves to achieve, in co-operation with the United Nations, the promotion of universal respect for and observance of human rights and fundamental freedoms,

Whereas a common understanding of these rights and freedoms is of the greatest importance for the full realization of this pledge,

Now, Therefore THE GENERAL ASSEMBLY proclaims THIS UNIVERSAL DECLARATION OF HUMAN RIGHTS as a common standard of achievement for all peoples and all nations, to the end that every individual and every organ of society, keeping this Declaration constantly in mind, shall strive by teaching and education to promote respect for these rights and freedoms and by progressive measures, national and international, to secure their universal and effective recognition and observance, both among the peoples of Member States themselves and among the peoples of territories under their jurisdiction.

Article 1. All human beings are born free and equal in dignity and rights. They are endowed with reason and conscience and should act towards one another in a spirit of brotherhood.

Article 2. Everyone is entitled to all the rights and freedoms set forth in this Declaration, without distinction of any kind, such as race, colour, sex, language, religion, political or other opinion, national or social origin, property, birth or other status. Furthermore, no distinction shall be made on the basis of the political, jurisdictional or international status of the country or territory to which a person belongs, whether it be independent, trust, non-self-governing or under any other limitation of sovereignty.

Article 3. Everyone has the right to life, liberty and security of person.

Article 4. No one shall be held in slavery or servitude; slavery and the slave trade shall be prohibited in all their forms.

Article 5. No one shall be subjected to torture or to cruel, inhuman or degrading treatment or punishment.

Article 6. Everyone has the right to recognition everywhere as a person before the law.

Article 7. All are equal before the law and are entitled without any discrimination to equal protection of the law. All are entitled to equal protection against any discrimination in violation of this Declaration and against any incitement to such discrimination.

Article 8. Everyone has the right to an effective remedy by the competent national tribunals for acts violating the fundamental rights granted him by the constitution or by law.

Article 9. No one shall be subjected to arbitrary arrest, detention or exile.

Article 10. Everyone is entitled in full equality to a fair and public hearing by an independent and impartial tribunal, in the determination of his rights and obligations and of any criminal charge against him.

Article 11. (1) Everyone charged with a penal offence has the right to be presumed innocent until proved guilty according to law in a public trial at which he has had all the guarantees necessary for his defence.

(2) No one shall be held guilty of any penal offence on account of any act or omission which did not constitute a penal offence, under national or international law, at the time when it was committed. Nor shall a heavier penalty be imposed than the one that was applicable at the time the penal offence was committed.

Article 12. No one shall be subjected to arbitrary interference with his privacy, family, home or correspondence, nor to attacks upon his honour and reputation. Everyone has the right to the protection of the law against such interference or attacks.

Article 13. (1) Everyone has the right to freedom of movement and residence within the borders of each state.

(2) Everyone has the right to leave any country, including his own, and to return to his country.

Article 14. (1) Everyone has the right to seek and enjoy in other countries asylum from persecution.

(2) This right may not be invoked in the case of prosecutions genuinely arising from non-political crimes or from acts contrary to the purposes and principles of the United Nations.

Article 15. (1) Everyone has the right to a nationality.

(2) No one shall be arbitrarily deprived of his nationality nor denied the right to change his nationality.

Article 16. (1) Men and women of full age, without any limitation due to race, nationality or religion, have the right to marry and to found a family. They are entitled to equal rights as to marriage, during marriage and at its dissolution.

(2) Marriage shall be entered into only with the free and full consent of the intending spouses.

(3) The family is the natural and fundamental group unit of society and is entitled to protection by society and the State.

Article 17. (1) Everyone has the right to own property alone as well as in association with others.

(2) No one shall be arbitrarily deprived of his property.

Article 18. Everyone has the right to freedom of thought, conscience and religion; this right includes freedom to change his religion or belief, and freedom, either alone or in community with others and in public or private, to manifest his religion or belief in teaching, practice, worship and observance.

Article 19. Everyone has the right to freedom of opinion and expression; this right includes freedom to hold opinions without interference and to seek, receive and impart information and ideas through any media and regardless of frontiers.

Article 20. (1) Everyone has the right to freedom of peaceful assembly and association.

(2) No one may be compelled to belong to an association.

Article 21. (1) Everyone has the right to take part in the government of his country, directly or through freely chosen representatives.

(2) Everyone has the right of equal access to public service in his country.

(3) The will of the people shall be the basis of the authority of government; this will shall be expressed in periodic and genuine elections which shall be by universal and equal suffrage and shall be held by secret vote or by equivalent free voting procedures.

Article 22. Everyone, as a member of society, has the right to social security and is entitled to realization, through national effort and international co-operation and in accordance with the organization and resources of the State, of the economic, social and cultural rights indispensable for his dignity and the free development of his personality.

Article 23. (1) Everyone has the right to work, to free choice of employment, to just and favourable conditions of work and to protection against unemployment.

(2) Everyone, without any discrimination, has the right to equal pay for equal work.

(3) Everyone who works has the right to just and favourable remuneration ensuring for himself and his family an existence worthy of human dignity, and supplemented, if necessary, by other means of social protection.

(4) Everyone has the right to form and to join trade unions for the protection of his interests.

Article 24. Everyone has the right to rest and leisure, including reasonable limitation of working hours and periodic holidays with pay.

Article 25. (1) Everyone has the right to a standard of living adequate for the health and well-being of himself and his family, including food, clothing, housing and medical care and necessary social services, and the right to security in the event of unemployment, sickness, disability, widowhood, old age or other lack of livelihood in circumstances beyond his control.

(2) Motherhood and childhood are entitled to special care and assistance. All children, whether born in or out of wedlock, shall enjoy the same social protection.

Article 26. (1) Everyone has the right to education. Education shall be free, at least in the elementary and fundamental stages. Elementary education shall be compulsory. Technical and professional education shall be made generally available and higher education shall be equally accessible to all on the basis of merit.

(2) Education shall be directed to the full development of the human personality and to the strengthening of respect for human rights and fundamental freedoms. It shall promote understanding, tolerance and friendship among all nations, racial or religious groups, and shall further the activities of the United Nations for the maintenance of peace.

(3) Parents have a prior right to choose the kind of education that shall be given to their children.

Article 27. (1) Everyone has the right freely to participate in the cultural life of the community, to enjoy the arts and to share in scientific advancement and its benefits.

(2) Everyone has the right to the protection of the moral and material interests resulting from any scientific, literary or artistic production of which he is the author.

Article 28. Everyone is entitled to a social and international order in which the rights and freedoms set forth in this Declaration can be fully realized.

Article 29. (1) Everyone has duties to the community in which alone the free and full development of his personality is possible.

(2) In the exercise of his rights and freedoms, everyone shall be subject only to such limitations as are determined by law solely for the purpose of securing due recognition and respect for the rights and freedoms of others and of meeting the just requirements of morality, public order and the general welfare in a democratic society.

(3) These rights and freedoms may in no case be exercised contrary to the purposes and principles of the United Nations.

Article 30. Nothing in this Declaration may be interpreted as implying for any State, group or person any right to engage in any activity or to perform any act aimed at the destruction of any of the rights and freedoms set forth herein.

United Nations System

Programmes and Funds. Social and economic development, aimed at achieving a better life for people everywhere, is a major part of the UN system of organizations. At the forefront of efforts to bring about such progress is the United Nations Development Programme (UNDP), the UN's global development network, advocating for change and connecting countries to knowledge, experience and resources to help people build a better life. In 2010–11, 150 countries were working with UNDP to implement 323 projects representing a total investment of US$5·2bn. in environmental and sustainable development priorities in these countries.

UNDP assistance is provided only at the request of governments and in response to their priority needs, integrated into overall national and regional plans. Its activities are funded mainly by voluntary contributions outside the regular UN budget. More than 80% of the UNDP's core programme funds go to countries with an annual per capita GNP of US$750 or less, which are home to 90% of the world's poorest peoples. Headquartered in New York, the UNDP is governed by a 36-member Executive Board, representing both developing and developed countries.

In addition to its regular programmes, UNDP administers various special-purpose funds, such as the *UN Capital Development Fund (UNCDF)*, which offers a unique combination of investment capital, capacity building and technical advisory services to promote microfinance and local development in the Least Developed Countries (LDCs) and the *United Nations Volunteers (UNV)*, which is the UN focal point for promoting and harnessing volunteerism for effective development. Together with the World Bank and the United Nations Environment Programme (UNEP), UNDP is one of the three implementing agencies of the Global Environment Facility (GEF), the world's largest fund for protecting the environment.

UNDP works with governments and local communities on their own solutions to global and national development challenges. In each country office, the UNDP Resident Representative normally also serves as the Resident Coordinator of development activities for the UN system as a whole.

Administrator: Helen Clark (New Zealand).

United Nations development and humanitarian agencies include the *United Nations Children's Fund (UNICEF)*. It was established in 1946 by the United Nations General Assembly as the United Nations International Children's Emergency Fund, to meet the emergency needs of children of post-war Europe. In 1953 the organization became a permanent part of the UN and its mandate was expanded to carry out long-term programmes to benefit children worldwide. Guided by the Convention on the Rights of the Child and its Optional Protocols and the Convention on the Elimination of All Forms of Discrimination Against Women, UNICEF supports low-cost community-based programmes to improve the wellbeing of children and women in more than 150 countries and territories. UNICEF also provides relief and rehabilitation assistance in emergencies.

UNICEF's medium-term strategic plan for 2006–13 focuses on young child survival and development, basic education and gender equality, HIV/AIDS and children, child protection from violence, exploitation and abuse, and policy advocacy and partnerships for children's rights. UNICEF contributes to interventions aimed at reducing under-five mortality rates through proper nutrition, immunization, quality health care, and clean water and sanitation. It is the largest supplier of vaccines to developing countries, reaching 56% of the world's children. The organization supported water, sanitation and hygiene (WASH) activities in nearly 100 countries in 2009. It also promotes universal access to primary education, and eliminating gender-based and other disparities in education.

UNICEF work to combat HIV and AIDS includes increased care and services for children orphaned and made vulnerable by HIV/AIDS, promoting expanded access to treatment for children and women and preventing infections among children and adolescents. In 2009 UNICEF supported prevention of mother-to-child transmission (PMTCT) and paediatric HIV-reduction programmes in 111 countries. UNICEF provides support in areas such as prevention and response to family separation, justice for children, birth registration, ending female genital mutilation/cutting and child marriage, and protecting children from child labour, trafficking and impact of armed conflict and natural disaster.

Executive Director: Anthony Lake (USA).

The *United Nations Population Fund (UNFPA)* became operational in 1969 and is the leading provider of United Nations assistance in the field of population. The Fund extends assistance to developing countries at their request to help them address reproductive health and population issues, and raises awareness of these issues in all countries.

In 2010 UNFPA provided assistance to some 155 countries, areas and territories, with special emphasis on increasing the quality of reproductive health services, ending gender discrimination and violence, formulating effective population policies and reducing the spread of HIV/AIDS.

UNFPA's main objectives are to expand access to comprehensive reproductive health care, including family planning and sexual health, skilled birth attendance and emergency obstetric care, to all couples and individuals in or before 2015. It also supports population and development strategies that enable capacity-building in population programming. UNFPA's strategy focuses on helping individual men and women exercise their right to health. Key to this approach is providing women with more choices through expanded access to education, health services and employment opportunities, and promoting the equal rights of women all over the world. UNFPA's *The State of World Population* report is published annually.

Executive Director: Dr Babatunde Osotimehin (Nigeria).

The *United Nations Environment Programme (UNEP)*, established in 1972, works to encourage sustainable development through sound environmental practices everywhere. UNEP has its headquarters in Nairobi, Kenya and regional offices in Bangkok, Geneva, Manama, Panama City and Washington, D.C. Its activities cover a wide range of issues, from atmosphere and terrestrial ecosystems, to the promotion of environmental science and information, to an early warning and emergency response capacity to deal with environmental disasters and emergencies. UNEP's present priorities include: environmental information, assessment and research; enhanced co-ordination of environmental conventions and development of policy instruments; fresh water; technology transfer and industry; and support to Africa. Information networks and monitoring systems established by the UNEP include: the Global Environment Information Exchange Network (INFOTERRA); Global Resource Information Database (GRID); the International Register of Potentially Toxic Chemicals (IRPTC); and the recent UNEP.net, a web-based interactive catalogue and multifaceted portal that offers access to environmentally relevant geographic, textual and pictorial information. In June 2000 the World Conservation and Monitoring Centre (WCMC) based in Cambridge, UK became UNEP's key biodiversity assessment centre. UNEP's latest state-of-the-environment report is the *UNEP Year Book* 2012.

Executive Director: Achim Steiner (Germany).

Other UN programmes working for development include: the *UN Conference on Trade and Development (UNCTAD)*, which promotes international trade, particularly by developing countries, in an attempt to increase their participation in the global economy; and the *World Food Programme (WFP)*, the world's largest international food aid organization, which is dedicated to both emergency relief and development programmes.

The *United Nations Human Settlements Programme (UN-Habitat)*, which assists over 600m. people living in health-threatening housing conditions, was established in 1978. The 58-member *UN Commission on Human Settlements (UNCHS)*, Habitat's governing body, meets every two years. The Centre serves as the focal point for human settlements action and the co-ordination of activities within the UN system.

The *United Nations Office on Drugs and Crime* (UNODC) educates the world about the dangers of drug abuse; strengthens international action against drug production, trafficking and drug related crime; promotes efforts to reduce drug abuse, particularly among the young and vulnerable; builds local, national and international partnerships to address drug issues; provides information, analysis and expertise on the drug issue; promotes international co-operation in crime prevention and control; supports the development of criminal justice systems; and assists member states in addressing the challenges and threats posed by the changing nature of transnational organized crime.

Executive Director: Yury Fedotov (Russia).

The UN work in crime prevention and criminal justice aims to lessen the human and material costs of crime and its impact on socio-economic development. The UN Congress on the Prevention of Crime and Treatment of Offenders has convened every five years since 1955 and provides a forum for the presentation of policies and progress. The Twelfth Crime Congress (Salvador, 2010) had as its theme 'Comprehensive strategies for global challenges: crime prevention and criminal justice systems and their development in a changing world'. The *Commission on Crime Prevention and Criminal Justice*, a functional body of ECOSOC, established in 1992, seeks to strengthen UN activities in the field, and meets annually in Vienna. The interregional research and training arm of the UN crime and criminal justice programme is the *United Nations Interregional Crime and Justice Research Institute (UNICRI)* in Rome.

Humanitarian assistance to refugees and victims of natural and man-made disasters is also an important function of the UN system. The main refugee organizations within the system are the *Office of the United Nations High Commissioner for Refugees (UNHCR)* and the *United Nations Relief and Works Agency for Palestine Refugees in the Near East (UNRWA).*

UNHCR was created in 1951 to resettle 1·2m. European refugees left homeless in the aftermath of the Second World War. It was initially envisioned as a temporary office with a projected lifespan of three years. However, in 2003, in a move to strengthen UNHCR's capacity to carry out its work more effectively, the General Assembly removed the time limitation on the organization's mandate and extended it indefinitely, until 'the refugee problem is solved'. Today, with some 20·8m. persons of concern across the globe, UNHCR has become one of the world's principal humanitarian agencies. Its Executive Committee currently comprises 87 member states. With its Headquarters in Geneva, UNHCR has a national and international staff of more than 7,600 working in 126 countries. The organization has twice been awarded the Nobel Peace Prize. UNHCR is a subsidiary organ of the United Nations General Assembly.

The work of UNHCR is humanitarian and non-political. International protection is its primary function. Its main objective is to promote and safeguard the rights and interests of refugees. In so doing UNHCR devotes special attention to promoting access to asylum and seeks to improve the legal, material and physical safety of refugees in their country of residence. Crucial to this status is the principle of *non-refoulement*, which prohibits the expulsion from or forcible return of refugees to a country where they may have reason to fear persecution. UNHCR pursues its objectives in the field of protection by encouraging the conclusion of intergovernmental legal instruments in favour of refugees, by supervising the implementation of their provisions and by encouraging governments to adopt legislation and administrative procedures for the benefit of refugees. UNHCR is often called upon to provide material assistance (e.g. the provision of food, shelter, medical care and essential supplies) while durable solutions are being sought. Durable solutions generally take one of three forms: voluntary repatriation, local integration or resettlement in another country.

UNHCR co-operates both multilaterally and bilaterally with a wide range of partners in order to fulfil its mandate for refugees and other people of concern to the Office. Partners include UN co-ordination bodies, other UN agencies and departments, intergovernmental organizations, non-governmental organizations (NGOs), universities and research institutes, regional organizations, foundations and corporate entities from the private sector, as well as governments, host communities and refugee and other displaced population representatives. In response to calls by the international community to improve the global humanitarian response capacity, today UNHCR is playing an active role in the inter-agency 'cluster leadership approach' with respect to protecting and assisting internally displaced persons. UNHCR's involvement is focused on conflict-generated situations of internal displacement, where it leads the protection 'cluster', the camp co-ordination and camp management 'cluster' and the emergency shelter 'cluster'. At present, UNHCR is funded almost entirely by voluntary contributions. In 2010 UNHCR's expenditure amounted to US$1·88bn.

High Commissioner: António Guterres (Portugal).

UNRWA was created by the General Assembly in 1949 as a temporary, non-political agency to provide relief to the nearly 750,000 people who became refugees as a result of the disturbances during and after the creation of the State of Israel in the former British Mandate territory of Palestine. 'Palestine refugees', as defined by UNRWA's mandate, are persons or descendants of persons whose normal residence was Palestine for at least two years prior to the 1948 conflict and who, as a result of the conflict, lost their homes and means of livelihood. UNRWA has also been called upon to help persons displaced by renewed hostilities in the Middle East in 1967. The situation of Palestine refugees in south Lebanon, affected in the aftermath of the 1982

Israeli invasion of Lebanon, was of special concern to the Agency in 1984. UNRWA provides education, health, relief and social services to eligible refugees among the 4·8m. registered Palestine refugees in its five fields of operation: Jordan, Lebanon, Syria, the West Bank and the Gaza Strip. Its mandate is renewed at intervals by the UN General Assembly, and has most recently been extended until 30 June 2014. The regular budget for 2011 amounted to US$624·7m.

Commissioner-General: Filippo Grandi (Italy).

The *Office of the High Commissioner for Human Rights (OHCHR)* represents the world's commitment to universal ideals of human dignity. The UN's activities in the field of human rights are the primary responsibility of the High Commissioner for Human Rights, a post established in 1993 under the direction and authority of the Secretary-General. The High Commissioner is nominated by the Secretary-General for a four-year term, renewable once. The principal co-ordinating human rights organ of the UN was until mid-2006 the 53-member Commission on Human Rights, set up by ECOSOC in 1946. On 15 March 2006 the UN General Assembly voted overwhelmingly to abolish the Commission after it was criticized for having member countries with poor human rights records. A 47-member *Human Rights Council* was established as its successor and held its first session in June 2006.

High Commissioner: Navanethem Pillay (South Africa).

Research and Training Institutes. There are four research and training institutes within the UN, all of them autonomous.

United Nations Institute for Disarmament Research (UNIDIR). Established in 1980 to undertake research on disarmament and security with the aim of assisting the international community in their disarmament thinking, decisions and efforts. Through its research projects, publications, small meetings and expert networks, UNIDIR promotes creative thinking and dialogue on both current and future security issues, through examination of topics as varied as tactical nuclear weapons, refugee security, computer warfare, regional confidence-building measures and small arms.

Address: Palais des Nations, 1211 Geneva 10, Switzerland.
Website: http://www.unidir.org

United Nations Institute for Training and Research (UNITAR). Founded in 1965, UNITAR is the leading UN institute offering training on global and strategic challenges. As an autonomous body within the UN system, UNITAR is led by an Executive Director, governed by a Board of Trustees and is supported by voluntary contributions from governments, intergovernmental organizations, foundations and the private sector. With 80,000 beneficiaries in 2008–09 the Institute provides short executive training to national and local government officials of UN member states and civil society representatives around the world. UNITAR aims to meet the growing demand, especially from the least developed countries, for capacity development in the fields of environment, peace, security and diplomacy, and governance.

Address: Palais des Nations, 1211 Geneva 10, Switzerland.
Website: http://www.unitar.org

United Nations Interregional Crime and Justice Research Institute (UNICRI). Established in 1967 to support countries worldwide in crime prevention and criminal justice, UNICRI offers technical co-operation, research and training at various levels for governments and the international community as a whole. The institute particularly focuses on security and counter-terrorism, counter-trafficking and preventing money laundering.

Address: 10 Viale Maestri del Lavoro, 10127 Turin, Italy.
Website: http://www.unicri.it

United Nations Research Institute for Social Development (UNRISD). Established in 1963 to conduct multidisciplinary research into the social dimensions of contemporary problems affecting development, it aims to provide governments, development agencies, grassroots organizations and scholars with a better understanding of how development policies and processes of economic, social and environmental change affect different social groups.

Address: Palais des Nations, CH-1211 Geneva 10, Switzerland.
Website: http://www.unrisd.org

Other UN Entities. In addition to the operational programmes and funds and the research and training institutes there are a number of other entities that fall within the UN system.

International Computing Centre (ICC). The Centre was established in 1971 as a common service, providing a wide range of Information and Communication Technology Services, on a cost recovery basis, to its users worldwide. More than 25 organizations, funds and programmes of the UN system currently use its services and participate in its governance.

Address: Palais des Nations, CH-1211 Geneva 10, Switzerland.
Website: http://www.unicc.org

Joint UN Programme on HIV/AIDS (UNAIDS). In 1996 the Assembly reviewed implementation of the global strategy for the prevention and control of AIDS, and progress of the Joint UN Programme on HIV/AIDS (UNAIDS), which became operational in 1996. The impact of the HIV/AIDS epidemic is seen to be expanding and intensifying, particularly in developing countries, and new resource mobilization mechanisms were called for to support countries in combating HIV/AIDS. UNAIDS brings together the HIV/AIDS responses of ten co-sponsor UN agencies, providing an overall framework for action and ensuring better co-ordination between its members. The co-sponsor agencies are: International Labour Organization (ILO), Office of the United Nations High Commissioner for Refugees (UNHCR), United Nations Children's Fund (UNICEF), United Nations Development Programme (UNDP), United Nations Educational, Scientific and Cultural Organization (UNESCO), United Nations Office on Drugs and Crime (UNODC), United Nations Population Fund (UNFPA), World Bank, World Food Programme (WFP) and World Health Organization (WHO). The proposed budget for 2012–13 amounted to US$484·8m.

Address: 20 avenue Appia, 1211 CH-Geneva 27, Switzerland.
Website: http://www.unaids.org

UN Office for Project Services (UNOPS). Established in 1995, the self-funding unit provides a range of services for other organizations in the UN system, the private sector, NGOs and academic institutions. Services offered include procurement, recruitment and human resources, and loan supervision.

Address: Midtermolen 3, PO Box 2695, 2100 Copenhagen, Denmark.
Website: http://www.unops.org

United Nations System Staff College (UNSSC). Established in 2002 as the pre-eminent learning arm of the UN, the College develops, co-ordinates and provides cross-organization training programmes with a view to strengthening collaboration within the UN system and increasing operational effectiveness. The UNSCC reaches an average of 7,000 beneficiaries worldwide every year.

Address: Viale Maestri del Lavoro 10, 10127 Turin, Italy.
Website: http://www.unssc.org/home

United Nations University (UNU). Sponsored jointly by the UN and UNESCO, UNU is guaranteed academic freedom by a charter approved by the General Assembly in 1973. It is governed by a 28-member Council of scholars and scientists, of whom 24 are appointed by the Secretary-General of the UN and the Director-General of UNESCO. Unlike a traditional university with a campus, students and faculty, it works through networks of collaborating institutions and individuals to undertake multidisciplinary research on problems of human survival, development and welfare; and to strengthen research and training capabilities in developing countries. It also provides postgraduate fellowships and PhD internships to scholars and scientists from developing countries. The University focuses its work within two programme areas: peace and governance, and environment and development.

Address: 5–53–70 Jingumae, Shibuya-ku, Tokyo 150-8925, Japan.
Website: http://www.unu.edu

United Nations Entity for Gender Equality and the Empowerment of Women (UN Women). Established in July 2010, UN Women supports international political negotiations to formulate globally agreed standards for gender equality and helps UN member states to implement those standards by providing expertise and financial support. It merges and builds on the work of four previously distinct parts of the UN system: the Division for the Advancement of Women (DAW), the International Research and Training Institute for the Advancement of Women (INSTRAW), the Office of the Special Adviser on Gender Issues and Advancement of Women (OSAGI) and the United Nations Development Fund for Women (UNIFEM).

Address: 405 East 42nd St., New York, NY 10017, USA.
Website: http://www.unwomen.org

Information. *The UN Statistics Division* in New York provides a wide range of statistical outputs and services for producers and users of statistics worldwide, facilitating national and international policy formulation, implementation and monitoring. It produces printed publications of statistics and statistical methods in the fields of international merchandise trade, national accounts, demography and population, gender, industry, energy, environment, human settlements and disability, as well as general statistics compendiums including the *Statistical Yearbook* and *World Statistics Pocketbook*. Many of its databases are available on CD-ROM and the internet.

Website: http://unstats.un.org

UN Information Centre. Public Inquiries Unit, Department of Public Information, Room GA-57, United Nations Plaza, New York, NY 10017. There are also 63 UN Information Centres in other parts of the world.

Website: http://www.un.org

Specialized Agencies of the UN

The intergovernmental agencies related to the UN by special agreements are separate autonomous organizations which work with the UN and each other through the co-ordinating machinery of the Economic and Social Council. Of these, 19 are 'Specialized Agencies' within the terms of the UN Charter, and report annually to ECOSOC.

Food and Agriculture Organization of the United Nations (FAO)

Origin. In 1943 the International Conference on Food and Agriculture, at Hot Springs, Virginia, set up an Interim Commission, based in Washington, with a remit to establish an organization. Its Constitution was signed on 16 Oct. 1945 in Quebec City. Today, membership totals 192 countries. The European Union was made a member as a 'regional economic integration organization' in 1991.

Aims and Activities. The aims of FAO are to raise levels of nutrition and standards of living; to improve the production and distribution of all food and agricultural products from farms, forests and fisheries; to improve the living conditions of rural populations; and, by these means, to eliminate hunger. Its priority objectives are to encourage sustainable agriculture and rural development as part of a long-term strategy for the conservation and management of natural resources; and to ensure the availability of adequate food supplies, by maximizing stability in the flow of supplies and securing access to food by the poor.

In carrying out these aims, FAO promotes investment in agriculture, better soil and water management, improved yields of crops and livestock, agricultural research and the transfer of technology to developing countries; and encourages the conservation of natural resources and rational use of fertilizers and pesticides; the development and sustainable utilization of marine and inland fisheries; the sustainable management of forest resources and the combating of animal disease. Technical assistance is provided in all of these fields, and in nutrition, agricultural engineering, agrarian reform, development communications, remote sensing for climate and vegetation, and the prevention of post-harvest food losses. In addition, FAO works to maintain global biodiversity with the emphasis on the genetic diversity of crop plants and domesticated animals; and plays a major role in the collection, analysis and dissemination of information on agricultural production and commodities. Finally, FAO acts as a neutral forum for the discussion of issues, and advises governments on policy, through international conferences like the *World Food Summit* in 1996, the *World Food Summit: five years later* in 2002 and the *World Summit on Food Security* in 2009, all held in Rome.

Special FAO programmes help countries prepare for, and provide relief in the event of, emergency food situations, in particular through the rehabilitation of agriculture after disasters. The *Special Programme for Food Security*, launched in 1994, is designed to assist target countries to increase food production and productivity as rapidly as possible, primarily through the widespread adoption by farmers of available improved production technologies, with the emphasis on high-potential areas. FAO provides support for the global co-ordination of the programme and helps attract funds. The *Emergency Prevention System for Transboundary Animal and Plant Pests and Diseases (EMPRES)*, established in 1994, strengthens FAO's existing contribution to the prevention, control and eradication of diseases and pests before they compromise food security, with locusts and rinderpest among its priorities. *The Global Information and Early Warning System (GIEWS)* provides current information on the world food situation and identifies countries threatened by shortages to guide potential donors. The interagency Food Insecurity and Vulnerability Information and Mapping System initiative (FIVIMS) was established in 1997, with FAO as its secretariat. Together with the UN, FAO sponsors the *World Food Programme (WFP)*.

Finance. The budget for the 2012–13 biennium is US$1,005·6m. FAO's Regular Programme budget, financed by contributions from member governments, covers the cost of its secretariat and Technical Co-operation Programme (TCP), and part of the costs of several special programmes.

FAO continues to provide technical advice and support through its field programmes in all areas of food and agriculture, fisheries, forestry and rural development. In 2009 FAO-assisted projects used US$647·1m. from donor agencies and governments for agricultural and rural development projects and emergencies.

Some 82·1% of Field Programme finances were taken from national trust funds. In the same year, FAO itself contributed 9·2% (or US$66·4m.), provided by the Regular Programme budget through its Technical Co-operation Programme and its national and regional programmes for food security. The FAO Investment Centre organizes more than 600 field missions for 140 investment projects in around 100 countries.

Organization. The FAO Conference, composed of all members, meets every other year to determine policy and approve the FAO's budget and programme. The 49-member Council, elected by the Conference, serves as FAO's governing body between conference sessions. Much of its work is carried out by dozens of regional or specialist commissions, such as the Asia-Pacific Fishery Commission, the European Commission on Agriculture and the Commission on Plant Genetic Resources. The Director-General is elected for a renewable six-year term.

> *Headquarters:* Viale delle Terme di Caracalla, 00153 Rome, Italy.
> *Website:* http://www.fao.org
> *Director-General*: José Graziano da Silva (Brazil).

Publications. Unasylva (quarterly), 1947 ff.; *The State of Food and Agriculture* (annual), 1947 ff.; *Animal Health Yearbook* (annual), 1957 ff.; *Statistical Yearbook* (annual), 2004 ff.; *FAO Commodity Review* (annual), 1961 ff.; *Yearbook of Forest Products* (annual), 1947 ff.; *Yearbook of Fishery Statistics* (in two volumes); *FAO Plant Protection Bulletin* (quarterly); *Environment and Energy Bulletin; Food Outlook* (monthly); *The State of World Fisheries and Aquaculture* (annual); *The State of the World's Forests; World Watch List for Domestic Animal Diversity; The State of Food Insecurity in the World.*

International Bank for Reconstruction and Development (IBRD) — The World Bank

Origin. Conceived at the UN Monetary and Financial Conference at Bretton Woods (New Hampshire, USA) in July 1944, the IBRD, frequently called the World Bank, began operations in June 1946, its purpose being to provide funds, policy guidance and technical assistance to facilitate economic development in its poorer member countries. The Group comprises four other organizations: the International Development Association (IDA), which provides interest-free loans and grants to governments of the poorest countries; the International Finance Corporation (IFC), which provides loans, equity and technical assistance to stimulate private sector investment in developing countries; the Multilateral Investment Guarantee Agency (MIGA), which provides guarantees against losses caused by non-commercial risks to investors in developing countries; and the International Centre for Settlement of Investment Disputes (ICSID), which provides international facilities for conciliation and arbitration of investment disputes.

Members. Afghanistan, Albania, Algeria, Angola, Antigua and Barbuda, Argentina, Armenia, Australia, Austria, Azerbaijan, Bahamas, Bahrain, Bangladesh, Barbados, Belarus, Belgium, Belize, Benin, Bhutan, Bolivia, Bosnia and Herzegovina, Botswana, Brazil, Brunei, Bulgaria, Burkina Faso, Burundi, Cambodia, Cameroon, Canada, Cape Verde, Central African Republic, Chad, Chile, China, Colombia, Comoros, Democratic Republic of the Congo, Republic of the Congo, Costa Rica, Côte d'Ivoire, Croatia, Cyprus, Czech Republic, Denmark, Djibouti, Dominica, Dominican Republic, Ecuador, Egypt, El Salvador, Equatorial Guinea, Eritrea, Estonia, Ethiopia, Fiji Islands, Finland, France, Gabon, Gambia, Georgia, Germany, Ghana, Greece, Grenada, Guatemala, Guinea, Guinea-Bissau, Guyana, Haiti, Honduras, Hungary, Iceland, India, Indonesia, Iran, Iraq, Ireland, Israel, Italy, Jamaica, Japan, Jordan, Kazakhstan, Kenya, Kiribati, South Korea, Kosovo, Kuwait, Kyrgyzstan, Laos, Latvia, Lebanon, Lesotho, Liberia, Libya, Lithuania, Luxembourg, Macedonia, Madagascar, Malaŵi, Malaysia, Maldives, Mali, Malta, Marshall Islands, Mauritania, Mauritius, Mexico, Micronesia, Moldova, Mongolia, Montenegro, Morocco, Mozambique, Myanmar, Namibia, Nepal, Netherlands, New Zealand, Nicaragua, Niger, Nigeria, Norway, Oman, Pakistan, Palau, Panama, Papua New Guinea, Paraguay, Peru, Philippines, Poland, Portugal, Qatar, Romania, Russia, Rwanda, St Kitts and Nevis, St Lucia, St Vincent and the Grenadines, Samoa, San Marino, São Tomé e Príncipe, Saudi Arabia, Senegal, Serbia, Seychelles, Sierra Leone, Singapore, Slovakia, Slovenia, Solomon Islands, Somalia, South Africa, South Sudan, Spain, Sri Lanka, Sudan, Suriname, Swaziland, Sweden, Switzerland, Syria, Tajikistan, Tanzania, Thailand, Timor-Leste, Togo, Tonga, Trinidad and Tobago, Tunisia, Turkey, Turkmenistan, Tuvalu, Uganda, Ukraine, United Arab Emirates, UK, USA, Uruguay, Uzbekistan, Vanuatu, Venezuela, Vietnam, Yemen, Zambia, Zimbabwe.

Activities. The Bank obtains its funds from the following sources: capital paid in by member countries; sales of its own securities; sales of parts of its loans; repayments; and net earnings. A resolution of the Board of Governors of 27 April 1988 provides that the paid-in portion of the shares authorized to be subscribed under it will be 3%.

The Bank is self-supporting, raising most of its money on the world's financial markets. In the fiscal year ending 30 June 2012 it achieved an operating income of US$783m. Income totalled US$4·3bn. and expenditure US$3·6bn.

In the fiscal year 2012 the Bank lent US$20·7bn. for 93 new operations. Cumulative lending had totalled US$400bn. by June 2012. In order to eliminate wasteful overlapping of development assistance and to ensure that the funds available are used to the best possible effect, the Bank has organized consortia or consultative groups of aid-giving nations for many countries. These include Bangladesh, Belarus, Bolivia, Bulgaria, Egypt, Ethiopia, Jordan, Kazakhstan, Kenya, Kyrgyzstan, Macedonia, Malaŵi, Mauritania, Moldova, Mozambique, Nicaragua, Pakistan, Peru, Romania, Sierra Leone, Tanzania, the [Palestinian] West Bank and Gaza Strip, Zambia, Zimbabwe and the Caribbean Group for Co-operation in Economic Development.

For the purposes of its analytical and operational work, in 2012 the IBRD characterized economies as follows: low income (average annual *per capita* gross national income of $1,025 or less); lower middle income (between $1,026 and $4,035); upper middle income (between $4,036 and $12,475); and high income ($12,476 or more).

A wide variety of technical assistance is at the core of IBRD's activities. It acts as executing agency for a number of pre-investment surveys financed by the UN Development Programme. Resident missions have been established in 64 developing member countries and there are regional offices for East and West Africa, the Baltic States and South-East Asia, which assist in the preparation and implementation of projects. The Bank maintains a staff college, the *World Bank Institute* in Washington, D.C., for senior officials of member countries.

Access to Information. Effective 1 July 2010, the World Bank Policy on Access to Information marked a pivotal shift in the World Bank's approach to making information available to the public. Now the public can obtain more information about projects under preparation, projects under implementation, analytic and advisory activities, and Board proceedings. The policy also includes a clear process for making information publicly available and a right to appeal if information seekers believe they were improperly or unreasonably denied access to information or there is a public interest case to override an exception that restricts certain information.

Organization. As of Feb. 2013 the Bank had 188 members, each with voting power in the institution, based on shareholding which in turn is based on a country's economic growth. The president is selected by the Bank's Board of Executive Directors. The Articles of Agreement do not specify the nationality of the president but by custom the US Executive Director makes a nomination, and by a long-standing, informal agreement, the president is a US national (while the managing director of the IMF is European). The initial term is five years, with a second of five years or less.

Current Leaders

Jim Yong Kim

Position
President

Introduction
Jim Yong Kim, a medical scientist and anthropologist, became president of the World Bank in July 2012. He is the first development professional to hold the post and has signalled his support for pursuing private enterprise growth alongside environmental sustainability and social protection.

Early Life
Jim Yong Kim was born on 8 Dec. 1959 in Seoul, South Korea, but moved to the USA aged five and became an American citizen. Transferring from Iowa University, he graduated in biology from Brown University in 1982. He received his medical degree in 1991 and a PhD in anthropology in 1993, both from Harvard.

In 1987 he co-founded Partners in Health (PIH), a non-profit organization pioneering community-based health care programmes in low-income areas and developing countries. Working initially in Boston and Haiti, Kim steered the expansion of PIH into Peru in the 1990s. Its success in treating a range of diseases was widely recognized and its model later adopted by the World Health Organization (WHO).

In 2003 Kim stepped down as executive director of PIH to become director of the WHO's HIV/AIDS department, where he led the '3 by 5' initiative to treat 3m. new HIV/AIDS patients in developing countries with antiretroviral drugs by 2005. The programme was credited with greatly accelerating the provision of effective HIV/AIDS treatment and reached its target in 2007.

Kim has held a range of teaching and research posts at Harvard since 1993. In 2006 the University appointed him chair of its department of social medicine (later renamed the department of global health and social medicine), where he developed a global health delivery programme. He has also served as head of the François-Xavier Bagnoud Center for Health and Human Rights at the Harvard School of Public Health, and as chief of the Brigham and Women's Hospital Division of Social Medicine and Health Inequalities.

In 2009 Kim left Harvard to become president of Dartmouth College, where he established a centre for health care delivery science. In March 2012 President Obama nominated him for the presidency of the World Bank and in April 2012 he beat two rivals in an election for the post. Both his opponents—the Nigerian finance minister, Ngozi Okonjo-Iweala, and a Colombian economist, José Antonio Ocampo—had argued that the president should be from a developing nation.

Career in Office
Kim took office on 1 July 2012. In Nov. 2012 he identified tackling climate change as crucial to ending poverty, announcing that the World Bank would integrate environmental sustainability into future development plans. Achieving this while promoting economic growth in emerging nations is set to be his key challenge.

European office: 66 avenue d'Iéna, 75116 Paris, France. *London office:* Milbank Tower, 12th Floor, 21–24 Milbank, London SW1P 4QP, England. *Tokyo office:* 10th Floor, Fukoku Seimei Building, 2-2-2 Uchisaiwai-cho, Chiyoda-ku, Tokyo 100-0011 Japan.

Headquarters: 1818 H St., NW, Washington, D.C., 20433, USA.
Website: http://www.worldbank.org
President: Jim Yong Kim (USA).

Publications. World Bank Annual Report; Summary Proceedings of Annual Meetings; The World Bank and International Finance Company, 1986; *The World Bank Atlas* (annual); *Catalog of Publications,* 1986 ff.; *World Development Report* (annual); *World Bank Economic Review* (thrice yearly); *World Bank and the Environment* (annual); *World Bank News* (weekly); *World Bank Research Observer; World Tables* (annual); *Social Indicators of Development* (annual); *ICSID Annual Report; ICSID Review: Foreign Investment Law Journal* (twice yearly); *Research News* (quarterly).

Further Reading

Phillips, David A., *Reforming the World Bank: Twenty Years of Trial – and Error.* 2011
The World Bank, A Guide to the World Bank. 3rd ed. 2011
Xu, Yi-Chong and Weller, Patrick, *Inside the World Bank: Exploding the Myth of the Monolithic Bank.* 2009

International Development Association (IDA)

A lending agency established in 1960 and administered by the IBRD to provide assistance on concessional terms to the poorest developing countries. Its resources consist of subscriptions and general replenishments from its more industrialized and developed members, special contributions and transfers from the net earnings of IBRD. Officers and staff of the IBRD serve concurrently as officers and staff of the IDA at the World Bank headquarters.

In fiscal year 2012 IDA commitments totalled US$14·8bn.; new commitments totalled 160 new operations. Since 1960 IDA has lent US$255bn. to 108 countries.

Headquarters: 1818 H St., NW, Washington, D.C., 20433, USA.
Website: http://www.worldbank.org/ida
President: Jim Yong Kim (USA).

International Finance Corporation (IFC)

Established in 1956 to help strengthen the private sector in developing countries, through the provision of long-term loans, equity investments, quasi-equity instruments, standby financing, and structured finance and risk management products. It helps to finance new ventures and assist established enterprises as they expand, upgrade or diversify. In partnership with other donors, it provides a variety of technical assistance and advisory services to public and private sector clients. To be eligible for financing, projects must be profitable for investors, must benefit the economy of the country concerned, and must comply with IFC's environmental and social guidelines.

The majority of its funds are borrowed from the international financial markets through public bond issues or private placements. Its authorized capital is US$2·58bn.; total capital at 30 June 2012 was US$20·6bn. IFC committed US$20·4bn. in total financing in fiscal year 2012 and committed 560 projects in 103 countries. It has 184 members.

Headquarters: 2121 Pennsylvania Ave., NW, Washington, D.C., 20433, USA.
Website: http://www.ifc.org
President: Jim Yong Kim (USA).

Publications. Annual Reports; Lessons of Experience (series); *Paths Out of Poverty.*

Multilateral Investment Guarantee Agency (MIGA)

Established in 1988 to encourage the flow of foreign direct investment to, and among, developing member countries, MIGA is the insurance arm of the World Bank. It provides investors with investment guarantees against non-commercial risk, such as expropriation and war, and gives advice to governments on improving climate for foreign investment. It may insure up to 90% of an investment, with a current limit of US$50m. per project. In March 1999 the Council of Governors adopted a resolution for a capital increase for the Agency of approximately US$850m. In addition US$150m. was transferred to MIGA by the World Bank as operating capital. In Feb. 2013 it had 175 member countries. Like IDA and ICSID, it is located at the World Bank headquarters in Washington (*see above*).

Headquarters: 1818 H St., NW, Washington, D.C., 20433, USA.
Website: http://www.miga.org
President: Jim Yong Kim (USA).

International Centre for Settlement of Investment Disputes (ICSID)

Founded in 1966 to promote increased flows of international investment by providing facilities for the conciliation and arbitration of disputes between governments and foreign investors. The Centre does not engage in such conciliation or arbitration. This is the task of conciliators and arbitrators appointed by the contracting parties, or as otherwise provided for in the Convention. Recourse to conciliation and arbitration by members is entirely voluntary.

In Feb. 2013 its Convention had 158 signatory countries. 251 cases had been concluded by it and 169 were pending. Disputes involved a variety of investment sectors: agriculture, banking, construction, energy, health, industrial, mining and tourism.

ICSID also undertakes research, publishing and advisory activities in the field of foreign investment law. Like IDA and MIGA, it is located at the World Bank headquarters in Washington (*see above*).

Headquarters: 1818 H St., NW, MSN U3-301, Washington, D.C., 20433, USA.
Website: http://www.worldbank.org/icsid
President: Jim Yong Kim (USA).
Secretary-General: Meg Kinnear (Canada).

Publications. ICSID Annual Report; News from ICSID; ICSID Review: Foreign Investment Law Journal; Investment Laws of the World; Investment Treaties.

Further Reading

Stone, D. L. and Wright, C., *The World Bank and Governance*. 2006
Woods, N., *The Globalizers: The IMF, the World Bank and Their Borrowers*. 2007

International Civil Aviation Organization (ICAO)

Origin. The Convention providing for the establishment of the ICAO was drawn up by the International Civil Aviation Conference held in Chicago in 1944. A Provisional International Civil Aviation Organization (PICAO) operated for 20 months until the formal establishment of ICAO on 4 April 1947. The Convention on International Civil Aviation superseded the provisions of the Paris Convention of 1919 and the Pan American Convention on Air Navigation of 1928.

Functions. It assists international civil aviation by establishing technical standards for safety and efficiency of air navigation and promoting simpler procedures at borders; develops regional plans for ground facilities and services needed for international flying; disseminates air-transport statistics and prepares studies on aviation economics; fosters the development of air law conventions and provides technical assistance to states in developing civil aviation programmes.

Organization. The principal organs of ICAO are an Assembly, consisting of all members of the Organization, and a Council, which is composed of 36 states elected by the Assembly for three years, which meets in virtually continuous session. In electing these states, the Assembly must give adequate representation to: (1) states of major importance in air transport; (2) states which make the largest contribution to the provision of facilities for the international civil air navigation; and (3) those states not otherwise included whose election would ensure that all major geographical areas of the world were represented. The budget approved for 2012 was $93·1m. CDN.

Headquarters: 999 University St., Montreal, PQ, Canada H3C 5H7.
Website: http://www.icao.int
President of the Council: Roberto Kobeh González (Mexico).
Secretary-General: Raymond Benjamin (France).

Publications. Annual Report of the Council; ICAO Journal (six yearly; quarterly in Russian); ICAO Training Manual; Aircraft Accident Digest; Procedures for Air Navigation Services.

International Fund for Agricultural Development (IFAD)

The idea for an International Fund for Agricultural Development arose at the 1974 World Food Conference. An agreement to establish IFAD entered into force on 30 Nov. 1977, and the agency began its operations the following month. IFAD is an international financial institution and a United Nations specialized agency dedicated to eradicating rural poverty in developing countries. It mobilizes resources from its 167 member countries to provide low-interest loans and grants to help middle and low-income member countries fight poverty in their poor rural communities. IFAD works with national partners to design and implement innovative initiatives that fit within national policies and systems. These enable poor rural people to access the assets, services, knowledge, skills and opportunities they need to overcome poverty. Since starting operations in 1978, IFAD has invested more than US$14·7bn. in 924 projects and programmes that have reached over 400m. people.

Organization. The highest body is the Governing Council, on which all 169 member countries are represented. Operations are overseen by an 18-member Executive Board (with 18 alternate members), which is responsible to the Governing Council. The Fund works with many partner institutions, including the World Bank, regional development banks and financial agencies, and other UN agencies; many of these co-finance IFAD programmes and projects.

Headquarters: Via Paolo di Dono 44, 00142 Rome, Italy.
Website: http://www.ifad.org
President: Kanayo F. Nwanze (Nigeria).

Publications. Annual Report; Polishing the Stone; What Meets the Eye: Images of Rural Poverty.

International Labour Organization (ILO)

Origin. The ILO was established in 1919 under the Treaty of Versailles as an autonomous institution associated with the League of Nations. An agreement establishing its relationship with the UN was approved in 1946, making the ILO the first Specialized Agency to be associated with the UN. An intergovernmental agency with a tripartite structure, in which representatives of governments, employers and workers participate, it seeks through international action to improve labour and living conditions, to promote productive employment and social justice for working people everywhere. On its fiftieth anniversary in 1969 it was awarded the Nobel Peace Prize. In Feb. 2013 it numbered 185 members.

Functions. The ILO's programme and budget set out four strategic objectives for the Organization at the turn of the century: i) to promote and realize fundamental principles and rights at work; ii) to create greater opportunities for women and men to secure decent employment and income; iii) to enhance the coverage and effectiveness of social protection for all; iv) to strengthen tripartism and social dialogue. The International Labour Conference 2011 adopted a budget of US$861·6m. for the 2012–13 biennium.

One of the ILO's principal functions is the formulation of international standards in the form of International Labour Conventions and Recommendations. Member countries are required to submit Conventions to their competent national authorities with a view to ratification. If a country ratifies a Convention it agrees to bring its laws into line with its terms and to report periodically how these regulations are being applied. More than 7,500 ratifications of 189 Conventions had been deposited by 30 June 2012. Procedures are in place to ascertain whether Conventions thus ratified are effectively applied. Recommendations do not require ratification, but member states are obliged to consider them with a view to giving effect to their provisions by legislation or other action. By 30 June 2012 the International Labour Conference had adopted 202 Recommendations.

In June 1998 delegates to the 86th International Labour Conference adopted an ILO Declaration in Fundamental Principles and Rights at Work, committing the Organization's member states to respect the principles inherent in a number of core labour standards: the right of workers and employers to freedom of association and the effective right to collective bargaining, and to work toward the elimination of all forms of forced or compulsory labour, the effective abolition of child labour and the elimination of discrimination in respect of employment and occupation.

Activities. In addition to its research and advisory activities, the ILO now conducts more than 1,000 technical co-operation programmes in over 80 countries with the help of some 60 donor institutions worldwide. The ILO has decentralized most such activities to its regional, area and branch offices in over 40 countries. Decent Work Country Programmes have been established as the main vehicle for delivery of ILO support to countries.

The ILO's standard-setting and technical co-operation are reinforced by an extensive research, training, education and publications programme. It has established two specialized educational institutions: the *International Institute for Labour Studies* in Geneva, and the *International Centre for Advanced Technical and Vocational Training* in Turin.

The *International Institute for Labour Studies* promotes the study and discussion of policy issues. The core theme of its activities is the interaction between labour institutions, development and civil society in a global economy. It identifies emerging social and labour issues by opening up new areas for research and action; and encourages systematic dialogue on social policy between the tripartite constituency of the ILO and the international academic community, and other public opinion-makers.

The *International Training Centre* was set up in 1965 to lead the training programmes implemented by the ILO as part of its technical co-operation activities. Member states and the UN system also call on its resources and experience, and a UN Staff College was established on the Turin Campus in 1996.

In June 2009 the International Labour Conference unanimously adopted a 'Global Jobs Pact' to address the social and employment impact of the recent international financial and economic crisis. The Pact promotes a productive recovery centred on investments, employment and social protection. It provides an internationally agreed basis for policy-making designed to reduce the time lag between economic recovery and a recovery with decent work opportunities.

Organization. The International Labour Conference is the supreme deliberative organ of the ILO; it meets annually in Geneva. National delegations are composed of two government delegates, one employers' delegate and one workers' delegate. The Governing Body, elected by the Conference, is the Executive Council. It is composed of 28 government members, 14 workers' members and 14 employers' members. Ten governments of countries of industrial importance hold permanent seats on the Governing Body. These are: Brazil, China, Germany, France, India, Italy, Japan, Russia, UK and USA. The remaining 18 government members are elected every three years. Workers' and employers' representatives are elected as individuals, not as national candidates.

Headquarters: International Labour Office, 4 route des Morillons, CH-1211 Geneva 22, Switzerland.
Website: http://www.ilo.org
Email: ilo@ilo.org
Director-General: Guy Ryder (United Kingdom).
Governing Body Chairman: Gilles de Robien (France).

Publications include: *International Labour Review; Bulletin of Labour Statistics; Official Bulletin* and *Labour Education; Yearbook of Labour Statistics* (annual); *World of Work Report* (annual); *Global Employment Trends* (annual); *Encyclopaedia of Occupational Health and Safety; Key Indicators of the Labour Market* (*KILM*).

International Maritime Organization (IMO)

Origin. The International Maritime Organization (formerly the InterGovernmental Maritime Consultative Organization) was established as a specialized agency of the UN by a convention drafted in 1948 at a UN maritime conference in Geneva. The Convention became effective on 17 March 1958 when it had been ratified by 21 countries, including seven with at least 1m. gross tons of shipping each. The IMCO started operations in 1959 and changed its name to the IMO in 1982.

Functions. To facilitate co-operation among governments on technical matters affecting merchant shipping, especially concerning safety and security at sea; to prevent and control marine pollution caused by ships; to facilitate international maritime traffic. The IMO is responsible for convening international maritime conferences and for drafting international maritime conventions. It also provides technical assistance to countries wishing to develop their maritime activities, and acts as a depositary authority for international conventions regulating maritime affairs. *The World Maritime University* (*WMU*), at Malmö, Sweden, was established in 1983; the *IMO International Maritime Law Institute* (*IMLI*), at Valletta, Malta and the *IMO International Maritime Academy*, at Trieste, Italy, both in 1989.

Organization. The IMO has 170 members and three associate members. The Assembly, composed of all member states, normally meets every two years. The 40-member Council acts as governing body between sessions. There are four principal committees (on maritime safety, legal matters, marine environment protection and technical co-operation), which submit reports or recommendations to the Assembly through the Council, and a Secretariat. The budget for 2010–11 amounted to £61,151,200.

Headquarters: 4 Albert Embankment, London SE1 7SR, UK.
Website: http://www.imo.org
Email: info@imo.org
Secretary-General: Koji Sekimizu (Japan).

Publication. IMO News.

International Monetary Fund (IMF)

Established in 1945 as an independent organization, the International Monetary Fund began financial operations on 1 March 1947. An agreement of mutual co-operation with the UN came into force on 15 Nov. 1947. The first amendment to the Articles of Agreement, creating the special drawing right (SDR), the IMF's reserve asset, took effect on 28 July 1969. The second amendment took effect on 1 April 1978, and established a new code of conduct for exchange arrangements in the wake of the collapse of the par value system. The third amendment came into force on 11 Nov. 1992; it allows for the suspension of voting and related rights of any member that fails to settle its outstanding obligations to the IMF. The fourth Amendment, which came into force on 10 Aug. 2009, provides for a special one-time allocation of SDRs.

Members. Afghanistan, Albania, Algeria, Angola, Antigua and Barbuda, Argentina, Armenia, Australia, Austria, Azerbaijan, Bahamas, Bahrain, Bangladesh, Barbados, Belarus, Belgium, Belize, Benin, Bhutan, Bolivia, Bosnia and Herzegovina, Botswana, Brazil, Brunei, Bulgaria, Burkina Faso, Burundi, Cambodia, Cameroon, Canada, Cape Verde, Central African Republic, Chad, Chile, China, Colombia, Comoros, Democratic Republic of the Congo, Republic of the Congo, Costa Rica, Côte d'Ivoire, Croatia, Cyprus, Czech Republic, Denmark, Djibouti, Dominica, Dominican Republic, Ecuador, Egypt, El Salvador, Equatorial Guinea, Eritrea, Estonia, Ethiopia, Fiji Islands, Finland, France, Gabon, Gambia, Georgia, Germany, Ghana, Greece, Grenada, Guatemala, Guinea, Guinea-Bissau, Guyana, Haiti, Honduras, Hungary, Iceland, India, Indonesia, Iran, Iraq, Ireland, Israel, Italy, Jamaica, Japan, Jordan, Kazakhstan, Kenya, Kiribati, South Korea, Kosovo, Kuwait, Kyrgyzstan, Laos, Latvia, Lebanon, Lesotho, Liberia, Libya, Lithuania, Luxembourg, Macedonia, Madagascar, Malaŵi, Malaysia, Maldives, Mali, Malta, Marshall Islands, Mauritania, Mauritius, Mexico, Micronesia, Moldova, Mongolia, Montenegro, Morocco, Mozambique, Myanmar, Namibia, Nepal, Netherlands, New Zealand, Nicaragua, Niger, Nigeria, Norway, Oman, Pakistan, Palau, Panama, Papua New Guinea, Paraguay, Peru, Philippines, Poland, Portugal, Qatar, Romania, Russia, Rwanda, St Kitts and Nevis, St Lucia, St Vincent and the Grenadines, Samoa, San Marino, São Tomé e Príncipe, Saudi Arabia, Senegal, Serbia, Seychelles, Sierra Leone, Singapore, Slovakia, Slovenia, Solomon Islands, Somalia, South Africa, South Sudan, Spain, Sri Lanka, Sudan, Suriname, Swaziland, Sweden, Switzerland, Syria, Tajikistan, Tanzania, Thailand, Timor-Leste, Togo, Tonga, Trinidad and Tobago, Tunisia, Turkey, Turkmenistan, Tuvalu, Uganda, Ukraine, United Arab Emirates, UK, USA, Uruguay, Uzbekistan, Vanuatu, Venezuela, Vietnam, Yemen, Zambia, Zimbabwe.

Aims. To promote international monetary co-operation, the expansion of international trade and exchange rate stability; to assist in the removal of exchange restrictions and the establishment of a multilateral system of payments; and to alleviate any serious disequilibrium in members' international balance of payments by making the financial resources of the IMF available to them, usually subject to economic policy conditions.

Activities. The IMF is mandated to oversee the international monetary system and monitor the economic and financial policies of its member countries. The IMF highlights possible risks to domestic and external stability and advises on policy adjustments.

Lending. A core responsibility of the IMF is to provide loans to member countries experiencing balance of payments problems. This financial assistance enables countries to rebuild their international reserves, stabilize their currencies, continue paying for imports and restore conditions for strong economic growth, while undertaking policies to correct underlying problems. Unlike development banks, the IMF does not lend for specific projects.

The IMF has various loan instruments, or 'facilities', that are tailored to address the specific circumstances of its diverse membership. Nonconcessional loans are provided mainly through Stand-By Arrangements (SBAs) and the Extended Fund Facility (which is useful primarily for longer-term needs). The Flexible Credit Line (FCL) was introduced in 2009, for countries with very strong fundamentals, policies and track records of policy implementation.

The IMF also offers special financing facilities for low-income countries. A new Poverty Reduction and Growth Trust, effective from Jan. 2010, incorporates: the Extended Credit Facility, which provides flexible medium-term support; the Standby Credit Facility, which addresses short-term and precautionary needs; and the Rapid Credit Facility, which offers emergency support with limited conditionality. The IMF also provides emergency assistance to support recovery from natural disasters and conflicts, in some cases at concessional interest rates.

A major reform of the IMF's lending facilities took place in March 2009. Conditions linked to IMF loan disbursements are to be better focused and more adequately tailored to the varying strengths of countries' policies and fundamentals. The flexibility of the SBA has been enhanced. In addition, access limits have been doubled, the cost and maturity structure of the Fund's lending has been simplified and its lending facilities have been streamlined.

Technical assistance. The IMF provides technical assistance in its areas of core expertise: macroeconomic policy, tax policy and revenue administration, expenditure management, monetary policy, the exchange rate system, financial sector sustainability, and macroeconomic and financial statistics. About 90% of IMF technical assistance goes to low and lower-middle income countries. The IMF operates eight regional technical assistance centres: in the Pacific (Fiji Islands), the Caribbean (Barbados), four in Africa (Gabon, Mali, Mauritius and Tanzania), the Middle East (Lebanon) and Central America (Guatemala).

Finances. Quota subscriptions from member countries are the IMF's main source of financing. A member's quota is largely determined by its economic position relative to other members; it is also linked to their drawing rights on the IMF, their voting power and their share of SDR allocations. Quotas are reviewed at least every five years, with the most recent review completed in Dec. 2010. The IMF can supplement its resources through borrowing if it believes that resources might fall short of members' needs.

The General Arrangements to Borrow (GAB) and New Arrangements to Borrow (NAB) are credit arrangements between the IMF and a group of member countries and institutions to provide supplementary resources to the IMF to deal with an exceptional situation that poses a threat to the stability of that system. The GAB, established in 1962, enables the IMF to borrow specified amounts of currencies from 11 industrial countries (or their central banks) under certain circumstances, at market-related rates of interest. The potential credit available to the IMF under the GAB totals SDR 17bn., with an additional SDR 1·5bn. available under an associated arrangement with Saudi Arabia. The NAB, which came into effect in 1998, is a set of credit arrangements between the IMF and 26 member countries and institutions. Importantly, the NAB is the facility of first and principal recourse vis-à-vis the GAB. The maximum amount of resources available to the IMF under the NAB and GAB is SDR 34bn.

In April 2009 the G20 agreed to increase the lending resources available to the IMF by up to US$500bn., thereby tripling total pre-crisis lending resources. The increase was to be made through immediate bilateral financing from IMF member countries and by subsequently incorporating this financing into an expanded and more flexible NAB increased by up to US$500bn. This objective was achieved by Sept. 2009.

Bilateral loans. Under such an agreement, the member normally commits to allow the Fund to make drawings up to a specified ceiling during the period for which drawings can be made. In 2009 the IMF signed a number of bilateral loan agreements.

IMF notes. Some official creditors may prefer to invest in paper or notes issued by the IMF. In 2009 the IMF's Executive Board approved a new framework for issuing notes to the official sector. China was the first country to have signed such a note purchase agreement.

SDR allocations. The IMF may allocate SDRs to members in proportion to their IMF quotas. Such an allocation provides each member with a costless asset. There have been three general SDR allocations, made in response to a long-term global need for reserve assets: (i) SDR 9·3bn., distributed in 1970–72; (ii) SDR 12·3bn., distributed in 1979–81; and (iii) SDR 162·1bn., distributed in Aug. 2009. A special one-off allocation of SDRs amounting to SDR 21·4bn. was implemented on 9 Sept. 2009. This allocation was for those countries that joined the Fund after 1981—more than one fifth of the IMF membership—and had never received an SDR allocation.

Governance reform. Implemented on 28 April 2008 this reform aims to make quotas more responsive to economic realities by increasing the representation of fast-growing economies while at the same time giving low-income countries more say in the IMF's decision making. The reform builds on an initial step agreed by the IMF's membership in Sept. 2006 to have *ad hoc* quota increases for four countries—China, South Korea, Mexico and Turkey. In Dec. 2010 the IMF Board of Governors approved a shift in quota share to dynamic emerging markets and developing countries of more than 6% using the current quota formula as the basis. However, the IMF membership failed to approve the reform at its annual meeting in Oct. 2012 and agreement on a revised quota formula has still to be reached.

Organization. The highest authority is the Board of Governors; each member government is represented. The Board of Governors has delegated many of its powers to the 24 executive directors in Washington, D.C., who are appointed or elected by individual member countries or groups of countries. The managing director is selected by the executive directors and serves as chairman of the Executive Board, but may not vote except in case of a tie. The term of office is for five years, but may be extended or terminated at the discretion of the executive directors. The managing director is responsible for the ordinary business of the IMF, under the direction of the executive directors, and supervises a staff of about 2,400. There are three deputy managing directors. As of Feb. 2013 the IMF had 188 members.

The IMF Institute is a specialized department providing training in macroeconomic analysis and policy, and related subjects, for officials of member countries. In addition to training offered in Washington, D.C., the IMF also offers training for country officials through a network of seven regional training institutes and programmes. These are: the IMF-Singapore Regional Training Institute; the Joint Africa Institute (in Tunisia); the Joint China-IMF Training Program (in Dalian, China); the Joint IMF-Arab Monetary Fund Regional Training Program (in the United Arab Emirates); the Joint India-IMF Training Program (in Pune, India); the Joint Regional Training Center for Latin America (in Brazil); and the Joint Vienna Institute (in Austria).

Current Leaders

Christine Lagarde

Position
Managing Director

Introduction
Christine Lagarde became managing director of the IMF in July 2011, against a backdrop of global financial instability and uncertainty. The 11th consecutive European to head the Fund, she promised to give more voting power to developing nations. One of her key challenges has been formulating the IMF response to the continuing sovereign debt crises.

Early Life
Born on 1 Jan. 1956 in Paris, Christine Lagarde attended school in Le Havre, France and Holton-Arms School in Bethesda, Maryland, USA before studying law at the Université de Paris X-Nanterre. After obtaining a masters degree from the Institut d'études politiques d'Aix-en-Provence, in 1981 she joined the Paris office of international law firm Baker & McKenzie, specializing in anti-trust law, employment law, and acquisitions and mergers.

Made a partner in 1987, she served on the executive committee from 1995–2004, moving to Chicago in 1999 when she became chairman of the global executive committee. As a member of a Washington-based think tank, the Centre for Strategic and International Studies, Lagarde headed the US–Poland defence industries working group from 1995–2002, promoting the interests of US and Polish companies.

She returned to France in 2005 to serve as minister for foreign trade in Dominique de Villepin's government, overseeing the growth of exports. She was minister of agriculture and fisheries from May–June 2007, before being appointed finance minister by President Sarkozy. With the onset of the global economic crisis in 2008 she gained a reputation as an astute negotiator, winning praise domestically for her representation of French interests on the international stage. In 2010 she was closely involved in the negotiations over IMF bailout loans for eurozone countries.

IMF managing director Dominique Strauss-Kahn resigned in May 2011, accused of sexual assault, and Lagarde emerged as the favourite to replace him, ahead of Agustín Carstens of Mexico. She was backed by the USA and, despite widely expressed concern over continued European dominance of the Fund, she also won support from the BRIC nations of Brazil, Russia, India and China.

Career in Office
Lagarde took office on 5 July 2011, promising to implement reforms to IMF governance, including more voting power for emerging nations. Her main preoccupation to date, however, has been the containment of sovereign debt crises in economically advanced countries around the world, and particularly in the eurozone where the IMF has contributed financial support for Greece, Ireland and Portugal in an international effort to sustain the viability of the European single currency. She has called for the eurozone to move swiftly towards a fiscal union.

Headquarters: 700 19th St., NW, Washington, D.C., 20431, USA; 1900 Pennsylvania Ave., NW, Washington, D.C., 20431, USA. European office in Paris and regional offices in Tokyo and Warsaw.
Website: http://www.imf.org/external/index.htm
Email: publicaffairs@imf.org
Managing Director: Christine Lagarde (France).

Publications: *Annual Report of the Executive Board; Annual Report on Exchange Arrangements and Exchange Restrictions; International Financial Statistics* (monthly); *IMF Survey* (online); *IMF Economic Review; World Economic Outlook; Global Financial Stability Report;* and *Finance & Development.* More publications information may be found online at: http://www.imf.org/external/pubind.htm.

Further Reading

Copelovitch, Mark S., *The International Monetary Fund in the Global Economy: Banks, Bonds, and Bailouts.* 2010

Dhingra, Vasudha, *The International Monetary Fund and the World Bank: Aspects of Convergence and Divergence.* 2010

Humphreys, N. K., *Historical Dictionary of the International Monetary Fund.* 1994

James, H., *International Monetary Cooperation since Bretton Woods.* 1996

Samans, Richard, Uzan, Marc and Lopez-Claros, Augusto, *The International Monetary System, the IMF, and the G-20: A Great Transformation in the Making?* 2007

Woods, N., *The Globalizers: The IMF, the World Bank and Their Borrowers.* 2007

International Telecommunication Union (ITU)

Origin. Founded in Paris in 1865 as the International Telegraph Union, the International Telecommunication Union took its present name in 1934 and became a specialized agency of the United Nations in 1947. Therefore, the ITU is the world's oldest intergovernmental body.

Functions. To maintain and extend international co-operation for the improvement and rational use of telecommunications of all kinds, and promote and offer technical assistance to developing countries in the field of telecommunications; to promote the development of technical facilities and their most efficient operation to improve the efficiency of telecommunication services, increasing their usefulness and making them, so far as possible, generally available to the public; to harmonize the actions of nations in the attainment of these ends.

Organization. The supreme organ of the ITU is the Plenipotentiary Conference, which normally meets every four years. A 48-member Council, elected by the Conference, meets annually in Geneva and is responsible for ensuring the co-ordination of the four permanent organs at ITU headquarters: the General Secretariat; Radiocommunication Sector; Telecommunication Standardization Sector; and Telecommunication Development Sector. The Secretary-General is also elected by the Conference. ITU has 193 member countries; a further 700 scientific and technical companies, public and private operators, broadcasters and other organizations are also ITU members.

Headquarters: Place des Nations, CH-1211 Geneva 20, Switzerland.
Website: http://www.itu.int
Secretary-General: Hamadoun Touré (Mali).

United Nations Educational, Scientific and Cultural Organization (UNESCO)

Origin. UNESCO's Constitution was signed in London on 16 Nov. 1945 by 37 countries and the Organization came into being in Nov. 1946 on the premise that: 'Since wars begin in the minds of men, it is in the minds of men that the defences of peace must be constructed'. In Feb. 2013 UNESCO had 195 members including the UK, which rejoined in 1997 having left in 1985, and the USA, which rejoined in 2003 having left in 1984. There are also eight associate members that are not members of the UN (Aruba, British Virgin Islands, Cayman Islands, Curaçao, the Faroe Islands, Macao, Sint Maarten and Tokelau).

Aims and Activities. UNESCO's primary objective is to contribute to peace and security in the world by promoting collaboration among the nations through education, science, communication, culture and the social and human sciences in order to further universal respect for justice, democracy, the rule of the law, human rights and fundamental freedoms, affirmed for all peoples by the UN Charter. Africa and gender equality are the Organization's two chief global priorities.

Education. Various activities support and foster national projects to renovate education systems and develop alternative educational strategies towards a goal of lifelong education for all. The World Development Forum in Dakar in 2000 set an agenda for progress towards this aim expressed as six goals. Two of these, attaining universal primary education by 2015 and gender parity in schooling by 2005, were also UN Millennium Development Goals. Three elements define the context for pursuing this purpose: promoting education as a fundamental right, improving the quality of education and stimulating experimentation, innovation and policy dialogue.

Science. UNESCO seeks to promote international scientific co-operation and encourages scientific research designed to improve living conditions and to protect ecosystems. Several international programmes to better understand the Earth's resources towards the advancement of sustainable development have been initiated, including the Man and the Biosphere (MAB) programme, the International Hydrological Programme (IHP), the Intergovernmental Oceanographic Commission (IOC) and the International Geoscience Programme (IGCP).

Culture. Promoting the preservation of heritage, both tangible and intangible, cultural diversity and intercultural dialogue is the principal priority of UNESCO's cultural programmes. UNESCO's World Heritage List, now covering 962 sites around the world, promotes the preservation of monuments, cultural landscapes and natural sites.

Communication. Activities are geared to promoting the free flow of information, freedom of expression, press freedom, media independence and pluralism. Another priority is to promote multilingualism on the internet, bridge the digital divide and help disadvantaged groups in North and South participate in the knowledge societies created through the information and communication technologies. To this end, UNESCO promotes access to public domain information, as well as encouraging the creation of local content.

Social and Human Sciences. UNESCO works to advance knowledge and intellectual co-operation in order to facilitate social transformations conducive to justice, freedom, peace and human dignity. It seeks to identify evolving social trends and develops and promotes principles and standards based on universal values and ethics, such as the *Universal Declaration on the Human Genome and Human Rights* (1997) and the *International Declaration on Human Genetic Data* (2003).

Organization. The General Conference, composed of representatives from each member state, meets biennially to decide policy, programme and budget. A 58-member Executive Board elected by the Conference meets twice a year and there is a Secretariat. In addition, national commissions act as liaison groups between UNESCO and the educational, scientific and cultural life of their own countries. The regular budget for the biennium 2012–13 is US$465m., with significant extra-budgetary contributions for specific programmes provided by both public and private bodies.

There are also ten separate UNESCO institutes and centres: the International Bureau of Education (IBE), in Geneva; the UNESCO Institute for Lifelong Learning (UIL), in Hamburg; the International Institute for Educational Planning (IIEP), in Paris and Buenos Aires; the International Institute for Capacity Building in Africa (IICBA), in Addis Ababa; the International Institute for Higher Education in Latin America and the Caribbean (IESALC), in Caracas; the Institute for Information Technologies in Education (IITE), in Moscow; the UNESCO Institute for Statistics (UIS), in Montreal; the UNESCO International Centre for Technical and Vocational Education and Training (UNEVOC), in Bonn; the UNESCO-IHE Institute for

Water Education (UNESCO-IHE), in Delft; and the International Centre for Theoretical Physics (ICTP), in Trieste.

Headquarters: UNESCO House, 7 Place de Fontenoy, 75352 Paris 07 SP, France; 1 rue Miollis, 75732 Paris Cedex 15, France.
Website: http://www.unesco.org
Director-General: Irina Bokova (Bulgaria).

Periodicals (published quarterly). *Museum International; International Social Science Journal; The UNESCO Courier; Prospects; Copyright Bulletin; World Heritage Review.*

United Nations Industrial Development Organization (UNIDO)

Origin. UNIDO was established by the UN General Assembly in 1966 and became a UN specialized agency in 1985.

Aims and Activities. UNIDO helps developing countries in the formulation of policies and programmes in the field of industrial development; analyses trends, disseminates information and co-ordinates activities in their industrial development; acts as a forum for consultations and negotiations directed towards the industrialization of developing countries; and provides technical co-operation to developing countries for the implementation of their development plans for sustainable industrialization in their public and private sectors.

UNIDO focuses its efforts on three thematic priority areas: poverty reduction through productive activities; trade capacity-building; and energy and the environment. Activities under the thematic priorities are reflected in UNIDO's medium-term programme frameworks and biennial programme documents. They are strictly aligned with the priorities of the current UN Decade for the Eradication of Poverty and relevant multilateral declarations.

Organization. As part of the United Nations common system, UNIDO has the responsibility for promoting industrialization throughout the developing world, in co-operation with its 174 member states. Its headquarters are in Vienna, Austria. In 2010 UNIDO's field network included ten regional offices, 19 country offices, 18 desks, five focal points and one regional centre, thus offering a UNIDO field presence in 53 countries. In addition to this official field structure there are a number of project-funded field units with UNIDO staff, including 13 Investment and Technology Promotion Offices, 43 National Cleaner Production Centres, 14 International Technology Centres, 32 Sub-contracting and Partnership Exchanges and two South–South Cooperation Centres.

The General Conference meets every two years to determine policy and approve the budget. The 53-member Industrial Development Board (membership according to constitutional lists) is elected by the General Conference. The General Conference also elects a 27-member Programme and Budget Committee for two years and appoints a Director-General for four years.

Finance. UNIDO's financial resources come from the regular and operational budgets, as well as voluntary contributions, budgeted for 2012–13 at US$153·2m., US$28·8m. and US$273·1m. respectively, totalling US$455·2m. The regular budget derives mainly from assessed contributions from member states with a marginal proportion provided from such other sources as interest income, sales publications and government contributions to the UNIDO field offices.

The Constitution of UNIDO provides for 6% of the net regular budget to be used for the Regular Programme of Technical Cooperation. These resources are primarily used for supporting the Organization's operational and normative activities. The operational budget derives mainly from support cost income

(of 5–13%) earned from the implementation of technical co-operation activities. Technical co-operation is funded mainly from voluntary contributions from donor countries and institutions as well as UNDP, the Multilateral Fund for the Implementation of the Montreal Protocol, the Global Environment Facility and the Common Fund for Communities.

Headquarters: Vienna International Centre, Wagramerstr. 5, POB 300, A-1400 Vienna, Austria.
Website: http://www.unido.org
Director-General: Kandeh Yumkella (Sierra Leone).

Publications. UNIDOScope (weekly internet newspaper); *UNIDO Annual Report; Industry for Growth into the New Millennium, African Industry 2000: The Challenge of Going Global; Using Statistics for Process Control and Improvement: An Introduction to Basic Concepts and Techniques; Guidelines for Project Evaluation; Practical Appraisal for Industrial Project Applications—Application of Social Cost-Benefit Analysis in Pakistan; Making It* (quarterly); *Manual for the Evaluation of Industrial Projects; Guide to Practical Project Appraisal—Social Benefit-Cost Analysis in Developing Countries; Manual for Small Industrial Businesses: Project Design and Appraisal; Manual for the Preparation of Industrial Feasibility Studies; Manual on Technology Transfer Negotiations; Guidelines for Infrastructure Development Through Build-Operate-Transfer (BOT) Projects; Gearing up for a New Development Agenda; Reforming the UN System: UNIDO's Need-Driven Model; UNIDO Times* (newsletter); *World Directory of Industrial Information Sources; Woodworking Machinery: A Manual on Selection Options; Competition and the World Economy; The International Yearbook of Industrial Statistics 2013; Industrial Development Report 2011.*

Universal Postal Union (UPU)

Origin. The UPU was established in 1875, when the Universal Postal Convention adopted by the Postal Congress of Berne on 9 Oct. 1874 came into force. It has 192 member countries.

Functions. The UPU provides co-operation between postal services and helps to ensure a universal network of up-to-date products and services. To this end, UPU members are united in a single postal territory for the reciprocal exchange of correspondence. A Specialized Agency of the UN since 1948, the UPU is governed by its Constitution, adopted in 1964 (Vienna), and subsequent protocol amendments (1969, Tokyo; 1974, Lausanne; 1984, Hamburg; 1989, Washington; 1994, Seoul; 1999, Beijing; 2004, Bucharest; 2008, Geneva).

Organization. It is composed of a Universal Postal Congress which meets every four years; a 41-member Council of Administration, which meets annually and is responsible for supervising the affairs of the UPU between Congresses; a 40-member Postal Operations Council; and an International Bureau which functions as the permanent secretariat, responsible for strategic planning and programme budgeting. A new UPU body, the Consultative Committee, was created at the Bucharest Congress. This committee represents the external shareholders of the postal sector as well as UPU member countries. The budget for 2011 was 37·2m. Swiss francs.

Headquarters: Weltpoststrasse 4, 3000 Berne 15, Switzerland.
Website: http://www.upu.int
Director-General: Bishar Abdirahman Hussein (Kenya).

Publications. Annual Report; ICTs, new services and transformation of the Post (2010); *Postal Economics in Developing Countries: Posts, Infrastructure of the 21st Century?* (2009); *Postal Statistics* (annual); *Union Postale* (quarterly); *POST*Code* (also in CD-ROM); *Bucharest World Postal Strategy* (2004).

World Health Organization (WHO)

Origin. An International Conference convened by the UN Economic and Social Council to consider a single health organization resulted in the adoption on 22 July 1946 of the Constitution of the World Health Organization, which came into force on 7 April 1948.

Functions. WHO's objective, as stated in the first article of the Constitution, is 'the attainment by all peoples of the highest possible level of health'. As the directing and co-ordinating authority on international health, it establishes and maintains collaboration with the UN, specialized agencies, governments, health administrations, professional and other groups concerned with health. The Constitution also directs WHO to assist governments to strengthen their health services; to stimulate and advance work to eradicate diseases; to promote maternal and child health, mental health, medical research and the prevention of accidents; to improve standards of teaching and training in the health professions, and of nutrition, housing, sanitation, working conditions and other aspects of environmental health. The Organization is also empowered to propose conventions, agreements and regulations, and make recommendations about international health matters; to develop, establish and promote international standards concerning foods, biological, pharmaceutical and similar substances; to revise the international nomenclature of diseases, causes of death and public health practices.

Methods of work. Co-operation in country projects is undertaken only on the request of the government concerned, through the six regional offices of the Organization. Worldwide technical services are made available by headquarters. Expert committees, chosen from the 47 advisory panels of experts, meet to advise the Director-General on a given subject. Scientific groups and consultative meetings are called for similar purposes. To further the education of health personnel of all categories, seminars, technical conferences and training courses are organized, and advisors, consultants and lecturers are provided. WHO awards fellowships for study to nationals of member countries.

Activities. The main thrust of WHO's activities in recent years has been towards promoting national, regional and global strategies for the attainment of the main social target of the member states: 'Health for All in the 21st Century', or the attainment by all citizens of the world of a level of health that will permit them to lead a socially and economically productive life. Almost all countries indicated a high level of political commitment to this goal; and guiding principles for formulating corresponding strategies and plans of action were subsequently prepared.

WHO has organized its responsibilities into four priorities: enhancing global health security, which includes preventing, detecting and containing disease outbreaks, preparing the world for controlling pandemic influenza, combating new diseases such as SARS, preparing for emergencies and responding quickly to minimize death and suffering; accelerating progress on the Millennium Development Goals (MDGs) by reducing maternal and child mortality, tackling the global epidemics of HIV/AIDS, tuberculosis and malaria, promoting safe drinking water and sanitation, promoting gender equality and increasing access to essential medicines; responding to non-communicable disease such as cardiovascular diseases, diabetes and cancers by reducing smoking, promoting a healthy diet and physical activity and reducing violence and road traffic crashes; promoting equity in health through strengthening health systems to reach everyone, particularly the most vulnerable people.

World Health Day is observed on 7 April every year. The 2013 theme for World Health Day was 'High blood pressure'; the theme for 2012 was 'Ageing and health'. World No-Tobacco Day is held on 31 May each year; International Day Against Drug Abuse on 26 June; World AIDS Day on 1 Dec.

The 50th World Health Assembly which met in 1997 adopted numerous resolutions on public health issues. *The World Health Report, 1997: Conquering suffering, enriching humanity* focused on 'non-communicable diseases'. It warned that the human and social costs of cancer, heart disease and other chronic diseases will rise unless confronted now.

The number of cancer cases is expected to increase by 37% between 2007 and 2030. The incidence of lung cancers in women and prostate cancers in men in the Western world is becoming far more prevalent. The incidence of other cancers is also rising rapidly, especially in developing countries. Heart disease and stroke, the leading causes of death in richer nations, will become more common in poorer countries. The number of people affected by diabetes has risen from 135m. in 1997 to 220m. in 2009, and is forecast to increase to over 360m. by 2030. There is likely to be a huge rise in some mental and neurological disorders, especially dementias and particularly Alzheimer's disease, which is projected to affect 34m. people by 2025. In 2003 an estimated 450m. people suffered from mental and neurological disorders. Dementia affected an estimated 24m. people in 2007 and some 50m. worldwide suffered from epilepsy.

These projected increases are reported to be owing to a combination of factors, not least population ageing and the rising prevalence of unhealthy lifestyles. Average life expectancy at birth globally reached 68 years in 2008. It is now well over 70 years in many countries and exceeds 80 years in some. In 2010 there were an estimated 524m. people over 65. By 2030 that number is expected to rise to 976m., representing nearly 12% of the world's population.

According to WHO, the top ten causes of death in the world in 2008 were: coronary heart disease, 7·3m. deaths; stroke and other cerebrovascular disease, 6·2m.; lower respiratory infections, 3·5m.; chronic obstructive pulmonary disease, 3·3m.; diarrhoeal diseases, 2·5m.; HIV/AIDS, 1·8m.; trachea, bronchus and lung cancers 1·4m.; tuberculosis, 1·3m.; diabetes mellitus, 1·3m.; road traffic accidents 1·2m. In total, tobacco use is responsible for the death of almost one in ten adults worldwide.

In response, WHO has called for an intensified and sustained global campaign to encourage healthy lifestyles and attack the main risk factors responsible for many of these diseases: unhealthy diet, inadequate physical activity, smoking and obesity.

The WHO Framework Convention on Tobacco Control (WHO FCTC) was developed in response to the globalization of the tobacco epidemic, and is the first global health treaty negotiated under the auspices of the World Health Organization. The provisions in the Treaty require countries to ban tobacco advertising, sponsorship and promotion; establish new packaging and labelling of tobacco products with prominent health warnings; establish smoking bans in public places, increase price and tax on tobacco products; and strengthen legislation to clamp down on tobacco smuggling, among other measures.

World Health Report, 2010: Health Systems Financing – The Path to Universal Coverage assesses the way health care is financed and what countries can do to modify their financing systems to ensure that all people can use health services, while being protected against financial hardship associated with paying for them. The report proposes ways the international community can better support efforts in low income countries to achieve universal coverage. Emphasis is placed on improving efficiency and moving away from direct payments towards prepayment and pooling.

Organization. The principal organs of WHO are the World Health Assembly, the Executive Board and the Secretariat. Each of the 194 member states has the right to be represented at the Assembly, which meets annually in Geneva. The 34-member Executive Board is composed of technically qualified health experts designated by as many member states as elected by the Assembly. The Secretariat consists of technical and administrative staff headed by a Director-General, who is appointed for not more than two five-year terms. Health activities in member countries are carried out through regional organizations which have been established in Africa (Brazzaville), South-East Asia (New Delhi), Europe (Copenhagen), Eastern Mediterranean (Cairo) and Western Pacific (Manila). The Pan American Sanitary Bureau in Washington serves as the regional office of WHO for the Americas. It is the oldest international health agency in the world and is the secretariat of the Pan American Health Organization (PAHO). Co-operation in country projects is undertaken only at the request of the government concerned, through the six regional offices.

Finance. The total two-year budget planned for 2010–11 was US$5·4bn.

Current Leaders
Margaret Chan Fung Fu-chun

Position
Director-General

Introduction
Dr Margaret Chan was appointed Director-General of WHO on 9 Nov. 2006 to secure a scheduled five-year term from 4 Jan. 2007 to 30 June 2012. In May 2012 she was reappointed for a second term until the end of June 2017.

Early Life
Chan was born in 1947 in Hong Kong. She graduated in medicine from the University of Western Ontario, Canada in 1977, and joined the Hong Kong department of health as a medical officer in Dec. 1978. In 1985 she gained an MSc. in public health from the National University of Singapore. In June 1994 she became the health department's first female director. Her nine-year tenure was marked by outbreaks of H5N1 avian influenza in 1997 and SARS in 2003.

In 2003 Chan joined WHO as director of the department for protection of the human environment. She was promoted to director of communicable diseases surveillance and response in June 2005 and also became the representative of the director-general for pandemic influenza. In Sept. that year she was appointed assistant director-general for communicable diseases. When Dr Lee Jong-wook died in May 2006, Chan was nominated by China to succeed him as Director-General.

Career in Office
Chan identified improvements in health of women and in Africa as key to her term. However, she courted early controversy when, in Feb. 2007, she was accused of favouring pharmaceutical companies over the sick in developing countries by humanitarian groups lobbying for cheaper generic drugs. Then in April she was criticized for her defence of WHO's refusal to extend membership to Taiwan. In June 2007 new international health regulations obliged governments to report potential pandemics to Chan as WHO Director-General immediately. In June 2009 WHO declared a global swine flu pandemic. The outbreak, first detected in March, had seemingly peaked in many countries by the end of the year, but Chan warned against complacency in tracking the evolution of the virus.

In Jan. 2011 Chan warned that WHO was overextended and facing serious funding shortfalls, necessitating administrative, budgetary and programmatic reform. She said that the organization needed to redirect resources in a more targeted manner, avoiding wasteful duplication with other health financiers and concentrating on areas where it could make the most impact.

Headquarters: 20 avenue Appia, CH-1211 Geneva 27, Switzerland.
Website: http://www.who.int
Director-General: Dr Margaret Chan Fung Fu-chun (China).

Publications. Annual Report on World Health; Bulletin of WHO (6 issues a year); *International Digest of Health Legislation* (quarterly); *Health and*

Safety Guides; International Statistical Classification of Diseases and Related Health Problems; WHO Technical Report Series; WHO AIDS Series; Public Health Papers; World Health Statistics Annual; Weekly Epidemiological Record; WHO Drug Information (quarterly).

World Intellectual Property Organization (WIPO)

Origin. The World Intellectual Property Organization (WIPO) was established in 1967 following the conclusion of the Convention Establishing the World Intellectual Property Organization in Stockholm in 1967. It was given a mandate by its member states to promote the protection of intellectual property (IP) through co-operation among states and in collaboration with other international organizations. The WIPO Convention entered into force on 26 April 1970 and WIPO became a specialized agency of the United Nations in 1974.

Aims and Activities. WIPO administers 24 treaties that deal with different legal and administrative aspects of intellectual property, notably the Paris Convention for the Protection of Industrial Property, the Patent Co-operation Treaty and the Bern Convention for the Protection of Literary and Artistic Works. WIPO is dedicated to developing a balanced and accessible international intellectual property (IP) system that rewards creativity, stimulates innovation and contributes to economic development while safeguarding the public interest.

In Dec. 2008 WIPO member states adopted a new strategic framework for the Organization comprising nine strategic goals that are designed to enable WIPO to more effectively respond to an evolving technological, cultural and geo-economic environment. In addition to goals relating to the balanced evolution of the international normative framework of IP, to facilitating the use of IP for development and to the provision of premier global IP services, WIPO's new goals include a focus on building respect for IP; on developing global IP infrastructure; on responsive communication; on becoming a world reference source for IP information; and on addressing IP in relation to global policy challenges such as climate change, public health and food security.

WIPO's activities fall broadly into three clusters of activities, namely: the progressive development of international IP law; IP capacity-building programmes to support the efficient use of IP, particularly in developing countries; and services to industry that facilitate the process of obtaining IP rights in multiple countries. Also, alternative dispute resolution options for private parties are available through the WIPO Arbitration and Mediation Center. The Center provides services under the Uniform Domain Name Dispute Resolution Policy (UDRP) designed to discourage and resolve the abusive registration of trademarks as domain names. WIPO also facilitates access to a number of IP-related databases and search services, including PATENSCOPE—a valuable technical resource that provides access to information about new technologies that are often disclosed for the first time as international patent applications.

Organization. As at Feb. 2013 WIPO had 185 member states. WIPO is unique among the family of UN organizations in that it is largely self-financing. The budget for the 2010–11 biennium was 618m. Swiss francs. Over 90% of the Organization's budget comes from earnings derived from the services that WIPO provides to industry and the private sector. The remainder of the budget is made up mainly of revenue generated by WIPO's Arbitration and Mediation Center, the sale of publications and contributions from member states.

Official languages: Arabic, Chinese, English, French, Russian and Spanish.
Headquarters: 34 chemin des Colombettes, CH-1211 Geneva 20, Switzerland.
Website: http://www.wipo.int
Director-General: Francis Gurry (Australia).

Periodicals. Industrial Property and Copyright (monthly, bi-monthly, in Spanish); *PCT Gazette* (weekly); *PCT Newsletter* (monthly); *International Designs Bulletin* (monthly); *WIPO Gazette of International Marks* (fortnightly); *Intellectual Property in Asia and the Pacific* (quarterly).

World Meteorological Organization (WMO)

Origin. A 1947 (Washington) Conference of Directors of the International Meteorological Organization (est. 1873) adopted a Convention creating the World Meteorological Organization. The WMO Convention became effective on 23 March 1950 and WMO was formally established. It was recognized as a Specialized Agency of the UN in 1951.

Functions. (1) To facilitate worldwide co-operation in the establishment of networks of stations for the making of meteorological observations as well as hydrological or other geophysical observations related to meteorology, and to promote the establishment and maintenance of meteorological centres charged with the provision of meteorological and related services; (2) to promote the establishment and maintenance of systems for the rapid exchange of meteorological and related information; (3) to promote standardization of meteorological and related observations and ensure the uniform publication of observations and statistics; (4) to further the application of meteorology to aviation, shipping, water problems, agriculture and other human activities; (5) to promote activities in operational hydrology and to further close co-operation between meteorological and hydrological services; and (6) to encourage research and training in meteorology and, as appropriate, to assist in co-ordinating the international aspects of such research and training.

Organization. WMO has 185 member states and six member territories responsible for the operation of their own meteorological services. Congress, which is its supreme body, meets every four years to approve policy, programme and budget, and adopt regulations. The Executive Council meets at least once a year to prepare studies and recommendations for Congress, and supervises the implementation of Congress resolutions and regulations. It has 37 members, comprising the President and three Vice-Presidents, as well as the Presidents of the six Regional Associations (Africa, Asia, South America, North America, Central America and the Caribbean, South-West Pacific, Europe), whose task is to co-ordinate meteorological activity within their regions, and 27 members elected in their personal capacity. There are eight Technical Commissions composed of experts nominated by members of WMO, whose remit includes the following areas: basic systems, climatology, instruments and methods of observation, atmospheric sciences, aeronautical meteorology, agricultural meteorology, hydrology, oceanography and marine meteorology. A permanent Secretariat is maintained in Geneva. There are three regional offices for Africa, Asia and the Pacific, and the Americas. The budget for 2008–11 was 269·8m. Swiss francs.

> *Headquarters:* 7 bis, avenue de la Paix, Case Postale 2300, CH-1211 Geneva 2, Switzerland.
> *Website:* http://www.wmo.int
> *Email:* wmo@wmo.int
> *Secretary-General:* Michel Jarraud (France).

Publications. WMO Bulletin (quarterly); *WMO Annual Report.*

World Tourism Organization (UNWTO)

Origin. Established in 1925 in The Hague as the International Congress of Official Tourist Traffic Associations. Renamed the International Union for Official Tourism Organizations after the Second World War when it moved to Geneva, it was renamed the World Tourism Organization in 1975 and moved its headquarters to Madrid the following year.

The World Tourism Organization became an executing agency of the United Nations Development Programme in 1976 and in 1977 a formal co-operation agreement was signed with the UN itself. With a UN resolution on 23 Dec. 2003 the World Tourism Organization became a specialized agency of the United Nations.

Aims. The World Tourism Organization exists to help nations throughout the world maximize the positive impacts of tourism, such as job creation, new infrastructure and foreign exchange earnings, while at the same time minimizing negative environmental or social impacts.

Membership. The World Tourism Organization has three categories of membership: full membership which is open to all sovereign states; associate membership which is open to all territories not responsible for their external relations; and affiliate membership which comprises a wide range of organizations and companies working either directly in travel and tourism or in related sectors. In Feb. 2013 the World Tourism Organization had 155 full members, seven associate members and more than 400 affiliate members.

Organization. The General Assembly meets every two years to approve the budget and programme of work and to debate topics of vital importance to the tourism sector. The Executive Council is the governing board, responsible for ensuring that the organization carries out its work and keeps within its budget. The World Tourism Organization has six regional commissions—Africa, the Americas, East Asia and the Pacific, Europe, the Middle East and South Asia—which meet at least once a year. Specialized committees of World Tourism Organization members advise on management and programme content.

> *Headquarters:* Capitán Haya 42, 28020 Madrid, Spain.
> *Website:* http://www.unwto.org
> *Secretary-General:* Dr Taleb Rifai (Jordan).

Publications. Yearbook of Tourism Statistics (annual); *Compendium of Tourism Statistics* (annual); *Travel and Tourism Barometer* (3 a year); *UNWTO News* (4 a year); *various others* (about 100 a year).

Other Organs Related to the UN

International Atomic Energy Agency (IAEA)

Origin. An intergovernmental agency, the IAEA was established in 1957 under the aegis of the UN and reports annually to the General Assembly. Its Statute was approved on 26 Oct. 1956 at a conference at UN Headquarters.

Functions. To enhance the contribution of atomic energy to peace, health and prosperity throughout the world; and to ensure that Agency assistance and activities are not used for any military purpose. In addition, under the terms of the Nuclear Non-Proliferation Treaty (NPT), non-nuclear-weapon states are required to allow the IAEA to verify that their nuclear activities are peaceful. Similar responsibilities are given to the IAEA as part of the nuclear-weapon-free zone treaties in Latin America, the South Pacific, Africa and Southeast Asia.

Activities. The IAEA gives advice and technical assistance to developing countries on a wide range of aspects of nuclear power development. In addition, it promotes the use of radiation and isotopes in agriculture, industry, medicine and hydrology through expert services, training courses and fellowships, grants of equipment and supplies, research contracts, scientific meetings and publications. In 2009 support for operational projects for technical co-operation involved 3,694 expert and lecturer assignments, 5,090 meeting participants, 2,493 participants in training courses and 1,532 fellows and scientific visitors.

The IAEA uses technical safeguards procedures to verify that nuclear equipment or materials are used exclusively for peaceful purposes. IAEA safeguards were applied in 2009 in 170 States at 1,128 nuclear facilities, with about 250 inspectors conducting nearly 2,000 inspections. The five nuclear-weapon states recognized by the NPT (China, France, Russia, UK and USA) are not required to accept safeguards but have concluded Voluntary Offer Agreements that permit the IAEA access to some of their civil nuclear activities.

Organization. The Statute provides for an annual General Conference, a 35-member Board of Governors and a Secretariat headed by a Director-General and currently staffed by 2,200 people from over 90 countries. The IAEA had 158 member states in Feb. 2013.

There are also research laboratories in Austria and Monaco. In addition, the *International Centre for Theoretical Physics* was established in Trieste, Italy, in 1964, and is operated jointly by UNESCO and the IAEA.

Headquarters: Vienna International Centre, PO Box 100, A-1400 Vienna, Austria.
Website: http://www.iaea.org
Director General: Yukiya Amano (Japan).

Publications. For a full list of IAEA publications, visit the website: http://www.iaea.org/Publications/index.html. Publications include: *Annual Report*; *IAEA Bulletin*; *INIS Reference Series*; *Legal Series*; *Nuclear Fusion*; *Nuclear Safety Review*; *International Nuclear Information System (INIS)*; *Technical Directories*; *Technical Reports Series*.

World Trade Organization (WTO)

Origin. The WTO came into being on 1 Jan. 1995. The bulk of the WTO's current work comes from the 1986–94 negotiations called the Uruguay Round and earlier negotiations under the General Agreement on Tariffs and Trade (GATT), which was created in 1948.

Aims and Activities. The WTO agreements have been negotiated and signed by the bulk of the world's trading nations and provide the legal ground rules for international commerce. They act as contracts, binding governments to keep their trade policies within agreed limits. The goal is to help producers of goods and services, exporters and importers conduct their business, while allowing governments to meet social and environmental objectives. The system's overriding purpose is to help trade flow as freely as possible.

The WTO agreements cover goods, services and intellectual property. They spell out the principles of liberalization and the permitted exceptions. They include individual countries' commitments to lower customs tariffs and other trade barriers, and to open and keep open services markets. They set procedures for settling disputes. The agreements are not static; they are renegotiated from time to time and new agreements can be added to the package. The WTO began new negotiations under the 'Doha Development Agenda' launched in Nov. 2001.

Governments are required to make their trade policies transparent by notifying the WTO about laws in force and measures adopted. Various WTO councils and committees seek to ensure that these requirements are being followed and that WTO agreements are being properly implemented. All WTO members must undergo periodic scrutiny of their trade policies and practices, each review containing reports by the country concerned and the WTO Secretariat.

The Dispute Settlement Understanding written into the WTO agreements provides a neutral procedure based on an agreed legal foundation when conflicts of interest arise between trading nations. Countries bring disputes to the WTO if they think their rights under the agreements are being infringed. Judgments by specially appointed independent experts are based on interpretations of the agreements and individual countries' commitments.

Special provision is provided for developing countries, including longer time periods to implement agreements and commitments, measures to increase their trading opportunities and support to help them build their trade capacity, to handle disputes and to implement technical standards. The WTO organizes hundreds of technical co-operation missions to developing countries annually. It also holds numerous courses each year in Geneva for government officials. Aid for Trade aims to help developing countries improve the skills and infrastructure needed to expand their trade.

The WTO maintains regular dialogue with non-governmental organizations, parliamentarians, other international organizations, the media and the general public on various aspects of the WTO and the ongoing Doha negotiations, with the aim of enhancing co-operation and increasing awareness of WTO activities.

Organization. As of Feb. 2013 the WTO had 158 members, accounting for around 95% of world trade. The WTO is run by its member governments and derives its income from annual contributions from its members. All major decisions are made by the membership as a whole, either by ministers (who usually meet at least once every two years) or by their ambassadors or delegates (who meet regularly in Geneva). Day-to-day work in between the ministerial conferences is handled by three bodies: the General Council, the Dispute Settlement Body and the Trade Policy Review Body. All three consist of all the WTO members. The previous GATT Secretariat now serves the WTO, which has no resources of its own other than its operating budget. The budget for 2012 was 196,003,900 Swiss francs.

Headquarters: Centre William Rappard, 154 rue de Lausanne, CH-1211 Geneva 21, Switzerland.
Website: http://www.wto.org
Email: enquiries@wto.org
Director-General: Pascal Lamy (France).

Publications include: *Annual Report*; *World Trade Report*; *International Trade Statistics*; *Trade Policy Reviews*.

Further Reading

Fulton, Richard and Buterbaugh, Kevin, *The WTO Primer: Tracing Trade's Visible Hand through Case Studies*. 2008
Narlikar, Amrita, Daunton, Martin and Stern, Robert M. (eds.) *The Oxford Handbook on the World Trade Organization*. 2012

Preparatory Commission for the Comprehensive Nuclear-Test-Ban Treaty Organization (CTBTO)

The Preparatory Commission for the Comprehensive Nuclear-Test-Ban Treaty Organization (CTBTO Preparatory Commission) is an international organization established by the States Signatories to the Treaty on 19 Nov. 1996. It carries out the necessary preparations for the effective implementation of the Treaty, and prepares for the first session of the Conference of the States Parties to the Treaty.

The Preparatory Commission consists of a plenary body composed of all the States Signatories, and the Provisional Technical Secretariat (PTS). Upon signing the Treaty a state becomes a member of the Commission. Member states oversee the work of the Preparatory Commission and fund its activities. The Commission's main task is the establishment of the 337 facility International Monitoring System and the International Data Centre, its provisional operation and the development of operational manuals. The Comprehensive Nuclear-Test-Ban

Treaty prohibits any nuclear weapon test explosion or any other nuclear explosion anywhere in the world. As of Feb. 2013 the Treaty had 183 States Signatories and 158 ratifications.

See also Nuclear Non-Proliferation Treaty (NPT) on page 73.

Headquarters: Vienna International Centre, PO Box 1200, A-1400 Vienna, Austria.
Website: http://www.ctbto.org
Executive Secretary: Tibor Tóth (Hungary).

Organization for the Prohibition of Chemical Weapons (OPCW)

The OPCW is responsible for the implementation of the Chemical Weapons Convention (CWC), which became effective on 29 April 1997. The principal organ of the OPCW is the Conference of the States Parties, composed of all the members of the Organization.

Given the relative simplicity of producing chemical warfare agents, the verification provisions of the CWC are far-reaching. The routine monitoring regime involves submission by States Parties of initial and annual declarations to the OPCW and initial visits and systematic inspections of declared weapons storage, production and destruction facilities. Verification is also applied to chemical industry facilities which produce, process or consume dual-use chemicals listed in the convention. The OPCW also when requested by any State Party conducts short-notice challenge inspections at any location under its jurisdiction or control of any other State Party.

The OPCW also co-ordinates assistance to any State Party that falls victim of chemical warfare as it fosters international co-operation in the peaceful application of chemistry.

By Feb. 2013 a total of 188 countries and territories were States Parties to the Chemical Weapons Convention.

Headquarters: Johan de Wittlaan 32, 2517 JR The Hague, Netherlands.
Website: http://www.opcw.org
Director General: Ahmet Üzümcü (Turkey).

European Union (EU)

Origin. The Union is founded on the existing European communities set up by the Treaties of Paris (1951) and Rome (1957), supplemented by revisions, the Single European Act in 1986, the Maastricht Treaty on European Union in 1992, the Treaty of Amsterdam in 1997, the Treaty of Nice in 2000 and the Treaty of Lisbon in 2009.

Members. (27). Austria, Belgium, Bulgaria, Cyprus (Greek-Cypriot sector only), the Czech Republic, Denmark, Estonia, Finland, France, Germany, Greece, Hungary, Ireland, Italy, Latvia, Lithuania, Luxembourg, Malta, the Netherlands, Poland, Portugal, Romania, Slovakia, Slovenia, Spain, Sweden and the UK.

History. European disillusionment with nationalism after the Second World War fostered a desire to bind key European states—France and (West) Germany—to each other and prevent future conflict. In 1946 Winston Churchill called for a moral union in the form of a 'united states of Europe'. Unsupported by the British Government, Churchill chaired the 1948 European Congress of The Hague, a meeting of 800 Europeanists that resulted in the creation of the *Council of Europe*, a European assembly of nations whose aim (Art. 1 of the Statute) was: 'to achieve a greater unity between its members for the purpose of

safeguarding and realizing the ideals and principles which are their common heritage'.

The formation of the *Benelux Economic Union* in 1948 provided a model for a regional customs union; the free movement of goods, people, capital and services was achieved by 1960. Further European integration and the eradication of tariff trade barriers were encouraged by the US-financed European Recovery Program (Marshall Plan) and the *Organisation for European Economic Co-operation* (later the *OECD*). Western European co-operation was also spurred on by distrust of Soviet power in the East, leading to the Brussels Treaty of 1948; the collective defence pact that established the *Western European Union (WEU)*. The North Atlantic Treaty of 1949 cemented Western European (and American) security co-operation.

Jean Monnet, a French economic advisor, suggested joint development to solve Franco-German tensions over the industrial power of the Ruhr and Saarland. Monnet's plan was championed by the French foreign minister, Robert Schuman, whose Declaration of 9 May 1950 (now celebrated as Europe Day) proposed the pooling of coal and steel production. Belgium, France, the Federal Republic of Germany, Italy, Luxembourg and the Netherlands signed the Treaty of Paris establishing the *European Coal and Steel Community (ECSC)*, regarded as a first step towards a united Europe. However, the *European Defence Community (EDC)* of 1952 was rejected by the French Parliament, ending hopes for a *European Political Community*. Encouraged by the success of the ECSC, European integrationists pressed for further economic co-operation. The *European Economic Community (EEC)* and the *European Atomic Energy Community (EAEC or Euratom)* were subsequently created under separate treaties signed in Rome on 25 March 1957. The treaties provided for the establishment by stages of a common market with a customs union at its core, the development of common transport and agricultural policies, and the promotion of growth and research in the nuclear industries for peaceful purposes. Euratom was awarded monopoly powers of acquisition of fissile materials for civil purposes (it is not concerned with the military uses of nuclear power).

The executives of the three communities (ECSC, Euratom, EEC) were amalgamated by a treaty signed in Brussels in 1965, forming a single Council and single Commission of the European Communities, today the core of the EU. The Commission is advised on matters relating to Euratom by a Scientific and Technical Committee.

Enlargement. On 30 June 1970 membership negotiations began between the European Community and the UK, Denmark, Ireland and Norway. On 22 Jan. 1972 all four countries signed a Treaty of Accession but Norway rejected membership in a referendum in Nov. The UK, Denmark and Ireland became full members on 1 Jan. 1973 (though Greenland exercised its autonomy under the Danish Crown to secede in 1985). Greece joined on 1 Jan. 1981; Spain and Portugal on 1 Jan. 1986. The former German Democratic Republic entered into full membership on reunification with Federal Germany in Oct. 1990 and, following referenda in favour, Austria, Finland and Sweden became members on 1 Jan. 1995. In a referendum in Nov. 1994 Norway again rejected membership. On 1 May 2004 a further ten countries became members—Cyprus, the Czech Republic, Estonia, Hungary, Latvia, Lithuania, Malta, Poland, Slovakia and Slovenia. On 1 Jan. 2007 Bulgaria and Romania also became members.

Single European Act. The enlarging of the Community resulted in renewed efforts to promote European integration, culminating in the signing in 1986 of the Single European Act. The SEA represented the first major revision of the Treaty of Rome. It provided for greater involvement of the European Parliament in the decision-making process and it extended Qualified Majority

Voting (QMV). The SEA also removed barriers within the EEC to movement and transnational business.

Maastricht Treaty on European Union. Following German reunification, closer European integration was pursued in the political as well as economic spheres. The Maastricht Summit of Dec. 1991 produced a new framework—a European Union based on three 'pillars': a central pillar of the existing European Communities and two supporting pillars based on formal intergovernmental co-operation. One pillar comprised a Common Foreign and Security Policy (CFSP) and the other focused on justice and home affairs, including policing, immigration and law enforcement. Signed in Feb. 1992, the Treaty on European Union laid down a timetable for the creation of a common currency (subject to specific conditions, including an opt-out clause for the UK). The Community Charter of Fundamental Social Rights for Workers, signed in 1989 by all members except the UK, was strengthened by a protocol, allowing member states to use EC institutions to co-ordinate social policy. The UK agreed to abide to the protocol in 1998. Ratification by member states of the Maastricht Treaty proved controversial. In June 1992 it was rejected in a Danish referendum but approved in a second referendum in May 1993. Ratification was finally completed during 1993, with the UK ratifying on 2 Aug. The European Union (EU) came into being officially on 1 Nov. that year.

Treaty of Amsterdam. The Turin Inter-Governmental Conference (IGC) of 1996 failed to advance the reform programme, in part because of the British Conservative Government's opposition to extending EU powers. The election in 1997 of a more Europeanist Labour Government led to the adoption of most of the IGC's proposals at the Amsterdam summit in 1997. Designed to further political integration, the Treaty did little more than adjust the institutions to prepare for EU enlargement. Strengthened policies included police co-operation, freedom of movement and the promotion of employment. Elements of the justice and home affairs 'pillar' were transferred to the Communities. The treaty also allows for member states to progress with selected areas of policy at different rates.

Treaty of Nice. Many of the problems unanswered at Amsterdam were left until the Dec. 2000 IGC at Nice. However, the tense summit failed to find consensus on key institutional reforms. The Treaty included a reweighting of votes in the Council of Ministers, adjustments to the composition of the Commission and several extensions to QMV. Ireland rejected the Treaty in a referendum in June 2001; this was reversed in the referendum in Oct. 2002. The Treaty came into effect on 1 Feb. 2003.

Charter of Fundamental Rights. The Charter, based on the Universal Declaration of Human Rights (UDHR), contains several provisions such as workers' rights and the right to good administration that are not included in the political and civil rights of the European Convention on Human Rights (ECHR). The Treaty was proclaimed by the European Parliament, the Commission and the Council—at the Nice IGC in 2000—but was not incorporated in the Treaty of Nice.

European Convention. In Dec. 2001 the Laeken European Conference adopted the Declaration on the Future of the European Union, committing the EU to becoming more democratic, transparent and effective, while opening the way to a constitution for the people of Europe. The European Council set up a convention, comprising 105 members—chaired by Valéry Giscard d'Estaing, a former French president—to draft the Treaty establishing a Constitution for Europe (the EU constitution), to be ratified by all member states. The Treaty was to confer legal personality on the European Union, giving it the right to represent itself as a single body under international law. The constitution included provision for a President of the European Council, to replace the current six-month rotating presidency, to

be elected by member states for 2½-year terms. The European Parliament was to be granted co-legislative powers in all policy areas with the Council. A new position of Union Minister of Foreign Affairs would have merged the responsibilities of the external relations Commissioner and the High Representative for the CFSP. The constitution incorporated the Charter of Fundamental Rights. Member states would have had reduced powers of veto, although the veto was to have remained in key areas including taxation, defence and foreign policy. Plans for the new constitution to be ready for EU governments to sign after the ten new members joined on 1 May 2004 were dropped when the Brussels summit of Dec. 2003 ended in stalemate over the weighting of voting rights in the Council of Ministers. On 29 Oct. 2004 the Treaty was approved for ratification by the 25 member countries, either by referendum or parliamentary vote. On 29 May 2005, after nine countries had ratified the Treaty, France became the first to reject it; the Netherlands followed suit three days later. Subsequently a further seven countries ratified the Treaty. In addition, Bulgaria and Romania ratified the constitution as part of their preparations for joining the EU. With ratification in the Czech Republic, Denmark, Ireland, Poland, Portugal, Sweden and the UK delayed indefinitely, the treaty's progress stalled.

In June 2007 the European Council initiated talks on a replacement Reform Treaty (the Treaty of Lisbon). Drafting was completed in Oct. 2007, with the intention that the treaty should be signed and ratified by all member governments in time for the European parliamentary elections of June 2009. The treaty removes much of the constitutional terminology in the draft constitution text, reduces the reach of the European charter of human rights and allows for individual states to opt out of certain legislative areas. It does, nonetheless, retain many of the draft constitution's provisions, including a reformed EU presidency, a representative for EU foreign affairs and security and the recognition of the EU as a full legal personality. The treaty was subject to ratification in the parliaments of member countries. There were calls in several countries for the reform to be put to a public vote, but only Ireland was required by its constitution to hold a referendum. The treaty was rejected in the Irish referendum held on 13 June 2008 but accepted in a second referendum on 2 Oct. 2009. On 3 Nov. 2009 the Czech Republic was the final nation to ratify the treaty, which came into effect on 1 Dec. 2009. On 1 Jan. 2010 Herman Van Rompuy, formerly Belgium's prime minister, was sworn in as the first president of the European Council. On 1 Dec. 2009 Catherine Ashton, a British politician, became high representative of the Union for foreign affairs and security policy.

Recent and Future Enlargement. On 15 July 1997 the European Commission adopted *Agenda 2000*, which included a detailed strategy for consolidating the Union through enlargement as far eastwards as Ukraine, Belarus and Moldova. It recommended the early start of accession negotiations with the Czech Republic, Estonia, Hungary, Poland and Slovenia under the provision of Article O of the Maastricht Treaty, whereby 'any European State may apply to become a member of the Union' (subject to the Copenhagen Criteria set by the European Council at its summit in 1993).

In 2002 it was announced that ten countries would be ready to join in 2004: Cyprus, the Czech Republic, Estonia, Hungary, Latvia, Lithuania, Malta, Poland, Slovakia and Slovenia. Following a series of referenda held in 2003 they all became members on 1 May 2004. Bulgaria and Romania signed an accession treaty in April 2005 and became members on 1 Jan. 2007. Entry talks with Croatia began in Oct. 2005—it was recognized as an official candidate country in June 2004—in response to greater co-operation with the International Criminal Tribunal for the former Yugoslavia. Turkey is also hoping to join, but although it became an official candidate country in Dec. 1999 talks on membership that began in Oct. 2005 may take up to 15 years. Switzerland applied for membership in May 1992 but this was rejected by a Swiss referendum later that year. The first Switzerland-EU summit, in 2004, brought Switzerland closer to the EU with a series of bilateral agreements. A referendum in June 2005 approved joining the Schengen Accord (*see* below). The former Yugoslav Republic of Macedonia (FYROM) applied for membership in March 2004. Montenegro applied for membership in Dec. 2008, as did Albania in April 2009, Iceland in July 2009 and Serbia in Dec. 2009. Croatia signed the accession treaty in Dec. 2011 and was scheduled to be the 28th country to join the EU on 1 July 2013. Macedonia became an official candidate country in Dec. 2005, as did Iceland in June 2010, Montenegro in Dec. 2010 and Serbia in March 2012.

Objectives. The Maastricht Treaty claimed the ultimate goal of the EU is 'an ever closer union among the peoples of Europe, in which decisions are taken as closely as possible to the citizen'. However, there are competing views over what that 'union' should be: political confederation or federation or primarily economic union. Priorities include: economic and monetary union; further expansion of the scope of the Communities; implementation of a common foreign and security policy; and development in the fields of justice and home affairs. The Lisbon Strategy, presented in 2000, strives to turn the EU into 'the most competitive and dynamic knowledge-based economy in the world'. Yet tensions remain over how to balance economic growth measures and social welfare provisions.

Structure. The EU's main institutions are: the European Commission, an independent policy-making executive with powers of proposal; the Council of the European Union (known informally as the Council of Ministers), a decision-making body drawn from the national governments (headed by a permanent president since ratification of the Treaty of Lisbon); the European Council, which defines the general political direction and priorities of the EU; the European Parliament, which has joint legislative powers in most policy areas and final say over the EU budget; the Court of Justice of the European Union, including the EU's supreme court; the European Court of Auditors, which checks the financing of the EU's activities; and the European Central Bank, which is responsible for maintaining price stability in the euro area.

Defence. At the 1999 Helsinki European Conference plans were drawn up for the formation of a rapid response capability that could be deployed at short notice. The success of Operation Artemis in 2003 (when an EU peacekeeping force, spearheaded by France, intervened in the humanitarian crisis in the Democratic Republic of the Congo) led to the statement at the Franco-British summit in Nov. 2003 that the EU should be able and willing to deploy forces within 15 days in response to a UN request. The 'EU Battlegroup Concept' was approved in 2004, reached its initial operational capability in Jan. 2005 and its full operational capability in Jan. 2007. The battlegroups are considered to be the smallest self-sufficient military unit capable of stand-alone operations or deployment in the initial phase of larger operations. The battlegroups are particularly suited for tasks such as conflict prevention, evacuation and humanitarian operations; however, they have yet to be deployed. There are currently 18 battlegroups, typically consisting of around 1,500 personnel. Two battlegroups remain on standby at any one time, rotating every six months. The European Union's first ever peacekeeping force (EUFOR) officially started work in Macedonia on 1 April 2003.

The European Union Institute for Security Studies (EUISS) was created by a Council Joint Action in July 2001 with the status of an autonomous agency. It contributes to the development of the Common Foreign and Security Policy (CFSP) through research and debate on major security and defence issues. The European Defence Agency, headed by Catherine Ashton, the EU's High Representative, was founded by the Council in July 2004 to improve defence co-ordination, especially crisis management.

Major Policy Areas. The major policy areas of the EU were laid down in the 1957 Treaty of Rome, which guaranteed certain rights to the citizens of all member states. Economic discrimination by nationality was outlawed, and member states were bound to apply 'the principle that men and women should receive equal pay for equal work'.

The Single Internal Market. The core of the process of economic integration is characterized by the removal of obstacles to the four fundamental freedoms of movement for persons, goods, capital and services. Under the Treaty, individuals or companies from one member state may establish themselves in another country (for the purposes of economic activity) or sell goods or services there on the same basis as nationals of that country. With a few exceptions, restrictions on the movement of capital have also been ended. Under the Single European Act the member states bound themselves to achieve the suppression of all barriers to free movement of persons, goods and services by 31 Dec. 1992. Since then the economies of the member states have expanded to such an extent that in 2007 the EU replaced the USA as the world's largest economy.

The *Schengen Accord* abolished border controls on persons and goods between certain EU states plus Norway and Iceland. It came into effect on 26 March 1995 and was signed by Austria, Belgium, Denmark, Finland, France, Germany, Greece, Iceland, Italy, Luxembourg, the Netherlands, Norway, Portugal, Spain and Sweden. The ten countries that joined the EU in 2004 signed the treaty, which came into force in nine of the countries (the Czech Republic, Estonia, Hungary, Latvia, Lithuania, Malta, Poland, Slovakia and Slovenia) on 21 Dec. 2007. Only Cyprus has yet to implement it. Switzerland became the 25th country to join the Schengen area on 12 Dec. 2008 and Liechtenstein the 26th on 19 Dec. 2011. Other signatory countries still to implement it are Bulgaria and Romania, which signed as part of their accession terms.

Economic and Monetary Union. The establishment of the single market provided for the next phase of integration: economic and monetary union. The *European Monetary System (EMS)* was founded in March 1979 to control inflation, protect European trade from international disturbances and ultimately promote convergence between the European economies. At its heart was the *Exchange Rate Mechanism (ERM)*. The ERM is run by the finance ministries and central banks of the EU countries on a day-to-day basis; monthly reviews are carried out by the EU Monetary Committee (finance ministries) and the EU Committee of Central Bankers. Sweden is not in the ERM; the UK suspended its membership on 17 Sept. 1992. In Jan. 1995 Austria joined the ERM. Finland followed in 1996, and in Nov. that year the Italian lira, which had been temporarily suspended, was readmitted.

With the introduction of the euro, exchange rates have been fixed for all member countries. The member countries are Austria, Belgium, Cyprus, Estonia, Finland, France, Germany, Greece, Ireland, Italy, Luxembourg, Malta, the Netherlands, Portugal, Slovakia, Slovenia and Spain.

European Monetary Union (EMU). The single European currency with 11 member states came into operation in Jan. 1999, although it was not until 2002 that the currency came into general circulation. Greece subsequently joined in Jan. 2001, Slovenia in Jan. 2007, Cyprus and Malta in Jan. 2008, Slovakia in Jan. 2009 and Estonia in Jan. 2011. Latvia is hoping to become the 18th European Union member to adopt the euro on 1 Jan. 2014. The euro became legal tender from 1 Jan. 2002 across the region (apart from in Cyprus, Estonia, Malta, Slovakia and Slovenia). National currencies were phased out in the 12 countries that were using the euro from 1 Jan. 2002 by the end of Feb. 2002. The eurozone is the world's second largest economy after the USA in terms of output and the largest in terms of trade. EMU currency consists of the euro of 100 cents. EU member countries not in EMU will select a central rate for their currency in consultation with members of the euro bloc and the European Central Bank. The rate is set according to an assessment of each country's chances of joining the eurozone.

An agreement on the legal status of the euro and currency discipline, the Stability and Growth Pact, was reached by all member states at the Dublin summit on 13 Dec. 1996. Financial penalties are meant to be applied to member states running a GDP deficit (negative growth) of up to 0·75%. If GDP falls between 0·75% and 2%, EU finance ministers have discretion as to whether to apply penalties. France and Germany exceeded their deficit limits repeatedly despite the efforts of the Commission to penalize them, resulting in the Council of Ministers' decision to relax the regulations in 2005 in an effort to make the law more enforceable. Members running an excessive deficit are automatically exempt from penalties in the event of a natural disaster or if the fall in GDP is at least 2% over one year.

In Jan. 2012 the Treaty on Stability, Coordination and Governance in the Economic and Monetary Union (also known as a 'fiscal compact') was signed by 25 of the 27 EU member states. It was designed to co-ordinate budget policy across the eurozone in a bid to prevent further unmanageable sovereign debt. The move followed a widespread debt crisis resulting from the global economic downturn, with Ireland, Greece and Portugal all requiring bailouts. The Czech Republic and the UK, both outside the eurozone, declined to sign the treaty.

Environment. The Single European Act made the protection of the environment an integral part of economic and social policies. Public support for EU environmental activism is strong, as evinced by the success of Green parties in the parliamentary elections. Community policy aims to prevent pollution (the Prevention Principle), rectify pollution at source, impose the costs of prevention or rectification on the polluters themselves (the Polluter Pays Principle), and promote sustainable development. The European Environment Agency (*see* below) was established to ensure that policy was based on reliable scientific data.

In 2002 the then 15 EU member states agreed to the 1997 Kyoto Protocol to the United Nations Framework Convention on Climate Change, which committed the EU to reduce its emissions of greenhouse gases by 8% of 1990 levels between 2008–12. By 2010 EU emissions had fallen by 16·8%, although ten of the 15 nations failed to achieve their national targets without the use of the agreement's 'flexible mechanisms' (such as carbon credits).

The Common Agricultural Policy (CAP). The objectives set out in the Treaty of Rome are to increase agricultural productivity, to ensure a fair standard of living for the agricultural community, to stabilize markets, to assure supplies, and to ensure reasonable consumer prices. In Dec. 1960 the Council laid down the fundamental principles on which the CAP is based: a single market, which calls for common prices, stable currency parities and the harmonizing of health and veterinary legislation; Community preference, which protects the single Community market from imports; common financing which seeks to improve agriculture and to stabilize markets against world price fluctuations through market intervention, with levies and refunds on exports. The CAP has made the EU virtually self-sufficient in food.

Following the disappearance of stable currency parities, artificial currency levels have been applied in the CAP. This factor, together with over-production owing to high producer prices, meant that the CAP consumed about two-thirds of the Community budget. In May 1992 it was agreed to reform CAP and to control over-production by reducing the price supports to farmers by 29% for cereals, 15% for beef and 5% for dairy products. In June 1995 the guaranteed intervention price for beef

was decreased by 5%. In July 1996 agriculture ministers agreed a reduction in the set-aside rate for cereals from 10% to 5%. Fruit and vegetable production subsidies were fixed at no more than 4% of the value of total marketed production, rising to 4·5% in 1999. Compensatory grants are made available to farmers who remove land from production or take early retirement. The CAP reform aims to make the agricultural sector more responsive to supply and demand.

Customs Union and External Trade Relations. Goods or services originating in one member state have free circulation within the EU, which implies common arrangements for trade with the rest of the world. Member states can no longer make bilateral trade agreements with third countries; this power has been ceded to the EU. The Customs Union was achieved in July 1968.

In Oct. 1991 a treaty forming the *European Economic Area (EEA)* was approved by the member states of the then EC and European Free Trade Association (EFTA). The EEA consists of the 27 EU members plus Iceland, Liechtenstein and Norway; a Swiss referendum rejected ratification of the EEA in Dec. 1992. Association agreements, which could lead to customs union, have been made with Algeria, Central America, Egypt, Israel, Jordan, Lebanon, Morocco, the Palestinian Authority and Tunisia. Stabilisation and association agreements have also been signed with Albania, Croatia, Macedonia and Montenegro. Customs unions came into force with Andorra in 1991, with Turkey in 1995 and with San Marino in 2002. In 1976 Canada signed a framework agreement for co-operation in industrial trade, science and natural resources, and a transatlantic pact was signed with the USA in 1995. Interim agreements on trade and trade related matters are in place with Bosnia and Herzegovina and Serbia, and the EU has signed free trade agreements with Chile, Colombia, the Faroe Islands, Mexico, Peru, South Africa and South Korea. An economic and commercial agreement has been signed with the Association of South East Asian Nations (ASEAN). Partnership and co-operation agreements exist with a number of eastern European and central Asian countries. In the Development Aid sector, the EU has an agreement (the Cotonou Agreement, signed in 2000, the successor of the Lomé Convention, originally signed in 1975 but renewed and enlarged in 1979, 1984 and 1989) with 77 African, Caribbean and Pacific (ACP) countries that removes customs duties without reciprocal arrangements for most of their imports to the Community. Since 2002, under the terms of the Cotonou Agreement, the EU and all participating ACP nations have been in talks to establish Economic Partnership Agreements (EPAs) designed to end non-reciprocal trade agreements that conflict with WTO rules. The EPAs were scheduled to come into force before the end of 2008 but by Feb. 2013 the only signatories to a formal (as opposed to interim) agreement were 15 Caribbean nations.

The application of common duties has been conducted mainly within the framework of the *General Agreement on Tariffs and Trade (GATT),* which was succeeded in 1995 by the establishment of the World Trade Organization.

Fisheries. The Common Fisheries Policy (CFP) came into effect in Jan. 1983. All EU fishermen have equal access to the waters of member countries (a zone extending up to 200 nautical miles from the shore), with the total allowable catch for each species being set and shared out between member countries according to pre-established quotas. In some cases 'historic rights' apply, as well as special rules to preserve marine biodiversity and ensure sustainable fishing.

A number of agreements are operating with third countries (with Cape Verde, Comoros, Côte d'Ivoire, the Faroe Islands, Greenland, Iceland, Kiribati, Madagascar, Mozambique, São Tomé e Príncipe and the Seychelles) allowing reciprocal fishing rights. When Greenland withdrew from the Community in 1985 EU boats retained their fishing rights subject to quotas and limits, which were revised in 1995 owing to concern about the over-fishing of Greenland halibut.

Transport. Failure to create a common transport policy, as expected by the Treaty of Rome, led in 1982 to parliamentary proceedings against the Council at the Court of Justice. Under the Maastricht Treaty, the Community must contribute to the establishment and development of Trans-European Networks (TENs) in the areas of transport, telecommunications and energy infrastructures. The TEN budget for 2000–06 was €4·6bn. Enlargement into Central and Eastern Europe necessitates a much larger budget; for 2007–13, €8·0bn. has been allocated. Common transport policy includes safety agreements, such as lorry weight and driver hours limits, the easing of border crossings for commercial vehicles and progress towards a common transport market. Rail plans include a 35,000 km high-speed train (HST) network, incorporating France's TGV and Germany's ICE networks.

Competition. The Competition (anti-trust) law of the EU is based on two principles: that businesses should not seek to nullify the creation of the common market by the erection of artificial national (or other) barriers to the free movement of goods; and that there should not be any abuse of dominant positions in any market. These two principles have led to the outlawing of prohibitions on exports to other member states, of price-fixing agreements and of refusal to supply; and to the refusal by the Commission to allow mergers or takeovers by dominant undertakings in specific cases.

Funds. There are two Structural Funds: the *European Regional Development Fund* (promoting economic and social cohesion through the reduction of imbalances between regions or social groups) and the *European Social Fund* (combating unemployment, developing human resources and promoting integration into the labour market). These Structural Funds along with the Cohesion Fund are the financial instruments of EU regional policy, which is intended to narrow the development disparities among regions and member states.

Finances. Around 35% of the EU budget for the period 2007–13 (€348bn.) has been allocated to regional policy; comprising €278bn. for the Structural Funds and €70bn. for the Cohesion Fund (which provides assistance in the fields of the environment and trans-European transport networks to member states with a gross national income below 90% of the EU average).

EU revenue in €1m.:

	2011
Customs duties	22,194·7
GNI-based own resource	88,414·3
VAT-based own resource	14,798·9
Miscellaneous	4,592·1
Total	130,000·0

Expenditure for 2011 was €129,349·9m., of which the expenditure on agriculture markets accounted for €43,817·9m. (33·9% of the total).

The budget for 2014–20 totals €960bn., representing a real terms cut of 3·3% on the 2007–13 budget (the first cut in EU history). The reduction, opposed by many members, was championed by major net-contributors including Finland, Germany, the Netherlands, Sweden and the UK.

The resources of the Community (the levies and duties mentioned above, and up to a 1·4% VAT charge) have been agreed by Treaty. The Budget is made by the Council and the Parliament acting jointly as the Budgetary Authority. The Parliament has control, within a certain margin, of non-obligatory expenditure (where the amount to be spent is not set out in the

legislation concerned), and can also reject the Budget. Otherwise, the Council decides.

Official languages: Bulgarian, Czech, Danish, Dutch, English, Estonian, Finnish, French, German, Greek, Hungarian, Irish, Italian, Latvian, Lithuanian, Maltese, Polish, Portuguese, Romanian, Slovak, Slovenian, Spanish and Swedish.
Website: http://europa.eu

EU Institutions

European Commission

The European Commission consists of 27 members. The Commission President is selected by a consensus of member state heads of government and serves a five-year term. The Commission acts as the EU executive body and as guardian of the Treaties. In this it has the right of initiative (putting proposals to the Council of Ministers for action) and of execution (once the Council has decided). It can take the other institutions or individual countries before the European Court of Justice should any of these fail to comply with European Law. Decisions on legislative proposals made by the Commission are taken in the Council of the European Union. Members of the Commission swear an oath of independence, distancing themselves from partisan influence from any source. The Commission operates through 44 Directorates-General and services.

At the European Summit held in Nice in Dec. 2000 it was decided that from 2005 each EU member state would have one commissioner until there are 27 members (which there have now been since 1 Jan. 2007). According to the Treaty of Lisbon, in 2014 the Commission will comprise representatives from two-thirds of the member states at the time on a rotating basis. However, in Dec. 2008 it was agreed that if the Treaty of Lisbon enters force (which it did on 1 Dec. 2009) a decision will be taken to allow the Commission to retain one national from each member state.

The current Commission took office in Feb. 2010. Members, their nationality and political affiliation (S-Socialist/Social Democrat; C-Christian Democrat/Conservative; L-Liberal; G-Green; Ind-Independent) in Feb. 2013 were as follows:
President: José Manuel Barroso (Portugal, S).
The commissioners are:
Vice-president: Catherine Ashton (UK, S); high representative of the Union for foreign affairs and security policy.
Vice-president: Viviane Reding (Luxembourg, C); responsible for justice, fundamental rights and citizenship.
Vice-president: Joaquín Almunia (Spain, S); responsible for competition.
Vice-president: Siim Kallas (Estonia, L); responsible for transport.
Vice-president: Neelie Kroes (Netherlands, L); responsible for digital agenda.
Vice-president: Antonio Tajani (Italy, C); responsible for industry and entrepreneurship.
Vice-president: Maroš Šefčovič (Slovakia, Ind); responsible for inter-institutional relations and administration.
Vice-president: Olli Rehn (Finland, L); responsible for economic and monetary affairs and the Euro.
Agriculture and Rural Development: Dacian Cioloş (Romania, C).
Climate Action: Connie Hedegaard (Denmark, C).
Development: Andris Piebalgs (Latvia, L).
Education, Culture, Multilingualism and Youth: Androulla Vassiliou (Cyprus, L).

Employment, Social Affairs and Inclusion: László Andor (Hungary, Ind).

Energy: Günther Oettinger (Germany, C).

Enlargement and European Neighbourhood Policy: Štefan Füle (Czech Republic, Ind).

Environment: Janez Potočnik (Slovenia, Ind).

Financial Programming and Budget: Janusz Lewandowski (Poland, C).

Health and Consumer Policy: Tonio Borg (Malta, C).

Home Affairs: Cecilia Malmström (Sweden, L).

Internal Market and Services: Michel Barnier (France, C).

International Co-operation, Humanitarian Aid and Crisis Response: Kristalina Georgieva (Bulgaria, C).

Maritime Affairs and Fisheries: Maria Damanaki (Greece, S).

Regional Policy: Johannes Hahn (Austria, C).

Research, Innovation and Science: Máire Geoghegan-Quinn (Ireland, L).

Taxation and Customs Union, Audit and Anti-Fraud: Algirdas Šemeta (Lithuania, C).

Trade: Karel De Gucht (Belgium, L).

Current Leaders

José Manuel Barroso

Position

President of the European Commission

Introduction

Former prime minister of Portugal and leader of the right-wing Partido Social Democrata (Social Democrats, PSD), José Manuel Barroso was nominated in June 2004 to succeed Romano Prodi as president of the European Commission and his appointment was approved by the European Parliament. He was re-elected for a second term in Sept. 2009. Since taking up office on 23 Nov. 2004 he has been confronted with controversial EU constitutional and membership issues, the fall-out from the global financial crisis and, more recently, debt-induced upheaval in the eurozone.

Early Life

Born on 23 March 1956, Barroso studied law at the Universidade de Lisboa and gained a masters degree in political science from the Université de Genève. He lectured at universities in Geneva and the USA, as well as working for the department of international relations at Universidade Lusíada, Lisbon.

In 1980 Barroso joined the PSD and was elected as a parliamentary deputy from 1985. In the 1990s he held foreign ministry posts before his election in 1999 as PSD leader. In elections that year the Socialist Party under incumbent prime minister António Guterres retained power, while the PSD came second. Two years later, with increasing criticism of the government's heavy public spending, Guterres resigned and elections were brought forward to March 2002. The PSD won a narrow victory and Barroso was appointed prime minister.

He reduced public spending, which affected local authorities' budgets and civil service recruitment, and plans were made to streamline or dissolve numerous state bodies. He also planned to accelerate privatization and introduce labour reforms and imposed an unpopular wage freeze.

In Jan. 2003 Barroso declared support for the efforts of the United States to disarm Iraq but made clear that Portugal would not take part in military action.

In June 2004 EU member states invited Barroso to succeed the incumbent, Romano Prodi, as president of the European Commission. Following his appointment, he resigned as prime minister of Portugal.

Career in Office

Rejection of the proposed new EU constitution in mid-2005 by the French and Dutch electorates in referendums posed a Union-wide dilemma. Barroso sought, particularly in 2007, to reopen the

debate in language more acceptable to national sensibilities. As a result of further negotiations, EU leaders reached an outline agreement in June that year on streamlining the institutional structure and operation of the enlarged Union and gathered in Lisbon in Dec. to sign a new treaty. However, in June 2008 the treaty was rejected by voters in a referendum in Ireland, which again threw the future of the agreement into question. Eventually the Lisbon Treaty was endorsed in a second Irish referendum in Oct. 2009 and entered into force on 1 Dec. that year. Meanwhile, Turkey's EU membership aspirations have continued to generate controversy, with accession negotiations hampered by the country's attitude to the issue of Cypriot sovereignty.

From the second half of 2008 European attention was focused increasingly on combating the global financial crisis and ensuing economic downturn. The fall-out from the crisis persisted, and was compounded in 2010 and 2011 by the growing sovereign debt problem in the eurozone, which prompted often controversial EU intervention in support of some eurozone member states. However, EU action failed to convince the international financial markets of the sustainability of the euro and the turbulence continued into 2012. Nevertheless, Barroso claimed by the end of the year that the 'existential threat against the euro has essentially been overcome'. Earlier, in 2011, he had stressed the need for progress on economic governance and policy co-ordination in the EU, describing it as a 'pragmatic need' rather than 'an ideological dream', and in Sept. 2012 he called for the Union to evolve into a federation of nation states.

In Dec. 2012 Barroso, together with the presidents of the European Council and European Parliament, accepted the Nobel Peace Prize on behalf of the European Union, which had received the annual award for its advancement of peace and reconciliation, democracy and human rights in Europe.

Headquarters: 200 rue de la Loi/Wetstraat, B-1049 Brussels, Belgium.
Secretary-General: Catherine Day (Ireland).

Statistical Office of the European Communities (Eurostat)
Eurostat is a directorate-general of the European Commission. Its mission is to provide the EU with a high-quality statistical service. It receives data collected according to uniform rules from the national statistical institutes of member states, then consolidates and harmonizes the data, before making them available to the public. The data are available from the Eurostat website.

Address: Joseph Bech Building, 5 Rue Alphonse Weicker, L-2721 Luxembourg.
Website: http://epp.eurostat.ec.europa.eu

Council of the European Union (Council of Ministers)

The Council of Ministers consists of ministers from the 27 national governments and is the only institution which directly represents the member states' national interests. It is the Union's principal decision-making body. Here, members legislate for the Union, set its political objectives, co-ordinate their national policies and resolve differences between themselves and other institutions. The presidency rotates every six months. Ireland had the presidency during the first half of 2013 and Lithuania has the presidency during the second half of 2013; Greece will have it during the first half of 2014 and Italy during the second half of 2014. Ireland, Lithuania and Greece are co-operating in a triple presidency over an 18-month period. There is only one Council, but it meets in different configurations depending on the items on the agenda. The meetings are held in Brussels, except in April, June and Oct. when all meetings are in Luxembourg. Around 100 formal ministerial sessions are held each year.

Decisions are taken either by qualified majority vote or by unanimity. Since the entry into force of the Single European Act in 1987 an increasing number of decisions are by majority vote, although some areas such as taxation and social security, immigration and border controls are reserved to unanimity. At the Nice Summit in Dec. 2000 agreement was reached that a further 39 articles of the EU's treaties would move to qualified majority voting. 26 votes were then needed to veto a decision (blocking minority), and member states were allocated the following number of votes: France, Germany, Italy and the UK, 10; Spain, 8; Belgium, Greece, the Netherlands and Portugal, 5; Austria and Sweden, 4; Denmark, Finland and the Republic of Ireland, 3; Luxembourg, 2. During a six-month transitional period that followed the accession of the ten new member states on 1 May 2004 these vote weightings remained unchanged, while the new members were allocated the following number of votes: Poland, 8; Czech Republic and Hungary, 5; Estonia, Latvia, Lithuania, Slovakia and Slovenia, 3; Cyprus and Malta, 2. Since Bulgaria and Romania joined the EU on 1 Jan. 2007 the allocation of vote weightings has been: France, Germany, Italy and the UK, 29; Poland and Spain, 27; Romania, 14; the Netherlands, 13; Belgium, the Czech Republic, Greece, Hungary and Portugal, 12; Austria, Bulgaria and Sweden, 10; Denmark, Finland, the Republic of Ireland, Lithuania and Slovakia, 7; Cyprus, Estonia, Latvia, Luxembourg and Slovenia, 4; Malta 3. A qualified majority will be reached if a majority of member states approve a proposal, the countries supporting the proposal represent at least 62% of the EU's population and a minimum of 255 votes is cast in favour of the proposal. Each member state has a national delegation in Brussels known as the Permanent Representation, headed by Permanent Representatives, senior diplomats whose committee (Coreper) prepares ministerial sessions. Coreper meets weekly and its main task is to ensure that only the most difficult and sensitive issues are dealt with at ministerial level.

The General Secretariat of the Council provides the practical infrastructure of the Council at all levels and prepares the meetings of the Council and the European Council by advising the Presidency and assisting the Coreper and the various committees and working groups of the Council.

Legislation. The Community's legislative process starts with a proposal from the Commission (either at the suggestion of its services or in pursuit of its declared political aims) to the Council, or in the case of co-decision, to both the Council and the European Parliament. The Council generally seeks the views of the European Parliament on the proposal, and the Parliament adopts a formal Opinion after consideration of the matter by its specialist Committees. The Council may also (and in some cases is obliged to) consult the Economic and Social Committee and the Committee of the Regions which similarly deliver an opinion. When these opinions have been received, the Council will decide. Most decisions are taken on a majority basis, but will take account of reservations expressed by individual member states.

Provisions of the Treaties and secondary legislation may be either directly applicable in member states or only applicable after member states have enacted their own implementing legislation. Community law, adopted by the Council (or by Parliament and the Council in the framework of the co-decision procedure) may take the following forms: (1) *Regulations*, which are of general application and binding in their entirety and directly applicable in all member states; (2) *Directives*, which are binding upon each member state as to the result to be achieved within a given time, but leave to the national authorities the choice of form and method of achieving this result; and (3) *Decisions*, which are binding in their entirety on their addressees. In addition the Council and Commission can issue recommendations, opinions, resolutions and conclusions which are essentially political acts and not legally binding.

Transparency. In order to make its decision-making process more transparent to the European citizens, the Council has, together with the European Parliament and the Commission, introduced a set of rules concerning public access to the documents of the three institutions. A considerable number of Council documents can be accessed electronically via the Council's public register of documents, whereas other documents which may not be directly accessible can be released to the public upon request. With a view to ensure the widest possible access to its decision-making process, some Council debates and deliberations are open to the public. The Council systematically publishes votes and explanations of votes and minutes of its meetings when it is acting as legislator.

Headquarters: 175 rue de la Loi, B-1048 Brussels, Belgium.
Website: http://www.consilium.eu.int
Secretary-General: Uwe Corsepius (Germany).

The European Council

Since 1974 Heads of State or Government have met at least twice a year (until the end of 2002 in the capital of the member state currently exercising the presidency of the Council of the European Union, since 2003 primarily in Brussels) in the form of the European Council or European Summit as it is commonly known. Its membership includes the President of the European Commission, and the President of the European Parliament is invited to make a presentation at the opening session. The European Council has become an increasingly important element of the Union, setting priorities, giving political direction, providing the impetus for its development and resolving contentious issues that prove too difficult for the Council of the European Union. It has a direct role to play in the context of the Common Foreign and Security Policy (CFSP) when deciding upon common strategies, and at a more general level, when deciding upon the establishing of closer co-operation between member states within certain policy areas covered by the EU-treaties. Moreover, during recent years, the European Council has played a preponderant role in defining the general political guidelines within key policy areas with a bearing on growth and employment and in the context of the strengthening of the EU as an area of freedom, security and justice. With the entry into force of the Treaty of Lisbon on 1 Dec. 2009, it has become a full EU institution.

Current Leaders

Herman Van Rompuy

Position
President of the European Council

Introduction
Former Belgian Prime Minister Herman Van Rompuy became the first permanent president of the European Council in Nov. 2009, a post created by the Treaty of Lisbon. He overcame a long list of rival candidates, including former British prime minister Tony Blair, to secure the consensus of the 27 member states. He was re-elected for a second term in March 2012.

Early Life
Van Rompuy was born in Oct. 1947 in Etterbeek, in the Brussels-Capital Region. In 1968 he graduated in philosophy from the Catholic University of Leuven before studying for a masters degree in economic science.

He began his political career as vice-president of the Young Christian People's Party after two years at the Belgian Central Bank. In 1978 he joined the national bureau of the Christian People's Party (CVP). By the end of the 1970s he was serving in the cabinet of Léo Tindemans and in 1980 was appointed director of the CVP Study Centre. In 1988 he was elected to the Senate and became president of the CVP, serving until his promotion to budget minister and deputy prime minister in 1993. During six years as budget minister Van Rompuy significantly reduced the national debt, which had stood at 130% of GDP when he took office. In 1995 he left the Senate to take up a seat in the Chamber of Representatives.

The heavy defeat of the CVP at the 1999 election, which followed a scandal concerning the contamination of feedstock with dioxins, precipitated a party crisis that saw it renamed as the Christian Democratic and Flemish (CD&V). After eight years out of power, the 2007 general election returned the CD&V to government, with Van Rompuy as speaker of the Chamber.

On 19 Dec. 2008 the Supreme Court announced 'strong indications' that the government had attempted to influence a court decision on the break-up of the Fortis financial group. King Albert accepted the resignation of Prime Minister Yves Leterme's administration and on 28 Dec. asked Van Rompuy to form a government. Van Rompuy was reportedly reluctant to assume the premiership but bowed to pressure from colleagues eager to prevent the return of ex-prime minister Guy Verhofstadt.

Van Rompuy sought to rebuild confidence in the political system, while confronting the financial crisis and attempting to diffuse tensions between the Dutch- and French-speaking communities. In Nov. 2009 he was selected for the post of president of the European Council, taking office in Jan. 2010.

Career in Office

Van Rompuy's tenure has been dominated by the continuing fall-out from the global financial crisis and the pressing need, against a background of volatility in international financial markets, for EU governments to devise mechanisms to deal with sovereign debt problems in a number of eurozone member states. Nevertheless, in a speech in Jan. 2013 he claimed that confidence in the eurozone was returning and that the economy was likely to resume growth during the coming year despite record unemployment. He has also sought to co-ordinate EU positions on the world stage, such as at G8 and G20 meetings and bilateral summits.

In Dec. 2012 Van Rompuy, together with the presidents of the European Commission and European Parliament, accepted the Nobel Peace Prize on behalf of the European Union, which had received the annual award for its advancement of peace and reconciliation, democracy and human rights in Europe.

Headquarters: 175 rue de la Loi, B-1048 Brussels, Belgium.
Website: http://www.european-council.europa.eu
President: Herman Van Rompuy (Belgium).

European Parliament

The European Parliament consists of 754 members (785 prior to the elections that took place in June 2009). All EU citizens may stand or vote in their adoptive country of residence. Germany returned 99 members in 2009 (and returned 99 in 2004), France, Italy and the UK 72 each (78 each in 2004), Poland and Spain 50 each (54 each in 2004), the Netherlands 25 (27 in 2004), Belgium, Czech Republic, Greece, Hungary and Portugal 22 each (24 each in 2004), Sweden 18 (19 in 2004), Austria 17 (18 in 2004), Denmark, Finland and Slovakia 13 each (14 each in 2004), Ireland and Lithuania 12 each (13 each in 2004), Latvia 8 (9 in 2004), Slovenia 7 (7 in 2004), Cyprus, Estonia and Luxembourg 6 each (6 each in 2004) and Malta 5 (5 in 2004). Romania and Bulgaria only joined the European Union in 2007; they did not vote in the 2004 elections. Romania returned 33 members in 2009 (35 in 2007) and Bulgaria 17 (18 in 2007). Turnout has declined from 62% at the first direct elections in 1979 to 42% in 2009. Although only 736 members were elected in June 2009, the Parliament has had 754 members since 1 Dec. 2011 under the terms of the Treaty of Lisbon. The 18 extra members are distributed as follows: four from Spain, two each from Austria, France and Sweden, and one each from Bulgaria, Italy, Latvia, Malta, the Netherlands, Poland, Slovenia and the UK.

Political groupings. Following the 2009 elections to the European Parliament the European People's Party (EPP) had 265 seats, Progressive Alliance of Socialists and Democrats (S&D) 184, Alliance of Liberals and Democrats for Europe (ALDE) 84, Greens/European Free Alliance (Greens/EFA) 55, European Conservatives and Reformists (ECR) 54, European United Left/Nordic Green Left (EUL/NGL) 35, Europe of Freedom and Democracy (EFD) 32, Non-attached members (NI) 27.

The Parliament has a right to be consulted on a wide range of legislative proposals and forms one arm of the Community's Budgetary Authority. Under the Single European Act, it gained greater authority in legislation through the 'concertation' procedure under which it can reject certain Council drafts in a second reading procedure. Under the Maastricht Treaty, it gained the right of 'co-decision' on legislation with the Council of Ministers on a restricted range of domestic matters. The President of the European Council must report to the Parliament on progress in the development of foreign and security policy. It also plays an important role in appointing the President and members of the Commission. It can hold individual commissioners to account and can pass a motion of censure on the entire Commission, a prospect that was realized in March 1999 when the Commission, including the President, Jacques Santer, was forced to resign following an investigation into mismanagement and corruption. Parliament's seat is in Strasbourg where the one-week plenary sessions are held each month. In the Chamber, members sit in political groups, not as national delegations. All the activities of the Parliament and its bodies are the responsibility of the Bureau, consisting of the President and 14 Vice-Presidents elected for a two-and-a-half year period.

Location: Brussels, but meets at least once a month in Strasbourg.
Website: http://www.europarl.europa.eu
President: Martin Schulz (Germany, S&D).

Court of Justice of the European Union

The Court of Justice of the European Union consists of three courts: the Court of Justice (the highest court in the European Union in matters of EU law, created in 1952 and composed of 27 Judges and eight Advocates General), the General Court (created in 1988 as the Court of First Instance) and the Civil Service Tribunal (created in 2004). Since their establishment, approximately 15,000 judgments have been delivered by the three courts.

Address: Blvd Konrad Adenauer, Kirchberg, L-2925 Luxembourg.
Website: http://curia.europa.eu
President of the Court of Justice: Vassilios Skouris (Greece).
President of the General Court: Marc Jaeger (Luxembourg).
President of the Civil Service Tribunal: Sean Van Raepenbusch (Belgium).

European Court of Auditors

The European Court of Auditors was established by a treaty of 22 July 1975 which took effect on 1 June 1977. It consists of 27 members (one from each member state) and was raised to the status of a full EU institution by the 1993 Maastricht Treaty. It audits the accounts and verifies the implementation of the budget of the EU.

Address: 12, rue Alcide De Gasperi, L-1615 Luxembourg.
Website: http://eca.europa.eu
Email: eca-info@eca.europa.eu
President: Vítor Manuel da Silva Caldeira (Portugal).

European Central Bank

The European System of Central Banks (ESCB) is composed of the European Central Bank (ECB) and 27 National Central Banks (NCBs). The NCBs of the member states not participating in the euro area are members with special status; while they are allowed to conduct their respective national monetary policies, they do not take part in decision-making regarding the single monetary policy for the euro area and the implementation of these policies. The Governing Council of the ECB makes a distinction between the ESCB and the 'Eurosystem' which is composed of the ECB and the 17 fully participating NCBs.

Members. The 17 fully participating National Central Banks are from: Austria, Belgium, Cyprus, Estonia, Finland, France, Germany, Greece, Ireland, Italy, Luxembourg, Malta, Netherlands, Portugal, Slovakia, Slovenia and Spain. The other ten EU members (those which do not use the euro as their currency) have special status.

Functions. The primary objective of the ESCB is to maintain price stability. Without prejudice to this, the ESCB supports general economic policies in the Community with a view to contributing to the achievement of the objectives of the Community. Tasks to be carried out include: i) defining and implementing the monetary policy of the Community; ii) conducting foreign exchange operations; iii) holding and managing the official foreign reserves of the participating member states; iv) promoting the smooth operation of payment systems; v) supporting the policies of the competent authorities relating to the prudential supervision of credit institutions and the stability of the financial system.

The ECB has the exclusive right to issue banknotes within the Community.

Organization. The ESCB is governed by the decision-making bodies of the ECB: the 23-member Governing Council and the Executive Board. The Governing Council is the supreme decision-making body and comprises all members of the Executive Board plus the governors of the NCBs forming the Eurosystem. The Executive Board comprises the president, vice-president and four other members, appointed by common accord of the heads of state and government of the participating member states. There is also a General Council which will exist while there remain members with special status.

Address: Eurotower, Kaiserstrasse 29, 60311 Frankfurt am Main, Germany.
Website: http://www.ecb.int
President: Mario Draghi (Italy).

Other EU Structures

European Investment Bank (EIB)

The EIB is the financing institution of the European Union, created by the Treaty of Rome in 1958 as an autonomous body set up to finance capital investment furthering European integration. To this end, the Bank raises its resources on the world's capital markets where it mobilizes significant volumes of funds on favourable terms. It directs these funds towards capital projects promoting EU economic policies. Outside the Union the EIB implements the financial components of agreements concluded under European Union development aid and co-operation policies. The members of the EIB are the member states of the European Union, who have all subscribed to the Bank's capital. Its governing body is its Board of Governors consisting of the ministers designated by each of the member states, usually the finance ministers.

Address: 98–100 Blvd Konrad Adenauer, L-2950 Luxembourg.
Website: http://www.eib.org
President and Chairman of the Board: Werner Hoyer (Germany).

European Investment Fund Founded in 1994 as a subsidiary of the European Investment Bank and the European Union's specialized financial institution. It has a dual mission that combines the pursuit of objectives such as innovation, the creation of employment and regional development with maintaining a commercial approach to investments. It particularly provides venture capital and guarantee instruments for the growth of small and medium-sized enterprises (SMEs). In 2002 it began advising entities in the setting up of financial enterprise and venture capital and SME guarantee schemes. A team has been created to structure and expand its advisory services.

Address: 96 Blvd Konrad Adenauer, L-2968 Luxembourg.
Website: http://www.eif.org

European Data Protection Supervisor

The European Data Protection Supervisor protects those individuals whose data are processed by the EU institutions and bodies by advising on new legislation and implications, processing and investigating complaints, and promoting a 'data protection culture' and awareness. The present incumbent is Peter Hustinx (Netherlands).

Address: Rue Wiertz, 60-MO 63, B-1047 Brussels, Belgium.
Website: http://www.edps.europa.eu/EDPSWEB
Email: edps@edps.europa.eu

European Ombudsman

The Ombudsman was inaugurated in 1995 and deals with complaints from citizens, companies and organizations concerning maladministration in the activities of the institutions and bodies of the European Union. The present incumbent is Paraskevas Nikiforos Diamandouros (Greece).

Address: 1 avenue du Président Robert Schuman, CS 30403, F-67001 Strasbourg Cedex, France.
Website: http://www.ombudsman.europa.eu

Advisory Bodies There are two main consultative committees whose members are appointed in a personal capacity and are not bound by any mandatory instruction.

1. *European Economic and Social Committee.* The 344-member committee is consulted by the Council of Ministers or by the European Commission, particularly with regard to agriculture, free movement of workers, harmonization of laws and transport. It is served by a permanent and independent General Secretariat, headed by a Secretary-General.

Address: Rue Belliard 99, B-1040 Brussels, Belgium.
Website: http://www.eesc.europa.eu
Secretary-General: Martin Westlake (UK).

2. *Committee of the Regions.* A political assembly which provides representatives of local, regional and city authorities with a voice at the heart of the European Union. Established by the Maastricht Treaty, the Committee consists of 344 full members and an equal number of alternates appointed for a four-year term. It must be consulted by the European Commission and Council of Ministers whenever legislative proposals are made in areas which have repercussions at the regional or local level. The Committee can also draw up opinions on its own initiative, which enables it to put issues on the EU agenda.

Address: Bâtiment Jacques Delors, Rue Belliard 99–101, B-1040 Brussels, Belgium.
Website: http://cor.europa.eu
President: Ramon Luis Valcárcel Siso (Spain).

Main EU Agencies

Agency for the Cooperation of Energy Regulators Launched in 2011 to assist and coordinate the work of national regulatory authorities, ensuring the proper functioning of the single European market in gas and electricity.

Address: Trg republike 3, 1000 Ljubljana, Slovenia.
Website: http://www.acer.europa.eu

Body of European Regulators for Electronic Communications Established in 2009 to help the European Commission and national regulatory authorities implement EU regulations on electronic communications.

Address: Zigfrīda Annas Meierovica Bulvaris 14, 2nd Floor, Riga, LV-1050, Latvia.
Website: http://berec.europa.eu

Community Plant Variety Office Launched in 1995 to administer a system of plant variety rights. The system allows Community Plant Variety Rights (CPVRs), valid throughout the European Union, to be granted for new plant varieties as sole and exclusive form of Community intellectual property rights.

Address: 3 Blvd Maréchal Foch, BP 10121, F-49101 Angers Cédex 02, France.
Website: http://www.cpvo.europa.eu

European Agency for Large-scale IT Systems Established in 2012 to provide a viable, long-term solution for the management of large-scale information systems in the areas of freedom, security and justice.

Address: EU House, 6th Floor, Rävala pst 4, 10143 Tallinn, Estonia.
Website: http://ec.europa.eu/dgs/home-affairs/what-we-do/policies/borders-and-visas/agency/index_en.htm

European Agency for Safety and Health at Work Founded in 1996 in order to serve the information needs of people with an interest in occupational safety and health.

Address: Gran Via 33, E-48009 Bilbao, Spain.
Website: http://osha.europa.eu

European Agency for the Management of Operational Cooperation at the External Borders Generally known as 'Frontex', it was established in 2004 to facilitate co-operation between member states in managing external borders. It assists in the training of national border guards, provides risk analyses and where necessary offers technical and operational assistance at external borders. It works in conjunction with other relevant EU partners, such as Europol.

Address: Rondo ONZ 1, 00 124 Warsaw, Poland.
Website: http://www.frontex.europa.eu

European Aviation Safety Agency Established in 2002 to establish and maintain a high uniform level of civil aviation safety in Europe. It offers technical expertise to the European Commission, such as assisting in the drafting of aviation safety regulations, and carries out executive tasks including the certification of aeronautical products and organizations involved in their design, production and maintenance.

Address: Ottoplatz 1, 50679 Cologne, Germany.
Website: http://www.easa.europa.eu

European Centre for Disease Prevention and Control Established in 2004 to help strengthen Europe's defences against infectious diseases, such as influenza, SARS and HIV/AIDS. It co-operates with a network of partners across the EU and the EEA/EFTA member states to strengthen and develop continent-wide disease surveillance and early warning systems. By pooling Europe's health knowledge, it provides analyses of risks posed by new and emerging infectious diseases.

Address: Tomtebodavägen 11A, 171 83 Stockholm, Sweden.
Website: http://ecdc.europa.eu

European Centre for the Development of Vocational Training Generally known as 'Cedefop', it was set up to help policy-makers and practitioners of the European Commission, the member states and social partner organizations across Europe make informed choices about vocational training policy.

Address: PO Box 22427, Thessaloniki 55102, Greece.
Website: http://www.cedefop.europa.eu

European Chemicals Agency Established in 2007 to manage the Registration, Evaluation, Authorisation and Restriction of Chemicals (REACH) system of the EU. There is a staff of around 200.

Address: Annankatu 18, PO Box 400, 00121 Helsinki, Finland.
Website: http://echa.europa.eu

European Defence Agency Established in 2004 to improve the EU's defence capabilities in the field of crisis management and to sustain the European Security and Defence Policy.

Address: Rue des Drapiers/Lakenweversstraat 17–23, BE-1050 Brussels, Belgium.
Website: http://www.eda.europa.eu

European Environment Agency Launched in 1993 to orchestrate and put to strategic use information of relevance to the protection and improvement of Europe's environment. It has a mandate to ensure objective, reliable and comprehensive information on the environment at European level. The Agency carries out its tasks through the European Information and Observation Network (EIONET). Membership is open to countries outside the EU and currently includes all EU countries, Iceland, Liechtenstein, Norway, Switzerland and Turkey.

Address: Kongens Nytorv 6, 1050 Copenhagen K, Denmark.
Website: http://eea.europa.eu

European Fisheries Control Agency Established in 2005 as the Community Fisheries Control Agency to oversee compliance with the 2002 reforms of the common fisheries policy. It was renamed the European Fisheries Control Agency from 1 Jan. 2012.

Address: Apartado de correos 771, E-36200 Vigo, Spain.
Website: http://cfca.europa.eu

European Food Safety Authority Founded in 2002 to provide independent scientific advice on all matters with a direct or indirect impact on food safety.

Address: Via Carlo Magno 1A, 43126 Parma, Italy.
Website: http://efsa.europa.eu

European Foundation for the Improvement of Living and Working Conditions Launched in 1975 to contribute to the planning and establishment of better living and working conditions. The Foundation's role is to provide findings, knowledge and advice from comparative research managed in a European perspective, which respond to the needs of the key parties at the EU level.

Address: Wyatville Road, Loughlinstown, Dublin 18, Ireland.
Website: http://www.eurofound.europa.eu

European GNSS Agency Established in 2004 as the European GNSS Supervisory Authority to manage the public interests related to and to be the regulatory authority for the European global navigation satellite system, taking over tasks previously assigned to the Galileo Joint Undertaking. It was renamed the European GNSS Agency in 2010.

Address: Janovského 438/2, Holešovice, 170 00 Prague 7, Czech Republic.
Website: http://gsa.europa.eu

European Institute for Gender Equality Established in 2007 to support the EU and its member states in their efforts to promote gender equality, to fight discrimination based on sex and to raise awareness about gender equality issues.

Address: Švitrigailos g. 11M, LT-03228 Vilnius, Lithuania.
Website: http://eige.europa.eu

European Maritime Safety Agency Established in 2003 to enhance the EU's pre-existing range of legal tools to deal with incidents resulting in serious casualties or pollution in European waters and coastlines. The agency gives advice to member states and offers support to the directorate-general of energy and transport. As well as the 27 member states, it also covers Norway and Iceland. The Agency is active across issues including maritime safety controls, classification societies and port reception facilities for hazardous substances. It plays a major role in harmonizing member states' methodologies in post-accident investigations.

Address: Cais do Sodré, 1249-206 Lisbon, Portugal.
Website: http://emsa.europa.eu

European Medicines Agency Founded in 1995 (as European Agency for the Evaluation of Medicinal Products) to evaluate the quality and effectiveness of health products for human and veterinary use.

Address: 7 Westferry Circus, Canary Wharf, London E14 4HB, UK.
Website: http://www.ema.europa.eu

European Monitoring Centre for Drugs and Drug Addiction Established in 1993 to provide the European Union and its member states with objective, reliable and comparable information on a European level concerning drugs and drug addiction and their consequences.

Address: Cais do Sodré, 1249-289 Lisbon, Portugal.
Website: http://emcdda.europa.eu

European Network and Information Security Agency Established in 2004 to serve as a centre of expertise on the digital economy, providing advice to member states and EU institutions in matters of network and information security.

Address: Science and Technology Park of Crete (ITE), Vassilika Vouton, 700 13 Heraklion, Greece.
Website: http://enisa.europa.eu

European Police College (CEPOL) Established in 2005 to encourage cross-border cooperation in the fight against crime, maintenance of public security and law and order by bringing together senior police officers across Europe.

Address: CEPOL House, Bramshill, Hook, Hampshire RG27 0JW, UK.
Website: http://cepol.europa.eu

European Police Office (Europol) Founded in 1994 to exchange criminal intelligence between EU countries. Its precursor was the Europol Drug Unit; Europol took up its activities in 1999. Europol's current mandate includes the prevention and combat of illicit drug trafficking, illegal immigration networks, vehicle trafficking, trafficking in human beings including child pornography, forgery of money, terrorism and associated money laundering activities. There are about 700 staff members from all member states. Of these, 130 are Europol Liaison Officers working for their national police, gendarmerie, customs or immigration services. The 2010 budget was €80·1m.

Address: Eisenhowerlaan 73, 2517 KK The Hague, Netherlands.
Website: http://www.europol.europa.eu

European Railway Agency Established in 2004 to reinforce safety and interoperability of railways throughout Europe. It has a staff of 100, most of whom come from the European railway sector.

Address: 120 rue Marc Lefrancq, 59300 Valenciennes, France.
Website: http://era.europa.eu

European Training Foundation Launched in 1995 to contribute to the process of vocational education and training reform that is currently taking place within the EU's partner countries and territories.

Address: Villa Gualino, viale Settimio Severo 65, I-10133 Turin, Italy.
Website: http://etf.europa.eu

European Union Agency for Fundamental Rights In 2007 the FRA succeeded the European Monitoring Centre on Racism and Xenophobia, which was established in 1997. It seeks to provide assistance and expertise to relevant institutions and authorities in relation to fundamental rights when implementing Community law.

Address: Schwarzenbergplatz 11, A-1040 Vienna, Austria.
Website: http://fra.europa.eu

European Union Institute for Security Studies Established in 2001 to help support and develop the EU's common foreign and security policy (CFSP) by offering analyses and forecasting to the High Representative for Foreign Affairs and Security Policy.

Address: 100 avenue de Suffren, 75015 Paris Cedex 16, France.
Website: http://www.iss.europa.eu

European Union Satellite Centre Set up in 2002 to support the EU's common foreign and security policy (CFSP) by providing analyses of satellite imagery and collateral data.

Address: Avda. de Cadiz - Ed. 457, Base Aérea de Torrejón, Madrid, Spain.
Website: http://www.eusc.europa.eu

Office for Harmonization in the Internal Market The Office was established in 1994, and is responsible for registering Community trade marks and designs. Both Community trade marks and Community designs confer their proprietors a uniform right, which covers all member states of the EU by means of one single application and one single registration procedure.

Address: Avenida de Europa 4, Apartado de Correos 77, E-03080 Alicante, Spain.
Website: http://oami.europa.eu

The European Union's Judicial Cooperation Unit Generally known as 'Eurojust', it was established in 2002 to stimulate and improve the coordination of investigations and prosecutions among the judicial authorities of the EU member states when dealing with serious cross-border and organized crime.

Address: Maanweg 174, 2516 AB The Hague, Netherlands.
Website: http://eurojust.europa.eu

Translation Centre for Bodies of the European Union Established in 1994, the Translation Centre's mission is to meet the translation needs of the other decentralized Community agencies. It also participates in the Interinstitutional Committee for Translation and Interpretation.

Address: Bâtiment Nouvel Hémicycle 1, rue du Fort Thüngen, L-1499 Luxembourg Kirchberg, Luxembourg.
Website: http://cdt.europa.eu

EU general information. The Office for Official Publications of the European Communities is the publishing house of the institutions and other bodies of the European Union. It is responsible for producing and distributing EU publications on all media and by all means.

Address: 2 rue Mercier, L-2985 Luxembourg.
Website: http://publications.eu.int

Further Reading

Official Journal of the European Communities.—General Report on the Activities of the European Communities (annual, from 1967).—*The Agricultural Situation in the Community* (annual).—*The Social Situation in the Community* (annual).—*Report on Competition Policy in the European Community* (annual).—*Bulletin of the European Community* (monthly).—*Register of Current Community Legal Instruments* (biannual).

Buonanno, Laurie and Nugent, Neill, *Policies and Policy Processes of the European Union.* 2013

Chang, Michele, *Monetary Integration in the European Union.* 2009

Christiansen, Thomas and Reh, Christine, *Constitutionalizing the European Union.* 2009

Cini, Michelle and McGowan, Lee, *Competition Policy in the European Union.* 2nd ed. 2008

Davies, N., *Europe: A History.* 1997

Déloye, Yves and Bruter, Michael, (eds.) *Encyclopaedia of European Elections.* 2007

Dinan, D., *Europe Recast: A History of European Union.* 2004.—*Ever Closer Union? An Introduction to the European Union.* 4th ed. 2010

Dod's European Companion. Occasional

Gänzle, Stefan and Sens, Allen G. (eds.) *The Changing Politics of European Security: Europe Alone?* 2007

Grabbe, Heather, *The EU's Transformative Power: Europeanization through Conditionality in Central and Eastern Europe.* 2005

Greenwood, Justin, *Interest Representation in the European Union.* 3rd ed. 2011

Hantrais, Linda, *Social Policy in the European Union.* 2007

Holland, Martin and Doidge, Mathew, *Development Policy of the European Union.* 2013

Howorth, Jolyon, *Security and Defence Policy in the European Union.* 2007

Judge, David and Earnshaw, David, *The European Parliament.* 2nd ed. 2008

Keukeleire, Stephan and MacNaughtan, Jennifer, *The Foreign Policy of the European Union.* 2008

Lea, Ruth, *The Essential Guide to the European Union.* 2004

Lewis, D. W. P., *The Road to Europe: History, Institutions and Prospects of European Integration, 1945–1993.* 1994

Mancini, Judge G. F., *Democracy and Constitutionalism in the European Union.* 2000

Mazower, M., *Dark Continent: Europe's 20th Century.* 1998

McCormick, John, *Understanding the European Union.* 5th ed. 2011.—*The European Superpower.* 2006

McGuire, Steven and Smith, Michael, *The European Union and the United States: Competition and Convergence in the Global Arena.* 2008

Menon, Anand, *Europe: the State of the Union.* 2008

Nugent, N., *The European Commission.* 2000.—*European Union Enlargement.* 2004.—*The Government and Politics of the European Union.* 7th ed. 2010

Wallace, Helen, Pollack, Mark and Young, Alasdair, (eds.) *Policy-Making in the European Union.* 6th ed. 2010

Council of Europe

Origin and Membership. In 1948 the Congress of Europe, bringing together at The Hague nearly 1,000 influential Europeans from 26 countries, called for the creation of a united Europe, including a European Assembly. This proposal, examined first by the Ministerial Council of the Brussels Treaty Organization, then by a conference of ambassadors, was at the origin of the Council of Europe, which is, with its 47 member States, the widest organization bringing together all European democracies. The Statute of the Council was signed at London on 5 May 1949 and came into force two months later.

The founder members were Belgium, Denmark, France, Ireland, Italy, Luxembourg, the Netherlands, Norway, Sweden and the UK. Turkey and Greece joined in 1949, Iceland in 1950, the Federal Republic of Germany in 1951 (having been an associate since 1950), Austria in 1956, Cyprus in 1961, Switzerland in 1963, Malta in 1965, Portugal in 1976, Spain in 1977, Liechtenstein in 1978, San Marino in 1988, Finland in 1989, Hungary in 1990, Czechoslovakia (after partitioning, the Czech Republic and Slovakia rejoined in 1993) and Poland in 1991, Bulgaria in 1992, Estonia, Lithuania, Romania and Slovenia in 1993, Andorra in 1994, Albania, Latvia, Macedonia, Moldova and Ukraine in 1995, Croatia and Russia in 1996, Georgia in 1999, Armenia and Azerbaijan in 2001, Bosnia and Herzegovina in 2002, Serbia in 2003 (as Serbia and Montenegro until 2006), Monaco in 2004 and Montenegro in 2007.

Membership is limited to European states which 'accept the principles of the rule of law and of the enjoyment by all persons within [their] jurisdiction of human rights and fundamental freedoms'. The Statute provides for both withdrawal (Article 7) and suspension (Articles 8 and 9). Greece withdrew during 1969–74.

Aims and Achievements. Article 1 of the Statute states that the Council's aim is 'to achieve a greater unity between its members for the purpose of safeguarding and realizing the ideals and principles which are their common heritage and facilitating their economic and social progress'; 'this aim shall be pursued ... by discussion of questions of common concern and by agreements and common action'. The only limitation is provided by Article 1 (d), which excludes 'matters relating to national defence'.

The main areas of the Council's activity are: human rights, the media, social and socio-economic questions, education, culture and sport, youth, public health, heritage and environment, local and regional government, and legal co-operation. 209 Conventions and Agreements have been concluded covering such matters as social security, cultural affairs, conservation of European wildlife and natural habitats, protection of archaeological heritage, extradition, medical treatment, equivalence of degrees and diplomas, the protection of television broadcasts, adoption of children and transportation of animals.

Treaties in the legal field include the adoption of the European Convention on the Suppression of Terrorism, the European Convention on the Legal Status of Migrant Workers and the Transfer of Sentenced Persons. The Committee of Ministers adopted a European Convention for the protection of individuals with regard to the automatic processing of personal data (1981), a Convention on the compensation of victims of violent crimes (1983), a Convention on spectator violence and misbehaviour at sports events and in particular at football matches (1985), the European Charter of Local Government (1985), and a Convention for the Prevention of Torture and Inhuman or Degrading Treatment or Punishment (1987). The European Social Charter of 1961 sets out the social and economic rights which all member governments agree to guarantee to their citizens.

European Social Charter. The Charter defines the rights and principles which are the basis of the Council's social policy, and guarantees a number of social and economic rights to the citizen, including the right to work, the right to form workers' organizations, the right to social security and assistance, the right of the family to protection and the right of migrant workers to protection and assistance. Two committees, comprising independent and government experts, supervise the parties' compliance with their obligations under the Charter. A revised charter, incorporating new rights such as protection for those without jobs and opportunities for workers with family responsibilities, was opened for signature on 3 May 1996 and entered into force on 1 July 1999.

Human rights. The promotion and development of human rights is one of the major tasks of the Council of Europe. The European Convention on Human Rights, signed in 1950, set up special machinery to guarantee internationally fundamental rights and freedoms. The European Commission of Human Rights which was set up has now been abolished and has been replaced by the new European Court of Human Rights, which came into operation on 1 Nov. 1998. The European Court of Human Rights in Strasbourg, set up under the European Convention on Human Rights as amended, is composed of a number of judges equal to that of the Contracting States (currently 46). There is no restriction on the number of judges of the same nationality. Judges are elected by the Parliamentary Assembly of the Council of Europe for a term of six years. The terms of office of one half of the judges elected at the first election expired after three years, so as to ensure that the terms of office of one half of the judges are renewed every three years. Any Contracting State (State application) or individual claiming to be a victim of a violation of the Convention (individual application) may lodge directly with the Court in Strasbourg an application alleging a breach by a Contracting State of one of the Convention rights.

> *President of the European Court of Human Rights*: Dean Spielmann (Luxembourg).

The Development Bank, formerly the Social Development Fund, was created in 1956. The main purpose of the Bank is to give financial aid in the spheres of housing, vocational training, regional planning and development.

The *European Youth Foundation* provides money to subsidize activities by European youth organizations in their own countries.

Structure. Under the Statute, two organs were set up: an intergovernmental *Committee of [Foreign] Ministers* with powers of decision and recommendation to governments, and an interparliamentary deliberative body, the *Parliamentary Assembly* (referred to in the Statute as the Consultative Assembly)—both served by the Secretariat. A Joint Committee acts as an organ of co-ordination and liaison between the two and gives members an opportunity to exchange views on matters of important European interest. In addition, a number of committees of experts have been established. On municipal matters the Committee of Ministers receives recommendations from the Congress of Local and Regional Authorities of Europe. The Committee meets at ministerial level once a year; the ministers' deputies meet once a week. The chairmanship of the Committee is rotated on a six-monthly basis.

The *Parliamentary Assembly* consists of 318 parliamentarians elected or appointed by their national parliaments (Albania 4, Andorra 2, Armenia 4, Austria 6, Azerbaijan 6, Belgium 7, Bosnia and Herzegovina 5, Bulgaria 6, Croatia 5, Cyprus 3, the Czech Republic 7, Denmark 5, Estonia 3, Finland 5, France 18, Georgia 5, Germany 18, Greece 7, Hungary 7, Iceland 3, Ireland 4, Italy 18, Latvia 3, Liechtenstein 2, Lithuania 4, Luxembourg 3, Macedonia 3, Malta 3, Moldova 5, Monaco 2, Montenegro 3, Netherlands 7, Norway 5, Poland 12, Portugal 7, Romania 10, Russia 18, San Marino 2, Serbia 7, Slovakia 5, Slovenia 3, Spain 12, Sweden 6, Switzerland 6, Turkey 12, Ukraine 12, UK 18). It meets three times a year for approximately a week. The work of the Assembly is prepared by parliamentary committees. Since June 1989 representatives of a number of central and East European countries have been permitted to attend as non-voting members ('special guests'). Armenia and Azerbaijan have subsequently become full members.

Although without legislative powers, the Assembly acts as the powerhouse of the Council, initiating European action in key areas by making recommendations to the Committee of Ministers. As the widest parliamentary forum in Western Europe, the Assembly also acts as the conscience of the area by voicing its opinions on important current issues. These are embodied in Resolutions. The Ministers' role is to translate the Assembly's recommendations into action, particularly as regards lowering the barriers between the European countries, harmonizing their legislation or introducing, where possible, common European laws, abolishing discrimination on grounds of nationality, and undertaking certain tasks on a joint European basis.

> *Official languages:* English and French.
> *Headquarters:* Council of Europe, F-67075 Strasbourg Cedex, France.
> *Website:* http://www.coe.int
> *Secretary-General:* Thorbjørn Jagland (Norway).

Publications. European Yearbook, The Hague; *Yearbook on the Convention on Human Rights,* Strasbourg; *Catalogue of Publications* (annual); *Activities Report* (annual). Information on other bulletins and documents is available on the Council of Europe's website.

Further Reading

Bond, Martyn, *The Council of Europe: Structure, History and Issues in European Politics.* 2011
Cook, C. and Paxton, J., *European Political Facts of the Twentieth Century.* 2000

Organization for Security and Co-operation in Europe (OSCE)

The OSCE is a pan-European security organization of 57 participating states. It has been recognized under the UN Charter as a primary instrument in its region for early warning, conflict prevention, crisis management and post-conflict rehabilitation.

Origin. Initiatives from both NATO and the Warsaw Pact culminated in the first summit Conference on Security and Co-operation in Europe (CSCE) attended by heads of state and government in Helsinki on 30 July–1 Aug. 1975. It adopted the *Helsinki Final Act* laying down ten principles governing the behaviour of States towards their citizens and each other, concerning human rights, self-determination and the interrelations of the participant states. The CSCE was to serve as a multilateral forum for dialogue and negotiations between East and West.

The Helsinki Final Act comprised three main sections: 1) politico-military aspects of security: principles guiding relations between and among participating States and military confidence-building measures; 2) co-operation in the fields of economics, science and technology and the environment; 3) co-operation in humanitarian and other fields.

From CSCE to OSCE. The Paris Summit of Nov. 1990 set the CSCE on a new course. In the Charter of Paris for a New Europe, the CSCE was called upon to contribute to managing the historic change in Europe and respond to the new challenges of the post-Cold War period. At the meeting, members of NATO and the Warsaw Pact signed an important Treaty on Conventional Armed Forces in Europe (CFE) and a declaration that they were 'no longer adversaries' and did not intend to 'use force against the territorial integrity or political independence of any state'. All 34 participants adopted the Vienna Document comprising Confidence and Security-Building Measures (CSBMs), which pertain to the exchange of military information, verification of military installations, objection to unusual military activities etc., and signed the Charter of Paris. The Charter sets out principles of human rights, democracy and the rule of law to which all the signatories undertake to adhere, and lays down the basis for East-West co-operation and other future action. The 1994 Budapest Summit recognized that the CSCE was no longer a conference and on 1 Jan. 1995 the CSCE changed its name to the Organization for Security and Co-operation in Europe (OSCE). The 1996 Lisbon Summit elaborated the OSCE's key role in fostering security and stability in all their dimensions. It also stimulated the development of an OSCE Document-Charter on European Security.

Members. Albania, Andorra, Armenia, Austria, Azerbaijan, Belarus, Belgium, Bosnia and Herzegovina, Bulgaria, Canada, Croatia, Cyprus, Czech Republic, Denmark, Estonia, Finland, France, Georgia, Germany, Greece, Holy See, Hungary, Iceland, Ireland, Italy, Kazakhstan, Kyrgyzstan, Latvia, Liechtenstein, Lithuania, Luxembourg, Macedonia, Malta, Moldova, Monaco, Mongolia, Montenegro, Netherlands, Norway, Poland, Portugal, Romania, Russian Federation, San Marino, Serbia, Slovak Republic, Slovenia, Spain, Sweden, Switzerland, Tajikistan, Turkey, Turkmenistan, Ukraine, UK, USA and Uzbekistan. *Partners for co-operation:* Afghanistan, Algeria, Australia, Egypt, Israel, Japan, Jordan, South Korea, Morocco, Thailand and Tunisia.

Organization. The OSCE's regular body for political consultation and decision-making is the Permanent Council. Its members, the Permanent Representatives of the OSCE participating States, meet weekly in the Hofburg Congress Center in Vienna to discuss and take decisions on all issues pertinent to the OSCE. The Forum for Security Co-operation (FSC), which deals with arms control and confidence- and security-building measures, also meets weekly in Vienna. Summits—periodic meetings of Heads of State or Government of OSCE participating States—set priorities and provide orientation at the highest political level. In the years between these summits, decision-making and governing power lies with the *Ministerial Council*, which is made up of the Foreign Ministers of the OSCE participating States. In addition, a Senior Council also meets once a year in special session as the Economic Forum. The Chairman-in-Office has overall responsibility for executive action and agenda-setting. The Chair rotates annually. The Secretary-General acts as representative of the Chairman-in-Office and manages OSCE structures and operations.

The Secretariat is based in Vienna and includes a *Conflict Prevention Centre* which provides operational support for OSCE field missions. There are some 400 staff employed in OSCE institutions, and about 1,000 professionals, seconded by OSCE-participating states, work at OSCE missions and other field operations, together with another 2,500 local staff.

The *Office for Democratic Institutions and Human Rights* is located in Warsaw. It is active in monitoring elections and developing national electoral and human rights institutions, providing technical assistance to national legal institutions, and promoting the development of the rule of law and civil society.

The *Office of the Representative on Freedom of the Media* is located in Vienna. Its main function is to observe relevant media developments in OSCE participating States with a view to providing an early warning on violations of freedom of expression.

The *Office of the High Commissioner on National Minorities* is located in The Hague. Its function is to identify and seek early resolution of ethnic tensions that might endanger peace, stability or friendly relations between the participating States of the OSCE.

The budget for 2012 was €148·1m.

Headquarters: Wallnerstrasse 6, A-1010 Vienna, Austria.
Website: http://www.osce.org
Email: info@osce.org
Chairman-in-Office: Leonid Kozhara (Ukraine).
Secretary-General: Lamberto Zannier (Italy).

Further Reading

Freeman, J., *Security and the CSCE Process: the Stockholm Conference and Beyond.* 1991
Galbreath, David J., *The Organization for Security and Co-operation in Europe.* 2007

European Bank for Reconstruction and Development (EBRD)

History. The European Bank for Reconstruction and Development was established in 1991 when communism was collapsing in central and eastern Europe and ex-Soviet countries needed support to nurture a new private sector in a democratic environment.

Activities. The EBRD is the largest single investor in the region and mobilizes significant foreign direct investment beyond its own financing. It is owned by 61 countries and two intergovernmental institutions. But despite its public sector shareholders, it invests mainly in private enterprises, usually together with commercial partners. Today the EBRD uses the tools of investment to help build market economies and democracies in 30 countries from Central Europe to Central Asia.

It provides project financing for banks, industries and businesses, for both new ventures and investments in existing companies. It also works with publicly-owned companies, to support privatization, restructuring of state-owned firms and improvement of municipal services. The EBRD uses its close relationship with governments in the region to promote policies that will bolster the business environment.

The mandate of the EBRD stipulates that it must only work in countries that are committed to democratic principles. Respect for the environment is part of the strong corporate governance attached to all EBRD investments.

Organization. All the powers of the EBRD are vested in a Board of Governors, to which each member appoints a governor, generally the minister of finance or an equivalent. The Board of Governors delegates powers to the Board of Directors, which is responsible for the direction of the EBRD's general operations and policies. The President is elected by the Board of Governors and is the legal representative of the EBRD. The President conducts the current business of the Bank under the guidance of the Board of Directors.

Headquarters: One Exchange Square, London EC2A 2JN, UK.
Website: http://www.ebrd.com
President: Sir Suma Chakrabarti (India).
Secretary-General: Enzo Quattrociocche (Italy).

European Free Trade Association (EFTA)

History and Membership. The Stockholm Convention establishing the Association entered into force on 3 May 1960. Founder members were Austria, Denmark, Norway, Portugal, Sweden, Switzerland and the UK. With the accession of Austria, Denmark, Finland, Portugal, Sweden and the UK to the EU, EFTA was reduced to four member countries: Iceland, Liechtenstein, Norway and Switzerland. In June 2001 the Vaduz Convention was signed. It liberalizes trade further among the four EFTA States in order to reflect the Swiss—EU bilateral agreements.

Activities. Free trade in industrial goods among EFTA members was achieved by 1966. Co-operation with the EU began in 1972 with the signing of free trade agreements and culminated in the establishment of a *European Economic Area (EEA)*, encompassing the free movement of goods, services, capital and labour throughout EFTA and the EU member countries. The Agreement was signed by all members of the EU and EFTA on 2 May 1992, but was rejected by Switzerland in a referendum on 6 Dec. 1992. The agreement came into force on 1 Jan. 1994.

The main provisions of the EEA Agreement are: free movement of products within the EEA from 1993 (with special arrangements to cover food, energy, coal and steel); EFTA to assume EU rules on company law, consumer protection, education, the environment, research and development, and social policy; EFTA to adopt EU competition rules on anti-trust matters, abuse of a dominant position, public procurement, mergers and state aid; EFTA to create an EFTA Surveillance Authority and an EFTA Court; individuals to be free to live, work and offer services throughout the EEA, with mutual recognition of professional qualifications; capital movements to be free with some restrictions on investments; EFTA countries not to be bound by the Common Agricultural Policy (CAP) or Common Fisheries Policy (CFP).

The EEA-EFTA states have established a Surveillance Authority and a Court to ensure implementation of the Agreement among the EFTA-EEA states. Political direction is given by the EEA Council which meets twice a year at ministerial level, while ongoing operation of the Agreement is overseen by the EEA Joint Committee. Legislative power remains with national governments and parliaments.

EFTA has formal relations with several other states. Free trade agreements have been signed with Turkey (1991), Israel and Czechoslovakia (1992, with protocols on succession with the Czech Republic and Slovakia in 1993), Poland and Romania (1992), Bulgaria and Hungary (1993), Estonia, Latvia, Lithuania and Slovenia (1995), Morocco (1997), the Palestine Liberation Organization on behalf of the Palestinian Authority (1998), the former Yugoslav Republic of Macedonia and Mexico (2000), Jordan and Croatia (2001), Singapore (2002), Chile (2003), Lebanon and Tunisia (2004), South Korea (2005), the Southern African Customs Union (2006), Egypt (2007), Canada and Colombia (2008), Albania, the Gulf Co-operation Council and Serbia (2009), Peru and Ukraine (2010), and Hong Kong and Montenegro (2011). Agreements with the countries that have joined the European Union in the meantime have been replaced by the relevant arrangements between the EFTA states and the EU. Negotiations on free trade agreements are ongoing with Algeria, Belarus, Bosnia and Herzegovina, Central American States, India, Indonesia, Kazakhstan, Malaysia, Russia, Thailand and Vietnam. There are currently Joint Declarations on Co-operation with Mercosur (2000), Mongolia (2007), Mauritius (2009), and Georgia and Pakistan (2012).

Organization. The operation of the free trade area among the EFTA states is the responsibility of the EFTA Council which meets regularly at ambassadorial level in Geneva. The Council is assisted by a Secretariat and standing committees. Each EFTA country holds the chairmanship of the Council for six months. For EEA matters there is a separate committee structure.

Brussels Office (EEA matters, press and information): 12–16 Rue Joseph II, B-1000 Brussels.

>*Headquarters:* 9–11 rue de Varembé, CH-1211 Geneva 20, Switzerland.
>*Website:* http://www.efta.int
>*Email:* mail.gva@efta.int
>*Secretary-General:* Kristinn F. Árnason (Iceland).

Publications. Convention Establishing the European Free Trade Association; EFTA Annual Report; EFTA Fact Sheets: Information Papers on Aspects of the EEA; EFTA Bulletin.

European Space Agency (ESA)

History. Established in 1975, replacing the European Space Research Organization (ESRO) and the European Launcher Development Organization (ELDO).

Members. Austria, Belgium, Czech Republic, Denmark, Finland, France, Germany, Greece, Ireland, Italy, Luxembourg, the Netherlands, Norway, Poland, Portugal, Romania, Spain, Sweden, Switzerland, United Kingdom. Canada takes part in some projects under a co-operation agreement.

Activities. ESA is the intergovernmental agency in Europe responsible for the exploitation of space science, research and technology for exclusively peaceful purposes. Its aim is to define and put into effect a long-term European space policy that allows Europe to remain competitive in the field of space technology. It has a policy of co-operation with various partners on the basis that pooling resources and sharing work will boost the effectiveness of its programmes. Its space plan covers the fields of science, Earth observation, telecommunications, navigation, space segment technologies, ground infrastructures, space transport systems and microgravity research.

>*Headquarters:* 8–10 rue Mario Nikis, 75738 Paris Cedex 15, France.
>*Website:* http://www.esa.int
>*Director-General:* Jean-Jacques Dordain (France).

CERN – The European Organization for Nuclear Research

Founded in 1954, CERN is the world's leading particle physics research centre. By studying the behaviour of nature's fundamental particles, CERN aims to find out what our Universe is made of and how it works. CERN's biggest accelerator, the Large Hadron Collider (LHC), became operational in Sept. 2008. One of the beneficial byproducts of CERN activity is the Worldwide Web, developed at CERN to give particle physicists easy access to shared data. One of Europe's first joint ventures, CERN now has a membership of 20 member states: Austria, Belgium, Bulgaria, Czech Republic, Denmark, Finland, France, Germany, Greece, Hungary, Italy, the Netherlands, Norway, Poland, Portugal, Slovak Republic, Spain, Sweden, Switzerland, United Kingdom. Some 6,500 scientists, half of the world's particle physicists, use CERN's facilities. They represent 500 institutions and 85 nationalities.

>*Address:* CH-1211 Geneva 23, Switzerland.
>*Website:* http://www.home.web.cern.ch
>*Director-General:* Rolf-Dieter Heuer (Germany).

Central European Initiative (CEI)

In Nov. 1989 Austria, Hungary, Italy and the then Yugoslavia met on Italy's initiative to form an economic and political co-operation group in the region.

Members. Albania, Austria, Belarus, Bosnia and Herzegovina, Bulgaria, Croatia, Czech Republic, Hungary, Italy, Macedonia, Moldova, Montenegro, Poland, Romania, Serbia, Slovakia, Slovenia, Ukraine.

>*Address:* Executive Secretariat, Via Genova 9, 34132 Trieste, Italy.
>*Website:* http://www.cei.int
>*Email:* cei@cei.int

Nordic Council

Founded in 1952 as a co-operative link between the parliaments and governments of the Nordic states. The co-operation focuses on Intra-Nordic co-operation, co-operation with Europe/EU/EEA and co-operation with the adjacent areas. The Council consists of 87 elected MPs and the committees meet several times a year, as required. Every year the Nordic Council grants prizes for literature, music, nature and environment.

Members. Denmark (including the Faroe Islands and Greenland), Finland (including Åland), Iceland, Norway, Sweden.

Address: Ved Stranden 18, DK-1061 Copenhagen K, Denmark.
Website: http://www.norden.org
Email: nordisk-rad@norden.org
President: Marit Nybakk (Norway).

Nordic Development Fund (NDF)

NDF is a multilateral development finance institution established by the five Nordic countries, Denmark, Finland, Iceland, Norway and Sweden. Since operations started in 1989, the Fund has provided soft loans to 190 projects of Nordic interest in developing countries. It entered a new phase in 2009 and changed its focus to grant aid for climate change related projects.

Address: Fabianinkatu 34, PO Box 185, FIN-00171 Helsinki, Finland.
Website: http://www.ndf.fi
Email: info.ndf@ndf.fi
Managing Director: Pasi Hellman (Finland).

Nordic Investment Bank (NIB)

The Nordic Investment Bank, which commenced operations in 1976, is a multilateral financial institution owned by Denmark, Estonia, Finland, Iceland, Latvia, Lithuania, Norway and Sweden. It finances public and private projects both within and outside the Nordic area. Priority is given to projects furthering economic co-operation between the member countries or improving the environment. Focal points include the neighbouring areas of the member countries.

Address: Fabianinkatu 34, PO Box 249, FI-00171 Helsinki, Finland.
Website: http://www.nib.int
Email: info@nib.int
President: Henrik Normann (Denmark).

Council of the Baltic Sea States

Established in 1992 in Copenhagen following a conference of ministers of foreign affairs.

Members. Denmark, Estonia, Finland, Germany, Iceland, Latvia, Lithuania, Norway, Poland, Russia, Sweden and the European Commission.

Aims. To promote co-operation in the Baltic Sea region in the field of trade, investment and economic exchanges, combating organized crime, civil security, culture and education, transport and communication, energy and environment, human rights and assistance to democratic institutions.

The Council meets at ministerial level once a year, chaired by rotating foreign ministers; it is the supreme decision-making body. Between annual sessions the Committee of Senior Officials and three working groups meet at regular intervals. In 1999 ministers of energy of the CBSS member states agreed to achieve the goal of creating effective, economically and environmentally sound and more integrated energy systems in the Baltic Sea region. Nine summits at the level of heads of government of CBSS member states and the President of the European Commission have taken place; in 1996 and then every other year since up to and including 2012. The Baltic Sea Region Energy Cooperation (BASREC) is made up of energy ministers from the region and is chaired by the energy minister from the chair country of the CBSS.

Official language: English.
CBSS Secretariat: Strömsborg, PO Box 2010, 103 11 Stockholm, Slussplan 9, Sweden.
Website: http://www.cbss.org
Email: cbss@cbss.org
Director of the Secretariat: Jan Lundin (Sweden).

European Broadcasting Union (EBU)

Founded in 1950 by western European radio and television broadcasters, the EBU is the world's largest professional association of national broadcasters, with 74 active members in 56 countries of Europe, North Africa and the Middle East, and 35 associate members worldwide.

The EBU merged with the OIRT, its counterpart in eastern Europe, in 1993. The EBU's Eurovision Operations Department has a permanent network offering 50 digital channels on five satellites. Two satellite channels also relay radio concerts, operas, sports fixtures and major news events for Euroradio.

Headquarters: Ancienne Route 17A, CH-1218 Grand-Saconnex, Geneva, Switzerland.
Website: http://www.ebu.ch
Email: ebu@ebu.ch
Director-General: Ingrid Deltenre (Netherlands).

Black Sea Economic Cooperation (BSEC)

Founded in 1992 to promote economic co-operation in the Black Sea region. Priority areas of interest include: trade and economic development; banking and finance; communications; energy; transport; agriculture and agro-industry; healthcare and pharmaceutics; environmental protection; tourism; science and technology; exchange of statistical data and economic information; combating organized crime, illicit trafficking of drugs, weapons and radioactive materials, all acts of terrorism and illegal immigration.

Members. Albania, Armenia, Azerbaijan, Bulgaria, Georgia, Greece, Moldova, Romania, Russia, Serbia, Turkey, Ukraine.

Observers. Austria, Belarus, Black Sea Commission, Commission of the European Communities, Croatia, Czech Republic, Egypt,

Energy Charter Secretariat, France, Germany, International Black Sea Club, Israel, Italy, Poland, Slovakia, Tunisia, USA.

The *Parliamentary Assembly of the Black Sea Economic Cooperation* is the BSEC parliamentary dimension. The *BSEC Business Council* is composed of representatives from the business circles of the member states. The *Black Sea Trade and Development Bank* is considered as the financial pillar of the BSEC. There is also an *International Center for Black Sea Studies* and a *Coordination Center for the Exchange of Statistical Data and Economic Information.*

Headquarters: Sakıp Sabancı Caddesi, Müşir Fuad Paşa Yalısı, Eski Tersane 34467, İstanbul, Turkey.
Website: http://www.bsec-organization.org
Secretary-General: Victor Ţvircun (Moldova).

Danube Commission

History and Membership. The Danube Commission was constituted in 1949 according to the Convention regarding the regime of navigation on the Danube signed in Belgrade on 18 Aug. 1948. The Belgrade Convention, amended by the Additional Protocol of 26 March 1998, declares that navigation on the Danube from Kelheim to the Black Sea (with access to the sea through the Sulina arm and the Sulina Canal) is equally free and open to the nationals, merchant shipping and merchandise of all states as to harbour and navigation fees as well as conditions of merchant navigation. The Commission holds annual sessions and is composed of one representative from each of its 11 member countries: Austria, Bulgaria, Croatia, Germany, Hungary, Moldova, Romania, Russia, Serbia, Slovakia and Ukraine.

Functions. To ensure that the provisions of the Belgrade Convention are carried out; to establish a uniform buoying system on all navigable waterways; to establish the basic regulations for navigation on the river and ensure facilities for shipping; to co-ordinate the regulations for river, customs and sanitation control as well as the hydrometeorological service; to collect relevant statistical data concerning navigation on the Danube; to propose measures for the prevention of pollution of the Danube caused by navigation; and to update its recommendations regularly with a view to bringing them in line with European Union regulations on inland waterway navigation.

Official languages: German, French, Russian.
Headquarters: Benczúr utca 25, H-1068 Budapest, Hungary.
Website: http://www.danubecommission.org
Email: secretariat@danubecom-intern.org
President: Biserka Benisheva (Bulgaria).
Director-General: István Valkár (Hungary).

European Trade Union Confederation (ETUC)

Established in 1973, the ETUC is recognized by the EU, the Council of Europe and EFTA as the only representative cross-sectoral trade union organization at a European level. It has grown steadily with a membership of 85 National Trade Union Confederations from 36 countries and ten European Industry Federations with a total of 60m. members. The Congress meets every four years; the 11th Statutory Congress took place in Athens in May 2011.

Address: 5 Blvd Roi Albert II, B-1210 Brussels, Belgium.
Website: http://www.etuc.org
Email: etuc@etuc.org
General Secretary: Bernadette Ségol (France).

Amnesty International (AI)

Origin. Founded in 1961 by British lawyer Peter Benenson as a one-year campaign for the release of prisoners of conscience, Amnesty International has grown to become a worldwide organization, winning the Nobel Peace Prize in 1977.

Activities. AI is a worldwide movement of people campaigning for human rights. It acts independently and impartially to promote respect for internationally recognized human rights standards.

Historically, the focus of AI's campaigning has been: to free all prisoners of conscience (a term coined by Peter Benenson); to ensure a prompt and fair trial for all political prisoners; to abolish the death penalty, torture and other cruel, inhuman or degrading punishments; to end extrajudicial executions and 'disappearances'; to fight impunity by working to ensure perpetrators of such abuses are brought to justice. AI is independent of any government or political ideology, and neither supports nor opposes the views of the victim it seeks to protect.

AI has over 3m. members, subscribers and regular donors in more than 150 countries. Major policy decisions are taken by an International Council comprising representatives from all national sections. AI's national sections, members and supporters are primarily responsible for funding the movement. During the financial year 1 April 2007–31 March 2008 AI's total income was £35,224,000.

Every year AI produces a global report detailing human rights violations in all regions of the world.

International Secretariat: Peter Benenson House, 1 Easton St., London WC1X 0DW, UK.
Website: http://www.amnesty.org
Secretary-General: Salil Shetty (India).

Further Reading

Power, Jonathan, *Like Water on Stone: The Story of Amnesty International.* 2001

Bank for International Settlements (BIS)

Origin. Founded on 17 May 1930, the Bank for International Settlements is the world's oldest international financial organization.

Activities. The BIS fosters international monetary and financial co-operation and serves as a bank for central banks. It acts as a forum to promote discussion and policy analysis among central banks and within the international financial community, a centre for economic and monetary research and an agent or trustee in connection with international financial operations.

As its customers are central banks and international organizations, the BIS does not accept deposits from, or provide financial services to, private individuals or corporate entities.

The head office is in Basle, Switzerland, and there are representative offices in Hong Kong and Mexico City.

Representative Office for Asia and the Pacific: 78th Floor, Two International Finance Centre, 8 Finance St., Central, Hong Kong SAR, People's Republic of China.
Representative Office for the Americas: Torre Chapultepec, Rubén Dario 281, Col. Bosque de Chapultepec, 11580 México, D. F., Mexico.
Headquarters: Centralbahnplatz 2, CH-4002 Basle, Switzerland.
Website: http://www.bis.org
Email: email@bis.org
Chairman of the Board of Directors: Christian Noyer (France).

Further Reading

Yago, Kazuhiko, *The Financial History of the Bank for International Settlements.* 2012

Commonwealth

The Commonwealth is a free association of sovereign independent states. It numbered 54 members in Feb. 2013. With a membership of over 2bn. people, it represents around 30% of the world's population. There is no charter, treaty or constitution; the association is expressed in co-operation, consultation and mutual assistance for which the Commonwealth Secretariat is the central co-ordinating body.

Origin. The Commonwealth was first defined by the Imperial Conference of 1926 as a group of 'autonomous Communities within the British Empire, equal in status, in no way subordinate one to another in any aspect of their domestic or external affairs, though united by a common allegiance to the Crown, and freely associated as members of the British Commonwealth of Nations'. The basis of the association changed from one owing allegiance to a common Crown, and the modern Commonwealth was born in 1949 when the member countries accepted India's intention of becoming a republic at the same time as continuing 'her full membership of the Commonwealth of Nations and her acceptance of the King as the symbol of the free association of its independent member nations and as such the Head of the Commonwealth'. In Feb. 2013 the Commonwealth consisted of 33 republics and 21 monarchies, of which 16 are Queen's realms. All acknowledge the Queen symbolically as Head of the Commonwealth. The Queen's legal title rests on the statute of 12 and 13 Will. III, c. 3, by which the succession to the Crown of Great Britain and Ireland was settled on the Princess Sophia of Hanover and the 'heirs of her body being Protestants'.

A number of territories, formerly under British jurisdiction or mandate, did not join the Commonwealth: Egypt, Iraq, Transjordan, Myanmar (then Burma), Palestine, Sudan, British Somaliland and Aden. Five countries, Ireland in 1948, South Africa in 1961, Pakistan in 1972, Fiji Islands (then Fiji) in 1987 and Zimbabwe in 2003 have left the Commonwealth. Pakistan was readmitted to the Commonwealth in 1989, South Africa in 1994, Fiji Islands in 1997. Nigeria was fully suspended in 1995 for violation of human rights but was fully reinstated on 29 May 1999. Pakistan was suspended from the Commonwealth's councils following a coup in Oct. 1999 but was readmitted in May 2004. It was again suspended in Nov. 2007 after President Musharraf declared emergency rule but was readmitted in May 2008. The Fiji Islands were suspended from the Commonwealth's councils in June 2000 following a coup there but were readmitted in Dec. 2001 following the restoration of democracy. They were again suspended from the councils following the coup of Dec. 2006 and fully suspended from membership in Sept. 2009. Zimbabwe was suspended from the Commonwealth's councils for a year on 19 March 2002 for a 'high level of politically motivated violence' during the vote that saw President Robert Mugabe re-elected. In March 2003 it was suspended for a further nine months. The suspension was extended at the Abuja meeting in Dec. 2003. Mugabe responded by withdrawing Zimbabwe from the Commonwealth. Mozambique, admitted in Nov. 1995, was the first member state not to have been a member of the former British Commonwealth or Empire.

Member States of the Commonwealth

The 54 member states, with year of admission:

	Year of admission		Year of admission
Antigua and Barbuda	1981	Namibia	1990
Australia[1]	1931	Nauru[2]	1968
Bahamas	1973	New Zealand[1]	1931
Bangladesh	1972	Nigeria[3]	1960
Barbados	1966	Pakistan[4]	1989
Belize	1981	Papua New Guinea	1975
Botswana	1966	Rwanda	2009
Brunei[5]	1984	St Kitts and Nevis	1983
Cameroon	1995	St Lucia	1979
Canada[1]	1931	St Vincent and the	1979
Cyprus	1961	Grenadines	
Dominica	1978	Samoa	1970
Fiji Islands[6]	1997	Seychelles	1976
Gambia	1965	Sierra Leone	1961
Ghana	1957	Singapore	1965
Grenada	1974	Solomon Islands	1978
Guyana	1966	South Africa[7]	1994
India	1947	Sri Lanka	1948
Jamaica	1962	Swaziland	1968
Kenya	1963	Tanzania	1961
Kiribati	1979	Tonga[5]	1970
Lesotho	1966	Trinidad and Tobago	1962
Malawi	1964	Tuvalu	1978
Malaysia	1957	Uganda	1982
Maldives	1982	United Kingdom	1931
Malta	1964	Vanuatu	1980
Mauritius	1968	Zambia	1964
Mozambique	1995		

[1]Independence given legal effect by the Statute of Westminster 1931. [2]Nauru joined as a special member on independence in 1968. It became a full member in 1999 but its status was changed back to that of a special member in 2006. [3]Nigeria was suspended in 1995 but readmitted as a full member in 1999. [4]Left 1972, rejoined 1989. [5]Brunei and Tonga had been sovereign states in treaty relationship with Britain. [6]Fiji left in 1987 but rejoined in 1997. It changed its name to Fiji Islands in 1998. The Fiji Islands were suspended in Sept. 2009 although technically they remain a member. [7]Left 1961, rejoined 1994.

Aims and Conditions of Membership. Membership involves acceptance of certain core principles, as set out in the Harare Declaration of 1991, and is subject to the approval of other member states. The Harare Declaration charted a course to take the Commonwealth into the 21st century affirming members' continued commitment to the Singapore Declaration of 1971, by which members committed themselves to the pursuit of world peace and support of the UN.

The core principles defined by the Harare Declaration are: political democracy, human rights, good governance and the rule of law, and the protection of the environment through sustainable development. Commitment to these principles was made binding as a condition of membership at the 1993 Heads of Government meeting in Cyprus.

The Millbrook Action Programme of 1995 aims to support countries in implementing the Harare Declaration, providing assistance in constitutional and judicial matters, running elections, training and technical advice. Violations of the Harare

Declaration will provoke a series of measures by the Commonwealth Secretariat, including: expression of disapproval, encouragement of bilateral actions by member states, appointment of fact-finders and mediators, stipulation of a period for the restoration of democracy, exclusion from ministerial meetings, suspension of all participation and aid and finally punitive measures including trade sanctions. A nine-member *Commonwealth Ministerial Action Group on the Harare Declaration* (CMAG) may be convened by the Secretary-General as and when necessary to deal with violations. The Group held its first meeting in Dec. 1995. Its terms of reference are as set out in the Millbrook Action Programme.

The *Commonwealth Parliamentary Association* was founded in 1911. As defined by its constitution, its objectives are to 'promote knowledge of the constitutional, legislative, economic, social and cultural aspects of parliamentary democracy'. It meets these objectives by organizing conferences, meetings and seminars for members, arranging exchange visits between members, publishing books, newsletters, reports, studies and a quarterly journal and providing an information service. Its principal governing body is the General Assembly, which meets annually during the Commonwealth Parliamentary Conference and is composed of members attending that Conference as delegates. The Association elects an Executive Committee comprising a Chair, President, Vice-President, Treasurer and 27 regional representatives, which meets twice a year. The Chair is elected for three-year terms.

Commonwealth Secretariat. The Commonwealth Secretariat is an international body at the service of all 54 member countries. It provides the central organization for joint consultation and co-operation in many fields. It was established in 1965 by Commonwealth Heads of Government as a 'visible symbol of the spirit of co-operation which animates the Commonwealth', and has observer status at the UN General Assembly.

The Secretariat disseminates information on matters of common concern, organizes and services meetings and conferences, co-ordinates many Commonwealth activities, and provides expert technical assistance for economic and social development through the multilateral Commonwealth Fund for Technical Co-operation. The Secretariat is organized in divisions and sections which correspond to its main areas of operation: political affairs, economic affairs, human rights, gender affairs, youth affairs, education, information, law, health and a range of technical assistance and advisory services. Within this structure the Secretariat organizes the biennial meetings of Commonwealth Heads of Government (CHOGMs), annual meetings of Finance Ministers of member countries, and regular meetings of Ministers of Education, Law, Health, Gender Affairs and others as appropriate. To emphasize the multilateral nature of the association, meetings are held in different cities and regions within the Commonwealth. Heads of Government decided that the Secretariat should work from London as it has the widest range of communications of any Commonwealth city, as well as the largest assembly of diplomatic missions from Commonwealth member countries.

Commonwealth Heads of Government Meetings (CHOGMs). Outside the UN, the CHOGM remains the largest inter-governmental conference in the world. Meetings are held every two years. The 2002 CHOGM in Coolum, Australia, scheduled for Oct. 2001 but postponed following the attacks on the United States of 11 Sept. 2001, was dominated by the Zimbabwe issue, as was the meeting held in Abuja, Nigeria in Dec. 2003. The last meeting was held in Oct. 2011 in Australia. The next CHOGM is scheduled to be held in Sri Lanka in Nov. 2013. A host of Commonwealth organizations and agencies are dedicated to enhancing inter-Commonwealth relations and the development of the potential of Commonwealth citizens. They are listed in the *Commonwealth Yearbook*, which is published by the Secretariat.

Commonwealth Day is celebrated on the second Monday in March each year. The theme for 2013 was 'Opportunity through Enterprise'.

Overseas Territories and Associated States. There are 14 United Kingdom overseas territories (*see* pages 1318–34), seven Australian external territories (*see* pages 155–8), two New Zealand dependent territories and two New Zealand associated states (*see* pages 925–8). A dependent territory is a territory belonging by settlement, conquest or annexation to the British, Australian or New Zealand Crown.

United Kingdom Overseas Territories administered through the Foreign and Commonwealth Office comprise, in the Indian Ocean: British Indian Ocean Territory; in the Mediterranean: Gibraltar, the Sovereign Base Areas of Akrotiri and Dhekelia in Cyprus; in the Atlantic Ocean: Bermuda, Falkland Islands, South Georgia and South Sandwich Islands, British Antarctic Territory, St Helena and Dependencies (Ascension and Tristan da Cunha); in the Caribbean: Montserrat, British Virgin Islands, Cayman Islands, Turks and Caicos Islands, Anguilla; in the Western Pacific: Pitcairn Group of Islands.

The Australian external territories are: Ashmore and Cartier Islands, Australian Antarctic Territory, Christmas Island, Cocos (Keeling) Islands, Coral Sea Islands, Heard and McDonald Islands and Norfolk Island. The New Zealand external territories are: Tokelau Islands and the Ross Dependency. The New Zealand associated states are: Cook Islands and Niue.

Headquarters: Marlborough House, Pall Mall, London SW1Y 5HX, UK.
Website: http://www.thecommonwealth.org
Email: info@commonwealth.int
Secretary-General: Kamalesh Sharma (India).

Selected publications. Commonwealth Yearbook; Commonwealth Today (biannual); *The Commonwealth at the Summit: Communiqués of Commonwealth Heads of Government Meetings.*

Further Reading

The Cambridge History of the British Empire. 8 vols. 1929 ff.
Chan, S., *Twelve Years of Commonwealth Diplomatic History: Summit Meetings, 1979–1991.* 1992
Judd, D. and Slinn, P., *The Evolution of the Modern Commonwealth.* 1982
Madden, F. and Fieldhouse, D., (eds.) *Selected Documents on the Constitutional History of the British Empire and Commonwealth.* 1994
Mayall, James, (ed.) *The Contemporary Commonwealth.* 2009
McIntyre, W. D., *The Significance of the Commonwealth, 1965–90.* 1991
Moore, R. J., *Making the New Commonwealth.* 1987

Commonwealth of Independent States (CIS)

The Commonwealth of Independent States, founded on 8 Dec. 1991 in Viskuli, a government villa in Belarus, is a community of independent states which proclaimed itself the successor to the Union of Soviet Socialist Republics in some aspects of international law and affairs. The member states are the founders, Russia, Belarus and Ukraine, and seven subsequent adherents: Armenia, Azerbaijan, Kazakhstan, Kyrgyzstan, Moldova, Tajikistan and Uzbekistan. Turkmenistan withdrew its permanent member status on 26 Aug. 2005 and became an associate member. Georgia withdrew on 18 Aug. 2008, with effect from 17 Aug. 2009.

History. Extended negotiations in the Union of Soviet Socialist Republics (USSR) in 1990 and 1991 sought to establish a 'renewed federation' or, subsequently, to conclude a new union treaty that would embrace all the 15 constituent republics of the USSR at that date. In Sept. 1991 the three Baltic republics—Estonia, Latvia and Lithuania—were recognized as independent states by the USSR State Council, and subsequently by the international community. Most of the remaining republics reached agreement on the broad outlines of a new 'union of sovereign states' in Nov. 1991, which would have retained a directly elected President and an all-union legislature, but which would have limited central authority to those powers specifically delegated to it by the members of the union.

A referendum in Ukraine in Dec. 1991, however, showed overwhelming support for full independence, and following this Russia, Belarus and Ukraine concluded the Minsk Agreement on 8 Dec. 1991, establishing a Commonwealth of Independent States (CIS), headquartered in Minsk. Each of the three republics individually renounced the 1922 treaty through which the USSR had been established.

In Dec. 1991 a further declaration was signed with eight other republics: Armenia, Azerbaijan, Kazakhstan, Kyrgyzstan, Moldova, Tajikistan, Turkmenistan and Uzbekistan. The declaration committed signatories to recognize the independence and sovereignty of other members, to respect human rights including those of national minorities, and to the observance of existing boundaries. Relations among the members of the CIS were to be conducted on an equal, multilateral, interstate basis, but it was agreed to endorse the principle of unitary control of strategic nuclear arms and the concept of a 'single economic space'. In a separate agreement the heads of member states agreed that Russia should take up the seat at the United Nations formerly occupied by the USSR, and a framework of interstate and intergovernment consultation was established. On 26 Dec. the USSR Supreme Soviet voted a formal end to the 1922 Treaty of Union, and dissolved itself. Georgia decided to join on 9 Dec. 1993 and on 1 March 1994 the national parliament ratified the act.

The Charter, adopted on 22 Jan. 1993 in Minsk, proclaims that the Commonwealth is based on the principles of the sovereign equality of all members. It is not a state and does not have supranational authority.

Activities and Institutions. The principal organs of the CIS, according to the agreement concluded in Alma-Ata on 21 Dec. 1991, are the *Council of Heads of States*, which meets twice a year, and the *Council of Heads of Government*, which meets every three months. Both councils may convene extraordinary sessions, and may hold joint sittings. There is also a *Council of Defence Ministers*, established in Feb. 1992, and a *Council of Foreign Ministers* (Dec. 1993). The Secretariat is the standing working organ.

At a summit meeting of heads of states (with the exception of Azerbaijan) in July 1992, agreements were reached on a way to divide former USSR assets abroad; on the legal cessionary of state archives of former Soviet states; on the status of an Economic Court; and on collective security. In 1992 an *Inter-Parliamentary Assembly* was established by seven member states (Armenia, Belarus, Kazakhstan, Kyrgyzstan, Russia, Tajikistan and Uzbekistan).

At a subsequent meeting in Jan. 1993 Armenia, Belarus, Kazakhstan, Kyrgyzstan, Russia, Tajikistan and Uzbekistan agreed on a charter to implement co-operation in political, economic, ecological, humanitarian, cultural and other spheres; thorough and balanced economic and social development within the common economic space; interstate co-operation and integration; and to ensure human rights and freedoms. Heads of State established an *Inter-State Bank* and adopted a Provision on it. Its charter was signed by ten Heads of State (Armenia, Belarus, Kazakhstan, Kyrgyzstan, Moldova, Russia, Tajikistan, Turkmenistan, Ukraine, Uzbekistan) on 22 Dec. 1993.

In accordance with the Agreement on Armed Forces and Border Troops, concluded on 30 Dec. 1991, it was decided to consider and solve the issue on the transference of the management of the General-Purpose Armed Forces in accordance with the national legislation of member states. On 14 Feb. 1992 the *Council of Defence Ministers* was established. In 1993 the Office of Commander-in-Chief of CIS Joint Armed Forces was reorganized in a Staff for Coordinating Military Co-operation.

On 24 Sept. 1993 Armenia, Azerbaijan, Belarus, Kazakhstan, Kyrgyzstan, Moldova, Russia, Tajikistan and Uzbekistan signed an agreement to form an *Economic Union*. Georgia and Turkmenistan signed later (14 and 23 Jan. 1994). Ukraine became an associated member on 15 April 1994. In Oct. 1994 a summit meeting established the *Inter-State Economic Committee (MEK)* to be based in Moscow. Members include all CIS states except Turkmenistan. A *Customs Union* to regulate payments between member states with nonconvertible independent currencies and a regulatory *Economic Court* have also been established.

On 29 March 1996 Belarus, Kazakhstan, Kyrgyzstan and Russia signed an agreement increasing their mutual economic and social integration by creating a *Community of Integrated States* (Tajikistan signed in 1998). The agreement established a Supreme Inter-Governmental Council comprising heads of state and government and foreign ministers, with, an integration committee of Ministers and an Inter-Parliamentary Committee. On 2 April 1996 the Presidents of Belarus and Russia signed a treaty providing for political, economic and military integration, creating the nucleus of a *Community of Russia and Belarus*. A further treaty was signed on 22 May 1997, instituting common citizenship, common deployment of military forces and the harmonization of the two economies with a view to the creation of a common currency. The Community was later renamed the *Union of Belarus and Russia* and signed subsequent agreements on equal rights for its citizens and equal conditions for state and private entrepreneurship. In March 1994 the CIS was accorded observer status in the UN.

Headquarters: 220030 Minsk, Kirova 17, Belarus.
Website (Russian only): http://www.cis.minsk.by
Executive Secretary: Sergei Lebedev (Russia).

Further Reading

Brzezinski, Z. and Sullivan, P. (eds.) *Russia and the Commonwealth of Independent States: Documents, Data and Analysis.* 1996

Eurasian Economic Community (EurAsEC)

EurAsEC was formed in 2000 to develop economic co-operation and trade and to establish a customs union. The UN General Assembly awarded the organization observer status in 2003. Between 2007 and 2010 a customs union between Belarus, Kazakhstan and Russia was created.

Membership. EurAsEC has five member states (Belarus, Kazakhstan, Kyrgyzstan, Russia, Tajikistan). Uzbekistan joined in 2006 but withdrew in 2008. Armenia, Moldova and Ukraine have observer status, as do the Interstate Aviation Committee and the Eurasian Development Bank.

Organization. The Interstate Council is the supreme governing body, comprising heads of state and government. The Integration Committee has offices in Almaty and Moscow, while the Inter-Parliamentary Assembly is based in St Petersburg and the Community Court has its headquarters in Minsk.

Headquarters: 6 1st Basmanny Pereulok, Bldg 4, Moscow 105066, Russia.
Website: http://www.evrazes.com
Secretary-General: Tair Mansurov (Kazakhstan).

International Air Transport Association (IATA)

Founded in 1945 for inter-airline co-operation in promoting safe, reliable, secure and economical air services, IATA has approximately 240 members from 115 nations worldwide. IATA is the successor to the International Air Traffic Association, founded in The Hague in 1919, the year of the world's first international scheduled services.

Main offices: IATA Centre, Route de l'Aéroport 33, PO Box 416, CH-1215 Geneva, Switzerland. 800 Place Victoria, PO Box 113, Montreal, Quebec, Canada H4Z 1M1. 111 Somerset Road, #14-05 Somerset Wing, Singapore 238164.
Website: http://www.iata.org
Director-General: Tony Tyler (UK).

International Committee of the Red Cross (ICRC)

The International Committee of the Red Cross (ICRC) is a Swiss-based impartial, neutral and independent organization ensuring humanitarian protection and assistance for victims of war and other situations of violence.

Established in 1863, the ICRC is a founding member of the International Red Cross and Red Crescent Movement and of international humanitarian law, notably the Geneva Conventions.

The ICRC is mandated by the international community to be the guardian and promoter of international humanitarian law. It has a permanent mandate under international law to take impartial action for prisoners, the wounded and sick, and civilians affected by conflict.

The ICRC aims to ensure that civilians not taking part in hostilities are spared and protected; to visit prisoners of war and security detainees and ensure that they are treated humanely and according to recognized international standards that forbid torture and other forms of abuse; to transmit messages to and reunite family members separated by armed conflict; to help find missing persons; to offer or facilitate access to basic health care facilities; to provide food, safe drinking water, sanitation and shelter in emergencies; to promote respect for, monitor compliance with and contribute to the development of international humanitarian law; to help reduce the impact of mines and explosive remnants of war on people; and to support national Red Cross and Red Crescent Societies to prepare for and respond to armed conflict and situations of violence.

The ICRC is a global presence with offices in over 80 countries and some 12,000 staff worldwide. Its HQ is in Geneva, Switzerland.

Headquarters: 19 avenue de la Paix, CH-1202 Geneva, Switzerland.
Website: http://www.icrc.org
President: Peter Maurer (Switzerland).

Further Reading

Forsythe, David P., *The Humanitarians: The International Committee of the Red Cross.* 2005
Forsythe, David P. and Rieffer-Flanagan, Barbara Ann J., *The International Committee of the Red Cross: A Neutral Humanitarian Actor.* 2007
Moorehead, Caroline, *Dunant's Dream: War, Switzerland and the History of the Red Cross.* 1998

International Criminal Court (ICC)

Origin. As far back as 1946 an international congress called for the adoption of an international criminal code prohibiting crimes against humanity and the prompt establishment of an international criminal court, but for more than 40 years little progress was made. In 1989 the end of the Cold War brought a dramatic increase in the number of UN peacekeeping operations and a world where the idea of establishing an International Criminal Court became more viable. The United Nations Conference of Plenipotentiaries on the Establishment of an International Criminal Court took place from 15 June–17 July 1998 in Rome, Italy.

Aims and Activities. The International Criminal Court is a permanent court for trying individuals who have been accused of committing genocide, war crimes and crimes against humanity, and is thus a successor to the *ad hoc* tribunals set up by the UN Security Council to try those responsible for atrocities in the former Yugoslavia and Rwanda. Ratification by 60 countries was required to bring the statute into effect. The court began operations on 1 July 2002 with 139 signatories and after ratification by 76 countries. By Feb. 2013 the number of ratifications had increased to 121. Its first trial, with Thomas Lubanga facing war crimes charges for his role in the Democratic Republic of the Congo's civil war, opened on 26 Jan. 2009 and was not concluded until 14 March 2012. Lubanga was found guilty of conscripting and enlisting children under the age of 15 and using them to participate in hostilities.

Judges. The International Criminal Court's first 18 judges were elected in Feb. 2003, with six serving for three years, six for six years and six for nine years. Every three years six new judges are elected. Anthony Carmona of Trinidad and Tobago was elected in Dec. 2011 and became a judge in March 2012 but resigned in Feb. 2013 to become the country's president. At the time of going to print a successor had yet to be elected. At present the 17 judges, with the year in which their term of office is scheduled to end, are: Joyce Aluoch (Kenya, 2018); Chile Eboe-Osuji (Nigeria, 2021); Silvia Fernández de Gurmendi (Argentina, 2018); Robert Fremr (Czech Republic, 2021); Olga Venecia Herrera Carbuccia (Dominican Republic, 2021); Hans-Peter Kaul (Germany, 2015); Erkki Kourula (Finland, 2015); Akua Kuenyehia (Ghana, 2015); Sanji Mmasenono Monageng (Botswana, 2018); Howard Morrison (United Kingdom, 2021); Kuniko Ozaki (Japan, 2018); Song Sang-hyun (South Korea, 2015); Miriam Defensor Santiago (Philippines, 2021); Cuno Tarfusser (Italy, 2018); Ekaterina Trendafilova (Bulgaria, 2015); Anita Ušacka (Latvia, 2015); Christine Van Den Wyngaert (Belgium, 2018).

Prosecutor. Fatou Bensouda (Gambia) was unanimously elected the second prosecutor of the Court on 12 Dec. 2011 and succeeded Luis Moreno-Ocampo (Argentina) on 16 June 2012.

Headquarters: Maanweg 174, 2516 AB The Hague, Netherlands.
Website: http://www.icc-cpi.int
President: Song Sang-hyun (South Korea).

Further Reading

Baker, Michael N. (ed.) *International Criminal Court: Developments and U.S. Policy.* 2012
Macedo, Stephen, (ed.) *Universal Jurisdiction: National Courts and the Prosecution of Serious Crimes Under International Law.* 2003
Mendes, Errol, *Peace and Justice at the International Criminal Court: A Court of Last Resort.* 2010
Schabas, William A., *An Introduction to the International Criminal Court.* 4th ed. 2011
Struett, Michael J., *The Politics of Constructing the International Criminal Court: NGOs, Discourse, and Agency.* 2008

International Institute for Democracy and Electoral Assistance (IDEA)

Created in 1995, International IDEA is an intergovernmental organization that supports sustainable democratic change through providing comparative knowledge, assisting in democratic reform, and influencing policies and politics. International IDEA focuses on the ability of democratic institutions to deliver a political system marked by public participation and inclusion, representative and accountable government, responsiveness to citizens' needs and aspirations, and the rule of law and equal rights for all citizens.

Aims and Activities. International IDEA undertakes work through three activity areas: providing comparative knowledge derived from practical experience on democracy-building processes—elections and referendums, constitutions, political parties, women's political empowerment and democracy self-assessments—from diverse contexts around the world; assisting political actors in reforming democratic institutions and processes, and engaging in political processes when invited to do so; influencing democracy-building policies and assistance to political actors.

Membership. The International IDEA had 27 full member states and one observer state in Feb. 2013.

Organization. IDEA has regional operations in Latin America, Africa, the Middle East, Asia and the Pacific, and has a staff of over 70 worldwide.

> *Headquarters:* Strömsborg, 103 34 Stockholm, Sweden.
> *Website:* http://www.idea.int
> *Secretary-General:* Vidar Helgesen (Norway).

International Mobile Satellite Organization (IMSO)

Founded in 1979 as the International Maritime Satellite Organization (Inmarsat) to establish a satellite system to improve maritime communications for distress and safety and commercial applications. Its competence was subsequently expanded to include aeronautical and land mobile communications. Privatization, which was completed in April 1999, transferred the business to a newly created company and the Organization remains as a regulator to ensure that the company fulfils its public services obligations. The company has taken the Inmarsat name and the Organization uses the acronym IMSO. In Feb. 2013 the Organization had 97 member parties.

Organization. The Assembly of all Parties to the Convention meets every two years.

> *Headquarters:* 99 City Road, London EC1Y 1AX, UK.
> *IMSO Website:* http://www.imso.org
> *Email:* info@imso.org
> *Inmarsat Website:* http://www.inmarsat.com
> *Director of the Secretariat, IMSO:* Esteban Pachá Vicente (Spain).
> *Chief Executive, Inmarsat Ltd:* Andrew Sukawaty (USA).

International Olympic Committee (IOC)

Founded in 1894 by French educator Baron Pierre de Coubertin, the International Olympic Committee is an international non-governmental, non-profit organization whose members act as the IOC's representatives in their respective countries, not as delegates of their countries within the IOC. The Committee's main responsibility is to supervise the organization of the summer and winter Olympic Games. It owns all rights to the Olympic symbols, flag, motto, anthem and Olympic Games.

Aims. 'To contribute to building a peaceful and better world by educating youth through sport, practised without discrimination of any kind and in the Olympic Spirit, which requires mutual understanding with a spirit of friendship, solidarity and fair play.'

Finances. The IOC receives no public funding. Its only source of funding is from private sectors, with the substantial part of these revenues coming from television broadcasters and sponsors.

> *Address:* Château de Vidy, Case Postale 356, CH-1007 Lausanne, Switzerland.
> *Website:* http://www.olympic.org
> *President:* Jacques Rogge (Belgium).

International Organisation of La Francophonie

The International Organisation of La Francophonie represents 77 countries and provinces/regions (including 20 with observer status) using French as an official language. It estimates that there are 220m. French speakers worldwide. Objectives include the promotion of peace, democracy, and economic and social development, through political and technical co-operation. The Secretary-General is based in Paris.

Members. Albania, Andorra, Armenia, Belgium, Benin, Bulgaria, Burkina Faso, Burundi, Cambodia, Cameroon, Canada, Canada—New Brunswick, Canada—Quebec, Cape Verde, Central African Republic, Chad, Comoros, Democratic Republic of the Congo, Republic of the Congo, Côte d'Ivoire, Djibouti, Dominica, Egypt, Equatorial Guinea, France, French Community of Belgium, Gabon, Greece, Guinea, Guinea-Bissau, Haiti, Laos, Lebanon, Luxembourg, Macedonia, Madagascar, Mali, Mauritania, Mauritius, Moldova, Monaco, Morocco, Niger, Romania, Rwanda, St Lucia, São Tomé e Príncipe, Senegal, Seychelles, Switzerland, Togo, Tunisia, Vanuatu, Vietnam. *Associate Members.* Cyprus, Ghana, Qatar. *Observers.* Austria, Bosnia and Herzegovina, Croatia, Czech Republic, Dominican Republic, Estonia, Georgia, Hungary, Latvia, Lithuania, Montenegro, Mozambique, Poland, Serbia, Slovakia, Slovenia, Thailand, Ukraine, United Arab Emirates, Uruguay.

> *Headquarters:* 19—21 avenue Bosquet, 75326 Paris, France.
> *Website (limited English):* http://www.francophonie.org
> *Secretary-General:* Abdou Diouf (Senegal).

International Organization for Migration (IOM)

Established in 1951, the International Organization for Migration (IOM) is the principal intergovernmental organization in the field of migration.

Members (149 as of Feb. 2013). Afghanistan, Albania, Algeria, Angola, Antigua and Barbuda, Argentina, Armenia, Australia, Austria, Azerbaijan, Bahamas, Bangladesh, Belarus, Belgium, Belize, Benin, Bolivia, Bosnia and Herzegovina, Botswana, Brazil,

Bulgaria, Burkina Faso, Burundi, Cambodia, Cameroon, Canada, Cape Verde, Central African Republic, Chad, Chile, Colombia, Comoros, Democratic Republic of the Congo, Republic of the Congo, Costa Rica, Côte d'Ivoire, Croatia, Cyprus, Czech Republic, Denmark, Djibouti, Dominican Republic, Ecuador, Egypt, El Salvador, Estonia, Ethiopia, Finland, France, Gabon, Gambia, Georgia, Germany, Ghana, Greece, Guatemala, Guinea, Guinea-Bissau, Guyana, Haiti, Holy See, Honduras, Hungary, India, Iran, Ireland, Israel, Italy, Jamaica, Japan, Jordan, Kazakhstan, Kenya, South Korea, Kyrgyzstan, Latvia, Lesotho, Liberia, Libya, Lithuania, Luxembourg, Madagascar, Maldives, Mali, Malta, Mauritania, Mauritius, Mexico, Micronesia, Moldova, Mongolia, Montenegro, Morocco, Mozambique, Myanmar, Namibia, Nauru, Nepal, Netherlands, New Zealand, Nicaragua, Niger, Nigeria, Norway, Pakistan, Panama, Papua New Guinea, Paraguay, Peru, Philippines, Poland, Portugal, Romania, Rwanda, St Vincent and the Grenadines, Senegal, Serbia, Seychelles, Sierra Leone, Slovakia, Slovenia, Somalia, South Africa, South Sudan, Spain, Sri Lanka, Sudan, Swaziland, Sweden, Switzerland, Tajikistan, United Republic of Tanzania, Thailand, Timor-Leste, Togo, Trinidad and Tobago, Tunisia, Turkey, Uganda, Ukraine, UK, USA, Uruguay, Vanuatu, Venezuela, Vietnam, Yemen, Zambia and Zimbabwe. A further 12 countries have observer status.

Activities. IOM works to help ensure the orderly and humane management of migration, to promote international co-operation on migration issues, to assist in the search for practical solutions to migration problems and to provide humanitarian assistance to migrants in need, be they refugees, displaced persons or other uprooted people. The IOM Constitution gives explicit recognition to the link between migration and economic, social and cultural development, as well as to the right of freedom of movement of persons. IOM works in the four broad areas of migration management: migration and development, facilitating migration, regulating migration and addressing forced migration. IOM's programme budget for 2010 exceeded US$1bn., funding over 2,820 active programmes and more than 7,000 staff members serving in over 420 field offices in more than 100 countries.

Official languages: English, French, Spanish.
Headquarters: Route des Morillons 17, POB 71, CH-1211 Geneva 19, Switzerland.
Website: http://www.iom.int
Director-General: William Lacy Swing (USA).

International Organization for Standardization (ISO)

Established in 1947, the International Organization for Standardization is a non-governmental federation of national standards bodies from 161 countries worldwide, one from each country. ISO's work results in international agreements which are published as International Standards. The first ISO standard was published in 1951 with the title 'Standard reference temperature for industrial length measurement'.

Some 19,000 ISO International Standards are available on subjects in such diverse fields as information technology, textiles, packaging, distribution of goods, energy production and utilization, building, banking and financial services. ISO standardization activities include the widely recognized ISO 9000 family of quality management system and standards and the ISO 14000 series of environmental management system standards. Standardization programmes are now being developed in

completely new fields, such as food safety, security, social responsibility and the service sector.

Mission. To promote the development of standardization and related activities in the world with a view to facilitating the international exchange of goods and services, and to developing co-operation in the spheres of intellectual, scientific, technological and economic activity.

Headquarters: 1 chemin de la Voie-Creuse, Case postale 56, CH-1211 Geneva 20, Switzerland.
Website: http://www.iso.org
Secretary-General: Rob Steele (New Zealand).

International Road Federation (IRF)

The IRF is a non-profit, non-political service organization whose purpose is to encourage better road and transportation systems worldwide and to help apply technology and management practices to give maximum economic and social returns from national road investments.

Founded following the Second World War, over the years the IRF has led major global road infrastructure developments, including achieving 1,000 km of new roads in Mexico in the 1950s, and promoting the Pan-American Highway linking North and South America. It publishes *World Road Statistics*, as well as road research studies, including a 140-country inventory of road and transport research in co-operation with the US Bureau of Public Roads.

Headquarters: 2 chemin de Blandonnet, CH-1214 Vernier, Geneva, Switzerland.
Website: http://www.irfnet.org
Chairman: Kiran K. Kapila (India).

International Seabed Authority (ISA)

The ISA is an autonomous international organization established under the UN Convention on the Law of the Sea (UNCLOS) of 1982 and the 1994 Agreement relating to the implementation of Part XI of the UNCLOS. It came into existence on 16 Nov. 1994 and became fully operational in June 1996.

The administrative expenses are met from assessed contributions from its members. Membership numbered 165 in Feb. 2013; the budget for the biennium 2011–12 was US$13,014,700.

The UNCLOS covers almost all ocean space and its uses: navigation and overflight, resource exploration and exploitation, conservation and pollution, fishing and shipping. It entitles coastal states and inhabitable islands to proclaim a 12-mile territorial sea, a contiguous zone, a 200-mile exclusive economic zone and an extended continental shelf (in some cases). Its 320 Articles and nine Annexes constitute a guide for behaviour by states in the world's oceans, defining maritime zones, laying down rules for drawing sea boundaries, assigning legal rights, duties and responsibilities to States, and providing machinery for the settlement of disputes.

Organization. The Assembly, consisting of representatives from all member states, is the supreme organ. The 36-member Council, elected by the Assembly, includes the four largest importers or consumers of seabed minerals, four largest investors in seabed minerals, four major exporters of the same, six developing countries representing special interests and 18 members from all the geographical regions. The Council is the executive organ of the Authority. There are also two subsidiary bodies: the Legal and Technical Commission (currently 25 experts) and the Finance Committee (currently 15 experts). The Secretariat serves all the bodies of the Authority and under the 1994 Agreement is performing functions of the Enterprise (until such time as it starts to operate independently of the Secretariat). The Enterprise is the organ through which the ISA carries out deep seabed activities directly or through joint ventures.

Activities. In July 2000 the ISA adopted the Regulations for Prospecting and Exploration for Polymetallic Nodules in the Area. Pursuant thereto, it signed exploration contracts with eight contractors who have submitted plans of work for deep seabed exploration. These are: Institut Français de Recherche pour l'Exploitation de la Mer (IFREMER) and Association Française pour l'Etude de la Recherche des Nodules (AFERNOD), France; Deep Ocean Resources Development Co. Ltd (DORD), Japan; State Enterprise Yuzhmorgeologiya, Russian Federation; China Ocean Minerals Research and Development Association (COMRA); Interoceanmetal Joint Organization (IOM), a consortium sponsored by Bulgaria, Cuba, Czech Republic, Poland, Russia and Slovakia; the government of the Republic of Korea; the Republic of India; and the Federal Institute for Geosciences and Natural Resources, Germany.

Between 1998 and 2010 the ISA organized 12 workshops on a range of topics, including: the development of guidelines for the assessment of the possible environmental impacts arising from exploration for polymetallic nodules; a standardized system of data interpretation; and prospects for international collaboration in marine environmental research. While continuing to develop a database on polymetallic nodules (POLYDAT), the Authority has also made significant progress towards the establishment of a central data repository for all marine minerals in the deep seabed.

Headquarters: 14−20 Port Royal St., Kingston, Jamaica.
Website: http://www.isa.org.jm
Secretary-General: Nii Allotey Odunton (Ghana).

Publications. Handbook 2012; Selected Decisions and Documents from the Authority's Sessions; various others.

International Telecommunications Satellite Organization (ITSO)

Founded in 1964 as Intelsat, the organization was the world's first commercial communications satellite operator. Today, with capacity on a fleet of geostationary satellites and expanding terrestrial network assets, Intelsat continues to provide connectivity for telephony, corporate network, broadcast and internet services.

Organization. In 2001 the member states of the organization implemented restructuring by transferring certain assets to Intelsat Ltd, a new commercial company under the supervision of the International Telecommunications Satellite Organization, now known as ITSO. In 2009 Intelsat Ltd moved its corporate headquarters to Luxembourg and became Intelsat S.A. ITSO's mission is to ensure that Intelsat provides public tele-communications services, including voice, data and video, on a global and non-discriminatory basis. The governing body of ITSO is the Assembly of Parties, which normally meets every other year. The Executive Organ is headed by the Director-General and is responsible to the Assembly of Parties. The Director-General supervises and monitors Intelsat's provision of public tele-communications services. There were 150 member countries in Feb. 2013.

Headquarters: 3400 International Drive, NW, Suite 3M−100, Washington, D.C., 20008−3006, USA.
Website: http://www.itso.int
Director-General: José Manuel Toscano (Portugal).

International Trade Union Confederation (ITUC)

Origin. Founded in Nov. 2006, the ITUC was formed through a unification process that included the merger of the International Confederation of Free Trade Unions (ICFTU) and the World Confederation of Labour (WCL) with the addition of several national centres that had not been affiliated with either organization. The WCL was established in 1920 as the International Federation of Christian Trade Unions, but went briefly out of existence in 1940 owing to the suppression of affiliated unions by the Nazi and Fascist regimes. It reconstituted in 1945 and became the WCL in 1968. The founding congress of the ICFTU took place in London in Dec. 1949 following the withdrawal of some Western trade unions from the World Federation of Trade Unions (WFTU), which was founded in 1945 but had come under Communist control.

In Feb. 2013 the ITUC represented 175m. members of 315 affiliates in 156 countries and territories.

Aims. The ITUC aims to defend and promote the rights of workers, particularly the right to trade union organization and collective bargaining; to combat discrimination at work and in society; to ensure that social concerns are put at the centre of global economic, trade and finance policies; to support young people's rights at work; and to promote the involvement of women in trade unions. In 2006 it also ran campaigns against child labour and to promote the prevention of HIV/AIDS.

Organization. The Congress meets every four years to set policies and to elect the General Secretary and the General Council, composed of 70 members, which is the main decision-making body between congresses. The President and Deputy Presidents are appointed by the General Council. The Founding Congress was held in Vienna in Nov. 2006. Its second Congress was held in 2010 in Vancouver.

The ITUC has regional organizations for Africa, the Americas, Asia-Pacific and Europe. It has offices that deal with the International Labour Organization (Geneva), and the United Nations, the World Bank and the International Monetary Fund (Washington, D.C.). There are also offices in Amman, Hong Kong, Moscow, Sarajevo and Vilnius.

The ITUC is a member of the Council of Global Unions, which was created in 2006 as a tool for structured co-operation and co-ordination.

Headquarters: Bd du Roi Albert II, N°5, bte 1, Brussels 1210, Belgium.
Website: http://www.ituc-csi.org
General Secretary: Sharan Burrow (Australia).
President: Michael Sommer (Germany).

International Tribunal for the Law of the Sea (ITLOS)

The International Tribunal for the Law of the Sea (ITLOS), founded in Oct. 1996 and based in Hamburg, adjudicates on disputes relating to the interpretation and application of the United Nations Convention on the Law of the Sea. The Convention gives the Tribunal jurisdiction to resolve a variety of international law of the sea disputes such as the delimitation of maritime zones, fisheries, navigation and the protection of the marine environment. Its Seabed Disputes Chamber has compulsory jurisdiction to resolve disputes amongst States, the International Seabed Authority, companies and private individuals, arising out of the exploitation of the deep seabed. The Tribunal also has compulsory jurisdiction in certain instances to protect the rights of parties to a dispute or to prevent serious harm to the marine environment, and over the prompt release of arrested vessels and their crews upon the deposit of a security. The jurisdiction of the Tribunal also extends to all matters specifically provided for in any other agreement which confers jurisdiction on the Tribunal. The Tribunal is composed of 21 judges, elected by signatories from five world regional blocs: five each from Africa and Asia; four from Western Europe and other States; four from Latin America and the Caribbean; and three from Eastern Europe. The judges serve a term of nine years, with one third of the judges' terms expiring every three years.

Headquarters: Am Internationalen Seegerichtshof 1, 22609 Hamburg, Germany.
Website: http://www.itlos.org
Registrar: Philippe Gautier (Belgium).

International Union Against Cancer (UICC)

Founded in 1933, the UICC is an international non-governmental association of 770 member organizations in 155 countries.

Objectives. The UICC is the only non-governmental organization dedicated exclusively to the global control of cancer. Its objectives are to advance scientific and medical knowledge in research, diagnosis, treatment and prevention of cancer, and to promote all other aspects of the campaign against cancer throughout the world. Particular emphasis is placed on professional and public education.

Membership. The UICC is made up of voluntary cancer leagues, patient organizations, associations and societies as well as cancer research and treatment centres and, in some countries, ministries of health.

Activities. The UICC creates and carries out programmes around the world in collaboration with several hundred volunteer experts, most of whom are professionally active in UICC member organizations. It promotes co-operation between cancer organizations, researchers, scientists, health professionals and cancer experts, with a focus in four key areas: building and enhancing cancer control capacity, tobacco control, population-based cancer prevention and control, and transfer of cancer knowledge and dissemination. The next UICC World Cancer Congress is scheduled to take place in Melbourne, Australia in 2014.

Address: 62 route de Frontenex, CH-1207 Geneva, Switzerland.
Website: http://www.uicc.org
President: Eduardo Cazap (Argentina).
Executive Director: Cary Adams (UK).

Inter-Parliamentary Union (IPU)

Founded in 1889 by William Randal Cremer (UK) and Frédéric Passy (France), the Inter-Parliamentary Union was the first permanent forum for political multilateral negotiations. The Union is a centre for dialogue and parliamentary diplomacy among legislators representing every political system and all the main political leanings in the world. It was instrumental in setting up what is now the Permanent Court of Arbitration in The Hague.

Activities. The IPU fosters contacts, co-ordination and the exchange of experience among parliaments and parliamentarians of all countries; considers questions of international interest and concern, and expresses its views on such issues in order to bring about action by parliaments and parliamentarians; contributes to the defence and promotion of human rights—an essential factor of parliamentary democracy and development; contributes to better knowledge of the working and development of representative institutions and to the strengthening of representative democracy.

Membership. The IPU had 162 members and ten associate members in Feb. 2013.

Headquarters: Chemin du Pommier 5, C.P. 330, CH-1218 Le Grand Saconnex, Geneva 19, Switzerland.
Website: http://www.ipu.org
President: Abdelwahad Radi (Morocco).
Secretary-General: Anders B. Johnsson (Sweden).

Interpol (International Criminal Police Organization)

Organization. Interpol was founded in 1923, disbanded in 1938 and reconstituted in 1946. The International Criminal Police Organization—Interpol was created to ensure and promote the widest possible mutual assistance between all criminal police authorities within the limits of the law existing in the different countries worldwide and the spirit of the Universal Declaration of Human Rights, and to establish and develop all institutions likely to contribute effectively to the prevention and suppression of ordinary law crimes.

Aims. Interpol provides a co-ordination centre (General Secretariat) for its 190 member countries. Its priority areas of activity concern criminal organizations, public safety and terrorism, drug-related crimes, financial crime and high-tech crime, trafficking in human beings and tracking fugitives from justice. Interpol centralizes records and information on international offenders; it operates a worldwide communication network.

Interpol's General Assembly is held annually. The General Assembly is the body of supreme authority in the organization. It is composed of delegates appointed by the members of the organization.

Interpol's Executive Committee, which meets four times a year, supervises the execution of the decisions of the General Assembly. The Executive Committee is composed of the president of the organization, the three vice-presidents and nine delegates.

Interpol's General Secretariat is the centre for co-ordinating the fight against international crime. Its activities, undertaken in response to requests from the police services and judicial authorities in its member countries, focus on crime prevention and law enforcement.

As of Feb. 2013 Interpol's Sub-Regional Bureaus were located in Abidjan, Buenos Aires, Harare, Nairobi, San Salvador and Yaoundé. Interpol's Liaison Office for Asia is located in Bangkok.

Headquarters: 200 Quai Charles de Gaulle, 69006 Lyon, France.
Website: http://www.interpol.int
President: Mireille Ballestrazzi (France).

Further Reading

Martha, Rutsel Silvestre J., *The Legal Foundations of INTERPOL*. 2010

Islamic Development Bank

The Agreement establishing the IDB (Banque islamique de développement) was adopted at the Second Islamic Finance Ministers' Conference held in Jeddah, Saudi Arabia in Aug. 1974. The Bank, which is open to all member countries of the Organization of the Islamic Conference, commenced operations in 1975. Its main objective is to foster economic development and social progress of member countries and Muslim communities individually as well as jointly in accordance with the principles of the Sharia. It is active in the promotion of trade and the flow of investments among member countries, and maintains a Special Assistance Fund for member countries suffering natural calamities. The Fund is also used to finance health and educational projects aimed at improving the socio-economic conditions of Muslim communities in non-member countries. A US$1·5bn. IDB Infrastructure Fund was launched in 1998 to invest in projects such as power, telecommunications, transportation, energy, natural resources, petro-chemical and other infrastructure-related sectors in member countries.

Members (56 as of Feb. 2013). Afghanistan, Albania, Algeria, Azerbaijan, Bahrain, Bangladesh, Benin, Brunei, Burkina Faso, Cameroon, Chad, Comoros, Côte d'Ivoire, Djibouti, Egypt, Gabon, Gambia, Guinea, Guinea-Bissau, Indonesia, Iran, Iraq, Jordan, Kazakhstan, Kuwait, Kyrgyzstan, Lebanon, Libya, Malaysia, Maldives, Mali, Mauritania, Morocco, Mozambique, Niger, Nigeria, Oman, Pakistan, Palestine, Qatar, Saudi Arabia, Senegal, Sierra Leone, Somalia, Sudan, Suriname, Syria, Tajikistan, Togo, Tunisia, Turkey, Turkmenistan, Uganda, United Arab Emirates, Uzbekistan, Yemen.

Official language: Arabic. *Working languages:* English, French.
Headquarters: PO Box 5925, Jeddah 21432, Saudi Arabia.
Website: http://www.isdb.org
President: Ahmed Mohamed Ali Al-Madani (Saudi Arabia).

Médecins Sans Frontières/Doctors Without Borders (MSF)

Origin. Médecins Sans Frontières/Doctors Without Borders was founded in 1971 by a small group of doctors and journalists who believed that all people have a right to emergency relief.

Functions. MSF was one of the first non-governmental organizations to provide both urgently needed medical assistance and to publicly bear witness to the plight of the people it helps. Today MSF is an international medical humanitarian organization with 19 sections and several additional offices around the world. Every year MSF sends around 3,000 volunteer doctors, nurses, other medical professionals, logistical experts, water and sanitation engineers, and administrators to join approximately 25,000 locally hired staff to provide medical aid in over 60 countries. MSF was awarded the 1999 Nobel Peace Prize.

Headquarters: MSF International Office, Rue de Lausanne 78, CP 116, CH-1211 Geneva 21, Switzerland.
Website: http://www.msf.org
Secretary-General: Jérôme Oberreit (France).
International Council President: Unni Karunakara (India).

Nobel Prizes

When the scientist, industrialist and inventor Alfred Nobel died in 1896, he made provision in his will for his fortune to be used for prizes in Physics, Chemistry, Physiology or Medicine, Literature and Peace. The Norwegian Nobel Committee awards the Nobel Peace Prize, and the Nobel Foundation in Stockholm (founded 1900; Mailing address: Box 5232, SE-10245, Stockholm, Sweden) awards the other four prizes plus the Sveriges Riksbank Prize in Economic Sciences in Memory of Alfred Nobel (often referred to as the Nobel Memorial Prize in Economic Sciences). The Prize Awarding Ceremony takes place on 10 Dec., the anniversary of Nobel's death. The last ten recipients of the Nobel Peace Prize, worth 8m. Sw. kr. in 2012 (down from 10m. Sw. kr. for the previous 11 years), are:

2003 – Shirin Ebadi (Iran) for her work fighting for democracy and the rights of women and children.

2004 – Wangari Maathai (Kenya) for her contribution to sustainable development, democracy and peace.

2005 – Mohamed ElBaradei and the IAEA for their efforts to prevent nuclear energy from being used for military purposes and to ensure that nuclear energy for peaceful purposes is used in the safest possible way.

2006 – Muhammad Yunus and Grameen Bank of Bangladesh for their efforts to create economic and social development from below.

2007 – the Intergovernmental Panel on Climate Change and Al Gore for their efforts to build up and disseminate greater knowledge about man-made climate change.

2008 – Martti Ahtisaari (Finland) for his important efforts, on several continents and over more than three decades, to resolve international conflicts.

2009 – Barack Obama (USA) for his extraordinary efforts to strengthen international diplomacy and co-operation between peoples.

2010 – Liu Xiaobo (China) for his long and non-violent struggle for fundamental human rights in China.

2011 – Ellen Johnson Sirleaf (Liberia), Leymah Gbowee (Liberia) and Tawakkul Karman (Yemen) for their non-violent struggle for the safety of women and for women's rights to full participation in peace-building work.

2012 – the European Union (EU) for its contribution for over six decades to the advancement of peace and reconciliation, democracy and human rights. The Nobel Committee highlighted the EU's role in bringing together historical enemies Germany and France, introducing democracy to Greece, Spain, Portugal and to the former socialist countries and advancing democracy and human rights in Turkey. The EU has played a stabilizing part in transforming Europe 'from a continent of war to a continent of peace'.

Sveriges Riksbank Prize in Economic Sciences in Memory of Alfred Nobel

The Sveriges Riksbank Prize in Economic Sciences in Memory of Alfred Nobel was set up by the Swedish central bank in 1968. The last ten recipients of the prize, worth 8m. Sw. kr. in 2012 (down from 10m. Sw. kr. for the previous 11 years), are:

2003 – Robert F. Engle III (USA) for methods of analyzing economic time series with time-varying volatility (ARCH), and Clive W. J. Granger (UK) for methods of analyzing economic time series with common trends (cointegration).

2004 – Finn E. Kydland (Norway) and Edward C. Prescott (USA) for their contributions to dynamic macroeconomics: the time consistency of economic policy and the driving forces behind business cycles.

2005 – Robert J. Aumann (Israel/USA) and Thomas C. Schelling (USA) for having enhanced our understanding of conflict and cooperation through game-theory analysis.

2006 – Edmund S. Phelps (USA) for his analysis of intertemporal tradeoffs in macroeconomic policy.

2007 – Leonid Hurwicz (USA), Eric S. Maskin (USA) and Roger B. Myerson (USA) for having laid the foundations of mechanism design theory.

2008 – Paul Krugman (USA) for his analysis of trade patterns and location of economic activity.

2009 – Elinor Ostrom (USA) for her analysis of economic governance, especially the commons, and Oliver E. Williamson (USA) for his analysis of economic governance, especially the boundaries of the firm.

2010 – Peter A. Diamond (USA), Dale T. Mortensen (USA) and Christopher A. Pissarides (Cyprus) for their analysis of markets with search frictions.

2011 – Thomas J. Sargent (USA) and Christopher A. Sims (USA) for their empirical research on cause and effect in macroeconomy.

2012 – Alvin E. Roth (USA) and Lloyd S. Shapley (USA) for the theory of stable allocations and the practice of market design. The two researchers worked independently of one another. Shapley used co-operative game theory to study and compare different matching methods and Roth built on these to make real-world changes to existing markets, including organ transplants and school choice. The combination of their work has generated a flourishing field of research and improved the performance of many markets.

North Atlantic Treaty Organization (NATO)

Origin. On 4 April 1949 the foreign ministers of Belgium, Canada, Denmark, France, Iceland, Italy, Luxembourg, the Netherlands, Norway, Portugal, the UK and the USA signed the North Atlantic Treaty, establishing the *North Atlantic Alliance*. In 1952 Greece and Turkey acceded to the Treaty; in 1955 the Federal Republic of Germany; in 1982 Spain; in 1999 the Czech Republic, Hungary and Poland; in 2004 Bulgaria, Estonia, Latvia, Lithuania, Romania, Slovakia and Slovenia; and in 2009 Albania and Croatia, bringing the total to 28 member countries.

Functions. The Alliance was established as a defensive political and military alliance of independent countries in accordance with the terms of the UN Charter. Its fundamental role is to safeguard the freedom and security of its members by political and military means. It also encourages consultation and co-operation with non-NATO countries in a wide range of security-related areas to help prevent conflicts within and beyond the frontiers of its member countries. NATO promotes democratic values and is committed to the peaceful resolution of disputes. If diplomatic efforts fail, it has the military capacity needed to undertake crisis-management operations alone or in co-operation with other countries and international organizations.

Reform and Transformation of the Alliance. Following the demise of the Warsaw Pact in 1991, and the improved relations with Russia, NATO established security dialogue and co-operation with the states of Central and Eastern Europe and those of the former USSR. These changes were reflected in the publication in 1991 of a new Strategic Concept for the Alliance outlining NATO's enduring purpose and nature, and its fundamental security tasks. Further changes in the security environment during the 1990s led to the development of the current Strategic Concept, published in 1999 to address new risks such as terrorism, ethnic conflict, human rights abuses, political instability, economic fragility, and the spread of nuclear, biological and chemical weapons and their means of delivery. A new Strategic Concept, which reflects new and emerging security threats, was published at a NATO summit meeting in Nov. 2010.

The Euro-Atlantic Partnership. In 1991 the North Atlantic Co-operation Council (NACC) was established as a forum for dialogue with former Warsaw Pact countries. The NACC was replaced in 1997 by the Euro-Atlantic Partnership Council (EAPC), which brings together all 50 NATO and partner countries in the Euro-Atlantic area for dialogue and consultation on political and security-related issues. It provides the overall political framework for NATO's co-operation with partner countries and the bilateral relationships developed between NATO and individual partner countries under the Partnership for Peace (PfP) programme, which was launched in 1994.

Since its launch, the PfP programme has been adapted to expand and intensify political and military co-operation throughout Europe. Core objectives are: the facilitation of transparency in national defence planning and budgeting processes; democratic control of defence forces; members' maintenance of capability and readiness to contribute to operations under the authority of the UN; development of co-operative military relations with NATO (joint planning, training and exercises) in order to strengthen participants' ability to undertake missions in the fields of peacekeeping, search and rescue, and humanitarian operations; development, over the longer term, of forces better able to operate with those of NATO member forces. NATO will consult with any active partner which perceives a direct threat to its territorial integrity, political independence or security.

One of the most tangible aspects of co-operation between partner countries and NATO has been their individual participation in NATO-led peace-support operations. PfP has done much to facilitate this and has also been a key factor in promoting a spirit of practical co-operation and commitment to the democratic principles that underpin the Alliance. In Feb. 2013 NATO had 22 PfP partners: Armenia, Austria, Azerbaijan, Belarus, Bosnia and Herzegovina, Finland, Georgia, Ireland, Kazakhstan, Kyrgyzstan, the former Yugoslav Republic of Macedonia, Malta, Moldova, Montenegro, Russia, Serbia, Sweden, Switzerland, Tajikistan, Turkmenistan, Ukraine and Uzbekistan. Many of these countries have accepted the Alliance's invitation to send liaison officers to permanent facilities at NATO Headquarters in Brussels and to the Partnership Co-ordination Cell in Mons, Belgium, where the Supreme Headquarters Allied Powers Europe (SHAPE) is located.

Country Relations. On 27 May 1997 in Paris, NATO and Russia signed the Founding Act on Mutual Relations, Co-operation and Security, committing themselves to build together a lasting peace in the Euro-Atlantic area, and establishing a new forum for consultation and co-operation called the NATO-Russia Permanent Joint Council. In May 2002 the Permanent Joint Council was replaced by a new NATO-Russia Council which brings together all NATO member countries and Russia in a forum in which they work as equal partners, identifying and pursuing opportunities for joint action in areas of common concern.

At the meeting in Sintra, Portugal in May 1997 a NATO-Ukraine Charter on a Distinctive Partnership was drawn up and signed in Madrid in July, establishing the NATO-Ukraine Commission (NUC). Dialogue and co-operation has become well-established with NATO and individual allies supporting Ukraine's ongoing reform efforts, particularly in the defence and security sectors.

NATO launched the Mediterranean Dialogue with six countries of the Mediterranean (Egypt, Israel, Jordan, Mauritania, Morocco and Tunisia) in 1995. In 1997 allied foreign ministers agreed to enhance the Dialogue. A new committee, the Mediterranean Co-operation Group, was established to take the Mediterranean Dialogue forward and Algeria joined in March 2000. Later, in 2004, NATO also launched the İstanbul Co-operation Initiative that aims to develop bilateral co-operation with countries in the broader Middle East. Bahrain, Kuwait, Qatar and the United Arab Emirates have since joined the Initiative.

Relations with other international organizations. NATO is gradually developing a strategic partnership with the European Union. Efforts to strengthen the security and defence role of NATO's European allies were initially organized through the Western European Union (WEU) during the 1990s, when there was a growing realization of the need for European countries to further develop defence capabilities and to assume greater responsibility for their common security. In 2000 the crisis management responsibilities of the WEU were increasingly assumed by the EU. Institutionalized relations between NATO and the EU were launched in 2001. The political principles underlying the NATO-EU relationship were set out in the Dec. 2002 NATO-EU Declaration on ESDP (European Security and Defence Policy). These decisions paved the way for the two organizations to work out the modalities for the transfer of responsibilities to the EU for the NATO-led military operations in the former Yugoslav Republic of Macedonia in 2003 and, from Dec. 2004, in Bosnia and Herzegovina.

NATO and the UN share a commitment to maintaining international peace and security and have been co-operating in this area since the early 1990s with consultations established between NATO and UN specialized bodies on a range of issues including crisis management, combating human trafficking, mine action and the fight against terrorism.

NATO and the OSCE work together to build security and promote stability in the Euro-Atlantic area, co-operating at both the political and the operational level in areas such as conflict prevention, crisis management and addressing new security threats.

Operations. One of the most significant aspects of NATO's transformation has been the decision to undertake peace-support and crisis-management operations in the Euro-Atlantic area and further afield.

In the wake of the disintegration of the former Yugoslavia, the Alliance has focused much of its attention on the Balkans. NATO first committed itself to peacekeeping in Bosnia and Herzegovina in Dec. 1995, through the NATO-led Implementation Force (IFOR), which was replaced by the Stabilization Force (SFOR) in 1996. Improvements in the security situation allowed NATO to hand over its operation to the EU in Dec. 2004.

Since 1999, following a 78-day air campaign against the Yugoslav regime to bring an end to the violent repression of ethnic Albanians, NATO has led a peacekeeping mission in Kosovo (the Kosovo Force, or KFOR). NATO also intervened in the former Yugoslav Republic of Macedonia at the request of the government to help avoid a civil war in 2001, and maintained a small peacekeeping presence there until March 2003, when the operation was handed over to the EU.

Following the attacks on New York and Washington, D.C. on 11 Sept. 2001, NATO invoked article five of the Washington Treaty (the collective defence clause) for the first time in its history, declaring it considered the attack on the USA as an attack against all members of the Alliance. It subsequently launched a series of initiatives aimed at curtailing terrorist activity. Operation *Active Endeavour* is a maritime operation led by NATO's naval forces to detect and deter terrorist activity in the Mediterranean. Operation *Eagle Assist* was one of the measures requested by the USA in the aftermath of the attacks in Sept. 2001. Aircraft from NATO's Airborne Warning and Control System (AWACS) patrolled American airspace from mid-Oct. 2001 to mid-May 2002.

In Aug. 2003 NATO took over responsibility for the International Security Assistance Force (ISAF) in Afghanistan. Currently NATO's largest and most robust peace-support operation, ISAF's presence was gradually expanded out from the capital Kabul into the provinces, covering the entire country by Oct. 2006. Provincial Reconstruction Teams, combining both civilian and military personnel, have been set up to help provide security and promote reconstruction and development. ISAF is also assisting the Afghan government in developing reliable security structures and training Afghan security forces.

NATO is also helping train Iraqi military personnel and supporting the development of security institutions in Iraq, as well as providing logistical support to the African Union's mission in Darfur, Sudan.

Since late 2008 NATO has been conducting counter-piracy operations in the Gulf of Aden and off the Horn of Africa, where piracy is threatening to undermine international humanitarian efforts in Africa, as well as the safety of commercial maritime routes and international navigation.

From March until Oct. 2011 a coalition of NATO allies and partners conducted Operation Unified Protector in Libya. Using only air and sea resources, its aim was to protect civilians, enforce an air embargo and maintain a no-fly zone. It was widely regarded as instrumental in the downfall of the Libyan leader, Col. Gaddafi.

Since the establishment in 1998 of the Euro-Atlantic Disaster Response Co-ordination Centre to serve as a focal point for co-ordinating the disaster-relief efforts of NATO member states and partner countries, NATO plays an increasingly important role in humanitarian relief. Most notably, in 2005 some NATO capabilities and forces were deployed to support relief efforts following Hurricane Katrina in the USA and the devastating earthquake in Pakistan and in 2010 after the Haitian earthquake and flooding in Pakistan.

Defence capabilities. This widened scope of NATO military operations is radically transforming the military requirements of the Alliance. The large defence forces of the past are being replaced by more flexible, mobile forces which are able to deploy at significant distances from their normal operating bases and to engage in the full range of missions, ranging from high-intensity combat to humanitarian support. A modernization process was launched at the Prague Summit in 2002 to ensure that NATO could effectively deal with the security challenges of the 21st century and measures to enhance the Alliance's military operational capabilities were agreed. A new capabilities initiative (the Prague Capabilities Commitment) and a NATO Response Force were created and the Alliance's military command structure streamlined. In addition, steps were taken to increase efforts in the areas of intelligence sharing and crisis response arrangements, as well as greater co-operation with partner countries. Five nuclear, biological and chemical (NBC) weapons defence initiatives were also endorsed, as well as the creation of a multinational chemical, biological, radiological and nuclear battalion. At subsequent NATO summit meetings in İstanbul in 2004 and Riga in 2006, further initiatives were taken to promote the Alliance's ongoing transformation.

Organization. The North Atlantic Council (NAC) is the highest decision-making body and forum for consultation within the Atlantic Alliance. Composed of Permanent Representatives of all the member countries, it meets at least once a week and also meets at higher levels involving foreign ministers, defence ministers or heads of state or government. The authority and powers of decision-making and status and validity of its decisions remain the same at whatever level it meets. All decisions are taken on the basis of consensus, reflecting the collective will of all member governments. The NAC is the only body within the Atlantic Alliance which derives its authority explicitly from the North Atlantic Treaty.

The Military Committee is responsible for making recommendations to the NAC and the Defence Planning Committee on military matters and for supplying guidance to the Allied Commanders. Composed of the Chiefs-of-Staff of member countries (Iceland, which has no military forces, may be represented by a civilian), the Committee is assisted by an International Military Staff. It meets at Chiefs-of-Staff level at least twice a year but remains in permanent session at the level of national military representatives. The military command structure of the Alliance is divided into two strategic commands, one based in Europe and the other based in the USA.

Finance. The greater part of each member country's contribution to NATO, in terms of resources, comes indirectly through its expenditure on its own national armed forces and on its efforts to make them interoperable with those of other members so that they can participate in multinational operations. Member countries usually incur the deployment costs involved whenever they volunteer forces to participate in NATO-led operations, although in 2006 agreement was reached on using common funding for some aspects of deployments on a trial basis.

Member countries make direct contributions to three budgets managed directly by NATO: namely the Civil Budget, the Military Budget and the Security Investment Programme. Member countries pay contributions to each of these budgets in accordance with agreed cost-sharing formulae broadly calculated in relation to their ability to pay.

Under the terms of the Partnership for Peace strategy, partner countries undertake to make available the necessary personnel, assets, facilities and capabilities to participate in the programme, and share the financial cost of any military exercises in which they participate.

Current Leaders

Anders Fogh Rasmussen

Position
Secretary General

Introduction
Anders Fogh Rasmussen took office as Secretary General on 1 Aug. 2009, having served as prime minister of Denmark for the preceding eight years. Fogh Rasmussen is the highest-ranked official to be appointed Secretary General.

Early Life
Fogh Rasmussen was born on 26 Jan. 1953 in Ginnerup, Jutland. He joined the Young Liberals (Venstres Ungdom) in 1970 and stood unsuccessfully as the Liberal parliamentary candidate for Viborg in Jan. 1973. In 1978 he graduated with a masters degree in economics from Aarhus University. In the same year he joined the Folketing as a replacement member for Viborg County.

From 1981–86 Fogh Rasmussen was vice chairman of the Folketing's housing committee and in 1985 was appointed deputy chairman of the Liberal Party (Venstre, or V). After re-election in 1987 he became taxation minister, adding the role of finance minister in 1990. In 1998 he was elected Liberal chairman. In Sept. 2001 incumbent Social Democrat prime minister Poul Nyrup Rasmussen called a snap election. The campaign was fought largely on the issue of immigration, with Fogh Rasmussen gaining popular support for his proposed hard line. Having defeated the Social Democrats he took office on 27 Nov. 2001 but needed to form a coalition with the Conservatives. Although

espousing a centre-right line, his government was also supported by the far-right Danish People's Party. Following the 2005 and 2007 general elections, he became the first Liberal leader to win a second, and then a third, consecutive term of office.

In opposition to public opinion Fogh Rasmussen supported the USA's invasion of Iraq in 2003, pledging 500 troops to the war effort. In early 2006 the republication in western European newspapers of cartoon caricatures of the Prophet Muhammad, which first appeared in Denmark in Sept. 2005, sparked mass protests and unofficial boycotts of Danish exports across the Muslim world. While arguing that the issue was one of freedom of expression, Fogh Rasmussen's government sought more effective engagement with Islamic opinion.

Fogh Rasmussen resigned as prime minister shortly after his appointment as NATO Secretary General was announced on 4 April 2009.

Career in Office
Despite having the backing of the major NATO members, Fogh Rasmussen's appointment as Secretary General was contested by Turkey, which only relented after the personal intervention of President Barack Obama at the NATO summit in Strasbourg in April 2009.

He came to power at a difficult time for the alliance as the situation in Afghanistan deteriorated. In July 2009, the month preceding his assumption of the post, NATO suffered record casualties. Nevertheless, on 4 Dec. 2009 Fogh Rasmussen declared that NATO would commit an extra 7,000 troops to the country.

He described his mandate as being to 'modernize, transform and reform so that NATO adapts to the security environment for the 21st century', and in Nov. 2010 a summit of heads of government agreed a new strategic concept for the alliance. The summit confirmed the military commitment to Afghanistan, while progressively handing security responsibility to Afghan forces, and also heralded a new relationship with Russia. Fogh Rasmussen said that the former foes 'have agreed together on which challenges NATO nations and Russia actually face today. What's most significant is what's not on the list: each other.'

In 2011 he oversaw the NATO military campaign in Libya. The mission was authorized by the United Nations to protect civilians from the Gaddafi regime, which had resisted widespread popular revolt from Feb. onwards but which collapsed with the violent death of the dictator in Oct.

In Nov. 2012 he told the annual NATO parliamentary assembly that investment in security remained essential despite the prevailing economic strictures, and urged legislators to halt the slide in defence spending and to increase it once the economic climate improved. He said: 'Freedom does not come for free and any decisions taken to improve our economy must not lead us into a different sort of crisis—a security crisis.' He also called on the European member states to take on greater responsibility for Alliance expenditure.

Headquarters: NATO, Blvd Leopold III, 1110 Brussels, Belgium.
Website: http://www.nato.int
Secretary General: Anders Fogh Rasmussen (Denmark).

Publications. For a full list of NATO publications, visit the website: http://www.nato.int/cps/en/natolive/publications.htm.

Further Reading

Cook, D., *The Forging of an Alliance.* 1989
Cottey, Andrew, *Security in 21st Century Europe.* 2nd ed. 2012
Heller, F. H. and Gillingham, J. R. (eds.) *NATO: the Founding of the Atlantic Alliance and the Integration of Europe.* 1992
Sloan, Stanley R., *Permanent Alliance?: NATO and the Transatlantic Bargain from Truman to Obama.* 2010
Smith, J. (ed.) *The Origins of NATO.* 1990

Organisation for Economic Co-operation and Development (OECD)

Origin. Founded in 1961 to replace the Organisation for European Economic Co-operation (OEEC), which was established in 1948 and linked to the Marshall Plan. The change of title marks the Organisation's altered status and functions: it ceased to be a European body with the accession of Canada and USA as full members and became a forum of global influence adding development to its list of core priorities. The Organisation aims to promote policies designed to achieve the highest sustainable economic growth and employment, as well as raising standards of living in member countries, while maintaining financial stability, thereby contributing to the development of the world economy; to contribute to sound economic expansion in member as well as non-member economies in the process of economic development; and to contribute to the expansion of world trade on a multilateral, non-discriminatory basis in accordance with international obligations.

Members. Australia, Austria, Belgium, Canada, Chile, Czech Republic, Denmark, Estonia, Finland, France, Germany, Greece, Hungary, Iceland, Ireland, Israel, Italy, Japan, South Korea, Luxembourg, Mexico, Netherlands, New Zealand, Norway, Poland, Portugal, Slovakia, Slovenia, Spain, Sweden, Switzerland, Turkey, UK and USA. Discussions that began in May 2007 on the possible future accession of Russia are ongoing.

Activities. The OECD's main fields of work are: economic policy; statistics; energy; development co-operation; sustainable development; public governance and territorial development; international trade; financial and enterprise affairs; tax policy and administration; food, agriculture and fisheries; environment; science, technology and industry; biotechnology and biodiversity; education; employment, labour and social affairs; entrepreneurship, small and middle-sized enterprises; and local development.

Relations with non-members. In order to ensure its continuing relevance as a hub for dialogue and action on globally significant policy issues, the OECD has developed an active global relations strategy and maintains co-operative relations with many economies outside the OECD. Officials from non-member economies increasingly discuss policy with their counterparts from member countries and conduct peer assessments while sharing each other's policy experiences. In 2007 the OECD launched a process of enhanced engagement with five countries whose engagement in the work of the OECD is particularly important for the fulfilment of the Organisation's mandate to promote policy convergence and global economic development: Brazil, China, India, Indonesia and South Africa. It aims to bring these countries closer to the OECD by engaging them actively in the OECD's analytical and policy development work, while supporting their own reform processes.

Other activities with non-OECD members are grouped around Global Forums in 12 policy areas. Created by Committees as stable, active networks of policy makers in OECD member and non-member economies, as well as other stakeholders, the Global Forums focus on: agriculture, biotechnology, competition, development, education, environment, finance, international investment, the knowledge economy, public governance, taxation and trade. A regional approach provides for targeted co-operation with non-OECD economies in Europe, Asia, Latin America, the Middle East and Northern Africa (MENA), and in Africa more generally, where the OECD supports the objectives of the New Partnership for Africa's Development (NEPAD) and the initiatives of the African Partnership Forum. The Centre for Co-operation with Non-Members develops the overall architecture of co-operation and general liaison with non-members, while the substantive directorates implement the programmes and activities in each policy area.

Relations with developing countries. Developing countries participate in many of the OECD's above-mentioned activities with non-members. The principal body dealing with issues related to development co-operation is the Development Assistance Committee (DAC). Its 23 members are major aid donors, collectively accounting for over 90% of total official development assistance (ODA) worldwide amounting to approximately US$128·7bn. in 2010. The DAC largely focuses on how to spend and invest this aid so as to help its partners achieve the Millennium Development Goals and produces analysis and guidance on a range of topics, including aid for trade, aid effectiveness, capacity development, poverty reduction, environment, conflict and fragility, gender equality, good governance, evaluation and aid architecture.

The OECD Development Centre links OECD member countries and developing countries in Africa, Asia and Latin America by helping policy makers in OECD and developing countries find solutions to the challenges of development, poverty alleviation and the curbing of inequality through recommendations designed to promote constructive policy change. Annual publications include the *Latin American Economic Outlook* and, jointly with the African Development Bank, the *African Economic Outlook*.

The OECD's policy dialogue is also developing at a regional level, particularly through the work of the Sahel and West Africa Club (SWAC) which acts as an interface between West African actors and OECD member countries. Administratively attached to the OECD, the SWAC is led by a secretariat based in Paris and combines direct field involvement with analyses of West African realities. The SWAC works with regional institutions, governments, business and civil society organizations to promote the regional dimension of development, support the formulation and implementation of joint or intergovernmental policies and thereby contribute to mobilizing and strengthening West African capacities.

Relations with other international organizations. Under a protocol signed at the same time as the OECD Convention, the European Commission takes part in the work of the OECD. EFTA may also send representatives to attend OECD meetings. Formal relations also exist with the Asian Development Bank, Inter-American Development Bank, World Bank, IMF, UNCTAD, WHO and the Parliamentary Assemblies of the Council of Europe and NATO.

Relations with civil society. Consultations with civil society organizations (CSOs) take place across the whole range of the OECD's work. The Business and Industry Advisory Committee to the OECD (BIAC) and the Trade Union Advisory Committee to the OECD (TUAC) have consultative status enabling them to discuss subjects of common interest and be consulted in a particular field by the relevant OECD Committee or its officers. Since 2000 the OECD has organized the annual OECD Forum, an international public conference offering business, labour and civil society the opportunity to discuss key issues of the 21st century with government ministers and leaders of international organizations.

Organization. The governing body of the OECD is the Council, comprising representatives of each member country and in which the European Commission participates. It usually meets once a year at the level of government ministers, with a rotating chairmanship at ministerial level among member governments. The Council also meets regularly, under the chairmanship of the Secretary-General at the level of Permanent Representatives to OECD (ambassadors who head resident diplomatic missions). It is responsible for all questions of general policy and may establish subsidiary bodies as required to achieve the aims of the

Organisation. Decisions and recommendations of the Council are adopted by consensus of all its members.

An Executive Committee, a Budget Committee and an External Relations Committee assist the Council although they have limited decision-making power within their fields of competence. In addition, the Executive Committee in Special Session meets, usually twice a year, and is attended by senior government officials. Most of the work of the OECD is prepared and carried out by about 250 specialized bodies (Committees, Working Parties, etc.). All members are normally represented on these bodies, except a few which have a more restricted membership. Funding is by member state contributions based on a formula related to their size and economy.

The International Energy Agency (IEA) and the Nuclear Energy Agency (NEA) are also part of the OECD system.

Headquarters: 2 rue André Pascal, 75775 Paris Cedex 16, France.
Website: http://www.oecd.org
Secretary-General: Angel Gurría (Mexico).
Deputy Secretaries-General: Richard A. Boucher (USA), Yves Leterme (Belgium), Pier Carlo Padoan (Italy), Rintaro Tamaki (Japan).

Publications include: *OECD Factbook*; *Economic, Environmental and Social Statistics* (annual); *OECD Policy Briefs* (20 a year); *OECD Economic Surveys* (by country); *Environmental Performance Reviews* (by country); *OECD Economic Outlook* (twice a year); *Economic Policy Reform: Going for Growth* (annual); *OECD-FAO Agricultural Outlook* (annual); *Education at a Glance* (annual); *OECD Employment Outlook* (annual); *OECD Science, Technology and Industry Outlook* (biennial); *International Migration Outlook* (annual); *Health at a Glance* (biennial); *Society at a Glance* (biennial); *OECD Health Data* (CD-ROM; annual); *Financial Market Trends* (twice a year); *Statistics of International Trade* (monthly); *International Trade by Commodity Statistics* (annual); *Main Economic Indicators* (monthly); *Energy Balances* (annual); *World Energy Outlook* (annual); *National Accounts* (quarterly and annual); *African Economic Outlook* (annual); *OECD Observer* (6 a year); *Quarterly Labour Force Statistics*; *Model Tax Convention*; *Development Co-operation Report* (annual); *Development Centre Policy Briefs*. For a full list of OECD publications, visit the website: http://www.oecdbookshop.org.

Organization of the Islamic Conference (OIC)

Founded in 1969, the objectives of the OIC are to promote Islamic solidarity among member states; to consolidate co-operation among member states in the economic, social, cultural, scientific and other vital fields of activities, and to carry out consultations among member states in international organizations; to endeavour to eliminate racial segregation, discrimination and to eradicate colonialism in all its forms; to take the necessary measures to support international peace and security founded on justice; to strengthen the struggle of all Muslim peoples with a view to safeguarding their dignity, independence and national rights; to create a suitable atmosphere for the promotion of co-operation and understanding among member states and other countries.

Members (57 as of Feb. 2013). Afghanistan, Albania, Algeria, Azerbaijan, Bahrain, Bangladesh, Benin, Brunei, Burkina Faso, Cameroon, Chad, Comoros, Côte d'Ivoire, Djibouti, Egypt, Gabon, Gambia, Guinea, Guinea-Bissau, Guyana, Indonesia, Iran, Iraq, Jordan, Kazakhstan, Kuwait, Kyrgyzstan, Lebanon, Libya, Malaysia, Maldives, Mali, Mauritania, Morocco, Mozambique, Niger, Nigeria, Oman, Pakistan, Palestine, Qatar, Saudi Arabia, Senegal, Sierra Leone, Somalia, Sudan, Suriname, Syria*, Tajikistan, Togo, Tunisia, Turkey, Turkmenistan, Uganda, United Arab Emirates, Uzbekistan, Yemen. *Observers.* Bosnia and

Herzegovina, Central African Republic, Russia, Thailand, Turkish Republic of Northern Cyprus. *Suspended since Aug. 2012.

Headquarters: PO Box 178, Jeddah 21411, Saudi Arabia.
Website: http://www.oic-oci.org
Secretary-General: Dr Ekmeleddin İhsanoğlu (Turkey).

Unrepresented Nations and Peoples Organization (UNPO)

UNPO is an international organization created by nations and peoples around the world who are not represented in the world's principal international organizations, such as the UN. Founded in 1991, UNPO had 42 members as at Feb. 2013 representing nearly 250m. people worldwide.

Membership. Open to all nations and peoples unrepresented, subject to adherence to the five principles that form the basis of UNPO's charter: equal right to self-determination of all nations and peoples; adherence to internationally accepted human rights standards; to the principles of democracy; promotion of non-violence; and protection of the environment. Applicants must show that they constitute a 'nation or people' as defined in the Covenant.

Functions and Activities. UNPO offers an international forum for occupied nations, indigenous peoples, minorities and oppressed majorities, who struggle to regain their lost countries, preserve their cultural identities, protect their basic human and economic rights, and safeguard their environment.

It does not represent those peoples; rather it assists and empowers them to represent themselves more effectively. To this end, it provides professional services and facilities as well as education and training in the fields of diplomacy, human rights law, democratic processes, conflict resolution and environmental protection. Members, private foundations and voluntary contributions fund the Organization.

In total six former members of UNPO (Armenia, Belau, Estonia, Georgia, Latvia and Timor-Leste) subsequently achieved full independence and gained representation in the UN. Belau is now called Palau. Current members Bougainville and Kosovo have achieved a degree of political autonomy. Kosovo has declared itself an independent state, although both Serbia and Russia oppose its sovereignty.

Headquarters: Laan van Meerdervoort 70, 2517 AN The Hague, Netherlands.
Website: http://www.unpo.org
Email: unpo@unpo.org
General Secretary: Marino Busdachin (Italy).

Publication. UNPO News (quarterly).

World Council of Churches

The World Council of Churches was formally constituted on 23 Aug. 1948 in Amsterdam. In Feb. 2013 member churches numbered 349 from more than 110 countries.

Origin. The World Council was founded by the coming together of Christian movements, including the overseas mission groups gathered from 1921 in the International Missionary Council, the Faith and Order Movement, and the Life and Work Movement. On 13 May 1938, at Utrecht, a provisional committee was appointed to prepare for the formation of a World Council of Churches.

Membership. The basis of membership (1975) states: 'The World Council of Churches is a fellowship of Churches which confess the Lord Jesus Christ as God and Saviour according to the Scriptures and therefore seek to fulfil together their common calling to the glory of the one God, Father, Son and Holy Spirit.' Membership is open to Churches which express their agreement with this basis and satisfy such criteria as the Assembly or Central Committee may prescribe. Today, more than 340 Churches of Protestant, Anglican, Orthodox, Old Catholic and Pentecostal confessions belong to this fellowship. The Roman Catholic Church is not a member of the WCC but works closely with it.

Activities. The WCC's Central Committee comprises the Programme Committee and the Finance Committee. Within the Programme Committee there are advisory groups on issues relating to communication, women, justice, peace and creation, youth, ecumenical relations and inter-religious relations. Following the WCC's 8th General Assembly in Harare, Zimbabwe in 1998 the work of the WCC was restructured. Activities were grouped into four 'clusters'—Relationships; Issues and Themes; Communication; and Finance, Services and Administration. The Relationships cluster comprises four teams (Church and Ecumenical Relations, Regional Relations and Ecumenical Sharing, Inter-Religious Relations and International Relations), as well as two programmes (Action by Churches Together and the Ecumenical Church Loan Fund). The Issues and Themes cluster comprises four teams (Faith and Order; Mission and Evangelism; Justice, Peace and Creation; and Education and Ecumenical Formation).

In Aug. 1997 the WCC launched a Peace to the City campaign, as the initial focus of a programme to overcome violence in troubled cities. The Decade to Overcome Violence was launched in Feb. 2001 during the meeting of the WCC Central Committee in Berlin.

Organization. The governing body of the World Council, consisting of delegates specially appointed by the member Churches, is the Assembly, which meets every seven or eight years to frame policy. It has no legislative powers and depends for the implementation of its decisions upon the action of member Churches. The 9th General Assembly, held in Porto Alegre, Brazil in Feb. 2006, had as its theme 'God, in your grace, transform the world'. A 159-member Central Committee meets annually to carry out the Assembly mandate, with a smaller 26-member Executive Committee meeting twice a year.

Headquarters: PO Box 2100, 150 route de Ferney, CH-1211 Geneva 2, Switzerland.
Website: http://www.oikoumene.org
General Secretary: Rev. Dr Olav Fykse Tveit (Norway).

Publications. Annual Reports; Dictionary of the Ecumenical Movement, Geneva, 1991; *Directory of Christian Councils,* 1985; *A History of the Ecumenical Movement,* Geneva, 1993; *Ecumenical Review* (quarterly); *Ecumenical News International* (weekly); *International Review of Mission* (quarterly).

Further Reading

Castro, E., *A Passion for Unity.* 1992
Raiser, K., *Ecumenism in Transition.* 1994
Van Elderen, M. and Conway, M., *Introducing the World Council of Churches.* 1991

World Customs Organization

Established in 1952 as the Customs Co-operation Council, the World Customs Organization is an intergovernmental body with worldwide membership, whose mission it is to enhance the effectiveness and efficiency of customs administrations throughout the world. It has 177 member countries or territories.

Headquarters: Rue du Marché, 30, B-1210 Brussels, Belgium.
Website: http://www.wcoomd.org
Secretary-General: Kunio Mikuriya (Japan).

World Federation of Trade Unions (WFTU)

Origin and History. The WFTU was founded on a worldwide basis in 1945 at the international trade union conferences held in London and Paris, with the participation of all the trade union centres in the countries of the anti-Hitler coalition. The aim was to reunite the world trade union movement at the end of the Second World War. The acute political differences among affiliates, especially the east–west confrontation in Europe on ideological lines, led to a split. A number of affiliated organizations withdrew in 1949 and established the ICFTU. The WFTU now draws its membership from the industrially developing countries like India, Vietnam and other Asian countries, Brazil, Peru, Cuba and other Latin American countries, Syria, Lebanon, Kuwait and other Arab countries, and it has affiliates and associates in more than 20 European countries. It has close relations with the International Confederation of Arab Trade Unions, the Organization of African Trade Union Unity as well as the All-China Federation of Trade Unions. The 16th Congress was held in Athens, Greece in April 2011. Its Trade Unions Internationals (TUIs) have affiliates in Russia, the Czech Republic, Poland and other East European countries, Portugal, France, Spain, Japan and other OECD countries.

The headquarters of the TUIs are situated in Helsinki, New Delhi, Budapest, Mexico, Paris and Moscow. The WFTU has 120m. members in 80 countries. It has regional offices in Dakar, Damascus, Havana, Johannesburg, Moscow, New Delhi and Nicosia and Permanent Representatives accredited to the UN in New York, Geneva, Paris and Rome.

Headquarters: 40 Zan Moreas St., 117 45 Athens, Greece.
Website: http://www.wftucentral.org
Email: info@wftucentral.org
President: Mohammad Assouz (Syria).
General Secretary: George Mavrikos (Greece).

Publications. Flashes From the Trade Unions (fortnightly, published in English, French, Spanish and Arabic), reports of Congresses, etc.

World Wide Fund for Nature (WWF)

Origin. WWF was officially formed and registered as a charity on 11 Sept. 1961. The first National Appeal was launched in the United Kingdom on 23 Nov. 1961, shortly followed by the United States and Switzerland.

Organization. WWF is the world's largest and most experienced independent conservation organization with over 4·7m. supporters and a global network of 27 National Organizations, five Associates and 24 Programme Offices.

The National Organizations carry out conservation activities in their own countries and contribute technical expertise and funding to WWF's international conservation programme. The Programme Offices implement WWF's fieldwork, advise national and local governments, and raise public understanding of conservation issues.

Mission. WWF has as its mission preserving genetic, species and ecosystem diversity; ensuring that the use of renewable natural resources is sustainable now and in the longer term, for the benefit of all life on Earth; promoting actions to reduce to a minimum pollution and the wasteful exploitation and consumption of resources and energy. WWF's ultimate goal is to stop, and eventually reverse, the accelerating degradation of our planet's natural environment, and to help build a future in which humans live in harmony with nature.

Address: Avenue du Mont-Blanc, CH-1196 Gland, Switzerland.
Website: http://wwf.panda.org
Director General: James P. Leape (USA).
President Emeritus: HRH The Prince Philip, Duke of Edinburgh.
President: Yolanda Kakabadse (Ecuador).

African Development Bank

Established in 1964 to promote economic and social development in Africa.

Regional Members. (53) Algeria, Angola, Benin, Botswana, Burkina Faso, Burundi, Cameroon, Cape Verde, Central African Republic, Chad, Comoros, Democratic Republic of the Congo, Republic of the Congo, Côte d'Ivoire, Djibouti, Egypt, Equatorial Guinea, Eritrea, Ethiopia, Gabon, Gambia, Ghana, Guinea, Guinea-Bissau, Kenya, Lesotho, Liberia, Libya, Madagascar, Malaŵi, Mali, Mauritania, Mauritius, Morocco, Mozambique, Namibia, Niger, Nigeria, Rwanda, São Tomé e Príncipe, Senegal, Seychelles, Sierra Leone, Somalia, South Africa, Sudan, Swaziland, Tanzania, Togo, Tunisia, Uganda, Zambia, Zimbabwe.

Non-regional Members. (24) Argentina, Austria, Belgium, Brazil, Canada, China, Denmark, Finland, France, Germany, India, Italy, Japan, South Korea, Kuwait, Netherlands, Norway, Portugal, Saudi Arabia, Spain, Sweden, Switzerland, UK, USA.

Within the ADB Group are the African Development Fund (ADF) and the Nigerian Trust Fund (NTF). The ADF, established in 1972, provides development finance on concessional terms to low-income Regional Member Countries which are unable to borrow on the non-concessional terms of the African Development Bank. Membership of the Fund is made up of 24 non-African State Participants and the African Development Bank. The NTF is a special ADB fund created in 1976 by agreement between the Bank Group and the Government of the Federal Republic of Nigeria. Its objective is to assist the development efforts of low-income Regional Member Countries whose economic and social conditions and prospects require concessional financing.

Official languages: English, French.
Headquarters: Rue Joseph Anoma, 01 BP 1387, Abidjan 01, Côte d'Ivoire.
Website: http://www.afdb.org
Email: afdb@afdb.org
President: Donald Kaberuka (Rwanda).

African Export–Import Bank (Afreximbank)

Established in 1987 under the auspices of the African Development Bank to facilitate, promote and expand intra-African and extra-African trade. Membership is made up of three categories of shareholders: Class 'A' Shareholders consisting

of African governments, African central banks and sub-regional and regional financial institutions and economic organizations; Class 'B' Shareholders consisting of African public and private financial institutions; and Class 'C' Shareholders consisting of international financial institutions, economic organizations and non-African states, banks, financial institutions and public and private investors.

Official languages: English, French, Arabic, Portuguese.
Headquarters: 72B El Maahad El Eshteraky St., Heliopolis, Cairo 11341, Egypt.
Website: http://www.afreximbank.com
President and Chairman of the Board: Jean-Louis Ekra (Côte d'Ivoire).

African Union (AU)

History. The Fourth Extraordinary Session of the Assembly of the Heads of State and Government of the Organization of African Unity (OAU) held in Sirté, Libya on 9 Sept. 1999 decided to establish an African Union. At Lomé, Togo on 11 July 2000 the OAU Assembly of the Heads of State and Government adopted the Constitutive Act of the African Union, which was later ratified by the required two-thirds of the member states of the Organization of African Unity (OAU); it came into force on 26 May 2001. The Lusaka Summit, in July 2001, gave a mandate to translate the transformation of the Organization of African Unity into the African Union, and on 9 July 2002 the Durban Summit, in South Africa, formally launched the African Union.

Members. Algeria, Angola, Benin, Botswana, Burkina Faso, Burundi, Cameroon, Cape Verde, Central African Republic, Chad, Comoros, Democratic Republic of the Congo, Republic of the Congo, Côte d'Ivoire, Djibouti, Egypt, Equatorial Guinea, Eritrea, Ethiopia, Gabon, Gambia, Ghana, Guinea, Guinea-Bissau*, Kenya, Lesotho, Liberia, Libya, Madagascar**, Malaŵi, Mali, Mauritania, Mauritius, Mozambique, Namibia, Niger, Nigeria, Rwanda, Sahrawi Arab Democratic Republic (Western Sahara), São Tomé e Príncipe, Senegal, Seychelles, Sierra Leone, Somalia, South Africa, South Sudan, Sudan, Swaziland, Tanzania, Togo, Tunisia, Uganda, Zambia, Zimbabwe. *Membership suspended since the coup in April 2012. **Membership suspended since the change of government in March 2009.

Aims. The African Union aims to promote unity, solidarity, cohesion and co-operation among the peoples of Africa and African states, and at the same time to co-ordinate efforts by African people to realize their goals of achieving economic, political and social integration.

Activities. The African Union became fully operational in July 2002, and is working towards establishing the organs stipulated in the constitutive act. These include a Pan-African parliament, an Economic, Social and Cultural Council (ECOSOCC) and a Peace and Security Council (which have now been inaugurated), plus a Central Bank and a Court of Justice.

Official languages: Arabic, English, French, Ki-Swahili, Portuguese and Spanish.
Headquarters: POB 3243, Roosevelt Street (Old Airport Area), W21K19 Addis Ababa, Ethiopia.
Website: http://www.au.int
Chairman: Hailemariam Desalegn (Ethiopia).
Chair of the African Union Commission: Nkosazana Dlamini-Zuma (South Africa).

Further Reading

Makinda, Samuel M., and Okumu, F. Wafula, *The African Union: Challenges of Globalization, Security, and Governance.* 2007

Miller-Jones, Edward R., *The African Union: Aiming to Unify the Continent.* 2010

Muthri, Tim, Akopari, John and Ndinga-Mavumba, Angela, (eds.) *The African Union and its Institutions.* 2008

Welz, Martin, *Integrating Africa: Decolonization's Legacies, Sovereignty and the African Union.* 2012

Bank of Central African States (BEAC)

The Bank of Central African States (Banque des Etats de l'Afrique Centrale) was established in 1973 when a new Convention of Monetary Co-operation with France was signed. The five original members, Cameroon, Central African Republic, Chad, Republic of the Congo and Gabon, were joined by Equatorial Guinea in 1985. Under its Convention and statutes, the BEAC is declared a 'Multinational African institution in the management and control of which France participates in return for the guarantee she provides for its currency'.

Official language: French.
Headquarters: 736 avenue Monseigneur Vogt, 1917 Yaoundé, Cameroon.
Website (French only): http://www.beac.int
Governor: Lucas Abaga Nchama (Equatorial Guinea).

Publications. Etudes et Statistiques (monthly bulletins); *Annual Report; Directory of Banks and Financial Establishments of BEAC Monetary Area* (annual); *Bulletin du Marché Monétaire* (monthly bulletins); *Annual Report of the Banking Commission.*

Central Bank of West African States (BCEAO)

Established in 1962, the Central Bank of West African States (Banque Centrale des Etats de l'Afrique de l'Ouest) is the common central bank of the eight member states that form the West African Monetary Union (WAMU). It has the sole right of currency issue throughout the Union territory and is responsible for the pooling of the Union's foreign exchange reserve; the management of the monetary policy of the member states; the keeping of the accounts of the member states treasury; and the definition of the banking law applicable to banks and financial establishments.

Members. Benin, Burkina Faso, Côte d'Ivoire, Guinea-Bissau, Mali, Niger, Senegal, Togo.

Official language: French.
Headquarters: Avenue Abdoulaye Fadiga, Dakar, Senegal.
Website: http://www.bceao.int
Governor: Tiémoko Meyliet Koné (Côte d'Ivoire).

Publications. Rapport annuel (annual); *Annuaire des Banques* (annual); *Bilan des Banques U.M.O.A.* (annual); *Notes d'information et statistiques* (monthly bulletin).

Common Market for Eastern and Southern Africa (COMESA)

COMESA is an African economic grouping of 20 member states who are committed to the creation of a Common Market for Eastern and Southern Africa. It was established in 1994 as a building block for the African Economic Community and replaced the Preferential Trade Area for Eastern and Southern Africa, which had been in existence since 1981.

Members. Burundi, Comoros, Democratic Republic of the Congo, Djibouti, Egypt, Eritrea, Ethiopia, Kenya, Libya, Madagascar, Malawi, Mauritius, Rwanda, Seychelles, South Sudan, Sudan, Swaziland, Uganda, Zambia and Zimbabwe.

Objectives. To facilitate the removal of the structural and institutional weaknesses of member states so that they are able to attain collective and sustainable development.

Activities. COMESA's Free Trade Area (FTA) was launched on 31 Oct. 2000 at a Summit of Heads of States and Government in Lusaka, Zambia. The FTA participating states have zero tariff on goods and services produced in these countries.

In addition to creating the policy environment for freeing trade, COMESA has also created specialized institutions like the Eastern and Southern African Trade and Development Bank (PTA Bank), the PTA Reinsurance Company (ZEP-RE), the Clearing House and the COMESA Court of Justice, to provide the required financial infrastructure and service support. COMESA has also promoted a political risk guarantee scheme, the Africa Trade Insurance Agency (ATI), a Leather and Leather Products Institute (LLPI), as well as a cross-border insurance scheme, the COMESA Yellow Card.

Official languages: English, French, Portuguese.
Headquarters: COMESA Secretariat, COMESA Centre, Ben Bella Road, PO Box 30051, 10101 Lusaka, Zambia.
Website: http://www.comesa.int
Secretary General: Sindiso Ngwenya (Zimbabwe).

East African Community (EAC)

The East African Community (EAC) was formally established on 30 Nov. 1999 with the signing in Arusha, Tanzania of the Treaty for the Establishment of the East African Community. The Treaty envisaged the establishment of a Customs Union, as the entry point of the Community, a Common Market, subsequently a Monetary Union and ultimately a Political Federation of the East African States. In Nov. 2003 the EAC partner states signed a Protocol on the Establishment of the East African Customs Union, which came into force on 1 Jan. 2005. The Common Market came into force on 1 July 2010.

Members. Burundi, Kenya, Rwanda, Tanzania, Uganda.

Headquarters: Arusha International Conference Centre, 5th Floor, Kilimanjaro Wing, PO Box 1096, Arusha, Tanzania.
Website: http://www.eac.int
Secretary General: Dr Richard Sezibera (Rwanda).

East African Development Bank (EADB)

Established originally under the Treaty for East African Co-operation in 1967 with Kenya, Tanzania and Uganda as signatories, a new Charter for the Bank (with the same signatories) came into force in 1980. Rwanda was admitted as a member in 2008. Under the original Treaty the Bank was confined to the provision of financial and technical assistance for the promotion of industrial development in member states but with the new Charter its remit was broadened to include involvement in agriculture, forestry, tourism, transport and the development of infrastructure, with preference for projects which promote regional co-operation.

Official language: English.
Headquarters: 4 Nile Ave., PO Box 7128, Kampala, Uganda.
Website: http://www.eadb.org
Chair of the Governing Council: Syda Bbumba (Uganda).

Economic Community of Central African States (CEEAC)

The Economic Community of Central African States (Communauté Economique des Etats de l'Afrique Centrale) was established in 1983 to promote regional economic co-operation and to establish a Central African Common Market. There are plans for both a common market and a single currency.

Members. Angola, Burundi, Cameroon, Central African Republic, Chad, Democratic Republic of the Congo, Republic of the Congo, Equatorial Guinea, Gabon, São Tomé e Príncipe.

Headquarters: BP 2112, Libreville, Gabon.
Website (French only): http://www.ceeac-eccas.org
President: Lieut.-Gen. Idriss Déby (Chad).
Secretary General: Nassour Guelengdoukssia Ouaidou (Chad).

Economic Community of West African States (ECOWAS)

Founded in 1975 as a regional common market, ECOWAS later also became a political forum involved in the promotion of a democratic environment and the pursuit of fundamental human rights. In July 1993 it revised its treaty to assume responsibility for the regulation of regional armed conflicts, acknowledging the inextricable link between development and peace and security. Thus it now has a new role in conflict management and prevention through its Mediation and Security Council, which monitors the moratorium on the export, import and manufacture of light weapons and ammunition. However, it still retains a military arm, the Economic Community of West African States Monitoring Group (generally known as ECOMOG). It is also involved in the war against drug abuse and illicit drug trafficking. There are plans to introduce a single currency, the *eco*, by 2020.

Members. Benin, Burkina Faso, Cape Verde, Côte d'Ivoire, Gambia, Ghana, Guinea, Guinea-Bissau, Liberia, Mali, Niger, Nigeria, Senegal, Sierra Leone, Togo.

Organization. The institutions of ECOWAS are: the Commission, the Community Parliament, the Community Court of Justice and the ECOWAS Bank for Investment and Development.

Official languages: English, French, Portuguese.
Headquarters: 101 Yakubu Gowon Crescent, Asokoro, Abuja, Nigeria.
Website: http://www.ecowas.int
Email: info@ecowas.int
ECOWAS Commission President: Kadré Désiré Ouedraogo (Burkina Faso).

Further Reading
Jaye, Thomas, Garuba, Dauda and Amadi, Stella, (eds.) *ECOWAS and the Dynamics of Conflict and Peace-building.* 2011

Intergovernmental Authority on Development

The Intergovernmental Authority on Development was created on 21 March 1996 and has its origins in the Intergovernmental Authority on Drought and Development, which had been established in 1986. It has three priority areas of co-operation: conflict prevention, management and humanitarian affairs; infrastructure development; food security and environment protection.

Members. Djibouti, Ethiopia, Kenya, Somalia, South Sudan, Sudan, Uganda. Eritrea was formerly a member but withdrew in April 2007.

Headquarters: Ave. Georges Clemenceau, PO Box 2653, Djibouti, Republic of Djibouti.
Website: http://igad.int
Executive Secretary: Mahboub Maalim (Kenya).

Lake Chad Basin Commission

Established by a Convention and Statute signed on 22 May 1964 by Cameroon, Chad, Niger and Nigeria, and later by the Central African Republic (Sudan has also been admitted as an observer), to regulate and control utilization of the water and other natural resources in the Basin; to initiate, promote and co-ordinate natural resources development projects and research within the Basin area; and to examine complaints and promote settlement of disputes, with a view to promoting regional co-operation.

In Dec. 1977, at Enugu in Nigeria, the 3rd summit of heads of state of the commission signed the protocol for the Harmonization of the Regulations Relating to Fauna and Flora in member countries, and adopted plans for the multi-donor approach towards major integrated development for the conventional basin. An international campaign to save Lake Chad following a report on the environmental degradation of the conventional basin was launched by heads of state at the 8th summit of the Commission in Abuja in March 1994. The 10th summit, held in N'Djaména in 2000, saw agreement on a US$1m. inter-basin water transfer project.

The Commission's budget for 2011 was 3·3bn. francs CFA. It also receives assistance from various international and donor agencies including the FAO, and UN Development and Environment Programmes.

Official languages: English, French.
Headquarters: BP 727, CBLT Siège, Rond Point de l'Etoile, N'Djaména, Chad.
Executive Secretary: Sanusi Imran Abdullahi (Nigeria).

Niger Basin Authority

As a result of a special meeting of the Niger River Commission (established in 1964), to discuss the revitalizing and restructuring of the organization to improve its efficiency, the Niger Basin Authority was established in 1980. Its responsibilities cover the harmonization and co-ordination of national development policies; the formulation of the general development policy of the Basin; the elaboration and implementation of an integrated development plan of the Basin; the initiation and monitoring of an orderly and rational regional policy for the utilization of the waters of the Niger River; the design and conduct of studies, researches and surveys; the formulation of plans, the construction, exploitation and maintenance of structure, and the elaboration of projects.

Members. Benin, Burkina Faso, Cameroon, Chad, Côte d'Ivoire, Guinea, Mali, Niger, Nigeria.

Official languages: English, French.
Headquarters: BP 729, Niamey, Niger.
Website: http://www.abn.ne
Executive Secretary: Maj.-Gen. (retd) Collins Remy Umunakwe Ihekire (Nigeria).

Southern African Customs Union (SACU)

Established by the Customs Union Convention between the British Colony of Cape of Good Hope and the Orange Free State Boer Republic in 1889, the Southern African Customs Union was extended in 1910 to include the then Union of South Africa and British High Commission Territories in Africa and remained unchanged after these countries gained independence. South Africa was the dominant member with sole-decision making power over customs and excise policies until the 2002 SACU Agreement which created a permanent Secretariat, a Council of Ministers headed by a minister from one of the member states on a rotational basis, a Customs Union Commission, Technical Liaison Committees, a SACU tribunal and a SACU tariff board.

Members. Botswana, Lesotho, Namibia, South Africa, Swaziland.

Aims. To promote economic development through regional co-ordination of trade.

Headquarters: Private Bag 13285, Windhoek, Namibia.
Website: http://www.sacu.int
Email: info@sacu.int
Executive Secretary: Tswelopele Cornelia Moremi (Botswana).

Southern African Development Community (SADC)

The Southern African Development Co-ordination Conference (SADCC), the precursor of the Southern African Development Community (SADC), was formed in Lusaka, Zambia on 1 April 1980, following the adoption of the Lusaka Declaration—*Southern Africa: Towards Economic Liberation*—by the nine founding member states.

Members. The nine founder member countries were Angola, Botswana, Lesotho, Malaŵi, Mozambique, Swaziland, Tanzania, Zambia and Zimbabwe. The Democratic Republic of the Congo, Madagascar, Mauritius, Namibia, the Seychelles and South Africa have since joined. The Seychelles left in July 2004 but rejoined in Aug. 2007. As a result there are now 15 members.

Aims and Activities. SADC's Common Agenda includes the following: the promotion of sustainable and equitable economic growth and socio-economic development that will ensure poverty alleviation with the ultimate objective of its eradication; the promotion of common political values, systems and other shared values that are transmitted through institutions that are democratic, legitimate and effective; and the consolidation and maintenance of democracy, peace and security.

In contrast to the country-based co-ordination of sectoral activities and programmes, SADC has now adopted a more centralized approach through which the 21 sectoral programmes are grouped into four clusters; namely: Trade, Industry, Finance and Investment; Infrastructure and Services; Food, Agriculture and Natural Resources; Social and Human Development and Special Programmes.

SADC has made significant progress in implementing its integration agenda since the 1992 Treaty came into force. Since then, more than 20 Protocols to spearhead the sectoral programmes and activities have been signed. Those Protocols that have entered into force include: Immunities and Privileges; Combating Illicit Drugs; Energy; Transport, Communications and Meteorology; Shared Watercourse Systems; Mining; Trade; Education and Training; Tourism; and Health.

Official languages: English, French, Portuguese.
Headquarters: SADC House, Plot No. 54385, Central Business District, Private Bag 0095, Gaborone, Botswana.
Website: http://www.sadc.int
Email: registry@sadc.int
Executive Secretary: Tomaz Augusto Salomão (Mozambique).

West African Development Bank (BOAD)

The West African Development Bank (Banque Ouest Africaine de Développement) was established in Nov. 1973 by an Agreement signed by the member states of the West African Monetary Union (UMOA), now the West African Economic and Monetary Union (UEMOA).

Aims. To promote balanced development of the States of the Union and to achieve West African economic integration.

Members. Benin, Burkina Faso, Côte d'Ivoire, Guinea-Bissau, Mali, Niger, Senegal, Togo.

Official language: French.
Headquarters: 68 avenue de la Libération, Lomé, Togo.
Website (French only): http://www.boad.org
Email: boadsiege@boad.org
President: Christian Adovèlandé (Benin).

West African Economic and Monetary Union (UEMOA)

Founded in 1994, the UEMOA (Union Economique et Monétaire Ouest Africaine) aims to reinforce the competitiveness of the economic and financial activities of member states in the context of an open and rival market and a rationalized and harmonized juridical environment; to ensure the convergence of the macro-economic performances and policies of member states; to create a common market among member states; to co-ordinate the national sector-based policies; and to harmonize the legislation, especially the fiscal system, of the member states.

Members. Benin, Burkina Faso, Côte d'Ivoire, Guinea-Bissau, Mali, Niger, Senegal, Togo.

> *Headquarters:* 01 B.P. 543, Ouagadougou 01, Burkina Faso.
> *Website (French only):* http://www.uemoa.int
> *Email:* commission@uemoa.int
> *President:* Cheikh Hadjibou Soumaré (Senegal).

Agency for the Prohibition of Nuclear Weapons in Latin America and the Caribbean (OPANAL)

The Agency (Organismo para la Proscripción de las Armas Nucleares en la América Latina y el Caribe) was established following the Cuban missile crisis to guarantee implementation of the world's first Nuclear-Weapon-Free-Zone (NWFZ) in the region. Created by the Treaty of Tlatelolco (1967), OPANAL is an inter-governmental agency responsible for ensuring that the requirements of the Treaty are enforced. OPANAL has played a major role in establishing other NWFZs throughout the world.

Organization. The Agency consists of three main bodies: the General Conference which meets for biennial sessions and special sessions when deemed necessary; the Council of OPANAL consisting of five member states which meet every two months plus special meetings when necessary; and the Secretariat General.

Members of the Treaty. Antigua and Barbuda, Argentina, Bahamas, Barbados, Belize, Bolivia, Brazil, Chile, Colombia, Costa Rica, Cuba, Dominica, Dominican Republic, Ecuador, El Salvador, Grenada, Guatemala, Guyana, Haiti, Honduras, Jamaica, Mexico, Nicaragua, Panama, Paraguay, Peru, St Kitts and Nevis, St Lucia, St Vincent and the Grenadines, Suriname, Trinidad and Tobago, Uruguay, Venezuela.

> *Headquarters:* Schiller No. 326, 5th Floor, Col. Chapultepec Morales, México, D. F. 11570, Mexico.
> *Website:* http://www.opanal.org
> *Email:* info@opanal.org
> *Secretary-General:* Gioconda Úbeda Rivera (Costa Rica).

Andean Community

On 26 May 1969 an agreement was signed by Bolivia, Chile, Colombia, Ecuador and Peru establishing the Cartagena Agreement (also referred to as the Andean Pact or the Andean Group). Chile withdrew from the Group in 1976. Venezuela, which was actively involved, did not sign the agreement until 1973. In 1997 Peru announced its withdrawal for five years. In 2006 Venezuela left as a result of Colombia and Peru signing bilateral trade agreements with the USA.

The Andean Free Trade Area came into effect on 1 Feb. 1993 as the first step towards the creation of a common market. Bolivia, Colombia, Ecuador and Peru have fully liberalized their trade. A Common External Tariff for imports from third countries has been in effect since 1 Feb. 1995.

In March 1996 at the Group's 8th summit in Trujillo in Peru, the then member countries (Bolivia, Colombia, Ecuador, Peru, Venezuela) set up the Andean Community, to promote greater economic, commercial and political integration between member countries under a new Andean Integration System (SAI).

The member countries and bodies of the Andean Integration System are working to establish an Andean Common Market and to implement a Common Foreign Policy, a social agenda, a Community policy on border integration, and policies for achieving joint macroeconomic targets.

Organization. The Andean Presidential Council, composed of the presidents of the member states, is the highest-level body of the Andean Integration System (SAI). The Commission and the Andean Council of Foreign Ministers are legislative bodies. The General Secretariat is the executive body and the Andean Parliament is the deliberative body of the SAI. The Court of Justice, which began operating in 1984, resolves disputes between members and interprets legislation. The SAI has other institutions: Andean Development Corporation (CAF), Latin American Reserve Fund (FLAR), Simón Bolívar Andean University, Andean Business Advisory Council, Andean Labour Advisory Council and various Social Agreements.

Further to the treaty signed by 12 South American countries in May 2008, it is anticipated that the Andean Community will gradually be integrated into the new Union of South American Nations.

> *Official language:* Spanish.
> *Headquarters:* Avda Paseo de la República 3895, San Isidro, Lima 17, Peru.
> *Website:* http://www.comunidadandina.org
> *Email:* contacto@comunidadandina.org
> *Secretary-General:* Adalid Contreras Baspineiro (Bolivia).

Association of Caribbean States (ACS)

The Convention establishing the ACS was signed on 24 July 1994 in Cartagena de Indias, Colombia, with the aim of promoting consultation, co-operation and concerted action among all the countries of the Caribbean, comprising 25 full member states and three (now five) associate members. A total of eight other non-independent Caribbean countries are eligible for associate membership.

Members. Antigua and Barbuda, Bahamas, Barbados, Belize, Colombia, Costa Rica, Cuba, Dominica, Dominican Republic, El Salvador, Grenada, Guatemala, Guyana, Haiti, Honduras, Jamaica, Mexico, Nicaragua, Panama, St Kitts and Nevis, St Lucia, St Vincent and the Grenadines, Suriname, Trinidad and Tobago, Venezuela.

Associate members. Aruba, Curaçao, France (on behalf of French Guiana, Guadeloupe and Martinique), Sint Maarten and the Turks and Caicos Islands.

The CARICOM Secretariat, the Latin American Economic System (SELA), the Central American Integration System (SICA) and the Permanent Secretariat of the General Treaty on Central American Economic Integration (SIECA) were declared Founding Observers of the ACS in 1996. The United Nations Economic Commission for Latin America and the Caribbean (ECLAC) and the Caribbean Tourism Organization (CTO) were admitted as Founding Observers in 2000 and 2001 respectively.

Functions. The objectives of the ACS are enshrined in the Convention and are based on the following: the strengthening of the regional co-operation and integration process, with a view to creating an enhanced economic space in the region; preserving the environmental integrity of the Caribbean Sea which is regarded as the common patrimony of the peoples of the region; and promoting the sustainable development of the Greater

Caribbean. Its current focal areas are trade, transport, sustainable tourism and natural disasters.

Organization. The main organs of the Association are the Ministerial Council and the Secretariat. There are Special Committees on: Trade Development and External Economic Relations; Sustainable Tourism; Transport; Natural Disasters; Budget and Administration. There is also a Council of National Representatives of the Special Fund responsible for overseeing resource mobilization efforts and project development.

Headquarters: ACS Secretariat, 5–7 Sweet Briar Road, St Clair, PO Box 660, Port of Spain, Trinidad and Tobago.
Website: http://www.acs-aec.org
Email: mail@acs-aec.org
Secretary-General: Alfonso Múnera (Colombia).

Caribbean Community (CARICOM)

Origin. The Treaty of Chaguaramas establishing the Caribbean Community and Common Market was signed by the prime ministers of Barbados, Guyana, Jamaica and Trinidad and Tobago at Chaguaramas, Trinidad, on 4 July 1973.

Six additional countries and territories (Belize, Dominica, Grenada, St Lucia, St Vincent and the Grenadines, Montserrat) signed the Treaty on 17 April 1974, and the Treaty came into effect for those countries on 1 May 1974. Antigua acceded to membership on 4 July that year; St Kitts and Nevis on 26 July; the Bahamas on 4 July 1983 (not Common Market); Suriname on 4 July 1995.

Members. Antigua and Barbuda, Bahamas, Barbados, Belize, Dominica, Grenada, Guyana, Haiti, Jamaica, Montserrat, St Kitts and Nevis, St Lucia, St Vincent and the Grenadines, Suriname, and Trinidad and Tobago. Anguilla, Bermuda, the British Virgin Islands, Cayman Islands and Turks and Caicos Islands are associate members.

Objectives. The Caribbean Community has the following objectives: improved standards of living and work; full employment of labour and other factors of production; accelerated, co-ordinated and sustained economic development and convergence; expansion of trade and economic relations with third States; enhanced levels of international competitiveness; organization for increased production and productivity; the achievement of a greater measure of economic leverage and effectiveness of member states in dealing with third States, groups of States and entities of any description; enhanced co-ordination of member states' foreign and foreign economic policies; enhanced functional co-operation.

At its 20th Meeting in July 1999 the Conference of Heads of Government of the Caribbean Community approved for signature the agreement establishing the Caribbean Court of Justice. They mandated the establishment of a Preparatory Committee comprising the Attorneys General of Barbados, Guyana, Jamaica, St Kitts and Nevis, St Lucia and Trinidad and Tobago assisted by other officials, to develop and implement a programme of public education within the Caribbean Community and to make appropriate arrangements for the inauguration of the Caribbean Court of Justice prior to the establishment of the CARICOM Single Market and Economy. To this end at its 23rd Meeting in July 2002 the Heads of Government agreed on immediate measures to inaugurate the Court by the second half of 2003, although delays meant it was not inaugurated until April 2005. Among the measures adopted was the establishment of a Trust Fund with a one-time settlement of US$100m. to finance the

Court. The President of the Caribbean Development Bank was authorized to raise the funds on international capital markets, so that member states could access these funds to meet their assessed contributions towards the financing of the Court. The agreement establishing the Regional Justice Protection Programme was also approved for signature.

Structure. The Conference of Heads of Government is the principal organ of the Community, and its primary responsibility is to determine and provide the policy direction for the Community. It is the final authority on behalf of the Community for the conclusion of treaties and for entering into relationships between the Community and international organizations and States. It is responsible for financial arrangements to meet the expenses of the Community.

The Community Council of Ministers is the second highest organ of the Community and consists of Ministers of Government responsible for Community Affairs. The Community Council has primary responsibility for the development of Community strategic planning and co-ordination in the areas of economic integration, functional co-operation and external relations.

The Secretariat is the principal administrative organ of the Community. The Secretary-General is appointed by the Conference (on the recommendation of the Community Council) for a term not exceeding five years, and may be reappointed. The Secretary-General is the Chief Executive Officer of the Community and acts in that capacity at all meetings of the Community Organs.

Associate Institutions. Caribbean Development Bank (CDB); University of Guyana (UG); University of the West Indies (UWI); Caribbean Law Institute (CLI)/Caribbean Law Institute Centre (CLIC); Organisation of Eastern Caribbean States; Anton de Kom University of Suriname.

Official language: English.
Headquarters: Bank of Guyana Building, PO Box 10827, Georgetown, Guyana.
Website: http://www.caricom.org
Secretary-General: Irwin LaRocque (Dominica).

Publications. CARICOM Perspective (annual); *Annual Report; Treaty Establishing the Caribbean Community; Caribbean Trade and Investment Report 2010.*

Further Reading

Payne, Anthony, *The Political History of CARICOM.* 2007

Caribbean Development Bank (CDB)

Established in 1969 by 16 regional and two non-regional members. Membership is open to all states and territories of the region and to non-regional states that are members of the UN or its Specialized Agencies or of the International Atomic Energy Agency.

Members—regional countries and territories: Anguilla, Antigua and Barbuda, Bahamas, Barbados, Belize, British Virgin Islands, Cayman Islands, Dominica, Grenada, Guyana, Haiti, Jamaica, Montserrat, St Kitts and Nevis, St Lucia, St Vincent and the Grenadines, Trinidad and Tobago, Turks and Caicos Islands. *Other regional countries:* Colombia, Mexico, Venezuela. *Non-regional countries:* Canada, China, Germany, Italy, United Kingdom.

Function. To contribute to the economic growth and development of the member countries of the Caribbean and promote economic co-operation and integration among them, with particular regard to the needs of the less developed countries.

Headquarters: PO Box 408, Wildey, St Michael, Barbados.
Website: http://www.caribank.org
Email: info@caribank.org
President: William Warren Smith (Jamaica).

Publications. Annual Report; Basic Information; Caribbean Development Bank: Its Purpose, Role and Functions; Summary of Proceedings of Annual Meetings of Board of Governors; Statements by the President; Financial Policies; Guidelines for Procurement; Procedures for the Selection and Engagement of Consultants by Recipients of CDB Financing; Special Development Fund Rules; Sector Policy Papers; CDB News (newsletter).

Central American Bank for Economic Integration (CABEI)

Established in 1960, the Bank is the financial institution created by the Central American Economic Integration Treaty and aims to implement the economic integration and balanced economic growth of the member states.

Regional members. Costa Rica, El Salvador, Guatemala, Honduras, Nicaragua.

Non-regional members. Argentina, Colombia, Dominican Republic, Mexico, Panama, Spain, Taiwan.

Official languages: Spanish, English.
Headquarters: Apartado Postal 772, Tegucigalpa, DC, Honduras.
Website: http://www.bcie.org
President: Nick Rischbieth (Honduras).

Central American Integration System (SICA)

The Central American Integration System (SICA) was established in Dec. 1991 and became operational in Feb. 1993. SICA is the successor body to the Organization of Central American States, which was suspended in 1973 after a war between El Salvador and Honduras (which were both member countries). It aims to achieve political, economic, social, cultural and ecological integration in Central America and transform the area into a region of peace, liberty, democracy and development.

Members. Belize, Costa Rica, El Salvador, Guatemala, Honduras, Nicaragua and Panama.
    The Framework Treaty on Democratic Security in Central America, signed in 1995, seeks to achieve a proper 'balance of forces' in the region, intensify the fight against trafficking of drugs and arms, and reintegrate refugees and displaced persons.

Headquarters: Final Bulevar Cancillería, Distrito El Espino, Ciudad Merliot, Antiguo Cuscatlán, La Libertad, El Salvador.
Website: http://www.sica.int
Secretary-General: Juan Daniel Alemán (Guatemala).

Eastern Caribbean Central Bank (ECCB)

The Eastern Caribbean Central Bank was established in 1983, replacing the East Caribbean Currency Authority (ECCA). Its purpose is to regulate the availability of money and credit; to promote and maintain monetary stability; to promote credit and exchange conditions and a sound financial structure conducive to the balanced growth and development of the economies of the territories of the participating governments; and to actively promote, through means consistent with its other objectives, the economic development of the territories of the participating governments.

Members. Anguilla, Antigua and Barbuda, Dominica, Grenada, Montserrat, St Kitts and Nevis, St Lucia, St Vincent and the Grenadines.

Official language: English.
Headquarters: PO Box 89, Bird Rock, Basseterre, St Kitts and Nevis.
Website: http://www.eccb-centralbank.org
Email: info@eccb-centralbank.org
Governor: Sir Dwight Venner (St Vincent and the Grenadines).

Inter-American Development Bank (IDB)

The IDB, the oldest and largest regional multilateral development institution, was established in 1959 to help accelerate economic and social development in Latin America and the Caribbean. The Bank's original membership included 19 Latin American and Caribbean countries and the USA. Today, membership totals 48 nations, including non-regional members.

Members. Argentina, Austria, Bahamas, Barbados, Belgium, Belize, Bolivia, Brazil, Canada, Chile, China, Colombia, Costa Rica, Croatia, Denmark, Dominican Republic, Ecuador, El Salvador, Finland, France, Germany, Guatemala, Guyana, Haiti, Honduras, Israel, Italy, Jamaica, Japan, South Korea, Mexico, the Netherlands, Nicaragua, Norway, Panama, Paraguay, Peru, Portugal, Slovenia, Spain, Suriname, Sweden, Switzerland, Trinidad and Tobago, UK, USA, Uruguay, Venezuela.
    The Bank's total lending up to 2011 was US$208bn. for projects with a total cost of over US$438bn. Its lending increased dramatically from the US$294m. approved in 1961 to US$10·9bn. in 2011.
    Current lending priorities include poverty reduction and social equity, modernization and integration, and the environment. The Bank has a Fund for Special Operations for lending on concessional terms for projects in countries classified as economically less developed. An additional facility, the Multilateral Investment Fund (MIF), was created in 1992 to help promote and accelerate investment reforms and private-sector development throughout the region.
    The Board of Governors is the Bank's highest authority. Governors are usually Ministers of Finance, Presidents of Central Banks or officers of comparable rank. The IDB has country offices in each of its borrowing countries, and in Paris and Tokyo.

Official languages: English, French, Portuguese, Spanish.
Headquarters: 1300 New York Ave., NW, Washington, D.C., 20577, USA.
Website: http://www.iadb.org
President: Luis Alberto Moreno (Colombia).

Latin American Economic System (SELA)

Established in 1975 by the Panama Convention, SELA (Sistema Económico Latinoamericano) promotes co-ordination on economic issues and social development among the countries of Latin America and the Caribbean.

Members. Argentina, Bahamas, Barbados, Belize, Bolivia, Brazil, Chile, Colombia, Costa Rica, Cuba, Dominican Republic, Ecuador, El Salvador, Grenada, Guatemala, Guyana, Haiti, Honduras, Jamaica, Mexico, Nicaragua, Panama, Paraguay, Peru, Suriname, Trinidad and Tobago, Uruguay, Venezuela.

Official languages: English, French, Portuguese, Spanish.
Headquarters: Torre Europa, Pisos 4 y 5, Avenida Francisco de Miranda, Urb. Campo Alegre, Caracas 1060, Venezuela.
Website: http://www.sela.org
Permanent Secretary: José Rivera Banuet (Mexico).

Publications. Capitulos (in Spanish and English, published thrice yearly); *SELA Antenna in the United States* (quarterly bulletin); *Integration Bulletin on Latin America and the Caribbean* (monthly).

Latin American Integration Association (ALADI/LAIA)

The ALADI was established to promote freer trade among member countries in the region.

Members. (14) Argentina, Bolivia, Brazil, Chile, Colombia, Cuba, Ecuador, Mexico, Nicaragua, Panama, Paraguay, Peru, Uruguay and Venezuela.

Observers. (28) Andean Development Corporation (CAF), China, Commission of the European Communities, Costa Rica, Dominican Republic, El Salvador, Guatemala, Honduras, Ibero-American General Secretariat (SEGIB), Inter-American Development Bank, Inter-American Institute for Cooperation on Agriculture (IICA), Italy, Japan, South Korea, Latin American Economic System (SELA), Nicaragua, Organization of American States (OAS), Pan American Health Organization (PAHO), Panama, Portugal, Romania, Russia, Spain, Switzerland, Ukraine, UN Development Programme, UN Economic Commission for Latin America and the Caribbean (ECLAC), World Health Organization (WHO).

Official languages: Portuguese, Spanish.
Headquarters: Calle Cebollatí 1461, Barrio Palermo, Casilla de Correos 20005, 11200 Montevideo, Uruguay.
Website: http://www.aladi.org
Secretary-General: Carlos Álvarez (Argentina).

Latin American Reserve Fund

Established in 1991 as successor to the Andean Reserve Fund, the Latin American Reserve Fund assists in correcting payment imbalances through loans with terms of up to four years and guarantees extended to members, to co-ordinate their monetary, exchange and financial policies and to promote the liberalization of trade and payments in the Andean sub-region.

Members. Bolivia, Colombia, Costa Rica, Ecuador, Peru, Uruguay, Venezuela.

Official language: Spanish.
Headquarters: Avenida 82, N° 12-18, piso 7, Bogotá, DC, Colombia.
Website: http://www.flar.net
Executive President: Ana María Carrasquilla (Colombia).

Organisation of Eastern Caribbean States (OECS)

Founded in 1981 when seven eastern Caribbean states signed the Treaty of Basseterre agreeing to co-operate with each other to promote unity and solidarity among the members.

Members. Antigua and Barbuda, Dominica, Grenada, Montserrat, St Kitts and Nevis, St Lucia, St Vincent and the Grenadines. The British Virgin Islands and Anguilla have associate membership.

Functions. As set out in the Treaty of Basseterre: to promote co-operation among the member states and to defend their sovereignty and independence; to assist member states in the realization of their obligations and responsibilities to the international community with due regard to the role of international law as a standard of conduct in their relationships; to assist member states in the realization of their obligations and responsibilities to the international community with due regard to the role of international issues; to establish and maintain, where possible, arrangements for joint overseas representation and common services; to pursue these through its respective institutions by discussion of questions of common concern and by agreement on common action.

The Authority is the highest decision-making body of the OECS, comprising the heads of government of the member countries. The OECS is administered by a Central Secretariat (based in Castries, St Lucia), headed by a director general who is responsible to the Authority. The secretariat is divided into four principal divisions: the division of the office of the director general; social and sustainable development division; corporate services division; and economic affairs division.

In June 2010 a Revised Treaty of Basseterre was signed. It came into effect in Jan. 2011, establishing the OECS Economic Union. Under its terms, member states agreed to the removal of trade barriers, the free movement of labour and capital, the establishment of a regional assembly of parliamentarians and the implementation of a common external tariff. Membership of the Economic Union initially comprised Antigua and Barbuda, Dominica, Grenada, St Vincent and the Grenadines, St Kitts and Nevis, and St Lucia.

Official language: English.
Headquarters: Morne Fortune, PO Box 179, Castries, St Lucia.
Website: http://www.oecs.org
Email: oecss@oecs.org
Director-General: Dr Len Ishmael (St Lucia).

Organization of American States (OAS)

Origin. On 14 April 1890 representatives of the American republics, meeting in Washington at the First International Conference of American States, established an International Union of American Republics and, as its central office, a Commercial Bureau of American Republics, which later became the Pan-American Union. This international organization's object was to foster mutual understanding and co-operation among the nations of the western hemisphere. This led to the adoption on 30 April 1948 by the Ninth International Conference of American States, at Bogotá, Colombia, of the Charter of the Organization of American States. This co-ordinated the work of all the former independent official entities in the inter-American system and defined their mutual relationships. The Charter of 1948 was subsequently amended by the Protocol of Buenos Aires (1967) and the Protocol of Cartagena de Indias (1985).

Members. Antigua and Barbuda, Argentina, Bahamas, Barbados, Belize, Bolivia, Brazil, Canada, Chile, Colombia, Costa Rica, Cuba (suspended 1962*), Dominica, Dominican Republic, Ecuador, El Salvador, Grenada, Guatemala, Guyana, Haiti, Honduras, Jamaica, Mexico, Nicaragua, Panama, Paraguay, Peru, St Kitts and Nevis, St Lucia, St Vincent and the Grenadines, Suriname, Trinidad and Tobago, USA, Uruguay, Venezuela. *In June 2009 the OAS voted to lift Cuba's suspension, although Cuba had stated that it did not wish to rejoin the organization.

Permanent Observers. Algeria, Angola, Armenia, Austria, Azerbaijan, Belgium, Benin, Bosnia and Herzegovina, Bulgaria, China, Croatia, Cyprus, Czech Republic, Denmark, Egypt, Equatorial Guinea, Estonia, EU, Finland, France, Georgia, Germany, Ghana, Greece, Holy See, Hungary, Iceland, India, Ireland, Israel, Italy, Japan, Kazakhstan, South Korea, Latvia, Lebanon, Luxembourg, Morocco, the Netherlands, Nigeria, Norway, Pakistan, Philippines, Poland, Portugal, Qatar, Romania, Russia, Saudi Arabia, Serbia, Slovakia, Slovenia, Spain, Sri Lanka, Sweden, Switzerland, Thailand, Tunisia, Turkey, UK, Ukraine, Vanuatu, Yemen.

Aims and Activities. To strengthen the peace and security of the continent; promote and consolidate representative democracy; promote by co-operative action economic, social and cultural development; and achieve an effective limitation of conventional weapons.

In Sept. 2001 an Inter-American Democratic Charter was adopted, declaring: 'The peoples of the Americas have a right to democracy and their governments have an obligation to promote and defend it.' The Charter compels the OAS to take action against any member state that disrupts its own democratic institutions.

Organization. Under its Charter the OAS accomplishes its purposes by means of:

(a) The General Assembly, which meets annually. The Secretary-General is elected by the General Assembly for five-year terms. The General Assembly approves the annual budget which is financed by quotas contributed by the member governments. The budget in 2012 amounted to US$85·35m.

(b) The Meeting of Consultation of Ministers of Foreign Affairs, held to consider problems of an urgent nature and of common interest.

(c) The Councils: The Permanent Council, which meets on a permanent basis at OAS headquarters and carries out decisions of the General Assembly, assists the member states in the peaceful settlement of disputes, acts as the Preparatory Committee of that Assembly, submits recommendations with regard to the functioning of the Organization, and considers the reports to the Assembly of the other organs. The Inter-American Council for Integral Development (CIDI) directs and monitors OAS technical co-operation programmes.

(d) The Inter-American Juridical Committee which acts as an advisory body to the OAS on juridical matters and promotes the development and codification of international law. 11 jurists, elected for four-year terms by the General Assembly, represent all the American States.

(e) The Inter-American Commission on Human Rights which oversees the observance and protection of human rights. Seven members elected for four-year terms by the General Assembly represent all the OAS member states.

(f) The General Secretariat, which is the central and permanent organ of the OAS.

(g) The Specialized Conferences, meeting to deal with special technical matters or to develop specific aspects of inter-American co-operation.

(h) The Specialized Organizations, intergovernmental organizations established by multilateral agreements to discharge specific functions in their respective fields of action, such as women's affairs, agriculture, child welfare, Indian affairs, geography and history, and health.

Headquarters: 17th St. and Constitution Ave., NW, Washington, D.C., 20006, USA.
Website: http://www.oas.org
Secretary-General: José Miguel Insulza (Chile).

Publications. Charter of the Organization of American States. 1948.—*As Amended by the Protocol of Buenos Aires in 1967 and the Protocol of Cartagena de Indias in 1985; The OAS and the Evolution of the Inter-American System; Annual Report of the Secretary-General; Status of Inter-American Treaties and Conventions* (annual).

Secretariat for Central American Economic Integration (SIECA)

SIECA (Secretaría de Integración Económica Centroamericana) was created by the General Treaty on Central American Economic Integration in Dec. 1960. The General Treaty incorporates the Agreement on the Regime for Central American Integration Industries. In Oct. 1993 the Protocol to the General Treaty on Central Economic Integration, known as the Guatemala Protocol, was signed.

Members. Costa Rica, El Salvador, Guatemala, Honduras, Nicaragua. *Observer:* Panama.

Official language: Spanish.
Headquarters: 4a Avenida 10–25, Zona 14, Ciudad de Guatemala, Guatemala.
Website: http://www.aic.sieca.int
Secretary-General: Ernesto Torres Chico (El Salvador).

Southern Common Market (MERCOSUR)

Founded in March 1991 by the Treaty of Asunción between Argentina, Brazil, Paraguay and Uruguay, MERCOSUR committed the signatories to the progressive reduction of tariffs culminating in the formation of a common market on 1 Jan. 1995. This duly came into effect as a free trade zone affecting 90% of commodities. A common external tariff averaging 14% applies to 80% of trade with countries outside MERCOSUR. Details were agreed at foreign minister level by the Protocol of Ouro Preto signed on 17 Dec. 1994.

In 1996 Chile negotiated a free-trade agreement with MERCOSUR which came into effect on 1 Oct. Subsequently Bolivia, Chile, Colombia, Ecuador and Peru have all been granted associate member status. Bolivia became an accessing member in 2012. Mexico has observer status. Venezuela, which had associate membership between 2004 and 2006, became the fifth member of MERCOSUR in July 2006, although it was not going to have full voting rights until all the other full members had ratified its entry into the organization. Paraguay was the only country still to approve Venezuela's full membership, but it was suspended from MERCOSUR in June 2012 following the impeachment of its president, Fernando Lugo. With Paraguay suspended, Venezuela was then formally admitted in July 2012.

Organization. The member states' foreign ministers form a Council responsible for leading the integration process, the chairmanship of which rotates every six months. The permanent executive body is the Common Market Group of member states, which takes decisions by consensus. There is a Trade Commission and Joint Parliamentary Commission, an arbitration tribunal whose decisions are binding on member countries, and a secretariat in Montevideo.

Further to the treaty signed by 12 South American countries in May 2008, it is anticipated that MERCOSUR will gradually be integrated into the new Union of South American Nations.

Headquarters: Dr Luis Piera 1992, Piso 1, 11200 Montevideo, Uruguay.
Website (Spanish and Portuguese only):
   http://www.mercosur.org.uy
Executive Director of the Secretariat: Agustín Colombo Sierra (Argentina).

Union of South American Nations (UNASUR)

History. Established in May 2008 in Brazil, it is anticipated that the Union of South American Nations will eventually supersede MERCOSUR and the Andean Community, creating an enlarged customs union with a single market, parliament, secretariat and central bank, based on the European Union structure. UNASUR is the successor body to the now defunct South American Community of Nations (CSN/SACN), founded in 2004. Despite initial problems, progress was made at UNASUR's fourth Summit in Nov. 2010 culminating in the 'Georgetown Declaration', with the attending heads of state and government and foreign ministers highlighting their commitment to working together to achieve a better South America. The Treaty establishing UNASUR became effective on 11 March 2011.

Organization. There is a permanent secretariat based in Quito, Ecuador. A proposed South American parliament is planned for Cochabamba, Bolivia. The heads of state of member nations meet annually.

Members. Argentina, Bolivia, Brazil, Chile, Colombia, Ecuador, Guyana, Paraguay (suspended following the impeachment of President Fernando Lugo in June 2012), Peru, Suriname, Uruguay, Venezuela.

Official languages: Portuguese, Spanish, Dutch and English.
Headquarters: Av. 6 de Diciembre N24-04 y Wilson, Quito, Ecuador.
Website: http://www.unasursg.org
Email: secretaria.general@unasursg.org
Secretary-General: Alí Rodríguez Araque (Venezuela).

Asian Development Bank

A multilateral development finance institution established in 1966 to promote economic and social progress in the Asian and Pacific region, the Bank's strategic objectives are to foster economic growth, reduce poverty, improve the status of women, support human development (including population planning) and protect the environment.

The bank's capital stock is owned by 67 member countries, 48 regional and 19 non-regional. The bank makes loans and equity investments, and provides technical assistance grants for the preparation and execution of development projects and programmes; promotes investment of public and private capital for development purposes; and assists in co-ordinating

development policies and plans in its developing member countries (DMCs).

The bank gives special attention to the needs of smaller or less developed countries, giving priority to projects that contribute to the economic growth of the region and promote regional co-operation. Loans from ordinary capital resources on non-concessional terms account for about 80% of cumulative lending. Loans from the bank's principal special fund, the Asian Development Fund, are made on highly concessional terms almost exclusively to the poorest borrowing countries.

Regional members. Afghanistan, Armenia, Australia, Azerbaijan, Bangladesh, Bhutan, Brunei, Cambodia, China, Cook Islands, Fiji Islands, Georgia, Hong Kong, India, Indonesia, Japan, Kazakhstan, Kiribati, South Korea, Kyrgyzstan, Laos, Malaysia, Maldives, Marshall Islands, Micronesia, Mongolia, Myanmar, Nauru, Nepal, New Zealand, Pakistan, Palau, Papua New Guinea, Philippines, Samoa, Singapore, Solomon Islands, Sri Lanka, Taiwan, Tajikistan, Thailand, Timor-Leste, Tonga, Turkmenistan, Tuvalu, Uzbekistan, Vanuatu and Vietnam.

Non-regional members. Austria, Belgium, Canada, Denmark, Finland, France, Germany, Ireland, Italy, Luxembourg, Netherlands, Norway, Portugal, Spain, Sweden, Switzerland, Turkey, UK, USA.

Organization. The bank's highest policy-making body is its Board of Governors, which meets annually. Its executive body is the 12-member Board of Directors (each with an alternate), eight from the regional members and four non-regional.

The ADB also has resident missions: in Afghanistan, Armenia, Azerbaijan, Bangladesh, Cambodia, China, Georgia, India, Indonesia, Kazakhstan, Kyrgyzstan, Laos, Mongolia, Nepal, Pakistan, Papua New Guinea, Philippines, Sri Lanka, Tajikistan, Thailand, Timor-Leste, Turkmenistan, Uzbekistan, Vietnam; a Pacific Liaison and Co-ordination Office in Sydney; and a South Pacific Subregional Office in Suva, Fiji Islands. There are also three representative offices: in Tokyo, Frankfurt and Washington, D.C.

Official language: English.
Headquarters: 6 ADB Avenue, Mandaluyong, Metro Manila, Philippines.
Website: http://www.adb.org
President: Takehiko Nakao (Japan).

Asia-Pacific Economic Co-operation (APEC)

Origin and Aims. APEC was originally established in 1989 to take advantage of the interdependence among Asia-Pacific economies, by facilitating economic growth for all participants and enhancing a sense of community in the region. Begun as an informal dialogue group, APEC is the premier forum for facilitating economic growth, co-operation, trade and investment in the Asia-Pacific region. APEC has a membership of 21 economic jurisdictions that together account for 40% of the world population, 43% of world trade and 55% of world GDP. APEC is working to achieve what are referred to as the 'Bogor Goals' of free and open trade and investment in the Asia-Pacific area.

Members. Australia, Brunei, Canada, Chile, China, Hong Kong, Indonesia, Japan, South Korea, Malaysia, Mexico, New Zealand, Papua New Guinea, Peru, Philippines, Russia, Singapore, Taiwan, Thailand, USA and Vietnam.

Activities. APEC works in three broad areas to meet the Bogor Goals. These three broad work areas, known as APEC's 'Three Pillars', are: Trade and Investment Liberalisation—reducing and eliminating tariff and non-tariff barriers to trade and investment, and opening markets; Business Facilitation—reducing the costs of business transactions, improving access to trade information and co-ordinating policy and business strategies to facilitate growth, and free and open trade; Economic and Technical Co-operation—assisting member economies build the necessary capacities to take advantage of global trade and the new economy. In 2012 Russia hosted APEC meetings under the theme 'Integrate to Grow, Integrate to Prosper'. The host for 2013 is Indonesia, using the theme 'Resilient Asia-Pacific, Engine of Global Growth'.

Official language: English.
Headquarters: 35 Heng Mui Keng Terrace, Singapore 119616.
Website: http://www.apec.org
Executive Director: Allan Bollard (New Zealand).

Association of South East Asian Nations (ASEAN)

History and Membership. ASEAN is a regional intergovernmental organization formed by the governments of Indonesia, Malaysia, the Philippines, Singapore and Thailand through the Bangkok Declaration which was signed by their foreign ministers on 8 Aug. 1967. Brunei joined in 1984, Vietnam in 1995, Laos and Myanmar in 1997 and Cambodia in 1999. Papua New Guinea also has observer status. The ASEAN Charter, signed in Nov. 2007, established the group as a legal entity and created permanent representation for members at its secretariat in Jakarta.

Objectives. The main objectives are to accelerate economic growth, social progress and cultural development, to promote active collaboration and mutual assistance in matters of common interest, to ensure the political and economic stability of the South East Asian region, and to maintain close co-operation with existing international and regional organizations with similar aims.

Activities. Principal projects concern economic co-operation and development, with the intensification of intra-ASEAN and global trade; joint research and technological programmes; co-operation in transportation and communications; promotion of tourism, South East Asian studies, cultural, scientific, educational and administrative exchanges. An *ASEAN Free Trade Area (AFTA)* agreement was signed in 1992. ASEAN member countries have in the meantime made significant progress in the lowering of intra-regional tariffs through the Common Effective Preferential Tariff (CEPT) Scheme for AFTA. The *ASEAN Charter* of 2007 established a schedule for the elimination of non-tariff barriers and other restrictions on trade. On 1 Jan. 2010 ASEAN signed a free trade agreement with China, creating the world's largest free trade area by population (encompassing 1·9bn. people) and the third largest by economic value.

Heads of government who met in Bangkok in Dec. 1995 established a South-East Asia Nuclear-Free Zone, which was extended to cover offshore economic exclusion zones. Individual signatories were to decide whether to allow port visits or transportation of nuclear weapons by foreign powers through territorial waters. The first formal meeting of the *ASEAN Regional Forum (ARF)* to discuss security issues in the region took place in July 1994 and was attended by the then six members (Brunei, Indonesia, Malaysia, Philippines, Singapore and Thailand). Also in attendance were ASEAN's dialogue partners (Australia, Canada, the EU, Japan, South Korea, New Zealand and the USA), consultative partners (China and Russia) and observers (Laos, Papua New Guinea and Vietnam). In 2012 the participants in the ARF were the ten ASEAN members, Australia, Bangladesh,

Canada, China, the EU, India, Japan, North Korea, South Korea, Mongolia, New Zealand, Pakistan, Papua New Guinea, Russia, Sri Lanka, Timor-Leste and the USA.

ASEAN is committed to resolving the dispute over sovereignty of the Spratly Islands, a group of more than 100 small islands and reefs in the South China Sea. Some or all of the largely uninhabited islands have been claimed by Brunei, China, Malaysia, the Philippines, Taiwan and Vietnam. The disputed areas have oil and gas resources.

Organization. The highest authority is the meeting of Heads of Government, which takes place twice annually. The highest policy-making body is the annual Meeting of Foreign Ministers, commonly known as AMM, the ASEAN Ministerial Meeting, which convenes in each of the member countries on a rotational basis in alphabetical order. The AEM (ASEAN Economic Meeting) meets each year to direct ASEAN economic co-operation. The AEM and AMM report jointly to the heads of government at summit meetings. The central secretariat in Jakarta is headed by the Secretary-General, a post that revolves among the member states in alphabetical order every five years.

Official language: English.
Headquarters: 70A Jl. Sisingamangaraja, Jakarta 12110, Indonesia.
Website: http://www.asean.org
Secretary-General: Le Luong Minh (Vietnam).

ASEAN-Mekong Basin Development Co-operation (Mekong Group)

The ministers and representatives of Brunei, Cambodia, China, Indonesia, Laos, Malaysia, Myanmar, Philippines, Singapore, Thailand and Vietnam met in Kuala Lumpur on 17 June 1996 and agreed the following objectives for the Group: to co-operate in the economic and social development of the Mekong Basin area and strengthen the link between it and ASEAN member countries, through a process of dialogue and common project identification.

Priorities include: development of infrastructure capacities in the fields of transport, telecommunications, irrigation and energy; development of trade and investment-generating activities; development of the agricultural sector to enhance production; sustainable development of forestry resources and development of mineral resources; development of the industrial sector, especially small to medium enterprises; development of tourism; human resource development and support for training; co-operation in the fields of science and technology.

Further Reading

Beeson, Mark, *Regionalism & Globalization in East Asia: Politics, Security & Economic Development.* 2006.—*Contemporary Southeast Asia.* 2nd ed. 2008.—*Institutions of the Asia-Pacific: ASEAN, APEC and Beyond.* 2008
Broinowski, A., *Understanding ASEAN.* 1982.—(ed.) *ASEAN into the 1990s.* 1990
Jarvis, Darryl S. L. and Welch, Anthony, (eds.) *ASEAN Industries and the Challenge from China.* 2011
Jones, Lee, *ASEAN, Sovereignty and Intervention in Southeast Asia.* 2011
Lee, Yoong Yoong, *ASEAN Matters! Reflecting on the Association of Southeast Asian Nations.* 2011

Colombo Plan

History. Founded in 1950 to promote the development of newly independent Asian member countries, the Colombo Plan has grown from a group of seven Commonwealth nations into an organization of 25 countries. Originally the Plan was conceived for a period of six years but the Consultative Committee gave the Plan an indefinite life span in 1980.

Members. Afghanistan, Australia, Bangladesh, Bhutan, Fiji Islands, India, Indonesia, Islamic Republic of Iran, Japan, South Korea, Lao People's Democratic Republic, Malaysia, Maldives, Mongolia, Myanmar, Nepal, New Zealand, Pakistan, Papua New Guinea, Philippines, Singapore, Sri Lanka, Thailand, USA and Vietnam. *(Provisional member country)* Brunei.

Aims. The aims of the Colombo Plan are: (1) to provide a forum for discussion, at local level, of development needs; (2) to facilitate development assistance by encouraging members to participate as donors and recipients of technical co-operation; and (3) to execute programmes to advance development within member countries. The Plan currently has the following programmes: Programme for Public Administration (PPA); South-South Technical Co-operation Data Bank Programme (SSTC/DB); Drug Advisory Programme (DAP); Programme for Private Sector Development (PPSD); Colombo Plan Staff College for Technician Education (CPSC).

Structure. The Consultative Committee is the principal policy-making body of the Colombo Plan. Consisting of all member countries, it meets every two years to review the economic and social progress of members, exchange views on technical co-operation programmes and review the Plan's activities. The Colombo Plan Council represents each member government and meets several times a year to identify development issues, recommend measures to be taken and ensure implementation.

Headquarters: PO Box 596, 31 Wijerama Road, Colombo 7, Sri Lanka.
Website: http://www.colombo-plan.org
Email: info@colombo-plan.org
Secretary-General: Adam Maniku (Maldives).

Publications. Consultative Committee Meeting—Proceedings and Conclusions (biennial); *Report of the Colombo Plan Council* (annual); *The Colombo Plan Brochure* (annual); *The Colombo Plan Focus* (quarterly newsletter); *South-South Technical Co-operation in Selected Member Countries.*

Economic Co-operation Organization (ECO)

The Economic Co-operation Organization (ECO) is an intergovernmental regional organization established in 1985 by Iran, Pakistan and Turkey and the successor of the Regional Co-operation for Development (RCD). ECO was expanded in 1992 to include seven new members: Afghanistan, Azerbaijan, Kazakhstan, Kyrgyzstan, Tajikistan, Turkmenistan and Uzbekistan. The organization's objectives, stipulated in its Charter, the Treaty of İzmir, include the promotion of conditions for sustained economic growth in the region. Transport and communications, trade and investment, and energy are the high priority areas in ECO's scheme of work although industry, agriculture, health, science and education, drug control and human development are also on the agenda.

The Council of Ministers (COM) remains the highest policy and decision-making body of the organization, meeting at least once a year and chaired by rotation among the member states.

ECO Summits were instituted with the First Summit held in Tehran in 1992. A further 11 Summits have been held since then, most recently in Baku in 2012.

The long-term perspectives and priorities of ECO are defined in the form of two Action Plans: the Quetta Plan of Action and the İstanbul Declaration and Economic Co-operation Strategy.

ECO enjoys observer status with the United Nations, World Trade Organization and the Organization of Islamic Conference.

Headquarters: 1 Goulbou Alley, Kamranieh, PO Box 14155-6176, Tehran, Islamic Republic of Iran.
Website: http://www.ecosecretariat.org
Email: Registry@ECOsecretariat.org
Secretary-General: Shamil Aleskerov (Azerbaijan).

Pacific Islands Forum (PIF)

In Oct. 2000 the South Pacific Forum changed its name to the Pacific Islands Forum. As the South Pacific Forum it held its first meeting of Heads of Government in New Zealand in 1971. The Agreement Establishing the Forum Secretariat defines the membership of the Forum and the Secretariat. Decisions are reached by consensus. The administrative arm of the Forum, known officially as the Pacific Islands Forum Secretariat, is based in Suva, Fiji Islands. In Oct. 1994 the Forum was granted observer status to the UN.

Members. Australia, Cook Islands, Fiji Islands*, Kiribati, Marshall Islands, Micronesia, Nauru, New Zealand, Niue, Palau, Papua New Guinea, Samoa, Solomon Islands, Tonga, Tuvalu and Vanuatu. *Associate Members.* French Polynesia, New Caledonia. *Observers.* Asian Development Bank, the Commonwealth, Timor-Leste, Tokelau, Wallis and Futuna, the World Bank. *Membership suspended since May 2009 after calls for fresh elections by a set date were ignored.

Functions. The Secretariat's mission is to provide policy options to the Pacific Islands Forum, and to promote Forum decisions and regional and international co-operation. The organization seeks to promote political stability and regional security; enhance the management of economies and the development process; improve trade and investment; and efficiently manage the resources of the Secretariat.

Activities. The Secretariat has four core divisions: Trade and Investment; Political and International Affairs; Development and Economic Policy; Corporate Services. It provides policy advice to members on social, economic and political issues. Since 1989 the Forum has held Post Forum Dialogues with key dialogue partners at ministerial level. There are currently 14 partners: Canada, China, EU, France, India, Indonesia, Italy, Japan, South Korea, Malaysia, the Philippines, Thailand, UK and USA.

Organization. Established in 1972, the South Pacific Bureau for Economic Co-operation (SPEC) began as a trade bureau before being reorganized as the South Pacific Forum Secretariat in 1988. The Secretariat is headed by a Secretary-General and Deputy Secretary-General who form the Executive. The governing body is the Forum Officials Committee, which acts as an intermediary between the Secretariat and the Forum. The Secretariat operates four Trade Offices in Auckland, Beijing, Sydney and Tokyo.

The Secretary-General is the permanent Chair of the Council of Regional Organisations in the Pacific (CROP), which brings together ten main regional organizations in the Pacific region: Fiji School of Medicine (FSMed); Pacific Aviation Safety Office (PASO); Pacific Islands Development Programme (PIDP); Pacific Islands Forum Fisheries Agency (PIFFA); Pacific Islands Forum Secretariat (PIFS); Pacific Power Association (PPA); Secretariat for the Pacific Community (SPC); Secretariat of the Pacific Regional

Environment Programme (SPREP); South Pacific Tourism Organisation (SPTO); and University of the South Pacific (USP).

Official language: English.
Headquarters: Ratu Sukuna Road, Suva, Fiji Islands.
Website: http://www.forumsec.org.fj
Secretary-General: Tuiloma Neroni Slade (Samoa).

Secretariat of the Pacific Community (SPC)

Until Feb. 1998 known as the South Pacific Commission, this is a regional intergovernmental organization founded in 1947 under an Agreement commonly referred to as the Canberra Agreement. It is funded by assessed contributions from its 26 members and by voluntary contributions from member and non-member countries, international organizations and other sources.

Members. American Samoa, Australia, Cook Islands, Fiji Islands, France, French Polynesia, Guam, Kiribati, Marshall Islands, Federated States of Micronesia, Nauru, New Caledonia, New Zealand, Niue, Northern Mariana Islands, Palau, Papua New Guinea, Pitcairn Islands, Samoa, Solomon Islands, Tokelau, Tonga, Tuvalu, USA, Vanuatu, and Wallis and Futuna.

Functions. The SPC has three main areas of work: land resources, marine resources and social resources. It conducts research and provides technical assistance and training in these areas to member Pacific Island countries and territories of the Pacific.

Organization. The Conference of the Pacific Community is the governing body of the Community. Its key focus is to appoint the Director-General, to consider major national or regional policy issues and to note changes to the Financial and Staff Regulations approved by the CRGA, the Committee of Representatives of Governments and Administrations. It meets every two years. The CRGA meets once a year and is the principal decision-making organ of the Community. There are also regional offices in the Fiji Islands and Micronesia.

Headquarters: BP D5, 98848 Nouméa Cedex, New Caledonia.
Website: http://www.spc.int
Email: spc@spc.int
Director-General: Dr Jimmie Rodgers (Solomon Islands).

South Asian Association for Regional Co-operation (SAARC)

SAARC was established to accelerate the process of economic and social development in member states. The foreign ministers of the seven member countries met for the first time in New Delhi in Aug. 1983 and adopted the Declaration on South Asian Regional Co-operation whereby an Integrated Programme of Action (IPA) was launched. The charter establishing SAARC was adopted at the first summit meeting in Dhaka in Dec. 1985.

Members. Afghanistan, Bangladesh, Bhutan, India, Maldives, Nepal, Pakistan, Sri Lanka. *Observers.* Australia, China, EU, Iran, Japan, South Korea, Mauritius, Myanmar, USA.

Objectives. To promote the welfare of the peoples of South Asia; to accelerate economic growth, social progress and cultural development; to promote and strengthen collective self-reliance among members; to promote active collaboration and mutual assistance in the economic, social, cultural, technical and scientific fields; to strengthen co-operation with other developing countries and among themselves. Co-operation within the framework is based on respect for the principles of sovereign equality, territorial integrity, political independence, non-interference in the internal affairs of other states and mutual benefit. Agreed areas of co-operation under the *Integrated Programme of Action (IPA)* include agriculture and rural development; human resource development; environment, meteorology and forestry; science and technology; transport and communications; energy; and social development.

A SAARC Preferential Trading Arrangement (SAPTA) designed to reduce trade tariffs between SAARC member states was signed in April 1993, entering into force in Dec. 1995. In 1998 at the Tenth Summit in Colombo, the importance of achieving a South Asian Free Trade Area (SAFTA) as mandated by the Malé Summit in 1997 was reiterated and it was decided to set up a Committee of Experts to work on drafting a comprehensive treaty regime for creating a free trade area. The Colombo Summit agreed that the text of this regulatory framework would be finalized by 2001.

Organization. The highest authority of the Association rests with the heads of state or government, who meet annually at Summit level. The Council of Foreign Ministers, which meets twice a year, formulates policy, reviews progress and decides on new areas of co-operation. The Council is supported by a Standing Committee of Foreign Secretaries, by the Programming Committee and by 11 Technical Committees which are responsible for individual areas of SAARC's activities. There is a secretariat in Kathmandu, headed by a Secretary-General, who is assisted in his work by seven Directors, appointed by the Secretary-General upon nomination by member states for a period of three years which may in special circumstances be extended.

Official language: English.
Headquarters: PO Box 4222, Kathmandu, Nepal.
Website: http://www.saarc-sec.org
Secretary-General: Ahmed Saleem (Maldives).

Arab Fund for Economic and Social Development (AFESD)

Established in 1968, the Fund commenced operations in 1974.

Functions. AFESD is an Arab regional financial institution that assists the economic and social development of Arab countries through: financing development projects, with preference given to overall Arab development and to joint Arab projects; encouraging the investment of private and public funds in Arab projects; and providing technical assistance services for Arab economic and social development.

Members. Algeria, Bahrain, Djibouti, Egypt, Iraq, Jordan, Kuwait, Lebanon, Libya, Mauritania, Morocco, Oman, Palestine, Qatar, Saudi Arabia, Somalia, Sudan, Syria, Tunisia, United Arab Emirates, Republic of Yemen.

Headquarters: PO Box 21923, Safat 13080, Kuwait.
Website: http://www.arabfund.org
Director General and Chairman of the Board of Directors: Abdulatif Y. Al Hamad (Kuwait).

Publications. Annual Report; Joint Arab Economic Report.

Arab Monetary Fund (AMF)

Origin. The Agreement establishing the Arab Monetary Fund was approved by the Economic Council of the League of Arab States in April 1976 and the first meeting of the Board of Governors was held on 19 April 1977.

Aims. To assist member countries in eliminating payments and trade restrictions, in achieving exchange rate stability, in developing capital markets and in correcting payments imbalances through the extension of short- and medium-term loans; the co-ordination of monetary policies of member countries; and the liberalization and promotion of trade and payments, as well as the encouragement of capital flows among member countries.

Members. Algeria, Bahrain, Comoros, Djibouti, Egypt, Iraq, Jordan, Kuwait, Lebanon, Libya, Mauritania, Morocco, Oman, Palestine, Qatar, Saudi Arabia, Somalia, Sudan, Syria, Tunisia, United Arab Emirates, Republic of Yemen.

Headquarters: PO Box 2818, Abu Dhabi, United Arab Emirates.
Website: http://www.amf.org.ae
Director General and Chairman of the Board of Directors: Jassim A. Al-Mannai (Bahrain).

Publications (in English and Arabic): *Annual Report; The Articles of Agreement of the Arab Monetary Fund; Money and Credit in Arab Countries* (annual); *National Accounts of Arab Countries* (annual); *Foreign Trade of Arab Countries* (annual); *Cross Exchange Rates of Arab Currencies* (annual); *Arab Countries: Economic Indicators* (annual); *Balance of Payments and External Public Debt of Arab Countries* (annual); *AMF Publications Catalogue* (annual). (In Arabic only): *The Joint Arabic Economic Report* (annual); *AMF Economic Bulletin; Developments in Arab Capital Markets* (quarterly).

Arab Organization for Agricultural Development (AOAD)

The AOAD was established in 1970 and commenced operations in 1972. Its aims are to develop natural and human resources in the agricultural sector and improve the means and methods of exploiting these resources on scientific bases; to increase agricultural productive efficiency and achieve agricultural integration between the Arab States and countries; to increase agricultural production with a view to achieving a higher degree of self-sufficiency; to facilitate the exchange of agricultural products between the Arab States and countries; to enhance the establishment of agricultural ventures and industries; and to increase the standards of living of the labour force engaged in the agricultural sector.

Organization. The structure comprises a General Assembly consisting of ministers of agriculture of the member states, an Executive Council, a Secretariat General, seven technical departments—Food Security, Human Resources Development, Water Resources, Studies and Research, Projects Execution, Technical Scientific Co-operation, and Financial Administrative Department—and two centres—the Arab Center for Agricultural Information and Documentation, and the Arab Bureau for Consultation and Implementation of Agricultural Projects.

Members. Algeria, Bahrain, Comoros, Djibouti, Egypt, Iraq, Jordan, Kuwait, Lebanon, Libya, Mauritania, Morocco, Oman, Palestine, Qatar, Saudi Arabia, Somalia, Sudan, Syria, Tunisia, United Arab Emirates, Republic of Yemen.

Official languages: Arabic (English and French used in translated documents and correspondence).
Headquarters: Street No. 7, Al-Amarat, Khartoum, Sudan.
Website: http://www.aoad.org
Director General: Dr Tariq Moosa Al-Zadjali.

Gulf Co-operation Council (GCC)

Origin. Also referred to as the Co-operation Council for the Arab States of the Gulf (CCASG), the Council was established on 25 May 1981 on signature of the Charter by Bahrain, Kuwait, Oman, Qatar, Saudi Arabia and the United Arab Emirates.

Aims. To assure security and stability of the region through economic and political co-operation; promote, expand and enhance economic ties on solid foundations, in the best interests of the people; co-ordinate and unify economic, financial and monetary policies, as well as commercial and industrial legislation and customs regulations; achieve self-sufficiency in basic food-stuffs.

Organization. The Supreme Council formed by the heads of member states is the highest authority. Its presidency rotates, based on the alphabetical order of the names of the member states. It holds one regular annual session in addition to a mid-year consultation session. Attached to the Supreme Council are the Commission for the Settlement of Disputes and the Consultative Commission. The Ministerial Council is formed of the Foreign Ministers of the member states or other delegated ministers and meets quarterly. The Secretariat-General is composed of Secretary-General, Assistant Secretaries-General and a number of staff as required. The Secretariat consists of the following sectors: Political Affairs, Military Affairs, Legal Affairs, Human and Environment Affairs, Information Centre, Media Department, Gulf Standardization Organization (GSO), GCC Patent Office, Secretary-General's Office, GCC Delegation in Brussels, Technical Telecommunications Bureau in Bahrain. In Jan. 2003 it launched a customs union, introducing a 5% duty on foreign imports across the trade bloc.

Finance. The annual budget of the GCC Secretariat is shared equally by the six member states.

Headquarters: PO Box 7153, Riyadh-11462, Saudi Arabia.
Website: http://www.GCC-SG.org
Secretary-General: Abdul Latif bin Rashid al-Zayani (Bahrain).

Publications. Attaawun (quarterly, in Arabic); *GCC Economic Bulletin* (annual); *Statistical Bulletin* (annual); *Legal Bulletin* (quarterly, in Arabic).

Further Reading

Twinam, J. W., *The Gulf, Co-operation and the Council: an American Perspective.* 1992

League of Arab States

Origin. The League of Arab States (often referred to as the Arab League) is a voluntary association of sovereign Arab states, established by a Pact signed in Cairo on 22 March 1945 by the representatives of Egypt, Iraq, Saudi Arabia, Syria, Lebanon, Jordan and Yemen. It seeks to promote closer ties among member states and to co-ordinate their economic, cultural and security policies with a view to developing collective co-operation, protecting national security and maintaining the independence and sovereignty of member states, in order to enhance the potential for joint Arab action across all fields.

Members. Algeria, Bahrain, Comoros, Djibouti, Egypt, Iraq, Jordan, Kuwait, Lebanon, Libya, Mauritania, Morocco, Oman, Palestine, Qatar, Saudi Arabia, Somalia, Sudan, Syria*, Tunisia, United Arab Emirates and Republic of Yemen. *Observers.* Brazil, Eritrea, India and Venezuela. *Membership suspended since Nov. 2011 after calls for the government to end violence against civilian protesters by a set date were ignored.

Joint Action. In the political field, the League is entrusted with defending the supreme interests and national causes of the Arab world through the implementation of joint action plans at regional and international levels. It examines any disputes that may arise between member states with a view to finding a peaceful resolution. The Joint Defence and Economic Co-operation Treaty signed in 1950 provided for the establishment of a Joint Defence Council as well as an Economic Council (renamed the Economic and Social Council in 1977). Economic, social and cultural activities constitute principal and vital elements of the joint action initiative.

Against the backdrop of the 2011 Arab Spring, the League backed a UN resolution authorizing action in Libya against Col. Gaddafi's air defences and suspended Syria for its government's oppression of the opposition movement. Also in 2011 the League supported a Palestinian bid for UN recognition.

Arab Common Market. An Arab Common Market came into operation on 1 Jan. 1965. Initial plans to abolish customs duties on agricultural products, natural resources and industrial products by incremental reductions never came to fruition although the concept remains an ambition shared by many people in the Arab world.

Organization. The machinery of the League consists of a Council, 11 specialized ministerial committees entrusted with drawing up common policies for the regulation and advancement of co-operation in their fields (information, internal affairs, justice, housing, transport, social affairs, youth and sports, health, environment, telecommunications and electricity), and a permanent secretariat.

The League is considered to be a regional organization within the framework of the United Nations at which its Secretary-General is an observer. It has permanent delegations in New York and Geneva for the UN and in Addis Ababa for the African Union (AU), as well as offices in a number of cities throughout the world.

> *Headquarters:* Al Tahrir Square, Cairo, Egypt.
> *Website (Arabic only):* http://www.arableagueonline.org
> *Secretary-General:* Nabil el-Araby (Egypt).

Further Reading

Bouhamidi, Soumia, *The Role of the League of Arab States: Mediating and Resolving Arab-Arab Conflicts.* 2011
Gomaa, A. M., *The Foundation of the League of Arab States.* 1977
Salem, Ahmed Ali, *International Relations Theories and Organizations: Realism, Constructivism, and Collective Security in the League of Arab States.* 2008

Organization of Arab Petroleum Exporting Countries (OAPEC)

Established in 1968 to promote co-operation and close ties between member states in economic activities related to the oil industry; to determine ways of safeguarding their legitimate interests, both individual and collective, in the oil industry; to unite their efforts so as to ensure the flow of oil to consumer markets on equitable and reasonable terms; and to create a favourable climate for the investment of capital and expertise in their petroleum industries.

Members. Algeria, Bahrain, Egypt, Iraq, Kuwait, Libya, Qatar, Saudi Arabia, Syria, Tunisia*, United Arab Emirates. *Tunisia's membership was made inactive in 1986.

> *Headquarters:* PO Box 20501, Safat 13066, Kuwait.
> *Website:* http://www.oapecorg.org
> *Secretary-General:* Abbas Ali Naqi (Kuwait).

Publications. Secretary General's Annual Report (Arabic and English editions); *Oil and Arab Co-operation* (quarterly; Arabic with English abstracts and bibliography); *OAPEC Monthly Bulletin* (Arabic and English editions); *Energy Resources Monitor* (Arabic); *OAPEC Annual Statistical Report* (Arabic/English).

Organization of the Petroleum Exporting Countries (OPEC)

Origin and Aims. Founded in Baghdad in 1960 by Iran, Iraq, Kuwait, Saudi Arabia and Venezuela. The principal aims are: to unify the petroleum policies of member countries and determine the best means for safeguarding their interests, individually and collectively; to devise ways and means of ensuring the stabilization of prices in international oil markets with a view to eliminating harmful and unnecessary fluctuations; and to secure a steady income for the producing countries, an efficient, economic and regular supply of petroleum to consuming nations, and a fair return on their capital to those investing in the petroleum industry. It is estimated that OPEC members possess 75% of the world's known reserves of crude petroleum, of which about two-thirds are in the Middle East. OPEC countries account for about 43% of world oil production (55% in the mid-1970s).

Members. (Feb. 2013) Algeria, Angola, Ecuador, Iran, Iraq, Kuwait, Libya, Nigeria, Qatar, Saudi Arabia, United Arab Emirates and Venezuela. Membership applications may be made by any other country having substantial net exports of crude petroleum, which has fundamentally similar interests to those of member countries. Gabon became an associated member in 1973 and a full member in 1975, but in 1996 withdrew owing to difficulty in meeting its percentage contribution. Ecuador joined the Organization in 1973 but left in 1992; it then rejoined in Oct. 2007. Indonesia joined in 1962 but left in 2008 as it had ceased to be an oil exporter.

Organization. The main organs are the Conference, the Board of Governors and the Secretariat. The Conference, which is the supreme authority meeting at least twice a year, consists of delegations from each member country, normally headed by the respective minister of oil, mines or energy. All decisions, other than those concerning procedural matters, must be adopted unanimously.

> *Headquarters:* Helferstorferstrasse 17, A-1010 Vienna, Austria.
> *Website:* http://www.opec.org
> *Secretary-General:* Abdullah Salem al-Badri (Libya).

Publications. Annual Statistical Bulletin; Annual Report; OPEC Bulletin (monthly); *OPEC Review* (quarterly); *OPEC General Information; Monthly Oil Market Report; OPEC Statute.*

Further Reading

Parra, Francisco, *Oil Politics: A Modern History of Petroleum.* 2010
Skeet, I., *OPEC: 25 Years of Prices and Policies.* 1988
Yergin, Daniel, *The Quest: Energy, Security and the Remaking of the Modern World.* 2011

OPEC Fund for International Development

The OPEC Fund for International Development (OFID) was established in 1976 as the OPEC Special Fund, with the aim of providing financial aid on concessional terms to developing countries (other than OPEC member states) and international development agencies whose beneficiaries are developing countries. In 1980 the Fund was transformed into a permanent autonomous international agency and renamed the OPEC Fund for International Development. It is administered by a Ministerial Council and a Governing Board. Each member country is normally represented on the Council by its finance minister, or if not then by another designated person.

The initial endowment of the fund amounted to US$800m. By the end of April 2010 OFID's total approved commitments (including public sector operations, private sector operations, trade finance operations, grants and contributions to other institutions) stood at US$11,926m. OFID had approved 2,656 operations by the end of April 2010, including US$7,326m. for project financing, US$724m. for balance-of-payments support, US$333m. for programme funding and US$270m. for debt relief under the *Highly Indebted Poor Countries Initiative*. In addition, and through its private sector window, OFID had approved financing worth a total of US$1,190m. in 144 operations in support of private sector entities in Africa, Asia, Latin America, the Caribbean and Europe. Through its grants programme OFID had also committed a total of US$483m. in support of a wide range of initiatives, ranging from technical assistance, research and emergency aid to dedicated operations to combat HIV/AIDS and hardship in Palestine.

Headquarters: Parkring 8, POB 995, A-1011 Vienna, Austria.
Website: http://www.ofid.org
Email: info@ofid.org
Director-General: Suleiman Jasir al-Herbish (Saudi Arabia).

Antarctic Treaty

Antarctica is an island continent some 15·5m. sq. km in area which lies almost entirely within the Antarctic Circle. Its surface is composed of an ice sheet over rock, and it is uninhabited except for research and other workers in the course of duty. It is in general ownerless: for countries with territorial claims, *see* ARGENTINA; AUSTRALIA: Australian Antarctic Territory; CHILE; FRANCE: Southern and Antarctic Territories; NEW ZEALAND: Ross Dependency; NORWAY: Queen Maud Land; UNITED KINGDOM: British Antarctic Territory.

12 countries which had maintained research stations in Antarctica during International Geophysical Year, 1957–58 (Argentina, Australia, Belgium, Chile, France, Japan, New Zealand, Norway, South Africa, the USSR, the UK and the USA) signed the Antarctic Treaty (Washington Treaty) on 1 Dec. 1959. Austria, Belarus, Brazil, Bulgaria, Canada, China, Colombia, Cuba, Czech Republic, Denmark, Ecuador, Estonia, Finland, Germany, Greece, Guatemala, Hungary, India, Italy, North Korea, South Korea, Malaysia, Monaco, the Netherlands, Pakistan, Papua New Guinea, Peru, Poland, Portugal, Romania, Slovakia, Spain, Sweden, Switzerland, Turkey, Ukraine, Uruguay and Venezuela subsequently acceded to the Treaty. The Treaty reserves the Antarctic area south of 60° S. lat. for peaceful purposes, provides for international co-operation in scientific investigation and research, and preserves, for the duration of the Treaty, the *status quo* with regard to territorial sovereignty, rights and claims. The Treaty entered into force on 23 June 1961. The 50 nations party to the Treaty (28 consultative or voting members and 22 non-consultative parties) meet biennially.

An agreement reached in Madrid in April 1991 and signed by all 39 parties in Oct. imposes a ban on mineral exploitation in Antarctica for 50 years, at the end of which any one of the 28 voting parties may request a review conference. After this the ban may be lifted by agreement of three quarters of the nations then voting, which must include the present 28.

Headquarters: Av. Leandro Alem 884–4° Piso, C1001AAQ, Buenos Aires, Argentina.
Website: http://www.ats.aq
Email: ats@ats.aq
Executive Secretary: Manfred Reinke (Germany).

Further Reading

Elliott, L. M., *International Environmental Politics: Protecting the Antarctic.* 1994
Jørgensen-Dahl, A. and Østreng, W., *The Antarctic Treaty System in World Politics.* 1991
Triggs, Gillian D. (ed.) *The Antarctic Treaty Regime: Law, Environment and Resources.* 2009

Nuclear Non-Proliferation Treaty (NPT)

The Treaty on the Non-Proliferation of Nuclear Weapons opened for signatories on 1 July 1968. It came into force on 5 March 1970. A review meeting takes place every five years. The initial treaty was limited to a 25-year term but it was extended indefinitely in 1995.

The treaty aims to prevent the spread of nuclear weapons and weapons technology, to promote co-operation in the peaceful uses of nuclear energy and to further the goal of achieving nuclear disarmament and general and complete disarmament. The International Atomic Energy Agency (*see* page 24) is responsible for setting safeguards to ensure compliance.

Of the treaty's 190 members only five have nuclear weapon capabilities: China, France, Russia, UK and USA. Three states known or believed to have developed nuclear weapons have not ratified the treaty: India, Israel and Pakistan. North Korea withdrew from the treaty in 2003, the only state to have done so.

See also Preparatory Commission for the Comprehensive Nuclear-Test-Ban Treaty Organization (CTBTO) on page 25.

Website:
http://www.un.org/disarmament/WMD/Nuclear/NPT.shtml

United Nations Framework Convention on Climate Change

The convention was produced at the 1992 UN Conference on Environment and Development with the stated aim of reducing global greenhouse gas emissions to 'a level that would prevent dangerous anthropogenic (human induced) interference with the climate system'. Signatories agreed to take account of climate change in their domestic policy and to develop national programmes that would slow its progress. However, no mandatory targets were established for the reduction of emissions so the treaty remained legally non-binding. Instead it operates as a 'framework' document, with provisions for regular updates and amendments.

The first of these additions was the Kyoto Protocol in 1997. Under the protocol, 37 developed countries are committed to reducing their collective emissions of six greenhouse gases to at least 5% below 1990 levels. These targets were scheduled to be met in the period 2008–12. By 2010 results were mixed. The EU had reduced emissions by 16·8% and Russia by 54·7%, while the USA's had risen by 8·6%, Canada's by 46·4%, New Zealand's by 59·5% and Turkey's by 147·5%. In Dec. 2011 Canada announced it would be the first signatory to formally withdraw from the agreement. A second commitment period of the Kyoto Protocol began on 1 Jan. 2013. By Feb. 2013, 191 countries plus the European Union had signed and ratified the treaty. The USA has not ratified the protocol. China and India, also amongst the world's top five producers of emissions, are exempt from the protocol's constraints by virtue of their status as developing countries.

The members of the UNFCCC meet on an annual basis. The conference in Indonesia in 2007 led to the creation of the 'Bali Roadmap', which timetables negotiations for a protocol to succeed Kyoto, a process continued at the 2008 conference in Poland. The subsequent Copenhagen Accord of 2009 was not legally binding and failed to set out concrete measures for tackling climate change. A package of decisions to encourage all governments to work towards a low-emissions future was adopted at the 16th conference held in 2010 in Cancún, Mexico. The 2011 conference, held in Durban, South Africa, advanced negotiations on the implementation of the Kyoto Protocol, the Bali Action Plan and the Cancún Agreements. At the 18th conference in Doha, Qatar in 2012 plans were laid for the development of a successor protocol by 2015 to be implemented by 2020.

Headquarters: United Nations Framework Convention on Climate Change, Haus Carstanjen, Martin-Luther-King-Strasse 8, 53175 Bonn, Germany.
Website: http://unfccc.int
Email: secretariat@unfccc.int
Executive Secretary: Christiana Figueres (Costa Rica).

Leading Think Tanks

Adam Smith Institute

Founded 1977. Independent, non-profit libertarian think tank that engineers policies to increase Britain's economic competitiveness, inject choice into public services and create a freer, more prosperous society. Research issues: tax and economy; education policy; health policy; justice and liberties; welfare and pensions; regulation and industry.

Address: 23 Great Smith St., London SW1P 3BL, UK.
Website: http://www.adamsmith.org
Director: Dr Eamonn Butler.

African Economic Research Consortium

Founded 1988. Non-profit organization that seeks to strengthen local capacity for conducting independent research into management problems of economies in sub-Saharan Africa. Two programme components: research; training.

Address: 3rd Floor, Middle East Bank Towers Building, Milimani Road, PO Box 62882 00200, Nairobi, Kenya.
Website: http://www.aercafrica.org
Chairman of the Board: Mthuli Ncube.

American Enterprise Institute (for Public Policy Research)

Founded 1943. Private, non-partisan think tank based around principles of private liberty, individual opportunity and free enterprise. Six principal research areas: economics; foreign and defence policy; health; legal and constitutional studies; political and public opinion studies; social and cultural studies.

Address: 1150 Seventeenth St., NW, Washington, D.C., 20036, USA.
Website: http://www.aei.org
President: Arthur C. Brooks.

Brookings Institution

Founded 1916. Independent, frequently cited as the world's best think tank. Goals are to strengthen American democracy; foster the economic and social welfare, security and opportunity of all Americans; and secure a more open, safe, prosperous and co-operative international system. Priority research areas include energy and climate, growth through innovation, managing global change, and opportunity and wellbeing.

Address: 1775 Massachusetts Ave., NW, Washington, D.C., 20036, USA.
Website: http://www.brookings.edu
President: Strobe Talbott.

Bruegel

Founded 2004. Independent European think tank working in the field of international economics. Research areas: emerging powers and global governance structures; Europe's macroeconomic and structural challenges; competitiveness, innovation and financial regulation; climate change and energy.

Address: Rue de la Charité 33, B-1210 Brussels, Belgium.
Website: http://www.bruegel.org
Director: Jean Pisani-Ferry.

Carnegie Endowment for International Peace

Founded 1910. Independent think tank specializing in international affairs with particular focus on Russia and Eurasia, China, the Indian subcontinent/South Asia, globalization, non-proliferation and security affairs. Aims to advance co-operation between nations and promote active international engagement by the USA and become 'the first truly multinational—ultimately global—think tank'. Offices in Washington, D.C., Moscow, Beijing, Beirut and Brussels.

Address: 1779 Massachusetts Ave., NW, Washington, D.C., 20036-2103, USA.
Website: http://www.carnegieendowment.org
President: Jessica T. Mathews.

Carnegie Middle East Center

Founded in 2006 as part of the Carnegie Endowment for International Peace's Middle East programme. Public policy think tank and research centre that aims to better inform the process of political change in the Arab Middle East and deepen understanding of the complex security and economic issues that affect it. Programmes: Middle East economies; Arab politics; regional relations; security.

Address: Lazarieh Tower, Building No 2026 1210, Fifth Floor, Emir Bechir St., Beirut, 11-1061 Riad El Solh, Lebanon.
Website: http://www.carnegie-mec.org
Director: Paul Salem.

Carnegie Moscow Center

Founded in 1994 as a subdivision of the Carnegie Endowment for International Peace. Analyzes the most important issues in international affairs and Russian domestic and foreign policy, as well as the regions they affect. Programmes: economic policy; foreign and security policy; non-proliferation; religion, society and security; Russian domestic politics and political intuitions; society and regions; the east–east: partnerships beyond borders.

Address: 16/2 Tverskaya, Moscow 125009, Russia.
Website: http://www.carnegie.ru
Director: Dmitry Trenin.

Cato Institute

Founded 1977. Non-profit public policy research foundation based on the principles of the American Revolution—limited government, free markets, individual liberty and peace. Comprises Centers for Constitutional Studies, Educational Freedom, Global Liberty and Prosperity, Representative Government and Trade Policy Studies.

Address: 1000 Massachusetts Ave., NW, Washington, D.C., 20001-5403, USA.
Website: http://www.cato.org
President: Edward H. Crane.

Center for American Progress

Founded 2003. Organization dedicated to improving the lives of Americans through progressive ideas and action. Research issues: domestic; economy; national security; energy and environment; media and progressive values.

Address: 1333 H St., NW, 10th Floor, Washington, D.C., 20005, USA.
Website: http://www.americanprogress.org
Chair: John Podesta.

Center for International Governance Innovation

Founded 2001. Independent, non-profit and non-partisan think tank that focuses on international governance. Research programmes: the global economy; the environment and energy; development; global security.

Address: 57 Erb St. West, Waterloo, Ontario, Canada N2L 6C2.
Website: http://www.cigionline.org
Executive Director: Thomas Bernes.

Center for Social and Economic Research

Founded 1991. Non-profit, independent economic and public policy research institution that aims to provide objective economic analysis and to foster the quality of policy-making to improve the lives of Europeans and their neighbours.

Address: Al. Jana Pawła II 61, Office 212, 01-031 Warsaw, Poland.
Website: http://www.case-research.eu
President of Management Board: Luca Barbone.

Center for Strategic and International Studies

Founded in 1962 during the Cold War to find ways for the USA to sustain its prominence and prosperity as a force for good in the world. Bipartisan, non-profit organization that conducts research and analysis and develops policy initiatives that look into the future and anticipate change. Research focuses on defence and security, energy and climate change, global health, global trends and forecasting, governance, human rights, technology, and trade and economics.

Address: 10 1800 K St., NW, Washington, D.C., 20006, USA.
Website: http://www.csis.org
President: John J. Hamre.

Centre for Economic Policy Research

Founded 1983. Non-profit, educational research network that promotes independent, objective analysis and public discussion of open economies and the relations among them. Programmes: development economics; financial economics; industrial organization; international macroeconomics; international trade and regional economics; labour economics; public policy.

Address: 77 Bastwick St., London EC1V 3PZ, UK.
Website: http://www.cepr.org
Director: Mathias Dewatripont.

Centre for European Policy Studies

Founded 1983. Independent institute specializing in European affairs. Research programmes: economic and social welfare policies; energy, climate change and sustainable development; EU neighbourhood, foreign and security policy; financial markets and institutions; justice and home affairs; politics and European institutions; regulatory policy; trade developments and agricultural policy.

Address: 1 Place du Congrès, B-1000 Brussels, Belgium.
Website: http://www.ceps.eu
Director: Daniel Gros.

Centre for European Reform

Pro-European think tank focusing on political, economic and social challenges facing Europe. Research topics: Britain and the EU; EU budget; EU foreign policy; EU institutions; justice and home affairs; economics and finance; energy and environment; education and research; the euro; security and defence policy; enlargement and Turkey; neighbourhood policy; transatlantic relations; Russia; China; the Middle East; Arab Reform initiative.

Address: 14 Great College St., London SW1P 3RX, UK.
Website: http://www.cer.org.uk
Director: Charles Grant.

Centre of Public Studies (Centro de Estudios Públicos)

Founded 1980. Private, non-profit academic foundation engaged in cultivating, analyzing and disseminating the values and principles of a free and democratic order. Plays a leading role in fostering a national debate on issues ranging from new developments in the social sciences to the concepts and values that support a free social order.

Address: Monseñor Sótero Sanz 162, C. Postal 7500011, Providencia, Santiago, Chile.
Website: http://www.cepchile.cl
President: Eliodoro Matte.

Chatham House (Royal Institute of International Affairs)

Founded 1920. Leading independent think tank whose research centres on three areas: energy, environment and resource governance; international economics; regional and security studies. Established the 'Chatham House Rule' that aids free and open debate by allowing for anonymity of speakers at meetings. Sister organization of Council on Foreign Relations in New York.

Address: 10 St James' Square, London SW1Y 4LE, UK.
Website: http://www.chathamhouse.org.uk
Director: Dr Robin Niblett.

Chinese Academy of Social Sciences

Founded 1977. China's highest academic research organization in the fields of philosophy and social sciences. Comprises 32 research institutes, three research centres and a graduate school. Research covers 120 key areas.

Address: 5 Jianguomennei Dajie, Beijing 100732, China.
Website (Chinese only): http://www.cass.net.cn
President: Chen Kuiyuan.

Civitas (Institute for the Study of Civil Society)

Founded 2000. Independent think tank that also provides primary education for children and provides teaching materials and speakers for schools. Research areas: crime; constitution; education; Europe; family; health reform; immigration.

Address: First Floor, 55 Tufton St., London SW1P 3QL, UK.
Website: http://www.civitas.org.uk
Director: Dr David G. Green.

Council on Foreign Relations

Founded 1921. Independent think tank that seeks to foster better understanding of the world and the foreign policy choices facing the USA and other countries. Research by the David Rockefeller Studies Program centres on major geopolitical areas but also covers global health, international institutions and global governance, national security, science and technology and US foreign policy. Sister organization of Chatham House in London.

Address: The Harold Pratt House, 58 East 68th St., New York, NY 10065, USA.
Website: http://www.cfr.org
Board of Directors Chairs: Carla A. Hills; Robert E. Rubin.
President: Richard N. Haass.

Danish Institute for International Studies

Founded 2002 by the Danish parliament. Independent institution engaged in research in international affairs in order to assess the security and foreign policy situation of Denmark. Research units: defence and security; foreign policy and EU studies; global economy, regulation and development; holocaust and genocide; migration; natural resources and poverty; politics and governance; the Middle East.

Address: Strandgade 56, 1401 Copenhagen, Denmark.
Website: http://www.diis.dk
Director: Nanna Hvidt.

European Council on Foreign Relations

Founded 2007. Pan-European think tank that conducts European foreign policy research and promotes a more integrated European foreign policy in support of shared European interests and values. Main programmes: Russia and wider Europe; China; democracy, human rights and the rule of law.

Address: 35 Old Queen St., London SW1H 9JA, UK.
Website: http://www.ecfr.eu
President: Asger Aamund.

Fraser Institute

Founded 1974. Independent non-partisan research and educational organization that aims to measure, study and communicate the impact of competitive markets and government interventions on the welfare of individuals. Research covers taxation, government spending, health care, school performance and trade.

Address: 4th Floor, 1770 Burrard St., Vancouver, BC, Canada V6J 3G7.
Website: http://www.fraserinstitute.org
President: Niels Veldhuis.

French Institute of International Relations (IFRI)

Founded 1979. The Institut Français des Relations Internationales is an independent research and debate institution dedicated to international affairs. Research centres on geographic regions as well as economy; energy; Franco–German relations; health/environment; migration; identities and citizenship; security and defence; space; and sport.

Address: 27 rue de la Procession, 75740 Paris Cedex 15, France.
Website: http://www.ifri.org
Secretary-General: Valérie Genin.

Friedrich Ebert Foundation (Friedrich-Ebert-Stiftung; FES)

Founded in 1925 as a political legacy of Germany's first democratically elected president, Friedrich Ebert. Non-profit foundation committed to the advancement of public policy issues in the spirit of the basic values of social democracy. Focuses on democracy promotion and international dialogue on the central topics of international politics, globalization, and economic, social and political development in the world.

Address: Berliner Haus, Hiroshimastrasse 17, 10785 Berlin, Germany; Bonner Haus, Godesberger Allee 149, 53175 Bonn, Germany.
Website: http://www.fes.de
President: Kurt Beck.

Fundação Getulio Vargas

Founded 1944. Higher education establishment dedicated to social sciences research to develop the socio-economic position of Brazil. Research covers business, citizenship, education, finance, justice, health, history, law, macro and microeconomics, politics, pollution, poverty and unemployment, sustainable development and welfare.

Address: Praia de Botafogo 190, Rio de Janeiro, 22250-900 Brazil.
Website: http://www.fgv.br
President: Carlos Ivan Simonsen Leal.

German Council on Foreign Relations

Founded 1945. Independent, non-partisan and non-profit membership organization and think tank that promotes public debate on foreign policy. Research programmes focus on: China; energy policy; European integration; global economics; international security policy; Middle East; Russia/Eurasia; transatlantic relations.

Address: Rauchstrasse 17–18, D-10787 Berlin, Germany.
Website: https://dgap.org
President: Arend Oetker.

German Institute for International and Security Affairs (Stiftung Wissenschaft und Politik; SWP)

Founded 1962. Independent scientific establishment that conducts practically oriented research on the basis of which it then advises the Bundestag and the German federal government on foreign and security policy issues. Research divisions: EU integration; EU external relations; international security; the Americas; Russian Federation/CIS; Middle East and Africa; Asia; global issues.

Address: Ludwigkirchplatz 3–4, 10719 Berlin, Germany.
Website: http://www.swp-berlin.org
Director: Prof. Dr Volker Perthes.

Heritage Foundation

Founded 1973. Conservative think tank aiming to formulate and promote public policies based on the principles of free enterprise, limited government, individual freedom, traditional American values and a strong national defence. Target audience includes members of Congress, key congressional staff members, policymakers in the executive branch, the news media, and the academic and public policy communities.

Address: 214 Massachusetts Ave., NE, Washington, D.C., 20002-4999, USA.
Website: http://www.heritage.org
President: Edwin J. Feulner.

Hoover Institution (on War, Revolution and Peace)

Founded in 1959 at Stanford University. Originated from research network built up by the Hoover War Library. Aims to secure and safeguard peace, improve the human condition and limit government intrusion into the lives of individuals. Three overarching research programmes focus on American institutions and economic performance, democracy and free markets, and international rivalries and global co-operation.

Address: 434 Galvez Mall, Stanford University, Stanford, CA 94305-6010, USA.
Website: http://www.hoover.org
Director: John Raisian.

Human Rights Watch

Founded 1978. Non-profit, non-governmental organization dedicated to protecting the human rights of people around the world. Research topics include: arms; children's rights; counterterrorism; disability rights; health; international justice; migrants; press freedom; refugees; terrorism; torture; women's rights.

Address: 2nd Floor, 2–12 Pentonville Road, London N1 9HF, UK.
Website: http://www.hrw.org
Executive Director: Kenneth Roth.

Institute for Government

Founded 2008. Independent charity that works with all the main political parties at Westminster and with senior civil servants in Whitehall to increase government effectiveness. Undertakes research, provides development opportunities for senior decision makers and organizes events for leading international experts to share new thinking on best government. Areas of work: a more effective Whitehall; better policy making; new models of governance and public services; parliament and the political process; leadership for government.

Address: 2 Carlton Gdns, London SW1Y 5AA, UK.
Website: http://www.instituteforgovernment.org.uk
Director: Rt Hon Peter Riddell CBE.

Institute of World Economy and International Relations

Founded 1956. Non-profit organization that carries out applied socio-economic, political and strategic research. Research areas include: current global problems; economic theory; economic, social and political problems of the transition period in Russia; forecasting and analysis of world economy dynamics and socio-political developments; international politics; military and strategic problems; theory of international relations; theory of social and political processes.

Address: 23 Profsoyuznaya St., Moscow 117997, Russia.
Website: http://www.imemo.ru
Director: Alexander A. Dynkin.

International Crisis Group

Founded 1995. Independent, non-profit organization committed to preventing and resolving deadly conflict. Combines field-based analysis, policy advice and high-level advocacy to highlight potential future conflicts, resolve peace negotiations and advise governments and intergovernmental bodies.

Address: 149 avenue Louise, Level 24, B-1050 Brussels, Belgium.
Website: http://www.crisisgroup.org
President: Louise Arbour.

International Institute for Strategic Studies (IISS)

Founded 1958. Independent organization, considered the world's leading authority on political-military conflict. Research programme themes: conflict; defence and military analysis; economics and conflict resolution; non-proliferation and disarmament; transnational threats and international political risk; transatlantic dialogue on climate change and security.

Address: Arundel House, 13–15 Arundel St., Temple Place, London WC2R 3DX, UK.
Website: http://www.iiss.org
Director-General: Dr John Chipman.

Kiel Institute for the World Economy (Institut für Weltwirtschaft an der Universität Kiel; IfW)

Founded 1914. Independent, international centre for research in global economic affairs, economic policy consulting, economic education and documentation. Research areas: the global division of labour; knowledge creation and growth; the environment and natural resources; poverty reduction, equity and development; monetary policy under market imperfections; financial markets and macroeconomic activity; reforming the welfare society.

Address: Hindenburgufer 66, 24105 Kiel, Germany.
Website: http://www.ifw-kiel.de
President: Prof. Dennis Snower.

Konrad Adenauer Foundation (Konrad-Adenauer-Stiftung)

Founded in 1955 as the Society for Christian Democratic Civic Education and renamed in 1964. Political foundation that focuses on consolidating democracy, the unification of Europe and the strengthening of transatlantic relations, as well as on development co-operation.

Address: Klingelhöferstrasse 23, 10785 Berlin, Germany; Rathausallee 12, 53757 Sankt Augustin, Germany.
Website: http://www.kas.de
Chairman: Dr Hans-Gert Pöttering.

Lowy Institute for International Policy

Founded 2003. Independent think tank that seeks to generate new ideas and dialogue on international developments and Australia's role in the world. Research ranges across all the dimensions of international policy debate in Australia—economic, political and strategic—and is not limited to a particular geographic region.

Address: 31 Bligh St., Sydney, NSW 2000, Australia.
Website: http://www.lowyinstitute.org
Executive Director: Michael Fullilove.

National Bureau of Economic Research (NBER)

Founded 1920. Private, non-profit, non-partisan research organization dedicated to promoting a greater understanding of how the economy works. Concentrates on four types of empirical research: developing new statistical measurements, estimating quantitative models of economic behaviour, assessing the economic effects of public policies and projecting the effects of alternative policy proposals.

Address: 1050 Massachusetts Ave., Cambridge, Massachusetts 02138-5398, USA.
Website: http://www.nber.org
President: James Poterba.

Netherlands Institute of International Relations 'Clingendael'

Non-profit, independent think tank for international relations. Identifies and analyses emerging political and social developments for the benefit of government and the general public. Programme concentrations: diplomatic studies; European studies; security and conflict; international energy.

Address: Clingendael 7, 2597 VH The Hague, Netherlands.
Website: http://www.clingendael.nl
Director: Prof. Dr Jaap (J. W.) de Zwaan.

Overseas Development Institute

Founded 1960. Independent think tank that focuses on international development and humanitarian issues. It aims to inform policy and practice that lead to the reduction of poverty, the alleviation of suffering and the achievement of sustainable livelihoods in developing countries.

Address: 11 Westminster Bridge Road, London SE1 7JD, UK.
Website: http://www.odi.org.uk
Director: Alison Evans.

Peace Research Institute Oslo (PRIO)

Founded 1959. Independent research institute that aims to promote peace through conflict resolution, dialogue and reconciliation, public information and policy-making activities. Research programmes: conflict resolution and peacebuilding;

ethics, norms and identities; security programme. Two research centres: Centre for the Study of Civil War; Cyprus Centre.

Address: Hausmanns gate 7, 0186 Oslo, Norway.
Website: http://www.prio.no
Director: Kristian Berg Harpviken.

Peterson Institute for International Economics

Founded 1981. Private, non-profit, non-partisan research institution devoted to the study of international economic policy. Research encompasses country and regional studies, debt and development, globalization, international finance/macroeconomics, international trade and investment, and US economic policy.

Address: 1750 Massachusetts Ave., NW, Washington, D.C., 20036-1903, USA.
Website: http://www.iie.com
Chairman: Peter G. Peterson.

Pew Research Center

Founded 2004. Non-partisan, non-profit fact tank that provides information on the issues, attitudes and trends shaping America and the world through public opinion polling, demographic studies, media content analysis and other empirical social science research.

Address: 1615 L St., NW, Suite 700, Washington, D.C., 20036, USA.
Website: http://pewresearch.org
President: Andrew Kohut.

Polish Institute of International Affairs

Founded 1999. Independent think tank that conducts original, policy-focused research and promotes the flow of ideas that inform and enhance the foreign policy of Poland. Programmes: international security; eastern and southeastern Europe; international economic relations and global issues; European Union; non-proliferation and arms control.

Address: 1A Warecka St., 00-950 Warsaw, Poland.
Website: http://www.pism.pl
Director: Marcin Zaborowski.

RAND Corporation

Founded in 1948 out of US military research and development during World War II by Douglas Aircraft. Independent, non-profit organization dedicated to promoting scientific, educational and charitable purposes for the public welfare. Research areas include health, education, national security, international affairs, law and business, and the environment. Houses three federally funded research and development centres sponsored by the US defence department: the RAND Arroyo Center, providing research and analysis for the army; the RAND National Defense Research Institute; and RAND Project Air Force.

Address: 1776 Main St., Santa Monica, CA 90401-3208, USA.
Website: http://www.rand.org
President: James A. Thomson.

Stockholm International Peace Research Institute (SIPRI)

Founded in 1966 by the Swedish parliament. Independent international institute dedicated to research into conflict, armaments, arms control and disarmament. Compiles detailed studies on multilateral peace operations, military expenditure, arms transfers and arms embargoes.

Address: Signalistgatan 9, SE-169 70 Solna, Sweden.
Website: http://www.sipri.org
Director: Dr Bates Gill.

Transparency International

Founded 1993. Non-partisan global civil society organization seeking to create change towards a world free of corruption. Global priorities: combating corruption in politics, public contracting and the private sector; international anti-corruption conventions; poverty and development.

Address: Alt-Moabit 96, 10559 Berlin, Germany.
Website: http://www.transparency.org
Director: Huguette Labelle.

Woodrow Wilson International Center for Scholars

Founded in 1968 by an act of Congress as a memorial to former US president Woodrow Wilson. Non-partisan, it promotes and develops relations between policy-makers and academic scholars. Research covers most public policy areas, specializing in the field of international affairs.

Address: One Woodrow Wilson Plaza, 1300 Pennsylvania Ave., NW, Washington, D.C., 20004-3027, USA.
Website: http://www.wilsoncenter.org
Director: Lee H. Hamilton.

PART II

COUNTRIES OF THE WORLD
A—Z

AFGHANISTAN

© Research Machines plc 2005

Da Afganistan Islami Jomhoriyat—
Jamhuri-ye Islami-ye Afganistan
(Islamic Republic of Afghanistan)

Capital: Kabul
Population projection, 2015: 36·74m.
GNI per capita, 2011: (PPP$) 1,416
HDI/world rank: 0·398/172
Internet domain extension: .af

KEY HISTORICAL EVENTS

Excavations near Kandahar in southern Afghanistan have revealed Neolithic settlements dating to 5000 BC. Later settlements indicate trade links with cities in the Indus Valley and Mesopotamia. Aryan tribes that settled in Afghanistan's northern plains near Balkh (Bactria) from around 1500 BC gave way to successive Persian dynasties, including the Acaemanid Empire (550 BC–330 BC) ruled by Darius the Great from Persepolis after 515 BC.

Alexander the Great conquered Bactria in 328 BC but his rule was short-lived and the region fell first to the Seleucid, then to the Parthian empires. Buddhism was spread by the Yuechi, who invaded from the northeast and established the Kushan dynasty at Peshawar in around 200 BC. The famous Buddha statues at Bamiyan, destroyed in 2001 by the Taliban, dated from the 3rd and 5th centuries AD. Scythians, White Huns and Turkish Tu-Kuie invaded in the first half of the first millennium. The Muslim conquest of Afghanistan began with the arrival of Arab settlers in AD 642, though it was during the reign of Mahmud of Ghazni (998–1030) that the region became a thriving cultural centre. Mahmud was succeeded by smaller dynasties, all of which fell to Genghis Khan's Mongol invasion in 1219.

Following Genghis Khan's death in 1227, chiefs and princes vied for supremacy until, late in the 14th century, one of his descendants, Timur-i Lang, incorporated Afghanistan into his central Asian empire. Babur, a descendant of Timur and founder of India's Moghul dynasty at the beginning of the 16th century, made Kabul his capital, although power later transferred to Delhi and Agra. The early sixteenth century also saw the rise of the Safavid dynasty in Iran which ruled over western Afghanistan, while the Shaibanid Uzbeks controlled northern Afghanistan and territory stretching northward across central Asia.

The early 18th century was marked by Afghan tribes revolting against foreign occupation. Ahmad Shah Durrani, a Pashtun, became the founder of modern Afghanistan following the death of the Persian ruler, Nadir Shah, in 1747. The subsequent rule by Dost Muhammad, who became Amir in 1826, was overshadowed by the power struggle between Britain, dominant in India, and the expanding Russian empire. The British attempt to oust Dost Muhammad sparked the first Afghan War (1838–42), during which the British were forced to retreat by an armed rebellion in Kabul in 1841.

Dost Muhammad returned from exile and governed for the next 20 years. British attempts to delineate India's northwest frontier and prevent Russian advances led to another invasion of Afghanistan in 1878, when British-Indian troops seized the Khyber Pass. Abd Ar-Rahman Khan, who became the Amir of Afghanistan in 1880, abolished the traditional regional centres of power and consolidated his government in Kabul. Border treaties were signed with Russia and British India. Following the Anglo-Russian agreement of 1907 Afghanistan's independence was guaranteed, although foreign affairs remained under British control until the 1919 Treaty of Rawalpindi. King Amanullah introduced Western-style reforms in the 1920s but development was restricted by tribal wars and banditry.

Záhir Shah took power in 1933 and ruled for 40 years, bringing stability and the expansion of education. In 1964 he established parliamentary democracy. In 1973 his cousin and brother-in-law, and a former prime minister, Mohammed Daoud, led a military coup and abolished the constitution to declare a republic. In April 1978 President Daoud was killed in a further coup which installed a pro-Soviet government. The new president, Noor Mohammad Taraki, was overthrown in Sept. 1979, whereupon the USSR invaded in Dec. to install Babrak Karmal in power.

In Dec. 1986 Sayid Mohammed Najibullah became president with civil war continuing between government and rebel Muslim forces. The USSR gave military support to the authorities while the USA backed the rebels. In the mid-1980s the UN began negotiating the withdrawal of Soviet troops and the establishment of a national unity government. The first Soviet troops withdrew in early 1988. After talks in Nov. 1991 with Afghan opposition movements ('mujahideen'), the USSR transferred its support from the Najibullah regime to an 'Islamic Interim Government'. As mujahideen insurgents closed in on Kabul on 16 April 1992 President Najibullah stepped down but fighting continued.

In 1994 a newly-formed militant Islamic movement, the 'Taliban' ('students of religion'), took Kabul, apparently with Pakistani support. The Taliban, most of whose leaders were Pashtuns, were in turn defeated by the troops of President Rabbani. However, in Sept. 1996 Taliban forces recaptured Kabul and set up an interim government under Mohamed Rabbani. Afghanistan was declared an Islamic state under Sharia law. Government forces counter-attacked but a new Taliban offensive, launched in Dec. 1996, took most of the country. The opposition Northern Alliance controlled the northeast of the country. Under the Taliban, irregular forces were disarmed and roads cleared of bandits. Rebuilding of towns and villages started. Only three countries—Pakistan, Saudi Arabia and the United Arab Emirates—recognized the Taliban as the legal government.

In March 2001 Afghanistan was widely condemned for destroying the ancient Buddha statues at Bamiyan. In May 2001 the Taliban refused to extradite Osama bin Laden, a Saudi militant, to the USA to face charges connected to the bombing of American embassies in Kenya and Tanzania in 1998. In Sept. 2001 Ahmed Shah Masood, leader of the Northern Alliance, was killed by two suicide bombers.

Following the attacks on the USA on 11 Sept. 2001, Saudi Arabia and the United Arab Emirates broke off relations with Afghanistan and the USA unsuccessfully put pressure on the Taliban to hand over bin Laden. Consequently the USA launched air strikes on 7 Oct. On 13 Nov. the Northern Alliance took the capital Kabul, effectively bringing an end to Taliban rule. With the surrender of Kandahar the Taliban lost control of their last stronghold. On 27 Nov. representatives of rival factions, but excluding the Taliban, joined UN-sponsored talks on Afghanistan's future. Hamid Karzai, a Pashtun tribal leader, was chosen to head an interim power-sharing council which took office in Dec. He was appointed president of the transitional government in June 2002. He survived an assassination attempt in Sept. 2002 and won re-election in 2004 and 2009 (the latter in a disputed contest). Following a meeting of international donors in London in 2006, Afghanistan was pledged US$10bn. in reconstruction aid over a five-year period.

Since Aug. 2003 the government has attempted to assert its authority with the help of the NATO-controlled International Security Assistance Force (ISAF), operating under a UN mandate. By 2007 the international force was battling a resurgent Taliban supported by around a quarter of the population in the south of the country. At the London Conference in Jan. 2010 it was agreed to increase ISAF personnel numbers from 96,000 to 135,000 in the course of 2010.

In May 2012 NATO endorsed plans to withdraw foreign combat troops by 2014. In Sept. 2012 the USA suspended the training of police recruits after a spate of attacks apparently carried out by members of the military and police forces with links to the Taliban.

TERRITORY AND POPULATION

Afghanistan is bounded in the north by Turkmenistan, Uzbekistan and Tajikistan, east by China, east and south by Pakistan and west by Iran.

The area is 652,230 sq. km (251,830 sq. miles). The last census was in 1979. Estimate, 2010, 31·41m.; density, 48·2 per sq. km. In 2011, 22·9% of the population lived in urban areas.

The UN gives a projected population for 2015 of 36·74m. Afghanistan's population has doubled since the early 1990s.

Successive wars in Afghanistan resulted in one of the world's largest refugee crises. Prior to the withdrawal of Soviet troops in 1989 there were more than 6m. Afghan refugees, mainly in Pakistan, Iran and to a lesser extent western Europe, Australia and North America. In the first half of the 1990s large numbers began to return, but after the Taliban came to power in 1996 repatriation slowed. As a consequence of the US-led war in Afghanistan beginning in Oct. 2001, numbers of refugees to Pakistan and Iran increased dramatically. In the meantime more than 5m. Afghans have returned to their country since a UN-sponsored programme began in early 2002, but at the end of 2011 there were still 2·7m. Afghan refugees living abroad—more than any other nationality and 26% of the global total. According to humanitarian agencies in Jan. 2002 there were almost 1·2m. internally displaced persons in Afghanistan. Approximately half of the internally displaced persons in Afghanistan moved prior to the events of Sept. 2001, for reasons such as drought and food scarcity. There were still 448,000 internally displaced persons at the end of 2011.

The country is divided into 34 regions (velayat). Area and estimated population in 2009:

Region	Area (sq. km)	Population (1,000)	Region	Area (sq. km)	Population (1,000)
Badakhshan	44,059	860	Daikondi	18,200	417
Badghis	20,591	449	Farah	48,471	459
Baghlan	21,118	819	Faryab	20,293	900
Balkh	17,249	1,169	Ghazni	22,915	1,111
Bamyan	14,175	405	Ghowr	36,479	625

Region	Area (sq. km)	Population (1,000)	Region	Area (sq. km)	Population (1,000)
Helmand	58,584	836	Nimroz	41,005	149
Herat	54,778	1,676	Nurestan	9,225	134
Jawzjan	11,798	485	Paktika	19,482	499
Kabul	4,462	3,569	Paktiya	6,432	394
Kandahar	54,022	1,080	Panjshir	3,610	139
Kapisa	1,842	400	Parwan	5,974	600
Khost	4,152	520	Samangan	11,262	350
Konar	4,942	408	Saripul	15,999	505
Kondoz	8,040	900	Takhar	12,333	886
Laghman	3,843	404	Uruzgan	12,600	317
Logar	3,880	355	Vardak	8,938	540
Nangarhar	7,727	1,358	Zabul	17,343	275

The capital, Kabul, had an estimated population of 2·94m. in 2009. Other towns (with population estimates, 2009): Herat (395,000), Kandahar (363,000), Mazar i Sharif (334,000), Jalalabad (188,000).

Main ethnic groups: Pashtuns, 38%; Tajiks, 25%; Hazaras, 19%; Uzbeks, 6%; others, 12%. The official languages are Pashto and Dari.

SOCIAL STATISTICS

Based on 2008 estimates: birth rate, 46·5 per 1,000 population; death rate, 19·6 per 1,000. Infant mortality (2010), 103 per 1,000 live births. Life expectancy at birth, 2007, was 43·5 years for women and 43·6 years for men (the lowest life expectancy for females and the second lowest overall, ahead of only Zimbabwe). Fertility rate, 2008, 6·6 births per woman.

The maternal mortality rate is among the highest in the world with some 16,000 pregnancy-related deaths every year.

CLIMATE

The climate is arid, with a big annual range of temperature and very little rain, apart from the period Jan. to April. Winters are very cold, with considerable snowfall, which may last the year round on mountain summits. Kabul, Jan. 27°F (–2·8°C), July 76°F (24·4°C). Annual rainfall 13" (338 mm).

CONSTITUTION AND GOVERNMENT

UN sanctions were imposed in 1999 but were withdrawn following the collapse of the Taliban regime. Following UN-sponsored talks in Bonn, Germany in Nov. 2001, on 22 Dec. 2001 power was handed over to an Afghan Interim Authority, designed to oversee the restructuring of the country until a second stage of government, the Transitional Authority, could be put into power. This second stage resulted from a *Loya Jirga* (Grand Council), which convened between 10–16 June 2002. The Loya Jirga established the Transitional Islamic State of Afghanistan. A constitutional commission was established, with UN assistance, to help the Constitutional Loya Jirga prepare a new constitution. A draft constitution was produced for public scrutiny in Nov. 2003 and was approved by Afghanistan's *Loya Jirga* on 4 Jan. 2004. The new constitution creates a strong presidential system, providing for a *President* and two *Vice-Presidents*, and a bicameral parliament. The constitution imposes a limit of two five-year terms for a president The lower house is the 249-member House of the People (*Wolesi Jirga*), directly elected for a five-year term, and the upper house the 102-member House of Elders (*Meshrano Jirga*). The upper house is elected in three divisions. The provincial councils elect one third of its members for a four-year term. The district councils elect the second third of the members for a three-year term. The president appoints the remaining third for a five-year term. At least one woman is elected to the *Wolesi Jirga* from each of the country's 32 regions, and half of the president's appointments to the *Meshrano Jirga* must be women. The constitution reserves 25% of the seats in the *Wolesi Jirga* for women. The president appoints ministers, the attorney general and central bank governor with the approval of the *Wolesi Jirga*. Cabinet ministers must be university graduates. Presidential and parliamentary elections, the first in 25 years, were scheduled for June 2004 but were put back to Oct. 2004. The parliamentary elections were subsequently delayed again and were set to be held in April 2005, but were postponed a further time until Sept. 2005. In Dec. 2005 an elected parliament sat for the first time since 1973.

National Anthem

'Daa watan Afghanistan di' ('This land is Afghanistan'); words by Abdul Bari Jahani' tune by Babrak Wassa.

RECENT ELECTIONS

In presidential elections held on 20 Aug. 2009 initial counts suggested that Hamid Karzai would be re-elected president with 54% of the vote compared to his closest rival Abdullah Abdullah's 28%. However, doubt as to the integrity of the elections was widespread in both the local and international communities with all candidates accused of ballot-rigging. A second round run-off was scheduled for 7 Nov. after Afghanistan's Independent Election Commission (IEC) found that Karzai had not obtained a 50% majority and only took 48·2% of the vote. On 2 Nov. Abdullah pulled out of the second round after calling for the resignation of the head of the IEC and disputing its impartiality. The run-off was subsequently cancelled and Karzai declared the winner. He was sworn in on 19 Nov. 2009 although legal experts questioned the legitimacy of his re-election.

Elections for the 249-member House of the People took place on 18 Sept. 2010; all elected representatives were non-partisan. There were widespread voting irregularities and at least 14 people died in election-related violence. Nearly a quarter of the ballot was declared void owing to fraud. Turnout was approximately 40%.

CURRENT GOVERNMENT

In March 2013 the government was composed as follows:

President: Hamid Karzai; b. 1957 (Pashtun; sworn in 19 June 2002 and re-elected 9 Oct. 2004 and 2 Nov. 2009).

Vice Presidents: Mohammad Qasim Fahim (Tajik); Karim Khalili (Hazara Shia).

Minister of Agriculture, Irrigation and Livestock: Mohammad Asif Rahimi. *Anti-Narcotics:* Ahmad Moqbel Zarar. *Border and Tribal Affairs:* Vacant. *Commerce and Industry:* Anwar Ul-Haq Ahady. *Communications and Technology:* Amirzai Sangin. *Defence:* Bismillah Mohammadi. *Economy:* Abdul Hadi Arghandiwal. *Education:* Ghulam Farooq Wardak. *Energy and Water:* Ismail Khan. *Finance:* Omar Zakhilwal. *Foreign Affairs:* Zalmai Rasul. *Hajj (Pilgrimage) and Awqaf:* Mohammad Yousuf Neyazi. *Higher Education:* Obaidullah Obaid. *Information and Culture:* Sayed Makhdum Rahin. *Justice:* Habibullah Ghaleb. *Labour, Social Affairs, Disabled and Martyrs:* Amina Afzali. *Mines:* Wahidullah Sharani. *Public Health:* Suraya Dalil. *Public Works:* Najibullah Awzhang. *Refugees and Repatriation:* Jamahir Anwari. *Rural Development:* Wais Barmak. *Transport and Civil Aviation:* Daoud Ali Najafi. *Urban Development:* Hassa Abdullahi. *Women's Affairs:* Hasan Ghazanfar.

Government Website: http://www.afghangovernment.com

CURRENT LEADERS

Hamid Karzai

Position
President

Introduction
Hamid Karzai was sworn in as chairman of the interim government of Afghanistan in Dec. 2001 at a conference in Bonn, Germany before taking the position permanently in June 2002. He was appointed by the United Nations in consultation with the Northern Alliance and the *Loya Jirga*, a group of elected delegates. His main aim has been to try and bring stability to the country, but there has been a resurgence of resistance by the Taliban since 2006 particularly in the south and east. His other major challenges have been tackling widespread corruption and drug-trafficking. He was re-elected controversially in autumn 2009.

Early Life
Karzai was born on 24 Dec. 1957 into the Popolzai tribe, one of southern Afghanistan's most powerful factions. His father, who was chief of the Popolzai clan, was assassinated in 1999 in what was widely believed to be a Taliban attack.

Karzai believes in a system of broad-based government called the *Loya Jirga*, with an integrated approach intended to reduce violence between tribal warlords. He first entered politics in the early 1980s during the Soviet occupation and organized the Pashtun Popolzai against Moscow. He spent time in Pakistan, during which time he developed his nationalist philosophy. Karzai returned to Afghanistan in 1992 and linked up with the leader of the Northern Alliance, Burhanuddin Rabbani. When Rabbani formed the first mujahideen government, Karzai served as the deputy foreign minister, but left the government because of infighting.

Karzai initially supported the Taliban when it was created in 1994 but in 1995 he rejected a government post, disillusioned by increasing foreign interference. Karzai left the country in 1996 but secretly re-entered in 2001 during the USA's post-11 Sept. air strikes. He co-ordinated Pashtun resistance to the Taliban and only narrowly evaded capture.

Career in Office
Hamid Karzai was sworn in as chairman of the interim administration in Dec. 2001, taking the title of president in June 2002. He has since enjoyed the support of a majority of the main tribal leaders. However, his lack of military strength has necessitated alliances with armed regional factions and his rule has remained tenuous outside the capital. In Sept. 2002 he survived an assassination attempt in Kandahar. Despite the precarious security situation, Karzai won outright the country's first-ever democratic presidential election on 9 Oct. 2004 with 55·4% of votes cast.

In May–June 2006 Afghanistan experienced the worst insurgent violence and casualties since the US invasion and toppling of the Taliban in 2001. Having taken over the leadership of military operations in the south from July 2006, NATO then assumed responsibility for security across the whole of the country from Oct., taking command in the east from a US-led coalition force. NATO and Afghan forces have since sought to contain a Taliban resurgence, although Karzai has increasingly expressed his concern over the high and ongoing civilian casualty rate in military operations.

In Feb. 2009 US President Barack Obama announced that a further 17,000 US troops would be deployed to Afghanistan, partly to train and support the Afghan army and police service, and in July NATO and Afghan forces pursued a new operation against Taliban strongholds in Helmand province. In Dec. 2009 Obama ordered a further deployment over six months of 30,000 US troops and other NATO members promised to provide another 7,000. This policy of reinforcement heralded the launch in Feb. 2010, again in Helmand, of the biggest coalition offensive in the country since the defeat of Taliban government in 2001. In July 2010 an international conference endorsed Karzai's timetable for control of security to be transferred from foreign to Afghan forces by 2014.

In 2011 Ahmad Wali Karzai, a half-brother of the president and a powerful political figure based in Kandahar, and former president Rabbani were both assassinated in Taliban attacks in July and Sept., respectively. In Oct. the government claimed to have foiled another plot to kill Karzai himself. However, despite the continuing violence, there were tentative moves towards opening channels of communication with the Taliban.

In Jan. 2012 the Taliban agreed to open a political office in Qatar for the purpose of holding peace talks with the Afghan government and international coalition, although Karzai demanded the declaration of a ceasefire before any negotiations could begin. In May 2012 NATO endorsed the planned withdrawal of foreign combat troops by the end of 2014, with Afghan forces assuming increased security responsibilities in the meantime. However, doubts emerged about the level of US logistic support that will be available to Afghanistan after 2014, without which Afghan forces would likely struggle against any Taliban resurgence.

Meanwhile, in Jan. 2011 Karzai had made the first official visit to Russia by an Afghan leader since the end of the Soviet occupation in 1989. In Oct. he signed a strategic partnership with India which further unsettled Afghanistan's already fragile relationship with Pakistan.

On the domestic political front, Karzai had sought re-election as president in 2009 in a campaign tainted by alleged widespread fraud. In the first round of voting in Aug. Karzai claimed to be ahead of rival candidate Abdullah Abdullah with 54% of the vote. In the face of domestic and international concern over the evidential scale of vote-rigging, Karzai subsequently conceded that the elections should go to a second round which was scheduled for early Nov. However, Abdullah Abdullah then withdrew his candidacy and Karzai was declared the winner as the only remaining contender. The president subsequently struggled in Jan. and Feb. 2010 to form a new government as a hostile parliament rejected many of his nominees for cabinet posts. He also remained under pressure from the international community to confront Afghanistan's pervasive corruption.

In late Feb. 2010 Karzai provoked further criticism when he assumed exclusive power to appoint all five members of the independent Electoral Complaints Commission (three of whom had previously been UN nominees), which had earlier rejected his claims to a first-round victory in the 2009 presidential elections. Parliamentary elections were then held in Sept. that year, but reports of widespread voting fraud threatened to undermine the validity of the poll. Official results were not released until Nov. and indicated that the Independent Electoral Commission had disqualified many candidates and approximately 25% of the votes cast owing to fraud. The political influence of Karzai's majority ethnic Pashtun community was significantly reduced in the new parliament, which the president reluctantly inaugurated in Jan. 2011.

DEFENCE

In 2008 military expenditure totalled US$180m. (US$7 per capita), representing 1·5% of GDP.

Since the fall of the Taliban, Afghanistan has had an all-volunteer professional-army. A UN-mandated international force, ISAF, assists the government in the maintenance of security throughout the country. It has been led by NATO since Aug. 2003 and comprises approximately 102,000 troops from 50 countries (of which 68,000 from the USA). US troop numbers peaked at around 100,000 in 2011.

Army

The decimation of the Taliban's armed forces left Afghanistan without an army. A multi-ethnic Afghan National Army, under the command of President Hamid Karzai, has been established, currently numbering approximately 200,000 (up from 12,000 in 2004).

Air Force

Afghanistan's air forces were severely damaged with all planes destroyed by US military operations in 2001, but the Afghan Air Force (the Afghan National Army Air Corps until June 2010) is being rebuilt with a planned strength of 8,000 by 2016.

INTERNATIONAL RELATIONS

Afghanistan was the world's largest recipient of foreign aid in both 2010 and 2011 (US$6·4bn. and US$6·7bn. respectively).

ECONOMY

In 2007 agriculture accounted for 37% of GDP, industry 25% and services 38%.

Afghanistan was rated the joint most corrupt country in the world in a 2012 survey of 176 countries carried out by the anti-corruption organization *Transparency International*.

Overview

Afghanistan is one of the world's poorest countries with formidable development, humanitarian and governance challenges. After the removal of the Taliban regime in 2001, work began on reconstructing an economy impoverished by more than 20 years of conflict, periodic earthquakes and drought. By 2011 the country remained among the bottom 5% of nations on the UN Human Development Index. Unemployment is widespread and the poverty rate is around 36%. The presence of large and active militant groups, poor governance, endemic corruption and low levels of human capacity restrict short- and long-term prospects.

Nevertheless, the authorities have taken steps towards establishing stability and growth. Real GDP growth averaged over 10% annually in the five years to 2010–11 and revenue collection has increased. However, the near-collapse and subsequent bailout of Kabul Bank in 2010 highlighted the country's corruption problems, with the majority of the Bank's loans (totalling around US$900m.) made to just 19 recipients. The financial sector was weakened as a result of the scandal, holding back private sector-led growth. The economy remains dependent on the opium trade, prompting the UN Office on Drugs and Crime to call for action.

Achieving self-sustainability and reaching development objectives pose a major challenge. The government's problems are exacerbated by the prospect of rising security costs, with foreign troops expected to withdraw by the end of 2014.

Currency

The *afghani* (AFN) was introduced in Oct. 2002 with one of the new notes worth 1,000 old *afghani* (AFA). The old *afghani* had been trading at around 46,000 to the US$. Following inflation of 26·8% in 2008 there was deflation of 12·2% in 2009, but then inflation again of 7·7% in 2010 and 11·8% in 2011.

Budget

The fiscal year begins on 21 March. Revenues in 2009–10 were 132,698m. afghani and expenditures 215,880m. afghani. Grants accounted for 60·4% of revenues in 2009–10; capital expenditure accounted for 55·1% of expenditures.

Performance

Real GDP growth was 21·0% in 2009–10 but only 8·4% in 2010–11 and 5·8% in 2011–12. Average annual per capita income has risen from US$180 in 2001 to US$355 in 2006. Total GDP in 2011 was US$19·2bn.

Banking and Finance

Da Afghanistan Bank undertakes the functions of a central bank, holding the exclusive right of note issue. Founded in 1939, its Governor is Noorullah Delawari. The banking sector has undergone major reconstruction since the removal of the Taliban government in 2001, with a number of new private banks having opened in the meantime.

In 2010 the banking system came under scrutiny after Kabul Bank, a leading commercial bank used by the government to pay civil servants and members of the security forces, was unable to account for US$910m. amid allegations of fraud and mismanagement. It was subsequently bailed out by the central bank to avert a collapse.

Total external debt in 2010 was US$2,297m., which was equivalent to 21% of GNI.

ENERGY AND NATURAL RESOURCES

Environment
Carbon dioxide emissions from the consumption and flaring of fossil fuels were the equivalent of less than 0·1 tonnes per capita in 2008.

Electricity
Installed capacity was an estimated 0·49m. kW in 2007–08. Production was approximately 999m. kWh in 2007–08 with consumption per capita about 42 kWh.

Oil and Gas
Proven natural gas reserves were 50bn. cu. metres in 2007. Production in 2007–08 was 3m. cu. metres.

Minerals
There are deposits of barite, coal, copper, emerald, lapis lazuli, salt and talc. Mining, particularly of copper, is considered to be the country's best prospect for economic growth. In spite of the instability, Afghanistan is still one of the world's leading producers of lapis lazuli.

Agriculture
The greater part of Afghanistan is mountainous but there are many fertile plains and valleys. In 2007 there were an estimated 8·53m. ha. of arable land and 0·13m. ha. of permanent cropland; 2·25m. ha. were irrigated in 2007. The agricultural population was approximately 16·0m. in 2007, of whom around 5·08m. were economically active.

Output, 2003, in 1,000 tonnes: wheat, 4,361; rice, 433; barley, 410; grapes, 365; maize, 310; potatoes, 240. Opium production in 2001 was just 185 tonnes (down from 4,565 tonnes in 1999), but in 2002 it went up to 3,400 tonnes, and further to 3,600 tonnes in 2003 and 4,200 tonnes in 2004. There was a slight decline in 2005, to 4,100 tonnes, but production then rose to 6,100 tonnes in 2006 and reached a record 8,200 tonnes in 2007. It has fallen since then, to 3,700 tonnes in 2012. The area under cultivation in 2012 was 154,000 ha., down from the record high of 193,000 ha. in 2007 although up from just 7,606 ha. in 2001. Afghanistan accounts for more than 90% of the world's opium. In Feb. 2001 the United Nations Drug Control Programme reported that opium production had been almost totally eradicated after the Taliban outlawed the cultivation of poppies. As a result in 2001 Myanmar became the largest producer of opium, but since then Afghanistan has again been the leading producer.

Livestock (2003): sheep, 8·8m.; goats, 7·3m.; cattle, 3·7m.; asses, 920,000; camels, 175,000; horses, 104,000; chickens, 6m.

Forestry
In 2010 forests covered 1·35m. ha., or 2% of the total land area. Timber production in 2007 was 3·29m. cu. metres.

Fisheries
In 2010 the total catch was estimated to be 1,000 tonnes, exclusively from inland waters.

INDUSTRY
Major industries include natural gas, fertilizers, cement, coalmining, small vehicle assembly plants, building, carpet weaving, cotton textiles, clothing and footwear, leather tanning and fruit canning.

Labour
The estimated economically active workforce was 9,445,000 in 2010 (73% male). In 2008 the unemployment rate was an estimated 40% (up from 8% in 1995).

INTERNATIONAL TRADE
In April 2003 the transitional government applied for membership of the WTO. Afghanistan is hopeful of joining before the end of 2014.

Imports and Exports
Total imports (2010), US$2,218m.; exports US$388·5m. Main imports: mineral fuels, lubricants and related materials; food and live animals; manufactured goods; machinery and transport equipment. Main exports: opium (illegal trade); food and live animals; manufactured goods; inedible crude materials (excluding fuels). The illegal trade in opium is the largest source of export earnings and accounts for half of Afghanistan's GDP. Main import sources in 2010 were Uzbekistan (21·1%), China (13·7%), Pakistan (11·6%) and Germany (8·2%) The leading export destination in 2010 was Pakistan (38·9%), followed by India (16·8%), Turkey (9·0%) and Iran (8·2%).

Imports and exports were largely unaffected by the sanctions imposed during the Taliban regime.

COMMUNICATIONS

Roads
There were 42,150 km of roads in 2006, of which 29·3% were paved. A large part of the road network is in a poor state of repair as a result of military action, but rebuilding is under way. In Jan. 2003 women regained the right to drive after a ten-year ban. 431,600 passenger cars (15 per 1,000 inhabitants in 2007) and 153,600 lorries and vans were in use in 2008.

Rail
Historically, Afghanistan has lacked its own railway system although two short stretches of railway extend inside the country from the Uzbek and Turkmen networks. In Feb. 2012 the first major Afghan-run railway opened to commercial traffic at a cost of US$170m., covering 75 km from Hairatan, a town on the border with Uzbekistan, to Mazar i Sharif. It is hoped it will be integrated into a wider network being developed as part of a Central Asia Regional Economic Co-operation programme.

Civil Aviation
There is an international airport at Kabul (Khwaja Rawash Airport). The national carrier is Ariana Afghan Airlines, which in 2010 operated direct flights from Kabul to Amritsar, Baku, Delhi, Dubai, Dushanbe, Frankfurt, Islamabad, İstanbul, Jeddah, Kuwait, Mashad, Moscow, Riyadh, Tehran and Urumqi, as well as domestic services. In 1999 scheduled airline traffic of Afghanistan-based carriers flew 2·7m. km, carrying 140,000 passengers (36,000 on international flights). The UN sanctions imposed on 14 Nov. 1999 included the cutting off of Afghanistan's air links to the outside world. In Jan. 2002 Ariana Afghan Airlines resumed services and Kabul airport was reopened. The airport was heavily bombed during the US campaign and although it is now functioning with some civilian flights it is still being used extensively by the military authorities. Afghanistan's first private airline, Kam Air, was launched in Nov. 2003.

Shipping
There are practically no navigable rivers. A port has been built at Qizil Qala on the Oxus and there are three river ports on the Amu Darya, linked by road to Kabul. The container port at Kheyrabad on the Amu Darya river has rail connections to Uzbekistan.

Telecommunications
There were 13,500 landline telephone subscriptions in 2011 (0·4 per 1,000 inhabitants) and 17,558,000 mobile phone subscriptions (542·6 per 1,000 inhabitants). In 2008, 1·5% of households had a computer and 1·2% of households had internet access at home. In March 2012 there were 257,000 Facebook users.

SOCIAL INSTITUTIONS

Justice

A Supreme Court was established in June 1978. It retained its authority under the Taliban regime.

Under the Taliban, a strict form of Sharia law was followed. This law, which was enforced by armed police, included prohibitions on alcohol, television broadcasts, internet use and photography, yet received its widest condemnation for its treatment of women. Public executions and amputations were widely used as punishment under the regime.

The Judicial Reform Commission is in the process of establishing a civil justice system in accordance with Islamic principles, international standards, the rule of law and Afghan legal traditions. The death penalty is still in force. There were 14 confirmed judicial executions in 2012 (two in 2011).

Education

Adult literacy was 28·1% in 2004.

In 2007 there were 4,718,077 pupils at primary schools (37% female) with 110,312 teaching staff and 1,035,782 at secondary schools (26% female) with 32,817 teaching staff. There were around 9,000 schools in 2008. Schools are often burned down and insurgents have attacked both teachers and pupils.

In 2009 there were 19 public universities and higher education institutions. There are also a number of private universities including the American University of Afghanistan. In 2008–09 there were 95,000 students (18% female) in tertiary education and 3,000 academic staff. Kabul University had some 8,000 students in 2006. Formerly one of Asia's finest educational institutes, Kabul University lost many of its staff during the Taliban regime, and following the US bombing attacks it was closed down, although it has since reopened.

In areas controlled by the Taliban education was forbidden for girls. Boys' schools taught only religious education and military training. Female teachers and pupils returned to education after five years of exclusion in 2002, and now nearly 40% of pupils and 30% of teachers are female.

Health

Afghanistan is one of the least successful countries in the battle against undernourishment. A survey carried out in 2004–05 found that 54% of children under five were chronically undernourished.

The bombing of Afghanistan beginning in Oct. 2001 severely disrupted the supply of aid to the country and left much of the population exposed to starvation, but progress since has resulted in more than 60% of the population now having access to basic health services compared to under 10% in 2001.

In 2005 there were 5,970 physicians, 900 dentistry personnel and 14,930 nursing and midwifery personnel in Afghanistan.

In 2008, 48% of the population were using improved drinking water sources.

RELIGION

The predominant religion is Islam. An estimated 86% of the population are Sunni Muslims and 9% Shias.

The Taliban provoked international censure in May 2001 by forcing the minority population of Afghan Hindus and Sikhs to wear yellow identification badges.

CULTURE

World Heritage Sites

There are two UNESCO sites in Afghanistan: the Minaret and Archaeological Remains of Jam (inscribed in 2002), a 12th century minaret; the Cultural Landscape and Archaeological Remains of the Bamiyan Valley (2003), including the monumental Buddha statues destroyed by the Taliban in 2001.

Press

Afghanistan had approximately 540 newspapers in 2008 including 16 paid-for dailies. The main dailies were *Hewad*, *Anis* and the English language publications *Daily Outlook Afghanistan* and *Kabul Times*.

Tourism

Owing to the political situation the tourism industry has been negligible since 2001. It is estimated that around 3,000–4,000 tourists visit the country annually.

Calendar

In 2002 the Afghan Interim Authority replaced the lunar calendar with the traditional Afghan solar calendar. The solar calendar had previously been used in Afghanistan until 1999 when it was changed by the Taliban authorities who wanted the country to adopt the system used in Saudi Arabia. The change back to the solar calendar means the current year is 1392.

DIPLOMATIC REPRESENTATIVES

Of Afghanistan in the United Kingdom (31 Prince's Gate, London, SW7 1QQ)
Ambassador: Dr Mohammed Daud Yaar.

Of the United Kingdom in Afghanistan (15th St., Roundabout Wazir Akbar Khan, PO Box 334, Kabul)
Ambassador: Sir Richard Stagg, KCMG.

Of Afghanistan in the USA (2341 Wyoming Ave., NW, Washington, D.C., 20008)
Ambassador: Eklil Ahmad Hakimi.

Of the USA in Afghanistan (Great Masood Rd between Radio Afghanistan and Ministry of Public Health, Kabul)
Ambassador: James Cunningham.

Of Afghanistan to the United Nations
Ambassador: Zahir Tanin.

Of Afghanistan to the European Union
Ambassador: Homayoun Tandar.

FURTHER READING

Cowper-Coles, Sherard, *Cables from Kabul: The Inside Story of the West's Afghanistan Campaign.* 2011

Dalrymple, William, *The Return of a King: Shah Shuja and the First Battle for Afghanistan, 1839–42.* 2013

Ewans, Martin, *Afghanistan, A New History.* 2001

Goodson, Larry, *Afghanistan's Endless War: State Failure, Regional Politics and the Rise of the Taliban.* 2001

Griffiths, John, *Afghanistan: A History of Conflict.* 2001

Hopkins, B. D., *The Making of Modern Afghanistan.* 2008

Hyman, A., *Afghanistan under Soviet Domination, 1964–1991.* 3rd ed. 1992

Magnus, Ralph H. and Naby, Eden, *Afghanistan: Mullah, Marx and Mujahid.* Revised ed. 2002

Maley, William, *The Afghanistan Wars.* 2nd ed. 2009

Margolis, Eric, *War at the Top of the World: The Struggle for Afghanistan, Kashmir and Tibet.* 2001

Nojumi, Neamatollah, *The Rise of the Taliban in Afghanistan.* 2001

Steele, Jonathan, *Ghosts of Afghanistan: The Haunted Battleground.* 2011

Vogelsang, Willem, *The Afghans.* 2002

National Statistical Office: Central Statistics Office, Ansar-i-Watt, Kabul.
Website: http://www.cso.gov.af

ALBANIA

© Research Machines plc 2006

Republika e Shqipërisë
(Republic of Albania)

Capital: Tirana
Population projection, 2015: 3·26m.
GNI per capita, 2011: (PPP$) 7,803
HDI/world rank: 0·739/70
Internet domain extension: .al

KEY HISTORICAL EVENTS

Albania was originally part of Illyria which stretched along the eastern coastal region of the Adriatic. By 168 BC the Romans, having conquered Illyria, administered it as a province (Illyricum). From AD 395 Illyria, as part of the eastern Byzantine empire, submitted to waves of Slavic invasions. During the middle ages the name Albania gained currency, possibly deriving from Albanoi, the name of an Illyrian tribe. Ottoman intrusion began in the 14th century and, despite years of resistance under the leadership of national hero Gjergj Kastrioti, Turkish suzerainty was imposed from 1478. During the 15th and 16th centuries, many Albanians fled to southern Italy to escape Ottoman rule and

conversion to Islam. After the Russo-Turkish war of 1877–78 there were demands for independence from Turkey. With the defeat of Turkey in the Balkan war of 1912, Albanian nationalists proclaimed independence and set up a provisional government.

In the First World War Albania was a battlefield for opposing occupation forces. Albania was admitted to the League of Nations on 20 Dec. 1920. In Nov. 1921 its 1913 frontiers were confirmed with minor alterations. Although declared a republic in 1925, Albania became a monarchy from Sept. 1928 until April 1939 when Italy's dictator, Mussolini, invaded and set up a puppet state. During the Second World War Albania suffered first Italian and then German occupation. Resistance was led by royalist, nationalist republican and Communist movements, often at odds with each other. The Communists enjoyed the support of Tito's partisans, who were instrumental in forming the Albanian Communist Party on 8 Nov. 1941. Communists dominated the Anti-Fascist National Liberation Committee which became the Provisional Democratic Government on 22 Oct. 1944 after the German withdrawal, with Enver Hoxha, a French-educated school teacher and member of the Communist Party Central Committee, at its head. Large estates were broken up and the land distributed, with collectivization imposed from 1955–59. Close ties were forged with the USSR. However, following Khrushchev's reconciliation with Tito in 1956, China replaced the Soviet Union as Albania's patron until the end of the Maoist phase in 1977. The regime then adopted a policy of 'revolutionary self-sufficiency'.

Following the collapse of the USSR, the People's Assembly legalized opposition parties. The Communists won the first multi-party elections in April 1991, but soon resigned from office following a general strike. They were replaced by a coalition government, which collapsed in Dec. 1991, and then by an interim technocratic administration. A new, non-Communist government was elected in March 1992.

In 1997 Albania was disrupted by financial crises caused by the collapse of fraudulent pyramid finance schemes. A period of violent anarchy led to the fall of the administration and to fresh elections which returned a Socialist led government. A UN peacekeeping force withdrew in Aug. 1997, but sporadic violence continued.

In April 1999 the Kosovo crisis which led to NATO air attacks on Yugoslavian military targets released a flood of refugees into Albania.

Having won a decisive victory in the 2001 elections, the ruling Socialist Party lost power to the opposition Democratic Party in July 2005.

TERRITORY AND POPULATION

Albania is bounded in the north by Montenegro and Serbia, east by Macedonia, south by Greece and west by the Adriatic. The area is 28,703 sq. km (11,082 sq. miles). The population at the census of Oct. 2011 was 2,821,138, giving a density of 98·3 per sq. km. The United Nations population estimate for 2011 was 3,216,000.

The UN gives a projected population for 2015 of 3·26m.

In 2011, 52·9% of the population lived in urban areas. The capital is Tirana (provisional population in 2011, 421,286); other large towns (population in 2011) are Durrës (115,550), Vlorë (79,948), Elbasan (79,810), Shkodër (74,867), Fier (57,198) and Korçë (51,683).

The country is administratively divided into 12 prefectures, 36 districts, 308 communes and 65 municipalities.

Prefectures	Area (sq. km)	2011 census population (provisional)	Prefectures	Area (sq. km)	2011 census population (provisional)
Berat	1,802	141,944	Durrës	827	262,785
Dibër	2,507	137,047	Elbasan	3,278	295,827

Prefectures	Area (sq. km)	2011 census population (provisional)	Prefectures	Area (sq. km)	2011 census population (provisional)
Fier	1,887	310,331	Lezhë	1,581	134,027
Gjirokastër	2,883	72,176	Shkodër	3,562	215,347
Korçë	3,711	220,357	Tirana	1,586	749,365
Kukës	2,373	85,292	Vlorë	2,706	175,640

In most cases prefectures are named after their capitals. The one exception is Dibër, where the capital is Peshkopi.

Albanians account for 91·7% of the population, Aromanians 3·6%, Greeks 2·3% and others 2·4%.

The official language is Albanian.

SOCIAL STATISTICS

2007: births, 33,163; deaths, 14,528. Rates in 2007 (per 1,000): births, 10·5; deaths, 4·6. Infant mortality, 2010, was 16 per 1,000 live births. Fertility rate (number of births per woman), 1·9 in 2008. Annual population growth rate, 2008–10, 0·4%. Life expectancy at birth, 2007, was 73·4 years for men and 79·8 years for women. Abortion was legalized in 1991.

CLIMATE

Mediterranean-type, with rainfall mainly in winter, but thunderstorms are frequent and severe in the great heat of the plains in summer. Winters in the highlands can be severe, with much snow. Tirana, Jan. 44°F (6·8°C), July 75°F (23·9°C). Annual rainfall 54" (1,353 mm). Shkodër, Jan. 39°F (3·9°C), July 77°F (25°C). Annual rainfall 57" (1,425 mm).

CONSTITUTION AND GOVERNMENT

A new constitution was adopted on 28 Nov. 1998. The supreme legislative body is the single-chamber *People's Assembly* of 140 deputies. As from April 2009 all members are elected through proportional representation, for four-year terms. Where no candidate wins an absolute majority, a run-off election is held. The *President* is elected by parliament for a five-year term.

National Anthem

'Rreth Flamurit të përbashkuar' ('The flag that united us in the struggle'); words by A. S. Drenova, tune by C. Porumbescu.

RECENT ELECTIONS

Parliamentary elections took place on 28 June 2009. The Democratic Party of Albania (PD) won 68 of the 140 seats with 40·0% of votes cast, the Socialist Party of Albania 65 with 40·8%, the Socialist Movement for Integration 4 with 4·8%. Three other parties won one seat each. The PD and its allies received a total of 70 seats, one short of a majority, and the Socialist Party and its allies 66. Turnout was an estimated 50%. The Socialist Movement for Integration joined the PD and its allies to form a working coalition in Sept. 2009.

Parliament elected Bujar Nishani (PD) president on 11 June 2012 in a fourth round after votes on 30 May, 4 June and 8 June had failed to result in the required three-fifths majority.

Parliamentary elections were scheduled to take place on 23 June 2013.

CURRENT GOVERNMENT

President: Bujar Nishani; b. 1966 (PD; in office since 24 July 2012).

The government is formed by a coalition of the Democratic Party of Albania (PD), the Republican Party of Albania (PR) and the Party for Justice and Integration (PDI). The Socialist Movement for Integration (LSI) left the government in April 2013. After the resulting reshuffle it comprised:

Prime Minister: Sali Berisha; b. 1944 (PD; sworn in 11 Sept. 2005, having previously been president from April 1992–July 1997).

Deputy Prime Minister and Minister for Education and Science: Myqerem Tafaj.

Minister for Agriculture, Food and Consumer Protection: Genc Ruli. *Defence:* Arben Imami. *Economy, Trade and Energy:* Florian Mima. *Environment:* Fatmir Mediu. *European Integration:* Majlinda Bregu. *Finance:* Ridvan Bode. *Foreign Affairs:* Aldo Bumçi. *Health:* Halim Kosova. *Innovation and ICT:* Genc Pollo. *Interior:* Flamur Noka. *Justice:* Eduard Halimi. *Labour, Social Affairs and Equal Opportunities:* Spiro Ksera. *Public Works, Transportation and Telecommunications:* Sokol Olldashi. *Tourism, Culture, Youth and Sports:* Visar Zhiti.

Albanian Parliament: http://www.parlament.al

CURRENT LEADERS

Bujar Nishani

Position
President

Introduction
Bujar Nishani was elected president in June 2012, although the main opposition Socialist Party boycotted the vote. A former justice and interior minister, he is close to Sali Berisha—the current prime minister, former president and leader of the Democratic Party.

Early Life
Bujar Nishani was born in Durrës on 29 Sept. 1966. He attended the Skënderbej Military Academy in Tirana, taking a teaching post there in 1988. He joined the newly established centre-right Democratic Party of Albania (PD) in 1991 headed by Sali Berisha, who led his party to victory in the country's first free parliamentary election in March 1992.

The following year Nishani served as director of foreign affairs in the ministry of defence. While the PD made progress in opening up the economy, reforming institutions and promoting foreign investment, the administration became embroiled in allegations of corruption while much of the country remained in poverty. The collapse of government pyramid investment schemes in 1997 led to violent protests, almost sparking civil war in the south. The government resigned and a snap election was won comfortably by a Socialist-led coalition.

Nishani took a law degree at Tirana University before turning to municipal politics. In 2001 he was elected secretary of the PD's Tirana branch and two years later he was elected to the capital's municipal council. At the 2005 parliamentary election, he won a seat in Tirana as the PD returned to power.

Appointed interior minister in March 2007, he sought to modernize the country's institutions and championed an electronic ID card scheme. As justice minister from Sept. 2009, after the PD had secured re-election in July that year, he attempted to tackle corruption in the justice system against a political backdrop of discontent fuelled by allegations of electoral fraud. Albania's request for EU candidate status was rejected in Nov. 2010.

In April 2011 Nishani was reappointed minister of the interior, a post he kept until June 2012 when he was elected by the national assembly to replace Bamir Topi as president. Nishani was the only candidate put forward by the ruling coalition led by Berisha, with the Socialists boycotting the vote after several failures to agree an opposition candidate.

Career in Office
On taking office, Nishani prioritized reforming the justice system and tackling the political impasse. His government received a boost in mid-Dec. 2012 when the European Parliament agreed to grant Albania EU candidate status dependent on the implementation of key reforms to the judiciary, public administration and the functioning of parliament.

Sali Berisha

Position
Prime Minister

Introduction
Dr Sali Berisha returned as prime minister on 11 Sept. 2005 after eight years in opposition. He was a leading opponent of the communist regime in the late 1980s and served as Albania's first elected post-communist president from 1992–97. Breathing life into Albania's ailing economy, reforming its institutions and tackling corruption and organized crime have remained his policy priorities. Following parliamentary elections in June 2009, Berisha retained the premiership at the head of a new coalition formed in Sept. between his Democratic Party of Albania and its allies and the Socialist Movement for Integration.

Early Life
Sali Ram Berisha was born in Vuçidol in northern Albania on 15 Oct. 1944, the year in which the communist leader, Enver Hoxha, seized power and established a hard-line Stalinist regime. Berisha graduated in medicine from the University of Tirana in 1967, subsequently specializing in cardiology and publishing numerous textbooks and scientific papers. He joined the communist Party of Labour in 1971.

In the late 1980s Berisha was one of a group of intellectuals who called for democratic reforms. Following student protests at the University of Tirana in Dec. 1990, he founded the PD and was elected a member of parliament in the country's first multi-party elections on 31 March 1991. During the PD's first Congress in Sept. 1991 Berisha was voted chairman and led the party to victory in the general election of 22 March 1992. Elected president of Albania on 9 April 1992, he set out to open up the economy, promote foreign investment and reform the country's institutions. However, the administration was marred by corruption and Albania remained mired in poverty. Tens of thousands emigrated and Berisha faced growing opposition to his increasingly authoritarian rule, particularly in the south of the country.

Support for Berisha was further eroded by the collapse of various pyramid investment schemes in 1997. An uprising in the south threatened to spill over into civil war and tensions remained high when Berisha refused to step down after the electoral victory of a socialist-led coalition in June 1997. He eventually resigned under international pressure on 23 July 1997, to be replaced by the head of the Socialist Party, Rexhep Meidani. Fatos Nano, a fellow socialist and arch-rival of Berisha, became prime minister.

Tensions between the two main parties remained. Berisha accused Fatos Nano's administration of corruption and incompetence and withdrew from parliament between 1998 and early 2002. The arrival of hundreds of thousands of ethnic Albanian refugees from Kosovo in 1999 placed further strain on the faltering economy and infrastructure. Berisha fought the general election on 3 July 2005 on an anti-corruption platform and the PD claimed victory, although foreign monitors criticized the vote as falling short of international standards. Having formed a coalition with other centre-right groups to take control of 81 of the 140 seats in the legislature, Berisha was sworn in as prime minister on 11 Sept. 2005.

Career in Office
Berisha promised to build a 'social state' by streamlining the government and purging it of corrupt elements, reducing taxes and enabling private enterprise to flourish. He also declared his aim of eventual Albanian membership of NATO and the EU. In June 2006 his government signed a Stabilization and Association Agreement with the EU after three years of negotiations, and in April 2009 Albania joined NATO and applied for EU membership.

Berisha has presided over an improvement in the economy, a revival of agriculture and the attraction of greater foreign investment. However, Albania's reputation for lawlessness has endured and criticisms of the country's ineffective judicial system remain. In March 2008 Berisha removed the defence minister, Fatmir Mediu, from office following a series of explosions at an ammunition depot near Tirana airport which killed 26 people and injured 250 more.

A new coalition government comprising the PD, the Republican Party of Albania, the Party for Justice and Integration and the Socialist Movement for Integration, with Berisha retaining the premiership, was sworn in on 17 Sept. 2009 following earlier parliamentary elections. However, the opposition Socialist Party maintained that the election result was fraudulent. The party's consequent campaign of demonstrations, civil disobedience and boycotting parliament resulted in political paralysis and prompted the EU's refusal in Nov. 2010 to grant Albania candidate membership status. The political climate deteriorated further in Jan. 2011 when an anti-government rally in the capital resulted in the killing of four protesters by security forces and led Berisha to accuse the opposition leader of trying to instigate a coup.

DEFENCE
Since 1 Jan. 2010 Albania has had an all-volunteer professional army. In 2008 defence expenditure totalled US\$254m. (US\$70 per capita), representing 1·9% of GDP.

Army
Strength of the Land Forces (Army) Command in 2007 was 6,200. There is an internal security force, and frontier guards number 500.

Navy
Navy Forces Command personnel in 2007 totalled 1,100. The fleet comprises 35 vessels including five torpedo craft. There are naval bases at Durrës and Vlorë.

Air Force
In 2007 the Air Forces Command had 1,370 personnel and operated 13 helicopters (including seven Agusta Bell 206s).

INTERNATIONAL RELATIONS
Albania applied to join the EU in April 2009, although membership is not expected until 2015 at the earliest.

ECONOMY
In 2009 agriculture accounted for 20·4% of GDP (down from 29·1% in 2000), industry 19·4% (up from 19·0% in 2000) and services 60·2% (up from 51·9% in 2000).

Overview
Following the dissolution of the USSR and economic collapse, the government embarked on a programme of privatization and economic liberalization. Privatization of land, small businesses and housing was achieved between 1991–93 and a privatization programme for large enterprises was initiated in 1995. The Tirana Stock Exchange was established in 1996 during a period of strong growth, with GDP annual growth averaging 9% between 1993–95. However, the absence of banking sector reform led to informal credit arrangements such as pyramid schemes, many of which collapsed in 1997, sparking widespread social unrest and the downfall of the government.

The agricultural sector remains important, although its share of GDP dropped from around 35% in 1990 to 21% in 2007. Recent economic growth has been stimulated chiefly by the manufacturing and service sectors, especially construction. Annual remittances from abroad account for about 15% of GDP.

Prior to the 2009 global economic crisis, Albania had experienced six years of growth averaging over 5·5% and unemployment was in prolonged decline. From 2002–08 the number of people living in poverty fell from 25% to 12%. It was one of the few European countries to maintain positive growth in

2009, owing to limited exposure to international markets and an expansionary fiscal policy. The chief impact of the crisis was a fall in remittances and exports.

Currency
The monetary unit is the *lek* (ALL), notionally of 100 *qindars*. In Sept. 1991 the lek (plural, *lekë* or leks) was pegged to the ecu at a rate of 30 leks = one ecu. In June 1992 it was devalued from 50 to 110 to US$1. There was inflation of 3·6% in 2010 and 3·4% in 2011, extending the economic stabilization that followed several years of high inflation (225% in 1992).

Foreign exchange reserves were US$1,180m. in July 2005, total money supply was 184,249m. leks and gold reserves totalled 69,000 troy oz.

Budget
The fiscal year is the calendar year. In 2008 general government revenue was 291,191m. leks (251,317m. leks in 2007) and expenditure 248,273m. leks (222,061m. leks in 2007).

Principal sources of revenue in 2007 were: taxes on goods and services, 130,349m. leks; social security contributions, 41,393m. leks; taxes on income, profits and capital gains, 38,159m. leks. Main items of expenditure by economic type in 2007 were: compensation of employees, 67,510m. leks; use of goods and services, 29,098m. leks; interest, 25,594m. leks.

VAT is 20%.

Performance
Total GDP in 2011 was US$13·0bn. Since the economy contracted by 10·2% in 1997 following the collapse of pyramid finance schemes, Albania has enjoyed consistently strong growth and managed to avoid recession in 2009. Every year between 1998 and 2001 and again from 2003 to 2008 real GDP growth was in excess of 5%. The economy then grew by 3·3% in 2009, 3·5% in 2010 and 3·0% in 2011.

Banking and Finance
The central bank and bank of issue is the Bank of Albania, founded in 1925 with Italian aid as the Albanian State Bank and renamed in 1993. Its *Governor* is Ardian Fullani. The Savings Bank of Albania, which serves around 75% of the Albanian market, was sold by the government to Raiffeisen Zentralbank Österreich AG in Dec. 2003. In 2002 it had total assets of US$1·4bn., approximately 60% of the total assets of the Albanian banking sector. In 2002 there were six other banks: American Bank of Albania; Arab-Albanian Islamic Bank; Fefad Bank; Italian-Albanian Bank; National Commercial Bank of Albania; and Tirana Bank SA.

Albania's external debt amounted to US$4,736m. in 2010, up from US$2,054m. in 2005 and US$1,070m. in 2000—this was equivalent to 40·5% of GNI (up from 24·1% in 2005).

A stock exchange opened in Tirana in 1996.

ENERGY AND NATURAL RESOURCES

Environment
Albania's carbon dioxide emissions from the consumption and flaring of fossil fuels in 2008 were the equivalent of 1·3 tonnes per capita (compared to the European average of 7·8 tonnes per capita).

Electricity
Albania is rich in hydro-electric potential. Although virtually all of the electricity is generated by hydro-electric power plants only 30% of potential hydro-electric sources are currently being used. Power cuts are common. Electricity capacity was an estimated 1·67m. kW in 2007. Production was 2·86bn. kWh in 2007 and consumption per capita 1,799 kWh. Albania imported 2,828m. kWh of electricity in 2007.

Oil and Gas
Offshore exploration began in 1991. Oil has been produced onshore since 1920. Oil reserves in 2007 were 165m. bbls. Crude oil production in 2007, 570,000 tonnes. Natural gas is extracted. Reserves in 2007 totalled 2bn. cu. metres; output in 2007 was 11m. cu. metres.

Minerals
Mineral wealth is considerable and includes lignite, chromium, copper and nickel. Output (in 1,000 tonnes): lignite (2007), 92; copper ore (2005), 73; chromite (2005), 66. Nickel reserves are 60m. tonnes of iron containing 1m. tonnes of nickel, but extraction had virtually ceased by 1996. A consortium of British and Italian companies is modernizing the chrome industry with the aim of making Albania the leading supplier of ferrochrome to European stainless steel producers.

Agriculture
In 2007 the agricultural population was an estimated 1·37m., of whom around 620,000 were economically active. The country is mountainous, except for the Adriatic littoral and the Korçë Basin, which are fertile. Only 24% of the land area is suitable for cultivation; 15% of Albania is used for pasture. In 2007 there were 578,000 ha. of arable land and 120,000 ha. of permanent cropland. 106,530 ha. were irrigated in 2007.

A law of Aug. 1991 privatized co-operatives' land. Families received allocations, according to their size, from village committees. In 2007 there were 369,598 agricultural holdings. Since 1995 owners have been permitted to buy and sell agricultural land. In 2007 there were 10,833 tractors in use and 1,643 harvester-threshers.

Production (in 1,000 tonnes), 2007: total grains, 494 (including wheat 250 and maize 216); watermelons, 190; tomatoes, 160; potatoes, 155; grapes, 147; onions, 73; cucumbers and gherkins, 50.

Livestock, 2007: sheep, 1,853,000; goats, 876,000; cattle, 577,000; pigs, 147,000; horses, 46,000; chickens, 4,712,000.

Livestock products, 2007 (in 1,000 tonnes): beef, 83; mutton, lamb and goat, 46; pork, 16; poultry, 13; milk, 1,016; eggs, 736m. units.

Forestry
Forests covered 776,000 ha. in 2010 (28% of the total land area), mainly oak, elm, pine and birch. Timber production in 2007 was 296,000 cu. metres.

Fisheries
The total catch in 2010 amounted to 6,144 tonnes (3,103 tonnes from sea fishing and 3,041 tonnes from inland fishing), up from 3,320 tonnes in 2000.

INDUSTRY
Output is small, and the principal industries are agricultural product processing, textiles, oil products and cement. Closures of loss-making plants in the chemical and engineering industries built up in the Communist era led to a 60% decline in the two years after the collapse of the USSR and the subsequent political changes in Albania. Output in 2007 unless otherwise indicated (in 1,000 tonnes): cement (2008), 737; rolled steel (2008), 194; wheat and blended flour, 191; distillate fuel oil, 82; residual fuel oil, 32; sawnwood (estimate), 97,000 cu. metres; beer, 36·6m. litres.

Labour
In 2003 the workforce was 1,089,000, of which 926,000 were employed (745,000 in the private sector). Unemployment was 15·0% at the end of 2003.

The average monthly wage in 2007 was 27,350 leks; the official minimum wage in 2010 was 16,120 leks. Minimum wages may not fall below one-third of maximum.

INTERNATIONAL TRADE

Imports and Exports

The following table shows the value of Albania's foreign trade (in US$1m.):

	2005	2006	2007	2008	2009
Imports (c.i.f.)	2,614·3	3,057·4	4,200·9	5,250·5	4,548·3
Exports (f.o.b.)	658·2	792·6	1,077·7	1,354·9	1,087·9

Principal imports in 2009: manufactured goods, 25·3%; machinery and transport equipment, 22·6%; food and livestock, 12·1%; mineral fuels, lubricants and related materials (including petroleum and petroleum products), 11·8%. Leading exports in 2009: clothing and apparel, 26·8%; footwear, 19·4%; petroleum and petroleum products, 8·3%; manufactures of metal, 4·8%.

Main import suppliers in 2009 were: Italy, 26·1%; Greece, 15·5%; China, 7·2%; Germany, 6·5%; Turkey, 6·4%. Main export destinations in 2009 were: Italy, 62·8%; Greece, 7·4%; Slovakia, 5·5%; China, 4·8%; Germany, 3·4%.

COMMUNICATIONS

Roads

In 2002 there were 3,220 km of main roads, 4,300 km of secondary roads and 10,480 km of other roads. There were 237,932 passenger cars in 2007, as well as 29,506 buses and coaches and 59,645 lorries and vans. There were 384 fatalities in road accidents in 2007.

Rail

Total length in operation in 2005 was 447 km. Passenger-km travelled in 2008 came to 41m. and freight tonne-km in 2007 to 53m.

Civil Aviation

The national carrier, Albanian Airlines, ceased operations in Nov. 2011. The only operating Albania-based airline is Belle Air, a low-cost carrier founded in 2005. It has international flights to a number of destinations in Europe as well as some charter flights. In 2006 scheduled airline traffic of Albania-based carriers flew 2m. km, carrying 213,000 passengers (all on international flights). The main airport is Mother Teresa International Airport at Rinas, 25 km from Tirana, which handled 1,394,688 passengers in 2009 (1,267,041 in 2008).

Shipping

In Jan. 2009 there were 46 ships of 300 GT or over registered, totalling 61,000 GT. The main port is Durrës, with secondary ports being Vlorë, Sarandë and Shëngjin.

Telecommunications

In 2010 there were 333,000 landline telephone subscriptions (equivalent to 103·9 per 1,000 inhabitants) and 2,692,000 mobile phone subscriptions (or 840·2 per 1,000 inhabitants). In 2009, 41·2% of the population were internet users. Fixed internet subscriptions totalled 105,000 in 2009 (32·9 per 1,000 inhabitants). In March 2012 Albania had 1·06m. Facebook users.

SOCIAL INSTITUTIONS

Justice

A new criminal code was introduced in June 1995. The administration of justice (made up of First Instance Courts, The Courts of Appeal and the Supreme Court) is presided over by the *Council of Justice*, chaired by the President of the Republic, which appoints judges to courts. A Ministry of Justice was re-established in 1990 and a Bar Council set up. In Nov. 1993 the number of capital offences was reduced from 13 to six and the death penalty was abolished for women. In 2000 the death penalty was abolished for peacetime offences; it was abolished for all crimes in 2007.

The prison population in Sept. 2009 was 4,482 (141 per 100,000 of national population).

Education

Primary education is free and compulsory from six to 14 years. Secondary education is also free and lasts four years. Pupils who fail exams at 14 are required to remain in school until the age of 16.

Secondary education is divided into three categories: general; technical and professional; vocational. There were, in 2008–09, 75,000 children enrolled in nursery schools and 4,000 teachers; 12,000 primary school teachers with 236,000 pupils; and 355,000 pupils and 24,000 teachers at secondary schools. In 2008–09 there were nine public universities, two academies (Academy of Arts and Academy of Sports), the Inter-University Center of Albanological Studies, the Academy of Public Order and the Military University. In Sept. 2008 there were 21 private universities. Adult literacy in 2008 was 95·9% (97·3% among males and 94·7% among females).

In 2007 public expenditure on education came to 3·3% of GDP and represented 11·1% of total government expenditure.

Health

Medical services are free, though medicines are charged for. In 2008 there were 41 public hospitals providing 29·1 beds per 10,000 population. There were 2,039 GPs, 1,587 specialized physicians and 12,746 nurses in 2008.

The budget for expenditure on health in 2007 was 24,104m. leks (8·4% of the total budget).

Welfare

The retirement age is 65 (men) and 60 (women). To be eligible for a full state pension contributions over 35 years are required. There is a partial pension with between 15 and 35 years of contributions. Old-age benefits consist of a basic pension and an earnings-related increment. The basic pension is indexed according to price changes of selected commodities.

Unemployment benefit was 5,240 leks per month as of 2007.

RELIGION

In 2001, 39% of the population were Muslims, mainly Sunni with some Bektashi, 17% Roman Catholic, 10% Albanian Orthodox and the remainder other religions or atheist. The Albanian Orthodox Church is autocephalous; it is headed by an Exarch, Anastasios, Archbishop of Tirana, Durrës and All Albania, and three metropolitans. In 2001 there were 118 priests. The Roman Catholic cathedral in Shkodër has been restored and the cathedral in Tirana has been rebuilt, opening in 2002. In 2000 there were one Roman Catholic archbishop and three bishops.

CULTURE

World Heritage Sites

In 1992 Butrint was added to the UNESCO World Heritage List (reinscribed in 1999 and 2007). Butrint is a settlement in the southwest of the country, near the port of Sarandë, which was inhabited from 800 BC and is now a major site of archaeological investigation. The site was extended for protection after looting in 1997. The historic town of Gjirokastër was added in 2005 (reinscribed in 2008) and is a rare example of a well-preserved Ottoman town built around the 13th century.

Press

In 2008 there were 28 paid-for dailies (combined circulation of 77,000) and 82 paid-for non-dailies. The leading newspaper in terms of circulation is *Shekulli*.

Tourism

In 2007 there were 1,127,000 non-resident tourists staying at hotels in Albania (including 1,061,000 from other European countries and 52,000 from the Americas), bringing revenue of US$1,479m.

Festivals

The main festivals are Tirana International Contemporary Art Biennial (Sept.–Oct.), Tirana Autumn (Nov.), a music and arts festival, and the Tirana International Film Festival (Nov.–Dec.).

DIPLOMATIC REPRESENTATIVES

Of Albania in the United Kingdom (33 St George's Drive, London, SW1V 4DG)
Ambassador: Vacant.
Chargé d'Affaires a.i.: Mal Berisha.

Of the United Kingdom in Albania (Rruga Skenderbeg 12, Tirana)
Ambassador: Nicholas Cannon, OBE.

Of Albania in the USA (2100 S St., NW, Washington, D.C., 20008)
Ambassador: Gilbert Galanxhi.

Of the USA in Albania (Tirana Rruga Elbasanit 103, Tirana)
Ambassador: Alexander Arvizu.

Of Albania to the United Nations
Ambassador: Ferit Hoxha.

Of Albania to the European Union
Ambassador: Mimoza Halimi.

FURTHER READING

Bezemer, Dirk J. (ed.) *On Eagle's Wings: The Albanian Economy in Transition.* 2009
Biberaj, Elez, *Albania in Transition: The Rocky Road to Democracy.* 1999
Hutchings, R., *Historical Dictionary of Albania.* 1997
Vickers, M., *The Albanians: a Modern History.* 1997
Vickers, M. and Pettifer, J., *Albania: from Anarchy to a Balkan Identity.* 1997—*The Albanian Question: Reshaping the Balkans.* 2009

National Statistical Office: Albanian Institute of Statistics, Blv. Zhan d'Ark, Nr. 3, Tirana. *Director General:* Ines Nurja.
Website: http://www.instat.gov.al

ALGERIA

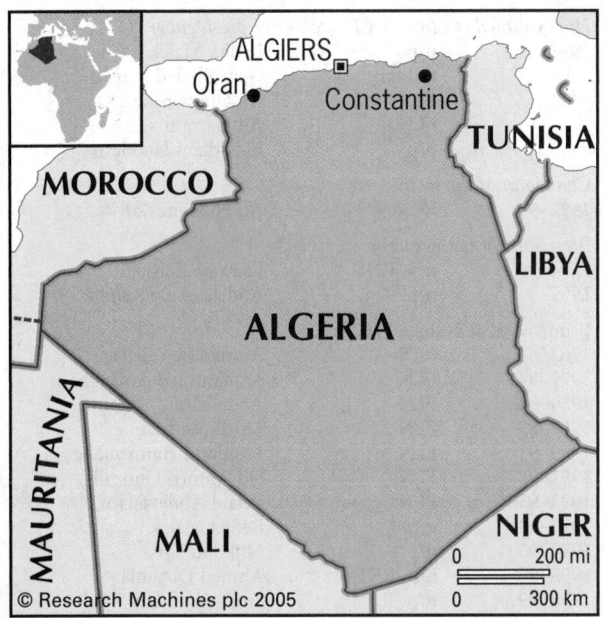

ALGIERS
Oran
Constantine
TUNISIA
MOROCCO
LIBYA
ALGERIA
MAURITANIA
MALI
NIGER
0   200 mi
0   300 km
© Research Machines plc 2005

Jumhuriya al-Jazairiya ad-Dimuqratiya ash-Shabiya
(People's Democratic Republic of Algeria)

Capital: Algiers
Population projection, 2015: 37·95m.
GNI per capita, 2011: (PPP$) 7,658
HDI/world rank: 0·698/96
Internet domain extension: .dz

KEY HISTORICAL EVENTS

Algeria came under French control in the 1850s. French settlers assumed political and economic power at the expense of the indigenous Muslim population. In Nov. 1954 the *Front de Libération Nationale* (FLN), representing the Muslim majority, declared open warfare against the French administration. Fierce fighting continued unabated until March 1962 when a ceasefire was agreed between the French government and the nationalists. Against the wishes of the French in Algeria, Gen. de Gaulle conceded Algerian independence on 3 July 1962.

The Political Bureau of the FLN took over the functions of government, a National Constituent Assembly was elected and the Republic was declared on 25 Sept. 1962. One of the founders of the FLN, Ahmed Ben Bella, became prime minister, and president the following year. On 15 June 1965 the government was overthrown by a junta of army officers, who established a Revolutionary Council under Col. Houari Boumedienne. After ten years of rule, Boumedienne proposed elections for a president and a National Assembly. A new constitution was accepted in a referendum in Nov. 1976 and Boumedienne was elected president unopposed. With all parties except the FLN banned from participating, a National Assembly was elected in Feb. 1977.

On the death of the president in Dec. 1978 the Revolutionary Council again took over the government. The Islamic Salvation Front (FIS) was banned in March 1992. The head of state, Mohamed Boudiaf, was assassinated on 29 July 1992 and a campaign of terrorism by fundamentalists has continued to the present day. It is estimated that over 100,000 lives have been lost,

although Algeria has emerged from the worst of the war, with guerrilla activity now restricted to the countryside. Unrest among Berbers, Algeria's main ethnic community, erupted into violence in May 2001, resulting in 60 deaths in the Berber region of Kabylie. In March 2002 President Bouteflika agreed to grant the Berber language official status alongside Arabic. The state of emergency declared in Feb. 1992 was not lifted until Feb. 2011, following a series of strikes and protests.

In Jan. 2013 Islamist militants seized a gas complex in the Sahara Desert populated by both Algerian and international workers. Algerian special forces stormed the complex after a four-day siege but the operation ended in heavy loss of life. France had used Algerian airspace earlier in the month to launch counter-insurgency operations against Islamist militants active in neighbouring northern Mali.

TERRITORY AND POPULATION

Algeria is bounded in the west by Morocco and Western Sahara, southwest by Mauritania and Mali, southeast by Niger, east by Libya and Tunisia, and north by the Mediterranean Sea. It has an area of 2,381,741 sq. km (919,595 sq. miles) and is the largest country in Africa. Population (census 2008) 34,080,030 (16,847,283 female); density, 14·3 per sq. km. In 2011, 67·1% of the population lived in urban areas.

The UN gives a projected population for 2015 of 37·95m.

2·5m. Algerians live in France.

86% of the population speak Arabic, 14% Berber; French is widely spoken. A law of Dec. 1996 made Arabic the sole official language, but in March 2002 Tamazight, the Berber language, was given official status and also made a national language.

The 2008 census populations of the 48 *wilayat* (provincial councils) were as follows:

Adrar	399,714	Laghouat	455,602
Aïn Defla	766,886	Mascara	784,073
Aïn Témouchent	371,239	Médéa	819,932
Algiers (El Djazaïr)	2,988,145	Mila	766,886
Annaba	609,499	Mostaganem	737,118
Batna	1,119,791	M'Sila	990,591
Béchar	270,061	Naâma	192,891
Béjaia	912,577	Oran (Ouahran)	1,454,078
Biskra	721,356	Ouargla	558,558
Blida	1,002,937	Oum El Bouaghi	621,612
Bordj Bou Arreridj	628,475	Relizane	726,180
Bouira	695,583	Saida	330,641
Boumerdès	802,083	Sétif	1,489,979
Chlef	1,002,088	Sidi-bel-Abbès	604,744
Constantine		Skikda	898,680
(Qacentina)	938,475	Souk Ahras	438,127
Djelfa	1,092,184	Tamanrasset	176,637
El Bayadh	228,624	Tébessa	648,703
El Oued	647,548	Tiaret	846,823
El Tarf	408,414	Tindouf	49,149[1]
Ghardaia	363,598	Tipaza	591,010
Guelma	482,430	Tissemsilt	294,476
Illizi	52,333	Tizi-Ouzou	1,127,607
Jijel	636,948	Tlemcen	949,135
Khenchela	386,683		

[1]Excluding Saharawi refugees in camps.

The capital is Algiers (2008 population estimate, 2·16m.). Other major towns (with 2008 census populations over 200,000): Oran, 759,645; Constantine, 448,374; Annaba, 376,197; Blida, 292,335; Batna, 289,504; Djelfa, 265,833; Sétif, 252,127; Chlef, 224,154; Sidi-bel-Abbès, 210,146; Biskra, 204,661.

SOCIAL STATISTICS

2007 estimates: births, 783,000; deaths, 149,000; marriages, 325,000. Rates (2007 estimates): births, 23·0 per 1,000; deaths, 4·4 per 1,000. Infant mortality in 2010 was 31 per 1,000 live births. Expectation of life (2007), 73·6 years for females and 70·8 years for males. Annual population growth rate, 1998–2008, 1·5%. Fertility rate, 2008, 2·4 births per woman.

CLIMATE

Coastal areas have a warm temperate climate, with most rain in winter, which is mild, while summers are hot and dry. Inland, conditions become more arid beyond the Atlas Mountains. Algiers, Jan. 54°F (12·2°C), July 76°F (24·4°C). Annual rainfall 30" (762 mm). Biskra, Jan. 52°F (11·1°C), July 93°F (33·9°C). Annual rainfall 6" (158 mm). Oran, Jan. 54°F (12·2°C), July 76°F (24·4°C). Annual rainfall 15" (376 mm).

CONSTITUTION AND GOVERNMENT

A referendum was held on 28 Nov. 1996. The electorate was 16,434,527; turnout was 79·6%. The electorate approved by 85·8% of votes cast a new constitution which defines the fundamental components of the Algerian people as Islam, Arab identity and Berber identity. It was signed into law on 7 Dec. 1996. Political parties are permitted, but not if based on a separatist feature such as race, religion, sex, language or region. There is no limit to the number of presidential terms after parliament voted in favour of abolishing the two-term limit in Nov. 2008, allowing the current president, Abdelaziz Bouteflika, to run for a third term. The President appoints the prime minister and cabinet ministers. Parliament is bicameral: a 462-member *National People's Assembly* elected by direct universal suffrage using proportional representation (389 prior to the elections of May 2012), and a 144-member *Council of the Nation*, one-third nominated by the President and two-thirds indirectly elected by the 48 local authorities. The Council of the Nation debates bills passed by the National Assembly which become law if a three-quarters majority is in favour.

In a referendum on 16 Sept. 1999 voters were asked 'Do you agree with the president's approach to restore peace and civilian concord?' Turnout was 85·1% and 98·6% of the votes cast were in favour.

National Anthem

'Qassaman bin nazilat Il-mahiqat' ('We swear by the lightning that destroys'); words by M. Zakaria, tune by Mohamed Fawzi.

GOVERNMENT CHRONOLOGY

(FLN = National Liberation Front; PRS = Revolutionary Socialist Party; RND = National Rally for Democracy; n/p = non-partisan)

Heads of State since 1962.

President of the Provisional Executive
| 1962 | FLN | Abderrahmane Farès |

Chairman of the National Constituent Assembly
| 1962 | FLN | Ferhat Abbas |

President of the Republic
| 1962–65 | FLN | Ahmed Ben Bella |

Chairman of the Revolutionary Council
| 1965–76 | military/FLN | Houari Boumedienne |

Presidents of the Republic
| 1976–78 | FLN | Houari Boumedienne |
| 1979–92 | FLN | Chadli Bendjedid |

Chairman of the Constitutional Council
| 1992 | FLN | Abdelmélik Benhabilès |

High Council of State (HCE) (collective presidency)
1992	military	Gen. Khaled Nezzar
	FLN	Ali Hussain Kafi
	FLN	Ali Haroun
	n/p	El-Tidjani Haddam

Chairman of the HCE
| 1992 | PRS | Mohamed Boudiaf |

High Council of State (HCE) (collective presidency)
1992	n/p	Redha Malek
	military	Gen. Khaled Nezzar
	FLN	Ali Hussain Kafi
	FLN	Ali Haroun
	n/p	El-Tidjani Haddam

Chairman of the HCE
| 1992–94 | FLN | Ali Hussain Kafi |

Presidents of the Republic
| 1994–99 | n/p, RND | Liamine Zéroual |
| 1999– | n/p | Abdelaziz Bouteflika |

Prime Ministers since 1962.
1962–63	FLN	Ahmed Ben Bella
1979–84	FLN	Mohammed Abdelghani
1984–88	FLN	Abdelhamid Brahimi
1988–89	FLN	Kasdi Merbah
1989–91	FLN	Mouloud Hamrouche
1991–92	FLN	Sid Ahmed Ghozali
1992–93	FLN	Belaid Abdessalam
1993–94	n/p	Redha Malek
1994–95	n/p	Mokdad Sifi
1995–98	n/p, RND	Ahmed Ouyahia
1998–99	n/p	Smail Hamdani
1999–2000	n/p	Ahmed Benbitour
2000–03	FLN	Ali Benflis
2003–06	RND	Ahmed Ouyahia
2006–08	FLN	Abdelaziz Belkhadem
2008–12	RND	Ahmed Ouyahia
2012–	n/p	Abdelmalek Sellal

RECENT ELECTIONS

In presidential elections on 9 April 2009 Abdelaziz Bouteflika won a third term of office, gaining 90·2% of the votes cast; Louisa Hanoune (Parti du Travail/Workers' Party) received 4·5% of votes cast; Moussa Touati (Front National Algérien/Algerian National Front), 2·0%; Djahid Younsi, 1·5%; Ali Fawzi Rebaine, 0·9%; and Mohammed Said, 0·9%. Turnout was 74·6%.

Parliamentary elections were held on 10 May 2012. The Front de Libération Nationale/National Liberation Front won 208 out of 462 seats; Rassemblement National Démocratique/National Rally for Democracy took 68 seats; the Green Algeria Alliance, 49; Front des Forces Socialistes/Front of Socialist Forces, 27; Parti des Travailleurs/Workers' Party, 24. Independents took 18 seats and the remainder went to minor parties. Turnout was 43·1%.

CURRENT GOVERNMENT

President and Minister of National Defence: Abdelaziz Bouteflika; b. 1937 (ind.; sworn in 27 April 1999; re-elected 8 April 2004 and 9 April 2009). In March 2013 the government comprised:

Prime Minister: Abdelmalek Sellal; b. 1948 (ind.; sworn in 3 Sept. 2012).

Minister of Agriculture and Rural Development: Rachid Benaïssa. *Commerce:* Mustapha Benbada. *Communication:* Mohand Oussaïd Belaid. *Country, Environmental and City Planning:* Amara Benyounès. *Culture:* Khalida Toumi. *Energy and Mines:* Youcef Yousfi. *Finance:* Karim Djoudi. *Fisheries and Marine Resources:* Sid Ahmed Ferroukhi. *Foreign Affairs:* Mourad Medelci. *Health, Population and Hospital Reform:* Abdelaziz Ziari. *Higher Education and Scientific Research:* Rachid Harraoubia. *Housing and Urban Planning:* Abdelmadjid Tebboune. *Industry, Small and Medium-Sized Businesses, and Promotion of Investments:* Chérif Rahmani.

Interior and Local Communities: Daho Ould Kablia. *Justice and Keeper of the Seals:* Mohamed Charfi. *Labour, Employment and Social Security:* Tayeb Louh. *Moudjahidine (War Veterans):* Mohamed Chérif Abbes. *National Education:* Abdelatif Baba Ahmed. *National Solidarity and Families:* Souad Bendjaballah. *Postal Services, Information and Communication Technologies:* Moussa Benhamadi. *Public Works:* Amar Ghoul. *Relations with Parliament:* Mahmoud Khedri. *Religious Affairs:* Bouabdellah Ghlamallah. *Tourism and Handicrafts:* Mohamed Benmeradi. *Transport:* Amar Tou. *Vocational and Educational Training:* Mohamed Mebarki. *Water Resources:* Hocine Necib. *Youth and Sports:* Mohamed Tahmi.

President's Website (Arabic and French only):
http://www.elmouradia.dz

CURRENT LEADERS

Abdelaziz Bouteflika

Position
President

Introduction
Abdelaziz Bouteflika became president in April 1999 following disputed elections, vowing to improve Algeria's weak economy and end civil discord. Improvements in the economy and state reform have since been slow. However, the significant reduction in Islamist rebel violence following an amnesty in 1999 was a key factor in Bouteflika's re-election in 2004, making him the first Algerian leader to be returned to power in a democratic vote since the country's independence. His charter for peace and national reconciliation was approved in a national referendum in Sept. 2005. He won a further term in the election of April 2009, having secured earlier parliamentary approval in Nov. 2008 of constitutional changes allowing him to run for a third consecutive term of office. In the wake of social unrest and anti-government protests in early 2011, Bouteflika lifted the longstanding state of emergency in Feb. and in April set up a committee to consider constitutional and democratic reform.

Early Life
Bouteflika was born on 2 March 1937 in Morocco. In 1956 he joined the Armée de Libération Nationale—a wing of the Front de Libération Nationale (FLN; National Liberation Front). Stationed in southern Algeria in 1960, he was involved in secret talks with the French authorities, which eventually led to Algerian independence two years later.

Bouteflika joined the government of Ahmed Ben Bella as minister of youth, sport and tourism and was appointed foreign minister in 1963. He retained the position in the government of Houari Boumedienne. Having failed to secure military support to succeed Boumedienne, he was pushed out of the political mainstream. In 1981 he was charged with corruption and forced into exile for seven years. With the charges dropped, he re-entered Algeria in Jan. 1987.

In Oct. 1988 he protested against human rights violations by government troops against young demonstrators. He rejoined the FLN congress the following year and was elected to the central committee. In Dec. 1998 he announced his intention to contest the presidency. At the elections, he won almost three quarters of the vote after his six opponents all stood down from the race the day before polling, accusing him of vote rigging.

Career in Office
Having become president and commanding army support, Bouteflika declared that his primary aim was to end Algeria's many years of civil unrest. In July 1999 parliament passed the National Harmony Law, which offered an amnesty to all rebels who had not been directly responsible for loss of life. A national referendum in Sept. 1999 approved the amnesty scheme, with nearly 99% in favour. Violence by some Islamist groups has since continued but at a greatly reduced rate. In foreign policy, he has striven to improve relations with neighbouring Morocco, while on the economic front he has encouraged the further development of the country's large oil and gas reserves.

Bouteflika's relations with his prime ministers have been turbulent. Ahmed Benbitour, prime minister from Dec. 1999 to Aug. 2000, resigned over divergent attitudes on economic recovery. His successor, Ali Benflis, was a personal friend and the leader of the largest party, the FLN. However, Benflis' reformist agenda was too radical for Bouteflika, who feared violent insurrection. Since Benflis' resignation in May 2003, the two have been at political loggerheads. Bouteflika appointed Ahmed Ouyahia to succeed Benflis.

Benflis refused to support the president's bid for re-election in 2004, instead announcing his own candidacy, supported by the FLN, in Oct. 2003. However, Bouteflika's bid was supported by a coalition of the Rassemblement National Démocratique (RND; National Rally for Democracy), the Islamic Mouvement de la Société pour la Paix (MSP; Movement of the Society for Peace) and also renegade members of the FLN. The elections on 8 April 2004 gave Bouteflika a resounding victory with 85% of the vote in a 58% turnout—against just 6% for Benflis. International observers declared the elections free and fair, despite opposition claims of electoral fraud. His dramatic win was ascribed to the much improved security situation and a steadily growing economy, even though unemployment remained around 30%.

Bouteflika pledged to investigate the disappearance of around 7,000 Algerians, allegedly killed or imprisoned by the security forces during the 1990s. Improving relations with the Berber community was also a priority (election disturbances in 2004 having been mainly limited to the Berber Kabylie region), and he promised reform of Algeria's family law, which he described as unfair to women. In Sept. 2005 his charter for peace and national reconciliation to end 13 years of civil war, envisaging a limited amnesty and compensation for some victims of violence, was approved in a national referendum. A six-month amnesty from March–Aug. 2006 resulted in the release of some 2,200 Islamist militants (except those accused of the most serious crimes) from prison. However, incidents of terrorism increased again, particularly in Aug. 2008 when almost 60 people were killed in a series of bombings in towns to the east of Algiers.

In May 2006 Bouteflika appointed Abdelaziz Belkhadem as prime minister in place of Ahmed Ouyahia, but reinstated the latter in June 2008. In Nov. Algeria's parliament overwhelmingly approved constitutional amendments (by 500 votes to 12, with eight abstentions) that abolished presidential term limits, paving the way for Bouteflika to retain the presidency in the election held on 9 April 2009. However, discontent over housing shortages, unemployment and low wages was reflected in strikes and demonstrations in 2009–10. Further violent disorder in early 2011, echoing widespread anti-government protests across much of the Arab world, led Bouteflika to promise democratic constitutional amendments. In the May 2012 parliamentary elections the ruling coalition headed by the FLN won a majority of seats in the National Assembly, and in Sept. Bouteflika named Abdelmalek Sellal as the new prime minister.

In Jan. 2013, in the wake of conflict in neighbouring Mali, about 30 militant Islamists attacked an isolated gas installation in southeast Algeria and took the foreign workers hostage. The Algerian government responded by ordering special forces to storm the plant, which they retook after four days. Most of the Islamists were killed, but at least 37 foreign workers also died.

Abdelmalek Sellal

Position
Prime Minister

Introduction
Abdelmalek Sellal became prime minister in Sept. 2012. He was appointed by President Abdelaziz Bouteflika following the resignation of Ahmed Ouyahia.

Early Life
Sellal was born in Aug. 1948 in Constantine, Algeria. He finished his secondary education at the National College of Administration and became an adviser to the prefect of Guelma province. He then took a post at the ministry of education and subsequently served as prefect of several provinces including Tamanrasset, Adrar, Laghouat, Boumerdès, Sidi-Bel-Abbès and Oran.

Sellal was Algeria's ambassador to Hungary when, in 1998, he was named minister of the interior by President Abdelaziz Bouteflika. In 1999 he was moved to the ministry of youth and sports, then to the ministry of public works in 2001, the ministry of transport in 2002 and the ministry of water in 2004 where he stayed until 2012.

Sellal guided Bouteflika's presidential re-election campaigns in 2004 and 2009. Following the resignation of Prime Minister Ahmed Ouyahia in 2012 after four years in office, Bouteflika appointed Sellal to the premiership.

Career in Office
Seen as a technocrat, Sellal does not belong to any political party despite his close allegiance to Bouteflika. He maintained much of his predecessor's cabinet and pledged to continue the reform programme laid out by Bouteflika. Sellal's challenges are to revitalize the economy, reduce inflation and lessen dependency on hydrocarbons. He has called for reforms to the financial sector and is expected to streamline banking. However, he has to cope with large-scale corruption and social unrest.

DEFENCE

Conscription is for 18 months (six months basic training and 12 months civilian tasks).

Military expenditure totalled an estimated US$5,172m. in 2008 (up by 18% in real terms on 2007), representing 20% of Africa's total military spending. In 2008 defence spending amounted to US$153 per capita (equivalent to 3·0% of GDP).

Army
There are six military regions. The Army had a strength of 127,000 (approximately 80,000 conscripts) in 2007. The Directorate of National Security maintains National Security Forces of 16,000. The Republican Guard numbers 1,200 personnel and the Gendarmerie 20,000. There were in addition legitimate defence groups (self-defence militia and communal guards) numbering around 150,000.

Navy
Naval personnel in 2007 totalled about 6,000. The Navy's 44 vessels included two submarines and three frigates. There are naval bases at Algiers, Annaba, Jijel and Mers el Kebir.

Air Force
The Air Force in 2007 had some 14,000 personnel; equipment included 141 combat-capable aircraft and 33 attack helicopters.

ECONOMY

In 2009 petroleum and natural gas (excluding refined petroleum) accounted for 31·0% of GDP; transport, communications, trade, restaurants, finance, real estate and services, 25·1%; public administration and defence, 10·9%; public utilities and construction, 10·9%.

Overview
The economy is heavily dependent on public spending and the sale of oil and gas. From 1994 an IMF-sponsored programme for economic reconstruction reduced inflation and created a free-market economy. The downside was large-scale unemployment. Privatization got off the ground in 1994, as demanded by the IMF, but failed to make significant strides until 2003.

The economy has achieved growth in every year since 1995, with rising oil prices prompting more robust growth from 2003. The global economic crisis reduced hydrocarbon exports in 2009 but Algeria weathered the crisis well. A primary challenge is to diversify the economy, as hydrocarbon revenues represent 98% of exports.

While the hydrocarbon sector continues to dominate, services and construction are making a stronger contribution. Non-hydrocarbon growth averaged about 6% over the decade from the early 2000s. Unemployment is considerably lower than its peak of 27% at the beginning of the new century, registering 10·2% at the end of 2009 as a result of government initiatives and robust growth. Nonetheless, the figure remains acute among the young.

Currency
The unit of currency is the *Algerian dinar* (DZD) of 100 *centimes*. Foreign exchange reserves were US$146,130m. in Sept. 2009, with gold reserves 5·58m. troy oz. Total money supply was 4,071·5bn. dinars in June 2009. Inflation rates (based on IMF statistics):

2002	2003	2004	2005	2006	2007	2008	2009	2010	2011
1·4%	2·6%	3·6%	1·6%	2·3%	3·6%	4·9%	5·7%	3·9%	4·5%

The dinar was devalued by 40% in April 1994.

Budget
The fiscal year starts on 1 Jan. In 2009 budgetary central government revenue totalled 3,740,500m. dinars and expenditure 2,556,900m. dinars. Principal sources of revenue in 2009 were: taxes on income, profits and capital gains, 2,231,700m. dinars; taxes on goods and services, 1,048,900m. dinars. Main items of expenditure by economic type in 2009: compensation of employees, 860,500m. dinars; social benefits, 555,600m. dinars.

VAT is 17%.

Performance
Real GDP growth rates (based on IMF statistics):

2002	2003	2004	2005	2006	2007	2008	2009	2010	2011
4·7%	6·9%	5·2%	5·1%	2·0%	3·0%	2·4%	2·4%	3·3%	2·4%

Total GDP was US$188·7bn. in 2011.

Banking and Finance
The central bank and bank of issue is the Banque d'Algérie. The *Governor* is Mohammed Laksaci. In 2002 it had total reserves of US$23·5bn. Private banking recommenced in Sept. 1995. In 2002 there were five state-owned commercial banks, four development banks, nine private banks and two foreign banks.

Foreign debt fell from US$25,388m. in 2000 to US$16,871m. in 2005 and further to US$5,276m. in 2010 (representing just 3·4% of GNI).

There is a stock exchange in Algiers.

ENERGY AND NATURAL RESOURCES

Environment
Algeria's carbon dioxide emissions from the consumption and flaring of fossil fuels in 2008 were the equivalent of 3·1 tonnes per capita.

Electricity

Installed capacity was 8·1m. kW in 2007 (3·0% is hydro-electric). Production in 2007 was 37·20bn. kWh, with consumption per capita 1,091 kWh.

Oil and Gas

A law of Nov. 1991 permits foreign companies to acquire up to 49% of known oil and gas reserves. Oil and gas production accounted for 48·0% of GDP in 2007. Oil production in 2008 was 85·6m. tonnes; oil reserves (2008) totalled 12·2bn. bbls. Production of natural gas in 2008 was 86·5bn. cu. metres (the sixth highest in the world); proven reserves in 2008 were 4,500bn. cu. metres.

Minerals

Output in 2004 unless otherwise indicated (in 1,000 tonnes): iron ore, 1,554; gypsum, 1,058; phosphate rock, 1,017; salt, 183; lead (2002), 1·1. There are also deposits of mercury, silver, gold, copper, antimony, kaolin, marble, onyx, salt and coal.

Agriculture

Much of the land is unsuitable for agriculture. The northern mountains provide grazing. There were 7·47m. ha. of arable land in 2007 and 0·92m. ha. of permanent crops. 0·91m. ha. were irrigated in 2007. In 1987 the government sold back to the private sector land which had been nationalized on the declaration of independence in 1962; a further 0·5m. ha., expropriated in 1973, were returned to some 30,000 small landowners in 1990. In 2007 the agricultural population was an estimated 7·41m. There were 135 tractors and 11 harvester-threshers per 10,000 ha. in 2006.

The chief crops in 2003 were (in 1,000 tonnes): wheat, 2,970; potatoes, 1,300; barley, 1,220; tomatoes, 820; melons and watermelons, 465; onions, 450; dates, 420; oranges, 360; olives, 300; grapes, 230; chillies and green peppers, 165; carrots, 158.

Livestock, 2003: sheep, 17·3m.; goats, 3·2m.; cattle, 1·5m.; camels, 245,000; asses, 170,000; horses, 44,000; mules, 43,000; chickens, 115m. Livestock products, 2003 (in 1,000 tonnes): poultry meat, 244; lamb and mutton, 165; beef and veal, 121; eggs, 110; cow's milk, 1,160; sheep's milk, 200; goat's milk, 155.

Forestry

Forests covered 1·49m. ha. in 2010, or 1% of the total land area. The greater part of the state forests are brushwood, but there are large areas with cork-oak trees, Aleppo pine, evergreen oak and cedar. The dwarf-palm is grown on the plains, alfalfa on the tableland. Timber is cut for firewood and industrial purposes, and bark for tanning. Timber production in 2007 was 7·97m. cu. metres.

Fisheries

The total catch in 2010 amounted to 93,607 tonnes, exclusively from marine waters.

INDUSTRY

Output (2007 unless otherwise specified, in 1,000 tonnes): cement, 15,886; distillate fuel oil, 6,388; residual fuel oil, 5,518; petrol, 2,100; clay bricks, 1,384; wheat and blended flour (2005), 1,377; crude steel, 1,278; pig iron, 1,193; jet fuel, 1,034; refined sugar (2004), 135; iron or steel tubes and pipes (2004), 122; woven cotton fabrics (2004), 14·8m. metres; woven woollen fabrics (2004), 1·2m. metres; beer (2004), 12·4m. litres. Production in units (2004): refrigerators and freezers, 215,000; TV sets, 205,800; tractors, 3,092; lorries, 2,698.

Labour

In 2005 there were 6,889,000 employed persons. The main areas of activity were: community, social and personal services, 1,826,000; agriculture, forestry and fishing, 1,440,000; transport, storage and communications, 1,268,000; construction, 902,000; manufacturing, 772,000. In 2006 unemployment was 17·3%.

INTERNATIONAL TRADE

Imports and Exports

Trade in US$1m.:

	Imports (c.i.f.)	Exports (f.o.b.)
2006	21,455·9	54,612·7
2007	27,631·2	60,163·2
2008	39,474·7	79,297·6
2009	39,258·3	45,193·9
2010	40,999·9	57,051·0

Principal import suppliers in 2006 were: France, 20·3%; Italy, 8·8%; China, 8·0%; Germany, 6·9%; USA, 6·6%; Spain, 4·8%. Principal export markets in 2006 were: USA, 27·2%; Italy, 17·1%; Spain, 11·0%; France, 8·4%; Canada, 6·6%; Netherlands, 5·2%.

Leading imports in 2006 were: machinery and transport equipment, 37·5%; manufactured goods, 22·6%; food and live animals, 16·9%. Leading exports in 2006 were: petroleum and petroleum products, 63·0%; natural and manufactured gas, 35·1%.

COMMUNICATIONS

Roads

There were, in 2008, 111,261 km of roads including 29,146 km of highways and main roads. There were 2,042,800 passenger cars (58 cars per 1,000 inhabitants in 2005) and 1,166,200 lorries and vans in use in 2006.

Rail

In 2005 there were 2,889 km of 1,435 mm route (283 km electrified) and 1,085 km of 1,055 mm gauge. The railways carried 6·9m. tonnes of freight and 24·7m. passengers in 2008.

A metro system opened in Algiers in Nov. 2011, 28 years after work on the project first began.

Civil Aviation

The main airport is Algiers International, which opened a new terminal in July 2006 to allow for more international air traffic; some international services also use Annaba, Constantine and Oran. The national carrier is the state-owned Air Algérie, which in 2010 operated direct flights to Abidjan, Amman, Bamako, Barcelona, Beijing, Beirut, Bordeaux, Brussels, Cairo, Casablanca, Dakar, Damascus, Dubai, Frankfurt, Geneva, İstanbul, Jeddah, Lille, London, Lyon, Madrid, Marseille, Metz/Nancy, Milan, Montreal, Moscow, Niamey, Nice, Nouakchott, Ouagadougou, Paris, Rome, Toulouse, Tripoli and Tunis, as well as domestic services. There were direct international flights in 2010 with other airlines to Alicante, Barcelona, Basel/Mulhouse, Cairo, Casablanca, Damascus, Doha, Frankfurt, İstanbul, Jeddah, Lille, London, Lyon, Madrid, Marseille, Palma de Mallorca, Paris, Rome, Toulouse, Tripoli and Tunis. In 2001 Houari Boumedienne International Airport handled 3,397,867 passengers (1,871,052 on domestic flights) and 16,191 tonnes of freight. In 2005 scheduled airline traffic of Algerian-based carriers flew 30·4m. km, carrying 2,760,700 passengers.

Shipping

In Jan. 2009 there were 39 ships of 300 GT or over registered, totalling 723,000 GT. Skikda, the leading port, handled 23,203,000 tonnes of cargo in 2008.

Telecommunications

In 2008 there were 3,314,000 main (fixed) telephone lines. In the same year mobile phone subscribers numbered 31,871,000 (927·2 per 1,000 persons). There were 4·1m. internet users in 2008. In June 2012 there were 3·6m. Facebook users.

Government plans to privatize Algérie Télécom, the major state-owned telecommunications company, were rejected in Feb. 2009.

SOCIAL INSTITUTIONS

Justice

The judiciary is constitutionally independent. Judges are appointed by the Supreme Council of Magistrature chaired by the President of the Republic. Criminal justice is organized as in France. The Supreme Court is at the same time Council of State and High Court of Appeal. The death penalty is in force for terrorism.

The population in penal institutions in 2010 was 58,000 (164 per 100,000 of national population).

Education

Adult literacy in 2006 was 72·6% (81·3% among males and 63·9% among females). In 2007 there were 171,000 children in pre-primary education. There were 4,079,000 pupils in primary schools in 2007, and 4,585,000 pupils in secondary schools in 2009.

The leading institute of higher education is the University of Algiers, founded in 1909 although Algerians were not admitted until 1946. In 2007 there were 901,562 students in tertiary education and 31,683 academic staff.

In 2008 expenditure on education came to 4·3% of GDP and represented 20·3% of total government expenditure.

Health

In 2004 there were 37,720 physicians, 8,842 dentists and 6,082 pharmacists. There were 17 hospital beds per 10,000 population in 2004. Health spending accounted for 4·4% of GDP in 2007.

Welfare

Welfare payments to 7·4m. beneficiaries on low incomes were introduced in 1992.

RELIGION

The 1996 constitution made Islam the state religion, established a consultative *High Islamic Council*, and forbids practices 'contrary to Islamic morality'. Over 99% of the population are Sunni Muslims. There are also around 180,000 Ibadiyah Muslims and 90,000 others. The Armed Islamic Group (GIA) vowed in 1994 to kill 'Jews, Christians and polytheists' in Algeria. Hundreds of foreign nationals, including priests and nuns, have since been killed. Signalling an increasing tolerance amongst the Muslim community, the Missionaries of Africa's house at Ghardaia Oasis was reopened in 2000.

CULTURE

World Heritage Sites

There are seven UNESCO sites in Algeria: Al Qal'a of Beni Hammad (inscribed in 1980), the 11th century ruined capital of the Hammadid emirs; Tassili n'Ajjer (1982), a group of over 15,000 prehistoric cave drawings; the M'zab Valley (1982), a tenth century community settlement of the Ibadites; Djémila (1982), or Cuicul, a mountainous Roman town; Tipasa (1982), an ancient Carthaginian port; Timgad (1982), a military colony founded by the Roman emperor Trajan in AD 100; the Kasbah of Algiers (1992), the medina of Algiers.

Press

Algeria had 52 paid-for daily newspapers in 2007, with a combined average daily circulation of 1·5m.

Tourism

In 2009 there were a record 2,070,000 foreign visitors, up from 1,912,000 in 2009 and 1,772,000 in 2008.

DIPLOMATIC REPRESENTATIVES

Of Algeria in the United Kingdom (54 Holland Park, London, W11 3RS)
Ambassador: Amar Abba.

Of the United Kingdom in Algeria (3 Chemin Capitaine Hocine Slimane [ex Chemin des Glycines], Hydra, Algiers)
Ambassador: Martyn Roper.

Of Algeria in the USA (2118 Kalorama Rd, NW, Washington, D.C., 20008)
Ambassador: Abdallah Baali.

Of the USA in Algeria (5 Chemin Cheich Bachir Ibrahimi, Algiers)
Ambassador: Henry S. Ensher.

Of Algeria to the United Nations
Ambassador: Mourad Benmehidi.

Of Algeria to the European Union
Ambassador: Amar Bendjama.

FURTHER READING

Ageron, C.-R., *Modern Algeria: a History from 1830 to the Present.* 1991
Eveno, P., *L'Algérie.* 1994
Heggoy, A. A. and Crout, R. R., *Historical Dictionary of Algeria.* 1995
Roberts, Hugh, *The Battlefield: Algeria 1998–2002, Studies in a Broken Polity.* 2003
Ruedy, J., *Modern Algeria: the Origins and Development of a Nation.* 1992
Stora, B., *Histoire de l'Algérie depuis l'Indépendance.* 1994
Volpi, Frédéric, *Islam and Democracy: The Failure of Dialogue in Algeria, 1998–2001.* 2003
Willis, M., *The Islamist Challenge in Algeria: A Political History.* 1997

National Statistical Office: Office National des Statistiques, 8–10 rue des Moussebilines, Algiers.
Website (French only): http://www.ons.dz

ANDORRA

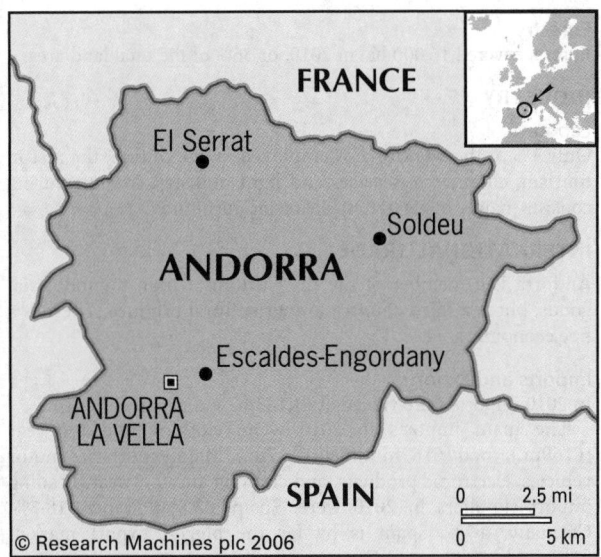

Principat d'Andorra
(Principality of Andorra)

Capital: Andorra la Vella
Population, 2010: 85,000
GNI per capita, 2010: US$41,138
HDI/world rank: 0·838/32
Internet domain extension: .ad

KEY HISTORICAL EVENTS

The Andosini, a tribe subdued by Hannibal in 218 BC, are the first recorded inhabitants of the Pyrenean state of Andorra. In the 9th century the Holy Roman Emperor, Charles II, reputedly made the bishop of Seo de Urgel overlord of Andorra. The *Paréage* of 1278 placed Andorra under the joint suzerainty of the bishop of Seo de Urgel and the Comte de Foix. The rights vested in the house of Foix passed by marriage to that of Bearn and, on the accession of Henri IV, to the French crown.

In the 19th century the *Consell General* (parliament) was strengthened, but the constitution remained traditional and unwritten until 8 Sept. 1993, when political parties and labour unions were legalized and Andorra joined the UN.

TERRITORY AND POPULATION

The co-principality of Andorra is situated in the eastern Pyrenees on the French–Spanish border. The country is mountainous and has an average altitude of 1,996 metres. Area, 464 sq. km. In lieu of a census, a register of population is kept. The estimated population at 31 Dec. 2010 was 85,015; density, 183 per sq. km.

In 2007, 89% of the population lived in urban areas.

The chief towns are Andorra la Vella, the capital (estimated population, 20,436 in 2009) and Escaldes-Engordany (16,861); other towns are Encamp, Sant Julià de Lòria and La Massana. In 2010, 38·8% of the residential population were Andorran, 31·4% Spanish, 15·4% Portuguese and 6·0% French.

Catalan is the official language, but Spanish and French are widely spoken.

SOCIAL STATISTICS

Births in 2006 numbered 843 (10·4 per 1,000 inhabitants) and deaths 260 (3·2). Life expectancy (2006): males, 78 years; females, 85 years. Annual population growth rate, 2000–05, 3·5%. Fertility rate, 2008, 1·3 births per woman (one of the lowest rates in the world). Infant mortality in 2010 was three per 1,000 live births.

CLIMATE

Escaldes-Engordany, Jan. 35·8°F (2·1°C), July 65·8°F (18·8°C). Annual rainfall 34·9" (886 mm).

CONSTITUTION AND GOVERNMENT

The joint heads of state are the co-princes—the President of the French Republic and the Bishop of Urgel.

A new democratic constitution was approved by 74·2% of votes cast at a referendum on 14 March 1993. The electorate was 9,123; turnout was 75·7%. The new constitution, which came into force on 4 May 1993, makes the co-princes a single constitutional monarch and provides for a parliament, the unicameral *General Council of the Valleys*, with 28 members, two from each of the seven parishes and 14 elected by proportional representation from the single national constituency, for four years. In 1982 an *Executive Council* was appointed and legislative and executive powers were separated. The General Council elects the President of the Executive Council, who is the head of the government.

There is a *Constitutional Court* of four members who hold office for eight-year terms, renewable once.

National Anthem

'El Gran Carlemany, mon pare' ('Great Charlemagne, my father'); words by D. J. Benlloch i Vivò, tune by Enric Marfany Bons.

RECENT ELECTIONS

Elections to the General Council were held on 3 April 2011. The Democrats for Andorra (DA) won 20 seats (55·1% of the vote) and the Social Democratic Party (PS) 6 (34·8%). The Lauredian Union won 2 seats. Although Andorra for Change gained 6·7% of the vote and Greens of Andorra 3·4% neither party won any seats. Turnout was 74·1%. As at 31 Oct. 2012, of the 28 Members of Parliament there were 14 men and 14 women (50%), the second highest percentage of women in a parliament after that of Rwanda (where 24 of 80 seats are set aside for women).

CURRENT GOVERNMENT

In March 2013 the government comprised:

President, Executive Council: Antoni Martí; b. 1963 (Democrats for Andorra; sworn in 12 May 2011).

Minister for Finance and Public Service: Jordi Cinca. *Economy and Land Management:* Jordi Alcobé. *Foreign Affairs:* Gilbert Saboya. *Justice and Interior:* Marc Vila. *Health and Wellbeing:* Cristina Rodríguez. *Education, Youth and Sports:* Roser Suñé. *Tourism and Environment:* Francesc Camp. *Culture:* Albert Esteve.

Government Website (Catalan only): http://www.govern.ad

CURRENT LEADERS

Antoni Martí Petit

Position
President of the Executive Council

Introduction
Antoni Martí Petit became the president of the executive council (*de facto* prime minister) after his party, the Democrats for

Andorra, gained an absolute majority in parliamentary elections held on 3 April 2011. An architect and the former mayor of Escaldes-Engordany, Andorra's second city, Martí has pledged to open up the state to foreign investment.

Early Life
Born in Escaldes-Engordany in 1963, Martí studied architecture at the University of Toulouse. He was elected to parliament in 1994. In 2003 he became mayor of Escaldes-Engordany, winning re-election to the post four years later. Towards the end of his second term he was elected to lead the recently-formed, centre-right Democrats for Andorra.

Early parliamentary elections were called for April 2011 after the dissolution of Andorra's General Council following deadlock over budget legislation. Martí's party defeated the incumbent Social Democrats.

Career in Office
Martí was elected head of government by parliament on 11 May and sworn in the following day. He campaigned on opposition to the introduction of income tax. He also promised to end the protectionism that has previously discouraged inward investment.

INTERNATIONAL RELATIONS

The 1993 constitution empowers Andorra to conduct its own foreign affairs, with consultation on matters affecting France or Spain.

ECONOMY

Overview
The mainstays of the economy are trade and tourism (accounting for 60% of GDP) and finance (16%). In 2007 there were 11m. visitors, attracted by Andorra's duty-free status and its thriving summer and winter resorts. However, increased openness in the economies of neighbouring Spain and France has eroded Andorra's advantage.

As an offshore financial centre, the banking sector plays a strong role in the economy. With only 2% of the country classified as arable land, most food is imported. While not a member of the EU, Andorra is a member of the EU Customs Union and is treated as an EU member for trade in manufactured goods (no tariff) but a non-EU member for trade in agricultural products.

The OECD's Committee on Fiscal Affairs removed Andorra from its List of Uncooperative Tax Havens in May 2009, following commitments to implement the OECD standards of transparency and exchange of information.

Currency
Since 1 Jan. 2002 Andorra has been using the euro (EUR). Inflation was 1·6% in 2010 (0·0% in 2009 and 2·0% in 2008).

Budget
In 2005 central government revenue was €516,000,000 and expenditure €484,900,000.

VAT at 4·5% was introduced on 1 Jan. 2013.

Performance
The economy shrank by 4·3% in 2008 and 2·9% in 2009. Total GDP was US$3·7bn. in 2008.

Banking and Finance
The banking sector, with its tax haven status, contributes substantially to the economy. Leading banks include: Andbane-Grup Agricol Reig; Banc Internacional d'Andorra SA; Banca Mora SA; Banca Privada d'Andorra SA; CaixaBank SA; and Crèdit Andorrà.

ENERGY AND NATURAL RESOURCES

Electricity
Installed capacity was 32,000 kW in 2007. Production in 2007 was 76m. kWh. 60% of Andorra's electricity comes from Spain.

Agriculture
In 2001 there were some 1,000 ha. of arable land (2% of total) and 1,000 ha. of permanent crops. Tobacco and potatoes are principal crops. The principal livestock activity is sheep raising.

Forestry
Forests covered 16,000 ha. in 2010, or 36% of the total land area.

INDUSTRY

Labour
Only 1% of the workforce is employed in agriculture, the rest in tourism, commerce, services and light industry. Manufacturing consists mainly of cigarettes, cigars and furniture.

INTERNATIONAL TRADE

Andorra is a member of the EU Customs Union for industrial goods, but is a third country for agricultural produce. There is a free economic zone.

Imports and Exports
In 2010 imports were valued at €1,143m. and exports at €41m.

The main imports in 2010 were: clothing and footwear (€169m.); food (€167m.); fuel (€117m.). Main exports are motor vehicles, electronic products, and clothing and footwear. Leading import suppliers in 2010 were: Spain, 63·8%; France, 18·7%; Germany, 4·7%. Spain is by far the biggest export market, followed by France and Germany.

COMMUNICATIONS

Roads
In 1999 there were 269 km of roads (198 km paved). There were 76,316 motor vehicles in 2009.

Civil Aviation
The nearest airport is Seo de Urgel, over the border in Spain 12 km to the south of Andorra.

Telecommunications
In 2010 there were 38,171 landline telephone subscriptions, equivalent to 449·8 per 1,000 inhabitants. There were 65,500 mobile phone subscriptions in 2010 (771·8 per 1,000 inhabitants). Andorra had 785·3 internet users per 1,000 inhabitants in 2009. Fixed internet subscriptions totalled 32,000 in 2009 (382·6 per 1,000 inhabitants).

SOCIAL INSTITUTIONS

Justice
Justice is administered by the High Council of Justice, comprising five members appointed for single six-year terms. The independence of judges is constitutionally guaranteed. Judicial power is exercised in civil matters in the first instance by Magistrates' Courts and a Judge's Court. Criminal justice is administered by the *Corts*, consisting of the judge of appeal, a general attorney and an attorney nominated for five years alternately by each of the co-princes. There is also a *raonador* (ombudsman) elected by the General Council of the Valleys.

Education
Free education in French- or Spanish-language schools is compulsory: six years primary starting at six years, followed by four years secondary. A Roman Catholic school provides education in Catalan. In 2007 there were 4,427 pupils in primary schools and 3,819 in secondary schools.

Health
In 2010 there was one public hospital; in 2009 there were 41 general practitioners, 220 medical specialists, 51 dentists and stomatologists, 309 nurses and 78 pharmacists.

RELIGION

The Roman Catholic is the established church, but the 1993 constitution guarantees religious liberty. In 2001 around 88% of the population were Catholics.

CULTURE

World Heritage Sites
There is one UNESCO site in Andorra: Madriu-Perafita-Claror Valley (entered on the list in 2004 and 2006).

Press
In 2006 there were three daily newspapers with a combined circulation of about 32,000. *Diari d'Andorra* and *El Periodic d'Andorra* are paid-for, while *Bondia* is free. *El 9 Esportiu*, a daily Catalan-language sports paper, is included in *Diari d'Andorra*.

Tourism
Tourism is the main industry, accounting for 80% of GDP. In 2010 there were 1,808,000 international tourist arrivals (excluding same-day visitors).

Festivals
In March the Big Snow festival in the ski resort of Arinsal combines music and winter sports. A jazz festival, the International Colours of Music Festival, is held in Escaldes-Engordany every July.

DIPLOMATIC REPRESENTATIVES

Of Andorra in the United Kingdom (63 Westover Rd, London, SW18 2RF)
Ambassador: Vacant.
Chargé d'Affaires a.i.: Eva Descàrrega Garcia.

Of the United Kingdom in Andorra
Ambassador: Giles Paxman, LVO (resides in Madrid).

Of Andorra in the USA (2 United Nations Plaza, 25th Floor, N.Y. 10017)
Ambassador: Narcís Casal de Fonsdeviela.

Of USA in Andorra
Ambassador: Alan D. Solomont (resides in Madrid).

Of Andorra to the United Nations
Ambassador: Narcís Casal de Fonsdeviela.

Of Andorra to the European Union
Ambassador: Eva Descàrrega Garcia.

FURTHER READING

A Strategic Assessment of Andorra. 2000

National Statistical Office: Servie d'Estudis, Ministeri de Finances, c/Doctor Vilanova, núm. 13, Edifici Davi, Esc. c, 5è, Andorra la Vella.
Website: http://www.estadistica.ad

ANGOLA

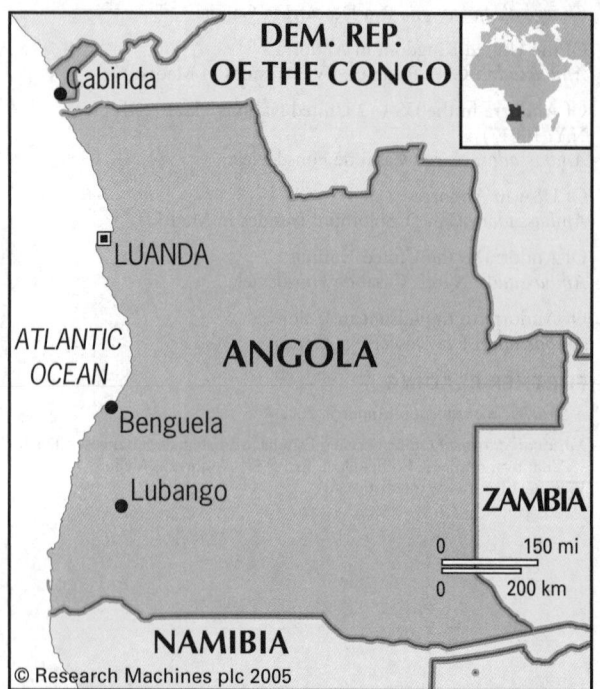

DEM. REP. OF THE CONGO

Cabinda

LUANDA

ATLANTIC OCEAN

ANGOLA

Benguela

Lubango

ZAMBIA

0   150 mi

0   200 km

NAMIBIA

© Research Machines plc 2005

República de Angola
(Republic of Angola)

Capital: Luanda
Population projection, 2015: 21·84m.
GNI per capita, 2011: (PPP$) 4,874
HDI/world rank: 0·486/148
Internet domain extension: .ao

KEY HISTORICAL EVENTS

Khoisan hunter-gatherers were in southern Angola at least 23,000 years ago. From AD 1000 Bantu-speaking groups moved south from the borders of present-day Cameroon and Nigeria. Substantial numbers of settlers arrived in present-day Angola during the 13th century, displacing Khoisan groups. Several powerful Bantu kingdoms developed, including the Bakongo, across northern Angola, parts of present-day Democratic Republic of the Congo, Republic of the Congo and Gabon, and the Ndongo kingdom further south. The name Angola is derived from Ngola a Kiluanje, the traditional title of the ruler of the Ndongo.

Portuguese vessels commanded by Diogo Cão reached Angola in 1482. Relations between Portugal and the Bakongo were initially cordial and a few Portuguese missionaries, craftsmen and merchants settled on Angola's coast. However, tensions rose in the 16th century as Portugal's colonial ambitions grew. Slaves were transported to São Tomé e Príncipe and later, Brazil, the Caribbean and North America.

Centred on Luanda, and, from 1617, Benguela, the slave trade used local clans as intermediaries. In 1665 Portuguese forces and their African allies defeated the Bakongo king, Antonio I, at the Battle of Mbwila, after which it splintered into small chieftaincies. The Ndongo kingdom went the same way in 1671. Nonetheless, several central and eastern Angolan kingdoms remained into the 18th century but were eventually undermined by the slave trade and Portuguese exploration of the interior. Portuguese control of Angola and its other African colonies was weakened after 1807, when Napoleon's armies forced the Portuguese court into exile in Brazil.

Political instability in Portugal between 1820 and 1834 spread to Angola, with uprisings and an army mutiny that toppled the colony's governor. But slave traders prospered from new markets in Brazil, Cuba and southern America, even after the progressive Portuguese prime minister, the Marquês de Sá da Bandeira, outlawed the trade in 1836. It is estimated that by 1850 four million Angolans had been enslaved and shipped overseas.

From the 1830s the Portuguese expanded to the south and east from Luanda. British forces held back Portuguese control of Cabinda and the mouth of the River Congo but both Cabinda and Massabi were annexed in 1883. Portugal's claims to Angola were outlined at the 1884 Berlin Conference and treaties with rival colonial powers were agreed by 1900.

By this time, a massive forced labour system had replaced slavery as the basis for the plantation economy and, later, the mining sector. Forced labour and British financing helped construct major rail routes, including one linking the port of Lobito with the copper zones of the Belgian Congo and what is now Zambia.

Economic growth did not translate into social development for native Angolans. The Portuguese encouraged white immigration, especially after 1950, which intensified racial antagonisms. As decolonization progressed elsewhere in Africa, Portugal rejected calls for independence. Nonetheless, three main militant movements had emerged by 1961: the Popular Movement for the Liberation of Angola (MPLA) led by Agostinho Neto; the National Front for the Liberation of Angola (FNLA), led by Holden Roberto; and the National Union for the Total Independence of Angola (UNITA), led by Jonas Savimbi.

A 1974 coup d'état in Portugal established a leftist military government in Lisbon that agreed, in the Alvor Accords of 1975, to hand power in Angola to a coalition of the three movements. The ideological differences between the three led to armed conflict, with FNLA and UNITA forces attempting to wrest control of Luanda from the MPLA. South African involvement and civil war continued until the New York Agreement of 1988 saw South Africa withdraw from Angola and Namibia. Cuba also agreed to the phased withdrawal of its troops.

A peace agreement was signed by all sides on 31 May 1991, allowing for a national army to be formed and multi-party elections to be held. In Sept. 1992 the MPLA won elections and José Eduardo dos Santos was re-elected president. UNITA's Jonas Savimbi rejected the result and resumed the war, seizing 70% of the country.

On 20 Nov. 1994 a peace agreement was signed in Lusaka, allowing for UNITA to share in government. In Jan. 1998 Savimbi met with President dos Santos but talks foundered and fighting resumed. Meanwhile, Angolan troops fought in the Democratic Republic of the Congo alongside the forces of President Kabila in his efforts to quash a Rwandan-backed rebellion in the east of his country. They remained after the assassination of Kabila in Jan. 2001 but renewed hopes of peace resulted in withdrawal in Jan. 2002. In Feb. 2002 Savimbi was killed in fighting with Angolan government troops. On 4 April 2002 the Angolan army and UNITA agreed a ceasefire. More than half a million Angolans had died in the unrest.

Following the ceasefire oil production increased and investment flowed, notably from China, despite some remaining pockets of unrest. In Sept. 2008 the MPLA, still under dos Santos, won a landslide in the first multiparty elections for 16 years.

The expected presidential election was repeatedly delayed until direct presidential elections were abolished by a new constitution in Jan. 2010.

TERRITORY AND POPULATION

Angola is bounded in the north by the Republic of the Congo, north and northeast by the Democratic Republic of the Congo, east by Zambia, south by Namibia and west by the Atlantic Ocean. The area is 1,246,700 sq. km (481,350 sq. miles) including the province of Cabinda, an exclave of territory separated by 30 sq. km of the Democratic Republic of the Congo's territory. The population at census, 1970, was 5,646,166, of whom 14% were urban. Estimate, 2012, 20·61m.; density, 16·5 per sq. km. In 2010, 58·5% of the population were living in urban areas. Population figures are rough estimates because the civil war led to huge movements of population. More than 300,000 Angolan refugees have returned to the country since the civil war ended in 2002. The number of Portuguese in Angola rose from 45,000 in 2007–08 to 92,000 a year later.

The UN gives a projected population for 2015 of 21·84m.

Area, population and chief towns of the provinces:

Province	Area (in sq. km)	Population estimate, 2012 (in 1,000)	Chief town
Bengo	31,371	351·5	Caxito
Benguela	31,788	1,985·4	Benguela
Bié	70,314	1,143·7	Kuito
Cabinda	7,270	441·1	Cabinda
Cunene	88,342	648·4	Ondjiva
Huambo	34,274	1,624·0	Huambo
Huíla	75,002	2,098·0	Lubango
Kuando-Kubango	199,049	394·4	Menongue
Kwanza Norte	24,110	376·9	Ndalatando
Kwanza Sul	55,660	1,353·8	Sumbe
Luanda	2,418	5,851·2	Luanda
Lunda Norte	102,783	770·3	Lucapa
Lunda Sul	56,985	387·1	Saurimo
Malanje	87,246	754·6	Malanje
Moxico	223,023	565·0	Luena
Namibe	58,137	368·0	Namibe
Uíge	58,698	1,101·2	Uíge
Zaire	40,130	394·8	Mbanza Congo

The most important towns are Luanda, the capital (2012 population estimate, 5·85m.), Huambo, Lobito, Benguela, Kuito, Lubango, Malanje and Namibe.

The main ethnic groups are Umbundo (Ovimbundo), Kimbundo, Bakongo, Chokwe, Ganguela, Luvale and Kwanyama.

Portuguese is the official language. Bantu and other African languages are also spoken.

SOCIAL STATISTICS

Life expectancy at birth, 2007, 44·6 years for males and 48·5 years for females. 2008 births (estimates), 775,000; deaths, 306,000. Estimated birth rate in 2008 was 43 per 1,000 population; estimated death rate, 17. Annual population growth rate, 2000–08, 2·9%. Fertility rate, 2008, 5·8 births per woman; infant mortality, 2010, 98 per 1,000 live births.

CLIMATE

The climate is tropical, with low rainfall in the west but increasing inland. Temperatures are constant over the year and most rain falls in March and April. Luanda, Jan. 78°F (25·6°C), July 69°F (20·6°C). Annual rainfall 13" (323 mm). Lobito, Jan. 77°F (25°C), July 68°F (20°C). Annual rainfall 14" (353 mm).

CONSTITUTION AND GOVERNMENT

Under the Constitution adopted at independence, the sole legal party was the MPLA. In Dec. 1990, however, the MPLA announced that the Constitution would be revised to permit opposition parties. The supreme organ of state is the 220-member *National Assembly.* For the 2008 elections 30% of seats were guaranteed for women. There is an executive *President*, elected for renewable terms of five years, who appoints a *Council of Ministers.*

In Dec. 2002 Angola's ruling party and the UNITA party of former rebels agreed on a new constitution. The president would keep key powers, including the power to name and to remove the prime minister. The president will also appoint provincial governors, rather than letting voters elect them, but the governor must be from the party that received a majority of votes in that province. A draft constitution was submitted to the constitutional commission of the Angolan parliament for consideration in Jan. 2004.

A new constitution was adopted on 21 Jan. 2010 and came into effect on 5 Feb. although the opposition party UNITA boycotted the vote. Direct presidential elections were abolished. Instead the party with the majority in parliament will choose the president. A two five-year term limit was introduced although it did not take effect until after the parliamentary elections in Aug. 2012, allowing President dos Santos to remain in power until 2022. The president was also made responsible for judicial appointments while the office of prime minister was replaced by that of a vice-president to be appointed by the president.

National Anthem

'O Pátria, nunca mais esqueceremos' ('Oh Fatherland, never shall we forget'); words by M. R. Alves Monteiro, tune by R. A. Dias Mingas.

GOVERNMENT CHRONOLOGY

Presidents since 1975. (MPLA = Popular Movement for the Liberation of Angola)

| 1975–79 | MPLA | António Agostinho Neto |
| 1979– | MPLA | José Eduardo dos Santos |

RECENT ELECTIONS

At the presidential elections of 29–30 Sept. 1992 the electorate was 4,862,748. Turnout was about 90%. José Eduardo dos Santos (Popular Movement for the Liberation of Angola/MPLA) was re-elected as president with 49·6% of votes cast against 40·1% for Jonas Savimbi (National Union for the Total Independence of Angola/UNITA). There were nine other candidates. Savimbi refused to accept the result.

In parliamentary elections held on 31 Aug. 2012 the electorate was 9,757,671. Turnout was 62·8%. The MPLA gained 175 seats in the National Assembly with 71·8% of votes cast, UNITA 32 with 18·7%, Broad Convergence for the Salvation of Angola–Electoral Coalition/CASA–CE 8 with 6·0%, Social Renewal Party 3 with 1·7% and the National Front for the Liberation of Angola 2 with 1·1%.

CURRENT GOVERNMENT

President: José Eduardo dos Santos; b. 1942 (MPLA; since 10 Sept. 1979; re-elected 9 Dec. 1985 and 29–30 Sept. 1992).

In March 2013 the government comprised:

Vice-President: Manuel Vicente.

Minister of Agriculture: Afonso Pedro Canga. *Commerce:* Rosa Escórcio Pacavira de Matos. *Construction:* Fernando Fonseca. *Culture:* Rosa Maria Martins da Cruz e Silva. *Economy:* Abraão Gourgel. *Education:* Mpinda Simão. *Energy and Water:* João Baptista Borges. *Environment:* Maria de Fátima Monteiro Jardim. *Family and Women's Affairs:* Maria Filomena Delgado. *Finance:* Carlos Alberto Lopes. *Fisheries:* Vitória de Barros Neto. *Foreign Affairs:* George Rebelo Pinto Chicoty. *Geology and Mines:* Francisco Manuel Queirós. *Health:* José Vieira Dias Van-Dúnem. *Higher Education:* Adão do Nascimento. *Hotels and Tourism:* Pedro Mutinde. *Industry:* Bernarda Henriques da Silva. *Interior:*

Ângelo de Barros Veiga Tavares. *Justice and Human Rights:* Rui Jorge Carneiro Mangueira. *National Defence:* Gen. Candido Pereira dos Santos Van-Dúnem. *Oil:* José Maria Botelho de Vasconcelos. *Parliamentary Affairs:* Rosa Luís de Sousa Micolo. *Planning and Territorial Development:* Job Graça. *Public Administration, Employment and Social Security:* António Domingos Pitra da Costa Neto. *Science and Technology:* Maria Cândida Teixeira. *Social Communication:* José Luís de Matos. *Telecommunications and Information Technology:* José Carvalho da Rocha. *Territorial Administration:* Bornito de Sousa Baltazar Diogo. *Transport:* Augusto da Silva Tomás. *Urban Development and Housing:* José António da Conceição e Silva. *War Veterans:* Kundi Paihama. *Welfare and Social Reintegration:* João Baptista Kussumua. *Youth and Sports:* Gonçalves Manuel Muandumba.

Government Website (Portuguese only):
   http://www.angola-portal.ao/PortaldoGoverno

CURRENT LEADERS

José Eduardo dos Santos

Position
President

Introduction
José Eduardo dos Santos, one of Africa's longest-serving leaders, has been president of Angola since the death of the country's first post-colonial president Agostinho Neto in 1979. He is also head of the ruling Movimento Popular de Libertação de Angola (MPLA; Popular Movement for the Liberation of Angola), and was prime minister from 1999–2002.

Early Life
Dos Santos was born on 28 Aug. 1942 in Luanda. In 1961 he joined Neto's MPLA rebel movement, fighting for Angolan independence. The movement was forced into exile in neighbouring Zaïre (now the Democratic Republic of the Congo). As his party standing increased, dos Santos founded the MPLA youth movement before being sent to Moscow to study telecommunications and petroleum engineering. He returned to fight for Angolan independence, which finally came in 1975. Under Neto's presidency, dos Santos served first as prime minister (1975–78) and then planning minister (1978–79). After Neto's death in Sept. 1979, dos Santos assumed the leadership as Angola's second post-independence president.

Career in Office
During the first ten years dos Santos upheld the MPLA's traditional Marxist doctrine and the government's single party rule while continuing the war against the UNITA rebels begun under his predecessor. The government received Cuban military help in the conflict and the Soviet Union supplied funds. The USA and South Africa meanwhile backed UNITA's leader Jonas Savimbi.

A rapprochement began in 1988 when both Cuba and South Africa withdrew their forces. In 1990, following the collapse of communism, dos Santos moved away from Marxism to adopt 'democratic socialism'. This allowed for the introduction of a free market economy and multi-party elections. The following year, a peace agreement signed in Lisbon culminated in Angola's first nationwide elections in 1992. In a turnout of 91% of registered voters, the MPLA won 54% compared to UNITA's 34%. In the presidential poll dos Santos secured 49·6%, while Savimbi polled 40·1%. Before a second round run-off, Savimbi rejected the election, claiming the first round results had been fraudulent. The civil war resumed and elections scheduled for 1997 were postponed indefinitely.

Attempts to resolve the conflict through amnesties, military action and peace talks all failed. In 1999 dos Santos assumed the role of prime minister and took over control of the armed forces. In Feb. 2002 Savimbi was killed by government soldiers and two

months later a ceasefire was signed between the government and the rebels. In 2001 dos Santos announced his intention to step down from the presidency at the next elections, although these would not take place until there was free movement of people and goods in the country and the many Angolans displaced by the conflict had returned home. Since the end of hostilities in most of the country in 2002, his government has committed substantial resources, financed by oil exports and diamonds, to reconstruction. Nevertheless, much of the population still lives in extreme poverty. Also, it was not until Aug. 2006 that a ceasefire agreement was achieved with separatists fighting for independence of the northern enclave of Cabinda, where much of Angola's oil wealth lies. The final stage of a United Nations refugee repatriation scheme, involving some 60,000 Angolans, began in Oct. 2006.

In Jan. 2007 dos Santos oversaw Angola's accession to the Organization of the Petroleum Exporting Countries (OPEC), and in Feb. he declared that parliamentary elections would be held in 2008 and presidential polls in 2009. The parliamentary polls took place in Sept. and resulted in a landslide victory for the ruling MPLA. Despite some criticisms of the poll by an observer mission from the European Union and the rejection of opposition demands for a rerun of voting in the capital Luanda, UNITA leader Isaias Samakuva accepted his party's defeat. In mid-2009 presidential elections scheduled for Sept. were postponed, reportedly to allow more time for the drafting of a new constitution that came into effect in Feb. 2010. Under the constitution the party with a parliamentary majority chooses the president, who is subject to a two-term limit, is responsible for judicial appointments and appoints a vice president (replacing the post of prime minister). Parliamentary elections in Aug. 2012 resulted in another comfortable victory for the MPLA and so secured another presidential term for dos Santos.

DEFENCE

Conscription is for two years. Defence expenditure totalled US$2,425m. in 2008 (US$194 per capita), representing 3·1% of GDP.

Army
In 2007 the Army had 42 regiments. Total strength was 100,000. In addition the paramilitary Rapid Reaction Police numbered 10,000.

Navy
Naval personnel in 2007 totalled about 1,000 with nine operational vessels. There is a naval base at Luanda.

Air Force
The Angolan People's Air Force (FAPA) was formed in 1976 and has 6,000 personnel. In 2007 there were 90 combat-capable aircraft and 16 attack helicopters.

ECONOMY

In 2006 agriculture accounted for 8·9% of GDP, industry 69·7% and services 21·4%.

Overview
Following four decades of civil war, Angola has emerged as Africa's second largest oil exporter and third largest economy. Oil revenue accounts for three-quarters of budgetary revenue. The economy grew by 15% per year on average between 2004 and 2007 as oil production increased and the country's infrastructure was rebuilt. Although Angola has a wealth of natural resources, including diamonds, coffee, fish and timber, the economy remains vulnerable to oil price shocks. Poverty is widespread.

As a result of the oil price collapse in 2008–09, Angola experienced a sharp contraction in its oil revenues, a one-third decline in international reserves and macroeconomic instability. In 2009 the government put in place a stabilization programme supported by a three-year stand-by arrangement with the IMF, which has led to a stronger fiscal position, a more comfortable level of international reserves, a stable exchange rate and lower inflation. Medium-term challenges include tackling social sector issues, putting in place a comprehensive fiscal framework and supporting economic diversification.

Currency

The unit of currency is the *kwanza* (AOA), introduced in Dec. 1999, replacing the *readjusted kwanza* at a rate of 1 kwanza = 1m. readjusted kwanzas. Foreign exchange reserves were US$2,017m. in July 2005 and money supply was 103,304m. kwanzas. Inflation, which reached 4,146% in 1996, was 14·5% in 2010 and 13·5% in 2011.

Budget

Revenues in 2009 were 1,848bn. kwanzas and expenditures 2,363bn. kwanzas. Petroleum tax revenue accounted for 63·1% of revenues in 2009; current expenditure accounted for 68·6% of expenditures.

Performance

Total GDP was US$104·3bn. in 2011. The civil war meant GDP growth in 1993 was negative, at –24·0%, but a recovery followed and in 2007 and 2008 it was 22·6% and 13·8% respectively, mainly thanks to booming diamond exports and post-war rebuilding. Angola's growth in both 2007 and 2008 was among the highest in the world. Growth fell to 2·4% in 2009 but was 3·4% in 2010 and 3·9% in 2011.

Banking and Finance

The Banco Nacional de Angola is the central bank and bank of issue (*Governor*, José Massano). All banks were state-owned until the sector was reopened to commercial competition in 1991. The largest banks in Dec. 2010 were: Banco Africano de Investimentos (BAI), with assets of US$8,373m.; Banco Espirito Santo Angola (BESA), with assets of US$7,892m.; Banco de Fomento Angola, with assets of US$6,435m.

Total external debt in 2010 was US$18,562m., equivalent to 24·6% of GNI.

Angola received US$9·9bn. in foreign direct investment in 2010 (the most of any African country, although down from a record US$16·6bn. in 2008).

ENERGY AND NATURAL RESOURCES

Environment

In 2008 Angola's carbon dioxide emissions from the consumption and flaring of fossil fuels were the equivalent of 1·9 tonnes per capita. An *Environmental Performance Index* compiled in 2008 ranked Angola 148th in the world out of 149 countries analysed, with 39·5%. The index examined various factors in six areas—air pollution, biodiversity and habitat, climate change, environmental health, productive natural resources and water resources.

Electricity

Installed capacity was 1·2m. kW in 2007. Production in 2007 was 3·17bn. kWh, with consumption per capita 259 kWh.

Oil and Gas

Oil is produced mainly offshore and in the Cabinda exclave. Oil production and supporting activities contribute more than half of Angolan GDP and provide the government with 80% of revenues. The oil industry is expected to invest US$3·5bn. a year in offshore Angola in the early part of the 21st century. Angola's oil production ranks among the fastest-growing in the world. There are plans for a new US$8bn. oil refinery near Lobito which is scheduled to be completed by 2015. Only Nigeria among sub-Saharan African countries produces more oil. It is believed that there are huge oil resources yet to be discovered. Proven oil reserves in 2008 were 13·5bn. bbls. Total production (2008) 92·2m. tonnes. Proven natural gas reserves (2007) 57bn. cu. metres; production, 2007, 844m. cu. metres.

Minerals

Mineral production in Angola is dominated by diamonds and 90% of all workers in the mining sector work in the diamond industry. Production in 2006 was an estimated 9·2m. carats. Angola has billions of dollars worth of unexploited diamond fields. In 2000 the government regained control of the nation's richest diamond provinces from UNITA rebels. Other minerals produced (2006 estimates) include granite, 1·5m. cu. metres; marble, 100,000 cu. metres; salt, 35,000 tonnes. Iron ore, phosphate, manganese and copper deposits exist.

Agriculture

In 2007 there were an estimated 3·3m. ha. of arable land and 0·3m. ha. of permanent crops. 75,000 ha. were irrigated in 2002. The agricultural population in 2007 was approximately 12·29m., of whom an estimated 5·41m. were economically active. Although more than 70% of the economically active population are engaged in agriculture it only accounts for 8% of GDP. There were 31 tractors per 10,000 ha. of arable land in 2006. Principal crops (with 2003 production, in 1,000 tonnes): cassava (5,699); maize (545); sweet potatoes (439); sugarcane (360); bananas (300); millet (97); citrus fruits (78); dry beans (66).

Livestock (2003): 4·1m. cattle, 340,000 sheep, 2·05m. goats, 780,000 pigs.

Forestry

In 2010, 58·48m. ha., or 47% of the total land area, was covered by forests. Timber production in 2007 was 4·84m. cu. metres.

Fisheries

Total catch in 2008 came to 305,762 tonnes, mainly from sea fishing.

INDUSTRY

The principal manufacturing branches are foodstuffs, textiles and oil refining. Output, 2007 unless otherwise indicated (in 1,000 tonnes): cement (2008 estimate), 1,780; residual fuel oil, 601; distillate fuel oil, 513; jet fuel, 351; petrol, 56; plywood (estimate), 10,000 cu. metres; beer (2003), 160m. litres.

Labour

In 2010 the estimated economically active population numbered 8,533,000 (53% males), up from 6,238,000 in 2000.

INTERNATIONAL TRADE

Imports and Exports

In 2007 imports (c.i.f.) totalled US$15,048m. (US$9,586m. in 2006); exports (f.o.b.), US$38,997m. (US$31,862m. in 2006).

Main exports, 2007 (in US$1m.): crude oil, 36,417; diamonds, 1,552; manufactures, 359. Chief import suppliers are Portugal, China, the USA and Brazil. Principal export markets are China, the USA, India and Taiwan.

COMMUNICATIONS

Roads

There were 51,429 km of roads in 2001 (7,944 km highways; 10·4% of all roads surfaced) and 671,100 vehicles in use in 2007. Many roads remain mined as a result of the civil war; a programme of de-mining and rehabilitation is under way.

Rail

Prior to the civil war there was in excess of 2,900 km of railway (predominantly 1,067 mm gauge track), but much of the network was damaged during the war. However, restoration and redevelopment of the network is now under way, notably the Benguela Railway, linking the port city of Lobito with Huambo in Angola's rich farmlands and neighbouring Democratic Republic of the Congo and Zambia.

Civil Aviation

There is an international airport at Luanda (Fourth of February). The national carrier is Linhas Aéreas de Angola (TAAG), which operated direct flights in 2010 to Bangui, Brazzaville, Cape Town, Douala, Harare, Johannesburg, Lisbon, Lusaka, Praia, Rio de Janeiro, São Paulo, São Tomé and Windhoek, as well as domestic services. There were direct flights in 2010 with other airlines to Abidjan, Addis Ababa, Bamako, Beijing, Brussels, Douala, Dubai, Frankfurt, Johannesburg, Lisbon, London, Maputo, Moscow, Paris, Pointe-Noire, Port Harcourt, Port-Gentil and Windhoek. In 2006 scheduled airline traffic of Angola-based carriers flew 7m. km, carrying 263,000 passengers (136,000 on international flights).

Shipping

There are ports at Luanda, Lobito and Namibe, and oil terminals at Malongo, Lobito and Soyo. In Jan. 2009 there were 28 ships of 300 GT or over registered, totalling 21,000 GT.

Telecommunications

In 2010 there were 303,200 main (fixed) telephone lines but mobile phone subscribers numbered 8·91m. There were 32·8 internet users per 1,000 inhabitants in 2009. Fixed internet subscriptions totalled 320,000 in 2009 (17·2 per 1,000 inhabitants). In June 2012 there were 433,000 Facebook users.

SOCIAL INSTITUTIONS

Justice

The Supreme Court and Court of Appeal are in Luanda. The death penalty was abolished in 1992. In 2002–03 the US government's Agency for International Development assisted in the modernization of the judicial system. Measures including the introduction of a court case numbering system were intended to reduce legal costs and attract foreign investment.

The population in penal institutions in Sept. 2009 was 16,183 (87 per 100,000 of national population).

Education

The education system provides three levels of general education totalling eight years, followed by schools for technical training, teacher training or pre-university studies. In 2000–01 there were 1,178,485 pupils and 33,478 teachers at primary schools, 399,712 pupils and 19,798 teachers at secondary schools and (1999–2000) 7,845 students with 796 academic staff in tertiary education institutions. There is one state university, Agostinho Neto University (formerly the University of Luanda and later the University of Angola), with campuses at Benguela, Cabinda, Dundo, Huambo, Luanda, Lubango, Namibe, Saurimo, Sumbe and Uíge. The Institute of International Relations is also publicly funded. Private schools have been permitted since 1991. There were five recognized private higher education institutes in 2005. The adult literacy rate was an estimated 70·0% in 2009 (82·9% among males and 57·6% among females).

In 2006 expenditure on education came to 3·0% of GNI.

Health

In 2004 there were 1,165 physicians, 18,485 nursing and midwifery personnel, 222 dentistry personnel and 919 pharmaceutical personnel. In 2005 there were eight hospitals beds per 10,000 population.

In 2008, 50% of the population were using improved drinking water sources and 57% were using improved sanitation facilities. 41% of the population were undernourished in the period 2005–07, down from 61% in 1995–97.

RELIGION

In 2001 there were 6·44m. Roman Catholics, 1·55m. Protestants and 710,000 African Christians, and most of the remainder follow traditional animist religions. In Feb. 2013 there was one cardinal.

CULTURE

Press

The government-owned *Jornal de Angola* (circulation of 41,000) was the only daily newspaper in 2006. The *Diário da República* is the official gazette. There are 12 private weekly publications and four smaller regional weeklies.

Tourism

In 2007 there were 195,000 non-resident tourists (including 89,000 from European countries and 38,000 from the Americas), bringing revenue of US$236m.

Festivals

The Luanda International Jazz Festival was first held in 2009.

DIPLOMATIC REPRESENTATIVES

Of Angola in the United Kingdom (22 Dorset St., London, W1U 6QY)
Ambassador: Miguel Gaspar Fernandes Neto.

Of the United Kingdom in Angola (Rua 17 de Setembro, Caixa Postal 1244, Luanda)
Ambassador: Richard Wildash, LVO.

Of Angola in the USA (2108 16th St., NW, Washington, D.C., 20009)
Ambassador: Alberto do Carmo Bento Ribeiro.

Of the USA in Angola (32 rua Houari Boumedienne, Miramar, Luanda)
Ambassador: Christopher McMullen.

Of Angola to the United Nations
Ambassador: Ismael Gaspar Martins.

Of Angola to the European Union
Ambassador: Maria Elizabeth Simbrão de Carvalho.

FURTHER READING

Anstee, M. J., *Orphan of the Cold War: the Inside Story of the Collapse of the Angolan Peace Process, 1992–93.* 1996
Brittain, Victoria, *Death of Dignity: Angola's Civil War.* 1999
Guimarães, Fernando Andersen, *The Origins of the Angolan Civil War: Foreign Intervention and Domestic Political Conflict.* 2001
Hodges, Tony, *Angola: Anatomy of an Oil State.* 2004

National Statistical Office: Instituto Nacional de Estatística, Rua Ho-Chi-Min, C.P. 1215, Luanda.
Website (Portuguese only): http://www.ine-ao.com

ANTIGUA AND BARBUDA

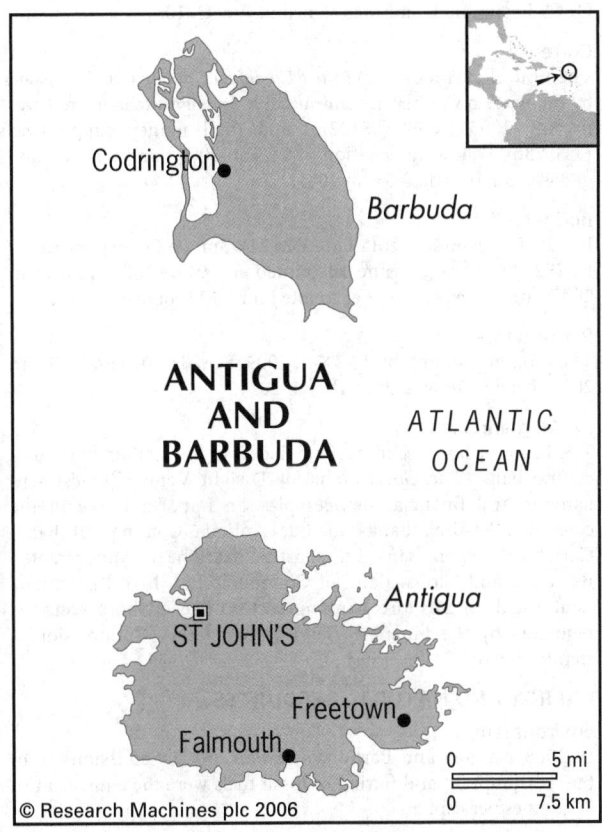

© Research Machines plc 2006

Capital: St John's
Population, 2011: 88,000
GNI per capita, 2011: (PPP$) 15,521
HDI/world rank: 0·764/60
Internet domain extension: .ag

KEY HISTORICAL EVENTS

Antigua and Barbuda were populated by Arawak-speaking people from at least 1000 BC. By 1493, when Columbus passed Antigua, it was occupied by Carib Indians. English settlers arrived in 1632. Sugar plantations, using slave labour, were established in the 1650s. As British colonies, Antigua and Barbuda formed part of the Leeward Islands Federation from 1871 until 1956 when they became a separate Crown Colony. This was merged into the West Indies Federation from Jan. 1958 until May 1962 and became an Associated State of the UK on 27 Feb. 1967. Antigua and Barbuda gained independence on 1 Nov. 1981.

TERRITORY AND POPULATION

Antigua and Barbuda comprises three islands of the Lesser Antilles situated in the eastern Caribbean with a total land area of 442 sq. km (171 sq. miles); it consists of Antigua (280 sq. km), Barbuda, 40 km to the north (161 sq. km) and uninhabited Redonda, 40 km to the southwest (one sq. km). The provisional population at the census of May 2011 was 87,774 (1,615 on Barbuda); density, 199 per sq. km. In 2011, 30·4% of the population lived in urban areas.

The chief town is St John's, the capital, on Antigua (24,451 inhabitants in 2001). Codrington (914) is the only settlement on Barbuda.

English is the official language; local dialects are also spoken.

SOCIAL STATISTICS

Expectation of life, 2009: males, 73 years, females, 76. Annual population growth rate, 2000–05, 2·7%. 2007: births, 1,240; deaths, 504. Infant mortality in 2010 was 7 per 1,000 live births; fertility rate, 2008, 2·1 births per woman.

CLIMATE

A tropical climate, but drier than most West Indies islands. The hot season is from May to Nov., when rainfall is greater. Mean annual rainfall is 40" (1,000 mm).

CONSTITUTION AND GOVERNMENT

H.M. Queen Elizabeth, as Head of State, is represented by a Governor-General appointed by her on the advice of the Prime Minister. There is a bicameral legislature, comprising a 17-member Senate appointed by the Governor-General and a 19-member House of Representatives (with 17 members elected by universal suffrage for a five-year term plus the Attorney General and the Speaker). The Governor-General appoints a Prime Minister and, on the latter's advice, other members of the Cabinet.

Barbuda is administered by a nine-member directly-elected council.

National Anthem

'Fair Antigua and Barbuda'; words by N. H. Richards, tune by W. G. Chambers.

RECENT ELECTIONS

At the elections to the House of Representatives of 12 March 2009 the United Progressive Party (UPP) won 9 seats, the Antigua Labour Party (ALP) 7 and the Barbuda People's Movement (BPM) 1.

CURRENT GOVERNMENT

Governor-General: Louise Lake-Tack, GCMG; b. 1944 (in office since 17 July 2007).

In March 2013 the UPP government comprised:

Prime Minister and Minister of Foreign Affairs: Baldwin Spencer; b. 1948 (UPP; in office since 24 March 2004).

Deputy Prime Minister and Minister of Health, Social Transformation and Consumer Affairs: Wilmoth Daniel. *Attorney General and Legal Affairs:* Justin Simon. *National Security:* Errol Cort. *Finance, the Economy and Public Administration:* Harold Lovell. *Agriculture, Lands, Housing and the Environment:* Hilson Baptiste. *Education, Sports, Youth and Gender Affairs:* Jacqui Quinn-Leandro. *Tourism, Civil Aviation and Culture:* John Maginley. *Works, Transport and Labour:* Trevor Walker.

Government Website: http://www.ab.gov.ag

CURRENT LEADERS

Baldwin Spencer

Position
Prime Minister

Introduction
Baldwin Spencer is leader of the United Progressive Party (UPP) and took office as prime minister in March 2004, defeating the Antigua Labour Party (ALP), which had held power continuously since 1976.

Early Life
Baldwin Spencer was born on 8 Oct. 1948 in Grays Green, Antigua. After secondary school, he studied social leadership at

St Francis Xavier University's Coady International Institute in Nova Scotia. He also obtained a diploma in labour and economic studies from Ruskin College (at Oxford in the UK) and in labour and industrial relations from Oslo University.

In the 1970s Spencer worked as a trade unionist, serving as vice-president and, later, assistant general secretary of the Antigua and Barbuda Workers' Union (AWU). He also served as president of the Caribbean Maritime and Aviation Council.

In 1989 Spencer entered parliament as the United Democratic Party (UNDP) representative for St John's Rural West constituency. In 1991 he became leader of the UNDP and, as leader of the opposition in parliament, formed an alliance with the two other main opposition parties, the Antigua Caribbean Liberation Movement and the Progressive Labour Movement. They merged in 1992 to form the UPP.

During the 1990s Spencer regularly accused Prime Minister Vere Bird, Sr and his ALP of corruption. The campaign helped exploit rifts within the government and Bird's own son, Lester Bird, called for his father's resignation. Lester Bird took over as prime minister shortly before the general election of March 2004. Spencer led the UPP into the election promising more transparent government, and the party won 12 of 17 seats.

Career in Office

Spencer vowed to combat corruption, develop tourism and foster economic co-operation with other countries. In the first year he introduced legislation to improve government accountability and took steps to de-politicize the government-owned media. However, corruption investigations were hampered by the loss of government files and in 2005 Spencer set up a task-force to tackle organized crime and corruption among officials.

In Oct. 2004 an IMF report concluded that Antigua and Barbuda's economy suffered from high levels of public debt and over-reliance on the government for jobs (accounting for 40% of total employment). Spencer responded in 2005 by launching a drive to expand the tourism industry, taking measures to cut the public service salary bill, and reintroducing income tax. The tourism drive was undermined, however, in July 2008 by the murder of a visiting British couple. Spencer's UPP won a second term by taking 9 of the 17 seats in the parliamentary elections of March 2009.

He has pursued closer ties with Brazil, China, India, Russia and neighbouring Caribbean countries (serving as chair of CARICOM in 2004).

DEFENCE

The Antigua and Barbuda Defence Force numbers 170. There are some 75 reserves. A coastguard service has been formed.

In 2008 defence expenditure totalled US$7m. (US$79 per capita), representing 0·6% of GDP.

Army

The strength of the Army section of the Defence Force was 125 in 2007.

Navy

There was a naval force of 45 operating three patrol craft in 2007.

ECONOMY

In 2009 agriculture accounted for 1·7% of GDP, industry 22·0% and services 76·3%.

Overview

Antigua's main industries include tourism (accounting for nearly 60% of GDP and 40% of investment), construction and light manufacturing. There was strong economic growth from 2003–07, owing mainly to a boom in construction connected to the Cricket World Cup. However, with the global economic crisis the country's largest financial institution collapsed. Tourism also

declined, along with FDI inflows and remittances. National debt remains high.

Antigua and Barbuda is a member of the Organisation of Eastern Caribbean States (OECS). The treaty establishing the OECS Economic Union was signed in June 2010.

Currency

The unit of currency is the *East Caribbean dollar* (XCD), issued by the Eastern Caribbean Central Bank. Foreign exchange reserves in July 2005 were US$122m. and total money supply was EC$573m. Following deflation of 0·6% in 2009 there was inflation of 3·4% in 2010 and 3·5% in 2011.

Budget

In 2007 revenues totalled EC$718·3m. and expenditures EC$923·8m. Tax revenue accounted for 91·4% of revenues in 2007; current expenditure accounted for 78·3% of expenditures.

Performance

The economy shrank by 10·7% in 2009, 8·5% in 2010 and 5·5% in 2011. Total GDP was US$1·1bn. in 2011.

Banking and Finance

The East Caribbean Central Bank based in St Kitts functions as a central bank. The *Governor* is Sir Dwight Venner. Investment banking and financial services play an important role in the economy. Leading banks include Antigua Commercial Bank, Caribbean Union Bank Ltd, Eastern Caribbean Amalgamated Bank Ltd and Global Bank of Commerce Ltd. In 1981 Antigua established an offshore banking sector. The offshore sector is regulated by the Financial Services Regulatory Commission, a statutory body.

ENERGY AND NATURAL RESOURCES

Environment

In 2008 Antigua and Barbuda's carbon dioxide emissions from the consumption and flaring of fossil fuels were the equivalent of 8·1 tonnes per capita.

Electricity

Capacity in 2007 was 27,000 kW. Production was estimated at 118m. kWh in 2007 and consumption per capita about 1,391 kWh.

Water

There is a desalination plant with a capacity of 0·6m. gallons per day, sufficient to meet the needs of the country.

Agriculture

In 2007 there were around 8,000 ha. of arable land and 1,000 ha. of permanent crops. Production (2003) of fruits and vegetables, 13,000 tonnes (notably melons and mangoes).

Livestock (2003): goats, 36,000; sheep, 19,000; cattle, 14,000; pigs, 6,000.

Forestry

Forests covered 10,000 ha., or 22% of the total land area, in 2010.

Fisheries

Total catch in 2010 came to 2,293 tonnes, exclusively from sea fishing.

INDUSTRY

Manufactures include beer, cement, toilet tissue, stoves, refrigerators, blenders, fans, garments and rum (molasses imported from Guyana).

Labour

The unemployment rate in 1998 was the lowest in the Caribbean, at 4·5%. Between 1994 and 1998, 2,543 jobs were created. The average annual salary in 1998 was US$8,345 per head of population.

INTERNATIONAL TRADE

Imports and Exports

Imports in 2010 (c.i.f.) totalled US$361·3m. and exports (f.o.b.) US$34·8m. The main trading partners were the USA, China and the UK for imports; and the USA, the UK and Panama for exports.

COMMUNICATIONS

Roads

In 2002 there were 1,165 km of roads of which 33·0% were paved. 23,700 passenger cars and 5,200 commercial vehicles were in use in 2002. More than EC$64m. was spent to rebuild major roads and highways in the three years following damage caused by hurricanes Luis and Marilyn in 1995.

Civil Aviation

V. C. Bird International Airport is near St John's. There were flights in 2010 to Anguilla, Atlanta, Barbados, Charlotte, Curaçao, Dominica, Dominican Republic, Frankfurt, Georgetown, Grenada, Hartford, Kingston, London, Miami, Milan, Montserrat, New York, Pointe-à-Pitre, Puerto Rico, St Croix, St Kitts and Nevis, St Lucia, St Maarten, St Vincent, Toronto, Trinidad and Tobago and the British and US Virgin Islands. A domestic flight links the airports on Antigua and Barbuda.

Shipping

The main port is St John's Harbour. In Jan. 2009 there were 1,166 ships of 300 GT or over registered, totalling 9,620,000 GT.

Telecommunications

There were 41,700 fixed telephone lines in 2010, or 470·5 per 1,000 inhabitants. Mobile phone subscribers numbered 163,900 in 2010. There were 742·0 internet users per 1,000 inhabitants in 2009. Fixed internet subscriptions totalled 15,600 in 2009 (177·7 per 1,000 inhabitants).

SOCIAL INSTITUTIONS

Justice

Law is based on UK common law as exercised by the Eastern Caribbean Supreme Court (ECSC) on St Lucia. There are Magistrates' Courts and a Court of Summary Jurisdiction. Appeals lie to the Court of Appeal of ECSC, or ultimately to the UK Privy Council. Antigua and Barbuda was one of ten countries to sign an agreement in Feb. 2001 establishing a Caribbean Court of Justice to replace the British Privy Council as the highest civil and criminal court. In the meantime the number of signatories has risen to 12. The court was inaugurated at Port-of-Spain, Trinidad on 16 April 2005 although Antigua and Barbuda has yet to accept it as its court of last resort.

The population in penal institutions in Dec. 2010 was 295 (equivalent to 330 per 100,000 of national population).

Education

Adult literacy was an estimated 98·9% in 2009. In 2007 there were 11,569 pupils at primary schools and 7,838 pupils at secondary schools. The University of Health Sciences Antigua is a private medical school, the University of the West Indies School of Continuing Studies offers adult training and the Antigua State College offers technical and teacher training. Antigua is a partner in the regional University of the West Indies.

In 2009 public expenditure on education came to 2·8% of GDP.

Health

There is one general hospital, a private clinic, seven health centres and 17 associated clinics. A new medical centre at Mount St John's opened in Feb. 2009. In the period 2000–06 there were two physicians per 10,000 inhabitants. In 2005 there were 16 dentists, 175 nurses, and 15 pharmacists and pharmacy assistants.

Welfare

The state operates a Medical Benefits Scheme providing free medical attention, and a Social Security Scheme, providing age and disability pensions and sickness benefits.

RELIGION

In 2001 there were 30,000 Protestants, 23,000 Anglicans and 8,000 Roman Catholics.

CULTURE

Press

The main newspapers are *The Antigua Sun* and *The Daily Observer*, with a combined circulation of 9,000 in 2006.

Tourism

Tourism is the main industry, contributing about 70% of GDP and 80% of foreign exchange earnings and related activities. In 2007 there were 261,786 tourist arrivals by air and 672,788 cruise passengers arrivals.

Festivals

Of particular interest are the International Sailing Week (April–May), the Annual Tennis Championship (May) and the Mid-Summer Carnival (July–Aug.).

DIPLOMATIC REPRESENTATIVES

Of Antigua and Barbuda in the United Kingdom (2nd Floor, 45 Crawford Place, London, W1H 4LP)
High Commissioner: Carl Roberts.

Of the United Kingdom in Antigua and Barbuda
High Commissioner: Paul Brummell (resides in Bridgetown, Barbados).

Of Antigua and Barbuda in the USA (3216 New Mexico Ave., NW, Washington, D.C., 20016)
Ambassador: Deborah Mae Lovell.

Of the USA in Antigua and Barbuda
Ambassador: Larry Palmer (resides in Bridgetown, Barbados).

Of Antigua and Barbuda to the United Nations
Ambassador: John W. Ashe.

Of Antigua and Barbuda to the European Union
Ambassador: Vacant.

FURTHER READING

Dyde, Brian, *The Unsuspected Isle: A History of Antigua.* 2000
Nicholson, Desmond, *Antigua, Barbuda and Redonda: A Historical Sketch.* 1991

ARGENTINA

República Argentina
(Argentine Republic)

Capital: Buenos Aires
Population projection, 2015: 42·18m.
GNI per capita, 2011: (PPP$) 14,257
HDI/world rank: 0·797/45
Internet domain extension: .ar

KEY HISTORICAL EVENTS

Before European colonization two main indigenous American groups and numerous nomadic tribes peopled the region that is now Argentina, constituting a population of some 300,000. Both groups—the Diaguita people in the northwest, and the Guarani people in the south and east—created the basis for a permanent agricultural civilization. The Diaguita also prevented the powerful Inca from expanding their empire from Bolivia into Argentina.

Europeans first came to Argentina in the early 16th century. The explorer Sebastian Cabot established the first Spanish settlement in 1526. He reported Argentina's natural silver resources, possibly inspiring the name *Argentina* ('of silver'). Ten years later Pedro de Mendoza founded Buenos Aires; however, it was not until 1580 that the region's indigenous peoples, weakened by European diseases as much as European military campaigns, were finally defeated and Spanish rule established.

Largely neglected as Spain looked instead to the riches of Peru, most settlers in Argentina came from the neighbouring colonies of Chile, Peru and Paraguay. Missions established by the Roman Catholic Church played an important part in the colonizing process.

In 1776 Buenos Aires, known throughout the 18th century as a smuggler's haunt, was made a free port at the centre of a viceroyalty comprising Argentina, Uruguay, Paraguay and Bolivia. As trade with Europe increased, Buenos Aires adopted European enlightenment and was seen as more cosmopolitan than its rivals.

When Spain came under Napoleonic control the British attacked Buenos Aires, first in 1806 and then again in 1807. On both occasions the city was able to repel the invasions without help from Spanish forces. This was a spur to the independence movement.

Independence
Having separated from Paraguay in 1814, Argentina gained its independence from Spain in 1816. Unable to control its outlying regions, it lost Bolivia in 1825 and Uruguay in 1828. In the early years of independence, the country was embroiled in internal struggles between the Unitarists and the Federalists. Unitarists wanted a strong central government, particularly as Britain agreed to recognize Argentinian independence only if it could devise a government representing the whole country. Federalists, on the other hand, advocated regional control, as each province had formed its own political regime, based on local interests and reinforced by military leaders dominant since the war.

In 1827 Argentinians joined forces with Uruguay to repel a Brazilian invasion, thereby securing independence for Uruguay and encouraging Argentinian unification. From 1835–52, the Federalists held power under Gen. Juan Manuel de Rosas, an important landowner and commander of a rural militia. Governor of Buenos Aires from 1829, Rosas proved a formidable leader who used a secret police force, the Mazorca, to defeat his opponents. He also consolidated church support, compelling priests to display his portrait at the altar.

Britain seized the Falkland Islands (Islas Malvinas) in 1833, while Bolivia, Paraguay and Uruguay continued to isolate the federation. In 1838, following a trade dispute with Uruguay, Argentinian political exiles gained French support in an attempt to overthrow Rosas. But he remained in power until 1853 when he was ousted by Gen. Justo José de Urquiza. The Unitarists inaugurated a new constitution to achieve a more stable government, although Urquiza's overthrow by Santiago Derquai led to another civil war. An agreement between Urquiza and Gen.

Bartolomé Mitre, governor of Buenos Aires, saw a return to stability and established the city as the seat of government.

In a period of strong economic growth, schools were built, public works started and liberal reforms instituted. From 1865–70 Argentina was involved in the War of the Triple Alliance, joining with Uruguay and Brazil in a campaign against Paraguay. This ultimately strengthened the newly centralized Argentina. From 1880–86 Argentina thrived under the leadership of Gen. Julio Roca. A Federalist, Roca nevertheless retained Buenos Aires as the capital. During his second term of office, Roca made peace with Chile after years of dispute over territory.

As a magnet for European immigration, the political system came under pressure to broaden its representation. Italian and Spanish immigrants established the new Socialist, Anarchist and Unión Cívica Radical parties. This latter became the main political force and under the leadership of Hipólito Irigoyen won their first presidential election in 1916, following Roque Sáenz Peña's electoral reforms of 1910–14. The conservatives regained power in 1930, supported by the military, and the activities of radicals were restricted until Gen. Augustín Pedro Justin, heading a coalition of conservatives, radicals and independent socialists, was elected in 1931. Political and economic reforms resulted in trade agreements with Britain which strengthened the economy.

Perón

Although Argentina remained neutral at the outbreak of the Second World War, another coup in 1943 brought Gen. Juan Domingo Perón to power. He chose to side with the allies and declared war on the axis powers.

Perón led a regime that was autocratic but populist and nationalistic, winning presidential elections in 1946 and 1951 with the support of the urban working class. His success was reinforced by his second wife Eva Duarte de Perón, 'Evita'. Acting as *de facto* minister of health and labour, she awarded wage increases that led to inflation. Her death in 1952 combined with Perón's increasing authoritarianism and his excommunication from the church, led to a fall in his popularity. In 1955 a coup by the armed forces sent him into exile.

In 1957 Argentina reverted to the constitution of 1853, and a year later Dr Arturo Frondizi was elected president. With US financial aid, Frondizi attempted to stabilize the economy but faced heavy criticism from left-wing parties and from the Peronists. Frondizi also fell out of favour with the military, whose intervention continued to overshadow Argentinian politics. When the Peronists achieved the highest number of votes in elections in 1962 the military took control, banning the Peronists, along with the Communist party, before elections in 1963. Dr Arturo Illia, a moderate liberal, was elected and many political prisoners were released. An attempted return by Perón in 1964 drove the military to install Gen. Carlos Onganía as president. Responding to popular resistance, the military eventually allowed the re-election of Perón in 1973.

After Perón's death in 1974 his third wife, María Estela Martínez de Perón, 'Isabelita', succeeded, becoming the Americas' first woman chief of state. She was deposed by military coup two years later and the army's commander-in-chief, Gen. Jorge Videla, became president. Once in power, Videla dissolved Congress, banned trade unions and imposed military control. Censorship and military curfews were imposed and the secret police was active. The savagely repressive regime implemented what became known as the 'Dirty War' when up to 15,000 of Videla's opponents disappeared, almost certainly tortured and executed.

Falklands War

Videla was succeeded by Gen. Leopoldo Galtieri, the army commander-in-chief. In April 1982, in an effort to distract attention from internal tension, Galtieri invaded the Falkland Islands. The subsequent defeat helped to precipitate Galtieri's fall in July 1982. Decades of state intervention, regulation, inward looking policies and special interest subsidies had caused economic chaos. Presidential elections held in Oct. 1983 restored civilian rule under Raúl Alfonsín, leader of the middle-class Unión Cívica Radical. Despite attempting to redress the finances of the bloated public sector, growing unemployment and four-figure inflation led to a Peronist victory in the 1989 elections and Carlos Menem became Argentina's new president.

After an attempted military coup in Dec. 1990, Menem met allegations of corruption in the country's privatization programme by reshuffling his government. Economy Minister Domingo Cavallo's plan to stabilize the economy allowed Menem to alter the constitution in 1994 to permit his re-election for a second term.

By 1999, with economic recession and high unemployment, Menem's popularity had plummeted. In the election that year, Fernando de la Rúa of the centrist Alliance became the first president in a decade from outside of the Peronist party. Menem was later accused of illegal arms deals but was released by a federal court and announced his intention to return to politics.

A state of emergency was introduced in Dec. 2001 as Argentina verged on bankruptcy. De la Rúa resigned on 20 Dec. 2001 after days of rioting and looting. Three interim presidents held power over a period of just 11 days before another Peronist, Eduardo Duhalde, was elected president by Congress. With the economy still in crisis, Duhalde held office until the election of Néstor Kirchner in May 2003. He was succeeded by his wife, Cristina Fernández de Kirchner, four years later.

TERRITORY AND POPULATION

The second largest country in South America, the Argentine Republic is bounded in the north by Bolivia, in the northeast by Paraguay, in the east by Brazil, Uruguay and the Atlantic Ocean, and the west by Chile. The republic consists of 23 provinces and one federal district with the following areas and populations in 2010 (in 1,000):

Provinces	Area (sq. km)	Population (census 2010)	Capital	Population (census 2010)
Buenos Aires	307,571	15,625	La Plata	642
Catamarca	102,602	368	Catamarca	159
Chaco	99,633	1,055	Resistencia	293
Chubut	224,686	509	Rawson	26
Córdoba	165,321	3,309	Córdoba	1,312
Corrientes	88,199	993	Corrientes	343
Entre Ríos	78,781	1,236	Paraná	251
Formosa	72,066	530	Formosa	221
Jujuy	53,219	673	San Salvador de Jujuy	258
La Pampa	143,440	319	Santa Rosa	103
La Rioja	89,680	334	La Rioja	178
Mendoza	148,827	1,739	Mendoza	115
Misiones	29,801	1,102	Posadas	289
Neuquén	94,078	551	Neuquén	233
Río Negro	203,013	639	Viedma	53
Salta	155,488	1,214	Salta	524
San Juan	89,651	681	San Juan	109
San Luis	76,748	432	San Luis	185
Santa Cruz	243,943	274	Río Gallegos	97
Santa Fé	133,007	3,195	Santa Fé	396
Santiago del Estero	136,351	874	Santiago del Estero	252
Tierra del Fuego	21,571	127	Ushuaia	57
Tucumán	22,524	1,448	San Miguel de Tucumán	548
Federal Capital	200	2,890	Buenos Aires	2,890

Argentina also claims territory in Antarctica.

The area is 2,780,400 sq. km (excluding the claimed Antarctic territory) and the population at the 2010 census 40,117,096, giving a density of 14 per sq. km.

The UN gives a projected population for 2015 of 42·18m.

In 2011, 92·6% of the population were urban.

In April 1990 the National Congress declared that the Falklands and other British-held islands in the South Atlantic were part of the new province of Tierra del Fuego formed from the former National Territory of the same name. The 1994 constitution reaffirms Argentine sovereignty over the Falkland Islands.

The population of the main metropolitan areas in 2010 was: Buenos Aires, 13,361,000; Córdoba, 1,444,600; Rosario, 1,240,400; Mendoza, 931,300; Tucumán, 788,800; La Plata, 786,300.

97% speak the national language, Spanish, while 2% speak Italian and 1% other languages. The 2010 census population included 1,805,957 persons born outside Argentina (550,713 born in Paraguay, 345,272 in Bolivia, 191,147 in Chile, 157,514 in Peru and 147,499 in Italy).

SOCIAL STATISTICS

2010 births, 756,176; deaths, 318,602. Rates, 2010 (per 1,000 population): birth, 18·7; death, 7·9. Infant mortality, 2010, 12 per 1,000 live births. Life expectancy at birth, 2007, 71·5 years for males and 79·0 years for females. Annual population growth rate, 2005–10, 0·9%; fertility rate, 2008, 2·2 births per woman. Argentina legalized same-sex marriage in July 2010.

CLIMATE

The climate is warm temperate over the pampas, where rainfall occurs in all seasons, but diminishes towards the west. In the north and west, the climate is more arid, with high summer temperatures, while in the extreme south conditions are also dry, but much cooler. Buenos Aires, Jan. 74°F (23·3°C), July 50°F (10°C). Annual rainfall 37" (950 mm). Bahía Blanca, Jan. 74°F (23·3°C), July 48°F (8·9°C). Annual rainfall 21" (523 mm). Mendoza, Jan. 75°F (23·9°C), July 47°F (8·3°C). Annual rainfall 8" (190 mm). Rosario, Jan. 76°F (24·4°C), July 51°F (10·6°C). Annual rainfall 35" (869 mm). San Juan, Jan. 78°F (25·6°C), July 50°F (10°C). Annual rainfall 4" (89 mm). San Miguel de Tucumán, Jan. 79°F (26·1°C), July 56°F (13·3°C). Annual rainfall 38" (970 mm). Ushuaia, Jan. 50°F (10°C), July 34°F (1·1°C). Annual rainfall 19" (475 mm).

CONSTITUTION AND GOVERNMENT

On 10 April 1994 elections were held for a 230-member constituent assembly to reform the 1853 constitution. The Justicialist National Movement (Peronist) gained 39% of votes cast and the Radical Union 20%. On 22 Aug. 1994 this assembly unanimously adopted a new constitution. This reduces the presidential term of office from six to four years, but permits the President to stand for two terms. The President is no longer elected by an electoral college, but directly by universal suffrage. A presidential candidate is elected with more than 45% of votes cast, or 40% if at least 10% ahead of an opponent; otherwise there is a second round. The Constitution reduces the President's powers by instituting a *Chief of Cabinet*. The bicameral *National Congress* consists of a Senate and a Chamber of Deputies. The Senate comprises 72 members (one-third of the members elected every two years to six-year terms). The Chamber of Deputies comprises 257 members (one-half of the members elected every two years to four-year terms) directly elected by universal suffrage. Voting is compulsory for citizens aged 18 to 70 and—with effect from mid-term elections in Oct. 2013—optional for those aged 16 and 17.

National Anthem

'Oíd, mortales, el grito sagrado: Libertad' ('Hear, mortals, the sacred cry of Liberty'); words by V. López y Planes, 1813; tune by J. Blas Parera.

GOVERNMENT CHRONOLOGY

Presidents since 1944. (FREJULI = Justicialista Liberation Front; FV = Front for Victory; PJ = Justicialist Party; PL = Labour Party; PP = Peronist Party; UCR = Radical Civic Union; UCRI = Radical Intransigent Civic Union; UCRP = People's Radical Civic Union)

1944–46	military	Edelmiro Julián Farrell Plaul
1946–55	military/PL/PP	Juan Domingo Perón Sosa
1955	military	Eduardo A. Lonardi Doucet
1955–58	military	Pedro Eugenio Aramburu Cilveti
1958–62	UCRI	Arturo Frondizi Ercoli
1962–63	UCRI	José María Guido
1963–66	UCRP	Arturo Umberto Illia Francesconi
1966–70	military	Juan Carlos Onganía Carballo
1970–71	military	Roberto Marcelo Levingston Laborda
1971–73	military	Alejandro Agustín Lanusse Gelly
1973	FREJULI	Héctor José Cámpora Demaestre
1973	FREJULI	Raúl Alberto Lastiri
1973–74	PJ	Juan Domingo Perón Sosa
1974–76	PJ	María Estela Martínez de Perón
1976–81	military	Jorge Rafael Videla
1981	military	Roberto Eduardo Viola
1981–82	military	Leopoldo Fortunato Galtieri
1982–83	military	Reynaldo Benito Bignone
1983–89	UCR	Raúl Ricardo Alfonsín
1989–99	PJ	Carlos Saúl Menem
1999–2001	UCR	Fernando de la Rúa
2002–03	PJ	Eduardo Duhalde
2003–07	PJ	Néstor Carlos Kirchner
2007–	FV	Cristina Fernández de Kirchner

RECENT ELECTIONS

In the presidential elections held on 23 Oct. 2011 incumbent Cristina Fernández de Kirchner (Front for Victory) won 54·0% of the vote, followed by Hermes Binner (Broad Progressive Front) with 16·9%, Ricardo Alfonsín (Union for the Social Development) with 11·1%, Alberto Rodríguez Saá (Federal Commitment) with 8·0%, Eduardo Duhalde (Popular Front) with 5·6%, Jorge Altamira (Workers Left Front) with 2·3% and Elisa Carrió (Civic Coalition) with 1·8%. Turnout was 78·9%.

In elections to the Chamber of Deputies also held on 23 Oct. 2011, 130 of the 257 seats that were not contested at the previous elections in June 2009 were at stake. Following the elections the Front for Victory and its allies held 135 seats (of which the Front for Victory itself had 116), the Radical Civic Union and its allies 42 (of which the Radical Civic Union itself had 38), Federal Peronism and its allies 23, the Broad Progressive Front 22, Republican Proposal and its allies 13 and others 22.

CURRENT GOVERNMENT

President: Cristina Fernández de Kirchner; b. 1953 (Front for Victory; sworn in 10 Dec. 2007 and re-elected in Oct. 2011).

Vice-President: Amado Boudou.

In March 2013 the cabinet comprised:

Chief of the Cabinet: Juan Manuel Abal Medina. *Minister of Agriculture*: Norberto Yauhar. *Defence*: Arturo Puricelli. *Economy*: Hernán Lorenzino. *Education*: Alberto Sileoni. *Federal Planning, Public Investment and Services*: Julio De Vido. *Foreign Affairs and Worship*: Héctor Timerman. *Health*: Juan Luis Manzur. *Interior*: Florencio Randazzo. *Justice, Security and Human Rights*: Julio Alak. *Labour, Employment and Social Security*: Carlos Tomada. *Production*: Débora Giorgi. *Science, Technology and Innovative Production*: Lino Barañao. *Security*: Nilda Garré. *Social Development*: Alicia Kirchner. *Tourism*: Enrique Meyer.

Office of the President (Spanish only):
http://www.presidencia.gov.ar

CURRENT LEADERS

Cristina Fernández de Kirchner

Position
President

Introduction
Cristina Fernández de Kirchner was sworn in as president on 10 Dec. 2007, representing the ruling Front for Victory party (FV). She succeeded her husband, Néstor Kirchner, and became the first elected female president of Argentina, though not the first female president. She was re-elected for a second term in Oct. 2011.

Early Life
Cristina Kirchner, also called Fernández, was born in 1953 in the La Plata region of Buenos Aires. Her father was a businessman and unionist of Spanish heritage. Her mother was a civil servant in the ministry for the economy and member of the Peronists of German decent. Fernández was educated at secondary level in Buenos Aires and at the Colegio Nuestra Señora de la Misericordia, a private college run by nuns.

During the 1970s she studied at La Plata National University, reading psychology before converting to law. In 1974 she met Néstor Kirchner, a law student, and they married after six months. Both were members of the Tendencia Revolucionaria faction of the Peronist Justicialista Party (PJ). In 1975, following Juan Perón's death, the Kirchners moved to Rio Gallegos, the capital of Santa Cruz. They set up a law practice and distanced themselves from politics while the military junta that had ended the Peronist government in 1976 with a coup d'état held power until 1983. Whether Fernández completed her law degree remains disputed.

Her political career began shortly after her husband's. In 1985 she became a member of the PJ and was elected provincial representative of Santa Cruz in 1989, 1993 and 1995. Her husband became mayor of Rio Gallegos in 1987 and governor of Santa Cruz in 1991. In 1995 Fernández resigned from provincial politics to represent Santa Cruz in the Senate, where she built her national profile. In 1997 and 2001 she was elected to represent Santa Cruz in the Chamber of Deputies. In 2003 she helped her husband win the presidential election and became first lady. Two years later she secured the senatorship for Buenos Aires, representing the FV. In 2007, despite good poll ratings, Néstor Kirchner decided not to run for re-election as president and Fernández was elected in his place.

Career in Office
Fernández pledged to further her husband's economic policies. However, in the early stages of her tenure she faced budget restrictions and other challenges, including an energy shortage, rising inflation, large public debt and dependency on GM soya exports. She also faced continuing accusations from the IMF that the national statistics institute manipulate official inflation figures. In July 2008, in a politically damaging defeat, Fernández cancelled tax increases on agricultural exports that had provoked months of protests by farmers. This was followed by another controversial government plan to nationalize private pension funds, ostensibly to protect pensioners' assets during the erupting global financial crisis, which was approved by parliament in Nov. As the economy deteriorated rapidly from late 2008 and her popularity slumped, Fernández brought forward partial congressional elections from Oct. 2009 to June. However, her party supporters still lost their absolute majorities in both parliamentary houses in the polling and her husband failed to gain election to the Chamber of Deputies.

In Oct. 2009 the government indicated its willingness to negotiate with holders of US$20bn. of bonds (on which Argentina had defaulted in 2001) in a debt restructuring initiative aimed at restoring the country's access to international credit. In April 2010 the government announced the terms of its offer to creditors, and by June about 90% of defaulted bonds had been exchanged.

Earlier, in Jan. 2010, Fernández had secured the removal of the governor of the Central Bank, Martín Redrado, for opposing her proposal to use the Bank's hard-currency reserves to repay debt. In Nov. that year she announced that the Paris Club of creditor countries had agreed to negotiate Argentina's debt default repayment of about US$7·5bn. without any intervention from the IMF (whose policies were rejected by both Fernández and her predecessor). In May 2012 the government announced that it would nationalize a majority stake in the energy firm YFP, the former state oil company owned by Spain's Repsol. The move incurred trading retaliation by the Spanish government and a referral to the World Trade Organization by the European Union.

Regarding social policy, in July 2010 Argentina became the first Latin American country to legalize same-sex marriage and give same-sex couples inheritance and adoption rights.

Internationally, Fernández sought to raise Argentina's profile, and she has represented Argentina at the G20 forum of wealthy nations. In 2010 the issue of disputed sovereignty over the Falkland Islands threatened to reignite tensions with the United Kingdom as the arrival of a British oil exploration rig in surrounding South Atlantic waters prompted diplomatic protests from the Argentinian government and the imposition of restrictions on shipping. Rancour over the disputed territory continued and in March 2013 the Falklands government held a referendum on the islands' political status in a bid to reaffirm UK sovereignty. 99·8% of votes cast were in favour of retaining the current status.

In Oct. 2010 Fernández was widowed as Néstor Kirchner died suddenly of a heart attack, depriving her not only of her husband but her principal political adviser.

Reflecting a revival in her popularity in the wake of stronger economic growth (but also rising inflation), Fernández won a second presidential term in elections in Oct. 2011 with 54% of the vote.

DEFENCE

Conscription was abolished in 1994. In 2008 defence expenditure totalled US$2,031m. (US$50 per capita), representing 0·6% of GDP (compared to over 8% in 1981).

Army
In 2007 the Army was 41,400 strong. There are no reserves formally established or trained.

There is a paramilitary gendarmerie of 18,000 run by the Ministry of Interior.

Navy
In 2007 the Argentinian Navy included three diesel submarines, five destroyers and nine frigates. Total personnel was 20,000 including 2,000 in Naval Aviation and 2,500 marines. Main bases are at Puerto Belgrano, Mar del Plata and Ushuaia.

The Naval Aviation Service had 28 combat-capable aircraft in 2007, including Super-Etendard strike aircraft, and 19 helicopters, including Agusta/Sikorsky ASH-3H Sea Kings.

Air Force
The Air Force is organized into Air Operations, Air Regions, Logistics and Personnel Commands. There were (2007) 14,600 personnel and 119 combat-capable aircraft including A-4 Skyhawk, Mirage 5 and Mirage III jet fighters.

ECONOMY

Agriculture contributed 7·5% of GDP in 2009, industry 31·8% and services 60·7%.

In Jan. 2006 the government repaid the country's entire US$9·57bn. debt to the IMF ahead of schedule.

Overview
By the late 1980s macroeconomic mismanagement had caused hyperinflation. The economy contracted at an average annual rate

of 0·7% over the decade. Structural reforms and a 1991 convertibility plan (establishing a currency peg to the US dollar) helped stabilize the economy for most of the 1990s. Between 1991 and 1997 GDP grew by 6·2% per year on average. The fixed exchange-rate regime survived the Mexican and Asian financial crises but the balance of payments was unable to withstand the pressure caused by subsequent shocks. By the fourth quarter of 1998 the economy was in recession which, combined with a low fiscal surplus and weak restraint in the provinces, threatened the country's ability to pay its foreign debt. The growing strength of the US dollar and the devaluation of the Brazilian *real* in 1999 increased pressure on the pegged peso.

The slowdown of the global economy in 2001 added to Argentina's plight. In Dec. 2001 it recorded the largest sovereign debt default in history and in Jan. 2002 abandoned convertibility. From 1999–2002 the economy contracted by 18·4% and poverty grew dramatically. Strong recovery between 2003 and 2007 was underpinned by a firm fiscal policy, successful debt restructuring and favourable international market conditions including high commodity prices, low interest rates and strong world growth. Unemployment fell and the urban poverty rate dropped from 48% in 2003 to 10% in 2010.

Argentina weathered the global financial crisis, with strong growth in 2010 thanks largely to high agricultural production and exports. Soybean products are the country's leading export. In June 2010 Argentina concluded an US$18·3bn. debt swap process, in which bad debts stemming from the 2001 default were exchanged for new bonds, with a participation rate of 66% among the government's creditors. This combined with the 2005 restructuring enabled the government to settle 92% of the US$100bn. default debt. However, those who rejected the debt swap offers in 2005 and 2010 have won a series of judgments in US courts ordering Argentina to pay.

Inflation was officially just below 10% in 2011 although independent consulting companies suggest a rate of more than double this. Argentina's reported Consumer Price Index and GDP data are not aligned with international statistical guidelines and the IMF approved a decision in 2012 calling on Argentina to address the quality of its reported data.

Currency
The monetary unit is the *peso* (ARS), which replaced the *austral* on 1 Jan. 1992 at a rate of one peso = 10,000 australs. For nearly a decade the peso was pegged at parity with the US dollar, but it was devalued by nearly 30% in Jan. 2002 and floated in Feb. 2002. Inflation rates (based on IMF statistics)[1]:

2002	2003	2004	2005	2006	2007	2008	2009	2010	2011
25·9%	13·4%	4·4%	9·6%	10·9%	8·8%	8·6%	6·3%	10·5%	9·8%

[1]The IMF has questioned the accuracy of Argentinian inflation figures since 2007.

Gold reserves were 1·76m. troy oz in Sept. 2009; foreign exchange reserves were US$43,111m. Total money supply was 107,615m. pesos in Aug. 2009.

Budget
In 2005 revenues totalled 82,106m. pesos and expenditures 77,531m. pesos. Tax revenue accounted for 77·4% of revenues in 2005; current expenditure accounted for 88·2% of expenditures.

VAT is 21% (reduced rate, 10·5%).

Performance
Real GDP growth rates (based on IMF statistics)[1]:

2002	2003	2004	2005	2006	2007	2008	2009	2010	2011
−10·9%	9·0%	8·9%	9·2%	8·5%	8·7%	6·8%	0·9%	9·2%	8·9%

[1]The IMF has questioned the accuracy of Argentinian GDP growth figures since 2008.

Total GDP was US$446·0bn. in 2011.

In March 2001 the economy minister, José Luis Machinea, resigned after a turbulent 15 months in which he had failed to revive a stagnant economy. As the economic situation deteriorated Argentina had a further five economy ministers in the space of just over a year. In Nov. 2001 the government tried to persuade creditors to accept a restructuring of the US$132bn. public debt, but on 23 Dec. 2001 interim President Adolfo Rodríguez Saá announced that Argentina would default on the debt payments—the biggest debt default in history.

Banking and Finance
The total assets of the Argentine Central Bank (BCRA) in Dec. 2007 were 228·92bn. pesos. The *President* of the Central Bank is Mercedes Marcó del Pont. In early 2002 banks and financial markets were temporarily closed as an emergency measure in response to the economic crisis that made the country virtually bankrupt. In 2002 there were 16 government banks, 24 private commercial banks, four co-operative banks, one other national bank (Banco Hipotecario Nacional) and 17 foreign banks.

In 2010 total external debt was US$127,849m., representing 36·1% of GNI.

There is a main stock exchange at Buenos Aires and there are others in Córdoba, Rosario, Mendoza and La Plata.

ENERGY AND NATURAL RESOURCES
Environment
Argentina's carbon dioxide emissions from the consumption and flaring of fossil fuels in 2008 were the equivalent of 4·3 tonnes per capita. An *Environmental Performance Index* compiled in 2008 ranked Argentina 38th in the world, with 81·8%. The index examined various factors in six areas—air pollution, biodiversity and habitat, climate change, environmental health, productive natural resources and water resources.

Electricity
Installed capacity in 2007 was 28·2m. kW. Electric power production (2007) was 115,296m. kWh (7,217m. kWh nuclear); consumption per capita in 2007 was 3,124 kWh. In 2010 there were two nuclear reactors operational.

Oil and Gas
Oil production (2008) was 34·1m. tonnes. Reserves were 2·6bn. bbls in 2008. The oil industry was privatized in 1993. Natural gas extraction in 2008 was 44·1bn. cu. metres. Reserves were 440bn. cu. metres in 2008. The main area in production is the Neuquen basin in western Argentina, with over 40% of the total oil reserves and nearly half the gas reserves. Natural gas accounts for approximately 45% of all the energy consumed in Argentina.

Minerals
Minerals (with production in 2005) include clays (6·4m. tonnes), salt (1·8m. tonnes), borates (632,792 tonnes), aluminium (270,714 tonnes), bentonite (247,101 tonnes), copper (187,317 tonnes), coal (110,000 tonnes in 2007), zinc (30,227 tonnes of metal), lead (10,683 tonnes of metal), silver (264 tonnes), gold (27,904 kg), granite, marble and tungsten. Production from the US$1·1bn. Alumbrera copper and gold mine, the country's biggest mining project, in Catamarca province in the northwest, started in 1997. In 1993 the mining laws were reformed and state regulation was swept away, creating a more stable tax regime for investors. In 1997 Argentina and Chile signed a treaty laying the legal and tax framework for mining operations straddling the 5,000 km border, allowing mining products to be transported out through both countries.

Agriculture
In 2007 there were around 32·5m. ha. of arable land and 1·0m. ha. of permanent crops. 1·56m. ha. were irrigated in 2006. The agricultural population was an estimated 3·18m. in 2008, of whom

1·42m. were economically active. In 2009 there were 4·40m. ha. of organic agricultural land (the second largest area after Australia), representing 3·3% of the total area under agriculture.

Livestock (2009 estimates): cattle, 50,750,000; sheep, 12,450,000; goats, 4,250,000; horses, 3,680,000; pigs, 2,270,000. In 2008–09 greasy wool production was 54,000 tonnes; milk (in 2007), 9,527m. litres; eggs (in 2007), 696m. dozen.

Crop production (in 1,000 tonnes) in 2006–07: soybeans, 47,500; sugarcane (2008 estimate), 29,950; maize, 21,800; wheat, 14,500; sunflower seeds, 3,500; potatoes (2008 estimate), 1,950. Cotton, vine, citrus fruit, olives and *yerba maté* (Paraguayan tea) are also cultivated. Argentina is the world's leading producer of sunflower seeds, and is now the fifth largest wine producer (15,046,000 hectolitres in 2007) after Italy, France, Spain and the USA; it ranked seventh in the world in 2007 for wine exports and eighth for wine consumption.

Forestry
The forest area was 29·40m. ha., or 11% of the total land area, in 2010. Production in 2006 included 1·75m. cu. metres of sawn wood, 8·80m. tonnes of rough timber, 1·43m. tonnes of paper and cardboard and 585,000 cu. metres of chipboard.

Fisheries
Fish landings in 2010 amounted to 811,749 tonnes (down from 921,797 tonnes in 2000), almost exclusively from sea fishing.

INDUSTRY
Production, 2007 unless otherwise indicated (in 1,000 tonnes): distillate fuel oil, 10,970; cement, 9,602; crude steel, 5,388; petrol, 4,846; pig iron, 4,389; residual fuel oil, 4,267; sugar (2006), 2,312; paper (2006), 1,721; jet fuel, 1,283; polyethylene, 575; synthetic rubber, 54. Motor vehicles produced in 2007 totalled 350,735; car tyres, 12,079,000; motorcycles, 225,397.

Labour
In 2005 the economically active population totalled 15·79m., of which 12·55m. were employed and 3·24m. were unemployed. The urban unemployment rate, which had been 12·9% in 1998, rose to a record 21·5% by May 2002 at the height of the economic crisis before falling to 15·6% in May 2003.

INTERNATIONAL TRADE
Imports and Exports
Imports (c.i.f.) in 2007 totalled US$44,707m. (US$34,154m. in 2006); exports (f.o.b.), US$55,780m. (US$46,546m. in 2006).

Principal imports in 2007 (in US$1m.) were machinery and transport equipment (21,201); chemicals and related products (8,252); petroleum and petroleum products (1,835); iron and steel (1,498). Principal exports in 2007 (in US$1m.) were food and livestock (18,270); machinery and transport equipment (7,656); fixed vegetable fats and oils (5,208); petroleum and petroleum products (4,699).

In 2007 imports (in US$1m.) were mainly from Brazil (14,660); USA (5,342); China (5,093); Germany (2,131). Exports went mainly to Brazil (10,486); China (5,167); USA (4,344); Chile (4,176).

COMMUNICATIONS
Roads
In 2003 there were 231,374 km of roads, of which 30·0% were paved. The four main roads constituting Argentina's portion of the Pan-American Highway were opened in 1942. Vehicles in use in 2007 totalled 12,399,900. In 2005, 3,443 people were killed in road accidents.

Rail
Much of the 34,000 km state-owned network (on 1,000 mm, 1,435 mm and 1,676 mm gauges) was privatized in 1993–94. 30-year concessions were awarded to five freight operators; long-distance passenger services are run by contractors to the requirements of local authorities. Metro, light rail and suburban railway services are also operated by concessionaires.

The rail company carrying the most passengers is Trenes de Buenos Aires (190m. in 2008); Ferrosur Roca carries the most freight (5·1m. tonnes in 2005–06).

The metro and light rail network in Buenos Aires extended to 75 km in 2005. A light railway opened in Mendoza in 2012, with a total length of 12·5 km.

Civil Aviation
The main international airport is Buenos Aires Ezeiza, which handled 7,910,048 passengers in 2009 (7,461,727 passengers on international flights). The second busiest airport is Buenos Aires Aeroparque Jorge Newbery, which handled 6,449,344 passengers in 2009. It is much more important as a domestic airport, with only 524,934 passengers on international flights in 2009. The national carrier, Aerolíneas Argentinas, was privatized in 1990 but renationalized in Sept. 2008. In 2010 it operated direct flights to Asunción, Auckland, Barcelona, Bogotá, Caracas, Florianópolis, Lima, Madrid, Miami, Montevideo, Porto Alegre, Punta del Este, Rio de Janeiro, Rome, Salvador, Santa Cruz, Santiago, São Paulo and Sydney, as well as domestic services. There were direct international flights with other airlines in 2010 to Asunción, Atlanta, Barcelona, Belo Horizonte, Bogotá, Cape Town, Cayo Coco, Cochabamba, Curitiba, Dallas, Florianópolis, Frankfurt, Guayaquil, Havana, Houston, Johannesburg, Kuala Lumpur, La Paz, Lima, London, Los Angeles, Madrid, Mexico City, Miami, Montevideo, New York, Panama City, Paris, Porto Alegre, Punta Cana, Punta del Este, Quito, Recife, Rio de Janeiro, Salvador, San José (Costa Rica), Santa Cruz, Santiago, São Paulo, Sydney, Toronto and Washington, D.C.

In 2006 scheduled airline traffic of Argentinian-based carriers flew 101m. km, carrying 6,636,000 passengers (1,818,000 on international flights).

Shipping
In Jan. 2009 there were 59 ships of 300 GT or over registered, totalling 500,000 GT. The leading ports are Buenos Aires (which handled 12,745,000 tonnes of cargo in 2008) and Bahía Blanca (12,676,000 tonnes of cargo in 2008).

Telecommunications
The telephone service Entel was privatized in 1990. The sell-off split Argentina into two monopolies, operated by Telefónica Internacional de España, and a holding controlled by France Télécom and Telecom Italia. In 2000 the industry was opened to unrestricted competition. In 2008 there were 9·7m. main (fixed) telephone lines. In the same year mobile phone subscribers numbered 46·5m. (1,166·1 per 1,000 persons). There were 11·2m. internet users in 2008. In June 2012 there were 19·0m. Facebook users.

SOCIAL INSTITUTIONS
Justice
Justice is administered by federal and provincial courts. The former deal only with cases of a national character, or those in which different provinces or inhabitants of different provinces are parties. The chief federal court is the Supreme Court, with five judges whose appointment is approved by the Senate. Other federal courts are the appeal courts, at Buenos Aires, Bahía Blanca, La Plata, Córdoba, Mendoza, Tucumán and Resistencia. Each province has its own judicial system, with a Supreme Court (generally so designated) and several minor chambers. The death penalty was reintroduced in 1976—for the killing of government, military police and judicial officials, and for participation in terrorist activities—but was abolished in 2008. The population in penal institutions in Dec. 2006 was 60,621 (154 per 100,000 of national population). In 2008 there were a total of 1,310,977 crimes reported.

The police force is centralized under the Federal Security Council.

Education

Adult literacy was 98% in 2008. In 2005, 1,324,529 children attended pre-school institutions, 6,510,382 pupils were in basic general education, 1,545,992 in 'multimodal' secondary schooling and 509,134 in higher non-universities.

In 2006, in the public sector, there were 39 universities (including one technical university and one art institute) and university institutes of aeronautics, military studies, naval and maritime studies, and police studies. In the private sector, there were 40 universities (including seven Roman Catholic universities) and ten university institutes. In 2006 there were 1,304,003 students attending public universities and 279,373 at private universities.

In 2006 public expenditure on education came to 4·6% of GDP and 14·0% of total government spending.

Health

Free medical attention is obtainable from public hospitals. In 2001 there were 7,833 public health care institutions with an average of 75,075 available beds. In 2002 there were 99,400 physicians.

Welfare

Until the end of 1996 trade unions had a monopoly in the handling of the compulsory social security contributions of employees, but private insurance agencies are now permitted to function alongside them.

Unique Social Security System Expenditure (in 1m. pesos):

	1999	2000
Retirement and pensions	17,508	17,386
Healthcare assistance and other forms of social insurance	5,249	5,440
Family allowances	1,879	1,920
Unemployment insurance, employment and training programmes	519	484
Work risks insurance	323	367
Other	2,201	2,361
Total	27,679	27,958

RELIGION

The Roman Catholic religion is supported by the State; affiliation numbered 29·92m. in 2001. There were three cardinals in April 2013. Jorge Mario Bergoglio was a cardinal from 2001 until March 2013, when he was selected to succeed Benedict XVI as Pope. There were 2·04m. Protestants of various denominations in 2001, 730,000 Muslims and 500,000 Jews. There were 275,000 Latter-day Saints (Mormons) in 1998.

CULTURE

World Heritage Sites

Argentina's heritage sites as classified by UNESCO (with year entered on list) are: Los Glaciares national park (1981), the Iguazu National Park (1984), and the Ischigualasto and Talampaya Natural Parks (2000). The Cueva de las Manos (Cave of Hands, 1999), in Patagonia, contains cave art that is between 1,000 and 10,000 years old. The Península Valdés (1999) in Patagonia protects several endangered species of marine mammal. The Jesuit Block and Estancias of Córdoba (2000) are the principal buildings of the Jesuit community from the 17th and 18th century. The Quebrada de Humahuaca (2003) is a valley on the Camino Inca trade route. Shared with Brazil, the Jesuit Missions of the Guaranis (1984) encompasses the ruins of five Jesuit missions.

Press

In 2008 there were 197 daily newspapers with a combined average daily circulation of 1·5m. The main newspapers are *Clarín*, *La Nación* and *Diario Popular*.

Tourism

In 2009, 4,329,000 tourists visited Argentina (excluding same-day visitors), down from a record 4,700,000 in 2008 and 4,562,000 in 2007. Of the 4,329,000 tourists in 2009, 3,413,000 were from elsewhere in the Americas and 722,000 were from Europe.

Festivals

Carnival is celebrated in Feb.–March. The Grape Harvest Festival (Fiesta Nacional de la Vendimia) takes place in Mendoza at the end of Feb. Mar del Plata holds an annual National Sea Festival (Dec.–Jan.) and International Film Festival (Nov.). Cosquín Rock (Feb.) and, in Buenos Aires, Pepsi music festival (Sept.–Oct.), Personal Fest (Oct.–Dec.) and Bue Festival (Nov.), are the leading pop and rock festivals. Buenos Aires Tango Festival takes place in Feb.–March.

DIPLOMATIC REPRESENTATIVES

Of Argentina in the United Kingdom (65 Brook St., London, W1K 4AH)
Ambassador: Alicia Castro.

Of the United Kingdom in Argentina (Dr Luis Agote 2412, 1425 Buenos Aires)
Ambassador: John Freeman.

Of Argentina in the USA (1600 New Hampshire Ave., NW, Washington, D.C., 2009)
Ambassador: Jorge Argüello.

Of the USA in Argentina (Av. Colombia 4300, 1425 Buenos Aires)
Ambassador: Vilma S. Martinez.

Of Argentina to the United Nations
Ambassador: Pablo Martin.

Of Argentina to the European Union
Ambassador: Jorge Remes Lenicov.

FURTHER READING

Bethell, L. (ed.) *Argentina since Independence.* 1994
Levitsky, Steven, *Argentine Democracy: The Politics of Institutional Weakness.* 2006
Pion-Berlin, David, *Broken Promises? The Argentine Crisis and Argentine Democracy.* 2006
Powers, Nancy R., *Grassroots Expectations of Democracy and Economy: Argentina in Comparative Perspective.* 2001
Romero, Luis Alberto, *A History of Argentina in the Twentieth Century;* translated from Spanish. 2002

National Statistical Office: Instituto Nacional de Estadística y Censos (INDEC). Av. Julio A. Roca 615, PB (1067) Buenos Aires. *Director:* Ana María Edwin.
Website: http://www.indec.gov.ar

ARMENIA

GEORGIA
Gyumri Vanadzor
ARMENIA
AZERBAIJAN
Echmiatsin Lake Sevan
YEREVAN
TURKEY
AZERBAIJAN
IRAN

0   35 mi
0   50 km
© Research Machines plc 2006

Hayastani Hanrapetoutiun
(Republic of Armenia)

Capital: Yerevan
Population projection, 2015: 3·13m.
GNI per capita, 2011: (PPP$) 5,188
HDI/world rank: 0·716/86
Internet domain extension: .am

KEY HISTORICAL EVENTS

Some of the world's earliest settlements, dating from around 6200 BC, were built on the Armenian plateau, part of the 'fertile crescent' encompassing Anatolia, Mesopotamia and the Levant. Independent culture and language developed and trade flourished with neighbouring civilizations. By 900 BC the kingdom of Urartu had taken root, centred on Lake Van, until falling to the Assyrian and Scythian empires in the sixth century BC.

In 190 BC a Hellenistic Armenian state emerged from the remnants of Alexander the Great's short-lived empire. Led by King Artaxias, Greater Armenia reached its peak around 70 BC, stretching from central Anatolia to the Levant, Syria and northwestern Persia. The imperialistic ambitions of King Tigranes ended in defeat against Rome, with Armenia becoming a tributary kingdom. Christianity was adopted as the state religion in AD 301. The Sassanid Persians overran Armenia in AD 328, although the persecution of Christians kindled nationalism, particularly after the partition of the kingdom in AD 387 between Persia and Rome.

Despite successive domination by Persian, Byzantine and Arab forces between the fifth and ninth centuries, Armenian identity was maintained by nobles ('nakharars') with highly militarized fiefdoms. In 884 sovereignty was restored under the Bagrunti dynasty, which oversaw prosperity and urban development for over 150 years.

Byzantine forces recaptured Armenia in 1040 but were overwhelmed by Seljuk Turks at the battle of Manzikert in 1071. The expansion of the Mongol empire forced Armenians, led by Prince Reuben, westward to the former Byzantine province of Cilicia. A new state was founded, centred on Sis, and prospered on trade between the Mediterranean, the Middle East and Central Asia, until conquered by Mamlik Turks in 1337.

There were further Mongol attacks until, following the death of the great Turkic warrior Tamerlane in 1405, the western part of the Armenian plateau came under Ottoman Turkish control, while its eastern reaches were dominated by Persia. Led by the merchants, many Armenians migrated, establishing communities in Constantinople and Tiflis.

Russian expansion into the Caucasus in the early 19th century resulted in eastern Armenia becoming a Russian province in 1828. Following Russia's victory in the Russo–Turkish War in 1878, the province incorporated Kars, Ardahan and Batumi. Western Armenia remained under the waning Ottoman Empire. The brutal reign of Sultan Abdul-Hamid II led to an uprising in 1894, with between 100,000 and 300,000 dying in reprisal massacres over the following two years.

The ultranationalist Committee of Union and Progress, which led the Ottoman government from 1913, has been blamed for the systematic killing in 1915 of between 600,000 and 1·5m. Armenians. Turkey and Azerbaijan deny this amounted to genocide, arguing that the number of deaths was inflated and resulted from inter-ethnic violence during the First World War.

On 28 May 1918 the Democratic Republic of Armenia became an independent state. Violent clashes with the newly established republics of Georgia and Azerbaijan followed. Hopes that its former provinces in eastern Anatolia would be returned to Armenia as outlined in the 1920 Treaty of Sèvres were dashed when opposing Turkish and Red Army forces invaded the country. The Treaty of Kars, which ended Bolshevik–Turkish hostilities in Oct. 1921, ceded Mount Ararat and the ancient city of Ani to Turkey. The Armenian Soviet Socialist Republic became one of the 15 constituent republics of the USSR in Dec. 1922.

The collapse of the Soviet Union prompted a declaration of independence in Sept. 1991. The new president, Levon Ter-Petrosyan, led the country into the Commonwealth of Independent States. Hostilities with Azerbaijan over sovereignty of Nagorno-Karabakh erupted in 1988, intensifying into full-scale war in 1992. A ceasefire was agreed in 1994, with 30,000 having lost their lives and up to 1m. Armenians and Azeris displaced. A new constitution in July 1995 led to elections. Ter-Petrosyan was re-elected in 1996, though OSCE observers noted 'serious irregularities' in electoral procedures.

Ter-Petrosyan was succeeded by Robert Kocharyan in 1998. On 27 Oct. 1999 gunmen invaded parliament and killed the premier, Vazgen Sargsyan, and seven other officials. Aram Sargsyan was named his brother's successor. Kocharyan won a second presidential term in 2003, amid renewed complaints of voting irregularities. Thousands marched against the president in Yerevan in April 2004. Serzh Sargsyan, whose right-wing Republican Party took a third of seats in the 2007 parliamentary election, became president in Feb. 2008. When tens of thousands of opposition supporters took to the streets, a state of emergency was declared.

The historic visit of the Turkish president, Abdullah Gül, in Sept. 2008 seemed to herald a rapprochement between the two nations, although Sargsyan's calls to establish diplomatic relations and open borders have gone unheeded.

TERRITORY AND POPULATION

Armenia covers an area of 29,743 sq. km (11,484 sq. miles). It is bounded in the north by Georgia, in the east by Azerbaijan and in the south and west by Iran and Turkey.

The 2011 census population (provisional) was 3,018,854; population density, 101 per sq. km. Armenians account for 97·9%, Kurds 1·3% and Russians 0·5%—in 1989, prior to the Nagorno-Karabakh conflict, 2·6% of the population were Azeris. Approximately 64% lived in urban areas in 2009.

The UN gives a projected population for 2015 of 3·13m.

There are an estimated 8m. Armenians worldwide, mainly living in Russia, the USA and Georgia as well as in Armenia itself.

The capital is Yerevan (estimated population of 1,116,600 in 2010). Other large towns are Gyumri (formerly Leninakan) (146,300, 2010 estimate) and Vanadzor (formerly Kirovakan) (104,800, 2010 estimate).

The official language is Armenian.

SOCIAL STATISTICS

2010 births, 44,825; deaths, 27,921; marriages, 17,984; divorces, 2,097. Rates, 2010 (per 1,000 population): birth, 13·8; death, 8·6; marriage, 5·5; divorce, 0·9. Infant mortality, 2010, 18 per 1,000 live births. Annual population growth rate, 2008–10, 0·2%. Life expectancy at birth, 2007, 70·1 years for men and 76·7 years for women; fertility rate, 2008, 1·7 births per woman.

CLIMATE

Summers are very dry and hot although nights can be cold. Winters are very cold, often with heavy snowfall. Yerevan, Jan. –9°C, July 28°C. Annual rainfall 318 mm.

CONSTITUTION AND GOVERNMENT

The constitution was adopted by a nationwide referendum on 5 July 1995. The head of state is the *President*, directly elected for five-year terms. Parliament is a 131-member *Azgayin Zhoghov* (National Assembly), with 90 deputies elected by party list and 41 chosen by direct election. The government is nominated by the President.

National Anthem

'Mer Hayrenik, azat ankakh' ('Land of our fathers, free and independent'); words by M. Nalbandyan, tune by B. Kanachyan.

RECENT ELECTIONS

In presidential elections held on 18 Feb. 2013 incumbent Serzh Sargsyan received 58·6% of votes cast, ahead of Raffi Hovannisian with 36·8% and five other candidates. Turnout was 60·1%. OSCE observes noted improvements over previous elections but concluded that the election was not genuinely competitive. There was subsequently a series of protests by the opposition against alleged electoral fraud.

Parliamentary elections were held on 6 May 2012. The Republican Party of Armenia won 69 seats (64 in 2007); Prosperous Armenia, 36 (25 in 2007); Armenian National Congress, 7 (none in 2007); Armenian Revolutionary Federation, 6 (16 in 2007); Rule of Law, 6; Heritage Party, 5. Two seats went to independents. Turnout was 62·3%.

CURRENT GOVERNMENT

President: Serzh Sargsyan; b. 1954 (HHK; in office since 9 April 2008).

In March 2013 the coalition government comprised:

Prime Minister: Tigran Sargsyan; b. 1960 (ind.; appointed 9 April 2008).

Deputy Prime Minister and Minister of Territorial Administration: Armen Gevorgyan.

Minister of Foreign Affairs: Edvard Nalbandyan. *Defence:* Seyran Ohanyan. *Justice:* Hrayr Tovmasyan. *Education and Science:* Armen Ashotyan. *Health:* Derenik Dumanyan. *Culture:* Hasmik Poghosyan. *Transport and Communication:* Gagik Beglaryan. *Agriculture:* Sergo Karapetyan. *Environmental Protection:* Aram Harutyunyan. *Economy:* Tigran Davtyan. *Finance:* Vache Gabrielyan. *Energy and Natural Resources:* Armen Movsissyan. *Urban Development:* Samvel Tadevosyan. *Labour and Social Affairs:* Artyom Asatryan. *Sport and Youth Affairs:* Hrachya Rostomyan. *Emergency Situations:* Armen Yeritsyan. *Diaspora Affairs:* Hranush Hacobyan.

Government Website: http://www.gov.am

CURRENT LEADERS

Serzh Sargsyan

Position
President

Introduction
Serzh Sargsyan was sworn into office on 9 April 2008, nearly two months after he was declared the winner of the disputed presidential election. Although said to be fair by international observers, the election sparked violent protests. A 20-day state of emergency was declared by the government before Sargsyan took office. He won re-election in Feb. 2013.

Early Life
Sargsyan was born on 30 June 1954 in Stepanakert, the capital of the Nagorno-Karabakh region. He enrolled at Yerevan State University in 1971 but served in the Soviet armed forces from 1972–74. He then worked as a metal turner from 1975–79 before graduating from the philological department of Yerevan State University in 1979. In the same year he became the divisional head of the Young Communist Union for Stepanakert.

Rising through the ranks of the Communist Party, Sargsyan was the leader of the Nagorno-Karabakh Republic self-defence forces committee during the Nagorno-Karabakh conflict of 1989–93, masterminding many battles. From 1993–95 he served as the Armenian minister of defence and was appointed head of the state security department in 1995. He was later promoted to minister of national security. In 1996 he took over the ministry of interior portfolio when it merged with the ministry of national security. When the offices separated in 1999, Sargsyan retained the ministry of national security. In the same year he joined the president's office as chief of staff and was selected as secretary of Armenia's national security council. In 2000 he returned to the defence ministry, where he served until 2007.

Sargsyan joined the conservative Republican Party of Armenia (HHK) in 2006 and from July 2006 to Nov. 2007 chaired the party council. He became party chairman in Nov. 2007. On 4 April 2007 President Robert Kocharyan appointed Sargsyan prime minister after the sudden death of incumbent Andranik Margaryan. On 19 Feb. 2008 Sargsyan won the presidential election by a landslide.

Career in Office
Responding to the violent protests that followed his election victory, Sargsyan called for unity and co-operation among all political factions. He appointed the non-partisan Tigran Sargsyan (no relation), former chairman of the central bank, as prime minister. Serzh Sargsyan has aimed to raise living standards and promote economic growth.

He has also sought to improve international relations, especially with Armenia's neighbours. Turkey's rejection of charges of genocide in Armenia during the First World War has previously posed a stumbling block to bilateral reconciliation, but in Oct. 2009 the two governments agreed on a framework to normalize relations, subject to parliamentary ratification by both sides. However, in April 2010 the Armenian parliament suspended the ratification process after the government accused Turkey of imposing preconditions, including a resolution of Armenia's territorial dispute with Azerbaijan. Azerbaijan has refused to concede its claim on Nagorno-Karabakh, although at several Russian-brokered meetings Sargsyan and the Azeri president have agreed to intensify their efforts to find a political settlement over the territory.

In parliamentary elections in May 2012 the governing pro-Sargsyan Republican Party retained its majority in the National Assembly.

Sargsyan was re-elected president in Feb. 2013, winning almost 59% of the vote. However, the Council of Europe expressed concern at the absence of major opposition candidates, with several boycotting the poll over fears of vote-rigging. The build-up to voting was marred by an apparent assassination attempt on the National Self-Determination Union's candidate.

DEFENCE

There is conscription for 24 months. Total active forces numbered 42,080 in 2007, including 25,105 conscripts.

Defence expenditure in 2008 totalled US$396m. (US$133 per capita), representing 3·3% of GDP.

There is a Russian military base in Armenia with 3,170 personnel in 2007.

Army

Current troop levels are 38,950, plus air and defence aviation forces of 2,220, air/air defence joint command forces of 915 and paramilitary forces of 4,750. There are approximately 210,000 Armenians who have received some kind of military service experience within the last 15 years.

INTERNATIONAL RELATIONS

There is a dispute over the mainly Armenian-populated enclave of Nagorno-Karabakh, which lies within Azerbaijan's borders—Armenia and Azerbaijan are technically still at war.

ECONOMY

In 2009 agriculture contributed 18·9% of GDP, industry 35·8% and services 45·3%.

Overview

After independence in 1991 most agricultural land was privatized. By the end of 2000, with support from the IMF and the World Bank, over 80% of medium and large enterprises and 90% of small enterprises had been privatized. Following reforms in 2001 to strengthen the business environment and promote exports and investment, there were several years of double-digit growth, driven mainly by construction, mining and services.

Poverty was reduced by 50% but rose again when the global economic crisis led to a fall in exports and remittances. Inflation rose steeply in late 2009 and real GDP declined by 14·1% in the year as a whole, with mining and construction hardest hit. Anti-crisis measures began to stabilize the economy in 2010.

Currency

In Nov. 1993 a new currency unit, the *dram* (AMD) of 100 *lumma*, was introduced to replace the rouble. Inflation, which had been 5,273% in 1994, was 3·5% in 2009 but increased to 7·3% in 2010 and 7·7% in 2011. Foreign exchange reserves were US$658m. in July 2005 and total money supply was 164,589m. drams.

Budget

Budgetary central government revenue totalled 568,918m. drams in 2007 and expenditure 415,630m. drams. Tax revenues in 2007 were 505,535m. drams. Main items of expenditure by economic type in 2007 were: use of goods and services (235,630m. drams) and grants (50,532m. drams).

VAT is 20%.

Performance

The economy contracted by 14·1% in 2009 but there was real GDP growth of 2·1% in 2010 and 4·6% in 2011. Total GDP in 2011 was US$10·2bn.

Banking and Finance

The *Chairman* of the Central Bank (founded in 1993) is Arthur Javadyan. In 2008 there were 22 commercial banks. There are commodity and stock exchanges in Yerevan and Gyumri.

In 2010 total foreign debt was US$6,103m., representing 64·8% of GNI.

ENERGY AND NATURAL RESOURCES

Environment

Armenia's carbon dioxide emissions from the consumption and flaring of fossil fuels in 2008 were the equivalent of 3·7 tonnes per capita.

Electricity

Output of electricity in 2007 was 5·9bn. kWh. Capacity was 3·2m. kW in 2007. Consumption per capita was 1,820 kWh in 2007. A nuclear plant closed in 1989 was reopened in 1995 because of the blockade of the electricity supply by Azerbaijan.

Minerals

There are deposits of copper, zinc, aluminium, molybdenum, marble, gold and granite.

Agriculture

The chief agricultural area is the valley of the Arax and the area round Yerevan. Here there are cotton plantations, orchards and vineyards. Almonds, olives and figs are also grown. In the mountainous areas the chief pursuit is livestock raising. In 2007 there were an estimated 406,000 ha. of arable land and 54,000 ha. of permanent crops Major agricultural production (in tonnes in 2003): potatoes, 508,000; tomatoes, 226,000; wheat, 217,000; cabbage, 98,000; grapes, 82,000; barley, 68,000. Livestock (2003): sheep, 553,000; cattle, 514,000; pigs, 111,000; goats, 50,000; chickens, 3m.

Forestry

In 2010 forests covered 0·26m. ha., or 9% of the total land area. Timber production in 2007 was 44,000 cu. metres.

Fisheries

The estimated total catch in 2008 came to 601 tonnes, exclusively from inland waters.

INDUSTRY

Among the chief industries are chemicals, producing mainly synthetic rubber and fertilizers, the extraction and processing of building materials, ginning- and textile-mills, carpet weaving, and food processing (including wine-making).

Labour

In 2006 the population of working age was 2·04m., of whom 1·09m. were employed: 46% in agriculture, hunting and forestry, 10% in manufacturing. The registered unemployment rate was 7·4% of the workforce in 2006. The average monthly salary in 2008 was 92,759 drams.

INTERNATIONAL TRADE

Imports and Exports

Imports and exports for calendar years in US$1m.:

	2005	2006	2007	2008	2009
Imports c.i.f.	1,691·5	2,194·4	3,052·6	4,101·2	3,174·6
Exports f.o.b.	937·0	1,004·0	1,121·2	1,055·0	684·0

The main import suppliers in 2009 were Russia (24·8%), China (9·0%), Ukraine (6·4%) and Turkey (5·6%). Principal export markets were Germany (16·8%), Russia (15·6%), USA (9·7%) and Bulgaria (8·8%). Machinery and transport equipment account for 24% of Armenia's imports, manufactured goods 21%, and mineral fuels and lubricants 15%. Manufactured goods (in particular non-ferrous metals and iron and steel) account for 45% of Armenia's exports, crude materials (excluding fuels) 20%, and beverages and tobacco 13%.

COMMUNICATIONS

Roads
There were 7,515 km of road network in 2007, of which 89·8% were paved. In 2007 there were 289,800 passenger cars and 25,679 buses and coaches. There were 371 fatalities as a result of road accidents in 2007.

Rail
Total length in 2005 was 845 km of 1,520 mm gauge. Passenger-km travelled in 2006 came to 27m. and freight tonne-km to 354m.
   There is a metro and a tramway in Yerevan.

Civil Aviation
There is an international airport at Yerevan (Zvartnots), which handled 1,443,557 passengers and 8,323 tonnes of freight in 2009. The Armenian flag carrier is Armavia, which was founded in 1996; in 2008 it carried 646,447 passengers. In 2010 there were direct flights from Yerevan to over 40 international destinations.

Telecommunications
There were 589,900 fixed telephone lines in 2010 (190·8 per 1,000 inhabitants). Mobile phone subscribers numbered 3·87m. in 2010. There were 153·0 internet users per 1,000 inhabitants in 2009. Fixed internet subscriptions totalled 96,000 in 2010 (31·1 per 1,000 inhabitants). In March 2012 there were 283,000 Facebook users.

SOCIAL INSTITUTIONS

Justice
In 2006, 9,757 crimes were reported, including 96 murders or attempted murders. The population in penal institutions in Jan. 2007 was 3,520 (109 per 100,000 of national population). The death penalty was abolished in 2003.

Education
Armenia's literacy rate was over 99% in 2008. In 2007, 48,015 children attended pre-school institutions. There were 127,546 pupils in primary schools with 6,606 teaching staff; and 336,877 pupils in secondary schools with 43,372 teaching staff. At the tertiary level there were 107,398 students in 2007 and 12,521 academic staff. The National Academy of Sciences of the Republic of Armenia (NAS RA), located in Yerevan, comprises more than 50 institutions and organizations with a staff of over 3,700.
   Public expenditure on education in 2006 came to 2·6% of GNI.

Health
In 2006 there were 12,388 physicians, 18,574 paramedics and 140 hospitals with 14,276 beds.

Welfare
In 2008 there were 523,839 pensioners. The average monthly pension was 21,370 drams in 2008.

RELIGION
Armenia adopted Christianity in AD 301, thus becoming the first Christian nation in the world. The Armenian Apostolic Church is headed by its Catholicos (Karekin II, b. 1951) whose seat is at Echmiatsin, and who is head of all the Armenian (Gregorian) communities throughout the world. In 1995 it numbered 7m. adherents (4m. in diaspora). The Catholicos is elected by representatives of parishes. The Catholicos of Cilicia is Aram I (b. 1947), with seat at Antelias. In 2001, 65% of the population belonged to the Armenian Apostolic Church.

CULTURE

World Heritage Sites
There are three UNESCO sites in Armenia: the Monasteries of Haghpat and Sanahin (inscribed in 1996 and 2000); the Monastery of Geghard and the Upper Azat Valley (2000); the Cathedral and Churches of Echmiatsin and the Archaeological Site of Zvartnots (2000).

Press
In 2008 there were 11 paid-for daily newspapers and 49 paid-for non-dailies with a combined circulation of 116,000.

Tourism
In 2010 there were 684,000 international tourist arrivals (excluding same-day visitors), up from 575,000 in 2009.

DIPLOMATIC REPRESENTATIVES
Of Armenia in the United Kingdom (25A Cheniston Gdns, London, W8 6TG)
Ambassador: Karine Kazinian.

Of the United Kingdom in Armenia (34 Baghramyan Ave., Yerevan 375019)
Ambassadors: Jonathan Aves and Katherine Leach.

Of Armenia in the USA (2225 R St., NW, Washington, D.C., 20008)
Ambassador: Tatoul Markarian.

Of the USA in Armenia (1 American Ave., Yerevan 375082)
Ambassador: John Heffern.

Of Armenia to the United Nations
Ambassador: Garen Nazarian.

Of Armenia to the European Union
Ambassador: Avet Adonts.

FURTHER READING
Brook, S., *Claws of the Crab: Georgia and Armenia in Crisis.* 1992
De Waal, Thomas, *Black Garden: Armenia and Azerbaijan Through Peace and War.* 2003
Hovannisian, R. G., *The Republic of Armenia.* 4 vols. 1996
Libaridian, Gerard J., *The Challenge of Statehood: Armenian Political Thinking Since Independence.* 1999

National Statistical Office: National Statistical Service of the Republic of Armenia, Republic Square, 3 Government House, Yerevan 375010.
   President: Stepan L. Mnatsakanyan.
Website: http://www.armstat.am

AUSTRALIA

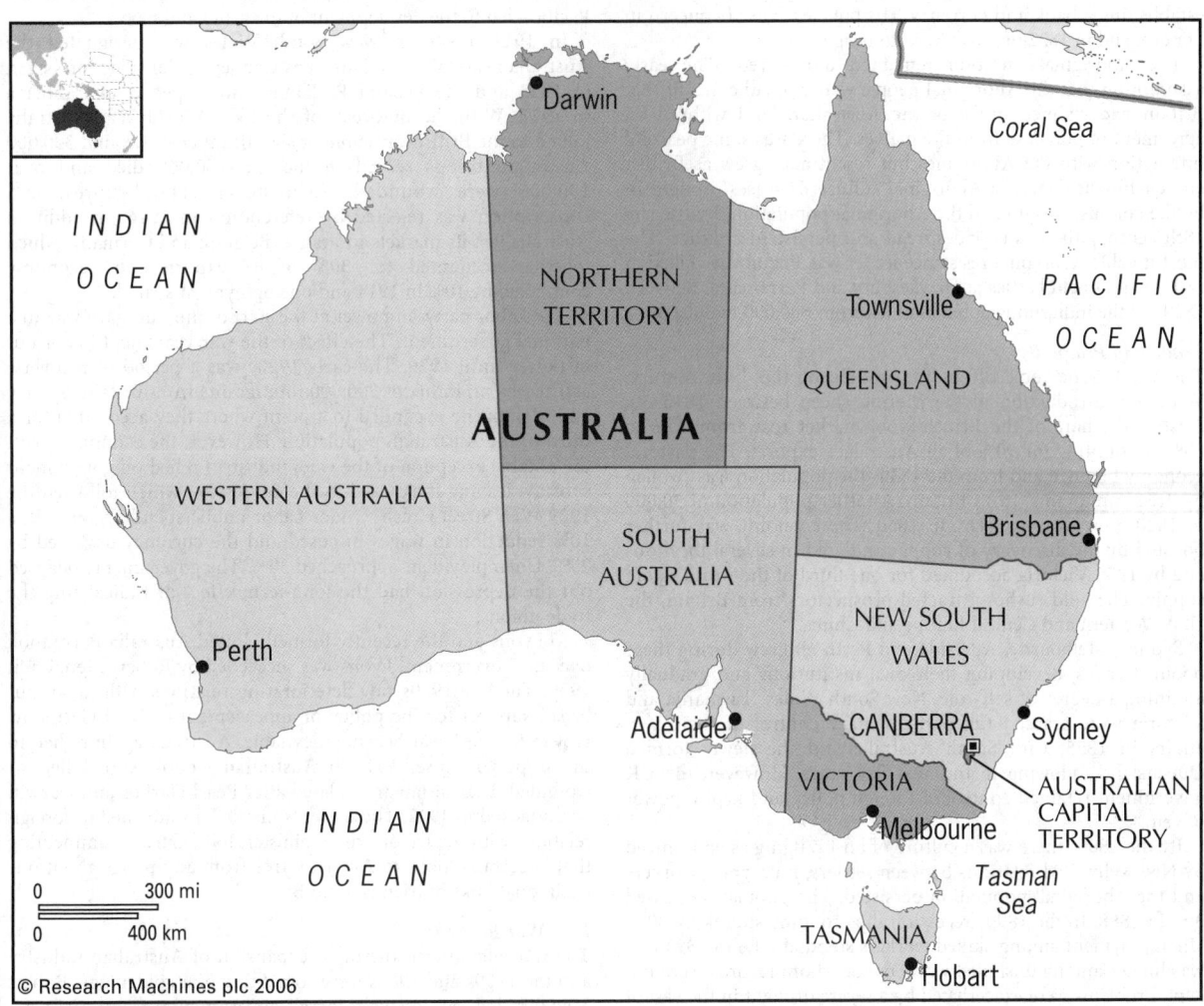

Darwin

Coral Sea

INDIAN
OCEAN

NORTHERN
TERRITORY

Townsville

PACIFIC
OCEAN

QUEENSLAND

AUSTRALIA

WESTERN AUSTRALIA

Brisbane

SOUTH
AUSTRALIA

NEW SOUTH
WALES

Perth

CANBERRA

Sydney

Adelaide

AUSTRALIAN
CAPITAL
TERRITORY

VICTORIA

Melbourne

Tasman
Sea

INDIAN

OCEAN

0   300 mi

0   400 km

TASMANIA

Hobart

© Research Machines plc 2006

Commonwealth of Australia

Capital: Canberra
Population projection, 2015: 23·79m.
GNI per capita, 2011: (PPP$) 34,431
HDI/world rank: 0·929/2
Internet domain extension: .au

KEY HISTORICAL EVENTS

The Australian landmass, reaching northwards to Papua New Guinea and including Tasmania in the south, was inhabited in prehistoric times until adverse climatic conditions led to an exodus between 15,000 and 25,000 years ago. Stone tools date to 2000–1000 BC. The Aboriginal society was based on extended family groups. At maximum there were 1m. Aborigines, using 200 different languages. By the early 18th century contact was made with traders from the area of modern Indonesia and Papua New Guinea.

Australia was sighted in 1522 by compatriot explorers of the Portuguese Ferdinand Magellan and in 1642 the Dutch explorer Abel Tasman mapped what is now Tasmania and part of New Zealand's east coast. By the middle of the century the Dutch had charted the western part of Australia, calling it New Holland.

But while the Dutch, Portuguese and Spanish made the early running in charting the continent, it was the discovery of the east coast by Capt. James Cook in 1770 that prompted colonization. Over several voyages he charted the Torres Strait and 8,000 kilometres of coastline. Having lost their penal settlements in the American War of Independence, the British decided to send convicts to Australia. Botany Bay was selected as the first settlement.

By 1800 convicts had established legal rights as crown subjects. Many freed men were able to make a successful living. However, there were several uprisings against penal rule culminating in the Rum Rebellion of 1808, in which John Macarthur led a troop of New South Wales officers against Gov. William Bligh. The response of the British government was to appoint Lachlan Macquarie, who promoted reform.

His tenure began a period of development in which Australia ceased to be primarily a penal settlement. The crossing of the Blue Mountains in 1813 was the first of many expeditions which led to discovery of vast areas of grazing land, although sealing and whaling were more important than agriculture until the 1830s. Macquarie's benevolent despotism rewarded freed men and several of the more talented were appointed to official posts. He did much to develop Sydney, instigating public works and establishing a bank and currency. His policies caused concern in London and in 1822 he was forced to resign.

The Aboriginal question remained unresolved. The early assumption that the Aboriginal people were nomadic meant that Britain had claimed much of the Australian land without an agreement of purchase from the natives. There was some peaceful interaction with the Aborigines but resentment grew as British encroachment damaged Aboriginal culture. Diseases brought in by the colonists decimated the Aboriginal population. By the late 18th century there was widespread and persistent conflict. The most notable Aboriginal resistance leader was Pemulwuy (killed in 1802) who fought battles at Hawkesbury and Parramatta. Between 1820–50 the indigenous population fell from 600,000 to 300,000.

Trade and Prosperity

The wool trade took off in the middle of the 19th century. With the introduction of the Merino sheep between 1830–50, Australia's share of the British wool market rose from 10% to 50%, accounting for 90% of all Australia's exports. The pastoral economy boomed and from the 1830s the population was swollen by 'free immigrants' from Britain. A settler population of 30,000 in 1820 grew to over 1·1m. in 1860. The economy was further boosted by the discovery of copper and gold in several locations and by 1850 Victoria accounted for one third of the world's gold supply. The gold rushes attracted prospectors from Britain, the USA, Western and Central Europe and China.

Sydney, Melbourne, Adelaide and Perth all grew during these boom decades, developing their own institutions and gradually attaining a degree of self-rule. New South Wales, Tasmania and Victoria were granted full parliamentary control of their own affairs in 1855, with South Australia and the newly formed Queensland following in the next five years. However, the UK government retained control of foreign policy and kept a power of veto.

By the 1860s there was a culture of bushranging as epitomized by Ned Kelly. Bad relations between settlers, immigrant workers and the Aboriginal population persisted. The population passed 3m. in 1888. In the 1890s recession the economy shrank by 30%. Unemployment among skilled workers stood at 30% in 1893 and was higher among unskilled workers, for whom records were not kept. Problems were exacerbated by a severe drought in the east of the country.

Social and Constitutional Reform

Trade unionism grew from the 1870s, gaining strength in the 1890s. Between 1899, when a first Labor government took power in Queensland, and the outbreak of the First World War, Labor was in government in every state. Union membership included a third of all workers. The 1890s witnessed the emergence of the federalist movement, with Sydney hosting conventions in 1891 and 1897–98. On 1 Jan. 1901 the six separately constituted colonies of New South Wales, Victoria, Queensland, South Australia, Western Australia and Tasmania were federated under the Commonwealth of Australia, the designation of 'colonies' being at the same time changed into that of 'states'—except in the case of Northern Territory, which was transferred from South Australia to the Commonwealth as a 'territory' on 1 Jan. 1911. A bicameral parliament was established while the states retained certain powers. Foreign policy continued to be guided by London and in 1907 Australia gained dominion status. By the following year there was female suffrage for the national and all state legislatures.

The Labor party formed its first national government in 1904 and their old allies in the protectionist parties forged an alliance with free trade parties, forming the Liberal party to challenge Labor's growth. A period in which the government switched between Labor and the Liberals saw pension and welfare reforms. The growing fear of an Asian (and especially Japanese) threat—Britain was reluctant to keep a strong military presence in the Pacific—led to the development of the army and navy.

In 1911 a site in New South Wales was designated the Australian capital, to be known as Canberra. Construction began in 1923 and the Federal Parliament was opened in Canberra in 1927. With the outbreak of the First World War Australia rallied to the British cause but at significant cost. Around 330,000 Australian troops served in the war—60,000 died and over 150,000 were wounded. Such losses caused unrest and conscription was rejected by referendum in 1916. In addition Australia lost its markets in France, Belgium and Germany, which together accounted for 30% of its exports. The economy contracted by 10% in 1914 and unemployment soared.

The Labor party, in power at the start of the war, gave way to a national government. The effect of the war years kept Labor out of power until 1929. The early 1920s was a period of recovery, assisted by an influx of 200,000 immigrants from Britain. Sydney and Melbourne expanded to a point where they accounted for a third of the Australian population. However, the economy (with the notable exception of the wool industry) relied on government subsidy, leaving it exposed in the depression which followed the 1929 Wall Street Crash. Under Labor public spending was cut, a 10% reduction in wages imposed and the currency devalued by 25%. Unemployment approached 30%. The government collapsed but the depression had the long-term effect of radicalizing the trade unions.

Jo Lyons and the recently formed United Australia Party took over the government. Lyons was succeeded by Robert Menzies in 1939. The late 1930s saw deteriorating relations with Japan but broad support for the policy of appeasement of Nazi Germany. However, when war became inevitable, Australia again rallied to the imperial cause. 100,000 Australian troops were killed or wounded. Fear of Japan escalated after Pearl Harbor and Darwin was attacked in 1942. These events marked a watershed in foreign relations with the Labor prime minister, John Curtin, commenting that 'Australia looks to America, free from any pangs about our traditional links of friendship to Britain.'

Post-War Recovery

The war encouraged the rapid expansion of Australian industry and the 1950s and 1960s were something of a golden age. Robert Menzies led successive Liberal governments from 1949–66. The population nearly doubled as immigration from Britain and continental Europe was encouraged. Unemployment was consistently low and the economy tripled in size during the two decades. Aware of its 'junior partner' status in the relationship with America, Australia undertook nuclear development with Britain. In 1951 it entered the ANZUS group with New Zealand and the USA and three years later joined the South-East Asian Treaty Organization. Australia's new found confidence was reflected in the success of the 1956 Melbourne Olympics.

The growing number of non-English speaking immigrants accentuated racial problems, with a succession of governments holding to a monocultural policy. The future of the Aboriginal population was one of assimilation. The movement for Aboriginal rights grew after the war, with a strike by Aboriginal workers at Pilbara in 1946 marking a new phase in the conflict. It climaxed in 1966 when an Aboriginal demand for equal pay in Northern Australia turned into demands for land. There was a swathe of moderate reforms favouring the Aboriginal population between 1959–67 but at the same time the assimilation policy allowed for the forced removal of large numbers of children from their families. The last of the Aboriginal reserves was taken over in the 1960s. The policy of forced removal of children did not prompt a formal apology from the Australian government until 2008.

When Menzies retired in 1966 he was followed by a succession of leaders who weakened the standing of the Liberals. Australia's participation in the Vietnam War also drew criticism. Gough Whitlam led the Labor party to power in 1972 and oversaw a radical administration. He withdrew Australian forces from Vietnam, set about modernizing the education and health programmes and funded extensive urban renewal. Government expenditure doubled over his three years in office and Australia was ill-prepared when the global oil crisis struck in 1974.

Whitlam's Liberal opponents, many of whom regarded him as a dangerous maverick, led a parliamentary revolt. A failure to win approval for the national budget led to a constitutional crisis in which the Governor-General John Kerr (himself recommended for the post by Whitlam) dismissed the prime minister and invited Malcolm Fraser to form an administration. Fraser believed that Australian society had become overly dependent on the state. He authorized cuts in public spending but was unable to counter rising unemployment and inflation and was voted out of government in 1983.

Free Market Politics
Bob Hawke took power at the head of a Labor government, assisted by his finance minister (and successor as prime minister), Paul Keating. Their terms of office, spanning 1983–96, saw a shift in Labor's stance on state control and economic planning to allow for an ambitious programme of privatization and financial deregulation. Trade with Asia took on increasing importance and Hawke stood fully behind US foreign policy. In March 1986 the Australia Act abolished the remaining legislative, executive and judicial controls of the British Parliament. By the end of the decade unemployment stood at a respectable 6% but the Australian dollar had suffered a 40% loss of value in 1986 and foreign debt stood at around 30% of GNP. Keating described the recession of the late 1980s as 'necessary'.

Keating took over the premiership in 1991 and mounted a programme of economic reform. He was replaced in 1996 by the Liberal, John Howard, who pressed on with economic reforms and won re-election two years later. The Aboriginal question continued to test every government. There were some symbolic gestures such as the return of Ayers Rock (with its Aboriginal name Uluru restored) in 1988. The High Court's Mabo ruling of 1992, which overturned a previous ruling that the Aboriginal title to land had not survived British settlement, raised expectations. Keating officially acknowledged injustices to the Aboriginal population when the 'Native Title' legislation was passed in Dec. 1993. Howard's tenure, however, saw disputes over indigenous land rights following a 1996 court ruling against Aboriginal access to cultural sites owned by non-Aboriginals.

A referendum to decide if Australia should become a republic was held on 6 Nov. 1999. 54·87% were in favour of the monarchy with Queen Elizabeth II as head of state, against 45·13% for a republic with a president chosen by parliament. In foreign policy Howard agreed to military involvement in UN peacekeeping in Timor-Leste and NATO action against Serbia. His government's approach to immigration came under the spotlight in July 2001, when a refugee-laden Norwegian cargo ship was caught in a diplomatic gridlock between Australia, the UN and Norway. Its passengers were eventually diverted to Papua New Guinea, with Howard assuming a firm and populist stance against asylum seekers. He won a further term of office at the elections of Nov. 2001. The Liberals also won the elections of Oct. 2004. Howard pursued an interventionist foreign policy in the Pacific region—with notable success in the peacekeeping mission to the Solomon Islands—and committed Australia to the US-led war in Iraq in 2003. Labor returned to power with victory in the elections of Nov. 2007, led by Kevin Rudd. After a steep decline in popularity in early 2010, he was succeeded by Australia's first female prime minister, Julia Gillard, in June that year. Following the election in

Aug. 2010 she remained prime minister in Australia's first hung parliament since 1940.

TERRITORY AND POPULATION

Australia, excluding external territories, covers a land area of 7,692,024 sq. km, extending from Cape York (10° 41′ S) in the north some 3,680 km to South East Cape, Tasmania (43° 39′ S), and from Cape Byron, New South Wales (153° 39′ E) in the east some 4,000 km west to Steep Point, Western Australia (113° 9′ E). External territories under the administration of Australia comprise the Ashmore and Cartier Islands, Australian Antarctic Territory, Christmas Island, the Cocos (Keeling) Islands, the Coral Sea Islands, the Heard and McDonald Islands and Norfolk Island. For these *see below*.

Growth in census population has been:

1901	3,774,310	1966	11,599,498	1991	16,852,258
1911	4,455,005	1971	12,755,638	1996	17,752,829
1921	5,435,734	1976	13,915,500	2001	18,769,249
1947	7,579,358	1981	15,053,600	2006	19,855,288
1961	10,508,186	1986	15,763,000	2011	21,507,719

The UN gives a projected population for 2015 of 23·79m.

At the census of 9 Aug. 2011 density was 2·8 per sq. km. In 2010, 89·1% of the population lived in urban areas.

Areas and populations of the States and Territories at the 2011 census:

States and Territories	Area (sq. km)	Population	Per sq. km
Australian Capital Territory (ACT) including Jervis Bay Territory	2,431	357,600	147·1
Northern Territory (NT)	1,349,129	211,944	0·2
New South Wales (NSW)	800,642	6,917,658	8·6
Queensland (Qld)	1,730,648	4,332,737	2·5
South Australia (SA)	983,482	1,596,570	1·6
Tasmania (Tas.)	68,401	495,350	7·2
Victoria (Vic.)	227,416	5,354,040	23·5
Western Australia (WA)	2,529,875	2,239,170	0·9

Resident population in the state capitals (2011 census figures):

Capital	State	Population	Capital	State	Population
Canberra	ACT	356,586	Adelaide	SA	1,225,235
Darwin	NT	120,585	Hobart	Tas.	211,656
Sydney	NSW	4,391,674	Melbourne	Vic.	3,999,982
Brisbane	Qld.	2,065,996	Perth	WA	1,728,867

The median age of the 2006 census population was 37 years.

Australians born overseas (census 2006), 4,416,037 (22·2%—one of the highest proportions in the industrialized world), of whom 856,939 (4·3%) were from England.

Aboriginals have been included in population statistics only since 1967. At the 2011 census 548,370 people identified themselves as being of indigenous origin (2·5% of the total population). A 1992 High Court ruling that the Meriam people of the Murray Islands had land rights before the European settlement reversed the previous assumption that Australia was *terra nullius* before that settlement. The Native Title Act setting up a system for deciding claims by Aborigines came into effect on 1 Jan. 1994.

Overseas arrivals and departures:

	Permanent arrival numbers	Permanent departure numbers	Net permanent gain
2005–06	131,593	67,853	63,740
2006–07	140,148	72,103	68,045
2007–08	149,365	76,923	72,442
2008–09	158,021	81,018	77,003

The Migration Act of Dec. 1989 sought to curb illegal entry and ensure that annual immigrant intakes were met but not exceeded.

Provisions for temporary visitors to become permanent were restricted. According to the 2006 census, 73% of the population born overseas have become Australian citizens.

The national language is English.

SOCIAL STATISTICS

Life expectancy at birth, 2007, 79·0 years for males and 83·7 years for females.

Statistics for years ended 30 June:

	Births	Deaths	Marriages	Divorces
2005	259,791	130,714	109,323	52,399
2006	265,949	133,739	114,222	51,375
2007	285,213	137,854	116,322	47,963
2008	296,621	143,946	118,756	47,209
2009	295,738	140,760	120,118	49,448

In 2007 the median age for marrying was 31·6 years for males and 29·3 for females. Infant mortality, 2007, was 4·2 per 1,000 live births. Population growth rate in the year ended 31 March 2008, 1·6%; fertility rate, 2007, 1·9 births per woman.

Suicide rates (per 100,000 population, 2009): 10·4 (men, 16·0; women, 4·9).

A UNICEF report published in 2010 showed that 10·9% of children in Australia live in relative poverty (living in a household in which disposable income—when adjusted for family size and composition—is less than 50% of the national median income), compared to just 4·7% in Iceland.

In the Human Development Index, or HDI (measuring progress in countries in longevity, knowledge and standard of living), Australia was ranked second (behind Norway) in the 2011 rankings published in the annual Human Development Report.

CLIMATE

Over most of the continent, four seasons may be recognized. Spring is from Sept. to Nov., summer from Dec. to Feb., autumn from March to May and winter from June to Aug., but because of its great size there are climates that range from tropical monsoon to cool temperate, with large areas of desert as well. In northern Australia there are only two seasons, the wet one lasting from Nov. to March, but rainfall amounts diminish markedly from the coast to the interior. Central and southern Queensland are subtropical, north and central New South Wales are warm temperate, as are parts of Victoria, Western Australia and Tasmania, where most rain falls in winter. Canberra, Jan. 68°F (20°C), July 42°F (5·6°C). Annual rainfall 25" (635 mm). Adelaide, Jan. 73°F (22·8°C), July 52°F (11·1°C). Annual rainfall 21" (528 mm). Brisbane, Jan. 77°F (25°C), July 58°F (14·4°C). Annual rainfall 45" (1,153 mm). Darwin, Jan. 83°F (28·3°C), July 77°F (25°C). Annual rainfall 59" (1,536 mm). Hobart, Jan. 62°F (16·7°C), July 46°F (7·8°C). Annual rainfall 23" (584 mm). Melbourne, Jan. 67°F (19·4°C), July 49°F (9·4°C). Annual rainfall 26" (659 mm). Perth, Jan. 74°F (23·3°C), July 55°F (12·8°C). Annual rainfall 35" (873 mm). Sydney, Jan. 71°F (21·7°C), July 53°F (11·7°C). Annual rainfall 47" (1,215 mm).

CONSTITUTION AND GOVERNMENT

Federal Government

Under the Constitution legislative power is vested in a Federal Parliament, consisting of the Queen, represented by a Governor-General, a Senate and a House of Representatives. Under the terms of the constitution there must be a session of parliament at least once a year.

The Senate (Upper House) comprises 76 Senators (12 for each State voting as one electorate and, as from Aug. 1974, two Senators respectively for the Australian Capital Territory and the Northern Territory). Senators representing the States are chosen for six years. The terms of Senators representing the Territories expire at the close of the day next preceding the polling day for the general elections of the House of Representatives. In general, the Senate is renewed to the extent of one-half every three years, but in case of disagreement with the House of Representatives, it, together with the House of Representatives, may be dissolved, and an entirely new Senate elected. Elections to the Senate are on the single transferable vote system; voters list candidates in order of preference. A candidate must reach a quota to be elected, otherwise the lowest-placed candidate drops out and his or her votes are transferred to other candidates.

The House of Representatives (Lower House) consists, as nearly as practicable, of twice as many Members as there are Senators, the numbers chosen in the several States being in proportion to population as shown by the latest statistics, but not less than five for any original State. The 150 membership is made up as follows: New South Wales, 48; Victoria, 37; Queensland, 30; South Australia, 11; Western Australia, 15; Tasmania, 5; ACT, 2; Northern Territory, 2. Elections to the House of Representatives are on the alternative vote system; voters list candidates in order of preference, and if no one candidate wins an overall majority, the lowest-placed drops out and his or her votes are transferred. The first Member for the Australian Capital Territory was given full voting rights as from the Parliament elected in Nov. 1966. The first Member for the Northern Territory was given full voting rights in 1968. The House of Representatives continues for three years from the date of its first meeting, unless sooner dissolved.

Every Senator or Member of the House of Representatives must be a subject of the Queen, be of full age, possess electoral qualifications and have resided for three years within Australia. The franchise for both Houses is the same and is based on universal (males and females aged 18 years) suffrage. Compulsory voting was introduced in 1925. If a Member of a State Parliament wishes to be a candidate in a federal election, he must first resign his State seat.

Executive power is vested in the Governor-General, advised by an Executive Council. The Governor-General presides over the Council, and its members hold office at his pleasure. All Ministers of State, who are members of the party or parties commanding a majority in the lower House, are members of the Executive Council under summons. A record of proceedings of meetings is kept by the Secretary to the Council. At Executive Council meetings the decisions of the Cabinet are (where necessary) given legal form, appointments made, resignations accepted, proclamations, regulations and the like made.

The policy of a ministry is, in practice, determined by the Ministers of State meeting without the Governor-General under the chairmanship of the Prime Minister. This group is known as the Cabinet. There are 11 Standing Committees of the Cabinet comprising varying numbers of Cabinet and non-Cabinet Ministers. In Labor governments all Ministers have been members of Cabinet; in Liberal and National Country Party governments, only the senior ministers. Cabinet meetings are private and deliberative, and records of meetings are not made public. The Cabinet does not form part of the legal mechanisms of government; the decisions it takes have, in themselves, no legal effect. The Cabinet substantially controls, in ordinary circumstances, not only the general legislative programme of Parliament but the whole course of Parliamentary proceedings. In effect, though not in form, the Cabinet, by reason of the fact that all Ministers are members of the Executive Council, is also the dominant element in the executive government of the country.

The legislative powers of the Federal Parliament embrace trade and commerce, shipping, etc.; taxation, finance, banking, currency, bills of exchange, bankruptcy, insurance, defence, external affairs, naturalization and aliens, quarantine, immigration and emigration; the people of any race for whom it is deemed necessary to make special laws; postal, telegraph and like services; census and statistics; weights and measures; astronomical and meteorological observations; copyrights; railways; conciliation and arbitration in disputes extending beyond the limits of any one

State; social services; marriage, divorce, etc.; service and execution of the civil and criminal process; recognition of the laws, Acts and records, and judicial proceedings of the States. The Senate may not originate or amend money bills. Disagreement with the House of Representatives may result in dissolution and, in the last resort, a joint sitting of the two Houses. The Federal Parliament has limited and enumerated powers, the several State parliaments retaining the residuary power of government over their respective territories. If a State law is inconsistent with a Commonwealth law, the latter prevails.

The Constitution also provides for the admission or creation of new States. Proposed laws for the alteration of the Constitution must be submitted to the electors, and they can be enacted only if approved by a majority of the States and by a majority of all the electors voting.

The Australia Acts 1986 removed residual powers of the British government to intervene in the government of Australia or the individual states.

In Feb. 1998 an Australian Constitutional Convention voted for Australia to become a republic. In a national referendum, held on 6 Nov. 1999, 54·9% voted against Australia becoming a republic.

State Government
In each of the six States (New South Wales, Victoria, Queensland, South Australia, Western Australia, Tasmania) there is a State government whose constitution, powers and laws continue, subject to changes embodied in the Australian Constitution and subsequent alterations and agreements, as they were before federation. The system of government is basically the same as that described above for the Commonwealth—i.e. the Sovereign, her representative (in this case a Governor), an upper and lower house of Parliament (except in Queensland, where the upper house was abolished in 1922), a cabinet led by the Premier and an Executive Council. Among the more important functions of the State governments are those relating to education, health, hospitals and charities, law, order and public safety, business undertakings such as railways and tramways, and public utilities such as water supply and sewerage. In the domains of education, hospitals, justice, the police, penal establishments, and railway and tramway operation, State government activity predominates. Care of the public health and recreative activities are shared with local government authorities and the Federal government; social services other than those referred to above are now primarily the concern of the Federal government; the operation of public utilities is shared with local and semi-government authorities.

Administration of Territories
Since 1911 responsibility for administration and development of the Australian Capital Territory (ACT) has been vested in Federal Ministers and Departments. The ACT became self-governing on 11 May 1989. The ACT House of Assembly has been accorded the forms of a legislature, but continues to perform an advisory function for the Minister for the Capital Territory.

On 1 July 1978 the Northern Territory of Australia became a self-governing Territory with expenditure responsibilities and revenue-raising powers broadly approximating those of a State.

National Anthem
'Advance Australia Fair' (adopted 19 April 1984; words and tune by P. D. McCormick). The 'Royal Anthem' (i.e. 'God Save the Queen') is used in the presence of the Royal Family.

GOVERNMENT CHRONOLOGY

Prime Ministers since 1945. (ALP = Australian Labor Party; LP = Liberal Party; CP = Australian Country Party)

1945	ALP	Francis Michael (Frank) Forde
1945–49	ALP	Joseph Benedict (Ben) Chifley
1949–66	LP	Robert Gordon Menzies
1966–67	LP	Harold Edward Holt
1967–68	CP	John (Jack) McEwen (acting)
1968–71	LP	John Grey Gorton
1971–72	LP	William (Bill) McMahon
1972–75	ALP	(Edward) Gough Whitlam
1975–83	LP	(John) Malcolm Fraser
1983–91	ALP	Robert James Lee (Bob) Hawke
1991–96	ALP	Paul John Keating
1996–2007	LP	John Winston Howard
2007–10	ALP	Kevin Michael Rudd
2010–	ALP	Julia Eileen Gillard

RECENT ELECTIONS
The 43rd Parliament was elected on 21 Aug. 2010.

House of Representatives
Australian Labor Party (ALP), 72 seats and 38·0% of votes cast; Liberal Party, 44 seats and 30·5% of votes cast; Liberal National Party of Queensland (LNP), 21 (9·1%); National Party of Australia (NP), 6 (3·4%); Australian Greens, 1 (11·7%); National Party of Western Australia, 1 (0·3%); Country Liberal Party, 1 (0·3%). Independents took four seats. The opposition Liberal/National coalition took 73 seats in total with 43·7%.

Senate
As at Feb. 2013 the make-up of the Senate was Australian Labor Party, 31; Liberal Party, 28; Australian Greens, 9; National Party of Australia/Country Liberal Party, 6; Democratic Labor Party, 1; Independent, 1.

CURRENT GOVERNMENT
Governor-General: Quentin Bryce, AC, b. 1942 (took office on 5 Sept. 2008).

In March 2013 the cabinet comprised:

Prime Minister: Julia Gillard; b. 1961 (ALP; in office since 24 June 2010).

Deputy Prime Minister and Treasurer: Wayne Swan.

Minister for Agriculture, Fisheries and Forestry: Joe Ludwig. *Attorney-General and Minister for Emergency Management:* Mark Dreyfus. *Broadband, Communications and the Digital Economy:* Stephen Conroy. *Climate Change and Energy Efficiency, and Industry and Innovation:* Greg Combet. *Defence:* Stephen Smith. *Families, Community Services and Indigenous Affairs, and Disability Reform:* Jenny Macklin. *Finance and Deregulation:* Penny Wong. *Financial Services and Superannuation, and Employment and Workplace Relations:* Bill Shorten. *Foreign Affairs:* Bob Carr. *Health:* Tanya Plibersek. *Immigration and Citizenship:* Brandan O'Connor. *Infrastructure and Transport:* Anthony Albanese. *Regional Australia, Regional Development and Local Government, and the Arts:* Simon Crean. *Resources and Energy, and Tourism:* Martin Ferguson. *School Education, Early Childhood and Youth:* Peter Garrett. *Social Inclusion, Mental Health and Ageing, and Housing and Homelessness:* Mark Butler. *Sustainability, Environment, Water, Population and Communities:* Tony Burke. *Tertiary Education, Skills, Science and Research, and Small Business:* Christopher Bowen. *Trade and Competitiveness:* Craig Emerson.

The *Speaker* is Anna Burke (ALP).

The *President* of the Senate is John Hogg (ALP).

Leader of the Opposition: Tony Abbott (LP).

Government: http://www.gov.au

CURRENT LEADERS

Julia Gillard

Position
Prime Minister

Introduction
Julia Gillard became prime minister in June 2010 when she successfully challenged incumbent Kevin Rudd for the leadership of the ruling Australian Labor Party. Considered a left-of-centre consensus politician, she has chiefly focused on domestic issues.

After Labor's weakened showing at elections in Aug. 2010, she agreed an alliance with Green and independent MPs to give herself a working majority of one seat.

Early Life

Julia Gillard was born in Barry, Wales on 29 Sept. 1961 and moved to Adelaide, Australia with her family in 1965. Educated at Mitcham Demonstration School and Unley High School, she went to Adelaide University in 1981 to study arts and law. While there she campaigned with the Labor Party against federal education budget cuts. She became vice president of the Australian Union of Students (AUS) in 1982, necessitating a move to Melbourne University where the AUS was based. In 1984 she was voted AUS president.

After graduating in 1986 she worked as a solicitor in Melbourne, specializing in industrial and employment law, and became a partner at the firm Slater & Gordon in 1990. From 1996–98 she was chief of staff to John Brumby, opposition leader in the state of Victoria. She was adopted as a parliamentary Labor candidate at her second attempt, successfully contesting the seat of Lalor at the 1998 election. From 1998–2001 she served on several committees, including the House of Representatives standing committee on employment, education and workplace relations.

Following Labor's defeat at the 2001 election, Gillard was appointed shadow minister for population and immigration, where she argued for a controlled and planned approach to immigration. In 2003 she was given the additional portfolio for reconciliation and indigenous affairs. From 2003–06 she was shadow minister for health, attracting notice for her combative challenges to her government opposite number, Tony Abbott. In Dec. 2006 she supported Kevin Rudd's bid for the Labor leadership, announcing herself as a candidate for the deputy leadership. Following Rudd's defeat of incumbent Kim Beazley, Gillard was elected unopposed as his deputy. She was named shadow minister for employment, industrial relations and social inclusion.

In Dec. 2007, following Labor's election victory, Gillard became deputy prime minister. She was given a wide-ranging portfolio, as minister for education, employment and workplace relations and social inclusion. She launched a programme of reforms including a $A16bn. investment in schools infrastructure and measures to gauge academic performance in national tests. The Howard government's controversial industrial relations legislation was replaced by a collective bargaining framework, overseen by a single government body.

In the aftermath of the 2008 global financial crisis, when Prime Minister Rudd's popularity suffered, Gillard was viewed as a credible alternative leader of the Labor Party. Although she shared responsibility for some of the government's more unpopular moves, including delaying expenditure on schools and postponing the introduction of a long-promised emissions trading programme, she maintained her personal appeal. In June 2010, having won the backing of the Australian Workers' Union and many factions within her party, she challenged Rudd for the leadership. He stood aside without a vote and on 24 June 2010 Gillard was sworn in as Australia's first female prime minister.

Career in Office

Gillard declared her intention to manage Australia's natural resources boom so that revenues could be invested for future growth. She opened negotiations with the mining industry over a proposed profits tax which, despite industry resistance, was approved by parliament in March 2012. She also promised a planned approach to solving the shortage of skilled labour, introducing tax reforms to encourage retired and part-time workers into fuller participation. She also signalled her intention to seek sustainable and controlled immigration rather than the 'big Australia' vision of her predecessor. Her controversial

proposal to process the rising numbers of asylum seekers through Malaysia was rejected in parliament in Oct. 2011 but in Sept. 2012 the government instead adopted an independent panel recommendation to re-establish holding centres in Nauru and Papua New Guinea. Other pressing challenges have included the environment, in particular rebuilding after the Queensland floods of early 2011, and devising controversial legislation on pricing carbon emissions, which completed its parliamentary passage in Nov. 2011 and came into force in July 2012. Gillard also promised to deliver a $A36bn. national broadband system and to take over from individual states the majority of funding responsibility for public hospitals. In Feb. 2012 Rudd resigned as foreign minister and challenged Gillard's leadership. The prime minister won the backing of her Labor party colleagues in the leadership ballot, by 71 votes to 31. In April 2012 Gillard announced that the withdrawal of most Australian troops from Afghanistan would be completed in 2013. In early 2013 the aim was still to achieve this by the end of the year.

DEFENCE

The Minister for Defence has responsibility under legislation for the control and administration of the Australian Defence Force (ADF). The Chief of the Defence Force is vested with command of the ADF. He is the principal military adviser to the Minister. The Chief of Navy, the Chief of Army and the Chief of Air Force command the Royal Australian Navy, Australian Army and Royal Australian Air Force respectively. They have delegated authority from the Chief of Defence Force Staff and the Secretary to administer matters relating to their particular Service. Conscription was abolished in 1972.

2008 defence expenditure was US$18,399m., amounting to US$876 per capita. In 2007 defence spending represented 1·9% of GDP.

Having contributed to the 2003 US-led invasion of Iraq, the last Australian troops left the country in July 2009. There were 1,100 Australian troops serving with the International Security Assistance Force (ISAF) in Afghanistan in Feb. 2013.

Army

The strength of the Army was 25,259 in 2007. The effective strength of the Army Reserve was 17,200.

Women have been eligible for combat duties since 1993.

Navy

The all-volunteer Navy had 12,784 personnel in 2007, with a reserve of 1,850. Equipment in 2007 included six diesel-powered submarines and 12 frigates.

The main naval bases are at Sydney and Perth with further bases at Cairns and Darwin.

Air Force

The Royal Australian Air Force (RAAF) operated 120 combat-capable aircraft including F-111s and F-18 'Hornets' in 2007. Personnel in 2007 numbered 13,250. There is also an Australian Air Force Reserve, 2,400-strong.

ECONOMY

In 2008 agriculture contributed 3% of GDP, industry 29% and services 68%.

According to the anti-corruption organization *Transparency International*, Australia ranked equal seventh in the world in a 2012 survey of the countries with the least corruption in business and government. It received 85 out of 100 in the annual index.

Australia gave US$5·0bn. in international aid in 2011, equivalent to 0·34% of GNI (compared to the UN target of 0·7%).

Overview

Australia has abundant natural resources, including coal, iron, copper, gold, natural gas and uranium. Prior to the global financial crisis the economy had experienced 17 years of robust expansion. Reforms in the 1980s and early 1990s liberalized the previously heavily-protected and regulated economy. Increased competition and technological advances have helped drive productivity. The Economist Intelligence Unit estimates that average productivity growth over 1991–2000 at 1·67%, six times greater than that for the previous decade. Reform and good management facilitated growth even in the face of the Asian crisis of 1997–98 and the US downturn of 2001–02.

The buoyant domestic economy withstood the impact of a severe drought in 2003 and a global slowdown. A fall in the housing market from its 2004 peak brought down growth rates but a commodities boom helped cushion the economy. The terms of trade improved by 30% from 2005–08 to register their highest level for over 50 years. Near full employment in 2007 saw unemployment at its lowest since the mid-1970s. A National Reform Agenda (NRA) agreed in 2006 aimed at lifting long-term productivity and workforce participation. Per capita GDP compares favourably with that of Western European countries and services account for a high proportion of output.

Australia has promoted itself to global and regional businesses (especially in the financial sector) as an ideal location for administrative and support operations. According to the World Bank, Australia ranked 10th out of 185 economies for the ease of doing business in 2013. The export sector is driven by mining and agriculture although there are concerns about the long-term dependency on natural resources.

The global financial crisis of the late 2000s hit the economy but real activity did not suffer as much as in other advanced nations thanks to limited higher-tech manufacturing, robust commodity exports and an effective policy response. The government implemented a US$50bn.-plus stimulus package and cut interest rates. Continued demand from China and India helped Australia escape recession in 2009, with the second best economic performance in the OECD (after Poland) that year. With a low risk of serious contraction, the Reserve Bank of Australia adopted a tighter monetary policy and raised interest rates. In 2010 the fiscal stimulus was reduced and the mining sector experienced a boom, with strong demand from emerging economies in Asia.

In Jan. 2011 severe flooding in Queensland, which contributes about 19% of output and produces 80% of the country's coking coal, was followed by a cyclone. The floods led to reduced output and a larger budget deficit. Exports of coal and iron ore were particularly disrupted. Several mines closed and agriculture, tourism, retail and manufacturing were also affected. The cost of clean-up and rebuilding was estimated at $A30bn., prompting the government to announce a levy on middle- and high-income earners, budget cuts and delays to other infrastructure projects.

The unemployment rate is low by international standards. While growth has accelerated since 2011, it is chiefly driven by a boom in mining investment, leaving the economy vulnerable to changes in the terms of trade. With the economy also susceptible to the effects of the European debt crisis and to lower growth rates in Asia (particularly China), the government is making efforts to move towards a more services-oriented economy.

Over the medium term, an increase in the number of older people and rising health care costs are expected to see public expenditure increase by over 3% of GDP by 2030.

Currency
On 14 Feb. 1966 Australia adopted a system of decimal currency. The currency unit, the Australian dollar (AUD), is divided into 100 *cents*.

Foreign exchange reserves were US$34,857m. in Sept. 2009 and gold reserves 2·57m. troy oz. Total money supply was $A376·5bn. in July 2009.

Inflation rates (based on OECD statistics):

2002	2003	2004	2005	2006	2007	2008	2009	2010	2011
3·0%	2·7%	2·3%	2·7%	3·6%	2·4%	4·3%	1·8%	2·9%	3·4%

Budget
The fiscal year is 1 July–30 June. In Aug. 1998 the Commonwealth government introduced a tax reform package including, from 2000, the introduction of a Goods and Services Tax (GST) at a 10% rate, with all the revenues going to the states in return for the abolition of a range of other indirect taxes; the abolition of Financial Assistance Grants to states; the abolition of wholesale sales tax (which is levied by the Commonwealth government); cuts in personal income tax; and increases in social security benefits, especially for families. In the 2007–08 Mid-Year Economic and Fiscal Outlook an underlying cash surplus of $A14·8bn. (1·3% of GDP) was anticipated in 2007–08, an increase of $A4·2bn. since the 2007–08 budget.

The Australian Government levies income taxes. State expenditure is backed by federal grants. Australian Government General Government Sector expenses and revenue outcomes (in $A1m.):

	2006–07	2007–08[1]
Total expenses (by function)	219,362	235,410
including		
General public services	14,615	15,774
Defence	16,854	19,243
Public order and safety	3,318	3,772
Education	16,898	17,386
Health	39,948	43,357
Social security and welfare	92,075	95,132
Housing and community amenities	2,909	3,345
Recreation and culture	2,561	2,861
Fuel and energy	4,635	5,310
Agriculture, fisheries and forestry	2,831	3,685
Mining, manufacturing and construction	1,920	2,046
Transport and communication	3,296	4,484
Other economic affairs	5,165	6,123
Other purposes	12,338	12,893
Total revenue (by source)		
including	237,008	251,885
Income tax		
Individuals and other withholding	117,614	121,650
Fringe benefits tax	3,754	3,970
Superannuation funds	7,879	10,130
Company tax	58,538	65,250
Petroleum resource rent tax	1,594	2,060
Indirect tax		
Total excise duty	22,734	23,090
Excise duty—petroleum products and crude oil	16,479	14,830
Other excise	6,255	8,260
Customs duty	5,644	6,010
Other taxes	3,748	3,710
Non-tax revenue	15,504	16,013

[1]Estimate.

Performance
Real GDP growth rates (based on OECD statistics):

2002	2003	2004	2005	2006	2007	2008	2009	2010	2011
3·8%	3·4%	3·8%	3·3%	2·6%	4·9%	2·2%	1·5%	2·4%	2·3%

In fiscal year 2012 (July 2011–June 2012) the GDP growth rate was 3·4%. During much of the 1990s the real GDP growth rate, along with lower inflation, made the Australian economic performance one of the best in the OECD area. In 2011 total GDP was US$1,379·4bn.

Banking and Finance
From 1 July 1998 a new financial regulatory framework based on three agencies was introduced by the Australian government, following recommendations by the Financial System Inquiry. The framework included changes in the role of the Reserve Bank of Australia and creation of the Australian Prudential Regulation Authority (APRA) with responsibility for the supervision of deposit-taking institutions (comprising banks, building societies

and credit unions), friendly societies, life and general insurance companies and superannuation funds. It further involved replacement of the Australian Securities Commission with the Australian Securities and Investments Commission (ASIC) with responsibility for the regulation of financial services and Australia's 1·2m. companies.

The banking system comprises:

(a) The Reserve Bank of Australia is the central bank. It has two broad responsibilities—monetary policy and the maintenance of financial stability, including stability of the payments system. It also issues Australia's currency notes and provides selected banking and registry services to Commonwealth Government customers and some overseas official institutions. Within the Reserve Bank there are two Boards: the Reserve Bank Board and the Payments System Board; the *Governor* (present incumbent, Glenn Stevens) is the Chairman of each.

At 30 June 2006 total assets of the Reserve Bank of Australia were $A109,932m., including gold and foreign exchange, $A76,889m.; and Australian dollar securities, $A31,102m. At 30 June 2006 capital and reserves were $A6,325m. and main liabilities were Australian notes on issue, $A38,069m.; and deposits, $A46,212m.

A wholly owned subsidiary of the Reserve Bank (Note Printing Australia Limited) manufactures currency notes and other security documents for Australia and for export.

(b) Four major banks: (i) The Commonwealth Bank of Australia; (ii) the Australia and New Zealand Banking Group Ltd; (iii) Westpac Banking Corporation; (iv) National Australia Bank.

(c) The Commonwealth Bank of Australia has a subsidiary—Commonwealth Development Bank. There are nine other Australian-owned banks—Adelaide Bank Ltd, AMP Bank Ltd, Bank of Queensland Ltd, Bendigo Bank Ltd, Elders Rural Bank (50% owned by Bendigo Bank Ltd), Macquarie Bank Ltd, Members Equity Pty Ltd, St George Bank Ltd and Suncorp-Metway Ltd.

(d) There are ten banks incorporated in Australia which are owned by foreign banks and 30 branches of foreign banks (these figures include five foreign banks which have both a subsidiary and a branch presence in Australia).

(e) According to the Australian Prudential Regulation Authority (APRA), as at 30 June 2006 there were 52 authorized banks with Australian banking assets of $A1,363·5bn., with 5,147 branches, 24,616 reported ATMs and 540,189 reported EFTPOS terminals. As at 30 June 2006 there were 14 building societies with assets of $A17·9bn. and 144 credit unions with assets of $A35·7bn.

Australia's gross external debt amounted to US$1,331,026m. in June 2012.

There is an Australian Stock Exchange (ASX) in Sydney.

ENERGY AND NATURAL RESOURCES

Environment

Australia's carbon dioxide emissions from the consumption and flaring of fossil fuels were the equivalent of 20·8 tonnes per capita in 2008. An *Environmental Performance Index* compiled in 2008 ranked Australia 46th in the world, with 79·8%. The index examined various factors in six areas—air pollution, biodiversity and habitat, climate change, environmental health, productive natural resources and water resources.

Electricity

Electricity supply is the responsibility of the State governments. 2007–08 total production was 228,600m. kWh. Coal is the main fuel source for electricity generation, accounting for 76% of the total, with gas accounting for 15% and hydro 5%. In 2007–08 consumption stood at 201,307m. kWh including 57,868m. kWh by residential customers.

Oil and Gas

The main fields are located in the Gippsland Basin (Vic.) and the North West Shelf (WA). In 2006, three new offshore fields (Enfield, Vincent and Stybarrow) in the Carnarvon Basin (WA) were expected to supply about 280,000 bbls of crude oil production per day. Crude oil and condensate production was 25,870m. litres in 2007–08, a reduction of 7·2% over the previous year. The ex-mine value of crude oil and natural gas production in 2008–09 totalled $A15·62bn. Oil reserves at the end of 2008 totalled 4·2bn. bbls and natural gas reserves 2,510bn. cu. metres. Natural gas production (2007–08) was 39·3bn. cu. metres.

Minerals

Australia is the world's largest producer of bauxite and alumina. It is the world's largest producer of diamonds, ranking first for industrial-grade diamonds and second for gem-grade diamonds, after Botswana. It is also the third largest gold and uranium producer. Black coal is Australia's major source of energy. Reserves are large (Dec. 2009: 44bn. economically recoverable tonnes) and easily worked. The main fields are in New South Wales and Queensland. Brown coal (lignite) reserves are mined principally in Victoria and South Australia. In 2005–06 raw coal production was 398m. tonnes; lignite production, 71m. tonnes; and iron ore and concentrates, 250m. tonnes.

Estimated production of other major minerals in 2005–06 (in tonnes): bauxite, 61m.; alumina, 17·8m.; salt, 11·8m.; manganese, 4·1m.; zinc, 1·4m.; nickel, 183,000; uranium, 9,974; silver, 2,218; gold, 250. Diamond production, 2005–06: 25·4m. carats.

Agriculture

In 2009–10 there were an estimated 134,184 establishments mainly engaged in agriculture; the estimated total area of land under agricultural use in the year ending 30 June 2010 was 398·6m. ha. (about 52% of total land area). Gross value of agricultural production in the same year, $A39·6bn., including (in $A1bn.) cattle and calves slaughtering, 7·3; sheep and lamb slaughtering, 2·6; wheat, 4·8; milk, 3·4; wool, 1·9. In year ending 30 June 2010 there were 26·0m. ha. of crops. An estimated 1·84m. ha. of crops and pastures were irrigated in 2009–10. Important crops (year ending 30 June 2010): wheat (21·8m. tonnes from 13·9m. ha.); barley (7·9m. tonnes from 4·4m. ha.); grain sorghum (1·5m. tonnes from 0·5m. ha.); oats (1·2m. tonnes from 0·85m. ha.); canola (1·9m. tonnes from 1·7m. ha.); sugarcane (31·2m. tonnes from 0·39m. ha.). In the year ended 30 June 2007 an estimated 1·53m. tonnes of grapes were harvested from 173,776 ha. of vines.

Beef cattle farming represents the largest sector, accounting for 25% of farming establishments. Livestock totals at June 2010: beef cattle and calves, 24·0m.; dairy cattle, 2·5m.; sheep and lambs, 68·1m.; pigs, 2·3m. Livestock products (in 1,000 tonnes) in 2009–10: beef and veal, 2,109; lamb and mutton, 574; pigmeat, 331; poultry meat, 834; wool, 353. Milk in the same year, 9,023m. litres.

Estimated fruit and vegetable production in year ending 30 June 2010 (in 1,000 tonnes): potatoes, 1,278; tomatoes, 472; oranges, 391; bananas, 302; carrots, 267; apples, 264; onions, 260.

Wine production (2006): 14,263,000 hectolitres. Australia was the sixth largest wine producer in the world in 2006 (5·0% of the global total) and the fourth largest wine exporter (9·1% of the global total).

In 2009 organic crops were grown in an area covering 12·0m. ha. (the largest area of any country in the world), representing 2·9% of all farmland.

Australia is the world's leading wool producer; only China has more sheep.

Forestry

The Federal government is responsible for forestry at the national level. Each State is responsible for the management of publicly owned forests. Total forest cover was 149·30m. ha. in 2010 (19% of Australia's land area). The major part of wood supplies derives from coniferous plantations, of which there were 1,001,100 ha. in

2006. Australia also had 807,400 ha. of broadleaved plantation in 2006. Timber production in 2007 was 32·26m. cu. metres.

Fisheries

The Australian Fishing Zone covers an area 16% larger than the Australian land mass and is the third largest fishing zone in the world, but fish production is insignificant by world standards owing to low productivity of the oceans. The major commercially exploited species are prawns, rock lobster, abalone, tuna, other fin fish, scallops, oysters and pearls. Estimated total fisheries production in 2005–06 came to 240,988 tonnes with a gross value of $A2·13bn. In the same year aquaculture production was an estimated 54,076 tonnes with a gross value of $A748·32m., which represented 35% of the total value of fisheries production.

INDUSTRY

The leading companies by market capitalization as at March 2012 were: BHP Billiton Ltd (Australian/British), a resources company (US$179·5bn.); Rio Tinto (Australian/British), a mining company (US$107·2bn.); and the Commonwealth Bank of Australia (US$82·1bn.).

In 2007–08 the manufacturing industry contributed 9·8% to Australia's GDP. In May 2006 almost 1·1m. people were employed, 10% of Australia's total employed.

Manufacturing by sector in 2007–08:

	Labour costs in $A1m.	Sales and service income in $A1m.
Food, beverages and tobacco	13,119	83,878
Textiles, clothing, footwear and leather products	1,970	9,905
Wood and paper products	4,383	23,003
Printing, publishing and recorded media	2,784	9,429
Petroleum, coal, chemical and associated products	7,946	82,893
Non-metallic mineral products	2,988	16,840
Primary and fabricated metal products	12,378	94,347
Transport equipment	7,042	34,939
Machinery and equipment	7,346	32,516
Furniture and other manufacturing	1,567	8,069

Manufactured products in 2007–08 included: clay bricks, 1,459m.; portland cement, 9·8m. tonnes; ready-mixed concrete, 27m. cu. metres; newsprint, 456,000 tonnes; raw steel, 8·1m. tonnes; pig iron, 6·3m. tonnes; automotive gasoline, 17,079m. litres; automotive diesel oil, 12,177m. litres; aviation turbine fuel, 5,182m. litres; beer, 1,677m. litres.

According to the World Bank's *Doing Business 2012* Australia is the second easiest country in which to start a business, after New Zealand.

Labour

In 2005–06 the total labour force (persons aged 15 and over) numbered 10,605,300 (4,770,700 females). In 2005–06 there were 10,065,800 employed persons (44·9% females) with 2,885,400 in part-time employment (72% females). The majority of wage and salary earners have had their minimum wages and conditions of work prescribed in awards by the Industrial Relations Commission. In Oct. 1991 the Commission decided to allow direct employer-employee wage bargaining, provided agreements reached are endorsed by the Commission. In some States, some conditions of work (e.g. weekly hours of work, leave) are set down in State legislation. Average weekly wage, May 2006, $A1,041·60 (men, $A1,101·20; women, $A932·90). Average weekly hours worked by full-time employed person, 2008–09: 39·8 hours. Four weeks annual leave is standard. In 2010–11 part-time work accounted for 30% of all employment in Australia. Foreign-born workers made up 26·5% of the labour force in 2008, the highest share of any major industrialized nation.

Employees in all States are covered by workers' compensation legislation and by certain industrial award provisions relating to work injuries.

In 2010 there were 227 industrial disputes recorded, which accounted for 126,600 working days lost (196,500 in 2008). In these disputes 54,800 workers were involved.

As at Dec. 2010 health care and social assistance (11·2% of employed persons) and retail trade (10·8%) were the largest employers, ahead of the construction industry (9·1%) and manufacturing (8·9%).

The following table shows the percentage distribution of employed persons in 2008–09 according to the *Australian and New Zealand Standard Classification of Occupations*:

	Employed persons (%)
Professionals	20·8
Clerical and administrative workers	15·4
Technicians and trades workers	15·2
Managers	12·9
Labourers	10·7
Sales workers	9·4
Community and personal service workers	8·9
Machinery operators and drivers	6·7

In 2010–11, 606,900 persons were unemployed, of whom 19% had been unemployed for more than one year. The unemployment rate in Dec. 2012 was 5·4% (up slightly from 5·1% in 2011 as a whole).

INTERNATIONAL TRADE

In 1990 Australia and New Zealand completed a Closer Economic Relations agreement (initiated in 1983) which establishes free trade in goods. Net foreign debt was $A600·0bn. as at 30 June 2008 (an increase of 9·6% on the previous year). In 1998 the effect of the Asian meltdown on exports resulted increasingly in shipments of commodities and exports of manufactures and some services being redirected to other destinations, notably the USA and Europe. Merchandise imports increased by 8% in 2006–07 against the previous year while exports rose by 10%.

Imports and Exports

The Australian customs tariff provides for preferences to goods produced in and shipped from certain countries as a result of reciprocal trade agreements. These include the UK, New Zealand, Canada and Ireland.

In 2009–10 merchandise imports totalled $A203,587m. ($A219,485m. in 2008–09); merchandise exports in 2009–10 totalled $A200,069m. ($A230,829m. in 2008–09).

Leading commodity imports, 2010 (in $A1m.): crude petroleum oils, 16,218; passenger motor vehicles, 15,917; refined petroleum, 9,970; medicaments (including veterinary), 7,896; telecommunications equipment and parts, 7,534; computers, 6,526. Most valuable commodity exports, 2010 (in $A1m.): iron ore and concentrates, 49,376; coal, 42,967; gold, 15,005; crude petroleum, 10,502; natural gas, 9,425; aluminium ores and concentrates (including alumina), 5,293.

Australia is the world's largest exporter of iron ore, black coal, bauxite, lead, diamonds, alumina, beef, barley and wool.

Trade by bloc or country in 2010 (in $A1m.):

	Imports	Exports
APEC	181,600	208,900
ASEAN	50,600	29,900
EU	51,100	27,000
China	41,000	64,400
Japan	20,400	45,700
Korea, South	7,700	22,400
New Zealand	9,900	11,400
Singapore	14,100	7,500
Thailand	13,000	6,900
UK	10,300	12,400
USA	35,300	14,500

COMMUNICATIONS

Roads

At 30 June 2007 there was an estimated 810,028 km of roads, of which 42·6% were paved.

As at 31 March 2006 registration totals were: 11,101,441 passenger vehicles, 2,114,333 light commercial vehicles, 475,519 trucks, 75,375 buses, 41,520 campervans and 463,057 motorcycles.

In 2008, 1,464 persons were killed in road accidents (less than half the 1985 total of 2,941).

Rail

Privatization of government railways began in Victoria in 1994 with West Coast Railway and Hoys Transport being granted seven-year franchises. Specialised Container Transport (SCT) won the first private rail freight franchise in 1995 followed by TNT (now Toll Holdings). Australian National Railway Commission was sold by the Commonwealth Government in Nov. 1997 and in Feb. 1999 V/Line Freight Corporation, owned by the Victorian Government, was sold to Freight Australia. Rail passenger services in Victoria were franchised in mid-1999. The Australian Railroad Group acquired Western Australia's government rail freight operation, Westrail, in Nov. 2000. In Jan. 2002 Toll Holdings and Lang acquired the rolling stock of the National Rail Corporation (NRC) and New South Wales freight carrier, FreightCorp. These two sales left Queensland Rail (QR) as the only government-owned rail freight operator in Australia. In July 2010 it was split into two separate companies with Queensland Rail as the State government-owned corporation responsible for passenger services and QR National, which was privatized in Nov. 2010, for freight.

In 2012 Australia had 33,299 route-km of open track of which 17,034 km were standard gauge (1,435 mm), 12,595 km were narrow gauge (1,067 mm), 3,281 km were narrow gauge (610 mm) and 389 km dual gauge; a total of 3,300 route-km were electrified. In 2005–06 a total of 684·9m. tonnes of freight were carried; passengers carried totalled 634m. urban (including train and tram); 9m. non-urban.

Under various Commonwealth–State standardization agreements, all the State capitals are now linked by standard gauge track. The 'AustralAsia Rail Project', which involved the construction of 1,420 km standard gauge railway between Alice Springs and Darwin, has been completed and passenger services from Adelaide through Alice Springs and on to Darwin commenced in Feb. 2004.

There are also private industrial and tourist railways, and tramways in Adelaide, Melbourne and Sydney. In the latter two cities there are also metro systems.

Civil Aviation

Qantas Airways is Australia's principal international airline. In 1992 Qantas merged with Australian Airlines, and in 1993, 25% of the company was purchased by British Airways. After it was floated on the stock exchange in the mid-1990s it was 55% Australian-owned and 45% foreign-owned. Under current law 51% must be Australian-owned. A total of 54 international airlines operated scheduled air services to and from Australia in the year ended June 2009. There are 13 international airports, the main ones being Adelaide, Brisbane, Cairns, Darwin, Melbourne, Perth and Sydney. In 2008–09 passenger movements totalled 122,024,912 (an increase of 1·6% on the previous financial year); domestic passenger numbers totalled 88,122,838, international 23,486,506 and regional 10,405,568; international freight decreased by 9·2% to 709,374 tonnes; international mail decreased by 6·0% to 36,581 tonnes.

Sydney (Kingsford Smith) handled the most traffic (26·5%) in Australia in 2008–09 (32,345,887 passengers, of which 20,095,054 on domestic flights), followed by Melbourne International (20·0%) and Brisbane (15·3%).

Internal airlines (domestic and regional) carried 43·7m. passengers in the year ended 31 July 2006. Domestic airlines were deregulated in Oct. 1990.

In 2008–09 there were 299 licensed, registered and certified aerodromes in Australia and its external territories. At 31 Dec. 2008 there were 13,459 registered aircraft in the Australian Civil Aircraft Register including 1,619 helicopters and 338 balloons.

Shipping

The chief ports are Brisbane, Dampier, Fremantle, Gladstone, Hay Point, Melbourne, Newcastle, Port Hedland, Port Kembla, Port Walcott, Sydney and Weipa. Port Hedland overtook Dampier as Australia's busiest port in 2008–09, handling 159,391,000 tonnes of cargo (158,382,000 tonnes loaded and 1,009,000 tonnes discharged) compared to 140,824,000 tonnes for Dampier (140,122,000 tonnes loaded and 702,000 tonnes discharged). Iron ore exports to China are the principal factor behind the rapid growth of Dampier and Port Hedland (both of which are in Western Australia) during the 2000s.

In Jan. 2009 there were 98 ships of 300 GT or over registered, totalling 1,258,000 GT. Of the 98 vessels registered, 45 were passenger ships, 26 general cargo ships, 15 bulk carriers, seven oil tankers, four liquid gas tankers and there was one container ship.

Telecommunications

In 1989 the domestic market became a regulated monopoly with Telstra as the government-owned company providing all services and, in 1991, a duopoly (with Optus) in fixed network services. In 1993 Vodafone joined Telstra and Optus in the provision of mobile phone services. A new regulatory regime was created by the introduction of the Telecommunications Act 1997 and both markets were opened to wholesale and retail competition. There is no limit to the number of carriers that can hold licences under the new arrangements and at 30 June 2010 there were 177 licensed carriers. The Australian Communications and Media Authority (ACMA) and the Australian Competition and Consumer Commission (ACCC) are the primary regulators with responsibility for the industry's development. The privatization of Telstra was completed in Nov. 2006.

In 2008 there were 9,370,000 main (fixed) telephone lines, down from 10,460,000 in 2003. Mobile phone subscribers numbered 22,120,000 in 2008 (1,049·6 per 1,000 persons). There were 15,170,000 internet users in 2008. The fixed broadband penetration rate in Dec. 2010 was 24·1 subscribers per 100 inhabitants. In Dec. 2011 there were 10·7m. Facebook users (48% of the population).

Three telecommunications satellites are in orbit covering the entire continent.

SOCIAL INSTITUTIONS

Justice

The judicial power of the Commonwealth of Australia is vested in the High Court of Australia (the Federal Supreme Court), in the Federal courts created by the Federal Parliament (the Federal Court of Australia and the Family Court of Australia) and in the State courts invested by Parliament with Federal jurisdiction.

High Court

The High Court consists of a Chief Justice and six other Justices, appointed by the Governor-General in Council. The Constitution confers on the High Court original jurisdiction, *inter alia*, in all matters arising under treaties or affecting consuls or other foreign representatives, matters between the States of the Commonwealth, matters to which the Commonwealth is a party and matters between residents of different States. Federal Parliament may make laws conferring original jurisdiction on the High Court, *inter alia*, in matters arising under the Constitution or under any laws made by the Parliament. It has in fact conferred jurisdiction on the High Court in matters arising under the Constitution and in matters arising under certain laws made by Parliament.

The High Court may hear and determine appeals from its own Justices exercising original jurisdiction, from any other Federal Court, from a Court exercising Federal jurisdiction and from the

Supreme Courts of the States. It also has jurisdiction to hear and determine appeals from the Supreme Courts of the Territories. The right of appeal from the High Court to the Privy Council in London was abolished in 1986.

Other Federal Courts
Since 1924, four other Federal courts have been created to exercise special Federal jurisdiction, i.e. the Federal Court of Australia, the Family Court of Australia, the Australian Industrial Court and the Federal Court of Bankruptcy. The Federal Court of Australia was created by the Federal Court of Australia Act 1976 and began to exercise jurisdiction on 1 Feb. 1977. It exercises such original jurisdiction as is invested in it by laws made by the Federal Parliament including jurisdiction formerly exercised by the Australian Industrial Court and the Federal Court of Bankruptcy, and in some matters previously invested in either the High Court or State and Territory Supreme Courts. The Federal Court also acts as a court of appeal from State and Territory courts in relation to Federal matters. Appeal from the Federal Court to the High Court will be by way of special leave only. The State Supreme Courts have also been invested with Federal jurisdiction in bankruptcy.

State Courts
The general Federal jurisdiction of the State courts extends, subject to certain restrictions and exceptions, to all matters in which the High Court has jurisdiction or in which jurisdiction may be conferred upon it.

Industrial Tribunals
The chief federal industrial tribunal is the Australian Conciliation and Arbitration Commission, constituted by presidential members (with the status of judges) and commissioners. The Commission's functions include settling industrial disputes, making awards, determining the standard hours of work and wage fixation. Questions of law, the judicial interpretation of awards and imposition of penalties in relation to industrial matters are dealt with by the Industrial Division of the Federal Court.

At 30 June 2008 the prison population was 27,615 (129 per 100,000 of national population).

Each State has its own individual police service which operates almost exclusively within its State boundaries. State police investigations include murder, robbery, street-level drug dealing, kidnapping, domestic violence and motor vehicle offences.

The role of the Australian Federal Police (AFP) is to enforce Commonwealth criminal law and protect Commonwealth and national interests from crime in Australia and overseas. Responsibilities include combating organized crime, trans-national crime, money laundering, illicit drug trafficking, e-crime, the investigation of fraud against the Australian Government and handling special references from Government. The AFP also provides a protection service to dignitaries and crucial witnesses as well as community policing services to the people of the Australian Capital Territory, Jervis Bay and Australia's External Territories.

Total Australian Federal Police personnel as at 30 June 2010 was 6,715, of which 3,056 were sworn employees (police officers) and 2,453 unsworn employees. In 2007–08 there were 55,387 sworn police officers in total in Australia.

Education
The governments of the Australian States and Territories have the major responsibility for education, including the administration and substantial funding of primary, secondary, and technical and further education. In most States, a single education department is responsible for these three levels but in Queensland, Western Australia and the Northern Territory, separate departments deal with school-based and technical and further education issues.

School attendance is compulsory between the ages of six (five in Tasmania) and 17 years (with the exception of the Northern Territory where the leaving age is 15 years), at either a government school or a recognized non-government educational institution. Between the ages of 16 and 17 in some states students have the option of undertaking vocational training, apprenticeships or other government approved learning programmes. Many children attend pre-schools for a year before entering school (usually in sessions of two–three hours, for two–five days per week). Government schools are usually co-educational and comprehensive. Non-government schools have been traditionally single-sex, particularly in secondary schools, but there is a trend towards co-education. Tuition is free at government schools, but fees are normally charged at non-government schools.

In Aug. 2006 there were 6,902 government (and 2,710 non-government) primary and secondary schools with 2,248,229 (1,119,807) full-time pupils and 158,194 (81,445) full-time teachers.

Vocational education and training (VET) is essentially a partnership between the Commonwealth, the States and Territories and industry. The Commonwealth is involved in VET through an agreed set of national arrangements for sharing responsibility with the States and Territories. The current mechanism for giving effect to this is the Australian National Training Authority (ANTA) Agreement, which sets out the roles and responsibilities for VET: they provide two-thirds of the funding and have all of the regulatory responsibility for the sector. They are also the 'owners' of the network of public Technical and Further Education (TAFE) institutes. In 2010 publicly-funded VET programmes were offered by some 170 TAFEs and other government institutions. A further 481 community education providers and over 2,000 other providers (mainly private providers) delivering VET were at least partly publicly funded. In 2010 there were 1·8m. people enrolled in publicly funded VET courses.

In 2011 there was a total of 128 higher education providers approved to receive Commonwealth Government funding through the Higher Education Support Act 2003, of which 38 were publicly-funded universities and 90 were other education institutions with students on accredited higher education courses. Most of these institutions operate under State and Territory legislation although several operate under Commonwealth legislation. There is also a completely privately-funded university and numerous private higher education providers in a range of specialist fields. Institutions established by appropriate legislation are autonomous and have the authority to accredit their own programmes and are primarily responsible for their own quality assurance. There were 1·19m. students in higher education in 2010; fields of study with the largest number of students in 2010 were management and commerce (28·5%); society and culture (21·6%); health (13·8%); and education (9·2%).

The higher education sector contributes a significant proportion of the research and research training undertaken in Australia. The Australian Research Council provides advice on research issues and administers the allocation of some research grants to higher education sector researchers and institutions.

The Commonwealth Government offers a number of programmes which provide financial assistance to students. The Youth Allowance is available for eligible full-time students aged 16 to 24, depending on the circumstances of study. Austudy is available to eligible full-time students aged over 25. Abstudy provides financial assistance for eligible Aboriginal and Torres Strait Islanders who undertake full-time or part-time study. AIC—the Assistance for Isolated Children scheme—provides special support to families whose children are isolated from schooling or who have physical disabilities.

Most students contribute to the cost of their higher education through the Higher Education Contribution Scheme (HECS). They can choose to make an upfront contribution (with a 25% discount) or to defer all or part of their payment until their income reaches a certain level when they must begin repaying their contribution through the taxation system. Overseas students generally pay full tuition fees. Universities are also able to offer full-fee places in postgraduate courses and to a limited number of domestic undergraduate students.

International education is increasingly important, with 335,273 enrolments by overseas students at Australian educational institutions in 2010, an increase of 4·5% on the previous year; 28·1% of higher education students were foreign, the highest proportion of any country.

Total operating expenses of Australian Government on education in 2007–08 were $A55,473m. Private expenditure on education in 2007–08 amounted to $A24,552m. The figures include government grants to the private sector which are also included in the operating expenses of Australian governments.

The adult literacy rate is at least 99%.

Health

In 2006–07 there were 758 public hospitals (including 19 psychiatric hospitals) and 543 private hospitals (including acute and psychiatric hospitals); there were 82,662 hospital beds (4·0 per 1,000 population). In 2006–07 there were 169,800 registered nurses, 37,000 general medical practitioners, 15,000 physiotherapists and 14,100 midwives. The Royal Flying Doctor Service serves remote areas. Total government expenditure on health goods and services (public and private sectors) in 2007–08 was $A103·6bn. ($A94·9bn. in the previous year), representing 9·1% of GDP. In 2007–08, 53% of Australians aged 15 years and over had private health insurance.

At 31 Dec. 2006 there were estimated to be 26,268 HIV cases, 10,119 AIDS diagnoses and 6,723 deaths following AIDS.

In 2007–08, 61% of the adult population (men, 68%; women, 55%) aged 18 years and over were considered overweight or obese, compared to 57% in 1995.

Welfare

All Commonwealth government social security pensions, benefits and allowances are financed from the Commonwealth government's general revenue. In addition, assistance is provided for welfare services.

Age Pensions—age pensions are payable to men 65 years of age or more who have lived in Australia for a specified period and, unless permanently blind, also satisfy an income and assets test. The minimum age for women's eligibility was raised by six months to 62 years on 1 July 2001 and has been lifted in six-monthly increments every two years until 1 July 2013 when it was scheduled to rise to 65 years. The qualifying age at 1 July 2011 was 64 years and six months. From 2017 the Age Pension qualifying age for both men and women will gradually increase from 65 to 67 years by 2023. In the year ending 30 June 2010, 2,158,303 age pensioners received a total of $A29,384·5m.

Disability Support Pension (DSP)—payable to persons aged 16 years or over with a physical, intellectual or psychiatric impairment of at least 20%, assessed as being unable to work for at least 15 hours a week. DSP for those of 21 years or over is paid at the same rate as Age Pensions and is subject to the same means test except for those who are permanently blind. In the year ending 30 June 2010, 792,581 disability support pensioners received a total of $A11,859·7m.

Carer Payment—payable to a person unable to support themselves owing to providing constant care and attention at home for a severely disabled person aged 16 or over, or a person who is frail aged, either permanently or for an extended period. Since 1 July 1998 Carer Payment has been extended to carers of children under 16 years of age with profound disabilities. Subject to income and assets tests, the rate of Carer Payment is the same as for other pensions. In the year ending 30 June 2010, 168,913 carers received a total of $A2,269·4m.

Carer Allowance—supplementary payment to a person providing constant care and attention at home for an adult or child with a disability or severe medical condition. The allowance is not income or assets tested. In the year ending 30 June 2010, 495,733 carers received a total of $A1,477·7m.

Sickness Allowance—paid to those over school-leaving age but below Age Pension age who are unable to work or continue full-time study temporarily owing to illness or injury. Eligibility rests on the person having a job or study course to which they can return. In the year ending 30 June 2010 a total of $A83·7m. was paid to 6,703 beneficiaries.

Family Tax Benefit (FTB)—replaced *Family Allowance* and *Family Tax Payment* on 1 July 2000. Family Tax Benefit Part A is paid to assist families with children under 21 years of age or dependent full-time students aged 21–24 years; Family Tax Benefit Part B provides additional assistance to families with only one income earner and children under 16 years of age or dependent full-time students aged 16–18 years. Both benefits are subject to an income and assets test. In the year ending 30 June 2010 FTB Part A and Part B payments were made to a total of 3·1m. families.

Parenting Payment (Single) and (Partnered)—is paid to assist those who care for children under 16, with income and assets under certain amounts, and have been an Australian resident for at least two years or a refugee or have become a lone parent while an Australian resident. Parenting Payment (Single) is paid to lone parents under pension rates and conditions; Parenting Payment (Partnered) is paid to one of the parents in the couple. [Since 1 July 2000 the basic component of Parenting Payment (Partnered) was incorporated into Family Tax Benefit with 375,233 beneficiaries transferring to Family Tax Benefit Part B.] In the year ending 30 June 2010, 333,512 Parenting Payment (Single) beneficiaries and 124,910 Parenting Payment (Partnered) beneficiaries received a total of $A5,467·1m.

Baby Bonus—recognizing the costs associated with a new baby, all families with a child born (including still born) or adopted are eligible for the payment; an income test applies. In the year ended 30 June 2010 a total of $A1,398·4m. was paid.

Newstart Allowance (NSA)—payable to those who are unemployed and are over 21 years of age but less than Age Pension age. Eligibility is subject to income and assets tests and recipients must satisfy the 'activity test' whereby they are actively seeking and willing to undertake suitable paid work, including casual and part-time work. To be eligible for benefit a person must have resided in Australia for at least 12 months preceding his or her claim or intend to remain in Australia permanently; unemployment must not be as a result of industrial action by that person or by members of a union to which that person is a member. In the year ended 30 June 2010 a total of $A6,136·8m. was paid to 553,893 NSA beneficiaries.

Youth Allowance—replaced five former schemes for young people, including the Youth Training Allowance. In the year ending 30 June 2010 a total of $A2,789·4m. was paid to YA beneficiaries.

Austudy—means tested payment for students or apprentices aged 25 years and over. In the year ending 30 June 2010 a total of $A343·4m. was paid.

Service Pensions—are paid by the Department of Veterans' Affairs. Male veterans who have reached the age of 60 years or are permanently unemployable, and who served in a theatre of war, are eligible subject to an income and assets test. The minimum age for female veterans' eligibility has been lifted in six-monthly increments every two years until 1 July 2013 when it was scheduled to rise to 60 years. The qualifying age at 1 July 2012 was 59 years 6 months. Wives of service pensioners are also eligible, provided that they do not receive a pension from the Department of Social Security. Disability pension is a compensatory payment in respect of incapacity attributable to war service. It is paid at a rate commensurate with the degree of incapacity and is free of any income test. In the year ended 30 June 2010, $A2,979·4m. for service pensions and $A3,292·9m. for disability and war widows' dependants' pensions were paid out; at 30 June 2010 there were 179,242 service pensioners and 222,876 disability and war widow(er)s' pensioners.

In addition to cash benefits, welfare services are provided, either directly or through State and local government authorities and voluntary agencies, for people with special needs.

Medicare—covers: automatic entitlement under a single public health fund to medical and optometrical benefits of 75% of the Medical Benefits Schedule fee, with a maximum patient payment for any service where the Schedule fee is charged; access without direct charge to public hospital accommodation and to inpatient and outpatient treatment by doctors appointed by the hospital; the restoration of funds for community health to approximately the same real level as 1975; a reduction in charges for private treatment in shared wards of public hospitals, and increases in the daily bed subsidy payable to private hospitals.

The Medicare programme is financed in part by a 1·5% levy on taxable incomes, with low income cut-off points, which were $A18,839 p.a. for a single person in 2009–10 and $A31,789 for a family; there is an extra income threshold for each child (lower threshold in 2010–11 of $A2,919). A levy surcharge of 1% was introduced from 1 July 1997 for single individuals with taxable incomes in excess of $A50,000 p.a. and couples and families with combined taxable incomes in excess of $A100,000 who do not have private hospital cover through private health insurance.

Medicare benefits are available to all persons ordinarily resident in Australia. Visitors from the UK, New Zealand, Italy, Sweden, the Netherlands and Malta have immediate access to necessary medical treatment, as do all visitors staying more than six months.

RELIGION

Under the Constitution the Commonwealth cannot make any law to establish any religion, to impose any religious observance or to prohibit the free exercise of any religion. The following percentages refer to those religions with the largest number of adherents at the census of 2006. Answering the census question on religious adherence was not obligatory, however.

Christian, 63·9% of population: Catholic, 25·8%; Anglican, 18·7%; Uniting Church, 5·7%; Presbyterian and Reformed, 3·0%; Orthodox, 2·7%; Baptist, 1·6%; Lutheran, 1·3%; Pentecostal, 1·1%; Jehovah's Witnesses, 0·4%; Salvation Army, 0·3%; Churches of Christ, 0·3%; other Christian, 3·0%. Religions other than Christian, 6·2%: Buddhism, 2·1%; Islam, 1·7%; Hinduism, 0·7%; Judaism, 0·4%; other religions, 1·3%; no religion, 18·7%; no statement, 11·2%.

The Anglican Synod voted for the ordination of ten women in Nov. 1992. In Feb. 2013 the Roman Catholic church had three cardinals.

Jupp, James, (ed.) *The Encyclopedia of Religion in Australia.* 2009
Thompson, R. C., *Religion in Australia, a History.* 1995

CULTURE

World Heritage Sites
There are 19 sites under Australian jurisdiction that appear on the UNESCO World Heritage List. They are (with year entered on list): Great Barrier Reef (1981), Kakadu National Park (1981, 1987 and 1992), Willandra Lakes Region (1981), Tasmanian Wilderness (1982 and 1989), Lord Howe Island Group (1982), Gondwana Rainforests of Australia (1986 and 1994), Uluru-Kata Tjuta National Park (1987 and 1994), Wet Tropics of Queensland (1988), Shark Bay (1991), Fraser Island (1992), Australian Fossil Mammal Sites (Riversleigh/Naracoorte) (1994), Heard and McDonald Islands (1997), Macquarie Island (1997), the Greater Blue Mountains Area (2000), Purnululu National Park (2003),

Royal Exhibition Building and Carlton Gardens in Melbourne (2004), Sydney Opera House (2007), Australian Convict Sites in New South Wales, Tasmania, Western Australia and Norfolk Island (2010), and Ningaloo Coast (2011).

Press
There were 52 English daily metropolitan newspapers in 2007 (two national, 13 metropolitan, 36 regional and one suburban). There are also 11 metropolitan Sunday newspapers. The papers with the largest circulations are the *Sunday Telegraph* (New South Wales), with an average of 671,500 per issue in 2007; the *Sunday Herald Sun* (Victoria), with an average of 620,000 per issue; and the *Sunday Mail* (Queensland), with an average of 592,440 per issue. In 2007 there were two free dailies, *mX* (with three editions, published in Brisbane, Melbourne and Sydney) and *Manly Daily*.

Tourism
In 2006 the total number of overseas visitors for the year stood at 5·5m. (the same number as in the previous year). The top source countries for visitors in 2006 were New Zealand (1,075,800); UK (734,200); Japan (651,000); USA (456,100); China (308,500); and South Korea (260,800). Tourism is Australia's largest single earner of foreign exchange.

Festivals
Among the largest events are the Sydney Festival (running for three weeks each Jan., with over 500 performers from across the arts), the National Multicultural Festival in Canberra (Feb.), the Perth International Arts Festival (Feb.), the Brisbane Festival (July) and the Melbourne International Arts Festival (Oct.). Adelaide and Darwin have their own large events too. Each June the Dreaming Festival in Woodford celebrates Indigenous culture. Among the biggest rock and pop festivals are the Big Day Out (Jan.–Feb.) and Good Vibrations (Feb.) concerts held across different locations in Sydney, Melbourne, Brisbane, Adelaide and Perth. Other music festivals include Tamworth Country Music Festival (Jan.), Canberra's National Folk Festival (Easter) and Bluesfest at Byron Bay (also Easter). The Melbourne International Film Festival (July–Aug.) is the largest film festival in the southern hemisphere with over 190,000 annual admissions.

DIPLOMATIC REPRESENTATIVES

Of Australia in the United Kingdom (Australia House, Strand, London, WC2B 4LA)
High Commissioner: John Dauth.

Of the United Kingdom in Australia (Commonwealth Ave., Yarralumla, A.C.T. 2600)
High Commissioner: Paul Madden.

Of Australia in the USA (1601 Massachusetts Ave., NW, Washington, D.C., 20036)
Ambassador: Kim Beazley.

Of the USA in Australia (Moonah Pl., Yarralumla, A.C.T. 2600)
Ambassador: Jeffrey Bleich.

Of Australia to the United Nations
Ambassador: Gary Quinlan.

Of Australia to the European Union
Ambassador: Duncan Lewis.

FURTHER READING

Australian Bureau of Statistics (ABS). *Year Book Australia,* since 1901.—*Australia at a Glance,* since 1994.—*Australian Economic Indicators* (absorbed *Monthly Summary of Statistics*) since 1994.—*Australian Social Trends,* since 1994. ABS also provide numerous online specialized statistical summaries.

Arthur, Bill and Morphy, Frances, *The Macquarie Atlas of Indigenous Australia.* 2006
Australian Encyclopædia. 12 vols. 1983

Blainey, G., *A Short History of Australia*. 1996

The Cambridge Encyclopedia of Australia. 1994

Concise Oxford Dictionary of Australian History. 2nd ed. 1995

Davison, Graeme, *et al.*, (eds.) *The Oxford Companion to Australian History*. 2nd ed. 2002

Docherty, J. D., *Historical Dictionary of Australia*. 1993

Foster, S. G., Marsden, S. and Russell, R. (compilers) *Federation. A guide to records*. 2000

Gilbert, A. D. and Inglis, K. S. (eds.) *Australians: a Historical Library*. 5 vols. 1988

Hirst, John, *The Sentimental Nation: The Making of the Australian Commonwealth*. 2000.—*Australia's Democracy: A Short History*. 2002

Irving, H. (ed.) *The Centenary Companion to Australian Federation*. 2000

Knightley, Phillip, *Australia: A biography of a Nation*. 2000

Macintyre, S., *A Concise History of Australia*. 2000

Oxford History of Australia. vol 2: 1770–1860. 1992. vol 5: 1942–88. 1990

Peel, Mark and Twomey, Christina, *A History of Australia*. 2011

Ward, Stuart, *Australia and the British Embrace: The Demise of the Imperial Ideal*. 2002

A more specialized title is listed under RELIGION, *above*.

National library: The National Library, Parkes Place, Parkes, Canberra, ACT 2600.

Website: http://www.nla.gov.au

National Statistical Office: Australian Bureau of Statistics (ABS), ABS House, 45 Benjamin Way, Belconnen, ACT 2617. The statistical services of the states are integrated with the Bureau.

ABS Website: http://www.abs.gov.au

AUSTRALIAN TERRITORIES AND STATES

Australian Capital Territory

KEY HISTORICAL EVENTS

The area that is now the Australian Capital Territory (ACT) was explored in 1820 by Charles Throsby who named it Limestone Plains. Settlement commenced in 1824. In 1901 the Commonwealth constitution stipulated that a land tract of at least 260 sq. km in area and not less than 160 km from Sydney be reserved as a capital district. The Canberra site was adopted by the Seat of Government Act 1908. The present site was surrendered by New South Wales and accepted by the Commonwealth in 1909. By subsequential proclamation the Territory became vested in the Commonwealth from 1 Jan. 1911. The Jervis Bay Territory was acquired by the Commonwealth of Australia from New South Wales in 1915 in order that the national seat of government at Canberra would have access to the sea. In 1911 an international competition for the city plan had been won by W. Burley Griffin of Chicago but construction was delayed by the First World War. It was not until 1927 that Canberra became the seat of government. Located on the Molonglo River surrounding an artificial lake, it was built as a compromise capital to stop squabbling between Melbourne and Sydney following the 1901 Federation of Australian States.

In Dec. 1988 self-government was proclaimed and in May 1989 the first ACT assembly was elected.

TERRITORY AND POPULATION

The total area is 2,431 sq. km (including Jervis Bay Territory). Around 60% is hilly or mountainous. Timbered mountains are located in the south and west, and plains and hill country in the north. The ACT lies within the upper Murrumbidgee River catchment, in the Murray-Darling Basin. The Murrumbidgee flows throughout the Territory from the south, and its tributary, the Molonglo, from the east. The Molonglo was dammed in 1964 to form Lake Burley Griffin. Population at the 2011 census excluding Jervis Bay Territory was 357,222 (2006: 324,034).

The Jervis Bay Territory (73 sq. km) is independent and administered by central government although the laws of the ACT apply. Population at the 2011 census was 378 (2006: 368).

SOCIAL STATISTICS

2007: births, 4,753; deaths, 1,597; marriages, 1,610; divorces, 1,333. Infant mortality rate (per 1,000 live births), 3·8. Expectation of life, 2007: males, 80·3 years; females, 84·0 years.

CLIMATE

ACT has a continental climate, characterized by a marked variation in temperature between seasons, with warm to hot summers and cold winters.

CONSTITUTION AND GOVERNMENT

The ACT became self-governing on 11 May 1989. It is represented by two members in the Commonwealth House of Representatives and two senators.

The parliament of the ACT, the *Legislative Assembly*, consists of 17 members elected for a three-year term. Its responsibilities are at State and Local government level. The Legislative Assembly elects a Chief Minister and a four-member cabinet.

Electors enrolled (30 June 2006) numbered 226,576.

RECENT ELECTIONS

At the elections of 20 Oct. 2012 the Labor Party and the Liberal party both won eight seats with 38·9% of the vote. The Green Party took three seats (10·7%).

CURRENT GOVERNMENT

In March 2013 the Labor-Green government comprised the following:

Chief Minister, Minister for Health, Regional Development, and Higher Education: Katy Gallagher.

Deputy Chief Minister, Treasurer, Minister for Economic Development, Community Services, Sport and Recreation, and Tourism and Events: Andrew Barr. *Education and Training, Disability, Children and Young People, Arts, Women, Multicultural Affairs, and Racing and Gaming*: Joy Burch. *Attorney General, Minister for Environment and Sustainable Development, Police and Emergency Services, and Workplace Safety and Industrial Relations*: Simon Corbell. *Territory and Municipal Services, Corrections, Housing, Aboriginal and Torres Strait Islander Affairs, and Ageing*: Shane Rattenbury.

Speaker: Vicki Dunne.

ACT Government Website: http://www.act.gov.au

ECONOMY

Budget

The ACT fully participates in the federal-state model underpinning the Australian Federal System. As a city-State, the ACT Government reflects State and local (municipal) government responsibilities, which is unique within the federal system.

However, the ACT is treated equitably with the other Australian jurisdictions regarding the distribution of federal funding.

In 2006 the Territory recorded total income of $A3,436m. (including the Territory's share of operating surplus from Joint Ventures accounted for using the Equity Method) and incurred expenditure of $A3,065m. achieving a surplus of $A372m. on accrual basis.

Banking and Finance
According to the Australian Prudential Regulation Authority, as at 30 June 2006 there were ten authorized banks with 66 branches; one building society with four branches and seven credit unions with 31 branches.

ENERGY AND NATURAL RESOURCES
Electricity
See NEW SOUTH WALES.

Agriculture
Sheep and/or beef cattle farming is the main agricultural activity. In 2005–06 there were 99 farming establishments with an estimated total area of 44,910 ha. Gross value of agricultural production in 2009–10: $A10·7m.

Forestry
There are about 10,000 ha. of plantation forest in the ACT (approximately 3·96% of the land area). Most of the area is managed for the production of softwood timber. The established pine forests, such as Kowen, Stromlo, Uriarra and Pierces Creek, are in the northern part of the Territory. After harvesting, 500–1,000 ha. of land are planted with new pine forest each year. No native forests or woodlands have been cleared for plantation since the mid-1970s.

INDUSTRY
Manufacturing industries in the year ended 30 June 2007 employed 4,967 persons and generated sales and service income of $A1,162m.

Labour
In June 2007 there were 187,900 employed persons and 6,000 unemployed persons, an unemployment rate of 3·1%.

In the year ending Feb. 2006, 25·7% of the ACT labour force was employed in public administration and defence; 14·9% in property and business services; 11·6% in retail trade. The average weekly wage in Nov. 2010 was $A1,468·20 (males $A1,550·80, females $A1,370·50).

INTERNATIONAL TRADE
Imports and Exports
In 2009–10 foreign merchandise imports were valued at $A5m.; exports at $A4m. Major sources of imports for 2009–10 (figures in $A1,000): Austria, 1,737; USA, 810; France, 660. Major export destinations (figures in $A1,000): Germany, 2,380; Saudi Arabia, 729; Hong Kong, 600. Principal commodity imports in 2009–10 were (with values in $A1,000): artwork and antiques, 1,558; telecommunications equipment and parts, 771; electrical machinery and parts, 518. Principal commodity exports (with values in $A1,000): gold coins and legal tender coins, 1,129; heating and cooling equipment, 729; optical instruments, 71.

COMMUNICATIONS
Roads
In 2007 there were an estimated 3,023 km of roads. At 31 March 2006 there were 224,076 vehicles registered in the ACT. In 2007 there were 14 road accident fatalities.

Civil Aviation
In 2009–10 Canberra International Airport handled 3,258,396 passengers.

Telecommunications
In 2005–06 there were 101,000 households with home computer access (82% of all households). In 2005–06, 89,000 households had home internet access (72%).

SOCIAL INSTITUTIONS
Justice
In the year ending 30 June 2011 there were 26,087 criminal incidents recorded by police. During the same year there were 740 full-time sworn police officers in the ACT and 206 unsworn police staff.

Education
In Feb. 2006 there were 221 schools comprising 82 pre-schools, 139 primary and secondary schools (including colleges) and five special schools. Of these 177 were government schools. There was a total of 60,142 full-time students. There were four higher education institutions in 2006: the Signadou Campus of the Australian Catholic University (ACU) had 596 students enrolled; the Australian National University, 14,476 students; the University of Canberra, 11,632; and the Australian Defence Force Academy, 2,136.

Health
The ACT is serviced by three public and 12 private hospitals (nine of the private hospitals are free-standing day hospital facilities only). At 30 June 2006 there were 2,056 medical practitioners, 4,469 registered nurses and 264 dentists.

Welfare
In 2005–06 there were 18,468 age pensioners (5·7% of ACT population); 7,281 persons received disability support pension (2·2%).

RELIGION
At the 2006 census, 60·2% of the population were Christian. Of these, 46·5% were Roman Catholic and 27·8% Anglican. Non-Christian religions accounted for 6·2%, the largest groups being Buddhists, Muslims and Hindus.

CULTURE
Tourism
In the year ending March 2009, 156,885 international visitors came to the ACT. Of these, the largest proportion (19%) was from the UK. At Dec. 2008 there were 56 hotels, guest houses and serviced apartments employing 2,467 persons.

FURTHER READING
Statistical Information: The State office of the Australian Bureau of Statistics (ABS) is at Level 5, QBE Insurance Building, 33–35 Ainslie Ave., Canberra City. Publications include: *Australian Capital Territory in Focus.* Annual (from 1994); *Australian Capital Territory at a Glance,* Annual (from 1995).

Sources: ACT in Focus 1307.8, Labour Force, Australia 6203.0 and *Labour Force, New South Wales and Australian Capital Territory 6201.1.*

Northern Territory

KEY HISTORICAL EVENTS
The Northern Territory, after forming part of New South Wales, was annexed on 6 July 1863 to South Australia. After the agreement of 7 Dec. 1907 for the transfer of the Northern Territory to the Commonwealth, it passed to the control of the Commonwealth government on 1 Jan. 1911. On 1 Feb. 1927 the Northern Territory was divided into two territories but in 1931 it was again administered as a single territory. The Legislative

Council for the Northern Territory, constituted in 1947, was reconstituted in 1959. In that year, citizenship rights were granted to Aboriginal people of 'full descent'. On 1 July 1978 self-government was granted.

TERRITORY AND POPULATION

The Northern Territory's total area is 1,349,129 sq. km and includes adjacent islands. It has 5,100 km of mainland coastline and 2,100 km of coast around the islands. The greater part of the interior consists of a tableland with excellent pasturage. The southern part is generally sandy and has a small rainfall.

The 2011 census population was 211,944 (2006: 192,898). The capital, seat of government and principal port is Darwin, on the north coast; 2006 census population, 105,991. Other main areas of population (2006 totals of local government areas) are Alice Springs (23,893); Palmerston (23,719); and Katherine (8,194). There are also a number of large self-contained Aboriginal communities. People identifying themselves as indigenous numbered 53,662 at the 2006 census.

SOCIAL STATISTICS

2007 totals: births, 3,894; deaths, 1,001; marriages, 779; divorces, 417. Infant mortality rate per 1,000 live births, 8·5. Life expectancy, 2007: 72·4 years for males, 78·4 for females. The annual rates per 1,000 population in 2007 were: births, 18·1; deaths, 4·7; marriages, 3·6; divorces, 1·9.

CLIMATE

See AUSTRALIA: Climate.

The highest temperature ever recorded in the NT was 118·9°F (48·3°C) at Finke in 1960, while the lowest recorded temperature was 18·5°F (–7·5°C) at Alice Springs in 1976.

CONSTITUTION AND GOVERNMENT

The Northern Territory (Self-Government) Act 1978 established the Northern Territory as a body politic as from 1 July 1978, with Ministers having control over and responsibility for Territory finances and the administration of the functions of government as specified by the Federal government. Regulations have been made conferring executive authority for the bulk of administrative functions.

The Northern Territory has federal representation, electing one member to the House of Representatives and two members to the Senate.

The *Legislative Assembly* has 25 members, directly elected for a period of four years. The *Administrator* (Tom Pauling) appoints Ministers on the advice of the Leader of the majority party.

Electors enrolled (30 June 2006) numbered 111,254.

RECENT ELECTIONS

In parliamentary elections held on 25 Aug. 2012 the opposition Country Liberal Party won 16 seats against eight for the Australian Labor Party. One independent was elected. Turnout was 76·9%.

CURRENT GOVERNMENT

Administrator: Sally Thomas, AM.

The Country Liberal Party Cabinet was as follows in March 2013:

Chief Minister, Minister for Police, Fire and Emergency Services, Corporate and Information Services, Trade, Economic Development, Asian Engagement, and Transport: Adam Giles.

Deputy Chief Minister, Treasurer, Minister for Business, Employment and Training, and Defence Liaison and Defence Industry Support: David Tollner. *Attorney General, and Minister for Justice, Public Employment, Correctional Services, and Statehood:* John Elferink. *Health, and Alcohol Rehabilitation:* Robyn Lambley. *Education, Housing, and Lands, Planning and the Environment:* Peter Chandler. *Primary Industry and Fisheries,*

Mines and Energy, Land Resource Management, and Essential Services: Willem Westra van Holthe. *Children and Families, Regional Development, Local Development, and Women's Policy:* Alison Anderson. *Central Australia, Tourism and Major Events, Sport and Recreation, Racing, Parks and Wildlife, and Arts and Museums:* Matthew Conlan. *Infrastructure, Multicultural Affairs, Senior Territorians and Young Territorians:* Peter Styles.

NT Government Website: http://www.nt.gov.au

ECONOMY

Budget

Revenue and expenditure in \$A1m.:

	2005–06	2006–07	2007–08[1]
Revenue	3,037	3,298	3,506
Expenditure	3,015	3,146	3,449

[1]Latest estimate from 2007–08 Mid-Year Report.

In 2007–08 total revenue was expected to be \$A3,506m. of which \$A2,770m. grants to the Northern Territory from the Commonwealth, \$A386m. in state-like taxes and \$A349m. Northern Territory Government own-source revenue.

Estimated expenditure in 2007–08 included \$A772m. for health; \$A741m. for education; \$A432m. for public order and safety.

Banking and Finance

According to the Australian Prudential Regulation Authority, as at 30 June 2006 there were seven authorized banks with 39 branches; and six credit unions with 24 branches.

ENERGY AND NATURAL RESOURCES

Environment

There are 93 parks and reserves covering 43,709 sq. km. Twelve of the parks are classified as national parks, including the Kakadu and Uluru-Kata Tjuta National Park which are included on the World Heritage List.

Electricity

The Power and Water Corporation supplies power to 72 indigenous and remote communities as well as the major centres. In the year ended 30 June 2009 total electricity generated was 1,525 GWh; total consumption was 1,748m. kWh. Total installed capacity at 30 June 2009 was 473 MW.

Oil and Gas

The Timor Sea is a petroleum producing province with five fields and more than 22m. cu. ft of known gas reserves. Gas is currently supplied from the Palm Valley and Mereenie fields in the onshore Amadeus Basin to the Channel Island Power Station in Darwin via one of Australia's longest onshore gas pipelines. The estimated total value of oil and gas production in 2009–10 was \$A2,401·4m., a decrease of \$A64·4m. on 2008–09. The Northern Territory produced 1,789 megalitres of crude oil and 509m. cu. metres of natural gas in 2006–07.

Water

The Power and Water Corporation (PAWC) is responsible for providing water supply (also electric power and sewerage services) throughout the NT. PAWC's subsidiary, Indigenous Essential Services Pty Ltd, maintains water supply to the 72 remote communities and Aboriginal outstations.

Minerals

Mining is the major contributor to the Territory's economy. Compared to 2005–06 the overall value of production in the mining industry rose by 51·2% in 2006–07. Value of major mineral commodities production in 2006–07 (in \$A1m.):

manganese, 982; zinc/lead concentrate, 566; alumina, 483; gold, 421; uranium oxide, 273; bauxite, 167; crushed rock, 13.

Agriculture

In the year ending June 2007 there were 640 agricultural establishments with a total area under holding of 61·2m. ha. Gross value of agricultural production in the year ending June 2007 rose by 21·8% to $A373m. Beef cattle production constitutes the largest farming industry. Total value of livestock slaughter and products in the year ending June 2007 was $A243m., an increase of 8·4% on the previous year. In the same year fruit production totalled (in tonnes): mangoes, 13,937; bananas, 1,701; table grapes, 1,327. Estimated value of the Northern Territory crocodile industry in 2010–11 was $A10·1m. with 64,835 kg of meat produced.

Forestry

In 2006 there were 27m. ha. of native forest, accounting for 20·2% of the Territory's total land area. As at June 2006 the total native forest cover consisted of 406,000 ha. closed forest, 7,139,000 ha. open forest and 25,290,000 ha. woodland. Total area of plantation forest was 26,000 ha., consisting mainly of softwoods. Hardwood plantations of fast-growing Acacia Mangium have been established on the Tiwi Islands for the production of woodchip for paper pulp. In addition, a number of operations for the production of sandalwood oil and neem products have been established near Batchelor.

Fisheries

Estimated total fisheries production in 2005–06 came to 8,257 tonnes with a gross value of $A98·0m. In the same year, aquaculture had a gross value of $A26·0m.

INDUSTRY

In the year ended 30 June 2007 the sales and service income generated by manufacturing industry was $A2,765m.; 4,634 persons were employed and salaries totalled $A283m. In Nov. 2006, 15,300 persons were employed in the wholesale and retail trade.

Labour

The labour force totalled 104,700 in Jan. 2007, of whom 102,100 were employed. The unemployment rate was 2·5%, down from 6·4% in Jan. 2006. The average weekly wage in Nov. 2006 was $A847·90 (males $A975·50, females $A724·80).

INTERNATIONAL TRADE

Imports and Exports

In 2009–10 foreign merchandise imports were valued at $A3,051m.; exports at $A4,980m. Major sources of imports for 2009–10 (figures in $A1m.): Singapore, 550; Kuwait, 331; Japan, 169; USA, 131; Thailand, 93. Major export destinations (figures in $A1m.): Japan, 1,952; China, 1,225; USA, 446; Indonesia, 339; Oman, 187. Principal commodity imports in 2009–10 were (with values in $A1m.): natural gas, 1,219; refined petroleum, 529; non-electric engines and motors, 138. Principal exports (with values in $A1m.): natural gas, 1,771; manganese ores and concentrates, 902; zinc ores and concentrates, 235.

COMMUNICATIONS

Roads

In 2007 there were an estimated 22,187 km of roads. The number of registered motor vehicles (excluding tractors and trailers) at 31 March 2006 was 114,015, including 73,302 passenger vehicles, 28,872 light commercial vehicles, 4,725 trucks, 2,989 buses and 3,950 motorcycles. There were 75 road accident fatalities in 2008. The death rate on the Northern Territory's roads in 2008 was, at 34·1 per 100,000 people, nearly five times higher than the rate for Australia as a whole.

Rail

In 1980 Alice Springs was linked to the Trans-continental network by a standard (1,435 mm) gauge railway to Tarcoola in South Australia (830 km). A 1,410 km standard gauge line operates between Darwin and Alice Springs. This $A1·3bn. AustralAsia Railway project links Darwin and Adelaide. The first train to complete the journey of 1,860 miles arrived in Darwin on 3 Feb. 2004.

Civil Aviation

Darwin and most regional centres in the Territory are serviced by daily flights to all State capitals and major cities. In 2010 there were direct international services connecting Darwin to Indonesia (Bali), Singapore, Timor-Leste and Vietnam. In 2009–10 Darwin airport handled 1,569,007 passengers (1,191,208 on domestic airlines); and Alice Springs 681,295 passengers (681,255 domestic).

Shipping

In 2008–09, 1,578 commercial vessels called at Northern Territory ports. General cargo imported in 2008–09 was 333,249 mass tonnes and general cargo exported was 247,324 mass tonnes.

Telecommunications

In 2005–06 there were 41,000 households with home computer access (70% of all households). In 2005–06, 35,000 households had home internet access (60%).

SOCIAL INSTITUTIONS

Justice

Voluntary euthanasia for the terminally ill was legalized in 1995 but the law was overturned by the Federal Senate on 24 March 1997. The first person to have recourse to legalized euthanasia died on 22 Sept. 1996.

Police personnel (sworn and unsworn) at 30 June 2007, 1,703 including 78 Aboriginal community police officers. In 2006–07 the Territory had two prisons with a daily average of 833 prisoners held.

Education

Education is compulsory from the age of six to 15 years. There were (Aug. 2006) 28,506 full-time students enrolled in 151 government schools and 9,074 enrolled in 35 non-government schools. Teaching staff in government and non-government schools totalled 3,205. The proportion of Indigenous students in the Territory is high, comprising 38·9% of all primary and secondary students at Aug. 2006. Bilingual programmes operate in some Aboriginal communities where traditional Aboriginal culture prevails.

The Northern Territory University (NTU), founded in 1989 by amalgamating the existing University College of the Northern Territory and the Darwin Institute of Technology, joined with the Alice Springs' Centralian College in 2004 to form the Charles Darwin University. In 2011, 8,744 students were enrolled in higher education courses of whom 4·7% were identified as Indigenous. The Batchelor Institute of Indigenous Tertiary Education, which provides higher and vocational education and training for Aboriginal and Torres Straits Islanders, had 355 students enrolled in higher education courses in 2011 and 2,550 in Vocational Education and Training (VET) courses. In 2010–11 there were 24,054 enrolments in the Northern Territory's Vocational Education and Training activities.

Health

In 2006–07 there were five public hospitals with a total of 600 beds and one private hospital. Community health services are provided from urban and rural Health Centres including mobile units. Remote communities are served by resident nursing staff, aboriginal health workers and in larger communities, resident

GPs. Emergency services are supported by the Aerial Medical Services throughout the Territory.

Welfare

Total social security and welfare expenditure for 2006–07 was $A175,331m. In the same year payments for welfare services for the aged totalled $A17,276m. and disability welfare services expenditure was $A50,811m.

RELIGION

Religious affiliation at the 2006 census: Roman Catholic, 21·1%; Anglican, 12·3%; Uniting Church, 7·0%; Lutheran, 3·9%; Baptist, 2·4%; other Christian, 8·0%; non-Christians, 5·1%; no religion, 23·1%; not stated, 17·1%.

CULTURE

Tourism

In 2009 a total of 1·4m. people visited the Northern Territory, a decrease of 1·9% from the previous year. In the same year tourist expenditure was $A1·8bn. Tourism is the second largest revenue earner after the mining industry.

FURTHER READING

Statistical Information: The State office of the Australian Bureau of Statistics (ABS) is at 7th Floor, AANT House, 81 Smith St., Darwin. Publications include: *Northern Territory at a Glance.* Annual (from 1994); *Regional Statistics, Northern Territory.* Annual (with exception of 1996) from 1995; and *National Regional Profile: Northern Territory* (from 2002).

The Northern Territory: Annual Report. Dept. of Territories, Canberra, from 1911. Dept. of the Interior, Canberra, from 1966–67. Dept. of Northern Territory, from 1972

Australian Territories, Dept. of Territories, Canberra, 1960 to 1973. Dept. of Special Minister of State, Canberra, 1973–75. Department of Administrative Services, 1976

Donovan, P. F., *A Land Full of Possibilities: A History of South Australia's Northern Territory 1863–1911.* 1981.—*At the Other End of Australia: The Commonwealth and the Northern Territory 1911–1978.* 1984

Heatley, A., *Almost Australians: the Politics of Northern Territory Self-Government.* 1990

Powell, A., *Far Country: A Short History of the Northern Territory.* 1996

State library: Northern Territory Library, Parliament House, State Sq., Darwin.

Website: http://www.ntl.nt.gov.au

New South Wales

KEY HISTORICAL EVENTS

The name New South Wales was applied to the entire east coast of Australia when Capt. James Cook claimed the land for the British Crown on 23 Aug. 1770. The separate colonies of Tasmania, South Australia, Victoria and Queensland were proclaimed in the 19th century. In 1911 and 1915 the Australian Capital Territory around Canberra and Jervis Bay was ceded to the Commonwealth. New South Wales was thus gradually reduced to its present area. The first settlement was made at Port Jackson in 1788 as a penal settlement. A partially elective council was established in 1843 and responsible government in 1856.

Gold discoveries from 1851 brought an influx of immigrants, and responsible government was at first unstable, with seven ministries holding office in the five years after 1856. Bitter conflict arose from land laws enacted in 1861. Lack of transport hampered agricultural expansion.

New South Wales federated with the other Australian states to form the Commonwealth of Australia in 1901.

TERRITORY AND POPULATION

New South Wales (NSW) is situated between the 29th and 38th parallels of S. lat. and 141st and 154th meridians of E. long., and comprises 800,642 sq. km, inclusive of Lord Howe Island, 17 sq. km, but exclusive of the Australian Capital Territory (2,352 sq. km) and 67 sq. km at Jervis Bay.

The population at the 2011 census was 6,917,658 (6,549,177 at 2006 census, of which 3,320,726 were female). In 2011 there were nine people per sq. km. Although NSW comprises only 10·4% of the total area of Australia, 33·0% of the Australian population live there. During the year ended June 2010, 42,300 permanent settlers arrived in New South Wales (47,000 in the year to June 2009).

The state is divided into 12 *Statistical Divisions.* The preliminary estimated population of these (in 1,000) at 30 June 2006 was: Sydney, 4,284·4; Hunter, 617·5; Illawarra, 414·5; Mid-North Coast, 297·0; Richmond-Tweed, 229·9; South Eastern, 207·2; Northern, 179·8; Central West, 178·5; Murrumbidgee, 154·2; North Western, 115·8; Murray, 115·6; Far West, 22·9. At June 2006 the preliminary estimated population of the Statistical Subdivisions Newcastle (within Hunter) and Wollongong (within Illawarra) was 517·5 and 278·1 respectively.

Lord Howe Island, 31° 33' 4" S., 159° 4' 26" E., which is part of New South Wales, is situated about 702 km northeast of Sydney; area, 1,654 ha., of which only about 120 ha. are arable; resident population (2006 census), 347 (175 females). The Island, which was discovered in 1788, is of volcanic origin. Mount Gower, the highest point, reaches a height of 866 metres.

The Lord Howe Island Board manages the affairs of the Island and supervises the Kentia palm-seed industry.

SOCIAL STATISTICS

Statistics for calendar years:

	Live births	Deaths	Marriages	Divorces
2004	85,894	46,440	37,431	15,007
2005	86,589	44,894	35,927	15,172
2006	87,336	46,034	38,071	14,482
2007	89,495	46,759	37,982	13,726

The annual rates per 1,000 of mean estimated resident population in 2007 were: births, 13·0; deaths, 6·8; marriages, 5·5; divorces, 2·0; infant mortality, 4·3 per 1,000 live births. Expectation of life in 2007: males, 79·1 years, females, 83·8.

CLIMATE

See AUSTRALIA: Climate.

CONSTITUTION AND GOVERNMENT

Within the State there are three levels of government: the Commonwealth government, with authority derived from a written constitution; the State government with residual powers; the local government authorities with powers based upon a State Act of Parliament, operating within incorporated areas extending over almost 90% of the State.

The Constitution of New South Wales is drawn from several diverse sources; certain Imperial statutes such as the Commonwealth of Australia Constitution Act (1900); the Australian States Constitution Act (1907); an element of inherited English law; amendments to the Commonwealth of Australia Constitution Act; the (State) Constitution Act; the Australia Acts of 1986; the Constitution (Amendment) Act 1987 and certain other State Statutes; numerous legal decisions; and a large amount of English and local convention.

The Parliament of New South Wales may legislate for the peace, welfare and good government of the State in all matters not specifically reserved to the Commonwealth government. The State Legislature consists of the Sovereign, represented by the Governor, and two Houses of Parliament, the *Legislative Council* (upper house) and the *Legislative Assembly* (lower house). Australian citizens aged 18 and over, and other British subjects who were enrolled prior to 26 Jan. 1984, men and women aged 18 years and over, are entitled to the franchise. Enrolment and voting is compulsory. The optional preferential method of voting is used for both houses. The Legislative Council has 42 members elected for a term of office equivalent to two terms of the Legislative Assembly, with 21 members retiring at the same time as the Legislative Assembly elections. The whole State constitutes a single electoral district. The Legislative Assembly has 93 members elected in single-seat electoral districts for a maximum period of four years.

Electors enrolled (30 June 2006) numbered 4,299,510.

RECENT ELECTIONS

In elections held on 26 March 2011 the Liberal-National Coalition won 69 of 93 seats (the Liberals 51 and the Nationals 18), the ruling Australian Labor Party 20, the Greens 1 and ind. 3.

CURRENT GOVERNMENT

In March 2013 the Legislative Council consisted of the following parties: Australian Labor Party, 14; Liberal Party of Australia, 12; National Party, 7; Greens, 5; Shooters and Fishers Party, 2; Christian Democratic Party (Fred Nile Group), 2.

The Legislative Assembly, which was elected in 2011, consisted of the following parties in March 2013: Liberal Party of Australia, 51 seats; Australian Labor Party, 20; National Party, 18; Greens, 1; ind., 3.

Governor: Prof. Marie Bashir, AC.

The New South Wales Liberal-National Ministry was as follows in March 2013:

Premier and Minister for Western Sydney: Barry O'Farrell (b. 1959).

Deputy Premier, Minister for Trade and Investment, and Regional Infrastructure and Services: Andrew Stoner. *Health, and Medical Research:* Jillian Skinner. *Education:* Adrian Piccoli. *Police and Emergency Services, and the Hunter:* Michael Gallacher. *Roads and Ports:* Duncan Gay. *Planning and Infrastructure:* Brad Hazzard. *Resources and Energy, Special Minister for State, and Minister for the Central Coast:* Chris Hartcher. *Transport:* Gladys Berejiklian. *Tourism, Major Events, Hospitality and Racing, and the Arts:* George Souris. *Treasurer and Minister for Industrial Relations:* Mike Baird. *Finance and Services, and the Illawarra:* Gregory Pearce. *Primary Industries, and Small Business:* Katrina Hodgkinson. *Ageing, and Disability Services:* Andrew Constance. *Attorney General and Minister for Justice:* Greg Smith. *Local Government, and the North Coast:* Donald Page. *Family and Community Services, and Women:* Pru Goward. *Fair Trading:* Anthony Roberts. *Mental Health, Healthy Lifestyles, and Western New South Wales:* Kevin Humphries. *Environment, and Heritage:* Robyn Parker. *Citizenship and Communities, and Aboriginal Affairs:* Victor Dominello. *Sports and Recreation:* Graham Annesley.

Speaker of the Legislative Assembly: Shelley Hancock.

NSW Government Website: http://www.nsw.gov.au

ECONOMY

Budget

Government sector revenue and expenses ($A1m.):

	2005–06	2006–07	2007–08[1]
Revenue	41,220	42,196	44,068
Expenditure	40,576	42,892	43,690
		[1]Forward estimate.	

In 2006–07 State government revenue from taxes amounted to $A16,719m.; grants and subsidies totalled $A17,625m.

Performance

In 2007–08 the gross state product of New South Wales represented 31·8% of Australia's total GDP.

Banking and Finance

According to the Australian Prudential Regulation Authority, as at 30 June 2006 there were 27 authorized banks with 1,525 branches; seven building societies with 153 branches; and 87 credit unions with 397 branches.

ENERGY AND NATURAL RESOURCES

Electricity

In 2008–09, 79,750 GWh were produced, of which 7·2% was from renewable sources. Black coal is the main fuel source for electricity generation in the state, producing 88·8% of the total output in 2008–09. Total installed capacity in 2010 was around 18,000 MW.

Oil and Gas

No natural gas is produced in NSW. Almost all gas is imported from the Moomba field in South Australia plus, since 2001, a small amount from Bass Strait.

Water

Ground water represents the largest source with at least 130 communities relying on it for drinking water.

Minerals

New South Wales contains extensive mineral deposits. In 2008–09 there were 60 coal mines directly employing 16,914 people. The value of metallic minerals produced in 2008–09 was $A2·59bn.; construction materials, $A346m. Output of principal products, 2008–09 (in tonnes): coal, 182·0m.; copper, 158,000; zinc, 122,000; lead, 73,000; zircon, 56,000; silver, 71; gold, 28.

Agriculture

NSW accounts for around 21% of the value of Australia's total agricultural production with a gross value of $A8,359·2m. in 2009–10. In the year ending 30 June 2010 there were 43,115 farming establishments with a total area under holding of 59m. ha. of which 6·9m. ha. were under crops.

Principal crops in 2009–10 with production in 1,000 tonnes: wheat for grain, 5,350; barley, 1,236; sorghum, 581; canola, 281. Value of crops, 2009–10, came to $A4·2bn. with wheat totalling $A1·2bn. and cotton $A453m.

The total area under vines in 2010 was 42,621 ha; winegrape production totalled 442,608 tonnes.

Orange production in 2009–10 was 192,500 tonnes (49% of the Australian total); apple production in 2007–08 was 44,900 tonnes.

2009–10 gross value of livestock products was $A1·3bn., including wool produced, $A641m.; and milk, $A522m. In the year ended 31 Aug. 2010 production (in tonnes) of beef and veal, 471,947; mutton and lamb, 133,637; pork, 63,691.

Forestry

The area of native and planted forests managed by State Forests of NSW in 2008–09 totalled 2·41m. ha.; there were 204,828 ha. of softwood and 46,483 ha. of hardwood plantation. In 2008–09, 2·66m. cu. metres of sawlogs and veneer logs were harvested.

Fisheries

Estimated total fisheries production in 2005–06 came to 27,916 tonnes with a gross value of $A143·3m. In the same year aquaculture production was an estimated 5,212 tonnes with a gross value of $A45·03m.

INDUSTRY

A wide range of manufacturing is undertaken in the Sydney area, and there are large iron and steel works near the coalfields at Newcastle and Port Kembla. Around one-third of Australian manufacturing takes place in NSW.

Manufacturing establishments' operations in the year ended 30 June 2007:

Industry	No. of persons employed (1,000)	Wages and salaries ($A1m.)	Sales and service income ($A1m.)
Food, beverages and tobacco	69·6	3,238	25,074
Primary and fabricated metal products	50·2	2,701	20,720
Chemicals, and chemical and rubber product manufacturing	28·2	1,655	12,463
Machinery and equipment	41·5	2,326	12,111
Wood and paper products	19·0	898	6,468
Transport equipment	17·4	946	4,745
Non-metallic mineral products	13·4	780	4,139
Printing, publishing and recorded media	21·9	944	4,068
Textiles, clothing, footwear and leather products	17·2	561	3,276
Petroleum and coal product manufacturing	1·9	—	—
Furniture and other manufacturing	11·0	—	—
Total manufacturing	291·3	14,712[1]	111,442[1]

[1]Although the totals include figures for both petroleum and coal product manufacturing and furniture and other manufacturing the individual figures are not available for those categories.

Labour

In 2007 the labour force totalled 3,493,200 persons, of whom 3,319,300 were employed: 476,100 in retail trade; 429,800 in property and business services; 347,300 in health and community services; 325,300 in manufacturing; 287,200 in construction; and 222,100 in education. There were 173,900 unemployed (a rate of 5·0%) in June 2007. The average weekly wage in Nov. 2005 was $A1,089·00 (males $A1,157·40, females $A973·90).

Industrial tribunals are authorized to fix minimum rates of wages and other conditions of employment. Their awards may be enforced by law, as may be industrial agreements between employers and organizations of employees, when registered.

INTERNATIONAL TRADE

Imports and Exports

In 2009–10 foreign merchandise imports were valued at $A76,004m.; exports at $A31,173. Major sources of imports for 2009–10 (figures in $A1m.): China, 18,370; USA, 8,948; Japan, 5,856; Germany, 4,134; Malaysia, 2,953. Major export destinations (figures in $A1m.): Japan, 8,306; China, 3,225; South Korea, 3,112; USA, 2,275; New Zealand, 2,191. Principal commodity imports in 2009–10 were (with values in $A1m.): medicaments (including veterinary), 6,322; telecommunications equipment and parts, 5,421; passenger motor vehicles, 4,882; computers, 4,806; crude petroleum, 3,954. Principal commodity exports (with values in $A1m.): coal, 8,460; copper ores and concentrates, 1,797; aluminium, 1,624; medicaments (including veterinary), 1,325; refined petroleum, 1,114.

COMMUNICATIONS

Roads

At 30 June 2006 there were 183,120 km of public roads in total. The Roads and Traffic Authority of New South Wales is responsible for the administration and upkeep of major roads. In 2006 there were 20,699 km of roads under its control, comprising 4,250 km of national highways, 13,503 km of state roads and 2,946 km of regional and local roads.

The number of registered motor vehicles (excluding tractors and trailers) at 31 March 2006 was 4,268,631, including 3,395,905 passenger vehicles, 587,713 light commercial vehicles, 133,662 trucks, 20,733 buses and 122,211 motorcycles. There were 510 road accident fatalities in 2006.

Rail

Rail Corporation New South Wales (RailCorp) was formed in 2004 following a merger of the State Rail Authority and Rail Infrastructure Corporation. RailCorp owns, operates and maintains the rail tracks and related infrastructure, and provides metropolitan (CityRail) and long distance (CountryLink) passenger services and access to rail freight operators. In the year ended 30 June 2009, 304·8m. passengers were carried on CityRail and 1·7m. on Countrylink. Also open for traffic are 325 km of Victorian government railways which extend over the border, 68 km of private railways (mainly in mining districts) and 53 km of Commonwealth government-owned track.

A tramway opened in Sydney in 1996. There is also a small overhead railway in the city centre.

Civil Aviation

Sydney Airport (Kingsford Smith) is the major airport in New South Wales and Australia's principal international air terminal. In 2002 it was sold to Macquarie Airports. As well as domestic flights, in 2010 there were direct international services connecting Sydney with nearly 50 destinations. In 2009–10 it handled a total of 34,462,117 passengers (21,394,601 on domestic airlines). It is also the leading airport for freight, handling 377,896 tonnes of international freight in 2009–10.

Shipping

The main ports are at Sydney, Newcastle and Port Kembla. In 2008–09, 4,664 commercial vessels called at New South Wales ports. General cargo imported in 2008–09 was 8,418,728 mass tonnes and general cargo exported was 7,177,701 mass tonnes.

Telecommunications

In the year ended 30 June 2011 there were 5·4m. mobile telephone subscribers aged 14 and over, a penetration rate of 85·6%[1]. In 2005–06 there were 1·82m. households with home computer access (69% of all households). In 2005–06, 1·57m. households had home internet access (60%).

[1]Source: Roy Morgan Single Source July 2010–June 2011

SOCIAL INSTITUTIONS

Justice

Legal processes may be conducted in Local Courts presided over by magistrates or in higher courts (District Court or Supreme Court) presided over by judges. There is also an appellate jurisdiction. Persons charged with more serious crimes must be tried before a higher court.

Children's Courts remove children as far as possible from the atmosphere of a public court. There are also a number of tribunals exercising special jurisdiction, e.g. the Industrial Commission and the Compensation Court.

As at 30 June 2006 there was a daily average of 9,911 persons held in prison. Police personnel (sworn) at 30 June 2006, 14,634.

Education

The State government maintains a system of free primary and secondary education, and attendance at school is compulsory from the age of six years of age. Since Jan. 2010 it has been compulsory for young people aged 15–17 to remain in school or, by arrangement, take part in an approved education and training pathway such as an apprenticeship or traineeship. Non-government schools are subject to government inspection.

In Aug. 2006 there were 2,187 government schools with 740,415 pupils (434,366 primary, 306,049 secondary) and 51,385 teachers, and 912 non-government schools with 369,902 pupils (185,963 primary, 183,939 secondary) and 26,775 teachers. There were 297,200 students in higher education in 2005, with the largest numbers enrolled in management and commerce (29·1% of total enrolments) and society and culture (23·7%). Student enrolments in 2006: University of Sydney (founded 1850), 45,039;

University of New England at Armidale (incorporated 1954), 17,854; University of New South Wales (founded 1949), 37,836; University of Newcastle (granted autonomy 1965), 22,997; University of Wollongong (founded 1951), 22,754; Macquarie University in Sydney (founded 1964), 31,660; University of Technology, Sydney, 32,708; University of Western Sydney, 35,061; Charles Sturt University, 34,261; Southern Cross University (founded 1994), 14,092. Colleges of advanced education were merged with universities in 1990. Post-school technical and further education is provided at State TAFE colleges. Enrolments in 2010 totalled 583,200.

Health

In 2006–07 there were 228 public and 173 private hospitals. In 2005–06 there were 27,918 medical practitioners, 4,358 dentists and 82,740 registered nurses.

Welfare

The number of age and disability pensions at 30 June 2007 was: age, 633,312; disability support, 226,955. There were 146,720 newstart allowance, 110,014 youth allowance and 130,038 single parent payments current at 30 June 2007.

Direct State government social welfare services are limited, for the most part, to the assistance of persons not eligible for Commonwealth government pensions or benefits, and the provision of certain forms of assistance not available from the Commonwealth government. The State also subsidizes many approved services for needy persons.

RELIGION

At the 2006 census 28·2% of the population were Roman Catholic and 21·8% Anglican. These two religions combined had nearly 3·3m. followers.

CULTURE

Tourism

In 2009, 2·7m. overseas visitors arrived for short-term visits, a 2·1% decrease on the previous year. The UK represented the major source of international visitors with 13·9%; followed by New Zealand, 13·2%; USA, 11·2%.

FURTHER READING

Statistical Information: The NSW Government Statistician's Office was established in 1886, and in 1957 was integrated with the Commonwealth Bureau of Census and Statistics (now called the Australian Bureau of Statistics). The state office of: the Australian Bureau of Statistics is at 5th Floor, St Andrews House. Sydney Sq., Sydney). Publications include: *New South Wales in Focus.* Replaces *New South Wales Yearbook.* Annual, since 2005.—*Regional Statistics, New South Wales.*

State library: The State Library of NSW, Macquarie St., Sydney.
Website: http://www.sl.nsw.gov.au

Queensland

KEY HISTORICAL EVENTS

Queensland was discovered by Capt. Cook in 1770. From 1778 it was part of New South Wales and was made a separate colony, with the name of Queensland, by letters patent of 8 June 1859, when responsible government was conferred. Although by 1868 gold had been discovered, wool was the colony's principal product. The first railway line was opened in 1865. Queensland federated with the other Australian states to form the Commonwealth of Australia in 1901. In 1982 Brisbane hosted the Commonwealth Games. Severe flooding in Dec. 2010 and Jan.

2011 killed at least 30 people and affected over 200,000, with damage estimated to have cost $A30bn.

TERRITORY AND POPULATION

Queensland comprises the whole northeastern portion of the Australian continent, including the adjacent islands in the Pacific Ocean and in the Gulf of Carpentaria. Area, 1,730,648 sq. km.

At the 2011 census the population was 4,332,737 (3,904,532 at 2006 census). At the 2006 census there were 127,578 Aboriginals and Torres Strait Islanders. Statistics on birthplaces from the 2006 census are as follows: Australia, 75·2%; England, 4·1%; New Zealand, 3·8%; South Africa, 0·6%; Scotland, 0·6%.

Brisbane, the capital, had at the time of the 2006 census a resident population of 1,763,131 (Statistical Division). The resident populations of the other major centres (Statistical Districts) at 30 June 2005 were: Gold Coast-Tweed (Queensland component), 432,000; Sunshine Coast, 213,000; Townsville, 149,000; Cairns, 128,000; Mackay, 82,000; Rockhampton, 69,000; Bundaberg, 61,000; Gladstone, 43,000.

SOCIAL STATISTICS

Statistics (including Aboriginals) for calendar years:

	Births	Deaths	Marriages	Divorces
2004	49,593	24,657	24,312	13,279
2005	51,700	23,584	24,303	12,383
2006	52,665	24,473	25,043	12,175
2007	61,249	25,801	25,808	11,058

The annual rates per 1,000 population in 2007 were: births, 14·6; deaths, 6·2; marriages, 6·2; divorces, 2·6. The infant mortality rate in 2007 was 5·0 per 1,000 live births. Life expectancy, 2007: 78·9 years for males, 83·6 for females.

CLIMATE

A typical subtropical to tropical climate. High daytime temperatures during Oct. to March give a short spring and long summer. Centigrade temperatures in the hottest inland areas often exceed the high 30s before the official commencement of summer on 1 Dec. Daytime temperatures in winter are quite mild, in the low- to mid-20s. Average rainfall varies from about 150 mm in the desert in the extreme southwestern corner of the State to about 4,000 mm in parts of the sugar lands of the wet northeastern coast, the latter being the wettest part of Australia.

CONSTITUTION AND GOVERNMENT

Queensland, formerly a portion of New South Wales, was formed into a separate colony in 1859, and responsible government was conferred. The power of making laws and imposing taxes is vested in a parliament of one house—the *Legislative Assembly*—which comprises 89 members, returned from four electoral zones for three years, elected from single-member constituencies by compulsory ballot.

Queensland elects 26 members to the Commonwealth House of Representatives.

The Elections Act, 1983, provides franchise for all males and females, 18 years of age and over, qualified by six months' residence in Australia and three months in the electoral district. Electors enrolled (30 June 2006) numbered 2,458,457.

RECENT ELECTIONS

In Legislative Assembly elections on 24 March 2012 the opposition Liberal National Party (LNP) won 78 seats, the Australian Labor Party (ALP) 7 and Katter's Australian Party 2.

CURRENT GOVERNMENT

Governor of Queensland: Penelope Wensley, AO (took office on 29 July 2008).

In March 2013 the LNP administration was as follows:

Premier: Campbell Newman (took office on 26 March 2012).

Deputy Premier, and Minister for State Development, Infrastructure and Planning: Jeff Seeney. *Treasurer and Minister for Trade:* Tim Nicholls. *Health:* Lawrence Springborg. *Education, Training and Employment:* John-Paul Langbroek. *Police and Community Safety:* Jack Dempsey. *Attorney General and Minister for Justice:* Jarrod Bleijie. *Transport and Main Roads:* Scott Emerson. *Housing and Public Works:* Tim Mander. *Agriculture, Fisheries and Forestry:* John McVeigh. *Environment and Heritage Protection:* Andrew Powell. *Natural Resources and Mines:* Andrew Cripps. *Energy and Water Supply:* Mark McArdle. *Local Government, Community Recovery and Resilience:* David Crisafulli. *Communities, Child Safety and Disability Services:* Tracy Davis. *Science, Information Technology, Innovation and the Arts:* Ian Walker. *National Parks, Recreation, Sport and Racing:* Steven Dickson. *Tourism, Major Events, Small Business and the Commonwealth Games:* Jann Stuckey. *Aboriginal and Torres Straight Islander and Multicultural Affairs, and Minister Assisting the Premier:* Glen Elmes.

Government Website: http://www.qld.gov.au

ECONOMY

Budget
In 2006–07 general government expenses by the state were expected to total $A28,825m.; revenue and grants received were expected to be $A29,070m.

Performance
Queensland's 2007–08 gross state product represented 18·9% of Australia's total GDP.

Banking and Finance
According to the Australian Prudential Regulation Authority, as at 30 June 2006 there were 16 authorized banks with 1,151 branches; six building societies with 151 branches; and 24 credit unions with 148 branches.

ENERGY AND NATURAL RESOURCES

Electricity
In 2008, 81% of electricity came from coal-fired power stations, 15% from gas and 4% from renewable energy. As at 31 Dec. 2008 total installed capacity was 12,487 MW.

Water
In the western portion of the State water is comparatively easily found by sinking artesian bores. Monitoring of water quality in Queensland is carried out by the Department of the Environment (estuarine and coastal waters) and the Department of Natural Resources (fresh water).

Minerals
There are large reserves of coal, bauxite, gold, copper, silver, lead, zinc, nickel, phosphate rock and limestone. The state is the largest producer of black coal in Australia. Most of the coal produced comes from the Bowen Basin coalfields in central Queensland. Copper, lead, silver and zinc are mined in the northwest and the State's largest goldmines are in the north. The total value of metallic minerals in 2008–09 was $A6·24bn. In 2008–09 there were 55 coal mines in operation producing 190·55m. tonnes of saleable coal (an increase of 5·6% on the previous year); and at 30 June 2009, 37,000 persons were employed in mining.

Agriculture
Queensland is Australia's leading beef-producing state and its chief producer of fruit and vegetables. In the year ending 30 June 2005 there were 26,955 agricultural establishments farming 143·8m. ha. of which 2·7m. ha. were under crops. Livestock numbered (at 30 June 2005) 11,380,000 beef cattle; 4,949,000 sheep and lambs; and 666,000 pigs. Total value of wool production, 2007–08: $A103m. The gross value of agricultural production in 2004–05 was $A8·3bn. which comprised crops, $A3·7bn.; livestock disposals, $4·2bn.; and livestock products, $A411m.

Forestry
Of a total of 56m. ha. of forests and woodlands in 2007, 9% was in national parks, World Heritage areas and other conservation reserves, while 5% was native forests in multiple use such as timber production. Queensland plantation forests comprise around 0·4% (225,000 ha.) of the state's total forest cover and supply around 10% of Australia's wood and paper products. The forestry industry is an important part of the state's economy, employing more than 19,000 people with an annual turnover of $A2·7bn.

Source: *Australia's Forests at a Glance 2007.* Australian Government Dept. of Agriculture, Fisheries and Forestry, Bureau of Rural Sciences publication, 2007.

Fisheries
Estimated total fisheries production in 2005–06 came to 34,417 tonnes with a gross value of $A322·6m. In the same year aquaculture production was an estimated 5,290 tonnes with a gross value of $A66·12m.

INDUSTRY

In the year ended 30 June 2005 manufacturing industry sales and service income was $A59,239m. with a total of 191,400 people employed. The largest manufacturing sector was food, beverages and tobacco (2004–05 sales and service income: $A14,995m.).

Labour
At Nov. 2006 the labour force numbered 2,163,900, of whom 2,082,900 (947,100 females) were employed. In June 2007 unemployment stood at 3·4%, down from 4·5% in June 2006. The average weekly wage in Nov. 2005 was $A959·60 (males $A1,014·00, females $A866·70).

INTERNATIONAL TRADE

Imports and Exports
In 2009–10 foreign merchandise imports were valued at $A31,046m.; exports at $A43,277m. Major sources of imports for 2009–10 (figures in $A1m.): China, 3,629; Japan, 3,625; USA, 3,356; Papua New Guinea, 2,081; Malaysia, 1,899. Major export destinations (figures in $A1m.): Japan, 9,763; China, 6,922; India, 5,254; South Korea, 4,724; Taiwan, 2,461. Principal commodity imports in 2009–10 were (with values in $A1m.): crude petroleum, 4,247; passenger motor vehicles, 3,204; refined petroleum, 2,425; goods vehicles, 1,757; gold, 1,459. Principal commodity exports (with values in $A1m.): coal, 20,527; beef, 2,492; copper, 1,444; aluminium, 1,001; lead ores and concentrates, 878.

COMMUNICATIONS

Roads
In 2007 there were an estimated 180,818 km of roads open to the public. The number of registered motor vehicles (excluding tractors and trailers) at 31 March 2006 was 2,897,867, including 2,138,364 passenger vehicles, 520,070 light commercial vehicles, 103,967 trucks, 16,516 buses and 110,501 motorcycles. There were 360 road accident fatalities in 2007.

Rail
In July 2010 Queensland Rail was divided into two separate companies with Queensland Rail as the State government-owned corporation responsible for passenger services and QR National, which was privatized in Nov. 2010, for freight. Total length of line as at 30 June 2009 was 8,038 km. In 2008–09, 66·0m. passengers and 247·7m. tonnes of freight were carried.

Civil Aviation

Queensland is well served with a network of air services, with overseas and interstate connections. Subsidiary companies provide planes for taxi and charter work, and the Flying Doctor Service operates throughout western Queensland. In 2009–10 Brisbane handled 18,897,115 passengers (13,670,871 on domestic airlines); Cairns, 3,549,828 passengers (2,795,344 on domestic airlines). As well as domestic flights, there were direct international services in 2010 from Brisbane to Brunei, Fiji Islands, Hong Kong, India, Indonesia, Japan, Malaysia, Nauru, New Caledonia, New Zealand, Papua New Guinea, Philippines, Samoa, Singapore, Solomon Islands, South Korea, Taiwan, Thailand, UAE, USA and Vanuatu, and from Cairns to Hong Kong, Guam, Japan, New Zealand, Papua New Guinea and Singapore. The number of aircraft registered at 31 Dec. 2005 was 2,715.

Shipping

Queensland has 14 modern trading ports, two community ports and a number of non-trading ports. In 2008–09 general cargo imported through Queensland ports was 5,152,027 mass tonnes and general cargo exported was 5,296,748 mass tonnes. There were 6,553 commercial ship calls during 2008–09.

Telecommunications

In 2005–06 there were 1·9m. households with home computer access (72% of all households). In 2005–06, 937,000 households had home internet access (61%).

SOCIAL INSTITUTIONS

Justice

Justice is administered by Higher Courts (Supreme and District), Magistrates' Courts and Children's Courts. The Supreme Court comprises the Chief Justice and 21 judges; the District Courts, 34 district court judges. Stipendiary magistrates preside over the Magistrates' and Children's Courts, except in the smaller centres, where justices of the peace officiate. A parole board may recommend prisoners for release.

Total police personnel (sworn and unsworn) at 30 June 2007 was 13,548. As at 30 June 2005 the average daily number of prisoners stood at 5,354.

Education

Education is compulsory between the ages of six and 16 years and is provided free in government schools. From 2008 students aged 15–17 must remain in school or, by arrangement, take part in an approved education and training pathway such as an apprenticeship or traineeship.

Primary and secondary education comprises 12 years of full-time formal schooling, and is provided by both the government and non-government sectors. In Aug. 2008 the State administered 1,250 schools with 308,771 primary students and 171,079 secondary students. In 2008 there were 40,189 teachers in government schools. There were 463 private schools in Aug. 2008 with 123,795 primary students and 102,817 secondary students. Educational programmes at private schools were provided by 17,320 teachers in 2008. In 2010 there were 303,000 subject enrolments in Vocational Education and Training activities. The one private and eight publicly-funded universities had approximately 190,000 full-time students in 2007.

Health

In 2006–07 there were 173 public acute hospitals and four public psychiatric hospitals. There were 51 private free-standing day hospital facilities and 57 private acute and psychiatric hospitals in 2006–07. In 2009 the percentage of Queensland adults considered overweight or obese rose to 55·3% of the State's population.

Welfare

Welfare institutions providing shelter and social care for the aged, the handicapped and children are maintained or assisted by the State. A child health service is provided throughout the State. Age, invalid, widows', disability and war service pensions, family allowances, and unemployment and sickness benefits are paid by the Federal government. The number of age and disability pensions at 30 June 2007 was: age, 344,648; disability support, 135,863. There were 75,144 newstart allowance, 58,024 youth allowance and 85,710 single parent payments current at 30 June 2007.

RELIGION

Religious affiliation at the 2006 census: Roman Catholic, 24·0%; Anglican, 20·4%; Uniting Church, 7·2%; Presbyterian and Reformed, 3·7%; Lutheran, 2·0%; Baptist, 1·9%; other Christian, 7·2%; non-Christian, 3·3%; no religion, 18·6%; not stated, 11·7%.

CULTURE

Tourism

Overseas visitors to Queensland in the year ending March 2011 totalled 2·0m., the main source being from New Zealand (accounting for 20·0% of tourists), the UK (11·0%), Japan (10·4%) and China (9·8%).

FURTHER READING

Statistical Information: The State office of the Australian Bureau of Statistics is at Level 3, 639 Wickham St., Brisbane. *A Queensland Official Year Book* was issued in 1901, the annual *ABC of Queensland Statistics* from 1905 to 1936 with exception of 1918 and 1922. Present publications include: *Queensland at a Glance.* Annual since 2000 with exception of 2001.—*Qld Stats.* Selected statistics available at *website:* http://www.abs.gov.au

Johnston, W. R., *A Bibliography of Queensland History.* 1981.—*The Call of the Land: A History of Queensland to the Present Day.* 1982
Johnston, W. R. and Zerner, M., *Guide to the History of Queensland.* 1985

State library: The State Library of Queensland, Queensland Cultural Centre, PO Box 3488, South Bank, South Brisbane.
Website: http://www.slq.qld.gov.au
Local Statistical Office: Office of Economic and Statistical Research, PO Box 15037, City East, Qld 4002.
Website: http://www.oesr.qld.gov.au

South Australia

KEY HISTORICAL EVENTS

South Australia was surveyed by Tasman in 1644 and charted by Flinders in 1802. It was made into a British province by letters of patent of Feb. 1836, and a partially elective legislative council was established in 1851. From 6 July 1863 the Northern Territory was placed under the jurisdiction of South Australia until the establishment of the Commonwealth of Australia in 1911.

TERRITORY AND POPULATION

The total area of South Australia is 983,482 sq. km. The settled part is divided into counties and hundreds. There are 49 counties proclaimed, and 536 hundreds, covering 23m. ha., of which 19m. ha. are occupied. Outside this area there are extensive pastoral districts, covering 76m. ha., 49m. of which are under pastoral leases.

The 2011 census population was 1,596,570. The 2006 census totalled 1,514,337 (25,557 Aboriginal and Torres Strait Islanders).

At the 2006 census the Adelaide Statistical Division had a population of 1,105,839 persons (73·0% of South Australia's total population) in 25 councils and four municipalities and other districts. Urban centres outside this area (with estimated populations at 30 June 2007) are Mount Gambier (24,640), Whyalla (22,612), Port Pirie (17,869), Port Lincoln (14,298) and Port Augusta (14,215).

SOCIAL STATISTICS

Statistics for calendar years:

	Live Births	Deaths	Marriages	Divorces
2004	17,140	11,629	7,883	4,147
2005	17,800	11,984	7,630	3,669
2006	18,260	11,921	7,841	3,913
2007	19,662	12,345	8,095	3,534

The rates per 1,000 population in 2007 were: births, 12·4; deaths, 7·8; marriages, 5·1; divorces, 2·2. The infant mortality rate in 2007 was 4·5 per 1,000 live births. Life expectancy for 2007 was 78·8 years for men and 83·9 years for women.

CLIMATE

Most of the state has an arid or semi-arid climate, with as little as 150 mm of rainfall per annum in desert areas. Oodnadatta, a town north of Adelaide on the outskirts of the Simpson Desert, recorded Australia's highest reliably-measured temperature at 50·7°C (123·3°F) on 2 Jan. 1960.

CONSTITUTION AND GOVERNMENT

The present Constitution dates from 24 Oct. 1856. It vests the legislative power in an elected Parliament, consisting of a *Legislative Council* and a *House of Assembly*. The former is composed of 22 members. Eleven members are elected at alternate elections for a term of at least six years and are elected on the basis of preferential proportional representation with the State as one multi-member electorate. The House of Assembly consists of 47 members elected by a preferential system of voting for the term of a Parliament (four years). Election of members of both Houses takes place by secret ballot. Voting is compulsory for those on the Electoral Roll. The qualifications of an elector are to be an Australian citizen, or a British subject who was, at some time within the period of three months commencing on 26 Oct. 1983, enrolled under the Repealed Act as an Assembly elector or enrolled on an electoral roll maintained under the Commonwealth or a Commonwealth Territory, must be at least 18 years of age and have lived in the subdivision for which the person is enrolled for at least one month. By the Constitution Act Amendment Act, 1894, the franchise was extended to women, who voted for the first time at the general election of 25 April 1896. Certain persons are ineligible for election to either House.

The executive power is vested in a Governor appointed by the Crown and an Executive Council, consisting of the Governor and the Ministers of the Crown. The Governor has the power to dissolve the House of Assembly but not the Legislative Council, unless that Chamber has twice consecutively with an election intervening defeated the same or substantially the same Bill passed in the House of Assembly by an absolute majority.

Electors enrolled (30 June 2006) numbered 1,058,029.

RECENT ELECTIONS

The House of Assembly, elected on 20 March 2010, consisted of the following members: Australian Labor Party (ALP), 26; Liberal Party (LP), 18; Independent (ind.), 3.

CURRENT GOVERNMENT

Governor: Kevin Scarce, AC, CSC.

In March 2013 the Labor Ministry was as follows:

Premier, Treasurer, Minister for State Development, the Public Sector, and the Arts: Jay Weatherill.

Deputy Premier, Attorney General, Minister for Planning, Industrial Relations, and Businesses Services and Consumers: John Rau. *Agriculture, Food and Fisheries, Forests, Regional Development, Status of Women, and State/Local Government Relations:* Gail Gago. *Health and Ageing, Mental Health and Substance Abuse, Defence Industries, and Veterans' Affairs:* John Snelling. *Education and Child Development, and Multicultural Affairs:*

Jennifer Rankine. *Transport and Infrastructure, Mineral Resources and Energy, and Housing and Urban Development:* Anastasious Koutsantonis. *Finance, Police, Correctional Services, Emergency Services, and Road Safety:* Michael O'Brien. *Employment, Higher Education and Skills, and Science and Information Economy:* Grace Portolesi. *Manufacturing, Innovation and Trade, and Small Business:* Tom Kenyon. *Transport Services and Minister Assisting the Minister for the Arts:* Chloe Fox. *Sustainability, Environment and Conservation, Water and the River Murray, and Aboriginal Affairs and Reconciliation:* Ian Hunter. *Communities and Social Inclusion, Social Housing, Disabilities, Youth, and Volunteers:* Antonio Piccolo. *Tourism, and Recreation and Sport:* Leon Bignell.

Speaker: Michael Atkinson (ALP).
President: John Gazzola (ALP).

SA Government Website: http://www.sa.gov.au

ECONOMY

Budget

Estimated government sector revenue and expenses ($A1m.):

	2004–05	2005–06	2006–07
Revenue	10,592	11,088	11,264
Expenditure	10,368	10,942	11,173

In 2006–07 State government revenue from taxes amounted to $A3,086m.; grants and subsidies totalled $A5,792m.

Performance

South Australia's 2007–08 gross state product represented 6·5% of Australia's total GDP.

Banking and Finance

According to the Australian Prudential Regulation Authority, as at 30 June 2006 there were 14 authorized banks with 439 branches; one building society with one branch; and 14 credit unions with 75 branches.

ENERGY AND NATURAL RESOURCES

Electricity

In 2010 installed capacity was 4,508 MW. In 2008–09 electricity generated from renewable sources constituted 14·8% of electricity production and 16·4% of consumption.

Minerals

The principal metallic minerals produced are copper, iron ore, uranium oxide, gold and silver. The total value of metallic minerals produced in 2005–06 was $A2,077m. including copper, $A1,444m.; uranium oxide, $A288m.; iron ore, $A191m. Total value of opal production (2005–06), $A25·1m. In 2005–06 there were 4,141 persons employed in mining.

Agriculture

In the year ending 30 June 2005 there were 14,111 establishments mainly engaged in agriculture with a total area under holding of 54·1m. ha. of which 4·4m. ha. were under crops. The gross value of agricultural production in 2004–05 was $A3·9bn. Total value of wool production, $A267·0m. Value of chief crops in 2004–05: wheat, $A531m.; barley, $A304m.; potatoes, $A12m. Production of grapes (2005) was 861,518 tonnes with virtually all being used for winemaking (vineyards' total area, 71,413 ha., including 4,434 ha. not yet bearing). Fruit culture is extensive with citrus and orchard fruits. The most valuable vegetable crops are potatoes, onions and carrots.

Livestock, 30 June 2005: cattle, 1,223,000; sheep and lambs, 12,476,000; pigs, 335,000. Gross value of livestock slaughtered, 2004–05, $A839·9m.

Forestry
Total area of plantations at 31 Dec. 2006 totalled 172,000 ha. Production of sawn timber in 2005–06 was 493,900 cu. metres.

Fisheries
Estimated total fisheries production in 2005–06 came to 69,400 tonnes with a gross value of $A466·8m. In the same year aquaculture production was an estimated 16,935 tonnes with a gross value of $A214·5m.

INDUSTRY
Sales and service income for manufacturing industries for 2004–05 was $A27,182m.; wages and salaries totalled $A4,315m.

Industry sub-division	No. of persons employed (1,000)	Wages and salaries ($A1m.)	Sales and service income ($A1m.)	Industry value added ($A1m.)
Food, beverages and tobacco	19·2	845	5,993	1,987
Textile, clothing, footwear and leather manufacturing	2·7	78	475	118
Wood and paper products manufacturing	6·7	315	1,623	736
Printing, publishing and recorded media	7·8	296	1,332	608
Chemical, petroleum, coal and associated products	8·2	385	2,307	719
Non-metallic mineral products	3·2	162	1,299	422
Metal products manufacturing	12·7	569	3,616	1,024
Machinery and equipment	31·4	1,531	9,825	2,367
Other manufacturing	5·1	133	712	250
Total	97·1	4,315	27,182	8,140

Labour
In Aug. 2007 the labour force stood at 799,700. There were 38,100 unemployed in June 2007, a rate of 4·8%. The average weekly wage in Nov. 2005 was $A950·60 (males $A982·40, females $A891·00).

INTERNATIONAL TRADE
Imports and Exports
In 2009–10 foreign merchandise imports were valued at $A6,446m.; exports at $A8,135m. Major sources of imports for 2009–10 were (figures in $A1m.): Singapore, 977; China, 964; Japan, 756; Thailand, 512; USA, 504. Major export destinations (figures in $A1m.): China, 1,251; USA, 956; Japan, 680; India, 599; UK, 531. Principal commodity imports in 2009–10 were (with values in $A1m.): refined petroleum, 972; passenger motor vehicles, 609; goods vehicles, 303; vehicle parts and accessories, 228. Principal commodity exports (with values in $A1m.): alcoholic beverages, 1,344; iron ore and concentrates, 624; copper ores and concentrates, 619; copper, 560.

COMMUNICATIONS
Roads
In 2007 there were an estimated 92,027 km of roads. The number of registered motor vehicles (excluding tractors and trailers) at 31 March 2006 was 1,137,957, including 915,059 passenger vehicles, 145,643 light commercial vehicles, 34,994 trucks, 4,413 buses and 33,772 motorcycles. In 2007 there were 124 road accident fatalities.

Rail
In Aug. 1997 the passenger operations of Australian National Railways were sold to Great Southern Railway and the freight operations to Australian Southern Railroad. Australian National Railways operates 4,415 km of railway in country areas. TransAdelaide operates 120 km of railway in the metropolitan area of Adelaide. In the year ended 31 March 2001, 19m. tonnes of freight were carried.

There is a tramway in Adelaide that runs from the city centre to the coast. A joint South Australia and Northern Territory project, The AustralAsia Rail Project between Alice Springs and Darwin, carried its first passenger train in Feb. 2004.

Civil Aviation
The main airport is Adelaide International Airport, which handled 7,015,509 passengers (5,993,091 on domestic airlines) in 2009–10. In 2010 there were direct international services to China (Hong Kong), Fiji Islands, Indonesia (Bali), Malaysia, New Zealand and Singapore.

Shipping
There are ten state and five private deep-sea ports. In 2008–09, 1,501 commercial vessels arrived in South Australia. General cargo imported in 2008–09 was 1,090,330 mass tonnes and general cargo exported was 2,233,279 mass tonnes.

Telecommunications
In the year ended 30 June 2011, 77·7% of households had a fixed telephone connected (93·1% in 2004)[1]. In 2005–06 there were 429,000 households with home computer access (67% of all households). In 2005–06, 356,000 households had home internet access (56%).

[1]Source: Roy Morgan Single Source July 2010–June 2011

SOCIAL INSTITUTIONS
Justice
There is a Supreme Court, which incorporates admiralty, civil, criminal, land and valuation, and testamentary jurisdiction; district criminal courts, which have jurisdiction in many indictable offences; and magistrates courts, which include the Youth Court. Circuit courts are held at several places. At 30 June 2009 the police force numbered 5,762. The average daily number of prisoners in 2007 was 1,780.

Education
Education is compulsory for children between the ages of six and 16 years although most children are enrolled at age five or soon after. Since Jan. 2009 it has been compulsory for young people to remain in full-time education or training until they reach the age of 17. Primary and secondary education at government schools is secular and free. In March 2006 there were 805 schools operating, of which 602 were government and 203 non-government schools. In that year there were 107,611 children in government and 49,767 in non-government primary schools, and 63,576 children in government and 35,424 in non-government secondary schools. In 2010 there were 123,900 enrolments in Vocational Education and Training activities. There were 32,266 students enrolled at the University of South Australia in 2005; University of Adelaide, 19,224; and Flinders University of South Australia, 15,110.

Health
In 2006–07 there were 79 public hospitals and 54 private hospitals. Beds available in public and private hospitals totalled 7,188.

Welfare
The number of age and disability pensions at 30 June 2007 was: age, 178,184; disability support, 68,159. There were 37,133 newstart allowance, 28,176 youth allowance and 31,923 single parent payments current at 30 June 2007.

RELIGION
Religious affiliation at the 2006 census: Catholic, 305,205; Anglican, 207,715; Uniting Church, 151,553; Lutheran, 71,251; Orthodox, 44,912; Baptist, 26,146; Presbyterian and Reformed, 21,030; other Christians, 78,252; non-Christians, 59,557; no religion, 367,161; not stated, 181,555.

CULTURE

Tourism

In the year ended 30 June 2007 international visitors totalled 375,500 (around 48% from Europe), an increase of 8·5% on the previous year. At 30 June 2007 there were 253 hotels, motels, guest houses and serviced apartments with 11,547 rooms.

FURTHER READING

Statistical Information: The State office of the Australian Bureau of Statistics (ABS) is at 7th Floor East, Commonwealth Centre, 55 Currie St., Adelaide. Although the first printed statistical publication was the *Statistics of South Australia, 1854*, with the title altered to *Statistical Register* in 1859, there is a manuscript volume for each year back to 1838. These contain simple records of trade, demography, production, etc. and were prepared only for the information of the Colonial Office; one copy was retained in the State.

ABS publications include the *South Australian Year Book* (now discontinued)—*South Australia at a Glance*, annual since 2005.—*SA Stats*, a quarterly bulletin of economic, social and environment statistics.

Gibbs, R. M., *A History of South Australia: from Colonial Days to the Present*. 3rd ed. revised. 1995

Prest, Wilfred, Round, Kerrie and Fort, Carol, (eds.) *The Wakefield Companion to South Australian History*. 2002

State library: The State Library of S.A., North Terrace, Adelaide.
Website: http://www.slsa.sa.gov.au/site/page.cfm

Tasmania

KEY HISTORICAL EVENTS

Abel Janszoon Tasman discovered Van Diemen's Land (Tasmania) on 24 Nov. 1642. The island became a British settlement in 1803 as a dependency of New South Wales. In 1825 its connection with New South Wales was terminated and in 1851 a partially elected Legislative Council was established. In 1856 a fully responsible government was inaugurated. On 1 Jan. 1901 Tasmania was federated with the other Australian states into the Commonwealth of Australia.

TERRITORY AND POPULATION

Tasmania is a group of islands separated from the mainland by Bass Strait with an area (including islands) of 68,401 sq. km, of which 63,447 sq. km form the area of the main island. The population at the 2011 census was 495,350 (476,481 at 2006 census, including 396,655 born in Australia, 18,918 in England and 4,158 in New Zealand).

The largest cities and towns (with populations at the 2006 census) are: Hobart (200,525), Launceston (103,325), Devonport (23,392) and Burnie (19,701).

SOCIAL STATISTICS

Statistics for calendar years:

	Births	Deaths	Marriages	Divorces
2004	5,809	3,892	2,648	1,404
2005	6,308	3,867	2,644	1,346
2006	6,475	3,934	2,664	1,233
2007	6,662	4,132	2,791	1,127

The annual rates per 1,000 of the mean resident population in 2007 were: births, 13·5; deaths, 8·4; marriages, 5·7; divorces, 2·3. Infant mortality rate, 2007, 4·2 per 1,000 live births. Expectation of life, 2007: males, 77·7 years; females, 82·4 years.

CLIMATE

Mostly a temperate maritime climate. The prevailing westerly airstream leads to a west coast and highlands that are cool, wet and cloudy, and an east coast and lowlands that are milder, drier and sunnier.

CONSTITUTION AND GOVERNMENT

Parliament consists of the Governor, the *Legislative Council* and the *House of Assembly*. The Council has 15 members, elected by adults with six months' residence. Members sit for six years, with either two or three retiring annually. There is no power to dissolve the Council. The House of Assembly has 25 members; the maximum term for the House of Assembly is four years. Women received the right to vote in 1903. Proportional representation was adopted in 1907, the method now being the single transferable vote in five member constituencies. Electors enrolled (30 June 2006) numbered 343,494.

A Minister must have a seat in one of the two Houses.

RECENT ELECTIONS

At the elections of 20 March 2010 the opposition Liberal Party won 10 seats in the House of Assembly, the ruling Australian Labor Party also 10 and the Tasmanian Greens 5.

CURRENT GOVERNMENT

Governor: Peter Underwood, AO; b. 1937 (took office on 2 April 2008).

In March 2013 the Labor-Green government comprised:
Premier, Treasurer, and Minister for the Arts: Lara Giddings (took office on 24 Jan. 2011).

Deputy Premier, and Minister for Primary Industries and Water, Energy and Resources, Local Government, Planning, and Racing: Bryan Green. *Tourism, Finance, Veterans' Affairs, and Hospitality:* Scott Bacon. *Education and Skills, Corrections and Consumer Protection, and Sustainable Transport:* Nick McKim. *Economic Development, Infrastructure, Innovation, Science and Technology, Police and Emergency Management, and Workplace Relations:* David O'Byrne. *Health, Children, and Sport and Recreation:* Michelle O'Byrne. *Human Services, Community Development, Climate Change, and Aboriginal Affairs:* Cassy O'Connor. *Attorney General and Minister for Justice, and Environment, Parks and Heritage:* Brian Wightman. *Leader of the Government in the Legislative Council:* Craig Farrell.

Speaker of the House of Assembly: Michael Polley.

TAS Government Website: http://www.tas.gov.au

ECONOMY

Budget

Consolidated Revenue Fund receipts and expenditure, in $A1m., for financial years ending 30 June:

	2005–06	2006–07
Revenue	3,572	3,695
Expenditure	3,453	3,680

In 2007–08 estimated State government revenue from taxes amounted to $A752m.; grants and subsidies, $A2,391m.

Banking and Finance

According to the Australian Prudential Regulation Authority, as at 30 June 2006 there were seven authorized banks with 126 branches; one building society with nine branches; and three credit unions with 22 branches.

ENERGY AND NATURAL RESOURCES

Electricity

As at 30 June 2008 installed capacity was 2,510 MW; total electricity generated in 2007–08 was 8,269 GWh.

Minerals

Output of principal metallic minerals in 2004–05 was (in 1,000 tonnes): iron ore and concentrate, 2,174; zinc, 95; lead, 33; copper, 29; tin, 1·7.

Agriculture

There were 3,877 agricultural establishments at 30 June 2005 occupying a total area of 1·8m. ha. Principal crops in 2004–05 with estimated production in 1,000 tonnes: potatoes, 321; apples, 46; barley, 28; wheat, 30; oats, 9. Gross value of recorded production in 2004–05 was: crops, $A385m.; livestock products, $A259m.; total gross value of agricultural production, $A903m. Livestock, 2005–06: meat cattle, 501,000; sheep and lambs, 2,963,000; pigs, 17,000. Wool produced during 2007–08 was 9,900 tonnes and had a value of $A71·2m.

Forestry

Indigenous forests, which cover a considerable part of the State, support sawmilling and woodchipping industries. Production of sawn timber in 2005–06 was 364,300 cu. metres. Newsprint and paper are produced from native hardwoods.

Fisheries

Estimated total fisheries production in 2005–06 came to 39,298 tonnes with a gross value of $A435·0m. In the same year aquaculture production was an estimated 22,756 tonnes with a gross value of $A247·2m.

INDUSTRY

The most important manufactures for export are refined metals, woodchips, newsprint and other paper manufactures, pigments, woollen goods, fruit pulp, confectionery, butter, cheese, preserved and dried vegetables, sawn timber and processed fish products. The electrolytic-zinc works at Risdon produce zinc, sulphuric acid, superphosphate, sulphate of ammonia, cadmium and other by-products. At George Town, large-scale plants produce refined aluminium and manganese alloys. In the year ending 30 June 2006 employment in manufacturing establishments was 21,400; wages and salaries totalled $A941m.

Labour

In 2006–07 the labour force stood at 238,000. In the same year there were 13,500 unemployed, a rate of 5·7%. The average weekly wage in Nov. 2005 was $A911·40 (males $A955·90, females $A827·60).

INTERNATIONAL TRADE

Imports and Exports

In 2009–10 foreign merchandise imports were valued at $A715m.; exports at $A3,005m. Major sources of imports for 2009–10 were (figures in $A1m.): Singapore, 131; USA, 72; Peru, 71; China, 67; Indonesia, 39. Major export destinations (figures in $A1m.): China, 457; Japan, 421; South Korea, 299; Taiwan, 294; Hong Kong, 284. Principal commodity imports in 2009–10 were (with values in $A1m.): refined petroleum, 114; cocoa, 65; zinc ores and concentrates, 39; animal feed, 28. Principal commodity exports (with values in $A1m.): zinc, 578; aluminium, 383; wood in chips or particles, 263; copper ores and concentrates, 201.

COMMUNICATIONS

Roads

In 2007 there were an estimated 19,618 km of roads open to general traffic. The number of registered motor vehicles (excluding tractors and trailers) at 31 March 2006 was 374,846, including 271,365 passenger vehicles, 74,586 light commercial vehicles, 12,639 trucks, 2,219 buses and 10,488 motorcycles. In 2007 there were 45 road accident fatalities.

Rail

Tasmania's rail network, incorporating 867 km of railways, is primarily a freight system with no regular passenger services.

There are some small tourist railways, notably the newly rebuilt 34 km Abt Wilderness Railway on the west coast.

Civil Aviation

Regular passenger and freight services connect the south, north and northwest of the State with the mainland. The main airports (Hobart and Launceston) handled 1,855,849 and 1,131,326 passengers respectively in 2009–10.

Shipping

There are four major commercial ports: Burnie, Devonport, Launceston and Hobart. In 2008–09, 1,905 commercial vessels called at Tasmanian ports. General cargo imported in 2008–09 was 2,412,126 mass tonnes and general cargo exported was 2,775,214 mass tonnes. Passenger ferry services connect Tasmania with the mainland and offshore islands.

Telecommunications

In 2005–06 there were 121,000 households with home computer access (60% of all households). In 2006–07, 112,000 households had home internet access (56%).

SOCIAL INSTITUTIONS

Justice

The Supreme Court of Tasmania is a superior court of record, with both original and appellate jurisdiction, and consists of a Chief Justice and five puisne judges. There are also inferior civil courts with limited jurisdiction.

In 2008–09 there were 31,615 recorded offences, including 25,635 against property; 4,884 against the person; and 547 fraud and similar offences. Total police personnel (sworn and unsworn) at 30 June 2009 was 1,723. There are five prisons, which in 2005–06 had a daily average of 504 prisoners held.

Education

Education is controlled by the State and is free, secular and compulsory between the ages of five and 16. Since Jan. 2008 it has been compulsory for students aged 16–17 to remain in school or, by arrangement, take part in an approved education and training pathway such as an apprenticeship or traineeship. In 2007, 212 government schools had a total enrolment of 58,926 pupils; 67 private schools had a total enrolment of 22,933 pupils.

In 2010 there were 49,600 students in Vocational Education and Training activities.

Tertiary education is offered at the University of Tasmania and the Australian Maritime College. In 2007 the University (established 1890) had 20,019 students and the Australian Maritime College 1,726 students.

Health

In 2006–07 there were 27 public hospitals with 1,353 beds and eight private hospitals with 948 beds, a total of 4·7 beds per 1,000 population.

Welfare

The number of age and disability pensions at 30 June 2007 was: age, 54,481; disability support, 24,948. There were 15,087 newstart allowance, 10,830 youth allowance and 11,631 single parent payments current at 30 June 2007.

RELIGION

At the census of 2006 the following numbers of adherents of the principal religions were recorded:

Anglican Church	139,379	Other Christian	30,614
Roman Catholic	87,784	Not stated	58,135
Uniting Church	27,507	No religion	102,577
Presbyterian and		Non-Christian	9,696
Reformed	12,125		
Baptist	8,664	Total	476,481

CULTURE
Tourism
In 2007, 815,200 adult visitors arrived in Tasmania (748,500 in 2004).

FURTHER READING
Statistical Information: The State Government Statistical Office (200 Collins St., Hobart), established in 1877, became in 1924 the Tasmanian Office of the Australian Bureau of Statistics, but continues to serve State statistical needs as required.
Main publications: *Tasmanian Year Book.* Annual (from 1967; biennial from 1986), now discontinued.—*Tasmania at a Glance.* Annual (from 1994).—*Regional Statistics, Tasmania.* Annual (from 1999).—*Statistics–Tasmania.* Annual.
Email address: Sales and Inquiries: client.services@abs.gov.au
Website: http://www.abs.gov.au

Robson, L., *A History of Tasmania. Vol. 1: Van Diemen's Land from the Earliest Times to 1855.* 1983.—*A History of Tasmania. Vol. 2: Colony and State from 1856 to the 1980s.* 1990

State library: The State Library of Tasmania, 91 Murray St., Hobart, TAS 7000.
Website: http://www.statelibrary.tas.gov.au

Victoria

KEY HISTORICAL EVENTS
The first permanent settlement was formed at Portland Bay in 1834. A government was established in 1839. Victoria, formerly a portion of New South Wales, was proclaimed a separate colony in 1851 at much the same time as gold was discovered. A new constitution giving responsible government to the colony was proclaimed on 23 Nov. 1855. This event had far-reaching effects, as the population increased from 76,162 in 1850 to 589,160 in 1864. By this time the impetus for the search for gold had waned and new arrivals made a living from pastoral and agricultural holdings and from the development of manufacturing industries. Victoria federated with the other Australian states to form the Commonwealth of Australia in 1901.

TERRITORY AND POPULATION
The State has an area of 227,416 sq. km. The 2011 census population was 5,354,040 (4,932,422 at 2006 census). Victoria has the greatest proportion of people from non-English-speaking countries of any State or Territory, with (2006 census) 1·7% from Italy, 1·2% from Vietnam and 1·1% from China.

2006 census population, within 11 'Statistical Divisions': Melbourne, 3,592,590; Barwon, 259,015; Goulburn, 195,239; Loddon, 168,840; Gippsland, 159,485; Central Highlands, 142,210; Western District, 98,854; Ovens-Murray, 92,587; Mallee, 88,598; East Gippsland, 80,114; Wimmera, 48,441.

SOCIAL STATISTICS
Statistics for calendar years:

	Births	Deaths	Marriages	Divorces
2004	62,417	32,522	25,587	12,544
2005	63,287	32,605	25,266	12,512
2006	65,236	33,311	26,564	12,110
2007	70,313	33,930	26,967	11,833

The annual rates per 1,000 of the mean resident population in 2007 were: births, 13·5; deaths, 6·5; marriages, 5·2; divorces, 2·3. Infant mortality rate, 2007, 3·8 per 1,000 live births. Expectation of life, 2007: males, 79·3 years; females, 83·8 years.

CLIMATE
See AUSTRALIA: Climate.

CONSTITUTION AND GOVERNMENT
Victoria, formerly a portion of New South Wales, was, in 1851, proclaimed a separate colony, with a partially elective Legislative Council. In 1856 responsible government was conferred, the legislative power being vested in a parliament consisting of a *Legislative Council* (Upper House) and a *Legislative Assembly* (Lower House). At present the Council consists of 44 members who are elected for two terms of the Assembly, with half of the seats up for renewal at each election. The Assembly consists of 88 members, elected for four years from the date of its first meeting unless sooner dissolved by the Governor. Members and electors of both Houses must be aged 18 years and Australian citizens or those British subjects previously enrolled as electors, according to the Constitution Act 1975. Single voting (one elector one vote) and compulsory preferential voting apply to Council and Assembly elections. Enrolment for Council and Assembly electors is compulsory.

In the exercise of the executive power the Governor is advised by a Cabinet of responsible Ministers. Section 50 of the Constitution Act 1975 provides that the number of Ministers shall not at any one time exceed 22, of whom not more than six may sit in the Legislative Council and not more than 17 may sit in the Legislative Assembly.

Electors enrolled (30 June 2006) numbered 3,324,691.

RECENT ELECTIONS
In elections to the Legislative Assembly on 27 Nov. 2010 the Australian Labor Party (ALP) won 43 seats with 36·3% of votes cast; the Liberal Party (LP) 35 (38·0%); the National Party (NP), 10 (6·8%). The Greens took 11·2% of the vote but no seats. Turnout was approximately 91%.

In the simultaneous elections to the Legislative Council the LP won 18 seats, the ALP 16, the Greens 3 and the NP 3.

CURRENT GOVERNMENT
Governor: Alex Chernov, AO, QC.
The Liberal-National Cabinet was as follows in March 2013:
Premier, and Minister for Regional Cities and Racing: Denis Napthine.
Deputy Premier, and Minister for Regional and Rural Development, and State Development: Peter Ryan.
Attorney-General, Minister for Finance, and Industrial Relations: Robert Clark. *Treasurer:* Michael O'Brien. *Minister for Aboriginal Affairs, and Local Government:* Jeanette Powell. *Ageing, and Health:* David Davis. *Agriculture and Food Security, and Water:* Peter Walsh. *Arts, Consumer Affairs, and Women's Affairs:* Heidi Victoria. *Bushfire Response, and Police and Emergency Services:* Kim Wells. *Children and Early Childhood Development, and Housing:* Wendy Lovell. *Community Services, Disability Services and Reform, and Mental Health:* Mary Wooldridge. *Corrections, Crime Prevention, and Liquor and Gaming Regulation:* Edward O'Donohue. *Education:* Martin Dixon. *Employment and Trade, Innovation, Services and Small Business, and Tourism and Major Events:* Louise Asher. *Energy and Resources, and Multicultural Affairs and Citizenship:* Nicholas Kotsiras. *Environment and Climate Change, and Youth Affairs:* Ryan Smith. *Higher Education and Skills:* Peter Hall. *Major Projects, Manufacturing, and Ports:* David Hodgett. *Planning:* Matthew Guy. *Public Transport, and Roads:* Terry Mulder. *Sport and Recreation, and Veterans' Affairs:* Hugh Delahunty. *Technology:* Gordon Rich-Phillips.

VIC Government Website: http://www.vic.gov.au

ECONOMY

Budget
In 2007–08 general government expenses by the state were estimated to total $A33,315·4m.; revenue and grants received were expected to increase by 2·9% to $A33,701·2m. ($A32,749·1m. in 2006–07).

Performance
In 2007–08 Victoria's gross state product represented 23·7% of Australia's total GDP.

Banking and Finance
The major trading banks in Victoria are the Commonwealth Bank of Australia, the Australia and New Zealand Banking Group, the Westpac Banking Corporation, the National Australia Bank, the St George Bank, the Bank of Queensland, Suncorp-Metway Bank, HSBC Bank Australia, Elders Rural Bank and the Bendigo Bank. According to the Australian Prudential Regulation Authority, as at 30 June 2006 there were 21 authorized banks with 1,282 branches; two building societies with five branches; and 45 credit unions with 139 branches.

ENERGY AND NATURAL RESOURCES

Electricity
In 1993 the State government began a major restructure of the government-owned electricity industry along competitive lines. The distribution sector was privatized in 1995, and four generator companies in 1997.

Total installed capacity in 2009 was 9,577 MW, of which brown coal 6,555 MW (69%), natural gas 1,943 MW (20%) and renewable sources 1,079 MW (11%). Over 90% of electricity generated is supplied by brown coal-fired generating stations in the Latrobe Valley.

Oil and Gas
Crude oil in commercially recoverable quantities was first discovered in 1967 in two large fields offshore, in East Gippsland in Bass Strait, between 65 and 80 km from land. These fields, with 20 other fields since discovered, have been assessed as containing initial recoverable oil reserves of 4,063m. bbls. Production of crude oil peaked at 450,000 bbls per day in 1985 but declined to 83,000 bbls per day in the fiscal year 2005–06, representing 19·5% of Australia's total production. Estimated remaining oil reserves as at May 2009 was 400·0m. bbls.

Natural gas was discovered offshore in East Gippsland in 1965. The initial recoverable gas reserves were 272·0m. cu. metres. Estimated remaining gas reserves (30 June 2005), 88·68m. cu. metres. Production of natural gas (2008–09), 9,374m. cu. metres.

Liquefied petroleum gas is produced after extraction of the propane and butane fractions from the untreated oil and gas.

Brown Coal
Major deposits of brown coal are located in the Latrobe Valley in the Central Gippsland region and comprise approximately 89% of the total resources in Victoria. In 2005 the resource was estimated to be 41,500 megatonnes, of which about 37,400 megatonnes were economically recoverable.

The primary use of these reserves is to fuel electricity generating stations. Production of brown coal in 2005–06 was 71·2m. tonnes.

Minerals
Production: limestone (2006–07), 1,565,433 tonnes; kaolin (2007–08), 151,669 tonnes. In 2007–08, 5,632 kg of gold were produced.

Agriculture
In the year ending 30 June 2007 there were 37,410 agricultural establishments (excluding those with an estimated value of agricultural operations less than $A5,000) with a total area of 13·3m. ha. of which around 3·4m. ha. were under crops. Gross value of agricultural production, 2006–07, $A8·7bn. Principal crops in 2006–07 with estimated production in 1,000 tonnes: wheat, 879; barley, 605; oats, 134; canola, 42.

Gross value of livestock production in 2006–07 totalled $A2·6bn., including wool production $A498m.

Grape growing, particularly for winemaking, is an important activity. In 2006–07, 308,501 tonnes of winegrapes were produced from 38,650 ha. of vineyards (including 1,904 ha. not yet bearing).

Forestry
Commercial timber production is an increasingly important source of income. As at Dec. 2006 there were 396,000 ha. of plantation. Of Victoria's 7·9m. ha. of native forest (Dec. 2005), 6·6m. ha. (83·4%) were publicly owned (3·1m. ha. in conservation reserves).

Fisheries
Estimated total fisheries production in 2005–06 came to 18,903 tonnes with a gross value of $A127·2m. In the same year aquaculture production was an estimated 3,034 tonnes with a gross value of $A21·6m.

INDUSTRY
Total sales and service income in manufacturing industry in the year ended 30 June 2005 was $A100,248m. In the same year there were 323,400 persons employed in the manufacturing sector with wages and salaries totalling $A15,359m.

Labour
At Aug. 2007 there were 2,706,600 persons in the labour force of whom 2,583,700 were employed: wholesale and retail trade, 503,600; finance, insurance, property and business services, 431,400; manufacturing, 330,600; health and community services, 264,000; construction, 211,900; education, 193,900; culture, recreation, personal and other services, 180,300; accommodation, cafes and restaurants, 118,100; transport and storage, 105,900; government administration and defence, 82,400; agriculture, forestry and fishing, 78,400; communication services, 50,100; electricity, gas and water supply, 20,400; mining, 12,600. There were 122,900 unemployed persons in Aug. 2007, a rate of 4·5%. The average weekly wage in Nov. 2005 was $A1,010·10 (males $A1,055·70, females $A915·00).

INTERNATIONAL TRADE

Imports and Exports
In 2009–10 foreign merchandise imports were valued at $A53,121m.; exports at $A18,427m. Major sources of imports for 2009–10 were (figures in $A1m.): China, 10,477; USA, 6,024; Japan, 5,108; Germany, 3,655; Thailand, 2,727. Major export destinations (figures in $A1m.): China, 2,380; New Zealand, 1,989; Japan, 1,561; USA, 1,514; Saudi Arabia, 1,069. Principal commodity imports in 2009–10 were (with values in $A1m.): passenger motor vehicles, 4,582; crude petroleum, 3,291; refined petroleum, 1,916; goods vehicles, 1,502; vehicle parts and accessories, 1,362. Principal commodity exports (with values in $A1m.): passenger motor vehicles, 1,560; aluminium, 1,106; medicaments (including veterinary), 1,072; wool and other animal hair, 884; milk and cream, 876.

COMMUNICATIONS

Roads
In 2007 there were an estimated 151,000 km of roads open to general traffic. The number of registered motor vehicles (excluding tractors and trailers) at 31 March 2006 was 3,740,726, including 2,997,856 passenger vehicles, 483,097 light commercial vehicles, 119,533 trucks, 16,508 buses and 114,438 motorcycles. There were 333 road accident fatalities in 2006.

Rail

The Victorian rail network is owned by the Victorian RailTrack Corporation (VicTrack), a Victorian Government corporation. Three interlinked rail networks operate in Victoria, comprising: 1,213 km of standard gauge interstate and intrastate non-urban track leased to the Australian Rail Track Corporation; 400 km of urban broad gauge track franchised to Metro Trains Melbourne, the metropolitan network access manager and passenger service operator; and 3,278 km of intrastate, non-urban rail network franchised to V/Line Passenger, the network access provider and regional passenger operator.

The regional instrastate rail network is leased to and managed by V/Line Passenger. Approximately 3m. tonnes of freight is transported on the instrastate rail network annually. There were more than 12m. passenger trips taken on regional train services in 2008–09.

Two new operators commenced operation from 30 Nov. 2009 following a competitive worldwide tender. The new train franchise agreement was awarded to Metro Trains Melbourne (Metro) and the new tram franchise to Keolis Downer EDI (KDR). The contracts operate for eight years with a possible extension of seven years based on good performance.

Melbourne's 249 km tramway and light rail network is operated by KDR (branded as Yarra Trams) across 28 main routes and had approximately 178·1m. passenger boardings in 2008–09. Melbourne's metropolitan passenger rail network, operated by Metro, comprises 15 lines and approximately 2,000 daily metropolitan rail services and had approximately 213·9m. passenger boardings in 2008–09.

Civil Aviation

Melbourne (Tullamarine) airport, Australia's second busiest airport after Sydney, handled 25,917,963 passengers in 2009–10 (19,828,176 on domestic airlines). Total international freight handled in 2009–10 was 199,033 tonnes (also the second highest total after Sydney). In 2010 there were direct international services to Abu Dhabi, Auckland, Bali, Bangkok, Beijing, Chicago, Christchurch, Doha, Dubai, Guangzhou, Hanoi, Ho Chi Minh City, Hong Kong, Honolulu, Jakarta, Johannesburg, Kuala Lumpur, London, Los Angeles, Macau, Manila, Mauritius, Nadi, Phuket, Queenstown, Seoul, Shanghai, Singapore and Wellington.

Shipping

The four major commercial ports are at Melbourne, Geelong, Portland and Hastings. In 2008–09, 4,189 commercial vessels called at Victorian ports. General cargo imported in 2008–09 was 12,452,210 mass tonnes and general cargo exported was 12,885,409 mass tonnes.

Telecommunications

In the year ended 30 June 2011, 83·8% of households had a fixed telephone connected; and in the same period 88·3% aged 14 and over had a mobile phone[1]. In 2005–06 there were 1·36m. households with home computer access (69% of all households) and 1·16m. households with home internet access (59%).

[1]Source: Roy Morgan Single Source July 2010–June 2011

SOCIAL INSTITUTIONS

Justice

There is a Supreme Court with a Chief Justice and 21 puisne judges. There are a county court, magistrates' courts, a court of licensing and a bankruptcy court.

In 1996–97 the State's prisons were upgraded with three new facilities developed, owned and operated by the private sector. In 2006–07 approximately 36% of Victoria's prison population was accommodated in the two private prisons still operating (one of the prisons having been returned to public ownership and operation in Nov. 2000). There are 12 public prisons remaining.

At 30 June 2007 the number of prisoners held stood at 4,183. Police personnel at 30 June 2007, 14,078.

Education

In Feb. 2006 there were 1,606 government schools with 307,577 pupils in primary schools and 222,827 in secondary schools. In the same year there were 694 non-government schools with 140,683 pupils at primary schools; and 153,039 pupils at secondary schools.

All higher education institutions, excluding continuing education and technical and further education (TAFE), now fall under the Unified National System, and can no longer be split into universities and colleges of advanced education. In addition, a number of institutional amalgamations and name changes occurred in the 12 months prior to the commencement of the 1992 academic year. In 2010 there were 520,000 enrolments in Vocational Education and Training activities.

There are ten publicly funded higher education institutions including eight State universities, Marcus Oldham College and the Australian Catholic University (partly privately funded), and the Melbourne University Private, established in 1998. In 2005 there were 292,761 students in higher education.

Health

In 2006–07 there were 144 public hospitals with 12,434 beds, and 144 private hospitals with 6,675 beds. Total government outlay on health in 2006–07 was $A6,716m.

Welfare

Victoria was the first State of Australia to make a statutory provision for the payment of Age Pensions. The Act came into operation on 18 Jan. 1901, and continued until 1 July 1909, when the Australian Invalid and Old Age Pension Act came into force. The Social Services Consolidation Act, which came into operation on 1 July 1947, repealed the various legislative enactments relating to age and invalid pensions, maternity allowances, child endowment, unemployment and sickness benefits and, while following in general the Acts repealed, considerably liberalized many of their provisions.

The number of age and disability pensions at 30 June 2007 was: age, 493,728; disability support, 169,694. There were 105,264 newstart allowance, 93,007 youth allowance and 90,551 single parent payments current at 30 June 2007.

RELIGION

There is no State Church, and no State assistance has been given to religion since 1875. At the 2006 census the following were the enumerated numbers of the principal religions: Catholic, 1,355,904; Anglican, 671,774; Uniting Church, 274,056; Orthodox, 224,038; Presbyterian and Reformed, 143,146; other Christian, 316,887; Buddhist, 132,633; Muslim, 109,369; Hindu, 42,310; Jewish, 41,108; no religion, 1,007,415; not stated, 550,309.

CULTURE

Tourism

For the year ending June 2007 the number of short-term overseas visitors to Australia who specified Victoria as their main destination was 1·4m. (28·4% of total overseas visitors to Australia), with 1·02m. nominating 'holiday' or 'visiting friends/relatives' as purpose of their trip. The UK represented the major source of international visitors with 16·1%; followed by New Zealand (16·0%), China (9·9%), USA (8·6%), Singapore (4·5%) and Japan (4·2%).

Source: International Visitor Survey, year ending June 2007

FURTHER READING

Statistical Information: The State office of the Australian Bureau of Statistics is at 5th Floor, Commercial Union Tower, 485 LaTrobe Street, Melbourne. Publications: *Victorian Year Book.* Annual.—*State and Regional Indicators.* Quarterly (from Sept. 2001).

State library: The State Library of Victoria, 328 Swanston St., Melbourne 3000.

Website: http://www.slv.vic.gov.au

Western Australia

KEY HISTORICAL EVENTS

In 1791 the British navigator George Vancouver took possession of the country around King George Sound. In 1826 the government of New South Wales sent 20 convicts and a detachment of soldiers to form a settlement then called Frederickstown. The following year, Capt. James Stirling surveyed the coast from King George Sound to the Swan River, and in May 1829 Capt. Charles Fremantle took possession of the territory. In June 1829 Capt. Stirling founded the Swan River Settlement (now the Commonwealth State of Western Australia) and the towns of Perth and Fremantle. He was appointed Lieut.-Governor.

Grants of land were made to the early settlers until, in 1850, with the colony languishing, they petitioned for the colony to be made a penal settlement. Between 1850 and 1868 (in which year transportation ceased), 9,668 convicts were sent out. In 1870 partially representative government was instituted. Western Australia federated with the other Australian states to form the Commonwealth of Australia in 1901.

In the 1914–18 war Western Australia provided more volunteers for overseas military service in proportion to population than any other State. The worldwide depression of 1929 brought unemployment (30% of trade union membership), and in 1933 over two-thirds voted to leave the Federation. While there were modest improvements in the standard of living through the 1930s, it was the 1939–45 war which brought full employment. Japanese aircraft attacked the Western Australia coast in 1942. Talk of a 'Brisbane line', which would abandon the West to invasion, only served to reinforce Western Australia's sense of isolation from the rest of the nation. The post-war years saw increasing demand for wheat and wool but the 1954–55 decline in farm incomes led to diversification. Work began in the early 1950s on steel production and oil processing. Oil was discovered in 1953 but it was not until 1966 that it was commercially exploited. The discovery of deposits of iron ore in the Pilbara, bauxite in the Darling scarp, nickel in Kambalda and ilmenite from sand led to the State becoming a major mineral producer by 1965.

TERRITORY AND POPULATION

Western Australia has an area of 2,529,875 sq. km and 12,500 km of coastline.

The population at the 2011 census was 2,239,170 (1,959,088 at 2006 census). In 2006, 65·3% of its population were born in Australia. Perth, the capital, had a 2006 census population of 1,445,078.

Principal local government areas outside the metropolitan area, with estimated resident population at 30 June 2007: Mandurah, 60,560; Albany, 33,545; Bunbury, 31,638; Kalgoorlie-Boulder, 30,903; Busselton, 27,500; Geraldton, 20,333; Roebourne, 18,240; Port Hedland, 13,060.

SOCIAL STATISTICS

Statistics for calendar years[1]:

	Births	Deaths	Marriages	Divorces
2004	25,295	11,184	10,601	4,337
2005	26,253	11,297	11,124	5,265
2006	27,776	11,643	11,602	5,544
2007	29,164	12,283	12,290	4,932

[1]Figures are on state of usual residence basis.

The annual rates per 1,000 of the mean resident population in 2007 were: births, 13·8; deaths, 5·8; marriages, 5·8; divorces, 2·3.

Infant mortality rate, 2007, 2·4 per 1,000 live births. Expectation of life, 2007: males, 79·2 years; females, 84·0 years.

CLIMATE

Western Australia is a region of several climate zones, ranging from the tropical north to the semi-arid interior and Mediterranean-style climate of the southwest. Most of the State is a plateau between 300 and 600 metres above sea level. Except in the far southwest coast, maximum temperatures in excess of 40°C have been recorded throughout the State. The normal average number of sunshine hours per day is 8·0.

CONSTITUTION AND GOVERNMENT

The *Legislative Council* consists of 36 members elected for a term of four years. There are six electoral regions for Legislative Council elections. Each electoral region returns six members. Each member represents the entire region.

There are 59 members of the *Legislative Assembly*, each member representing one of the 59 electoral districts of the State. Members are elected for a period of up to four years. A system of proportional representation is used to elect members.

Electors enrolled (30 June 2006) numbered 1,259,528.

RECENT ELECTIONS

In elections to the Legislative Assembly on 9 March 2013 the Liberal Party (LP) won 31 seats with 47·1% of votes cast; the Labor Party (ALP) won 21 with 33·1%; and the National Party (NP), 7 with 6·1%. Turnout was 89·2%.

CURRENT GOVERNMENT

Governor: Malcolm McCusker, AO, CVO, QC.

Lieut.-Governor: Wayne Martin, AC.

In March 2013 the Cabinet comprised:

Premier, and Minister for State Development, and Science: Colin Barnett (Liberal Party).

Deputy Premier, and Minister for Health, and Tourism: Kim Hames. *Regional Development, and Lands:* Brendon Grylls. *Education, Aboriginal Affairs, and Electoral Affairs:* Peter Collier. *Treasurer, and Minister for Transport, and Fisheries:* Troy Buswell. *Planning, and Culture and the Arts:* John Day. *Police, Road Safety, Small Business, and Women's Interests:* Liza Harvey. *Training and Workforce Development, Water, and Forestry:* Terry Redman. *Mental Health, Disability Services, and Child Protection:* Helen Morton. *Attorney General, and Minister for Commerce:* Michael Mischin. *Mines and Petroleum, and Housing:* Bill Marmion. *Sport and Recreation, and Racing and Gaming:* Terry Waldron. *Agriculture and Food:* Ken Baston. *Energy, Finance, and Citizenship and Multicultural Interests:* Mike Nahan. *Local Government, Community Services, Seniors and Volunteering, and Youth:* Tony Simpson. *Environment, and Heritage:* Albert Jacob. *Emergency Services, Corrective Services, and Veterans:* Joe Francis.

Speaker of the Legislative Assembly: Michael Sutherland.

WA Government Website: http://wa.gov.au

ECONOMY

Budget

Revenue and expenditure (in $A1m.) in years ending 30 June:

	2005–06	2006–07	2007–08[1]
Revenue	16,123	16,510	17,593
Expenditure	14,141	15,234	16,141

[1]Projected.

A general government net operating surplus of $A1,853m. was projected for 2006–07, a decrease of $A812m. on the 2005–06 outcome.

Performance

Western Australia's 2007–08 gross state product represented 13·8% of Australia's total GDP, up from 11·0% in 2002–03. The economy grew at the fastest rate of any of the States and Territories in 2005–06, 2006–07 and again in 2007–08.

Banking and Finance

According to the Australian Prudential Regulation Authority, as at 30 June 2006 there were 15 authorized banks with 519 branches; one building society with 19 branches; and 16 credit unions with 41 branches.

ENERGY AND NATURAL RESOURCES

Electricity

In 2008–09 electricity consumption was 31,343 GWh of which 979 GWh (3·1%) was from renewable sources.

Oil and Gas

Petroleum continued to be the State's largest resource sector with sales increasing by 33% to $A22·3bn. in 2008. Owing to low oil prices in 2008–09 the value of crude oil sales decreased by 12% to $A7·7bn.; output was 81·4m. bbls. The State accounts for 66% of Australia's oil and condensate production.

Western Australia has significant natural gas resources and, with a $A2·4bn. expansion of the North West Shelf liquefied natural gas (LNG) project, exports are forecast to rise by around $A1bn. Total natural gas production, 2005–06: 25,887 gigalitres.

Source: Western Australian Department of Mineral and Petroleum Resources.

Minerals

Mining is a significant contributor to the Western Australia economy. The State is the world's third largest producer of iron ore and accounts for almost 88% of Australia's iron ore production.

Principal minerals produced in 2005–06 were: gold, 165 tonnes; iron ore and concentrate, 253·4m. tonnes; diamonds, 25·3m. carats. Most of the State's coal production (an estimated 6·2m. tonnes in 2004–05) is used by Western Power's electricity generation.

Agriculture

In the year ending 30 June 2005 there were 11,745 establishments mainly engaged in agriculture with a total area of 104·6m. ha. of which 8·33m. ha. were under crops. Gross value of agricultural production in 2004–05 totalled $A5·1bn., a reduction of 18% on the previous year.

Principal crops in 2004–05 with estimated production in 1,000 tonnes: wheat, 8,619; barley, 2,489; lupins for grain, 792; oats, 460; canola, 488.

Value of livestock products in 2004–05 totalled $A625m. Total value of wool produced in 2004–05 was $A490m.

Forestry

The area of State forests and timber reserves at 30 June 2010 was 1,427,954 ha. In 2009–10 sawlog production totalled 769,780 cu. metres, of which plantation softwood sawlogs and veneer logs 570,556 cu. metres and Jarrah and Karri hardwoods 183,816 cu. metres.

Fisheries

Estimated total fisheries production in 2007–08 came to 29,301 tonnes with a gross value of $A448·4m., of which $A216·9m. from rock lobster. In the same year aquaculture production was an estimated 1,013 tonnes with a gross value of $A122·8m. Pearling is the most valuable form of aquaculture in the State with the Pearl Oyster Fishery producing an estimated $A113m. worth of pearls from wild captured and hatchery produced oysters in 2007–08.

INDUSTRY

Heavy industry is concentrated in the southwest, and is largely tied to export-orientated mineral processing, especially alumina and nickel.

The following table shows manufacturing industry statistics for the year ended 30 June 2005:

Industry sub-division	Persons employed 1,000	Wages and salaries $A1m.	Sales and service income $A1m.
Food, beverages and tobacco	16·5	631	5,538
Textiles, clothing and leather products	3·8	107	539
Wood and paper products	4·6	170	966
Printing and publishing and recorded media	8·0	347	1,445
Petroleum, coal, chemical products	8·3	464	7,484
Non-metallic mineral products	6·0	276	1,799
Metal products	24·0	1,117	14,503
Machinery and equipment	18·3	749	4,171
Other manufacturing	8·8	225	1,263

Labour

The labour force comprised 1,054,800 employed and 44,500 unemployed persons in 2005 (an unemployment rate of 4·1%). The average weekly wage in Nov. 2005 was $A1,054·80 (males $A1,148·60, females $A863·40).

INTERNATIONAL TRADE

Imports and Exports

In 2009–10 foreign merchandise imports were valued at $A27,939m.; exports at $A83,357m. Major sources of imports for 2009–10 were (figures in $A1m.): Thailand, 4,528; USA, 2,822; China, 2,820; Singapore, 2,543; Japan, 2,244. Major export destinations (figures in $A1m.): China, 30,801; Japan, 14,245; India, 8,584; South Korea, 6,694; UK, 4,494. Principal commodity imports in 2009–10 were (with values in $A1m.): gold, 6,200; crude petroleum, 3,229; refined petroleum, 1,966; pumps (excluding liquid pumps) and parts, 1,747; passenger motor vehicles, 1,528. Principal commodity exports (with values in $A1m.): iron ore and concentrates, 34,125; gold, 13,659; crude petroleum, 8,810; natural gas, 6,016; wheat, 1,677.

COMMUNICATIONS

The Public Transport Authority of Western Australian (PTA), which replaced the Western Australia Government Railways Commission in July 2003, is responsible for rail, bus and ferry services in the metropolitan area (Transperth); public transport services in regional centres; coach and rail passenger services in regional areas (Transwa); and school buses.

Roads

In 2007 there were an estimated 152,262 km of roads open to general traffic. The number of registered motor vehicles (excluding tractors and trailers) at 31 March 2006 was 1,600,566, including 1,205,266 passenger vehicles, 254,164 light commercial vehicles, 63,316 trucks, 11,051 buses and 59,675 motorcycles. In 2008 there were 209 road accident fatalities.

Rail

Transperth is responsible for metropolitan rail services and Transwa for rail passenger services to regional areas. In 2005–06, 34·1m. passenger journeys were made on urban services. In the year ended 31 March 2001, 196m. tonnes of freight were carried.

Civil Aviation

An extensive system of regular air services operates for passengers, freight and mail. In 2009–10 Perth International Airport handled 9,992,583 passengers (6,606,335 on domestic airlines). In 2010 there were direct international services to Brunei, China (Hong Kong), Indonesia, Japan, Malaysia, Mauritius, New Zealand, Singapore, South Africa, Thailand and the UAE.

Shipping

In 2008–09, 5,869 commercial vessels called at Western Australian ports (1,789 at Dampier and 1,774 at Fremantle). General cargo imported in 2008–09 was 10,334,358 mass tonnes and general cargo exported was 10,968,803 mass tonnes.

Telecommunications

In 2005–06 there were 560,000 households with home computer access (71% of all households). In 2005–06, 484,000 households had home internet access (62%).

SOCIAL INSTITUTIONS

Justice

Justice is administered by a Supreme Court, consisting of a Chief Justice, 16 other judges and two masters; a District Court comprising a chief judge and 20 other judges; a Magistrates Court, a Chief Stipendiary Magistrate, 37 Stipendiary Magistrates and Justices of the Peace. All courts exercise both civil and criminal jurisdiction except Justices of the Peace who deal with summary criminal matters only. Juvenile offenders are dealt with by the Children's Court. The Family Court also forms part of the justice system.

At 30 June 2005 there was a daily average of 3,482 prisoners held. At 30 June 2006 police personnel stood at 6,875.

Education

School attendance is compulsory from the age of six until the end of the year in which the child attains 16 years. Since Jan. 2008 young people aged 15–17 have been obliged to remain in school or, by arrangement, take part in an approved education and training pathway such as an apprenticeship or traineeship.

In Aug. 2006 there were 777 government primary and secondary schools (with 16,737 full-time equivalent teaching staff) providing free education to 131,294 primary and 83,064 secondary students; in the same year there were 303 non-government schools for 51,577 primary and 53,828 secondary students.

Higher education is available through four state universities and one private (Notre Dame). In 2010 there were 166,000 enrolments in Vocational Education and Training activities. In 2006 there were approximately 98,745 students in tertiary education at the University of Western Australia, Murdoch University, the University of Notre Dame Australia, Curtin University of Technology and the Edith Cowan University.

Health

In 2006–07 there were 94 acute public hospitals and one public psychiatric hospital; and 23 acute and psychiatric private hospitals and 20 private free-standing day hospitals.

Welfare

The Department for Community Development is responsible for the provision of welfare and community services throughout the State. The number of age and disability pensions at 30 June 2007 was: age, 166,239; disability support, 58,467. There were 27,650 newstart allowance, 26,001 youth allowance and 37,900 single parent payments current at 30 June 2007.

RELIGION

At the census of 2006 the principal denominations were: Catholic, 464,005; Anglican, 400,480; Uniting Church, 74,333; Presbyterian and Reformed, 43,807; Baptist, 32,730; other Christian, 147,171. There were 97,916 persons practising non-Christian religions and 448,435 persons had no religion.

CULTURE

Tourism

In the year ending March 2007 there were 634,000 overseas visitors. Of these, 26·9% were from the UK, 9·1% from Singapore, 7·5% from New Zealand and 7·1% from Japan.

FURTHER READING

Statistical Information: The State Government Statistician's Office was established in 1897 and now functions as the Western Australian Office of the Australian Bureau of Statistics (Level 15, Exchange Plaza, Sherwood Court, Perth). Publications include: *Western Australia at a Glance,* Annual (from 1997). *Western Australia Statistical Indicators,* Quarterly (from Sept. 2000).

Crowley, F. K., *Australia's Western Third: A History of Western Australia from the First Settlements to Modern Times.* (Rev. ed.) 1970
Stannage, C. T. (ed.) *A New History of Western Australia.* 1980

State library: Alexander Library Building, Perth.
Website: http://www.slwa.wa.gov.au

AUSTRALIAN EXTERNAL TERRITORIES

Ashmore and Cartier Islands

By Imperial Order in Council of 23 July 1931, Ashmore Islands (known as Middle, East and West Islands) and Cartier Island, situated in the Indian Ocean, some 320 km off the northwest coast of Australia (area, 5 sq. km), were placed under the authority of the Commonwealth. Under the Ashmore and Cartier Islands Acceptance Act, 1933, the islands were accepted by the Commonwealth as the Territory of Ashmore and Cartier Islands. It was the intention that the Territory should be administered by the State of Western Australia but owing to administrative difficulties the Territory was deemed to form part of the Northern Territory of Australia (by amendment to the Act in 1938). On 16 Aug. 1983 Ashmore Reef was declared a National Nature Reserve. The islands are uninhabited but Indonesian fishing boats fish within the Territory and land to collect water in accordance with an agreement between the governments of Australia and Indonesia. It is believed that the islands and their waters may house considerable oil reserves.

Australian Antarctic Territory

An Imperial Order in Council of 7 Feb. 1933 placed under Australian authority all the islands and territories other than Adélie Land situated south of 60° S. lat. and lying between 160° E. long. and 45° E. long. The Order came into force with a Proclamation issued by the Governor-General on 24 Aug. 1936 after the passage of the Australian Antarctic Territory Acceptance Act 1933. The boundaries of Adélie Land were definitively fixed by a French Decree of 1 April 1938 as the islands and territories south of 60° S. lat. lying between 136° E. long. and 142° E. long. The Australian Antarctic Territory Act 1954 declared that the laws in force in the Australian Capital Territory are, so far as they are applicable and are not inconsistent with any ordinance made under the Act, in force in the Australian Antarctic Territory.

The area of the territory is estimated at 6,119,818 sq. km (2,362,875 sq. miles).

There is a research station on MacRobertson Land at lat. 67° 37' S. and long. 62° 52' E. (Mawson), one on the coast of Princess Elizabeth Land at lat. 68° 34' S. and long. 77° 58' E. (Davis), and one at lat. 66° 17' S. and long. 110° 32' E. (Casey). The Antarctic Division also operates a station on Macquarie Island.

Christmas Island

GENERAL DETAILS

Christmas Island is an isolated peak in the Indian Ocean, lat. 10° 25' 22" S., long. 105° 39' 59" E. It lies 360 km S. 8° E. of Java Head, and 970 km N. 79° E. from Cocos (Keeling) Islands, 1,310 km from Singapore and 2,650 km from Perth. Area: 136·7 sq. km. The climate is tropical with temperatures varying little over the year at 27° C. The wet season lasts from Nov. to April with an average rainfall of 1,930 mm. The island was formally annexed by the UK on 6 June 1888, placed under the administration of the Governor of the Straits Settlements in 1889, and incorporated with the Settlement of Singapore in 1900. Sovereignty was transferred to the Australian government on 1 Oct. 1958. The population at the 2006 census was 1,347. In 2010 there were also some 2,700 asylum seekers in a centre for boat people.

The legislative, judicial and administrative systems are regulated by the Christmas Island Act, 1958. The first Island Assembly was elected in Sept. 1985, and is now replaced by the elected members of the Shire of Christmas Island. The Territory underwent major changes to its legal system when the Federal Parliament passed the Territories Law Reform Bill of 1992; Commonwealth and State laws applying in the state of Western Australia now apply in the Territory as a result, although some Western Australia laws have been disallowed to take into account the unique status of the Territory.

Extraction and export of rock phosphate dust is the main industry. The government is also encouraging the private sector development of tourism.

CONSTITUTION AND GOVERNMENT

The Shire of Christmas Island has nine members elected for terms of four years. Every two years half of the positions are elected. The last elections were in Oct. 2009.

CURRENT GOVERNMENT

Administrator: Jon Stanhope (appointed 5 Oct. 2012).

ECONOMY

Currency
The Australian dollar is legal tender.

ENERGY AND NATURAL RESOURCES

Electricity
Electricity consumption in 2005–06 was 24·6m. kWh.

COMMUNICATIONS

Roads
The Shire of Christmas Island has responsibility for approximately 140 km of roads with the remaining 100 km of haul roads and tracks maintained by Christmas Island Phosphates and Park Australia North. As at Oct. 2006 there were 1,527 registered vehicles on the island.

Civil Aviation
Virgin Australia operates four scheduled flights weekly to and from Perth.

Shipping
There are up to eight phosphate ships a month, approximately 11 general cargo and seven fuel tankers visiting the island per year.

SOCIAL INSTITUTIONS

Education
As at Oct. 2006 there were 335 students at the Christmas Island District High School, which caters for education under the Western Australian curriculum for kindergarten through to Year 12.

Health
There is a nine-bed hospital staffed (as at Oct. 2006) by two doctors, one dentist, a Director of Nursing and 32 other on-island staff. A range of visiting medical specialists provides services to the island.

RELIGION

About 50% of the population are Buddhists or Taoists, 16% Muslims and 30% Christians.

Cocos (Keeling) Islands

GENERAL DETAILS

The Cocos (Keeling) Islands are two separate atolls comprising some 27 small coral islands with a total area of about 14·2 sq. km, and are situated in the Indian Ocean at 12° 05' S. lat. and 96° 53' E. long. They lie 2,950 km northwest of Perth. The islands are low-lying, flat and thickly covered by coconut palms, and surround a lagoon in which ships drawing up to seven metres may be anchored. There is an equable and pleasant climate, affected for much of the year by the southeast trade winds. Temperatures range over the year from 68° F (20° C) to 88° F (31° C) and rainfall averages 80" (2,000 mm) a year. Most of this rain falls between Nov. and May. Feb. and March are usually the wettest months.

The main islands are: West Island (the largest, about 14 km long), home to most of the European community; Home Island, occupied by the Cocos Malay community; Direction, South and Horsburgh Islands, and North Keeling Island, 24 km to the north of the group. The population of the Territory (2006 Census) was 571, distributed between Home Island (75%) and West Island (25%).

The islands were discovered in 1609 by Capt. William Keeling but remained uninhabited until 1826. In 1857 the islands were annexed to the Crown; the governments of Ceylon and Singapore held jurisdiction over the islands at different periods until they were placed under the authority of the Australian government as the Territory of Cocos (Keeling) Islands on 23 Nov. 1955. An *Administrator*, appointed by the Governor-General, is the government's representative in the Territory and is responsible to the Minister for Territories and Local Government.

In 1978 and 1993 the Australian government purchased the various interests of the Clunies-Ross family, who had been granted the land in its entirety by Queen Victoria. A Cocos Malay co-operative was established to engage in local businesses.

CONSTITUTION AND GOVERNMENT

The Shire of Cocos (Keeling) Islands has seven members elected for terms of four years. Every two years half of the elected members' terms expire. The last Local Government elections were in Oct. 2009.

CURRENT GOVERNMENT

Administrator: Jon Stanhope (appointed 5 Oct. 2012).

ECONOMY

Currency
The Australian dollar is legal tender.

ENERGY AND NATURAL RESOURCES

Electricity
Electricity consumption in 2005–06 was 5.3m. kWh.

COMMUNICATIONS

Roads
The Shire of Cocos (Keeling) Islands has responsibility for approximately 10 km of sealed roads and approximately 12 km of unsealed roads. As at Oct. 2006 there were 170 registered motor vehicles.

Civil Aviation
Virgin Australia operates three scheduled flights weekly to and from Perth via Christmas Island.

Shipping
There is approximately one general cargo ship every month and a tanker twice per year.

SOCIAL INSTITUTIONS

Education
As at Oct. 2006 there were two school campuses catering for 128 students. The campus on Home Island caters for students up to secondary level and the campus on West Island for students from kindergarten to Year 10.

Health
There is a health centre on both West and Home Islands. As at Oct. 2006 there were one doctor, one nurse manager and nine staff across the two locations. The Christmas Island dentist visits the islands at regular intervals and a range of visiting medical specialists provides services to the island.

RELIGION
About 85% of the population are Muslim and 15% Christian.

Coral Sea Islands

GENERAL DETAILS
The Coral Sea Islands, which became a Territory of the Commonwealth of Australia under the Coral Sea Islands Act 1969, comprises scattered reefs and islands over a sea area of about 1m. sq. km. The Territory is uninhabited apart from a meteorological station on Willis Island.

FURTHER READING
Australian Department of Arts, Sport, the Environment, Tourism and Territories. *Christmas Island: Annual Report.—Cocos (Keeling) Islands: Annual Report.—Norfolk Island: Annual Report.*

Heard and McDonald Islands

These islands, about 2,500 miles southwest of Fremantle, were transferred from British to Australian control from 26 Dec. 1947.

Heard Island is about 43 km long and 21 km wide; Shag Island is about 8 km north of Heard. The total area is 412 sq. km (159 sq. miles). The McDonald Islands are 42 km to the west of Heard. Heard is an active stratovolcano that has erupted eight times since 1910, most recently in 1993. A volcano on McDonald Island had been dormant for 75,000 years before it erupted in 1992. The most recent eruption was in 2005. In 1985–88 a major research programme was set up by the Australian National Antarctic Research Expeditions to investigate the wildlife as part of international studies of the Southern Ocean ecosystem. Subsequent expeditions followed from June 1990 through to 1992.

Norfolk Island

KEY HISTORICAL EVENTS
The island was formerly part of the colony of New South Wales and then of Van Diemen's Land (now known as Tasmania). A penal colony between 1788–1814 and 1825–55, it was separated from the state of Tasmania in 1856 and placed under the jurisdiction of the Australian State of New South Wales. Following the Norfolk Island Act 1913 (Cth), the Island was accepted as a Territory of Australia with the Australian Federal Government having jurisdiction for the Island.

TERRITORY AND POPULATION
Situated 29° 02′ S. lat. 167° 57′ E. long.; area 3,455 ha.; permanent population (Aug. 2011 census), 1,795.

Descendants of the *Bounty* mutineer families constitute the 'original' settlers and are known locally as 'Islanders', while later settlers, mostly from Australia and New Zealand, are identified as 'mainlanders'. According to the Aug. 2011 census, 79% of the Island's permanent population are Australian citizens with 13% being New Zealand citizens. Descendants of the Pitcairn Islanders make up about 45·0% of the permanent resident population. Over the years the Islanders have preserved their own lifestyle and customs, and their language remains a mixture of West Country English, Gaelic and Tahitian.

SOCIAL STATISTICS
Births in 2004–05 totalled 25 and deaths 15.

CLIMATE
Sub-tropical. Summer temperatures (Dec.–March) average about 75°F (25°C), and 65°F (18°C) in winter (June–Sept.). Annual rainfall is approximately 50" (1,200 mm), most of which falls in winter.

CONSTITUTION AND GOVERNMENT
An *Administrator*, appointed by the Governor-General and responsible to the Minister for Territories and Local Government, is the senior government representative in the Territory. The seat of administration is Kingston.

The Norfolk Island Act 1979 gives Norfolk Island responsible legislative and executive government to enable it to run its own affairs. Wide powers are exercised by the Norfolk Island Legislative Assembly of nine members, elected for a period of three years, and by an Executive Council. The Norfolk Island Act also provides for consultation with the Federal Government in respect of certain types of laws proposed by Norfolk Island's Legislative Assembly.

RECENT ELECTIONS
At the last elections, on 13 March 2013, non-partisans won eight out of nine seats and the Norfolk Liberals one.

CURRENT GOVERNMENT

Administrator: Neil Pope (since 1 April 2012).
  Chief Minister: Lisle Snell (since 20 March 2013).

ECONOMY

The office of the Administrator is financed from Commonwealth expenditure which in 2004–05 was $A1,195,000; local revenue for 2004–05 totalled $A24,551,220; expenditure, $A23,553,196.

Currency

Australian notes and coins are the legal currency.

Banking and Finance

There are two banks, Westpac and the Commonwealth Bank of Australia.

COMMUNICATIONS

Roads

There are 100 km of roads (53 km paved), some 2,800 passenger cars and 200 commercial vehicles.

Civil Aviation

In 2010 there were scheduled flights to Brisbane, Melbourne, Newcastle and Sydney (operated by Our Airline), and Auckland (Air New Zealand).

Telecommunications

In 2006 there were 2,057 telephone lines in service.

SOCIAL INSTITUTIONS

Justice

The Island's Supreme Court sits as required and a Court of Petty Sessions exercises both civil and criminal jurisdiction. Appeals from decisions of the Norfolk Island Supreme Court are heard by the Federal Court of Australia and by the High Court of Australia.

Education

A school is run by the New South Wales Department of Education for children aged 5 to 12. It had 310 pupils in Aug. 2006. The Island also has a pre-school facility.

Health

In 2006 the Norfolk Island Hospital Enterprise had three doctors, one dentist, one physiotherapist, one radiographer, a pharmacist, a medical scientist, a generalist counsellor and registered nursing staff. The hospital had 24 beds. Visiting specialists attend the Island on a regular basis.

RELIGION

40% of the population are Anglicans.

CULTURE

Press

There is one weekly with a circulation of 1,200.

Tourism

In the year ended 30 June 2007, 34,318 tourists visited Norfolk Island.

AUSTRIA

© Research Machines plc 2006

Republik Österreich
(Austrian Republic)

Capital: Vienna
Population projection, 2015: 8·46m.
GNI per capita, 2011: (PPP$) 35,719
HDI/world rank: 0·885/19
Internet domain extension: .at

KEY HISTORICAL EVENTS

The oldest historical site in Austria is at the Gudenus caves in the Kremstal valley, where hunters' stone implements and bones dating from the Paleolothic Age have been found. The Early Iron Age Hallstatt culture prevailed from around 750–400 BC, covering the area north of the Alps, large parts of Slovakia, the north Balkans and Hungary. This area became renowned for its ceramics, ornamental Hallstatt burial grounds and the development of salt mining.

In the fifth century BC Celtic tribes stormed the eastern Alps, their culture named after the site in Switzerland (La-Tène) where tools, weapons and other artefacts were found. La-Tène culture showed Greek and Etruscan influences. Around the middle of the second century BC some of these tribes united to found Noricum, the first recognizable state on Austrian territory. In 113 BC a treaty of friendship was signed between Rome and Noricum. Around 15 BC Noricum was incorporated into the Roman Empire. Present day Austria (with parts of Germany, Switzerland, Slovenia and Hungary) was eventually divided into the three Roman provinces of Raetia, Noricum and Pannonia.

Germanic tribes, in particular the Marcomanni and Quadi, later known as Bavarians, invaded in AD 166–80. Emperor Marcus Aurelius campaigned against them from Vindobona (Vienna), where he died in battle in AD 180. His successors fought unsuccessfully against the Alemanni and other invading tribes. Brigantium (Bregenz) became a border town of the Roman Empire after a peace agreement was signed with the Alemanni.

Charlemagne conquered the Bavarian duke Tassilo III and the Avars at the end of the 8th century and established a territory in the Danube Valley known as the Ostmark in 803. The church of Salzburg (Roman Juvavum) was founded at the end of the 8th century and consequently became the Bavarian-Frankish spiritual centre and archbishopric. Christianity spread throughout the region, led by the Slav apostles Cyril and Methodius.

The Magyar invasions culminated in the battle for Vienna in 881 and the loss of Lower Austrian territories. The Babenbergs took over the margravate of Bavaria in 976 under Leopold I. The name 'Ostarichi', originating from Old High German, first appeared in a document of Emperor Otto III in 996. In 1156 the margravate of Austria became a separate duchy.

From 1160–1200 Vienna, the most important trading town on the Danube, became the residence of the art-loving Babenberger dukes, receiving its town charter in 1198. In 1246 the last Babenberger, Friedrich II, fell in the battle of Leitha against King Bela IV of Hungary. Count Rudolf IV of Habsburg was elected German King Rudolf I in 1273 and set about conquering the former Babenberg lands, naming his son Duke Albert I the sole ruler in 1283. From the rule of Albert's son and successor Frederick I onwards, the territory of the Habsburgs was known as the *dominium Austriae*.

Holy Roman Empire

In 1452 Frederick III became Holy Roman Emperor. In one of the first of a series of dynastic marriages to expand the Habsburg realm, his son Maximilian I was married to Mary, the heiress of Burgundy. The marriage of their son, Philip the Handsome, to Juana, the heiress of the Spanish crowns, forged a massive and disparate Habsburg inheritance, encompassing Spain and its empire, Hungary, Bohemia, the Burgundian Netherlands and territory in Italy. Austria was harried by the Ottoman Turks, whose armies had advanced across southeast Europe as far as Hungary under Sultan Süleyman the Magnificent. A truce was agreed after the Ottoman defeat at Vienna in 1529.

The Reformation brought Protestantism to Austria but was resisted by Rudolf II. Brought up a strict Catholic, he embraced the Counter-Reformation in 1576. Tensions between Protestants and Catholics came to a head with the Defenestration of Prague in 1618, which led to the Thirty Years War. Peace was restored by the Treaty of Westphalia in 1648.

Philip, a grandson of Louis XIV of France, was designated heir to the last of the Spanish Habsburgs, Charles II. During the subsequent War of the Spanish Succession, Holland and England joined the coalition forces. Peace negotiations at Rastatt awarded Austria Spanish territory in Italy and the Netherlands. Austrian boundaries reached their furthest limit in 1720.

In 1740 Emperor Charles VI died without a male heir. By pragmatic sanction, his daughter Maria Theresa was allowed to succeed him as the first female Habsburg ruler, though she was denied the Imperial crown. Challenged by Prussia, the War of the Austrian Succession divided the two alliances of Bavaria, France, Spain and Prussia against Austria, Holland and Great Britain. Maria Theresa introduced a number of reforms aimed at strengthening the Habsburg monarchy. The army was doubled in size, a public education system was introduced and administrative and financial structures were overhauled, centralizing government and laying the foundations of a modern state. During the Seven Years' War (1756–63) Maria Theresa aimed to reconquer Silesia, lost in the War of the Austrian Succession. However, the Austrians, bereft of allies, were defeated at Burkersdorf in July 1762. Silesia was settled on Prussia by the Peace of Hubertusburg.

Enlightened Despotism

In 1772 Austria, Prussia and Russia carried out the first partition of Poland, with Austria gaining Galicia. Maria Theresa's youngest daughter of 16 children, Marie-Antoinette, married Louis XVI, thus achieving a closer political alliance with France. Maria Theresa was succeeded by Joseph, her eldest son. Also a reformist, and heavily influenced by the Enlightenment, Joseph II proclaimed the Edict of Toleration in 1781, giving more rights to faiths other than Catholicism. He also abolished serfdom and homogenized land tax laws. Not all of these changes were popular, such as the introduction of German as the official language in the Hungarian government. His heavy-handed reforms provoked resistance, leading to a revolt in the Austrian Netherlands opposing absolutist rule. In the face of widespread opposition to his reforms, Joseph revoked a number of them in 1790.

From 1792–1815 the Habsburgs were involved in almost continuous warfare. Austria and Prussia voiced their disapproval of the ideals of the French Revolution in the Declaration of Pillnitz. Ill-received by the French government, this led to a declaration of war followed by 23 years of hostilities and five separate wars. During the first, second and third coalition wars Vienna was occupied twice by French troops. Napoleon defeated Austria in the Battle of Austerlitz in 1805, forcing Francis to surrender his title of Holy Roman Emperor and to hand over one of his daughters, Archduchess Marie Louise, in marriage. After Napoleon was defeated and exiled to Elba in 1814, the Congress of Vienna met to re-establish Europe's internal borders. Leading Austria's foreign policy, Prince von Metternich created the German confederation of 35 states and four free cities to succeed the Holy Roman Empire.

With the growth of the cities in the first half of the 19th century came an expansion of the markets for agricultural goods. The first Austrian railway, the Kaiser-Ferdinand Nordbahn, was built between Linz and Budweis and steam navigation started on the Danube. An uprising in Vienna in 1848 forced the Habsburgs to flee the city and Metternich to resign. Ferdinand I abdicated in Dec. and was succeeded by his 18-year-old nephew, Francis Joseph I. The pan-European wave of revolution spread to Hungary, where the fight for emancipation was led by Lajos Kossuth. The Habsburgs refused to accept Hungary's independence and enlisted Russian aid to quell the uprising. Italian and Slavic revolts followed. In 1859, during the Austro-Italian war, the Habsburgs were defeated and Italy was unified.

The German unification campaign was headed by Otto von Bismarck, who expanded Prussian influence by isolating Austria from her allies. This led to the Austro-Prussian war of 1866. Austria's defeat at the battle of Königgrätz stripped it of all presiding powers over Germany. The loss of Venetia in 1867 added a further blow to the shrinking Habsburg empire. Forced to compromise with Hungary, Emperor Francis Joseph I negotiated the 'Ausgleich' in 1867, giving Hungary its own constitution and quasi-independent status. Francis Joseph was henceforth recognized as the Apostolic king of Hungary and emperor of Austria within the Austro-Hungarian monarchy, also referred to as the Dual Monarchy. Foreign policy, the army and finances were administered jointly.

German Alliance

In 1879 Austria entered the Dual Alliance with the German Reich and the two states pledged mutual support in the eventuality of Russian aggression. Italy joined in 1882, making a Triple Alliance. Tensions between Austria and Russia heightened over territory in the Balkans, with an uprising in Macedonia in 1903. King Alexander of Serbia was assassinated while the Serbs were fighting for the unification of the Southern Slavs. The Habsburg empire responded with a livestock embargo, known as the Pig War. Austria-Hungary then annexed the two provinces of Bosnia and Herzegovina in 1908, prompting a Serb revolt and protests by the pan-Slav movement. Austria tried to gain control over Serbia during the Balkan wars of 1912–13. In June 1914 the heir to the Habsburg throne, Archduke Franz Ferdinand, and his wife were assassinated in Sarajevo by a Bosnian nationalist. The refusal of Serbia to accept the blame led to an Austrian declaration of war on Serbia in July. German backing for Austria encompassed a settling of its own scores with France and Russia. Austria was obliged to support Germany against France and Russia. Germany declared war on Russia and France in early Aug., beginning the First World War. The Austro-Hungarian army suffered major setbacks and the monarchy began to crumble after Francis Joseph I's death in 1916. His great-nephew, Charles I of Austria, succeeded him, trying but failing to achieve a secret truce with the Allies. A series of strikes, mutiny in the army and navy, and food shortages were among the factors that defeated Austro-Hungarian forces in the spring and summer of 1918. About 1·2m. soldiers from Austria-Hungary died during the war.

Emperor Charles I issued a manifesto on 17 Oct. 1918 guaranteeing the independence of the non-German speaking states. Each province established a national council which then developed into a government. The Poles declared themselves an independent unified state in Warsaw on 7 Oct. 1918; the Czechs founded an independent republic in Prague and the Southern Slavs merged with Serbia. An armistice was agreed on 3 Nov. and the Hungarian government announced its complete separation from Austria. Within days Austria and Hungary declared republics. The Habsburg monarchy was formally dissolved on 11 Nov. with the abdication of Charles I. The Treaty of Saint Germain in 1919 stipulated that the German–Austrian lands were not permitted to unite (*Anschluss*) with Germany without the consent of the League of Nations. Austria was declared a federal state on 1 Oct. 1920.

Nazism

The economic crises of the 1920s led to the rise of a nationalist movement much influenced by Germany's adoption of National Socialism. When Engelbert Dollfuss of the conservative Christian Socialists became chancellor in 1932 he faced Nazi and Marxist opposition. He dissolved parliament and governed by emergency decree, founding the conservative Fatherland Front in 1933. The anti-Marxist Heimwehr (home defence forces) supported him. When the Social Democrats fought back they were defeated by Dollfuss with the backing of the Heimwehr. All political parties except the Fatherland Front were subsequently banned. On 25 July 1934 a Nazi gang murdered Dollfuss but was forced to surrender and the coup leaders were executed. Dollfuss' successor, Kurt Schuschnigg, signed the Austrian–German agreement whereby Germany recognized Austria's sovereignty in return for Austria calling itself a German state. German Nazis pressurized Schuschnigg to allow them more influence, Hitler demanding that leading Nazis take on top positions in the Austrian cabinet. Schuschnigg planned a plebiscite to decide upon Anschluss but on 12 March 1938 German forces marched into Austria, establishing a Nazi government led by Arthur Seyss-Inquart. Renamed 'Ostmark', Austria was put under the central authority of the German Third Reich. After anti-Nazis were rounded up a plebiscite held on 10 April 1938 showed over 99% support for Hitler.

The Allied Moscow conference of 1943 agreed Austrian independence along the demarcation lines of 1937. The borders were finally set at the Yalta conference of Feb. 1945. On 13 April 1945 Vienna was liberated by Soviet troops. Two weeks later, Dr Karl Renner proclaimed a provisional government and Austria a republic, recognized officially by western powers at the Potsdam conference. The first elections were held in Nov., with former Nazis excluded from voting. Leopold Figl of the Austrian People's Party (Christian Socialists) became the first chancellor of the Second Republic, with Karl Renner as its first president. The Paris treaty of Sept. 1946 granted autonomy for South Tyrol within Italy, the Allies refusing to return it to Austria. Financial support from the United Nations and the Marshall Plan enabled Austria to begin rebuilding its economy.

The Soviet Union, Britain, France and the USA occupied Austria until 1955, when the country reaffirmed its neutrality. Full sovereignty was restored by the State Treaty in Vienna. Anschluss between Austria and Germany or restoration of the Habsburgs were expressly prohibited. The rights of ethnic minorities were guaranteed and former German assets confiscated by the Western allies were returned. The Soviet Union, however, demanded reparations including US$150m. for former German businesses. Austria became a member of the United Nations in 1955 and subsequently joined EFTA in 1959 and the OECD in 1960.

The first all-socialist cabinet was formed in 1970 under Chancellor Bruno Kreisky, who established economic stability and prosperity throughout the 1970s. The late 1970s saw the emergence of the Green party, with the planned construction of a nuclear power station becoming a central political issue. Kreisky's achievements were followed by political scandals which led to his resignation in 1983.

Kurt Waldheim (elected as secretary general of the UN in 1971) became Austrian president in 1986 despite allegations of a Nazi past. Although this was never proved, the affair caused him to be placed on the USA's list of undesirable aliens. In 1995 Austria became a member of the EU. In Jan. 2000 the Freedom Party, headed by the far-right leader Jörg Haider, joined the government. Although Haider, described as a 'dangerous extremist' by EU leaders, did not take a post in the government, he continued to exercise political influence and remained popular with the Austrian electorate. Sanctions against Austria were imposed by the European Union in Feb. 2000 but were lifted in Sept. 2000. In the parliamentary elections of Oct. 2002 the Freedom Party, with a new leader, saw its support fall by half. Nevertheless the Freedom Party was again invited to join a coalition government.

TERRITORY AND POPULATION

Austria is bounded in the north by Germany and the Czech Republic, east by Slovakia and Hungary, south by Slovenia and Italy, and west by Switzerland and Liechtenstein. It has an area of 83,879 sq. km (32,386 sq. miles), including 1,444 sq. km (558 sq. miles) of inland waters. Population (2011, provisional) 8,430,558; density, 102·3 per sq. km. Austria has now adopted a register-based method of calculating the population rather than a traditional census, and had a full register-based census in 2011 for the first time. Previous population censuses: (1923) 6·53m., (1934) 6·76m., (1951) 6·93m., (1971) 7·49m., (1981) 7·56m., (1991) 7·96m, (2001) 8·03m. In 2011, 67·8% of the population lived in urban areas.

The UN gives a projected population for 2015 of 8·46m.

In 2011, 88·8% of residents were of Austrian nationality. The countries of origin of the principal minorities in Jan. 2012 were: Germany (153,000); Serbia and Montenegro combined, including Kosovo (136,000); Turkey (114,000); Croatia (57,000); Romania (48,000). Since the mid-1980s the number of foreigners living in Austria has more than doubled, from 4% to 11%.

The areas, populations and capitals of the nine federal states:

Federal States	Area (sq. km)	(2001)	Population at censuses (2011, provisional)	State capitals
Vienna (Wien)	415	1,550,261	1,724,381	Vienna
Lower Austria (Niederösterreich)	19,186	1,545,794	1,617,444	St Pölten
Burgenland	3,962	277,558	286,029	Eisenstadt
Upper Austria (Oberösterreich)	11,980	1,376,607	1,416,102	Linz
Salzburg	7,156	515,454	533,247	Salzburg
Styria (Steiermark)	16,401	1,183,246	1,212,415	Graz
Carinthia (Kärnten)	9,538	559,346	557,671	Klagenfurt
Tyrol	12,640	673,543	712,077	Innsbruck
Vorarlberg	2,601	351,048	371,192	Bregenz

Provisional populations of the main towns at the census of 2011 (and 2001 final figures): Vienna, 1,724,381 (1,550,261); Graz, 264,351 (226,241); Linz, 190,802 (183,614); Salzburg, 148,236 (142,808); Innsbruck, 121,076 (113,457); Klagenfurt, 94,683 (90,145); Villach, 59,458 (57,492); Wels, 58,709 (56,481); St Pölten, 52,091 (49,117).

The official language is German. For orthographical changes agreed in 1996 *see* GERMANY: Territory and Population.

SOCIAL STATISTICS

Statistics, 2008: live births, 77,752 (rate of 9·3 per 1,000 population); deaths, 75,083 (rate of 9·0 per 1,000 population); infant deaths, 287; stillborn, 258; marriages, 35,223; divorces, 19,701. In 2007 there were 1,280 suicides (rate of 15·4 per 100,000 population), of which 965 males and 315 females. Average annual population growth rate, 2007–11, 0·4%. Life expectancy at birth, 2011, 83·4 years for women and 78·1 years for men. In 2006 the most popular age range for marrying was 30–34 for males and 25–29 for females. Infant mortality, 2010, was four per 1,000 live births; fertility rate, 2008, 1·4 children per woman. In 2011, 130,208 people immigrated into Austria and 94,604 emigrants left Austria. Net immigration in 2011, at 35,604, was the highest since 2005. Asylum applications totalled 15,821 in 2009, up from 12,841 in 2008.

CLIMATE

The climate is temperate and from west to east in transition from marine to more continental. Depending on the elevation, the climate is also predominated by alpine influence. Winters are cold with snowfall. In the eastern parts summers are warm and dry.

Vienna, Jan. 0·0°C, July 20·2°C. Annual rainfall 624 mm. Graz, Jan. –1·0°C, July 19·4°C. Annual rainfall 825 mm. Innsbruck, Jan. –1·7°C, July 18·1°C. Annual rainfall 885 mm. Salzburg, Jan. –0·9°C, July 18·6°C. Annual rainfall 1,174 mm.

CONSTITUTION AND GOVERNMENT

The constitution of 1 Oct. 1920 was revised in 1929 and restored on 1 May 1945. Austria is a democratic federal republic comprising nine states *(Länder)*, with a federal *President (Bundespräsident)* directly elected for not more than two successive six-year terms, and a bicameral National Assembly which comprises a National Council and a Federal Council.

The National Council *(Nationalrat)* comprises 183 members directly elected for a five-year term by proportional representation in a three-tier system by which seats are allocated at the level of 43 regional and nine state constituencies, and one federal constituency. Any party gaining 4% of votes cast nationally is represented in the National Council. In 2007 Austria's voting age was reduced to 16—the lowest for national elections in the EU.

The Federal Council *(Bundesrat)* has 62 members appointed by the nine states for the duration of the individual State Assemblies' terms; the number of deputies for each state is proportional to that state's population. In March 2013 the ÖVP held 27 of the 62 seats, the SPÖ 22, the FPÖ 8, the Greens 3, the BZÖ 1 and Liste Fritz Dinkhauser (FRITZ) 1.

The head of government is a *Federal Chancellor*, who is appointed by the President (usually the head of the party winning the most seats in National Council elections). The *Vice-Chancellor*, the *Federal Ministers* and the *State Secretaries* are appointed by the President at the Chancellor's recommendation.

National Anthem

'Land der Berge, Land am Strome' ('Land of mountains, land on the river'); words by Paula Preradovic; tune attributed to Mozart.

GOVERNMENT CHRONOLOGY

Presidents since 1945. (ÖVP = Austrian People's Party; SPÖ = Social Democratic Party)

1945–50	SPÖ	Karl Renner
1951–57	SPÖ	Theodor Körner
1957–65	SPÖ	Adolf Schärf
1965–74	SPÖ	Franz Joseph Jonas
1974–86	SPÖ	Rudolf Kirchschläger
1986–92	ÖVP	Kurt Josef Waldheim

| 1992–2004 | ÖVP | Thomas Klestil |
| 2004– | SPÖ | Heinz Fischer |

Federal Chancellors since 1945.

1945	SPÖ	Karl Renner
1945–53	ÖVP	Leopold Figl
1953–61	ÖVP	Julius Raab
1961–64	ÖVP	Alfons Gorbach
1964–70	ÖVP	Josef Klaus
1970–83	SPÖ	Bruno Kreisky
1983–86	SPÖ	Alfred (Fred) Sinowatz
1986–97	SPÖ	Franz Vranitzky
1997–2000	SPÖ	Viktor Klima
2000–07	ÖVP	Wolfgang Schüssel
2007–08	SPÖ	Dr Alfred Gusenbauer
2008–	SPÖ	Werner Faymann

RECENT ELECTIONS

Elections were held on 28 Sept. 2008. The Social Democratic Party (SPÖ) won 57 seats with 29·3% of votes cast (68 with 35·3% in 2006); the Austrian People's Party (ÖVP), 51 with 26·0% (66 with 34·3%); the Freedom Party (FPÖ), 34 with 17·5% (21 with 11·0%); the Alliance for the Future of Austria (BZÖ), 21 with 10·7% (7 with 4·1%); the Greens, 20 with 10·4% (21 with 11·0%). Turnout was 78·8%.

In the presidential election held on 25 April 2010 incumbent Heinz Fischer (SPÖ) won 78·9% of the vote against 15·6% for Barbara Rosenkranz (FPÖ) and 5·4% for Rudolf Gehring (Christian Party of Austria). Turnout was 49·2%.

Parliamentary elections are scheduled to take place on 29 Sept. 2013.

European Parliament
Austria has 19 (18 in 2004) representatives. At the June 2009 elections turnout was 46·0% (42·4% in 2004). The ÖVP won 6 with 30·0% of votes cast (political affiliation in European Parliament: European People's Party); the SPÖ 4 with 23·7% (Progressive Alliance of Socialists and Democrats); Liste Martin, 3 with 17·7% (non-attached); the FPÖ 2 with 12·7% (non-attached); the Greens, 2 with 9·9% (Greens/European Free Alliance).

CURRENT GOVERNMENT

President: Dr Heinz Fischer; b. 1938 (SPÖ; took office on 8 July 2004 and re-elected 25 April 2010).

Following the elections of Sept. 2008 the SPÖ and the ÖVP agreed in Nov. 2008 to form a grand coalition, with Werner Faymann (SPÖ) as chancellor. In March 2013 the government comprised:

Chancellor: Werner Faymann; b. 1960 (SPÖ; sworn in 2 Dec. 2008).

Deputy Chancellor and Minister for European and International Affairs: Michael Spindelegger (ÖVP).

Minister for Finance: Maria Fekter (ÖVP). Defence and Sports: Gerald Klug (SPÖ). Economy, Family and Youth: Reinhold Mitterlehner (ÖVP). Agriculture, Forestry, Environment and Water Management: Nikolaus Berlakovich (ÖVP). Health: Alois Stöger (SPÖ). Interior: Johanna Mikl-Leitner (ÖVP). Education, Art and Culture: Claudia Schmied (SPÖ). Science and Research: Karlheinz Töchterle (ÖVP). Justice: Beatrix Karl (ÖVP). Labour, Social Affairs and Consumer Protection: Rudolf Hundstorfer (SPÖ). Transport, Innovation and Technology: Doris Bures (SPÖ). Women and Public Administration: Gabriele Heinisch-Hosek (SPÖ). State Secretary in the Federal Chancellery: Josef Ostermayer (SPÖ). State Secretary in the Ministry of Finance: Andreas Schieder (SPÖ). State Secretary in the Ministry of Foreign Affairs: Wolfgang Waldner (ÖVP). State Secretary for Integration in the Ministry of Interior: Sebastian Kurz (ÖVP).

Government Website: http://www.austria.gv.at

CURRENT LEADERS

Dr Heinz Fischer

Position
President

Introduction
Following his electoral victory on 25 April 2004 at the age of 65, Dr Heinz Fischer took office as Austria's first socialist federal president for 18 years on 8 July 2004. He is committed to maintaining the country's neutral foreign policy and the welfare state. Critics have labelled him a Berufspolitiker ('professional politician') who has tended to avoid controversy and conflict.

Early Life
Heinz Fischer was born into a political family in Graz on 9 Oct. 1938. His father was state secretary in the ministry of trade from 1954–56. Fischer attended the Humanistisches Gymnasium in Vienna and went on to study law and political science at the University of Vienna, attaining a PhD in 1961. He entered politics two years later, becoming secretary to the Social Democratic Party (SPÖ) in the Austrian parliament, a position he held until 1975. Despite being elected as a member of parliament in 1971, Fischer continued his academic career. He was appointed associate professor of political science at the University of Innsbruck in 1978 and was made a full professor in 1994.

Fischer served as federal minister of science and research from 1983–86, under a coalition government headed by Fred Sinowatz of the SPÖ. In 1986 the SPÖ joined the Austrian People's Party (ÖVP) in a 'grand coalition' that retained control of the government through the 1990s. Fischer was elected president of the National Council in Nov. 1990, holding the office for 12 years until Dec. 2002. He also served as a member of the national security council and the foreign affairs council.

He was elected federal president on 25 April 2004 as the SPÖ candidate, polling 52·4% of the vote to defeat Benita Ferrero-Waldner, foreign minister in the ruling ÖVP-led conservative coalition.

Career in Office
On 8 July 2004 Fischer was sworn in for a six-year term. Although a largely ceremonial post, the president is commander-in-chief of the military and has the constitutional power to reject nominations for cabinet ministers and to remove them from office. In his opening address, Fischer recalled how many Austrians had grown up 'sensitive to war and peace' and aware that 'peace and the politics to promote peace... must have a central role in our political efforts'. The consolidation of the basic values of democracy is another of his priorities: '...Consensus is very important to me. But consensus means to build bridges. Bridges between solid shores.'

Following the collapse of the governing coalition in July 2008, early parliamentary elections were held in Sept. Fischer subsequently asked SPÖ leader Werner Faymann, as head of the largest party, to form a new government. Faymann renewed the coalition with the ÖVP, excluding the resurgent far-right parties that had made gains in the elections, and his government took office in Dec. Fischer stressed that a stable and competent administration was in the national interest to deal with serious challenges confronting Austria, including the global financial crisis. In April 2010 Fischer was re-elected for a further six-year presidential term, taking almost 80% of the vote.

He has written numerous books and publications on law and political science. He is also co-editor of the Austrian Zeitschrift für Politikwissenschaft (Journal of Political Science) and Journal für Rechtspolitik (Journal of Law Policy).

Werner Faymann

Position
Chancellor

Introduction
A career politician, Werner Faymann became leader of the Social Democratic Party (SPÖ) in June 2008. He led the party to a narrow victory at the Sept. 2008 general election and heads a coalition government.

Early Life
Werner Faymann was born on 4 May 1960 in Vienna. He attended a grammar school in Vienna and studied law at the University of Vienna before joining the youth branch of the Vienna SPÖ. There he led campaigns and protests, becoming the group's chairman in 1985. From 1985–88 he also worked as a consultant at the Zentralsparkasse (now Bank of Austria). In 1985 he was elected to the Vienna state parliament and from 1988–94 led the board for tenants' rights. He was appointed councillor for housing and urban development in Vienna in 1995, serving until 2007 when Chancellor Gusenbauer appointed him federal minister for transport, technology and development.

In June 2008, with the 'grand coalition' between the SPÖ and the Austrian People's Party (ÖVP) under strain, the SPÖ separated the role of chancellor from that of party leader. Faymann was chosen to head the party. Faymann and Gusenbauer then published an open letter that reversed earlier SPÖ policy by promising to put all amendments to EU treaties affecting the national interest to a referendum. The ÖVP consequently abandoned the coalition, triggering a general election.

Faymann fought the election on a platform of social investment and populist scepticism about the EU. On 28 Sept. 2008 the SPÖ won the most seats, although with a reduced share of the vote (just under 30%). Refusing to consider a coalition with the right-wing Alliance for the Future of Austria or the anti-immigration Freedom Party, he entered into an agreement in Nov. with the ÖVP under its new leader Josef Pröll. On 2 Dec. 2008 Faymann was sworn in as chancellor.

Career in Office
Faymann faced immediate challenges from the global economic crisis and in early 2009 introduced tax cuts to boost the economy. He has advocated EU aid to support struggling European economies, in part to safeguard large loans Austria has made to eastern European countries.

DEFENCE

The Federal President is C.-in-C. of the armed forces. Conscription is for a six-month period, with liability for at least another 30 days' reservist refresher training spread over eight to ten years. Conscientious objectors can instead choose to undertake nine months' civilian service. In Jan. 2013 a referendum was held on whether to continue with conscription; 59·8% of votes cast were in favour of maintaining compulsory military service. Since 2005 the total 'on mobilization strength' of the forces has been reduced from 110,000 to 55,000 troops. In 2010 approximately 1,200 personnel were deployed in peace support operations in various parts of the world.

Defence expenditure in 2008 totalled €2,560m. (€307 per capita), representing 0·9% of GDP.

Army

The Army is structured in four brigades and nine provincial military commands. Two brigades are mechanized and two are infantry brigades. The mechanized brigades are equipped with Leopard 2/A4 main battle tanks. One of the infantry brigades is earmarked for airborne operations and the other is specialized in mountain operations. M-109 armoured self-propelled guns equip the artillery battalions. Active personnel, 2010, 50,000 including 25,000 conscripts. Women started to serve in the armed forces in 1998.

Air Force

The Air Force Command comprises two brigades (one air support and one air defence), approximately 150 aircraft and a number of fixed and mobile radar stations. Some 15 Eurofighter interceptors equip a surveillance wing responsible for the defence of the Austrian air space and a fighter-bomber wing operates SAAB 105s. Helicopters including the S-70 Black Hawk equip six squadrons for transport/support, communication, observation, and search and rescue duties. Fixed-wing aircraft including PC-6s, PC-7s, and C-130 Hercules are operated as trainers and for transport.

ECONOMY

In 2007 agriculture accounted for 2% of GDP, industry 31% and services 67%.

Overview

Austria has had one of the OECD's strongest productivity growth rates since the late 1990s while maintaining an unemployment rate below the OECD average. The economy has benefited from increasing integration with Central and Eastern European Countries (CEECs), with its market share expanding as FDI into the region has grown. Membership of the EU and the EMU has added momentum and encouraged market liberalization, with many companies coming under foreign (particularly German) ownership. Social partnership in the labour market has ensured wage moderation and low unemployment. Tourism is important, reflecting Austria's central location and Alpine appeal.

GDP growth accelerated in 2004 to its highest rate since 2000 and was higher still in the next three years, averaging 2·8% over four years until the global financial crisis and rising unemployment led to recession. Austria has high dependency on exports, which amount to about 60% of GDP, and is vulnerable to fluctuating conditions among its CEEC partners.

Government debt stood at 71·8% of GDP in 2010, with the IMF expecting debt to remain above 70% in the medium term. However, private consumption and employment rates fared better than in many European counterparts. When external demand picked up in 2010, the economy rebounded. Services and tourism performed well. At 4·1% in Dec. 2011, the unemployment rate is among the lowest in Europe.

Government spending accounts for about half of GDP, one of the highest levels in the OECD. 11% of GDP in 2009 was spent on health, well above the 9·5% OECD average. With one of the most expensive public pension systems in Europe, reforms were introduced in 2003 and 2005 to improve long-term financial sustainability.

Currency

On 1 Jan. 1999 the euro (EUR) became the legal currency in Austria at the irrevocable conversion rate of 13·7603 schillings to one euro. The euro, which consists of 100 cents, has been in circulation since 1 Jan. 2002. On the introduction of the euro there was a 'dual circulation' period before the schilling ceased to be legal tender on 28 Feb. 2002.

Inflation rates (based on OECD statistics):

2002	2003	2004	2005	2006	2007	2008	2009	2010	2011
1·7%	1·3%	2·0%	2·1%	1·7%	2·2%	3·2%	0·4%	1·7%	3·6%

Foreign exchange reserves were US$5,269m. in Sept. 2009 and gold reserves 9·00m. troy oz. Total money supply was €96,422m. in Aug. 2009.

Budget

Central government revenue and expenditure in €1m.:

	2007	2008	2009[1]
Revenue	102,734	106,113	101,081
Expenditure	105,024	108,551	108,757

[1]Provisional.

Austria's budget deficit in 2011 was 2·6% of GDP (2010, 4·5%; 2009, 4·1%). The required target set by the EU is a deficit of no more than 3%.

VAT is 20% (reduced rates, 12% and 10%).

Performance

Real GDP growth rates (based on OECD statistics):

2002	2003	2004	2005	2006	2007	2008	2009	2010	2011
1·6%	0·9%	2·3%	2·7%	3·6%	3·7%	1·1%	−3·5%	2·2%	2·7%

Total GDP was US$417·7bn. in 2011.

Banking and Finance

The Oesterreichische Nationalbank, central bank of Austria, opened on 1 Jan. 1923 but was taken over by the German Reichsbank on 17 March 1938. It was re-established on 3 July 1945. Its *Governor* is Ewald Nowotny. At 31 Dec. 2011 it had assets of €99,348m.

In 2011 banking and insurance accounted for 4·6% of gross domestic product at current prices. In 2010 an average of 124,789 individuals were engaged in banking and insurance (78,309 in banking and credit institutions, 28,101 in insurance companies and pension funds and 18,379 in other financial and insurance services).

In 2011 there were 824 bank and credit institution head offices and 4,441 branch offices. The leading banks with total assets in Dec. 2010 (in €1m.) were: Erste Group, 205,938; Bank Austria, 193,049; Raiffeisen Zentralbank Österreich, 136,497.

Gross external debt amounted to US$779,544m. in June 2012.

There is a stock exchange in Vienna (VEX). It is one of the oldest in Europe and one of the smallest.

ENERGY AND NATURAL RESOURCES

In 2008, 28·5% of energy consumption came from renewables (wind power, solar power, hydro-electric power, tidal power, geothermal energy and biomass), compared to the European Union average of 10·3%. A target of 34% has been set by the EU for 2020.

Environment

Austria's carbon dioxide emissions from the consumption and flaring of fossil fuels were the equivalent of 8·6 tonnes per capita in 2008. An *Environmental Performance Index* compiled in 2008 ranked Austria sixth in the world, with 89·4%. The index examined various factors in six areas—air pollution, biodiversity and habitat, climate change, environmental health, productive natural resources and water resources.

Austria is one of the world leaders in recycling. In 2008 an estimated 69% of all municipal waste was composted or recycled.

Electricity

The Austrian electricity market was fully liberalized on 1 Oct. 2001. Installed capacity was 19·5m. kW in 2007. Production in 2007 was 63·36bn. kWh. Consumption per capita, 2007: 8,415 kWh. Renewable sources accounted for 62% of electricity consumption in 2008—the highest share in the EU.

Oil and Gas

The commercial production of petroleum began in the early 1930s. Production of crude oil, 2006: 856,274 tonnes. Crude oil reserves, 2007, were 50m. bbls.

The Austrian gas market was fully liberalized on 1 Oct. 2002. Production of natural gas, 2008: 1,532m. cu. metres. Natural gas reserves in 2007 amounted to 16bn. cu. metres.

Minerals

The most important minerals are sand and gravel (2011 production, 25,063,350 tonnes), limestone (2011 production, 21,570,972 tonnes), dolomite (2011 production, 3,710,729 tonnes) and granite and granulite (2011 production, 3,034,265 tonnes).

Agriculture

In 2010 the agricultural workforce numbered 413,755 (520,984 in 2005). There were 173,317 farms in 2010. There were 1·38m. ha. of arable land in 2007 and 68,000 ha. of permanent crops. In 2009 Austria set aside 519,000 ha. (18·5% of its agricultural land—the highest percentage in the European Union) for the growth of organic crops. Food and live animals accounted for 4·2% of exports and 5·3% of imports in 2007.

The chief products in 2010 (area in 1,000 ha.; yield in tonnes) were as follows: barley (168·9; 777,961); grain maize (179·8; 1,692,450); potatoes (22·0; 671,722); silo maize (81·2; 3,557,330); sugar beets (44·8; 3,131,666); wheat (302·9; 1,517,805). Other important agricultural products include apples (197,413 tonnes in 2010) and pears (8,185 tonnes in 2010). Wine production in 2007 totalled 2,628,021 hectolitres.

Livestock in 2011: cattle, 1,976,527; pigs, 3,004,907; sheep, 361,183; poultry, 14,644,413.

Forestry

Forested area in 2010, 3·89m. ha. (47% of the land area), around three-quarters of which was coniferous. Felled timber, in 1,000 cu. metres: 2008, 21,795·4; 2009, 16,727·4; 2010, 17,831·0.

Fisheries

Total catch in 2010 came to 375 tonnes, exclusively from inland waters.

INDUSTRY

The leading companies by market capitalization in Austria in March 2012 were: OMV Group, an oil and gas company (US$11·6bn.); Erste Group, a banking group (US$9·1bn.); and Raiffeisen International Bank (US$6·9bn.).

Production (in 1,000 tonnes): pig iron (2007), 5,808; paper and paperboard (2007), 5,199; cement (2008), 5,309; distillate fuel oil (2007), 3,461; petrol (2007), 1,702; residual fuel oil (2007), 880; sawnwood (2007), 11·26m. cu. metres; beer (2010), 883·4m. litres.

Labour

Austria has the lowest unemployment rate in the European Union (4·3% in Dec. 2012).

Of 3,421,755 employees in 2011 (annual average), 573,571 worked in the manufacturing and production of goods; 529,976 in public administration and defence, and compulsory social security; 518,188 in wholesale and retail trade, and the repair of motor vehicles; 233,994 in human health and social work; 184,548 in accommodation and food service activities. In 2011 there were an average of 73,800 job vacancies.

There were no recorded strikes between 2005 and 2009.

Austria has one of the lowest average retirement ages but reforms passed in 1997 now make it less attractive to retire before 60. Only 15% of men and 6% of women in the 60–65 age range work, although the legal retirement ages are 60 for women and 65 for men.

INTERNATIONAL TRADE

Imports and Exports

In 2011 imports were valued at €131,008m. (€113,652m. in 2010) and exports €121,774m. (€109,373m. in 2010).

Leading import suppliers in 2011 (% of total imports) were: Germany, 38·2%; Italy, 6·5%; Switzerland, 5·4%; China, 4·9%;

Czech Republic, 3·7%; USA, 2·9%. Principal export markets in 2011 (% of total exports) were: Germany, 31·2%; Italy, 7·7%; USA, 5·2%; Switzerland, 4·9%; France, 4·1%; Czech Republic 3·9%.

In 2011 chemicals, manufactured goods classified chiefly by material and miscellaneous manufactured articles accounted for 42·5% of Austria's imports and 47·7% of exports; machinery and transport equipment 31·8% of imports and 37·8% of exports; food, live animals, beverages and tobacco 6·3% of imports and 6·7% of exports; mineral fuels, lubricants and related materials 12·0% of imports and 3·4% of exports; inedible crude materials (except fuels), and animal and vegetable oil and fats 5·5% of imports and 3·5% of exports.

Trade Fairs

Vienna ranks as the fourth most popular convention city in the world (behind Singapore, Brussels and Paris) according to the Union des Associations Internationales (UAI), hosting 2·3% of all meetings held in 2010.

COMMUNICATIONS

Roads

In 2007 the road network totalled 107,262 km (Autobahn, 1,677 km; highways, 10,408 km; secondary roads, 23,657 km). In 2007 passenger cars in use numbered 4,245,600, lorries and vans 372,600, buses and coaches 9,300, and motorcycles and mopeds 642,800. There were 691 fatalities in road accidents in 2007.

Rail

The Austrian Federal Railways (ÖBB) has been restructured and was split up into ten new companies, which became operational on 1 Jan. 2005. Length of route in 2011, 5,500 km, of which 3,763 km were electrified. There are also a number of private railways. In 2011, 244·0m. passengers and 107·6m. tonnes of freight were carried by Federal Railways. There is a metro and tramway in Vienna, and tramways in Gmunden, Graz, Innsbruck and Linz.

Civil Aviation

There are international airports at Vienna (Schwechat), Graz, Innsbruck, Klagenfurt, Linz and Salzburg. The national airline is Austrian Airlines, which was privatized after a takeover by Lufthansa in Sept. 2009. In 2010, 65 other airlines had scheduled flights to and from Vienna. In 2011, 312,502 commercial aircraft and 25,704,655 passengers arrived and departed; 227,938 tonnes of freight and 13,551 tonnes of mail were handled. In 2011 Vienna handled 21,069,398 passengers and 218,835 tonnes of freight. Austrian Airlines carried 11,261,000 passengers in 2011. In 2010 there were direct international services to more than 140 destinations from Vienna, 30 from Salzburg, 24 from Graz, 17 from Linz and 16 from Innsbruck.

Shipping

The Danube is an important waterway. Goods traffic (in 1,000 tonnes): 12,084 in 2005; 11,782 in 2006; 12,107 in 2007; 11,209 in 2008 (including the Rhine-Main-Danube Canal). There were four vessels of 300 GT or over registered in Jan. 2009, totalling 14,000 GT.

Telecommunications

Österreichische Industrie Holding AG, the Austrian investment and privatization agency, holds a 28·4% stake in Telekom Austria. In 2008 there were 3,285,000 main (fixed) telephone lines. In the same year mobile phone subscribers numbered 10,816,000 (1,297·3 per 1,000 persons). There were 5·9m. internet users in 2008. In March 2012 there were 2·8m. Facebook users.

SOCIAL INSTITUTIONS

Justice

The Supreme Court of Justice (Oberster Gerichtshof) in Vienna is the highest court in civil and criminal cases. In addition, in 2012 there were four Courts of Appeal (Oberlandesgerichte),

20 High Courts (Landesgerichte) and 141 District Courts (Bezirksgerichte). There is also a Supreme Constitutional Court (Verfassungsgerichtshof) and a Supreme Administrative Court (Verwaltungsgerichtshof), both seated in Vienna. In 2011 a total of 540,007 criminal offences were reported to the police and 36,461 people were convicted of offences. The population in penal institutions in Aug. 2008 was 7,909 (95 per 100,000 of national population).

Education

In 2006–07 there were 5,011 general compulsory schools (including special education) with 71,796 teachers and 639,433 pupils. Secondary schools totalled 1,676 in 2006–07, with 571,651 pupils. Secondary technical and vocational colleges numbered 317 in 2006–07, with 134,609 pupils.

The dominant institutions of higher education are the 16 public universities and six colleges of arts. Student fees were introduced in 2001, but in 2008 the government decided to eliminate fees for Austrian nationals who complete their studies within the minimum time.

In the winter term 2007–08 there were 224,597 students (46,102 foreign) enrolled at the public universities and 9,853 (5,934 foreign) at the colleges of arts. In 1994 Higher Technical Study Centres (Fachhochschul-Studiengänge, FHS) were established, which are private, but government-dependent, institutions. A federal law was passed in 1999 to allow the accreditation of private universities. In 2007 there were 11 accredited private universities.

In 2009 public expenditure on education came to 6·0% of GDP and 11·4% of total government spending. The adult literacy rate is at least 99%.

Health

In 2006 there were 30,068 doctors, 4,467 dentists and 53,782 nurses and midwives. In 2009 there were 267 hospitals and 64,069 hospital beds. In 2008 Austria spent 10·5% of its GDP on health.

Welfare

Maternity/paternity leave is until the child's second birthday. A new parenting allowance was introduced on 1 Jan. 2002, replacing the maternity/paternity allowance. The latest formula is based on family benefit financed from the Family Fund. The basic allowance is €436 per month for a maximum of three years. In June 2003 a reform of the pensions system was approved involving the reduction of pension benefits by 10%, the raising of the retirement age to 65 by increasing the workers' contribution period from 40 to 45 years and the abolition of early retirement by 2017. These reforms provoked Austria's first general strike in over 50 years. There were 2,705,480 pensioners in Dec. 2011.

RELIGION

In 2001 there were 5,915,000 Roman Catholics (73·6%), 376,000 Evangelical Lutherans (4·7%), 339,000 Muslims (4·2%), 963,000 without religious allegiance (12·0%) and 439,000 others (5·5%). The Roman Catholic Church has two archbishoprics and seven bishoprics. In Feb. 2013 there was one cardinal.

CULTURE

World Heritage Sites

There are nine UNESCO sites in Austria. They are: the historic centre of the city of Salzburg (inscribed in 1996); the Palace and gardens of Schönbrunn (1996); Hallstatt-Dachstein Salzkammergut cultural landscape (1997); Semmering Railway (1998); the historic centre of the city of Graz (1999); the Wachau cultural landscape (2000); and the historic centre of the city of Vienna (2001).

Austria shares the Cultural Landscape of Fertö/Neusiedlersee site (2001) with Hungary and the Prehistoric Pile dwellings around the Alps (2011) with France, Germany, Italy, Slovenia and Switzerland.

Press

There were 19 daily newspapers and 223 non-daily newspapers in 2008. The most popular newspaper is the tabloid *Kronen Zeitung*, with an average daily circulation of 949,000 in 2008. In the 2011–12 *World Press Freedom Index* compiled by Reporters Without Borders, Austria ranked fifth out of 179 countries.

Tourism

In 2011, 13,359 hotels and boarding houses had a total of 594,357 beds available; in the same year 23,012,000 non-resident tourists stayed in holiday accommodation and international tourist spending came to €14·3bn. Of 126,002,551 overnight stays in tourist accommodation in 2011, 35,296,997 were by Austrians and 47,389,531 by Germans.

Festivals

The main festivals are the Salzburg Festival, held every July–Aug. (213,592 visitors in 2011), Operafestival St Margarethen, also held in July–Aug. (209,800 visitors in 2010) and the Mörbisch Festival on the Lake (184,779 visitors in 2011), again held in July–Aug. Popular music festivals include Snowbombing music and snow sports festival in Mayrhofen (April), Vienna Jazz Fest (May–July), Donauinselfest in Vienna (June) and Nova Rock Festival at Nickelsdorf (June).

DIPLOMATIC REPRESENTATIVES

Of Austria in the United Kingdom (18 Belgrave Mews West, London, SW1X 8HU)
Ambassador: Emil Brix.

Of the United Kingdom in Austria (Jaurèsgasse 12, 1030 Vienna)
Ambassador: Susan le Jeune d'Allegeershecque.

Of Austria in the USA (3524 International Court, NW, Washington, D.C., 20008)
Ambassador: Hans Peter Manz.

Of the USA in Austria (Boltzmanngasse 16, 1090 Vienna)
Ambassador: William C. Eacho.

Of Austria to the United Nations
Ambassador: Martin Sajdik.

Of Austria to the European Union
Permanent Representative: Walter Grahammer.

FURTHER READING

Austrian Central Statistical Office. *Main publications: Statistisches Jahrbuch für die Republik Österreich.* New Series from 1950. Annual.—*Statistische Nachrichten.* Monthly.—*Beiträge zur österreichischen Statistik.—Statistik in Österreich 1918–1938.* [Bibliography] 1985.—*Veröffentlichungen des Österreichischen Statistischen Zentralamtes 1945–1985.* [Bibliography] 1990.—*Republik Österreich, 1945–1995.*

Bischof, Günter, Pelinka, Anton and Gehler, Michael, (eds.) *Austria in the European Union.* 2002.—*Austrian Foreign Policy in Historical Context.* 2005

Brook-Shepherd, G., *The Austrians: a Thousand-Year Odyssey.* 1997

Bruckmüller, Ernst, *Austrian Nation: Cultural Consciousness and Socio-Political Processes.* 2003

Pick, Hella, *Guilty Victim: Austria from the Holocaust to Haider.* 2000

Steininger, Rolf, *Austria in the Twentieth Century.* 2002

Wolfram, H. (ed.) *Österreichische Geschichte.* 10 vols. 1994

National library: Österreichische Nationalbibliothek, Josefsplatz, 1015 Vienna.

National Statistical Office: Austrian Central Statistical Office, Guglgasse 13, A-1110 Vienna.

Website: http://www.statistik.at

AZERBAIJAN

© Research Machines plc 2005

Azarbaijchan Respublikasy
(Republic of Azerbaijan)

Capital: Baku
Population projection, 2015: 9·75m.
GNI per capita, 2011: (PPP$) 8,666
HDI/world rank: 0·700/91
Internet domain extension: .az

KEY HISTORICAL EVENTS

Rock art at Gobustan, close to Azerbaijan's Caspian Sea coast, dates from 20,000 BC. Part of southern Azerbaijan came under the influence of the Assyrian Empire around 800 BC and was later subsumed into the ancient kingdoms of Manue, Urartu and Medea. During the 6th century BC the Persian Akhemenid dynasty held sway in what was known as Caucasian Albania, fortified by the Zoroastrian religion. Persian influence continued in the form of the Parthian Empire from around 200 BC, followed by periods of Roman rule. The Arshakid dynasty, installed by the Romans to control much of the Caucasus, survived until the Persian Sassanid Empire in the 4th century AD.

Sassanid power reached its zenith under Khosrau II around AD 620, with a sphere of influence stretching from Egypt to the Caucasus, central Asia and northwest India. Arab tribes, newly united around Islam, made incursions into Azerbaijan from the mid-7th century. From 750 the territory came under the rule of Caliph Abu Abbas, whose Abbasid dynasty was centred on Baghdad.

Arab control of Azerbaijan was waning by the late 10th century and the region dissolved into numerous fiefdoms, including Shirvan (centred on Shemakha) and Arran (based at Terter). The 11th century saw an influx of nomadic Oghuz Turks under the Seljuk dynasty from Central Asia. This began a process of Turkification that continued for the next three centuries, punctuated by Mongol incursions including the Golden Horde in 1319 and Timur in 1380.

At the end of the 15th century Azerbaijan became the power base of the Safavid dynasty, centred on Ardabil (now in northern Iran). Shah Ismail I forged a kingdom that controlled all Persia by 1520 and which promoted the Shia branch of Islam as the state religion. There was conflict with Ottoman Turkey, which followed Sunni Muslim traditions, and the Safavids shifted their focus eastwards from Azerbaijan, eventually establishing their capital in Esfahan.

The overthrow of Safavid control in Azerbaijan in 1722 preceded increasing Russian influence in the region, which led to the annexation of Georgia by Tsar Alexander I and the control of the khanates in Azerbaijan. By 1807 Nakhichevan remained the only independent khanate, although Russian forces had to fight off Persian challenges for control. The territory of the present Azerbaijan was acquired by Russia from Persia through the treaties of Gulistan (1813) and Turkmenchai (1828). This period of Russian rule was marked by an influx of Armenians into western Azerbaijan, fleeing persecution from Ottoman Turkey and Persia.

Azerbaijan's oil industry, centred on Baku, began to develop in the 1870s. By the early 20th century half the world's production was supplied from Baku. Tensions between Azeris and Armenians rose against the backdrop of revolutions in Russia (1905–07) and Iran (1906–11), prompting the rise of Azeri nationalism. Following the Bolshevik revolution in 1917, Azerbaijan joined with Armenia and Georgia to form the short-lived Transcaucasian Federation. On 28 May 1918 the Azerbaijani Democratic Republic was declared, with Gandja as its capital. The fledgling nation was soon occupied, first by Ottoman troops and then by British forces until they withdrew in Aug. 1919. After the Red Army achieved victory in Russia's civil war in early 1920, it moved swiftly to secure control of oil-rich Azerbaijan, reaching Baku on 28 April 1920.

In 1922 Azerbaijan joined the Transcaucasian Soviet Federal Socialist Republic, led by Nariman Narimanov. Stalin's administrative reorganization of 1936 made Azerbaijan a separate Soviet Republic, forcing it to break links with other Turkic and Islamic states. Mass killings, deportations and imprisonments followed, reaching their height in the Stalinist purges of 1937. Baku's oilfields were an objective for Nazi forces in the Second World War but their defeat at Stalingrad meant that Azerbaijan remained under Soviet control.

Conflict with Armenia over the enclave of Nagorno-Karabakh escalated in 1988, leading to violent expulsions of Armenians in Azerbaijan and Azeris in Armenia. In 'Black January' 1990 Soviet tanks moved into Baku following rioting and over 100 civilians were killed. War broke out between the two countries in 1992 after the Armenian population of Nagorno-Karabakh declared independence from Azerbaijan. A ceasefire was agreed in 1994 but the dispute over territory remains unsettled, despite the efforts of the Organization for Security and Co-operation in Europe's Minsk Group (co-chaired by France, Russia and the USA) to broker a peace.

On 18 Aug. 1991 the Supreme Soviet of Azerbaijan declared independence from the crumbling Soviet Union. Following an attempted coup in June 1993, Heidar Aliyev was appointed president by the national assembly and then confirmed in a general election in Oct. Parliament ratified association with the Commonwealth of Independent States on 20 Sept. 1993. A treaty of friendship and co-operation was signed with Russia on 3 July 1997 and Aliyev was re-elected in Oct. 1998, although the administration of the election was criticized by international observers. Following serious illness, Aliyev stood down from the presidency in Oct. 2003. He controversially appointed his son, Ilham, as the party's sole presidential candidate. Ilham Aliyev duly won the presidential election of 15 Oct. 2003 but international observers again criticized the contest as falling below accepted standards. He was re-elected on 15 Oct. 2008 with 88·7% of the vote, although several major parties boycotted the election.

TERRITORY AND POPULATION

Azerbaijan is bounded in the west by Armenia, in the north by Georgia and the Russian Federation (Dagestan), in the east by the Caspian sea and in the south by Turkey and Iran. Its area is 86,600 sq. km (33,430 sq. miles), and it includes the Nakhichevan Autonomous Republic and the largely Armenian-inhabited Nagorno-Karabakh.

The population at the 2009 census was 8,922,447 (50·5% females); density, 103 per sq. km. In 2011, 52·1% of the population lived in urban areas. The population breaks down into 91·6% Azerbaijanis, 2·0 Lezgis, 1·3% Armenians and 1·3% Russians (2009 census).

The UN gives a projected population for 2015 of 9.75m.

Chief cities (estimates of Jan. 2012): Baku, 1,184,000; Gandja 320,700; Sumgait 284,600. There are 66 districts and 13 cities.

The official language is Azeri. On 1 Aug. 2001 Azerbaijan abolished the use of the Cyrillic alphabet and switched to using Latin script.

SOCIAL STATISTICS

In 2009: births, 152,139; deaths, 52,514; marriages, 78,072; divorces, 7,784. Rates, 2009 (per 1,000 population): births, 17·2; deaths, 5·9; infant mortality (2010, per 1,000 live births), 39. Life expectancy in 2007: 72·3 years for females and 67·6 years for males. Annual population growth rate, 2000–05, 0·8%; fertility rate, 2008, 2·1 children per woman.

CLIMATE

The climate is almost tropical in summer and the winters slightly warmer than in regions north of the Caucasus. Cold spells do occur, however, both on the high mountains and in the enclosed valleys. There are nine climatic zones. Baku, Jan. –6°C, July 25°C. Annual rainfall 318 mm.

CONSTITUTION AND GOVERNMENT

Parliament is the 125-member *Melli-Majlis*, with all seats elected from single-member districts. A constitutional referendum and parliamentary elections were held on 12 Nov. 1995. Turnout for the referendum was 86%. The new constitution was approved by 91·9% of votes cast. As a result of a referendum held on 24 Aug. 2002 a number of changes were made to the constitution, including the distribution of the *Melli-Majlis* seats—previously, 25 seats were distributed proportionally among political parties. The validity of the outcome of the referendum was questioned by international observers. In a referendum on 18 March 2009 a measure to abolish presidential term limits was approved, with 91·8% of votes cast in favour.

National Anthem

'Azerbaijan! Azerbaijan!'; words by A. Javad, tune by U. Hajibeyov.

RECENT ELECTIONS

At elections on 15 Oct. 2008 Ilham Aliyev of the New Azerbaijan Party (YAP) was re-elected president with 88·7% of votes cast. Igbal Aghazade of the Azerbaijan Hope Party (AUP) won 2·9%, Fazil Mustafayev of the Great Order Party (BQP) won 2·5%, Gudrat Hasanguliyev of the All-Azerbaijan Popular Front Party (BAKJP) won 2·3% and Gulamhuseyn Alibayli (ind.) won 2·2%. There were two other candidates who received less than 1% of the vote each. Turnout was 75·6%.

At the parliamentary elections held on 7 Nov. 2010 the YAP gained 72 seats, Civic Solidarity 3, Motherland Party 2, and ind. and others 48. Turnout was 50·1%.

CURRENT GOVERNMENT

President: Ilham Aliyev; b. 1961 (YAP; sworn in 31 Oct. 2003).

In March 2013 the government comprised:

Prime Minister: Artur Rasizade; b. 1935 (YAP; in office since 6 Aug. 2003, until 4 Nov. as acting prime minister, having previously been prime minister from 20 July 1996 to 4 Aug. 2003).

First Deputy Prime Minister: Yagub Eyyubov. *Deputy Prime Ministers:* Elchin Efendiyev; Ali Hasanov; Abid Sharifov.

Minister of Foreign Affairs: Elmar Mamedyarov. *Interior:* Ramil Usubov. *Culture and Tourism:* Abulfaz Garayev. *Education:* Misir Mardanov. *Emergency Situations:* Kamaladdin Heydarov. *National Security:* Eldar Mahmudov. *Defence:* Col.-Gen. Safar Abiyev. *Defence Industry:* Yavar Jamalov. *Communications and Information Technologies:* Ali Abbasov. *Agriculture:* Ismat Abbasov. *Justice:* Fikret Mamedov. *Health:* Ogtay Shiraliyev. *Finance:* Samir Sharifov. *Labour and Social Protection:* Fizuli Alakbarov. *Youth and Sport:* Azad Rahimov. *Economic Development:* Shahin Mustafayev. *Ecology and Natural Resources:* Huseyngulu Bagirov. *Industry and Energy:* Natig Aliyev. *Taxation:* Fazil Mamedov. *Transport:* Ziya Mammadov.

Speaker of the National Assembly (Melli-Majlis): Ogtay Asadov.

Office of the President: http://www.president.az

CURRENT LEADERS

Ilham Aliyev

Position
President

Introduction
Ilham Aliyev succeeded his father, Heidar Aliyev, as president in Oct. 2003. The Moscow-educated politician has presided over a rapidly-growing economy but also increasing repression of political opposition. He was re-elected president in Oct. 2008, although most leading opposition candidates dismissed the poll as neither free nor fair.

Early Life
Ilham Heidar oglu Aliyev was born in Baku, capital of the Soviet Socialist Republic of Azerbaijan, on 24 Dec. 1961. His father was Heidar Aliyev, who became deputy prime minister of the Soviet Union under Mikhail Gorbachev and, in Oct. 1993, president of Azerbaijan. Ilham Aliyev graduated in history from the Moscow State Institute for International Relations in 1982. He subsequently gained a PhD in history and began teaching at the Institute. Plans to enter the diplomatic service were curtailed by the collapse of the Soviet Union in 1991 and he established 'business interests' in Moscow and İstanbul. Between 1991 and 1994 his flamboyant lifestyle attracted media attention and he was accused of accumulating large gambling debts.

In May 1994 Ilham Aliyev was appointed vice-president of Azerbaijan's state oil company. Four months later President Aliyev signed a 30-year deal valued at more than US$7bn. with eight foreign companies to develop the country's substantial oil reserves. In 1995 Ilham Aliyev was elected to parliament and was subsequently appointed president of the national Olympic committee and head of the Azerbaijan delegation to the Council of Europe. In Dec. 1999 he became a deputy of the ruling New Azerbaijan Party (YAP) and, in 2001, was appointed party vice president.

Following the surprise resignation of Prime Minister Artur Rasizade in Aug. 2003, Heidar Aliyev appointed his son as prime minister. The move was approved by a 101–1 vote in the National Assembly but opposition parties boycotted the election. Critics took the move as proof that the increasingly frail president planned to hand over power to his son (an amendment to the constitution in Aug. 2002 providing for the prime minister to become interim president in the event that the president dies in office or is incapacitated). A few weeks before the presidential elections of Oct. 2003 Heidar Aliyev pulled out of the running, leaving Ilham as the YAP's candidate.

Official results gave Aliyev victory with 76·8% of the vote but opposition parties staged mass protests over alleged intimidation and fraud, charges backed by international observers. Aliyev was sworn in as the president on 31 Oct. 2003. His father died on 12 Dec. 2003.

Career in Office
Aliyev has presided over a rapidly expanding economy, a consequence of the discovery of offshore gas fields, high international oil prices and the completion of pipelines crossing from the Caucasus to Turkey. Unemployment and poverty levels nonetheless remain high, especially in rural areas.

Aliyev released a number of opposition figures from prison in March 2005 following international pressure, particularly from the Council of Europe. However, his administration was criticized for intimidating opposition activists in the run-up to parliamentary elections in Nov. 2005. The YAP, led by Aliyev since March 2005, won 56 of 125 parliamentary seats but the Organization for Security and Co-operation in Europe's international election observer mission reported harassment and vote buying. The opposition Azadlig party refused to accept the election results and organized mass protests in Nov. and Dec. 2005. In Nov. 2006 an independent broadcaster was closed down and a newspaper evicted from offices in Baku. These measures were denounced by the opposition as a clampdown by the regime on freedom of speech. In Oct. 2008 Western observers judged that Aliyev's overwhelming re-election as president was an improvement on the conduct of previous polls but still fell short of fully-democratic standards. A constitutional amendment to abolish presidential term limits was approved in a referendum in March 2009 with nearly 92% of the vote, allowing Aliyev to stand for election a third time. In Nov. 2010 the YAP won a majority of seats in parliamentary elections, although international monitors again cited widespread voting irregularities. There were repeated demonstrations against Aliyev's regime in 2011 but protesters' demands for democratic reforms were rejected and a security crackdown imposed.

In Feb. 2006 Aliyev met Armenia's then president, Robert Kocharian, in Paris, France, but the two failed to agree on a 'declaration of principles' on the disputed territory of Nagorno-Karabakh. Azerbaijan has continued to reject any move towards independence for the Armenian enclave, although Aliyev and current Armenian president Serzh Sargsyan have agreed at several Russian-brokered meetings to intensify their efforts to find a political settlement over the territory. Sporadic border fighting has nevertheless continued, notably in March 2008 and mid-2012.

Artur Rasizade

Position
Prime Minister

Introduction
Artur Rasizade, an oil engineer-turned-politician, has been prime minister since 1996. He was appointed by President Heidar Aliyev and has served under his son, Ilham Aliyev, since Oct. 2003.

Early Life
Artur Tahir oglu Rasizade was born on 26 Feb. 1935 in Gandja in the Transcaucasian Soviet Federated Socialist Republic. Educated at the Azerbaijan Institute of Industry in Baku, Azerbaijan Soviet Socialist Republic, Rasizade began work as an engineer at the Institute of Oil Machine Construction in 1957. He served as chief engineer at Trust Soyuzneftemash from 1973–77, before taking the post of deputy head of the Azerbaijan state planning committee.

In 1986, after five years as bureau chief of the central committee of the Communist Party of Azerbaijan, Rasizade became first deputy prime minister under Kamran Baghirov, who has been widely blamed for the Republic's economic stagnation and the escalating tension with Armenia over Nagorno-Karabakh.

Following the break-up of the Soviet Union and Azerbaijan's declaration of independence in Aug. 1991, Rasizade became an adviser to the foundation for economic reforms. He served as an assistant to President Heidar Aliyev in early 1996 and was then appointed first deputy prime minister. He was appointed prime minister when Fuad Kuliev resigned following accusations by Aliyev of economic mismanagement. The National Assembly endorsed Rasizade's appointment and he took office on 26 Dec. 1996.

Career in Office
Heidar Aliyev won the presidential election in Oct. 1998 and retained Rasizade (a fellow member of the New Azerbaijan Party) as prime minister until 4 Aug. 2003, when the premier unexpectedly resigned. Rasizade's departure, ostensibly for health reasons, paved the way for Ilham Aliyev to assume office. Ilham Aliyev contested the presidential election of 15 Oct. 2003 and emerged victorious, although the opposition staged mass protests, alleging intimidation and fraud. On 4 Nov. 2003 Rasizade was formally reinstated as prime minister. He was reappointed again in Oct. 2008 following Aliyev's re-election as president.

DEFENCE
Conscription is for 18 months, or 12 in the case of university graduates. Defence expenditure in 2008 totalled US$1,585m. (US$194 per capita), representing 3·2% of GDP.

Army
Personnel, 2007, 56,840. In addition there is a reserve force of 300,000 Azerbaijanis who have received some kind of military service experience within the last 15 years. There is also a paramilitary Ministry of the Interior militia of more than 10,000 and a border guard of approximately 5,000.

Navy
The flotilla is based at Baku on the Caspian Sea. Equipment includes six patrol craft. Personnel numbered about 2,000 in 2007.

Air Force
How many ex-Soviet aircraft are usable is not known but there are 47 combat-capable aircraft and 15 attack helicopters. Personnel, 7,900 in 2007.

INTERNATIONAL RELATIONS
There is a dispute with Armenia over the status of the chiefly Armenian-populated Azerbaijani enclave of Nagorno-Karabakh. A ceasefire was negotiated from 1994 with 20% of Azerbaijan's land in Armenian hands and with 1m. Azeri refugees and displaced persons.

ECONOMY
In 2010 agriculture accounted for 5·8% of GDP, industry 64·7% and services 29·5%.

Overview
After Azerbaijan gained independence in 1991, the economy suffered from corruption, distortion and weak social infra-structure. In 1994 the government signed a production-sharing agreement with a consortium of foreign companies to develop oil deposits in the Caspian Sea and transport oil from Azerbaijan to Turkey via Georgia through the Baku-Tbilisi-Ceyhan and Shah Deniz pipelines. In 1999 the State Oil Fund was created to manage the oil boom. Its assets were valued at US$11·2bn. by the end of 2008. Oil and gas make up 90% of exports and over 50% of GDP.

At 25% in 2007, Azerbaijan has had one of the world's fastest growth rates. The economy was equipped to weather the global economic crisis, with fiscal stimulus and a stable exchange rate preventing non-oil growth falling below 3%. The central bank provided support to state enterprises, including provision for losses on overseas investments.

Strong oil-based growth combined with increased wages and well-targeted social programmes to reduce the poverty rate from 50% in 2001 to 7·6% in 2011. However, in the long term the authorities must seek to reduce dependence on oil and gas revenues. According to the IMF, diversification is dependent on stimulating private sector growth by improving governance and the business environment.

Currency
The *manat* (AZM) of 100 *gyapiks* replaced the *rouble* in Jan. 1994. It was in turn replaced in Jan. 2006 by the *new manat* (AZN), also of 100 *gyapiks*, at 1 new manat = 5,000 manats. Inflation was 20·8% in 2008, falling to 1·5% in 2009 but increased to 5·7% in 2010 and 7·9% in 2011. Foreign exchange reserves were US$1,028m. in July 2005 and total money supply was 3,846·1bn. manats.

Budget
In 2007 revenues totalled 7,949m. manats and expenditures 7,356m. manats. Tax revenue accounted for 70·9% of revenues in 2007; current expenditure accounted for 62·5% of expenditures.

Performance
Total GDP was US$63·4bn. in 2011. Azerbaijan has one of the fastest-growing economies in the world. Real GDP growth was 26·4% in 2005, 34·5% in 2006 and 25·0% in 2007—the highest rates of any country in each of these years. This was largely thanks to the oil boom. Growth slowed to 5·0% in 2010 and just 0·1% in 2011.

Banking and Finance
The central bank and bank of issue is the National Bank (*Chairman*, Dr Elman Rustamov). The largest of the 44 commercial banks in Dec. 2011 was the International Bank of Azerbaijan (the only state-owned bank), with assets of US$6·2bn. The largest private bank is Bank Standard.

In 2010 total external debt was US$6,974m., representing 14·9% of GNI.

ENERGY AND NATURAL RESOURCES
Environment
Azerbaijan's carbon dioxide emissions from the consumption and flaring of fossil fuels were the equivalent of 4·7 tonnes per capita in 2008.

Electricity
Output was 21·8bn. kWh in 2007; consumption per capita in 2007 was 2,518 kWh. Capacity in 2007 was an estimated 5·7m. kW.

Oil and Gas
The most important industry is crude oil extraction. Baku is at the centre of oil exploration in the Caspian. Partnerships with Turkish, western European and US companies have been forged.

In 2008 oil reserves totalled 7·0bn. bbls. A century ago Azerbaijan produced half of the world's oil, but output today is just 1% of the total (although production started to increase in the late 1990s, rising every year between 1997 and 2008). Oil production in 2008 was 44·7m. tonnes. In July 1999 BP Amoco announced a major natural gas discovery in the Shah Deniz offshore field, with reserves of at least 700bn. cu. metres and perhaps as much as 1,000bn. cu. metres. There were proven reserves of 1,200bn. cu. metres in 2008. Natural gas production in 2008 amounted to 14·7bn. cu. metres—an increase of 50% on 2007 and up from just 4·5bn. cu. metres in 2004.

Accords for the construction of an oil pipeline from Baku, the Azerbaijani capital, on the Caspian Sea through Georgia to Ceyhan in southern Turkey (the BTC pipeline) were signed in

Nov. 1999. Work on the pipeline began in Sept. 2002 and it was officially opened in May 2005.

A gas pipeline from Baku through Georgia to Erzurum in Turkey (the South Caucasus pipeline) was commissioned in 2006, allowing Azerbaijan to become a net exporter of natural gas.

Minerals
The republic is rich in natural resources: iron, bauxite, manganese, aluminium, copper ores, lead, zinc, precious metals, sulphur pyrites, nepheline syenites, limestone and salt. Cobalt ore reserves have been discovered in Dashkasan, and Azerbaijan has the largest iodine-bromine ore reserves of the former Soviet Union (the Neftchala region has an iodine-bromine mill).

Agriculture
In 2008 the total area devoted to agriculture was 4·8m. ha. In 2007 there were 1·85m. ha. of arable land and 0·22m. ha. of permanent crops. 1·43m. ha. were irrigated in 2007. In 2009, 38% of the economically active population was engaged in agriculture, hunting and forestry. Principal crops include grain, cotton, rice, grapes, citrus fruit, vegetables, tobacco and silk.

Output of main agricultural products (in 1,000 tonnes) in 2008: wheat, 1,646; potatoes, 1,077; barley, 606; tomatoes, 438; watermelons, 408; apples, 205.

Livestock (2009): cattle, 2·28m.; goats, 591,000; sheep, 7·69m.; poultry, 22m. Livestock products (2008, in 1,000 tonnes): beef and veal, 77; poultry, 52; mutton and goat meat, 46; cow's milk, 1,382; eggs, 1·0bn. units.

Forestry
In 2010 forests covered 1·04m. ha., or 12·0% of the total land area. Timber production in 2007 was 7,000 cu. metres.

Fisheries
Total fish catch in 2010 came to 1,081 tonnes, exclusively from inland waters.

INDUSTRY
There are oil extraction and refining, oil-related machinery, iron and steel, aluminium, copper, chemical, cement, building materials, timber, synthetic rubber, salt, textiles, food and fishing industries. Production (2007) in 1,000 tonnes: residual fuel oil, 2,341; distillate fuel oil, 2,109; cement (2009), 1,283; petrol, 1,129; jet fuel, 793; bread and bakery products (2003), 686. Output of other products: footwear (2003), 455,900 pairs.

Labour
In 2009 the economically active workforce numbered 4,331,800. The main areas of activity were: agriculture, hunting and forestry, 1,562,400; wholesale and retail trade/repair of motor vehicles, motorcycles and personal and household goods, 661,500; education, 346,900; public administration and defence/social security, 277,200. The unemployment rate in 2009 was 6·0%. The average monthly salary in 2009 was 298 manats.

INTERNATIONAL TRADE
Imports and Exports
In 2007 imports (c.i.f.) were valued at US$5,712·2m. and exports (f.o.b.) at US$6,058·3m.

Principal imports in 2007 were machinery and transport equipment, manufactured goods, food and live animals, chemicals and related products, miscellaneous manufactured articles, and beverages and tobacco. Petroleum and related products accounted for 81% of exports. Manufactured goods, machinery and transport equipment, inedible crude materials (except fuels), and chemicals and related products are also important exports.

Leading import suppliers in 2007 were Russia (17·6%), Turkey (10·9%), Germany (8·2%), Ukraine (8·2%), UK (7·2%). The main export markets were Turkey (17·4%), Italy (15·5%), Russia (8·7%), Iran (7·2%), Indonesia (6·4%).

COMMUNICATIONS

Roads

There were 59,141 km of roads (6,928 km highways and main roads) in 2006. Passenger cars in use in 2006 totalled 548,979 (57 per 1,000 inhabitants in 2005). In addition, there were 9,916 lorries and vans, and 27,474 buses and coaches. There were 1,107 fatalities as a result of road accidents in 2007.

Rail

Total length in 2008 was 2,099 km of 1,524 mm gauge (1,244 km electrified). Passenger-km travelled in 2008 came to 1,047m. and freight tonne-km to 10·0bn.

There is a metro and tramway in Baku and a tramway in Sumgait.

Civil Aviation

There is an international airport at Baku. Azerbaijan Airlines, the national airline, had international flights in 2010 to Aktau, Ankara, Astrakhan, Dubai, İstanbul, London, Milan, Moscow, Paris, Rostov, St Petersburg, Tel Aviv and Urumqi. There were direct flights in 2010 with other airlines to Almaty, Ashgabat, Atyrau, Dnepropetrovsk, Donetsk, Dushanbe, Ekaterinburg, Frankfurt, İstanbul, Kabul, Khanty-Mansiysk, Kyiv, London, Mineralnye Vody, Minsk, Moscow, Nizhnevartovsk, Nizhny Novgorod, Novosibirsk, Odesa, Riga, St Petersburg, Samara, Surgut, Tashkent, Tblisi, Tehran, Tyumen, Urumqi and Vienna. In 2005 scheduled airline traffic of Azerbaijan-based carriers flew 14·8m. km, carrying 494,800 passengers.

Shipping

In 2008 merchant shipping totalled 423,000 GRT (including oil tankers 229,000 GRT). In 2005 vessels totalling 1,627,000 NRT entered ports and vessels totalling 1,578,000 NRT cleared.

Telecommunications

In 2011 there were 1,684,000 landline telephone subscriptions (180·9 per 1,000 inhabitants) and 10,120,000 mobile phone subscriptions (1,087·5 per 1,000 inhabitants). There were 274·0 internet users per 1,000 inhabitants in 2009. Fixed internet subscriptions totalled 871,000 in 2010 (94·8 per 1,000 inhabitants). In March 2012 there were 782,000 Facebook users.

SOCIAL INSTITUTIONS

Justice

The number of recorded crimes in 2009 was 22,830, including 236 murders or attempted murders (604 in 1995); there were 259 crimes per 1,000 inhabitants.

The population in penal institutions in Dec. 2006 was 19,559 (233 per 100,000 of national population).

The death penalty was abolished in 1998.

Education

In 2009–10 there were 7,048 pupils at 368 primary schools, 79,069 pupils at 853 secondary schools and 1,272,361 pupils at 3,299 joint primary and secondary schools. There were a total of 173,299 teachers in primary and secondary schools. 107,954 children were enrolled at pre-school institutions. In 2009–10 there were 117,934 students and 12,813 teaching staff at 37 state higher educational institutions and 21,260 students and 2,120 teaching staff at 16 non-state higher educational institutions. There were 43 institutes of higher education in Baku, with 110,131 students in 2009–10. The Azerbaijan Academy of Sciences, founded in 1945, has 31 research institutes. Adult literacy is 98·8%.

In 2007 public expenditure on education came to 2·9% of GNI and represented 12·6% of total government expenditure.

Health

In 2009 there were 752 state and private hospitals with 67,800 beds. There were 32,503 physicians in 2009 (377 per 100,000 population). In 2006 there were 71,265 nursing and midwifery personnel, 2,431 dentistry personnel and 1,074 pharmaceutical personnel.

Welfare

In Jan. 2004 there were 751,000 age pensioners and 576,000 other pensioners.

RELIGION

In 2003 the population was 92% Muslim (mostly Shia), the balance being mainly Russian Orthodox, Armenian Apostolic and Jewish.

CULTURE

World Heritage Sites

There are two UNESCO World Heritage sites in Azerbaijan: the Walled City of Baku with the Shirvanshah's Palace and Maiden Tower (2000), which was damaged by an earthquake in 2000, and the Gobustan rock art cultural landscape (2007).

Press

In 2008 Azerbaijan published 32 paid-for daily newspapers with a combined circulation of 120,000. The leading paid-for daily is *Yeni Müsavat*, with an average daily circulation of 25,000 in 2008.

Tourism

In 2007 there were 1,010,000 non-resident tourists; spending by tourists totalled US$317m. in 2007.

DIPLOMATIC REPRESENTATIVES

Of Azerbaijan in the United Kingdom (4 Kensington Court, London, W8 5DL)
Ambassador: Fakhraddin Gurbanov.

Of the United Kingdom in Azerbaijan (45 Khagani St., AZ-1010 Baku)
Ambassador: Peter Bateman.

Of Azerbaijan in the USA (2741 34th St., NW, Washington, D.C., 20008)
Ambassador: Elin Suleymanov.

Of the USA in Azerbaijan (83 Azadliq Prospect, AZ-1007 Baku)
Ambassador: Richard Morningstar.

Of Azerbaijan to the United Nations
Ambassador: Agshin Mehdiyev.

Of Azerbaijan to the European Union
Ambassador: Fuad Eldar oglu Isgandarov.

FURTHER READING

Azerbaijan. A Country Study. 2004

Chorbajian, Levon, *The Making of Nagorno-Karabagh: From Secession to Republic.* 2001
De Waal, Thomas, *Black Garden: Armenia and Azerbaijan Through Peace and War.* 2003
Swietochowski, T., *Russia and a Divided Azerbaijan.* 1995
Van Der Leeuw, C., *Azerbaijan.* 1999

National Statistical Office: The State Statistical Committee of the Republic of Azerbaijan, Inshaatchilar Av., Baku AZ1136.
Website: http://www.azstat.org

Nakhichevan

This territory, on the borders of Turkey and Iran, forms part of Azerbaijan although separated from it by the territory of Armenia. In 1999 its population was 99·1% Azerbaijani. It was annexed by Russia in 1828. In June 1923 it was constituted as an Autonomous Region within Azerbaijan. On 9 Feb. 1924 it was elevated to the status of Autonomous Republic. The 1996 Azerbaijani Constitution defines it as an Autonomous State within Azerbaijan.

Area, 5,500 sq. km (2,120 sq. miles); population (2009 census, provisional), 398,400. Capital, Nakhichevan (2009 provisional census population, 73,900).

Chairman of the Supreme Council: Vasif Talybov.

Prime Minister: Alovsat Bakhshiyev.

Approximately 70% of the economically active population are engaged in agriculture of which the main branches are cotton and tobacco growing. Fruit and grapes are also produced.

In 2009–10 there were 221 primary and secondary schools with 51,846 pupils, and 5,052 students in three higher educational institutions.

Nakhichevan had 43 hospitals and 2,990 hospital beds in 2009; there were 800 physicians and 2,701 paramedic staff.

Nagorno-Karabakh

Established on 7 July 1923 as an Autonomous Region within Azerbaijan, in 1989 the area was placed under a 'special form of administration' subordinate to the USSR government. In Sept. 1991 the regional Soviet and the Shaumyan district Soviet jointly declared a Nagorno-Karabakh republic, which declared itself independent with a 99·9% popular vote (only the Armenian community took part in this vote as the Azeri population had already been expelled from Nagorno-Karabakh) in Dec. 1991. The autonomous status of the region was meanwhile abolished by the Azerbaijan Supreme Soviet in Nov. 1991, and the capital renamed Khankendi. A presidential decree of Jan. 1992 placed the region under direct rule. Azeri-Armenian fighting for possession of the region culminated in its occupation by Armenia in 1993 (and the occupation of seven other Azerbaijani regions outside it), despite attempts at international mediation. Since May 1994 there has been a ceasefire. Negotiations on settlements are conducted within the OSCE Minsk Group. International pressure on Azerbaijan and Armenia to find a resolution to the conflict increased in 2005, following an OSCE fact-finding mission in the occupied provinces, but bilateral talks held in Aug. of that year proved to be inconclusive. In a referendum held on 10 Dec. 2006, 99% of votes cast were in favour of a constitution declaring a sovereign state with 1% against. Azerbaijan, the USA, the EU and the OSCE all refused to recognize the results. Following renewed violence in the region in March 2008, Armenia and Azerbaijan signed a joint agreement in Nov. 2008 aimed at reinvigorating efforts to end the dispute but talks in Nov. 2009 failed to yield significant progress. A summit of the presidents of the two countries, hosted by Russia's President Medvedev in June 2010, ended in deadlock. There were further skirmishes between Armenian and Azeri forces in the disputed region over the following months.

Area, 4,400 sq. km (1,700 sq. miles); population (Dec. 2008 est.), 139,900. Capital, Khankendi (Dec. 2008 est., 51,600). It is mainly populated by Armenians (an estimated 95% in 2001), with some Assyrians, Azerbaijanis, Greeks and Kurds.

In presidential elections held on 19 July 2012 Bako Sahakyan was re-elected with 66·7% of votes cast against 32·5% for Vitaly Balasanyan and 0·8% for Arkady Soghomonyan. With Nagorno-Karabakh being *de jure* part of Azerbaijan, the election did not receive international support. Legislative elections were held on 23 May 2010. The Free Motherland Party won 14 seats; the Democratic Artsakh Party 7; the Armenian Revolutionary Federation 6; and ind. 6.

President: Bako Sahakyan.

Prime Minister: Araik Arutyunyan.

Main industries are silk, wine, dairy farming and building materials. Cotton, grapes and winter wheat are grown.

In 2008–09, 20,813 pupils were studying in 236 general education schools with 4,525 teachers, 1,365 pupils in four specialized secondary schools and 6,900 students in seven higher educational institutions.

BAHAMAS

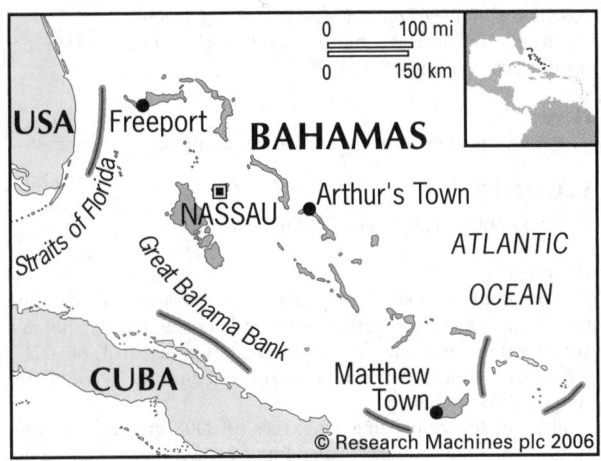

Commonwealth of The Bahamas

Capital: Nassau
Population projection, 2015: 363,000
GNI per capita, 2010: US$21,663
HDI/world rank: 0·771/53
Internet domain extension: .bs

KEY HISTORICAL EVENTS

First inhabited in the 9th century by the Lucayans, a branch of the Arawaks, The Bahamas received their name 'Baja Mar' (low sea) from Christopher Columbus who landed on San Salvador in 1492. Colonized by English puritans from Bermuda during the 17th century, The Bahamas were later plagued by notorious pirates such as Blackbeard, until they were driven out by Governor Woodes Rogers in 1718. The Bahamas played an important part in the American Civil War—blockaded by the Union navy in 1861, the islanders traded Confederate cotton with Britain and supplied military equipment to the Confederacy. During Prohibition The Bahamas prospered as a rum-smuggling base but experienced a severe economic downturn when the Prohibition law was repealed in 1933. An important Atlantic base during WWII, the tourist industry benefited greatly from Cuba's closure to western visitors in the 1950s. Internal self-government with cabinet responsibility was introduced on 7 Jan. 1964 and full independence achieved on 10 July 1973.

TERRITORY AND POPULATION

The Commonwealth of The Bahamas consists of over 700 islands and inhabited cays off the southeast coast of Florida extending for about 260,000 sq. miles. Only 22 islands are inhabited. Land area, 5,382 sq. miles (13,939 sq. km).

The areas and populations of the 19 divisions used for the most recent census in 2010 were as follows:

	Area (in sq. km)	Population
New Providence	207	246,329
Grand Bahama	1,373	51,368
Abaco	1,681	17,224
Eleuthera	484	8,202
Andros	5,957	7,490
Exuma and Cays	290	6,928
Long Island	596	3,094
Biminis	23	1,988
Harbour Island	8	1,762

	Area (in sq. km)	Population
Spanish Wells	26	1,551
Cat Island	388	1,522
San Salvador	163	940
Inagua	1,551	913
Berry Islands	31	807
Acklins	497	565
Crooked Island	241	330
Mayaguana	285	277
Rum Cay	78	99
Ragged Island	36	72

Total census population for 2010 was 351,461.

The UN gives a projected population for 2015 of 363,000.

In 2011, 84·3% of the population were urban. The capital is Nassau on New Providence Island (246,329 in 2010). Other large towns are Freeport (on Grand Bahama), West End (also on Grand Bahama) and Coopers Town (on Abaco).

English is the official language. Creole is spoken among Haitian immigrants.

SOCIAL STATISTICS

2008 estimated births, 5,600; deaths, 2,000. Rates, 2008 estimates (per 1,000 population): birth, 16·7; death, 6·0; infant mortality (per 1,000 live births), 2010, 14. Expectation of life was 70·4 years for males and 76·0 years for females in 2007. Annual population growth rate, 2000–08, 1·3%; fertility rate, 2008, 2·0 children per woman.

CLIMATE

Winters are mild and summers pleasantly warm. Most rain falls in May, June, Sept. and Oct., and thunderstorms are frequent in summer. Rainfall amounts vary over the islands from 30" (750 mm) to 60" (1,500 mm). Nassau, Jan. 71°F (21·7°C), July 81°F (27·2°C). Annual rainfall 47" (1,179 mm).

CONSTITUTION AND GOVERNMENT

The Commonwealth of The Bahamas is a free and democratic sovereign state. Executive power rests with Her Majesty the Queen, who appoints a Governor-General to represent her, advised by a Cabinet whom he appoints. There is a bicameral legislature. The *Senate* comprises 16 members all appointed by the Governor-General for five-year terms, nine on the advice of the Prime Minister, four on the advice of the Leader of the Opposition, and three after consultation with both of them. The *House of Assembly* consists of 38 members elected from single-member constituencies for a maximum term of five years.

National Anthem

'Lift up your head to the rising sun, Bahamaland'; words and tune by T. Gibson.

RECENT ELECTIONS

In parliamentary elections held on 7 May 2012 the Progressive Liberal Party (PLP) won 48·6% of votes cast and 29 out of 38 seats (18 in 2007) against the ruling Free National Movement (FNM) with 42·1% and 9 seats (23 in 2007).

CURRENT GOVERNMENT

Governor-General: Sir Arthur Foulkes; b. 1928 (sworn in 14 April 2010).

In March 2013 the cabinet was composed as follows:

Prime Minister and Minister of Finance: Perry Christie; b. 1943 (PLP; took office on 8 May 2012, having previously been prime minister from May 2002–May 2007).

Deputy Prime Minister and Minister of Works and Urban Development: Philip Davis.

Minister of Agriculture, Marine Resources and Local Government: Alfred Gray. *Education, Science and Technology:* Jerome Fitzgerald. *Environment and Housing:* Kendred Dorsett. *Financial Services:* Ryan Pinder. *Foreign Affairs and Immigration:* Frederick Mitchell. *Grand Bahama:* Michael Darville. *Health:* Dr Perry Gomez. *Investments in the Office of the Prime Minister:* Khaalis Rolle. *Labour and National Insurance:* Shane Gibson. *Legal Affairs and Attorney General:* Allyson Maynard Gibson. *National Security:* Bernard Nottage. *Social Services:* Melanie Griffin. *Tourism:* Obediah Wilchcombe. *Transport and Aviation:* Glenys Hanna Martin. *Youth, Sports and Culture:* Daniel Johnson.

Government Website: http://www.bahamas.gov.bs

CURRENT LEADERS

Perry Christie

Position
Prime Minister

Introduction
Perry Christie was elected prime minister in May 2012. He had previously served as premier from 2002–07.

Early Life
Perry Gladstone Christie was born in Nassau on 21 Aug. 1943. He was schooled in New Providence before moving to the UK, where he graduated from Birmingham University in 1969 and was called to the Bar at London's Inner Temple. He represented the Bahamas in the triple jump at the 1962 Central American and Caribbean Games.

Christie was appointed to the Bahamian senate in Nov. 1974 by Prime Minister Lynden Pindling. He became head of the national gaming board in Jan. 1977 and in the same year successfully stood in the general election as the Progressive Liberal Party (PLP) candidate for the Centreville constituency. He was subsequently named minister of health and national insurance and after the 1982 election took over the tourism portfolio. However, he left the government two years later amid allegations that several of his party colleagues had taken bribes. Reclaiming his parliamentary seat in 1987 as an independent, he rejoined the PLP in March 1990 and was appointed minister for agriculture, trade and industry.

At the general election of 1992 the PLP lost power for the first time since independence, giving way to the Free National Movement. In Jan. 1993 Christie was elected PLP deputy leader. He succeeded Pindling as leader in April 1997 and led the party to a landslide victory at the polls in May 2002.

In government the PLP was dogged by scandal, including the resignation of the immigration minister, Shane Gibson, over charges that he expedited the residency application of the actress and model Anna Nicole Smith. At parliamentary elections on 4 May 2007 the PLP was defeated by the Free National Movement but Christie remained at the head of the party. As leader of the opposition he proposed a referendum on the inauguration of a national lottery as a means of tackling mounting national debt.

He campaigned for the 2012 election on a platform of lowering the cost of electricity, helping struggling homeowners and championing national health insurance.

Career in Office
Sworn into office on 8 May 2012, Christie pledged to tackle unemployment (which had risen to 15%), address the escalating crime rate, and diversify the economy to reduce its dependency on tourism and the country's status as an offshore financial centre. He also appointed a constitutional commission to advise on possible amendments to the constitution, which has remained unchanged since the country gained independence in 1973.

DEFENCE

The Royal Bahamian Defence Force is a primarily maritime force tasked with naval patrols and protection duties in the extensive waters of the archipelago. Personnel in 2007 numbered 860. The base is at Coral Harbour on New Providence Island.

In 2008 defence expenditure totalled US$49m. (US$161 per capita), representing 0·7% of GDP.

Navy
The Navy operates 14 patrol craft and four aircraft.

ECONOMY

Services contributed 79% of GDP in 2006.

Overview
The economy is heavily dependent on tourism and offshore banking. Tourism together with related activities, notably construction, account for approximately one-third of GDP. Financial services, including offshore banking, account for roughly 20% of GDP.

Prior to the global financial crisis of 2008 growth had been robust, with the economy expanding by 5% in 2005, boosted by construction on new resorts and second homes. The global crisis saw a collapse in tourism. Domestic economic activity contracted sharply, unemployment rose and central government debt increased to nearly 49% of GDP. Output, however, began to grow in 2010 and gained strength by 2011. Short-term growth remains over-reliant on tourism and on the economic health of the USA, which accounts for 80% of visitors.

Currency
The unit of currency is the *Bahamian dollar* (BSD) of 100 *cents*. American currency is generally accepted. Inflation was 1·0% in 2010 and 2·5% in 2011. Foreign exchange reserves were US$731m. in July 2005 and total money supply was B$1,144m.

Budget
The fiscal year is 1 July–30 June.

In 2009–10 revenues were B$1,400·1m. and expenditures B$1,639·3m. Tax revenue accounted for 88·4% of revenues in 2009–10; education accounted for 17·4% of expenditures, health 16·9% and general administration 14·5%.

Performance
The Bahamas experienced a recession during the period 1988–94; this was mainly owing to the recession in the USA leading to a fall in the number of American tourists. The economy has generally been growing since (until the global economic crisis of 2008–09), and there are continuing efforts to diversify. Freeport's tax-free status was extended by 25 years in 1995, and import duties were reduced in the 1996–97 budget.

The economy contracted by 4·9% in 2009 but grew by 0·2% in 2010 and 1·6% in 2011. Total GDP in 2011 was US$7·8bn.

Banking and Finance
The Central Bank of The Bahamas was established in 1974. Its *Governor* is Wendy Craigg. The Bahamas is an important centre for offshore banking. Financial business produced 11·3% of GDP in 2004. In 2004, 259 banks and trust companies were licensed, about half being branches of foreign companies. Leading Bahamian-based banks include The Bank of Bahamas Ltd, The Commonwealth Bank Ltd and Private Investment Bank Ltd. There is also a Development Bank.

Gross external debt amounted to US$1,287m. in June 2012.

A stock exchange, the Bahamas International Securities Exchange (BISX) based in Nassau, was inaugurated in May 2000.

ENERGY AND NATURAL RESOURCES

Environment
The carbon dioxide emissions of The Bahamas from the consumption and flaring of fossil fuels were the equivalent of 17·5 tonnes per capita in 2008.

Electricity
In 2007 installed capacity was 0·5m. kW, all thermal. Output in 2007 was approximately 2·11bn. kWh; consumption per capita in 2007 was about 6,317 kWh.

Oil and Gas
The Bahamas does not have reserves of either oil or gas, but oil is refined in The Bahamas. The Bahamas Oil Refining Company (BORCO), in Grand Bahama, operates as a terminal which trans-ships, stores and blends oil.

Minerals
Aragonite is extracted from the seabed.

Agriculture
In 2007 there were some 8,000 ha. of arable land and 4,000 ha. of permanent crops. Production (in 1,000 tonnes), 2003: sugarcane, 56; fruit, 28 (notably grapefruit, lemons and limes); vegetables, 24.
    Livestock (2003): cattle, 1,000; sheep, 5,000; goats, 14,000; pigs, 6,000; chickens, 3m.

Forestry
In 2010 forests covered 0·52m. ha. or 51% of the total land area. Timber production in 2007 was 50,000 cu. metres.

Fisheries
Total catch in 2008 amounted to 9,117 tonnes, exclusively from sea fishing.

INDUSTRY
Tourism and offshore banking are the main industries. Two industrial sites, one in New Providence and the other in Grand Bahama, have been developed as part of an industrialization programme. The main products are pharmaceutical chemicals, salt and rum.

Labour
A total of 176,330 persons were in employment in April 2004 (excluding armed forces). The main areas of activity were: wholesale and retail trade, restaurants and hotels, 32%; community, social and personal services, 30%; construction, 11%; financing, insurance, real estate and business services, 11%. Unemployment was 10·2% in 2004.

INTERNATIONAL TRADE
There is a free trade zone on Grand Bahama. Although a member of CARICOM, The Bahamas is not a signatory to its trade protocol.

Imports and Exports
Imports and exports for calendar years in US$1m.:

	2006	2007	2008	2009	2010
Imports c.i.f.	2,984	3,103	3,230	2,699	2,862
Exports f.o.b.	509	670	702	585	620

    In 2010 the principal imports were (in US$1m.): mineral fuels, lubricants and related materials, 687; machinery and transport equipment, 494; food and live animals, 427. The principal exports were (in US$1m.): mineral fuels, lubricants and related materials, 160; food and live animals, 75; machinery and transport equipment, 74. In 2010 the USA was the source of 91% of imports; the main export markets were the USA (76%) and the UK (5%).

COMMUNICATIONS

Roads
There were about 2,717 km of roads in 2002 (57·4% paved). In 2007 there were around 27,100 vehicles in use.

Civil Aviation
There are international airports at Nassau and Freeport (Grand Bahama Island). The national carrier is the state-owned Bahamasair, which in 2010 flew to Fort Lauderdale, Havana, Miami, Orlando and the Turks and Caicos Islands, as well as providing services between different parts of The Bahamas. There were direct international flights in 2010 with other airlines to Atlanta, Atlantic City, Baltimore, Birmingham (USA), Boston, Buffalo, Calgary, Cayman Islands, Charlotte, Cleveland, Dallas/Fort Worth, Detroit, Fort Lauderdale, Frankfurt, Havana, Holguín, Houston, Kingston, Little Rock, London, Louisville, Miami, Montego Bay, Montreal, Myrtle Beach, New Orleans, New York, Orlando, Philadelphia, Pittsburgh, Toronto, the Turks and Caicos Islands and Washington, D.C. In 2006 scheduled airline traffic of Bahamas-based carriers flew 8m. km, carrying 1,033,000 passengers (456,000 on international flights).

Shipping
In Jan. 2009 there were 1,240 ships of 300 GT or over registered, totalling 43·93m. GT (representing 5·6% of the world total and a figure exceeded only by the fleets of Panama and Liberia). Of the 1,240 vessels registered, 439 were general cargo ships, 273 oil tankers, 225 bulk carriers, 146 passenger ships, 78 liquid gas tankers, 62 container ships and 17 chemical tankers.

Telecommunications
There were 129,300 fixed telephone lines in 2010 (377·1 per 1,000 inhabitants) and mobile phone subscribers numbered 428,400. There were 338·8 internet users per 1,000 inhabitants in 2009. Fixed internet subscriptions totalled 38,600 in 2009 (114·0 per 1,000 inhabitants). In June 2012 there were 164,000 Facebook users.

SOCIAL INSTITUTIONS

Justice
English Common Law is the basis of the Bahamian judicial system, although there is a large volume of Bahamian Statute Law. The highest tribunal in the country is the Court of Appeal. New Providence has 14 Magistrates' Courts, Grand Bahama has three and Abaco one.
    The strength of the police force (2004) was 3,352 officers.
    There were 44 murders in 2004 (a rate of 14·5 per 100,000 population). The death penalty is in force, the most recent execution being carried out in 2000. The population in penal institutions in 2010 was 1,322 (382 per 100,000 of national population).

Education
Education is compulsory between five and 16 years of age. The adult literacy rate in 2001 was 95·5% (94·6% among males and 96·3% among females). In 2004 there were 188 schools (30 independent). In 2004–05 there were 62,110 pupils with 1,365 teachers in primary schools and 26,185 pupils with 1,792 teachers in secondary education. Courses lead to The Bahamas General Certificate of Secondary Education (BGCSE). Independent schools provide education at primary, secondary and high school levels.
    The four institutions offering higher education are: the government-sponsored College of The Bahamas, established in 1974; the University of the West Indies (regional), affiliated with The Bahamas since 1960; The Bahamas Hotel Training College, sponsored by the Ministry of Education and the hotel industry; and The Bahamas Technical and Vocational Institute, established

to provide basic skills. Several schools of continuing education offer secretarial and academic courses.

Health

In 2003 there was a government general hospital (423 beds) and a psychiatric/geriatric care centre (515 beds) in Nassau, and a hospital in Freeport (88 beds). The Family Islands, comprising 33 health districts, had 21 health centres and 66 main clinics in 2003. There were two private hospitals (495 beds) in New Providence in 2003. In 2004 there were 720 physicians, 76 dentists and 1,323 nurses.

Welfare

Social Services are provided by the Department of Social Services, a government agency which grants assistance to restore, reinforce and enhance the capacity of the individual to perform life tasks, and to provide for the protection of children in The Bahamas.

The Department's divisions include: community support services, child welfare, family services, senior citizens, disability affairs, Family Island and research planning, training and community relations.

RELIGION

In 2001, 44% of the population were Protestant, 16% Roman Catholic, 10% Anglican and the remainder other religions.

CULTURE

Press
There were four paid-for dailies in 2008.

Tourism
Tourism is the most important industry, accounting for about 60% of GDP. In 2007 there were 1,527,726 non-resident overnight tourists and 2,970,659 cruise ship visitors. Tourist expenditure was US$2,198m. in 2007.

Festivals
Junkanoo is the quintessential Bahamian celebration, a parade or 'rush-out' characterized by colourful costumes, goatskin drums, cowbells, horns and a brass section. It is staged in the early hours of 26 Dec. and the early hours of 1 Jan.

DIPLOMATIC REPRESENTATIVES

Of The Bahamas in the United Kingdom (10 Chesterfield St., London, W1J 5JL)
High Commissioner: Paul H. Farquharson.

Of the United Kingdom in The Bahamas
High Commissioner: Howard Drake, OBE (resides in Kingston, Jamaica).

Of The Bahamas in the USA (2220 Massachusetts Ave., NW, Washington, D.C., 20008)
Ambassador: Cornelius A. Smith.

Of the USA in The Bahamas (Mosmar Bldg, Queen St., Nassau)
Ambassador: Vacant.
Chargé d'Affaires a.i.: John Dinkelman.

Of The Bahamas to the United Nations
Ambassador: Paulette Bethel.

Of The Bahamas to the European Union
Ambassador: Paul H. Farquharson.

FURTHER READING

Cash, P., *et al.*, *Making of Bahamian History.* 1991
Craton, M. and Saunders, G., *Islanders in the Stream: a History of the Bahamian People.* 2 vols. 1998
Storr, Virgil Henry, *Enterprising Slaves and Master Pirates: Understanding Economic Life in the Bahamas.* 2004

National Statistical Office: Department of Statistics, Clarence A. Bain Building, Thompson Blvd, PO Box N-3904, Nassau.
Website: http://statistics.bahamas.gov.bs

BAHRAIN

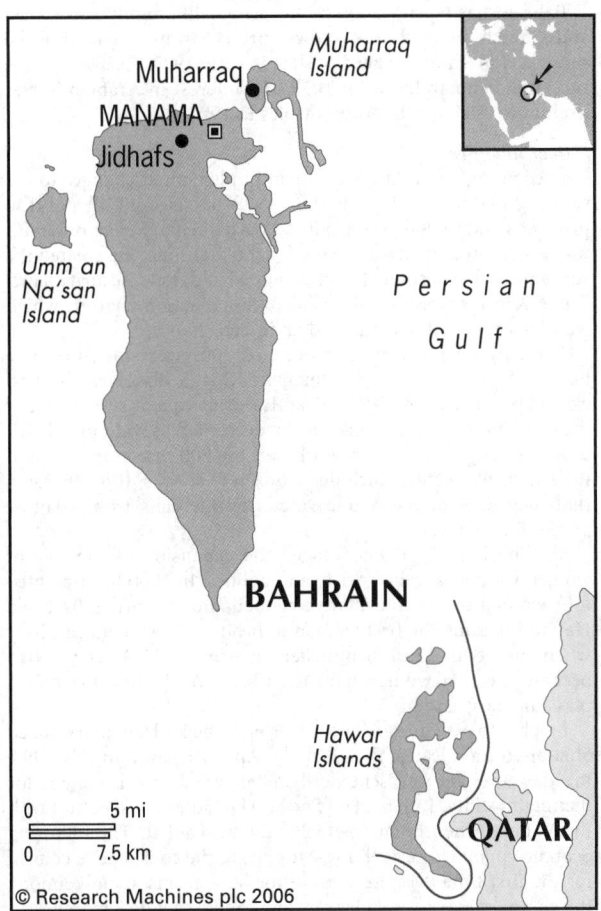

Al-Mamlaka Al-Bahrayn
(Kingdom of Bahrain)

Capital: Manama
Population projection, 2015: 1·40m.
GNI per capita, 2011: (PPP$) 28,169
HDI/world rank: 0·806/42
Internet domain extension: .bh

KEY HISTORICAL EVENTS

Bahrain was controlled by the Portuguese from 1521 until 1602. The Khalifa family gained control in 1783 and has ruled since that date. British assistance was sought to retain independence and from 1861 until 1971 Bahrain was in all but name a British protectorate. Bahrain declared its independence in 1971. Sheikh Isa bin Salman Al-Khalifa became the Amir. A constitution was ratified in June 1973 providing for a National Assembly of 30 members, popularly elected for a four-year term, together with all members of the cabinet (appointed by the Amir). However, in 1975 the National Assembly was dissolved and the Amir began ruling by decree. In 1987 the main island was joined to the Saudi mainland by a causeway. In Feb. 2002 Bahrain became a kingdom, with the Amir proclaiming himself king. In Feb. and March 2011 there were popular anti-government protests in the capital, Manama, echoing similar demonstrations in Tunisia, Egypt and Libya. A state of emergency, imposed in March 2011 at the height of the protests, was lifted seven weeks later.

TERRITORY AND POPULATION

The Kingdom of Bahrain forms an archipelago of 36 low-lying islands in the Persian Gulf, between the Qatar peninsula and the mainland of Saudi Arabia. The total area is 720 sq. km.

The island of Bahrain (578 sq. km) is connected by a 2·4 km causeway to the second largest island, Muharraq to the northeast, and by a causeway with the island of Sitra to the east. A causeway links Bahrain with Saudi Arabia. From Sitra, oil pipelines and a causeway carrying a road extend out to sea for 4·8 km to a deep-water anchorage.

In March 2001 the International Court of Justice ruled on a long-standing dispute between Bahrain and Qatar over the boundary between the two countries and ownership of certain islands. Both countries accepted the decision.

Total census population in 2001 was 650,604. Population (2008 est.) 1,106,509 (males, 677,999; females, 428,510) of which 537,719 were Bahraini and 568,790 non-Bahraini. The population density was 1,537 per sq. km in 2008. In 2011, 88·7% of the population were urban.

The UN gives a projected population for 2015 of 1·40m.

There are five governorates: Capital, Central, Muharraq, Northern, Southern. Manama, the capital and commercial centre, had a 2001 census population of 143,035. Other towns (2001 census population) are Muharraq (91,307), Rifa'a (79,550), Hamad Town (52,718), Al-Ali (47,529) and Isa Town (36,833).

Arabic is the official language. English is widely used in business.

SOCIAL STATISTICS

Statistics, 2009: births, 17,841; deaths, 2,387. Rates (per 1,000 population) in 2009: birth, 15·1; death, 2·0. Infant mortality (per 1,000 live births), 9 (2010). Life expectancy at birth, 2007, was 74·2 years for men and 77·4 years for women. Annual population growth rate, 2000–05, 2·6%; fertility rate, 2008, 2·3 children per woman. In 2006 there were 4,714 marriages and 1,130 divorces.

CLIMATE

The climate is pleasantly warm between Dec. and March but from June to Sept. the conditions are very hot and humid. The period June to Nov. is virtually rainless. Bahrain, Jan. 66°F (19°C), July 97°F (36°C). Annual rainfall 5·2" (130 mm).

CONSTITUTION AND GOVERNMENT

The ruling family is the Al-Khalifa who have been in power since 1783.

The constitution changing Bahrain from an Emirate to a Kingdom dates from 14 Feb. 2002. The new constitutional hereditary monarchy has a bicameral legislature, inaugurated on 14 Dec. 2002. National elections for a legislative body took place on 24 and 31 Oct. 2002 (the first since the National Assembly was adjourned 27 years earlier). One chamber *(Council of Representatives)* is a directly elected assembly while the second (upper) chamber, a *Shura* consultative council of experts, is appointed by the King. Both chambers have 40 members. All Bahraini citizens over the age of 21—men and women—are able to vote for the elected assembly. In the Oct. 2002 national elections women stood for office for the first time.

National Anthem

'Bahrain ona, baladolaman' ('Our Bahrain, secure as a country'); words by M. S. Ayyash, tune anonymous.

GOVERNMENT CHRONOLOGY

Heads of State since 1942.

Hakims
1942–61	Sheikh Salman bin Hamad Al-Khalifa
1961–71	Sheikh Isa bin Salman Al-Khalifa

Amirs
1971–99	Sheikh Isa bin Salman Al-Khalifa
1999–2002	Sheikh Hamad bin Isa Al-Khalifa

King
2002–	Sheikh Hamad bin Isa Al-Khalifa

RECENT ELECTIONS

Parliamentary elections, held on 23 and 30 Oct. 2010, were dominated by Shia candidates. The Shia Al-Wefaq party took 18 seats, while the Sunni Al-Asala and Al-Menbar groupings won 3 and 2 seats respectively. Independents won 17.

CURRENT GOVERNMENT

The present king (formerly Amir), HH Sheikh Hamad bin Isa Al-Khalifa, KCMG (b. 1950), succeeded on 6 March 1999 and became king on 14 Feb. 2002.

In March 2013 the cabinet was composed as follows:

Prime Minister: Sheikh Khalifa bin Salman Al-Khalifa; b. 1936. He is currently the longest-serving prime minister of any sovereign country, having been Bahrain's prime minister since it became independent in Aug. 1971.

Deputy Prime Ministers: Jawad Al-Arrayed; Sheikh Ali bin Khalifa Al-Khalifa; Sheikh Mohammed bin Mubarak Al-Khalifa; Sheikh Khalid bin Abdulla Al-Khalifa.

Minister of Cabinet Affairs: Kamal Bin Ahmed Mohammed. *Culture:* Shaikha Mai bin Mohammad Al-Khalifa. *Education:* Majid Ali Al-Nuaymi. *Energy:* Abdulhussain Mirza. *Finance:* Sheikh Ahmad bin Muhammad Al-Khalifa. *Foreign Affairs:* Sheikh Khalid bin Ahmed bin Mohammed Al-Khalifa. *Health:* Sadik bin Abdulkarim Al-Shehabi. *Housing:* Basim bin Yacub Al-Hamar. *Human Rights and Social Development:* Fatima Ahmed Al-Beloushi. *Industry and Commerce:* Dr Hassan bin Abdullah Fakhro. *Interior:* Gen. Rashed bin Abdullah bin Ahmed Al-Khalifa. *Justice and Islamic Affairs:* Sheikh Khalid bin Ali Al-Khalifa. *Labour:* Jameel Humaidan. *Municipal Affairs and Urban Planning:* Dr Juma'a bin Ahmed Al-Ka'abi. *Transport:* Kamal bin Ahmed Mohammed. *Works:* Essam bin Abdulla Khalaf.

Government Website: http://www.e.gov.bh

CURRENT LEADERS

HH Sheikh Hamad bin Isa Al-Khalifa

Position
King

Introduction
HH Sheikh Hamad bin Isa Al-Khalifa became Amir in March 1999. In Feb. 2001 his national action charter, encompassing a range of reforms, was approved by popular referendum. The state became a kingdom and Sheikh Hamad's title was changed to King. The king is the supreme authority, with members of the ruling family holding the majority of senior political and military positions. Bahrain was not immune to the popular disaffection with the political status quo that spread across much of the Arab world in 2011 but Sheikh Hamad has maintained a repressive response to pro-democracy demands.

Early Life
Sheikh Hamad bin Isa Al-Khalifa was born on 28 Jan. 1950 in Ar-Rifa', Bahrain. He was educated in Bahrain and the UK before pursuing a military career. He attended Mons Officer Cadet School in Aldershot, UK, and continued his military training at the US Army Command and Staff College in Fort Leavenworth, Kansas. In 1968 he founded the Bahrain Defence Force (BDF) and served as the minister of defence from 1971–88.

Bahrain has been headed by the Al-Khalifa family since 1783 and Sheikh Hamad was crown prince from 1964 until he succeeded his father as head of state in early 1999. He also became supreme commander of the BDF. His interest in Arabian horses led him to establish the Amiri Stables in 1977.

Career in Office
On becoming Amir, Sheikh Hamad implemented changes to the running of the country. In June 1999 he released all political prisoners and in 2002 reintroduced parliamentary elections, with women granted the right to vote for the first time. He created the supreme judicial council and scrapped old state security laws. These reforms were based on his national action charter, which won 98% approval in a referendum in Feb. 2001.

The parliamentary elections of Oct. 2002 were the first to be held since Sheikh Isa bin Salman Al-Khalifa dissolved the first elected parliament in 1975. Sheikh Isa's subsequent suspension of the constitution and ensuing rule by decree led to widespread civil unrest amongst the Shia majority but the 2002 elections ensured the House of Deputies included a dozen Shia MPs. It is estimated that over 50% of the public voted, despite calls from Islamist parties for a boycott.

Sheikh Hamad has encouraged the expansion of the role of women within Bahraini society and politics. In 2000 he appointed four women to the consultative council and in April 2004 Nada Haffadh became the first woman to head a government ministry when she became health minister. In April 2005 Alees Samoan became the first woman, and the first non-Muslim, to chair a parliamentary session.

In a bid to ease inter-religious tensions, Sheikh Hamad pardoned Shia opposition leader Sheikh Abdel Amir Al-Jamin in July 1999, the day after he was sentenced to ten years imprisonment for inciting hostility. In Jan. 2000 Sheikh Hamad established ties with the Vatican when he met with Pope John Paul II. The following Sept. he appointed several non-Muslims to the consultative council for the first time. Despite support for his reforms, mainly among the Sunni minority population, thousands attended marches in 2005 to demand a fully elected government. Parliamentary elections in Nov. and Dec. 2006 resulted in a stronger showing by the opposition Shia Al-Wefaq group, campaigning against broad social grievances such as unemployment, poor services and corruption.

Shia discontent continued, provoking frequent public protests in poorer areas outside the predominantly Sunni capital of Manama in the early months of 2008. In the run-up to the Oct. 2010 parliamentary elections Shia opposition leaders were arrested, accused of instigating violent protests to overthrow the government. However, the Shia Al-Wefaq party won 18 seats in the elections to become the single largest group in the House of Deputies. The incumbent prime minister was meanwhile reappointed and asked to form a new cabinet. In the early months of 2011 political discontent across the Arab world spread to Bahrain. Pro-democracy demonstrations prompted Sheikh Hamad to initially make conciliatory gestures to the Shia opposition before declaring martial law and calling in Saudi military support in March to forcefully suppress the protests. Sporadic unrest has nevertheless continued, prompting the government to announce a ban in Oct. 2012 on all demonstrations and gatherings.

DEFENCE

The Crown Prince is C.-in-C. of the armed forces. An agreement with the USA in Oct. 1991 gave port facilities to the US Navy and provided for mutual manoeuvres.

Military expenditure totalled US$553m. in 2008 (US$768 per capita), representing 2·8% of GDP.

Army

The Army consists of one armoured brigade, one infantry brigade, one artillery brigade, one special forces battalion and one air defence battalion. Personnel, 2007, 6,000. In addition there is a National Guard of approximately 2,000 and a paramilitary police force of 9,000.

Navy

The Naval force based at Mina Sulman numbered 700 in 2007.

Air Force

Personnel (2007), 1,500. Equipment includes 33 combat-capable aircraft and 24 attack helicopters.

ECONOMY

Finance and real estate accounted for 28·5% of GDP in 2009, crude petroleum and natural gas 23·1% and manufacturing 14·7%.

Overview

Among the first countries in the Middle East to exploit its oil, the government has sought to diversify the economy to match the prosperity of neighbours with larger oil reserves. Leading sectors include tourism and financial services, aluminium production, and oil-related industries such as refining. In Aug. 2006 Bahrain became the first Gulf state to sign a free trade agreement with the USA.

Civil unrest in early 2011 forced banks and shops to close and damaged tourism and real estate. High oil prices supported a subsequent recovery, although questions remain over Bahrain's attractiveness to investors while political tensions remain.

Currency

The unit of currency is the *Bahraini dinar* (BHD), divided into 1,000 *fils*. In Feb. 2005 foreign exchange reserves were US$1,742m. Total money supply was BD948m. in April 2005 and gold reserves were 150,000 troy oz. Inflation was 2·0% in 2010 but there was deflation of 0·4% in 2011.

In 2001 the six Gulf Arab states—Bahrain, along with Kuwait, Oman, Qatar, Saudi Arabia and the United Arab Emirates—signed an agreement to establish a single currency by 2010. In June 2009 it was agreed to postpone the implementation of the new currency, the *khaleeji*, until 2013. It has since been put back further still. Both Oman and the United Arab Emirates have now withdrawn from the scheme, in 2007 and 2009 respectively.

Budget

Budgetary central government revenue and expenditure (in BD1m.):

	2008	2009	2010
Revenue	2,677·7	1,708·0	2,175·6
Expenditure	1,567·8	1,704·5	1,882·2

Performance

Total GDP in 2010 was US$22·9bn. Real GDP growth was 3·2% in 2009, 4·7% in 2010 and 2·1% in 2011.

Banking and Finance

The Central Bank of Bahrain (*Governor*, Rasheed Mohammed Al-Maraj) has central banking powers. In 2003 Bahrain had 50 offshore banking units. Offshore banking units may not engage in local business. In Sept. 2010 there were 15 locally incorporated commercial banks, 15 foreign commercial banks, 76 wholesale banks and 27 representative offices; assets totalled US$216·4bn.

There is a stock exchange in Manama linked with those of Kuwait and Oman.

ENERGY AND NATURAL RESOURCES

Environment

Bahrain's carbon dioxide emissions from the consumption and flaring of fossil fuels in 2008 were the equivalent of 43·2 tonnes per capita, among the highest in the world.

Electricity

In 2006 installed capacity was 2·5m. kW; 10·91bn. kWh were produced in 2007. Electricity consumption per capita was 14,349 kWh in 2007.

Oil and Gas

In 1931 oil was discovered. Operations were at first conducted by the Bahrain Petroleum Co. (BAPCO) under concession. In 1975 the government assumed a 60% interest in the oilfield and related crude oil facilities of BAPCO. Oil reserves in 2005 were 125m. bbls. Production (2007) was 9·2m. tonnes. Refinery distillation output amounted to 12·9m. tonnes in 2007.

There were known natural gas reserves of 90bn. cu. metres in 2008. Production in 2008 was 13·4bn. cu. metres. Gas reserves are government-owned.

Water

Water is obtained from artesian wells and desalination plants and there is a piped supply to Manama, Muharraq, Isa Town, Rifa'a and most villages.

Minerals

Aluminium is Bahrain's oldest major industry after oil and gas; production in 2009 was 847,738 tonnes.

Agriculture

In 2007 there were some 2,000 ha. of arable land and 4,000 ha. of permanent crops. There are about 900 farms and smallholdings (average 2·5 ha.) operated by about 2,500 farmers who produce a wide variety of fruits (22,000 tonnes in 2003) including dates (17,000 tonnes). In 2003 an estimated 10,000 tonnes of vegetables were produced. The major crop is alfalfa for animal fodder.

Livestock (2003): sheep, 18,000; goats, 16,000; cattle, 13,000; camels, 1,000.

In 2003 an estimated 8,000 tonnes of lamb and mutton, 6,000 tonnes of poultry meat, 2,000 tonnes of eggs and 14,000 tonnes of fresh milk were produced.

Fisheries

The total catch in 2010 was 13,490 tonnes, exclusively from sea fishing.

INDUSTRY

The largest company by market capitalization in Bahrain in April 2012 was Ahli United Bank (US$3·8bn.).

Industry is being developed with foreign participation: aluminium smelting (and ancillary industries), shipbuilding and repair, petrochemicals, electronics assembly and light industry.

Traditional crafts include boatbuilding, weaving and pottery.

Labour

The workforce (estimate 2006) was 359,500 of which 76% were male. The unemployed rate was 3·7% in Feb. 2010.

INTERNATIONAL TRADE

Bahrain, along with Kuwait, Oman, Qatar, Saudi Arabia and the United Arab Emirates entered into a customs union in Jan. 2003.

Imports and Exports

In 2008 imports totalled US$12,530m. and exports US$18,865m. In 2007 crude petroleum accounted for 50·9% of imports; refined petroleum made up 79·1% of exports. In 2008 the main import sources were Saudi Arabia, Japan and the USA; the main export markets were India, Saudi Arabia and the UAE.

COMMUNICATIONS

Roads

A 25-km causeway links Bahrain with Saudi Arabia. In 2008 there were 3,942 km of roads, including 475 km of main roads and 563 km of secondary roads. Bahrain has one of the densest road networks in the world. In 2008 there were 310,200 passenger cars in use (404 per 1,000 inhabitants in 2007). In 2007 there were 91 fatalities in road accidents.

Civil Aviation

The national carrier is Gulf Air, now fully owned by the government of Bahrain after the other three former partners, Qatar, Abu Dhabi and Oman, withdrew in 2002, 2006 and 2007 respectively. In 2010 Gulf Air flew to about 40 international destinations. In 2001 Bahrain International Airport handled 3·44m. passengers (all on international flights) and 152,100 tonnes of freight. In 2003 scheduled airline traffic of Bahrain-based carriers flew 40m. km, carrying 1,850,000 passengers (all on international flights).

Shipping

In Jan. 2009 there were 15 ships of 300 GT or over registered, totalling 338,000 GT. The port of Mina Sulman is a free transit and industrial area.

Telecommunications

Bahrain's telecommunications industry was fully liberalized on 1 July 2004. In 2011 there were 276,500 landline telephone subscriptions (equivalent to 208·9 per 1,000 inhabitants) and 1,694,000 mobile phone subscriptions (1,279·6 per 1,000 inhabitants). There were 519·5 internet users per 1,000 inhabitants in 2008. Fixed internet subscriptions totalled 79,400 in 2009 (67·9 per 1,000 inhabitants). In March 2012 there were 346,000 Facebook users.

SOCIAL INSTITUTIONS

Justice

The new constitution which came into force in Feb. 2002 included the creation of an independent judiciary. The State Security Law and the State Security Court were both abolished in the lead-up to the change to a constitutional monarchy.

The population in penal institutions at 2010 was 1,100 (136 per 100,000 of national population). The death penalty is still in force, though rarely used. The last execution was in 2010.

Education

Adult literacy was 91% in 2008. Government schools provide free education from primary to technical college level. Schooling is in three stages: primary (six years), intermediate (three years) and secondary (three years). Secondary education may be general or specialized. In 2005–06 there were 67,528 primary school pupils, 32,359 intermediate school pupils and 29,223 secondary school pupils; there were a total of 203 schools and 10,836 teachers in 2005–06 including two religious institutes with 1,197 male students and 114 teachers.

In the private sector there were 56 schools with 34,378 pupils and 2,629 teachers in 2005–06.

There were 16 universities and similar institutions in 2005–06 with 29,678 students; and 2,392 persons attending adult education centres.

In 2006 expenditure on education came to 10·8% of total government spending.

Health

There is a free medical service for all residents. In 2003 there were 1,295 physicians, 186 dentists, 3,156 nurses and 158 pharmacists. In 2003 there were ten general hospitals (four government; six private), 21 health centres and five maternity hospitals.

Welfare

In 1976 a pensions, sickness benefits and unemployment, maternity and family allowances scheme was established. Employers contribute 7% of salaries and Bahraini employees 11%. In 2009, 27,503 persons received state welfare payments totalling BD13,904,790. A total of BD6,548,490 was paid out to pensioners, and BD2,183,460 to families.

RELIGION

Islam is the state religion. In 2001, 87% of the population were Muslim (65% Shia and 22% Sunni). There are also Christian, Jewish, Bahai, Hindu and Parsee minorities.

CULTURE

World Heritage Sites

In 2005 Qal'at Al-Bahrain archaeological site was added to the UNESCO World Heritage List (reinscribed in 2008). The site was an area of human occupation from about 2300 BC to the 16th century. It is now a site of major excavation. Pearling was added to the list in 2012. The site, consisting of urban properties in Muharraq City, a fort, the seashore and oyster beds, is an exceptional illustration of the cultural tradition of pearling, which brought prosperity to the Persian Gulf between the 2nd and early 20th centuries.

Press

There were eight daily newspapers in 2008 with a combined average daily circulation of 155,000.

Tourism

In 2007 there were 7,834,000 foreign visitors (up from 7,289,000 in 2006 and 6,313,000 in 2005).

DIPLOMATIC REPRESENTATIVES

Of Bahrain in the United Kingdom (30 Belgrave Square, London, SW1X 8QB)
Ambassador: Alice Thomas Samaan.

Of the United Kingdom in Bahrain (21 Government Ave., Manama 306, PO Box 114, Bahrain)
Ambassador: Iain Lindsay.

Of Bahrain in the USA (3502 International Dr., NW, Washington, D.C., 20008)
Ambassador: Houda Nonoo.

Of the USA in Bahrain (Building No. 979, Rd No. 3119, Block 331, Zinj District, Manama)
Ambassador: Thomas Krajeski.

Of Bahrain to the United Nations
Ambassador: Jamal Fares Al-Ruwae.

Of Bahrain to the European Union
Ambassador: Ahmed Mohamed Yousif Aldoseri.

FURTHER READING

Bahrain Monetary Authority. *Quarterly Statistical Bulletin.*
Central Statistics Organization. *Statistical Abstract.* Annual

Al-Khalifa, A. and Rice, M. (eds.) *Bahrain through the Ages.* 1993
Al-Khalifa, H. bin I., *First Light: Modern Bahrain and its Heritage.* 1995
Moore, Philip, *Bahrain: A New Era.* 2001

National Statistical Office: Central Statistics Organization, Council of Ministers, Manama.
Website: http://www.cio.gov.bh

BANGLADESH

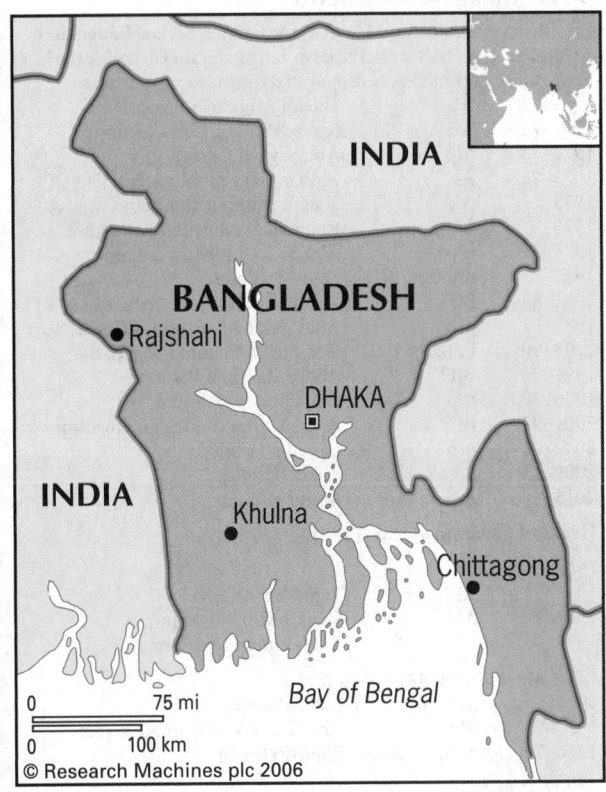

INDIA

BANGLADESH
• Rajshahi

DHAKA ▣

INDIA

Khulna

Chittagong

Bay of Bengal

0   75 mi

0   100 km

© Research Machines plc 2006

Gana Prajatantri Bangladesh
(People's Republic of Bangladesh)

Capital: Dhaka
Population projection, 2015: 158·32m.
GNI per capita, 2011: (PPP$) 1,529
HDI/world rank: 0·500/146
Internet domain extension: .bd

KEY HISTORICAL EVENTS

India's Maurya Empire established Buddhism in Bengal (*Bangla*) in the 3rd century BC. The Buddhist Pala Dynasty ruled Bengal and Bihar independently from AD 750, exporting Buddhism to Tibet. Hinduism regained dominance under the Sena Dynasty in the 11th century until the Muslim invasions in 1203–04. Rule from Delhi was broken in the 14th century by local Bengali kings. The Afghan adventurer Sher Shah conquered Bengal in 1539 and defeated the Mughal Emperor Humayun, creating an extensive administrative empire in North India. However, Mughal power was re-established by Akbar in 1576.

The Portuguese arrived in the 15th century, drawn to the rich Bengali cotton trade. They were followed by the Dutch and the British, whose East India Company was centred at Calcutta. The Nawab of Bengal was defeated by Robert Clive's army at the Battle of Plassey in 1757. British rule was maintained through the mainly Hindu *zamindar* land-owners and the Company was replaced by the Crown in 1858.

The partition of Bengal in 1905 was an attempt to undermine the nationalist influence of the *bhadralok*, the Hindu middle-classes, by forming a Muslim-dominated eastern province. Religious violence increased and political interests were represented by newly formed parties, including the All-India Muslim League. The partition was reversed in 1912. Tensions between the Muslim and Hindu communities escalated in the 1930s. The League suffered electoral defeat in 1936 but calls for a Muslim state were strengthened by the Pakistan Resolution of 1940. An agreement between Hindu and Muslim leaders to create an independent, secular Bengal was resisted by Mahatma Gandhi. Further violence, such as the Great Calcutta Killing in 1946, put pressure on the administration and India was hastily partitioned—East Bengal was united with the northwestern Muslim provinces as Pakistan. East Pakistan (as East Bengal became under the 1956 constitution) received 0·7m. Muslims, mainly from Bihar and West Bengal, while over 2·5m. Hindus left for India.

Relations with West Pakistan were strained from the outset. Bengali demands for recognition of their language and resentment of preferential investment in their western partner led to the formation of the Awami League in 1949 to represent Bengali interests. Led by Sheikh Mujibur Rahman (Mujib), the Awami League triumphed as part of the 'United Front' in elections in 1954 but its government was dismissed by Governor-General Ghulam Mohammad after two months. A military government was installed from 1958–62 and in 1966 Mujib was arrested. Civilian government was again suspended in 1969 and in elections in 1970–71 the Awami League won all East Pakistani seats. While talks to form a government foundered, President Yahya Khan sent troops to the East and suspended the assembly, provoking a civil disobedience campaign. On 25 March 1971 the army began a crackdown. Members of the Awami League were arrested or fled to Calcutta, where they declared a provisional Bengali government. 10m. Bengalis fled to India to escape the bloody repression. India, with Soviet support, invaded on 3 Dec. 1971, forcing the surrender of the Pakistani army on 16 Dec. Mujib was released to become prime minister of independent Bangladesh.

Bangladesh suffered famine in 1974 and disorder led to Mujib assuming the presidency with dictatorial powers. Assassinated in Aug. 1975, a coup brought to power Maj.-Gen. Ziaur Rahman, who turned against the former ally, India. Rahman was assassinated in 1981. Hussain Mohammad Ershad became martial law administrator in 1982 and president in 1983. Ershad's National Party triumphed in parliamentary elections in May 1986 but presidential elections in Oct. were boycotted by opposition parties. A campaign of demonstrations and national strikes forced Ershad's resignation in 1990. Rahman's widow, Khaleda, became prime minister after her Bangladesh Nationalist Party won nearly half the seats. The Bangladesh Awami League's victory in 1996 brought Mujib's daughter, Sheikh Hasina Wajed, to the premiership but Khaleda Zia returned to power in 2001.

A state of emergency was declared in Jan. 2007 and elections were postponed following several weeks of violence that claimed more than 40 lives. The Bangladesh Awami League had previously announced that it would boycott the elections, maintaining that they were not free and fair. Sheikh Hasina Wajed became prime minister for a second time when the Bangladesh Awami League won elections held in Dec. 2008, at which time the state of emergency was lifted.

TERRITORY AND POPULATION

Bangladesh is bounded in the west and north by India, east by India and Myanmar and south by the Bay of Bengal. The area is 147,570 sq. km (56,977 sq. miles). In 1992 India granted a 999-year lease of the Tin Bigha corridor linking Bangladesh with its enclaves of Angarpota and Dahagram. The most recent census took place in March 2011; population, 144,043,697 (72,109,796 males), giving a density of 976 persons per sq. km.

The UN gives a projected population for 2015 of 158·32m.

In 2011, 28·6% of the population lived in urban areas. The country is administratively divided into seven divisions, subdivided into 21 *anchal* and 64 *zila*. Area (in sq. km) and population (in 1,000) in 2011 of the seven divisions:

	Area	Population
Barisal division	13,297	8,326
Chittagong division	33,771	28,423
Dhaka division	31,120	47,424
Khulna division	22,272	15,688
Rajshahi division	18,197	18,485
Rangpur division	16,317	15,788
Sylhet division	12,596	9,910

The populations of the chief cities (2011 census) were as follows:

Dhaka[1]	7,033,075	Tongi	476,350
Chittagong	2,592,439	Rajshahi	449,756
Khulna	663,342	Bogra	350,397
Narayanganj	543,090	Barisal	328,278
Sylhet	479,837	Comilla	326,386

[1]Metropolitan area, 11,086,309.

The official language is Bengali. English is also in use for official, legal and commercial purposes.

SOCIAL STATISTICS

2008 estimated births, 3,429,000; deaths, 1,054,000. In 2008 the birth rate was an estimated 21·4 per 1,000 population; death rate, 6·6; infant mortality, 2010, 38 per 1,000 live births (down from 99 per 1,000 in 1990). Life expectancy at birth, 2007, 66·7 years for females and 64·7 years for males. Annual population growth rate, 2000–08, 1·6%; fertility rate, 2008, 2·3 births per woman (down from 4·4 in 1990). Bangladesh has made some of the best progress in recent years in reducing child mortality. The number of deaths per 1,000 live births among children under five was reduced from 143 in 1990 to 48 in 2010.

CLIMATE

A tropical monsoon climate with heat, extreme humidity and heavy rainfall in the monsoon season, from June to Oct. The short winter season (Nov.–Feb.) is mild and dry. Rainfall varies between 50" (1,250 mm) in the west to 100" (2,500 mm) in the southeast and up to 200" (5,000 mm) in the northeast. Dhaka, Jan. 66°F (19°C), July 84°F (28·9°C). Annual rainfall 81" (2,025 mm). Chittagong, Jan. 66°F (19°C), July 81°F (27·2°C). Annual rainfall 108" (2,831 mm).

CONSTITUTION AND GOVERNMENT

Bangladesh is a unitary republic. The Constitution came into force on 16 Dec. 1972 and provides for a parliamentary democracy. The head of state is the *President*, elected by parliament every five years, who appoints a *Vice-President*. A referendum of Sept. 1991 was in favour of abandoning the executive presidential system and opted for a parliamentary system. Turnout was low. An amendment to the constitution in 1996 allowed for a caretaker government, which the president may install to supervise elections should the parliament be dissolved. There is a *Council of Ministers* to assist and advise the President. The President appoints the government ministers.

Following a constitutional amendment made in June 2011 parliament has one chamber of 350 members, 300 directly elected every five years by citizens over 18 and 50 reserved for women, elected by the 300 MPs based on proportional representation in parliament.

National Anthem

'Amar Sonar Bangla, ami tomay bhalobashi' ('My Bengal of gold, I love you'); words and tune by Rabindranath Tagore.

GOVERNMENT CHRONOLOGY

Presidents since 1971. (AL = Awami League; BAL = Bangladesh Awami League; BJD = Bangladesh Jatiyatabadi Dal/Bangladesh Nationalist Party; JD = National Party; n/p = non-partisan)

1971–72	AL	Sheikh Mujibur Rahman
1971–72	AL	Sayeed Nazrul Islam (acting)
1972–73	n/p	Abu Sayeed Chowdhury
1973–75	AL	Mohammad Mohammadullah
1975	AL	Sheikh Mujibur Rahman
1975	AL	Khandakar Mushtaq Ahmed
1975–77	n/p	Abu Sadat Mohammad Sayem
1977–81	military/BJD	Ziaur Rahman
1981–82	BJD	Abdus Sattar
1982–83	n/p	Abul Fazal Ahsanuddin Chowdhury
1983–90	military/JD	Hossain Mohammad Ershad
1991–96	BJD	Abdur Rahman Biswas
1996–2001	n/p	Shahabuddin Ahmed
2001–02	BJD	A. Q. M. Badruddoza Chowdhury
2002–09	n/p	Iajuddin Ahmed
2009–13	BAL	Zillur Rahman
2013–	AL	Abdul Hamid

Heads of government since 1971.

Prime Ministers

1971–72	AL	Tajuddin Ahmed
1972–75	AL	Sheikh Mujibur Rahman
1975		Mohammad Mansoor Ali

Chief Martial Law Administrators

1975	military	Ziaur Rahman
1975–76	n/p	Abu Sadat Mohammad Sayem
1976–79	military/BJD	Ziaur Rahman

Prime Ministers

1979–82	BJD	Shah Azizur Rahman
1984–86	JD	Ataur Rahman Khan
1986–88	JD	Mizanur Rahman Chowdhury
1988–89	JD	Moudud Ahmed
1989–90	JD	Kazi Zafar Ahmed
1991–96	BJD	Khaleda Zia
1996	n/p	Mohammad Habibur Rahman
1996–2001	BAL	Sheikh Hasina Wajed
2001	n/p	Latifur Rahman
2001–06	BJD	Khaleda Zia
2006–07	n/p	Iajuddin Ahmed
2007–09	n/p	Fakhruddin Ahmed
2009–	BAL	Sheikh Hasina Wajed

RECENT ELECTIONS

Abdul Hamid was elected president unopposed on 22 April 2013 after incumbent Zillur Rahman died on 20 March 2013.

In parliamentary elections of 29 Dec. 2008 the Bangladesh Awami League (BAL) and its coalition partners gained 57·2% of votes cast. The BAL itself gained 230 seats, with allies the National Party (JD) gaining 27, Jatiyo Samajtantrik Dal (JSD) 3, the Workers Party of Bangladesh (WPB) 2 and the Liberal Democratic Party (LDP) 1. The Bangladesh Jatiyatabadi Dal/Bangladesh Nationalist Party (BJD) gained 30 seats (33·2%), the Jamaat-e-Islami Bangladesh (JIB) 2 and the Bangladesh Jatiya Party (BJP) 1 with remaining seats going to independents. Turnout was an estimated 87%.

CURRENT GOVERNMENT

President: Abdul Hamid; b. 1944 (since 14 March 2013—acting until 24 April 2013).

In March 2013 the government comprised:

Prime Minister and Minister of Public Administration, Defence, and Power, Energy and Mineral Resources: Sheikh Hasina Wajed; b. 1947 (BAL; since 6 Jan. 2009, having previously held office from June 1996–July 2001).

Minister for Agriculture: Matia Chowdhury. *Civil Aviation and Tourism:* Lieut.-Col. (retd) Faruq Khan. *Commerce:* G. M. Quader. *Communications:* Obaidul Quader. *Cultural Affairs:* Abul Kalam Azad. *Disaster Management and Relief:* Abul Hasan Mahmud Ali. *Education:* Nurul Islam Naheed. *Environment and Forests:* Hasan Mahmud. *Expatriates' Welfare and Overseas Employment:* Khandaker Mosharraf Hossain. *Finance:* Abul Maal Abdul Muhith. *Fisheries and Livestock:* Abdul Latif Biswas. *Food:* Dr Abdur Razzak. *Foreign Affairs:* Dr Dipu Moni. *Health and Family Welfare:* Dr A. F. M. Ruhul Huq. *Home Affairs:* Mahiuddin Khan Alamgir. *Industries:* Dilip Barua. *Information:* Hassanul Haque Inu. *Information and Communication Technology:* Mostafa Faruque Mohammad. *Labour and Employment:* Rajiuddin Ahmed Raju. *Land:* Rezaul Karim Hira. *Law, Justice and Parliamentary Affairs:* Shafique Ahmed. *Local Government, Rural Development and Co-operatives:* Syed Ashraful Islam. *Planning:* A. K. Khandker. *Posts and Telecommunications:* Sahara Khatun. *Primary and Mass Education:* Dr Afsarul Amin. *Railways:* Mujubul Haque. *Shipping:* Shahjahan Khan. *Social Welfare:* Enamul Huq Mostafa Shaheed. *Textiles and Jute:* Abdul Latif Siddique. *Water Resources:* Ramesh Chandra Sen.

Government Website: http://www.bangladesh.gov.bd

CURRENT LEADERS

Sheikh Hasina Wajed

Position
Prime Minister

Introduction
Sheikh Hasina, leader of the Bangladesh Awami League (BAL), was elected prime minister in Dec. 2008 after two years of military government. It is her second term as premier, having previously served from 1996–2001.

Early Life
Sheikh Hasina was born on 28 Sept. 1947 in Tungipara, in the Gopalganj district of Bangladesh. She is the oldest of five children of Bangabandhu Sheikh Mujibar Rahman, former leader of the AL (which later became the BAL) and the nation's first head of state following the 1971 liberation war with Pakistan.

Politically active during her years at Eden College, Dhaka in the 1960s, Hasina was vice-president of the student union at the Government Intermediate College from 1966–67. She was a member of the AL's student Chhatra League while at Dhaka University, from where she graduated in 1973. In 1981 she was elected head of the BAL despite being in self-imposed exile, six years after the assassinations of her father, mother and three of her siblings.

Her chief political rival was Bangladesh Jatiyatabadi Dal/ Bangladesh Nationalist Party (BJD) leader Khaleda Zia, widow of the assassinated military president Ziaur Rahman, who had taken power following the death of Hasina's father. Throughout the 1980s and early 1990s Hasina worked to remove the military government, including a BAL boycott of Zia's government in 1994. In 1996 a civilian caretaker government was established ahead of elections, which Hasina won with a landslide to become prime minister. As premier, she secured a 30-year Ganges Water Sharing Treaty with India and signed the Chittagong Hill Tract Peace Accords with rebel tribes in the southeast of the country. Nonetheless, Bangladesh descended into chaos and in 2001 Transparency International ranked it the most corrupt country in the world (although the situation has since improved and in 2010 it was ranked 134th out of 178 countries). Also in 2001, the BJD won a landslide election victory.

Political violence involving supporters of the BJD and BAL escalated from 2004–07. In 2007 a military-backed caretaker government seized control after failed elections and implemented emergency rule. Hasina was arrested on charges of corruption, extortion and murder. She was briefly jailed before being released to seek medical care in the USA. Despite attempts to prevent Hasina from returning, she arrived in Bangladesh in time to lead the BAL to election victory on 29 Dec. 2008.

Career in Office
Hasina's primary challenge was to re-establish national stability. Internationally, relations with India have been consolidated by trade and co-operation pacts and, more recently, by an agreement to establish transit routes across Bangladesh to India's remote northeastern states. However, in domestic affairs the Nov. 2010 eviction, backed by a High Court order, of Khaleda Zia from her home in Dhaka—indicative of the personal hostility between Hasina and her principal political rival—sparked strikes and protests by BJD supporters. Opposition boycotts of parliament have since been commonplace and Hasina has been accused of increasingly autocratic rule. In Jan. 2012 the army claimed to have uncovered a coup plot by officers hostile to the government, and stability has been further undermined by street demonstrations and communal violence.

The economy, meanwhile, has been adversely affected by a fall in remittances from migrant Bangladeshi workers in the Middle East in the wake of political upheaval across much of the Arab world.

DEFENCE

The supreme command of defence services is vested in the president. Defence expenditure in 2008 totalled US$1,195m. (US$8 per capita), representing 1·5% of GDP.

As at 31 Jan. 2013 Bangladesh had 8,781 personnel, including 6,927 troops, serving in UN peacekeeping operations (the largest contingent of any country).

Army
Strength (2007) 120,000. There is also a 5,000-strong specialized police unit (the Armed Police), 20,000 security guards (Ansars) and the Bangladesh Rifles (border guard) numbering 38,000.

Navy
Naval bases are at Chittagong, Dhaka, Kaptai, Khulna and Mongla. The fleet comprises four frigates, nine missile craft, four torpedo craft and 25 patrol craft. Personnel in 2007 were estimated at 16,000.

Air Force
Personnel strength (2007) 14,000. There were 76 combat-capable aircraft (mainly Chinese F-7s and A-5s) and 30 helicopters in 2007.

ECONOMY

In 2006 agriculture accounted for 19·6% of GDP, industry 27·9% and services 52·5%.

Overview
The economy has achieved impressive and steady growth in recent years, remaining resilient throughout the global financial crisis. Growth declined only moderately to 5·9% in 2009 and remained broad-based, reflecting the economy's trade openness and limited financial exposure. The export-generating textiles and garments sector fared particularly well.

Although the service sector is the dominant component of GDP, 48·4% of the population is employed in agriculture, hunting and fishing. Despite political upheaval, international reserves have strengthened as a result of strong remittances and exports, and weak imports. The balance of payments current account recorded a surplus of near 3% of GDP in 2009, up from less than 1% the previous year.

Progress on structural reforms has been mixed. Tax reforms have gained momentum but power and gas shortages continue to hold back growth while the development of Public Private Partnerships (PPP) has been slow. Low skill levels, poor governance and a weak infrastructure remain obstacles to sustained growth and poverty reduction. FDI is stagnant at US$1bn. per year but could rise if the government awards contracts for gas exploration.

Currency
The unit of currency is the *taka* (BDT) of 100 *poisha*, which was floated in 1976. Foreign exchange reserves in July 2005 were US$2,780m., gold reserves were 113,000 troy oz and total money supply was Tk.384,290m. Inflation was 8·1% in 2010 and 10·7% in 2011.

Budget
The fiscal year ends on 30 June. Budget, 2009–10: revenue, Tk.794·8bn.; expenditure, Tk.1,105·2bn.
VAT is 15%.

Performance
Real GDP growth was 5·9% in 2009, 6·4% in 2010 and 6·5% in 2011. Total GDP was US$111·9bn. in 2011.

Banking and Finance
Bangladesh Bank is the central bank (*Governor*, Dr Atiur Rahman). There were 47 commercial banks in 2010 of which four were state-owned. In 2008 the Bangladesh Bank had deposits of Tk.211,805m. Sonali Bank Limited is the largest of the nationalized commercial banks, with deposits of Tk.406,152m. at 31 Dec. 2009.

In 2010 total foreign debt was US$24,963m., representing 22·8% of GNI.

There are stock exchanges in Dhaka and Chittagong.

ENERGY AND NATURAL RESOURCES
Environment
Bangladesh's carbon dioxide emissions from the consumption and flaring of fossil fuels in 2008 were the equivalent of 0·3 tonnes per capita.

Electricity
Installed capacity was an estimated 5·2m. kW in 2006–07. Electricity generated, 2006–07, 24·38bn. kWh; consumption per capita in 2006–07 was 169 kWh.

Oil and Gas
In 2008 Bangladesh had proven natural gas reserves of 370bn. cu. metres in about 20 mainly onshore fields. Total natural gas production in 2008 amounted to 17·3bn. cu. metres.

Minerals
The principal minerals are lignite, limestone, china clay and glass sand. There are reserves of good-quality coal of 300m. tonnes. Production, 2001–02: limestone, 32,000 tonnes; kaolin, 8,100 tonnes.

Agriculture
In 2007 the agricultural population was approximately 76·11m., of whom 35·54m. were economically active. There were 7·96m. ha. of arable land in 2005 and some 0·5m. ha. of permanent crops. 5·60m. ha. were irrigated in 2005. Bangladesh is a major producer of jute: production, 2003, 801,000 tonnes. Rice is the most important food crop; production in 2003 (in 1m. metric tonnes), 38·06. Other major crops (1m. tonnes): sugarcane, 6·84; potatoes, 3·69; wheat, 1·55; bananas, 0·65. Livestock in 2003: goats, 34·50m.; cattle, 24·50m.; sheep, 1·24m.; buffalo, 850,000; chickens, 140m. Livestock products in 2003 (tonnes): beef and veal, 180,000; goat meat, 130,000; poultry meat, 115,000; goat's milk, 1·31m.; cow's milk, 797,000; sheep's milk, 25,000; buffalo's milk, 23,000; hen's

eggs, 134,000. Bangladesh is the second largest producer of goat's milk, after India.

Forestry
In 2010 the area under forests was 1·44m. ha., or 11% of the total land area. Timber production in 2007 was 27·79m. cu. metres.

Fisheries
Bangladesh is a major producer of fish and fish products. The total catch in 2010 amounted to 1,726,586 tonnes, of which 1,119,094 tonnes came from inland waters. Only China and India have larger annual catches of freshwater fish. Aquaculture production totalled 1,308,515 tonnes in 2010 (the fifth highest in the world).

INDUSTRY
Manufacturing contributes around 11% of GDP. The principal industries are jute and cotton textiles, tea, paper, newsprint, cement, chemical fertilizers and light engineering. Production, in 1,000 tonnes: cement (2007–08 estimate), 5,100; nitrogenous fertilizer (2001), 1,875; jute goods (2001–02), 536; sugar (2002), 229. Output of other products: cotton woven fabrics (2000–01), 63m. sq. metres; cigarettes (2002–03), 22·5bn. units; television sets (2001), 133,000 units; bicycles (2000–01), 13,000 units.

Labour
In 2000 the economically active workforce totalled 51,764,000 over the age of 15 years (32,369,000 males). The main areas of activity (in 1,000) were as follows: agriculture, hunting, forestry and fishing, 32,171; wholesale and retail trade, restaurants and hotels, 6,275; manufacturing, 3,783; community, social and personal services, 2,969; transport, storage and communication, 2,509; construction, 1,099. On average, wage rates are among the lowest of developing countries. In 1999–2000, 3·3% of the workforce aged 15 or over were unemployed.

INTERNATIONAL TRADE
Imports and Exports
In 2006 imports (c.i.f.) were valued at US$15,688·5m. (US$12,630·5m. in 2005) and exports (f.o.b.) at US$11,696·7m. (US$9,331·6m. in 2005). In 2006, 16·4% of imports came from China, 12·0% from India, 9·3% from Kuwait and 5·7% from Japan. 26·7% of exports in 2006 went to the USA, 15·0% to Germany, 9·0% to the UK and 6·6% to China. The main imports are machinery and transport equipment, manufactured goods, mineral fuels and lubricants, and chemicals and related products; the main exports are miscellaneous manufactured articles, manufactured goods, food and live animals, and inedible crude materials (except fuels). Since the early 1980s the garment industry has developed from virtually nothing to earn some 70% of the country's hard currency. Exports of garments and clothing accessories in 2006 earned US$8·3bn.

COMMUNICATIONS
Roads
In 2003 the total road network covered 239,226 km, including 22,378 km of national roads and 81,670 km of secondary roads. In 2007 there were 158,100 passenger cars, 168,600 vans and lorries, 31,600 buses and coaches, and 653,500 motorcycles and mopeds. There were 3,160 fatalities as a result of road accidents in 2006.

Rail
In 2005 there were 2,855 km of railways, comprising 660 km of 1,676 mm gauge, 1,830 km of metre gauge and 365 km of dual gauge. Passenger-km travelled in 2008 came to 402·4m. and freight tonne-km to 952m.

Civil Aviation
There are international airports at Dhaka (Zia) and Chittagong, and eight domestic airports. Biman Bangladesh Airlines was

state-owned until July 2007 when it became a public limited company. In addition to domestic routes, in 2010 it operated international services to Abu Dhabi, Bangkok, Delhi, Dubai, Hong Kong, Jeddah, Karachi, Kathmandu, Kolkata, Kuala Lumpur, Kuwait, London, Muscat, Riyadh, Rome and Singapore. There were direct international flights in 2010 with other airlines to Abu Dhabi, Bahrain, Bangkok, Beijing, Dammam, Delhi, Doha, Dubai, Guangzhou, Hong Kong, Jeddah, Karachi, Kathmandu, Khartoum, Kolkata, Kuala Lumpur, Kunming, Kuwait, London, Madinah, Mumbai, Muscat, Paro, Riyadh, Sharjah, Singapore and Tripoli. In 2001 Dhaka's Zia International Airport handled 2,863,575 passengers (2,322,743 on international flights) and 106,291 tonnes of freight. In 2003 scheduled airline traffic of Bangladesh-based carriers flew 29m. km, carrying 1,579,000 passengers (1,205,000 on international flights).

Shipping
In Jan. 2009 there were 197 ships of 300 GT or over registered, totalling 407,000 GT. The main port is Chittagong, which handled 27,026,000 tonnes of cargo in 2006 (3,090,000 tonnes loaded and 23,936,000 tonnes unloaded). There is also a seaport at Mongla. There are 8,000 km of navigable inland waterways.

Telecommunications
There were 1,522,900 fixed telephone lines in 2009 (10·4 per 1,000 inhabitants). Mobile phone subscribers numbered 68·7m. in 2010. There were 37·0 internet users per 1,000 inhabitants in 2010. Fixed internet subscriptions totalled 150,000 in 2006 (1·1 per 1,000 inhabitants). In March 2012 there were 2·5m. Facebook users.

SOCIAL INSTITUTIONS

Justice
The Supreme Court comprises an Appellate and a High Court Division, the latter having control over all subordinate courts. Judges are appointed by the President and retire at 65. There are benches at Comilla, Rangpur, Jessore, Barisal, Chittagong and Sylhet, and courts at District level.

The population in penal institutions in Sept. 2008 was 83,000 (51 per 100,000 of national population). The death penalty is still in force. Amnesty International reported that there was one execution in 2012 (at least five in 2011).

Education
In 2007 there were 16·3m. pupils and 364,494 teaching staff at primary schools; 10·4m. pupils and 413,746 teaching staff at secondary schools; 1·1m. students and 60,915 academic staff in tertiary education. In 2006 there were 78 universities of which 27 were public, 42 medical colleges (15 public), 171 polytechnic institutes (42 public) and 3,197 general colleges.

Adult literacy was an estimated 55·9% in 2009 (60·7% among males and 51·0% among females).

In 2007 public expenditure on education came to 2·4% of GNI and 15·8% of total government spending.

Health
In 2008 there were 2,860 hospitals and health complexes with 74,400 beds. There were 49,994 registered physicians in 2008, and 2,344 dentistry personnel and 39,471 nursing and midwifery personnel in 2005.

RELIGION
Islam is the state religion. In 2001 the population was 87% Muslim and 12% Hindu.

CULTURE

World Heritage Sites
There are three UNESCO sites in Bangladesh: the Historic Mosque City of Bagerhat (inscribed in 1985); the Ruins of the Buddhist Vihara at Paharpur (1985); the Sundarbans (1997), 140,000 ha. of mangrove forest.

Press
In 2008 there were 430 paid-for daily newspapers with a combined circulation of 1·5m.

Tourism
In 2010 there were 303,000 non-resident tourists, spending US$81m.

DIPLOMATIC REPRESENTATIVES
Of Bangladesh in the United Kingdom (28 Queen's Gate, London, SW7 5JA)
High Commissioner: Mohammad Sayeedur Rahman Khan.

Of the United Kingdom in Bangladesh (United Nations Rd, Baridhara, PO Box 6079, Dhaka 1212)
High Commissioner: Robert Winnington Gibson, CMG.

Of Bangladesh in the USA (3510 International Drive, NW, Washington, D.C., 20007)
Ambassador: Akramul Qader.

Of the USA in Bangladesh (Madani Ave., Baridhara, Dhaka 1212)
Ambassador: Dan W. Mozena.

Of Bangladesh to the United Nations
Ambassador: Dr Abdul Momen.

Of Bangladesh to the European Union
Ambassador: Ismat Jahan.

FURTHER READING
Bangladesh Bureau of Statistics. *Statistical Yearbook of Bangladesh.— Statistical Pocket Book of Bangladesh.*

Bakshi, S. R., *Bangladesh: Government and Politics.* 2002
Baxter, Craig, *Historical Dictionary of Bangladesh.* 2003
Karlekar, Hiranmay, *Bangladesh: The Next Afghanistan?* 2006
Lewis, David, *Bangladesh: Politics, Economy and Civil Society.* 2011
Muhith, A. M. A., *Issues of Governance in Bangladesh.* 2000
Rashid, H. U., *Foreign Relations of Bangladesh.* 2001
Riaz, Ali, *God Willing: The Politics of Islamism in Bangladesh.* 2004.— *Unfolding State: The Transformation of Bangladesh.* 2005
Tajuddin, M., *Foreign Policy of Bangladesh: Liberation War to Sheikh Hasina.* 2001
Umar, Badruddin, *Emergence of Bangladesh.* 2004
Van Schendel, Willem, *A History of Bangladesh.* 2009

National Statistical Office: Bangladesh Bureau of Statistics, Ministry of Planning, E-27/A, Agargaon, Sher-e-banglanagar, Dhaka 1207.
Website: http://www.bbs.gov.bd

BARBADOS

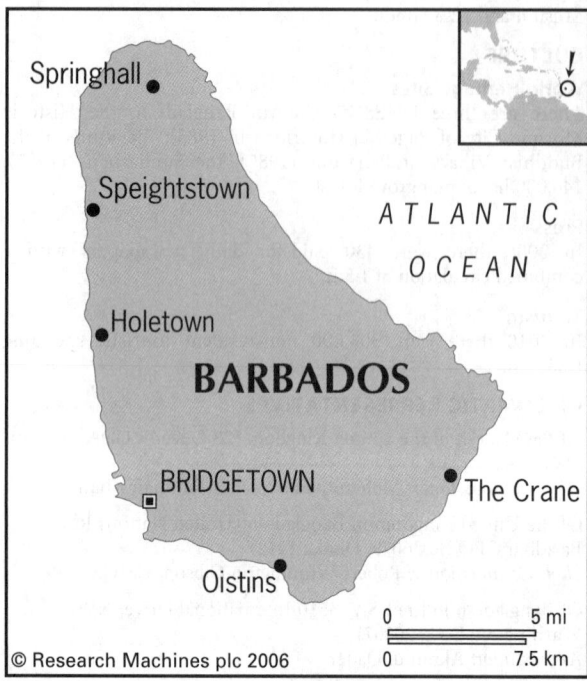

Capital: Bridgetown
Population projection, 2015: 276,000
GNI per capita, 2011: (PPP$) 17,966
HDI/world rank: 0·793/47
Internet domain extension: .bb

KEY HISTORICAL EVENTS

Archaeological evidence suggests that Barbados was inhabited by Barrancoid Indians from at least 1000 BC, and by Arawak people for about 400 years from around AD 1000. Portuguese mariners who landed on the island in 1536 reported that it was uninhabited. An Englishman, William Courteen, established Jamestown in 1627. Sugar plantations were developed in the 1640s, using imported slave labour from Africa until the practice was abolished in 1834. In 1951 universal suffrage was introduced, followed in 1954 by cabinet government. Full internal self-government was attained in Oct. 1961. On 30 Nov. 1966 Barbados became an independent sovereign state within the British Commonwealth.

TERRITORY AND POPULATION

Barbados lies to the east of the Windward Islands. Area 430 sq. km (166 sq. miles). In 2000 the census population was 268,792; density, 625·1 per sq. km.

The UN gives a projected population for 2015 of 276,000.

In 2011, 45·1% of the population were urban. Bridgetown is the principal city: population (including suburbs), 133,000 in 1999. The country is divided into 11 parishes.

The official language is English.

SOCIAL STATISTICS

In 2007: births registered, 3,537; deaths registered, 2,213; birth rate, 12·9 per 1,000 population; death rate, 8·1 per 1,000 population; infant mortality (2010), 17 per 1,000 live births. Expectation of life,

2007, males 74·0 years and females 79·7. Population growth rate, 2005, 0·3%; fertility rate, 2008, 1·5 children per woman.

CLIMATE

An equable climate in winter, but the wet season, from June to Nov., is more humid. Rainfall varies from 50" (1,250 mm) on the coast to 75" (1,875 mm) in the higher interior. Bridgetown, Jan. 76°F (24·4°C), July 80°F (26·7°C). Annual rainfall 51" (1,275 mm).

CONSTITUTION AND GOVERNMENT

The head of state is the British sovereign, represented by an appointed Governor-General. The bicameral Parliament consists of a Senate and a House of Assembly. The *Senate* comprises 21 members appointed by the Governor-General, 12 being appointed on the advice of the Prime Minister, two on the advice of the Leader of the Opposition and seven at the Governor-General's discretion. The *House of Assembly* comprises 30 members elected every five years. In 1963 the voting age was reduced to 18.

The *Privy Council* is appointed by the Governor-General after consultation with the Prime Minister. It consists of 12 members and the Governor-General as chairman. It advises the Governor-General in the exercise of the royal prerogative of mercy and in the exercise of his disciplinary powers over members of the public and police services.

National Anthem

'In plenty and in time of need'; words by Irvine Burgie, tune by V. R. Edwards.

RECENT ELECTIONS

In the general election of 21 Feb. 2013 the ruling Democratic Labour Party (DLP) won 16 of 30 seats (51·3% of the total vote) and the Barbados Labour Party (BLP) 14 (48·3%). Three smaller parties and independents failed to win any seats.

CURRENT GOVERNMENT

Governor-General: Elliot Belgrave; b. 1931.

In March 2013 the government comprised:

Prime Minister and Minister of National Security, Public Service and Urban Development: Freundel Stuart; b. 1951 (DLP; appointed 23 Oct. 2010).

Minister of Agriculture, Food, Fisheries and Water Resources: Dr David Estwick. *Attorney General and Minister of Home Affairs:* Adriel Brathwaite. *Commerce and Trade:* Haynesley Benn. *Education and Human Resource Development:* Ronald Jones. *Environment and Drainage:* Dennis Lowe. *Family, Culture, Sports and Youth:* Stephen Lashley. *Finance and Economic Affairs:* Christopher Sinckler. *Foreign Affairs and Foreign Trade:* Maxine McClean. *Health:* Donville Inniss. *Housing and Lands:* Michael Lashley. *Industry, Small Business and Rural Development:* Denis Kellman. *International Business and International Transport:* George Hutson. *Labour:* Dr Esther Byer-Suckoo. *Social Care, Constituency Empowerment and Community Development:* Steve Blackett. *Tourism:* Richard Sealy. *Transport and Works:* John Boyce.

Government of Barbados Information Network: http://www.gov.bb

CURRENT LEADERS

Freundel Stuart

Position
Prime Minister

Introduction
Freundel Stuart was sworn into office in Oct. 2010 after the death of David Thompson, having served as acting prime minister since

May that year when Thompson took a leave of absence. In Feb. 2013 he led his Democratic Labour Party to electoral victory.

Early Life

Freundel Jerome Stuart was born on 27 April 1951 in the parish of St Philip. He was educated at Christ Church Boys' Foundation School. He later taught history and Spanish at the Princess Margaret Secondary School in St Philip. In 1970 he joined the Democratic Labour Party (DLP) but did not stand for election until the 1990s.

Stuart graduated in history and political science from the University of the West Indies at Cave Hill in 1975. He gained further degrees in law and was called to the Barbados Bar in 1984. Beginning in criminal law, he later switched to the civil code.

In 1994 Stuart won the seat of St Philip South which he subsequently lost at the 1999 election. From 2003–07 he served in the Senate but in 2008 returned to the lower house after winning the constituency of St Michael South. He was also admitted to the Inner Bar and appointed Queen's Counsel. Stuart joined the Thompson administration as minister of home affairs and was later appointed attorney general and deputy prime minister.

In May 2010 Stuart was named acting prime minister when Thompson took leave because of ill health. On 23 Oct. 2010, after Thompson's death from cancer, Stuart was selected by the DLP as his successor.

Career in Office

Upon taking office, Stuart made few changes to the cabinet except for the appointment of Adriel Brathwaite as attorney general and minister of home affairs. Stuart has largely followed his predecessor's programme, which aims to tackle poverty, unemployment and housing shortages. On 20 Jan. 2011 the government was boosted when Mara Thompson, David Thompson's widow, won the by-election for her late husband's St John seat by an overwhelming majority.

At the general election of 21 Feb. 2013 the DLP won 16 of the 30 parliamentary seats, securing Stuart a mandate for a full term in office. Polls prior to the election had suggested the DLP were running behind the opposition Barbados Labour Party, which took 14 seats.

DEFENCE

The Barbados Defence Force has a strength of about 610. In 2008 defence expenditure totalled US$30m. (US$106 per capita), representing 0·8% of GDP.

Army

Army strength was 500 with reserves numbering 430 in 2007.

Navy

A small maritime unit numbering 110 (2007) operates nine patrol vessels. The unit is based at St Ann's Fort Garrison, Bridgetown.

ECONOMY

In 2009 agriculture accounted for 3% of GDP, industry 23% and services 74%.

According to the anti-corruption organization *Transparency International*, Barbados ranked 15th in the world in a 2012 survey of the countries with the least corruption in business and government. It received 76 out of 100 in the annual index.

Overview

Since independence in 1966, Barbados has become an upper middle-income economy with one of the highest per capita incomes in the region. Once dependent on sugar production, agriculture has played a diminishing role while tourism and financial services now account for around 75% of GDP.

For several years prior to the global financial crisis, real GDP growth averaged 4% per year as a result of strong tourist receipts and construction activity, which contributed to record employment. The government began to liberalize the capital account in 2008 as a means of developing Barbados as a centre for tourism and financial services.

However, the global crisis saw growth contract by more than 4% in 2009 as all key economic activities, including tourism and financial services, were affected. The unemployment rate has risen steadily, while foreign direct investment inflows came to a near standstill in 2009. Lower government revenues and higher expenditure have contributed to persistently large fiscal imbalances, with public debt reaching 110% of GDP in the 2009–10 financial year. While measures to raise VAT rates and public transportation tariffs are praised, the IMF has recommended broadening the tax base, streamlining government operations and reducing public spending to place public debt on a sustainable path.

Currency

The unit of currency is the *Barbados dollar* (BBD), usually written as BDS$, of 100 *cents*, which is pegged to the US dollar at BDS$2=US$1. Inflation was 5·8% in 2010 and 9·4% in 2011. Total money supply was BDS$1,691m. in June 2005. Foreign exchange reserves were US$575m. in July 2005.

Budget

The financial year runs from 1 April. Budgetary central government revenue and expenditure (in BDS$1m.):

	2008–09	2009–10	2010–11
Revenue	2,662·5	2,531·7	2,410·4
Expenditure	2,957·0	3,035·6	3,112·3

Tax revenue accounted for 92·8% of revenues in 2008–09. Main sources of expenditure in 2008–09 were subsidies, 31·3%; compensation of employees, 28·2%; use of goods and services, 14·7%.

VAT is 17·5% (reduced rate, 7·5%).

Performance

Total GDP in 2011 was US$3·7bn. Real GDP contracted by 4·1% in 2009 but grew by 0·2% in 2010 and 0·6% in 2011.

Banking and Finance

The central bank and bank of issue is the Central Bank of Barbados (*Governor*, Dr DeLisle Worrell), which had total assets of BDS$1,248·5m. in Dec. 2003. The provisional figures for the total assets of commercial banks in Dec. 2003 were BDS$6,812·6m. and savings banks' deposits BDS$5,493·8m. In 2003 there were 4,635 international business companies, 413 exempt insurance companies and 51 offshore banks. In addition, there were three commercial banks, one regional development bank, one National Bank (privatized in the meantime), three foreign banks and seven trust companies.

External debt amounted to BDS$2,224m. in 2004.

There is a stock exchange which participates in the regional Caribbean exchange.

ENERGY AND NATURAL RESOURCES

Environment

Carbon dioxide emissions from the consumption and flaring of fossil fuels in Barbados in 2008 were the equivalent of 5·0 tonnes per capita.

Electricity

Production in 2007, 948m. kWh. Capacity in 2007 was 0·2m. kW. Consumption per capita was 3,455 kWh in 2007.

Oil and Gas

Crude oil production in 2006 was 51,000 tonnes and reserves in 2007 were 3·0m. bbls. Output of natural gas (2007) 30·7m. cu. metres, and reserves (2008) 141m. cu. metres. Production of Liquid Petroleum Gas (LPG) was 1·2m. cu. metres in 2007.

Agriculture

The agricultural sector accounted for 4·4% of GDP in 2003 (24% in 1967). Of the total labour force in 2003, 4·6% were employed in agriculture. Of the total area of Barbados (42,995 ha.), about 16,000 ha. are arable land, which is intensively cultivated. In 2003, 7,515 ha. were under sugarcane cultivation. Production, 2003 (in tonnes): sugarcane, 48,500; sweet potatoes, 2,610; cucumbers, 2,018; okra, 1,446; tomatoes, 1,234; yams, 1,234; carrots, 1,012; cabbage, 640.

Meat and dairy products, 2003 (in tonnes): poultry, 11,458; cow's milk, 7,017; pork, 1,756; eggs, 1,620; beef, 346.

Livestock (2003): sheep, 27,000; pigs, 17,000; cattle, 14,000; chickens, 3m.

Forestry

Timber production in 2007 was 11,000 cu. metres. In 2010 forests covered 8,000 ha. or 19% of the total land area.

Fisheries

The total catch in 2008 was 3,551 tonnes (the highest annual catch since 1998), exclusively from sea fishing.

INDUSTRY

Industry has traditionally been centred on sugar, but there is also light manufacturing and component assembly for export. In 2003, 36,300 tonnes of raw sugar were produced.

Labour

In 2003 the workforce was 145,500, of whom 129,500 were employed. Unemployment stood at 11·0%, down from 24·5% in 1993.

INTERNATIONAL TRADE

Imports and Exports

Trade in US$1m.:

	Imports (c.i.f.)	Exports (f.o.b.)
2006	1,628·6	441·2
2007	1,298·7	314·2
2008	1,744·3	454·2
2009	1,340·7	322·7
2010	1,196·2	313·7

In 2010 imports (in US$1m.) were mainly from the USA (525·7), Trinidad and Tobago (85·9), the UK (64·2) and China (57·8). Principal export markets in 2010 were (in US$1m.) the USA (78·2), the UK (52·6), Trinidad and Tobago (26·5) and St Lucia (18·2).

The main imports are machinery and transport equipment; food and live animals; manufactured goods. The main exports are chemicals and petroleum products; manufactured articles; beverages.

COMMUNICATIONS

Roads

There were 1,600 km of roads in 2004. In 2007 there were 103,500 passenger cars, 15,200 lorries and vans, and 630 buses and coaches. There were 38 deaths as a result of road accidents in 2007.

Civil Aviation

Grantley Adams International Airport is 16 km from Bridgetown. In 2009 it handled 1,939,059 passengers (down from 2,165,125 in 2008) and 21,098 tonnes of freight (up from 19,479 in 2008). There were flights in 2010 to Antigua, Atlanta, Beef Island, Boston, Canouan Island, Charlotte, Curaçao, Dominica, Fort de France, Frankfurt, Georgetown, Grenada, Kingston, London, Manchester, Miami, Montreal, New York, Philadelphia, Port of Spain, St Kitts, St Lucia, St Maarten, St Vincent, San Juan, Tobago and Toronto.

Shipping

There is a deep-water harbour at Bridgetown. In Jan. 2009 there were 91 ships of 300 GT or over registered, totalling 642,000 GT.

Telecommunications

In 2011 there were 141,000 landline telephone subscriptions (equivalent to 513·5 per 1,000 inhabitants) and 348,000 mobile phone subscriptions (1,270·1 per 1,000 inhabitants). Fixed internet subscriptions totalled 61,000 in 2009 (223·5 per 1,000 inhabitants). In Dec. 2011 Barbados had 118,000 Facebook users.

SOCIAL INSTITUTIONS

Justice

Justice is administered by the Supreme Court and Justices' Appeal Court, and by magistrates' courts. All have both civil and criminal jurisdiction. There is a Chief Justice, three judges of appeal, five puisne judges of the Supreme Court and nine magistrates. The death penalty is authorized. Barbados was one of ten countries to sign an agreement in Feb. 2001 establishing a Caribbean Court of Justice (CCJ) to replace the British Privy Council as the highest civil and criminal court. In the meantime the number of signatories has risen to 12. The court was inaugurated at Port-of-Spain, Trinidad on 16 April 2005. Barbados is one of only three countries, along with Belize and Guyana, to have accepted it as its court of final appeal.

The population in penal institutions in Feb. 2008 was 1,030 (379 per 100,000 of national population).

Education

The adult literacy rate was 99·7% in 2003. There were 22,584 pupils at primary schools in 2007 with 1,553 teaching staff and 20,855 pupils at secondary schools in 2006 with 1,430 teaching staff. There were 23 public and eight private secondary schools in 2003. Education is free in all government-owned and government-maintained institutions from primary to university level.

In 2005 public expenditure on education came to 6·9% of GDP and 16·4% of total government spending.

In 2007 there were 11,405 students in higher education and 786 academic staff. One of the three main campuses of the University of the West Indies is in Barbados, at Cave Hill.

Health

In 2001 there was one general hospital, one psychiatric hospital, five district hospitals, eight health centres and two private hospitals with 35 beds. There were 2,049 hospital beds and 420 doctors in the same year.

Welfare

The National Insurance and Social Security Scheme provides contributory sickness, age, maternity, disability and survivors benefits. Sugar workers have their own scheme.

RELIGION

In 2001, 63% of the population were Protestants, 5% Roman Catholics and the remainder other religions.

CULTURE

World Heritage Sites

Historic Bridgetown and its Garrsion, an outstanding example of British colonial architecture, was inscribed on the UNESCO World Heritage List in 2011.

Press

In 2008 there were two daily newspapers, the *Barbados Advocate* (est. 1895) and the *Daily Nation* (est. 1973). The *Daily Nation* has an average daily circulation of 33,000; the *Barbados Advocate*, 15,000.

Tourism

There were 574,533 non-resident overnight tourists in 2007 (including 303,404 from elsewhere in the Americas and 250,924 from Europe) plus 647,636 cruise passenger arrivals, up from 526,558 and 539,092 respectively in 2006.

Festivals
The National Cultural Foundation organizes three annual national festivals: the nine-day Congaline Carnival which begins in the last week of April; Crop Over, a three-week festival held from mid-July until Aug.; the National Independence Festival of Creative Arts (NIFCA) which runs throughout Nov. The biggest music festivals are the Barbados Jazz Festival (Jan.) and Gospelfest (May). Other festivals include the Holetown Festival (Feb.), which commemorates the anniversary of the first settlement of Barbados, Holders Season (March), an arts festival at St James, and the Oistins Fish Festival (March–April).

DIPLOMATIC REPRESENTATIVES

Of Barbados in the United Kingdom (1 Great Russell St., London, WC1B 3ND)
High Commissioner: Hugh Anthony Arthur.

Of the United Kingdom in Barbados (Lower Collymore Rock, PO Box 676, Bridgetown)
High Commissioner: Paul Brummell.

Of Barbados in the USA (2144 Wyoming Ave., NW, Washington, D.C. 20008)
Ambassador: John Beale.

Of the USA in Barbados (Wildey Business Park, Wildey, St Michael, BB 14006, Bridgetown)
Ambassador: Larry Palmer.

Of Barbados to the United Nations
Ambassador: Joseph Goddard.

Of Barbados to the European Union
Ambassador: Samuel Jefferson Chandler.

FURTHER READING

Beckles, H., *A History of Barbados: from Amerindian Settlement to Nation-State.* 1990
Carmichael, Trevor A. (ed.) *Barbados: Thirty Years of Independence.* 1998
Carter, R. and Downes, A. S., *Analysis of Economic and Social Development in Barbados: A Model for Small Island Developing States.* 2000
Hoyos, F. A., *Tom Adams: a Biography.* 1988.—*Barbados: A History from the Amerindians to Independence.* 2nd ed. 1992

National Statistical Office: Barbados Statistical Service, Fairchild Street, Bridgetown.
Website: http://www.barstats.gov.bb

The world in focus at
www.statesmansyearbook.com

BELARUS

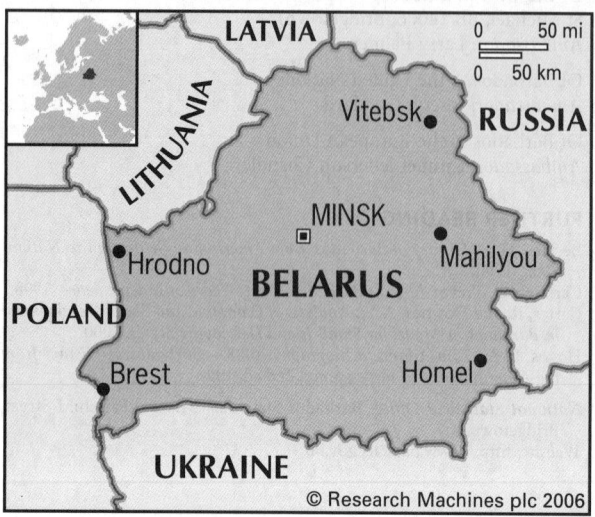

© Research Machines plc 2006

Respublika Belarus
(Republic of Belarus)

Capital: Minsk
Population projection, 2015: 9·44m.
GNI per capita, 2011: (PPP$) 13,439
HDI/world rank: 0·756/65
Internet domain extension: .by

KEY HISTORICAL EVENTS

There is evidence of Neolithic settlement dating to 4000 BC at Asaviec in northern Belarus. During the third and second millennia BC, parts of present-day Belarus were settled by Finno-Ugric tribes, as well as groups migrating along the Dniepr river and from the southern Baltic. Slavic tribes, notably the Kryvichans, gained dominance between the 5th and 8th centuries AD, seeing off Huns and Avars from the eastern steppes. The territory came under the sway of Kyivan Rus during the 9th century, although the Duchy of Polotsk challenged Kyiv and Novgorod for control of the northern reaches until its decline in the late 12th century. The Duchy of Turau and Pinsk to the south also became powerful during this period.

Kyiv's influence waned in the 13th century and raids by Mongol horsemen in the 1220s and 1230s left the area weak and open to conquest by the Dukes of Lithuania, initially under Mindaugas. Grand Duke Ladislaus Jagiełło's marriage to a Polish princess, Jadwiga, in 1385 made him king of Poland and a powerful rival to Muscovy, and several wars ensued between the two over access to the Baltic Sea, with Belarus one of the battlegrounds.

Jewish settlers arrived in Belarus in the late 14th century, initially in the cities of Brest and Hrodna. Belarusian culture flourished until the Polish–Lithuanian union in 1569, which put Polish rulers in the ascendency. The Commonwealth reached its zenith in the early 17th century then gradually declined until partitions in 1772, 1793 and 1795 saw control of all Belarusian territory pass to the Russian empire. By the 1830s Russification had taken hold, with the Polish language outlawed in schools and the Russian Orthodox church dominant.

Poverty and hardship under Tsarist rule led to significant emigration, notably by Jews to the USA, and fuelled rebellion. In 1863 anti-Russian uprisings erupted across the eastern Baltic region, led in Belarus by Kastus Kalinouski. By the 1880s several organizations championed Belarusian independence, although the movement gained little ground until the Russian Revolution of 1905, when bans on a free press and non-Russian languages were lifted. The outbreak of the First World War brought a million Russian troops into Belarus, with as many of the local population fleeing or being evacuated.

In March 1918 the first all-Belarusian Congress in Minsk declared an independent republic, but nine months later the Bolshevik government in Smolensk proclaimed the Byelorussian Soviet Socialist Republic (BSSR). A battlefield in the Russo–Polish war, it was divided by the Treaty of Riga in 1921, with 40% of the territory coming under Polish rule.

In Sept. 1939 the Red Army annexed western Belarus and western Ukraine, setting the stage for the region's devastation. Up to 300,000 people were deported under Stalin prior to the German offensive in June 1941. During the three years of Nazi occupation 2·2m. people died, including almost all the Jewish population, and over 1m. buildings were destroyed. The Red Army liberated Minsk in July 1944 and Belarus was fully integrated into the Soviet Union at the end of the war, becoming the most militarized Soviet Republic and a centre for light industry.

Gorbachev's reforms of the mid-1980s encouraged demands for greater freedom. On 25 Aug. 1991 Belarus declared independence, with the reformist Stanislau Shushkevich becoming head of state. Alyaksandr Lukashenka who, in 1991, had opposed the dissolution of the USSR, was elected president in July 1994. By 1996 only 11% of state enterprises had been privatized and the pro-Russian government worked to establish a Russia–Belarus Union. A referendum held in Nov. 1996 extended the president's term of office from three to five years and increased his powers.

The government maintained its political stranglehold at the 2008 parliamentary election, winning all seats. Lukashenka remained defiant in the face of domestic and international pressure for reform. In Dec. 2010 he secured a further presidential term, although the opposition and international observers alleged vote rigging while the EU reinstated a travel ban and froze his assets.

TERRITORY AND POPULATION

Belarus is situated along the western Dvina and Dnieper. It is bounded in the west by Poland, north by Latvia and Lithuania, east by Russia and south by Ukraine. The area is 207,600 sq. km (80,155 sq. miles). The capital is Minsk. Other important towns are Homel, Vitebsk, Mahilyou, Bobruisk, Hrodno and Brest. On 2 Nov. 1939 western Belorussia was incorporated with an area of over 108,000 sq. km and a population of 4·8m. Census population, 2009, 9,503,807; density, 45·8 per sq. km. Estimate, Jan. 2012: 9,465,150.

The UN gives a projected population for 2015 of 9·44m.

In 2011, 75·2% of the population lived in urban areas. Major ethnic groups: 81·2% Belarusians, 11·4% Russians, 3·9% Poles, 2·4% Ukrainians.

Belarus comprises six regions and one municipality. Areas and populations:

Province	Area sq. km	Population (2009 census)	Capital	Population (2009 census)
Brest	32,300	1,401,177	Brest	309,764
Homel	40,400	1,440,718	Homel	482,652
Hrodno	25,000	1,072,381	Hrodno	327,540
Mahilyou	29,000	1,099,374	Mahilyou	358,279
Minsk	40,500	1,422,528	Minsk	—
Minsk City	300	1,836,808		
Vitebsk	40,100	1,230,821	Vitebsk	347,928

Belarusian and Russian are both official languages.

SOCIAL STATISTICS

2007 births, 103,626 (rate of 10·7 per 1,000 population); deaths, 132,993 (rate of 13·7 per 1,000 population); marriages, 90,444; divorces, 36,146. Annual population growth rate, 2000–05, –0·5%. Life expectancy at birth, 2007, was 63·1 years for men and 75·2 years for women. Only Russia has a bigger difference between its male and female life expectancy. Infant mortality, 2010, four per 1,000 live births; fertility rate, 2008, 1·3 children per woman (one of the lowest rates in the world).

CLIMATE

Moderately continental and humid with temperatures averaging 20°F (–6°C) in Jan. and 64°F (18°C) in July. Annual precipitation is 22–28" (550–700 mm).

CONSTITUTION AND GOVERNMENT

A new constitution was adopted on 15 March 1994. It provides for a *President* who must be a citizen of at least 35 years of age, have resided for ten years in Belarus and whose candidacy must be supported by the signatures of 70 deputies or 100,000 electors. At a referendum held on 17 Oct. 2004, 86·2% of votes cast were in favour of the abolition of the two-term limit on the presidency. The vote was widely regarded as fraudulent.

There is an 11-member *Constitutional Court*. The chief justice and five other judges are appointed by the president.

Four referendums held on 14 May 1995 gave the president powers to dissolve parliament; work for closer economic integration with Russia; establish Russian as an official language of equal status with Belarusian; and introduce a new flag.

At a further referendum of 24 Nov. 1996 turnout was 84%. 79% of votes cast were in favour of the creation of an upper house of parliament nominated by provincial governors and 70% in favour of extending the presidential term of office by two years to five years. The Supreme Soviet was dissolved and a 110-member lower *House of Representatives* established, whose members are directly elected by universal adult suffrage every four years. The upper chamber is the *Council of the Republic* (64 seats; 56 members elected by regional councils and eight members appointed by the president, all for four-year terms). In practice, since 1996 the Belarusian parliament has only had a ceremonial function.

National Anthem

'My Bielarusy' ('We, the Belarusians'); words by M. Klimkovich and U. Karyzna, tune by Nester Sakalouski.

GOVERNMENT CHRONOLOGY

Heads of State since 1991.

Chairmen of the Supreme Council
1991–94   Stanislau Stanislavavich Shushkevich
1994   Myechyslau Ivanavich Hryb

President
1994–   Alyaksandr Rygorovich Lukashenka

RECENT ELECTIONS

Parliamentary elections were held on 23 Sept. 2012. 109 deputies were elected; one of the 110 seats went to a second round. The two main opposition parties boycotted the election. Independent candidates won 104 seats, the Communist Party 3, the Agrarian Party 1 and the Republican Party of Labour and Justice 1. All the elected deputies are supporters of President Lukashenka. Turnout was 74·2%. According to international observers the elections were neither free nor impartial.

Presidential elections were held on 19 Dec. 2010. Alyaksandr Lukashenka was re-elected with 79·7% of votes cast against 2·4% for Andrei Sannikov and eight other opponents. The election took place amid accusations of vote rigging. Turnout was 90·4%.

CURRENT GOVERNMENT

President: Alyaksandr Lukashenka; b. 1954 (sworn in 20 July 1994 and re-elected in Sept. 2001, March 2006 and Dec. 2010).

Prime Minister: Mikhail Myasnikovich; b. 1950 (took office on 28 Dec. 2010).

In March 2013 the government comprised:

First Deputy Prime Minister: Vladimir Semashko. *Deputy Prime Ministers:* Anatol Kalinin; Mikhail Rusyi; Anatol Tozik.

Minister for Agriculture and Food: Leonid Zayats. *Architecture and Construction:* Anatoly Nichkasov. *Communications and Information Technology:* Nikolai Pantelei. *Culture:* Boris Svetlov. *Defence:* Yuri Zhadobin. *Economy:* Nikolai Snopkov. *Education:* Sergei Maskevich. *Emergencies:* Vladimir Vashchenko. *Energy:* Alexander Ozerets. *Finance:* Andrei Kharkovets. *Foreign Affairs:* Vladimir Makei. *Forestry:* Mikhail Amelyanovich. *Health:* Vasily Zharko. *Housing and Communal Services:* Andrei Shorets. *Industry:* Dmitry Katerinich. *Information:* Oleg Proleskovsky. *Internal Affairs:* Anatoly Kuleshov. *Justice:* Oleg Slizhevsky. *Labour and Social Protection:* Marianna Shchetkina. *Natural Resources and Environmental Protection:* Vladimir Tsalko. *Sports and Tourism:* Aleksandr Shamko. *Taxes and Duties:* Vladimir Poluyan. *Trade:* Valentin Chekanov. *Transport and Communications:* Anatoly Sivak.

Government Website: http://www.government.by

CURRENT LEADERS

Alyaksandr Rygorovich Lukashenka

Position
President

Introduction
Alyaksandr Lukashenka has been president since 1994. A communist from an early age, he has strengthened his powers within the constitution but has drawn international condemnation for his autocratic leadership and human rights abuses against political opponents and journalists.

Early Life
Lukashenka was born on 30 Aug. 1954 in Kepys. After studying history and agricultural economics, he taught at the Mahilyou Teaching Institute and the Belarusian SSR Agro-economics Academy. He was a member of Komsomol, a young communists' group, before working on collective farms. After working in local politics, he became a deputy on the Belarusian Supreme Council in 1990. When Belarus gained independence the following year, he opposed the formation of the CIS, set up by Russia, Belarus and the Ukraine, as well as the nationalist and free market tendencies of Stanislau Shushkevich (the first post-independence leader). By the time of the 1994 presidential elections, Shushkevich had been forced to stand down. Campaigning on a pro-Russian manifesto, Lukashenka won in a second round run-off with former prime minister (1990–94) Vyachaslau Kebich, receiving 80·1% of votes.

Career in Office
Following his election, Lukashenka set about increasing presidential powers and strengthening state control. Countering Shushkevich's efforts to promote Belarusian culture, he reinstated Russian as the official language and sought closer ties, not always successfully, with the Russian Federation. Trade agreements were made with Russia and several other former Soviet states. In 1996 he pressed through constitutional changes to give himself more power and to extend his term of office by two years to 2001. He reversed reforms made by his predecessor after the collapse of the Soviet Union and secured control over the state-owned media and security services. He also rejected Shushkevich's previous moves towards privatization.

Lukashenka used his presidential decree to impose restrictions on opponents, and journalists in particular were subject to harassment for criticizing his regime. He was condemned by the international community for human rights abuses and disregard for democracy.

In the 2000 parliamentary elections, Lukashenka's supporters won 81 of 110 seats. The following year he won a second presidential term, taking 75·6% of the vote against 15·4% for Uladzimir Hancharyk. Both elections were criticized by international observers as undemocratic, with many opposition politicians either boycotting them or remaining in exile. International organizations continued to criticize Lukashenka's repressive regime during 2002 (when the authorities expelled an OSCE delegation) and 2003. The USA meanwhile passed legislation allowing the provision of financial support to the democratic opposition in Belarus.

Further parliamentary elections and a referendum were held on 17 Oct. 2004. No opposition candidates won a parliamentary seat in the poll, while Lukashenka claimed overwhelming support in the referendum for his intention to change the constitution and run for another presidential term. The results were dismissed as fraudulent by most international observers, human rights activists and opposition politicians.

Lukashenka was re-elected for a further term in March 2006, again amid accusations of vote rigging. From April 2006 the EU and the USA imposed travel bans and asset freezes on various Belarusian officials, including Lukashenka, deemed responsible for electoral fraud and civil repression.

In 2007 Belarus was in dispute with Russia over oil supplies, price rises and unpaid debts, and in May that year Belarus' bid for a seat on the United Nations Human Rights Council was rejected.

In the spring of 2008 relations with the USA deteriorated further, as Lukashenka's government expelled US diplomats for criticizing Belarus' human rights record. In Sept. 2008 candidates loyal to the president were again returned to every seat in parliamentary elections considered by European monitors to be only marginally less flawed than previous polls.

Following the lifting of the EU's travel ban in Oct. 2008, Lukashenka visited the Vatican in April 2009 for talks with Pope Benedict XVI in his first official visit to Western Europe since the mid-1990s.

Tensions with Russia over energy trading continued. In Jan. 2010 the Belarus government threatened to cut electricity deliveries to the Russian Baltic enclave of Kaliningrad because of an unresolved dispute over Russian oil supplies to Belarus, and there was a further dispute in June over Belarus' unpaid debts and transit fees. Russia began reducing gas supplies and Lukashenka threatened to disrupt Russian gas and oil deliveries to Europe before the situation was resolved.

The EU reinstated the travel ban on Lukashenka in response to accusations of human rights violations ahead of the presidential election in Dec. 2010, in which Lukashenka claimed an overwhelming victory. Subsequent protests in Minsk were broken up by force and there were mass arrests of opposition activists. In April 2011 a terrorist attack on a metro station in the capital was cited by Lukashenka as a plot to destabilize the country. Against a background of increasing social unrest, the economy continued to suffer in 2011, with rising inflation, a sharp devaluation of the currency, declining financial support from Russia and debt-servicing difficulties. In Jan. 2012 new legislation imposed restrictions on access to foreign websites, while in parliamentary elections in Sept., which were boycotted by opposition groups, loyal Lukashenka supporters again secured every assembly seat.

DEFENCE

Conscription is for 18 months, or 12 in the case of university and college graduates. A treaty with Russia of April 1993 co-ordinates their military activities. All nuclear weapons had been transferred to Russia by Dec. 1996. Total active armed forces in 2007 numbered 72,940. In addition there are Ministry of Interior paramilitary troops numbering 110,000.

Defence expenditure in 2008 totalled US$674m. (US$70 per capita), representing 1·1% of GDP.

Army

In 2007 Army personnel numbered 29,600. In addition there were 289,500 reserves. Equipment in 2007 included 1,586 main battle tanks (T-55s, T-72s and T-80s).

Air Force

In 2007 the Air Force and Air Defence Forces operated 175 combat-capable aircraft, including MiG-29s, Su-24s, Su-25s and Su-27s, and 50 attack helicopters. Personnel, 2007, 18,170.

INTERNATIONAL RELATIONS

A treaty of friendship with Russia was signed on 21 Feb. 1995. A further treaty signed by the respective presidents on 2 April 1997 provided for even closer integration.

ECONOMY

In 2006 agriculture contributed 9·3% of GDP, industry 42·0% and services 48·7%. In 2006, 57·4% of economic output was being produced by the private sector.

Overview

Following the collapse of the Soviet Union, Belarus' economy underperformed relative to other transition economies. From 1991 to 1995 the economy contracted at an average annual rate of 8·1% compared to the 0·6% contraction averaged by eastern European economies. In 1995 President Lukashenka reinstated price controls, state intervention in private enterprise and the obstruction of foreign investment. In spite of this, the economy grew at an annual average rate of 6·9% in the decade 1996–2005.

The economy is largely sustained by Russian oil subsidies and the processing and re-export of Russian oil at Belarus' two refining centres, Mozyr and Naftan. Growth averaged 8% per year from 2001–11 thanks to improved terms of trade, under-priced energy from the Russian Federation and strong growth among Belarus' main trading partners. Poverty decreased seven-fold from a peak of 30·5% in 2002.

However, the global financial crisis led to a loss of external financing and a sharp decline in exports, exacerbated by an already high current account deficit and low reserves. In 2011 there was a balance of payments crisis as a result of unsustainably loose macroeconomic policies at the end of 2010. Annual inflation soared and the poverty headcount rose from 5·2% in 2010 to 7·3% in 2011. Stabilization measures implemented from mid-2011 included a more flexible exchange rate regime, fiscal consolidation and constrained credit growth. In addition, oil and gas agreements with Russia saw lower priced imports. Belarus' foreign exchange markets rebounded and the current account deficit was reduced.

Currency

The *rouble* was retained under an agreement of Sept. 1993 and a treaty with Russia on monetary union of April 1994. Foreign currencies ceased to be legal tender in Oct. 1994. Only banknotes are issued—there are no coins in circulation. In Jan. 2000 the Belarusian rouble was revalued at 1 new rouble (BYR) = 1,000 old roubles (BYB). In Nov. 2000 President Lukashenka and President Putin of Russia agreed the introduction of a single currency, but plans to introduce the Russian rouble to Belarus have since been postponed indefinitely. The inflation rate in 1994 was 2,434% but has since fallen, and was inflation at 7·7% in 2010. Inflation then climbed to 53·2%, the highest rate of any country in 2011. Foreign exchange reserves in June 2005 were US$1,151m., total money supply was 4,025·7bn. roubles and gold reserves were 100,000 troy oz.

Budget
Central government revenue and expenditure (in 1bn. roubles):

	2008	2009	2010
Revenue	50,478	48,453	51,692
Expenditure	44,064	45,148	51,493

In 2010 tax revenue totalled 27,823bn. roubles (including social security contributions, 19,249bn. roubles). Main items of expenditure by economic type in 2009 were: social benefits, 21,204bn. roubles; grants, 9,791bn. roubles; compensation of employees, 6,147bn. roubles.

VAT is 20%.

Performance
Real GDP growth was 0·2% in 2009, 7·7% in 2010 and 5·3% in 2011. Total GDP in 2011 was US$55·1bn.

Banking and Finance
The central bank is the National Bank (*Chairman*, Nadezhda Ermakova). In 2003 there were 28 commercial banks. The largest banks are Belagroprombank, Belarusbank, Belinvestbank and BPS-Bank.

In 2010 total external debt was US$25,726m., equivalent to 47% of GNI.

There is a stock exchange in Minsk.

ENERGY AND NATURAL RESOURCES
Environment
Carbon dioxide emissions from the consumption and flaring of fossil fuels in Belarus were the equivalent of 7·0 tonnes per capita in 2008.

Electricity
Installed capacity was an estimated 8·0m. kW in 2007. Production was 31·83bn. kWh in 2007. Consumption per capita in 2007 was 3,733 kWh. A joint Belarusian–Russian project envisages the construction of the country's first nuclear power plant at Ostrovetsk, with a first reactor scheduled to become operational in 2017 and a second in 2018.

Oil and Gas
In 2007 output of crude petroleum totalled 1·76m. tonnes; reserves in 2007 were 198m. bbls. Natural gas production in 2007 was 202m. cu. metres; reserves were 2·8bn. cu. metres in 2007.

Minerals
Particular attention has been paid to the development of the peat industry with a view to making Belarus as far as possible self-supporting in fuel. There are over 6,500 peat deposits. There are rich deposits of rock salt and of iron ore.

Agriculture
Belarus is hilly, with a general slope towards the south. It contains large tracts of marshland, particularly to the southwest.

Agriculturally, it may be divided into three main sections— Northern: growing flax, fodder, grasses and breeding cattle for meat and dairy produce; Central: potato growing and pig breeding; Southern: good natural pasture land, hemp cultivation and cattle breeding for meat and dairy produce. In 2007 the agricultural population was around 972,000, of whom approximately 488,000 were economically active.

Output of main agricultural products (in 1m. tonnes) in 2007: potatoes, 8·74; sugar beets, 3·63; barley, 1·91; wheat, 1·40; rye, 1·31; oats, 0·58; milk, 5·91; eggs, 3,230m. units. In 2007 there were 3·99m. cattle; 3·64m. pigs; 156,200 horses; and 28·7m. poultry.

Since 1991 individuals may own land and pass it to their heirs, but not sell it. In 2006 there were 5·54m. ha. of arable land and 121,000 ha. of permanent crops. There were 4,723 farms in 2003. The private and commercial sectors accounted for 49% of the value of agricultural output in 2003, but only 20% of the total agricultural land. Agricultural output grew by 12·9% in 2004, the fifth successive year of growth. In 2004 state support for the agricultural sector accounted for 4% of GDP.

Forestry
Forests occupied 8·63m. ha., or 42% of the land area, in 2010. There are valuable reserves of oak, elm, maple and white beech. Timber production in 2007 was 8·76m. cu. metres.

Fisheries
The total catch in 2010 came to 897 tonnes, exclusively from inland waters. Aquaculture production totalled 16,265 tonnes in 2010 (95% of total fish production).

INDUSTRY
There are food-processing, chemical, textile, artificial silk, flax-spinning, motor vehicle, leather, machine-tool and agricultural machinery industries. Output in 1,000 tonnes (2007 unless otherwise indicated): distillate fuel oil, 6,679; residual fuel oil, 6,195; fertilizers (2004), 5,403; cement (2009), 4,350; petrol, 3,181; crude steel (2009), 2,449. Output of other products: 18,699m. cigarettes (2007); refrigerators (2003), 886,000; TV sets (2003), 690,000; tractors (2004), 34,000; lorries (2004), 21,500; beer (2006), 332m. litres; footwear (2002), 12·7m. pairs; cotton woven fabrics (2003), 57m. sq. metres; silk fabrics (2003), 40m. sq. metres; linen fabrics (2003), 30m. sq. metres. Machine-building equipment and chemical products are also important. Most industry is still state-controlled.

Labour
In 2007 the labour force totalled 4,525,200. In 2007, out of 4,476,600 economically active people, 1,183,400 were in industry; 638,700 in trade and public catering, material and technical supply and sale; 453,000 in education; and 441,900 in agriculture. In 2007 according to official statistics there were 48,600 unemployed persons, or 1·1% of the workforce, although some analysts maintain that unemployment is around 15%. The minimum unemployment benefit is the minimum wage and the maximum benefit is twice the minimum wage. The minimum wage in 2007 was 179,050 roubles a month.

INTERNATIONAL TRADE
In Jan. 2010 Belarus joined the newly established Customs Union, together with Kazakhstan and Russia.

Imports and Exports
In 2006 imports were valued at US$22,237m. and exports at US$19,838m. The main import suppliers in 2006 were Russia (58·6%), Germany (7·5%), Ukraine (5·5%), Poland (3·4%) and China (2·5%). Principal export markets were Russia (34·7%), the Netherlands (17·7%), United Kingdom (7·5%), Ukraine (6·3%) and Poland (5·2%). Main import commodities are petroleum, natural gas, rolled metal and coal. Export commodities include machinery and transport equipment, diesel fuel, synthetic fibres and consumer goods.

COMMUNICATIONS
Roads
In 2005 there were 97,536 km of roads (88·0% paved), including 15,432 km of national roads. There were 2,329,200 passenger cars in use in 2007 (240 per 1,000 inhabitants). In 2007 public transport totalled 9,375m. passenger-km and freight 19,200m. tonne-km. There were 1,517 fatalities as a result of road accidents in 2007.

Rail
In 2005 there were 5,507 km of 1,520 mm gauge railways (874 km electrified). Passenger-km travelled in 2008 came to 8·2bn. and freight tonne-km to 49·0bn. There is a metro in Minsk.

Civil Aviation

The main airport is Minsk International 2, which handled 421,000 passengers (all international) and 2,700 tonnes of freight in 2001. The national carrier is Belavia. In 2003 Belavia flew on domestic routes and operated international services to Adler/Sochi, Baku, Berlin, Frankfurt, Hurghada, İstanbul, Kaliningrad, Kyiv, Larnaca, London, Moscow, Paris, Prague, Rome, Shannon, Stockholm, Tashkent, Tbilisi, Tel Aviv, Vienna, Warsaw and Yerevan. In 2003 scheduled airline traffic of Belarus-based carriers flew 7m. km, carrying 234,000 passengers (232,000 on international flights).

Telecommunications

In 2011 there were 4,208,000 landline telephone subscriptions (equivalent to 440·2 per 1,000 inhabitants) and 10,694,900 mobile phone subscriptions (or 1,118·8 per 1,000 inhabitants). In 2011, 39·6% of the population were internet users. In March 2012 there were 409,000 Facebook users.

SOCIAL INSTITUTIONS

Justice

The death penalty is retained following the constitutional referendum of Nov. 1996; there were two confirmed executions in both 2011 and 2012. Belarus is the only European country that still uses the death penalty.

180,427 crimes were reported in 2007. There were 39,552 prisoners in 2007, at a rate of 408 prisoners per 100,000 population.

Education

Adult literacy rate in 2008 was over 99%. In 2007 there were 365,298 children and 50,568 teachers at pre-school institutions, 362,282 pupils and 22,990 teachers at primary schools, 772,584 pupils and 99,011 teachers in secondary schools, and 576,679 students and 42,603 academic staff at institutions of tertiary education.

In 2007 there were 56 higher educational establishments (46 public and ten private) of which eight were general universities and 24 specialized universities (including four medical, three agricultural, three economics and three technical). There were also seven academies and 14 institutes. In 2007 there were 354,988 people enrolled at state higher education establishments.

Public expenditure on education in 2007 came to 5·3% of GNI.

Health

In 2007 there were 46,965 doctors (48·5 per 10,000 population), 1,964 dentists, 79,941 nurses, 4,921 midwives and 4,179 pharmacists. There were 112·4 hospital beds per 10,000 population in 2007.

Welfare

To qualify for an old-age pension men must be age 60 with 25 years of insurance coverage and women must be 55 with 20 years of insurance coverage. Minimum old-age pension is 25% of the average per capita subsistence budget. The maximum pension is 75% of wage base. Benefits are adjusted periodically according to changes in the minimum wage, which in 2009 was 229,000 roubles a month.

RELIGION

The Orthodox is the largest church. There is a Roman Catholic archdiocese of Minsk and Mahilyou, and five dioceses embracing 455 parishes. In 2001, 32% of the population were Belarusian Orthodox and 18% Roman Catholics.

CULTURE

World Heritage Sites

There are four UNESCO sites in Belarus: the Mir Castle Complex, begun in the 15th century (inscribed on the list in 2000); the Radziwill Family complex at Nesvizh (2005); the Belovezhskaya Pushcha/Białowieża Forest site (1979 and 1992), shared with Poland; and the Struve Geodetic Arc (2005). The Arc is a chain of survey triangulations spanning from Norway to the Black Sea that helped establish the exact shape and size of the earth and is shared with nine other countries.

Press

There were two state-owned daily newspapers in Jan. 2006. The only independent daily newspaper, *Narodnaya Volya*, has been published in Russia since Oct. 2005. There is also a Belarusian edition of the Russian daily *Komsomolskaya Pravda*. The most widely read paper is *Sovetskaya Belarussiya*, with a daily circulation of 390,000 in 2006.

Tourism

In 2007 there were 105,400 foreign tourists on organized trips. Spending by tourists totalled US$479m. in 2007.

DIPLOMATIC REPRESENTATIVES

Of Belarus in the United Kingdom (6 Kensington Court, London, W8 5DL)
Ambassador: Vacant.
Chargé d'Affaires a.i.: Valery Dougan.

Of the United Kingdom in Belarus (37 Karl Marx St., Minsk 220030)
Ambassador: Bruce Bucknell.

Of Belarus in the USA (1619 New Hampshire Ave., NW, Washington, D.C., 20009)
Ambassador: Vacant.
Chargé d'Affaires a.i.: Oleg Kravchenko.

Of the USA in Belarus (46 Starovilenskaya, Minsk 220002)
Ambassador: Vacant.
Chargé d'Affaires a.i.: Ethan Goldrich.

Of Belarus to the United Nations
Ambassador: Andrei Dapkiunas.

Of Belarus to the European Union
Ambassador: Andrei Yeudachenka.

FURTHER READING

Balmaceda, Margarita M., *Independent Belarus: Domestic Determinants, Regional Dynamics and Implications for the West*. 2003
Korosteleva, Elena, *Contemporary Belarus: Between Democracy and Dictatorship*. 2002
Marples, D. R., *Belarus: from Soviet Rule to Nuclear Catastrophe*. 1996
White, Stephen, *Postcommunist Belarus*. 2004
Zaprudnik, J., *Belarus at the Crossroads in History*. 1993

National Statistical Office: Ministry of Statistics and Analysis of the Republic of Belarus, 12 Partizansky Avenue, Minsk 220070.
Website: http://www.belstat.gov.by

BELGIUM

© Research Machines plc 2006

Royaume de Belgique Koninkrijk België
(Kingdom of Belgium)

Capital: Brussels
Population projection, 2015: 10·87m.
GNI per capita, 2011: (PPP$) 33,357
HDI/world rank: 0·886/18
Internet domain extension: .be

KEY HISTORICAL EVENTS

The Neanderthal Mousterian culture of the Ardennes region produced flint tools between 80–35,000 years ago. Omalien tribes of the early Neolithic period (5–4000 BC) developed settled agricultural practices, sophisticated tools and decorated black pottery. The Bronze Age Hilversum culture left evidence of contact overseas in Wessex, southern England. During the pre-Roman period Celtic and Germanic tribes moved across the region. Much of northern Gaul and southern Britain was settled by a Celtic group known as the Belgae, forming numerous tribes including the seafaring Morini and Menapii in what is now Flanders and the bellicose Nervii in Artois. In alliance with Germanic and other Belgic tribes, the Nervii led resistance to the invasion of Julius Caesar in 59 BC, succumbing five years later. Roman control was extended as far north as the Rhine and the area divided into the provinces of Gallia Belgica and Germania Inferior. Several tribes survived as Roman administrative *civitates*.

The network of Roman power, based around wealthy *villae* (country estates), declined from the mid-3rd century AD, despite the campaigns of Julian, who became emperor in AD 361. The massive influx of Germanic tribes (known commonly as the Barbarian invasions) over the Rhine in 406/7 effectively brought Roman rule in Belgium to an end. Chief among the German tribes were the Franks, who settled in Toxandria (modern Brabant). The Frankish Merovingian Empire, established by Childeric I, was based at Tournai and extended by Childeric's son, Clovis. The successors to the Merovingians, the Pippins (or Carolingians), ruled all but in name from Austrasia in the Ardennes. A partnership between the nobility and the church allowed the expansion of Frankish power north and east across the Rhine, bringing Christianity to the Low Countries by the 7th century, under the sees of Arras, Tournai, Cambrai and, from 720, Liège.

The death of Louis the Pious in 840 precipitated the fragmentation of Charlemagne's huge empire. Viking attacks on the Low Countries came at the end of the 8th century, intensifying in the period 841–75. Resistance began under Charlemagne; provincial princes capitalized on the fortification process and land reclamation to increase their own power. Baldwin 'Iron Arm', count of Flanders, fortified Ghent around 867, cementing his authority over the Flemish towns. His successors extended control into Artois (and to Hainault by personal union) in defiance of the French kings. Philippe IV of France was defeated in 1302 at the Battle of the Golden Spurs at Kortrijk, Flanders and formally recognized Flemish independence. Flanders' alliance with England during the Hundred Years War created an advantageous trading relationship, especially the importation of English wool for the textile industry. Bruges (Brugge), Ypres and Ghent flourished and in the 14th century had to be forcibly restrained from becoming city-states by Philip of Burgundy.

Other principalities emerged in the wake of the Carolingian empire, most notably the duchies of Brabant and Limburg and the prince-bishopric of Liège. Their union was forged under the dukes of Burgundy. The marriage of Philip the Bold, duke of Burgundy, to Margaret of Flanders in 1369 was the first step towards what became the Burgundian *Kreis* (lands) under Emperor Charles V. The addition of Hainault-Holland, Namur and Luxembourg encouraged Burgundian ambitions of centralization and even a unitary empire, vainly attempted by Duke Charles the Bold in the 1470s. The provinces and towns jealously guarded their imperial and local privileges and resisted the high taxation imposed by their Burgundian lord. Burgundian authority was reinforced by the 1477 marriage of Charles' daughter (and heir), Mary, to the Habsburg Maximilian of Austria, later Holy Roman Emperor, beginning over three centuries of Habsburg rule in the Low Countries.

Trade Centres

Bruges became the principal market of northwest Europe in the 14th century but after the silting of its waterways ceded its role in the 1490s to Antwerp. The volume of Portuguese and English traders and Italian financiers testified to the importance of Antwerp as a mercantile and financial centre; the *Antwerpen beurs* (stock exchange) was established in 1531. Antwerp's population reached 100,000 in the mid-16th century. The artistic achievements of the 15th century were supported by court patronage. Artists such as Rogier van der Weyden in Brussels and Jan van Eyck in Ghent formed part of a Flemish school that greatly influenced northern European art. The University of Louvain (Leuven), founded in 1425, was a centre of Dutch Humanism made famous by scholars such as Erasmus.

Dynastic pressures increased with the marriage of Archduke Philip the Handsome to the heiress of the Spanish crowns, Juana the Mad, leaving the Netherlands (the Burgundian Low Countries, including Belgium) under the supervision of governors-general in Brussels. Centralization continued under Philip's son, Charles of Ghent (Holy Roman Emperor Charles V), who regulated the succession to his Burgundian territories by pragmatic sanction. However, it was the imposition of a new ecclesiastical hierarchy by Charles' son, Philip II of Spain, that unified opposition in the Netherlands.

The Netherlands was highly receptive to non-conformist ideas, most notably those of Jean Calvin, whose influence had extended to Antwerp by 1545. Appealing to the urban middle classes and the lower nobility, Calvinism suffered repression, especially after the radical iconoclasm of 1566. The following year Philip sent the duke of Alba to stamp out the religious and political uprisings, sparking revolution in Holland and other northern provinces. The execution of the counts of Hoorne and Egmond in Brussels in 1568 reinforced opposition to Spanish rule.

The secession of the northern Netherlands was partly the result of the inability of the Spanish armies to penetrate the marshes and dendritic waterways of Holland and Zeeland. Although Alba managed to reassert Philip's authority in the south, his armies never retook the provinces north of the Rhine after 1574. Separation was also caused by the extreme demands of the northern Calvinists, who alienated the more Catholic southern provinces. However, a measure of unity was achieved at the Pacification of Ghent after the atrocities of the 'Spanish Fury'— Spanish troops massacred 7,000 people in Antwerp in 1576.

In 1578 Philip appointed as governor-general Alessandro Farnese, duke of Parma, who ejected Protestants from the governments of the southern provinces and waged successful campaigns against the revolutionaries. The Union of Arras (1579), led by Flanders and Hainault, accepted the sovereignty of the Spanish king, supported Catholicism and ended the revolt of the southern provinces. In reaction the northern provinces drew up the Union of Utrecht, thus marking the birth of the 'Dutch Republic', though Philip was not rejected as sovereign in the north until 1581. Farnese took Antwerp in 1585, effectively creating the boundaries of the renegade Dutch state. Although fighting resumed after the Twelve Year Truce (1609–21), the Habsburg government accepted the independence of the Dutch Netherlands at the Peace of Westphalia in 1648.

The Spanish Netherlands was governed autonomously for much of the 17th century—Liège remained neutral in the revolt and separate until 1795. Antwerp was at first eclipsed as the principal trading centre of the region by Dutch Amsterdam, which attracted many of the south's skilled artisans and merchants. Economic recovery was helped by new industries such as linen production and diamond processing in Antwerp. Art was dominated by the Baroque style, patronized by the Catholic Church and adopted by Rubens, Van Dyck and Jordaens.

War of Succession
The death of Charles II of Spain without an heir in 1700 caused a constitutional crisis and the War of the Spanish Succession. Philip of Anjou, the heir-designate and a grandson of Louis XIV of France, was urged to hand over the Spanish Netherlands to France. The intervention of England and the Dutch was motivated by the fear of either Franco-Spanish union or the reuniting of the Austrian and Spanish Habsburgs. The Spanish Netherlands were finally settled on Emperor Charles VI after the Treaty of Utrecht in 1713, thus bringing the Netherlands under the sway of the Austrian Habsburgs.

Dynastic succession was again the cause of war after the death of Charles VI in 1740. By pragmatic sanction, his daughter, Archduchess Maria Theresa, succeeded to the Habsburg territories but was barred from the Imperial crown (it was secured for her husband, Francis Stephen of Lorraine). Supported by Great Britain, the Dutch and the Hungarian diet, Maria Theresa's armies repelled the French from the Austrian Netherlands, establishing her authority by the Treaty of Aix-la-Chapelle in 1748 (though she lost Silesia to Prussia). Her reign saw great economic gains in the Netherlands and the beginning of industrial capitalism thanks to good trading relations with Great Britain.

The Austrian regime lost popular support under Maria Theresa's successor, Emperor Joseph II, whose abolition of local privileges and his attempts to exchange the Austrian Netherlands for Bavaria made him highly unpopular. Coupled with his attacks on the power of the Catholic Church, his 'Belgian' subjects— conservatives and progressives alike—revolted in 1789. The bulk of Joseph's forces being engaged on the Ottoman frontier, the Austrian army was easily routed at Turnhout. Conservative elements were victorious in the 'Brabant Revolution' and proclaimed the United States of Belgium in 1790.

Although Joseph's brother, Leopold II, reasserted Austrian authority, the seeds of revolution had been sown, encouraged by events in France. Republican France invaded Belgium in 1795, ending the independence and religious rule of Liège. Discontent with French rule was immediate, partly in reaction to military conscription, persecution of the Church and the denial of autonomy. After peasant uprisings in 1798 the Napoleonic consulate agreed a compromise with the papacy and the Belgian church, ending persecution.

Towards Independence
Napoleon's defeat in 1814 left Belgium in the hands of the Great Powers. Belgium was reunified with the Dutch Netherlands as the Kingdom of the Netherlands at the Congress of Vienna. Despite cultural and linguistic affinities, the relationship between the two Netherlands was uneasy. Belgium, ruled by the Dutch William of Orange, was under-represented in the States General and its French-speaking elite alienated by the declaration of Dutch as the sole legal language. The two economies were in contrast, the Dutch being primarily mercantile while the Belgian was geared towards mechanized industry. Belgium was one of the first areas of continental Europe to adopt the innovations of the Industrial Revolution, most notably in the Ghent textile industry and coal mining in Hainaut. The loss of the French market in 1814 and William's refusal to increase tariffs to protect Belgian industry impacted heavily on the Belgian economy.

Dissent paved the way for the Belgian Revolution of 1830. The secession of Belgium was secured by the intervention of France and Britain, which recognized Belgian independence in 1831. Repelled by the French, William accepted the loss of Belgium in 1838. Limburg and Luxembourg were partitioned and a liberal constitution implemented. The great powers insisted on a Belgian monarch and Prince Leopold of Saxe-Coburg was duly installed.

A Liberal government came to power in 1847 after three devastating harvests. The Liberal prime minister, Walthère Frère-Orban, championed the removal of church control of the schools. The Schools War dominated the political agenda and saw a conservative counter-offensive with the establishment of a network of independent Catholic schools. A conservative Catholic victory in 1884 brought Auguste Beernaert to the premiership. Closely associated with the Flemish revivalists, Beernaert managed the Flemish Equality Law, giving the Flemish language the same rights as French. Changes in the electoral system brought in full male suffrage (over 25 years of age) in 1893.

The Congo
Belgian foreign policy was bound by recognition (and imposition) of neutrality. King Leopold II, set on expanding his kingdom, looked to Africa. The Belgian Congo, acquired as his personal possession in 1885, was soon infamous for colonial abuse and exploitation. After widespread international condemnation, the 'Congo Free State' was formally annexed by Belgium in 1908, thus curbing Leopold's inhuman regime. In Europe, threats were perceived to the west and east. Attempts by Albert I (reigned 1909–34) to arm Belgium against French and German aggression were frustrated by the domestic pacifist movement led by Beernaert. Belgian neutrality was violated by Germany in 1914 after Albert's refusal to allow German free passage. Albert remained with his army on the Yser River throughout the First World War, supported by Allied troops. Neutrality, seen by many Belgians as a hindrance, was abolished under the Treaty of Versailles in 1919; Belgium was awarded the provinces of Eupen and Malmédy and jurisdiction over Ruanda-Urundi from the defeated Germany.

Economic reconstruction after the First World War included an economic union with Luxembourg in 1921. The Belgian Congo was exploited for its minerals. Constitutional changes allowed for equal suffrage of all men over 21 (women were denied the vote until 1948) and the formal linguistic separation of Flanders and Wallonia (excluding Brussels).

The 1930s was a time of rising unemployment and concern over German ambitions. Germany invaded Belgium on 10 May 1940. Capitulation after just 18 days made King Leopold III unpopular, despite his refusal to flee to France (and later to London) with the government. Resistance was active for much of the Second World War. Insurgents managed to protect the port of Antwerp, crucial for Allied support, during the liberation of Belgium in Sept. 1944.

The post-war economy made a speedy recovery but political unity foundered on the royal question. A referendum on the return of the king from imprisonment in Austria caused violent protest in Wallonia and in 1951 Leopold abdicated in favour of his son, Baudouin.

European Union
Belgium embarked on international co-operation under the leadership of Prime Minister Paul-Henri Spaak. Economic union with Luxembourg was re-established and extended to include the Netherlands, forming the Benelux Economic Union. In 1949 Belgium joined the North Atlantic Treaty Organization (NATO) and in 1951 the European Coal and Steel Community (ECSC). Encouraged by the success of the ECSC, plans were laid for the establishment of two more communities. The European Economic Community (EEC) and the European Atomic Energy Community (Euratom) were subsequently created under separate treaties signed in Rome on 25 March 1957.

The administration of the Belgian Congo resisted political reform and demands for greater participation until the 1950s. After violent protest and agitation, local government reform was passed in 1957 but was too late to quell the independence movement, led by Patrice Lumumba. In 1959 the Belgian government rushed through a decolonization programme, leaving the Congo abruptly in 1960. Rwanda and Burundi became independent in 1962.

A milestone in domestic politics was reached in 1958 with the School Pact, ending a century of conflict between secularists and conservative Catholics. Prime Minister Gaston Eyskens negotiated a guarantee of funding for state secondary schools and private religious schools. Relations between Walloon and Flemish society became difficult as a result of the decline of Walloon industry. Strikes and discontent with government subsidies set Belgium on the course of federalization. After the division of Brabant along linguistic lines Belgium officially became a federal state in 1993. King Baudouin, who died in 1993, was respected for his even-handed approach to Belgium's divided society and seen as an important symbol of unity. He was succeeded by his brother, Albert II.

Following elections in June 2007 in which the Christian Democratic and Flemish/New Flemish Alliance won most seats but fell short of an overall majority, no new government was formed for 282 days. After a period of interim administration a coalition took office in March 2008. However, Prime Minister Yves Leterme resigned in Dec. 2008 after criticism of the government's role in the bailing-out of Fortis Bank. He was succeeded by Herman Van Rompuy but returned for another term in Nov. 2009 when Van Rompuy left to become the first president of the European Council. Leterme's coalition collapsed six months later after a dispute over francophone voting rights.

With the parliamentary election of June 2010 resulting in no party having an absolute majority, attempts to form a coalition followed. Fundamental differences between the Flemish separatist New Flemish Alliance and the Francophone Socialist Party led to breakdowns in negotiations. After a series of failed talks, it was only in Dec. 2011 that Elio Di Rupo, leader of the Socialist Party, formed a government comprising six parties. Without an administration for 541 days, Belgium has the record for failing to form a new government following an election.

TERRITORY AND POPULATION

Belgium is bounded in the north by the Netherlands, northwest by the North Sea, west and south by France, and east by Germany and Luxembourg. Its area is 30,528 sq. km. Population (at 1 Jan. 2012), 11,035,948 (5,622,147 females); density, 361·5 per sq. km. The Belgian exclave of Baarle-Hertog in the Netherlands has an area of seven sq. km and a population (2010) of 2,504. There were 971,448 resident foreign nationals as at 1 Jan. 2008. In 2011, 97·4% of the population lived in urban areas.

The UN gives a projected population for 2015 of 10·87m.

Dutch (Flemish) is spoken by the Flemish section of the population in the north, French by the Walloon south. The linguistic frontier passes south of the capital, Brussels, which is bilingual. Some German is spoken in the east. Each language has official status in its own community. (Bracketed names below signify French/Dutch alternatives.)

Area, population and chief towns of the ten provinces on 1 Jan. 2012:

Province	Area (sq. km)	Population	Chief Town
Flemish Region			
Antwerp	2,867	1,781,904	Antwerp (Antwerpen/ Anvers)
East Flanders	2,982	1,454,716	Ghent (Gent/Gand)
West Flanders	3,144	1,169,990	Bruges (Brugge)
Flemish Brabant	2,106	1,094,751	Leuven (Louvain)
Limburg	2,422	849,404	Hasselt
Walloon Region			
Hainaut (Henegouwen)	3,786	1,328,196	Mons (Bergen)
Liège (Luik)	3,862	1,083,400	Liège (Luik)
Namur (Namen)	3,666	480,105	Namur (Namen)
Walloon Brabant	1,091	385,990	Wavre (Waver)
Luxembourg	4,440	273,638	Arlon (Aarlen)

Population of the regions on 1 Jan. 2012: Brussels-Capital Region, 1,138,854; Flemish Region, 6,350,765; Walloon Region, 3,546,329.

The most populous towns, with population on 1 Jan. 2012:

Brussels (Brussel/ Bruxelles)[1]	1,138,854	La Louvière	78,774
		Kortrijk (Courtrai)	75,219
Antwerp (Antwerpen/ Anvers)	502,604	Hasselt	74,588
		St Niklaas (St Nicolas)	72,883
Ghent (Gent/Gand)	248,242	Ostend (Oostende/ Ostende)	70,284
Charleroi	203,871		
Liège (Luik)	195,576	Tournai (Doornik)	69,593
Bruges (Brugge)	117,170	Genk	65,264
Namur (Namen)	110,096	Seraing	63,575
Leuven (Louvain)	97,656	Roeselare (Roulers)	58,823
Mons (Bergen)	93,072	Mouscron (Moeskroen)	56,011
Mechelen (Malines)	82,325	Verviers	55,936
Aalst (Alost)	81,853		

[1] 19 communes.

SOCIAL STATISTICS

Statistics for calendar years:

	Births	Deaths	Marriages	Divorces
2005	118,002	103,278	43,141	30,840
2006	121,382	101,587	44,813	29,189
2007	120,663	100,658	45,561	30,081
2008	128,049	104,587	45,613	35,366
2009	127,297	104,509	43,303	32,606

In 2010 Belgium received 19,941 asylum applications, equivalent to 1·9 per 1,000 inhabitants. Annual population growth rate, 2005–10, 0·7%. Life expectancy at birth, 2009, was 77·2 years for men and 82·4 years for women. 2009 birth rate (per 1,000 population): 11·8; death rate: 9·7. Infant mortality, 2008, 3·8 per 1,000 live births; fertility rate, 2008, 1·8 children per woman.

In 2003 Belgium became the second country to legalize same-sex marriage.

CLIMATE

Cool temperate climate influenced by the sea, giving mild winters and cool summers. Brussels, Jan. 36°F (2·2°C), July 64°F (17·8°C). Annual rainfall 33" (825 mm). Ostend, Jan. 38°F (3·3°C), July 62°F (16·7°C). Annual rainfall 31" (775 mm).

CONSTITUTION AND GOVERNMENT

According to the constitution of 1831, Belgium is a constitutional, representative and hereditary monarchy. The legislative power is vested in the King, the federal parliament and the community and regional councils. The King convokes parliament after an election or the resignation of a government, and has the power to dissolve it in accordance with Article 46 of the Constitution.

The reigning King is **Albert II**, born 6 June 1934, who succeeded his brother, Baudouin, on 9 Aug. 1993. Married on 2 July 1959 to Paola Ruffo di Calabria, daughter of Don Fuleo and Donna Luisa Gazelli de Rossena. *Offspring*: Prince Philippe, Duke of Brabant, b. 15 April 1960; Princess Astrid, b. 5 June 1962; Prince Laurent, b. 19 Oct. 1963. Prince Philippe married Mathilde d'Udekem d'Acoz, 4 Dec. 1999. *Offspring*: Princess Elisabeth, b. 25 Oct. 2001; Prince Gabriel, b. 20 Aug. 2003; Prince Emmanuel, b. 4 Oct. 2005; Princess Eléonore, b. 16 April 2008. Princess Astrid married Archduke Lorenz of Austria, 22 Sept. 1984. *Offspring*: Prince Amedeo, b. 21 Feb. 1986; Princess Maria Laura, b. 26 Aug. 1988; Prince Joachim, b. 9 Dec. 1991; Princess Luisa Maria, b. 11 Oct. 1995; Princess Laetitia Maria, b. 23 April 2003. Prince Laurent married Claire Coombs, 12 April 2003. *Offspring*: Princess Louise, b. 6 Feb. 2004; Prince Nicolas, b. 13 Dec. 2005; Prince Aymeric, b. 13 Dec. 2005.

A constitutional amendment of June 1991 permits women to accede to the throne.

The King received an allowance of €10,338,000 for 2011; Queen Fabiola received €1,441,000; Prince Philippe, €922,000; Princess Astrid €319,000; and Prince Laurent, €307,000. Owing to the country's financial troubles, the King agreed that allowances for the members of the Royal Family should be frozen in 2012–13.

Constitutional reforms begun in Dec. 1970 culminated in May 1993 in the transformation of Belgium from a unitary into a 'federal state, composed of communities and regions'. The communities are three in number and based on language: Flemish, French and German. The regions also number three, and are based territorially: Flemish, Walloon and the Brussels-Capital Region.

Since 1995 the federal parliament has consisted of a 150-member *Chamber of Representatives*, directly elected by obligatory universal suffrage from 20 constituencies on a proportional representation system for four-year terms; and a *Senate* of 71 members (excluding senators by right, i.e. certain members of the Royal Family). 25 senators are elected by a Dutch-speaking, and 15 by a French-speaking, electoral college; 21 are designated by community councils (ten Flemish, ten French and one German). These senators co-opt a further ten senators (six Dutch-speaking and four French-speaking).

The federal parliament's powers relate to constitutional reform, federal finance, foreign affairs, defence, justice, internal security, social security and some areas of public health. The Senate is essentially a revising chamber, though it may initiate certain legislation, and is equally competent with the Chamber of Representatives in matters concerning constitutional reform and the assent to international treaties.

The number of ministers in the federal government is limited to 15. The Council of Ministers, apart from the Prime Minister, must comprise an equal number of Dutch- and French-speakers. Members of parliament, if appointed ministers, are replaced in parliament by the runner-up on the electoral list for the minister's period of office. Community and regional councillors may not be members of the Chamber of Representatives or Senate.

National Anthem

'La Brabançonne'; words by A. Dechet, tune by F. van Campenhout. The Flemish version is 'O dierbaar België, O heilig land der vaad'ren' ('Noble Belgium, for ever a dear land').

GOVERNMENT CHRONOLOGY

Prime Ministers since 1939. (BSP/PSB = Belgian Socialist Party; Christian Democratic and Flemish = CD&V; CVP = Christian People's Party; CVP/PSC = Christian People's/Social Christian Party; PS = Socialist Party; VLD = Flemish Liberals and Democrats)

1939–45	CVP/PSC	Hubert Pierlot
1945–46	BSP/PSB	Achille Van Acker
1946	BSP/PSB	Paul-Henri Spaak
1946	BSP/PSB	Achille Van Acker
1946–47	BSP/PSB	Camille Huysmans
1947–49	BSP/PSB	Paul-Henri Spaak
1949–50	CVP/PSC	Gaston Eyskens
1950	CVP/PSC	Jean Pierre Duvieusart
1950–52	CVP/PSC	Louis Marie Joseph Pholien
1952–54	CVP/PSC	Jean Marie Van Houtte
1954–58	BSP/PSB	Achille Van Acker
1958–61	CVP/PSC	Gaston Eyskens
1961–65	CVP/PSC	Théodore Lefèvre
1965–66	CVP/PSC	Pierre Charles Harmel
1966–68	CVP/PSC	Paul Vanden Boeynants
1968–73	CVP	Gaston Eyskens
1973–74	BSP/PSB	Edmond Jules Leburton
1974–78	CVP	Léo Tindemans
1978–79	CVP	Paul Vanden Boeynants
1979–81	CVP	Wilfried Martens
1981	CVP	Mark Eyskens
1981–92	CVP	Wilfried Martens
1992–99	CVP	Jean-Luc Dehaene
1999–2008	VLD	Guy Verhofstadt
2008	CD&V	Yves Leterme
2008–09	CD&V	Herman Van Rompuy
2009–11	CD&V	Yves Leterme
2011–	PS	Elio Di Rupo

RECENT ELECTIONS

Elections to the 150-member Chamber of Representatives were held on 13 June 2010. The New Flemish Alliance (N-VA) won 27 seats with 17·4% of votes cast; the Socialist Party (PS) won 26 seats (13·7%); Reformist Movement (MR) won 18 seats (9·3%); Christian Democratic and Flemish (CD&V) won 17 seats (10·9%); Socialist Party Alternative (sp.a) won 13 seats (9·2%); Open Flemish Liberals and Democrats (Open Vld) won 13 seats (8·6%); Flemish Interest (VB) won 12 seats (7·8%); the Humanist Democratic Centre (CDH) won 9 seats (5·5%); Ecolo won 8 seats (4·8%); Groen! won 5 seats (4·4%); List Dedecker won 1 seat (2·3%); and the Popular Party won 1 seat (1·3%). Turnout was 89·2%.

Voting for the 40 electable seats in the Senate took place on the same day. N-VA won 9 seats; PS, 7; CD&V, sp.a, MR and Open Vld, 4 each; VB, 3; Ecolo and Humanist Democratic Centre, 2 each; and Groen!, 1. There are also 31 indirectly elected senators.

European Parliament

Belgium has 22 (24 in 2004) representatives. At the June 2009 elections turnout was 90·4% (90·8% in 2004). The CD&V won 3 seats with 14·4% of the vote (political affiliation in European Parliament: European People's Party); the Open Vld, 3 with 12·8% (Alliance of Liberals and Democrats for Europe); the PS, 3 with 10·9% (Progressive Alliance of Socialists and Democrats); the VB, 2 with 9·9% (non-attached); the MR, 2 with 9·7% (Alliance of Liberals and Democrats for Europe); Ecolo, 2 with 8·6% (Greens/

European Free Alliance); sp.a, 2 with 8·2% (Progressive Alliance of Socialists and Democrats); the N-VA, 1 with 6·1% (Greens/European Free Alliance); the CDH, 1 with 5·0% (European People's Party); Groen!, 1 with 4·9% (Greens/European Free Alliance); the LDD, 1 with 4·5% (European Conservatives and Reformists); the Christian-Social Party, 1 with 0·2% (European People's Party).

CURRENT GOVERNMENT

In March 2013 the government comprised:

Prime Minister: Elio Di Rupo; b. 1960 (PS; sworn in 6 Dec. 2011).

Deputy Prime Ministers: Pieter De Crem (CD&V; also *Minister of Defence*); Didier Reynders (MR; also *Minister of Foreign and European Affairs, and Foreign Trade*); Johan Vande Lanotte (sp.a; also *Minister of Economy, Consumer Affairs and the North Sea*); Alexander De Croo (Open Vld; also *Minister of Pensions*); Jöelle Milquet (CDH; also *Minister of Interior*); Laurette Onkelinx (PS; also *Minister of Social and Public Health, in Charge of Beliris and Federal Cultural Institutions*).

Minister for Budget and Administrative Simplification: Olivier Chastel (MR). *Employment:* Monica De Coninck (sp.a). *Finance, in Charge of the Civil Service:* Koen Geens (CD&V). *Justice:* Annemie Turtelboom (Open Vld). *Public Enterprises and Development Co-operation, in Charge of Cities:* Jean-Pascal Labille (PS). *Middle Classes, Small and Medium-Sized Enterprises, the Self-Employed and Agriculture:* Sabine Laruelle (MR).

Government Website: http://www.belgium.be

CURRENT LEADERS

Elio Di Rupo

Position

Prime Minister

Introduction

Elio Di Rupo became prime minister in Dec. 2011 at the head of a six-party coalition, bringing to a conclusion talks that lasted 541 days. A French speaker from the Walloon region, Di Rupo is regarded as a pragmatist and a skilled negotiator.

Early Life

Born on 18 July 1951 in Morlanwelz to Italian migrant parents, Elio Di Rupo was awarded a PhD in chemistry from the University of Mons-Hainault (UMH). He taught at Leeds University in the UK from 1977 before returning to Belgium the following year to work at UMH and the Institute for Scientific Research in Industry and Agriculture.

He joined the Socialist Party (PS) and was elected a municipal councillor in Mons in 1982. In 1987 he entered parliament as MP for Mons-Borinage, serving until 1989 when he was elected a Member of the European Parliament. In 1991 he became a senator and from 1992–94 was minister of education for the French-speaking community, as well as minister for audiovisual policy from 1993–94. In 1994 Di Rupo was appointed deputy prime minister and minister for communications and public enterprises. He remained deputy prime minister following the 1995 parliamentary election but was named minister for the economy and telecommunications. In 1998 he was also put in charge of foreign trade.

After the 1999 election Di Rupo helped negotiate the 'rainbow coalition' of Liberals, Socialists and Greens. As minister-president of the Walloon region, he initiated a ten-year programme designed to lift the region out of economic decline. In Oct. 1999 he was elected president of the PS and resigned his minister-president post. He also became vice-president of the Socialist International group.

While serving as mayor of Mons from 2000–05, he was appointed minister of state, representing Belgium in talks on a European constitution. As president of the PS, Di Rupo oversaw a review to widen the party's appeal and, following its strong performance in the 2003 election, brought it into the 'Purple Coalition' of Liberals and Socialists. After updating his development plan for the Walloon region, he was reappointed its minister-president in Oct. 2005, holding the post until July 2007 when he again became mayor of Mons.

In the June 2010 general election the PS became the second largest party in parliament, one seat behind the nationalist New-Flemish Alliance (N-VA). Di Rupo participated in negotiations with N-VA to form a coalition government but differences over increased autonomy for the regions resulted in deadlock. In Oct. 2011 he instigated talks with five other parties to form a government. In Nov. 2011, after Belgium's credit status was downgraded, the parties agreed a package of budget cuts and constitutional reforms, including increased fiscal autonomy for the regions.

Career in Office

Di Rupo took office on 6 Dec. 2011 at the head of a coalition of Liberal, Socialist and Christian Democrat parties, promising to cut €11·3bn. from the national budget. In early 2012 his austerity measures provoked strikes and his centre-right coalition partners resisted planned tax increases. His principal task is to maintain a coalition of diverse partners while addressing Belgium's economic difficulties. In addition, he faces opposition from Flemish nationalists.

DEFENCE

Conscription was abolished in 1994 and the Armed Forces were restructured, with the aim of progressively reducing the size and making more use of civilian personnel. Since 1 Jan. 2002 they have been organized into one unified structure consisting of four main components: the Land Component (Army), Naval Component (Navy), Air Component (Air Force) and Medical Component.

In 2008 defence expenditure totalled US$5,551m. (US$534 per capita), representing 1·1% of GDP.

Army

The Land Component (formerly Army) has five 'capacities': the command capacity, the combat capacity, the support capacity, the services capacity and the training capacity. Total strength (2007) 12,571. In addition there are 2,040 reserves. All tracked vehicles are in the process of being phased out in favour of wheeled vehicles. The transition is scheduled to be completed by 2015.

Navy

The Naval Component (formerly Navy), based at Ostend and Zeebrugge, includes two frigates. Personnel (2007) totalled 1,605.

The naval air arm comprises three general utility helicopters.

Air Force

The Air Component (formerly Belgian Air Force) has a strength of (2007) 7,470 personnel. There are two tactical wings, based at Florennes and Kleine Brogel. Equipment in 2007 included 71 combat-capable aircraft (Lockheed Martin F-16s), plus 44 helicopters.

ECONOMY

Services contributed 75% of GDP in 2007, with industry accounting for 24% and agriculture 1%.

According to the anti-corruption organization *Transparency International*, Belgium ranked 16th in the world in a 2012 survey of the countries with the least corruption in business and government. It received 75 out of 100 in the annual index.

Overview

Belgium's open economy is closely linked with those of Germany, France and the Netherlands, which jointly accounted for 49·5% of Belgian exports in 2009.

Economic activity slowed at the end of 2007 and into 2008, the result of low private consumption and investment as well as negative export growth stemming from the US sub-prime mortgage crisis. Bankruptcies increased by 12·5% year-on-year in Dec. 2008. The end of 2008 also saw a downturn in the manufacturing industry, although consumer confidence and activity in the trade and construction sectors recovered slowly. In Oct. 2008 the French bank BNP Paribas bailed out the Belgian part of Fortis, the Belgian–Dutch bank conglomerate, the Dutch part having been nationalized the same month. In Feb. 2009 Fortis shareholders voted against the sale of the bank to BNP Paribas and the Dutch government. A revised deal giving BNP Paribas a 75% stake in Fortis was approved by the Belgian government in March 2009.

A slow recovery in GDP growth began in the third quarter of 2009 and continued through 2010, when GDP expanded by 2·4% thanks to a modest increase in private consumption and exports. However, growth slowed to 1·8% in 2011 as confidence weakened in tandem with a significant slowdown across the rest of the eurozone. The eurozone crisis and the drying up of liquidity in those financial markets led to the restructuring of the Dexia Group, with its retail operations in Belgium nationalized at a cost of 1% of GDP.

Inflexibilities in the labour market, including low labour mobility and wage inflexibility, have contributed to long-term unemployment. Labour utilization is low by OECD standards, especially for older workers (employment stood at 32·8% among people aged between 55 and 64 in 2008), younger workers (26·9% among those aged between 15 and 24 in 2008) and minorities. Employment for the rest of the prime-age population is close to international rates. The unemployment rate of ethnic minorities is three times that of native Belgians because of poor education, language barriers and ineffective anti-discrimination legislation.

The OECD estimates that the old-age ratio will double by 2050, reducing economic growth and putting pressure on public finances. Since 2000 fiscal discipline has enabled a steady reduction in the public debt ratio. Public debt fell below 100% of GDP at the end of 2004 for the first time in 30 years and declined further to 84·1% in 2007. However, loans and capital injections into financial institutions in 2008 prompted public debt to increase to 97% of GDP in 2011. A Stability Programme implemented in the same year seeks to achieve a balanced budget by 2015.

After more than 540 days under a caretaker government, a federal coalition government assumed office in Dec. 2011 with the aim of achieving fiscal sustainability in the medium-term, containing risks in the financial sector and increasing employment and growth. The 2012 budget included a fiscal consolidation package of 2·5% of GDP in order to bring the fiscal deficit below 3% of GDP in 2012. Authorities have committed to strengthening banking supervision through implementation of the Basel III and Solvency II regulatory frameworks, and the minimum age for early retirement is expected to rise from 60 to 62 years by 2016. However, continued eurozone uncertainty and slow growth across the continent contribute to an uncertain outlook for the economy.

Currency

On 1 Jan. 1999 the euro (EUR) became the legal currency in Belgium at the irrevocable conversion rate of BEF40·3399 to EUR1. The euro, which consists of 100 cents, has been in circulation since 1 Jan. 2002. On the introduction of the euro there was a 'dual circulation' period before the Belgian franc ceased to be legal tender on 28 Feb. 2002. Euro banknotes in circulation on 1 Jan. 2002 had a total value of €24·0bn.

Inflation rates (based on OECD statistics):

2002	2003	2004	2005	2006	2007	2008	2009	2010	2011
1·6%	1·5%	1·9%	2·5%	2·3%	1·8%	4·5%	0·0%	2·3%	3·5%

In Sept. 2009 gold reserves were 7·32m. troy oz and foreign exchange reserves US$8,060m. Total money supply was €97,656m. in Aug. 2009.

Budget

Central government revenue and expenditure in €1m. for calendar years:

	2008	2009	2010[1]
Revenue	142,438	136,829	144,655
Expenditure	146,861	153,666	156,383

[1]Provisional.

Principal sources of revenue in 2008: taxes on income, profits and capital gains, €53,007m.; social security contributions, €45,899m.; taxes on goods and services, €33,643m. Main items of expenditure by economic type in 2008: social benefits, €70,391m.; grants, €41,512m.; interest, €12,261m.

Belgium's budget deficit in 2011 was 3·7% of GDP (2010, 3·8%; 2009, 5·6%). The required target set by the EU is a budget deficit of no more than 3%.

VAT is 21% (reduced rates, 12% and 6%).

Performance

Real GDP growth rates (based on OECD statistics):

2002	2003	2004	2005	2006	2007	2008	2009	2010	2011
1·4%	0·8%	3·2%	1·8%	2·7%	2·9%	1·0%	−2·7%	2·4%	1·8%

Total GDP in 2011 was US$513·7bn.

Banking and Finance

The National Bank of Belgium was established in 1850. The *Governor*—Luc Coene—was appointed in 2011 for a five-year period. Its shares are listed on Euronext (Brussels).

The law of 22 Feb. 1998 has adapted the status of the National Bank of Belgium in view of the realization of the Economic and Monetary Union.

The National Bank of Belgium is within the ESCB-framework in charge of the issue of banknotes, the execution of exchange rate policy and monetary policy. Furthermore, it is the Bank of banks and the cashier of the federal state.

The law of 4 Dec. 1990 on financial transactions and financial markets defines the legal framework for collective investment institutions, the sole object of which is the collective investment of capital raised from the public. It transposes into Belgian legislation the European Directive of 20 Dec. 1985 on the co-ordination of laws, regulations and administrative provisions relating to undertakings for collective investment in transferable securities.

The law of 6 April 1995 relating to secondary markets, status and supervision of investment firms, intermediaries and investment consultants, provides the credit institutions with direct access to securities' stock exchanges. Stock exchange legislation was also subject to an important reform. The law fundamentally modifies the competitive environment and strengthens exercise conditions for securities' dealers.

On 30 June 2006, 103 credit institutions with a balance sheet totalling €1,150bn. were established in Belgium: 52 governed by Belgian law and 51 by foreign law. 372 collective investment institutions (152 Belgian and 220 foreign) were marketed in Belgium and supervised by the Banking, Finance and Insurance Commission; and 80 investment firms were operating in Belgium with the approval of the Banking, Finance and Insurance Commission.

Gross external debt totalled US$1,379,070m. in June 2012.

There is a stock exchange (a component of Euronext) in Brussels. Euronext was created in Sept. 2000 through the merger of the Amsterdam, Brussels and Paris bourses.

ENERGY AND NATURAL RESOURCES

In 2008, 3·3% of energy consumption came from renewables (wind power, solar power, hydro-electric power, tidal power, geothermal energy and biomass), compared to the European Union average of 10·3%. A target of 13% has been set by the EU for 2020.

Environment

Belgium's carbon dioxide emissions from the consumption and flaring of fossil fuels were the equivalent of 14·9 tonnes per capita in 2008.

Electricity

The production of electricity amounted to 88·6bn. kWh in 2007; consumption per capita (2007) was 8,981 kWh. Installed capacity (2007) was 16·4m. kW. 54% of production in 2007 was nuclear-produced. Belgium had seven operational nuclear reactors in 2010.

Minerals

Belgium's mineral resources are very limited; the most abundantly occurring mineral is calcite.

Agriculture

There were, in 2006, 1,382,390 ha. under cultivation, of which 841,666 ha. were arable land. There were 49,850 farms in 2006.

Chief crops	Area (ha.) 2005	2006	Production (tonnes) 2005	2006
Barley	39,965	49,008	301,647	367,348
Beet (fodder)	3,750	3,423	371,694	330,290
Beet (sugar)	85,527	82,912	5,983,173	5,666,621
Chicory	15,649	8,210	704,821	371,238
Maize (fodder)	163,825	161,178	7,745,553	6,600,738
Maize grain	54,256	56,500	634,088	575,898
Potatoes	64,952	67,267	2,780,865	2,592,820
Wheat	204,209	201,330	1,737,552	1,661,958

At the May 2006 agricultural census there were 6,294,904 pigs, 2,663,076 cattle, 153,976 sheep, 34,799 horses, 27,985 goats and 32,866,650 poultry.

Forestry

In 2010 forest covered 0·68m. ha. (22% of the total land area). Timber production in 2007 was 4·95m. cu. metres.

Fisheries

In 2011 the Belgian fishing fleet numbered 89 vessels of 15,800 gross tonnes. Total catch in 2010 was 22,418 tonnes, almost entirely from marine waters.

INDUSTRY

The leading companies by market capitalization in March 2012 were: Anheuser-Busch InBev, a beverages company (US$117·2bn.); Groupe Bruxelles Lambert (GBL), a financial services company (US$12·5bn.); and Belgacom, a telecommunications company (US$10·9bn.).

Output, 2007 unless otherwise indicated, in 1,000 tonnes: distillate fuel oil, 12,737; crude steel (2006), 11,238; cement (2006), 8,192; residual fuel oil, 7,391; petrol, 5,041; sugar (2005), 3,394; beer (2006), 1,784·1m. litres; mineral water (2006), 1,066·8m. litres.

Labour

In 2010 (Labour Force Survey), 60,686 persons worked in the primary sector (agriculture, fishing and mining), 1,049,239 in the secondary sector (industry and construction) and 3,298,598 in the tertiary sector (services). The unemployment rate was 7·5% in Dec. 2012. In French-speaking Wallonia the rate is more than double that in Dutch-speaking Flanders.

INTERNATIONAL TRADE

In 1922 the customs frontier between Belgium and Luxembourg was abolished; their foreign trade figures are amalgamated.

Imports and Exports

Trade in €1m.:

	Imports	Exports
2007	300,298·2	314,449·4
2008	317,044·0	320,806·0
2009	253,344·5	265,609·8

Leading imports and exports (in €1m.):

	Imports 2008	2009	Exports 2008	2009
Chemicals and pharmaceutical products	62,438·6	59,436·0	71,991·1	71,047·5
Machinery and appliances	43,127·1	34,190·4	38,929·6	30,623·7
Vehicles and transport equipment	36,280·9	28,441·7	35,631·2	26,805·3
Mineral products	52,198·0	31,730·0	32,099·4	20,356·2
Base metals	27,684·2	17,318·5	32,022·4	20,364·1
Plastics and rubber	15,943·8	12,968·4	25,575·6	20,913·3
Food industry	10,416·9	10,065·6	12,715·3	12,490·4
Textile and textile articles	12,202·7	9,122·2	12,107·5	10,558·9
Precious stones and precious metals	12,053·8	8,368·5	12,992·5	9,866·5
Vegetable products	9,094·6	8,104·3	7,680·4	6,927·8

Trade by selected countries (in €1m.):

	Imports from 2008	2009	Exports to 2008	2009
China	5,805·9	5,014·4	2,788·4	3,382·2
France	28,294·5	24,183·4	38,931·9	31,601·4
Germany	37,860·9	30,747·0	38,302·6	29,933·4
India	2,668·8	1,992·9	4,756·0	4,250·5
Italy	7,436·3	6,359·2	9,818·4	7,573·1
Japan	5,660·9	5,102·2	1,187·5	1,058·5
Luxembourg	2,714·8	1,681·9	5,479·6	4,126·5
Netherlands	57,334·6	41,667·0	30,339·9	24,573·4
Poland	2,445·4	2,143·3	3,958·6	3,103·4
Russia	4,759·0	3,552·6	2,976·1	1,668·3
Spain	5,277·4	4,469·3	7,064·9	5,488·0
Sweden	5,195·6	3,169·3	3,332·1	2,420·8
Switzerland	2,382·8	2,163·1	3,307·8	2,305·8
UK	14,594·9	10,444·9	15,676·1	12,805·6
USA	9,955·0	8,017·7	8,567·8	7,588·8

In 2009 fellow EU-member countries accounted for 54·0% of imports and 50·5% of exports.

Trade Fairs

Brussels ranks as the second most popular convention city behind Singapore according to the Union des Associations Internationales (UAI), hosting 4·4% of all international meetings held in 2010.

COMMUNICATIONS

Roads

Length of roads, 2006: motorways, 1,763 km; national roads, 12,585 km; secondary roads, 1,349 km; local roads, 136,559 km. Belgium has one of the densest road networks in the world. In 2007 there were 5,006,300 passenger cars in use, 29,000 buses and coaches, 696,700 lorries and vans, and 371,500 motorcycles and mopeds. Road accidents caused 994 fatalities in both 2008 and 2009 (1,470 in 2000).

Rail

The main Belgian lines were a State enterprise from their inception in 1834. In 1926 the Société Nationale des Chemins de Fer Belges (SNCB) was formed to take over the railways. In 2005 SNCB was divided into separate operating and infrastructure companies. The length of railway operated in 2005 was 3,696 km (electrified, 3,110 km). In 2008, 217m. passengers and 55·5m. tonnes of freight were carried.

The regional transport undertakings Société Régionale Wallonne de Transport and Vlaamse Vervoermaatschappij operate tramways around Charleroi (20 km) and from De Panne to Knokke (55 km). There is also a metro and tramway in Brussels (175 km), and tramways in Antwerp (57 km) and Ghent (30 km). An urban commuter rail network around Brussels is expected to become operational by 2015–16. A tram link between Hasselt and Maastricht in the Netherlands is currently under construction; it is expected to open in 2017.

Civil Aviation
The former national airline SABENA (*Société anonyme belge d'exploitation de la navigation aérienne*) was set up in 1923. However, in Nov. 2001 it filed for bankruptcy. Its successor, Delta Air Transport (DAT), a former SABENA subsidiary, was given a new identity in Feb. 2002 as SN Brussels Airlines. In Nov. 2006 SN Brussels Airlines merged with Virgin Express and since March 2007 has been trading under the name Brussels Airlines.

The busiest airport is Brussels National Airport (Zaventem), which handled 18,710,388 passengers in 2008 and 658,743 tonnes of freight. Charleroi is the second busiest airport in terms of passenger numbers and Liège the third busiest.

Shipping
In Jan. 2009 there were 81 ships of 300 GT or over registered, totalling 3·93m. GT. Of the 81 vessels registered, 23 were general cargo ships, 19 liquid gas tankers, 18 bulk carriers, ten oil tankers, six container ships, four passenger ships and there was one chemical tanker. Antwerp is Europe's second busiest port in terms of both total cargo handled and container traffic after Rotterdam. In 2008, 189,390,000 tonnes of cargo were handled at the port of Antwerp (84,371,000 tonnes loaded and 105,018,000 tonnes discharged), with total container throughput 8,663,000 TEUs (twenty-foot equivalent units).

The length of navigable inland waterways was 1,516 km in 2004; 108·2m. tonnes of freight were carried on inland waterways in 2009.

Telecommunications
In 2008 there were 4,457,000 main (fixed) telephone lines. In the same year mobile phone subscribers numbered 11,822,000 (1,116·3 per 1,000 persons). There were 7·3m. internet users in 2008. The fixed broadband penetration rate in Dec. 2010 was 30·8 subscribers per 100 inhabitants. In March 2012 there were 4·6m. Facebook users.

SOCIAL INSTITUTIONS
Justice
Judges are appointed for life. There is a court of cassation, five courts of appeal and assize courts for political and criminal cases. There are 27 judicial districts, each with a court of first instance. In each of the 222 cantons is a justice and judge of the peace. There are also various special tribunals. There is trial by jury in assize courts. The death penalty, which had been in abeyance for 45 years, was formally abolished in 1991.

The Gendarmerie ceased to be part of the Army in Jan. 1992.

The population in penal institutions in June 2008 was 10,002 (93 per 100,000 of national population). Owing to overcrowding in Belgian prisons some inmates have since Feb. 2010 been accommodated at Tilburg prison in the Netherlands.

In Aug. 2003 a new act reformed war crimes legislation introduced in 1993 which allowed for charges to be brought against foreign nationals accused of abuses committed outside Belgian jurisdiction. The amendment requires that either accuser or defendant be a citizen of or resident in Belgium.

Education
Following the constitutional reform of 1988, education is the responsibility of the three Communities (the French Community, the Flemish Community and the German-speaking Community). Education is free and compulsory from the age of six to 18, although from 16 to 18 it may be part-time.

In 2007 there were 411,951 children and 29,550 teaching staff in pre-primary schools; 732,411 pupils and 65,378 teaching staff in primary schools; 825,293 pupils and (2006) 81,873 teaching staff in secondary schools; and 393,687 students and 26,298 academic staff in tertiary education. There were 16 universities and 73 non-university colleges and institutes in 2006–07. There are five royal academies of fine arts and six royal conservatoires at Brussels (one Flemish and one French), Liège, Ghent, Antwerp and Mons.

Public expenditure on education in 2005 amounted to 6·0% of GNI and represented 12·1% of total government expenditure.

The adult literacy rate is at least 99%.

Health
On 31 Dec. 2006 there were 42,326 physicians, 8,423 dentists and 12,109 pharmacists. There were 210 hospitals with 55,000 beds in 2007. Total health spending accounted for 9·4% of GDP in 2007. In Jan. 2000 the Belgian government agreed to decriminalize the use of cannabis. Euthanasia became legal on 24 Sept. 2002. The Belgian Chamber of Representatives had given its approval on 16 May 2002 to a measure adopted by the Senate on 26 Oct. 2001. Belgium was the second country to legalize euthanasia, after the Netherlands.

Welfare
Expenditure on social security, 2006: wage earners, €50,774·6m.; self employed, €3,858·9m. Expenditure on pensions, 2006: wage earners, €15,324·55m.; self employed, €2,192·28m.

The retirement age for men and women is 65. In 2005 the government increased the early retirement age from 58 to 60 years of age in a move that met with opposition from trade unions. A full pension is 60% of average lifetime earnings (75% for married couples). Based on size and sustainability of payments, a report by Aon Consulting in Nov. 2005 rated Belgium as the country with the worst pensions system of the 15 pre-expansion EU countries.

RELIGION
There is full religious liberty, and part of the income of the ministers of all denominations is paid by the State. In 2001 there were 8·31m. Roman Catholics. Numbers of clergy, 1996: Roman Catholic, 3,899; Protestant, 84; Anglican, 9; Jews, 26; Greek Orthodox, 39. There are eight Roman Catholic dioceses subdivided into 260 deaneries. In Feb. 2013 there were one cardinal. The Protestant (Evangelical) Church is under a synod. There is also a Central Jewish Consistory, a Central Committee of the Anglican Church and a Free Protestant Church.

CULTURE
World Heritage Sites
Belgium has 11 sites which have been included on the UNESCO world heritage list. They are: the Flemish Béguinages (1998); the four lifts on the Canal du Centre and their environs (1998); La Grand Place in Brussels (1998); the historic centre of Bruges (2000); the major town houses of the architect Victor Horta in Brussels (2000); the Neolithic flint mines at Spiennes (2000); Notre Dame cathedral in Tournai (2000); the Plantin-Moretus Museum, a Renaissance printing and publishing house (2005); Stoclet House in Brussels, a house designed by artists of the Vienna Secession movement (2009); and Major Mining Sites of Wallonia, the four best-preserved mining sites of the country dating back to the late 17th century (2012).

Belgium shares the belfries of Belgium and France (1999 and 2005) with France.

Press

In 2008 there were 23 daily newspapers (21 paid-for and two free) with a combined circulation of 1,669,000, at a rate of 190 per 1,000 adult population. Belgium's biggest-selling national daily is *Het Laatste Nieuws*, with an average daily circulation of 287,000 copies in 2008.

Tourism

In 2006, 29,372,011 tourist nights were spent in 3,485 establishments in accommodation for 367,866 persons. The number of overnight stays in 2006 accounted for by leisure, holiday and recreation was 22,611,797, with 3,622,212 for congresses and conferences, and 3,138,002 for other business purposes. Total number of tourists was 11,800,974 (8,372,294 leisure, 1,938,469 conference and 1,490,211 for other business purposes).

Festivals

The country's largest festival is Gentse Feesten (July), which has been held annually in Ghent since 1843 and combines music and theatre. The main rock and pop festivals are Rock Werchter (July) at Werchter near Leuven, Klinkers (July–Aug.) in Bruges, Pukkelpop (Aug.) in Hasselt and Marktrock in Leuven (Aug.).

DIPLOMATIC REPRESENTATIVES

Of Belgium in the United Kingdom (17 Grosvenor Cres., London, SW1X 7EE)
Ambassador: Johan Verbeke.

Of the United Kingdom in Belgium (Av. d'Auderghem 10, Oudergemlaan, 1040 Brussels)
Ambassador: Jonathan Brenton.

Of Belgium in the USA (3330 Garfield St., NW, Washington, D.C., 20008)
Ambassador: Jan Matthysen.

Of the USA in Belgium (Blvd du Régent 27, 1000 Brussels)
Ambassador: Howard. W. Gutman.

Of Belgium to the United Nations
Ambassador: Jan Grauls.

Of Belgium to the European Union
Permanent Representative: Dirk Wouters.

FURTHER READING

The Institut National de Statistique. *Statistiques du commerce extérieur* (monthly). *Bulletin de Statistique.* Bi-monthly. *Annuaire Statistique de la Belgique* (from 1870).—*Annuaire statistique de poche* (from 1965).
Service Fédéral d'Information. *Guide de l'Administration Fédérale.* Occasional

Arblaster, Paul, *A History of the Low Countries.* 2005
Blom, J. C. H. and Lamberts, E. (eds.) *History of the Low Countries.* Revised ed. 2006
Deprez, K., and Vos, L., *Nationalism in Belgium—Shifting Identities, 1780–1995.* 1998
Deschouwer, Kris, *The Politics of Belgium: Governing a Divided Society.* 2nd ed. 2012
Fitzmaurice, J., *The Politics of Belgium: a Unique Federalism.* 1996
Hermans, T. J., *et al.,* (eds.) *The Flemish Movement: a Documentary History.* 1992
Witte, Els, *Political History of Belgium from 1830 Onwards.* 2000

National Statistical Office: Institut National de Statistique, Rue de Louvain 44, 1000 Brussels.
Service Fédérale d'Information: POB 3000, 1040 Brussels 4.
Website: http://statbel.fgov.be

BELIZE

MEXICO
Corozal
BELIZE
San Pedro
Belize City
⊡ BELMOPAN
Middlesex
Caribbean Sea
GUATEMALA
Punta Gorda
HONDURAS
© Research Machines plc 2006
0   25 mi
0   50 km

Capital: Belmopan
Population projection, 2015: 344,000
GNI per capita, 2011: (PPP$) 5,812
HDI/world rank: 0·699/93
Internet domain extension: .bz

KEY HISTORICAL EVENTS

Evidence of farming settlements at Cuello in northern Belize dates to around 2000 BC. Over the following centuries Mayan towns and villages encompassed Belize, Mexico's Yucatán peninsula and much of Guatemala. The city of Caracol, near Belize's border with Guatemala, is estimated to have spread over 140 sq. km, with some 180,000 residents at its height in the 7th century AD. However, the Mayan civilization declined rapidly after AD 900 for reasons that are still unclear. Construction of the great pyramid temples ceased, literacy was abandoned and subsistence farming returned.

By 1525 the Spanish adventurer, Hernán Cortés, established a base in Honduras. From 1543 Melchor and Alonso Pacheco took control of land around Tipu in southern Belize, which became part of the Spanish Empire. However, Spanish control was limited and Tipu became a centre of Mayan resistance in the late 1630s. By the 1640s British buccaneers were attacking Spanish ships from havens along Belize's coast. Some established settlements and traded logwood, then used in dyes. After the capture of Jamaica from Spain in 1655 they were joined by demobilized British soldiers and sailors.

Spanish attempts to expel the British settlers ended with the defeat of the commander of Yucatán, Arturo O'Neill, at the Battle of St George's Caye in 1798. Belize was part of Britain's 'informal empire' during the 18th century. Logwood and mahogany were harvested by slaves with British Honduras, as it was known, a focal point for Central American trade until the Panama railway was completed in 1855. In 1862 British Honduras was declared a British colony with a legislative assembly and a lieutenant-governor under the governor of Jamaica. The administrative connection with Jamaica was severed in 1884.

Belize's small economy was hit by the depression in the 1930s and the capital, Belize City, was laid waste by a hurricane in Sept. 1931. Widespread protests over unemployment and poor living conditions in the mid-1930s were led by Antonio Soberanis Gómez. The establishment of the General Workers' Union became one of the foundations of Belize's nationalist movement after the Second World War. The People's United Party (PUP), formed in 1950, became the dominant political force under George Price. Universal suffrage was introduced in 1964 and thereafter the majority of the legislature were elected rather than appointed.

The road to independence from Britain was complicated by turbulent relations with Guatemala, which had long claimed Belize as its territory. Price rejected calls for an 'associated state' of Guatemala and full independence was achieved on 21 Sept. 1981, prompting Guatemala to threaten war. Price served as Belize's first prime minister until his party was defeated by the United Democratic Party (UDP) under Manuel Esquivel in Dec. 1984. Guatemala officially recognized Belize as an independent sovereign nation in Sept. 1991. But a border dispute rumbled on, remaining unresolved at the time of Dean Barrow's election as prime minister in Feb. 2008, following a landslide victory for the UDP.

TERRITORY AND POPULATION

Belize is bounded in the north by Mexico, west and south by Guatemala and east by the Caribbean. Fringing the coast there are three atolls and some 400 islets (cays) in the world's second longest barrier reef (140 miles), which was declared a world heritage site in 1996. Area, 22,966 sq. km.

There are six districts as follows, with area, 2010 provisional census population and chief city:

District	Area (in sq. km)	Population	Chief City	Population
Belize	4,307	89,247	Belize City	53,532
Cayo	5,196	73,202	San Ignacio/ Santa Elena	16,977
Corozal	1,860	40,324	Corozal	9,871
Orange Walk	4,636	45,419	Orange Walk	13,400
Stann Creek	2,554	32,166	Dangriga	9,096
Toledo	4,413	30,538	Punta Gorda	5,205

Provisional population at the 2010 census, 312,971; density, 13·6 per sq. km.

The UN gives a projected population for 2015 of 344,000.

The capital is Belmopan (2010 provisional census population, 13,654). In 2007, 51·0% of the population were urban.

English is the official language. Spanish is widely spoken. In 2010 the main ethnic groups were Mestizo (Spanish-Maya), 49·7%; Creole (African descent), 20·8%; Mayans, 9·9%; and Garifuna (Caribs), 4·6%.

SOCIAL STATISTICS

2009 births (est.), 8,000; deaths (est.), 1,000. In 2009 the estimated birth rate per 1,000 was 25 and the death rate 4; infant mortality in 2010 was 14 per 1,000 live births; there were 2,020 marriages in 2004. Life expectancy in 2007 was 74·2 years for males and 78·0 for females. Annual population growth rate, 2000–05, 3·1%; fertility rate, 2008, 2·9 children per woman.

CLIMATE

A tropical climate with high rainfall and small annual range of temperature. The driest months are Feb. and March. Belize City, Jan. 74°F (23·3°C), July 81°F (27·2°C). Annual rainfall 76" (1,890 mm).

CONSTITUTION AND GOVERNMENT

The head of state is the British sovereign, represented by an appointed Governor-General. The Constitution, which came into force on 21 Sept. 1981, provided for a National Assembly, with a five-year term, comprising a 32-member *House of Representatives* (31 elected by universal suffrage plus the Speaker), and a *Senate* consisting of 13 members, six appointed by the Governor-General on the advice of the Prime Minister, three on the advice of the Leader of the Opposition, one on the advice of the Belize Council of Churches and the Evangelical Association of Churches, one on the advice of the Belize Chamber of Commerce and Industry and the Belize Business Bureau and one on the advice of the National Trade Union Congress of Belize and the Civil Society Steering Committee plus the Senate President.

National Anthem

'O, Land of the Free'; words by S. A. Haynes, tune by S. W. Young.

RECENT ELECTIONS

In elections to the House of Representatives held on 7 March 2012 the ruling United Democratic Party won 17 of 31 seats with 50·4% of votes cast and the People's United Party 14 with 47·5%. Turnout was 73·2%.

CURRENT GOVERNMENT

Governor-General: Sir Colville Young, GCMG; b. 1932 (sworn in 17 Nov. 1993).

In March 2013 the cabinet comprised as follows:

Prime Minister and Minister for Finance and Economic Development: Dean Barrow; b. 1951 (UDP; sworn in 8 Feb. 2008).

Deputy Prime Minister and Minister of Natural Resources and Agriculture: Gaspar Vega.

Attorney General and Minister of Foreign Affairs: Wilfred Elrington. *Education, Youth and Sports:* Patrick Faber. *Energy, Science and Technology, and Public Utilities:* Joy Grant. *Forestry, Fisheries, Sustainable Development and Indigenous People:* Liselle Alamilla. *Health:* Pablo Marin. *Housing and Urban Development:* Michael Finnegan. *Human Development, Social Transformation and Poverty Alleviation:* Anthony Martinez. *Labour, Local Government and Rural Development:* Godwin Hulse. *National Security:* John Saldivar. *Public Service, and Elections and Boundaries:* Charles Gibson. *Tourism and Culture:* Manuel Heredia. *Trade, Investment, Private Sector Development and Consumer Protection:* Erwin Contreras. *Works and Transport:* Rene Montero.

Government Website: http://www.belize.gov.bz

CURRENT LEADERS

Dean Barrow

Position
Prime Minister

Introduction
Dean Barrow, leader of the United Democratic Party (UDP), won a landslide victory in the Feb. 2008 general election to become Belize's first black prime minister. He took office promising to root out corruption, reduce crime and reform the faltering economy.

Early Life
Dean Barrow was born on 2 March 1951 in Belize City and was educated at St Michael's College. He studied law at the University of the West Indies, Barbados, and the Norman Manley Law School in Kingston, Jamaica. He entered the legal profession in 1975, joining his uncle Dean Lindo's Church Street Chambers in Belize City and becoming a partner in 1977. In the early 1980s he completed a masters degree in international relations at the University of Miami.

In 1983 Barrow was elected to Belize City council. He successfully contested the Dec. 1984 general election for the UDP, winning the Queen's Square division. He was appointed attorney general and minister of foreign affairs, serving until 1989 when the UDP lost office. In 1989 Barrow set up with Rodwell Williams the law firm Barrow & Williams which has acted for influential clients including the Belize Bank and Belize Telecommunications Ltd (BTL). In 1990 Barrow became deputy leader of the UDP.

With the UDP victory at the 1993 general election, Barrow returned to his previous posts of attorney general and minister of foreign affairs but also took on responsibility for the national security, immigration and nationality, and media portfolios. This concentration of power attracted some criticism. In 1998, after the UDP lost all but three seats in the general election, Barrow became party head and oversaw the securing of seven seats in the 2003 election.

In opposition, Barrow argued for greater transparency in public finances and advocated building public infrastructure. In the Feb. 2008 general election the UDP upset predictions of a close result to win 25 out of 31 seats. Barrow took office as prime minister on 8 Feb. 2008 and on 11 Feb. announced his new cabinet with himself as minister of finance.

Career in Office
Barrow's early months in office were dominated by investigations into a financial scandal inherited from the previous administration, involving the alleged misuse of US$20m. of overseas grants. With politicians, the Belize Bank and a health care company all implicated, Barrow requested US assistance to set up an audit. Among his other major challenges were the revitalization of the economy, and implementing a programme to build houses, roads and health centres. The ongoing problem of crime was highlighted in Sept. 2011 as the US government added Belize to a blacklist of countries considered to be producers of or transit routes for illegal drugs. The UDP won the parliamentary election of March 2012 albeit with a reduced majority, giving Barrow a second term in office.

DEFENCE

The Belize Defence Force numbers around 1,050 (2007) with 700 reservists. There is an Air Wing and a Maritime Wing.

In 2006 defence expenditure totalled US$18m. (US$63 per capita), representing 1·5% of GDP.

ECONOMY

In 2006 agriculture accounted for 14·0% of GDP, industry 21·0% and services 65·0%.

Overview

Belize is classified as an upper middle-income country by the World Bank. The largest industries by output are garment production, food processing, tourism, construction and oil. Since oil was discovered in 2005, crude petroleum has become the country's major export. Other exports include marine products, sugar, citrus fruits, bananas and papayas. A Commercial Free Zone established in 1994 has seen FDI substantially increase since 2000. Two tropical storms in 2008 caused damage to agriculture and infrastructure valued at 4·8% of GDP.

Belize weathered the global financial crisis relatively well compared to other Caribbean Community countries. Increased exports of petroleum, citrus fruits and bananas, allied with reduced imports, narrowed the external current account deficit from 6·1% of GDP in 2009 to 3·1% in 2010.

In Feb. 2007 the government restructured 98% of its external debt, worth US$900m., although debt remains a concern. The government re-nationalized Belize Telemedia (BTL) and Belize Electricity Limited (BEL) in 2011, prompting Standard & Poor's (S&P) to downgrade Belize's sovereign credit rating from 'B' to 'B-'. In Feb. 2012 S&P further downgraded the rating to 'CCC+'.

Annual GDP growth rates averaged 7% in the 1980s but declined to less than 4% in the period 2000–10. The overall poverty rate rose from 34% in 2002 to 41% in 2009, with unemployment remaining above double digits. The IMF highlights improving the business environment as key to stronger growth.

Currency
The unit of currency is the *Belize dollar* (BZD) of 100 *cents*. Since 1976 $B2 has been fixed at US$1. Total money supply was $B463m. in July 2005 and foreign exchange reserves were US$101m. There was deflation of 1·1% in 2009 but inflation of 0·9% in 2010 and 1·5% in 2011.

Budget
Revenues in 2006–07 were $B598·0m. and expenditures $B667·9m. Tax revenues accounted for 85·9% of total revenues; current expenditure accounted for 84·1% of total expenditures.

A Goods and Services Tax (GST) was introduced in July 2006, initially of 10% and since April 2010 of 12·5%.

Performance
Real GDP growth was 3·5% in 2008 but there was zero growth in 2009. The economy then grew by 2·7% in 2010 and 2·0% in 2011. Total GDP in 2011 was US$1·4bn.

Banking and Finance
A Central Bank was established in 1981 (*Governor*, Glenford Ysaguirre) and in 2001 had deposits of $B148m. There were (2001) one development bank and six other banks.

External debt was US$1,057m. in 2007.

ENERGY AND NATURAL RESOURCES

Environment
Carbon dioxide emissions from the consumption and flaring of fossil fuels in Belize were the equivalent of 3·3 tonnes per capita in 2008.

Electricity
Installed capacity in 2007 was 74,000 kW. Production was 197m. kWh in 2007 and consumption per capita 1,508 kWh.

Oil and Gas
After several years of exploration oil was discovered in 2005 by Belize Natural Energy. It is the only company producing oil in Belize. Production of crude oil was 4,300 bbls per day in 2010.

Agriculture
In 2005 there were 70,000 ha. of arable land and 32,000 ha. of permanent crops. Output, 2006 (in 1,000 tonnes): sugarcane, 1,192; oranges, 212; bananas, 87. Livestock (2001): cattle, 56,000; pigs, 28,000; horses, 5,000; mules, 4,000; chickens, 1m.

Forestry
In 2010, 1·40m. ha. (61% of the total land area) were under forests. Timber production in 2007 was 711,000 cu. metres.

Fisheries
In 2007 there were five registered fishing co-operatives. The total catch in 2007 amounted to 23,902 tonnes, exclusively from sea fishing.

INDUSTRY
Manufacturing is mainly confined to processing agricultural products and timber. There is also a clothing industry. Sugar production in 2006 was 113,000 tonnes; molasses, 42,000 tonnes.

Labour
In 2004 the economically active labour force totalled 180,030; the unemployment rate in 2004 was 11·6%.

INTERNATIONAL TRADE

Imports and Exports
Imports (c.i.f.) in 2007 totalled US$684·3m.; exports (f.o.b.) in 2007 amounted to US$266·6m. Main imports in 2007 were: machinery and transport equipment (18%); mineral fuels and lubricants (16%); manufactured goods (12%); and food and live animals (10%). Main exports were: food and live animals (63%), in particular orange juice and sugar; mineral fuels and lubricants (27%); and miscellaneous manufactured articles (5%). The USA is by far the leading trading partner, in 2007 accounting for 33·9% of imports and 26·8% of exports.

COMMUNICATIONS

Roads
In 2006 there were 575 km of main roads and 2,432 km of other roads. There were 40,000 passenger cars in use in 2006 and 14,800 trucks and vans. In 2006 there were 68 deaths as a result of road accidents.

Civil Aviation
There is an international airport (Philip S. W. Goldson) in Belize City. The national carrier is Maya Island Air, which in 2003 operated domestic services and international flights to Flores (Guatemala). There were direct flights in 2003 with other airlines to Boston, Charlotte, Dallas, Houston, Indianapolis, Las Vegas, McAllen, Miami, Montego Bay, New York, Raleigh, San Pedreo Sula, San Salvador and Washington, D.C. In 2001 Philip S. W. Goldson International handled 497,464 passengers (364,711 on international flights).

Shipping
The main port is Belize City, with a modern deep-water port able to handle containerized shipping. There are also ports at Commerce Bight and Big Creek. In Jan. 2009 there were 246 ships of 300 GT or over registered, totalling 950,000 GT. Nine cargo shipping lines serve Belize, and there are coastal passenger services to the offshore islands and Guatemala.

Telecommunications
In 2011 there were 28,800 landline telephone subscriptions (equivalent to 90·7 per 1,000 inhabitants) and 203,100 mobile phone subscriptions (or 638·7 per 1,000 inhabitants). Fixed internet subscriptions totalled 9,400 in 2010 (30·1 per 1,000 inhabitants).

SOCIAL INSTITUTIONS

Justice
Each of the six judicial districts has summary jurisdiction courts (criminal) and district courts (civil), both of which are presided over by magistrates. There is a Supreme Court, a Court of Appeal and a Family Court. There is a Director of Public Prosecutions, a Chief Justice and two Puisne Judges. Belize was one of ten countries to sign an agreement in Feb. 2001 establishing a Caribbean Court of Justice (CCJ) to replace the British Privy Council as the highest civil and criminal court. In the meantime the number of signatories has risen to 12. The court was inaugurated at Port-of-Spain, Trinidad on 16 April 2005. Belize became the third country to abolish appeals to the Privy Council and accept the CCJ as its court of last resort in June 2010.

The population in penal institutions in 2008 was 1,334 (455 per 100,000 of national population). Belize's prison population rate ranks in the top ten in the world.

Education

The adult literacy rate was 76·9% in 2003 (76·7% among males and 77·1% among females). Education is in English. State education is managed jointly by the government and the Roman Catholic and Anglican Churches. It is compulsory for children between five and 14 years and primary education is free, although there are plans to raise the leaving age to 16 by 2015. In 2008–09 there were 53,000 pupils at primary schools and 32,000 at secondary schools. There are two government-maintained schools for children with special needs. There is a teacher training college. The University College of Belize opened in 1986. The University of the West Indies maintains an extramural department in Belize City.

In 2009 public expenditure on education came to 6·1% of GDP. Expenditure on education was 18·7% of total government spending in 2008.

Health

In 2006 there were 11 hospitals with 12 beds per 10,000 persons. There were 263 physicians, 12 dentists, 441 nurses and 46 pharmacists. Medical services in rural areas are provided by health care centres and mobile clinics.

RELIGION

In 2001, 58% of the population was Roman Catholic and 34% Protestant.

CULTURE

World Heritage Sites

The Belize Barrier Reef Reserve System was inscribed on the UNESCO World Heritage List in 1996.

Press

There are no daily newspapers although there were eight non-dailies in 2008, the largest of which were *Belize Times*, *The Amandala Press* and *The Reporter*.

Tourism

In 2008 there were 842,396 visitors of which 245,026 stayed overnight and 597,370 arrived on cruise ships. There were 561 hotels and 5,789 hotel rooms in 2006.

Festivals

The Belize Carnival is celebrated in the week before Lent begins, with particularly extravagant celebrations in San Pedro. There is also a series of Lobsterfests (at which lobsters prepared in many different ways are consumed) held at different venues throughout June and July. 10 Sept. is St George's Caye Day, held in memory of the defeat of Spanish forces in 1798 and including a battle re-enactment. National Independence Day follows on 21 Sept.

DIPLOMATIC REPRESENTATIVES

Of Belize in the United Kingdom (3rd Floor, 45 Crawford Place, London, W1H 4LP)
High Commissioner: Perla Maria Perdomo.

Of the United Kingdom in Belize (PO Box 91, Belmopan, Belize)
High Commissioner: Pat Ashworth.

Of Belize in the USA (2535 Massachusetts Ave., NW, Washington, D.C., 20008)
Ambassador: Nestor Mendez.

Of the USA in Belize (Floral Park Rd, Belmopan, Cayo)
Ambassador: Vinai Thummalapally.

Of Belize to the United Nations
Ambassador: Lois Michele Young.

Of Belize to the European Union
Ambassador: Audrey Joy Grant.

FURTHER READING

Leslie, Robert, (ed.) *A History of Belize: Nation in the Making.* 2nd ed. 1995
Shoman, Assad, *Thirteen Chapters of a History of Belize.* 1994
Sutherland, Anne, *The Making of Belize: Globalization in the Margins.* 1998
Twigg, Alan, *Understanding Belize: A Historical Guide.* 2006

National Statistical Office: Statistical Institute of Belize, 1902 Constitution Drive, Belmopan.
Website: http://www.statisticsbelize.org.bz

BENIN

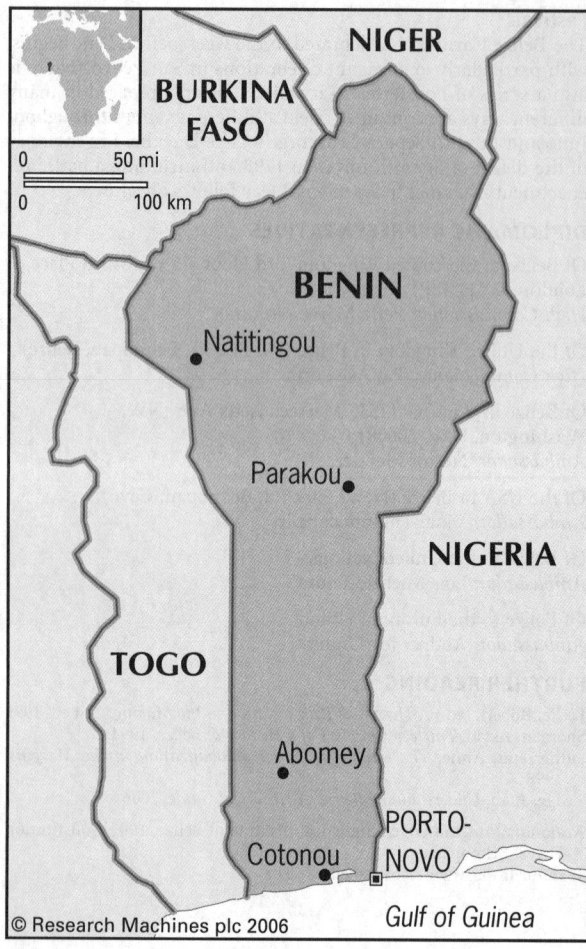

© Research Machines plc 2006

République du Bénin
(Republic of Benin)

Capital: Porto-Novo
Population projection, 2015: 10·13m.
GNI per capita, 2011: (PPP$) 1,364
HDI/world rank: 0·427/167
Internet domain extension: .bj

KEY HISTORICAL EVENTS

The People's Republic of Benin is the former Republic of Dahomey. Dahomey was a powerful, well-organized state from the 17th century, trading extensively in slaves through the port of Whydah with the Portuguese, British and French. On the coast an educated African elite was established in the 19th century.

After the defeat of Dahomey and the abolition of the monarchy in 1894, the French occupied territory inland up to the River Niger and absorbed Dahomey into French West Africa. Protest against French rule grew after the Second World War and in 1946 Dahomey was designated a French overseas territory.

After becoming independent on 1 Aug. 1960 civilian government was interrupted by long periods of military rule. In Oct. 1972 Gen. Mathieu Kérékou seized power and installed a left-wing regime. In 1975 Kérékou renamed the country Benin. A constitution was adopted in 1977, based on a single Marxist-Leninist party, the *Parti de la Révolution Populaire du Bénin* (PRPB). Benin is beset with economic problems, factional fighting and frequent plots to overthrow the regime.

TERRITORY AND POPULATION

Benin is bounded in the east by Nigeria, north by Niger and Burkina Faso, west by Togo and south by the Gulf of Guinea. The area is 112,622 sq. km, and the population (census 2002) 6,769,914; density, 60·1 per sq. km.

The UN gives a projected population for 2015 of 10·13m.

In 2011, 42·5% of the population were urban.

The areas and populations of the 12 departments are as follows:

Department	Sq. km	Census 2002	Department	Sq. km	Census 2002
Alibori	25,683	521,093	Donga	10,691	350,062
Atacora	20,459	549,417	Littoral	79	665,100
Atlantique	3,233	801,683	Mono	1,396	360,037
Borgou	25,310	724,171	Ouémé	2,835	730,772
Collines	13,561	535,923	Plateau	1,865	407,116
Couffo	2,404	524,586	Zou	5,106	599,954

Major towns, with 2002 census population: Cotonou, 665,100; Porto-Novo, 223,552; Parakou, 149,819; Bohicon, 65,974; Abomey, 59,672.

In 1992 the main ethnic groups numbered (in 1,000): Fon, 1,930; Yoruba, 590; Adja, 540; Aizo, 420; Bariba, 420; Somba, 320; Fulani, 270. The official language is French. Over half the people speak Fon.

SOCIAL STATISTICS

2006 (estimates) births, 303,000; deaths, 79,000. Rates, 2006 estimates (per 1,000 population): births, 38·7; deaths, 10·1. Infant mortality, 2010 (per 1,000 live births), 73. Expectation of life in 2007 was 59·8 years for males and 62·1 for females. Annual population growth rate, 1994–2004, 3·2%. Fertility rate, 2008, 5·4 children per woman.

CLIMATE

In coastal parts there is an equatorial climate, with a long rainy season from March to July and a short rainy season in Oct. and Nov. The dry season increases in length from the coast, with inland areas having rain only between May and Sept. Porto-Novo, Jan. 82°F (27·8°C), July 78°F (25·6°C). Annual rainfall 52" (1,300 mm). Cotonou, Jan. 81°F (27·2°C), July 77°F (25°C). Annual rainfall 53" (1,325 mm).

CONSTITUTION AND GOVERNMENT

The Benin Party of Popular Revolution (PRPB) held a monopoly of power from 1977 to 1989.

In Feb. 1990 a 'National Conference of the Active Forces of the Nation' proclaimed its sovereignty and appointed Nicéphore Soglo prime minister of a provisional government. At a referendum in Dec. 1990, 93·2% of votes cast were in favour of the new constitution, which introduced a presidential regime. The *President* is directly elected for renewable five-year terms. Parliament is the unicameral *National Assembly* of 83 members elected by proportional representation for four-year terms.

A 30-member advisory *Social and Economic Council* was set up in 1994. There is a *Constitutional Court*.

National Anthem

'L'Aube Nouvelle' ('The Dawn of a New Day'); words and tune by Gilbert Dagnon.

RECENT ELECTIONS

Presidential elections were held on 13 March 2011. Incumbent Yayi Boni (ind.) was re-elected with 53·2% of the vote, ahead of Adrien Houngbédji (Democratic Renewal Party) with 35·7% and Abdoulaye Bio-Tchané (ind.) with 6·3%. There were a further 11 candidates who each received less than 1% of the vote.

Parliamentary elections were held on 30 April 2011. The Cauri Forces for an Emerging Benin, a coalition supporting President Yayi Boni, won 41 of 83 seats; Unite the Nation, 30; Alliance G13 Baobab, 2; Amana Alliance, 2; Cauris 2 Alliance, 2; Hope Force–Union for Relief, 2; Strength in Unity Alliance, 2; and Union for Benin, 2.

CURRENT GOVERNMENT

President and Minister of Defence: Yayi Boni; b. 1952 (ind.; sworn in 6 April 2006 and re-elected 13 March 2011).

In March 2013 the government comprised:

Prime Minister, Responsible for Co-ordination of Government Action, Evaluation of Public Policies, Privatization and Social Dialogue: Pascal Irénée Koupaki; b. 1951 (in office since 29 May 2011).

Deputy Minister to the President of the Republic in Charge of the Maritime Economy and Port Infrastructure: Valentin Djènontin.

Minister of State for Presidential Affairs: Issifou Kogui N'Douro.

Minister of Administrative and Institutional Reform: Martial Sounton. *Agriculture, Husbandry and Fisheries:* Katé Sadaï. *Communication and Information Technology:* Max Awékè. *Culture, Literacy, Handicrafts and Tourism:* Jean-Michel Abimbola. *Decentralization, Local Government, Administration and Land Management:* Raphaël Édou. *Economic Analysis, Development and Planning:* Marcel de Souza. *Economy and Finance:* Jonas Gbian. *Energy, Petroleum and Mineral Research, Water, and Renewable Energy Development:* Barthélémy Dahoga Kassa. *Environment, Housing and Town Planning:* Blaise Ahanhanzo Glele. *Family, Social Affairs, Solidarity, Senior Citizens and the Disabled:* Fatouma Amadou Djibril. *Foreign Affairs, African Integration, Francophone Affairs and Beninese Abroad:* Nassirou Arifari Bako. *Health:* Akoko Kindé Gazar. *Higher Education and Scientific Research:* François Abiola. *Industry, Commerce and Small and Medium-Sized Businesses:* Marie-Élise Gbèdo. *Interior and Public Security:* Benoit Dégla. *Justice, Legislation and Human Rights, and Keeper of the Seals:* Reckya Madougou. *Labour and Civil Service:* Maïmouna Kora Zaki. *Microfinance, and Youth Employment:* Onfadé Baba Moussa. *Public Works and Transport:* Lambert Koty. *Pre-school and Primary Education:* Eric N'Dah. *Relations with Institutions and Religion:* Safiatou Bassari. *Secondary Education, Technical Training and Professional Training:* Alassane Soumanou. *Youth, Sports and Leisure:* Didier Aplogan Djibodé.

Government Website (French only): http://www.gouv.bj

CURRENT LEADERS

Yayi Boni

Position
President

Introduction
Yayi Boni, a former banker with little political experience and no party backing, won a run-off for the presidency by a landslide and was sworn in on 6 April 2006. He succeeded Gen. Mathieu Kérékou, who led the country for 30 of the 34 years following independence. In March 2011 he was re-elected for a second term.

Early Life
Yayi Boni was born in 1952 in Tchaourou, northern Dahomey, then part of French West Africa. He attended schools in Tchaourou and Parakou, before studying economics at the National University of Benin and then banking and finance at the University of Dakar, Senegal. Boni later read politics and economics at the University of Orléans, France, and received a PhD in economics from Université Paris Dauphine in 1991.

Having worked at the Commercial Bank of Benin for two years, Boni joined the Central Bank of the States of West Africa in 1977. By the time he left in 1989, he was the organization's deputy director. Following a three-year spell as deputy director for professional development at the West African Centre for Banking Studies in Dakar, Boni returned to Benin as an adviser to President Nicéphore Soglo on banking and monetary policy.

Boni was appointed president of the Togo-based West African Development Bank in 1994 and oversaw a programme of modernization. Resigning in 2005 to contest Benin's presidential election, he campaigned on a platform of economic reforms aimed at reducing the country's dependence on cotton exports. Twenty-six candidates contested the first round of voting held on 5 March 2006, with Boni polling 36%. A run-off was held on 19 March between Boni and Adrien Houngbédji of the Democratic Renewal Party. Boni won with 74·5% of the vote and was sworn into office in Porto-Novo on 6 April 2006.

Career in Office
Boni promised sweeping reforms to tackle poverty and corruption and to strengthen the institutions of democracy. In March 2007 the pro-Boni Cauri Forces for an Emerging Benin won control of parliament in elections, and in April 2008 parties supporting the president won a majority of seats in polling for local councils. In July 2010 Boni replaced interior minister Armand Zinzindohoué, accusing him of involvement in a financial scandal defrauding thousands of small investors, but was himself accused in Aug. by a majority of members of parliament of involvement in the swindle. He secured a further presidential term in March 2011, although his main rival Adrien Houngbédji claimed the result was fraudulent. In April 2011 Boni's allies increased their parliamentary representation in legislative elections. An alleged assassination plot against Boni led to the arrest of three people, including one of the president's relatives, in Oct. 2012, Also in 2012 Boni served as chair of the African Union for a year-long term.

DEFENCE

There is selective conscription for 18 months. Defence expenditure totalled US$67m. in 2008 (US$8 per capita), representing 1·0% of GDP.

Army

The Army strength (2007) was 4,300, with an additional 2,500-strong paramilitary gendarmerie.

Navy

Personnel in 2007 numbered about 100; the force is based at Cotonou.

Air Force

The Air Force has suffered a shortage of funds and operates no combat aircraft. Personnel, 2007, 350.

ECONOMY

Agriculture and fisheries accounted for 32·7% of GDP in 2009; trade and restaurants, 17·1%; finance, 10·7%; public administration, defence and services, 10·1%.

Overview

Over the first decade of the century the economy grew an average 3·9% per year. Agriculture is a key sector and cotton exports account for 25–40% of GDP. The West Africa Agricultural Program aims to improve crop productivity while an Agricultural Productivity and Diversification Program seeks to diversify production to insulate the economy from external shocks. In 2010

GDP growth was 2·6%, lower than predicted, as a result of flooding and weak exports.

Benin's geographical location makes it an important transit point for neighbouring countries including Burkina Faso and Nigeria. Transit-related revenues generate a fifth of GDP. However, corruption and inefficiency are impediments to long-term growth.

Currency

The unit of currency is the *franc CFA* (XOF) with a parity of 655·957 francs CFA to one euro. Total money supply was 394,434m. francs CFA in June 2005 and foreign exchange reserves were US$692m. Inflation was 2·1% in 2010 and 2·7% in 2011.

Budget

The fiscal year is the calendar year. In 2008 budgetary central government revenues were 608·5bn. francs CFA and expenditures 446·5bn. francs CFA.

VAT is 18%.

Performance

Real GDP growth was 2·7% in 2009, 2·6% in 2010 and 3·5% in 2011. Total GDP was US$7·3bn. in 2011.

Banking and Finance

The bank of issue and the central bank is the regional Central Bank of West African States (BCEAO). The *Governor* is Tiémoko Meyliet Koné. In Dec. 2001 it had total assets of 5,517,700m. francs CFA. In 2009 there were 13 banks and one financial institute. The Caisse Autonome d'Amortissement du Bénin manages state funds.

Total foreign debt was US$1,221m. in 2010, representing 18% of GNI.

ENERGY AND NATURAL RESOURCES

Environment

Benin's carbon dioxide emissions from the consumption and flaring of fossil fuels in 2008 were the equivalent of 0·4 tonnes per capita.

Electricity

Installed capacity in 2007 was an estimated 60,000 kW. In 2007 production was 132m. kWh; Benin also imported 588m. kWh. Consumption per capita in 2007 was 80 kWh.

Oil and Gas

The Sémé oilfield, located 15 km offshore, was discovered in 1968. Production commenced in 1982 but ceased in 2004. Crude petroleum reserves in 2007 were 8m. bbls.

Agriculture

Benin's economy is underdeveloped, and is dependent on subsistence agriculture. In 2002, 3·69m. persons depended on agriculture, of whom 1·55m. were economically active. Small independent farms produce about 90% of output. In 2007 an estimated 2·7m. ha. were arable and 0·27m. ha. permanent crops; about 12,000 ha. was irrigated in 2002. There were 185 tractors in 2002. The chief agricultural products, 2002 (in 1,000 tonnes) were: cassava, 2,452; yams, 1,875; maize, 622; seed cotton, 486; cottonseed, 267; sorghum, 195; groundnuts, 146; tomatoes, 141.

Livestock, 2003 estimates: cattle, 1,600,000; goats, 1,300,000; sheep, 670,000; pigs, 550,000; chickens, 10m.

Livestock products, 2003 (estimates, in 1,000 tonnes): beef and veal, 19; pork, bacon and ham, 7; goat meat, 4; poultry meat, 12; eggs, 7; milk, 31.

Forestry

In 2010 there were 4·56m. ha. of forest (41% of the total land area), mainly in the north. Timber production in 2007 was 6·57m. cu. metres.

Fisheries

Total catch in 2010 was 39,791 tonnes, mainly freshwater fish.

INDUSTRY

Only about 2% of the workforce is employed in industry. The main activities include palm-oil processing, brewing and the manufacture of cement, sugar and textiles. Also important are cigarettes, food, construction materials and petroleum. Production (2007 unless otherwise specified, in 1,000 tonnes): cement (2005), 1,100; palm oil (estimate), 40; cottonseed oil, 22; groundnut oil (estimate), 10; raw sugar, 10; sawnwood, 84,000 cu. metres.

Labour

The estimated labour force numbered 3,825,000 in 2010 (54% males), up from 3,212,000 in 2005. Approximately half of the economically active population is engaged in agriculture, fishing and forestry.

INTERNATIONAL TRADE

Commercial and transport activities, which make up 36% of GDP, are extremely vulnerable to developments in neighbouring Nigeria, with which there is a significant amount of illegal trade.

Imports and Exports

Imports (c.i.f.) in 2006 totalled US$1,003·3m.; exports (f.o.b.), US$224·6m.

Principal import suppliers, 2006: France, 17·2%; China, 8·5%; Côte d'Ivoire, 6·9%; Ghana, 6·8%; UK, 6·3%; Togo, 5·3%. Principal export markets, 2006: China, 24·0%; Nigeria, 8·7%; India, 8·6%; Niger, 7·2%; Côte d'Ivoire, 5·7%; Thailand, 4·3%.

Main imports in 2005 were: petroleum and petroleum products (13·8%); machinery and transport equipment (12·2%); rice (11·2%); chemicals and related products (6·7%). The main exports were: cotton (58·0%); cashew nuts (6·9%); cigarettes (6·7%); cement (4·1%).

COMMUNICATIONS

Roads

There were 19,000 km of roads in 2004, of which 9·5% were paved. Passenger cars in use in 2007 totalled 149,300, buses and coaches 1,100, and lorries and vans 35,700.

Rail

In 2005 there were 438 km of metre-gauge railway. In 2007 railways carried 0·1m. tonnes of freight. Passenger services were suspended in 2007 but have since resumed on a limited basis.

Civil Aviation

The international airport is at Cotonou (Cadjehoun), which in 2001 handled 227,000 passengers (all on international flights) and 3,200 tonnes of freight. In 2001 scheduled airline traffic of Benin-based carriers flew 1m. km, carrying 46,000 passengers (all on international flights). In 2003 Trans African Airlines flew to Abidjan, Bamako, Brazzaville, Dakar, Lomé and Pointe-Noire; Trans Air Benin operated services to Abidjan, Brazzaville and Lomé; and Aero Benin flew to Bamako, Brazzaville, Johannesburg, Libreville and Ouagadougou.

Shipping

There is a port at Cotonou, which handled 6,307,000 tonnes of cargo in 2008 (714,000 tonnes loaded and 5,593,000 tonnes discharged), up from 5,889,000 tonnes in 2007.

Telecommunications

In 2008 there were 103,200 main (fixed) telephone lines; mobile phone subscribers numbered 3,625,400 in 2008 (41·9 per 100 persons). There were 160,000 internet users in 2008. In June 2012 there were 143,000 Facebook users.

SOCIAL INSTITUTIONS

Justice

The Supreme Court is at Cotonou. There are Magistrates Courts and a *tribunal de conciliation* in each district. The legal system is based on French civil law and customary law.

The population in penal institutions in May 2006 was 5,834 (75 per 100,000 of national population). The death penalty was abolished in Aug. 2011.

Education

Adult literacy rate was 41% in 2008. In 2006 there were 1,356,818 pupils in primary schools with 31,103 teaching staff; and, in 2004, 344,890 pupils in secondary schools with 14,410 teaching staff. There were 42,603 students in higher education in 2006. The leading institution in the tertiary sector is the National University of Benin (Université Nationale du Bénin), located in Cotonou.

In 2006 public expenditure on education came to 3·9% of GNI and 18·0% of total government spending.

Health

In 2006 there were 1,088 physicians, 3,563 nurses and 999 midwives. There were three hospital beds per 10,000 inhabitants in 2006.

RELIGION

Some 51% of the population follow traditional animist beliefs. Voodoo became an official religion in 1996. In 2001 there were 1·37m. Roman Catholics and 1·32m. Muslims.

CULTURE

World Heritage Sites

The Royal Palaces of Abomey joined the World Heritage List in 1985 (reinscribed in 2007). They preserve the remains of the palaces of 12 kings who ruled the former kingdom of Abomey between 1625 and 1900.

Press

Benin has dozens of newspapers and periodicals but in 2007 only around 20 dailies were published regularly. The main newspapers are *Le Matinal, Les Echos du Jour* and the government-controlled *La Nation*.

Tourism

In 2007 there were an estimated 186,000 non-resident tourists. Tourist spending totalled US$124m. in 2007.

DIPLOMATIC REPRESENTATIVES

Of Benin in the United Kingdom
Ambassador: Albert Agossou (resides in Paris).
Honorary Consul: Lawrence Landau (Millennium House, Humber Rd, London, NW2 6DW).

Of the United Kingdom in Benin
Ambassador: Dr Andrew Pocock, CMG (resides in Abuja, Nigeria).

Of Benin in the USA (2124 Kalorama Rd, NW, Washington, D.C., 20008)
Ambassador: Segbe Cyrille Oguin.

Of the USA in Benin (Rue Caporal Bernard Anani, Cotonou)
Ambassador: Michael Raynor.

Of Benin to the United Nations
Ambassador: Jean-Francis Régis Zinsou.

Of Benin to the European Union
Ambassador: Charles Borromée Todjinou.

FURTHER READING

Bay, E., *Wives of the Leopard: Gender, Politics, and Culture in the Kingdom of Dahomey.* 1998

National Statistical Office: Institut National de la Statistique et de l'Analyse Économique, 01 BP 323, Cotonou.
Website (French only): http://www.insae-bj.org

BHUTAN

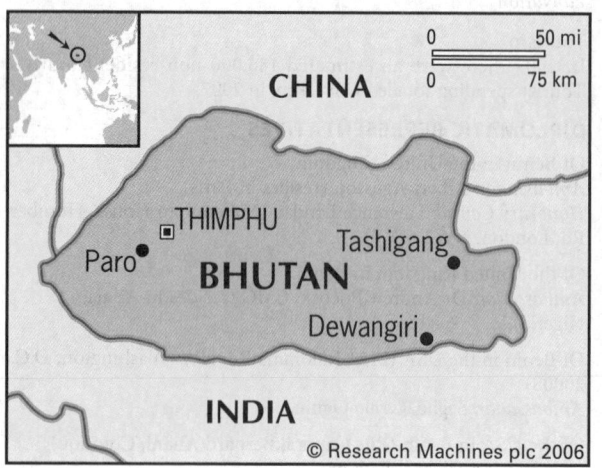

0 ____ 50 mi
0 ____ 75 km

CHINA

THIMPHU

Paro · BHUTAN · Tashigang

Dewangiri

INDIA

© Research Machines plc 2006

Druk-yul
(Kingdom of Bhutan)

Capital: Thimphu
Population projection, 2015: 784,000
GNI per capita, 2011: (PPP$) 5,293
HDI/world rank: 0·522/141
Internet domain extension: .bt

KEY HISTORICAL EVENTS

Indigenous Monpa clans established settlements in the eastern Himalayas by around 2000 BC. Buddhism was brought to Bhutan in the 7th century AD when Tibetan lamas (monks) founded monasteries at Bumthang and Kyichi, although animist beliefs persisted among the scattered villages. It was the arrival in 1616 of a monk, Zhabdrung Nawang Namgyal, fleeing persecution in Tibet, which led to the foundation of the kingdom of Bhutan. Over a period of 35 years Zhabdrung and his followers built fortresses and monasteries and established the Drukpa sect of Buddhism as well as a dual system of governance known as the Chhoesid. Power was split between the Deb Raja, the head of secular affairs (responsible for four regional governors) and the Dharma Raja, the spiritual head who was charged with enacting laws. In 1720 the Ch'ing dynasty took control of Tibet, claiming suzerainty of it and neighbouring Bhutan.

Tensions between Bhutan and Bengal to the south culminated in a Bhutanese invasion of Cooch Behar in 1772. This prompted the governor of the province to seek military assistance from the British who had defeated the Nawab of Bengal in 1757. Skirmishes continued for two years until peace was brokered by Tashi Lama, then Regent of Tibet. British attempts to develop trade with Bhutan in the 1780s were unsuccessful and new tensions emerged when British India took control of neighbouring Assam in 1826. In 1864 British India claimed ownership of a strip of southern Bhutan known as the Duars. It was formally ceded to India the following year, although the Treaty of Sinchula provided for an annual subsidy to Bhutan as compensation.

In 1907 the office of Dharma Raja came to an end. The governor of Tongsa, Ugyen Wangchuck, was then chosen as Maharajah of Bhutan, the first of a hereditary line (the title is now King of Bhutan). He concluded a treaty with British India in 1910 which allowed for internal autonomy with British control of foreign policy in return for doubling Britain's annual subsidy.

After Indian independence, a treaty of 1949 returned the Duars to Bhutan. The kingdom continued to manage its internal affairs while India inherited control of Bhutan's defence and foreign affairs.

When Communist Chinese forces invaded Tibet in 1950, Bhutan was claimed as part of 'Greater Tibet'. In response, India closed the Bhutan–Tibet border and upgraded roads linking Bengal with Bhutan. Amid mounting Indo-Sino tensions in the early 1960s, Bhutan established a small army, trained and equipped by India. The reign of the third king, Jigme Dorji Wangchuk (1952–72), was marked by gradual economic development and an opening up to the outside world. The 151-seat National Assembly was established in 1952 and Bhutan became a member of the United Nations in 1971, since when relations with China have slowly thawed.

In the early 1990s, tens of thousands of 'illegal immigrants', mostly Nepali-speaking Hindus, were forcibly expelled from southern and western Bhutan. Twenty years on, there are still more than 85,000 people claiming to be Bhutanese refugees in camps set up by the UNHCR in eastern Nepal, although around 23,000 have been resettled in the USA and Europe since March 2008. Bhutan was ruled from 1972–2006 by King Jigme Singye Wangchuck. He abdicated in Dec. 2006 in favour of his son, Jigme Kesar Namgyel Wangchuck. In Dec. 2007 and March 2008 Bhutan held its first democratic parliamentary elections. The leader of the Druk Phuensum Tshogpa (Bhutan Peace and Prosperity Party), Jigme Thinley, took office as the first elected prime minister of Bhutan on 9 April 2008.

TERRITORY AND POPULATION

Bhutan is situated in the eastern Himalayas, bounded in the north by Tibet and on all other sides by India. In 1949 India retroceded 83 sq. km of Dewangiri, annexed in 1865. Area 46,650 sq. km (18,012 sq. miles); 2005 census population, 672,425 (364,482 males), giving a density of 14 per sq. km.

The UN gives a projected population for 2015 of 784,000.

In 2011, 35·5% of the population lived in urban areas. A Nepalese minority makes up 30–35% of the population, mainly in the south. The capital is Thimphu (2005 population, 79,185). The country is divided into 20 districts (*dzongkhag*).

The official language is Dzongkha.

SOCIAL STATISTICS

2008 (estimates) births, 14,800 (rate of 21·5 per 1,000 population); deaths, 4,900 (rate of 7·1 per 1,000 population). Life expectancy at birth, 2007, was 64·0 years for men and 67·6 years for women. Infant mortality, 2010, 44 per 1,000 live births. Annual population growth rate, 2000–08, 2·5%; fertility rate, 2008, 2·6 children per woman.

CLIMATE

The climate is largely controlled by altitude. The mountainous north is cold, with perpetual snow on the summits, but the centre has a more moderate climate, though winters are cold, with rainfall under 40" (1,000 mm). In the south, the climate is humid sub-tropical and rainfall approaches 200" (5,000 mm).

CONSTITUTION AND GOVERNMENT

Bhutan's first formal constitution came into force on 18 July 2008, after a period of almost seven years of planning. There is a bicameral parliament. The lower house is the *National Assembly* (with a maximum of 55 members but currently with 47, all elected) and the upper house the 25-member *National Council* (with 20 members elected and five appointed by the king). Executive power is vested in the *Council of Ministers*.

The reigning King is Jigme Kesar Namgyel Wangchuck (b. 1980), who succeeded his father King Jigme Singye Wangchuck (abdicated 14 Dec. 2006). He was crowned on 6 Nov. 2008. With the introduction of democratic elections in 2007–08, the King's role became more ceremonial. Nonetheless, all leading political parties have affirmed their loyalty to the monarchy, which remains central to political life.

In 1907 the Tongsa Penlop (the governor of the province of Tongsa in central Bhutan), Sir Ugyen Wangchuck, GCIE, KCSI, was elected as the first hereditary Maharaja of Bhutan. The Bhutanese title is Druk Gyalpo, and his successors are addressed as King of Bhutan. The stated goal is to increase Gross National Happiness.

National Anthem

'Druk tsendhen koipi gyelknap na' ('In the Thunder Dragon Kingdom'); words by Gyaldun Dasho Thinley Dorji, tune by A. Tongmi.

RECENT ELECTIONS

Bhutan's first ever elections were held on 31 Dec. 2007 when 15 members (all independents) were elected to the National Council (the non-partisan upper house of Bhutan's bicameral parliament). A further five members were elected on 29 Jan. 2008. In the second National Council elections, 20 members (again, all independents) were elected in single-member constituencies on 23 April 2013. A further five members were appointed by the king.

On 22 March 2008 elections were held for Bhutan's National Assembly. The 47 seats were contested by two parties—the Druk Phuensum Tshogpa (DPT; Bhutan Peace and Prosperity Party), led by former prime minister Jigme Thinley, and the People's Democratic Party (PDP), led by another former prime minister, Sangay Ngedup. The DPT won 45 of the 47 seats with 67·0% of the vote. The PDP gained only two seats, with 33·0% of the vote.

CURRENT GOVERNMENT

In March 2013 the government comprised:

Prime Minister: Jigme Thinley; b. 1952 (took office for the third time on 9 April 2008, having previously been prime minister from July 1998–July 1999 and Aug. 2003–Aug. 2004).

Minister for Economy: Khandu Wangchuk. *Information and Communication:* Nandalal Rai. *Education:* Thakur Singh Powdyel. *Finance:* Wangdi Norbu. *Health:* Zangley Dukpa. *Labour and Human Resources:* Dorji Wangdi. *Home and Cultural Affairs:* Minjur Dorji. *Agriculture:* Pema Gyamtsho. *Foreign Affairs:* Ugyen Tshering. *Works and Human Settlement:* Yeshey Zimba. *Chief Justice:* Sonam Tobgye.

Speaker: Jigme Tshultim.

Government Website: http://www.cabinet.gov.bt

CURRENT LEADERS

Jigme Thinley

Position
Prime Minister

Introduction
Jigme Thinley, a civil servant and former government minister, became Bhutan's first ever democratically elected prime minister on 24 March 2008.

Early Life
Jigme Yoser Thinley was born in 1952 in Bumthang, northern Bhutan and educated at Dr Graham's Homes in Kalimpong, northeastern India. He graduated from St Stephen's College at the University of Delhi and subsequently earned a masters degree in public administration from Penn State University in the USA. He later studied manpower planning and management at Manchester University in the UK.

Having joined Bhutan's civil service in 1974 as a trainee officer in the ministry of home affairs, Thinley went on to hold a range of posts including, in 1990, administrator of Bhutan's six eastern districts. He became a secretary in the ministry of home affairs in 1992 and was promoted to deputy minister in 1994. In the same year he was appointed as Bhutan's permanent representative to the UN and other international organizations.

Career in Office
From July 1998 to July 1999 and again from Aug. 2003 to Aug. 2004, Thinley was the royal appointee as prime minister. He also served as minister of foreign affairs between 1998 and 2003. In March 2008, in the run-up to Bhutan's first multi-party elections, Thinley stood as leader of the new DPT. The party won 45 of the 47 seats in the National Assembly, making Thinley the country's first democratically elected premier. He took office on 9 April 2008, promising to make democracy a success and to provide a transparent and corruption-free government.

DEFENCE

In 2003 defence spending totalled US$22m. (US$25 per capita), representing 3·3% of GDP.

Army

In 2007 the Royal Bhutan Army had a strength of 9,021. It is lightly armed, mainly with weapons supplied by India. There is also an Air Wing of around 80 personnel.

ECONOMY

Agriculture accounted for 18·7% of GDP in 2009, with industry accounting for 43·2% and services 38·1%.

Overview

One of the smallest economies in the world, 23·2% of the population lived in poverty in 2007. However, Bhutan has seen steady improvement in social indicators including poverty reduction. Growth averaged 8% in the decade 2000–10, underpinned by development of the hydropower sector and strong ties with India. Earnings from the hydropower sector accounted for 23% of total revenue in 2009–10.

43% of the labour force is employed in agriculture, although agriculture's share of GDP has steadily declined in recent years. Public expenditure, guided by the government ideology of Gross National Happiness, is focused on health and education.

India is the destination for 82% of Bhutan's exports and source of 75% of its imports. The currency, the *ngultrum*, is pegged to the Indian rupee and the Indian government finances a large part of Bhutan's budget expenditures.

Total debt stood at 80% of GDP in 2011, half of which was incurred through hydropower projects. Bhutan was expected to reduce electricity exports to India from 6,500m. units in 2010–11 to 5,500m. units in 2012–13 because of reduced surplus power. The IMF projects growth to remain at between 8% and 9% in the near term.

Other economic activities are underdeveloped, partly owing to a poor business climate. In the World Bank's *Doing Business* 2011 report, Bhutan ranked 142nd of 183 economies.

Currency

The unit of currency is the *ngultrum* (BTN) of 100 *chetrum*, at parity with the Indian rupee. Indian currency is also legal tender. Foreign exchange reserves were US$425m. in May 2005 and total money supply was Nu8,088m. Inflation was 7·0% in 2010 and 8·9% in 2011.

Budget

Budgetary central government revenue and expenditure (in Nu1m.):

	2006–07	2007–08	2008–09
Revenue	16,083·1	18,277·9	20,624·1
Expenditure	9,092·2	12,352·5	14,257·0

Grants accounted for 31·9% of revenues in 2008–09 and taxes 27·4%; compensation of employees accounted for 35·0% of expenditures and use of goods and services 32·7%.

Performance

Real GDP growth was 17·9% in 2007 (one of the highest rates for the year in the world), 4·7% in 2008, 6·7% in 2009, 11·8% in 2010 and 5·3% in 2011. Total GDP in 2011 was US$1·7bn.

Banking and Finance

The Royal Monetary Authority (founded 1982; *Governor,* Daw Tenzin) acts as the central bank. Assets (June 2010) Nu38·25bn. Financial sector total assets were Nu55·29bn. in June 2010 with the country's four commercial banks accounting for 90·6% of assets. The oldest and largest commercial bank is the Bank of Bhutan, which was established in 1968 and operated as the central bank until 1982. The headquarters are at Phuentsholing with 26 branches throughout the country. It is 80%-owned by the government of Bhutan and 20%-owned by the Indian government. There is also a development bank (the Bhutan Development Finance Corporation) and a stock exchange in Thimphu.

In 2010 Bhutan's total external debt amounted to US$898m., equivalent to 63·3% of GNI.

ENERGY AND NATURAL RESOURCES

Environment

Bhutan's carbon dioxide emissions from the consumption and flaring of fossil fuels in 2008 were the equivalent of 0·5 tonnes per capita.

Electricity

Installed capacity in 2006–07 was 1·5m kW (nearly all hydro-electric). Production (2006–07) was 6·6bn. kWh. Consumption per capita in 2006–07 was about 1,731 kWh. Bhutan exports around 80% of the electricity that it produces, mainly to India.

In March 2007 the Tala hydro-electric plant was commissioned. A joint project between Bhutan and India, it is designed to generate 4·9bn. kWh annually.

Minerals

Large deposits of limestone, marble, dolomite, slate, graphite, lead, copper, coal, talc, gypsum, beryl, mica, pyrites and tufa have been found. Most mining activity (principally limestone, coal, slate and dolomite) is on a small-scale. Output, 2010: dolomite, 1,192,000 tonnes; limestone, 716,000 tonnes; gypsum, 344,000 tonnes; coal, 88,000 tonnes.

Agriculture

The agricultural area in 2007 was an estimated 562,000 ha. In 2007 there were an estimated 128,000 ha. of arable land and 27,000 ha. of permanent crops. The chief products (2007 estimated production in 1,000 tonnes) are rice (74), maize (62), potatoes (61), oranges (37), dried chillies and peppers (11) and ginger (10).

Livestock (2005): cattle, 381,000; goats (estimate), 30,000; pigs, 28,000; horses, 25,000; sheep, 18,000.

Forestry

In 2010, 3·25m. ha. (69% of the land area) were forested. Timber production in 2007 was 4·80m. cu. metres.

Fisheries

The total catch in 2010 amounted to an estimated 160 tonnes, exclusively from inland waters.

INDUSTRY

Industries in Bhutan include cement, wood products, processed fruits, alcoholic beverages and calcium carbide. In 2009 manufacturing accounted for 8·4% of GDP. 2001 production: cement, 160,000 tonnes; veneer sheets, 16,000 cu. metres; particle board, 12,000 cu. metres; plywood, 4,000 cu. metres. In 2001 there were 12,878 licensed industrial establishments, of which 8,536 were construction, 3,773 service and 569 manufacturing industries. The latter included 317 forest-based companies, 116 agriculture-based and 46 mineral-based.

Labour

In 2005 the economically active population was 256,895. Of those in employment, 63·5% were males. Unemployment in 2005 was 3·1%.

INTERNATIONAL TRADE

Financial support is received from India, the UN and other international aid organizations.

Imports and Exports

The following table shows the value of Bhutan's foreign trade (in US$1m.):

	Imports (c.i.f.)	Exports (f.o.b.)
2006	418·9	413·9
2007	498·1	674·5
2008	543·3	521·4
2009	529·4	495·8
2010	853·8	413·5

India is by far the biggest trading partner. In 2010 the leading import suppliers (in US$1m.) were: India (640·8); South Korea (43·8); Thailand (21·6). The main export markets in 2010 (in US$1m.) were: India (340·8); Hong Kong (47·8); Bangladesh (19·8). Major imports are machinery and transport equipment, manufactured goods, and mineral fuels and lubricants. Hydropower, ferrosilicon and cement are the main exports.

COMMUNICATIONS

Roads

In 2006 there were about 4,153 km of roads, of which 1,577 km were highways. In 2007 there were 19,600 passenger cars, 180 buses and coaches, 5,400 lorries and vans, and 7,500 motorcycles and mopeds. A number of sets of traffic lights were installed during the late 1990s but all have subsequently been removed as they were considered to be eyesores. There had previously been just one set. There were 111 fatalities in road accidents in 2007.

Rail

Bhutan does not currently have a railway network but there are plans for a line funded by India that would link the town of Toribari with Hasimara in India.

Civil Aviation

In 2003 Druk-Air flew from Paro to Bangkok, Delhi, Dhaka, Kathmandu, Calcutta and Rangoon (Yangon). In 2003 Druk-Air flew 2m. km, carrying 36,000 passengers (all on international flights).

Telecommunications

There were 26,300 fixed telephone lines in 2010 (36·2 per 1,000 inhabitants). Mobile phone subscribers numbered 394,300 in 2010. There were 136·0 internet users per 1,000 inhabitants in 2010. Fixed internet subscriptions totalled 6,700 in 2009 (9·3 per 1,000 inhabitants).

SOCIAL INSTITUTIONS

Justice
The High Court consists of eight judges appointed by the King. There is a Magistrate's Court in each district, under a *Thrimpon*, from which appeal is to the High Court at Thimphu. The death penalty, which had not been used for 40 years, was abolished in 2004.

Education
In 2008–09 there were 109,000 pupils and 4,000 teachers in primary schools and 57,000 pupils with 3,000 teachers in secondary schools. In 2006 there were 4,141 students in tertiary education and 375 academic staff. Adult literacy was 52·8% in 2005.

In 2005 public expenditure on education came to 7·1% of GNI and 17·2% of total government spending.

Health
In 2006 there were 29 hospitals with 1,133 beds, 176 basic health units and 514 outreach clinics. There were two doctors per 10,000 population in 2005. Free health facilities are available to 90% of the population.

RELIGION

The state religion of Bhutan is the Drukpa Kagyupa, a branch of Mahayana Buddhism. There are also Hindu and Muslim minorities.

CULTURE

Press
Until 2006 there was only one newspaper, the government-controlled *Kuensel*, which is published in English, Dzongkha and Nepali. Two private weeklies were launched in 2006. The country's first daily paper, the English-language *Bhutan Today*, was launched in 2008 and had an average daily circulation of 18,000 that year.

Tourism
Bhutan was not formally opened to foreign tourists until 1974, but tourism is now the largest source of foreign exchange. In 2009, 23,000 tourists visited Bhutan; revenue totalled US$32m.

Festivals
Bhutan's most famous festival is Tshechu, a religious celebration of Guru Padsambhava held throughout the country at the end of the harvest. The Thimphu Tshechu takes place around mid-Sept. and includes spectacular masked dances performed by monks. Other important festivals include Dromche (to honour the protective deities of the Bhutanese people) and Jambay Lhakhan Drup (including a fire dance to promote fertility).

DIPLOMATIC REPRESENTATIVES

Of Bhutan in the United Kingdom
Honorary Consul: Michael Rutland (2 Windacres, Warren Rd, Guildford, Surrey, GU1 2HG).

Of Bhutan to the United Nations
Ambassador: Lhato Wangchuk.

Of Bhutan to the European Union
Ambassador: Sonam Tshong.

FURTHER READING

Crossette, B., *So Close to Heaven: The Vanishing Buddhist Kingdoms of the Himalayas.* 1995
Parmanand, Parashar, *The Politics of Bhutan: Retrospect and Prospect.* 2002
Rahul, Ram, *Royal Bhutan: A Political History.* 1997
Savada, A. M. (ed.) *Nepal and Bhutan: Country Studies.* 1993
Sinha, A. C., *Bhutan: Ethnic Identity and National Dilemma.* 1998.— *Himalayan Kingdom Bhutan: Tradition, Transition and Transformation.* 2002

National Statistical Office: National Statistics Bureau, Thimphu.
Website: http://www.nsb.gov.bt

BOLIVIA

Estado Plurinacional de Bolivia
(Plurinational State of Bolivia)

Capital: Sucre
Seat of government: La Paz
Population projection, 2015: 10·74m.
GNI per capita, 2011: (PPP$) 4,054
HDI/world rank: 0·663/108
Internet domain extension: .bo

KEY HISTORICAL EVENTS

Bolivia was part of the Inca Empire until conquered by the Spanish in the 16th century. Independence was won and the Republic of Bolivia was proclaimed on 6 Aug. 1825. During the first century and a half of its independence, Bolivia had frequent changes of government. Many of these were as a result of coups of which there have been some 190 in total. The most recent was in 1980. In the 1960s the Argentinian revolutionary and former minister of the Cuban government, Ernesto 'Che' Guevara, was killed in Bolivia while fighting with a left-wing guerrilla group. In 1971 Bolivia was governed briefly by the revolutionary Popular Assembly. Repression under Gen. Hugo Banzer took a heavy toll on the left-wing parties. Banzer was followed by a succession of military-led governments until civilian rule was restored in Oct. 1982 when Dr Siles Zuazo became president. He introduced economic reforms embracing free markets and open trade which succeeded in restoring stability but also widened the gap between rich and poor. Amid growing discontent, in Dec. 2005 Evo Morales Ayma was elected as the country's first indigenous president.

TERRITORY AND POPULATION

Bolivia is a landlocked state bounded in the north and east by Brazil, south by Paraguay and Argentina, and west by Chile and Peru, with an area of some 1,098,581 sq. km (424,165 sq. miles).

A coastal strip of land on the Pacific passed to Chile after a war in 1884. In 1953 Chile declared Arica a free port and Bolivia has certain privileges there.

Population (2012 census, provisional): 10,389,913 (50·1% female); density, 9·5 per sq. km. In 2011, 67·0% of the population lived in urban areas.

The UN gives a projected population for 2015 of 10·74m.

Area and population of the departments (capitals in brackets) at the 2001 and 2012 censuses:

Departments	Area (sq. km)	Census (2001)	Census (2012, provisional)
Beni (Trinidad)	213,564	362,521	425,780
Chuquiaca (Sucre)	51,524	531,522	600,728
Cochabamba (Cochabamba)	55,631	1,455,711	1,938,401
La Paz (La Paz)	133,985	2,350,466	2,741,554
Oruro (Oruro)	53,588	391,870	490,612
Pando (Cobija)	63,827	52,525	109,173
Potosí (Potosí)	118,218	709,013	798,664
Santa Cruz (Santa Cruz de la Sierra)	370,621	2,029,471	2,776,244
Tarija (Tarija)	37,623	391,226	508,757
Total	1,098,581	8,274,325	10,389,913

Population (2010 estimates, in 1,000) of the principal towns: Santa Cruz de la Sierra, 1,616; El Alto, 953; La Paz, 835; Cochabamba, 618; Sucre, 284; Oruro, 217; Tarija, 194; Sacaba, 156; Potosí, 155.

Spanish along with the Amerindian languages Quechua and Aymará are all official languages; Tupi Guaraní is also spoken. Indigenous peoples account for 50% of the population.

SOCIAL STATISTICS

In 2008 births totalled an estimated 263,000 (birth rate of 27·1 per 1,000 population); deaths totalled an estimated 73,000 (rate, 7·5 per 1,000); infant mortality (2010), 42 per 1,000 live births, the highest in South America. Expectation of life (2007) was 63·3 years for men and 67·5 years for women. Annual population growth rate, 2000–08, 1·9%. Fertility rate, 2008, 3·5 children per woman (the highest in South America).

CLIMATE

The varied geography produces different climates. The low-lying areas in the Amazon Basin are warm and damp throughout the year, with heavy rainfall from Nov. to March; the Altiplano is generally dry between May and Nov. with sunshine but cold nights in June and July, while the months from Dec. to March are the wettest. La Paz, Jan. 55·9°F (13·3°C), July 50·5°F (10·3°C). Annual rainfall 20·8" (529 mm). Sucre, Jan. 58·5°F (14·7°C), July 52·7°F (11·5°C). Annual rainfall 20·1" (510 mm).

CONSTITUTION AND GOVERNMENT

Bolivia's first constitution was adopted on 19 Nov. 1826. The present constitution, the fifteenth, came into effect following its acceptance in a referendum on 25 Jan. 2009 and defined Bolivia as 'a United Social State of Plurinational Communitarian Law'. Under its terms, running to 411 articles, the majority indigenous population has been granted increased rights (including recognition of indigenous systems of justice), state control is extended over the exploitation of natural resources and regional autonomy is enhanced. The separation of church and state is recognized and land reforms in favour of indigenous populations enshrined. A new 'plurinational legislative assembly', consisting of a 130-member *Chamber of Deputies* and a 36-member *Senate*, took office following elections in Dec. 2009. The constitution also allows for the president to serve a maximum of two consecutive terms.

National Anthem

'Bolivianos, el hado propicio' ('Bolivians, a favourable destiny'); words by José Ignacio Sanjinés, tune by Benedetto Vincenti.

GOVERNMENT CHRONOLOGY

Heads of State since 1943. (ADN = Nationalist Democratic Action; FRB = Front of the Bolivian Revolution; MAS = Movement Towards Socialism; MIR = Movement of Revolutionary Left; MNR = Nationalist Revolutionary Movement; MNRI = Nationalist Revolutionary Movement of the Left; PSD = Social Democratic Party; PURS = Party of the Republican Socialist Union; n/p = non-partisan)

President of the Republic

1943–	military	Gualberto Villarroel López

Presidents of the Provisional Junta of Government

1946	n/p	Néstor Guillén Olmos
1946–47	n/p	Tomás Monje Gutiérrez

Presidents of the Republic

1947–49	PURS	José Enrique Hertzog Garaizábal
1949–51	PURS	Mamerto Urriolagoitia Harriague

President of the Military Junta of Government

1951–52	military	Hugo Ballivián Rojas

Presidents of the Republic

1952	MNR	Hernán Siles Zuazo
1952–56	MNR	Ángel Víctor Paz Estenssoro
1956–60	MNR	Hernán Siles Zuazo
1960–64	MNR	Ángel Víctor Paz Estenssoro

Presidents of the Military Junta of Government

1964	military	Alfredo Ovando Candía
1964–65	military	René Barrientos Ortuño
1965–66	military	René Barrientos Ortuño; Alfredo Ovando Candía
1966	military	Alfredo Ovando Candía

Presidents of the Republic

1966–69	FRB	René Barrientos Ortuño
1969	PSD-FRB	Luis Adolfo Siles Salinas
1969–70	military	Alfredo Ovando Candía
1970–71	military	Juan José Torres González
1971–78	military	Hugo Banzer Suárez
1978	military	Juan Pereda Asbún

President of the Military Junta of Government

1978–79	military	David Padilla Arancibia

Presidents of the Republic

1979	military	Alberto Natusch Busch
1980–81	military	Luis García Meza Tejada
1981–82	military	Celso Torrelio Villa
1982	military	Guido Vildoso Calderón
1982–85	MNRI	Hernán Siles Zuazo
1985–89	MNR	Ángel Víctor Paz Estenssoro
1989–93	MIR	Jaime Paz Zamora
1993–97	MNR	Gonzalo Sánchez de Lozada
1997–2001	ADN	Hugo Banzer Suárez
2001–02	ADN	Jorge Fernando Quiroga Ramírez
2002–03	MNR	Gonzalo Sánchez de Lozada
2003–05	MNR	Carlos Diego Mesa Gisbert
2005–06	n/p	Eduardo Rodríguez Veltzé
2006–	MAS	Evo Morales Ayma

RECENT ELECTIONS

Presidential elections were held on 6 Dec. 2009. Evo Morales Ayma (Movement Towards Socialism) won 64·1% of votes cast against 26·6% for Manfred Reyes Villa (Progress Plan for Bolivia), 5·7% for Samuel Doria Medina (National Unity Front) and 2·3% for René Joaquino (Social Alliance). There were four other candidates.

In elections to the Chamber of Deputies, also held on 6 Dec. 2009, the Movement Towards Socialism won 88 seats with 64·2% of the vote, Progress Plan for Bolivia 37 with 26·5%, the National Unity Front 3 with 5·7% and the Social Alliance 2 with 2·3%. In Senate elections of the same day Movement Towards Socialism won 26 seats and Progress Plan for Bolivia 10.

CURRENT GOVERNMENT

President: Evo Morales Ayma; b. 1959 (Movement Towards Socialism; sworn in 22 Jan. 2006 and re-elected 6 Dec. 2009).

Vice-President: Álvaro García Linera.

In March 2013 the cabinet was composed as follows:

Minister of Autonomy: Claudia Peña. *Communication:* Amanda Dávila. *Culture:* Pablo Groux. *Economy and Public Finance:* Luis Alberto Arce Catacora. *Education:* Roberto Aguilar. *Environment and Water:* Felipe Quisipe. *Foreign Relations and Worship:* David Choquehuanca Céspedes. *Health and Sport:* Juan Carlos Calvimontes. *Hydrocarbons and Energy:* Juan José Sosa. *Institutional Transparency and the Fight Against Corruption:* Nardy Suxo. *Interior:* Carlos Romero Bonifaz. *Justice:* Cecilia Ayllón. *Labour, Employment and Social Security:* Daniel Santalla Tórrez. *Mining and Metals:* Mario Virreyra. *National Defence:* Rubén Saavedra. *Planning and Development:* Elba Viviana Caro Hinojosa. *Presidency:* Juan Ramón Quintana. *Productive Development:* Ana Teresa Morales Olivera. *Public Works, Services and Housing:* Vladimir Sánchez. *Rural Affairs and Lands:* Nemecia Achacollo Tola.

Office of the President (Spanish only):
  http://www.presidencia.gob.bo

CURRENT LEADERS

Juan Evo Morales Ayma

Position
President

Introduction
Evo Morales Ayma became Bolivia's first indigenous president in Jan. 2006, promising to transform the fortunes of South America's poorest country. The former coca farmer and left-wing activist has sought to embed a state-led socialist economy and strengthen presidential powers, and has been a leading critic of US foreign policies. He was re-elected in Dec. 2009.

Early Life
Evo Morales Ayma was born in the mining town of Orinoca, Oruro on 26 Oct. 1959. When the mines began to close in the early 1980s, his family moved to become coca farmers in Chapare. Morales became a leader of the *cocaleros* and during the 1990s clashed with successive governments, particularly that of the former right-wing military dictator, Hugo Banzer Suárez. Banzer joined the US 'war on drugs' and introduced a five-year plan to eradicate coca cultivation and pressurize the *cocaleros* into growing alternative crops. In 1997 Morales, by then a member of the Movement Towards Socialism (MAS), was elected to Congress where he continued to fight the government's policy, arguing that coca consumption was an accepted part of daily life for workers and that the West had a responsibility to suppress cocaine demand.

In Jan. 2002 he was removed from Congress on a terrorism charge related to riots in Sacaba over coca eradication, though many claimed his dismissal followed pressure from the US embassy. Morales nevertheless declared his candidacy for the congressional and presidential elections, held in June 2002, on a platform of nationalizing strategic industries, providing basic services for all, land reform and tackling corruption. The MAS came second with 20·9% of the vote, with Morales crediting much of his success to inflammatory comments made against him by the then US ambassador. Refusing to join a coalition with the Nationalist Revolutionary Movement (MNR), led by President Gonzalo Sánchez de Lozada, the MAS became the leading opposition party.

Following a general strike, Morales was involved in an uprising that led to the ousting of de Lozada in Oct. 2003. He was also a key figure in mass protests in La Paz in April 2005 calling for an end to poverty and a greater share of profits from the country's gas reserves. A blockade led to food and fuel shortages in La Paz and when clashes erupted between the police and protesters, Carlos Mesa, the new president, fled the city. In fresh elections in Dec. 2005, Morales defeated the conservative former president, Jorge Quiroga, with 53·7% of the vote, and was sworn in as president on 22 Jan. 2006.

Career in Office
Morales promised 'equality and justice' for the poor and marginalized, and a new constitution providing greater legal representation and more rights for indigenous people. His choice of inexperienced left-wing activists to fill most of the cabinet led to comparisons with Venezuela's president, Hugo Chávez, with whom Morales swiftly signed an energy co-operation accord. Close links were also forged with communist Cuba, but relations with the USA remained strained (leading to the expulsion of the US ambassador and the suspension of operations of the US drug enforcement agency in Bolivia in Oct.–Nov. 2008).

On 1 May 2006 Morales announced the nationalization of Bolivia's oil and gas industry. The decree required foreign energy companies already operating in the market, including Brazil's Petrobras, to accept harsh new contracts within six months. However, a temporary suspension was announced in Aug. owing to shortage of finance, and the terms accepted by foreign companies by the end of Oct. were less draconian than in the May decree.

In July 2006 Morales and the MAS won elections for a new Constituent Assembly to revise the constitution, but fell well short of the two-thirds majority needed to rewrite it at will. This led MAS delegates to introduce a controversial measure in Sept. allowing the Assembly to approve individual clauses by simple majority vote. Opponents feared the loss of regional autonomy and the greater centralization of political power in the president's hands. There was also resistance from large landowners to Morales' land redistribution reform in favour of the indigenous majority, which passed into law in Nov. 2006. Although Morales retained popular support, he remained locked in a power struggle over the proposed constitution with the wealthier eastern departments, four of which declared autonomy in protest. In response, Morales staged a recall referendum on his leadership in Aug. 2008, which he won convincingly. In Oct. Congress approved the text of the new constitution, which was accepted in a referendum in Jan. 2009.

Presidential and parliamentary elections in Dec. 2009 returned Morales to office in a landslide victory, while his MAS won majorities in the Chamber of Deputies and Senate. Morales resumed the nationalization programme, taking over four electricity companies in May 2010, but he also faced strikes and protests—the first such opposition since he took office in 2006—against his government's restrictive wage policy. In Oct. 2010 Morales and the Peruvian leader signed an accord allowing landlocked Bolivia to build its own Pacific port on a small stretch of Peru's coastline.

There were further mass demonstrations in early 2011 against a government proposal to raise fuel prices sharply, and again in Aug.–Sept. against a controversial plan to build a highway through a rainforest reserve. Both initiatives were subsequently dropped. In mid-2012 Morales announced another electricity nationalization, this time of the Bolivian subsidiary of the Spanish

power company REE. He also faced strikes by government workers, particularly police officers, over pay.

DEFENCE
In 2008 defence expenditure totalled US$250m. (US$26 per capita), representing 1·5% of GDP.

Army
There are six military regions. Strength (2009): 34,800 (25,000 conscripts), including a Presidential Guard infantry regiment under direct headquarters command.

Navy
A small force exists for river and lake patrol duties. Despite being landlocked, Bolivia's naval personnel totalled 4,800 in 2009, including 1,700 marines. There were six Naval Districts in 2009. The Navy had 73 vessels in 2009, including 54 patrol craft.

Air Force
The Air Force, established in 1923, had 33 combat-capable aircraft and 15 armed helicopters in 2009. Personnel strength (2009): 6,500 (including conscripts).

ECONOMY
In 2007 agriculture accounted for 12·9% of GDP, industry 36·4% and services 50·7%.

Bolivia's 'shadow' (black market) economy is estimated to constitute approximately 63% of the country's official GDP, one of the highest percentages of any country in the world.

Overview
Bolivia has large natural gas reserves, with gas exports accounting for 46·6% of total exports. The country also exports minerals and agricultural goods, and is among the world's largest producers of coca (the raw material for cocaine). Hydrocarbon revenues account for 31·6% of total revenue, making Bolivia vulnerable to fluctuations in global commodity prices.

Growth has been volatile over the decades. The 1960s and 1970s were boom years owing to a flourishing mining sector and capital inflows, while the 1980s were marked by hyperinflation, high debt and political crises. In the 1990s consumption-driven expansion led to an average annual investment growth rate of 4–5%. The state privatized some utilities but by the end of the decade growth was slowed by the knock-on effect of the economic crises in Brazil and Argentina, both major trading partners. Real GDP growth averaged 3·3% per year over the period 1996–2005.

Growth rates have improved in recent years led by domestic services, construction and rising prices for hydrocarbons and mining products. Exports of goods tripled between 2005 and 2012, with GDP growth averaging 4·7% per year from 2006–11. External debt was significantly reduced and unemployment reached a record low 5·5% in 2012.

Social indicators have also improved. In 2005 more than 60% of the population was living in poverty. By 2011 the figure had fallen to just above 48%. Nonetheless, Bolivia remains among the poorest countries in Latin America.

Currency
The unit of currency is the *boliviano* (BOB) of 100 *centavos*, which replaced the *peso* on 1 Jan. 1987 at a rate of one boliviano = 1m. pesos. Inflation was 10·3% in 2008, falling to 2·5% in 2010 but it increased again to 9·9% in 2011. In July 2005 foreign exchange reserves were US$922m., total money supply was 7,481m. bolivianos and gold reserves totalled 911,000 troy oz.

Budget
Central government revenue was 26,797m. bolivianos in 2007 (35,733m. bolivianos in 2006) and expenditure 22,467m. bolivianos (21,963m. bolivianos in 2006).

Performance

Real GDP growth was 3·4% in 2009, 4·1% in 2010 and 5·2% in 2011. Total GDP was US$23·9bn. in 2011.

Banking and Finance

The Central Bank (*President*, Marcelo Zabalaga) is the bank of issue. In 2000 there were eight commercial banks and five foreign banks.

In 2010 total external debt was US$5,267m., representing 27·8% of GNI.

There is a stock exchange in La Paz.

ENERGY AND NATURAL RESOURCES

Environment

In 2008 Bolivia's carbon dioxide emissions from the consumption and flaring of fossil fuels were the equivalent of 1·4 tonnes per capita.

Electricity

Installed capacity was 1·5m. kW in 2007. Production from all sources (2007), 5·73bn. kWh; consumption per capita was 583 kWh in 2007.

Oil and Gas

There are petroleum and natural gas deposits in the Tarija, Santa Cruz and Cochabamba areas. Production of oil in 2005 was 15,416,919 bbls. Reserves in 2005 were 465m. bbls. The US$2·2bn. Bolivia–Brazil pipeline was completed in 2000 and is the longest natural gas pipeline in South America. The 3,150 km pipeline connects Bolivia's gas sources with the southeast regions of Brazil. Natural gas output in 2008 was 13·9bn. cu. metres, with proven reserves of 710bn. cu. metres. In May 2006 Bolivia nationalized the oil and natural gas industries and six months later, in Nov., Bolivia signed up 40 new contracts with oil and natural gas enterprises.

Minerals

Mining accounted for 4% of GDP in 2005. Tin-mining had been the mainstay of the economy until the collapse of the international tin market in 1985. Estimated production in 2005 (preliminary, in tonnes): zinc, 157,019; tin, 18,694; lead, 11,093; antimony, 5,225; wolfram, 658; silver, 420; gold, 8,906 fine kg.

Agriculture

The agricultural population was 3·17m. in 2002, of whom 1·56m. were economically active. There were 2·35m. ha. of arable land in 2004–05. Output in 1,000 tonnes in 2004–05 was: sugarcane, 5,328; soybeans, 1,689; maize, 817; potatoes, 762; rice, 479; bananas, 443; cassava, 370; sorghum, 246; alfalfa, 167. In 2004, 49,000 tonnes of coca (the source of cocaine) were grown. Since 1987 Bolivia has received international (mainly US) aid to reduce the amount of coca grown, with compensation for farmers who co-operate. Bolivia (not Brazil) is the largest producer of brazil nuts (39,000 tonnes in 2009).

Livestock, 2005: cattle, 7,314,000; sheep, 8,816,000; pigs, 2,390,000; goats, 1,896,000; llamas, 2,130,000; asses (2003), 635,000; horses (2003), 323,000; alpacas, 255,000; chickens (2003), 75m.

Forestry

Forests covered 57·20m. ha. (53% of the land area) in 2010. Tropical forests with woods ranging from the 'iron tree' to the light balsa are exploited. Timber production in 2007 was 3·10m. cu. metres.

Fisheries

In 2010 the total catch was 6,946 tonnes, exclusively from inland waters.

INDUSTRY

In 2007 industry accounted for 36·4% of GDP, with manufacturing contributing 14·7%. The principal manufactures are mining, petroleum, smelting, foodstuffs, tobacco and textiles.

Labour

Out of 3,884,251 people (54·8% male) in employment in 2001, 44·1% were in agriculture, ranching and hunting, 14·0% in retail and repair, 9·2% in industrial manufacturing and 4·9% in construction. The unemployment rate in 2004 was 8·7%. In 2006 the minimum wage was 500 bolivianos a month.

INTERNATIONAL TRADE

An agreement of Jan. 1992 with Peru gives Bolivia duty-free transit for imports and exports through a corridor leading to the Peruvian Pacific port of Ilo from the Bolivian frontier town of Desaguadero, in return for Peruvian access to the Atlantic via Bolivia's roads and railways.

Imports and Exports

In 2005 imports (f.o.b.) amounted to US$2,191·8m.; exports (f.o.b.) US$2,810·4m. Main import commodities are road vehicles and parts, machinery for specific industries, cereals and cereal preparations, general industrial machinery, chemicals, petroleum, food, and iron and steel. Main exports in 2005 (provisional, in US$1m.): natural gas, 984·0; soybeans and products, 340·1; fuels, 314·8; zinc, 200·1; metallic tin, 102·0; silver ore, 88·3; metallic gold, 78·6; nuts, 74·4; wood and wood products, 67·4.

Main import suppliers in 2005 (provisional, in US$1m.): Brazil, 486·0; Argentina, 370·2; USA, 294·2; Chile, 154·3; Peru, 145·6; Japan, 135·9. Main export markets in 2005 (provisional, in US$1m.): Brazil, 1,011·6; USA, 385·3; Argentina, 260·5; Colombia, 180·5; Venezuela, 154·3; Japan, 134·3.

COMMUNICATIONS

Roads

The total length of the road system was 62,479 km in 2004, of which 14,336 km were national roads. Total passenger cars in use in 2007 numbered 174,900, lorries and vans 468,800, and buses and coaches 7,000. There were 1,073 road accident fatalities in 2007.

Rail

In 2007 the railway network totalled 2,866 km of metre gauge track. Passenger-km travelled in 2007 came to 313m. and freight tonne-km in 2005 to 1,027m. In July 2007 President Morales announced plans to nationalize the railways.

Civil Aviation

The three international airports are La Paz (El Alto), Santa Cruz de la Sierra (Viru Viru) and Cochabamba (Jorge Wisterman). The main airline is Aerosur, which in 2007 ran scheduled services to Asunción, Buenos Aires, Lima, Salta and São Paulo, as well as internal services. The operations of Lloyd Aéreo Boliviano, for many years the national airline, were suspended in April 2007 owing to financial problems but charter flights were resumed in Dec. In Oct. 2007 the government announced the creation of a new airline, Boliviana de Aviación, as a successor flag carrier. It began commercial flights in March 2009 and in March 2013 ran scheduled services to Buenos Aires, Madrid and São Paulo as well as internal services. In 2005 scheduled airline traffic of Bolivian-based carriers flew 17·1m. km, carrying 1,396,400 passengers.

Shipping

Lake Titicaca and about 19,000 km of rivers are open to navigation. In Jan. 2009 there were 43 ships of 300 GT or over registered, totalling 78,000 GT.

Telecommunications

In 2011 there were 879,800 landline telephone subscriptions (equivalent to 87·2 per 1,000 inhabitants) and 8,353,300 mobile phone subscriptions (or 828·0 per 1,000 inhabitants). Fixed internet subscriptions totalled 114,000 in 2010 (11·5 per 1,000 inhabitants). In June 2012 there were 1·6m. Facebook users.

SOCIAL INSTITUTIONS

Justice

Justice is administered by the Supreme Court, superior department courts (of five or seven judges) and courts of local justice. The Supreme Court, with headquarters at Sucre, is divided into two sections, civil and criminal, of five justices each, with the Chief Justice presiding over both. Members of the Supreme Court are chosen on a two-thirds vote of Congress. The death penalty was abolished for ordinary crimes in 1997.

The population in penal institutions in Oct. 2006 was 7,682 (82 per 100,000 of national population).

Education

Adult literacy was 90·7% in 2008 (male, 95·9%; female, 86·8%). Primary instruction is free and obligatory between the ages of six and 14 years. In 2007–08 there were 1,508,000 pupils and (2006–07) 62,000 teachers in primary schools, 1,060,000 pupils and (2006–07) 58,000 teachers in secondary schools; there were an estimated 353,000 students and 16,000 academic staff in tertiary education in 2006–07.

In 2005 there were ten universities, two technical universities, one Roman Catholic university, one musical conservatory, and colleges in the following fields: business, six; teacher training, four; industry, one; nursing, one; technical teacher training, one; fine arts, one; rural education, one; physical education, one. Bolivia's largest university is the Universidad Mayor de San Andrés, in La Paz, founded in 1830. There were 37 private universities in 2006–07.

In 2006 public expenditure on education came to 6·3% of GDP.

Health

In 2001 there were 1,999 doctors and 4,025 nurses. There were 241 hospitals with 9,886 hospital beds (one per 954 persons) in 2005.

Welfare

The pensions and social security systems in Bolivia were reformed in 2010. Instead of a defined-contribution system based on privately managed individual capitalization accounts (introduced in 1996), a defined-benefit publicly managed pension system was reintroduced. In 2008 the solidarity bonus (Bonosol) was replaced by a new universal pension (Renta Dignidad). The new universal pension—worth approximately US$340 a year—is paid to all Bolivians over the age of 60.

RELIGION

The Roman Catholic church was disestablished in 1961. It is under a cardinal (in Sucre), an archbishop (in La Paz), six bishops and vicars apostolic. It had 7·54m. adherents in 2001. In 2001, 78% of the population were Roman Catholics and 16% Protestants. In Feb. 2013 there was one cardinal.

CULTURE

World Heritage Sites

There are six UNESCO sites in Bolivia: the City of Potosí (inscribed in 1987), the largest industrial mining complex of the 16th century; the Jesuit Missions of the Chiquitos (1990), six settlements for converted Indians built between 1696 and 1760; the Historic City of Sucre (1991), containing 16th century colonial architecture; El Fuerte de Samaipata (1998), a pre-Hispanic sculptured rock and political and religious centre; Noel Kempff Mercado National Park (2000), a 1,523,000 ha. park in the Amazon Basin; and Tiwanaku: Spiritual and Political Centre of the Tiwanaku Culture (2000), monumental remains from AD 500 to 900.

Press

There were 19 paid-for daily newspapers in 2007 with a combined circulation of 140,000. The top-selling daily is the tabloid *El Deber*, with an average daily circulation of 15,000 (30,000 on Sundays).

Tourism

In 2005 there were 504,000 foreign tourists; total revenue from tourism was US$346m. in 2005.

DIPLOMATIC REPRESENTATIVES

Of Bolivia in the United Kingdom (106 Eaton Sq., London, SW1W 9AD)
Ambassador: Vacant.
Chargé d'Affaires a.i.: Eduardo Daza Sandoval.

Of the United Kingdom in Bolivia (Avenida Arce 2732, Casilla 694, La Paz)
Ambassador: Ross Denny.

Of Bolivia in the USA (3014 Massachusetts Ave., NW, Washington, D.C., 20008)
Ambassador: Vacant.
Chargé d'Affaires a.i.: Freddy Bersatti Tudela.

Of the USA in Bolivia (Avenida Arce 2780, Casilla 425, La Paz)
Ambassador: Vacant.
Chargé d'Affaires a.i.: Larry Memmott.

Of Bolivia to the United Nations
Ambassador: Sacha Sergio Llorenti Solíz.

Of Bolivia to the European Union
Ambassador: René Fernández Revollo.

FURTHER READING

Jemio, Luis Carlos, *Debt, Crisis and Reform in Bolivia: Biting the Bullet.* 2001
Klein, Herbert S., *A Concise History of Bolivia.* 2003
Morales, Waltraud Q., *Bolivia.* 2004
Muñoz-Pogossian, Betilde, *Electoral Rules and the Transformation of Bolivian Politics: The Rise of Evo Morales.* 2008

National Statistical Office: Instituto Nacional de Estadística, Av. José Carrasco 1391, CP 6129, La Paz.
Website (Spanish only): http://www.ine.gob.bo

BOSNIA AND HERZEGOVINA

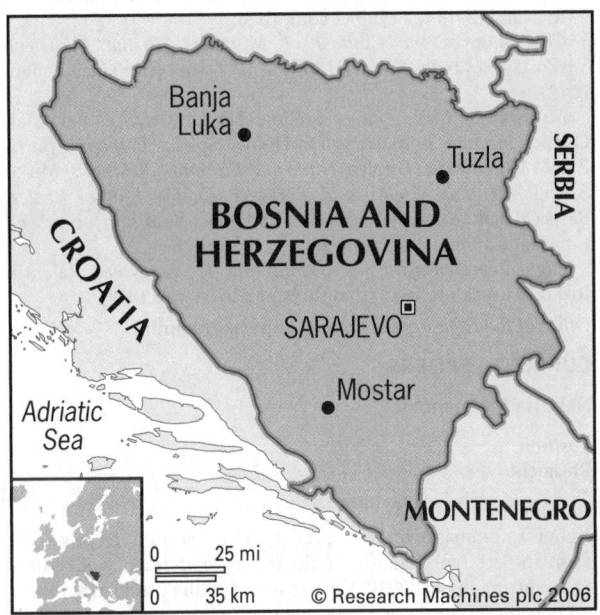

Republika Bosna i Hercegovina
(Republic of Bosnia and Herzegovina)

Capital: Sarajevo
Population projection, 2015: 3·72m.
GNI per capita, 2011: (PPP$) 7,664
HDI/world rank: 0·733/74
Internet domain extension: .ba

KEY HISTORICAL EVENTS

Settled by Slavs in the 7th century, Bosnia was conquered by the Turks in 1463 when much of the population was gradually converted to Islam. At the Congress of Berlin (1878) the territory was assigned to Austro-Hungary under nominal Turkish suzerainty. Austria-Hungary's outright annexation in 1908 contributed to the outbreak of the First World War. After 1918 Bosnia and Herzegovina became part of a new Kingdom of Serbs, Croats and Slovenes under the Serbian monarchy. Its name was changed to Yugoslavia in 1929. (See SERBIA and MONTENEGRO for developments up to and beyond the Second World War.)

On 15 Oct. 1991 the National Assembly adopted a 'Memorandum on Sovereignty' supporting Bosnian autonomy within a Yugoslav federation. Though boycotted by Serbs, a referendum in March 1992 supported independence. In March 1992 an agreement was reached by Muslims, Serbs and Croats to set up three autonomous ethnic communities under a central Bosnian authority.

Bosnia and Herzegovina declared independence on 5 April 1992. Fighting broke out between the Serb, Croat and Muslim communities, with heavy casualties and destruction in Sarajevo, Muslim territorial losses and an exodus of refugees. UN-sponsored ceasefires were repeatedly violated.

On 13 Aug. 1992 the UN Security Council authorized force to ensure the delivery of humanitarian aid to besieged civilians. Internationally sponsored peace talks were held in Geneva in 1993, but Serb-Muslim-Croat fighting continued. In April 1993 the UN established havens for Muslim civilians in Sarajevo, Srebrenica and Goražde.

In Dec. 1994 Bosnian Serbs and Muslims signed a countrywide interim ceasefire. Bosnian Croats also signed in Jan. 1995. However, Croatian Serbs and the Muslim secessionist forces under Fikret Abdić continued fighting. On 16 June 1995 Bosnian government forces launched an attack to break the Bosnian Serb siege of Sarajevo. On 11 July Bosnian Serb forces began to occupy UN security zones despite retaliatory NATO air strikes, and on 28 Aug. shelled Sarajevo. In July Srebrenica was the scene of the worst massacre of the war, when Bosnian Serb troops killed over 7,000 Muslim men and boys. Dutch peacekeeping forces whose duty was to protect the area failed to prevent the massacre.

To stop the shelling of UN safe areas, more than 60 NATO aircraft attacked Bosnian Serb military installations on 30–31 Aug. On 26 Sept. in Washington the foreign ministers of Bosnia, Croatia and Yugoslavia (the latter negotiating for the Bosnian Serbs) agreed a draft Bosnian constitution under which a central government would handle foreign affairs and commerce and a Serb Zone, while a Muslim-Croat Federation administered internal affairs. A ceasefire came into force on 12 Oct. 1995.

In Dayton (Ohio) on 21 Nov. 1995 the prime ministers of Bosnia, Croatia and Yugoslavia initialled a US-brokered agreement to end hostilities. The Bosnian state was divided into a Muslim-Croat Federation containing 51% of Bosnian territory and a Serb Republic containing 49%. A central government authority representing all ethnic groups with responsibility for foreign and monetary policy and citizenship issues was established and free elections held. On 20 Dec. 1995 a NATO contingent (IFOR) took over from UN peacekeeping forces to enforce the Paris peace agreements and set up a 4 km separation zone between the Serb and Muslim-Croat territories. After a year IFOR was replaced by SFOR, a 'Stabilization Force'. On 2 Dec. 2004 a 7,000-strong European Union force 'EUFOR' took over from SFOR.

TERRITORY AND POPULATION

The republic is bounded in the north and west by Croatia, in the east by Serbia and in the southeast by Montenegro. It has a coastline of only 20 km with no harbours. Its area is 51,210 sq. km, including 210 sq. km of inland waters. The capital is Sarajevo (estimated population, 2005: 380,000).

Population at the 1991 census: 4,377,033, of which the predominating ethnic groups were Muslims (1,905,829), Serbs (1,369,258) and Croats (755,892). Population of the principal cities in 1991: Sarajevo, 415,631 (est. 1999, 522,000); Banja Luka, 142,644; Zenica, 96,238. By 1996, following the civil war, 1,319,250 Bosnians had taken refuge abroad, including 0·45m. in Serbia and Montenegro, 0·32m. in Germany, 0·17m. in Croatia and 0·12m. in Sweden. Population estimate, 2010, 3·76m.; density, 74 per sq. km.

The UN gives a projected population for 2015 of 3·72m.

In 2011, 49·2% of the population lived in urban areas.

In accordance with the Dayton Agreement the country is divided into two entities: the Federation of Bosnia and Herzegovina (51% of the territory) and Republika Srpska, sometimes referred to as the the Serb Republic (49% of the territory). In addition there is a self-governing unit, Brčko District, which is a *de facto* third entity in the northeast of the country. *See also* CONSTITUTION AND GOVERNMENT *below*.

The official languages are Bosnian, Croatian and Serbian.

SOCIAL STATISTICS

2010 births, 33,779; deaths, 34,633. Rates per 1,000, 2010: birth, 8·8; death, 9·0. Annual population growth rate, 2000–08, 0·3%. Life expectancy at birth, 2007, was 72·4 years for men and 77·7 years for women. Infant mortality, 2010, eight per 1,000 live births; fertility rate, 2008, 1·2 children per woman (the joint lowest rate in the world).

CLIMATE

The climate is generally continental with steady rainfall throughout the year, although in areas nearer the coast it is more Mediterranean.

CONSTITUTION AND GOVERNMENT

On 18 March 1994, in Washington, Bosnian Muslims and Croats reached an agreement for the creation of a federation of cantons with a central government responsible for foreign affairs, defence and commerce. It was envisaged that there would be a president elected by a two-house legislature alternating annually between the nationalities.

On 31 May 1994 the National Assembly approved the creation of the Muslim Croat federation (the Federation of Bosnia and Herzegovina). Alija Izetbegović remained the unitary states' President. An interim government with Hasan Muratović as Prime Minister was formed on 30 Jan. 1996.

The Dayton Agreement including the new constitution was signed and came into force on 14 Dec. 1995. The government structure was established in 1996 as follows:

Heading the state is a three-member *Presidency* (one Croat, one Muslim, one Serb) with a rotating president. The Presidency is elected by direct universal suffrage, and is responsible for foreign affairs and the nomination of the prime minister. There is a two-chamber parliament: the *House of Representatives* (which meets in Sarajevo) comprises 42 directly elected deputies, two-thirds Croat and Muslim and one-third Serb; and the *House of Peoples* (which meets in Lukavica) comprises five Croat, five Muslim and five Serb delegates.

Below the national level the country is divided into two self-governing entities along ethnic lines.

The Bosniak-Croat Federation of Bosnia and Herzegovina (Federacija Bosna i Hercegovina) is headed by a President and Vice-President, alternately Croat and Muslim, a 98-member Chamber of Representatives and a 58-member Chamber of Peoples. The Serb Republic (Republika Srpska) is also headed by an elected President and Vice-President, and there is a National Assembly of 83 members, elected by proportional representation.

Central government is conducted by a *Council of Ministers*, which comprises Muslim and Serb Co-Prime Ministers and a Croat Deputy Prime Minister. The Co-Prime Ministers alternate in office every week.

In Nov. 2005 leaders of the three main ethnic groups agreed on a series of constitutional reforms aimed at enhancing the authority of the central government, reducing the powers of the Federation of Bosnia and Herzegovina and the Serb Republic, and streamlining the parliament and the office of the presidency.

National Anthem

'Intermezzo'; tune by Dusan Sestić; no words.

RECENT ELECTIONS

Elections were held on 3 Oct. 2010 for the Presidium and the federal parliament. Elected to the Presidency were: Bakir Izetbegović (Muslim; Party of Democratic Action—SDA); Željko Komšić (Croat; Social Democratic Party of Bosnia and Herzegovina—SDP BiH); and Nebojša Radmanović (Serb; Alliance of Independent Social Democrats—SNSD). In the parliamentary elections the SDP BiH won 8 seats with 284,358 votes, the SNSD 8 with 277,817 votes, the SDA 7 with 214,261 votes, the Union for a Better Future of BiH 4, the Serbian Democratic Party 4, the Croatian Democratic Union of Bosnia and Herzegovina 3, HDZ 1990 2 and the Party for Bosnia and Herzegovina 2. The Party of Democratic Progress, the People's Party Work for Betterment, the Democratic People's Alliance and the Democratic People's Community won a single seat each.

CURRENT GOVERNMENT

Presidency Chairman: Nebojša Radmanović (Serb, SNSD), took rotating presidency on 10 Nov. 2012). *Presidency Members:* Željko Komšić (Croat, SDP BiH); Bakir Izetbegović (Muslim, SDA).

In March 2013 the cabinet comprised:

Chairman of the Council of Ministers (Prime Minister): Vjekoslav Bevanda (Croat, HDZ); b. 1956 (in office since 12 Jan. 2012).

Minister of Civil Affairs: Sredoje Nović. *Defence:* Zekerijah Osmić. *Finance and Treasury:* Nikola Špirić. *Foreign Affairs:* Zlatko Lagumdžija. *Foreign Trade and Economic Relations:* Mirko Šarović. *Human Rights and Refugees:* Damir Ljubić. *Justice:* Bariša Čolak. *Security:* Fahrudin Radončić. *Transportation and Communications:* Damir Hadžić.

High Representative for Bosnia and Herzegovina: Valentin Inzko (Austria); b. 1949 (in office since 26 March 2009).

Office of the High Representative: http://www.ohr.int

CURRENT LEADERS

Nebojša Radmanović

Position
President

Introduction
Nebojša Radmanović was elected to his second term as the Bosnian-Serb representative to the tripartite presidency of Bosnia-Herzegovina in Oct. 2010. Chairmanship of the presidency rotates every eight months to ensure equality of representation among the country's three predominant ethnic groups. The president is chiefly responsible for foreign policy and finance.

Early Life
Radmanović, a Serb by nationality, was born in 1949 in Gracanica, northeastern Bosnia, in the then Socialist Federal Republic of Yugoslavia. He was educated in Banja Luka and read philosophy at the University of Belgrade.

After the collapse of communism in eastern Europe and the break-up of Yugoslavia, Radmanović worked in education and the arts. In the 1990s he was director of the Bosanska Krajina archives (based in Banja Luka) and of the Republic of Srpska Archives, as well as director of the National Theatre in Banja Luka and editor of *Glas Javnosti*, a Belgrade-based newspaper.

In the late 1990s Radmanović served as president of the executive board of the city of Banja Luka. He was a member of the Socialist Party of Serbia (SPRS) until April 2000 when he and other members formed the splinter Socialist Democratic Party (DSP) and allied themselves with the Independent Social Democrats (SNSD). Radmanović became leader of the DSP and ran unsuccessfully for the presidency in 2002 representing the SNSD–DSP alliance.

In 2003 the SNSD–DSP formed the Alliance of Independent Social Democrats (SNSD). Radmanović was appointed the Bosnian-Serb representative to the rotating presidential council in Oct. 2006, a post to which he was reappointed four years later after securing a narrow victory over the more moderate Mladen Ivanić.

Career in Office
Radmanović has repeatedly stated his opposition to any moves to amend the terms of the 1995 Dayton peace accords that divided the state into its two constituent entities—a Serb Republic and a Bosniak-Croat Federation. On succeeding Bosniak representative Haris Silajdžić as chairman of the presidency in Nov. 2008, Radmanović declared that his priorities would be to advance his country's progress towards membership of the European Union and NATO and to improve relations with neighbouring Serbia, Croatia and Montenegro.

With a reputation as a hardline Serb nationalist, Radmanović's re-election in 2010 was perceived as an obstacle to centralizing power and creating a more united tripartite presidency. In contrast to his moderate Croat and Muslim counterparts, Radmanović has argued in favour of Bosnian-Serb secession from the country. He was criticized in some quarters for his re-election campaign and its avowedly nationalist rhetoric.

On assuming the chairmanship of the presidency in Nov. 2012, Radmanović said that he hoped to make progress on Bosnia and Herzegovina's accession to the European Union. He has voiced concern that the EU's internal divisions will make it more difficult to gain membership, particularly in the absence of changes to the country's constitution demanded by a 2009 European Court of Human Rights ruling.

DEFENCE

Defence expenditure in 2008 totalled US$244m. (US$53 per capita), representing 1·3% of GDP.

An EU-led peacekeeping contingent 'EUFOR' took over military operations from the NATO-led 'SFOR' on 2 Dec. 2004. Its mission is to focus on the apprehension of indicted war criminals and counter-terrorism, and provide advice on defence reform. Initially numbering 7,000 personnel, by early 2013 the strength had been reduced to 600.

Army

Reform of the defence forces began in 2003 and resulted, in 2006, in the establishment of a single state army (in place of the previously separate armed forces of the Federation of Bosnia and Herzegovina and the Serb Republic). Its personnel—including that of the Air Force and Anti-Air Defence Brigade—totalled 9,007 in 2007. The Army had 325 main battle tanks in 2007 plus 193 armoured personnel carriers and 132 armoured infantry fighting vehicles.

Air Force

The Air Force and Anti-Aircraft Defence was established in 2006. It had 17 combat-capable aircraft and 21 helicopters in 2007.

INTERNATIONAL RELATIONS

The Serb Republic and the then Yugoslavia signed an agreement on 28 Feb. 1997 establishing 'special parallel relations' between them. The agreement envisages co-operation in cultural, commercial, security and foreign policy matters, allows visa-free transit of borders and includes a non-aggression pact. A customs agreement followed on 31 March.

ECONOMY

In 2008 agriculture accounted for 9·1% of GDP, industry 28·5% and services 62·4%.

Overview

Infrastructure and output were damaged by war in the early 1990s but growth has been continuous since the Dayton Accords ended fighting in 1995, although the economy is yet to reach pre-war levels. In the post-war decade GDP tripled and exports expanded ten-fold. Two-thirds of the economy is generated by services. The financial sector plays an important role, with the banking sector dominated by foreign banks. It is estimated that donor commitment from international assistance programmes in the first decade of the century exceeded US$5bn. Full membership of the Central European Free Trade Agreement came in Sept. 2007.

Prior to the global financial crisis, growth relied on domestic demand financed from abroad. However, the crisis triggered a collapse and the economy fell into recession in 2009. The government agreed a stand-by arrangement with the IMF to cushion the impact of the deteriorating external environment and adopted policies to redress fiscal imbalances. These included gradual fiscal consolidation accompanied by structural fiscal reforms, with substantial financing from the IMF, World Bank

and EU. The economy stabilized and registered marginal growth in 2010 but business conditions remain challenging and further stabilization measures have been hampered by slow progress in implementing the IMF programme.

Currency

A new currency, the *konvertibilna marka* (BAM) consisting of 100 *pfennig*, was introduced in June 1998. Initially trading at a strict 1-to-1 against the Deutsche Mark, it is now pegged to the euro at a rate of 1·95583 convertible marks to the euro. Inflation was 2·1% in 2010 and 3·7% in 2011. Total money supply was 4,193m. convertible marks in July 2005.

Budget

Budgetary central government revenue in 2009 was 5,509m. convertible marks; expenditure was 5,931m. convertible marks. VAT of 17% was introduced on 1 Jan. 2006.

Performance

Real GDP contracted by 2·9% in 2009 but grew by 0·7% in 2010 and 1·3% in 2011. Total GDP was US$18·1bn. in 2011.

Banking and Finance

There is a Central Bank (*Governor*, Kemal Kozarić). In 2005 there were 28 commercial banks (19 in the Federation and 9 in the Serb Republic).

In 2010 total external debt was US$8,457m., representing 48·8% of GNI.

There are stock exchanges in Banja Luka and Sarajevo.

ENERGY AND NATURAL RESOURCES

Environment

Bosnia and Herzegovina's carbon dioxide emissions from the consumption and flaring of fossil fuels in 2008 were the equivalent of 4·1 tonnes per capita.

Electricity

Bosnia and Herzegovina is rich in hydro-electric potential and is a net exporter of electricity. Installed capacity was estimated at 4·3m. kW in 2007. Production in 2007 was 11·82bn. kWh. In 2007 consumption per capita was 2,920 kWh.

Minerals

Output in 2008: lignite, 11·2m. tonnes; crushed stone, 4·4m. tonnes; iron ore, 2·7m. tonnes; gravel, 1·5m. tonnes; bauxite, 1·0m. tonnes; aluminium, 156,000 tonnes.

Agriculture

In 2007 there were 1·02m. ha. of arable land and 95,000 ha. of permanent crops. 2009 production (in 1,000 tonnes): maize, 963; potatoes, 414; wheat, 256; plums, 156; cabbage and kale, 82; barley, 77; apples, 72; tomatoes, 46. Livestock in 2009: cattle, 458,000; pigs, 529,000; sheep, 1,055,000; poultry, 18·7m.

Source: Agency for Statistics of Bosnia and Herzegovina

Forestry

In 2010 forests covered 2·19m. ha., or 43% of the total land area. Timber production in 2007 was 3·75m. cu. metres.

Fisheries

Estimated total fish catch in 2010: 2,000 tonnes (almost exclusively freshwater).

INDUSTRY

Industry accounted for 29% of GDP in 2008, with manufacturing contributing 14%. Output in 2008 (in 1,000 tonnes): cement, 1,406; crude steel, 608; lime, 216.

Labour

The active labour force totalled 1,157,940 in April 2010 (62% males). Unemployment in April 2010 was 27·2% (25·6% for men and 29·9% for women). Among 15–24 year olds it was 57·5%.

Source: Agency for Statistics of Bosnia and Herzegovina

INTERNATIONAL TRADE

Imports and Exports

2007 external trade (in US$1m.): imports (c.i.f.), 9,720·1; exports (f.o.b.), 4,152·0. Principal import sources in 2007 were Croatia, 17·6%; Germany, 12·5%; Serbia, 10·2%; Italy, 9·0%; Slovenia, 6·4%. Main export markets in 2007 were Croatia, 18·4%; Serbia, 13·7%; Italy, 13·1%; Germany, 12·8%; Slovenia, 10·9%.

COMMUNICATIONS

Roads

In 2005 there were an estimated 22,419 km of roads (4,104 km main roads). Passenger cars numbered 473,076 in 2007 (123 per 1,000 inhabitants). There were 428 road accident fatalities in 2007.

Source: Agency for Statistics of Bosnia and Herzegovina

Rail

There were 1,017 km of railways in 2008 (771 km electrified). It is estimated that up to 80% of the rail network was destroyed in the civil war, and it was not until July 2001 that the first international services were resumed. There are two state-owned rail companies —the Railway of the Federation of Bosnia and Herzegovina (ŽFBH) and the Railway of the Serb Republic (ŽRS). In 2008 ŽFBH carried 528,000 passengers and 8·1m. tonnes of freight while ŽRS carried 727,000 passengers and 5·0m. tonnes of freight.

Civil Aviation

There are airports at Sarajevo (Butmir), Tuzla, Banja Luka and Mostar. In 2005 there were direct flights to Belgrade, Cologne/Bonn, Düsseldorf, Frankfurt, İstanbul, İzmir, Ljubljana, Milan, Prague, Stockholm, Stuttgart, Vienna, Zagreb and Zürich. In 2001 Sarajevo handled 313,000 passengers (all international) and 1,300 tonnes of freight.

Telecommunications

In 2011 there were 955,900 landline telephone subscriptions (equivalent to 254·8 per 1,000 inhabitants) and 3,171,300 mobile phone subscriptions (or 845·2 per 1,000 inhabitants). In 2010, 52·0% of the population were internet users. Three companies run the telephone networks in different parts of the country, the largest of which is the Sarajevo-based BH Telecom.

SOCIAL INSTITUTIONS

Justice

The population in penal institutions in the Federation of Bosnia and Herzegovina in April 2008 was 1,750; in the Serb Republic the prison population was 928 in Sept. 2007.

Police

The European Union Police Mission (EUPM) in Bosnia and Herzegovina, the EU's first civilian crisis management operation, took over from the UN's International Police Task Force on 1 Jan. 2003. It aims to help the authorities develop their police forces to the highest European and international standards.

Education

The adult literacy rate was 98% in 2008. In 2008–09 there were 174,000 pupils in primary schools, 334,000 in secondary schools and 105,000 students in tertiary education. The largest university is the University of Sarajevo, established in 1949 but with its origins dating back to 1531.

Health

In 2005 there were 5,540 physicians, 629 dentistry personnel, 18,332 nursing and midwifery personnel and 308 pharmaceutical personnel. In 2005 there were 30 hospital beds per 10,000 inhabitants.

Welfare

There were 509,000 pensioners in 2006. Only around 30% of the population over 65 receives an old-age pension. The average monthly pension was 246·6 convertible marks in 2006.

RELIGION

In 2001 there were estimated to be 1,690,000 Sunni Muslims, 1,180,000 Serbian Orthodox, 710,000 Roman Catholics and 350,000 followers of other religions. In Feb. 2013 the Roman Catholic church had one cardinal.

CULTURE

World Heritage Sites

There are two UNESCO World Heritage sites in Bosnia and Herzegovina: the Old Bridge area of the Old City of Mostar (inscribed on the list in 2005), an important Ottoman frontier town, and the Mehmed Paša Sokolović Bridge in Višegrad (2007), constructed in the 16th century.

Press

There were seven paid-for daily newspapers in 2008 with a combined circulation of 75,000 and 46 paid-for non-dailies.

Tourism

In 2010, 365,000 non-resident tourists stayed in holiday accommodation (up from 311,000 in 2009 and 171,000 in 2000).

DIPLOMATIC REPRESENTATIVES

Of Bosnia and Herzegovina in the United Kingdom (5–7 Lexham Gdns, London, W8 5JJ)
Ambassador: Mustafa Mujezinović.

Of the United Kingdom in Bosnia and Herzegovina (8 Tina Ujevića, 71000 Sarajevo)
Ambassador: Nigel Casey, MVO.

Of Bosnia and Herzegovina in the USA (2109 E St., NW, Washington, D.C., 20037)
Ambassador: Jadranka Negodić.

Of the USA in Bosnia and Herzegovina (1 Robert C. Frasure St., 71000 Sarajevo)
Ambassador: Patrick Moon.

Of Bosnia and Herzegovina to the United Nations
Ambassador: Mirsada Čolaković.

Of Bosnia and Herzegovina to the European Union
Ambassador: Igor Davidović.

FURTHER READING

Bieber, Florian, *Post-War Bosnia: Ethnicity, Inequality and Public Sector Governance.* 2005
Burg, Steven L. and Shoup, Paul S., *The War in Bosnia-Herzegovina.* 1999
Cigar, N., *Genocide in Bosnia: the Policy of Ethnic Cleansing.* 1995
Fine, J. V. A. and Donia, R. J., *Bosnia-Hercegovina: a Tradition Betrayed.* 1994
Friedman, F., *The Bosnian Muslims: Denial of a Nation.* 1996
Hoare, Marko Attila, *The History of Bosnia: From the Middle Ages to the Present Day.* 2006
Malcolm, N., *Bosnia: a Short History.* 3rd ed. 2002

National Statistical Office: Agency for Statistics of Bosnia and Herzegovina, Zelenih beretki 26, 71000 Sarajevo. *Director:* Zdenko Milinović.
Website: http://www.bhas.ba

BOTSWANA

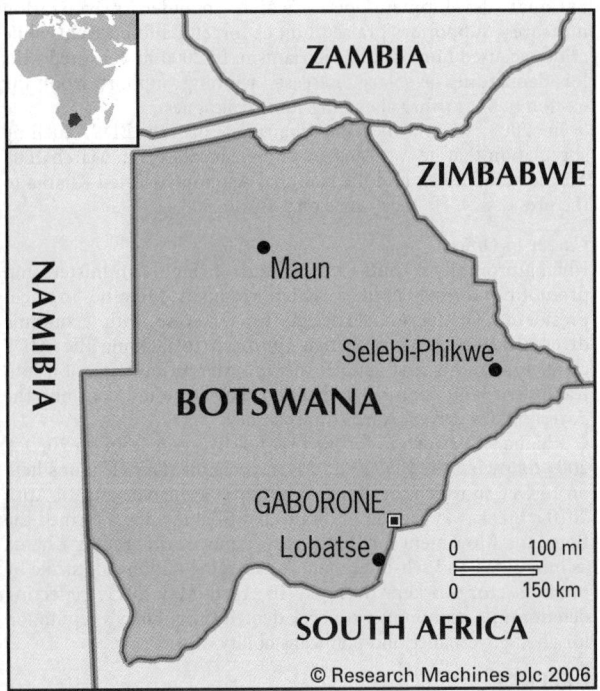

ZAMBIA

ZIMBABWE

NAMIBIA

Maun

Selebi-Phikwe

BOTSWANA

GABORONE

Lobatse

0   100 mi
0   150 km

SOUTH AFRICA

© Research Machines plc 2006

**Lefatshe la Botswana
(Republic of Botswana)**

Capital: Gaborone
Population projection, 2015: 2·12m.
GNI per capita, 2011: (PPP$) 13,049
HDI/world rank: 0·633/118
Internet domain extension: .bw

KEY HISTORICAL EVENTS

The Tswana or Batswana people are the principal inhabitants of the country formerly known as Bechuanaland. The territory was declared a British protectorate in 1895 administered by the High Commissioner in South Africa until the post was abolished in 1964. Proposals for the merging of Bechuanaland and the other two High Commission Territories into South Africa were strongly opposed. Economically, however, the country was closely tied to that of South Africa and has remained so. In Dec. 1960 Bechuanaland received its first constitution. Further constitutional change brought self-government in 1965 and independence on 30 Sept. 1966. Botswana had difficulties with the neighbouring settler regime in Rhodesia, until that country became Zimbabwe in 1980. Relations with South Africa were also strained until the ending of apartheid. Today Botswana is one of the few African countries to enjoy stability and (2009 apart) a fast-growing economy.

TERRITORY AND POPULATION

Botswana is bounded in the west and north by Namibia, northeast by Zambia and Zimbabwe, and east and south by South Africa. The area is 581,730 sq. km. 2011 census population, 2,024,904; density, 3·5 per sq. km.

The UN gives a projected population for 2015 of 2·12m.

In 2011, 61·8% of the population were urban.

The country is divided into nine districts (Central, Ghanzi, Kgalagadi, Kgatleng, Kweneng, North East, North West, South East and Southern).

The main towns (2011 census population) are Gaborone (231,592), Francistown (98,961), Molepolole (66,466), Maun (60,263), Mogoditshane (58,079), Serowe (50,820), Selebi-Phikwe (49,411), Kanye (47,007), Mochudi (44,815) and Mahalapye (43,289).

The official languages are Setswana and English. Setswana is spoken by over 90% of the population and English by approximately 40%. More than ten other languages, including Herero, Hottentot, Kalanga, Mbukushu, San and Sekgalagadi are spoken in various tribal areas. The main ethnic groups are the Tswana (67%), Kalanga (15%) and Ndebele (2%).

SOCIAL STATISTICS

2008 (estimates) births, 47,000; deaths, 23,000. Rates, 2008 estimates (per 1,000 population): births, 24·5; deaths, 12·1. Infant mortality, 2010 (per 1,000 live births), 36. Expectation of life in 2007 was 53·2 years for males and 53·3 for females. In 2007, 23·9% of all adults between 15 and 49 were infected with HIV. Annual population growth rate, 2000–08, 1·4%. Fertility rate, 2008, 2·9 children per woman.

CLIMATE

In winter, days are warm and nights cold, with occasional frosts. Summer heat is tempered by prevailing northeast winds. Rainfall comes mainly in summer, from Oct. to April, while the rest of the year is almost completely dry with very high sunshine amounts. Gaborone, Jan. 79°F (26·1°C), July 55°F (12·8°C). Annual rainfall varies from 650 mm in the north to 250 mm in the southeast. The country is prone to droughts.

CONSTITUTION AND GOVERNMENT

The Constitution was adopted in March 1965 and became effective on 30 Sept. 1966. It provides for a republican form of government headed by the President with three main organs: the Legislature, the Executive and the Judiciary. The executive rests with the President who is responsible to the National Assembly. The President is elected for five-year terms by the National Assembly.

The *National Assembly* consists of 63 members, of which 57 are elected by universal suffrage, four are specially elected members and two, the President and the Speaker, are *ex officio*.

Elections are held every five years. Voting is on the first-past-the-post system.

There is also a *House of Chiefs* to advise the government. It consists of the Chiefs of the eight tribes who were autonomous during the days of the British protectorate, plus four members elected by and from among the sub-chiefs in four districts; these 12 members elect a further three politically independent members.

National Anthem

'Fatshe leno la rona' ('Blessed be this noble land'); words and tune by K. T. Motsete.

GOVERNMENT CHRONOLOGY

Presidents since 1966. (BDP = Botswana Democratic Party)

1966–80	BDP	Seretse Khama
1980–98	BDP	Quett Ketumile Joni Masire
1998–2008	BDP	Festus Gontebanye Mogae
2008–	BDP	Lieut.-Gen. Seretse Khama Ian Khama

RECENT ELECTIONS

In National Assembly elections held on 16 Oct. 2009 the Botswana Democratic Party (BDP) gained 45 seats with 53·3% of the vote, the

Botswana National Front 6 with 21·9%, the Botswana Congress Party 4 with 19·2%, the Botswana Alliance Movement 1 with 2·3% and ind. 1 with 1·9%. Turnout was 76·7%.

CURRENT GOVERNMENT

President: Lieut.-Gen. Seretse Khama Ian Khama; b. 1953 (BDP; sworn in 1 April 2008).

Vice-President: Ponatshego Kedikilwe.

In March 2013 the cabinet was as follows:

Minister of Presidential Affairs and Public Administration: Mokgweetsi Masisi. *Defence, Justice and Security:* Dikgakgamatso Seretse. *Foreign Affairs and International Co-operation:* Phandu Skelemani. *Finance and Development Planning:* Kenneth Matambo. *Infrastructure, Science and Technology:* Johnnie Swartz. *Lands and Housing:* Lebonamang Mokalake. *Labour and Home Affairs:* Edwin Batshu. *Minerals, Energy and Water Resources:* Kitso Mokaila. *Youth, Sports and Culture:* Shaw Kgathi. *Trade and Industry:* Baledzi Gaolathe. *Local Government:* Peter Siele. *Agriculture:* Christian de Graaf. *Transport and Communications:* Nonofo Molefi. *Education and Skills Development:* Pelonomi Venson-Moitoi. *Environment, Wildlife and Tourism:* Tshekedi Khama. *Health:* John Seakgosing.

Government Website: http://www.gov.bw

CURRENT LEADERS

Lieut.-Gen. Seretse Khama Ian Khama

Position
President

Introduction
Lieut.-Gen. (retd) Seretse Khama Ian Khama succeeded Festus Mogae as president in 2008. Formerly head of the army and a paramount chief of the Bamangwato people, Khama retains a reputation as a military man rather than a career politician. Since entering politics in 1998 he has been closely associated with the government's programme to diversify the economy and promote transparency.

Early Life
Seretse Khama Ian Khama was born in the UK on 27 Feb. 1953, the son of Seretse Khama (who later became Botswana's first president) and his English wife. Following the family's return to Botswana in 1956, Khama grew up in Serowe and attended the local school. He pursued further studies in Zimbabwe (then Rhodesia), Swaziland and Switzerland, before enrolling at Sandhurst Military Academy in the UK. After graduating, he joined the Botswana Defence Force (BDF) and rose rapidly through the ranks. In 1977 he was promoted to brigadier and became deputy commander of the BDF under Lieut.-Gen. Mompati Merafhe.

In 1979 he was made paramount chief of the Bamangwato people, Botswana's largest tribal group. However, Khama devoted his time to the army and in 1989, when Merafhe retired to enter politics, became its commander. Under Khama the BDF developed into a professional force, participating in international peacekeeping, disaster relief and anti-poaching missions.

On 1 April 1998 Khama, newly retired from the army, was named vice-president by the incoming president Festus Mogae. The appointment was widely seen as an attempt to inject new blood into the ruling Botswana Democratic Party (BDP) and to tap into Khama's reputation and influence. Initially unable to take up his post because he did not hold a seat in the National Assembly, Khama was sworn in on 13 July 1998 after winning a by-election in Serowe North. He was put in charge of presidential administration and public affairs and became an arbiter of complaints against government ministers, which made him some political enemies.

Following the BDF's victory in the 1999 general election, Khama was controversially allowed a year's sabbatical to fulfil his duties as chief of the Bamangwato. On his return to the national scene he oversaw the implementation of the government's national development plan, which included privatization measures. Supporters praised him as forceful and efficient while critics accused him of authoritarianism. In 2000 he censured MPs for demanding a salary increase, winning support from the electorate but further alienating some colleagues.

In 2003 Khama became chairman of the BDP, fuelling speculation that he was Mogae's chosen successor. In March 2008 Mogae stood down and the National Assembly elected Khama to the presidency. He took office on 1 April 2008.

Career in Office
Khama promptly reshuffled the cabinet, firing five ministers and promoting former military chief Mompati Merafhe to vice-president. Khama was expected to persevere with economic diversification and a proactive approach to tackling the AIDS crisis. One of his first actions was to order an investigation into fraudulent land deals, an indication that he would continue the government's drive towards transparency.

Khama was sworn in for his first full five-year term on 20 Oct. 2009 following the BDP's success in parliamentary elections held on 16 Oct. in which they took 45 of the 57 available seats. In April 2010 a breakaway faction of disaffected BDP members formed the Botswana Movement for Democracy, a move criticized by Khama as misguided and self-interested. A damaging nationwide strike by public sector workers over pay in April–May 2011, reflecting deteriorating economic prospects, dented the country's reputation for good governance and political stability.

DEFENCE

In 2008 defence expenditure totalled US$293m. (US$150 per capita), representing 2·2% of GDP.

Army

The Army personnel (2007) numbered 8,500. There is also a 1,500-strong paramilitary force.

Air Force

The Air Wing operated 31 combat-capable aircraft in 2007 and numbered 500.

ECONOMY

Industry accounted for 55·2% of GDP in 2006, services 42·9% and agriculture 1·9%.

Overview

From the 1980s until the early 2000s Botswana recorded 8% annual growth, fuelled by the diamond industry. From being one of the world's poorest countries in the 1960s, it achieved middle income status and today holds the second highest sovereign credit rating in Africa according to the ratings agency Standard & Poor's. This is a result of healthy management of resources and prudent government expenditure.

When Botswana gained independence in 1966 mining accounted for only 1% of GDP but today accounts for 33%. Nickel and copper exports are also important. Agricultural development is limited by the dry landscape. The government is attempting to diversify the economy by developing the financial services sector and safari tourism. In 1994 Botswana was the first country to be removed from the UN's list of Least Developed Countries (LDCs).

Alongside a slowdown in diamond production, the global economic crisis of 2008–09 saw the economy shrink by 4·7% in 2009 although sub-Saharan Africa as a whole experienced growth of nearly 3%. However, a rebound in diamond exports and strong growth in the non-mining sector helped the economy return to solid growth in 2010. Growth since the latter part of 2010 has been among the strongest of any middle-income country.

Nonetheless, the fragile global economy and a possible reduction in mineral export demand continue to pose threats.

The Botswana Core Welfare Indicators Survey for 2009/10 showed a decline in the number of people below the poverty line from 30·6% in 2003 to 20·7% in 2010. However, 116,000 people were still living below one dollar a day in 2009–10 and income inequality is high. Furthermore, the HIV/AIDS rate was estimated to be the second highest in the world in 2009. Government policy is focused on poverty eradication and tackling high unemployment.

Currency

The unit of currency is the *pula* (BWP) of 100 *thebe*. The pula was devalued by 7·5% in Feb. 2004 and 12·5% in May 2005. Inflation was 6·9% in 2010 and 8·5% in 2011. Foreign exchange reserves were US$5,770m. in June 2005 and total money supply was P3,878m.

Budget

The fiscal year begins in April. Revenue in 2006–07 was P27,397·7m.; expenditure was P19,737·4m. Tax revenue accounted for 92·1% of total revenues; social services accounted for 46·2% of expenditures and general services including defence 29·2%.

VAT is 12%.

Performance

Real GDP contracted by 4·7% in 2009 but grew by 7·0% in 2010 and 5·1% in 2011. Total GDP in 2011 was US$17·3bn.

Banking and Finance

There were five commercial banks in 2004. Total assets were P24,718·2m. at July 2004. The Bank of Botswana (*Governor*, Linah Mohohlo), established in 1976, is the central bank. The National Development Bank, founded in 1964, has six regional offices, and agricultural, industrial and commercial development divisions. The Botswana Co-operative Bank is banker to co-operatives and to thrift and loan societies. The government-owned Post Office Savings Bank (Botswana Post) operates throughout the country.

In 2010 external debt totalled US$1,709m., representing 11·6% of GNI.

There is a stock exchange in Gaborone.

ENERGY AND NATURAL RESOURCES

Environment

Botswana's carbon dioxide emissions from the consumption and flaring of fossil fuels in 2008 were the equivalent of 2·4 tonnes per capita.

Electricity

Installed capacity was an estimated 217,000 kW in 2007. Production in 2010 was 430m. kWh. The coal-fired power station at Morupule supplies cities and major towns.

Minerals

Botswana is the world's biggest diamond producer in terms of value, accounting for over a quarter of world diamond revenue; in 2003 the total value was estimated to be US$2·5bn. Diamonds were first discovered in Botswana in 1967. Debswana, a partnership between the government and De Beers, runs three mines producing around 30m. carats a year, with plans to double the capacity of the largest mine from 6m. to 12m. carats a year. Coal reserves are estimated at 17bn. tonnes. There is also copper, salt and soda ash. Estimated diamond production in 2005, 31·9m. carats (the third largest quantity after Australia and Russia). Other mineral production, 2003: coal, 823,000 tonnes; salt, 30,000 tonnes; copper, 8,000 tonnes; gold, 8 kg.

Agriculture

70% of the total land area is desert. 80% of the population is rural, 71% of all land is 'tribal', protected and allocated to prevent over-grazing, maintain small farmers and foster commercial ranching. Agriculture provides a livelihood for over 70% of the population, but accounts for only 2·4% of GDP (2003). In 2003, 360,000 ha. were arable and 3,000 ha. permanent crops. There were 7,000 tractors in 2004 and 102 harvester-threshers. Cattle-rearing is the chief industry after diamond-mining, and the country is more a pastoral than an agricultural one, crops depending entirely upon the rainfall. In 2007 an estimated 295,000 persons were economically active in agriculture. In 2004 there were: cattle, 2·2m.; goats, 1·6m.; asses, 330,000 (estimate); sheep, 244,000; chickens, 4·5m. (estimate). A serious outbreak of cattle lung disease in 1995–96 led to the slaughter of around 300,000 animals.

Production in 2004 (in 1,000 tonnes) included: sorghum, 12; maize, 8; millet, 3; beans and pulses, 2.

17% of the land is set aside for wildlife conservation and 20% for wildlife management areas, with four national parks and game reserves.

Forestry

Forests covered 11,351,000 ha., or 20% of the total land area, in 2010. There are forest nurseries and plantations. Concessions have been granted to harvest 7,500 cu. metres in Kasane and Chobe Forestry Reserves, and up to 2,500 cu. metres in the Masame area. In 2007, 774,000 cu. metres of roundwood were cut.

Fisheries

In 2010 the total catch was 60 tonnes, exclusively from inland waters.

INDUSTRY

The most important sector is the diamond industry. A diamond-processing plant opened in Gaborone in March 2008. Meat is processed, and beer, soft drinks, textiles and foodstuffs manufactured. Rural technology is being developed and traditional crafts encouraged. In June 2003 there were 16,773 enterprises operating in Botswana, of which a third were in the wholesale and retail trade.

Labour

In 2005–06, 787,962 persons were economically active, of which 539,150 persons were in employment; 29·9% worked in agriculture, 14·4% in wholesale and retail trade, 8·0% in education and 6·9% in public administration. In March 2005 there were an estimated 288,800 paid employees (including 86,700 in central government and 24,700 in local government). Botswana's biggest individual employer is the Debswana Diamond Company, with a workforce of 4,000. In 2005–06 the unemployment rate was 17·5%.

INTERNATIONAL TRADE

Botswana is a member of the Southern African Customs Union (SACU) with Lesotho, Namibia, South Africa and Swaziland. There are no foreign exchange restrictions.

Imports and Exports

In 2007 imports (c.i.f.) totalled US$3,986·9m. More than 80% of all imports are from the SACU countries, the main commodities being machinery and transport equipment, manufactured goods, mineral fuels and lubricants, food and live animals, and chemicals and related products.

In 2007 exports (f.o.b.) totalled US$5,072·5m., including diamonds, nickel, garments and clothing accessories, copper and meat. Diamond exports totalled US$3,172·3m. in 2007.

Principal import sources in 2007 were South Africa, 83·5%; China, 1·8%; Belgium, 1·6%; UK, 1·4%. Main export markets were the UK, 65·0%; South Africa, 10·2%; Norway, 8·1%; Zimbabwe, 7·3%.

COMMUNICATIONS

Roads

In 2005 the total road network was estimated to be 25,798 km (32·6% paved). In Dec. 2008 there were 256,498 motor vehicles registered. There were 497 deaths in road accidents in 2007.

Rail

The main line from Mafeking in South Africa to Bulawayo in Zimbabwe traverses Botswana. The total length of the rail system was 888 km in 2005, including two branch lines. In 2006, 426,894 passengers and 1,712,607 tonnes of freight were carried.

Civil Aviation

There are international airports at Gaborone (Sir Seretse Khama) and at Maun and six domestic airports. The national carrier is the state-owned Air Botswana. In 2003 direct flights were operated to Harare and Johannesburg. In 2003 scheduled airline traffic of Botswana-based carriers flew 3m. km, carrying 189,000 passengers (131,000 on international flights). In Oct. 1999 an Air Botswana pilot who had been suspended two months earlier crashed an empty passenger plane into the airline's two serviceable aeroplanes at Gaborone Airport, killing himself and destroying the airline's complete fleet in the process. In 2006 Gaborone handled 289,550 passengers.

Telecommunications

In 2011 there were 149,600 landline telephone subscriptions (equivalent to 73·7 per 1,000 inhabitants) and 2,900,300 mobile phone subscriptions (or 1,428·2 per 1,000 inhabitants). In 2011, 7·0% of the population were internet users. In June 2012 there were 224,000 Facebook users.

SOCIAL INSTITUTIONS

Justice

Law is based on the Roman-Dutch law of the former Cape Colony, but judges and magistrates are also qualified in English common law. The Court of Appeal has jurisdiction in respect of criminal and civil appeals emanating from the High Court, and in all criminal and civil cases and proceedings. Magistrates' courts and traditional courts are in each administrative district. As well as a national police force there are local customary law enforcement officers. The death penalty is still in force and was used in 2012. The population in penal institutions in May 2007 was 5,917 (329 per 100,000 of national population).

Education

Adult literacy rate in 2008 was 83%. Basic free education, introduced in 1986, consists of seven years of primary and three years of junior secondary schooling. In 2005 enrolment in primary schools was 326,500 with 13,472 teaching staff, and 168,720 pupils at secondary level with 12,371 teaching staff. There were 10,950 students is higher education in 2005 with 529 academic staff. 'Brigades' (community-managed private bodies) provide lower-level vocational training. The Department of Non-Formal Education offers secondary-level correspondence courses and is the executing agency for the National Literacy Programme.

In 2009 total expenditure on education came to 8·2% of GNI and 16·2% of total government spending.

Health

In 2004 there were 16 primary hospitals, one mental hospital, three referral hospitals, 15 health centres, 257 clinics and 366 health posts. There were also 761 stops for mobile health teams. In 2004 there were 89 doctors and 2,129 nurses in government health facilities. There are other private health facilities with more personnel.

RELIGION

Freedom of worship is guaranteed under the Constitution. About 43% of the population is Christian. Non-Christian religions include Bahais, Muslims and Hindus.

CULTURE

World Heritage Sites

Tsodilo was created a UNESCO World Heritage Site in 2001. It is the site of over 4,500 prehistoric rock paintings in the Kalahari Desert.

Press

The government-owned *Daily News* is distributed free (circulation, 2006: 65,000). There is one other daily, the independent *Mmegi* ('The Reporter'), and 11 non-dailies.

Tourism

There were 1·7m. non-resident tourists in 2005 with tourism receipts totalling US$561m.

DIPLOMATIC REPRESENTATIVES

Of Botswana in the United Kingdom (6 Stratford Pl., London, W1C 1AY)
High Commissioner: Roy Warren Blackbeard.

Of the United Kingdom in Botswana (Private Bag 0023, Gaborone)
High Commissioner: Nicholas Pyle OBE, MBE.

Of Botswana in the USA (1531–1533 New Hampshire Ave., NW, Washington, D.C., 20036)
Ambassador: Tebelelo Seretse.

Of the USA in Botswana (PO Box 90, Gaborone)
Ambassador: Michelle Gavin.

Of Botswana to the United Nations
Ambassador: Charles Thembani Ntwaagae.

Of Botswana to the European Union
Ambassador: Samuel Otsile Outlule.

FURTHER READING

Central Statistics Office. *Statistical Bulletin* (Quarterly).
Ministry of Information and Broadcasting. *Botswana Handbook.— Kutlwano* (Monthly).
Molomo, M. G. and Mokopakgosi, B. (eds.) *Multi-Party Democracy in Botswana.* 1991
Perrings, C., *Sustainable Development and Poverty Alleviation in Sub-Saharan Africa: the Case of Botswana.* 1995

National Statistical Office: Central Statistics Office, Private Bag 0024, Gaborone.
Website: http://www.cso.gov.bw

BRAZIL

República Federativa do Brasil
(Federative Republic of Brazil)

Capital: Brasília (Federal District)
Population projection, 2015: 203·29m.
GNI per capita, 2011: (PPP$) 10,162
HDI/world rank: 0·718/84
Internet domain extension: .br

KEY HISTORICAL EVENTS

Evidence of human habitation in Brazil dates to 9000 BC. Before the Portuguese discovery and occupation of Brazil there was a large indigenous population fragmented into smaller tribes. The largest was the Tupi-Guarani, which survived the sub-tropical environment by clearing land for crops.

The first Europeans to encounter the indigenous peoples were exiled criminals, or *degredados*, and Jesuit missionaries. The first Portuguese contact with Brazil was Pedro Alvares Cabral who left Lisbon in 1500 with orders to travel along the Cape of Good Hope route discovered by the Portuguese navigator Vasco da Gama in 1497–98. In an attempt to avoid storms he set a course more westerly than da Gama's and was carried still farther westward, landing in a place he named *Terra da Vera Cruz* (Land of the True Cross) and later renamed *Terra do Brasil* (Land of Brazil).

Although the official motive for Portuguese exploration was religious—the conversion of the natives to the Catholic faith—the greater incentive was to find a direct all-water trade route with Asia and thereby break Italy's commercial domination. Early Portuguese economic activity in Brazil revolved around the exploitation of the huge timber (Brazilwood) resources. This was soon superseded by sugarcane and, to a lesser extent, tobacco, harvested on plantations that sprang up in the interior in the 16th and 17th centuries. As these industries came to dominate the economy the need for large-scale labour became more pressing. Where the indigenous people proved unsuitable or unavailable, largely owing to ill health from newly introduced European diseases, millions of Africans were enslaved and shipped to the region.

The first attempt to establish a working government came in 1533 when the Portuguese divided the land into 15 captaincies, subdivided into leagues and ruled by selected governors (*donatários*). In 1549 John III sought to establish a more centralized power structure and appointed Tomé de Sousa as governor general, ruling from the newly founded capital, Salvador

(Bahia). In 1567 Governor-General Mem da Sá founded Rio de Janeiro to protect its harbour from French incursions. During the Union of Portugal and Spain (1580–1640), Brazil became subject to attacks from Spanish enemies, notably the Netherlands, whose forces were not expelled until 1654.

The Portuguese settlers had set about conquering the vast Brazilian interior by the late 17th century. Early excursions were made by *bandeirantes*, men pursuing private enterprise and dreams of personal wealth. They discovered the first gold in the region at Minas Gerais in 1695. This opened up a new resource for exploitation by the Portuguese crown. Rio de Janeiro benefited greatly and in 1763 became the colonial capital in place of Salvador. Recife and Ouro Preto were the other major colonial urban centres.

The third quarter of the 18th century saw Spain accept many of Portugal's claims in the region. Portuguese Prime Minister Sebastião José de Carvalho e Mello ended the rule of the *donatários,* expelled the Jesuits, gave new freedoms to the native population and established two companies to regulate Brazilian trade. As Brazilian government became increasingly centralized a burgeoning nationalism emerged, most famously in the failed rebellion against the Portuguese led by Joaquim José da Silva Xavier (Tiradentes) in 1789.

In 1807 an invasion of Portugal by Napoleon Bonaparte forced the royal family to flee Portugal and take refuge in Rio, declaring it the temporary capital of the Portuguese Empire. In 1816 King João VI ascended the Portuguese throne but for five years refused to return to Lisbon. While in Brazil he initiated reforms which ended Portugal's commercial monopoly and in 1815 granted Brazil equal status with Portugal when he established the United Kingdom of Portugal, Brazil and the Algarves. On his return to Portugal João's son, Pedro, became Brazil's regent.

Independence

Pedro's regency ran into trouble when the Cortes (the Portuguese parliamentary body) demanded his return to Portugal. The Cortes repealed many of João's reforms for the former colony and sought to reduce it to its earlier colonial status. When, in Sept. 1822, the Cortes decided to reduce Pedro's powers he called for Brazilian independence. On 1 Dec. 1822 he was crowned Constitutional Emperor and Perpetual Defender of Brazil. The United States recognized Brazil's independence in May 1824 followed by Portugal itself in 1825.

Pedro was forced to abdicate in 1831 following a disastrous war with Argentina and a financial crisis deepened by his promise to free the slaves. He left his five-year old son Pedro II as the ruler in waiting. In 1840 Pedro II succeeded after nine years of weak rule and civil strife. By 1847, Pedro was established as a leader free of political influences. He ruled for nearly 50 years and despite uprisings Brazil remained relatively stable and its economy strong.

Pedro was instrumental in the overthrow of Juan Manuel de Rosas in Argentina in the 1850s and became involved in Uruguay's civil war in the 1860s. In the 1870s the three nations united to repel the advances of the Paraguayan forces of Francisco Solano López. Pedro outlawed the slave trade in 1854. Emancipation was achieved in 1888 when three quarters of a million slaves were freed without compensation to their owners. In 1889 Gen. Manuel Deodoro da Fonseca led a military revolt which forced Pedro's abdication. In Feb. 1891 Brazil became a Federal Republic with Fonseca elected as its first president. Forced to resign when he attempted to bypass congress, he was succeeded by Floriano Peixoto who used the military to restore order. In 1894 he was replaced by Brazil's first civilian head of state, Prudente de Morais. He and his immediate successors enjoyed relative peace as Brazil grew rich on coffee exports.

Brazil underwent significant territorial expansion in the early years of the 20th century during the Baron of Rio Branco's tenure as foreign minister. As well as winning 900,000 sq. km of land from other South American nations, he pursued close relations with the USA and UK, which led to a declaration of war against Germany in 1917. However, by the 1920s there was growing internal resentment at the wealth of the coffee barons and in 1922 a failed military coup initiated eight years of civil strife. Amid economic crisis in 1930 Getúlio Vargas lost the presidential election but was swept to power by a military junta which dismissed the legitimately elected government.

The constitution of 1934 provided for universal suffrage and three years later a new constitution, drafted in the aftermath of a failed coup, gave Vargas greatly extended power. During his period of rule some areas, including São Paulo, saw considerable industrial development, helping Brazil to create a modern economy. In the 1940s the first steel plant was built in the state of Rio de Janeiro at Volta Redonda with US financing. In 1942 Brazil followed the lead of the USA and declared war against the Axis powers. In Oct. 1945 the military staged a coup and Vargas was forced to step down.

Economic Problems

The subsequent election was won by Eurico Gaspar Dutra, a favourite of Vargas, whose government set presidential terms at five years and reduced the power of central government. Vargas was returned to power in 1950 but failed to dominate as he had previously. Brazil's economic problems spiralled and in 1954 Vargas was implicated in the attempted murder of a journalist. When the High Command demanded his resignation Vargas shot himself.

Juscelino Kubitschek, popularly known as JK, was elected president in 1956. He launched road and hydroelectric schemes and built a new capital, Brasília. It was hoped these programmes would be the catalyst for the development of Brazil's huge interior but instead brought uncontrollable inflation. The presidency of Jânio Quadros in 1961 was marked by his decoration of Che Guevara in a public ceremony. Antagonizing the right wing military, he resigned after six months in office. Vice-president João Goulart took over but his leftist policies led to his overthrow by the military in 1964.

There followed 20 years of single party rule and censored press. Humberto Castelo Branco was installed as president in April 1964. Chosen by the military to enforce fundamental political and economic reforms, Branco instead sought to introduce change through democratic channels. He narrowly survived a coup in 1965 but was forced by the military powerbrokers to take a radical line. Laws passed in Oct. 1965 suspended political parties and gave the president emergency powers. An ostensibly two-party state was created, consisting of the government-backed National Renewal Alliance (ARENA) and the Brazilian Democratic Movement (MDB). The MDB declined to field a candidate at the presidential elections of 1966 and ARENA's Costa e Silva took the presidency.

Brazil's military regime was not as brutal as those of Chile or Argentina, but at its height, around 1968 and 1969 when Costa e Silva awarded himself emergency powers, the use of torture was widespread. Costa e Silva had a stroke in Aug. 1970 and was replaced by Gen. Emílio Garrastazú Médici in Oct. He was succeeded in early 1974 by Gen. Ernesto Geisel. Geisel promoted measures to reduce censorship and increase political freedom but reverted to political oppression when electoral victory was in doubt. In April 1977 he dismissed congress when it failed to pass judicial reforms and governed using emergency powers. He resigned in 1979 to be replaced by his favoured successor, Gen. João Baptista de Oliveira Figueiredo. The generals benefited from the Brazilian economic miracle in the late 1960s and '70s, when the economy was growing by more than an annual 10%. However, uncoordinated growth led to rampant bureaucracy, corruption and inflation.

Return to Democracy

Unable to contain hyperinflation, the government authorized the restitution of political rights. In 1980 a militant working-class movement sprang up under the charismatic leadership of a worker, Luiz Inácio Lula da Silva (better known as Lula). Popular opposition, together with economic problems, forced Figueiredo to adopt the *abertura* (opening)—a slow process of returning to democratic government.

Tancredo Neves, leader of the Partido do Movimento Democratico Brasiliero (PMDB), the main opposition party, surprised his military opponents by winning the 1985 elections, but died shortly before taking power. José Sarney, his vice-president, guided the country through the transition from military to civilian rule as well as overseeing the drafting and implementation of a new democratic constitution. But the country drifted into the economic chaos afflicting the whole continent, with finance ministers changing frequently and foreign debt reaching CR$115,000m. Price and wage freezes set out in Sarney's Cruzado Plan succeeded only briefly in bringing down inflation. In the presidential run-off of Dec. 1989 voters backed two of Sarney's most vociferous critics, with Fernando Collor de Mello narrowly defeating Labour Party candidate, Lula.

Collor, of the National Reconstruction party, promised reductions in inflation and corruption. In March 1990 he confiscated 80% of every bank account worth more than US$1,200, promising to release them 18 months later with interest. He also announced the privatization of state-owned companies and the opening of Brazilian markets to foreign competition and capital. By 1992 Collor's government had failed to reach many of its targets and was embroiled in scandals and corruption, some of which were linked directly to his family. Inflation was again spiralling. Parliament, under public pressure, forced an impeachment and Itamar Franco, Collor's vice-president, took office until elections were held in Oct. 1994. Under Franco inflation leapt towards 3,000% but his fourth finance minister, Fernando Henrique Cardoso, introduced successful economic reforms.

In 1994 Cardoso was elected president for the Partido da Social Democracia Brasiliera, formed in 1990 by PMDB dissidents. He oversaw an economic revolution that included a radical privatization programme, the lowering of trade barriers and the introduction of a new currency, the *real*. A constitutional amendment in 1997 provided for consecutive presidential terms and the following year Cardoso won re-election at a time when Brazil was feeling the effects of the economic turbulence in the Far East. In Jan. 1999 the *real* was devalued, losing 35% of its value against the dollar in two months. In Cardoso's second term the public debt reached US$260bn. and the government was forced to reduce spending on health and welfare while increasing taxes. His government failed to address the problems of inequality and corruption and at the 2002 presidential elections Cardoso's successor, José Serra, was defeated by the left-wing leader, Luiz Inácio Lula da Silva.

Lula, the country's first elected socialist president, pledged to combat Brazil's widespread poverty while co-operating with the business sector and international community. In May 2003 he invited the Democratic Movement into government to ensure the passage of key economic reforms. In 2011 Dilma Rousseff, also of the Workers' Party, succeeded him as president.

TERRITORY AND POPULATION

Brazil is bounded in the east by the Atlantic and on its northern, western and southern borders by all the South American countries except Chile and Ecuador. The total area (including inland waters) is 8,514,877 sq. km. It is the world's fifth largest country and occupies 47·8% of South America. Area and population as at the census of 2010:

Federal Unit and Capital	Area (sq. km)	Census 2010
North	3,853,327	
Rondônia (Porto Velho)	237,576	1,562,409
Acre (Rio Branco)	164,165	733,559

Federal Unit and Capital	Area (sq. km)	Census 2010
Amazonas (Manaus)	1,559,161	3,483,985
Roraima (Boa Vista)	224,299	450,479
Pará (Belém)	1,247,690	7,581,051
Amapá (Macapá)	142,815	669,526
Tocantins (Palmas)	277,621	1,383,445
North-East	1,554,257[1]	
Maranhão (São Luís)	331,983	6,574,789
Piauí (Teresina)	251,529	3,118,360
Ceará (Fortaleza)	148,826	8,452,381
Rio Grande do Norte (Natal)	52,797	3,168,027
Paraíba (João Pessoa)	56,440	3,766,528
Pernambuco (Recife)	98,312	8,796,448
Alagoas (Maceió)	27,768	3,120,494
Sergipe (Aracaju)	21,910	2,068,017
Bahia (Salvador)	564,693	14,016,906
South-East	924,511	
Minas Gerais (Belo Horizonte)	586,528	19,597,330
Espírito Santo (Vitória)	46,078	3,514,952
Rio de Janeiro (Rio de Janeiro)	43,696	15,989,929
São Paulo (São Paulo)	248,209	41,262,199
South	576,410	
Paraná (Curitiba)	199,315	10,444,526
Santa Catarina (Florianópolis)	95,346	6,248,436
Rio Grande do Sul (Porto Alegre)	281,749	10,693,929
Central West	1,606,372	
Mato Grosso (Cuiabá)	903,358	3,035,122
Mato Grosso do Sul (Campo Grande)	357,125	2,449,024
Goiás (Goiânia)	340,087	6,003,788
Distrito Federal (Brasília)	5,802	2,570,160
Total	8,514,877	190,755,799

[1]Including disputed areas between states of Piauí and Ceará.

Population density, 2010, 22·4 per sq. km. The 2010 census showed 93,406,990 males and 97,348,809 females. The urban population comprised 84·3% of the population in 2010.

The UN gives a projected population for 2015 of 203·29m.

The official language is Portuguese.

Population of principal cities (2010 census):

São Paulo	11,152,344	Nova Iguaçu	787,563
Rio de Janeiro	6,320,446	Campo Grande	776,242
Salvador	2,674,923	Teresina	767,557
Brasília	2,482,210	São Bernardo do	
Fortaleza	2,452,185	Campo	752,658
Belo Horizonte	2,375,151	João Pessoa	720,785
Manaus	1,792,881	Santo André	676,407
Curitiba	1,751,907	Osasco	666,740
Recife	1,537,704	Jaboatão dos	
Porto Alegre	1,409,351	Guararapes	630,595
Belém	1,381,475	São José dos Campos	617,106
Goiânia	1,297,076	Ribeirão Preto	602,966
Guarulhos	1,221,979	Contagem	601,400
Campinas	1,061,540	Uberlândia	587,266
São Gonçalo	998,999	Sorocaba	580,655
São Luís	958,522	Aracaju	571,149
Maceió	932,129	Cuiabá	540,814
Duque de Caxias	852,138	Feira de Santana	510,635
Natal	803,739	Juiz de Fora	510,378

The principal metropolitan areas (census, 2010) were São Paulo (11,152,344), Rio de Janeiro (6,320,446), Salvador (2,674,923), Brasília (2,482,210), Fortaleza (2,452,185), Belo Horizonte (2,375,151), Manaus (1,792,881), Curitiba (1,751,907), Recife (1,537,704) and Porto Alegre (1,409,351).

Approximately 54% of the population of Brazil is White, 40% mixed White and Black, and 5% Black. There are some 260,000 native Indians.

SOCIAL STATISTICS

The total number of registered live births in 2006 was 2,799,128 (rate of 15·4 per 1,000 population); deaths, 1,020,211 (5·6); marriages, 889,828 (4·9); divorces 162,244 (0·9). The average age at first marriage in 2006 was 28·3 years for men and 25·4 for women. Life expectancy in 2006 was 68·5 years for males and 76·1 for females. Annual population growth rate, 2000–05, 1·5%; infant mortality, 2010, 17 per 1,000 live births (down from 50 per 1,000 in 1990); fertility rate, 2006, 2·0 children per woman. The number of deaths per 1,000 live births among children under five was reduced from 59 in 1990 to 19 in 2010. Brazil's recent economic advances enabled 6m. people to move out of poverty in 2006.

CLIMATE

Because of its latitude, the climate is predominantly tropical, but factors such as altitude, prevailing winds and distance from the sea cause certain variations, though temperatures are not notably extreme. In tropical parts, winters are dry and summers wet, while in Amazonia conditions are constantly warm and humid. The northeast *sertão* is hot and arid, with frequent droughts. In the south and east, spring and autumn are sunny and warm, summers are hot, but winters can be cold when polar air-masses impinge. Brasília, Jan. 72°F (22·3°C), July 68°F (19·8°C). Annual rainfall 60" (1,512 mm). Belém, Jan. 78°F (25·8°C), July 80°F (26·4°C). Annual rainfall 105" (2,664 mm). Manaus, Jan. 79°F (26·1°C), July 80°F (26·7°C). Annual rainfall 92" (2,329 mm). Recife, Jan. 80°F (26·6°C), July 77°F (24·8°C). Annual rainfall 75" (1,907 mm). Rio de Janeiro, Jan. 83°F (28·5°C), July 67°F (19·6°C). Annual rainfall 67" (1,758 mm). São Paulo, Jan. 75°F (24°C), July 57°F (13·7°C). Annual rainfall 62" (1,584 mm). Salvador, Jan. 80°F (26·5°C), July 74°F (23·5°C). Annual rainfall 105" (2,669 mm). Porto Alegre, Jan. 75°F (23·9°C), July 62°F (16·7°C). Annual rainfall 59" (1,502 mm).

CONSTITUTION AND GOVERNMENT

The present Constitution came into force on 5 Oct. 1988, the eighth since independence. The *President* and *Vice-President* are elected for a four-year term. To be elected candidates must secure 50% plus one vote of all the valid votes, otherwise a second round of voting is held to elect the President between the two most voted candidates. Voting is compulsory for men and women between the ages of 18 and 70 apart from illiterates (for whom it is optional); it is also optional for persons from 16 to 18 years old and persons over 70. A referendum on constitutional change was held on 21 April 1993. Turnout was 80%. 66·1% of votes cast were in favour of retaining a republican form of government, and 10·2% for re-establishing a monarchy. 56·4% favoured an executive presidency, 24·7% parliamentary supremacy.

A constitutional amendment of June 1997 authorizes the re-election of the President for one extra term of four years.

Congress consists of an 81-member *Senate* (three Senators per federal unit plus three from the Federal District of Brasília) and a 513-member *Chamber of Deputies*. The Senate is directly elected (two-thirds of it and one-third of it elected for eight years in rotation every four years). The Chamber of Deputies is elected by universal franchise for four years. There is a *Council of the Republic* which is convened only in national emergencies.

Baaklini, A. I., *The Brazilian Legislature and Political System*. 1992
Kingstone, Peter and Power, Timothy, (eds.) *Democratic Brazil Revisited.* 2008
Martinez-Lara, J., *Building Democracy in Brazil: the Politics of Constitutional Change.* 1996
Montero, Alfred, *Brazilian Politics: Reforming a Democratic State in a Changing World.* 2006
Skidmore, Thomas E., *Brazil: Five Centuries of Change.* 2009

National Anthem

'Ouviram do Ipiranga às margens plácidas de um povo heróico o brado retumbante' ('The peaceful banks of the Ipiranga heard the resounding cry of an heroic people'); words by J. O. Duque Estrada, tune by F. M. da Silva.

GOVERNMENT CHRONOLOGY

Presidents since 1930. (ARENA = National Renewal Alliance; PMDB = Brazilian Democratic Movement Party; PRN = Party for National Reconstruction; PSD = Social Democratic Party; PSDB = Brazilian Social Democracy Party; PT = Workers' Party; PTB = Brazilian Labour Party; n/p = non-partisan)

1930–45	n/p	Getúlio Dornelles Vargas
1945–46	n/p	José Linhares
1946–51	military/PSD	Eurico Gaspar Dutra
1951–54	PTB	Getúlio Dornelles Vargas
1954–56	PTB	João Fernandes de Campos Café (Filho)
1956–61	PSD	Juscelino Kubitschek de Oliveira
1961	n/p	Jânio da Silva Quadros
1961–64	PTB	João Belchior Marques Goulart
1964–67	military	Humberto de Alencar Castelo Branco
1967–69	military	Artur da Costa e Silva
1969	Triumvirate (military)	Augusto Hamann Rademaker Grünewald, Aurélio de Lyra Tavares, Márcio de Souza e Mello
1969–74	military/ARENA	Emílio Garrastazú Médici
1974–79	military/ARENA	Ernesto (Beckmann) Geisel
1979–85	military/ARENA/PDS	João Baptista de Oliveira Figueiredo
1985–90	PMDB	José Sarney Costa
1990–92	PRN	Fernando Affonso Collor de Mello
1992–95	n/p	Itamar Augusto Cautiero Franco
1995–2003	PSDB	Fernando Henrique Silva Cardoso
2003–11	PT	Luiz Inácio Lula da Silva
2011–	PT	Dilma Rousseff

RECENT ELECTIONS

In presidential elections held on 3 Oct. 2010, Dilma Rousseff of the Workers' Party (PT) won 46·9% of votes cast, José Serra of the Brazilian Social Democracy Party (PSDB) 32·6% and Marina Silva of the Green Party (PV) 19·3%. There were six other candidates. In the second round run-off on 31 Oct., Rousseff won with 56·1% of the vote against 43·9% for Serra.

Parliamentary elections were also held on 3 Oct. 2010 for both the Chamber of Deputies and the Senate.

In the elections to the 513-seat Chamber of Deputies, the Workers' Party (PT) won 88 seats; Brazilian Democratic Movement Party (PMDB) 79; Brazilian Social Democracy Party (PSDB), 53; Democrats (DEM), 43; Progressive Party (PP), 41; Republic Party (PR), 41; Brazilian Socialist Party (PSB), 34; Democratic Labour Party (PDT), 28; Brazilian Labour Party (PTB), 21; Christian Social Party (PSC), 17; Communist Party of Brazil (PCdoB), 15; Green Party (PV), 15; Socialist People's Party (PPS), 12; Brazilian Republican Party (PRB), 8; Party of National Mobilization (PMN), 4. Other parties won three seats or fewer. The ruling alliance 'For Brazil to Keep on Changing' led by the PT and also including PCdoB, PDT, PMDB, PR, PRB, PSB, PSC, Christian Labour Party (PTC) and National Labour Party (PTN) won 311 seats in total. The opposition coalition 'Brazil Can Do More' formed by the DEM, PMN, PPS, PSDB, PTB and Labour Party of Brazil (PTdoB) won 136 seats.

Following the Senate elections of 3 Oct. 2010 the PMDB had 20 seats; PT, 15; PSDB, 11; DEM, 6; PTB, 6; PDT, 4; PP, 4; PR, 4; PSB, 3; PCdoB, 2; Socialism and Liberty Party (PSOL), 2; PMN, 1; PPS, 1; PRB, 1; PSC, 1.

CURRENT GOVERNMENT

President: Dilma Rousseff; b. 1947 (Workers' Party; sworn in 1 Jan. 2011).

Vice-President: Michel Temer.

In March 2013 the six-party coalition government was composed as follows:

Minister of Agrarian Development: Pepe Vargas. *Agriculture, Livestock and Food Supply:* Mendes Ribeiro Filho. *Cities:* Aguinaldo Ribeiro. *Communications:* Paulo Bernardo. *Culture:* Marta Suplicy. *Defence:* Celso Amorim. *Development, Industry and Foreign Trade:* Fernando Pimentel. *Education:* Aloizio Mercadante Oliva. *Environment:* Izabella Teixeira. *External Relations:* Antônio Patriota. *Finance:* Guido Mantega. *Fisheries and Aquaculture:* Marcello Crivella. *Health:* Alexandre Padilha. *Justice:* José Eduardo Cardozo. *Labour and Employment:* Carlos Daudt Brizola. *Mines and Energy:* Edison Lobão. *National Integration:* Fernando Bezerra Coelho. *Planning, Budget and Management:* Miriam Belchior. *Science, Technology and Innovation:* Marco Antônio Raupp. *Social Development and Hunger Alleviation:* Tereza Campelo. *Social Security:* Garibaldi Alves. *Sport:* Aldo Rebelo. *Tourism:* Gastão Vieira. *Transport:* Paulo Sérgio Passos. *Chief of Staff:* Gleisi Hoffmann.

Government Website: http://www.brasil.gov.br

CURRENT LEADERS

Dilma Rousseff

Position
President

Introduction
Dilma Rousseff became Brazil's first female president on 1 Jan. 2011. A left-winger from the Workers' Party, she succeeded Luiz Inácio Lula da Silva, whom she had served as chief of staff for five years and with whose policies she had been closely associated.

Early Life
Rousseff was born on 14 Dec. 1947 in Belo Horizonte, the state capital of Minas Gerais, to a Bulgarian émigré and his teacher wife. Her father had fled Europe in the 1920s, fearing persecution because of his involvement with the Bulgarian Communist Party. In Brazil he became a successful businessman.

Dilma Rousseff was schooled first at a francophone nunnery and then at Central State High School. In 1967 she joined the Worker's Politics faction of the Brazilian Socialist Party. Committed to armed struggle as the best means of achieving socialism's aims, Rousseff became involved in the militant Colina ('National Liberation Command'). She was also editor of a Marxist newspaper, *The Piquet*, and married fellow militant, Cláudio Galeno Linhares, in 1968. Fearing arrest, Rousseff abandoned her studies at the Minas Gerais Federal University School of Economics and moved to Rio de Janeiro in 1969. Her first marriage fell apart when she began an affair with another militant, Carlos Araújo, whom she later married.

Rousseff became a leading member of the VAR Palmares (Palmares Armed Revolutionary Vanguard), a military-political organization. In 1970 she was arrested in São Paulo and subsequently claimed that she suffered torture for 22 days. Following her release from jail at the end of 1972 she moved to Porto Alegre, the state capital of Rio Grande do Sul. She graduated with a degree in economics from its Federal University.

In the late 1970s she co-founded the Democratic Labour Party (PDT) of Rio Grande do Sul and was involved in the Direct Elections Now movement, credited with helping bring an end to Brazil's military regime. Having made the transition to mainstream politics, Rousseff was appointed municipal secretary of treasury in Porto Alegre in 1986. She later led the national economics and statistics foundation as well as serving as secretary of energy and communications for Rio Grande do Sul. In the

latter role she oversaw a major increase in power production and distribution that ensured the state avoided the power cuts that routinely blighted much of the rest of Brazil.

In 2001 Rousseff joined the Workers' Party led by Luiz Inácio Lula da Silva. In 2002 she was named minister of mines and energy by the newly-elected president. The following year she became director of Brazilian oil conglomerate Petrobras. On 21 June 2005 she was appointed Lula's chief of staff following the resignation of José Dirceu in the wake of the Mensalão scandal that rocked the Lula administration. In April 2010 she resigned her position at Petrobras. In June 2010 she announced she was standing for the presidency. Promising to continue the populist policies of Lula, she claimed 56% of the vote in a run-off against her centre-right opponent, José Serra.

Career in Office
Never before elected to public office, Rousseff campaigned on a broad commitment to continue the left-leaning policies of her predecessor. She was also expected to forge close ties with other leftist South American leaders. As a former head of Petrobras, Rousseff supports state involvement in key areas of the economy including banking, the oil industry and energy.

Growth figures released in March 2011 indicated that Brazil had overtaken Italy to become the world's seventh largest economy, but Rousseff's government had to introduce unpopular measures to forestall economic overheating and reduce inflation. She was weakened politically in 2011 by the resignations of her chief of staff and the ministers of transport, agricultural and sports, together with her dismissal of the defence minister, over corruption allegations that continued to cast a shadow over her administration through 2012. Despite antipathy towards privatization, she announced plans in Aug. 2012 to encourage private involvement in the building and running of new and essential national infrastructure, while resisting the pay demands of public sector workers.

In May 2012 Rousseff established a truth commission to investigate the abuses committed during the military dictatorship between 1964 and 1985.

DEFENCE

Conscription is for 12 months.

In 2008 defence expenditure totalled US$23,302m. (US$120 per capita), a 5% real-term year-on-year increase. In 2007 defence spending represented 1·5% of GDP. Brazil is responsible for 48% of South America's military spending.

As at 31 Jan. 2013, 2,202 personnel (including 2,170 troops) were deployed in UN peacekeeping operations.

Army

There are seven military commands and 12 military regions. Strength, 2007, 238,200 (including 89,000 conscripts). Equipment in 2007 included 224 main battle tanks.

Navy

The principal ship of the Navy and Brazil's only aircraft carrier is the 32,700-tonne *São Paulo* (formerly the French *Foch*), commissioned in 1963 and purchased in 2000. There are also five diesel submarines and ten frigates including three bought from the UK in 1995 and 1996.

Naval bases are at Rio de Janeiro, Salvador, Natal, Belém, Rio Grande and São Paulo, with river bases at Brasília, Ladário and Manaus.

Active personnel, 2007, totalled 62,261 (more than 11,000 conscripts), including 14,500 Marines and 1,300 in Naval Aviation.

Air Force

The Air Force has four commands: COMGAR (operations), COMDABRA (aerospace defence), COMGAP (logistics) and COMGEP (personnel). There are seven air regions. Personnel strength, 2007, 67,440. There were 309 combat aircraft in 2007, including Mirage F-2000s and F-5Es.

ECONOMY

Agriculture accounted for 6·1% of GDP in 2009, industry 25·4% and services 68·5%.

In Dec. 2005 the government repaid the country's entire US$15·5bn. debt to the IMF two years ahead of schedule.

Overview

Brazil is South America's largest economy and among the world's top ten largest economies. In 2007 the government launched a Growth Acceleration Plan to stimulate public and private investment and to provide tax incentives for faster growth, spurring growth of over 5% that year and again in 2008. In mid-2008 Brazil achieved an investment grade rating by Standard & Poor's and Fitch as a result of its strong growth prospects and continued sound macroeconomic policies. However, the global economic crisis saw Brazil enter recession in 2009. Strong private consumption and domestic investment, along with a robust export environment and a healthy financial system, prompted GDP to grow by 7·5% in 2010, its fastest rate in 25 years. In 2011 Brazil became the sixth largest economy in the world on the back of high prices for oil and agricultural products, although growth slowed considerably in 2012. The limited impact of the global financial turmoil has underlined the economy's ability to withstand external shocks.

Brazil has abundant natural resources and in 2011 became the ninth largest oil exporter. Reserves are estimated at 12·86bn. bbls, with further natural gas and oil reserves lying off shore. Meanwhile, the diverse industrial sector recorded growth of 0·3% in 2011, driven by textiles, cement, iron ore, tin, steel, automobiles and aircraft part manufactures. China is Brazil's major trading partner.

Nonetheless, high labour prices and lack of infrastructure contribute to the so-called 'Custo Brasil', the rising cost of doing business in the country that threatens long-term economic aspirations. It is hoped that preparations for the soccer World Cup in 2014 and the Olympic Games in 2016 will see investment channelled into essential infrastructure projects. President Dilma Rousseff announced a US$50bn. five-year investment programme in 2012, in part driven by the selling-off of 14,000 km of railways and roads and the privatization of some ports.

Though still a major problem, poverty has reduced. The number of those living on US$2 a day dropped from 21% to 11% between 2003 and 2009, while those living on US$1·25 per day dropped from 10·0% in 2004 to 2·2% in 2009. The growth in income among the poorest averaged 7% per year from 2001 to 2009, while the richest 10% enjoyed growth of 1·7%. Income inequality reached a 50-year low in 2010, while unemployment fell to an historic low of 6·4% in May 2011. To build on this success, Brazil must invest in its education system and particularly the provision of English teaching to take advantage of that language's position as the gold standard of business communication.

Currency

The unit of currency is the *real* (BRL) of 100 *centavos*, which was introduced on 1 July 1994 to replace the former *cruzeiro real* at a rate of 1 real (R$1) = 2,750 cruzeiros reais (CR$2,750). The *real* was devalued in Sept. 1994, March 1995, June 1995 and Jan. 1999, when it was allowed to float. Inflation rates (based on IMF statistics):

2002	2003	2004	2005	2006	2007	2008	2009	2010	2011
8·4%	14·8%	6·6%	6·9%	4·2%	3·6%	5·7%	4·9%	5·0%	6·6%

In 1990 inflation had been 2,948%.

In Sept. 2009 foreign exchange reserves were US$215,978m. (US$32,434m. in 2000); gold reserves totalled 1·08m. troy oz. Total money supply in Aug. 2009 was R$201,761m.

Budget

Central government revenue and expenditure (in R$1m.):

	2007	2008	2009
Revenue	617,232	714,504	736,939
Expenditure	645,232	723,080	815,568

Principal sources of revenue in 2009 were: taxes on goods and services, R$243,260m.; taxes on income, profits and capital gains, R$222,838m.; social security contributions, R$191,248m. Main items of expenditure by economic type in 2009: social benefits, R$270,749m.; interest, R$152,709m.; compensation of employees, R$151,653m.

Performance

Real GDP growth rates (based on IMF statistics):

2002	2003	2004	2005	2006	2007	2008	2009	2010	2011
2·7%	1·1%	5·7%	3·2%	4·0%	6·1%	5·2%	−0·3%	7·5%	2·7%

Brazil experienced a recession with negative growth in the fourth quarter of 2008 and the first quarter of 2009, but the economy then recovered to grow throughout the rest of 2009. Total GDP in 2011 was US$2,476·7bn.

Banking and Finance

On 31 Dec. 1964 the Banco Central do Brasil (*Governor*, Alexandre Tombini) was founded as the national bank of issue and at May 2010 had total assets of R$1,192·4bn.

The Banco do Brasil/Bank of Brazil (founded in 1853 and reorganized in 1906) is a state-owned commercial bank; it had 4,048 branches in 2007 throughout the country. In Nov. 2008 Banco Itaú and Unibanco agreed to merge in a move that created South America's biggest bank, Itaú Unibanco, with assets in 2010 of R$651bn. (US$366bn.). In March 2007 total deposits in banks were R$386·8bn.

In Nov. 1998 the IMF announced a US$41·5bn. financing package to help shore up the Brazilian economy. In Aug. 2001 it gave approval for a new US$15bn. stand-by credit, and in Aug. 2002 granted an additional US$30bn. loan to try to prevent a financial meltdown that was threatening to devastate the region. In Dec. 2005 Brazil repaid all its IMF debts.

Brazil received a record US$66·7bn. worth of foreign direct investment in 2011, up from US$48·5bn. in 2010 and US$25·9bn. in 2009.

In 2010 total external debt amounted to US$346,978m. (equivalent to 16·9% of GNI), up from US$187,526m. in 2005.

There is a stock exchange in São Paulo.

ENERGY AND NATURAL RESOURCES

Environment

Brazil's carbon dioxide emissions from the consumption and flaring of fossil fuels in 2008 were the equivalent of 2·2 tonnes per capita. An *Environmental Performance Index* compiled in 2008 ranked Brazil 35th in the world, with 82·7%. The index examined various factors in six areas—air pollution, biodiversity and habitat, climate change, environmental health, productive natural resources and water resources.

Brazil has the world's biggest river system and about a quarter of the world's primary rainforest. Current environmental issues are deforestation in the Amazon Basin, air and water pollution in Rio de Janeiro and São Paulo (the world's fourth largest city), and land degradation and water pollution caused by improper mining activities. Contaminated drinking water causes 70% of child deaths.

Electricity
Hydro-electric power accounts for over 80% of Brazil's total electricity output. Although Brazil was only the ninth largest electricity producer overall in the world in 2007, it was the second largest producer of hydro-electric power (behind only China). Installed electric capacity (2007) was 100·4m. kW, of which 76·9m. kW hydro-electric. Production cannot meet demand, making Brazil a net importer of electricity. There were two nuclear power plants in 2007, supplying some 2·8% of total output. A third nuclear plant, currently under construction, is scheduled for completion in 2015. There are plans for a further four plants to be built. Production (2007) 444,583 GWh. Consumption per capita in 2007 was 2,521 kWh.

Oil and Gas
There are 13 oil refineries, of which 11 are state-owned. Oil production in 2008 was a record 93·9m. tonnes. Production rose every year between 2004 and 2008. Proven oil reserves were 12·6bn. bbls in 2008. Brazil began to open its markets in 1999 by inviting foreign companies to drill for oil, and in 2000 the monopoly of the state-owned Petrobras on importing oil products was removed. In Nov. 2007 the discovery of a huge offshore oilfield was announced off Brazil's southeastern Atlantic coast that could increase the country's oil reserves by as much as 40%. The Tupi field, which contains reserves estimated at between 5bn. and 8bn. bbls of oil, is the largest new oilfield discovered since Kazakhstan's Kashagan field in 2000. Oil now contributes 12% of total GDP. It has been suggested that by 2020 Brazil could be the world's fifth largest oil producer.

Natural gas production in 2008 was 13·9bn. cu. metres (up from 11·3bn. cu. metres in 2007) with reserves of 330bn. cu. metres. One of the most significant developments has been the construction of the 3,150-km Bolivia–Brazil gas pipeline, one of Latin America's biggest infrastructure projects, costing around US$2bn. (£1·2bn.). The pipeline runs from the Bolivian interior across the Brazilian border at Puerto Suárez-Corumbá to the far southern port city of Porto Alegre. Gas from Bolivia began to be pumped to São Paulo in 1999.

Ethanol
Brazil is the second largest producer of ethanol (after the USA) and by far the largest exporter. Production, almost exclusively from sugarcane, totalled 17·0bn. litres in 2006. Ethanol accounts for half of all the transport fuel used in Brazil.

Minerals
The chief minerals are bauxite, gold, iron ore, manganese, nickel, phosphates, platinum, tin and uranium. Output figures, 2004 (in 1,000 tonnes): phosphate rock, 35,000; bauxite (2003), 17,363; salt, 6,648; hard coal, 5,077; asbestos (crude ore), 3,950; manganese ore, 3,143; aluminium (2003), 1,381; magnesite, 1,339; graphite, 650; chrome (crude ore, 2003), 404; zinc, 159; barytes, 64; nickel ore, 52; zirconium, 35; copper (2003), 26; lead (lead content in concentrate), 15; tin (tin content), 12. Deposits of coal exist in Rio Grande do Sul, Santa Catarina and Paraná. Total reserves were estimated at 9,920m. tonnes in 2005.

Iron is found chiefly in Minas Gerais, notably the Cauê Peak at Itabira. Proven reserves of iron ore amounted to around 15,800m. tonnes in 2005. Total output of iron ore, 2004 was 261·7m. tonnes. Brazil is the second largest producer of iron ore after China.

Gold is chiefly from Pará, Mato Grosso and Minas Gerais; total production (2004), 47·6 tonnes. Silver output (2004), 35·5 tonnes. Diamond output in 2005 was an estimated 900,000 carats, mainly from Minas Gerais and Mato Grosso.

Agriculture
In 2007 the agricultural population was an estimated 23·06m. There were 5·2m. farms in 2006. There were some 59·5m. ha. of arable land in 2007 and 7·0m. ha. of permanent crops.

Production (in tonnes):

	2004	2005
Apples	980,203	850,535
Bananas	6,583,564	6,703,400
Beans	2,967,007	3,021,641
Cassava	23,926,553	25,872,015
Coconut (1,000 fruits)	2,078,226	2,079,291
Coffee	2,465,710	2,140,169
Cotton	3,801,382	3,668,283
Grapes	1,291,382	1,232,564
Maize	41,787,558	35,113,312
Onions	1,157,562	1,137,684
Oranges	18,313,717	17,853,443
Pineapples (1,000 fruits)	1,477,299	1,528,313
Potatoes	3,047,083	3,130,174
Rice	13,277,008	13,192,863
Soya	49,549,941	51,182,074
Sugarcane	415,205,835	422,956,646
Tomatoes	3,515,567	3,452,973
Wheat	5,818,846	4,658,790

Brazil is the world's leading producer of sugarcane, oranges and coffee (and the second largest consumer of coffee after the USA). Harvested coffee area, 2005, 2,325,920 ha., principally in the states of Minas Gerais, Espírito Santo, São Paulo and Paraná. Harvested cocoa area, 2005, 625,384 ha. Bahia furnished 65% of the output in 2005. Two crops a year are grown. Brazil accounts for more than a quarter of annual coffee production worldwide. Harvested castor-bean area, 2005, 230,911 ha. Tobacco is grown chiefly in Rio Grande do Sul and Santa Catarina.

Rubber is produced chiefly in the states of São Paulo, Mato Grosso, Bahia, Espírito Santo and Minas Gerais. Output, 2005, 172,847 tonnes.

Livestock, 2005: cattle, 207·2m.; pigs, 34·1m.; sheep, 15·6m.; goats, 10·3m.; horses, 5·8m.; mules, 1·4m.; asses, 1·2m.; chickens and other poultry, 999·0m.

Livestock products, 2005 (in 1,000 tonnes): beef and veal, 6,346; pork, bacon and ham, 2,157; poultry meat, 7,866; milk, 24·6bn. litres; hen's eggs, 2·8bn. dozen; wool, 11; honey, 34.

Forestry
With forest lands covering 519,522,000 ha. in 2010, only Russia had a larger area of forests. Brazil's forests account for 13% of global forest cover. In 2010, 62·4% of the total land area of Brazil was under forests. The annual loss of 2,194,000 ha. of forests between 2005 and 2010 was the biggest in any country in the world over the same period. Of the total area under forests 92% was primary forest in 2010, 7% other naturally regenerated forest and 1% planted forest.

In 1996 the government ruled that Amazonian landowners could log only 20% of their holdings, instead of 50%, as had previously been permitted. Timber production in 2007 totalled 244·96m. cu. metres, a figure exceeded only in the USA, India and China. By Dec. 2003 the government had seized illegally-cut mahogany to the value of US$60m. Between Aug. 2004 and Aug. 2006 the government's environmental agency, Ibama, issued fines worth R$4·97bn. for illegal logging, but only 2·5% of these were actually collected. The government has pledged to end net deforestation by 2015, although Amazon deforestation increased in 2007–08 for the first time since 2003–04.

Fisheries
In 2009 the fishing industry had a catch of 825,412 tonnes, of which 585,919 tonnes came from sea fishing and 239,493 tonnes from inland fishing. Aquaculture production totalled 479,399 tonnes in 2010. Imports of fishery commodities were valued at US$690m. in 2008 and exports at US$274m.

INDUSTRY
The leading companies by market capitalization in Brazil in March 2012 were: Petrobras (US$170·8bn.); Vale, the world's

largest iron ore producer (US$124·5bn.); and Ambev, a beverages company (US$117·1bn.).

The main industries are textiles, shoes, chemicals, cement, lumber, iron ore, tin, steel, aircraft, motor vehicles and parts, and other machinery and equipment. The National Iron and Steel Co. at Volta Redonda, State of Rio de Janeiro, furnishes a substantial part of Brazil's steel. Production (in 1,000 tonnes): cement (2008), 51,970; pig iron (2008), 34,871; distillate fuel oil (2007), 34,035; crude steel (2008), 33,716; sugar (2007), 28,226; rolled steel (2008), 24,693; petrol (2007), 15,733; residual fuel oil (2007), 15,707; cellulose (2008), 12,697; paper (2008), 9,409. Output of other products in 2007: 68·43m. mobile phones; 37·54m. rubber tyres for cars; 12·84m. TV sets; 2·55m. motor vehicles (2008); soft drinks, 12,642·2m. litres; beer, 10,020·2m. litres.

Labour
The economically active population of Brazil totalled 99,500,000 in 2008. In 2004 a total of 84,596,000 persons were in employment (49,242,000 males), including: 17,330,000 engaged in agriculture, hunting and forestry; 14,653,000 in wholesale and retail trade; 11,724,000 in manufacturing; 6,472,000 in private households with employed persons; 5,354,000 in construction. A constitutional amendment of Oct. 1996 prohibits the employment of children under 14 years. In 1992, 13% of children between five and 14 were working but by 2008 this had fallen to 5%, helped by monetary incentives for families to send children to school. There is a minimum monthly wage, which was increased from R$545 to R$622 with effect from 1 Jan. 2012. In Dec. 2010, 5·3% of the workforce was unemployed (down from 10·9% in Dec. 2003 and 7·5% in Dec. 2007).

INTERNATIONAL TRADE

In 1990 Brazil repealed most of its protectionist legislation. Import tariffs on some 13,000 items were reduced in 1995. In 1991 the government permitted an annual US$100m. of foreign debt to be converted into funds for environmental protection.

Imports and Exports
Imports and exports for calendar years (in US$1m.):

	2005	2006	2007	2008
Imports (c.i.f.)	77,628	95,836	126,564	182,361
Exports (f.o.b.)	118,529	137,807	160,649	197,942

Principal imports in 2004 were: machinery and transport equipment, 34·8%; chemicals and related products, 22·1%; mineral fuels, lubricants and related materials, 18·8%; manufactured goods, 9·9%; and miscellaneous manufactured articles, 5·7%.

Principal exports in 2004 were: machinery and transport equipment, 25·5%; manufactured goods, 19·6%; food and live animals, 19·4% (including coffee, 2·2%); crude materials (excluding fuels), 16·4%; chemicals and related products, 6·0%. Brazil is the world's largest exporter of a number of commodities including coffee, sugar and orange juice.

The leading import suppliers in 2004 were: USA, 18·3%; Argentina, 9·0%; Germany, 7·5%; China, 6·2%; Nigeria, 5·5%; Japan, 4·6%. The principal export destinations in 2004 were: USA, 21·4%; Argentina, 7·8%; Netherlands, 6·2%; China, 5·7%; Germany, 4·2%; Mexico, 4·2%. China has in the meantime become Brazil's largest export market and its largest trading partner overall.

COMMUNICATIONS

Roads
In 2004 there were 1,751,868 km of roads, of which 93,071 km were highways, national and main roads. In 2007 there were 37,978,000 vehicles in use, including 30,283,000 passenger cars. In 2006, 407,685 persons were injured in road accidents and 35,155 were killed.

Rail
The Brazilian railways have largely been privatized: all six branches of the large RFFSA network are now under private management. The largest areas of the network are now run by América Latina Logística (12,883 km of metre gauge in 2007) and Ferrovia Centro-Atlântica (7,080 km of metre-gauge). The Rio de Janeiro suburban network is run by SuperVia (128m. passengers in 2008) and the São Paulo network by Cia Paulista de Trens Metropolitanos (390m. passengers in 2005). There are plans for South America's first high-speed rail service to be introduced in 2016 (although possibly not in time for the Olympic Games), linking Rio de Janeiro and São Paulo.

Several other freight routes operate independently and are mainly used by the mining industry. There are metros in Belo Horizonte (30 km), Brasília (40 km), Fortaleza (24 km), Porto Alegre (31 km), Recife (39 km), Rio de Janeiro (46 km), Salvador (7 km), São Paulo (57 km) and Teresina (15 km).

Civil Aviation
There are major international airports at Rio de Janeiro-Galeão (Antonio Carlos Jobim International) and São Paulo (Guarulhos) and some international flights from Brasília, Porto Alegre, Recife and Salvador. The main airlines are LATAM (created in June 2012 when the Brazilian carrier TAM merged with LAN Airlines, Chile's largest airline) and Gol (a low-cost airline launched in 2001). In 2005 TAM carried 17,109,193 passengers and Varig 13,268,869 passengers. Varig was previously Brazil's biggest airline but was surpassed by TAM and was bought by Gol in April 2007. LATAM controls 40% of the Latin American market and serves 150 destinations in 22 countries.

Brazil's busiest airport is Guarulhos (São Paulo), which handled 18,795,596 passengers in 2007, followed by Congonhas (São Paulo) with 15,244,401 passengers (all on domestic flights) and Brasília International (Presidente Juscelino Kubitschek International Airport) with 11,119,872 passengers. In Feb. 2012 Guarulhos was one of three Brazilian airports to be privatized.

Shipping
Inland waterways, mostly rivers, are open to navigation over some 43,000 km. Tubarão and Itaqui are the leading ports. In 2008 Santos, the leading container port, handled 2·68m. TEUs (twenty-foot equivalent units). In Jan. 2009 there were 166 ships of 300 GT or over registered, totalling 2·02m. GT. Of the 166 vessels registered, 52 were oil tankers, 44 general cargo ships, 24 passenger ships, 19 bulk carriers, 14 liquid gas tankers, nine container ships and four chemical tankers. In 2007 the cargo moved through Brazilian ports and terminals totalled 754·7m. tonnes of which 35·44% was iron ore.

Telecommunications
The state-owned telephone system was privatized in 1998. There were 41,141,000 main (fixed) telephone lines in 2008. Mobile phone services were opened to the private sector in 1996. In 2008 there were 150,641,000 mobile phone subscribers. There were 72·0m. internet users in 2008. In June 2012 Brazil had 51·2m. Facebook users, the second highest total after the USA (26% of the population).

SOCIAL INSTITUTIONS

Justice
There is a Supreme Federal Court of Justice at Brasília composed of 11 judges, and a Supreme Court of Justice; all judges are appointed by the President with the approval of the Senate. There are also Regional Federal Courts, Labour Courts, Electoral Courts and Military Courts. Each state organizes its own courts and judicial system in accordance with the federal Constitution.

In Dec. 1999 then President Cardoso created the country's first intelligence agency (the Brazilian Intelligence Agency) under civilian rule. It replaced informal networks which were a legacy of the military dictatorship, and helps authorities crack down on organized drug gangs.

The prison population was 440,013 in June 2008 (227 per 100,000 of national population). Brazil's annual murder rate fell slightly between 2006 and 2008 but remains in excess of 25 per 100,000 population, around five times that of the USA.

Education
Elementary education is compulsory from seven to 14. Adult literacy in 2008 was 90·0% (male, 89·8%; female, 90·2%). In 2006 there were 107,375 pre-primary schools, with 5,588,153 pupils and 310,241 teachers; 159,016 elementary schools, with 33,282,663 pupils and 1,665,341 teachers; 24,131 secondary schools, with 8,906,820 pupils and 519,935 teachers. In 2003 there were 1,637 higher education institutions, with 3,479,913 students and 227,844 teachers. In 2006, 97·6% of children between the ages of seven and 14 were enrolled at schools.

There were 1,859 universities in Brazil in 2003, of which 207 were public and 1,652 were private. Of the 207 public universities, 83 were federal, 65 were state and 59 were municipal institutions.

In 2008 total expenditure on education came to 5·4% of GDP. Expenditure on education was 16·1% of total government spending in 2007.

Health
In 2005 there were 62,483 hospitals, clinics and health centres (18,496 private). There were a total of 443,210 beds at hospitals, clinics and health centres in 2005 (294,244 private). In 2007, 329,041 doctors, 219,827 dentists, 178,546 nurses and 104,098 pharmacists were registered with Brazil's national health system. In 2006 there were 115 physicians per 100,000 population.

Brazil has been one of the most successful countries in the developing world in the campaign against AIDS. The death rate from HIV/AIDS fell from 9·6 deaths per 100,000 people in 1996 to 6·0 in 2006.

Welfare
Old-age pensions begin at 65 years (men) or 60 years (women) for employees and the urban self-employed, and ages 60 (men) or 55 (women) for the rural self-employed. To qualify there must be at least 35 years contributions for men or 30 years contributions for women. The maximum monthly pension was R$1,869·34 in June 2003.

Unemployment benefits vary depending on insurance but, as a general rule, cover 50% of average earnings in the last three months of employment, up to three times the minimum wage. The minimum benefit is 100% of the minimum monthly wage (R$622 in Jan. 2012).

Family allowances are granted to low-income families with one or more children under the age of 14 or with disabled children attending school. In 2003, R$13·48 a month was provided for each child.

RELIGION
In 2000 there were 124,980,000 Roman Catholics (including syncretic Afro-Catholic cults having spiritualist beliefs and rituals) and 26,185,000 Evangelical Protestants, with 5,274,000 followers of other religions. Roman Catholic estimates in 1991 suggest that 90% were baptized Roman Catholic but only 35% were regular attenders. In 1991 there were 338 bishops and some 14,000 priests. In Feb. 2013 there were nine Roman Catholic cardinals. There are numerous sects, some evangelical, some African-derived (e.g. Candomble).

CULTURE

World Heritage Sites
The sites under Brazilian jurisdiction entered on the UNESCO World Heritage List (with year entered) are: the Historic Town of Ouro Preto (1980), the centre of the gold rush founded at the end of the 17th century; the Historic Centre of Olinda (1982), founded by the Portuguese in the 16th century and largely rebuilt in the 18th century; the centre of Salvador de Bahia, Brazil's first capital (1549–1763), with its early mix of European, African and Amerindian cultures and many colonial buildings; and the Sanctuary of Bom Jesus do Congonhas, an ornate church dating to the late 18th century (both 1985); the Iguaçu National Park (1986), with its 2,700 metre waterfall and an impressive range of flora and fauna, shared with the Iguazu National Park in Argentina; Brasília (1987), Brazil's purpose built capital city; Serra da Capivara National Park (1991) including cave paintings over 25,000 years old; the Historic Centre of São Luís (1997), which has examples of late 17th-century architecture; the Historic Centre of Diamantina, a colonial village inhabited by diamond prospectors in the 18th century; the Discovery Coast Atlantic Forest Reserves, incorporating eight separate areas protecting 100,000 ha. of Atlantic forest; and Atlantic Forest Southeast Reserves, covering 470,000 ha. over 25 protected areas (all 1999); the Pantanal Conservation Area (2000), incorporating four protected areas covering 188,000 ha. and offering access to a major freshwater wetland ecosystem; the Central Amazon Conservation Complex, covering over 6m. ha. of Amazon basin (2000 and 2003); the Cerrado Protected Areas, comprising Chapada dos Veadeiros and Emas National Parks, home to a diverse tropical ecosystem; Brazilian Atlantic Islands, comprising Fernando de Noronha and Atol das Rocas Reserves, protecting important marine flora and fauna such as dolphins and turtles and a large concentration of tropical seabirds; the Historic Centre of Goiás, established by colonizing powers in the 18th and 19th centuries (all 2001); São Francisco Square in the town of São Cristóvão, a Franciscan complex dating from the late 17th century (2010); and Rio de Janeiro: Carioca Landscapes between the Mountain and the Sea, key natural elements that have shaped the city including the peaks of the Tijuca mountains, the Botanical Gardens, Corcovado Mountain and its statue of Christ, the hills around Guanabara Bay and the extensive designed landscapes along Copacabana Bay (2012).

Brazil shares the Jesuit Missions of the Guaranis (1983 and 1984), the ruins of five Jesuit missions in the tropical rainforest dating from the 17th and 18th centuries, with Argentina.

Press
In 2008 there were 673 daily newspapers (including six free papers) with a combined circulation of 8,992,000. Average circulation of dailies increased by 37·9% between 2004 and 2008. The dailies with the highest circulation are Folha de S. Paulo (daily average of 311,000 in 2008), the tabloid Super Notícia, Extra and O Globo.

Tourism
In 2007, 5,026,000 tourists visited Brazil (up from 5,017,000 in 2006 although down from 5,358,000 in 2005). In 2007 the largest number of tourists came from elsewhere in the Americas (2,775,000); 1,942,000 European tourists visited the country. Receipts in 2007 totalled US$5·28bn. (US$4·58bn. in 2006).

Festivals
New Year's Eve in Rio de Janeiro is always marked with special celebrations, with a major fireworks display at Copacabana Beach. Immediately afterwards, preparations start for Carnival, which in 2014 will be held from 28 Feb.–4 March. Other notable cultural festivals include the Parintins Folk Festival, held at the end of June and attracting over 40,000 people, and the Bahia Carnival, held in Salvador in Feb. to celebrate African influences in the region.

DIPLOMATIC REPRESENTATIVES
Of Brazil in the United Kingdom (14–16 Cockspur St., London, SW1Y 5BN)
Ambassador: Roberto Jaguaribe.

Of the United Kingdom in Brazil (Setor de Embaixadas Sul, Quadra 801, Lote 8, CEP 70408-900, Brasília, DF)
Ambassador: Alan Charlton.

Of Brazil in the USA (3006 Massachusetts Ave., NW, Washington, D.C. 20008)
Ambassador: Mauro Luiz Iecker Vieira.

Of the USA in Brazil (Av. das Nações, Quadra 801, Lote 03, CEP: 70403-900, Brasília, D.F.)
Ambassador: Thomas A. Shannon, Jr.

Of Brazil to the United Nations
Ambassador: Maria Luiza Ribeiro Viotti.

Of Brazil to the European Union
Ambassador: Ricardo Neiva Tavares.

FURTHER READING

Banco Central do Brasil: *Boletim do Banco Central do Brasil* (monthly).
Instituto Brasileiro de Geografia e Estatística: *Anuário Estatístico do Brasil.* —*Indicadores IBGE* (monthly).

Baer, W., *The Brazilian Economy: Growth and Development.* 6th ed. 2007

Brainard, Lael and Martinez-Diaz, Leonardo, (eds.) *Brazil as an Economic Superpower?: Understanding Brazil's Changing Role in the Global Economy.* 2009
Burns, Bradford E., *A History of Brazil.* 1993
Eakin, Marshall C., *Brazil: The Once and Future Country.* 1997
Falk, P. S. and Fleischer, D. V., *Brazil's Economic and Political Future.* 1988
Fausto, Boris, *A Concise History of Brazil.* 1999
Klein, Herbert, *Brazil Since 1980.* 2006
Levine, Robert M., *History of Brazil.* 2003
Love, Jospeh L. and Baer, Werner, (eds.) *Brazil Under Lula: Economy, Politics, and Society Under the Worker-President.* 2009
Montero, Alfred, *Brazilian Politics: Reforming a Democratic State in a Changing World.* 2006
Riordan, Roett, *The New Brazil.* 2011
Rohter, Larry, *Brazil on the Rise: The Story of a Country Transformed.* 2010
Skidmore, Thomas E., *Brazil: Five Centuries of Change.* 2009

For other more specialized titles see under CONSTITUTION AND GOVERNMENT *above.*

National library: Biblioteca Nacional, Avenida Rio Branco 219, 22040-008 Rio de Janeiro, RJ.
National Statistical Office: Instituto Brasileiro de Geografia e Estatística (IBGE), Avenida Franklin Roosevelt 166, 20021-120 Rio de Janeiro, RJ.
Website: http://www.ibge.gov.br

BRUNEI

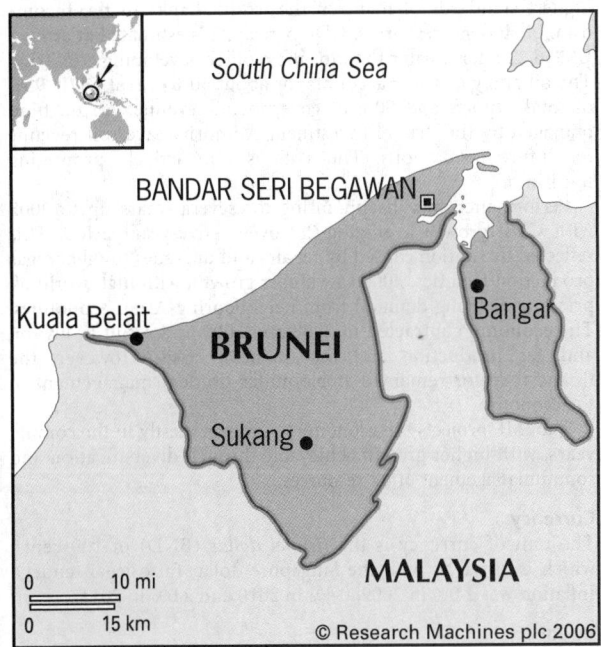

South China Sea

BANDAR SERI BEGAWAN

Kuala Belait

BRUNEI

Bangar

Sukang

MALAYSIA

| 0 | 10 mi |
| 0 | 15 km |

© Research Machines plc 2006

**Negara Brunei Darussalam
(State of Brunei Darussalam)**

Capital: Bandar Seri Begawan
Population projection, 2015: 433,000
GNI per capita, 2011: (PPP$) 45,753
HDI/world rank: 0·838/33
Internet domain extension: .bn

KEY HISTORICAL EVENTS

Brunei became an independent Sultanate in the 15th century controlling most of Borneo, its neighbouring islands and the Suhi Archipelago. By the end of the 16th century, however, the power of Brunei was on the wane. By the middle of the 19th century the state had been reduced to its present limits. Brunei became a British protectorate in 1888. The discovery of major oilfields in the west of the country in the 1920s brought economic prosperity. Brunei was occupied by the Japanese in 1941 and liberated by the Australians in 1945. Self-government was introduced in 1959 but Britain retained responsibility for foreign affairs. In 1965 constitutional changes were made which led to direct elections for a new Legislative Council. Full independence and sovereignty were gained in Jan. 1984.

TERRITORY AND POPULATION

Brunei, on the coast of Borneo, is bounded in the northwest by the South China Sea and on all other sides by Sarawak (Malaysia), which splits it into two parts, the smaller portion forming the Temburong district. Area, 5,765 sq. km (2,226 sq. miles). Population (2011 census, provisional) 393,162 (202,668 males), giving a density of 68·2 per sq. km.

The UN gives a projected population for 2015 of 433,000.

In 2011, 76·1% of the population lived in urban areas. The four districts are Brunei/Muara (2011 census, provisional: 279,842), Belait (60,609), Tutong (43,855) and Temburong (8,856). The

capital is Bandar Seri Begawan (estimate 2001: 27,285); other large towns are Kuala Belait (2001: 27,975) and Seria (2001: 15,819). Ethnic groups include Malays 67% and Chinese 11%.

The official language is Malay but English is in use.

SOCIAL STATISTICS

2005 births, 6,933; deaths, 1,072. Rates, 2005: birth per 1,000 population, 18·7; death, 2·9. There were 2,018 marriages in 2005. Life expectancy in 2007: males, 74·9 years; females, 79·6. Annual population growth rate, 1995–2005, 2·5%. Infant mortality, 2010, six per 1,000 live births; fertility rate, 2005, 2·1 children per woman.

CLIMATE

The climate is tropical marine, hot and moist, but nights are cool. Humidity is high and rainfall heavy, varying from 100" (2,500 mm) on the coast to 200" (5,000 mm) inland. There is no dry season. Bandar Seri Begawan, Jan. 80°F (26·7°C), July 82°F (27·8°C). Annual rainfall 131" (3,275 mm).

CONSTITUTION AND GOVERNMENT

The Sultan and Yang Di Pertuan of Brunei Darussalam is HM Sultan Haji Hassanal Bolkiah Mu'izzadin Waddaulah. He succeeded on 5 Oct. 1967 at his father's abdication and was crowned on 1 Aug. 1968. On 10 Aug. 1998 his son, Oxford-graduate Prince Al-Muhtadee Billah, was inaugurated as Crown Prince and heir apparent.

On 29 Sept. 1959 the Sultan promulgated a constitution, but parts of it have been in abeyance since Dec. 1962 under emergency powers assumed by the Sultan. Since 1984 the Legislative Council (*Majlis Masyuarat Megeri*) has been effectively replaced by a Council of Cabinet Ministers appointed and presided over by the Sultan. The constitution was amended in Sept. 2004, allowing for the Legislative Council to be reconvened, but with no independent executive powers and its 21 members chosen by the Sultan. The amendment allowed for the first elections since 1962, with a third of the members of a new 45-member parliament to be directly elected. However, no date has been set for elections and in Sept. 2005 a 29-member Legislative Council including five indirectly-elected members was appointed. The Sultan is both the head of state and head of government.

National Anthem

'Ya Allah, lanjutkan lah usia' ('God bless His Majesty'); words by P. Rahim, tune by I. Sagap.

CURRENT GOVERNMENT

In March 2013 the Council of Ministers was composed as follows:
Prime Minister, Minister of Defence and of Finance: HM Sultan Haji Hassanal Bolkiah Mu'izzadin Waddaulah (the Sultan).
Minister of Communications: Dato Seri Setia Awang Haji Abdullah bin Begawan Mudim Dato Paduka Haji Bakar. *Culture, Youth and Sports:* Dato Paduka Haji Hazair bin Haji Abdullah. *Development:* Dato Seri Setia Haji Suyoi bin Haji Osman. *Education:* Dato Seri Setia Haji Awang Abu Bakar bin Haji Apong. *Energy:* Dato Seri Paduka Haji Mohammad Yasmin bin Haji Umar. *Finance (No. 2):* Pehin Dato Haji Abdul Rahman bin Haji Ibrahim. *Foreign Affairs and Trade:* Prince Haji Mohammad Bolkiah. *Foreign Affairs and Trade (No. 2):* Pehin Dato Seri Paduka Lim Jock Seng. *Health:* Dato Seri Setia Awang Haji Adanan bin Begawan Pehin Siraja Khatib Dato Seri Setia Awang Haji Mohd Yussof. *Home Affairs:* Dato Paduka Seri Setia Ustaz Haji Awang Badaruddin bin Pengarah Dato Paduka Haji Othman. *Industry and Primary Resources:* Pehin Dato Seri Setia Haji Yahya

Begawan Mudim Dato Paduka Haji Bakar. *Religious Affairs:* Dato Seri Dr Haji Mohammad bin Pengiran Haji Abd Rahman. *Attorney General:* Datin Paduka Hajah Hayati binti Pehin Orang Kaya Shahbandar Dato Seri Paduka Haji Mohd Salleh.

Government Website: http://www.brunei.gov.bn

CURRENT LEADERS

Sultan Sir Hassanal Bolkiah (Sultan of Brunei)

Position
Head of State

Introduction
The Sultan was crowned Brunei's 29th head of state on 1 Aug. 1968 following the abdication of his father. He is among the world's richest men.

Early Life
The Sultan was born on the 15 July 1946 in Bandar Seri Begawan. He was educated in Darussalam, Brunei and Malaysia. He became the Crown Prince of Brunei in 1961 and in 1966–67 he enlisted as an officer cadet at the Royal Military Academy in Sandhurst in the UK.

In 1978 he led the mission to London which paved the way for Brunei to become a sovereign state. On 1 Jan. 1984 a treaty of friendship ended British control over Brunei's foreign affairs and defence.

Career in Office
The Sultan is head of government as well as head of state. He is prime minister, minister of defence and minister of finance. Under the 1959 constitution and the Malay Muslim Monarchy tradition, he is assisted by a council of cabinet ministers, a privy council, council of succession and a religious council. A legislative council (a third of whose 45 members will, following a constitutional amendment in Sept. 2004, be directly elected) is to be revived, although elections have yet to be scheduled.

In May 2010 the Sultan reshuffled non-royal members of the cabinet, appointing the first female minister and raising the position of attorney general to cabinet rank.

Brunei's wealth originates from the country's large oil and gas reserves, although earnings from overseas investments have exceeded those from exports. The people of Brunei enjoy high subsidies and pay no taxes.

DEFENCE

In 2008 military expenditure totalled US$360m. (US$945 per capita), representing 2·5% of GDP.

Army
The armed forces are known as the Task Force and contain the naval and air elements. Only Malays are eligible for service. Strength (2007): 4,900.

There is a Gurkha reserve unit of 400–500.

Navy
The Royal Brunei Armed Forces Flotilla includes three fast missile-armed attack craft. Personnel in 2007 numbered 1,000.

Air Force
The Royal Brunei Air Force (formerly known as the Air Wing) was formed in 1965. Personnel (2007), 1,100. There are no combat aircraft.

ECONOMY

In 2008 petroleum, natural gas and mining accounted for 57·2% of GDP, manufacturing 13·7%, public administration and defence 10·6%, and finance and real estate 8·1%. The fall in oil prices in 1997–98 led to the setting up of an Economic Council to advise the Sultan on reforms. In 1998 an investigation was mounted into the Amedeo Corporation, Brunei's largest private company, run by Prince Jefri, the Sultan's brother. Amadeo collapsed with large debts.

Overview
Since independence in 1984, Brunei has boasted one of the highest standards of living in the world thanks to the income from oil and gas resources. GDP per capita is estimated at around US$36,225, far greater than for most other developing countries. The oil and gas sector accounts for about 50% of real GDP, 95% of total exports and 90% of government revenue. An oil fund managed by the Brunei Investment Authority saves oil revenue for future generations. The state is also actively promoting tourism.

Performance was disappointing for several years up to 2005 with GDP growth averaging 2% over a five-year period. This reflected disruption caused by repairs and upgrades to oil and gas production facilities. 2006 saw higher growth, with high world oil prices and strong demand from neighbouring Asian economies. The economy contracted in 2008 and 2009 as a result of low oil and gas production and the financial crisis. However, the financial sector remained stable under prudent management of the economy.

The IMF projects the economy to grow modestly in the coming years, with higher growth achievable through diversification and sound management of oil resources.

Currency
The unit of currency is the *Brunei dollar* (BND) of 100 cents, which is at parity with the Singapore dollar (also legal tender). Inflation was 1·0% in 2009, 0·4% in 2010 and 2·0% in 2011.

Budget
The financial year runs from 1 April–31 March. Revenues in 2008–09 totalled B$11,378m.; expenditures were B$5,974m. Tax revenues accounted for 65·3% of revenues in 2008–09; current expenditure accounted for 83·0% of total expenditures.

Performance
The economy shrank by 1·9% in 2008 and 1·8% in 2009 but grew by 2·6% in 2010 and 2·2% in 2011. Total GDP in 2011 was US$16·4bn.

Banking and Finance
The Brunei Currency Board is the note-issuing monetary authority. In 2002 there were three commercial banks, six foreign banks and one off-shore bank. Total bank assets in 2005 were B$16,971m.

The International Brunei Exchange Ltd (IBX) established an international securities exchange in May 2002.

ENERGY AND NATURAL RESOURCES

Environment
Brunei's carbon dioxide emissions from the consumption and flaring of fossil fuels were the equivalent of 27·3 tonnes per capita in 2008.

Electricity
Installed capacity was 0·8m. kW in 2005. Production in 2005 was 2·91bn. kWh and consumption per capita 7,280 kWh.

Oil and Gas
The Seria oilfield, discovered in 1929, has passed its peak production. The high level of crude oil production is maintained through the increase of offshore oilfields production. Output was 8·5m. tonnes in 2008. The crude oil is exported directly, and only a small amount is refined at Seria for domestic uses. There were proven oil reserves of 1·1bn. bbls in 2008.

Natural gas is produced (12·1bn. cu. metres in 2008) at one of the largest liquefied natural gas plants in the world and is exported to Japan. There were proven reserves of 350bn. cu. metres in 2008.

Agriculture

In 2007 there were about 3,000 ha. of arable land and 5,000 ha. of permanent crops. The main crops produced in 2005 were (estimates, in 1,000 tonnes): vegetables, 11; fruit, 5 (notably bananas and pineapples); rice, 1.

Livestock in 2005: cattle, 1,079; buffaloes, 4,790; goats, 2,578; pigs (2003), 1,000; chickens (2003), 11m.

Livestock products (2003 estimates, in 1,000 tonnes): beef and veal, 3; poultry meat, 24; eggs, 6.

Forestry

Forests covered 0·38m. ha., or 72% of the total land area, in 2010. Most of the interior is under forest, containing large potential supplies of serviceable timber. In 2010, 69% of forest area was primary forest. Timber production in 2007 was 124,000 cu. metres.

Fisheries

The 2008 catch totalled 2,357 tonnes, exclusively from sea fishing.

INDUSTRY

Although Brunei has long been dependent on its oil and gas industry, the government has begun a programme of diversification, recognizing oil and gas as non-renewable resources. Greater emphasis is now placed on other sectors such as manufacturing, services, tourism and high technology, but oil still accounted for nearly two-thirds of total GDP in 2004.

Labour

The labour force totalled 169,200 in 2005 (60·4% males). Unemployment in 2005 was 4·3%.

INTERNATIONAL TRADE

Imports and Exports

Imports (c.i.f.) totalled US$1,676m. in 2006 and exports (f.o.b.) US$7,636m.

Principal imports (c.i.f.) in 2006 (in US$1m.) were: road vehicles (175); iron and steel (127); power-generating machinery and equipment (100); textile yarn, fabrics and finished articles (88). Main exports (f.o.b.) in 2006 (in US$1m.) were: petroleum, petroleum products and related materials (5,141); liquefied natural gas (2,215); articles of apparel and clothing accessories (131); metalworking machinery (23).

Of imports in 2006, 21·6% came from Malaysia, 17·4% from Singapore, 12·8% from Japan, 9·0% from the USA and 7·9% from China. Of exports in 2006, 30·6% went to Japan, 19·8% to Indonesia, 15·1% to South Korea, 12·2% to Australia and 6·7% to the USA.

COMMUNICATIONS

Roads

There were an estimated 3,560 km of roads in 2005; 77·2% of all roads were paved in 2005. The main road connects Bandar Seri Begawan with Kuala Belait and Seria. In 2007 there were 252,700 passenger cars in use (649 per 1,000 inhabitants—one of the highest rates in the world), 16,700 vans and lorries, 1,500 buses and coaches, and 12,200 motorcycles and mopeds. There were 38 fatalities in road accidents in 2005.

Civil Aviation

Brunei International Airport (Bandar Seri Begawan) handled 1,262,343 passengers (all international) in 2005. The national carrier is the state-owned Royal Brunei Airlines (RBA). In 2006 RBA operated services to Auckland, Bangkok, Brisbane, Darwin, Denpasar Bali, Dubai, Frankfurt, Ho Chi Minh City, Hong Kong, Jakarta, Jeddah, Kota Kinabalu, Kuala Lumpur, London, Manila, Perth, Shanghai, Sharjah, Singapore, Surabaya and Sydney. In 2003 RBA flew 28m. km, carrying 956,000 passengers (all on international flights).

Shipping

Regular shipping services operate from Singapore, Hong Kong, Sarawak and Sabah to Bandar Seri Begawan, and there is a daily passenger ferry between Bandar Seri Begawan and Labuan. In 2005 merchant shipping totalled 2·4m. GRT. In 2005 vessels totalling 1,066,381 NRT entered ports and vessels totalling 1,061,339 NRT cleared.

Telecommunications

There is a telephone network linking the main centres. Brunei had 381,600 telephone subscribers in 2006 (or 999·2 per 1,000 inhabitants), including 301,400 mobile phone subscribers. There were 446·8 internet users per 1,000 inhabitants in 2007. Fixed internet subscriptions totalled 100,000 in 2009 (255·6 per 1,000 inhabitants). In March 2012 there were 234,000 Facebook users.

SOCIAL INSTITUTIONS

Justice

The Supreme Court comprises a High Court and a Court of Appeal and the Magistrates' Courts. The High Court receives appeals from subordinate courts in the districts and is itself a court of first instance for criminal and civil cases. The Judicial Committee of the Privy Council in London is the final court of appeal. Sharia Courts deal with Islamic law. 4,552 crimes were reported in 2005.

The population in penal institutions in mid-2010 was 379 (93 per 100,000 of national population)

Education

The government provides free education to all citizens from pre-school up to the highest level at local and overseas universities and institutions. In 2005 there were 12,999 children and 701 teachers in pre-primary education; 46,012 pupils and 4,548 teachers in primary education; 41,107 pupils and 3,907 teachers in secondary education; 4,154 students and 658 academic staff in tertiary education. The University of Brunei Darussalam was founded in 1985; in 2006 there were also eight technical and vocational colleges, one teacher training college and an institute of advanced education.

Estimated adult literacy rate, 2009, 95·3% (male, 96·8%; female, 93·7%). Public expenditure on education in 2010 came to 2·0% of GDP.

Health

Medical and health services are free to citizens and those in government service and their dependants. Citizens are sent overseas, at government expense, for medical care not available in Brunei. Flying medical services are provided to remote areas. In 2005 there were four government hospitals and the Jerudong Park private hospital with a total of 1,154 beds; there were 390 physicians, 73 dentists, 1,789 nurses, 748 midwives and 41 pharmacists.

RELIGION

The official religion is Islam. In 2001, 75% of the population were Muslim (mostly Malays). There are Buddhist and Christian minorities.

CULTURE

Press

In 2006 there were three daily newspapers with an average circulation of 35,000. The *Borneo Bulletin* and the *Brunei Times* are English-language papers, while *Media Permata* is a Malay paper.

Tourism

In 2010, 214,290 non-resident tourists (excluding same-day visitors) arrived by air—up from 157,474 in 2009 but down from 225,757 in 2008.

Festivals

The national day is celebrated on 23 Feb.

DIPLOMATIC REPRESENTATIVES

Of Brunei in the United Kingdom (19/20 Belgrave Sq., London, SW1X 8PG)
High Commissioner: Mohd Aziyan Abdullah.

Of the United Kingdom in Brunei (2.01, 2nd Floor, Block D, Kompleks Yayasan Sultan Haji Hassanal Bolkiah, Bandar Seri Begawan, BS 8711)
High Commissioner: Rob Fenn.

Of Brunei in the USA (3520 International Court, NW, Washington, D.C., 20008)
Ambassador: Dato Yusoff Abd Hamid.

Of the USA in Brunei (Simpang 336-52-16-9 Jalan Kebangsaan, Bandar Seri Begawan)
Ambassador: Daniel L. Shields.

Of Brunei to the United Nations
Ambassador: Latif bin Tuah.

Of Brunei to the European Union
Ambassador: Dato Paduka Serbini Ali.

FURTHER READING

Department of Economic Planning and Development, Prime Minister's Office. *Brunei Darussalam Statistical Yearbook.*

Cleary, M. and Wong, S. Y., *Oil, Economic Development and Diversification in Brunei.* 1994
Saunders, G., *History of Brunei.* 1996
Sidhu, Jatswan S, *Historical Dictionary of Brunei Darussalam.* 2010

National Statistical Office: Department of Statistics, Department of Economic Planning and Development, Prime Minister's Office, Block 2A, Jalan Ong Sum Ping, Bandar Seri Begawan, BA 1311.
Website: http://www.depd.gov.bn/home.html

BULGARIA

© Research Machines plc 2006

Republika Bulgaria
(Republic of Bulgaria)

Capital: Sofia
Population projection, 2015: 7·25m.
GNI per capita, 2011: (PPP$) 11,412
HDI/world rank: 0·771/55
Internet domain extension: .bg

KEY HISTORICAL EVENTS

Neolithic settlements in central Bulgaria dating from 6000 BC are among the oldest man-made structures yet discovered. Thracian tribes entered the Balkans from the east around 1500 BC, establishing fortified villages. By 500 BC Thracian rulers were under Hellenic influence and Greek colonists founded settlements on the Black Sea coast. Thrace was conquered by Philip of Macedonia in 324 BC and later absorbed into the eastern (Byzantine) Roman Empire, along with Moesia to the north.

Semi-nomadic Turkic clans entered the region—among them Bulgars who united under Khan Kublat in 632. They formed alliances with Slavic tribes during the sixth and seventh centuries and defeated the Byzantines in 681. The first Bulgarian state, centred on Pliska, was characterized by intermarriage between various ethnic groups and the rise of the Slavic language and Orthodox Christianity (adopted by Boris I in 864). Tsar Simeon moved the capital to Preslav in 893. Subsequent tsars battled with Byzantium and the expanding Kievan Rus. Preslav fell in 971, followed by a westward shift and defeat to the Byzantines in 1014.

Ivan Asen I's defeat of the Byzantines in 1187 heralded the Second Bulgarian Empire, centred on Tarnovo. It held sway until the late 13th century, when invading Mongol Hordes divided the territory into rival principalities. Following battles at Kosovo (1389) and Nikopol (1396), Bulgaria was absorbed into the Ottoman Empire. With fertile plains and located on a key European trade route, it became an important imperial province.

The Bulgarian population was forced to work the land although their Christian faith was broadly tolerated.

Uprisings against feudal rule occurred at Tarnovo in 1598 and 1668. The failed siege of Vienna in 1683 plus the strengthening Austrian and Russian influence contributed to a gradual decline in Ottoman dominance in the Balkans. Meanwhile, a growing merchant class demanded reform.

Bulgaria's 'national awakening' began in 1762 with the publication of Paisi Khilandarski's *History of the Bulgarian Slavs*. Calls for a separate Bulgarian church in the mid-19th century developed into broader revolutionary movements in the 1860s. The Russo–Turkish wars of 1877 and the subsequent Treaty of San Stefano paved the way for autonomous Bulgaria, which was reduced to territory south of the Danube by the 1878 Treaty of Berlin. Present-day southern Bulgaria became a separate autonomous province known as Eastern Rumelia until annexation in 1885.

After Austria annexed Bosnia in 1908, Prince Ferdinand Saxe-Coburg-Gotha became Tsar of independent Bulgaria. To block Austrian expansion into the Balkans, Russia encouraged Greece, Serbia, Montenegro and Bulgaria to attack Turkey (First Balkan War, 1912). The dispute that followed Bulgaria's claims to Macedonia led to the Second Balkan War and the entry of Bulgaria into the First World War in 1915 on the side of Germany and Austria-Hungary.

When post-war economic decline produced social unrest, Ferdinand I abdicated in favour of his son, Boris III. Parliamentary government was ended by a military coup in 1934. A year later Boris established a royal dictatorship.

Bulgaria entered the Second World War in March 1941 allied to Germany. In Sept. 1944 the USSR declared war. An alliance of Communists, Agrarians and Pro-Soviet army officers led by Kimon Georgiev seized power and a referendum in 1946 abolished the monarchy and established Bulgaria as a republic, with Georgi Dimitrov as the premier of a one-party state. The government was aligned with Moscow under the authoritarian Todor Zhivkov from 1954. The regime's attempts to persuade ethnic minorities to replace their traditional names with Bulgarian ones sparked violence in 1984 and led to a mass exodus of Bulgarian Turks in 1989.

After demonstrations in Sofia in Nov. 1989, reformers within the Communist Party forced Zhivkov's resignation. Multi-party elections in Oct. 1991 were won by the centre-right Union of Democratic Forces, which formed a coalition with the predominantly ethnic-Turkish Movement for Rights and Freedom. Zhelyu Zhelev became the first directly elected president in Jan. 1992.

Petar Stoyanov, elected as an anti-Communist pro-reform president in 1996, failed to contain political and economic unrest. In June 2001 the former King, Simeon Saxe-Coburg-Gotha, became premier. Bulgaria joined NATO in 2004 and in 2005 the Socialist leader, Sergey Stanishev, replaced Simeon as prime minister and led Bulgaria into the European Union on 1 Jan. 2007. The general election of July 2009 was won by the Citizens for the European Development of Bulgaria (GERB), led by Boyko Borisov. He resigned amid popular anti-government protests in Feb. 2013 and was replaced by Marin Raikov at the head of a caretaker administration.

TERRITORY AND POPULATION

The area of Bulgaria is 111,002 sq. km (42,858 sq. miles). It is bounded in the north by Romania, east by the Black Sea, south by Turkey and Greece, and west by Serbia and the Republic of Macedonia. The country is divided into 28 districts.

Area and population in 2011 (census):

District	Area (sq. km)	Population	District	Area (sq. km)	Population
Blagoevgrad	6,449	323,552	Shumen	3,390	180,528
Bourgas	7,748	415,817	Silistra	2,846	119,474
Dobrich	4,720	189,677	Sliven	3,544	197,473
Gabrovo	2,023	122,702	Smolyan	3,193	121,752
Haskovo	5,533	246,238	Sofia (city)	1,349	1,291,591
Kardzhali	3,209	152,808	Sofia (district)	7,062	247,489
Kyustendil	3,052	136,686	Stara Zagora	5,151	333,265
Lovech	4,129	141,422	Targovishte	2,559	120,818
Montana	3,636	148,098	Varna	3,819	475,074
Pazardzhik	4,457	275,548	Veliko		
Pernik	2,394	133,530	Tarnovo	4,662	258,494
Pleven	4,335	269,752	Vidin	3,033	101,018
Plovdiv	5,973	683,027	Vratsa	3,938	186,848
Razgrad	2,640	125,190	Yambol	3,355	131,447
Rousse	2,803	235,252	Total	111,002	7,364,570

The capital, Sofia, has district status.

The population of Bulgaria at the census of 2011 was 7,364,570 (females, 3,777,999); population density 66·3 per sq. km. Bulgaria's population has been declining since the mid-1980s. It has been falling at such a rate that by 2012 it was the same as it had been in the mid-1950s. The United Nations predicts that it will lose more than a quarter of its population by 2050. In 2011, 71·7% of the population were urban.

The UN gives a projected population for 2015 of 7·25m.

Population of principal towns (2011 census): Sofia, 1,202,761; Plovdiv, 338,153; Varna, 334,870; Bourgas, 200,271; Rousse, 149,642; Stara Zagora, 138,272; Pleven, 106,954; Sliven, 91,620; Dobrich, 91,030.

Ethnic groups at the 2011 census: Bulgarians, 5,664,624; Turks, 588,318; Roma, 325,343.

Bulgarian is the official language.

SOCIAL STATISTICS

2008: live births, 77,712; deaths, 110,523; marriages, 27,722; divorces, 14,104. Rates per 1,000 population, 2008: birth, 10·2; death, 14·5; marriage, 3·6; divorce, 1·9; infant mortality, 11 per 1,000 live births (2010). There were 37,272 reported abortions in 2006. In 2005 the most popular age range for marrying was 25–29 for males and 20–24 for females. Expectation of life in 2007 was 69·6 years among males and 76·7 years among females. The annual population growth rate for the period 2000–05 was −1·1%, giving Bulgaria one of the fastest declining populations of any country. Fertility rate, 2008, 1·4 children per woman.

CLIMATE

The southern parts have a Mediterranean climate, with winters mild and moist and summers hot and dry, but further north the conditions become more Continental, with a larger range of temperature and greater amounts of rainfall in summer and early autumn. Sofia, Jan. 28°F (−2·2°C), July 69°F (20·6°C). Annual rainfall 25·4" (635 mm).

CONSTITUTION AND GOVERNMENT

A new constitution was adopted at Tarnovo on 12 July 1991. The *President* is directly elected for not more than two five-year terms. Candidates for the presidency must be at least 40 years old and have lived for the last five years in Bulgaria. American-style primary elections were introduced in 1996; voting is open to all the electorate.

The 240-member *National Assembly* is directly elected by proportional representation. The *President* nominates a candidate from the largest parliamentary party as *Prime Minister*.

National Anthem

'Gorda stara planina' ('Proud and ancient mountains'); words and tune by T. Radoslavov.

GOVERNMENT CHRONOLOGY

(BKP = Bulgarian Communist Party; BSP = Bulgarian Socialist Party; GERB = Citizens for the European Development of Bulgaria; NMS = National Movement Simeon II; SDS = Union of Democratic Forces; Zveno = People's League Zveno; n/p = non-partisan)

Heads of State since 1943.

King

1943–46		Simeon II (Simeon Sakskoburggotski)

Chairman of Provisional Presidency

1946–47	BKP	Vasil Petrov Kolarov

Chairmen of the Presidium of the National Assembly

1947–50	BKP	Mincho Kolev Neychev
1950–58	BKP	Georgi Parvanov Damyanov
1958–64	BKP	Dimitar Ganev Varbanov
1964–71	BKP	Georgi Traykov Girovski

Chairmen of the Council of State

1971–89	BKP	Todor Khristov Zhivkov
1989–90	BKP	Petar Toshev Mladenov

Presidents of the Republic

1990	n/p	Petar Toshev Mladenov
1990–97	SDS	Zhelyu Mitev Zhelev
1997–2002	SDS	Petar Stefanov Stoyanov
2002–12	BSP	Georgi Sedefchov Parvanov
2012–	GERB	Rosen Asenov Plevneliev

Prime Ministers since 1944.

1944–46	Zveno	Kimon Gheorgiev Stoyanov
1946–49	BKP	Georgi Mihaylov Dimitrov
1949–50	BKP	Vasil Petrov Kolarov
1950–56	BKP	Vŭlko Velov Chervenkov
1956–62	BKP	Anton Tanev Yugov
1962–71	BKP	Todor Khristov Zhivkov
1971–81	BKP	Stanko Todorov Georgiev
1981–86	BKP	Grisha Stanchev Filipov
1986–90	BKP	Georgi Ivanov Atanasov
1990	BSP	Andrey Karlov Lukanov
1990–91	n/p	Dimitar Popov
1991–92	SDS	Filip Dimitrov Dimitrov
1992–94	n/p	Lyuben Borisov Berov
1995–97	BSP	Zhan Vasilev Videnov
1997–01	SDS	Ivan Yordanov Kostov
2001–05	NMS	Simeon Borisov Sakskoburggotski
2005–09	BSP	Sergey Dimitrievich Stanishev
2009–13	GERB	Boyko Metodiev Borisov
2013–	n/p	Marin Raikov (interim)

RECENT ELECTIONS

Presidential elections were held in two rounds on 23 and 30 Oct. 2011. Rosen Plevneliev won the first round with 40·1% of votes cast, ahead of Ivailo Kalfin with 29·0%, Meglena Kuneva with 14·0%, Volen Siderov with 3·6% and Stefan Solakov with 2·5%. There were 13 other candidates, each receiving less than 2% of the vote. In the second round Plevneliev won 52·6% of the vote and Kalfin 47·4%. Turnout was 52·3% in the first round and 48·2% in the second.

At the elections of 5 July 2009 Citizens for the European Development of Bulgaria (GERB) won 116 of 240 seats with 39·7% of the vote; the Coalition for Bulgaria (headed by the Bulgarian Socialist Party) won 40 seats with 17·7% of the vote; the Movement for Rights and Freedoms won 38 seats with 14·5%; the Attack coalition 21 with 9·4%; the Blue Coalition 15 with 6·8%; and Order, Law and Justice 10 with 4·1%. Turnout was 60·2%.

Parliamentary elections were scheduled to take place on 12 May 2013.

European Parliament

Bulgaria has 18 (18 in 2007) representatives. At the June 2009 elections turnout was 39·0% (29·2% in 2007). GERB won 5 seats with 24·4% of the vote (political affiliation in European Parliament: European People's Party); the Coalition for Bulgaria, 4 with 18·5% (Progressive Alliance of Socialists and Democrats); the Movement for Rights and Freedoms, 3 with 14·1% (Alliance of Liberals and Democrats for Europe); the Attack coalition, 2 with 12·0% (non-attached); the National Movement for Stability and Progress, 2 with 8·0% (Alliance of Liberals and Democrats for Europe); the Blue Coalition, 1 with 8·0% (European People's Party).

CURRENT GOVERNMENT

President: Rosen Plevneliev; b. 1964 (Citizens for the European Development of Bulgaria; in office since 22 Jan. 2012).

Vice-President: Margarita Popova.

In March 2013 the caretaker government comprised:

Prime Minister and Minister of Foreign Affairs: Marin Raikov; b. 1960 (ind.; sworn in 13 March 2013).

Deputy Prime Ministers: Ekaterina Zaharieva (also *Minister of Regional Development and Public Works*); Deyana Kostadinova (also *Minister of Labour and Social Policy*); Iliyana Tsanova (*in Charge of EU funds*).

Minister of Agriculture and Food Industry: Ivan Stankov. *Culture:* Vladimir Penev. *Defence:* Todor Tagarev. *Economy, Energy and Tourism:* Assen Vassilev. *Education, Youth and Science:* Nikolai Miloshev. *E-government:* Roman Vassilev. *Environment and Water:* Yulian Popov. *Finance:* Kalin Hristov. *Health:* Nikolai Petrov. *Interior:* Petya Purvanova. *Justice:* Dragomir Yordanov. *Physical Education and Sports:* Petar Stoychev. *Transport, Communications and Information Technologies:* Kristian Krustev.

Government Website: http://www.government.bg

CURRENT LEADERS

Rosen Plevneliev

Position
President

Introduction
Rosen Plevneliev was sworn in as president on 22 Jan. 2012, having been elected in Oct. 2011 with the support of the centre-right Citizens for the European Development of Bulgaria (GERB). He succeeded the Socialist Georgi Parvanov, who had held the presidency for the maximum two terms. Prior to the election Plevneliev had never run for public office, having worked as a non-partisan member of the cabinet.

Early Life
Rosen Plevneliev was born in Gotse Delchev, southern Bulgaria in 1964. Having graduated in engineering from the Sofia Technical University in 1989, he founded a building firm, Iris International. From 1991 he worked in Germany as the co-owner of a construction firm subcontracting for the German conglomerate Lindner. In 1998 he returned to Bulgaria to head the Lindner Bulgaria group.

Plevneliev entered politics in June 2009 as part of the GERB economic policy team. The following month he was appointed minister of regional development and public works in Boyko Borisov's cabinet. Plevneliev swiftly established himself as a popular and competent member of government, delivering several long-delayed, large-scale infrastructure projects.

On 4 Sept. 2011 he was nominated as GERB's candidate for the presidency. In a second round of polling, Plevneliev and his vice presidential candidate, former justice minister Margarita Popova, gained 52·6% of the vote to defeat the Socialist candidate, Ivailo Kalfin, and his running mate, Stefan Danailov.

Career in Office
In his electoral campaign Plevneliev argued that his presidency would promote cross-governmental unity. Although the position is largely ceremonial, his predecessor Georgi Parvanov had been criticized for obstructing legislation and civil service appointments already approved by parliament.

Plevneliev, a pro-European, pledged to oversee the reduction of the budget deficit and implement further austerity measures. He is expected to push for the development of regional and international trade. The diversification of energy supplies is among his most pressing challenges.

Marin Raikov

Position
Prime Minister

Introduction
Marin Raikov was sworn in as prime minister on 13 March 2013, heading a caretaker administration charged with overseeing early elections following the resignation of the former incumbent, Boyko Borisov, on 20 Feb. Raikov was previously Bulgaria's ambassador to France and also served as deputy foreign minister under Borisov.

Early Life
Marin Raikov was born on 17 Dec. 1960 in Washington, D.C., USA, where his diplomat father, Rayko Nikolov, was then based. As a result of his father's various postings, Raikov was school in Paris and Belgrade. In 1984 he graduated in international relations from the Karl Marx Higher Institute in Economics, Sofia (later renamed the University of National and World Economy).

He followed his father into the diplomatic corps, joining the foreign ministry in 1987 in the Balkan department. From 1992 until 1995 he worked at the Bulgarian embassy in Belgrade, then moved to Strasbourg to take a position at the Council of Europe. In 1998 he joined the centre-right government of Ivan Kostov, serving as deputy minister for foreign affairs until 2001. He left to become Bulgaria's ambassador to France but resumed the foreign affairs post in 2009 and 2010 in the administration of Boyko Borisov before again leaving to become ambassador to France.

In Feb. 2013 nationwide protests against rising electricity prices escalated into demonstrations against the government. Clashes with police in Sofia prompted Borisov to resign and withdraw his centre-right Citizens for the European Development of Bulgaria (GERB) from government. The resulting power vacuum led President Rosen Plevneliev to appoint Raikov to head an interim government staffed by technocrats. With his links to Europe and loose affiliation to the centre-right, Raikov was considered to be the least controversial choice.

Career in Office
As well as prime minister, Raikov is also foreign minister. On taking office, he named a cabinet that included three deputy prime ministers. The primary challenges of his caretaker administration are to reform the country's energy supplies, manage its European funds and quell popular unrest. Raikov pledged to ensure that elections scheduled for May 2013 would be free and fair.

DEFENCE

Since 1 Jan. 2008 Bulgaria has had an all-volunteer professional army. Following restructuring the total strength of the armed forces has been reduced from more than 68,000 in 2002 to less than 41,000 in 2007.

Defence expenditure in 2008 totalled US$1,315m. (US$181 per capita), representing 2·6% of GDP.

Army

In 2007 the Army had a strength of 18,773. In addition there are reserves of 250,500, 12,000 border guards and 18,000 railway and construction troops.

Navy

The Navy, mostly ex-Soviet or Soviet-built, includes one old diesel submarine and two small frigates. The Naval Aviation Wing operates six armed helicopters. The naval headquarters are at Varna (Northern Command) and Bourgas (Southern Command), and there are further bases at Atiya, Vidin, Balchik and Sozopol. Personnel in 2007 totalled 4,100, with 7,500 reservists.

Air Force

The Air Force had (2007) 9,344 personnel, with 45,000 reservists. There were 80 combat-capable aircraft in 2007, including MiG-21s, MiG-29s and Su-25s, and 18 attack helicopters.

INTERNATIONAL RELATIONS

At the European Union's Helsinki Summit in Dec. 1999 Bulgaria, along with five other countries, was invited to begin full negotiations for membership in Feb. 2000. Bulgaria joined the EU on 1 Jan. 2007. It became a member of NATO on 29 March 2004.

ECONOMY

Agriculture accounted for 8·5% of GDP in 2006, industry 31·2% and services 60·3%.

Overview

After communism, Bulgaria was slow to adapt to the market economy. In early 1997 the collapse of the banking system led to a currency crisis and a sharp decline in GDP. Government reforms later that year privatized state-owned enterprises and liberalized prices. The creation of a privatized financial sector inspired high growth rates.

Further market reforms to prepare the economy for EU entry were undertaken from 2001–05. When Bulgaria joined the EU in Jan. 2007, increased capital inflows and a credit boom generated GDP growth of over 6% per annum. Nonetheless, the country remains the poorest in the EU.

During the global financial crisis the economy has maintained stability through prudent macroeconomic policing, including an extensive 2011 pension reform. However, growth has been weak and unemployment high, running at over 12% in the second quarter of 2012. Strong economic ties to the eurozone, and especially Greece, pose a serious risk to growth, while reliance on energy imports leaves the country vulnerable to oil price increases.

Currency

The unit of currency is the *lev* (BGN) of 100 *stotinki*. In May 1996 the lev was devalued by 68%. A new *lev* was introduced on 5 July 1999, at 1 new *lev* = 1,000 old *leva*. Runaway inflation (123·0% in 1996 rising to 1,061% in 1997) forced the closure of 14 banks in 1996. However, by 1999 the rate had slowed to 2·6%. Inflation was 12·0% in 2008, falling to 2·5% in 2009 before rising to 3·0% in 2010 and 3·4% in 2011. In June 1997 the new government introduced a currency board financial system which stabilized the lev and renewed economic growth. Under it, the lev is pegged to the euro at one euro = 1·95583 new leva. Foreign exchange reserves were US$7,792m. in July 2005, gold reserves were 1·28m. troy oz and total money supply was 11,494m. leva.

Budget

The fiscal year is the calendar year.

Central government revenue and expenditure (in 1m. new leva):

	2007	2008	2009
Revenue	22,162	25,018	23,292
Expenditure	18,081	20,624	21,631

Bulgaria's budget deficit in 2011 was 2·1% of GDP (2010, 3·1%; 2009, 4·3%). In 2008 it had a surplus of 1·7%. The required target set by the EU is a budget deficit of no more than 3%.

VAT is 20%. There is a flat income tax rate of 10%.

Performance

Total GDP in 2011 was US$53·5bn. The economy contracted by 5·5% in 2009 but grew by 0·4% in 2010 and 1·7% in 2011.

Banking and Finance

The National Bank (*Governor*, Ivan Iskrov) is the central bank and bank of issue. There is also a Currency Board, established in 1997. The largest Bulgarian bank in terms of assets is UniCredit Bulbank (assets of 11·52bn. leva in Dec. 2009). Other major banks are DSK Bank (the last state bank to be privatized, in 2003), United Bulgarian Bank and Raiffeisenbank Bulgaria.

Foreign direct investment totalled US$2,170m. in 2010 (down from US$3,351m. in 2009).

In 2010 total external debt was US$48,077m., representing 104·8% of GNI.

There is a stock exchange in Sofia.

ENERGY AND NATURAL RESOURCES

In 2008, 9·4% of energy consumption came from renewables (wind power, solar power, hydro-electric power, tidal power, geothermal energy and biomass), compared to the European Union average of 10·3%. A target of 16% has been set by the EU for 2020.

Environment

Bulgaria's carbon dioxide emissions from the consumption and flaring of fossil fuels were the equivalent of 7·3 tonnes per capita in 2008.

Electricity

In 2012 there were two nuclear reactors in use, at the country's sole nuclear power plant in Kozloduy (dating from the 1970s). The two oldest of the plant's six reactors closed in Dec. 2002. A further two closed in Dec. 2006, as a condition of Bulgaria's accession to the EU. To compensate, the government had approved plans to complete a nuclear plant in Belene, started in the 1980s but suspended in 1990. However, the project was abandoned by the Borisov government in March 2012. Installed electrical capacity was 12·0m. kW in 2007. Output, 2007, 43·3bn. kWh (59% thermal, 34% nuclear and 7% hydro-electric). Consumption per capita: 5,081 kWh (2007).

Oil and Gas

Oil is extracted in the Balchik district on the Black Sea coast, in an area 100 km north of Varna, and at Dolni Dubnik near Pleven. There are refineries at Bourgas (annual capacity 5m. tonnes) and Dolni Dubnik (7m. tonnes). Crude oil production (2007) was 26,000 tonnes.

Minerals

Output in 2007 (in tonnes): lignite, 28·42m.; coal, 215,000. There are also deposits of gold, silver and copper.

Agriculture

In 2002 the total area of land in agricultural use was 5,796,208 ha. (52·2% of the overall territory of the country); there were 3,080,829 ha. under crops (including 2,217,560 ha. for cereals), 2,502,723 ha. of permanent grassland (including meadows and orchards), and 212,656 ha. of perennial plantations. There were 32,000 tractors in use in 2002 and 9,000 harvester-threshers.

Legislation of 1991 and 1992 provided for the redistribution of collectivized land to its former owners up to 30 ha. In 2002 there were 37,836 registered agricultural producers, including 33,633 individual farmers. The agricultural workforce was 222,000 in 2004.

Production in 2002 (in 1,000 tonnes): wheat, 4,123; maize, 1,288; sunflower seeds, 645; potatoes, 627; barley, 569; grapes, 409; tomatoes, 390; watermelons, 200; chillies and green peppers, 167; cucumbers and gherkins, 125; cabbage, 109. Bulgaria is a leading producer of attar of roses (rose oil). Bulgaria produced 205,000 tonnes of wine in 2002. Other products (in 1,000 tonnes) in 2002: meat, 488; cow's milk, 1,306; goat's milk, 105; sheep's milk, 93; eggs, 91.

Livestock (2002), in 1,000: cattle, 635; sheep, 1,571; pigs, 1,014; goats, 899; chickens, 18,000.

Forestry
In 2010 forests covered 3·93m. ha., or 36% of the total land area. Timber production in 2007 totalled 5·70m. cu. metres.

Fisheries
In 2010 total catch was 10,769 tonnes, mainly from sea fishing.

INDUSTRY
In 2005 the total output of industrial enterprises was 35,260m. new leva, of which 31,459m. new leva (89·2%) came from the private and 3,801m. new leva (10·8·%) from the public sector.

Output in 1,000 tonnes (2007 unless otherwise indicated): cement (2005), 3,618; distillate fuel oil, 2,376; crude steel, 1,909; rolled steel (2005), 1,817; residual fuel oil, 1,550; petrol, 1,466; pig iron, 1,069; sulphuric acid (2001), 620; nitrogenous fertilizers (2002), 193; paper (2002), 171. Production of other products: cotton fabrics (2005), 67·7m. sq. metres; silk fabrics (2005), 18·3m. sq. metres; woollen fabrics (2005), 9·9m. sq. metres; 17·4bn. cigarettes (2006); 145,000 refrigerators (2001).

Labour
There is a 40-hour five-day working week. The average annual salary in 2006 was 4,225 new leva. In 2001 the labour force numbered 3,412,600. A total of 2,940,300 persons were in employment in 2001 (excluding the armed forces), with the leading areas of activity as follows: agriculture, fishing, forestry and hunting, 774,100; manufacturing, 591,800; wholesale and retail trade/repair of motor vehicles, motorcycles and personal and household goods, 355,200; transport, storage and communications, 214,200; and health and social work, 138,300. The unemployment rate was 12·3% in June 2012, up from 11·2% in 2011 as a whole and 5·6% in 2008. The monthly minimum wage was raised from 270 leva to 290 leva in May 2012.

INTERNATIONAL TRADE
Legislation in force as of Feb. 1992 abolished restrictions imposed in 1990 on the repatriation of profits and allows foreign nationals to own and set up companies in Bulgaria. Western share participation in joint ventures may exceed 50%.

Imports and Exports
Imports (c.i.f.) and exports (f.o.b.) for calendar years in US$1m.:

	2007	2008	2009	2010
Imports	30,085·4	37,015·4	23,340·8	25,359·9
Exports	18,575·1	22,485·5	16,502·2	20,608·0

Leading import commodities are mineral products, machinery and apparatus, electrical equipment and parts, textile materials and articles, and transportation facilities. Leading export commodities are non-precious metals and articles, textile materials and articles, mineral products and chemical industry produce.

Leading import suppliers in 2005: Russia, 15·6%; Germany, 13·6%; Italy, 9·0%; Turkey, 6·1%; Greece, 5·0%; France, 4·7%. Main export markets in 2005: Italy, 12·0%; Turkey, 10·5%; Germany, 9·8%; Greece, 9·4%; Belgium, 6·0%; France, 4·6%. Trade with the European Union grew steadily in the years prior to Bulgaria becoming a member in Jan. 2007, with imports from the EU rising from 35% of the total in 1996 to 50% in 2005 and exports to the EU increasing from 39% of all exports in 1996 to 56% in 2005.

COMMUNICATIONS
Roads
In 2005 Bulgaria had 40,231 km of roads, including 331 km of motorways and 2,961 km of main roads. In 2007 there were 1,971,500 passenger cars (257 per 1,000 inhabitants), 262,900 lorries and vans, 26,300 buses and coaches, and 78,900 motorcycles and mopeds. In 2005 public transport totalled 13·7bn. passenger-km. In 2007, 9,827 persons were injured in road accidents and 1,006 were killed.

Rail
In 2005 there were 5,040 km of 1,435 mm gauge railway (3,900 km electrified) and 245 km of 760 mm gauge. Passenger-km travelled in 2008 came to 2·34bn. and freight tonne-km to 4·67bn.

There is a tramway and a 31-km long metro in Sofia.

Civil Aviation
There is an international airport at Sofia (Vrazhdebna), which handled 1,101,734 passengers (1,049,738 on international flights) and 7,395 tonnes of freight in 2001. The bankrupt former state-owned Balkan Bulgarian Airlines was replaced by Bulgaria Air (initially named Balkan Air Tour) in 2002 as the new national flag carrier. In 2003 Bulgaria Air operated direct services to Berlin, Brussels, Budapest, Copenhagen, Frankfurt, Lisbon, London, Madrid, Moscow, Paris, Prague, Rome, Stockholm, Tel Aviv, Vienna, Warsaw and Zürich. Balkan Air Tour has in the meantime been renamed Bulgaria Air. The independent Hemus Air operated services in 2003 to Athens, Beirut, Bucharest, Damascus, Dubai, Larnaca, Tirana and Tripoli. In 2003 scheduled airline traffic of Bulgarian-based carriers flew 7m. km, carrying 311,000 passengers (270,000 on international flights).

Shipping
In Jan. 2009 there were 86 ships of 300 GT or over registered, totalling 890,000 GT. Bourgas is a fishing and oil-port. Varna is the other important port.

Telecommunications
The Bulgarian Telecommunications Company was privatized in Jan. 2004. In 2011 there were 2,310,800 landline telephone subscriptions (equivalent to 310·3 per 1,000 inhabitants) and 10,475,100 mobile phone subscriptions (or 1,406·8 per 1,000 inhabitants). In 2011, 51·0% of the population were internet users. In March 2012 there were 2·4m. Facebook users.

SOCIAL INSTITUTIONS
Justice
A law of Nov. 1982 provides for the election (and recall) of all judges by the National Assembly. There are a Supreme Court, 28 provincial courts (including Sofia) and regional courts. Jurors are elected at the local government elections. The Prosecutor General and judges are elected by the Supreme Judicial Council established in 1992.

The population in penal institutions in Jan. 2008 was 10,271 (134 per 100,000 of national population). The maximum term of imprisonment is 20 years. The death penalty was abolished for all crimes in 1998.

Education
Adult literacy rate in 2008 was 98%. Education is free, and compulsory for children between the ages of 7 and 16.

In 2003–04 there were 6,648 educational establishments: 3,278 kindergartens, 2,823 general and special schools, 496 vocational schools and 51 higher education institutions. There were 122,986 teaching staff (22,532 in higher education) and 1,451,284 pupils and students (228,468 in higher education); 114 schools (with

8,721 pupils) and 14 higher institutions (with 32,802 students) were private. There are eight state universities, four private universities and several specialized higher education institutions, some of which have university status. The Academy of Sciences was founded in 1869.

In 2006 public expenditure on education came to 4·3% of GNI and 11·6% of total government spending.

Health

All medical services are free. Private medical services were authorized in Jan. 1991. In 2003 there were 249 hospitals; Bulgaria had 64 beds per 10,000 inhabitants in 2007. There were 28,111 physicians, 6,512 dentists and 35,028 nurses and midwives in 2006. In 2007 health spending represented 7·3% of GDP.

Welfare

In 2000 Bulgaria's official retirement age remained unchanged since the Soviet era at 60 years (men) and 55 years (women). However, as part of EU accession reform, the age level increased gradually until 2009 when retirement ages were 63 (men) and 60 (women). There are plans to increase the ages still further to 65 for both sexes by 2022. The minimum old-age pension is 116 leva a month.

The family allowance is 35 leva a month for each child below age 16 (or age 18 if the child attends secondary school).

In order for a person to be eligible for unemployment benefits they must have been working and making social insurance contributions for at least nine of the previous 15 months.

RELIGION

'The traditional church of the Bulgarian people' (as it is officially described) is that of the Eastern Orthodox Church. It was disestablished under the 1947 constitution. In 1953 the Bulgarian Patriarchate was revived. The Patriarch is Neofit (enthroned Feb. 2013). The seat of the Patriarch is at Sofia. There are 11 dioceses (each under a Metropolitan), ten bishops, 2,600 parishes, 1,500 priests, 120 monasteries (with about 400 monks and nuns), 3,700 churches and chapels, one seminary and one theological college.

In 2002 there were some 80,000 Roman Catholics with 51 priests and 54 parishes in three bishoprics. At the 2001 census, 6,638,870 Christians were recorded and 966,978 Muslims (Pomaks). There is a Chief Mufti elected by regional muftis.

CULTURE

World Heritage Sites

There are nine Bulgarian sites that appear on the UNESCO World Heritage List. They are (with year entered on list): Boyana Church (1979); Madara Rider (1979), an 8th century sculpture carved into a rockface; Rock-hewn Churches of Ivanovo (1979); Thracian Tomb of Kazanlak (1979); Ancient City of Nessebar (1983);

Srebarna Nature Reserve (1983); Pirin National Park (1983); Rila Monastery (1983 and 2008); Thracian tomb of Sveshtari (1985).

Press

In 2007 there were 64 daily newspapers with a combined daily circulation of 629,000. The two biggest circulation paid-for dailies are *Telegraph* (which was only launched in 2005) and *Trud*, the only title from the socialist era that survived after 1989. A total of 6,432 book titles were published in 2004, including 2,047 in sociology and politics and 1,756 literary texts for adults.

Tourism

There were 5,151,000 non-resident tourists in 2007. Earnings from tourism were US$3,975m. in 2007.

Festivals

The Surva International Festival of Masquerade Games in Pernik takes place every two years in Jan. and is one of the most important celebrations of traditional folkloric culture in the country. Classical music is showcased at the March Music Days festival in Rousse while the Spirit of Bourgas pop and dance music festival takes place in Aug.

DIPLOMATIC REPRESENTATIVES

Of Bulgaria in the United Kingdom (186–188 Queen's Gate, London, SW7 5HL)
Ambassador: Konstantin Stefanov Dimitrov.

Of the United Kingdom in Bulgaria (9 Moskovska St., Sofia 1000)
Ambassador: Jonathan Allen.

Of Bulgaria in the USA (1621 22nd St., NW, Washington, D.C., 20008)
Ambassador: Elena Poptodorova.

Of the USA in Bulgaria (16 Kozyak St., 1407 Sofia)
Ambassador: Marcie B. Ries.

Of Bulgaria to the United Nations
Ambassador: Stefan Tafrov.

Of Bulgaria to the European Union
Permanent Representative: Dimiter Tzantchev.

FURTHER READING

Central Statistical Office. *Statisticheski Godishnik.—Statisticheski Spravochnik* (annual).—*Statistical Reference Book of Republic of Bulgaria* (annual).

Crampton, Richard J., *A Concise History of Bulgaria.* 2nd ed. 2005
Melone, A., *Creating Parliamentary Government: The Transition to Democracy in Bulgaria.* 1998

National Statistical Office: Natsionalen Statisticheski Institut, 2 P. Volov St., 1038 Sofia. *President:* Reneta Indjova.
Website: http://www.nsi.bg

BURKINA FASO

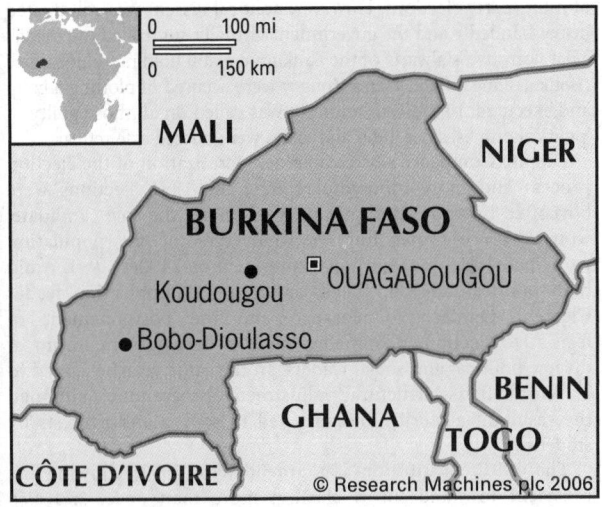

République Démocratique du Burkina Faso
(Democratic Republic of Burkina Faso)

Capital: Ouagadougou
Population projection, 2015: 19·12m.
GNI per capita, 2011: (PPP$) 1,141
HDI/world rank: 0·331/181
Internet domain extension: .bf

KEY HISTORICAL EVENTS

Formerly known as Upper Volta, the country's name was changed in 1984 to Burkina Faso, meaning 'the land of honest men'. The area it covers was settled by farming communities until an invasion by the Mossi people in the 11th century. The Mossi successfully resisted Islamic crusades and attacks by neighbouring empires for seven centuries until conquered by the French between 1895 and 1903.

France made Upper Volta a colony in 1919, only to abolish it as such in 1932, dividing its territory between the Ivory Coast (now Côte d'Ivoire), French Sudan (now Mali) and Niger. In 1947 the territory of Upper Volta was reconstituted as a territory within French West Africa. In 1958 it was granted the status of autonomous republic within the French Community before winning full independence two years later.

Upper Volta remained a desperately poor country often hit by drought, particularly in 1972–74 and again in 1982–84. The military has held power for most of the period after independence. In Aug. 1983 a coup was led by Capt. Thomas Sankara, a leading radical, who headed a left-wing regime. Sankara was overthrown and killed in a coup on 15 Oct. 1987, the fifth since 1960, mounted by his former friend Capt. Blaise Compaoré.

TERRITORY AND POPULATION

Burkina Faso is bounded in the north and west by Mali, east by Niger and south by Benin, Togo, Ghana and Côte d'Ivoire. Area: 270,764 sq. km; 2006 census population, 14,017,262, giving a density of 51·8 per sq. km. In 2011 the population was 26·5% urban.

The UN gives a projected population for 2015 of 19·12m.

The largest cities in 2006 were Ouagadougou, the capital (1,475,223), Bobo-Dioulasso (489,967), Koudougou (88,184), Banfora (75,917), Ouahigouya (73,153) and Pouytenga (60,618).

The country is administratively divided into 13 regions (with capitals): Boucle du Mouhoun (Dédougou), Cascades (Banfora), Centre (Ouagadougou), Centre-Est (Tenkodogo), Centre-Nord (Kaya), Centre-Ouest (Koudougou), Centre-Sud (Manga), Est (Fada N'Gourma), Hauts-Bassins (Bobo-Dioulasso), Nord (Ouahigouya), Plateau-Central (Ziniaré), Sahel (Dori), Sud-Ouest (Gaoua).

The principal ethnic groups are the Mossi (48%), Fulani (10%), Bobo (7%), Lobi (7%), Mandé (7%), Grosi (5%), Gurma (5%), Sénoufo (5%) and Tuareg (3%).

French is the official language.

SOCIAL STATISTICS

2008 births (estimates), 719,000; deaths, 198,000. Estimated birth rate in 2008 was 47·2 per 1,000 population; estimated death rate, 13·0. Burkina Faso has one of the youngest populations of any country, with 73% of the population under the age of 30 and 45% under 15. Annual population growth rate, 2000–08, 3·3%. Expectation of life at birth, 2007, 54·0 years for females and 51·4 for males. Infant mortality, 2010 (per 1,000 live births), 93. Fertility rate, 2008, 5·9 children per woman.

CLIMATE

A tropical climate with a wet season from May to Nov. and a dry season from Dec. to April. Rainfall decreases from south to north. Ouagadougou, Jan. 76°F (24·4°C), July 83°F (28·3°C). Annual rainfall 36" (894 mm).

CONSTITUTION AND GOVERNMENT

At a referendum in June 1991 a new constitution was approved; there is an executive presidency and a multi-party system. Parliament consists of the 127-member *National Assembly*, elected by universal suffrage. The *Chamber of Representatives*, a consultative body representing social, religious, professional and political organizations, was abolished in 2002. There is also an *Economic and Social Council*. In April 2000 parliament passed a law reducing presidential terms from seven to five years, with a maximum of two terms. The new law did not affect President Blaise Compaoré's seven-year term which was to expire in Nov. 2005, and he has now been elected for a further term.

National Anthem

'Contre la férule humiliante' ('Against the shameful fetters'); words by T. Sankara, tune anonymous.

RECENT ELECTIONS

At the presidential elections of 21 Nov. 2010 Blaise Compaoré was re-elected with 80·2% of votes cast, defeating six other candidates. Turnout was 54·9%.

Parliamentary elections were held on 2 Dec. 2012. The Congress for Democracy and Progress (CDP) won 70 out of 127 seats; the Alliance for Democracy and Federation–African Democratic Rally (ADF–RDA), 19; the Union for Progress and Reform, 19; the Union for Rebirth/Sankarist Movement, 4; Union for the Republic, 4; Convention of Democratic Forces, 3; Party for Democracy and Socialism–Builders' Party, 2. Six other parties won one seat each. Turnout was 76·0%.

CURRENT GOVERNMENT

President and Minister of Defence and War Veterans: Capt. Blaise Compaoré; b. 1951 (CDP; in office since 1987, most recently re-elected 21 Nov. 2010).

In March 2013 the government comprised:

Prime Minister: Luc-Adolphe Tiao; b. 1954 (CDP; sworn in 18 April 2011).

Minister of State and Minister of Relations with Parliament and Political Reforms: Bongnessan Arsène Yé. *Minister of State and Minister of Foreign Affairs and Regional Co-operation:* Yipènè Djibril Bassolet. *Minister of State at the Presidency:* Assimi Koanda.

Minister of Agriculture and Food Safety: Mahama Zoungrana. *Animal Resources and Fisheries:* Jérémie Ouédraogo. *Civil Service, Labour and Social Security:* Vincent Zakané. *Communication and Government Spokesman:* Alain Edouard Traoré. *Culture and Tourism:* Baba Hama. *Digital Economy Development and Post:* Jean Koulidiati. *Economy and Finance:* Lucien Marie Noël Bembamba. *Environment and Sustainable Development:* Salif Ouédraogo. *Health:* Léné Sebgo. *Housing and Town Planning:* Yacouba Barry. *Human Rights and Civic Education:* Prudence Julie Nigna-Somda. *Industry, Trade and Handicrafts:* Patiendé Arthur Kafando. *Infrastructure, Rural Public Works Development and Transport:* Jean Bertin Ouédraogo. *Justice and Keeper of the Seals:* Dramane Yaméogo. *Land Management and Decentralization:* Toussaint Abel Coulibaly. *Mines and Energy:* Lamoussa Salif Kaboré. *National Education and Literacy:* Koumba Boly-Barry. *Promotion of Women:* Nestorine Sangaré-Compaoré. *Scientific Research and Innovation:* Gnissa Isaïe Konaté. *Secondary and Higher Education:* Moussa Ouattara. *Social Action and National Solidarity:* Alain Zoubga. *Sports and Leisure:* Yacouba Ouédraogo. *Territorial Administration and Security:* Jérôme Bougouma. *Water, Water Resources and Sanitation:* Mamounata Belem-Ouédraogo. *Youth, Professional Training and Employment:* Basga Emile Dialla.

Office of the Prime Minister (French only):
   http://www.primature.gov.bf

CURRENT LEADERS

Blaise Compaoré

Position
President

Introduction
Blaise Compaoré came to power in 1987 after the assassination of Thomas Sankara, the president of the Conseil National de la Révolution (CNR; National Revolutionary Council). Compaoré has attempted to deregulate the economy and improve relations with the West. Politically, however, his tenure has been marked by strikes, unrest and political murders. His human rights record has also attracted international condemnation.

Early Life
Born on 3 Feb. 1951 in Ouagadougou, Compaoré received his early education in Burkina Faso and became a secondary school teacher. In 1971 he joined the army and in 1975 went to Cameroon and France for military training. His friendship with Sankara began in Morocco in 1978 while he was serving as a parachute instructor. By 1981 Compaoré had achieved the rank of captain.

Although involved in the establishment in 1982 of the Conseil de Salut du Peuple (CSP; People's Salvation Council, led by Jean-Baptiste Ouédraogo), Compaoré and Sankara broke with the CSP in 1983 to form the more left-wing CNR. When Sankara was later arrested, Compaoré led an anti-government revolt that overthrew Ouédraogo and brought Sankara and the CNR to power. Compaoré was appointed vice-premier. However, growing dissatisfaction with Sankara's increasingly autocratic leadership led in 1987 to his assassination by soldiers loyal to Compaoré, who replaced his former ally as head of state.

Career in Office
Compaoré took office promising a continuation of the CNR's guiding principles, but with 'rectification'. He restored links with the business community, traditional chiefs and the army, and

sought to reassure the West on whom he relied for aid. A new party—the Organization for Popular Democracy/Labour Movement (ODP/MT; Organisation pour la Démocratie Populaire/Mouvement du Travail)—was created in 1989 and provision made for the return of multi-party elections. However, political dissent was still treated heavy-handedly and the government virtually controlled the media. That year, two stalwarts of the Sankara era still holding senior office (Boukari Lingani and Henri Zongo) were accused of plotting a coup and executed. In 1991 an amnesty was called on all those guilty of 'political crimes' since 1960 and exiles were offered safe return.

Despite Compaoré's ostensible democratization of the election process and more moderate regime, the 1991 elections were boycotted by opposition groups. Compaoré, the sole candidate, won 90·4% of votes but less than 25% of the population participated. He was sworn in as president on 24 Dec. 1991. Amid high political tension, the assassination of opposition leader Clément Oumarou Ouédraogo and the postponement of legislative elections, Compaoré called a development forum of diverse political and social leaders. In the same year he agreed to a World Bank structural adjustment programme, although the resultant austerity measures led to strikes and protests by students.

The 1991 constitution was amended in 1997, allowing the president to stand for re-election more than once, and also restructuring parliament and provincial government. In addition the national anthem and the flag were modified to break with the revolutionary past. Compaoré was re-elected president in Nov. 1998 with more than 87% of votes (in a 56·1% turnout), although doubt was cast on the legitimacy of the electoral process. On 13 Dec. 1998 a journalist critical of Compaoré, Norbert Zongo, and three of his colleagues were murdered. There followed public protests and the arrest of opposition leaders.

A report of May 1999 suggested that the presidential bodyguard was behind the murders of Zongo and his colleagues. The conclusions led to student protests in the capital. Compaoré's human rights record fell under further scrutiny as more opposition leaders and independent journalists were arrested. Despite the offer of compensation to the victims' families and the release of several political prisoners, tensions remained high, and strikes and other protests persisted.

In parliamentary elections in May 2002, despite a stronger opposition performance, the pro-Compaoré Congress for Democracy and Progress (CDP) won 57 of the 111 National Assembly seats, and it took 73 of the 111 seats in the May 2007 elections. Meanwhile, although deemed unconstitutional by opposition politicians, Compaoré was re-elected for further presidential terms in Nov. 2005 and Nov. 2010, winning an overwhelming share of the vote on each occasion. In April 2009 the parliament adopted legislation requiring at least 30% of political party candidates in future elections to be women. A serious challenge to Compaoré's authority was posed in April 2011 by a military revolt over unpaid allowances, following popular protests over rising prices, which prompted his appointment of a new prime minister and cabinet.

In foreign policy, his regime has been accused of destabilizing interference in civil wars in Liberia and Sierra Leone.

DEFENCE

There are three military regions. Defence expenditure totalled US$112m. in 2008 (US$7 per capita), representing 1·2% of GDP.

Army

Strength (2007), 6,400 with a paramilitary Gendarmerie of 4,200. In addition there is a People's Militia of 45,000.

Air Force

Personnel total (2007), 200 with five combat-capable aircraft.

ECONOMY

In 2006 agriculture accounted for 33·3% of GDP, industry 22·4% and services 44·4%.

Overview

Ranked 181st of 187 countries on the UNDP's Human Development Index, Burkina Faso is among the world's poorest nations. Although there are large gold deposits, the economy is dependent on cotton exports that are susceptible to droughts and fluctuating world prices. In 2009 gold surpassed cotton as the main export commodity.

Macroeconomic performance has been sound since the late 1990s, with real GDP growth averaging 5·5% per year. Growth slowed with the global economic downturn affecting the cotton sector in particular. The government has instigated structural reforms, including modernization and reform of the tax system, although further fiscal consolidation may be required to restore debt sustainability.

Currency

The unit of currency is the *franc CFA* (XOF) with a parity of 655·957 francs CFA to one euro. Foreign exchange reserves were US$520m. in June 2005 and total money supply was 375,711m. francs CFA. There was deflation of 0·6% in 2010 but inflation of 2·7% in 2011.

Budget

Budgetary central government revenues in 2009 totalled 771·5bn. francs CFA (630·7bn. francs CFA in 2008) and expenses were 499·1bn. francs CFA (453·6bn. francs CFA in 2008). Taxes accounted for 64·1% of revenues in 2009 and compensation of employees 45·8% of expenses.

VAT is 18%.

Performance

Real GDP growth was 3·0% in 2009, 7·9% in 2010 and 4·2% in 2011. Total GDP was US$10·2bn. in 2011.

Banking and Finance

The bank of issue which functions as the central bank is the regional Central Bank of West African States (BCEAO; *Governor*, Tiémoko Meyliet Koné). There are seven other banks and three credit institutions.

In 2010 external debt totalled US$2,053m., representing 23·3% of GNI.

There is a stock exchange in Ouagadougou.

ENERGY AND NATURAL RESOURCES

Environment

Burkina Faso's carbon dioxide emissions from the consumption and flaring of fossil fuels in 2008 were the equivalent of 0·1 tonnes per capita. An *Environmental Performance Index* compiled in 2008 ranked Burkina Faso 144th in the world out of 149 countries analysed, with 44·3%. The index examined various factors in six areas—air pollution, biodiversity and habitat, climate change, environmental health, productive natural resources and water resources.

Electricity

Production of electricity (2007) was 612m. kWh. Thermal capacity in 2007 was approximately 217,000 kW; hydro-electric capacity in 2007 was around 32,000 kW. Total installed capacity was 249,000 kW in 2007. Consumption per capita was 50 kWh in 2007.

Minerals

There are deposits of manganese, zinc, limestone, phosphate and diamonds. Gold production was 5,482 kg in 2008.

Agriculture

In 2007 there were about 5·2m. ha. of arable land and 60,000 ha. of permanent crops. 25,000 ha. were irrigated in 2001. There were four tractors per 10,000 ha. of arable land in 2006. The agricultural population in 2007 totalled approximately 13·56m., of whom 6·12m. were economically active. Production (2003, in 1,000 tonnes): sorghum, 1,519; millet, 1,214; maize, 738; seed cotton, 500; sugarcane, 420; groundnuts, 301; cottonseed, 250; cotton lint, 207; rice, 97.

Livestock (2003 estimates): goats, 8·8m.; sheep, 6·8m.; cattle, 5·0m.; pigs, 640,000; asses, 531,000; chickens, 24m. Livestock products, 2003 estimates (in 1,000 tonnes): beef and veal, 62; goat meat, 24; poultry meat, 28; cow's milk, 180; goat's milk, 54; eggs, 18.

Forestry

In 2010 forests covered 5·65m. ha., or 21% of the total land area. Timber production in 2007 was 13·41m. cu. metres.

Fisheries

In 2010 total catch was 14,520 tonnes, exclusively from inland waters.

INDUSTRY

In 2002 manufacturing contributed 14·5% of GDP, primarily food-processing and textiles. Industry is underdeveloped and employs only 1% of the workforce. The country's manufactures are mainly restricted to basic consumer goods and processed foods. Output of major products, in 1,000 tonnes: vegetable oil (2000), 31; sugar (2002), 35; flour (2002), 10; soap (2002), 10; beer (2003), 55·0m. litres; printed fabric (2000), 275,000 sq. metres.

Labour

In 2003 the labour force was 5,918,880 (54% males). Over 90% of the economically active population are engaged in agriculture, fishing and forestry.

INTERNATIONAL TRADE

Imports and Exports

In 2010 imports totalled US$2,048·2m. and exports US$1,288·1m. Principal import suppliers, 2010: Côte d'Ivoire, 16·0%; France, 10·3%; China, 9·7%; Togo, 4·5%. Principal export markets, 2010: Switzerland, 63·5%; South Africa, 11·2%; Singapore, 4·9%; United Kingdom, 3·0%. Machinery and transport equipment accounted for 22·9% of the country's imports in 2010; gold (including gold plated with platinum) accounted for 68·6% of exports in 2010.

COMMUNICATIONS

Roads

The road system comprised 92,495 km in 2004 (including 15,271 km of main roads). There were 97,100 passenger cars (seven per 1,000 inhabitants), 55,700 lorries and vans, and 356,400 motorcycles and mopeds in use in 2007.

Rail

The railway from Abidjan in Côte d'Ivoire to Kaya (600 km of metre gauge within Burkina Faso) is operated by the mixed public-private company Sitarail, a concessionaire to both governments. Across both countries Sitarail carried 156,569 passengers in 2004 and 759,957 tonnes of freight in 2005.

Civil Aviation

The international airports are Ouagadougou (which handled 337,000 passengers in 2007) and Bobo-Dioulasso. The national carrier is Air Burkina, which in 2003 flew to Abidjan, Bamako, Cotonou, Dakar, Lomé and Niamey in addition to operating on domestic routes. In 2003 scheduled airline traffic of Burkina Faso-based carriers flew 1m. km, carrying 54,000 passengers.

Telecommunications

In 2011 there were 141,500 landline telephone subscriptions (equivalent to 8·3 per 1,000 inhabitants) and 7,628,100 mobile phone subscriptions (or 452·7 per 1,000 inhabitants). Fixed internet subscriptions totalled 28,700 in 2010 (1·7 per 1,000 inhabitants). In June 2012 there were 116,000 Facebook users.

SOCIAL INSTITUTIONS

Justice
Civilian courts replaced revolutionary tribunals in 1993. A law passed in April 2000 split the supreme court into four separate entities—a constitutional court, an appeal court, a council of state and a government audit office.

The population in penal institutions in 2010 was 5,238 (32 per 100,000 of national population).

Education
In 2007 adult literacy was 28·7%, among the lowest in the world. The 1994–96 development programme established an adult literacy campaign, and centres for the education of 10–15-year-old non-school attenders. In 2007 there were 1,561,258 pupils at primary schools with 32,760 teaching staff. During the period 1990–95 only 24% of females of primary school age were enrolled in school but by 2006 this had increased to 42%. In 2007 there were 352,376 pupils and 12,498 teaching staff in secondary schools, and 33,459 students in higher education with 1,886 academic staff.

In 2006 public expenditure on education came to 4·5% of GNI and 15·4% of total government expenditure.

Health
In 2007 there were three national hospitals, nine regional hospitals and 75 medical centres. There were 441 physicians, 38 dentists, 4,262 nurses, 604 midwives and 58 pharmacists in the public sector in 2007.

RELIGION

In 2001 there were 5·96m. Muslims and 2·04m. Christians (mainly Roman Catholic). Many of the remaining population follow traditional animist religions.

CULTURE

World Heritage Sites
The Ruins of Loropéni were inscribed on the UNESCO World Heritage List in 2009. At least 1,000 years old, Loropéni is the best preserved of ten similar fortresses in the Lobi area and is part of a larger group of around 100 stone-built enclosures with ties to the trans-Saharan gold trade.

Press
There were five dailies (two government-owned) with a combined circulation of 36,000 in 2008. The leading newspaper in terms of circulation is *Le Pays*.

Tourism
In 2009, 269,000 foreign tourists stayed in hotels or similar accommodation.

DIPLOMATIC REPRESENTATIVES

Of Burkina Faso in the United Kingdom
Ambassador: Vacant (resides in Brussels).
Honorary Consul: Colin Seelig (The Lilacs, Stane St., Ockley, Surrey, RH5 5LU).

Of the United Kingdom in Burkina Faso
Ambassador: Peter Jones (resides in Accra, Ghana).

Of Burkina Faso in the USA (2340 Massachusetts Ave., NW, Washington, D.C., 20008)
Ambassador: Seydou Bouda.

Of the USA in Burkina Faso (Secteur 15, Ouaga 2000, Ave. Sembène Ousmane, rue 15.873, Ouagadougou)
Ambassador: Thomas Dougherty.

Of Burkina Faso to the United Nations
Ambassador: Der Kogda.

Of Burkina Faso to the European Union
Ambassador: Vacant.

FURTHER READING

Nnaji, B. O., *Blaise Compaoré: Architect of the Burkina Faso Revolution.* 1991

National Statistical Office: Institut National de la Statistique et de la Démographie (INSD), Ave. Pascal Zagre, Ouaga 2000, 01 BP 374, Ouagadougou.
Website (French only): http://www.insd.bf

**The world in focus at
www.statesmansyearbook.com**

BURUNDI

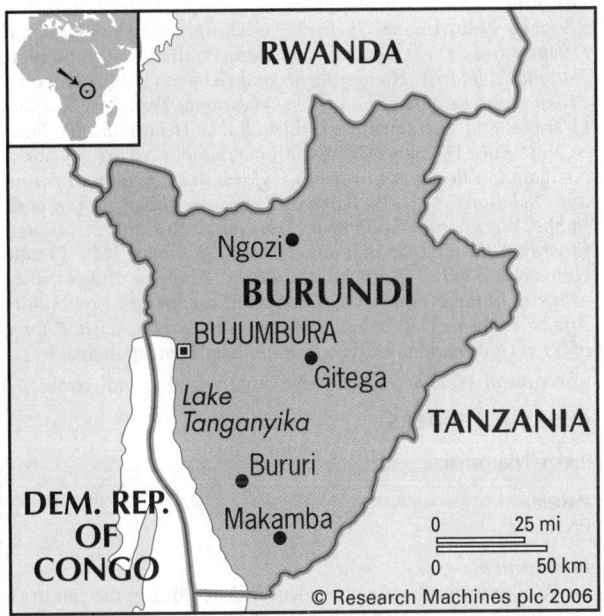

Republika y'Uburundi
(Republic of Burundi)

Capital: Bujumbura
Population projection, 2015: 9·23m.
GNI per capita, 2011: (PPP$) 368
HDI/world rank: 0·316/185
Internet domain extension: .bi

KEY HISTORICAL EVENTS

From 1890 Burundi was part of German East Africa and from 1919 part of Ruanda-Urundi, administered by Belgium as a League of Nations mandate. Internal self-government was granted on 1 Jan. 1962, followed by independence on 1 July 1962. In April 1972 fighting broke out between rebels from both Burundi and neighbouring countries and the ruling Tutsi, apparently with the intention of destroying the Tutsi hegemony. Up to 120,000 died. On 1 Nov. 1976 President Micombero was deposed by the Army, as was President Bagaza on 3 Sept. 1987. Maj. Pierre Buyoya assumed the presidency on 1 Oct. 1987.

On 1 June 1993 President Buyoya was defeated in elections by Melchior Ndadaye, who thus became the country's first elected president and the first Hutu president, but on 21 Oct. 1993 President Ndadaye and six ministers were killed in an attempted military coup. A wave of Tutsi-Hutu violence cost thousands of lives. On 6 April 1994 the new president, Cyprien Ntaryamira, was also killed, possibly assassinated, together with the president of Rwanda.

On 25 July 1996 the army seized power, installing Maj. Pierre Buyoya, a Tutsi, as president for the second time. In June 1998 Buyoya drew up a settlement for a power-sharing transitional government and the replacement of the prime minister by two vice-presidents, one Hutu and one Tutsi. Extremists on both sides denounced the agreement. An attempted coup in April 2001 failed. In July 2001 it was agreed that a three-year transitional government should be installed with Buyoya as president and Domitien Ndayizeye, a Hutu, as vice-president for the first 18

months, after which the roles would be reversed. A further attempted coup shortly after the announcement of the agreement also failed, although fighting continued. A ceasefire was eventually signed in Dec. 2002 by the government and the Forces for the Defense of Democracy (FDD), the country's principal rebel movement. In Oct. 2003 the FDD and the government sealed a peace deal to implement the ceasefire agreed in 2002.

In March 2005 a new power-sharing constitution won the popular vote and Pierre Nkurunziza became president in Aug. 2005. In April 2007 Burundi joined with Rwanda and the Democratic Republic of the Congo to relaunch the Great Lakes Countries Economic Community and in March 2009 the country had US$134m. of debt wiped out by the Paris Club of creditors. The following month the Forces for National Liberation, the last remaining rebel group, gave up its campaign of violence and became a legitimate political party. An estimated 300,000 have died in civil conflict since 1993. In June 2010 Nkurunziza won a new term as president.

TERRITORY AND POPULATION

Burundi is bounded in the north by Rwanda, east and south by Tanzania and west by the Democratic Republic of the Congo, and has an area of 27,830 sq. km (10,745 sq. miles) including 2,150 sq. km of inland water (830 sq. miles). The population at the 2008 census was 8,053,574 (4,088,668 females); density, 314 per sq. km. In 2011, 11·3% of the population lived in urban areas (the smallest proportion of any country in the world).

The UN gives a projected population for 2015 of 9·23m.

There are 17 regions, all named after their chief towns. They are (with 2008 provisional populations): Bubanza (348,188), Bujumbura Mairie (478,155), Bujumbura Rural (564,070), Bururi (570,929), Cankuzo (221,391), Cibitoke (460,626), Gitega (715,080), Karusi (433,061), Kayanza (586,096), Kirundo (636,298), Makamba (428,917), Muramvya (294,891), Muyinga (632,346), Mwaro (269,048), Ngozi (661,310), Rutana (336,394), Ruyigi (400,818).

The capital, Bujumbura, had a population (provisional) of 478,155 in 2008. Other large towns are Gitega, Ngozi and Bururi.

There are four ethnic groups—Hutu (Bantu, forming 81% of the total); Tutsi (Nilotic, 16%); Lingala (2%); Twa (pygmoids, 1%). The local language, Kirundi, and French are both official languages. Kiswahili is spoken in the commercial centres.

SOCIAL STATISTICS

2008 estimates: births, 278,000; deaths, 112,000. Rates, 2008 estimates (per 1,000 population): birth, 34·5; death, 13·9. Life expectancy at birth, 2007, was 48·6 years for men and 51·4 years for women. Infant mortality, 2010, 88 per 1,000 live births. Annual population growth rate, 2000–08, 2·8%; fertility rate, 2008, 4·6 children per woman.

CLIMATE

An equatorial climate, modified by altitude. The eastern plateau is generally cool, the easternmost savanna several degrees hotter. The wet seasons are from March to May and Sept. to Dec. Bujumbura, Jan. 73°F (22·8°C), July 73°F (22·8°C). Annual rainfall 33" (825 mm).

CONSTITUTION AND GOVERNMENT

The constitution of 1981 provided for a one-party state. In Jan. 1991 the government of President Maj. Pierre Buyoya, leader of the sole party, the Union for National Progress (UPRONA), proposed a new constitution which was approved by a referendum in March 1992 (with 89% of votes cast in favour), legalizing parties not based on ethnic group, region or religion and providing for presidential elections by direct universal suffrage. On 28 Feb. 2005 citizens voted overwhelmingly to adopt a new constitution laying the foundations for the end of a 12-year civil war, with 92% of votes cast in favour of the constitution. The constitution gives Tutsis (who have traditionally held power in Burundi but only make up 15% of the population) 40% of seats in the National Assembly, while the Hutus, who constitute 83% of the population, are given 60% of the seats.

Burundi has a bicameral legislature, consisting of the *National Assembly* of 106 members, with 100 members elected to serve five-year terms and six appointed to ensure that ethnic and gender quotas are met, and the *Senate* of 41 members (34 elected and seven appointed, including four former presidents).

In July 2001 agreement was reached on President Buyoya's presidency for the first 18 months of a three-year transition period of multi-ethnic broad-based government. In accordance with the terms of the Arusha peace accord, initially he was being assisted by Hutu Vice-President Domitien Ndayizeye, after which the roles were to be reversed for the second 18 months. The transitional government was established on 1 Nov. 2001. On 30 April 2003 Ndayizeye became president but Alphonse Marie Kadege, like Buyoya a Tutsi from the Party of Unity and National Progress, became the vice-president. In Oct. 2004 the transitional government was extended for a further six months, with elections scheduled for 22 April 2005. In April 2005 the transitional period was extended for a further four months and a new deadline of 19 Aug. 2005 set for elections. The success of the referendum in Feb. 2005 held under 1993 electoral laws was seen as proof that presidential elections need not be postponed further. Parliamentary elections that were generally deemed free and fair were held in July 2005, with presidential elections following in Aug.

National Anthem

'Burundi Bwacu' ('Dear Burundi'); words by a committee, tune by M. Barengayabo.

RECENT ELECTIONS

Pierre Nkurunziza was re-elected president on 28 June 2010 with 91·6% of votes cast. Several opposition parties boycotted the election. In contrast to when Nkurunziza was elected president by parliament in 2005, in 2010 there was to be a direct election although as a result of the withdrawals Nkurunziza was the only candidate. Turnout was 77·0%.

At the parliamentary elections of 23 July 2010 the National Council for the Defense of Democracy–Forces for the Defense of Democracy (CNDD–FDD) won 80 of 100 available seats with 81·2%% of the vote, the Union for National Progress (UPRONA) 16 with 11·1% and Sahwanya Frodebu-Nyakuri 4 with 5·9%. In addition, three ethnic Twa members were appointed to the National Assembly along with one member from the three other parties. Turnout was 66·7%. In indirect Senate elections held on 28 July 2010 the National Council for the Defense of Democracy–Forces for the Defense of Democracy won 32 of 34 available seats and the Union for National Progress 2. Three ethnic Twa members were appointed to the Senate as were four former presidents. The opposition boycotted both elections.

CURRENT GOVERNMENT

President: Pierre Nkurunziza; b. 1963 (CNDD–FDD; sworn in 26 Aug. 2005).

In March 2013 the government comprised:

First Vice-President in Charge of Political, Administrative, Legal and Security Issues: Thérence Sinunguruza.

Second Vice-President in Charge of Economic and Social Issues: Gervais Rufyikiri.

Minister of Agriculture and Livestock: Odette Kaytesi. *Civil Service, Labour and Social Security:* Anonciate Sendazirasa. *Commerce, Industry, Posts and Tourism:* Victoire Ndikumana.

Communal Development: Jean Claude Ndihokubwayo. *Energy and Mining:* Côme Manirakiza. *External Relations and International Co-operation:* Laurent Kavakure. *Finance and Economic Development Planning:* Tabu Abdallah Manirakisa. *Higher Education and Scientific Research:* Joseph Butore. *Interior:* Edouard Nduwimana. *Justice:* Pascal Barandagiye. *National Defence and War Veterans:* Maj.-Gen. Pontien Gaciyubwenge. *National Solidarity, Human Rights and Gender:* Clotilde Niragira. *Primary and Secondary Education, Vocational Training, Teaching of Trades and Literacy:* Rose Gahiru. *Public Health and the Fight Against Aids:* Dr Sabine Ntakarutimana. *Public Security:* Gabriel Nizigama. *Telecommunications, Information, Communication and Relations with Parliament:* Léocadie Nihazi. *Transport, Public Works and Equipment:* Déogratias Rurimunzu. *Water, Environment, and Spatial and Urban Planning:* Jean Claude Nduwayo. *Youth, Sports and Culture:* Adolphe Rukenkanya. *Minister in the Presidency in Charge of East African Community Affairs:* Léontine Nzeyimana. *Minister in the Presidency in Charge of Good Governance and Privatization:* Issa Ngendakumana.

Government Website (French only): http://www.burundi-gov.bi

CURRENT LEADERS

Pierre Nkurunziza

Position
President

Introduction
Pierre Nkurunziza became president in Aug. 2005 in the country's first democratic elections since the start of the civil war in 1993 and was re-elected in 2010. The key challenges facing the former leader of Burundi's largest ethnic Hutu rebel group have been the rebuilding of the economy, overseeing the repatriation of tens of thousands of refugees and maintaining relations with the Tutsi minority.

Early Life
Pierre Nkurunziza was born in Burundi's capital, Bujumbura, on 18 Dec. 1963. He attended primary school in Ngozi where, in 1972, his father was killed in ethnic violence that claimed over 100,000 lives. The family moved to Gitega, where Nkurunziza attended secondary school. He studied physical education at the University of Burundi in Bujumbura and became involved with the New Sporting football club as player and coach. Having graduated in 1990, he combined teaching at Muramvya High School with further studies in psychology and pedagogy. A year later he began lecturing in physical education at the country's leading military academy and at the University of Burundi.

Civil war followed the assassination in Oct. 1993 of Burundi's first ethnic Hutu president, Melchior Ndadaye, and spilled on to the University of Burundi campus in 1995 when 200 Hutu students were killed by Tutsi militia. Nkurunziza was reportedly shot at but escaped and joined the National Council for the Defense of Democracy–Forces for the Defense of Democracy (CNDD–FDD) as a soldier. The group was one of several rebel Hutu groups that fought the Tutsi-dominated army, in a conflict that killed thousands and had created an estimated 700,000 refugees by the late 1990s. In 1998 Nkurunziza was promoted to deputy secretary-general and co-ordinated the activities of the armed and political wings of the CNDD–FDD. In the same year he was sentenced to death by a Burundian court for alleged involvement in a series of ambushes, but was granted immunity during peace talks that culminated in the Arusha Peace Accord of Aug. 2000.

Elected chairman of the CNDD–FDD at its first congress in 2001, Nkurunziza began negotiations with Burundi's transitional government. In Nov. 2003 he signed a ceasefire accord, winning official recognition of the CNDD–FDD as a political party. Nkurunziza became state minister of good governance in the transitional government led by Domitien Ndayizeye. As such, he was a key figure in forging a power-sharing agreement and setting a timetable for democratic elections, ratified by heads of state from the Great Lakes region in Aug. 2004.

Following a series of CNDD–FDD victories in elections held in June and July 2005, Nkurunziza was nominated as the party's presidential candidate. He won an overwhelming victory in a vote by members of parliament (acting as an electoral college) on 19 Aug. and was sworn in as president on 26 Aug. 2005.

Career in Office
President Nkurunziza called for the Palipehutu–FNL opposition to lay down arms and rejoin negotiations. He appointed a cabinet of 20 ministers comprising 11 Hutus and nine Tutsis in accordance with the fixed quotas stipulated in the constitution. He announced that free primary education would be available to all children with immediate effect. He also promised to aid the return of Burundian refugees from Tanzania and Rwanda. In Aug. 2006 Nkurunziza's government claimed to have foiled a coup attempt, which prompted a crackdown on opposition and the arrest of several prominent political and military figures. The following month the government signed a ceasefire agreement with the Palipehutu–FNL rebel group which finally entered the peace process. In 2007, despite some residual violence attributed to the FNL, the United Nations completed its peacekeeping mandate in Burundi. In April 2008 a renewal of fighting between government forces and the FNL left around 100 people dead before another ceasefire was negotiated the following month. Then, in April 2009, the FNL officially disarmed and transformed into a legal political party.

Nkurunziza was re-elected overwhelmingly at presidential elections on 28 June 2010, but opposition members dismissed the legitimacy of the poll. In parliamentary elections in July that year, which were again boycotted by the opposition, the CNDD–FDD was returned with a large majority, prompting an increase in political tensions and a sporadic resurgence of violence across the country in 2011 and 2012.

DEFENCE

A new National Defence Force combining government forces and rebels from the FDD was created following the Oct. 2003 peace settlement.

Defence expenditure totalled US$83m. in 2008 (US$9 per capita), representing 7·5% of GDP.

Army
The Army had a strength (2007) of 35,000 including an air wing.

Air Force
There were 200 air wing personnel in 2007 with two combat-capable aircraft and two attack helicopters.

ECONOMY

Agriculture accounted for 42% of GDP in 2009; transport and communications, 15%; finance, public administration, defence and services, 12%; trade, 10%.

Overview
Being landlocked and among the poorest countries, Burundi is dependent on foreign aid. 90% of the population relies on agriculture for subsistence living. The economy is driven by coffee exports and, to a lesser extent, tea, sugar, cotton and hides. Ethnic tensions exacerbated by the civil war that ended in 2005 persist, particularly in light of the Tutsi minority effectively controlling the coffee trade.

GDP increased by around 4% per year from 2006–10, with the tertiary sector increasingly important. Under the Heavily Indebted Poor Countries Initiative, Burundi received US$832m. in debt relief in 2009. Vulnerability to shocks (especially food and fuel prices), a high debt burden, poor governance and poor infrastructure pose long-term challenges.

Currency
The unit of currency is the *Burundi franc* (BIF) of 100 *centimes*. Inflation was 26·0% in 2008, falling to 4·1% in 2010 before increasing to 14·9% in 2011. In July 2005 gold reserves were 1,000 troy oz and foreign exchange reserves US$88m. Total money supply was 139,353m. Burundi francs in Sept. 2004.

Budget
Total revenues in 2008 were 595bn. Burundi francs and expenditures 594bn. Burundi francs. Grants accounted for 50·9% of revenues in 2008; current expenditure accounted for 59·4% of expenditures.

VAT of 18% was introduced in July 2009, replacing the 17% transaction tax.

Performance
Real GDP growth was 3·8% in 2010 and 4·2% in 2011. Total GDP in 2011 was US$2·3bn.

Banking and Finance
The Bank of the Republic of Burundi is the central bank and bank of issue. Its *Governor* is Gaspard Sindayigaya. The largest commercial banks in 2007 were Interbank Burundi (assets of US$126m.), Banque de Crédit de Bujumbura (US$110m.) and Banque Commerciale du Burundi (US$67m.). Consolidated assets of financial institutions totalled 42·5bn. Burundi francs in 2009. In 2010 foreign debt totalled US$537m., representing 33·8% of GNI.

ENERGY AND NATURAL RESOURCES

Environment
Burundi's carbon dioxide emissions from the consumption and flaring of fossil fuels in 2008 were the equivalent of less than 0·1 tonnes per capita.

Electricity
Installed capacity was 33,000 kW in 2007. Production was 119m. kWh in 2007. Consumption per capita in 2007 was 24 kWh.

Minerals
Gold is mined on a small scale. Deposits of nickel (280m. tonnes) and vanadium remain to be exploited. There are proven reserves of phosphates of 17·6m. tonnes.

Agriculture
The main economic activity is agriculture, which contributed 35% of GDP in 2005. In 2007, 0·96m. ha. were arable and 0·35m. ha. permanent crops. 74,000 ha. were irrigated in 2001. There were 170 tractors in 2001. Beans, cassava, maize, sweet potatoes, groundnuts, peas, sorghum and bananas are grown according to the climate and the region.

The main cash crop is coffee, of which about 95% is arabica. It accounts for 90% of exports, and taxes and levies on coffee constitute a major source of revenue. Production (2003) 36,000 tonnes. The main agricultural crops (2002 production, in 1,000 tonnes) are bananas (1,603), sweet potatoes (833), cassava (750), dry beans (245), sugarcane (176) and maize (127).

Livestock (2002): 750,000 goats, 324,000 cattle, 230,000 sheep, 70,000 pigs and 4m. chickens.

Forestry
Forests covered 0·17m. ha., or 7% of the total land area, in 2010. Timber production in 2007 was 9·16m. cu. metres, the majority of it for fuel.

Fisheries
In 2008 the total catch was 17,766 tonnes, exclusively from inland waters.

INDUSTRY

In 2003 production of sugar totalled 20,000 tonnes. Other major products are (2003 output): beer (87·5m. litres), soft drinks (12·1m. litres), cigarettes (354m. units) and blankets (123,000 units).

Labour

In 2004 there were 3,335,000 employed persons.

INTERNATIONAL TRADE

With Rwanda and the Democratic Republic of the Congo, Burundi forms part of the Economic Community of the Great Lakes.

Imports and Exports

Imports and exports for calendar years in US$1m.:

	2006	2007	2008	2009	2010
Imports c.i.f.	433·6	423·0	315·2	344·8	832·5
Exports f.o.b.	228·5	156·2	141·8	112·9	275·5

Main exports are coffee, gold and tea. Leading imports are road vehicles, medicinal and pharmaceutical products, and cement. In 2010 main import sources (in US$1m.) were: Belgium (101·0); China (101·0); Japan (78·0). Leading export markets in 2010 (in US$1m.) were: Switzerland (74·2); UK (36·8); Belgium (35·8).

COMMUNICATIONS

Roads

In 2004 there were 12,322 km of roads of which 10·4% were paved. There were 15,500 passenger cars (two per 1,000 inhabitants) and 32,700 lorries and vans in use in 2007.

Civil Aviation

There were direct flights to Addis Ababa, Douala, Entebbe/ Kampala, Kigali and Nairobi in 2003. In 1998 scheduled airline traffic of Burundi-based carriers flew 800,000 km, carrying 12,000 passengers (all on international flights). Bujumbura International airport handled 86,353 passengers and 2,240 tonnes of freight in 2003.

Shipping

There are lake services from Bujumbura to Kigoma (Tanzania) and Kalémie (Democratic Republic of the Congo). The main route for exports and imports is via Kigoma, and thence by rail to Dar es Salaam.

Telecommunications

There were 32,600 fixed telephone lines in 2010 (3·9 per 1,000 inhabitants) and mobile phone subscribers numbered 1·15m. There were 21·0 internet users per 1,000 inhabitants in 2010. Fixed internet subscriptions totalled 5,000 in 2009 (0·6 per 1,000 inhabitants).

SOCIAL INSTITUTIONS

Justice

There is a Supreme Court, an appeal court and a court of first instance at Bujumbura, and provincial courts in each provincial capital.

The death penalty was abolished in April 2009. The population in penal institutions in May 2008 was 9,114 (104 per 100,000 of national population).

Education

Adult literacy rate was 66% in 2008. In 2007 there were 1,490,844 pupils in primary schools with 28,671 teachers and 209,945 pupils in secondary schools with 7,501 teachers. There were 15,623 students in higher education in 2007 and 1,007 academic staff. The leading institution in the tertiary sector is the Université du Burundi, located in Bujumbura.

In 2005 public expenditure on education came to 5·2% of GNI.

Health

In 2004 there were 200 physicians and 1,348 nursing and midwifery personnel. In 2006 there were seven hospital beds per 10,000 inhabitants (up from less than one per 10,000 in 1996).

RELIGION

In 2001 there were 4·05m. Roman Catholics with an archbishop and three bishops. About 3% of the population are Pentecostal, 1% Anglican and 1% Muslim, while the balance follow traditional tribal beliefs.

CULTURE

Press

There was one state-controlled daily newspaper (*Le Renouveau*) in 2008 with a circulation of 20,000.

Tourism

There were 201,000 foreign tourists in 2006 (148,000 in 2005).

DIPLOMATIC REPRESENTATIVES

Of Burundi in the United Kingdom (Uganda House, Second Floor, 58–59 Trafalgar Square, London, WC2N 5DX)
Ambassador: Vacant.
Chargé d'Affaires a.i.: Bernard Ntahiraja.

Of the United Kingdom in Burundi
Ambassador: Benedict Llewellyn Jones, OBE (resides in Kigali, Rwanda).

Of Burundi in the USA (2233 Wisconsin Ave., NW, Suite 212, Washington, D.C., 20007)
Ambassador: Angele Niyuhire.

Of the USA in Burundi (PO Box 1720, Ave. des Etats-Unis, Bujumbura)
Ambassador: Dawn M. Liberi.

Of Burundi to the United Nations
Ambassador: Herménégilde Niyonzima.

Of Burundi to the European Union
Ambassador: Balthazar Bigirimana.

FURTHER READING

Lemarchand, R., *Burundi: Ethnic Conflict and Genocide.* 1996
Melson, Robert, *Genocide and Crisis in Central Africa: Conflict Roots, Mass Violence and Regional War.* 2001

National Statistical Office: Service des Études et Statistiques, Ministère du Plan, B. P. 1156, Bujumbura.

CAMBODIA

© Research Machines plc 2006

Preah Reach Ana Pak Kampuchea (Kingdom of Cambodia)

Capital: Phnom Penh
Population projection, 2015: 15·02m.
GNI per capita, 2011: (PPP$) 1,848
HDI/world rank: 0·523/139
Internet domain extension: .kh

KEY HISTORICAL EVENTS

Neolithic communities, probably linked to migration from southeast China, were established by 1000 BC in the Kompong Cham province of eastern Cambodia. From around 300 BC the Indianized Funan kingdom held sway across much of present-day Cambodia with trading links to China, India, the Middle East and Rome. The state of Chenla broke away from Funan control during the 6th century and over the next 300 years its influence spread to western Cambodia, central Laos and northern Thailand. Cambodia's southern coast came under Javanese control in the eighth century, forcing Khmer-speaking groups inland. The crowning of Jayavarman II as a deva-raja (or god king) in 802 heralded a long period of regional Khmer domination centred around Angkor.

The Indian-influenced civilization prospered with the development of agriculture and trade, notably under King Yasovarman I around AD 900 and during the reign of Suryavarman II (1113–50), when the temple complex at Angkor Wat was constructed. The empire was attacked by the Islamic Kingdom of Champa (now in central Vietnam) in the 12th and 13th centuries and was subsequently vulnerable to Thai incursions from Ayudhya. Thai forces sacked Angkor in 1432, ushering in a period of dominance by the kings of Siam on Cambodia's western frontier and pressure from Annam to the east. Cambodian nobility established a capital at Phnom Penh in the 16th century, although its influence was limited and the city was sacked by Rama I of Siam in 1772. In 1811–12 Siamese and Vietnamese forces fought for control of Cambodia, with the kingdom coming under Vietnam's control in 1835–40. King Ang Duong wrested back control in 1848 and appealed for military assistance from

France, a major naval force in the region with bases in southern Vietnam. Under King Norodom I, a French protectorate was established in 1863, although resentment grew as French control extended beyond foreign affairs and defence. Rebellions were suppressed in the 1870s and 1880s and Cambodia became part of the Union of Indochina in 1887.

Anti-French feeling strengthened in 1940–41 when the Vichy French submitted to Japanese demands for bases in Cambodia. Son Ngoc Thanh was one of the leaders of a nascent nationalist movement after French control was reasserted in 1945. Having returned to Cambodia in 1951, he joined the Khmer Issarak guerrillas to fight the French. Independence was achieved on 9 Nov. 1953, the result of deft diplomacy by King Norodom Sihanouk and France's increasingly weak position in a war against Vietnam's communist revolutionaries. Sihanouk abdicated in 1955 to form the popular Socialist Party which dominated the general election of that year. He served as the country's leader until 1970 when he was deposed by the US-backed Lon Nol amid growing communist incursions from North Vietnam, a rapidly deteriorating economy and political corruption.

The country's name was changed to the Khmer Republic in Oct. 1970. US forces attacked Cambodia's communist strongholds, which led to the rise of the Khmer Rouge. In 1973 direct US involvement came to an end, precipitating a civil war between the Khmer Republic and the United National Cambodian Front (including the Khmer Rouge), supported by North Vietnam and China. After unsuccessful attempts to capture Phnom Penh in 1973 and 1974, the Khmer Rouge, led by Pol Pot, overthrew Lon Nol's government in April 1975.

Renaming the country Democratic Kampuchea, Pol Pot instituted a harsh and highly centralized regime. All cities and towns were forcibly evacuated and citizens set to work in the fields. Intellectuals, members of the professional classes and people identified as enemies of the Khmer Rouge were murdered in their hundreds of thousands and many more died as a result of disease, malnutrition and overwork, resulting in the loss of an estimated 2m. Cambodian lives from 1975–79. In response to repeated border attacks, Vietnam invaded Cambodia in 1978. On 7 Jan. 1979 Phnom Penh was captured by the Vietnamese and Pol Pot fled. During the 1980s the country was destabilized by warring factions fighting both the Vietnamese and the Khmer Rouge. On 23 Oct. 1991 an agreement was signed in Paris by the warring factions and 19 countries, instituting a UN-monitored ceasefire.

A new constitution was promulgated in 1993 restoring parliamentary monarchy. The Khmer Rouge continued hostilities, refusing to take part in the 1993 elections, and by 1996 had split into two factions. The leader of one faction, Ieng Sary, had been sentenced to death *in absentia* for genocide but was pardoned in Sept. 1996. In Nov. 1996 Ieng Sary and 4,000 of his troops joined with government forces. Prince Norodom Ranariddh, styled as First Prime Minister of the Royal Government, was exiled in July 1997 and coup-leader Hun Sen appointed himself prime minister. In March 1998 Ranariddh returned with a Japanese-brokered plan to ensure 'fair and free' elections, which took place in July 1998. Against a background of violence Hun Sen's Cambodian People's Party (KPK) declared victory.

King Norodom Sihanouk abdicated in Oct. 2004 for health reasons and was succeeded by his son, Norodom Sihamoni. In July 2007 UN-backed tribunals investigated allegations of genocide by the Khmer Rouge. The KPK claimed victory in the parliamentary elections of July 2008 and announced that it would remain in coalition with the depleted Royalist FUNCINPEC

(National United Front for an Independent, Neutral, Peaceful and Co-operative Cambodia).

TERRITORY AND POPULATION

Cambodia is bounded in the north by Laos and Thailand, west by Thailand, east by Vietnam and south by the Gulf of Thailand. It has an area of about 181,035 sq. km (69,898 sq. miles).

Population, 13,395,682 (2008 census), of whom 6,879,628 were females; density, 74·0 per sq. km. In 2011, 20·4% of the population lived in urban areas.

The UN gives a projected population for 2015 of 15·02m.

The capital, Phnom Penh, had a population of 1,242,992 in 2008. Other cities are Battambang and Siem Reap. Khmers make up around 90% of the population. There are also Vietnamese, Chinese and ethnic hill tribes.

Khmer is the official language.

SOCIAL STATISTICS

2008 estimated births, 360,000; deaths, 121,000. Rates, 2008 estimates (per 1,000 population): births, 24·7; deaths, 8·3. Infant mortality, 2010 (per 1,000 live births), 43. Expectation of life in 2007 was 58·6 years for males and 62·3 for females. Annual population growth rate, 2000–08, 1·7%. Fertility rate, 2008, 2·9 children per woman. Cambodia has had one of the largest reductions in its fertility rate of any country in the world over the past quarter of a century, having had a rate of 5·8 births per woman in 1990.

CLIMATE

A tropical climate, with high temperatures all the year. Phnom Penh, Jan. 78°F (25·6°C), July 84°F (28·9°C). Annual rainfall 52″ (1,308 mm).

CONSTITUTION AND GOVERNMENT

A parliamentary monarchy was re-established by the 1993 constitution. King Norodom Sihamoni (b. 14 May 1953; appointed 14 Oct. 2004 and sworn in on 29 Oct. 2004) was chosen in the first ever meeting of the nine-member Throne Council following the abdication of his father King Norodom Sihanouk (b. 31 Oct. 1922) on health grounds. As the Cambodian constitution allowed for a succession only in the event of the monarch's death, a new law had to be approved after King Norodom Sihanouk announced his abdication.

Cambodia has a bicameral legislature. There is a 123-member *National Assembly*, which on 14 June 1993 elected Prince Sihanouk head of state. On 21 Sept. it adopted a constitution (promulgated on 24 Sept.) by 113 votes to five with two abstentions making him monarch of a parliamentary democracy. Its members are elected by popular vote to serve five-year terms. There is also a 61-member *Senate*, established in 1999.

National Anthem

'Nokoreach' ('Royal kingdom'); words by Chuon Nat, tune adapted from a Cambodian folk song.

RECENT ELECTIONS

Parliamentary elections were held on 27 July 2008; turnout was 75·2%. The Cambodian People's Party (KPK) won 90 seats with 58·1% of the vote, the party of government critic Sam Rainsy won 26 seats with 21·9%, the Human Rights Party won three seats with 6·6%, the party of Prince Norodom Ranariddh two seats with 5·6% and the royalist FUNCINPEC party two seats with 5·1%. In the meantime the Sam Rainsy Party and the Human Rights Party have merged to form one large opposition party, known as the National Salvation Party.

Parliamentary elections were scheduled to take place on 28 July 2013.

CURRENT GOVERNMENT

In March 2013 the government comprised:

Prime Minister: Hun Sen; b. 1951 (KPK; sworn in 30 Nov. 1998 and reappointed 14 July 2004 and 25 Sept. 2008 having first become prime minister in 1985).

Deputy Prime Ministers: Men Sam An; Sar Kheng (also *Minister of Internal Affairs*); Sok An (also *Minister in Charge of the Office of the Council of Ministers*); Gen. Tea Banh (also *Minister of Defence*); Hor Nam Hong (also *Minister of Foreign Affairs and International Co-operation*); Keat Chhon (also *Minister of Economy and Finance*); Bin Chhin; Nhek Bunchhay; Yim Chhai Ly.

Senior Ministers: Cham Prasidh (also *Minister of Commerce*); Dr Mok Mareth (also *Minister of Environment*); Chhay Than (also *Minister of Planning*); Im Chhun Lim (also *Minister of Land Management, Urban Affairs and Construction*); Nhim Vanda; Tao Seng Huor; Khun Haing; Ly Thuch; Kol Pheng; Sun Chanthol; Veng Sereyvuth; Nuth Sokhom; Om Yentieng; Ieng Mouly; Va Kimhong; Yim Nol La.

Minister of Agriculture, Forestry and Fisheries: Chan Sarun. *Industry, Mines and Energy:* Suy Sem. *Social Affairs, War Veterans and Youth Rehabilitation:* Ith Sam Heng. *Water Resources and Meteorology:* Lim Kean Hor. *Information:* Khieu Kanharith. *Justice:* Ang Vong Vathana. *Post and Telecommunications:* So Khun. *Health:* Mam Bunheng. *Culture and Fine Arts:* Him Chhem. *Tourism:* Thong Khon. *Women's Affairs:* Ing Kantha Phavi. *Labour and Vocational Training:* Vong Sauth. *Rural Development:* Chea Sophara. *Parliamentary Affairs and Inspection:* Sam Kim Suor. *Religious Affairs:* Min Khin. *Education, Youth and Sports:* Im Sithy. *Public Works and Transport:* Tram Iv Tek.

Government Website: http://www.pressocm.gov.kh

CURRENT LEADERS

Hun Sen

Position
Prime Minister

Introduction
Hun Sen has been the dominant figure in Cambodian politics since becoming prime minister in 1985. His tenure has coincided with national recovery from the rule of the Khmer Rouge in the 1970s and the subsequent occupation by Vietnamese forces. He has been criticized for the repressive tactics he has used to retain power. The economy has suffered from endemic corruption and a failure to attract foreign investment. However, Hun Sen has moved the economy away from state socialism towards one based on free market principles. A UN-backed tribunal to investigate charges of genocide against prominent Khmer Rouge figures may conclude that turbulent period in the nation's history.

Early Life
Hun Sen was born into a peasant family in Kompang in 1951 and was educated in Phnom Penh by Buddhist monks. In 1970 he joined the Khmer Rouge, losing an eye in battle in 1975, but in 1977 he joined anti-Khmer Rouge forces operating out of Vietnam. After Vietnamese troops invaded Kampuchea in 1979, Hun Sen was appointed foreign minister in the newly-established People's Republic of Kampuchea. He became prime minister in 1985.

Career in Office
On assuming office, Hun Sen was confronted with a country in turmoil as pro- and anti-Vietnamese forces waged guerrilla war. In 1989 Vietnamese forces withdrew from the country, which was renamed the State of Cambodia. Buddhism was re-established as the state religion and Hun Sen announced the end of state socialism. In 1991 the UN brokered a peace treaty and a transitional government was installed, with Prince Sihanouk as head of state.

FUNCINPEC, the royalist party of Prince Norodom Ranariddh, was victorious at the 1993 elections but Hun Sen refused to relinquish power. A compromise government was formed with Ranariddh as first prime minister and Hun Sen as his deputy. In 1997 Hun Sen received international condemnation and saw Cambodia expelled from ASEAN for deposing Ranariddh while he was absent from the country. The Cambodian People's Party (KPK) won the 1998 elections but did not gain enough seats to form a government. Hun Sen agreed to head a coalition that included FUNCINPEC. Ranariddh was found guilty *in absentia* of arms smuggling but received a royal pardon and was named president of the National Assembly. Pol Pot died that year, having been sentenced to life imprisonment for his crimes the previous year.

In 2001 Prince Norodom Sirivudh became leader of the opposition FUNCINPEC party, having received a royal pardon in 1999 for a ten-year prison sentence he received *in absentia* for his alleged involvement in an assassination plot against Hun Sen. In the same year the east and west of the country were connected for the first time by a bridge across the Mekong River.

In 2002 Thailand agreed to extradite Sok Yoeun, a leading figure in the Cambodian Sam Rainsy Party, for alleged involvement in an assassination plot against Hun Sen. Sok Yoeun declared that the charges were part of a political attack and Amnesty International classified him as a prisoner of conscience.

In 2002 the KPK was the dominant party at the country's first multi-party local elections, although opposition groups claimed ballots were rigged. At the general elections of 2003 the KPK emerged as the single biggest party but fell short of the two thirds majority required to form the government. Coalition talks were frosty and relations were not improved when Hun Sen removed 17 FUNCINPEC politicians from senior posts, claiming neglect of duty. The opposition FUNCINPEC and Sam Rainsy parties agreed to join a ruling coalition but not under Hun Sen. The king finally confirmed Hun Sen as head of government in July 2004, after almost a year without a properly functioning administration.

After winning senate approval, Cambodia began talks with the UN on a tribunal to try former Khmer Rouge leaders for genocide. Negotiations were protracted and in 2003 the UN concluded that the tribunal was unlikely to function alongside Cambodia's existing judicial system, which claimed precedence over international law. The issue was further complicated by Khmer Rouge leaders transferring allegiance to the government in the years following Pol Pot's fall from power. Hun Sen himself was a soldier in the Khmer Rouge, though he denied claims that he ever held a senior position. Nevertheless, the UN-backed tribunal, run by Cambodian and international judges, finally held its first public hearing in Nov. 2007 and achieved its first conviction (of a notorious prison commandant) in July 2010. Preliminary hearings against three senior Khmer Rouge figures began in June 2011 and their trial started in Nov.

In 2003 a Thai celebrity suggested that the religious complex of Angkor Wat had been stolen by Cambodia from Thailand. The comments caused outrage in Cambodia and led to a siege of the Thai embassy in Phnom Penh. Observers accused Hun Sen of aggravating the situation. In 2004 Hun Sen threatened to boycott a joint Asian-European summit if, as the European Union was requesting, Myanmar was banned from attending. In the same year, with economic growth and foreign investment falling, Hun Sen outlined proposals to cut business costs, reduce red tape and counter corruption. He also promised to increase civil service salaries. However, he did not offer a timeframe for these developments. Entry into the WTO was ratified in Aug. 2004 after long delays.

Other significant problems confronting Hun Sen's government have included deforestation and the spread of AIDS. In Aug.–Sept. 2006 he pushed two controversial measures through the National Assembly. The first curtailed the right of parliamentarians to speak openly without fear of prosecution and the second imposed a ban on adultery.

Following public criticism by Hun Sen, Prince Ranariddh resigned as president of the National Assembly and in March 2007 was sentenced in absentia to 18 months imprisonment for breach of trust over the sale of the FUNCINPEC headquarters.

On 26 Sept. 2008 Hun Sen was reappointed prime minister after the KPK had secured over two-thirds of the seats in the National Assembly in elections in July that were criticized by EU monitors as falling short of international standards.

In Oct. 2008 tensions between Cambodia and Thailand worsened as border skirmishes resulted in the deaths of two Cambodian soldiers in a disputed area around Preah Vihear, an ancient temple and (since July 2008) a UNESCO World Heritage site. Exchanges of gunfire on the border were reported in the spring of 2009 and again in Feb. and April 2011 before the International Court of Justice ruled in July that both sides should withdraw their forces from the area. There had earlier been diplomatic tensions when a former Thai prime minister, Thaksin Shinawatra, who was accused of corruption by the Thai government, became an economic adviser to Hun Sen's administration. However, in Aug. 2010 it was reported that Thaksin had relinquished his role and that the two governments had normalized their relations.

In Sept. 2010 exiled opposition leader Sam Rainsy was sentenced *in absentia* to ten years imprisonment by a Cambodian court for disinformation and manipulating public documents.

DEFENCE

The King is C.-in-C. of the Royal Cambodian Armed Forces (RCAF). Conscription has not been implemented since 1993 although it is authorized. Defence expenditure in 2008 totalled US$225m. (US$18 per capita), representing 2·3% of GDP.

Army

Strength in 2007 was an estimated 75,000. There are also provincial forces numbering some 45,000 and paramilitary local forces organized at village level.

Navy

Naval personnel in 2007 totalled about 2,800 including a naval infantry of 1,500. In 2007 there were two coastal fast patrol craft, two riverine patrol craft and six patrol boats.

Air Force

Aviation operations were resumed in 1988, initially under the aegis of the Army and since 1993 as part of the RCAF. Personnel (2007), 1,500. There are 24 combat-capable aircraft but serviceability is in doubt.

ECONOMY

Agriculture accounted for 36% of GDP in 2009, industry 23% and services 41%.

Overview

Cambodia is among the poorest countries outside of Africa, with per capita income of US$760 in 2010. Efforts to reduce poverty have been frustrated by low agricultural productivity, inadequate infrastructure and a stifling regulatory environment.

The economy was devastated by the Khmer Rouge regime of 1975–79, when trade collapsed to an agrarian barter system, the industrial base was destroyed, and the banking system and domestic currency abolished. Economic progress remained slow until the Paris Peace Accord of 1991 brought about a ceasefire in the civil war. At the time of signing Cambodia faced rapid inflation, exchange rate depreciation, monetary instability, negative real interest rates and high fiscal deficits. Since the Accord, and with the aid of the World Bank and the IMF, the government has tried to restore monetary stability and improve fiscal performance. Privatization of state-owned enterprises was completed by 1996 and the trade regime was liberalized as Cambodia joined ASEAN and prepared for WTO accession.

From 2000–07 growth averaged over 9%, thanks to expansion in agriculture, buoyant exports and a healthy urban economy underpinned by construction, services and the financial sector. The overall poverty rate declined from 47% in 1993 to 30% in 2007 although inequality increased, largely because of the failure to address persistent rural poverty. The global financial crisis resulted in minimal growth in 2009, but the economy rebounded in 2010 with growth of 6%, led by textiles and tourism. Garment exports to the USA and Europe account for 40% of total exports, leaving Cambodia exposed to fluctuations in the global economy. To secure long term prosperity the government must address high levels of corruption, low fiscal revenues, the excessively dollarized economy and a narrow export base.

Currency

The unit of currency is the *riel* (KHR) of 100 *sen*. In July 2005 total money supply was 1,222·4bn. riels, foreign exchange reserves were US$937m. and gold reserves 400,000 troy oz. Inflation was 25·0% in 2008 but there was deflation of 0·7% in 2009. There was then inflation of 4·0% in 2010 and 5·5% in 2011.

Budget

In 2007 revenues were 3,280bn. riels and expenditures 3,295bn. riels. Tax revenue accounted for 58·3% of revenues in 2007; current expenditure accounted for 59·7% of expenditures.

VAT is 10%.

Performance

The economy grew by 0·1% in 2009, 6·1% in 2010 and 7·1% in 2011. Total GDP in 2011 was US$12·8bn.

Banking and Finance

Banking and money were banned during the rule of the Khmer Rouge from 1975–79. The National Bank of Cambodia (*Governor*, Chea Chanto), which was founded in 1954 and re-established in Oct. 1979 after the fall of the Khmer Rouge, is the bank of issue. In 2011 there were 31 commercial banks of which 22 were local incorporated banks and nine were foreign bank branches, seven specialized banks and 32 licensed microfinance institutions. The largest banks in Dec. 2011 were Acleda Bank (with assets of US$1·49bn.), Canadia Bank (assets of US$1·30bn.) and Cambodian Public Bank (assets of US$1·01bn.). The largest specialized bank is the Rural Development Bank.

In 2010 external debt totalled US$4,676m., equivalent to 43·4% of GNI.

A stock exchange opened in Phnom Penh in July 2011 but only started trading in April 2012.

ENERGY AND NATURAL RESOURCES

Environment

Carbon dioxide emissions from the consumption and flaring of fossil fuels were the equivalent of less than 0·3 tonnes per capita in 2008.

Electricity

Installed capacity was an estimated 382,000 kW in 2007. Production (2007) was 1,349m. kWh; consumption per capita was 106 kWh.

Oil and Gas

Oil was discovered in 2005 off the coast of Cambodia. The reserves are estimated to total at least 400m. bbls. Construction of the first oil refinery is expected to begin in the course of 2013.

Minerals

There are phosphates and high-grade iron-ore deposits. Some small-scale gold panning and gem (mainly zircon) mining is carried out.

Agriculture

The majority of the population is engaged in agriculture, fishing or forestry. Before the spread of war in the 1970s the high productivity provided for a low but well-fed standard of living for the peasant farmers, the majority of whom owned the land they worked before agriculture was collectivized. A relatively small proportion of the food production entered the cash economy. The war and unwise pricing policies led to a disastrous reduction in production, so much so that the country became a net importer of rice. Private ownership of land was restored by the 1989 constitution. In 2007 there were around 3·80m. ha. of arable land and 155,000 ha. of permanent crops.

A crop of 3·82m. tonnes of rice was produced in 2002. Production of other crops, 2002 (in 1,000 tonnes): sugarcane, 209; maize, 149; bananas, 146; cassava, 122; coconuts, 70; oranges, 63.

Livestock (2002): cattle, 2·92m.; pigs, 2·11m.; buffaloes, 626,000; poultry, 17m.

Livestock products, 2003 (in 1,000 tonnes): pork, bacon and ham, 105; beef and veal, 54; poultry meat, 28; buffalo meat, 13; eggs, 17; milk, 20.

Forestry

Some 10·09m. ha., or 57% of the land area, were covered by forests in 2010. Timber exports have been banned since Dec. 1996. There are substantial reserves of pitch pine. Rubber plantations are a valuable asset with production at around 40,000 tonnes per year. Timber production in 2007 was 9·00m. cu. metres.

Fisheries

2010 catch was approximately 490,000 tonnes (mainly from inland waters).

INDUSTRY

Some development of industry had taken place before the spread of open warfare in 1970, but little was in operation by the 1990s except for rubber processing, sea-food processing, jute sack making and cigarette manufacture. Garment manufacture, rice milling, wood and wood products, rubber, cement and textiles production are the main industries. In the private sector small family concerns produce a wide range of goods. Light industry is generally better developed than heavy industry.

Labour

In 2004 the labour force was 7,496,000. Females constituted 51·8% of the labour force in 2001—among the highest proportions of women in the workforce in the world. More than 66% of employed persons are engaged in agriculture, fishing and forestry.

INTERNATIONAL TRADE

Foreign investment has been encouraged since 1989. Legislation of 1994 exempts profits from taxation for eight years, removes duties from various raw and semi-finished materials and offers tax incentives to investors in tourism, energy, the infrastructure and labour-intensive industries.

Imports and Exports

In 2010 imports (c.i.f.) totalled US$4,902·5m. (US$3,905·7m. in 2009); exports (f.o.b.), US$5,590·1m. (US$4,992·0m. in 2009).

Main imports are: textile yarn, fabrics and finished articles; road vehicles; specialized machinery; gold. Main exports are: textile yarn, fabrics and finished articles; printed matter; specialized machinery; road vehicles.

Major import sources, 2010: China (24·2%), Thailand (14·1%) and Hong Kong (11·3%). Principal export destinations, 2010: USA (34·1%), Hong Kong (24·8%) Singapore (7·7%).

COMMUNICATIONS

Roads

There were 8,257 km of roads in 2004, of which 6·3% were paved. In 2005 there were 195,300 passenger cars in use plus 3,200 buses and coaches, 32,100 lorries and vans and 566,300 motorcycles and mopeds. There were 1,545 fatalities in road accidents in 2007.

Rail
All official rail services had been suspended by 2009 owing to the dilapidated state of the 600-km metre gauge network. However, a rehabilitation project began in 2006 and freight services were resumed in Oct. 2010 between Phnom Penh and Touk Meas. Some passenger services have been restored and plans are under way to build a 255-km extension to link the country to Vietnam by 2015.

Civil Aviation
Phnom Penh International Airport airport handled 1,587,986 passengers in 2009 and Siem Reap International Airport 1,255,166. The flag carrier is Cambodia Angkor Air (51% state-owned), which began services in 2009. The former national airline, Royal Air Cambodge, had gone bankrupt in 2001.

Shipping
There is an ocean port at Kompong Som; the port of Phnom Penh can be reached by the Mekong (through Vietnam) by ships of between 3,000 and 4,000 tonnes. In Jan. 2009 there were 730 ships of 300 GT or over registered, totalling 1,966,000 GT.

Telecommunications
In 2009 Cambodia had just 54,200 main (fixed) telephone lines, but there were 6,268,000 mobile phone subscribers (448·4 for every 1,000 persons). Cambodia has among the highest ratios of mobile phone subscriptions to fixed telephone lines. There were 74,000 internet users in 2008 (5·1 for every 1,000 persons). In March 2012 there were 449,000 Facebook users.

SOCIAL INSTITUTIONS

Justice
The population in penal institutions in 2007 was 10,337 (71 per 100,000 of national population). The death penalty was abolished in 1989.

Education
In 2004–05 there were 6,990 schools of which 6,180 were primary, 578 junior high and 232 senior high. There were 111,090 pupils in pre-primary schools and 4,415 teaching staff in 2007; 2,479,644 pupils and 48,736 teaching staff in primary schools; and in general secondary education 30,258 teaching staff for 875,120 pupils. There were 92,340 students in tertiary education in 2007 and (2006) 3,261 academic staff. Adult literacy in 2008 was 78%.

In 2007 public expenditure on education came to 1·7% of GNI and 12·4% of total government spending.

Health
In 2000 there were 2,047 physicians, 209 dentists, 8,085 nurses and 3,040 midwives. Only 30% of the population had access to safe drinking water in 2000.

RELIGION
The constitution of 1989 reinstated Buddhism as the state religion; it had 10·8m. adherents in 2001. About 2,800 monasteries were active in 1994. There are small Roman Catholic and Muslim minorities.

CULTURE

World Heritage Sites
There are two UNESCO world heritage sites in Cambodia: Angkor (inscribed in 1992), an archaeological park that was the site of various capitals of the Khmer Empire from the 9th to the 15th centuries containing the Temple of Angkor Wat and the Bayon Temple at Angkor Thom; and the Temple of Preah Vihear (2008), an outstanding example of Khmer architecture that dates back to the first half of the 11th century and is composed of a series of sanctuaries linked by a system of pavements and staircases.

Press
There were 22 paid-for daily newspapers in 2008 with a combined circulation of 60,000, including the English-language *Cambodia Daily*.

Tourism
In 2010 there were 2,399,000 international tourist arrivals (excluding same-day visitors), up from 2,046,000 in 2009 and 2,001,000 in 2008.

DIPLOMATIC REPRESENTATIVES

Of Cambodia in the United Kingdom (64 Brondesbury Park, London, NW6 7AT)
Ambassador: Hor Nambora.

Of the United Kingdom in Cambodia (27–29 St. 75, Phnom Penh)
Ambassador: Mark Gooding.

Of Cambodia in the USA (4530 16th St., NW, Washington, D.C., 20011)
Ambassador: Hem Heng.

Of the USA in Cambodia (1 St. 96, Phnom Penh)
Ambassador: William E. Todd.

Of Cambodia to the United Nations
Ambassador: Sea Kosal.

Of Cambodia to the European Union
Ambassador: Hem Saem.

FURTHER READING
Chandler, D. P., *A History of Cambodia*. 4th ed. 2007
Etcheson, Craig, *After the Killing Fields: Lessons from the Cambodian Genocide*. 2005
Gottesman, Evan R., *Cambodia After the Khmer Rouge: Inside the Politics of Nation Building*. 2004
Peschoux, C., *Le Cambodge dans la Tourmente: le Troisième Conflit Indochinois, 1978–1991*. 1992.—*Les 'Nouveaux' Khmers Rouges*. 1992
Short, Philip, *Pol Pot: The History of a Nightmare*. 2004

National Statistical Office: National Institute of Statistics, Ministry of Planning, 386 Preah Monivong Blvd, Boeung Keng Kong 1, Phnom Penh.
Website: http://www.nis.gov.kh

CAMEROON

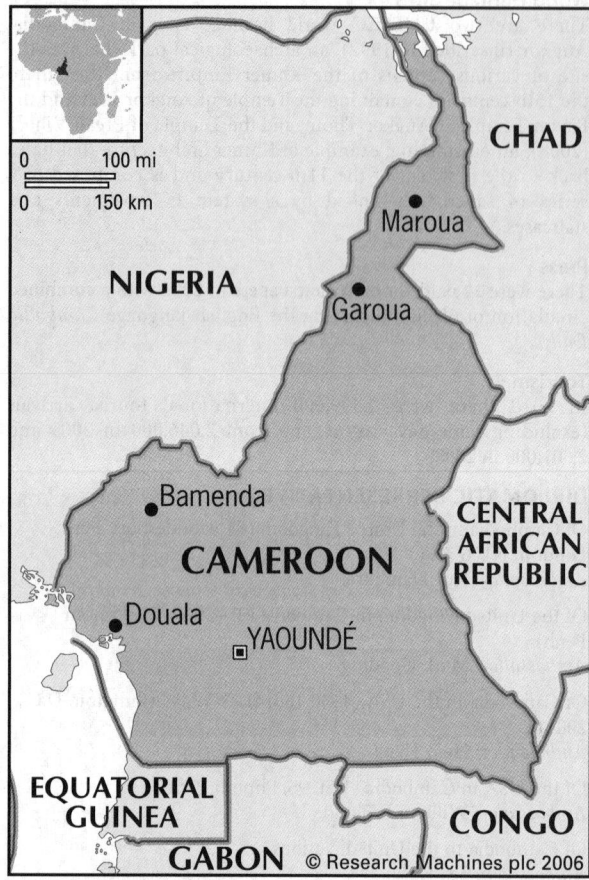

République du Cameroun
(Republic of Cameroon)

Capital: Yaoundé
Population projection, 2015: 21·82m.
GNI per capita, 2011: (PPP$) 2,031
HDI/world rank: 0·482/150
Internet domain extension: .cm

KEY HISTORICAL EVENTS

Neolithic settlements at Shum Laka and Abeke date from between 5000 and 4000 BC. From 500 BC Bantu-speaking farmers originating in central Cameroon migrated to the east and south and, over the next 2,000 years, dispersed across the African continent. The Sao civilization developed in the far north, close to Lake Chad, from around 600 AD. It developed ties with the Kanem kingdom, which for a time controlled trans-Saharan trade.

By the 14th century Sao was reduced to small, scattered settlements under the control of the Kotoko state, itself comprised of several kingdoms across northern Cameroon. The Duala, mainly located in the coastal lowlands, made contact with Portuguese, Dutch and British mariners during the 15th and 16th centuries, with Fernando Gomes reaching the Wouri river in 1472 and naming it Rio de Cameroes (or 'shrimp river').

The Duala acted as middlemen between European explorers and tribes in the interior, trading ivory, palm oil and slaves. The port of Bimbia and its environs became a slave-trading centre, with hundreds of thousands shipped to plantations in São Tomé e Príncipe, Equatorial Guinea, the Caribbean and the Americas.

By the early 1800s the British controlled much of the coastal trade, though its influence inland was limited. The north was increasingly Islamic following the arrival of Fulani settlers from the northwest, led by Uthman Dan Fodio. In 1858 an English missionary and engineer, Alfred Saker, founded the colony of Victoria (Limbe), reliant on gold, ivory and palm oil after the demise of the slave trade.

British colonial operations centred on Lagos were supplanted by German influence in 1884 when a treaty between the Duala and Gustav Nachtigal (on behalf of Kaiser Wilhelm) created the protectorate of Kamerun. The port of Kamerunstadt (Douala) was developed and roads, bridges and settlements (including the future capital, Yaoundé) were established inland using forced labour.

In 1915 a combined force of British, French and Belgian troops defeated the German administrators. After the First World War the territory was governed briefly by Britain and France, before partition. The British sector encompassed two small provinces along the border with Nigeria (the North and South Cameroons), administered from Lagos, while the French sector (Cameroun) aligned with most of the former German colony. Both areas became mandated territories under the League of Nations in 1922 and, after the Second World War, Trust Territories of the newly established UN.

In French Cameroun the Union des Populations du Cameroun (UPC) was founded in 1948 to call for independence and 'reunification' with British Cameroons. In 1955 uprisings organized by the UPC broke out in several towns. From 1958 more moderate parties, such as l'Union Camerounaise of Ahmadou Ahidjo, were permitted to work towards independence within the French administration. This was achieved on 1 Jan. 1960 when Ahidjo was inaugurated as president. In a referendum in Feb. 1961 British Southern Cameroons voted to join the new federal republic of Cameroon, with British Northern Cameroons joining Nigeria.

Re-elected in 1965 and 1970, Ahidjo became increasingly authoritarian and by 1976 opposition parties were outlawed. He stepped down, apparently for health reasons, in 1982, to be succeeded by Paul Biya, a Christian from the south. Cameroon faced economic hardship in the late 1980s linked to the depletion of oil reserves, sparking demands for democracy.

Biya narrowly won the first multi-party elections in Oct. 1992. Tensions with Nigeria over the potentially oil-rich Bakassi peninsula erupted in 1994, with flare-ups continuing until Nigeria retracted its claims in Aug. 2008. Biya won presidential elections in 1997, 2004 and 2011, although opposition parties rejected the results on each occasion amid claims of widespread fraud.

TERRITORY AND POPULATION

Cameroon is bounded in the west by the Gulf of Guinea, northwest by Nigeria, east by Chad and the Central African Republic, and south by the Republic of the Congo, Gabon and Equatorial Guinea. The total area is 475,650 sq. km (land area, 466,050 sq. km). In 1994 Cameroon asked the International Court of Justice to confirm its sovereignty over the oil-rich Bakassi peninsula, occupied by Nigerian troops. The dispute continued for eight years, with Equatorial Guinea also becoming involved. In Oct. 2002 the Court rejected Nigeria's claims and awarded Bakassi to Cameroon. All parties agreed to accept the Court's judgment. At the 2005 census the population was 17,463,836 (50·6% female); Jan. 2010 estimate, 19,406,100, giving a density of 41·6 per sq. km.

The UN gives a projected population for 2015 of 21·82m.

In 2011, 59·2% of the population were urban.

The areas, estimated populations and chief towns of the ten provinces are:

Province	Sq. km	Census 2005	Chief town	Census 2005
Adamaoua	63,701	884,289	Ngaoundéré	152,698
Centre	68,953	3,098,044	Yaoundé	1,817,524
Est	109,002	771,755	Bertoua	88,462
Extrême-Nord	34,263	3,111,792	Maroua	201,371
Littoral	20,248	2,510,263	Douala	1,906,962
Nord (Bénoué)	66,090	1,687,959	Garoua	235,996
Nord-Ouest	17,300	1,728,953	Bamenda	269,530
Ouest	13,892	1,720,047	Bafoussam	239,287
Sud	47,191	634,655	Ebolowa	64,980
Sud-Ouest	25,410	1,316,079	Buéa	90,090

The population is composed of Sudanic-speaking people in the north (Fulani, Sao and others) and Bantu-speaking groups, mainly Bamileke, Beti, Bulu, Tikar, Bassa and Duala, in the rest of the country. The official languages are French and English.

SOCIAL STATISTICS

2008 estimates: births, 704,000; deaths, 271,000. Rates, 2008 estimates (per 1,000 population): birth, 36·9; death, 14·2. Annual population growth rate, 2000–08, 2·3%. Infant mortality, 2010, 84 per 1,000 live births. Life expectancy in 2007: males, 50·3 years; females, 51·4. Fertility rate, 2008, 4·6 children per woman.

CLIMATE

An equatorial climate, with high temperatures and plentiful rain, especially from March to June and Sept. to Nov. Further inland, rain occurs at all seasons. Yaoundé, Jan. 76°F (24·4°C), July 73°F (22·8°C). Annual rainfall 62" (1,555 mm). Douala, Jan. 79°F (26·1°C), July 75°F (23·9°C). Annual rainfall 160" (4,026 mm).

CONSTITUTION AND GOVERNMENT

The constitution was approved by referendum on 20 May 1972 and became effective on 2 June; it was amended in Jan. 1996. It provides for a *President* as head of state and government. The President is directly elected for a seven-year term, and there is a *Council of Ministers* whose members must not be members of parliament. A constitutional bill removing a two-term presidential limit was adopted in April 2008.

The *National Assembly*, elected by universal adult suffrage for five years, consists of 180 representatives. After 1966 the sole legal party was the Cameroon People's Democratic Movement (RDPC), but in Dec. 1990 the National Assembly legalized opposition parties. The 1996 amendment to the constitution established a 100-seat *Senate*, although elections to it were not held until April 2013.

National Anthem

'O Cameroon, Thou Cradle of our Fathers'/'O Cameroun, Berceau de nos Ancêtres'; words by R. Afame, tune by R. Afame, S. Bamba and M. Nko'o.

GOVERNMENT CHRONOLOGY

Presidents since 1960. (UC = Cameroonian Union; UNC = Cameroonian National Union; RDPC = Cameroonian People's Democratic Rally)

1960–82	UC, UNC	Ahmadou Babatoura Ahidjo
1982–	UNC, RDPC	Paul Biya

RECENT ELECTIONS

Presidential elections were held on 9 Oct. 2011. Incumbent Paul Biya was re-elected with 78·0% of the votes ahead of John Fru Ndi with 10·7% and Garga Haman Adji with 3·2%. There were 20 other candidates. Turnout was 68·3%.

The most recent National Assembly elections were held on 22 July and 30 Sept. 2007. The ruling Cameroon People's Democratic Movement (Rassemblement Démocratique du Peuple Camerounais; RDPC) won 153 seats, Social-Democratic Front (Front Social-Démocratique; SDF) 16, National Union for Democracy and Progress (Union Nationale pour la Démocratie et le Progrès; UNDP) 6, Democratic Union of Cameroon (Union Démocratique du Cameroun; UDC) 4 and Progressive Movement (Mouvement Progressiste; MP) 1.

First elections to the Senate took place on 14 April 2013. The Cameroon People's Democratic Movement (RDPC) won 56 seats of the 70 contested with 73% of the vote; the opposition Social-Democratic Front (SDF) obtained 14 seats with 17%. The remaining 30 seats are appointed.

CURRENT GOVERNMENT

President: Paul Biya; b. 1933 (RDPC; assumed office 6 Nov. 1982, elected 14 Jan. 1984, re-elected 24 April 1988, also 10 Oct. 1992, 12 Oct. 1997, 11 Oct. 2004 and once again re-elected 9 Oct. 2011).

In March 2013 the cabinet comprised:

Prime Minister: Philémon Yang; b. 1947 (RDPC; in office since 30 June 2009).

Deputy Prime Minister and *Minister Delegate at the Presidency in Charge of Relations with the Assembly:* Amadou Ali.

Minister for Agriculture and Rural Development: Essimi Menye. *Arts and Culture:* Ama Tutu Muna. *Commerce:* Luc Magoire Mbarga Atangana. *Communication:* Issa Tchiroma Bakari. *Economy, Planning and Regional Development:* Emmanuel Nganou Djoumessi. *Employment and Professional Training:* Zacharie Perevet. *Energy and Water Resources:* Basile Atangana Kouna. *Environment and Nature Protection:* Pierre Hélé. *External Relations:* Pierre Moukoko Mbonjo. *Finance:* Alamine Ousmane Mey. *Forests and Wildlife:* Ngolle Philip Ngwese. *Health:* Andre Mama Fouda. *Higher Education:* Jacques Fame Ndongo. *Industry, Mines and Technological Development:* Emmanuel Bonde. *Labour and Social Insurance:* Gregoire Owona. *Lands, Survey and State Property:* Koum Abissike Jacqueline. *Livestock, Fisheries and Animal Industries:* Dr Taiga. *Posts and Telecommunications:* Jean-Pierre Biyiti Bi Essam. *Promotion of Women and Family Affairs:* Marie Thérèse Abena Ondoa. *Public Service and Administrative Reforms:* Angouen Michel Ange. *Public Works:* Patrice Ambassala. *Scientific Research and Innovation:* Madeleine Tchuenté. *Small and Medium-Sized Enterprises, and Social Economy:* Laurent Etoundi Ngoa. *Sports and Physical Education:* Adoum Garoua. *Territorial Administration and Decentralization:* Rene Sadi. *Transport:* Robert Nkili. *Youth Affairs and Civic Education:* Pierre Ismael Bidoung Mkpatt.

Ministers of State: Laurent Esso (also *Minister of Justice and Keeper of the Seals*); Bello Bouba Maigari (also *Minister of Tourism and Leisure*).

Minister Delegate at the Presidency in Charge of Defence: Edgard Alain Mebe Ngo'o. *Minister Delegate at the Presidency in Charge of Public Contracts:* Abba Sadou. *Minister Delegate at the Presidency in Charge of Supreme State Control:* Henri Eyebe Ayissi. *Secretary General at the Presidency:* Ferdinand Ngo Ngo.

Office of the President: http://www.prc.cm

CURRENT LEADERS

Paul Biya

Position
President

Introduction
President since 1982, Biya has kept tight political control over Cameroon. Despite some concessions to democracy, his 30-year-old regime has been dogged by allegations of widespread corruption. Meanwhile his international profile has been low, although he has fostered close relations with France.

Early Life

Born on 13 Feb. 1933 in Mvomeka'a, Sud Province, Biya attended a Catholic mission school and, in the early 1950s, a seminary. He specialized in philosophy at the Lycée Général Leclerc in Yaoundé before studying law at the Université de la Sorbonne, Paris. His postgraduate studies included a diploma in public law from the Institut des Hautes Etudes d'Outre-Mer.

Biya's political career began in 1962 as chargé de mission at the Presidency of the Republic. He became secretary-general of the ministry of national education in 1965. In 1970 he was made a minister of state, serving as secretary-general to the Presidency. On 30 June 1975 Biya was appointed prime minister by President Ahmadou Ahidjo. An amendment to the constitution in 1979, designating the prime minister as successor to the president in case of vacancy, allowed Biya to assume the presidency on 6 Nov. 1982 following Ahidjo's resignation.

Career in Office

The succession, although constitutional, was not peaceful. In Aug. 1983 Biya forced Ahidjo into exile, and then consolidated his position by replacing Ahidjo's northern supporters with fellow southerners. Direct presidential elections by universal suffrage were instituted in Jan. 1984, which Biya won. Despite his initial democratic and modernizing aspirations, freedom of speech and the press were soon curtailed, largely as a result of problems with the old regime. The Republican Guard revolt of April 1984 provoked Biya to reform the sole political party, the Cameroon National Union (UNC), which was seen as Ahidjo's personal support base. Transformed into the Cameroon People's Democratic Movement (RDPC; Rassemblement Démocratique du Peuple Camerounais), it elected Biya as party president in March 1985.

In the early 1980s Cameroon's economy suffered from a downturn in the commodity export trade. Biya only admitted the severity of the economic crisis in 1987, submitting the national economy to scrutiny and assistance from the World Bank. Despite popular dissatisfaction, he was re-elected in April 1988.

Opposition political parties were legalized in Dec. 1990, although the delay in organizing multi-party elections and a ban on opposition party meetings caused rioting and a general strike in 1991. Biya relented in Oct. promising elections, which took place in March 1992. The RDPC was forced into coalition with the Movement for the Defence of the Republic (MDR; Mouvement pour la Défense de la République) to attain a majority in the National Assembly. Biya himself was re-elected in Oct. by a narrow majority.

Conflict over the Bakassi Peninsula, on the Nigerian border, and its oil and fishing rights began when the Nigerian leader Gen. Abacha sent troops to claim the area. Biya responded with military force and appealed to the International Court of Justice, which ruled in Cameroon's favour in 2002 (with Cameroon eventually taking control of the territory in Aug. 2008).

The presidential elections of Oct. 1997 were boycotted by the three main opposition parties after a year of popular unrest. The removal of elected mayors after the 1996 municipal elections and the Supreme Court's controversial rulings concerning the May 1997 National Assembly elections were two of the more prominent causes of the boycott. Biya was re-elected with a large majority of the vote.

By that time, Biya's presidency was marked by anglophone separatist unrest and allegations of corruption and human rights abuses. The secessionist Southern Cameroon National Council (SCNC), claiming to represent the country's 5m. English speakers, was targeted by the government and its leaders charged with treason. In 1998 Transparency International, the Berlin-based anti-corruption organization, classed Cameroon as the most corrupt country of the 85 covered by their survey. Amnesty International meanwhile claimed that extrajudicial executions and

politically-motivated detentions were continuing despite international pressure.

In parliamentary elections in 2002 the RDPC won 149 of the 180 seats in National Assembly. On 11 Oct. 2004 Biya was re-elected for a further presidential term with over 70% of the vote, although opposition parties alleged widespread fraud and international observers said the poll lacked credibility in key areas. By 2006 new measures against corruption were in place, including a law requiring the declaration of assets by public officials and the establishment of an anti-corruption commission. In the 2007 legislative elections the RDPC retained its overwhelming majority and in April 2008 parliament approved a controversial constitutional amendment enabling Biya to run for a third term of office. He again secured re-election in Oct. 2011 by a landslide margin, claiming 78% of the vote.

DEFENCE

The President of the Republic is C.-in-C. of the armed forces. Defence expenditure totalled US$306m. in 2008 (US$17 per capita), representing 1·3% of GDP.

Army

Total strength (2007) is 12,500 and includes a Presidential Guard; there is a Gendarmerie 9,000 strong.

Navy

Personnel in 2007 numbered about 1,300. There are bases at Douala (HQ), Limbe and Kribi.

Air Force

Aircraft availability is low because of funding problems. Personnel (2007), 300. There are 15 combat-capable aircraft.

ECONOMY

In 2006 agriculture accounted for 19·9% of GDP, industry 31·4% and services 48·7%.

Overview

Cameroon became an oil exporter in 1977, with production peaking in 1985. The country came to be regarded as an African success story in the early 1980s. However, an economic reversal was triggered in 1985 by a sharp fall in prices for primary products including cocoa and oil. Underlying the external trade shock were weak economic and political structures, a fiscal crisis and an overvalued exchange rate. In 1994 the currency was devalued.

A programme of privatization began in 1995, taking in the export, infrastructure, banking and insurance sectors. In 1996, supported by the IMF and World Bank, the government adopted structural reforms and macroeconomic policies encompassing forestry, banking, transportation and privatization of public utilities. Despite these reforms, corruption and poor resource management remain.

The Heavily Indebted Poor Countries Initiative and the Multilateral Debt Relief Initiative helped reduce the country's debt burden from 51·8% of GDP in 2005 to 9·5% in 2008. However, the debt level has subsequently increased, standing at 13·7% in 2011.

Average real GDP growth was 3% over the period 2007–11 though this has not been reflected in social indicators. The poverty rate remains close to 40%. A further downturn in the fortunes of the eurozone, Cameroon's main trading partner, threatens to hold back economic activity.

Currency

The unit of currency is the *franc CFA* (XAF) with a parity of 655·957 francs CFA to one euro. In June 2005 foreign exchange reserves were US$759m. (negligible in 1997), total money supply was 757,006m. francs CFA and gold reserves were 30,000 troy oz. Inflation was 1·3% in 2010 and 2·9% in 2011.

Budget

The financial year used to end on 30 June but since 2003 has been the calendar year. In 2008 revenues totalled 2,214bn. francs CFA and expenditures 1,966bn. francs CFA.

VAT is 19·25%.

Performance

Real GDP growth was 2·0% in 2009, 2·9% in 2010 and 4·2% in 2011. Total GDP in 2011 was US$25·2bn.

Banking and Finance

The Banque des États de l'Afrique Centrale (*Governor*, Lucas Abaga Nchama) is the sole bank of issue. The largest commercial banks are Afriland First Bank, Banque internationale du Cameroun pour l'épargne et le crédit (BICEC) and Société générale des banques au Cameroun (SGBC).

External debt totalled US$2,964m. in 2010 and represented 13·5% of GNI.

The Douala Stock Exchange was opened in 2003.

ENERGY AND NATURAL RESOURCES

Environment

Cameroon's carbon dioxide emissions from the consumption and flaring of fossil fuels in 2008 were the equivalent of 0·4 tonnes per capita.

Electricity

Installed capacity in 2007 was an estimated 1·1m. kW. Total production in 2007 was 5·75bn. kWh (67% hydro-electric), with consumption per capita 308 kWh. Cameroon is rich in hydro-electric potential, of which only around 5% is currently exploited.

Oil and Gas

Oil production (2008), mainly from Kole oilfield, was 4·3m. tonnes. In 2007 there were proven reserves of 400m. bbls. In June 2000 the World Bank approved funding for a 1,000-km US$4bn. pipeline to run from 300 new oil wells in Chad through Cameroon to the Atlantic Ocean. Oil started pumping in July 2003.

Minerals

Tin ore and limestone are extracted. There are deposits of aluminium, bauxite, uranium, nickel, gold, cassiterite and kyanite. Aluminium production in 2008 was 89,700 tonnes.

Agriculture

In 2002, 57·1% of the economically active population were engaged in agriculture. In 2001 there were 5·96m. ha. of arable land and 1·20m. ha. of permanent crops. 33,000 ha. were irrigated in 2001. There were 500 tractors in 2001. Main agricultural crops (with 2003 production in 1,000 tonnes): cassava, 2,619; sugarcane, 1,350; plantains, 1,200; maize, 1,040; bananas, 689; tomatoes, 419; sorghum, 331; yams, 311; groundnuts, 295; seed cotton, 244; sweet potatoes, 234; millet, 196; dry beans, 182; potatoes, 162; palm oil, 144; cocoa beans, 125; pumpkins and squash, 122; cucumbers and gherkins, 115.

Livestock (2003): 5·9m. cattle; 4·4m. goats; 3·8m. sheep; 1·3m. pigs; 24m. chickens.

Livestock products (in 1,000 tonnes), 2003: beef and veal, 95; lamb and mutton, 16; pork, bacon and ham, 16; goat meat, 15; poultry meat, 29; cow's milk, 125; goat's milk, 42; sheep's milk, 17; eggs, 13.

Forestry

Forests covered 19·92m. ha. in 2010 (42% of the total land area), ranging from tropical rain forests in the south (producing hardwoods such as mahogany, ebony and sapele) to semi-deciduous forests in the centre and wooded savannah in the north. Timber production in 2007 was 11·45m. cu. metres.

Fisheries

In 2007 the total catch was 138,612 tonnes (mainly from inland waters).

INDUSTRY

Manufacturing is largely small-scale. Output in 1,000 tonnes (2007 unless otherwise indicated): cement (2005), 1,026; distillate fuel oil, 694; kerosene, 404; petrol, 390; residual fuel oil, 387. In 2005, 444m. litres of beer and 1·8bn. cigarettes were produced. There are also factories producing shoes, soap, oil and food products.

Labour

In 2001 there were 5,465,000 employed persons, of whom 58·5% were occupied in agriculture.

INTERNATIONAL TRADE

Imports and Exports

In 2010 total imports (c.i.f.) amounted to US$5,133·3m. and exports (f.o.b.) to US$3,878·4m. Principal imports, 2006: petroleum and petroleum products, 31·6%; cereal and cereal preparations, 9·0%; road vehicles, 5·8%; non-metallic mineral manufactures, 3·1%. Principal exports, 2006: petroleum and petroleum products, 61·6%; sawn wood, 9·5%; cocoa, 7·3%; aluminium, 4·5%.

Main import suppliers, 2006: Nigeria, 23·3%; France, 17·2%; China, 6·3%. Main export markets, 2006: Spain, 25·9%; Italy, 23·1%; France, 10·7%.

COMMUNICATIONS

Roads

There were about 51,300 km of roads in 2004, of which 8·4% were paved. In 2005 there were 174,900 passenger cars, 56,200 lorries and vans, 15,600 buses and coaches, and 65,600 motorcycles and mopeds. In 2007 there were 990 deaths in road accidents.

Rail

Cameroon Railways (Camrail), 987 km in 2005, link Douala with Nkongsamba and Ngaoundéré, with branches from M'Banga to Kumba and Makak to M'Balmayo. In 2008 railways carried 1·2m. passengers and 1·6m. tonnes of freight.

Civil Aviation

There are 45 airports including three international airports at Douala, Garoua and Yaoundé (Nsimalen). In 2001 Douala handled 485,000 passengers (386,000 on international flights). In 2003 Cameroon Airlines (Camair), the then national carrier, operated on domestic routes and provided international services to Abidjan, Bamako, Bangui, Brazzaville, Bujumbura, Cotonou, Dakar, Johannesburg, Kigali, Kinshasa, Lagos, Libreville, Malabo, N'Djaména, Paris and Pointe-Noire. It ceased operations in March 2008. In 2003 scheduled airline traffic of Cameroon-based carriers flew 9m. km, carrying 315,000 passengers (225,000 on international flights).

Shipping

In Jan. 2009 there were four vessels of 300 GT or over registered, totalling 2,000 GT. The main port is Douala; other ports are Bota, Campo, Garoua (only navigable in the rainy season), Kribi and Limbo-Tiko.

Telecommunications

There were 496,500 fixed telephone lines in 2010 (25·3 per 1,000 inhabitants) and mobile phone subscribers numbered 8·16m. There were 40·0 internet users per 1,000 inhabitants in 2010. Fixed internet subscriptions totalled 25,000 in 2006 (1·4 per 1,000 inhabitants). In June 2012 there were 494,000 Facebook users.

SOCIAL INSTITUTIONS

Justice

The Supreme Court sits at Yaoundé, as does the High Court of Justice (consisting of nine titular judges and six surrogates all appointed by the National Assembly). There are magistrates' courts situated in the provinces.

The population in penal institutions in Dec. 2009 was 23,368 (119 per 100,000 of national population).

Education

In 2007 there were 217,284 children and 12,349 teaching staff at pre-primary schools. There were 3,120,357 pupils in primary schools in 2007 with 70,230 teaching staff and 750,777 secondary level pupils with (2006) 43,193 teaching staff.

In 2007, 132,134 students were in tertiary education with 3,040 academic staff. There were six public universities and five private universities in 2007. The adult literacy rate in 2008 was 76%.

In 2007 public expenditure on education came to 3·9% of GNI and 17·0% of total government spending.

Health

In 2004 there were 3,124 physicians, 147 dentists, 26,042 nurses and 700 pharmacists. There were 15 hospital beds per 10,000 population in 2006.

RELIGION

In 2001 there were 4·18m. Roman Catholics, 3·35m. Muslims and 3·27m. Protestants. Some of the population follow traditional animist religions. In Feb. 2013 there was one cardinal.

CULTURE

World Heritage Sites

The Dja Faunal Reserve was inscribed on the UNESCO World Heritage List in 1987. Surrounded by the Dja River, it is one of Africa's largest rainforests. Sangha Trinational, a transboundary conservation complex situated in Cameroon, Central African Republic and the Republic of the Congo, was added to the UNESCO World Heritage List in 2012. The site consists of three contiguous national parks totalling around 750,000 ha. Much of it is unaffected by human activity and features a wide range of humid tropical forest ecosystems with rich flora and fauna.

Press

In 2008 there was one national government-owned daily newspaper with a circulation of 25,000, four privately-owned dailies and about 200 other privately owned-newspapers that appeared at irregular intervals.

Tourism

In 2005, 176,000 non-resident tourists stayed in hotels or similar accommodation (including 89,000 from other African countries and 68,000 from Europe). Receipts from tourism totalled US$229m. in 2005.

Festivals

The Cameroon National Festival takes place on 20 May. The Ngonda festival in Douala takes place over ten days ending on the first Sunday of Dec. and involves water ceremonies of the jengu cult.

DIPLOMATIC REPRESENTATIVES

Of Cameroon in the United Kingdom (84 Holland Park, London, W11 3SB)
High Commissioner: Nkwelle Ekaney.

Of the United Kingdom in Cameroon (Ave. Winston Churchill, BP 547, Yaoundé)
High Commissioner: Bharat Joshi.

Of Cameroon in the USA (2349 Massachusetts Ave., NW, Washington, D.C., 20008)
Ambassador: Joseph Foe-Atangana.

Of the USA in Cameroon (Ave. Rosa Parks, BP 817, Yaoundé)
Ambassador: Robert P. Jackson.

Of Cameroon to the United Nations
Ambassador: Michel Tommo Monthe.

Of Cameroon to the European Union
Ambassador: Daniel Evina Abe'e.

FURTHER READING

Ardener, E., *Kingdom on Mount Cameroon: Studies in the History of the Cameroon Coast 1500–1970.* 1996
Gros, Jean-Germain, *Cameroon: Politics and Society in Critical Perspective.* 2003

National Statistical Office: Institut National de la Statistique du Cameroun, Ministère de l'Economie, de la Planification et de l'Aménagement du Territoire, Yaoundé.
Website: http://www.statistics-cameroon.org

CANADA

ARCTIC OCEAN

GREENLAND
(DENMARK)

ALASKA
(U.S.)

Beaufort
Sea

Baffin
Bay

YUKON
Whitehorse

NORTHWEST
TERRITORIES

NUNAVUT

Iqaluit

ATLANTIC
OCEAN

Yellowknife

CANADA

NEWFOUNDLAND
AND LABRADOR

BRITISH
COLUMBIA

ALBERTA

Edmonton

SASKATCHEWAN

Hudson
Bay

MANITOBA

QUÉBEC

St John's

NEW
BRUNSWICK

Vancouver

Calgary

Regina

Winnipeg

ONTARIO

Québec

Montréal

PRINCE
EDWARD
ISLAND

UNITED STATES

OTTAWA
Toronto

NOVA
SCOTIA

© Research Machines plc 2006

Capital: Ottawa
Population projection, 2015: 35·62m.
GNI per capita, 2011: (PPP$) 35,166
HDI/world rank: 0·908/6
Internet domain extension: .ca

KEY HISTORICAL EVENTS

The first habitation in Canada dates from the last stages of the Pleistocene Ice Age up to 30,000 years ago. Mongoloid tribes from Asia crossed the Bering Strait by a land bridge in search of mammoth, bison and elk. These hunter-gatherers were the forefathers of some of Canada's native people referred to today as the First Nations. There are currently two other Aboriginal groups; the Inuit (Arctic people, formerly known as Eskimos) and the Métis. The Inuit were one of the last groups to arrive, around 1000 BC, whereas the Métis evolved from the union of natives and Europeans (mostly French).

The numerous tribes that made up the First Nations consisted of 12 major language groups with a number of sub groups with diverse spiritual beliefs, laws and customs. Around 6000 BC, during the Boreal Archaic age, the glaciers of the Canadian Shield

melted and lakes were formed. The Iroquois speaking tribes, including the Mohawks and the Huron, settled along the St Lawrence River and the Great Lakes. Excellent farmers, they lived in large communities. Trade flourished but tribal wars were common. By 1000 BC the Early Woodland Culture had developed in the east. Among the eight tribes were the Algonquin, one of the largest language groups, who spread west to the Plains to hunt buffalo along with the Blackfoot, Sioux and Cree. The tribes of the Pacific Coast, such as the Tlingit and Salish, made a living from whaling and salmon fishing and enjoyed a more elaborate social structure. The peoples in the north around Yukon and Mackenzie River basins and the Inuit around the Arctic were nomadic hunters foraging for limited food in small family groups. However, one factor common to all was that they were self-governing and politically independent.

In 1963 remains of a Viking settlement were found at L'Anse aux Meadows in Newfoundland and Labrador dating from AD 1000. Trade had been established between the Norse men and the Inuit but settlements were abandoned when the Norse withdrew from Greenland. John Cabot, an Italian navigator, commissioned by King Henry VII of England in 1497, charted the coasts around Labrador and Newfoundland and found large resources of fish.

The Frenchman, Jacques Cartier, discovered the Gulf of St Lawrence in 1534 and claimed it for the French crown. In the following years fisheries were set up by the English and French with Indians bringing valuable furs, mostly beaver, to trade for iron and other goods. Realising the potential, the French sent Samuel de Champlain in 1604 to establish a fur trade and organize a settlement. This he achieved in 1605 in an area called Acadia (now New Brunswick, Nova Scotia and Prince Edward Island). The French traded with the Algonquin and Huron and supported them during fierce raids by the Iroquois. In retaliation the Iroquois later became the fur trading allies of the Dutch and then the English. Champlain, having founded Quebec City, went on to explore Huron territory, now central Ontario, and is considered by many to be the father of New France.

English–French Rivalry

In opposition to French expansion, England sent explorers such as Martin Frobisher, William Baffin and Henry Hudson to claim new territory. Colonies sprang up along the English coast and the Hudson Bay Company was formed in 1670 to gain a fur-trading monopoly over the area. Rivalry between the English and French for trade at Hudson Bay persisted throughout the 17th century. In 1713 the Treaty of Utrecht, signed by Queen Anne of England and Louis XIV, gave England complete control of the Hudson Bay territory, Acadia and Newfoundland. France, however, retained Cape Breton Island, the St Lawrence Islands and fishing rights in Newfoundland. Led by Gen. Wolfe, Britain's victory over France at the Battle of the Plains of Abraham in 1759 gained Quebec, and in 1760 Montreal too was taken. This brought an end to the Seven Years' War (1756–63) confirmed in 1763 by the Treaty of Paris by which all French Canadian territory was ceded to the British.

Relations with Indian tribes formerly allied to the French were strained and there was much resentment at the invasion of their lands by white settlers. The Royal Proclamation of 1763, administered by the Indian Department, ruled that aboriginal peoples could only sell land to crown representatives. Numerous treaties were signed over the following decades, redistributing thousands of acres of land. To soothe the British rule of a French speaking colony, the British government passed the Quebec Act in 1774 allowing French Canadians religious and linguistic freedom, the right to collect tithes and recognition of French civil law.

The attack on Montreal in 1775, during the American War of Independence, failed and Americans, loyal to Britain, sought refuge in Canada. Around 50,000 emigrated to Nova Scotia, from which New Brunswick was created in 1784. In an attempt to keep the peace the Constitutional Act of 1791 divided Quebec into Lower Canada (mostly French) and Upper Canada (mostly British from America).

Exploration continued on the Pacific Coast and into the Plains and the Northwest. Capt. James Cook charted the Pacific Coast from Vancouver to Alaska in 1778. From trading posts set up by the Hudson Bay Company (HBC) expeditions were made by traders including Samuel Hearne who, in 1771, was the first man to reach the Arctic Ocean by land. A rival company, the North West Company (NWC), was set up in Montreal in 1783. In 1793 Alexander Mackenzie, from the NWC, crossed the Rocky Mountains and reached the coast making him the first man to cross the continent. Competition between the two companies erupted into violence between new settlers and established traders including the Métis who were hired, mainly by NWC, to transport furs and supply food. To resolve the conflict the British government pressed for the merger of the two companies. This was achieved in 1821.

After the American war against Britain in Canada in 1812 (which ended in stalemate), large numbers of English, Scottish and Irish settlers swelled the English-speaking population. By 1837 radical reformers were seeking accountable government with a broader electorate. Rebellions led by William Lyon Mackenzie in Upper Canada and by Louis-Joseph Papineau in Lower Canada were quashed by government troops. Following a report by Lord Durham, who was sent from England as governor-general to conduct an enquiry, the two colonies were united under one central government in 1841 with both enjoying equal representation. Vancouver Island was acknowledged to be British by the Oregon Boundary Treaty of 1846. The 1850s saw a significant period of growth. Railways were built and industry and commerce thrived, enhanced by the Reciprocity Treaty of 1854 with the USA.

Dominion Status

However, by the 1860s ethnic clashes had made Canada almost ungovernable. The American Civil War also posed a threat. Three political leaders, George Étienne Cartier (Conservative–Canada East), George Brown (Reform Movement–Canada West) and John A. Macdonald (Conservative–Canada West) formed a coalition government in 1864. Nova Scotia, New Brunswick and the Canadas (now Ontario and Quebec) were united in 1867 as the Dominion of Canada. What became known as the Constitution Act, confirmed the language and legal rights of the French and provided for the division of power between the federal government and the provinces. John Macdonald was elected prime minister.

One of the first actions of the new federal government was to purchase the Northwest Territories from the HBC, a move that led to rebellion by white settlers and the Métis, under Louis Riel. The result was the creation of Manitoba in 1870 with political power divided between the French and English. British Columbia joined the federation in 1871 and Prince Edward Island in 1873. The former agreed to join on the promise of a federally financed railway. To make way for new settlers from 1868–77, treaties were negotiated with Indians from Ontario to the Rocky Mountains. In return for moving to reserves the Indians were to receive financial support and other concessions. They were also to be assimilated into Christian society. In the years following the government failed in its obligations and many Plains Indians suffered poverty, starvation and disease. Later legislation even outlawed traditional practices such as the Sun Dance and the potlatch (exchange of gifts). Many Métis had moved out of Manitoba and settled further west but were not awarded the same rights as those of the Aboriginal Indians. Resistance grew, and Riel, who had emigrated to Montana, was urged to lead a revolt. In response, troops were rushed in by rail and the rebellion was crushed. The importance of the railway became evident and money was found for its completion across the Rockies. The new railway was opened in 1885, the same year in which Riel was executed.

Prosperity and Reform

Following the death of Macdonald in 1891, the Liberal leader Wilfrid Laurier came to power in 1896 and there followed a period of growth and stability. Mineral resources were found in British Columbia and Ontario as well as gold which precipitated the Klondike gold rush of 1897. The Yukon Territory was established in 1898 to ensure Canadian jurisdiction over the exploitation of gold. The provinces of Saskatchewan and Alberta were created in 1905, each with its own premier and elected assembly. By 1911 the population in the provinces had doubled and there was a powerful business sector. When reformers called for action to alleviate conditions in overcrowded cities, health and welfare programmes were introduced. A new women's movement campaigned for equal rights and women's suffrage. Anti-monopoly legislation was passed in 1910. However, much industrial investment was backed by American money and many French Canadians began to agitate for autonomy. In 1910 Laurier founded the Canadian Navy with the provision that in time of war it would be placed under British command. This further angered the French Canadians. Laurier's decision to negotiate a new trade agreement with the USA, coupled with the navy issue, lost him the 1911 election to the Conservatives. Robert Laird Borden became prime minister.

When the First World War broke out in 1914 thousands of British-born Canadians volunteered to fight. At first troops were under British command, but by the time conscription had been introduced in 1917 they were under Canadian leadership. Around 60,000 men lost their lives in the battles of Ypres, Vimy Ridge and Passchendaele, and another 173,000 were wounded. Having won recognition for its contribution to the war effort, Canada participated as an independent state at the Paris Peace Conference and joined the League of Nations. French Canadians had been bitterly opposed to conscription and to counteract this Borden formed a joint government of Liberals and Conservatives. This was split into the English speaking Unionists and the French speaking Liberals. At the election in 1917 the Unionists won every province but Quebec.

Women's suffrage was granted in 1918. 1921 saw the Liberals back in power under William Lyon Mackenzie King who strove to unify the nation and gain autonomy. This was achieved in 1931 by the Statute of Westminster in which Canada was granted complete independence. In the same year Norway formally recognized the Canadian title to the Sverdrup group of Arctic islands. Canada thus holds sovereignty in the whole Arctic sector north of the Canadian mainland.

Following the Wall Street Crash of 1929, the country chose the Conservatives under the leadership of Richard Bedford Bennett in the election of 1930. Despite measures to alleviate the effects of the depression and the severe drought in the prairies, Bennett was unsuccessful and in 1935 King was re-elected. King introduced a new Reciprocity Treaty (1936) with the USA, nationalized the Bank of Canada, created the Canadian Broadcasting Corporation and made available federal money to provide social services.

Post-War Politics

Canada's contribution in the Second World War was even more extensive than that in the First World War although casualties were lower. A post-war plan introduced unemployment insurance, family allowances, veterans' benefits, subsidized housing, health plans and improved pensions. Industrial controls were lifted and trade was encouraged. Canada became a founder member of the United Nations in 1945 and has been active in a peacekeeping role ever since. King retired in 1948 to be succeeded by Louis St Laurent, a Quebec lawyer, who won an overwhelming victory in 1949. In the same year, Newfoundland—including Labrador—became a Canadian province thus completing the Confederation. Also in that year, Canada joined NATO.

An amendment to the Indian Act in 1959 increased opportunities for Indians to influence decisions affecting them, and in 1960 the federal government granted the franchise to all Indians, with several provinces following suit.

For 20 years after 1950 Canada enjoyed growth, prosperity and a 'baby boom'. The face of industrial Canada changed with the discovery of radium, petroleum and natural gas. Canada took a more active part in foreign affairs, especially in the 1956 Suez war when Lester B. Pearson, external affairs minister, won the Nobel Peace Prize. The North American Air Defense Command (NORAD) was formed with the USA in 1957. In the same year, after 22 years of Liberal rule, the Conservatives won the election with John Diefenbaker as leader. However, internal struggles and an economic recession saw the return of the Liberals in 1963 under Pearson. During his five years as prime minister Canada gained a national flag, a social security system and medical care for all its citizens.

Having chosen Pierre Trudeau to succeed Pearson, the Liberals won the 1968 election. At the same time there was a revival of French nationalism, especially in Quebec, with the formation of the Parti Québécois (PQ) led by René Levasque. Keen to preserve national unity, Trudeau passed the Official Languages Act in 1969 which affirmed the equality of French and English in all governmental activities. However, in 1970 he had to send troops into Quebec following the murder of the Labour minister, Pierre Laporte, by the separatist Front de Libération du Québec. In 1976 a pledge on separatism won the PQ the provincial election and French became the official language of Quebec. Despite these milestones, a referendum to make the province an independent country was rejected by Quebec voters in 1980.

Indian Rights

In the sixties and seventies Indians sought special rights and settlement of their outstanding treaty claims. The National Indian Brotherhood was formed in 1968 to represent the interest of Indians at federal level. By 1973 the Department of Indian Affairs and Northern Development was instructed to resolve these claims. In 1982, with the exception of Quebec, the country agreed to a new constitution giving Canada, as opposed to the British Parliament, prerogative over all future constitutional changes. At the same time a charter of Rights and Freedoms was introduced recognizing the nation's multi-cultural heritage, affirming the existing rights of native peoples and the principle of equality of benefits to the provinces.

Retiring in 1984, Trudeau was succeeded by John Turner who was ousted the same year by the Conservative leader Brian Mulroney. In 1987 at a meeting in Meech Lake, a series of constitutional amendments were drawn up to win Quebec's acceptance of the new constitution. English Canadians objected to the Meech Lake Accord and it was rejected by Newfoundland and Manitoba. This failure sparked another separatist revival in Quebec leading to the drafting of the Charlottetown Accord incorporating extensive amendments, including recognition of Quebec as a 'distinct society', offering better representation in parliament, and self-government for indigenous peoples. This was defeated in a national referendum in 1992.

Mulroney negotiated a free trade agreement with the United States which went into effect in 1989. This was followed in 1994 by North American Free Trade Agreement (NAFTA) with the USA and Mexico. Voter opposition coupled with a recession in the early nineties forced Mulroney to resign in 1993. He was replaced by Kim Campbell, Canada's first female prime minister. In the Oct. election of that year, Campbell and the Conservatives suffered a major defeat retaining only two of their 154 seats. Led by Jean Chrétien, the Liberals won 177 seats and the Reform Party 52 seats. The PQ became the major opposition party with 54 seats.

Another referendum on Quebec's independence from Canada failed narrowly in 1995. Again in 1997 leaders of all provinces and territories (apart from Quebec) met in Calgary and signed a declaration recognizing the 'unique character' of Quebec's society. The following year the Supreme Court ruled that Quebec was prohibited from declaring itself independent without first negotiating an agreement with the federal government and other provinces.

In the 1990s further protests were made by native peoples anxious to claim their territory. In 1997 the Supreme Court ruled that two aboriginal groups had title to 22,000 square miles of ancestral lands in British Columbia. The following year a formal apology was issued by the government for the treatment Indian and Inuit peoples had received since the arrival of the Europeans. After decades of complex negotiations, the Inuit were granted their own territory of Nunavut in 1999, an area once part of the Northwest Territories. In the same year the government agreed that Indians and the Inuit should have the right of self-government.

Following a period of economic growth Chrétien's government made major tax cuts in 2000. In an attempt to form a more effective opposition, the Reform Party accepted the 'united alternative' proposed by party leader Preston Manning which resulted in the Canadian Alliance (Canadian Reform Conservative Alliance), a broad-based conservative party with Stockwell Day as its first leader. However, with the Progressive Conservative Party declining to join forces, Chrétien was returned for a third term in Nov. 2000 with an increased majority.

Chrétien stood down from the premiership in late 2003 and was replaced by his Liberal colleague, Paul Martin. Martin and his party soon became embroiled in a financial scandal over misuse of government money for advertising. In the Jan. 2006 general election the Conservative Party, led by Stephen Harper, defeated the Liberals, taking power for the first time in 12 years.

TERRITORY AND POPULATION

Canada is bounded in the northwest by the Beaufort Sea, north by the Arctic Ocean, northeast by Baffin Bay, east by the Davis Strait, Labrador Sea and Atlantic Ocean, south by the USA and west by the Pacific Ocean and USA (Alaska). The area is 9,984,670 sq. km, of which 891,163 sq. km are fresh water. 2011 census population, 33,476,688 (51·0% female), giving a density of 3·7 per sq. km. In 2011, 81·1% of the population were urban.

The UN gives a projected population for 2015 of 35·62m.

Population at previous censuses:

1861	3,229,633	1931	10,376,786	1981	24,343,181
1871	3,689,257	1941	11,506,655	1986[1]	25,309,331
1881	4,324,810	1951	14,009,429	1991	27,296,859[2]
1891	4,833,239	1961	18,238,247	1996[1]	28,848,761[2]
1901	5,371,315	1971	21,568,311	2001	30,007,094
1911	7,206,643	1976[1]	22,992,604	2006[1]	31,612,897
1921	8,787,949				

[1]It became a statutory requirement to conduct a census every five years in 1971. [2]Excludes data from incompletely enumerated Indian reserves and Indian settlements.

Figures for the 2006 census population according to ethnic origin (leading categories), were[1]:

Canadian	10,066,290	Chinese	1,346,510
English	6,570,015	North American Indian	1,253,260
French	4,941,210	Ukrainian	1,209,090
Scottish	4,719,850	Dutch	1,035,965
Irish	4,354,155	Polish	984,565
German	3,179,425	East Indian	962,670
Italian	1,445,330	Russian	500,500

[1]Census respondents who reported multiple ethnic origins are counted for each origin they reported.

The aboriginal population (those persons identifying with at least one aboriginal group, and including North American Indian, Métis or Inuit) numbered 1,172,785 in 2006. In 2006, 57·2% of the population gave their mother tongue as English and 21·8% as French (English and French are both official languages); Chinese was reported as the third most common language, accounting for 3·2% of the total population. In 2006, 1·2m. residents were immigrants who had arrived between 2001 and 2006, accounting for 3·8% of the total population; 40·8% of all immigrants in 2006 were from Asia (including the Middle East), 36·8% from Europe, 11·3% from the Caribbean, Central and South America, 6·1% from Africa, 4·0% from the USA and 1·0% from Oceania and other countries.

Of the total population in 2006, 80·4% were Canadian-born. The percentage of the population in each of the provinces and territories in 2006 born outside Canada was as follows: Ontario, 27·9%; British Columbia, 27·2%; Alberta, 16·0%; Manitoba, 13·2%; Quebec, 11·3%; Yukon, 9·9%; Northwest Territories, 6·8%; Saskatchewan, 5·0%; Nova Scotia, 4·9%; New Brunswick, 3·6%; Prince Edward Island, 3·5%; Newfoundland and Labrador, 1·7%; Nunavut, 1·5%.

Yukon had the biggest population increase between 2006 and 2011 with 11·6%, while the Northwest Territories had the smallest population growth at 0·0%.

Populations of Census Metropolitan Areas (CMA) and Cities (proper), 2011 census:

	CMA	City proper		CMA	City proper
Toronto	5,583,064	2,615,060	Halifax	390,328	390,096
Montreal	3,824,221	1,649,519	Oshawa	356,177	149,607
Vancouver	2,313,328	603,502	Victoria	344,615	80,017
Ottawa-			Windsor	319,246	210,891
Gatineau	1,236,324	—	Saskatoon	260,600	222,189
Ottawa	—	883,391	Regina	210,556	193,100
Gatineau	—	265,349	Sherbrooke	201,890	154,601
Calgary	1,214,839	1,096,833	St John's	196,966	106,172
Edmonton	1,159,869	812,201	Barrie	187,013	135,711
Quebec	765,706	516,622	Kelowna	179,839	117,312
Winnipeg	730,018	663,617	Abbotsford	170,191	133,497
Hamilton	721,053	519,949	Greater		
Kitchener	477,160	366,151	Sudbury	160,770	160,274
London	474,786	219,153	Kingston	159,561	123,363
St Catharines-			Saguenay	157,790	144,746
Niagara	392,184	—	Trois-Rivières	151,773	131,338
St Catharines	—	131,400	Guelph	141,097	121,688
Niagara Falls	—	82,997			

SOCIAL STATISTICS

Statistics for period from July–June:

	Live births	Deaths
2006–07	360,916	233,825
2007–08	373,695	236,525
2008–09	380,767	239,930
2009–10	383,585	244,677

Average annual population growth rate, 2000–10, 1·1%. Birth rate, 2009–10 (per 1,000 population), 11·2; death rate, 7·2. Marriages, 2008, numbered 147,288; divorces, 2008, 70,226. Suicides, 2007, 3,611 (11·0 per 100,000 population). Life expectancy at birth, 2007, was 78·2 years for men and 82·9 years for women. Infant mortality, 2010, five per 1,000 live births; fertility rate, 2008, 1·7 children per woman. Canada legalized same-sex marriage in 2005.

CLIMATE

The climate ranges from polar conditions in the north to cool temperate in the south, but with considerable differences between east coast, west coast and the interior, affecting temperatures, rainfall amounts and seasonal distribution. Winters are very severe over much of the country, but summers can be very hot inland. See individual provinces for climatic details.

CONSTITUTION AND GOVERNMENT

In Nov. 1981 the Canadian government agreed on the provisions of an amended constitution, to the end that it should replace the British North America Act and that its future amendment should be the prerogative of Canada. These proposals were adopted by the Parliament of Canada and were enacted by the UK Parliament as the Canada Act of 1982. This was the final act of the UK Parliament in Canadian constitutional development. The Act gave to Canada the power to amend the Constitution according to procedures determined by the Constitutional Act 1982. The latter added to the Canadian Constitution a charter of Rights and Freedoms, and provisions which recognize the nation's multi-cultural heritage, affirm the existing rights of native peoples, confirm the principle of equalization of benefits among the provinces, and strengthen provincial ownership of natural resources.

Under the Constitution legislative power is vested in Parliament, consisting of the Queen, represented by a Governor-General, a Senate and a House of Commons. The members of the *Senate* are appointed until age 75 by summons of the Governor-General under the Great Seal of Canada. Members appointed before 2 June 1965 may remain in office for life. The Senate consists of 105 senators: 24 from Ontario, 24 from Quebec, 10 from Nova Scotia, 10 from New Brunswick, 6 from Manitoba, 6 from British Columbia, 6 from Alberta, 6 from Saskatchewan, 6 from Newfoundland and Labrador, 4 from Prince Edward Island, 1 from Yukon, 1 from the Northwest Territories, and 1 from Nunavut. Each senator must be at least 30 years of age and reside in the province for which he or she is appointed. The *House of Commons*, currently of 308 members, is elected by universal secret suffrage, by a first-past-the-post system. Legislation that came into force in May 2007 stipulates that elections will be held on the third Monday of Oct. in the fourth calendar year following the previous general election, except when a government loses a vote of confidence. Representation is based on the population of all the provinces taken as a whole with readjustments made after each census. State of the parties in the Senate (March 2013): Conservatives, 64; Liberals, 36; ind., 4; vacant, 1.

The First Nations have representation in the *Assembly of First Nations* (National Chief: Shawn A-in-chut Atleo, elected July 2009).

The office and appointment of the Governor-General are regulated by letters patent of 1947. In 1977 the Queen approved the transfer to the Governor-General of functions discharged by the Sovereign. The Governor-General is assisted by a *Privy Council* composed of Cabinet Ministers.

Canada: The State of the Federation. Queen's Univ., annual
Canadian Parliamentary Guide. Annual.
Bejermi, J., *Canadian Parliamentary Handbook*. 2008
Cairns, A. C., *Charter versus Federalism: the Dilemmas of Constitutional Reform*. 1992
Courtney, John and Smith, David, *The Oxford Handbook of Canadian Politics*. 2010
Fox, P. W. and White, G., *Politics Canada*. 8th ed. 1995
Hogg, P. W., *Constitutional Law of Canada*. 2001
Kernaghan, K., *Public Administration in Canada: a Text*. 1991
Mahler, G., *Contemporary Canadian Politics, 1970–1994: an Annotated Bibliography*. 2 vols. 1995
Reesor, B., *The Canadian Constitution in Historical Perspective*. 1992
Tardi, G., *The Legal Framework of Government: a Canadian Guide*. 1992

National Anthem

'O Canada, our home and native land'/'O Canada, terre de nos aïeux'; words by A. Routhier, tune by C. Lavallée.

GOVERNMENT CHRONOLOGY

Prime Ministers since 1935. (CPC = Conservative Party of Canada; LP = Liberal Party; PC = Progressive Conservative Party)

1935–48	LP	William Lyon Mackenzie King
1948–57	LP	Louis Stephen Saint Laurent
1957–63	PC	John George Diefenbaker
1963–68	LP	Lester Bowles Pearson
1968–79	LP	Pierre Elliott Trudeau
1979–80	PC	Charles Joseph (Joe) Clark
1980–84	LP	Pierre Elliott Trudeau
1984	LP	John Napier Turner
1984–93	PC	Martin Brian Mulroney
1993	PC	Avril Phaedra (Kim) Campbell
1993–2003	LP	Joseph Jacques Jean Chrétien
2003–06	LP	Paul Joseph Martin (Jr)
2006–	CPC	Stephen Joseph Harper

RECENT ELECTIONS

At the elections of 2 May 2011 the incumbent Conservative Party won 167 of 308 seats (143 in 2008) with 39·6% (37·6% in 2008) of votes cast; the New Democratic Party 102 with 30·6% (37 in 2008 with 18·2%); the Liberal Party 34 with 18·9% (76 in 2008 with 26·2%); the Bloc Québécois 4 with 6·0% (50 in 2008 with 10·0%); and the Green Party 1 with 3·9% (no seats in 2008 with 6·8%). Turnout was 61·4% (59·1% in 2008).

CURRENT GOVERNMENT

Governor-General: David Johnston (b. 1941; sworn in 1 Oct. 2010).

In March 2013 the Conservative cabinet comprised:

Prime Minister: Stephen Harper; b. 1959 (Conservative Party; took office on 6 Feb. 2006).

Minister of Aboriginal Affairs and Northern Development: Bernard Valcourt. *Agriculture and Agri-Food, and the Canadian Wheat Board:* Gerry Ritz. *Canadian Heritage and Official Languages:* James Moore. *Citizenship, Immigration and Multiculturalism:* Jason Kenney. *Environment:* Peter Kent. *Finance:* James Flaherty. *Fisheries and Oceans, and the Atlantic Gateway:* Keith Ashfield. *Foreign Affairs:* John Baird. *Health, and the Canadian Northern Economic Development Agency:* Leona Aglukkaq. *Human Resources and Skills Development:* Diane Finley. *Industry:* Christian Paradis. *International Co-operation:* Julian Fantino. *International Trade and the Asian-Pacific Gateway:* Edward Fast. *Justice and Attorney General:* Robert Nicholson. *Labour:* Lisa Raitt. *National Defence:* Peter MacKay. *National Revenue:* Gail Shea. *Natural Resources:* Joe Oliver. *Public Safety:* Vic Toews. *Public Works and Government Services, and Status of Women:* Rona Ambrose. *Transport, Infrastructure and Communities, and the Economic Development Agency of Canada for the Regions of Quebec:* Denis Lebel. *Veterans' Affairs:* Steven Blaney. *Leader of the Government in the House of Commons:* Peter Van Loan. *Leader of the Government in the Senate:* Marjory LeBreton. *President of the Queen's Privy Council for Canada, Minister of Intergovernmental Affairs:* Peter Penashue. *President of the Treasury Board and Minister for the Federal Economic Development Initiative in Northern Ontario:* Tony Clement.

The *Leader of the Opposition* is Thomas Mulcair.

Office of the Prime Minister: http://www.pm.gc.ca

CURRENT LEADERS

Stephen Harper

Position
Prime Minister

Introduction
Stephen Harper's victory in federal elections in Jan. 2006 represented a shift to the right after 12 years of Liberal government overshadowed by allegations of corruption. The free-market economist and leader of the Conservative Party has cast himself as a moderate, progressive, centre-right politician promising to tackle corruption, reduce taxes and lead a more efficient government. He retained power following the Conservative victory in further elections in Oct. 2008 and May 2011.

Early Life
Stephen Harper was born on 30 April 1959 in Toronto, Canada. He graduated from Richview Collegiate Institute in 1978 and moved to Edmonton, Alberta, where he worked as a computer programmer in the oil and gas industry. While studying economics at the University of Calgary in the early 1980s, Harper was influenced by the right-wing monetarist ideas espoused by Ronald Reagan in the USA and Margaret Thatcher in the UK. He graduated with a BA in economics in 1985 and began working for a Conservative member of parliament, Jim Hawkes.

Disillusioned with the Progressive Conservatives (PC) and the government of Brian Mulroney, Harper joined the newly established Reform Party of Canada in 1987, led by the economist Preston Manning. As chief policy officer, Harper helped draft the party's manifesto for the elections of 1988. He became legislative assistant to the Reform Party MP, Deborah Gray, after she won a by-election to represent Beaver River, Alberta in March 1989.

At the elections of Oct. 1993 Harper beat Jim Hawkes to win Calgary West for the Reform Party and became the party's spokesman on finance and national unity. In a run-up to a referendum in Oct. 1995 on the status of Quebec, Harper argued to maintain but decentralize the federation. Disagreements with

Manning led to Harper's decision in late 1996 not to stand in the next election. He resigned his seat in Jan. 1997 and was appointed vice president of the conservative lobby group, the National Citizens Coalition. He also worked as a regular political commentator for the Canadian Broadcasting Corporation.

Harper rejected invitations to run for the PC leadership but returned to politics in March 2002 when he was elected to succeed Stockwell Day as leader of the Canadian Alliance party. He successfully contested a by-election for Calgary Southwest two months later and returned to the House of Commons as leader of the opposition. Following protracted negotiations, Harper reached agreement with the PC leader, Peter MacKay, on a merger between the two parties to form the Conservative Party of Canada in Dec. 2003.

Harper won the new party's leadership election in March 2004 and fought the Liberal prime minister, Paul Martin, in the 2004 election. After taking an early poll lead, the Conservatives lost ground with Harper criticized for supporting the US-led war on Iraq in March 2003. The election in June 2004 saw a victory for the Liberals, who took 135 seats against 99 for the Conservatives.

When the Liberals became mired in a corruption scandal in April 2005, Harper argued that the government had 'lost the moral authority to govern'. He introduced a motion of no confidence in Paul Martin's administration on 24 Nov. 2005, which was passed by 171–133. Parliament was dissolved and elections were scheduled for 23 Jan. 2006. Harper's campaign presented him as head of a modernizing centre-right party that would stimulate economic growth by lowering taxes. Having held a comfortable lead in the opinion polls, the Conservatives won the elections with 36% of the vote, though short of a parliamentary majority. Harper was sworn in as prime minister on 6 Feb. 2006.

Career in Office
Harper promised a smaller 'more focused and effective' government. Eager to pursue closer relations with the USA, he announced a settlement in April of Canada's long-running dispute over softwood timber exports to its neighbour. He also promised more respect for provincial autonomy, particularly in relation to French-speaking Quebec. In Nov. 2006 his proposal to recognize Quebec 'as a nation within a united Canada' was approved by the House of Commons. In foreign policy, he maintained the previous administration's position on Afghanistan, but brought Canadian combat involvement in the country to an end in July 2011. In June 2006 police arrested 17 Islamic extremists allegedly plotting to kill the prime minister.

In Oct. 2007 Harper set out a new government programme focusing on the environment, tax cuts, the fiscal imbalance between the provinces and the federal government, and reform of political institutions (to include an elected Senate rather than an appointed body). He also promised a parliamentary vote on any extension of the Canadian military mission in Afghanistan.

Despite the shadow of global financial turmoil and opposition accusations of economic complacency, Harper sought a more secure mandate for his administration by calling n early general election for Oct. 2008 (although he had previously pushed through legislation fixing the normal life of a parliament at four years). The Conservatives improved their representation in the House of Commons but again failed to achieve a parliamentary majority. In Dec. the opposition parties sought to bring down the minority government over its response to the economic situation, but Harper asked the Governor-General to suspend parliament until Jan. 2009 thereby postponing a no-confidence vote. When parliament resumed in late Jan. the opposition alliance backed down and in Feb. the government secured approval of its 2009 budget, including a two-year stimulus package for the economy, with conditional Liberal Party support which ensured Harper's political survival. However, in Oct. 2009 the Liberals tabled a no-confidence motion accusing the government of having lost control

of the public finances. The motion was defeated by 144 votes to 117 with the help of the New Democratic Party and Bloc Québécois, and in Nov. the Conservatives gained two seats in four parliamentary by-elections, suggesting some improvement in the government's standing with the electorate.

At the end of Dec. 2009 Harper again successfully sought the prorogation of parliament until the beginning of March 2010, a move that attracted considerable criticism from opposition leaders who accused him of seeking to avoid political debate, particularly on Canada's military actions in Afghanistan.

At the general election in May 2011 Harper led the Conservatives to a decisive victory to head a majority government for the first time since taking office.

In Dec. 2011 Canada became the first country to formally withdraw from the 1997 Kyoto Protocol on climate change, a decision which was widely criticized internationally.

Relations with Iran deteriorated from late 2011 as Canada supported a tightening of international sanctions over Tehran's nuclear development programme. In Sept. 2012 the Harper government broke off diplomatic relations, claiming Iran was a threat to world security.

DEFENCE

The armed forces have been unified since 1968 as the Canadian Armed Forces (usually referred to as the Canadian Forces). The three commands are the Canadian Army (known until Aug. 2011 as Land Force Command), the Royal Canadian Navy (Maritime Command until Aug. 2011) and the Royal Canadian Air Force (Air Command until Aug. 2011). In 2007 the armed forces numbered 64,000; reserves, 65,800.

Military expenditure totalled US$19,290m. in 2008 (US$581 per capita). In 2007 defence spending represented 1·2% of GDP.

Army
The Land Force Command (now the Canadian Army) numbered 33,300 in 2007; reserves include a Militia of 24,700.

Navy
The naval combatant force is headquartered at Halifax (Nova Scotia), and includes four diesel submarines, three destroyers and 12 helicopter-carrying frigates. Naval personnel in 2007 numbered about 11,100, with 4,200 reserves. The main bases are Halifax, where about two-thirds of the fleet is based, and Esquimalt (British Columbia).

Air Force
The air forces numbered 19,600 in 2007 with 107 combat-capable aircraft (mainly CF-18s).

ECONOMY

Services accounted for 66% of GDP in 2008, industry 32% and agriculture 2%.

According to the anti-corruption organization *Transparency International*, Canada ranked equal ninth in the world in a 2012 survey of the countries with the least corruption in business and government. It received 84 out of 100 in the annual index.

Overview
Canada has prodigious natural resources, including the second largest oil reserves in the world. In 2010 Canada produced 2·2m. bbls of oil per day and 94·7bn. cubic metres of natural gas.

Canada has a trading surplus with the USA, which buys three-quarters of exports (mainly energy, including oil, natural gas, uranium and electrical power supplies). The next biggest trading partner is the UK, accounting for just 3% of exports. The 1989 US-Canada Free Trade Agreement and the 1994 North American Free Trade Agreement greatly increased economic integration between Canada, the USA and Mexico. However, over-reliance on the fortunes of the US economy contributed to the economy contracting by 2·8% in 2009. Nonetheless, the economy bounced back to grow by 3·2% in 2010, 2·6% in 2011 and a provisional 1·8% in 2012. Growth has been buoyed by rising employment, vibrant consumer spending and increased demand for industrial goods.

The banking sector weathered the global crisis far better than its US counterpart and did not require a taxpayer bailout. With a largely prime mortgage market, well-capitalized banks and solid deposit institutions anchoring investment banks, the World Economic Forum ranked Canada as the most sound banking system in the world for five years in a row to 2012. In Aug. 2012, only a third of mortgages were in arrears. Strict fiscal enforcement and regulation by the Office of Superintendent of Financial Institutions and the Financial Consumer Agency of Canada maintains the reputation of the banking system.

The labour force is well-skilled, the economy technologically well-equipped and the economy diversified. The service sector accounts for two-thirds of GDP and employs approximately three-quarters of the workforce. The primary sector accounts for roughly 2% of GDP but contributes a quarter of total export earnings and is the leading source of income in several provinces.

Canada joined the G8 in 1997. Its GNI per capita ranks among the highest of any OECD nation. However, an ageing population and increasing health care spending present long-term challenges.

Currency
The unit of currency is the *Canadian dollar* (CAD) of 100 *cents*. In Sept. 2009 gold reserves were 0·11m. troy oz and foreign exchange reserves totalled US$46,327m. Total money supply was $434,519m. CDN in Dec. 2008.

Inflation rates (based on OECD statistics):

2002	2003	2004	2005	2006	2007	2008	2009	2010	2011
2·3%	2·7%	1·8%	2·2%	2·0%	2·1%	2·4%	0·3%	1·8%	2·9%

Budget
Consolidated federal, provincial, territorial and local government revenue and expenditure for fiscal years ending 31 March (in $1m. CDN):

	2005–06	2006–07	2007–08	2008–09
Revenue	533,031	561,238	600,575	585,799
Expenditure	516,669	545,533	580,922	594,594

In 2008–09 revenue included (in $1m. CDN): income taxes, 248,655; consumption taxes, 107,150; property and related taxes, 54,862; investment income, 54,068. Expenditure included: social services, 151,869; health, 121,577; education, 95,732; debt charges, 45,384.

On 1 Jan. 1991 a 7% Goods and Services Tax (GST) was introduced, superseding a 13·5% Manufacturers' Sales Tax. This was reduced to 6% from 1 July 2006 and 5% from 1 Jan. 2008. A Harmonized Sales Tax (HST) that combines the GST and the regional Provincial Sales Tax (PST) into a single value added sales tax applies in five provinces: New Brunswick (13%), Newfoundland and Labrador (13%), Nova Scotia (15%), Ontario (13%) and Prince Edward Island (14%). Following a referendum held in Aug. 2011, British Columbia—which had been using HST—reinstated the GST/PST system on 1 April 2013.

Performance
Real GDP growth rates (based on OECD statistics):

2002	2003	2004	2005	2006	2007	2008	2009	2010	2011
2·8%	2·0%	3·2%	3·1%	2·7%	2·1%	1·1%	−2·8%	3·2%	2·6%

Real GDP grew by 1·8% in 2012 according to Statistics Canada. Total GDP was US$1,736·1bn. in 2011.

Banking and Finance
The Bank of Canada (established 1935) is the central bank and bank of issue. The *Governor* is appointed by the Bank's directors for seven-year terms. Mark Carney was set to step down on 1 June 2013 to be succeeded by Stephen Poloz. The Minister of Finance owns the capital stock of the Bank on behalf of Canada. Banks in Canada are chartered under the terms of the Bank Act, which imposes strict conditions on capital reserves, returns to the federal government, types of lending operations, ownership and other matters. As of June 2008 there were 20 domestic banks, 24 foreign bank subsidiaries, 22 full service foreign bank branches and seven foreign bank lending branches operating in Canada; these manage over $2·7trn. CDN in assets between them. Chartered banks accounted collectively for over 70% of the total assets of the Canadian financial services sector, with the six largest domestic banks (Canadian Imperial Bank of Commerce, Bank of Nova Scotia, Bank of Montreal, National Bank of Canada, TD Canada Trust and Royal Bank of Canada) accounting for over 90% of the total assets held by the banking industry. In 2006 Canada had the highest number of automated bank machines per capita in the world (at 1,735 per 1m. inhabitants). The First Nations Bank was founded in Dec. 1996 to provide finance to Inuit and Indian entrepreneurs.

The activities of banks are monitored by the federal Office of the Superintendent of Financial Institutions (OSFI), which reports to the Minister of Finance. Canada's federal financial institutions legislation is reviewed at least every five years. Significant legislative changes were made in 1992, updating the regulatory framework and removing barriers separating the activities of various types of financial institutions. In 1999 legislation was passed allowing foreign banks to establish operations in Canada without having to set up Canadian-incorporated subsidiaries. In 2001 Bill C-8, establishing the Financial Consumer Agency of Canada (the FCAC), was implemented. It aimed to foster competition in the financial sector and provide a holding company option allowing additional organizational flexibility to banks and insurance companies. The FCAC is responsible for enforcing consumer-related provisions of laws governing federal financial institutions.

In June 2012 Canada's gross external debt amounted to US$1,202,858m.

There are stock exchanges at Calgary (Alberta Stock Exchange), Montreal, Toronto, Vancouver and Winnipeg.

ENERGY AND NATURAL RESOURCES
Environment
Canada's carbon dioxide emissions from the consumption and flaring of fossil fuels in 2008 were the equivalent of 17·3 tonnes per capita.

An *Environmental Performance Index* compiled in 2008 ranked Canada 12th in the world, with 86·6%. The index examined various factors in six areas—air pollution, biodiversity and habitat, climate change, environmental health, productive natural resources and water resources.

Electricity
Generating capacity, 2007, 124·7m. kW. Production, 2007, 639·84bn. kWh (368·52bn. kWh hydro-electric, 174·75bn. kWh thermal and 93·49bn. kWh nuclear); consumption per capita was 18,636 kWh in 2007. In 2010 there were 18 nuclear reactors in use. Canada is one of the world's leading exporters of electricity, with 44·7bn. kWh in 2007.

Oil and Gas
Oil reserves in 2011 were 175·2bn. bbls (6·0bn. bbls of conventional crude oil and condensate reserves and 169·2bn. bbls of oil sands reserves in Alberta), ranking Canada third after Venezuela and Saudi Arabia. Natural gas reserves in 2008 were 1,630bn. cu. metres. Production of oil, 2008, 156·7m. tonnes; natural gas, 175·2bn. cu. metres.

Canada is the third largest producer of natural gas, after the USA and Russia. Canada's first off-shore field, 250 km off Nova Scotia, began producing in June 1992.

Minerals
Mineral production in 1,000 tonnes (in 2008): sand and gravel, 241,591; coal, 67,750; iron ore, 32,102; salt, 14,224; potash, 10,379 (the highest of any country in the world); lignite, 9,900; gypsum, 5,819; aluminium, 3,127; lime, 2,046; peat, 1,246; zinc, 705; copper, 584; nickel, 246; asbestos, 160; lead, 87; uranium (2010), 9·8; cobalt, 4·8. Silver production in 2008 was 709 tonnes; gold, 95 tonnes; diamonds, 14·5m. carats.

Agriculture
Grain growing, dairy farming, fruit farming, ranching and fur farming are all practised. In 2006 over 346,000 people were engaged in agriculture (2% of the labour force).

According to the 2006 census the total land area was 9,220,770 sq. km, of which 675,867 sq. km were on farms. There were 229,373 farms in 2006 (246,923 in 2001); average size, 294·6 ha. Total farm cash receipts (2006), $36,949,543,000 CDN. There were 733,182 tractors in 2006 and 102,924 combine harvesters.

The following table shows the value of receipts for selected agricultural commodities in 2009 (in $1m. CDN):

Crops	22,971	Livestock and products	17,903
Barley	763	Beef	5,278
Canola (rapeseed)	5,037	Dairy	5,456
Corn	1,300	Hogs	2,892
Soybeans	1,339	Poultry	2,383
Wheat	4,034		

Output (in 1,000 tonnes) and harvested area (in 1,000 ha.) of crops:

	Output		Harvested Area	
	2008	2009	2008	2009
Wheat	28,611	26,848	10,032	9,638
Rapeseeds	12,643	11,825	6,494	6,104
Maize	10,592	9,561	1,169	1,142
Barley	11,781	9,517	3,502	2,918
Potatoes	4,724	4,581	151	146
Soybeans	3,336	3,504	1,195	1,382
Peas	3,571	3,379	1,582	1,487
Oats	4,273	2,798	1,448	948
Lentils	1,043	1,510	700	963
Linseeds	861	930	625	623
Sugar beets	345	658	7	11
Tomatoes	770	458	8	6
Carrots and turnips	277	359	8	8
Rye	316	280	131	115
Beans	266	220	125	113
Cabbage, broccoli, etc.	176	189	8	9
Sunflower seeds	112	102	69	63
Cucumbers and gherkins	212	54	2	2

Canada is the world's second largest producer of rapeseeds and oats.

Livestock
In parts of Saskatchewan and Alberta, stockraising is still carried on as a primary industry, but the livestock industry of the country at large is mainly a subsidiary of mixed farming The following table shows the numbers of livestock (in 1,000) by provinces in July 2009:

Provinces	Total cattle and calves (including dairy cows)	Dairy cows	Sheep and lambs	Pigs
Newfoundland and Labrador	11·8	6·2	4·0	1·8
Prince Edward Island	68·5	13·0	4·1	48·8
Nova Scotia	88·8	22·2	26·2	16·4
New Brunswick	82·1	18·9	8·1	81·2
Quebec	1,385·0	366·0	285·0	3,870·0

Provinces	Total cattle and calves (including dairy cows)	Dairy cows	Sheep and lambs	Pigs
Ontario	1,828·8	321·0	315·0	3,064·8
Manitoba	1,430·0	45·5	71·0	2,530·0
Saskatchewan	3,370·0	29·5	114·0	810·0
Alberta	5,870·0	88·5	177·0	1,530·0
British Columbia	705·0	72·0	58·0	112·0
Total	14,840·0	982·8	1,062·4	12,065·0

Other livestock totals (2007 estimates): chickens, 165m.; turkeys, 5·6m.

Livestock Products
Slaughterings in 2008: pigs, 21·69m.; cattle, 3·84m.; sheep, 0·72m. Production, 2009 (in 1,000 tonnes): pork, bacon and ham, 1,945; beef and veal, 1,255; poultry meat, 1,212; horse meat, 18; lamb and mutton, 17; cow's milk, 8,213; hens' eggs, 422; cheese, 390; honey, 29; hides, 96.

Fruit production in 2007, in 1,000 tonnes: apples, 446; blueberries, 78; grapes, 76; cranberries, 71; peaches, 34; strawberries, 24; raspberries, 12; pears, 12.

Forestry
Forests and other wooded land make up nearly half of Canada's landmass. Forestry is of great economic importance, and forestry products (pulp, newsprint, building timber) constitute Canada's most valuable exports. In 2010 Canada had 310·13m. ha. of forest land (8% of the world's forest cover) and 91·95m. ha. of other wooded land. 1·7m. ha. were burned by forest fires in 2007. 195·91m. cu. metres of roundwood was produced in 2007.

Fur Trade
In 2006, 1,047,400 wildlife pelts (valued at $25,777,200 CDN) and 1,652,200 ranch-raised pelts (valued at $90,176,000 CDN) were produced.

Fisheries
In 2007 landings of commercial fisheries totalled 1,019,224 tonnes with a value of $1,951m. CDN. More than 96% of the total catch in 2007 was from sea fishing; Atlantic landings totalled 815,903 tonnes and Pacific landings 171,019 tonnes, with freshwater fish totalling 32,303 tonnes. Value of sea fish landed in 2007 was $1,888m. CDN and of freshwater fish $64m. CDN.

INDUSTRY

The leading companies by market capitalization in Canada in March 2012 were: Royal Bank of Canada (US$83·4bn.); Toronto-Dominion Bank (US$76·7bn.); and Bank of Nova Scotia (US$63·5bn.).

Value of manufacturing shipments for all industries in 2008 was $598,217·1m. CDN. Principal manufactures in 1,000 tonnes: petrol (2007), 32,630; distillate fuel oil (2007), 31,223; wood pulp (2007), 22,381; paper and paperboard (2007), 18,113; crude steel (2007), 15,569; cement (producers' shipments, 2008), 13,672; pig iron (2007), 8,577; residual fuel oil (2007), 8,407; newsprint (2007), 6,640; sulphuric acid (2007), 4,328; jet fuel (2007), 4,038; kerosene (2007), 1,544. Output of other products: 2·05m. motor vehicles (2008); 19bn. cigarettes and equivalent (2009; down from 73bn. in 1981); sawn lumber (2008), 57·25m. cu. metres; chipboard (2007), 10·04m. cu. metres; plywood (2007), 2·07m. cu. metres.

According to the World Bank's *Doing Business 2010* Canada is the second easiest country in which to start a business, after New Zealand, and the eighth easiest country in which to do business.

Labour
In 2008 there were (in 1,000), 17,125·8 (8,104·5 females) in employment, with principal areas of activity as follows: trade, 2,678·8; manufacturing, 1,970·3; health care and social assistance,

1,903·4; construction, 1,232·2; educational services, 1,192·8; finance, insurance, real estate and leasing, 1,075·4; accommodation and food services, 1,073·5; public administration, 925·7; transport and warehousing, 857·7. In Dec. 2012 the unemployment rate was 7·1%.

In 2006, 813,000 working days were lost in industrial disputes, a sharp decline from 4·1m. in 2005. Between 1996 and 2005 strikes cost Canada an average of 208 days per 1,000 employees a year. Canada's figure was one of the highest in the industrialized world.

INTERNATIONAL TRADE

A North American Free Trade Agreement (NAFTA) between Canada, Mexico and the USA was signed on 7 Oct. 1992 and came into force on 1 Jan. 1994.

Imports and Exports
Trade in $1m. CDN:

	2004	2005	2006	2007	2008
Imports	363,158	387,838	404,346	415,229	442,988
Exports	429,006	450,210	453,951	463,127	489,857

Canada is heavily dependent on foreign trade. In 2006 imports of goods and services were equivalent to 34% of GDP and exports equivalent to 36%.

Leading import suppliers, 2007: USA, 54·2%; China, 9·4%; Mexico, 4·2%; Japan, 3·8%; Germany, 2·8%; UK, 2·8%. Leading export markets, 2007: USA, 79·0%; UK, 2·8%; China, 2·1%; Japan, 2·0%; Mexico, 1·1%; Germany, 0·9%.

Main categories of imports, 2008 (in $1m. CDN): machinery and equipment, 122,628·3 (industrial and agricultural machinery, 34,251·8; aircraft and other transportation equipment, 17,544·2); industrial goods and materials, 91,573·8 (metals and metal ores, 32,573·2; chemicals and plastics, 31,561·7); automotive products, 71,959·0 (motor vehicle parts, 30,847·2). Exports, 2008 (in $1m. CDN): energy products, 125,792·2 (crude petroleum, 60,969·7; natural gas, 33,046·0); industrial goods and materials, 111,511·5 (metals and alloys, 39,976·7; chemicals, plastics and fertilizers, 35,910·1); machinery and equipment, 92,994·4 (industrial and agricultural machinery, 23,442·6; aircraft and other transport equipment, 20,339·6); automotive products, 61,082·6 (passenger cars and chassis, 34,069·2; motor vehicle parts, 19,749·5); forestry products, 25,659·2 (newsprint and other paper and paperboard products, 10,092·0; lumber and sawmill products, 9,169·0).

COMMUNICATIONS

Roads
In 2007 there were 1,409,000 km of public roads, including 17,000 km of motorways, 86,000 km of main roads and 115,000 km of secondary roads. The National Highway System, spanning almost 25,000 km, includes the Trans-Canada Highway and other major east–west and north–south highways. While representing only 3% of total road infrastructure, the system carries about 30% of all vehicle travel in Canada.

Registered road motor vehicles totalled 21,387,132 in 2009; they comprised 19,876,990 passenger cars and light vehicles, 829,695 trucks and truck tractors (weighing at least 4,500 kg), 85,579 buses and 594,866 motorcycles and mopeds.

There were 2,209 fatalities (a rate of 6·6 deaths per 100,000 population) in road accidents in 2009, the lowest annual total since records began.

Rail
Canada has two great trans-continental systems: the Canadian National Railway system (CN), a body privatized in 1995 which operated 25,185 km of routes in 2005, and the Canadian Pacific Railway (CP), operating 22,370 km. A government-funded organization, VIA Rail, operates passenger services in all regions of Canada; 4·6m. passengers were carried in 2008. There are

several provincial and private railways operating 12,705 km (2005).

There are metros in Montreal and Toronto, and tram/light rail systems in Calgary, Edmonton, Ottawa, Toronto and Vancouver.

Civil Aviation
Civil aviation is under the jurisdiction of the federal government. The technical and administrative aspects are supervised by Transport Canada, while the economic functions are assigned to the Canadian Transportation Agency.

The busiest Canadian airport is Toronto Pearson International, which in 2008 handled 32,335,000 passengers (18,439,000 on international flights), ahead of Vancouver International, with 17,852,000 passengers (9,345,000 on domestic flights) and Montreal (Pierre Elliot Trudeau International), with 12,813,000 passengers (5,278,000 on domestic flights). Toronto is also the busiest airport for freight, handling 483,975 tonnes in 2008.

Air Canada (privatized in July 1989) took over its main competitor, Canadian Airlines, in April 2000. In 2005 Air Canada carried 23·5m. passengers (11·9m. on international flights); passenger-km totalled 71·0bn. Other major Canadian airlines are Air Transat and WestJet.

Shipping
In Jan. 2009 there were 171 ships of 300 GT or over registered, totalling 1·29m. GT. Of the 171 vessels registered, 114 were passenger ships, 26 oil tankers, 18 general cargo ships, nine bulk carriers, two chemical tankers and two container ships.

In 2009 the total tonnage handled by Canadian ports was 409·1m. tonnes (251·0m. loaded and 158·1m. unloaded). Canada's leading port in terms of cargo handled is Vancouver. Other major ports are Saint John, Fraser River, Montreal and Quebec.

The major canals are those of the St Lawrence Seaway. Main commodities moved along the seaway are grain, iron ore, coal, other bulk and steel. The St Lawrence Seaway Management Corporation was established in 1998 as a non-profit making corporation to operate the Canadian assets of the seaway for the federal government under a long-term agreement with Transport Canada.

In 2010 total traffic on the Montreal-Lake Ontario (MLO) section of the seaway was 26,920,000 tonnes; on the Welland Canal section it was 29,180,000 tonnes. There were 3,925 vessel transits in 2010, generating $60·7m. CDN in toll revenue.

Telecommunications
In 2008 there were 18,250,000 main (fixed) telephone lines. In the same year mobile phone subscribers numbered 22,093,000 (664·2 per 1,000 persons). A 2010 survey found that 78% of households had a mobile phone and 13% only had a mobile phone; 17% only had a landline. Canada had 25·1m. internet users in 2008. The fixed broadband penetration rate in Dec. 2010 was 30·7 subscribers per 100 inhabitants. In Dec. 2011 there were 17·1m. Facebook users (49% of the population).

SOCIAL INSTITUTIONS

Justice
The courts in Canada are organized in a four-tier structure. The Supreme Court of Canada, based in Ottawa, is the highest court, having general appellate jurisdiction in civil and criminal cases throughout the country. It is comprised of a Chief Justice and eight puisne judges appointed by the Governor-in-Council, with a minimum of three judges coming from Quebec. The second tier consists of the Federal Court of Appeal and the various provincial courts of appeal. The third tier consists of the Federal Court (which replaced the Exchequer Court in 1971), the Tax Court of Canada and the provincial and territorial superior courts (which include both a court of general trial jurisdiction and a provincial court of appeal). The majority of cases are heard by the provincial courts, the fourth tier in the hierarchy. They are generally divided within each province into various divisions defined by the subject

matter of their respective jurisdictions (for example a Traffic Division, a Small Claims Division, a Family Division and a Criminal Division).

There were 2,194,968 Criminal Code Offences (excluding traffic) reported in 2008 (down from 2,440,650 in 1998), including 310,398 crimes of violence (up from 300,058 in 1998). In 2010 there were 554 homicides in Canada, giving a rate of 1·6 homicides per 100,000 population (the lowest rate since 1966). In 2007–08 the average daily population in penal institutions (including pre-trial detainees) was 38,348 (116 per 100,000 of national population). The death penalty was abolished for all crimes in 1998.

Police
Total police officers in Canada in 2008 numbered 65,283. There were 12,207 female police officers, up from 3,573 in June 1990. Policing costs in 2007 totalled $10·54bn. CDN.

Royal Canadian Mounted Police (RCMP)
The RCMP is Canada's national police force maintained by the federal government. Established in 1873 as the North-West Mounted Police, it became the Royal Northwest Mounted Police in 1904. Its sphere of operations was expanded in 1918 to include all of Canada west of Thunder Bay, Ontario. In 1920 the force absorbed the Dominion Police and its headquarters was transferred from Regina, Saskatchewan to Ottawa, Ontario. Its title also changed to Royal Canadian Mounted Police. The RCMP is responsible to the Minister of Public Safety and Emergency Preparedness Canada and is controlled by a Commissioner who is empowered to appoint peace officers in all the provinces and territories of Canada.

The responsibilities of the RCMP are national in scope. The administration of justice within the provinces, including the enforcement of the Criminal Code of Canada, is the responsibility of provincial governments, but all the provinces except Ontario and Quebec have entered into contracts with the RCMP to enforce criminal and provincial laws under the direction of the respective Attorneys-General. In these eight provinces the RCMP is under agreement to provide police services to municipalities as well. The RCMP is also responsible for all police work in the three territories—Yukon, Northwest Territories and Nunavut— enforcing federal law and territorial ordinances. The 16 Divisions, alphabetically designated, make up the strength of the RCMP across Canada; they comprise 740 detachments containing varying numbers of police officers. Headquarters Division, as well as the Office of the Commissioner, is located in Ottawa.

Supporting Canada's law enforcement agencies, the RCMP's National Police Services includes seven diverse service lines providing a broad range of programmes and services throughout Canada. It comprises Information and Identification Services, Forensic Laboratory Services, Canadian Police College, Criminal Intelligence Service Canada, Technical Operations, National Child Exploitation Coordination Centre and Chief Information Officer.

In 2005 the Force had a total strength of over 22,000 including regular members, special constables, civilian members and public service employees. It maintained 10,385 motor vehicles, 112 police dog teams across Canada and 195 horses.

The Force has 16 divisions actively engaged in law enforcement, one Headquarters Division and one training division. Marine services are divisional responsibilities and the Force currently has 308 boats at various points across Canada. The Air Services Branch has offices throughout the country and maintains a fleet of 34 operational aircraft.

Education
Under the Constitution the provincial legislatures have powers over education. These are subject to certain qualifications respecting the rights of denominational and minority language schools. School board revenues derive from local taxation on real property, and government grants from general provincial revenue.

In 2006–07 there were 5,162,363 pupils enrolled in elementary and secondary public schools and 376,294 educators (including teaching support staff and administrators).

The federal government shares responsibility with First Nations for the provision of education to children ordinarily resident on reserve and attending provincial, federal or band-operated schools. In 2006–07 Indian and Northern Affairs Canada (since renamed Aboriginal Affairs and Northern Development Canada) supported the elementary and secondary education of 120,000 First Nations children living on reserves. About 60% of First Nations children attend schools on reserves, while 40% go off reserve to schools under provincial authority. Funding is also provided for postsecondary tuition, books and living allowances for students residing on or off reserve.

The Association of Universities and Colleges of Canada represents 94 public and private not-for-profit universities and degree-level colleges. In 2005–06 there were 781,440 full-time and 266,070 part-time students enrolled in universities and 461,589 full-time and 151,911 part-time students enrolled in colleges. Full-time faculty staff in universities numbered 40,800 (13,400 women) in 2006. According to 2006 census data, 3,985,745 adults aged between 25 and 64 had a university degree (up 24% from 3,207,440 in 2001). Six out of every ten adults between 25 and 64 (10,541,900 people according to the census) had completed some form of post-secondary education in 2006, with a university degree, college diploma or post-secondary certificate.

The adult literacy rate is at least 99%.

In 2006–07 public education expenditure represented 3·3% of GDP.

Health
Constitutional responsibility for health care services rests with the provinces and territories. Accordingly, Canada's national health insurance system consists of an interlocking set of provincial and territorial hospital and medical insurance plans conforming to certain national standards rather than a single national programme. The Canada Health Act (which took effect from April 1984 and consolidated the original federal health insurance legislation) sets out the national standards that provinces and territories are required to meet in order to qualify for full federal health contributions, including: provision of a comprehensive range of hospital and medical benefits; universal population coverage; access to necessary services on uniform terms and conditions; portability of benefits; and public administration of provincial and territorial insurance plans. From 1996–97 the federal government's contribution to provincial health and social programmes was consolidated into a single block transfer—the Canada Health and Social Transfer (CHST). However, following reforms introduced by the 2003 Accord on Health Care Renewal to improve health spending accountability it was split into the Canada Health Transfer (CHT) for health and the Canada Social Transfer (CST) for post-secondary education, social service and social assistance in April 2004. The CHT is the federal government's largest transfer and funding is provided to provinces as a combination of cash contributions and tax transfers. In 2012–13 the provinces and territories were scheduled to receive a CHT cash transfer of $29·0bn. CDN and CHT tax transfers of $14·7bn. CDN. Over and above these health transfers, the federal government also provides financial support for such provincial and territorial extended health care service programmes as nursing-home care, certain home care services, ambulatory health care services and adult residential care services.

The approach taken by Canada is one of state-sponsored health insurance. The advent of insurance programmes produced little change in the ownership of hospitals, almost all of which are owned by non-government non-profit corporations, or in the rights and privileges of private medical practice. Patients are free to choose their own general practitioner. Except for a small percentage of the population whose care is provided for under other legislation (such as serving members of the Canadian Armed Forces and inmates of federal penitentiaries), all residents are eligible, regardless of whether they are in the workforce. Benefits are available without upper limit so long as they are medically necessary, provided any registration obligations are met.

In addition to the benefits qualifying for federal contributions, provinces and territories provide additional benefits at their own discretion. Most fund their portion of health costs out of general provincial and territorial revenues. Most have charges for long-term chronic hospital care geared, approximately, to the room and board portion of the OAS–GIS payment mentioned under Welfare *below*. Health spending accounted for 10·1% of GDP in 2007.

In 2006 there were: 62,307 physicians, giving a rate of 19 per 10,000 population; 38,310 dentists (12 per 10,000 population); and 327,224 nursing and midwifery personnel (101 per 10,000 population).

Welfare

The social security system provides financial benefits and social services to individuals and their families through programmes administered by federal, provincial and municipal governments and voluntary organizations. Federally, Human Resources and Skills Development is responsible for research into the areas of social issues, provision of grants and contributions for various social services and the administration of income security programmes, including the Old Age Security (OAS) programme, the Guaranteed Income Supplement, the Allowance and the Canada Pension Plan (CPP).

The Old Age Security pension is payable to persons 65 years of age and over who satisfy the residence requirements stipulated in the Old Age Security Act. The amount payable, whether full or partial, is also governed by stipulated conditions, as is the payment of an OAS pension to a recipient who absents himself from Canada. OAS pensioners with little or no income apart from OAS may, upon application, receive a full or partial supplement known as the Guaranteed Income Supplement (GIS). Entitlement is normally based on the pensioner's income in the preceding year, calculated in accordance with the Income Tax Act (which excepts OAS income). The spouse or common-law partner (same sex or opposite sex since the Modernization of Benefits and Obligations Act of 2000) of an OAS pensioner, meeting specific citizenship/residency requirements and aged 60 to 64, may be eligible for a full or partial benefit called the Allowance (introduced in 1975 and formerly known as the Spouse's Allowance). The Allowance is payable, on application, to spouses/partners of individuals receiving OAS and GIS, where the couple's joint income is classified as low. An Allowance for the Survivor (introduced in 1985) is available to persons aged 60–64 whose spouse or common-law partner has died and where the survivor has not entered into a subsequent marriage or common law relationship. The Allowance and Allowance for the Survivor stop when the recipient becomes eligible for the OAS at age 65.

As of 1 July 2011 the maximum OAS pension was $533·70 CDN monthly; the maximum Guaranteed Income Supplement was $723·65 CDN monthly for a single pensioner or a married pensioner whose spouse or common law partner was not receiving an OAS pension, and $479·84 CDN monthly for each spouse of a married couple/common law partner where both were pensioners.

The Canada Pension Plan is designed to provide workers with a basic level of income protection in the event of retirement, disability or death. Benefits may be payable to a contributor, a surviving spouse or an eligible child. Actuarially adjusted retirement benefits may begin as early as age 60 or as late as age 70. Benefits are determined by the contributor's earnings and contributions made to the Plan. Contribution is compulsory for most employed and self-employed Canadians aged 18 to 65. The CPP does not operate in Quebec, which has exercised its

constitutional prerogative to establish a similar plan. In 2011 the maximum monthly retirement pension payable under CPP was $960·00 CDN; the maximum disability pension was $1,153·37 CDN; and the maximum surviving spouse's pension was $576·00 CDN (for survivors 65 years of age and over). The survivor pension payable to a surviving spouse under 65 (maximum of $529·09 CDN in 2011) is composed of two parts: a flat-rate component and an earnings-related portion.

The CPP is financed through mandatory contributions from employers, employees and the self-employed, and through revenue generated from investments. The combined contribution rate is 9·9% of earnings between the year's basic exemption ($3,500 CDN in 2010) and the year's maximum pensionable earnings ($47,200 CDN in 2010). The contribution rate (maximum rate of $4,326·30 CDN in 2010) is split equally between employers and employees, while the self-employed pay both shares. Changes to the CPP are being introduced from 2011 to 2016 whereby the monthly pension amount increases by a larger percentage if taken after age 65 and decreases by a larger percentage if taken before 65.

The Canada Pension Plan Investment Board, an independent investment organization separate from the CPP, was established to invest excess CPP funds in a diversified portfolio of securities, beginning operations in April 1998. A total of 4·3m. Canadians received Canada Pension Plan benefits amounting to $27·5bn. CDN in fiscal year 2007–08. Social security agreements co-ordinate the operation of the Old Age Security and the CPP with the comparable social security programmes of certain other countries.

Canada Child Tax Benefit (CCTB) is a tax-free monthly payment made to eligible families to help them with the cost of raising children under 18. Included with the CCTB is the National Child Benefit Supplement (NCBS), a monthly benefit for low-income families with children.

RELIGION

Membership of religious denominations (according to census analysis):

	1991	2001	% change 1991–2001
Anglican Church of Canada	2,188,110	2,035,500	−7·0
Canadian Baptist Ministries	663,360	729,470	10·0
Christian Orthodox	387,395	479,620	23·8
Lutheran Church	636,205	606,590	−4·7
Pentecostal Assemblies of Canada	436,435	369,475	−15·3
Presbyterian Church	636,295	409,830	−35·6
Roman Catholic Church	12,203,625	12,793,125	4·8
United Church of Canada	3,093,120	2,839,125	−8·2

Membership of other denominations in 2001 (census figures): Jehovah's Witnesses, 154,745; Jews, 329,995; Latter-day Saints (Mormons), 104,750; Mennonites, 191,465; Muslims, 579,640; Salvation Army, 87,785. In Feb. 2013 the Roman Catholic church had three cardinals.

CULTURE

World Heritage Sites

Sites under Canadian jurisdiction which appear on UNESCO's world heritage list are (with year entered on list): L'Anse aux Meadows National Historic Site (1978), the remains of an 11th-century Viking settlement in Newfoundland and Labrador; Nahanni National Park (1978), containing canyons, waterfalls and a limestone cave system—fauna includes wolves, grizzly bears, caribou, Dall's sheep and mountain goats; Dinosaur Provincial Park (1979), in Alberta, a major area for fossil discoveries, including 35 species of dinosaur; SGaang Gwaii (Anthony Island) (1981), illustrating the Haida people's art and way of life; Head-Smashed-In Buffalo Jump (1981), in southwest Alberta, incorporating an aboriginal camp—the name relates to the aboriginal custom of killing buffalo by chasing them over a precipice; Wood Buffalo National Park (1983), in north-central Canada, home to America's largest population of wild bison as well as important fauna; Canadian Rocky Mountain Parks (1984 and 1990), incorporating the neighbouring parks of Banff, Jasper, Kootenay and Yoho, as well as the Mount Robson, Mount Assiniboine and Hamber provincial parks, and the Burgess Shale fossil site; Historic District of Québec (1985), retaining aspects of its French colonial past; Gros Morne National Park (1987), in Newfoundland and Labrador; Old Town Lunenburg (1995), a well-preserved British colonial settlement established in 1753; Miguasha Park (1999), among the world's most important fossil sites for fish species of the Devonian age, forerunners of the first four-legged, air-breathing terrestrial vertebrates (tetrapods); Rideau Canal (2007), a 19th century monumental canal running from Ottawa to Kingston Harbour; Joggins Fossil Cliffs (2008), a 689 ha. palaeontological site along the coast of Nova Scotia; and Landscape of Grand Pré in Nova Scotia (2012), a symbolic reference landscape for the Acadian way of life and deportation.

Two UNESCO World Heritage Sites fall under joint Canadian and US jurisdiction: Kluane/Wrangell-St Elias/Glacier Bay/ Tatshenshini-Alsek (1979, 1992 and 1994), parks in Yukon, British Columbia and Alaska, include the world's largest non-polar icefield, glaciers and high peaks, and important fauna; Waterton Glacier International Peace Park (1995), in Alberta and Montana, with collections of plant and mammal species as well as prairie, forest, and alpine and glacial features.

Press
In 2008 there were 104 daily papers with a total average circulation of 5·92m.; *The Toronto Star* had the largest circulation at 431,000 in 2008, then *The Globe and Mail* with 330,000. *Le Journal de Montréal* is the largest francophone daily with an average daily circulation in 2008 of 266,000.

Tourism
In 2010 foreign visitors made 15,864,000 overnight trips to Canada of which 11,749,000 were made by Americans. The next biggest tourist markets are the UK, France, Germany, Japan and Australia. Tourism expenditure by staying visitors amounted to $11,902m. CDN in 2010. In 2010, 617,300 were employed in tourism.

Festivals
Canada Day, held each July, is marked nationwide with firework displays, parades and parties. The Quebec Winter Festival is held each Feb. The Calgary Stampede (the world's largest rodeo, incorporating a series of concerts and a carnival) is in July. Also in July are the Québec Festival d'Été/Summer Festival (featuring music and art performances) and the Montréal Juste Pour Rire/ Just for Laughs comedy festival. The Toronto international film festival takes place in Sept. and the Vancouver international film festival is the following month. Popular music festivals include the Montreal International Jazz Festival and North by Northeast in Toronto in June, the Ottawa International Jazz Festival in June–July, Ottawa Bluesfest and Halifax Rocks in July and Osheaga in Montreal in July–Aug. Les FrancoFolies de Montréal, a festival dedicated to francophone music, is held every summer.

DIPLOMATIC REPRESENTATIVES
Of Canada in the United Kingdom (Macdonald House, 1 Grosvenor Sq., London, W1K 4AB)
High Commissioner: Gordon Campbell.

Of the United Kingdom in Canada (80 Elgin St., Ottawa, K1P 5K7)
Acting High Commissioner: Corin Robertson.

Of Canada in the USA (501 Pennsylvania Ave., NW, Washington, D.C., 20001)
Ambassador: Gary Doer.

Of the USA in Canada (490 Sussex Drive, Ottawa, K1N 1G8)
Ambassador: David Jacobson.

Of Canada to the United Nations
Ambassador: Guillermo Rishchynski.

Of Canada to the European Union
Ambassador: David Plunkett.

FURTHER READING
Canadian Annual Review of Politics and Public Affairs. From 1960
Canadian Encyclopedia. 2000

Brown, R. C., *An Illustrated History of Canada.* 1991
Dawson, R. M. and Dawson, W. F., *Democratic Government in Canada.* 5th ed. 1989
Fierlbeck, Katherine, *The Development of Political Thought in Canada: An Anthology.* 2005
Jackson, R. J. and Jackson, D., *Politics in Canada: Culture, Institutions, Behaviour and Public Policy.* 7th ed. 2009
O'Reilly, Marc, *Handbook of Canadian Foreign Policy.* 2006
Silver, A. I. (ed.) *Introduction to Canadian History.* 1994

Other more specialized titles are listed under CONSTITUTION AND GOVERNMENT *above.*

Library and Archives Canada: 395 Wellington Street, Ottawa, K1A ON4. *Librarian and Archivist of Canada:* Daniel J. Caron.
National Statistical Office: Statistics Canada, 150 Tunney's Pasture Driveway, Ottawa, Ontario, K1A 0T6.
Website: http://www.statcan.gc.ca

CANADIAN PROVINCES

GENERAL DETAILS
The ten provinces each have a separate parliament and administration, with a Lieut.-Governor, appointed by the Governor-General in Council at the head of the executive. They have full powers to regulate their own local affairs and dispose of their revenues, provided that they do not interfere with the action and policy of the central administration. Among the subjects assigned exclusively to the provincial legislatures are: the amendment of the provincial constitution, except as regards the office of the Lieut.-Governor; property and civil rights; direct taxation for revenue purposes; borrowing; management and sale of Crown lands; provincial hospitals, reformatories, etc.; shop, saloon, tavern, auctioneer and other licences for local or provinciaql purposes; local works and undertakings, except lines of ships, railways, canals, telegraphs, etc., extending beyond the province or connecting with other provinces, and excepting also such works as the Canadian Parliament declares are for the general good; marriages, administration of justice within the province; education. On 18 July 1994 the federal and provincial governments signed an agreement easing inter-provincial barriers

on government procurement, labour mobility, transport licences and product standards. Federal legislation of Dec. 1995 grants provinces a right of constitutional veto.

For the administration of the three territories *see* Northwest Territories, Nunavut and Yukon *below*.

Areas of the ten provinces and three territories (Northwest Territories, Nunavut and Yukon) (in sq. km) and population at recent censuses:

Province	Land area	Total land and fresh water area	Population, 2001	Population, 2006	Population, 2011
Newfoundland and Labrador (Nfld.)	373,872	405,212	512,930	505,469	514,536
Prince Edward Island (PEI)	5,660	5,660	135,294	135,851	140,204
Nova Scotia (NS)	53,338	55,284	908,007	913,462	921,727
New Brunswick (NB)	71,450	72,908	729,498	729,997	751,171
Quebec (Que.)[1]	1,365,128	1,542,056	7,237,479	7,546,131	7,903,001
Ontario (Ont.)[1]	917,741	1,076,395	11,410,046	12,160,282	12,851,821
Manitoba (Man.)[1]	553,556	647,797	1,119,583	1,148,401	1,208,268
Saskatchewan (Sask.)[1]	591,670	651,036	978,933	968,157	1,033,381
Alberta (Alta.)[1]	642,317	661,848	2,974,807	3,290,350	3,645,257
British Columbia (BC)[1]	925,186	944,735	3,907,738	4,113,487	4,400,057
Nunavut (Nvt.)	1,936,113	2,093,190	26,745	29,474	31,906
Northwest Territories (NWT)	1,183,085	1,346,106	37,360	41,464	41,462
Yukon (YT)	474,391	482,443	28,674	30,372	33,897

[1]Excludes population data from incompletely enumerated Indian reserves and Indian settlements.

Local Government

Under the terms of the British North America Act the provinces are given full powers over local government. All local government institutions are, therefore, supervised by the provinces, and are incorporated and function under provincial acts.

The acts under which municipalities operate vary from province to province. A municipal corporation is usually administered by an elected council headed by a mayor or reeve, whose powers to administer affairs and to raise funds by taxation and other methods are set forth in provincial laws, as is the scope of its obligations to, and on behalf of, the citizens. Similarly, the types of municipal corporations, their official designations and the requirements for their incorporation vary between provinces. The following table sets out the classifications as at the 2011 census:

	Economic regions	Census divisions
Nfld.	4	11
PEI	1	3[1]
NS	5	18[1]
NB	5	15[1]
Que.	17	98[2]
Ont.	11	49[3]
Man.	8	23
Sask.	6	18
Alta.	8	19
BC	8	29[4]
Nvt.	1	3[5]
NWT	1	6[5]
YT	1	1[6]

[1]Counties. [2]81 municipalités régionales de comté, 12 territoires équivalents, 5 divisions de recensement. [3]20 counties, 10 districts, 9 census divisions, 6 regional municipalities, 3 united counties, 1 district municipality. [4]28 regional districts, 1 region. [5]Regions. [6]Territory.

SOCIAL INSTITUTIONS

Justice

The administration of justice within the provinces, including the enforcement of the Criminal Code of Canada, is the responsibility of provincial governments, but all the provinces except Ontario and Quebec have entered into contracts with the Royal Canadian Mounted Police (RCMP) to enforce criminal and provincial law. In addition, in these eight provinces the RCMP is under agreement to provide police services to municipalities.

Alberta

KEY HISTORICAL EVENTS

The southern half of Alberta was administered from 1670 as part of Rupert's land by the Hudson's Bay Company. Trading posts were set up after 1783 when the North West Company took a share in the fur trade. In 1869 Rupert's Land was transferred from the Hudson's Bay Company (which had absorbed its rival in 1821) to the new Dominion and in the following year this land was combined with the former Crown land of the North Western Territories to form the Northwest Territories. In 1882 'Alberta' first appeared as a provisional 'district', consisting of the southern half of the present province. In 1905 the Athabasca district to the north was added when provincial status was granted to Alberta.

TERRITORY AND POPULATION

The area of the province is 661,848 sq. km, 642,317 sq. km being land area and 16,531 sq. km water area. The population at the 2011 census was 3,645,257. Alberta had the fastest-growing population of any Canadian province between 1996 and 2006, with a 25·2% increase over the ten-year period. The urban population (2006), centres of 1,000 or over, was 82·1% and the rural 17·9%. Population (2006 census) of the 16 cities, as well as the two largest specialized municipalities: Calgary, 988,193; Edmonton, 730,372; Red Deer, 82,772; Lethbridge, 74,637; St Albert, 57,719; Medicine Hat, 56,997; Grande Prairie, 47,076; Airdrie, 28,927; Spruce Grove, 19,496; Leduc, 16,967; Lloydminster (Alberta portion), 15,910; Camrose, 15,620; Fort Saskatchewan, 14,957; Brooks, 12,498; Cold Lake, 11,991; Wetaskiwin, 11,673; Specialized Municipality of Strathcona County (Sherwood Park), 82,551; Specialized Municipality of Wood Buffalo (Fort McMurray), 51,496.

SOCIAL STATISTICS

Births in 2007–08 numbered 49,568 (a rate of 13·9 per 1,000 population) and deaths 20,699 (rate of 5·8 per 1,000 population). There were 18,632 marriages in 2006 and 8,075 divorces in 2005.

CLIMATE

Alberta has a continental climate of warm summers and cold winters—extremes of temperature. For the capital city, Edmonton, the hottest month is usually July (mean 17·5°C), while the coldest are Dec. and Jan. (–12°C). Rainfall amounts are greatest between May and Sept. In a year, the average precipitation is 461 mm (19·6") with about 129·6 cm of snowfall.

CONSTITUTION AND GOVERNMENT

The constitution of Alberta is contained in the British North America Act of 1867, and amending Acts; also in the Alberta Act of 1905, passed by the Parliament of the Dominion of Canada, which created the province out of the then Northwest Territories. The province is represented by six members in the Senate and 28 in the House of Commons of Canada.

The executive is vested nominally in the *Lieut.-Governor*, who is appointed by the federal government, but actually in the *Executive Council* or the Cabinet of the legislature. Legislative power is vested in the Assembly in the name of the Queen.

Members of the 87-member *Legislative Assembly* are elected by the universal vote of adults, 18 years of age and older.

RECENT ELECTIONS

In elections on 23 April 2012 the ruling Progressive Conservative Party won 44·0% of the vote (taking 61 of 87 seats), the Wildrose Party 34·3% (17), the Liberal Party 9·9% (5) and the New Democratic Party 9·8% (4). Turnout was 57%.

CURRENT GOVERNMENT

Lieut.-Governor: Donald Ethell (sworn in 11 May 2010).

As of March 2013 the members of the Executive Council were as follows:

Premier and President of Executive Council: Alison Redford; b. 1965 (Progressive Conservative; sworn in 7 Oct. 2011).

Deputy Premier and Minister of Enterprise and Advanced Education: Thomas Lukaszuk. *President of the Treasury Board and Minister of Finance:* Doug Horner. *Human Services:* David Hancock. *International and Intergovernmental Relations:* Cal Dallas. *Environment and Sustainable Resource Development:* Diana McQueen. *Health:* Fred Horne. *Energy:* Ken Hughes. *Education:* Jeff Johnson. *Agriculture and Rural Development:* Verlyn Olson. *Justice and Solicitor General:* Jonathan Denis. *Municipal Affairs:* Doug Griffiths. *Aboriginal Relations:* Robin Campbell. *Culture:* Heather Klimchuk. *Service Alberta:* Manmeet Bhullar. *Infrastructure:* Wayne Drysdale. *Transportation:* Ric McIver. *Tourism, Parks and Recreation:* Richard Starke.

Office of the Premier: http://premier.alberta.ca

ECONOMY

GDP per person in 2009 was $67,321 CDN.

Budget

Total revenue in 2008–09 was $51,388m. CDN (own source revenue, $47,005m. CDN; special purpose transfers, $3,159m. CDN; general purpose transfers, $1,224m. CDN). Total expenditures in 2008–09 amounted to $49,201m. CDN (including: health, $13,119m. CDN; education, $11,509m. CDN; transport and communications, $5,469m. CDN).

Performance

Real GDP growth was negative in 2009, at –4·5%. The economy then began to recover in 2010, achieving growth of 3·8%.

ENERGY AND NATURAL RESOURCES

Electricity

Electricity generation totalled 66·0bn. kWh in 2009, representing 11·1% of Canada's production.

Oil and Gas

Oil sands underlie some 60,000 sq. km of Alberta, the four major deposits being: the Athabasca, Cold Lake, Peace River and Buffalo Head Hills deposits. Some 7% (3,250 sq. km) of the Athabasca deposit can be recovered by open-pit mining techniques. The rest of the Athabasca, and all the deposits in the other areas, are deeper reserves which must be developed through *in situ* techniques. These reserves reach depths of 760 metres. In 2004 Alberta produced 773,300 bbls per day of crude oil and 962,300 bbls per day of synthetic crude oil and bitumen. The 2004 value of Albertan producers' sales of crude oil, condensate and pentanes was $40·94bn. CDN. Sales of oilsands were worth $14·94bn. CDN. Alberta produced 67% of Canada's crude petroleum output in 2004.

Natural gas is found in abundance in numerous localities. In 2004, 4,923bn. cu. ft valued at $31·1bn. CDN were produced in Alberta.

Minerals

Coal production in 2004 was 27·2m. tonnes with 1·7m. tonnes of coal being exported.

The preliminary value of mineral production in 2004 (excluding oil and gas) was $1,200·0m. CDN.

Agriculture

There were 49,431 farms in Alberta in 2006 with a total area of 21,095,393 ha.; 9,621,606 ha. were land in crop in 2006. The majority of farms are made up of cattle, followed by grains and oilseed, and wheat. For particulars of livestock *see* CANADA: Agriculture.

Farm cash receipts in 2004 totalled $8,043·4m. CDN of which crops contributed $2,616·8m. CDN, livestock and products $3,993·3m. CDN and direct payments $1,433·3m. CDN.

Forestry

Forest and other wooded land in 2001 covered some 36,388,000 ha. In 2003–04 Alberta had a regulated harvest of 24,819,100 cu. metres of net merchantable forest.

Fisheries

The largest catch in commercial fishing is whitefish. Perch, tullibee, walleye, pike and lake trout are also caught in smaller quantities. Commercial fish production in 2003–04 was 2,127 tonnes, value $3·46m. CDN.

INDUSTRY

The leading manufacturing industries are food and beverages, petroleum refining, metal fabricating, wood industries, primary metal, chemical and chemical products and non-metallic mineral products.

Manufacturing shipments had a total value of $52,965·8m. CDN in 2004. Greatest among these shipments were (in $1m. CDN): refined petroleum and coal products, 10,018; chemicals and chemical products, 9,645; food, 9,087; fabricated metal products, 4,135; machinery, 4,062; wood products, 3,753; primary metal, 2,137; paper and allied products, 1,784; non-metal mineral products, 1,695; and computer and electronic products, 1,471.

Total retail sales in 2004 were $43,703m. CDN, as compared to 2003 with $38,925m. CDN in sales. Main sales in 2004 were (in $1m. CDN): automobiles, 10,007·8; food, 7,650·9; general merchandise, 5,032·2; fuel 4,072·8; and pharmacies and personal care, 2,111·8.

Labour

In 2004 the labour force was 1,843,400 (837,200 females), of whom approximately 1,757,900 (797,700) were employed. In 2004 a total of 40,000 new jobs were created. Alberta's unemployment rate dropped to 4·6% in 2004, compared to the national average of 7·2%.

INTERNATIONAL TRADE

Imports and Exports

Alberta's exports were valued at a record $80·6bn. CDN in 2007, an increase of 3·7% on 2006. The largest export markets were the USA, China, Japan, Mexico and the Netherlands, which together accounted for 93% of Alberta's international exports. Energy accounted for 68% of exports in 2007.

COMMUNICATIONS

Roads

In 2005 there were 30,800 km of provincial highways and 153,500 km of local roads.

On 31 March 2005 there were 2,459,926 motor vehicles registered.

Rail
In 2003 the length of main railway lines was 7,136 km. There are light rail networks in Edmonton (12·3 km) and Calgary (35·7 km).

Civil Aviation
Calgary International is Canada's fourth busiest airport (handling 11,775,000 passengers in 2010) and Edmonton the fifth busiest (5,981,000 passengers in 2010).

Telecommunications
In Dec. 2008, 84·5% of households had at least one mobile phone (the highest rate of any of the Canadian provinces) and 11·5% of households only had a mobile phone (also the highest rate). 83% of households had home internet access in 2010.

SOCIAL INSTITUTIONS

Justice
The Supreme Judicial authority of the province is the Court of Appeal. Judges of the Court of Appeal and Court of Queen's Bench are appointed by the Federal government and hold office until retirement at the age of 75. There are courts of lesser jurisdiction in both civil and criminal matters. The Court of Queen's Bench has full jurisdiction over civil proceedings. A Provincial Court which has jurisdiction in civil matters up to $2,000 CDN is presided over by provincially appointed judges. Youth Courts have power to try boys and girls 12–17 years old inclusive for offences against the Young Offenders Act.

The jurisdiction of all criminal courts in Alberta is enacted in the provisions of the Criminal Code. The system of procedure in civil and criminal cases conforms as nearly as possible to the English system. In 2009–10, 53,375 Criminal Code cases were disposed of in an adult criminal court. In 2010 there were 77 homicides (a rate of 2·1 per 100,000 population).

Education
Schools of all grades are included under the term of public school (including those in the separate school system, which are publicly supported). The same board of trustees controls the schools from kindergarten to university entrance. In 2001–02 there were approximately 546,961 pupils enrolled in grades 1–12, including private schools and special education programmes. The University of Alberta (in Edmonton), founded in 1907, had, in 2004–05, 35,666 students; the University of Calgary had 28,306 students; Athabasca University had 29,542 students; the University of Lethbridge had 7,086 students. Alberta has 34 post-secondary institutions including four universities and two technical colleges.

CULTURE

Tourism
Alberta attracted more than 4,661,000 visitors from outside the province in 2003. It is known for its mountains, museums, parks and festivals. Total tourism receipts in 2003 were $4·3bn. CDN.

FURTHER READING
Savage, H., Kroetsch, R., Wiebe, R., *Alberta.* 1993

Statistical office: Alberta Finance, Statistics, Room 259, Terrace Bldg, 9515–107 St., Edmonton, AB T5K 2C3.
Websites: http://www.albertacanada.com;
   http://www.finance.alberta.ca/aboutalberta/index.html

British Columbia

KEY HISTORICAL EVENTS

British Columbia, formerly known as New Caledonia, was first administered by the Hudson's Bay Company. In 1849 Vancouver Island was given crown colony status and in 1853 the Queen Charlotte Islands became a dependency. The discovery of gold on the Fraser river and the subsequent influx of population resulted in the creation in 1858 of the mainland crown colony of British Columbia, to which the Strikine Territory (established 1862) was later added. In 1866 the two colonies were united.

TERRITORY AND POPULATION

British Columbia has an area of 944,735 sq. km of which land area is 925,186 sq. km. The capital is Victoria. The province is bordered westerly by the Pacific Ocean and Alaska Panhandle, northerly by the Yukon and Northwest Territories, easterly by the Province of Alberta and southerly by the USA along the 49th parallel. A chain of islands, the largest of which are Vancouver Island and the Queen Charlotte Islands, affords protection to the mainland coast.

The population at the 2011 census was 4,400,057. The principal metropolitan areas and cities and their population census for 2006 are as follows: Metropolitan Vancouver, 2,116,581; Metropolitan Victoria, 330,088; Abbotsford (amalgamated with Matsqui), 123,864; Kelowna, 106,707; Kamloops, 80,376; Nanaimo, 78,692; Prince George, 70,981; Chilliwack, 69,217; Vernon, 35,944; Penticton, 31,909; Campbell River, 29,572; Courtenay, 21,940; Cranbrook, 18,267; Port Alberni, 17,548; Fort St John, 17,402; Salmon Arm, 16,012.

SOCIAL STATISTICS

Births in 2007–08 numbered 44,087 (a rate of 10·1 per 1,000 population) and deaths 31,789 (rate of 7·3 per 1,000 population). There were 20,665 marriages in 2006 and 9,954 divorces in 2005. Life expectancy, at 81·4 years in 2006, is the highest in Canada.

CLIMATE

The climate is cool temperate, but mountain influences affect temperatures and rainfall considerably. Driest months occur in summer. Vancouver, Jan. 36°F (2·2°C), July 64°F (17·8°C). Annual rainfall 58" (1,458 mm).

CONSTITUTION AND GOVERNMENT

The British North America Act of 1867 provided for eventual admission into Canadian Confederation, and on 20 July 1871 British Columbia became the sixth province of the Dominion.

British Columbia has a unicameral legislature of 85 elected members. Government policy is determined by the *Executive Council* responsible to the Legislature. The *Lieut.-Governor* is appointed by the Governor-General of Canada, usually for a term of five years, and is the head of the executive government of the province.

The *Legislative Assembly* is elected for a maximum term of five years. Every Canadian citizen 18 years and over, having resided a minimum of six months in the province, duly registered, is entitled to vote. The province is represented in the Federal Parliament by 36 members in the House of Commons and six Senators.

RECENT ELECTIONS

At the Legislative Assembly elections of 12 May 2009 the Liberal Party won 46·0% of the vote and 49 of the 85 available seats, the New Democratic Party won 42·1% and 35 seats, and the Green Party 8·1%; an independent took one seat. Turnout was 51%.

CURRENT GOVERNMENT

Lieut.-Governor: Judith Guichon (sworn in 2 Nov. 2012).
   The Liberal Executive Council comprised in March 2013:
   Premier and President of the Executive Council: Christy Clark.
   Deputy Premier, Minister of Energy, Mines and Natural Gas, and Minister Responsible for Housing: Rich Coleman.
   Minister of Aboriginal Relations and Reconciliation: Ida Chong.
   Advanced Education, Innovation and Technology: John Yap.
   Agriculture: Norm Letnick. *Children and Family Development:*

Stephanie Cadieux. *Citizens' Services and Open Government:* Ben Stewart. *Community, Sport and Cultural Development:* Bill Bennett. *Education:* Don McRae. *Environment:* Terry Lake. *Finance:* Michael de Jong. *Forests, Lands and Natural Resources Operations:* Steve Thomson. *Health:* Margaret MacDiarmid. *Jobs, Tourism and Skills Training:* Pat Bell. *Justice and Attorney General:* Shirley Bond. *Social Development:* Moira Stilwell. *Transportation and Infrastructure:* Mary Polak.

Office of the Premier: http://www.gov.bc.ca/prem

ECONOMY

GDP per person in 2009 was $42,827 CDN.

Budget
Total revenue in 2005–06 was $34,070m. CDN (own source revenue, $28,333m. CDN; general purpose transfers, $2,086m. CDN; special purpose transfers, $3,651m. CDN). Total expenditures in 2005–06 came to $32,910m. CDN (including: health, $12,468m. CDN; education, $6,562m. CDN; social services, $4,923m. CDN; debt charges, $2,602m. CDN; transport and communication, $1,671m. CDN; resource conservation and industrial development, $1,452m. CDN; protection of persons and property, $1,430m. CDN).

Performance
Real GDP growth was negative in 2009, at −1·8%. The economy then began to recover strongly in 2010, achieving growth of 4·0%.

ENERGY AND NATURAL RESOURCES

Electricity
Generation in 2007 totalled 71,820 GWh (64,337 GWh from hydro-electric sources), of which 10,984 GWh were delivered outside the province. Available within the province were 68,863 GWh (with imports of 8,027 GWh).

Oil and Gas
In 2004 natural gas production, from the northeastern part of the province, was valued at $5·83bn. CDN.

Minerals
Coal, copper, gold, zinc, silver and molybdenum are the most important minerals produced but natural gas amounts to approximately half of the value of mineral and fuel extraction. The value of mineral production in 2006 was estimated at $5·99bn. CDN. Coal production (from the northeastern and southeastern regions) was valued at $2·11bn. CDN. Copper was the most valuable metal with production totalling $2·19bn. CDN; gold production amounted to $344m. CDN.

Agriculture
Only 3% of the total land area is arable or potentially arable. Farm holdings (19,844 in 2006) cover 2·8m. ha. with an average size of 143 ha. Farm cash receipts in 2007 were $2·4bn. CDN, led by dairy products valued at $424m. CDN, floriculture and nursery products valued at $404m. CDN and poultry and eggs valued at $382m. CDN. For particulars of livestock *see* CANADA: Agriculture.

Forestry
Around 49·9m. ha. are considered productive forest land of which 48·0m. ha. are provincial crown lands managed by the Ministry of Forests. Approximately 96% of the forested land is coniferous. The total timber harvest in 2007 was 72·9m. cu. metres. Output of forest-based products, 2007: lumber, 3·7m. cu. metres; plywood (2006), 1·6m. cu. metres; pulp, 4·7m. tonnes; newsprint, paper and paperboard, 2·5m. tonnes.

Fisheries
In 2006 the total landed value of the catch was $786m. CDN; wholesale value $1·3bn. CDN. Salmon (wild and farmed) generated 53% of the wholesale value of seafood products, followed by groundfish and shellfish. In 2006, 7,800 people worked in the commercial fishery, aquaculture and fish processing industries.

INDUSTRY
The value of shipments from all manufacturing industries reached $42·2bn. CDN in 2004, including wood ($13·0bn. CDN), paper ($5·9bn. CDN), food ($5·0bn. CDN) and primary metals ($2·4bn. CDN).

Labour
In 2007 the labour force averaged 2,366,000 persons with 2,266,000 employed (47% female) and 100,000 unemployed (4·2%). Of the employed workforce 1·77m. were in service industries and 496,000 in goods production. There were 365,000 employed in trade, 240,000 in health care and social assistance, 205,000 in manufacturing, 197,000 in construction and 173,000 in accommodation and food industries.

INTERNATIONAL TRADE

Imports and Exports
Imports in 2007 totalled $38,687m. CDN in value, while exports amounted to $31,459m. CDN. The USA is the largest market for products exported through British Columbia customs ports ($19,009m. CDN in 2007), followed by Japan ($4,102m. CDN) and People's Republic of China, excluding Hong Kong ($1,744m. CDN).

Wood products accounted for 22·7% of exports in 2007, energy products 19·6%, pulp and paper products 16·2%, machinery and equipment 11·0% and metallic mineral products 10·9%.

COMMUNICATIONS

Roads
In 2001 there were 42,440 km of provincial highway, of which 23,710 km were paved. In 2007, 1,995,000 passenger cars and 664,000 commercial vehicles were registered.

Rail
The province is served by two transcontinental railways, the Canadian Pacific Railway and the Canadian National Railway. Passenger service is provided by VIA Rail, a Crown Corporation, and the publicly owned British Columbia Railway. In 1995 the American company Amtrak began operating a service between Seattle and Vancouver after a 14-year hiatus. British Columbia is also served by the freight trains of the B.C. Hydro and Power Authority, the Northern Alberta Railways Company and the Burlington Northern and Southern Railways Inc. The combined route-mileage of mainline track operated by the CPR, CNR and BCR totals 6,800 km. The system also includes CPR and CNR wagon ferry connections to Vancouver Island, between Prince Rupert and Alaska, and interchanges with American railways at southern border points. There is a light rail system in Vancouver, opened in 1985 (50 km). A commuter rail service linking Vancouver and the Fraser Valley was established in 1995 (69 km).

Civil Aviation
Vancouver International Airport is Canada's second busiest airport. It handled 16,254,000 passengers in 2010, up from 15,660,000 in 2009. The second busiest in British Columbia is Victoria International Airport. It handled 1,464,000 passengers in 2010, down from 1,491,000 in 2009.

Shipping
The major ports are Vancouver (the largest dry cargo port on the North American Pacific coast), Prince Rupert and ports on the Fraser River. Other deep-sea ports include Nanaimo, Port Alberni, Campbell River, Powell River, Kitimat, Stewart and Squamish. Total cargo shipped through the port of Vancouver in 2007 was 82·7m. tonnes. 961,000 cruise passengers visited Vancouver in 2007.

British Columbia Ferries—one of the largest ferry systems in the world—connect Vancouver Island with the mainland and also provide service to other coastal points; in 2006–07, over 21m. passengers and more than 8·5m. vehicles were carried. Service by other ferry systems is also provided between Vancouver Island and the USA. The Alaska State Ferries connect Prince Rupert with centres in Alaska.

Telecommunications
In Dec. 2008, 77·8% of households had at least one mobile phone and 10·3% of households only had a mobile phone. 84% of households had home internet access in 2010.

SOCIAL INSTITUTIONS

Justice
The judicial system is composed of the Court of Appeal, the Supreme Court, County Courts and various Provincial Courts, including Magistrates' Courts and Small Claims Courts. The federal courts include the Supreme Court of Canada and the Federal Court of Canada.

In 2009–10, 40,467 Criminal Code cases were disposed of in an adult criminal court. In 2010 there were 83 homicides (a rate of 1·8 per 100,000 population).

Education
Education, free up to Grade XII level, is financed jointly from municipal and provincial government revenues. Attendance is compulsory from the age of five to 16. There were 582,691 pupils enrolled in 1,634 public schools from kindergarten to Grade 12 in Sept. 2007.

The universities had a full-time enrolment of 114,536 for 2006–07. Enrolment at the six universities (2006–07): the University of British Columbia, 48,293; Simon Fraser University, 24,842; University of Victoria, 19,372; Thompson Rivers University, 14,711; University of Northern British Columbia, 3,672; Royal Roads University, 3,646. There were three university-colleges in 2006: Kwantlen University College, Surrey; Malaspina University-College, Nanaimo; University College of the Fraser Valley, Abbotsford. British Columbia also had 12 colleges and five institutes in 2006: Camosun College, Victoria; Capilano College, North Vancouver; College of New Caledonia, Prince George; College of the Rockies, Cranbrook; Douglas College, New Westminster; Langara College, Vancouver; North Island College, Courtenay; Northern Lights College, Dawson Creek; Northwest Community College, Terrace; Okanagan College, Kelowna; Selkirk College, Castlegar; Vancouver Community College, Vancouver; British Columbia Institute of Technology, Burnaby; Emily Carr Institute of Art and Design, Vancouver; Institute of Indigenous Government, Burnaby; Justice Institute of British Columbia, New Westminster; Nicola Valley Institute of Technology, Merritt.

Televised distance education and special programmes through KNOW, the Knowledge Network of the West, are also provided.

Health
The government operates a hospital insurance scheme giving universal coverage after a qualifying period of three months' residence in the province. The province has come under a national medicare scheme which is partially subsidized by the provincial government and partially by the federal government. In March 2003 there were approximately 8,400 acute care and rehabilitation hospital beds. The provincial government spent an estimated $12·1bn. CDN on health programmes in 2006–07. 38% of the government's total expenditure was for health care in 2005–06.

CULTURE

Tourism
British Columbia's greatest attractions are Vancouver, and the provincial parks and ecological reserves that make up the Protected Areas System. There were more than 11,000 campsites and over 6,000 km of hiking trails in 2007. In 2003, 21·87m. tourists spent $8·95n. CDN in the province.

FURTHER READING
Barman, J., *The West beyond the West: a History of British Columbia.* 1991

Statistical office: BC STATS, Ministry of Finance and Corporate Relations, P. O. Box 9410, Stn. Prov. Govt., Victoria V8W 9V1.
Website: http://www.bcstats.gov.bc.ca

Manitoba

KEY HISTORICAL EVENTS

Manitoba was known as the Red River Settlement before it entered the dominion in 1870. During the 18th century its only inhabitants were fur-trappers, but a more settled colonization began in the 19th century. The area was administered by the Hudson's Bay Company until 1869 when it was purchased by the new dominion. In 1870 it was given provincial status. It was enlarged in 1881 and again in 1912 by the addition of part of the Northwest Territories.

TERRITORY AND POPULATION

The area of the province is 647,797 sq. km (250,114 sq. miles), of which 553,556 sq. km are land and 94,241 sq. km water. From north to south it is 1,225 km, and at the widest point it is 793 km.

The population at the 2011 census was 1,208,268. The 2006 census showed the following figures for areas of population of over 10,000 people: Winnipeg, the province's capital and largest city, 694,668; Brandon, 41,511; Thompson, 13,446; Portage la Prairie, 12,773; Steinbach, 11,066.

SOCIAL STATISTICS

Births in 2007–08 numbered 15,417 (a rate of 12·8 per 1,000 population) and deaths 10,137 (rate of 8·4 per 1,000 population). There were 5,722 marriages in 2006 and 2,429 divorces in 2005.

CLIMATE

The climate is cold continental, with very severe winters but pleasantly warm summers. Rainfall amounts are greatest in the months May to Sept. Winnipeg, Jan. –3°F (–19·3°C), July 67°F (19·6°C). Annual rainfall 21" (539 mm).

CONSTITUTION AND GOVERNMENT

The provincial government is administered by a *Lieut.-Governor* assisted by an *Executive Council* (Cabinet), which is appointed from and responsible to a *Legislative Assembly* of 57 members elected for five years. Women were enfranchised in 1916. The province is represented by six members in the Senate and 14 in the House of Commons of Canada.

RECENT ELECTIONS

In elections to the Legislative Assembly held on 4 Oct. 2011 the New Democratic Party of Manitoba won 37 out of 57 seats (46·0% of the vote), the Progressive Conservative Party 19 seats (43·9%) and the Liberal Party 1 seat (7·8%). Turnout was 57·2%.

CURRENT GOVERNMENT

Lieut.-Governor: Philip S. Lee; b. 1944 (took office on 4 Aug. 2009).

The members of the New Democratic Party of Manitoba government in March 2013 were:
Premier, President of the Executive Council, Minister of Federal-Provincial Relations: Greg Selinger; b. 1951.

Minister of Aboriginal and Northern Affairs: Eric Robinson. *Advanced Education and Literacy:* Erin Selby. *Agriculture, Food* and Rural Initiatives: Ron Kostyshyn. *Children and Youth Opportunities:* Kevin Chief. *Conservation and Water Stewardship:* Gord Mackintosh. *Culture, Heritage and Tourism:* Flor Marcelino. *Education:* Nancy Allan. *Entrepreneurship, Training and Trade:* Peter Bjornson. *Family Services and Labour:* Jennifer Howard. *Finance:* Stan Struthers. *Health:* Theresa Oswald. *Healthy Living, Seniors and Consumer Affairs:* Jim Rondeau. *Housing and Community Development:* Kerri Irvin-Ross. *Immigration and Multiculturalism:* Christine Melnick. *Infrastructure and Transportation:* Steve Ashton. *Innovation, Energy and Mines:* Dave Chomiak. *Justice and Attorney General:* Andrew Swan. *Local Government:* Ron Lemieux.

Manitoba Government Website: http://www.gov.mb.ca

ECONOMY

GDP per capita was $41,809 CDN in 2009.

Budget

Total revenue in 2008–09 was $11,791m. CDN (own source revenue, $8,004m. CDN; general purpose transfers, $2,458m. CDN; special purpose transfers, $1,328m. CDN). Total expenditures in 2008–09 amounted to $11,713m. CDN (including: health, $3,979m. CDN; education, $1,983m. CDN; social services, $1,859m. CDN).

Performance

Real GDP growth was 3·2% in 2006 (2·7% in 2005).

ENERGY AND NATURAL RESOURCES

Electricity

Electricity generation amounted to 34·8bn. kWh in 2009 (34·2bn. kWh hydro-electric), representing 5·9% of Canada's total production.

Oil and Gas

Oil production in 2010 was a record 11·7m. bbls. The estimated value of oil sold in 2010 was $907m. CDN, up 45·3% from 2009.

Minerals

Principal minerals mined are nickel, zinc, copper, gold and small quantities of silver. The value of mineral production increased 24·0% in 2010 to $1,663·5m. CDN. At $674m. CDN, nickel is Manitoba's most important mineral product, accounting for 26·2% of the province's total value of mineral production. Copper accounted for 24·6% of the value of mineral production in 2010, with a large increase in price and an 8·0% gain in Manitoba's output. Zinc, which accounts for 10·3% of Manitoba's mineral production value, also saw a jump in price, while its volume of production was steady in 2010. Gold was 10·8% of mineral production value, with strong growth in both price and production in Manitoba.

Agriculture

Rich farmland is the main primary resource, although the area in farms is only about 18% of the total land area. In 2010 total farm cash receipts increased by 0·1% to $4·85bn. CDN. Crop receipts fell 2·2% to $2·74bn. CDN but livestock receipts increased 6·2% to $1·76bn. CDN. Crop receipts accounted for 61% of total market receipts while livestock accounted for 39%. Within crops, the oilseeds share increased compared to wheat during the 2000s, illustrating its growing importance to Manitoba agriculture. For particulars of livestock *see* CANADA: Agriculture.

Fisheries

From about 57,000 sq. km of rivers and lakes, the value of fisheries production to fishers was about $27·0m. CDN in 2008–09 representing about 12,130 tonnes of fish. Pickerel, pike, sauger and whitefish are the principal varieties of fish caught.

INDUSTRY

Manitoba's diverse manufacturing sector is the province's largest industry, accounting for approximately 10·7% of total GDP. The value of manufacturing shipments declined 1·6% in 2010 to $14·4bn. CDN.

Labour

Manitoba's total employment rose by 11,500 in 2010 (5,700 full-time and 5,900 part-time jobs), a 1·9% increase, bringing employment to a record high level of 619,800. Manitoba had the second lowest unemployment rate among the Canadian provinces at 5·4%. It also had the second lowest youth unemployment rate in the country at 11·1%.

INTERNATIONAL TRADE

Imports and Exports

In 2002 Manitoba merchandise exports to the US rose 1% to $7·6bn. CDN. Manitoba is the only province in Canada to record higher exports to the USA in 2001 and 2002. Merchandise exports to the USA comprise 82% of Manitoba's total foreign merchandise exports. In 2002 merchandise exports to Japan (Manitoba's second-most important foreign market) increased 3·2% while exports declined to Mexico, Hong Kong, Belgium and China. Manufacturing industries' exports, which account for about two-thirds of Manitoba's total foreign exports, increased by 1%. Gains were posted by four of the five largest manufacturing industry categories. Leading growth export industries include machinery, printing and wood products.

COMMUNICATIONS

Roads

Highways and provincial roads total 18,500 km, with 2,800 bridges and other structures. In 2003 there were 498,880 passenger vehicles (including taxis), 118,823 trucks, 51,122 farm trucks, 27,978 off-road vehicles and 9,138 motorcycles registered in the province.

Rail

The province has about 5,650 km of commercial track, not including industrial track, yards and sidings. Most of the track belongs to the country's two national railways. Canadian Pacific owns about 1,950 km and Canadian National about 2,400 km. The Hudson Bay Railway, operated by Denver-based Omnitrax, has about 1,300 km of track. Fort Worth-based Burlington Northern's railcars are moved in Manitoba on CN and CP tracks and trains.

Civil Aviation

Winnipeg (James Armstrong Richardson International Airport) was Canada's eighth busiest airport in 2010, handling 3,385,000 passengers (3,372,000 in 2009).

Telecommunications

In Dec. 2008, 73·8% of households had at least one mobile phone and 10·0% of households only had a mobile phone. 73% of households had home internet access in 2010.

SOCIAL INSTITUTIONS

Justice

In 2009–10, 16,849 Criminal Code cases were disposed of in an adult criminal court. In 2010 there were 45 homicides (a rate of 3·6 per 100,000 population)

Education

Education is controlled through locally elected school divisions. There were 181,446 students enrolled in the province's public schools in the 2007–08 school year. Student teacher ratios (including all instructors but excluding school-based administrators) averaged one teacher for every 17·6 students in 2006–07.

Manitoba has four universities with a total full- and part-time undergraduate and graduate enrolment for the 2003–04 academic year of 39,500. They are the University of Manitoba, founded in 1877; the University of Winnipeg; Brandon University; and the Collège universitaire de Saint Boniface.

Community colleges in Brandon, The Pas and Winnipeg offer two-year diploma courses in a number of fields, as well as specialized training in many trades. They also give a large number and variety of shorter courses, both at their campuses and in many communities throughout the province. Provincial government expenditure on education and training for the 2003–04 fiscal year was budgeted at $1·59bn. CDN.

CULTURE

Tourism

Between 2009 and 2010 Manitoba recorded a 14·1% increase in total visitor numbers. In 2009 visitors spent $1·2bn. CDN (representing 2·8% of GDP). The province received a total of 6,910,000 visits in 2009 of which 371,000 were from the USA and 82,000 from other foreign countries. 24,800 people were employed in tourism-related jobs in 2009.

FURTHER READING

General Information: Inquiries may be addressed to Manitoba Government Inquiry. *Email:* mgi@gov.mb.ca

---

New Brunswick

KEY HISTORICAL EVENTS

Visited by Jacques Cartier in 1534, New Brunswick was first explored by Samuel de Champlain in 1604. With Nova Scotia, it originally formed one French colony called Acadia. It was ceded by the French in the Treaty of Utrecht in 1713 and became a permanent British possession in 1759. It was first settled by British colonists in 1764 but was separated from Nova Scotia, and became a province in June 1784 as a result of the influx of United Empire Loyalists. Responsible government from 1848 consisted of an executive council, a legislative council (later abolished) and a House of Assembly. In 1867 New Brunswick entered the Confederation.

TERRITORY AND POPULATION

The area of the province is 72,908 sq. km (28,150 sq. miles), of which 71,450 sq. km (27,587 sq. miles) is land area. The 2011 census counted 751,171 people in New Brunswick. At the time of the 2006 census, the most frequently reported ethnic origin, whether reported alone or in combination with other origins, was Canadian (53%). French was the second most frequently reported ancestry (27%), followed by English (25%), Irish (21%) and Scottish (20%). A total of 36,015 persons in New Brunswick identified themselves as Aboriginal (that is, as a North American Indian, Métis or Inuit) in 2006.

The seven urban centres of the province and their respective populations based on 2006 census figures are: Moncton, 126,424; Saint John, 122,389; Fredericton (capital), 85,688; Bathurst, 31,424; Miramichi, 24,737; Edmundston, 21,442; Campbellton (part only), 14,826. The official languages are English and French.

SOCIAL STATISTICS

Births in 2007–08 numbered 7,120 (a rate of 9·5 per 1,000 population) and deaths 6,277 (rate of 8·4 per 1,000 population). There were 3,497 marriages in 2006 and 1,444 divorces in 2005.

CLIMATE

A cool temperate climate, with rain in all seasons but temperatures modified by the influence of the Gulf Stream. Annual average total precipitation in Fredericton: 1,131 mm. Warmest month, July (average high) 25·6°C.

CONSTITUTION AND GOVERNMENT

The government is vested in a *Lieut.-Governor*, appointed by the Queen's representative in New Brunswick, and a *Legislative Assembly* of 55 members, each of whom is individually elected to represent the voters in one constituency or riding. The political party with the largest number of elected representatives, after a Provincial election, forms the government.

The province has ten appointed members in the Canadian Senate and elects ten members in the House of Commons.

RECENT ELECTIONS

Elections to the provincial assembly were held on 27 Sept. 2010. The opposition Progressive Conservative Party won 42 seats (with 48·8% of the vote), the Liberal Party 13 seats (with 34·4%) and the New Democratic Party no seats (10·4%).

CURRENT GOVERNMENT

Lieut.-Governor: Graydon Nicholas; b. 1946 (took office on 30 Sept. 2009).

The members of the Progressive Conservative government were as follows in March 2013:

Premier and President of Executive Council Office: David Alward; b. 1959.

Minister of Economic Development: Paul Robichaud. *Justice and Attorney General:* Marie-Claude Blais. *Agriculture, Aquaculture and Fisheries:* Mike Olscamp. *Education and Early Childhood Development:* Jody Carr. *Energy and Mines:* Craig Leonard. *Environment and Local Government:* Bruce Fitch. *Finance:* Blaine Higgs. *Government Services:* Sue Stultz. *Health:* Hugh Flemming. *Healthy and Inclusive Communities:* Dorothy Shephard. *Human Resources:* Troy Lifford. *Natural Resources:* Bruce Northrup. *Post-Secondary Education, Training and Labour:* Danny Soucy. *Public Safety and Solicitor General:* Robert Trevors. *Social Development:* Madeleine Dube. *Tourism, Heritage and Culture:* Trevor Holder. *Transportation and Infrastructure:* Claude Williams.

Government of New Brunswick Website: http://www.gnb.ca

ECONOMY

GDP per capita in 2009 was $36,663 CDN.

Budget

The ordinary budget (in $1m. CDN) is shown as follows (financial years ended 31 March):

	2001–02	2002–03	2003–04	2004–05	2005–06
Gross revenue	5,251·4	5,261·1	5,479·8	5,994·1	6,325·5
Gross expenditure	5,072·9	5,370·5	5,571·0	5,848·7	6,202·9

Funded debt and capital loans outstanding (exclusive of Treasury Bills) as of 31 March 2006 was $6,685·1m. CDN.

ENERGY AND NATURAL RESOURCES

Electricity

Hydro-electric, thermal and nuclear generating stations of NB Power had an installed capacity of 3,142 MW at 31 March 2010, consisting of 14 generating stations. Electricity generation amounted to 11·2bn. kWh in 2009.

Oil and Gas

In 2002 Enbridge Gas New Brunswick continued developing the natural gas distribution system in the province, which is now available in Fredericton, Moncton, St John, St George and Oromocto.

Minerals
The total value of minerals produced in 2006 was $1,538·6m. CDN. The top four contributors to mineral production are zinc, lead, silver and peat, accounting for 79·7% of total value in 2006. In 2007 New Brunswick ranked first in Canada for the production of zinc, bismuth and lead, second for silver and sixth for copper.

Agriculture
The total area under crops was 151,996 ha. in 2006. Farms numbered 2,776 and averaged 142 ha. (census 2006). Potatoes account for 33% of total farm cash receipts and dairy products 11%. New Brunswick is self-sufficient in fluid milk and supplies a processing industry. For particulars of livestock see CANADA: Agriculture. Net farm income in 2005 was $25·4m. CDN.

Forestry
New Brunswick contains some 6m. ha. of productive forest lands. The value of shipments of forest products in 2008 was $1·25bn. CDN; the value of manufacturing shipments for the wood-related industries was $871m. CDN. In 2008 nearly 13,000 people were employed in all aspects of the forest industry.

Fisheries
Commercial fishing is one of the most important primary industries of the province, employing 6,366 in 2007. Landings in 2005 (117,295 tonnes) amounted to $197m. CDN. In 2005 molluscs and crustaceans ranked first with a value of $178m. CDN, 90% of the total landed value. Exports in 2005, totalling $653·4m. CDN, went mainly to the USA and Japan.

INDUSTRY
Important industries include food and beverages, paper and allied industries, and timber products.

Labour
New Brunswick's labour force increased by 2·4% in 2002 to 385,700 while employment increased to 345,000. Goods producing industries employed 79,700 and the service-producing industries employed 253,700. Nearly 20% of the industrial labour force work in Saint John. In 2002 unemployment was 10·4%.

INTERNATIONAL TRADE

Imports and Exports
New Brunswick's location, with deepwater harbours open throughout the year and container facilities at Saint John, makes it ideal for exporting. The main exports include lumber, wood pulp, newsprint, refined petroleum products and electricity. In 2009 the major trading partners of the province were the USA with 86% of total exports, followed by the Netherlands with 3%. Imports totalled $9,396m. CDN while exports reached $9,902m. CDN in 2009.

COMMUNICATIONS

Roads
There are 21,423 km of roads in the Provincial Highway system, of which 8,333 km consists of arterial, collector and local roads that provide access to most areas. The main highway system, including approximately 964 km of the Trans-Canada Highway, links the province with the principal roads in Quebec, Nova Scotia and Prince Edward Island, as well as the Interstate Highway System in the eastern seaboard states of the USA. At 31 March 2009 total road motor vehicle registrations numbered 524,300 of which 489,507 were vehicles weighing less than 4,500 kilograms, 7,662 were vehicles weighing 4,500 kilograms to 14,999 kilograms, 4,643 were vehicles weighing 15,000 kilograms or more, 3,086 were buses and 19,399 were motorcycles and mopeds.

Rail
New Brunswick is served by the Canadian National Railways, Springfield Terminal Railway, New Brunswick Southern Railway, New Brunswick East Coast Railway, Le Chemin de fer de la Matapédia et du Golfe and VIA Rail. The Salem-Hillsborough rail is popular with tourists.

Civil Aviation
There are three major airports at Fredericton, Moncton and Saint John. There are also a number of small regional airports.

Shipping
New Brunswick has five major ports. The Port of Saint John handles approximately 20m. tonnes of cargo each year including forest products, steel, potash and petroleum. The Port of Belledune is a deep-water port and open all year round. Other ports are Dalhousie, Bayside/St Andrews and Miramichi.

Telecommunications
In Dec. 2008, 66·0% of households had at least one mobile phone and 4·3% of households only had a mobile phone. 70% of households had home internet access in 2010.

SOCIAL INSTITUTIONS

Justice
In 2009, 47,923 Criminal Code offences were reported, including 12 homicides.

Education
Public education is free and non-sectarian.

There were, in Sept. 2009, 106,394 students (including kindergarten) and 7,896 full-time equivalent/professional educational staff in the province's 322 schools.

There are four universities. The University of New Brunswick at Fredericton (founded 13 Dec. 1785 by the Loyalists, elevated to university status in 1823 and reorganized as the University of New Brunswick in 1859) had 9,007 full-time students at the Fredericton campus and 3,017 full-time students at the Saint John campus (2002–03); the Université de Moncton at Moncton, 5,089 full-time students; St Thomas University at Fredericton, 2,897 full-time students; Mount Allison University at Sackville had 2,199 full-time students.

CULTURE

Press
In 2009 New Brunswick had four daily newspapers (one in French) and 17 weekly newspapers (seven in French and one bilingual).

Tourism
New Brunswick has a number of historic buildings as well as libraries, museums and other cultural sites. Tourism is one of the leading contributors to the economy. In 2007 tourism revenues reached $1·4bn. CDN.

FURTHER READING
Industrial Information: Dept. of Business New Brunswick, Fredericton. *Economic Information:* Dept. of Finance, New Brunswick Statistics Agency, Fredericton. *General Information:* Communications New Brunswick, Fredericton.

Newfoundland and Labrador

KEY HISTORICAL EVENTS
Archaeological finds at L'Anse aux Meadows in northern Newfoundland show that the Vikings established a colony here in about AD 1000. This site is the only known Viking colony in North America. Newfoundland was discovered by John Cabot on 24 June 1497, and was soon frequented in the summer months by the Portuguese, Spanish and French for its fisheries. It was

formally occupied in Aug. 1583 by Sir Humphrey Gilbert on behalf of the English Crown but various attempts to colonize the island remained unsuccessful. Although British sovereignty was recognized in 1713 by the Treaty of Utrecht, disputes over fishing rights with the French were not finally settled until 1904. By the Anglo-French Convention of 1904, France renounced her exclusive fishing rights along part of the coast, granted under the Treaty of Utrecht, but retained sovereignty of the offshore islands of St Pierre and Miquelon. Self-governing from 1855, the colony remained outside of the Canadian confederation in 1867 and continued to govern itself until 1934, when a commission of government appointed by the British Crown assumed responsibility for governing the colony and Labrador. This body controlled the country until union with Canada in 1949.

TERRITORY AND POPULATION

Area, 405,212 sq. km (156,452 sq. miles), of which freshwater, 31,340 sq. km (12,100 sq. miles). In March 1927 the Privy Council decided the boundary between Canada and Newfoundland in Labrador. This area, now part of the Province of Newfoundland and Labrador, is 294,330 sq. km (113,641 sq. miles) of land area.

Newfoundland island's coastline is punctuated with numerous bays, fjords and inlets, providing many good deep water harbours. Approximately one-third of the area is covered by water. Grand Lake, the largest body of water, has an area of about 530 sq. km. Good agricultural land is generally found in the valleys of the Terra Nova River, the Gander River, the Exploits River and the Humber River, which are also heavily timbered. The Strait of Belle Isle separates the island from Labrador to the north. Bordering on the Canadian province of Quebec, Labrador is a vast, pristine wilderness and extremely sparsely populated (approximately 10 sq. km per person). Labrador's Lake Melville is 2,934 sq. km and its highest peak, Mount Caubvick, is 1,700 metres.

The population at the 2011 census was 514,536. The capital of the province is the City of St John's (2006 census population, 100,646). The other cities are Mt Pearl (24,671 in 2006) and Corner Brook (20,083); important towns are Conception Bay South (21,966), Grand Falls-Windsor (13,558), Paradise (12,584), Gander (9,951), Happy Valley-Goose Bay (7,572), Labrador City (7,240), Stephenville (6,588), Portugal Cove-St Philip's (6,575), Torbay (6,281), Marystown (5,436), Bay Roberts (5,414) and Clarenville (5,274).

SOCIAL STATISTICS

Births in 2007–08 numbered 4,521 (a rate of 8·9 per 1,000 population) and deaths 4,656 (rate of 9·2 per 1,000 population). Newfoundland and Labrador was the only province in which deaths exceeded births in 2007–08. There were 2,722 marriages in 2006 and 789 divorces in 2005.

CLIMATE

The cool temperate climate is marked by heavy precipitation, distributed evenly over the year, a cool summer and frequent fogs in spring. St. John's, Jan. −4°C, July 15·8°C. Annual rainfall 1,240 mm.

CONSTITUTION AND GOVERNMENT

Until 1832 Newfoundland was ruled by a British Governor. In that year a Legislature was brought into existence, but the Governor and his Executive Council were not responsible to it. Under the constitution of 1855, the government was administered by the *Governor* appointed by the Crown with an *Executive Council* responsible to the House of Assembly.

Parliamentary government was suspended in 1933 on financial grounds and Government by Commission was inaugurated on 16 Feb. 1934. Confederation with Canada was approved by a referendum in July 1948. In the Canadian Senate on 18 Feb. 1949 Royal Assent was given to the terms of union of Newfoundland and Labrador with Canada, and on 23 March 1949, in the House of Lords, London, Royal Assent was given to an amendment to the British North America Act, made necessary by the inclusion of Newfoundland and Labrador as the tenth Province of Canada. Since April 1949 Newfoundland and Labrador has had a *Lieut.-Governor* rather than a Governor, and its House of Assembly (comprised of 48 members) was reconstituted.

The province is represented by six members in the Senate and by seven members in the House of Commons of Canada.

RECENT ELECTIONS

Elections were held on 11 Oct. 2011. The ruling Progressive Conservative Party (PC) won 37 of the 48 seats in the House of Assembly with 56·1% of the vote; the Liberal Party (Lib.), 6 (19·1%); and the New Democratic Party (NDP), 5 (24·6%).

CURRENT GOVERNMENT

Lieut.-Governor: Frank Fagan; b. 1945 (assumed office 19 March. 2013).

In March 2013 the Progressive Conservative cabinet was composed as follows:

Premier: Kathy Dunderdale; b. 1952 (sworn in 3 Dec. 2010).

Minister of Advanced Education and Skills: Joan Shea. *Child, Youth and Family Services:* Charlene Johnson. *Education:* Clyde Jackman. *Environment and Conservation:* Thomas J. Hedderson. *Finance and President of the Treasury Board:* Jerome Kennedy, QC. *Fisheries and Aquaculture:* Derrick Dalley. *Health and Community Services:* Susan Sullivan. *Innovation, Business and Rural Development:* Keith Hutchings. *Intergovernmental Affairs and Aboriginal Affairs:* Felix Collins. *Service Newfoundland and Labrador:* Nick McGrath. *Justice:* Darin King. *Municipal Affairs and Registrar General:* Kevin O'Brien. *Natural Resources and Attorney General:* Thomas Marshall, QC. *Tourism, Culture and Recreation:* Terry French. *Transportation and Works:* Paul Davis.

Speaker of the House of Assembly: Ross Wiseman.

Office of the Premier: http://www.premier.gov.nl.ca/premier

ECONOMY

GDP per capita was $49,067 CDN in 2009. Real GDP growth in 2010 was 6·0%.

Budget

Government budget in $1,000 CDN in fiscal years ending 31 March:

	2004–05	2005–06	2006–07[1]
Gross Revenue	4,153,356	4,928,642	4,909,628
Gross Expenditure	4,009,775	4,229,153	4,599,079

[1]Estimate.

ENERGY AND NATURAL RESOURCES

Electricity

Newfoundland and Labrador is served by two large physically independent electrical systems with a total of 7,412 MW of operational electrical generating capacity and 23 small isolated systems primarily supplied by diesel-fuelled internal combustion generators. In 2011 total provincial electricity generation equalled 41·2bn. kWh, of which about 97% was from hydro-electric sources. Approximately 73% of total electricity generation was exported outside the province. Electricity service for a total of 274,000 retail customers is provided by two utilities and regulated by the Board of Commissioners of Public Utilities.

Oil and Gas

Newfoundland and Labrador is home to three active offshore oil projects: Hibernia, Terra Nova and White Rose. In 2011 oil production from Hibernia reached 56·3m. bbls. The Terra Nova development started producing oil in Jan. 2002 and in 2003 produced more than 40m. bbls, although production declined in 2011 to 15·7m. bbls. Production at the White Rose development started in Nov. 2005 and reached 25·2m. bbls in 2011. Oil production from a fourth development, Hebron, is expected to begin in 2017.

Minerals

The mineral resources are vast but only partially documented. Large deposits of iron ore, with an ore reserve of over 5,000m. tonnes at Labrador City, Wabush City and in the Knob Lake area, are supplying approximately half of Canada's production. Other large deposits of iron ore are known to exist in the Julienne Lake area. The Central Mineral Belt, which extends from the Smallwood Reservoir to the Atlantic coast near Makkovik, holds uranium, copper, beryllium and molybdenite potential.

The percentage share of mineral shipment value in 2009 stood at 59% for iron ore and 20% for nickel. Other major mineral products were copper, zinc and gold. The value of mineral shipments in 2011 totalled $4·6bn. CDN, representing a 22% increase over 2010.

Agriculture

Farm receipts in 2011 were $124·8m. CDN, an increase of 5·8% on 2010. In 2011 dairy products contributed $43·3m. CDN, crops $18·2m. CDN, eggs $17·1m. CDN and furs $15·8m. CDN. For particulars of livestock *see* CANADA: Agriculture.

Forestry

The forestry economy in the province is mainly dependent on the operation of three newsprint mills—Corner Brook Pulp and Paper and Abitibi-Consolidated (which operates two mills). In 2005 the value of newsprint exported totalled $563m. CDN, an increase of 7·5% over 2004. Lumber mills and saw-log operations produced 125m. flat bd ft in 2005.

Fisheries

Closure of the northern cod and other groundfish fisheries has switched attention to secondary seafood production and aquaculture. The total catch in 2009 was 301,496 tonnes valued at $423m. CDN. Shellfish accounted for 55·3% of total landings and 83·1% of landed value. 10,300 people were employed in the fishing industry in 2009.

INDUSTRY

The total value of manufacturing shipments in 2005 was $2·98bn. CDN, a 2·8% reduction on 2004. This consisted largely of fish products, refined petroleum and newsprint.

Labour

In 2005 those in employment numbered 214,100 with 16,800 workers employed in manufacturing. The unemployment rate was 15·2% in 2005 (15·7% in 2004).

COMMUNICATIONS

Roads

In 2007 there were 19,250 km of roads, of which 10,595 km were paved. In 2005 there were 266,716 motor vehicles registered.

Rail

The Quebec North Shore and Labrador Railway operated both freight and passenger services on its 588 km main line from Sept-Iles, Quebec, to Schefferville, Quebec and its 58 km spur line from Ross Bay Junction to Labrador City, Newfoundland. In 2006 iron ore freight totalled 20·0m. tonnes.

Civil Aviation

The province is linked to the rest of Canada by regular air services provided by Air Canada and a number of smaller air carriers.

Shipping

At Jan. 2006 there were 1,851 ships on register in Newfoundland and Labrador. Marine Atlantic, a federal crown corporation, provides a freight and passenger service all year round from Channel-Port aux Basques to North Sydney, Nova Scotia; and seasonal ferries connect Argentia with North Sydney, and Lewisporte with Goose Bay, Labrador.

Telecommunications

In Dec. 2008, 68·2% of households had at least one mobile phone and 4·1% of households only had a mobile phone. 74% of households had home internet access in 2010.

SOCIAL INSTITUTIONS

Justice

In 2009–10, 4,846 Criminal Code cases were disposed of in an adult criminal court. In 2010 there were four homicides (a rate of 0·8 per 100,000 population).

Education

In 2005–06 total enrolment for elementary and secondary education was 76,763; full time teachers numbered 5,485; total number of schools was 294. Memorial University, offering courses in arts, science, engineering, education, nursing and medicine, had 15,000 full-time students in 2004–05.

CULTURE

Tourism

In 2005, 469,600 non-resident tourists (449,300 in 2004) spent approximately $336·4m. CDN in the province.

FURTHER READING

Statistical office: Newfoundland & Labrador Statistics Agency, POB 8700, St John's, NL A1B 4J6.
Website: http://www.stats.gov.nl.ca

Nova Scotia

KEY HISTORICAL EVENTS

Nova Scotia was visited by John and Sebastian Cabot in 1497–98. In 1605 a number of French colonists settled at Port Royal. The old name of the colony, Acadia, was changed in 1621 to Nova Scotia. The French were granted possession of the colony by the Treaty of St-Germain-en-Laye (1632). In 1654 Oliver Cromwell sent a force to occupy the settlement. Charles II, by the Treaty of Breda (1667), restored Nova Scotia to the French. It was finally ceded to the British by the Treaty of Utrecht in 1713. In the Treaty of Paris (1763) France resigned all claims and in 1820 Cape Breton Island united with Nova Scotia. Representative government was granted as early as 1758 and a fully responsible legislative assembly was established in 1848. In 1867 the province entered the dominion of Canada.

TERRITORY AND POPULATION

The area of the province is 55,284 sq. km (21,345 sq. miles), of which 53,338 sq. km are land area and 1,946 sq. km water area. The population at the 2011 census was 921,727.

Population of the major urban areas (2006 census): Halifax Regional Municipality, 372,679; Cape Breton Regional Municipality, 102,250. Principal towns (2006 census): Truro, 11,765; Amherst, 9,505; New Glasgow, 9,455; Bridgewater, 7,944; Yarmouth, 7,162; Kentville, 5,815.

SOCIAL STATISTICS

Births in 2007–08 numbered 8,848 (a rate of 9·5 per 1,000 population) and deaths 8,401 (rate of 9·0 per 1,000 population). There were 4,513 marriages in 2006 and 1,961 divorces in 2005.

CLIMATE

A cool temperate climate, with rainfall occurring evenly over the year. The Gulf Stream moderates the temperatures in winter so that ports remain ice-free. Halifax, Jan. 23·7°F (−4·6°C), July 63·5°F (17·5°C). Annual rainfall 54" (1,371 mm).

CONSTITUTION AND GOVERNMENT

Under the British North America Act of 1867 the legislature of Nova Scotia may exclusively make laws in relation to local matters, including direct taxation within the province, education and the administration of justice. The legislature of Nova Scotia consists of a *Lieut.-Governor*, appointed and paid by the federal government, and holding office for five years, and a *House of Assembly* of 52 members, chosen by popular vote at least every five years. The province is represented in the Canadian Senate by ten members, and in the House of Commons by 11.

RECENT ELECTIONS

At the provincial elections of 9 June 2009 the New Democratic Party won 31 seats (45·3% of the vote), the Liberals 11 (27·2%) and the Progressive Conservatives 10 (24·5% of the vote). Turnout was 58%.

CURRENT GOVERNMENT

Lieut.-Governor: John James Grant; b. 1936 (assumed office 12 April 2012).

The members of Nova Scotia's first ever New Democratic cabinet in March 2013 were:

Premier, President of the Executive Council, and Minister of Intergovernmental Affairs, Policy and Priorities, and Aboriginal Affairs: Darrell Dexter.

Deputy Premier, Deputy President of the Executive Council, and Minister of the Public Service Commission, Communications Nova Scotia, and Information Management: Frank Corbett. *Finance:* Maureen MacDonald. *Agriculture, and Service Nova Scotia and Municipal Relations:* John MacDonell. *Labour and Advanced Education, and Immigration:* Marilyn More. *Environment, and Fisheries and Aquaculture:* Sterling Belliveau. *Economic and Rural Development and Tourism, and African Nova Scotian Affairs:* Percy Paris. *Community Services, and Seniors:* Denise Peterson-Rafuse. *Education:* Ramona Jennex. *Attorney General and Minister of Justice:* Ross Landry. *Natural Resources, and Energy:* Charlie Parker. *Health and Wellness, and Acadian Affairs:* David Wilson. *Communities, Culture and Heritage:* Leonard Preyra. *Transportation and Infrastructure Renewal, and Gaelic Affairs:* Maurice Smith.

Speaker of the House of Assembly: Gordie Gosse.

Government of Nova Scotia Website: http://www.gov.ns.ca

ECONOMY

GDP per capita was $36,460 CDN in 2009.

Budget

Summary of operations and net funding requirements for the consolidated entity (in $1m. CDN) for fiscal years ending 31 March:

	2004[1]	2005[2]	2006[3]
Revenues	5,857·7	6,267·7	6,587·3
Net Programme Expenditures/Expenses	5,192·7	5,588·4	5,987·2
Net Debt Servicing Costs	890·3	872·1	884·6
Pension Valuation Adjustment	6·3	38·3	33·2
Total Net Expenditures/Expenses	6,089·4	6,498·7	6,905·0
Consolidation Adjustment	47·5	35·5	54·5
Net Income from Government Business Enterprises	349·5	346·6	335·1
Surplus (Deficit)	166·3	151·0	71·9

[1]Actual. [2]Forecast. [3]Estimate.

Performance

GDP (market prices) was $31,344m. CDN in 2005, an increase of 5·0% on 2004. GDP per person in 2005 was $33,487 CDN.

ENERGY AND NATURAL RESOURCES

Electricity

Electricity generation amounted to 11·2bn. kWh in 2009, mostly from fossil fuels.

Oil and Gas

Significant finds of offshore natural gas are currently under development. Gas is flowing to markets in Canada and the USA (the pipeline was completed in 1999). Total marketable gas receipts for 2005 was 3·9bn. cu. metres.

Minerals

Principal minerals in 2005 were: gypsum, 6·8m. tonnes, valued at $81·1m. CDN; stone, 10·8m. tonnes, valued at $73·0m. CDN. Total value of mineral production in 2005 was $286·1m. CDN.

Agriculture

In 2006 there were 3,795 farms in the province with 116,609 ha. of land under crops. Dairying, poultry and egg production, livestock and fruit growing are the most important branches. Farm cash receipts for 2005 were $453·4m. CDN. Cash receipts from sale of dairy products were $107·0m. CDN, with total milk and cream sales of 166·7m. litres. The production of poultry meat in 2005 was 37,030 tonnes, of which 33,508 tonnes were chicken and 3,522 tonnes were turkey. Egg production in 2005 was 17·9m. dozen. For particulars of livestock *see* CANADA: Agriculture.

The main fruit crops in 2005 were apples, 39,372 tonnes; blueberries, 16,239 tonnes; strawberries, 1,769 tonnes.

Forestry

The estimated forest area of Nova Scotia is 15,830 sq. miles (40,990 sq. km), of which about 28% is owned by the province. Softwood species represented 89·4% of the 6,254,716 cu. metres of the forest round products produced in 2005. Employment in the forest sector was 4,100 persons in 2005.

Fisheries

The fisheries of the province in 2005 had a landed value of $647m. CDN of sea fish; including lobster fishery, $339m. CDN; and crab fishery, $73m. CDN. Aquaculture production in 2005 was 8,917 tonnes with a value of $40·4m. CDN; finfish accounted for 64% of total value while shellfish made up the remainder.

INDUSTRY

The number of employees in manufacturing establishments was 38,055 in 2004; wages and salaries totalled $1,444m. CDN. The value of shipments in 2005 was $10,596m. CDN, and the leading industries were food, paper production, and plastic and rubber products.

Labour

In 2005 the labour force was 483,900 (232,100 females), of whom 443,100 (214,500) were employed. The provincial unemployment rate stood at 8·4% while the participation rate was 63·6%.

INTERNATIONAL TRADE

Imports and Exports

Total of imports and exports to and from Nova Scotia (in $1m. CDN):

	2002	2003	2004	2005
Imports	5,140	5,816	6,590	6,989
Exports	5,345	5,477	5,859	5,815

The main exports in 2005 included fish and fish products, natural gas and paper. Major trading partners were the USA with 80·1% of total exports, followed by Japan and the United Kingdom.

COMMUNICATIONS

Roads

In 2005 there were 26,000 km of highways, of which 13,600 km were paved. The Trans Canada and 100 series highways are limited access, all-weather, rapid transit routes. The province's first toll road opened in Dec. 1997. In 2005 total road vehicle registrations numbered 561,325 and over 600,000 persons had road motor vehicle operators licences.

Rail

The province has an 805-km network of mainline track operated predominantly by Canadian National Railways. The Cape Breton and Central Nova Scotia Railway operates between Truro and Cape Breton Island. The Windsor and Hantsport Railway operates in the Annapolis Valley region. VIA Rail operates the Ocean for six days a week, a transcontinental service between Halifax and Montreal.

Civil Aviation

Halifax (Robert L. Stanfield International Airport) was Canada's seventh busiest airport in 2010, handling 3,509,000 passengers. There are other major airports at Sydney and Yarmouth.

Shipping

Ferry services connect Nova Scotia to the provinces of Newfoundland and Labrador, Prince Edward Island and New Brunswick as well as to the USA. The deep-water, ice-free Port of Halifax handles about 14m. tonnes of cargo annually.

Telecommunications

In Dec. 2008, 72·3% of households had at least one mobile phone and 6·6% of households only had a mobile phone. 77% of households had home internet access in 2010.

SOCIAL INSTITUTIONS

Justice

The Supreme Court (Trial Division and Appeal Division) is the superior court of Nova Scotia and has original and appellate jurisdiction in all civil and criminal matters unless they have been specifically assigned to another court by Statute. An appeal from the Supreme Court, Appeal Division, is to the Supreme Court of Canada.

In 2009–10, 11,760 Criminal Code cases were disposed of in an adult criminal court. In 2010 there were 21 homicides (a rate of 2·2 per 100,000 population).

Education

Public education in Nova Scotia is free, compulsory and undenominational through elementary and high school. Attendance is compulsory to the age of 16. In 2004–05 there were 438 elementary-secondary public schools, with 9,581 full-time teachers and 145,396 pupils. The province has 11 degree-granting institutions. The Nova Scotia Agricultural College is located at Truro. The Technical University of Nova Scotia, which grants degrees in engineering and architecture, amalgamated with Dalhousie University and is now known as DalTech.

The Nova Scotia government offers financial support and organizational assistance to local school boards for provision of weekend and evening courses in academic and vocational subjects, and citizenship for new Canadians.

Health

A provincial retail sales tax of 8% provides funds for free hospital in-patient care up to ward level and free medically required services of physicians. The Queen Elizabeth II Hospital in Halifax is the overall referral hospital for the province and, in many instances, for the Atlantic region. The Izaak Walton Killam Hospital provides similar regional specialization for children.

Welfare

General and specialized welfare services in the province are under the jurisdiction of the Department of Community Services. The provincial government funds all of the costs.

RELIGION

The population is predominantly Christian. In 2001, 36·6% were Roman Catholic, 15·9% were United Church, 13·4% Anglicans, 10·6% Baptist and 2·5% Presbyterian.

CULTURE

Press

Nova Scotia has approximately 50 newspapers, including eight dailies. Daily newspapers with the largest circulations are *The Chronicle Herald* and *Mail Star* of Halifax, *The Daily News* of Dartmouth and *The Cape Breton Post* of Sydney.

Tourism

Tourism revenues were $1·3bn. CDN in 2005. Total number of visitors in 2005 was 2,114,000.

FURTHER READING

Nova Scotia Statistical Review. 2007
Nova Scotia at a Glance. 2006

Statistical office: Statistics Division, Department of Finance, POB 187, Halifax, Nova Scotia B3J 2N3.
Website: http://www.gov.ns.ca/finance/statistics/agency/index.asp

---

Ontario

KEY HISTORICAL EVENTS

The French explorer Samuel de Champlain explored the Ottawa River from 1613. The area was governed by the French, first under a joint stock company and then as a royal province, from 1627 and was ceded to Great Britain in 1763. A constitutional act of 1791 created the province of Upper Canada, largely to accommodate loyalists of English descent who had immigrated after the United States war of independence. Upper Canada entered the Confederation as Ontario in 1867.

TERRITORY AND POPULATION

The area is 1,076,395 sq. km (415,596 sq. miles), of which some 917,741 sq. km (354,340 sq. miles) are land area and some 158,654 sq. km (61,256 sq. miles) are lakes and fresh water rivers. The province extends 1,690 km (1,050 miles) from east to west and 1,730 km (1,075 miles) from north to south. It is bounded in the north by the Hudson and James Bays, in the east by Quebec, in the west by Manitoba, and in the south by the USA, the Great Lakes and the St Lawrence Seaway.

The census population in 2011 was 12,851,821. Population of the principal cities (2006 census):

Toronto[1]	2,503,281	Windsor	216,473	Oshawa	141,590
Ottawa	812,129	Kitchener	204,668	St Catharines	131,989
Mississauga	668,549	Oakville	165,613	Barrie	128,430
Hamilton	504,559	Burlington	164,415	Cambridge	120,371
Brampton	433,806	Richmond		Kingston	117,207
London	352,395	Hill	162,704	Guelph	114,943
Markham	261,573	Greater		Whitby	111,184
Vaughan	238,866	Sudbury	157,857	Thunder Bay	109,140

[1]The new City of Toronto was created on 1 Jan. 1998 through the amalgamation of seven municipalities: Metropolitan Toronto and six local area municipalities of Toronto, North York, Scarborough, Etobicoke, East York and York.

There are over 1m. French-speaking people and 0·25m. native Indians. An agreement with the Ontario government of Aug. 1991 recognized Indians' right to self-government.

SOCIAL STATISTICS

Births in 2007–08 numbered 138,985 (a rate of 10·8 per 1,000 population) and deaths 89,141 (a rate of 6·9 per 1,000 population). There were 63,151 marriages in 2006 and 28,805 divorces in 2005.

CLIMATE

A temperate continental climate, but conditions can be quite severe in winter, though proximity to the Great Lakes has a moderating influence on temperatures. Ottawa, average temperature, Jan. –10·8°C, July 20·8°C. Annual rainfall (including snow) 911 mm. Toronto, average temperature, Jan. –4·5°C, July 22·1°C. Annual rainfall (including snow) 818 mm.

CONSTITUTION AND GOVERNMENT

The provincial government is administered by a *Lieut.-Governor*, a cabinet and a single-chamber 107-member *Legislative Assembly* elected by a general franchise for a period of no longer than five years. The minimum voting age is 18 years. The province is represented by 24 members in the Senate and 106 in the House of Commons of Canada.

RECENT ELECTIONS

At the elections on 6 Oct. 2011 to the Legislative Assembly the governing Liberal Party won 53 of a possible 107 seats (with 37·6% of the vote), the Progressive Conservative Party 37 seats (35·4%) and the New Democratic Party 17 seats (22·7%). Turnout was a record low of 49·2%.

CURRENT GOVERNMENT

Lieut.-Governor: David Onley, O.Ont.; b. 1950 (in office since 5 Sept. 2007).

In March 2013 the Executive Council comprised:

Premier, and Minister of Agriculture and Food: Kathleen Wynne; b. 1953 (sworn in 11 Feb. 2013).

Deputy Premier and Minister of Health and Long-Term Care: Deb Matthews. *Aboriginal Affairs:* David Zimmer. *Attorney General:* John Gerretsen. *Children and Youth Services:* Teresa Piruzza. *Citizenship and Immigration:* Michael Coteau. *Community and Social Services:* Ted McMeekin. *Community Safety and Correctional Services, and Minister Responsible for Francophone Affairs:* Madeleine Meilleur. *Consumer Services:* Tracy McCharles. *Economic Development, Trade and Employment:* Eric Hoskins. *Education:* Liz Sandals. *Energy:* Bob Chiarelli. *Environment:* Jim Bradley. *Finance:* Charles Sousa. *Government Services:* Harinder Takhar. *Intergovernmental Affairs:* Laurel Broten. *Labour:* Yasir Naqvi. *Municipal Affairs and Housing:* Linda Jeffrey. *Natural Resources:* David Orazietti. *Northern Development and Mines:* Michael Gravelle. *Rural Affairs:* Jeff Leal. *Research and Innovation:* Reza Moridi. *Tourism, Culture, Sport and Minister Responsible for the 2015 Pan/Parapan American Games:* Michael Chan. *Training, Colleges and Universities:* Brad Duguid. *Transportation and Infrastructure:* Glen Murray. *Minister Responsible for Seniors:* Mario Sergio. *Government House Leader:* John Milloy.

Office of the Premier: http://www.premier.gov.on.ca

ECONOMY

GDP per person in 2009 was $44,228 CDN.

Budget

Provincial revenue and expenditure (in $1m. CDN) for years ending 31 March:

	2001–02	2002–03	2003–04	2004–05	2005–06
Gross revenue	71,013	71,585	72,432	80,247	86,811
Gross expenditure	68,626	72,106	77,807	83,747	89,061

Performance

Real GDP growth was negative in 2009, at –3·6%. The economy then began to recover in 2010, achieving growth of 3·4%.

ENERGY AND NATURAL RESOURCES

Electricity

Electricity generation totalled 140·4bn. kWh in 2009 (of which 86·8bn. kWh were nuclear and 34·7bn. kWh hydro-electric), representing 23·7% of Canada's production. Ontario ranks second after Quebec for electricity production.

Oil and Gas

Ontario is Canada's leading petroleum refining region. The province's five refineries have an annual capacity of 170m. bbls (27m. cu. metres).

Minerals

The total value of mineral production in 2002 was $5·7bn. CDN. In 2003 the most valuable commodities (production in $1m. CDN) were: gold, 1,253; nickel, 1,192; cement, 614; stone, 506; sand and gravel, 410; copper, 393. Total direct employment in the mining industry was 14,000 (9,000, metals) in 2002.

Agriculture

In 2006, 57,211 census farms operated on 5,386,453 ha.; total gross farm receipts in 2005 (excluding forest products sold) were $10·3bn. CDN. Net farm income in 2005 totalled $341·8m. CDN. For particulars of livestock *see* CANADA: Agriculture.

Forestry

The forested area totals 69·1m. ha., approximately 65% of Ontario's total area. Composition of Ontario forests: conifer, 56%; mixed, 26%; deciduous, 18%. The total growing stock (62% conifer, 38% hardwood) equals 5·3bn. cu. metres with an annual harvest level of 23m. cu. metres.

INDUSTRY

Ontario is Canada's most industrialized province, with GDP in 2007 of $532,842m. CDN, or 40·5% of the Canadian total. In 2006 manufacturing accounted for 19·1% of Ontario's GDP.

Leading manufacturing industries include: motor vehicles and parts; office and industrial electrical equipment; food processing; chemicals; and steel.

In 2006 Ontario was responsible for about 43% ($177·4bn. CDN) of Canada's merchandise exports, and for 96% ($69·2bn. CDN) of exports of motor vehicles and motor vehicle parts.

Labour

In 2006 the labour force was 6,928,000, of whom 6,493,000 were employed (4,388,000 in the private sector, 1,170,000 in the public sector and 935,000 self-employed). The major employers (2006 in thousands) were: trade, 1,016; manufacturing, 1,007; health care and social assistance, 638; finance, insurance, real estate and leasing, 477; professional, scientific and technical services, 454; educational services, 445; construction, 405. The unemployment rate in 2006 was 6·3%.

INTERNATIONAL TRADE

Imports and Exports

Ontario's imports were $241·7bn. CDN in 2008, up from $240·3bn. CDN in 2007. Exports were $188·8bn. CDN in 2008, down from $202·5bn. CDN.

COMMUNICATIONS

Roads

Almost 40% of the population of North America is within one day's drive of Ontario. There were, in 1998, 159,456 km of roads (municipal, 143,000). Motor licences (on the road) numbered (2004) 10,360,891, of which 6,218,458 were passenger cars, 1,267,244 commercial vehicles, 31,038 buses, 2,100,511 trailers, 158,103 motorcycles and 306,479 snow vehicles.

Rail

In 2007 there were 17 short lines (eight provincially-licensed freight railways, four provincially-licensed tourist railways and five federally-licensed railways), plus the provincially-owned Ontario Northland Railway. The Canadian National and Canadian Pacific Railways operate in Ontario. Total track length, approximately 11,800 km. There is a metro and tramway network in Toronto.

Civil Aviation

Toronto's Lester B. Pearson International Airport is Canada's busiest, handling 30,911,000 passengers in 2010 (29,326,000 in 2009).

Shipping

The Great Lakes/St Lawrence Seaway, a 3,747 km system of locks, canal and natural water connecting Ontario to the Atlantic Ocean, has 95,000 sq. miles of navigable waters and serves the water-borne cargo needs of four Canadian provinces and 17 American States.

Telecommunications

In Dec. 2008, 76·8% of households had at least one mobile phone and 7·4% of households only had a mobile phone. 81% of households had home internet access in 2010.

SOCIAL INSTITUTIONS

Justice

In 2009 there were 4,964 criminal code violations per 100,000 population, compared to a national average of 6,840 per 100,000 population.

Education

There is a provincial system of publicly financed elementary and secondary schools as well as private schools. Publicly financed elementary and secondary schools had a total enrolment of 2,118,544 pupils in 2005–06 (1,411,011 elementary and 707,533 secondary) and 114,191 teachers in 2004–05. In 2001–02, of the $64,270m. CDN total expenditure, 18·5% was on education.

There are 18 publicly funded universities (Brock, Carleton, Dominican, Guelph, Lakehead, Laurentian, McMaster, Nipissing, Ottawa, Queen's, Ryerson, Toronto, Trent, Waterloo, Western Ontario, Wilfred Laurier, Windsor and York), the Royal Military College of Canada and the University of Ontario Institute of Technology as well as one institute of equivalent status (Ontario College of Art and Design) with full-time enrolment for 2004–05 of 333,219. All receive operating grants from the Ontario government. There are also 24 publicly financed Colleges of Applied Arts and Technology, with a full-time enrolment of 150,000 in 2005–06.

Government funding for education in Ontario in 2005–06 was $17·2bn. CDN.

Health

Ontario Health Insurance Plan health care services are available to eligible Ontario residents at no cost. The Ontario Health Insurance Plan (OHIP) is funded, in part, by an Employer Health Tax.

FURTHER READING

Statistical Information: Annual publications of the Ontario Ministry of Finance include: Ontario Statistics; Ontario Budget; Public Accounts; Financial Report.

Prince Edward Island

KEY HISTORICAL EVENTS

The first recorded European visit was by Jacques Cartier in 1534, who named it Isle St-Jean. In 1719 it was settled by the French, but was taken from them by the English in 1758, annexed to Nova Scotia in 1763, and constituted a separate colony in 1769. Named Prince Edward Island in honour of Prince Edward, Duke of Kent, in 1799, it joined the Canadian Confederation on 1 July 1873.

TERRITORY AND POPULATION

The province lies in the Gulf of St Lawrence, and is separated from the mainland of New Brunswick and Nova Scotia by Northumberland Strait. The area of the island is 5,660 sq. km (2,185 sq. miles). The population at the 2011 census was 140,204. Population of the principal cities (2006): Charlottetown (capital), 32,174; Summerside, 14,500.

SOCIAL STATISTICS

Births in 2007–08 numbered 1,388 (a rate of 10·0 per 1,000 population) and deaths 1,217 (rate of 8·8 per 1,000 population). There were 844 marriages in 2006 and 283 divorces in 2005.

CLIMATE

The cool temperate climate is affected in winter by the freezing of the St Lawrence, which reduces winter temperatures. Charlottetown, Jan. –3°C to –11°C, July 14°C to 23°C. Annual rainfall 853·5 mm.

CONSTITUTION AND GOVERNMENT

The provincial government is administered by a Lieut.-Governor-in-Council (Cabinet) and a Legislative Assembly of 27 members who are elected for up to five years. The province is represented by four members in the Senate and four in the House of Commons of Canada.

RECENT ELECTIONS

At provincial elections on 3 Oct. 2011 the governing Liberal Party won 22 of the available 27 seats (with 51·4% of the vote) and the Progressive Conservatives took five seats (40·2%). The Greens and the New Democratic Party won no seats (4·3% and 3·2% respectively). Turnout was 76·4%.

CURRENT GOVERNMENT

Lieut.-Governor: Frank Lewis, O.PEI (sworn in 15 Aug. 2011).

The Liberal Party Executive Council was composed as follows in March 2013:

Premier, President of the Executive Council, and Minister Responsible for Intergovernmental Affairs, Acadian and Francophone Affairs, and Aboriginal Affairs: Robert Ghiz; b. 1974.

Deputy Premier and Minister of Agriculture and Forestry: George Webster. Community Services and Seniors, and Minister Responsible for the Status of Women: Valerie Docherty. Education and Early Childhood Development: Alan McIsaac. Environment, Labour and Justice: Janice Sherry. Finance, Energy and Municipal Affairs: Wesley Sheridan. Fisheries, Aquaculture and Rural Development: Ronald MacKinley. Health and Wellness: Doug Currie. Innovation and Advanced Learning: Allen Roach. Tourism and Culture: Robert Henderson. Transportation and Infrastructure Renewal: Robert Vessey.

Office of the Premier: http://www.gov.pe.ca/premier

ECONOMY

GDP per person was $33,640 CDN in 2009.

Budget

Total revenue in 2005–06 was $1,196m. CDN (own source revenue, $746m. CDN; general purpose transfers, $319m. CDN; special purpose transfers, $132m. CDN). Total expenditures in 2005–06 amounted to $1,213m. CDN (including: health, $365m. CDN; education, $232m. CDN; debt charges, $124m. CDN; resource conservation and industrial development, $112m. CDN; social services, $107m. CDN).

ENERGY AND NATURAL RESOURCES

Electricity

Prince Edward Island's electricity generation in 2009 totalled 500m. kWh, almost exclusively from wind power.

Oil and Gas

In 2006 Prince Edward Island had more than 400,000 ha. under permit for oil and natural gas exploration.

Agriculture

Total area of farmland occupies approximately half of the total land area of 566,177 ha. Farm cash receipts in 2003 were $353m. CDN, with cash receipts from potatoes accounting for about 50% of the total. Cash receipts from dairy products, hogs and cattle followed in importance. For particulars of livestock, *see* CANADA: Agriculture.

Forestry

Total forested area is 280,000 ha. Of this 87% is owned by 12,000 woodlot owners. Most of the harvest takes place on private woodlots. The forest cover is 23% softwood, 29% hardwood and 48% mixed wood. In 2003 the volume of wood harvested reached 677,031 cu. metres, an increase of 5·2% on the previous year. The total value of wood industry shipments was $50·6m. CDN in 2003.

Fisheries

The total catch of 114·8m. lb in 2008 had a landed value of $155·2m. CDN. Lobsters accounted for $100·7m. CDN; other shellfish, $44·0m. CDN; pelagic and estuarial, $9·7m. CDN; groundfish, $0·4m. CDN; seaplants, $0·4m. CDN.

INDUSTRY

Value of manufacturing shipments for all industries in 2003 was $1,356·1m. CDN. In 2003 (provisional) provincial GDP in constant prices for manufacturing was $394·9m. CDN; construction, $160·7m. CDN. In 2003 the total value of retail trade was $1,318·0m. CDN.

Labour

The average weekly wage (industrial aggregate) rose from $540·77 CDN in 2002 to $547·04 CDN in 2003. The labour force averaged 78,500 in 2004, with employment averaging 69,600. The unemployment rate was 11·1% in 2003.

COMMUNICATIONS

Roads

In 2006 there were 3,700 km of paved highway and 1,900 km of unpaved road as well as 1,200 bridge structures. The Confederation Bridge, a 12·9-km two-lane bridge that joins Borden-Carleton with Cape Jourimain in New Brunswick, was opened in June 1997. A bus service operates twice daily to the mainland.

Civil Aviation

Prince Edward Island's busiest airport is Charlottetown, which handled 285,000 passengers in 2010 (271,000 in 2009). There were direct flights in 2010 to Edmonton, Halifax, Montreal, Ottawa and Toronto.

Shipping

Car ferries link the Island to New Brunswick year-round, with ice-breaking ferries during the winter months. Ferry services are operated to Nova Scotia from late April to mid-Dec. A service to the Magdalen Islands (Quebec) operates from 1 April to 31 Jan. The main ports are Summerside and Charlottetown, with additional capacity provided at Souris and Georgetown.

Telecommunications

In Dec. 2008, 73·5% of households had at least one mobile phone and 6·1% of households only had a mobile phone. 73% of households had home internet access in 2010.

SOCIAL INSTITUTIONS

Justice

In 2009–10, 1,215 Criminal Code cases were disposed of in an adult criminal court. There were no homicides in 2010.

Education

In 2003–04 there were 10,731 elementary students and 12,352 secondary students in both private and public schools. There is one undergraduate university (3,294 full-time and 599 part-time students), a veterinary college (237 students), and a Master of Science programme (33 students), all in Charlottetown. Holland College provides training for employment in business, applied arts and technology, with approximately 2,500 full-time students in post-secondary and vocational career programmes. The college offers extensive academic and career preparation programmes for adults.

Estimated government expenditure on education, 2000–01, $183·4m. CDN.

CULTURE

Tourism

The value of the tourist industry was estimated at $350m. CDN in 2003, with 1·1m. visitors in that year.

FURTHER READING

Baldwin, D. O., *Abegweit: Land of the Red Soil.* 1985

Quebec—Québec

KEY HISTORICAL EVENTS

Quebec was known as New France from 1534 to 1763; as the province of Quebec from 1763 to 1790; as Lower Canada from 1791 to 1846; as Canada East from 1846 to 1867, and when, by the union of the four original provinces, the Confederation of the Dominion of Canada was formed, it again became known as the province of Quebec (Québec).

The Quebec Act, passed by the British Parliament in 1774, guaranteed to the people of the newly conquered French territory in North America security in their religion and language, their customs and tenures, under their own civil laws. In a referendum on 20 May 1980, 59·5% voted against 'separatism'. At a further referendum on 30 Oct. 1995, 50·6% of votes cast were against Quebec becoming 'sovereign in a new economic and political partnership' with Canada. The electorate was 5m.; turnout was 93%. On 20 Aug. 1998 Canada's supreme court ruled that Quebec was prohibited by both the constitution and international law from seceding unilaterally from the rest of the country but that a clear majority in a referendum would impose a duty on the Canadian government to negotiate. Both sides claimed victory. On 27 Nov. 2006 Canada's parliament passed a motion recognizing that 'the Québécois form a nation within a united Canada'.

TERRITORY AND POPULATION

The area of Quebec (as amended by the Labrador Boundary Award) is 1,542,056 sq. km (595,388 sq. miles), of which 1,365,128 sq. km is land area (including the Territory of Ungava, annexed in 1912 under the Quebec Boundaries Extension Act). The population at the 2011 census was 7,903,001.

Principal cities (2006 census populations): Montreal, 1,620,693; Quebec (capital), 491,142; Laval, 368,709; Gatineau, 242,124;

Longueuil, 229,330; Sherbrooke, 147,427; Saguenay, 143,692; Lévis, 130,006; Trois-Rivières, 126,323; Terrebonne, 94,703; Saint-Jean-sur-Richelieu, 87,492; Repentigny, 76,237; Brossard, 71,154; Drummondville, 67,392; Saint-Jérôme, 63,729; Shawnigan, 51,904; Sainte-Hyacinthe, 51,616.

SOCIAL STATISTICS

Births in 2007–08 numbered 85,608 (a rate of 11·1 per 1,000 population) and deaths 56,200 (rate of 7·3 per 1,000 population). There were 21,956 marriages in 2006 and 15,423 divorces in 2005.

CLIMATE

Cool temperate in the south, but conditions are more extreme towards the north. Winters are severe and snowfall considerable, but summer temperatures are quite warm. Quebec, Jan. −12·5°C, July 19·1°C. Annual rainfall 1,123 mm. Montreal, Jan. −10·7°C, July 20·2°C. Annual rainfall 936 mm.

CONSTITUTION AND GOVERNMENT

There is a *Legislative Assembly* consisting of 125 members, elected in 125 electoral districts for four years. The province is represented by 24 members in the Senate and 75 in the House of Commons of Canada.

RECENT ELECTIONS

At the elections of 4 Sept. 2012 the Parti Québécois won 54 seats with 32·0% of votes cast, the Quebec Liberal Party 50 seats with 31·2%, the Coalition Avenir Québec 19 seats with 27·1% and Québec Solidaire 2 seats with 6·0%. Turnout was 74·6%.

CURRENT GOVERNMENT

Lieut.-Governor: Pierre Duchesne (took office on 7 June 2007).
　　Members of the Parti Québécois cabinet in March 2013:
　　Premier: Pauline Marois; b. 1949.
　　Deputy Premier and Minister of Agriculture, Fisheries and Food: François Gendron. *Chair of the Conseil du trésor:* Stéphane Bédard. *Culture and Communications:* Maka Kotto. *Education, Recreation and Sports:* Marie Malavoy. *Families:* Nicole Léger. *Finance and the Economy:* Nicolas Marceau. *Health and Social Services:* Réjean Hébert. *Higher Education, Research, Science and Technology:* Pierre Duchesne. *Immigration and Cultural Communities:* Diane De Courcy. *International Relations, La Francophonie and External Trade:* Jean-François Lisée. *Justice:* Bertrand St-Arnaud. *Labour, and Employment and Social Solidarity:* Agnès Maltais. *Natural Resources:* Martine Ouellet. *Public Security:* Stéphane Bergeron. *Sustainable Development, Environment, Wildlife and Parks:* Yves-François Blanchet. *Transport, and Municipal Affairs, Regions and Land Occupancy:* Sylvain Gaudreault.

Government of Quebec Website: http://www.gouv.qc.ca

ECONOMY

GDP per person in 2009 was $38,808 CDN.

Budget

Revenue and expenditure (in $1,000 CDN) for fiscal years ending 31 March:

	2002–03	2003–04	2004–05	2005–06
Revenue	58,186,000	60,808,000	64,439,000	69,311,000
Expenditure	60,469,000	63,135,000	66,833,000	69,832,000

The total gross debt at 31 March 2009 was $151,385m. CDN.

Performance

Real GDP growth was negative in 2009, at −0·3%. The economy then began to recover strongly in 2010, achieving growth of 2·7%.

ENERGY AND NATURAL RESOURCES

Electricity

Water power is one of the most important natural resources of Quebec. At the end of 2011 the installed generating capacity was 36,971 MW. Production, 2009, was 191,510m. kWh. Quebec produces around a third of Canada's electricity and ranks among the largest producers of hydroelectric power in the world.

Water

There are 4,500 rivers and 500,000 lakes in Quebec, which possesses 3% of the world's freshwater resources.

Minerals

For 2006 the value of mineral production was $4,826m. CDN. Chief minerals: iron ore (confidential); nickel, $628·1m. CDN; gold, $508·3m. CDN; zinc, $319·0m. CDN; copper, $144·6m. CDN. Non-metallic minerals produced include: asbestos, titanium-dioxide, peat and quartz (silica). Among the building materials produced were: stone, $367·8m. CDN; cement, $329·0m. CDN; sand and gravel, $90·0m. CDN.

Agriculture

In 2011 the agricultural area was 3,341,333 ha. The production of the principal crops was (2011 in 1,000 tonnes):

Crops	Production	Crops	Production
Tame hay	3,900	Barley	196
Corn for grain	3,125	Wheat	116
Fodder corn	2,000	Mixed grains	42
Soya	800	Canola	36
Oats	223		

There were 29,437 farms operating in 2011. Cash receipts, 2011, $7,968m. CDN (livestock and livestock products, 61·6%; crops, 30·2%; direct payments, 8·2%). Quebec was a net importer of food and agricultural produce in 2011. For particulars of livestock see CANADA: Agriculture.

Forestry

Forests cover an area of 591,549 sq. km. 424,114 sq. km are classified as productive forests, of which 355,004 sq. km are provincial forest land and 66,198 sq. km are privately owned. Quebec leads the Canadian provinces in paper production, having nearly half of the Canadian estimated total.

In 2006 production of pulp, paper and cardboard was 9,832,000 tonnes.

Fisheries

The principal fish and seafood are Greenland halibut, Atlantic halibut, herring, snow crab, lobster and shrimp. The landed value in 2011 of fish, seafood and shellfish was $149m. CDN.

INDUSTRY

In 2001 there were 15,191 industrial establishments in the province; employees, 567,999; salaries and wages, $20,691m. CDN; value of shipments, $141,537m. CDN. Among the leading industries are petroleum refining, pulp and paper mills, smelting and refining, dairy products, slaughtering and meat processing, motor vehicle manufacturing, women's clothing, sawmills and planing mills, iron and steel mills, and commercial printing.

Labour

In 2009 there were 3,844,200 persons (1,854,000 female) in employment.

INTERNATIONAL TRADE

Imports and Exports

In 2008 Quebec's imports were valued at $87·6bn. CDN; value of exports, $71·0bn. CDN.

COMMUNICATIONS

Roads
In 2007 there were 30,086 km of roads and (2006) 5,402,353 registered motor vehicles.

Rail
There were (2008) 9,728 km of railway. There is a metro system in Montreal (71 km of which 65 km for passenger use).

Civil Aviation
Quebec's busiest airports are Montreal (Pierre Elliott Trudeau International), which is the third busiest in Canada and handled 12,700,000 passengers in 2010, and Quebec City (Jean Lesage International Airport).

Telecommunications
In Dec. 2008, 65·5% of households had at least one mobile phone and 6·7% of households only had a mobile phone. 73% of households had home internet access in 2010.

SOCIAL INSTITUTIONS

Justice
In 2009, 434,084 Criminal Code offences were reported; there were 88 homicides.

Education
Education is compulsory for children aged 6–16. Pre-school education and elementary and secondary training are free in some 2,520 public schools. In July 1998 the number of school boards was reduced to 72. These were organized along linguistic lines, 60 French, nine English and three special school boards that served native students in the Côte-Nord and Nord-du-Québec regions. Around 12% of the student population attends private schools: in 2004–05, 348 establishments were authorized to provide pre-school, elementary and secondary education. After six years of elementary and five years of secondary school education, students attend Cegep, a post-secondary educational institution. In 2004–05 college, pre-university and technical training for young and adult students was provided by 52 Cegeps, 11 government schools and 60 private establishments.

In 2005–06 in pre-kindergartens there were 14,808 pupils; in kindergartens, 74,123; in primary schools, 510,340; in secondary schools, 489,054; in general education for adults, 257,443; in colleges (post-secondary, non-university), 188,549; and in universities, 264,240.

The operating expenditures of education institutions totalled $14,663·0m. CDN in 2004–05. This included $4,163·2m. CDN for universities, $8,000·3m. CDN for public primary and secondary schools, $873·6m. CDN for private primary and secondary schools and $1,625·9m. CDN for colleges.

In 2004–05 the province had 19 universities and affiliated schools of which seven were major universities: four French-language universities—Université Laval (Quebec, founded 1852), Université de Montréal (opened 1876 as a branch of Laval, independent 1920), Université de Sherbrooke (founded 1954), Université du Québec (founded 1968); and three English-language universities—McGill University (Montreal, founded 1821), Bishop's University (Lennoxville, founded 1843) and Concordia University (Montreal, granted a charter 1975). Université de Montréal has two affiliated schools: HEC Montréal (École des Hautes Études Commerciales), a business school founded in 1907; and École Polytechnique de Montréal, an engineering school founded in 1873. In 2004–05 there were 164,644 full-time university students and 97,045 part-time.

Health
Quebec's socio-health network consisted of 294 public and private establishments in 2009, of which 191 were public.

CULTURE

Press
In 2007 there were five major French-language newspapers (*La Presse, Le Devoir, Le Journal de Québec, Le Journal de Montréal* and *Le Soleil*) and one major English-language newspaper (*The Gazette*).

FURTHER READING
Dickinson, J. A. and Young, B., *A Short History of Quebec.* 4th ed. 2008

Statistical office: Institut de la statistique du Québec, 200 chemin Sainte-Foy, Québec, G1R 5T4.
Website: http://www.stat.gouv.qc.ca

Saskatchewan

KEY HISTORICAL EVENTS

Saskatchewan derives its name from its major river system, which the Cree Indians called 'Kis-is-ska-tche-wan', meaning 'swift flowing'. It officially became a province when it joined the Confederation on 1 Sept. 1905.

In 1670 Charles II granted to Prince Rupert and his friends a charter covering exclusive trading rights in 'all the land drained by streams finding their outlet in the Hudson Bay'. This included what is now Saskatchewan. The trading company was first known as The Governor and Company of Adventurers of England; later as the Hudson's Bay Company. In 1869 the Northwest Territories was formed, and this included Saskatchewan. The North-West Mounted Police Force was inaugurated four years later. In 1882 the District of Saskatchewan was formed and in 1885 the Canadian Pacific Railway's transcontinental line was completed, bringing a stream of immigrants to southern Saskatchewan. The Hudson's Bay Company surrendered its claim to territory in return for cash and land around the existing trading posts.

TERRITORY AND POPULATION

Saskatchewan is bounded in the west by Alberta, in the east by Manitoba, in the north by the Northwest Territories and in the south by the USA. The area of the province is 651,036 sq. km (251,365 sq. miles), of which 591,670 sq. km is land area and 59,366 sq. km is water. The population at the 2011 census was 1,033,381. Population of cities, 2006 census: Saskatoon, 202,340; Regina (capital), 179,246; Prince Albert, 34,138; Moose Jaw, 32,132; Yorkton, 15,038; Swift Current, 14,946; North Battleford, 13,190; Estevan, 10,084; Weyburn, 9,433; Lloydminster, 8,118; Melfort, 5,192; Humboldt, 4,998; Melville, 4,149.

SOCIAL STATISTICS

Births in 2007–08 numbered 13,438 (a rate of 13·3 per 1,000 population) and deaths 9,295 (rate of 9·2 per 1,000 population). There were 5,030 marriages in 2006 and 1,922 divorces in 2005.

CLIMATE

A cold continental climate, with severe winters and warm summers. Rainfall amounts are greatest from May to Aug. Regina, Jan. 0°F (−17·8°C), July 65°F (18·3°C). Annual rainfall 15″ (373 mm).

CONSTITUTION AND GOVERNMENT

The provincial government is vested in a *Lieut.-Governor*, an *Executive Council* and a *Legislative Assembly* of 58 seats, elected for five years. Women were given the franchise in 1916. The province is represented by six members in the Senate and 14 in the House of Commons of Canada.

RECENT ELECTIONS

In elections on 7 Nov. 2011 the Saskatchewan Party (SP) won 49 of 58 seats (64·2% of the vote) and the New Democrats (NDP), 9 (32·0%).

CURRENT GOVERNMENT

Lieut.-Governor: Vaughn Schofield (took office 22 March 2012).

The Saskatchewan Party ministry comprised as follows in March 2013:

Premier, President of the Executive Council, and Minister of Intergovernmental Affairs: Brad Wall.

Deputy Premier and Minister of Finance: Ken Krawetz. *Economy:* Bill Boyd. *Social Services:* June Draude. *Highways and Infrastructure:* Don McMorris. *Advanced Education, Labour Relations and Workplace Safety:* Don Morgan, QC. *Crown Investments:* Donna Harpauer. *Health:* Dustin Duncan. *Justice and Attorney General:* Gordon Wyant, QC. *Environment:* Ken Cheveldayoff. *Government Relations:* Jim Reiter. *Education:* Russ Marchuk. *Agriculture:* Lyle Stewart. *Central Services:* Nancy Heppner. *Parks, Culture and Sport:* Kevin Doherty.

Office of the Premier: http://www.gov.sk.ca/premier

ECONOMY

GDP per capita in 2009 was $54,943 CDN.

Budget

Budget and net assets (years ending 31 March) in $1,000 CDN:

	2002–03	2003–04	2004–05	2005–06
Budgetary revenue	6,094,300	6,228,000	6,590,500	7,006,800
Budgetary expenditure	6,319,255	6,561,383	6,761,533	7,151,731

ENERGY AND NATURAL RESOURCES

Electricity

SaskPower (the Saskatchewan Power Corporation) generated 20,969m. kWh in 2011, 55% of which came from coal-fired power plants.

Minerals

In 2005 mineral sales were valued at $13,050m. CDN, including (in $1m. CDN): petroleum, 6,671·5; potash, 2,697·8; natural gas, 2,095·9; coal and others, 889·9; salt, 24·6. Other major minerals included copper, zinc, potassium sulphate, ammonium sulphate, bentonite, uranium, gold and base metals.

Agriculture

Agriculture accounted for 11·4% of the provincial GDP in 2011, second to the mining sector which accounted for 13·4%. Saskatchewan normally produces about two-thirds of Canada's wheat. Wheat production in 2005 (in 1,000 tonnes) was 13,742 (12,261 in 2004) from 14·0m. acres; barley, 5,345 from 4·8m. acres; canola, 4,633 from 6·6m. acres; oats, 1,672 from 2·0m. acres; flax, 881 from 1·6m. acres; rye, 184 from 195,000 acres. Livestock (1 July 2005): cattle and calves, 3·7m.; swine, 1·4m.; sheep and lambs, 145,000. Poultry in 2005: chickens, 22·0m.; turkeys, 674,000. Cash income from the sale of farm products in 2005 was $6,355m. CDN. At the 2006 census there were 44,329 farms in the province covering an area of 26,002,605 ha. (with 14,960,103 ha. of land under crops).

The South Saskatchewan River irrigation project, the main feature of which is the Gardiner Dam, was completed in 1967. It will ultimately provide for an area of 0·2m. to 0·5m. acres of irrigated cultivation in Central Saskatchewan. As of 2006, 247,158 acres were intensively irrigated. Total irrigated land in the province, 339,583 acres.

Forestry

Half of Saskatchewan's area is forested, but only 115,000 sq. km are of commercial value at present. Forest products valued at $612m. CDN were produced in 2003–04.

Fisheries

In 2010 commercial freshwater fisheries landings had a total weight of 2,731 tonnes (of which whitefish 1,137 tonnes) and a value of $3,192,000 CDN.

INDUSTRY

In 2005 there were 1,008 manufacturing establishments, employing 20,293 persons. In 2005 manufacturing contributed $2,336·1m. CDN and construction $1,776·0m. CDN to total GDP at basic prices of $31,575·0m. CDN.

Labour

In 2005 the labour force was 509,400 (234,700 females), of whom 483,500 (224,000) were employed.

COMMUNICATIONS

Roads

In 2005 there were 26,168 km of provincial highways and 198,375 km of municipal roads (including prairie trails). Motor vehicles registered totalled 750,640 (2005). Bus services are provided by two major lines.

Rail

In 2005 there were approximately 9,513 km of railway track.

Civil Aviation

Saskatchewan's busiest airports are Saskatoon and Regina, which handled 1,196,000 and 1,102,000 passengers in 2010 respectively.

Telecommunications

In Dec. 2008, 78·0% of households had at least one mobile phone and 7·2% of households only had a mobile phone. 76% of households had home internet access in 2010.

SOCIAL INSTITUTIONS

Justice

In 2009–10, 21,665 Criminal Code cases were disposed of in an adult criminal court. In 2010 there were 34 homicides (a rate of 3·3 per 100,000 population).

Education

The Saskatchewan education system in 2005–06 consisted of 28 school divisions, of which nine are Roman Catholic, serving 107,331 elementary pupils, 59,801 high-school students and 1,577 students enrolled in special classes. In addition, the Saskatchewan Institute of Applied Science and Technology (SIAST) had approximately 13,200 full-time and 29,000 part-time and extension course registration students in 2005–06. There are also eight regional colleges with an enrolment of approximately 19,700 students in 2004–05.

The University of Saskatchewan was established at Saskatoon in 1907. In 2005–06 it had 15,300 full-time students, 4,000 part-time students and 961 full-time academic staff. The University of Regina, established in 1974, had 9,678 full-time and 2,977 part-time students and 425 full-time academic staff in 2005–06.

CULTURE

Tourism

An estimated 1·8m. out-of-province tourists spent $513·3m. CDN in 2004.

FURTHER READING

Archer, J. H., *Saskatchewan: A History.* 1980

Statistical office: Bureau of Statistics, 9th Floor, 2350 Albert St., Regina, SK, S4P 4A6.

Website: http://www.stats.gov.sk.ca

The Northwest Territories

KEY HISTORICAL EVENTS

The Territory was developed by the Hudson's Bay Company and the North West Company (of Montreal) from the 17th century. The Canadian government bought out the Hudson's Bay Company in 1869 and the Territory was annexed to Canada in 1870. The Arctic Islands lying north of the Canadian mainland were annexed to Canada in 1880.

A plebiscite held in March 1992 approved the division of the Northwest Territories into two separate territories. (For the territory of Nunavut *see* CONSTITUTION AND GOVERNMENT *below*, and NUNAVUT on page 298).

TERRITORY AND POPULATION

The Northwest Territories comprises all that portion of Canada lying north of the 60th parallel of N. lat. except those portions within Nunavut, Yukon and the provinces of Quebec and Newfoundland and Labrador. The total area of the Territories was 3,426,320 sq. km, but since the formation of Nunavut is now 1,346,106 km. Of its five former administrative regions—Fort Smith, Inuvik, Kitikmeot, Keewatin and Baffin—only Fort Smith and Inuvik remain in the Northwest Territories.

The population at the 2011 census was 41,462. The capital is Yellowknife, population (2006); 18,700. Other main centres (with population in 2006): Hay River (3,648), Inuvik (3,484), Fort Smith (2,364), Behchokò (1,894). Iqaluit and Rankin Inlet, formerly in the Northwest Territories, are now in Nunavut. In Aug. 2003 an agreement was reached for the Tlicho First Nation to assume control over 39,000 sq. km of land in the Northwest Territories (including Canada's two diamond mines), creating the largest single block of First Nation-owned land in Canada.

SOCIAL STATISTICS

Births in 2007–08 numbered 727 (a rate of 16·7 per 1,000 population) and deaths 193 (rate of 4·4 per 1,000 population). There were 129 marriages in 2006 and 65 divorces in 2005.

CLIMATE

Conditions range from cold continental to polar, with long hard winters and short cool summers. Precipitation is low. Yellowknife, Jan. mean high –24·7°C, low –33°C; July mean high 20·7°C, low 11·8°C. Annual rainfall 26·7 cm.

CONSTITUTION AND GOVERNMENT

The Northwest Territories is governed by a *Premier*, with a cabinet (the *Executive Council*) of eight members including the *Speaker*, and a *Legislative Assembly*, who choose the premier and ministers by consensus. There are no political parties. The Assembly is composed of 19 members elected for a four-year term of office. A *Commissioner* of the Northwest Territories is the federal government's senior representative in the Territorial government. The seat of government was transferred from Ottawa to Yellowknife when it was named Territorial Capital on 18 Jan. 1967. On 10 Nov. 1997 the governments of Canada and the Northwest Territories signed an agreement so that the territorial government could assume full responsibility to manage its elections.

Legislative powers are exercised by the Executive Council on such matters as taxation within the Territories in order to raise revenue, maintenance of justice, licences, solemnization of marriages, education, public health, property, civil rights and generally all matters of a local nature. The territory is represented by one member in the Senate and one in the House of Commons of Canada.

The Territorial government has assumed most of the responsibility for the administration of the Northwest Territories but political control of Crown lands. In a Territory-wide plebiscite in April 1982, a majority of residents voted in favour of dividing the Northwest Territories into two jurisdictions, east and west. Constitutions for an eastern and western government have been under discussion since 1992. A referendum was held in Nov. 1992 among the Inuit on the formation of a third territory, **Nunavut** ('Our Land'), in the eastern Arctic. Nunavut became Canada's third territory on 1 April 1999.

RECENT ELECTIONS

On 3 Oct. 2011, 19 members (MLAs) were returned to the 17th Legislative Assembly. There were 47 candidates, three of whom had been previously returned unopposed. All members are independent.

CURRENT GOVERNMENT

Commissioner: George L. Tuccaro; b. 1950 (took office in May 2010).

Members of the Executive Council of Ministers in March 2013:

Premier, and Minister of Executive, and Aboriginal Affairs and Intergovernmental Relations: Bob McLeod.

Deputy Premier, and Minister of Education, Culture and Employment: Jackson Lafferty. *Finance, and Environment and Natural Resources:* Michael Miltenberger. *Municipal and Community Affairs:* Robert McLeod. *Industry, Tourism and Investment, and Transportation:* David Ramsay. *Justice, Human Resources, and Public Works and Services:* Glen Abernethy. *Health and Social Services:* Tom Beaulieu.

Speaker: Jackie Jacobson.

Government of the Northwest Territories Website: http://www.gov.nt.ca

ECONOMY

GDP per person in 2009 was $93,280 CDN, the highest of any Canadian province or territory.

Budget

Total revenue in 2005–06 was $1,259m. CDN (own source revenue, $268m. CDN; general purpose transfers, $801m. CDN; special purpose transfers, $190m. CDN). Total expenditures in 2005–06 amounted to $1,333m. CDN (including: health, $264m. CDN; education, $251m. CDN; social services, $137m. CDN).

Performance

In 2005 real GDP grew at a rate of 3·5% but there was then a recession in 2006, with the economy shrinking by 0·4%.

ENERGY AND NATURAL RESOURCES

Oil and Gas

Crude petroleum production was 1,373,000 cu. metres in 2003; natural gas production was 780m. cu. metres in 2003.

Minerals

Mineral production in 2006: diamonds, 12,976,000 carats ($1,567,019,000 CDN); sand and gravel (including Nunavut), 362,000 tonnes ($2,157,000 CDN); stone, 461,000 tonnes ($4,500,000 CDN); tungsten, 2,500 tonnes ($64,497,000 CDN). Total mineral production in 2006 was valued at $1·64bn. CDN.

Forestry

Forest land area in the Northwest Territories consists of 61·4m. ha., about 18% of the total land area. The principal trees are white and black spruce, jack-pine, tamarack, balsam poplar, aspen and birch. In 2004, 26,000 cu. metres of timber were produced.

Trapping and Game

Wildlife harvesting is the largest economic activity undertaken by aboriginal residents in the Northwest Territories. The value of the subsistence food harvest is estimated at $28m. CDN annually in terms of imports replaced. Fur-trapping (the most valuable pelts being white fox, wolverine, beaver, mink, lynx and red fox) was once a major industry, but has been hit by anti-fur campaigns. In 2009–10, 27,489 pelts worth $830,921 CDN were sold.

Fisheries

Fish marketed through the Freshwater Fish Marketing Corporation in 2005–06 totalled 734,000 kg at a value of $705,000 CDN, principally whitefish, northern pike and trout.

INDUSTRY

Co-operatives

There are 37 active co-operatives, including two housing co-operatives and two central organizations to service local co-operatives, in the Northwest Territories. They are active in handicrafts, furs, fisheries, retail stores, hotels, cable TV, post offices, petroleum delivery and print shops. Total revenue in 2000 was about $97m. CDN.

COMMUNICATIONS

Roads

The Mackenzie Route connects Grimshaw, Alberta, with Hay River, Pine Point, Fort Smith, Fort Providence, Rae-Edzo and Yellowknife. The Mackenzie Highway extension to Fort Simpson and a road between Pine Point and Fort Resolution have both been opened.

Highway service to Inuvik in the Mackenzie Delta was opened in spring 1980, extending north from Dawson, Yukon as the Dempster Highway. The Liard Highway connecting the communities of the Liard River valley to British Columbia opened in 1984.

In 2005 there were a total of 28,212 vehicle registrations, including 23,184 road motor vehicles and 3,864 trailers.

Rail

There is one small railway system in the north which runs from Hay River, on the south shore of Great Slave Lake, 435 miles south to Grimshaw, Alberta, where it connects with the Canadian National Railways, but it is not in use.

Civil Aviation

The busiest airport is Yellowknife, which handled 319,000 passengers in both 2009 and 2010.

Shipping

A direct inland-water transportation route for about 1,700 miles is provided by the Mackenzie River and its tributaries, the Athabasca and Slave rivers. Subsidiary routes on Lake Athabasca, Great Slave Lake and Great Bear Lake total more than 800 miles. Communities in the eastern Arctic are resupplied by ship each summer via the Atlantic and Arctic Oceans or Hudson Bay.

Telecommunications

In 2003, 13,000 households (95·5%) had telephones. Those few communities without a telephone service have high frequency or very high frequency radios for emergency use.

SOCIAL INSTITUTIONS

Justice

In 2009–10, 1,706 Criminal Code cases were disposed of in an adult criminal court. In 2010 there was one homicide (a rate of 2·3 per 100,000 population).

Education

The Education System in the Northwest Territories is comprised of eight regional bodies (boards) that have responsibilities for the K–12 education programme. Three of these jurisdictions are located in Yellowknife; a public school authority, a catholic school authority and a Commission scolaire francophone that oversees a school operating in Yellowknife, and one in Hay River.

For the 2010–11 school year there were 46 public plus three (Catholic) private schools operating in the NWT. Within this system there were 758 teachers, excluding those at Aurora College and community learning centres, for 8,576 students. 98% of students have access to high school programmes in their home communities. There is a full range of courses available in the school system, including academic, French immersion, Aboriginal language, cultural programmes, technical and occupational programmes.

A range of post secondary programmes are available through the Northwest Territories' Aurora College. The majority of these programmes are offered at the three main campus locations: Inuvik, Yellowknife and Fort Smith.

Health

In 2011 there were eight health authorities. Expenditure on health and social services totalled $349·7m. CDN in 2009–10.

Welfare

Welfare services are provided by professional social workers. Facilities included (2006): family violence services in seven communities, eight helplines, seven group homes or shelters and one residential treatment centre.

FURTHER READING

Northwest Territories—2011: By the Numbers. Online only

Zaslow, M., *The Opening of the Canadian North 1870–1914.* 1971

Statistical office: Bureau of Statistics, Government of the Northwest Territories, PO Box 1320, Yellowknife, NWT X1A 2L9.
Website: http://www.statsnwt.ca

Nunavut

KEY HISTORICAL EVENTS

Inuit communities were to be found in what is now the Canadian Arctic between 4500 BC and AD 1000. By the 19th century these communities were under the jurisdiction of the Northwest Territories. In 1963 the Canadian government first introduced legislation to divide the territory, a proposal that failed at the order paper stage. In 1973 the Comprehensive Land Claims Policy was established which sought to define the rights and benefits of the Aboriginal population in a land claim settlement agreement. The Northwest Territories Legislative Assembly voted in favour of dividing the territory in 1980, and in a public referendum of 1982, 56% of votes cast were also for the division. In 1992 the proposed boundary was ratified in a public vote and the Inuit population approved their land claim settlement. A year later, the Nunavut Act (creating the territory) and the Nunavut Land Claim Agreement Act were passed by parliament. Iqaluit was selected as the capital in 1995.

On 15 Feb. 1999 Nunavut held elections for its Legislative Assembly. The territory was officially designated and the government inaugurated on 1 April 1999.

TERRITORY AND POPULATION

The total area of the region is 2,093,190 sq. km or about 21% of Canada's total mass, making Nunavut Canada's largest territory. It contains seven of Canada's 12 largest islands and two-thirds of the country's coastline. The territory is divided into three regions: Qikiqtaaluk (Baffin), Kivalliq (Keewatin) and Kitikmeot. The total population at the 2011 census was 31,906. The population is divided up into 25 communities of which the largest is in the capital Iqaluit, numbering 6,184.

The native Inuit language is Inuktitut.

SOCIAL STATISTICS

Births in 2007–08 numbered 797 (a rate of 25·3 per 1,000 population) and deaths 136 (rate of 4·3 per 1,000 population).

Nunavut's birth rate is the highest in Canada and is more than twice the national average of 11·2 per 1,000 births. It also has the lowest death rate.

CLIMATE

Conditions range from cold continental to polar, with long hard winters and short cool summers. In Iqaluit there can be as little as four hours sunshine per day in winter and up to 21 hours per day at the summer solstice. Iqaluit, Jan. mean high –22°C; July mean high, 15°C.

CONSTITUTION AND GOVERNMENT

Government is by a *Legislative Assembly* of 19 elected members, who then choose a leader and ministers by consensus. There are no political parties. Government is highly decentralized, consisting of ten departments spread over 11 different communities. By 2020 Inuktitut is intended to be the working language of government but government agencies will also offer services in English and French. Although the Inuits are the dominant force in public government, non-Inuit citizens have the same voting rights. The territory is represented by one member in the Senate and one in the House of Commons of Canada.

RECENT ELECTIONS

Legislative Assembly elections were held on 27 Oct. 2008 (3 Nov. 2008 in one district). There were 50 non-partisan candidates; 18 were elected (one woman) with the election in one district cancelled.

CURRENT GOVERNMENT

Commissioner: Edna Elias (took office in May 2010).

In March 2013 the cabinet was as follows:

Premier, Minister of Executive and Intergovernmental Affairs, and Education: Eva Aariak (took office on 19 Nov. 2008).

Deputy Premier and Minister of Economic Development and Transportation: Peter Taptuna. *Finance, and Health and Social Services:* Keith Peterson. *Justice:* Daniel Shewchuk. *Human Resources:* Monica Ell. *Community and Government Services:* Lorne Kusugak. *Culture, Language, Elders and Youth, Environment, and Languages:* James Arreak.

Speaker: Hunter Tootoo.

Government of Nunavut Website: http://www.gov.nu.ca

ECONOMY

GDP per person was $47,360 CDN in 2009.

Currency
The Canadian dollar is the standard currency.

Budget
Total revenue in 2005–06 was $1,181m. CDN (own source revenue, $111m. CDN; general purpose transfers, $877m. CDN; special purpose transfers, $192m.). Total expenditures in 2005–06 amounted to $1,119m. CDN (including: health, $256m. CDN; education, $192m. CDN; housing, $145m. CDN).

Performance
Real GDP growth was 3·4% in 2006, the second highest among Canada's provinces and territories after Alberta.

ENERGY AND NATURAL RESOURCES

Minerals
There are two lead and zinc mines operating in the High Arctic region. There are also known deposits of copper, gold, silver and diamonds.

Hunting and Trapping
Most communities still rely on traditional foodstuffs such as caribou and seal. The Canadian government now provides meat inspections so that caribou and musk ox meat can be sold across the country.

Fisheries
Fishing is still very important in Inuit life. The principal catches are shrimp, scallops and arctic char.

INDUSTRY

The main industries are mining, tourism, fishing, hunting and trapping and arts and crafts production.

Labour
The unemployment rate was 15·6% in May 2006.

COMMUNICATIONS

Roads
There is one 21-km government-maintained road between Arctic Bay and Nanisivik. There are a few paved roads in Iqaluit and Rankin Inlet, but most are unpaved. Some communities have local roads and tracks but Kivalliq has no direct land connections with southern Canada.

Civil Aviation
The busiest airport is Iqaluit, which handled 129,000 passengers in 2010.

Shipping
There is an annual summer sea-lift by ship and barge for transport of construction materials, dry goods, non-perishable food, trucks and cars.

Telecommunications
In 2010, 85% of households had telephones. Because of the wide distances between communities, there is a very high rate of internet use in Nunavut. However, line speeds are slow and there is a problem with satellite bounce.

SOCIAL INSTITUTIONS

Justice
In 2009–10, 1,494 Criminal Code cases were disposed of in an adult criminal court. In 2010 there were six homicides (a rate of 18·1 per 100,000 population—by some way the highest of any Canadian province or territory).

Education
Approximately one quarter of Nunavut's population aged over 15 have less than Grade 9 schooling. Training and development is seen as central to securing a firm economic foundation for the province.

Courses in computer science, business management and public administration may be undertaken at Arctic College. In 2010–11 there were 42 schools with 8,855 students.

Health
There is one hospital in Iqaluit. 26 health centres provide nursing care for communities. For more specialized treatment, patients of Qikiqtaaluk may be flown to Montreal, patients in Kivalliq to Churchill or Winnipeg and patients in Kitikmeot to Yellowknife's Stanton Regional Hospital.

CULTURE

Tourism
Nunavut has four national parks—Auyuittuq, Quttinirpaaq, Sirmilik and Ukkusiksalik—offering the opportunity of seeing Inuit life first-hand. Tourism generates $30m. CDN per year for the Territory, with 20% of visitors arriving on cruise ships. During the 2008 high season (June–Oct.) there were 33,000 visitors.

FURTHER READING

The Nunavut Handbook. 2004

Yukon

KEY HISTORICAL EVENTS

The territory owes its fame to the discovery of gold in the Klondike at the end of the 19th century. Formerly part of the Northwest Territories, the Yukon joined the Dominion as a separate territory on 13 June 1898.

Yukon First Nations People lived a semi-nomadic subsistence long before the region was established as a territory. The earliest evidence of human activity was found in caves containing stone tools and animal bones estimated to be 20,000 years old. The Athapaskan cultural linguistic tradition to which most Yukon First Nations belong is more than 1,000 years old. The territory's name comes from the native 'Yu-kun-ah' for the great river that drains most of this area.

The Yukon was created as a district of the Northwest Territories in 1895. The Klondike Gold Rush in the late 1890s saw thousands pouring into the gold fields of the Canadian northwest. Population at the peak of the rush reached 40,000. This event spurred the federal government to set up basic administrative structures in the Yukon. The territory was given the status of a separate geographical and political entity with an appointed legislative council in 1898. In 1953 the capital was moved south from Dawson City to Whitehorse, where most of the economic activity was centred. The federal government granted the Yukon responsible government in 1979.

The *Yukon Act* of 1 April 2003 gave the territory more control over its own governance and changed its name from the Yukon Territory to Yukon.

TERRITORY AND POPULATION

The territory consists of one city, three towns, four villages, two hamlets, 13 unincorporated communities and eight rural communities. It is situated in the northwestern region of Canada and comprises 482,443 sq. km of which 8,052 sq. km is fresh water.

The population at the 2011 census was 33,897.

Principal centres in 2006 were Whitehorse, the capital, 20,461; Dawson City, 1,327; Watson Lake, 846; Haines Junction, 589; Carmacks, 425.

Yukon represents 4·8% of Canada's total land area.

SOCIAL STATISTICS

Births in 2007–08 numbered 355 (a rate of 10·8 per 1,000 population) and deaths 189 (rate of 5·7 per 1,000 population). There were 152 marriages in 2006 and 109 divorces in 2005.

CLIMATE

Temperatures in Yukon are usually more extreme than those experienced in the southern provinces of Canada. A cold climate in winter with moderate temperatures in summer provide a considerable annual range of temperature and moderate rainfall. Whitehorse, Jan. –18·7°C (–2·0°F), July 14°C (57·2°F). Annual precipitation 268·8 mm. Dawson City, Jan. –30·7°C (–23·3°F), July 15·6°C (60·1°F). Annual precipitation 182·7 mm.

CONSTITUTION AND GOVERNMENT

Yukon was constituted a separate territory on 13 June 1898. The *Yukon Legislative Assembly* consists of 19 elected members and functions in much the same way as a provincial legislature. The seat of government is at Whitehorse. It consists of an *Executive Council* with parliamentary powers similar to those of a provincial cabinet. The Yukon government consists of 14 departments, as well as a Workers' Compensation Health and Safety Board and four Crown corporations. The territory is represented by one member in the Senate and one in the House of Commons of Canada.

RECENT ELECTIONS

At elections held on 11 Oct. 2011 the Yukon Party took 11 of the available 19 seats; the New Democratic Party 6; and the Liberal Party 2.

CURRENT GOVERNMENT

Commissioner: Douglas Phillips (took office on 17 Dec. 2010).

In March 2013 the Yukon Party Ministry comprised:

Premier, Minister Responsible for Executive Council Office, and Minister of Finance: Darrell Pasloski.

Deputy Premier, and Minister of Community Services: Elaine Taylor. *Energy, Mines and Resources:* Brad Cathers. *Health and Social Services:* Doug Graham. *Education:* Scott Kent. *Economic Development, and Environment:* Currie Dixon. *Highways and Public Works:* Wade Istchenko. *Justice, and Tourism and Culture:* Mike Nixon.

Speaker: David Laxton.

Government of Yukon Website: http://www.gov.yk.ca

ECONOMY

GDP per person was $63,332 CDN in 2009.

Budget

Total revenue in 2005–06 was $776m. CDN (own source revenue, $136m. CDN; general purpose transfers, $545m. CDN; special purpose transfers, $95m. CDN). Total expenditures in 2005–06 amounted to $784m. CDN (including: education, $129m. CDN; health, $125m. CDN; transport and communication, $124m. CDN).

Performance

GDP at basic prices in 2005 for all industries was $1,176m. CDN (at market prices, GDP was $1,521m. CDN). Mining, oil and gas production was estimated at $55·5m. CDN in 2005 and shipments in the manufacturing sector were valued at $24·6m. CDN. Revenue from agriculture, forestry, hunting and fishing was estimated at $3·5m. CDN.

ENERGY AND NATURAL RESOURCES

Electricity

Electricity generation totalled 400m. kWh in 2009, almost exclusively hdyro-electric.

Oil and Gas

In 1997 the Yukon Oil and Gas Act was passed, replacing the federal legislation. This Act provides for the transfer of responsibility for oil and gas resources to Yukon jurisdiction. Five unexplored oil and gas basins with rich potential exist. Current net production is about 1·7m. cu. metres of natural gas per day.

Minerals

Gold and silver are the chief minerals. There are also deposits of lead, zinc, copper, tungsten and iron ore. Gold deposits, both hard rock and placer, are being mined.

Estimates for 2006 mineral production: gold, $37·5m. CDN; and silver, $0·2m. CDN. Total: $37·7m. CDN.

Agriculture

Many areas have suitable soils and climate for the production of forages, vegetables, domestic livestock and game farming. The greenhouse industry is Yukon's largest horticulture sector.

In 2006 there were 148 farms operating full- and part-time. The total area of farms was 10,125 ha. of which 2,658 ha. were in crops.

Gross farm receipts in 2005 were estimated at $4·1m. CDN. Total farm capital at market value in 2006 was $66·1m. CDN.

Forestry

The forests, covering 281,000 sq. km of the territory, are part of the great Boreal forest region of Canada, which covers 58% of Yukon.

Fur Trade

The fur-trapping industry is considered vital to rural and remote residents and especially First Nations people wishing to maintain a traditional lifestyle. Preliminary fur production in 2006 (mostly marten, lynx, wolverine, wolf, beaver and muskrat) was valued at $428,810 CDN.

Fisheries

Commercial fishing concentrates on chinook salmon, chum salmon, lake trout and whitefish.

INDUSTRY

The key sectors of the economy are tourism and government.

Labour

The 2006 labour force average was 16,200, of whom 15,500 were employed.

INTERNATIONAL TRADE

Imports and Exports

In 2003 exports made up 15·6% of Yukon goods and services produced. In 2003 exports were valued at $353m. CDN.

COMMUNICATIONS

Roads

The Alaska Highway and branch highway systems connect Yukon's main communities with Alaska, the Northwest Territories, southern Canada and the United States. The 733-km Dempster Highway north of Dawson City connects with Inuvik, in the Northwest Territories. In 2006 there were 4,902·5 km of roads maintained by the government of Yukon: 3,780·8 km is primary (including the 998·1 km Alaska Highway); 1,121·7 km is secondary. Vehicles registered in 2006 totalled 29,003 (excluding buses, motorcycles and trailers), including 26,621 passenger vehicles.

Rail

The 176-km White Pass and Yukon Railway connected Whitehorse with year-round ocean shipping at Skagway, Alaska, but was closed in 1982. A modified passenger service was restarted in 1988 to take cruise ship tourists from Skagway to Carcross, Yukon, over the White Pass summit.

Civil Aviation

The busiest airport is Whitehorse (Erik Nielsen Whitehorse International Airport), which handled 230,000 passengers in 2010.

Shipping

The majority of goods are shipped into the territory by truck over the Alaska and Stewart-Cassiar Highways. Some goods are shipped through the ports of Skagway and Haines, Alaska, and then trucked to Whitehorse for distribution throughout the territory.

Telecommunications

All telephone and telecommunications, including internet access in most communities, are provided by Northwestel, a subsidiary of Bell Canada Enterprises. In 2003, 10,000 households (93·4%) had telephones.

SOCIAL INSTITUTIONS

Justice

In 2009–10, 1,050 Criminal Code cases were disposed of in an adult criminal court. In 2010 there was one homicide (a rate of 2·9 per 100,000 population).

Education

The Yukon Department of Education operates (with the assistance of elected school boards) the territory's 29 schools, both public and private, from kindergarten to grade 12. In May 2006 there were 5,148 pupils. There is also one French First Language school and three Roman Catholic schools. The total enrolment figure for Yukon College in 2005–06 was 5,057. The Whitehorse campus is the administrative and programme centre for 13 other campuses located throughout the territory. In 2005–06 a total of 600 full-time and 4,457 part-time students enrolled in programmes and courses.

Health

In 2006 there were two hospitals with 71 staffed beds, 14 health centres, 74 resident doctors and 16 resident dentists.

CULTURE

Tourism

In 2006, 396,377 visitors came to Yukon. Tourism is the largest private sector employer. In 2005 approximately 80% of employees in Yukon worked for businesses that reported some level of tourism revenue. In 2005, 15% of businesses generated more than a third of gross revenues from tourism.

FURTHER READING

Annual Report of the Government of Yukon.
Yukon Executive Council, *Annual Statistical Review.*

Berton, P., *Klondike.* (Rev. ed.) 1987
Coates, K. and Morrison, W., *Land of the Midnight Sun: A History of the Yukon.* 1988

Statistical office: Bureau of Statistics, Executive Council Office, Box 2703, Whitehorse, Yukon Y1A 2C6. There is also a Yukon Archive at Yukon College, Whitehorse.
Website: http://www.eco.gov.yk.ca/stats

The world in focus at
www.statesmansyearbook.com

CAPE VERDE

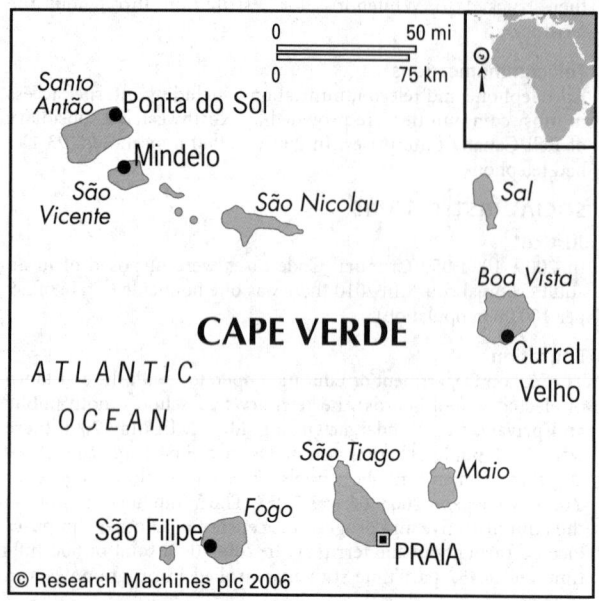

República de Cabo Verde
(Republic of Cape Verde)

Capital: Praia
Population projection, 2015: 520,000
GNI per capita, 2011: (PPP$) 3,402
HDI/world rank: 0·568/133
Internet domain extension: .cv

KEY HISTORICAL EVENTS

During centuries of Portuguese rule the islands were gradually peopled with Portuguese, slaves from Africa and people of mixed African-European descent who formed the majority. While retaining some African culture, the Cape Verdians spoke Portuguese or the Portuguese-derived Crioulo (Creole) language and became Catholics. In 1956 nationalists from Cape Verde and Portuguese Guinea founded the *Partido Africano da Independência da Guiné e Cabo Verde* (PAIGC). In the 1960s the PAIGC waged a successful guerrilla war. On 5 July 1975 Cape Verde became independent, ruled by the PAIGC, which was already the ruling party in the former Portuguese colony of Guinea-Bissau. But resentment at Cape Verdians' privileged position in Guinea-Bissau led to the end of the ties between the two countries' ruling parties. Although the PAIGC retained its name in Guinea-Bissau, in Jan. 1981 it was renamed the *Partido Africano da Independência do Cabo Verde* (PAICV) in Cape Verde. The constitution of 1981 made the PAICV the sole legal party but in Sept. 1990 the National Assembly abolished its monopoly and free elections were permitted.

TERRITORY AND POPULATION

Cape Verde is situated in the Atlantic Ocean 620 km off west Africa and consists of ten islands (Boa Vista, Brava, Fogo, Maio, Sal, Santa Luzia, Santo Antão, São Nicolau, São Tiago and São Vicente) and five islets. The islands are divided into two groups, named Barlavento (windward) and Sotavento (leeward). The total area is 4,033 sq. km (1,557 sq. miles). The 2010 census population

was 491,875 (248,282 female), giving a density of 122 per sq. km. In 2010, 62% of the population lived in urban areas.

The UN gives a projected population for 2015 of 520,000.

Over 600,000 Cape Verdeans live abroad (more than still live in the country), mainly in the USA.

Areas and populations of the islands:

Island	Area (sq. km)	Population census 2000	Population census 2010
Santo Antão	779	47,389	43,915
São Vicente[1]	227	67,511	76,140
São Nicolau	388	13,735	12,817
Sal	216	14,892	25,779
Boa Vista	620	4,225	9,162
Barlavento	*2,230*	*147,752*	*167,813*
Maio	269	6,788	6,952
São Tiago	991	237,828	274,044
Fogo	476	37,617	37,071
Brava	64	6,838	5,995
Sotavento	*1,803*	*289,071*	*324,062*

[1]Including Santa Luzia island, which is uninhabited.

The main towns are Praia, the capital, on São Tiago (127,832, 2010 provisional census population) and Mindelo on São Vicente (70,468, 2010 provisional census population). Ethnic groups in 2000 included: Mixed, 70%; Fulani, 12%; Balanta, 10%; Mandyako, 5%. The official language is Portuguese; a creole (Crioulo) is in ordinary use.

SOCIAL STATISTICS

2008 estimates: births, 12,000; deaths, 2,000. Rates, 2008 estimates (per 1,000 population): birth, 24·1; death, 5·0. Annual population growth rate, 2000–08, 1·6%. Annual emigration varies between 2,000 and 10,000. Life expectancy at birth, 2007, was 68·2 years for men and 73·5 years for women. Infant mortality, 2010, 29 per 1,000 live births; fertility rate, 2008, 2·7 children per woman. Cape Verde has had one of the largest reductions in its fertility rate of any country in the world over the past quarter of a century, having had a rate of 5·3 births per woman in 1990.

CLIMATE

The climate is arid, with a cool dry season from Dec. to June and warm dry conditions for the rest of the year. Rainfall is sparse, rarely exceeding 5" (127 mm) in the northern islands or 12" (304 mm) in the southern ones. There are periodic severe droughts. Praia, Jan. 72°F (22·2°C), July 77°F (25°C). Annual rainfall 10" (250 mm).

CONSTITUTION AND GOVERNMENT

The Constitution was adopted in Sept. 1992 and was revised in 1995 and 1999.

A constitutional referendum was held on 28 Dec. 1994; turnout was 45%. 82·06% of votes cast favoured a reform extending the powers of the presidency and strengthening the autonomy of local authorities. The *President* is elected for five-year terms by universal suffrage.

The 72-member *National Assembly* (*Assembleia Nacional*) is elected for five-year terms.

National Anthem

'Cântico da Liberdade' ('Song of Freedom'); words by A. S. Lopes, tune by A. H. T. Silva.

RECENT ELECTIONS

Elections for the National Assembly of 72 members were held on 6 Feb. 2011. Turnout was 75·5%. The ruling African Party for the

Independence of Cape Verde (PAICV) won 37 seats with 50·9% of votes cast, the Movement for Democracy (MPD) won 33 seats with 41·9% and the Democratic and Independent Cape Verdean Union won 2 with 4·9%. Two smaller parties failed to win any seats.

In the first round of presidential elections held on 7 Aug. 2011 Jorge Carlos Fonseca won 37·3% of the vote against Manuel Inocêncio Sousa with 32·0%, Aristides Lima with 27·4% and Joaquim Jaime Monteiro with 2·0%. Turnout was 53·1%. In the run-off held on 21 Aug. 2011, Fonseca won 54·5% of the vote and Sousa 45·5%. Turnout was 59·7%.

CURRENT GOVERNMENT

President: Jorge Carlos Fonseca; b. 1950 (MPD; sworn in 9 Sept. 2011).

In March 2013 the government comprised:

Prime Minister and Minister for State Reform: José Maria Neves; b. 1959 (PAICV; sworn in 1 Feb. 2001).

Minister for the Presidency of the Council of Ministers and National Defence: Jorge Tolentino. *Minister for Health:* Cristina Fontes Lima. *Finance and Planning:* Cristina Duarte. *Foreign Affairs:* Jorge Borges. *Parliamentary Affairs:* Rui Semedo. *Home Affairs:* Marisa Morais. *Justice:* José Carlos Lopes Correia. *Infrastructure and Marine Economy:* José Maria Veiga. *Environment, Housing and Spatial Planning:* Sara Lopes. *Youth, Employment and Human Resources Development:* Janira Hopffer Almada. *Education and Sport:* Fernanda Marques. *Rural Development:* Eva Ortet. *Higher Education, Science and Innovation:* António Leão Correia e Silva. *Tourism, Industry and Energy:* Humberto Brito. *Communities:* Fernanda Fernandes. *Culture:* Mário Lúcio Sousa.

Government Website: http://www.governo.cv

CURRENT LEADERS

Jorge Carlos Fonseca

Position
President

Introduction
Jorge Carlos Fonseca was elected president in Aug. 2011, having previously served as foreign minister. As leader of the main opposition party, the Movement for Democracy (MPD), he pledged co-operation with the ruling African Party for the Independence of Cape Verde (PAICV) to boost economic growth and to strengthen representative democracy.

Early Life
Born on 20 Oct. 1950 in São Velenti, Jorge Carlos de Almeida Fonseca was educated in Praia and Mindelo, before studying law at the University of Lisbon. He returned to Cape Verde in 1975 when the island gained independence from Portugal. From 1975–77 he served under the ruling Partido Africano da Independência da Guiné et Cabo Verde (PAIGC) as director general of emigration. From 1977–79 he was secretary general at the ministry of foreign affairs.

Returning to academic life, he worked at the University of Lisbon from 1982 before becoming a professor of criminal law at Lisbon's National Institute of Legal Medicine in 1987. From 1989–90 he was a resident director and associate professor of law and public administration at the University of East Asia, Macau. As a founder member of the centrist MPD in 1990, he helped negotiate Cape Verde's first multi-party elections in Jan. 1991, in which he was elected to parliament. He served as minister of foreign affairs in the MPD government from 1991–93, before leaving the party to co-found the Democratic Convergence Party (PCD) in 1994.

Representing a coalition of minor parties, he stood unsuccessfully for the presidency in 2001. For the next decade he pursued a career as an academic and political activist. He co-founded the Law and Justice Foundation in 2004 and was appointed president of the Higher Institute of Legal and Social Sciences in Praia. He fought the Aug. 2011 presidential election as the MPD candidate, promising to sustain economic growth and build the island's infrastructure. He won in the second round with close to 55% of the vote.

Career in Office
Fonseca took office on 9 Sept. 2011, pledging to work with the governing PAICV to maintain political stability and attract inward investment. His main challenge is to boost the economy and reduce high unemployment in an adverse global economic climate.

DEFENCE

There is selective conscription for 12 to 14 months. The President of the Republic is C.-in-C. of the armed forces. Defence expenditure totalled US$9m. in 2008 (US$21 per capita), representing 0·5% of GDP.

Army

The Army is composed of two battalions and had a strength of 1,000 in 2007.

Navy

There is a coast guard of about 100 (2007) with two patrol craft.

Air Force

The Air Force had under 100 personnel and no combat aircraft in 2007.

ECONOMY

Agriculture accounted for 8·9% of GDP in 2009, industry 19·7% and services 71·4%.

Overview

Despite poor natural resources and its small island status, Cape Verde has achieved remarkable growth over the past two decades. Real per capita GDP has expanded on average by 7% per year, a figure greater than other small island economies and countries within sub-Saharan Africa, while absolute poverty fell from 49% in 1988–89 to 24% in 2010. In 2007 Cape Verde became only the second country to graduate from the UN's list of Least Developed Countries (LDCs) to 'developing country' status.

The economy's success derives from good governance and sound macroeconomic management. Notable measures include tax reforms to reduce domestic debt, the attraction of tourism-related foreign direct investment, trade openness and integration into the global economy, and enhanced private sector involvement in the economy. Remittances from the diaspora also make a strong contribution, accounting for around 9% of GDP in recent years. Cape Verde is a special partner of the EU and joined the WTO in 2008, marking its accession to middle-income country status.

As a result of the global financial crisis, real GDP growth fell from 8·6% in 2007 to 3·7% in 2009, with FDI and remittances particularly weakened. The government responded with a package of counter-cyclical policies, including lower household taxes and an aggressive public investment programme. The economy rebounded in 2010, led primarily by tourism and construction. However, government stimulus resulted in a sharp deterioration in the public finance deficit from 1·2% of GDP in 2008 to over 9% in 2009.

Currency

The unit of currency is the *Cape Verde escudo* (CVE) of 100 *centavos*, which is pegged at 110·6521 to the euro. Foreign exchange reserves were US$164m. in June 2005 and total money supply was 25,156m. escudos. There was inflation of 2·1% in 2010 and 4·5% in 2011.

Budget
Revenues in 2008 were 40,129m. escudos (tax revenue, 73·7%) and expenditures 41,304m. escudos (current expenditure, 60·6%).

VAT is 15% (reduced rate, 6%).

Performance
Real GDP growth was 3·7% in 2009, 5·2% in 2010 and 5·0% in 2011. Total GDP in 2011 was US$1·9bn.

Banking and Finance
The Banco de Cabo Verde is the central bank (*Governor*, Carlos Burgo) and bank of issue, and was also previously a commercial bank. Its latter functions have been taken over by the Banco Comercial do Atlântico, mainly financed by public funds. The Caixa Econômica de Cabo Verde (CECV) has been upgraded into a commercial and development bank. Two foreign banks have also been established there. In addition, the Fundo de Solidariedade Nacional acts as the country's leading savings institution while the Fundo de Desenvolvimento Nacional administers public investment resources and the Instituto Caboverdiano channels international aid.

Foreign debt was US$599m. in 2007.

There is a stock exchange in Mindelo.

ENERGY AND NATURAL RESOURCES

Environment
Cape Verde's carbon dioxide emissions from the consumption and flaring of fossil fuels in 2008 were the equivalent of 0·7 tonnes per capita.

Electricity
Installed capacity was 75,000 kW in 2007. Production was 269m. kWh in 2007. Consumption per capita in 2007 was 548 kWh.

Minerals
Salt is obtained on the islands of Sal, Boa Vista and Maio. Volcanic rock (pozzolana) is mined for export. There are also deposits of kaolin, clay, gypsum and basalt.

Agriculture
In 2002, 21·7% of the economically active population were engaged in agriculture. Some 10–15% of the land area is suitable for farming. In 2003, 46,000 ha. were arable and 3,000 ha. permanent crops, mainly confined to inland valleys. 3,000 ha. were irrigated in 2003. The chief crops (production, 2002, in 1,000 tonnes) are: sugarcane, 14; bananas, 6; coconuts, 6; maize, 5; cabbage, 4; mangoes, 4; sweet potatoes, 4; tomatoes, 4.

Livestock (2003): 200,000 pigs, 112,000 goats, 22,000 cattle, 14,000 asses.

Forestry
In 2010 the forest area was 85,000 ha., or 21% of the total land area.

Fisheries
In 2009 the total catch was 16,828 tonnes, exclusively from marine waters.

INDUSTRY
The main industries are the manufacture of paint, beer, soft drinks, rum, flour, cigarettes, canned tuna and shoes.

Labour
In 2010 the estimated economically active population was 221,000 (57% males).

INTERNATIONAL TRADE

Imports and Exports
Imports and exports (f.o.b.) for calendar years in US$1m.:

	2001	2002	2003	2004
Imports	233·8	313·7	355·0	421·5
Exports (including re-exports)	32·1	40·9	49·9	53·2

In 2004 machinery and transport equipment constituted 30·8% of imports, with food and live animals accounting for 22·3% and manufactured goods 16·4%; clothing made up 57·4% of exports, footwear 31·2% and seafood 8·2%.

Main import suppliers, 2004: Portugal, 42·5%; Netherlands, 13·7%; USA, 13·2%. Leading export markets, 2004: Portugal, 78·3%; USA, 19·4%; Guinea-Bissau, 0·6%.

COMMUNICATIONS

Roads
In 2002 there were an estimated 1,100 km of roads (78% paved); in 2005 there were 2,244 registered vehicles.

Civil Aviation
Amilcar Cabral International Airport, at Espargos on Sal, is a major refuelling point on flights to Africa and Latin America. A new international airport, Praia International Airport, has been built at Praia on São Tiago, and was opened in 2005. Transportes Aéreos de Cabo Verde (TACV), the national carrier, provided services to most of the other islands in 2003, and internationally to Abidjan, Amsterdam, Bamako, Bissau, Conakry, Dakar, Fortaleza, Las Palmas, Lisbon, Madrid, Milan, Munich, Paris and Zürich. In 2006 Amilcar Cabral International Airport handled 562,972 passengers and 1,415 tonnes of freight. In 2003 scheduled airline traffic of Cape Verde-based carriers flew 5m. km, carrying 253,000 passengers (86,000 on international flights).

Shipping
The main ports are Mindelo and Praia. In Jan. 2009 there were 19 vessels of 300 GT or over registered, totalling 19,000 GT. There is a state-owned ferry service between the islands.

Telecommunications
In 2011 there were 74,500 landline telephone subscriptions (equivalent to 148·8 per 1,000 inhabitants) and 396,400 mobile phone subscriptions (or 791·9 per 1,000 inhabitants). There were 296·7 internet users per 1,000 inhabitants in 2009. Fixed internet subscriptions totalled 12,900 in 2009 (26·3 per 1,000 inhabitants).

SOCIAL INSTITUTIONS

Justice
There is a network of People's Tribunals, with a Supreme Court in Praia. The Supreme Court is composed of a minimum of five Judges, of whom one is appointed by the President, one elected by the National Assembly and the other by the Supreme Council of Magistrates.

The population in penal institutions in 2010 was approximately 1,300 (253 per 100,000 of national population). The death penalty was abolished in 1981.

Education
Adult literacy in 2009 was estimated at 84·8% (90·1% among males and 80·2% among females). Primary schooling is followed by lower (13–15 years) and upper (16–18 years) secondary education options. In 2005–06 there were 3,196 primary school teachers for 81,162 pupils; and 2,363 teachers for 52,969 pupils at secondary schools. There are two universities: the Jean Piaget University of Cape Verde and the University of Cape Verde.

In 2007 public expenditure on education came to 5·9% of GNI.

Health
Cape Verde's main hospitals are located in Praia and Mindelo. There is a network of health centres and clinics. In 2005 there

were 21 hospital beds per 10,000 inhabitants. There were 217 physicians and 471 nurses in 2006.

RELIGION

In 2001, 83% of the population were Roman Catholic and 8% were followers of other religions.

CULTURE

World Heritage Sites

Cidade Velha, the historic centre of Ribeira Grande, was inscribed on the UNESCO World Heritage List in 2009. It was the first European colonial town to be built in the tropics.

Press

In 2008 there were 12 non-daily newspapers although no dailies. The most popular newspaper is the weekly *A Semana*.

Tourism

Tourism has experienced huge growth in the past few years. In 2009 there were 287,183 non-resident tourists staying at hotels and similar establishments, compared to 267,188 in 2007 and 197,844 in 2005.

Festivals

Festivals include the São Vicente Creole Carnival (Feb.) and the music festivals, Gamboa Festival in São Tiago (May) and São Vicente Baia das Gatas Festival (Aug.).

DIPLOMATIC REPRESENTATIVES

Of Cape Verde in the United Kingdom
Ambassador: Vacant.
Chargé d'Affaires a.i.: Maria de Jesus Veiga Miranda Mascarenhas (resides in Brussels).
Honorary Consul: Jonathan Lux (7a The Grove, London, N6 6JU).

Of the United Kingdom in Cape Verde
Ambassador: John Marshall (resides in Dakar, Senegal).

Of Cape Verde in the USA (3415 Massachusetts Ave., NW, Washington, D.C., 20007)
Ambassador: Maria de Fatima Lima da Veiga.

Of the USA in Cape Verde (Rua Abilio Macedo 6, Praia)
Ambassador: Adrienne S. O'Neal.

Of Cape Verde to the United Nations
Ambassador: António Pedro Monteiro Lima.

Of Cape Verde to the European Union
Ambassador: Maria de Jesus Veiga Miranda Mascarenhas.

FURTHER READING

Lobban, Richard, *Historical Dictionary of the Republic of Cape Verde.* 1995.—*Cape Verde: Crioulo Colony to Independent Nation.* 1998

National Statistical Office: Instituto Nacional de Estatística, Praia.
Website (Portuguese only): http://www.ine.cv

CENTRAL AFRICAN REPUBLIC

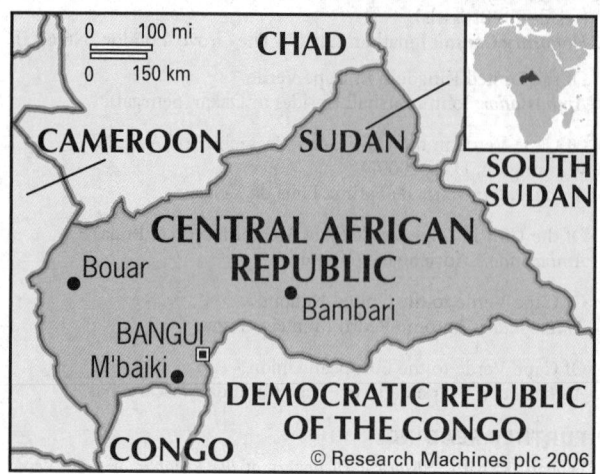

Prefecture	Sq. km	2003 census	Capital
Bamingui-Bangoran	58,200	43,229	Ndele
Bangui[1]	67	622,771	Bangui
Basse-Kotto	17,604	249,150	Mobaye
Haute-Kotto	86,650	90,316	Bria
Haut-M'bomou	55,530	57,602	Obo
Kemo	17,204	118,420	Sibut
Lobaye	19,235	246,875	M'baiki
Mambere Kadéi	30,203	364,795	Berbérati
M'bomou	61,150	164,009	Bangassou
Nana Grebizi[2]	19,996	117,816	Kaga-Bandoro
Nana-Mambere	26,600	233,666	Bouar
Ombella-M'poko	31,835	356,725	Bimbo
Ouaka	49,900	276,710	Bambari
Ouham	50,250	369,220	Bossangoa
Ouham-Pendé	32,100	430,506	Bozoum
Sangha M'baéré[2]	19,412	101,074	Nola
Vakaga	46,500	52,255	Birao

[1]Autonomous commune. [2]Economic prefecture.

The capital, Bangui, had a census population in 2003 of 622,771. Other main towns, with 2003 census populations, are Bimbo (124,176), Bebérati (76,918), Carnot (45,421), Bambari (41,356) and Bouar (40,353).

There are a number of ethnic groups, the largest being Gbaya (34%), Banda (27%) and Mandja (21%).

Sango and French are the official languages.

République Centrafricaine

Capital: Bangui
Population projection, 2015: 4·85m.
GNI per capita, 2011: (PPP$) 707
HDI/world rank: 0·343/179
Internet domain extension: .cf

KEY HISTORICAL EVENTS

One of the four territories of French Equatorial Africa, Central African Republic became independent on 13 Aug. 1960. A constitution of 1976 provided for a parliamentary democracy to be known as the Central African Empire. President Bokassa became Emperor Bokassa I. He was overthrown in 1979. In 1981 Gen. André Kolingba took power, initiating a gradual return to constitutional rule.

On 5 June 1996, following an army mutiny, President Patassé accepted an agreement brokered by France which led to the formation of a government of national unity. But mutineers demanded the replacement of President Patassé. France chaired a mediation committee of various neighbouring French-speaking states. An agreement to end the mutiny was signed in 1997 and a peacekeeping force from neighbouring states, MISAB, was set up. Conflicts between the mutineers and MISAB continued until a ceasefire was concluded on 2 July 1997. There was an attempted coup on 28 May 2001, allegedly led by Gen. Kolingba, who had been the country's military ruler from 1981 to 1993. However, it failed following several days of fighting in and around the capital, Bangui. Fighting erupted once more in Oct. 2002 after another coup attempt. In March 2003 a further coup saw Gen. François Bozizé, a former army chief, seize power.

TERRITORY AND POPULATION

The republic is bounded in the north by Chad, northeast by Sudan, east by South Sudan, south by the Democratic Republic of the Congo and the Republic of the Congo, and west by Cameroon. Area, 622,984 sq. km (240,534 sq. miles). The population at the 2003 census was 3,895,139, giving a density of 6 per sq. km. In 2011, 39·2% of the population were urban.

The UN gives a projected population for 2015 of 4·85m.

The areas, populations and capitals of the prefectures are as follows:

There are a number of ethnic groups, the largest being Gbaya (34%), Banda (27%) and Mandja (21%).

Sango and French are the official languages.

SOCIAL STATISTICS

2008 births (estimates), 154,000; deaths, 74,000. Estimated birth rate in 2008 was 35·4 per 1,000 population; estimated death rate, 17·0. Infant mortality, 2010 (per 1,000 live births), 106. Expectation of life in 2007 was 45·1 years for males and 48·2 for females. Annual population growth rate, 2000–08, 1·8%. Fertility rate, 2008, 4·8 children per woman.

CLIMATE

A tropical climate with little variation in temperature. The wet months are May, June, Oct. and Nov. Bangui, Jan. 31·9°C, July 20·7°C. Annual rainfall 1,289·3 mm. Ndele, Jan. 36·3°C, July 30·5°C. Annual rainfall 203·6 mm.

CONSTITUTION AND GOVERNMENT

Under the Constitution adopted by a referendum on 21 Nov. 1986, the sole legal political party was the *Rassemblement Démocratique Centrafricain*. In Aug. 1992 the Constitution was revised to permit multi-party democracy. Further constitutional reforms followed a referendum in Dec. 1994, including the establishment of a *Constitutional Court*. Following the coup of March 2003 Gen. François Bozizé suspended the constitution and dissolved parliament. However, at a referendum on 5 Dec. 2004, 90·4% of voters approved the adoption of a new constitution; voter participation was 77·4%. The new constitution resembles the previous one but permits the *President* to serve not more than two terms of five years. The President appoints the *Prime Minister* and leads the *Council of Ministers*. There is a 105-member *National Assembly*, with members elected in single-member constituencies for a five-year term.

National Anthem

'La Renaissance' ('Rebirth'); words by B. Boganda, tune by H. Pepper.

RECENT ELECTIONS

At the presidential elections held on 23 Jan. 2011, incumbent president Gen. François Bozizé won 66·1% of the vote, Ange-Félix

Patassé (ind.) 20·1%, Martin Ziguélé (Movement for the Liberation of the Central African People) 6·5%, Emile Gros Raymond Nakombo (Central African Democratic Rally) 4·6% and Jean-Jacques Démafouth (New Alliance for Progress/People's Army for the Restoration of Democracy) 2·7%. Turnout was 54·0%.

Parliamentary elections were held on 23 Jan. and 27 March 2011. The National Convergence 'Kwa Na Kwa' won a total of 61 seats, independents 26, candidates of the presidential majority 11, the Movement for the Liberation of the Central African People 1 and the Central African Democratic Rally 1.

CURRENT GOVERNMENT

Former army chief Gen. François Bozizé seized power on 15 March 2003 in a coup and the following day declared himself president, saying that he had dissolved the National Assembly and government. A transitional government was formed comprising representatives of civil society and all political parties. Gen. Bozizé said that a transition period would last between one and three years, after which elections would be held to decide on a new government. The period of transitional government ended with the 2005 elections. Bozizé was in turn deposed in March 2013. In April 2013 the government comprised the following:

Interim President and Minister of National Defence, Restructuring of the Army, Veterans and War Victims: Michel Djotodia; b. 1949 (since 24 March 2013).

Prime Minister: Nicolas Tiangaye; b. 1956 (took office on 17 Jan. 2013).

Minister of State in Charge of Communication, Promotion of Civic Culture and National Reconciliation: Christophe Gazzam Betty. *Minister of State for Mines, Oil and Water:* Djono Ahaba. *Minister of State for Equipment and Spokesman for the Government:* Crépin Mboli Goumba. *Minister of State for Public Security, Emigration, Immigration and Public Order:* Noureddine Adam. *Minister of State for Water, Forests, Hunting and Fishing, in Charge of the Environment:* Mohamed Moussa Daffhane.

Minister of Commerce and Industry: Amalas Amias. *Economy, Planning and International Co-operation:* Abdallah Hassan Kadre. *Education and Higher Education:* Marcel Loudegue. *Finance and Budget:* Georges Bozanga. *Foreign Affairs, Integration, Francophonie and Central Africans Abroad:* Charles Armel Doubane. *Health, Population and the Fight against HIV/AIDS:* Dr Aguide Soubouk. *Housing and Habitat:* Marie-Madeleine Nkouet. *Justice and Judicial Reform:* Arsene Sendé. *Post and Telecommunications:* Henri Pouzere. *Public Service, Labour, Employment and Social Security:* Sabin Kpokolo. *Rural Development:* Jérémie Tchimanguere. *Small and Medium-Sized Enterprises and Business Climate Improvement:* Maurice Yondo. *Social Affairs, National Solidarity and Gender Promotion:* Marie Madeleine Moussa Yadouma. *Territorial Administration:* Aristide Sokambi. *Tourism:* Mahamat Ataib Yacoub. *Transport and Civil Aviation:* Arnaud Djoubaye Abazene. *Urban Development, Reconstruction of Public Buildings and Land Reform:* Resigala Ramadan. *Youth and Sports, and Arts and Culture:* Abdoulaye Hissene. *Secretary General of the Government, in Charge of Relations with Institutions:* Harold Ahamat Deya.

CURRENT LEADERS

Michel Moana Djotodia

Position
President

Introduction
Michel Djotodia declared himself president of the Central African Republic (CAR) on 24 March 2013 after ousting decade-long president François Bozizé, who himself came to power in 2003 on the back of a coup before winning two subsequent elections.

Early Life
Djotodia was born in 1949 in Vakaga, in the northeast of the CAR. He studied economics in the former Soviet Union, where he lived for ten years. After returning to the CAR, he twice unsuccessfully attempted to win a seat in parliament in Vakaga Prefecture in the 1980s. He became a civil servant in the administration of Ange Félix-Patassé that began in 1993, working variously for the ministry of planning, the foreign ministry and as a diplomatic consul to Nyala in Sudan. When Félix-Patassé's government was overthrown by François Bozizé in 2003, Djotodia helped establish the rebel Union of Democratic Forces for Unity (UFDR).

In Oct. 2006 the UFDR captured the town of Birao in northern CAR. Djotodia, in Benin at the time, was subsequently arrested and imprisoned at Bozizé's request but was released in Feb. 2008 as part of a peace agreement between Bozizé and the rebels. Djotodia moved to South Sudan and reputedly cultivated links with Chadian and Sudanese fighters who would form the backbone of Séléka, a rebel coalition that included the UFDR. In Dec. 2012 Séléka came close to capturing the CAR capital, Bangui.

In Jan. 2013 a regionally-brokered peace treaty was signed, with Séléka forming a unity government alongside Bozizé to serve until elections in 2016. As part of the agreement Djotodia became first deputy minister for national defence. In March 2013 Séléka pulled out of the administration, accusing Bozizé of running a parallel administration and failing to release political prisoners. The group, boasting a force of 3,000, overran the capital a week later and Bozizé fled into exile. Djotodia declared himself president on 24 March 2013.

Career in Office
Djotodia established an interim government and pledged to hold elections in 2016, saying that he hopes 'to be the last rebel chief president of Central Africa'. A National Transitional Council confirmed his presidency on 13 April. Nicolas Tiangaye, prime minister in the power-sharing government under Bozizé, was asked to remain in his post while 34 ministries were shared between former opposition figures and members of Séléka. Djotodia himself heads the ministry of defence. His government faces regional and international criticism and lacks recognition by, among others, the US government. Bozizé has claimed that the coup was carried out with the backing of neighbouring Chad.

DEFENCE

Selective national service for a two-year period is in force.

Defence expenditure totalled US$20m. in 2008 (US$5 per capita), representing 1·0% of GDP.

Army
The Army consisted (2007) of about 2,000 personnel. There is a territorial defence regiment, a combined arms regiment and a support/HQ regiment. In addition there are some 1,000 personnel in the paramilitary Gendarmerie.

Navy
The Army includes a small naval wing operating a handful of patrol craft.

Air Force
Personnel strength (2007) was 150. There are no combat aircraft.

ECONOMY

Agriculture accounted for 56·5% of GDP in 2009, industry 14·8% and services 28·7%.

Overview
Despite being rich in natural resources including diamonds, gold, timber and uranium, and having favourable agricultural conditions, Central African Republic remains one of the world's poorest countries. Political turmoil and armed conflict, along

with droughts, a landlocked geography, a poor transport system and government mismanagement have hampered growth over decades.

A coup d'état in 2003 preceded two years of transition culminating in legislative and presidential elections in May 2005. The return to representative government prompted the strongest economic growth for a decade. GDP growth was around 4% in 2006 owing to increased private consumption following the resumption of regular salary payments to civil servants, a pick-up in investment, a recovery in diamond and timber exports, and an upturn in agriculture (the economy's largest sector). A series of external shocks, including the breakdown of a major hydro-power plant, a price surge in imported commodities and the global economic downturn, led to slower growth and accelerated inflation from 2008. However, the current account deficit narrowed to 7·7% of GDP in 2009 from 10·3% the previous year, reflecting improvements in terms of trade as a result of lower oil prices and rising diamond prices.

The economy receives international aid from the IMF-backed Poverty Reduction and Growth Facility (approved in Dec. 2006) and from the Heavily Indebted Poor Countries Initiative (since June 2009).

Currency
The unit of currency is the *franc CFA* (XAF) with a parity of 655·957 francs CFA to one euro. Total money supply in June 2005 was 98,664m. francs CFA, with foreign exchange reserves US$132m. and gold reserves 11,000 troy oz. Inflation was 9·3% in 2008, falling to 1·5% in 2010 and 1·2% in 2011.

Budget
In 2006 revenue totalled 176,300m. francs CFA and expenditure 107,200m. francs CFA.

VAT is 19%.

Performance
Total GDP in 2011 was US$2·2bn. Real GDP growth was 3·0% in 2010 and 3·3% in 2011.

Banking and Finance
The Banque des États de l'Afrique Centrale (BEAC) acts as the central bank and bank of issue. The *Governor* is Lucas Abaga Nchama. There are three commercial banks, a development bank and an investment bank.

In 2010 external debt totalled US$385m., representing 19% of GNI.

ENERGY AND NATURAL RESOURCES

Environment
The Central African Republic's carbon dioxide emissions from the consumption and flaring of fossil fuels in 2008 were the equivalent of 0·1 tonnes per capita.

Electricity
Installed capacity was an estimated 40,000 kW in 2007. Production in 2007 totalled 160m. kWh (around 81% hydro-electric). Consumption per capita in 2007 was about 37 kWh.

Minerals
In 2005 an estimated 265,000 carats of gem diamonds and 88,000 carats of industrial diamonds were mined; and, in 2007, 10 kg of gold. There are also oil, uranium and other mineral deposits which are for the most part unexploited.

Agriculture
In 2002 the agricultural population numbered 2·21m. persons, of whom 1·27m. were economically active. In 2002 about 1·93m. ha. were arable and 94,000 ha. permanent crops. The main crops (production 2002, in 1,000 tonnes) are cassava, 563; yams, 350; groundnuts, 128; bananas, 115; maize, 113; taro, 100; sugarcane, 90; plantains, 82; sorghum, 48.

Livestock, 2002: cattle, 3·27m.; goats, 2·92m.; pigs, 738,000; sheep, 246,000; chickens, 5m.

Forestry
There were 22·61m. ha. of forest in 2010, or 36% of the total land area. The extensive hardwood forests, particularly in the southwest, provide mahogany, obeche and limba. Timber production in 2007 was 2·83m. cu. metres.

Fisheries
The catch in 2010 was estimated at 35,000 tonnes, exclusively from inland waters.

INDUSTRY
The small industrial sector includes factories producing wood products, cotton fabrics, footwear, beer and radios. Output: sugar (2001), 13,000 tonnes; oils and fats (2000), 7,000 tonnes; beer (2003), 12·2m. litres; cotton fabrics (1992), 5·32m. metres; sawnwood (2001), 150,000 cu. metres.

Labour
In 2003 there were 1,162,000 employed persons (53% males).

INTERNATIONAL TRADE

Imports and Exports
Imports (c.i.f.) in 2005 totalled US$186·3m. (US$70·5m. in 2000); exports (f.o.b.) in 2005 totalled US$116·2m. (US$79·3m. in 2000).

The main import suppliers are Cameroon, France, Democratic Republic of the Congo and Japan. The main export markets are Belgium, France, Switzerland and Cameroon. Main imports include food, textiles, petroleum products, machinery, electrical equipment and motor vehicles. Main exports are diamonds, timber, coffee and motor vehicles.

COMMUNICATIONS

Roads
There were 23,417 km of roads in 2002, including 5,200 km of highways or main roads. In 2007 there were 1,200 passenger cars, 58 lorries and vans, and 4,500 motorcycles and mopeds. There were 583 road accident deaths in 2007.

Civil Aviation
There is an international airport at M'Poko, near Bangui, which handled 44,000 passengers (41,000 on international flights) in 2001. In 2003 there were direct services operating to Douala, Khartoum, Nyala, Paris and Yaoundé. In 2001 scheduled airline traffic of Central African Republic-based carriers flew 1m. km, carrying 46,000 passengers (all on international flights).

Shipping
Timber and barges are taken to Brazzaville (Republic of the Congo).

Telecommunications
There were 12,000 fixed telephone lines in 2010 (2·7 per 1,000 inhabitants). Mobile phone subscribers numbered 1·02m. in 2010. In 2008, 1·0% of the population were internet users. In June 2012 there were 144,000 Facebook users.

SOCIAL INSTITUTIONS

Justice
The Criminal Court and Supreme Court are situated in Bangui. There are 16 high courts throughout the country. The population in penal institutions in Oct. 2007 was 1,233 (29 per 100,000 of national population).

Education
Adult literacy rate was an estimated 55·2% in 2009 (69·1% among males and 42·1% among females).

In 2002–03 there were 284,869 pupils at the *fondamental 1* (lower primary) level; 49,922 pupils at *fondamental 2*; and 9,945 pupils at secondary schools. The University of Bangui, founded in 1969, is the leading institution in the tertiary sector. In 2006 there were 4,462 students in total in higher education.

Public expenditure on education came to 1·3% of GNI in 2006.

Health
In 2006 there were 12 hospital beds per 10,000 population. In 2004 there were 331 physicians and 1,613 nursing and midwifery personnel. Expenditure on health in 2005 came to 4·0% of GDP and represented 10·9% of total government expenditure.

RELIGION

In 2001 there were 660,000 Roman Catholics, 560,000 Muslims and 520,000 Protestants. Traditional animist beliefs are still widespread.

CULTURE

World Heritage Sites
The Manovo-Gounda St Floris National Park was inscribed on the UNESCO World Heritage List in 1988. Poaching and violence closed the park to tourism in 1997. Sangha Trinational, a transboundary conservation complex situated in Cameroon, Central African Republic and the Republic of the Congo, was added to the UNESCO World Heritage List in 2012. The site consists of three contiguous national parks totalling around 750,000 ha. Much of it is unaffected by human activity and features a wide range of humid tropical forest ecosystems with rich flora and fauna.

Press
In 2008 there were 30 newspapers, of which six were dailies with a circulation of 5,000.

Tourism
In 2009, 52,000 non-resident tourists—excluding same-day visitors—arrived by air (up from 31,000 in 2008).

DIPLOMATIC REPRESENTATIVES

Of Central African Republic in the United Kingdom
Ambassador: Jean Willybiro Sako (resides in Paris).

Of the United Kingdom in Central African Republic
Ambassador: Bharat Joshi (resides in Yaoundé, Cameroon).

Of Central African Republic in the USA (2704 Ontario Rd., NW, Washington, D.C., 20009)
Ambassador: Stanislas Moussa-Kembe.

Of the USA in Central African Republic (Ave. David Dacko, Bangui)
Ambassador: Laurence D. Wohlers.

Of Central African Republic to the United Nations
Ambassador: Charles-Armel Doubane.

Of Central African Republic to the European Union
Ambassador: Armand-Guy Zounguere-Sokambi.

FURTHER READING

Kalck, Pierre, *Historical Dictionary of the Central African Republic.* 3rd ed. 2004
Titley, B., *Dark Age: The Political Odyssey of Emperor Bokassa.* 1997

National Statistical Office: Division des Statistiques, des Études Économiques et Sociales, BP 696, Bangui.
Website (French only): http://www.stat-centrafrique.com

Explore the world at
www.statesmansyearbook.com

CHAD

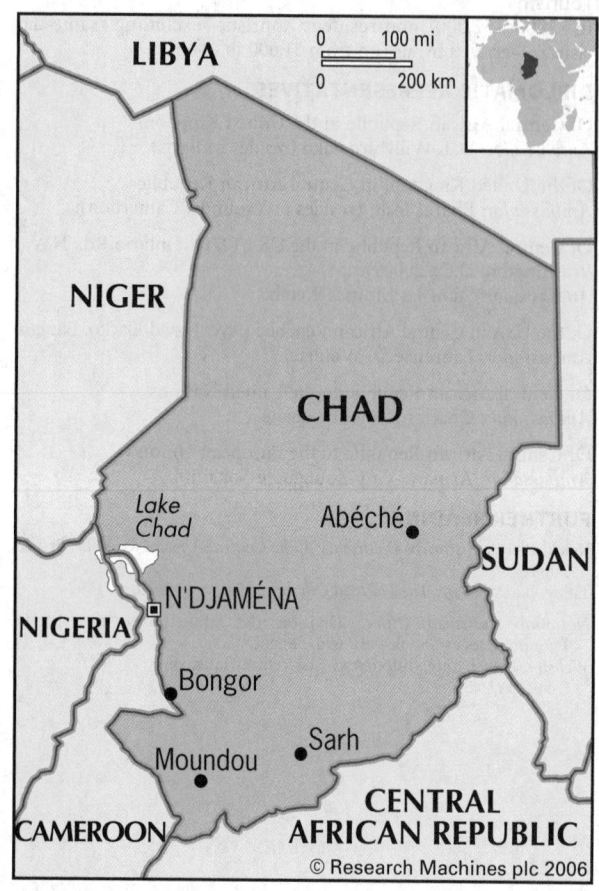

République du Tchad
(Republic of Chad)

Capital: N'Djaména
Population projection, 2015: 12·78m.
GNI per capita, 2011: (PPP$) 1,105
HDI/world rank: 0·328/183
Internet domain extension: .td

KEY HISTORICAL EVENTS

France proclaimed a protectorate over Chad in 1900 and in July 1908 the territory was incorporated into French Equatorial Africa. It became a separate colony in 1920, and in 1946 one of the four constituent territories of French Equatorial Africa. It achieved full independence on 11 Aug. 1960. Conflicts between the government and secessionist groups, particularly in the Muslim north and centre, from 1965 led to civil war. In 1982 Hissène Habré gained control of the country. In June 1983 Libyan-backed forces reoccupied some territory but a ceasefire took effect in Sept. 1987. Rebel forces of the Popular Salvation Movement led by Idriss Déby entered Chad from Sudan in Nov. 1990. On 4 Dec. 1990 Déby declared himself President. In Feb. 2000 Hissène Habré was charged with torture and barbarity and put under house arrest in Senegal.

TERRITORY AND POPULATION

Chad is bounded in the west by Cameroon, Nigeria and Niger, north by Libya, east by Sudan and south by the Central African Republic. In Feb. 1994 the International Court of Justice ruled that the Aozou Strip along the Libyan border, occupied by Libya since 1973, was part of Chad. Area, 1,284,000 sq. km. The provisional population at the 2009 census was 11,175,915. In 1993, 94% of the population were settled (of whom 22% were urban and 6% nomadic).

The UN gives a projected population for 2015 of 12·78m.

In 2011, 28·2% of the population were urban. The capital is N'Djaména with 993,492 inhabitants (2009 census, provisional), other large towns being (2009 provisional census figures) Moundou (132,411) and Sarh (99,099).

Following administrative reforms of 2002 and 2008, Chad's 14 prefectures were divided into 22 regions, including the City of N'Djaména (which is a commune governed by a special statute). The 22 regions are (with 2009 provisional census population and capital): Barh el Ghazel, population 260,865 (Moussoro); Batha, 527,031 (Ati); Bourkou, 97,251 (Faya); Chari-Baguirmi, 621,785 (Massénya); Ennedi, 173,606 (Fada); Guéra, 553,795 (Mongo); Hadjer-Lamis, 562,957 (Massakory); Kanem, 354,603 (Mao); Lac, 451,369 (Bol); Logone Occidental, 683,293 (Moundou); Logone Oriental, 796,453 (Doba); Mandoul, 637,086 (Koumra); Mayo-Kebbi Est, 769,178 (Bongor); Mayo-Kebbi Ouest, 565,087 (Pala); Moyen-Chari, 598,284 (Sarh); N'Djaména, 993,492 (N'Djaména); Ouaddaï, 731,679 (Abéché); Salamat, 308,605 (Am-Timan); Sila, 289,776 (Goz Beïda); Tandjilé, 682,817 (Laï); Tibesti, 21,970 (Bardaï); Wadi Fira, 494,933 (Biltine).

The official languages are French and Arabic, but more than 100 different languages and dialects are spoken. The largest ethnic group is the Sara of southern Chad (27·7% of the total population), followed by the Sudanic Arabs (11·5%).

SOCIAL STATISTICS

2008 estimates: births, 499,000; deaths, 182,000. Rates, 2008 estimates (per 1,000 population): births, 45·7; deaths, 16·7. Chad has one of the youngest populations of any country, with 73% of the population under the age of 30 and 45% under 15. Annual rate of growth, 2000–08, 3·3%. Expectation of life in 2007 was 47·3 years among males and 49·9 among females. Infant mortality, 2010 (per 1,000 live births), 99. Fertility rate, 2008, 6·2 children per woman.

CLIMATE

A tropical climate, with adequate rainfall in the south, though Nov. to April are virtually rainless months. Further north, desert conditions prevail. N'Djaména, Jan. 75°F (23·9°C), July 82°F (27·8°C). Annual rainfall 30" (744 mm).

CONSTITUTION AND GOVERNMENT

After overthrowing the regime of Hissène Habré, Idriss Déby proclaimed himself *President* and was sworn in on 4 March 1991.

A law of Oct. 1991 permits the formation of political parties provided they are not based on regionalism, tribalism or intolerance. There were over 70 political parties in 2008.

At a referendum on 31 March 1996 a new constitution was approved by 63·5% of votes cast. It defines Chad as a unitary state. The head of state is the *President*, elected by universal suffrage. On 26 May 2004 the *National Assembly* passed an amendment scrapping the two-term limit on the presidency, replacing it with an age limit of 70. The amendment was approved by referendum in June 2005.

The National Assembly has 188 members, elected for a four-year term. A *Senate* was stipulated in the 1996 constitution, but has yet to be created.

National Anthem

'Peuple tchadien, debout et à l'ouvrage' ('People of Chad, arise and take up the task'); words by L. Gidrol, tune by P. Villard.

RECENT ELECTIONS

Presidential elections were held on 25 April 2011. Turnout was 64·2%. Incumbent Idriss Déby won re-election, with 88·7% of the vote, against 6·0% for Albert Pahimi Padacké and 5·3% for Nadji Madou.

In parliamentary elections held on 13 Feb. 2011 (the first since April 2002) the Patriotic Salvation Movement (MPS) of President Idriss Déby and allied parties won 131 seats, the National Union for Democracy and Renewal (UNDR) 10, the Union for Renewal and Democracy (URD) 8, the National Rally for Democracy in Chad (RNDT) 6 and the Federation Action for the Republic (FAR) 4. A number of smaller parties each took one or two seats. Turnout was 56·6%.

CURRENT GOVERNMENT

President: Lieut.-Gen. Idriss Déby; b. 1952 (MPS; in office since Dec. 1990 and re-elected 3 July 1996, 20 May 2001, 3 May 2006 and 25 April 2011).

In March 2013 the government comprised:

Prime Minister: Djimrangar Dadnadji; b. 1954 (MPS; in office since 21 Jan. 2013).

Minister at the Presidency in Charge of National Defence and Veterans: Benaindo Tatola. *Secretary General of the Government Responsible for Relations with Parliament:* Samir Adam Annour.

Minister of Agriculture and Irrigation: Dangdé Laoubélé Damaye. *Basic Education and Literacy:* Hassan Tcholnaye. *Civil Service and Labour:* Abdoulaye Abakar. *Commerce and Industry:* Hamid Mahamat Dahalop. *Communication, and Government Spokesman:* Hassane Sylla Bakari. *Culture, Arts and Heritage Conservation:* Dayang Menwa Enoch. *Economy, Planning and International Co-operation:* Issa Ali Taher. *Energy and Oil Resources:* Djérassem Le Bémadjiel. *Environment and Fisheries:* Mahamat Issa Halikimi. *Finance and Budget:* Atteib Habib Doutoum. *Foreign Affairs and African Integration:* Moussa Faki Mahamat. *Higher Education, Research and Vocational Training:* Adoum Goudja. *Human Rights and Fundamental Freedoms:* Amina Kodjiyana. *Infrastructure and Equipment:* Gata Ngoulou. *Interior and Public Security:* Ali Mahamat Zen Ali Fadel. *Justice and Keeper of the Seals:* Abdoulaye Sabre Fadoul. *Land Management, Decentralization and Local Freedoms:* Yokabdjim Mandigui. *Livestock and Animal Resources:* Amir Adoudou Artine. *Micro-credit and Promotion of Women and Youth:* Martin Bagrim Kibassim. *Mines and Geology:* Oumar Adoum Sini. *Posts, and Information and Communication Technology:* Rosine Amane Benaïwa Djibergui. *Public Health:* Ahmat Djidda Mahamat. *Public Sanitation and Good Governance:* Hinsou Hara. *Secondary Education and Professional Training:* Abdelkerim Seid Bauche. *Social Action, Family Affairs and National Solidarity:* Sadie Goukouni Weddeye. *Tourism and Handicrafts:* Abderahim Younous Ali. *Transport and Civil Aviation:* Dillo Adoum. *Urban and Rural Water Resources:* Ali Mahamat Abdoulaye. *Urban Planning, Housing and Land Affairs:* Jean-Bernard Padaré. *Youth and Sports:* Mahamat Adoum.

Office of the President (Arabic and French only):
http://www.presidencetchad.org

CURRENT LEADERS

Idriss Déby

Position

President

Introduction

Lieut.-Gen. Idriss Déby became president in Feb. 1991 after participating in a coup to overthrow Hissène Habré. He oversaw multi-party elections, but opponents cited electoral irregularities after Déby's victories at the 1996 and 2001 polls and largely boycotted the 2006 and 2011 presidential contests. His political survival has been attributed in large part to the presence of French forces in the country, while his tenure has been marked by civil war and an overspill of fighting from Darfur in neighbouring Sudan, hindering attempts at reducing Chad's extreme poverty.

Early Life

Déby was born in 1952 into the Bidyate clan of the Zaghawa peoples. While serving in the army he helped Hissène Habré take power in 1982, overthrowing Goukouni Oueddei in a coup. His relationship with Habré declined and in 1989 Déby was accused of involvement in an alleged coup and went into exile in Sudan. In Dec. 1990, as head of the Patriotic Salvation Movement (MPS) and with the support of Libya, he removed Habré from power. Déby was proclaimed president in Feb. 1991 and in 1993 was appointed interim head of a transitional government charged with preparing democratic elections to be held within a year.

Career in Office

Déby went on to establish a multi-party constitution and triumphed at presidential elections held in 1996. The MPS won elections to the legislative assembly the following year.

Déby has had to cope with tensions between the largely Arab-Muslim north and the mainly Christian and animist south. In 1998 there was a surge in rebel activity in the north, spearheaded by the Movement for Democracy and Justice in Chad (MDJT), led by Déby's former defence minister Youssouf Togoimi. In early 2002 Libyan leader Col. Gaddafi, formerly a supporter of Chadian rebel movements, brokered a peace agreement which included provision for an amnesty for MDJT members. It soon failed but in Jan. 2003 the government reached a peace agreement with the rebel National Resistance Army in the east and in Dec. that year a new accord was signed with the MDJT.

Déby was re-elected in 2001 although the electoral commission discounted results from 25% of polling stations for electoral irregularities. Six of Déby's defeated rivals were subsequently arrested for 'inciting violence and civil disobedience' but later released as human rights organizations and trades unions pressed for a general strike. Déby was sworn into office in Aug. 2001. In 2001 the Senegalese judicial system concluded that it lacked the authority to try Habré, the deposed former president in exile in Senegal, on charges of authorizing torture. A constitutional amendment in June 2005 permitted Déby to stand for a third term of office, which he did successfully in May 2006. He was re-elected for a fourth term in April 2011, claiming 89% of the vote, although the main opposition parties continued to boycott the polls.

Since 2003 fighting has spilled over the border from Darfur in neighbouring Sudan, prompting a large influx of refugees to Chad. From 2005 Chad accused Sudan of supporting Chadian rebels, who made unsuccessful assaults on the capital, N'Djaména, in April 2006 and Feb. 2008. The government meanwhile declared a state of emergency in eastern provinces in Nov. 2006 and again in Oct. 2007. In March 2008 Déby and Sudan's president signed an accord aimed at stopping hostilities. However, alleged Chadian involvement in a rebel attack on Omdurman in Sudan in May prompted Sudan to sever diplomatic relations and Déby to cut economic links in retaliation. In Jan. 2009 several rebel groups united to form the Union of Resistance Forces (UFR), which in May launched a major offensive in eastern Chad from bases in Sudan. The offensive was repelled, but the action further exacerbated tensions between the Chadian and Sudanese governments until Jan. 2010 when both sides stated their readiness to normalize relations. In April the Chad–Sudan border reopened seven years after the Darfur conflict had forced its closure.

DEFENCE

There are seven military regions. Total armed forces personnel numbered 25,350 in 2007, including republican guards. Defence expenditure totalled US$145m. in 2008 (US$14 per capita), representing 1·8% of GDP.

Army

In 2007 the strength was about 17,000 although it is being reorganized. In addition there was a paramilitary Gendarmerie of 4,500 and a Republican Guard of 5,000.

Air Force

Personnel (2007), 350 including four combat-capable aircraft and two attack helicopters.

ECONOMY

Agriculture accounted for 20·5% of GDP in 2006, industry 54·8% and services 24·7%.

Overview

After 30 years of civil war, Chad is dependent on external aid from the IMF, the World Bank and the EU. The agrarian sector supports 80% of the population and accounts for an estimated 52% of GDP. Chad is ranked 183rd out of 187 countries on the Human Development Index.

GDP per capita was US$220 per annum in 2002, with development hindered by political instability, droughts and primitive infrastructure. However, oil-related investments and the completion of the Chad–Cameroon oil pipeline in July 2003 saw real GDP growth peak at over 30% in 2004. GDP per capita had risen to around US$755 by 2008. The majority of oil revenues have been earmarked for priority sectors including education, health care, infrastructure and rural development.

After the global economic crisis in 2008–09, a recovery in international oil prices and growth in the agricultural sector helped to improve the economic position in 2010. However, reducing the country's over-dependence on oil reserves is a major long-term challenge.

Chad became a member of CAEMU (the Central African Economic and Monetary Union) in 2004. The currency is pegged to the euro and managed by the Banque des États de l'Afrique Centrale.

Currency

The unit of currency is the *franc CFA* (XAF) with a parity of 655·957 francs CFA to one euro. There was inflation of 10·1% in 2009 but deflation of 2·1% in 2010. In 2011 inflation stood at 1·9%. Foreign exchange reserves were US$214m. in June 2005, total money supply was 213,920m. francs CFA and gold reserves were 11,000 troy oz.

Budget

Revenues in 2007 were 764·9bn. francs CFA (petroleum revenue, 73·6%) and expenditures 709·3bn. francs CFA (current expenditure, 65·4%).

VAT is 18%.

Performance

In 2009 the economy shrank by 1·2%. However, Chad's growth in 2010 was among the highest in the world at 13·0%. In 2004 there had been growth of 33·6%, thanks mainly to the construction of an oil pipeline from landlocked Chad to neighbouring Cameroon's Atlantic coast. Chad became the world's newest oil producer at the time in 2003, and as a result in 2004 recorded the highest economic growth of any country. In 2011 the economy grew by 1·8%; total GDP was US$9·5bn.

Banking and Finance

The Banque des États de l'Afrique Centrale (*Governor*, Lucas Abaga Nchama) is the bank of issue. Other leading banks include: Banque Agricole du Soudan au Tchad; Banque Commerciale du

Chari; Banque Internationale de l'Afrique au Tchad; Commercial Bank Tchad; Financial Bank Tchad; and Société Générale Tchadienne de Banque.

In 2010 foreign debt totalled US$1,733m., representing 25·7% of GNI.

ENERGY AND NATURAL RESOURCES

Environment

Carbon dioxide emissions from the consumption and flaring of fossil fuels were the equivalent of less than 0·1 tonnes per capita in 2008. An *Environmental Performance Index* compiled in 2008 ranked Chad 143rd in the world out of 149 countries analysed, with 45·9%. The index examined various factors in six areas—air pollution, biodiversity and habitat, climate change, environmental health, productive natural resources and water resources.

Electricity

Installed capacity was estimated at 32,000 kW in 2007. Production in 2007 amounted to about 105m. kWh. Consumption per capita was an estimated 10 kWh in 2007.

Oil and Gas

The oilfield in Kanem prefecture has been linked by pipeline to a new refinery at N'Djaména but production has remained minimal. There is a larger oilfield in the Doba Basin. In June 2000 the World Bank approved funding for a 1,070-km US$4bn. pipeline to run from 300 new oil wells in Chad through Cameroon to the Atlantic Ocean. Oil started pumping in July 2003. Oil production in 2008 was 6·7m. tonnes. Proven reserves totalled 0·9bn. bbls in 2008. Revenue in 2008 was US$1·9bn.

Minerals

Salt (about 4,000 tonnes per annum) is mined around Lake Chad, and there are deposits of uranium, gold, iron ore and bauxite. There are small-scale workings for gold and iron.

Agriculture

Some 80% of the workforce is involved in subsistence agriculture and fisheries. In 2001, 3·60m. ha. were arable and 30,000 ha. permanent crops. There were 175 tractors in 2001. Cotton growing (in the south) and animal husbandry (in the animal zone) are the most important branches. Production, 2002 (in 1,000 tonnes): sorghum, 481; groundnuts (2001), 477; millet, 357; sugarcane, 355; cassava (2001), 306; yams, 230; seed cotton, 170; rice, 135; cottonseed, 100; maize, 84; dry beans, 78.

Livestock, 2001: cattle, 5,992,000; goats, 5,304,000; sheep, 2,431,000; camels, 725,000; chickens, 5m.

Forestry

In 2010 the area under forests was 11·53m. ha., or 9% of the total land area. Timber production in 2007 was 7·47m. cu. metres.

Fisheries

Total catches, from Lake Chad and the Chari and Logone rivers, were an estimated 40,000 tonnes in 2010.

INDUSTRY

Output in 1,000 tonnes (2007): raw sugar, 35; groundnut oil, 29; cottonseed oil, 9.

Labour

In 2003 the labour force was 4,171,710 (55% males). In 2003 approximately 72% of the economically active population were engaged in agriculture.

INTERNATIONAL TRADE

Imports and Exports

Imports (c.i.f.) in 2009 totalled US$1,950m.; exports (f.o.b.) in 2009 totalled US$2,800m.

Main import suppliers are France, USA and Germany. Main export markets are Portugal, Germany and USA. The principal imports are machinery and transportation equipment, industrial goods and tobacco. Mineral fuels and oils account for the majority of exports.

COMMUNICATIONS

Roads
In 2006 there were around 40,000 km of roads. 18,900 passenger cars were in use in 2006, plus 3,300 buses and coaches, 35,400 lorries and vans, and 63,000 motorcycles and mopeds. In 2007 there were 840 deaths in road accidents.

Civil Aviation
There is an international airport at N'Djaména, from which there were direct flights in 2003 to Addis Ababa, Bamako, Bangui, Douala, Garoua, Kano, Paris, Tripoli and Yaoundé. In 2001 scheduled airline traffic of Chad-based carriers flew 1m. km, carrying 46,000 passengers (all on international flights). In 2000 N'Djaména handled 17,000 passengers and 2,300 tonnes of freight.

Telecommunications
There were 51,200 fixed telephone lines in 2010 (4·6 per 1,000 inhabitants). Mobile phone subscribers numbered 2·61m. in 2010. There were 17·0 internet users per 1,000 inhabitants in 2010. Fixed internet subscriptions totalled 4,600 in 2009 (0·4 per 1,000 inhabitants).

SOCIAL INSTITUTIONS

Justice
There are criminal courts and magistrates courts in N'Djaména, Moundou, Sarh and Abéché, with a Court of Appeal situated in N'Djaména.

The population in penal institutions in 2010 was 4,775 (43 per 100,000 of national population).

The death penalty is still in force, although it has not been used since 2003 when there were nine executions in the space of four days.

Education
In 2007 there were 1,324,298 pupils in primary schools with 21,933 teaching staff and 314,470 pupils in secondary schools with 9,555 teaching staff. In 2005 there were 10,468 students with 1,100 academic staff at institutes of tertiary education. Adult literacy rate was 33% in 2005.

In 2005 public expenditure on education came to 2·3% of GNI and 10·1% of total government spending.

Health
In 2001 there were 4,105 hospital beds. There were 205 doctors, 1,220 nurses, 161 midwives and 38 pharmacists in 2001.

Chad has made significant progress in the reduction of undernourishment in the past 20 years. Between the period 1990–92 and 2001–03 the proportion of undernourished people declined from 58% of the population to 33%.

RELIGION
The northern and central parts of the country are predominantly Muslim. In 2001 there were estimated to be 4,690,000 Muslims, 1,770,000 Roman Catholics and 1,250,000 Protestants. Traditional beliefs are still widespread.

CULTURE

World Heritage Sites
The Lakes of Ounianga were added to the UNESCO World Heritage List in 2012. The site includes 18 permanent interconnected lakes in a desert setting, a remarkable natural phenomenon that is still to be fully understood.

Press
There are no daily newspapers; there were five non-dailies in 2008, including the government-owned *Info-Tchad*. Combined circulation was 4,000.

Tourism
In 2009, 31,000 non-resident tourists (including 16,000 from Europe) stayed in hotels or similar accommodation.

DIPLOMATIC REPRESENTATIVES
Of Chad in the United Kingdom
Ambassador: Ahmat Awad Sakine (resides in Brussels).

Of the United Kingdom in Chad
Ambassador: Bharat Joshi (resides in Yaoundé, Cameroon).

Of Chad in the USA (2002 R. St., NW, Washington, D.C., 20009)
Ambassador: Maitine Djoumbe.

Of the USA in Chad (Ave. Felix Eboue, N'Djaména)
Ambassador: Mark Boulware.

Of Chad to the United Nations
Ambassador: Ahmad Alaam-mi.

Of Chad to the European Union
Ambassador: Ahmat Awad Sakine.

FURTHER READING
National Statistical Office: Direction de la Statistique des Études Économiques et Démographiques, Ministère du Plan, de l'Économie et de la Coopération Internationale, BP 453 N'Djaména.

CHILE

© Research Machines plc 2006

República de Chile
(Republic of Chile)

Capitals: Santiago (Administrative), Valparaíso (Legislative)
Population projection, 2015: 17·87m.
GNI per capita, 2011: (PPP$) 13,329
HDI/world rank: 0·805/44
Internet domain extension: .cl

KEY HISTORICAL EVENTS

Archaeological evidence suggests the earliest settlements of hunter-gatherers in Chile date from around 10,500 BC. They were probably the descendents of Paleo-Indians who crossed from Siberia by way of the Bering Strait (at various times a land bridge). Prior to the arrival of Europeans, the indigenous peoples included the Atacameno, living in small settlements in the northern deserts, the Araucanians, farmers in the more temperate valleys of central Chile, and the Chono, Alacaluf and Yahgan tribes from the mountainous southern areas.

Ferdinand Magellan was the first European to catch sight of what is now Chile when, in 1520, he sailed through the bleak archipelago at the tip of South America en route for the Pacific Ocean. Fifteen years later a Spanish expeditionary force, led by Diego de Almagro, set off from the newly-captured Inca city of Cusco to explore land to the south. Almagro travelled as far as the Itata river but came under repeated attacks from hostile Araucanians and was unable to establish a foothold. He returned to Peru in 1536, with news only of 'a cursed land without gold, inhabited by savages of the worst kind.' Five years passed before the next Spanish expedition to Chile left Cusco, headed by Pedro de Valdivia. After months of hardship, battles with Araucanians and internal divisions, Valdivia's forces established the settlement of Santiago in early 1541. In the next ten years the Spanish built fortified towns at Concepción, La Serena, Valdivia and Villarrica. North of Concepción, they began to convert the Araucanians to Christianity and established mines run on forced labour. Subjugating the indigenous people south of Concepción, however, proved difficult. In that region, Araucanians known as Mapuche fought hard and quickly adapted their weapons and tactics to become effective guerrilla fighters. Fifty years after Valdivia's forces arrived in Chile, the colony remained a frontier, dependent on the Viceroyalty of Peru and governed by military officers based in dispersed fort towns. Gold was discovered but the wealth it generated was minimal compared with the riches that poured out of Mexico and Peru.

Opposition to Spain

Following further military defeats at the hands of the Mapuche and the destruction of Concepción in an earthquake in 1570, King Felipe II of Spain named a veteran conquistador, Rodrigo de Quiroga, as governor of Chile. After 1575 Quiroga attempted to quell rebellion with a brutal campaign against the Araucanians, capturing them for forced labour and mutilating their feet to prevent escape. A subsequent governor, Garcia Onez de Loyola, attempted to moderate these abuses, but was killed at the battle of Curalaba in 1598. Subsequently, all major Spanish settlements south of the Bíobío river were destroyed or abandoned. In 1600 the king of Spain granted a permanent military subsidy to fund the war in Chile and in 1608 he signed a royal decree legalizing the enslavement of 'rebellious' Indians. At around this time, coastal settlements such as Valparaíso came under attack from English and Dutch adventurers and pirates in search of wealth and as part of a prolonged effort to force Spain to allow other nations to trade with its colonies. The relative lack of mineral wealth in Chile led the 5,000 or so Spanish settlers to develop a pastoral and agricultural society; they grew a wide range of cereals and raised livestock. North of the Bíobío river, there was considerable intermarriage and the rapid growth of a mestizo (mixed Amerindian and European) group. The social status of mestizos was determined by the extent to which they were Hispanicized and by their kinship ties with the landed class.

In 1664 the governorship of Chile passed to Francisco de Meneses, an opportunist who took advantage of the warfare economy, taxing ships unless they carried his merchandise. He demanded bribes and accumulated vast wealth from the slave trade. An earthquake that shook Lima in 1687 disrupted the supply of food to the Peruvian city for some years. Chilean merchants cashed in by shipping wheat from the country's central belt and there was a rapid expansion in wheat production. The wheat 'boom' continued until 1700 with trade controlled by a clique of merchants who colluded with corrupt officials. Attempts by successive governors to make peace with the Mapuche (the Pact of Quillin) ended in failure, and sporadic fighting continued throughout the 17th and 18th centuries. Many thousands of Mapuche migrated eastwards across the Andes to Argentina.

Bourbon Rule

The Habsburg dynasty's rule over Spain ended in 1700. They were succeeded by the Bourbons who gave the *audiencia* of Chile (based in Santiago) greater independence from the Viceroyalty of Peru. One of the most charismatic governors of the Bourbon era was the Irish-born Ambrosio O'Higgins, who presided over increased economic production and strengthened the military. In 1791 he also outlawed forced labour. Economic links with Argentina increased after it became the Viceroyalty of the Río de la Plata in 1776 and by the end of the 18th century Chile was engaging in direct trade with Europe. Freer trade brought with it knowledge of politics abroad, particularly the spread of liberalism in Europe and American independence. The Royal University of San Felipe was established at Santiago in 1758 but most educated Chileans followed the traditional ideology of the Spanish crown and the Roman Catholic Church, while the majority of mestizos and Araucanians remained illiterate and subordinate.

The French Revolution and Napoleon Bonaparte's subsequent invasion of Spain in 1807 eventually led to greater autonomy and independence. On 18 Sept. 1810 the Santiago elite, employing the town council as a junta, announced their intention to govern the colony until Fernando VII was reinstated. They remained loyal to the ousted Spanish king but insisted they had the right to rule and immediately relaxed trade restrictions. Chile's first government was led by José Miguel Carrera Verdugo. Carrera and his brothers, as well as Bernardo O'Higgins (son of former governor Ambrosio O'Higgins) soon saw the opportunity to replace temporary self-rule with permanent independence, although others remained loyal to Spain. 1814 saw the start of the Reconquest (*La Reconquista*), and the Spanish authorities managed to reassert control of Chile by winning the Battle of Rancagua. O'Higgins and many of the Chilean rebels escaped to Argentina, from where they plotted to liberate their country. O'Higgins won the support of the revolutionary government in Buenos Aires under José de San Martín. Their joint forces freed Chile in 1817, defeating the Spaniards and their supporters at the Battle of Chacabuco.

Independence

Bernardo O'Higgins ruled Chile from 1817–23, formally proclaiming independence on 12 Feb. 1818 at Talca. He founded schools and expelled the remaining Spaniards but his authoritarian style and attempted reforms of the land tenure system caused unrest among the powerful landowners. A succession of poor harvests forced him to abdicate in 1823. Civil conflict continued throughout the 1820s, owing largely to a split between the Chilean oligarchs and the army. The harmful effects on the economy prompted conservatives to seize control in 1830. Diego Portales reached a compromise between the oligarchs and promulgated a constitution in 1833, beginning a prolonged period of political stability and economic revival. A free port was created at Valparaíso to encourage trade with foreign, especially British, merchants. Chilean landowners and merchants profited from new markets in California and Australia in the 1850s. Economic improvement was underpinned by discoveries of silver and copper in northern Chile and significant coal deposits around Concepción. The period after 1860, known as the 'Liberal Republic', saw the emergence of rival political groups influenced by economic, scientific and literary ideas from Europe. Great

Britain became the main trading partner and British entrepreneurs invested in the railways and the modernization of the ports.

Access to valuable saltpetre (nitrate) deposits in the far north, bordering Peru and Bolivia, led to the War of the Pacific (1879–83), in which the Chilean army and navy prevailed. During the presidency of José Manuel de Balmaceda (1886–91) the government attempted to use revenue from mineral extraction to strengthen its administration, a policy opposed by the oligarchs who forced Balmaceda's abdication. Thereafter Chile's presidential republic was transformed into a parliamentary republic. The next 20 years saw the emergence of new political parties representing the emerging working and middle classes. The Democratic Party was formed in 1887 to represent artisans and urban workers, while the Radical Party was backed by the middle class. Marxist ideology spread among workers in the late 1890s and the Socialist Party was established in 1901. By the start of the 20th century Chile was becoming increasingly urbanized as workers poured into the cities from rural areas. Society was polarized, with parts of Santiago and Valparaíso mirroring prosperous and elegant European cities, while the masses remained largely poverty stricken and illiterate.

The outbreak of the First World War, with Britain and Germany on opposite sides, brought disaster to the Chilean economy. Demand for saltpetre fell away and thousands of workers lost their jobs. The reformist president Arturo Alessandri Palma was elected in 1920 but his initiatives were blocked by the legislature and he resigned. The army intervened and returned Alessandri to power in 1925, after which the constitution was amended to strengthen the executive at the expense of the legislature. It established a presidential republic, separated church and state, and enshrined new labour and welfare legislation. Attempts to reduce the power of the oligarchs failed. Alessandri resigned for a second time and was replaced by Carlos Ibáñez del Campo in 1927. His military dictatorship led to improvements in education and public services but also failed to address the economic power of the oligarchs. The world depression of the 1930s was hard on Chile as demand for mineral exports plummeted. A democratic-leftist coalition, the Popular Front, took power following the elections of 1938. Chile remained neutral in the Second World War until 1942 when President Juan Antonio Ríos declared war on Germany, Italy and Japan.

Conflict with the USA

The Radical Party joined with the Communists to field a left-wing Radical, Gabriel González Videla, for president in the 1946 election. Once in office, González Videla (president, 1946–52) turned against his Communist allies, expelling them from his cabinet and banning the party in 1948. He also severed relations with the Soviet Union, prompting accusations of a Cold War agreement with the United States. The early 1950s were characterized by slow economic growth, spiralling inflation and increasing social demands. By 1952 Chileans were alienated by multi-party politics and reacted by turning to two symbols of the past: first, the 1920s dictator Ibáñez and, following the 1958 elections, the son of former president Alessandri. The Christian Democrat party, under the leadership of Eduardo Frei Montalva, undertook a 'Chileanization programme', wresting back control from the US-owned copper mines and making progress in land reform by establishing peasant co-operatives. There were also advances in education and housing. The agrarian reforms increased support for the various Socialist and Communist parties. In 1969 they formed the Popular Unity coalition, headed by Salvador Allende Gossens, who was elected president in 1970. Allende nationalized many private companies, attempted to improve conditions for the working classes and established ties with other socialist states. The first year was heralded a success, but in 1971–72 Chile was afflicted by rapid inflation and shortages of foods and consumer goods. The United States, which had become by far the largest foreign investor in the decades that followed the Second World War, withdrew much of its backing.

In Sept. 1973, with covert American support, the armed forces staged a coup and Allende died during an assault on the presidential palace in Santiago. Gen. Augusto Pinochet Ugarte was installed as president. The military closed Congress, censored the media, purged the universities and banned Marxist parties and union activities. It is estimated that over 3,000 of Allende's supporters lost their lives, over 30,000 were forced into exile and more than 130,000 were arrested over a three-year period. The return to market capitalism led to steady economic improvement from 1976 but falling copper prices and mounting foreign debt led to spiralling inflation and growing unemployment in the early 1980s. In 1981 a new constitution was approved, guaranteeing an eight-year extension to Pinochet's rule but also allowing a transition to civilian government by the end of the decade. The first free elections since the 1973 coup took place in Dec. 1989 and Patricio Aylwin Azócar emerged victorious, heading a coalition of left and centrist parties. The 1990s saw a rapid strengthening of the economy, underpinned by large inflows of foreign investment.

Pinochet remained head of the military until 1998, after which he claimed his constitutional right to become a senator for life (and hence immune from prosecution). While visiting Britain for medical treatment in 1998, Pinochet was arrested and held on human rights charges instigated by Spain. In early 2000 he returned to Chile after the British government ruled he was too ill to be extradited to Spain to face charges. Stripped of his immunity from prosecution, he died in Dec. 2006 without facing trial. On his election in Jan. 2000, Ricardo Lagos Escobar, leader of the Coalition of Parties for Democracy (CPD) pledged to reform the labour code, increase the minimum wage, introduce unemployment insurance and provide better health care, education and housing. In 2003 allegations of corruption damaged investor confidence.

In Jan. 2006 Michelle Bachelet, of the centre-left Concertación coalition, became Chile's first female president. In March 2010 she was succeeded by Sebastián Piñera, leader of the rightist Coalition for Change.

TERRITORY AND POPULATION

Chile is bounded in the north by Peru, east by Bolivia and Argentina, and south and west by the Pacific Ocean. The area is 756,096 sq. km (291,928 sq. miles) excluding the claimed Antarctic territory. Many islands to the west and south belong to Chile: the Islas Juan Fernández (147 sq. km with 633 inhabitants in 2002) lie about 600 km west of Valparaíso, and the volcanic Isla de Pascua (Easter Island or Rapa Nui, 164 sq. km with 3,791 inhabitants in 2002), lies about 3,000 km west-northwest of Valparaíso. Small uninhabited dependencies include Sala y Goméz (400 km east of Easter Is.), San Félix and San Ambrosio (1,000 km northwest of Valparaíso, and 20 km apart) and Islas Diego Ramírez (100 km southwest of Cape Horn).

In 1940 Chile declared, and in each subsequent year has reaffirmed, its ownership of the sector of the Antarctic lying between 53° and 90° W. long., and asserted that the British claim to the sector between the meridians 20° and 80° W. long. overlapped the Chilean by 27°. There are ten Chilean bases in Antarctica. A law of 1955 put the governor of Magallanes in charge of the 'Chilean Antarctic Territory' which has an area of 1,250,000 sq. km and a population (2002) of 2,392.

The provisional population at the census of April 2012 was 16,572,475 (8,513,327 females and 8,059,148 males); density, 21·9 per sq. km. 89·2% of the population lived in urban areas in 2011.

The UN gives a projected population for 2015 of 17·87m.

Area, population and capitals of the 15 regions:

Region	Sq. km	2012 census population (provisional)	Capital
Aisén del Gral. Carlos Ibáñez del Campo	108,494	98,413	Coihaique
De Antofagasta	126,049	542,504	Antofagasta
De La Araucanía	31,842	907,333	Temuco
De Arica-Parinacota	16,873	213,595	Arica
De Atacama	75,176	290,581	Copiapó
Del Bíobío	37,063	1,965,199	Concepción
De Coquimbo	40,580	704,908	La Serena
Del Libertador Gral. B. O'Higgins	16,387	872,510	Rancagua
De Los Lagos	48,584	785,169	Puerto Montt
De Los Ríos	18,430	363,887	Valdivia
De Magallanes y de la Antártica Chilena	132,297	159,102	Punta Arenas
Del Maule	30,296	963,618	Talca
Metropolitana de Santiago	15,403	6,683,852	Santiago
De Tarapacá	42,226	298,257	Iquique
De Valparaíso	16,396	1,723,547	Valparaíso

Other large towns (2002 census populations) are: Viña del Mar (286,931), Arica (175,441), Coquimbo (148,434), Chillán (146,701), Osorno (132,245), Calama (126,135) and Los Ángeles (117,972). 69·7% of the population is mixed or mestizo, 20% are of European descent and 10·3% declared themselves to be indigenous Amerindians of the Mapuche, Aymara, Atacameño and Quechua groups. Language and culture remain of European origin, with 604,349 Mapudungun-speaking (mainly Mapuche) Indians the only sizeable minority.

The official language is Spanish.

SOCIAL STATISTICS

2007 births, 240,569; deaths, 93,000; marriages, 57,792. Rates, 2007 (per 1,000 population): birth, 14·6; death, 5·6; marriage, 3·5. Divorce was only made legal in 2004; abortion remains illegal. Annual population growth rate, 2008–10, 0·9%. Infant mortality, 2010 (per 1,000 live births), 8. In 2009 the most popular age range for marrying was 25–29 for both males and females. Expectation of life at birth (2007): males 75·5 years, females 81·6 years. Chile has the highest life expectancy in South America. Fertility rate, 2008, 1·9 children per woman.

CLIMATE

With its enormous range of latitude and the influence of the Andean Cordillera, the climate of Chile is very complex, ranging from extreme aridity in the north, through a Mediterranean climate in Central Chile, where winters are wet and summers dry, to a cool temperate zone in the south, with rain at all seasons. In the extreme south, conditions are very wet and stormy. Santiago, Jan. 67°F (19·5°C), July 46°F (8°C). Annual rainfall 15″ (375 mm). Antofagasta, Jan. 69°F (20·6°C), July 57°F (14°C). Annual rainfall 0·5″ (12·7 mm). Valparaíso, Jan. 64°F (17·8°C), July 53°F (11·7°C). Annual rainfall 20″ (505 mm).

CONSTITUTION AND GOVERNMENT

A new constitution was approved by 67·5% of the voters on 11 Sept. 1980 and came into force on 11 March 1981. It provided for a return to democracy after a minimum period of eight years. Gen. Pinochet would remain in office during this period after which the government would nominate a single candidate for President. At a plebiscite on 5 Oct. 1988 President Pinochet was rejected as a presidential candidate by 54·6% of votes cast. The Constitution has been amended on a number of occasions since then.

The *President* is directly elected for a non-renewable four-year term. Parliament consists of a 120-member *Chamber of Deputies* and a *Senate* of 38 members. In March 2006 the Senate became fully elected, by abolishing non-elected senators and eliminating life seats for former presidents. Senators are elected for an eight-year term.

Santiago is the administrative capital of Chile, but since 11 March 1990 Valparaíso has been the legislative capital.

National Anthem

'Dulce patria, recibe los votos' ('Sweet Fatherland, receive the vows'); words by E. Lillo, tune by Ramón Carnicer.

GOVERNMENT CHRONOLOGY

Heads of State since 1942. (APL = Popular Liberating Alliance; FP = Popular Front; PC = Conservative Party; PDC = Christian Democratic Party; PS = Socialist Party; RN = National Renewal)

Presidents of the Republic

1942–46	FP	Juan Antonio Ríos Morales
1946–52	FP	Gabriel González Videla
1952–58	APL	Carlos Ibáñez del Campo
1958–64	PC	Jorge Alessandri Rodríguez
1964–70	PDC	Eduardo Nicanor Frei Montalva
1970–73	PS	Salvador Allende Gossens

Military Junta

1973–74		Gen. Augusto J. R. Pinochet (chair); Gen. César Raúl Benavides Escobar; Admr. José Toribio Merino Castro; Gen. Gustavo Leigh Guzmán; Gen. Fernando Matthei Aubel; Gen. César Mendoza Durán; Gen. Rodolfo Stange Oelckers

Presidents of the Republic

1974–90	military	Augusto J. R. Pinochet
1990–94	PDC	Patricio Aylwin
1994–2000	PDC	Eduardo Frei Ruiz-Tagle
2000–06	PS	Ricardo Froilán Lagos
2006–10	PS	Michelle Bachelet
2010–	RN	Sebastián Piñera Echenique

RECENT ELECTIONS

In the presidential run-off held on 17 Jan. 2010 the centre-right candidate Sebastián Piñera Echenique (National Renewal) polled 51·6%, defeating the ruling leftist 'Concertación' candidate Eduardo Frei Ruiz-Tagle (Christian Democratic Party), with 48·4%. Two other candidates had participated in the first round of voting on 13 Dec. 2009.

In elections to the Chamber of Deputies on 13 Dec. 2009 the Coalition for Change won 58 seats with 43·4% of the vote (Independent Democratic Union, 37 and 23·0%; National Renewal, 18 and 17·8%; ind. 3 and 2·3%) against 57 seats (44·4%) for the Coalition of Parties for Democracy/Concertación (Christian Democratic Party, 19 and 14·2%; Party for Democracy, 18 and 12·7%; Socialist Party, 11 and 9·9%; Radical Social Democratic Party, 5 and 3·8%; Communist Party, 3 and 2·0%; ind., 1 and 1·8%) and 3 for Clean Chile Vote Happy (5·4%). The remaining two seats went to independent candidates. After partial elections to the Senate on the same day the composition in the Senate was: Coalition of Parties for Democracy, 19 seats; Coalition for Change, 16; Clean Chile Vote Happy, 1; ind., 2.

Presidential and parliamentary elections are scheduled to take place on 17 Nov. 2013.

CURRENT GOVERNMENT

President: Sebastián Piñera; b. 1949 (National Renewal; sworn in 11 March 2010).

In March 2013 the government comprised:

Minister of Agriculture: Luis Mayol. *Culture and the Arts:* Luciano Cruz-Coke. *Economy, Development and Tourism:* Pablo Longueira. *Education:* Harald Beyer. *Energy:* Jorge Bunster. *Environment:* María Ignacia Benitez. *Finance:* Felipe Larraín. *Foreign Affairs:* Alfredo Moreno. *Health:* Jaime Mañalich. *Housing and Urban Development, and National Heritage:* Rodrigo Pérez.

Interior and Public Security: Andrés Chadwick. *Justice:* Patricia Pérez. *Labour and Social Security:* Evelyn Matthei. *Mining:* Hernán de Solminihac. *National Defence:* Rodrigo Hinzpeter. *National Women's Service:* Carolina Schmidt. *Public Works:* María Loreto Silva. *Social Development:* Joaquín Lavín. *Transport and Telecommunications:* Pedro Pablo Errázuriz. *General Secretary of the Government:* Cecilia Pérez. *General Secretary of the Presidency:* Cristián Larroulet.

Government Website (Spanish only): http://www.gobiernodechile.cl

CURRENT LEADERS

Sebastián Piñera Echenique

Position
President

Introduction
Sebastián Piñera became president in March 2010 in the aftermath of a severe earthquake. A billionaire and right-of-centre politician, his victory ended 22 years of centre-left rule. He was expected to focus on strengthening an economy emerging from recession and increasing employment, while maintaining many existing social policies.

Early Life
Miguel Juan Sebastián Piñera Echenique was born on 1 Dec. 1949 in Santiago, the son of a diplomat. He grew up in Belgium, New York and Chile, graduated in economics from the Pontifical Catholic University of Chile in 1971 and then worked as a university tutor for two years. He undertook postgraduate studies at Harvard University from 1973–76, gaining a doctorate. Returning to Chile, he continued to teach economics while amassing personal wealth by establishing a credit card business and, from 1977–80, serving as general manager of the Bank of Talca.

In 1982 allegations of fraud relating to the bank led to the issue of an arrest warrant but the case against him was dropped. In 1989 Piñera entered politics to lead the presidential campaign of Hernán Büchi, a former finance minister. From 1990–98 Piñera was senator for East Santiago, representing the centre-right National Renewal party. During this period he served on the senate financial committee and in 1992 made an unsuccessful attempt to win the party's presidential candidacy. In 1993 he created the Future Foundation, later renamed the Foundation for Culture and Society, to develop policies on social justice, human rights and the environment.

From 2001–04 he was party president and in 2005 contested the presidential election, coming second behind Michelle Bachelet of the governing centre-left coalition. He made another run for the presidency in 2009, promising to increase economic growth, create a million new jobs and privatize 20% of state-owned Codelco (the world's largest copper-producing company). With a lead from the first round of voting on 13 Dec. 2009, he won the run-off with 52% of the vote on 17 Jan. 2010.

Career in Office
Piñera took office on 11 March 2010 as Chile struggled to cope with the impact of the previous month's earthquake when hundreds of people died and about half a million homes were destroyed. Restoring the social fabric was his immediate challenge—he unveiled an US$8·4bn. reconstruction plan in April —together with rebuilding the economy and returning to a balanced budget. Internationally, he signalled his intention to distance Chile from Venezuela and Cuba and to strengthen bonds with Colombia and other right-of-centre neighbours.

In Oct. 2010 he gained worldwide media exposure and considerable credit for his close involvement in a successful operation to rescue 33 Chilean miners trapped deep underground for over two months in a collapsed copper mine. However, his popularity declined sharply in 2011 and 2012, particularly in the face of mass student protests demanding more investment in state education. There has also been opposition to government plans for a hydroelectric project in Patagonia and calls for better state pension provision and labour law reform. In municipal elections in Oct. 2012 Piñera's ruling conservative coalition lost substantial ground to the centre-left opposition, although the voter turnout was low at 40%.

DEFENCE

Conscription is compulsory, but only when there are not enough voluntary recruits. It is currently a maximum of 14 months in the Army and in the Air Force and a maximum of two years in the Navy.

In 2008 defence expenditure totalled US$5,561m. (US$338 per capita), representing 3·3% of GDP. In 1985 defence spending had accounted for 10·0% of GDP.

Army
Strength (2005): 41,000 (18,366 conscripts) with 50,000 reserves. There is a 36,800-strong force of Carabineros.

Navy
The principal ships of the Navy are three ex-British destroyers, three diesel submarines and three frigates. There is a Naval Air Service numbering 600 personnel with 13 combat aircraft.

Naval personnel in 2005 totalled 20,092 (1,030 conscripts) including 3,800 marines and 1,300 Coast Guard. There are HQs at Iquique, Valparaíso, Talcahuano and Punta Arenas.

Air Force
Strength (2005) was 9,971 personnel (950 conscripts). There are 76 combat aircraft made up largely of Mirage jets.

ECONOMY

Agriculture accounted for 3·2% of GDP in 2010, industry 42·9% and services 53·9%.

According to the anti-corruption organization *Transparency International*, Chile ranked equal 20th in the world in a 2012 survey of the countries with the least corruption in business and government. It received 72 out of 100 in the annual index.

Overview
Chile's economy was liberalized under the Pinochet regime (1973–90) and reform continued under the democratic government of the 1990s. It had the highest foreign direct investment to GDP ratio in Latin America and strong growth for most of the decade. By 2011 foreign direct investment was US$15bn. and GDP growth has averaged 3·9% since the early 1990s.

In 2011 the government introduced a seven-point plan aimed at reducing poverty. The World Bank has helped fund public sector modernization, job creation and equality, along with the promotion of sustainable investments. Strong in mining, the country is the world's leading copper and iodine producer and, increasingly, a source of gold and non-metallic minerals. The iron processing, steel, timber and textiles industries are also important.

While manufacturing has gradually declined, fruit, fish, poultry, beef, wine and methanol production account for an increasing share of GDP, with exports making up a third of the total. In 2011, 22·8% of exports went to the USA, 11·1% to Japan and 5·1% to Brazil. Chile is signatory to 59 bilateral trading agreements with, amongst others, China, India, South Korea, Mexico and the EU.

In 1999 a floating exchange rate was introduced and a counter-cyclical fiscal policy implemented. In late 2007 growth slowed as the effects of the international financial crisis, along with falling copper prices, hit. Growth slowed in 2008 to its lowest level since 2002 before the economy entered recession in 2009. After an earthquake contributed to a further contraction of –3·1 % in the first quarter of 2010, President Sebastián Piñera initiated an ambitious reconstruction programme. The economy grew by around 6% in both 2010 and 2011 but poverty remained high at 15·1%.

Currency
The unit of currency is the *Chilean peso* (CLP) of 100 *centavos*. The peso was revalued 3·5% against the US dollar in Nov. 1994. In Sept. 1999 the managed exchange-rate system was abandoned and the peso allowed to float. Inflation rates (based on OECD statistics):

2002	2003	2004	2005	2006	2007	2008	2009	2010	2011
2·5%	2·8%	1·1%	3·1%	3·4%	4·4%	8·7%	0·4%	1·4%	3·3%

In Aug. 2009 gold reserves were 8,000 troy oz and foreign exchange reserves US$23,711m. Total money supply was 9,687·2bn. pesos in May 2009.

Budget
The fiscal year is the calendar year.

Budgetary central government revenue and expenditure (in 1bn. pesos):

	2005	2006	2007
Revenue	15,304·0	19,382·0	22,784·1
Expenditure	11,481·0	12,701·0	14,175·9

VAT is 19%.

Performance
Real GDP growth rates (based on OECD statistics):

2002	2003	2004	2005	2006	2007	2008	2009	2010	2011
2·2%	4·0%	6·8%	6·3%	5·8%	5·2%	3·1%	–0·9%	6·1%	5·9%

Total GDP in 2011 was US$248·6bn.

Banking and Finance
The Superintendencia de Bancos e Instituciones Financieras, affiliated to the finance ministry, is the banking supervisory authority. There is a Central Bank and a State Bank. The Central Bank was made independent of government control in March 1990. The *Governor* is Rodrigo Vergara. There were 21 domestic and six foreign banks in 2005. In Jan. 2005 deposits in domestic banks totalled 25,779,053m. pesos; in foreign banks, 2,163,183m. pesos, and in other finance companies, 4,829,090m. pesos.

External debt totalled US$86,349m. in 2010 and represented 45·9% of GNI.

There are stock exchanges in Santiago and Valparaíso.

ENERGY AND NATURAL RESOURCES

Environment
Chile's carbon dioxide emissions from the consumption and flaring of fossil fuels in 2008 were the equivalent of 3·9 tonnes per capita.

Electricity
Installed capacity was 15·9m. kW in 2007. Production of electricity was 58·51bn. kWh in 2007, of which 40% was hydro-electric. Consumption per capita in 2007 was 3,623 kWh.

Oil and Gas
Production of crude oil, 2007, was 123,000 tonnes. Natural gas production, 2007, was 1·3bn. cu. metres. Chile imports nearly all of its oil and around 70% of the natural gas that it consumes.

Minerals
The wealth of the country consists chiefly in its minerals. Chile is the world's largest copper producer; copper is the most important source of foreign exchange and government revenues. Production, 2004, 5,418,800 fine tonnes. Coal is low-grade and mining is

difficult, made possible by state subsidies. Production, 2004, 238,307 tonnes.

Output of other minerals, 2004 (in tonnes): iron ore, 8,003,491; limestone, 6,653,343; salt, 4,938,928; molybdenum, 41,883; zinc, 27,635; manganese, 7,188; silver, 1,360. Gold (39,986 kg in 2004), lithium, nitrate, iodine and sodium sulphate are also produced.

Agriculture

In 2007, 1·29m. ha. were arable land and 0·46m. ha. permanent crops. 0·96m. ha. were irrigated in 2007. There were 400 tractors and 66 harvester-threshers per 10,000 ha. of arable land in 2006.

Principal crops were as follows:

Crop	Area harvested, 1,000 ha 2004	Production, 1,000 tonnes 2004	Crop	Area harvested, 1,000 ha 2004	Production, 1,000 tonnes 2004
Sugar beets	31	2,598	Tomatoes	7	470
Wheat	420	1,852	Oats	77	357
Maize	134	1,508	Onions	6	290[1]
Potatoes	56	1,116	Rice	25	117

[1]2003.

Fruit production, 2004 (in 1,000 tonnes): apples, 1,300; grapes, 1,150; peaches and nectarines, 311; plums, 250; pears, 210; lemons and limes, 165; oranges, 140. Wine production in 2005 totalled 7,886,000 hectolitres.

Livestock, 2003: sheep, 4·1m.; cattle, 3·9m.; pigs, 3·2m.; goats, 1·0m.; horses, 700,000; poultry, 98m. Livestock products, 2003 (in 1,000 tonnes): pork, bacon and ham, 386; beef and veal, 190; poultry meat, 457; milk, 2,180; eggs, 116.

Since 1985 agricultural trade has been consistently in surplus. Wine exports rose from US$52m. in 1990 to US$844m. in 2004.

Forestry

In 2010, 16·23m. ha., or 22% of the total land area, was under forests. Of the total area under forests 27% was primary forest in 2010, 58% other naturally regenerated forest and 15% planted forest.

Fisheries

The catch in 2010 was 2,679,736 tonnes, the ninth highest in the world. Exports of fishery commodities in 2008 were valued at US$3·93bn., against imports of US$251m. Chile is one of the leading aquaculture producers, with 701,062 tonnes in 2010. Only Norway produces more farmed salmon. Chile is also a major producer of fishmeal, with 2009 production second only to that of Peru at 641,000 tonnes.

INDUSTRY

The leading companies by market capitalization in Chile in March 2012 were: Falabella, a retail company (US$23·3bn.); and Copec, an oil and gas producer (US$21·7bn.).

Output of major products in 2003 unless otherwise indicated (in 1,000 tonnes): cement (2009), 3,876; distillate fuel oil (2007), 3,623; sulphuric acid, 2,866; residual fuel oil (2007), 2,445; petrol (2007), 2,349; cellulose, 1,430; fishmeal, 580; iron or steel plates, 403; sugar (2004), 401; newsprint, 177; paper and cardboard, 113. Output of other products: soft drinks, 796m. litres; beer (2007), 550m. litres; 20,136 motor vehicles (2001); 4·74m. motor tyres.

Labour

In 2005 there were 5,779,660 people in employment (1,997,690 women). In Sept. 2005, 1,678,880 persons were employed in social or personal services, 1,117,900 in trade, 765,580 in manufacturing, 683,100 in agriculture, forestry and fisheries, 470,110 in transport and communications and 447,480 in building. In 2005 there was a monthly minimum wage of 127,500 pesos. In Nov. 2012, 6·5% of the civilian workforce was unemployed (compared to 7·1% in 2011 and 8·2% in 2010).

INTERNATIONAL TRADE

Imports and Exports

Trade in US$1m.:

	2004	2005	2006	2007	2008
Imports c.i.f.	24,794	32,735	38,406	47,164	61,903
Exports f.o.b.	32,520	41,267	58,680	67,666	66,456

In 2004 the principal exports were (in US$1m.): minerals, 16,633·6 (of which copper, 14,358·4, equivalent to 87·2% of all exports); manufactures, 11,928·7; and agricultural products, 2,339·3. Principal imports in 2004 were (in US$1m.): manufactures, 17,928·7; minerals, 3,919·6; and agricultural products, 416·9. Major import suppliers (as % of total), 2004: Argentina, 18·5; USA, 15·1; Brazil, 12·4; China, 8·3; Germany 3·7. Major export markets, 2004: USA, 14·8; Japan, 12·0; China, 10·4; South Korea, 5·8; Netherlands, 5·4.

COMMUNICATIONS

Roads

In 2004 there were 80,505 km of roads, but only 20·8% were hard-surfaced. There were 2,414 km of motorways and 16,785 km of main roads. In 2007 there were 1,701,036 passenger cars, 849,282 trucks and vans, 170,217 buses and coaches and 63,257 motorcycles and mopeds. In 2006 there were 2,280 road accident fatalities.

Rail

The total length of railway lines was (2004) 5,775 km, including 1,051 km electrified, of broad- and metre-gauge. The state railway (EFE) transported 11·3m. passengers in 2005. Freight operations are in the hands of the semi-private companies Ferronor, Pacifico and the Antofagasta (Chili) and Bolivia Railway (973 km, metre-gauge) which links the port of Antofagasta with Bolivia and Argentina. Passenger-km travelled in 2008 came to 759m. and freight tonne-km in 2006 to 3,660m.

There are metro systems in Santiago (46·2 km) and Valparaíso (42·5 km).

Civil Aviation

There are 389 airports, with nine international airports at Antofagasta, Arica, Coihaique, Concepción, Easter Island (Isla de Pascua), Iquique, Puerto Montt, Punta Arenas and Santiago (Comodoro Arturo Merino Benítez). The largest airline is LATAM, created in June 2012 when the Chilean carrier LAN Airlines (formerly LAN-Chile) merged with TAM, Brazil's largest airline. LATAM controls 40% of the Latin American market and serves 150 destinations in 22 countries. In 2004 Santiago handled 6,057,279 passengers (3,603,267 on international flights) and 269,660 tonnes of freight.

Shipping

In Jan. 2009 there were 89 ships of 300 GT or over registered, totalling 632,000 GT. The leading ports are Antofagasta, Arica, Iquique, Puerto Ventanas, San Antonio, Talcahuano/San Vicente and Valparaíso.

Telecommunications

In 2008 there were 3,252,000 main (fixed) telephone lines. In the same year mobile phone subscribers numbered 14,797,000 (880·5 per 1,000 persons). There were 5·7m. internet users in 2006. In June 2012 there were 9·4m. Facebook users.

SOCIAL INSTITUTIONS

Justice

There is a High Court of Justice in the capital, 16 courts of appeal distributed over the republic, courts of first instance in the departmental capitals and second-class judges in the sub-delegations. There were 642 public prosecutors, 782 judges and 417 defence lawyers in 2002.

The population in penal institutions in April 2011 was 52,563 (305 per 100,000 of national population).

The death penalty for ordinary crimes was abolished in 2001.

Education

In 2007–08 there were 402,000 children at pre-primary schools, 1·66m. primary school pupils and 1·59m. pupils at secondary level. Adult literacy rate in 2008 was 99%.

In 2004 there were 567,114 students in higher education. There were 162 universities with 403,370 students, 140 professional institutes with 101,674 students and 211 technical education centres with 62,070 students. The number of students at higher education institutions has doubled since 1990.

In 2007 public expenditure on education came to 3·8% of GNI and represented 18·2% of total government expenditure.

Health

There were 418 hospitals in 2008. In 2003 there were 15,006 doctors, 2,846 dentists and 6,900 university nurses in the public sector. In 2004 there were 20,776 junior doctors.

Welfare

In 1981 Chile abolished its state-sponsored pension plan and became the first country to establish private mandatory retirement savings. The system is managed by competitive private companies called AFPs (Pension Fund Administrators). Employees are required to save 13% of their pay. In April 2005 it had 7,132,983 members and assets of 35,051,470m. pesos. In 2005 about 65% of the population over the age of 14 had private health insurance.

RELIGION

At the 2002 census Chile had 7,853,428 Roman Catholics. In Jan. 2002 there were five archbishops, 25 bishops and two vicars apostolic. There were two cardinals in Feb. 2013. In 2002 there were 1,699,725 Evangelical Christians, 119,455 Jehovah's Witnesses, 103,735 Latter-day Saints, 14,976 Jews, 6,959 Orthodox Christians and 2,894 Muslims.

CULTURE

World Heritage Sites

Chile's five UNESCO protected sites are the Rapa Nui National Park, the Churches of Chiloé, the Historic Quarter of the Seaport of Valparaíso, the Humberstone and Santa Laura Saltpeter Works, and the Sewell Mining Town. Entered on the list in 1995, the Rapa Nui National Park encompasses much of the coastline of Easter Island and protects the shrines and statues (*Moai*) carved between the 10th–16th centuries. On the island of Chiloé off the Región de Los Lagos coastline, wooden churches were built by Jesuit missionaries at the turn of the 17th century. The churches were entered on the list in 2000. The Valparaíso site was inscribed on the list in 2003 as a model of urban and architectural development in 19th-century Latin America. The Humberstone and Santa Laura Saltpeter Works were inscribed in 2005 and are where workers from Peru, Chile and Bolivia formed a distinctive communal pampinos culture. The Sewell Mining Town, inscribed in 2006, is an example of the company towns that existed in the early 20th century in many remote parts of the world.

Press

In 2008 there were 57 daily newspapers (55 paid-for and two free) and 41 non-dailies (37 paid-for and four free). The dailies had a combined average daily circulation of 859,000 in 2008. The papers with the highest circulation are *La Tercera*, *El Mercurio* and *La Cuarta*.

Tourism

There were 2,749,913 non resident overnight tourists in 2009. Tourist receipts were US$2,172m. in 2007.

DIPLOMATIC REPRESENTATIVES

Of Chile in the United Kingdom (37–41 Old Queen St., London, SW1H 9JA)
Ambassador: Tomás Müller Sproat.

Of the United Kingdom in Chile (Av. El Bosque Norte 0125, Piso 2, Las Condes, Santiago)
Ambassador: Jon Benjamin.

Of Chile in the USA (1732 Massachusetts Ave., NW, Washington, D.C., 20036)
Ambassador: Felipe Bulnes.

Of the USA in Chile (Av. Andrés Bello 2800, Las Condes, Santiago)
Ambassador: Alejandro. D. Wolff.

Of Chile to the United Nations
Ambassador: Octavio Errazuriz Guilisasti.

Of Chile to the European Union
Ambassador: Carlos Appelgren Balbontín.

FURTHER READING

Bizzarro, Salvatore, *Historical Dictionary of Chile*. 2005
Collier, S. and Sater, W. F., *A History of Chile, 1808–1994*. 1996
Hojman, D. E., *Chile: the Political Economy of Development and Democracy in the 1990s*. 1993.—(ed.) *Change in the Chilean Countryside: from Pinochet to Aylwin and Beyond*. 1993
Oppenheim, L. H., *Politics in Chile: Democracy, Authoritarianism and the Search for Development*. 1993
Rector, John L., *The History of Chile*. 2006

National Statistical Office: Instituto Nacional de Estadísticas (INE), Paseo Bulnes 418, Santiago.
Website: http://www.ine.cl

CHINA

© Research Machines plc 2006

Zhonghua Renmin Gonghe Guo (People's Republic of China)

Capital: Beijing (Peking)
Population projection, 2015: 1,369·74m.
GNI per capita, 2011: (PPP$) 7,476
HDI/world rank: 0·687/101
Internet domain extension: .cn

KEY HISTORICAL EVENTS

An embryonic Chinese state emerged in the fertile Huang He (Yellow River) basin before 4000 BC. Chinese culture reached the Chang Jiang (Yangtze) basin by 2500 BC and within 500 years the far south was also within the Chinese orbit. Four thousand years ago the Xia dynasty ruled in the Huang He basin. About 1500 BC it was supplanted by the Shang dynasty, the cultural ancestor of modern China.

Shang civilization spread out from the Huang He region. In the west, the Shang came into conflict with the Zhou, whose rulers replaced the Shang dynasty around 1000 BC. Under the Zhou, a centralized administration developed. In about 500 BC one court official, Kongfuzi (Confucius), outlined his vision of society. Confucianism, which introduced a system of civil service recruitment through examination, remained dominant until the mid-20th century.

The Zhou expanded the Chinese state south beyond the Chang Jiang. Dependent territories periodically rebelled against the central authority. In 221 BC the ruler of Qin became the first emperor of China. He built an empire extending from the South China Sea to the edge of Central Asia where work was begun on the Great Wall of China, a massive fortification to keep threatening nomads at bay. The Qin dynasty standardized laws, money and administration throughout the empire but it was short-lived. By 206 BC the state had divided into three.

Reunification came gradually under the Han dynasty (202 BC–AD 200) with its efficient, centralized bureaucracy. A nation with boundaries similar to those of modern China was created. But peripheral territories proved too distant to hold and the Han empire fell to rebellion and invasion. It was followed by the Jin (265–316) and Sui (589–612) dynasties, interspersed by internecine war and anarchy. Reunification was achieved by the Tang dynasty which brought prosperity to China from 618–917. Eventually the Tang empire fell victim to separatism.

Under the Song (960–1127), the balance of power shifted south. The Song state lost control of the area north of the Chang Jiang in 1126 when nomads from Manchuria invaded. A declining Song empire survived in the south until 1279.

Genghis Khan

The northern invaders were overthrown by the Mongols, led by Genghis Khan (c. 1162–1227), who went on to claim the rest of China. In 1280 Kublai Khan (1251–94), who had founded the Yuan dynasty in 1271, swept into southern China. The Mongol Yuan dynasty adopted Chinese ways but was overthrown by a nationalist uprising in 1368, led by Hongwu (1328–98), a former beggar who established the Ming dynasty.

The Ming empire collapsed in a peasants' revolt in 1644. The capital, Beijing (Peking), was only 64 km from the Great Wall and vulnerable to attack from the north. Within months the peasants' leader was swept aside by the invasion of the Manchus, whose Qing dynasty ruled China until 1911. Preoccupied with threats from the north, China neglected its southern coastal frontier where European traders were attempting to open up the country. The Portuguese, who landed on the Chinese coast in 1516, were followed by the Dutch in 1622 and the English in 1637.

The Qing empire expanded into Mongolia, Tibet, Vietnam and Kazakhstan. But by the 19th century, under pressure from rural revolts ignited by crippling taxation and poverty, the Qing dynasty was crumbling. Two Opium Wars (1838–42; 1856–58) forced China to allow the import of opium from India into China, while Britain, France, Germany and other European states gained concessions in 'treaty ports' that virtually came under foreign rule.

The Taiping Rebellion (1851–64) set up a revolutionary egalitarian state in southern China. The European powers intervened to crush the rebellion and in 1860 British and French forces invaded Beijing and burnt the imperial palace. Further trading concessions were demanded. A weakened China was defeated by Japan in 1895 and lost both Taiwan and Korea.

The xenophobic Boxer Rebellion, led by members of a secret society called the Fists of Righteous Harmony, broke out in 1900. The Guangxu emperor (1875–1908) attempted modernization in the Hundred Days Reform, but was taken captive by the conservative dowager empress who harnessed the Boxer Rebellion to her own ends. The rebellion was put down by European troops in 1901. China was then divided into zones of influence between the major European states and Japan.

With imperial authority so weakened, much of the country was ungovernable and ripe for rebellion. The turning point came in 1911 when a revolution led by the Kuomintang (Guomintang or Nationalist movement) of Sun Yet-sen (Sun Zhong Shan; 1866–1925) overthrew the emperor and the imperial system. The authoritarian Yuan Shih-kai ruled as president from 1913 to 1916. Following the overthrow of Yuan, China disintegrated into warlord anarchy.

In 1916 Sun founded a republic in southern China but the north remained beyond his control. Reorganizing the Nationalist party on Soviet lines, Sun co-operated with the Communists to re-establish national unity. But rivalry between the two parties increased, particularly after the death of Sun in 1925.

Nationalism and Communism

After Sun's death the nationalist movement was taken over by his ally Chiang Kai-shek (Jiang Jie Shi; 1887–1976). As commander in chief of the Nationalist army from 1925, Chiang's power grew. In April 1927 he tried to suppress the Chinese Communist Party in a bloody campaign in which thousands of Communists were slaughtered. The survivors fled to the far western province of Jiangxi. In 1928 Chiang's army entered Beijing. With the greater part of the country reunited under Chiang's rule, he made Nanjing the capital of China.

In 1934 the Communists were forced to retreat from Jiangxi province. Led by Mao Zedong (Mao Tse-tung; 1893–1976) they trekked for more than a year on the 5,600-mile Long March. Harried by the Nationalists they eventually took refuge in Shaanxi province.

In 1931 the Japanese invaded Manchuria. By 1937 they had seized Beijing and most of coastal China. The Nationalists and Communists finally co-operated against the invader but were unable to achieve much against the superior Japanese forces.

During the Second World War (1939–45), a Nationalist government ruled unoccupied China ineffectually from a temporary capital in Chongqing. At the end of the war, Nationalist-Communist co-operation was short-lived. The Soviet Union sponsored the Communist Party, which marched into Manchuria in 1946. So began the civil war which lasted until 1949. Although the Nationalist forces of Chiang Kai-shek received support from some western countries, particularly the United States, the Communists were victorious. On 1 Oct. 1949 Mao declared the People's Republic of China in Beijing.

Chiang fled with the remains of his Nationalist forces to the island of Taiwan, where he established a government that claimed to be a continuation of the Republic of China. At first, that administration was recognized as the government of China by most Western countries and Taiwan kept China's Security Council seat at the United Nations until 1971. Chiang's authoritarian regime was periodically challenged by Red China, which bombed Taiwan's small offshore islands near the mainland. In the 1960s and 1970s, Taiwan gradually lost recognition as the legitimate government of China and in 1978 the USA recognized the People's Republic of China.

Expansionism

In 1950 China invaded Tibet, independent since 1916. Chinese rule quickly alienated the Tibetans who rebelled in 1959. The Tibetan religious leader, the Dalai Lama, was forced to flee to India. Since then, the settlement of large numbers of ethnic Chinese in the main cities of Tibet has threatened to swamp Tibetan culture.

During the 1950s and 1960s China was involved in a number of border disputes and wars in neighbouring states. The Communists posted 'volunteers' to fight alongside Communist North Korea during the Korean War (1950–53). There were clashes on the Soviet border in the 1950s and the Indian border in the 1960s, when China occupied some Indian territory.

From the establishment of the Peoples' Republic of China, Communist China and the Soviet Union were allies. Communist China initially depended upon Soviet assistance for economic development. A Soviet-style five-year plan was put into action in 1953, but the relationship with Moscow was already showing signs of strain. The two Communist powers fell out over interpretations of Marxist orthodoxy. By the end of the 1950s the Soviet Union and China were rivals, spurring the Chinese arms race. Chinese research into atomic weapons culminated in the testing of the first Chinese atomic bomb in 1964.

Mao introduced rapid collectivization of farms in 1955. The plan was not met with universal approval in the Communist Party but its implementation demonstrated Mao's authority over the fortunes of the nation. In 1956 he launched the doctrine of letting a 'hundred flowers bloom', encouraging intellectual debate. However, the new freedoms took a turn Mao did not expect and led to the questioning of the role of the party. Strict controls were reimposed and free-thinkers were sent to work in the countryside to be 're-educated'.

In May 1958 Mao launched another ill-fated policy, the Great Leap Forward. To promote rapid industrialization and socialism, the collectives were reorganized into larger units. Neither the resources nor trained personnel were available for this huge task. Backyard blast furnaces were set up to increase production of iron and steel. The Great Leap Forward was a disaster. It is believed that 30m. died from famine. Soviet advice against the project was ignored and a breakdown in relations with Moscow came in 1963, when Soviet assistance was withdrawn. A rapprochement with the United States was achieved in the early 1970s.

Cultural Revolution

Having published his 'Thoughts' in the 'Little Red Book' in 1964, Mao set the Cultural Revolution in motion. Militant students were organized into groups of Red Guards to attack the party hierarchy. Anyone perceived to lack enthusiasm for Mao Zedong Thought was denounced. Thousands died as the students lost control and the army was eventually called in to restore order.

After Mao's death in 1976 the Gang of Four, led by Mao's widow Jiang Qing, attempted to seize power. These hard-liners were denounced and arrested. China effectively came under the control of Deng Xiaoping, despite the fact that he held none of the great offices of state. Deng pursued economic reform. The country was opened to Western investment. Special Economic Zones and 'open cities' were designated and private enterprise gradually returned, on a small scale at first.

Improved standards of living and a thriving economy increased expectations for civil liberties. The demand for political change climaxed in demonstrations by workers and students in April 1989, following the funeral of Communist Party leader Hu Yaobang. Protests were held in several major cities. In Beijing where demonstrators peacefully occupied Tiananmen Square, they were evicted by the military who opened fire, killing more than 1,500. Hard-liners took control of the government, and martial law was imposed from May 1989 to Jan. 1990.

Since 1989 the leadership has concentrated on economic development. Hong Kong was returned to China from British rule in 1997 and Macao from Portuguese rule in 1999. The late 1990s saw a cautious extension of civil liberties but Chinese citizens are still denied most basic political rights.

Beijing was chosen for the 2008 Olympic Games. China's treatment of Tibet came under the international spotlight in the build-up to the games, following violent protests in Tibet's capital city, Lhasa.

The arrest by Japan of a Chinese trawler in disputed waters in 2010 marked the beginning of heightened tensions between the two nations in the East and South China Seas. In 2011 China became the world's second largest national economy. In Nov. 2012 the Communist Party congress selected Xi Jinping to succeed Hu Jintao as president.

For the background to the handover of Hong Kong in 1997, *see* page 333.

TERRITORY AND POPULATION

China is bounded in the north by Russia and Mongolia; east by North Korea, the Yellow Sea and the East China Sea, with Hong Kong and Macao as enclaves on the southeast coast; south by Vietnam, Laos, Myanmar, India, Bhutan and Nepal; west by India, Pakistan, Afghanistan, Tajikistan, Kyrgyzstan and Kazakhstan. The total area (including Taiwan, Hong Kong and Macao) is estimated at 9,572,900 sq. km (3,696,100 sq. miles). A law of Feb. 1992 claimed the Spratly, Paracel and Diaoyutasi Islands. An agreement of 7 Sept. 1993 at prime ministerial level settled Sino-Indian border disputes which had first emerged in the war of 1962.

China's sixth national census was held on 1 Nov. 2010. According to preliminary results, the total population of the 31 provinces, autonomous regions and municipalities and of servicemen on the mainland was 1,339,724,852 (652,872,280 females, representing 48·73%); density, 140 per sq. km. China's population in 2010 represented 19% of the world's total population. The population rose by 73,899,804 (or 5·84%) since the census in 2000. There were 665,575,306 urban residents, accounting for 49·68% of the population; compared to the 2000 census, the proportion of urban residents rose by 13·46% (reflecting the increasing migration from the countryside to towns and cities since the economy was opened up in the late 1970s).

China has a fast-growing ageing population. Whereas in 1980 only 5·2% of the population was aged 65 or over and by 2010 this had increased to 8·9%, by 2030 it is expected to rise to 16·5%. Long-term projections suggest that in 2050 as much as 23·8% of the population will be 65 or older. The population is expected to peak around 2025 and then begin to decline to such an extent that by around 2043 it will be back to the 2010 level. China is set to lose its status as the world's most populous country to India in about 2026.

The UN gives a projected population for 2015 of 1,369·74m.

1979 regulations restricting married couples to a single child, a policy enforced by compulsory abortions and economic sanctions, have been widely ignored, and it was admitted in 1988 that the population target of 1,200m. by 2000 would have to be revised to 1,270m. Since 1988 peasant couples have been permitted a second child after four years if the first born is a girl, a measure to combat infanticide. In 1999 China started to implement a more widespread gradual relaxation of the one-child policy.

An estimated 55m. persons of Chinese origin lived abroad in 2005.

A number of widely divergent varieties of Chinese are spoken. The official 'Modern Standard Chinese' is based on the dialect of North China. Mandarin in one form or another is spoken by 885m. people in China, or around 70% of the population of mainland China. The Wu language and its dialects has some 77m. native speakers and Cantonese 66m. The ideographic writing system of 'characters' is uniform throughout the country, and has undergone systematic simplification. In 1958 a phonetic alphabet (*Pinyin*) was devised to transcribe the characters, and in 1979 this was officially adopted for use in all texts in the Roman alphabet. The previous transcription scheme (Wade) is still used in Taiwan and Hong Kong.

Mainland China is administratively divided into 22 provinces, five autonomous regions (originally entirely or largely inhabited by ethnic minorities, though in some regions now outnumbered by Han immigrants) and four government-controlled municipalities. These are in turn divided into 332 prefectures, 658 cities (of which 265 are at prefecture level and 393 at county level), 2,053 counties and 808 urban districts.

Government controlled municipalities	Area (1,000 sq. km)	2010 census population (1,000, provisional)	Density per sq. km (2010, provisional)	Capital
Beijing	16·8	19,612	1,167	—
Chongqing	82·0	28,846	352	—
Shanghai	6·2	23,019	3,713	—
Tianjin	11·3	12,938	1,145	—
Provinces				
Anhui	139·9	59,501	425	Hefei
Fujian	123·1	36,894	300	Fuzhou
Gansu[1]	366·5	25,575	70	Lanzhou
Guangdong[1]	197·1	104,303	529	Guangzhou
Guizhou[1]	174·0	34,746	200	Guiyang
Hainan[1]	34·3	8,672	253	Haikou
Hebei[1]	202·7	71,854	354	Shijiazhuang
Heilongjiang[1]	463·6	38,312	83	Haerbin
Henan	167·0	94,024	563	Zhengzhou
Hubei[1]	187·5	57,238	305	Wuhan
Hunan[1]	210·5	65,684	312	Changsha
Jiangsu	102·6	78,660	767	Nanjing
Jiangxi	164·8	44,567	270	Nanchang
Jilin[1]	187·0	27,462	147	Changchun
Liaoning[1]	151·0	43,746	290	Shenyang
Qinghai[1]	721·0	5,627	8	Xining
Shaanxi	195·8	37,327	191	Xian
Shandong	153·3	95,793	625	Jinan
Shanxi	157·1	35,712	227	Taiyuan
Sichuan[1]	487·0	80,418	165	Chengdu
Yunnan[1]	436·2	45,966	105	Kunming
Zhejiang[1]	101·8	54,427	535	Hangzhou
Autonomous regions				
Guangxi Zhuang	220·4	46,027	209	Nanning
Inner Mongolia	1,177·5	24,706	21	Hohhot
Ningxia Hui	66·4	6,301	95	Yinchuan
Tibet[2]	1,221·6	3,002	2	Lhasa
Xinjiang Uighur	1,646·9	21,813	13	Urumqi

[1]Also designated minority nationality autonomous area.
[2]See also Tibet below.

Population of largest cities in 2000: Shanghai, 14·23m.; Beijing (Peking), 10·30m.; Guangzhou (Canton), 7·55m.; Tianjin, 6·84m.; Wuhan, 6·79m.; Shenzhen, 6·48m.; Chongqing, 5·09m.; Shenyang, 4·60m.; Chengdu, 4·27m.; Foshan, 4·01m.; Xian, 3·87m.; Dongguan, 3·87m.; Nanjing, 3·78m.; Haerbin, 3·63m.; Hangzhou, 3·24m.; Shantou, 3·07m.; Dalian, 2·87m.; Jinan, 2·80m.; Changchun, 2·75m.; Qingdao, 2·72m.; Kunming, 2·55m.; Taiyuan, 2·54m.; Zhengzhou, 2·50m.; Changsha, 2·12m.; Fuzhou, 2·03m.; Shijiazhuang, 1·94m.; Zibo, 1·93m.; Lanzhou, 1·91m.; Guiyang, 1·89m.; Wuxi, 1·87m.; Suzhou, 1·75m.; Urumqi (Wulumuqi), 1·73m.; Ningbo, 1·70m.; Nanchang, 1·68m.; Nanning, 1·67m.; Tangshan, 1·66m.; Wenzhou, 1·58m.; Hefei, 1·55m.; Changzhou, 1·51m.

China has 56 ethnic groups. According to the 2010 census 1,225,932,641 people (91·51%) were of Han nationality and 113,792,211 (8·49%) were from national minorities (including Zhuang, Manchu, Hui, Miao, Uighur, Yi, Tujia, Mongolian and Tibetan). Compared with the 2000 census, the Han population increased by 66,537,177 (5·74%), while the ethnic minorities increased by 7,362,627 (6·92%). Non-Han populations predominate in the autonomous regions, most notably in Tibet where national minorities accounted for 96·8% of the population in 2006.

Chang, Chiung-Fang, Lee, Che-Fu, McKibben, Sherry L., Poston, Dudley L. and Walther, Carol S. (eds.) *Fertility, Family Planning and Population Policy in China*. 2009

Li Chengrui, *A Study of China's Population*. 1992

Zhao, Zhongwei and Guo, Fei, *Transition and Challenge: China's Population at the Beginning of the 21st Century*. 2007

Tibet

After the 1959 revolt was suppressed, the Preparatory Committee for the Autonomous Region of Tibet (set up in 1955) took over the functions of local government, led by its Vice-Chairman, the Panchen Lama, in the absence of its Chairman, the Dalai Lama, who had fled to India in 1959. In Dec. 1964 both the Dalai and Panchen Lamas were removed from their posts and on 9 Sept. 1965 Tibet became an Autonomous Region. 301 delegates were elected to the first People's Congress, of whom 226 were Tibetans. The senior spiritual leader, the Dalai Lama (who announced in March 2011 that he was relinquishing his political role as head of the Tibetan government-in-exile), has remained abroad. He was awarded the Nobel Peace Prize in 1989. Following the death of the 10th Panchen Lama (Tibet's second most important spiritual leader) in Jan. 1989, the Dalai Lama announced Gendun Choekyi Nyima (b. 1989) as the 11th Panchen Lama in May 1995. Beijing rejected the choice and appointed Gyaltsen Norbu (b. 1989) in his place. Gendun Choekyi Nyima has been missing since 1995. The borders were opened for trade with neighbouring countries in 1980. In July 1988 Tibetan was reinstated as a 'major official language', competence in which is required of all administrative officials. Monasteries and shrines have been renovated and reopened. There were some 46,000 monks and nuns in 2007. In 1984 a Buddhist seminary in Lhasa opened with 200 students. A further softening of Beijing's attitude towards Tibet was shown during President Bill Clinton's visit to China in June 1998. Jiang Zemin, China's then president, said he was prepared to meet the Dalai Lama provided he acknowledged Chinese sovereignty over Tibet and Taiwan. In Sept. 2002 direct contact between the exiled government and China was re-established after a nine-year gap.

In March 2008 anti-Chinese protests in Lhasa, the regional capital, ended in violence, with dozens reportedly killed by the Chinese authorities. The episode focused international attention on China's human rights record ahead of the 2008 Olympic Games in Beijing.

The estimated population of Tibet in 2007 was 2·74m., of which 95% were Tibetans and the remainder from other ethnic groups. At the 2010 census the population had risen to 3·00m. The average population density was 2·26 persons per sq. km in 2008, although the majority of residents live in the southern and eastern parts of the region. Birth rate (per 1,000), 2006, 17·4; death rate,

5·7. Estimated population of the Lhasa (capital) region in 2007 was 460,470.

About 80% of the population is engaged in the dominant industries of farming and animal husbandry. In 2006 the total sown area was 233,000 ha. (including 171,700 ha. of grain crops). Output in 2006: total grain crops, 923,700 tonnes; vegetables, 450,000 tonnes. In 2005 there were 10·7m. sheep, 6·3m. cattle and buffaloes, 6·3m. goats and 0·4m. horses.

Tibet has over 2,000 mineral ore fields. Minerals output, 2006: copper, 1,260,000 tonnes; vanadium, 415,000 tonnes; chrome, 121,800 tonnes. Cement production, 2006: 1·66m. tonnes. Timber output, 2007: 356,800 cu. metres.

In 2006 there were 44,813 km of roads (21,842 km in 1990). There are airports at Lhasa, Bangda and (since 2006) Nyingchi providing external links. In 2006, 154,800 foreign tourists visited Tibet. In July 2006 a 1,142-km railway linking Lhasa with the town of Golmud opened. It is the highest railway in the world. Direct services have subsequently been introduced between Lhasa and a number of major Chinese cities, including Beijing and Shanghai. An extension from Lhasa to Shigatse, Tibet's second largest city, is under construction and scheduled for completion in 2014.

In Dec. 2006 Tibet had 890 elementary schools and 1,568 teaching centres (with 329,500 pupils), 93 junior middle schools (127,900 pupils), 13 senior middle schools (37,700 pupils) and ten secondary vocational schools (14,775 pupils). There were also six higher education institutes (including Tibet University, Tibet Nationalities Institute, Tibet Agriculture and Animal Husbandry College, and Institute of Tibetan Medicines) with 23,327 enrolled students. The illiteracy rate of young and middle-aged people fell from 39% in 2000 to below 10% in 2006.

In 2006 there were 10,746 medical personnel (including 4,310 doctors and 2,000 registered nurses) and 1,349 medical institutions, with a total of 7,496 beds.

Lixiong, Wang and Shakya, Tsering, *The Struggle for Tibet*. 2009

Margolis, Eric, *War at the Top of the World: The Struggle for Afghanistan, Kashmir and Tibet*. 2001

Shakya, Tsering, *The Dragon in the Land of Snows: The History of Modern Tibet since 1947*. 1999

Smith, W. W., *A History of Tibet: Nationalism and Self-Determination*. 1996

Van Schaik, Sam, *Tibet: A History*. 2011

SOCIAL STATISTICS—CHINA

Births, 2005, 16,210,000; deaths, 8,510,000. 2005 birth rate (per 1,000 population), 12·40; death rate, 6·51. In 2005 the birth rate rose for the first time since 1987. There were 9,450,000 marriages and 1,893,000 divorces in 2006. In April 2001 parliament passed revisions to the marriage law prohibiting bigamy and cohabitation outside marriage. The World Health Organization estimated in 2005 that the suicide rate in China was about 17·4 per 100,000 population. China is the only major country in which the suicide rate is higher among females—over half the world's female suicides occur in China. Life expectancy at birth, 2007, was 71·3 years for men and 74·7 years for women. Infant mortality, 2010, 16 per 1,000 live births. Fertility rate, 2008, 1·8 births per woman. Annual population growth rate, 2000–10, 0·6%. According to the World Bank, the number of people living in poverty (less than US$1·25 a day) at purchasing power parity declined from 835m. in 1981 to 173m. in 2009.

CLIMATE

Most of China has a temperate climate but, with such a large country, extending far inland and embracing a wide range of latitude as well as containing large areas at high altitude, many parts experience extremes of climate, especially in winter. Most rain falls during the summer, from May to Sept., though amounts decrease inland. Monthly average temperatures and annual rainfall (2006): Beijing (Peking), Jan. 28·6°F (−1·9°C), July 78·6°F (25·9°C). Annual rainfall 12·5" (318 mm). Chongqing, Jan. 46·0°F

(7·8°C), July 87·8°F (31·0°C). Annual rainfall 33·1" (840 mm). Shanghai, Jan. 42·3°F (5·7°C), July 84·9°F (29·4°C). Annual rainfall 45·3" (1,150 mm). Tianjin, Jan. 27·1°F (−2·7°C), July 78·6°F (25·9°C). Annual rainfall 16·3" (415 mm).

CONSTITUTION AND GOVERNMENT

On 21 Sept. 1949 the *Chinese People's Political Consultative Conference* met in Beijing, convened by the Chinese Communist Party. The Conference adopted a 'Common Programme' of 60 articles and the 'Organic Law of the Central People's Government' (31 articles). Both became the basis of the Constitution adopted on 20 Sept. 1954 by the 1st National People's Congress, the supreme legislative body. The Consultative Conference continued to exist after 1954 as an advisory body. Three further constitutions have been promulgated under Communist rule—in 1975, 1978 and 1982 (currently in force). The latter was partially amended in 1988, 1993 and 1999, endorsing the principles of a socialist market economy and of private ownership.

The unicameral *National People's Congress* is the highest organ of state power. Usually meeting for one session a year, it can amend the constitution and nominally elects and has power to remove from office the highest officers of state. There are a maximum of 3,000 members of the Congress (and currently 2,987), who are elected to serve five-year terms by municipal, regional and provincial people's congresses. The Congress elects a *Standing Committee* (which supervises the *State Council*) and the *President* and *Vice-President* for a five-year term. When not in session, Congress business is carried on by the Standing Committee.

The State Council is the supreme executive organ and comprises the Prime Minister, Deputy Prime Ministers and State Councillors.

The *Central Military Commission* is the highest state military organ.

National Anthem

'March of the Volunteers'; words by Tien Han, tune by Nie Er.

GOVERNMENT CHRONOLOGY

Leaders of the Communist Party of China since 1935.

Chairmen
1935–76	Mao Zedong
1976–81	Hua Guofeng
1981–82	Hu Yaobang

General Secretaries
1956–57	Deng Xiaoping
1980–87	Hu Yaobang
1987–89	Zhao Ziyang
1989–2002	Jiang Zemin
2002–12	Hu Jintao
2012–	Xi Jinping

De facto ruler
1978–97	Deng Xiaoping

Heads of State since 1949.

Chairman of the Central People's Government
1949–54	Mao Zedong

Chairmen (Presidents)
1954–59	Mao Zedong
1959–68	Liu Shaoqi
1968–75	Dong Biwu

Chairmen of the Standing Committee of the National People's Congress
1975–76	Zhu De
1978–83	Ye Jianying

Presidents of the Republic
1983–88	Li Xiannian
1988–93	Yang Shangkun

1993–2003	Jiang Zemin
2003–13	Hu Jintao
2013–	Xi Jinping

Prime Ministers since 1949.
1949–76	Zhou Enlai
1976–80	Hua Guofeng
1980–87	Zhao Ziyang
1987–1998	Li Peng
1998–2003	Zhu Rongji
2003–13	Wen Jiabao
2013–	Li Keqiang

RECENT ELECTIONS

Elections of delegates to the 12th National People's Congress were held between Oct. 2012 and Feb. 2013 by municipal, regional and provincial people's congresses. At its annual session in March 2013 the Congress elected Xi Jinping as President and Li Yuanchao as Vice-President.

CURRENT GOVERNMENT

President and Chairman of Central Military Commission: Xi Jinping; b. 1953 (Chinese Communist Party; elected 14 March 2013).

Deputy President: Li Yuanchao.

In March 2013 the government comprised:

Premier of the State Council (Prime Minister): Li Keqiang; b. 1955 (Chinese Communist Party; appointed 15 March 2013).

Deputy Prime Ministers: Zhang Gaoli; Liu Yandong; Wang Yang; Ma Kai.

Minister of Agriculture: Han Changfu. *Civil Administration:* Li Liguo. *Commerce:* Gao Hucheng. *Culture:* Cai Wu. *Education:* Yuan Guiren. *Environmental Protection:* Zhou Shengxian. *Finance:* Lou Jiwei. *Foreign Affairs:* Wang Yi. *Health:* Chen Zhu. *Housing, and Urban and Rural Development:* Jiang Weixin. *Human Resources and Social Security:* Yin Weimin. *Industry and Information:* Miao Wei. *Justice:* Wu Aiying. *Land and Resources:* Jiang Daming. *National Defence:* Chang Wanquan. *Public Security:* Guo Shengkun. *Science and Technology:* Wan Gang. *State Security:* Geng Huichang. *Supervision:* Huang Shuxian. *Transport:* Yang Chuantang. *Water Resources:* Chen Lei.

Ministers heading State Commissions: *Ethnic Affairs*, Wang Zhengwei. *National Development and Reform*, Xu Shaoshi. *National Population and Family Planning*, Li Bin.

De facto power is in the hands of the Communist Party of China, which had 80·27m. members in 2010. There are eight other parties, all members of the Chinese People's Political Consultative Conference.

The members of the Standing Committee of the Politburo in March 2013 were: Xi Jinping (*General Secretary*); Li Keqiang; Zhang Dejiang; Yu Zhengsheng; Liu Yunshan; Wang Qishan; Zhang Gaoli.

Government Website: http://www.gov.cn

CURRENT LEADERS

Xi Jinping

Position
President

Introduction
Xi Jinping succeeded Hu Jintao as president in March 2013 at the 12th National People's Congress. Tipped for the role since his appointment as secretary general of the Communist Party and chairman of the Central Military Commission in Nov. 2012, Xi is known for his tough stance on corruption and is likely to be open to political and market reforms. He is expected to complete two five-year terms in office.

Early Life

Xi Jinping was born on 15 June 1953 in Beijing, son of Xi Zhongxun, one of the first generation of communist leaders. In 1969, following his father's purge and as part of the 'Down to the Countryside Movement' in which urban youth were exposed to rural life, Xi was sent to Shaanxi. He joined the Communist Party (CCP) in 1974 and left Shaanxi a year later. He graduated from Tsinghua University in 1979 with a degree in chemical engineering and became secretary to one of his father's allies, Geng Biao, the vice-premier and secretary-general of the Central Military Commission.

Xi became the Zhengding County Committee deputy secretary in Hebei province in 1982 and the following year was promoted to secretary. In 1985 he was made deputy mayor of Xiamen City, Fujian province. He undertook various party roles in the province and in 1999 became deputy governor of Fujian. Appointed governor a year later, he gained renown for attracting Taiwanese investment to boost the local economy.

In 2002 he moved to Zhejiang, another of China's economically successful provinces, and made his first inroads into national politics when he was made a member of the 16th Central Committee. From 2003–07 he was party secretary of Fujian, overseeing economic growth averaging 14% a year and earning a reputation as an opponent of corruption.

In March 2007 Xi transferred to Shanghai to take the role of party secretary following the dismissal of the incumbent, Chen Liangyu, on corruption charges. Xi's appointment to such an important regional post was seen as a sign of confidence from the central government and he became a member of the Politburo standing committee at the 17th Party Congress in Oct. 2007. He was also made a high-ranking member of the central secretariat. On 15 March 2008 he was elected vice-president at the 11th National People's Congress and took on a number of high profile portfolios including the presidency of the Central Party School. He was also Beijing's senior representative for Hong Kong and Macao and headed up preparations for the 2008 Olympic Games in Beijing.

On 18 Oct. 2010 Xi was appointed vice-chairman of the CCP and Central Military Commission, marking him as Hu's successor. He was elected general-secretary of the CCP and chairman of the Central Military Commission by the 18th Central Committee on 15 Nov. 2012 and was sworn in as president on 14 March 2013.

Career in Office

Ahead of his presidency, Xi said little about his policy ambitions but there was hope in the international community that he would champion political and social reform and take a firm line against corruption. He must deal with a widening wealth gap between rich and poor, and between urban and rural communities. He also faces the conundrum of how to provide adequate healthcare to a rapidly aging population while maintaining strong economic growth.

Li Keqiang

Position

Premier of the State Council

Introduction

Li Keqiang took office as premier of the State Council, a role equivalent to prime minister, in March 2013. He succeeded Wen Jiabao and is expected to serve two five-year terms.

Early Life

Li Keqiang was born on 1 July 1955 in Dingyuan County, Anhui province. His father was a county-level Communist Party (CCP) official. Following Li's high school graduation in 1974, he moved to an agricultural commune in Fengyang County, Anhui province, as part of the government's 'Down to the Countryside Movement'

that sent urban youth to be educated in rural life. Li joined the CCP and from 1976–78 served as party secretary in a local production brigade.

In 1982 he graduated in law from Peking University, serving as head of the Students' Federation from 1978–82. He went on to earn a masters degree and doctorate in economics and headed the University's Communist Youth League of China (CYLC) committee. Over the following two decades he rose through the CYLC ranks, joining the secretariat of the league's central committee in the 1980s and serving as its first secretary in the 1990s. At this time he built up his power base and forged close ties with Hu Jintao, a fellow CYLC committee member and future Chinese president.

In 1998 Li became deputy party secretary for Henan province and a year later was appointed Henan's governor. Despite strong economic growth, his term was marred by three major fires and the spread of HIV/AIDS, which his administration failed to curb. In Dec. 2004 he was named party secretary for Liaoning province where he spearheaded a major coastal infrastructure project, the '5 Points and One Line' highway development. In 2009 this template was adopted at the national level to rejuvenate industrial northeast China. He also oversaw the rehousing of 1·5m. shanty-town residents into new apartment blocks over a three-year period.

Li advanced to national level politics when he was elected to the Politburo standing committee in Oct. 2007. He was appointed vice-premier of the State Council in March 2008, leading a medical and health care reform programme aimed at creating an accessible public healthcare service. He also chaired an affordable housing programme and introduced reform plans to replace turnover tax with value-added tax. However, his image abroad suffered during a visit to Hong Kong in Aug. 2011 when demonstrators were restrained by security forces. Li was re-elected in Nov. 2012 as a member of the Politburo standing committee and on 15 March 2013 became premier of the State Council at the 12th National People's Congress.

Career in Office

Based on his political track record, Li is expected to focus on provision of basic national healthcare, increases in affordable housing, employment growth, regional development and a push towards cleaner energy technology. He will also attempt to secure China's long-term economic expansion.

DEFENCE

The Chinese president is chairman of the State and Party's Military Commissions. China is divided into seven military regions. The military commander also commands the air, naval and civilian militia forces assigned to each region.

China's armed forces, totalling more than 2·2m. in 2006, are the largest of any country.

Conscription is compulsory, but for organizational reasons, is selective: only some 10% of potential recruits are called up. Service is for two years. A military academy to train senior officers in modern warfare was established in 1985.

Defence expenditure in 2008 was estimated at 590bn. yuan (US$84,900m., equivalent to US$63 per capita). China's military spending during the 2000s more than trebled. Defence spending in 2007 represented an estimated 2·0% of GDP. Only the USA spent more on defence in 2008, but China's defence expenditure totalled around a seventh of that of the USA. In the period 2006–10 China's expenditure on major conventional weapons was the fourth highest in the world (after that of India, South Korea and Pakistan) at US$6·3bn. In March 2007 it was announced that defence budget for the year would rise by 17·8%.

Nuclear Weapons

Having carried out its first test in 1964, there have been 45 tests in all at Lop Nur, in Xinjiang (the last in 1996). The nuclear arsenal consisted of approximately 240 operational warheads in Jan. 2012 according to the Stockholm International Peace Research

Institute. China has been helping Pakistan with its nuclear efforts. Despite China's official position, *Deadly Arsenals*, published by the Carnegie Endowment for International Peace, alleges that the Chinese government is secretly pursuing chemical and biological weapons programmes.

Army

The Army (PLA: 'People's Liberation Army') is divided into main and local forces. Main forces, administered by the seven military regions in which they are stationed, but commanded by the Ministry of Defence, are available for operation anywhere and are better equipped. Local forces concentrate on the defence of their own regions. Ground forces are divided into infantry, armour, artillery, air defence, aviation, engineering, chemical defence and communications service arms. There are also specialized units for electronic counter-measures, reconnaissance and mapping. In 2009 there were 18 group armies covering seven military regions. They included: 17 armoured divisions and brigades; 15 mechanized infantry divisions, brigades and regiments; 28 motorized infantry divisions and brigades; seven special operations units; 18 artillery divisions and brigades; 11 surface-to-surface missile brigades and regiments; 22 air defence brigades and regiments; 17 engineering brigades and regiments; five electronic warfare regiments; 12 aviation brigades and regiments; and two guard divisions. Total strength in 2009 was 1·60m. including some 800,000 conscripts. Reserve forces are undergoing major reorganization on a provincial basis but are estimated to number some 510,000.

In Sept. 2003 it was announced that the strength of the PLA was to be reduced by 200,000 as part of a move to modernize the military.

There is a paramilitary People's Armed Police force estimated at 660,000 under PLA command.

Navy

The naval arm of the PLA comprises one aircraft carrier, one nuclear-powered ballistic missile armed submarine, four nuclear-propelled fleet submarines, one diesel-powered cruise missile submarine and some 51 patrol submarines. Surface combatant forces include 27 missile-armed destroyers, 44 frigates and some 52 missile craft. Sea trials of China's first aircraft carrier, *Liaoning* (a former Soviet warship purchased from Ukraine), began in Aug. 2011. It entered service in Sept. 2012, initially only to be used for training.

There is a land-based naval air force of about 792 combat aircraft, primarily for defensive and anti-submarine service. The force includes H-5 torpedo bombers, Q-5 fighter/ground attack aircraft, J-6 (MiG-19) and J-7 (MiG-21) fighters.

The naval arm is split into a North Sea Fleet, an East Sea Fleet and a South Sea Fleet.

In 2006 naval personnel were estimated at 255,000, including 26,000 in the naval air force and 40,000 conscripts.

Air Force

There are 32 air divisions. Up to four squadrons make up an air regiment and three air regiments form an air division. The Air Force has an estimated 2,600 combat aircraft.

Equipment includes J-7 (MiG-21) interceptors and fighter-bombers, H-5 (Il-28) jet bombers, H-6 Chinese-built copies of Tu-16 strategic bombers, Q-5 fighter-bombers (evolved from the MiG-19) and Su-27 fighters supplied by Russia. About 165 of a locally-developed fighter designated J-8 (known in the West as 'Finback') are in service.

Total strength (2006) was 400,000 (150,000 conscripts), including 210,000 in air defence organization. The Air Force headquarters are in Beijing.

ECONOMY

In 2010 agriculture accounted for 10·1% of GDP, industry 46·8% and services 43·1%.

Overview

China's economic performance has been marked by high rates of growth for over 25 years. GDP growth in the early 2000s consistently exceeded 10% until a slump followed the global financial crisis. China holds the world's largest foreign reserves, at more than US$3trn. in mid-2012. It is among the top recipients of foreign direct investment and is the world's largest producer and consumer of coal. In 2005 China made the transition from receiver of foreign aid to donor and is a key player in Africa's development. According to the ministry of commerce, China's cumulative FDI in Africa at the end of 2011 exceeded US$14·7bn., up 60% from 2009.

The first steps towards a more market-oriented economy were taken by Deng Xiaoping in the late 1970s. He opened the economy to foreign trade and investment, decentralized industrial management and allowed the private sector to flourish. In 2001 China became a member of the WTO, establishing trade relations with many countries. Much of China's recent dynamism has come from 'collective' enterprises, particularly those at township and village level run by managers under the auspices of local government. Local governments have an incentive to see enterprises run efficiently as officials are allowed to keep surplus revenues. However, this has led to abuses. In 2010 the government recovered about US$84·3bn. in misdirected funds.

Private entrepreneurs and foreign investors have played an important role in manufacturing. Even before 1978 the economy was heavily skewed towards manufacturing but thereafter output increased and there was a structural shift away from large state-owned enterprises (SOEs), although SOEs remain a significant part of the economy. Many new enterprises are labour-intensive as distinct from the capital-intensive SOEs. Growth has been fuelled by low added value and labour-intensive exports. But the country has moved up the added value curve and Chinese firms are predicted to become increasingly competitive with higher added value producers, such as South Korea. The Twelfth Five-Year Plan (2011–15) aims to shift the economy toward a consumer-based growth model.

There are threats to continued growth. Inefficient production and outmoded equipment have led to a deterioration of the environment, especially in the north. Air pollution, soil erosion and a declining water table are particular problems. The government aims to diversify its energy sources, relying less on coal and more on nuclear and alternative energy sources.

The global economic crisis reduced the rate of growth and the inflow of FDI but China's recovery was among the earliest and most spectacular. GDP growth was 7·9% in the second quarter of 2009, up from a two-decade low of 6·1% in the first quarter of that year. Growth was rooted in a stimulus package of 4trn. yuan (US$586bn. or 13% of 2008 GDP), including fiscal spending and interest rate cuts, as well as expansionary monetary policy. Central government committed 1·18trn. yuan, with the rest coming from local government, banks and SOEs. Although exports declined by around 17% in 2009, other countries fared worse and China's share of world exports increased to nearly 10% (up from 3% in 1999), making it the world's largest merchandise exporter in 2010. Growth in 2010 stood at more than 10%, which saw China overtake Japan as the world's second largest economy.

However, efforts to restructure the economy away from investment and export-led growth towards private consumption were interrupted by the global crisis. Nonetheless, structural reforms to redirect the export-oriented economy include increased worker mobility and improved public sector efficiency. SOEs continue to dominate 'strategic' industries and remain burdened by excess labour. China also faces the growing burden of an ageing population. Those aged 60 and over accounted for 14·3% of the total population in 2012, up from 13·7% the previous year. This is projected to increase to about 31% of the total population by 2050. For continued long term growth, there is a need for further job creation.

Other challenges include rising property costs, high levels of local government debt, the lack of enforcement of intellectual property rights and endemic corruption. Corruption is concentrated in sectors with a high degree of state involvement. The number of government officials caught embezzling more than 1m. yuan (US$146,000) increased by 19% in 2009, while in 2007 the US-based Carnegie Endowment for International Peace put the direct cost of corruption at US$86bn. per year. Inflation has been an important social and economic issue, with the consumer price index increasing by 5·4% in 2011 and food prices increasing by 11·8%. Inflation peaked at 6·5% in July 2011 but stood at 2·5% in Dec. 2012.

The World Bank estimates that the number of people living below the poverty line declined from 65% in 1981 to 4% in 2007. Those who still have consumption levels below a dollar a day are located mainly in remote and resource-poor regions, particularly in the west and the interior. In 2012 the income of urban households was more than three times as high as that of rural households. The number of migrant workers stood at 236m. in 2012.

For further developments *see* www.statesmansyearbook.com.

Currency
The currency is called Renminbi (i.e. People's Currency). The unit of currency is the *yuan* (CNY) which is divided into ten *jiao*, the *jiao* being divided into ten *fen*. The yuan was floated to reflect market forces on 1 Jan. 1994 while remaining state-controlled. For 11 years the People's Bank of China maintained the yuan at about 8·28 to the US dollar, allowing it to fluctuate but only by a fraction of 1% in closely supervised trading. In July 2005 it was revalued and pegged against a 'market basket' of currencies central parities of which were determined every night. In July 2008, after three years of sharp appreciation, it was repegged at around 6·83 yuan to the dollar, leading to claims from some international observers that it was being kept unfairly low to boost exports. In June 2010 the government announced that the yuan would be allowed to move freely against the dollar as long as a rise or fall does not exceed 0·5% within a single day. In Aug. 2009 total money supply was 20,039·5bn. yuan, gold reserves were 33·89m. troy oz and foreign exchange reserves US$2,210·8bn. (US$75·4bn. in 1995). China's reserves are the highest of any country, having overtaken those of Japan in 2006.

Inflation rates (based on IMF statistics):

2002	2003	2004	2005	2006	2007	2008	2009	2010	2011
−0·8%	1·2%	3·9%	1·8%	1·5%	4·8%	5·9%	−0·7%	3·3%	5·4%

China's economy overheated in the early 1990s, leading to inflation rates of 14·7% in 1993, 24·1% in 1994 and 17·1% in 1995. The 2008 rate was the highest since 1996.

Budget
Total revenue and expenditure (in 1bn. yuan):

	2006	2007	2008	2009	2010
Revenue	3,876·0	5,132·2	6,133·0	6,851·8	8,310·2
Expenditure	4,042·3	4,978·1	6,259·3	7,630·0	8,987·4

Total revenue in the central budget for 2006 was 2,123·2bn. yuan, comprising 2,045·0bn. yuan in revenue collected by central government and 78·3bn. yuan transferred to central government from local authorities. Total expenditure in the central budget amounted to 2,348·2bn. yuan, of which 999·2bn. yuan of expenditure for the central government and 1,349·1bn. yuan in the form of subsidies for local authorities. Local government revenue in 2006 came to 3,177·2bn. yuan (1,828·1bn. yuan in revenue collected by local authorities and 1,349·1bn. yuan in central government subsidies) and expenditure amounted to

3,100·4bn. yuan (3,022·2bn. yuan of expenditure in local budgets and 78·3bn. yuan transferred to central government). The deficit in the central budget in 2006 was 275·0bn. yuan.

Performance
GDP totalled US$7,318·5bn. in 2011, the second highest behind the USA. China replaced Japan as the second largest economy in 2010. It is forecast that around 2020 China will overtake the USA to become the world's largest economy. As recently as 2000 the US economy was around eight times larger than China's. Real GDP growth rates (based on IMF statistics):

2002	2003	2004	2005	2006	2007	2008	2009	2010	2011
9·1%	10·0%	10·1%	11·3%	12·7%	14·2%	9·6%	9·2%	10·4%	9·2%

GDP growth in 2012 was 7·8% (provisional) according to the National Bureau of Statistics, the lowest rate since 1999. In spite of high growth in recent years, China's GDP per capita at purchasing power parity was $5,383 in 2007 compared to the very high human development average of $37,272.

Banking and Finance
The People's Bank of China is the central bank and bank of issue (*Governor*, Zhou Xiaochuan). There are three state policy banks—the State Development Bank, Export and Import Bank of China and Agricultural Development Bank of China—and four state-owned commercial banks (the Bank of China, Industrial and Commercial Bank of China, Agricultural Bank of China and China Construction Bank). The four state-owned commercial banks have all sold minority stakes to foreign investors. The Bank of China is responsible for foreign banking operations. In April 2003 the China Banking Regulatory Commission was launched, taking over the role of regulating and supervising the country's banks and other deposit-taking financial institutions from the central bank. Legislation of 1995 permitted the establishment of commercial banks; credit co-operatives may be transformed into banks, mainly to provide credit to small businesses. In Oct. 2005 there were 30,438 rural credit co-operatives and 626 urban credit co-operatives. In Sept. 2007 deposits in rural credit co-operatives amounted to 4,400bn. yuan and loans reached 3,200bn. yuan. Insurance is handled by the People's Insurance Company.

Savings deposits in various forms in all banking institutions totalled 30,020·9bn. yuan at the end of 2005; loans amounted to 20,683·8bn. yuan.

There are stock exchanges in the Shenzhen Special Economic Zone and in Shanghai. A securities trading system linking six cities (Securities Automated Quotations System) was inaugurated in 1990 for trading in government bonds.

China received US$114·7bn. worth of foreign direct investment in 2010 and US$124·0bn. in 2011.

External debt totalled US$548,551m. in 2010 (up from US$145,339m. in 2000) and represented 9·3% of GNI.

ENERGY AND NATURAL RESOURCES

Environment
China's carbon dioxide emissions from the consumption and flaring of fossil fuels in 2008 accounted for 21·5% of the world total (the biggest emissions producer having overtaken the USA in 2007) and were equivalent to 4·9 tonnes per capita. An *Environmental Performance Index* compiled in 2008 ranked China 105th in the world out of 149 countries analysed, with 65·1%. The index examined various factors in six areas—air pollution, biodiversity and habitat, climate change, environmental health, productive natural resources and water resources. Pollution is estimated to cost China about 10% of GDP annually.

Electricity
Installed generating capacity in 2008 was 849m. kW, compared with 254m. kW in 1998. In 2009 electricity output was 3,714,950 GWh, up from 2,865,726 GWh in 2006. Consumption per capita

was 2,476 kWh in 2007, up from 1,688 kWh in 2004. Rapidly increasing demand has meant that more than half of China's provinces have had to ration power. Sources of electricity in 2005 as percentage of total production: coal, 81·5%; hydro-electric power, 16·0%; nuclear, 2·1%; others, 0·4%. In 2010 there were 13 nuclear reactors in use and 25 under construction. Generating electricity is not centralized; local units range between 30 and 60 MW of output. In Dec. 2002 China formally broke up its state power monopoly, creating instead five generating and two transmission firms. The Three Gorges dam project on the Yangtze river was launched in 1993 and is intended to produce abundant hydro-electricity (as well as helping flood control). The first three 700,000-kW generators in service at the project's hydro-power station began commercial operation in July 2003. The original specification was completed in Oct. 2008, although six more generators have been added in the meantime (bringing the total to 32). The final two generators become operational in July 2012, giving the dam an overall capacity of 22·5 GW.

Oil and Gas

On-shore oil reserves are found mainly in the northeast (particularly the Daqing and Liaohe fields) and northwest. There are off-shore fields in the continental shelves of east China. Oil production was 189·7m. tonnes in 2008. China is the second largest consumer of oil after the USA. Ever-growing demand has meant that increasing amounts of oil are having to be imported. A 964-km pipeline from Skovorodino in Russia to Daqing in the northeast of China was inaugurated in Jan. 2011, allowing China to increase significantly its imports of oil from the world's largest producer. The 1,833-km Turkmenistan–China gas pipeline, bringing natural gas to Xinjiang in China via Kazakhstan and Uzbekistan, was inaugurated in Dec. 2009. This connects with China's Second West–East gas pipeline. Only the USA and Japan import more oil. Domestic production now accounts for only 55% of consumption, compared to nearly 85% in 1998. Proven reserves in 2008 were 15·5bn. bbls, but they are expected to be exhausted by 2019.

The largest natural gas reserves are located in the western and north-central regions. Production was 76·1bn. cu. metres in 2008, with proven reserves of 2,460bn. cu. metres.

Wind

China is one of the largest producers of wind-power. In 2010 total installed capacity amounted to 44,733 MW, the highest of any country and 22·7% of the world total.

Minerals

In 2006 there were 158 varieties of proven mineral deposits in China, making it the third richest in the world in total reserves. Recoverable deposits of coal totalled 3,334·8bn. tonnes, mainly distributed in north China (particularly Shanxi province and the Inner Mongolia Autonomous Region). Coal production was 2,320m. tonnes in 2008. Annual coal production has increased every year since 2000. Growing domestic demand meant that China became a net importer of coal in 2009, and now consumes three times more coal than any other country.

The iron ore reserve base was 46bn. tonnes in 2005. Deposits are abundant in the anthracite field of Shanxi, in Hebei and in Shandong, and are found in conjunction with coal and worked in the northeast. Production in 2006 was 601m. tonnes, making China the world's leading iron ore producer.

Tin ore is plentiful in Yunnan, where the tin-mining industry has long existed. Tin production was 126,000 tonnes in 2006.

China is a major producer of wolfram (tungsten ore). There is mining of wolfram in Hunan, Guangdong and Yunnan.

Salt production was 56·6m. tonnes in 2006; gold production was 245 tonnes. Output of other minerals (in 1,000 tonnes) in 2006: bauxite, 21,000; aluminium, 13,700; zinc, 2,840; lead, 1,330; copper, 873. Estimated diamond production in 2005, 1,060,000 carats. Other minerals produced: nickel, barite, bismuth, graphite,

gypsum, mercury, molybdenum, silver. Reserves (in tonnes) of salt, 402,400m.; phosphate ore, 15,766m.; sylvite, 458m. China's gold production rose to 276 tonnes in 2007, in the process surpassing South Africa as the world's leading gold producer.

Agriculture

Agriculture accounted for approximately 11·7% of GDP in 2006, compared to over 50% in 1949 at the time of the birth of the People's Republic of China and over 30% in 1980. In 2009 areas harvested for major crops were (in 1m. ha.): maize, 31·20; rice, 29·88; wheat, 24·29; soybeans, 9·19; rapeseed, 7·28; sweet potatoes, 3·56. Intensive agriculture and horticulture have been practised for millennia. Present-day policy aims to avert the traditional threats from floods and droughts by soil conservancy, afforestation, irrigation and drainage projects, and to increase the 'high stable yields' areas by introducing fertilizers, pesticides and improved crops. In Aug. 1998 more than 21m. ha., notably in the Yangtze valley, were under water as China experienced its worst flooding since the 1950s. The 1998 flood season claimed over 4,100 lives.

'Township and village enterprises' in agriculture comprise enterprises previously run by the communes of the Maoist era, co-operatives run by rural labourers and individual firms of a certain size. Such enterprises employed 146·8m. people in 2006. There were 1,896 state farms in 2006 with 3·29m. employees. In 2005 there were 252·22m. rural households. The rural workforce in 2005 was 503·87m., of whom 299·76m. were employed in agriculture, fishing or land management. Net per capita annual peasant income, 2006: 3,587 yuan. Around 40% of the total workforce is engaged in agriculture, down from 68% in 1980. In 2006 rural residents accounted for 56·1% of the population (1996, 69·5%).

In 2009 there were an estimated 110·0m. ha. of arable land and 14·3m. ha. of permanent cropland; 64·5m. ha. were irrigated.

There were 1·4m. large/medium-sized tractors in 2006 and 550,000 combine harvesters.

Agricultural production of main crops (in 1m. tonnes), 2009: rice, 196·68; maize, 164·11; sugarcane, 116·25; wheat, 115·12; melons and watermelons, 77·23; sweet potatoes, 76·77; potatoes, 73·28; tomatoes, 45·37; cucumbers and gherkins, 44·25; cabbage, broccoli, etc., 30·22; aubergines, 25·91; onions, dry, 21·05; seed cotton, 19·13; garlic, 17·97; soybeans, 14·98; groundnuts, 14·76; chillies and green peppers, 14·52; pears, 14·42; rapeseeds, 13·66; tangerines and mandarins, 9·75. Tea production in 2009 was 1,376,000 tonnes. China is the world's leading producer of a number of agricultural crops, including rice, sweet potatoes, wheat, potatoes, watermelons, groundnuts and honey. The gross value of agricultural output in 2006 was 4,242,400m. yuan.

Livestock, 2009: pigs, 451,178,000; goats, 152,458,000; sheep, 128,557,000; cattle, 92,132,000 (estimate); buffaloes, 23,704,000 (estimate); horses, 6,752,000 (estimate); chickens, 4·70bn. (estimate); ducks, 771m. (estimate). China has more pigs, goats, sheep, horses and chickens than any other country. It is also home to nearly two-thirds of the world's ducks. Meat production in 2009 was estimated at 78·21m. tonnes; milk, 40·55m. tonnes; eggs, 27·90m. tonnes; honey, 367,000 tonnes. China is the world's leading producer of meat and eggs.

Gale, Fred, (ed.) *China's Food and Agriculture: Issues for the 21st Century*. 2012

Powell, S. G., *Agricultural Reform in China: from Communes to Commodity Economy, 1978–1990*. 1992

Forestry

In 2010 the area under forests was 206·86m. ha., or 21·9% of the total land area. The average annual increase in forest cover of 2,763,000 ha. between 2005 and 2010 was the highest of any country in the world. Total roundwood production in 2007 was 294·40m. cu. metres, making China the world's third largest timber producer (8·2% of the world total in 2007). It is the world's leading importer of roundwood, accounting for 28·2% of world timber imports in 2007.

Fisheries
Total catch, 2008: 14,791,163 tonnes, of which 12,542,986 tonnes were from marine waters. China's annual catch is the largest in the world, and currently accounts for approximately 16% of the world total. In 1989 the annual catch had been just 5·3m. tonnes. China's aquaculture production is also the largest in the world, at 36,734,215 tonnes in 2010.

INDUSTRY
The leading companies by market capitalization in China in March 2013 were PetroChina (US$248·9bn.); China Mobile (Hong Kong), a telecommunications company (US$219·0bn.) and China Construction Bank (US$199·8bn.). In Nov. 2007 PetroChina was briefly the world's largest company after its flotation on the Shanghai stock market, with a market capitalization in excess of US$1trn.; in March 2013 it ranked fifth after the USA's Apple Inc., Exxon Mobil, Google Inc. and Berkshire Hathaway Inc.

Industry accounted for 46·8% of GDP in 2010, up from 21% in 1949 when the People's Republic of China came into existence. Cottage industries persist into the 21st century. Industrial output grew by 15·7% in 2010. Modern industrial development began with the manufacture of cotton textiles and the establishment of silk filatures, steel plants, flour mills and match factories. In 2006 there were 287,406 non-state-owned industrial enterprises with an annual revenue of more than 5m. yuan, and a combined gross industrial output value of 28,586·1bn. yuan. Of these enterprises, 226,534 were domestically funded, 31,691 were foreign funded and 29,181 were dependent on funds from Hong Kong, Macao and Taiwan. There were 14,555 state-owned industrial enterprises in total, with a gross output value of 3,072·8bn. yuan.

Output of major products, 2009 (in tonnes): cement, 1,644·0m.; rolled steel, 694·1m.; crude steel, 572·2m.; pig iron, 552·8m.; distillate fuel oil (2008), 134·1m.; paper and paperboard, 89·7m.; chemical fertilizers, 63·9m.; petrol (2008), 62·9m.; sulphuric acid, 59·6m.; yarn, 23·9m.; residual fuel oil (2008), 17·4m.; sugar, 13·4m. Also produced in 2009: cloth, 75,342m. metres; beer, 41,621·8m. litres; 619·2m. mobile telephones; 150·1m. notebook PCs; 119·2m. watches (2008); 99·0m. colour TV sets; 84·6m. cameras; 80·8m. air conditioners; 59·3m. home refrigerators; 57·6m. bicycles; 49·7m. washing machines; 27·6m. motorcycles; 13·8m. motor vehicles. China is the world's leading cement, steel and pig iron manufacturer (producing 53% of the world's cement, 46% of crude steel and 58% of pig iron); since 2000 output of cement has doubled and production of crude steel and pig iron has quadrupled.

Labour
The employed population at the 1990 census was 647·2m. (291·1m. female). By 2005 it had risen to 758·3m. (6·3m. more than in 2004), of whom 484·9m. worked in rural areas (2·3m. fewer than in 2004) and 273·3m. in urban areas (8·6m. more than in 2004). By 2015 China's working age population will begin to decline as a consequence of the country's one-child policy. In Dec. 2010 China's registered urban jobless was 4·1%, with 9·08m. registered unemployed in the country's cities. In 2005 there were 312·06m. people working in agriculture, forestry and fisheries; 80·80m. in manufacturing; 49·66m. in wholesale and retail trade, restaurants and hotels; 45·22m. in community, social and personal services; and 40·77m. in construction.

In 2006 China had 149,736 private industrial enterprises employing almost 20m. people. It was not until the late 1970s that the private sector even came into existence in China.

The average non-agricultural annual wage in 2005 was 18,364 yuan: 11,283 yuan, urban collectives; 19,313 yuan, state-owned enterprises; 18,244 yuan, other enterprises. There is a 6-day

48-hour working week. Minimum working age was fixed at 16 in 1991. There were 693,000 labour disputes in 2008 involving 1,214,000 workers, up from 350,000 disputes and 650,000 workers in 2007. Strikes over pay have become ever more frequent in China, particularly at foreign-owned facilities.

INTERNATIONAL TRADE
There are five Special Economic Zones at Shenzhen, Xiamen, Zhuhai, Shantou and Hainan in which concessions are made to foreign businessmen. The Pudong New Area in Shanghai is also designated a special development area. Since 1979 joint ventures with foreign firms have been permitted. A law of April 1991 reduced taxation on joint ventures to 33%. There is no maximum limit on the foreign share of the holdings; the minimum limit is 25%.

In May 2000 the USA granted normal trade relations to China, a progression after a number of years when China was accorded 'most favoured nation' status. China subsequently joined the World Trade Organization on 11 Dec. 2001.

Saee, John, *China and the Global Economy in the 21st Century.* 2011

Imports and Exports
Trade in US$1m.:

	2006	2007	2008	2009	2010
Imports	791,461	956,115	1,132,562	1,005,555	1,394,200
Exports	968,936	1,220,060	1,430,693	1,201,647	1,578,193

China is the second largest trading nation in the world after the USA, in 2009 accounting for 7·9% of global merchandise imports by value and 9·6% of global merchandise exports (up from 4·3% when it joined the WTO in 2001). It overtook Germany as the largest exporter of goods in 2009 and its imports are second only to those of the USA.

Main imports in 2010 (in US$1bn.): machinery and transport equipment, 550·0; inedible crude materials (except fuels), and animal and vegetable oil and fats, 219·7; mineral fuels, lubricants and related materials, 188·5; chemicals, 149·3. Major exports in 2010 (in US$1bn.): machinery and transport equipment, 781·3; miscellaneous manufactured articles, 376·9; manufactured goods classified chiefly by material, 249·2; chemicals, 87·6. Chinese imports and exports both increased fivefold between 2001 and 2010.

Main import suppliers, 2009: Japan, 13·0%; South Korea, 10·2%; USA, 7·7%; Germany, 5·5%. Main export markets in 2009: USA, 18·4%; Hong Kong, 13·8%; Japan, 8·1%; South Korea, 4·5%. Customs duties with Taiwan were abolished in 1980. Trade with the European Union is fast expanding, having increased from €258·7bn. in 2006 to €395·1bn. in 2010.

COMMUNICATIONS

Roads
The total road length in 2005 was 1,931,000 km, including 41,000 km of motorways (there had not been any motorways as recently as the mid-1980s). 14,663m. tonnes of freight and 18,605m. persons were transported by road in 2006. The number of civil motor vehicles was 31·60m. in 2005, including 21·32m. buses and cars and 9·56m. trucks. China is the world's fastest-growing car market. There were 378,871 traffic accidents in 2006, with 89,455 fatalities.

Rail
In 2010 there were 91,000 km of railway. The high-speed network, at 9,676 km in June 2011, is the longest in the world. The high-speed line linking Beijing and Guangzhou, which opened in Dec. 2012, is the longest in the world at 2,293 km. The railways carried 1·53bn. passengers in 2009 and 3·22bn. tonnes of freight. China's railways are the busiest in the world, carrying 24% of global rail traffic. There are metro systems in Beijing, Chengdu, Dalian,

Guangzhou, Hangzhou (where the first line opened in June 2012), Kunming (where the first line opened in June 2012), Nanjing, Shanghai, Shenyang, Shenzhen, Suzhou (where the first line opened in April 2012), Tianjin, Wuhan and Xian.

Civil Aviation

There are major international airports at Beijing (Capital), Guangzhou (Baiyun), Hong Kong (Chek Lap Kok) and Shanghai (Hongqiao and Pudong). In 2006 there were 142 civil airports for regular flights. The national and major airlines are state-owned, except Shanghai Airlines and Shenzhen Airlines. The leading Chinese airlines operating scheduled services in 2006 were China Southern Airlines (49·2m. passengers), China Eastern Airlines (35·0m.) and Air China (34·0m.). Other Chinese airlines include Changan Airlines, Hainan Airlines, Shandong Airlines, Shanghai Airlines, Shanxi Airlines, Shenzhen Airlines, Sichuan Airlines and Xiamen Airlines. In Feb. 2010 Shanghai Airlines merged with China Eastern Airlines but they have both retained their brand and livery.

In 2007 the busiest airport was Beijing (Capital International), with 53,611,747 passengers; followed by Hong Kong International (Chek Lap Kok), with 47,042,419 passengers; Guangzhou (Baiyun), with 30,958,467 passengers; and Shanghai (Pudong), with 28,920,432 passengers. Shanghai (Pudong) is the busiest airport for freight, with 2,559,246 tonnes of cargo handled in 2007. In the meantime numbers have increased at Beijing Capital to such an extent that in 2011 it was the second busiest airport in the world, with a total of 77,403,668 passengers. As recently as 2003 it had not featured among the world's 20 busiest airports. In 2006 China had a total of 1,336 scheduled flight routes, of which 1,068 were domestic air routes and 268 were international air routes. Total passenger traffic in 2006 reached 159·68m.; freight traffic totalled 3·49m. tonnes.

Regular direct flights between mainland China and Taiwan resumed in July 2008 for the first time since 1949.

Shipping

In Jan. 2009 there were 2,495 ships of 300 GT or over registered, totalling 25·36m. GT. Of the 2,495 vessels registered, 1,022 were general cargo ships, 482 bulk carriers, 476 oil tankers, 178 passenger ships, 174 container ships, 83 liquid gas tankers and 80 chemical tankers.

Mainland China's busiest port in 2007 was Shanghai (561·5m. tonnes of cargo handled, up from 537·5m. tonnes in 2006), followed by Ningbo (344·0m. tonnes, up from 309·7m. tonnes), Guangzhou (Canton) (343·3m. tonnes, up from 302·9m. tonnes), Tianjin (309·5m. tonnes), Qingdao (265·0m. tonnes), Qinhuangdao (248·9m. tonnes) and Dalian (222·9m. tonnes). Shanghai is the second busiest container port in the world after Singapore, handling 26·2m. 20-ft equivalent units (TEUs) in 2007 (19·6m. in 2006). Shenzhen, mainland China's second busiest port for container traffic and the world's fourth busiest, handled 21·1m. 20-ft equivalent units (TEUs) in 2007. Hong Kong handled 24·0m. 20-ft equivalent units (TEUs) in 2007.

In Jan. 2001 the first legal direct shipping links between the Chinese mainland and Taiwanese islands in more than 50 years were inaugurated.

Inland waterways totalled 123,400 km in 2006; 2,487·0m. tonnes of freight and 220·47m. passengers were carried. In June 2003 the Three Gorges Reservoir on the Chang Jiang River, the largest water control project in the world, reached sufficient depth to support the resumption of passenger and cargo shipping.

Telecommunications

In 2008 there were 340,810,000 main (fixed) telephone lines. In the same year mobile phone subscribers numbered 641,230,000 (479·5 per 1,000 persons), making China the biggest market for mobile phones in the world. The two main mobile operators are China Mobile and China Unicom. The main landline operators are China Telecom and China Netcom. In 1998 there were around 500,000 internet users, but by 2008 this had risen to 298·0m. China has by far the most internet users of any country. In March 2012 there were only 447,000 Facebook users in mainland China (less than 0·1% of the population).

SOCIAL INSTITUTIONS

Justice

Six new codes of law (including criminal and electoral) came into force in 1980, to regularize the legal unorthodoxy of previous years. There is no provision for *habeas corpus*. The death penalty has been extended from treason and murder to include rape, embezzlement, smuggling, fraud, theft, drug-dealing, bribery and robbery with violence. Capital punishment applies to a total of 55 offences including 31 non-violent crimes. China does not divulge figures on its use of the death penalty; however, Amnesty International reported that thousands of people were executed in China in 2012—more than the rest of the world put together. 'People's courts' are divided into some 30 higher, 200 intermediate and 2,000 basic-level courts, and headed by the Supreme People's Court. The latter, the highest state judicial organ, tries cases, hears appeals and supervises the people's courts. It is responsible to the National People's Congress and its Standing Committee. People's courts are composed of a president, vice-presidents, judges and 'people's assessors' who are the equivalent of jurors. 'People's conciliation committees' are charged with settling minor disputes. There are also special military courts. Procuratorial powers and functions are exercised by the Supreme People's Procuracy and local procuracies.

The number of sentenced prisoners in Dec. 2005 was 1,565,771 (119 per 100,000 of national population).

Education

An educational reform of 1985 brought in compulsory nine-year education consisting of six years of primary schooling and three years of secondary schooling, to replace a previous five-year system.

In mainland China the 2010 population census revealed the following levels of educational attainment: 119·63m. people had finished university education; 187·99m. had received senior secondary education; 519·66m. had received junior secondary education; and 358·76m. had had primary education. 54·66m. people over 15 years of age or 4·08% of the population were illiterate, although this compared favourably with a 15·88% rate of illiteracy in the 1990 census and a 6·72% rate in 2000. In 2009 adult literacy was estimated at 94·0% and youth literacy at 99·4%. In 2006 there were 130,495 kindergartens with 22·64m. children and 776,500 teachers; 396,567 primary schools with 109·77m. pupils and 5·63m. teachers; 94,116 secondary schools (of which: 16,992 senior secondary; 62,431 junior secondary; 6,048 specialized; 5,765 vocational; and 2,880 technical) with 103·50m. pupils and 5·67m. teachers. There were also 363,000 children at 1,605 special education schools. Institutes of higher education, including universities, numbered 1,867 in 2006, with 17·39m. undergraduates and 1·10m. postgraduate level students, and 1·08m. teaching staff. In 2009, 25% of the population of tertiary age were in post-secondary education, compared to 3% at the beginning of the 1990s. A national system of student loans was established in 1999. The number of Chinese students studying abroad has exceeded 100,000 annually since 2002 (making the country the largest exporter of students in the world); in 2006 the figure was 134,000.

There are more than 1,300 non-governmental private higher education institutions (including 12 private universities) with 1·5m. students, or 39% of the total college and university students nationwide.

There is an Academy of Sciences with provincial branches. An Academy of Social Sciences was established in 1977.

In 2005 total expenditure on education came to 841,884m. yuan; government appropriation was 516,108m. yuan.

Health

Medical treatment is free only for certain groups of employees, but where costs are incurred they are partly borne by the patient's employing organization.

In 2006 there were 308,969 health institutions throughout China, comprising 60,037 hospitals and health centres, 264 sanatoriums, 212,243 clinics, 1,402 specialized prevention and treatment centres, 3,548 centres for disease control and prevention, 3,003 maternity and child care centres, 248 medical research institutions and 28,224 other institutions; the number of beds totalled 3·51m.

China's first AIDS case was reported in 1985. Approximately 740,000 Chinese were HIV-infected in 2009.

In the first half of 2003 China was struck by an epidemic of a pneumonia-type virus identified as SARS (severe acute respiratory syndrome). The virus was first detected in southern China and was subsequently reported in over 30 other countries. According to the Ministry of Health, by the time the outbreak had been contained a total of 5,327 cases had been reported on the Chinese mainland; 4,959 patients were cured and discharged from hospital, and 349 died.

In 2002 some 67% of males and 4% of females smoked; one in every three cigarettes consumed worldwide in 2002 was smoked in China, and about 3,000 people died each day due to smoking.

In the period 2004–06, 10% of the population were undernourished compared to 22% in 1979.

Welfare

In 2006 there were 30,199 social welfare enterprises with 1,512,000 inmates. Numbers (in 1,000) of beneficiaries of relief funds: persons receiving minimum living allowance and traditional relief in rural areas, 29,878; persons receiving minimum living allowance in urban areas, 22,401; persons receiving temporary relief in poor rural households, 9,638; persons receiving temporary relief in poor urban households, 1,230; persons in rural households entitled to 'the five guarantees' (food, clothing, medical care, housing, education for children or funeral expenses), 5,033. The official retirement age for men is 60 and for women 50 (or 55 in the case of civil servants and professionals).

RELIGION

The government accords legality to five religions only: Buddhism, Islam, Protestantism, Roman Catholicism and Taoism. Confucianism, Buddhism and Taoism have long been practised. Confucianism has no ecclesiastical organization and appears rather as a philosophy of ethics and government. Taoism—of Chinese origin—copied Buddhist ceremonial soon after the arrival of Buddhism two millennia ago. Buddhism in return adopted many Taoist beliefs and practices. A more tolerant attitude towards religion had emerged by 1979, and the government's Bureau of Religious Affairs (since renamed the State Administration for Religious Affairs) was reactivated.

Ceremonies of reverence to ancestors have been observed by the whole population regardless of philosophical or religious beliefs.

A new quasi-religious movement, Falun Gong, was founded in 1992, but has since been banned by the authorities. The movement has claimed some 100m. adherents, although the Chinese government has disputed this.

Muslims are found in every province of China, being most numerous in the Ningxia-Hui Autonomous Region, Yunnan, Shaanxi, Gansu, Hebei, Henan, Shandong, Sichuan, Xinjiang and Shanxi.

Roman Catholicism has had a footing in China for more than three centuries. Two Christian organizations—the Chinese Patriotic Catholic Association, which declared its independence from Rome in 1958, and the Protestant Three-Self Patriotic Movement—are sanctioned by the Chinese government.

According to estimates (by the state-approved Xinhua news agency, the Chinese Academy of Social Sciences and the State Administration for Religious Affairs) there were 100m. Buddhists, 23m. Christians and more than 21m. Muslims in the country in 2009. Other official figures indicate that there are 5·3m. Catholics, although unofficial estimates are much higher.

Legislation of 1994 prohibits foreign nationals from setting up religious organizations.

CULTURE

World Heritage Sites

There are 43 sites in the People's Republic of China that appear on the UNESCO World Heritage List. They are (with year entered on list): the Great Wall of China (1987), Zhoukoudian, the Peking Man site (1987), Imperial Palaces of the Ming and Qing Dynasties in Beijing and Shenyang (1987 and 2004), mausoleum of first Qing dynasty emperor, Beijing (1987), Taishan mountain (1987), Mogao Caves (1987), Mount Huangshan (1990), Huanglong Scenic Reserve (1992), Jiuzhaigou National Reserve (1992), Wulingyuan Scenic Reserve (1992), Chengde mountain resort and temples (1994), Potala palace, Lhasa (1994, 2000 and 2001), ancient building complex in the Wudang Mountains (1994), Qufu temple, cemetery and mansion of Confucius (1994), Mount Emei Scenic Reserve, including the Leshan Buddha (1996), Lushan National Park (1996), Lijiang old town (1997), Ping Yao old town (1997), Suzhou classical gardens (1997 and 2000), Summer Palace, Beijing (1998), Temple of Heaven, Beijing (1998), Mount Wuyi (1999), Dazu rock carvings (1999), Mount Qincheng and Dujiangyan irrigation system (2000), Xidi and Hongcun ancient villages, Anhui (2000), Longmen grottoes (2000), Ming and Qing dynasty tombs (2000, 2003 and 2004), the Yungang Grottoes (2001), the Three Parallel Rivers of Yunnan Protected Areas (2003), the Capital Cities and Tombs of the Ancient Koguryo Kingdom (2004), the historic centre of Macao (2005), the Sichuan Giant Panda sanctuaries (2006), Yin Xu (2006), Kaiping Diaolou and villages (2007), South China Karst (2007), Fuijan Tulou (2008), Mount Sanqingshan National Park (2008), Mount Wutai (2009), China Danxia, six subtropical areas of erosional landforms (2010), the Historic Monuments of Dengfeng (2010), the West Lake Cultural Landscape of Hangzhou (2011), Chengjiang Fossil Site, one of the earliest records of a complex marine ecosystem (2012), and Xanadu, the remains of the summer capital of the Yuan Dynasty (2012).

Press

China has two news agencies: Xinhua (New China) News Agency (the nation's official agency) and China News Service. In 2006 there were 1,938 newspapers and 9,468 magazines; 42,500m. copies of newspapers and 2,850m. copies of magazines were published. In 1980 there were fewer than 400 newspapers. The Communist Party newspaper is *Renmin Ribao* (People's Daily), which had a daily circulation of 2·8m. in 2006. The most widely read newspaper is *Cankao Xiaoxi*, with a daily circulation of 3·2m. in 2006. China has the highest circulation of daily newspapers in the world, with an average daily total of 107·51m. in 2007. In the 2010 *World Press Freedom Index* compiled by Reporters Without Borders, China ranked 171st out of 178 countries.

In 2006, 6,410m. volumes of books were produced.

Tourism

In 2010 tourist numbers totalled 55·7m. The World Tourism Organization predicts that China will overtake France as the world's most visited destination by 2020. It is currently the third most visited destination after France and the USA. Income from tourists in 2010 was US$45·8bn., ranking it fourth behind the USA, Spain and France. Expenditure by Chinese travellers outside of mainland China for 2010 was US$54·9bn., third only to spending by German and US travellers in foreign countries.

Festivals

The lunar New Year, also known as the 'Spring Festival', is a time of great excitement for the Chinese people. The festivities get under way 22 days prior to the New Year date and continue for

15 days afterwards. Dates of the lunar New Year: Year of the Snake, 10 Feb. 2013; Year of the Horse, 31 Jan. 2014. Lantern Festival, or Yuanxiao Jie, is an important, traditional Chinese festival, which is on the 15th of the first month of the Chinese New Year. Guanyin's Birthday is on the 19th day of the second month of the Chinese lunar calendar. Guanyin is the Chinese goddess of mercy. Tomb Sweeping Day, as the name implies, is a day for visiting and cleaning the ancestral tomb and usually falls on 5 April. Dragon Boat Festival is called Duan Wu Jie in Chinese. The festival is celebrated on the 5th of the 5th month of the Chinese lunar calendar. The Moon Festival is on the 15th of the 8th lunar month. It is sometimes called Mid-Autumn Festival. The Moon Festival is an occasion for family reunion. China's largest rock festivals include the Midi Modern Music Festival in Beijing (May), Beijing Pop Festival (Sept.) and Modern Sky Festival, also in Beijing (Oct.).

DIPLOMATIC REPRESENTATIVES

Of China in the United Kingdom (49–51 Portland Pl., London, W1B 1JL)
Ambassador: Liu Xiaoming.

Of the United Kingdom in China (11 Guang Hua Lu, Jian Guo Men Wai, Beijing 100600)
Ambassador: Sebastian Wood, CMG.

Of China in the USA (3505 International Place, NW, Washington, D.C., 20008)
Ambassador: Zhang Yesui.

Of the USA in China (55 An Jia Lou Lu, 100600 Beijing)
Ambassador: Gary Locke.

Of China to the United Nations
Ambassador: Li Baodong.

Of China to the European Union
Ambassador: Wu Hailong.

FURTHER READING

State Statistical Bureau. *China Statistical Yearbook*
China Directory [in Pinyin and Chinese]. Annual

Adshead, S. A. M., *China in World History*. 1999
Baum, R., *Burying Mao: Chinese Politics in the Age of Deng Xiaoping*. 1994
Becker, Jasper, *The Chinese*. 2000
Breslin, Shaun, *China and the Global Political Economy*. 2007
Brown, Kerry, *Contemporary China*. 2013
The Cambridge Encyclopaedia of China. 2nd ed. 1991
The Cambridge History of China. 14 vols. 1978 ff.
Chang, David Wen-Wei and Chuang, Richard Y., *The Politics of Hong Kong's Reversion to China*. 1999
Chang, Jung and Halliday, Jon, *Mao: The Unknown Story*. 2005
Cook, Sarah, Yao, Shujie and Zhuang, Juzhong, (eds.) *The Chinese Economy Under Transition*. 1999
De Crespigny, R., *China This Century*. 2nd ed. 1993
Dikötter, Frank, *Mao's Great Famine: The History of China's Most Devastating Catastrophe, 1958–62*. 2010
Dillon, Michael, *China: A Modern History*. 2006
Dittmer, Lowell, *China's Deep Reform: Domestic Politics in Transition*. 2006
Dixin, Xu and Chengming, Wu, (eds.) *Chinese Capitalism, 1522–1840*. 1999
Evans, R., *Deng Xiaoping and the Making of Modern China*. 1993
Fairbank, J. K., *The Great Chinese Revolution 1800–1985*. 1987.—*China: a New History*. 1992
Glassman, R. M., *China in Transition: Communism, Capitalism and Democracy*. 1991
Goldman, M., *Sowing the Seeds of Democracy in China: Political Reform in the Deng Xiaoping Era*. 1994
Guo, Jian, *Historical Dictionary of the Chinese Cultural Revolution*. 2006
Hsü, Immanuel C. Y., *The Rise of Modern China*. 6th ed. 2000
Huang, R., *China: a Macro History*. 2nd ed. 1997
Jisheng, Yang, *Tombstone: The Untold Story of Mao's Great Famine*. 2012
Kissinger, Henry, *On China*. 2011
Kruger, Rayne, *All Under Heaven: A Complete History of China*. 2004
Lam, Willy Wo-Lap, *Chinese Politics in the Hu Jintao Era: New Leaders, New Challenges*. 2006
Lynch, Michael, *Modern China*. 2006

Ma, Jun, *Chinese Economy in the 1990s*. 1999
MacFarquhar, R. (ed.) *The Politics of China: the Eras of Mao and Deng*. 2nd ed. 1997.—*The Origins of the Cultural Revolution*. 3 vols. 1998
McGregor, Richard, *The Party: the Secret World of China's Communist Rulers*. 2010
Mok, Ka-Ho, *Social and Political Development in Post-Reform China*. 1999
Roberts, J. A. G., *A History of China*. 3rd ed. 2011
Saich, Tony, *Governance and Politics of China*. 3rd ed. 2010
Schram, S. (ed.) *Mao's Road to Power: Revolutionary Writings 1912–1949*. 4 vols. 1998
Shenkar, Oded, *The Chinese Century: The Rising Chinese Economy and Its Impact on the Global Economy, the Balance of Power, and Your Job*. 2004
Short, Philip, *Mao: A Life*. 2000
Spence, Jonathan, D., *The Chan's Great Continent: China in Western Minds*. 1998.—*Mao Zedong*. 2000
Suyin, H., *Eldest Son, Zhou Enlai and The Making of Modern China*. 1995
Tseng, Wanda and Cowen, David, *India's and China's Recent Experience with Reform and Growth*. 2007
Tubilewicz, Czeslaw, *Critical Issues in Contemporary China*. 2006
Zha, Jianying, *Tide Players: The Movers and Shakers of a Rising China*. 2011

Other more specialized titles are listed under TERRITORY AND POPULATION; TIBET; AGRICULTURE; INTERNATIONAL TRADE.

National Statistical Office: National Bureau of Statistics, 57 Yuetan Nanjie, Sanlihe, Xicheng District, Beijing 100826.
Website: http://www.stats.gov.cn

Hong Kong

Xianggang

Population projection, 2015: 7·43m.
GNI per capita, 2011: (PPP$) 44,805
HDI/world rank: 0·898/13

KEY HISTORICAL EVENTS

Hong Kong island and the southern tip of the Kowloon peninsula were ceded in perpetuity to the British Crown in 1841 and 1860 respectively. The area lying immediately to the north of Kowloon known as the New Territories was leased to Britain for 99 years in 1898. Talks began in Sept. 1982 between Britain and China over the future of Hong Kong after the lease expiry in 1997. On 19 Dec. 1984 the two countries signed a Joint Declaration by which Hong Kong became, with effect from 1 July 1997, a Special Administrative Region of the People's Republic of China, enjoying a high degree of autonomy and vested with executive, legislative and independent judicial power, including that of final adjudication. The existing social and economic systems were to remain unchanged for another 50 years. This 'one country, two systems' principle, embodied in the Basic Law, became the constitution for the Hong Kong Special Administrative Region of the People's Republic of China.

TERRITORY AND POPULATION

Hong Kong ('Xianggang' in Mandarin *Pinyin*) island is situated off the southern coast of the Chinese mainland 32 km east of the mouth of the Pearl River. The area of the island is 81 sq. km. It is separated from the mainland by a fine natural harbour. On the opposite side is the peninsula of Kowloon (47 sq. km). The 'New Territories' include the mainland area lying to the north of Kowloon together with over 200 offshore islands (976 sq. km). Total area of the Territory is 1,104 sq. km, a large part of it being steep and unproductive hillside. Country parks and special areas cover over 40% of the land area. Since 1945 the government has reclaimed over 5,400 ha. from the sea, principally from the seafronts of Hong Kong and Kowloon, facing the harbour.

Based on the results of the 2011 population census Hong Kong's resident population in March 2011 was 7,071,576 and the population density 6,405 per sq. km. In 2006, 60·3% of the population were born in Hong Kong, 33·5% in other parts of China and 6·2% in the rest of the world.

In 2011, 100% of the population lived in urban areas.

The UN gives a projected population for 2015 of 7·43m.

The official languages are Chinese and English.

SOCIAL STATISTICS

Annual population growth rate, 2006–11, 0·7%. Vital statistics, 2010: known births, 88,600; known deaths, 42,200; registered marriages, 52,600. Rates (per 1,000): birth, 12·5; death, 6·0; marriage, 7·4; infant mortality, 2010, 1·7 per 1 000 live births (one of the lowest rates in the world). Expectation of life at birth, 2010: males, 80·0 years; females, 85·9. The median age for marrying in 2010 was 33·2 years for males and 29·8 for females. Total fertility rate, 2010, 1·1 child per woman.

CLIMATE

The climate is sub-tropical, tending towards temperate for nearly half the year, the winter being cool and dry and the summer hot and humid, May to Sept. being the wettest months. Normal temperatures are Jan. 60°F (15·8°C), July 84°F (28·8°C). Annual rainfall 87" (2,214·3 mm).

THE BRITISH ADMINISTRATION

Hong Kong used to be administered by the Hong Kong government. The Governor was the head of government and presided over the *Executive Council*, which advised the Governor on all important matters. The last British Governor was Chris Patten. In Oct. 1996 the Executive Council consisted of three *ex officio* members and ten appointed members, of whom one was an official member. The chief functions of the *Legislative Council* were to enact laws, control public expenditure and put questions to the administration on matters of public interest. The Legislative Council elected in Sept. 1995 was, for the first time, constituted solely by election. It comprised 60 members, of whom 20 were elected from geographical constituencies, 30 from functional constituencies encompassing all eligible persons in a workforce of 2·9m., and ten from an election committee formed by members of 18 district boards. A president was elected from and by the members.

At the elections on 17 Sept. 1995 turnout for the geographical seats was 35·79%, and for the functional seats (21 of which were contested), 40·42%. The Democratic Party and its allies gained 29 seats, the Liberal Party 10 and the pro-Beijing Democratic Alliance 6. The remaining seats went to independents.

CONSTITUTION AND GOVERNMENT

In Dec. 1995 the Standing Committee of China's National People's Congress set up a Preparatory Committee of 150 members (including 94 from Hong Kong) to oversee the retrocession of Hong Kong to China on 1 July 1997. In Nov. 1996 the Preparatory Committee nominated a 400-member Selection Committee to select the *Chief Executive of Hong Kong* and a provisional legislature to replace the Legislative Council. The Selection Committee was composed of Hong Kong residents, with 60 seats reserved for delegates to the National People's Congress and appointees of the People's Political Consultative Conference. On 11 Dec. 1996 Tung Chee Hwa was elected Chief Executive by 80% of the Selection Committee's votes.

On 21 Dec. 1996 the Selection Committee selected a provisional legislature which began its activities in Jan. 1997 while the Legislative Council was still functioning. In Jan. 1997 the provisional legislature started its work by enacting legislation which would be applicable to the Hong Kong Special Administrative Region and compatible with the Basic Law.

Constitutionally Hong Kong is a Special Administrative Region of the People's Republic of China. The Basic Law enables Hong Kong to retain a high degree of autonomy. It provides that the legislative, judicial and administrative systems which were previously in operation are to remain in place. The Special Administrative Region Government is also empowered to decide on Hong Kong's monetary and economic policies independent of China.

In July 1997 the first-past-the-post system of returning members from geographical constituencies to the Legislative Council was replaced by proportional representation. There were 20 directly elected seats out of 60 for the first elections to the Legislative Council following Hong Kong's return to Chinese sovereignty, increasing in accordance with the Basic Law to 24 for the 2000 election with 36 indirectly elected. In the Sept. 2004 Legislative Council election (and that of Sept. 2008) 30 of the 60 seats were directly elected. For the election in Sept. 2012 the number of seats was increased to 70, with 35 directly elected and 30 indirectly elected by functional constituencies. There were also five new functional constituency seats nominated by elected District Council members. The Chief Executive is chosen by a Beijing-backed 1,200-member election committee (800 prior to the March 2012 election), although it has been stated that universal suffrage is the ultimate aim—potentially in 2017 for the Chief Executive and 2020 for all Legislative Council seats.

In July 2002 a new accountability or 'ministerial' system was introduced, under which the Chief Executive nominates for appointment 14 policy secretaries, who report directly to the Chief Executive. The Chief Executive is aided by the *Executive Council*, consisting of the three senior Secretaries of Department (the Chief Secretary, the Financial Secretary and the Secretary for Justice) and eleven other secretaries plus five non-officials.

RECENT ELECTIONS

In the Legislative Council election held on 9 Sept. 2012 turnout was 53%, up from 45% at the 2008 vote. 35 of the 70 seats were directly elected, the other 35 being returned by committees and professional associations in 'functional constituencies'. Pro-Beijing parties won 43 of the 70 seats (35 of 60 in 2008); pro-democracy parties won 27 (23 of 60 in 2008).

Leung Chun-ying was elected chief executive on 25 March 2012, receiving 689 of 1,132 votes in the Election Committee.

CURRENT GOVERNMENT

In March 2013 the government of the Hong Kong Special Administrative Region comprised:

Chief Executive: Leung Chun-ying; b. 1954 (since 1 July 2012).

Chief Secretary for Administration: Stephen Lam Sui-lung. *Financial Secretary:* John Tsang Chun-wah. *Secretary for Justice:* Wong Yan Lung. *Education:* Michael Suen Ming-yeung. *Constitutional and Mainland Affairs:* Raymond Tam Chi-yuen. *Security:* Ambrose Lee Siu-kwong. *Food and Health:* Dr York Chow. *Civil Service:* Denise Yue Chung-yee. *Home Affairs:* Tsang Tak-sing. *Labour and Welfare:* Matthew Cheung Kin-chung. *Financial Services and the Treasury:* Prof. K. C. Chan. *Development:* Carrie Lam Cheng Yuet-ngor. *Environment:* Edward Yau Tang-wah. *Transport and Housing:* Eva Cheng. *Commerce and Economic Development:* Gregory So Kam-leung.

Government Website: http://www.gov.hk

ECONOMY

Services accounted for 93% of GDP in 2010 and industry 7%.

According to the anti-corruption organization *Transparency International*, Hong Kong ranked 14th in the world in a 2012 survey of the countries and regions with the least corruption in business and government. It received 77 out of 100 in the annual index.

Hong Kong adopted a flat tax rate in 1948. Income tax is a flat 16% and only 25% of the population pay any tax at all. 6% of the population pays 80% of the total income tax bill. Hong Kong represents 20% of China's total worth.

Overview

A major financial centre, Hong Kong has a per capita GDP that compares favourably with other OECD countries. Its economic rise was founded on its role as an international trade emporium. Mainland China, the USA and Japan are its major export partners, accounting for 48·7%, 13·7% and 4·5% of exports respectively in 2007. The island is dependent on imports of food and other resources, importing 46·3% of goods from mainland China and 10% from Japan.

In 2004 and 2005 the economy grew strongly on the back of a rise in Chinese tourism, healthy global demand for exports and improving domestic consumer confidence. However, the financial crisis and the slowdown in growth of the Chinese economy saw the economy shrink by 2·6% in 2009 before rebounding with 7% growth the following year. It then expanded by 5% in 2011.

The service sector dominates the economy. Retail, restaurants, hotels and related services makes up 41·1% of employment, with real estate, financial services and insurance accounting for 12·5%. The manufacturing and construction sectors employ 4·3% and 2·4% respectively.

Foreign direct investment levels are high, reaching US$68·9bn. in 2011. The World Economic Foundation ranks Hong Kong as the 11th most competitive economy in the world.

Currency

The unit of currency is the *Hong Kong dollar* (HKD) of 100 *cents*. It is pegged at a rate of HK$7·8 to the US dollar. Banknotes are issued by the Hongkong and Shanghai Banking Corporation and the Standard Chartered Bank, and, from May 1994, the Bank of China. Total money supply was HK$529,161m. in July 2009. In Aug. 2009 gold reserves were 67,000 troy oz and foreign exchange reserves were US$223,211m.

Inflation rates (based on IMF statistics):

2002	2003	2004	2005	2006	2007	2008	2009	2010	2011
–3·0%	–2·6%	–0·4%	0·9%	2·0%	2·0%	4·3%	0·6%	2·3%	5·3%

Budget

In 2007–08 revenue totalled HK$358·5bn. and expenditure HK$252·4bn. Earnings and profits taxes accounted for 37·3% of revenues in 2007–08 and indirect taxes 26·9%; education accounted for 21·3% of expenditures and social welfare 13·8%.

Performance

Total GDP was US$248·6bn. in 2011. Real GDP growth rates (based on IMF statistics):

2002	2003	2004	2005	2006	2007	2008	2009	2010	2011
1·8%	3·0%	8·5%	7·1%	7·0%	6·4%	2·3%	–2·6%	7·1%	5·0%

In the 2012 *World Competitiveness Yearbook*, compiled by the International Institute for Management Development, Hong Kong came first in the world ranking. The annual publication ranks and analyzes how a nation's environment creates and sustains the competitiveness of enterprises.

Banking and Finance

The Hong Kong Monetary Authority acts as a central bank. The *Chief Executive* is Norman Chan. As at Dec. 2009 there were 145 banks licensed under the Banking Ordinance, of which 23 were locally incorporated. There were also 26 restricted licence banks, 28 deposit-taking companies and 71 representative offices of foreign banks. Licensed bank deposits were HK$5,193,003m. in July 2007; restricted licence bank deposits were HK$22,065m. There are three banks of issue: Bank of China (Hong Kong); The Hongkong and Shanghai Banking Corporation; and Standard Chartered Bank.

Gross external debt amounted to US$1,029,927m. in June 2012.

The principal regulator of Hong Kong's securities and futures markets is the Securities and Futures Commission. Hong Kong Exchanges and Clearing (HKEx), which was created in March 2000, owns and operates the only stock and futures exchange in Hong and their related clearing houses.

ENERGY AND NATURAL RESOURCES

Environment

Hong Kong's carbon dioxide emissions from the consumption and flaring of fossil fuels in 2008 were the equivalent of 12·0 tonnes per capita.

Electricity

Installed capacity was 12·6m. kW in 2007. Production in 2007 was 38·95bn. kWh. Hong Kong is a net importer of electricity. Consumption in 2007 was 45·87bn. kWh.

Agriculture

The local agricultural industry is directed towards the production of high quality fresh food through intensive land use and modern farming techniques. Out of the territory's total land area of 1,103 sq. km, only 60 sq. km is currently farmed. In 2006 local production accounted for 55% of live poultry consumed, 23% of live pigs and 4% of fresh vegetables. The gross value of local agricultural production totalled HK$1,184m. in 2006, with pig production valued at HK$585m., poultry production (including eggs) at HK$340m., and vegetable and flower production at HK$254m.

Fisheries

In 2006 the capture and mariculture fisheries supplied about 21% of seafood consumed in Hong Kong and pond fish farms produced about 5% of the freshwater fish consumed. The capture fishing fleet comprises some 3,900 fishing vessels, almost all mechanized. In 2006 the industry produced 155,000 tonnes of fisheries produce, valued at HK$1·6bn. There are 26 fish culture zones occupying a total sea area of 209 ha. with some 1,080 licensed operators. The estimated production in 2006 was 1,488 tonnes. The inland fish ponds, covering a total of 1,024 ha., produced 1,943 tonnes of freshwater fish in 2006.

INDUSTRY

The leading companies by market capitalization in Hong Kong in March 2013 were: China Mobile (Hong Kong), a mobile telecommunications company (US$219·0bn.); CNOOC, an oil and natural gas company (US$85·4bn.); and AIA Group Limited, a life insurance organization (US$53·3bn.).

Industry is mainly export-oriented. In Sept. 2001 there were 19,801 manufacturing establishments employing 209,329 persons. Other establishment statistics by product type (and persons engaged) were: printing, publishing and allied industries, 4,778 (42,963); textiles and clothing, 3,696 (58,821); plastics, 973 (5,938); electronics, 748 (20,939); watches and clocks, 347 (2,945); shipbuilding, 325 (3,173); electrical appliances, 49 (390).

Labour

In 2011 the size of the labour force (synonymous with the economically active population) was 3,703,100 (1,760,400 females). The persons engaged in June 2012 included 1,090,059 people in wholesale, retail and import/export trades, accommodation and food services, 664,652 in finance, insurance, real estate, professional and business services, 159,217 in the civil service, 107,637 in manufacturing and 71,721 in construction sites (manual workers only). A minimum wage of HK$28 per hour was introduced for the first time on 1 May 2011.

Unemployment stood at 3·1% in the period Sept.–Dec. 2011.

EXTERNAL ECONOMIC RELATIONS

Imports and Exports

In 2009 the total value of imports was HK$2,692,356m. and total exports HK$2,469,089m. The main suppliers of imports in 2009 were mainland China (46·4%), Japan (8·8%), Taiwan (6·5%), Singapore (6·5%) and USA (5·3%). In 2009, 51·2% of total exports went to mainland China, 11·6% to the USA, 4·4% to Japan, 3·2% to Germany and 2·4% to the United Kingdom.

The chief import items in 2009 were: electrical machinery, apparatus and appliances, etc. (26·8%); telecommunications, sound recording and reproducing equipment (13·7%); office machines and automatic data processing machines (9·2%); articles of apparel and clothing accessories (4·5%). The main exports in 2009 were: electrical machinery, apparatus and appliances, etc. (26·4%); telecommunications, sound recording and reproducing equipment (16·8%); office machines and automatic data processing machines (10·1%); articles of apparel and clothing accessories (7·2%).

Hong Kong has a free exchange market. Foreign merchants may remit profits or repatriate capital. Import and export controls are kept to the minimum, consistent with strategic requirements.

COMMUNICATIONS

Roads
In 2011 there were 2,086 km of roads, over 50% of which were in the New Territories. There are 16 road tunnels, including three under Victoria Harbour. In 2011 there were 435,000 private cars, 111,000 goods vehicles, 20,000 buses and coaches, and 39,000 motorcycles and mopeds. There were 15,541 road accidents in 2011, of which 128 were fatal. A total of 26·7m. tonnes of cargo were transported by road in 2011.

A 50-km bridge linking Hong Kong, Zhuhai in Guangdong Province in mainland China and Macao is currently under construction and is expected to be finished in 2016.

Hong Kong was ranked third for its road infrastructure in the World Economic Forum's *Global Competitiveness Report 2009–2010*.

Rail
The railway network covers around 229 km. The electrified Kowloon-Canton Railway, East Rail, runs for 53·9 km from the terminus at East Tsim Sha Tsui in Kowloon to border points at Lo Wu and Lok Ma Chau. Ma On Shan Rail branches off the main East Rail at Tai Wai and runs to Wu Kai Sha. East Rail and Ma On Shan Rail together carried 337m. passenger in 2006; cargo transported in 2006 totalled 184,000 tonnes. Another passenger rail service, West Rail, runs for 30·5 km from Tuen Mun in the New Territories to Nam Cheong in West Kowloon. It carried 72m. passengers in 2006. A light rail system (36 km and 58 stops) is operated by the Kowloon-Canton Railway Corporation in Tuen Mun, Yuen Long and Tin Shui Wai; it carried 136m. passengers in 2006.

The electric tramway on the northern shore of Hong Kong Island commenced operating in 1904 and has a total track length of 16 km. The Peak Tram, a funicular railway connecting the Peak district with the lower levels in Victoria, has a track length of 1·4 km and two tramcars (each with a capacity of 120 passengers per trip).

A metro, the Mass Transit Railway system, comprises 91 km with 53 stations and carried 867m. passengers in 2006.

The Airport Express Line (35 km) opened in 1998 and carried a total of 9·6m. passengers in 2006.

In 2006 a total of 4·1bn. passenger journeys were made on public transport (including local railways, buses, etc.).

In the World Economic Forum's *Global Competitiveness Report 2009–2010* Hong Kong ranked third for quality of rail infrastructure.

Civil Aviation
The new Hong Kong International Airport (generally known as Chek Lap Kok), built on reclaimed land off Lantau Island to the west of Hong Kong, was opened on 6 July 1998 to replace the old Hong Kong International Airport at Kai Tak, which was situated on the north shore of Kowloon Bay. More than 100 airlines now operate scheduled services to and from Hong Kong. In 2012 Cathay Pacific Airways, the largest Hong Kong-based airline, operated approximately 105,000 passenger and cargo services to 172 destinations in 41 countries and territories around the world. Cathay Pacific carried 21,146,492 passengers and 1·4m. tonnes of cargo in 2012. Dragonair, a Cathay Pacific subsidiary, provided scheduled services to 41 cities in mainland China and Asia in 2012. In 2012 Air Hong Kong, an all-cargo operator, provided scheduled services to Bangkok, Beijing, Ho Chi Minh City, Manila, Nagoya, Osaka, Penang (via Bangkok), Seoul, Shanghai, Singapore, Taipei and Tokyo. Hong Kong International Airport handled more international freight in 2009 than any other airport. In 2011, 334,000 aircraft arrived and departed and 54m. passengers and 3·94m. tonnes of freight were carried on aircraft.

Hong Kong was second, behind only Singapore, in the rankings for air transport infrastructure in the World Economic Forum's *Global Competitiveness Report 2011–2012*.

Shipping
The port of Hong Kong handled 23·5m. 20-ft equivalent units (TEUs) in 2006, making it the world's second busiest container port after Singapore. The Kwai Chung Container Port has 24 berths with 7,694 metres of quay backed by 275 ha. of cargo handling area. Merchant shipping in 2004 totalled 25,562,000 GRT, including oil tankers 5,416,000 GRT. In 2004, 35,900 ocean-going vessels, 117,540 river cargo vessels and 71,980 river passenger vessels called at Hong Kong. In 2004, 221m. tonnes of freight were handled. In 2004 vessels totalling 399,031,000 NRT entered ports and vessels totalling 399,025,000 NRT cleared.

Only Singapore ranked ahead of Hong Kong for quality of port facilities in the World Economic Forum's *Global Competitiveness Report 2009–2010*.

Telecommunications
In 2008 there were 4,099,900 main (fixed) telephone lines. The local fixed telecommunications network services (FTNS) market in Hong Kong was liberalized in 1995. In Oct. 2007 there were five mobile network operators in Hong Kong. There were ten wireline-based local FTNS operators in Oct. 2007 and one wireless-based FTNS operator. There were only 687,600 mobile phone subscribers in 1995, since when the sector has expanded substantially. In 2008 there were 11,580,100 mobile phone subscribers (1,658·5 per 1,000 population). The internet market has also seen considerable growth. In 2008 there were 4·7m. internet users, up from 1·9m. in 2000. There were 26·1 broadband subscribers per 100 inhabitants in June 2007. In March 2012 there were 3·8m. Facebook users.

The external telecommunications services market has been fully liberalized since 1 Jan. 1999, and the external telecommunications facilities market was also liberalized starting from 1 Jan. 2000.

SOCIAL INSTITUTIONS

Justice
The Hong Kong Act of 1985 provided for Hong Kong ordinances to replace English laws in specified fields.

The courts of justice comprise the Court of Final Appeal (inaugurated 1 July 1997), which hears appeals on civil and criminal matters from the High Court; the High Court (consisting of the Court of Appeal and the Court of First Instance); the Lands Tribunal, which determines on statutory claims for compensation over land and certain landlord and tenant matters; the District Court (which includes the Family Court); the Magistracies (including the Juvenile Court); the Coroner's Court; the Labour Tribunal, which provides a quick and inexpensive method of settling disputes between employers and employees; the Small Claims Tribunal, which deals with monetary claims involving amounts not exceeding HK$50,000; and the Obscene Articles Tribunal.

While the High Court has unlimited jurisdiction in both civil and criminal matters, the District Court has limited jurisdiction. The maximum term of imprisonment it may impose is seven years. Magistracies exercise criminal jurisdiction over a wide range of offences, and the powers of punishment are generally restricted to a maximum of two years' imprisonment or a fine of HK$100,000.

After being in abeyance for 25 years, the death penalty was abolished in 1992.

75,936 crimes were reported in 2011, of which 13,100 were violent crimes. 38,327 people were arrested in 2011, of whom 8,962 were for violent crimes. The population in penal institutions was 9,067 at 31 Dec. 2011 (127 per 100,000 population).

Education

In 2003 adult literacy was 97·1% among males and 90·5% among females. Universal basic education is available to all children aged from six to 15 years. In around three-quarters of the ordinary secondary day schools teaching has been in Cantonese since 1998–99, with about a quarter of ordinary secondary day schools still using English. In 2010 there were 148,940 pupils in 951 kindergartens, 331,112 in 572 primary schools (including 40 international schools) and 458,131 in 565 secondary schools (including 27 international schools).

The Hong Kong Technical Institutes and the Hong Kong Technical Colleges were renamed the Hong Kong Institute of Vocational Education in 1999. In the academic year 2005–06 there were 55,531 students enrolled at the Hong Kong Institute of Vocational Education.

The University of Hong Kong (founded 1911) had 12,916 full-time and 736 part-time students in 2010–11; the Chinese University of Hong Kong (founded 1963), 13,260 full-time and 654 part-time students; the Hong Kong University of Science and Technology (founded 1991), 7,208 full-time and 26 part-time students; the Hong Kong Polytechnic University (founded 1972 as the Hong Kong Polytechnic), 13,925 full-time and 807 part-time students; the City University of Hong Kong (founded 1984 as the City Polytechnic of Hong Kong), 10,221 full-time and 11 part-time students; the Hong Kong Baptist University (founded 1956 as the Hong Kong Baptist College), 5,050 full-time and 506 part-time students; the Lingnan University (founded 1967 as the Lingnan College), 2,287 full-time and five part-time students; and the Hong Kong Institute of Education (founded 1994), 3,270 full-time and 3,706 part-time students.

Estimated total government expenditure on education in 2011–12 was HK$68·3bn. (18·6% of total government spending and 3·6% of GDP).

Health

The Department of Health (DH) is the Government's health adviser and regulatory authority. The Hospital Authority (HA) is an independent body responsible for the management of all public hospitals. In 2009 there were 12,424 registered doctors, equivalent to 1·8 doctors per 1,000 population. In 2009 there were 2,126 dentists, 38,641 nurses and 4,525 midwives. The total number of hospital beds in 2009 was 35,062, including 26,872 beds in 38 public hospitals under the HA and 3,818 beds in 13 private hospitals. The bed-population ratio was 5·0 beds per thousand population.

The Chinese Medicine Ordinance was passed by the Legislative Council in July 1999 to establish a statutory framework to accord a professional status for Chinese medicine practitioners and ensure safety, quality and efficacy of Chinese medicine. In 2009 there were 6,048 registered Chinese medicine practitioners.

Total expenditure on health in 2006–07 amounted to HK$75,048m., an increase of 6·1% in real terms over that in 2005–06.

Welfare

Social welfare programmes include social security, family services, child care, services for the elderly, medical social services, youth and community work, probation, and corrections and rehabilitation. 171 non-governmental organizations are subsidized by public funds.

The government gives non-contributory cash assistance to needy families, unemployed able-bodied adults, the severely disabled and the elderly. Caseload as at Aug. 2011 totalled 280,358. Victims of natural disasters, crimes of violence and traffic accidents are financially assisted. Estimated recurrent government expenditure on social welfare for 2011–12 was HK$42·2bn.

RELIGION

In 2001 there were 4,970,000 Buddhists and Taoists, 290,000 Protestants and 280,000 Roman Catholics. The remainder of the population are followers of other religions. Joseph Zen Ze-kiun became Hong Kong's first cardinal in 2006. In Feb. 2013 the Roman Catholic church had two cardinals.

CULTURE

Press

In 2006 there were 48 newspapers of which 21 were Chinese-language dailies, 14 English dailies, eight bilingual dailies and five Japanese dailies. The newspapers with the highest circulation figures are all Chinese-language papers—*Oriental Daily News*, *Apple Daily* and *The Sun*. Circulation of dailies (including free papers) in 2008 was 3·6m. (2·0m. paid-for and 1·6m. free). Daily newspapers reached 80% of the population in 2007. A number of news agency bulletins are registered as newspapers.

Tourism

There were a record 36,030,300 visitor arrivals in 2010. Expenditure associated to inbound tourism totalled HK$209,983·0m. in 2010.

Festivals

The Hong Kong Arts Festival takes place in Feb.–March and features music, theatre, dance and opera. The Hong Kong International Film Festival (Aug.–Sept.) has been running annually since 1977.

FURTHER READING

Statistical Information: The Census and Statistics Department is responsible for the preparation and collation of government statistics. These statistics are published mainly in the *Hong Kong Monthly Digest of Statistics*. The Department also publishes monthly trade statistics, economic indicators and an annual review of overseas trade, etc. Website: http://www.censtatd.gov.hk

Hong Kong [various years] Hong Kong Government Press
Brown, J. M. (ed.) *Hong Kong's Transitions, 1842–1997.* 1997
Buckley, R., *Hong Kong: the Road to 1997.* 1997
Cottrell, R., *The End of Hong Kong: the Secret Diplomacy of Imperial Retreat.* 1993
Courtauld, C. and Holdsworth, M., *The Hong Kong Story.* 1997
Flowerdew, J., *The Final Years of British Hong Kong: the Discourse of Colonial Withdrawal.* 1997
Keay, J., *Last Post: the End of Empire in the Far East.* 1997
Lo, S.-H., *The Politics of Democratization in Hong Kong.* 1997
Lok, Sang Ho and Ash, Robert, *China, Hong Kong and the World Economy.* 2006
Roberts, E. V., *et al.*, *Historical Dictionary of Hong Kong and Macau.* 1993
Shipp, S., *Hong Kong, China: a Political History of the British Crown Colony's Transfer to Chinese Rule.* 1995
Welsh, F., *A History of Hong Kong.* 3rd ed. 1997

Macao

Região Administrativa Especial de Macau (Macao Special Administrative Region)

Population projection, 2015: 601,000
GNI per capita, 2010: US$47,022

KEY HISTORICAL EVENTS

Macao was visited by Portuguese traders from 1513 and became a Portuguese colony in 1557. Initially sovereignty remained vested in China, with the Portuguese paying an annual rent. In 1848–49 the Portuguese declared Macao a free port and established jurisdiction over the territory. On 6 Jan. 1987 Portugal agreed to return Macao to China on 20 Dec. 1999 when it would become a special administrative zone of China, with considerable autonomy.

TERRITORY AND POPULATION

The Macao Special Administrative Region, which lies at the mouth of the Pearl River, comprises a peninsula (9·3 sq. km) connected by a narrow isthmus to the People's Republic of China, on which is built the city of Santa Nome de Deus de Macao, the islands of Taipa (6·5 sq. km), linked to Macao by a 2-km bridge, Colôane (7·6 sq. km) linked to Taipa by a 2-km causeway, and Cotai, a strip of reclaimed land between Colôane and Taipa (5·2 km). The total area of Macao in 2006 was 28·6 sq. km. Additional land continues to be reclaimed from the sea. The population at the 2011 census was 552,503 (287,359 females); density, 19,318 people per sq. km. According to UN estimates, the entire population lived in urban areas in 2007. The official languages are Chinese and Portuguese, with the majority speaking the Cantonese dialect. Only about 2,000 people speak Portuguese as their first language.

The UN gives a projected population for 2015 of 601,000.

In 2009, 9,489 foreigners were legally registered for residency in Macao. There were 3,121 legal immigrants from mainland China.

SOCIAL STATISTICS

2009: births, 4,764 (8·8 per 1,000 population); deaths, 1,664 (3·1); marriages, 3,035 (5·6); divorces, 782 (1·4). Infant mortality, 2009, 2·1 per 1,000 live births. Life expectancy at birth (2005–08), 82·1 years.

CLIMATE

Sub-tropical tending towards temperate, with an average temperature of 23·0°C. The number of rainy days is around a third of the year. Average annual rainfall varies from 47–87" (1,200–2,200 mm). It is very humid from May to Sept.

CONSTITUTION AND GOVERNMENT

Macao's constitution is the 'Basic Law', promulgated by China's National People's Congress on 31 March 1993 and in effect since 20 Dec. 1999. It is a Special Administrative Region (SAR) of the People's Republic of China, and is directly under the Central People's Government while enjoying a high degree of autonomy.

RECENT ELECTIONS

At the elections held on 20 Sept. 2009 the Union for Development won two of 12 elected seats with 14·9% of votes cast, the Association of United Citizens of Macau two with 12·0% and the Democratic Prosperous Macau Association two with 11·6%. Six other parties won a single seat each. Turnout was 59·9%.

Fernando Chui Sai-on was elected chief executive on 26 July 2009, receiving 282 out of 296 votes in the Election Committee.

CURRENT GOVERNMENT

Chief Executive: Fernando Chui Sai-on; b. 1957 (sworn in 20 Dec. 2009).

Government Website: http://www.gov.mo

ECONOMY

Gaming is of major importance to the economy of Macao. It accounted for 37·2% of total GDP in 2008 and provides billions of dollars in taxes. In 2009, 19·7% of the workforce was directly employed by the casinos. In 2008 Macao overtook Nevada as the world's largest gaming market; in 2011 gaming industry total revenue was US$33,483m. (more than double the 2009 total and more than three times Nevada's 2011 total).

Overview

After its transfer of sovereignty to the People's Republic of China in 1999, Macao achieved high growth thanks to tourism and gambling. China's relaxation of travel restrictions in 1999 resulted in an increase in mainland visitors, reaching nearly 27m. in 2007 (up from 7m. in 1999). The cessation of Stanley Ho's monopoly of the local gaming industry in 2001 and its opening up to foreign competition led to an influx of foreign investment that made Macao the world's biggest gaming centre in 2008. Casino and gambling revenue totalled US$23·5bn. in 2010, up 57% on the previous year.

The economy grew an average 15·5% per year from 2003–09 according to the Economist Intelligence Unit. Over-dependence on the gaming sector leaves the economy susceptible to plans for casinos in other countries in the region. Macao's traditional manufacturing industries have virtually disappeared following the transfer of much of the textile industry to the Chinese mainland and, in 2005, the termination of the Multifibre Arrangement, which had governed international textile trade flows for three decades.

Currency

The unit of currency is the *pataca* (MOP) of 100 *avos* which is tied to the Hong Kong dollar at parity. Inflation was 8·6% in 2008 and 1·2% in 2009. Foreign exchange reserves were US$18,350m. in 2009. Total money supply was 30,608m. patacas in 2009.

Budget

In 2008 revenues totalled 51,077m. patacas; expenditures, 25,943m. patacas. Revenues from gaming tax accounted for 82·0% of total revenue in 2008; current expenditure accounted for 85·7% of expenditure.

Performance

Real GDP growth was 6·9% in 2005, rising to 16·6% in 2006. Total GDP in 2011 was US$36·4bn.

Banking and Finance

There are two note-issuing banks in Macao—the Macao branch of the Bank of China and the Macao branch of the Banco Nacional Ultramarino. The Monetary Authority of Macao functions as a central bank (*Chairman,* Teng Lin Seng). Commercial business is handled (2009) by 26 banks, 11 of which are local and 15 foreign. Total deposits, 2009 (including non-resident deposits), 290,534·0m. patacas. There are no foreign-exchange controls within Macao.

ENERGY AND NATURAL RESOURCES

Environment

Macao's carbon dioxide emissions from the consumption and flaring of fossil fuels in 2008 were the equivalent of 4·4 tonnes per capita.

Electricity

Installed capacity was 0·47m. kW in 2007; production, 1·52bn. kWh. Macao imported 1,683m. kWh of electricity in 2007.

Oil and Gas

202,688,000 litres of fuel oil were imported in 2009.

Fisheries

The catch in 2010 was estimated at 1,500 tonnes.

INDUSTRY

Although the economy is based on gaming and tourism there is a light industrial base of textiles and garments. In 2009 the number of manufacturing establishments was 1,002 (textiles and clothing, 304; food products and beverages, 234; publishing, printing and recorded media, 126).

Labour

In 2009 a total of 317,500 people were in employment, including 62,700 (19·7%) in gaming (up from 12,500 in 1999); 43,700 (13·8%), hotels, restaurants and similar activities; 41,500 (13·1%), wholesale and retail trade, repair of motor vehicles, motorcycles and personal and household goods; 32,700 (10·3%), construction; 25,600 (8·1%), real estate, renting and business activities; 20,300 (6·4%), public administration, defence and compulsory social security. Employment in 2009 was 96·5% of the labour force; unemployment rate stood at 3·6% (3·0% in 2008).

INTERNATIONAL TRADE

Imports and Exports

In 2009 imports (c.i.f.) were valued at US$4,750·9m., of which the main products were telecommunications, sound recording and reproducing equipment; petroleum and petroleum products; and gold, silverware, jewellery and articles of precious materials. In 2009 the chief import sources (in US$1m.) were: mainland China (1,451·6); Hong Kong (505·4); Japan (381·9).

2009 exports (f.o.b.) were valued at US$960·7m., of which the leading products were articles of apparel and clothing accessories; gold, silverware, jewellery and articles of precious materials; and petroleum oils and oils obtained from bituminous minerals. In 2009 the main export markets (in US$1m.) were: Hong Kong (377·5); USA (163·8); mainland China (139·9).

COMMUNICATIONS

Roads

In 2007 there were 401 km of roads. In 2007 there were 68,800 passenger cars in use (143 cars per 1,000 inhabitants), 2,100 buses and coaches, 4,600 lorries and vans, and 85,400 motorcycles and mopeds. There were 17 fatalities in road accidents in 2007.

A 50-km bridge linking Macao, Zhuhai in Guangdong Province in mainland China and Hong Kong is currently under construction and is expected to be finished in 2016.

Civil Aviation

An international airport opened in Dec. 1995. In 2009 Macau International Airport handled 3,643,970 passengers and 52,464 tonnes of freight (including transit cargo). In 2003 Air Macau flew to Bangkok, Beijing, Chengdu, Guilin, Haikou, Kaohsiung, Kota Kinabalu, Kuala Lumpur, Kunming, Manila, Nanjing, Ningbo, Shanghai, Singapore, Taipei and Xiamen.

Shipping

Regular services connect Macao with Hong Kong, 65 km to the northeast.

Telecommunications

In 2011 there were 165,500 landline telephone subscriptions (equivalent to 297·9 per 1,000 inhabitants) and 1,353,200 mobile phone subscriptions (or 2,435·0 per 1,000 inhabitants). In 2010, 53·8% of the population were internet users. In March 2012 there were 205,000 Facebook users.

SOCIAL INSTITUTIONS

Justice

There is a judicial district court, a criminal court and an administrative court with 24 magistrates in all.

In 2009 there were 12,406 crimes, of which 6,462 were against property. There were 930 persons in prison in Dec. 2009.

Education

There are three types of schools: public, church-run and private. In 2008–09 there were 142 schools and colleges. Number of students at the end of the 2008–09 academic year (with number of teachers): pre-primary, 9,270 (548); primary, 27,481 (1,727); secondary, 39,463 (2,572). There were nine special schools with

482 pupils and 108 teachers in 2008–09. In 2008–09 there were 12 higher education institutions with student enrolment of 20,917. There were also 137 adult education establishments with a total of 129,146 students enrolled and a teaching faculty of 1,628.

Expenditure on education came to 2·0% of GDP in 2008 and 13·0% of total government spending in 2009.

Health

In 2009 there were 723 doctors, 108 dentists and 450 nurses working in primary health care, and 560 doctors, 14 dentists and 1,169 nurses working in hospitals. In 2009 there were 1,294 hospital beds; there were 2·4 doctors per 1,000 population.

RELIGION

Non-religious persons account for 62% of the population. About 17% are Buddhists and 7% Roman Catholics.

CULTURE

World Heritage Sites

The historic centre of Macao was inscribed on the UNESCO World Heritage List in 2005.

Press

In 2009 there were 14 daily newspapers (nine in Chinese, three in Portuguese and two in English) and 11 weekly newspapers (ten in Chinese and one in Portuguese).

Tourism

Tourism is one of the mainstays of the economy. In 2009 there were 21·8m. tourists (of which 11·0m. were from mainland China, 6·7m. from Hong Kong and 1·3m. from Taiwan), down 5% on the 2008 total but nearly three times the 1999 total. Receipts in 2009 totalled US$17,843m.

Festivals

The government-run Macao International Music Festival featuring a wide range of Chinese and Western music takes place in Oct.–Nov.

FURTHER READING

Direcção dos Serviços de Estatística e Censos. *Anuário Estatístico/Yearbook of Statistics Macau in Figures.* Annual

Porter, J., *Macau, the Imaginary City: Culture and Society, 1557 to the Present.* 1996
Roberts, E. V., *Historical Dictionary of Hong Kong and Macau.* 1993

Statistics and Census Service Website: http://www.dsec.gov.mo

---

Taiwan[1]

Zhonghua Minguo
('Republic of China')

Capital: Taipei
Population, 2011: 23·2m.
GNI per capita: not available

KEY HISTORICAL EVENTS

Taiwan, christened Ilha Formosa ('beautiful island') by the Portuguese, was ceded to Japan by China by the Treaty of Shimonoseki in 1895. After the Second World War the island was surrendered to Gen. Chiang Kai-shek who made it the headquarters for his crumbling Nationalist Government. Until 1970 the USA supported Taiwan's claims to represent all of China. Only in 1971 did the government of the People's Republic

[1]See note on transcription of names in CHINA: Territory and Population.

of China manage to replace that of Chiang Kai-shek at the UN. In Jan. 1979 the USA established formal diplomatic relations with the People's Republic of China, breaking off formal ties with Taiwan. Taiwan itself has continued to reject attempts at reunification, and although there have been frequent threats of direct action from mainland China (including military manoeuvres off the Taiwanese coast) the prospect of confrontation with the USA supports the status quo.

In July 1999 President Lee Teng-hui repudiated Taiwan's 50-year-old 'One China' policy—the pretence of a common goal of unification—arguing that Taiwan and China should maintain equal 'state to state' relations. This was a rejection of Beijing's view that Taiwan is no more than a renegade Chinese province which must be reunited with the mainland, by force if necessary. In the presidential election of 18 March 2000 Chen Shui-bian, leader of the Democratic Progressive Party, was elected, together with Annette Lu Hsiu-lien as his Vice President. Both support independence although Chen Shui-bian has made friendly gestures towards China and has distanced himself from colleagues who want an immediate declaration of independence. Following his wife's indictment on embezzlement charges in Nov. 2006, President Chen survived three parliamentary attempts to impeach him.

TERRITORY AND POPULATION

Taiwan lies between the East and South China Seas about 160 km from the coast of Fujian. The territories currently under the control of the Republic of China include Taiwan, Penghu (the Pescadores), Kinmen (Quemoy) and Lienchiang (the Matsu Islands), as well as the archipelagos in the South China Sea. Off the Pacific coast of Taiwan are Green Island and Orchid Island. To the northeast of Taiwan are the Tiaoyutai Islets. The total area of Taiwan Island, the Penghu Archipelago and the Kinmen area (including the fortified offshore islands of Quemoy and Matsu) is 36,193 sq. km (13,974 sq. miles). Population (2011), 23,224,912. The ethnic composition is 84% native Taiwanese (including 15% of Hakka), 14% of Mainland Chinese, and 2% aborigine of Malayo-Polynesian origin. There were also 519,984 aboriginals of Malay origin in Dec. 2011. Population density: 642 per sq. km.

Taiwan's administrative units comprise (with 2011 populations): five special municipalities: Kaohsiung (2,774,470), New Taipei (3,916,451), Taichung (2,664,394), Tainan (1,876,960), Taipei, the capital (2,650,968); three provincial cities: Chiayi (271,526), Hsinchu (420,052), Keelung (379,927); 12 counties (*hsien*) in Taiwan Province: Changhwa (1,303,039), Chiayi (537,942), Hsinchu (517,641), Hualien (336,838), Ilan (459,061), Miaoli (562,010), Nantou (522,807), Penghu (97,157), Pingtung (864,529), Taitung (228,290), Taoyuan (2,013,305), Yunlin (713,556); two counties in Fujian Province: Kinmen (103,883), Lienchiang (10,106).

SOCIAL STATISTICS

In 2006 the birth rate was 9·0 per 1,000 population; death rate, 6·0 per 1,000. Population growth rate, 2006, 0·5%. Life expectancy, 2006: males, 74·1 years; females, 80·2 years. Infant mortality, 2006, 5·8 per 1,000 live births.

CLIMATE

The climate is subtropical in the north and tropical in the south. The typhoon season extends from July to Sept. The average monthly temperatures of Jan. and July in Taipei are 59·5°F (15·3°C) and 83·3°F (28·5°C) respectively, and average annual rainfall is 84·99" (2,158·8 mm). Kaohsiung's average monthly temperatures of Jan. and July are 65·66°F (18·9°C) and 83·3°F (28·5°C) respectively, and average annual rainfall is 69·65" (1,769·2 mm).

CONSTITUTION AND GOVERNMENT

The ROC Constitution is based on the Principles of Nationalism, Democracy and Social Wellbeing formulated by Dr Sun Yat-sen,

the founding father of the Republic of China. The ROC government is divided into three main levels: central, provincial/municipal and county/city, each of which has well-defined powers.

The central government consists of the *Office of the President*, the *National Assembly*, which is specially elected only for constitutional amendment, and five governing branches called '*yuan*', namely the Executive Yuan, the Legislative Yuan, the Judicial Yuan, the Examination Yuan and the Control Yuan. Beginning with the elections to the seventh Legislative Yuan held on 12 Jan. 2008 the Legislative Yuan has 113 members (formerly 225).

From 5 May to 23 July 1997 the *Additional Articles of the Constitution of the Republic of China* underwent yet another amendment. As a result a resolution on the impeachment of the President or Vice President is no longer to be instituted by the Control Yuan but rather by the Legislative Yuan. The Legislative Yuan has the power to pass a no-confidence vote against the premier of the Executive Yuan, while the president of the Republic has the power to dissolve the Legislative Yuan. The premier of the Executive Yuan is now directly appointed by the president of the Republic. Hence the consent of the Legislative Yuan is no longer needed.

In Dec. 2003 a law came into effect allowing for referendums to be held.

National Anthem

'San Min Chu I'; words by Dr Sun Yat-sen, tune by Cheng Mao-yun.

RECENT ELECTIONS

Presidential elections took place on 14 Jan. 2012. Ma Ying-jeou (Nationalist Party/Kuomintang) won 51·6% of the vote and Tsai Ing-wen (Democratic Progressive Party) 45·6%.

Elections to the Legislative Yuan were also held on 14 Jan. 2012. The Nationalist Party won 64 seats with 44·5% of votes cast; the Democratic Progressive Party, 40 (34·6%); the People First Party, 3; the Taiwan Solidarity Union, 3; the Non-Partisan Solidarity Union, 2; ind., 1.

Elections for an *ad hoc* National Assembly charged with amending the constitution were held on 14 May 2005. The Democratic Progressive Party took 127 of 300 seats (with 42·5% of the vote), the Nationalist Party 117 (38·9%), the Taiwan Solidarity Union 21 (7·1%), the People First Party 18 (6·1%) and the Jhang Ya Jhong Union 5 (1·7%). Turnout was 23·4%.

CURRENT GOVERNMENT

President: Ma Ying-jeou; b. 1950 (Nationalist Party/Kuomintang; sworn in 20 May 2008 and re-elected in Jan. 2012).

Vice President: Wu Den-yih.

Prime Minister and *President of the Executive Yuan:* Jiang Yi-huah; b. 1960 (Kuomintang; sworn in 18 Feb. 2013). There are nine ministries under the Executive Yuan: Interior; Foreign Affairs; National Defence; Finance; Education; Justice; Economic Affairs; Transport and Communications; Culture.

Vice President of the Executive Yuan (Deputy Premier): Mao Chi-kuo. *President, Control Yuan:* Wang Chien-shien. *President, Examination Yuan:* John Kuan. *President, Judicial Yuan:* Rai Hau-min. *President, Legislative Yuan:* Wang Jin-pyng. *Secretary General, Executive Yuan:* Chen Wei-zen. *Minister of Interior:* Lee Hong-yuan. *Foreign Affairs:* Lin Yung-lo. *National Defence:* Kao Hua-chu. *Finance:* Chang Sheng-ford. *Education:* Chiang Wei-ling. *Justice:* Tseng Yung-fu. *Economic Affairs:* Chang Chia-juch. *Transport and Communications:* Yeh Kuang-shih. *Culture:* Lung Ying-tai. *Ministers without Portfolio:* Lin Junq-tzer; Luo Ying-shay (also Minister of the *Mongolian and Tibetan Affairs Commission*); Huang Kuang-nan; Chang San-cheng; Yang Chiu-hsing; Kuan Chung-ming (also Minister of the *Council for Economic Planning and Development*); Chern Jenn-chuan (also Minister of the *Public Construction Commission*); Steven S. K. Chen; Schive Chi.

A number of commissions and subordinate organizations have been formed with the resolution of the Executive Yuan Council and the Legislature to meet new demands and handle new affairs. Examples include the Mongolian and Tibetan Affairs Commission; the Mainland Affairs Council; the Fair Trade Commission; the Public Construction Commission; and the Financial Supervisory Commission. These commissions, councils and agencies are headed by:

Council of Agriculture: Chen Bao-ji. *Atomic Energy Council:* Tsai Chuen-horng. *Directorate General of Budget, Accounting and Statistics:* Shih Su-mei. *Central Election Commission:* Chang Po-ya. *Coast Guard Administration:* Wang Ginn-wang. *Environmental Protection Administration:* Stephen Shu-hung Shen. *Fair Trade Commission:* Wu Shiow-ming. *Financial Supervisory Commission:* Chen Yuh-chang. *Council for Hakka Affairs:* Huang Yu-cheng. *Department of Health:* Chiu Wen-ta. *Council of Indigenous Peoples:* Sun Ta-chuan. *Council of Labour Affairs:* Pan Shih-wei. *Mainland Affairs Council:* Wang Yu-chi. *National Communications Commission:* Shyr Shyr-hau. *National Science Council:* Cyrus Chu. *Overseas Chinese Affairs Council:* Wu Ying-yih. *Directorate General of Personnel:* Huang Fu-yuan. *Research, Development and Evaluation Commission:* Sung Yu-hsieh. *Veterans' Affairs Commission:* Tseng Jing-ling.

Government Website: http://www.ey.gov.tw

DEFENCE

Conscription was reduced from 14 months to 12 months in 2009. The government has announced its intention to move towards a volunteer professional force—a process that was scheduled to start in 2011 and end in 2014. However, the costs involved mean that this transformation is unlikely to be viable. Defence expenditure in 2008 totalled US$10,495m. (US$458 per capita), representing 2·8% of GDP.

Army

The Army was estimated to number about 190,000 in 2000, including military police. Army reserves numbered 2·7m. In addition the Ministry of Justice, Ministry of Interior and the Ministry of Defence each command paramilitary forces totalling 25,000 personnel in all. The Army consists of Army Corps, Defence Commands, Airborne Cavalry Brigades, Armoured Brigades, Motorized Rifle Brigades, Infantry Brigades, Special Warfare Brigades and Missile Command.

Navy

Active personnel in the Navy in 2000 totalled 50,000. There are 425,000 naval reservists. The operational and land-based forces consist of four submarines, 16 destroyers and 21 frigates. There is a naval air wing operating 31 combat aircraft and 21 armed helicopters.

Air Force

Units in the operational system are equipped with aircraft that include locally developed IDF, F-16, Mirage 2000-5 and F-5E fighter-interceptors. There were 50,000 Air Force personnel in 2000 and 334,000 reservists.

INTERNATIONAL RELATIONS

By a treaty of 2 Dec. 1954 the USA pledged to defend Taiwan, but this treaty lapsed one year after the USA established diplomatic relations with the People's Republic of China on 1 Jan. 1979. In April 1979 the Taiwan Relations Act was passed by the US Congress to maintain commercial, cultural and other relations between USA and Taiwan through the American Institute in Taiwan and its Taiwan counterpart, the Co-ordination Council for North American Affairs in the USA, which were accorded quasi-diplomatic status in 1980. The People's Republic took over the China seat in the UN from Taiwan on 25 Oct. 1971. In May 1991 Taiwan ended its formal state of war with the People's Republic. Taiwan became a member of the World Trade Organization on 1 Jan. 2002.

In March 2013 Taiwan had formal diplomatic ties with 23 countries after Malawi agreed to recognize the People's Republic of China instead. In Aug. 2007, 15 of the diplomatic allies sponsored an unsuccessful proposal for Taiwan to join the UN.

ECONOMY

Overview

Taiwan has made a successful transition from an agricultural economy to one based on high-tech electronics. Economic growth averaged 8% during the last three decades, driven primarily by high value added manufacturing and exports, especially in electronics and computers. China has overtaken the USA as the prime export market.

Government-owned enterprises, including banks, have been privatized. Though largely escaping the impact of the 1997 Asian financial crisis, the economy went into recession in 2001 with the first year of negative growth ever recorded and unemployment reaching record highs. Strong export performance stimulated a recovery, with annual GDP growth above 3% since 2004, unemployment falling below 4% in 2007 and inflation consistently low.

Owing to a heavy dependence on exports, Taiwan suffered a severe downturn as a result of the global financial crisis. Major export industries such as semiconductors and memory chips declined, unemployment reached its highest levels since 2003 and, in 2008, the economy went into recession. However, a US$5·6bn. stimulus package was unveiled in late 2008 and by 2010 the economy was recording its highest growth rates for nearly three decades.

Currency

The unit of currency is the *New Taiwan dollar* (TWD) of 100 *cents*. Gold reserves were 13·62m. oz in Dec. 2010. There was deflation of 0·9% in 2009 but inflation of 1·0% in 2010 and 1·4% in 2011. Foreign exchange reserves were US$382·0bn. in Dec. 2010.

Budget

In 2006 general government revenues totalled NT$2,172,436m. and expenditures NT$2,261,958m. Tax revenue accounted for 71·7% of revenues in 2006; education, science and culture accounted for 21·6% of expenditures, economic development 17·0% and general administration 15·3%.

Performance

Taiwan sustained rapid economic growth at an annual rate of 9·2% from 1960 up to 1990. The rate slipped to 6·4% in the 1990s and 5·9% in 2000; Taiwan suffered from the Asian financial crisis, though less than its neighbours. Consumer prices showed increasing stability, rising at an average annual rate of 6·3% from 1960 to 1989, 2·9% in the 1990s and 1·3% in 2000. In 2001 global economic sluggishness and the events of 11 Sept. in the USA severely affected Taiwan's economy, which contracted by 2·2%. Per capita GNP stood at US$12,876, while consumer prices remained almost unchanged. Subsequent economic recovery led to growth of 5·4% in 2006 and 6·0% in 2007. Although there was negative growth of 1·8% in 2009, the economy grew by 10·7% in 2010 and by 4·0% in 2011.

Banking and Finance

The Central Bank of The Republic of China (Taiwan), reactivated in 1961, regulates the money supply, manages foreign exchange and issues currency. The *Governor* is Perng Fai-nan. The Bank of Taiwan is the largest commercial bank and the fiscal agent of the government. There are seven domestic banks, 38 commercial banks and 36 foreign banks.

There are two stock exchanges in Taipei.

ENERGY AND NATURAL RESOURCES

Environment
Taiwan's carbon dioxide emissions from the consumption and flaring of fossil fuels in 2008 were the equivalent of 13·3 tonnes per capita.

Electricity
Output of electricity in 2011 was 238·6m. MWh; total installed capacity was 41,401 MW. There were six units in three nuclear power stations in 2010.

Oil and Gas
Crude oil production in 2010 was 91,000 bbls; natural gas, 290m. cu. metres. Taiwan imports most of the oil and natural gas that it consumes.

Agriculture
In 2010 the cultivated area was 813,126 ha., of which 410,832 ha. were paddy fields. Rice production totalled 1,451,011 tonnes. Livestock production was valued at NT$144,614m., accounting for 34% of Taiwan's total agricultural production value.

Forestry
Forest area, 2010: 2,102,000 ha. Forest reserves: trees, 357,492,000 cu. metres; bamboo, 1,109m. poles. Timber production, 19,468 cu. metres.

Fisheries
The catch in 2010 was 851,384 tonnes, almost exclusively from sea fishing.

INDUSTRY

The largest companies in Taiwan by market capitalization in March 2012 were: Taiwan Semiconductor Manufacturing (US$74·6bn.), Hon Hai Precision Industry, an electronics manufacturer (US$41·5bn.); and Formosa Petrochemical (US$29·7bn.).

Output (in tonnes) in 2010: crude steel; 20·5m.; cement, 16·3m.; cotton fabrics, 270·5m. sq. metres; integrated circuit packages, 50·5trn. units; Global Positioning System (GPS) sets, 20·9bn. units.

Labour
In 2010 the average total labour force was 11·07m., of whom 10·49m. were employed. Of the employed population, 27·3% worked in manufacturing; 16·6% in wholesale and retail trade; 7·6% in construction; 6·9% in accommodation and food services; 5·9% in education; 5·2% in agriculture, forestry and fisheries. The unemployment rate was 5·2%.

INTERNATIONAL TRADE

Restrictions on the repatriation of investment earnings by foreign nationals were removed in 1994.

Imports and Exports
Imports in 2010 totalled US$251,236m. and exports US$274,601m.

In 2010 the main import suppliers were Japan (20·7%), mainland China (14·3%), USA (10·1%), South Korea (6·4%) and Saudi Arabia (4·7%). The main export markets were mainland China (28·0%), Hong Kong (13·8%), USA (11·5%), Japan (6·6%) and Singapore (4·4%).

Principal imports (2010), in US$1bn.: machinery and electrical equipment, 90·8; mineral fuels and lubricants, 51·6; chemicals, 34·5; manufactured goods (classified chiefly by material), 30·0; miscellaneous manufactured articles, 19·1; crude materials (inedible) except fuels, 12·7.

Principal exports (2010), in US$1bn.: machinery and transport equipment, 135·2; manufactured goods (classified chiefly by material), 41·7; miscellaneous manufactured articles, 38·5; chemicals, 35·6; mineral fuels and lubricants, 14·4.

COMMUNICATIONS

Roads
In 2006 there were 39,286 km of roads. In 2007, 5·7m. passenger cars, 117,100 buses and coaches, 1·0m. lorries and vans, and 13·9m. motorcycles and mopeds were in use. 1,007m. passengers and 594m. tonnes of freight were transported in 2006. There were 3,140 fatalities in road accidents in 2006.

Rail
In 2010 freight traffic amounted to 14·5m. tonnes and passenger traffic to 864m. Total route length was 1,741 km. There are metro systems in Taipei (opened in 1996) and in Kaohsiung (opened in 2008).

Civil Aviation
There are currently two international airports: Taiwan Taoyuan International Airport at Taoyuan near Taipei, and Kaohsiung International in the south. In addition there are 14 domestic airports: Taipei, Hualien, Taitung, Taichung, Tainan, Chiayi, Pingtung, Makung, Chimei, Orchid Island, Green Island, Wangan, Kinmen and Matsu (Peikan). A second passenger terminal at Taiwan Taoyuan International Airport opened in July 2000 as part of a US$800m. expansion project, which included aircraft bays, airport connection roads, a rapid transit link with Taipei, car parks and the expansion of air freight facilities, begun in 1989. In 2010 Taiwan Taoyuan International Airport handled 25,114,418 passengers, up from 18,681,462 in 2000.

The top airlines serving Taiwan (by capacity) as of Sept. 2011 were China Airlines (CAL), EVA Air, Cathay Pacific Airways, UNI Airways, TransAsia Airways (TNA) and Mandarin Airlines (MDA; CAL's subsidiary). In 2010, 37·5m. passengers and 1·2m. tonnes of freight were flown.

Regular direct flights between Taiwan and mainland China resumed in July 2008 for the first time since 1949.

Shipping
Maritime transportation is vital to the trade-oriented economy of Taiwan. In Jan. 2009 there were 154 ships of 300 GT or over registered, totalling 2·59m. GT. Of the 154 vessels registered, 50 were general cargo ships, 39 bulk carriers, 30 oil tankers, 24 container ships, eight passenger ships and three chemical tankers. There are six international ports: Kaohsiung, Keelung, Taichung, Hualien, Anping and Suao. The first three are container centres, Kaohsiung handling 9·68m. 20-ft equivalent units in 2008, making it the world's 12th busiest container port in terms of number of containers handled. Suao port is an auxiliary port to Keelung. In Jan. 2001 the first legal direct shipping links between Taiwanese islands and the Chinese mainland in more than 50 years were inaugurated.

Telecommunications
In 2011 there were 16,907,300 landline telephone subscribers (726·8 per 1,000 inhabitants). Taiwan's biggest telecommunications firm, the state-owned Chunghwa Telecom, lost its fixed-line monopoly in Aug. 2001. In 2011 there were 28,861,800 mobile phone subscribers, equivalent to 1,240·7 per 1,000 persons. There were 5·88m. fixed internet subscriptions in 2010 (253·3 per 1,000 inhabitants). In June 2007 there were 20·9 broadband subscribers per 100 inhabitants. In March 2012 there were 11·9m. Facebook users.

SOCIAL INSTITUTIONS

Justice
The Judicial Yuan is the supreme judicial organ of state. Comprising 15 grand justices, since 2003 these have been nominated and, with the consent of the Legislative Yuan, appointed by the President of the Republic. The grand justices hold meetings to interpret the Constitution and unify the interpretation of laws and orders. There are three levels of judiciary: district courts and their branches deal with civil and

criminal cases in the first instance; high courts and their branches deal with appeals against judgments of district courts; the Supreme Court reviews judgments by the lower courts. There is also the Supreme Administrative Court, high administrative courts and a Commission on the Disciplinary Sanctions of Public Functionaries. Criminal cases relating to rebellion, treason and offences against friendly relations with foreign states are handled by high courts as the courts of first instance.

The death penalty is still in force. There were six executions in 2012 (five in 2011). The population in penal institutions in Oct. 2008 was 63,370 (276 per 100,000 of national population).

Education
Since 1968 there has been compulsory education for six to 15-year-olds with free tuition. The illiteracy rate dropped from 7·1% in 1989 to 2·5% by 2006. There were 2,654 primary schools, 1,061 secondary schools and 156 vocational schools in 2008; and 102 universities, 45 colleges and 15 junior colleges. In 2005–06 there were 1,831,913 pupils with 101,682 teaching staff at elementary schools; 951,236 pupils and 48,816 teaching staff at junior high schools; 420,608 pupils and 34,112 teaching staff at senior high schools; and 331,604 students and 15,590 teaching staff at senior vocational schools. There were 1,259,490 students in universities and colleges in 2005–06 with 48,047 academic staff.

Health
In 2011 there were 40,002 physicians (one for every 510 persons), 5,570 doctors of Chinese medicine, 133,336 nurses, 12,032 dentists and assistants, and 31,300 pharmacists and assistants.

In 2010 there were 20,691 medical facilities serving 1,119 persons per facility; there were 158,922 beds and 68·6 beds per 10,000 persons.

In 2010 cancers, heart diseases, cerebrovascular diseases, diabetes and accidents were the first five leading causes of death.

Welfare
A universal health insurance scheme came into force in 1995 as an extension to 13 social insurance plans that cover only 59% of Taiwan's population. Premium shares among the government, employer and insured are varied according to the insured statuses. By the end of 2010, 23·07m. people or 99% of the population were covered by the National Health Insurance programme.

RELIGION
The Religious Affairs section of the Ministry of the Interior estimated that, in 2006, 35% of the population considered itself Buddhist and 33% Taoist. Smaller percentages adhered to various Christian denominations and to Islam.

CULTURE
Press
There were 23 daily newspapers in 2008 with a circulation of 4·2m. and 21 non-dailies with a circulation of 3·8m. The biggest circulation dailies are *The Liberty Times* and *Apple Daily*.

Tourism
In 2006, 3,520,000 international tourists visited Taiwan. Receipts totalled US$5,136m.

Festivals
The pop festival, Spring Scream, is held in April in Kenting.

FURTHER READING
Statistical Yearbook of the Republic of China. Annual. *The Republic of China Yearbook.* Annual. *Taiwan Statistical Data Book.* Annual. *Annual Review of Government Administration, Republic of China.* Annual.

Arrigo, L. G., et al., *The Other Taiwan: 1945 to the Present Day.* 1994
Cooper, J. F., *Historical Dictionary of Taiwan.* 1993
Hughes, C., *Taiwan and Chinese Nationalism: National Identity and Status in International Society.* 1997
Lary, Diana, *China's Republic.* 2006
Tsang, S. (ed.) *In the Shadow of China: Political Developments in Taiwan since 1949.* 1994

National library: National Central Library, Taipei (established 1986).
National Statistics Website: http://www.stat.gov.tw

COLOMBIA

Caribbean Sea

Barranquilla
Cartagena

PANAMA

VENEZUELA

PACIFIC
OCEAN

Medellín

□BOGOTÁ

Cali

COLOMBIA

Pasto

ECUADOR

BRAZIL

PERU

© Research Machines plc 2006

0   125 mi

0   200 km

República de Colombia
(Republic of Colombia)

Capital: Bogotá
Population projection, 2015: 49·37m.
GNI per capita, 2011: (PPP$) 8,315
HDI/world rank: 0·710/87
Internet domain extension: .co

KEY HISTORICAL EVENTS

Cundinamarca in central Colombia shows evidence of human habitation from 10,500 BC. The central highlands became the stronghold of the Chibcha people from 550 BC, with settlements based on the cultivation of corn, potatoes and beans. Much of the north, including the Caribbean coast, was associated with the Tairona civilization whose stone dwellings and ceremonial sites date from AD 1000.

Columbus explored South America's Caribbean coast in 1498, heralding the European influx. The port of Cartagena, founded in 1533 by Pedro de Heredia, became a leading administrative and commercial centre, its wealth derived from gold mined in the Cauca valley, Antioquia and the Choco. The Spanish explorer, Gonzalo Jiménez de Quesada, founded Bogotá in 1538 and within a decade the Chibcha were conquered and the New Kingdom of Granada declared. In 1564 the Spanish Crown appointed a president, loosely attached to the viceroyalty of Peru and centred on Bogotá, with jurisdiction over Colombia, Ecuador, Panama and Venezuela. Cartagena and the districts near the gold mines

prospered, while other parts of the colony remained undeveloped. Disputes with Lima led to the creation of the Viceroyalty of New Granada in 1719.

An independence movement followed the Napoleonic invasion of Spain in 1808. Antonio Nariño was a leading revolutionary figure, along with Francisco de Paula Santander and Simón Bolívar, who became an officer in the Patriot forces in 1811. Forced to flee to Jamaica when Gen. Pablo Morilla launched the Spanish reconquest in 1815, Bolívar settled in the Llanos (Colombia's eastern plains) and assembled a new army. Victory at Boyaca in 1819 secured Greater Colombia as an independent state. Following the Spanish surrender of Cartagena in 1821, Bolívar became president, ruling from Bogotá what is now Venezuela, Panama and (from 1822) Ecuador.

Separatist movements soon emerged and by 1826 Venezuela was autonomous. Ecuador withdrew three years later, with Colombia and Panama emerging as the Republic of New Granada, renamed the United States of Colombia following the signing of the Treaty of Rionegro in 1863. Politics was dominated by the reformist Liberal Party (PL) and the Conservatives, who sought to preserve Spanish and Roman Catholic heritage. Free trade led to the growth of exports, notably tobacco, quinine and coffee. Concerns over government weakness brought about a new Nationalist movement under Rafael Núñez. President from 1879, he introduced a constitution in 1886 that reversed the federalist trend and brought strong centralist control to the new Republic of Colombia. A Liberal revolt in 1899 sparked off the War of a Thousand Days, which had claimed over 100,000 lives by the time peace was achieved in 1902. The department of Panama became independent in 1903, backed by the USA, after plans for a canal across the isthmus were rejected by Colombia.

The rule of Gen. Rafael Reyes from 1904 was characterized by centralized power and industrial growth. Collapsing coffee prices caused economic hardship in the early 1930s but conditions improved under the Liberal governments of Enrique Olaya Herrera (1930–34) and Alfonso López Pumarejo (1934–38). After re-election in 1942, López Pumarejo faced a split in the PL, which culminated in an aborted coup and his resignation in 1945. The Conservatives took power the following year under Mariano Ospina Pérez, setting the stage for a civil war (*La Violencia*) that claimed 300,000 lives by 1957.

Conservatives and Liberals agreed to unite under the National Front in 1958, while other parties were banned. Several leftist guerrilla organizations emerged in the 1960s, including the Colombian Revolutionary Armed Forces (FARC), which aimed to install a Marxist regime. Generating wealth from the illegal drugs trade, it was especially influential in the rural south and east. Additionally, the National Liberation Army (ELN) was behind waves of kidnapping and violence in the mid-1990s.

In 1998 peace talks were started when a Conservative, Andrés Pastrana Arango, was elected president. FARC was granted a haven in the southeast. In 2000 Pastrana's 'Plan Colombia' initiative was launched, backed with US$1bn. from Washington to fight drug-trafficking. In 2001 FARC was accused of preparing attacks and conducting drug deals, and in Feb. 2002, following the kidnapping of a senator, Pastrana broke off peace talks. Three months later Álvaro Uribe Vélez, an independent, became president but within days declared a state of emergency after 20 died in bomb blasts in Bogotá. Relations with several rebel groups subsequently thawed, with talks with the rightist United Self-Defence Forces of Colombia (AUC) beginning in 2004 and with the ELN in 2005.

Parties loyal to Uribe secured victory in elections in 2006 and the president won a new term. In July 2007 hundreds of thousands of Colombians staged protests demanding the release of 3,000 people held hostage. The government made inroads into FARC territory in

2008 and secured the release of several high-profile hostages, including former presidential candidate Ingrid Betancourt. In July 2009 relations with Venezuela deteriorated over Colombia's plans to allow US military access to its bases and claims by Colombia that Caracas supplied arms to FARC. In June 2010 Juan Manuel Santos, a former defence minister, became president. In Dec. 2010 a state of emergency was declared following floods and landslides that killed almost 300 and affected over 2m. people.

TERRITORY AND POPULATION

Colombia is bounded in the north by the Caribbean Sea, northwest by Panama, west by the Pacific Ocean, southwest by Ecuador and Peru, northeast by Venezuela and southeast by Brazil. The estimated area is 1,141,748 sq. km (440,829 sq. miles). Population census (2005), 42,888,592; density, 37·6 per sq. km. More than 3·6m. Colombians protected and assisted by the Office of the United Nations High Commissioner for Refugees are displaced within the country as a consequence of continuing conflict involving left-wing guerrillas since the 1960s and right-wing paramilitaries and drug gangs since the 1980s.

The UN gives a projected population for 2015 of 49·37m.

In 2011, 75·4% lived in urban areas. Bogotá, the capital (census 2005): 6,824,510.

The following table gives census populations for departments and their capitals for 2005:

Departments	Area (sq. km)	Population	Capital	Population
Amazonas	109,665	67,726	Leticia	23,811
Antioquia	63,612	5,682,276	Medellín	2,175,681
Arauca	23,818	232,118	Arauca	62,634
Atlántico	3,388	2,166,156	Barranquilla	1,142,312
Bogotá[1]	1,587	6,840,116	—	
Bolívar	25,978	1,878,993	Cartagena	842,228
Boyacá	23,189	1,255,311	Tunja	146,621
Caldas	7,888	968,740	Manizales	353,312
Caquetá	88,965	420,337	Florencia	121,898
Casanare	44,640	295,353	Yopal	90,218
Cauca	29,308	1,268,937	Popayán	226,978
Cesar	22,905	903,279	Valledupar	299,065
Chocó	46,530	454,030	Quibdó	101,134
Córdoba	25,020	1,467,929	Montería	286,575
Cundinamarca	22,623	2,280,037	Bogotá	—
Guainía	72,238	35,230	Puerto Inírida	10,793
Guaviare	42,327	95,551	San José del Guaviare	34,863
Huila	19,890	1,011,418	Neiva	295,961
La Guajira	20,848	681,575	Riohacha	136,183
Magdalena	23,188	1,149,917	Santa Marta	385,122
Meta	85,635	783,168	Villavicencio	356,464
Nariño	33,268	1,541,956	Pasto	312,377
Norte de Santander	21,658	1,243,975	Cúcuta	567,664
Putumayo	24,885	310,132	Mocoa	25,751
Quindío	1,845	534,552	Armenia	273,114
Risaralda	4,140	897,509	Pereira	371,239
San Andrés y Providencia	44	70,554	San Andrés	48,421
Santander	30,537	1,957,789	Bucaramanga	509,216
Sucre	10,917	772,010	Sincelejo	219,639
Tolima	23,562	1,365,342	Ibagué	468,647
Valle del Cauca	22,140	4,161,425	Cali	2,083,171
Vaupés	65,268	39,279	Mitú	13,066
Vichada	100,242	55,872	Puerto Carreño	10,032

[1]Capital District.

Ethnic divisions (2000): Mestizo 47%, Mulatto 23%, White 20%, Black 6%, Indian 3%, mixed Black-Indian 1%.

The official language is Spanish.

SOCIAL STATISTICS

2008 estimates: births, 918,000; deaths, 248,000. Rates, 2008 estimates (per 1,000 population): births, 20·4; deaths, 5·5. Annual population growth rate, 2000–08, 1·5%. Life expectancy at birth, 2007, was 69·1 years for men and 76·5 years for women. Infant mortality, 2010, 17 per 1,000 live births; fertility rate, 2008, 2·4 children per woman. Abortion is illegal.

CLIMATE

The climate includes equatorial and tropical conditions, according to situation and altitude. In tropical areas, the wettest months are March to May and Oct. to Nov. Bogotá, Jan. 58°F (14·4°C), July 57°F (13·9°C). Annual rainfall 42" (1,052 mm). Barranquilla, Jan. 80°F (26·7°C), July 82°F (27·8°C). Annual rainfall 32" (799 mm). Cali, Jan. 75°F (23·9°C), July 75°F (23·9°C). Annual rainfall 37" (915 mm). Medellín, Jan. 71°F (21·7°C), July 72°F (22·2°C). Annual rainfall 64" (1,606 mm).

CONSTITUTION AND GOVERNMENT

Simultaneously with the presidential elections of May 1990, a referendum was held in which 7m. votes were cast for the establishment of a special assembly to draft a new constitution. Elections were held on 9 Dec. 1990 for this 74-member 'Constitutional Assembly' which operated from Feb. to July 1991. The electorate was 14·2m.; turnout was 3·7m. The Liberals gained 24 seats, M-19 (a former guerrilla organization), 19. The Assembly produced a new constitution which came into force on 5 July 1991. It stresses the state's obligation to protect human rights, and establishes constitutional rights to health care, social security and leisure. Indians are allotted two Senate seats. Congress may dismiss ministers, and representatives may be recalled by their electors.

The *President* is elected by direct vote. In Oct. 2005 the constitution was amended to allow a president to be re-elected for a second term. A vice-presidency was instituted in July 1991.

The legislative power rests with a *Congress* of two houses, the *Senate*, of 102 members (including two elected from a special list set aside for American Indian communities), and the *House of Representatives*, of 166 members, both elected for four years by proportional representation. Congress meets annually at Bogotá on 20 July.

National Anthem

'O! Gloria inmarcesible' ('Oh unfading Glory!'); words by R. Núñez, tune by O. Síndici.

GOVERNMENT CHRONOLOGY

Heads of State since 1945. (Partido de la U = Social National Unity Party; PLC = Colombian Liberal Party; PSC = Colombian Conservative Party/Colombian Social Conservative Party; n/p = non-partisan)

Presidents
1945–46	PLC	Alberto Lleras Camargo
1946–50	PSC	Luis Mariano Ospina Pérez
1950–51	PSC	Laureano Eleuterio Gómez Castro
1951–53	PSC	Roberto Urdaneta Arbeláez
1953	PSC	Laureano Eleuterio Gómez Castro
1953–57	military	Gustavo Rojas Pinilla

Military Junta
1957–58		Gabriel Paris Gordillo (chair); Rubén Piedrahíta Arango; Deogracias Fonseca Espinosa; Luis Ernesto Ordóñez Castillo; Rafael Navas Pardo

Presidents
1958–62	PLC	Alberto Lleras Camargo
1962–66	PSC	Guillermo León Valencia Muñoz
1966–70	PLC	Carlos Lleras Restrepo
1970–74	PSC	Misael Eduardo Pastrana Borrero
1974–78	PLC	Alfonso López Michelsen
1978–82	PLC	Julio César Turbay Ayala
1982–86	PSC	Belisario Betancur Cuartas

1986–90	PLC	Virgilio Barco Vargas
1990–94	PLC	César Augusto Gaviria Trujillo
1994–98	PLC	Ernesto Samper Pizano
1998–2002	PSC	Andrés Pastrana Arango
2002–10	n/p	Álvaro Uribe Vélez
2010–	Partido de la U	Juan Manuel Santos

RECENT ELECTIONS

Presidential elections were held on 30 May 2010, in which Juan Manuel Santos (Social National Unity Party) won with 46·7% of votes cast, against 21·5% for Antanas Mockus Šivickas (Green Party), 10·1% for Germán Vargas Lleras (Radical Change Party), 9·1% for Gustavo Petro (Alternative Democratic Pole) and 6·1% for Noemí Sanín (Colombian Conservative Party). Four other candidates received less than 5% of the vote each. A second round run-off was held between Santos and Mockus on 20 June 2010 in which Santos took 69·1% of the vote compared to Mockus' 27·5%. Turnout was 49·2% in the first round and 44·5% in the second.

Congressional elections were held on 14 March 2010. In elections to the House of Representatives the Social National Unity Party won 47 seats, the Colombian Conservative Party 38, the Colombian Liberal Party 37, the Radical Change Party 15, the National Integration Party 12, Alternative Democratic Pole 4 and the Green Party 3, with smaller parties accounting for the remainder. In the elections to the Senate the Social National Unity Party won 28 seats, the Colombian Conservative Party 22, the Colombian Liberal Party 17, the National Integration Party 9, the Radical Change Party 8, Alternative Democratic Pole 8 and the Green Party 5, again with smaller parties accounting for the remainder.

CURRENT GOVERNMENT

President: Juan Manuel Santos; b. 1951 (Social National Unity Party/Partido de la U; sworn in 7 Aug. 2010).

Vice President: Angelino Garzón.

In March 2013 the government comprised:

Minister of Agriculture and Rural Development: Juan Camilo Restrepo. *Culture:* Mariana Garcés Córdoba. *Defence:* Juan Carlos Pinzón Bueno. *Education:* María Fernanda Campo Saavedra. *Environment and Sustainable Development:* Juan Gabriel Uribe. *Finance and Public Credit:* Mauricio Cárdenas. *Foreign Relations:* María Ángela Holguín Cuéllar. *Health and Social Welfare:* Alejandro Gaviria Uribe. *Housing and Territorial Development:* Germán Vargas Lleras. *Information Technologies and Communications:* Diego Molano Vega. *Interior:* Fernando Carillo. *Justice:* Ruth Stella Correa. *Labour:* Rafael Pardo Rueda. *Mines and Energy:* Federico Renjifo. *Trade, Industry and Tourism:* Sergio Díaz-Granados. *Transport:* Cecilia Álvarez-Correa Glenn.

Office of the President (limited English):
   http://www.presidencia.gov.co

CURRENT LEADERS

Juan Manuel Santos Calderón

Position
President

Introduction
Juan Manuel Santos became president in Aug. 2010. He succeeded Álvaro Uribe, under whom he had served as defence minister. He is the head of the Partido de la U.

Early Life
Santos was born in Aug. 1951 in Bogotá into a prestigious family. His great uncle, Eduardo Santos, was president from 1938–42. His father, Enrique Castillo, was editor of a national newspaper, *El Tiempo*, for 50 years. Santos attended the private Colegio San Carlos and the Cartagena Naval School before graduating in

economics and business administration from Kansas University in the USA in 1972.

He then returned to Colombia and worked for the Colombian Coffee Delegation, which he represented in London. During this time he completed further studies at the London School of Economics, Harvard University and the Fletcher School of Law and Diplomacy. In 1981 he joined *El Tiempo*.

He served as minister of foreign trade in César Gaviria's Liberal Party (PLC) government from 1990–94, resigning from *El Tiempo* in 1992 to dedicate himself to politics. Although not invited to join the 1994 administration of Ernesto Samper, he was involved in peace talks with FARC and was elected to the PLC's directorial committee in 1995. In 1999 Santos returned to *El Tiempo* and became head of the UN Economic Commission for Latin America and the Caribbean (ECLAC).

After the 2002 presidential election, Santos became a dissident within the Liberal ranks when he came out in support of the non-partisan President Uribe. He expressed particular support for Uribe's tough 'democratic security' policy in the face of calls for further talks with FARC. In 2005 Santos left the PLC and founded the Social National Unity Party (the Partido de la U), a pro-Uribe party of which Uribe himself was not a member.

Uribe named Santos his defence minister in 2006. He took a tough line against FARC while strengthening US–Colombian military ties. In 2008 he oversaw the rescue of several hostages held by FARC, including former presidential candidate Ingrid Betancourt. In the same year the defence forces came under scrutiny for the killing of FARC's second-in-command, Raúl Reyes, in an unauthorized military operation within Ecuador's northern border. There was also a scandal after evidence surfaced that the armed forces had carried out extrajudicial executions to boost the number of guerrillas killed.

In 2010 Santos stood for president on a platform of maintaining the legacy of Uribe and fighting 'narco-terrorist' violence. He also promised ambitious land reforms and fairer wealth distribution. In all he outlined 109 policy proposals, framed around the ideas of 'democratic security' and 'democratic prosperity'. He won the election by the largest margin in Colombian history.

Career in Office
Santos pledged the creation of 2·4m. extra jobs to bring unemployment below 9% and the construction of 1m. new homes over four years. He also promised a cut in income and corporate taxes, and reforms to the revenue system, land possession and health imbalances. His government was expected to invest 10% of its revenues from natural resources into science and technology. In addition, he has created a ministry of justice, raised the minimum wage and re-established diplomatic relations with Venezuela.

In May 2011 Congress approved a law, described as historic by Santos, to compensate the many victims who have suffered in the country's internal armed conflicts. In Aug.–Sept., following a resurgence of FARC guerrilla activity, Santos replaced his defence minister and appointed new military commanders, and in Nov. 2011 Alfonso Cano, the FARC leader, was killed in a gun battle with government forces. Santos subsequently pursued a policy of dialogue with FARC and exploratory contacts led to the launch of the first substantive round of peace negotiations in Nov. 2012 in Cuba.

In May 2012 a free trade agreement with the USA, which had been signed in 2006, came into force.

Santos underwent reportedly successful surgery in Bogotá in Oct. 2012, having earlier been diagnosed with cancer.

DEFENCE

There is selective conscription for 12 to 24 months. In 2008 defence expenditure totalled US$9,546m. (US$221 per capita), representing 3·9% of GDP. Colombia is the second largest military spender in South America, after Brazil.

Army
Personnel (2007) 217,000 (conscripts, 173,000); reserves number 54,700. The national police numbered (2007) 136,100.

Navy
The Navy has two diesel powered submarines, two midget submarines and four small frigates. Naval personnel in 2007 totalled 27,600. There are also two brigades of marines numbering 14,000. An air arm operates light reconnaissance aircraft.

The Navy's main ocean base is Cartagena with Pacific bases at Buenaventura and Málaga. There are in addition numerous river bases.

Air Force
The Air Force has been independent of the Army and Navy since 1943, when its reorganization began with US assistance. It has 115 combat-capable aircraft and 31 attack helicopters. There are six combat air commands plus a command responsible for air operations in a specific geographical area. Total strength (2006), 8,600 personnel, including 1,900 conscripts.

INTERNATIONAL RELATIONS
It was announced in Aug. 2000 that Colombia would receive US$1·3bn. in anti drug-trafficking aid (mostly of a military nature) from the USA as part of 'Plan Colombia', a five-year long series of projects intended to serve as a foundation for stability and peace of which the focal point is the fight against drugs. By May 2005 the USA had given aid amounting to US$4·5bn.

In March 2008 Colombian forces killed a high-level member of the rebel FARC movement during an unsanctioned raid over the Ecuadorian border. A week-long diplomatic incident ensued, in which both Ecuador and Venezuela massed military personnel on their respective borders with Colombia.

ECONOMY
In 2010 agriculture accounted for 7·1% of GDP, industry 36·3% and services 56·6%.

Overview
Colombia's economy has been blighted by guerrilla insurgencies, drug cartels, human rights violations and an unsustainable fiscal deficit. In 2008 the World Bank estimated that conflict had reduced growth by 2% per year since 1980, reducing GDP per capita by two-thirds. The poverty rate stood at 37·2% in 2010, down from 47·4% in 2009. Youth unemployment is a long-term problem, while school enrolment rates in 2011 had dropped to their lowest level in a decade.

Increased internal security has helped stabilize the country though corruption in government and the judicial system pose obstacles to long-term development. Substantial oil reserves and deposits of gold, silver, emeralds, platinum and coal have driven growth. Improvements in regulatory frameworks and more open market policies along with increased government spending and private sector investment helped GDP grow by above the Latin American average in 2011 at 5·9%.

The country is the largest producer of cocaine in the world and it is estimated that trade in illegal drugs accounts for 3% of GDP, though production is declining. Colombia is also among the largest exporters of coffee. However, a combination of the rising value of the peso, a fall in the coffee price and unfavourable weather conditions resulted in failure to reach 2011–12 production targets, which were already at their lowest level for 33 years. Improving agricultural productivity remains a key goal for the government, with 560,000 farmers owning an average of 13 acres each. However, more efficient production of crops, including coffee, tobacco, rice, bananas and sugarcane, remains elusive.

Annual foreign direct investment (FDI), focused on the oil sector, reached a record US$13bn. in 2011. A Free Trade Agreement with the USA was approved by the Colombian Congress in 2007 but was not ratified by the US Congress until Oct. 2011 and only came into force in May 2012. Colombia also has trading agreements with Mexico, Chile, Canada, the EU and Switzerland. The USA remains its chief trading partner, accounting for 34% of all trade in 2010.

Currency
The unit of currency is the *Colombian peso* (COP) of 100 *centavos*. Inflation rates (based on IMF statistics):

2002	2003	2004	2005	2006	2007	2008	2009	2010	2011
6·3%	7·1%	5·9%	5·0%	4·3%	5·5%	7·0%	4·2%	2·3%	3·4%

In Sept. 2009 gold reserves were 221,000 troy oz and foreign exchange reserves US$22,926m. Total money supply was 41,403bn. pesos in Aug. 2009.

Budget
In 2007 revenues totalled 103,986bn. pesos and expenditures 110,014bn. pesos. Tax revenue accounted for 56·4% of revenues in 2007; grants accounted for 37·9% of expenditures and interest 25·1%.

The standard rate of VAT is 16%.

Performance
Real GDP growth rates (based on IMF statistics):

2003	2004	2005	2006	2007	2008	2009	2010	2011
3·9%	5·3%	4·7%	6·7%	6·9%	3·5%	1·7%	4·0%	5·9%

Total GDP in 2011 was US$333·4bn.

Banking and Finance
In 1923 the Bank of the Republic (*Governor*, José Darío Uribe Escobar) was inaugurated as a semi-official central bank, with the exclusive privilege of issuing banknotes. Its note issues must be covered by a reserve in gold of foreign exchange of 25% of their value. Interest rates of 40% plus are imposed.

There are 24 commercial banks, of which 18 are private or mixed, and six official. There is also an Agricultural, Industrial and Mining Credit Institute, a Central Mortgage Bank and a Social Savings Bank. Demand deposits totalled 11,256bn. pesos in Dec. 2002. The Superintendencia Bancaria acts as a supervising body.

External debt totalled US$63,064m. in 2010 and represented 22·8% of GNI.

There are stock exchanges in Bogotá, Medellín and Cali.

ENERGY AND NATURAL RESOURCES

Environment
In 2008 Colombia's carbon dioxide emissions from the consumption and flaring of fossil fuels were the equivalent of 1·5 tonnes per capita. An *Environmental Performance Index* compiled in 2008 ranked Colombia ninth in the world, with 88·3%. The index examined various factors in six areas—air pollution, biodiversity and habitat, climate change, environmental health, productive natural resources and water resources.

Electricity
Installed capacity of electric power (2007) was 13·3m. kW. In 2007 production was 55·31bn. kWh and consumption per capita 1,146 kWh. Colombia exported 877m. kWh of electricity in 2007.

Oil and Gas
Oil production (2011) 46·0m. tonnes, up from 29·6m. tonnes in 2008. Colombia's oil output has been growing at one of the fastest rates of any producer in recent years. Natural gas production in 2008 totalled 9·1bn. cu. metres. In 2008 there were proven oil reserves of 1·4bn. bbls and proven natural gas reserves of 110bn. cu. metres.

Minerals

Production (2005): gold, 35,785 kg; silver, 7,142 kg; platinum, 1,082 kg. Other important minerals include: copper, lead, mercury, manganese, nickel and emeralds (of which Colombia accounts for over half of world production).

Coal production (2007): 69·90m. tonnes; iron ore (2005): 498,623 tonnes; salt production (2005): 473,996 tonnes.

Agriculture

There is a wide range of climate and, consequently, crops. In 2002 there were 2·29m. ha. of arable land and 1·56m. ha. of permanent crops.

In 2002, 19·3% of the economically active population were engaged in agriculture. Production, 2002 (in 1,000 tonnes): sugarcane, 35,800; plantains, 2,921; potatoes, 2,841; rice, 2,347; cassava, 1,768; bananas, 1,424; maize, 1,189; coffee, 691. Coca was cultivated in 2005 on approximately 144,000 ha., up from 34,000 ha. in 1988. Estimated coca leaf production in 2005 totalled 170,730 tonnes, making Colombia the world's largest producer of coca leaves, the raw material for cocaine. Production of cocaine was estimated at 640 tonnes.

Livestock (2002): 24·76m. cattle; 2·65m. horses; 2·26m. sheep; 2·23m. pigs; 1·10m. goats; 723,000 asses; 115m. chickens.

Livestock products, 2002: beef and veal, 6·76m. tonnes; poultry meat, 629,000 tonnes; pork, bacon and ham, 110,000 tonnes; milk, 6,021,000 tonnes; eggs, 314,000 tonnes.

Forestry

In 2010 the area under forests was 60·50m. ha., or 55% of the total land area. Timber production in 2007 was 10·44m. cu. metres.

Fisheries

The estimated total catch (2008) was 135,000 tonnes, of which 85% was from marine waters.

INDUSTRY

The largest companies by market capitalization in Colombia in April 2012 were Ecopetrol (US$119·7bn.); GrupoAval, a holding company engaged in a variety of financial activities (US$13·7bn.); and Bancolombia, the country's largest commercial bank (US$13·3bn.).

Production, 2007 unless otherwise indicated (in 1,000 tonnes): cement (2005), 9,959; distillate fuel oil, 4,395; residual fuel oil, 3,318; petrol, 3,164; raw sugar, 2,277; crude steel and semi-finished products, 1,245; jet fuel, 537; sawnwood, 382,000 cu. metres; passenger cars, 68,000 units.

Labour

The economically active workforce in 2001 was 18·65m., of which 16·62m. were employed. The main areas of activity in 2001 were: wholesale and retail trade, restaurants and hotels (employing 4·19m. persons); community, social and personal services (3·74m.); and agriculture, hunting, forestry and fishing (3·49m.). The unemployment rate in 2002 was 15·7%.

INTERNATIONAL TRADE

Foreign companies are liable for basic income tax of 30% and surtax of 7·5%. Since 1993 tax on profit remittance has started at 12%, reducing (except for oil companies) to 7% after three years.

The Group of Three (G-3) free trade pact with Mexico and Venezuela came into effect on 1 Jan. 1995. In Nov. 2006 Colombia and the USA signed a free trade agreement that eliminates tariffs on each other's goods; it was ratified in Oct. 2011 and came into effect in May 2012.

Imports and Exports

In US$1m.:

	2007	2008	2009	2010	2011
Imports c.i.f.	32,897	39,669	32,898	40,683	54,675
Exports f.o.b.	29,991	37,626	32,853	39,820	56,954

Imports and exports have both doubled since 2006. Major import suppliers, 2010: USA (25·9%), China (13·5%), Mexico (9·5%), Brazil (5·8%). Major export markets, 2010: USA (43·1%), China (4·9%), Ecuador (4·6%), Netherlands (4·1%).

Main imports in 2006 were (in US$1m.): machinery and transport equipment (10,508·9), chemicals and related products (5,230·2), food and live animals (1,890·2), iron and steel (1,257·5), textile yarn and fabrics (863·5). Main exports in 2006 were (in US$1m.): petroleum and petroleum products (6,309·8), coal (2,807·2), chemicals and related products (2,024·4), coffee and coffee substitutes (1,633·7), machinery and transport equipment (1,519·8).

COMMUNICATIONS

Roads

Total length of roads was 164,278 km in 2006 (including 14,143 km of main roads). In 2005 there were 2,686,000 vehicles in use, including 1,607,000 passenger cars. There were 5,486 road accident fatalities in 2006.

Rail

The National Railways (2,532 km of route, 914 mm gauge) went into liquidation in 1990. There are currently two concessions operating—Ferrocarril del Oeste and Red Férrea del Atlántico. Ferrocarril del Oeste carried 360,000 tonnes of freight in 2007 and Red Férrea del Atlántico 22m. tonnes in 2006. Passenger services are very limited. Total length in 2007 was 1,663 km. A metro system operates in Medellín.

Civil Aviation

There are international airports at Barranquilla, Bogotá (Eldorado), Cali, Cartagena, Medellín and San Andrés. The main Colombian airline is Avianca. In 2005 scheduled traffic of Colombian-based carriers flew 132·5m. km and carried 9,933,100 passengers. The busiest airport is Bogotá, which in 2000 handled 7,154,312 passengers (5,234,807 on domestic flights) and 372,957 tonnes of freight.

Shipping

In Jan. 2009 there were 31 ships of 300 GT or over registered, totalling 46,000 GT. The chief port is Cartagena, which handled 25·6m. tonnes of foreign cargo in 2008. The Magdalena River is subject to drought, and navigation is always impeded during the dry season, but it is an important artery of passenger and goods traffic. The river is navigable for 1,400 km; steamers ascend to La Dorada, 953 km from Barranquilla.

Telecommunications

In 2008 there were 8,054,000 main (fixed) telephone lines. In the same year mobile phone subscribers numbered 41,365,000 (919·0 per 1,000 persons). There were 17,330,000 internet users in 2008. In June 2012 there were 16·8m. Facebook users.

SOCIAL INSTITUTIONS

Justice

The July 1991 constitution introduced the offices of public prosecutor and public defence. The Supreme Court, at Bogotá, of 20 members, is divided into three chambers—civil cassation (six), criminal cassation (eight), labour cassation (six). Each of the 61 judicial districts has a superior court with various sub-dependent tribunals of lower juridical grade. In Dec. 1997 the constitution was amended to allow the extradition of Colombian nationals.

In 2010 there were 15,459 murders, continuing a downward trend since the high of 28,837 in 2002. Colombia's murder rate, at 33 per 100,000 persons, although declining is still among the highest in the world. In 2011 the reported number of kidnappings totalled 298, down from 3,572 in 2000 although up from 213 in 2009.

Colombia abolished the death penalty in 1997. The population in penal institutions in Sept. 2008 was 69,689 (149 per 100,000 of national population).

Education
Education between the ages of five and 15 became compulsory in 1991. Schools are both state and privately controlled. In 2007 there were 49,538 teaching staff for 1,081,343 children in pre-primary schools; 187,821 teaching staff for 5,298,567 pupils in primary schools; and 4,657,360 pupils with 164,484 teaching staff in secondary schools. In 2008 there were 44,757 pre-primary schools, 57,711 primary schools and 18,412 secondary schools.

There were 32 public universities in 2007. The National University of Colombia (Universidad Nacional de Colombia), founded in 1867, is the leading institution in the tertiary sector. Colombia also has many private universities and colleges of art and music. In 2007 there were 1,372,674 students in total in higher education and 88,337 academic staff.

Adult literacy in 2008 was 93%.

In 2007 public expenditure on education came to 5·1% of GNI and represented 12·6% of total government expenditure.

Health
In 2003 there were 1,165 hospitals with 49,000 beds. Medical personnel (2002) was as follows: doctors, 58,761; dentists, 33,951; nurses and midwives, 103,158.

Welfare
The retirement age is 60 (men), or 55 (women) although it is set to be raised to 62 for men and 60 for women after 2014; to be eligible for a state pension 1,000 weeks of contributions are required. The minimum social insurance pension is equal to the minimum wage. If a private pension is less than the minimum pension set by law, the government makes up the difference.

Unemployment benefit is a month's wage for every year of employment.

RELIGION
The religion is Roman Catholic (41·02m. adherents in 2007), with the Cardinal Archbishop of Bogotá as Primate of Colombia and 20 other archbishoprics. There are also 94 bishops, 3,994 parishes and 8,312 priests. In Feb. 2013 there were three cardinals. Other forms of religion are permitted so long as their exercise is 'not contrary to Christian morals or the law'. In addition to Roman Catholics there are also some agnostics, atheists, Bahais, Ethnoreligionists and Spiritists.

CULTURE

World Heritage Sites
Colombia's heritage sites as classified by UNESCO (with year entered on list) are: the Port, Fortresses and Group of Monuments, Cartagena (1984)—on the Caribbean coast, Cartagena was one of the first cities to be founded in South America and has the most extensive fortifications in the continent; Los Katíos National Park (1994) covers 72,000 ha. in northwest Colombia; founded 200 km inland on the River Magdalena in the 1530s, the Historic Centre of Santa Cruz de Mompox (1995), or Mompós, was a focal point for colonization and a vital trade post between the Caribbean coast and the interior—its colonial aspect has been preserved; the National Archaeological Park of Tierradentro (1995) in the southwest contains statues and elaborately decorated underground tombs dating from the 6th–10th centuries; the San Agustín Archaeological Park (1995) protects religious monuments and sculptures from the 1st–8th centuries; the Malpelo Fauna and Flora sanctuary (2006) provides a critical habitat for internationally threatened marine species; and the Coffee Cultural Landscape of Colombia (2011) reflects a tradition of coffee growing in small plots in the high forests.

Press
There were 44 daily newspapers in 2008 (42 paid-for and two free), with daily circulation totalling 1·78m.

Tourism
In 2007 there were 1,195,000 non-resident visitors (including 944,000 from elsewhere in the Americas and 220,000 from Europe), bringing revenue of US$2,262m.

DIPLOMATIC REPRESENTATIVES

Of Colombia in the United Kingdom (Flat 3a, 3 Hans Cres., London, SW1X 0LN)
Ambassador: Dr José Mauricio Rodríguez Múnera.

Of the United Kingdom in Colombia (Edificio Ing. Barings, Carrera 9 No 76–49, Piso 8, Bogotá)
Ambassador: Lindsay Croisdale-Appleby.

Of Colombia in the USA (2118 Leroy Pl., NW, Washington, D.C., 20008)
Ambassador: Carlos Urrutia.

Of the USA in Colombia (Carrera 45 # 22D-45, Bogotá)
Ambassador: P. Michael McKinley.

Of Colombia to the United Nations
Ambassador: Néstor Osorio Londoño.

Of Colombia to the European Union
Ambassador: Rodrigo Rivera Salazar.

FURTHER READING

Departamento Administrativo Nacional de Estadística. *Boletín de Estadística.* Monthly.

Dudley, Steven, *Walking Ghosts: Murder and Guerrilla Politics in Colombia.* 2004
Hylton, Forrest, *Evil Hour in Colombia.* 2006
Palacios, Marco, *Between Legitimacy and Violence: A History of Colombia, 1875–2002.* 2006

National Statistical Office: Departamento Administrativo Nacional de Estadística (DANE), AA 80043, Zona Postal 611, Bogotá.
Website: http://www.dane.gov.co

Explore the world at
www.statesmansyearbook.com

COMOROS

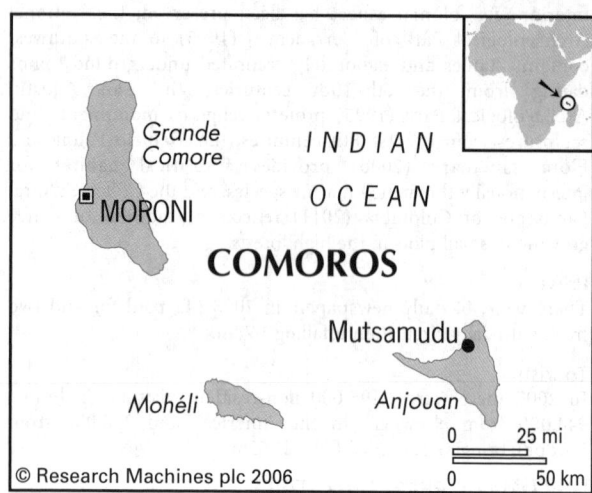

© Research Machines plc 2006

Union des Comores
(Union of the Comoros)

Capital: Moroni
Population projection, 2015: 832,000
GNI per capita, 2011: (PPP$) 1,079
HDI/world rank: 0·433/163
Internet domain extension: .km

KEY HISTORICAL EVENTS

The three islands forming the present state became French protectorates at the end of the 19th century and were proclaimed colonies in 1912. With neighbouring Mayotte they were administratively attached to Madagascar from 1914 until 1947 when the four islands became a French Overseas Territory, achieving internal self-government in Dec. 1961. In referendums held on each island on 22 Dec. 1974, the three western islands voted overwhelmingly for independence, while Mayotte voted to remain French.

There have been dozens of coups or attempted takeovers since independence. In April 1999 an agreement was brokered on a federal structure for the three main islands—Grande Comore, Anjouan and Mohéli—to be known as the Union of the Comoros. However, the delegates from Anjouan did not sign the agreement. Violence broke out in the capital, Moroni, against Anjouans living there. A military coup followed on 30 April 1999, led by Col. Azaly Assoumani. He subsequently dissolved the government and the constitution, declaring a transitional government.

A new constitution in 2001 kept the union in place but with greater autonomy for each constituent island. Col. Azaly Assoumani was elected president of the federation in April 2002, holding office until May 2006 when he was replaced by Ahmed Abdallah Mohamed Sambi. In May 2007 Mohamed Bascar, president of Anjouan, refused to stand down and held local elections in defiance of the federal government, prompting an African Union-led blockade. In March 2009 African Union and Comoran forces seized control of the island.

TERRITORY AND POPULATION

The Comoros consists of three islands in the Indian Ocean between the African mainland and Madagascar with a total area of 1,862 sq. km (719 sq. miles). The population at the 2003 census was 575,660 (285,590 males), giving a density of 309 per sq. km.

The UN gives a projected population for 2015 of 832,000.

In 2011, 28·3% of the population were urban.

The areas, populations and chief towns of the islands are:

	Area (sq. km)	Population (2003 census)	Chief town
Njazídja (Grande Comore)	1,148	296,177	Moroni
Nzwani (Anjouan)	424	243,732	Mutsamudu
Mwali (Mohéli)	290	35,751	Fomboni

Estimated population of the chief towns (2002): Moroni, 40,275; Mutsamudu, 21,558; Domoni, 13,254; Fomboni, 13,053.

The indigenous population are a mixture of Malagasy, African, Malay and Arab peoples; the vast majority speak Comorian, an Arabized dialect of Swahili and one of the three official languages, but a small proportion speak one of the other official languages, French and Arabic, or Makua (a Bantu language).

SOCIAL STATISTICS

2008 births (estimates), 21,000; deaths, 4,000. Estimated birth rate in 2008 was 32·4 per 1,000 population; estimated death rate, 6·7. Annual population growth rate, 2000–08, 2·2%. Infant mortality, 63 per 1,000 live births (2010). Expectation of life in 2007 was 62·8 years among men and 67·2 among females. Fertility rate, 2008, 4·0 children per woman.

CLIMATE

There is a tropical climate, affected by Indian monsoon winds from the north, which gives a wet season from Nov. to April. Moroni, Jan. 81°F (27·2°C), July 75°F (23·9°C). Annual rainfall, 113" (2,825 mm).

CONSTITUTION AND GOVERNMENT

At a referendum on 23 Dec. 2001, 77% of voters approved a new constitution that keeps the three islands as one country while granting each one greater autonomy.

The *President of the Union* is Head of State. The presidency rotates every four years among the three main islands.

There used to be a *Federal Assembly* comprised of 42 democratically elected officials and a 15-member *Senate* chosen by an electoral college, but these were dissolved after the 1999 coup. A new 33-member *Federal Parliament* was established following the elections of April 2004. In 2004 there were 15 deputies selected by the individual islands' parliaments and 18 by universal suffrage but this was changed to nine selected by the individual islands' parliaments and 24 by universal suffrage for the 2009 elections.

National Anthem

'Udzima wa ya Masiwa' ('The union of the islands'); words by S. H. Abderamane, tune by K. Abdallah and S. H. Abderamane.

RECENT ELECTIONS

In the first round of presidential elections held on 7 Nov. 2010 on Mohéli, Ikililou Dhoinine (at the time one of the two Vice Presidents) won 28·2% of the vote, Mohamed Said Fazul 22·9%, Abdou Djabir 9·9% and Bianrifi Tarmidhi 9·3%. There were six other candidates. Turnout was 67·1%. A second round was held on 26 Dec. 2010. Ikililou Dhoinine won 61·1% of the vote, Mohamed Said Fazul 32·7% and Abdou Djabir 6·2%. Turnout in the second round was 52·8%.

Parliamentary elections were held on 6 and 20 Dec. 2009. The pro-presidential party and its allies won 20 of 33 seats while the opposition won four. The remaining nine were allocated to regional assembly representatives.

CURRENT GOVERNMENT

President of the Union: Ikililou Dhoinine; b. 1962 (sworn in 26 May 2011).

In March 2013 the government comprised:

Vice President for Production, Environment, Energy, Industry and Handicrafts: Fouad Mohadji. *Vice President for Finance, Economics, Investment and Foreign Trade, and Privatization:* Mohamed Ali Soilihi. *Vice President for Land Management, Urban Planning and Housing:* Nourdine Bourhane.

Chief of the Cabinet, in Charge of Defence: M'Madi Ali. *Minister for External Relations and Co-operation, Diaspora, Francophonie and the Arab World:* Mohamed Bakri Ben Abdoulfatah Charif. *Posts and Telecommunications, Promotion of New Information Technologies, and Communication, Transport and Tourism:* Rastami Mouhidine. *Keeper of the Seals, Justice, Public Service, Administrative Reform, Human Rights and Islamic Affairs:* Anliane Ahmed. *Education, Research, Arts and Culture, Youth and Sports:* Mohamed Issimaila. *Health, Solidarity, Social Cohesion and Gender Promotion:* Moinafouraha Ahmed. *Employment, Work, Vocational Training and Female Entrepreneurship, and Government Spokesman:* Siti Kassim. *Interior, Information, Decentralization and Institutional Relations:* Hamada Abdallah.

Office of the President (French only): http://www.beit-salam.km

CURRENT LEADERS

Ikililou Dhoinine

Position
President

Introduction
Ikililou Dhoinine was sworn in as the president on 26 May 2011, assuming power five months after securing electoral victory. Dhoinine had previously served for five years as vice president to the departing head of state, Ahmed Abdallah Mohamed Sambi. The 2001 constitution demands that the presidency rotates every four years between representatives of the three islands of which the archipelago is composed. Dhoinine is the first president to hail from the federation's smallest island, Mohéli, and his appointment was only the second peaceful handover of power.

Early Life
Ikililou Dhoinine was born in Djoièzi, a village on Mohéli, on 14 Aug. 1962. Before entering government he qualified as a pharmacist. As vice president, he took responsibility for the budget and women's entrepreneurship. He also served for five days in March 2008 as provisional president of Anjouan (one of the Comorian islands) after a diplomatic crisis there resulted in an armed intervention by the federal government.

Following the second round of presidential elections on 26 Dec. 2010 the electoral commission announced that Dhoinine had won 61·1% of the popular vote. His closest rival, Mohamed Said Fazul, alleged electoral fraud but the election monitoring group concluded that any breaches of protocol had not been sufficient to alter the result. The opposition later complained that the five-month gap between polling and the new leader's inauguration amounted to an unconstitutional extension of office for the outgoing president. Sambi's tenure had exceeded its mandate by several months even before the election.

Career in Office
Dhoinine's peaceful assumption of the presidency suggested a degree of political stability that has hitherto eluded the country. He pledged to consolidate national unity while tackling the endemic corruption that has blighted the federation's governance

since it gained independence. He campaigned specifically on a policy of much-needed infrastructure investment.

DEFENCE

Army
The Army numbered an estimated 1,060 in 2008.

Navy
There is no navy. The Army operates two small patrol boats that were supplied by Japan in 1982.

ECONOMY

Agriculture accounted for 46·3% of GDP in 2009, industry 12·1% and services 41·6%.

Overview
Economic development has been hampered by recurrent political crises. Agriculture, fishing, hunting and forestry contribute 40% to GDP and employ 80% of the labour force. Remittances supplement GDP. Dependent on food imports, especially rice, the main staple, the Comoros is vulnerable to global food prices. The primary sector is geared towards high-value export crops, chiefly vanilla, cloves and ylang-ylang, which provide up to 95% of export earnings.

In April 2010 a preliminary assessment by the World Bank and the IMF qualified the Comoros for assistance under the Enhanced Heavily Indebted Poor Countries Initiative.

Currency
The unit of currency is the *Comorian franc* (KMF) of 100 *centimes.* It is pegged to the euro at 491·96775 *Comorian francs* to the euro. Foreign exchange reserves were US$92m. in July 2005, total money supply was 24,096m. Comorian francs and gold reserves were 1,000 troy oz. Inflation was 3·9% in 2010 and 6·8% in 2011.

Budget
Revenues in 2007 were 33·9bn. Comorian francs and expenditures 37·3bn. Comorian francs.

VAT is 10%.

Performance
Real GDP growth was 1·8% in 2009, 2·1% in 2010 and 2·2% in 2011. In 2011 total GDP was US$0·6bn.

Banking and Finance
The Central Bank of the Comoros (*Governor,* Mzé Abdou Mohamed Chanfiou) is the bank of issue. Chief commercial banks include the Banque Internationale des Comores, the Banque de Développement des Comores and the Banque pour l'Industrie et le Commerce-Comores.

Total foreign debt was US$291m. in 2007.

ENERGY AND NATURAL RESOURCES

Environment
Carbon dioxide emissions from the consumption and flaring of fossil fuels were the equivalent of 0·2 tonnes per capita in 2008.

Electricity
In 2007 estimated installed capacity was 6,000 kW. Production was approximately 50m. kWh in 2007; consumption per capita was an estimated 70 kWh in 2007.

Agriculture
80% of the economically active population depends upon agriculture, which (including fishing, hunting and forestry) contributed 41% to GDP in 2002. There were about 80,000 ha. of arable land in 2002 and 52,000 ha. of permanent crops. The chief product was formerly sugarcane, but now vanilla, copra, maize and other food crops, cloves and essential oils (citronella, ylang-ylang, lemongrass) are the most important products. Production

(2002 in 1,000 tonnes): coconuts, 77; bananas, 61; cassava, 55; rice, 17; taro, 9; copra, 8; sweet potatoes, 5.

Livestock (2003): goats, 115,000; cattle, 52,000; sheep, 21,000; asses, 5,000.

Forestry
In 2010 the area under forest was 3,000 ha., or 2% of the total land area. The forested area has been severely reduced because of the shortage of cultivable land and ylang-ylang production. In 2007, 9,000 cu. metres of timber were cut.

Fisheries
The catch totalled 19,676 tonnes in 2007.

INDUSTRY
Branches include perfume distillation, textiles, furniture, jewellery, soft drinks and the processing of vanilla and copra.

Labour
The estimated economically active population in 2010 was 342,000 (53% males).

INTERNATIONAL TRADE
Imports and Exports
In 2009 imports amounted to US$181·5m. (up from US$163·0m. in 2008) and exports to US$12·6m. (up from US$5·4m. in 2008).

Main import suppliers, 2006: United Arab Emirates, 30·7%; France, 21·3%; South Africa, 9·5%. Main export markets, 2006: France, 53·8%; India, 15·4%; Germany, 11·5%. The principal imports are machinery and transport equipment (35·0% of total imports in 2006), food and live animals (25·1% in 2006), cement (6·6% in 2006), petroleum and petroleum products, chemicals and related products, and iron and steel. Main exports are vanilla (53·8% in 2006) and cloves (30·8% in 2006).

COMMUNICATIONS
Roads
In 2002 there were 880 km of roads, of which 76·5% were paved.

Civil Aviation
There is an international airport at Moroni (International Prince Said Ibrahim). In 2009 it handled 149,071 passengers (98,638 international) and 627 tonnes of freight.

Shipping
In Jan. 2009 there were 184 ships of 300 GT or over registered, totalling 625,000 GT.

Telecommunications
In 2011 there were 23,600 landline telephone subscriptions (equivalent to 31·3 per 1,000 inhabitants) and 216,400 mobile phone subscriptions (or 287·1 per 1,000 inhabitants). Fixed internet subscriptions totalled 1,600 in 2009 (2·3 per 1,000 inhabitants).

SOCIAL INSTITUTIONS
Justice
French and Muslim law is in a new consolidated code. The Supreme Court comprises seven members, two each appointed by the President and the Federal Assembly, and one by each island's Legislative Council. The death penalty is authorized for murder. The last execution was in 1996.

Education
After two pre-primary years at Koran school, which 50% of children attend, there are six years of primary schooling for seven- to 13-year-olds followed by a four-year secondary stage attended by 25% of children. Some 5% of 17- to 20-year-olds conclude schooling at *lycées*. There were 106,700 pupils with 3,050 teaching staff in primary schools in 2005 and 43,349 pupils at secondary schools with 3,138 teaching staff. At the tertiary level there were approximately 3,000 students in 2009.

The adult literacy rate in 2009 was an estimated 74·2% (79·7% among males and 68·7% among females).

In 2008 public expenditure on education came to 7·6% of GDP.

Health
In 2004 there were 115 physicians, 29 dentistry personnel and 588 nursing and midwifery personnel. In 2006 there were 17 hospital beds per 10,000 inhabitants.

RELIGION
Islam is the official religion: 98% of the population are Muslims; there is a small Christian minority. Following the coup of April 1999 the federal government discouraged the practice of religions other than Islam, with Christians especially facing restrictions on worship.

CULTURE
Press
There has not been a daily newspaper since *Le Matin des Comores* ceased publication in 2006. *Le Canal*, which is published in Mayotte, is distributed in the Comoros. There were five non-dailies in 2008. *Al-Watwan* is published four days a week in French and one day a week in Arabic.

Tourism
In 2006 there were 29,000 non-resident tourists, bringing revenue of US$27m.

DIPLOMATIC REPRESENTATIVES
Of the United Kingdom in the Comoros
Ambassador: Nick Leake (resides in Port Louis, Mauritius).

Of the Comoros in the USA (Temporary: c/o the Permanent Mission of the Union of the Comoros to the United Nations, 420 E 50th St., N.Y. 10022)
Ambassador: Roubani Kaambi.

Of the USA in the Comoros
Ambassador: Vacant (resides in Antananarivo, Madagascar).
Chargé d'Affaires a.i.: Eric M. Wong.

Of the Comoros to the United Nations
Ambassador: Roubani Kaambi.

Of the Comoros to the European Union
Ambassador: Ali Said Mdahoma.

FURTHER READING
Ottenheimer, M. and Ottenheimer, H. J., *Historical Dictionary of the Comoro Islands.* 1994

CONGO, DEMOCRATIC REPUBLIC OF THE

République Démocratique du Congo

Capital: Kinshasa
Population projection, 2015: 75·19m.
GNI per capita, 2011: (PPP$) 280
HDI/world rank: 0·286/187
Internet domain extension: .cd

KEY HISTORICAL EVENTS

Bantu tribes migrated to the Congo basin from the northwest in the first millennium AD, forming several kingdoms and many smaller forest communities. Congo emerged as a kingdom on the Atlantic coast in the 14th century. King Nzinga Mbemba entered into diplomatic relations with Portugal after 1492. The Luba kingdom was centred on the marshy Upemba depression in the southeast. Expansion began in the late 18th century under Ilungu Sungu. In central Congo the Kuba kingdom was established in the 17th century as a federation of Bantu groups. Agriculture became the mainstay of the Kuba economy, strengthened by the introduction of American crops by Europeans. Trade made the Kuba elite, especially the Bushoong group, wealthy and encouraged the development of art and decorated cloth. Kuba thrived until the incursions of the Nsapo in the late 19th century.

King Leopold II of the Belgians claimed the Congo Basin as a personal possession in 1885. Exploitation of the native population provoked international condemnation. The Belgian government responded by annexing the Congo in 1908. Political representation was denied the Congolese until 1957 when the colonial administration introduced the *statut des villes* in response to the revolutionary demands of the *Alliance des BaKongo* (Abako). Violence increased, instigated by the *Mouvement National Congolais* (MNC), led by Patrice Lumumba. Local elections were held in Dec. 1959 and in Jan. 1960 the Belgian government announced a rapid independence programme. After general elections in May, the Republic of the Congo became independent on 1 June 1960, with Lumumba as prime minister and Joseph Kasavubu, the Abako leader, as president.

Independence and Anarchy

The country descended into anarchy, with the mineral-rich Katanga region declaring independence. Lumumba was ousted and in 1961 assassinated. Only in 2002 did Belgium admit to participating in his murder. Lieut.-Gen. Joseph-Désiré Mobutu (later Sese Seko) seized power in 1965. At first he was seen as a strongman who could hold together a huge, unstable country comprising hundreds of tribes and language groups. He changed the country's name to Zaïre in 1971. In the 1970s he was feted by the USA, which used Zaïre as a springboard for operations into neighbouring Angola where western-backed Unita rebels were locked in civil war with a Cuban and Soviet-backed government. Because Mobutu was useful in the fight against Communism the brutality and repressiveness of his regime was ignored.

After armed insurrection by Tutsi rebels in the province of Kivu, the government alleged pro-Tutsi intervention by the armies of Burundi and Rwanda and on 25 Oct. 1996 declared a state of emergency. By Dec. the secessionist forces of Laurent-Désiré Kabila, the *Alliance des Forces Démocratiques pour la Libération du Congo-Zaïre* (AFDL), had begun to drive the regular Zaïrean army out of Kivu and an attempt was made to establish a rebel administration, called 'Democratic Congo'. In the face of continuing rebel military successes and the disaffection of the army, the Government accepted a UN resolution demanding the immediate cessation of hostilities. The Security Council asked the rebels to make a public declaration of their acceptance. However, they continued in their advance westwards, capturing Kisangani on 15 March 1997, then Kasai and Shaba, giving Kabila control of eastern Zaïre, and crucially the country's mineral wealth. After a futile attempt to deploy Serbian mercenaries, Mobutu succumbed to pressure from the USA and South Africa to meet Kabila, an occasion that had all the trappings of a symbolic surrender. Mobutu fled on the night of 15–16 May 1997. One of the most destructive tyrants of the African independence era, he died of cancer four months later.

On coming to power Kabila changed the name of the country to the Democratic Republic of the Congo. Hopes for democratic and economic renewal were soon disappointed. The Kabila regime relied too closely on its military backup, mainly Rwandans and eastern Congolese from the Tutsi minority who seemed more interested in eliminating tribal enemies in eastern border areas than in establishing democracy. As a result, Rwanda and Uganda switched support to rebel forces. When Zimbabwe and Angola sent in troops to help President Kabila, full-scale civil war threatened. A ceasefire was negotiated at a Franco-African summit in Nov. 1998 but the military build-up continued into the new year and violence intensified.

A ceasefire was signed by leaders from more than a dozen African countries in July 1999. Rival factions of the *Rassemblement Congolais pour la Démocratie* (RCD), the main rebel group opposed to the president, also signed the accord, but not until Sept.

Violence and Collapse

On 16 Jan. 2001 President Kabila was assassinated, allegedly by one of his own bodyguards. He was succeeded by his son, Joseph. Prospects for peace improved dramatically in Feb. 2001 when the UN Security Council approved the deployment of 3,000 UN-supported peacekeepers. In early 2002 talks between the government and rebels on how to end the conflict ended without agreement. However, in July 2002 the presidents of the Democratic Republic of the Congo and neighbouring Rwanda signed a peace deal that was expected to be the first stage towards ending a war that has claimed more than 3m. lives. In Oct. 2002 Rwanda completed the withdrawal of its forces. The Democratic Republic of the Congo and Uganda also signed a peace agreement.

The conflict, described as Africa's first continental war, had drawn in Zimbabwe, Angola and Namibia (and, for a time, Sudan and Chad) on the side of the government, which controls the west of the country, while Rwanda and Uganda backed other rival factions. The RCD, which controls areas in the east, was backed by Rwanda, while Uganda supported the *Mouvement de Libération du Congo* (MLC), based in the north and northeast of the country. Burundi also had troops in the country, allied to the Rwandans, although they stayed close to the border with Burundi. In addition to the huge death toll, large numbers of people were displaced and sought asylum in Tanzania and Zambia. In Dec. 2002 the government and leading rebel forces reached an agreement on power-sharing. Its terms allowed for Joseph Kabila to remain as president until elections which, after several postponements, were held in July 2006. Kabila was elected president in a second round run-off in Oct. 2006 that signalled the end of the period of transitional government.

In April 2007 the Democratic Republic of the Congo resurrected the Economic Community of the Great Lakes alongside Rwanda and Burundi. Political instability within the country continued, particularly in eastern areas. In Jan. 2008 the government signed a peace deal with militia groups, including that of the renegade Gen. Laurent Nkunda. However, fighting between the army and Rwandan Hutu militias had resumed by April. In Oct. the government accused Rwanda of secretly backing Nkunda. Fighting intensified and Goma, the provincial capital of Nord-Kivu, was paralysed as rebel forces encroached and some 200,000 people were displaced. In Jan. 2009 Rwandan forces arrested Nkunda, signalling a new diplomatic direction.

TERRITORY AND POPULATION

The Democratic Republic of the Congo, sometimes referred to as Congo (Kinshasa), is bounded in the north by the Central African Republic, northeast by South Sudan, east by Uganda, Rwanda, Burundi and Lake Tanganyika, south by Zambia, southwest by Angola and northwest by the Republic of the Congo. There is a 37-km stretch of coastline that gives access to the Atlantic Ocean, with the Angolan exclave of Cabinda to the immediate north, and Angola itself to the south. The area is 2,344,860 sq. km (905,360 sq. miles), including 77,810 sq. km (30,040 sq. miles) of inland waters. A census has not been held since 1984, when the population was 29,916,800. The United Nations gave an estimated population for 2012 of 69·58m.; density, 15 per sq. km. 35·9% of the population was urban in 2011.

The UN gives a projected population for 2015 of 75·19m.

There were 141,000 refugees in the country in Sept. 2012, including 72,000 from Angola. Around 62,000 refugees who escaped the fighting between Hutus and Tutsis in Rwanda and Burundi in 1994 are still in the Democratic Republic of the Congo (out of 1m. who came originally).

The country is administratively divided into ten provinces plus Kinshasa city, as follows (with capitals): Bandundu (Bandundu), Bas-Congo (Matadi), Équateur (Mbandaka), Kasai-Occidental (Kananga), Kasai-Oriental (Mbuji-Mayi), Katanga (Lubumbashi), Kinshasa (Kinshasa), Maniema (Kindu), Nord-Kivu (Goma), Orientale (Kisangani), Sud-Kivu (Bukavu).

The capital is Kinshasa (2010 population estimate, 8,415,000). Other main cities (with 2010 population estimates) are: Lubumbashi (1,486,000); Mbuji-Mayi (1,433,000); Kananga (846,000); Kisangani (783,000).

The population is Bantu, with minorities of Sudanese (in the north), Nilotes (northeast), Pygmies and Hamites (in the east). French is the official language, but of more than 200 languages spoken, four are recognized as national languages: Kiswahili, Tshiluba, Kikongo and Lingala. Lingala has become the *lingua franca* after French.

SOCIAL STATISTICS

2008 estimates: births, 2,883,000; deaths, 1,090,000. Rates (2008 estimates, per 1,000 population); birth, 44·9; death, 17·0. Annual population growth rate, 2000–08, 2·9%. Infant mortality in 2010 was 112 per 1,000 live births (the second highest in the world after Sierra Leone). Expectation of life in 2007 was 46·1 years for men and 49·2 for females. Fertility rate, 2008, 6·0 children per woman.

CLIMATE

The climate is varied, the central region having an equatorial climate, with year-long high temperatures and rain at all seasons. Elsewhere, depending on position north or south of the Equator, there are well-marked wet and dry seasons. The mountains of the east and south have a temperate mountain climate, with the highest summits having considerable snowfall. Kinshasa, Jan. 79°F (26·1°C), July 73°F (22·8°C). Annual rainfall 45" (1,125 mm). Kananga, Jan. 76°F (24·4°C), July 74°F (23·3°C). Annual rainfall 62" (1,584 mm). Kisangani, Jan. 78°F (25·6°C), July 75°F (23·9°C). Annual rainfall 68" (1,704 mm). Lubumbashi, Jan. 72°F (22·2°C), July 61°F (16·1°C). Annual rainfall 50" (1,237 mm).

CONSTITUTION AND GOVERNMENT

A new constitution was adopted by the transitional parliament on 16 May 2005. It limits the powers of the president, who may now serve a maximum of two five-year terms and lowers the minimum age for presidential candidates from 35 to 30. It allows a greater degree of federalism and recognises as citizens all ethnic groups at the time of independence in 1960. It also called for presidential elections by June 2006. In a referendum held on 18–19 Dec. 2005, 83% of voters approved the constitution in the country's first free vote in 40 years. The constitution was promulgated on 18 Feb. 2006.

The 240-member *Constituent and Legislative Assembly* was appointed in Aug. 2000 by former President Laurent Désiré Kabila. In Aug. 2003 a new bicameral parliament of 500 members and 120 senators met in Kinshasa. The representatives were chosen from the groups that comprised the newly-formed transitional government. In accordance with the constitution of the time, for the 2006 presidential election the *President* was elected by direct popular vote to serve a five-year term. In the *National Assembly*, 60 members were elected by majority vote in single-member constituencies and 440 members by open list proportional representation in multi-member constituencies to serve five-year terms. The 108 members of the *Senate* were elected by indirect vote, by provincial deputies, to serve five-year terms.

National Anthem

'Debout Congolais' ('Stand up, Congolese'); words and tune by J. Lutumba and S. Boka di Mpasi Londi.

RECENT ELECTIONS

Presidential elections, only the second since the Democratic Republic of the Congo's became independent in 1960, were held on 28 Nov. 2011. In the presidential election incumbent Joseph Kabila received 49·0% of votes cast and Étienne Tshisekedi 32·3%. There were nine other candidates. Owing to a change in election laws, a potential run-off on 26 Feb. 2012 was shelved. Elections were held under difficult conditions, with incidents of violence. Tshisekedi rejected the result and declared himself president raising fears of civil unrest.

In the parliamentary elections held on the same day Kaliba's People's Party for Reconstruction and Democracy (PPRD) won 62 of 500 seats and Tshisekedi's Union for Democracy and Social Progress (UDPS) 41. A total of 98 parties won seats, of which 46 only took one seat.

CURRENT GOVERNMENT

President: Joseph Kabila; b. 1971 (in office since 17 Jan. 2001, elected on 29 Oct. 2006 and re-elected in Nov. 2011).

In March 2013 the government comprised:

Prime Minister: Augustin Matata Ponyo; b. 1964 (People's Party for Reconstruction and Democracy; since 18 April 2012).

Deputy Prime Ministers: Daniel Mukoko Samba (also *Minister of the Budget*); Alexandre Lubal Tamu (also *Minister of Defence*).

Minister of Agriculture and Rural Development: Jean-Chrysostome Wahamiti. *Civil Service:* Jean-Claude Kibala. *Communication Channels and Transport:* Justin Kalumba. *Economy and Trade:* Jean-Paul Nomayato. *Employment, Labour and Social Welfare:* Modeste Bahati Lukwebo. *Environment, Nature Conservation and Tourism:* Louis Bavon Mputu. *Foreign Affairs, International Co-operation and Francophone:* Raymond Tshibanda. *Gender and Family:* Genevieve Inagosi. *Higher Education, University and Scientific Research:* Kyelo Lotama. *Hydrocarbons:* Crispin Atama. *Industry, and Small and Medium Enterprises:* Rémy Musungay. *Justice and Human Rights:* Wivine Mumba. *Land Affairs:* Robert Mbuinga Bila. *Land Management, Public Works, Urban Planning and Housing:* Fridolin Kaswesi Kusoka. *Media, and Relations with Parliament:* Lambert Mende. *Mines:* Martin Kabwelulu. *Parastatals:* Louise Munga. *Plan Monitoring and Implementation of the Revolution of Modernity:* Célestin Vunabandi. *Post, Telecommunications, and New Information and Communication Technologies:* Kin-Kiey Mulumba. *Primary, Secondary and Vocational Education:* Maker Mwangu. *Public Health:* Félix Kabange Numbi. *Social Affairs, Humanitarian Action and National Solidarity:* Charles Nawej. *Water Resources and Electricity:* Bruno Kasanji Kalalasanji Kalala. *Youth, Sports, Culture and Arts:* Banza Mukalay.

CURRENT LEADERS

Joseph Kabila

Position
President

Introduction
Joseph Kabila is the son of the former president Laurent Kabila who was assassinated on 16 Jan. 2001. Despite being little known outside his own group, Joseph Kabila was appointed president. He inherited a country divided by civil war which had spanned his father's tenure. Although promising unification in Congo, Kabila has not been able to restore lasting peace to a country split along tribal lines and, despite its rich natural resources, suffering from poverty and economic instability. He was elected president following two rounds of voting held in July and Oct. 2006 and re-elected in Nov. 2011.

Early Life
Joseph Kabila was born on 4 Dec. 1971 at Laurent Kabila's anti-Mobutu guerrilla movement's headquarters (Hewa Bora) in the Fizi territory of Sud-Kivu. He was educated at a French-language school in Tanzania and then studied at the Makerere University in Uganda. He also did military training in China.

Career in Office
Following his father's death, Kabila replaced him as president. In his inaugural speech in 2001, he promised a ceasefire with rebel forces, an end to corruption and an improvement in living standards. He pledged to lead the country into multi-party democracy, and in May 2001 he lifted restrictions on political parties. Subsequently, over 200 parties were registered.

Peace negotiations brokered by South Africa in 2002 between the Kabila government and rebel factions led to agreement in Dec. on a power-sharing accord. Kabila was to remain as president pending future democratic elections, while his supporters, the civilian political opposition and two main rebel groups would each appoint a vice-president and seven ministers to serve under him on an interim basis. On 30 June 2003 Kabila announced the composition of the new transitional government. Although this was a major breakthrough, members of the new government feared for their safety in Kinshasa amid the mistrust between the various factions. The new government was inaugurated in Kinshasa in July 2003, but subsequent political progress was very slow and lawlessness and human rights abuses continued, particularly in the east and northeast of the country. Following a referendum on a new constitution in Dec. 2005, elections were scheduled for mid-2006. In the first round of the presidential poll in July, Kabila won 45% of the vote while his main rival, vice-president and former rebel Jean-Pierre Bemba, took 20%. In the Oct. 2006 run-off, Kabila claimed victory with 58%. The result was ratified by the Supreme Court the following month. Although Bemba denounced the outcome as rigged and his supporters reacted violently, he ultimately conceded defeat and in Dec. 2006 Kabila was sworn in as the country's first freely-elected president in more than 40 years.

Nevertheless, violence has persisted between government troops and rebel groups, resulting in the continuing displacement and abuse of many thousands of civilians. The east of the country has suffered particularly, with lawlessness dominating despite the presence of the world's second largest UN peacekeeping mission. Notably, instances of mass rape allegedly perpetrated by army troops and rebels have been reported in Nord-Kivu province. The brief occupation in late 2012 of the eastern city of Goma by rebel forces, allegedly with Rwandan and Ugandan backing, emphasized Kabila's tenuous authority over swathes of the country.

In Jan. 2011 constitutional changes, strongly criticized by opposition figures, provided for presidential elections to take place in one round rather than two and for a candidate to win with a simple majority. In the elections that followed in Nov., Kabila retained the presidency with 49% of the vote.

DEFENCE

Following the overthrow of the Mobutu regime in May 1997, the former Zaïrean armed forces were in disarray. In June 2003 command of ground forces and naval forces were handed over to the RCD-Goma and MLC factions respectively as part of the power-sharing transitional government. Supreme command of the armed forces will remain in the hands of the former government faction.

A UN mission, MONUSCO, has been in the Democratic Republic of the Congo since 1999 (under the name of MONUC until June 2010). With 19,130 uniformed personnel in Jan. 2013 it is the second largest UN peacekeeping force in the world.

Defence expenditure totalled US$168m. in 2008 (US$3 per capita), representing 1·5% of GDP.

Army

The total strength of the Army was estimated at 125,000 (2007), including some 14,000 republican guards. There is an additional paramilitary National Police Force of unknown size. There are thought to be 14 infantry brigades, one mechanized infantry brigade and two commando regiments.

Navy

Naval strength (2007), 6,700.

Air Force

Personnel (2007), 2,500.

ECONOMY

Agriculture accounted for 43% of GDP in 2009 (one of the highest percentages of any country), industry 24% and services 33%.

Overview

The Democratic Republic of the Congo is a resource-rich economy that has suffered socio-political unrest since the mid-1980s. It is ranked last of the 187 countries on the UN Human Development Index. Per capita income has dropped steadily since independence, from US$250 in 1960 to US$139 in 2006.

Ceasefire agreements in 1999 and 2001 led to economic and social reforms. Between 2003 and 2006 the transitional government introduced prudent macroeconomic policies and structural reforms, bringing hyperinflation under control and helping recovery. After a decade of contraction, growth increased to over 6·0% in 2008 before falling to 2·8% in 2009 in the wake of the global financial crisis.

Despite IMF debt relief, external debt represents 93% of GDP. Mineral smuggling, particularly of gold, is a heavy drain on government income.

Currency
The unit of currency is the *Congolese franc* (CDF) which replaced the former *zaïre* in July 1998. The value of the new currency fell by two-thirds in the six months following its launch. Total money supply was 120,605m. Congolese francs in July 2005. Inflation, which reached 23,760% in 1994, had declined to 4·0% by 2004 before rising to 46·2% in 2009 (the highest rate of any country that year). Inflation has remained high, with rates of 23·5% in 2010 and 15·5% in 2011. In May 2001 the franc was floated in an effort to overcome the economic chaos caused by three years of state control and inter-regional war.

Budget
In 2008 revenues totalled 1,326·8bn. Congolese francs and expenditures 1,478·7bn. Congolese francs.

Performance
Real GDP growth was 2·8% in 2009, 7·2% in 2010 and 6·9% in 2011. In Feb. 1998 GDP was reported to be 65% lower than it was in 1960, when the country gained independence. Total GDP in 2011 was US$15·7bn.

Banking and Finance
The central bank, the Banque Centrale du Congo (*Governor*, Jean-Claude Masangu), achieved independence in May 2002. In 2010 there were 23 commercial banks, two specialized financial institutions, one savings bank and 120 co-operative banks. The largest bank in 2010 was the Banque Commerciale du Congo, in which the state has a 25·5% stake.

Foreign debt totalled US$5,774m. in 2010 and represented 47·1% of GNI.

ENERGY AND NATURAL RESOURCES
Environment
Carbon dioxide emissions from the consumption and flaring of fossil fuels were the equivalent of less than 0·1 tonnes per capita in 2008. An *Environmental Performance Index* compiled in 2008 ranked Democratic Republic of the Congo 142nd in the world out of 149 countries analysed, with 47·3%. The index examined various factors in six areas—air pollution, biodiversity and habitat, climate change, environmental health, productive natural resources and water resources.

Electricity
Production (2007), 8·3bn. kWh. Installed capacity was an estimated 2·4m. kW in 2007. The Democratic Republic of the Congo is rich in untapped hydro-electric potential, but the unstable climate and poor infrastructure have made it difficult to attract investors. Consumption per capita was 114 kWh in 2007.

Oil and Gas
Production of oil in 2007 was 836,000 tonnes; reserves in 2007 were 180m. bbls. There is an oil refinery at Kinlao-Muanda.

Minerals
Production, 2004 (in 1,000 tonnes): coal, 108; copper, 70; cobalt (estimate), 20; gold (estimate), 10,500 kg. Diamond production,

2004: 29·5m. carats. Only Australia, Russia and Botswana produce more diamonds. The country holds an estimated 80% of the world's coltan (columbite-tantalite) reserves. Coal, tin and silver are also found. The most important mining area is in the province of Katanga.

Agriculture
There were, in 2001, 6·70m. ha. of arable land and 1·18m. ha. of permanent crops. 11,000 ha. were irrigated in 2001. There were 2,430 tractors in 2001. The main agricultural crops (2002 production in 1,000 tonnes) are: cassava, 14,929; sugarcane, 1,650; plantains (2001), 1,216; maize, 1,154; groundnuts, 355; yams, 320; rice, 315; bananas, 313. Livestock (2002): goats, 4,004,000; pigs, 953,000; sheep, 897,000; cattle, 761,000; poultry, 20m.

Forestry
Forests covered 154·14m. ha. in 2010, or 68% of the land area. Timber production in 2007 was 77·66m. cu. metres.

Fisheries
The catch for 2007 was approximately 236,000 tonnes, almost entirely from inland waters.

INDUSTRY
Output in 1,000 tonnes: cement (2001), 192; steel (2001), 80; sugar (2002), 65; soap (1995), 47; tyres (1995), 50,000 units; printed fabrics (1995), 15·73m. sq. metres; shoes (1995), 1·6m. pairs; beer (2003), 149·6m. litres.

Labour
The estimated economically active population in 2010 was 25·77m. (59% males), up from 21·79m. in 2005. Agriculture employs around 65% of the total labour force.

INTERNATIONAL TRADE
Imports and Exports
Imports in 2007 were US$5,257m.; exports were US$6,143m. Main commodities for import are consumer goods, foodstuffs, mining and other machinery, transport equipment and fuels; and for export: diamonds, copper, coffee, cobalt and crude oil. Main import sources in 2004 were South Africa, 18·5%; Belgium, 15·6%; France, 10·9%; USA, 6·2%. Main export markets (2004) were Belgium, 42·5%; Finland, 17·8%; Zimbabwe, 12·2%; USA, 9·2%.

COMMUNICATIONS
Roads
In 2004 there were 153,497 km of roads (1·8% paved). In 2007 there were around 312,000 vehicles in use.

Rail
Total route length was 4,499 km on three gauges in 2004, of which 858 km was electrified. However, the length of track actually in use has been severely reduced by the civil conflict and only amounted to 3,641 km. In 2008 the Société Nationale des Chemins de Fer du Congo (SNCC) carried 169,000 passengers and in 2007 the Office National des Transports carried 39,000 passengers. The SNCC carried 921,000 tonnes of freight in 2008.

Civil Aviation
There is an international airport at Kinshasa (Ndjili). Other major airports are at Lubumbashi (Luano), Bukavu, Goma and Kisangani. The main carrier is Hewa Bora Airways. In 2009 Kinshasa handled 672,347 passengers (385,923 international) and 67,544 tonnes of freight.

Shipping
The River Congo and its tributaries are navigable to 300-tonne vessels for about 14,500 km. Regular traffic has been established between Kinshasa and Kisangani as well as Ilebo, on the Lualaba (i.e. the river above Kisangani), on some tributaries and on the lakes. The Democratic Republic of the Congo has only 37 km of

sea coast. In Jan. 2009 there were two ships of 300 GT or over registered, totalling 2,000 GT. Matadi, Kinshasa and Kalemie are the main seaports.

Telecommunications

In 2009 the Democratic Republic of the Congo had just 42,300 main (fixed) telephone lines (0·7 for every 1,000 persons), but there were 9,459,000 mobile phone subscribers (147·3 for every 1,000 persons). The Democratic Republic of the Congo has among the highest ratios of mobile phone subscriptions to fixed telephone lines. In 2010, 0·7% of the population were internet users. In June 2012 there were 808,000 Facebook users.

SOCIAL INSTITUTIONS

Justice

There is a Supreme Court at Kinshasa, 11 courts of appeal, 36 courts of first instance and 24 'peace tribunals'. The death penalty is in force.

The population in penal institutions in 2010 was approximately 22,000 (33 per 100,000 of national population).

Education

In 2007 there were 230,834 teaching staff in primary schools for 8·8m. pupils, and 2·8m. pupils in secondary schools with 179,635 teaching staff. In 2007 there were 237,836 students in higher education and 16,913 academic staff. The largest public universities are the University of Kinshasa, the University of Kisangani and the University of Lubumbashi.

Adult literacy rate was 67% in 2008.

Health

In 2004 there were 5,827 physicians, 159 dentistry personnel and 28,789 nursing and midwifery personnel.

The Democratic Republic of the Congo has been one of the least successful countries in the battle against undernourishment in the past 20 years. The proportion of the population classified as undernourished increased from 29% in 1990–92 to 75% by 2004–06.

RELIGION

In 2001 there were 21·99m. Roman Catholics, 16·95m. Protestants, 7·17m. Kimbanguistes (African Christians) and 0·75m. Muslims. Animist beliefs persist. In Feb. 2013 there was one Roman Catholic cardinal.

CULTURE

World Heritage Sites

(With year entered on list). Virunga National Park (1979); Kahuzi-Biega National Park (1980); Garamba National Park (1980); Salonga National Park (1984); and Okapi Wildlife Reserve (1996).

Press

In 2008 there were 12 daily newspapers with a combined circulation of 50,000.

Tourism

In 2007 there were 47,000 non-resident tourist arrivals by air.

DIPLOMATIC REPRESENTATIVES

Of the Democratic Republic of the Congo in the United Kingdom (45–47 Great Portland St., London, W1W 7LT)
Ambassador: Dr Barnabe Kikaya Bin Karubi.

Of the United Kingdom in the Democratic Republic of the Congo (83 Ave. du Roi Baudouin, Kinshasa)
Ambassador: Neil Wigan.

Of the Democratic Republic of the Congo in the USA (1726 M St., Suite 601, NW, Washington, D.C., 20009)
Ambassador: Faida Mitifu.

Of the USA in the Democratic Republic of the Congo (310 Ave. des Aviateurs, Kinshasa)
Ambassador: James F. Entwistle.

Of the Democratic Republic of the Congo to the United Nations
Ambassador: Ignace Gata Mavita wa Lufuta.

Of the Democratic Republic of the Congo to the European Union
Ambassador: Henri Mova Sakany.

FURTHER READING

Gondola, Didier, *The History of Congo*. 2003
Hochschild, Adam, *King Leopold's Ghost: A Study of Greed, Terror and Heroism in Colonial Africa*. 1999
Melson, Robert, *Genocide and Crisis in Central Africa: Conflict Roots, Mass Violence and Regional War*. 2001
Renton, David, *The Congo: Plunder and Resistance*. 2006
Stearns, Jason, *Dancing in the Glory of Monsters: the Collapse of the Congo and the Great War of Africa*. 2011

CONGO, REPUBLIC OF THE

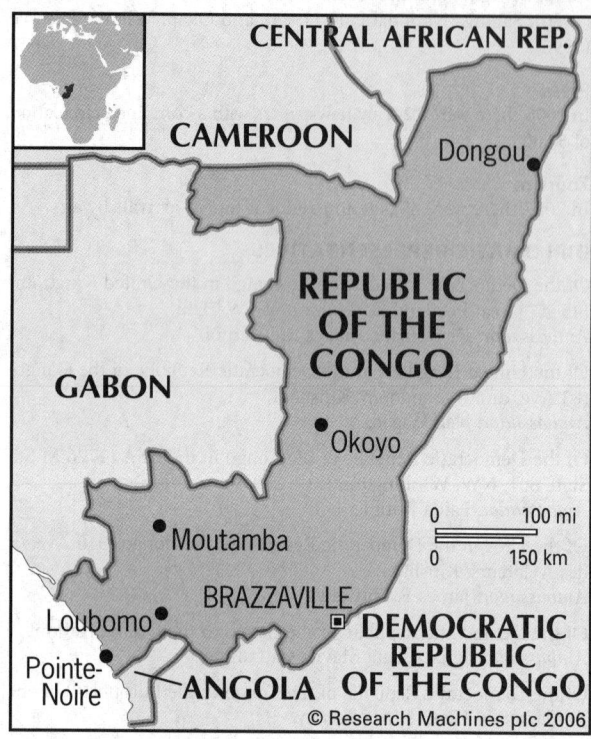

© Research Machines plc 2006

République du Congo

Capital: Brazzaville
Population projection, 2015: 4·51m.
GNI per capita, 2011: (PPP$) 3,066
HDI/world rank: 0·533/137
Internet domain extension: .cg

KEY HISTORICAL EVENTS

First occupied by France in 1882, the Congo became a territory of French Equatorial Africa from 1910–58, and then a member state of the French Community. Between 1940 and 1944, thanks to Equatorial Africa's allegiance to Gen. de Gaulle, he named Brazzaville the capital of the Empire and Liberated France. Independence was granted in 1960. A Marxist-Leninist state was introduced in 1970. Free elections were restored in 1992 but violence erupted when, in June 1997, President Lissouba tried to disarm opposition militia ahead of a fresh election. There followed four months of civil war with fighting concentrated on Brazzaville which became a ghost town. In Oct. Gen. Sassou-Nguesso proclaimed victory, having relied upon military support from Angola. President Lissouba went into hiding in Burkina Faso. A peace agreement signed in Nov. 1999 between President Sassou-Nguesso and the 'Cocoye' and 'Ninja' militias brought a period of relative stability.

TERRITORY AND POPULATION

The Republic of the Congo, sometimes referred to as Congo (Brazzaville), is bounded by Cameroon and the Central African Republic in the north, the Democratic Republic of the Congo to the east and south, Angola and the Atlantic Ocean to the southwest and Gabon to the west, and covers 342,000 sq. km.

At the census of 2007 the population was 3,697,490 (1,876,133 females); density, 11 per sq. km.

The UN gives a projected population for 2015 of 4·51m.

In 2011, 62·5% of the population were urban.

Area, census population and county towns of the departments in 2007 were:

Region	Sq. km	Population	County town
Bouenza	12,266	309,073	Madingou
Capital District	100	1,373,382	Brazzaville
Cuvette	47,650	156,044	Owando
Cuvette Ouest	27,200	72,999	Ewo
Kouilou[1]	13,694	807,289	Pointe-Noire
Lékoumou	20,950	96,393	Sibiti
Likouala	66,044	154,115	Impfondo
Niari	25,941	231,271	Loubomo (Dolisie)
Plateaux	38,400	174,591	Djambala
Pool	33,955	236,595	Kinkala
Sangha	55,800	85,738	Ouesso

[1]A new department, Pointe-Noire (formerly part of Kouilou), has been created since the 2007 census.

Census population of major cities in 2007: Brazzaville, the capital, 1,373,382; Pointe-Noire, 715,334; Loubomo (Dolisie), 83,798; N'Kayi, 71,620; Impfondo, 33,911; Ouesso, 28,179. Main ethnic groups are: Kongo (48%), Sangha (20%), Teke (17%) and M'Bochi (12%).

French is the official language. Kongo languages are widely spoken. Monokutuba and Lingala serve as *lingua francas*.

SOCIAL STATISTICS

2008 estimates: births, 125,000; deaths, 46,000. Rates, 2008 estimates (per 1,000 population): births, 34·5; deaths, 12·9. Infant mortality, 2010 (per 1,000 live births), 61. Expectation of life in 2007 was 52·5 years for males and 54·4 for females. Annual population growth rate, 2000–08, 2·2%. Fertility rate, 2008, 4·4 children per woman.

CLIMATE

An equatorial climate, with moderate rainfall and a small range of temperature. There is a long dry season from May to Oct. in the southwest plateaux, but the Congo Basin in the northeast is more humid, with rainfall approaching 100" (2,500 mm). Brazzaville, Jan. 78°F (25·6°C), July 73°F (22·8°C). Annual rainfall 59" (1,473 mm).

CONSTITUTION AND GOVERNMENT

A new constitution was approved in a referendum held in Jan. 2002. Under the new constitution the president's term of office is increased from five to seven years. The constitution provides for a new two-chamber assembly consisting of a house of representatives and a senate. The president may also appoint and dismiss ministers. 84·3% of voters were in favour of the draft constitution and 11·3% against. Turnout was 78%, despite calls from opposition parties for a boycott. The new constitution came into force in Aug. 2002.

There is a 139-seat *National Assembly*, with members elected for a five-year term in single-seat constituencies, and a 72-seat *Senate*, with members elected for a six-year term (one third of members every two years).

National Anthem

'La Congolaise'; words by Levent Kimbangui, tune by Français Jacques Tondra.

RECENT ELECTIONS

Presidential elections were held on 12 July 2009. Incumbent Denis Sassou-Nguesso won with 78·6% of votes cast, against 7·5% for Joseph Kignoumbi Kia Mboungou and 7·0% for Nicéphore Antoine Fylla de Saint-Eudes. The turnout was 66·4%. There were ten other candidates. Six opposition candidates boycotted the elections, claiming that the electoral lists were flawed.

Parliamentary elections were held on 15 July and 5 Aug. 2012. President Denis Sassou-Nguesso's Congolese Labour Party won 89 out of 139 seats; Congolese Movement for Democracy and Integral Development, 7; Pan-African Union for Social Democracy, 7; Rally for Democracy and Social Progress, 5; Action Movement for Renewal, 4; Citizen Rally, 3; Movement for Unity, Solidarity and Work, 2; Patriotic Union for Democracy and Progress, 2. Five parties won one seat each, 12 went to independents and three were vacant.

CURRENT GOVERNMENT

President: Denis Sassou-Nguesso; b. 1943 (Congolese Labour Party; sworn in 25 Oct. 1997 for a second time and re-elected in March 2002 and July 2009, having previously held office 1979–92).

In March 2013 the government comprised:

Minister at the Presidency in Charge of Defence: Charles Richard Mondjo. *Minister at the Presidency in Charge of Planning and Large-Scale Works Management:* Jean-Jacques Bouya. *Minister at the Presidency in Charge of Special Economic Zones:* Alain Akouala Atipault.

Minister of State for the Economy, Finances, Planning, Public Portfolio and Integration: Gilbert Ondongo. *Justice, Guardian of the Seals and Human Rights:* Aimé Emmanuel Yoka. *Transport, Civil Aviation and the Merchant Marine:* Rodolphe Adada. *Work and Social Security:* Florent Tsiba. *Industrial Development and Promotion of the Private Sector:* Isidore Mvouba.

Minister of Agriculture and Livestock: Rigobert Maboundou. *Civic Education and Youth:* Anatole Collinet Makosso. *Civil Service and State Reform:* Guy Brice Parfait Kolélas. *Commerce and Supplies:* Claudine Munari. *Communications and Relations with Parliament:* Bienvenu Okiemy. *Construction, Town Planning and Housing:* Claude Alphonse Nsilou. *Culture and the Arts:* Jean-Claude Gakosso. *Energy and Water Resources:* Henri Ossebi. *Equipment and Public Works:* Émile Ouosso. *Fisheries and Aquaculture:* Bernard Tchibambelela. *Foreign Affairs and Co-operation:* Basile Ikouébé. *Forest Economy and Sustainable Development:* Henri Djombo. *Health and Population:* François Ibovi. *Higher Education:* Georges Moyen. *Hydrocarbons:* André Raphaël Loemba. *Interior and Decentralization:* Raymond Zéphirin Mboulou. *Land Affairs and Public Territory:* Pierre Mabiala. *Mines, Mining Industry and Geology:* Pierre Oba. *Posts and Telecommunications:* Thierry Moungala. *Primary and Secondary Education and Literacy:* Hellot Mampouya Matson. *Promotion of Women and the Involvement of Women in Development:* Catherine Embondza. *Scientific Research and New Technologies:* Bruno Jean-Richard Itoua. *Small and Medium-Sized Businesses, and Handicrafts:* Adélaïde Mougany. *Social Affairs, Humanitarian Affairs and Solidarity:* Émilienne Raoul. *Sports and Physical Education:* Léon Alfred Opimbat. *Technical and Vocational Training, and Employment:* Serge-Blaise Zoniaba. *Tourism and Environment:* Josué Rodrigue Ngouonimba.

President's Website (French only): http://www.presidence.cg/accueil

CURRENT LEADERS

Denis Sassou-Nguesso

Position
President

Introduction
A military leader from the 1960s, Denis Sassou-Nguesso ruled the Republic of the Congo between 1979–92, regaining power in 1997

in a coup. He maintained Congo's one-party Marxist-Leninist state—through the Congolese Labour Party (PCT; Parti Congolais du Travail)—until 1992, when he introduced free elections in which he was defeated. Much of his tenure since 1997 has been marred by civil war and economic crises.

Early Life
Sassou-Nguesso was born in 1943 in Edou. Joining the army in 1960, he trained at the General Leclerc military school in the Congo and the Saint-Maixent military school in France. In 1963 he was involved in the overthrow of Fulbert Youlou, the first president of independent Congo from 1959, and then of his successor Alphonse Massemba-Débat (1963–68). When Marien Ngouabi took power in 1969 he introduced a Marxist-Leninist state with the newly-created PCT at its core. Sassou-Nguesso joined the PCT in the same year.

In 1977 Sassou-Nguesso formed part of a military junta which took power following the assassination of Ngouabi by forces supposedly loyal to Massemba-Débat, before serving as defence minister under the leadership of Col. Joachim Yhombi-Opango (1977–79). At the same time he achieved the rank of colonel and became vice president of the military committee. Two years later Sassou-Nguesso replaced Yhombi-Opango as president.

Career in Office
On taking power, Sassou-Nguesso introduced a new constitution with the PCT continuing to lead a one-party government. Although reinforcing the Marxist-Leninist state introduced by Ngouabi, and despite signing a co-operation treaty with the USSR in 1981, Sassou-Nguesso sought a rapprochement with France and international investment in Congo's oil resources. In 1990 the party abandoned Marxism for democratic socialism and in 1992, following a referendum, free elections were introduced. The first multi-party elections saw the defeat of Sassou-Nguesso by Pascal Lissouba of the Pan-African Union for Social Development. Sassou-Nguesso came third with 16·9% of votes.

Thereafter, tension between the governing and opposition parties led to recurring violence between their respective militias, including the 'Cobra' militia loyal to Sassou-Nguesso. Violence increased ahead of elections scheduled for 1997, accelerating from June 1997 into four months of civil war. Sassou-Nguesso, with the aid of Angolan soldiers, finally ousted Lissouba. The latter went into hiding in Burkina Faso and Sassou-Nguesso was sworn in as president again in Oct. 1997. Between 10,000–15,000 people were killed in the conflict. Fighting erupted again in Jan. 1999 as 'Cocoye' rebels loyal to Lissouba attacked Brazzaville, and in April there was an attack on Pointe-Noire by the 'Ninja' rebels loyal to former prime minister Bernard Kolélas. Then in Aug. 1999 peace talks were held by Sassou-Nguesso, Lissouba and Kolélas at which a ceasefire was agreed.

In Jan. 2001 Sassou-Nguesso passed a new constitution which increased the president's term from five to seven years and strengthened presidential powers. In presidential elections held in March 2002, he won 89·4% of votes. The new constitution deemed Lissouba and Kolélas ineligible to stand, while another opposition candidate, Andre Milongo, refused to participate claiming 'irregularities'. The elections were marred by violence between the 'Ninja' rebels and government forces in the Pool area of the country, which caused the displacement of around 66,000 people. Legislative elections which followed in May and June 2002 resulted in a large parliamentary majority for the PCT and its allies, although the results were criticized by international observers and provoked further widespread militia violence. A peace agreement reached in March 2003 has remained fragile. Following the parliamentary elections in mid-2007 the PCT remained the largest party in the National Assembly. In July 2009 Sassou-Nguesso was re-elected for a further seven-year presidential term, claiming 78·6% of the vote. The conduct of the poll was criticized by a European Commission delegation and the results were disputed, provoking street protests in Brazzaville that were dispersed by riot police.

In a cabinet reshuffle in Sept. 2009, Sassou-Nguesso abolished the post of prime minister and took over those duties himself under the mantle of the presidency. The PCT maintained its parliamentary dominance in legislative elections in July–Aug. 2012.

In Aug. 2010 Sassou-Nguesso presided over national celebrations marking the 50th anniversary of the country's independence from France.

DEFENCE

In 2008 military expenditure totalled US$112m. (US$29 per capita), representing 0·9% of GDP.

Army

Total personnel (2007) 8,000. There is a Gendarmerie of 2,000.

Navy

Personnel in 2007 totalled about 800. The Navy is based at Pointe-Noire.

Air Force

The Air Force had (2007) about 1,200 personnel and 24 aircraft although none are combat-capable and their serviceability is questionable.

ECONOMY

Agriculture produced 4·5% of GDP in 2009, industry 71·1% and services 24·4%.

Overview

The economy is dominated by the oil sector (more than 60% of nominal GDP, 85% of exports and 75% of revenues) and is thus vulnerable to external shocks. The closing of one of the country's largest oil fields following an accident led to a recession in 2007, while the global financial crisis prompted a sharp decline in oil prices (although prices subsequently recovered). Sectors with strong growth potential include transport services, agriculture, forestry and mining.

Civil war and poor governance during the 1990s damaged the health and education systems as well as the transport infra-structure. In 2005, 50·7% of the population was below the poverty line. In 2010 the World Bank and the IMF approved debt relief of US$1·9bn. (reducing the debt burden by 34%), Brazzaville having implemented policies recommended by the Heavily Indebted Poor Countries and the Multilateral Debt Relief Initiatives over a four-year period.

Currency

The unit of currency is the *franc CFA* (XAF) with a parity of 655·957 francs CFA to one euro. Total money supply in June 2005 was 322,269m. francs CFA and foreign exchange reserves were US$226m. Gold reserves were 11,000 troy oz in July 2005. Inflation was 5·0% in 2010 and 1·8% in 2011.

Budget

In 2008 revenues were 2,465·9bn. francs CFA and expenditures 1,227·5bn. francs CFA. Petroleum revenue accounted for 85·9% of revenues in 2008; current expenditure accounted for 63·8% of expenditures.

Performance

Total GDP in 2011 was US$14·4bn. Real GDP growth was 7·5% in 2009, 8·8% in 2010 and 3·4% in 2011.

Banking and Finance

The Banque des États de l'Afrique Centrale (*Governor*, Lucas Abaga Nchama) is the bank of issue. There are four commercial banks and a development bank, in all of which the government has majority stakes. There is also a co-operative banking organization (Mutuelle Congolaise de l'Épargne et de Crédit).

In 2010 external debt totalled US$3,781m. and represented 43·9% of GNI.

ENERGY AND NATURAL RESOURCES

Environment

Carbon dioxide emissions from the consumption and flaring of fossil fuels in 2008 were the equivalent of 1·4 tonnes per capita.

Electricity

Installed capacity was an estimated 112,000 kW in 2007. Total production in 2007 was 407m. kWh and consumption per capita 227 kWh.

Oil and Gas

Oil was discovered in the mid-1960s when Elf Aquitaine was given exclusive rights to production. Elf still has the lion's share but Agip Congo is also involved in oil exploitation. In 2008 production was 12·9m. tonnes. Proven reserves in 2008 were 1·9bn. bbls, including major off-shore deposits. Oil provides about 90% of government revenue and exports. There is a refinery at Pointe-Noire, the second largest city. There were proven natural gas reserves of 91bn. cu. metres in 2007.

Minerals

A government mine produces several metals; gold and diamonds are extracted by individuals. There are reserves of potash (4·5m. tonnes), iron ore (1,000m. tonnes), and also clay, bituminous sand, phosphates, zinc and lead.

Agriculture

In 2001 there were 175,000 ha. of arable land and 45,000 ha. of permanent crops. There were some 700 tractors and 85 thresher-harvesters in use in 2001. Production (2002, in thousand tonnes): cassava, 862; sugarcane, 459; bananas, 84; plantains, 71; mangoes, 25; groundnuts, 24; palm oil, 17; yams, 11; maize, 7.

Livestock (2002): goats, 294,000; sheep, 98,000; cattle, 93,000; pigs, 46,000; poultry, 2m.

Forestry

In 2010 equatorial forests covered 22·41m. ha. (66% of the total land area). In 2007, 3·71m. cu. metres of timber were produced, mainly okoumé from the south and sapele from the north. Timber companies are required to replant, and to process at least 60% of their production locally. Before the development of the oil industry, forestry was the mainstay of the economy.

Fisheries

The catch for 2009 was 61,218 tonnes, of which 32,833 tonnes were from marine waters.

INDUSTRY

There is a growing manufacturing sector, located mainly in the four major towns, producing processed foods, textiles, cement, metal goods and chemicals. Industry produced 65·2% of GDP in 2001, including 4·1% from manufacturing. Production (2004): residual fuel oil, 295,000 tonnes; distillate fuel oil, 120,000 tonnes; petrol, 49,000 tonnes; kerosene, 19,000 tonnes; cigarettes (1994), 655m. cartons; beer (2003), 66·0m. litres; veneer sheets (2001), 12,000 cu. metres; cotton textiles (1993), 1·8m. metres.

Labour

In 2010 the estimated economically active population was 1,637,000 (56% males), up from 1,256,000 in 2000. More than 50% of the labour force in 2010 were engaged in agriculture.

INTERNATIONAL TRADE

Imports and Exports

Imports and exports for calendar years in US$1m.:

	2001	2002	2003	2004	2005
Imports f.o.b.	681	691	831	969	1,356
Exports f.o.b.	2,055	2,289	2,637	3,433	4,730

Principal imported commodities are intermediate manufactures, capital equipment, construction materials, foodstuffs and petroleum products. Apart from crude oil, other significant commodities for export are lumber, plywood, sugar, cocoa, coffee and diamonds. In 2003 the main import suppliers were the Netherlands, France, USA, Italy and Germany. The main export markets were China, South Korea, USA, North Korea and France.

COMMUNICATIONS

Roads
In 2004 there were 17,289 km of roads, of which 5·0% were surfaced. Passenger cars in use in 2007 numbered 56,000 (15 per 1,000 inhabitants). There were 214 deaths in road accidents in 2007.

Rail
A railway connects Brazzaville with Pointe-Noire via Loubomo and Bilinga, and a branch links Mont-Belo with Mbinda on the Gabon border. Total length in 2005 was 797 km (1,067 mm gauge). In 2006 passenger-km totalled 167m. and freight tonne-km 264m.

Civil Aviation
The principal airports are at Brazzaville (Maya Maya) and Pointe-Noire. In 2003 Trans Air Congo operated services to Abidjan, Cotonou and Lomé, as well as domestic services. Trans African Airlines flew to Abidjan, Bamako, Cotonou, Dakar and Lomé. In 2001 Brazzaville handled 433,000 passengers (341,000 on domestic flights) and 27,000 tonnes of freight.

Shipping
The only seaport is Pointe-Noire. There are some 5,000 km of navigable rivers, and river transport is an important service for timber and other freight as well as passengers.

Telecommunications
There were 9,800 fixed telephone lines in 2010 (2·4 per 1,000 inhabitants). Mobile phone subscribers numbered 3·80m. in 2010. There were 42·9 internet users per 1,000 inhabitants in 2008.

SOCIAL INSTITUTIONS

Justice
The Supreme Court, Court of Appeal and a criminal court are situated in Brazzaville, with a network of *tribunaux de grande instance* and *tribunaux d'instance* in the regions.

Education
In 2007 there were 10,631 teaching staff for 621,702 pupils at primary schools and (2004) 6,965 secondary school teaching staff for 232,026 pupils. There were 12,456 students at tertiary level in 2003 with 894 academic staff. Adult literacy rate in 2004 was 84·7%.

In 2005 public expenditure on education came to 1·8% of GDP and 8·1% of total government spending.

Health
In 2001 there were 103 hospitals with 5,195 beds. In 2000 there were 540 physicians, 75 pharmacists, 1,439 nurses and 579 midwives.

RELIGION
In 2001 there were 1·43m. Roman Catholics, 0·49m. Protestants and 0·36m. Kimbanguistes (African Christians). Traditional animist beliefs are still widespread.

CULTURE

World Heritage Sites
Sangha Trinational, a transboundary conservation complex situated in Cameroon, Central African Republic and the Republic of the Congo, was added to the UNESCO World Heritage List in 2012. The site consists of three contiguous national parks totalling around 750,000 ha. Much of it is unaffected by human activity and features a wide range of humid tropical forest ecosystems with rich flora and fauna.

Press
In 2008 there were five daily newspapers with a combined circulation of 8,000.

Tourism
In 2009, 85,000 non-resident tourists stayed in hotels and similar accommodation (up from 63,000 in 2008 and 54,000 in 2007).

DIPLOMATIC REPRESENTATIVES

Of the Republic of the Congo in the United Kingdom
Ambassador: Henri Marie Joseph Lopes (resides in Paris).
Honorary Consul: Louis Muzzu (The Arena, 24 Southwark Bridge Rd, London, SE1 9HF).

Of the United Kingdom in the Republic of the Congo
Ambassador: Neil Wigan (resides in Kinshasa, Democratic Republic of the Congo).

Of the Republic of the Congo in the USA (4891 Colorado Ave., NW, Washington, D.C., 20011)
Ambassador: Serge Mombouli.

Of the USA in the Republic of the Congo (70–83 Section D, Maya-Maya Blvd, Brazzaville)
Ambassador: Christopher W. Murray.

Of the Republic of the Congo to the United Nations
Ambassador: Raymond Serge Balé.

Of the Republic of the Congo to the European Union
Ambassador: Roger Julien Menga.

FURTHER READING
Thompson, V. and Adloff, R., *Historical Dictionary of the People's Republic of the Congo.* 2nd ed. 1984

National Statistical Office: Centre National de la Statistique et des Études Économiques, BP 2031, Brazzaville.
Website (French only): http://www.cnsee.org

COSTA RICA

© Research Machines plc 2006

República de Costa Rica
(Republic of Costa Rica)

Capital: San José
Population projection, 2015: 4·99m.
GNI per capita, 2011: (PPP$) 10,497
HDI/world rank: 0·744/69
Internet domain extension: .cr

KEY HISTORICAL EVENTS

Discovered by Columbus in 1502 on his last voyage, Costa Rica (Rich Coast) was part of the Spanish viceroyalty of New Spain from 1540 to 1821, then of the Central American Federation until 1838 when it achieved full independence. Coffee was introduced in 1808 and became a mainstay of the economy, helping to create a peasant land-owning class. In 1948 accusations of election fraud led to a six-week civil war, at the conclusion of which José Figueres Ferrer won power at the head of a revolutionary junta. A new constitution abolished the Army. In 1986 Oscar Arias Sánchez was elected president. He promised to prevent Nicaraguan anti-Sandinista (*contra*) forces using Costa Rica as a base. In 1987 he received the Nobel Peace Prize as recognition of his Central American peace plan, agreed to by the other Central American states. Costa Rica was beset with economic problems in the early 1990s when several politicians, including President Calderón, were accused of profiting from drug trafficking.

TERRITORY AND POPULATION

Costa Rica is bounded in the north by Nicaragua, east by the Caribbean, southeast by Panama, and south and west by the Pacific. The area is estimated at 51,100 sq. km (19,730 sq. miles). The population at the census of May 2011 was 4,301,712 (2,195,649 females); density, 84·2 per sq. km. In 2011, 64·9% of the population were urban.

The UN gives a projected population for 2015 of 4·99m.

There are seven provinces (with 2011 census population): Alajuela (848,146); Cartago (490,903); Guanacaste (326,953); Heredia (433,677); Limón (386,862); Puntarenas (410,929); San

José (1,404,242). The largest cities, with estimated 2000 populations, are San José (346,600); Limón (62,000); and Alajuela (53,900). Main ethnic groups (2011): White or Mestizo 84%, Mulatto 7%, Amerindian 2%, Black or Afro-Caribbean 1%.

Spanish is the official language.

SOCIAL STATISTICS

Statistics for calendar years:

	Marriages	*Births*	*Deaths*
2005	25,631	71,548	16,139
2006	26,575	71,291	16,766
2007	26,010	73,144	17,071
2008	25,034	75,187	18,021

2008 rates per 1,000 population: births, 16·9; deaths, 4·0. Annual population growth rate, 2005–10, 1·6%. Life expectancy at birth, 2007, was 76·4 years for men and 81·3 years for women. Infant mortality, 2008, 9·0 per 1,000 live births; fertility rate, 2008, 2·0 children per woman.

CLIMATE

The climate is tropical, with a small range of temperature and abundant rain. The dry season is from Dec. to April. San José, Jan. 66°F (18·9°C), July 69°F (20·6°C). Annual rainfall 72" (1,793 mm).

CONSTITUTION AND GOVERNMENT

The Constitution was promulgated on 7 Nov. 1949. The legislative power is vested in a single-chamber *Legislative Assembly* of 57 deputies elected for four years. The *President* and two *Vice-Presidents* are elected for four years; the candidate receiving the largest vote, provided it is over 40% of the total, is declared elected, but a second ballot is required if no candidate gets 40% of the total. Since 2003 former presidents have been permitted to stand again. Elections are normally held on the first Sunday in Feb.

The President may appoint and remove members of the cabinet.

National Anthem

'Noble patria, tu hermosa bandera' ('Noble fatherland, thy beautiful banner'); words by J. M. Zeledón Brenes, tune by M. M. Gutiérrez.

GOVERNMENT CHRONOLOGY

Presidents since 1944. (PLN = National Liberation Party; PRD = Party of the Democratic Renewal; PRN = National Republican Party; PUN = National Union Party; PUSC = Social Christian Unity Party)

1944–48	PRN	Teodoro Picado Michalski
1948–49	military	José María Figueres Ferrer
1949–53	PUN	Luis Otilio Ulate Blanco
1953–58	PLN	José María Figueres Ferrer
1958–62	PUN	Mario José Echandi Jiménez
1962–66	PLN	Francisco José Orlich Bolmarcich
1966–70	PUN	José Joaquín Trejos Fernández
1970–74	PLN	José María Figueres Ferrer
1974–78	PLN	Daniel Oduber Quirós
1978–82	PRD	Rodrigo José Carazo Odio
1982–86	PLN	Luis Alberto Monge Álvarez
1986–90	PLN	Óscar Rafael Arias Sánchez
1990–94	PUSC	Rafael Ángel Calderón Fournier
1994–98	PLN	José María Figueres Olsen
1998–2002	PUSC	Miguel Ángel Rodríguez Echeverría
2002–06	PUSC	Abel Pacheco de la Espriella

| 2006–10 | PLN | Óscar Rafael Arias Sánchez |
| 2010– | PLN | Laura Chinchilla Miranda |

RECENT ELECTIONS

In presidential elections held on 7 Feb. 2010 vice-president Laura Chinchilla of the National Liberation Party (PLN) won 46·8% of the vote, Ottón Solís of the Citizens' Action Party (PAC) 25·2%, Otto Guevara of the Libertarian Movement (ML) 20·8% and Luis Fishman of the Social Christian Unity Party (PUSC) 3·9%. There were five other candidates. Turnout was 69·1%. In parliamentary elections held on the same day the PLN won 24 of 57 seats with 37·3% of votes cast, the PAC 11 (17·6%), the ML 9 (14·5%), the PUSC 6 (8·2%), the PASE (Accessibility Without Exclusion Party) 4 (9·0%), with three parties each winning one seat.

CURRENT GOVERNMENT

President: Laura Chinchilla Miranda; b. 1959 (PLN; sworn in 8 May 2010).
　First Vice-President: Alfio Piva Mesén. *Second Vice-President:* Luis Liberman Ginsburg.
　In March 2013 the government comprised:
　Minister of Agriculture and Livestock: Gloria Abraham Peralta. *Communications:* Francisco Chacón. *Culture and Youth:* Manuel Obregón López. *Decentralization and Local Development:* Juan Marín Quirós. *Economy, Industry and Commerce:* Mayi Antillón Guerrero. *Environment, Energy and Telecommunications:* René Castro Salazar. *Finance:* Edgar Ayales. *Foreign Relations and Religion:* Enrique Castillo. *Foreign Trade:* Anabel González Campabadal. *Housing and Human Settlements:* Guido Monge. *Interior, Police and Public Security:* Mario Zamora Cordero. *Justice and Peace:* Fernando Ferraro Castro. *Labour and Social Security:* Olman Segura. *National Planning and Economic Policy:* Roberto Gallardo. *Presidency:* Carlos Ricardo Benavides Jiménez. *Public Education:* Leonardo Garnier Rímolo. *Public Health:* Daisy Corrales Díaz. *Public Works and Transportation:* Pedro Castro. *Science and Technology:* Alejandro Cruz Molina. *Social Wellbeing:* Fernando Alfonso Marín Rojas. *Sport:* William Corrales. *Tourism:* Allan Flores. *Women's Status:* Maureen Clarke.

Costa Rican Parliament (Spanish only): http://www.asamblea.go.cr

CURRENT LEADERS

Laura Chinchilla

Position
President

Introduction
Laura Chinchilla was elected Costa Rica's first female president in Feb. 2010, continuing the centre-right National Liberation Party's (PLN) hold on government. A former justice minister, she has won praise for her attempts to tackle drug networks and improve public security.

Early Life
Laura Chinchilla Miranda was born on 28 March 1959 in Desamparados, a suburb of the capital, San José. She graduated in political science from the University of Costa Rica in 1981 and went on to earn a masters in public administration from Georgetown University in the USA. Returning to Costa Rica, she worked as a consultant to various international organizations including the US Agency for International Development, the UN Development Program and the Inter-American Development Bank, on issues such as public safety, border security, human rights and judicial reform.
　Consultancy for the ministry of national planning and economic policy in the early 1990s led to Chinchilla being offered a senior role in government. In May 1994 she was appointed deputy minister of public security in the administration of José María Figueres Olsen, whose PLN had won the general election

three months earlier. Chinchilla also led the national council of migration and foreign affairs and was a member of the board of the national drug council. She was promoted to minister for public security in 1996 and gained a reputation for her tough stance on criminal drugs activities.
　Chinchilla was elected to the legislative assembly as a deputy for the province of San José in 2002 and served as vice-president under Oscar Arias of the PLN. Arias, who had served as president in the late 1980s and was awarded the Nobel Peace Prize for his contribution to democracy and peace in central America, enjoyed popular support and won a further term in Jan. 2006. Chinchilla served as both vice-president and justice minister until she left office in Oct. 2008 to mount a presidential campaign. She emerged victorious after the first round on 7 Feb. 2010, taking 47% of the vote, well ahead of nearest rival Ottón Solís of the left-leaning Citizens' Action Party on 25%.

Career in Office
Chinchilla was sworn in on 8 May 2010. She pledged to make Costa Rica the first developed country in Central America, confirmed her commitment to carbon neutrality by 2021 and promised to improve health care, safety and security.
　In Nov. 2010 a dispute with neighbouring Nicaragua over a river border area led to the deployment of security forces by both countries. In March 2011 the International Court of Justice ruled that all troops should be withdrawn from the area in a judgment viewed as favouring Costa Rica.

DEFENCE

In 2008 defence expenditure totalled US$156m. (US$37 per capita), representing 0·5% of GDP.

Army

The Army was abolished in 1948 and replaced by a Civil Guard numbering 4,500 in 2007. In addition there is a Border Security Police of 2,500 and a Rural Guard, 2,000-strong.

Navy

The paramilitary Coast Guard Unit numbered (2007) 400.

Air Wing

There is a 400-strong Air Surveillance Unit attached to the ministry of public security, equipped with ten light planes and two helicopters.

ECONOMY

Agriculture accounted for 7·3% of GDP in 2009, industry 27·3% and services 65·4%.

Overview

The economy suffered a crisis in the 1980s as a result of falling prices of key exports including bananas, pineapples and coffee. Since then growth has been healthy, despite dips in the mid-1990s and early 2000s. There has been a large-scale diversification into manufacturing, with microprocessors, medical equipment, plastics and construction materials the main industrial products.
　Tax-free trade zones were also established to attract foreign companies, with foreign direct investment (FDI) secured initially for assembly plants producing clothing for the US market. After Intel opened a computer chip assembly and testing plant in 1998, other US manufacturing companies followed because of Costa Rica's proximity, relative political stability and well-educated, English-speaking population. The USA is the biggest trading partner, accounting for 43% of foreign trade in 2010, followed by China and Mexico.
　The economy achieved peak growth of 8·8% in 2006 but contracted by 1·0% in 2009 as a result of the global economic crisis. In response, central government expenditure increased and the World Bank approved a US$500m. credit line in April 2009, enabling domestic investment and higher private consumption.

In addition, on 1 Jan. 2009 the US Central American Free Trade Agreement came into force. Real GDP growth exceeded 4% in both 2010 and 2011. Poverty nonetheless remained high, with 24·2% falling below the poverty line in 2010, up from 21·7% in 2009.

Currency

The unit of currency is the *Costa Rican colón* (CRC) of 100 *céntimos*. The official rate is used for all imports on an essential list and by the government and autonomous institutions, and a free rate is used for all other transactions. In July 2005 total money supply was 1,101·6bn. colones, foreign exchange reserves were US$2,215m. and gold reserves were 2,000 troy oz. Inflation was at 13·4% in 2008 but fell to 5·7% in 2010 and 4·9% in 2011.

Budget

In 2010 central government revenues were 4,534·6bn. colones (4,146·8bn. colones in 2009) and expenditures 5,066·0bn. colones (4,361·6bn. colones in 2009).

There is a sales tax of 13%.

Performance

Costa Rica is considered to be amongst the most stable countries in Central America. Although the economy contracted by 1·0% in 2009 it grew by 4·7% in 2010 and 4·2% in 2011. Total GDP in 2011 was US$40·9bn.

Banking and Finance

The bank of issue is the Central Bank (founded 1950) which supervises the national monetary system, foreign exchange dealings and banking operations. The bank has a board of seven directors appointed by the government, including *ex officio* the Minister of Finance and the Planning Office Director. The *President* is Dr Rodrigo Bolaños Zamora.

There are three state-owned banks (Banco de Costa Rica, Banco Nacional de Costa Rica and Banco Popular y de Desarrollo Comunal), 17 private banks and one credit co-operative.

External debt totalled US$8,849m. in 2010 and represented 26·8% of GNI.

There is a stock exchange in San José.

ENERGY AND NATURAL RESOURCES

Environment

Costa Rica's carbon dioxide emissions from the consumption and flaring of fossil fuels in 2008 were the equivalent of 1·7 tonnes per capita. An *Environmental Performance Index* compiled in 2008 ranked Costa Rica fifth in the world, with 90·5%. The index examined various factors in six areas—air pollution, biodiversity and habitat, climate change, environmental health, productive natural resources and water resources.

Electricity

Installed capacity was an estimated 2·1m. kW in 2007. Production was 9·05bn. kWh in 2007; consumption per capita in 2007 was 2,074 kWh.

Minerals

In 2007 production of gold was 1,036 kg; crushed rock and rough stone, 9,260,000 tonnes; diatomite, 1,712,000 tonnes.

Agriculture

Agriculture is a key sector, with an estimated 327,000 people being economically active in 2007. There were about 0·2m. ha. of arable land in 2007 and 0·3m. ha. of permanent crops. The principal agricultural products are coffee, bananas and sugar. Cattle are also of great importance. Production figures for 2003 (in 1,000 tonnes): sugarcane, 3,924; bananas, 1,863; pineapples, 725; oranges, 367; melons and watermelons, 292; rice, 180; coffee, 132; palm oil, 131; cassava, 94; potatoes, 82; plantains, 70.

Livestock (2003): cattle, 1·15m.; pigs, 500,000; horses, 115,000; chickens, 18m.

Forestry

In 2010 forests covered 2·61m. ha., or 51% of the land area. Timber production in 2007 was 4·61m. cu. metres.

Fisheries

Total catch in 2007 amounted to 21,735 tonnes, mostly from sea fishing.

INDUSTRY

The main manufactured goods are foodstuffs, palm oil, textiles, fertilizers, pharmaceuticals, furniture, cement, tyres, canning, clothing, plastic goods, plywood and electrical equipment.

Labour

In July 2001 there were 1,552,920 people in employment. In July 2001 there were 100,397 unemployed persons, or 6·1% of the workforce. The main area of employment is transport, storage and communications (303,000 people in 2002), followed by agriculture, hunting, forestry and fisheries (243,000 in 2002).

INTERNATIONAL TRADE

A free trade agreement was signed with Mexico in March 1994. Some 2,300 products were freed from tariffs, with others to follow over ten years. In 2007 a national referendum approved the adoption of the Central America-Dominican Republic-United States Free Trade Agreement (CAFTA-DR), which establishes a free trade zone with the Dominican Republic, El Salvador, Guatemala, Honduras, Nicaragua and the USA. It was approved by the Legislative Assembly in Nov. 2008 and subsequently entered into force on 1 Jan. 2009.

Imports and Exports

The value of imports and exports in US$1m. was:

	2006	2007	2008	2009	2010
Imports c.i.f.	11,070·5	12,757·8	15,289·4	11,550·5	13,920·2
Exports f.o.b.	7,254·9	8,927·6	9,744·5	8,836·3	9,044·8

Principal imports: machinery and apparatus, chemicals and chemical products, mineral fuels, food. Chief exports: manufactured goods and other products, coffee, bananas, sugar, cocoa. Main import suppliers, 2006: USA, 39·6%; Japan, 5·3%; Venezuela, 5·3%; Mexico, 5·2%. Major export markets, 2006: USA, 42·5%; China, 7·7%; Hong Kong, 7·2%; Netherlands, 6·9%.

COMMUNICATIONS

Roads

In 2007 there were 36,654 km of roads, including 7,640 km of main roads. On the Costa Rica section of the Inter-American Highway it is possible to drive to Panama during the dry season. The Pan-American Highway into Nicaragua is metalled for most of the way and a new highway between San José and Caldera opened in Jan. 2010. Passenger cars in use in 2007 numbered 525,400, buses and coaches 12,300, vans and lorries 139,600 and motorcycles and mopeds 100,100. There were 339 fatalities as a result of road accidents in 2007.

Rail

The nationalized railway system (Incofer) was closed in 1995 following an earthquake in 1991. Freight services and some commuter services have now been resumed. In 2007 passenger-km totalled 872,000 and freight tonne-km 230,000.

Civil Aviation

There are international airports at San José (Juan Santamaria) and Liberia (Daniel Oduber Quirós). The national carrier is Líneas Aéreas Costarriquenses (LACSA). In 2003 scheduled airline traffic of Costa Rican-based carriers flew 20m. km, carrying 750,000 passengers (584,000 on international flights). In 2001 San José handled 2,108,713 passengers (1,972,606 on international flights) and 67,858 tonnes of freight.

Shipping

The chief ports are Limón on the Atlantic and Caldera on the Pacific. In Jan. 2009 there were two ships of 300 GT or over registered, totalling 2,000 GT.

Telecommunications

In 2011 there were 1,490,600 landline telephone subscriptions (equivalent to 315·4 per 1,000 inhabitants) and 4,358,100 mobile phone subscriptions (or 922·0 per 1,000 inhabitants). There were 343·3 internet users per 1,000 inhabitants in 2009. Fixed internet subscriptions totalled 271,500 in 2009 (59·1 per 1,000 inhabitants). In Dec. 2011 there were 1·6m. Facebook users.

SOCIAL INSTITUTIONS

Justice

Justice is administered by the Supreme Court and five appeal courts divided into five chambers—the Court of Cassation, the Higher and Lower Criminal Courts, and the Higher and Lower Civil Courts. There are also subordinate courts in the separate provinces and local justices throughout the republic. There is no capital punishment.

The population in penal institutions in Oct. 2008 was 8,924 (198 per 100,000 of national population).

Education

The adult literacy rate in 2005 was 97·4% (96·0% among males and 96·2% among females). Primary instruction is compulsory and free from six to 14 years; secondary education (since 1949) is also free. Primary schools are provided and maintained by local school councils, while the national government pays the teachers, besides making subventions in aid of local funds. In 2006 there were 4,026 public and private primary schools with 35,413 teachers and administrative staff and 521,505 enrolled pupils, and 752 public and private secondary schools with 24,445 teachers and 338,508 pupils. In 2004 there were 166,417 university students. The largest of the four public universities is the University of Costa Rica (Universidad de Costa Rica). There are also a number of private universities.

In 2006 public expenditure on education came to 4·9% of GNI, representing 20·6% of total government expenditure.

Health

In 2006 there were 6,987 doctors, 2,800 dentists, 6,943 nurses, 3,058 pharmacists and 29 hospitals. There were 14 beds per 10,000 inhabitants in 2006.

RELIGION

Roman Catholicism is the state religion; it had 3·38m. adherents in 2001. There is entire religious liberty under the constitution. The Archbishop of Costa Rica has six bishops at Alajuela, Ciudad Quesada, Limón, Puntarenas, San Isidro el General and Tilarán. There were 360,000 Protestants in 2001. The remainder of the population are followers of other religions.

CULTURE

World Heritage Sites

Costa Rica has three sites on the UNESCO World Heritage List: Cocos Island National Park (1997 and 2002); and the Area de Conservación Guanacaste (1999 and 2004), an important dry forest habitat.

Costa Rica shares a UNESCO site with Panama: the Talamanca Range-La Amistad Reserves (1983 and 1990), an important cross-breeding site for North and South American flora and fauna.

Press

There were six daily newspapers in 2008 with a combined circulation of 255,000, and 40 non-dailies. The most widely read dailies are *La República*, *Diario Extra* and *La Nación*.

Tourism

In 2009 there were 1,923,000 non-resident tourists (excluding same-day visitors), down from 2,089,000 in 2008.

DIPLOMATIC REPRESENTATIVES

Of Costa Rica in the United Kingdom (Flat 1, 14 Lancaster Gate, London, W2 3LH)
Ambassador: Pilar Saborío Rocafort.

Of the United Kingdom in Costa Rica (Edificio Centro Colón, 11th Floor, Apartado 815, San José 1007)
Ambassador: Sharon Campbell.

Of Costa Rica in the USA (2114 S St., NW, Washington, D.C., 20008)
Ambassador: Meta Shanon Figueres Boggs.

Of the USA in Costa Rica (Calle 120 Avenida 0, Pavas, San José)
Ambassador: Anne Slaughter Andrew.

Of Costa Rica to the United Nations
Ambassador: Eduardo Ulibarri.

Of Costa Rica to the European Union
Ambassador: Francisco Tomás Dueñas.

FURTHER READING

Creedman, T. S., *Historical Dictionary of Costa Rica*. 2nd ed. 1991
Cruz, Consuelo, *Political Culture and Institutional Development in Costa Rica and Nicaragua: World Making in the Tropics*. 2005

National Statistical Office: Instituto Nacional de Estadística y Censos, San José.
Website (Spanish only): http://www.inec.go.cr

Explore the world at
www.statesmansyearbook.com

CÔTE D'IVOIRE

MALI
BURKINA FASO
Boundiali
GUINEA
Korhogo
CÔTE D'IVOIRE
Man
GHANA
YAMOUSSOUKRO
Daloa
0   100 mi
Adzopé
0   150 km
LIBERIA
Abidjan
Gulf of Guinea
© Research Machines plc 2006

**République de la Côte d'Ivoire
(Republic of Côte d'Ivoire)**

Capital: Yamoussoukro
Seat of government: Abidjan
Population projection, 2015: 22·02m.
GNI per capita, 2011: (PPP$) 1,387
HDI/world rank: 0·400/170
Internet domain extension: .ci

KEY HISTORICAL EVENTS

Evidence of Neolithic settlement has been found at Boundali in northern Côte d'Ivoire but the country's forested central and southern regions are believed to have been thinly populated. From the ninth century the northwest fell under the Ghana empire and prospered from Trans-Saharan trade. From the 13th to the 17th centuries the north came under the control of the Islamic Mali empire.

Malinké ethnic groups moved southwards during the 16th and 17th centuries, founding kingdoms including Kong and Kabadugu. The central and southern regions became associated with two ethnic groups, the Kru and the Akan, the latter principally emigrants fleeing the Ashanti empire (present-day Ghana) after 1720. Of the Akan groups, the Baoulé dominated the Ivorian centre. The Gyaman kingdom, centred on Bondoukou in the east, became famous for its Muslim scholars in the 18th and 19h centuries. Various Anyi kingdoms, notably the Sanwi, held sway in the southeast.

Lacking natural harbours, Côte d'Ivoire was neglected by European traders in the 15th and 16th centuries. The first French attempt to establish a missionary station in 1637 at Assinie was abandoned for almost a century but coastal settlements developed in the 18th century, although with less slave trading than elsewhere in West Africa. In 1842 the French Admiral Bouët-Willaumez signed treaties with local chiefs establishing a French protectorate, beginning with Assinie and Grand Bassam and extending along the coast over the next 25 years.

In the early 1870s the French unsuccessfully offered Côte d'Ivoire to the British in exchange for the Gambia, which bisected the French colony of Senegal. However, rumours of gold rekindled French interest in the region and expeditions were mounted to sign treaties with inland kingdoms and the trading centres of Kong and Bondoukou. In 1893 Côte d'Ivoire was declared a French colony, with the explorer Louis-Gustave Binger as governor. He negotiated boundary treaties with Liberia and with the UK's Gold Coast colony (now Ghana), as well as exiling the Malinké chief, Samory Touré, in 1898. In 1904 Côte d'Ivoire became part of the Federation of French West Africa. Governor Gabriel Angoulvant embarked on a full military conquest in 1908, brutally suppressing revolts by the Baoulé, Dan and Bété.

Sweeping reforms to the governance of French West Africa after the Second World War saw the establishment of Côte d'Ivoire's first political party, the Democratic Party of Côte d'Ivoire. Its leader, Félix Houphouët-Boigny, eventually adopted a policy of co-operation with the French and by the mid-1950s the country was the wealthiest in French West Africa. In 1958 it became an autonomous republic within the French Community and achieved full independence on 7 Aug. 1960, with Houphouët-Boigny as president. He developed ties with the West and oversaw economic development. Revenues from agriculture (principally coffee and cocoa, but also bananas, sugar, palm oil and rubber) were invested in infrastructure, education and industry.

Resentment against one-party rule grew from the late 1980s. Under Henri Konan Bédié, who became president after Houphouët-Boigny's death in 1993, Côte d'Ivoire faced economic challenges triggered by corruption and falling prices for cash crops. In Dec. 1999 Bédié was ousted in a coup led by Gen. Robert Guéï, the country's military chief from 1990 to 1995. After Guéï declared victory in the presidential election in Oct. 2000, 2,000 people died in an uprising and Guéï fled to Benin. The veteran opposition candidate, Laurent Gbagbo, was declared the rightful winner.

In Sept. 2002 a failed coup claimed over 20 lives, including those of Gen. Guéï and the interior minister. The country descended into civil war between the rebel-held north and the government-held south. In April 2005 government, rebel and opposition leaders signed a deal to end the war but promised elections were repeatedly postponed. Under a power-sharing peace deal agreed in March 2007 between the government and rebels from the New Forces movement, the latter's leader, Guillaume Soro, became prime minister. In Dec. 2010 the electoral commission declared Gbagbo had lost the presidency to Alassane Ouattara in a run-off held the previous month. Despite international pressure to leave office, Gbagbo ignored the ruling, leading to rising tensions as both candidates claimed power. Captured in April 2011, Gbagbo was handed over to the International Criminal Court. In the Dec. 2011 parliamentary elections, Ouattara and his allies secured a majority, although the political scene remains unstable.

TERRITORY AND POPULATION

Côte d'Ivoire is bounded in the west by Liberia and Guinea, north by Mali and Burkina Faso, east by Ghana, and south by the Gulf of Guinea. It has an area of 322,463 sq. km (including 4,460 sq. km of inland water). The population at the 1998 census was 15,366,672; density, 48·3 per sq. km. The population was 51·3% urban in 2011.

The UN gives a projected population for 2015 of 22·02m.

Since 2000 the country has been divided into 19 regions comprising 58 departments.

Areas, populations (1998 census) and capitals of the regions are:

Region	Area (in sq. km)	Population	Capital
Agnéby	9,200	525,211	Agboville
Bafing	8,900	139,251	Touba
Bas-Sassandra	26,400	1,395,251	San-Pédro
Denguélé	21,000	222,446	Odienné
Dix-Huit Montagnes	16,800	936,510	Man
Fromager	6,900	542,992	Gagnoa
Haut-Sassandra	15,200	1,071,977	Daloa
Lacs	8,900	476,235	Yamoussoukro
Lagunes	13,300	3,733,413	Abidjan
Marahoué	8,700	554,807	Bouaflé
Moyen-Cavally	14,300	508,733	Guiglo
Moyen-Comoé	6,900	394,761	Abengourou
N'zi-Comoé	19,200	633,927	Dimbokro
Savanes	40,200	929,673	Korhogo
Sud-Bandama	10,700	682,021	Divo
Sud-Comoé	7,300	459,487	Aboisso
Vallée du Bandama	28,500	1,080,509	Bouaké
Worodougou	22,200	378,463	Séguéla
Zanzan	38,300	701,005	Bondoukou

In 2000 the population of Abidjan stood at 3,790,000. Other major towns (with 1998 census population): Bouaké, 461,618; Yamoussoukro, 299,243; Daloa, 173,107; Korhogo, 142,093.

There are about 60 ethnic groups, the principal ones being the Baoulé (23%), the Bété (18%) and the Sénoufo (15%). A referendum held in July 2000 on the adoption of a new constitution set eligibility conditions for presidential candidates (the candidate and both his parents had to be Ivorian). This excluded a northern Muslim leader and in effect made foreigners out of millions of Ivorians. The north of the country is predominantly Muslim and the south predominantly Christian and animist.

Approximately 30% of the population are immigrants, in particular from Burkina Faso, Mali, Guinea and Senegal.

French is the official language.

SOCIAL STATISTICS

2008 estimates: births, 720,000; deaths, 223,000. Rates (2008 estimates, per 1,000 population); birth, 35·0; death, 10·8. Expectation of life in 2007 was 55·7 years for males and 58·3 for females. Annual population growth rate, 2000–08, 2·2%. Infant mortality, 2010, 86 per 1,000 live births; fertility rate, 2008, 4·6 births per woman. 29% of the population are migrants.

CLIMATE

A tropical climate, affected by distance from the sea. In coastal areas, there are wet seasons from May to July and in Oct. and Nov., but in central areas the periods are March to May and July to Nov. In the north, there is one wet season from June to Oct. Abidjan, Jan. 81°F (27·2°C), July 75°F (23·9°C). Annual rainfall 84" (2,100 mm). Bouaké, Jan. 81°F (27·2°C), July 77°F (25°C). Annual rainfall 48" (1,200 mm).

CONSTITUTION AND GOVERNMENT

The 1960 constitution was amended in 1971, 1975, 1980, 1985, 1986, 1990, 1998 and 2000. The sole legal party was the Democratic Party of Côte d'Ivoire, but opposition parties were legalized in 1990. There is a 255-member *National Assembly* elected by universal suffrage for a five-year term. The *President* is also directly elected for a five-year term (renewable). He appoints and leads a Council of Ministers.

In Nov. 1990 the National Assembly voted that its Speaker should become President in the event of the latter's incapacity, and created the post of Prime Minister to be appointed by the President. Following the death of President Houphouët-Boigny on

7 Dec. 1993, the speaker, Henri Konan Bédié, proclaimed himself head of state till the end of the presidential term in Sept. 1995.

Following the coup of Dec. 1999 a referendum was held on 23 July 2000 on the adoption of a new constitution, which set eligibility conditions for presidential candidates (the candidate and both his parents must be Ivorian), reduced the voting age from 21 to 18, and abolished the death penalty. It also offered an amnesty to soldiers who staged the coup and the junta, but committed the junta to hand over power to an elected civilian head of state and parliament within six months of the proclamation of the text. Approximately 87% of votes cast were in favour of the new constitution. This was subsequently adopted on 4 Aug. 2000.

National Anthem

'L'Abidjanaise' ('Song of Abidjan'); words by M. Ekra, J. Bony and P. M. Coty, tune by P. M. Pango.

GOVERNMENT CHRONOLOGY

Presidents since 1960. (PDCI-RDA = Democratic Party of Ivory Coast-African Democratic Rally; FPI = Ivorian Popular Front; RDR = Rally of the Republicans)

1960–93	PDCI-RDA	Félix Houphouët-Boigny
1993–99	PDCI-RDA	Aimé Henri Konan Bédié
1999–2000	military	Robert Guéï
2000–11	FPI	Laurent Gbagbo
2010–	RDR	Alassane Ouattara

RECENT ELECTIONS

In the first round of presidential elections on 31 Oct. 2010 incumbent Laurent Gbagbo won 38·0%, Alassane Ouattara 32·1% and former president Henri Konan Bédié 25·2%. There were 11 other candidates and voter turnout was 83·7%. In the second round on 28 Nov. Ouattara was declared the winner by the Independent Electoral Commission with 54·1% of the vote against 45·9% for Gbagbo. Turnout was 81·1%. However, Gbagbo refused to concede and the result was then overturned on the grounds of alleged fraud by the Constitutional Council. Gbagbo was credited with 51·5% and was sworn in for a new term on 4 Dec. Ouattara, with broad international support, nevertheless continued to claim victory.

In parliamentary elections held on 11 Dec. 2011, Ouattara's Rally of the Republicans (RDR) won 127 of 255 National Assembly seats and his ally Henri Konan Bédié's Democratic Party of Côte d'Ivoire 77 seats. Independents took 35 seats, the Union for Democracy and Peace 7, the Rally of Houphouetists for Democracy and Peace 4, the Movement of Forces of the Future 3 and the Union for Côte d'Ivoire 1. One seat was vacant. Turnout was 36·7%. The Ivorian Popular Front of ousted President Laurent Gbagbo boycotted the election.

CURRENT GOVERNMENT

President and Minister of Defence: Alassane Dramane Ouattara; b. 1942 (RDR; assumed office 4 Dec. 2010, although former incumbent Laurent Gbagbo refused to surrender the office before being captured by Ouattara's forces on 11 April 2011).

In March 2013 the government comprised:

Prime Minister, and Minister of Economy and Finance: Daniel Kablan Duncan; b. 1943 (Democratic Party of Côte d'Ivoire-African Democratic Rally/PDCI-RDA; took office 21 Nov. 2012).

Minister of State for Employment, Social Affairs and Solidarity: Moussa Dosso. *Foreign Affairs:* Charles Koffi Diby. *Interior:* Hamed Bakayoko. *Planning and Development:* Albert Mabri Toikeusse.

Minister of African Integration and Ivorians Abroad: Ally Coulibaly. *Agriculture:* Mamadou Sangafowa Coulibaly. *Animal Husbandry and Fisheries:* Kobenan Adjoumani. *Commerce, Handicrafts and the Promotion of Small and Medium-Sized Businesses:* Jean-Louis Billon. *Communication:* Affoussiata Bamba

Lamine. *Construction, Housing, Sanitation and Urban Affairs:* Mamadou Sanogo. *Culture and Francophonie:* Maurice Bandaman. *Economic Infrastructure:* Patrick Achi. *Environment, Sustainable Development and Urban Hygiene:* Remi Kouadio Allah. *Family, Women and Children:* Anne Désirée Ouloto. *Health and the Fight Against AIDS:* Raymonde Goudou Coffie. *Higher Education and Scientific Research:* Ibrahima Cissé. *Industry:* Jean-Claude Brou. *Justice, Human Rights and Civil Liberties:* Gnenema Coulibaly. *Mines, Oil and Energy:* Adama Toungara. *National Education and Technical Education:* Kandia Camara. *Post and Information and Communication Technology, and Government Spokesman:* Bruno Nabagné Koné. *Promotion of Youth, Sports and Leisure:* Alain Lobognon. *Public Services and Administrative Reform:* Konan Gnamien. *Tourism:* Roger Kacou. *Transport:* Gaoussou Touré. *Water and Forestry:* Mathieu Babaud Darret. *Minister at the Presidency in Charge of Defence:* Paul Koffi Koffi. *Minister of State at the Presidency:* Jeannot Ahoussou.

CURRENT LEADERS

Alassane Ouattara

Position
President

Introduction
Alassane Ouattara was elected president in Nov. 2010 but was prevented from taking office when the incumbent, Laurent Gbagbo, disputed the result. An economist who formerly worked for the IMF, Ouattara draws his support largely from the Muslim north of the country. He spent the early months of his term under UN guard as conflict waged between his supporters and those of Gbagbo before the former president's capture in April 2011.

Early Life
Alassane Ouattara was born on 1 Jan. 1942 in Dimbokro to a family with links to Burkina Faso. After graduating in business administration from the Drexel Institute of Technology in the USA, he gained an MA in 1967 and a PhD in economics in 1972, both from the University of Pennsylvania. He worked as an economist at the IMF from 1968–73, before moving to the Paris office of the Central Bank of West African States (BCEAO), becoming vice governor in 1983. From 1984–88 he was director of the African department at the IMF and served as counsellor to the managing director from 1987–88. From 1988–93 he was governor of the BCEAO.

In April 1990 Côte d'Ivoire's president, Félix Houphouët-Boigny, appointed Ouattara chairman of an inter-ministerial committee responsible for the national economic recovery programme. From Nov. 1990–Dec. 1993 he served as prime minister, acting as president for the last nine months of that period when Houphouët-Boigny fell ill. Ouattara then returned to the IMF as deputy managing director until 1999. In July 1995 he was nominated by the opposition Rally of the Republicans (RDR) as its presidential candidate but was banned from standing by Gbagbo on the grounds that one of his parents was a foreign national.

In 2000, after becoming leader of the RDR, he was again banned on nationality grounds from contesting the presidency, leading to increased tensions between the Muslim north and predominantly Christian south that helped to precipitate the 2002 civil war. Ouattara finally became the RDR's presidential candidate in 2008, promising to rebuild the country and unite its people. He emerged victorious in the UN-supervised presidential election of Nov. 2010. However, Gbagbo disputed the result and refused to leave office; on 4 Dec., Gbagbo and Ouattara took rival presidential oaths.

Career in Office
The international community refused to recognize Gbagbo's presidency and on 8 Dec. 2010 the UN Security Council declared Ouattara the winner. He was also endorsed by the USA, European Union, African Union and ECOWAS. Gbagbo nonetheless remained in power, with escalating violence between the RDR and Gbagbo's supporters resulting in tens of thousands of civilians fleeing to neighbouring Liberia. By March 2011 Ouattara's forces had gained supremacy in much of the country and on 11 April captured Gbagbo and his inner circle after surrounding the presidential compound in Abidjan. Ouattara was inaugurated in May, and in Sept. a truth and reconciliation commission was set up with the aim of restoring national unity in the wake of the post-election violence.

In parliamentary elections in Dec. 2011 Ouattara's RDR and its Democratic Party ally won an overwhelming majority of seats in the National Assembly. Supporters of Gbagbo boycotted the poll and were implicated in June 2012 in an alleged anti-government plot. Ouattara subsequently dissolved the government led by Jeannot Ahoussou-Kouadio in Nov. over a controversial proposed change to marriage legislation and appointed foreign minister Daniel Kablan Duncan as prime minister of a new administration.

In Feb. 2012 Ouattara took over the chair of the Economic Community of West African States for a year-long term.

DEFENCE

Defence expenditure totalled US$336m. in 2008 (US$17 per capita), representing 1·5% of GDP.

About 11,000 uniformed UN personnel are in the country to maintain a ceasefire.

Army

Total strength (2007), 6,500. In addition there is a Presidential Guard of 1,350, a Gendarmerie of 7,600 and a Militia of 1,500.

Navy

Personnel in 2007 totalled about 900 with the force based at Locodjo (Abidjan).

Air Force

There are six combat-capable aircraft, although their serviceability is in doubt. Personnel (2007) 700.

ECONOMY

Agriculture accounted for 23·1% of GDP in 2006, industry 26·8% and services 50·2%.

Côte d'Ivoire's 'shadow' (black market) economy is estimated to constitute approximately 47% of the country's official GDP.

Overview

After a burst of economic expansion following independence from France in the 1960s, Côte d'Ivoire has seen little growth. The civil war that began in 2002 held growth rates at around 1% per year for most of the 2000s. The economy began to recover in 2008, spurred by the Ouagadougou Political Accord of 2007 that reunified the country. However, the political crisis following disputed elections in 2010 led to an economic slowdown, as well as the destruction of much of the country's social infrastructure.

The crisis also had a negative impact on the West African Economic and Monetary Union (UEMOA), in which Côte d'Ivoire is the leading economy. Nonetheless, economic activity has gradually recovered since 2011, and was further buoyed in June 2012 when the IMF and World Bank announced more than US$4bn. in debt relief.

Export industries include cocoa beans, crude oil and manufactures. Côte d'Ivoire is the world's largest producer of cocoa but, since 2006, oil and gas production have taken precedence.

Currency

The unit of currency is the *franc CFA* (XOF) with a parity of 655·957 francs CFA to one euro. Foreign exchange reserves were US$1,379m. in June 2005 and total money supply was 1,225·0bn. francs CFA. Inflation was 1·4% in 2010 and 4·9% in 2011.

Budget

Revenues in 2008 were 2,156bn. francs CFA (tax revenue, 76·0%) and expenditures 2,217bn. francs CFA (current expenditure, 84·8%).

VAT is 18%.

Performance

Real GDP growth was 2·4% in 2010 but the economy then contracted by 4·7% in 2011 as a consequence of the political crisis that paralysed the country for the early part of the year. However, there was a strong recovery in 2012 with growth estimated to be in the region of 8%. Total GDP in 2011 was US$24·1bn.

Banking and Finance

The regional Banque Centrale des États de l'Afrique de l'Ouest is the central bank and bank of issue. The *Governor* is Tiémoko Meyliet Koné. In 2002 there were 13 commercial banks and six credit institutions. The African Development Bank is based in Abidjan.

External debt amounted to US$11,430m. in 2010 (equivalent to 52·6% of GNI), down from US$12,138m. in 2000.

ENERGY AND NATURAL RESOURCES

Environment

Carbon dioxide emissions from the consumption and flaring of fossil fuels in 2008 were the equivalent of 0·3 tonnes per capita.

Electricity

Installed capacity was an estimated 1·5m. kW in 2007. Production in 2007 amounted to 5·63bn. kWh, with consumption per capita 240 kWh.

Oil and Gas

Petroleum has been produced (offshore) since 1977. Production (2007), 2·4m. tonnes. Oil reserves, 2007, 100m. bbls. Natural gas reserves, 2007, 28bn. cu. metres; production (2007), 1,220m. cu. metres.

Minerals

Côte d'Ivoire has large deposits of iron ores, bauxite, tantalite, diamonds, gold, nickel and manganese, most of which are untapped. Gold production totalled 6·9 tonnes in 2009. Estimated diamond production, 2005: 300,000 carats.

Agriculture

In 2002 the agricultural population was 9·09m., of whom 3·13m. were economically active. In 2002 agriculture accounted for 58% of exports. There were 3·10m. ha. of arable land in 2001 and 4·40m. ha. of permanent crops. 73,000 ha. were irrigated in 2001. There were 3,800 tractors in 2001 and 70 harvester-threshers. Côte d'Ivoire is the world's largest producer and exporter of cocoa beans, with an output of 1·23m. tonnes in 2003 (more than 37% of the world total). It is also a leading coffee producer, although production dropped from 365,000 tonnes in 2000 to 160,000 tonnes in 2003. The cocoa and coffee industries have for years relied on foreign workers, but tens of thousands have left the country since the 1999 coup resulting in labour shortages. Other main crops, with 2001 production figures in 1,000 tonnes, are: yams (2,938), cassava (1,688), plantains (1,410), rice (1,212), sugarcane (1,155), maize (573).

Livestock, 2002: 1·52m. sheep, 1·48m. cattle, 1·19m. goats, 356,000 pigs and 33m. chickens.

Forestry

In 2010 forests covered 10·40m. ha., or 33% of the total land area. Of the total area under forests 6% was primary forest in 2010, 91% other naturally regenerated forest and 3% planted forest. Products include teak, mahogany and ebony. In 2007, 10·25m. cu. metres of roundwood were produced.

Fisheries

The total catch in 2010 amounted to 71,812 tonnes, of which 64,749 tonnes came from sea fishing and 7,063 tonnes from inland fishing. Imports of fishery commodities were valued at US$398m. in 2008 and exports at US$199m.

INDUSTRY

Industrialization has developed rapidly since independence, particularly food processing, textiles and sawmills. Output in 2007 (in 1,000 tonnes): distillate fuel oil, 1,089; kerosene, 933; petrol, 564; residual fuel oil, 500; cement (2008 estimate), 360; sawnwood (2010), 456,000 cu. metres; veneer sheets (2008), 396,000 cu. metres.

Labour

In 2003 the workforce was 7·3m. (60% males).

INTERNATIONAL TRADE

Imports and Exports

In 2007 imports (c.i.f.) totalled US$6·68bn. (US$5·82bn. in 2006); exports (f.o.b.), US$8·07bn. (US$8·15bn. in 2006). Main imports, 2007: petroleum and petroleum products, 30·0%; machinery and transport equipment, 19·7%; food and live animals, 15·3%; chemicals and related products, 11·0%; arms and ammunition, 3·9%. Principal exports, 2007: petroleum and petroleum products, 32·6%; cocoa, 26·4%; machinery and transport equipment, 4·9%; natural rubber, 4·4%; arms and ammunition, 4·0%. Main import suppliers, 2007: Nigeria, 24·1%; France, 21·7%; China, 6·6%; Venezuela, 3·2%. Main export markets, 2007: France, 20·5%; Netherlands, 9·1%; Nigeria, 8·0%; USA, 6·8%.

COMMUNICATIONS

Roads

In 2007 there were 81,996 km of roads, including 142 km of motorways. There were 314,165 passenger cars, 78,575 vans and trucks, 38,105 motorcycles and mopeds and 17,512 buses and coaches in use in 2007.

Rail

From Abidjan a metre-gauge railway runs to Ouangolodougou near the border with Burkina Faso (660 km), and thence through Burkina Faso to Ouagadougou and Kaya. Operation of the railway in both countries is franchised to the mixed public-private company Sitarail. Across both countries Sitarail carried 156,569 passengers in 2004 and 759,957 tonnes of freight in 2005.

Civil Aviation

There is an international airport at Abidjan (Félix Houphouët-Boigny Airport), which in 2001 handled 912,000 passengers (all on international flights) and 19,100 tonnes of freight. The national carrier is the state-owned Air Ivoire. It provides domestic services and in 2003 operated international flights to Accra, Bamako, Conakry, Cotonou, Dakar, Douala, Libreville, Lomé, Niamey and Ouagadougou. There were direct flights in 2003 with other airlines to Addis Ababa, Banjul, Beirut, Bobo Dioulasso, Brazzaville, Brussels, Cairo, Casablanca, Freetown, Johannesburg, Lagos, Monrovia, Nairobi, Nouakchott, Paris, Pointe-Noire and Tripoli. In 2001 scheduled airline traffic of Côte d'Ivoire-based carriers flew 1m. km, carrying 46,000 passengers (all on international flights).

Shipping

The main port is Abidjan, which handled 22·1m. tonnes of foreign cargo in 2008. In Jan. 2009 there were two ships of 300 GT or over registered, totalling 1,000 GT.

Telecommunications

There were 223,200 fixed telephone lines in 2010 (11 per 1,000 inhabitants). Mobile phone subscribers numbered 14·91m. in 2010. In 2011 an estimated 2% of the population were internet users.

SOCIAL INSTITUTIONS

Justice
There are 28 courts of first instance and three assize courts in Abidjan, Bouaké and Daloa, two courts of appeal in Abidjan and Bouaké, and a supreme court in Abidjan. Côte d'Ivoire abolished the death penalty in 2000.

The population in penal institutions under government control was 10,621 (55 per 100,000 of national population) in Dec. 2007.

Education
The adult literacy rate in 2009 was estimated at 55·3% (64·7% among males and 45·3% among females). There were, in 2007, 2,179,801 pupils with 53,161 teaching staff in primary schools and (2000–01) 663,636 pupils with 23,184 teachers at secondary schools. In 2007 there were 156,772 students in higher education. There were three state universities and six other public higher education institutions in 2003, plus seven private institutions of higher education.

In 2008 expenditure on education came to 4·6% of GDP and 24·6% of total government spending.

Health
In 2006 there were four hospital beds per 10,000 inhabitants. In 2004 there were 2,081 physicians, 339 dentistry personnel and 10,180 nursing and midwifery personnel.

RELIGION

In 2001 there were 6·3m. Muslims (mainly in the north) and 4·3m. Christians (chiefly Roman Catholics in the south). Although Christians are in the majority among Ivorians, when Côte d'Ivoire's large immigrant population is taken into account Muslims are in the majority. Traditional animist beliefs are also practised. In Feb. 2013 the Roman Catholic church had one cardinal.

CULTURE

World Heritage Sites
UNESCO world heritage sites in Côte d'Ivoire are: Taï National Park (inscribed on the list in 1982); the Comoé National Park (1983); the Mount Nimba Strict Nature Reserve (1981 and 1982), shared with Guinea; and the Historic Town of Grand-Bassam (2012).

Press
In 2008 there were 23 paid-for daily newspapers with an estimated combined circulation of 200,000.

Tourism
There were 479,000 foreign tourists in 2002; spending by tourists in 2007 (excluding passenger transport) totalled US$104m.

DIPLOMATIC REPRESENTATIVES

Of Côte d'Ivoire in the United Kingdom (2 Upper Belgrave St., London, SW1X 8BJ)
Ambassador: Claude Bouah-Kamon.

Of the United Kingdom in Côte d'Ivoire (Cocody Quartier, Ambassades, Rue d'Impasse du Belier, 01 BP 2581, Abidjan 01)
Ambassador: Simon Tonge.

Of Côte d'Ivoire in the USA (2424 Massachusetts Ave., NW, Washington, D.C., 20008)
Ambassador: Daouda Diabaté.

Of the USA in Côte d'Ivoire (Riviera Golf, 01 B.P. 1712, Abidjan)
Ambassador: Philip Carter III.

Of Côte d'Ivoire to the United Nations
Ambassador: Youssoufou Bamba.

Of Côte d'Ivoire to the European Union
Ambassador: Jean-Vincent Zinsou.

FURTHER READING

Direction de la Statistique. *Bulletin Mensuel de Statistique.*

Hellweg, Joseph, *Hunting the Ethical State: The Benkadi Movement of Côte d'Ivoire.* 2011

McGovern, Mike, *Making War in Côte d'Ivoire.* 2006

Mundt, Robert J., *Historical Dictionary of Côte d'Ivoire.* 1995

National Statistical Office: Institut National de la Statistique, BP V 55, Abidjan 01.

CROATIA

Republika Hrvatska
(Republic of Croatia)

Capital: Zagreb
Population projection, 2015: 4·36m.
GNI per capita, 2011: (PPP$) 15,729
HDI/world rank: 0·796/46
Internet domain extension: .hr

KEY HISTORICAL EVENTS

Croatia was united with Hungary in 1091 and remained under Hungarian administration until the end of the First World War. On 1 Dec. 1918 Croatia became a part of the new Kingdom of Serbs, Croats and Slovenes, which was renamed Yugoslavia in 1929. During the Second World War an independent fascist (Ustaša) state was set up under the aegis of the German occupiers. During the Communist period Croatia became one of the six 'Socialist Republics' constituting the Yugoslav federation led by Marshal Tito. With the collapse of Communism, an independence movement gained momentum.

In a referendum on 19 May 1991, 94·17% of votes cast were in favour of Croatia becoming an independent sovereign state with the option of joining a future Yugoslav confederation. The Krajina and other predominantly Serbian areas of Croatia wanted union with Serbia and seized power. Croatian forces and Serb insurgents backed by federal forces became embroiled in a conflict throughout 1991 until the arrival of a UN peacekeeping mission at the beginning of 1992 and the establishment of four UN peacekeeping zones ('pink zones'). In early May 1995 Croatian forces retook Western Slavonia from the Serbs and opened the Zagreb-Belgrade highway. In a 60-hour operation mounted on 4 Aug. 1995 the former self-declared Serb Republic of Krajina was occupied, provoking an exodus of 180,000 Serb refugees. Croats who had left the area in 1991 began to return. On 12 Nov. 1995 the Croatian government and Bosnian Serbs reached an agreement to place Eastern Slavonia, the last Croatian territory still in Bosnian Serb control, under UN administration.

TERRITORY AND POPULATION

Croatia is bounded in the north by Slovenia and Hungary, in the east by Serbia and Bosnia and Herzegovina and in the southeast by Montenegro. It includes the areas of Dalmatia, Istria and Slavonia, which no longer have administrative status. Its area is 56,542 sq. km. 2011 census population, 4,284,889; population density, 75·8 per sq. km. The United Nations population estimate for 2011 was 4·40m.

The UN gives a projected population for 2015 of 4·36m.

In 2011, 58·0% of the population lived in urban areas.

The area, 2011 census population and capital of the 20 counties and one city:

County	Area (sq. km)	Population	Capital
Bjelovarska-Bilogorska	2,638	119,764	Bjelovar
Brodsko-Posavska	2,027	158,575	Slavonski Brod
Dubrovačko-Neretvanska	1,782	122,568	Dubrovnik
Istarska	2,813	208,055	Pazin
Karlovačka	3,622	128,899	Karlovac
Koprivničko-Križevačka	1,734	115,584	Koprivnica
Krapinsko-Zagorska	1,230	132,892	Krapina
Ličko-Senjska	5,350	50,927	Gospić
Međimurska	730	113,804	Čakovec
Osječko-Baranjska	4,149	305,032	Osijek
Požeško-Slavonska	1,821	78,034	Požega
Primorsko-Goranska	3,590	296,195	Rijeka
Šibensko-Kninska	2,994	109,375	Šibenik
Sisačko-Moslavačka	4,448	172,439	Sisak
Splitsko-Dalmatinska	4,524	454,798	Split
Varaždinska	1,260	175,951	Varaždin
Virovitičko-Podravska	2,021	84,836	Virovitica
Vukovarsko-Srijemska	2,448	179,521	Vukovar
Zadarska	3,643	170,017	Zadar
Zagrebačka	3,078	317,606	Zagreb
Zagreb (city)	640	790,017	Zagreb

Zagreb, the capital, had a 2011 population of 688,163. Other major towns (with 2011 census population): Split (167,121), Rijeka (128,384) and Osijek (84,104).

The official language is Croatian.

SOCIAL STATISTICS

2007: births, 41,910 (9·4 per 1,000 population); deaths, 52,367 (11·8); marriages, 23,140 (5·2); divorces, 4,785 (1·1); suicides, 776 (17·5 per 100,000). Infant mortality, 2010, five per 1,000 live births. Annual population growth rate, 2004–07, 0·0%. In 2006 the most popular age range for marrying was 25–29 for both males and females. Life expectancy at birth, 2007, was 72·6 years for males and 79·4 years for females. Fertility rate, 2008, 1·4 children per woman.

CLIMATE

Inland Croatia has a central European type of climate, with cold winters and hot summers, but the Adriatic coastal region experiences a Mediterranean climate with mild, moist winters and hot, brilliantly sunny summers with less than average rainfall. Average annual temperature and rainfall: Dubrovnik, 16·6°C and 1,051 mm. Zadar, 15·6°C and 963 mm. Rijeka, 14·3°C and 1,809 mm. Zagreb, 12·4°C and 1,000 mm. Osijek, 11·3°C and 683 mm.

CONSTITUTION AND GOVERNMENT

A new constitution was adopted on 22 Dec. 1990 and was revised in both 2000 and 2001. The *President* is elected for renewable five-year terms. There is a unicameral Parliament (*Hrvatski Sabor*), consisting of 151 deputies; 140 members are elected from multi-seat constituencies for a four-year term, eight seats are reserved for national minorities and three members representing Croatians abroad are chosen by proportional representation. The upper house, the *Chamber of Counties*, was abolished in 2001.

National Anthem

'Lijepa nasva domovino' ('Beautiful our homeland'); words by A. Mihanović, tune by J. Runjanin.

GOVERNMENT CHRONOLOGY

(HDZ = Croatian Democratic Union; SDP = Social Democratic Party of Croatia; n/p = non-partisan)

Presidents since 1990.

1990–99	HDZ	Franjo Tuđman
2000–10	n/p	Stjepan (Stipe) Mesić
2010–	SDP	Ivo Josipović

Prime Ministers since 1990.

1990	HDZ	Stjepan (Stipe) Mesić
1990–91	HDZ	Josip Manolić
1991–92	HDZ	Franjo Gregurić
1992–93	HDZ	Hrvoje Šarinić
1993–95	HDZ	Nikica Valentić
1995–2000	HDZ	Zlatko Mateša
2000–03	SDP	Ivica Račan
2003–09	HDZ	Ivo Sanader
2009–11	HDZ	Jadranka Kosor
2011–	SDP	Zoran Milanović

RECENT ELECTIONS

Presidential elections were held on 27 Dec. 2009. Ivo Josipović (Social Democratic Party of Croatia/SDP) received 32·4% of the vote, Milan Bandić (ind.) 14·8%, Andrija Hebrang (Croatian Democratic Union/HDZ) 12·0% and Nadan Vidošević (ind.) 11·3%. There were eight other candidates. Turnout was 44·0%. As a result a second round was required. In the run-off on 10 Jan. 2010 Ivo Josipović received 60·3% of votes cast, against 39·7% for Milan Bandić.

Elections to the Sabor were held on 4 Dec. 2011. The Kukuriku coalition (comprising the SDP, Croatian People's Party–Liberal Democrats, Istrian Democratic Assembly and Croatian Party of Pensioners) won 81 seats (40·0% of the vote), the HDZ 47 (23·5%), the Croatian Labourists–Labour Party 6 (5·1%), the Croatian Democratic Alliance of Slavonija and Baranja 6 (2·9%) and the independent list Ivan Grubišić 2 (2·8%). Two other parties took one seat each. Domestic turnout was 61·8%. In the national minority electoral district the Independent Democratic Serb Party won three seats and other national minority representatives four.

CURRENT GOVERNMENT

President: Ivo Josipović; b. 1957 (SDP; sworn in 18 Feb. 2010).

In March 2013 the government comprised:

Prime Minister: Zoran Milanović; b. 1966 (SDP; sworn in 23 Dec. 2011).

Deputy Prime Ministers: Vesna Pusić (also *Minister of Foreign and European Affairs*); Milanka Opačić (also *Minister of Social Welfare and Youth*); Neven Mimica (for *Internal, Foreign and European Policies*); Branko Grčić (also *Minister of Regional Development and EU Funds*).

Minister of Administration: Arsen Bauk. *Agriculture:* Tihomir Jakovina. *Construction and Physical Planning:* Anka Mrak-Taritaš. *Culture:* Andrea Violić. *Defence:* Ante Kotromanović. *Economy:* Ivan Vrdoljak. *Entrepreneurship and Trade:* Gordan Maras. *Environment and Nature Protection:* Mihael Zmajlović. *Finance:* Slavko Linić. *Health:* Rajko Ostojić. *Interior:* Ranko Ostojić. *Justice:* Orsat Miljenić. *Labour and Pension System:* Mirando Mrsić. *Maritime Affairs, Transport and Infrastructure:* Siniša Hajdaš Dončić. *Science, Education and Sports:* Željko Jovanović. *Tourism:* Veljko Ostojić. *War Veterans:* Predrag Matić.

Government Website: http://www.vlada.hr

CURRENT LEADERS

Ivo Josipović

Position
President

Introduction
Ivo Josipović was sworn in as president on 18 Feb. 2010 after winning the second round of elections on 10 Jan. 2010 on an anti-corruption platform. His term in the largely ceremonial role is for five years. He was expected to preside over Croatia's entry into the European Union in July 2013 (negotiations for which were completed in June 2011), and has sought to improve relations with the country's ex-Yugoslav neighbours.

Early Life
Josipović was born on 28 Aug. 1957 in Zagreb. He studied law at the University of Zagreb, qualifying for the bar in 1980. He returned as a lecturer in 1984, specializing in criminal procedure and international crime. In 1985 he gained an MA in criminal law and in 1994 received his PhD in criminal sciences. Josipović also pursued musical interests, graduating from the composition department of the Zagreb Music Academy in 1983. From 1987–2004 he taught at the Academy and has written over 50 compositions and won several awards for his work.

He began his political career in 1980 when he joined the League of Communists of Croatia (SKH). The party rebranded its image in the early 1990s, with Josipović helping to write the first statutes of the new Social Democratic Party of Croatia (SDP). In 1994 he retired from politics to work as an international law specialist in cooperation with the International Criminal Tribunal for the former Yugoslavia in The Hague. He was a key author of Croatia's genocide case against Serbia before the International Court of Justice.

In 2003 Josipović returned to politics. Elected to parliament as an independent MP, he was selected as vice-president of the SDP Representatives' Group in parliament. In 2005 he became a representative in the City of Zagreb assembly and was re-elected to parliament in 2007, formally rejoining the SDP a year later. On 12 July 2009 he was selected as the SDP presidential candidate and on 27 Dec. 2009 won the first round of voting with 32% of the vote. He received 60% in the second round to secure the presidency.

Career in Office
Josipović's top priority has been to fight corruption, a prerequisite to Croatia's membership of the EU. He has also faced the challenge of mending ties with ex-Yugoslav neighbours in the wake of the Balkan wars of the 1990s—most notably Serbia. In April 2010 Josipović went to the Bosnian parliament and expressed regret over Croatia's part in the Bosnian conflict. In July that year he also visited the Serbian capital of Belgrade, heralding an improvement in bilateral relations that was further encouraged by the Serbian president's visit in Nov. to the Croatian city of Vukovar, the site of wartime civilian killings by Serb forces.

Zoran Milanović

Position
Prime Minister

Introduction
Zoran Milanović became prime minister in Dec. 2011. A lawyer and diplomat, he promised sweeping reforms after leading a centre-left bloc to a landslide victory in parliamentary elections.

Early Life
Zoran Milanović was born on 30 Oct. 1966 in Zagreb, the son of an economist and a teacher. He spent his childhood in Trnje, a suburb of Zagreb, and graduated in law from Zagreb University in

1991. He trained in the city's commercial court for two years, before joining the then recently independent Croatia's ministry of foreign affairs under future prime minister Ivo Sanader. Milanović served on a UN mission to the disputed former Soviet territory of Nagorno-Karabakh before moving to Brussels in 1996 to serve on Croatia's mission to the EU and NATO.

Returning to Croatia in 1999, which was then under the premiership of Franjo Tuđman and his right-wing Croatian Democratic Union (HDZ) administration, Milanović joined the opposition Social Democratic Party of Croatia (SDP). Following the party's victory (in coalition with the Croatian Social Liberal Party and four minor parties) at the Jan. 2000 parliamentary election, Milanović worked in various diplomatic roles within the ministry of foreign affairs. After the ruling coalition lost at elections in Nov. 2003, Milanović joined the SDP's executive in 2004 and grew close to the party's founder and leader, Ivica Račan.

When Račan resigned because of ill health in early 2007, Milanović ran for the party leadership and unexpectedly won, defeating the interim leader, Željko Antunović, and other more experienced candidates. The closely fought parliamentary elections of Nov. 2007 eventually resulted in Sanader forming an HDZ-led coalition, with Milanović leading the opposition. Milanović was re-elected president of the SDP at the party's convention in May 2008.

Milanović led a centre-left opposition bloc known as Kukuriku to victory in the legislative elections of 4 Dec. 2011, securing 81 seats in the 151-member parliament. The HDZ took 47 seats after presiding over a series of corruption scandals, a stagnating economy and soaring unemployment.

Career in Office

Milanović has been committed to reforms and public spending cuts to deter downgrading by international credit rating agencies ahead of Croatia's accession to the EU, which was scheduled for 1 July 2013. He has urged Croats to work 'more, harder, longer' to turn the economy around.

DEFENCE

Conscription was abolished on 1 Jan. 2008. Defence expenditure in 2008 totalled US$1,090m. (US$243 per capita), representing 1·6% of GDP.

Army

Personnel, 2007, 12,300 (around 1,300 conscripts). Paramilitary forces include an armed police of 3,000. There are 95,000 reserves.

Navy

In 2007 the fleet included two tactical submarines and two missile-armed corvettes. Total personnel in 2007 numbered 1,700 (250 conscripts), including two companies of marines.

Air Force

Personnel, 2007, 1,800 (including Air Defence and 200 conscripts). There were 12 combat-capable aircraft (MiG-21s) in 2007.

INTERNATIONAL RELATIONS

Croatia applied for European Union membership in 2003 and was accepted as an official candidate country in 2004. Having signed the accession treaty in Dec. 2011 a referendum was held on 22 Jan. 2012 in which 66·7% of votes cast were in favour of membership. Croatia was expected to become a member on 1 July 2013.

ECONOMY

Agriculture contributed 5% of GDP in 2010, industry 27% and services 68%.

Overview

In the 1990s the economy was rocked by war and international isolation. Increased political stability from 2000 onwards saw the economy expanding at 4–5% annually prior to the 2008 global financial crisis, with per capita income reaching about 63% of the EU average. Structural reforms included a privatization programme and modernization of the bankruptcy, company and labour laws. There was also strong investment in fixed capital that had been neglected in the previous decade.

Tourism potential has attracted investment. Household consumption has expanded at rates slightly below economic growth but has acted as an anchor for domestic demand. The banking sector is almost entirely owned by foreign banks that have benefited from borrowing at low external interest rates for lending at higher rates in Croatia.

However, the global financial crisis led to declines in investment and private consumption, resulting in a fall in GDP. Public debt increased to more than 52% of GDP. The economy continued to decline in early 2010 and an Economic Recovery Programme was adopted in April 2010. Croatia completed EU accession negotiations in June 2011 and was set to become the 28th member in July 2013.

Currency

On 30 May 1994 the *kuna* (HRK; a name used in 1941–45) of 100 *lipa* replaced the Croatian dinar at one kuna = 1,000 dinars. Foreign exchange reserves were US$8,529m. in July 2005 and total money supply was 38,305m. kuna. Inflation was 6·1% in 2008 but fell to 1·0% in 2010 before rising to 2·3% in 2011.

Budget

Budgetary central government revenue totalled 110,258m. kuna in 2009 (115,773m. kuna in 2008) and expenditure 117,924m. kuna (115,293m. kuna in 2008).

Principal sources of revenue in 2009 were: taxes on goods and services, 49,238m. kuna; social security contributions, 39,995m. kuna. Main items of expenditure by economic type in 2009: social benefits, 56,149m. kuna; compensation of employees, 31,289m. kuna.

VAT is 25%.

Performance

The economy shrank by 6·9% in 2009 and 1·4% in 2010. There was then zero growth in 2011. Total GDP was US$62·5bn. in 2011.

Banking and Finance

The National Bank of Croatia (*Governor,* Boris Vujčić) is the bank of issue. In 2010 there were 32 registered banks. The largest banks are Zagrebačka Banka, with assets in Dec. 2009 of US$18·2bn., and Privredna Banka Zagreb. There is a stock exchange in Zagreb.

Gross external debt amounted to US$59,028m. in June 2012.

Total foreign direct investment in 2010 was US$583m., down from US$2,911m. in 2009.

ENERGY AND NATURAL RESOURCES

Environment

Croatia's carbon dioxide emissions from the consumption and flaring of fossil fuels in 2008 were the equivalent of 5·4 tonnes per capita.

Electricity

Installed capacity in 2007 was 3·9m. kW. Output was 12·24bn. kWh in 2007, with consumption per capita 4,141 kWh in 2007.

Oil and Gas

In 2007, 768,000 tonnes of crude oil were produced. Natural gas output totalled 2·9bn. cu. metres in 2007; reserves were 30bn. cu. metres in 2007.

Minerals

Production (in 1,000 tonnes): salt (2006), 30.

Agriculture

Agriculture and fishing generate approximately 6% of GDP. Some 4·6% of the workforce were engaged in agriculture, hunting and forestry in 2008. Agricultural land totalled 1·3m. ha. in 2009; there were 863,000 ha. of arable land and 87,900 ha. of permanent crops. Production (in 1,000 tonnes, 2008): maize, 2,505; sugar beets, 1,270; wheat, 858; potatoes, 256; grapes, 185; sunflower seeds, 120; soybeans, 108.

Livestock, 2008: pigs, 1,104,000; sheep, 643,000; cattle, 454,000; chickens, 6·7m. Livestock products, 2009: milk, 838,000 tonnes; meat, 227,000 tonnes; eggs, 48,000 tonnes; cheese, 28,000 tonnes.

Forestry

Forests covered 1·92m. ha. in 2010. In 2007, 4·21m. cu. metres of roundwood were produced.

Fisheries

Total catch in 2010 was 52,828 tonnes, almost exclusively from sea fishing. The annual catch has doubled since 2003.

INDUSTRY

The largest company in Croatia in April 2012 by market capitalization was INA (Industrija nafte), the national oil company of Croatia (US$6·3bn.).

In 2009 industrial production fell 9·2% in comparison with 2008. Output in 2009: cement, 2,823,000 tonnes; distillate fuel oil, 1,536,000 tonnes; petrol, 1,211,000 tonnes; residual fuel oil, 1,050,000 tonnes; paper and paperboard (2007), 545,000 tonnes; cigarettes, 11,382m. units; cotton woven fabrics, 17m. sq. metres; beer (2007), 381m. litres.

Labour

In 2009 the labour force numbered 1,762,000 people of whom 1,498,800 were employed. In 2009 the unemployment rate was 9·1%; youth unemployment among 15 to 24-year-olds was 25·1%. The main areas of activity in 2009 were manufacturing (employing 272,800 persons), wholesale and retail trade/repair of motor vehicles and motorcycles (243,300), construction (140,700) and public administration and defence/compulsory social security (113,500).

INTERNATIONAL TRADE

Imports and Exports

Imports in 2008 came to US$30,728m. Exports were valued at US$14,112m. in 2008.

Principal imports in 2004 were: machinery and transport equipment, 34·9%; manufactured goods, 19·6%; mineral fuels, 12·0%; miscellaneous manufactured articles, 11·9%; chemicals, 11·2%; food and live animals, 7·2%. Main exports in 2004 were: machinery and transport equipment, 32·3%; miscellaneous manufactured articles, 17·8%; manufactured goods, 14·8%; mineral fuels, 11·3%; chemicals, 9·4%; food and live animals, 6·3%.

In 2004 the main import suppliers were (in US$1m.): Italy (2,819); Germany (2,569); Russia (1,206); Slovenia (1,179); Austria (1,131). Main export markets (in US$1m.): Italy (1,834); Bosnia and Herzegovina (1,154); Germany (895); Austria (757); Slovenia (601).

COMMUNICATIONS

Roads

There were 29,333 km of roads in 2010 (including 1,126 km of motorways and 6,929 km of highways, national and main roads). In 2010 there were 1,515,449 passenger cars, 4,877 buses and coaches, and 157,731 goods vehicles. 56m. passengers and 75m. tonnes of freight were carried by road transport in 2010. There were 426 deaths in road accidents in 2010.

Rail

There were 2,722 km of 1,435 mm gauge rail in 2009 (985 km electrified). In 2009 railways carried 74m. passengers and 12m. tonnes of freight. In Sept. 2010 the state-owned railway companies of Croatia, Serbia and Slovenia announced the creation of a joint venture called Cargo 10 to improve the management of freight trains along the route known as Corridor 10 that passes through all three countries.

Civil Aviation

The biggest international airports are Zagreb (Pleso), Split and Dubrovnik. The national carrier is Croatia Airlines. In 2004 scheduled airline traffic of Croatian-based carriers flew 12m. km, carrying 1,336,411 passengers (886,215 on international flights). In 2004 Zagreb handled 1,389,537 passengers (926,011 on international flights) and 7,692 tonnes of freight, Dubrovnik 860,672 passengers (704,331 on international flights) and Split 768,706 passengers (575,019 on international flights).

Shipping

The main ports in 2010 (ports that had total traffic of goods greater than or equal to 1m. tonnes) are Bakar (2·4m. tonnes), Omišalj (5·9m. tonnes), Ploče (4·5m. tonnes), Raša (1·9m. tonnes), Rijeka (2·0m. tonnes) and Split (2·7m. tonnes). Figures for 2010 show that 27·5m. passengers and 24·3m. tonnes of cargo were transported. In 2010 merchant shipping (passenger and cargo ships) totalled 1,625,210 GT, including liquid bulk carriers 554,805 GT.

Telecommunications

There were 1·87m. fixed telephone lines in 2010 (423·7 per 1,000 inhabitants). Mobile phone subscribers numbered 6·36m. in 2010. There were 603·2 internet users per 1,000 inhabitants in 2010. Fixed internet subscriptions totalled 1·50m. in 2009 (339·7 per 1,000 inhabitants). In March 2012 there were 1·5m. Facebook users.

SOCIAL INSTITUTIONS

Justice

The population in penal institutions in July 2007 was 4,127 (93 per 100,000 of national population).

Education

In 2009–10 there were 1,444 pre-school institutions with 121,433 children and 16,216 childcare workers; 2,131 primary schools with 361,052 pupils and 32,083 teachers; 713 secondary schools with 180,582 pupils and 24,004 teachers. In 2009–10 there were 132 institutes of higher education with 149,853 students and 15,863 academic staff. In 2009–10 there were seven universities (Dubrovnik, Osijek, Pula, Rijeka, Split, Zadar and Zagreb). Adult literacy rate in 2009 was estimated at 98·8% (male, 99·5%; female, 98·1%).

In 2008 public spending on education came to 4·3% of GDP.

Health

In 2010 there were 69 hospitals with 25,017 beds; and 12,341 physicians, 3,121 dentists, 23,447 nurses, 2,851 pharmacists and 1,553 midwives.

Welfare

The official retirement age is 65 years (men) and 60 years (women). It was gradually increased in six-month increments over ten years from the 1999 levels of 60 (men) and 55 (women). The old-age pension is dependent on wages earned in relation to the average wage of all employed persons. The minimum old-age pension is defined for every year of the qualifying period as 0·825% of the average gross salary of all employees in 1998, adjusted until the date of entitlement. This amount (56·59 kuna at July 2009) is adjusted for inflation. The average unemployment benefit was 1,477·54 kuna in 2009.

RELIGION

In 2001 there were 3,890,000 Roman Catholics, 250,000 Serbian Orthodox and 100,000 Sunni Muslims. The remainder of the population were followers of other religions. In Feb. 2013 there was one cardinal.

CULTURE

World Heritage Sites

Croatia has seven UNESCO protected sites: the Old City of Dubrovnik (entered on the List in 1979 and 1994), known as the 'Pearl of the Adriatic'; the Historic Complex of Split with the Palace of Diocletian (1979), including Roman Emperor Diocletian's mausoleum, now the cathedral; Plitvice Lakes National Park (1979 and 2000), a series of lakes and waterfalls and a habitat for bears and wolves; the Episcopal Complex of the Euphrasian Basilica in the Historical Centre of Poreč (1997); the Historic City of Trogir (1997), a Venetian city based on a Hellenistic plan; the Cathedral of St James in Šibenik (2000), built in the Gothic and Renaissance styles between 1431–1535; and Stari Grad Plain (2008), agricultural land on the island of Hvar that was first cultivated by the Greeks in the 4th century BC.

Press

In 2008 there were 16 daily newspapers, of which 14 were in Croatian, one in English and one in Italian. The newspapers with the highest circulation are *24sata* (185,000 copies in 2008), *Večerni list* (daily average of 135,000 copies) and *Jutarnji list* (daily average of 120,000 copies in 2006).

Tourism

In 2010, 9·11m. non-resident tourists stayed in holiday accommodation (up from 8·69m. in 2009 and 7·74m. in 2005).

Festivals

Croatia has a number of cultural and traditional festivals, including the International Folk Dance Festival in Zagreb (July); the Zagreb Summer Festival (July–Aug.); Dubrovnik Summer Festival (July–Aug.); Split Summer Festival (July–Aug.); Alka Festival (traditional medieval tilting), Sinj (Aug.). The best-known music festivals include Mars Festival in Zagreb (May), INmusic Festival near Zagreb (June) and Garden Festival near Zadar (July).

DIPLOMATIC REPRESENTATIVES

Of Croatia in the United Kingdom (21 Conway St., London, W1T 6BN)
Ambassador: Ivan Grdešić.

Of the United Kingdom in Croatia (Ivana Lučića 4, 10000 Zagreb)
Ambassador: David Slinn, CMG, OBE.

Of Croatia in the USA (2343 Massachusetts Ave., NW, Washington, D.C., 20008)
Ambassador: Josip Paro.

Of the USA in Croatia (Thomasa Jeffersona 2, 10010 Zagreb)
Ambassador: Kenneth H. Merten.

Of Croatia to the United Nations
Ambassador: Ranko Vilović.

Of Croatia to the European Union
Ambassador: Vladimir Drobnjak.

FURTHER READING

Central Bureau of Statistics. *Statistical Yearbook, Monthly Statistical Report, Statistical Information, Statistical Reports.*

Fisher, Sharon, *Political Change in Post-Communist Slovakia and Croatia: From Nationalist to Europeanist.* 2006
Jovanovic, Nikolina, *Croatia: A History.* Translated from Croatian. 2000
Stallaerts, Robert, *Historical Dictionary of the Republic of Croatia.* 2nd ed. 2003
Tanner, M. C., *A Nation Forged in War.* 1997
Uzelac, Gordana, *The Development of the Croatian Nation: An Historical and Sociological Analysis.* 2006

National Statistical Office: Central Bureau of Statistics, 3 Ilica, 10000 Zagreb. *Director:* Ivan Kovač.
Website: http://www.dzs.hr

CUBA

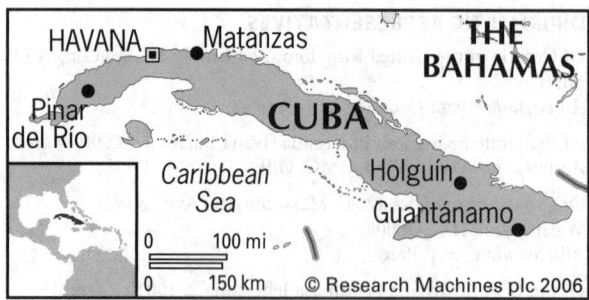

República de Cuba
(Republic of Cuba)

Capital: Havana
Population projection, 2015: 11·23m.
GNI per capita, 2010: US$5,621
HDI/world rank: 0·776/51
Internet domain extension: .cu

KEY HISTORICAL EVENTS

Cuba's first inhabitants were the Taíno, Ciboney and Guanahatabey tribes. Christopher Columbus landed in 1492 and a permanent settlement was established by Diego Velázquez in 1511. Within 50 years oppression and European diseases virtually exterminated the indigenous population and African slaves were imported as replacements. In 1607 Havana was declared the capital.

Resistance to Spanish rule grew after the removal of Cuban delegates from the Spanish *Cortes* in 1837. Repeated offers by the USA to buy Cuba were rejected. Slavery was suppressed from the 1850s though not abolished until 1886. The first rebellion, the Ten-Year War, broke out in 1868 and was led by Gen. Máximo Gómez. An assembly was granted in 1869. However, José Martí y Pérez created the Cuban Revolutionary Party from New York and launched an invasion of Cuba in 1895 with Gómez and Antonio Maceo. The USA intervened in 1898, winning control of Cuba from Spain at the Treaty of Paris. Municipal elections in 1900 rejected annexationist policies and Cuba achieved independence in 1902. The Platt Amendment allowed for US intervention to preserve independence and stability and awarded the USA control of Guantánamo Bay. At the request of President Estrada Palma, US forces were installed on the island between 1906 and 1909. In 1912 and 1917 there were further interventions by American forces.

Gerardo Machado's dictatorial presidency began in 1925 and was ended by a coup in 1933. The Revolt of the Sergeants brought Fulgencio Batista y Zaldívar to power. The Platt Amendment was revoked and in 1940 a socially progressive constitution was inaugurated. Batista was returned at disputed elections in 1940 but was voted out of office four years later. He ran for re-election in 1952 but, with little chance of victory, led a bloodless coup before elections could be held, suspending the constitution and instigating a repressive and corrupt regime.

Fidel Castro, imprisoned in 1953 after a failed revolt, arrived from Mexico with 80 supporters in 1956. Castro and Che Guevara led a guerrilla war from the Sierra Maestra mountains and, despite US financial support, Batista fled after revolutionaries seized Havana in Jan. 1959. The USA recognized the new regime but relations soon deteriorated. Castro, as prime minister, launched a nationalization programme, seizing American assets and outlawing foreign land ownership. In Oct. 1960 the USA's trade embargo began and diplomatic relations were broken in Jan. 1961. Close relations between Cuba and the USSR provoked covert US support for the doomed Bay of Pigs invasion in April 1961, in which an offensive by a group of exiled Cubans was defeated by Castro's troops. Castro declared Cuba to be a socialist state. In 1962 the USA and USSR neared nuclear conflict during the Cuban Missile Crisis, with the US Navy imposing a blockade on Cuba from 22 Oct. until 22 Nov. to force the USSR to withdraw Soviet missile bases. In return, the USA guaranteed not to invade Cuba and, in a secret deal, to withdraw missiles from Turkey. Between 1965 and 1973, 250,000 Cubans left for America on Freedom Flights agreed between the two nations. In 1976 a new constitution consolidated Castro's power as head of state, government and the armed forces.

Cuba continued to receive financial aid and technical advice from the USSR until the early 1990s when subsidies were suspended. This led to a 40% drop in GDP between 1989 and 1993. The USA has maintained an economic embargo against the island and relations between Cuba and the USA have remained embittered, although contact between the two countries has been growing in recent years. From Jan. 2002 suspected al-Qaeda and Taliban prisoners were brought from Afghanistan to the military prison at the American naval base at Guantánamo Bay.

In July 2006 Castro was hospitalized and temporarily handed power to his brother, Raúl. When Fidel failed to appear at a series of high-profile presidential appointments, there was speculation that the power transfer would become permanent. In Feb. 2008 Fidel announced his resignation and was immediately replaced by Raúl.

TERRITORY AND POPULATION

The island of Cuba forms the largest and most westerly of the Greater Antilles group and lies 215 km (135 miles) south of the tip of Florida, USA. The area is 109,886 sq. km, and comprises the island of Cuba (104,341 sq. km); the Isle of Youth (Isla de la Juventud, formerly the Isle of Pines; 2,419 sq. km); and some 1,600 small isles ('cays'; 3,126 sq. km). Population, census (2002), 11,177,743, giving a density of 100·8 per sq. km. Estimate, Dec. 2010: 11,241,000. In 2011, 75·2% of the population were urban. The United Nations estimates that the population peaked in 2007 and has been slowly falling since then.

The UN gives a projected population for 2015 of 11·23m.

At the 2002 census the area and population of the 14 provinces and the special Municipality of the Isle of Youth (Isla de la Juventud) were as follows:

	Area (sq. km)	Population
Ciudad de La Habana	721	2,201,610
Santiago de Cuba	6,156	1,036,281
Holguín	9,293	1,021,321
Villa Clara	8,412	817,395
Granma	8,375	822,452
Camagüey	15,615	784,178
Pinar del Río	10,904	726,574
La Habana	5,732	711,066
Matanzas	11,803	670,427
Las Tunas	6,588	525,485
Guantánamo	6,168	507,118
Sancti Spíritus	6,737	460,328
Ciego de Avila	6,783	411,766
Cienfuegos	4,180	395,183
Isla de la Juventud	2,419	86,559

Two new provinces, Artemisa and Mayabeque, were created in Jan. 2011 and La Habana was abolished. Ciudad de la Habana was renamed La Habana. The capital city, Havana, had a population in 2002 of 2,201,610. Other major cities (2002 census populations in 1,000): Santiago de Cuba (423), Camagüey (302), Holguín (270), Santa Clara (210), Guantánamo (208), Bayamo (145), Las Tunas (144), Cienfuegos (141), Pinar del Río (139) and Matanzas (127).

The official language is Spanish.

SOCIAL STATISTICS

2008 births, 122,569; deaths, 86,357; marriages, 61,852; divorces, 35,882; suicides, 1,357. Rates, 2008 (per 1,000 population): birth, 10·9; death, 7·7; marriage, 5·5; divorce, 3·2; suicide, 12·1 per 100,000 population. Infant mortality rate, 2008, 4·7 per 1,000 live births. Annual population growth rate, 2005–10, 0·0%. Life expectancy, 2005–07: 76·0 years for males and 80·0 for females. The fertility rate in 2008 was 1·6 births per woman.

CLIMATE

Situated in the sub-tropical zone, Cuba has a generally rainy climate, affected by the Gulf Stream and the N.E. Trades, although winters are comparatively dry after the heaviest rains in Sept. and Oct. Hurricanes are liable to occur between June and Nov. Havana, Jan. 72°F (22·2°C), July 82°F (27·8°C). Annual rainfall 48" (1,224 mm).

CONSTITUTION AND GOVERNMENT

A Communist Constitution came into force on 24 Feb. 1976. It was amended in July 1992 to permit direct parliamentary elections and in June 2002 to make the country's socialist system 'irrevocable'.

Legislative power is vested in the *National Assembly of People's Power*, which meets twice a year and consists of 612 deputies elected for a five-year term by universal suffrage. Citizens are entitled to vote at the age of 16. Lists of candidates are drawn up by mass organizations (trade unions, etc.). The National Assembly elects a 31-member *Council of State* as its permanent organ. The Council of State's President, who is head of state and of government, nominates and leads a Council of Ministers approved by the National Assembly.

National Anthem

'Al combate corred bayameses' ('Run, Bayamans, to the combat'); words and tune by P. Figueredo.

RECENT ELECTIONS

Elections to the National Assembly were held on 3 Feb. 2013. 50% of candidates must come from municipal assemblies, with the other 50% being candidates of national or provincial importance. All 612 candidates received the requisite 50% of votes for election.

CURRENT GOVERNMENT

President: Gen. Raúl Castro Ruz (b. 1931) became *President* of the Council of State and of the Council of Ministers on 24 Feb. 2008. He is also First Secretary of the Cuban Communist Party.

In March 2013 the government comprised:

First Vice-President of the Council of State: Miguel Díaz-Canel Bermúdez.

First Vice-President of the Council of Ministers: José Ramón Machado.

Vice-Presidents of the Council of Ministers: Miguel Díaz-Canel Bermúdez, Ulises Rosales del Toro, Ramiro Valdés Menéndéz, Ricardo Cabrisas Ruiz, Marino Murillo Jorge, Antonio Enrique Lussón Batlle, Adel Yzquierdo Rodríguez. *Secretary of the Council of Ministers:* Brig. Gen. José Amado Ricardo Guerra.

Minister of Agriculture: Gustavo Rodríguez Rollero. *Basic Industries:* Alfredo López Valdés. *Construction:* Rene Mesa Villafaña. *Culture:* Rafael Bernal Alemany. *Defence:* Gen. Leopoldo Cintra Frías. *Domestic Trade:* Mary Blanca Ortega Barredo. *Economy and Planning:* Adel Izquierdo Rodríguez. *Education:* Ena Elsa Velázquez Cobiella. *Energy and Mines:* Alfredo López Valdés.

Finance and Prices: Lina Pedraza Rodríguez. *Food Industry:* María del Carmen Concepción González. *Foreign Relations:* Bruno Rodríguez Parrilla. *Foreign Trade and Investment:* Rodrigo Malmierca Díaz. *Higher Education:* Rodolfo Alarcón Ortíz. *Industries:* Brig. Gen. Salvador Pardo Cruz. *Information and Communications:* Maimir Mesa Ramos. *Interior:* Gen. Abelardo Colomé Ibarra. *Justice:* María Esther Reus. *Labour and Social Security:* Margarita Marlene González Fernández. *Light Industry:* Damar Maceo. *Public Health:* Roberto Morales. *Science, Technology and Environment:* Elba Rosa Pérez Montoya. *Tourism:* Manuel Marrero Cruz. *Transport:* César Ignacio Arocha Masid.

The Congress of the Cuban Communist Party (PCC) elects a Central Committee of 150 members, which in turn appoints a Political Bureau comprising 25 members.

Government Website (Spanish only): http://www.cubagob.cu

CURRENT LEADERS

Raúl Modesto Castro Ruz

Position
President

Introduction
Raúl Castro was elected president by the National Assembly of People's Power on 24 Feb. 2008. His five-year term began when he replaced his brother, Fidel, who had held office for 49 years. While less charismatic than his brother, he has preserved the essence of Cuba's brand of socialism.

Early Life
Raúl Castro was born on 3 June 1931 in Birán, in the Oriente province (now Santiago de Cuba) in eastern Cuba. His father was a sugar plantation owner of Spanish origin and his mother a housemaid. After expulsion from his first school Raúl attended the Jesuit-run Colegio Dolores in Santiago and the Belén School in Havana, before graduating as a sergeant from military college. He attended the University of Havana until 1953, when his involvement in politics cut short his studies.

Raúl was a member of the Socialist Youth group, affiliated to the Moscow-orientated Popular Socialist Party. Although the party supported President Fulgencio Batista, Raúl's travels to the Soviet bloc in 1953 prompted him to turn against Batista which culminated that year in the failed 26 July attack on the Moncada barracks alongside brother Fidel. Both were imprisoned but then exiled to Mexico 22 months later, where Raúl reputedly introduced Che Guevara into his brother's circle. Raúl subsequently helped organize the 26 July Revolutionary Movement and the failed coup attempt of 1956.

Raúl found refuge in the Sierra Maestra mountains before taking charge of the military campaign in the east of the country. Military victory was achieved in 1959 and Batista went into exile. Raúl secured his place as Fidel's right hand man, politically and militarily. He was appointed minister for the Revolutionary Armed Forces (FAR) and played a key role during the Bay of Pigs invasion of 1961 and in the Cuban missile crisis the following year.

In 1965 he was promoted to the Cuban Communist Party (PCC) Politburo and made second secretary of the central committee behind Fidel. In 1972 he became first vice premier and four years later the new National Assembly of People's Power elected him vice president. The Castro hierarchy was endorsed with total support at each of the congressional party sessions from 1975–97. As vice president he was pivotal in the close relationship with the Eastern European communist bloc until the 1990s when he engineered the economic shift away from former Soviet dependency and introduced some freedom in the agricultural sector. From 2000–02 his public profile grew as he stood in for Fidel on diplomatic tours of China and South East Asia.

When Fidel underwent abdominal surgery in July 2006, Raúl became acting president. On 19 Feb. 2008 Fidel announced his formal resignation and Raúl was elected his successor by the National Assembly nine days later.

Career in Office

Raúl's key challenges have been economic and he promised to reduce red tape and to revise the dual-currency system. Early signs of greater market freedom included allowing Cubans to stay in tourist hotels and rent cars and the lifting of bans on ownership of consumer products such as mobile phones, computers and DVD players. However, such products remained unaffordable to average citizens.

In mid-2008 the government relaxed restrictions on the amount of idle state-owned land available to private farmers and announced plans to abandon salary equality in a radical departure from Marxist principles. In the wake of two hurricanes in 2008 that devastated homes and crops in Cuba, a US offer of emergency aid was rejected by Raúl, who instead demanded a lifting of the longstanding US trade embargo.

In March 2009 and again in Jan. 2011 the Obama administration eased the US embargo, lifting restrictions on remittances and visits to Cuba by Cuban-Americans. Despite this slight thaw in relations and Cuba's acute economic problems, Raúl continued to crack down on political dissent. Nevertheless, in Sept. 2010 he unveiled limited plans to further liberalize the economy by laying off thousands of state employees while legalizing self-employment in many areas and allowing more private enterprise. In Nov. he also announced the convening of a long-overdue PCC congress to be held in April 2011 (the first since 1997) to approve the plans, at which he was elected first secretary of the party. The Organization of American States had meanwhile voted in June 2009 to end Cuba's diplomatic suspension dating back to 1962.

Further significant shifts in domestic policy were evident from late 2011. In Nov. legislation enabling individuals to buy and sell property for the first time since the revolution was passed. A subsequent papal visit to Cuba in March 2012 prompted a government amnesty for some 2,500 prisoners, including political detainees, and the restoration of recognition for a religious holiday. Then in Oct. the government abolished the requirement that most citizens, other than certain professionals, acquire exit visas to travel abroad.

DEFENCE

The National Defence Council is headed by the president of the republic. Conscription is for two years.

In 2008 defence expenditure totalled US$2,296m. (US$201 per capita), representing 4·0% of GDP.

Army

The strength was estimated at 38,000 in 2006, with 39,000 reservists. Border Guard and State Security forces total 26,500. The Territorial Militia is estimated at 1m. (reservists), all armed. In addition there is a Youth Labour Army of 70,000 and a Civil Defence Force of 50,000.

Navy

Personnel in 2006 totalled about 3,000 including some 550 marines. The Navy has five patrol and coastal combatants, five mine warfare vessels and one support vessel. Main bases are at Cabañas and Holguín. The USA still occupies the Guantánamo naval base.

Air Force

In 2006 the Air Force had a strength of some 8,000 and about 130 combat aircraft of which only around 25 are thought to be operational. They include MiG-29, MiG-23 and MiG-21 jet fighters.

ECONOMY

Services accounted for 75% of GDP in 2009, industry 20% and agriculture 5%.

Overview

Cuba is classified as a middle income country with a GDP per capita of US$2,500. After the withdrawal of annual subsidies worth US$4–6bn. from the USSR in 1990, the economy fell into deep recession and had to cope with falling tourism, low export prices and hurricane damage. The USA has imposed a trade embargo since 1963 and has effectively blocked access to funds from the IMF and the World Bank, although the UN continues to operate in Cuba. The black market is bigger than the legal economy and basic economic activities (such as the sale of milk and bread) take place in the informal sector.

Following the collapse of the Soviet Union, the government helped stimulate growth through legalization of the US dollar in shops and other businesses. However, commercial transactions in dollars were banned in Nov. 2004 in response to tighter US sanctions. The economy has received substantial aid from Venezuela.

Damage from Hurricanes Gustav and Ike in 2008 resulted in damage valued at 20% of GDP. The global economic crisis also had a significant impact. A combination of higher food prices, a decline in the value of exports (particularly nickel, Cuba's principal export) and tighter conditions for external financing, led to a significant drop in growth during 2009. Fiscal revenues declined following the crisis, leading to policy-tightening measures from the government, including rationing of electricity. A US$600m. loan from China helped ease external debt pressures.

Raúl Castro has overseen limited reforms since assuming power in 2006, including expansion of the private sector to take in 40% of the workforce by 2015, and the legalization of private property sales. In 2011 Richard Feinberg, a former adviser to President Clinton, urged the USA to back IMF and World Bank engagement with Havana. However, relations between the two countries remain strained.

Currency

There are two currencies in Cuba. The official currency ('moneda nacional') is the Cuban peso (CUP) of 100 centavos. The Convertible peso (CUC), introduced in 1994 (pegged since April 2005 at 1 Convertible peso = US$1·08), is the 'tourist' currency. The US dollar ceased to be legal tender in 2004. 9,710m. pesos were in circulation in 1998. Inflation is low, averaging 1·8% between 1995–2002.

Budget

In 2008 revenues were 42,056m. pesos (tax revenue, 61·5%) and expenditure 46,256m. pesos (current expenditure, 90·3%).

Performance

Cuba's economic growth was officially put as 4·3% in 2008, down from 7·5% in 2007 and 12·5% in 2006.

Banking and Finance

The Central Bank of Cuba (President, Ernesto Medina Villaveirán) replaced the National Bank of Cuba as the central bank in 1997. On 14 Oct. 1960 all banks were nationalized. Changes to the banking structure beginning in 1996 divested the National Bank of its commercial functions, and created new commercial and investment institutions. The Grupa Nueva Banca has majority holdings in each institution of the new structure. There were eight commercial banks in 2002 and 18 local non-banking financial institutions. In addition, there were 13 representative offices of foreign banks and four representative offices of non-banking financial institutions. All insurance business was nationalized in 1964. A National Savings Bank was established in 1983.

ENERGY AND NATURAL RESOURCES

Environment
Cuba's carbon dioxide emissions from the consumption and flaring of fossil fuels were the equivalent of 2·3 tonnes per capita in 2008.

Electricity
Installed capacity was 5·4m. kW in 2007. Production was 17·6bn. kWh in 2007; consumption per capita in 2007 was 1,564 kWh.

Oil and Gas
Crude oil production (2007), 2·91m. tonnes. There were known natural gas reserves of 71bn. cu. metres in 2005. Natural gas production (2007), 1·2bn. cu. metres.

Minerals
Iron ore abounds, with deposits estimated at 3,500m. tonnes; output (2006), 7·8m. tonnes. The output of salt was 180,000 tonnes in 2006; nickel, 74,000 tonnes (2005); chromite, 34,000 tonnes (2005); lime, 34,000 tonnes (2005). Other minerals are cobalt and silica. Nickel is Cuba's second largest foreign exchange earner, after tourism. Gold is also worked.

Agriculture
In 1959 all land over 30 *caballerías* was nationalized and eventually turned into state farms. In 2007 there were 3·0m. ha. of arable land and 1·8m. ha. of permanent crops. Under legislation of 1993, state farms were reorganized as 'units of basic co-operative production'. Unit workers select their own managers and are paid an advance on earnings. In 1963 private holdings were reduced to a maximum of five *caballerías*. In 1994 farmers were permitted to trade on free market principles after state delivery quotas had been met. In 2008 private farmers were granted access to underused government land in the form of ten-year leases in an effort to boost food production and curb dependence on imported produce.

During the 1980s average annual sugar production totalled 7·5m. tonnes making it one of the country's most important crops. However, the negative economic impact of the collapse of the Soviet Union in 1991 followed by a series of weather disasters in subsequent years resulted in production shrinking to 3·2m. tonnes in 1998, the smallest crop for 50 years. In 2002 the government closed over half of the 156 sugar mills and by 2008 production had fallen to 1·4m. tonnes with the country relying on imports to meet domestic demand. Production of other crops in 2008 was (in 1,000 tonnes): bananas, 758; tomatoes, 576; rice, 436; sweet potatoes, 375; cassava, 340; maize, 326; oranges, 200; potatoes, 196; pomelos, 166.

In 2008 livestock included 3·8m. cattle; 1·9m. pigs; 1·1m. goats; 565,000 horses; 277,500 sheep; 29m. chickens.

Forestry
Cuba had 2·87m. ha. of forests in 2010, representing 26% of the land area. These forests contain valuable cabinet woods. Cedar is used locally for cigar boxes, and mahogany is exported. In 2007, 2·35m. cu. metres of roundwood were produced.

Fisheries
The total catch was 23,951 tonnes in 2010, of which 21,923 tonnes were from marine waters. The catch in 2010 was less than half the 2001 total and less than a third of the 1997 total.

INDUSTRY
In 2008 manufacturing accounted for 14% of GDP. All industrial enterprises used to be state-controlled, but in 1995 the economy was officially stated to comprise state property, commercial property based on activity by state enterprises, joint co-operative and private property. Production in 2007 (in 1,000 tonnes): cement, 1,812; raw sugar (2008), 1,446; residual fuel oil, 940; distillate fuel oil, 464; sulphuric acid, 428; petrol, 392; wheat and blended flour, 391; crude steel and steel semi-finished products,

247; televisions, 118,300 units; clay building bricks, 24·9m. units; cigarettes, 13·8bn. units; shoes, 721,000 pairs; soft drinks, 350·6m. litres; beer, 250·4m. litres; spirits and liqueurs, 95·1m. litres; mineral water, 33·2m. litres; cotton fabrics, 6·1m. sq. metres.

Labour
In 2005 the labour force was 6,679,900, with 4,722,500 in employment. Self-employment was legalized in 1993. Under legislation of 1994 employees made redundant must be assigned to other jobs or to strategic social or economic tasks; failing this, they are paid 60% of former salary.

INTERNATIONAL TRADE
Foreign debt to non-communist countries was US$12·3bn. in 2000. Since July 1992 foreign investment has been permitted in selected state enterprises, and Cuban companies have been able to import and export without seeking government permission. Foreign ownership is recognized in joint ventures. A free-trade zone opened at Havana in 1993. In 1994 the productive, real estate and service sectors were opened to foreign investment. Legislation of 1995 opened all sectors of the economy to foreign investment except defence, education and health services. 100% foreign-owned investments and investments in property are now permitted.

The Helms-Burton Law of March 1996 gives US nationals the right to sue foreign companies investing in Cuban estate expropriated by the Cuban government.

Imports and Exports
In 2006 imports (c.i.f.) totalled US$10,173·6m. and exports (f.o.b.) US$2,980·2m. The principal exports are nickel, tobacco and medical products. Sugar used to account for more than half of Cuba's export revenues, but revenues have been gradually declining and constituted less than 1% of the total in 2006. In 2005 the chief import sources (in US$1m.) were: Venezuela (2,021·5); China (926·2); Spain (702·9); USA (520·7). The chief export markets (in US$1m.) were: Netherlands (645·1); Canada (471·4); Venezuela (434·2); Spain (159·5).

COMMUNICATIONS

Roads
In 2002 there were estimated to be 60,856 km of roads (including 638 km of motorways), of which 29,819 km were paved. Vehicles in use in 2008 included 236,881 passenger cars and 171,081 trucks and vans. There were 1,403 fatalities as a result of road accidents in 2008.

Rail
There were 4,066 km of public railway (1,435 mm gauge) in 2005, of which 140 km was electrified. Passenger-km travelled in 2007 came to 1,285m. and freight tonne-km to 783m. In addition, the large sugar estates have 7,162 km of lines in total on 1,435 mm, 914 mm and 760 mm gauges.

Civil Aviation
There is an international airport at Havana (Jose Martí). The state airline Cubana operates all services internally, and in 2003 had international flights from Havana to Bogotá, Buenos Aires, Cancún, Caracas, Curaçao, Fort de France, Guatemala City, Guayaquil, Kingston, Las Palmas, London, Madrid, Mexico City, Montego Bay, Montreal, Moscow, Panama City, Paris, Pointe-à-Pitre, Quito, Rome, San José (Costa Rica), Santiago, Santo Domingo, São Paulo and Toronto. In 2003 scheduled airline traffic of Cuban-based carriers flew 19m. km, carrying 664,000 passengers. In 2001 Havana Jose Martí International handled 2,472,300 passengers and 19,302 tonnes of freight.

Shipping
There are 11 ports, the largest being Havana, Cienfuegos and Mariel. In Jan. 2009 there were 15 ships of 300 GT or over registered, totalling 32,000 GT.

Telecommunications

In 2011 there were 1,193,400 landline telephone subscriptions (equivalent to 106·0 per 1,000 inhabitants) and 1,315,100 mobile phone subscriptions (or 116·9 per 1,000 inhabitants). In 2011, 23·2% of the population were internet users.

SOCIAL INSTITUTIONS

Justice

There is a Supreme Court in Havana and seven regional courts of appeal. The provinces are divided into judicial districts, with courts for civil and criminal actions, and municipal courts for minor offences. The civil code guarantees aliens the same property and personal rights as those enjoyed by nationals.

The 1959 Agrarian Reform Law and the Urban Reform Law passed on 14 Oct. 1960 have placed certain restrictions on both. Revolutionary Summary Tribunals have wide powers.

The death penalty is still in force. There were three executions in 2003, but none since then.

The population in penal institutions in Nov. 2006 was approximately 60,000 (531 per 100,000 of national population).

Education

Education is compulsory (between the ages of six and 14), free and universal. In 2007 there were 883,132 pupils in primary school and 91,530 teaching staff; and 898,833 secondary school pupils with 93,311 teaching staff. There were 864,846 students and 135,800 academic staff in higher education in 2007.

There are four universities, plus ten teacher training, two agricultural, four medical and ten other higher educational institutions.

The adult literacy rate was estimated at 99·8% in 2009.

In 2007 public expenditure on education came to 13·6% of GNI and 20·6% of total government spending.

Health

At 31 Dec. 2008 there were 74,552 physicians, 107,761 nurses and 2,962 pharmacists. There were 217 hospitals in 2008 with 43,434 beds. An additional 23,834 beds were provided by other health care units.

Free medical services are provided by the state polyclinics, though a few doctors still have private practices.

Welfare

The official retirement age is 60 (men) or 55 (women). However, the qualifying age falls to 55 (men) or 50 (women) if the last 12 years of employment or 75% of employment was in dangerous or arduous work. In 2008 plans were announced to raise the retirement ages to 65 (men) and 60 (women) by 2015. The minimum pension in 2003 was 59 pesos a month, or 79 pesos a month, or 80% of wages, depending on average earnings and the number of years of employment. The maximum pension is 90% of average earnings.

Cuba has a sickness and maternity support programme.

RELIGION

Religious liberty was constitutionally guaranteed in July 1992. 40% of the population was estimated to be Roman Catholic in 2001. In 1994 Cardinal Jaime Ortega (b. 1936) was nominated Primate by Pope John Paul II. In Feb. 2013 there was one cardinal. In 2002 there were 180 Roman Catholic priests, approximately half of them foreign nationals. There is a seminary in Havana which had 61 students in 1996. There is a bishop of the American Episcopal Church in Havana; there are congregations of Methodists in Havana and in the provinces as well as Baptists and other denominations. Cults of African origin (mainly Santería) still persist.

CULTURE

World Heritage Sites

There are nine UNESCO sites in Cuba: Old Havana and its fortifications (inscribed in 1982), Trinidad and the Valley de los Ingenios (19th century sugar mills; 1988), San Pedro de la Roca Castle in Santiago de Cuba (1997), Desembarco del Granma National Park (marine terraces; 1999), Viñales Valley (1999), 19th-century coffee plantations at Sierra Maestra (2000), Alejandro de Humboldt National Park (2001), the urban historic centre of Cienfuegos (2005) and the historic centre of Camagüey (2008).

Press

There were (2008) four national daily newspapers and 15 regional and local dailies with a combined circulation of 1·8m. The most widely read newspaper is the Communist Party's *Granma*.

Tourism

Tourism is Cuba's largest foreign exchange earner. Between 2007 and 2008 tourist spending increased by 6%. There were 2,316,000 foreign tourists in 2008 (2,119,000 in 2007). Total receipts from tourism in 2008 amounted to 2,099bn. convertible pesos.

DIPLOMATIC REPRESENTATIVES

Of Cuba in the United Kingdom (167 High Holborn, London, WC1 6PA)
Ambassador: Esther Armenteros Cárdenas.

Of the United Kingdom in Cuba (Calle 34, No. 702/4, entre 7ma Avenida y 17, Miramar, Havana)
Ambassador: Tim Cole.

Of Cuba to the United Nations
Ambassador: Rodolfo Reyes Rodríguez.

Of Cuba to the European Union
Ambassador: Mirtha Hormilla Castro.

The USA broke off diplomatic relations with Cuba on 3 Jan. 1961 but Cuba has an Interests Section in the Swiss Embassy in Washington, D.C., and the USA has an Interests Section in the Swiss Embassy in Havana.

FURTHER READING

Bethell, L. (ed.) *Cuba: a Short History.* 1993
Bunck, J. M., *Fidel Castro and the Quest for a Revolutionary Culture in Cuba.* 1994
Cabrera Infantye, G., *Mea Cuba;* translated into English from Spanish. 1994
Dosal, Paul J., *Cuba Libre: A Brief History of Cuba.* 2006
Fursenko, A. and Naftali, T., *'One Hell of a Gamble': Khrushchev, Castro and Kennedy, 1958–1964.* 1997
Gott, Richard, *Cuba: A New History.* 2004
Levine, Robert, *Secret Missions to Cuba: Fidel Castro, Bernardo Benes, and Cuban Miami.* 2002
May, E. R. and Zelikow, P. D., *The Kennedy Tapes: Inside the White House during the Cuban Missile Crisis.* 1997
Mesa-Lago, C. (ed.) *Cuba: After the Cold War.* 1993
Sweig, Julia, *Inside the Cuban Revolution.* 2002

National Statistical Office: Oficina Nacional de Estadísticas, Paseo No. 60e/ 3ra y 5ta, Vedado, Plaza de la Revolución, Havana, CP 10400.
Website (Spanish only): http://www.one.cu

CYPRUS

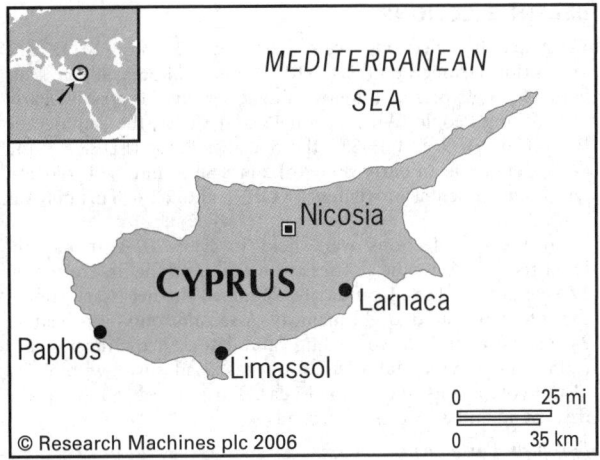

© Research Machines plc 2006

Kypriaki Dimokratia—Kibris Çumhuriyeti
(Republic of Cyprus)

Capital: Nicosia
Population projection, 2015: 1·17m.
GNI per capita, 2011: (PPP$) 24,841
HDI/world rank: 0·840/31
Internet domain extension: .cy

KEY HISTORICAL EVENTS

Cyprus has been settled since the early Neolithic period in the 9th millennium BC. During the 2nd millennium BC Greek colonies were established and Greek culture predominated. Cyprus then came under Assyrian and Egyptian rule, subsequently becoming part of the Persian Empire. In the 2nd and 1st centuries BC the island was Hellenized. It became a Roman province in 58 BC and remained within the Roman Empire for most of the next three centuries, when Christianity was adopted.

The island became part of the Byzantine Empire in AD 364. In 488 the Church of Cyprus was confirmed as independent. Successive Arab invasions in the 7th century led to joint rule by Arabs and Byzantines until 965, when Byzantium regained sole rule. From 1191 Cyprus was attacked by Crusaders and in 1193 it became a Frankish kingdom under Guy de Lusignan. The Venetians ruled it as a dependency from 1489–1571, when it was taken by the Ottoman Turks. Under the Ottomans, the population was divided between Muslim Turks and Christian Greeks. After Greece achieved independence in 1829, some Greek Cypriots called for a union between Cyprus and Greece. On 4 June 1878 Great Britain and the Ottomans signed the Cyprus Convention, which gave Cyprus to Britain as a protectorate, in return for Britain supporting Turkey against Russia. In 1914, in response to Turkey's alliance with Germany, Britain annexed Cyprus. On 1 May 1925 it became a Crown Colony.

In the 1930s Greek Cypriots began campaigning for union with Greece (*enosis*). In 1955 the EOKA (National Organization of Cypriot Fighters), an anti-British guerrilla movement, was formed, led by Archbishop Makarios, head of the Greek Orthodox Church in Cyprus. In 1959 Greek and Turkish Cypriots agreed a constitution for an independent Cyprus and Makarios was elected president. On 16 June 1960 Cyprus became an independent state.

In Dec. 1963 conflict over proposed changes to the power-sharing mechanisms led to the Turkish Cypriots withdrawing from government. Fighting broke out and a UN peacekeeping force was deployed in 1964. Further fighting occurred in 1967–68, after which the Turkish Cypriots formed a provisional administration to run their community's affairs. Some Turkish Cypriots began to call for *taksim*, division of the island. On 15 July 1974 the Greek Cypriot National Guard staged a coup, with backing from Athens, and deposed President Makarios. On 20 July 1974 Turkey invaded. Turkish forces rapidly occupied the northern third of the island and, as the coup collapsed, enforced partition between north and south. During the fighting, and later as part of a UN-supervised population transfer, 200,000 Greek Cypriots moved south, while an estimated 65,000 Turkish Cypriots migrated north.

In Dec. 1974 Makarios returned as president. In 1975 a Turkish Cypriot Federated State was proclaimed in the north, with Rauf Denktaş its president. In 1983 the Turkish state unilaterally proclaimed itself the 'Turkish Republic of Northern Cyprus' ('TRNC'). UN-sponsored peace talks continued without success throughout the 1980s. In 1991 the UN rejected Rauf Denktaş' demands for recognition of the 'TRNC', including the right to secession. In 1998 the Greek and Cypriot governments rejected Denktaş' proposal that the Greek and Turkish communities should join in a federation recognizing 'the equal and sovereign status of Cyprus' Greek and Turkish parts'.

In 2002 Cyprus was one of ten countries to be granted EU membership starting in 2004; the 'TRNC' would be included only if UN-brokered talks to reunify the country succeeded. In Nov. 2002 the UN presented a peace plan to the Greek Cypriot and Turkish Cypriot leaders for a 'common' state with two 'component' states and a rotating presidency. In March 2003 UN-brokered talks to pave the way for reunification collapsed. However, as a goodwill measure the 'TRNC' opened the 'Green Line' separating the island's two sectors in April 2003. In a referendum held in the Greek-speaking and the Turkish-speaking areas on 24 April 2004, Turkish Cypriots voted in favour of a UN plan to reunite the island but Greek Cypriots voted overwhelmingly against. This result meant that after EU accession, EU benefits and laws would apply only to Greek Cypriots. On 1 May 2004, under these conditions, Cyprus became a member of the EU.

In Dec. 2004 Turkey extended its EU customs union agreement to Cyprus but declared that this did not amount to a recognition of Cyprus. In 2006 at UN-sponsored talks, a series of confidence building measures were agreed by Greek and Turkish Cypriots. However, in Nov. 2006 EU-Turkish talks on Cyprus broke down. Hopes of finding a solution were boosted with the election in March 2008 of President Dimitris Christofias who vowed to resurrect negotiations with Turkey. A month later the Ledra Street crossing that had symbolized the island's division since its closure 44 years earlier was reopened.

TERRITORY AND POPULATION

The island lies in the Mediterranean, about 60 km off the south coast of Turkey and 90 km off the coast of Syria. Area, 9,251 sq. km (3,572 sq. miles). The Turkish-occupied area is 3,355 sq. km. Population of Cyprus by ethnic group:

Ethnic group	1960 census	1973 census	1992	2011
Greek Cypriot	452,291	498,511	599,200	681,000
Turkish Cypriot	104,942	116,000	94,500[1]	90,100[1]
Others	16,333	17,267	20,000	181,000[2]
Total	573,566	631,778	713,700[1]	952,100[1]

[1]Excluding Turkish settlers and troops.
[2]Including foreign workers and residents.

The 2011 census population (government-controlled area only) was 856,857.

The UN gives a projected population for 2015 of 1·17m.

70·3% of the population lived in urban areas in 2010. Principal towns with populations (2011 census): Nicosia (the capital), 239,277; Limassol, 180,201; Larnaca, 84,591; Paphos, 62,122.

As a result of the Turkish occupation of the northern part of Cyprus, 0·2m. Greek Cypriots were displaced and forced to find refuge in the south. The urban centres of Famagusta, Kyrenia and Morphou were completely evacuated. *See below* for details on the 'Turkish Republic of Northern Cyprus'. (The 'TRNC' was unilaterally declared as a 'state' in 1983 in the area of the Republic of Cyprus, which has been under Turkish occupation since 1974, when Turkish forces invaded the island. The establishment of the 'TRNC' was declared illegal by UN Security Resolutions 541/83 and 550/84. The 'TRNC' is not recognized by any country in the world except Turkey). Nicosia is a divided city, with the UN-patrolled Green Line passing through it.

Greek and Turkish are official languages. English is widely spoken.

SOCIAL STATISTICS

2009 births, 9,608; deaths, 5,182; marriages, 12,769; divorces, 1,738. Rates, 2009 (per 1,000 population): birth, 12·0; death, 6·5; marriage (residents of Cyprus only), 7·9; divorce, 2·2. Life expectancy at birth, 2007, was 77·3 years for males and 81·9 years for females. Population growth rate, 2009, 0·8%; infant mortality, 2009, 3·3 per 1,000 live births; fertility rate, 2008, 1·5 children per woman. In 2009 the average age of first marriage (residents of Cyprus only) was 29·5 years for men and 27·4 years for women.

CLIMATE

The climate is Mediterranean, with very hot, dry summers and variable winters. Maximum temperatures may reach 112°F (44·5°C) in July and Aug., but minimum figures may fall to 22°F (−5·5°C) in the mountains in winter, when snow is experienced. Rainfall is generally between 10" and 27" (250 and 675 mm) and occurs mainly in the winter months, but it may reach 48" (1,200 mm) in the Troodos mountains. Nicosia, Jan. 50°F (10·0°C), July 83°F (28·3°C). Annual rainfall 19·6" (500 mm).

CONSTITUTION AND GOVERNMENT

Under the 1960 constitution executive power is vested in a *President* elected for a five-year term by universal suffrage, and exercised through a Council of Ministers appointed by him or her. The *House of Representatives* exercises legislative power. It is elected by universal suffrage for five-year terms, and consists of 80 members, of whom 56 are elected by the Greek Cypriot and 24 by the Turkish Cypriot community. As from Dec. 1963 the Turkish Cypriot members have ceased to attend, and the 24 seats allocated to the Turkish Cypriot community are no longer contested. Voting is compulsory, and is by preferential vote in a proportional representation system with reallocation of votes at national level.

National Anthem

'Imnos eis tin Eleftherian' ('Hymn to Freedom'); words by Dionysios Solomos, tune by N. Mantzaros.
(Same as Greece.)

GOVERNMENT CHRONOLOGY

Presidents since 1960. AKEL = Progressive Party of the Working People; DIKO = Democratic Party; DISI = Democratic Rally; EOKA = National Organization of Cypriot Fighters; n/p = non-partisan)

1960–74	n/p	Makarios III
1974	EOKA	Nikolaos (Nikos) Sampson
1974–77	n/p	Makarios III
1977–88	DIKO	Spyros Achilleos Kyprianou
1988–93	n/p	Georgios Vasou Vasiliou
1993–2003	DISI	Glafcos Ioannou Clerides
2003–08	DIKO	Tassos Nikolaou Papadopoulos
2008–13	AKEL	Dimitris Christofias
2013–	DISI	Nicos Anastasiades

RECENT ELECTIONS

Parliamentary elections were held on 22 May 2011. The opposition Democratic Rally (DISI) won 20 of 56 available seats with 34·3% of votes cast, followed by the ruling Progressive Party of Working People (AKEL) with 19 and 32·7%, the Democratic Party (DIKO) 9 and 15·8%, the Socialist Party (EDEK) 5 and 8·9%, the European Party (Evroko) 2 and 3·9% and the Ecological and Environmental Movement (KOP) 1 and 2·2%. Turnout was 78·7%.

Presidential elections were held in Feb. 2013. Incumbent Dimitris Christofias did not seek re-election. In the first round on 17 Feb. none of the 11 candidates obtained the necessary 50% of the vote to win an outright majority. A second round was held on 24 Feb. between the two leading candidates, Nicos Anastasiades (DISI) and Stavros Malas (AKEL). Anastasiades won with 57·5% of the vote against 42·5% for Malas. Turnout was 83·1% in the first round and 81·6% in the second.

European Parliament

Cyprus has six representatives. At the June 2009 elections turnout was 59·4% (72·5% in 2004). The DISI won 2 seats with 35·7% of votes cast (political affiliation in European Parliament: European People's Party), AKEL 2 seats with 34·9% (European United Left/Nordic Green Left), DIKO 1 seat with 12·3% (Progressive Alliance of Socialists and Democrats) and EDEK 1 seat with 9·9% (Progressive Alliance of Socialists and Democrats).

CURRENT GOVERNMENT

President: Nicos Anastasiades; b. 1946 (Democratic Rally; sworn in 28 Feb. 2013).

In March 2013 the Council of Ministers consisted of:

Minister of Foreign Affairs: Ioannis Kasoulides. *Interior:* Socratis Hasikos. *Agriculture, Natural Resources and Environment:* Nicos Kouyialis. *Defence:* Fotis Fotiou. *Commerce, Industry and Tourism:* Giorgios Lakkotrypis. *Health:* Petros Petrides. *Communications and Works:* Tasos Mitsopoulus. *Finance:* Haris Georgiades. *Education and Culture:* Kyriacos Kenevezos. *Labour and Social Insurance:* Zeta Emilianidou. *Justice and Public Order:* Ionas Nicolaou.

Government Website: http://www.cyprus.gov.cy

CURRENT LEADERS

Nicos Anastasiades

Position
President

Introduction
Nicos Anastasiades was sworn into office on 28 Feb. 2013 following his victory in the second round of a presidential poll. He is faced with an economy in crisis.

Early Life
Nicos Anastasiades was born on 27 Sept. 1946 in Pera Pedi, Limassol District. In 1969 he graduated in law from the National and Kapodistrian University of Athens. Two years later he completed his postgraduate studies in shipping law at University College, London, and in 1972 established his own firm, Nicos Chr. Anastasiades & Partners.

Anastasiades became a founding member of Democratic Rally (DISI) in 1976. He was active in its youth wing (NEDISI) from that date until 1990, serving as both its vice-president and president. At the 1981 general election he was voted into parliament as the representative for Limassol. In 1990 he became DISI's vice-president, a year later was named deputy parliamentary spokesman for the DISI–Democratic Party coalition and from

1993–97 served as its parliamentary spokesman. He was appointed DISI deputy president in 1995 and took the party leadership two years later, a role he has kept ever since.

In March 2012 Anastasiades was named DISI's presidential nominee after receiving 86·7% backing from the party executive. The campaign trail was dominated by the country's financial crisis, with Anastasiades pledging to finalize a bailout programme and rebuild Cyprus' economic credibility on the international stage. On 17 Feb. 2013 he won 45% of the first round vote and in a run-off a week later claimed the presidency with 57·5% support.

Career in Office

Considered a safe pair of hands by the IMF and EU, Anastasiades began his tenure working with foreign lenders to seal a bailout package worth €17bn. (equivalent to almost 75% of the country's GDP). In return, he will be expected to implement financial reforms and austerity measures more severe than those that his predecessor, Dimitris Christofias, had been willing to put in place.

The recent discovery of natural gas off Cyprus' coast is an opportunity to secure its energy future and generate much needed revenue. However, Anastasiades must decide on how best to exploit the discovery without escalating tensions with Turkey. He is also under pressure to restart talks with the 'Turkish Republic of Northern Cyprus' ('TRNC') based on the guidelines laid out in the Annan Plan, which recommends the unification of Cyprus and the 'TRNC' within a federation. He has previously supported the plan despite a majority of his party rejecting it.

DEFENCE

Conscription is for 24 months. Defence expenditure in 2008 totalled US$537m. (US$503 per capita), representing 2·2% of GDP. In 1998 the president cancelled a US$450m. contract with Russia for the deployment of S-300 anti-aircraft missiles on the island and negotiated to place them on Crete instead.

National Guard

Total strength (2007) 10,000 (8,700 conscripts). There is also a paramilitary force of some 500 armed police.

There are two British bases (Army and Royal Air Force) and some 3,250 personnel. Greek (1,150) and UN peacekeeping (909 uniformed personnel; UNFICYP) forces are also stationed on the island.

There are approximately 36,000 Turkish troops stationed in the occupied area of Cyprus. The Turkish Cypriot Security Force amounts to around 5,000 troops, with 26,000 reservists and a paramilitary armed police of approximately 150.

Navy

The Maritime Wing of the National Guard operates six vessels. In the Turkish-occupied area of Cyprus the Coast Guard operates six patrol craft.

Air Force

The Air Wing of the National Guard operates a handful of aircraft and 16 attack helicopters, including 12 Mi-35s.

ECONOMY

Overview

The traditional leading economic sectors are tourism, financial services and real estate. The southern part of Cyprus joined the EU in 2004 and adopted the euro in Jan. 2008. It accounts for 0·2% of the eurozone's GDP. Major regulatory and institutional reform implemented after 2004 included low tax incentives to attract international corporations. The economy expanded for 35 years before the global downturn prompted shrinkage of 1·9% in 2009, followed by slow growth in 2010 and 2011 and an estimated contraction of 1·2% in 2012. The EU predicted a further contraction of 8·7% in 2013. The budget deficit reached 6·3% of GDP in 2011 and public debt stood at 72%. Unemployment was 10% in 2012.

Attempts to bring together the economies of the north and south of the island have largely failed. The 'Turkish Republic of Northern Cyprus' ('TRNC') lacks an independent monetary policy and is vulnerable to economic and monetary fluctuations in Turkey. Nonetheless, economic activity in the 'TRNC' accelerated after the partial lifting of obstacles to free movement in 2003, especially in the tourism and construction sectors. Between 2001 and 2005 GDP in constant prices increased by more than 50%.

In June 2012 Cyprus became the fifth eurozone country to request a bailout, having grown vulnerable to a combination of light regulation, rapid credit expansion, a property market bubble, exposure to toxic Greek debt and the knock-on effects of the eurozone crisis. After an initial deal was rejected by parliament, a bailout valued at €10bn. was agreed in March 2013 with the banking sector close to collapse and access to the international capital markets severely curtailed. The IMF reported bank assets equivalent to 835% of GDP in 2011, over double the eurozone average.

With fellow EU states concerned at the scale of Russian-origin deposits in the banking system, Cyprus came under pressure to reform the sector. The bailout demanded a levy on all deposits over €100,000 and wound up Cyprus' second largest bank, Laiki Bank, with the Bank of Cyprus scheduled to take over its healthy assets. The economy faces a precarious future.

Currency

On 1 Jan. 2008 the euro (EUR) replaced the Cyprus pound (CYP) as the legal currency of Cyprus at the irrevocable conversion rate of £C0·585274 to one euro. Inflation was 2·6% in 2010 and 3·5% in 2011.

In July 2005 gold reserves were 465,000 troy oz, foreign exchange reserves were US$3,655m. and total money supply was £C1,504m.

Budget

Revenue in 2009 (2008) was €6·7bn. (€7·5bn.) and expenditure €7·8bn. (€7·3bn.). Tax revenue in 2009 totalled €4,452m. and non-tax revenue €2,291m.

Main divisions of expenditure in 2009 (in €1m.): wages and salaries, 2,644; social security payments, 2,251; other goods and services, 952. Capital expenditure for 2009 totalled €849m., of which €690m. was investment expenditure.

The outstanding domestic debt at 31 Dec. 2009 was €5,841·9m. and the foreign debt was €3,984·5m.

Cyprus' budget deficit in 2011 was 6·3% of GDP (2010, 5·3%; 2009, 6·1%). In 2008 it had a surplus of 0·9%. The required target set by the EU is a budget deficit of no more than 3%.

VAT is 18%. There are reduced rates of 8·0% and 5·0%.

Performance

There was negative growth of 1·9% in 2009 but the economy grew by 1·1% in 2010 and 0·5% in 2011. Total GDP in 2011 was US$24·7bn.

Banking and Finance

The Central Bank of Cyprus, established in 1963, is the bank of issue. It regulates money supply, credit and foreign exchange and supervises the banking system. The *Governor* is Panicos Demetriades.

In 2012 there were six domestic banks, 35 International Banking Units and one representative office of a foreign bank. The leading banks are Bank of Cyprus, Cyprus Popular Bank and Hellenic Bank. At 31 Oct. 2012 banks' total deposits and lending amounted to €70,306m. and €71,415m. respectively.

There is a stock exchange in Nicosia.

ENERGY AND NATURAL RESOURCES

In 2008, 4·1% of energy consumption came from renewables (wind power, solar power, hydro-electric power, tidal power, geothermal energy and biomass), compared to the European

Union average of 10·3%. A target of 13% has been set by the EU for 2020.

Environment
Carbon dioxide emissions from the consumption and flaring of fossil fuels in Cyprus were the equivalent of 9·5 tonnes per capita in 2008.

Electricity
Installed capacity was 1·1m. kW in 2007. Production in 2007 was 4·87bn. kWh and consumption per capita 6,213 kWh.

Minerals
The principal minerals extracted in 2010 were (in 1,000 tonnes): crushed aggregate, 4,589; sand, 4,571; gravel, 4,253; crushed limestone, 250; gypsum, 240; clay, 210.

Agriculture
21% of the government-controlled area is cultivated. There were 102,400 ha. of arable land in 2008 and 37,500 ha. of permanent crops. 24,400 ha. were irrigated in 2008. About 6·3% (2008) of the economically active population were engaged in agriculture.

Chief agricultural products in 2008 (1,000 tonnes): milk, 194·9; potatoes, 115·0; citrus fruit, 111·8; meat, 100·5; fresh fruit, 52·2; grapes, 32·4; olives, 15·6; eggs, 9·9; carobs, 6·5; cereals (barley, wheat and oats), 6·3; carrots, 1·9; almonds, 0·4; other vegetables, 71·9. Livestock in 2008: cattle, 55,800; goats, 318,400; pigs, 464,900; sheep, 267,300; poultry, 2·9m.

Forestry
Total forest area in 2010 was 0·17m. ha. (19% of the land area). In 2007, 20,000 cu. metres of timber were produced.

Fisheries
Catches in 2008 totalled 2,011 tonnes; aquaculture production that year amounted to 3,403 tonnes. Imports of fishery commodities were valued at US$100m. in 2008.

INDUSTRY
The most important manufacturing industries in 2008 were: food products, other non-metallic mineral products, fabricated metal products, beverages, wood products, printing, pharmaceutical products and plastic products. The manufacturing industry in 2008 contributed 7·5% of the GDP.

Labour
Out of an average of 381,300 people in employment in 2009, 71,800 were in wholesale and retail trade/repair of motor vehicles and motorcycles; 44,300 in construction; and 34,800 in manufacturing. The unemployment rate was 7·1% in Sept. 2010. There were a total of 1,211 working days lost to strike action in 2009, up from 1,034 in 2008 but down from 10,289 in 2007.

INTERNATIONAL TRADE
Imports and Exports
In 2011 imports (c.i.f.) totalled €6,311m. (€6,517m. in 2010); exports (f.o.b.), €1,404m. (€1,137m. in 2010).

Chief imports, 2009 (in €1,000): machinery, electrical equipment, sound and television recorders, 1,018,788; mineral products, 1,007,937; prepared foodstuffs, beverages, spirits and vinegar, tobacco, 542,945; products of chemical or allied industries, 514,872; vehicles, aircraft, vessels and associated transport equipment, 478,516. Chief domestic exports, 2009 (in €1,000): pharmaceutical products, 108,019; photosensitive semiconductor devices, 57,421; haloumi cheese, 41,027; potatoes, 38,094; waste and scrap, 29,900.

Main import suppliers, 2007: Greece, 17·5%; Italy, 10·1%; UK, 9·7%; Germany, 9·3%. Main export markets, 2007: Greece, 19·8%; Bunkers and ship's stores, 15·7%; UK, 13·9%; Germany, 6·1%.

COMMUNICATIONS
Roads
In 2007 the total length of roads in the government-controlled area was 12,246 km, of which 64·0% were paved. In 2007 there were 410,936 passenger cars, 117,498 trucks and vans, 3,292 buses and coaches and 41,211 motorcycles and mopeds. There were 71 deaths as a result of road accidents in 2009.

The area controlled by the government of the Republic and that controlled by the 'TRNC' are now served by separate transport systems, and there are no services linking the two areas.

Civil Aviation
Nicosia airport has been closed since the Turkish invasion in 1974. It is situated in the UN controlled buffer zone. There are international airports at Larnaca (the main airport) and Paphos. In 2009, 7,068,080 passengers, 59,092 aircraft and 38,502 tonnes of commercial freight went through these airports. Both airports have been expanded considerably to increase capacity, with new terminals at Paphos and Larnaca coming into operation in Nov. 2008 and Nov. 2009 respectively. In 2009 Larnaca handled 5,360,369 passengers and 38,092 tonnes of freight, and Paphos handled 1,707,711 passengers and 410 tonnes of freight. In 2009, 2,318,107 passengers arrived at the two airports on scheduled flights and 2,306,850 passengers departed. The national carrier is Cyprus Airways, which is 69·57% state-owned.

Shipping
The two main ports are Limassol and Larnaca. In 2009, 3,915 ships of 22,897,408 net registered tons entered Cyprus ports carrying 7,859,269 tonnes of cargo from, to and via Cyprus. In Jan. 2009 there were 867 ships of 300 GT or over registered, totalling 20·03m. GT. Among the 867 vessels registered were 286 bulk carriers, 216 general cargo ships and 192 container ships. The port in Famagusta has been closed to international traffic since the Turkish invasion in 1974.

Telecommunications
In 2011 there were 405,000 landline telephone subscriptions (equivalent to 362·7 per 1,000 inhabitants) and 1,090,900 mobile phone subscriptions (or 977·1 per 1,000 inhabitants). There were 529·9 internet users per 1,000 inhabitants in 2010. Fixed internet subscriptions totalled 190,900 in 2009 (175·1 per 1,000 inhabitants). In March 2012 there were 554,000 Facebook users. CYTA (short for Cyprus Telecommunications Authority) is the national telecommunications provider. It has developed an extensive submarine fiber optic cable network linking Cyprus with neighbouring countries such as Greece, Israel and Egypt and—by extension—with the rest of the world.

SOCIAL INSTITUTIONS
Justice
There is a Supreme Court, Assize Courts and District Courts. The Supreme Court is composed of 13 judges, one of whom is the President of the Court. The Assize Courts have unlimited criminal jurisdiction. The District Courts exercise civil and criminal jurisdiction, the extent of which varies with the composition of the Bench.

A Supreme Council of Judicature, consisting of the President and Judges of the Supreme Court, is entrusted with the appointment, promotion, transfers, termination of appointment and disciplinary control over all judicial officers, other than the Judges of the Supreme Court. The Attorney-General (Petros Clerides) is head of the independent Law Office and legal adviser to the President and his Ministers.

The population in penal institutions in the government-controlled area in Aug. 2008 was 671 (83 per 100,000 of national population).

The death penalty was abolished for all crimes in 2002.

Education

Greek-Cypriot Education. Primary education is compulsory and is provided free in six grades to children between 5 years 8 months and 11 years 8 months. A one-year pre-primary education is compulsory and is provided free by the public kindergartens. There are also schools for the deaf and blind, and nine schools for disabled children. In 2008–09 the Ministry of Education and Culture ran 250 kindergartens for children in the age group 3–5 years 8 months; there were also 60 communal and 154 private kindergartens. There were 371 primary schools in 2008–09 with 55,552 pupils and 4,686 teachers.

Secondary education is also free and attendance for the first cycle (gymnasium) is compulsory. The secondary school lasts six years—three years at the gymnasium followed by three years at the *lykeion* (lyceum) or three years at one of the technical schools that provide technical and vocational education for industry. In 2008–09 there were 165 secondary schools with 7,652 teachers and 65,445 pupils.

Tertiary education is provided at nine public and 34 private institutions. There are three public universities: the University of Cyprus, the Open University of Cyprus and the Cyprus University of Technology, which admitted their first students in Sept. 1992, Sept. 2006 and Sept. 2007 respectively and had a combined total of 7,527 students in 2008–09. The other public institutions are the Higher Technical Institute, which provides courses lasting three to four years for technicians in civil, electrical, mechanical and marine engineering; the Cyprus Forestry College (administered by the Ministry of Agriculture, Natural Resources and Environment); the Higher Hotel Institute (Ministry of Labour and Social Insurance); the Mediterranean Institute of Management (Ministry of Labour and Social Insurance); the School of Nursing (Ministry of Health), which runs courses lasting two to three years; the Cyprus Police Academy, which provides a three-year training programme. Among the 34 private institutions are three private universities: the European University Cyprus, the University of Nicosia and the Frederick University. All three universities opened in Sept. 2007; they had a combined total of 10,367 students in 2008–09.

There are also various public and private institutions that provide non-formal education. These include the Apprenticeship Training Scheme and Evening Technical Classes, and other vocational and technical courses organized by the Human Resources Development Authority.

In 2008 the adult literacy rate was 98%. The percentage of the population aged 20 years and over that has attended school was 93·0% in 2008.

In 2008 public expenditure on education came to 7·5% of GDP and 16·9% of total government spending.

Health

In 2010 there were 2,442 physicians, 772 dentists, 3,930 nurses and midwives and 181 pharmacists. There were 87 registered private hospitals/clinics, six government hospitals and two rural hospitals in 2009.

Welfare

Cyprus has a compulsory earnings-related Social Insurance Scheme financed by tripartite contributions, which covers all the gainfully employed population. Employees in the broader public sector are covered by supplementary mandatory pension schemes or provident funds. A large proportion of the private sector's employees have supplementary coverage under non-statutory provident funds established by collective agreements.

RELIGION

The Greek Cypriots are predominantly Greek Orthodox Christians, and almost all Turkish Cypriots are Muslims (mostly Sunnis of the Hanafi sect). There are also small groups of the Armenian Apostolic Church, Roman Catholics (Maronites and Latin Rite) and Protestants (mainly Anglicans). *See also* CYPRUS: Territory and Population.

CULTURE

World Heritage Sites

There are three sites under Cypriot jurisdiction in the World Heritage List: Paphos (entered on the list in 1980); the churches of the Troodos region (1985, 2001); and Choirokoitia (1998). Paphos, inhabited since the Neolithic age, was a site of worship of the goddess Aphrodite and prehistoric fertility deities. The site has the remains of villas, palaces, theatres, fortresses and tombs. The Troodos region has one of the largest groups of Byzantine churches and monasteries. Decorated with murals and frescoes, they include the Church of St John Lampadistis, built between the 11th and 15th centuries. The Neolithic settlement of Choirokhoitia, between Larnaca and Limassol and dating from the 7th to the 4th millennium BC, is one of the most important prehistoric sites in the eastern Mediterranean.

Press

In 2008 there were 22 paid-for dailies daily newspapers with a circulation of 103,000. The most widely read daily is *Phileleftheros*.

Tourism

There were 2,173,000 international tourist arrivals in 2010 (excluding same-day visitors). Most tourists in 2010 were from the UK (45·8%), followed by Russia (10·3%), Germany (6·4%) and Greece (5·9%). Tourist spending in 2010 totalled US$2,371m.

Festivals

The capital, Nicosia, is the focal point for Independence Day festivities each Oct. Limassol hosts many of the country's largest festivals, including a ten-day Carnival as part of the nationwide Apokreo festival that precedes Easter. Each May Anthestiria (a flower festival) is held, and there are annual festivals of Shakespearean and Ancient Greek theatre.

DIPLOMATIC REPRESENTATIVES

Of Cyprus in the United Kingdom (13 St James' Sq., London, SW1Y 4LB)
High Commissioner: Alexandros N. Zenon.

Of the United Kingdom in Cyprus (Alexander Pallis St., Nicosia)
High Commissioner: Mathew Kidd.

Of Cyprus in the USA (2211 R St., NW, Washington, D.C., 20008)
Ambassador: Pavlos Anastasiades.

Of the USA in Cyprus (Metochiou and Ploutarchou Streets, Engomi, Nicosia)
Ambassador: John M. Koenig.

Of Cyprus to the United Nations
Ambassador: Nicholas Emiliou.

Of Cyprus to the European Union
Permanent Representative: Kornelios Korneliou.

FURTHER READING

Calotychos, V., *Cyprus and Its People: Nation, Identity and Experience in an Unimaginable Community 1955–1997.* 1999
Faustmann, Hubert and Ker-Lindsay, James, *The Government and Politics of Cyprus.* 2008
Hakkı, Murat Metin, *The Cyprus Issue: A Documentary History, 1878–2006.* 2007
Hitchens, Christopher, *Hostage to History: Cyprus from the Ottomans to Kissinger.* 1997
Joseph, Joseph S., *Cyprus: Ethnic Conflict and International Politics: From Independence to the Threshold of the European Union.* 1999
Ker-Lindsay, James, *The Cyprus Problem: What Everyone Needs to Know.* 2011
Mallinson, William, *Cyprus: A Modern History.* 2005

Michael, Michális Stavrou, *Resolving the Cyprus Conflict: Negotiating History.* 2009

Pace, Roderick, *The European Union's Mediterranean Enlargement: Cyprus and Malta.* 2006

Papadakis, Yiannis, *Divided Cyprus: Modernity, History and an Island in Conflict.* 2006

Richter, Heinz, *A Concise History of Modern Cyprus: 1878–2009.* 2010

Sepos, Angelos, *The Europeanization of Cyprus: Polity, Policies and Politics.* 2008

Statistical Information: Statistical Service of the Republic of Cyprus, Michalakis Karaolis Street, 1444 Nicosia.

Website: http://www.mof.gov.cy/mof/cystat/statistics.nsf

'Turkish Republic of Northern Cyprus (TRNC)'

Kuzey Kıbrıs Türk Cumhuriyeti

KEY HISTORICAL EVENTS

See CYPRUS: Key Historical Events.

TERRITORY AND POPULATION

The Turkish Republic of Northern Cyprus occupies 3,355 sq. km (about 33% of the island of Cyprus) and its population at the 2011 census was 286,257. Distribution of population by districts (2011 census): Lefkoşa (Nicosia), 94,824; Girne (Kyrenia), 69,163; Mağusa (Famagusta), 69,741; Güzelyurt (Morphou), 30,037; İskele (Trikomo), 22,492.

CONSTITUTION AND GOVERNMENT

The Turkish Republic of Northern Cyprus was proclaimed on 15 Nov. 1983. The 50 members of the *Legislative Assembly* are elected under a proportional representation system.

RECENT ELECTIONS

Presidential elections were held on 18 April 2010. Prime Minister Derviş Eroğlu (National Unity Party/UBP) won 50·4% against incumbent president, Mehmet Ali Talat (ind.), who claimed 42·9%. Turnout was 76·4%.

In parliamentary elections on 19 April 2009 the opposition National Unity Party won 44·1% of the vote (26 of 50 seats), the ruling Republican Turkish Party-United Forces 29·2% (15), the Democrat Party 10·7% (5), the Communal Democracy Party 6·9% (2) and the Freedom and Reform Party 6·2% (2). Turnout was 81·4%.

CURRENT GOVERNMENT

President: Derviş Eroğlu; b. 1938 (sworn in 23 April 2010).

In March 2013 the government consisted of:

Prime Minister: İrsen Küçük; b. 1940 (UBP; since 17 May 2010).

Minister for Agriculture and Natural Resources: Ali Çetin Amcaoğlu. *Economy and Energy:* Sunat Atun. *Education, Youth and Sport:* Mutlu Atasayan. *Finance:* Erşin Tatar. *Foreign Affairs:* Hüseyin Özgürgün. *Health:* Ertugrul Hasipoğlu. *Interior and Local Administration:* Nazim Çavuşoğlu. *Labour and Social Security:* Şerife Ünverdi. *Public Works and Communications:* Hamza Ersan Saner. *Tourism, Environment and Culture:* Ünal Üstel.

Government Website (Turkish only): http://www.cm.gov.nc.tr

DEFENCE

In 2007 around 36,000 members of Turkey's armed forces were stationed in the TRNC with eight main battle tanks (plus 441 tanks for training purposes). TRNC forces comprise seven infantry battalions with an estimated total personnel strength of 5,000. Conscription is for 15 months.

INTERNATIONAL RELATIONS

In April 2004 the European Union pledged to release almost US$310m. as a reward for approval of a UN plan to reunify the island. The TRNC voted in favour of reunification in a referendum but it was rejected by the Greek Cypriot south, which joined the EU separately on 1 May 2004. The aid package was approved in Feb. 2006 and aimed to improve economic growth and development in order to encourage reunification.

ECONOMY

Currency

The Turkish lira is used.

Budget

Revenue in 2008 (in US$1m.) was 1,457·9 (of which local revenues 1,235·9 and foreign aid and loans 222·1); expenditure, 1,787·5.

Performance

After shrinking by 5·6% in 2009 the economy grew by 3·6% in 2010.

Banking and Finance

46 banks, including 21 offshore banks, were operating in 2004. Control is exercised by the Central Bank of the TRNC.

ENERGY AND NATURAL RESOURCES

Agriculture

Agriculture accounted for 6·4% of GDP in 2009 (provisional).

INTERNATIONAL TRADE

Exports earned US$96·4m. in 2010. Imports cost US$1,604·2m. Customs tariffs with Turkey were reduced in July 1990. There is a free port at Famagusta.

COMMUNICATIONS

Civil Aviation

There is an international airport at Ercan. In 2004 there were flights to Adana, Ankara, Antalya, Dalaman, İstanbul and İzmir with Turkish Airlines and Cyprus Turkish Airlines.

SOCIAL INSTITUTIONS

Education

In 2007–08 there were 18,101 pupils in 96 primary schools; 16,048 pupils in 55 junior and senior high schools; 440 students in technical and vocational schools; and 43,021 students in higher education. There are seven universities.

Health

In 2008 there were 557 doctors, 140 dentists, 138 pharmacists and 1,211 beds in state hospitals and private clinics.

CULTURE

Press

In 2008 there were 11 daily newspapers. The most widely read paper is *Kıbrıs*.

Tourism

There were 715,749 tourists in 2006, of which 572,633 were from Turkey and 143,116 from other countries.

FURTHER READING

Dodd, C. H. (ed.) *The Political, Social and Economic Development of Northern Cyprus.* 1993

Hanworth, R., *The Heritage of Northern Cyprus.* 1993

Ioannides, C. P., *In Turkey's Image: the Transformation of Occupied Cyprus into a Turkish Province.* 1991

CZECH REPUBLIC

Česká Republika

Capital: Prague
Population projection, 2015: 10·63m.
GNI per capita, 2011: (PPP$) 21,405
HDI/world rank: 0·865/27
Internet domain extension: .cz

KEY HISTORICAL EVENTS

The area that is today the Czech Republic was originally inhabited by Celts around the 4th century BC. The Celtic Boii tribe gave the country its Latin name—Boiohaemum (Bohemia)—but was driven out by Germanic tribes. Slav tribes migrated to central Europe during the period known as the Migration of Peoples and were well established by the 6th century. The first half of the 7th century saw allied Slavonic tribes defending their territory from the Avar Empire in the Hungarian lowlands and from Frank attackers to the West.

Mojmír established The Great Moravian Empire in 830, comprising Bohemia, Moravia and Slovakia. The Empire reached its height under Moravian ruler Svatopluk but was engulfed and destroyed by the Magyars around 903–07. In 1041, after the defeat of Prince Břetislav the Restorer by the German Emperor Henry III, Bohemia became a fief of the Holy Roman Empire. Dynastic squabbles, exacerbated by German interference, weakened the power of the dukes, but in 1212 Otakar I (1197–1230) was granted a hereditary kingship from the Holy Roman Emperor who also declared the indivisibility of Bohemia as a key independent state within the realm.

A period of prosperity followed, aided by the immigration of German miners and merchants. Bohemia expanded under the last Přemysl kings. Wenceslas I (1230–53) seized Austria in 1251, though it passed to the Habsburgs when Otakar II was killed at the battle of Marchfeld in 1278. His son Wenceslas II was elected king of Poland in 1300. Wenceslas III was assassinated in 1306, thus ending the Přemysl line. After four years of struggle John of Luxemburg succeeded the throne in 1310.

John's son Charles (1346–78) became Holy Roman Emperor as Charles IV (Charles I of Bohemia) in 1355. He declared Prague the capital of the German Empire, designating the Crownlands of Bohemia to include parts of modern Germany and Poland. The Golden Bull of Nürnberg in 1356 granted the king of Bohemia first place among the empire's electors.

In what became known as the Golden Age, Charles fostered the commercial and cultural development of Bohemia. In 1348 he founded Prague University, the first university in Central Europe. Work started on the building of St Vitus cathedral, Charles Bridge and the Czech castle of the Grail in Karlstejn during his reign.

Hussite Revolution

Dissatisfaction with the Catholic church in the 14th and 15th centuries climaxed with the Hussite Revolution, the clerical reform movement associated with Jan Hus. The movement had undertones of anti-German Czech nationalism and found support amongst the urban middle classes and lesser rural gentry as well as the urban poor and peasantry. Hus was condemned as a heretic and burnt at the stake in Constance in 1415. Anti-Hussite rulings by Wenceslas IV led to the first 'Defenestration of Prague' in 1419, when Catholic councillors were thrown from the Town Hall windows. In the ensuing Hussite wars Sigismund, the Hungarian Holy Roman Emperor, failed to recover the Bohemian crown. Five crusades were launched against the Hussites in the years 1420–31, all of which were defeated.

The Hussites were eventually weakened by divisions between moderates, the Utraquists, and radicals, the Taborites (originating from the town of Tabor in South Bohemia). The latter's militant leader, Jan Žižka, was defeated in the battle of Lipany in 1434 by the Prague faction supported by Sigismund. This victory allowed for a temporary agreement between Hussite Bohemia and Catholic Europe, known as the Compacts of Basle, which reunited the Utraquist faction with Rome in 1436. A degree of post-war recovery followed under the moderate Hussite king, George of Poděbrady (1457–71), who crushed the radical Taborite dissidents.

In 1471 Vladislav Jagellonský, son of King Casimir IV of Poland, was elected King of Bohemia. His son Louis inherited the throne but was killed at Mohás fighting the Turks in 1526. From 1490, Hungary and Bohemia were both ruled by the Jagiellon Dynasty. Under their rule, the provincial diet of three estates (nobility, gentry and burgesses) enhanced the power of the nobility at the expense of the burgesses. Animosity between the Hussite church and the minority Catholic Church continued.

In 1526, after the extinction of the Jagiellon line, Czech nobles elected Archduke Ferdinand. The advent of the Habsburgs saw Roman Catholicism reintroduced. When Rudolf II (1576–1611) left Vienna to make Prague the capital of the German Empire and the seat of a papal nuncio, the city became a centre of European culture. However, religious divisions with the beginnings of the Counter-Reformation led to the second Defenestration of Prague. Two Catholic governors and their secretary were thrown from a window of Prague Castle, an action which sparked off the Thirty Years' War (1618–48). This brought political disorder and economic devastation to the country with the Habsburg forces wiping out a third of the Bohemian population. The estates deposed Emperor Ferdinand II in favour of the Calvinist Frederick V but the latter's forces were defeated at the battle of White Mountain in 1620. A period of Habsburg hegemony ensued.

After the suppression of the Taborites, the only remaining Protestant Church in Bohemia was the Unity of Czech Brethren, to which Catholics, Utraquists and Lutherans were opposed. The Czech nobility was replaced by German-speaking adventurers, the burgesses lost their rights, the peasantry suffered hardships and citizens were forced to embrace the Catholic faith or emigrate. The throne of Bohemia was made hereditary in the Habsburg Dynasty and the most important offices transferred to Vienna.

Opposition was savagely repressed. The next two centuries became known as the Dark Ages.

Enlightened Despotism
In the Enlightenment of the late 18th century, Empress Maria Theresa and her son Joseph II granted freedom of worship and movement to the peasantry in 1781, a precondition for the industrial revolution of the next century which made Bohemia the most developed economy within the Empire. Bohemia and Moravia each became independent parts of the Habsburg Monarchy although, conversely, the reforms led to greater Germanization and centralization of power, threatening the Slavic identities of the Empire's subjects.

The revival of the Czech nation, which began as a cultural movement, soon progressed into a struggle for political emancipation. Calls for the promotion of the Czech language encouraged Czech nationalists to campaign for the formation of a new Czechoslovakia. It was a concept fostered by Tomáš Garrigue Masaryk, a philosophy professor at the University of Prague and Professor for Slavonic Studies at King's College, London, who was to become the first president of the Czechoslovak Republic.

Male suffrage was granted in 1906 but the chamber of deputies was constantly bypassed by the emperor. The First World War brought estrangement between the Czechs and the Germans, the latter supporting the war effort, the former seeing it as a clash of monarchy versus democracy. Masaryk went into exile in London, where he committed himself to enlisting the support of Britain, France, Russia and the United States in founding a post-war independent Czechoslovak state. He worked with other exiled Czechs and Slovaks, including Dr Edvard Beneš, who later also became president of Czechoslovakia. In 1916 a Czechoslovak National Council was set up in Paris under Masaryk's chairmanship.

In 1918 Masaryk secured the support of US president Woodrow Wilson for Czech and Slovak unity and in May the Pittsburgh agreement was signed in the USA by exiles of both lands. On 18 Oct. 1918 the National Council transformed itself into a provisional government, and was recognized by the Allies.

Creation of the State
Austria accepted President Wilson's terms on 27 Oct. 1918, and the next day a republic was proclaimed with Masaryk, almost 70, as president, and Beneš as foreign minister. In drawing up the frontiers of the new state the principles of Wilsonian self-determination were defeated by the ethnic mix; other criteria employed were the partial restoration of the historic provinces and the need to establish an economically viable and defensible state. Among the minorities were 3·25m. Sudeten Germans. Borders were confirmed in the Treaty of Versailles in June 1919 (although Hungary subsequently called for these to be reformed) along with the official recognition of the state of Czechoslovakia by the international community.

The constitution of 1920 provided for a two-chamber parliament with adult suffrage. The French Constitution and the American Declaration of Rights were both used as models. The electoral system worked so that all governments were coalitions. Slovakia was granted an assembly in 1927 but the state was basically centralist and the Slovaks maintained their own parties. Between the wars Czechoslovakia became one of the ten most developed and stable countries of the world, although it suffered severe economic depression in the 1930s.

In Nov. 1935 Masaryk was succeeded by Beneš. Meanwhile in Germany, Hitler's designs on expanding the Third Reich stirred up nationalist agitation among the Sudeten Germans. The 1930 census showed 22·3% of the Czechoslovak population were ethnic Germans. In 1933 the German National Socialist Worker's Party in Czechoslovakia, directly affiliated to the Nazi party in Germany, was banned. The Sudeten German Party, led by Konrad Henlein, was formed in its place and won 67% of the German vote in the parliamentary elections of 1935. Czechoslovakia had relied on its 1925 pact with France to defend it against the threat of German aggression, but in the Munich Conference of 29 Sept. 1938 France sided with Britain and Italy in stipulating that all districts with a German population of more than 50% should be ceded to Germany. This legitimized the annexing of the Czech Sudetenland to Germany.

On 5 Oct. 1938 Beneš resigned and went into exile in Britain, and on 15 March 1939 Hitler's troops invaded Prague, contravening the Munich agreement. Slovakia, though allied to the Germans, declared independence under the fascist leadership of Jozef Tiso. The Czech territory became the German Protectorate of Bohemia-Moravia.

Czechoslovakia suffered further territorial losses to Poland and Hungary under the Vienna Arbitration of 2 Nov. 1938. In all, it lost 30% of its territory and almost 34% of its population. Over 1m. Czechs, Slovaks and Ukrainians came under German, Polish and Hungarian rule. Hitler's declared aim was to drive the Czechs out of Central Europe. The Protectorate became a centre for arms production.

Nazi Occupation
Initial attempts at resistance were brutally crushed. On 17 Nov. 1939 German occupiers closed down all Czech universities, executed nine of the leaders of the student movement and transported scores of other students to concentration camps. In 1940 the Gestapo transformed the town of Terezín (Theresienstadt) near Prague into a concentration camp, evacuating the pre-war population to accommodate 140,000 Jews from all parts of the Reich, the majority from the Protectorate of Bohemia-Moravia. 85,000 were then transported to death camps in the East, chiefly to Auschwitz, and over 30,000 prisoners were held in the fortress. Many Communists and Czech resistance fighters also met their deaths there.

Mass expulsions were a regular feature of the Protectorate. In 1942, 30,000 people were forced to leave their homes in order to make way for the military. In another case around 5,000 families were expelled from around 30 villages in Moravia in an attempt to create an ethnic German enclave. Growing resistance led to the appointment of SS General Reinhard Heydrich, head of the Reich's Security Office, who launched a savage offensive against underground organizations and executed general Alois Elias, head of the Protectorate's government, for his connections with the Beneš government in Britain. The latter, along with the Allies, were in turn behind the assassination of Heydrich, carried out on 27 May 1942 by two paratroopers. The Nazis responded with a frenzy of terror known as the 'Heydrichiade'. Revenge murders of the entire populations of Lidice (on 10 June 1942) and Lezáky (24 June 1942) were carried out.

The Beneš government in exile in London, with Jan Šrámek as prime minister, was officially recognized by Britain and the USSR on 18 June 1941. A 20-year treaty of alliance with the USSR was signed on 12 Dec. 1943 and, in March 1945, Beneš went to Moscow to prepare for post-war government in the wake of the Soviet Army advance. Liberation by the Soviet Army and US forces was completed in 1945 following an insurrection on 5 May. Territories taken by Germans, Poles and Hungarians were restored to Czechoslovakia. Subcarpathian Ruthenia (now in Ukraine) was transferred to the USSR, creating a common border with the Soviet state.

The Sudeten Germans suffered brutal expulsions, during which up to 250,000 died, 6,000 of whom were murdered. The 'resettlement' of ethnic Germans was officially approved at the Potsdam Conference on 1 Aug. 1945, when the USA and Britain insisted on humane transfer, subsequently supervised by the Allies and the Red Cross. German or Hungarian Czechs had their citizenship taken away unless they were naturalized Czechs or Slovaks. In order to stay, German or Hungarian Czechs had to prove that they had remained faithful to the Czechoslovak Republic, had fought in the resistance or had personally suffered at the hands of fascists. Further decrees expropriated property and agricultural land. A total of 2,700,000 Germans were expelled, and in the Czech census of March 1991 only 47,000 claimed German as their nationality. A Czech–German Declaration of 21 Jan. 1997 saw both sides admit to and apologize for their atrocities during the period.

Soviet Domination

Beneš once again became president of Czechoslovakia but, under pressure from the USSR, measures were taken to confiscate and redistribute property and to take key industries into public ownership. Before 1939 the Communist party of Czechoslovakia (CPCz) had never gained more than 13% of the vote, but its patriotic stance in the late 1930s as well as its clear affiliation with the country's main liberator, the USSR, led to a sharp increase in popularity. In the elections of 26 May 1946 the Communists won almost 40% of the vote in Czech areas and 30% in Slovakia. This made the party the largest group in the new Constituent National Assembly, with 114 of the 300 seats and Klement Gottwald as premier. Its leaders pledged commitment to democratic traditions while pursuing a 'specific Czechoslovak road to socialism'.

The party's independence was first put to the test when Czechoslovakia was pressured by Stalin into withdrawing from the American Marshall Plan for European economic reconstruction, seen from Moscow as a threat to its own influence. Stalin's encouragement of the CPCz grew in the autumn of 1947 when a people's militia was formed and non-communist parties in the governing coalition found their influence eroded. In protest, 12 non-communist ministers handed in their resignations. The CPCz adroitly handled this situation to its own advantage, forcing President Beneš to appoint a predominantly communist government on 25 Feb. 1948.

A new constitution, declaring Czechoslovakia a 'people's democracy' was approved on 9 May 1948, and elections were held on 30 May with a single list of candidates, resulting in an 89% majority for the government. Beneš resigned on 2 June after refusing to ratify the Communist Constitution. 12 days later Gottwald succeeded him as President. Civil rights were severely restricted; from 1950 monasteries and convents were nationalized, with 219 monasteries being taken over by the People's Militia on the night of 13 April alone; monastic orders were abolished.

The secret service became one of the most systematic and omnipresent in the Soviet bloc. Unsubstantiated charges of treason and resistance to the communist cause led to a series of show trials and executions, with many victims from the CPCz itself. It is estimated that between 200,000 and 280,000 suffered death or persecution during the Stalin era. Stalin died on 5 March 1953, and Gottwald just a week later, but the communists maintained their grip. Workers' demonstrations in Plzeň and other Czech towns against price rises and currency reform, which devalued savings, were brutally suppressed by the military in June that year. Political trials of 'Slovak nationalists' in 1954 had many Slovak Communists imprisoned, including Gustáv Husák, who later went on to become secretary-general of the CPCz and president of Czechoslovakia.

The Soviet five-year economic plans emphasized engineering, arms production and heavy industry. A third of Czechoslovak output came from the arms industry. The service and consumer goods industries were virtually abolished and all farms collectivized. The founding of COMECON on 1 Jan. 1949 and the signing of the Warsaw Pact in 1955 limited Czechoslovakia's trading partners to the Eastern bloc.

A new constitution, introduced in 1960, reinforced the power of the Communist Party and changed the country's name to the Socialist Republic of Czechoslovakia (ČSSR).

Prague Spring

The failure of the third five-year plan to meet its targets gave a push to economic reform. A mixed economy was introduced which led to a period of cultural liberalization. Support for political reform grew within party ranks. Antonín Novotný was persuaded to resign as CPCz leader on 5 Jan. 1968. He was replaced by Alexander Dubček, leader of the Communist Party of Slovakia. On 22 March, Novotný resigned as president in favour of Gen. Ludvík Svoboda, who was elected by the National Assembly. Precipitating what became known as the Prague Spring, Dubček introduced many reforms in his pursuit of 'socialism with a human face'. New political bodies were formed, breaking up the monopoly of the Communist Party. Press censorship was abolished and restraints on freedom of expression relaxed.

The reforms caused consternation in Moscow. Brezhnev unsuccessfully put pressure on Czechoslovakia between May and Aug. He then ordered the invasion of Czechoslovakia by troops of the Warsaw Pact countries (with the exception of Romania) on 20–21 Aug. The Soviet intention of replacing Dubček and his government with more hard-line Communists failed when President Svoboda turned down Soviet nominees. Dubček and other party leaders were then abducted to Moscow and forced to sign an agreement to keep Soviet troops stationed in Czechoslovakia. The Prague Spring all but withered. The only surviving measure of reform was the introduction of the federal system on 1 Jan. 1969. Separate Czech and Slovak states came into force within a Czechoslovak federation, a move that satisfied the Slovaks who had been seeking autonomy for some time. Each state was awarded its own administration and a national council, and the National Assembly divided into two chambers. All other reforms of the Dubček administration were stamped out or reversed and the Soviet policy of 'normalization' took hold.

In protest at the Soviet repression, student Jan Palach set fire to himself in Wenceslas Square on 16 Jan. 1969. On 28 March when the Czechoslovak national ice hockey team beat the Soviets, celebrations turned into anti-Soviet demonstrations and led to many arrests. On 17 April Dubček was forced from office and replaced with the hard-liner, Gustáv Husák. Repression led to mass emigrations. So-called 'enemies of the state' were put under constant surveillance, blacklisted for jobs and their children denied university places. Freedom to travel abroad was no longer granted to ordinary citizens. Protest resurfaced again in 1977, with the signing of Charter 77 and founding of the Committee for the Defence of the Rights of the Unjustly Persecuted (VONS). These two organizations, led by artists and intellectuals, alerted the public to civil rights abuses and campaigned for civil and political rights.

Velvet Revolution

The advent of Mikhail Gorbachev's *glasnost* and *perestroika* in the mid-1980s initially changed little. Husák was replaced as leader of the CPCz by Miloš Jakeš, who had been responsible for Party purges in 1970. But there were renewed demonstrations in Aug. 1988 on the 20th anniversary of the Soviet invasion, in Oct. on the 70th anniversary of the founding of Czechoslovakia and in Jan. 1989 on the 20th anniversary of Jan Palach's suicide when the crowd was brutally dispersed and leading dissidents, including Václav Havel, arrested and imprisoned. Protests, petitions and further demonstrations followed in May, Aug. and Oct. of 1989. The fall of the Berlin Wall on 9 Nov. 1989 galvanized the pro-democracy movement. Another major demonstration by students on 17 Nov. led to further protests until the entire CPCz leadership resigned on 24 Nov. Civic Forum (OF) in the Czech Republic and the Public Against Violence (VPN) in Slovakia were formed to co-ordinate all pro-democratic forces, and an interim broad coalition 'Government of National Understanding' with a minority of Communists took over. This became known as the Velvet Revolution. Václav Havel was unanimously elected president of Czechoslovakia by the Federal Assembly on 29 Dec. 1989.

The first free elections since the Second World War were held in June 1990, with a turnout of 96·4%. The Communists were roundly defeated, and Civic Forum won 52% of the votes. Second parliamentary elections were fixed for mid-1992, by which time

Civic Forum (OF) and VPN had been replaced by fully-fledged parties across the political spectrum. In the Czech lands, the right wing emerged as the strongest element, with the Civic Democratic party (ODS) the largest coalition party. Its leader, Václav Klaus, who was also finance minister, became prime minister and stayed in the post until Nov. 1997.

By contrast, in Slovakia, Vladimír Mečiar's Democratic Slovakia party and other post-communists were successful in the elections. A continuing Slovak desire for independence from Prague, along with differences in opinion on economic policy and the role of the state, strained relations between the two federal partners. An agreement to a 'velvet divorce' was reached with the two devolving into separate sovereign states from 1 Jan. 1993. Economic property was divided in accordance with a federal law of 13 Nov. 1992 and real estate became the property of the republic in which it was located. Other property was divided by specially-constituted commissions in the proportion of 2 (Czech Republic) to 1 (Slovakia) on the basis of population. Military material was also divided on the 2:1 principle, and regular military personnel were invited to choose in which army they would serve.

The Czech Republic showed the greater eagerness to become westernized. Although many feared a precipitous move towards a market economy, Klaus argued that danger lay in delaying reform, and that the creation of a market economy would lead the 'return to Europe' and the opening up of new markets.

Klaus immediately embarked on a series of radical reforms, impressing Western investors with his Thatcherite rhetoric, policies and publications. A programme of mass privatization was implemented, at first successfully, and the early to mid-nineties were a time of economic boom, with the Czech Republic a leading contender for Western trade and investment. However, many of the tough policies were not fully implemented and an increasing number of financial scandals were associated with the Klaus administration. 1997 also saw a currency crisis, where the crown devalued 12% against the dollar. Klaus was forced to resign as prime minister on 18 Nov. 1997.

Josef Tošovský formed a caretaker government until 1998 when Miloš Zeman formed a minority Social Democratic government, the country's first left-wing government since the fall of socialism. The Czech Republic joined NATO in 1999 and in 2001 Vladimír Špidla succeeded Zeman as party leader and prime minister. The Czech Republic became a member of the EU on 1 May 2004.

Following an inconclusive election in June 2006 there were seven months of political deadlock when the country was without a government. A new centre-right coalition was formed under Mirek Topolánek in Jan. 2007 but collapsed two years later. An interim government took office until July 2010 when Petr Nečas formed a coalition. Austerity measures prompted mass protests. In Dec. 2011 Václav Havel died and was given a state funeral.

TERRITORY AND POPULATION

The Czech Republic is bounded in the west by Germany, north by Poland, east by Slovakia and south by Austria. Minor exchanges of territory to straighten their mutual border were agreed between the Czech Republic and Slovakia on 4 Jan. 1996, but the Czech parliament refused to ratify them on 24 April 1996. Its area is 78,867 sq. km (30,451 sq. miles), including 1,620 sq. km (625 sq. miles) of inland waters. The population at the 2011 census was 10,436,560; density, 135·1 per sq. km. In 2011, 73·6% of the population lived in urban areas.

The UN gives a projected population for 2015 of 10·63m.

There are 14 administrative regions (Kraj), one of which is the capital, Prague (Praha).

Region	Chief city	Area in sq. km	2011 census population
Jihočeský	České Budějovice	10,057	628,336
Jihomoravský	Brno	7,196	1,163,508
Karlovarský	Karlovy Vary	3,315	295,595

Region	Chief city	Area in sq. km	2011 census population
Královéhradecký	Hradec Králové	4,758	547,916
Liberecký	Liberec	3,163	432,439
Moravskoslezský	Ostrava	5,427	1,205,834
Olomoucký	Olomouc	5,267	628,427
Pardubický	Pardubice	4,519	511,627
Plzeňský	Pilsen (Plzeň)	7,561	570,401
Prague (Praha)	—	496	1,268,796
Středočeský	Prague (Praha)	11,015	1,289,211
Ústecký	Ústí nad Labem	5,335	808,961
Vysočina	Jihlava	6,796	505,565
Zlínský	Zlín	3,964	579,944

The census population of the principal towns in 2011 (in 1,000):

Prague (Praha)	1,269	Olomouc	101	Havířov	77
Brno	386	České Budějovice	94	Zlín	75
Ostrava	296	Hradec Králové	94	Kladno	68
Pilsen (Plzeň)	170	Ústí nad Labem	93	Most	65
Liberec	103	Pardubice	91	Opava	58

At the 2001 census 90·4% of the population was Czech, 3·7% Moravian and 1·9% Slovak. There were also (in 1,000): Poles, 52; Germans, 39; Roma (Gypsies), 12; Silesians, 11. Although only 12,000 people described themselves as being Roma in the 2001 census they are in reality estimated to number up to 300,000.

The official language is Czech.

SOCIAL STATISTICS

2009 births, 118,667; deaths, 107,421; marriages, 47,862; divorces, 29,133. Rates (per 1,000 population), 2009: birth, 11·3; death, 10·0; marriage, 4·6; divorce, 2·8. Life expectancy at birth, 2007, 73·2 years for males and 79·4 years for females. In 2009 the most popular age range for marrying was 30–34 for males and 25–29 for females. Annual population growth rate, 2005–10, 0·5%. Infant mortality, 2010, three per 1,000 live births; fertility rate, 2008, 1·4 children per woman.

CLIMATE

A humid continental climate, with warm summers and cold winters. Precipitation is generally greater in summer, with thunderstorms. Autumn, with dry clear weather, and spring, which is damp, are each of short duration. Prague, Jan. 29·5°F (−1·5°C), July 67°F (19·4°C). Annual rainfall 19·3" (483 mm). Brno, Jan. 31°F (−0·6°C), July 67°F (19·4°C). Annual rainfall 21" (525 mm).

CONSTITUTION AND GOVERNMENT

The constitution of 1 Jan. 1993 provides for a parliament comprising a 200-member *Chamber of Deputies*, elected for four-year terms by proportional representation, and an 81-member *Senate* elected for six-year terms in single-member districts, 27 senators being elected every two years. The main function of the Senate is to scrutinize proposed legislation. Senators must be at least 40 years of age, and are elected on a first-past-the-post basis, with a run-off in constituencies where no candidate wins more than half the votes cast. For the House of Representatives there is a 5% threshold; votes for parties failing to surmount this are redistributed on the basis of results in each of the eight electoral districts.

There is a *Constitutional Court* at Brno, whose 15 members are nominated by the President and approved by the Senate for ten-year terms.

Following a constitutional amendment that took effect in Oct. 2012, the *President* of the Republic is directly elected for a five-year term. Candidates standing for office must be 40 years of age. In the event of no candidate winning an absolute majority, a second round is held between the two most successful candidates.

A president may not serve more than two consecutive five-year terms.

National Anthem

'Kde domov můj?' ('Where is my homeland?'); words by J. K. Tyl, tune by F. J. Škroup.

GOVERNMENT CHRONOLOGY

(ČSSD = Czech Social Democratic Party; ODS = Civic Democratic Party; SPOZ = Party of Civic Rights–Zemanovci; n/p = non-partisan)

Presidents since 1993.

1993–2003	n/p	Václav Havel
2003–13	ODS	Václav Klaus
2013–	SPOZ	Miloš Zeman

Prime Ministers since 1993.

1993–97	ODS	Václav Klaus
1997–98	n/p	Josef Tošovský
1998–2002	ČSSD	Miloš Zeman
2002–04	ČSSD	Vladimír Špidla
2004–05	ČSSD	Stanislav Gross
2005–06	ČSSD	Jiří Paroubek
2006–09	ODS	Mirek Topolánek
2009–10	n/p	Jan Fischer
2010–	ODS	Petr Nečas

RECENT ELECTIONS

In 2012 legislation was passed changing the presidential election from indirect to direct. The first round of direct elections took place on 11–12 Jan. 2013. Incumbent president Václav Klaus was barred from seeking a third term. None of the nine candidates secured enough votes in the first round and a run-off between Miloš Zeman (Party of Civic Rights–Zemanovci) and Karel Schwarzenberg (Tradition Responsibility Prosperity 09/TOP 09) was held on 25–26 Jan. 2013. Zeman was elected president with 54·8% of the vote to Schwarzenberg's 45·2%. Turnout was 61·3% in the first round and 59·1% in the second.

Elections to the National Assembly were held on 28 and 29 May 2010; turnout was 62·6%. The Czech Social Democratic Party (ČSSD) gained 56 seats with 22·1% of votes cast; Civic Democratic Party (ODS) gained 53 with 20·2%; TOP 09, 41 with 16·7%; the Communist Party of Bohemia and Moravia (KSČM), 26 with 11·3%; Public Affairs (VV), 24 with 10·9%.

Elections for a third of the seats in the Senate were held on 12, 13, 19 and 20 Oct. 2012. As a result ČSSD had 48 seats in the Senate; ODS 15; the TOP 09–Mayors and Independents coalition 4; the Christian and Democratic Union–Czechoslovak People's Party coalition (KDU–ČSL) 4. Independents and other parties held three seats or fewer.

European Parliament.

The Czech Republic has 22 (24 in 2004) representatives. At the June 2009 elections turnout was 28·2% (28·3% in 2004). The ODS won 9 seats with 31·5% of votes cast (political affiliation in European Parliament: European Conservatives and Reformists); ČSSD 7 with 22·4% (Progressive Alliance of Socialists and Democrats); KSČM, 4 with 14·2% (European United Left/Nordic Green Left); KDU–ČSL, 2 with 7·6% (European People's Party).

CURRENT GOVERNMENT

President: Miloš Zeman; b. 1944 (SPOZ; sworn in 8 March 2013).
    Prime Minister: Petr Nečas; b. 1964 (ODS; sworn in 13 July 2010).

In March 2013 the government comprised:
    Deputy Prime Minister and Minister of Foreign Affairs: Karel Schwarzenberg. Deputy Prime Minister: Karolína Peake.
    Minister of Agriculture: Petr Bendl. Culture: Alena Hanáková.
    Defence: Vlastimil Picek. Education, Youth and Sports: Petr Fiala.
    Environment: Tomáš Chalupa. Finance: Miroslav Kalousek.

Health: Leoš Heger. Industry and Trade: Martin Kuba. Interior: Jan Kubice. Justice: Pavel Blažek. Labour and Social Affairs: Ludmila Müllerová. Local Development: Kamil Jankovský. Transport: Zbynek Stanjura. Minister without Portfolio: Petr Mlsna.

Government Website: http://www.vlada.cz

CURRENT LEADERS

Miloš Zeman

Introduction

Miloš Zeman, leader of the Party of Civic Rights–Zemanovci (SPOZ), was sworn in as president on 8 March 2013. He served as prime minister from 1998–2002, representing the centre-left Czech Social Democratic Party (ČSSD).

Early Life

Miloš Zeman was born on 28 Sept. 1944 in Kolin, a town 55 km east of Prague. Having completed his schooling he was refused entry to university because of his political beliefs, but graduated from the Prague University of Economics in 1969 after undertaking external studies. After his attempts to join the Social Democratic Party were frustrated when the party was banned, he joined the Czechoslovakian Communist Party. This was in 1968 at the zenith of Dubček's Prague Spring, when the country experienced rapid liberal reform. After the Soviets crushed the movement, Zeman was expelled from the party for his opposition to Soviet 'normalization'.

After working in several sports and fitness organizations, Zeman established a forecasting centre, which was closed in 1984 by the communist regime. He then worked for an agricultural organization specializing in social systems forecasting but was dismissed from his post in 1989 after writing an article critical of the authorities.

Joining the Civic Forum, an alliance of anti-communist reformist groups, Zeman became a leading voice for the centre-left. Following the success of the Velvet Revolution and the over-throw of the communist regime in 1989, he entered parliament and became chairman of the budget and planning committee in 1990. After the Civic Forum split, he joined ČSSD in 1992 and won re-election to parliament. In 1993 he was elected party chairman, a position he retained in 1995 and 1997. Following the general elections of 1996 he was elected speaker of the parliament.

In 1997 the administration of Václav Klaus collapsed. After elections the following year in which the ČSSD emerged victorious, Zeman replaced him as prime minister. Over his four-year term Zeman reduced the pace of market reforms and privatization and advocated increased public spending. Leading a minority government, he secured an alliance with Klaus' Civic Democrats, a move that led to widespread criticism. Popular discontent resulted in demonstrations in Prague in late 1999 calling for the resignations of Zeman and Klaus to make room for a new generation of politicians. Zeman's relationship with the Czech media was strained and in 2000 national television journalists went on strike in protest at what they claimed was the politically-motivated appointment of a new director general.

On the international scene, Zeman oversaw the Czech Republic's entry into NATO in 1999. After relations with Austria became strained over a nuclear power plant dispute, Zeman resigned both his party leadership and the premiership in 2001. In 2003 he ran unsuccessfully for president and went into semi-retirement. In 2007 he left the ČSSD after a number of hostile public exchanges and established SPOZ. Despite SPOZ's shaky performance at its first elections in 2010, the party endorsed Zeman as leader and in 2012 he announced his candidature for the presidency.

Career in Office

On 8 March 2013 Zeman took office as the first president to be directly elected by universal suffrage in Czech history. He ran on a social-democratic pro-Europe platform in a heated campaign against Karel Schwarzenberg, winning support from poorer and older voters. He faced a fraught post-election political atmosphere as he sought to establish a working relationship with a politically opposed right-of-centre coalition government led by Petr Nečas. Previously Zeman had questioned the legitimacy of the coalition and continues to refuse to acknowledge LIDEM–Liberal Democrats, one of the participating parties. The president's primary role is to represent the Czech Republic abroad and make appointments to the constitutional court and central bank. Zeman champions European integration and has pledged to fight corruption.

Petr Nečas

Position
Prime Minister

Introduction
Petr Nečas, leader of the Civic Democratic Party, was sworn in as prime minister on 13 July 2010 at the head of a coalition government. He had previously served as deputy minister for defence and deputy prime minister.

Early Life
Nečas was born in Nov. 1964 in Uherské Hradiště in the southeast of the country, near the current Slovak border. From 1983 he studied physics at the University of Brno, undertaking post-graduate studies in natural sciences. In 1988 he became an engineering researcher for Tesla Rožnov, an electronics manufacturer.

In 1992 he joined Václav Klaus' recently formed Civic Democratic Party (ODS), serving on committees for defence, intelligence and foreign/EU relations. His first government post came in 1995 when Klaus appointed him deputy minister of defence. In 2006, after a series of electoral defeats for the party, Mirek Topolánek defeated Nečas in the race to succeed Klaus as ODS leader. Topolánek formed a minority government in which Nečas served as his deputy prime minister.

Topolánek's government collapsed after a confidence vote in 2009. He resigned as ODS chairman in 2010 and Nečas was elected in his place. Despite the ODS suffering a 15% drop in support at the May 2010 elections, Nečas was able to negotiate a centre-right conservative coalition with the Public Affairs party (VV) and TOP 09 to form a government in July 2010.

Career in Office
Having campaigned on a platform of reforming social benefit and fighting corruption, Nečas aimed for fiscal consolidation to reduce the public deficit and fulfil the criteria necessary to join the eurozone. He undertook a programme of austerity measures, including cutting public salaries by 10% and reducing expenditure on the health sector and pensions, and sought to save money by streamlining defence procurement procedures. In Dec. 2010 he also joined in criticism, led by the UK, of EU plans to raise its own budget. The coalition has since become more fractious, undermined by a corruption scandal that led to a split in Public Affairs (VV) and by popular discontent over the austerity measures. Nevertheless, Nečas' government narrowly survived a parliamentary no-confidence vote in April 2012 and formed a new coalition with TOP 09 and LIDEM–Liberal Democrats, which was founded in May 2012 by former members of VV.

DEFENCE

Conscription ended in Dec. 2004 when the armed forces became all-volunteer. Defence expenditure in 2008 totalled US$3,165m. (US$310 per capita), representing 1·5% of GDP.

Army
Strength (2007) 16,960. There are also paramilitary Border Guards (3,000-strong) and Internal Security Forces (100).

Air Force
The Air Force has a strength of 6,130 including Air Defence Forces. There were 50 combat-capable aircraft in 2007 (L-159s and JAS 39s) and 38 attack helicopters.

INTERNATIONAL RELATIONS

In 1974 the Federal Republic of Germany and Czechoslovakia annulled the Munich agreement of 1938. On 14 Feb. 1997 the Czech parliament ratified a declaration of German–Czech reconciliation, with particular reference to the Sudeten German problems.

The Czech Republic became a member of the EU on 1 May 2004. A referendum held on 13–14 June 2003 approved accession, with 77·3% of votes cast for membership and 22·7% against. In 2000 a visa requirement for Russians entering the country was introduced as one of the conditions for EU membership.

The Czech Republic's Senate approved the European Union's Treaty of Lisbon on 6 May 2009, after the Chamber of Deputies had done so on 18 Feb. 2009. However, President Klaus stated that he would not sign it until Ireland had ratified it (which happened after a second referendum held on 2 Oct. 2009). Klaus subsequently gave his presidential assent on 3 Nov. 2009, making the Czech Republic the last country to ratify it.

ECONOMY

Agriculture accounted for 2% of GDP in 2008, industry 38% and services 60%.

Overview
Until 1996 the Czech Republic was viewed as the most successful European transition economy. Industrial production and employment declined significantly after communism but job losses were contained by rising service sector employment and soft loans given to loss-making enterprises by state-banks. After four years of 3% average annual growth, the economy fell into recession in 1997–98. In May 1997 large current account deficits fuelled a speculative attack on the *koruna*, forcing the country to adopt a tight monetary policy and fiscal austerity. Foreign direct investment began flowing from the West in 1999 and solid growth resumed in the 2000s without high inflation.

With strength in engineering, low labour costs, good infrastructure and a favourable geographical position, the diversified industrial export sector has been a key engine of growth. Integration with the EU and large-scale FDI further boosted export-oriented growth prior to the global financial crisis in 2008. The economy was severely hit by the crisis and, in particular, the downturn in Germany, the country's largest trading partner. Exports and investment declined, unemployment rose and the trade deficit widened while debt increased.

Exports drove post-recession expansion but the recovery stalled in 2011 as exports lost momentum and domestic demand remained stagnant. Nonetheless, fiscal consolidation measures have improved overall deficit levels and public debt, at 41·5% of GDP, reflects the economy's strengthening fiscal position. However, a slowing of the EU economy threatens to further depress exports, while the impact of fiscal consolidation suggests sluggish domestic demand over the short-term.

Currency
The unit of currency is the *koruna* (CZK) or crown of 100 *haler*, introduced on 8 Feb. 1993 at parity with the former Czechoslovakian koruna. Gold reserves were 415,000 troy oz in Sept. 2009 and foreign exchange reserves were US$39,101m. Total money supply was Kč. 1,736·0bn. in Aug. 2009.

Inflation rates (based on OECD statistics):

2002	2003	2004	2005	2006	2007	2008	2009	2010	2011
1·8%	0·1%	2·8%	1·9%	2·6%	3·0%	6·3%	1·0%	1·5%	1·9%

The koruna became convertible on 1 Oct. 1995. In May 1997 the koruna was devalued 10% and allowed to float.

Budget

Budgetary central government revenue in 2008 totalled Kč. 1,024·16bn. (Kč. 974·33bn. in 2007) and expenditure Kč. 1,083·83bn. (Kč. 1,019·36bn. in 2007).

Principal sources of revenue in 2008 were: social security contributions, Kč. 368·12bn.; taxes on goods and services, Kč. 299·81bn.; taxes on income, profits and capital gains, Kč. 222·13bn. Main items of expenditure by economic type in 2008: social benefits, Kč. 400·92bn.; grants, Kč. 304·95bn.; compensation of employees, Kč. 100·80bn.

The Czech Republic's budget deficit in 2011 was 3·1% of GDP (2010, 4·8%; 2009, 5·8%). The required target set by the EU is a budget deficit of no more than 3%.

VAT is 21% (reduced rate, 15%).

Performance

Real GDP growth rates (based on OECD statistics):

2002	2003	2004	2005	2006	2007	2008	2009	2010	2011
2·1%	3·8%	4·7%	6·8%	7·0%	5·7%	3·1%	−4·5%	2·5%	1·9%

Total GDP was US$217·0bn. in 2011.

Banking and Finance

The central bank and bank of issue is the Czech National Bank (*Governor*, Miroslav Singer), which also acts as banking supervisor and regulator. Decentralization of the banking system began in 1991, and private banks began to operate. The only legal form of domestically operating banks are joint stock companies and branches of foreign banks. The Commercial Bank and Investment Bank are privatized nationwide networks with a significant government holding. Specialized banks include the Czech Savings Bank and the Czech Commercial Bank (for foreign trade payments). Private banks tend to be on a regional basis, many of them agricultural banks. In 1997 the cabinet agreed to sell off large stakes in three of the largest state-held banks to individual foreign investors through tenders, in preparation for European Union entry. In 2000 the country's fourth largest bank, Československá obchodní banka (ČSOB), acquired the operations of the third largest bank, IPB (Investiční a Poštovní banka). The newly-formed institution (which retained the name ČSOB) is the largest bank in central and eastern Europe, with assets of US$50·2bn. in Sept. 2010. Other major banks are Česká Spořitelna (assets of US$49·3bn. in Sept. 2010) and Komerční banka (assets of US$38·5bn. in Sept. 2010). Other capital market participants are subject to the supervision of the Czech Securities Commission. Savings deposits were Kč. 2,698,236m. in 2007.

In June 2012 gross external debt amounted to US$93,069m.

Foreign direct investment was US$6·8bn. in 2010, up from US$2·9bn. in 2009.

A stock exchange was founded in Prague in 1992.

ENERGY AND NATURAL RESOURCES

In 2008, 7·2% of energy consumption came from renewables (wind power, solar power, hydro-electric power, tidal power, geothermal energy and biomass), compared to the European Union average of 10·3%. A target of 13% has been set by the EU for 2020.

Environment

The Czech Republic's carbon dioxide emissions from the consumption and flaring of fossil fuels in 2008 were the equivalent of 9·9 tonnes per capita.

Electricity

Installed capacity was 17·6m. kW in 2007. Production in 2007 was 88·2bn. kWh. 67% of electricity was produced by thermal power stations (mainly using brown coal) and 30% was nuclear. In 2010 there were six nuclear reactors in operation. Consumption per capita in 2007 was 6,940 kWh.

Oil and Gas

Natural gas reserves in 2007 totalled 4·0bn. cu. metres. Production in the same year was 170m. cu. metres. In 2007 crude petroleum reserves were 15m. bbls; production was 330,000 tonnes that year.

Minerals

There are hard coal and lignite reserves (chief fields: Most, Chomutov, Kladno, Ostrava and Sokolov). Lignite production in 2007 was 55·0m. tonnes; hard coal production in 2007 was 7·7m. tonnes.

Agriculture

In 2002 there were 4,273,000 ha. of agricultural land. In 2002 there were 3·07m. ha. of arable land and 0·97m. ha. of permanent crops. Approximately 19,900 ha. were irrigated in 2007. Agriculture employs just 4·9% of the workforce—the smallest proportion of any of the ex-Communist countries in eastern Europe. A law of May 1991 returned land seized by the Communist regime to its original owners, to a maximum of 150 ha. of arable to a single owner. Main agricultural production figures, 2002 (1,000 tonnes): wheat, 3,867; sugar beets, 3,833; barley, 1,793; potatoes, 901; rapeseed, 710; apples, 339; maize, 304; rye, 119. Livestock, 2003: cattle, 1·47m.; pigs, 3·36m.; sheep, 103,000; poultry, 27m. In 2002 production of meat was 787,000 tonnes; cheese, 146,000 tonnes; milk, 2,728m. litres; 1,829m. eggs.

Forestry

In 2010 forests covered 2·66m. ha., or 34% of the total land area. Timber production in 2007 was 18·51m. cu. metres.

Fisheries

Freshwater aquaculture (particularly carp) is an integral part of the Czech agriculture sector; production in 2010 totalled 20,420 tonnes (84% of total fish production). Fish landings in 2010 amounted to 3,990 tonnes, entirely from inland waters.

INDUSTRY

The leading companies in the Czech Republic in March 2012 were ČEZ (České Energetické Závody a.s.), with a market capitalization of US$23·0bn.; Komerční banka (US$7·5bn.); and Telefónica O2 Czech Republic (US$6·7bn.).

In 2008 there were 1,747,020 small private businesses (of which 17,448 were incorporated), 311,309 companies and partnerships (of which 22,700 were joint-stock companies), 15,338 co-operatives and 526 state enterprises. Output (2008 unless otherwise indicated) includes: crude steel, 6·4m. tonnes; cement (2007), 4·9m. tonnes; pig iron, 4·7m. tonnes; 940,300 cars; soft drinks, 2,935·4m. litres; beer, 1,904·4m. litres.

Labour

In the fourth quarter of 2010 the economically active population numbered 5,281,800; 1·28m. persons worked in manufacturing; 589,400 in trade; 445,500 in construction; 338,300 in human health and social work activities; and 327,500 in public administration and defence. In Dec. 2012 the unemployment rate was 7·5% (up from 6·7% in 2011 as a whole). The average monthly gross wage was Kč. 23,004 in 2010. Pay increases are regulated in firms where wages grow faster than production. Fines are levied if wages rise by more than 15% over four years. There is a minimum wage of Kč. 8,000 a month.

INTERNATIONAL TRADE

Imports and Exports

Trading with EU and EFTA countries has increased significantly while trading with all post-communist states has fallen.

Trade, 2010, in US$1m. (2009 in brackets): imports c.i.f., 125,691 (104,850); exports f.o.b., 132,141 (112,884). In 2010 main import sources (in US$1m.) were: Germany (32,057); China (15,332); Poland (8,041); Russia (6,813); Slovakia (6,505). Main export markets in 2010 (in US$1m.) were: Germany (42,213); Slovakia (11,596); Poland (8,129); France (7,101); UK (6,443).

Principal imports in 2010 (in US$1m.) were: machinery and transport equipment (53,253); manufactured goods (21,457); chemicals and related products (12,719). Principal exports in 2010 (in US$1m.) were: machinery and transport equipment (70,592); manufactured goods (22,238); miscellaneous manufactured articles (14,131).

COMMUNICATIONS

Roads

In 2007 there were 657 km of motorways, 6,191 km of highways and main roads, 48,736 km of secondary roads and 72,927 km of other roads, forming a total network of 128,511 km. Passenger cars in use in 2007 numbered 4,280,100 (414 per 1,000 inhabitants), and there were also 555,200 lorries and vans and 20,400 buses and coaches. Motorcycles and mopeds numbered 860,100. There were 832 deaths as a result of road accidents in 2009.

Rail

In 2008 Czech State Railways had a route length of 9,486 km (9,464 km on 1,435 mm gauge), of which 3,078 km were electrified. Passenger-km travelled in 2008 came to 6·76bn. and freight tonne-km to 14·36bn. There is a metro (44 km) and tram/light rail system (496 km) in Prague, and also tram/light rail networks in Brno, Liberec, Most, Olomouc, Ostrava and Plzeň.

Civil Aviation

There are international airports at Prague (Ruzyně), Ostrava (Mošnov) and Brno (Turany). The national carrier is Czech Airlines, 95·69% of which is owned by the state. In 2007 it flew 82·9m. km and carried 5,492,200 passengers (5,379,500 on international flights). In 2007 Prague handled 12,436,254 passengers; there were a total of 174,662 take-offs and landings.

Shipping

In 2009, 804,000 tonnes of freight were carried by inland waterways.

Telecommunications

In 2008 there were 2,264,000 main (fixed) telephone lines. In the same year mobile phone subscribers numbered 13,780,000 (1,335·4 per 1,000 persons). Český Telecom was sold to the Spanish telecommunications firm Telefónica in April 2005. It has since become Telefónica O2 Czech Republic. There were 6·0m. internet users in 2008. In March 2012 there were 3·5m. Facebook users.

SOCIAL INSTITUTIONS

Justice

The post-Communist judicial system was established in July 1991. This provides for a unified system of civil, criminal, commercial and administrative courts. Commercial courts arbitrate in disputes arising from business activities. Administrative courts examine the legality of the decisions of state institutions when appealed by citizens. In addition, there are military courts which operate under the jurisdiction of the Ministry of Defence. There is a Supreme Court, and a hierarchy of courts under the Ministry of Justice at republic, region and district level. District courts are courts of first instance. Cases are usually decided by senates comprising a judge and two associate judges, though occasionally by a single judge. (Associate judges are citizens in good standing over the age of 25

who are elected for four-year terms). Regional courts are courts of first instance in more serious cases and also courts of appeal for district courts. Cases are usually decided by a senate of two judges and three associate judges, although occasionally by a single judge. There is also a Supreme Administrative Court. The Supreme Court interprets law as a guide to other courts and functions also as a court of appeal. Decisions are made by senates of three judges. Judges are appointed for life by the National Council.

There is no death penalty. In 2008, 343,799 crimes were reported (33·0 per 1,000 inhabitants) of which 37·2% were solved. The population in penal institutions at 31 Dec. 2007 was 18,901.

Education

Elementary education up to age 15 is compulsory. 52% of children continue their education in vocational schools and 48% move on to secondary schools. In 2007 there were 462,820 children in primary schools with 24,713 teaching staff and 937,026 secondary school pupils with (2006) 91,622 teaching staff.

In 2009 there were 73 universities, of which 26 were public, 45 private and two state—the University of Defence and the Police Academy. Private universities have only existed since 1990. There were 389,231 students at the public and private universities in 2009.

In 2006 public expenditure on education came to 4·8% of GNI and 10·5% of total government spending.

The adult literacy rate is at least 99%.

Health

At 31 Dec. 2008 there were 192 hospitals with 63,263 beds. Another 54,472 beds were available in other health establishments. There were 34,437 physicians, 6,595 dentists, 77,956 nurses, 5,583 pharmacists and 3,842 midwives in 2007. In 2008 the Czech Republic spent 7·1% of its GDP on health.

Welfare

Since 1 Jan. 1996 the retirement age has been gradually increasing by two months per year for men and by four months per year for women. The retirement age, as of Sept. 2008, was 61 years 10 months (men) and 56 to 60 (women), according to the number of children. The target age was to be 63 by the end of 2013 for men and by the end of 2028 for childless women, and between 59 and 62 for women depending on the number of children. However, in Jan. 2010 the retirement age for anyone born after 1968 was increased to 65 for all men and between 62 and 65 for women depending on the number of children.

The old-age pension is calculated as a flat-rate basic amount of Kč. 1,310 plus an earnings-related percentage calculated on personal assessment and the number of years of insurance. As at 31 Dec. 2008 the average monthly old-age pension was Kč. 9,638. Early pensions are available up to three years before the standard date of retirement, with the claimant required to have at least 25 years of contributions.

To qualify for unemployment benefit the applicant must have been in employment for at least 12 months in the previous three years. The maximum unemployment benefit in 2006 was Kč. 11,050 per month.

RELIGION

In 2009 there were 25 registered churches and religious societies. In 2001 church membership was estimated to be: Roman Catholic, 2,740,800; Evangelical Church of the Czech Brethren, 117,200; Hussites, 99,100; Eastern Orthodox, 23,000; Silesian Evangelicals, 14,000. 6,040,000 persons were classified as atheist or non-religious, and there were 331,000 adherents of other religions.

Dominik Duka (b. 1943) was installed as Archbishop of Prague and Primate of Bohemia in April 2010. The national Czech church, created in 1918, took the name 'Hussite' (Czechoslovak Hussite Church) in 1972. It has a patriarch, five bishops and over 260 pastors (49% women). There are also around a dozen other Protestant churches, the largest being the Evangelical Church of

Czech Brethren, which unites Calvinists and Lutherans and has about 115,000 members. In Feb. 2013 the Roman Catholic church had two cardinals.

CULTURE

World Heritage Sites

Sites under Czech jurisdiction which appear on UNESCO's World Heritage List are (with year entered on list): Historic Centre of Prague (1992); Historic Centre of Český Krumlov (1992); Historic Centre of Telč (1992); Pilgrimage Church of St John of Nipomuk at Zelená Hora in Žďár nad Sázavou (1994); Kutná Hora—the Historical Town Centre with the Church of Saint Barbara and the Cathedral of our Lady at Sedlec (1995); Lednice-Valtice Cultural Landscape (1996); Holašovice Historical Village Reservation (1998); Gardens and Castle at Kroměříž (1998); Litomyšl Castle (1999); Holy Trinity Column in Olomouc (2000); Tugendhat Villa in Brno (2001); and the Jewish Quarter and St Procopius' Basilica in Třebíč (2003).

Press

There were 85 daily newspapers in 2008 (82 paid-for and three free) with a combined average daily circulation of 2,075,000. There were also 490 non-dailies in 2008. The newspaper with the highest circulation is *Blesk* (daily average of 436,000 copies in 2008).

Tourism

In 2010, 6,334,000 non-resident tourists stayed in holiday accommodation. Of these, 1,350,000 were from Germany, 414,000 from Russia, 368,000 from the UK, 351,000 from Poland and 333,000 from Italy.

Festivals

The leading classical music festivals are the Prague Spring International Music Festival (May–June) and Smetana's Litomyšl International Opera Festival (June–July). The main popular music festivals are Rock for People in Hradec Králové (July) and Trutnov Open Air Music Festival (Aug.).

DIPLOMATIC REPRESENTATIVES

Of the Czech Republic in the United Kingdom (26 Kensington Palace Gdns, London, W8 4QY)
Ambassador: Michael Žantovský.

Of the United Kingdom in the Czech Republic (Thunovská 14, 118 00 Prague 1)
Ambassador: Sian MacLeod, OBE.

Of the Czech Republic in the USA (3900 Spring of Freedom St., NW, Washington, D.C., 20008)
Ambassador: Petr Gandalovič.

Of the USA in the Czech Republic (Tržiste 15, 118 01 Prague 1)
Ambassador: Norman Eisen.

Of the Czech Republic to the United Nations
Ambassador: Edita Hrdá.

Of the Czech Republic to the European Union
Permanent Representative: Martin Povejšil.

FURTHER READING

Czech Statistical Office. *Statistical Yearbook of the Czech Republic.*

Krejčí, Jaroslav and Machonin, Pavel, *Czechoslovakia 1918–1992: A Laboratory for Social Change.* 1996

Leff, C. S., *National Conflict in Czechoslovakia: The Making and Remaking of a State, 1918–1987.* 1988

Simmons, M., *The Reluctant President: a Political Life of Vaclav Havel.* 1992

National Statistical Office: Czech Statistical Office, Na Padesátém 81, 100 82 Prague 10.
Website: http://www.czso.cz

Explore the world at
www.statesmansyearbook.com

DENMARK

**Kongeriget Danmark
(Kingdom of Denmark)**

Capital: Copenhagen
Population projection, 2015: 5·65m.
GNI per capita, 2011: (PPP$) 34,347
HDI/world rank: 0·895/16
Internet domain extension: .dk

KEY HISTORICAL EVENTS

Evidence of habitation exists from the Bølling period (12500–12000 BC). By 7700 BC reindeer hunters were settled on the Jutland Peninsula and around 3900 BC agriculture developed. Metal tools and weapons were imported in the Dagger Period (*c.* 2000 BC) but trading stations on the coast did not appear until around AD 300. The first towns developed in the Germanic Iron Age (AD 400–750). The first trading market was held in the 8th century in Hedeby. Denmark was converted to Christianity in 860 when Ansgar built churches in Hedeby and Ribe.

Danish Vikings first attacked England's northeast coast in 793. In about 900 Harold Bluetooth became the first king of Denmark and Skåne. His grandson, Canute the Great, fought successfully to incorporate England into his North Sea Empire and from 1018–35, Denmark, England and Norway were one nation. However, civil war broke out and in 1146 the kingdom was divided between Magnus the Strong and Knud Lavard. In 1157 Knud's son Valdemar was recognized as the ruler of Denmark. By 1200 Skåne, Halland and Blekinge in the South of Sweden were part of the Danish kingdom. The southern border of Denmark extended to the Eider in what is today northern Germany. In 1219 Valdemar conquered Estonia. He also established a code of law and a land register (Jordebog). The first written constitution was a coronation charter signed by Erik V in 1282.

In the 13th century, agriculture was supplemented by the expansion of fishing to supply inland Europe. Other industries also benefited and this brought with it a passion for building, particularly cathedrals and churches. Economic growth strengthened German influence. The Hanseatic League of German entrepreneurs was granted trade concessions for herring, salt and grain and also played a leading role in the country's political affairs. Valdemar IV Atterdag was crowned king in 1340. He challenged the privileges of the Hanseatic League, and was brought into conflict with Sweden over the southern provinces of

Skåne, Halland and Blekinge. In 1361 Valdemar Atterdag took Gotland in one of the bloodiest of Nordic battles. When the king died in 1375, his daughter Margaret (married to King Håkon of Norway) claimed the throne on behalf of her five-year-old son Olav. After Håkon's death in 1388 she also became regent of Norway. While resisting the Hanseatic League, she succeeded in defeating her opponent Albrecht of Mecklenburg, king of Sweden, thus clearing the way to a Nordic union. In 1397, after Olav's death, Margaret's nephew Erik of Pomerania became king of Denmark, Norway and Sweden. In 1412 Erik was opposed by the Swedish nobles who resented being taxed to finance Danish wars in northern Germany. When Erik abdicated, Christian I was elected king of Denmark and Norway in 1448.

By the 16th century Scandinavia was divided between Denmark–Norway (including Iceland and Greenland) and Sweden–Finland. In 1520 a power struggle in Sweden made the country vulnerable to a Danish invasion. Christian II was crowned king of Sweden in 1520 (having assumed the thrones of Norway and Denmark in 1513) but was soon challenged by Gustav Vasa, who replaced him in Sweden in 1521. In 1523 Christian was succeeded in Denmark and Norway by Frederick I, who ended the union with Sweden. Following the Lutheran Reformation, the monarchy enhanced its power by confiscating the property of the Roman Catholic Church.

Imperial Rise and Fall

Christian IV (1577–1648) is regarded as one of Denmark's greatest rulers. Around this time overseas colonies were established, including Tranquebar (India), Danish Gold Coast (Ghana) and the Danish West Indies (the US Virgin Islands). However, in 1626 Denmark was defeated in the Thirty Years' War. Denmark lost Gotland and the Norwegian territories of Jämtland and Härjedalen to Sweden. In 1660 Sweden gained Skåne, Halland and Blekinge. A new constitution proclaimed the Danish king absolute sovereign. In 1661 the Supreme Court was established and in 1683 the law was codified.

In the Great Northern War, Denmark allied itself with Russia, the Netherlands and France, a policy which lasted for the rest of the 18th century. In the Napoleonic Wars, Denmark, smarting under the British bombardment of Copenhagen, allied itself to Napoleon. The price Denmark had to pay was signing away its rights to Norway, which it did by the treaty of Kiel in 1814. Danish possessions were now reduced to Iceland, Greenland, the Faroes and Schleswig-Holstein. Holstein was lost to Germany in 1863 and Schleswig a year later. The surrender of so much rich agricultural land, with nearly 1m. inhabitants, brought Denmark to the edge of bankruptcy. But within a few years the country had pulled itself back from one of the lowest points in its history. The economy benefited from a land-reclamation programme in Jutland. Socially, Bishop Grundtvig (founder of the folk high-schools), who reconciled patriotism with a reduced status for Denmark in European affairs, had a great influence. There were demands for a liberal constitution. In 1846 Anton Frederik Tscherning founded the Society of the Friends of the Peasant (Bondevennernes Selskab), which later became the Liberal Party (Venstre).

Social Reform

In 1901 the Left Reform Party (Venstrereformparti) came to power to introduce free-trade, popular education and changes in the revenue system to make income rather than land the criterion for taxation. The First World War gave neutral Denmark an improved export market but there was a shortage of raw materials. In 1929 a Social Democrat government, with Thorvald Stauning as prime minister, combined rural and urban interests in one of the most ambitious programmes of social reforms ever mounted. The 1930s Great Depression led to unemployment made worse when Britain favoured Commonwealth food imports over those from Denmark. In the late 1930s trade improved and industry expanded.

In 1939 when the Second World War broke out, Denmark again declared neutrality. On 9 April 1940 German troops entered and occupied the country. The Germans permitted Danish self-government until growing resistance led to direct rule. After the liberation a Liberal government was elected with Knud Kristensen as prime minister. Kristensen's campaign for the return of southern Schleswig from Germany brought down his government in 1947. Denmark joined NATO in 1949.

With its share of the Marshall Plan, Denmark entered on a new industrial revolution. By the mid-1950s the value of manufacturing equalled that of agriculture. However, there was a high rate of inflation. In 1953 the Social Democrats came back to power where they remained until the mid-1960s. By then the rate of inflation was higher than in any comparable country. In 1968 a centre-right coalition was elected, led by Hilmar Baunsgaard. But the change of government did not signify a change in strategy. Taxes were kept high and the budget expanded to increase social welfare. After the 1971 election, the Social Democrat leader Jens Otto Krag negotiated entry into the European Union, making Copenhagen the bridge between the Nordic capitals and Brussels. In 1982 a Conservative-led minority government was formed, led by Poul Schlüter, the first Conservative prime minister since 1901. He remained in power until 1993 when a Social Democratic coalition led by Poul Nyrup Rasmussen took office. Following the 2001 election, a right-wing government came to power under Anders Fogh Rasmussen, whose campaign for entry into the EMU (European Monetary Union) was rejected in a referendum in 2000. In Nov. 2008 Greenland approved a referendum in favour of greater autonomy from Denmark. Anders Fogh Rasmussen left office in April 2009 to become Secretary General of NATO. He was replaced by Lars Løkke Rasmussen (no relation) as prime minister.

TERRITORY AND POPULATION

Denmark is bounded in the west by the North Sea, northwest and north by the Skagerrak and Kattegat straits (separating it from Norway and Sweden), and south by Germany. A 16-km long fixed link with Sweden was opened in July 2000 when the Øresund motorway and railway bridge between Copenhagen and Malmö was completed.

Regions	Area (sq. km)	Population 1 Jan. 2011	Population per sq. km 2011
Capital Region (Hovedstaden)	2,561	1,699,387	663·6
Central Jutland (Midtjylland)	13,124	1,260,993	96·1
North Jutland (Nordjylland)	7,933	579,829	73·1
Zealand (Sjælland)	7,273	819,763	112·7
South Denmark (Syddanmark)	12,206	1,200,656	98·4
Total	43,098	5,560,628	129·0

Denmark has not used a traditional census since 1970, but instead uses a register-based method of calculating the population. It was the first country in the world to implement such a change.

The UN gives a projected population for 2015 of 5·65m.

In 2010 an estimated 86·7% of the population lived in urban areas. In 2010, 91·4% of the inhabitants were born in Denmark, including the Faroe Islands and Greenland.

On 1 Jan. 2011 the population of the capital, Copenhagen (comprising Copenhagen, Frederiksberg and Gentofte municipalities), was 1,199,224; Aarhus, 249,709; Odense, 167,615; Aalborg, 103,545; Esbjerg, 71,576; Randers, 60,656; Kolding, 57,197; Horsens, 53,807; Vejle, 51,341.

The official language is Danish.

SOCIAL STATISTICS

Statistics for calendar years:

	Live births	Marriages	Divorces	Deaths	Emigration	Immigration
2005	64,282	36,148	15,300	54,962	45,869	52,458
2006	64,984	36,452	14,343	55,477	46,786	56,750
2007	64,082	36,576	14,066	55,604	41,566	64,656
2008	65,038	37,376	14,695	54,591	43,490	72,749
2009	62,818	32,934	14,940	54,872	44,874	67,161

2009 rates per 1,000 population: birth, 11·4; death, 9·9. Births outside marriage: 2006, 46·4%; 2007, 46·1%; 2008, 46·2%; 2009, 46·5%. Average annual population growth rate, 2005–09, 0·5%. Suicide rate, 2006 (per 100,000 population) was 11·9 (men, 17·5; women, 6·4). Life expectancy at birth, 2008–09, was 76·5 years for males and 80·8 years for females. In 2007 the most popular age range for marrying was 30–34 for males and 25–29 for females. Denmark was the first country to legalize same-sex unions, in 1989. Infant mortality, 2009, 3·0 per 1,000 live births. Fertility rate, 2009, 1·8 births per woman. In 2009 Denmark received 3,855 asylum applications, equivalent to 0·7 per 1,000 inhabitants. In July 2002 a controversial new immigration law was introduced in an attempt to deter potential asylum seekers. Denmark legalized same-sex marriage in June 2012 (although the legislation does not apply in the Faroe Islands and Greenland).

CLIMATE

The climate is much modified by marine influences and the effect of the Gulf Stream, to give winters that may be both cold or mild and often cloudy. Summers may be warm and sunny or chilly and rainy. Generally the east is drier than the west. Long periods of calm weather are exceptional and windy conditions are common. Copenhagen, Jan. 33°F (0·5°C), July 63°F (17°C). Annual rainfall 650 mm. Esbjerg, Jan. 33°F (0·5°C), July 61°F (16°C). Annual rainfall 800 mm. In general 10% of precipitation is snow.

CONSTITUTION AND GOVERNMENT

The present constitution is founded upon the Basic Law of 5 June 1953. The legislative power lies with the Queen and the *Folketing* (parliament) jointly. The executive power is vested in the monarch, who exercises authority through the ministers.

The reigning Queen is **Margrethe II**, b. 16 April 1940; married 10 June 1967 to Prince Henrik, b. Count de Monpezat. She succeeded to the throne on the death of her father, King Frederik IX, on 14 Jan. 1972. *Offspring:* Crown Prince Frederik, b. 26 May 1968, married 14 May 2004 Mary Elizabeth Donaldson, b. 5 Feb. 1972 (*offspring:* Prince Christian Valdemar Henri John, b. 15 Oct. 2005; Princess Isabella Henrietta Ingrid Margrethe, b. 21 April 2007; Prince Vincent Frederik Minik Alexander, b. 8 Jan. 2011; Princess Josephine Sophia Ivalo Mathilda, b. 8 Jan. 2011); Prince Joachim, b. 7 June 1969, married 18 Nov. 1995 Alexandra Manley, b. 30 June 1964, divorced 8 April 2005 (*offspring:* Prince Nikolai William Alexander Frederik, b. 28 Aug. 1999; Prince Felix Henrik Valdemar Christian, b. 22 July 2002), married 24 May 2008 Marie Cavallier, b. 6 Feb. 1976 (*offspring:* Prince Henrik Carl Joachim Alain, b. 4 May 2009; Princess Athena Marguerite Françoise Marie, b. 24 Jan. 2012).

Sisters of the Queen. Princess Benedikte, b. 29 April 1944; married 3 Feb. 1968 to Prince Richard of Sayn-Wittgenstein-Berleburg; Princess Anne-Marie, b. 30 Aug. 1946; married 18 Sept. 1964 to King Constantine of Greece.

The crown was elective from the earliest times but became hereditary by right in 1660. The direct male line of the house of Oldenburg became extinct with King Frederik VII on 15 Nov. 1863. In view of the death of the king, without direct heirs, the Great Powers signed a treaty at London on 8 May 1852, by the terms of which the succession to the crown was made over to Prince Christian of Schleswig-Holstein-Sonderburg-Glücksburg, and to the direct male descendants of his union with the Princess Louise of Hesse-Cassel. This became law on 31 July 1853. Linked to the constitution of 5 June 1953, a new law of succession, dated 27 March 1953, has come into force, which restricts the right of succession to the descendants of King Christian X and Queen Alexandrine, and admits the sovereign's daughters to the line of succession, ranking after the sovereign's sons.

The Queen receives a tax-free annual sum from the state. This was 71·1m. kroner in 2011.

The judicial power is with the courts. The monarch must be a member of the Evangelical-Lutheran Church, the official Church of the State, and may not assume major international obligations without the consent of the Folketing. The Folketing consists of one chamber. All men and women of Danish nationality of more than 18 years of age and permanently resident in Denmark possess the franchise, and are eligible for election to the Folketing, which is at present composed of 179 members; 135 members are elected by the method of proportional representation in 17 constituencies. In order to attain an equal representation of the different parties, 40 additional seats are divided among such parties which have not obtained sufficient returns at the constituency elections. Two members are elected for the Faroe Islands and two for Greenland. The term of the legislature is four years, but a general election may be called at any time. The Folketing convenes every year on the first Tuesday in Oct. Besides its legislative functions, every six years it appoints judges who, together with the ordinary members of the Supreme Court, form the *Rigsret*, a tribunal which can alone try parliamentary impeachments.

National Anthem

'Kong Kristian stod ved højen mast' ('King Christian stood by the lofty mast'); words by J. Ewald, tune by D. L. Rogert.

GOVERNMENT CHRONOLOGY

Prime Ministers since 1945. (KF = Conservative Party; RV = Radical Liberal Party; SD = Social Democratic Party; V = Liberal Party)

1945	SD	Vilhelm Buhl
1945–47	V	Knud Kristensen
1947–50	SD	Hans Hedtoft
1950–53	V	Erik Eriksen
1953–55	SD	Hans Hedtoft
1955–60	SD	Hans Christian Hansen
1960–62	SD	Viggo Kampmann
1962–68	SD	Jens Otto Krag
1968–71	RV	Hilmar Baunsgaard
1971–72	SD	Jens Otto Krag
1972–73	SD	Anker Jørgensen
1973–75	V	Poul Hartling
1975–82	SD	Anker Jørgensen
1982–93	KF	Poul Holmskov Schlüter
1993–2001	SD	Poul Nyrup Rasmussen
2001–09	V	Anders Fogh Rasmussen
2009–11	V	Lars Løkke Rasmussen
2011–	SD	Helle Thorning-Schmidt

RECENT ELECTIONS

Parliamentary elections were held on 15 Sept. 2011; turnout was 87·7%. The Liberal Party (V) won 47 seats, with 26·7% of mainland votes cast (46 seats with 26·3% in 2007); the Social Democratic Party (SD) 44 with 24·8% (45 with 25·5%); the Danish People's Party (DF) 22 with 12·3% (25 with 13·8%); the Social Liberal Party (RV) 17 with 9·5% (9 with 5·1%); the Socialist People's Party (SF) 16 with 9·2% (23 with 13·0%); the Unity List— the Red-Greens (EL) 12 with 6·7% (4 with 2·2%); the Liberal Alliance (I) 9 with 5·0% (5 with 2·8% as the New Alliance); and the Conservative Party (KF) 8 with 4·9% (18 with 10·4%). Although the Liberal Party won the most seats, the Social

Democratic Party-led 'Red bloc' took 89 seats compared to 86 for the Liberal party-led 'Blue bloc'. Four remaining seats go to representative parties from the Faroe Islands and Greenland.

European Parliament
Denmark has 13 (14 in 2004) representatives. At the June 2009 elections turnout was 59·5% (47·9% in 2004). The SD won 4 seats with 20·9% of votes cast (political affiliation in European Parliament: Progressive Alliance of Socialists and Democrats); V, 3 with 19·6% (Alliance of Liberals and Democrats for Europe); the SF, 2 with 15·4% (Greens/European Free Alliance); DF, 2 with 14·8% (Europe of Freedom and Democracy); KF, 1 with 12·3% (European People's Party); the People's Movement Against the EU, 1 with 7·0% (European United Left/Nordic Green Left).

CURRENT GOVERNMENT

Following the elections of Sept. 2011 a three-party centre-left coalition was formed consisting of the Social Democratic Party (SD), the Social Liberal Party (RV) and the Socialist People's Party (SF). In March 2013 it comprised:

Prime Minister: Helle Thorning-Schmidt; b. 1966 (SD; took office on 3 Oct. 2011).

Minister for Economy and Interior: Margrethe Vestager (RV). *Foreign Affairs:* Villy Søvndal (SF). *Finance:* Bjarne Corydon (SD). *Justice:* Morten Bødskov (SD). *Defence:* Nick Hækkerup (SD). *Culture:* Marianne Jelved (RV). *Taxation:* Holger K. Nielsen (SF). *Research, Innovation and Higher Education:* Morten Østergaard (RV). *Business and Growth:* Annette Vilhelmsen (SF). *City, Housing and Rural Affairs:* Carsten Hansen (SD). *Employment:* Mette Frederiksen (SD). *Children and Education:* Christine Antorini (SD). *Integration and Social Affairs:* Karen Hækkerup (SD). *Development Co-operation:* Christian Friis Bach (RV). *Food, Agriculture and Fisheries:* Mette Gjerskov (SD). *Trade and Investment:* Pia Olsen Dyhr (SF). *Climate and Energy:* Martin Lidegaard (RV). *Transport:* Henrik Dam Kristensen (SD). *Health and Prevention:* Astrid Krag (SF). *European Affairs:* Nicolai Wammen (SD). *Environment:* Ida Auken (SF). *Equality, Church and Nordic Co-operation:* Manu Sareen (RV).

Office of the Prime Minister: http://www.stm.dk

CURRENT LEADERS

Helle Thorning-Schmidt

Position
Prime Minister

Introduction
Helle Thorning-Schmidt took office on 3 Oct. 2011, heading a Social Democrat-led coalition. She is the country's first female prime minister.

Early Life
Born on 14 Dec. 1966 in Rødovre, Helle Thorning-Schmidt completed her early education at Ishøj Gymnasium, Copenhagen. From 1987–94 she studied political science at the University of Copenhagen. In 1992 she was selected by the ministry of foreign affairs to attend the College of Europe in Bruges, Belgium, where she obtained a masters degree in European studies, specialising in policy and public administration. While in Belgium she became involved with the Social Democrats there and joined the Danish Social Democrats in 1993.

From 1994–97 Thorning-Schmidt headed the party's secretariat in the European Parliament. She then joined the Danish Confederation of Trade Unions as a consultant before being elected to the European Parliament as a member of the Party of European Socialists in 1999. She subsequently sat on the employment and social committee and co-founded the Campaign for Parliament Reform (CPR).

In Feb. 2005 Thorning-Schmidt won a seat in the Folketing. On 12 April 2005 she was elected leader of the Social Democrats following Mogens Lykketoft's resignation. At the 2007 general election Thorning-Schmidt campaigned on relaxing immigration regulations, increasing welfare spending and ensuring that 45% of Denmark's energy requirements are met by renewable sources by 2025.

Despite electoral defeat, Thorning-Schmidt kept her post and the Social Democrats remained the largest opposition party. In 2011 she campaigned to raise taxes for high earners, increase public spending and liberalize immigration policies. Her centre-left coalition secured a narrow victory at the general election.

Career in Office

Thorning-Schmidt's primary aim on taking office was to revitalize the economy by creating jobs via education, green energy and infrastructure projects. However, building consensus within her coalition has been a challenge. Attempts to review aspects of the welfare state, including a tax reform package agreed with the centre-right opposition, led to a slump in public support for the Social Democrats in the first year of her premiership. Meanwhile, Denmark's six-month presidency of the European Union from Jan.–June 2012 was dominated by the debt crisis in the eurozone, of which Denmark is not a member.

DEFENCE

Pursuant to the Defence Agreement covering 2005–09 the Danish defence system is being completely restructured. A new security mechanism is being built from scratch to make it relevant in today's security environment. The entire transformation process centres on increasing Denmark's deployable capabilities. The composition of armed forces personnel has changed to 60:40 in favour of operational elements. Change has been accomplished by establishing so-called functional services in a number of areas formerly run by each individual service, and by reducing the staff- and support structure.

Denmark will be able simultaneously to deploy 2,000 soldiers on international missions and offer considerable High Readiness forces to NATO or coalition partners. The Armed Forces as a whole have been professionalized. Formal conscription still exists, but no specific combat training takes place. New conscripts train for four months in basic Homeland Defence, for example fire fighting, relief work during *force majeure* and basic weapons handling. The 20% of conscripts expected to sign up for additional time undergo military training focusing on participation in international operations for eight months and then deploy on an international assignment for six months.

The overall organization of the Danish Armed Forces includes the Ministry of Defence (MoD), the Danish Defence Command, the Army, the Navy, the Air Force, Functional Services and Schools and several joint service institutions and authorities; to this should be added the Home Guard, which is an integral part of Danish military defence. The Chief of Defence (CHOD), answering to the Minister of Defence, is in full command of the Army, the Navy and the Air Force.

Denmark has a compulsory military service with mobilization based on the constitution of 1849. This states that it is the duty of every fit man to contribute to the national defence. In 2007 defence expenditure totalled US$3,788m. (US$692 per capita), representing 1·3% of GDP.

Army

The Danish Army is comprised of field army formations and local defence forces. The strength of the Danish Army is approximately 10,600. The Danish Army is organized in two brigades, the first made up of professional soldiers and the second functioning as a training structure for conscripts.

Navy

The strength of the Royal Danish Navy is approximately 3,500. The two main naval bases are located at Frederikshavn and Korsør.

Air Force

The strength of the Royal Danish Air Force is approximately 3,500. The Royal Danish Air Force consists of Tactical Air Command Denmark and the Danish Air Materiel Command.

Home Guard (Hjemmeværnet)

The overall Home Guard organization comprises the Home Guard Command, the Army Home Guard, the Naval Home Guard, the Air Force Home Guard and supporting institutions. The personnel are recruited on a voluntary basis. The personnel establishment of the Home Guard is approximately 50,000 soldiers.

INTERNATIONAL RELATIONS

In a referendum in June 1992 the electorate voted against ratifying the Maastricht Treaty for closer political union within the EU. Turnout was 82%. 50·7% of votes were against ratification, 49·3% in favour. However, a second referendum on 18 May 1993 reversed this result, with 56·8% of votes cast in favour of ratification and 43·2% against. Turnout was 86·2%. In a referendum held on 28 Sept. 2000 Danish voters rejected their country's entry into the common European currency, 53·2% opposing membership of the euro against 46·8% voting in favour. Turnout was 87·6%.

ECONOMY

In 2010 agriculture accounted for 1·2% of GDP, industry 21·8% and services 77·0%.

According to the Berlin-based organization *Transparency International*, Denmark ranked equal first in the world in a 2012 survey of countries with the least corruption in business and government. It received 90 out of 100 in the corruption perceptions index.

Denmark gave US$2·9bn. in international aid in 2011, equivalent to 0·85% of GNI (making Denmark one of only five countries to exceed the UN target of 0·7%).

Overview

Until the 1960s the economy was heavily based on a competitive agricultural sector. 57% of the labour force was employed in agriculture, fisheries, manufacturing and industry in 1960. By 2009 that figure had declined to 22%. Almost three-quarters of the workforce are now employed in services, although the sector with the greatest proportional gain in employment in the last half-century is public services.

From 1993–2003 total research and development (R&D) expenditure nearly doubled, with expenditure in the private sector accounting for over two-thirds of total R&D spending. Since 2002 R&D expenditure as a share of GDP has been around 2·5% annually, well above both the EU and OECD averages. It is particularly high in manufacturing, knowledge services (such as information and communications technologies) and, increasingly, financial services. Denmark is ranked fourth in the world in terms of the Gini coefficient (a measure of income equality) and 12th in the World Economic Forum's *Global Competitiveness Report* for 2012–13.

After weak growth during the Scandinavian banking crisis of 1991–93 with unemployment at 12·3% in Jan. 1994, the economy grew robustly from 1994–2007, although unemployment remained above 6% until 1998. From 2001–03 Denmark avoided recession in the face of a global economic slowdown, although annual growth (driven by private consumption) averaged less than 2%. A thriving real estate market, in which house prices increased by 22%, boosted consumer demand, with higher investment and a strong export performance also supporting growth.

In early 2007 the housing boom ended and the economy began to slow, exacerbated by the global financial crisis. A 20% correction in house prices contributed to the economy shrinking by 5·7% in 2009, before rebounding by 1·6% in 2010. However, private consumption and business investment were subdued and the recovery faltered in late 2011. Domestic demand remained weak and output fell throughout most of 2012.

The economy is tightly linked with the eurozone and its currency is pegged to the euro. Germany and Sweden are Denmark's largest trading partners, accounting for about a third of foreign trade. Reflecting Europe's economic woes, export growth has been stunted since the late 2000s and renewed recession in the eurozone poses a significant threat to the longer-term economic outlook.

Currency
The monetary unit is the *Danish krone* (DKK) of 100 øre. Inflation rates (based on OECD statistics):

2002	2003	2004	2005	2006	2007	2008	2009	2010	2011
2·4%	2·1%	1·2%	1·8%	1·9%	1·7%	3·4%	1·3%	2·3%	2·8%

In Aug. 2009 foreign exchange reserves were US$67,774m., gold reserves were 2·14m. troy oz and total money supply was 832·2bn. kroner.

While not participating directly in EMU, the Danish krone is pegged to the euro in ERM-2, the successor to the exchange rate mechanism.

Budget
The following shows the actual revenue and expenditure in central government accounts for the calendar years 2006 and 2007, the approved budget figures for 2008 and the budget for 2009 (in 1,000 kroner):

	2006	2007	2008	2009
Revenue[1]	562,987,500	649,559,500	647,946,900	669,496,400
Expenditure[1]	459,623,500	535,301,500	556,050,500	597,606,700

[1]Receipts and expenditures of special government funds and expenditures on public works are included.

The 2009 budget envisaged revenue of 335,266·7m. kroner from income and property taxes and 279,872·0m. kroner from consumer taxes. The central government debt on 31 Dec. 2007 amounted to 255,074m. kroner.

In 2007 tax revenues were 48·9% of GDP (the highest percentage of any developed country).

Denmark's budget deficit in 2011 was 1·8% of GDP (2010, 2·5%; 2009, 2·7%). In 2008 it had a surplus of 3·2%. The required target set by the EU is a budget deficit of no more than 3%.

VAT is 25%.

Performance
Real GDP growth rates (based on OECD statistics):

2002	2003	2004	2005	2006	2007	2008	2009	2010	2011
0·5%	0·4%	2·3%	2·4%	3·4%	1·6%	−0·8%	−5·7%	1·6%	1·1%

Total GDP was US$333·6bn. in 2011.

Banking and Finance
In 2009 the accounts of the National Bank (*Governor*, Lars Rohde) balanced at 550,151m. kroner. The assets included official net foreign reserves of 370,861m. kroner. The liabilities included notes and coins totalling 60,761m. kroner. On 31 Dec. 2008 there were 138 commercial banks and savings banks, with deposits of 1,438,028m. kroner.

The two largest commercial banks are Danske Bank and Nordea Bank Danmark. The supervisory boards of all banks must include public representation.

Gross external debt totalled US$584,152m. in June 2012.

There is a stock exchange in Copenhagen.

ENERGY AND NATURAL RESOURCES

In 2008, 18·8% of energy consumption came from renewables (wind power, solar power, hydro-electric power, tidal power,

geothermal energy and biomass), compared to the European Union average of 10·3%. A target of 30% has been set by the EU for 2020.

Environment
Denmark's carbon dioxide emissions from the consumption and flaring of fossil fuels in 2008 were the equivalent of 9·9 tonnes per capita.

Electricity
Installed capacity was 13·0m. kW in 2007. Production (2008), 34,737m. kWh. Consumption per capita in 2007 was 6,505 kWh. In 2007 some 5,212 wind turbines produced 19·7% of output.

Oil and Gas
Oil production was (2008) 14·0m. tonnes with 800m. bbls of proven reserves. Production of natural gas was (2008) 10·1bn. cu. metres with 60bn. cu. metres of proven reserves.

Wind
Denmark is one of the world's largest wind-power producers, with an installed capacity of 3,734 MW at the end of 2010. Denmark generated 19·3% of its electricity from wind in 2009, the highest proportion of any country.

Agriculture
Agriculture accounted for 9·7% of exports and 2·5% of imports in 2009. Land ownership is widely distributed. In 2008 there were 43,413 holdings with at least 5 ha. of agricultural area (or at least a production equivalent to that from 5 ha. of barley). There were 10,214 small holdings (with less than 10 ha.), 18,465 medium-sized holdings (10–50 ha.) and 14,734 holdings with more than 50 ha. Approximately 5·1% of all agricultural land is used for organic farming. There were 24,675 agricultural workers in 2007. In 2007 Denmark had 2·30m. ha. of arable land and 8,000 ha. of permanent cultures.

In 2008 the cultivated area was (in 1,000 ha.): grain, 1,505; green fodder and grass, 705; root crops, 84; set aside, 71; other crops, 298; pulses, 5; total cultivated area, 2,668.

	Area (1,000 ha.)				Production (in 1,000 tonnes)			
Chief crops	2006	2007	2008	2009	2006	2007	2008	2009
Wheat	686	689	638	739	4,802	4,519	5,019	5,940
Barley	679	632	717	593	3,270	3,104	3,396	3,394
Potatoes	39	41	41	39	1,361	1,626	1,705	1,618
Oats	69	66	84	67	274	312	322	315
Rye	28	30	29	44	130	135	152	238
Other root crops	46	43	41	43	2,585	3,518	2,525	2,278

Livestock, 2008 (in 1,000): pigs, 12,738; cattle, 1,564; sheep, 136; horses, 60; poultry, 15,106.

Production (in 1,000 tonnes) in 2009: pork, 1,898; beef, 137; milk, 4,733; cheese, 324; eggs, 73; butter, 37.

On 1 Jan. 2010 tractors numbered 99,700.

Forestry
The area under forests in 2010 was 0·54m. ha., or 13% of the total land area. Timber production in 2007 was 2·57m. cu. metres.

Fisheries
The total value of the fish caught was (in 1m. kroner): 1970, 854; 1980, 2,888; 1985, 3,542; 1990, 3,485; 1995, 3,020; 2000, 3,141; 2005, 2,781; 2008, 2,487; 2009, 2,154.

In 2009 the total catch was 738,094 tonnes, almost exclusively from sea fishing. Denmark is one of the leading fishing nations in the EU.

INDUSTRY

The leading companies by market capitalization in Denmark in March 2012 were: Novo Nordisk A/S, a health care company (US$65·3bn.); A. P. Møller-Mærsk, a shipping company (US$33·1bn.); and Danske Bank (US$15·8bn.).

The following table is of gross value added by kind of activity (in 1m. kroner; 2000 constant prices):

	2007	2008	2009
Total	1,229,577	1,227,087	1,169,778
Agriculture, fishing and quarrying	55,028	50,296	52,120
Manufacturing	190,818	198,674	174,528
Electricity, gas and water supply	21,252	22,931	21,895
Construction	62,040	58,498	49,900
Wholesale and retail trade	174,600	169,413	147,085
Transport, post and telecommunication	111,394	99,815	91,999
Financial and business activities	310,271	319,820	321,501
Public and personal services	304,173	307,641	310,749

In the following table 'number of jobs' refers to 18,779 local business enterprises including self-employed businesses with no employees (Nov. 2006):

Branch of industry	Number of jobs
Food, beverages and tobacco	72,508
Textiles, clothing and footwear, and leather	8,963
Wood and wood products	15,537
Paper products	44,541
Refined petroleum products	1,008
Chemicals and man-made fibres	27,940
Rubber and plastic products	21,157
Non-metallic mineral products	16,585
Basic metals	52,902
Machinery and equipment	62,976
Electrical and optical equipment	47,873
Transport equipment	15,040
Furniture and other manufactures	26,510
Total manufacturing	413,540

Labour

In 2008 the labour force was 2,917,400. 34·7% of the working population in 2008 were in public and personal services; 18·8% in wholesale and retail trade, hotels and restaurants; 15·7% in financial intermediation, commerce, etc.; 14·0% in manufacturing; 6·8% in construction; 6·1% in transport, storage and telecommunications; 3·1% in agriculture, fisheries and quarrying; and 0·5% in electricity, gas and water supply. In 2008, 399,600 persons were employed in manufacturing. In Dec. 2012 the unemployment rate was 8·0% (up from 7·6% in 2011 as a whole). Between 2005 and 2009 strikes cost Denmark an average of 159 days per 1,000 employees a year, the highest total for any EU-member country.

INTERNATIONAL TRADE

Imports and Exports

In 2009 imports totalled 437,998m. kroner and exports 495,577m. kroner.

Imports and exports (in 1m. kroner) for calendar years:

	2008		2009	
Leading commodities	Imports	Exports	Imports	Exports
Live animals, meat and meat preparations	7,911	33,199	7,519	31,756
Dairy products and eggs	4,297	13,438	3,267	12,346
Fish, crustaceans, etc. and preparations	10,188	16,331	8,767	14,276
Cereals and cereal preparations	6,532	5,412	4,448	5,344
Fodder for animals	7,344	5,348	6,649	5,192
Wood and cork	4,616	1,060	3,666	790
Textile fibres, yarns, fabrics, etc.	8,452	7,448	6,222	5,926
Mineral fuels, lubricants, etc.	41,598	56,804	27,974	37,592
Chemicals and plastics, etc.	31,387	27,202	24,041	26,004
Medicine and pharmaceutical products	17,119	40,630	17,654	42,503
Metals, manufacture of metals	54,553	40,348	32,100	29,658
Machinery, electrical, equipment, etc.	126,803	131,770	99,186	107,296
Transport equipment	59,488	24,566	50,049	19,243
Furniture, etc.	8,989	13,916	7,243	11,432
Clothing and clothing accessories	24,131	20,884	21,171	19,022

Distribution of foreign trade (in 1m. kroner) according to countries of origin and destination for 2009:

Countries	Imports	Exports
Austria	4,242·5	3,661·4
Belgium	15,145·2	8,033·3
Canada	2,641·3	4,931·3
China	28,780·1	11,475·4
Finland	7,547·9	11,844·6
France	15,203·1	20,631·1
Germany	92,689·5	85,607·0
Greece	1,060·1	3,556·7
Greenland	2,082·6	2,764·0
Hong Kong	1,104·9	4,844·8
Ireland	4,787·2	5,839·2
Italy	15,274·4	14,991·9
Japan	2,283·0	10,038·3
Netherlands	30,796·1	22,860·9
Norway	23,228·0	31,423·4
Poland	11,345·1	12,228·4
Russia	4,754·6	8,266·8
South Korea	2,928·9	3,103·1
Spain	6,511·3	12,847·5
Sweden	57,758·3	63,666·1
Switzerland	4,721·6	4,715·2
Turkey	4,021·7	2,879·6
United Kingdom	24,296·7	41,970·1
United States of America	15,175·8	31,018·5

In 2006 fellow European Union member countries accounted for 72·6% of imports and 69·1% of exports.

COMMUNICATIONS

Roads

Denmark proper had (1 Jan. 2009) 1,128 km of motorways, 3,790 km of other state roads and 69,500 km of other commercial roads. Motor vehicles registered at 1 Jan. 2010 comprised 2,120,322 passenger cars, 32,300 trucks, 462,359 vans, 14,509 buses and 147,373 motorcycles. There were 5,250 casualties in road accidents in 2009, resulting in 303 fatalities.

Rail

In 2008 there were 2,132 km of State railways of 1,435 mm gauge (619 km electrified), which carried 164m. passengers (2007) and 7·14m. tonnes of freight. There were also 514 km of private railways. A metro system was opened in Copenhagen in 2002.

Civil Aviation

The main international airport is at Copenhagen (Kastrup), and there are also international flights from Aalborg, Aarhus, Billund and Esbjerg. The Scandinavian Airlines System (SAS) resulted from the 1950 merger of the three former Scandinavian airlines. SAS Denmark A/S is the Danish partner (SAS Norge ASA and SAS Sverige AB being the other two). Denmark and Norway each hold 14·3% of the capital of SAS and Sweden 21·4%. The remaining 50% of SAS shares are listed on the stock exchanges of Copenhagen, Oslo and Stockholm.

On 1 Jan. 2009 Denmark had 1,122 aircraft with a capacity of 19,077 seats. Copenhagen (Kastrup) handled 9,848,000 departing passengers in 2009, Billund 1,151,000, Aalborg 561,000 and Aarhus 255,000.

Shipping

On 1 Jan. 2010 the merchant fleet consisted of 462 vessels (above 100 GT) totalling 10·7m. GT. In 2009, 40m. tonnes of cargo were unloaded and 29m. tonnes were loaded in Danish ports; traffic by passenger ships and ferries is not included.

Telecommunications

In 2009 there were 2,062,000 main (fixed) telephone lines. In the same year mobile phone subscribers numbered 7,424,000 (134·1 per 100 persons). In 2010, 86% of the population had access to the internet at home and 88% had access to a computer at home. Denmark has one of the highest fixed broadband penetration rates, at 37·7 subscribers per 100 inhabitants in Dec. 2010. In March 2012 there were 2·8m. Facebook users.

SOCIAL INSTITUTIONS

Justice

The lowest courts of justice are organized in 24 tribunals (*byretter*), where minor cases are dealt with by a single judge. The tribunal at Copenhagen has one president and 42 other judges; and Aarhus one president and 13 other judges; the other tribunals have one to 11 judges. Cases of greater consequence are dealt with by the two High Courts (*Landsretterne*); these courts are also courts of appeal for minor cases. The Eastern High Court in Copenhagen has one president and 60 other judges; and the Western in Viborg one president and 39 other judges. From these an appeal lies to the Supreme Court in Copenhagen, composed of a president and 19 other judges. Judges under 65 years of age can be removed only by judicial sentence.

In 2009, 487,851 penal code offences were reported, including 56 homicides. In 2008 the daily average population in penal institutions was 3,679 (67·0 per 100,000 of national population).

Education

Education has been compulsory since 1814. The first stage of the Danish education system is the basic school (education at first level). This starts with a pre-school year (education preceding the first level), which has been compulsory since the beginning of the 2009–10 school year, and continues up to and including the optional 10th year in the *folkeskole* (municipal primary and lower secondary school). In 2006, 649,000 pupils attended education at first level and second level, first stage. In 2010 the number of pupils beginning their education at pre-school was 67,174.

Of all students leaving basic school in 2007–08, 79% had commenced further education within three months. Over half of the students (53%) had elected to attend general upper-secondary education (general programmes of education at secondary level, second stage), while 25% opted for vocational education and training at secondary level, second stage.

Education that qualifies students for tertiary level education is called general upper-secondary education and comprises general upper-secondary education (general programmes of education at secondary level, second stage), such as *gymnasium* (upper-secondary school), higher preparatory examination and adult upper-secondary level courses as well as general/vocational upper-secondary education at the vocational education institutions. In 2008, 119,000 students attended general upper-secondary education and 125,000 students attended upper-secondary vocational education and training.

Higher education is divided into three levels: short-cycle higher education involves two years of training, sometimes practical, after completion of upper-secondary education (19,000 students in 2008); medium-cycle higher education involves two–four years of mainly theoretical training (64,000 students in 2008); long-cycle higher education requires more than four years of education, mainly theoretical, divided between a bachelor's degree, candidate programme and PhD programme (bachelor's students in 2008: 62,000; master's: 53,000; PhD: 6,800).

Universities have been reorganized as a result of several mergers in Jan. 2007. The universities ranked by student population are: the University of Copenhagen (founded 1479), 36,600 students; the University of Aarhus (1928), 30,100; the University of Southern Denmark (1964), 14,300; the Copenhagen Business School, 13,900; the University of Aalborg (1974), 11,300; Roskilde University (1972), 8,100; the Technical University of Denmark, 6,200; the IT University of Copenhagen, 1,300.

Other types of post-secondary education have also been reorganized through mergers of institutions. Eight university colleges have been formed with student numbers ranging from 3,000 to 13,000. The university colleges encompass: schools of nursing; schools of midwifery education; colleges of physiotherapy; social education colleges; teacher training colleges; engineering colleges. There are also post-secondary educational institutions in the cultural sector in areas such as music, architecture, media and the visual arts.

In 2008 public expenditure on education was 15·0% of total government spending.

The adult literacy rate is at least 99%.

Health

In 2005 there were 17,350 doctors (321 per 100,000 persons), 4,634 dentists, 52,843 nurses, 27,072 auxiliary nurses and 1,304 midwives. There were 59 hospitals in 2005 (20,058 beds). In 2007 Denmark spent 9·8% of its GDP on health. In 2006 an estimated 26% of men and 23% of women smoked.

Welfare

The main body of Danish social welfare legislation is consolidated in seven acts concerning: (1) public health security, (2) sick-day benefits, (3) social pensions (for early retirement and old age), (4) employment injuries insurance, (5) employment services, unemployment insurance and activation measures, (6) social assistance including assistance to handicapped, rehabilitation, child and juvenile guidance, daycare institutions, care of the aged and sick, and (7) family allowances.

Public health security, covering the entire population, provides free medical care, substantial subsidies for certain essential medicines together with some dental care, and a funeral allowance. Hospitals are primarily municipal and treatment is normally free. All employed workers are granted daily sickness allowances; others can have limited daily sickness allowances. Daily cash benefits are granted in the case of temporary incapacity because of illness, injury or childbirth to all persons in paid employment. The benefit is paid up to the rate of 100% of the average weekly earnings. There is, however, a maximum rate of 3,415 kroner a week.

Social pensions cover the entire population. Entitlement to the old-age pension at the full rate is subject to the condition that the beneficiary has been ordinarily resident in Denmark for 40 years. For a shorter period of residence, the benefits are reduced proportionally. The basic amount of the old-age pension in Jan. 2007 was 174,720 kroner a year to married couples and 119,244 to single persons. Various supplementary allowances, depending on age and income, may be payable with the basic amount. The retirement age is 65, or 67 for those born before 1 July 1939. Depending on health and income, persons aged 60–64 (60–66 for those born before 1 July 1939) may apply for an early retirement pension. Persons over 65 (or 67) years of age are entitled to the basic amount. The pensions to a married couple are calculated and paid to the husband and the wife separately. Early retirement pension to a disabled person is payable at ages 18–64 (or 66) years, at a rate of 177,636 kroner to a single person. Early retirement pensions may be subject to income regulation. The same applies to the old-age pension.

Employment injuries insurance provides for disability or survivors' pensions and compensations. The scheme covers practically all employees.

Employment services are provided by regional public employment agencies. Insurance against unemployment provides daily allowances and covers about 85% of the unemployed. The unemployment insurance system is based on state subsidized insurance funds linked to the trade unions. The unemployment insurance funds had a membership of 2,065,700 in Jan. 2010.

The *Social Assistance Act* comprises three acts (the act on active social policy, the act on social service and the act on integration of foreigners). From these acts individual benefits are applied, in contrast to the other fields of social legislation which apply to fixed benefits. Total social expenditure, including hospital and health services, statutory pensions, etc. amounted in the financial year 2008 to 515,935·0m. kroner.

RELIGION

There is complete religious liberty. The state church is the Evangelical-Lutheran to which 80·9% of the population belonged in 2010. There are ten dioceses, each with a Bishop. The Bishop together with the Chief Administrative Officer of the county make up the diocesan-governing body, responsible for all matters of ecclesiastical local finance and general administration. Bishops are appointed by the Crown after an election by the clergy and parish council members. Each diocese is divided into a number of deaneries (107 in the whole country), each with its own Dean and Deanery Committee, who have certain financial powers. 81% of church finance derives from a voluntary tax paid by members, at a rate between 0·4–1·5% of income depending upon location. A further 12% comes from state subsidiaries and 7% from other sources, such as church lands.

CULTURE

World Heritage Sites
Denmark has four sites on the UNESCO World Heritage List: the burial mounds, runic stones and church at Jelling (inscribed on the list in 1994); Roskilde Cathedral (1995); Kronborg Castle (2000); and Ilulissat Icefjord (2004), the sea mouth of Sermeq Kujalleq in Greenland.

Press
In 2009 there were 37 daily newspapers with a combined circulation of 1·66m. The newspaper with the largest average circulation in the period Jan.–June 2009 was *MetroXpress* (a free paper; 228,000 on weekdays), followed by *24timer* (also a free paper; 180,000 on weekdays) and *Urban* (again a free paper; 162,000 on weekdays).

Tourism
In 2009, 8,457,000 overnight tourists visited Denmark; foreign tourists spent some 35,482m. kroner in the same year. Foreigners spent 11,164,000 nights in holiday cottages, 4,258,000 nights in hotels and 2,750,000 nights at camping sites in 2009.

Festivals
Roskilde, one of Europe's largest music festivals, is held annually in July. Other festivals include the Winter Jazz Festival in late Jan./early Feb. which takes place across the country, Carnival in Aalborg, held every May, the Copenhagen Jazz Festival in July, Skanderborg festival (music) in Aug., Tønder Festival (folk music) in Aug., Aarhus Festival (performing arts) in Aug./Sept. and the Copenhagen International Film Festival in Sept./Oct.

DIPLOMATIC REPRESENTATIVES

Of Denmark in the United Kingdom (55 Sloane St., London, SW1X 9SR)
Ambassador: Anne Hedensted Steffensen.

Of the United Kingdom in Denmark (Kastelsvej 36–40, DK-2100, Copenhagen Ø)
Ambassador: Vivien Life.

Of Denmark in the USA (3200 Whitehaven St., NW, Washington, D.C., 20008)
Ambassador: Peter Taksoe-Jensen.

Of the USA in Denmark (Dag Hammarskjölds Allé 24, DK-2100, Copenhagen Ø)
Ambassador: Laurie S. Fulton.

Of Denmark to the United Nations
Ambassador: Carsten Staur.

Of Denmark to the European Union
Permanent Representative: Jeppe Tranholm-Mikkelsen.

FURTHER READING

Statistical Information: Danmarks Statistik was founded in 1849 and reorganized in 1966 as an independent institution; it is administratively placed under the Minister of Economic Affairs. Its main publications are: *Statistisk Årbog* (Statistical Yearbook). From 1896: *Statistiske Efterretninger* (Statistical News). *Konjunkturstatistik* (Main indicators); *Statistisk Tiårsoversigt* (Statistical Ten-Year Review).

Dania polyglotta. Annual Bibliography of Books . . . in Foreign Languages. Annual
Kongelig Dansk Hof og Statskalender. Annual
Jespersen, Knud J. V., *A History of Denmark.* 2nd ed. 2011
Larsen, Henrik, *Analysing Small State Foreign Policy in the EU: The Case of Denmark.* 2005

National library: Det kongelige Bibliotek, POB 2149, DK-1016 Copenhagen K. *Director:* Erland Kolding Nielsen.
National Statistical Office: Statistics Denmark, Sejrøgade 11, DK-2100 Copenhagen Ø. *Director General:* Jan Plovsing.
Website: http://www.dst.dk

The Faroe Islands

Føroyar/Færøerne

KEY HISTORICAL EVENTS

A Norwegian province until the peace treaty of 14 Jan. 1814, the islands have been represented by two members in the Danish parliament since 1851. In 1852 they were granted an elected parliament which in 1948 secured a degree of home-rule. The islands are not part of the EU. Recently, negotiations for independence were given a push by the prospect of exploiting offshore oil and gas.

TERRITORY AND POPULATION

The archipelago is situated due north of Scotland, 300 km from the Shetland Islands, 675 km from Norway and 450 km from Iceland, with a total land area of 1,399 sq. km (540 sq. miles). There are 17 inhabited islands (the main ones being Streymoy, Eysturoy, Vágoy, Suðuroy, Sandoy and Borðoy) and numerous islets, all mountainous and of volcanic origin. Population in Jan. 2010 was 48,650; density, 34·8 per sq. km. In 2007 an estimated 59·1% of the population lived in rural areas. The capital is Tórshavn (12,375 residents in Jan. 2010) on Streymoy.

The official languages are Faroese and Danish.

SOCIAL STATISTICS

Birth rate per 1,000 inhabitants (2011), 11·9; death rate, 7·5. Life expectancy at birth (2010–11): 81·8 years.

CONSTITUTION AND GOVERNMENT

There is a 33-member parliament, the *Løgting*, which is elected by proportional representation by universal suffrage at age 18. Parliament elects a government of at least three members that administers home rule. Denmark is represented in parliament by the *High Commissioner*. A referendum was to be held on 26 May

2001 on the government's plan to move towards full sovereignty, but it was called off after the Danish prime minister at the time Poul Nyrup Rasmussen stated that subsidies would cease after four years if the islanders voted for independence.

RECENT ELECTIONS

Parliamentary elections were held on 29 Oct. 2011: the Union Party (Sambandsflokkurin) won 8 seats with 24·7% of the vote; the People's Party (Fólkaflokkurin) 8 seats (22·5%); the Republic (Tjóðveldi) 6 (18·3%); the Social Democratic Party (Javnaðarflokkurin) 6 (17·7%); Progress (Framsókn) 2 (6·3%); the Centre Party (Miðflokkurin) 2 (6·2%); and the Self-Government Party (Sjálvstýrisflokkurin) 1 (4·2%). Turnout was 86·6%.

CURRENT GOVERNMENT

High Commissioner: Dan M. Knudsen (b. 1962; took office on 1 Jan. 2008).

Prime Minister: Kaj Leo Johannesen; b. 1964 (Union Party; took office on 26 Sept. 2008).

Office of the Prime Minister: http://www.tinganes.fo

ECONOMY

Currency

Since 1940 the currency has been the Faroese *króna* (kr.) which remains freely interchangeable with the Danish krone.

Budget

In 2008 revenues totalled 4,332m. kr. and expenditures 4,315m. kr. Tax revenue constituted 85·4% of revenues in 2008; current expenditure accounted for 95·2% of expenditures.

Performance

Total GDP in 2009 was US$2·2bn.

Banking and Finance

The largest bank is the state-owned Føroya Banki. There are four other banks.

ENERGY AND NATURAL RESOURCES

Environment

Carbon dioxide emissions from the consumption and flaring of fossil fuels in 2008 were the equivalent of 16·8 tonnes per capita.

Electricity

Installed capacity was approximately 100,000 kW in 2007. Total production in 2007 was 269m. kWh, of which 39% was hydro-electric. There are five hydro-electric stations at Vestmanna on Streymoy and one at Eiði on Eysturoy. Consumption per capita was an estimated 5,554 kWh in 2007.

Agriculture

Only 2% of the surface is cultivated; it is chiefly used for sheep and cattle grazing. Potatoes are grown for home consumption. Livestock (2009 estimate): sheep, 68,000; cattle, 2,000.

Fisheries

Fishery products, including farmed salmon, represent 85% of Faroese merchandise exports. Total catch in 2011 was 354,954 tonnes (down from a peak of 623,122 tonnes in 2006).

INTERNATIONAL TRADE

Imports and Exports

Trade, 2006, in US$1m. (2005 in brackets): imports c.i.f., 783·2 (746·6); exports f.o.b., 631·0 (601·8). In 2006 machinery and transport equipment accounted for 28·3% of imports, mineral fuels 18·8% and manufactured goods 14·1%. Chilled and frozen fish constituted 61·1% of exports in 2006, salted and smoked fish 15·2% and feeding stuff for animals 11·4%. Denmark supplied 30·1% of imports in 2006, Norway 20·3% and Germany 7·6%; the United Kingdom took 27·2% of exports in 2006, Denmark 12·2% and Norway 10·3%.

COMMUNICATIONS

Roads

In 2006 there were 463 km of highways. At 1 Jan. 2012 there were 20,050 private cars and 4,214 lorries and vans.

Civil Aviation

The airport is on Vágoy, from which there are regular services to Aberdeen, Billund, Copenhagen and Reykjavík.

Shipping

The chief port is Tórshavn, with smaller ports at Klaksvik, Vestmanna, Skálafjørður, Tvøroyri, Vágur and Fuglafjørður. In Dec. 2007 there were 257 vessels of 20 GT or over (including 150 fishing vessels), totalling 252,000 GT.

Telecommunications

There were 20,200 landline telephone subscriptions in 2010 (or 414·2 per 1,000 inhabitants) and 59,400 mobile phone subscriptions (1,220·5 per 1,000 inhabitants). In 2010, 75·2% of the population were internet users.

SOCIAL INSTITUTIONS

Education

In 2008–09 there were 5,113 primary and 1,899 secondary school pupils (total of 649 teachers).

Health

In 2011 there were 81 physicians, 40 dentists and 375 nurses; and three hospitals with 243 beds.

RELIGION

About 80% are Evangelical Lutherans and 20% are Plymouth Brethren, or belong to small communities of Roman Catholics, Pentecostalists, Adventists, Jehovah's Witnesses and Bahais.

CULTURE

Press

In 2008 there were two daily newspapers (*Dimmalætting* and *Sosialurin*) and five non-dailies (four paid-for and one free). The dailies had a combined circulation of 16,000.

FURTHER READING

Árbók fyri Føroyar. Annual.
Rutherford, G. K. (ed.) *The Physical Environment of the Fœroe Islands.* 1982
Wylie, J., *The Faroe Islands: Interpretations of History.* 1987

National Statistical Office: Hagstova Føroya, Glyvursvegur 1, PO Box 2068, FO-165 Argir.
Website: http://www.hagstova.fo

Greenland

Grønland/Kalaallit Nunaat

KEY HISTORICAL EVENTS

A Danish possession since 1380, Greenland became an integral part of the Danish kingdom on 5 June 1953. Following a referendum in Jan. 1979, home rule was introduced from 1 May 1979. Having joined the then European Economic Community (now European Union) along with Denmark in Jan. 1973, a referendum was held in Feb. 1982 on Greenland's continued membership. With the vote being in favour of withdrawal, Greenland officially left on 1 Jan. 1985. In June 2009 laws providing for an extension of Greenland's autonomy came into force, providing Greenland with increased control over its energy resources and the adoption of Greenlandic as the sole official language.

TERRITORY AND POPULATION

Area, 2,166,086 sq. km (840,000 sq. miles), made up of 1,755,437 sq. km of ice cap and 410,449 sq. km of ice-free land. The population at 1 Jan. 2009 was 56,194, of whom 50,023 were born in Greenland and 6,171 were born outside Greenland; density 0·03 per sq. km. The population of the four communes in Jan. 2009 was as follows: Kujalleq, 7,631; Qaasuitsup, 17,678; Qeqqata, 9,686; Sermersooq, 20,955; and 244 not belonging to any specific municipality. The capital is Nuuk (Godthåb), with a population in Jan. 2007 of 14,719. In Jan. 2007, 47,000 persons were urban (83%).

The predominant language is Kalaallisut (Greenlandic), which since June 2009 has been the sole official language. Most of the population also speak Danish, which also had official status until June 2009.

SOCIAL STATISTICS

Live births (2005), 887; deaths (2005), 466. Number of abortions (2005): 899. Birth rate per 1,000 population (2005), 15·7; death rate per 1,000 population (2005), 8·2. There were 49 suicides in 2005. Population growth rate (2005), –0·1%.

CONSTITUTION AND GOVERNMENT

There is a 31-member Home Rule Parliament, which is elected for four-year terms and meets two to three times a year. The seven-member cabinet is elected by parliament. Ministers need not be members of parliament. In accordance with the Home Rule Act, the Greenland Home Rule government is constituted by an elected parliament, *Landstinget* (The Greenland Parliament), and an administration headed by a local government, *Landsstyret* (The Cabinet).

Greenland elects two representatives to the Danish parliament (Folketing). Denmark is represented by an appointed High Commissioner. Following a referendum in Nov. 2008 approved by 75·5% of voters and ratification by the parliaments of Greenland and Denmark, a series of autonomous reforms came into operation on 21 June 2009. These include extending Greenland's authority over its police force, courts and coastguard as well as revenues derived from its natural resources. Greenlandic is the sole official language. Denmark, meanwhile, has cut its annual subsidies to the island. The moves are considered a major staging post to independence.

RECENT ELECTIONS

At parliamentary elections held on 12 March 2013 the opposition Siumut (Social Democratic) won 14 of 31 seats with 42·8% of votes cast, the ruling Inuit Ataqatigiit (leftist) 11 (34·4%), Solidarity 2 (8·1%), Partii Inuit 2 (6·4%) and the Democrats 2 (6·2%). Turnout was 74·2%.

CURRENT GOVERNMENT

A coalition government of Siumut, Atassut and Partii Inuit was formed in March 2013.

Prime Minister: Aleqa Hammond; b. 1965 (Siumut, in office since 5 April 2013).

High Commissioner: Mikaela Engell (appointed 2011).

Greenland Home Rule Website: http://www.nanoq.gl

ECONOMY

Currency

The unit of currency is the *Danish krone*.

Budget

The budget (*finanslovsforslag*) for the following year must be approved by the Home Rule Parliament (*Landstinget*) no later than 31 Oct.

The following table shows the actual revenue and expenditure as shown in Home Rule government accounts for the calendar years 2003–05 and the approved budget figures for 2006 and 2007. Figures are in 1m. kroner.

	2003	2004	2005	2006	2007
Revenue	5,315	5,395	5,648	5,834	5,808
Expenditure	5,184	5,294	5,341	5,473	5,670

Performance

In 2006 and 2007 the rate of economic growth quickened, primarily as a result of high growth rates in private consumption expenditure and investment in new buildings. However, owing to the global financial crisis growth then slowed and in 2009 the economy shrank by 5·4%. Total GDP was US$1·3bn. in 2009.

Banking and Finance

There are two private banks, Grønlandsbanken and Sparbank Vest.

ENERGY AND NATURAL RESOURCES

Environment

Greenland's carbon dioxide emissions from the consumption and flaring of fossil fuels in 2008 were the equivalent of 11·2 tonnes per capita.

Electricity

In 2005 the production of electricity in the cities totalled 331m. kWh.

Oil and Gas

Imports of fuel and fuel oil (2005), 275·5m. litres, with a value of 741·6m. kroner. There are indications of significant oil deposits off the west coast of Greenland, with exploration ongoing.

Agriculture

Livestock, 2003: sheep, 19,259; reindeer, 3,100. There are about 57 sheep-breeding farms in southwest Greenland.

Fisheries

Fishing and product-processing are the principal industry. The total catch in 2007 was 233,754 tonnes. In 2006, 84% of Greenland's total exports were derived from fish products, particularly prawns and halibut. In 2006, 192 large whales and 2,787 smaller cetacean mammals such as porpoise were caught (subject to the International Whaling Commission's regulations); plus 130,927 seals.

INDUSTRY

Six shipyards repair and maintain ships and produce industrial tanks, containers and steel constructions for building.

Labour

At 1 Jan. 2007 the potential labour force was 38,652.

INTERNATIONAL TRADE

Imports and Exports

In 2006 imports totalled 3,454m. kroner and exports 2,418m. kroner.

Principal import commodities in 2006 were machinery and vehicles (22%); mineral fuels (22%); and food, beverages and tobacco products (19%). Main export commodities were fish and fish products (84%), notably shrimp and halibut.

Principal import sources, 2006: Denmark, 59·7%; Sweden, 22·5%; Germany, 3·8%; Norway, 1·8%. Main export markets, 2006: Denmark, 86·6%; Spain, 7·0%; UK, 1·9%; Iceland, 1·6%.

COMMUNICATIONS

Roads

There are no roads between towns. Registered vehicles (2004): passenger cars, 2,861; commercial vehicles and trucks, 1,467; total (including others), 4,827.

Civil Aviation

Number of passengers to/from Greenland (2003): 98,118. Domestic flights—number of passengers (2003): aeroplanes, 177,554; helicopters, 39,457. Air Greenland operates domestic services and international flights to Denmark and Baltimore in the USA. There are international airports at Kangerlussuaq (Søndre Strømfjord), Narsarsuaq and Kulusuk and 18 local airports/heliports with scheduled services. There are cargo services to Denmark, Iceland and Canada.

Shipping

There are no overseas passenger services. In 2006, 43,448 passengers were carried on coastal services. There are cargo services to Denmark, Iceland and St John's (Canada).

Telecommunications

In 2008 there were 22,800 main (fixed) telephone lines; mobile phone subscribers numbered 55,800 in 2008 (97·4 per 100 persons). There were 628·3 internet users per 1,000 inhabitants in 2009. Fixed internet subscriptions totalled 12,200 in 2009 (212·8 per 1,000 inhabitants).

SOCIAL INSTITUTIONS

Justice

Cases in the High Court in Nuuk are led by one professional judge and two lay magistrates, while there are 18 district courts under lay assessors.

The population in penal institutions in Dec. 2009 was 194 (340 per 100,000 of national population).

Education

Education is compulsory from six to 15 years. A further three years of schooling are optional. Primary schools (2006–07) had 10,688 pupils and 1,216 teachers; secondary schools, 780 pupils.

Health

The medical service is free to all citizens. There is a central hospital in Nuuk and 15 smaller district hospitals. In 2006 there were 95 doctors.

Non-natural death occurred in approximately one-fifth of all deaths in 2005. Suicide is the most dominant non-natural cause of death.

Welfare

Pensions are granted to persons who are 63 or above. The right to maternity leave has been extended to two weeks before the expected birth and up to 20 weeks after birth against a total of 21 weeks in earlier regulations. The father's right to one week's paternity leave in connection with the birth was extended to three weeks from 1 Jan. 2000. Wage earners who are members of SIK (The National Workers' Union) receive financial assistance (unemployment benefit) according to fixed rates, in case of unemployment or illness.

RELIGION

About 80% of the population are Evangelical Lutherans. In 2006 there were 17 parishes with 92 churches and chapels, and 27 ministers.

CULTURE

World Heritage Sites

Greenland has one site on the UNESCO World Heritage List: the Ilulissat Icefjord, the sea mouth of Sermeq Kujalleq on the west coast of the island (inscribed on the list in 2004).

Press

In 2008 there were not any daily newspapers but there were two national non-dailies. *Atuagagdliutit/Grønlandsposten*, generally referred to as AG, appears twice a week and *Sermitsiaq* once a week. Local newspapers are also published.

Tourism

In 2006 visitors stayed 245,432 nights in hotels (including 104,012 Greenlandic citizens).

FURTHER READING

Statistics Greenland. *Statistical Yearbook 2010* in English. *Greenland in Figures 2012* in English. Online only

Gad, F., *A History of Greenland*. 2 vols. 1970–73

Greenland National Library, P. O. Box 1011, DK-3900 Nuuk
National Statistical Office: Statistics Greenland, PO Box 1025, DK-3900 Nuuk.
Website: http://www.stat.gl

DJIBOUTI

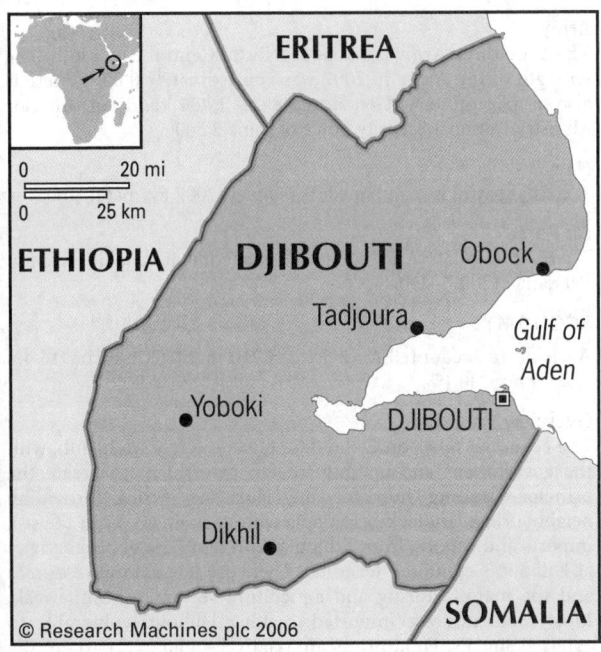

ERITREA

ETHIOPIA   DJIBOUTI   Obock

Tadjoura   Gulf of Aden

Yoboki   DJIBOUTI

Dikhil

SOMALIA

0 20 mi
0 25 km

© Research Machines plc 2006

Jumhouriyya Djibouti
(Republic of Djibouti)

Capital: Djibouti
Population projection, 2015: 975,000
GNI per capita, 2011: (PPP$) 2,335
HDI/world rank: 0·430/165
Internet domain extension: .dj

KEY HISTORICAL EVENTS

At a referendum held on 19 March 1967, 60% of the electorate voted for continued association with France rather than independence. France affirmed that the Territory of the Afars and the Issas was destined for independence but no date was fixed. Independence as the Republic of Djibouti was achieved on 27 June 1977. Afar rebels in the north, belonging to the Front for the Restoration of Unity and Democracy (FRUD), signed a 'Peace and National Reconciliation Agreement' with the government on 26 Dec. 1994, envisaging the formation of a national coalition government, the redrafting of the electoral roll and the integration of FRUD militants into the armed forces and civil service.

TERRITORY AND POPULATION

Djibouti is in effect a city-state surrounded by a semi-desert hinterland. It is bounded in the northwest by Eritrea, northeast by the Gulf of Aden, southeast by Somalia and southwest by Ethiopia. The area is 23,200 sq. km (8,958 sq. miles). The provisional population at the 2009 census was 818,159. In 2011, 76·3% of the population lived in urban areas. Around half the population in 2005 were Somali (Issa, Gadaboursi and Issaq), 35% Afar, with some Europeans (mainly French) and Arabs.

The UN gives a projected population for 2015 of 975,000.

There are five administrative regions, plus the city of Djibouti (areas in sq. km): Ali-Sabieh (2,200); Arta (1,800); Dikhil (7,200); Djibouti (200); Obock (4,700); Tadjourah (7,100). The capital is Djibouti (2009 provisional population, 475,322).

French and Arabic are official languages; Somali and Afar are also spoken.

SOCIAL STATISTICS

2008 estimates: births, 24,100; deaths, 9,300. Rates (2008 estimates, per 1,000 population); birth, 28·4; death, 11·0. 2006: marriages, 3,059; divorces, 723. Annual population growth rate, 2000–08, 1·9%. Infant mortality, 2010, 73 per 1,000 live births. Expectation of life, 2007: 53·7 years for men, 56·5 for women. Fertility rate, 2008, 3·9 children per woman.

CLIMATE

Conditions are hot throughout the year, with very little rain. Djibouti, Jan. 78°F (25·6°C), July 96°F (35·6°C). Annual rainfall 5" (130 mm).

CONSTITUTION AND GOVERNMENT

After a referendum at which turnout was 70%, a new constitution was approved on 4 Sept. 1992 by 96·63% of votes cast, which permits the existence of up to four political parties. Parties are required to maintain an ethnic balance in their membership. The *President* is directly elected for a renewable six-year term. Parliament is a 65-member *National Assembly* elected for five-year terms. In April 2010 the constitution was amended to allow the president to stand for a third consecutive term. It also provided for the creation of a *Senate*.

National Anthem

'Hinjinne u sara kaca' ('Arise with strength'); words by A. Elmi, tune by A. Robleh.

RECENT ELECTIONS

In the presidential election on 8 April 2011 Ismail Omar Guelleh was re-elected with 80·6% of the vote. Mohamed Warsama Ragueh took 19·4%. Turnout was 69·7%.

At the parliamentary elections of 22 Feb. 2013 the Union for a Presidential Majority won 43 seats with 61·5% of votes cast; the Union for National Salvation, 21 with 35·6%; the Centre of Unified Democrats, 1 with 3·0%. Turnout was 69·2%. Unlike in 2003 and 2008 the opposition did not boycott the election, but it did reject the result and claimed that the vote was rigged.

CURRENT GOVERNMENT

President: Ismail Omar Guelleh; b. 1947 (RPP; sworn in 8 May 1999 and re-elected in April 2005 and April 2011).

In April 2013 the Council of Ministers comprised:

Prime Minister: Abdoulkader Kamil Mohamed; b. 1951 (RPP; took office 1 April 2013).

Minister of Agriculture, Water, Livestock and Fisheries: Mohamed Ahmed Awaleh. *Budget:* Bodeh Ahmed Robleh. *Communication, in Charge of Posts and Telecommunications:* Ali Hassan Bahdon. *Defence:* Hassan Darar Houffaneh. *Economy and Finance, in Charge of Industry:* Ilyas Moussa Dawaleh. *Energy and Natural Resources:* Ali Yacoub Mahamoud. *Equipment and Transport:* Moussa Ahmed Hassan. *Foreign Affairs and International Co-operation, and Government Spokesman:* Mahamoud Ali Youssouf. *Health:* Dr Kassim Issak Osman. *Higher Education and Research:* Nabil Mohamed Ahmed. *Housing, Town Planning and Environment:* Mohamed Moussa Ibrahim Balala. *Interior:* Hassan Omar Mohamed. *Justice, in Charge of Human Rights:* Ali Farah Assoweh. *Labour, in Charge of Administrative Reform:* Abdi Hussein Ahmed. *Muslim Affairs, Culture and*

Awqaf: Aden Hassan Aden. *National Education and Vocational Training:* Dr Djama Elmi Okieh. *Promotion of Women and Family Planning, in Charge of Parliamentary Affairs:* Hasna Barkad Daoud.

Government Website (French only): http://www.presidence.dj

CURRENT LEADERS

Ismail Omar Guelleh

Position
President

Introduction
Ismail Omar Guelleh was elected for a third six-year term as president in April 2011. He succeeded his uncle in 1999, becoming the country's second president since independence in 1977.

Early Life
Ismail Omar Guelleh was born on 27 Nov. 1947 in Dire Dawa, Ethiopia. He is the grandson of Guelleh Batal, one of the chiefs of the Issa clan who signed the 1917 agreement placing the Issa territories under French administration. From 1974 Guelleh became increasingly involved in the fight for independence as a member of the African Popular League for Independence (LPAI). Following Djibouti's declaration of independence on 27 June 1977, Guelleh was appointed principal private secretary to the president, his uncle, Hassan Gouled Aptidon.

Guelleh became head of the security services and joined the People's Rally for Democracy (RPP) when it was established in March 1979. He became head of the party's cultural commission in 1981, the year in which Gouled made the Issa-dominated RPP the country's only legal political party, causing resentment among the Afar community. Civil war followed in 1991. When Gouled announced that he would not contest the April 1999 presidential elections, Guelleh stood as the RPP candidate. He was sworn in as president on 8 May 1999.

Career in Office
In Feb. 2000 Guelleh signed a peace agreement with the radical faction of the Afar party, the Front for the Restoration of Unity and Democracy (FRUD), ending nine years of civil war. During the multi-party elections of Jan. 2003 the coalition supporting him—the Union for a Presidential Majority—won all 65 seats, prompting opposition accusations of vote-rigging. In the run-up to the April 2005 presidential election, Guelleh pledged to reduce poverty and the country's dependence on food imports while boosting women's rights and institutional accountability. The election was boycotted by the opposition and he was sworn in for a second six-year term with 100% of votes cast. In Feb. 2008 the Union for a Presidential Majority again won all 65 seats in parliamentary elections boycotted by the main opposition parties. A constitutional amendment permitting the president to seek a third term was approved by parliament in April 2010, and in April 2011 Guelleh was re-elected with about 80% of the vote.

In Sept. 2002, in support of the US-led war on terror, Guelleh allowed 900 US troops to be based in Djibouti. Although his government has denied interference in neighbouring Somalia's affairs, US air strikes in Jan. 2007 on the retreating Islamist militias that had earlier taken control of the Somali capital, Mogadishu, and much of the south of the country, were launched from the US base in Djibouti. In June 2008 border clashes between troops from Djibouti and Eritrea led Guelleh to declare war with the neighbouring state. However, following the imposition of sanctions against Eritrea by the UN Security Council in Dec. 2009, both sides agreed in June 2010 to resolve their border issues peacefully. Guelleh won a third term in the presidential election of April 2011.

DEFENCE

France maintains a naval base and forces numbering 2,000 under an agreement renewed in Feb. 1991. Defence expenditure totalled US$15m. in 2008 (US$22 per capita), representing 1·5% of GDP.

Army
There are three Army commands: North, Central and South. The strength of the Army in 2007 was approximately 8,000. There is also a paramilitary Gendarmerie of 1,400, and an Interior Ministry National Security Force of some 2,500.

Navy
A coastal patrol is maintained. Personnel (2007 estimate), 200.

Air Force
There is a small Air Force with no combat-capable aircraft. Personnel (2007), 250.

ECONOMY

Agriculture accounted for 3·5% of GDP in 2006, industry 16·4% and services 80·1%.

Overview
The economy relies on Djibouti's status as a free trade hub, with the government encouraging foreign investment to create the principal trading hub for the Horn of Africa. Providing neighbouring, landlocked Ethiopia with its main access to the sea, imports and exports from Ethiopia represent 70% of port activity at Djibouti's container terminal. There are few natural resources and the manufacturing and agriculture sectors are both weak. Most foodstuffs are imported, making Djibouti vulnerable to external shocks. Food prices are relatively high and parts of the population depend on food aid.

Djibouti weathered the global economic crisis well, although real GDP growth slowed from 5·8% in 2008 to 5·0% in 2009 and FDI declined from 23·8% of GDP to 18·0%. Growth was driven mainly by expansion in construction, banking and shipping, as well as public investment. FDI and transshipment remain weakened.

Unemployment was around 60% in 2007 and recent growth has not translated into reduced unemployment rates. Pegged to the US dollar, the currency is overvalued. Other challenges to economic wellbeing include high population growth and a high risk of debt distress (with debt being sold on to third parties at below face value).

Currency
The currency is the *Djibouti franc* (DJF), notionally of 100 *centimes.* Foreign exchange reserves were US$90m. in July 2005 and total money supply was 49,822m. Djibouti francs. Inflation was 12·0% in 2008 but fell to 4·0% in 2010 before rising to 5·1% in 2011.

Budget
Revenues in 2009 were 67·7bn. Djibouti francs and expenditures 69·8bn. Djibouti francs.

Performance
Real GDP growth was 5·0% in 2009, 3·5% in 2010 and 4·5% in 2011. Total GDP in 2009 was US$1·0bn.

Banking and Finance
The Banque Nationale de Djibouti is the bank of issue. There are three commercial banks and a development bank. Foreign debt totalled US$472m. in 2007.

ENERGY AND NATURAL RESOURCES

Environment
Djibouti's carbon dioxide emissions from the consumption and flaring of fossil fuels in 2008 were the equivalent of 2·6 tonnes per capita.

Electricity
Installed capacity in 2007 was an estimated 118,000 kW. Production in 2007 was 292m. kWh; consumption per capita was 350 kWh in 2007.

Agriculture
There were around 1,000 ha. of arable land in 2002. Production is dependent on irrigation which in 2002 covered 1,000 ha. Vegetable production (2003) 24,000 tonnes. The most common crops are tomatoes, mangoes, papayas and melons. Livestock (2005 estimates): goats, 512,000; sheep, 466,000; cattle, 297,000; camels, 69,000. Livestock products, 2005 estimates: meat, 11,000 tonnes; milk, 14,000 tonnes.

Forestry
In 2010 the area under forests was 6,000 ha., or 0·2% of the total land area.

Fisheries
In 2009 the catch was 1,058 tonnes, entirely from sea fishing.

INDUSTRY
Labour
In 2007 the economically active population totalled 366,000. Unemployment in 2009 was estimated at 60%.

INTERNATIONAL TRADE
Imports and Exports
The main economic activity is the operation of the port. Exports are largely re-exports. In 2006 imports totalled US$335·7m. and exports US$55·2m. The chief imports are cotton goods, sugar, cement, flour, fuel oil and vehicles; the chief exports are hides, cattle and coffee (transit from Ethiopia).

Main import suppliers are Saudi Arabia, India, UAE and China. Main export markets are Saudi Arabia, Kenya, Egypt and Yemen.

COMMUNICATIONS
Roads
In 2002 there were estimated to be 2,890 km of roads, of which 12·6% were hard-surfaced. An estimated 15,700 passenger cars were in use in 2002 (23·5 per 1,000 inhabitants), plus 3,200 vans and trucks.

Rail
For the line from Djibouti to Addis Ababa, of which 97 km lie within Djibouti, see ETHIOPIA: Communications. Traffic carried is mainly in transit to and from Ethiopia.

Civil Aviation
There is an international airport at Djibouti (Ambouli), 5 km south of Djibouti. Djibouti-based carriers are Daallo Airlines and Djibouti Airlines. They operated flights in 2003 to Addis Ababa, Asmara, Borama, Bossaso, Burao, Dire Dawa, Dubai, Galcaio, Hargeisa, Jeddah, London, Mogadishu, Paris and Ta'iz.

Shipping
Djibouti is a free port and container terminal. In 2008, 5·82m. tonnes of cargo were handled (7·33m. tonnes in 2007).

Telecommunications
There were 18,500 fixed telephone lines in 2010 (20·8 per 1,000 inhabitants). Mobile phone subscribers numbered 165,600 in 2010. There were 22·6 internet users per 1,000 inhabitants in 2008. Fixed internet subscriptions totalled 11,900 in 2010 (13·4 per 1,000 inhabitants).

SOCIAL INSTITUTIONS
Justice
There is a Court of First Instance and a Court of Appeal in the capital. The judicial system is based on Islamic law. The death penalty was abolished for all crimes in 1994.

The population in penal institutions in 2010 was approximately 600 (68 per 100,000 of national population).

Education
Adult literacy in 2001 was 65·5% (76·1% of men; 55·5% of women). In 2007 there were 56,667 pupils and 1,597 teaching staff in primary schools, and 34,667 pupils and 1,021 teachers in secondary schools. In 2007 there were 2,192 students at tertiary education institutions and 121 academic staff.

In 2007 public expenditure on education came to 7·8% of GNI.

Health
In 2007 there were a total of 1,220 hospital beds. There were 85 physicians, nine dentists, 196 nurses, 90 midwives and eight pharmacists in 2007.

RELIGION
In 2001, 96% of the population were Muslim; there were small Roman Catholic, Protestant and Orthodox minorities.

CULTURE
Press
There are no daily newspapers; in 2008 the government-owned La Nation was published four times a week.

Tourism
There were 40,000 foreign tourists staying at hotels and similar establishments in 2006; tourist spending (excluding passenger transport) totalled US$9m.

DIPLOMATIC REPRESENTATIVES
Of Djibouti in the United Kingdom
Ambassador: Rachad Farah (resides in Paris).

Of the United Kingdom in Djibouti
Ambassador: Gregory Dorey, CVO (resides in Addis Ababa, Ethiopia).

Of Djibouti in the USA and to the United Nations (1156 15th St., NW, Suite 515, Washington, D.C., 20005)
Ambassador: Roble Olhaye.

Of the USA in Djibouti (B. P. 185, Lot no. 350-B, Lotissement Haramous, Djibouti)
Ambassador: Geeta Pasi.

Of Djibouti to the European Union
Ambassador: Badri Ali Bogoreh.

FURTHER READING
Direction Nationale de la Statistique. Annuaire Statistique de Djibouti
Alwan, Daoud A., Historical Dictionary of Djibouti. 2000

National Statistical Office: Direction Nationale de la Statistique, Ministère de l'Économie et des Finances, chargé de l'Industrie, BP 13, Djibouti.
Website (French only): http://www.ministere-finances.dj

DOMINICA

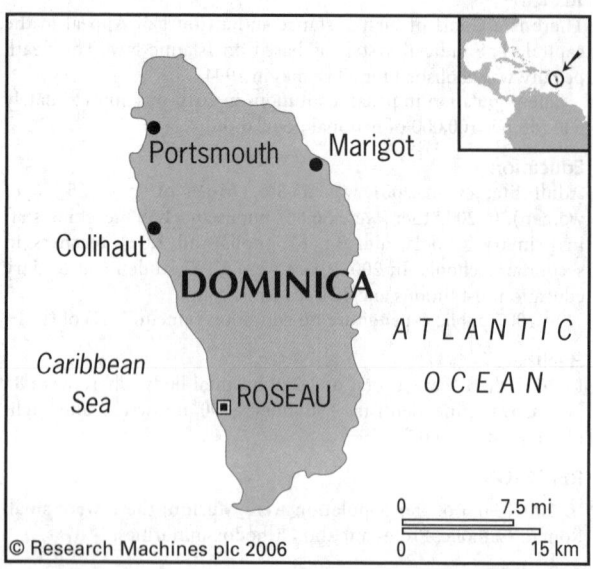

Commonwealth of Dominica

Capital: Roseau
Population, 2011: 71,000
GNI per capita, 2011: (PPP$) 7,889
HDI/world rank: 0·724/81
Internet domain extension: .dm

KEY HISTORICAL EVENTS

When Christopher Columbus sighted Dominica on 3 Nov. 1493 it was occupied by Carib Indians, who are thought to have overrun the previous inhabitants, the Arawak, from around 1300. Dominica remained a 'Carib Isle' until the 1630s, when French farmers and missionaries established sugar plantations. Control was contested between the British and French until it was awarded to the British by the Treaty of Versailles in 1783. In March 1967 Dominica became a self-governing state within the West Indies Associated States, with Britain retaining control of external relations and defence. The island became an independent republic, the Commonwealth of Dominica, on 3 Nov. 1978.

TERRITORY AND POPULATION

Dominica is an island in the Windward group of the West Indies situated between Martinique and Guadeloupe. It has an area of 750 sq. km (290 sq. miles) and a provisional population at the 2011 census of 71,293. The population density in 2011 was 95·1 per sq. km.

In 2010, 67·2% of the population were urban. The chief town, Roseau, had 14,725 inhabitants (provisional) in 2011.

The population is mainly of African and mixed origins, with small white and Asian minorities. There is a Carib settlement of about 500, almost entirely of mixed blood.

The official language is English, although 90% of the population also speak a French Creole.

SOCIAL STATISTICS

Births, 2006 estimates, 1,080 (rate of 14·9 per 1,000 population); deaths, 540 (rate of 7·5); marriages (2009), 250; divorces (2009), 85. Life expectancy, 2007: male, 72 years; female, 76 years. Annual population growth rate, 2000–08, –0·2%. Infant mortality rate, 2010, 11 per 1,000 live births. Fertility rate, 2008, 2·1 births per woman.

CLIMATE

A tropical climate, with pleasant conditions between Dec. and March, but there is a rainy season from June to Oct., when hurricanes may occur. Rainfall is heavy, with coastal areas having 70" (1,750 mm) but the mountains may have up to 225" (6,250 mm). Roseau, Jan. 76°F (24·2°C), July 81°F (27·2°C). Annual rainfall 78" (1,956 mm).

CONSTITUTION AND GOVERNMENT

The head of state is the *President*, nominated by the *Prime Minister* and the Leader of the Opposition, and elected for a five-year term (renewable once) by the House of Assembly. The *House of Assembly* has 32 members, of whom 21 members are elected and nine nominated by the President in addition to the Speaker and the Attorney General.

National Anthem

'Isle of beauty, isle of splendour'; words by W. Pond, tune by L. M. Christian.

RECENT ELECTIONS

Elections were held on 18 Dec. 2009. The ruling Dominica Labour Party (DLP) won 18 of 21 elected seats (12 in 2005), the United Workers Party (UWP) won 3 seats (8 in 2005). The Dominica Freedom Party (DFP) did not win any seats.

CURRENT GOVERNMENT

President: Eliud Williams; b. 1948 (sworn in 17 Sept. 2012).

Prime Minister, Minister of Finance, Foreign Affairs and Information Technology: Roosevelt Skerrit; b. 1972 (DLP; sworn in 8 Jan. 2004 and again on 21 Dec. 2009).

In March 2013 the cabinet comprised:

Minister of Agriculture and Forestry: Matthew Walter. *Carib Affairs:* Ashton Graneau. *Culture, Youth and Sports:* Justina Charles. *Education and Human Resource Development:* Petter Saint-Jean. *Employment, Trade, Industry and Diaspora Affairs:* Colin McIntyre. *Environment, Natural Resources, Physical Planning and Fisheries:* Kenneth Darroux. *Health:* Julius Timothy. *Information, Telecommunications and Constituency Empowerment:* Ambrose George. *Lands, Housing, Settlements and Water Resource Management:* Reginald Austrie. *National Security, Labour and Immigration:* Charles Savarin. *Public Works, Energy and Ports:* Rayburn Blackmore. *Social Services, Community Development and Gender Affairs:* Gloria Shillingford. *Tourism and Legal Affairs:* Ian Douglas. *Attorney General:* Levi Peter.

Government Website: http://www.dominica.gov.dm

CURRENT LEADERS

Eliud Williams

Position
President

Introduction
Eliud Williams was elected president by parliamentary vote on 17 Sept. 2012 after the resignation of his predecessor, Nicholas Liverpool, owing to ill health. Williams' election to the largely ceremonial post was boycotted by the main opposition party, the United Workers Party (UWP), on the grounds that his nomination had been unconstitutional.

Early Life

Eliud Thaddeus Williams was born on 21 Aug. 1948. After working as a financial controller, he graduated in accounting from the University of the West Indies in 1985. He joined the Dominican civil service, first as a commissioner of co-operatives and, from 1987, as permanent secretary in various ministries including health and social security, and agriculture and the environment.

In 1995 Williams received his masters degree in business administration from the University of the West Indies. The following year he became chairman of the government's rural enterprise project, holding the position until 2000. In 2004 he became managing director of the Eastern Caribbean Telecommunications Authority, based in St Lucia, but left in 2008 to take on a senior position with WHITCO Inc., a management consultancy.

On 17 Sept. 2012 Nicholas Liverpool resigned as president and Williams was named as the Labour-led government's nominee to succeed him. However, the UWP argued that the proper procedure had not been followed by the prime minister, who did not consult the leader of the opposition over the selection. Despite the UWP's boycott, Williams received the necessary support from parliament and was sworn into office.

Career in Office

Williams is scheduled to serve out the remainder of Liverpool's term, scheduled to end in Oct. 2013.

Roosevelt Skerrit

Position

Prime Minister

Introduction

Roosevelt Skerrit became Dominica's youngest ever prime minister in Jan. 2004 when he took office following the death of his predecessor, Pierre Charles. Appointed by parliamentary recommendation to lead the coalition government, he was returned as prime minister in the May 2005 general election when the Dominica Labour Party (DLP) won an outright majority, and again in Dec. 2009.

Early Life

Roosevelt Skerrit was born in 1972 and grew up in Vieille Case in northeast Dominica. From 1994–97 he studied psychology and English at the University of Mississippi and New Mexico State University. On his return to Dominica he worked as a teacher, first in a high school then at the Dominica Community College. In 1999 he entered politics and in 2000 was elected as a DLP representative to the House of Assembly.

In the coalition government of the DLP and the Dominica Freedom Party (DFP), Skerrit served as minister for sports and youth affairs and later also for education. When Pierre Charles died of a heart attack in Jan. 2004, then President Nicholas Liverpool appointed Skerrit as his replacement.

Career in Office

Skerrit inherited a small working majority and sought to maintain unity in the coalition government. Against a background of economic troubles, the government introduced unpopular austerity measures, with a combination of spending cuts and higher taxes prompting strikes. During his first year in office he pursued a Caribbean Community (CARICOM) initiative to raise US$50m. of aid and has subsequently been involved in developing a regional stabilization fund under the auspices of the Caribbean Development Bank.

In March 2004 Skerrit reversed Dominica's traditional policy of pursuing diplomatic ties with Taiwan in preference to relations with China, obtaining a six-year aid package from China worth US$117m. In an attempt to reduce Dominica's dependence on agriculture, the government has invested in tourism and major infrastructure projects.

In the election of 5 May 2005 Skerrit led the DLP to victory with 12 of 21 elected seats. In Sept. 2005 Dominica was one of several CARICOM countries to enter into an oil-purchasing deal with Venezuela and in Oct. 2005 he secured around US$2m. of direct US aid for public and private sector investment. The International Monetary Fund commended his government in Dec. 2006 for its successful economic programme, including significant progress in debt restructuring. In Feb. 2009 Skerrit announced that his government had secured US$49m. in grants from Venezuela within the framework of the ALBA (Alianza Bolivariana para los Pueblos de Nuestra América) trade group of leftist Latin American states. In the Dec. 2009 election Skerrit again led the DLP to victory, this time with 18 of the 21 elected seats.

Skerrit served a six-month tenure as chairman of CARICOM in the first half of 2010.

ECONOMY

In 2009 agriculture accounted for 14% of GDP, industry 15% and services 71%.

Overview

Dominica is highly vulnerable to external shocks and susceptible to a variety of natural disasters. Traditionally the economy has relied on bananas as its main export earner but has large potential for tourism. Having completed an IMF Poverty Reduction and Growth Facility (PRGF) in 2006, the economy's fiscal position stabilized and the public debt-to-GDP ratio fell to below 95% in 2007. In Aug. 2007 Dominica was struck by Hurricane Dean, causing damage estimated at 20% of GDP.

The global downturn has harmed tourism and caused a decline in foreign direct investment and remittance inflows. The state has responded by maintaining post-hurricane levels of capital spending and increasing social assistance, made possible by prudent fiscal management since 2005 and strong VAT returns since its introduction in 2006.

Currency

The East Caribbean dollar (XCD) and the US dollar are legal tender. Foreign exchange reserves were US$50m. in July 2005 and total money supply was EC$149m. Inflation was 2·8% in 2010 and 1·4% in 2011.

Budget

The fiscal year begins on 1 July. Revenues for the fiscal year 2009–10 were EC$467·7m. and expenditures EC$468·8m.

The standard rate of VAT is 15% (reduced rate, 10%).

Performance

The economy contracted by 1·3% in 2009 but grew by 1·2% in 2010 and 1·0% in 2011. In 2011 total GDP was US$0·5bn.

Banking and Finance

The East Caribbean Central Bank based in St Kitts and Nevis functions as a central bank. The Governor is Sir Dwight Venner. In 2001 there were five commercial banks (four foreign, one domestic), a development bank and a credit union. Dominica is affiliated to the Eastern Caribbean Securities Exchange in Basseterre, St Kitts and Nevis. Total foreign debt was US$290m. in 2007.

ENERGY AND NATURAL RESOURCES

Environment

Carbon dioxide emissions from the consumption and flaring of fossil fuels in 2008 were the equivalent of 1·7 tonnes per capita.

Electricity

Installed capacity was an estimated 24,000 kW in 2007. Production in 2007 was approximately 85m. kWh. Consumption per capita in 2007 was about 1,174 kWh. There is a hydro-electric power station.

Agriculture

Agriculture employs 26% of the labour force. In 2007 there were around 5,000 ha. of arable land and 16,000 ha. of permanent crops. Estimated production, 2003, in 1,000 tonnes): bananas, 29; grapefruit and pomelos, 17; coconuts, 12; taro, 11; yams, 8; oranges, 7; plantains, 6. Livestock (2003 estimates): cattle, 13,000; goats, 10,000; sheep, 8,000; pigs, 5,000.

Forestry

In 2010 forests covered 45,000 ha., or 60% of the total land area.

Fisheries

In 2009 fish landings were 790 tonnes (all from sea fishing).

INDUSTRY

Manufactures include soap (10,500 tonnes in 2001), coconut oil, copra, cement blocks, furniture and footwear.

Labour

Around 25% of the economically active population are engaged in agriculture, fishing and forestry. In 2006 the minimum wage was US$0·75 an hour. The unemployment rate in 2003 was 15·7%.

INTERNATIONAL TRADE

Imports and Exports

In 2010 imports (c.i.f.) totalled US$224·6m. and exports (f.o.b.) US$34·1m. Main import sources in 2010 (in US$1m.) were: USA (98·3); Trinidad and Tobago (34·3); Venezuela (16·3). Main export markets in 2010 (in US$1m.) were: St Kitts and Nevis (6·0); Jamaica (5·9); Trinidad and Tobago (4·7).

Principal imports in 2010 (in US$1m.) were: machinery and transport equipment (51·0); mineral fuels, lubricants and related materials (38·8); food (37·0). Principal exports in 2010 (in US$1m.) were: chemicals and related products (16·5); food (7·2); miscellaneous manufactured articles (5·0).

COMMUNICATIONS

Roads

In 2002 there were an estimated 788 km of roads, of which 50·4% were paved. Approximately 10,300 passenger cars and 3,500 commercial vehicles were in use in 2002.

Civil Aviation

There are international airports at Melville Hall and Cane Field. In 2003 there were direct flights to Anguilla, Antigua, Barbados, British Virgin Islands, Grenada, Guadeloupe, Martinique, Puerto Rico, St Kitts, St Lucia, St Maarten, St Vincent, Trinidad and the US Virgin Islands.

Shipping

There are deep-water harbours at Roseau and Woodbridge Bay. Roseau has a cruise ship berth. In Jan. 2009 there were 60 ships of 300 GT or over registered (including 15 bulk carriers and nine oil tankers), totalling 936,000 GT.

Telecommunications

There were 15,500 fixed telephone lines in 2010 (228·5 per 1,000 inhabitants). Mobile phone subscribers numbered 98,100 in 2010. There were 474·5 internet users per 1,000 inhabitants in 2010.

SOCIAL INSTITUTIONS

Justice

There is a supreme court and 14 magistrates courts. Law is based on UK common law as exercised by the Eastern Caribbean Supreme Court on St Lucia. Final appeal lies to the UK Privy Council. Dominica was one of 12 countries to sign an agreement establishing a Caribbean Court of Justice to replace the British Privy Council as the highest civil and criminal court. The court was inaugurated at Port-of-Spain, Trinidad on 16 April 2005 although Dominica has yet to accept it as its court of final appeal.

The police force has a residual responsibility for defence. The population in penal institutions in Aug. 2007 was 254 (equivalent to 348 per 100,000 of national population).

Education

In 1998 adult literacy was 94%. Education is free and compulsory between the ages of five and 16 years. In 2007 there were 499 teaching staff and 8,643 pupils in primary schools, and 469 teaching staff and 7,481 pupils in general secondary level education. The leading higher education institution is Ross University School of Medicine, established in 1978. In 2007 public expenditure on education came to 5·5% of GNI.

Health

In 2005 there were 39 hospital beds per 10,000 inhabitants. There were 38 physicians, ten dentists and 361 nurses in 2003. Large numbers of professional nurses take up employment abroad, especially in the USA, causing a shortage of health care workers in Dominica.

RELIGION

70% of the population was Roman Catholic in 2001.

CULTURE

World Heritage Sites

Dominica has one site on the UNESCO World Heritage List: Morne Trois Pitons National Park (1997), a tropical forest centred on the Morne Trois Pitons volcano.

Press

In 2008 there were no daily newspapers but there were four weeklies—*The Chronicle, The Sun, The Times* and *The Tropical Star*.

Tourism

During the 2007–08 cruise ship season a total of 366,692 passengers (233 cruise ship calls) visited Dominica. Tourism receipts in 2007 (excluding passenger transport) totalled US$71m.

DIPLOMATIC REPRESENTATIVES

Of Dominica in the United Kingdom (1 Collingham Gdns, South Kensington, London, SW5 0HW)
High Commissioner: Francine Baron.

Of the United Kingdom in Dominica
High Commissioner: Paul Brummell (resides in Bridgetown, Barbados).

Of Dominica in the USA (3216 New Mexico Ave., NW, Washington, D.C., 20016)
Ambassador: Hubert Charles.

Of the USA in Dominica
Ambassador: Larry Palmer (resides in Bridgetown, Barbados).

Of Dominica to the United Nations
Ambassador: Vince Henderson.

Of Dominica to the European Union
Ambassador: Shirley Skerritt-Andrew.

FURTHER READING

Baker, P. L., *Centring the Periphery: Chaos, Order and the Ethnohistory of Dominica.* 1994

Honychurch, L., *The Dominica Story: a History of the Island.* 2nd ed. 1995

National Statistical Office: Central Statistical Office, Kennedy Avenue, Roseau.

DOMINICAN REPUBLIC

ATLANTIC OCEAN

Santiago

DOMINICAN REPUBLIC

San Juan

La Romana

SANTO
DOMINGO

Barahona

HAITI

Caribbean Sea

0   50 mi
0   75 km

© Research Machines plc 2006

	Area (in sq. km)	Population
La Altagracia	3,010	273,210
Azua	2,532	214,311
Bahoruco	1,282	97,313
Barahona	1,739	187,105
Dajabón	1,021	63,955
Distrito Nacional (Santo Domingo area)	1,401	3,339,410
Duarte	1,605	289,574
Elías Piña	1,426	63,029
Espaillat	839	231,938
Hato Mayor	1,329	85,017
Independencia	2,006	52,589
María Trinidad Sánchez	1,272	140,925
Monseñor Nouel	992	165,224
Monte Cristi	1,924	109,607
Monte Plata	2,632	185,956
Pedernales	2,075	31,587
Peravia	792	184,344
Puerto Plata	1,853	321,597
La Romana	654	245,433
Salcedo	440	92,193
Samaná	854	101,494
Sánchez Ramírez	1,196	151,392
San Cristóbal	1,266	596,930
San José de Ocoa	855	59,544
San Juan	3,569	232,333
San Pedro de Macorís	1,255	290,458
Santiago	2,837	963,422
Santiago Rodríguez	1,111	57,476
El Seíbo	1,787	87,680
Valverde	823	163,030
La Vega	2,287	394,205

República Dominicana

Capital: Santo Domingo
Population projection, 2015: 10·55m.
GNI per capita, 2011: (PPP$) 8,087
HDI/world rank: 0·689/98
Internet domain extension: .do

KEY HISTORICAL EVENTS

In 1492 Columbus discovered the island of Hispaniola, which he called La Isla Española, and which for a time was also known as Santo Domingo. The city of Santo Domingo, founded by his brother, Bartholomew, in 1496, is the oldest city in the Americas. The western third of the island—now the Republic of Haiti—was later occupied and colonized by the French, to whom the Spanish colony of Santo Domingo was also ceded in 1795. In 1808 the Dominican population routed the French at the battle of Palo Hincado. Eventually, with the aid of a British naval squadron, the French were forced to return the colony to Spanish rule, from which it declared its independence in 1821. It was invaded and held by the Haitians from 1822 to 1844, when the Dominican Republic was founded and a constitution adopted.

Thereafter the rule was dictatorship interspersed with brief democratic interludes. Between 1916 and 1924 the country was under US military occupation. From 1930 until his assassination in 1961, Rafael Trujillo was one of Latin America's legendary dictators using puppet presidents to remain in control of the country after his legal terms in office ended. The conservative pro-American Joaquin Balaguer was president from 1966 to 1978. In 1986 Balaguer returned to power at the head of the Socialist Christian Reform Party, leading the way to economic reforms. But there was violent opposition to spending cuts and general austerity. The 1996 elections brought in a reforming government pledged to act against corruption.

TERRITORY AND POPULATION

The Dominican Republic occupies the eastern portion (about two-thirds) of the island of Hispaniola, the western division forming the Republic of Haiti. The area is 48,671 sq. km (18,792 sq. miles). The area and 2010 census populations of the provinces and National District (Santo Domingo area) were:

Census population 2010, 9,445,281. In 2011 the population was 69·8% urban.

The UN gives a projected population for 2015 of 10·55m.

Population of the main towns (2010 census, in 1,000): Santo Domingo, the capital, 2,582; Santiago de los Caballeros, 551; Los Alcarrizos, 245; La Romana, 225; San Pedro de Macorís, 185.

The population is mainly composed of a mixed race of European (Spanish) and African blood. The official language is Spanish; about 0·18m. persons speak a Haitian-French Creole.

SOCIAL STATISTICS

2009 estimates: births, 216,000; deaths, 59,000. Rates, 2009 estimates (per 1,000 population): birth, 22; death, 6. Annual population growth rate, 2005–10, 1·4%. Life expectancy, 2007: male, 69·8 years; female, 75·2 years. Infant mortality, 2010, 22 per 1,000 live births. Fertility rate, 2008, 2·6 children per woman.

CLIMATE

A tropical maritime climate with most rain falling in the summer months. The rainy season extends from May to Nov. and amounts are greatest in the north and east. Hurricanes may occur from June to Nov. Santo Domingo, Jan. 75°F (23·9°C), July 81°F (27·2°C). Annual rainfall 56" (1,400 mm).

CONSTITUTION AND GOVERNMENT

A new constitution came into force on 26 Jan. 2010, replacing one from 1966. The new constitution's provisions included the establishment of a Constitutional Court, Council of the Judiciary and Supreme Electoral Court. It also provides for recourse to instruments of direct democracy, including referenda and plebiscites. It outlaws same-sex marriages and abortion and defines Dominican nationals as the children of Dominican parents.

The *President*, who has executive power, is elected for four years by direct vote but is prohibited from serving consecutive terms. A second round of voting in a presidential election is authorized when no candidate secures an absolute majority in the first ballot. There is a bicameral legislature, the *Congress*, comprising a 32-member *Senate* (one member for each province and one for the National District of Santo Domingo) and a 183-member *Chamber of Deputies*, both elected for four-year terms. Citizens are entitled to vote at the age of 18.

National Anthem

'Quisqueyanos valientes, alcemos' ('Valiant Quisqueyans, Let us raise our voices'); words by E. Prud'homme, tune by J. Reyes.

GOVERNMENT CHRONOLOGY

Heads of State since 1942. (PD = Dominican Party; PLD = Dominican Liberation Party; PR = Reformist Party; PRD = Dominican Revolutionary Party; PRSC = Social Christian Reformist Party; REP = Republican Party; UCN = National Civic Union; n/p = non-partisan)

Presidents

1942–52	PD/military	Rafael Leonidas Trujillo Molina
1952–60	PD	Héctor Bienvenido Trujillo Molina
1960–62	PD	Joaquín Antonio Balaguer Ricardo
1962–63	REP	Rafael Filiberto Bonelly Fondeur
1963	PRD	Juan Emilio Bosch Gaviño

Chairmen of the Triumvirate

1963	n/p	Emilio de los Santos
1963–65	UCN	Donald Joseph Reid Cabral

Chairman of Military Junta of Government

1965	military	Pedro Bartolomé Benoit Vanderhorst

President of the Government of National Reconstruction

1965	military	Antonio Cosme Imbert Barrera

Presidents

1965	military	Francisco Alberto Caamaño Deñó
1965–66	PR	Héctor Federico García-Godoy Cáceres
1966–78	PR	Joaquín Antonio Balaguer Ricardo
1978–82	PRD	Silvestre Antonio Guzmán Fernández
1982	PRD	Jacobo Majluta Azar
1982–86	PRD	Salvador Jorge Blanco
1986–96	PRSC	Joaquín Antonio Balaguer Ricardo
1996–2000	PLD	Leonel Antonio Fernández Reyna
2000–04	PRD	Rafael Hipólito Mejía Domínguez
2004–12	PLD	Leonel Antonio Fernández Reyna
2012–	PLD	Danilo Medina Sánchez

RECENT ELECTIONS

Presidential elections were held on 20 May 2012. Danilo Medina Sánchez of the ruling Dominican Liberation Party (PLD) won 51·2% of the votes and Hipólito Mejía of the Dominican Revolutionary Party (PRD) 47·0%. The four other candidates received less than 1·5% of the vote each.

Parliamentary elections were held on 16 May 2010. In the election to the Chamber of the Deputies the PLD and its allies won 105 seats with 54·6% of the vote, the PRD and its allies 75 (41·9%) and the PRSC 3 (1·5%). In the Senate elections of the same day, the PLD and its allies won 31 seats and the PRSC 1. Turnout was 54·2%.

CURRENT GOVERNMENT

President: Danilo Medina Sánchez; b. 1951 (PLD; sworn in 16 Aug. 2012).

Vice-President: Margarita Cedeño de Fernández.

In March 2013 the government comprised:

Secretary of State for Agriculture: Luis Ramón Rodríguez.
Armed Forces: Adm. Sigfrido Pared Pérez. *Culture*: José Antonio Rodríguez. *Economy, Planning and Development*: Juan Temístocles Montas. *Education*: Josefina Pimentel. *Environment and Natural Resources*: Bautista Rojas Gómez. *Finance*: Simón Lizardo. *Foreign Relations*: Carlos Morales Troncoso. *Higher Education, Science and Technology*: Ligia Amada Melo. *Industry and Commerce*: José del Castillo. *Interior and Police*: José Ramón Fadul. *Labour*: Maritza Hernández. *Presidency*: Gustavo Montalvo. *Public Administration*: Ramón Ventura Camejo. *Public Health and Social Welfare*: Félix Hidalgo Ramírez. *Public Works and Communications*: Gonzalo Castillo. *Sport*: Jaime David Fernández Mirabal. *Tourism*: Francisco Javier García Fernández. *Women*: Alejandrina Germán. *Youth*: Jorge Minaya. *Ministers without Portfolio*: Miguel Mejía; Franklin Almeyda; Antonio Isa Conde.

Office of the President (Spanish only):
http://www.presidencia.gob.do

CURRENT LEADERS

Danilo Medina Sánchez

Position
President

Introduction
Danilo Medina was elected president in May 2012. He leads the Dominican Liberation Party (PLD), which has been in power since 2004, and he is scheduled to serve a four-year term.

Early Life
Medina was born on 10 Nov. 1951 in Bohechío, Dominican Republic, the eldest of eight brothers. After leaving school in 1965, he became involved in student associations affiliated with the Dominican Revolutionary Party (PRD). A year after moving to Santo Domingo in 1972 to further his studies, he followed PRD leader Juan Bosch to the newly founded Dominican Liberation Party (PLD). Having abandoned his initial studies, Medina enrolled at the Instituto Tecnológico de San Domingo in 1980 and graduated in economics in 1984.

Medina was elected to congress in 1986 and again in 1990. From 1994–95 he was Bosch's head campaign strategist and also served as president of the chamber of deputies. In 1995 he became party spokesman and was twice secretary of state to President Leonel Fernández Reyna, from 1996–99 and 2004–06.

Medina unsuccessfully contested the presidency in 2000 but waited until 2011, when Fernández withdrew from politics, before making a second—this time successful—bid.

Career in Office
Medina was sworn in as president in Aug. 2012. He has positioned himself away from the policies of his predecessor. He campaigned on a platform of tackling structural poverty, addressing public service deficiencies, fighting corruption and developing an economic plan focused on tourism and natural resources.

He has set an annual GDP growth target of 4·5%, with the aim of lifting 400,000 people out of extreme poverty. Despite strong congressional support, he faces significant public finance challenges and must contend with a bloated public sector and social tensions over corruption. In a bid to tackle the budget deficit, he raised VAT from 16% to 18% in Nov. 2012.

DEFENCE

In 2008 defence expenditure totalled US$278m. (US$29 per capita), representing 0·6% of GDP.

Army

There are five defence zones. The Army has a strength (2007) of 15,000 and includes a special forces unit and a Presidential Guard. There is a paramilitary National Police 15,000-strong.

Navy

The Navy is equipped with former US vessels. Personnel in 2007 totalled 4,000, based at Santo Domingo and Las Calderas.

Air Force

The Air Force, with HQ at San Isidoro, has 11 aircraft. Personnel strength (2007), 5,500.

ECONOMY

In 2009 agriculture accounted for 6·2% of GDP, industry 32·5% and services 61·3%.

Overview

The Dominican Republic is noted for its plantation crop exports of sugar, cocoa, coffee and tobacco, but their combined share of total exports has declined. The pillars of the economy are now tourism, worker remittances from the USA and the free trade zones where in-bond factories (known as *maquiladoras*) produce clothing for big name brands.

Economic growth was interrupted in 2003 when a banking crisis, triggered by corruption, sent the sovereign debt to the brink of default while the *peso* depreciated dramatically and inflation soared. Tight fiscal and monetary policy measures, as well as renewed strength in the US economy, helped lift the country out of crisis. Real growth averaged 6·6% between 2004 and 2010, peaking at 10·7% in 2006 before the global economic slowdown saw it decline to 3·5% in 2009. However, driven by agriculture, commerce and tourism, GDP growth increased to 3·8% in the first half of 2012. Nonetheless, unemployment remained high, at over 13% in 2011.

In March 2007 the Dominican Republic-Central America-United States Free Trade Agreement (DR-CAFTA) came into effect, providing free access to the US market. Over 50% of exports went to the USA in 2011, of which the chief components were optical and medical instruments, precious stones, electrical machinery, tobacco and knitwear.

Currency

The unit of currency is the *peso* (DOP), written as RD$, of 100 *centavos*. Gold reserves were 18,000 troy oz in July 2005, foreign exchange reserves US$1,537m. and total money supply was RD$78,490m. Inflation spiralled to 51·5% in 2004, second in the world only to Zimbabwe, but has since slowed with rates of 6·3% in 2010 and 8·5% in 2011.

Budget

Budgetary central government revenue totalled RD$192,577m. and expenditure RD$183,795m. in 2006. Tax revenues in 2006 were RD$176,581m. Main items of expenditure by economic type in 2006 were compensation of employees (RD$44,270m.) and grants (RD$31,240m.).

VAT was raised from 16% to 18% with effect from 1 Jan. 2013.

Performance

Real GDP growth was 3·5% in 2009, 7·8% in 2010 and 4·5% in 2011. Total GDP in 2011 was US$55·6bn.

Banking and Finance

In 1947 the Central Bank was established (*Governor*, Héctor Valdez Albizu). Its foreign assets were US$3,337·6m. at Sept. 2010. In 2010 there were 13 commercial banks (two foreign).

In 2010 external debt totalled US$13,045m., representing 26·2% of GNI.

The Santo Domingo Securities Exchange is a member of the Association of Central American Stock Exchanges (Bolcen).

ENERGY AND NATURAL RESOURCES

Environment

Carbon dioxide emissions from the consumption and flaring of fossil fuels in 2008 were the equivalent of 2·0 tonnes per capita.

Electricity

Installed capacity was 5·5m. kW in 2007. Production was 14·84bn. kWh in 2007; consumption per capita was 1,563 kWh. Power failures are frequent.

Minerals

Bauxite output in 1988 was 167,800 tonnes, but had declined to nil by 1992. Output: nickel (2005), 53,124 tonnes; gold (2009 estimate), 173 kg. Gold production had been in decline but in 2007 a new project launched by two Canadian companies revived the industry.

Agriculture

Agriculture and processing are the chief sources of income, sugar cultivation being the principal industry. In 2001 there were 1·1m. ha. of arable land and 500,000 ha. of permanent cropland. 275,000 ha. were irrigated in 2001.

Production, 2003 (in 1,000 tonnes): sugarcane, 5,036; rice, 609; bananas, 481; plantains, 192; mangoes, 186; coconuts, 178; tomatoes, 155; avocados, 150; cassava, 124; pineapples, 110; oranges, 88.

Livestock in 2003: 2·16m. cattle; 578,000 pigs; 342,000 horses; 188,000 goats; 46m. chickens. Livestock products, 2003 (in 1,000 tonnes): poultry meat, 186; beef and veal, 72; pork, bacon and ham, 65; eggs, 83; milk, 520.

Forestry

Forests and woodlands covered 2·41m. ha. in 2010, representing 50% of the total land area. In 2007, 903,000 cu. metres of timber were cut.

Fisheries

The total catch in 2010 was 14,490 tonnes, mainly from sea fishing.

INDUSTRY

Production, 2007 unless otherwise indicated (in 1,000 tonnes): cement, 4,100; residual fuel oil, 753; raw sugar, 488; petrol, 404; distillate fuel oil, 359; kerosene, 158; rum (2005), 49·9m. litres; beer (2006), 213·4m. litres; cigarettes (2006), 1·4bn. units.

Labour

In 2005 the economically active population was 4,026,000. The unemployment rate in 2005 was 19·3%.

INTERNATIONAL TRADE

On 1 March 2007 the Central America-Dominican Republic-United States Free Trade Agreement (CAFTA-DR) entered into force between the Dominican Republic and El Salvador, Guatemala, Honduras, Nicaragua and the USA. Costa Rica implemented the agreement on 1 Jan. 2009.

Imports and Exports

Trade, 2009, in US$1m.: imports (f.o.b.), 12,053·9; exports (f.o.b.), 4,690·9. Principal imports in 2009 were: machinery and transport equipment, 22·2%; mineral fuels and lubricants, 20·9%; manufactured goods, 18·2%; chemicals and related products, 11·5%. Main exports in 2009 were: miscellaneous manufactured articles, 39·4%; food, live animals, beverages and tobacco, 24·6%; manufactured goods, 16·4%; machinery and transport equipment, 11·5%. Main import suppliers, 2009: USA, 42·2%; China, 10·1%; Venezuela, 5·4%. Main export markets, 2009: USA, 61·9%; Haiti, 13·8%; Netherlands, 2·2%.

COMMUNICATIONS

Roads

In 2002 the road network covered an estimated 19,705 km, of which 51·2% were paved. In 2007 there were 602,700 passenger cars (62 per 1,000 inhabitants), 525,400 lorries and vans, and 64,200 buses and coaches. Motorcycles and mopeds numbered

1·04m. In 2008 there were 1,648 fatal road accidents resulting in 1,846 deaths.

Rail

The railway system has been closed down with the exception of 142 km line from Guayubin to the port of Pepillo, used primarily for the banana trade.

There is a metro in Santo Domingo.

Civil Aviation

The main airports are at Puerto Plata, Punta Cana and Santo Domingo (Las Américas). In 2009 Punta Cana was the busiest airport, handling 4,077,596 passengers, followed by Santo Domingo (2,887,175 passengers) and Puerto Plata (1,096,267). The largest airline operating in the Dominican Republic is the American low-cost airline JetBlue.

Shipping

The main ports are Santo Domingo, Puerto Plata, La Romana and Haina. In Jan. 2009 there were five ships of 300 GT or over registered, totalling 5,000 GT.

Telecommunications

In 2011 there were 1,044,200 landline telephone subscriptions (equivalent to 103·8 per 1,000 inhabitants) and 8,770,800 mobile phone subscriptions (or 872·2 per 1,000 inhabitants). In 2011, 35·5% of the population were internet users. In Dec. 2011 there were 2·5m. Facebook users.

SOCIAL INSTITUTIONS

Justice

The judicial power resides in the Supreme Court of Justice, the courts of appeal, the courts of first instance, the communal courts and other tribunals created by special laws, such as the land courts. The Supreme Court, consisting of a president and eight judges chosen by the Senate, and the procurator-general, appointed by the executive, supervises the lower courts. Each province forms a judicial district, as does the National District, and each has its own procurator fiscal and court of first instance; these districts are subdivided, in all, into 97 municipalities, each with one or more local justices. The death penalty was abolished in 1924.

The population in penal institutions in Sept. 2008 was 16,457 (165 per 100,000 of national population).

Education

Primary education is free and compulsory for children between five and 14 years of age; there are also secondary, normal, vocational and special schools, all of which are either wholly maintained by the State or state-aided. In 2007 there were 1,355,085 primary school pupils with 56,744 teaching staff and 920,494 pupils at secondary level with 31,710 teaching staff. The Universidad Autónoma de Santo Domingo, founded in 1914, is the leading public university; there were 158,534 students enrolled in 2005–06. The leading private university is the Universidad Tecnológica de Santiago, created in 1976. There were 293,565 students and 11,367 academic staff in tertiary education in 2004. Adult literacy was 88% in 2007.

In 2007 public expenditure on education came to 2·6% of GNI and 11·0% of total government spending.

Health

In 2007 there were 14,479 doctors. There were 136 hospitals in 2006 with a total of 9,517 beds. In 2005 there were 20 hospital beds per 10,000 inhabitants.

RELIGION

The religion of the state is Roman Catholic; there were 7·11m. adherents in 2001. Protestants numbered 560,000 in 2001. In Feb. 2013 there was one cardinal.

CULTURE

World Heritage Sites

The Dominican Republic has one site on the UNESCO World Heritage List: the Colonial City of Santo Domingo (1990)—founded in 1492, it is the site of the first cathedral and university in the Americas.

Press

In 2008 there were ten dailies (eight paid-for and two free) with a combined circulation of 465,000.

Tourism

In 2006 there were 3,965,055 non-resident air arrivals and 303,489 cruise ship visitors. Tourism receipts in 2006 totalled US$3,917m. In 2007 there were 65,106 hotel rooms (54,730 in 2002).

DIPLOMATIC REPRESENTATIVES

Of the Dominican Republic in the United Kingdom (139 Inverness Terrace, London, W2 6JF)
Ambassador: Federico Alberto Cuello Camilo.

Of the United Kingdom in the Dominican Republic (Edificio Corominas Pepin, Ave. 27 de Febrero 233, Santo Domingo)
Ambassador: Stephen Fisher.

Of the Dominican Republic in the USA (1715 22nd St., NW, Washington, D.C., 20008)
Ambassador: Aníbal de Castro.

Of the USA in the Dominican Republic (Calle César Nicolás Penson, Santo Domingo)
Ambassador: Raul Yzaguirre.

Of the Dominican Republic to the United Nations
Ambassador: Héctor Virgilio Alcántara Mejía.

Of the Dominican Republic to the European Union
Ambassador: Alejandro González Pons.

FURTHER READING

Gregory, Steven, *The Devil Behind the Mirror: Globalization and Politics in the Dominican Republic.* 2006

Hartlyn, Jonathan, *The Struggle for Democratic Politics in the Dominican Republic.* 1998

Peguero, Valentina, *The Militarization of Culture in the Dominican Republic, from the Captains General to General Trujillo.* 2004

Wucker, Michele, *Why the Cocks Fight: Dominicans, Haitians, and the Struggle for Hispaniola.* 2000

National Statistical Office: Oficina Nacional de Estadística, Av. México esq. Leopoldo Navarro, Edificio Oficinas Gubernamentales 'Juan Pablo Duarte' Pisos 8 y 9 Gazcue, Santo Domingo.

Website (Spanish only): http://www.one.gov.do

ECUADOR

República del Ecuador
(Republic of Ecuador)

Capital: Quito
Population projection, 2015: 15·45m.
GNI per capita, 2011: (PPP$) 7,589
HDI/world rank: 0·720/83
Internet domain extension: .ec

KEY HISTORICAL EVENTS

In 1532 the Spaniards founded a colony in Ecuador, then called Quito. In 1821 a revolt led to the defeat of the Spaniards at Pichincha and thus independence from Spain. On 13 March 1830, Quito became the Republic of Ecuador. Political instability was endemic. From the mid-1930s, President José Maria Velasco Ibarra was deposed by military coups from four of his five presidencies.

From 1963 to 1966 and from 1976 to 1979 military juntas ruled the country. The second of these juntas produced a new constitution which came into force on 10 Aug. 1979. Presidencies were more stable but civil unrest continued in the wake of economic reforms and attempts to combat political corruption.

In Jan. 2000 President Mahaud declared a state of emergency when protesters demanded his resignation over his handling of an economic crisis. There was a coup on 21 Jan. but, after five hours in control, the military junta handed power to the former vice-president, Gustavo Noboa.

In April 2005 President Lucio Gutiérrez was ousted by Ecuador's congress after public protest at his attempts to implement IMF-backed economic policies and his substitution of 27 out of 31 Supreme Court judges with his allies. The replacement judges promptly dropped corruption charges against two former presidents, increasing public outcry. Four days after being dismissed, Gutiérrez fled to Brazil. He was replaced by Alfredo Palacio, who immediately issued a warrant for Gutiérrez's arrest. Gutiérrez was arrested in Oct. 2005 and charged with endangering

national security but was freed in March 2006 when a judge dismissed the claims. Palacio was defeated in the elections of Nov. 2006, with Rafael Correa taking over as president. He won a second term in April 2009. In Sept. 2010 he was hospitalized after being hit by tear gas fired by police and military personnel unhappy at austerity measures, claiming that he was then held against his will in hospital. Correa decried the events as an opposition-sponsored coup attempt and imposed a state of emergency.

TERRITORY AND POPULATION

Ecuador is bounded in the north by Colombia, in the east and south by Peru and in the west by the Pacific ocean. The frontier with Peru has long been a source of dispute. It was delimited in the Treaty of Rio, 29 Jan. 1942, when, after being invaded by Peru, Ecuador lost over half her Amazonian territories. Ecuador unilaterally denounced this treaty in Sept. 1961. Fighting between Peru and Ecuador began again in Jan. 1981 over this border issue but a ceasefire was agreed in early Feb. Following a confrontation of soldiers in Aug. 1991 the foreign ministers of both countries signed a pact creating a security zone, and took their cases to the UN in Oct. 1991. On 26 Jan. 1995 further armed clashes broke out with Peruvian forces in the undemarcated mutual border area (*Cordillera del Cóndor*). On 2 Feb. talks were held under the auspices of the guarantor nations of the 1942 Protocol of Rio de Janeiro (Argentina, Brazil, Chile and the USA) but fighting continued. A ceasefire was agreed on 17 Feb., which was broken, and again on 28 Feb. On 25 July 1995 an agreement between Ecuador and Peru established a demilitarized zone along their joint frontier. The frontier was reopened on 4 Sept. 1995. Since 23 Feb. 1996 Ecuador and Peru have signed three further agreements to regulate the dispute. The dispute was settled in Oct. 1998. Confirming the Peruvian claim that the border lies along the high peaks of the Cóndor, Ecuador gained navigation rights on the Amazon within Peru.

No definite figure of the area of the country can yet be given. One estimate of the area of Ecuador is 272,045 sq. km, excluding the litigation zone between Peru and Ecuador, which is 190,807 sq. km, but including the Galápagos Archipelago (8,010 sq. km), situated in the Pacific ocean about 960 km west of Ecuador, and comprising 13 islands and 19 islets. These were discovered in 1535 by Fray Tomás de Berlanga and had a population of 18,640 in 2001. They constitute a national park, and had approximately 122,000 visitors in 2005.

The population is an amalgam of European, Amerindian and African origins. Some 41% of the population is Amerindian: Quechua, Shiwiar, Achuar and Zaparo. In May 1992 they were granted title to the 1m. ha. of land they occupy in Pastaza.

The official language is Spanish. Quechua and other languages are also spoken.

Census population in 2010, 14,483,449; density, 53 per sq. km. In 2011, 67·6% lived in urban areas.

The UN gives a projected population for 2015 of 15·45m.

The population was distributed by provinces as follows in 2010 (census figures):

Province	Sq. km	Population	Capital	Population
Azuay	7,995	712,127	Cuenca	329,928
Bolívar	3,926	183,641	Guaranda	23,874
Cañar	3,142	225,184	Azogues	33,848
Carchi	3,750	164,524	Tulcán	53,558
Chimborazo	6,470	458,581	Riobamba	146,324
Cotopaxi	5,985	409,205	Latacunga	63,842
El Oro	5,817	600,659	Machala	231,260
Esmeraldas	16,219	534,092	Esmeraldas	154,035
Guayas	16,803	3,645,483	Guayaquil	2,278,691

Province	Sq. km	Population	Capital	Population
Imbabura	4,615	398,244	Ibarra	131,856
Loja	10,995	448,966	Loja	170,280
Los Ríos	7,151	778,115	Babahoyo	90,191
Manabí	18,894	1,369,780	Portoviejo	206,682
Morona-Santiago	23,797	147,940	Macas	18,984
Napo	12,483	103,697	Tena	23,307
Orellana	21,675	136,396	Francisco de Orellana	40,730
Pastaza	29,325	83,933	Puyo	33,557
Pichincha	9,465	2,576,287	Quito	1,607,734
Santa Elena	3,763	308,693	Santa Elena	39,681
Santo Domingo de los Tsáchilas	3,805	368,013	Santo Domingo de los Colorados	270,875
Sucumbíos	18,008	176,472	Nueva Loja	48,562
Tungurahua	3,369	504,583	Ambato	165,185
Zamora-Chinchipe	10,456	91,376	Zamora	12,386
Galápagos	8,010	25,124	Puerto Baquerizo Moreno	6,672
Non-delimited zones	775	32,384		

SOCIAL STATISTICS

2008 estimates: births, 280,000; deaths, 70,000. Rates, 2008 estimates (per 1,000 population): birth, 20·8; death, 5·2. Life expectancy at birth, 2007, was 72·1 years for males and 78·0 years for females. Annual population growth rate, 2000–08, 1·1%. Infant mortality, 2010, 18 per 1,000 live births; fertility rate, 2008, 2·6 children per woman. In 2009 the most popular age for marrying was 20–24 for both men and women.

CLIMATE

The climate varies from equatorial, through warm temperate to mountain conditions, according to altitude, which affects temperatures and rainfall. In coastal areas, the dry season is from May to Dec., but only from June to Sept. in mountainous parts, where temperatures may be 20°F colder than on the coast. Quito, Jan. 59°F (15°C), July 58°F (14·4°C). Annual rainfall 44" (1,115 mm). Guayaquil, Jan. 79°F (26·1°C), July 75°F (23·9°C). Annual rainfall 39" (986 mm).

CONSTITUTION AND GOVERNMENT

An executive *President* and a *Vice-President* are directly elected by universal suffrage. The president appoints and leads a *Council of Ministers*, and determines the number and functions of the ministries that comprise the executive branch. Legislative power is vested in a *National Assembly* of 137 members, popularly elected by province. One seat is reserved for overseas voters. Citizens must be at least 16 years of age to vote. Voting is obligatory for all literate citizens of 18–65 years. It is optional for those aged 16 and 17 and other eligible voters.

A new constitution came into force on 20 Oct. 2008. It was drafted by a Constituent Assembly set up by President Correa in Nov. 2007 and was approved with 63·9% of the vote in a referendum on 28 Sept. 2008. It superseded the previous constitution that had been in place for ten years. The 2008 constitution, which includes 444 articles, allows a president to run for two consecutive four-year terms, dissolve parliament and call early elections, and set monetary policy. The *National Congress* was abolished and replaced by a new *National Assembly*. The constitution also allows for tighter control of key industries, the expropriation and redistribution of idle farm land, free health care for the elderly and the legalization of same-sex civil marriages. The government can also declare some foreign loans illegitimate.

In May 2011 a package of ten changes to the constitution was passed in a referendum. The proposals included giving the executive branch of government increased influence over the appointment of Supreme Court judges, creating a body to regulate media content, extending the legal period of detention without trial, outlawing 'unjustified' wealth and restricting gambling, requiring owners of banks and media organizations to declare other commercial holdings, and banning cock- and bull-fighting.

National Anthem

'Salve, Oh Patria, mil veces, Oh Patria' ('Hail, Oh Fatherland, a thousand times, Oh Fatherland'); words by J. L. Mera, tune by A. Neumane.

GOVERNMENT CHRONOLOGY

Heads of State since 1944. (AD = Democratic Alliance; Alianza PAIS = Proud and Sovereign Fatherland Alliance; CFP = Concentration of Popular Forces; CID = Democratic Institutional Coalition; DP–UDC = People's Democracy–Christian Democratic Union; FNV = National Velasquista Federation; FRA = Alfarista Radical Front; ID = Democratic Left; MCDN = Nacional Democratic Civic Movement; MSC = Social Christian Party [called PSC since 1967]; PRE = Ecuadorian Roldosist Party; PSC = Social Christian Party; PSP = Patriotic Society January 21; PUR = Republican Union Party; n/p = non-partisan)

Presidents

1944–47	AD	José María Velasco Ibarra
1947–48	n/p	Carlos Julio Arosemena Tola
1948–52	MCDN	Galo Plaza Lasso
1952–56	FNV	José María Velasco Ibarra
1956–60	MSC	Camilo Ponce Enríquez
1960–61	FNV	José María Velasco Ibarra
1961–63	FNV	Carlos Julio Arosemena Monroy

Military Junta

1963–66		Adm. Ramón Castro Jijón (chair); Gen. Luis Cabrera Sevilla; Col. Guillermo Freile Posso; Gen. Mario Gándara Enríquez

Presidents

1966	n/p	Clemente Yerovi Indaburu
1966–68	CID	Otto Arosemena Gómez
1968–72	FNV	José María Velasco Ibarra
1972–76	military	Gen. Guillermo Rodríguez Lara

Military Junta

1976–79		Admr. Alfredo Ernesto Poveda Burbano (chair); Gen. Luis Leoro Franco; Gen. Luis G. Durán Arcentales

Presidents

1979–81	CFP	Jaime Roldós Aguilera
1981–84	DP–UDC	Osvaldo Hurtado Larrea
1984–88	PSC	León Esteban Febres Cordero
1988–92	ID	Rodrigo Borja Cevallos
1992–96	PUR	Sixto Alfonso Durán-Ballén
1996–97	PRE	Abdalá Jaime Bucaram Ortiz
1997–98	FRA	Fabián Ernesto Alarcón Rivera
1998–2000	DP–UDC	Jorge Jamil Mahuad Witt
2000–03	DP–UDC	Gustavo Noboa Bejarano
2003–05	PSP	Lucio Edwin Gutiérrez Borbúa
2005–07	n/p	Luis Alfredo Palacio González
2007–	Alianza PAIS	Rafael Vicente Correa Delgado

RECENT ELECTIONS

In the presidential election held on 17 Feb. 2013 Rafael Vicente Correa Delgado was re-elected with 57·2% of the vote, against 22·7% for Guillermo Lasso, 6·7% for Lucio Gutiérrez, 3·9% for Mauricio Rodas, 3·7% for Alvaro Noboa and 3·3% for Alberto Acosta. There were two other candidates who each received less than 2% of the vote.

In the parliamentary election also held on 17 Feb. 2013 the Proud and Sovereign Fatherland Alliance (Alianza PAIS) won 100 of 137 seats with 52·3% of the vote; Creating Opportunities, 11 (11·4%); Social Christian Party, 6 (9·0%); Patriotic Society January

21, 5 with 5·6%; Plurinational Unity of the Lefts, 5 with 4·7%; Partido Avanza, 5 with 2·9%; Ecuadorian Roldosist Party, 1 with 4·5%; SUMA, 1 with 3·2%. Independents took three seats. Turnout was 81·1%.

CURRENT GOVERNMENT

President: Rafael Correa; b. 1963 (Alianza PAIS; sworn in 15 Jan. 2007, and re-elected 26 April 2009 and 17 Feb. 2013).

Vice-President: Lenín Moreno.

In March 2013 the cabinet comprised:

Minister of Agriculture, Livestock, Aquaculture and Fisheries: Javier Ponce. *Culture:* Érika Sylva Charvet. *Defence:* María Fernanda Espinosa. *Economic and Social Inclusion:* Doris Soliz. *Education:* Gloria Vidal. *Electricity and Renewable Energy:* Esteban Albornoz. *Environment:* Lorena Tapia. *Finance:* Patricio Rivera. *Foreign Relations, Trade and Integration:* Ricardo Patiño. *Human Talent and Knowledge:* Augusto Espinosa. *Industry and Competitiveness:* Verónica Sión. *Interior:* José Serrano. *Justice and Human Rights:* Johana Pesántez. *Labour Relations:* Francisco Vacas. *Non-Renewable Natural Resources:* Wilson Pástor Morris. *Public Health:* Carina Vance Mafla. *Social Development Co-ordination:* Richard Espinosa. *Sport:* José Francisco Cevallos. *Telecommunications and Information Society:* Jaime Guerrero Ruíz. *Tourism:* Freddy Ehlers. *Transport and Public Works:* María de los Ángeles Duarte. *Urban Development and Housing:* Pedro Jaramillo.

Parliament Website (Spanish only):
http://www.asambleanacional.gov.ec

CURRENT LEADERS

Rafael Correa

Position
President

Introduction
Rafael Correa started his four-year presidential term in Jan. 2007. An economist educated in Europe and the USA, he has mostly enjoyed strong popular support and was re-elected in April 2009 and again in Feb. 2013.

Early Life
Correa was born in April 1963 in Ecuador's second city, Guayaquil. He graduated in economics from the Catholic University of Guayaquil and worked for a year in an indigenous community in the Cotopaxi region. He went on to postgraduate studies at the Catholic University of Leuven (Belgium) and the University of Illinois at Urbana-Champaign (USA).

Correa embarked on an academic career, rising to become dean of economics at the private University San Francisco de Quito. An economic analyst known for his anti-neoliberal and nationalist stance, he worked as a consultant for the UN Development Programme and the Japanese Development Bank among others. He was a notable opponent of Ecuadorian dollarization in 2000.

When popular revolts forced Lucio Gutiérrez to resign as president in April 2005, his successor Alfredo Palacio appointed Correa as finance minister. Correa held the position for four months, during which time he was critical of the World Bank and IMF and advocated poverty reduction and economic sovereignty schemes. When the World Bank withheld a loan in protest at various economic policies, Correa resigned his office. In 2006 he founded the Alianza PAIS movement, which allied itself with the Socialist Party in the run-up to general elections.

Career in Office
Correa campaigned for sustainable socio-economic revolution and Latin American integration. On taking office, he pledged a referendum on the establishment of a constitutional assembly to draft a new constitution, aimed at maintaining good relations with the USA (although his rejection of a free trade agreement, Ecuador's refusal to extend US use of the Manta military base in the Pacific and, in 2011, US allegations of corruption in the Ecuador police have strained ties) and promised that dollarization would remain in place during his tenure. He also guaranteed Ecuador's non-involvement in Colombia's internal conflict. Other policies included engagement with Colombia and Brazil in trade negotiations and talks with Argentina to renegotiate the terms of Ecuador's multi-billion dollar debt. In addition, Correa announced that he would renegotiate 'entrapping' oil contracts with transnational companies, consider Ecuador's re-entry into OPEC (which happened in Nov. 2007) and limit debt repayment in favour of social spending (although quick repayment of IMF debts would remain on the agenda).

In April 2007 he won referendum approval to set up a constituent assembly to rewrite the constitution. Elections to the assembly in Sept. resulted in a majority victory for Correa's allies over opposition parties that had dominated the Congress since the Oct. 2006 polls. In Nov. 2007 the assembly voted to dissolve the Congress and proceeded to act as a legislature. A new draft constitution increasing presidential powers and tenure, banning foreign military bases and enhancing state influence over the economy was approved by the assembly in July 2008 and endorsed in a national referendum on 28 Sept.

In March 2008 Ecuador's relations with Colombia were seriously undermined by a cross-border Colombian strike against a FARC guerrilla target in Ecuadorian territory, prompting Correa to cut diplomatic ties and send troops to the border. Relations worsened in July 2009 when a FARC military commander claimed that his organization had contributed funding to Correa's 2006 election campaign, and have remained strained.

Correa won an outright victory in the April 2009 presidential elections, avoiding the need for a second round run-off. His party won a legislative majority at the same time. However, in Sept. 2010 Correa was assaulted and then besieged in hospital by rebellious police officers protesting against pay cuts. The incident was denounced as a coup attempt by Correa and prompted the government to declare an indefinite state of emergency. In a further constitutional referendum in May 2011 voters approved a range of amendments, including extra presidential powers over the judiciary and media.

In June 2012 Julian Assange, the Australian founder of the WikiLeaks website which publicized secret US diplomatic documents, took refuge in Ecuador's embassy in London, England to evade extradition to Sweden on sexual assault charges. In Aug. Correa granted Assange diplomatic asylum, prompting tensions with the UK government.

Correa won a third term of office in Feb. 2013, winning over 50% of the vote against seven opponents. Correa promised 'another four years of revolution', although critics accuse him of autocratic tendencies.

DEFENCE

Military service is selective, with a one-year period of conscription. The country is divided into four military zones, with headquarters at Quito, Guayaquil, Cuenca and Pastaza.

In 2008 defence expenditure totalled US$1,105m. (US$77 per capita), representing 2·0% of GDP.

Army
Strength (2007) 47,000.

Navy
Navy combatant forces include two diesel submarines (although their serviceability is in doubt) and two ex-UK frigates. The Naval Aviation has 12 aircraft but no combat-capable aircraft. Naval personnel in 2007 totalled 6,100 including 1,500 marines.

Air Force

The Air Force had a 2007 strength of 4,000 personnel and some 57 combat-capable aircraft, and includes Cessna A-37s, Mirage F-1s and Kfirs.

INTERNATIONAL RELATIONS

In March 2008 Ecuador responded to a Colombian incursion into its territory, during which a senior figure in the rebel FARC movement was killed, by sending troops to the border with Colombia. Venezuela made a similar gesture in sympathy before a diplomatic solution to the stand-off came into effect a week later.

ECONOMY

Agriculture accounted for 6·7% of GDP in 2008, industry 40·6% and services 52·7%.

Overview

In the 1980s and 1990s per capita income stagnated at a level above the Latin American average but below the world average. Heavily dependent on oil exports (oil and mining account for 14·2% of GDP), the fall in oil prices in the late 1990s combined with natural disasters to trigger a collapse in GDP and income levels. However, high oil prices throughout most of the 2000s reinvigorated growth.

In Dec. 2008 the government announced a default on its sovereign debt, representing 80% of its private external debt, which the government had labelled 'illegitimate'. It bought back the majority of its defaulted bonds in 2009.

The global financial crisis, along with a sharp drop in world oil prices and remittances, led to the slowest growth for ten years in 2009 although the economy has since recovered strongly. Oil output has faced regular industrial disruption and corruption remains endemic.

Currency

The monetary unit is the US dollar. Inflation came down from 8·4% in 2008 to 3·6% in 2010, although it then rose to 4·5% in 2011. In March 2000 the government passed a law to phase out the former national currency, the *sucre*, to be replaced by the US dollar, and in April bank cash machines began dispensing dollars instead of sucres. On 11 Sept. 2000 the dollar became the only legal currency. Foreign exchange reserves were US$1,398m. in July 2005 with gold reserves of 845,000 troy oz.

Budget

Revenues in 2009 totalled US$11,583m. and expenditures US$14,218m.

VAT is 12% and corporate tax 25%.

Performance

Real GDP growth was 0·4% in 2009, 3·6% in 2010 and 7·8% in 2011. Total GDP in 2011 was US$65·9bn.

Banking and Finance

The Central Bank of Ecuador (*President of the Directorate*, Diego Martínez), the bank of issue, with reserves of US$1,994·5m. in Dec. 2008, is modelled after the Federal Reserve Banks of the USA. There were 24 banks in Dec. 2006 with deposits of US$8,755m. Ecuador's largest private bank is Banco del Pichincha. All commercial banks must be affiliated to the Central Bank. The national monetary board is based in Quito.

In 2010 foreign debt totalled US$14,815m., representing 23·1% of GNI.

There are stock exchanges in Quito and Guayaquil.

ENERGY AND NATURAL RESOURCES

Environment

Ecuador's carbon dioxide emissions from the consumption and flaring of fossil fuels were the equivalent of 2·0 tonnes per capita in 2008.

Electricity

Installed capacity was 4·49m. kW in 2007. Production was 17·34bn. kWh in 2007; consumption per capita was 1,335 kWh.

Oil and Gas

Production of oil in 2008 was 26·2m. tonnes. Estimated reserves, 2008, 3·8bn. bbls. In 2007 natural gas production was 794m. cu. metres. Proven reserves (2007), 90bn. cu. metres.

Minerals

Main products are silver, gold, copper and zinc. The country also has some iron, uranium, lead, coal, cobalt, manganese and titanium.

Agriculture

There were 1·20m. ha. of arable land in 2007 and 1·22m. ha. of permanent crops. In 2007 the agricultural population was an estimated 2·86m., of which about 1·19m. were economically active.

Main crops, in 1,000 tonnes, in 2003: sugarcane, 5,691; bananas, 5,609; rice, 1,236; plantains, 860; maize, 677; potatoes, 427; palm oil, 244; oranges, 189; soybeans, 111.

Livestock, 2003: cattle, 4·98m.; pigs, 3·01m.; sheep, 2·65m.; horses, 530,000; asses, 280,000; goats, 279,000; chickens, 142m.

Forestry

Excepting the agricultural zones and a few arid spots on the Pacific coast, Ecuador is a vast forest. 9·87m. ha., or 36% of the land area, was forested in 2010, but much of the forest is not commercially accessible. In 2007, 6·08m. cu. metres of roundwood were produced.

Fisheries

Fish landings in 2008 were 434,239 tonnes (almost entirely from sea fishing). Exports of fishery commodities were valued at US$1·75bn. in 2008.

INDUSTRY

Industry produced 40·6% of GDP in 2008, including 9·7% from manufacturing. Manufacturing grew by 8·1% in 2008. Main products include (in 1,000 tonnes): cement (2007 estimate), 4,420; residual fuel oil (2007), 3,323; petrol (2007), 1,940; distillate fuel oil (2007), 1,603.

Labour

Out of 3,673,200 people in urban employment in 2001, 1,026,700 were in wholesale and retail trade/repair of motor vehicles, motorcycles and personal and household goods; 610,600 in manufacturing; 244,600 in transport, storage and communications; and 239,800 in agriculture, hunting and forestry. In June 2001, 10·4% of the workforce was unemployed, down from 14·1% in June 2000.

INTERNATIONAL TRADE

Most restrictions on foreign investment were removed in 1992 and the repatriation of profits was permitted.

Imports and Exports

In 2008 imports totalled US$18,692m. (US$13,565m. in 2007); exports, US$18,490m. (US$13,852m. in 2007). Main imports in 2004 were (in US$1m.): road vehicles, 835·2; telecommunications, sound recording and reproducing equipment, 615·3; iron and steel, 477·7; petroleum and petroleum products, 378·8; medicinal and pharmaceutical products, 377·1. Ecuador is the world's leading exporter of bananas (US$1,022·9m. in 2004), with approximately a third of world banana exports. Other major exports (2004, in US$1m.): petroleum and petroleum products, 4,233·8; fish and seafood, 734·6; cut flowers and foliage, 342·2; cocoa, 145·7. Main import suppliers, 2004: USA, 20·7%; Colombia, 14·6%; Venezuela, 6·8%; Brazil, 6·1%. Main export markets in 2004: USA, 42·9%; Peru, 7·9%; Italy, 4·6%; Colombia, 3·9%.

COMMUNICATIONS

Roads
In 2007 there were 43,670 km of roads. There were 507,500 passenger cars in 2007 (38 per 1,000 inhabitants) and 323,500 lorries and vans. There were 1,848 fatalities in road accidents in 2007.

In 1998 storms and floods on the coast, caused by El Niño, resulted in 2,000 km of roads being damaged or destroyed.

Rail
The railway network, once 971 km long, now has a total length of just 204 km. In 2002 passenger-km travelled came to 33m.

Civil Aviation
The Ecuadorian flag carrier is Tame. There are international airports at Quito (Mariscal Sucre) and Guayaquil (José Joaquín de Olmedo). In 2009 Quito handled 4,746,292 passengers and 143,767 tonnes of freight, and Guayaquil handled 3,382,554 passengers and 55,605 tonnes of freight.

Shipping
Ecuador has three major seaports, of which Guayaquil is the most important, and six minor ones. In Jan. 2009 there were 60 ships of 300 GT or over registered, totalling 222,000 GT.

Telecommunications
In 2011 there were 2,210,600 landline telephone subscriptions (equivalent to 150·7 per 1,000 inhabitants) and 15,332,700 mobile phone subscriptions (or 1,045·5 per 1,000 inhabitants). In 2011, 31·4% of the population were internet users. In June 2012 there were 4·7m. Facebook users.

SOCIAL INSTITUTIONS

Justice
The Supreme Court in Quito, consisting of a President and 30 Justices, comprises ten chambers each of three Justices. It is also a Court of Appeal. There is a Superior Court in each province, comprising chambers (as appointed by the Supreme Court) of three magistrates each. The Superior Courts are at the apex of a hierarchy of various tribunals. There is no death penalty.

The population in penal institutions in Aug. 2008 was 17,065 (126 per 100,000 of national population).

Education
There were an estimated 287,000 pre-primary pupils in 2007–08 and an estimated 19,000 pre-primary teachers in 2008–09. Primary education is free and compulsory. Private schools, both primary and secondary, are under some state supervision. In 2007 there were 2·04m. pupils and 90,366 teaching staff in primary schools; and 1·14m. pupils with 77,904 teaching staff in secondary schools. In the public sector in 2000–01 there were: 9 universities, 8 technical universities, 2 institutes of technology, 1 polytechnical university, 1 military polytechnic and 1 agricultural university; and in the private sector: 9 universities, 3 Roman Catholic universities, 4 institutes of technology, 2 polytechnic institutes and 1 technical university. There were 443,509 students in tertiary education in 2007 with 22,714 academic staff. Adult literacy was 84·2% in 2009 (male, 87·1%; female, 81·5%).

In 2000–01 total expenditure on education came to 1·7% of GNP and 8·0% of total government spending.

Health
In 2002 there were 3,496 hospitals and clinics with 14 beds per 10,000 inhabitants. There were 18,335 physicians, 2,062 dentists, 19,549 nurses and 1,037 midwives in 2000.

Welfare
Those who qualify for a pension must have at least 480 months of contributions (at any age), or be aged 60 with 360 months of contributions, 65 with 180 months of contributions or 70 with 120 months of contributions. In 2003 the minimum monthly pension was US$25, and the maximum pension was US$125.

Social insurance providing lump-sum benefits for unemployment is gradually being phased out in favour of a system linking payments to income and length of employment.

RELIGION
The state recognizes no religion and grants freedom of worship to all. In 2001 there were 11·91m. Roman Catholics. There were also small numbers of Protestants and followers of other faiths. In Feb. 2013 there was one Roman Catholic cardinal.

CULTURE

World Heritage Sites
Ecuador has four sites on the UNESCO World Heritage List: the Galápagos Islands (inscribed on the list in 1978 and 2001); the City of Quito (1978); Sangay National Park (1983); and the Historic Centre of Santa Ana de los Ríos de Cuenca (1999).

Press
There were 47 daily newspapers in 2008, with a circulation of 591,000.

Tourism
Foreign visitors numbered 968,000 in 2009, of whom 735,000 were from elsewhere in the Americas and 197,000 from Europe.

DIPLOMATIC REPRESENTATIVES

Of Ecuador in the United Kingdom (Flat 3b, 3 Hans Cres., London, SW1X 0LS)
Ambassador: Ana Albán Mora.

Of the United Kingdom in Ecuador (Citiplaza Bldg, Naciones Unidas Ave. and República de El Salvador, 14th Floor, Quito)
Ambassador: Patrick Mullee.

Of Ecuador in the USA (2535 15th St., NW, Washington, D.C., 20009)
Ambassador: Nathalie Cely.

Of the USA in Ecuador (Avenida Avigiras E12–170 y Avenida Eloy Alfaro, Quito)
Ambassador: Adam E. Namm.

Of Ecuador to the United Nations
Ambassador: Xavier Lasso Mendoza.

Of Ecuador to the European Union
Ambassador: Fernando Yépez Lasso.

FURTHER READING
Roos, W. and van Renterghem, O., *Ecuador in Focus: A Guide to the People, Politics and Culture.* 1997

Sawyer, Suzana, *Crude Chronicles: Indigenous Politics, Multinational Oil, and Neoliberalism in Ecuador.* 2004

Selverston-Scher, M., *Ethnopolitics in Ecuador: Indigenous Rights and the Strengthening of Democracy.* 2001

National Statistical Office: Instituto Nacional de Estadística y Censos (INEC), Juan Larrea N15-36 y José Riofrío, Quito.

EGYPT

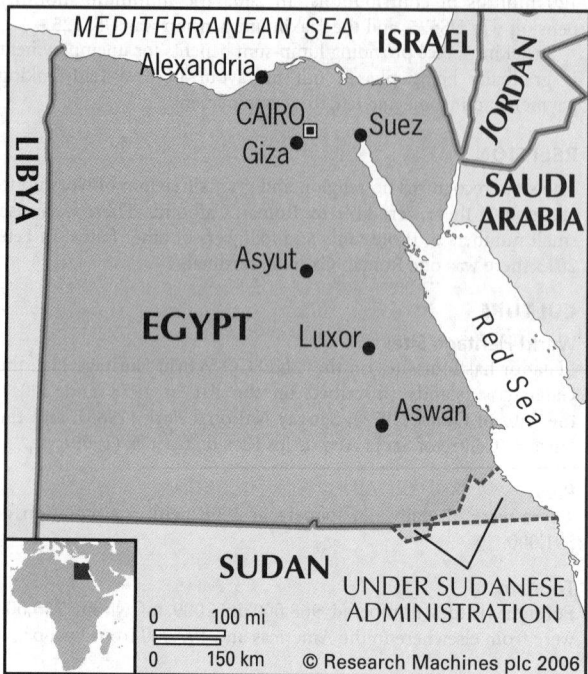

MEDITERRANEAN SEA
ISRAEL
Alexandria
CAIRO
Giza
Suez
JORDAN
SAUDI ARABIA
LIBYA
Asyut
EGYPT
Luxor
Red Sea
Aswan
SUDAN
UNDER SUDANESE ADMINISTRATION
0   100 mi
0   150 km
© Research Machines plc 2006

**Jumhuriyat Misr al-Arabiya
(Arab Republic of Egypt)**

Capital: Cairo
Population projection, 2015: 88·18m.
GNI per capita, 2011: (PPP$) 5,269
HDI/world rank: 0·644/113
Internet domain extension: .eg

KEY HISTORICAL EVENTS

There is evidence of pastoralism and the cultivation of cereals in southwest Egypt from as early as 7000 BC. Settlements grew along the Nile valley, though Upper and Lower Egypt only united around 3100 BC under Pharoah Menes. The subsequent Early Dynastic period was marked by flourishing trade with Sinai, the Levant and as far north as the Black Sea. The astonishing artistic and intellectual developments of the Old Kingdom began during the IVth dynasty (2575–2465 BC), when sun-worship took hold and temples and pyramids, including those at Giza, were constructed. Egypt was governed from the city of Memphis, south of modern Cairo, reaching its height during the VIth dynasty before losing power to local rulers from around 2200 BC.

Centralized power was restored at Thebes from 2134 BC. The Middle Kingdom (XIIth dynasty) saw Egypt expand south into Nubia under Amenemhat I. A cultural flowering included the invention of a writing system. The XIIth dynasty ended in 1786 BC when the region was invaded by the Hyksos, a nomadic Asiatic tribe. The New Kingdom came into being with the expulsion of the Hyksos around 1550 BC and lasted until 1050 BC. It was now that Egypt achieved its greatest territorial dominance, with Syria, Palestine and northern Iraq all under Egyptian jurisdiction. The conquests brought great prosperity and bold architecture, which reached its zenith under the XVIIIth dynasty pharaohs, Tuthmosis III and Tutankhamun.

Ancient Egyptian civilization began to fragment in conflict with Hittite invaders during the XIXth dynasty (around 1250 BC). The subsequent rise of Assyria to the northeast and Nubian conquests from the south hastened the decline. The last pharaoh was ousted by Persian invading forces led by Cambyses in 525 BC. The Persians remained in power until overrun by Alexander the Great in 332 BC. He founded the port of Alexandria, including its great lighthouse, and made the city the commercial and cultural centre of the Greek world. On his death in 305 BC, Ptolemy of Macedonia seized power, establishing a dynasty which lasted until 30 BC and the suicide of Cleopatra.

Egypt then became a province of the Roman Empire until Arabian forces invaded in AD 642, absorbing the Nile valley into the Ummayad Caliphate, centred on Damascus. The Arabic language became the official language of government in 706. The Abbasid defeat of the Ummayad dynasty in 750 brought a shift of Arab power to the new city of Baghdad. Abdullah bin Tahir sent a deputy to rule Egypt. The Fatimid caliphs, whose origins were in Tunisia, entered Egypt in 960 and founded Cairo as their capital, later establishing the Muslim university. Over the next 200 years they built an empire that stretched from Tunisia to Syria and Yemen, with extensive trade routes across the Mediterranean and into the Indian Ocean.

Weakened by the Christian Crusades, the Fatimid caliphate fell to the Ayyubid dynasty in 1169, led by Tikrit-born Saladin (Salahuddin al-Ayyubi). Operating from Damascus, he fought the Crusader States and by the end of the 12th century controlled much of the Eastern Mediterranean. Egypt was largely governed by his deputy, Karaksh. In 1250 Egypt was seized by Saif ad Din Qutuz, a Turkic former slave who founded the Mamluk Sultanate. There were frequent revolts and changes of leaders (*beys*, or princes), but the Sultanate survived until 1517 when Egypt was absorbed into the Ottoman empire.

Napoleonic forces seized the country between 1798 and 1801 but were forced out by a combined Anglo-Ottoman force. Muhammad Ali, appointed Egyptian pasha by the Ottoman emperor in 1805, swiftly destroyed the remnants of Mamluk power. He introduced sweeping political and social reforms and modernized agriculture. Like many of his predecessors he took control of Syria, Nubia and part of the Arabian peninsula. The opening of the Suez Canal during the reign of Muhammad Said Pasha in 1867 heralded an era of foreign intervention and domination, with Said selling his shares in the Suez Canal to the British in 1875. British forces occupied Alexandria and then Cairo in 1882, ruling the country through the consul general, Lord Cromer. During the First World War, Britain declared Egypt a British Protectorate, ousting the khedive, Abbas I, for supporting Germany. Calls for independence grew louder after the war ended and Fuad I ruled a partially independent Egypt from 1923.

Although Egypt was officially neutral until the last days of the Second World War, Britain was the dominant power. From 1940 it was the arena for the Desert War that saw Allied forces ultimately repel Axis attempts to occupy Egypt, take control of the Suez Canal and open up access to the oil fields of the Middle East. The decisive victory came when Gen. Montgomery's Eighth Army overpowered the German and Italian forces under Gen. Rommel at the Second Battle of El Alamein in July 1942.

Following a revolution in July 1952 led by Gen. Neguib, King Farouk abdicated in favour of his son but in 1953 the monarchy was abolished. Neguib became president but encountered opposition from the military when he attempted to move towards a parliamentary republic. Col. Gamal Abdel Nasser became head of state on 14 June 1954 (president from 1956). In 1956 Egypt nationalized the Suez Canal, a move which led Britain, France and

422

Israel to mount military attacks against Egypt until UN and US pressure forced a withdrawal.

The 1960s and 1970s were years of conflict with Israel, notably the Six-Day War in June 1967, when Egypt (together with Syria and Jordan) declared war on Israel but were defeated despite having more troops and armaments. After the war Egypt received economic and military aid from the Soviet Union. Following Nasser's death in Sept. 1970 Muhammad Anwar Sadat took over as president. Having launched a peace initiative during a visit to Jerusalem in 1977, Sadat secured a treaty with Israel in March 1979. Sadat was assassinated on 6 Oct. 1981 and was succeeded by his vice-president, Lieut.-Gen. Muhammad Hosni Mubarak of the National Democratic Party (NDP). He gained a fifth consecutive term after emerging victorious in presidential elections in Sept. 2005 (the first multi-candidate presidential poll) to become the country's longest-serving leader since Muhammad Ali. In Jan. and Feb. 2011 he faced popular protests demanding his resignation. Despite attempts to appease the protesters through various concessions, on the 18th day of unrest Mubarak resigned and handed control to the Armed Forces Supreme Council. In June 2012, Mohamed Morsy became the nation's first democratically elected president.

TERRITORY AND POPULATION

Egypt is bounded in the east by Israel and Palestine, the Gulf of Aqaba and the Red Sea, south by Sudan, west by Libya and north by the Mediterranean. The total area is 1,009,450 sq. km (including 6,000 sq. km of inland water), but the cultivated and settled area, that is the Nile Valley, Delta and oases, covers only 35,000 sq. km. A number of new desert cities are being developed to entice people away from the overcrowded Nile valley, where 99% of the population lives. The 2006 census population was 72,798,031; density 72·5 per sq. km. The United Nations population estimate for 2006 was 75,568,000. Estimate, Jan. 2012: 81,395,000. In 2011, 43·5% of the population were urban.

The UN gives a projected population for 2015 of 88·18m.

3·9m. Egyptians were living abroad in 2006.

Area, population and capitals of the governorates (1996 and 2006 censuses):

Governorate	Area (in sq. km)	Population (1996 census)	(2006 census)	Capital
Alexandria	2,300	3,339,076	4,123,869	Alexandria
Aswan	62,726	960,510	1,186,482	Aswan
Asyut	25,926	2,802,334	3,444,967	Asyut
Behera	9,826	3,994,297	4,747,283	Damanhur
Beni Suef	10,954	1,859,213	2,291,618	Beni Suef
Cairo[1]	3,085	6,800,991	6,758,581	Cairo
Dakahlia	3,716	4,223,338	4,989,997	Mansura
Damietta	910	913,555	1,097,339	Damietta
Fayum	6,068	1,989,772	2,511,027	Fayum
Gharbia	1,948	3,404,339	4,011,320	Tanta
Giza[2]	13,184	4,784,095	3,143,486	Giza
Helwan[1]	—	—	1,713,278	Helwan
Ismailia	5,067	714,828	953,006	Ismailia
Kafr El Shaikh	3,748	2,223,383	2,620,208	Kafr El Shaikh
Kalyubia	1,124	3,281,135	4,251,672	Benha
Luxor	2,410	361,138	457,286	Luxor
Matruh	166,563	212,001	323,381	Matruh
Menia	32,279	3,310,129	4,166,299	Menia
Menufia	2,499	2,760,429	3,270,431	Shibin Al Kom
New Valley	440,098	141,774	187,263	Al Kharija
Port Said	1,351	472,331	570,603	Port Said
Qena	10,798	2,442,016	3,001,681	Qena
Red Sea	119,099	157,314	288,661	El Gurdakah
Sharkia	4,911	4,281,068	5,354,041	Zagazig
North Sinai	27,564	252,160	343,681	Al Arish
6th October[2]	—	—	2,581,059	6th October City
South Sinai	31,272	54,806	150,088	At Tur
Suez	9,002	417,526	512,135	Suez
Suhag	11,022	3,123,114	3,747,289	Suhag

[1]Helwan was created in 2008 but was re-merged with Cairo in 2011.
[2]6th October was created in 2008 but re-merged with Giza in 2011.

The capital, Cairo, had a census population in 2006 of 7,740,018. Other major cities, with populations at the 2006 census (in 1,000): Alexandria, 4,085; Giza, 2,891; Shubra Al Khayma, 1,026; Helwan, 650; Port Said, 571; Suez (2005 estimate), 489.

Smaller cities, with 2006 populations (in 1,000): Mahalla Al Kubra, 443; Mansura, 439; Tanta, 423; Asyut, 389; Fayum, 316; Zagazig, 303; Ismailia, 293; Al Khusus, 291; Aswan, 266; Damanhur, 244; Menia, 236; Damietta, 207; Luxor (Uqsur), 202; Qena, 201.

The official language is Arabic, although French and English are widely spoken.

SOCIAL STATISTICS

Births (est.), 2009, 2,217,000 (28·8 per 1,000 population); deaths, 477,000 (6·2). Annual population growth rate, 2000–05, 2·3%. In 2010, 73% of the population was under 40 years old. Life expectancy at birth, 2007, was 68·2 years for males and 71·7 years for females. Fertility rate, 2008, 2·9 births per woman; infant mortality, 2010, 19 per 1,000 live births. Egypt has made some of the best progress in recent years in reducing child mortality. The number of deaths per 1,000 live births among children under five was reduced from more than 100 in 1990 to 26 in 2005.

CLIMATE

The climate is mainly dry, but there are winter rains along the Mediterranean coast. Elsewhere, rainfall is very low and erratic in its distribution. Winter temperatures are comfortable everywhere, but summer temperatures are very high, especially in the south. Cairo, Jan. 56°F (13·3°C), July 83°F (28·3°C). Annual rainfall 1·2" (28 mm). Alexandria, Jan. 58°F (14·4°C), July 79°F (26·1°C). Annual rainfall 7" (178 mm). Aswan, Jan. 62°F (16·7°C), July 92°F (33·3°C). Annual rainfall (trace). Giza, Jan. 55°F (12·8°C), July 78°F (25·6°C). Annual rainfall 16" (389 mm). Ismailia, Jan. 56°F (13·3°C), July 84°F (28·9°C). Annual rainfall 1·5" (37 mm). Luxor, Jan. 59°F (15°C), July 86°F (30°C). Annual rainfall (trace). Port Said, Jan. 58°F (14·4°C), July 78°F (27·2°C). Annual rainfall 3" (76 mm).

CONSTITUTION AND GOVERNMENT

Following the popular uprising that led to President Hosni Mubarak being deposed in Feb. 2011, a Provisional Constitution came into force on 30 March 2011 to supersede the previous constitution dating from 1971.

After elections in March 2012 parliament elected a constituent assembly to draft a new constitution. This assembly was dissolved the following month amid claims that it was unrepresentative. A second 100-member assembly was appointed, delivering a 234-article draft constitution in Nov. 2012.

Under its terms, Egypt is a democratic state within the Arab world, with the principles of Islamic law the chief source of its legislation. It provides for a multi-party political system and restricts the *President* to a maximum of two four-year terms. There is a bicameral parliament consisting of the *House of Representatives* (comprising 508 members of which 498 elected by direct, secret public balloting and ten appointed) and the *Shura Council* (comprising 270 members of which 180 directly elected and 90 appointed). The *House of Representatives* sits for five-year terms and the *Shura Council* for six years.

The military has the right to try civilians in military tribunals in the case of crimes damaging to the armed forces. Freedom of opinion and thought is guaranteed, along with the freedom to follow any of the Abrahamic religions (Christianity, Islam and Judaism), though 'insulting prophets and messengers is forbidden'. Torture is outlawed and citizens are equal before the law regardless of 'gender, origin, language, religion, belief, opinion, social status or disability'. Leaders of the former ruling National Democratic Party are banned from seeking office for ten years from the promulgation of the constitution.

Fifteen members of the constituent assembly (including all Christian members and all but four women) boycotted the assembly's final vote on the draft on 29 Nov. 2012, amid accusations that it did not sufficiently enshrine the rights and freedoms of woman and minority groups. Passed by the remaining members of the assembly, the constitution received 64% backing in a two-round referendum in Dec. 2012, though turnout was only around a third. It was passed into law by President Morsy on 26 Dec. 2012.

National Anthem
'Biladi' ('My homeland'); words and tune by S. Darwish.

GOVERNMENT CHRONOLOGY

Heads of State since 1953. (ASU = Arab Socialist Union; LR = Liberation Rally; NDP = National Democratic Party; NU = National Union)

President
| 1953–54 | military, LR | Muhammad Neguib |

Chairman of the Revolutionary Command Council
| 1954 | military, LR | Gamal Abdel Nasser |

President
| 1954 | military, LR | Muhammad Neguib |

Chairman of the Revolutionary Command Council
| 1954–56 | military, LR | Gamal Abdel Nasser |

Presidents
1956–70	NU, ASU	Gamal Abdel Nasser
1970–81	ASU, NDP	Muhammad Anwar Sadat
1981–2011	NDP	Muhammad Hosni Mubarak

Head of the Armed Forces Supreme Council
| 2011–12 | Military | Mohamed Hussein Tantawi |

President
| 2012– | ind. | Mohamed Morsy |

RECENT ELECTIONS

Parliamentary elections were held in three phases—a first phase on 28–29 Nov. 2011 (first round) and 5–6 Dec. (second round), a second phase on 14–15 Dec. (first round) and 21–22 Dec. (second round), and a third phase on 3–4 Jan. 2012 (first round) and 10–11 Jan. (second round). The Democratic Alliance of Egypt won 235 of 498 elected seats (including 213 for the Freedom and Justice Party), taking 37·5% of the vote; the Islamist Bloc 123 seats (including 107 for the Al-Nour Party), taking 27·8%; the New Wafd Party 38 (9·2%); the Egyptian Bloc 34 (8·9%); Al-Wasat Party 10 (3·7%); the Revolution Continues Alliance 9 (2·8%); and the Reform and Development Party 9 (2·2%). Eight other parties gained five seats or fewer and 21 went to independents. With ten seats reserved for presidential appointees, the total number of seats is 508.

Presidential elections took place on 23–24 May 2012. Mohamed Morsy (Freedom and Justice Party) came first with 24·8% of the vote, followed by Ahmed Shafik (ind.) with 23·7%, Hamdeen Sabahi (Dignity Party) 20·7%, Abdel Moneim Aboul Fotouh (ind.) 17·5% and Amr Moussa (ind.) 11·1%. There were eight other candidates. In the run-off held on 16–17 June 2012, Morsy won 51·7% of the vote and Shafik 48·3%. Turnout was 46·4% in the first round and 51·9% in the second.

CURRENT GOVERNMENT

President: Mohamed Morsy; b. 1951 (ind.; sworn in 30 June 2012).

In March 2013 the government comprised:

Prime Minister: Hisham Qandil; b. 1962 (ind.; sworn in 2 Aug. 2012).

Minister of Agriculture and Land Reclamation: Salah Abdel Momen. *Civil Aviation:* Wael El-Maadawi. *Communications and Information Technology:* Atef Helmi. *Culture:* Mohammed Saber Arab. *Defence:* Abdel Fattah al-Sisi. *Drinking Water and*

Sanitation Facilities: Abdel Kawi Khalifa. *Education:* Ibrahim Ahmed Ghoneim. *Electricity and Energy:* Ahmed Emam. *Environment:* Khlaed Fahmi. *Finance:* El-Morsi Hegazy. *Foreign Affairs:* Mohammed Kamel Amr. *Health:* Mohamed Mostafa Hamed. *Higher Education:* Mostafa al-Sayed Mosad. *Housing and Urban Communities:* Tarek Wafiq. *Industry and Foreign Trade:* Hatem Saleh. *Information:* Salah Abdel Maqsud. *Insurance and Social Affairs:* Nagwa Khalil. *Interior:* Gen. Mohamed Ibrahim. *Investment:* Osama Saleh. *Justice:* Ahmed Mekky. *Local Development:* Mohamed Ali Beshr. *Manpower and Migration:* Khaled al-Azhary. *Oil:* Osama Kamal. *Planning and International Co-operation:* Ashraf Abdel Fatah. *Religious Affairs (Awqaf):* Talaat Afify. *Scientific Research:* Nadia Zakhary. *Supplies and Domestic Trade:* Bassem Ouda. *Tourism:* Hisham Zaazou. *Transport:* Hatem Abdel-Latif. *Water Resources and Irrigation:* Mohamed Baha-Eddin Saad.

CURRENT LEADERS

Mohamed Morsy

Position
President

Introduction
Mohamed Morsy became Egypt's first democratically elected president on 30 June 2012. He succeeded Hosni Mubarak, who resigned on 11 Feb. 2011 in the wake of a popular uprising. A Sunni Muslim, Morsy previously served as chairman of the Freedom and Justice Party, a nominally independent Islamist party with strong ties to the Muslim Brotherhood.

Early Life
Mohamed Morsy was born on 20 Aug. 1951 in the northern Sharkia governorate. He received a masters degree in engineering from Cairo University in 1978. Four years later he completed a PhD at the University of Southern California, staying in the USA to work as a professor at the University of North Ridge, California. The eldest two of his five children were born in Los Angeles and hold American citizenship. Morsy returned to Egypt in 1985 to head the engineering faculty at Zagazig University, a post he held until 2010.

He was elected to the Egyptian People's Assembly in 2000. Although nominally an independent because the Muslim Brotherhood was barred from parliament under Mubarak, he was a member of the Brotherhood's executive office and spokesperson for the parliamentary bloc that aligned itself with the organization's Islamist policies. He failed to win re-election in 2005, a loss that the Muslim Brotherhood attributed to electoral fraud. Prominent in opposing state interference with the judiciary, Morsy spent seven months in jail in 2006. He was detained again in Jan. 2011 as protests against Mubarak swept the country.

In the wake of this uprising, the Muslim Brotherhood founded the Freedom and Justice Party on 30 April 2011, with Morsy as chairman. He was nominated as the party's presidential candidate after its first choice, Khairat al-Shater, was disqualified by the electoral commission. In the second round of presidential elections on 16–17 June 2012, Morsy took 51·7% of the vote, defeating the former prime minister, Ahmed Shafik. Morsy immediately resigned from the Muslim Brotherhood and the Freedom and Justice Party to take up the presidency.

Career in Office
Previously an outspoken critic of Israel, Morsy helped mediate a ceasefire with Hamas after Israel's incursion into Gaza in Nov. 2012. In the same month his tenure came under the international spotlight when he issued a constitutional declaration that effectively granted him unlimited legislative power. After mass protests, Morsy annulled the decree on 8 Dec. 2012.

DEFENCE

Conscription is selective, and for 12–36 months, depending on the level of education. Military expenditure totalled US$4,562m. in 2008 (US$59 per capita), representing 2·9% of GDP. According to *Deadly Arsenals*, published by the Carnegie Endowment for International Peace, Egypt has a chemical and biological weapons programme.

Army

Estimated strength (2007) 310,000 (around 205,000 conscripts). In addition there were 375,000 reservists, a Central Security Force of 325,000, a National Guard of 60,000 and 12,000 Border Guards.

Navy

Major surface combatants include one destroyer and ten frigates. A small shore-based naval aviation branch operates 20 helicopters. There are naval bases at Al Ghardaqah, Alexandria, Hurghada, Mersa Matruh, Port Said, Port Tewfik, Safaqa and Suez. Naval personnel in 2007 totalled an estimated 18,500 (including 10,000 conscripts and 2,000 coast guards).

Air Force

Until 1979 the Air Force was equipped largely with aircraft of USSR design, but subsequent re-equipment involves aircraft bought in the West, as well as some supplied by China. Strength (2007) is about 30,000 personnel (10,000 conscripts), 115 attack helicopters and 489 combat-capable aircraft including F-7s, F-16s and *Mirages*.

ECONOMY

In 2009 agriculture accounted for 13·7% of GDP, industry 37·3% and services 49·0%.

Overview

Egypt, one of the most economically diversified countries in the Middle East, is beginning to recover after the economic turmoil that accompanied the overthrow of Hosni Mubarak in 2011.

The 1980s was a decade of macroeconomic disorder in Egypt before an IMF-backed reform programme in the 1990s helped it achieve economic stability. However, the late 1990s saw a downturn and privatization efforts stalled in the early 2000s. In 2004 an economically liberal cabinet was appointed and the reform agenda was revived, with President Mubarak investing political capital in structural reforms to generate jobs and promote foreign investment.

Growth in Egypt averaged 6·4% per year between 2005 and 2008, underpinned by record levels of foreign direct investment and a favourable external environment. The economy held up relatively well during the global financial crisis, with a decline in remittances and external demand partially offset by resilient domestic demand and strong performances in the construction, communications and trade sectors.

In the wake of the Jan. 2011 revolution, the economy slumped and GDP fell by 9% in the first quarter. Revenues from tourism collapsed, triggering a slide in foreign reserves. Political uncertainty in the post-Mubarak era continues to hinder economic stability, while long-term challenges include reducing the budget deficit and addressing high levels of poverty and youth unemployment.

Currency

The monetary unit is the *Egyptian pound* (EGP) of 100 *piastres*. Inflation rates (based on IMF statistics) for fiscal years:

2002	2003	2004	2005	2006	2007	2008	2009	2010	2011
2·4%	3·2%	8·1%	8·8%	4·2%	11·0%	11·7%	16·2%	11·7%	11·1%

Faced with slowing economic activity, the country devalued the Egyptian pound four times in 2001. In Jan. 2003 the Egyptian pound was allowed to float against the dollar after years of a government-controlled foreign exchange regime. In June 2009 foreign exchange reserves were US$29,278m., gold reserves totalled 2·43m. troy oz and total money supply was £E182,991m.

Budget

The financial year runs from 1 July. Budgetary central government revenues in 2008–09 (provisional) were £E282,505m. and expenditures £E308,070m. Taxes on income, profits and capital gains accounted for 28·4% of revenue and taxes on goods and services 22·2%. Main items of expenditure were subsidies (30·5%) and compensation of employees (24·7%).

Performance

Real GDP growth rates (based on IMF statistics):

2002	2003	2004	2005	2006	2007	2008	2009	2010	2011
3·2%	3·2%	4·1%	4·5%	6·8%	7·1%	7·2%	4·7%	5·1%	1·8%

Total GDP in 2011 was US$229·5bn.

Banking and Finance

The Central Bank of Egypt (founded 1960) is the central bank and bank of issue. The *Governor* is Hisham Ramez.

In 2003, four major public-sector commercial banks accounted for some 77% of all banking assets: the National Bank of Egypt (the largest bank, with assets of £E299bn. in June 2010), the Banque Misr, the Bank of Alexandria and the Banque du Caïre. There were 40 banks in total in 2008. Foreign banks have only been allowed to operate since 1996.

Foreign direct investment inflows, which were just US$237m. in 2003, rose to US$11·6bn. in 2007, but declined to US$6·4bn. in 2010.

In 2010 external debt totalled US$34,844m., representing 16·2% of GNI.

There are stock exchanges in Cairo and Alexandria.

ENERGY AND NATURAL RESOURCES

Environment

Egypt's carbon dioxide emissions from the consumption and flaring of fossil fuels in 2008 were the equivalent of 2·1 tonnes per capita.

Electricity

Installed capacity was 21·0m. kW in 2006–07. Electricity generated in 2006–07 was 128·13bn. kWh. Consumption per capita was an estimated 1,732 kWh in 2006–07. The use of solar energy is expanding. There are plans to build a nuclear power station to help meet the growing demand for electricity.

Oil and Gas

Oil was discovered in 1909. Oil policy is controlled by the state-owned Egyptian General Petroleum Corporation, whole or part-owner of the production and refining companies. Oil production in 2008 was 34·6m. tonnes with 4·3bn. bbls of proven reserves.

As a result of a series of new discoveries in 1999 and 2000 gas reserves have been steadily increasing. Egypt began exporting natural gas in 2003 and is now a major exporter. 2008 total production amounted to 58·9bn. cu. metres. There were proven natural gas reserves of 2,170bn. cu. metres in 2008.

Minerals

Production (2003–04, in tonnes): phosphate, 2·08m.; iron ore, 1·98m.; salt, 1·55m.; kaolin, 221,000; aluminium (2002 estimate), 190,000; quartz, 19,000; asbestos (2002 estimate), 2,000.

Agriculture

There were 2·86m. ha. of arable land in 2001 and 0·48m. ha. of permanent crops. In 1996, of the total cultivated area 18·4% was reclaimed desert. Irrigation is vital to agriculture and is being developed by government programmes; it now reaches most cultivated areas and in 2001 covered 3·34m. ha. The Nile provides

85% of the water used in irrigation, some 55,000m. cu. metres annually. There were 102,584 tractors in 2007 and 2,451 harvester-threshers.

In 1994 there were 5,214 agricultural co-operatives. 0·71m. feddan of land had been distributed by 1991 to 0·35m. families under an agrarian reform programme. In 2001, 5·07m. persons were engaged in agriculture. Cotton, sugarcane and rice are subject to government price controls and procurement quotas.

Output (in 1,000 tonnes), 2007: sugarcane, 17,014; tomatoes, 8,695; wheat, 7,379; maize, 6,930; rice, 6,877; sugar beets, 5,458; potatoes, 2,760; melons and watermelons, 2,121; oranges, 2,055; dry onions, 1,756; grapes, 1,485; dates, 1,314; aubergines, 1,160; bananas, 945; sorghum, 844. Egypt is Africa's largest producer of a number of crops, including wheat, rice, tomatoes, potatoes and oranges. In spite of being a significant producer of wheat, Egypt is also the largest importer.

Livestock, 2000: sheep, 4·45m.; goats, 3·30m.; buffaloes, 3·20m.; cattle, 3·18m.; asses, 3·05m.; camels, 120,000; chickens, 88m. Livestock products in 2000 (in 1,000 tonnes): buffalo's milk, 2,079; cow's milk, 1,645; meat, 1,391; eggs, 170. 464,000 tonnes of cheese were produced in 2000, making Egypt the largest cheese producer in Africa.

Forestry
In 2010 forests covered 70,000 ha., representing 0·1% of the total land area. In 2007, 17·44m. cu. metres of roundwood were produced.

Fisheries
The catch in 2010 was 385,209 tonnes, of which 263,847 tonnes were freshwater fish. Aquaculture production totalled 919,585 tonnes in 2010 (70% of total fish production).

INDUSTRY
The largest company by market capitalization in Egypt in March 2012 was Orascom Construction Industries (US$8·8bn.).

Production, in 1,000 tonnes: cement (2007), 38,469; residual fuel oil (2007), 10,989; distillate fuel oil (2007), 8,803; crude steel (2008), 6,198; petrol (2005), 2,972; sugar (2001–02), 1,555; fertilizers (2002), 1,269; tobacco (1997–98), 595; paper and paperboard (2001), 460; cotton yarn (2000), 164. Motor vehicles (2002), 46,479 units; washing machines (1999), 252,000 units; cigarettes (2000), 53·0bn. units.

Labour
In 2002–03 the labour force was 19·9m. In 2003, 29·0% of employed persons were engaged in agriculture, hunting and forestry; 11·9% in wholesale and retail trade/repair of motor vehicles, motorcycles and personal and household goods; 11·2% in public administration, defence and compulsory social security; and 10·9% in manufacturing. Unemployment was 9·9% in 2003. The high birth rate of the 1980s has meant that there are now some 800,000 new entrants into the job market annually.

INTERNATIONAL TRADE
Imports and Exports
In 2010 imports (c.i.f.) totalled US$53,003m. (US$44,913m. in 2009); exports (f.o.b.), US$26,332m. (US$24,182m. in 2009).

Leading imports, 2004: machinery and apparatus, 15·9%; vegetable products, 11·4%; metal products, 10·0%; chemicals and chemical products, 9·1%. Leading exports, 2004: petroleum, 40·0%; finished goods, 27·2%; semi-manufactured goods, 14·1%; raw cotton, 6·3%.

Main import suppliers, 2004: Free zones, 11·8%; USA, 10·3%; Germany, 6·6%; China, 5·1%; Italy, 4·9%. Main export markets, 2004: Italy, 12·5%; Bunkers and ship's stores, 9·8%; USA, 7·5%; Free zones, 5·7%; Spain, 5·5%.

COMMUNICATIONS
Roads
In 2006 there were 99,672 km of roads, of which 81·0% were paved. Vehicles in use in 2006 (in 1,000): passenger cars, 2,372 (29 per 1,000 inhabitants in 2005); lorries and vans, 1,463; motorcycles and mopeds, 751; buses and coaches, 79. There were 12,295 fatalities as a result of road accidents in 2007.

Rail
In 2005 there were 5,063 km of state railways (1,435 mm gauge), of which 42 km were electrified. Passenger-km travelled in 2008 came to 40·8bn. and freight tonne-km to 3·8bn.

There are tramway networks in Cairo, Heliopolis and Alexandria, and a metro (63 km) opened in Cairo in 1987.

Civil Aviation
There are international airports at Cairo, Luxor, Alexandria, Hurghada and Sharm El-Sheikh. The national carrier is Egyptair. In 2003 scheduled airline traffic of Egyptian-based carriers flew 63m. km, carrying 4,181,000 passengers (2,916,000 on international flights). In 2003 Cairo handled 8,337,000 passengers and 187,280 tonnes of freight. Sharm El-Sheikh was the second busiest in 2003, with 3,423,000 passengers.

Shipping
In Jan. 2009 there were 100 ships of 300 GT or over registered, totalling 923,000 GT. Of the 100 vessels registered, 39 were general cargo ships, 34 oil tankers, 11 bulk carriers, eight passenger ships, six chemical tankers and two container ships. The Egyptian-controlled fleet comprised 113 vessels of 1,000 GT or over in Jan. 2009, of which 59 were under the Egyptian flag and 54 under foreign flags. The leading ports are Adabeya, Alexandria, Damietta, Dekheila, Port Said and Sokhna.

Suez Canal
The Suez Canal was opened for navigation on 17 Nov. 1869 and nationalized in June 1956. By the convention of Constantinople of 29 Oct. 1888, the canal is open to vessels of all nations and is free from blockade, except in time of war. It is 190 km long, connecting the Mediterranean with the Red Sea. It has a maximum depth of 22·5 metres and a maximum width of 365 metres. Vessels of up to 210,000 DWT fully laden are able to pass through the canal.

In 2004, 16,850 vessels (net tonnage, 621m.) went through the canal. In 2004, 521m. tonnes of cargo were transported. Toll revenue in 2004 was US$3,085m.

Telecommunications
In 2008 there were 11,936,000 main (fixed) telephone lines. In the same year mobile phone subscribers numbered 41,273,000 (506·2 per 1,000 persons). In Dec. 2005 the Egyptian government sold 20% of its holding in Telecom Egypt. There were 13,573,000 internet users in 2008. In June 2012 there were 11·3m. Facebook users.

SOCIAL INSTITUTIONS
Justice
The court system comprises: a Court of Cassation with a bench of five judges which constitutes the highest court of appeal in both criminal and civil cases; five Courts of Appeal with three judges; Assize Courts with three judges which deal with all cases of serious crime; Central Tribunals with three judges which deal with ordinary civil and commercial cases; Summary Tribunals presided over by a single judge which hear minor civil disputes and criminal offences. Contempt for religion and what is judged to be a false interpretation of the Koran may result in prison sentences.

The population in penal institutions in Dec. 2006 was 64,378 (87 per 100,000 of national population). The death penalty is in force; there was one execution in 2011 but none in 2012.

Education

The adult literacy rate in 2005 was 71·4%. Free compulsory education is provided in primary schools (eight years). Secondary and technical education is also free. In 2002–03, 53·9% of girls and 46·1% of boys were enrolled in the primary school system. In 2004–05 there were 5,845 pre-primary schools with 494,334 pupils. In 2004–05 there were 16,369 primary schools with 8,634,115 pupils, 8,757 preparatory schools with 2,889,212 pupils and 2,170 general secondary schools with 1,299,233 pupils. In 2004–05 there were 788,017 students in 841 commercial secondary schools, 1,050,970 in 855 industrial secondary schools and 251,021 in 172 agricultural secondary schools.

Al Azhar institutes educate students who intend enrolling at Al Azhar University, one of the world's oldest universities and Sunni Islam's foremost seat of learning. In 2003–04 there were 6,690 institutes in the Al Azhar system with 1,449,048 pupils.

In 2002–03 there were 12 state universities, the Al Azhar university and five private universities including a French and a German university. There were 2·0m. students enrolled in university and higher education.

Public education expenditure in 2007 was 3·7% of GNI and 12·6% of total government spending.

Health

At 1 Jan. 2002 there were 1,112 hospitals with 80,519 beds. There were 157,000 physicians and 188,000 nurses in 2005. In 2004–05 health expenditure represented 3·4% of GDP.

Welfare

In 2003–04 there were 18·7m. welfare beneficiaries including, in 2002–03, 7·4m. recipients of pensions.

RELIGION

Islam is constitutionally the state religion. In 2001 there were 58·1m. Sunni Muslims (84% of the population); some 9% of the population are Coptic Christians, the remainder being Roman Catholics, Protestants or Greek Orthodox, with a small number of Jews. A Patriarch heads the Coptic Church, and there are 25 metropolitans and bishops in Egypt; four metropolitans for Ethiopia, Jerusalem, Khartoum and Omdurman, and 12 bishops in Ethiopia. The Copts use the Diocletian (or Martyrs') calendar, which begins in AD 284. In Feb. 2013 there was one cardinal in the Roman Catholic church.

CULTURE

World Heritage Sites

There are seven sites under Egyptian jurisdiction that appear on the UNESCO World Heritage List. The first five were entered on the list in 1979. They are: Memphis and its Necropolis (the Pyramid Fields from Giza to Dahshur); Ancient Thebes with its Necropolis; Nubian Monuments from Abu Simbel to Philae; Historic Cairo; and Abu Mena. Memphis was considered one of the Seven Wonders of the World. Ancient Thebes was the capital of Egypt during the period of the middle (c. 2000 BC) and new (c. 1600 BC) kingdoms. The Nubian monuments include the temples of Ramses II in Abu Simbel and the Sanctuary of Isis in Philae. Historic Cairo, founded in the 10th century, became the centre of the Islamic world. Abu Mena was an early Christian holy city.

The Saint Catherine Area was added to the list in 2002. It included Saint Catherine's Monastery, an outstanding example of an Orthodox Christian monastic settlement, dating from the 6th century AD. The area is centred on Mount Sinai (Jebel Musa or Mount Horeb). In 2005 Wadi Al-Hitan (Whale Valley) was inscribed on the list. The site is an area rich in fossil remains in Egypt's western desert.

Press

In 2008 there were 18 dailies (17 paid-for and one free) with a total average circulation of 2·74m. The leading dailies are *Al-Ahram* and *Al-Gomhuriya*.

Tourism

There were 12,535,885 foreign visitors in 2009 (up from 8,607,067 in 2005). The main countries of origin in 2009 were: Russia (2,035,330); UK (1,346,724); Germany (1,202,339).

Festivals

Annual film festivals are held in Alexandria (Sept.) and Cairo (Nov.–Dec.). The Arab Music Festival at the Cairo Opera House takes place in Aug.

DIPLOMATIC REPRESENTATIVES

Of Egypt in the United Kingdom (26 South St., London, W1K 1DW)
Ambassador: Hatem Seif El-Nasr.

Of the United Kingdom in Egypt (7 Ahmed Ragheb St., Garden City, Cairo)
Ambassador: James Watt, CVO.

Of Egypt in the USA (3521 International Court, NW, Washington, D.C., 20008)
Ambassador: Mohamed Tawfik.

Of the USA in Egypt (8 Kamal el-Din Salah St., Garden City, Cairo)
Ambassador: Anne W. Patterson.

Of Egypt to the United Nations
Ambassador: Mootaz Ahmadein Khalil.

Of Egypt to the European Union
Ambassador: Fatma Elzahraa Etman.

FURTHER READING

CAPMAS, *Statistical Year Book, Arab Republic of Egypt*

Abdel-Khalek, G., *Stabilization and Adjustment in Egypt.* 2001
Baker, Raymond William, *Islam without Fear: Egypt and the New Islamists.* 2006
Bradley, John R., *Inside Egypt: The Land of The Pharaohs on the Brink of a Revolution.* Revised ed. 2011
Cook, Steven, *The Struggle for Egypt: From Nasser to Tahrir Square.* 2011
Daly, M. W. (ed.) *The Cambridge History of Egypt.* 2 vols. 2000
El-Mikawy, Noha and Handoussa, Heba, (eds.) *Institutional Reform and Economic Development in Egypt.* 2004
Hopwood, D., *Egypt: Politics and Society 1945–1990.* 3rd ed. 1992
Ibrahim, Fouad N. and Ibrahim, Barbara, *Egypt: An Economic Geography.* 2001
Khalil, Ashraf, *Liberation Square: Inside the Egyptian Revolution and the Rebirth of a Nation.* 2012
King, J. W., *Historical Dictionary of Egypt.* 2nd ed. Revised by A. Goldschmidt. 1995
Malek, J. (ed.) *Egypt.* 1993
Osman, Tarek, *Egypt on the Brink: From Nasser to Mubarak.* Revised ed. 2011
Raymond, André, *Cairo.* 2001
Rodenbeck, M., *Cairo—the City Victorious.* 1998
Rubin, Barry, *Islamic Fundamentalism in Egyptian Politics.* 2002
Turner, Barry, *Suez 1956: the Inside Story of the First Oil War.* 2007
Vatikiotis, P. J., *History of Modern Egypt: from Muhammad Ali to Mubarak.* 1991

National Statistical Office: Central Agency for Public Mobilization and Statistics (CAPMAS), Salah Salam Street, Nasr City, Cairo.
Website: http://www.capmas.gov.eg

EL SALVADOR

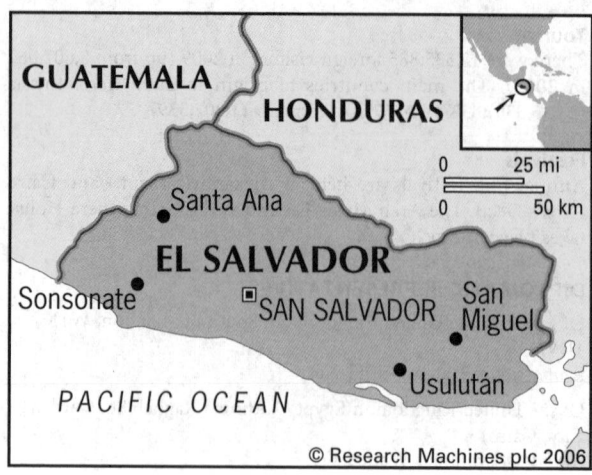

República de El Salvador
(Republic of El Salvador)

Capital: San Salvador
Population projection, 2015: 6·39m.
GNI per capita, 2011: (PPP$) 5,925
HDI/world rank: 0·674/105
Internet domain extension: .sv

KEY HISTORICAL EVENTS

Conquered by Spain in 1526, El Salvador remained under Spanish rule until 1821. Thereafter, El Salvador was a member of the Central American Federation comprising the states of El Salvador, Guatemala, Honduras, Nicaragua and Costa Rica until this federation was dissolved in 1839. In 1841 El Salvador declared itself an independent republic.

The country's history has been marked by political violence. The repressive dictatorship of President Maximiliano Hernandez Martínez lasted from 1931 to 1944 when he was deposed as were his successors in 1948 and 1960. The military junta that followed gave way to more secure presidential succession although left-wing guerrilla groups were fighting government troops in the late 1970s. As the guerrillas grew stronger, gaining control over a part of the country, the USA sent economic aid and assisted in the training of Salvadorean troops. A new constitution was enacted in Dec. 1983 but the presidential election was boycotted by the main left-wing organization, the Favabundo Marti National Liberation Front (FMLN). Talks between the government and the FMLN in April 1991 led to constitutional reforms in May, envisaging the establishment of civilian control over the armed forces and a reduction in their size. On 16 Jan. 1992 the government and the FMLN signed a peace agreement.

TERRITORY AND POPULATION

El Salvador is bounded in the northwest by Guatemala, northeast and east by Honduras and south by the Pacific Ocean. The area (including 247 sq. km of inland lakes) is 21,040 sq. km. Population (2007 census), 5,744,113 (female 53%), giving a population density of 273 per sq. km. The United Nations population estimate for 2007 was 6,101,000.

The UN gives a projected population for 2015 of 6·39m.

In 2007, 62·7% of the population were urban. Some 2·5m. Salvadoreans live abroad, mainly in the USA.

The republic is divided into 14 departments. Areas (in sq. km) and 2007 populations:

Department	Area	Population	Chief town	Population
Ahuachapán	1,240	319,503	Ahuachapán	63,981
Cabañas	1,104	149,326	Sensuntepeque	15,395
Chalatenango	2,017	192,788	Chalatenango	16,976
Cuscatlán	756	231,480	Cojutepeque	41,072
La Libertad	1,653	660,652	Santa Tecla	108,840
La Paz	1,224	308,087	Zacatecoluca	42,127
La Unión	2,074	238,217	La Unión	18,046
Morazán	1,447	174,406	San Francisco	15,307
San Miguel	2,077	434,003	San Miguel	158,136
San Salvador	886	1,567,156	San Salvador	316,090[1]
San Vicente	1,184	161,645	San Vicente	36,700
Santa Ana	2,023	523,655	Santa Ana	204,340
Sonsonate	1,225	438,960	Sonsonate	49,129
Usulatán	2,130	344,235	Usulután	51,496

[1]Greater San Salvador conurbation (2007), 1,566,629.

The official language is Spanish.

SOCIAL STATISTICS

2008 births (est.), 112,000; deaths (est.), 32,000. Rates (2008, per 1,000 population): births (est.), 18·3; deaths (est.), 5·2. Life expectancy at birth in 2007 was 66·4 years for males and 75·9 years for females. Annual population growth rate, 2000–05, 1·8%. Infant mortality, 2010, 14 per 1,000 live births; fertility rate, 2008, 2·3 births per woman. Abortion is illegal.

CLIMATE

Despite its proximity to the equator, the climate is warm rather than hot, and nights are cool inland. Light rains occur in the dry season from Nov. to April, while the rest of the year has heavy rains, especially on the coastal plain. San Salvador, Jan. 71°F (21·7°C), July 75°F (23·9°C). Annual rainfall 71" (1,775 mm). San Miguel, Jan. 77°F (25°C), July 83°F (28·3°C). Annual rainfall 68" (1,700 mm).

CONSTITUTION AND GOVERNMENT

A new constitution was enacted in Dec. 1983. Executive power is vested in a *President* and *Vice-President* elected for a non-renewable term of five years. There is a *Legislative Assembly* of 84 members elected by universal suffrage and proportional representation: 64 locally and 20 nationally, for a term of three years.

National Anthem

'Saludemos la patria orgullosos' ('We proudly salute the Fatherland'); words by J. J. Cañas, tune by J. Aberle.

GOVERNMENT CHRONOLOGY

Heads of State since 1944. (ARENA = Nationalist Republican Alliance; FMLN = Farabundo Martí National Liberation Front; PCN = National Conciliation Party; PDC = Christian Democratic Party; PRUD = Revolutionary Party of Democratic Unification; n/p = non-partisan)

Presidents
1944–45	military	Osmín Aguirre y Salinas
1945–48	military	Salvador Castaneda Castro

Military Junta
1948–50		Manuel de Jesús Córdova; Óscar Osorio Hernández; Reinaldo Galindo Pohl; Óscar A. Bolaños; Humberto Costa

Presidents

1950–56	military, PRUD	Óscar Osorio Hernández
1956–60	military, PRUD	José María Lemus López

Junta

1960–61	Miguel Ángel Castillo; César Yanes Urías; Rubén Alonso Rosales; Ricardo Falla Cáceres; Fabio Castillo Figueroa; Rene Fortín Magaña

Civic-Military Directory

1961–62	José Antonio Rodríguez Porth; José Francisco Valiente; Feliciano Avelar; Aníbal Portillo; Julio Adalberto Rivera Carballo; Mariano Castro Morán

Presidents

1962	n/p	Eusebio Rodolfo Cordón Cea
1962–67	military, PCN	Julio Adalberto Rivera Carballo
1967–72	military, PCN	Fidel Sánchez Hernández
1972–77	military, PCN	Arturo Armando Molina Barraza
1977–79	military, PCN	Carlos Humberto Romero Mena

Revolutionary Junta of Government (I)

1979–80	Adolfo Arnaldo Majano Ramos; Jaime Abdul Gutiérrez Avendaño; Román Mayorga Quirós; Guillermo Manuel Ungo Revelo; Mario Antonio Andino

Revolutionary Junta of Government (II)

1980	Adolfo Arnaldo Majano Ramos (military); Jaime Abdul Gutiérrez Avendaño (military); José Antonio Morales Ehrlich (PDC); Héctor Miguel Dada Hirezi (PDC); José Napoleón Duarte Fuentes (PDC); José Ramón Ávalos Navarrete (n/p)

Chairman of the Revolutionary Junta of Government

1980–82	PDC	José Napoleón Duarte Fuentes

Presidents

1982–84	n/p	Álvaro Alfredo Magaña Borja
1984–89	PDC	José Napoleón Duarte Fuentes
1989–94	ARENA	Alfredo Félix Cristiani Burkard
1994–99	ARENA	Armando Calderón Sol
1999–2004	ARENA	Francisco Guillermo Flores Pérez
2004–09	ARENA	Elías Antonio Saca González
2009–	FMLN	Carlos Mauricio Funes Cartagena

RECENT ELECTIONS

Presidential elections were held on 15 March 2009. Mauricio Funes (Farabundo Martí National Liberation Front; FMLN) received 51·3% of votes cast against Rodrigo Ávila (Nationalist Republican Alliance; ARENA) with 48·7%.

In parliamentary elections on 11 March 2012 the ARENA gained 33 of a possible 84 seats in the Legislative Assembly (39·8% of the vote), ahead of the ruling FMLN with 31 (36·8%), the Grand Alliance for National Unity (GANA) 11 (9·6%), the National Coalition (CN) 6 (7·2%), the Party of Hope (PES) 1 (2·8%) and Democratic Change (CD) 1 (2·1%). One seat went to a candidate running jointly for CN and PES.

CURRENT GOVERNMENT

President: Mauricio Funes; b. 1959 (FMLN; sworn in 1 June 2009).
In March 2013 the cabinet comprised:
Vice-President: Prof. Salvador Sánchez Cerén.

Minister of Agriculture and Livestock: Pablo Alcides Ochoa. *Defence:* José Atilio Benítez Parada. *Economy:* Armando Flores. *Education:* Hato Hasbún. *Environment and Natural Resources:* Hermán Humberto Rosa Chávez. *Finance:* Juan Ramón Carlos Enrique Cáceres Chávez. *Foreign Affairs:* Hugo Roger Martínez Bonilla. *Governance:* Ernesto Zelayandia. *Health:* Dr María Isabel Rodríguez. *Justice and Public Security:* Col. David Victoriano Munguía Payés. *Labour and Social Welfare:* Humberto Centeno. *Public Works:* Manuel Orlando Quinteros Aguilar. *Tourism:* José Napoleón Duarte Durán.

Office of the President (Spanish only):
http://www.presidencia.gob.sv

CURRENT LEADERS

Carlos Mauricio Funes Cartagena

Position
President

Introduction
Mauricio Funes took office in June 2009 at the head of the left-wing Farabundo Martí National Liberation Front (FMLN) administration, ending 20 years of Nationalist Republican Alliance (ARENA) government.

Early Life
Funes was born on 18 Oct. 1959 in San Salvador. Schooled at the Colegio Centroamérica and the private Colegio Externado San José, he went in 1975 to study literature at the city's José Simón Cañas University.

In 1980 civil war broke out between the authoritarian government and leftist organizations, from which the FMLN was created. Mauricio's older brother was killed by police during a student demonstration. In 1986 Mauricio became a reporter for state television news. In 1997 he became head of news at Canal 12. However, after several run-ins—notably over criticism of the government after a 2001 earthquake—he left the station in 2005.

His public popularity paved the way for a move into politics, where he established a relationship with the FMLN. In Sept. 2007 the party chose him as its presidential candidate, a month before Funes' son was murdered in Paris. Funes is the first FMLN leader not to have fought as a guerrilla. At the presidential elections in March 2009 Funes defeated Rodrigo Ávila of ARENA, winning over 50% of the vote.

Career in Office
Amid claims that an FMLN victory would jeopardize US relations and turn the country into a satellite of Venezuela, Funes declared he would maintain a good relationship with Washington and that El Salvador would remain part of the Central American Free Trade Agreement, signed by ARENA in 2006. Meanwhile, he re-established ties with Cuba, cut off since 1961.

He offered ministerial portfolios to supporters from a range of political backgrounds. His primary challenges have been to address the troubled economy and reduce the large deficit (aided by loans from the IMF and development banks) and to cut crime, while maintaining the co-operation of a congress where the FMLN failed to secure an overall majority of seats at the elections in Jan. 2009 and March 2012. In Nov. 2009 he was also faced with the humanitarian consequences of a hurricane that left about 140 people dead and thousands homeless. In Sept. 2011 the ongoing problem of crime was highlighted as the US government added El Salvador to a blacklist of countries considered major producers of or transit routes for illegal drugs.

DEFENCE

There is selective conscription for 12 months. In 2008 defence expenditure totalled US$115m. (US$16 per capita), representing 0·5% of GDP.

Army

Strength (2007): 13,850 (4,000 conscripts). The National Civilian Police numbers about 12,000.

Navy

There is a small coastguard force with 860 (2007) personnel including one company of Naval Infantry numbering 160 and one company of Commandos numbering 90.

Air Force

Estimated strength (2007): 950 personnel (200 conscripts). There are some 15 combat-capable aircraft and 33 helicopters.

ECONOMY

Agriculture accounted for 10·9% of GDP in 2006, industry 29·4% and services 59·7%.

Overview

El Salvador's economic performance in the early to mid-2000s was underpinned by wide-ranging reforms, calling for more open trade and changes to pension and fiscal policies. In 2006 the country joined the Central America–Dominican Republic–United States Free Trade Agreement (CAFTA–DR), along with the USA, Costa Rica, the Dominican Republic, Guatemala, Honduras and Nicaragua.

Key industries include coffee and textiles, making the economy vulnerable to weather conditions and fluctuating international markets. Remittances contributed 17% of GDP in 2011, with 21% of all households as recipients. The USA is the leading trade partner, accounting for 48% of exports, and the largest source of remittance inflows.

After achieving its highest growth rates in a decade in 2007, El Salvador was hit by the global financial crisis. Remittances decreased by nearly 10% while GDP contracted by 3·1% in 2009. A US$250m. assistance package was announced by the World Bank in July 2009. The economy has since shown modest growth, but lower than the average for Latin America and the Caribbean in both 2010 and 2011.

Currency

The *dollar* (USD) replaced the *colón* as the legal currency of El Salvador in 2003. Inflation was 1·2% in 2010 and 3·6% in 2011. Foreign exchange reserves were US$1,741m. and gold reserves 326,000 troy oz in May 2005.

Budget

Budgetary central government revenue totalled US$3,105·0m. in 2010 and expenditure US$3,556·3m.

VAT is 13%.

Performance

The economy contracted by 3·1% in 2009 but grew by 1·4% in both 2010 and 2011. Total GDP in 2011 was US$23·1bn.

Banking and Finance

The bank of issue is the Central Reserve Bank (*President*, Carlos Acevedo), formed in 1934 and nationalized in 1961. In 2008 there were ten commercial banks (eight foreign). Foreign debt was US$11,069m. in 2010, representing 53% of GNI.

There is a stock exchange in San Salvador, founded in 1992.

ENERGY AND NATURAL RESOURCES

Environment

El Salvador's carbon dioxide emissions from the consumption and flaring of fossil fuels were the equivalent of 0·9 tonnes per capita in 2008.

Electricity

Installed capacity in 2007 was 1,372,000 kW, of which 472,000 kW hydro-electric. Production in 2007 was 5·81bn. kWh; consumption per capita was 822 kWh in 2007.

Minerals

El Salvador has few mineral resources. In 2008 an estimated 1·2m. tonnes of limestone were produced. Annual marine salt production averages 30,000 tonnes.

Agriculture

27% of the land surface is given over to arable farming. There were about 682,000 ha. of arable land in 2007 and 237,000 ha. of permanent crops. In 2007 the agricultural population was an estimated 1,696,000, of which about 608,000 were economically active. Large landholdings have been progressively expropriated and redistributed in accordance with legislation initiated in 1980. Since the mid-19th century El Salvador's economy has been dominated by coffee. Output, in 1,000 tonnes (2003): sugarcane, 4,532; maize, 618; sorghum, 144; coffee, 92; melons and watermelons, 90; dry beans, 78; bananas, 65. Livestock (2003): 1·0m. cattle, 153,000 pigs, 96,000 horses, 8m. chickens.

Forestry

Forest area was 0·29m. ha. (14% of the land area) in 2010. Balsam trees abound: El Salvador is the world's principal source of this medicinal gum. In 2007, 4·89m. cu. metres of roundwood were cut.

Fisheries

The catch in 2010 was 30,527 tonnes (90% from marine waters).

INDUSTRY

Production in 1,000 tonnes: cement (2006), 1,311; sugar (2005), 633; residual fuel oil (2007), 477; distillate fuel oil (2007), 229; petrol (2007), 117; paper and paperboard (2006 estimate), 56. Traditional industries include food processing and textiles.

Labour

Out of 2,412,800 people in employment in 2002, 688,500 were in wholesale and retail trade/repair of motor vehicles, motorcycles and personal and household goods/hotels and restaurants; 458,400 in agriculture, forestry and hunting; 434,100 in manufacturing; and 155,400 in health and social work, and other community, social and personal service activities. There were 160,200 unemployed persons, or 6·2% of the workforce, in 2002.

INTERNATIONAL TRADE

Imports and Exports

Imports and exports in calendar years (in US$1m.):

	2006	2007	2008	2009	2010
Imports c.i.f.	7,763	8,821	9,818	7,306	8,485
Exports f.o.b.	3,730	4,015	4,641	3,866	4,499

In 2010 main import sources (in US$1m.) were: USA (3,130); Guatemala (803); Mexico (752); China (483). Main export markets in 2010 (in US$1m.) were: USA (2,176); Guatemala (629); Honduras (579); Nicaragua (244). Main import commodities are chemicals and chemical products, transport equipment, and food and beverages; main export commodities are coffee, paper and paper products, and clothing.

COMMUNICATIONS

Roads

In 2002 there were an estimated 10,029 km of roads, including 327 km of motorways. Vehicles in use in 2002: passenger cars, 112,700; trucks and vans, 234,500. There were 12,396 road accidents in 2009 resulting in 1,033 fatalities.

Rail

There are 555 km of 914 mm gauge railway. The railway was closed from 2002–06 but a limited service resumed in 2007 and continues to operate.

Civil Aviation

The international airport is El Salvador International in San Salvador. The national carrier is Taca International Airlines. In 2003 scheduled airline traffic of El Salvador-based carriers flew 34m. km, carrying 2,271,000 passengers (2,182,000 on international flights). It flies to various destinations in the USA, Mexico and all Central American countries. In 2001 El Salvador International handled 1,294,864 passengers on international flights and 26,276 tonnes of international freight.

Shipping

The main ports are Acajutla (which handled 5·86m. tonnes of cargo in 2008) and Cutuco.

Telecommunications

The telephone system has been privatized and is owned by two international telephone companies. In 2010 there were 1,000,900 landline telephone subscriptions (equivalent to 161·6 per 1,000 inhabitants) and 7,700,300 mobile phone subscriptions (or 1,243·4 per 1,000 inhabitants). In 2010, 15·9% of the population were internet users. In Dec. 2011 there were 1·3m. Facebook users.

SOCIAL INSTITUTIONS

Justice

Justice is administered by the Supreme Court (six members appointed for three-year terms by the Legislative Assembly and six by bar associations), courts of first and second instance, and minor tribunals. Following the disbanding of security forces in Jan. 1992 a new National Civilian Police Force was created that numbered 12,000 by 2007. The population in penal institutions in Dec. 2007 was 14,682 (208 per 100,000 of national population).

El Salvador had the second highest annual murder rate in the world in 2010 (after Honduras), at 65 per 100,000 people (4,004 homicides in total in 2010, although down slightly from 4,367 in 2009). The rate nearly doubled between 2002 and 2009.

Education

The adult literacy rate in 2009 was 84·1%. Education, run by the state, is free and compulsory. In 2007 there were 229,539 children in nursery schools, 1,075,041 in primary schools and 536,017 in secondary schools. The University of El Salvador (Universidad de El Salvador), founded in 1841, is the country's oldest and largest public university. There are also several private universities. In 2007 there were 132,246 students and 8,370 academic staff in tertiary education.

In 2007 public expenditure on education came to 3·1% of GNI and 13·1% of total government spending.

Health

In 2003 there were 30 hospitals with nine beds per 10,000 inhabitants. There were 8,171 physicians, 3,573 dentists and 11,777 nurses in 2002.

Welfare

There are old age, disability and survivors' pensions and sickness, maternity and work injury benefits. The official retirement age is 55 years (women) and 60 years (men). The minimum monthly old age pension is US$143·64. Maternity benefit is equal to 75% of average monthly earnings for up to 12 weeks.

RELIGION

In 2001 there were 4,880,000 Roman Catholics. Under the 1962 constitution, churches are exempted from the property tax; the Catholic Church is recognized as a legal person, and other churches are entitled to secure similar recognition. There is an archbishop in San Salvador and bishops at Santa Ana, San Miguel, San Vicente, Santiago de María, Usulután, Sonsonate and Zacatecoluca. There were about 1,070,000 Protestants in 2001 and 290,000 followers of other religions.

CULTURE

World Heritage Sites

El Salvador has one site on the UNESCO World Heritage List: the Joya de Cerén Archaeological Site (1993), a pre-Hispanic farming community preserved under volcanic ash.

Press

In 2005 there were five daily newspapers with a combined circulation of 250,000.

Tourism

There were 1,091,000 non-resident tourists in 2009 (excluding same-day visitors), down from 1,385,000 in 2008.

DIPLOMATIC REPRESENTATIVES

Of El Salvador in the United Kingdom (8 Dorset Sq., London, NW1 6PU)
Ambassador: Werner Matías Romero.

Of the United Kingdom in El Salvador (Torre Futura, Colonia Escalón, San Salvador)
Ambassador: Linda Cross, MBE.

Of El Salvador in the USA (1400 16th St., NW, Suite 100, Washington, D.C., 20036)
Ambassador: Francisco Altschul.

Of the USA in El Salvador (Urbanización Santa Elena, Antiguo Cuscatlán, San Salvador)
Ambassador: Mari Carmen Aponte.

Of El Salvador to the United Nations
Ambassador: Joaquín Alexander Maza Martelli.

Of El Salvador to the European Union
Ambassador: Edgar Hernán Varela.

FURTHER READING

Lauria-Santiago, Aldo and Binford, Leigh, (eds.) *Landscapes of Struggle: Politics, Society and Community in El Salvador.* 2004
Ladutke, Lawrence Michael, *Freedom of Expression in El Salvador: The Struggle for Human Rights and Democracy.* 2004
Tilley, Virginia Q., *Seeing Indians: A Study of Race, Nation, and Power in El Salvador.* 2005

National Statistical Office: Dirección General de Estadística y Censos, Av. Juan Bertis No. 79, Ciudad Delgado, San Salvador.
Website (Spanish only): http://www.digestyc.gob.sv

EQUATORIAL GUINEA

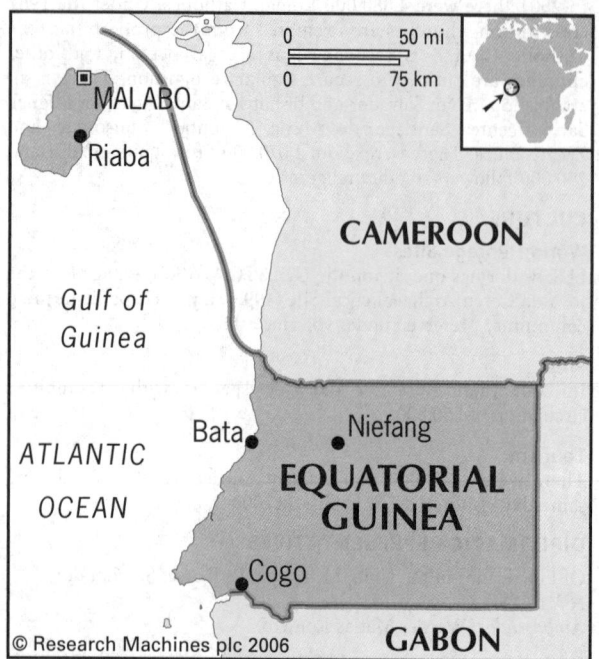

República de Guinea Ecuatorial
(Republic of Equatorial Guinea)

Capital: Malabo
Population projection, 2015: 802,000
GNI per capita, 2011: (PPP$) 17,608
HDI/world rank: 0·537/136
Internet domain extension: .gq

KEY HISTORICAL EVENTS

Equatorial Guinea consists of the island of Bioko, for centuries called Fernando Pó; other smaller islands and the mainland territory of Rio Muni. Fernando Pó was named after the Portuguese navigator Fernão do Pó. The island was ruled for three centuries by Portugal until 1778 when it was ceded to Spain. For some decades after taking possession, Spain did not have a strong presence. Britain established a naval base at Clarence (later Santa Isabel), which was central to the suppression of slave trading over a wide area. Spain asserted its rule from the 1840s when cocoa was cultivated on European-owned plantations using imported African labour. This traffic led to an international scandal in 1930 when Liberians were found to be held in virtual slavery. Later many Nigerians were employed, often in poor conditions.

African nationalist movements began in the 1950s. Internal self-government was granted in 1963 and in 1968 the colony gained independence. The two parts of Equatorial Guinea were united under Macías Nguema who established single-party rule. During his bloody dictatorship up to a third of the population was killed or left the country. Macías was declared President-for-Life in July 1972 but was overthrown by a military coup on 3 Aug. 1979.

A constitution approved by a referendum on 3 Aug. 1982 restored some institutions but a Supreme Military Council remained the sole political body until constitutional rule was resumed on 12 Oct. 1982.

TERRITORY AND POPULATION

The mainland part of Equatorial Guinea is bounded in the north by Cameroon, east and south by Gabon, and west by the Gulf of Guinea, in which lie the islands of Bioko (formerly Macías Nguema, and before that Fernando Pó) and Annobón (called Pagalu from 1973 to 1979). The total area is 28,051 sq. km (10,831 sq. miles). Although the population at the 2001 census was officially given as 1,014,999, the United Nations does not consider this to be an accurate figure.

The UN gives a projected population for 2015 of 802,000.

In 2011, 39·9% of the population were urban.

The seven provinces are grouped into two regions—Continental (C), chief town Bata; and Insular (I), chief town Malabo—with areas and 2001 census populations as follows:

	Area (sq. km)	Population	Chief town
Annobón (I)	17	5,008	San Antonio de Palea
Bioko Norte (I)	776	231,428	Malabo
Bioko Sur (I)	1,241	29,034	Luba
Centro Sur (C)	9,931	125,856	Evinayong
Kié-Ntem (C)	3,943	167,279	Ebebiyin
Litoral (C)	6,665[1]	298,414	Bata
Wele-Nzas (C)	5,478	157,980	Mongomo

[1] Including the adjacent islets of Corisco, Elobey Grande and Elobey Chico (17 sq. km).

In 2003 the capital, Malabo, had an estimated population of 92,900.

The main ethnic group on the mainland is the Fang, which comprises 85% of the total population; there are several minority groups along the coast and adjacent islets. On Bioko the indigenous inhabitants (Bubis) constitute 60% of the population there, the balance being mainly Fang and coast people. On Annobón the indigenous inhabitants are the descendants of Portuguese slaves and still speak a Portuguese patois. The official languages are French, Portuguese and Spanish.

SOCIAL STATISTICS

2008 estimates: births, 25,000; deaths, 10,000. Rates (2008 estimates, per 1,000 population); birth, 38·0; death, 15·0. Life expectancy (2007): male, 48·7 years; female, 51·1. Annual population growth rate, 2000–08, 2·8%. Infant mortality, 2010, 81 per 1,000 live births; fertility rate, 2008, 5·3 births per woman.

CLIMATE

The climate is equatorial, with alternate wet and dry seasons. In Rio Muni, the wet season lasts from Dec. to Feb.

CONSTITUTION AND GOVERNMENT

A Constitution was approved in a plebiscite in Aug. 1982 by 95% of the votes cast and was amended in Jan. 1995. It provided for an 11-member Council of State, and for a 41-member House of Representatives of the People. The President presides over a Council of Ministers.

On 12 Oct. 1987 a single new political party was formed as the *Partido Democrático de Guinea Ecuatorial*. A referendum on 17 Nov. 1991 approved the institution of multi-party democracy, and a law to this effect was passed in Jan. 1992. The electorate is restricted to citizens who have resided in Equatorial Guinea for at least ten years. A parliament created as a result, the *Cámara de Representantes del Pueblo* (House of People's Representatives), has 100 seats, with members elected for a five-year term by proportional representation in multi-member constituencies.

In Nov. 2011 further constitutional amendments were approved by referendum. Official results indicated 97·7% support and turnout of 91·8%, although opposition parties alleged fraud. The amendments relaxed restrictions on the number of terms the *President* can serve and on the age of incumbents (previously set at between 40 and 75 years old), and provide for the creation of a *Senate*. A new position of *Vice President* was established, to be appointed by the President.

National Anthem

'Caminemos pisando las sendas' ('Let us journey treading the pathways'); words by A. N. Miyongo, tune anonymous.

RECENT ELECTIONS

At parliamentary elections on 4 May 2008, boycotted by most opposition parties, the ruling Democratic Party of Equatorial Guinea (PDGE) won 89 of the 100 seats and its allies (the so-called 'democratic opposition') won 10. The Convergence for Social Democracy (CPDS) won the remaining seat.

Presidential elections were held on 29 Nov. 2009. President Nguema Mbasogo was re-elected with 95·4% of votes cast. His main rival, opposition candidate Plácido Micó Abogo of the CPDS, received 3·6% but refused to accept the result, citing fraud. Turnout was 93·5%.

Parliamentary elections were scheduled to take place on 26 May 2013.

CURRENT GOVERNMENT

President of the Supreme Military Council: Brig.-Gen. Teodoro Obiang Nguema Mbasogo; b. 1942 (PDGE; in office since 1979, most recently re-elected in 2009).

Vice President, and Head of Presidential Affairs: Ignacio Milam Tang. *Second Vice President, and Head of Defence and State Security:* Teodoro Nguema Obiang Mangue.

In March 2013 the government comprised:

Prime Minister: Vicente Ehate Tomi; b. 1968 (ind.; sworn in 21 May 2012).

First Deputy Prime Minister, Head of Political Affairs and Democracy, and Minister of Internal Affairs and Local Corporations: Clemente Engonga Nguema Onguene. *Second Deputy Prime Minister, in Charge of Social and Human Rights Affairs:* Alfonso Nsue Mokuy.

Minister of State to the Presidency of the Republic in Charge of Missions: Alejandro Evuna Owono Asangono. *Minister of State to the Presidency of the Republic in Charge of the Civil Cabinet:* Braulio Ncogo Abegue. *Minister of State to the Presidency of the Government in Charge of Relations with Parliament and Legal Affairs:* Angel Masie Mibuy. *Minister of State to the Presidency of the Government in Charge of Regional Integration:* Baltasar Engonga Edjo. *Minister of State for Justice, Religious Affairs and Penitentiary Institutions:* Francisco Javier Ngomo Mbengono. *Minister of State for National Defence:* Gen. Antonio Mba Nguema. *Minister-Secretary General of the Presidency of the Government:* Tomas Esono Ava.

Minister of Foreign Affairs and Co-operation: Agapito Mba Mokuy. *Economy, Trade and Business Promotion:* Celestino Bonifacio Bakale Obiang. *Finance and Budget:* Marcelino Owono Edu. *Planning, Economic Development and Public Investment:* Conrado Okenve Ndoho. *Transport, Technology, Postal Services and Communications:* Francisco Mba Olo Bahamonde. *National Security:* Nicolas Obama Nchama. *Education and Science:* Maria del Carmen Ekoro. *Health and Social Welfare:* Tomas Mecheba Fernandez. *Public Works and Infrastructures:* Juan Nko Mbula. *Mines, Industry and Energy:* Gabriel Mbega Obiang Lima. *Labour and Social Security:* Miguel Abia Biteo Borico. *Agriculture and Forestry:* Miguel Oyono Ndong Mifumu. *Information, Press and Radio:* Agustin Nse Nfumu. *Social Affairs and Women's Development:* Maria Leonor Epam Biribe. *Fisheries and the Environment:* Crescencio Tamarite Castaño. *Public Service and*

Administrative Reform: Purificacion Buari Lasaquero. *Youth and Sports:* Francisco Pascual Obama Asue.

Government Website: http://www.guineaecuatorialpress.com

CURRENT LEADERS

Brig.-Gen. Teodoro Obiang Nguema Mbasogo

Position
President

Introduction
Brig.-Gen. Teodoro Obiang Nguema Mbasogo became president of Equatorial Guinea in Aug. 1979, having led a coup d'état against the dictatorial regime of his uncle, Macías Nguema. After introducing some liberalizing reforms, Obiang himself adopted an authoritarian form of government, leading to widespread allegations of civil rights abuses and electoral fraud. The discovery of major fossil fuel reserves in the mid-1990s created an economic boom, although Obiang has been criticized for the government's lack of transparency in administering this new wealth.

Early Life
Obiang, an ethnic Fang, was born on 5 June 1942 into the Esangui clan of Acoacán. He undertook military training in Spain and, following the election of his uncle as president of the newly-independent country, was made a lieutenant. He had stints as governor of Bioko, presidential aide-de-camp and head of the military, while Macías Nguema's regime became increasingly repressive, with around a third of the population leaving the country during the 1970s.

In Aug. 1979 Obiang ousted his uncle, who was subsequently put on trial and executed.

Career in Office
On assuming the presidency on 3 Aug. 1979 it was hoped that Obiang would implement a more liberal and democratic approach to government. One of his first acts was to call an amnesty on refugees and to free 5,000 political prisoners. However, he retained many of the powers of his uncle and soon came under criticism for his style of government. Local and national political appointments have been blighted by nepotism, which has led to some interfamilial feuding within the political and military establishments.

The discovery of large oil and gas reserves off Bioko in the mid-1990s led to a massive upturn in the economy as Equatorial Guinea became one of sub-Saharan Africa's leading oil exporters. However, it has been widely claimed that the benefits of this oil money have failed to reach the population at large. The IMF and the World Bank demanded increased transparency concerning government oil revenues, which Obiang claimed were a state secret, and warned against an over-reliance on the limited oil reserves.

The country's first multi-party elections in 1993 were won by Obiang's Democratic Party of Equatorial Guinea (PDGE), but boycotted by most of the opposition parties. Then, in the presidential election in Feb. 1996, he was returned with a reported 99% of the vote. At the presidential election of Dec. 2002 he again claimed over 97% of the vote and opposition parties accused the government of vote rigging. A government-in-exile formed by Obiang's opponents was established in Spain. Obiang's treatment of opposition politicians has received condemnation from, among others, the EU and Amnesty International and the president has been accused of using torture on political prisoners. There have also been high-profile public trials such as that in 2002 which resulted in the one-year imprisonment of opposition leader Fabian Nseu Guema for insulting Obiang on a website. In Nov. 2009 Obiang was again re-elected overwhelmingly as president in polling that was denounced by opposition figures as fraudulent and manipulated.

In foreign policy, Obiang has been in dispute with Gabon over the latter's long-term occupation of Mbagne, an island in the Bay of Corisco thought to contain further significant oil supplies. In 2002 he signed an agreement with Nigeria for the development of the Zafiro-Ekanga oil field along their joint maritime border.

In 2004 a plane flying from Zimbabwe was intercepted after Obiang announced it was carrying mercenaries preparing a coup against him. Those accused of involvement included Mark Thatcher (son of former British prime minister Margaret Thatcher), who was arrested in South Africa and later fined and given a suspended prison term, and British mercenary Simon Mann, arrested in Zimbabwe. Obiang claimed that the arrests were evidence of a plot by the secret services of the USA, UK and Spain to overthrow him. In Feb. 2008 Mann was extradited from Zimbabwe to stand trial for his alleged role in the coup and in July was sentenced to 34 years' imprisonment. However, he was granted a presidential pardon in Nov. 2009 on humanitarian grounds.

The PDGE retained its dominance in parliamentary elections in April 2004, but most opposition parties boycotted the poll and foreign observers claimed that there were serious irregularities. In Aug. 2006 Obiang accepted the resignation of Abia Biteo's government, having accused it of corruption and poor leadership. Ricardo Mangue Obama Nfubea was appointed as the new prime minister, but he too submitted the resignation of his government in July 2008 after similar accusations by Obiang. He was replaced by Ignacio Milam Tang, who also tendered his administration's resignation in Jan. 2010 but was reappointed along with his key ministers. According to official results, voters in a referendum in Nov. 2011 overwhelmingly approved a new constitution based on the US presidential system, including a new vice-presidential office and Senate and a limitation on presidential terms of office. However, opposition figures dismissed the changes as a sham. In May 2012 Vicente Ehate Tomi replaced Ignacio Milam Tang as prime minister.

In 2011 Obiang served as chair of the African Union.

DEFENCE

In 2008 defence expenditure totalled US$11m. (US$18 per capita), representing 0·1% of GDP.

Army

The Army consists of three infantry battalions with (2007) 1,100 personnel. There is also a paramilitary Guardia Civil.

Navy

A small force, numbering an estimated 120 in 2007 and based at Malabo and Bata, operates five patrol and coastal combatants.

Air Force

There are no combat-capable aircraft or armed helicopters. Personnel (2007), 100.

ECONOMY

Agriculture accounted for 2% of GDP in 2008, industry 96% (the highest percentage of any country) and services 2%.

Overview

Equatorial Guinea has been one of the world's fastest-growing economies since the discovery of oil in the 1990s. Oil revenues grew from US$3m. in 1993 to an estimated US$8·4bn. in 2009, making it the third largest oil producer in sub-Saharan Africa (behind Nigeria and Angola). GDP growth performance has been exceptional since the late 1990s, reaching a high of 150·0% in 1997. However, a recent slowdown in hydrocarbon production, which accounts for 73% of GDP, has seen GDP growth fall from an average of 28·9% annually between 2001–05 to 9·7% over the period 2006–09. Non-oil GDP growth has remained positive, with large public infrastructure investment and private housing construction the main drivers of growth. The oil boom has generated inflationary pressures, averaging 6% since the early 2000s, but public investment in infrastructure has helped offset the problem.

Despite per capita income rising to that of middle-income countries, living standards remain poor. Poverty is widespread, life expectancy low and access to safe water among the worst in the world. The economy is equipped to make socio-economic progress but a stronger institutional capacity to direct resources to priority areas is needed. Corruption remains a problem, with government officials and their families owning most businesses. An Extractive Industries Transparency Initiative (EITI) and a national poverty reduction strategy are now in place.

Currency

On 2 Jan. 1985 the country joined the Franc Zone and the *ekpwele* was replaced by the *franc CFA* (XAF) which now has a parity value of 655·957 francs CFA to one euro. Foreign exchange reserves were US$1,416m. in June 2005 and total money supply was 209,768m. francs CFA. Inflation was 6·1% in 2010 and 6·3% in 2011.

Budget

In 2009 revenue was 2,368bn. francs CFA and expenditure 2,828bn. francs CFA. Oil revenue accounted for 94·3% of revenues in 2005; capital expenditure accounted for 63·9% of total expenditures in 2005.

Performance

Equatorial Guinea was one of the world's fastest-growing economies during much of the 1990s and 2000s thanks to the rapid expansion of its oil sector. The economy grew by a record 150·0% in 1997. This strong performance continued with growth in both 2007 and 2008 among the highest in the world, at 21·4% and 10·7% respectively. Growth slowed significantly in 2009 to 4·6%. The economy then contracted by 0·5% in 2010 but grew again by 7·8% in 2011. Total GDP in 2011 was US$19·8bn.

Banking and Finance

The Banque des États de l'Afrique Centrale (*Governor*, Lucas Abaga Nchama) became the bank of issue in Jan. 1985. There are two commercial banks (Caisse Commune d'Épargne et d'Investissement Guinée Equatoriale; Société Générale des Banques GE) and two development banks.

Foreign debt was US$260m. in 2002.

ENERGY AND NATURAL RESOURCES

Environment

Carbon dioxide emissions from the consumption and flaring of fossil fuels in 2008 were the equivalent of 7·4 tonnes per capita.

Electricity

Installed capacity was an estimated 31,000 kW in 2007. Production was around 95m. kWh in 2007; consumption per capita in 2007 was approximately 187 kWh.

Oil and Gas

Oil production started in 1992 and in 2005 totalled 17·6m. tonnes, up from 5·8m. tonnes in 2000. There were proven reserves of 1·8bn. bbls in 2005. Since oil in commercial quantities was discovered in 1995 the total stock of foreign direct investment has risen from US$0·2bn. to US$8·8bn. in 2011.

Natural gas reserves were 37bn. cu. metres in 2007.

Minerals

There is some small-scale alluvial gold production.

Agriculture

There were an estimated 130,000 ha. of arable land in 2007 and 90,000 ha. of permanent crops. Subsistence farming predominates. In 2007 the agricultural population was approximately 425,000 of which about 164,000 were economically active. The major crops (estimated production, 2003, in 1,000 tonnes) are: cassava, 45; sweet potatoes, 36; plantains, 31; bananas, 20; coconuts, 6; cocoa beans, 4; coffee, 4; palm oil, 4. Plantations in the hinterland have been abandoned by their Spanish former owners and, except for cocoa and coffee, commercial agriculture is in serious difficulties. Livestock, 2003 estimates: cattle, 5,000; goats, 9,000; pigs, 6,000; sheep, 38,000.

Forestry
In 2010 forests covered 1·63m. ha., or 58% of the total land area. Timber production in 2007 totalled 606,000 cu. metres.

Fisheries
The total catch in 2010 was estimated to be 7,700 tonnes (mainly from sea fishing).

INDUSTRY
The once-flourishing light industry collapsed under the Macías regime. Oil production is now the major activity. Production of veneer sheets, 2007 estimate, 15,000 cu. metres. Food processing is also being developed.

Labour
In 2010 the estimated economically active population was 270,000 (69% males). The wage-earning non-agricultural workforce is small.

INTERNATIONAL TRADE
Imports and Exports
Imports for 2007 came to US$3,098m.; exports in 2007 were valued at US$10,095m.

Main import suppliers, 2003: USA, 30·6%; UK, 16·0%; France, 15·0%; Côte d'Ivoire, 11·9%; Spain, 8·2%. Main export markets, 2003: USA, 33·2%; Spain, 25·4%; China, 14·2%; Canada, 12·7%; Italy, 6·3%. Principal import commodities are machinery and transport equipment, and petroleum and petroleum products; principal export commodities are petroleum, cocoa and timber.

COMMUNICATIONS
Roads
In 2002 the road network covered 2,880 km. Most roads are in a state of disrepair. There were 4,700 passenger cars (9·6 per 1,000 inhabitants) and 3,600 vans and trucks in 2002.

Civil Aviation
There is an international airport at Malabo. There were international flights in 2003 to Cotonou, Douala, Libreville, Madrid, Yaoundé and Zürich. In 1998 Malabo handled 54,000 passengers.

Shipping
Bata is the main port, handling mainly timber. The other ports are Luba, formerly San Carlos, in Bioko, and Malabo, Evinayong and Mbini on the mainland. In Jan. 2009 there were seven ships of 300 GT or over registered, totalling 5,000 GT.

Telecommunications
In 2010 there were 13,500 main (fixed) telephone lines. In the same year mobile phone subscribers numbered 399,000 (570·1 per 1,000 persons). There were 60·0 internet users per 1,000 inhabitants in 2010.

SOCIAL INSTITUTIONS
Justice
The Constitution guarantees an independent judiciary. The Supreme Tribunal, the highest court of appeal, is located at Malabo. There are Courts of First Instance and Courts of Appeal at Malabo and Bata. The death penalty is in force; there were four executions in 2010 but none in 2011 or 2012.

Education
In 2007–08 there were 2,000 teachers for 40,000 children in pre-primary schools; (2009–10) 3,000 teachers for 85,000 pupils in primary schools; and (1999–2000) 21,000 secondary pupils with 800 teachers. In 1993 there were 2 teacher training colleges, 2 post-secondary vocational schools and 1 agricultural institute. Adult literacy was an estimated 93·3% in 2009 (male, 97·0%; female, 89·8%). In 2003 public expenditure on education came to 1·3% of GNP.

Health
In 2005 there were 22 hospital beds per 10,000 inhabitants. There were 153 physicians, 15 dentistry personnel and 271 nursing and midwifery personnel in 2004.

RELIGION
Christianity was proscribed under President Macías but reinstated in 1979. In 2001 there were 390,000 Roman Catholics with the remainder of the population followers of other religions.

CULTURE
Press
There are no daily newspapers, although there are a number of periodicals that are published at varying degrees of regularity.

Tourism
Foreign tourists brought in revenue of US$14m. in 2001.

DIPLOMATIC REPRESENTATIVES
Of Equatorial Guinea in the United Kingdom (13 Park Place, London, SW1A 1LP)
Ambassador: Mari-Cruz Evuna Andeme.

Of the United Kingdom in Equatorial Guinea
Ambassador: Dr Andrew Pocock, CMG (resides in Abuja, Nigeria).

Of Equatorial Guinea in the USA (2020 16th St., NW, Washington, D.C., 20009)
Ambassador: Purificación Angue Ondo.

Of the USA in Equatorial Guinea (Carretera de Aeropuerto, K-3 El Paraiso, Apt 95, Malabo)
Ambassador: Mark L. Asquino.

Of Equatorial Guinea to the United Nations
Ambassador: Anatolio Ndong Mba.

Of Equatorial Guinea to the European Union
Ambassador: Carmelo Nvono-Nca.

FURTHER READING
Liniger-Goumaz, M., *Guinea Ecuatorial: Bibliografía General.* 1974–91.— *Small is Not Always Beautiful: The Story of Equatorial Guinea.* 1988.— *Historical Dictionary of Equatorial Guinea.* 2000
Molino, A. M. del, *La Ciudad de Clarence.* 1994

National Statistical Office: Dirección General de Estadísticas y Cuentas Nacionales.
Website (Spanish only): http://www.dgecnstat-ge.org

ERITREA

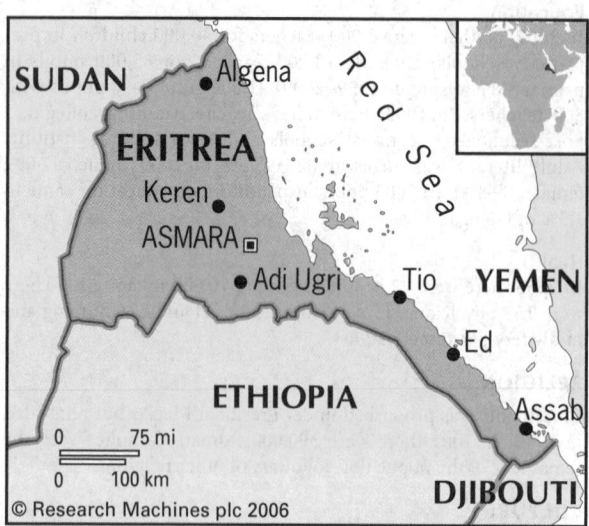

Hagere Ertra
(State of Eritrea)

Capital: Asmara
Population projection, 2015: 6·08m.
GNI per capita, 2011: (PPP$) 536
HDI/world rank: 0·349/177
Internet domain extension: .er

KEY HISTORICAL EVENTS

Italy was the colonial ruler from 1890 until 1941 when Eritrea fell to British forces. A British protectorate ended in 1952 when the UN sanctioned federation with Ethiopia. In 1962 Ethiopia became a unitary state and Eritrea was incorporated as a province. Eritreans began an armed struggle for independence under the leadership of the Eritrean People's Liberation Front (EPLF) which culminated successfully in the capture of Asmara on 24 May 1991. Thereafter the EPLF maintained a *de facto* independent administration recognized by the Ethiopian government. Sovereignty was proclaimed on 24 May 1993. In 1999 fighting broke out along the border with Ethiopia, following a series of skirmishes the previous year. After the failure of international mediation, the 13-month long-truce between Eritrea and Ethiopia ended in May 2000. Ethiopia launched a major offensive. In June both sides agreed to an Organization of African Unity peace deal.

In 2003 the UN border commission's award of the disputed town of Badame to Eritrea was rejected by Ethiopia, leading to a standoff between the two nations. UN helicopters were banned from Eritrean airspace in Nov. 2005. The following month Eritrea expelled all North American and European UN representatives, with the UN threatening sanctions against both countries unless they adhered to the 2000 peace plan. Meanwhile, the international commission in The Hague judged that Eritrea's 1998 attacks on Ethiopia breached international law.

Eritrean relations with the UN deteriorated further throughout 2006, with five UN staff expelled on charges of spying in Sept. In Nov. 2007 Eritrea agreed to a new judgment by the international boundary commission, which Ethiopia rejected. In July 2008 the UN Security Council voted to end its peacekeeping mission along the border. The previous month there were skirmishes between troops from Eritrea and Djibouti over the disputed borderlands of Ras Doumeira. In Dec. 2009 Eritrea was the subject of UN sanctions amid accusations that it was backing Islamist insurgents in Somalia. In June 2010 Djibouti and Eritrea pledged to resolve their dispute diplomatically.

TERRITORY AND POPULATION

Eritrea is bounded in the northeast by the Red Sea, southeast by Djibouti, south by Ethiopia and west by Sudan. Some 300 islands form the Dahlak Archipelago, most of them uninhabited. For the dispute with Yemen over the islands of Greater and Lesser Hanish *see* YEMEN: Territory and Population. Its area is 117,600 sq. km (45,410 sq. miles), including 16,600 sq. km (6,410 sq. miles) of inland waters. There has not been a census since Eritrea became independent in 1993. Population estimate, 2010, 5·25m.; density, 52 per sq. km. 22·1% of the population were urban in 2011.

The UN gives a projected population for 2015 of 6·08m.

In 2009 there were around 200,000 Eritreans living abroad, more than half of them as refugees in Sudan.

There are six regions: Anseba, Debub, Debubawi Keyih Bahri, Gash Barka, Maekel and Semenawi Keyih Bahri. The capital is Asmara (2002 estimated population, 500,600). Other large towns (with 2002 populations) are Keren (74,800) and Adi Ugri (25,700). An agreement of July 1993 gives Ethiopia rights to use the ports of Assab and Massawa.

49% of the population speak Tigrinya and 32% Tigré, and there are seven other indigenous languages. Arabic is spoken on the coast and along the Sudanese border, and English is used in secondary schools. Arabic and Tigrinya are the official languages.

SOCIAL STATISTICS

2008 births (estimates), 182,000; deaths, 42,000. Estimated birth rate in 2008 was 37·0 per 1,000 population; estimated death rate, 12·4. Annual population growth rate, 2000–08, 3·7%. Life expectancy at birth, 2007, was 56·8 years for males and 61·4 years for females. Infant mortality, 2010, 42 per 1,000 live births; fertility rate, 2008, 4·6 births per woman.

CLIMATE

Massawa, Jan. 78°F (25·6°C), July 94°F (34·4°C). Annual rainfall 8" (193 mm).

CONSTITUTION AND GOVERNMENT

A referendum to approve independence was held on 23–25 April 1993. The electorate was 1,173,506. 99·8% of votes cast were in favour.

The transitional government consists of the *President* and a 150-member *National Assembly*. It elects the President, who in turn appoints the *State Council* made up of 14 ministers and the governors of the ten provinces. The President chairs both the State Council and the National Assembly.

National Anthem

'Ertra, Ertra, Ertra' ('Eritrea, Eritrea, Eritrea'); words by S. Beraki, tune by I. Meharezghi and A. Tesfatsion.

RECENT ELECTIONS

In the presidential and legislative elections in May 1997, President Afewerki was re-elected to office. National Assembly elections, postponed in 1998, were set to take place before the end of 2003 but have been put back indefinitely. In the meantime several dissident politicians have been jailed.

CURRENT GOVERNMENT

President: Issaias Afewerki; b. 1945 (People's Front for Democracy and Justice, formerly the Eritrean People's Liberation Front; elected 22 May 1993 and re-elected in May 1997).

In March 2013 the ministers in the State Council were:

Minister of Agriculture: Arefaine Berhe. *Defence:* Sebhat Ephrem. *Education:* Semere Russom. *Energy and Mining:* Ahmed Haj Ali. *Finance:* Berhane Abrehe. *Foreign Affairs:* Osman Saleh. *Health:* Amina Nurhussein. *Information:* Ali Abdu. *Justice:* Fozia Hashim. *Labour and Human Welfare:* Salma Hassen. *Land, Water and Environment:* Tesfai Ghebreselassie. *Marine Resources:* Tewoldi Kelati. *National Development:* Giorgis Teklemikael. *Public Works:* Abraha Asfaha. *Tourism:* Askalu Menkerios. *Trade and Industry:* Estifanos Habte. *Transport and Communications:* Woldemikael Abraha.

CURRENT LEADERS

Issaias Afewerki

Position
President

Introduction
Issaias Afewerki has been president of Eritrea since it achieved independence from Ethiopia in 1993, having been a leading campaigner for secession since the mid-1960s. However, his tenure has been marked by civil rights abuses. In foreign policy he has overseen a bloody war with Ethiopia in 1999–2000, which has since threatened to reignite over an unresolved border dispute, as well as difficult relations with Djibouti and Sudan.

Early Life
Issaias Afewerki was born in 1945 in Asmara, Eritrea's capital, which was then under British administration. Eritrea became part of Ethiopia in 1962 and in 1966 Afewerki joined the secessionist Eritrean Liberation Front (ELF). Having received military training in China, he became a deputy divisional commander. In 1970 he helped found the Eritrean People's Liberation Front (EPLF), becoming its general secretary in 1987.

Following the collapse of the Mengistu military regime in Ethiopia in 1991, the new government agreed to a referendum on Eritrean independence. The referendum was held in 1993 and Eritrea declared independence in May of that year. Eritrea's National Assembly selected Afewerki as the country's first president.

Career in Office
The EPLF initially suggested a multi-party political system and in the early stages of his tenure Afewerki advocated close economic relations with Ethiopia. However, in Feb. 2002 the National Assembly, composed of EPLF representatives, refused to ratify a bill on the establishment of new political parties. Multi-party elections, previously scheduled for the end of 2001, were shelved.

In 2001 Afewerki authorized the arrest of critical journalists and political opponents. The move was condemned internationally and the Italian ambassador, who had voiced concerns over human rights violations, was expelled. International aid was consequently cut. In 2002 Afewerki set out his plans for the creation of a 'responsible' press, soon after an opposition party—the Eritrean People's Liberation Front Democratic Party—had emerged to challenge him. The party was believed to have been co-founded by Mesfin Hagos, Afewerki's former defence minister.

In 1999 border disputes escalated into full-scale war between Ethiopian and Eritrean forces which resulted in 70,000 deaths. A ceasefire was agreed in June 2000, with Ethiopia withdrawing its forces under UN supervision. A formal peace treaty was signed in Dec. 2000. Tensions remained, particularly concerning the ownership of the small border settlement of Badame, and in May 2001 the countries agreed to abide by the decision of an international boundary commission. The commission awarded Badame to Eritrea, but Ethiopia refused to accept the decision. Fears of a renewed conflict mounted in late 2005 after Eritrea expelled UN observers policing the militarized border region.

Relations with the UN deteriorated further during the autumn of 2006 as Eritrea expelled several UN peacekeeping staff for allegedly spying and moved troops into the buffer zone on the Ethiopian border in violation of the ceasefire. At the same time, the Eritrean government was accused of providing arms and supplies to the Islamist militias confronting the Ethiopian-backed transitional government in Somalia. In Nov. 2007 Eritrea accepted a border demarcation proposal by an independent boundary commission but Ethiopia rejected it. In Jan. 2008 the UN Security Council extended the mandate of its peacekeeping mission on the Eritrean-Ethiopian border for a further six months (despite Eritrean opposition), but brought it to a close at the end of July. Relations deteriorated again in March–April 2011 as Ethiopia accused Eritrea of terrorist infiltration and then declared openly that it would support Eritrean rebel forces aiming to overthrow Afewerki.

Afewerki meanwhile oversaw the restoration of diplomatic ties with Sudan and Djibouti, although relations with both countries remain unsettled. Eritrea has claimed that Islamic fundamentalist groups active within the country have received backing from Khartoum, and has also accused Djibouti of providing military support to Ethiopia, a claim denied by Djibouti. In June 2008 there were border clashes between troops from Djibouti and Eritrea following several weeks of rising tensions. However, following the imposition of sanctions against Eritrea by the UN Security Council in Dec. 2009, both sides agreed in June 2010 to resolve their border issues peacefully. In mid-2009 the African Union (AU) rebuked Afewerki's regime for continuing to aid Islamist insurgents in Somalia, so endangering civilians and AU peacekeeping forces, and the UN Security Council similarly acted against Eritrea with further sanctions in Dec. 2011.

DEFENCE

Conscripts (both male and female) are subjected to six months military training and 12 months work on national reconstruction. It has since been reduced to 16 months. The total strength of all forces was estimated at 201,750 in 2007.

Defence expenditure totalled US$65m. in 2005 (US$14 per capita and 6·3% of GDP).

Army
The Army had a strength of around 200,000 in 2007. There were also approximately 120,000 reservists available.

Navy
Most of the former Ethiopian Navy is now in Eritrean hands. The main bases and training establishments are at Massawa, Assab and Dahlak. Personnel numbered 1,400 in 2007.

Air Force
Personnel numbers were estimated at 350 in 2007. There were 18 combat-capable aircraft including MiG-29s, MiG-23s and MiG-21s.

INTERNATIONAL RELATIONS

A border dispute between Eritrea and Ethiopia broke out in May 1998. Eritrean troops took over the border town of Badame after a skirmish between Ethiopian police units and armed men from Eritrea. Ethiopia maintained that Badame and Sheraro, a nearby town, had always been part of Ethiopia and called Eritrea's action an invasion. An agreement ending hostilities was signed in June 2000, followed by a peace accord in Dec. A buffer zone has been created to separate the armies, but tensions do still arise from time to time, notably in late 2005 following a further dispute between the two countries over Badame.

ECONOMY

In 2007 agriculture accounted for 24·3% of GDP, industry 19·2% and services 56·5%.

Eritrea's resources are meagre, the population small and poorly-educated; communications are difficult and there is a shortage of energy.

Overview

In the half decade following Eritrea's declaration of independence in 1993 and after 30 years of war with Ethiopia, the average annual growth rate was 10·9%. However, renewed conflict with Ethiopia from 1998–2000 damaged much of the economic and social infrastructure and displaced a large part of the population. Eritrea is now one of the world's poorest countries, with an estimated two-thirds of the population living below the poverty line. Sustained real economic growth of above 7% is required to reduce the number living in long-term extreme poverty.

Subsistence agriculture is the main economic activity. A series of external shocks devastated the economy in 2008, including a drought that resulted in a harvest a quarter the size of the previous year, prompting emergency imports of food. As a net importer of food and oil products, Eritrea was also hit by a surge in international commodity prices. Inflation reached 33·0% in 2009 but is reported to be on a downward trend. Fiscal consolidation since the end of the border war has helped reduce the fiscal deficit though it remains high at 10% of GDP, with resulting macroeconomic imbalances managed through regulations and price controls.

Currency

The *nakfa* (ERN) replaced the Ethiopian currency, the *birr*, in 1997. However, its introduction led to tensions with Ethiopia, adversely affecting cross-border trade. Inflation was 33·0% in 2009, falling to 12·7% in 2010 before rising slightly to 13·3% in 2011. Total money supply was 8,063m. nakfa in May 2005.

Budget

Revenues in 2008 were 4·46bn. nafka and expenditures 9·84bn. nafka.

Performance

Total GDP in 2008 was US$1·7bn. The economy expanded by 8·8% in 2001 following the end of the conflict with neighbouring Ethiopia but 2003, 2006 and 2008 all saw negative growth. There was real GDP growth of 3·9% in 2009, 2·2% in 2010 and 8·7% in 2011. Total GDP in 2011 was US$2·6bn.

Banking and Finance

The central bank is the National Bank of Eritrea (*Acting Governor*, Kibreab Woldemariam). All banks and financial institutions are state-run. There is a Commercial Bank of Eritrea with 15 branches, an Eritrean Investment and Development Bank with 13 branches, a Housing and Commercial Bank of Eritrea with seven branches and an Insurance Corporation.

In 2010 external debt totalled US$1,010m., representing 48·2% of GNI.

ENERGY AND NATURAL RESOURCES

Environment

Carbon dioxide emissions from the consumption and flaring of fossil fuels were the equivalent of 0·1 tonnes per capita in 2008.

Electricity

Installed capacity was an estimated 167,000 kW in 2007. Electricity is provided to only some 20% of the population. Total production was 288m. kWh in 2007.

Minerals

There are deposits of gold, silver, copper, zinc, sulphur, nickel, chrome and potash. Basalt, limestone, marble, sand and silicates are extracted. Oil exploration is taking place in the Red Sea. Salt production totals 200,000 tonnes annually.

Agriculture

Agriculture engaged approximately 77% of the economically active population in 2002. Several systems of land ownership (state, colonial, traditional) co-exist. In 1994 the PFDJ proclaimed the sole right of the state to own land. There were 500,000 ha. of arable land in 2001 and 3,000 ha. of permanent crops. 21,000 ha. were irrigated in 2001. There were 463 tractors in 2001 and 125 harvester-threshers.

Main agricultural products, 2003 (in 1,000 tonnes): sorghum, 64; potatoes, 33; millet, 17; barley, 9; wheat, 5; maize, 4.

Livestock, 2003: sheep, 2·1m.; cattle, 1·9m.; goats, 1·7m.; camels, 75,000; chickens, 1m.

Livestock products, 2003 (in 1,000 tonnes): beef and veal, 14; goat meat, 6; lamb and mutton, 6; poultry meat, 2; milk, 51.

Forestry

In 2010 forests covered 1·53m. ha., or 15% of the total land area. Timber production in 2007 was 2·53m. cu. metres.

Fisheries

The total catch in 2010 was 3,286 tonnes, exclusively from marine waters.

INDUSTRY

Light industry was well developed in the colonial period but capability has declined. Processed food, textiles, leatherwear, building materials, glassware and oil products are produced. Industrial production accounted for 19·2% of GDP in 2007, with the manufacturing sector providing 5·5%.

Labour

In 2010 the estimated labour force was 2,230,000 (55% males).

INTERNATIONAL TRADE

Eritrea is dependent on foreign aid for most of its capital expenditure.

Imports and Exports

In 2009 imports (c.i.f.) were valued at US$540m. and exports (f.o.b.) at US$15m. The leading imports are machinery and transport equipment, basic manufactures, and food and live animals. The main exports are drinks, leather and products, textiles and oil products. Principal import suppliers, 2003: USA, 15·9%; UAE, 12·2%; Italy, 11·6%; Saudi Arabia, 10·5%; India, 6·3%. Principal export markets, 2003: Sudan, 19·7%; Italy, 12·1%; Singapore, 12·1%; Netherlands, 10·6%; India, 7·6%.

COMMUNICATIONS

Roads

There were some 4,010 km of roads in 2000, of which 21·8% were paved. A tarmac road links the capital Asmara with one of the main ports, Massawa. In 2007 there were 6·4 passenger cars per 1,000 inhabitants. About 500 buses operate regular services.

Rail

In 2000 the reconstruction of the 117 km Massawa–Asmara line reached Embatkala, thus opening up an 80 km stretch from Massawa on the coast. In 2003 the line was rebuilt right through to Asmara.

Civil Aviation

There is an international airport at Asmara (Yohannes IV Airport). In 2003 there were scheduled flights to Cairo, Djibouti, Dubai, Frankfurt, Jeddah, Milan, Nairobi and Sana'a. In 2001 Asmara handled 140,000 passengers (129,000 on international flights) and 3,200 tonnes of freight.

Shipping
Massawa is the main port; Assab used to be the main port for imports to Ethiopia. Both were free ports for Ethiopia until the onset of hostilities. In Jan. 2009 there were five ships of 300 GT or over registered, totalling 12,000 GT.

Telecommunications
In 2011 there were 58,000 landline telephone subscriptions (equivalent to 10·7 per 1,000 inhabitants) and 241,900 mobile phone subscriptions (or 44·7 per 1,000 inhabitants). In 2010, 5·4% of the population were internet users.

SOCIAL INSTITUTIONS
Justice
The legal system derives from a decree of May 1993.

Education
Adult literacy was 65% in 2008. In 2007 there were 331,855 pupils and 6,933 teaching staff in primary schools, and 218,369 pupils at secondary schools with 4,425 teaching staff. There were 10,000 students and 1,000 academic staff in tertiary education in 2009–10. In 2006 public expenditure on education came to 2·4% of GNI.

Health
In 2006 there were 12 hospital beds per 10,000 inhabitants. In 2004 there were 215 physicians, 16 dentistry personnel and 2,505 nursing and midwifery personnel.

Eritrea has one of the highest rates of undernourishment of any country. The proportion of the population classified as undernourished was 66% in the period 2004–06, down from 70% in 2000–02.

RELIGION
Half the population are Sunni Muslims (along the coast and in the north), and half Coptic Christians (in the south).

CULTURE
Press
In 2008 there were three government newspapers, one published three times a week and the others once a week. In Sept. 2001 the government closed down the country's eight independent newspapers. In the 2011–12 *World Press Freedom Index*, compiled by Reporters Without Borders, Eritrea ranked 179th and last out of the 179 countries covered. A number of journalists have been jailed.

Tourism
There were 79,000 foreign visitors in 2009, up from 70,000 in 2008.

DIPLOMATIC REPRESENTATIVES
Of Eritrea in the United Kingdom (96 White Lion St., London, N1 9PF)
Ambassador: Tesfamicael Gerahtu Ogbaghiorghis.

Of the United Kingdom in Eritrea (66–68 Mariam Ghimbi St., PO Box 5584, Asmara)
Ambassador: Dr Amanda Tanfield.

Of Eritrea in the USA (1708 New Hampshire Ave., NW, Washington, D.C., 20009)
Ambassador: Vacant.
Chargé d'Affaires a.i.: Berhane Gebrihiwet Solomon.

Of the USA in Eritrea (179 Alaa St., POB 211, Asmara)
Ambassador: Vacant.
Chargé d'Affaires a.i.: Sue Bremner.

Of Eritrea to the United Nations
Ambassador: Araya Desta.

Of Eritrea to the European Union
Ambassador: Mohammed Sulieman Ahmed.

FURTHER READING
Connel, D., *Against All Odds: a Chronicle of the Eritrean Revolution.* 1993
Henze, Paul, *Eritrea's War: Confrontation, International Response, Outcome, Prospects.* 2001
Mengisteab, Kidane, *Anatomy of an African Tragedy: Political, Economic and Foreign Policy Crisis in Post-Independence Eritrea.* 2005
Negash, Tekeste and Tronvoll, Kjetil, *Brothers at War: Making Sense of the Eritrean–Ethiopian War.* 2001
Wrong, Michaela, *I Didn't Do It For You: How the World Betrayed a Small African Nation.* 2005

ESTONIA

© Research Machines plc 2006

Eesti Vabariik
(Republic of Estonia)

Capital: Tallinn
Population projection, 2015: 1·34m.
GNI per capita, 2011: (PPP$) 16,799
HDI/world rank: 0·835/34
Internet domain extension: .ee

KEY HISTORICAL EVENTS

There is evidence of human habitation from around 11,000 BC. Remnants of a 'comb' pottery culture from 5000 BC show the arrival of the ancestors of the Eestii, one of the first known peoples to inhabit the Baltic's eastern shores and the forerunners of modern Estonians. Before the arrival of Christianity, animism was widespread and the cult of Tharapita (or Taara), a god of war, was popular in northern Estonia and the island of Saaremaa.

The failed Danish invasion of Saaremaa in 1206 under the Bishop of Lund marked the first attempt at the Christianization of Estonia. German invasions began in 1208 with the capture of Otepää in the southeast. Christianization came with Bishop Albert and his Sword Brothers, a military order established in 1202 that became the Livonian Order. Livonia (including southern Estonia) fell to them in 1217. Valdemar II of Denmark invaded the north in 1219, establishing Tanin Lidna (later Tallinn) at Reval on the Gulf of Finland. Danish forces had prevailed against indigenous forces by 1227 and took Saaremaa (Ösel).

In 1238 Wilhelm of Modena, the papal legate, negotiated the re-establishment of Danish power in the north after its seizure by the Sword Brothers, who shared the rest of the land with the prince-bishops of Ösel-Wiek and Dorpat (Tartu). The eastward expansion of the Sword Brothers was checked in 1242 at Lake Peipus by Alexander Nevsky, Prince of Novgorod.

Coastal Swedes arrived from the mid-13th century and a German merchant class emerged in the towns. In the St George's Night Uprising, beginning on 23 April 1343 and lasting until 1345, Estonians rebelled against Danish rule. The peasant army appealed to Swedish Finland for help but was defeated by the Danish vice-regents, who called on the Livonian Order. In 1346 Denmark's Valdemar IV sold his share of Estonia to the Teutonic Knights who gave control to the Livonian Order.

The Reformation reached Estonia's semi-autonomous cities in the 1520s but met resistance from the Livonian Order. In 1558 Tsar Ivan IV demanded taxes from the bishopric of Dorpat

(Tartu) as a pretext to invading the Livonian Confederation and gaining access to the Baltic Sea. The weakened Livonian Order disbanded in 1561. The treaties ending the Livonian War, which ran for 25 years, saw the division of Estonia between Sweden, Denmark and the Kingdom of Poland–Lithuania.

By 1600 Sweden, repelling the incursions of Muscovy, had strengthened its hold over Estland. Sweden's Gustavus Adolphus attempted to build a Reformist Baltic empire, winning all of Estonia in 1629 and reducing the power of German landowners while defending the country from Polish and Russian attacks. He also made education available to the peasants and established a university at Tartu. In 1645 the Danes ceded Ösel to the Swedes. Estonian territories were badly affected by the Great Famine of 1695–97, when over 70,000 people (around 20% of the population) died.

Russian ambitions to establish a Baltic base and Danish resentment of Swedish hegemony led to the Great Northern War (1700–21). Swedish forces won a string of victories under Charles XII, routing Peter the Great's army at Narva in 1700. However, Charles' over-ambitious attack on Poland led to a Swedish collapse. Peter seized Ingria (where he built St Petersburg) and ravaged Livonia. The Swedish defeat at Poltava in 1709 led to Estonia passing to Russia by the 1721 Treaty of Nystad.

Estonia remained dominated by German aristocrats after the reversal of Swedish land seizures and the reinforcement of serfdom. Estonian nationalism emerged in the mid-19th century despite Russian attempts to contain it. Baltic autonomy was severely curtailed, with the Russian language imposed in education and public administration, and Russian nationals parachuted into high office.

After nationalist success in the municipal elections of 1904, many Estonians joined the all-Russia workers' strikes of 1905, which were brutally suppressed. In the ensuing anti-landlord violence, about a fifth of the country's German-owned manors were destroyed by ethnic Estonians. Nicholas II's subsequent October Manifesto allowed for the establishment of political parties in Estonia, giving renewed momentum to the nationalist movement.

After the overthrow of the monarchy in 1917, Russia's provisional government amalgamated Estonia and Estonian-speaking northern Livonia. As the First World War continued into 1918, German forces pushed the Russians eastwards leading to their retreat from Estonia. In the brief period after the Soviet withdrawal and before the arrival of the German army, the Estonian Rescue Committee declared independence on 24 Feb. 1918. However, German troops took Tallinn the following day and the subsequent German occupation lasted until Nov.

The German withdrawal preceded Bolshevik attempts to regain control. However, a defensive campaign, referred to as the Estonian War of Independence, ejected the Russians in May 1919. The Treaty of Tartu on 2 Feb. 1920 saw Russia acknowledge Estonian independence.

In March 1934 this regime was, in turn, overthrown by a quasi-fascist coup. The secret protocol of the Soviet–German agreement of Aug. 1939 assigned Estonia to the Soviet sphere of interest. An ultimatum in June 1940 saw the formation of the Estonian Soviet Socialist Republic but a German occupation lasted from June 1941–Sept. 1944. An attempt to secure national independence quickly crumbled and the return to Soviet control saw the loss of 2,200 sq. km of Estonian territory to Pskov Oblast. There followed a Sovietization programme that lasted until the mid-1980s.

However, with crises emerging all over the Soviet Union, the Estonian Supreme Soviet unilaterally declared a sovereign republic in Nov. 1988. Pro-independence protests were brutally suppressed in Riga and Vilnius (in Latvia and Lithuania respectively) in 1990 on Kremlin orders but a popular referendum in Estonia the following year saw 77·8% vote in favour of independence. As Moscow was reeling from the August Coup of 1991, the Estonian Supreme Council passed a new independence resolution recognized by Moscow on 6 Sept. 1991. Estonia was admitted to the Council of Europe in 1993 and all Russian troops were withdrawn by Aug. 1994, though a sizeable Russian population remains (around a third of the total). Estonia became a member of NATO in March 2004 and the European Union in May 2004.

TERRITORY AND POPULATION

Estonia is bounded in the west and north by the Baltic Sea, east by Russia and south by Latvia. There are 1,521 offshore islands, of which the largest are Saaremaa and Hiiumaa, but only 12 are permanently inhabited. Area, 45,227 sq. km (17,462 sq. miles). The census population in Dec. 2011 was 1,294,455 (693,929 females), giving a density of 27·6 per sq. km.

The UN gives a projected population for 2015 of 1·34m.

In 2010, 69·5% of the population lived in urban areas. Of the whole population, Estonians accounted for 68·7% in 2011, Russians 24·8% and Ukrainians 1·7%. The capital is Tallinn (2011 population, 393,222 or 31·5%). Other large towns are Tartu (97,600), Narva (58,663), Pärnu (39,728) and Kohtla-Järve (37,201). In 2011 there were 15 counties, 47 cities and 193 rural municipalities.

The official language is Estonian.

SOCIAL STATISTICS

2009 registered births, 15,763; deaths, 16,081. Rates (per 1,000 population): birth, 11·8; death, 12·0. There were 9,394 reported abortions in 2006. Expectation of life in 2007 was 67·3 years for males and 78·3 for females. The annual population growth rate in the period 2005–10 was −0·1%. The suicide rate was 18·1 per 100,000 population in 2008 (rate among males, 30·6). The rate has more than halved in 13 years, having been 40·1 per 100,000 in 1995. Infant mortality in 2010 was four per 1,000 births. In 2008 total fertility rate was 1·7 births per woman.

CLIMATE

Because of its maritime location Estonia has a moderate climate, with cool summers and mild winters. Average daily temperatures in 2008: Jan. −1·5°C; July 17·0°C. Rainfall is heavy, 600–800 mm per year, and evaporation low.

CONSTITUTION AND GOVERNMENT

A draft constitution drawn up by a constitutional assembly was approved by 91·1% of votes cast at a referendum on 28 June 1992. Turnout was 66·6%. The constitution came into effect on 3 July 1992. It defines Estonia as a 'democratic state guided by the rule of law, where universally recognized norms of international law are an inseparable part of the legal system.' It provides for a 101-member national assembly (*Riigikogu*) elected for four-year terms. There are 12 electoral districts with eight to 12 mandates each. Candidates may be elected: a) by gaining more than 'quota', i.e. the number of votes cast in a district divided by the number of its mandates; b) by standing for a party which attracts for all of its candidates more than the quota, in order of listing; c) by being listed nationally for parties which clear a 5% threshold and eligible for the seats remaining according to position on the lists. The head of state is the *President*, elected by the Riigikogu for five-year terms. Presidential candidates must gain the nominations of at least 20% of parliamentary deputies. If no candidate wins a two-thirds majority in any of three rounds, the Speaker convenes an electoral college, composed of parliamentary deputies and local councillors. At this stage any 21 electors may nominate an additional candidate. The electoral college elects the President by a simple majority.

Citizenship requirements are two years residence and competence in Estonian for existing residents. For residents immigrating after 1 April 1995, five years qualifying residence is required.

National Anthem

'Mu isamaa, mu õnn ja rõõm' ('My native land, my pride and joy'); words by J. V. Jannsen, tune by F. Pacius (same as Finland).

GOVERNMENT CHRONOLOGY

Heads of State since independence.

Chairman of the Supreme Council

1991–92	Arnold Rüütel

Presidents

1992–2001	Lennart Georg Meri
2001–06	Arnold Rüütel
2006–	Toomas Hendrik Ilves

Prime Ministers since independence. (IERSP (Isamaaliit) = Pro Patria Union; KMÜ-K = Estonian Coalition Party; Rahvarinne = Popular Front of Estonia; RE (Reformierakond) = Estonian Reform Party; ResP = Union for the Republic-Res Publica; RK Isamaa = National Coalition Party Pro Patria; n/p = non-partisan)

1990–92	Rahvarinne	Edgar Savisaar
1992	n/p	Tiit Vähi
1992–94	RK Isamaa	Mart Laar
1994–95	n/p	Andres Tarand
1995–97	KMÜ-K	Tiit Vähi
1997–99	KMÜ-K	Mart Siimann
1999–2002	IERSP	Mart Laar
2002–03	RE	Siim Kallas
2003–05	ResP	Juhan Parts
2005–	RE	Andrus Ansip

RECENT ELECTIONS

Parliamentary elections were held on 6 March 2011; turnout was 63·5%. The Estonian Reform Party (Reform) won 33 of 101 seats (with 28·6% of the total votes); Estonian Centre Party (Kesk), 26 seats (23·3%); Union of Pro Patria and Res Publica (IRL), 23 seats (20·5%); Social Democratic Party (SDE), 19 seats (17·1%). Six other parties failed to win seats.

On 29 Aug. 2011 Toomas Ilves was re-elected as president, winning 73 votes to Indrek Tarand's 25 in the 101-seat parliament.

European Parliament

Estonia has six representatives. At the June 2009 elections turnout was 43·9% (26·8% in 2004). Kesk won 2 seats with 26·1% of the vote (political affiliation in European Parliament: Alliance of Liberals and Democrats for Europe); Reform, 1 with 15·3% (Alliance of Liberals and Democrats for Europe); IRL, 1 with 12·2% (European People's Party); SDE, 1 with 8·7% (Progressive Alliance of Socialists and Democrats). One independent was elected, with 25·8% (Greens/European Free Alliance).

CURRENT GOVERNMENT

President: Toomas Ilves; b. 1953 (sworn in 9 Oct. 2006 and re-elected 29 Aug. 2011).

In March 2013 the coalition government comprised:

Prime Minister: Andrus Ansip; b. 1956 (Reform; in office since 13 April 2005).

Minister of Agriculture: Helir-Valdor Seeder (IRL). *Culture:* Rein Lang (Reform). *Defence:* Urmas Reinsalu (IRL). *Economic Affairs and Communications:* Juhan Parts (IRL). *Education and Research:* Jaak Aaviksoo (IRL). *Environment:* Keit Pentus (Reform). *Finance:* Jürgen Ligi (Reform). *Foreign Affairs:* Urmas Paet (Reform). *Internal Affairs:* Ken-Marti Vaher (IRL). *Justice:* Hanno Pevkur (Reform). *Regional Affairs:* Siim-Valmar Kiisler (IRL). *Social Affairs:* Taavi Rõivas (Reform).

Government Website: http://www.valitsus.ee

CURRENT LEADERS

Toomas Hendrik Ilves

Position
President

Introduction
Toomas Ilves began his first five-year largely ceremonial term as president on 9 Oct. 2006, replacing Arnold Rüütel, and was re-elected in Aug. 2011.

Early Life
Toomas Hendrik Ilves was born on 26 Dec. 1953 in Sweden, his parents having fled Estonia during the Soviet occupation in the 1940s. Ilves studied psychology in the USA, first at Columbia University before completing his MA at Pennsylvania University. He then lectured in Vancouver on Estonian literature and linguistics before working as an analyst and researcher for Radio Free Europe.

He returned to Estonia in 1993, two years after the country regained independence. In the late 1990s he served as ambassador to the USA, Canada and Mexico. He also had two spells as minister for foreign affairs and was chairman of the North American Institute. From 2004–06 he was a member of the European Parliament, representing the Social Democratic Party.

Career in Office
Ilves' principal duty as president is to represent the country abroad. In Nov. 2006 George W. Bush became the first US president to visit Estonia. Ilves justified the participation of Estonian forces in Afghanistan as a necessary duty of NATO membership, and an increase in the Estonian deployment was approved by parliament in June 2009.

Ilves has sought greater integration of Estonia's large Russian-speaking minority. Relations between Estonia and Russia were tense during the early months of his tenure, particularly over the relocation by the Estonian government of a prominent Soviet war memorial out of the centre of Tallinn. Ilves was re-elected for a second presidential term by parliament in Aug. 2011.

Andrus Ansip

Position
Prime Minister

Introduction
When Andrus Ansip was sworn in as prime minister of Estonia on 13 April 2005, he took charge of the country's 12th government since its independence in 1991. A right-leaning former investment banker who was mayor of the second largest city, Tartu, for six years, Ansip pledged to implement policies that would attract investment and strengthen Estonia's position as a dynamic, post-industrial economy. His coalition government retained power in parliamentary elections in March 2007 and March 2011.

Early Life
Ansip was born in Tartu in the Soviet Republic of Estonia (ESSR) on 1 Oct. 1956. He attended local schools and graduated from the University of Tartu with a diploma in chemistry in 1979. He remained at the historic university to undertake further academic study, and later joined the municipal Committee of the Estonian Communist Party (ECP), which had been led, since 1978, by Karl Vaino, a Russian-born Estonian. The ESSR experienced increased Russification and 'Sovietization' in the early 1980s, in accordance with the policy of the Soviet leader, Leonid Brezhnev. By 1988 however, there was growing opposition to the communist leadership and, against a backdrop of gradual economic liberalization, Ansip joined Estkompexim, a 'joint-venture' specializing in the import and export of foodstuffs. He was head

of Estkompexim's Tartu office during 1991, when, following the collapse of the USSR, Estonia was internationally recognized as an independent nation. The following year Ansip attended a business management course at the University of York in Toronto, Canada.

On his return to Estonia in 1993, Ansip entered the rapidly evolving banking and investment sector, serving as a member of the board of directors of Rahvapank (the People's Bank) until 1995 and then chairman of the board of Livonia Privatization. In 1997 he was chief executive officer of the investment fund, Fondiinvesteeringu Maakler AS, as well as chairman of the board of Radio Tartu. The following year he was elected Mayor of Tartu as a candidate of the centre-right Estonian Reform Party (Reform), established in 1994 by Siim Kallas, a former governor of Estonia's central bank. A popular mayor, Ansip was credited with attracting investment to the country's second city and overseeing developments such as the Baltic Defence College and a new biomedical research institute which capitalized on Tartu's long-standing reputation as an academic centre.

On 13 Sept. 2004, shortly after Estonia joined the EU and NATO, Ansip was nominated to replace Meelis Atonen as the minister of economic affairs and communications. Two months later, he became chairman of Reform, which had formed part of the Res Publica-led coalition government under Juhan Parts since March 2003. His appointment followed the departure of Reform's leader (and former prime minister), Siim Kallas, to Brussels to become an EU commissioner. When, in March 2005, the Riigikogu (parliament) passed a vote of no confidence in the country's justice minister over proposed anti-corruption measures, Parts resigned as prime minister. On 31 March 2005 the president, Arnold Rüütel, asked Ansip to form a new government. He succeeded in forging a coalition with the Estonian Centre Party (Kesk) and the Estonian Peoples' Union (Rahvaliit). Ansip was backed by 53 out of 101 members of the Riigikogu, and was inaugurated as prime minister on 13 April 2005.

Career in Office
Ansip confirmed Estonia's aspiration to adopt the single European currency. However, he acknowledged that it would be tough to fulfil the criterion of holding inflation to no more than 1·5 percentage points above that of the three lowest-inflation EU countries, given that Estonia's economy was growing at that time at around 7% a year and its exports had risen by 20% in its first 12 months in the EU. He pledged to maintain the previous government's tax-cutting agenda, as well as increasing social welfare measures to bridge the gap between the relatively wealthy, young urban population and poorer rural citizens whose skills date from the Soviet period. He retained the premiership following legislative elections in March 2007, forming a new Reform coalition with Union of Pro Patria and Res Publica (IRL) and the Social Democratic Party.

Despite a sharp economic contraction in the wake of the global financial crisis that unfolded in the latter half of 2008, Estonia fared better than its Baltic neighbours in the ensuing downturn. The government acted quickly to implement spending cuts and adjustments to stem the rise in the budget deficit, while earlier prudent management of the public finances provided a buffer of fiscal reserves, with no requirement for support from the International Monetary Fund. Estonia became the 17th country to adopt the euro in Jan. 2011.

In May 2009 the government lost its parliamentary majority as Ansip dismissed the three Social Democratic ministers from the coalition cabinet in a dispute over economic policy, replacing them in June with IRL members and continuing on a minority basis.

In foreign affairs, Ansip's government formally signed a border treaty agreement with Russia in May 2005. However, Russia subsequently withdrew from the agreement because Estonia had attached an unacceptable preamble to the text referring to the Soviet occupation. Relations with Russia soured further in early 2007 when the Estonian parliament passed legislation banning monuments glorifying Soviet rule and the government approved the removal of a Soviet war memorial in Tallinn.

Ansip secured another term in office when Reform won the March 2011 parliamentary election.

DEFENCE

The President is the head of national defence. Conscription is eight to 11 months for men and voluntary for women. Conscientious objectors may opt for 16 months civilian service instead.

Defence expenditure in 2008 totalled US$450m. (US$344 per capita), representing 1·9% of GDP.

The Estonian Defence Forces (EDF) regular component is divided into the Army, the Air Force and the Navy.

Army

The Army consists of seven battalions (three infantry, one reconnaissance, one artillery, one guard and one peacekeeping). The total number of personnel in the Army in 2007 was 3,600 (1,200 conscripts). There is a Border Guard numbering 2,600.

Navy

The Navy consists of the Naval Staff (Naval HQ), the Naval Base, and the Mine Countermeasures (MCM) Squadron. The total number of personnel in the Navy in 2007 was 300 including a platoon-sized conscript unit. Estonia, Latvia and Lithuania have established a joint naval unit 'BALTRON' (Baltic Naval Squadron), with bases at Tallinn in Estonia, Liepāja, Riga and Ventspils in Latvia, and Klaipėda in Lithuania.

Air Force

The Air Force consists of an Air Force Staff, Air Force Base and Air Surveillance Wing. The total number of personnel in the Air Force in 2007 was 200.

INTERNATIONAL RELATIONS

Estonia became a member of NATO on 29 March 2004 and of the EU on 1 May 2004. Estonia held a referendum on EU membership on 14 Sept. 2003, in which 66·9% of votes cast were in favour of accession, with 33·1% against.

ECONOMY

Agriculture contributed 3% of GDP in 2007, industry 30% and services 67%.

Overview

Estonia is a gateway for trade with the Nordic countries, especially Finland, the destination for 18·3% of Estonian exports in 2008. When Estonia gained independence in 1991 it was among the most competitive of the former Soviet Union countries, boasting the highest per capital income, strong infrastructure and high education levels compared to its neighbours.

The cornerstones of economic reform were the introduction of a new currency, tight budgetary control, privatization programmes and trade liberalization. Since independence, Estonia has been among the fastest growing of the EU accession countries (having joined in 2004). Performance has been driven in large part by low unemployment, strong domestic demand and buoyant exports, with GDP growth averaging 8% per year from 2003–07. Unemployment dropped from 11·0% in 2003 to 4·7% in 2007.

However, the economy began to slow in 2007 and fell into recession as the global financial crisis hit. Growth contracted by 14·3% in 2009. Despite this sharp contraction, the government persevered with efforts to join the euro area with a fiscal adjustment of nearly 9% of GDP in 2009. The economy rebounded strongly in 2010, with growth resuming in the second quarter as a result of rising exports and increases in manufacturing activity. Estonia joined the euro on 1 Jan. 2011 and continued its impressive recovery. Unemployment fell from 19·8% in the first quarter of 2010 to 14·4% in the first quarter of 2011, while government debt at 6·6% of GDP is the lowest in the EU.

Currency

On 1 Jan. 2011 the euro (EUR) replaced the *kroon* (EEK) as the legal currency of Estonia at the irrevocable conversion rate of 15·6466 krooni to one euro. Foreign exchange reserves were US$1,768m. in July 2005, gold reserves 8,000 troy oz and total money supply was 42,396m. krooni. Inflation was 10·4% in 2008, fell to 2·9% in 2010 but went up to 5·1% in 2011.

Budget

Budgetary central government revenue and expenditure in 1m. krooni for calendar years:

	2007	2008	2009
Revenue	69,206	67,853	69,577
Expenditure	58,479	68,793	67,481

Tax revenue provided 37,776m. krooni in 2009; social benefits (25,591m. krooni) were the main item of expenditure. There is a flat income tax rate of 21%.

Estonia registered a budget surplus in 2011 of 1·0% of GDP. There was also a surplus in 2010 of 0·2% but a deficit in 2009 of 2·0%. The required target set by the EU is a budget deficit of no more than 3%.

Performance

Estonia suffered particularly badly in the global downturn with the economy shrinking by 14·3% in 2009 but there was growth of 2·3% in 2010 and 7·6% in 2011. Total GDP in 2011 was US$22·2bn.

Banking and Finance

A central bank, the Bank of Estonia, was re-established in 1990 (*Governor*, Ardo Hansson). The Estonian Investment Bank was established in 1992 to provide financing for privatized and private companies. As of 30 June 2011 there were seven licensed credit institutions and 11 affiliated branches of foreign credit institutions in Estonia. The four largest institutions control 89% of the market. Total assets of Estonian credit institutions as of 30 June 2011 were €19,105m. The Estonian Banking Association was founded in 1992. It has 12 member banks that represent approximately 98% of the Estonian banking sector's assets.

Gross external debt amounted to US$20,224m. in June 2012.

In 2010 Estonia received US$1·5bn. worth of foreign direct investment.

A stock exchange opened in Tallinn in 1996.

ENERGY AND NATURAL RESOURCES

In 2008, 19·1% of energy consumption came from renewables (wind power, solar power, hydro-electric power, tidal power, geothermal energy and biomass), compared to the European Union average of 10·3%. A target of 25% has been set by the EU for 2020.

Environment

Estonia's carbon dioxide emissions from the consumption and flaring of fossil fuels in 2008 were the equivalent of 15·8 tonnes per capita. Estonia's greenhouse gas emissions fell by 50·4% between 1990 and 2008, mainly owing to the decline of polluting industries from the Soviet era.

Electricity

Estonia is a net electricity exporter. In 2005 installed capacity was 2·7m. kW, with production of 9·1bn. kWh. Consumption per capita was 6,733 kWh in 2005. 91% of electricity was produced by burning oil shale. Production of hydro and wind energy accounted for about 0·7% of total electricity production. About 20% of net production was exported, mainly to Latvia.

Oil and Gas
Oil shale deposits were estimated at 4,898m. tonnes in 2006. A factory for the production of gas from shale and a 208 km-pipeline from Kohtla-Järve supplies shale gas to Tallinn, and exports to St Petersburg. Natural gas is imported from Russia.

Minerals
Oil shale is the most valuable mineral resource. Production volume has decreased (from 21m. tonnes in 1990 to 12m. tonnes in 2006) because of falls in exports and domestic electricity consumption, and an increase in the use of natural gas. Peatlands occupy about 22% of Estonia's territory; there are extensive deposits, totalling an estimated 1·64bn. tonnes in 2004. Phosphorites and super-phosphates are found and refined, and lignite (16·54m. tonnes in 2007), limestone, dolomite, clay, sand and gravel are mined.

Agriculture
Farming employed 4·3% of the population in 2010. In the same year there were 19,613 holdings (55,748 in 2001). In 2010 there were 640,000 ha. of arable land and 3,000 ha. of permanent crops. Total agricultural output in 2011 was valued at €811·6m., including: animal production €387·8m.; crop production, €336·3m.; agricultural services and other non-agricultural production, €86·5m.

Output of main agricultural products (in 1,000 tonnes) in 2011: wheat, 360; barley, 295; potatoes, 165; rapeseed, 144; oats, 63; rye, 31.

In 2011 there were 365,700 pigs, 238,300 cattle, 83,900 sheep and 1,973,300 chickens.

Livestock products, 2011: meat, 81,000 tonnes; milk, 693,000 tonnes; eggs, 184m. units.

Forestry
In 2010, 2·22m. ha. were covered by forests (52% of the total land area), which provide material for sawmills, furniture, and the match and pulp industries, as well as wood fuel. Private, municipal and state ownership of forests is allowed. In 2007 the annual timber cut was 5·90m. cu. metres.

Fisheries
In 2010 the Estonian fishing fleet numbered 947 vessels of 17,300 gross tonnes. The total catch in 2010 was 95,398 tonnes.

INDUSTRY
Important industries are engineering, metalworking, food products, wood products, furniture and textiles. In 2010 manufacturing accounted for 17·0% of GDP.

Labour
The workforce in 2010 totalled 686,800, of whom 570,900 were employed. The average monthly gross wage in the fourth quarter of 2011 was €865. The unemployment rate in Nov. 2012 was 9·9% (down from 12·6% in 2011 as a whole).

Retirement age was 63 years for men and 61 years for women in 2010 although the female retirement age is to increase gradually to 63 by 2016.

INTERNATIONAL TRADE
Imports and Exports
Imports (c.i.f.) in 2007 (and 2006) were valued at US$15,458·8m. (US$13,285·1m.); exports (f.o.b.), US$11,009·7m. (US$9,607·5m.).

Principal imports are road vehicles, petroleum and petroleum products, and electrical machinery and equipment; principal exports are petroleum and petroleum products, road vehicles, and electrical machinery and equipment.

Main import suppliers in 2007: Finland, 15·9%; Germany, 12·8%; Russia, 10·2%; Sweden, 10·1%; Latvia, 7·6%. Main export

markets, 2007: Finland, 18·0%; Sweden, 13·3%; Latvia, 11·4%; Russia, 8·9%; Lithuania, 5·8%.

COMMUNICATIONS
Roads
As of 1 Jan. 2009 there were 16,487 km of national roads (28·4% of the total Estonian road network of 58,034 km). In Dec. 2010 there were 552,684 registered passenger cars in use, plus 81,204 lorries, 4,167 buses and 19,671 motorcycles. There were 1,340 road accidents and 78 fatalities in 2010.

Rail
Length of railways in 2009 was 919 km (1,520 mm gauge), of which 131 km was electrified. In 2009, 4·9m. passengers and 45·9m. tonnes of freight were carried.

Civil Aviation
There is an international airport at Tallinn (Lennart Meri International Airport), which handled 1·3m. passengers (98% on international flights) and 21,000 tonnes of freight in 2009. The national carrier is Estonian Air, 97% state-owned. In 2010 Estonian Air operated services to Amsterdam, Athens, Berlin, Brussels, Copenhagen, Dublin, Hamburg, Kyiv, London, Milan, Minsk, Moscow, Nice, Oslo, Paris, Rome, St Petersburg, Stockholm and Vilnius as well as domestic flights. In 2007 Estonian-based airlines carried 1,141,600 passengers (1,120,900 on international flights).

Shipping
There were 11 commercial ports and five ports offering international passenger services in 2009. Tallinn handled 31·6m. tonnes of cargo traffic in 2009 (82% of total transport of freight through Estonian ports). The port of Tallinn makes most of its money by shipping out Russian oil and importing goods destined for Russia. In 2009, 7·26m. passengers travelled through the port of Tallinn (more than 80% on the Tallinn–Helsinki route). In Jan. 2009 the Estonian-controlled fleet comprised 108 vessels of 1,000 GT or over, of which 87 were under foreign flags.

Telecommunications
In 2011 there were 470,500 landline telephone subscriptions (equivalent to 351·0 per 1,000 inhabitants) and 1,863,000 mobile phone subscriptions (or 1,389·8 per 1,000 inhabitants). In 2011, 76·5% of the population were internet users. In March 2012 there were 448,000 Facebook users. In 2000 the Estonian parliament voted to guarantee internet access to its citizens.

SOCIAL INSTITUTIONS
Justice
A post-Soviet criminal code was introduced in 1992. There is a three-tier court system with the State Court at its apex, and there are both city and district courts. The latter act as courts of appeal. The State Court is the final court of appeal, and also functions as a constitutional court. There are also administrative courts for petty offences. Judges are appointed for life. City and district judges are appointed by the President; State Court judges are elected by Parliament.

In 2008, 50,977 crimes were recorded; there were 12 murders and four attempted murders. In Jan. 2008, 3,467 persons were in penal institutions (259 per 100,000 of national population).

The death penalty was abolished for all crimes in 1998.

Education
Adult literacy rate in 2009 was estimated at 99·8% (99·8% for both males and females). There are nine years of comprehensive school starting at age six, followed by three years secondary school. In 2010–11 there were 545 general education schools: 68 primary, 253 basic and 224 secondary/upper secondary. Of these, 454 were Estonian-language, 31 non Estonian-language and 60 mixed-language. There were 43 schools for children with special needs.

The total number of pupils at basic school level in general education was 112,600 in 2010–11, with 33,300 in secondary education. At the start of the 2010–11 academic year there were 69,100 higher education students studying at six public universities, three private universities, ten state higher schools and 12 private higher schools; two vocational educational institutions also provide higher education.

In 2006 central government expenditure on education came to 4,510·5m. krooni. In 2008 public expenditure on education came to 5·7% of GDP.

Health

Estonia had 60 hospitals (22 private) in 2008, down from 78 hospitals (28 private) in 1999. There were 7,662 hospital beds in 2008. In Dec. 2008 there were 4,444 doctors, 1,156 dentists, 937 pharmacists and 10,338 nurses.

Welfare

In 2012 there were 0·4m. pensioners. The average monthly pension in the third quarter of 2012 was €279·40. An official poverty line was introduced in 1993 (then 280 krooni—equivalent to €17·90—per month). Persons receiving less than the subsistence level (€76·70 per month in 2012) are entitled to state benefit. Unemployment allowance was €65·41 a month in 2012.

RELIGION

There is freedom of religion in Estonia and no state church, although most of the population is Lutheran. The Estonian Orthodox Church owed allegiance to Constantinople until it was forcibly brought under Moscow's control in 1940; a synod of the free Estonian Orthodox Church was established in Stockholm. Returning from exile, it registered itself in 1993 as the Estonian Apostolic Orthodox Church. By an agreement in 1996 between the Moscow and Constantinople Orthodox Patriarchates, there are now two Orthodox jurisdictions in Estonia. In 2000 there were 152,000 Lutherans and 144,000 Orthodox. Other Christian denominations, including Methodist, Baptist and Roman Catholic, are also represented.

CULTURE

World Heritage Sites

Estonia has two sites on the UNESCO World Heritage List: the Historic Centre (Old Town) of Tallinn (1997 and 2008) and the Struve Geodetic Arc (2005). The Arc is a chain of survey triangulations spanning from Norway to the Black Sea that helped establish the exact shape and size of the earth and is shared with nine other countries.

Press

In 2008 there were 15 daily newspapers (combined circulation of 333,000) and 28 non-dailies (248,000). *The Baltic Times* is an English-language weekly covering news from Estonia, Latvia and Lithuania.

Tourism

In 2011, 808,000 non-resident tourists and 918,000 Estonians stayed in holiday accommodation. Of the foreign tourists most were from Finland (841,000), followed by Russia (203,000), Germany (104,000), Sweden (86,000) and Latvia (85,000).

Festivals

Festivals include: International Folklore Festival, BALTICA, which is staged every three years; Festival of Baroque Music; Jazz festival, JAZZKAAR; Pärnu International Documentary and Anthropology Film Festival and the Viljandi Folk Music Festival. Estonia's Song Festival, which was first held in 1869, is held every five years and is next scheduled to take place in 2014.

Baltoscandal, an international theatre festival which takes place every two years, celebrated its 12th staging in July 2012.

DIPLOMATIC REPRESENTATIVES

Of Estonia in the United Kingdom (16 Hyde Park Gate, London, SW7 5DG)
Ambassador: Aino Lepik von Wirén.

Of the United Kingdom in Estonia (Wismari 6, 10136 Tallinn)
Ambassador: Christopher B. Holtby, OBE.

Of Estonia in the USA (2131 Massachusetts Ave., NW, Washington, D.C., 20036)
Ambassador: Marina Kaljurand.

Of the USA in Estonia (Kentmanni 20, 15099 Tallinn)
Ambassador: Jeffrey D. Levine.

Of Estonia to the United Nations
Ambassador: Margus Kolga.

Of Estonia to the European Union
Permanent Representative: Matti Maasikas.

FURTHER READING

Statistical Office of Estonia. *Statistical Yearbook.*
Ministry of the Economy. *Estonian Economy.* Annual

Hood, N., *et al.*, (eds.) *Transition in the Baltic States.* 1997
Kasekamp, Andres, *A History of the Baltic States.* 2010
Kolsto, Pal, *National Integration and Violent Conflict in Post-Soviet Societies: The Cases of Estonia and Moldova.* 2002
Lieven, A., *The Baltic Revolution: Estonia, Latvia, Lithuania and the Path to Independence.* 2nd ed. 1994
Misiunas, R.-J. and Taagepera, R., *The Baltic States: Years of Dependence 1940–1990.* 2nd ed. 1993
O'Connor, Kevin, *The History of the Baltic States.* 2003
Plakans, Andrejs, *A Concise History of the Baltic States.* 2011
Smith, David J., Purs, Aldis, Pabriks, Artis and Lane, Thomas, (eds.) *The Baltic States: Estonia, Latvia and Lithuania.* 2002
Taagepera, R., *Estonia: Return to Independence.* 1993

National Statistical Office: Statistical Office of Estonia, Endla 15, 15174 Tallinn.
Website: http://www.stat.ee

ETHIOPIA

Ye-Ityoppya Federalawi Dimokrasiyawi Ripeblik
(Federal Democratic Republic of Ethiopia)

Capital: Addis Ababa
Population projection, 2015: 92·00m.
GNI per capita, 2011: (PPP$) 971
HDI/world rank: 0·363/174
Internet domain extension: .et

KEY HISTORICAL EVENTS

From as early as 3000 BC Egyptian Pharaohs referred to northern Ethiopia as the Land of Punt, rich in precious resources including gold, myrrh and ivory. The region was in contact with southern Arabia by around 2000 BC, with settlers bringing Semitic languages and stone-building techniques. Early in the 1st century AD a prosperous and advanced civilization arose in the northern highlands, centred on Aksum. Christianity reached Aksum in the 4th century AD when King Ezana was converted by Frumentius of Tyre. At its height in the 6th century AD the Aksumite empire controlled much of the Red Sea coast and traded with the Mediterranean powers, as well as Persia and India.

Between the 8th–10th centuries the declining Aksumite realm shifted southwards, while northern Ethiopia increasingly came under Arabian influence. Christianity held sway in the highlands and was central to the culture of the post-Aksumite Zagwe kingdom, founded in 1137. The Zagwe were ousted around 1270 by Yekuno Amlak, an Amharic warrior who restored the Solomonic dynasty which claimed descent from Aksum, King Solomon and the Queen of Sheba. The kingdom expanded to the south, particularly under Zara Jakob (ruled 1434–68).

Portuguese mariners reached the Red Sea in the early 1500s and a diplomatic mission arrived in Ethiopia in 1508. Faced with raids from neighbouring Islamic states, Emperor Lebna Dengel sought an alliance with the Portuguese who, in 1543, assisted in defeating Ahmad ibn Ibrahim al Ghazi, the conqueror of much of southern and eastern Ethiopia. The Solomonic monarchy was reinstated in 1632 by Emperor Fasidas, who established a new capital at Gondar. The coastal provinces came largely under Ottoman rule.

Tigray and Amhara experienced sporadic civil wars until the emergence of Lij Kasa in the 19th century. Crowned Emperor Tewodros II in 1855, he set about unifying the country although his efforts to end the slave trade led to tensions with local rulers. Following disagreements with Britain, Tewodros arrested several British officials including the consul. Britain responded by sending 12,000 troops to Ethiopia under Robert Napier. When Napier overwhelmed the fortress at Magdala, Tewodros committed suicide, igniting civil war.

The opening of the Suez Canal in 1869 intensified the regional scramble for influence among the European powers. At the same time Ethiopia faced armed incursions from Egypt and the Madhists in Sudan. Menelik II, ruler of Shoa in central Ethiopia, increased his power base with Italian support and seized control of Ethiopia in 1889. He renounced Italian claims to Ethiopia and defeated the Italian army at Adwa in 1896 before founding a new capital at Addis Ababa. Menelik II centralized authority and developed the country's infrastructure. He was succeeded in 1913 by his grandson, Lij Iyasu, who was deposed three years later in favour of Empress Zawaditu, with Ras Tafari Makonnen as regent and heir apparent. Following Empress Zawaditu's death in 1930 Ras Tafari became emperor as Haile Selassie I, claiming direct descent from King Solomon and the Queen of Sheba.

Although Ethiopia was recognized as independent in 1923, the League of Nations was unable to prevent Benito Mussolini from launching a second Italian invasion from Eritrea on 3 Oct. 1935. When Addis Ababa was captured in May 1936 Haile Selassie fled to Britain, only returning when Allied forces defeated the Italians in 1941. He brought in social and political reforms and established a national assembly. In 1950 Eritrea, an Italian colony under British military administration since 1941, was handed over to Ethiopia. A secessionist movement, the Eritrean Peoples' Liberation Front (EPLF), began a guerrilla war for independence. Following famine and economic decline, a military government (the Dirgue) assumed power on 12 Sept. 1974 under Lieut. Col. Mengistu Haile Mariam. It deposed Haile Selassie (who was murdered in prison in 1975), abolished the monarchy and mounted an agricultural collectivization programme.

In 1977 Somalia invaded Ethiopia and took control of the Ogaden region. After a counter offensive, with Soviet and Cuban support, the area was recaptured. Droughts in the late 1970s and early 1980s led to a devastating famine, which received international attention in 1984 when the death toll had already reached 200,000. War-torn Tigray and Eritrea were again at conflict in 1989. In 1991 the Ethiopian People's Revolutionary Democratic Front (EPRDF), led by Meles Zenawi, defeated the Ethiopian army, forcing Mengistu to flee to Zimbabwe. In July 1991 a conference of 24 political groups, called to appoint a transitional government, agreed a democratic charter. Eritrea seceded and became independent on 24 May 1993.

In 1994 a new constitution established a bicameral legislature and a judicial system. Meles Zenawi was elected prime minister in May 1995 with Negasso Gidada as president. The ongoing conflict with Eritrea flared up in 1999 and thousands were killed before a peace deal was brokered in June 2000. Economic progress raised hopes of higher living standards until three successive years of drought left food resources seriously depleted. Widespread malnutrition was alleviated by international aid. Meles' EPRDF won contested elections in May 2005, paving the way for his third five-year stint as prime minister.

TERRITORY AND POPULATION

Ethiopia is bounded in the northeast by Eritrea, east by Djibouti and Somalia, south by Kenya and west by South Sudan and Sudan. It has a total area of 1,127,127 sq. km. The secession of Eritrea in 1993 left Ethiopia without a coastline. An

Eritrean–Ethiopian agreement of July 1993 gives Ethiopia rights to use the Eritrean ports of Assab and Massawa.

The first census was carried out in 1984: population, 42,616,876 (including Eritrea). The 2007 census population was 73,750,932 (36,533,802 females); density, 65·4 per sq. km. The United Nations population estimate for 2007 was 77·72m. Ethiopia is Africa's second most populous country, after Nigeria. In 2007, 83·9% of the population lived in rural areas.

The UN gives a projected population for 2015 of 92·00m.

Ethiopia has 11 administrative divisions—eight states (Afar, Amhara, Benshangul/Gumaz, Gambella, Oromia, the Peoples of the South, Somalia and Tigre) and three cities (Addis Ababa, Dire Dawa and Harar).

The population of the capital, Addis Ababa, was 2,739,551 in 2007. Other large towns (2007 populations): Dire Dawa, 233,224; Nazret, 220,212; Mekele, 215,914; Gonder, 207,044.

There are seven major ethnic groups (in % of total population in 2007): Oromo, 35%; Amhara, 27%; Somali, 6%; Tigrinya, 6%; Sidamo, 4%; Gurage, 3%; Welaita, 2%. The de facto official language is Amharic (which uses its own alphabet). Oromo is also widely spoken. In total there are around 80 local languages.

SOCIAL STATISTICS

Births, 2008 estimate, 3,086,000; deaths, 954,000. Rates per 1,000 population, 2008 estimates: births, 38·2; deaths, 11·8. Expectation of life at birth in 2007 was 53·3 years for males and 56·2 years for females. Annual population growth rate, 2000–08, 2·6%; infant mortality, 2010, 68 per 1,000 live births; fertility rate, 2008, 5·3 births per woman.

CLIMATE

The wide range of latitude produces many climatic variations between the high, temperate plateaus and the hot, humid lowlands. The main rainy season lasts from June to Aug., with light rains from Feb. to April, but the country is very vulnerable to drought. Addis Ababa, Jan. 59°F (15°C), July 59°F (15°C). Annual rainfall 50" (1,237 mm). Harar, Jan. 65°F (18·3°C), July 64°F (17·8°C). Annual rainfall 35" (897 mm). Massawa, Jan. 78°F (25·6°C), July 94°F (34·4°C). Annual rainfall 8" (193 mm).

CONSTITUTION AND GOVERNMENT

A 548-member constituent assembly was elected on 5 June 1994; turnout was 55%. The EPRDF gained 484 seats. On 8 Dec. 1994 it unanimously adopted a new federal constitution which became effective on 22 Aug. 1995. It provided for the creation of a federation of nine regions based (except the capital and the southern region) on a predominant ethnic group. These regions have the right of secession after a referendum. The President, a largely ceremonial post, is elected for a six-year term by both chambers of parliament (renewable once only). The lower house is the 547-member House of People's Representatives; the upper house the 135-member House of the Federation.

National Anthem

'Yazegennat keber ba-Ityop yachchen santo' ('In our Ethiopia our civic pride is strong'); words by D. M. Mengesha, tune by S. Lulu Mitiku.

RECENT ELECTIONS

Parliamentary elections were held on 23 May 2010. Although the violence of the 2005 elections was not repeated, the results were disputed by the opposition parties and the poll failed to meet international standards. The Ethiopian People's Revolutionary Democratic Front (EPRDF) won 499 seats. The EPRDF's allies took 46 seats: the Somali People's Democratic Party 24, the Benishengul Gumuz People's Democratic Party 9, the Afar National Democratic Party 8, the Gambella People's Unity Democratic Movement 3, the Harrari National League 1 and the Argoba Nationality Democratic Movement 1. The opposition

Ethiopian Federal Democratic Unity Forum took one seat and there was one independent. Turnout was 93·4%.

Girma Wolde-Giyorgis was re-elected president on 9 Oct. 2007. He received 430 votes with 88 against and 11 abstentions in the Council of People's Representatives, having already won the required two-thirds majority in the Federal Council.

CURRENT GOVERNMENT

President: Girma Wolde-Giyorgis; b. 1925 (elected on 8 Oct. 2001 and re-elected 9 Oct. 2007).

In March 2013 the government comprised:

Prime Minister: Hailemariam Desalegn; b. 1965 (in office since 20 Aug. 2012—acting until 21 Sept. 2012).

Deputy Prime Minister and Minister of Education: Demeke Mekonnen. Deputy Prime Minister and Minister of Communications and Information Technology: Debretsion Mikeal. Deputy Prime Minister and Minister of Civil Service: Ato Muktar Kedir.

Minister of Agriculture: Tefera Deribew. Culture and Tourism: Amin Abdulkadir. Defence: Siraj Fegeta. Federal Affairs: Dr Shiferaw Tekle-Mariam. Finance and Economic Development: Sufyan Ahmad. Foreign Affairs: Dr Tewodros Adhanom. Health: Dr Kesetebirhan Admassu. Industry: Mekonnen Manyazewal. Justice: Berhan Hailu. Labour and Social Affairs: Abdulfeta Abdul Ahmed. Mines: Sinkenesh Ejigu. Science and Technology: Dese Dalke. Trade: Kebede Chane. Transport: Diriba Kuma. Urban Development and Construction: Mekuria Haile. Water and Energy: Alemayehu Tegenu. Women's Affairs, Children and Youth Affairs: Zenebu Tadesse.

Ethiopian Parliament: http://www.ethiopar.net

CURRENT LEADERS

Hailemariam Desalegn

Position
Prime Minister

Introduction
Hailemariam Desalegn became prime minister, initially in an acting capacity after Meles Zenawi died of undisclosed illness, in Aug. 2012. He had previously served as deputy prime minister under Meles and was minister of foreign affairs from 2010–12. He is the first premier from the Ethiopian Apostolic denomination.

Early Life
Hailemariam was born on 19 July 1965 in what is now the Southern Nations, Nationalities, and Peoples' Region in the south of Ethiopia. At school, he joined a political youth group attached to the communist military junta of Mengistu Haile Mariam.

In 1988 Hailemariam graduated in civil engineering from Addis Ababa University, before taking up a post as a graduate assistant in the Arba Minch Water Technology Institute. In 1990 he won a scholarship to Tampere University of Technology in Finland to study for a masters degree in sanitation engineering. Returning to Ethiopia, he worked in various academic and administrative positions for over a decade, during which time he earned a masters in organizational leadership from Azusa Pacific University in California.

Hailemariam served in senior management positions at the Hawassa and Wolayta Soddo Universities, the Addis Ababa Water Supply and Sewerage Authority, the Construction Design Share Company, the Ethiopian Maritime and Transit Service Enterprise, the Privatization and Public Enterprise Supervising Agency, and the Walta Information and Public Relations Center.

From the late 1990s he was increasingly involved in politics, joining Ethiopia's ruling party, the Ethiopian People's Revolutionary Democratic Front (EPRDF). A member of the country's Southern Nations, Nationalities and People's Region (SNNPR) council between 1995 and 2008, he served as its vice-president from 2000–01 and as its president from 2001–06. He is also chairman of the Southern Ethiopian People's Democratic Movement and deputy chairman of the executive council of the EPRDF.

From 2006–08 Hailemariam was special adviser (with the rank of minister) to the prime minister on social affairs, civic organizations and partnership. From 2008–10 he was the government's chief whip in the House of People's Representatives. In Sept. 2010 Hailemariam was appointed deputy prime minister and minister of foreign affairs under the premiership of Meles Zenawi.

Career in Office
After Meles died in Aug. 2012, Hailemariam succeeded him as prime minister, serving in an interim capacity for a month. On 15 Sept. 2012 he was elected chairman of the EPRDF before being sworn in as the fully-mandated prime minister a week later.

DEFENCE

In 2006 defence expenditure totalled US$366m. (US$4 per capita), representing 1·2% of GDP.

Army
Following the overthrow of President Mengistu's government Ethiopian armed forces were constituted from former members of the Tigray People's Liberation Front. The strength of the Army was 135,000 in 2007.

Air Force
Owing to its role in the war with Eritrea aircraft operability has improved. There were 48 combat-capable aircraft in 2007, including MiG-21s and MiG-23s, and 25 attack helicopters. Personnel numbered 3,000 in 2007.

INTERNATIONAL RELATIONS

A border dispute between Ethiopia and Eritrea broke out in May 1998. Eritrean troops took over the border town of Badame after a skirmish between Ethiopian police units and armed men from Eritrea. Ethiopia maintained that Badame and Sheraro, a nearby town, had always been part of Ethiopia and called Eritrea's action an invasion. An agreement ending hostilities was signed in June 2000, followed by a peace accord in Dec. A buffer zone has been created to separate the armies but tensions do still arise from time to time, notably in late 2005 following a further dispute between the two countries over Badame.

ECONOMY

Agriculture accounted for 46·3% of GDP in 2007, industry 13·3% and services 40·4%.

Overview
Ethiopia is among the poorest countries in the world, with economic development stalled by frequent droughts, conflict with Eritrea since the late 1990s and the global financial crisis. Nonetheless, growth averaged 10·4% from 2005–11 and those living below the poverty line fell from 39% in 2006–07 to 29% in 2011–12.

Donor assistance increased significantly in 2004 under the IMF-World Bank Indebted Poor Country Initiative. The country also qualified for debt relief under the Multilateral Debt Relief Initiative of Dec. 2005. In 2010 the government launched a five-year growth and transformation plan.

Currency
The *birr* (ETB), of 100 *cents*, is the unit of currency. The birr was devalued in Oct. 1992. In April 2005 total money supply was 24,297m. birr. Foreign exchange reserves were US$1,444m. in May 2005. Inflation was 44·4% in 2008, fell to 8·1% in 2010 but increased again to 33·1% in 2011.

Budget
The fiscal year ends on 7 July. Revenue, 2006–07, 30,274m. birrs; expenditure, 35,564m. birrs. Tax revenue accounted for 57·3% of revenues in 2006–07; capital expenditure accounted for 51·7% of expenditures.

VAT of 15% was introduced in 2003.

Performance
After the economy contracted by 2·1% in 2003 there was a recovery with growth of 11·7% in 2004. This strong performance has continued since with real GDP growth of 10·0% in 2009, 8·0% in 2010 and 7·5% in 2011. Ethiopia ranked among Africa's top half-dozen performing economies every year between 2004 and 2010 and is one of the world's fastest-growing non-oil producing economies. Total GDP was US$30·2bn. in 2011.

Banking and Finance
The central bank and bank of issue is the National Bank of Ethiopia (founded 1964; *Governor*, Teklewold Atnafu). The country's largest bank is the state-owned Commercial Bank of Ethiopia. The complete monopoly held by the bank ended with deregulation in 1994, but it still commands about 90% of the market share. There are eight other banks.

In 2010 external debt totalled US$7,147m., equivalent to 24% of GNI.

ENERGY AND NATURAL RESOURCES

Environment
Carbon dioxide emissions from the consumption and flaring of fossil fuels were the equivalent of 0·1 tonnes per capita in 2008.

Electricity
Installed capacity in 2007 was 0·8m. kW. Production in 2007 was 3·50bn. kWh. Hydro-electricity accounts for 96% of generation. Consumption per capita was 46 kWh in 2007. In 2008 only 15% of the population had access to electricity, although the stated aim is for electricity to be available to the entire country by 2018.

Oil and Gas
The Calub gas field in the southeast of Ethiopia had proven reserves of 25bn. cu. metres in 2007.

Minerals
Gold and salt are produced. Lege Dembi, an open-pit gold mine in the south of the country, has proven reserves of over 62 tonnes and produces more than five tonnes a year.

Agriculture
Small-scale farmers make up about 85% of Ethiopia's population. There were 10·7m. ha. of arable land in 2001 and 750,000 ha. of permanent crops. 190,000 ha. were irrigated in 2001. There were 3,000 tractors in 2001 and 100 harvester-threshers.

Coffee is by far the most important source of rural income. Main agricultural products (2002, in 1,000 tonnes): maize, 2,968; sugarcane, 2,232; sorghum, 1,566; wheat, 1,478; barley, 1,093; broad beans, 447; potatoes, 385; millet, 308; yams, 300; papayas, 226; coffee, 220. Teff (*Eragrastis abyssinica*) and durra are also major products.

Livestock, 2002: cattle, 35·5m.; sheep, 11·4m.; goats, 9·6m.; asses, 3·4m.; horses, 1·3m.; camels, 326,000; chickens, 38m.

Forestry
In 2010 forests covered 12·30m. ha., representing 11% of the land area. Ethiopia is Africa's leading roundwood producer, with removals totalling 100·06m. cu. metres in 2007.

Fisheries
The catch in 2010 was 18,058 tonnes, entirely from inland waters.

INDUSTRY

Most public industrial enterprises are controlled by the state. Industrial activity is centred around Addis Ababa. Processed food, cement, textiles and drinks are the main commodities produced. Industrial production accounted for 13·3% of GDP in 2007, including 5·0% from manufacturing.

Labour

The estimated labour force in 2010 was 41,310,000 (52% males), up from 28,996,000 in 2000. Coffee provides a livelihood to a quarter of the population.

INTERNATIONAL TRADE

Imports and Exports

In 2010 imports (c.i.f.) were valued at US$8,601·8m.; exports (f.o.b.) totalled US$2,329·8m. Main imports (2006): machinery and transport equipment (35·6%); petroleum and petroleum products (19·8%); manufactured goods (16·1%). Main exports (2006): coffee and coffee substitutes (40·8%); sesame seeds (15·4%); crude vegetable materials (12·4%). Major import suppliers, 2006: Saudi Arabia, 17·9%; China, 12·3%; Italy, 7·7%. Major export markets, 2006: Germany, 12·6%; China, 9·7%; Japan, 8·4%.

COMMUNICATIONS

Roads

There were 44,359 km of roads in 2007. Passenger cars in use in 2007 numbered 70,900 (one per 1,000 inhabitants) and there were also 149,000 lorries and vans, and 17,100 buses and coaches. In 2007 there were 2,517 deaths in road accidents.

Rail

The Ethiopian-Djibouti Railway has a length of 781 km (metre-gauge), but much of the route is in need of renovation. Passenger-km travelled in 2005 came to 145m. and freight tonne-km to 118m.

Civil Aviation

There are international airports at Addis Ababa (Bole) and Dire Dawa. The national carrier is the state-owned Ethiopian Airlines. In 2003 it served 43 international and 25 domestic destinations. In 2003 scheduled airline traffic of Ethiopian-based carriers flew 35m. km, carrying 1,147,000 passengers (881,000 on international flights). In 2001 Addis Ababa (Bole) handled 1,096,500 passengers and 26,490 tonnes of freight.

Shipping

In Jan. 2009 there were nine ships of 300 GT or over registered, totalling 118,000 GT.

Telecommunications

There were 908,900 fixed telephone lines in 2010 (11·0 per 1,000 inhabitants). Mobile phone subscribers numbered 6·52m. in 2010. There were 7·5 internet users per 1,000 inhabitants in 2010. Fixed internet subscriptions totalled 74,600 in 2009 (0·9 per 1,000 inhabitants). In June 2012 there were 599,000 Facebook users.

SOCIAL INSTITUTIONS

Justice

The legal system is based on the Justinian Code. A new penal code came into force in 1958 and Special Penal Law in 1974. Codes of criminal procedure, civil, commercial and maritime codes have since been promulgated. Provincial and district courts have been established, and High Court judges visit the provincial courts on circuit. The Supreme Court at Addis Ababa is presided over by the Chief Justice. The death penalty is in force; there was one execution in 2007 but none since.

The population in penal institutions in Sept. 2007 was approximately 80,000 (98 per 100,000 of national population).

Education

The adult literacy rate in 2007 was 39·0%. Primary education commences at seven years and continues with optional secondary education at 13 years. Up to the age of 12, education is in the local language of the federal region. In 2007 there were 12·17m. pupils at primary schools and 3·43m. pupils at secondary schools. During the period 1990–95 only 19% of females of primary school age were enrolled in school but this had increased to 62% by 2006. There were 21 public universities in 2007, with many having just opened in the previous few years. There were 210,456 students in tertiary education in 2007 and 8,355 academic staff.

In 2007 public expenditure on education came to 5·5% of GNI and 23·3% of total government spending.

Health

In 2002 there were 1,971 physicians, 61 dentists, 13,018 nurses, 1,142 midwives and 125 pharmacists. In 2000 only 24% of the population had access to safe drinking water.

RELIGION

About 59% of the population are Christian, mainly belonging to the Ethiopian Orthodox Church, and 32% Sunni Muslims. Amhara, Tigreans and some Oromos are Christian. Somalis, Afars and some Oromos are Muslims. About 5% of the population follow traditional animist beliefs.

CULTURE

World Heritage Sites

There are nine sites in Ethiopia that appear on the UNESCO World Heritage List. They are (with the year entered on list): the Rock-hewn Churches at Laibela (1978), 11 monolithic 13th century churches; Simien National Park (1978); Fasil Ghebbi, Gondar Region (1979), a 16th century fortress city; Aksum (1980), the capital of the ancient Kingdom of Aksum, containing tombs and castles dating from the first millennium AD; the Lower Valley of the Awash (1980), an important palaeontological site; the Lower Valley of the Omo (1980), where *Homo gracilis* was discovered; Tiya (1980), a group of archaeological sites south of Addis Ababa; Harar Jugol (2006), a fortified historic town; and the Konso Cultural Landscape (2011), an area of stone walled terraces and fortified settlements in the Konso highlands.

Press

In 2008 there were three paid-for daily newspapers with a combined circulation of 92,000 and 54 paid-for non-dailies.

Tourism

In 2008 there were 330,000 non-resident tourist arrivals (excluding same-day visitors).

Calendar

The Ethiopian calendar is based on the ancient Coptic calendar; the year has 13 months (12 months with 30 days and one month with five or six, depending on the leap-year). It begins on 11 or 12 Sept. (Gregorian) and is seven or eight years behind the Gregorian calendar.

DIPLOMATIC REPRESENTATIVES

Of Ethiopia in the United Kingdom (17 Prince's Gate, London, SW7 1PZ)
Ambassador: Ato Berhanu Kebede.

Of the United Kingdom in Ethiopia (Comoros St., Addis Ababa)
Ambassador: Gregory Dorey, CVO.

Of Ethiopia in the USA (3506 International Drive, NW, Washington, D.C., 20008)
Ambassador: Girma Birru Geda.

Of the USA in Ethiopia (Entoto St., Addis Ababa)
Ambassador: Donald E. Booth.

Of Ethiopia to the United Nations
Ambassador: Tekeda Alemu.

Of Ethiopia to the European Union
Ambassador: Kassu Yilala.

FURTHER READING

Araia, G., *Ethiopia: the Political Economy of Transition.* 1995
Bigsten, Arne, Shimeles, Adebe and Kebede, Bereket, (eds.) *Poverty, Income Distribution and Labour Markets in Ethiopia.* 2005

Crummey, Donald, *Land and Society in the Christian Kingdom of Ethiopia: From the Thirteenth to the Twentieth Century.* 2000
Henze, Paul B., *Layers of Time: A History of Ethiopia.* 2000
Marcus, H. G., *A History of Ethiopia.* 1994
Negash, Tekeste and Tronvoll, Kjetil, *Brothers at War: Making Sense of the Eritrean–Ethiopian War.* 2001
Pankhurst, Richard, *The Ethiopians.* 1999
Woodward, Peter, *The Horn of Africa: Politics and International Relations.* 2002

National Statistical Office: Central Statistical Office, Addis Ababa.
Website: http://www.csa.gov.et

**The world in focus at
www.statesmansyearbook.com**

FIJI ISLANDS

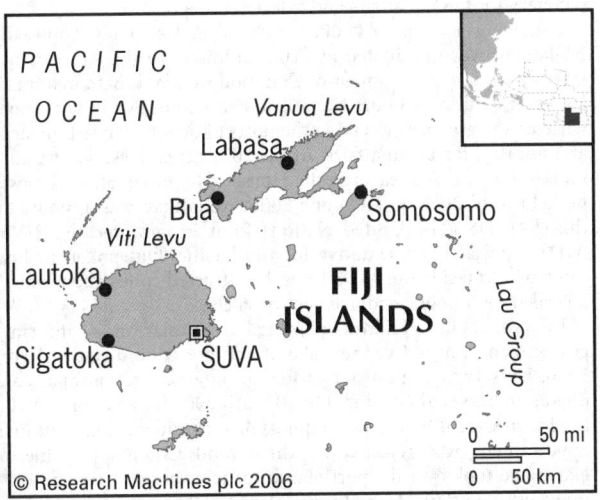

PACIFIC OCEAN

Vanua Levu
Labasa
Bua   Somosomo
Viti Levu
Lautoka
FIJI
ISLANDS
Sigatoka   SUVA
Lau Group

0   50 mi
0   50 km

© Research Machines plc 2006

Kai Vakarairai ni Fiji
(Republic of the Fiji Islands)

Capital: Suva
Population projection, 2015: 896,000
GNI per capita, 2011: (PPP$) 4,145
HDI/world rank: 0·688/100
Internet domain extension: .fj

KEY HISTORICAL EVENTS

The Fiji Islands were first recorded in detail by Capt. Bligh after the mutiny of the *Bounty* (1789). In the 19th century the demand for sandalwood attracted merchant ships. Deserters and shipwrecked sailors settled. Tribal wars were bloody and widespread until Fiji was ceded to Britain on 10 Oct. 1874. Fiji gained independence on 10 Oct. 1970. It remained an independent state within the Commonwealth with a Governor-General appointed by the Queen until 1987. In the general election of 12 April 1987 a left-wing coalition came to power with the support of the Indian population who outnumbered the indigenous Fijians by 50% to 44%. However, it was overthrown in a military coup. A month later, Fiji declared itself a Republic and Fiji's Commonwealth membership lapsed.

In 1990 a new coalition restored civilian rule but made it impossible for Fijian Indians to hold power. A rapprochement with Indian leaders led to an agreement to restore multi-racial government in 1998. Fiji rejoined the Commonwealth in 1997. On 27 July 1998 a new constitution changed the country's name from Fiji to Fiji Islands.

A coup was staged in May 2000 under the leadership of George Speight, a failed businessman. His main aim was to exclude Indians from the government. An interim government, excluding Speight supporters, was appointed on 3 July 2000 to rule for 18 months. On 26 July George Speight and 400 of his supporters were arrested. On 18 Feb. 2002 Speight was sentenced to death although this was subsequently commuted to life imprisonment.

On 5 Dec. 2006 the Fiji Islands suffered their fourth coup in less than 20 years when Commodore Frank Bainimarama ousted Prime Minister Laisenia Qarase, placing him under house arrest. Bainimarama assumed the powers of president and prime minister, although he subsequently restored his predecessor, Ratu Josefa Iloilo, to the presidency.

TERRITORY AND POPULATION

The Fiji Islands comprise 332 islands and islets (about one-third are inhabited) lying between 15° and 22° S. lat. and 174° E. and 178° W. long. The largest is Viti Levu, area 10,429 sq. km (4,027 sq. miles); followed by Vanua Levu, 5,556 sq. km (2,145 sq. miles). The island of Rotuma (47 sq. km, 18 sq. miles), about 12° 30' S. lat., 178° E. long., was added to the colony in 1881. Total area, 18,333 sq. km (7,078 sq. miles). Total population (2007 census), 837,271 (females, 410,095); ethnic groups: Fijian, 475,739; Indian, 313,798; other Pacific islanders, 15,311; part-European/European, 13,724; Rotuman, 10,335; Chinese, 4,704; other, 3,660. Population density (2007), 45·7 per sq. km; 50·7% of the population lived in urban areas in 2007.

The UN gives a projected population for 2015 of 896,000.

The population of the capital, Suva (including Nasinu), was 173,137 at the 2007 census. Other large towns are Lautoka (52,220), Nausori (47,604) and Nadi (42,284).

English, Fijian and Hindustani are all official languages.

SOCIAL STATISTICS

2009 estimates: births, 19,000; deaths, 6,000. Rates, 2009 estimates (per 1,000 population): birth, 22; death, 7. Annual population growth rate, 2000–05, 0·8%. Life expectancy at birth in 2007 was 66·5 years for males and 71·0 years for females. Infant mortality, 2010, 15 per 1,000 live births; fertility rate, 2008, 2·7 births per woman.

CLIMATE

A tropical climate, but oceanic influences prevent undue extremes of heat or humidity. The S. E. Trades blow from May to Nov., during which time nights are cool and rainfall amounts least. Suva, Jan. 80°F (26·7°C), July 73°F (22·8°C). Annual rainfall 117" (2,974 mm).

CONSTITUTION AND GOVERNMENT

The executive authority of the State is vested in the *President*. The *Prime Minister* is appointed by the President. The Prime Minister must establish a multi-party cabinet. The President's term of office is five years.

A new constitution unanimously passed by Parliament and assented to by the President came into force on 27 July 1998. The country's name was changed from Fiji to Fiji Islands and the people were to be known as Fiji Islanders instead of Fijians. The new constitution stated that it was no longer a condition to have an indigenous Prime Minister and established a 71-seat *House of Representatives* (Lower House), with 46 elected on a communal role and 25 from an open electoral roll. Of the 46, 23 are elected from a roll of voters registered as Fijians, 19 from a roll of voters registered as Indians, one from a roll of voters registered as Rotumans and three from a roll of voters registered who are none of these. The Upper House or *Senate* has 32 members.

Parliament was reopened in Oct. 2001, having been suspended following a coup in May 2000. In 2006 another coup brought Commodore Frank Bainimarama to power but on 9 April 2009 the court of appeal declared his government illegal and he stood down. The next day the president repealed the constitution and assumed all governing power. The court was disbanded and Bainimarama's government restored on 12 April for the next five years. Elections were delayed resulting in suspension from the Pacific Islands Forum. In May 2009 Bainimarama announced that the next elections would not be held before Sept. 2014. In March 2012 Bainimarama disbanded the Great Council of Chiefs, which had existed in name only since April 2007 when Bainimarama suspended its operations. It had previously been responsible for appointing the president and 14 members of the Senate.

National Anthem

'Meda Dau Doka' ('God Bless Fiji'); words and tune by M. Prescott.

RECENT ELECTIONS

Mahendra Chaudhry, the Fiji Labour Party leader, became the country's first Indian prime minister in 1999, but was ousted in the coup of May 2000 after just over a year in office. In parliamentary elections held between 6–13 May 2006, incumbent Prime Minister Laisenia Qarase's Fiji United Party won 36 of 71 available seats with 44·6% of votes cast, followed by the Fiji Labour Party with 31 (39·2%), the United People's Party 2 (0·8%) and ind. 2 (4·9%). Turnout was 87·7%.

CURRENT GOVERNMENT

President: Ratu Epeli Nailatikau; b. 1941 (sworn in 5 Nov. 2009, having been acting president since 30 July 2009; reappointed 12 Nov. 2012).

In March 2013 the interim government comprised:

Prime Minister, Minister for Finance, Strategic Planning, National Development and Statistics, Public Service, People's Charter for Change and Progress, Information, National Archives and Library Services of Fiji, I-Taukei, Sugar Industry, and Lands and Mineral Resources: Commodore Frank Bainimarama; b. 1954 (sworn in 5 Jan. 2007, then briefly ousted from 9–12 April 2009).

Attorney General and Minister for Justice, Anti Corruption, Public Enterprises, Communications, Civil Aviation, Tourism, Industry and Trade: Aiyaz Sayed-Khaiyum. *Women, Social Welfare and Poverty Alleviation:* Dr Jiko Luveni. *Foreign Affairs and International Co-operation:* Ratu Inoke Kubuabola. *Education, National Heritage, Culture and the Arts:* Filipe Bole. *Defence, National Security and Immigration:* Joketani Cokanasiga. *Public Utilities, Works and Transport:* Timoci Lesi Natuva. *Health:* Dr Neil Sharma. *Local Government, Urban Development, Housing and the Environment:* Col. Samu A. Saumatua. *Labour, Industrial Relations and Employment:* Jone Usamate. *Youth and Sports:* Viliame Naupoto. *Agriculture, Fisheries and Forests, and Provincial Development and National Disaster Management:* Lieut.-Col. Inia Seruiratu.

Fiji Islands Government Online: http://www.fiji.gov.fj

CURRENT LEADERS

Ratu Epeli Nailatikau

Position
President

Introduction
Ratu Epeli Nailatikau was sworn into office on 5 Nov. 2009, having been acting president for the previous three months following the resignation of Ratu Josefa Iloilo on 30 July 2009.

Early Life
Nailatikau was born on 5 July 1941 to a family of politically powerful chieftains. After completing his education in Fiji, he joined the armed forces and was sent for training in New Zealand. In 1966 he was posted with the 1st Battalion, Royal New Zealand Infantry Regiment to Sarawak, Malaysia during the Indonesia–Malaysia confrontation. On his return to Fiji, he joined the Fiji Infantry Regiment, rising steadily through the ranks. By 1987 he was a Brigadier-General and Commander of the Royal Fiji Military Forces but was ousted in a military coup headed by Sitiveni Rabuka.

Pursuing a new career in the diplomatic service, Nailatikau completed the foreign service programme at the University of Oxford and was appointed High Commissioner to the United

Kingdom, a portfolio that also included Denmark, Egypt, Germany, Israel and the Holy See. He went on to become Roving Ambassador and High Commissioner to the member states of the South Pacific Forum and, in 1999, was appointed permanent secretary for foreign affairs and external trade.

Following a coup in 2000, which he had strongly opposed, Nailatikau was nominated as prime minister in the subsequent interim military government of Commodore Frank Bainimarama. However, within 24 hours Nailatikau had withdrawn in favour of Laisenia Qarase, instead taking the posts of deputy prime minister and minister for Fijian affairs. At the 2001 general election he put himself forward as speaker of the House of Representatives, a post he held until 2006. On 14 June 2005, Nailatikau was appointed the UNAIDS (The United Nations Joint Programme on HIV/AIDS) Special Representative for the Pacific. Outspoken in his campaign to tackle the AIDS crisis, he attracted controversy when he called for a public endorsement of safe sex.

In Jan. 2007 Nailatikau joined Bainimarama's interim government, formed in the wake of another coup d'état in Dec. 2006. He served as minister of foreign affairs, international co-operation and civil aviation. On 10 April 2007 he was nominated as vice-president by the newly reinstalled President Iloilo but his appointment was rejected by the Grand Council of Chiefs. Nailatikau took over the portfolio for provincial development and multi-ethnic affairs from Prime Minister Bainimarama in Oct. 2008, with his previous duties reassigned to the premier.

Nailatikau became vice-president on 17 April 2009 following a constitutional crisis in which the appeal court ruled that the military regime formed in 2006 was illegal, prompting Iloilo to repeal the constitution and sack the appeal court judges. With Bainimarama restored to the premiership, Nailatikau took over the presidency on 30 July 2009 in an acting capacity, until 5 Nov. 2009 when he was sworn in as Iloilo's successor.

Career in Office
Nailatikau's appointment, made behind closed doors, signalled that the military would continue to play a pivotal in Fijian politics. On 29 Jan. 2010 he signed an extension of the public emergency regulation, in place since April 2009 to give power to the authorities to stop events they deem to be a threat to national security. He was reappointed as president for a further three years in Nov. 2012.

Josaia Voreqe Bainimarama

Position
Prime Minister

Introduction
Voreqe Bainimarama, military commander of the Fiji Islands, became prime minister after leading a coup in Dec. 2006 against the incumbent, Laisenia Qarase. Previously a Qarase ally, Bainimarama was a key figure in defeating a coup attempt by George Speight in 2000 and installing Qarase to the premiership. Bainimarama's relationship with Qarase deteriorated in 2006 and Bainimarama deposed him. He appointed himself as president before relinquishing that position to become prime minister. He is popularly known in the Fiji Islands as 'Frank' Bainimarama.

Early Life
An ethnic Fijian, Bainimarama was born on 27 April 1954 on Bau Island and educated at the Marist Brothers High School. He enlisted in the navy in July 1975 and received his first command post in the early 1980s. He was promoted to lieutenant commander in Feb. 1986 and then served with the Multinational Force and Observers peacekeeping force in the Sinai Peninsula until returning to Fiji in Sept. 1987. He was appointed commander of the navy in Oct. 1988.

During his term as commander Bainimarama undertook extensive training, including in maritime surveillance, in disaster management and in exclusive economic zone management. He became acting chief of staff in Nov. 1997, and was promoted to the rank of commodore and appointed commander of the armed forces in March 1999.

In May 2000 George Speight led a coup with the declared aim of promoting Fijian nationalism and excluding ethnic Indians from government. Prime Minister Chaudhry was deposed and when President Mara fled, Bainimarama declared martial law. On 30 May he appointed an interim military government and appointed Laisenia Qarase as prime minister on 4 July. On 6 July the interim government responded to a military mutiny by signing an accord with Speight granting him immunity from prosecution. On 13 July it installed Iloilo as president, together with a pro-Speight vice-president, Jope Seniloli.

On 27 July the interim government revoked Speight's immunity, with Bainimarama claiming that the accord had been signed under duress. Speight and 369 others were arrested. On 2 Nov. pro-Speight soldiers mutinied in Suva, forcing Bainimarama to flee. The mutiny was quelled and Bainimarama accused a former prime minister, Sitiveni Rabuka, of involvement. He persisted in attempts to prove the involvement of Rabuka and other alleged conspirators and vehemently opposed Prime Minister Qarase's proposals in 2006 to offer amnesty to the rebels.

In Oct. 2006 Qarase attempted unsuccessfully to replace Bainimarama as commander of the armed forces while he was out of the country. Bainimarama then staged a coup on 5 Dec. 2006, with the support of senior military figures. Claiming he wanted to end corruption and stop racial divisions threatening national unity, he dismissed Qarase's government and appointed himself acting president. He placed government ministries under the control of their chief executive officers and appointed Jona Senilagakali as acting prime minister. International pressure forced Bainimarama to return the presidency to Iloilo on 4 Jan. 2007 and on 5 Jan. he replaced Senilagakali as interim prime minister.

Career in Office
Bainimarama's political legitimacy has since come under intense scrutiny domestically and internationally. Australia and New Zealand called for him to relinquish power and refused visas to members of his government. At home political opponents and leading institutions, including the influential Methodist Church, condemned the coup alleging human rights abuses although talks in Feb. 2007 led to the Methodist Church declaring its support for the interim government. Internationally, particularly within the Pacific islands region, pressure continued to mount. In Feb. 2007 Bainimarama promised to work towards democratic elections, but initially would not commit to a time-frame. Then, in Oct. 2007 at the Pacific Islands Forum annual summit in Tonga, he agreed to hold a general election by March 2009. However, in July 2008 he reneged on this commitment on the grounds that electoral reforms could not be completed in time.

Having briefly reimposed a state of emergency from Sept.–Oct. 2007 he claimed the following month to have foiled a plot to assassinate him. On 9 April 2009 Bainimarama stood down following a court of appeal judgment that his government was illegal. However, President Iloilo reinstated the cabinet on 12 April for another five years after having dismissed the judiciary and annulled the constitution.

In Sept. 2009 the Commonwealth suspended the Fiji Islands' membership and cut off all aid because of the country's lack of progress in re-establishing democracy. Bainimarama incurred further diplomatic reproach in Nov. when he accused Australia and New Zealand of interfering in the Fiji Islands' internal affairs and expelled their high commissioners.

In Sept. 2011 the Bainimarama government introduced a controversial decree severely restricting the rights of trade unions to strike in any industry designated by the authorities and nullifying existing collective bargaining agreements. However, in an apparently more liberal move, he announced the lifting of martial law in Jan. 2012 and set the stage for consultations on a new constitution and democratic elections. His subsequent agreement to schedule elections for Sept. 2014 led Australia and New Zealand to restore diplomatic ties in July 2012.

DEFENCE

In 2008 defence expenditure totalled US$57m. (US$61 per capita), representing 1·8% of GDP.

Army
Personnel in 2007 numbered 3,200 including 300 recalled reserves. There is an additional reserve force of around 6,000.

Navy
A small naval division of the armed forces numbered 300 in 2007.

ECONOMY

Agriculture accounted for 13% of GDP in 2009, industry 18% and services 69%.

Overview
The growing tourism sector, sugar exports and remittances from citizens working abroad have been the main sources of foreign exchange for the Fiji Islands over the last decade. Following a military coup in 2000 the economy experienced reduced growth of 2·8%, falling to 1% in 2001. GDP growth in 2005 was 0·7%, partly reflecting the loss of preferential trade agreements with the USA and the EU in the garment and sugar industries. After another coup in Dec. 2006, tourist arrivals fell by 6% in 2007 and the business climate deteriorated. The 2008 global financial crisis, adverse weather conditions and a weak domestic investment climate caused economic contraction from 2007–09. Subsequent growth has been slow while public debt is rising.

At independence in 1970, sugar accounted for 70% of export earnings. In 1973 the government brought all four sugar mills under the Fiji Sugar Corporation (FSC), a majority state-owned company. The FSC has been operating at a loss since 2006 and during this period the government has provided guarantees on borrowing. Total guaranteed debt of the FSC stood at 3% of GDP at the end of 2010. Public debt rose to 56% of GDP in 2010. It is estimated that over 20% of the population still depend on the sugar industry for their livelihoods, though the industry's share of export earnings is declining.

Reducing public debt to below 50% of GDP, restructuring of the FSC and the sugar industry and improved debt management are key priorities. The weak business climate remains a major impediment to sustained growth.

Currency
The unit of currency is the *Fiji dollar* (FJD) of 100 *cents*. In June 2005 total money supply was $F1,082m., foreign exchange reserves were US$397m. and gold reserves 1,000 troy oz. Inflation was 5·5% in 2010 and 8·7% in 2011. The Fiji dollar was devalued by 20% in both Jan. 1998 and April 2009.

Budget
Revenues in 2008 totalled $F1,455m. and expenditures $F1,427m.

VAT of 10% was introduced in 1992 (increased to 12·5% in 2003 and 15% in 2011).

Performance
The economy contracted by 1·3% in 2009 and 0·2% in 2010 but grew by 2·1% in 2011. Total GDP in 2011 was US$3·8bn.

Banking and Finance
The financial system in the Fiji Islands comprises the central bank, banking, insurance and superannuation industries, non-bank financial institutions (NBFIs), restricted foreign exchange dealers and money changers, and the South Pacific Stock Exchange.

The central bank and bank of issue is the Reserve Bank of Fiji (*Governor*, Barry Whiteside). Total assets at 31 Dec. 2007 were $F1,170m.

Total assets of commercial banks were $F3,957m. at the end of 2007. Total assets for the Fiji National Provident Fund (the sole player in the superannuation industry) at 31 Dec. 2007 were $F3,437m. Non-regulated financial institutions include the Fiji Development Bank, Housing Authority and Unit Trust of Fiji. Their assets totalled $F738m. in Dec. 2007.

Foreign debt was US$387m. in 2007.

The South Pacific Stock Exchange is based in Suva.

ENERGY AND NATURAL RESOURCES

Environment

Carbon dioxide emissions from the consumption and flaring of fossil fuels in 2008 were the equivalent of 2·9 tonnes per capita.

Electricity

The Fiji Electricity Authority is responsible for the generation, transmission and distribution of electricity in most of the country. It operates 13 power stations, three of which are operated on hydro power, and one 10 MW wind farm. The largest energy project is one of hydro-electricity that is capable of generating 70% of the main island's electric needs. Two rural hydro schemes have been completed, one private generating 100 kW and the other—operated by the Authority—producing 800 kW.

Installed capacity in 2007 was about 184,000 kW. Production in 2007 was 836m. kWh with consumption per capita 999 kWh.

Minerals

The main gold-mine normally accounts for about a twelfth of the country's exports. Gold has for many years been one of the Fiji Islands' main exports. However, after an extended closure in 2006–07 gold production in 2007 was only 77 kg, valued at $F2·55m.

Agriculture

With a total land area of 1·8m. ha., only 16% is suitable for farming. In 2007 there were an estimated 170,000 ha. of arable land and 83,000 ha. of permanent crops. Arable land: 24% sugarcane, 23% coconut and 53% other crops. Production figures for 2003 (in 1,000 tonnes): sugarcane, 3,300; coconut, 170; taro, 38; cassava, 33; rice, 16; copra, 14; bananas, 6; sweet potatoes, 6. Ginger is becoming increasingly important.

Livestock (2003): cattle, 320,000; goats, 248,000; pigs, 139,000; chickens, 4m. Products, 2003 (in 1,000 tonnes): beef and veal, 9; poultry meat, 9; pork, bacon and ham, 4; eggs, 3. Total production of milk was 58,000 tonnes in 2003.

Forestry

Forests covered 1·01m. ha.—56% of the land area—in 2010. Forestry contributed about 0·5% of GDP in 2007. It is the sixth most important export commodity, valued at $F48m. in 2007, of which $F27m. was in the form of wood chips. Hardwood plantations covered over 67,000 ha. in 2007. About 42,000 ha. of softwood plantations were held by Fiji Pine Ltd in 2007. There was no export of unprocessed timber. Roundwood production in 2007 was 509,000 cu. metres.

Fisheries

The catch in 2007 was 45,963 tonnes, of which 43,775 tonnes came from sea fishing. In 2007 fisheries accounted for around 2% of GDP, valued at $F101·3m. Mainstay of export fisheries are skipjack and albacore tuna for canning.

INDUSTRY

The main industries are tourism, sugar, fish, mineral water, garments and gold which in 2007 accounted for 12·6%, 3·4%, 1·9%, 1·9%, 1·8% and 0·5% of GDP respectively.

Output, 2007 (in tonnes): sugar, 240,000; cement, 144,000; flour, 52,677; animal feed, 37,820; coconut oil, 9,657; soap, 5,556; soft drinks, 213·4m. litres; beer, 19·0m. litres; cigarettes, 401m. units.

Labour

The labour force was estimated at 371,400 in 2003. In 2002 there were 23,000 people out of work and seeking employment—the number of unemployed people doubled between 1996 and 2002.

INTERNATIONAL TRADE

The Tax Free Factory/Tax Free Zone Scheme was introduced in 1987 to stimulate investment and encourage export-oriented businesses.

Imports and Exports

In 2009 imports (c.i.f.) totalled US$1,437·0m. and exports (f.o.b.) US$628·7m. Chief exports are clothing and apparel, sugar, gold, prepared and preserved fish, beverages, and cereal and cereal preparations.

In 2009 the chief import sources (in US$1m.) were: Singapore (397·7); Australia (317·8); New Zealand (228·2). The chief export markets (in US$1m.) were: Singapore (103·1); Australia (100·0); United Kingdom (94·4).

COMMUNICATIONS

Roads

Total road length in 2002 was an estimated 3,440 km, of which almost half were surfaced. There were a total of 94,400 passenger cars and 48,000 lorries and vans in 2007. In 2006, 89 fatalities were caused by road accidents.

Rail

Fiji Sugar Cane Corporation runs 600 mm gauge railways at four of its mills on Viti Levu and Vanua Levu, totalling 597 km in 2005.

Civil Aviation

There are international airports at Nadi and Suva. The national carrier is Air Pacific (51% government-owned). In 2003 it provided services to Australia, Japan, New Zealand, USA and a number of Pacific island nations. Air Fiji only operates on domestic routes. In 2001 Nadi handled 911,000 passengers (808,000 on international flights).

Shipping

The three main ports are Suva, Lautoka and Levuka. The gross registered tonnage of ocean-going shipping entering the ports in 2007 totalled 8,361,785 GRT including liquid bulk carriers of 2,530,718 GRT. A total of 694 foreign vessels called into Suva port in 2007, 348 into Lautoka and 93 into Levuka. The inter-island shipping fleet is made up of private and government vessels.

Telecommunications

In 2008 there were 129,100 main (fixed) telephone lines; mobile phone subscribers numbered 530,000 in 2008 (63·2 per 100 persons). There were 148·2 internet users per 1,000 inhabitants in 2010. Fixed internet subscriptions totalled 13,800 in 2007 (16·5 per 1,000 inhabitants). In Dec. 2011 there were 163,000 Facebook users.

SOCIAL INSTITUTIONS

Justice

An independent Judiciary is guaranteed under the constitution. A High Court has unlimited original jurisdiction to hear and determine any civil or criminal proceedings under any law. The High Court also has jurisdiction to hear and determine constitutional and electoral questions including the membership of the House of Representatives. The Chief Justice of the Fiji Islands is appointed by the President after consultation with the Prime Minister. The substantive Chief Justice was removed from office following the Dec. 2006 military coup and was replaced by an Acting Chief Justice in Jan. 2007, who was made permanent in Dec. 2008. Following the abolition of the constitution and the dismissal of all judges in April 2009, the Chief Justice was reappointed to his post in May 2009.

The Fiji Islands' Court of Appeal, of which the Chief Justice is *ex officio* President, is formed by three specially appointed Justices of Appeal, appointed by the President after consultation with the Judicial and Legal Services Commission. Generally, any person convicted of an offence has a right of appeal from the High Court of Appeal. The final appellant court is the Supreme Court. Most matters coming before the Superior Courts originate in Magistrates' Courts.

The population in penal institutions in 2007 was 841 (100 per 100,000 of national population).

Police
In 2008 the Royal Fiji Police Force had a total strength of 2,655 established officers, 60 support staff and 1,600 Special Constables.

Education
Adult literacy rate was 93·2% in 2001 (95·2% among males and 91·2% among females). Total enrolment: pre-primary schools (2004), 8,628; primary schools (2005), 141,089 (with 5,006 teachers); secondary schools (2005), 68,521 (with 4,141 teachers); special schools (2005), 1,007 (with 103 teachers); teacher training (2005), 713 (with 87 teachers); technical/vocational education (2005), 2,115 (1,048 teachers in 2004). There were 531 pre-primary schools, 719 primary schools, 162 secondary schools, 17 special schools, 4 teacher training schools and 63 technical/vocational schools in 2005.

The University of the South Pacific, which is located in Suva, serves 12 countries in the South Pacific region. The Fiji Islands also has a college of agriculture, school of medicine and nursing, an institute of technology, a primary school teacher training college and an advanced college of education.

In 2005 public expenditure on education came to 6·2% of GDP.

Health
In 2007 there were 25 public hospitals with 1,727 beds, two private hospitals, 76 health centres and 101 nursing stations. There were 318 doctors, 196 dental staff and 1,820 nurses.

Through its national health service system, the government continues to provide the bulk of health services both in the curative and public health programmes.

RELIGION
In 2001 the population consisted of 53% Christians, 38% Hindus, 8% Muslims and 1% others.

CULTURE

Press
In 2008 there were three national dailies with a combined circulation of 40,000.

Tourism
There were 542,000 foreign tourists in 2009 (excluding same-day visitors), down from 585,000 in 2008.

DIPLOMATIC REPRESENTATIVES
Of the Fiji Islands in the United Kingdom (34 Hyde Park Gate, London, SW7 5DN)
High Commissioner: Ratu Naivakarurubalavu Solo Mara.

Of the United Kingdom in the Fiji Islands (Victoria House, 47 Gladstone Rd, Suva)
Acting High Commissioner: Martin Fidler.

Of the Fiji Islands in the USA (2233 Wisconsin Ave., NW, Washington, D.C., 20007)
Ambassador: Winston Thompson.

Of the USA in the Fiji Islands (31 Loftus St., Suva)
Ambassador: Frankie A. Reed.

Of the Fiji Islands to the United Nations
Ambassador: Peter Thomson.

Of the Fiji Islands to the European Union
Ambassador: Peceli Vuniwaga Vocea.

FURTHER READING
Bureau of Statistics. *Annual Report; Current Economic Statistics.* Quarterly Reserve Bank of Fiji. *Quarterly Review*

Belshaw, Cyril S., *Under the Ivi Tree: Society and Economic Growth in Rural Fiji.* 2004
Kelly, John D. and Kaplan, Martha, *Represented Communities: Fiji and World Decolonization.* 2001
Lal, B. J., *Broken Waves: a History of the Fiji Islands in the Twentieth Century.* 1992
Robertson, Robert and Sutherland, William, *Government by the Gun: Fiji and the 2000 Coup.* 2002

National Statistical Office: Bureau of Statistics, POB 2221, Government Buildings, Suva.
Website: http://www.statsfiji.gov.fj

FINLAND

NORWAY

RUSSIA

SWEDEN

• Rovaniemi

Oulu •

FINLAND

Gulf of
Bothnia

• Tampere

Turku
Espoo ■
■ HELSINKI

RUSSIA

0   50 mi
0   100 km

© Research Machines plc 2006

**Suomen Tasavalta—Republiken Finland
(Republic of Finland)**

Capital: Helsinki
Population projection, 2015: 5·45m.
GNI per capita, 2011: (PPP$) 32,438
HDI/world rank: 0·882/22
Internet domain extension: .fi

KEY HISTORICAL EVENTS

Finland's first inhabitants moved northwards at the end of the Ice Age. Further waves of settlement came in 4000 BC and 1000 BC and social groups began to develop. During the Viking era Finland's location on the trade route between Russia and Sweden brought prosperity and conflict in equal measure, with attacks made on Finnish trading posts by the Swedes and the Danes.

In the 12th century economic and religious rivalry between Sweden and Russia centred on Finland. The defeat of Birger Jarl in 1240 marked the end of the Swedish incursions but efforts at strengthening the Swedish presence in areas it already held were intensified. By 1323 Russia was forced to recognize a boundary marking off those parts of Finland that were under Swedish control including all of western and southern Finland. Finland remained a duchy of Sweden until 1581.

In the 18th century Russian forces conquered the southeast territory. The rest of the country was ceded to Russia by the treaty of Hamina in 1809 when Finland became an autonomous grand duchy, retaining its laws and institutions but owing allegiance to the tsar.

Throughout the 19th century Finland remained in Russia's shadow while building on its status as a grand duchy. By the 1880s Finland had its own army. This proved too much for the Russian military, who feared that moves towards Finnish separatism would make more difficult their task of defending the long western border. With the appointment of Gen. Bobrikov as governor general in 1898 a start was made on bringing Finland back into the imperial fold. The army was put under Russian command, the Russian language was made compulsory for the civil service and for schools and decision-making reverted to the tsar's appointees. In June 1904 Bobrikov was assassinated and, as the Russian revolutionary movement gathered pace, the Marxists won an absolute majority in parliamentary elections. It was a short-lived victory but the far left held its popular appeal.

Civil War

Following the Russian Revolution, on 6 Dec. 1917 Finland declared independence. This was recognized by the Russian Bolsheviks on 31 Dec. By now, however, the left- and right-wing parties were in open conflict. In Jan. 1918 the Whites (the government forces) took the western, Russian-controlled province of Ostrobothnia while the Reds (supported by the Bolsheviks) seized power in the south. At the end of Jan. the Reds staged a coup and the Whites were forced to abandon Helsinki, relocating to Vaasa. But government forces, led by Gen. Gustaf Mannerheim and aided by German troops, prevailed.

A new constitution by which a German prince, Friedrich Karl, would become regent was abandoned with the collapse of Germany at the end of the First World War. In the summer of 1919 Finland became a republic with K. J. Ståhlberg as its first president.

Throughout the 1920s and early 1930s class antagonism inherited from the civil war remained the dominant political issue. As a conciliatory measure the Social Democrats were brought into government and the party formed a minority government in 1926–27. In the early 1930s fascism entered domestic politics with the emergence of the Lapua Movement. After an unsuccessful coup attempt in 1932 the movement was banned. Finland was hit by the pre-war depression but cushioned by the dominant role of agriculture in the Finnish economy.

Winter War

As Europe was anticipating German aggression, the Finns were more fearful of Moscow's territorial demands. Outnumbered and outmatched in arms and equipment, their hopes were pinned on foreign involvement. When this failed to materialize, there was no option but to concede all the Russians demanded, including the Karelian Isthmus. The 1940 treaty, which ended the Winter War, required the resettlement of 12% of the Finnish population. Fearing worse to come from the Soviet Union, Helsinki opened up contacts with the Germans, allowing transit for military traffic in return for food and armaments.

There followed the German invasion of Russia, a campaign which Mannerheim, justifying active participation, described as a 'holy war' to restore Finnish borders. Having achieved this objective, the Finns wanted out, a desire which became all the more determined as the German advance ground to a halt at Stalingrad. But there was no basis for a settlement and the Finns could only wait for the inevitable Russian counter-attack. When it came, retreating Germans took revenge by devastating everything in their path.

Having fought first against Russia then against Germany, the country emerged from the Second World War defeated, demoralized and in political disarray. The terms of the 1944 armistice included the surrender of one-twelfth of Finnish territory and reparations in goods valued at US$300m.

Revered as a national hero by the right, Mannerheim became president. Carl Enckell, a close associate for many years, took charge of foreign affairs while Juho Paasikivi was appointed premier. The 1945 election confirmed Paasikivi and Enckell in their jobs and a cabinet was formed giving roughly equal representation to the social democrats, communists and the farmers' party. When Mannerheim, who had turned 78 and ailing fast, was persuaded to stand down in mid-term of his presidency, Paasikivi was the obvious successor.

A peace treaty with Russia signed in Feb. 1947 confirmed what had already been agreed by the 1944 armistice. Finland lost 12% of her border territory to the Soviet Union, including the country's second largest city, Viipuri, and the port and province of Petsamo on the Arctic coast. With a large part of the province of Karelia taken over by the Russians the frontier was moved back from a distance of only 31 km from Leningrad to a new line 180 km from the former Russian capital. 400,000 people had to be resettled. The Åland Islands were to remain demilitarized and limitations were imposed on the size of the Finnish armed forces and its weaponry.

Pacifying Russia
An 'invitation' to negotiate a mutual assistance agreement suggested that only an administration answerable to Moscow would satisfy the Russians. Paasikivi argued that the interests of the Soviet Union on her northwestern border (the only part of Finland that really mattered to the Russian military) could best be served by a sovereign Finland whose sympathetic relations with her eastern neighbour precluded her territory being used as a platform for attack. Skilfully, this shifted the emphasis away from Russian ambitions for making Finland an ally towards the prospect of the two countries' entering into a joint security arrangement. Finland promised to defend herself against an attack from Germany or an allied state, to confer with Russia in case of war or threat of war and, if necessary, to accept Russian aid. Great play was made of Finland's ambition 'to remain outside the conflicting interests of the great powers'.

The popular view in Europe was that Finland was as much under the control of Moscow as any of the communist satellites. Paasikivi tried to counteract this impression by reacting decisively to any hint of a threat to his authority. When in 1948 there were rumours that the communists were planning to seize power, he dismissed the powerful minister of internal affairs, Yrjö Leino. But the president was hyper-sensitive to Moscow's needs for reassurances of Finnish good faith. The press was told to tone down criticism of the Soviet Union. Most of the nation's productive capacity had survived the war, and the export demand for wood products was strong. But paying off reparations meant a transfer of resources to the engineering industry. All this had to be achieved without a share in Marshall Aid, though US loans totalling US$150m. were channelled in other ways. With skilled labour in short supply, wages and prices climbed steeply. In 1948 prices were eight times their pre-war level.

In the summer of 1949 the communists disrupted industry with a series of strikes, splitting the trade union movement and raising fears of an imminent coup. Paasikivi promptly replaced the social democrat government with one formed by the agrarians under the leadership of Urho Kekkonen. Like Paasikivi, Kekkonen worked hard to establish good personal relations with the USSR. In 1952 he put up a plan for 'a neutral alliance between the Scandinavian countries', which 'would remove even the theoretical threat of an

attack ... via Finland's territory'. In reality, a Scandinavian alliance, neutral or otherwise, was impracticable since Denmark and Norway had only recently joined NATO. But the gain to Kekkonen was approval from the Soviet Union.

Defending Neutrality
In 1955, two years after the death of Stalin had brought the first signs of an easing in the cold war, Finland negotiated the return of the Porkkala base near Helsinki, which had been leased to the Soviet Union for 50 years. This meant the departure of the last Soviet troops on Finnish territory. That same year Finland joined the United Nations but stayed out of the latest formation of Soviet defence, the Warsaw Pact. In 1956 Kekkonen succeeded Paasikivi as president. A succession of weak governments consolidated presidential power and confirmed Kekkonen as the only leader capable of handling the Russians.

His first move was to assert his country's freedom of action, by suggesting that Finland might come to a deal with the European Community. In response, the Soviet Union activated article 2 of the 1948 Treaty by demanding consultation on measures to ensure the defence of their frontiers. It was the most serious challenge yet to Finnish neutrality. It had been said that article 2 could be acted upon only when both parties agreed that a threat existed; it came as a shock to realize that a unilateral declaration of interest by the stronger partner was sufficient to start the process of military consultation. If the Soviet claim went uncontested Finnish independence would be seen as a sham.

Western observers expected the worst; nothing less than military bases on Finnish soil would satisfy Moscow. But Kekkonen called successfully for a postponement of military talks in favour of discussions aimed at reassuring the Kremlin that Finland would remain true to her foreign policy. This was accompanied by a warning that if military consultations went ahead there would be a war scare in Scandinavia, possibly leading to counter-measures by the West. When the Soviet Union backed down Kekkonen was feted as the country's saviour. He was elected for a second six-year term by an overwhelming majority on the first round of voting.

In 1981 the ailing Kekkonen was replaced by Mauno Koivisto. At first he adopted the foreign policy of his predecessor but with the collapse of the Soviet Union at the end of the 1980s he was able to move Finland towards closer ties with Western Europe. Koivisto played a major role in dismantling the 1948 Treaty and in the early 1990s fostered close relations with the EU. A referendum held in 1995 paved the way for Finland to join the EU.

TERRITORY AND POPULATION

Finland, a country of lakes and forests, is bounded in the northwest and north by Norway, east by Russia, south by the Baltic Sea and west by the Gulf of Bothnia and Sweden. At the most recent ten-yearly census on 31 Dec. 2010 the population was 5,375,276. Finland has used a register-based method of calculating the population since 1990. The areas, populations and population densities of Finland and its regions on 31 Dec. 2011 (Swedish names in brackets) were as follows:

Regions	Area (sq. km)[1]	Population	Population per sq. km
Ahvenanmaa (Åland)	1,552	28,354	18·3
Etelä-Karjala (Södra Karelen)	5,613	133,311	23·8
Etelä-Pohjanmaa (Södra Österbotten)	13,444	193,735	14·4
Etelä-Savo (Södra Savolax)	13,977	153,738	11·0
Kainuu (Kajanaland)	21,501	81,298	3·8
Kanta-Häme (Egentliga Tavastland)	5,200	175,230	33·7
Keski-Pohjanmaa (Mellersta Österbotten)	5,019	68,484	13·6
Keski-Suomi (Mellersta Finland)	16,704	274,379	16·4
Kymenlaakso (Kymmenedalen)	5,148	181,829	35·3
Lappi (Lappland)	92,662	183,330	2·0
Päijät-Häme (Päijänne-Tavastland)	5,125	202,236	39·4

Regions	Area (sq. km)[1]	Population	Population per sq. km
Pirkanmaa (Birkaland)	12,446	491,472	39·5
Pohjanmaa (Österbotten)	7,750	179,106	23·1
Pohjois-Karjala (Norra Karelen)	17,763	165,906	9·3
Pohjois-Pohjanmaa (Norra Österbotten)	35,507	397,887	11·2
Pohjois-Savo (Norra Savolax)	16,768	248,130	14·8
Satakunta	7,957	226,567	28·5
Uusimaa (Nyland)	9,096	1,549,058	170·3
Varsinais-Suomi (Egentliga Finland)	10,661	467,217	43·8
Total	303,893	5,401,267	17·8

[1]Excluding inland water area which totals 34,534 sq. km.

The semi-autonomous province of the **Åland Islands** (Ahvenanmaa) occupies a special position as a demilitarized area and is 91% Swedish-speaking. **Åland** elects a 30-member parliament (*Lagting*), which in turn elects the provincial government (*Landskapsstyrelse*). It has a population of 28,354. The capital is Mariehamn (Maarianhamina).

The growth of Finland's population, which was 421,500 in 1750, has been:

End of year	Urban[1]	Semi-urban[2]	Rural	Total	Percentage urban
1800	46,600	—	786,100	832,700	5·6
1900	333,300	—	2,322,600	2,655,900	12·5
1950	1,302,400	—	2,727,400	4,029,800	32·3
1970	2,340,300	—	2,258,000	4,598,300	50·9
1980	2,865,100	—	1,922,700	4,787,800	59·8
1990	2,846,220	803,224	1,349,034	4,998,500	56·9
2000	3,167,668	898,860	1,114,587	5,181,115	61·1
2005	3,294,777	896,181	1,064,622	5,255,580	62·7
2006	3,327,207	913,614	1,036,134	5,276,955	63·1
2007	3,444,620	852,225	1,003,639	5,300,484	65·0
2008	3,616,471	837,892	871,951	5,326,314	67·9
2009	3,644,491	851,259	855,677	5,351,427	68·1

The classification urban/rural has been revised as follows: [1]Urban—at least 90% of the population lives in urban settlements, or in which the population of the largest settlement is at least 15,000. [2]Semi-urban—at least 60% but less than 90% live in urban settlements, or the population of the largest settlement is more than 4,000 but less than 15,000.

The population on 31 Dec. 2009 by language spoken: Finnish, 4,852,209; Swedish, 290,392; Sami, 1,789; other languages, 207,037.

The UN gives a projected population for 2015 of 5·45m.

The principal towns with resident population, 31 Dec. 2011, are (Swedish names in brackets):

Helsinki (Helsingfors)		Mikkeli (St Michel)	48,907
—capital	595,384	Porvoo (Borgå)	48,833
Espoo (Esbo)	252,439	Kokkola (Karleby)	46,585
Tampere (Tammerfors)	215,168	Hyvinkää (Hyvinge)	45,527
Vantaa (Vanda)	203,001	Nurmijärvi	40,349
Turku (Åbo)	178,630	Rauma (Raumo)	39,820
Oulu (Uleåborg)	143,909	Lohja (Lojo)	39,726
Jyväskylä	132,062	Järvenpää	38,966
Lahti	102,308	Kajaani (Kajana)	38,045
Kuopio	97,433	Tuusula (Tusby)	37,667
Kouvola	87,567	Kirkkonummi	
Pori (Björneborg)	83,133	(Kyrkslätt)	37,192
Joensuu	73,758	Kerava (Kervo)	34,549
Lappeenranta		Nokia	32,056
(Villmanstrand)	72,133	Kaarina (St Karins)	31,081
Hämeenlinna		Ylöjärvi	30,942
(Tavastehus)	67,270	Kangasala	29,891
Rovaniemi	60,637	Riihimäki	29,018
Vaasa (Vasa)	60,398	Raasepori (Raseborg)	28,959
Seinäjoki	58,703	Vihti (Vichtis)	28,581
Salo	55,283	Imatra	28,472
Kotka	54,831	Savonlinna (Nyslott)	27,585

In 2009, 68·1% of the population lived in urban areas. Nearly one-fifth of the total population lives in the Helsinki metropolitan region.

Finnish and Swedish are the official languages. Three Sami languages are spoken in Lapland.

SOCIAL STATISTICS

Statistics in calendar years:

	Living births	Of which outside marriage	Still-born	Marriages	Deaths (exclusive of still-born)	Emigration
2005	57,745	23,319	182	29,283	47,928	12,369
2006	58,840	23,858	193	28,236	48,065	12,107
2007	58,729	23,824	204	29,497	49,077	12,443
2008	59,530	24,246	189	31,014	49,094	13,657
2009	60,430	24,697	205	29,836	49,883	12,151

In 2009 the rate per 1,000 population was: births, 11; deaths, 9; marriages, 6; infant deaths (per 1,000 live births), 2·6. Annual population growth rate, 1999–2009, 0·3%. In 2008 the suicide rate per 100,000 population was 30·7 among men and 8·5 among women, giving Finland one of the highest suicide rates in Europe. Life expectancy at birth, 2008, 76·3 years for males and 83·0 years for females. The most popular age range for marrying in 2008 was 25–29 for both males and females. Fertility rate, 2009, 1·9 births per woman. In 2008 Finland received 4,305 asylum applications, equivalent to 0·8 per 1,000 inhabitants.

A UNICEF report published in 2010 showed that 5·3% of children in Finland live in relative poverty (living in a household in which disposable income—when adjusted for family size and composition—is less than 50% of the national median income), the second lowest percentage of any country behind Iceland.

CLIMATE

A quarter of Finland lies north of the Arctic Circle. The climate is severe in winter, which lasts about six months, but mean temperatures in the south and southwest are less harsh, 21°F (−6°C). In the north, mean temperatures may fall to 8·5°F (−13°C). Snow covers the ground for three months in the south and for over six months in the far north. Summers are short but quite warm, with occasional very hot days. Precipitation is light throughout the country, with one third falling as snow, the remainder mainly as rain in summer and autumn. Helsinki (Helsingfors), Jan. 30·2°F (−1·0°C), July 68·4°F (20·2°C). Annual rainfall 27·9" (708·7 mm).

CONSTITUTION AND GOVERNMENT

Finland is a republic governed by the constitution of 1 March 2000 (which replaced the previous constitution dating from 1919). Although the president used to choose who formed the government, under the new constitution it is the responsibility of parliament to select the prime minister. The government is in charge of domestic and EU affairs with the president responsible for foreign policy 'in co-operation with the government'.

Parliament consists of one chamber (*Eduskunta*) of 200 members chosen by direct and proportional election by all citizens of 18 or over. The country is divided into 15 electoral districts, with a representation proportional to their population. Every citizen over the age of 18 is eligible for parliament, which is elected for four years, but can be dissolved sooner by the president.

The *president* is elected for six years by direct popular vote. In the event of no candidate winning an absolute majority, a second round is held between the two most successful candidates.

National Anthem

'Maamme'/'Vårt land' ('Our land'); words by J. L. Runeberg, tune by F. Pacius (same as Estonia).

GOVERNMENT CHRONOLOGY

(KESK = Centre Party; KOK = National Coalition Party; ML = Agrarian League; SDP = Social Democratic Party; SFP = Swedish People's Party; SKDL = Finnish People's Democratic League; VL = Liberal League; n/p = non-partisan)

Presidents of the Republic

1944–46	military	Carl Gustaf Emil Mannerheim
1946–56	KOK	Juho Kusti Paasikivi
1956–82	ML/KESK	Urho Kaleva Kekkonen
1982–94	SDP	Mauno Henrik Koivisto
1994–2000	SDP	Martti Oiva Kalevi Ahtisaari
2000–12	SDP	Tarja Kaarina Halonen
2012–	KOK	Sauli Väinämö Niinistö

Prime Ministers

1944–46	KOK	Juho Kusti Paasikivi
1946–48	SKDL	Mauno Pekkala
1948–50	SDP	Karl-August Fagerholm
1950–53	ML	Urho Kaleva Kekkonen
1953–54	VL	Sakari Severi Tuomioja
1954	SFP	Ralf Johan Gustaf Törngren
1954–56	ML	Urho Kaleva Kekkonen
1956–57	SDP	Karl-August Fagerholm
1957	ML	Väinö Johannes Sukselainen
1957–58	n/p	Berndt Rainer von Fieandt
1958	n/p	Reino Iisakki Kuuskoski
1958–59	SDP	Karl-August Fagerholm
1959–61	ML	Väinö Johannes Sukselainen
1961–62	ML	Martti Juhani Miettunen
1962–63	ML	Ahti Kalle Samuli Karjalainen
1963–64	n/p	Reino Ragnar Lehto
1964–66	ML/KESK	Johannes Virolainen
1966–68	SDP	Kustaa Rafael Paasio
1968–1970	SDP	Mauno Henrik Koivisto
1970	n/p	Teuvo Ensio Aura
1970–71	KESK	Ahti Kalle Samuli Karjalainen
1971–72	n/p	Teuvo Ensio Aura
1972	SDP	Kustaa Rafael Paasio
1972–75	SDP	Taisto Kalevi Sorsa
1975	n/p	Keijo Antero Liinamaa
1975–77	KESK	Martti Juhani Miettunen
1977–79	SDP	Taisto Kalevi Sorsa
1979–81	SDP	Mauno Henrik Koivisto
1982–87	SDP	Taisto Kalevi Sorsa
1987–91	KOK	Harri Hermanni Holkeri
1991–95	KESK	Esko Tapani Aho
1995–2003	SDP	Paavo Tapio Lipponen
2003	KESK	Anneli Tuulikki Jäätteenmäki
2003–10	KESK	Matti Taneli Vanhanen
2010–11	KESK	Mari Johanna Kiviniemi
2011–	KOK	Jyrki Tapani Katainen

RECENT ELECTIONS

In the first round of presidential elections held on 22 Jan. 2012 Sauli Niinistö of the National Coalition Party (KOK) came first with 37·0% of the vote, followed by Pekka Haavisto (Green League) with 18·8%, Paavo Väyrynen (Centre Party) 17·5% and Timo Soini (True Finns) 9·4%. There were four other candidates. Turnout was 72·8%. In the run-off held on 5 Feb. 2012 Sauli Niinistö won 62·6% of the vote against 37·4% for Pekka Haavisto. Turnout was 68·8%.

At the elections for the 200-member parliament on 17 April 2011, turnout was 70·5%. The KOK won 44 seats with 20·4% of the votes cast (50 seats in 2007), the Social Democratic Party (SDP) 42 with 19·1% (45 seats in 2007), the True Finns (PS) 39 with 19·1% (5 seats in 2007), the Centre Party (KESK) 35 with 15·8% (51), the Left Alliance (VAS) 14 with 8·1% (17), the Green League (VIHR) 10 with 7·3% (15), the Swedish People's Party (SFP) 9 with 4·3% (9) and the Christian Democrats (KD) 6 with 4·0% (7). One representative from the province of Åland was also elected. Following the April 2011 election 42·5% of the seats in parliament were held by women.

European Parliament

Finland has 13 (14 in 2004) representatives. At the June 2009 elections turnout was 40·3% (39·4% in 2004). KOK won 3 seats with 23·2% of votes cast (political affiliation in European Parliament: European People's Party); KESK, 3 with 19·0% (Alliance of Liberals and Democrats for Europe); SDP, 2 with 17·5% (Progressive Alliance of Socialists and Democrats); KD–PS, 2 with 14·0% (one with European People's Party and one with Europe of Freedom and Democracy); VIHR, 2 with 12·4% (Greens/European Free Alliance); SFP, 1 with 6·1% (Alliance of Liberals and Democrats for Europe).

CURRENT GOVERNMENT

President: Sauli Niinistö; b. 1948 (National Coalition Party; sworn in 1 March 2012).

Following the elections of April 2011 a six-party coalition was formed consisting of the National Coalition Party (KOK), the Social Democratic Party (SDP), the Left Alliance (VAS), the Green League (VIHR), the Swedish People's Party (SFP) and the Christian Democrats (KD). In March 2013 it comprised:

Prime Minister: Jyrki Katainen; b. 1971 (KOK; sworn in 22 June 2011).

Deputy Prime Minister and Minister of Finance: Jutta Urpilainen (SDP). *Foreign Affairs:* Erkki Tuomioja (SDP). *European Affairs and Foreign Trade:* Alexander Stubb (KOK). *International Development:* Heidi Hautala (VIHR). *Justice:* Anna-Maja Henriksson (SFP). *Interior:* Päivi Räsänen (KD). *Defence:* Carl Haglund (SFP). *Public Administration and Local Government:* Henna Virkkunen (KOK). *Education and Science:* Jukka Gustafsson (SDP). *Culture and Sport:* Paavo Arhinmäki (VAS). *Agriculture and Forestry:* Jari Koskinen (KOK). *Transport:* Merja Kyllönen (VAS). *Economic Affairs:* Jan Vapaavuori (KOK). *Labour:* Lauri Ihalainen (SDP). *Social Affairs and Health:* Paula Risikko (KOK). *Health and Social Services:* Maria Guzenina-Richardson (SDP). *Environment:* Ville Niinistö (VIHR). *Housing and Communications:* Krista Kiuru (SDP).

The *Speaker* is Eero Heinäluoma.

Government Website: http://www.valtioneuvosto.fi

CURRENT LEADERS

Sauli Niinistö

Position
President

Introduction
Sauli Niinistö became president in March 2012, the first member of the conservative National Coalition Party (KOK) to hold the post since 1956. A fiscal conservative and pro-European, he is regarded as a pragmatist. He intends to maintain Finland's membership stance within the European Union, advocating restraint on financial bailout packages to partner countries and on further expansion.

Early Life
Born on 24 Aug. 1948 in Salo, Sauli Niinistö graduated in law from the University of Turku in 1974 before establishing his own law firm in Salo. In 1977 he was elected to Salo's municipal council, where he served until 1992. In 1987 he entered the national parliament as a member of the KOK, becoming chairman of the committee on constitutional law in 1993. He was elected leader of the KOK in 1994.

In the coalition government led by social democrat Paavo Lipponen, he served as justice minister from 1995–96 and as finance minister from 1996–2003. Between 1995 and 2001 he was also deputy prime minister. As finance minister he tackled recession by cutting social spending and taxes, and took Finland into the single European currency in 1999. He was chair of the European Democratic Union (EDU) grouping from 1998 and became honorary president of the European's People Party (EPP), following its merger with the EDU in 2002.

After standing down as KOK leader in 2001, Niinistö left parliament in 2003 to become vice-chairman of the board of directors of the European Investment Bank, a post he held until 2007. In 2006 he unsuccessfully challenged for the Finnish presidency, narrowly losing to incumbent Tarja Halonen. He re-entered parliament in 2007 and served as speaker from 2007–11, supporting measures for administrative reform.

In the wake of the economic turmoil in Europe amid the global financial crisis, he advocated continuing Finnish membership of the EU but was sceptical on further expansion and critical of the financial bailouts proposed for struggling member states. In 2011 he again contested the presidency, arguing for fiscal discipline and measures to help young people into employment. He was elected in the second round, winning 63% of the vote to beat Pekka Haavisto of the Green League.

Career in Office
Niinistö took office on 1 March 2012, promising to use the presidency to consolidate Finland's place within the EU while strengthening relations with the USA and China. His chief domestic challenges arise from economic uncertainty, a growing youth unemployment rate and an ageing population. Internationally, he must respond to the continuing crises in the eurozone and aims to maintain good relations with Russia.

Jyrki Katainen

Position
Prime Minister

Introduction
Jyrki Katainen became prime minister in June 2011, heading a six-party coalition. Leader of the centre-right National Coalition Party (KOK) and a former finance minister, he is a fiscal conservative who, though broadly pro-European, has signalled caution on further EU financial bailouts for struggling member states.

Early Life
Born on 14 Oct. 1971 in Siilinjärvi, Jyrki Katainen studied social sciences at Tampere University. In 1993 he was elected to the Siilinjärvi municipal council as a KOK member and served as the council's second vice-chair from 1997–98. In 1998 he obtained a masters degree in social sciences and spent the 1990s working as a supply teacher and at the National Education Association.

In 1997 he became a member of the regional council of Northern Savo, serving as its first vice-chair from 2001–04. In 1999 he was elected to parliament for the KOK. He was vice-president of the European People's Party (EPP) youth organization from 1998–2000 and in 2005 was elected vice-president of the EPP as a whole. During the same period he rose within the KOK to become its deputy chair in 2001, before successfully challenging Ville Itälä for the leadership in 2004.

He served on the foreign affairs committee from 2004–07 in the government of Matti Vanhanen. Following the KOK's entry into coalition government with the Centre Party in 2007, Katainen was appointed finance minister and deputy prime minister. In the aftermath of the 2008 global economic crisis, he won praise for controlling Finland's deficit and retaining its top-grade international credit rating. He continued to embrace fiscal discipline during recession in 2009. In 2010 he called for stronger IMF surveillance of financial systems. In April 2011 the KOK emerged from general elections as the largest party and, after complex negotiations, Katainen formed a coalition government with the centre-left Social Democratic Party and four smaller parties from across the political spectrum.

Career in Office
Katainen took office on 22 June 2011, pledging to rebuild the economy while limiting government borrowing. In late 2011 he signalled that Finland might require guarantees or stricter conditions before agreeing to further EU bailout measures. His chief challenges remain economic uncertainty at home and abroad, which he must confront while maintaining unity across his coalition in the face of strong opposition from the nationalist True Finns party.

DEFENCE

Conscript service is 6–12 months. Total strength of trained and equipped reserves is about 490,000 (to be 350,000).

In 2007 defence expenditure totalled €2,203m. (€416 per capita), representing 1·2% of GDP.

Army
The Army consists of one armoured training brigade, three readiness brigades, three infantry training brigades, three jaeger regiments, one artillery brigade, three brigade artillery regiments, two air defence regiments, one engineer regiment (including ABC school), three brigade engineer battalions, one signals regiment, four brigade signals battalions and a reserve officer school. Total strength of 37,700 (26,000 conscripts).

Border Guard
This comes under the purview of the Ministry of the Interior, but is militarily organized to participate in the defence of the country. It is in charge of border surveillance and border controls. It is also responsible for conducting maritime search and rescue operations. The Border Guard's mobilization force can be utilized for border security tasks if needed. If necessary in the interests of defence capability, the border troops or parts thereof may be attached to the Defence Forces. Personnel, 2011, 2,800 (professional) with a potential mobilizational force of 12,000.

Navy
The organization of the Navy was changed on 1 July 1998. The Coastal Defence, comprising the coast artillery and naval infantry, was merged into the Navy.

About 50% of the combatant units are kept manned, with the others on short-notice reserve and reactivated on a regular basis. Naval bases exist at Upinniemi (near Helsinki), Turku and Kotka. Naval Infantry mobile troops are trained at Tammisaari. Total personnel strength (2006) was 6,600, of whom 4,300 were conscripts.

Air Force
Personnel (2006), 4,600 (1,500 conscripts). Equipment included 62 F-18 Hornets.

ECONOMY

Agriculture accounted for 3% of GDP in 2009, industry 28% and services 69%.

According to the Berlin-based organization *Transparency International*, Finland ranked equal first in a 2012 survey of countries with the least corruption in business and government. It received 90 out of 100 in the corruption perceptions index.

Overview
The economy, once based on basic metals and forestry, has evolved to become a leading force in knowledge-based, high-tech production. It is one of the world's leading information and communications technology (ICT) producers. The country's legal framework provides effective protection of intellectual property while open market policies encourage entrepreneurship and innovation.

In 2005–06 Finland was ranked first in the World Economic Forum's *Global Competitiveness Report* although it had dropped to seventh by 2010–11. The emergence of venture capital financing in the 1990s created opportunities for high-risk technology start-ups. Expenditure on research and development (R&D) made up 3·9% of GDP in 2010, one of the highest levels in the world.

Finland joined the EU in 1995 and subsequently adopted the euro. Growth prior to the 2008 financial crisis was well above the eurozone average but the country experienced the worst recession of any euro nation in 2009. Exports, on which the country is heavily reliant, declined by 20% and there was a sharp drop in investment. GDP fell by 8·5% in 2009 and unemployment rose from 6·5% to over 8%. In 2011 unemployment declined to 7·8% and in 2012 exports grew by 2·4%. Although the economy grew overall in 2010 and 2011, a fall in GDP in the second quarter of 2012 sparked fears of a return to recession. Foreign investment dropped by 4% in 2012, while debt as a percentage of GDP in 2011 was 49·2%.

With a fiscal policy designed to support growth, the budget position has deteriorated. A rapidly ageing population presents long-term problems.

Currency

On 1 Jan. 1999 the euro (EUR) became the legal currency in Finland at the irrevocable conversion rate of 5·94573 marks to one euro. The euro, which consists of 100 cents, has been in circulation since 1 Jan. 2002. On the introduction of the euro there was a 'dual circulation' period before the mark ceased to be legal tender on 28 Feb. 2002. Euro banknotes in circulation on 31 Oct. 2010 had a total value of €10·2bn.

Inflation rates (based on OECD statistics):

2002	2003	2004	2005	2006	2007	2008	2009	2010	2011
2·0%	1·3%	0·1%	0·8%	1·3%	1·6%	3·9%	1·6%	1·7%	3·3%

Foreign exchange reserves were US$7,002m. in Sept. 2009 and gold reserves were 1·58m. troy oz. Total money supply was €62,585m. in Aug. 2009.

Budget

Revenue and expenditure for the calendar years 2005–09 in €1m:

	2005	2006	2007	2008	2009
Revenue	39,023	40,979	43,212	44,292	47,798
Expenditure	41,247	40,871	43,252	44,923	46,897

Of the total revenue in 2009, 29% derived from value added tax, 22% from income and property tax, 22% from net loans, 11% from excise duties, 4% from other taxes and similar revenue and 12% from miscellaneous sources. Of the total expenditure, 2009, 21% went to health and social security, 13% to education, 6% to agriculture and forestry, 6% to defence, 5% to transport and 49% to other expenditure.

VAT is 24% (reduced rates, 14% and 10%).

At the end of Dec. 2009 the central government debt totalled €64,281m. Foreign debt amounted to €45m. at the end of 2006.

Finland's budget deficit in 2011 was 0·5% of GDP (2·5% in both 2010 and 2009). In 2008 it had a surplus of 4·3%. The required target set by the EU is a budget deficit of no more than 3%.

Performance

GDP growth rates (based on OECD statistics):

2002	2003	2004	2005	2006	2007	2008	2009	2010	2011
1·8%	2·0%	4·1%	2·9%	4·4%	5·3%	0·3%	−8·5%	3·3%	2·7%

Total GDP was US$263·0bn. in 2011.

Finland was ranked third in the Global Competitiveness Index in the World Economic Forum's *Global Competitiveness Report 2012–2013*. The index analyses 12 areas of competitiveness for over 100 countries including macroeconomy, higher education and training, institutions, innovation and infrastructure.

Banking and Finance

The central bank is the Bank of Finland (founded in 1811), operating under the guarantee and supervision of parliament. The Bank is a member of the European System of Central Banks. As a member of the euro area, the Bank issues euro banknotes and coins in Finland by permission of the European Central Bank. The *Governor* is Erkki Liikanen.

The most important groups of banking institutions in 2009 were:

	Number of institutions	Number of branches	Deposits (€1m.)	Loans (€1m.)
Commercial banks	15	530	71,030	100,583
Savings banks	35	166	5,744	5,326
Co-operative banks	263	418	33,611	35,705
Foreign banks	15	57	—	—

The three largest banks are Nordea Bank Finland, Pohjola Bank and Sampo Bank. In 2009, 72% of the population between the ages of 16 and 74 were using online banking (one of the highest percentages of any country).

Gross external debt amounted to US$594,188m. in June 2012.

In 2010 Finland received US$4·3bn. worth of foreign direct investment (down from a record US$12·4bn. in 2007).

There is a stock exchange in Helsinki.

ENERGY AND NATURAL RESOURCES

In 2008, 30·5% of energy consumption came from renewables (wind power, solar power, hydro-electric power, tidal power, geothermal energy and biomass), compared to the European Union average of 10·3%. A target of 38% has been set by the EU for 2020.

Environment

Finland's carbon dioxide emissions in 2009 were the equivalent of 9·7 tonnes per capita. An *Environmental Performance Index* compiled in 2008 ranked Finland fourth in the world, with 91·4%. The index examined various factors in six areas—air pollution, biodiversity and habitat, climate change, environmental health, productive natural resources and water resources.

Electricity

Installed capacity was 17·0m. kW at the beginning of 2009. Production was 68,711m. kWh. in 2009 (18% hydro-electric). Consumption per capita in 2008 was an estimated 16,420 kWh. In 2009 there were four nuclear reactors, which contributed 33% of production. In May 2002 parliament approved the construction of a fifth reactor, on the island of Olkiluoto, which was scheduled to become operational in 2009. However, a number of problems have now delayed the start of electricity production until 2016.

Water

Finland has abundant surface water and groundwater resources relative to its population and level of consumption. The total groundwater yield is estimated to be 10–30m. cu. metres a day, of which some 6m. is suitable for water supplies. Approximately 15% of this latter figure is made use of at the present time.

Minerals

Notable of the mines are Pyhäsalmi (zinc–copper), Pahtavaara (gold ore), Hitura (nickel) and Keminmaa (chromium). In 2008 the metal content (in tonnes) of the output of zinc ore was 27,800; of copper ore, 13,300; of nickel ore, 4,000; of chromium ore and concentrate, 614,500.

Agriculture

The cultivated area covers only 8% of the land, and of the economically active population 4·8% were employed in agriculture and forestry in 2009. In 2008 there were 2·28m. ha. of arable land. This arable area was divided into 65,802 farms (including 411 farms with under one hectare of arable land). The distribution of this area by the size of the farms was: less than 5 ha. cultivated, 5,401 farms; 5–20 ha., 22,750 farms; 20–50 ha., 23,379 farms; 50–100 ha., 10,952 farms; over 100 ha., 3,320 farms.

Agriculture accounted for 0·9% of exports and 2·8% of imports in 2009.

The principal crops (area in 1,000 ha., yield in 1,000 tonnes) were in 2009:

Crop	Area	Yield
Barley	600·7	2,171·0
Oats	342·6	1,114·7
Wheat	218·3	887·0
Potatoes	26·4	755·3
Hay	86·1	289·7

The total area under cultivation in 2008 was 2,275,282 ha. Approximately 6·6% of all agricultural land is used for organic farming. Production of dairy butter in 2007 was 48,283 tonnes; and of cheese, 94,156 tonnes.

Livestock (2008): pigs, 1,483,000; cattle, 915,000; reindeer, 193,000; horses, 69,000 (including trotting and riding horses, and ponies); poultry, 4,056,000.

Forestry
Forests covered 22·16m. ha. in 2010, or 73% of the total land area. The productive forest land covered 20·1m. ha. in 2008. Timber production in 2009 was 41·4m. cu. metres. Finland is one of the largest producers of roundwood in Europe. Finland's per capita consumption of roundwood is the highest in the world, at 13·7 cu. metres per person in 2008.

Fisheries
The catch in 2007 was 122,355 tonnes, of which 117,857 tonnes came from sea fishing. In 2007 there were 201 food fish production farms in operation, of which 61 were freshwater farms. Their total production amounted to 13,031 tonnes. In addition there were 108 fry-farms and 235 natural food rearers, most of these in fresh water.

INDUSTRY
The leading companies by market capitalization in Finland in March 2012 were: Fortum, an energy company (US$21·5bn.); Nokia, a leading mobile phone producer (US$20·4bn.); and Sampo, an insurance company (US$16·1bn.).

Forests are still Finland's most crucial raw material resource, although the metal and engineering industry has long been Finland's leading branch of manufacturing, both in terms of value added and as an employer. In 2008 there were 28,339 establishments in industry (of which 25,799 were manufacturing concerns) with 420,241 personnel (of whom 400,421 were in manufacturing). Gross value of industrial production in 2008 was €141,417m., of which manufacturing accounted for €132,235m.

Labour
In 2009 the labour force was 2,678,000 (51% males). Of this total, 71·5% of the economically active population worked in services (including 15·9% in trade and restaurants) and 15·4% in manufacturing. In Dec. 2012 unemployment was 7·7%.

INTERNATIONAL TRADE
At the start of the 1990s a collapse in trade with Russia led to the worst recession in the country's recent history. Today, exports to Russia are about 9% of the total.

Imports and Exports
In 1960 the wood and paper industry dominated exports with 69% of the total, but today the electronics industry/metal and engineering industry sector is the largest export sector.

Imports and exports for calendar years, in €1m.:

	2006	2007	2008	2009
Imports	55,253	59,616	62,402	43,654
Exports	61,489	65,688	65,580	45,063

Use of Goods	Imports 2009
Intermediate goods	33%
Investment goods	23%
Energy	17%
Durable consumer goods	8%
Other	19%

Industry	Exports 2009
Metal, engineering, electronics	47%
Forest industry	19%
Chemical industry	15%
Other	19%

Region	Imports 2009	Exports 2009
European Union	56%	56%
Other Europe	21%	16%
Developing countries	16%	17%
Other countries	7%	11%

Trade with principal partners in 2009 was as follows (in €1m.):

	Imports	Exports		Imports	Exports
Belgium	1,023	1,231	Netherlands	1,994	2,645
Brazil	466	599	Norway	1,007	1,345
China	3,475	1,857	Poland	897	1,428
Denmark	1,177	883	Russia	7,035	4,028
Estonia	989	1,025	South Korea	507	553
France	1,964	1,659	Spain	549	1,025
Germany	6,400	4,653	Sweden	4,271	4,403
Italy	1,214	1,365	UK	1,452	2,358
Japan	859	740	USA	1,502	3,518

COMMUNICATIONS
Roads
At 1 Jan. 2010 there were 78,161 km of public roads, of which 50,987 km were paved. At the end of 2009 there were 3,246,414 registered cars, 111,267 lorries, 332,645 vans and pick-ups, 13,017 buses and coaches and 12,821 special automobiles. Road accidents caused 279 fatalities in 2009.

Rail
In 2009 the total length of the line operated was 5,919 km (3,067 km electrified), all of it owned by the State. The gauge is 1,524 mm. In 2009, 67·6m. passengers and 32·9m. tonnes of freight were carried. There is a metro (21 km) and tram/light rail network (85 km) in Helsinki although extensions to both systems are scheduled to open between 2013–15.

Civil Aviation
The main international airport is at Helsinki (Vantaa), and there are also international airports at Turku, Tampere, Rovaniemi and Oulu. The national carrier is Finnair. Scheduled traffic of Finnish airlines covered 178m. km in 2007. The number of passengers was 11·0m. and the number of passenger-km 22,704m.; the air transport of freight and mail amounted to 508·6m. tonne-km. Helsinki-Vantaa handled 12,611,187 passengers in 2009 (10,238,302 on international flights) and 122,107 tonnes of freight and mail. Oulu is the second busiest airport, handling 688,860 passengers in 2009, and Tampere-Pirkkala the third busiest, with 628,105 in 2009.

Shipping
The total registered mercantile marine in 2009 was 644 vessels of 1,534,000 GRT. In 2009 the total number of vessels arriving in Finland from abroad was 30,238 and the goods discharged amounted to 45·1m. tonnes. The goods loaded for export from Finnish ports amounted to 37·5m. tonnes.

The lakes, rivers and canals are navigable for about 9,747 km. Timber floating is still practised; in 2010 bundle floating was about 0·5m. tonnes.

Finland was ranked fourth in the World Economic Forum's *Global Competitiveness Report 2009–2010* for the quality of its port facilities.

Telecommunications

In 2008 there were 1,650,000 main (fixed) telephone lines. In the same year mobile phone subscribers numbered 6,830,000 (1,287·6 per 1,000 persons). In Aug. 2010 around 99% of Finnish households owned at least one mobile phone. In Nov. 2008 approximately 69% of Finnish households only had a mobile phone and did not have a fixed-line phone at all. Finland has the lowest rates in Europe for both fixed and mobile phone calls.

There were 4·4m. internet users in 2008. In Dec. 2010 there were 84·8 wireless broadband subscribers per 100 inhabitants and 28·6 fixed broadband subscribers per 100. In March 2012 there were 2·1m. Facebook users.

According to the World Economic Forum's *Global Information Technology Report 2010–11* Finland is ranked third in the world in exploiting global information technology developments.

SOCIAL INSTITUTIONS

Justice

The lowest court of justice is the District Court. In most civil cases a District Court has a quorum of three legally qualified members. In criminal cases as well as in some cases related to family law the District Court has a quorum with a chair and three lay judges. In the preliminary preparation of a civil case and in a criminal case concerning a minor offence, a District Court is composed of the chair only. From the District Court an appeal lies to the courts of appeal in Turku, Vaasa, Kuopio, Helsinki, Kouvola and Rovaniemi. The Supreme Court sits in Helsinki. Appeals from the decisions of administrative authorities are in the final instance decided by the Supreme Administrative Court, also in Helsinki. Judges can be removed only by judicial sentence. Two functionaries, the Chancellor of Justice and the Ombudsman or Solicitor-General, exercise control over the administration of justice. The former acts also as counsel and public prosecutor for the government; the latter is appointed by Parliament.

At the end of 2008 the daily average number of prisoners was 3,526 of which 232 were women. The number of convictions in 2007 was 323,751, of which 26,671 carried a penalty of imprisonment. 10,423 of the prison sentences were unconditional.

Education

Number of institutions, teachers and students (2008):

Primary and Secondary Education

	Number of institutions	Teachers[1]	Students
First-level Education (Lower sections of the comprehensive schools, grades I–VI)			364,223[2]
Second-level Education General education (Upper sections of the comprehensive schools, grades VII–IX, and upper secondary general schools)	3,580	53,157	311,078
Vocational Education	219[3]	17,048[4]	275,376

[1]Data for teachers refers to 2007. [2]Including pre-primary education (13,128 pupils) in comprehensive schools. [3]Numbers of institutions for vocational education refer to secondary and tertiary education. [4]Number of teachers for vocational education refer to secondary and tertiary education.

Tertiary Education

Vocational education at tertiary education level was provided for 122 students in 2008. In 2008 polytechnic education was provided at 28 polytechnics with 132,501 students and 5,971 teachers

(2007). In 2008, 26·9% of the population aged 15 years or over had been through tertiary education.

University Education

Universities with the number of teachers and students in 2008:

	Founded[1]	Teachers	Students Total	Women
Universities				
Helsinki	1640	1,638	35,216	22,605
Turku (Swedish)	1918	331	6,119	3,636
Turku (Finnish)	1922	774	15,483	9,740
Tampere	1925	550	14,626	9,553
Jyväskylä	1934	722	12,896	8,069
Oulu	1958	740	15,254	7,328
Vaasa	1968	165	4,394	2,301
Joensuu	1969	374	7,465	4,586
Kuopio	1972	342	5,572	3,727
Lapland	1979	202	4,409	3,092
Universities of Technology				
Helsinki	1849	531	14,282	3,161
Tampere	1965	353	11,381	2,444
Lappeenranta	1969	181	5,579	1,640
Schools of Economics and Business Administration				
Helsinki (Swedish)	1909	100	2,127	947
Helsinki (Finnish)	1911	164	3,183	1,450
Turku (Finnish)	1950	125	2,258	1,084
Universities of Art				
Academy of Fine Arts	1848	35	246	137
University of Art and Design	1871	172	1,931	1,216
Sibelius Academy	1882	232	1,259	733
Theatre Academy	1943	54	388	227
Total		7,785	164,068	87,676

[1]Year when the institution was founded regardless of status at the time.

Adult Education

Adult education provided by educational institutions in 2008:

Type of institution	Participants[1]
General education institutions[2]	1,681,300
Vocational education institutions	411,900
Permanent polytechnics	76,700
Universities[3]	160,300
Summer universities	97,800
	2,428,000

[1]Participants are persons who have attended adult education courses run by educational institutions in the course of the calendar year. The same person may have attended a number of different courses and has been recorded as a participant in each one of them. [2]Including study centres. [3]Adult education at continuing education centres of universities.

In 2007 public expenditure on education came to 5·9% of GDP and 12·5% of total government spending.

The adult literacy rate in 2008 was almost 100%.

According to the OECD's 2009 PISA (Programme for International Student Assessment) study, 15-year-olds in Finland rank first in science and second in mathematics and reading among OECD countries. The three-yearly study compares educational achievement of pupils in the major industrialized countries. Although education is only compulsory until 16, 93% of pupils stay on at school to 18.

Health

In 2008 there were 19,114 physicians, 4,718 dentists and 30,097 hospital beds. In 2007, 26% of males and 17% of females aged 15 and over smoked on a daily basis. The average Finn aged 15 or over drank 10·5 litres of alcohol in 2007.

In 2008 Finland spent 8·4% of its GDP on health.

Welfare

The Social Insurance Institution administers old-age pensions (to all persons over 65 years of age and disabled younger persons)

and health insurance. There is also a system of special assistance for resident immigrants over 65. The universal old-age pension paid between €11·38 and €510·80 per month in 2006. An additional system of compulsory old-age pensions paid for by employers is in force and works through the Central Pension Security Institute. Reforms of 2005 provided for pensions to be linked to life expectancy, early retirement to be phased out and the pension age to rise to 63. Incentives to encourage workers to carry on after retirement age are a bid to counter the problems posed by an ageing population and a shortage of younger workers. Pensioners are predicted to account for 25% of the population by 2020. Systems for other public aid are administered by the communes and supervised by the National Social Board and the Ministry of Social Affairs and Health.

The total cost of social security amounted to €48,572m. in 2008. Of this €22,248m. (46%) was spent on old age and disability, €12,654m. (26%) on health, €7,005m. (14%) on family allowances and child welfare, €3,346m. (7%) on unemployment and €3,317m. (7%) on general welfare purposes and administration. Out of the total expenditure, 38·4% was financed by employers, 25·1% by the State, 18·6% by local authorities, 11·2% by the insured and 6·7% by property income.

RELIGION

Liberty of conscience is guaranteed to members of all religions. National churches are the Lutheran National Church and the Greek Orthodox Church of Finland. The Lutheran Church is divided into nine dioceses (Turku being the archiepiscopal see) and some 460 parishes. The Greek Orthodox Church is divided into three bishoprics (Kuopio being the archiepiscopal see) and 27 parishes, in addition to which there are a monastery and a convent. Percentage of the total population at the end of 2009: Lutherans, 79·9; Greek Orthodox, 1·1; others, 1·3; not members of any religion, 17·7.

CULTURE

World Heritage Sites

There are seven UNESCO sites in Finland: Old Rauma harbour (inscribed in 1991); the sea fortress of Suomenlinna (1991); the old church of Petäjävesi (1994); Verla groundwood and board mill (1996); the Bronze Age burial site of Sammallahdenmäki (1999); the Kvarken Archipelago and High Coast (2000 and 2006), shared with Sweden; and the Struve Geodetic Arc (2005). The Arc is a chain of survey triangulations spanning from Norway to the Black Sea that helped establish the exact shape and size of the earth and is shared with nine other countries.

Press

Finland has 53 newspapers that are published four to seven times a week, nine of which are in Swedish, and 151 with one to three issues per week. The total circulation of all newspapers is 3·2m. There are 4,801 registered periodicals with a total circulation of over 15m. The bestselling newspapers in 2008 were *Helsingin Sanomat* (average daily circulation, 412,421 copies), *Ilta-Sanomat* (161,615) and *Aamulehti* (139,130). In 2008 a total of 13,419 book titles were published.

Tourism

There were 2,220,267 foreign tourists in 2009; the income from tourism was €2,022m.

Major international tourist attractions include Uspensky Cathedral, Helsinki Cathedral and Suomenlinna (all in Helsinki). Helsinki's churches and Santa Park in Rovaniemi are particularly popular among foreigners, who account for the majority of their visitors.

Festivals

Major music festivals include the Lakeside Blues Festival in Järvenpää, Pori Jazz Festival, Kaustinen Folk Music Festival, Ankkarock in Vantaa, Provinssirock in Seinäjoki, Ruisrock on the island of Ruissalo and Savonlinna Opera Festival. Other main festivals are the Helsinki Festival Week, the Maritime Festival in Kotka, Tampere Theatre Festival and Seinäjoki's Tango Festival.

DIPLOMATIC REPRESENTATIVES

Of Finland in the United Kingdom (38 Chesham Pl., London, SW1X 8HW)
Ambassador: Pekka Huhtaniemi.

Of the United Kingdom in Finland (Itäinen Puistotie 17, 00140 Helsinki)
Ambassador: Matthew Lodge.

Of Finland in the USA (3301 Massachusetts Ave., NW, Washington, D.C., 20008)
Ambassador: Ritva Koukku-Ronde.

Of the USA in Finland (Itäinen Puistotie 14, Helsinki 00140)
Ambassador: Bruce Oreck.

Of Finland to the United Nations
Ambassador: Jarmo Viinanen.

Of Finland to the European Union
Permanent Representative: Jan Store.

FURTHER READING

Statistics Finland. *Statistical Yearbook of Finland* (from 1879).—*Bulletin of Statistics* (quarterly, from 1971).
Constitution Act and Parliament Act of Finland. 1999
Suomen valtiokalenteri—Finlands statskalender (State Calendar of Finland). Annual
Facts About Finland. Annual
Finland in Figures. Annual

Jussila, Osmo, Hentila, Seppo and Nevakivi, Jukka, *From Grand Duchy to a Modern State: A Political History of Finland since 1809.* 2000
Kirby, D. G., *A Concise History of Finland.* 2006
Klinge, M., *A Brief History of Finland.* 1987
Lewis, Richard D., *Finland, Cultural Lone Wolf.* 2004
Pesonen, Pertti and Riihinen, Olavi, *Dynamic Finland: The Political System and the Welfare State.* 2004
Raunio, Tapio and Tiilikainen, Teija, *Finland in the European Union.* 2003
Singleton, F., *A Short History of Finland.* 2nd ed. 1998

National Statistical Office: Statistics Finland, Työpajankatu 13, Helsinki, FIN-00022.
Website: http://www.stat.fi

FRANCE

UNITED KINGDOM
English Channel
BELGIUM
GERMANY
Lille
LUX.
PARIS
Strasbourg
Nantes
Dijon
FRANCE
SWITZ.
Lyons
ITALY
Bay of Biscay
Bordeaux
MONACO
Toulouse
Marseilles
Nice
SPAIN
ANDORRA
Mediterranean Sea
© Research Machines plc 2006

0   100 mi
0   150 km

République Française
(French Republic)

Capital: Paris
Population projection, 2015: 64·41m.
GNI per capita, 2011: (PPP$) 30,462
HDI/world rank: 0·884/20
Internet domain extension: .fr

KEY HISTORICAL EVENTS

The Dordogne has evidence of Mousterian industry from 40,000 BC and of Cro-Magnon man of the Upper Paleolithic period. With the end of the Ice Age, agricultural settlement appeared around 7000 BC. By the beginning of the 8th century BC, Celtic tribes from Central Europe were inhabiting the Rhône valley of Gaul (now France) while the Greeks were building cities such as Massalia (Marseille) along the southern coast. The Romans crossed the Alps into southern France in 121 BC and Gaul was conquered by Julius Caesar in 52 BC. The country benefited from protected trade routes and from Roman infrastructure, speech and government. Roman rule was consolidated by the reign of Augustus at the end of the 1st century AD. But the Empire was threatened by Germanic ('barbarian') incursions from the north and east. Many of these tribes were assimilated as *foederati* (treaty nations) into the Gallo-Roman Empire but they assumed authority in their domains as Roman government receded in the 4th and 5th centuries. After the repulse of Attila and his Huns in 451, the Salian Franks emerged as the strongest of the Germanic tribes—their leader, Merovius, was the progenitor of the Merovingian dynasty that ruled France until the beginning of the 8th century.

On the death of Merovius' grandson, Clovis, the kingdom was divided between his three sons. The Merovingians remained in power for two centuries but their rule, weakened by internecine warfare, gave way to the Carolingian dynasty in 751. Having extended his empire over Germany and Italy, Charlemagne was crowned emperor of the West by the Pope in 800. He moved his seat of government to Aix-la-Chapelle (Aachen) where he presided over a revival of learning and education.

Charlemagne died in 814 and his empire was fought over by his grandsons before the 843 Treaty of Verdun officially split the territories. Charles le Chauve (823–77) inherited the western territories, an area roughly corresponding to modern day France. But by 912 Vikings had settled in Rouen, having laid siege to Paris. Further threats came from Muslim Saracens in the south and Hungarian Magyars in the east. The Carolingians struggled to keep their power for another century but they were weakened by unrest and disunity. In 987 Hugh Capet, the duke of the Franks, ousted the legitimate claimant to the throne, Charles of Lorraine, and appointed himself king. To control a diverse country, power was centralized on Paris.

Between 1150 and 1300 France underwent a period of economic expansion, though the 12th century also saw the Holy Land crusades and the expulsion of the Jews, followed by the bloody Albigensian Crusade against the heretical Cathars of Languedoc in 1209. The last Capetian king, Charles IV, died in 1328 (leaving only daughters) and the Capetian dynasty gave way to the House of Valois. However, King Edward III of England disputed Philippe de Valois' claim to the French throne, prompting the start of the Hundred Years War (1337–1453). With his son the Black Prince, Edward III's successful invasion led to the Treaty of Brétigny in 1360, which ceded Aquitaine to England. Edward renounced all claims to the French throne but the warfare continued until Charles V (ruled 1364–80) won back most of their territories.

In 1415 Henry V of England, with the backing of the Burgundians, defeated the French at Agincourt. He married the daughter of Charles IV and obtained the right of succession to the French throne. The war continued between his son Henry VI and the dauphin Charles (VI) who enlisted the help of Joan of Arc (Jeanne d'Arc). After leading a series of successful campaigns against the English, she was captured, tried as a heretic by a court of Burgundian ecclesiastics and burnt at the stake in Rouen in 1431. Nevertheless, French successes continued and eventually the English were driven from all their French possessions except Calais.

Rising Power

The reign of Louis XI (1461–83) saw a change from a medieval social system to a more modern state. Provincial governments were set up in major cities and nobles wielding independent power were crushed. In 1494 Charles VIII, encouraged to pursue his claim to the crown of Naples by Ludovico Sforza, duke of Milan, invaded Italy. The speed of his advance shocked the Italian cities into an alliance to expel his army. The appearance of Spanish power in Naples began the Habsburg-Valois wars that used Italy as a battlefield until the Peace of Cateau-Cambrésis in 1559. François I is considered the first Renaissance French king. He patronized some of Italy's greatest artists, commissioning palaces such as the Château de Chambord and rebuilding the Louvre and Château de Fontainebleau. To finance his cultural interests and his military failures in Italy—he was captured by Spanish forces at the Battle of Pavia in 1525—François imposed huge tax rises, severely straining the French economy.

Between 1562–98 the Wars of Religion raged in France between the Protestant Huguenots and the Spanish-supported Catholic League. The civil war reached its peak with the 1572 St Bartholomew's Day massacre, in which 20,000 Huguenots were killed, before ending with Henry of Navarre's conversion to Catholicism. He did not abandon his Huguenot roots, however, and the 1598 Edict of Nantes guaranteed Protestants political and religious rights.

After Henry's assassination in 1610 the young Louis XIII took the throne with his mother, Marie de Médicis, acting as regent. Between 1624–42 Cardinal Richelieu held the reins of government and set about establishing absolute royal power in France, with the suppression of Protestant influences. This policy was continued for the next 20 years by his successor Cardinal Mazarin. On Mazarin's death Louis XIV (1643–1715) was able to govern alone. Louis, the 'Sun King', attempted to impose a centralized absolutism, gathering the aristocracy around him and thus denying it traditional regional power. The king formally revoked the Edict of Nantes, Protestant churches were destroyed and religious minorities persecuted. His successor, Louis XV, married Maria, the daughter of the deposed king of Poland, who drew France into the War of the Polish Succession. Further costly military disasters followed including the Seven Years' War, in which France lost her colonies in India, North America and the West Indies.

When Louis XVI succeeded to the throne in 1774, financial crises caused by prolonged military failure coupled with a succession of bad harvests led to grain riots in 1787–88 in Paris, Lyon, Nantes and Grenoble. The subsequent reforms were rejected by the aristocracy (les privilégiés), the upper ranks of the clergy (the First Estate) and the majority of the nobility (the Second Estate), who feared a reduction in their tax-levying privileges. Meanwhile, Louis XVI supported the American colonies in their struggle for independence from Britain, a policy that was financially disastrous and also did much to disseminate revolutionary and democratic ideals in France.

Revolution

The French Revolution erupted in 1789 when the Third Estate (the non-privilégiés) assumed power in the National Assembly and overthrew the government. Riots broke out across France, culminating in the storming of the Bastille in Paris on 14 July 1789. A new legislative assembly was formed and although the moderate Girondins held power at the start, the more extreme followers of Danton, Robespierre and Marat—the Jacobins— seized power and in 1792 declared a republic.

On 21 Jan. 1793 Louis XVI was guillotined in the Place de la Révolution. After his death a reign of terror led by Maximilien Robespierre followed in which thousands of people were guillotined. Despite the efforts of the royalists to re-establish a monarchy, in 1795 the 'Directory of Five' was appointed to run the country. As one of these five, Paul Barras had been responsible for the promotion of a young Corsican, Napoleon Bonaparte, to the rank of general. Over four years, Napoleon commanded the French troops in a series of successful campaigns against the Austrians and the British. On his return to Paris, he found the Directory in disarray and in 1799 overthrew the government and declared himself first consul. Napoleon immediately faced a hostile coalition of England, Austria and Russia. In 1805 he defeated Austria and Russia at the Battle of Austerlitz but the British naval victory at the Battle of Trafalgar earlier the same year gave Britain maritime supremacy. Napoleon's best troops were bogged down supporting his brother Joseph in the Peninsula War in Spain and his success at Borodino, Russia in 1812 was followed by the army's forced retreat from Moscow during the harsh winter months. The Prussian army retaliated at Leipzig, entered France and forced the surrender of Paris in March 1814. Napoleon abdicated at Fontainebleau on 20 April 1814 and retired to Elba. But when Louis XVIII returned from exile in England later that year, Napoleon left Elba to attempt to recover his empire. He marched north towards Paris, gathering support on the way. But his defeat in 1815 at Waterloo by the Allies led by the duke of Wellington ended his 'Hundred Days' reign. He was exiled to the island of St Helena where he died in 1821.

Second Empire

The monarchy was restored with the Bourbon family. A revolution in 1830 brought Louis Philippe, son of the duke of Orléans, to the throne as a constitutional monarch. This 'July Monarchy' was overthrown in 1848 and superseded by the Second Republic, with Louis Napoleon (nephew of Napoleon I) elected president. In 1852 he took the title of Emperor Napoleon III, and hence began the Second Empire. However, the defeat of France in the Franco-Prussian War (1870–71) led to Napoleon being deposed and the proclamation of the Third Republic in 1870. After a four-month siege, Paris capitulated in Jan. 1871. By Sept. 1873 the occupying troops had gone but Alsace and Lorraine had been lost and French politics, embittered by the Dreyfus Affair (1894–1906), in which forged evidence resulted in a Jewish general staff captain being falsely imprisoned for spying, went from crisis to crisis.

An entente cordiale was established between France and Britain in 1904, putting an end to colonial rivalry and paving the way for future co-operation. In 1905 the Church was separated from the State, a measure to counteract ecclesiastical influence over education.

European War

Although Paris was saved from occupation during the First World War, ten departments were overrun and four long years of trench warfare followed. The tide began to turn against Germany in 1916 with the Battle of the Somme, the French stand at Verdun and the arrival of the Americans in 1917; the Armistice was finally signed on 11 Nov. 1918. By the end of the war France had lost a total of 1·3m. men. The main thrust of France's efforts to rebuild her defences after the First World War was concentrated on the 'Maginot Line'—a supposedly impregnable barrier running along the German frontier, but which was sidestepped by the advancing German forces in 1939. Demoralized French troops, unable to resist the German advance, were forced to retreat towards Dunkerque (Dunkirk). The French government capitulated and a pro-German government presided over by Marshal Pétain (a hero of the Battle of Verdun) was established at Vichy. A truce was signed with Germany agreeing German occupation in the northern third of the country and collaborationist government control in the south. Gen. Charles de Gaulle established the Forces Françaises Libres (Free French Forces) and declared the Comité National Français to be the true French government-in-exile with its headquarters first in London and then in Algiers. With help from the Resistance in France, in Aug. 1944 de Gaulle returned at the head of the allied armies and liberated Paris. An armistice with Germany was signed in March 1945.

In Oct. 1946 the Fourth Republic, institutionally similar to the Third Republic, was established but during prolonged wrangling over the form of the new constitution Gen. de Gaulle retired. Despite frequent changes of government and defeat in Indo-China, France achieved economic recovery. In 1957 a European common market was established of which France, West Germany, Italy and the Benelux countries were founder members.

Fifth Republic

Between 1954–62 France was embroiled in a war of independence with Algeria that split public and political opinion. In 1958 de Gaulle prepared a new constitution and was persuaded to return first as prime minister and then, by popular election, as the first president of the newly declared Fifth Republic. The new constitution greatly enhanced the power of the president. The politics of the early Fifth Republic was dominated by the centre-right, with a succession of parties (including the Union of Democrats for the Republic, Union of Democrats for the V Republic, Union for the New Republic, Union for the French Republic-Democratic Union of Labour) working to a Gaullist agenda. There was an emphasis on national independence, government involvement in the economy and broadly conservative social policies.

In 1962 Algeria gained independence. De Gaulle continued to preside over a period of relative stability and economic growth but serious student riots in Paris in 1968 precipitated reforms to the authoritarian system of education. The students were joined by workers wanting better pay and conditions. The National Assembly was dissolved and, although the Gaullists were returned to power in the new election, de Gaulle's referendum proposing decentralization was defeated and in 1969 he resigned. Georges Pompidou, who had been de Gaulle's prime minister, succeeded him. Pompidou attempted to consolidate de Gaulle's legacy by concentrating on economic reform. When he died in office in 1974 he was succeeded by Valéry Giscard d'Estaing who continued right-wing policies, eventually precipitating a swing to the left. In 1981 the Socialist leader François Mitterrand was elected president. He immediately implemented widespread social reforms but a deep recession in 1983 forced him to take a series of unpopular deflationary measures.

When the ailing Mitterrand's term of office expired in 1995, Jacques Chirac was elected president with Alain Juppé as prime minister. After the Socialists won an assembly majority in 1997, Juppé resigned making way for the Socialist leader Lionel Jospin to take over as prime minister. The right and left-wing *cohabitation* lasted five years until Jospin retired after a disastrous result in the first round presidential elections. Chirac's second electoral success was consolidated by the moderate right taking an assembly majority in the 2002 legislative elections.

In Oct. 2005 the death of two youths of African origin led to several days of rioting in immigrant ghettoes, prompting the government to declare a state of emergency, which was lifted in Jan. 2006. In May 2007 Nicolas Sarkozy was elected president. In 2009 he led France back into NATO's integrated military command after a 43-year absence. François Hollande was elected president in May 2012 to end 17 years of centre-right rule.

TERRITORY AND POPULATION

France is bounded in the north by the English Channel (*La Manche*), northeast by Belgium and Luxembourg, east by Germany, Switzerland and Italy, south by the Mediterranean (with Monaco as a coastal enclave), southwest by Spain and Andorra, and west by the Atlantic Ocean. The total area of metropolitan France is 543,965 sq. km. More than 14% of the population of Paris are foreign and 19% are foreign born.

The population was 58,518,395 at the census of 1999 and 62,765,235 on 1 Jan. 2010 (density, 115·4 persons per sq. km).

The UN gives a projected population for 2015 of 64·41m.

In 2011, 85·9% of the population lived in urban areas.

The growth of the population has been as follows:

Census	Population	Census	Population	Census	Population
1801	27,349,003	1962	46,519,997	2007[2]	61,795,238
1861	37,386,313	1968	49,778,540	2008[2]	62,134,866
1901	38,961,945	1975	52,655,802	2009[2]	62,465,709
1921	39,209,518	1982	54,334,871	2010[2]	62,765,235
1931	41,834,923	1990	56,615,155	2011[3]	63,088,990
1946	40,506,639	1999	58,518,395		
1954	42,777,174	2006[1]	61,399,733		

[1]First recorded figure using the new 'rolling census' method of calculating the population that came into effect in Jan. 2004. [2]Calculated using the 'rolling census' method. [3]Provisional.

In 2004 there were 4·96m. people of foreign extraction in France (8·1% of the population). The largest groups of foreigners with residence permits in 2004 were: Algerians (679,000), Moroccans (625,000) and Portuguese (567,000). France's Muslim population, at an estimated 5m., is the highest in Europe.

Controls on illegal immigration were tightened in July 1991. Automatic right to citizenship for those born on French soil was restored in 1997 by the new left-wing coalition government. New immigration legislation, which came into force in 1998, brought in harsher penalties for organized traffic in illegal immigrants and

extended asylum laws to include people whose lives are at risk from non-state as well as state groups. It also extended nationality at the age of 18 to those born in France of non-French parents, provided they have lived a minimum of five years in France since the age of 11.

The areas, recorded populations and chief towns of the 22 metropolitan regions in Jan. 2009 were as follows:

Regions	Area (sq. km)	Population	Chief town
Alsace	8,280	1,843,053	Strasbourg
Aquitaine	41,308	3,206,137	Bordeaux
Auvergne	26,013	1,343,964	Clermont-Ferrand
Basse-Normandie	17,589	1,470,880	Caen
Bourgogne (Burgundy)	31,582	1,642,440	Dijon
Bretagne (Brittany)	27,208	3,175,064	Rennes
Centre	39,151	2,538,590	Orléans
Champagne-Ardenne	25,606	1,337,953	Reims
Corse (Corsica)	8,680	305,674	Ajaccio
Franche-Comté	16,202	1,168,208	Besançon
Haute-Normandie	12,317	1,832,942	Rouen
Île-de-France	12,012	11,728,240	Paris
Languedoc-Roussillon	27,376	2,610,890	Montpellier
Limousin	16,942	741,785	Limoges
Lorraine	23,547	2,350,112	Nancy
Midi-Pyrénées	45,348	2,862,707	Toulouse
Nord-Pas-de-Calais	12,414	4,033,197	Lille
Pays de la Loire	32,082	3,539,048	Nantes
Picardie	19,399	1,911,157	Amiens
Poitou-Charentes	25,810	1,760,575	Poitiers
Provence-Alpes-Côte d'Azur	31,400	4,889,053	Marseille
Rhône-Alpes	43,698	6,174,040	Lyon

The 22 regions are divided into 96 metropolitan *départements*, which in 2007 consisted of 36,569 communes.

Populations of the principal conurbations (in descending order of size) and towns in 2006:

	Conurbation	Town
Paris	10,142,977[1]	2,181,371
Marseille–Aix-en-Provence	1,418,481[2]	839,043
Lyon	1,417,463[3]	472,305
Lille	1,016,205[4]	226,014
Nice	940,017	347,060
Toulouse	850,873	437,715
Bordeaux	803,117	232,260
Nantes	568,743	282,853
Toulon	543,065	167,816
Douai–Lens	512,462	. . .[5]
Strasbourg	440,265	272,975
Grenoble	427,658	156,107
Rouen	388,798	107,904
Valenciennes	355,660	42,426
Nancy	331,279	105,468
Metz	322,946	124,435
Montpellier	318,225	251,634
Tours	306,974	136,942
Saint-Étienne	286,400	177,480
Rennes	282,550	209,613
Avignon	273,359	92,454
Orléans	269,283	113,130
Clermont-Ferrand	260,657	138,992
Béthune	259,293	26,472
Le Havre	238,776	182,580
Mulhouse	238,638	110,514
Dijon	238,088	151,504
Angers	227,771	152,337
Reims	212,021	183,837
Brest	206,394	144,548
Caen	196,323	110,399
Pau	193,991	83,903
Le Mans	192,910	144,016
Bayonne	189,836	44,406
Dunkerque	182,973	69,274
Perpignan	178,501	115,326
Limoges	177,439	136,539
Nîmes	161,565	144,092

	Conurbation	Town
Amiens	161,311	136,105
Annecy	144,682	51,023
Saint-Nazaire	143,106	68,838
Besançon	134,951	117,080
Troyes	131,039	61,344
Thionville	130,437	41,127
Poitiers	126,652	88,776
Valence	120,922	65,263
La Rochelle	119,702	77,196
Chambéry	119,266	57,543
Genève–Annemasse	118,554	...⁶
Lorient	116,764	58,547
Montbéliard	109,118	26,535
Angoulême	105,021	42,096
Calais	103,277	74,888
Creil	101,100	33,479

[1]Including Boulogne-Billancourt (110,251), Argenteuil (102,683), Montreuil (101,587), Versailles (87,549), Saint-Denis (97,875), Nanterre (88,316), Créteil (88,939), Aulnay-sous-Bois (81,600), Vitry-sur-Seine (82,902). [2]Including Aix-en-Provence (142,534). [3]Including Villeurbanne (136,473), Vénissieux (57,179). [4]Including Roubaix (97,952), Tourcoing (92,357). [5]Including Douai (42,766), Lens (35,583). [6]Including Annemasse (28,572).

France (including its overseas territories) has nine national parks, 45 regional nature parks and 164 national nature reserves.

Languages

The official language is French. Breton and Basque are spoken in their regions. The *Toubon* legislation of 1994 seeks to restrict the use of foreign words in official communications, broadcasting and advertisements (a previous such decree dated from 1975). The Constitutional Court has since ruled that imposing such restrictions on private citizens would infringe their freedom of expression.

SOCIAL STATISTICS

Statistics for calendar years (mainland France):

	Births	Deaths	Marriages	Divorces
2004	767,816	509,429	271,598	131,335
2005	774,355	527,533	276,303	152,020
2006	796,896	516,416	267,260	135,910
2007	785,985	521,016	267,194	131,316
2008	796,044	532,131	258,739	129,379
2009	793,420	538,116	245,151	127,578

Live birth rate (2009) was 12·7 per 1,000 population; death rate, 8·6; marriage rate, 3·9; divorce rate, 2·0. 52·9% of births in 2009 were outside marriage. In 2009 the average age at first marriage was 31·7 years for males and 29·8 years for females. Abortions were legalized in 1975; there were an estimated 209,300 in 2009. Life expectancy at birth, 2009, 77·7 years for males and 84·4 years for females. Annual population growth rate, 2005–10, 0·6%. In 2007 the suicide rate per 100,000 population was 16·3 (males, 24·7; females, 8·5). Infant mortality, 2010, three per 1,000 live births; fertility rate, 2008, 1·9 births per woman. In 2008 France received 35,404 asylum applications (59,768 in 2003), the second highest total after the USA. France legalized same-sex marriage in April 2013.

CLIMATE

The northwest has a moderate maritime climate, with small temperature range and abundant rainfall; inland, rainfall becomes more seasonal, with a summer maximum, and the annual range of temperature increases. Southern France has a Mediterranean climate, with mild moist winters and hot dry summers. Eastern France has a continental climate and a rainfall maximum in summer, with thunderstorms prevalent. Paris, Jan. 37°F (3°C), July 64°F (18°C). Annual rainfall 22·9" (573 mm). Bordeaux, Jan. 41°F (5°C), July 68°F (20°C). Annual rainfall 31·4" (786 mm). Lyon, Jan. 37°F (3°C), July 68°F (20°C). Annual rainfall 31·8" (794 mm).

CONSTITUTION AND GOVERNMENT

The Constitution of the Fifth Republic, superseding that of 1946, came into force on 4 Oct. 1958. It consists of a preamble, dealing with the Rights of Man, and 89 articles.

France is a decentralized republic, indivisible, secular, democratic and social; all citizens are equal before the law (Art. 1). National sovereignty resides with the people, who exercise it through their representatives and by referendums (Art. 3). Constitutional reforms of July 1995 widened the range of issues on which referendums may be called. Political parties carry out their activities freely, but must respect the principles of national sovereignty and democracy (Art. 4).

A constitutional amendment of 4 Aug. 1995 deleted all references to the 'community' (*communauté*) between France and her overseas possessions, representing an important step towards the constitutional dismantling of the former French colonial empire.

The head of state is the *President*, who must be a French citizen, have attained the age of 18 years and be qualified to vote. The President sees that the Constitution is respected; ensures the regular functioning of the public authorities, as well as the continuity of the state; is the protector of national independence and territorial integrity (Art. 5). As a result of a referendum held on 24 Sept. 2000 the President is elected for five years by direct universal suffrage (Art. 6). Previously the term of office had been seven years. The President appoints (and dismisses) a Prime Minister and, on the latter's advice, appoints and dismisses the other members of the government (*Council of Ministers*) (Art. 8); presides over the Council of Ministers (Art. 9); may dissolve the National Assembly, after consultation with the Prime Minister and the Presidents of the two Houses (Art. 12); appoints to the civil and military offices of the state (Art. 13). In times of crisis, the President may take such emergency powers as the circumstances demand; the National Assembly cannot be dissolved during such a period (Art. 16).

Parliament consists of the National Assembly and the Senate. The *National Assembly* is elected by direct suffrage by the second ballot system (by which candidates winning 50% or more of the vote in their constituencies are elected, candidates winning less than 12·5% are eliminated and other candidates go on to a second round of voting); the Senate is elected by indirect suffrage (Art. 24). Since 1996 the National Assembly has convened for an annual nine-month session. It comprises 577 deputies, elected by a two-ballot system for a five-year term from single-member constituencies (including 11 constituencies for French residents abroad for the June 2012 election), and may be dissolved by the President.

The *Senate* comprises 348 senators (343 prior to the elections of Sept. 2011) elected for six-year terms (one-half every three years) by an electoral college in each Department or overseas dependency, made up of all members of the Departmental Council or its equivalent in overseas dependencies, together with all members of Municipal Councils within that area. The *President* of the Senate deputizes for the President of the Republic in the event of the latter's incapacity. Senate elections were last held on 25 Sept. 2011.

The *Constitutional Council* is composed of nine members whose term of office is nine years (non-renewable), one-third every three years; three are appointed by the President of the Republic, three by the President of the National Assembly, three by the President of the Senate; in addition, former Presidents of the Republic are, by right, life members of the Constitutional Council (Art. 56). It oversees the fairness of the elections of the President (Art. 58) and Parliament (Art. 59), and of referendums (Art. 60), and acts as a guardian of the Constitution (Art. 61). Its *President* is Jean-Louis Debré (appointed 5 March 2007).

The *Economic, Social and Environmental Council* advises on Government and Private Members' Bills (Art. 69). It comprises representatives of employers', workers' and farmers' organizations in each Department and Overseas Territory.

Constitutional amendments of 25 March 2003 and 1 March 2005 added provisions for European Union arrest warrants and allowed for a referendum on the European Union constitution.

Ameller, M., *L'Assemblée Nationale*. 1994
Duhamel, O. and Mény, Y., *Dictionnaire Constitutionnel*. 1992

Local Government
The traditional system of centralized government was overhauled in 1982 to provide local government with greater power. There are three basic layers of local government: *régions* (of which there are 22), *départements* (of which there are 96) and *communes* (which number about 36,000). Paris, Lyon and Marseille each have special status.

Régions—Mainland France comprises 22 *régions*, which are principally responsible for economic development, town and country planning and education. Government is through regional councils, with members elected every six years. The council works with an economic and social committee, which includes representatives from business and commerce, trade unions, voluntary bodies and other organizations. The council elects a president every three years.

Départements—Government at the department level is principally concerned with health and social welfare, rural capital works, highways and the administration of colleges. A prefect, appointed by the government, oversees the work of the department administration. Decision-making rests with a general council, with members elected every six years. Each department is divided into *cantons* (of which there are around 4,000 in France) to serve as electoral constituencies. The council elects a chairman who holds executive power.

Communes—These municipalities can vary greatly in size. Around 80% of communes have fewer than 1,000 citizens. As a result, smaller communes will often group themselves together, either as urban communities (*communautés urbaines*) or as associations called *syndicats intercommunaux*. Municipal government consists of a decision-making municipal council and a mayor, elected by the council, who wields executive power and also acts as the state's representative. The size of a municipal council is dependent on the size of the commune and members are elected every six years. The council oversees management in areas such as schools and the environment. In addition, the mayor has jurisdiction in security, public health and crime, as well as responsibility for registering births, deaths and marriages.

National Anthem
'La Marseillaise'; words and tune by C. Rouget de Lisle.

GOVERNMENT CHRONOLOGY

(CD = Democratic Centre; CNIP = National Centre of Independents and Peasants; DL = Liberal Democracy; FNRI = National Federation of Independent Republicans; MRP = People's Republican Movement; PR = Republican Party; PS = Socialist Party; Rad. = Radical Party; RPR = Rally for the Republic; SFIO = French Section of the Workers International; UDF = Union for the French Democracy; UDR = Union of Democrats for the Republic; UDSR = Democratic and Social Union of the Resistance; UDT = Democratic Union of Labour; UDVe = Union of Democrats for the V Republic; UMP = Union for a Popular Movement; UNR = Union for the New Republic; UNR-UDT = Union for the French Republic-Democratic Union of Labour; n/p = non-partisan)

Presidents of the French Republic since the Second World War.

1947–54	SFIO	Vincent Auriol
1954–59	CNIP	René Coty

With the advent of the Fifth Republic the power of the president gained at the expense of the prime minister.

1959–69	UNR, UNR-UDT, UDVe, UDR	Charles de Gaulle
1969	CD	Alain Poher
1969–74	UDR	Georges Pompidou
1974	CD	Alain Poher
1974–81	FNRI, PR-UDF	Valéry Giscard d'Estaing
1981–95	PS	François Mitterrand
1995–2007	RPR, UMP	Jacques Chirac
2007–12	UMP	Nicolas Sarkozy
2012–	PS	François Hollande

Heads of Government since 1944.

Chairmen of the Provisional Government of the French Republic

1944–46	n/p	Charles de Gaulle
1946	SFIO	Félix Gouin
1946	MRP	Georges Bidault
1946–47	SFIO	Léon Blum

Chairmen of the Council of Ministers

1947	SFIO	Paul Ramadier
1947–48	MRP	Robert Schuman
1948	Rad.	André Marie
1948	MRP	Robert Schuman
1948–49	Rad.	Antoine Henri Queuille
1949–50	MRP	Georges Bidault
1950	Rad.	Antoine Henri Queuille
1950–51	UDSR	René Pleven
1951	Rad.	Antoine Henri Queuille
1951–52	UDSR	René Pleven
1952	Rad.	Edgar Faure
1952–53	CNIP	Antoine Pinay
1953	Rad.	René Mayer
1953–54	CNIP	Joseph Laniel
1954–55	Rad.	Pierre Mendès France
1955–56	Rad.	Edgar Faure
1956–57	SFIO	Guy Mollet
1957	Rad.	Maurice Bourgès-Maunoury
1957–58	Rad.	Félix Gaillard
1958	MRP	Pierre Pflimlin
1958–59	UNR	Charles de Gaulle

Prime Ministers

1959–62	UNR	Michel Debré
1962–68	UNR, UNR-UDT, UDVe	Georges Pompidou
1968–69	UDVe, UDR	Maurice Couve de Murville
1969–72	UDR	Jacques Chaban-Delmas
1972–74	UDR	Pierre Messmer
1974–76	UDR	Jacques Chirac
1976–81	n/p, UDF	Raymond Barre
1981–84	PS	Pierre Mauroy
1984–86	PS	Laurent Fabius
1986–88	RPR	Jacques Chirac
1988–91	PS	Michel Rocard
1991–92	PS	Édith Cresson
1992–93	PS	Pierre Bérégovoy
1993–95	RPR	Édouard Balladur
1995–97	RPR	Alain Juppé
1997–2002	PS	Lionel Jospin
2002–05	DL, UMP	Jean-Pierre Raffarin
2005–07	UMP	Dominique de Villepin
2007–12	UMP	François Fillon
2012–	PS	Jean-Marc Ayrault

RECENT ELECTIONS

At the first round of presidential elections on 22 April 2012 François Hollande, the Socialist Party candidate, gained the largest number of votes (28·63% of those cast) against nine opponents. His nearest rivals were the incumbent president and candidate of the Union for a Popular Movement, Nicolas Sarkozy, who came second with 27·18% of votes cast, and the National Front candidate Marine Le Pen, with 17·90%. In the second round of voting, held on 6 May 2012, Hollande was elected president with

51·63% of votes cast against 48·37% for Sarkozy. Turnout was 80·3% in the second round (79·5% in the first round).

Elections to the National Assembly were held on 10 and 17 June 2012. The election was won by the Socialist Party along with its Presidential Majority allies with a total of 331 seats, providing the new government with an absolute parliamentary majority. The Socialist Party (PS) won 280 of the 577 available seats; the Union for a Popular Movement (UMP), 194; Miscellaneous Left (DVG), 22; Europe Ecology–the Greens (EELV), 17; Miscellaneous Right (DVD), 15; Radical Party of the Left (PRG), 12; the New Centre (NC), 12; the Left Front (FDG), 10; the Radical Party (PRV), 6; and others, 9.

Following the indirect election held on 25 Sept. 2011, the Senate was composed of (by group, including affiliates): the Socialist Party, 140; Union for a Popular Movement, 132; the Centrist Group, 31; Républicain, Communiste et Citoyen (RCC), 21; Democratic and Social European Rally, 16; unattached, 8. In Oct. 2011 Jean-Pierre Bel (PS) was elected *President* of the Senate for a three-year term. He is the first left-of-centre Senate President since the Fifth Republic was founded in 1958.

European Parliament
France has 74 (78 in 2004) representatives. At the June 2009 elections turnout was 40·6% (42·8% in 2004). The UMP won 29 seats with 27·8% of votes cast (political affiliation in European Parliament: European People's Party); PS 14 with 16·5% (Progressive Alliance of Socialists and Democrats); Europe Ecologie 14 with 16·3% (Greens/European Free Alliance); MoDem, 6 with 8·4% (Alliance of Liberals and Democrats for Europe); Front de Gauche, 4 with 6·0% (European United Left/Nordic Green Left); Front National, 3 with 6·3% (non-attached); Libertas, 1 with 4·6% (Europe of Freedom and Democracy); Alliance des Outre-mers, 1 with 0·4% (European United Left/Nordic Green Left).

CURRENT GOVERNMENT

President: François Hollande; b. 1954 (Socialist Party; sworn in 15 May 2012).

In March 2013 the government comprised:

Prime Minister: Jean-Marc Ayrault; b. 1950 (Socialist Party; sworn in 16 May 2012).

Minister of Foreign Affairs: Laurent Fabius. *National Education:* Vincent Peillon. *Justice and Keeper of the Seals:* Christiane Taubira. *Economy and Finance:* Pierre Moscovici. *Social Affairs and Health:* Marisol Touraine. *Territorial Equality and Housing:* Cécile Duflot. *Interior:* Manuel Valls. *External Trade:* Nicole Bricq. *Industrial Recovery:* Arnaud Montebourg. *Ecology, Sustainable Development and Energy:* Delphine Batho. *Labour, Employment, Vocational Training and Social Dialogue:* Michel Sapin. *Defence:* Jean-Yves Le Drian. *Culture and Communication:* Aurélie Filippetti. *Higher Education and Research:* Geneviève Fioraso. *Women's Rights and Government Spokesperson:* Najat Vallaud-Belkacem. *Agriculture and the Food Processing Industry:* Stéphane Le Foll. *State Reform, Decentralization and the Civil Service:* Marylise Lebranchu. *Overseas Territories:* Victorin Lurel. *Crafts, Trade and Tourism:* Sylvia Pinel. *Sports, Youth, Popular Education and Community Life:* Valérie Fourneyron.

President of the National Assembly: Claude Bartolone.

Office of the President: http://www.elysee.fr

CURRENT LEADERS

François Hollande

Position
President

Introduction
François Hollande was elected the 24th president of the French Republic in May 2012. A Socialist Party veteran, Hollande's election returned the Left to the Élysée Palace for the first time since his one-time mentor, François Mitterrand, departed office in 1995. Renowned for his understated manner and dry wit, Hollande led the Socialist Party from 1997–2008 but had never served in a ministerial role. His first months in office were focused on economic issues in the wake of continuing eurozone instability. In Jan. 2013 foreign affairs took centre-stage as he ordered French military intervention in the north of the former African colony of Mali in support of its government against Islamist rebels.

Early Life
François Hollande was born on 12 Aug. 1954 in Rouen, Normandy. The son of a social worker and a doctor, he attended the Paris Institute of Political Studies ('Sciences Po') before continuing his studies at the HEC Paris business school. In 1978 he won a place at the École Nationale d'Administration (ENA), the elite graduate school for those aspiring to high office.

On leaving the ENA in 1980, Hollande joined François Mitterrand's second presidential election campaign. When Mitterrand was sworn in the following year as France's first socialist president under the Fifth Republic, Hollande was appointed an economic adviser. During his time at the Élysée Palace he worked as chief of staff to the foreign minister, Roland Dumas, who would later give influential support to Hollande's run to become the Socialist Party's presidential candidate.

In 1988 Hollande was elected to the National Assembly as the representative for Corrèze, a rural department in south-central France. He served as mayor of the department's capital, Tulle, from 2001 until 2008. Corrèze was regarded as a power base for the then president Jacques Chirac, the nemesis of the French Left, and Hollande's ability to build popularity there was considered a significant achievement.

Chirac's decision to call a snap parliamentary election in 1997 saw Lionel Jospin installed as prime minister, with Hollande selected to succeed him as leader of the Socialist Party. Hollande's long-term partner, Ségolène Royal—with whom he has four children—was appointed minister of schools by Jospin. Despite not being a member of the cabinet, Hollande was a close adviser to the premier at the same time as overseeing a programme of modernization in the Socialist Party comparable to that undertaken by the British Labour Party and the German Social Democrats.

In 2002 Jospin ran for the presidency, only to be defeated in the first round by the far-right candidate, Jean-Marie Le Pen. A committed pro-European, Hollande went on to secure his party's support for a 'yes' vote in the 2005 referendum to ratify the European Constitutional Treaty, although the electorate rejected the treaty. In 2007 Ségolène Royal stood as the Socialist Party's presidential candidate but was defeated by Nicolas Sarkozy. Hollande and Royal announced the end of their relationship soon afterwards. In 2008 Hollande stepped down as secretary-general of the Socialist Party, to be succeeded by Martine Aubry.

In 2011 the Socialist Party held its first ever open primary to select a presidential candidate. Dominique Strauss-Kahn, previously the strong favourite for the nomination, elected not to stand following his arrest in New York on charges of sexual assault. Hollande, who used the primary campaign to declare his intention to be a 'président normal', defeated Martine Aubry in a second round run-off on 16 Oct. 2011.

The contrast of his low-key approach with the belligerence of the presidential incumbent, Nicolas Sarkozy, was a cornerstone of Hollande's election campaign. In the first round of voting on 22 April 2012 Hollande took 28·6% of the vote, and in the second round on 6 May he defeated Sarkozy, taking 51·6% of the vote.

Career in Office
Hollande's electoral victory and his Socialist Party's subsequent success in parliamentary elections in June 2012 were quickly overshadowed by the ongoing debt crisis in the eurozone and growing concern over France's stalled economic growth and high

unemployment. Ratings agencies Standard & Poor's and Moody's downgraded France's triple-A status during the year. Hollande had also pledged to renegotiate the terms of the 2011 EU treaty setting strict limits on state deficits, arguing that the euro crisis was symptomatic of a wider 'failure of European governance'. The treaty was nevertheless ratified by the French parliament in Oct.

On the international stage, in Nov. 2012 France recognized the Syrian National Coalition for Revolutionary and Opposition Forces, representing the disparate anti-government groups seeking to topple the increasingly beleaguered regime of President Bashar al-Assad. In Jan. 2013 Hollande sent French troops and aircraft to Mali to oust militant Islamist insurgents who had taken control of the north of the country during 2012 and threatened regional stability across West Africa. By early Feb. most of the main northern towns had been recaptured by French and Malian forces.

Meanwhile, Hollande's government courted considerable domestic opposition over proposed legislation to legalize gay marriage and adoption, which began its passage through the National Assembly in Jan. 2013 and was legalized in April.

Jean-Marc Ayrault

Position
Prime Minister

Introduction
Jean-Marc Ayrault was appointed prime minister on 16 May 2012 by President François Hollande. Ayrault served as president of the cross-party Socialist group in the National Assembly from 1997 to 2012 and was mayor of Nantes between 1989 and 2012.

Early Life
Jean-Marc Ayrault was born on 25 Jan. 1950 in Maulévrier, Maine-et-Loire. In 1971 he graduated in German from Nantes University before gaining a teaching diploma and becoming a German teacher. He taught at Angevinière College in Saint-Herblain, a suburb of Nantes, from 1973 until 1986.

In 1971 Ayrault joined the Socialist Party (PS) and affiliated himself with Jean Poperen, who unsuccessfully challenged François Mitterrand for the party leadership that year. From 1976–82 he served as a regional councillor for Loire-Atlantique and in 1977 was elected mayor of Saint-Herblain, a post he held for 11 years. In 1979 he joined the PS national committee and in 1981 its executive committee. In 1986 he was elected to the National Assembly to represent Loire-Atlantique, a seat he held for the next six legislative elections.

In 1989 Ayrault became mayor of Nantes, winning re-election in 1995, 2001 and 2008. In 1997 he was elected president of the parliamentary Socialist group, the same year as he received a six-month suspended prison sentence after the Nantes government awarded a newspaper contract without a public tender. He was a special adviser to François Hollande on his successful presidential campaign in 2012 and was subsequently appointed premier.

Career in Office
Ayrault's principal challenge is to rejuvenate the economy and improve the public finances. Together with Hollande, he has pushed the European Commission to support the Greek economy. Domestically, he lowered the retirement age from 62 to 60 for some workers, cut ministerial salaries by up to 30% and announced plans to spend €2·5bn. by 2017 to help the poor. His government has introduced new capital gains and inheritance taxes, an exit tax for entrepreneurs, a temporary 75% tax on annual incomes over a €1m., an income tax hike from 41% to 45% on salaries over €150,000 and increased taxes on stock options, dividends and financial transactions.

DEFENCE

The President of the Republic is the supreme head of defence policy and exercises command over the Armed Forces. He is the only person empowered to give the order to use nuclear weapons. He is assisted by the Council of Ministers, which studies defence problems, and by the Defence Council and the Restricted Defence Committee, which formulate directives.

Legislation of 1996 inaugurated a wide-ranging reform of the defence system over 1997–2002, with regard to the professionalization of the armed forces (brought about by the ending of military conscription and consequent switch to an all-volunteer defence force), the modification and modernization of equipment and the restructuring of the defence industry. In 2008 defence expenditure totalled US$65,675m. (equivalent to US$1,061 per capita), ensuring France overtook the UK as the country with the world's third highest defence expenditure. Defence spending as a proportion of GDP has fallen from 3·9% in 1985 to 2·3% in 2008.

In Nov. 2010 France and the UK signed a Defence and Security Cooperation Treaty providing for the creation of a rapid reaction force, with troops from both nations called up as required after a joint political decision. Aircraft carriers may be jointly used under certain circumstances and there will be joint training exercises, pooling of resources for the maintenance and logistics of the A400M transport aircraft, and joint work on several other projects. A separate Joint Radiographic/Hydrodynamics Facilities treaty will see the sharing of testing facilities at atomic weapons establishments in France (Valduc) and the UK (Aldermaston), and the development of a new joint hydrodynamic facility.

France rejoined NATO as a full member in April 2009, having withdrawn from its integrated military structure in 1966. In 2009 French military personnel were stationed in a number of countries outside France, including Afghanistan, Chad, Côte d'Ivoire, Djibouti, Kosovo and Lebanon. President Hollande withdrew the last combat troops from Afghanistan in Dec. 2012. However, 1,500 non-combat troops stayed on to help repatriate French military equipment and train the Afghan Army.

Conscription was for ten months, but France officially ended its military draft on 27 June 2001 with a reprieve granted to all conscripts (barring those serving in civil positions) on 30 Nov. 2001.

Nuclear Weapons
Having carried out its first test in 1960, there have been 210 tests in all. The last French test was in 1996 (this compares with the last UK test in 1991 and the last US test in 1993). The nuclear arsenal consisted of approximately 300 warheads in Jan. 2012 according to the Stockholm International Peace Research Institute. In 2008 France announced a 33% reduction in the airborne component of its nuclear forces.

Army
The Army comprises the Logistic Force (CFLT), based in Montlhéry with two logistic brigades, and the Land Force Command (CFAT), based in Lille. Apart from the Franco-German brigade, there are 13 brigades, each made up of between four and seven battalions, including one airmobile brigade.

Personnel numbered (2006) 133,500 including 14,700 marines and a Foreign Legion of 7,700. There were 11,350 army reservists in 2006. Equipment levels in 2006 included 926 main battle tanks and 393 helicopters.

Gendarmerie
The paramilitary police force exists to ensure public security and maintain law and order, as well as participate in the operational defence of French territory as part of the armed forces. On 1 Jan. 2009 budgetary responsibility for the Gendarmerie was transferred from the ministry of defence to the ministry of the interior. It consisted in 2006 of 102,322 personnel including 1,953 civilians. It comprises a territorial force of 66,537 personnel throughout the country, a mobile force of 16,859 personnel and specialized formations including the Republican Guard, the Air Force and Naval Gendarmeries, and an anti-terrorist unit.

Navy

The missions of the Navy are to provide the prime element of the French independent nuclear deterrent through its force of strategic submarines; to assure the security of the French offshore zones; to contribute to NATO's missions; and to provide on-station and deployment forces overseas in support of French territorial interests and UN commitments. French territorial seas and economic zones are organized into two maritime districts (with headquarters in Brest and Toulon).

The strategic deterrent force comprises four nuclear-powered strategic-missile submarines of the *Triomphant* class (*Le Triomphant*, *Le Téméraire*, *Le Vigilant* and *Le Terrible*, which entered service in 1997, 1999, 2004 and 2010 respectively).

The principal surface ship is the 40,000-tonne nuclear-powered aircraft carrier *Charles de Gaulle*, which was launched at Brest in 1994 and commissioned in May 2001. Other surface combatants include 13 destroyers and 20 frigates. There are also six *Rubis* class nuclear-powered submarines.

The naval air arm, *Aviation Navale*, numbers some 6,400 personnel. Operational aircraft include Super-Étendard nuclear-capable strike aircraft, Étendard reconnaissance aircraft and maritime Rafale combat aircraft. A small Marine force of 1,550 *Fusiliers Marins* provides assault groups.

Personnel in 2006 numbered 43,995, including 10,265 civilians. There were 6,000 reserves in 2006.

Air Force

Created in 1934, the Air Force was reorganized in June 1994. The Conventional Forces in Europe (CFE) Agreement imposes a ceiling of 800 combat aircraft. In 2006 there were 304 combat aircraft, 157 transport aircraft and 342 aircraft for training purposes.

Personnel (2006) 63,600. Air Force reserves in 2006 numbered 4,300.

INTERNATIONAL RELATIONS

At a referendum in Sept. 1992 to approve the ratification of the Maastricht treaty on European Union of 7 Feb. 1992, 12,967,498 votes (50·8%) were cast for and 12,550,651 (49·2%) against. On 29 May 2005 France became the first European Union member to reject the proposed EU constitution, with 54·67% of votes cast in a referendum against the constitution and only 45·33% in favour.

France is the focus of the *Communauté Francophone* (French-speaking Community) which formally links France with many of its former colonies in Africa. A wide range of agreements, both with members of the Community and with other French-speaking countries, extend to economic and technical matters, and in particular to the disbursement of overseas aid.

ECONOMY

Agriculture accounted for 2% of GDP in 2008, industry 20% and services 78%.

Overview

France has high per capita income but GDP growth has lagged behind other major European economies. Growth averaged 2% from 2003–07, driven by private consumption resulting from a steady increase in real disposable income, although the failure of domestic supply to keep up with demand resulted in a net external drag on growth of roughly 0·5% of GDP per year over the period.

An estimated 3% of the workforce was employed in agriculture in 2010. Having secured the common agricultural policy (CAP), which subsidizes European agriculture, as a condition for establishing the EU, France is the largest beneficiary. In 2003 CAP reform saw most subsidies converted from price supports to direct income payments. While this creates less trade distortion, farm support remains a major point of contention in EU budget and WTO trade negotiations.

Labour force participation is particularly low and structural unemployment high. The re-emergence of government-aided jobs and labour market reforms has aided employment growth but continued rises in the minimum wage have held back job creation. Having reached a low of 7·6% in March 2008, unemployment rose steeply from late 2008 owing to the global economic downturn and stood at over 10% in 2012, its highest level since 1999. Youth unemployment reached 26·9% in Jan. 2013. The IMF suggests efforts to foster job creation, especially for young, low-skilled and senior workers, would aid efforts to boost growth and competitiveness, safeguard fiscal sustainability and reduce welfare spending.

The global financial crisis forced the economy into recession in 2008. Following four successive quarters of contraction, France emerged from recession in the second quarter of 2009 but GDP declined by 3·1% for the year as a whole. Policy response included a fiscal stimulus package comprising temporary investment expenditures and tax breaks, and measures to recapitalize banks to support liquidity. The eurozone debt crisis, shrinking private investment, loss of competitiveness and rising unemployment resulted in the economy once more falling into recession in the first half of 2012.

In 2002 France breached the 3% GDP budget deficit limit set by the eurozone's growth and stability pact, causing public debt to exceed 60% of GDP in 2003. It reduced its deficit to less than 3% of GDP in 2005, while health care spending was on target for the first time in recent history and real government spending remained constant for four years running. With the global financial crisis, the government deficit again increased, from 3·3% of GDP in 2008 to 7·5% in 2009 (although it has fallen since then), while public debt increased from 68·2% in 2008 to 89·3% of GDP in 2012. The Hollande government set about tackling the budget deficit largely through increased taxes, with the aim of reducing the overall fiscal deficit to 3% of GDP by 2013—although stalling growth has left this target subject to revision. The government has maintained its medium-term goal of a zero deficit by 2017. In Jan. 2012 Standard & Poor's stripped France of its triple-A credit rating. France's current account deteriorated from a surplus of 3·1% of GDP in 1999 to a deficit of 2·2 % of GDP in 2012, owing largely to weak exports. France's share of European exports dropped to 9·3% in 2011, down from 12·7% in 2000.

Efforts to address population ageing are key to long-term growth. The average legal age of retirement across the OECD is between 63 and 64 but in France most people stop working before they are 60, and only 40% of people over 55 were employed in 2011, compared to the OECD average of 54%. Major pension reforms were passed in 2010 aimed at gradually increasing the retirement age from 60 to 62 years, although in June 2012 the Hollande government announced it would lower the retirement age back to 60 years for a small group of workers.

The IMF has highlighted improving public finances, reforming the tax system and labour market, and introducing structural reforms to boost productivity as economic priorities.

For further developments *see* www.statesmansyearbook.com.

Currency

On 1 Jan. 1999 the euro (EUR) became the legal currency in France at the irrevocable conversion rate of 6·55957 francs to one euro. The euro, which consists of 100 cents, has been in circulation since 1 Jan. 2002. On the introduction of the euro there was a 'dual circulation' period before the franc ceased to be legal tender on 17 Feb. 2002. Euro banknotes in circulation on 1 Jan. 2002 had a total value of €84·2bn.

Foreign exchange reserves were US$26,170m. in Sept. 2009 and gold reserves 78·30m. troy oz. Total money supply was €431,879m. in Aug. 2009. Inflation rates (based on OECD statistics):

2002	2003	2004	2005	2006	2007	2008	2009	2010	2011
1·9%	2·2%	2·3%	1·9%	1·9%	1·6%	3·2%	0·1%	1·7%	2·3%

Franc Zone
13 former French colonies (Benin, Burkina Faso, Cameroon, Central African Republic, Chad, Comoros, the Republic of the Congo, Côte d'Ivoire, Gabon, Mali, Niger, Senegal and Togo), the former Spanish colony of Equatorial Guinea and the former Portuguese colony of Guinea-Bissau are members of a Franc Zone, the CFA (*Communauté Financière Africaine*). Comoros uses the Comorian franc. The *franc CFA* is pegged to the euro at a rate of 655·957 francs CFA to one euro. The franc CFP (*Comptoirs Français du Pacifique*) is the common currency of French Polynesia, New Caledonia and Wallis and Futuna. It is pegged to the euro at 119·3317422 francs CFP to the euro.

Budget
Central government revenue and expenditure in €1m.:

	2007	2008	2009
Revenue	798,126	816,535	722,353
Expenditure	841,194	871,631	908,581

Principal sources of revenue in 2009 were: social security contributions, €308·06bn.; taxes on goods and services, €178·92bn. Main items of expenditure by economic type in 2009: social benefits, €459·23bn.; compensation of employees, €187·65bn.

France's budget deficit in 2011 was 5·2% of GDP (2010, 7·1%; 2009, 7·5%). The required target set by the EU is a budget deficit of no more than 3%.

The standard rate of VAT is 19·6% (reduced rates, 7·0%, 5·5% and 2·1%). In 2013 the top rate of income tax was 45·0% and corporate tax was 33·3%.

Performance
Real GDP growth rates (based on OECD statistics):

2002	2003	2004	2005	2006	2007	2008	2009	2010	2011
0·9%	0·9%	2·3%	1·8%	2·6%	2·2%	−0·2%	−3·1%	1·6%	1·7%

The real GDP growth rate in 2012 according to INSEE, the French National Institute for Statistics and Economic Studies, was 0·0%. Total GDP in 2011 was US$2,773·0bn.

Banking and Finance
The central bank and bank of issue is the Banque de France (*Governor*, Christian Noyer, appointed 2003 and reappointed 2009), founded in 1800, and nationalized on 2 Dec. 1945. The Governor is appointed for a six-year term (renewable once) and heads the nine-member Council of Monetary Policy.

The National Credit Council, formed in 1945 to regulate banking activity and consulted in all political decisions on monetary policy, comprises 51 members nominated by the government; its president is the minister for the economy; its vice-president is the governor of the Banque de France.

In 2008 there were 722 banks and other credit institutions, including 304 financial companies, 290 commercial banks and 104 mutual or co-operative banks. Four principal deposit banks were nationalized in 1945, the remainder in 1982. The banking and insurance sectors underwent a flurry of mergers, privatizations, foreign investment, corporate restructuring and consolidation in 1997, in both the national and international fields. The largest banks in April 2009 by assets were BNP Paribas (US$2,888·73bn.) and Crédit Agricole (US$2,064·17bn.). In April 2010 the largest banks by market capitalization were BNP Paribas (US$68·44bn.) and Société Générale (US$34·25bn.).

The former state banks, the Caisses d'Épargne, became co-operative savings banks in 1999 although the group remains partly state-owned. There is a state-owned postal bank, La Banque Postale. Deposited funds are centralized by a non-banking body, the Caisse des Dépôts et Consignations, which finances a large number of local authorities and state-aided housing projects, and carries an important portfolio of transferable securities.

Gross external debt totalled US$4,838,186m. in June 2012.

France attracted US$40·9bn. worth of foreign direct investment in 2011 (down from US$96·2bn. in 2007 although up from US$24·2bn. in 2009).

There is a stock exchange (*Bourse*) in Paris; it is a component of Euronext, which was created in Sept. 2000 through the merger of the Paris, Brussels and Amsterdam bourses.

ENERGY AND NATURAL RESOURCES

In 2008, 11·0% of energy consumption in mainland France came from renewables (wind power, solar power, hydro-electric power, tidal power, geothermal energy and biomass), compared to the European Union average of 10·3%. A target of 23% has been set by the EU for 2020.

Environment
France's carbon dioxide emissions from the consumption and flaring of fossil fuels in 2008 were the equivalent of 6·5 tonnes per capita. An *Environmental Performance Index* compiled in 2008 ranked France tenth in the world, with 87·8%. The index examined various factors in six areas—air pollution, biodiversity and habitat, climate change, environmental health, productive natural resources and water resources.

Electricity
EDF is responsible for power generation and supply. It was privatized in Nov. 2005 when the government sold a 15·6% stake in the company. Installed capacity was 116·5m. kW in 2007. Electricity production in 2007: 569·84bn. kWh, of which 77·2% was nuclear. Hydro-electric power contributes about 11·2% of total electricity output. Consumption per capita in 2007 was 8,313 kWh. In 2007 France was the world's biggest exporter of electricity, with 67·6bn. kWh. EDF is Europe's leading electricity producer, generating 610·6bn. kWh in 2007.

France, not rich in natural energy resources, is at the centre of Europe's nuclear energy industry. In 2010 there were 58 nuclear reactors in operation—more than in any other country in the world apart from the USA—with a generating capacity of 63,236 MW. France has the highest percentage of its electricity generated through nuclear power of any country.

Oil and Gas
In 2007, 974,000 tonnes of crude oil were produced. The greater part came from the Parentis oilfield in the Landes. Reserves in 2007 totalled 122m. bbls. The importation and distribution of natural gas is the responsibility of GDF Suez, a company formed in July 2008 following a merger between Gaz de France and fellow utility group Suez. The French government has a 36·7% stake in the combined company.

Production of natural gas (2007) was 1·0bn. cu. metres. Natural gas reserves were 9·7bn. cu. metres in 2007.

Minerals
France is a significant producer of nickel, uranium, iron ore, bauxite, potash, pig iron, aluminium and coal. Société Le Nickel extracts in New Caledonia and is the world's third largest nickel producer.

France's last coal mine closed in April 2004. Production of other principal minerals and metals, in 1,000 tonnes (2006): salt, 9,371; aluminium, 442.

Agriculture
France has the highest agricultural production in Europe. In 2007 the agricultural sector employed about 1,020,000 people, down from 2,038,000 in 1988 and over 6m. in the mid-1950s. Agriculture accounts for 10·4% of exports and 8·3% of imports.

In 2007 there were 507,000 holdings (average size 54 ha.), down from over 1m. in 1988. There were 1,176,000 tractors and 80,000 harvester-threshers in 2005. Although the total number of tractors has been declining steadily in recent years, increasingly more powerful ones are being used. In 2005, 497,000 tractors in use were of 80 hp or higher, compared to 96,000 in 1979.

Of the total area of France (54·9m. ha.), the utilized agricultural area comprised 29·27m. ha. in 2009. 18·26m. ha. were arable, 9·87m. ha. were grassland and 1·06m. ha. were under permanent crops including vines (0·84m. ha.).

Area under cultivation and yield for principal crops:

	Area (1,000 ha.)			Production (1,000 tonnes)		
	2006	2007	2008	2006	2007	2008
Wheat	5,246	5,239	5,492	35,364	32,764	39,002
Sugar beets	379	393	349	29,871	33,230	30,306
Maize	1,465	1,484	1,702	12,775	14,357	15,819
Barley	1,667	1,699	1,799	10,401	9,474	12,171
Potatoes	158	158	156	6,363	7,183	6,808
Rapeseed	1,406	1,619	1,421	4,144	4,691	4,719
Triticale[1]	331	324	343	1,694	1,450	1,821
Sunflower seeds	645	520	630	1,440	1,311	1,608

[1]Cross between wheat and rye.

Production of principal fruit crops (in 1,000 tonnes) as follows:

	2006	2007	2008
Grapes	6,777	6,019	5,664
Apples	2,081	2,144	1,940
Peaches and nectarines	395	365	301
Melons	301	242	274
Pears	226	203	156
Plums and sloes	234	249	147

Total fruit and vegetable production in 2008 was 13,732,900 tonnes. Other important vegetables include tomatoes (714,683 tonnes in 2008), carrots (556,517 tonnes), sweetcorn (479,718 tonnes) and cauliflowers (392,648 tonnes). France is the world's leading producer of sugar beets. Total area under cultivation and yield of grapes from the vine (2008): 813,496 ha.; 5·7m. tonnes. Wine production (2008): 45,692,000 hectolitres. France was the second largest wine producer in the world in 2008 (16·1% of the world total) after Italy. Consumption in France has declined dramatically in recent times, from nearly 120 litres per person in 1966 to 53 litres per person in 2008.

In 2008 France set aside 583,800 ha. (2·1% of its agricultural land) for the growth of organic crops, compared to the EU average of 4·1% in 2007.

Livestock (2008, in 1,000): cattle, 19,887; pigs, 14,801; sheep, 8,171; goats, 1,228; horses, 420; chickens (estimate), 175,000; turkeys, 25,253; ducks, 22,848. Livestock products (2008, in 1,000 tonnes): pork, bacon and ham, 2,259; beef and veal, 1,496; lamb and mutton, 130; eggs, 11,618m. units. Milk production, 2008 (in 1,000 hectolitres): cow, 235,410; goat, 5,673; sheep, 2,442. Cheese production, 2008, 1,861,000 tonnes. France is the second largest cheese producer in the world after the USA.

Source: SCEES/Agreste

Forestry
Forestry is France's richest natural resource. In 2010 forests covered 15·95m. ha. (29% of the land area). In 1990 the area under forests had been 14·54m. ha. 48,000 ha. of land in France was reforested annually between 2005 and 2010. Growing stock was 2,584m. cu. metres in 2010 (1,647m. cu. metres broadleaved and 937m. cu. metres coniferous). Timber production in 2007 was 62·76m. cu. metres.

Fisheries
In 2008 there were 7,389 fishing vessels (of which 4,979 were in mainland France). Catch in 2010 was 426,514 tonnes, of which 424,014 tonnes were from marine waters.

INDUSTRY
The leading companies by market capitalization in France in March 2012 were: Total, an integrated oil company (US$120·4bn.); Sanofi, a pharmaceuticals company (US$104·0bn.); and LVMH, a luxury goods conglomerate (US$87·1bn.).

Chief industries: steel, chemicals, textiles, aircraft, machinery, electronic equipment, tourism, wine and perfume. In 2008 industry accounted for 20% of GDP, with manufacturing contributing 12%.

Industrial production (in 1,000 tonnes): distillate fuel oil (2007, including Monaco), 34,392; cement (2008), 21,443; crude steel (2007), 19,252; rolled steel (2006), 17,437; petrol (2007, including Monaco), 16,479; pig iron (2006), 13,013; residual fuel oil (2007, including Monaco), 11,441; jet fuel (2007, including Monaco), 5,536. France is one of the biggest producers of mineral water, with 12,159m. litres in 2005. In 2005 soft drinks production was 4,777m. litres, beer production 1,720m. litres and cigarette production 46·5bn. units.

Engineering production (in 1,000 units): cars (2009), 1,819; commercial vehicles (2009), 228; car tyres (2007), 54,000.

Labour
Of 25,628,000 people in employment in 2008, 46·9% were women. By sector, 74·5% worked in services (58·1% in 1980), 22·0% in industry and construction (33·1% in 1980) and 3·4% in agriculture (8·8% in 1980). Some 3·1m. people work in the public sector at national and local level.

A new definition of 'unemployed' was adopted in Aug. 1995, omitting persons who had worked at least 78 hours in the previous month. The unemployment rate was 10·6% in Dec. 2012 (up from 9·6% in 2011 as a whole). The rate among the under 25s is more than double the overall national rate.

Conciliation boards (conseils de prud'hommes) mediate in labour disputes. They are elected for five-year terms by two colleges of employers and employees. There were a total of 1,553,000 working days lost to strike action in 2007. Between 1996 and 2005 strikes cost France an average of 53 days per 1,000 employees a year. In Jan. 2013 the minimum wage (SMIC) was raised to €9·43 an hour (€1,430·22 month for a 35-hour week); it affected about 3·4m. wage-earners in July 2008. The average annual salary in the private and semi-public sectors was €31,932 in 2007. The minimum retirement age was lowered from 65 to 60 by then President Mitterrand in 1983. In Nov. 2010 then President Sarkozy signed a bill to increase it gradually to 62 by 2018. In Nov. 2011 it was announced that the retirement age would rise from 60 to 62 by 2017 instead of 2018. However, after taking office in May 2012 President François Hollande partially reversed the decision, reducing the retirement age from 62 to 60 for people who have completed a minimum 41 years of work. The average actual age for retirement was 59 in 2008. A five-week annual holiday is statutory.

In March 2005 the National Assembly voted by 350 to 135 to amend the working hours law restricting the legal working week to 35 hours, introduced by the former between 1998–2000. Under the new proposals employees can, in agreement with their employer, work up to 48 hours per week. There is no change in the legal working week: any increased hours are on a voluntary basis. The proposal also allows for the increase of overtime hours from 180 to 220 per year, payable at 125% of the normal hourly rate. To encourage the working of longer hours, payment of overtime work is now exempt from individual income tax, social contributions and social taxes (since 1 Oct. 2007).

INTERNATIONAL TRADE
Imports and Exports
In 2008 imports (c.i.f.) totalled US$696·65bn. (US$619·27bn. in 2007); exports (f.o.b.), US$596·10bn. (US$542·64bn. in 2007). Principal imports include: oil, machinery and equipment, chemicals, iron and steel, and foodstuffs. Major exports: metals,

chemicals, industrial equipment, consumer goods and agricultural products.

In 2007 chemicals and manufactured goods accounted for 41·4% of France's imports and 42·6% of exports; machinery and transport equipment 35·0% of imports and 40·1% of exports; food, live animals, beverages and tobacco 7·3% of imports and 10·7% of exports; mineral fuels, lubricants and related materials 13·5% of imports and 3·9% of exports; inedible crude materials (except fuels), and animal and vegetable oil and fats 2·7% of imports and 2·6% of exports.

In 2007 the chief import sources (as % of total imports) were as follows: Germany, 16·6%; Italy, 8·5%; Belgium, 8·3%; Spain, 7·0%; China, 6·3%. The chief export markets (as % of total) were: Germany, 14·4%; Spain, 9·6%; Italy, 9·2%; UK, 8·4%; Belgium, 7·5%. Imports from fellow European Union members accounted for 62·0% of all imports, and exports to fellow European Union members constituted 65·5% of the total.

Trade Fairs
Paris ranks as the third most popular convention city behind Singapore and Brussels according to the Union des Associations Internationales (UAI), hosting 3·6% of all international meetings held in 2010.

COMMUNICATIONS

Roads
In 2007 there were 951,125 km of road, including 11,010 km of motorway and 9,115 km of highways and main roads. France has the longest road network in the EU. Around 90% of all freight is transported by road. In 2007 there were 30·70m. passenger cars (498 per 1,000 inhabitants), 6·27m. lorries and vans, and 83,000 buses and coaches. Road passenger traffic in 2007 totalled 775bn. passenger-km. In 2007 there were 4,620 road deaths, down from 8,445 in 1997.

Only Singapore ranked ahead of France for quality of road infrastructure in the World Economic Forum's *Global Competitiveness Report 2009–2010*.

Rail
In 1938 all the independent railway companies were merged with the existing state railway system in a Société Nationale des Chemins de Fer Français (SNCF), which became a public industrial and commercial establishment in 1983. Legislation came into effect in 1997 which vested ownership of the railway infrastructure (track and signalling) in a newly established public corporation, the Réseau Ferré de France (RFF/French Rail Network). The RFF is funded by payments for usage from the SNCF, government and local subventions and authority capital made available by the state derived from the proceeds of privatization. The SNCF remains responsible for maintenance and management of the rail network. The legislation also envisages the establishment of regional railway services which receive funds previously given to the SNCF as well as a state subvention. These regional bodies negotiate with SNCF for the provision of suitable services for their area. SNCF is the most heavily indebted and subsidized company in France.

In 2010 the RFF-managed network totalled 29,473 km of track (15,424 km electrified). High-speed TGV lines link Paris to the southwest, southeast and east of France, and north from Paris and Lille to the Channel Tunnel (Eurostar). The high-speed TGV line appeared in 1981; it had 1,896 km of track in 2010, and another 2,000 km planned by 2020. Services from London through the Channel Tunnel began operating in 1994. Rail passenger traffic in 2008 totalled 98·3bn. passenger km and freight tonne-km came to 40·6bn.

The Paris transport network consisted in 2005 of 212 km of metro (297 stations), 115 km of regional express railways and 31 km of tramway. Outside Paris and the Île-de-France region there are metros in Lille (45 km), Lyon (30 km), Marseille (22 km),

Rennes (9 km) and Toulouse (27 km), and tram/light railway networks in Angers (12 km), Bordeaux (25 km), Brest (14 km), Caen (16 km), Clermont-Ferrand (14 km), Grenoble (28 km), Le Havre (13 km), Le Mans (15 km), Lille (22 km), Lyon (18 km), Marseille (12 km), Montpellier (12 km), Mulhouse (12 km), Nancy (11 km), Nantes (40 km), Nice (9 km), Orléans (18 km), Reims (11 km), Rouen (15 km), St Étienne (9 km), Strasbourg (25 km), Toulouse (11 km) and Valenciennes (9 km).

France was ranked fourth for rail infrastructure in the World Economic Forum's *Global Competitiveness Report 2009–2010*.

Civil Aviation
The main international airports are at Paris (Charles de Gaulle), Paris (Orly), Bordeaux (Mérignac), Lyon (Satolas), Marseille-Provence, Nice-Côte d'Azur, Strasbourg (Entzheim), Toulouse (Blagnac), Clermont-Ferrand (Aulunat) and Nantes (Atlantique). The following had international flights to only a few destinations in 2003: Brest, Caen, Carcassonne, Le Havre, Le Touquet, Lille, Pau, Rennes, Rouen and Saint-Étienne. The national airline, Air France, was 54·4% state-owned but merged in Oct. 2003 with the Dutch carrier KLM to form Air France-KLM. In the process the share owned by the French state fell to 44·2%. In Dec. 2004 the government sold off a further 18·4% to reduce its stake to 25·8%, and in the meantime the government's share has come down still further to 15·9%. In 2005 Air France carried 42·9m. passengers (25·9m. on international flights); passenger-km totalled 115·1bn. The main other French airline is Corsairfly. In 2008 Charles de Gaulle airport handled 60,874,681 passengers (55,825,413 on international flights) and 2,039,460 tonnes of freight. Only Heathrow handled more international passengers in 2008. Orly was the second busiest airport, handling 26,209,703 passengers (11,825,460 on domestic flights) and 95,770 tonnes of freight. Nice was the third busiest for passengers, with 10,382,566 (6,061,002 on international flights).

In April 2003 Air France announced that Concorde, the world's first supersonic jet which began commercial service in 1976, would be permanently grounded from Oct. 2003.

Shipping
In Jan. 2009 there were 232 ships of 300 GT or over registered, totalling 6,025,000 GT. Of the 232 vessels registered, 88 were passenger ships, 58 oil tankers, 43 general cargo ships, 26 container ships, 13 liquid gas tankers, three bulk carriers and there was one chemical tanker. The French-controlled fleet comprised 285 vessels of 1,000 GT or over in Jan. 2009, of which 164 were under foreign flags and 121 under the French flag. The chief ports are Marseille, Le Havre, Dunkerque, Calais and Saint-Nazaire.

France has extensive inland waterways. Canals are administered by the public authority France Navigable Waterways (VNF). In 2006 there were approximately 8,800 km of navigable rivers and canals (the longest network in the EU), with a total traffic in 2009 of 68·0m. tonnes.

Telecommunications
France Télécom became a limited company on 1 Jan. 1997. In 2007 there were 34·8m. main (fixed) telephone lines. In the same year mobile phone subscribers numbered 55·4m. (897·0 per 1,000 persons). The largest operators are Orange France, with a 46% share of the market, and SFR, with a 35% share. There were 42·3m. internet users in 2008. The fixed broadband penetration rate in Dec. 2010 was 33·7 subscribers per 100 inhabitants. In March 2012 there were 23·5m. Facebook users (37% of the population).

SOCIAL INSTITUTIONS

Justice
The system of justice is divided into two jurisdictions: the judicial and the administrative. Within the judicial jurisdiction are common law courts including 473 lower courts (*tribunaux*

d'instance, 11 in overseas departments), 181 higher courts (*tribunaux de grande instance*, 5 *tribunaux de première instance* in the overseas territories) and 454 police courts (*tribunaux de police*, 11 in overseas departments).

The *tribunaux d'instance* are presided over by a single judge. The *tribunaux de grande instance* usually have a collegiate composition, but may be presided over by a single judge in some civil cases. The *tribunaux de police*, presided over by a judge on duty in the *tribunal d'instance*, deal with petty offences (*contraventions*); correctional chambers (*chambres correctionelles*, of which there is at least one in each *tribunal de grande instance*) deal with graver offences (*délits*), including cases involving imprisonment up to five years. Correctional chambers normally consist of three judges of a *tribunal de grande instance* (a single judge in some cases). Sometimes in cases of *délit*, and in all cases of more serious *crimes*, a preliminary inquiry is made in secrecy by one of 569 examining magistrates (*juges d'instruction*), who either dismisses the case or sends it for trial before a public prosecutor.

Within the judicial jurisdiction are various specialized courts, including 191 commercial courts (*tribunaux de commerce*), composed of tradesmen and manufacturers elected for two years initially, and then for four years; 271 conciliation boards (*conseils de prud'hommes*), composed of an equal number of employers and employees elected for five years to deal with labour disputes; 437 courts for settling rural landholding disputes (*tribunaux paritaires des baux ruraux*, 11 in overseas departments); and 116 social security courts (*tribunaux des affaires de sécurité sociale*).

When the decisions of any of these courts are susceptible of appeal, the case goes to one of the 35 courts of appeal (*cours d'appel*), composed each of a president and a variable number of members. There are 104 courts of assize (*cours d'assises*), each composed of a president who is a member of the court of appeal, and two other magistrates, and assisted by a lay jury of nine members. These try crimes involving imprisonment of over five years. The decisions of the courts of appeal and the courts of assize are final. However, the Court of Cassation (*cour de cassation*) has discretion to verify if the law has been correctly interpreted and if the rules of procedure have been followed exactly. The Court of Cassation may annul any judgment, following which the cases must be retried by a court of appeal or a court of assizes.

The administrative jurisdiction exists to resolve conflicts arising between citizens and central and local government authorities. It consists of 36 administrative courts (*tribunaux administratifs*, of which eight are in overseas departments and territories) and 15 administrative courts of appeal (*cours administratives d'appel*, of which eight are in overseas departments and territories). The Council of State is the final court of appeal in administrative cases, though it may also act as a court of first instance.

Cases of doubt as to whether the judicial or administrative jurisdiction is competent in any case are resolved by a *Tribunal de conflits* composed in equal measure of members of the Court of Cassation and the Council of State. In 1997 the government restricted its ability to intervene in individual cases of justice.

Penal code
A revised penal code came into force on 1 March 1994, replacing the *Code Napoléon* of 1810. Penal institutions consist of: (1) *maisons d'arrêt*, where persons awaiting trial as well as those condemned to short periods of imprisonment are kept; (2) punishment institutions – (a) central prisons (*maisons centrales*) for those sentenced to long imprisonment, (b) detention centres for offenders showing promise of rehabilitation, and (c) penitentiary centres, establishments combining (a) and (b); (3) hospitals for the sick. Special attention is being paid to classified treatment and the rehabilitation and vocational re-education of prisoners including work in open-air and semi-free establishments. Juvenile delinquents go before special judges in 139 (11 in overseas departments and territories) juvenile courts

(*tribunaux pour enfants*); they are sent to public or private institutions of supervision and re-education.

A new post of Defender of Rights (*Défenseur des droits*) was created in March 2011 as a successor to the Ombudsman (*Médiateur*). The current Defender of Rights is Dominique Baudis.

Capital punishment was abolished in Aug. 1981. In metropolitan France the detention rate in 2006 was 98·4 prisoners per 100,000 population, up from 50 per 100,000 in 1975. The average period of detention in 2004 was 8·4 months. The principal offences committed were: theft, 21·1%; drink driving, 20·6%; assault (including rape), 16·8%; drug-related offences, 5·8%. The population of the 194 penal establishments (six for minors) in 2008 was 64,003, including 2,379 women.

Elliott, Catherine, Jeanpierre, Eric and Vernon, Catherine, *French Legal System*. 2nd ed. 2006
Weston, M., *English Reader's Guide to the French Legal System*. 1991

Education
The primary, secondary and higher state schools constitute the 'Université de France'. Its Supreme Council of 84 members has deliberative, administrative and judiciary functions, and as a consultative committee advises respecting the working of the school system; the inspectors-general are in direct communication with the Minister. For local education administration France is divided into 25 academic areas, each of which has an Academic Council whose members include a certain number elected by the professors or teachers. The Academic Council deals with all grades of education. Each is under a Rector, and each is provided with academy inspectors, one for each department.

Compulsory education is provided for children of 6–16. The educational stages are as follows:

1. Non-compulsory pre-school instruction for children aged 2–5, to be given in infant schools or infant classes attached to primary schools.

2. Compulsory elementary instruction for children aged 6–11, to be given in primary schools and certain classes of the *lycées*. It consists of three courses: preparatory (one year), elementary (two years) and intermediary (two years). Children with special needs are cared for in special institutions or special classes of primary schools.

3. Lower secondary education (*Enseignement du premier cycle du Second Degré*) for pupils aged 11–15, consists of four years of study in the *lycées* (grammar schools), *Collèges d'Enseignement Technique* (CES) or *Collèges d'Enseignement Général* (CEG).

4. Upper secondary education (*Enseignement du second cycle du Second Degré*) for pupils aged 15–18: (1) *Long, général* or *professionel* provided by the *lycées* and leading to the *baccalauréat* or to the *baccalauréat de technicien* after three years; and (2) *Court*, professional courses of three, two and one year are taught in the *lycées d'enseignement professionel*, or the specialized sections of the *lycées*, CES or CEG.

The following table shows the number of schools in 2006–07 and the numbers of teaching staff and pupils in 2007:

	Number of schools	Teaching staff	Pupils
Nursery	17,410	141,476	2,594,074
Primary	38,257	216,654	4,105,628
Secondary	11,410	490,955	5,940,366

Higher education is provided by the state free of charge in universities and in special schools, and by private individuals in the free faculties and schools. Legislation of 1968 redefined the activities and workings of universities. Bringing several disciplines together, 780 units for teaching and research (*UER—Unités d'Enseignement et de Recherche*) were formed which decided their own teaching activities, research programmes and procedures for checking the level of knowledge gained. In 1984 they were reclassified as units for training and research (*UFR—Unités de Formation et de Recherche*). They and the other parts of each university must respect the rules designed to maintain the national standard of qualifications. The UFRs form the basic units of the 78 state universities in mainland France and three national polytechnic institutes (with university status), which are grouped into 25 administrative Académies. Private universities include seven Catholic universities, in Angers, Lille, Lyon, Paris, Rennes, Toulouse and La Roche-sur-Yon in the Vendée region. There were 2,179,505 students in higher education in 2007.

Outside the university system, higher education (academic, professional and technical) is provided by over 400 schools and institutes, including the 177 *Grandes Écoles*, which are highly selective public or private institutions offering mainly technological or commercial curricula. These have an annual output of about 20,000 graduates, and in 2004–05 there were also 73,147 students in preparatory classes leading to the *Grandes Écoles*; 230,275 students were registered in the Sections de Techniciens Supérieurs and 100,899 in the Écoles d'Ingénieurs.

The adult literacy rate is at least 99%.

In 2006 public expenditure on education came to 5·6% of GNI and represented 10·6% of total government expenditure.

Health

Ordinances of 1996 created a new regional regime of hospital administration and introduced a system of patients' records to prevent abuses of public health benefits. In 2007 there were 972 public and 1,800 private health care establishments with 316,551 and 174,925 beds respectively. There were 208,191 physicians, 41,444 dentists, 483,380 nurses, 70,498 pharmacists and 17,483 midwives in 2007.

In 2008 France spent 11·2% of its GDP on health (the highest percentage in the EU), with public spending amounting to 77·8% of the total.

In 2009, 33% of the population aged 15 and over were smokers. The average French adult drinks 13·0 litres of alcohol a year.

Welfare

An order of 4 Oct. 1945 laid down the framework of a comprehensive plan of Social Security and created a single organization which superseded the various laws relating to social insurance, workmen's compensation, health insurance, family allowances, etc. All previous matters relating to Social Security are dealt with in the Social Security Code, 1956; this has been revised several times. The Chamber of Deputies and Senate, meeting as Congress on 19 Feb. 1996, adopted an important revision of the Constitution giving parliament powers to review annually the funding of social security (previously managed by the trade unions and employers' associations), and to fix targets for expenditure in the light of anticipated receipts.

In 2008 the welfare system accounted for €601bn., representing 31% of GDP. The Social Security budget had a deficit of some €20·2bn. in 2009.

Contributions. The general social security contribution (CSG) introduced in 1991 was raised by 4% to 7·5% in 1997 by the government of Lionel Jospin in an attempt to dramatically reduce the deficit on social security spending, effectively almost doubling the CSG. All wage-earning workers or those of equivalent status are insured regardless of the amount or the nature of the salary or earnings. The funds for the general scheme are raised mainly from professional contributions, these being fixed within the limits of a ceiling and calculated as a percentage of the salaries. The calculation of contributions payable for family allowances, old age and industrial injuries relates only to this amount; on the other hand, the amount payable for sickness, maternity expenses, disability and death is calculated partly within the limit of the 'ceiling' and partly on the whole salary. These contributions are the responsibility of both employer and employee, except in the case of family allowances or industrial injuries, where they are the sole responsibility of the employer.

Self-employed Workers. From 17 Jan. 1948 allowances and old-age pensions were paid to self-employed workers by independent insurance funds set up within their own profession, trade or business. Schemes of compulsory insurance for sickness were instituted in 1961 for farmers, and in 1966, with modifications in 1970, for other non-wage-earning workers.

Social Insurance. The orders laid down in Aug. 1967 ensure that the whole population can benefit from the Social Security Scheme; at present all elderly persons who have been engaged in the professions, as well as the surviving spouse, are entitled to claim an old-age benefit.

Sickness Insurance refunds the costs of treatment required by the insured and the needs of dependants.

Maternity Insurance covers the costs of medical treatment relating to the pregnancy, confinement and lying-in period; the beneficiaries being the insured person or the spouse.

Insurance for Invalids is divided into three categories: (1) those who are capable of working; (2) those who cannot work; (3) those who, in addition, are in need of the help of another person. According to the category, the pension rate varies from 30 to 50% of the average salary for the last ten years, with additional allowance for home help for the third category.

Old-Age Pensions for workers were introduced in 1910. Over the period 2003–08 the duration a private sector wage-earner or standard civil servant had to work in order to qualify for a pension was raised from 37½ years to 40 years. As a result, standard public sector workers were required to contribute for the same period as private sector workers. The contribution period rose to 41 years in 2012 and thereafter was scheduled to increase in line with rises in life expectancy, so maintaining the ratio of pension payment period to contribution period. Pensions are payable at 60 to anyone with at least 25% insurance coverage. A pension worth 50% of the adjusted average salary is provided to those who have paid a full 40 years (160 quarters) worth of insurance. The pension is proportionately reduced for coverage less than 160 quarters (or less than 150 quarters for those claiming their pension before 2004). The exception is the special retirement plan for which employees of several state-owned companies (notably the SNCF) and organizations including the military and the police are eligible, allowing them to claim a full pension after 37½ years' contribution. A law was passed in Nov. 2010 to raise the minimum retirement age from 60 to 62 by 2017 (its initial target of 2018 amended in Nov. 2011 as part of a package of austerity measures), with the fully pensionable retirement age rising from 65 to 67. However, in June 2012 the Hollande government lowered the minimum retirement age back to 60 years for people who have worked 41 years, to be financed by an increase in pension contributions.

There is also an allowance payable to low-income pensioners, who also receive an old-age supplement at 65. A child's supplement, worth 10% of the pension, is awarded to those who have three or more children. Citizens who do not qualify for a pension may be allowed to claim an old-age special allowance.

Family Allowances. A controversial programme of means-testing for Family Allowance was introduced in 1997 by the new administration. The Family Allowance benefit system comprises: (a) Family allowances proper, equivalent to 25·5% of the basic monthly salary for two dependent children, 46% for the third child, 41% for the fourth child, and 39% for the fifth and each subsequent child; a supplement equivalent to 9% of the basic monthly salary for the second and each subsequent dependent child more than ten years old, and 16% for each dependent child over 15 years. (b) Family supplement for persons with at least three children or one child aged less than three years. (c) Antenatal grants. (d) Maternity grant is equal to 260% of basic salary. Increase for multiple births or adoptions, 198%; increase for birth or adoption of third or subsequent child, 457%. (e) Allowance for specialized education of handicapped children. (f) Allowance for orphans. (g) Single parent allowance. (h) Allowance for opening of school term. (i) Allowance for accommodation, under certain circumstances. (j) Minimum family income for those with at least three children. Allowances (b), (g), (h) and (j) only apply to those whose annual income falls below a specified level.

Workmen's Compensation. The law passed by the National Assembly on 30 Oct. 1946 forms part of the Social Security Code and is administered by the Social Security Organization. Employers are invited to take preventive measures. The application of these measures is supervised by consulting engineers (assessors) of the local funds dealing with sickness insurance, who may compel employers who do not respect these measures to make additional contributions; they may, in like manner, grant rebates to employers who have in operation suitable preventive measures. The injured person receives free treatment, the insurance fund reimburses the practitioners, hospitals and suppliers chosen freely by the injured. In cases of temporary disablement, the daily payments are equal to half the total daily wage received by the injured. In case of permanent disablement, the injured person receives a pension, the amount of which varies according to the degree of disablement and the salary received during the past 12 months.

Unemployment Benefits vary according to circumstances (full or partial unemployment) which are means-tested.

Ambler, J. S. (ed.) *The French Welfare State: Surviving Social and Ideological Change.* 1992

RELIGION

A law of 1905 separated church and state. In 2005 there were 96 Roman Catholic dioceses in metropolitan France and 106 bishops. In Feb. 2013 there were eight cardinals. A survey conducted by the French Institute of Public Opinion in 2010 estimated that some 64% of the population was Roman Catholic, 28% non-religious/atheist, 3% Protestant and 5% belonged to other religions. There are generally estimated to be about 5m.–6m. Muslims in France. France has the third-largest Jewish population, after Israel and the USA.

CULTURE

Marseille is one of two European Capitals of Culture for 2013. The title attracts large European Union grants.

World Heritage Sites

There are 38 sites under French jurisdiction that appear on the UNESCO World Heritage List. They are (with year entered on list): Mont-Saint-Michel and its Bay, the Versailles Palace and Park, the church and hill at Vézelay, Burgundy (all 1979 and 2007); the prehistoric sites and decorated grottoes of the Vézère Valley (Dordogne) and Chartres Cathedral (both 1979); Fontainebleau Palace and Park, Amiens Cathedral, and the Roman and Romanesque monuments of Arles (all 1981); Fontenay's Cistercian Abbey and Orange's Roman theatre and Arc de Triomphe (both 1981 and 2007); from the Great Saltworks of Salins-les-Bains to the Royal Saltworks of Arc-et-Senans, Franche-Comté (1982 and 2009); the Place Stanislas, Place de la Carrière and Place d'Alliance in Nancy and the Gulf of Porto (the Gulf of Girolata, Scandola Nature Reserve and the Calanche of Piana) in Corsica (both 1983); the Church of Saint-Savin-sur-Gartempe in Poitou-Charentes (1983 and 2007); the Pont du Gard Roman aqueduct, Languedoc (1985 and 2007); Grande-Île, Strasbourg (1988); the Banks of the Seine and Reims' Notre Dame Cathedral, Abbey of Saint-Remi and Tau Palace (both 1991); Bourges Cathedral (1992); Avignon's historic centre (1995); the Canal du Midi, Languedoc (1996); the Historic Fortified City of Carcassonne (1997); Lyon's historic sites and the route of Santiago de Compostela (both 1998); Saint-Émilion Jurisdiction (1999); the Loire Valley between Sully-sur-Loire and Chalonnes-sur-Loire

(2000); Provins, the Town of Medieval Fairs (2001); the city of Le Havre (2005); the historic centre of Bordeaux (2007); fortifications of Vauban and the lagoons of New Caledonia: reef diversity and associated ecosystems (both 2008); the Episcopal City of Albi, Languedoc and the Pitons, cirques and remparts of Réunion (both 2010); the Causses and the Cévennes, Mediterranean agro-pastoral cultural landscapes (2011); and Nord-Pas de Calais Mining Basin (2012).

France shares the Pyrénées–Mount Perdu site (1997 and 1999) with Spain; the Belfries of Belgium and France (1999 and 2005) with Belgium; and the Preshistoric Pile dwellings around the Alps (2011) with Austria, Germany, Italy, Slovenia and Switzerland.

Press

There were 87 daily papers (18 nationals, 69 provincials) in 2008. The leading dailies are: *Ouest-France* (average circulation, 792,000), *Le Figaro* (average circulation, 333,000), *Le Monde* (average circulation, 332,000), *Le Parisien, L'Équipe, Sud Ouest, La Voix du Nord* and *Le Dauphiné Libéré.* The *Journal du Dimanche* is the only national Sunday paper. In 2008 total average daily press circulation was 10·3m. copies. In 2010 a total of 67,278 book titles were published.

Tourism

There were 76,800,000 foreign tourists in 2009; tourism receipts were US$49·4bn. France is the most popular tourist destination in the world, and receipts from tourism in 2009 were exceeded only in the USA and Spain. The most visited tourist attractions in 2007 were Disneyland Paris (14·5m.), the Louvre (8·2m.) and the Eiffel Tower (6·8m.). Around 17m. foreigners a year visit Paris. In 2009, 85·0% of tourists were from elsewhere in Europe and 7·1% from the Americas. Most visitors come from the UK, Belgium/Luxembourg, Germany, Italy, the Netherlands and Switzerland. There were 614,532 classified hotel rooms in 17,721 hotels in 2008.

Festivals
Religious Festivals
Ascension Day (40 days after Easter Sunday), Assumption of the Blessed Virgin Mary (15 Aug.) and All Saints Day (1 Nov.) are all public holidays.

Cultural Festivals
The Grande Parade de Montmartre, Paris (1 Jan.); the Carnival of Nice (Feb.–March); the Fête de la Victoire (8 May), celebrates victory in World War Two; the May Feasts take place in Nice regularly throughout May; the prestigious Cannes Film Festival, which has been running since 1946, lasts two weeks in mid-May; the Fête de la Musique (21 June); the Avignon Festival is a celebration of theatre that attracts average attendances of 140,000 each year and runs for most of July; Bastille Day (14 July) sees celebrations, parties and fireworks across the country. The Festival International d'Art Lyrique, focusing on classical music, opera and ballet, takes place in Aix-en-Provence every July. There are annual festivals of opera at Orange (July–Aug.) and baroque music at Ambronay (Sept.–Oct.). Celtic music and dance is celebrated at the Festival Interceltique de Lorient (Aug.). The biggest pop and rock festivals are Eurockéennes in Belfort (July), Les Vieilles Charrues in Carhaix (July), Rock en Seine in Paris (Aug.) and La Route du Rock in Saint-Malo (Aug.).

DIPLOMATIC REPRESENTATIVES

Of France in the United Kingdom (58 Knightsbridge, London, SW1X 7JT)
Ambassador: Bernard Emié.

Of the United Kingdom in France (35 rue du Faubourg St Honoré, 75363 Paris Cedex 08)
Ambassador: Sir Peter Ricketts, GCMG.

Of France in the USA (4101 Reservoir Rd, NW, Washington, D.C., 20007)
Ambassador: François Delattre.

Of the USA in France (2 Ave. Gabriel, 75382 Paris Cedex 08)
Ambassador: Charles H. Rivkin.

Of France to the United Nations
Ambassador: Gérard Araud.

Of France to the European Union
Permanent Representative: Philippe Étienne.

FURTHER READING

Institut National de la Statistique et des Études Économiques: *Annuaire statistique de la France* (from 1878); *Bulletin mensuel de statistique* (monthly); *Documentation économique* (bi-monthly); *Économie et Statistique* (monthly); *Tableaux de l'Économie Française* (biennially, from 1956); *Tendances de la Conjoncture* (monthly).

Agulhon, Maurice, *De Gaulle: Histoire, Symbole, Mythe.* 2000
Agulhon, M., and Nevill, A., *The French Republic, 1879–1992.* 1993
Ardagh, John, *France in the New Century: Portrait of a Changing Society.* 1999
Ardant, P., *Les Institutions de la Ve République.* 1992
Bell, David, *Presidential Power in Fifth Republic France.* 2000.—*Parties and Democracy in France: Parties under Presidentialism.* 2000
Brouard, Sylvain, Appelton, Andrew M. and Mazur, Amy G. (eds.) *The French Fifth Republic at Fifty: Beyond Stereotypes.* 2008
Chafer, Tony and Sackur, Amanda, (eds.) *French Colonial Empire and the Popular Front.* 1999
Cole, Alistair, Le Galès, Patrick and Levy, Jonah, (eds.) *Developments in French Politics 4.* 2008
Cubertafond, A., *Le Pouvoir, la Politique et l'État en France.* 1993
Culpepper, Pepper D., Hall, Peter A. and Palier, Bruno, (eds.) *Changing France: The Politics that Markets Make.* 2008
Drake, Helen, *Contemporary France.* 2011
L'État de la France. Annual

Friend, Julius W., *The Long Presidency: France in the Mitterrand Years, 1981–95.* 1999
Gildea, R., *France since 1945.* 1996
Guyard, Marius-François, (ed.) *Charles de Gaulle: Mémoires.* 2000
Hollifield, J. F. and Ross, G., *Searching for the New France.* 1991
Jack, A., *The French Exception.* 1999
Jones, C., *The Cambridge Illustrated History of France.* 1994
Kedward, Rod, *France and the French: A Modern History.* 2007
Knapp, Andrew, *Parties and the Party System in France: A Disconnected Democracy?* 2004
Lacoutre, Jean, *Mitterrand: Une histoire de Français.* 2 vols. 1999
Lewis-Beck, Michael S., *The French Voter: Before and After the 2002 Elections.* 2004
MacLean, Mairi, *The Mitterrand Years: Legacy and Evaluation.* 1999
McMillan, J. F., *Twentieth-Century France: Politics and Society in France, 1898–1991.* 2nd ed. [of *Dreyfus to De Gaulle*]. 1992
Milner, Susan and Parsons, Nick, (eds.) *Reinventing France: State and Society in the 21st Century.* 2004
Noin, D. and White, P., *Paris.* 1998
Peyrefitte, Alain, *C'était de Gaulle.* 2000
Popkin, J. D., *A History of Modern France.* 1994
Price, Roger, *A Concise History of France.* 1993
Sowerwine, Charles, *France since 1870.* 2nd ed. 2009
Stevens, Anne, *Government and Politics of France.* 2003
Tiersky, Ronald, *Mitterrand in Light and Shadow.* 1999.—*François Mitterrand: The Last French President.* 2000
Tippett-Spiritou, Sandy, *French Catholicism.* 1999
Zeldin, T., *The French.* 1997

(Also see specialized titles listed under relevant sections, above.)

National Statistical Office: Institut National de la Statistique et des Études Économiques (INSEE), 75582 Paris Cedex 12.
Website: http://www.insee.fr

DEPARTMENTS AND COLLECTIVITIES OVERSEAS

Départements (DOM) et collectivités d'outre-mer (COM)

GENERAL DETAILS

These fall into two main categories: *Overseas Departments and Regions* (French Guiana, Guadeloupe, Martinique, Mayotte, Réunion) and *Overseas Collectivities* (French Polynesia, St Barthélemy, St Martin, St Pierre and Miquelon, Wallis and Futuna). In addition there are two *Sui Generis Collectivities* (New Caledonia, Southern and Antarctic Territories) and one *Minor Territory* (Clipperton Island).

FURTHER READING

Aldrich, R. and Connell, J., *France's Overseas Frontier: Départements et Territoires d'Outre-Mer.* 1992

OVERSEAS DEPARTMENTS AND REGIONS

Départements et régions d'outre-mer

French Guiana

Guyane Française

KEY HISTORICAL EVENTS

A French settlement on the island of Cayenne was established in 1604 and the territory between the Maroni and Oyapock rivers finally became a French possession in 1817. Convict settlements were established from 1852, that on Devil's Island being the most notorious; all were closed by 1945. On 19 March 1946 the status of French Guiana was changed to that of an Overseas Department.

TERRITORY AND POPULATION

French Guiana is situated on the northeast coast of Latin America, and is bounded in the northeast by the Atlantic Ocean, west by Suriname, and south and east by Brazil. It includes the offshore Devil's Island, Royal Island and St Joseph, and has an area of 85,534 sq. km. Recorded population in 2010: 229,040; density: 2·7 per sq. km. The UN gives a projected population for 2015 of 262,000. In 2010 an estimated 76·2% lived in urban areas. The chief towns are (with 2010 recorded populations): the capital, Cayenne (55,753 inhabitants), Saint-Laurent-du-Maroni (38,367) and Kourou (25,189). About 58% of inhabitants are of African descent.

The official language is French.

SOCIAL STATISTICS

2008 births, 6,247; deaths, 762. 40% of the population are migrants. Average annual population growth rate, 2000–05, 4·0%.

CLIMATE

Equatorial type climate with most of the country having a main rainy season between April and July and a fairly dry period between Aug. and Dec. Both temperatures and humidity are high the whole year round. Cayenne, Jan. 26°C, July 29°C. Annual rainfall 3,202 mm.

CONSTITUTION AND GOVERNMENT

French Guiana is administered by a *General Council* of 19 members directly elected for five-year terms, and by a *Regional Council* of 31 members. It is represented in the National Assembly by two deputies and in the Senate by two senators, and on the Economic, Social and Environmental Council by one councillor. The French government is represented by a Prefect. There are two *arrondissements* (Cayenne and Saint Laurent-du-Maroni) subdivided into 22 communes and 19 cantons.

70·2% of voters rejected proposals for greater autonomy from France in a referendum on 10 Jan. 2010. However, a second referendum on 24 Jan. proposing a change in status from department to unique collectivity was passed with 57·5% of the vote and is scheduled to take effect in March 2014. The General Council and the Regional Council will be combined as a sole entity.

CURRENT GOVERNMENT

Prefect: Denis Labbé.

President of the General Council: Alain Tien-Long (Divers Gauche).

President of the Regional Council: Rodolphe Alexandre (Divers Gauche).

Government Website (French only):
http://www.guyane.pref.gouv.fr

ECONOMY

Currency

Since 1 Jan. 2002 the euro has been the official currency as in metropolitan France.

Performance

In 2007 GDP was €2,931m.; GDP per capita was €13,489. Real GDP growth was 3·4% in 2008.

Banking and Finance

In 2009 there were four commercial banks (Banque Nationale de Paris-Guyane, Banque Postale, Banque Française Commerciale Antilles/Guyane and Banque des Antilles Françaises), three cooperative or mutual savings banks and six other financial institutions.

ENERGY AND NATURAL RESOURCES

Electricity

Installed capacity was 0·1m. kW in 2007. Production in 2007 was 435m. kWh.

Minerals

Placer gold mining is the most important industry in French Guiana. In 2010, 1,140 kg of gold were produced.

Agriculture

There were 12,930 ha. of arable land in 2008. Principal crops (2006, in 1,000 tonnes): cassava, 30; rice, 15; passion fruit, 8; sugarcane, 7.

Livestock (2008 estimates): 11,000 pigs; 9,300 cattle; 2,700 sheep; 900 goats; 200,000 chickens.

Forestry

The country has immense forests that are rich in many kinds of timber. In 2010 forests covered 8·08m. ha., or 98% of the total land area. Roundwood production (2007), 177,000 cu. metres. The trees also yield oils, essences and gum products.

Fisheries

The catch in 2009 was 4,140 tonnes, mainly shrimp.

INDUSTRY

Important products include rum, rosewood essence and beer. The island has sawmills and one sugar factory.

Labour

In 2006 there were 47,500 people in paid work and 6,100 in unpaid work. In Jan. 2013 the minimum wage (SMIC) was raised to €9·43 an hour (€1,430·22 month for a 35-hour week). The unemployment rate was 20·2% in 2009; among the under 25s the rate was 37·6%.

INTERNATIONAL TRADE

Imports and Exports

Imports (2006), €2,504m.; exports (2006), €489m. Main import suppliers are France, Trinidad and Tobago and Germany. Leading export markets are France, Switzerland and Italy.

COMMUNICATIONS

Roads

There were (2005) 464 km of national and an estimated 400 km of departmental roads. In 2006, 5,866 new vehicles were registered. In 2009 there were 339 road accidents that caused 457 injuries and 28 deaths.

Civil Aviation

In 2007 Rochambeau International Airport (Cayenne) handled 383,168 passengers. There are smaller airports at Maripasoula and Saul for internal flights. The base of the European Space Agency (ESA) is located near Kourou and has been operational since 1979.

Shipping

212 merchant ships arrived and departed in 2008; 193,300 tonnes of petroleum products and 372,500 tonnes of other products were discharged, and 29,100 tonnes of freight loaded. Chief ports: Cayenne, St-Laurent-du-Maroni and Kourou. There are also inland waterways navigable by small craft.

Telecommunications

In 2008 there were 51,000 main (fixed) telephone lines. Mobile phone subscribers numbered 98,000 in 2004. In 2009 an estimated 26% of the population were internet users.

SOCIAL INSTITUTIONS

Justice

At Cayenne there is a *tribunal d'instance* and a *tribunal de grande instance*, from which appeal is to the regional *cour d'appel* in Martinique.

The population in penal institutions in Oct. 2007 was 746 (365 per 100,000 population).

Education

Primary education is free and compulsory. In 2008–09 there were 40,890 pupils at pre-elementary and primary schools, and 28,758 at secondary level. In 2004–05 there were 1,775 students at the French Guiana campuses of the University of Antilles-Guyana.

Health

In 2007 there were 804 hospital beds. There were (2007) 363 doctors, 41 dentists, 89 pharmacists, 70 midwives and 744 nursing personnel.

RELIGION

In 2001 approximately 55% of the population was Roman Catholic.

CULTURE

Press

There were two daily newspapers in 2008 with a combined circulation of 15,000.

Tourism

Total number of non-resident tourists (2005), 95,000; receipts totalled US$45m.

FURTHER READING

Redfield, Peter, *Space in the Tropics: From Convicts to Rockets in French Guiana.* 2000

Guadeloupe

KEY HISTORICAL EVENTS

The islands were discovered by Columbus in 1493. The Carib inhabitants resisted Spanish attempts to colonize. A French colony was established on 28 June 1635, and apart from short periods of occupancy by British forces, Guadeloupe has since remained a French possession. On 19 March 1946 Guadeloupe became an Overseas Department. In Feb. 2009 there was widespread unrest and a general strike that ran for over a month in protest at high prices and low wages.

TERRITORY AND POPULATION

Guadeloupe consists of a group of islands in the Lesser Antilles with a total area of 1,630 sq. km. The two main islands, Basse-Terre (to the west) and Grande-Terre (to the east), are joined by a bridge over a narrow channel. Adjacent to these are the islands of Marie-Galante (to the southeast), La Désirade (to the east), and the Îles des Saintes (to the south). The islands of St Martin (2010 population of 36,979) and St Barthélemy (2010 population of 8,938) seceded from Guadeloupe in Feb. 2007.

Island	Area (sq. km)	2010 populations	Chief town
Grande-Terre	590	197,620	Pointe-à-Pitre
Basse-Terre	848	189,713	Basse-Terre
Marie-Galante	158	11,561	Grand-Bourg
Îles des Saintes	13	2,882	Terre-de-Bas
La Désirade	20	1,579	Grande Anse

Recorded population in 2010, 403,355. An estimated 98·4% of the population were urban in 2010. Basse-Terre (2010 population, 11,915) is the seat of government, while larger Pointe-à-Pitre (2010 population, 16,427) is the department's main economic centre and port; Les Abymes (2010 population, 58,534) is a 'suburb' of Pointe-à-Pitre.

French is the official language, but Creole is spoken by the vast majority.

SOCIAL STATISTICS

2006: live births, 6,228; deaths, 2,763. Marriages (2007), 1,427. 2006 rates (per 1,000 population): birth, 15·5; death, 6·9. Annual population growth rate, 2000–05, 0·8%. Life expectancy at birth, 2005, 75·3 years for males and 81·6 years for females.

CLIMATE

Warm and humid. Pointe-à-Pitre, Jan. 74°F (23·4°C), July 80°F (26·7°C). Annual rainfall 71" (1,814 mm).

CONSTITUTION AND GOVERNMENT

Guadeloupe is administered by a *General Council* of 42 members directly elected for six-year terms (assisted by an Economic and Social Committee of 40 members) and by a *Regional Council* of 41 members. It is represented in the National Assembly by four deputies; in the Senate by three senators; and on the Economic, Social and Environmental Council by one councillor. There are four *arrondissements,* sub-divided into 42 cantons and 34 communes, each administered by an elected municipal council. The French government is represented by an appointed *Prefect.*

CURRENT GOVERNMENT

Prefect: Marcelle Pierrot.
    President of the General Council: Jacques Gillot.
    President of the Regional Council: Victorin Lurel.

Government Website (French only):
    http://www.guadeloupe.pref.gouv.fr

ECONOMY

Currency

Since 1 Jan. 2002 the euro has been the official currency as in metropolitan France.

Performance

In 2007 GDP was €8,147m.; GDP per capita was €18,244. Real GDP growth was 3·7% in 2007.

Banking and Finance

In Dec. 2009 there were ten banks and eight other financial institutions.

ENERGY AND NATURAL RESOURCES

Electricity

Total production (2009): 1·61bn. kWh. Installed capacity was 0·4m. kW in 2009.

Agriculture

Chief products (2007, in 1,000 tonnes): sugarcane, 789; bananas, 45; melons, 9; pineapples, 7; cucumbers, 6. Other fruits and vegetables are also grown for both export and domestic consumption.

Livestock (2008 estimates): cattle, 75,000; goats, 48,000; pigs, 30,000; chickens, 475,000.

Forestry

In 2010 forests covered 64,000 ha., or 39% of the total land area. Timber production in 2007 was 32,000 cu. metres.

Fisheries

Total catch in 2008 amounted to an estimated 10,100 tonnes, exclusively from sea fishing.

INDUSTRY

The main industries are sugar refining, food processing and rum distilling, carried out by small and medium-sized businesses. Other important industries are cement production and tourism.

Labour

The economically active population in 2005 was 160,000. In Jan. 2013 the minimum wage (SMIC) was raised to €9·43 an hour

(€1,430·22 month for a 35-hour week). The unemployment rate was 23·4% in 2009; among the under 25s the rate was 59·3%.

INTERNATIONAL TRADE

Imports and Exports

Total imports (2006): €2,310m.; total exports (2006): €164m. Main export commodities are bananas, sugar and rum. Main import sources in 2006 were France, 60·8%; Italy, 3·5%; Germany, 3·4%; Trinidad and Tobago, 3·0%. Main export markets in 2006 were France, 54·9%; Martinique, 30·7%; French Guiana, 2·9%; Venezuela, 2·6%.

COMMUNICATIONS

Roads

There were 416 km of national roads in 2009 and 619 km of departmental roads. In 2009, 16,876 new vehicles were registered. There were 779 road accidents with 56 fatalities in 2008.

Civil Aviation

Air France and six other airlines call at Guadeloupe airport. In 2007 there were 1,707,781 passengers on domestic flights and 155,582 passengers on international flights at Le Raizet (Pointe-à-Pitre) airport. There is a smaller airport at Marie-Galante only for internal flights. Most domestic services are operated by Air Caraibes.

Shipping

In 2008, 3,582,054 tonnes of freight passed through the Port autonome de la Guadeloupe (2,841,539 tonnes inbound and 740,515 tonnes outbound).

Telecommunications

In 2008 there were 246,000 main (fixed) telephone lines. Mobile phone subscribers numbered 314,700 in 2004. There were 79,000 internet users in 2004. In Dec. 2011 there were 138,000 Facebook users.

SOCIAL INSTITUTIONS

Justice

There are four *tribunaux d'instance* and two *tribunaux de grande instance* at Basse-Terre and Pointe-à-Pitre; there is also a court of appeal and a court of assizes.

The population in penal institutions in Oct. 2007 was 790 (174 per 100,000 population).

Education

Education is free and compulsory from six to 16 years. In 2008–09 there were 60,741 pupils at pre-elementary and primary schools, and 52,547 at secondary level. In 2004–05 there were 5,965 students at the Guadeloupe campuses of the University of Antilles-Guyana.

Health

As at 1 Jan. 2007 there were 1,783 hospital beds (560 in the private sector). In 2007 there were 2·3 hospital beds per 1,000 inhabitants. There were 1,014 doctors, 172 dentists, 2,439 nurses, 314 pharmacists and 162 midwives at 1 Jan. 2007.

RELIGION

The majority of the population are Roman Catholic.

CULTURE

Press

There were three daily newspapers in 2008. *Le Pélican* is the only paid-for daily, with a circulation of 2,000.

Tourism

Tourism is the chief economic activity. In 2009, 346,507 non-resident tourists—excluding same-day visitors—arrived by air (411,799 in 2008).

Martinique

KEY HISTORICAL EVENTS

Discovered by Columbus in 1502, Martinique became a French colony in 1635 and apart from brief periods of British occupation the island has since remained under French control. On 19 March 1946 its status was altered to that of an Overseas Department.

TERRITORY AND POPULATION

The island, situated in the Lesser Antilles between Dominica and St Lucia, occupies an area of 1,128 sq. km. Recorded population in 2010, 394,173; density, 349 per sq. km. The UN gives a projected population for 2015 of 411,000. An estimated 89·0% of the population were urban in 2010. Recorded population of principal towns in 2010: the capital and main port Fort-de-France, 87,216; Le Lamentin, 39,360; Le Robert, 23,918; Schoelcher, 20,814; Le François, 19,218; Sainte-Marie, 18,389. French is the official language but the majority of people speak Creole.

SOCIAL STATISTICS

2008: live births, 5,333; deaths, 2,793. 2008 estimates per 1,000 population: birth rate, 13·3; death rate, 7·0. Average annual population growth rate, 2000–05, 0·7%. Life expectancy at birth, 2006, 76·2 years for males and 84·5 years for females.

CLIMATE

The dry season is from Dec. to May, and the humid season from June to Nov. Fort-de-France, Jan. 74°F (23·5°C), July 78°F (25·6°C). Annual rainfall 72" (1,840 mm).

CONSTITUTION AND GOVERNMENT

The island is administered by a *General Council* of 45 members directly elected for six-year terms and by a *Regional Council* of 41 members. The French government is represented by an appointed *Prefect*. There are four *arrondissements*, sub-divided into 45 cantons and 34 communes, each administered by an elected municipal council. Martinique is represented in the National Assembly by four deputies, in the Senate by two senators and on the Economic, Social and Environmental Council by one councillor.

79·3% of voters rejected proposals for greater autonomy from France in a referendum on 10 Jan. 2010. However, a second referendum on 24 Jan. proposing a change in status from department to unique collectivity was passed with 68·3% of the vote and is scheduled to take effect in March 2014. The General Council and the Regional Council will be combined as a sole entity.

CURRENT GOVERNMENT

Prefect: Laurent Prévost; b. 1967 (took office on 30 March 2011).
   President of the General Council: Josette Manin.
   President of the Regional Council: Serge Letchimy.

Government Website: http://www.martinique.pref.gouv.fr

ECONOMY

Main sectors of activity: tradeable services, distribution, industry, building and public works, transport and telecommunications, agriculture and tourism.

Currency

Since 1 Jan. 2002 the euro has been the official currency as in metropolitan France.

Performance

In 2007 GDP was €7,893m.; GDP per capita in 2007 was €19,800. Real GDP growth was 0·9% in 2007.

Banking and Finance

The *Agence Française de Développement* is the government's vehicle for the promotion of economic development in the region. There were 12 banks and six other financial institutions in 2009.

ENERGY AND NATURAL RESOURCES

Electricity

A network of 4,510 km of cables covers 98% of Martinique and supplies some 183,000 customers. In 2009, 97·2% of electricity production was thermal energy and 2·8% was from renewable sources. Total production (2009): 1·55bn. kWh. Installed capacity (2009): 0·4m. kW.

Agriculture

Crops by area (2008, in ha.): bananas, 5,750; sugarcane, 4,150; vegetables, 2,450; citrus fruits, 300. Production (2007, in 1,000 tonnes): sugarcane, 226; bananas, 144; cucumbers, 7; tomatoes, 3.

Livestock (2008): 14,900 sheep; 14,800 pigs; 8,200 cattle; 8,100 goats; 310,000 poultry.

Forestry

In 2010 there were 49,000 ha. of forest, or 46% of the total land area. Timber production in 2007 was 26,000 cu. metres.

Fisheries

The catch in 2008 was an estimated 6,200 tonnes, exclusively from sea fishing.

INDUSTRY

Some food processing and chemical engineering is carried out by small and medium-size businesses. There were 33,063 businesses in 2006. There is an important cement industry; there are also 11 rum distilleries and an oil refinery, with an annual treatment capacity of 0·75m. tonnes. Martinique has five industrial zones.

Labour

In 2006, 4·6% of the employed population worked in agriculture, 12·0% in commerce, 6·2% in construction, 7·6% in industry and 69·6% in services. There were 120,900 people in paid work in 2006 and 11,100 in unpaid work. In Jan. 2013 the minimum wage (SMIC) was raised to €9·43 an hour (€1,430·22 a month for a 35-hour week). The unemployment rate was 21·8% in 2009; among the under 25s the rate was 57·6%.

INTERNATIONAL TRADE

Imports and Exports

Martinique has a structural trade deficit owing to the nature of goods traded. It imports high-value-added goods (foodstuffs, capital goods, consumer goods and motor vehicles) and exports agricultural produce (bananas) and refined oil.

In 2006 imports were valued at €2,504m.; exports at €489m. Main trading partners: France, the UK, Guadeloupe, the USA and Italy. Trade with France accounted for 55·6% of imports and 21·8% of exports in 2006.

COMMUNICATIONS

Roads

Martinique had 291 km of national roads and 7 km of motorway in 2008. In 2008, 38,029 vehicles were registered and 7,567 driving licences granted.

Civil Aviation

There is an international airport at Fort-de-France (Lamentin), which handled 1,482,465 passengers on internal flights and 119,919 on international flights in 2007.

Shipping

The island is visited regularly by French, American and other lines. The main sea links to and from Martinique are ensured by CGM Sud. In 2007, 838 merchant ships called at Martinique and discharged 2,252,000 tonnes of freight and embarked 891,000 tonnes of freight. In 2008, 109 cruise ships carrying 86,488 passengers called at Martinique. There were 150,301 passengers on inter-island crossings.

Telecommunications

In 2008 there were 172,000 main (fixed) telephone lines. Mobile phone subscribers numbered 295,400 in 2004. The main operator is France Télécom. There were 107,000 internet users in 2004. In Dec. 2011 there were 126,000 Facebook users.

SOCIAL INSTITUTIONS

Justice

Justice is administered by two lower courts (*tribunaux d'instance*), a higher court (*tribunal de grande instance*), a regional court of appeal, a commercial court and an administrative court.

The population in penal institutions on 1 Jan. 2008 was 760 (189 per 100,000 population).

Education

Education is compulsory between the ages of six and 16 years. In 2009–10 there were 46,124 pupils in nursery and primary schools, and 42,740 pupils in secondary schools. 8,942 students were in higher education in 2009–10. In 2004–05 there were 5,417 students at the Martinique campus of the University of Antilles-Guyana.

Health

As at 1 Jan. 2007 there were 1,795 hospital beds (170 in the private sector). In 2007 there were 2·2 hospital beds per 1,000 inhabitants. There were 1,013 doctors, 156 dentists, 2,689 nurses, 318 pharmacists and 180 midwives at 1 Jan. 2007.

RELIGION

In 2001, 87% of the population was Roman Catholic.

CULTURE

Press

In 2006 there was one daily newspaper with a circulation of 65,000.

Tourism

In 2006 there were 503,475 staying visitors and 96,089 cruise passenger arrivals. Tourism receipts totalled US$299m. in 2007. In 2006 there were 99 classified hotels, with 4,747 rooms.

Mayotte

KEY HISTORICAL EVENTS

Mayotte was a French colony from 1843 until 1914 when it was attached, with the other Comoros islands, to the government-general of Madagascar. The Comoro group was granted administrative autonomy within the French Republic and became an Overseas Territory. When the other three islands voted to become independent (as the Comoros state) in 1974, Mayotte voted against and remained a French dependency. In Dec. 1976, following a further referendum, it became a Territorial Collectivity. On 11 July 2001 Mayotte became a Departmental Collectivity—a constitutional innovation—as a result of a referendum. This was denounced by the Comorian authorities, who claim Mayotte as part of the Union of the Comoros. On 28 March 2003 it became an Overseas Collectivity. As a result of a further referendum held on 29 March 2009 the island became an Overseas Department on 31 March 2011.

TERRITORY AND POPULATION

Mayotte, southeast of the Comoros, had a total population at the 2012 census of 212,645 (population density of 569 persons per sq. km). The whole territory covers 374 sq. km (144 sq. miles). It consists of a main island (363 sq. km) with (2012 census) 188,442 inhabitants, containing the chief town, Mamoudzou (57,281 inhabitants in 2012); and the smaller island of Pamanzi (11 sq. km) lying 2 km to the east (24,203 in 2012) containing the old capital of Dzaoudzi (14,311 in 2012).

The UN gives a projected population for 2015 of 237,000.

The spoken language is Shimaoré (akin to Comorian, an Arabized dialect of Swahili), but French remains the official, commercial and administrative language.

SOCIAL STATISTICS

There were 7,643 births and 572 deaths in 2004.

CLIMATE

The dry and sunniest season is from May to Oct. The hot but rainy season is from Nov. to April. Average temperatures are 27°C from Dec. to March and 24°C from May to Sept.

CONSTITUTION AND GOVERNMENT

The island is administered by a *General Council* of 19 members, directly elected for a six-year term. The French government is represented by an appointed *Prefect*. In accordance with the legislation of 11 July 2001 executive powers were transferred from the prefect to the president of the General Council in March 2004. Mayotte is represented by two deputies in the National Assembly, by two senators in the Senate and by one councillor on the Economic, Social and Environmental Council. There are 17 communes, including two on Pamanzi.

CURRENT GOVERNMENT

Prefect: Jacques Witkowski.

President of the General Council: Daniel Zaïdani (Mouvement Départementaliste Mahorais/MDM).

Government Website (French only): http://www.mayotte.pref.gouv.fr

ECONOMY

Currency

Since 1 Jan. 2002 the euro has been the official currency as in metropolitan France.

Banking and Finance

The Institut d'Émission d'Outre-mer and the Banque Française Commerciale both have branches in Dzaoudzi and Mamoudzou.

ENERGY AND NATURAL RESOURCES

Agriculture

Agricultural land in Mayotte totalled an estimated 20,000 ha. in 2007. Mayotte is the world's second largest producer of ylang-ylang essence. Important cash crops include cinnamon, ylang-ylang, vanilla and coconut. The main food crops (2004) were bananas (11,500 tonnes) and cassava (9,000 tonnes). Livestock (2004): cattle, 17,235; goats, 22,811; sheep, 1,430.

Forestry

There were 14,000 ha. of forest in 2010 (37% of the land area).

Fisheries

A lobster and shrimp industry has been created. Fish landings in 2010 totalled 20,842 tonnes, up from 2,194 tonnes in 2005.

INDUSTRY

Labour

In 2007, 19% of the active population was engaged in education, health and social care; 17% in public administration; 13% in transport, real estate and services; and 12% in commerce. Unemployment rate, 2007, 26·4%.

INTERNATIONAL TRADE

Imports and Exports

In 2009 imports totalled €428·1m. and exports €5·1m. Main export commodities are fish and ylang-ylang. Main imports sources in 2009: France, 51%; China, 8%. Main export destinations, 2009: France, 40%; Comoros, 15%.

COMMUNICATIONS

Roads

In 2009 there were 90 km of national roads and 139 km of departmental roads.

Civil Aviation

There is an airport at Pamandzi, with scheduled services in 2010 provided to the Comoros, mainland France, Kenya, Madagascar and Réunion.

Shipping

The commercial port is situated at Longoni. There is a secondary port at Dzaoudzi and a marina at Mamoudzou. There are passenger ferry services between Mayotte and Anjouan (Comoros).

Telecommunications

In 2008 there were 10,000 main (fixed) telephone lines. There were 48,100 mobile phone subscribers in 2004.

SOCIAL INSTITUTIONS

Justice

There is a *tribunal de première instance* and a *tribunal supérieur d'appel*.

Education

In 2004 there were 39,807 pupils in nursery and primary schools, and 17,316 pupils at 16 *collèges* and four *lycées* at secondary level. There were also 3,011 pupils enrolled in pre-professional classes and professional *lycées*. There is a teacher training college.

Health

The main hospital is at Mamoudzou. There were four other hospitals and 15 health clinics in 2009. Hospital beds totalled 252 in 2009. There were 279 doctors, 13 dentists, 113 midwives and 516 nurses in 2009.

RELIGION

The population is 97% Sunni Muslim, with a small Christian (mainly Roman Catholic) minority.

CULTURE

Press

In 2008 there was one newspaper, *Kwezi*, which was published weekly and also distributed in the Comoros.

Tourism

In 2008 there were some 38,000 visitors of which 45·3% came from mainland France, 45·2% from Réunion and 9·6% from other countries. The average length of stay was 23 days.

Réunion

KEY HISTORICAL EVENTS

Réunion (formerly Île Bourbon) became a French possession in 1638 and remained so until 19 March 1946, when its status was altered to that of an Overseas Department.

TERRITORY AND POPULATION

The island of Réunion lies in the Indian Ocean, about 880 km east of Madagascar and 210 km southwest of Mauritius. It has an area of 2,512 sq. km. Recorded population in 2010: 821,136, giving a density of 327 per sq. km. An estimated 94·0% of the population were urban in 2010. The capital is Saint-Denis (population, 2010: 145,022); other large towns are Saint-Paul (103,346), Saint-Pierre (79,228) and le Tampon (73,365).

The UN gives a projected population for 2015 of 893,000.

French is the official language, but Creole is also spoken.

SOCIAL STATISTICS

2005: births, 14,610; deaths, 4,255; marriages, 3,115; divorces, 1,499. Rates per 1,000 population (2005): birth, 18·7; death, 5·5. Average annual population growth rate, 2000–05, 1·5%. Life expectancy at birth, 2005, 72·4 years for males and 80·0 years for females. Infant mortality, 2005, 7·9 per 1,000 live births; fertility rate, 2005, 2·4 births per woman.

CLIMATE

There is a sub-tropical maritime climate, free from extremes of weather, although the island lies in the cyclone belt of the Indian Ocean. Conditions are generally humid and there is no well-defined dry season. Saint-Denis, Jan. 80°F (26·7°C), July 70°F (21·1°C). Annual rainfall 56" (1,400 mm).

CONSTITUTION AND GOVERNMENT

Réunion is administered by a *General Council* of 47 members directly elected for six-year terms, and by a *Regional Council* of 45 members. Réunion is represented in the National Assembly in Paris by five deputies; in the Senate by four senators; and in the Economic, Social and Environmental Council by one councillor. There are four *arrondissements* sub-divided into 47 cantons and 24 communes, each administered by an elected municipal council. The French government is represented by an appointed *Prefect*.

CURRENT GOVERNMENT

Prefect: Jean-Luc Marx.
   President of the General Council: Nassimah Dindar.
   President of the Regional Council: Didier Robert.

Government Website (French only): http://www.reunion.pref.gouv.fr

ECONOMY

Currency

Since 1 Jan. 2002 the euro has been the official currency as in metropolitan France. Owing to its geographical location, Réunion was, by two hours, the first territory to introduce the euro.

Performance

GDP was €12,061m. in 2005; real GDP growth was 7·4% in 2004. GDP per capita in 2005 was €15,475.

Banking and Finance

The Institut d'Émission des Départements d'Outre-mer has the right to issue bank-notes. Banks operating in Réunion are the Banque de la Réunion (Caisse d'Épargne), the Banque Nationale de Paris Intercontinentale, the Crédit Agricole de la Réunion, the Banque Française Commerciale (BFC) CCP, Trésorerie Générale and the Banque de la Réunion pour l'Économie et le Développement (BRED).

ENERGY AND NATURAL RESOURCES

Electricity

Production in 2007 was about 1,792m. kWh. Estimated consumption per capita (2007), 2,223 kWh. Installed capacity (2007 estimate): 0·4m. kW.

Agriculture

There were 47,425 ha. of land used for agriculture in 2006 of which 25,569 ha. were planted with sugarcane. Main agricultural products (2006 in 1,000 tonnes): sugarcane, 1,864; pineapples, 16; bananas, 10; citrus fruits, 8; tomatoes, 8.

Livestock (2006): 73,000 pigs, 36,000 cattle, 36,000 goats, 14·9m. poultry (estimate). Meat production (2004, in tonnes): pork, 12,394; beef and veal, 1,723; poultry, 8,319. Milk production (2004), 238,470 hectolitres.

Forestry

There were 88,000 ha. of forest in 2010, or 35% of the total land area. Timber production in 2007 was 36,000 cu. metres.

Fisheries

In 2009 the catch was 3,050 tonnes, entirely from marine waters.

INDUSTRY

The major industries are electricity and sugar. Food processing, chemical engineering, printing and the production of perfume, textiles, leathers, tobacco, wood and construction materials are carried out by small and medium-sized businesses. In 2004 there were 9,018 craft businesses employing about 27,000 persons. Production of sugar was 220,470 tonnes in 2004; rum, 86,130 hectolitres.

Labour

In 2006 there were 197,800 people in paid work and 24,100 in unpaid work. In Jan. 2013 the minimum wage (SMIC) was raised to €9·43 an hour (€1,430·22 month for a 35-hour week). In 2009 the unemployment rate was 27·1%. Among the under 25s the unemployment rate is nearly 50%.

INTERNATIONAL TRADE

Imports and Exports

In 2006 imports totalled €3,912m. and exports €238m. The chief export is sugar, accounting for 41·0% of total exports in 2006. France provided 42·2% of imports in 2006 and took 59·6% of exports.

COMMUNICATIONS

Roads

There were, in 2001, 2,914 km of roads. In 2008 there were 327,800 registered passenger cars. There were 67 road-related fatalities in 2004.

Civil Aviation

In 2007 Roland Garros Saint-Denis airport handled 1,049,791 passengers on domestic flights and 470,171 passengers on international flights.

Shipping

In 2008, 3,639,100 tonnes of freight were unloaded and 647,900 tonnes loaded at Port-Réunion.

Telecommunications

In 2008 there were 440,000 main (fixed) telephone lines. Mobile phone subscribers numbered 579,200 in 2004. There were 200,000 internet users in 2004. In June 2012 there were 229,000 Facebook users.

SOCIAL INSTITUTIONS

Justice

There are three lower courts (*tribunaux d'instance*), two higher courts (*tribunaux de grande instance*), one appeal, one administrative court and one conciliation board.

The population in penal institutions in Jan. 2008 was 1,296 (162 per 100,000 population).

Education

In 2008–09 there were 122,298 pupils in primary schools and 101,262 in secondary schools. The Université Française de l'Océan Indien (founded 1971) had 12,267 students in 2007–08.

Health

In 2007 there were 1,359 hospital beds and 2,130 doctors. There were 105 general doctors per 100,000 inhabitants in 2007.

RELIGION

In 2001, 82% of the population was Roman Catholic.

CULTURE

World Heritage Sites

Réunion has one site on the UNESCO World Heritage List: its Pitons, cirques and remparts (inscribed on the list in 2010). At 100,000 ha. the site covers 40% of the island and incorporates a large area of the national park containing subtropical rainforests and cloud forests.

Press

There were three daily newspapers (*Quotidien*, *Journal de l'Île* and *Témoignages*) in 2008 with a combined circulation of 72,000.

Tourism

Tourism is a major resource industry. There were 381,000 non-resident tourists in 2007. Receipts in 2007 totalled US$446m. In 2008 there were 49 classified hotels with 2,127 rooms.

FURTHER READING

Institut National de la Statistique et des Études Économiques: *Tableau Économique de la Réunion.* Annual

Bertile, W., *Atlas Thématique et Régional.* 1990

OVERSEAS COLLECTIVITIES

Collectivités d'outre-mer

French Polynesia

Territoire de la Polynésie Française

KEY HISTORICAL EVENTS

French protectorates since 1843, these islands were annexed to France in 1880–82 to form 'French Settlements in Oceania', which opted in Nov. 1958 for the status of an overseas territory within the French Community. In March 2003 French Polynesia became an Overseas Collectivity.

TERRITORY AND POPULATION

The total land area of these five archipelagoes, comprising 121 volcanic islands and coral atolls (76 inhabited) scattered over a wide area in the eastern Pacific, is 3,521 sq. km. The population (2012 census) was 268,270; density, 76 per sq. km. In 2007 French forces stationed in Polynesia numbered 1,500 (based mostly on Tahiti). In 2007 an estimated 50·7% of the population lived in urban areas.

The UN gives a projected population for 2015 of 285,000.

The official languages are French and Tahitian.

The islands are administratively divided into five *subdivisions administratives* as follows:

Windward Islands (Îles du Vent) (200,881 inhabitants, 2012) comprise Tahiti with an area of 1,042 sq. km and 178,132 inhabitants in 2007; Mooréa with an area of 132 sq. km and 16,191 inhabitants in 2007; Maiao (Tubuai Manu) with an area of 9 sq. km and 299 inhabitants in 2007; and the smaller Mehetia (uninhabited) and Tetiaroa (population of one). The capital is Papeete, Tahiti (26,017 inhabitants in 2007, excluding suburbs; urban area, 131,693).

Leeward Islands (Îles sous le Vent) comprise the five volcanic islands of Raiatéa, Tahaa, Huahine, Bora-Bora and Maupiti, together with four small atolls (Tupai, Mopelia, Scilly, Bellinghausen), the group having a total land area of 404 sq. km and 34,622 inhabitants in 2012. The chief town is Uturoa on Raiatéa. The Windward and Leeward Islands together are called the Society Archipelago (Archipel de la Société). Tahitian, a Polynesian language, is spoken throughout the archipelago and used as a *lingua franca* in the rest of the territory.

Marquesas Islands 12 islands lying north of the Tuamotu Archipelago, with a total area of 1,049 sq. km and 9,264 inhabitants in 2012. There are six inhabited islands: Nuku Hiva, Ua Pou, Ua Uka, Hiva Oa, Tahuata, Fatu Hiva; and six smaller (uninhabited) ones; the chief centre is Taiohae on Nuku Hiva.

Austral or Tubuai Islands lying south of the Society Archipelago, comprise a 1,300 km chain of volcanic islands and reefs. There are five inhabited islands (Rimatara, Rurutu, Tubuai, Raivavae and, 500 km to the south, Rapa), with a combined area of 148 sq. km (6,839 inhabitants in 2012); the chief centre is Mataura on Tubuai.

Tuamotu and Gambier Islands comprise the Tuamotu Islands, two parallel ranges of 76 atolls (53 inhabited) lying north and east of the Society Archipelago with a total area of 690 sq. km and 15,510 inhabitants in 2007; and the Gambier Islands to the southeast of the Tuamotu Islands with a total area of 36 sq. km and 1,337 inhabitants in 2007. The most populous atolls are Rangiroa (3,210 inhabitants in 2007), Manihi (1,379 in 2007) and Hao (1,342 in 2007).

The Mururoa and Fangataufa atolls in the southeast of the group were ceded to France in 1964 by the Territorial Assembly, and were used by France for nuclear tests from 1966–96. The Pacific Testing Centre (CEP) was dismantled in 1998. A small military presence remains to ensure permanent radiological control.

SOCIAL STATISTICS

2005: births, 4,467; deaths, 1,239. Average annual population growth rate, 2000–05, 1·6%. Life expectancy at birth, 2005, 71·4 years for males and 76·4 years for females. Infant mortality, 2005, 6·3 per 1,000 live births; fertility rate, 2·2 births per woman.

CLIMATE

Papeete, Jan. 81°F (27·1°C), July 75°F (24°C). Annual rainfall 83" (2,106 mm).

CONSTITUTION AND GOVERNMENT

Under the 1984 constitution, the Territory is administered by a *Council of Ministers*, whose President is elected by the Territorial Assembly from among its own members; the President appoints a Vice-President and 14 other ministers. French Polynesia is represented in the French Assembly by two deputies, in the Senate by two senators and on the Economic, Social and Environmental Council by one councillor.

The French government is represented by a *High Commissioner*. The Territorial Assembly comprises 57 members elected every five years from five constituencies by universal suffrage, using the same proportional representation system as in metropolitan French regional elections. To be elected a party must gain at least 5% of votes cast.

In Dec. 2003 French Polynesia's status was changed to that of an Overseas Country within the French Republic. However, the designation has no legal consequences and the 2004 status of French Polynesia acknowledged that it belongs to the category of Overseas Collectivity.

RECENT ELECTIONS

Elections were held on 21 April and 5 May 2013. The Popular Rally (Tahoera'a Huiraatira) won 38 seats (with 45·1% of the vote), the Union for Democracy (UPD) 11 seats with 29·3% and the A Ti'a Porinetia 8 with 25·6%.

An indirect presidential election was held on 11 Feb. 2009 in the Territorial Assembly. In the first round Oscar Temaru (Tavini Huiraatira/People's Servant Party) received 24 votes against 20 for outgoing president Gaston Tong Sang (To Tatou Ai'a), 12 for Edouard Fritch (Tahoera'a Huiraatira) and 1 for Sandra Levy-Agami (ind.). Temaru defeated Tong Sang in the second round by 37 votes to 20.

CURRENT GOVERNMENT

High Commissioner: Jean-Pierre Laflaquière; b. 1947 (took office on 3 Sept. 2012).

President: Oscar Temaru; b. 1944 (Tavini Huiraatira; took office for a fifth time on 1 April 2011).

Government Website (French only):
  http://www.polynesie-francaise.pref.gouv.fr

ECONOMY

Currency
The unit of currency is the franc CFP (XPF). Up to 31 Dec. 1998, its parity was to the French franc: 1 franc CFP = 0·055 French francs; from 1 Jan. 1999 parity was linked to the euro: 119·3317422 francs CPF = one euro.

Budget
Revenues totalled 143·1bn. francs CPF in 2007 and expenditures 144·5bn. francs CPF.

Performance
Total GDP was US$6,172m. in 2007; GDP per capita was US$23,488. Real GDP growth was 3·0% in 2007.

Banking and Finance
There are four commercial banks: Banque de Tahiti, Banque de Polynésie, Société de Crédit et de Développement de l'Océanie and the Banque Westpac.

ENERGY AND NATURAL RESOURCES

French Polynesia is heavily dependent on external sources for its energy.

Environment
Carbon dioxide emissions from the consumption and flaring of fossil fuels in 2008 were the equivalent of 3·8 tonnes per capita.

Electricity
Production (2007) was 687m. kWh, of which approximately 24% was hydro-electric. Installed capacity in 2007 was an estimated 113,000 kW.

Agriculture
Agriculture used to be the primary economic sector but now accounts for only 2% of GDP. Important products are copra (coconut trees cover the coastal plains of the mountainous islands and the greater part of the low-lying islands) and the nono fruit, which has medicinal value. Production in tonnes (2007): copra, 9,508; fruit, 8,920; vegetables, 5,675; nono, 2,089; vanilla, 37. Tropical fruits, such as bananas, pineapples and oranges, are grown for local consumption.

Livestock (2006 estimates): cattle 12,000; goats 17,000; pigs 27,000; poultry 232,000.

Forestry
There were 0·16m. ha. of forest in 2010, or 42% of the total land area.

Fisheries
Polynesia has an exclusive zone of 5·0m. sq. km, one of the largest in the world. Catch (2010): 13,015 tonnes, almost exclusively from sea fishing.

INDUSTRY

Some 2,200 industrial enterprises employ 5,800 people. Principal industries include food and drink products, cosmetics, clothing and jewellery, furniture-making, metalwork and shipbuilding.

INTERNATIONAL TRADE

Imports and Exports
French Polynesia imports a great deal and exports very little. Trade, 2010, in US$1m. (2009 in brackets): imports (c.i.f.), 1,726 (1,717); exports (f.o.b.), 153 (148).

The chief exports are pearls, ships and boats, fruit and vegetables, fish and precious jewellery.

Polynesia is among the world's largest producers of pearls. While pearl production remains the second largest industry in Polynesia after tourism, the number of pearl farms has shrunk in recent years from over 2,500 in 2001 to just 524 in 2010 as a result of over-production and declining pearl quality.

The major trading partner overall is France, although Hong Kong is the leading export market. In 2010 France accounted for 28% of imports and Singapore 13%. Hong Kong accounted for 30% of exports in 2010 and Japan 20%.

COMMUNICATIONS

Roads
In 2009, 5,430 new cars were registered.

Civil Aviation
The main airport is at Papeete (Tahiti-Faa'a). Air France and nine other international airlines (including Air New Zealand and Qantas) connect Tahiti International Airport with Paris, Auckland, Honolulu, Los Angeles, Osaka, Santiago, Tokyo and many Pacific islands. In 2007, 1,788 flights landed at Papeete and 1,786 took off. 332,444 passengers arrived in 2007, 332,894 departed and 17,342 were in transit.

Shipping
Ten shipping companies connect France, San Francisco, New Zealand, Japan, Australia, southeast Asia and most Pacific locations with Papeete. In 2005 vessels totalling 19,749,378 GRT passed through Papeete's main port. Around 1·4m. people pass through the port each year.

Telecommunications
There were 54,900 fixed telephone lines in 2010 (202·9 per 1,000 inhabitants). Tikiphone, the sole provider of mobile phone services, had 222,800 subscribers in 2011 (813·9 per 1,000 inhabitants). In 2010, 49·0% of the population were internet users.

SOCIAL INSTITUTIONS

Justice

There is a *tribunal de première instance* and a *cour d'appel* at Papeete. The population in penal institutions in Oct. 2007 was 404 (153 per 100,000 population).

Education

In 2006–07 there were 165 primary schools and 99 secondary schools. 15,249 children attended pre-primary schools in 2006–07, 24,961 pupils primary schools and 33,845 pupils secondary schools. The University of French Polynesia (Université de la Polynésie Française) was formed from the Tahitian campus of the now-defunct French University of the Pacific (UFP) in 1999. In 2008–09 student enrolments numbered 2,664.

Health

In 2007 there were a total of 613 beds in public sector establishments and 260 in the private sector. Medical personnel in 2007 numbered 609 doctors, 1,123 nurses, 149 pharmacists, 128 midwives and 113 dentists.

RELIGION

In 2001 there were approximately 119,000 protestants (about 49% of the population) and 94,000 Roman Catholics (39%).

CULTURE

Press

In 2008 there were two daily newspapers with a combined circulation of 22,000.

Tourism

Tourism is the main industry. There were 218,241 tourist arrivals in 2007.

FURTHER READING

Local Statistical Office: Institut Statistique de Polynésie Française, Immeuble Uupa, 1st Floor, Rue Edouard Ahnne, Papeete.
Website (French only): http://www.ispf.pf

St Barthélemy

Saint-Barthélemy

KEY HISTORICAL EVENTS

There is evidence of habitation dating to 1000 BC. Columbus visited in 1493 but it was the French who settled St Barthélemy in 1648. The island came under Swedish rule in 1784, after which it prospered as a free port. The British briefly held sway in 1801–02 but the French regained control in 1878 and the island was administered by Guadeloupe until seceding on 22 Feb. 2007 to become a French overseas collectivity.

TERRITORY AND POPULATION

The island has an area of 21 sq. km and is situated 200 km northwest of Guadeloupe. 2007 recorded population, 8,450. The capital is Gustavia.

CONSTITUTION AND GOVERNMENT

The laws of France are enforceable. Government on the island is by a 19-seat *Territorial Council*, elected by popular vote every five years. St Barthélemy is represented in the Senate in Paris by one senator and on the Economic, Social and Environmental Council by one councillor. The council elects a *President* and the French government appoints a *Prefect*.

CURRENT GOVERNMENT

Prefect: Philippe Chopin.
President of the Territorial Council: Bruno Magras.

St Martin

Saint-Martin

KEY HISTORICAL EVENTS

Sighted by Columbus in 1493, St Martin was settled by the Dutch in 1631. The Spanish claimed dominion in 1633 but by 1648 the island was divided between Dutch and French interests. The French part of the island was administered by Guadeloupe until it seceded on 22 Feb. 2007 to become a French overseas collectivity (encompassing several neighbouring islets including Île Tintamarre).

TERRITORY AND POPULATION

St Martin is situated 300 km southeast of Puerto Rico. The French-run part of the island covers roughly the northern two-thirds while the southern third (Sint Maarten) is an autonomous country within the Kingdom of the Netherlands. The French area is 53 sq. km; the recorded population in 2007 was 35,925. Marigot is the capital.

CONSTITUTION AND GOVERNMENT

The laws of France are enforceable. Government on the island is by a 23-seat *Territorial Council*, elected by popular vote every five years. St Martin is represented in the Senate in Paris by one senator and on the Economic, Social and Environmental Council by one councillor. The council elects a *President* and the French government appoints a *Prefect*.

CURRENT GOVERNMENT

Prefect: Philippe Chopin.
President of the Territorial Council: Alain Richardson.

St Pierre and Miquelon

Saint-Pierre et Miquelon

KEY HISTORICAL EVENTS

The only remaining fragment of the once-extensive French possessions in North America, the archipelago was settled from France in the 17th century. It was a French colony from 1816 until 1976, an overseas department until 1985, and is now an Overseas Collectivity.

TERRITORY AND POPULATION

The archipelago consists of two islands off the south coast of Newfoundland, with a total area of 242 sq. km, comprising the Saint-Pierre group (26 sq. km) and the Miquelon-Langlade group (216 sq. km). The recorded population at 2007 was 6,099, representing a decrease of 217 from the 1999 census. Approximately 90% of the population lives on Saint-Pierre. The chief town is St Pierre.

The official language is French.

SOCIAL STATISTICS

2009: births, 64; deaths, 45.

CONSTITUTION AND GOVERNMENT

The Overseas Collectivity is administered by a *Territorial Council* of 19 members directly elected for a six-year term. It is represented in the National Assembly in Paris by one deputy, in the Senate by one senator and on the Economic, Social and Environmental Council by one councillor. The French government is represented by a *Prefect*.

RECENT ELECTIONS

At the Territorial Council elections on 18 March 2012, 14 seats went to Archipel Demain and five to Ensemble pour l'Avenir.

CURRENT GOVERNMENT

Prefect: Patrice Latron.
  President of the Territorial Council: Stéphane Artano.

Government Website (French only):
  http://www.saint-pierre-et-miquelon.pref.gouv.fr

ECONOMY

Currency
Since 1 Jan. 2002 the euro has been the official currency as in metropolitan France.

Budget
Receipts of the Territorial Council in 2009 were €46·3m. and expenditures €38·9m.

Banking and Finance
Banks include the Banque des Îles Saint-Pierre et Miquelon, the Crédit Saint-Pierrais and the Caisse d'Épargne.
  A Development Agency was created in 1996 to help with investment projects.

ENERGY AND NATURAL RESOURCES

Environment
Carbon dioxide emissions from the consumption and flaring of fossil fuels in 2008 were the equivalent of 12·9 tonnes per capita.

Electricity
Production (2009): 44,920 GWh. Consumption (2009): 46,584 GWh. Installed capacity (2007 estimate): 27,000 kW.

Agriculture
The islands, being mostly barren rock, are unsuited for agriculture, but some vegetables are grown and livestock is kept for local consumption.

Forestry
Forests extended over 3,000. ha. in 2010, representing 13% of the land area.

Fisheries
The 2010 catch amounted to 2,043 tonnes, the lowest since the mid-1990s.

INDUSTRY

In 2009 there were 528 businesses (including 122 commercial, 76 services, 64 construction, 63 real estate, 42 accommodation and restaurants, 37 primary sector and 28 transport). The main industry is fish processing. Diversification activities are in progress (such as aquaculture and sea products processing).

Labour
The economically active population in 2006 was 3,185. Of those in employment in 2006, 85·6% worked in the service sector. The unemployment rate was 7·7% in 2006. In Jan. 2013 the minimum wage (SMIC) was raised to €9·43 an hour (€1,430·22 month for a 35-hour week).

INTERNATIONAL TRADE

Imports and Exports
Trade in €1m. (2009): imports, 60·2; exports, 3·9.

COMMUNICATIONS

Roads
In 2009 there were 117 km of roads. There were 5,773 vehicles in use in 2009; 380 new vehicles were registered in the course of the year.

Civil Aviation
Air Saint-Pierre connects St Pierre with Halifax, Moncton (New Brunswick), Montreal, Sydney (Nova Scotia) and St John's (Newfoundland). In addition, a new airport capable of receiving medium-haul aeroplanes was opened in 1999.

Shipping
St Pierre has regular services to Fortune in Canada. In 2009, 25,286 tonnes of freight (excluding petroleum products) were handled by St Pierre's ports.

Telecommunications
There were 4,800 main (fixed) telephone lines in 2008.

SOCIAL INSTITUTIONS

Justice
There is a court of first instance and a higher court of appeal at St Pierre.

Education
Primary instruction is free. In 2009 there were 644 pupils in pre-elementary education and primary education and 624 in secondary education. There were 155 teachers, 62% of whom were employed in the public sector.

Health
In 2009 there was one hospital with 40 beds, one retirement home with 60 beds and one care centre for the disabled with 20 beds.

RELIGION

The population is chiefly Roman Catholic.

CULTURE

Tourism
In 2009 there were 11,767 visitors (14·9% from mainland France), down from 15,098 in 2008.

Wallis and Futuna

Wallis et Futuna

KEY HISTORICAL EVENTS

French dependencies since 1842, the inhabitants of these islands voted on 22 Dec. 1959 by 4,307 votes out of 4,576 in favour of exchanging their status to that of an overseas territory, which took effect from 29 July 1961. In March 2003 Wallis and Futuna became an Overseas Collectivity.

TERRITORY AND POPULATION

The territory comprises two groups of islands in the central Pacific (total area, 274 sq. km; census population, 13,445 in 2008). The Îles de Hoorn lie 255 km northeast of the Fiji Islands and consist of two main islands: Futuna (64 sq. km, 4,238 inhabitants) and uninhabited Alofi (51 sq. km). The Wallis Archipelago lies another 160 km further northeast, and has an area of 159 sq. km (9,207 inhabitants). It comprises the main island of Uvéa (60 sq.

km) and neighbouring uninhabited islands, with a surrounding coral reef. The capital is Mata-Utu (2008 census population of 1,126) on Uvéa. Wallisian and Futunian are distinct Polynesian languages.

SOCIAL STATISTICS

Estimates per 1,000 population, 2003: birth rate, 19·4; death rate, 5·9.

CONSTITUTION AND GOVERNMENT

A *Prefect* represents the French government and carries out the duties of head of the territory, assisted by a 20-member *Territorial Assembly* directly elected for a five-year term, and a six-member *Territorial Council*, comprising the three traditional chiefs and three nominees of the Prefect agreed by the Territorial Assembly. The territory is represented by one deputy in the French National Assembly, by one senator in the Senate, and by one member on the Economic, Social and Environmental Council. There are three districts: Singave and Alo (both on Futuna), and Wallis; in each, tribal kings exercise customary powers assisted by ministers and district and village chiefs.

CURRENT GOVERNMENT

Prefect: Michel Aubouin.
   President of the Territorial Assembly: Sosefo Suve.

ECONOMY

Currency
The unit of currency is the franc CFP (XPF), with a parity of 119·3317422 francs CPF to the euro.

Budget
The budget for 2005 balanced at 2,623m. francs CFP.

Banking and Finance
There is a branch of Banque Indosuez at Mata-Utu.

ENERGY AND NATURAL RESOURCES

Electricity
There is a thermal power station at Mata-Utu.

Agriculture
The chief products are bananas, coconuts, copra, cassava, yams and taro.

Livestock (2009 estimate): 25,000 pigs; 7,000 goats.

Forestry
There were 6,000 ha. of forest in 2010, or 42% of the total land area.

Fisheries
The catch in 2010 was an estimated 800 tonnes.

COMMUNICATIONS

Roads
There are about 100 km of roads on Uvéa.

Civil Aviation
There is an airport on Wallis, at Hihifo, and another near Alo on Futuna. Eight flights a week link Wallis and Futuna. Air Calédonie International operates two flights a week to Nouméa (three in the summer) and two flights a week to Nadi.

Shipping
There are ports at Mata-Utu (Wallis) and Leava (Futuna).

Telecommunications
There were 2,143 main telephone lines in 2006.

SOCIAL INSTITUTIONS

Justice
There is a court of first instance, from which appeals can be made to the Court of Appeal in New Caledonia.

Education
In 2005 there were 2,573 pupils in nursery and primary schools and 2,355 in secondary schools. The South Pacific University Institute for Teacher Training, founded in 1992 (part of the French University of the Pacific, UFP) has three colleges: in Wallis and Futuna, French Polynesia and Nouméa (New Caledonia), where it is headquartered.

Health
In 2005 there were two hospitals with 72 beds, and three dispensaries.

RELIGION
The majority of the population is Roman Catholic.

SUI GENERIS COLLECTIVITIES
Collectivités sui generis

New Caledonia
Nouvelle-Calédonie

KEY HISTORICAL EVENTS

From the 11th century Melanesians settled in the islands that now form New Caledonia and dependencies. Capt. James Cook was the first European to arrive on Grande Terre on 4 Sept. 1774. The first European settlers (English Protestants and French Catholics) came in 1840. In 1853 New Caledonia was annexed by France and was used as a penal colony, taking in 21,000 convicts by 1897. Nickel was discovered in 1863, the mining of which provoked revolt among the Kanak tribes. During the Second World War, New Caledonia was used as a military base by the USA. Having fought for France during the war, the Kanaks were awarded citizenship in 1946. Together with most of its former dependencies, New Caledonia was made an Overseas Territory in 1958. It became a Territorial Collectivity under the Nouméa Accord of May 1998, which agreed on a gradual handover of responsibilities and the creation of New Caledonian citizenship. New Caledonia became a *Sui Generis* Collectivity (one that does not conform to the normal administrative structure) in March 2003. A referendum on independence is set to be held between 2014 and 2018.

TERRITORY AND POPULATION

The territory comprises Grande Terre (New Caledonia mainland) and various outlying islands, all situated in the southwest Pacific (Melanesia) with a total land area of 18,575 sq. km (7,172 sq. miles). New Caledonia has the second biggest coral reef in the world. The population (2009 census, provisional) was 245,580; density, 13·2 per sq. km. The main ethnic groups are the native Melanesians (Kanaks) and the Europeans (mostly French). There are also Wallisians and Futunians, Tahitians and Vietnamese and smaller minorities. In 2004 an estimated 63% of the population lived in urban areas. The UN gives a projected population for 2015 of 270,000. The capital, Nouméa, had 7,579 inhabitants in 2009.

There are four main islands (or groups of):

Grande Terre An area of 16,372 sq. km (about 400 km long, 50 km wide) with a population (2004 census) of 205,939. A central mountain range separates a humid east coast and a drier temperate west coast. The east coast is predominantly Melanesian; the Nouméa region predominantly European; and the rest of the west coast is of mixed population.

Loyalty Islands 100 km (60 miles) east of New Caledonia, consisting of four large islands: Maré, Lifou, Uvéa and Tiga. It has a total area of 1,981 sq. km and a population (2004) of 22,080.

Isle of Pines A tourist and fishing centre 50 km (30 miles) to the southeast of Nouméa, with an area of 152 sq. km and a population (2004) of 1,840.

Bélep Archipelago About 50 km northwest of New Caledonia, with an area of 70 sq. km and a population (2004) of 930.

The remaining islands are very small and have no permanent inhabitants.

At the 1996 census there were 341 tribes (which have legal status under a high chief) living in 160 reserves, covering a surface area of 392,550 ha. (21% of total land), and representing about 28·7% of the population. 80,000 Melanesians belong to a tribe.

New Caledonia has a remarkable diversity of Melanesian languages (29 vernacular), divided into four main groups (Northern, Central, Southern and Loyalty Islands). There were 62,648 speakers in 2004. In 2006, eight Melanesian languages were taught in schools.

SOCIAL STATISTICS

2009: live births, 4,103; deaths, 1,261; marriages, 961; divorces, 243. Population growth rate, 2008, 1·1%. Life expectancy at birth, 2007, 75·9 years. Infant mortality, 2007, 6 per 1,000 live births; fertility rate, 2·2 births per woman.

CLIMATE

2006: Nouméa, Jan. 25·5°C, July 20·5°C (average temperature, 23·2°C; max. 33·8°C, min. 14·6°C). Annual rainfall 739 mm.

CONSTITUTION AND GOVERNMENT

Subsequent to the referendum law of 9 Nov. 1988, the organic and ordinary laws of 19 March 1999 define New Caledonia's new statute. In March 2003 New Caledonia became a *Sui Generis* Collectivity with specific status endowed with wide autonomy. New Caledonia's institutions comprise the congress, government, economic and social council (CES), the customary senate and customary councils. The congress is made up of 54 members called 'Councillors of New Caledonia' from the provincial assemblies. The 11-member government is elected by congress on a proportional ballot from party lists. The president is elected by majority vote of all members. Each member is allocated to lead and control a given sector in the administration. The government's mandate ends when the mandate of the Congress that elected it comes to an end. New Caledonia is represented by two deputies and two senators in the French parliament, and by two councillors on the Economic, Social and Environmental Council.

RECENT ELECTIONS

On 8 Nov. 1998 there was a referendum for the agreement of the Nouméa Accord. Nearly 72% of those who voted approved. Turnout was 74·2%. Voting was restricted to those people resident in New Caledonia before 1998.

In elections to the Territorial Congress on 10 May 2009 the conservative Rassemblement-UMP won 13 of 54 seats, Calédonie Ensemble 10, Union Calédonienne 8, Union Nationale pour l'Indépendance–Front de Libération Nationale Kanak et Socialiste (UNI–FLNKS) 8, Avenir Ensemble–Le Mouvement de la Diversité 6, FLNKS 3, Parti Travailliste 3, Rassemblement pour la Calédonie 2 and Libération Kanak Socialiste 1. Turnout was 72·3%.

CURRENT GOVERNMENT

High Commissioner: Jean-Jacques Brot.
   President of the Government: Harold Martin.
   President of the Congress: Gérard Poadja.

Government Website (French only):
   http://www.nouvelle-caledonie.gouv.fr

ECONOMY

Currency

The unit of currency is the *franc CFP* (XPF), with a parity of 1,000 francs CPF = 8·38 euros. 344,036m. francs CFP were in circulation in Dec. 2006.

Budget

New Caledonia's expenditures in 2008 totalled 163,834m. francs CFP (of which direct taxes 36·3% and indirect taxes 29·1%) and receipts 184,661m. francs CFP (of which current expenditure 93·3%).

Performance

Total GDP was 737bn. francs CFP in 2008.

Banking and Finance

In 2006 the banks were: Banque Calédonienne d'Investissement (BCI), Banque de Nouvelle-Calédonie (BNC), Banque Nationale de Paris/Nouvelle-Calédonie (BNP/NC), Société Générale Calédonienne de Banque (SGCB) and Caisse d'Épargne.

ENERGY AND NATURAL RESOURCES

Environment

Carbon dioxide emissions from the consumption and flaring of fossil fuels in 2008 were the equivalent of 12·8 tonnes per capita.

Electricity

Production (2007): 1,926m. kWh. Installed capacity was 0·4m. kW in 2007.

Minerals

A wide range of minerals has been found in New Caledonia including: nickel, copper and lead, gold, chrome, gypsum and platinum metals. The nickel deposits are of special value, being without arsenic, and constitute between 20–40% of the world's known nickel resources located on land.

Production of nickel ore (2006): 102,986 tonnes, of which saprolitic ore (80,794 tonnes) and lateritic ore (22,192).

Agriculture

In 2007 there were an estimated 9,000 ha. of arable land and 4,000 ha. of permanent crops. Estimated livestock, 2008: cattle, 90,000; pigs, 29,000; horses, 12,000; goats, 8,200; poultry, 600,000. The chief products are beef, pork, poultry, coffee, copra, maize, fruit and vegetables. Production (2008 estimates, in 1,000 tonnes): coconuts, 17; yams, 13; maize, 4; cassava, 3; sweet potatoes, 3.

Forestry

There were 0·84m. ha. of forest in 2010, or 46% of the total land area. Timber production (2007), 5,000 cu. metres.

Fisheries

Total catch in 2010 was 3,771 tonnes. Aquaculture (consisting mainly of saltwater prawns) provides New Caledonia's second highest source of export income after nickel.

INDUSTRY

Up until the end of the 1970s the New Caledonia economy was almost totally dependent on the nickel industry. Subsequently transformation or processing industries gained in importance to reach levels similar to those in metallurgic industries.

Labour

The employed population (2009) was 96,410. In Feb. 2010 the guaranteed monthly minimum wage was 132,000 francs CFP. In 2009 the unemployment rate stood at 13·8%.

INTERNATIONAL TRADE

Imports and Exports

Trade, 2007, in US$1m.: imports c.i.f., 2,427; exports f.o.b., 1,648. In 2007, 26·6% of imports came from France, 13·6% from Singapore and 10·7% from Australia. In 2007, 24·1% of exports went to Japan, 15·8% to France and 12·1% to China. In 2007 machinery and transport equipment accounted for 34·0% of imports, mineral fuels and lubricants 14·1% and manufactured goods 11·3%. Ferro-nickel accounted for 57·5% of exports, and nickel ores and concentrates 24·4%.

COMMUNICATIONS

Roads

In 2006 there were 5,622 km of roads and 110,000 vehicles. In 2006 road accidents injured 878 and killed 56 persons.

Civil Aviation

New Caledonia is connected by air routes with Australia, Japan, Vanuatu, Wallis and Futuna, Fiji Islands, New Zealand and French Polynesia. Regular domestic air services are provided by Air Calédonie from Magenta aerodrome in Nouméa. In 2006 there were 297,257 passengers recorded at Magenta Aerodrome. Internal services with Air Calédonie link Nouméa to a number of domestic airfields.

In 2006, 414,990 passengers and 5,440 tonnes of freight were carried via La Tontouta International Airport, near Nouméa.

Shipping

In 2005, 235 vessels unloaded 1,566,000 tonnes of freight in New Caledonia; 341 vessels loaded 3,643,000 tonnes of cargo.

Telecommunications

There were 76,200 landline telephone subscriptions in 2011 (or 299·0 per 1,000 inhabitants) and 227,300 mobile phone subscriptions (892·1 per 1,000 inhabitants). In 2009, 34·0% of the population were internet users.

SOCIAL INSTITUTIONS

Justice

The judicial system is based on a two-tier court structure. A dispute may be heard first by the relevant Court of the First Instance (*Tribunal de Première Instance*). The Court of Appeal (*Cour d'Appel*) sits in Nouméa, under the joint responsibility of the first President of the Court and the Public Prosecutor.

The population in penal institutions in 2010 was 406 (162 per 100,000 population).

Education

In 2011 there were 36,048 pupils at primary schools and 33,672 at secondary schools. There were a total of 235 public schools and 121 private schools in 2011. By decree of 1999 the New Caledonia campus of the French University of the Pacific (UFP), established in 1987, was separated from the campus, to become University of New Caledonia (UNC). In 2008 it had 2,800 students.

Health

In 2008 there were 545 doctors, 125 dentists, 141 pharmacists, 1,091 nurses and 106 midwives. There were 701 hospital beds for short-term care in 2007.

Welfare

There are two main forms of social security cover: Free Medical Aid, which provides total sickness cover for non-waged persons and low-income earners; and the Family Benefit, Workplace Injury and Contingency Fund for Workers in New Caledonia (CAFAT). There are also numerous mutual benefit societies. In 2006 Free Medical Aid had 57,873 beneficiaries; CAFAT had 214,638 beneficiaries.

RELIGION

There were about 130,000 Roman Catholics in 2001.

CULTURE

World Heritage Sites

New Caledonia has one site on the UNESCO World Heritage List: its lagoons—reef diversity and associated ecosystems (inscribed on the list in 2008), some of the most extensive in the world.

Press

In 2010 there was one daily newspaper, *Les Nouvelles Calédoniennes*.

Tourism

In 2011 there were 112,000 non-resident tourist arrivals, up from 99,000 in 2010.

FURTHER READING

Institut de la Statistique et des Études Économiques: *Tableaux de l'Économie Calédonienne* (TEC 2011); *Informations Statistiques Rapides de Nouvelle-Calédonie* (monthly).

Imprimerie Administrative, Nouméa: *Journal Officiel de la Nouvelle Calédonie.*

Local Statistical Office: Institut Territorial de la Statistique et des Études Économiques, BP 823, 98845 Nouméa.

Website: http://www.isee.nc

Southern and Antarctic Territories

Terres Australes et Antarctiques Françaises (TAAF)

GENERAL DETAILS

The Territory of the TAAF was created on 6 Aug. 1955. It comprises the Kerguelen and Crozet archipelagoes, the islands of Saint-Paul and Amsterdam (formerly Nouvelle Amsterdam) and the Scattered Islands group, all in the southern Indian Ocean, plus Terre Adélie. It has been classified as a *Sui Generis* Collectivity (one that does not conform to the normal administrative structure) since 2007. The Scattered Islands were incorporated into the TAAF in Feb. 2007. Since 2 April 1997 the administration has had its seat in Saint-Pierre, Réunion; before that it was in Paris. The Administrator is assisted by a seven-member consultative council which meets twice yearly in Paris; its members are nominated by the government for five years. The 15-member Polar Environment Committee, which in 1993 replaced the former Consultative Committee on the Environment (est. 1982), meets at least once a year to discuss all problems relating to the preservation of the environment.

The French Institute for Polar Research and Technology was set up to organize scientific research and expeditions in Jan. 1992. The staff of the permanent scientific stations of the TAAF (approximately 200 in 2005) is renewed every 6 or 12 months and forms the only population.

Prefect: Pascal Bolot.

Amsterdam and **Saint-Paul Islands** Situated 38–39° S. lat., 77° E. long. Amsterdam, with an area of 54 sq. km (21 sq. miles) was discovered in 1522 by Magellan's companions; Saint-Paul, lying about 100 km to the south, with an area of 7 sq. km (2·7 sq. miles), was probably discovered in 1559 by Portuguese sailors. Both were first visited in 1633 by the Dutch explorer, Van Diemen, and were annexed by France in 1843. They are both extinct volcanoes. The only inhabitants are at Base Martin de Vivies (est. 1949 on Amsterdam Island), including several scientific research stations, a hospital, communication and other facilities (ranging from 25–45 persons). Crayfish are caught commercially on Amsterdam.

Crozet Islands Situated 46° S. lat., 50–52° E. long.; consists of five larger and 15 tiny islands, with a total area of 505 sq. km (195 sq. miles). The western group includes Apostles, Pigs and Penguins islands; the eastern group, Possession and Eastern islands. The archipelago was discovered in 1772 by Marion Dufresne, whose first mate, Crozet, annexed it for Louis XV. A meteorological and scientific station (ranging from 25–45 persons) at Base Alfred-Faure on Possession Island was built in 1964.

Kerguelen Islands Situated 48–50° S. lat., 68–70° E. long.; consists of one large and 85 smaller islands, and over 200 islets and rocks, with a total area of 7,215 sq. km (2,786 sq. miles) of which Grande Terre occupies 6,675 sq. km (2,577 sq. miles). It was discovered in 1772 by Yves de Kerguelen, but was effectively occupied by France only in 1949. Port-aux-Français has several scientific research stations (ranging from 70–110 persons). Reindeer, trout and sheep have been acclimatized.

Scattered Islands Situated in the Indian Ocean around Madagascar between 11–22° S. lat., 39–54° E. long., comprising Bassas da India, Europa Island, the Glorieuses Islands, Juan de Nova Island and Tromelin Island. Formerly French minor territories, the islands—which have a total area of 39 sq. km (15 sq. miles)—were incorporated into the TAAF in Feb. 2007. Sovereignty of individual islands is disputed, with the Comoros, Madagascar, Mauritius and the Seychelles all making claims. The population numbers around 50. The islands are designated as nature reserves and support a range of meteorological stations, military garrisons and radio stations. About 12,000 tonnes of guano are mined annually on Juan de Nova Island.

Terre Adélie Comprises that section of the Antarctic continent between 136° and 142° E. long., south of 60° S. lat. The ice-covered plateau has an area of about 432,000 sq. km (166,800 sq. miles), and was discovered in 1840 by Dumont d'Urville. A research station (ranging from 30–110 persons) is situated at Base Dumont d'Urville, which is maintained by the French Institute for Polar Research and Technology.

MINOR TERRITORIES

Dépendances

Clipperton Island

Île Clipperton

In the 18th century the island was the hideout of a pirate, John Clipperton, for whom it was named. In 1855 it was claimed by France, and in 1897 by Mexico. It was awarded to France by international arbitration in 1935. Clipperton Island is a Pacific atoll, 3 km long, some 1,120 km southwest of the coast of Mexico. It covers an area of 7 sq. km and is uninhabited. The island is administered by the Minister of Overseas France. The island is occasionally visited by tuna fishermen.

GABON

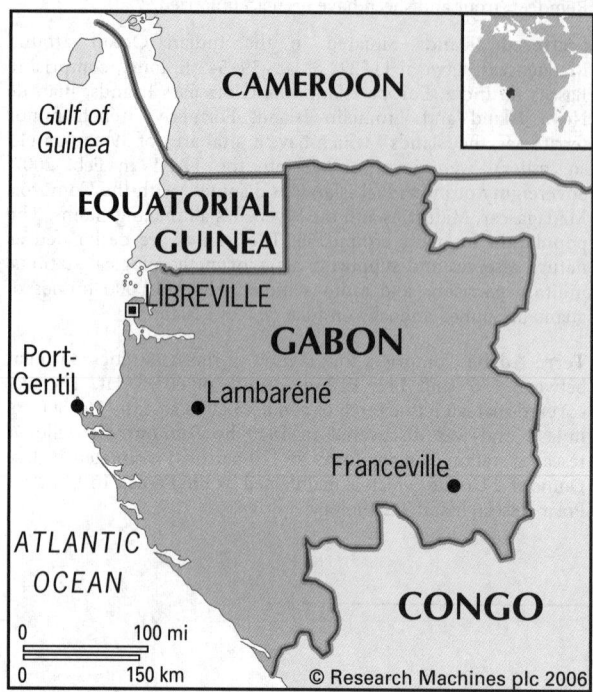

République Gabonaise
(Gabonese Republic)

Capital: Libreville
Population projection, 2015: 1·66m.
GNI per capita, 2011: (PPP$) 12,249
HDI/world rank: 0·674/106
Internet domain extension: .ga

KEY HISTORICAL EVENTS

The earliest inhabitants of Gabon were pygmy hunter-gatherers; Baka pygmies continue to inhabit the northern forests, while the Babongo remain in parts of the southeast. Bantu-speaking farmers originating in present-day Cameroon migrated south and east across the African continent from around 500 BC, leaving evidence of pottery and tools at Njole in central Gabon. Later Bantu migrations from the north included the Mpongwe in the 15th century and the Fang in the 18th century.

Portuguese mariners reached Equatorial Africa in 1472. The name Gabon derives from the Portuguese for a hooded coat, which the estuary of the Komo river apparently resembled. Settlement was slow to develop compared with the islands of São Tomé, Príncipe and Bioko, where sugar plantations were established. However, Portuguese and Dutch trade with the Vili group grew in the 17th century, with iron tools, cloth and tobacco exchanged for ivory, timber, rubber and, increasingly, slaves. The Ogooué river delta became a hub for Portuguese merchants trading with local Orungu chieftains for slaves from the interior, who were shipped mainly to Brazil and Cuba. The Mpongwe also profited from trade (including slaves) with Europeans around the Komo estuary.

By the early 19th century the British and French were vying for control of trade in the Gulf of Guinea. In 1839 Capt. Bouët-Willaumez negotiated treaties with two Mpongwe chiefs to end the slave trade and accept French sovereignty over their land. The current capital, Libreville, was founded by freed slaves in 1849 on the Komo estuary. Over the next two decades French agents made treaties with all of the coastal clans and from 1875 turned their attention inland.

Expeditions along the Ogooué river were led by Savorgnan de Brazza, who founded Franceville in 1880. In 1886 he established French Congo, which included Gabon and much of the present-day Republic of the Congo. The land rights of indigenous groups were denied and the region's resources depleted. Borders with German Cameroon and Spanish Guinea were fixed in the late 1890s and in 1910 Gabon became one of the four territories comprising French Equatorial Africa.

Conditions improved markedly after the Second World War, when the territory was incorporated into the French Fourth Republic with its own assembly and representation in Paris. Investment in infrastructure, industry, agriculture, education and healthcare invigorated the economy. Following the collapse of the Fourth Republic in 1958, Gabon became an autonomous republic within the French Community. Léon Mba, a prominent member of the Fang ethnic group and former mayor of Libreville, became the first prime minister under the new constitution of 1959 and president following independence on 17 Aug. 1960.

In Feb. 1964 Mba was toppled by the military in a bloodless coup, but was reinstated shortly afterwards with French assistance. His death in March 1967 ushered in Omar Bongo, the vice-president. Within months Bongo had established the Gabonese Democratic Party (PDG) as the sole political party. The economy developed steadily as minerals—including oil, manganese and uranium—were extracted. The stable political scene facilitated foreign investment in infrastructure and education, although freedoms were suppressed.

Financial difficulties emerged in the late 1970s caused by an over-reliance on petroleum and high interest rates on foreign loans. Bongo was also dogged by allegations of corruption and the misuse of public funds throughout the 1980s. Reports in 1989 that he and his associates had amassed vast personal fortunes, combined with the poor economic outlook and news of the break-up of Eastern Europe's autocratic regimes, spurred protests in Libreville in Jan. 1990. Bongo agreed to legalize opposition parties and in March 1991 the National Assembly adopted a constitution restoring a multi-party system.

Bongo narrowly won the presidential election in Dec. 1993, although opposition candidates alleged fraud. In the 1996 legislative election the PDG secured a landslide and Bongo was returned to the presidency in 1998 and 2005. His death in a Spanish hospital in June 2009 forced a presidential election that attracted 18 candidates. When Ali Ben Bongo was announced the winner, opposition leaders André Mba Obame and Pierre Mamboundou protested that the poll had been fixed to ensure dynastic succession. Unrest ensued in Libreville and Port-Gentil.

In 2010 Gabon signalled a shift in geopolitical focus away from France by announcing deals with India and Singapore for major infrastructure projects.

TERRITORY AND POPULATION

Gabon is bounded in the west by the Atlantic Ocean, north by Equatorial Guinea and Cameroon and east and south by the Republic of the Congo. The area covers 267,670 sq. km, including 10,000 sq. km of inland waters. Its population at the 2003 census was reported as 1,517,685, although this is believed to be an over-count. United Nations estimate, 2010, 1·51m.; density, 6 per sq. km. In 2011, 86·4% of the population were urban.

The UN gives a projected population for 2015 of 1·66m.

494

The capital is Libreville (538,195 inhabitants, 2003 census), other large towns (1993 census) being Port-Gentil (79,225), Franceville (31,183), Oyem (22,404) and Moanda (21,882).

Provincial areas, populations and capitals:

Province	Area in sq. km	Population 2003 census	Capital
Estuaire	20,740	662,028	Libreville
Haut-Ogooué	36,547	228,471	Franceville (Masuku)
Moyen-Ogooué	18,535	60,990	Lambaréné
Ngounié	37,750	101,415	Mouila
Nyanga	21,285	50,297	Tchibanga
Ogooué-Ivindo	46,075	64,163	Makokou
Ogooué-Lolo	25,380	64,534	Koulamoutou
Ogooué-Maritime	22,890	128,774	Port-Gentil
Woleu-Ntem	38,465	157,013	Oyem

The largest ethnic groups are the Fangs (25%) in the north and the Bapounou (24%) in the south. There are some 40 smaller groups. French is the official language.

SOCIAL STATISTICS

2008 estimates: births, 39,000; deaths, 14,000. Estimated rates, 2008 (per 1,000 population): births, 27·3; deaths, 9·7. Annual population growth rate, 2000–08, 2·0%. Expectation of life at birth, 2007, 58·7 years for males and 61·5 years for females. Infant mortality, 2010, 54 per 1,000 live births; fertility rate, 2008, 3·3 births per woman.

CLIMATE

The climate is equatorial, with high temperatures and considerable rainfall. Mid-May to mid-Sept. is the long dry season, followed by a short rainy season, then a dry season again from mid-Dec. to mid-Feb., and finally a long rainy season once more. Libreville, Jan. 80°F (26·7°C), July 75°F (23·9°C). Annual rainfall 99" (2,510 mm).

CONSTITUTION AND GOVERNMENT

On 21 March 1997 the government presented to the Parliament legislation aimed at reforming the constitution in a number of key areas: notably, the bill mandated the creation of a Vice-President of the Republic, the extension of the presidential term of office from five to seven years, and the transformation of the Senate into an Upper Chamber of Parliament. Gabon has a bicameral legislature, consisting of a 120-member *National Assembly* (with members elected by direct, popular vote to serve five-year terms) and a 102-member *Senate* (elected for six-year terms in single-seat constituencies by local and departmental councillors). At a referendum on electoral reform on 23 July 1995, 96·48% of votes cast were in favour; turnout was 63·45%. The 1991 constitution provides for an executive *President* directly elected for a five-year term (renewable once only). In July 2003 Gabon's parliament approved an amendment to the constitution that allows the president to seek re-election indefinitely. The head of government is the *Prime Minister*, who appoints a Council of Ministers.

National Anthem

'La concorde' ('The Concord'); words and tune by G. Damas Aleka.

RECENT ELECTIONS

Presidential elections were held on 30 Aug. 2009. The late president's son, Ali-Ben Bongo Ondimba, was elected with 41·7% of votes cast against 25·9% for Andre Mba Obame and 25·2% for Pierre Mamboundou. There were 15 other candidates. Bongo's victory was marred by accusations of vote-rigging and post-election violence.

Elections for the National Assembly were held on 17 Dec. 2011. The Gabonese Democratic Party (PDG) won 114 of 120 seats, the Rally for Gabon took three, and the Circle of Liberal Reformers, the Social Democratic Party and the Union for the New Republic one seat each. Turnout was 34·8%. A number of opposition parties withdrew from the election process and urged their supporters to boycott the elections.

Senate elections were held on 18 Jan. 2009. The PDG won 75 of 102 seats; Rally for Gabon, 6; Gabonese Union for Democracy and Development, 3. Independents claimed nine seats and the remaining nine went to minor parties.

CURRENT GOVERNMENT

President: Ali-Ben Bongo Ondimba; b. 1959 (PDG; sworn in 16 Oct. 2009).

Vice President: Didjob Divungi Di Ndinge.

In March 2013 the Council of Ministers comprised:

Prime Minister: Raymond Ndong Sima; b. 1955 (PDG; sworn in 27 Feb. 2012).

Minister Delegate to the Prime Minister, in Charge of State Reform: Calixte Isidore Nsie Edang.

Minister for Agriculture, Livestock, Fishing and Rural Development: Julien Nkoghé Bekale. *Budget, Public Finance and Civil Service:* Christiane Rose Ossoucah Raponda. *Defence:* Rufin Pacôme Ondzounga. *Digital Economy, Communications and Posts:* Blaise Louembé. *Economy, Employment and Sustainable Development:* Luc Oyoubi. *Family and Social Affairs:* Honorine Nzet Biteghe. *Foreign Affairs, International Co-operation and Francophonie Affairs:* Emmanuel Issozé Ngondet. *Health:* Léon Nzouba. *Industry and Mines:* Régis Immongault Tatagani. *Interior, Public Security, Immigration and Decentralization:* Jean François Ndongou. *Investment Promotion, Public Works, Transport, Housing and Tourism:* Magloire Ngambia. *Justice, Keeper of the Seals, Human Rights, Relations with Constitutional Institutions and Government Spokesperson:* Ida Reteno Assonouet. *National Education, Higher Education, Technical and Vocational Training, Culture, and Youth and Sports:* Séraphin Moundounga. *Oil, Energy and Hydraulics:* Etienne Ngoubou. *Small and Medium-Sized Enterprise, Handicrafts and Trade:* Fidèle Mengue M'engouang. *Water and Forests:* Gabriel Ntchango.

Gabonese Parliament (French only): http://www.assemblee.ga

CURRENT LEADERS

Ali-Ben Bongo Ondimba

Position
President

Introduction
Ali-Ben Bongo Ondimba was elected president in Aug. 2009 following the death of his father, Omar Bongo Ondimba (president from 1967–2009). Ali Bongo served as minister of foreign affairs under his father, as well as a deputy of the National Assembly, minister of defence and vice-president of the Gabonese Democratic Party (PDG).

Early Life
Ali-Ben Bongo Ondimba is the oldest son of Omar Bongo, the country's longest-serving president. He was born Alain-Bernard Bongo in Feb. 1959 in Brazzaville, Republic of the Congo. His mother, Gabonese singer Josephine Kama (later Patience Dabany), was 15 at the time of his birth.

The family moved to Gabon in 1960, just after its independence from France. Alain-Bernard spent most of his youth in France, studying at a protestant primary school in Cévennes and then the Catholic College Notre-Dame de Sainte-Croix, on the outskirts of Paris. In 1973 Albert-Bernard Bongo and Alain-Bernard Bongo changed their names to Omar and Ali-Ben, taking the second name Ondimba as part of their conversion to Islam. Ali graduated in law from the Sorbonne before returning to Gabon in 1981. He went straight into the upper ranks of the PDG, presided over by his father.

In 1983 Ali-Ben Bongo was elected a member of the PDG central committee and in 1984 became his father's personal spokesman. In 1989 he was appointed minister of foreign relations, replacing his cousin Martin Bongo, who had been minister since 1976. At the 1990 elections Bongo was returned as representative for the province of Haut-Ogooué. Since the constitution of 1991, which allowed for multi-party politics, also set a minimum age of 35 for a minister, Bongo was replaced at the department of foreign relations. Bongo was re-elected to represent Haut-Ogooué in 1996 and again in 2006. In 1999 he returned to the cabinet as minister for defence. He became party vice president in 2003 and in 2005 helped run his father's election campaign. After Omar Bongo died in Spain in June 2009, Bongo put himself at centre stage to run for the presidency in new elections in Aug.

In competition against 17 opponents, he won 42% of the vote. As with previous elections, the results were contested amid allegations of government electoral fraud and bribery. Thousands took to the streets in the capital, Libreville, and in the second city, Port-Gentil. A recount was held and the result upheld by the Supreme Court. Six newspapers were suspended for criticism of the government.

Career in Office
Bongo's first act as president was to reinstate Paul Biyoghé Mba as head of the cabinet and to cut down the number of ministers, having pledged to slash government expenditure. Ministers have undergone curbs on privileges and been subject to pay cuts. In Nov. 2009 Bongo travelled to Paris to meet the then French president Nicolas Sarkozy and in Dec. 2009 he made a state visit to the Vatican and Pope Benedict XVI.

Bongo pledged to fight corruption. Nonetheless, Transparency International and other NGOs have pushed for investigations into the Bongo family's finances.

In Dec. 2010 parliament adopted a controversial constitutional amendment allowing the president to extend his mandate in the case of an emergency. In early 2011 Bongo's main rival in the 2009 election, André Mba Obame, took refuge in the United Nations compound in Libreville. Mba Obame claimed to have been the rightful winner of the presidential poll, in response to which Bongo banned his party. In National Assembly elections in Dec. 2011 the PDG retained its substantial parliamentary majority. In Feb. 2012 Bongo appointed Raymond Ndong Sima as prime minister following Biyoghé Mba's resignation.

Raymond Ndong Sima

Position
Prime Minister

Introduction
Raymond Ndong Sima, a former minister of agriculture, was appointed prime minister by President Ali-Ben Bongo in Feb. 2012 following the victory of the ruling Gabonese Democratic Party (PDG) in parliamentary elections. Sima has combined careers in politics and private enterprise and is regarded as a key figure in Bongo's plan to diversify the economy. He is the first person from Gabon's less developed northern region to become prime minister.

Early Life
Sima was born on 23 Jan. 1955 in Oyem in the province of Woleu-Ntem in the north of Gabon. He was educated in Oyem, at Bessieux College in Libreville and in Algeria. In 1981 he received a masters degree in economics from the University of Paris IX-Dauphine in France. Returning to Gabon later that year, he worked as a researcher for the department of the economy.

In 1986, during an economic crisis, Sima joined the cabinet as minister for economy and planning. From 1992–94 he served as director general of the economy, managing Gabon's negotiations

with the IMF and the World Bank. He was director of Hévégab, the state-owned rubber company, from 1994–98 and in the late 1990s was made director of Gabon's state railway operator, C.E.C.F.T., serving until 2001. In 2003 he founded a private bus company.

In 2009, following the victory of Ali-Ben Bongo in presidential elections after the death of his father, Sima was appointed minister of agriculture, livestock, fisheries and rural development. He oversaw the introduction of a US$1·3m. agricultural development programme, backed by the European Union and the UN Food and Agriculture Organization. In Dec. 2011 he won a seat in parliament for the ruling PDG, representing the constituency of Kye in Woleu-Ntem province.

With the PDG retaining power at the legislative election, the incumbent prime minister, Paul Biyoghé Mba, stepped down. President Bongo appointed Sima as his successor on 27 Feb. 2012, citing Sima's combination of business expertise and administrative experience as key to the appointment.

Career in Office
On becoming prime minister Sima made substantial changes to the administration, bringing 14 newcomers into ministerial and deputy ministerial posts. One of his principal tasks is to implement Bongo's 'Emerging Gabon' project to diversify the economy away from oil and to build up the services, industrial and environmental sectors.

DEFENCE

In 2008 military expenditure totalled US$134m. (US$90 per capita), representing 0·9% of GDP.

Army
The Army totalled (2007) 3,200. A referendum of 23 July 1995 favoured the transformation of the Presidential Guard into a republican guard. There is also a paramilitary Gendarmerie of 2,000. France maintains 700 Army personnel in Gabon.

Navy
There is a small naval flotilla, about 500 strong in 2007. France maintains 1,560 Naval personnel in Gabon.

Air Force
Personnel (2007) 1,000. There are around 16 combat-capable aircraft (including nine Mirage 5s) and five attack helicopters.

ECONOMY

Agriculture accounted for 5% of GDP in 2009, industry 54% and services 41%.

Gabon's 'shadow' (black market) economy is estimated to constitute approximately 47% of the country's official GDP.

Overview
Gabon is rich in natural resources and one of the largest oil producers in sub-Saharan Africa. Income per capita is greater than the sub-Saharan Africa average but, although extreme poverty has been cut drastically over the years, 33% of the population live under the poverty line. Unemployment remains high.

Until the early 1970s when offshore oil deposits were discovered, principal exports were timber and manganese. Because of the role played by the oil industry, the agricultural sector accounts for a small percentage of total output relative to other low-income countries. After the oil rush there was volatile growth marked by fiscal mismanagement leading to a 50% currency devaluation in 1994. In the years following, an inflationary spike prompted French, US and IMF financial support. Among oil-exporting African nations, Gabon stands out as a growth laggard, averaging only 1·1% annual growth from 1996–2005 compared to a 9·1% average among the continent's top seven oil exporters.

Gabon has seen oil production drop in recent years, with few additional deposits predicted. The global economic crisis had a knock-on effect on the price of oil and other exports. Economic diversification is key to further growth.

Currency

The unit of currency is the *franc CFA* (XAF) with a parity of 655·957 francs CFA to one euro. Foreign exchange reserves were US$466m. in June 2005 and total money supply was 416,653m. francs CFA. Gold reserves were 13,000 troy oz in July 2005. Inflation was 1·4% in 2010 and 1·3% in 2011.

Budget

In 2008 revenue totalled 2,078·1bn. francs CFA and expenditure 1,296·3bn. francs CFA. Oil revenues accounted for 65·5% of total revenue in 2008; current expenditure accounted for 69·9% of total expenditure.

The standard rate of VAT is 18% (reduced rate, 10%).

Performance

The economy contracted by 1·4% in 2009 but there was growth of 6·6% in both 2010 and 2011. Total GDP in 2011 was US$17·1bn.

Banking and Finance

The Banque des États de l'Afrique Centrale (*Governor*, Lucas Abaga Nchama) is the bank of issue. There are five commercial banks. The largest are Banque Internationale pour le Commerce et l'Industrie du Gabon, BGFIBANK and Union Gabonaise de Banque, which between them had 80% of the market share in 2003.

In 2010 foreign debt amounted to US$2,331m., representing 20·3% of GNI.

ENERGY AND NATURAL RESOURCES

Environment

Gabon's carbon dioxide emissions from the consumption and flaring of fossil fuels in 2008 were the equivalent of 3·1 tonnes per capita.

Electricity

Installed capacity was 0·4m. kW in 2007. Production totalled 1·84bn. kWh in 2007 (approximately 57% thermal and 43% hydroelectric). Consumption per capita was an estimated 1,297 kWh in 2007.

Oil and Gas

Proven oil reserves (2008), 3·2bn. bbls. Production, 2008, 11·8m. tonnes. In 2007 exports of oil stood at 9·7m. tonnes. There were proven natural gas reserves of 28bn. cu. metres in 2007. Natural gas production (2007) was 158m. cu. metres.

Minerals

There are an estimated 200m. tonnes of manganese ore and 850m. tonnes of iron ore deposits. Gold, zinc and phosphates also occur. Output, 2006: manganese ore, 2·98m. tonnes.

Agriculture

In 2007 the agricultural population was approximately 415,000, of whom 192,000 were economically active. There were 325,000 ha. of arable land in 2001 and 170,000 ha. of permanent crops. 15,000 ha. were irrigated in 2001.

The major crops (estimated production, 2003, in 1,000 tonnes) are: plantains, 270; sugarcane, 235; cassava, 230; yams, 155; taro, 59; maize, 31; groundnuts, 20; bananas, 12; rubber, 11. Other important products include palm oil, sweet potatoes and soybeans.

Livestock (2002 estimates): 212,000 pigs; 195,000 sheep; 90,000 goats; 35,000 cattle; 3m. chickens.

In 2002 an estimated 32,000 tonnes of meat, 2,000 tonnes of eggs and 2,000 tonnes of fresh milk were produced.

Forestry

Forests covered 22·0m. ha. in 2010, or 85% of the total land area. Of the total area under forests 65% in 2010 was primary forest and 35% other naturally regenerated forest. Timber production in 2007 was 3·93m. cu. metres.

Since 2002 a tenth of the country has been transformed into 13 national parks covering nearly 30,000 sq. km.

Fisheries

The catch in 2006 was 41,521 tonnes, of which 32,162 tonnes were from marine waters and 9,359 tonnes from inland waters.

INDUSTRY

Most manufacturing is based on the processing of food (particularly sugar), timber and mineral resources, cement and chemical production and oil refining. Production figures in 1,000 tonnes: residual fuel oil (2007), 356; cement (2006), 260; distillate fuel oil (2007), 260; petrol (2007), 56; beer (2003), 75·4m. litres; soft drinks (2005), 60·7m. litres.

Labour

The economically active workforce in 2005 numbered 664,000 (59% males). The unemployment rate was 14·8% in 2005. In 2007 the legal minimum monthly wage was 80,000 francs CFA. There is a 40-hour working week.

INTERNATIONAL TRADE

The government retains the right to participate in foreign investment in oil and mineral extraction.

Imports and Exports

Imports and exports for calendar years in US$1m.:

	2002	2003	2004	2005	2006
Imports c.i.f.	745·2	770·0	964·9	1,471·9	1,724·9
Exports f.o.b.	2,411·1	319·9	2,780·0	5,068·5	6,015·2

Machinery and transport equipment accounted for 42·8% of imports in 2006, food and live animals 13·0%, chemicals and related products 9·2% and iron and steel 7·4%. Petroleum and petroleum products constituted 85·6% of exports in 2006, and cork and wood 6·3%.

Main import suppliers, 2006: France, 39·9%; Belgium, 14·2%; USA, 7·3%. Main export markets, 2006: USA, 58·4%; China, 10·6%; France, 7·1%.

COMMUNICATIONS

Roads

In 2004 there were an estimated 9,170 km of roads (10·2% of which were paved); and in 2002 some 25,600 passenger cars plus 17,000 trucks and vans. There were 293 deaths in road accidents in 2000.

Rail

The 669 km standard gauge Transgabonais railway runs from the port of Owendo to Franceville. Total length of railways, 2005, 649 km. In 2008 passenger-km travelled came to 95m. and freight tonne-km to 2,485m.

Civil Aviation

There are international airports at Libreville (Léon M'Ba Airport), Port-Gentil and Franceville (Masuku); scheduled internal services link these to a number of domestic airfields. Libreville, the main airport, handled 757,000 passengers and 17,700 tonnes of freight in 2001. In 2003 scheduled airline traffic of Gabonese-based carriers flew 8m. km, carrying 386,000 passengers (170,000 on international flights). Gabon Airlines was established in July 2006 as a successor to the bankrupt national carrier Air Gabon.

Shipping

In Jan. 2009 there were 12 ships of 300 GT or over registered, totalling 6,000 GT. Owendo (near Libreville), Mayumba and Port-Gentil are the main ports. Rivers are an important means of inland transport.

Telecommunications

In 2010 there were 30,400 landline telephone subscriptions (equivalent to 20·2 per 1,000 inhabitants) and 1,610,000 mobile phone subscriptions (or 1,069·4 per 1,000 inhabitants). Fixed internet subscriptions totalled 22,200 in 2010 (14·7 per 1,000 inhabitants).

SOCIAL INSTITUTIONS

Justice

There are *Tribunaux de grande instance* at Libreville, Port-Gentil, Lambaréné, Mouila, Oyem, Franceville (Masuku) and Koulamoutou, from which cases move progressively to a central Criminal Court, Court of Appeal and Supreme Court, all three located in Libreville. Civil police number about 900. The death penalty was abolished in Feb. 2010.

Education

The adult literacy rate in 2009 was estimated at 87·7% (91·4% among males and 84·1% among females). Education is compulsory between 6–16 years. In 2000–01 there were 265,714 pupils and 5,399 teachers in primary schools, and 101,681 pupils with 2,727 (1996–97) teachers at secondary schools; in 1996–97 there were 6,703 students in 11 technical and professional schools and 76 students in two teacher-training establishments.

In 2009 there was one university at Libreville (the Omar Bongo University) and one university of science and technology at Franceville (Masuku). In 2004 a university of health sciences (previously part of the Omar Bongo University) was created in Libreville.

In 2000–01 total expenditure on education came to 4·6% of GNP.

Health

In 2004 there were 395 physicians, 66 dentistry personnel, 63 pharmaceutical personnel and 6,778 nursing and midwifery personnel. In 2006 there were 20 hospital beds per 10,000 inhabitants.

RELIGION

In 2001 there were 0·69m. Roman Catholics, 0·22m. Protestants and 0·17m. followers of African Christian sects. The majority of the remaining population follow animist beliefs. There are about 12,000 Muslims.

CULTURE

World Heritage Sites

Gabon has one site on the UNESCO World Heritage List: the ecosystem and relict cultural landscape of Lopé-Okanda (inscribed on the list in 2007), an area of rainforest and savannah containing the remains of Neolithic and Iron Age settlements.

Press

In 2008 there was one government-controlled daily newspaper (*L'Union*) with a circulation of 20,000. In the 2010 *World Press Freedom Index*, compiled by Reporters Without Borders, Gabon was ranked 107th out of 178 countries.

Tourism

358,000 non-resident tourists arrived at Libreville airport in 2008, up from 169,000 in 2001.

Festivals

Gabao Hip Hop Festival takes place in Libreville in June.

DIPLOMATIC REPRESENTATIVES

Of Gabon in the United Kingdom (27 Elvaston Place, London, SW7 5NL)
Ambassador: Omer Piankali.

Of the United Kingdom in Gabon
Ambassador: Bharat Joshi (resides in Yaoundé, Cameroon).

Of Gabon in the USA (1630 Connecticut Ave., NW, Washington, D.C., 20009)
Ambassador: Michael Moussa-Adamo.

Of the USA in Gabon (Blvd du Bord de Mer, Libreville)
Ambassador: Eric Benjaminson.

Of Gabon to the United Nations
Ambassador: Noel Nelson Messone.

Of Gabon to the European Union
Ambassador: Félicité Ongouori Ngoubili.

FURTHER READING

Barnes, J. F. G., *Gabon: Beyond the Colonial Legacy*. 1992
Gardinier, David E., *Historical Dictionary of Gabon*. 3rd ed. 2006

National Statistical Office: Direction Générale de la Statistique, Ministère de l'économie, de l'emploi et du développement durable, BP 2119, Libreville.
Website (French only): http://www.stat-gabon.org

THE GAMBIA

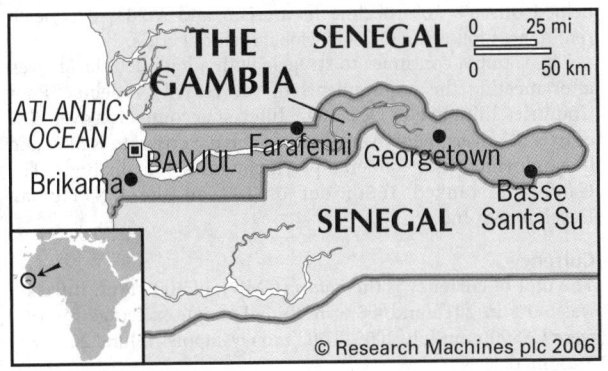

Republic of The Gambia

Capital: Banjul
Population projection, 2015: 1·98m.
GNI per capita, 2011: (PPP$) 1,282
HDI/world rank: 0·420/168
Internet domain extension: .gm

KEY HISTORICAL EVENTS

Stone circles thought to have been constructed by ancestors of the Jola people are estimated to date from AD 600. Kingdoms of Mandinka-speaking people were established near the Gambia River from around 1100. State-building by the Jolof and Serer groups gathered pace from around 1400. Portuguese mariners entered the Gambia River in 1455 but the first permanent European settlement was founded by traders from the Baltic Duchy of Courland (Latvia) in 1651. English and French merchants subsequently vied for control of the region (Senegambia). The British Captain, Alexander Grant, established Bathurst (Banjul) as a garrison in 1816 and it was controlled from the Freetown Colony (Sierra Leone). The Gambia became an independent member of the British Commonwealth on 18 Feb. 1965 and an independent republic on 24 April 1970.

TERRITORY AND POPULATION

The Gambia takes its name from the River Gambia, and consists of a strip of territory never wider than 10 km on both banks. It is bounded in the west by the Atlantic Ocean and on all other sides by Senegal. The area is 10,690 sq. km, including 2,077 sq. km of inland water. Population (census, 2003), 1,360,681; density, 158 per sq. km. In 2011, 58·9% of the population were urban.

The UN gives a projected population for 2015 of 1·98m.

The largest ethnic group is the Mandingo, followed by the Wolofs, Fulas, Jolas and Sarahuley. The country is administratively divided into eight local government areas (LGAs).

The eight LGAs, with their areas, populations and chief towns are:

Division	Area in sq. km	Population 2003 census	Chief town
Banjul	12	35,061	Banjul
Basse	2,070	182,586	Basse Santa Su
Brikama	1,764	389,594	Brikama
Janjangbureh	1,428	107,212	Janjangbureh
Kanifeng	76	322,735	Kanifeng
Kerewan	2,255	172,835	Kerewan
Kuntaur	1,467	78,491	Kuntaur
Mansakonko	1,618	72,167	Mansakonko

The official language is English.

SOCIAL STATISTICS

2008 estimates: births, 61,000; deaths, 19,000. Estimated birth rate in 2008 was 36·8 per 1,000 population; estimated death rate, 11·3. Annual population growth rate, 2000–08, 3·0%. Expectation of life, 2007, was 54·1 years for males and 57·3 for females. Fertility rate, 2008, 5·1 births per woman; infant mortality, 2010, 57 per 1,000 live births. The Gambia has made some of the best progress in recent years in reducing child mortality. The number of deaths per 1,000 live births among children under five was reduced from around 153 in 1990 to approximately 106 in 2008.

CLIMATE

The climate is characterized by two very different seasons. The dry season lasts from Nov. to May, when precipitation is very light and humidity moderate. Days are warm but nights quite cool. The SW monsoon is likely to set in with spectacular storms and produces considerable rainfall from July to Oct., with increased humidity. Banjul, Jan. 73°F (22·8°C), July 80°F (26·7°C). Annual rainfall 52" (1,295 mm).

CONSTITUTION AND GOVERNMENT

The 1970 constitution provided for an executive *President* elected directly for renewable five-year terms. The President appoints a *Vice-President* who is the government's chief minister. The single-chamber *National Assembly* has 53 members (48 elected by universal adult suffrage for a five-year term and five appointed by the President).

A referendum of 8 Aug. 1996 approved a new constitution by 70·4% of votes cast. It took effect in Jan. 1997 and thereby created the Second Republic. Under this, the ban on political parties imposed in July 1994 was lifted. Members of the ruling Military Council resigned from their military positions before joining the Alliance for Patriotic Reorientation and Construction (APRC).

National Anthem

'For the Gambia, our homeland'; words by V. J. Howe, tune traditional.

RECENT ELECTIONS

Presidential elections were held on 24 Nov. 2011. President Jammeh (Alliance for Patriotic Reorientation and Construction) was re-elected with 71·5% of the vote, against 17·4% for Ousainou Darboe (United Democratic Party) and 11·1% for Hamat Bah (ind.).

Parliamentary elections were held on 29 March 2012. The Alliance for Patriotic Reorientation and Construction (APRC) won 43 seats with 51·8% of votes cast, the National Reconciliation Party 1 (9·4%) and ind. 4 (38·8%). Turnout was 38·7%. Six parties boycotted the election.

CURRENT GOVERNMENT

President, C.-in-C. of the Armed Forces: Col. (retd) Yahya Jammeh; b. 1965 (APRC; seized power 22 July 1994; elected 26 Sept. 1996 and re-elected in 2001, 2006 and 2011).

In March 2013 the government comprised:

Vice-President and Minister of Women's Affairs: Isatou Njie Saidy.

Minister of Agriculture: Solomon Owens. *Basic and Secondary Education:* Fatou Faye. *Energy:* Teneng Mba Jaiteh. *Environment, Parks and Wildlife:* Fatou Gaye. *Finance and Economic Affairs:* Abdou Kolley. *Fisheries and Water Resources:* Mass Axi Gye. *Foreign Affairs, International Co-operation and Gambians Abroad:* Susan Waffa Ogoo. *Health and Social Welfare, Responsible for National Assembly Matters:* Bala Garba Jahumpa. *Higher Education,*

Research, Science and Technology: Dr Mamadou Tangara. *Information and Communication Infrastructure:* Nana Grey-Johnson. *Interior:* Ousman Sonko. *Regional Administration, Lands and Traditional Rulers:* Momodou Aki Bayo. *Tourism and Culture:* Fatou Mass Jobe-Njie. *Trade, Industry, Regional Integration and Employment:* Kebba S. Touray. *Works, Construction and Infrastructure:* Francis Liti Mboge. *Youth and Sports:* Alieu K. Jammeh. *Attorney General and Minister of Justice:* Lamin Jobarteh. *Secretary General and Head of the Civil Service, and Minister for Presidential Affairs:* Njogou L. Bah.

Office of the President: http://www.statehouse.gm

CURRENT LEADERS

Retd Col. Yahya Jammeh

Position
President

Introduction
Former army colonel Yahya Jammeh came to power in a military coup in July 1994. Leading the APRC, he was elected to office in 1996 and re-elected in 2001, 2006 and 2011.

Early Life
Yahya A. J. J. Jammeh was born in the Foni Kansala district of The Gambia on 25 May 1965, the year in which the country gained independence from the United Kingdom. He joined the army in 1984, rising to captain by 1992, and on 22 July 1994 led a successful coup against Sir Dawda Jawara, the president since 1970.

Career in Office
In 1996 a new constitution was approved by referendum. Jammeh was confirmed as president in Sept. that year and his Alliance for Patriotic Reorientation and Construction (APRC) secured a majority of seats in parliamentary elections in Jan. 1997. He was re-elected president in Oct. 2001. The 2002 parliamentary elections, in which the APRC won nearly all the seats, were boycotted by the main opposition party.

Legislation passed in 2004 set jail terms for journalists guilty of libel, and broadcast news has remained tightly controlled. Jammeh was re-elected in Sept. 2006. Although the poll was considered free and fair on the day by observers, the Commonwealth Secretariat noted 'abuses of incumbency' before the vote. Alleged coup attempts against the regime led to the imposition of death sentences on six military officials and two businessmen in July 2010 and of long prison terms on former army and navy chiefs in May 2011. Jammeh was overwhelmingly re-elected in Nov. 2011 and his APRC again won almost all of the seats at the March 2012 parliamentary elections, which six opposition parties boycotted claiming that it was rigged.

DEFENCE

The Gambian National Army, 800 strong, has two infantry battalions, one engineer squadron and one company of presidential guards.

The marine unit of the Army consisted in 2007 of approximately 70 personnel operating four inshore patrol craft, based at Banjul.

Defence expenditure totalled US$16m. in 2008 (US$9 per capita), representing 2·2% of GDP.

ECONOMY

Agriculture accounted for 29% of GDP in 2007, industry 15% and services 56%.

Overview

With few natural resources and a limited agricultural base, remittances from workers overseas are crucial. Tourism is a key driver of the economy and the most significant foreign exchange earner. Agriculture, which is dominated by groundnuts, accounts for about 75% of employment and 60% of domestically produced exports. The main export commodities are peanut products, fish, cotton lint and palm kernels. Good groundnut and rice harvests helped offset a 10% decline in tourism and a 20% decline in remittances following the 2008 global financial crisis.

The Gambia continues to struggle with a high debt level, even after meeting the requirements of the Heavily Indebted Poor Countries Initiative in Dec. 2007. Interest accounts for over 25% of overall government expenditures. The poverty rate is 58% and unemployment and underemployment rates remain high. The Gambia is ranked 168th out of 187 on the UN Human Development Index.

Currency

The unit of currency is the *dalasi* (GMD), of 100 *butut*. Inflation was 5·0% in 2010 and 4·8% in 2011. Foreign exchange reserves were US$90m. in July 2005. Total money supply in June 2005 was 3,256m. dalasis.

Budget

2007 revenues were 3,663·5m. dalasis (tax revenue, 82·9%) and expenditures 3,635·0m. dalasis (current expenditure, 71·1%).

VAT at 15% was introduced on 1 Jan. 2013.

Performance

Real GDP growth was 6·7% in 2009, 5·5% in 2010 and 3·3% in 2011. Total GDP was US$0·9bn. in 2011.

Banking and Finance

The Central Bank of The Gambia (founded 1971; *Governor*, Amadou Colley) is the bank of issue. There are six other banks: Standard Chartered, Trust Bank, Arab Gambian Islamic Bank, Continent Bank, First International and the International Bank for Commerce and Industry. In 2010 foreign debt totalled US$470m., representing 63·3% of GNI.

ENERGY AND NATURAL RESOURCES

Environment

Carbon dioxide emissions from the consumption and flaring of fossil fuels in 2008 were the equivalent of 0·2 tonnes per capita.

Electricity

Installed capacity was 53,000 kW in 2007. Production was 229m. kWh in 2007; consumption per capita in 2007 was 140 kWh.

Oil and Gas

Although The Gambia is not currently an oil-producing country, there is ongoing exploration both onshore and offshore.

Minerals

Heavy minerals, including ilmenite, zircon and rutile, have been discovered in Sanyang, Batokunku and Kartong areas.

Agriculture

About 68% of the population depend upon agriculture. There were 0·25m. ha. of arable land in 2001 and 5,000 ha. of permanent crops. Almost all commercial activity centres upon the marketing of groundnuts, which is the only export crop of financial significance. Cotton is also exported on a limited scale. Rice is of increasing importance for local consumption. Major products (2002, in 1,000 tonnes), are: millet, 85; groundnuts, 72; rice, 20; maize, 19; sorghum, 15; cassava, 8.

Livestock (2002): 327,000 cattle, 262,000 goats, 146,000 sheep and 1m. poultry.

Forestry

In 2010 forests covered 0·48m. ha., or 48% of the land area. Timber production in 2007 was 778,000 cu. metres.

Fisheries

The total catch in 2008 was estimated at 42,645 tonnes (mainly marine fish).

INDUSTRY

Labour

The labour force in 2010 totalled 751,000 (52·1% males). Around 70% of the economically active population are engaged in agriculture.

INTERNATIONAL TRADE

Imports and Exports

In 2009 imports (c.i.f.) were valued at US$303·9m. and exports (f.o.b.) at US$66·0m. Re-exports account for more than 70% of total exports. Chief imports in 2006 were: machinery and transport equipment, 25·2%; food and live animals, 22·8%; petroleum and petroleum products, 17·2%. In 2006 groundnuts accounted for 47·8% of domestic exports, fruit and vegetables 28·7% and machinery and transport equipment 9·6%. Main import sources, 2006: Denmark, 16·5%; USA, 12·0%; China, 9·3%; Côte d'Ivoire, 8·6%. Leading markets for domestic exports, 2006: UK, 48·7%; Senegal, 30·4%; France, 4·3%; Germany, 4·3%.

COMMUNICATIONS

Roads

There were some 3,742 km of roads in 2004, of which 19·3% were paved. Number of vehicles (2007): 8,800 passenger cars; 2,600 lorries and vans.

Civil Aviation

There is an international airport at Banjul (Yundum), which handled 313,173 passengers and 1,247 tonnes of freight in 2009. The then national carrier, Gambia International Airlines, ceased operations in 2007. A new national carrier, Gambia Bird, began scheduled services in Oct. 2012.

Shipping

The chief port is Banjul. Ocean-going vessels can travel up the Gambia River as far as Kuntaur. In Jan. 2009 there were five ships of 300 GT or over registered, totalling 32,000 GT.

Telecommunications

In 2010 there were 48,800 landline telephone subscriptions (equivalent to 28·2 per 1,000 inhabitants) and 1,478,300 mobile phone subscriptions (or 855·3 per 1,000 inhabitants). There were 92·0 internet users per 1,000 inhabitants in 2010. Fixed internet subscriptions totalled 3,500 in 2007 (2·2 per 1,000 inhabitants).

SOCIAL INSTITUTIONS

Justice

Justice is administered by a Supreme Court consisting of a chief justice and puisne judges. The High Court has unlimited original jurisdiction in civil and criminal matters. The Supreme Court is the highest court of appeal and succeeds the judicial committee of the Privy Council in London. There are Magistrates Courts in each of the divisions plus one in Banjul and two in nearby Kombo St Mary's Division—eight in all. There are resident magistrates in provincial areas. There are also Muslim courts, district tribunals dealing with cases concerned with customary law, and two juvenile courts.

The death penalty was abolished in 1993 but restored by decree in 1995. It was used in Aug. 2012—for the first time since 1985—when nine prisoners were executed. The population in penal institutions in Dec. 2009 was 780 (45 per 100,000 of national population).

Education

The adult literacy rate in 2008 was 45%. In 2007 there were 218,341 pupils with 5,341 teaching staff at primary schools and 101,670 pupils with 4,475 teachers at secondary schools. Higher education institutes include The Gambia College, a technical training institute, a management development institute, a multi-media training institute and a hotel training school. In 2008 there were 6,000 students and 400 academic staff in tertiary education.

In 2009 public expenditure on education came to 3·8% of GDP.

Health

In 2005 there were eight hospital beds per 10,000 inhabitants. There were 156 physicians, 43 dentistry personnel and 1,881 nursing and midwifery personnel in 2003.

RELIGION

More than 90% of the population is Muslim. Banjul is the seat of an Anglican and a Roman Catholic bishop. There is a Methodist mission. A few sections of the population retain their original animist beliefs.

CULTURE

World Heritage Sites

The Gambia has two sites on the UNESCO World Heritage List: James Island and Related Sites (inscribed on the list in 2003), containing important evidence of early Afro-European encounters and the slave trade.

The Gambia shares a UNESCO site with Senegal: the Stone Circles of Senegambia (added in 2006) is a collection of 93 stone circles, tumuli and burial mounds from between the 3rd century BC and the 16th century AD.

Press

In 2006 there were two daily newspapers—the *Daily Observer* and the government-owned *Gambia Daily*.

Tourism

Tourism is The Gambia's biggest foreign exchange earner. In 2009 there were 142,000 non-resident charter tourists (excluding same-day visitors).

DIPLOMATIC REPRESENTATIVES

Of The Gambia in the United Kingdom (57 Kensington Ct., London, W8 5DG)
High Commissioner: Elizabeth Ya Eli Harding.

Of the United Kingdom in The Gambia (48 Atlantic Rd, Fajara, Banjul)
High Commissioner: David Morley.

Of The Gambia in USA (Suite 240, Georgetown Plaza, 2233 Wisconsin Ave., NW, Washington, D.C., 20007)
Ambassador: Alieu Momodou Ngum.

Of the USA in The Gambia (Fajara (East), Kairaba Ave., Banjul)
Ambassador: Edward M. Alford.

Of The Gambia to the United Nations
Ambassador: David Daroll Tommy.

Of The Gambia to the European Union
Ambassador: Mamour A. Jagne.

FURTHER READING

Gailey, Harry A., *Historical Dictionary of the Gambia.* 1999
Hughes, A. and Perfect, D., *A Political History of The Gambia, 1816–1994.* 2006

National Statistical Office: The Gambia Bureau of Statistics, Kanifing Institutional Layout, Serekunda, P. O. Box 3504, Serekunda.

GEORGIA

Sakartvelos Respublika
(Republic of Georgia)

Capital: Tbilisi
Population projection, 2015: 4·23m.
GNI per capita, 2011: (PPP$) 4,780
HDI/world rank: 0·733/75
Internet domain extension: .ge

KEY HISTORICAL EVENTS

By the 1st millennium BC the Diakhi (Taokhi) and the Qolha (Colchis) tribal groups had developed bronze casting. A two-state confederation emerged as early as the 6th century BC, with Colchis (Egrisi) in the west and Kartli (Iberia) in the east. The Greeks established Black Sea colonies from the 6th century BC, including Phasis (present-day Poti), Gyenos (Ochamchire) and Dioscuras (Sukhumi). Parnavaz I ruled a united Kartli (or Georgia) from his citadel of Armaztsikhe from c. 302–237 BC.

The Romans under Pompey attacked Kartli in 65 BC, defeating King Artag and establishing a client state. In the first half of the 2nd century AD Kartli grew under Parsman II. Rome, realizing its value as an ally against Parthia (Iran), recognized Kartli's extended borders. In 298 the Sassanid Iranians acknowledged Roman jurisdiction over Kartli-Iberia and recognized King Mirian III (284–361), who adopted Christianity as the state religion. Christianity brought close ties with Rome's successor, Byzantium, though Persia controlled Kartli for much of the 4th–6th centuries.

Tbilisi fell to an Arab army in 645 but Kartli retained considerable autonomy under local Arab rulers. In 813 the Armenian prince Ashot I of the Bagrationi family took control of Georgia, beginning nearly a millennium of Bagratid rule. Bagrat V (1027–72) united west and east and David IV ('the Builder', 1099–1125) presided over a golden age, repulsing the Seljuk Turks and disseminating Georgian influence in the Caucasus.

The Mongols invaded in 1236 and the Turkic conqueror Timur destroyed Tbilisi in 1386. Turkish dominance was sealed when the Byzantine Empire collapsed. By the 18th century the Bagratids had regained autonomy under nominal Persian rule. In 1762 Herekle II took control of the eastern regions of Kartli and Kakhetia, reducing the powers of the Georgian nobility. Herekle opened channels with Russia to gain protection from the Turks, though Russo-Turkish rivalry soon afflicted Georgia. After the Persians sacked Tbilisi in 1795 Herekle again sought Russian protection, leading to annexation.

In 1801 Tsar Alexander I abolished the kingdom of Kartli-Kakhetia (Eastern Georgia). Western Georgia (Imeretia) was annexed in 1804 and the Georgian Orthodox Church lost its autocephalous status in 1811. Russification intensified in the second half of the 19th century but a national liberation movement emerged in the 20th century, with Russia declaring martial law in 1905 following peasant revolts and general strikes. Revolutionaries split between the gradualist Mensheviks and the radical Bolsheviks (led by Ioseb Jugashvili, later Joseph Stalin).

The Caucasus became a major battleground in 1915 when Russia invaded Turkey. In May 1918 Georgia declared independence under the protection of Germany (to prevent invasion by the Turks). Lenin gave recognition in May 1920 and the Menshevik-dominated government redistributed swathes of aristocratic landholdings. In 1921 the Bolshevik Red Army invaded Georgia. From 1922–36 Georgia was part of the Transcaucasian Soviet Federated Socialist Republic within the USSR. In 1936 the republic was divided into Armenia, Azerbaijan and Georgia.

Over 500,000 Georgians served in the Red Army in the Second World War and the autonomy of the Georgian Orthodox Church was restored in 1943. Stalin oversaw forced urbanization and industrialization and although he and Lavrenti Beria (chief of secret police) were Georgians, Georgia suffered greatly under his regime. Criticism of Stalin in Georgia gained momentum under Gorbachev's *glasnost* policy of the late 1980s. In 1972 Eduard Shevardnadze took control of the Georgian Communist Party. His clearout of officials prompted dissident nationalists (led by the academic, Zviad Gamsakhurdia) to warn of Russification. In 1978 leaders of the Abkhazian Autonomous Republic threatened to secede from Georgia. Shevardnadze took steps to diffuse the crisis, including an affirmative action programme for ethnic Abkhaz. When Shevardnadze became Soviet foreign minister in 1985 his successor, Jumber Patiashvili, removed some of the Shevardnadze appointees, forcing reformist leaders underground.

In April 1989 Soviet troops broke up peaceful demonstrations in Tbilisi and 20 Georgians were killed. Following the 'April Tragedy' Shevardnadze was sent to restore calm. Multiparty elections in Oct. 1990 saw Gamsakhurdia's Round Table/Free Georgia coalition win a solid majority. Supported by a referendum, parliament declared independence and declined to participate in the Commonwealth of Independent States.

In May 1991 Gamsakhurdia was elected president but the National Guard joined the opposition. When fighting in central Tbilisi in late Dec. forced Gamsakhurdia to flee to Chechnya, a military council took control and invited Shevardnadze to return in March 1992. Post-Soviet Georgia was dogged by separatism, especially in South Ossetia and Abkhazia. Relations with Russia were strained by its support for these insurgencies. In 1990 Gamsakhurdia had removed South Ossetia's autonomous status. When the South Ossetian regional legislature attempted to unite with Russian North Ossetia, Georgian forces invaded. Thousands died and tens of thousands were displaced before Yeltsin mediated a ceasefire in July 1992.

In Abkhazia the ethnic Abkhaz population feared cultural annihilation while the Georgian majority resented disproportionate allocation of political and administrative positions to the Abkhaz. In July 1992 the Abkhazian Supreme Soviet voted to separate from Georgia. In Aug. 1992 the National Guard seized the Abkhazian capital, Sukhumi. Hundreds died and refugees fled to Russia and other parts of Georgia. The Abkhazian government requested Russian intervention, retaking Sukhumi and forcing the Georgian army out in 1993. A ceasefire began in early 1994. When Gamsakhurdia invaded Mingrelia (his home region where he had retained support) Shevardnadze asked for Russian military assistance. Defeated, Gamsakhurdia committed suicide in Jan. 1994.

Opposition to Shevardnadze's increasingly corrupt rule grew in the late 1990s, despite advantageous relations with the West and a US$3bn. Baku-Tbilisi-Ceyhan oil pipeline deal. In Nov. 2003 Shevardnadze resigned after opposition forces, alleging electoral fraud, stormed parliament. Following the 'Rose Revolution' Mikheil Saakashvili became president in Jan. 2004. Saakashvili's presidency has been dominated by diplomatic sparring with Russia over Russian interference in Abkhazia and South Ossetia. Relations reached a new low in the months following Saakashvili's re-election in Jan. 2008. In March Abkhazia's separatist government requested the UN to recognize the region's independence and in April 2008 Moscow announced closer ties with both regions. In Aug. Saakashvili sent troops into South Ossetia to attack separatist forces. Russia claimed its citizens were under attack and responded by sending in several thousand troops. A week of fierce fighting was brought to an end by a French-brokered peace deal, with Russia recognizing independence for both South Ossetia and Abkhazia. Moscow was subsequently accused of delaying its troop withdrawal. Adjara, another autonomous region, was brought under Tbilisi's control with the ejection of its populist leader, Aslan Abashidze, in 2004.

TERRITORY AND POPULATION

Georgia is bounded in the west by the Black Sea and south by Turkey, Armenia and Azerbaijan. Area, 69,700 sq. km (26,900 sq. miles). Its census population in 2002 was 4,371,535 (excluding Abkhazia and South Ossetia). Jan. 2012 estimate (excluding Abkhazia and South Ossetia), 4,497,600; density (excluding Abkhazia and South Ossetia), 78·7 per sq. km.

The UN gives a projected population for 2015 of 4·23m.

In 2012, 53·2% of the population lived in urban areas. The capital is Tbilisi (Jan. 2012 population estimate, 1·17m.). Other principal towns (with Jan. 2012 population estimates in brackets): Kutaisi (196,800), Batumi (125,800), Rustavi (122,500). After Russian-backed Abkhaz forces and their allies took Sukhumi in 1993, non-Abkhaz residents were ejected from the region. Sukhumi's population fell from 121,000 in 1991 to an estimated 64,500 in 2011.

Georgians accounted for 83·8% of the 2002 census population; others included 6·5% Azerbaijanis, 5·7% Armenians and 1·5% Russians. Georgia includes the Autonomous Republics of Abkhazia and Adjara and the former Autonomous Region of South Ossetia.

Georgian is the official language. Armenian, Russian and Azeri are also spoken.

SOCIAL STATISTICS

Births, 2011, 58,014; deaths, 49,818. Rates, 2011: birth, 12·9 per 1,000 population; death, 11·1 per 1,000. Annual population growth rate, 2000–05, –0·3%. Life expectancy, 2007, 68·1 years for males and 75·0 years for females. Infant mortality, 2010, 20 per 1,000 live births; fertility rate, 2008, 1·6 births per woman.

CLIMATE

The Georgian climate is extremely varied. The relatively small territory covers different climatic zones, ranging from humid sub-tropical zones to permanent snow and glaciers. In Tbilisi summer is hot: 25–35°C. Nov. sees the beginning of the Georgian winter and the temperature in Tbilisi can drop to –8°C; however, average temperature ranges from 2–6°C.

CONSTITUTION AND GOVERNMENT

A new constitution of 24 Aug. 1995 defines Georgia as a presidential republic with federal elements. The head of state is the *President*, elected by universal suffrage for not more than two five-year terms. The 150-member *Supreme Council* is elected for four-year terms, with 73 members elected in single-seat constituencies and 77 by proportional representation. There is a 5% threshold.

Amendments limiting the power of the president in favour of the prime minister were passed in Oct. 2010 and are set to come into force after presidential elections scheduled for Oct. 2013. The prime minister will then have executive authority over domestic and foreign policy. The president will remain head of state but will not have the right to initiate laws, introduce a budget or hold an office in a political party.

National Anthem

'Tavisupleba' ('Freedom'); words by Dawit Magradse, tune by Zakaria Paliashvili.

RECENT ELECTIONS

At the presidential election held on 5 Jan. 2008 Mikheil Saakashvili of the United National Movement was re-elected president with 53·5% of the vote. Levan Gachechiladze took 25·7% and Badri Patarkatsishvili 7·1%. Turnout was 56·2%.

At the parliamentary elections of 1 Oct. 2012 the opposition Georgian Dream coalition won 83 of the 150 seats (with 54·9% of the party list vote and 53·4% of the constituency vote). The ruling United National Movement won 67 seats (40·4% of the party list vote and 46·6% of the constituency vote). Turnout was 59·8%.

CURRENT GOVERNMENT

President: Mikheil Saakashvili; b. 1967 (United National Movement; since 20 Jan. 2008, having previously been president from Jan. 2004 to Nov. 2007).

In March 2013 the government comprised:

Prime Minister: Bidzina Ivanishvili; b. 1956 (Georgian Dream; sworn in 25 Oct. 2012).

First Deputy Prime Minister and Minister of Education and Science: Giorgi Margvelashvili.

Deputy Prime Minister and Minister of Energy: Kakha Kaladze.

Minister of Agriculture: David Kirvalidze. *Corrections:* Sozar Subari. *Culture:* Guram Odisharia. *Defence:* Irakli Alasania. *Economic and Sustainable Development:* Giorgi Kvirikashvili. *Environmental Protection:* Khatuna Gogaladze. *Finance:* Nodar Khaduri. *Foreign Affairs:* Maya Panjikidze. *Interior:* Irakli Garibashvili. *Internally Displaced Persons from the Occupied Territories, Accommodation and Refugees of Georgia:* David Darakhvelidze. *Justice:* Tea Tsulukiani. *Labour, Health and Social Affairs:* David Sergeenko. *Regional Development and Infrastructure:* Davit Narmania. *Sports and Youth:* Levan Kipiani.

Minister of State for Diaspora Affairs: Kote Surguladze. *Employment:* Kakha Sakandelidze. *European Integration:* Alexi Petriashvili. *Reintegration:* Paata Zakareishvili.

Georgian Parliament: http://www.parliament.ge

CURRENT LEADERS

Mikheil Saakashvili

Position
President

Introduction
Mikheil Saakashvili was first elected president on 4 Jan. 2004. A former protégé of President Eduard Shevardnadze, he led opposition to the rigged parliamentary elections of Nov. 2003 and forced the president to resign. Saakashvili's peaceful management of the 'rose revolution' earned him respect at home and abroad and even the support of the ousted Shevardnadze. However, he has since faced increasing domestic opposition. He secured re-election as president in Jan. 2008, but with a much reduced majority, and was further undermined as he entered the final year of his second presidential term when an opposition coalition won control of parliament in elections in Oct. 2012. Relations with Russia, which were shattered in Aug. 2008 when Georgian military action against the separatist enclave of South Ossetia

provoked a retaliatory invasion by Russian forces, have remained fraught.

Early Life

Mikheil (Mikhail in Russian) Saakashvili was born on 21 Dec. 1967 in Tbilisi. He received law degrees from Kyiv University, Ukraine in 1992 and Columbia University, New York in 1994 and completed a doctorate in juridical science at George Washington University, Washington, D.C. He pursued further studies in Florence and Strasbourg.

While working for a New York law firm, Saakashvili was approached by Zurab Zhvania, speaker of the Georgian parliament. The Georgian leader, Eduard Shevardnadze, was seeking potential parliamentary candidates, unconnected to the Soviet system. Saakashvili returned home and was elected to parliament in Dec. 1995. As a member of Shevardnadze's Citizens' Union and a trained jurist, his political career developed rapidly. He served as chairman of the parliamentary committee on electoral reform and contributed to the drafting of the 1995 constitution. He led the Citizens' Union in parliament from 1998–99 and was appointed vice-president of the Council of Europe's parliamentary assembly in Jan. 2000.

Appointed justice minister in Oct. 2000, Saakashvili attempted an overhaul of the judicial system, condemned by international observers as highly corrupt. He also tried to reform the prison system and proposed a bill on illegal property confiscation, which was blocked by Shevardnadze. His programme was cut short when he openly accused the ministers for economics and state security and Tbilisi's head of police of corruption and profiteering. Shevardnadze, then president, refused to act on these charges, forcing Saakashvili to resign in Sept. 2001.

Having left the government, he continued his anti-corruption programme by forming a party, the United National Movement, to represent Georgia's reformist elements. Coalition partners included the small ideological Republican Party and the Union of National Forces. Support was widened by alliances with Zurab Zhvania's United Democrats and the Burjanadze-Democrats, led by Nino Burjanadze, Zhvania's replacement as speaker. Saakashvili was elected chairman of the Tbilisi Assembly in June 2002.

Saakashvili campaigned for the Nov. 2003 parliamentary elections on an anti-Shevardnadze platform, drawing large crowds with his energetic rhetoric, especially in Adjara, in the southwest, and Kvemo Kartli, a province in the southeast with a large Azeri minority. Despite OSCE sponsorship, the elections on 2 Nov. were chaotic and heavily rigged by the ruling Citizens' Union. The delayed announcement of provisional results provoked accusations of electoral fraud. Burjanadze and Zhvania agreed to form a coalition with Saakashvili—the United Opposition Front—and massive demonstrations were organized night after night in Tbilisi. After the results were announced, putting Saakashvili's United National Movement in third place, a boycott of parliament was declared by the coalition. Shevardnadze refused to compromise and opened parliament on 22 Nov. Saakashvili responded by summoning his national supporters to Tbilisi and demanded the president's resignation. During Shevardnadze's opening address Saakashvili burst into the assembly, brandishing a rose, the symbol of the peaceful demonstrations. His supporters occupied the chancellery, causing the president to declare a state of emergency.

Bereft of support from abroad and in his own government, Shevardnadze finally resigned on 23 Nov. His avoidance of a military solution was praised internationally and by the opposition leaders. In the wake of his resignation, Saakashvili, Burjanadze and Zhvania agreed to present a united front in immediate presidential elections, the former gaining the support from the other two in exchange for senior government positions. On 4 Jan. 2004 Saakashvili won the presidential election with just over 96% of the votes. His electoral promises were broad and ambitious: the abolition of taxes on small businesses, doubling of pensions and public sector salaries and swift punishment for the worst abuses of the previous regime. His campaign was strengthened by the co-operation of Shevardnadze, who voted for his successor.

Career in Office

Saakashvili declared his priorities in office as maintaining the territorial integrity of Georgia and his anti-corruption programme. He consolidated his political position in March 2004 when the National Movement–Democrats bloc won the parliamentary elections with a substantial majority of seats. Tension increased with Adjara's president, Aslan Abashidze, who declared a state of emergency in his jurisdiction and rejected the new central government's authority. Saakashvili reasserted direct control over Adjara in May 2004 after popular demonstrations in Batumi forced Abashidze to step down. Elsewhere, Saakashvili proposed giving greater autonomy—but not full independence—to the separatist regions of South Ossetia and Abkhazia.

Despite friction with Russia over its links with these regions, its military bases in Georgia and the conflict in Chechnya (where Russia has accused Georgia of aiding Chechen fighters), Saakashvili initially sought to improve relations between the two countries. Nevertheless, bilateral tensions resurfaced in 2006 as energy supplies from Russia were disrupted in Jan., Russia suspended imports of Georgian wine and mineral water on health grounds in March–May, and Georgia briefly detained and then expelled four Russian army officers on spying charges in Sept.–Oct. (in retaliation for which Russia imposed a transport blockade and adopted punitive measures against ethnic Georgians living in Russia). In Dec. 2006 the Georgian government accepted a doubling of the price of Russian natural gas supplies, but accused its powerful neighbour of political blackmail.

Tension continued through 2007, as Georgia accused Russia of violations of its airspace in Aug. and of orchestrating mass opposition demonstrations in Tbilisi against Saakashvili's government which in Nov. led to a police crackdown and temporary state of emergency. The protests were sparked particularly by allegations of corruption and murder against Saakashvili made by a former defence minister in Sept. Despite his apparent unpopularity and opposition claims of fraud and vote rigging, Saakashvili was re-elected as president in polling in Jan. 2008, and in May the ruling United National Movement won a landslide victory in parliamentary elections.

Meanwhile, suspicious of Georgia's aspirations to join NATO, Russia stated in April 2008 that it would strengthen its links with South Ossetia and Abkhazia. This prompted the Georgian government to accuse Russia of planning military intervention and annexation of the regions. In Aug. continuing tensions escalated into outright military conflict after Georgian troops had mounted an attack of separatist forces in South Ossetia. Russian forces occupied the enclaves and advanced deep into Georgian territory, destroying strategic targets, before agreeing a French-brokered ceasefire following diplomatic intervention by the European Union. Later in the month, Russia unilaterally recognized the independence of Abkhazia and South Ossetia, a move rejected by Saakashvili and the Western nations. In Sept. Russia announced that it would keep troops in Abkhazia and South Ossetia but completed the withdrawal from the rest of Georgia in Oct.

Also in Oct. 2008, Saakashvili's former ally, Nino Burjanadze, announced the creation of a new opposition group, claiming that the government was not capable of protecting the country and should face fresh elections. Opposition agitation against the president and his powers continued in 2009 and in May the authorities quelled a military mutiny. In May 2011 riot police violently broke up a rally in Tbilisi by opposition protesters calling on Saakashvili to resign, and in May 2012 a massive anti-government demonstration took place in the capital. In elections in Oct. 2012, the opposition Georgian Dream coalition led by Bidzina Ivanishvili won a parliamentary majority to oust the ruling pro-Saakashvili United National Movement administration.

Georgia's relations with Russia thawed in the first few months of 2010 as direct air traffic resumed in Jan. and a border crossing closed since 2006 was reopened in March. However, in Aug. Russia announced that it was deploying air defence systems in Abkhazia and South Ossetia in defiance of Georgian sovereignty.

DEFENCE

The total strength of the Armed Forces consisted of 21,150 personnel in 2007. Conscription is currently for 15 months, but there are plans to move to an all-professional army by 2017. The UN peacekeeping mission (United Nations Observer Mission in Georgia, or UNOMIG, which was established in Aug. 1993) ended in June 2009 owing to a lack of consensus among Security Council members on mandate extension. Following the collapse of the USSR in 1991 Russia maintained two bases in Georgia with some 4,000 personnel. The last Russian troops left Georgia in Nov. 2007. However, several thousand soldiers returned in Aug. 2008 when Moscow responded to Georgia's military attack on separatist forces in South Ossetia. Despite a subsequent withdrawal, some forces remain as part of a 'buffer zone' around South Ossetia and Abkhazia.

Georgia hopes to join NATO, although its bid to become a member is fiercely opposed by Russia.

Defence expenditure in 2008 totalled US$1,037m. (US$224 per capita), representing 8·1% of GDP (up from 4·4% of GDP in 2006).

Army

The Army totalled 17,767 (3,767 conscripts) in 2007. In addition there were 1,578 active reservists in the National Guard and 6,300 Ministry of the Interior troops. A paramilitary border guard exists, estimated at 5,400.

Navy

Former Soviet facilities at Poti have been taken over. The headquarters are at Tbilisi. Personnel, 2007, 495.

Air Force

Personnel, 2007, 1,310 (290 conscripts). There were nine combat-capable aircraft in 2007 (mainly Su-25 fighter-bombers) and nine attack helicopters.

ECONOMY

Agriculture accounted for 10·0% of GDP in 2008, industry 21·2% and services 68·8%.

Georgia's 'shadow' (black market) economy is estimated to constitute approximately 62% of the country's official GDP, one of the highest percentages of any country in the world.

Overview

Georgia is a small, lower-income transition economy. After independence in 1991, civil war and the loss of markets in the former Soviet Union led to economic collapse, with exports declining by 90% and output falling by 70%. Political tensions, declining living standards and low tax revenues undermined the funding of basic services. Progress towards macroeconomic stabilization was made in the late 1990s but was disrupted by internal fragmentation, drought and the 1998 financial crisis in Russia. IMF-supported programmes from 2001–04 aided growth recovery and price stability.

Prudent macroeconomic policies and structural reforms, notably a privatization programme, helped achieve strong growth and single-digit inflation. Anti-corruption initiatives and government efforts to create a more business-friendly environment saw the World Bank label the economy as the world's number one reformer in a 2006 'Doing Business' survey. Sanctions imposed by Russia, its major export market, have been offset by reform-led growth but uncertainty remains.

Military conflict with Russia in Aug. 2008 and the global financial crisis led to falling FDI, a sharp credit squeeze and a collapse in confidence. The economy contracted by nearly 4% in 2009. A strong fiscal policy response with 10% of GDP injected into the economy during 2008 and 2009 led to broad-based growth across all sectors, driven by a rebound of credit to the private sector and strong export demand. However, FDI inflows remain far below pre-crisis levels and public debt is high at 39% of GDP.

Currency

The unit of currency is the *lari* (GEL) of 100 *tetri*, which replaced coupons at 1 lari = 1m. coupons on 25 Sept. 1995. Inflation was 7·1% in 2010 and 8·5% in 2011, having been 163% in 1995 and 15,606% in 1994. Gold reserves are negligible. Total money supply was 882m. laris in July 2005.

Budget

Revenues in 2007 totalled 5,159m. laris and expenditures 5,237m. laris. Tax revenue accounted for 72·4% of total revenues; defence accounted for 28·6% of total expenditures, and social security and welfare 14·8%.

VAT is 18%.

Performance

The economy contracted by 3·8% in 2009 but there was growth of 6·3% in 2010 and 7·0% in 2011.

Total GDP was US$14·4bn. in 2011 (excluding Abkhazia and South Ossetia).

Banking and Finance

The *President* of the Central Bank is Giorgi Kadagidze. In 2005 there were 19 commercial banks. Two foreign banks had representative offices. The largest bank is Bank of Georgia, with a 36% market share by assets.

In 2010 external debt totalled US$9,238m., equivalent to 80% of GNI.

ENERGY AND NATURAL RESOURCES

Environment

Carbon dioxide emissions from the consumption and flaring of fossil fuels in Georgia were the equivalent of 1·2 tonnes per capita in 2008.

Electricity

The many fast-flowing rivers provide an important hydro-electric resource. Installed capacity was around 4·4m. kW in 2007. Production in 2007 was 8·33bn. kWh (82% hydro-electric); consumption per capita in 2007 was 1,853 kWh.

Oil and Gas

Output (2007) of crude petroleum, 57,000 tonnes. A 930 km long oil pipeline from an offshore Azerbaijani oilfield in the Caspian Sea across Azerbaijan and Georgia to a new oil terminal at Supsa, near Poti, on the Black Sea Coast started pumping oil in early 1999; the US$600m. pipeline allowed Georgia to create 25,000 new jobs. However, Georgia is still heavily dependent on Russia for natural gas. Accords for the construction of a second oil pipeline through Georgia were signed in Nov. 1999, to take oil from Azerbaijan to Turkey via Georgia. Work on the pipeline began in Sept. 2002 and it was officially opened in May 2005. Natural gas production was 16m. cu. metres in 2007.

A gas pipeline from Baku through Georgia to Erzurum in Turkey was commissioned in June 2006.

Minerals

Manganese deposits are calculated at 250m. tonnes. Other important minerals are barytes, clays, gold, diatomite shale, agate, marble, alabaster, iron and other ores, building stone, arsenic, molybdenum, tungsten and mercury.

Agriculture

Agriculture plays an important part in Georgia's economy, contributing 12·8% of GDP in 2006. In 2001 there were 795,000 ha. of arable land and 268,000 ha. of permanent crops. 469,000 ha. were irrigated in 2001.

Output of main agricultural products (in 1,000 tonnes) in 2003: maize, 450; potatoes, 425; wheat, 225; grapes, 200; tomatoes, 185; cabbage, 148; onions, 98; apples, 86; wine, 80; watermelons (including melons, pumpkins and squash), 60. Livestock, 2003: cattle, 1,216,000; sheep, 611,000; pigs, 446,000; chickens, 10m. Livestock products, 2003 (in 1,000 tonnes): meat, 109; milk, 762; eggs, 24.

Forestry

There were 2·74m. ha. of forest in 2010, or 39% of the total land area. Timber production in 2007 was 616,000 cu. metres.

Fisheries

The catch in 2008 was 26,512 tonnes, down from 147,688 tonnes in 1989.

INDUSTRY

Industry accounted for 21·2% of GDP in 2008. There is a metallurgical plant and a motor works. There are factories for processing tea, creameries and breweries. There are also textile and silk industries.

Production, 2008 unless otherwise specified (in 1,000 tonnes): cement, 1,351; flour, 218; refined sugar (2006), 123; footwear, 28,290 pairs; beer, 62·5m. litres; spirits (2007), 2·6m. litres; cigarettes, 5,156m. units.

Labour

The economically active workforce numbered 1,991,800 in 2009 (1,071,300 males). In 2007, 53·4% of employees were engaged in agriculture, 10·4% in industry and 36·0% in services. The unemployment rate was 16·9% in 2009. Approximately 500,000 Georgians, or a tenth of the population, work in Russia, but in Dec. 2000 Russia began requiring Georgians to have a visa to visit the country.

INTERNATIONAL TRADE

Imports and Exports

Imports and exports for calendar years in US$1m.:

	2004	2005	2006	2007	2008
Imports c.i.f.	1,846	2,490	3,678	5,217	6,066
Exports f.o.b.	647	865	993	1,240	1,507

Major commodities imported are machinery and transport equipment, food products and mineral fuels. Major commodities for export are iron and steel products, food and beverages, and machinery.

The leading import suppliers in 2004 were: Russia, 14·0%; Turkey, 10·9%; UK, 9·3%; Azerbaijan, 8·5%. Principal export markets in 2004 were: Turkey, 18·3%; Turkmenistan, 17·7%; Russia, 16·1%; Armenia, 8·4%.

COMMUNICATIONS

Roads

There were 20,329 km of roads in 2007 (94·1% hard-surfaced). Passenger cars in use in 2007 numbered 416,300, and there were also 51,500 lorries and vans and 42,800 buses and coaches. In 2007 there were 737 road deaths.

Rail

Total length in 2005 was 1,522 km of 1,520 mm gauge (1,481 km electrified). In 2005 railways carried 19m. tonnes of freight and 3·6m. passengers. There is a metro system in Tbilisi.

Civil Aviation

The main airport is Tbilisi International Airport. The main Georgian carrier is Georgian Airways. In 2007 it had flights to Amsterdam, Athens, Dubai, Frankfurt, Kyiv, Minsk, Moscow, Paris, Tel Aviv and Vienna. In 2009 Tbilisi handled 702,596 passengers (714,976 in 2008) and 12,245 tonnes of freight.

Shipping

In Jan. 2009 there were 199 ships of 300 GT or over registered, totalling 617,000 GT. The principal port is Poti, which handled 8·1m. tonnes of cargo in 2008 (7·7m. tonnes in 2007).

Telecommunications

In 2011 there were 1,342,400 landline telephone subscriptions (equivalent to 310·1 per 1,000 inhabitants) and 4,430,600 mobile phone subscriptions (or 1,023·5 per 1,000 inhabitants). There were 269·0 internet users per 1,000 inhabitants in 2010. Fixed internet subscriptions totalled 176,500 in 2009 (40·4 per 1,000 inhabitants). In March 2012 there were 908,000 Facebook users.

SOCIAL INSTITUTIONS

Justice

The population in penal institutions in Jan. 2008 was 18,170 (415 per 100,000 of national population). The death penalty was abolished in 1997.

Education

In 2005 there were 1,214 public pre-primary schools with 6,883 teachers for 76,416 pupils. In 2005–06 there were 2,744 public and private schools with 74,300 teachers and 634,700 pupils (326,600 at primary level and 308,100 at secondary). In 2005–06 there were 171 higher education institutions with 144,300 students and 11,280 academic staff; the largest is Tbilisi State University with 35,000 students in 2005–06. Adult literacy rate in 2009 was estimated at 99·7%.

Public spending on education in 2007 came to 2·6% of GNI and 7·8% of total government expenditure.

Health

Georgia had 13,600 hospital beds in 2009. In 2009 there were 20,609 physicians and 12,933 nurses; there were 1,269 dentistry personnel in 2006.

Welfare

In 2005 there were 549,900 old age and 352,200 other pensioners.

RELIGION

The Georgian Orthodox Church has its own organization under Catholicos (Patriarch) Ilia II who is resident in Tbilisi. In 2001 there were 1·8m. Georgian Orthodox, 550,000 Sunni Muslims, 280,000 Armenian Apostolic (Orthodox) and 130,000 Russian Orthodox.

CULTURE

World Heritage Sites

Georgia has three sites on the UNESCO World Heritage List: City-Museum Reserve of Mtskheta (inscribed on the list in 1994), churches of the former Georgian capital; Bagrati Cathedral and Gelati Monastery (1994); and Upper Svaneti (1996), a mountainous area of medieval villages.

Press

In 2008 there were ten dailies with a combined circulation of 45,000, as well as 73 other papers.

Tourism

Investment in tourism has increased substantially in recent years, and large numbers of hotels have been built. In 2010 there were 2,032,000 international visitors, up from 1,423,000 in 2009. Most visitors in 2010 were from Armenia (548,000), Turkey (536,000), Azerbaijan (498,000) and Russia (171,000).

DIPLOMATIC REPRESENTATIVES

Of Georgia in the United Kingdom (4 Russell Gdns, London, W14 8EZ)
Ambassador: Giorgi Badridze.

Of the United Kingdom in Georgia (51 Krtsanisi St., 0114 Tbilisi)
Ambassador: Judith Gough.

Of Georgia in the USA (2209 Massachusetts Ave., Washington, D.C., 20008)
Ambassador: Temuri Yakobashvili.

Of the USA in Georgia (11 George Balanchine St., 0131 Tbilisi)
Ambassador: Richard B. Norland.

Of Georgia to the United Nations
Ambassador: Alexander Lomaia.

Of Georgia to the European Union
Ambassador: Salome Samadashvili.

FURTHER READING

Areshidze, Irakly, *Democracy and Autocracy in Eurasia: Georgia in Transition.* 2007
Coppieters, Bruno and Legvold, Robert, (eds.) *Statehood and Security: Georgia after the Rose Revolution.* 2005
Gachechiladze, R., *The New Georgia: Space, Society, Politics.* 1995
Jones, Stephen, *Georgia: A Political History since Independence.* 2012
Mikaberidze, Alexander, *Historical Dictionary of Georgia.* 2007
Nodia, Ghia and Scholtbach, Alvaro Pinto, *The Political Landscape of Georgia: Political Parties, Achievements, Challenges, and Prospects.* 2007
Pelkmans, Mathijs, *Defending the Border: Identity, Religion, and Modernity in the Republic of Georgia.* 2006
Rayfield, Donald, *Edge of Empires: A History of Georgia.* 2012
Suny, R. G., *The Making of the Georgian Nation.* 2nd ed. 1994
Wheatley, Jonathan, *Georgia from National Awakening to Rose Revolution: Delayed Transition in the Former Soviet Union.* 2005

National Statistical Office: National Statistics Service of Georgia, 39 Guramishvili str., 0181 Tskneti, Tbilisi.
Website: http://www.geostat.ge

Abkhazia

GENERAL DETAILS

Area, 8,600 sq. km (3,320 sq. miles); population (Jan. 2004 est.), 178,600. Capital, Sukhumi (2002 population, 45,000). This area, the ancient Colchis, saw the establishment of a West Georgian kingdom in the 4th century and a Russian protectorate in 1810. In March 1921 a congress of local Soviets proclaimed it a Soviet Republic, and its status as an Autonomous Republic, within Georgia, was confirmed on 17 April 1930 and again by the Georgian Constitution of 1995.

Around 300,000 ethnic Georgians were displaced as a result of the 1992–94 war and ethnic Abkhazians are now thought to constitute the majority population followed by Armenians and Russians with Georgians in the minority.

In July 1992 the Abkhazian parliament declared sovereignty under the presidency of Vladislav Ardzinba and the restoration of its 1925 constitution. Fighting broke out as Georgian forces moved into Abkhazia. On 3 Sept. and on 19 Nov. ceasefires were agreed, but fighting continued into 1993 and by Sept. Georgian forces were driven out. On 15 May 1994 Georgian and Abkhazian delegates under Russian auspices signed an agreement on a ceasefire and deployment of 2,500 Russian troops as a peacekeeping force. On 26 Nov. 1994 parliament adopted a new constitution proclaiming Abkhazian sovereignty. CIS economic sanctions were imposed in Jan. 1996. Parliamentary elections were held on 23 Nov. 1996. Neither the constitution nor the elections

were recognized by the Georgian government or the international community. Fighting flared up between rival militia forces again in May 1998 after Abkhazian forces ejected thousands of ethnic Mingrelian and Georgian refugees who had returned to the southern Abkhazian region of Gali. After the fighting in 1998, the worst in five years, both sides declared a ceasefire. Up to 20,000 Georgians lost their homes. Abkhazia has expressed a desire to join the Russian Federation. In March 2008 Abkhazia appealed to the UN to have its independence recognized. Following fighting between Russian and Georgian forces in Aug. 2008, after Tbilisi sent troops into South Ossetia, Russia confirmed its recognition of Abkhazia as an independent state (as have Nauru, Nicaragua, Tuvalu, Vanuatu and Venezuela in the meantime).

President: Aleksandr Ankvab (in office since 29 May 2011—acting until 26 Sept. 2011). In elections held on 26 Aug. 2011 Aleksandr Ankvab took 54·9% of the vote, Sergey Shamba 21·0% and Raul Khadjimba 19·8%. Turnout was 71·9%. The USA, the European Union and Georgia refused to recognize the election.

Prime Minister: Leonid Lakerbaya (appointed on 27 Sept. 2011).

The unit of currency is the Russian *rouble* (RUB), of 100 *kopeks.* Imports, 2006: 3,270·2m. roubles; exports, 2006: 627·2m. roubles. Since 1997 the EU has provided over €20m. in funding for humanitarian projects. The European Commission provided €9·4m. for the restoration of the Inguri hydro power station, which was inaugurated in Oct. 2006 having fallen into severe disrepair in the early 1990s.

The republic has coal, electric power, building materials and light industries. Tourism is an important growth industry. In the first nine months of 2009, 88,865 tourists stayed in Abkhazian hotels and resorts (up from 68,905 for 2008 as a whole).

All agricultural land is owned by the state and private ownership is forbidden under still prevailing Soviet-era regulations. 90% of all households keep cattle. During the Soviet era Abkhazia provided 15–20% of tea for the USSR but since its collapse the industry has declined by around 80%. Tobacco production has also virtually ceased. Peak annual production levels were 120,000 tonnes of citrus fruits, 110,000 tonnes of tea and 14,000 of tonnes of tobacco. Current cash crops are citrus fruits and hazelnuts.

There is a university at Sukhumi.

FURTHER READING

Coppieters, Bruno, Darchiashvili, David and Akaba, Natella, *Federal Practice: Exploring Alternatives for Georgia and Abkhazia.* 2001

Adjara

Area, 2,900 sq. km (1,160 sq. miles); census population (2002): 376,016. Density, 129 per sq. km. Capital, Batumi (2002 population, 121,806, mostly Sunni Muslim). Adjara, which is bounded by Turkey to the south, fell under Turkish rule in the 17th century and was annexed to Russia (rejoining Georgia) after the Berlin Treaty of 1878. On 16 July 1921 the territory was constituted as an Autonomous Republic within the Georgian SSR, a status confirmed by the Georgian Constitution of 1995. In Jan. 2004 Adjaran leader Aslan Abashidze refused to acknowledge the central government of Mikheil Saakashvili and declared a state of emergency. Fearing a Georgian invasion, Abashidze destroyed road and rail links to the rest of Georgia but was forced to step down on 6 May after popular demonstrations in Batumi. Saakashvili imposed direct rule over Adjara. Elections to the *Supreme Council,* Adjara's parliament, were held on 1 Oct. 2012. The opposition Georgian Dream won 13 of 21 seats with 57·6% of the vote and Saakashvili's United National Movement 8 with

36·9%. The last Russian troops left the military base in Batumi in Nov. 2007.

Ethnic groups at the 2002 census: Georgians, 93·4%; Russians, 2·4%; Armenians, 2·3%.

Chairman of the Supreme Council: Avtandil Beridze.

Prime Minister: Archil Khabadze.

Adjara specializes in sub-tropical agricultural products. These include tea, citruses, bamboo, eucalyptus and tobacco. Livestock (Dec. 2003): 128,400 cattle, 1,900 pigs, 16,100 sheep and goats.

There is a port and a shipyard at Batumi, oil-refining, food-processing and canning factories, clothing, building materials and pharmaceutical factories. The reconstructed Batumi airport opened in May 2007 with flights to Turkey and Ukraine.

Approximately 166,300 persons were in paid employment in 2003; the unemployment rate was 12·1%.

South Ossetia

Area, 3,900 sq. km (1,505 sq. miles); population (Jan. 2004 est.), 49,200. Ethnic groups include Ossetians, Georgians, Russians and Armenians. The capital, Tskhinvali, had a population of approximately 30,000 before the 2008 conflict with Georgia. The UN High Commission for Refugees reported that some 30,000 people were displaced within South Ossetia as a result of the 2008 conflict.

This area was populated by Ossetians from across the Caucasus (North Ossetia), driven out by the Mongols in the 13th century. The region was set up within the Georgian SSR on 20 April 1922. Formerly an Autonomous Region, its administrative autonomy was abolished by the Georgian Supreme Soviet on 11 Dec. 1990, and it has been named the Tskhinvali Region.

Fighting broke out in 1990 between insurgents wishing to unite with North Ossetia and Georgian forces. By a Russo-Georgian agreement of July 1992 Russian peacekeeping forces moved into a seven-km buffer zone between South Ossetia and Georgia pending negotiations. An OSCE peacekeeping force has been deployed since 1992.

At elections not recognized by the Georgian government on 10 Nov. 1996, Lyudvig Chibirov was elected president. Though maintaining a commitment to independence, President Chibirov came to a political agreement with the Georgian government in 1996 that neither force nor sanctions should be applied. In July 2003 his successor, President Eduard Kokoyty, asked Vladimir Putin to let South Ossetia become a member of the Russian Federation. Georgian President Mikheil Saakashvili, who took office in Jan. 2004, has made clear his wish to revive the authority of the Georgian government in the regions. In Aug. 2008 he sent in troops against the region's separatist forces. Moscow responded by mobilizing troops, leading to a week of fierce fighting. Russia subsequently recognized South Ossetian independence, as have Nauru, Nicaragua, Tuvalu, Vanuatu and Venezuela in the meantime.

President: Leonid Tibilov (in office since 19 April 2012).

Prime Minister: Rostislav Khugayev (in office since 26 April 2012—acting until 15 May 2012).

In presidential elections held on 25 March 2012, Leonid Tibilov won 42·5% of the votes cast, David Sanakoyev 24·6%, Dmitriy Medoyev 23·8% and Stanislav Kochiyev 5·3%. Turnout was 70·3%. In the second round on 8 April, Tibilov won 54·1% of the vote against 42·7% for Sanakoyev. Turnout was 71·3%. Parliamentary elections were held on 31 May 2009. The ruling Unity Party won 17 of 34 seats, the People's Party of South Ossetia 9 and the Communist Party of South Ossetia 8. Neither Georgia nor NATO, the USA or the EU recognized either the presidential election of March/April 2012 or the parliamentary election of May 2009.

Main industries are mining, timber, electrical engineering and building materials.

GERMANY

0   40 mi
0   75 km

NETHER-
LANDS

SCHLESWIG-
HOLSTEIN

Kiel

Rostock
MECKLENBURG-
WEST POMERANIA

Hamburg
HAMBURG

BREMEN   Bremen
LOWER SAXONY

POLAND

Osnabrück

Hannover

BERLIN   BERLIN
BRANDENBURG

SAXONY-
ANHALT

Cottbus

Bielefeld

NORTH RHINE-

Essen

WESTPHALIA

Düsseldorf

Göttingen

GERMANY

Halle

Leipzig

Dresden

SAXONY

Cologne
Bonn

Kassel

HESSEN

Erfurt
THURINGIA

Koblenz

RHINELAND-
PALATINATE

Frankfurt
am Main

Mainz

CZECH
REPUBLIC

SAARLAND
Saarbrücken

Nuremberg

LUXEMBOURG

BAVARIA

FRANCE

Stuttgart

BADEN-
WÜRTTEMBERG

Ulm

Augsburg

Munich

Freiburg

AUSTRIA

SWITZERLAND

© Research Machines plc 2006

Bundesrepublik Deutschland
(Federal Republic of Germany)

Capital: Berlin
Seats of government: Berlin, Bonn
Population projection, 2015: 81·47m.
GNI per capita, 2011: (PPP$) 34,854
HDI/world rank: 0·905/9
Internet domain extension: .de

KEY HISTORICAL EVENTS

From the 8th century BC the Celtic peoples inhabited most of present-day Germany but by about 500 BC Germanic tribes had pushed their way north and settled in the Celtic lands. The expanding Roman Empire established its boundaries along the Rhine and the Danube rivers but attempts to move further east had to be abandoned after the Roman provincial Governor Varius was defeated in AD 9 by the Germanic forces under Arminius. For the next thousand years the towns of Trier, Regensburg, Augsburg, Mainz and Cologne, founded by the Romans, formed the main centres of urban settlement. Christianity was introduced under Emperor Constantine and the first bishopric north of the Alps was established in Trier in AD 314.

At the start of the 5th century the Huns forced the indigenous Saxons north, towards Britain. However, the Franks, who came from the lowlands and were to become the founders of a German state, gradually asserted themselves over all the other Germanic people. Towards the end of the 5th century a powerful Rhenish state was founded under King Clovis, a descendant of Merovech (Merovius), a Salian Frankish king. The Merovingian dynasty eventually gave way to the Carolingians, whose authority was strengthened by papal support.

Charlemagne succeeded to the throne in 768, founding what was later known as the First Reich (Empire). The Franks continued to thrive until their influence stretched from Rome to the North Sea and from the Pyrenees to the River Elbe. The Pope crowned Charlemagne emperor on Christmas Day 800, creating what was to become known as the Holy Roman Empire. But the empire was too unwieldy to survive Charlemagne. On his death in 814 it began to break up. The Treaty of Verdun in 843 divided the French and German people for the first time, creating a Germanic Central Europe and a Latin Western Europe. The first king of the newly formed eastern kingdom was Ludwig the German and under him a specific German race and culture began to take shape. The last of Charlemagne's descendants died in 911 and with it the Carolingian dynasty. Power shifted, via Conrad I, duke of Franconia, to Henry I, duke of Saxony. Henry's son Otto the Great crushed the power of the hereditary duchies and by making grants of land to the Church he strengthened ties with Rome. His coronation as emperor of the Romans in 962 was the first to associate German kingship with the office of the Holy Roman Emperor.

Over two centuries, powerful dynasties emerged to threaten the emperor. After an intense feud the Hohenstaufens (of Swabia) gained supremacy over the Guelfs (the counts of Bavaria; later denoting anti-imperial loyalties and the papal faction) and managed to keep the upper hand for well over a century. Frederick Barbarossa, descended from both dynasties, led several expeditions to subjugate Italy and died on the Third Crusade. The Knights of the Teutonic Order set about converting Eastern Europe to Christianity and by the 14th century they had conquered much of the Baltic. By controlling the lucrative grain trade Germany grew rich.

Habsburg Rule

The Golden Bull of 1356 established the method for electing an emperor by setting up an Electoral College composed of seven princes or *electors*. Three of these were drawn from the church (the archbishops of Cologne, Mainz and Trier), and four from the nobility (the king of Bohemia, the duke of Saxony, the margrave of Brandenburg and the count palatine of the Rhine), all of whom had the right to build castles, mint their own coinage, impose taxes and act as judges. The title of Holy Roman Emperor nearly always went to an outsider and increasingly to members of the Austrian Habsburg dynasty. In 1273 Count Rudolph IV was the first Habsburg to be crowned king of the Germans. The Great Schism of 1378–1417, which resulted in rival popes holding court in Rome and Avignon, effectively ended the church's residual power over German affairs.

The Hundred Years War between France and England benefited the growing number of Free Imperial Cities along the German trading routes. Merchants and craftsmen organized themselves into guilds, wresting control of civic life away from the nobility and laying the foundations for a capitalist economy. Founded as a defence and trading league at Lübeck, the Hanseatic League combated piracy and established Germanic economic and political domination of the Baltic and North Sea. German communities were founded in Scandinavia and along the opposite coast as far as Estonia.

In the 14th century the bubonic plague wiped out a quarter of the German population. Rather than put the onus on their own tradesmen returning from Asia, it was the Jews, living in tightly-knit segregated communities, who were blamed. Excluded from guilds and trades they took to money lending, an occupation forbidden to Christians, and one which engendered envy and suspicion.

In 1273 Count Rudolph IV was the first Habsburg to be crowned king of the Germans. Over two centuries the Habsburg dynasty became increasingly powerful, retaining the title of Holy Roman Emperor from 1432 until its abolition nearly four centuries later. Succeeding to the title in 1493 Maximilian I gained the Netherlands by his marriage to Mary of Burgundy and control of Hungary and Bohemia by other marital alliances. Spain was added to the Habsburg dominions by the marriage of Maximilian's son, Philip, to Juana the Mad.

Reformation

In the early part of the 15th century the unpopularity of the church was linked to the corrupt sale of indulgences. Land taxes were levied to pay for St Peter's Church and other sacred buildings in Rome. In 1517 an Augustinian monk named Martin Luther, professor of theology at the University of Wittenberg, made his famous protest with 95 Theses or arguments against indulgences. This open attack on the Church of Rome marked the beginning of the Reformation. Luther challenged the power of the pope, the privileged position of the priests and the doctrine of transubstantiation that had always been at the heart of Catholic dogma. But for the death of Maximilian I in 1519 and the subsequent power struggle for the title of Holy Roman Emperor, Luther might well have been executed as a heretic.

Following Maximilian's death, Francis I of France staked a claim to the succession in an attempt to avoid the concentration of power that would result in the election of the Habsburg candidate, Charles I of Spain. To placate the electors of Germany, the pope named Luther's patron, Frederick the Wise of Saxony, as a compromise candidate. This gained Luther only a temporary reprieve as, after much intrigue, the king of Spain was elected. Although Luther was excommunicated in 1520, he had the right to a hearing before an Imperial Diet (court). This was convened at Worms and although he was branded an outlaw and his books were ordered to be burned, he was given haven in Wartburg Castle where he translated the Bible into German. Thanks to the revolutionary system of printing invented by Johannes Gutenberg, Luther's ideas spread quickly throughout Germany. His doctrine of 'justification by faith alone', with its apparent invitation to resist the authority of the church, was one of the main causes of the Peasants' War of 1524–25 that led to wholesale destruction of monasteries and castles. To the surprise of the rebels, Luther aligned himself with the authorities and the uprising was crushed. The Reformation thus gained political authority. By 1555 so many of the small independent German states had joined the Protestant cause that Charles V admitted defeat and abdicated, retiring to a monastery in Spain. His brother Ferdinand succeeded him and signed the Peace of Augsburg. This agreement gave the secular rulers of each state the right to decide on religious practices (*cuius regio, eius religio*), so dividing Germany between Catholics and Lutherans.

Thirty Years War

Martin Luther died in 1546. The Catholics then launched a Counter-Reformation following the church reforms agreed at the Council of Trent. Bavaria's annexation of the mostly Protestant free city of Donauwörth in 1608 led to the formation of the Protestant Union, an armed alliance under the leadership of the Palatinate. The Catholic League was set up by the Bavarians the following year. Rudolf II, who reigned as emperor for 36 years, chose Prague as his power base, thus weakening his authority over his more distant territories. After he was deposed in 1611 a series of dynastic and religious conflicts set in train the Thirty Years' War. The countryside was laid waste, towns were pillaged and mass slaughter reduced the population by as much as a third. Although the Catholics were the early victors, Denmark and Sweden as well as Catholic France (who preferred the Protestants to the Habsburgs) backed the Protestants while Spain supported the Catholics. After repeated attempts to end the war, the Peace of Westphalia (signed in 1648) brought peace but deprived the emperor of much of his authority. Power was divided between 300 principalities and over 1,000 other territories.

In the 17th and 18th centuries the German princes consolidated their power. The Hanoverian branch of the Welf family inherited the British Crown in 1714; a royal union that was to last until 1837.

Meanwhile, the Habsburgs were struggling to hold on to the title of Holy Roman Emperor. The Turks reached Vienna in 1683 but after they were repulsed, the Austrians pushed eastwards and began to build up an empire in the Balkans. This left their western borders vulnerable where the French, who had long regarded the Rhine as the natural limit of their territory to their east, annexed Alsace and Strasbourg in 1688 and 1697. To the north, the presence of Brandenburg-Prussia under the Hohenzollern family was beginning to be felt. Throughout the 18th century Prussia was built up into a powerful independent state with its capital in Berlin. Strong militarism and a strict class-dominated society helped Prussia to become a major European power. When Frederick the Great came to the throne in 1740, he softened his country's military image by introducing reforms and by creating a cultured life at his court. His main preoccupation, however, continued to be expansion by force. His annexation of Silesia (under an old Brandenburg claim) provoked the Habsburgs to retaliate and, backed by Russia and France, they launched the Seven Years' War. Frederick had only the tacit support of Hanover and Britain to fall back on and, within three years, the Prussian armies were seriously overextended. But Frederick engineered a dramatic change in his fortunes by swelling the ranks of his armies with fresh recruits and in 1772, helped by the collapse of the alliance between Austria and Russia, he annexed most of Poland.

Unification

Revolutionary France had expanded east and when the left bank of the Rhine fell under French control during the War of the First Coalition of 1792–97 the way was paved for the unification of Germany. After Napoleon Bonaparte defeated Austria in 1802 he redrew the map of Germany. All but a few of the free German cities and all the ecclesiastical territories were stripped of their independence. In their place he created a series of buffer states. Bavaria, Württemberg and Saxony were raised to the status of

kingdoms, with Baden and Hesse-Darmstadt as duchies. In 1806 the Holy Roman Empire was ended. The Habsburgs promoted themselves from archdukes to emperors of Austria and set about consolidating their position in the Balkans. After the defeat of Prussia and the occupation of Berlin by Napoleon, the country was forced to sign away half its territory. In the aftermath, Prussia abolished serfdom and allowed the cities their own municipal governments. Prussia played a critical role in the defeat of Napoleon at Waterloo in 1815. The Congress of Vienna, which met to determine the structure of post-Napoleon Europe, established Prussian dominance in German affairs. Westphalia and the Rhineland were added to its territories and although there were still 39 independent states, much of Napoleon's original vision for the reorganization of the Holy Roman Empire was ratified. A German Confederation was established, each state was represented in the Frankfurt-based Diet and Austria held the permanent right to the presidency with Prussia holding the vice-presidency.

The dominant political forces in Germany after the Congress of Vienna were conservative but the rapid advance of the industrial revolution brought about the emergence of a new social order, with wage-earning workers and a growing bourgeoisie. The workers were quick to agitate for better working conditions and the middle classes for political representation. Adding to the social unrest was the peasant class, whose poor living standards were made worse by the failed harvests of the late 1840s. By 1848 violence had erupted all over Europe, forcing the Prussian king to allow elections to the National Assembly in Frankfurt. Although this presented an opportunity for the electorate to introduce liberal social reforms, the middle class members of the Assembly blocked all radical measures. When armed rebellions broke out in 1849, the National Assembly was disbanded and the Prussian army, backed by other German kingdoms and principalities, seized power. From the 1850s, Prussia was in an unassailable position. Realizing the importance of industrial might, Prussia became the driving force for creating a single German market.

Bismarck

In 1862 Wilhelm I appointed Otto von Bismarck as chancellor. Although a leading member of the Junker class, he set about introducing widespread reforms. In order to unite the liberal and conservative wings, he backed the demands for universal male suffrage and, in return for the Chancellor's support for a united Germany, the liberals supported his plans for modernizing the army. Bismarck persuaded Austria to back him in a war against Denmark, which resulted in the recapture of Schleswig and Holstein, and in a subsequent row with Austria over the spoils (the Seven Years' War) Austria was crushed by the superior strength of Prussian arms and military organization. Austria was forced out of German affairs and the previously neutral Hanover and Hesse-Kassel joined the other small German states under Prussia to form a North German Confederation. Bismarck still needed to bring the southern German states into the fold and, in 1870, he rallied all the German states to provoke a war with France. The outcome of the Franco-Prussian War of 1870–71 was the defeat of France and the creation of a united Germany (including the long disputed provinces of Alsace and Lorraine). Wilhelm I of Prussia was named kaiser and the empire was dubbed the Second Reich, commemorating the revival of German imperial tradition after a hiatus of 65 years.

At home, Bismarck continued with his liberal reforms. Uniform systems of law, currency, banking and administration were introduced nationwide, restrictions on trade and labour movements were lifted and the cities were given civic autonomy. These measures were designed to contain the liberals while he set about trying to undermine the influence of the Catholic Church. Although he forced the Catholics to support his agricultural policies designed to protect the interests of the Junker landowners, he had to back down on other issues. Despite the introduction of welfare benefits, opposition to Bismarck grew with the formation of the Social Democratic Party (SPD) in 1870.

Meanwhile, Bismarck was competing with Britain and France in the acquisition of colonies in Africa and the Pacific. In Germany he managed a political balancing act, on one hand creating an alliance of the three great imperial powers of Germany, Russia and Austria and on the other a Mediterranean alliance with Britain to prevent Russia from expanding into the Balkans. When Wilhelm I died in 1888, he was succeeded by his son, Friedrich III, who died after only a few months, and then by his grandson Kaiser Wilhelm II, a firm believer in the divine right of kings. After dismissing Bismarck from office in 1890, he appointed a series of 'yes men' to run his government, thereby seriously undermining the strength and stability that had been built up under Bismarck in the previous decade. Britain had long been an ally of Germany, bound by the ties of dynasty and common distrust of France but after the Kaiser came out in open support of the South African Boers, with whom Britain was in conflict, relations between the two countries plummeted and the European arms race accelerated. Bismarck's juggling of alliances collapsed and Europe was divided into two hostile camps. On one side, Germany was allied once more with Austria (who needed help in propping up her collapsing Eastern Empire) and with Italy. On the other, France and Russia, united in common mistrust of the German-speaking nations, drew Britain closer to them. The European war that was brewing was set in motion in 1914 when a Bosnian nationalist assassinated the Austrian Archduke Franz Ferdinand at Sarajevo. Austria sent a threatening memo falsely accusing Serbia of causing the assassination. This led Russia to mobilize in defence of her Slavic neighbours. Seeing this as an excuse to strike first, Germany attacked France. Belgium's neutrality (which had been guaranteed by Britain) was violated when the German armies marched through on their way to France and Britain declared war on Germany in 1914.

First World War

The German generals miscalculated the strength of the resistance from France and Russia. They had counted on a quick victory and when this did not happen, they were fighting on two fronts. A new form of warfare emerged with the digging of trenches all along Northern France and Belgium. The injury and loss of life suffered by both sides during the next four years was to devastate an entire generation of young men all over Europe. In 1917 the United States entered the war and although the Bolshevik revolution in Russia that same year gained Germany a reprieve, allowing the transfer of vast numbers of troops from the eastern to western fronts, the respite was short-lived. Troops returning from Russia agitated against the war. At the same time, the German lines were weakened by over-extension. On 8 Aug. 1918 the German defences were finally broken. In 1916 the Kaiser had handed over military and political power to Generals Paul von Hindenburg and Erich Ludendorff. As the threat of defeat came closer and in an attempt to minimize the potential damage of a harsh peace treaty, Ludendorff left peace negotiations to a parliamentary delegation. He felt they would be more likely to gain lenient terms and this might serve to nip a possible Bolshevik-style revolution in the bud. Two months of frenzied political activity followed which resulted in the abdication of the Kaiser and the announcement of a new German republic. The First World War ended on 11 Nov. 1918.

The Elections that followed confirmed the Social Democratic Party as the new political force in Germany. Friedrich Ebert, leader of the SPD, was made president, with Philipp Scheidemann as chancellor. In 1919 a new constitution was drawn up at Weimar and a republican government under Chancellor Ebert attempted to restore political and economic stability. But the Treaty of Versailles had exacted painful losses. The rich industrial regions of Saarland and Alsace-Lorraine were ceded to France and Upper Silesia was given to a resurrected Poland. A Polish corridor to the sea effectively cut off East Prussia from the rest of the country and all Germany's overseas colonies were confiscated. The Rhineland was declared a demilitarized zone and the size of the armed forces was severely limited. The German economy was burdened with a heavy reparation bill.

Feeling betrayed by what they saw as a harsh settlement, the German military fostered the 'stab in the back' excuse for failure, which was readily accepted by a disillusioned public. Scheidemann was forced to resign and in the elections of 1920 the SPD withdrew altogether, leaving power in the hands of minorities drawn from the liberal and moderate conservative parties. When, in 1923, reparation payments were withheld, France occupied the industrial region of the Ruhr. Passive resistance brought production to a halt and inflation ruined the middle class as the currency became worthless. Although the Weimar Republic seemed bound to fail, a new chancellor, Gustav Stresemann, ended passive resistance in the Ruhr and negotiated loans from the United States to help rebuild Germany's economy.

Rise of Hitler

By Oct. 1924 the currency was re-established at more or less its former value. Near full employment and general prosperity followed. Scheidemann went on to serve as foreign secretary when he re-established Germany as a world power. Reparation payments were scaled down and more US aid was negotiated. Even with the ailing and aged Hindenburg elected president, the German Republic seemed secure.

The National Socialist German Workers' Party was founded in 1918. Its first leader, a locksmith named Anton Drexler, was soon ousted by Adolf Hitler, a former soldier from Austria whose fanaticism had been fed by defeat in 1918. The party attracted political extremists and misfits whose views mixed the extremes of right and left wing opinion.

Putting his followers into uniform as the Brown Shirts or Storm Troopers (SA), Hitler led a failed *Putsch* in Bavaria in 1923. He was arrested and convicted of high treason. He served only nine months of his sentence and emerged having used his time in prison to write his political manifesto, *Mein Kampf*. Initially, sales of *Mein Kampf* were negligible and Hitler's views were treated as something of a joke. Only the power of his personality, and Joseph Goebbels' propaganda skills, sustained the National Socialists on the German political scene.

The recession of the late 1920s proved fertile ground for Hitler's ideas, which began to appeal to wounded national pride and seemed to offer an attractive solution to the growing economic crisis. Elections were held in 1930 and the Nazi party gained an astonishing 6·4m. votes, becoming the country's second largest party.

The young, the unemployed and the impoverished middle classes were Hitler's main supporters but it was the decision of the right wing traditionalists to back Hitler in order to gain control over his supporters that gave him respectability. Financial support from leading industrialists and giant corporations followed, enabling the Nazi party to fight a strong campaign in the 1932 presidential elections. Hindenburg, backed by the SPD and other democratic parties, scraped a victory. Appointed chancellor, Heinrich Brüning introduced a series of economic reforms, negotiated the end of reparation payments and regained Germany's right to arms equality. But his attempt to introduce land reform lost him the support of the landowners, who undermined his efforts to make the Republic work. Two short-term chancellors followed—Franz von Papen and Gen. Kurt von Schleicher. In 1932 one inconclusive election followed another. Hitler, greatly helped by a campaign of terror by his storm troopers, won increased support. Von Papen, who plotted to persuade Hindenburg to declare Hitler chancellor, mistakenly believed that his party's majority in the Reichstag would enable him to retain control. Hitler was sworn in as chancellor on 30 Jan. 1933.

No sooner had Hitler assumed power than he set about destroying all opposition, stepping up the campaign of terror, which was now backed by the apparatus of the state. He was greatly helped by Hermann Göring who, as Prussian minister of the interior, had control of the police. A month later, the Reichstag was burned down and Hindenburg was obliged to declare a state of emergency, giving Hitler the excuse to silence his opponents legally. Hitler was now the country's dictator, declaring himself president of the Third Reich in 1934.

Race to War

Hitler's policies embraced a theory of Aryan racial supremacy by which, during the following decade, millions of Jews, gypsies, and other non-Aryan 'undesirables' were persecuted, used as slave labour, shipped off to concentration camps, murdered and their assets confiscated. Hitler's expansionism led to his annexation of Austria (the *Anschluss*) and German-speaking Czechoslovakia (the Sudetenland) in 1938. The following year he declared all of Bohemia-Moravia a German protectorate and invaded Poland, attempting to restore the authority exercised there by Prussia before 1918. After the invasion of Czechoslovakia, Britain and France signed an agreement with Germany sacrificing Czech national integrity in return for what they believed would be world peace. Interpreting this as a sign of weakness, Hitler ordered the invasion of Poland, signed a non-aggression pact with Russia and expected a similar collapse of resistance on the part of other western powers. However, by now Britain and France had realized that the Munich agreement was a humiliating sham and that Hitler's invasion of Poland on 1 Sept. 1939 meant that he would pursue his expansionist policy of *Lebensraum* ('living space'). Two days after German tanks rolled into Poland, Britain and France declared war on Germany and the Second World War began.

Germany was well prepared for conflict and to begin with the war went well for Hitler. The fall of Poland was quickly followed by the defeat of the Low Countries and in 1940 France was forced to sign an armistice with Germany and to set up a puppet government in Vichy. Hitler bombarded Britain from the air but held back from invasion. Instead, he turned to the east, subduing the Balkans and, in 1941, planned an invasion of the Soviet Union. In Dec. 1941, after Hitler's Japanese allies attacked the United States naval base at Pearl Harbor, America declared war on the Axis powers. By this time, Germany was hopelessly overextended. Defeat in North Africa, in May 1943, was followed by the halt of the German advance on Russia. The Allies invaded France in June 1944, liberating Paris in Aug. while Russian troops advanced from the east. Hitler was faced with certain defeat but refused to surrender, ordering the German people to defend every square inch of German territory to the death. On 30 April 1945, as Soviet forces marched into Berlin, Hitler committed suicide in his bunker. Germany surrendered unconditionally on 7 May 1945, bringing the Third Reich to an end.

Post-War Period

The Allied forces occupied Germany—the UK, the USA and France holding the west and the USSR the east. By the Berlin Declaration of 5 June 1945 each was allocated a zone of occupation. The zone commanders-in-chief together made up the Allied Control Council in Berlin. The area of Greater Berlin was also divided into four sectors.

At the Potsdam Conference of 1945 northern East Prussia was transferred to the USSR. It was also agreed that, pending a final peace settlement, Poland should administer the areas east of the rivers Oder and Neisse, with the frontier fixed on the Oder and Western Neisse down to the Czechoslovak frontier.

By 1948 it had become clear that there would be no agreement between the occupying powers on the future of Germany. Accordingly, the western allies united their zones into one unit in March 1948. In protest, the USSR withdrew from the Allied Control Council, blockaded Berlin until May 1949, and consolidated control of eastern Germany, establishing the German Democratic Republic (East Germany).

A People's Council appointed in 1948 drew up a constitution for the German Democratic Republic (GDR) that came into force in Oct. 1949, providing for a communist state of five Länder with a centrally planned economy. In 1952 the government marked the division between its own territory and that of the Federal Republic of Germany (West Germany), with a three-mile cordon fenced along the frontier. Berlin was closed as a migration route by the construction of a concrete boundary wall in 1961. In 1953 there were popular revolts against food shortages and the pressure to collectivize. In 1954 the government eased economic restraints, the USSR ceased to collect reparation payments, and sovereignty was granted. The GDR signed the Warsaw Pact in 1955. Socialist policies were stepped up in 1958, leading to flight to the West of skilled workers.

Meanwhile, a constituent assembly met in Bonn in Sept. 1948. A Basic Law, which came into force in May 1949, created the Federal Republic of Germany. The occupation forces retained some powers, however, and the Republic did not become a sovereign state until 1955 when the Occupation Statute was revoked.

The Republic consisted of the states of Schleswig-Holstein, Hamburg, Lower Saxony, Bremen, North Rhine-Westphalia, Hessen, Rhineland-Palatinate, Baden-Württemberg, Bavaria and Saarland, together with West Berlin.

The first chancellor, Konrad Adenauer (1949–63), was committed to the ultimate reunification of Germany and refused to acknowledge the German Democratic Republic. It was not until 1972 that the two German states signed an agreement of mutual recognition and intent to co-operate, forged by West German Chancellor Willy Brandt.

The most marked feature of post-war West Germany was rapid population and economic growth. Immigration from the German Democratic Republic, about 3m. since 1945, stopped when the Berlin Wall was built in 1961; however, there was a strong movement of German-speaking people from German settlements in countries of the Soviet bloc. Industrial growth also attracted labour from Turkey, Yugoslavia, Italy and Spain.

Reunification

The Paris Treaty, which came into force in 1955, ensured the Republic's contribution to NATO and NATO forces were stationed along the Rhine in large numbers, with consequent dispute about the deployment of nuclear missiles on German soil. Even before sovereignty, the Republic had begun negotiations for a measure of European unity, and joined in creating the European Coal and Steel Community in 1951 and the European Economic Community in 1957. In Jan. 1957 the Saarland was returned to full German control. In 1973 the Federal Republic entered the UN.

In the autumn of 1989 movements for political liberalization in the GDR and reunification with the Federal Republic of Germany gathered strength. Erich Honecker and other long-serving Communist leaders were dismissed in Oct.–Nov. The Berlin Wall was breached on 9 Nov. Following the reforms in the GDR in Nov. 1989 the Federal Chancellor Helmut Kohl issued a plan for German confederation. The ambassadors of the four wartime allies met in Berlin in Dec. After talks with Chancellor Kohl on 11 Feb. 1990, Soviet General Secretary Mikhail Gorbachev said the USSR would raise no objection to German reunification. The Allies agreed a formula for reunification talks to begin after the GDR elections on 18 March. On 18 May the Federal Republic of Germany and the GDR signed a treaty extending the Federal Republic's currency, together with its economic, monetary and social legislation, to the GDR as of 1 July. On 23 Aug. the *Volkskammer* (the parliament of the GDR) by 294 votes to 62 'declared its accession to the jurisdiction of the Federal Republic as from 3 Oct. according to article 23 of the Basic Law', which provided for the Länder of pre-war Germany to accede to the

Federal Republic. On 12 Sept. the Treaty on the Final Settlement with Respect to Germany was signed by the Federal Republic of Germany, the GDR and the four wartime allies: France, the USSR, the UK and the USA.

The single most important event in German post-war history took place on 3 Oct. 1990 with the reunification of the Federal Republic and the former GDR. That it happened at all was remarkable enough but that it was achieved without major social and political disruption was a huge tribute to the strength of a still young democracy. That is not to say that reunification has been trouble free. Notwithstanding the injection of billions of Deutsche Marks of public subsidy which transformed the infrastructure and restored urban areas, the easterners found the transition from communism to capitalism more painful than they had anticipated. Part of the problem was the adoption of the Deutsche Mark which, by virtue of its strength as an international currency, inspired confidence but at the same time made it harder for export markets in central and eastern Europe to afford to buy German. The collapse of much traditional industry in the east was hastened by wage equalization deals, pushing up labour costs.

The Federal Assembly (*Bundestag*) moved from Bonn to the renovated *Reichstag* in Berlin in 1999. As a psychological factor, the government move to Berlin was calculated to do much to bring eastern Germany back into the centre of national life as an equal part of the country. Gerhard Schröder served as chancellor from 1998 until 2005, when he was replaced by Angela Merkel of the Christian Democratic Union. She headed a 'grand coalition' of the Christian Democratic Union and the Social Democratic Party until securing a further term in 2009 with the Free Democratic Party as coalition partners.

TERRITORY AND POPULATION

Germany is bounded in the north by Denmark and the North and Baltic Seas, east by Poland, east and southeast by the Czech Republic, southeast and south by Austria, south by Switzerland and west by France, Luxembourg, Belgium and the Netherlands. Area: 357,137 sq. km. Population estimate, 31 Dec. 2011: 81,844,000 (41,637,000 females); density 229 per sq. km. Of the total population of 81,752,000 in Dec. 2010, 65,426,000 lived in the former Federal Republic of Germany (excluding West Berlin), 12,865,000 in the five new states of the former German Democratic Republic and 3,461,000 in Berlin. In 2010, 73·8% of the population lived in urban areas. There were 40·30m. households in 2010 of which 16·20m. were single-person. Germany has an ageing population. The proportion of the population over 60 has been steadily rising, and that of the under 20s steadily declining. By the mid-1990s the number of over 60s had surpassed the number of under 20s and now stands at 25% of the total population.

The UN gives a projected population for 2015 of 81·47m.

On 14 Nov. 1990 Germany and Poland signed a treaty confirming Poland's existing western frontier and renouncing German claims to territory lost as a result of the Second World War.

The capital is Berlin; the Federal German government moved from Bonn to Berlin in 1999.

The Federation comprises 16 *Bundesländer* (states). Area and population:

Bundesländer	Area in sq. km	Population (in 1,000) 1987 census	Dec. 2011 estimate	Density per sq. km (2011)
Baden-Württemberg (BW)	35,751	9,286	10,814[1]	302[1]
Bavaria (BY)	70,550	10,903	12,596	179
Berlin (BE)[2]	892	—	3,502	3,926
Brandenburg (BB)[3]	29,484	—	2,496	85
Bremen (HB)	419	660	661	1,578
Hamburg (HH)	755	1,593	1,799	2,382
Hessen (HE)	21,115	5,508	6,092	289
Lower Saxony (NI)	47,614	7,162	7,914	166
Mecklenburg-West Pomerania (MV)[3]	23,194	—	1,635	70
North Rhine-Westphalia (NW)	34,098	16,712	17,842	523
Rhineland-Palatinate (RP)	19,854	3,631	3,999	201
Saarland (SL)	2,569	1,056	1,013	394
Saxony (SN)[3]	18,420	—	4,137	225
Saxony-Anhalt (ST)[3]	20,450	—	2,313	113
Schleswig-Holstein (SH)	15,800	2,554	2,838	180
Thuringia (TH)[3]	16,172	—	2,221	137

[1]June 2012. [2]1987 census population of West Berlin: 2,013,000.
[3]Reconstituted in 1990 in the Federal Republic.

On 31 Dec. 2010 there were 6,753,621 resident foreigners, including 1,629,480 Turks, 517,546 Italians, 419,435 Poles, 276,685 Greeks and 220,199 Croats. More than 1·6m. of these were born in Germany. In 2010 Germany received 41,332 asylum applications, up from 19,164 in 2007 but down from 438,200 in 1992. The main countries of origin in 2010 were Afghanistan (5,905), Iraq (5,555), Serbia (4,978), Iran (2,475) and Macedonia (2,466). Tighter controls on entry from abroad were applied as from 1993. At the end of 2010 there were 594,000 refugees, a figure exceeded only in Pakistan, Iran and Syria. 96,122 persons were naturalized in 2009, of whom 24,647 were from Turkey. In 2009 there were 733,800 emigrants and 721,000 immigrants. In 2008 emigration exceeded immigration for the first time since the reunification of Germany in 1990. New citizenship laws were introduced on 1 Jan. 2000, whereby a child of non-Germans will have German citizenship automatically if the birth is in Germany, if at the time of the birth one parent has made Germany his or her customary legal place of abode for at least eight years, and if this parent has had an unlimited residence permit for at least three years. Previously at least one parent had to hold German citizenship for the child to become a German national.

Populations of the 80 towns of over 100,000 inhabitants in Dec. 2010 (in 1,000):

Town (and Bundesland)	Population (in 1,000)	Ranking by population	Town (and Bundesland)	Population (in 1,000)	Ranking by population
Aachen (NW)	258·7	25	Fürth (BY)	114·6	67
Augsburg (BY)	264·7	24	Gelsenkirchen (NW)	258·0	26
Bergisch Gladbach (NW)	105·7	72	Göttingen (NI)	121·1	62
Berlin (BE)	3,460·7	1	Hagen (NW)	188·5	41
Bielefeld (NW)	323·3	19	Halle (ST)	233·0	32
Bochum (NW)	374·7	16	Hamburg (HH)	1,786·4	2
Bonn (NW)	324·9	18	Hamm (NW)	181·8	42
Bottrop (NW)	116·8	66	Hanover (NI)	522·7	13
Braunschweig (NI)	248·9	28	Heidelberg (BW)	147·3	53
Bremen (HB)	547·3	10	Heilbronn (BW)	122·9	59
Bremerhaven (HB)	113·4	68	Herne (NW)	164·8	45
Chemnitz (SN)	243·2	29	Hildesheim (NI)	102·8	78
Cologne/Köln (NW)	1,007·1	4	Ingolstadt (BY)	125·1	58
Cottbus (BB)	102·1	80	Jena (TH)	105·1	76
Darmstadt (HE)	144·4	55	Karlsruhe (BW)	294·8	21
Dortmund (NW)	580·4	8	Kassel (HE)	195·5	40
Dresden (SN)	523·1	11	Kiel (SH)	239·5	30
Duisburg (NW)	489·6	15	Koblenz (RP)	106·4	71
Düsseldorf (NW)	588·7	7	Krefeld (NW)	235·1	31
Erfurt (TH)	205·0	37	Leipzig (SN)	522·9	12
Erlangen (BY)	105·6	73	Leverkusen (NW)	160·8	49
Essen (NW)	574·6	9	Lübeck (SH)	210·2	36
Frankfurt am Main (HE)	679·7	5	Ludwigshafen am Rhein (RP)	164·4	46
Freiburg im Breisgau (BW)	224·2	34	Magdeburg (ST)	231·5	33
			Mainz (RP)	199·2	39
			Mannheim (BW)	313·2	20
			Moers (NW)	105·5	74

Town (and Bundesland)	Population (in 1,000)	Ranking by population	Town (and Bundesland)	Population (in 1,000)	Ranking by population
Mönchengladbach (NW)	258·0	26	Potsdam (BB)	156·9	51
Mülheim a. d. Ruhr (NW)	167·3	44	Recklinghausen (NW)	118·4	65
Munich/ München (BY)	1,353·2	3	Regensburg (BY)	135·5	56
Münster (NW)	279·8	22	Remscheid (NW)	110·6	70
Neuss (NW)	151·4	52	Reutlingen (BW)	112·5	69
Nuremberg/ Nürnberg (BY)	505·7	14	Rostock (MV)	202·7	38
Oberhausen (NW)	212·9	35	Saarbrücken (SL)	175·7	43
			Salzgitter (NI)	102·4	79
Offenbach am Main (HE)	120·4	63	Siegen (NW)	103·4	77
			Solingen (NW)	159·9	50
Oldenburg (NI)	162·2	48	Stuttgart (BW)	606·6	6
Osnabrück (NI)	164·1	47	Trier (RP)	105·3	75
Paderborn (NW)	146·3	54	Ulm (BW)	122·8	60
Pforzheim (BW)	119·8	64	Wiesbaden (HE)	276·0	23
			Wolfsburg (NI)	121·5	61
			Wuppertal (NW)	349·7	17
			Würzburg (BY)	133·8	57

The official language is German. Minor orthographical amendments were agreed in 1995. An agreement between German-speaking countries on 1 July 1996 in Vienna provided for minor orthographical changes and established a Commission for German Orthography in Mannheim. There have been objections within Germany, particularly in the North, and many *Bundesländer* are to decide their own language programmes for schools. Generally, both old and new spellings are acceptable.

SOCIAL STATISTICS

Calendar years:

	Marriages	Live births	Of these to single parents	Deaths	Divorces
2004	395,992	705,622	197,129	818,271	213,691
2005	388,451	685,795	200,122	830,227	201,693
2006	373,681	672,724	201,519	821,627	190,928
2007	368,922	684,862	211,053	827,155	187,072
2008	377,055	682,514	218,887	844,439	191,948
2009	378,439	665,126	217,758	854,544	185,817

The annual number of births declined every year between 1997 and 2006 before rising in 2007, but it then fell again and the 2009 figure was a post-war low. Of the 378,439 marriages in 2009, 26,540 involved a foreign male and 32,492 involved a foreign female. The average age of bridegrooms in 2009 was 37·2 years, and of brides 34·0. The average first-time marrying age for men was 33·1 and for women 30·2.

Rates (per 1,000 population), 2009: birth, 8·1; death, 10·4; marriage, 4·6; divorce, 2·3; infant mortality, 3·5 per 1,000 live births; stillborn rate, 3·5 per 1,000 live births. Life expectancy, 2007–09: men, 77·3 years; women, 82·5. Suicide rates, 2007, per 100,000 population, 11·4 (men, 17·4; women, 5·7). Annual population growth rate, 2005–10, −0·2%; fertility rate, 2010, 1·4 births per woman (one of the lowest rates in the world).

Since 1 Aug. 2001 same-sex couples have been permitted to exchange vows at registry offices. The law also gives them the same rights as heterosexual couples in inheritance and insurance law.

A UNICEF report published in 2010 showed that 8·5% of children in Germany live in relative poverty (living in a household in which disposable income—when adjusted for family size and composition—is less than 50% of the national median income).

CLIMATE

Oceanic influences are only found in the northwest where winters are quite mild but stormy. Elsewhere a continental climate is general. To the east and south, winter temperatures are lower, with bright frosty weather and considerable snowfall. Summer temperatures are fairly uniform throughout. Berlin, Jan. 31°F (−0·5°C), July 66°F (19°C). Annual rainfall 22·5" (563 mm). Cologne, Jan. 36°F (2·2°C), July 66°F (18·9°C). Annual rainfall 27" (676 mm). Dresden, Jan. 30°F (−0·1°C), July 65°F (18·5°C). Annual rainfall 27·2" (680 mm). Frankfurt, Jan. 33°F (0·6°C), July 66°F (18·9°C). Annual rainfall 24" (601 mm). Hamburg, Jan. 31°F (−0·6°C), July 63°F (17·2°C). Annual rainfall 29" (726 mm). Hanover, Jan. 33°F (0·6°C), July 64°F (17·8°C). Annual rainfall 24" (604 mm). Munich, Jan. 28°F (−2·2°C), July 63°F (17·2°C). Annual rainfall 34" (855 mm). Stuttgart, Jan. 33°F (0·6°C), July 66°F (18·9°C). Annual rainfall 27" (677 mm).

CONSTITUTION AND GOVERNMENT

The Basic Law (*Grundgesetz*) was approved by the parliaments of the participating *Bundesländer* and came into force on 23 May 1949. It is to remain in force until 'a constitution adopted by a free decision of the German people comes into being'. The Federal Republic is a democratic and social constitutional state on a parliamentary basis. The federation is constituted by the 16 *Bundesländer* (states). The Basic Law decrees that the general rules of international law form part of the federal law. The constitutions of the *Bundesländer* must conform to the principles of a republican, democratic and social state based on the rule of law. Executive power is vested in the *Bundesländer*, unless the Basic Law prescribes or permits otherwise. Federal law takes precedence over state law.

Legislative power is vested in the *Bundestag* (Federal Assembly) and the *Bundesrat* (Federal Council). The Bundestag is currently composed of 622 members and is elected in universal, free, equal and secret elections for a term of four years. A party must gain 5% of total votes cast in order to gain representation in the Bundestag, although if a party has three candidates elected directly, they may take their seats even if the party obtains less than 5% of the national vote. The electoral system combines relative-majority and proportional voting; each voter has two votes, the first for the direct constituency representative, the second for the competing party lists in the *Bundesländer*. All directly elected constituency representatives enter parliament, but if a party receives more 'indirect' than 'direct' votes, the first name in order on the party list not to have a seat becomes a member—the number of seats is increased by the difference ('overhang votes'). Thus the number of seats in the Bundestag varies, but is 598 regular members (for the 2009 election, the same as in 2005 and 2002, but down from 656 for the previous elections since reunification) plus the 'overhang votes' (24 at the 2009 election, giving a total of 622 members). The Bundesrat consists of 69 members appointed by the governments of the *Bundesländer* in proportions determined by the number of inhabitants. Each *Bundesland* has at least three votes.

The Head of State is the Federal *President*, who is elected for a five-year term by a *Federal Convention* specially convened for this purpose. This Convention consists of all the members of the Bundestag and an equal number of members elected by the *Bundesländer* parliaments in accordance with party strengths, but who need not themselves be members of the parliaments. No president may serve more than two terms. Executive power is vested in the Federal government, which consists of the Federal *Chancellor*, elected by the Bundestag on the proposal of the Federal President, and the Federal Ministers, who are appointed and dismissed by the Federal President upon the proposal of the Federal Chancellor.

The Federal Republic has exclusive legislation on: (1) foreign affairs; (2) federal citizenship; (3) freedom of movement, passports, immigration and emigration, and extradition; (4) currency, money and coinage, weights and measures, and regulation of time and calendar; (5) customs, commercial and navigation agreements, traffic in goods and payments with foreign countries, including customs and frontier protection; (6) federal railways and air traffic; (7) post and telecommunications; (8) the legal status of persons in the employment of the Federation and of public law corporations under direct supervision of the Federal government; (9) trade marks, copyright and publishing rights; (10) co-operation of the Federal Republic and the *Bundesländer* in the criminal police and in matters concerning the protection of the constitution, the establishment of a Federal Office of Criminal Police, as well as the combating of international crime; (11) federal statistics.

In the field of finance the Federal Republic has exclusive legislation on customs and financial monopolies and concurrent legislation on: (1) excise taxes and taxes on transactions, in particular, taxes on real-estate acquisition, incremented value and on fire protection; (2) taxes on income, property, inheritance and donations; (3) real estate, industrial and trade taxes, with the exception of the determining of the tax rates.

Federal laws are passed by the Bundestag and after their adoption submitted to the Bundesrat, which has a limited veto. The Basic Law may be amended only upon the approval of two-thirds of the members of the Bundestag and two-thirds of the votes of the Bundesrat.

Staatshandbuch Bund. Annual

National Anthem

'Einigkeit und Recht und Freiheit' ('Unity and right and freedom'); words by H. Hoffmann, tune by J. Haydn.

GOVERNMENT CHRONOLOGY

Federal Republic of Germany (prior to reunification).
Chancellors since 1949 (CDU = Christian Democratic Union; FDP = Free Democratic Party; SPD = Social Democratic Party)

1949–63	CDU	Konrad Adenauer
1963–66	CDU	Ludwig Erhard
1966–69	CDU	Kurt Georg Kiesinger
1969–74	SPD	Willy Brandt
1974	FDP	Walter Scheel
1974–82	SPD	Helmut Schmidt
1982–90	CDU	Helmut Kohl

German Democratic Republic = Presidents of the Republic (1949–60) then Leaders of the Council of State. (LDPD = Liberal Democratic Party of Germany; SED = Socialist Unity Party of Germany)

1949	LDPD	Johannes Dieckmann
1949–60	SED	Willhelm Pieck
1960	LDPD	Johannes Dieckmann
1960–73	SED	Walter Ulbricht
1973	SED	Friedrich Ebert
1973–76	SED	Willi Stoph
1976–89	SED	Erich Honecker
1989	SED	Egon Krenz
1989–90	LDPD	Manfred Gerlach

Federal Republic of Germany.
Chancellors since reunification. (CDU = Christian Democratic Union; FDP = Free Democratic Party; n/p = non-partisan; SPD = Social Democratic Party)

1990–98	CDU	Helmut Kohl
1998–2005	SPD	Gerhard Schröder
2005–	CDU	Angela Merkel

Presidents since 1949.

1949–59	FDP	Theodor Heuss
1959–69	CDU	Karl Heinrich Lübke
1969–74	SPD	Gustav Heinemann
1974–79	FDP	Walter Scheel
1979–84	CDU	Karl Carstens
1984–94	CDU	Richard von Weizsäcker
1994–99	CDU	Roman Herzog
1999–2004	SPD	Johannes Rau
2004–10	CDU	Horst Köhler
2010	SPD	Jens Böhrnsen (acting)
2010–12	CDU	Christian Wulff
2012	CSU	Horst Seehofer (acting)
2012–	n/p	Joachim Gauck

RECENT ELECTIONS

On 18 March 2012 Joachim Gauck was elected Federal President by the Federal Convention with 991 votes, against 126 for Beate Klarsfeld and 108 abstentions.

Bundestag elections were held on 27 Sept. 2009. The Christian Democratic Union/Christian Social Union of Chancellor Angela Merkel (CDU/CSU; the CSU is a Bavarian party where the CDU does not stand) won 239 seats with 33·8% of votes cast (226 with 35·2% in 2005); the Social Democratic Party (SPD) won 146 with 23·0% (222 seats with 34·2%); the Free Democratic Party (FDP), 93 with 14·6% (61 with 9·8%); the Left Party (former Party for Democratic Socialism), 76 with 11·9% (54 with 8·7%); the Greens, 68 with 10·7% (51 with 8·1%). Turnout was 70·8% (77·7% in 2005). The CDU formed a coalition with the FDP to create Germany's first centre-right government since 1998.

Parliamentary elections are scheduled to take place on 22 Sept. 2013.

European Parliament

Germany has 99 representatives. At the June 2009 elections turnout was 43·3% (43·0% in 2004). The CDU/CSU won 42 seats—CDU 34 and CSU 8—with 37·9% of votes cast (political affiliation in European Parliament: European People's Party); the SPD, 23 with 20·8% (Progressive Alliance of Socialists and Democrats); the Greens, 14 with 12·1% (Greens/European Free Alliance); the FDP, 12 with 11·0% (Alliance of Liberals and Democrats for Europe); the Left Party, 8 with 7·5% (European Left/Nordic Green Left).

CURRENT GOVERNMENT

Federal President: Joachim Gauck; b. 1940 (ind.; since 18 March 2012).

Chancellor: Angela Merkel; b. 1954 (CDU; sworn in 22 Nov. 2005 and re-elected in Oct. 2009). In March 2013 the cabinet comprised:

Vice Chancellor and Minister for Economics and Technology: Philipp Rösler (FDP). *Foreign Affairs:* Guido Westerwelle (FDP). *Interior:* Hans-Peter Friedrich (CSU). *Justice:* Sabine Leutheusser-Schnarrenberger (FDP). *Finance:* Wolfgang Schäuble (CDU). *Labour and Social Affairs:* Ursula von der Leyen (CDU). *Food, Agriculture and Consumer Protection:* Ilse Aigner (CSU). *Defence:* Thomas de Maizière (CDU). *Family Affairs, Senior Citizens, Women and Youth:* Kristina Schröder (CDU). *Health:* Daniel Bahr (FDP). *Transport, Building and Urban Development:* Peter Ramsauer (CSU). *Environment, Nature Conservation and Nuclear Safety:* Peter Altmaier (CDU). *Education and Research:* Johanna Wanka (CDU). *Economic Co-operation and Development:* Dirk Niebel (FDP). *Head of the Federal Chancellery and Minister for Special Tasks:* Ronald Pofalla (CDU).

President of the Bundestag: Norbert Lammert (CDU; elected Oct. 2005).

Government Website: http://www.bundesregierung.de

CURRENT LEADERS

Joachim Gauck

Position
President

Introduction
Joachim Gauck was elected president on 18 March 2012. His appointment alongside Chancellor Angela Merkel meant that the two most senior positions in government were held by East Germans for the first time since reunification. Gauck is best known as a pro-democracy activist who has exposed crimes

committed by the secret police (the Stasi) in the former German Democratic Republic (GDR).

Early Life

Gauck was born on 24 Jan. 1940 in the town of Rostock. In 1951 his father was arrested by Soviet forces on charges of espionage and served three years in a Soviet prison, an event that moulded Joachim's political beliefs.

Gauck refused to join the communist Free German Youth movement and was denied the opportunity of becoming a journalist. Instead, he studied theology and became a Lutheran pastor, a position which brought him into conflict with the ruling communist regime and to the attention of the Stasi.

In 1989 Gauck became spokesman for the New Forum, a pro-democratic opposition movement, taking part in the series of peaceful demonstrations that led the way to the regime's collapse. After the dissolution of the GDR in Oct. 1990, the new federal government appointed him as a special representative to research the Stasi archives. In this role, which he held until 2000, he investigated crimes perpetrated by the secret police in the period of communist rule.

Gauck ran for the presidency in 2010 backed by the opposition Social Democratic Party (SPD) and the Greens. He was narrowly defeated by the governing coalition candidate, Christian Wulff, after three parliamentary ballots. When Wulff resigned in Feb. 2012 in the wake of a corruption scandal, Gauck was again put forward by the SPD and Greens and this time gained the support of Merkel's coalition. He faced one opponent, 'Nazi hunter' Beate Klarsfeld, who was proposed by the Left Party. In a vote by the Federal Assembly on 18 March, Gauck took 991 votes against 126 for Klarsfeld, with 108 abstentions.

Career in Office

Gauck came to power as Germany was striving to negotiate a resolution to the eurozone debt crisis. A charismatic and popular figure, his election was seen by some commentators as threatening the authority of Chancellor Merkel, who had opposed his initial candidacy in 2010.

Angela Merkel

Position

Chancellor

Introduction

Angela Merkel became Germany's first female chancellor in Nov. 2005. Her appointment came after three weeks of negotiations following elections that failed to give a parliamentary majority to either her party, the Christian Democrats (CDU), or the Social Democrats (SPD) of incumbent Gerhard Schröder. Despite her initial success in re-energizing Germany's economy, global turmoil in financial markets led to recession in 2008, which highlighted considerable differences in her uneasy CDU–SPD coalition over the appropriate policy response. She was nevertheless re-elected Chancellor in the elections of Sept. 2009. Her second term has seen economic concerns at both domestic and international levels remain centre-stage.

Early Life

Angela Dorothea Kasner was born on 17 July 1954 in Hamburg, West Germany, the daughter of a Lutheran pastor and a teacher. Later in 1954 her father received a pastorship in East Germany (GDR) and the family moved to Templin, 50 km north of Berlin. Merkel was educated in Templin before studying physics at the University of Leipzig from 1973–78. Having married Ulrich Merkel in 1977, she continued her studies at the Academy of Sciences in East Berlin, receiving a doctorate in 1986, and subsequently combined research in quantum chemistry with lecturing.

Following the fall of the Berlin Wall in Nov. 1989 and the democratic elections in the GDR on 18 March 1990, Merkel became a member of the East German Christian Democratic Union (CDU) and deputy spokesperson of the new government under Lothar de Maizière. She was elected to the *Bundestag* in the first post-unification general elections in Dec. 1990, representing the united CDU in a Baltic coast constituency encompassing Rügen and the city of Stralsund. She was also appointed to Chancellor Helmut Kohl's cabinet as minister for women and youth, a position she held until being promoted to minister for the environment in 1994.

Merkel lost ministerial office in 1998 when the CDU was defeated in federal elections but later that year was appointed secretary-general of the party. She oversaw a string of CDU provincial election victories in 1999, although it was a funding scandal implicating the party's chairman, Wolfgang Schäuble, and Kohl himself that thrust Merkel into the limelight. She criticized Kohl (who was later stripped of his CDU honorary chairmanship), called for a fresh start for the party and was duly elected party president on 10 April 2000.

Unable to garner sufficient support to challenge Chancellor Schröder in the 2002 federal elections, Merkel ceded that role to Edmund Stoiber, leader of the CDU's sister party, the Bavarian Christian Social Union (CSU). Following Stoiber's narrow defeat, Merkel became leader of the conservative opposition in the *Bundestag*. She advocated institutional reform through simplifying the tax code and lowering taxes and overhauling health care and pensions. She also argued for a loosening of labour law and, in 2003, controversially backed the US-led invasion of Iraq.

In May 2005 Merkel won the CDU/CSU nomination to challenge Schröder in the 2005 elections. In the elections on 18 Sept. the CDU/CSU won 35·2% of the vote to the SPD's 34·2%. Both Merkel and Schröder claimed victory and weeks of wrangling ensued. A deal for a grand coalition was eventually reached whereby Merkel would become chancellor and the SPD would hold eight of the 14 cabinet posts. Merkel was elected chancellor by a majority of delegates in the *Bundestag* on 22 Nov. 2005.

Career in Office

Merkel won plaudits for brokering an EU budget deal between France's Jacques Chirac and Britain's Tony Blair within weeks of becoming chancellor. However, countering high unemployment and the sluggish German economy, and delivering health care and tax reforms, were likely to prove difficult. Nevertheless, by Aug. 2006 she claimed that the country had 'turned the corner' with a reduction in the budget deficit and lower unemployment, and in the first half of 2007 the public sector recorded a surplus for the first time since unification in 1990. Meanwhile, in Jan. 2007 Germany took over the six-month rotating presidency of the EU and Merkel was instrumental in facilitating agreement among the heads of government on a new draft treaty streamlining the Union's institutional structure and decision-making process (which, officially concluded as the Treaty of Lisbon in Dec. 2007, eventually entered into force in Dec. 2009).

In March 2008 Merkel made an historic address to the Israeli parliament, the first by a German head of government, during a visit marking the 60th anniversary of the founding of Israel.

As the global financial crisis began to bite through 2008, Germany's economy slipped into recession. In the autumn the government stepped in to prevent the collapse of one of the country's largest banks, Hypo Real Estate, and made €500bn. available in loan guarantees and capital to bolster the banking system. Merkel remained unconvinced, however, about the wisdom of the huge economic stimulus packages unveiled by many of her EU partners and said that Germany would not engage in a 'pointless race to spend billions'. This apparent inaction in the face of a worsening recession generated external and internal opposition, and Merkel subsequently endorsed a co-ordinated economic recovery package (and joint action on global warming) agreed at a summit of EU leaders in mid-Dec. 2008. Also in Dec., Merkel was re-elected overwhelmingly as the leader of the CDU.

The CDU emerged as the largest party in the German elections to the European Parliament in June 2009 and increased its majority in the Federal Assembly in general elections in Sept. In Oct. Merkel was sworn in as chancellor for a second term at the head of a new centre-right coalition with the Free Democratic Party (FDP). However, her political authority was subsequently undermined. In June 2010 her preferred candidate for the post of federal president, Christian Wulff, was elected by parliament but only after three rounds of voting, highlighting dissension within her government and party. Wulff then attracted widespread criticism at the end of 2011 after it emerged that he had tried to suppress media coverage about a personal loan that he had accepted while premier of Lower Saxony. He subsequently resigned in Feb. 2012 and a month later Joachim Gauck was elected the new federal president.

In addition, the growing debt crisis in the eurozone led to large and costly bailout packages for some partner states, which Merkel reluctantly endorsed. This has been unpopular with the German public, prompting CDU losses in regional elections in several states (including Baden-Württemberg, Hamburg, Mecklenburg-West Pomerania, North Rhine-Westphalia and Lower Saxony). Merkel strived throughout 2011 and 2012 to build a political consensus on containing Europe's debt problems, which in turn continued to impact negatively on German economic growth. She helped negotiate a treaty change proposing greater fiscal union in the eurozone (but not the EU as a whole following a British veto in Dec. 2011) and also, in 2012, a rescue strategy for struggling eurozone economies (through the European Stability Mechanism) in return for stricter budgetary discipline.

In contrast to British and French policy towards the popular uprising in Libya against the Gaddafi regime, Germany abstained in the United Nations Security Council vote in March 2011 sanctioning NATO military intervention to protect Libyan civilians.

Following the disaster in Japan at the Fukushima nuclear power plant, Merkel's government announced in May 2011 that all of Germany's nuclear power stations would be phased out by 2022.

Ahead of the federal election scheduled for Sept. 2013, she was re-elected overwhelmingly as party leader of the CDU in Dec. 2012.

DEFENCE

Germany officially ended its compulsory military service on 1 July 2011. In July 1994 the Constitutional Court ruled that German armed forces might be sent on peacekeeping missions abroad. Germany has increased the number of professionals available for military missions abroad and sent troops to Afghanistan as part of the international alliance against terrorism in the aftermath of 11 Sept. 2001. The first time that German armed forces were deployed in this way since the Second World War, the move provoked controversy in Germany. In 2006 there were over 8,000 German troops abroad. In addition to those in Afghanistan (nearly half of all German troops deployed abroad) there were German peacekeepers in various parts of the world including Kosovo, Lebanon, Bosnia and Sudan. Since Jan. 2001 women have been allowed to serve in all branches of the military on the same basis as men.

The total strength of the *Bundeswehr* (Federal Defence Forces of Germany) in Oct. 2012 was 195,893, including 7,132 in the vocational training service. In 2008 defence expenditure totalled US$46,759m. (US$568 per capita). In 2007 defence spending represented 1·3% of GDP. Military spending fell by 11·0% in real terms between 1999 and 2008.

Arms Trade
Germany was the world's third largest exporter after the USA and Russia in 2008, with sales worth US$2,900m. or 9·1% of the world total.

Army
The Army is organized in the Army Forces Command. The equipment of the former East German army is in store. Total strength was 70,062 in Oct. 2012.

The Territorial Army is organized into five Military Districts, under three Territorial Commands. Its main task is to defend rear areas and remains under national control even in wartime.

Navy
The Fleet Commander operates from a modern Maritime Headquarters at Glücksburg, close to the Danish border.

The fleet includes 14 diesel coastal submarines and 14 frigates. The main naval bases are at Wilhelmshaven, Olpenitz, Kiel, Eckernförde and Warnemünde.

The Naval Air Arm, 3,700 strong in 2006, is organized into two wings and includes 16 combat aircraft (Atlantics) and 29 armed helicopters.

Personnel in Oct. 2012 numbered 15,852.

Air Force
Since 1970 the *Luftwaffe* has comprised the following commands: German Air Force Tactical Command, German Air Force Support Command (including two German Air Force Regional Support Commands—North and South) and General Air Force Office. Personnel in Oct. 2012 was 33,449. There were 426 combat aircraft in 2006, including *Tornados*, F-4Fs, T-37Bs and T-38As.

ECONOMY

Services accounted for 71% of GDP in 2010, industry 28% and agriculture 1%. Manufacturing's share of total GDP was 21%.

According to the anti-corruption organization *Transparency International*, Germany ranked 13th in the world in a 2012 survey of the countries with the least corruption in business and government. It received 79 out of 100 in the annual index.

Germany gave US$14·1bn. in international aid in 2011, equivalent to 0·39% of GNI (compared to the UN target of 0·7%).

Overview
Germany is Europe's largest economy and the world's second largest exporter, behind China. However, the reunification of the two Germanies in 1989 proved costly for the more prosperous West Germans. GDP growth averaged 4·5% per annum in the 1960s but slowed to under 1% in the first half of the 2000s, the lowest in the eurozone alongside Italy. In 2006 the economy recorded its highest growth since 2000, spurred by an increase in exports and a burst of domestic demand prior to a 3% VAT increase implemented in Jan. 2007.

The principal manufacturing is in the auto and chemical industries, with telecommunications an increasingly important sector. Manufacturing (especially of automobiles) and related services rank higher in the economy than in other developed countries. This leaves Germany vulnerable to slowdowns and recessions in its principal export markets.

The post-Second World War economic miracle (*Wirtschaftswunder*) was marked by prudent fiscal and monetary policy, the growth of a globally competitive manufacturing sector and good industrial relations. The economy is described as a 'social market' in that it embraces enlightened company management and social welfare. Companies are managed on the 'stakeholder' concept, which means they are responsible not only to their shareholders but also to employees, customers, suppliers and local communities. However, the system is changing in response to globalization and the ending of large cross-shareholdings by companies and banks.

The global financial crisis put a strain on the economy and the government was required to intervene in the markets on several occasions. In Oct. 2008 the finance ministry agreed a €50bn. plan to rescue Hypo Real Estate, one of the country's biggest banks, while in Jan. 2009 Commerzbank became the first German bank to be partly nationalized via a government bailout fund. The government also announced a €50bn. stimulus package of public investment and tax cuts aimed at halting a descent into severe recession, alongside a €100bn. fund to underwrite fresh credit to companies starved of new loans.

In the period April–June 2009 the economy emerged from recession, growing by 0·4% on the back of increasing exports, government fiscal stimulus and stronger private consumption. Though there was an overall contraction of 5·1% in 2009, the economy rebounded strongly in 2010 with growth of 4·0% following a pick-up in global trade and strong household and government spending. The economy contracted once again in the final quarter of 2011 owing to financial market turbulence and weakening external demand. In 2012 falling exports led GDP to shrink by 0·6% in the fourth quarter, its sharpest contraction since the height of the financial crisis. The near-term outlook remains uncertain, reflecting the eurozone debt crisis and Germany's dependence on EU markets (which account for 40% of all exports).

Having achieved a balanced budget in 2008, Germany's budget deficit reached 4·3% of GDP in 2010, above the 3% target set by the EU's Stability and Growth Pact. The government committed to reducing it to less than 2% by 2013 and in June 2010 Chancellor Merkel announced €80bn. worth of cuts, including lower spending on welfare and unemployment benefits, a proposed slash to defence spending (of some 10% by 2015) and a flight tax on air travel. The measures aimed to reduce the structural deficit from 2·5% of GDP in 2010 to 0·35% by 2016, as required by an amendment to the constitution and to 'set an example' for the rest of Europe. The financial crisis led to an increase in public debt from 65% of GDP in 2007 to 81% in 2011.

Relatively low growth in the post-unification years has been attributed to weaknesses in the labour market and the high cost of restructuring the economy of the former GDR. However, reforms of the labour market prompted by the global economic crisis—including increased working-hour flexibility and reduced structural unemployment—have contributed to the lowest overall jobless rate in decades (5·3%) and the lowest youth unemployment in Europe. The normalization of working hours and significant one-off payments have pushed overall wage growth to near 3%, supporting the transition to domestic demand-led growth. Increasing full-time female labour participation—by lowering fiscal disincentives for second earners and improving the supply of childcare—is crucial in the medium term given an ageing population.

For further developments see www.statesmansyearbook.com.

Currency

On 1 Jan. 1999 the euro (EUR) became the legal currency in Germany at the irrevocable conversion rate of 1·95583 DM (Deutsche Mark) to one euro. The euro, which consists of 100 cents, has been in circulation since 1 Jan. 2002. It was still possible to make cash transactions in German marks until 28 Feb. 2002, although formally the mark had ceased to be legal tender on 31 Dec. 2001. Euro banknotes in circulation on 1 Jan. 2002 had a total value of €254·2bn.

Foreign exchange reserves were US$37,492m. in Sept. 2009 and gold reserves were 109·56m. troy oz. Only the USA, with 261·50m. troy oz, had more in Sept. 2009. Total money supply was €934,352m. in Aug. 2009.

Inflation rates (based on OECD statistics):

2002	2003	2004	2005	2006	2007	2008	2009	2010	2011
1·4%	1·0%	1·8%	1·9%	1·8%	2·3%	2·8%	0·2%	1·2%	2·5%

The inflation rate in 2012 according to Destatis, the Federal Statistical Office, was 2·0%.

Budget

After winning the 2009 election, Chancellor Merkel's CDU agreed upon a coalition agreement with the liberal FDP based on major tax cuts. In June 2011 the government announced plans to introduce tax cuts worth up to €10bn. within two years. In the space of ten years corporate taxes were cut from 51·6% in 2000 to 29·8% in 2010, a drop of almost 22%.

VAT is (since 1 Jan. 2007) 19% (reduced rate, 7%). In 2006 the federal government and the *Bundesländer* each received 42·5% of income tax and the local authorities 15%. Corporation tax was equally split between the federal government and the *Bundesländer*. In 2007 the federal government received approximately 55% of VAT, the *Bundesländer* around 43% and the local authorities about 2%.

Budget for 2010 (in €1m.):

	All public authorities	Federal portion
Revenue	*Current*	
Taxes	924,862	254,938
Economic activities	19,438	5,659
Interest	12,236	6,784
Current allocations and subsidies	328,004	28,118
Other receipts	42,787	10,398
minus equalising payments	304,410	13,153
	1,022,917	292,744
Revenue	*Capital*	
Sale of assets	14,454	7,979
Allocations for investment	27,157	277
Repayment of loans	8,840	2,153
Public sector borrowing	1,476	—
minus equalising payments	25,205	10
	26,723	10,398
Total revenues	1,049,640	303,143
Expenditure	*Current*	
Staff	215,625	41,888
Materials	282,171	23,026
Interest	64,578	38,408
Allocations and subsidies	779,029	225,986
Other current expenditures	240	—
minus equalising payments	304,410	13,153
	1,037,233	316,156
Expenditure	*Capital*	
Construction	30,875	6,790
Acquisition of property	10,386	2,028
Allocations and subsidies	51,057	21,254
Loans	11,391	2,694
Acquisition of shares	9,960	4,388
Repayments in the public sector	1,266	—
Other expenditures	819	—
minus equalising payments	25,205	10
	90,548	37,143
Total expenditures	1,127,781	353,299

Germany's budget deficit in 2011 was 1·0% of GDP (2010, 4·3%; 2009, 3·2%). The required target set by the EU is a budget deficit of no more than 3%.

Performance

Real GDP growth rates (based on OECD statistics):

2002	2003	2004	2005	2006	2007	2008	2009	2010	2011
0·0%	−0·4%	0·7%	0·8%	3·9%	3·4%	0·8%	−5·1%	4·0%	3·1%

In 2002 real GDP growth was 0·0%, the lowest since 1993, and in 2003 the economy contracted by 0·4%, although Germany came out of recession in the second half of the year. Germany had four quarters of negative growth from the second quarter of 2008 but emerged from recession in the second quarter of 2009. The real GDP growth rate in 2012 according to Destatis, the Federal Statistical Office, was 0·7%. Total GDP in 2011 was US$3,600·8bn., the fourth highest in the world.

Banking and Finance

The Deutsche Bundesbank (German Federal Bank) is the central bank and bank of issue. Its duty is to protect the stability of the currency. It is independent of the government but obliged to support the government's general policy. Its Governor is appointed by the government for eight years. The *President* is Jens Weidmann. Its assets were €671,259m. in Dec. 2010. The largest private banks are the Deutsche Bank, Commerzbank, Dresdner Bank and DZ Bank. In April 2001 Dresdner Bank accepted a takeover offer from Allianz, the country's largest insurance company. Commerzbank in turn agreed to buy Dresdner Bank from Allianz in Aug. 2008. In June 2005 Italy's UniCredit finalized an agreement to acquire HypoVereinsbank in Europe's biggest cross-border banking takeover.

In 2010 there were 2,093 credit institutes, including 300 banks, 429 savings banks, 23 mortgage lenders and 1,141 credit societies. They are represented in the wholesale market by nine public sector *Landesbanken*. Total assets, 2009, €7,509,829m. Savings deposits were €628,154m. in 2010. In 2007 approximately 39% of the German population were using e-banking.

A single stock exchange, the Deutsche Börse, was created in 1992, based on the former Frankfurt stock exchange in a union with the smaller exchanges in Berlin, Bremen, Düsseldorf, Hamburg, Hanover, Munich and Stuttgart. Frankfurt processes 90% of equities trading in the country.

Gross external debt amounted to US$5,617,751m. in June 2012.

Germany attracted US$40·40bn. worth of foreign direct investment in 2011, up from US$8·11bn. in 2008 but down from the record US$198·28bn. of 2000.

Gull, L., *et al.*, *The Deutsche Bank, 1870–1995*. 1996

ENERGY AND NATURAL RESOURCES

In 2008, 8·9% of energy consumption came from renewables (wind power, solar power, hydro-electric power, tidal power, geothermal energy and biomass), compared to the European Union average of 10·3%. A target of 18% has been set by the EU for 2020.

Environment

Germany's carbon dioxide emissions from the consumption and flaring of fossil fuels were the equivalent of 10·1 tonnes per capita in 2008. An *Environmental Performance Index* compiled in 2008 ranked Germany 13th in the world, with 86·3%. The index examined various factors in six areas—air pollution, biodiversity and habitat, climate change, environmental health, productive natural resources and water resources.

Germany is one of the world leaders in recycling. In 2004, 56% of all household waste was recycled.

Electricity

Installed capacity in 2009 was 118·62m. kW. In 2009 there were 17 nuclear reactors in operation. Production of electricity was 524·60bn. kWh in 2009, of which about 26% was nuclear. There is a moratorium on further nuclear plant construction, and the SPD–Green coalition government agreed in 2000 to begin phasing out nuclear power, with the final plant closure scheduled for 2022. After the CDU/CSU–FDP coalition came to power following the Sept. 2009 election this date was extended to 2036. However, in the wake of Japan's nuclear incidents in March 2011, Chancellor Angela Merkel suspended the extension and temporarily shut down seven of Germany's oldest reactors. Electricity consumption per capita was 7,543 kWh in 2007. In April 1998 the electricity market was liberalized, leading to huge cuts in bills for both

industrial and residential customers. In June 2000 Veba and Viag merged to form E.ON, which became the world's largest private energy service provider. Germany is the second largest exporter of electricity in the world (after France), with 62·5bn. kWh in 2007.

Oil and Gas

The chief oilfields are in Emsland (Lower Saxony). In 2010, 1·91m. tonnes of crude oil were produced. Natural gas production was 13·0bn. cu. metres in 2008. Natural gas reserves were 120bn. cu. metres in 2008; crude petroleum reserves were 367m. bbls in 2007.

Wind

Germany is one of the world's largest wind-power producers. By the end of 2010 there were 21,607 wind turbines with a total rated power of 27,215 MW (14·0% of the world total), ranking Germany third behind China and the USA. Production of wind-generated electricity in 2010 totalled 37·3bn. kWh.

Minerals

The main production areas are: North Rhine-Westphalia (for coal, iron and metal smelting-works), Central Germany (for lignite) and Lower Saxony (Salzgitter for iron ore; the Harz for metal ore).

Production (in 1,000 tonnes): lignite (2010), 174,847; salt (2009), 18,939; coal (2009), 13,766. In 2009 proved coal reserves were 6·7bn. tonnes. Germany is the world's largest lignite producer and the third largest salt producer after China and the USA.

Agriculture

In 2010 there were 11·85m. ha. of arable land. Sown areas in 2010 (in 1,000 ha.) included: wheat, 3,297·7; corn for silage, 1,828·9; barley, 1,641·4; rape, 1,461·2; rye and maslin, 627·1; maize, 466·6; triticale, 413·3; grassland, 386·9; sugar beets, 364·1; potatoes, 254·4; oats, 141·4. Crop production, 2008 (in 1,000 tonnes): fodder, 75,876; wheat, 26,001; sugar beets (2010), 22,441; barley, 11,972; potatoes (2010), 10,143; maize, 5,158; rapeseed, 5,154; rye, 3,743; oats, 793. Germany is the world's largest producer of hops (39,700 tonnes in 2008) and the second largest producer of rye.

In 2008, 5·4% of agricultural land was farmed organically in Germany. Organic food sales for Germany in 2008 were valued at €5·8bn. (the second highest in the world behind the USA).

In 2010 there were 299,100 farms, of which 27,400 were between two and five ha. and 33,600 over 100 ha. In 2010 there were 274,600 farmers assisted by 293,600 household members and 529,500 hired labourers (334,000 of them seasonal).

Wine production was 690·6m. litres in 2010 (down from 922·8m. litres in 2009).

Livestock, 2010 (in 1,000): pigs, 27,571·4; beef cattle, 12,534·5; dairy cows, 4,164·8; sheep, 2,088·5; horses (2007), 541·9; poultry, 128,899·8. Livestock products (in 1,000 tonnes): milk (2009), 29,199; meat (2010), 6,719; cheese (2008) (except soft cheese and cottage cheese), 1,487; eggs, 8,007m. units.

Forestry

Forest area in 2010 was 11·08m. ha., of which about half was owned by the State. Timber production was 54·42m. cu metres in 2010. In recent years depredation has occurred through pollution with acid rain.

Fisheries

The total catch in 2010 was 222,771 tonnes (207,761 tonnes from marine waters), down from 281,368 tonnes in 2005. In 2010 the fishing fleet consisted of 1,766 vessels totalling 68,200 GT. Total employment in the German fleet in 2010 was about 1,640. Imports of fishery commodities were valued at US$4,502m. in 2008 and exports at US$2,472m.

INDUSTRY

The leading companies by market capitalization in Germany in March 2012 were: Siemens, an engineering conglomerate including industry, energy and health care (US$92·0bn.); SAP, a software and computer services company (US$85·6bn.); and BASF, a chemicals company (US$80·2bn.).

In 2010 a total of 862,986 firms were registered (728,978 in 2001).

Output of major industrial products, 2010 unless otherwise indicated (in 1,000 tonnes): distillate fuel oil (2008), 48,709; crude steel, 43,830; rolled steel products, 36,827; cement, 29,661; petrol (2008), 24,822; pig iron (2009), 20,104; plastics, 17,747; paper, 12,832; residual fuel oil (2008), 12,023; flour, 5,365; jet fuel (2008), 4,760; sulphuric acid, 1,690; nitrogenous fertilizers, 1,373; synthetic fibre, 373; passenger cars, 6,065,000 units; household dishwashing machines, 3,024,000 units; refrigerators, 2,739,000 units; glass bottles, 8,535m. units; TV sets, 416,000 units; beer, 8,674m. litres; soft drinks (excluding milk-based beverages), 22,031m. litres.

Labour

Retirement age was traditionally 65 years, but is being raised gradually to 67 in a process that started at the beginning of 2012 and is to continue through to 2029. In 2010 the workforce was 43·32m., of whom 40·38m. were working and 3·24m. (1·48m. females) were registered as unemployed. In 2010 there were 35·96m. employees and 4·41m. self-employed (including those helping family members). Of the total workforce in 2010 the year average for the number of employees in each industry was as follows: 7,384,000 in the mining, processing and manufacturing industries; 5,196,000 in the vehicle trade and maintenance; 4,905,000 in real estate and corporate services (2009); 3,913,000 in health, veterinary and social services (2009); 2,654,000 in the civil service and armed forces (2009); 2,314,000 in education (2009); 2,067,000 in transport and communications; 1,767,000 in the construction industry; 1,604,000 in the hotel and catering industries; 1,042,000 in banking and insurance (2009); and 461,000 in agriculture, forestry and fisheries. In 2010 there were 359,038 job vacancies.

The standardized unemployment rate was 5·3% in Dec. 2012 (down from 6·0% in 2011 as a whole). Unemployment in 2012 was at its lowest level since the reunification of Germany in 1990. Youth unemployment (under 25) was at just 8·0% in June 2012 the lowest in the European Union, helped by the fact that a quarter of employers provide formal apprenticeship schemes for young people. Long-term unemployment is particularly high, with 47·4% of the labour force in 2010 having been out of work for more than a year. In Jan. 2005 the number of people out of work reached 5m., the highest total since the 1930s, although by Oct. 2010 it had fallen to below 3m. as Germany made a strong recovery from the recession. The gross annual earnings of full-time employees in the industry and services sector amounted to an average of €41,509 per person in 2008. There is no national minimum wage in Germany.

INTERNATIONAL TRADE

In 2007 Germany had its highest ever annual trade surplus at €195·3bn. for the year, although it has declined since then.

Imports and Exports

Trade in €1m.:

	2007	2008	2009	2010[1]
Imports	769,887	805,842	664,615	806,164
Exports	965,236	984,140	803,312	959,497

[1]Provisional.

Main import sources in 2010 (provisional trade figures in €1m.): China, 76,528; Netherlands, 68,767; France, 61,751; USA, 45,063; Italy, 43,667; UK, 38,594; Austria, 34,315; Belgium, 33,700; Switzerland, 32,485; Russia, 31,780. Main export markets in 2010 (provisional trade figures in €1m.): France, 90,694; USA, 65,570; Netherlands, 63,235; UK, 59,487; Italy, 58,477; Austria, 53,721;

China, 53,636; Belgium, 46,407; Switzerland, 41,712; Poland, 38,053.

Distribution of imports and exports by commodities in 2010 (provisional, in €1m.) includes: finished goods, 540,935 and 808,205; semi-finished goods, 71,699 and 52,273; foodstuffs, 50,422 and 41,888; raw materials, 80,864 and 9,671; alcohol and tobacco, 8,981 and 8,985; live animals, 1,275 and 959.

Germany is the third largest trading nation in the world after the USA and China. In 2003 it took over from the USA as the world's leading exporter and in 2009 accounted for 9·0% of global exports. However, China took over as the largest exporter of goods in 2009 in spite of Germany's share increasing from 8·1% in 2008.

Trade Fairs

Germany has a number of major annual trade fairs, among the most important of which are Internationale Grüne Woche Berlin (International Green Week Berlin—Exhibition for the Food Industry, Agriculture and Horticulture), held in Berlin in Jan.; Ambiente (for high quality consumer goods and new products), held in Frankfurt in Feb.; ITB Berlin (International Tourism Exchange), held in Berlin in March; CeBit (World Business Fair for Office Automation, Information Technology and Telecommunications), held in Hanover in March; Hannover Messe (the World's Leading Fair for Industry, Automation and Innovation), held in Hanover, in April; Internationale Funkausstellung Berlin (Your World of Consumer Electronics), held in Berlin in late Aug./early Sept.; and Frankfurter Buchmesse (Frankfurt Book Fair) held in Frankfurt in Oct. Hanover's trade fair site is the largest in Europe and Frankfurt's the second largest.

COMMUNICATIONS

Roads

In 2010 the total length of the road network was 230,969 km, including 12,813 km of motorway (Autobahn), 39,887 km of federal highways and 86,615 km of secondary roads. The motorway network is the largest in Europe. On 1 Jan. 2011 there were 50,902,131 motor vehicles, including: passenger cars, 42,301,563 (approximately one car for every two persons); lorries, 2,441,377; buses, 76,500; motorcycles, 3,827,900. In 2010, 8,892m. passengers were transported by scheduled road transport services. There were 288,297 accidents in 2010 resulting in injuries to passengers. Road casualties in 2010 totalled 374,818, with 371,170 injured and 3,648 killed. In 2010 there were 4·5 fatalities per 100,000 population.

Germany was ranked fifth for its road infrastructure in the World Economic Forum's *Global Competitiveness Report 2009–10*.

Rail

Legislation of 1993 provides for the eventual privatization of the railways, but the state-owned Deutsche Bahn still dominates the market. On 1 Jan. 1994 West German Bundesbahn and the former GDR Reichsbahn were amalgamated as the Deutsche Bahn, a joint-stock company in which track, long-distance passenger traffic, regional passenger traffic, goods traffic and railway stations/services are run as five separate administrative entities. These were intended after 3–5 years to become companies themselves, at first under a holding company, and ultimately independent. In 2009 the total length of railway track of all kinds was 41,104 km (nearly all 1,435 mm gauge track); 56% of the network was electrified. 2,370m. passengers and 355·7m. tonnes of freight were carried in 2010.

There are metros in Berlin (152 km), Hamburg (101 km), Munich (101 km) and Nuremberg (35 km), and tram/light rail networks in over 50 cities.

In the World Economic Forum's *Global Competitiveness Report 2009–10* Germany ranked fifth for quality of rail infrastructure.

Civil Aviation

Lufthansa, the largest carrier, was set up in 1953 and was originally 75% state-owned. The government sold its final shares in 1997. Other airlines include Air Berlin (Germany's second largest airline and Europe's third largest low-cost carrier), Condor, Eurowings, Germanwings and TUIfly. In 2005 Lufthansa carried 49·0m. passengers (35·7m. on international flights); passenger-km totalled 112·8bn. In 2010 civil aviation had 799 aircraft over 20 tonnes (772 jets).

In 2010 there were 94·97m. passenger arrivals and 95·04m. departures. Main international airports: Bremen, Cologne-Bonn, Düsseldorf, Frankfurt am Main, Frankfurt (Hahn), Hamburg (Fuhlsbüttel), Hanover, Munich, Nuremberg, Stuttgart, Weeze (Niederrhein) and two at Berlin (Tegel and Schönefeld). Secondary airports in terms of passenger numbers include Dortmund, Dresden, Karlsruhe, Leipzig, Münster and Paderborn.

In 2010 Frankfurt am Main handled 52·95m. passengers and 2,304,000 tonnes of freight. It is the busiest airport in Europe in terms of freight handled. Munich was the second busiest German airport in terms of passenger traffic in 2010 (34·6m.) but fourth for freight. Leipzig was the second busiest in 2010 for freight, with 663,000 tonnes, but only 14th for passenger traffic.

Shipping

At 31 Dec. 2010 the mercantile marine comprised 571 ocean-going vessels of 15,526,000 GRT. Sea-going ships in 2010 carried 276·0m. tonnes of cargo. The busiest port, Hamburg, handled 104·5m. tonnes of cargo in 2010 (42·6m. tonnes loaded and 61·9m. tonnes discharged), ranking it third in Europe behind Rotterdam and Antwerp. Hamburg is Europe's second busiest container port after Rotterdam. Navigable rivers and canals have a total length of 7,707 km. The inland-waterways fleet on 31 Dec. 2010 included 917 motor freight vessels totalling 1·17m. tonnes and 419 tankers of 759,454 tonnes. 229·6m. tonnes of freight were transported in 2010.

Germany was ranked fifth in the World Economic Forum's *Global Competitiveness Report 2009–10* for the quality of its port facilities.

Telecommunications

Telecommunications were deregulated in 1989. On 1 Jan. 1995, three state-owned joint-stock companies were set up: Deutsche Telekom, Postdienst and Postbank. The partial privatization of Deutsche Telekom began in Nov. 1996; in 2012 the German government held only 15·0% of shares directly, and a further 17·0% indirectly through the government bank KfW.

In 2010 there were 45·6m. main (fixed) telephone lines, down from 54·8m. in 2005. Mobile phone subscribers numbered 104·6m. in 2010 (1,270·4 per 1,000 persons), up from 79·3m. in 2005. T-Mobile and D2 Vodafone are the largest networks, with 36% and 32% of the market share respectively. Germany had 67·4m. internet users in Dec. 2011. The fixed broadband penetration rate in Dec. 2010 was 31·9 subscribers per 100 inhabitants. In March 2012 there were 22·1m. Facebook users (27% of the population).

SOCIAL INSTITUTIONS

Justice

Justice is administered by the federal courts and by the courts of the *Bundesländer*. In criminal procedures, civil cases and procedures of non-contentious jurisdiction the courts on the state level are the local courts (*Amtsgerichte*), the regional courts (*Landgerichte*) and the courts of appeal (*Oberlandesgerichte*). Constitutional federal disputes are dealt with by the Federal Constitutional Court (*Bundesverfassungsgericht*) elected by the Bundestag and Bundesrat. The *Bundesländer* also have constitutional courts. In labour law disputes the courts of the first and second instance are the labour courts and the *Bundesland* labour courts, and in the third instance the Federal Labour Court

(*Bundesarbeitsgericht*). Disputes about public law in matters of social security, unemployment insurance, maintenance of war victims and similar cases are dealt with in the first and second instances by the social courts and the *Bundesland* social courts and in the third instance by the Federal Social Court (*Bundessozialgericht*). In most tax matters the finance courts of the *Bundesländer* are competent, and in the second instance the Federal Finance Court (*Bundesfinanzhof*). Other controversies of public law in non-constitutional matters are decided in the first and second instance by the administrative and the higher administrative courts (*Oberverwaltungsgerichte*) of the *Bundesländer*, and in the third instance by the Federal Administrative Court (*Bundesverwaltungsgericht*).

For inquiries into maritime accidents the admiralty courts (*Seeämter*) are competent on the state level and in the second instance the Federal Admiralty Court (*Bundesoberseeamt*) in Hamburg.

The death sentence was abolished in the Federal Republic of Germany in 1949 and in the German Democratic Republic in 1987.

The population in penal institutions at 30 Nov. 2010 was 69,385 (of which 3,755 women) including 8,852 in open prisons. In 2009, 6,054,000 crimes were recorded by the police (6,633,000 in 2004). There were 690 murders in 2010, down from 757 in 2007.

Education

Education is compulsory for children aged six to 15, although between the ages of 15 and 18 young people are obliged to pursue at least part-time vocational secondary education. After the first four (or six) years at primary school (*Grundschulen*) children attend post-primary (*Hauptschulen*), secondary modern (*Real-schulen*), grammar (*Gymnasien*), or comprehensive schools (*Integrierte Gesamtschulen*). Secondary modern school lasts six years. Grammar school traditionally lasted nine years but this is in the process of being reduced to eight years (although Rhineland-Palatinate is retaining the nine-year system). Entry to higher education is by the final Grammar School Certificate (*Abitur—Higher School Certificate*). There are also schools for children with physical disabilities and those with other special needs (*Sonderschulen*).

In 2009–10 there were 1,568 kindergartens with 27,863 pupils and 2,517 teachers; 16,305 primary schools with 2,914,858 pupils and 189,465 teachers; 3,306 special schools with 387,792 pupils and 72,975 teachers; 7,614 secondary modern schools with 2,099,471 pupils and 144,551 teachers; 3,094 grammar schools with 2,475,371 pupils and 176,296 teachers; 999 comprehensive schools with 610,947 pupils and 47,151 teachers.

In 2009–10 there were 670,927 working teachers, of whom 470,284 were female.

The adult literacy rate is at least 99%.

In 2006 total expenditure on education came to €142·9bn. In 2008 public expenditure on education came to 4·6% of GDP and represented 10·4% of total government expenditure.

Vocational education is provided in part-time, full-time and advanced vocational schools (*Berufs-, Berufsaufbau-, Berufsfach-* and *Fachschulen,* including *Fachschulen für Technik* and *Schulen des Gesundheitswesens*). Occupation-related, part-time vocational training of six to 12 hours per week is compulsory for all (including unemployed) up to the age of 18 years or until the completion of the practical vocational training. Full-time vocational schools comprise courses of at least one year. They prepare for commercial and domestic occupations as well as specialized occupations in the field of handicrafts. Advanced full-time vocational schools are attended by pupils over 18. Courses vary from six months to three or more years.

In 2009–10 there were 8,935 full- and part-time vocational schools with 2,768,771 students and 124,306 teachers.

Higher Education. In the winter term of the 2010–11 academic year there were 418 institutes of higher education (*Hochschulen*) with 2,214,112 students, including 105 universities (1,441,921 students), six teacher training colleges (22,381), 16 theological seminaries (2,411), 51 schools of art (33,021), 211 technical colleges (684,856) and 29 management schools (29,522). Only 385,348 students (17·4%) were in their first year. It is for individual *Bundesländer* to decide whether to charge tuition fees. Some do but the majority offer tuition-free higher education (although they may charge registration fees).

Health

In 2009 there were 325,945 doctors, 67,157 dentists and 57,832 pharmacists. There were 2,084 hospitals in 2009 with 503,341 beds (61·5 for every 10,000 people). In 2008 Germany spent 10·5% of its GDP on health, with public spending amounting to 76·8% of the total. In 2009 total expenditure on health came to €278·4bn.

Welfare

Social Health Insurance (introduced in 1883). Wage-earners and apprentices, salaried employees with an income below a certain limit and social insurance pensioners are compulsorily insured within the state system. Voluntary insurance is also possible.

Benefits: medical treatment, medicines, hospital and nursing care, maternity benefits, death benefits for the insured and their families, sickness payments and out-patients' allowances. Economy measures of Dec. 1992 introduced prescription charges related to recipients' income.

As part of a series of measures to tackle a funding shortfall in the health service, a patient charge of €10 was introduced from Jan. 2004, payable for the first visit only per quarter to a doctor.

51·37m. persons were insured in 2010 (30·06m. compulsorily). Number of cases of incapacity for work (2009) totalled 33·66m., and the number of working days lost were 218·34m. (men) and 208·18m. (women). A total of €160,398m. was paid to beneficiaries in 2009.

Accident Insurance (introduced in 1884). Those insured are all persons in employment or service, apprentices and the majority of the self-employed and the unpaid family workers.

Benefits in the case of industrial injuries and occupational diseases: medical treatment and nursing care, sickness payments, pensions and other payments in cash and in kind, surviving dependants' pensions.

Number of insured in 2009, 61·43m.; number of current pensions, 984,092. A total of €10,225m. was paid to beneficiaries in 2009.

Workers' and Employees' Old-Age Insurance Scheme (introduced in 1889). All wage-earners and salaried employees, the members of certain liberal professions and—subject to certain conditions—self-employed craftsmen are compulsorily insured. The insured may voluntarily continue to insure when no longer liable to do so or increase the insurance.

Benefits: measures designed to maintain, improve and restore the earning capacity; pensions paid to persons incapable of work, old age and surviving dependants' pensions.

Number of current pensions in July 2010, 24·89m. (including old age pensions, 17·57m.; pensions to widows and widowers, 5·38m.). A total of €246,285m. was paid to beneficiaries in 2009.

There are also special retirement and unemployment pension schemes for miners and farmers, assistance for war victims and compensation payments to members of German minorities in East European countries expelled after the Second World War and persons who suffered damage because of the war or in connection with the currency reform.

Family Allowances. €33·53bn. were dispensed to 8·82m. recipients (1·08m. foreigners) in 2010 on behalf of 14·51m. children. Paid child care leave is available for three years to mothers or fathers.

Unemployment Allowances. In 2010, 1·02m. persons (0·43m. women) were receiving unemployment benefit and 6·71m. (3·37m. women) basic cost-of-living benefit for jobseekers. Total expenditure on these and similar benefits (e.g. short-working supplement, job creation schemes) was €45·21bn. in 2010. Unemployment assistance was abolished in Jan. 2005 and replaced with a new so-called 'Unemployment benefit II'. The new benefit is no longer tied to the former income of the recipient but is around the same flat-rate level as the social assistance benefit.

Public Welfare. In 2009, €23·03bn. were distributed to 2,270,000 recipients (1,156,000 women).

Public Youth Welfare. For supervision of foster children, official guardianship, assistance with adoptions and affiliations, social assistance in juvenile courts, educational assistance and correctional education under a court order. A total of €26·51bn. was spent on recipients in 2009.

Pension Reform. A major reform of the German pension system became law on 11 May 2001. The changes entail a cut in the value of the average state pension from 70% to approximately 67% of average final earnings by 2030. There will be incentives in the form of tax concessions and direct payments to encourage individuals to build up supplementary provision by contributing up to 4% of their earnings into private-sector personal pensions. In the long term these could supply up to 40% of overall pension income, with 60% coming from the state as opposed to 85% prior to the changes.

A survey by Aon Consulting in 2005 which ranked the pension systems of the 15 pre-expansion EU countries (based on size and sustainability of payments) placed Germany's system at 13th. The OECD notes that Germany has one of the smallest active workforces aged between 55 and 65 (with a participation rate of 40%). Future reforms are expected to increase employee contributions. Workers and employers currently contribute around 20% of salary, with the state contributing a further 10%.

The retirement age is being increased gradually from 65 to 67 in a process that started in Jan. 2012. By 2029 Germans will only be eligible for a state pension at the age of 67.

RELIGION

In 2009 there were 24,909,000 Roman Catholics in 12,000 parishes, 24,195,000 Protestants in 15,281 parishes; and in 2010, 104,024 Jews with 56 rabbis and 96 synagogues. The Federal Ministry of the Interior estimated in 2007 that there were between 3·1m. and 3·4m. Muslims resident in Germany, a number exceeded in the EU only in France.

There are seven Roman Catholic archbishoprics (Bamberg, Berlin, Cologne, Freiburg, Hamburg, Munich and Freising, Paderborn) and 20 bishoprics. Chairman of the German Bishops' Conference is Robert Zollitsch, Archbishop of Freiburg im Breisgau. A concordat between Germany and the Holy See dates from 10 Sept. 1933. In April 2005 Cardinal Joseph Ratzinger, former archbishop of Munich and Freising, was elected Pope as Benedict XVI. In Feb. 2013 he became the first Pope to resign in 600 years, citing age and declining health as the reasons for his decision. There were nine cardinals in Feb. 2013.

The Evangelical (Protestant) Church (EKD) consists of 22 member-churches comprising nine Lutheran Churches, 11 United-Lutheran-Reformed Churches and two Reformed Churches. Its organs are the Synod, the Church Conference and the Council under the chairmanship of Nikolaus Schneider. The Free Evangelical Church (BFeG) has some 420 communities.

CULTURE

World Heritage Sites

Germany has 37 sites on the UNESCO World Heritage List (date of inscription on the list in brackets): Aachen Cathedral (1978), begun in the 8th century under Charlemagne; Speyer Cathedral (1981), founded in 1030 and constructed in the Romanesque style;

Würzburg Residence, with the Court Gardens and Residence Square (1981), an 18th century Baroque palace; Pilgrimage Church of Wies (1983), an 18th century Baroque-Rococo church; Castles of Augustusburg and Falkenlust at Brühl (1984), early examples of 18th century Rococo architecture; St Mary's Cathedral and St Michael's Church at Hildesheim (1985 and 2008), Romanesque constructions from the 11th century; Roman Monuments in Trier (1986), a Roman colony from the 1st century, and the Cathedral of St Peter and Church of Our Lady; Hanseatic City of Lübeck (1987), founded in the 12th century; Palaces and Parks of Potsdam and Berlin (1990, 1992 and 1999), an eclectic mix of 150 buildings covering 500 hectares built between 1730 and 1916; Abbey and Altenmünster of Lorsch (1991), an example of Carolignian architecture; Mines of Rammelsberg and Historic Town of Goslar (1992 and 2008), with a well-preserved historic centre; Town of Bamberg (1993), the country's biggest intact historical city core; Maulbronn Monastery Complex (1993), a former Cistercian abbey over 850 years old; Collegiate Church, Castle, and Old Town of Quedlinburg (1994), capital of the East Franconian German Empire; Völklingen Ironworks (1994), a preserved 19th/20th centuries ironworks; Messel Pit Fossil site (1995), containing important fossils from 57m.–36m. BC; Cologne Cathedral (1996 and 2008), a Gothic masterpiece begun in 1248; Bauhaus and its sites in Weimar and Dessau (1996), buildings of the influential early-20th century architectural movement; Luther Memorials in Eisleben and Wittenberg (1996), including his birthplace, baptism church and religious sites; Classical Weimar (1998), a cultural epicentre during the 18th and early 19th centuries; Museumsinsel (Museum Island), Berlin (1999), including Altes Museum, Bodemuseum, Neues Museum and Pergamonmuseum; Wartburg Castle (1999), dating from the feudal period and rebuilt in the 19th century— Luther translated the New Testament here; Garden Kingdom of Dessau-Wörlitz (2000), an 18th century landscaped garden in the Enlightenment style; Monastic Island of Reichenau (2000), on Lake Constance, incorporating medieval churches and the remains of an 8th century Benedictine monastery; Zollverein Coal Mine Industrial Complex in Essen (2001), a 20th century mining complex with modernist buildings; Upper Middle Rhine Valley (2002), a 65 km-stretch of one of Europe's most important historical transport conduits; Historic Centres of Stralsund and Wismar (2002), Hanseatic towns; Town Hall and Roland on the Marketplace of Bremen (2004); Old Town of Regensburg with Stadtamhof (2006); Primeval Beech Forests of the Carpathians and the Ancient Beech Forests of Germany (2007); the Berlin Modernism Housing Estates (2008); the Fagus Factory in Alfeld (2011), a landmark in the development of modern architecture and industrial design; and the Margravial Opera House Bayreuth (2012), a stunning illustration of Baroque theatre architecture.

Germany and Poland are jointly responsible for Muskauer Park/Park Mużakowski (2004), a landscaped park astride the Neisse river. Germany and the United Kingdom share the Frontiers of the Roman Empire sites (1987, 2005 and 2008), which contain the border line of the Roman Empire at its greatest extent in the 2nd century AD. The Wadden Sea, the largest unbroken system of intertidal sand and mud flats in the world (2009), is located in Germany and the Netherlands. The Prehistorical Pile dwellings around the Alps (2011) are shared with Austria, France, Italy, Slovenia and Switzerland.

Press

The daily press is mainly regional. The dailies with the highest circulation are (average figures for July–Sept. 2008): the tabloid *Bild* (3·36m. copies per day); *Süddeutsche Zeitung* (Munich, 0·44m.); *Frankfurter Allgemeine Zeitung* (0·37m.); and *Die Welt* (which in 2007 made the first profit in its 60-year history, 0·27m.). Other important opinion leaders are the weeklies *Die Zeit*, *Die Woche* and *Rheinischer Merkur*. In the period April–June 2009 the total circulation figures for some 350 German daily newspapers

came to 25·3m. The entire range of popular magazines includes some 2,300 publications and has a total circulation of more than 120m. The total circulation of daily newspapers in Germany is the highest in Europe. 78% of the population over the age of 14 regularly read a daily newspaper. There were also 267 online daily newspapers in 2008. Among magazines the most widely read are *Der Spiegel* (1·04m. weekly) and *Stern* (1·02m. weekly). In 2010 a total of 95,838 book titles were published, up from 93,124 in 2009 although down on the record 96,479 in 2007.

Tourism

In 2010 there were 55,315 places of accommodation with 3,516,544 beds (including 13,487 hotels with 1,053,614 beds). 26,875,288 foreign visitors and 113,139,484 tourists resident in Germany spent a total of 380,334,025 nights in holiday accommodation. The most visited city is Berlin with 9,051,430 overnight visitors in 2010; Bavaria is the most visited *Bundesland* with 28,288,883 (5,572,955 visited Munich). In 2010 the Netherlands was the country of origin of the largest number of overnight visitors (3,917,640), ahead of the USA (2,206,339) and Switzerland (2,028,423). In 2010 tourism brought in €26·2bn. Expenditure by German travellers in foreign countries for 2010 was €58·9bn., the most of any nationality.

Festivals

The Munich Opera Festival takes place annually in June–July, and the Wagner Festspiele (the Wagner Festival) in Bayreuth is held from late July to the end of Aug. The Oberammergau Passion Play, which takes place every ten years, was last held in 2010. Karneval (Fasching in some areas), in Jan./Feb./March, is a major event in the annual calendar in cities such as Cologne, Munich, Düsseldorf and Mainz. The annual Berlin Film Festival (Berlinale) takes place over a two-week period in Feb. Oktoberfest, Munich's famous beer festival which first began in 1810, takes place each year in late Sept. and early Oct. and regularly attracts 7m. visitors.

DIPLOMATIC REPRESENTATIVES

Of Germany in the United Kingdom (23 Belgrave Sq./Chesham Pl., London, SW1X 8PZ)
Ambassador: Georg Boomgaarden.

Of the United Kingdom in Germany (Wilhelmstrasse 70, 10117 Berlin)
Ambassador: Simon McDonald.

Of Germany in the USA (4645 Reservoir Rd, NW, Washington, D.C., 20007)
Ambassador: Peter Ammon.

Of the USA in Germany (Pariser Platz 2, 10117 Berlin)
Ambassador: Philip D. Murphy.

Of Germany to the United Nations
Ambassador: Peter Wittig.

Of Germany to the European Union
Permanent Representative: Peter Tempel.

FURTHER READING

Statistisches Bundesamt. *Statistisches Jahrbuch für die Bundesrepublik Deutschland; Wirtschaft und Statistik* (monthly, from 1949).

Ardagh, J., *Germany and the Germans.* 3rd ed. 1995
Balfour, M., *Germany: the Tides of Power.* 1992
Bark, D. L. and Gress, D. R., *A History of West Germany, 1945–1991.* 2nd ed. 1993
Betz, H. G., *Postmodern Politics in Germany.* 1991
Blackbourn, D., *Fontana History of Germany, 1780–1918: The Long Nineteenth Century.* 1997
Blackbourn, D. and Eley, G., *The Peculiarities of German History.* 1985
Carr, W. and Allinson, M., *A History of Germany, 1815–2002.* 5th ed. 2010
Childs, D., *Germany in the 20th Century.* 1991.—*The Stasi: The East German Intelligence and Security Service.* 1999
Dennis, M., *The German Democratic Republic: Politics, Economics and Society.* 1987

Fulbrook, Mary, *A Concise History of Germany*. 1991.—*The Divided Nation: A History of Germany, 1918–1990*. 1992.—*The Fontana History of Germany: 1918–1990 The Divided Nation*. 1994.—*German National Identity After the Holocaust*. 1999.—*Interpretation of the Two Germanies, 1945–1997*. 1999

Fulbrook, Mary, (ed.) *Twentieth-Century Germany: Politics, Culture and Society, 1918–1990*. 2001

Glees, A., *Reinventing Germany: German Political Development since 1945*. 1996

Green, Simon, Hough, Dan, Miskimmon, Alister and Timmins, Graham, *The Politics of the New Germany*. Revised ed. 2007

Heneghan, Tom, *Unchained Eagle: Germany After the Wall*. 2000

Huelshoff, M. G., *et al.*, (eds.) *From Bundesrepublik to Deutschland: German Politics after Reunification*. 1993

Kielinger, T., *Crossroads and Roundabouts, Junctions in German-British Relations*. 1997

Kitchen, Martin, *A History of Modern Germany: 1800 to the Present*. 2011

Langewiesche, Dieter, *Liberalism in Germany*. 1999

Lees, Charles, *Party Politics in Germany*. 2005

Loth, W., *Stalin's Unwanted Child—The Soviet Union, the German Question and the Founding of the GDR*. 1998

Maier, C. S., *Dissolution: The Crisis of Communism and the End of East Germany*. 1997

Maull, Hanns W., *German Foreign Policy Since Reunification*. 2005

Merkl, Peter H. (ed.) *The Federal Republic of Germany at Fifty: The End of a Century of Turmoil*. 1999

Miskimmon, Alister, *Germany and the Common Foreign and Security Policy of the European Union: Between Europeanization and National Adaptation*. 2007

Miskimmon, Alister, Paterson, William E. and Sloam, James, (eds.) *Germany's Gathering Crisis: The 2005 Federal Election and the Grand Coalition*. 2008

Müller, Jan-Werner, *Another Country: German Intellectuals, Unification and National Identity*. 2000

Nicholls, A. J., *The Bonn Republic: West German Democracy, 1945–1990*. 1998

Olsen, Jonathan, *Nature and Nationalism: Right-wing Ecology and the Politics of Identity in Contemporary Germany*. 2000

Orlow, D., *A History of Modern Germany, 1871 to the Present*. 6th ed. 2009

Padgett, Stephen, Paterson, William E. and Smith, Gordon, (eds.) *Developments in German Politics 3*. 2003

Pulzer, P., *German Politics, 1945–1995*. 1995

Schulze, Hagen, *Germany: A New History*. 2001

Schwartz, H-P., translator, Willmot, L., *Konrad Adenauer Vol 1: From the German Empire to the Federal Republic, 1876–1952*. 1995.—*Konrad Adenauer Vol 2: The Statesman: 1952–1967*. 1997

Schweitzer, C.-C., Karsten, D., Spencer, R., Cole, R. T., Kommers, D. P. and Nicholls, A. J. (eds.) *Politics and Government in Germany, 1944–1994: Basic Documents*. 2nd ed. 1995

Sereny, Gitta, *The German Trauma: Experiences and Reflections, 1938–99*. 2000

Sinn, G. and Sinn, H.-W., *Jumpstart: the Economic Reunification of Germany*. 1993

Smyser, W. R., *The Economy of United Germany: Colossus at the Crossroads*. 1992.—*From Yalta to Berlin: The Cold War Struggle over Germany*. 1999

Speirs, Ronald and Breuilly, John, (eds.) *Germany's Two Unifications: Anticipations, Experiences, Responses*. 2005

Taylor, R., *Berlin and its Culture*. 1997

Thompson, W. C., *et al.*, *Historical Dictionary of Germany*. 1995

Turner, H. A., *Germany from Partition to Reunification*. 2nd ed. [of *Two Germanies since 1945*]. 1993

Tusa, A., *The Last Division – A History of Berlin, 1945–1989*. 1997

Watson, A., *The Germans: Who Are They Now?* 2nd ed. 1995

Watson, Peter, *The German Genius: Europe's Third Renaissance, the Second Scientific Revolution and the Twentieth Century*. 2010

Wende, Peter, *History of Germany*. 2004

Williams, C., *Adenauer: The Father of the New Germany*. 2000

Other more specialized titles are listed under CONSTITUTION AND GOVERNMENT *and* BANKING AND FINANCE, *above*.

National library: Deutsche Nationalbibliothek, Deutscher Platz 1, 04103 Leipzig; Adickesallee 1, 60322 Frankfurt am Main; Deutsches Musikarchiv, Gärtnerstrasse 25–32, 12207 Berlin. *Director General*: Elisabeth Niggemann.

National Statistical Office: Statistisches Bundesamt, 65189 Wiesbaden, Gustav Stresemann Ring 11. *President*: Roderich Egeler.

Website: http://www.destatis.de

THE BUNDESLÄNDER

Baden-Württemberg

KEY HISTORICAL EVENTS

The *Bundesland* is a combination of former states. Baden (the western part of the present *Bundesland*) became a united margravate in 1771, after being divided as Baden-Baden and Baden-Durlach since 1535; Baden-Baden was predominantly Catholic, and Baden-Durlach predominantly Protestant. The margrave became an ally of Napoleon, ceding land west of the Rhine and receiving northern and southern territory as compensation. In 1805 Baden became a grand duchy and in 1806 a member state of the Confederation of the Rhine, extending from the Main to Lake Constance. In 1815 it was a founder-state of the German Confederation. A constitution was granted by the grand duke in 1818, but later rulers were less liberal and there was revolution in 1848, put down with Prussian help. The Grand Duchy was abolished and replaced by a *Bundesland* in 1919.

In 1949 Baden was combined with Württemberg to form three states; the three joined as one in 1952.

Württemberg, having been a duchy since 1495, became a kingdom in 1805 and joined the Confederations as did Baden. A constitution was granted in 1819 and the state remained liberal. In

1866 the king allied himself with Austria against Prussia, but in 1870 joined Prussia in war against France. The monarchy came to an end with the abdication of William II in 1918, and Württemberg became a state of the German Republic. In 1945 the state was divided between Allied occupation authorities but the divisions ended in 1952.

TERRITORY AND POPULATION

Baden-Württemberg comprises 35,751 sq. km, with a population (at 30 June 2012) of 10,813,603 (5,473,789 females, 5,339,814 males).

The *Bundesland* is divided into four administrative regions, nine urban and 35 rural districts, and numbers 1,101 communes.

The capital is Stuttgart, with a population (30 June 2012) of 616,137.

SOCIAL STATISTICS

Statistics for calendar years:

	Live births	Marriages	Divorces	Deaths
2008	91,909	48,612	22,792	96,431
2009	89,678	48,378	22,100	97,556
2010	90,965	48,927	21,958	98,807
2011	88,823	48,991	23,113	97,732

CONSTITUTION AND GOVERNMENT

Baden-Württemberg is a merger of Baden, Württemberg-Baden and Württemberg-Hohenzollern, which were formed after 1945. The merger was approved by a plebiscite held on 9 Dec. 1951, when 70% of the population voted in its favour. It has six votes in the Bundesrat.

RECENT ELECTIONS

At the elections to the 138-member Diet of 27 March 2011, turnout was 66·3%. The Christian Democrats won 60 seats with 39·0% of the vote, the Greens 36 with 24·2%, the Social Democrats 35 with 23·1% and the Free Democrats 7 with 5·3%. The Left received 2·8% but won no seats.

CURRENT GOVERNMENT

Winfried Kretschmann (Greens) is *Prime Minister.*

Government Website: http://www.baden-wuerttemberg.de

ECONOMY

Performance

GDP in 2011 was €376,285m., which amounted to 14·6% of Germany's total GDP. Industries *(Produzierendes Gewerbe)* provided around 36% of GDP in 2010 (44·6% in 1991). Real GDP growth in 2011 was 4·7%. Service enterprises accounted for 63% of GDP in 2010.

Banking and Finance

There is a stock exchange in Stuttgart. Turnover of shares and bonds in 2011 was €108·6bn.

ENERGY AND NATURAL RESOURCES

Agriculture

Area and yield of the most important crops:

	Area (in 1,000 ha.)			Yield (in 1,000 tonnes)		
	2009	2010	2011	2009	2010	2011
Wheat	238·3	238·5	236·1	1,744·0	1,638·5	1,640·6
Sugar beet	17·9	15·7	17·8	1,303·7	1,080·0	1,429·7
Barley	179·3	158·5	157·8	1,101·2	941·3	900·1
Potatoes	5·9	5·4	5·7	218·6	185·3	247·5
Oats	29·1	25·3	23·2	171·8	125·4	110·8
Rye	10·8	10·6	9·7	64·7	59·2	46·3

Livestock in Nov. 2012 (in thousands): cattle, 995·8 (including dairy cows, 340·4); pigs, 1,952·1; sheep 221·7; poultry (2010), 4,566·8.

Forestry

Total area covered by forests is 13,693 sq. km or 38·3% of the total area.

INDUSTRY

Baden-Württemberg is one of Germany's most industrialized states. In 2011, 8,102 establishments (with 20 or more employees) employed 1,192,238 persons; of these, 289,405 were employed in machine construction; 210,815 in car manufacture; 159,454 in the manufacture of computer, electronic and optical products and of electrical equipment; 20,869 in the textile and clothing industry.

Labour

Economically active persons totalled 5,506,800 at the 1%-EU-sample survey of 2011: 4·90m. were employees and 605,100 were self-employed (including family workers); 1,906,300 were engaged in power supply, mining, manufacturing and building; 1,260,600 in commerce and transport; 68,500 in agriculture and forestry; 2,271,300 in other industries and services. There were 205,500 unemployed in 2011, a rate of 4·0%.

INTERNATIONAL TRADE

Imports and Exports

Total imports (2011): €143,409m. Total exports: €171,910m., of which €88,294m. went to the EU. Automotive exports totalled €39,407m. and machinery exports €38,061m.

COMMUNICATIONS

Roads

On 1 Jan. 2012 there were 27,449 km of 'classified' roads, comprising 1,059 km of Autobahn, 4,382 km of federal roads, 9,929 km of first-class and 12,079 km of second-class highways. Motor vehicles, at 1 Jan. 2012, numbered 7,173,076, including 5,897,054 passenger cars, 8,561 buses, 298,037 lorries, 351,879 tractors and 587,642 motorcycles.

Civil Aviation

The largest airport in Baden-Württemberg is at Stuttgart, which in 2011 handled 9,534,000 passengers and 31,521 tonnes of freight. There are two further airports, Karlsruhe/Baden-Baden and Friedrichshafen.

Shipping

The harbour in Mannheim is the largest in Baden-Württemberg. In 2011 it handled 6·6m. tonnes of freight, compared to 6·0m. tonnes in Karlsruhe.

SOCIAL INSTITUTIONS

Justice

There are a constitutional court *(Staatsgerichtshof)*, two courts of appeal, 17 regional courts, 108 local courts, a *Bundesland* labour court, nine labour courts, a *Bundesland* social court, eight social courts, a finance court, a higher administrative court *(Verwaltungsgerichtshof)* and four administrative courts.

Education

In 2011–12 there were 2,680 primary schools *(Grund- und Werkreal-/Hauptschulen)* with 34,167 teachers and 524,056 pupils; 582 special schools with 11,832 teachers and 52,822 pupils; 494 intermediate schools with 14,257 teachers and 245,006 pupils; 451 high schools with 23,907 teachers and 344,002 pupils; 57 *Freie Waldorf* schools with 1,669 teachers and 23,635 pupils. Other general schools had 672 teachers and 10,478 pupils in total; there were also 770 vocational schools with 425,935 pupils. There were 44 universities of applied sciences *(Hochschulen für angewandte Wissenschaft)* with 93,779 students in winter term 2011–12.

In the winter term 2011–12 there were nine universities (Freiburg, 22,205 students; Heidelberg, 26,958; Hohenheim, 8,808; Karlsruhe, 22,062; Konstanz, 10,176; Mannheim, 10,636; Stuttgart, 21,608; Tübingen, 24,047; Ulm 8,628); six teacher training colleges with 22,500 students; five colleges of music and three colleges of fine arts with a total of 4,424 students.

Health

In 2011 the 285 hospitals in Baden-Württemberg had 56,910 beds and treated 2,059,083 patients. The average occupancy rate was 77·1%.

RELIGION

In 2008, 36·9% of the population were Roman Catholics and 33·0% were Protestants.

CULTURE

Tourism

In 2011, 17,853,382 visitors spent a total of 45,616,399 nights in Baden-Württemberg. Only Bavaria of the German *Bundesländer* recorded more overnight stays.

FURTHER READING

Statistical Information: Statistisches Landesamt Baden-Württemberg (70158 Stuttgart) (*President:* Dr Carmina Brenner), publishes: *Statistisches Monatsheft* (monthly); *Statistisches Taschenbuch* (latest issue 2012).
Website (German only): http://www.statistik.rlp.de
State libraries: Württembergische Landesbibliothek, Konrad-Adenauer-Str. 8, 70173 Stuttgart. Badische Landesbibliothek Karlsruhe, Erbprinzenstr. 15, 76133 Karlsruhe.

Bavaria
Bayern

KEY HISTORICAL EVENTS

Bavaria was ruled by the Wittelsbach family from 1180. The duchy remained Catholic after the Reformation, which made it a natural ally of Austria and the Habsburg Emperors.

The present boundaries were set during the Napoleonic wars, and Bavaria became a kingdom in 1806. Despite the granting of a constitution and parliament, radical feeling forced the abdication of King Ludwig I in 1848. Maximilian II was followed by Ludwig II who allied himself with Austria against Prussia in 1866, but was reconciled with Prussia and entered the German Empire in 1871. In 1918 the King Ludwig III abdicated. The first years of republican government were filled with unrest, attempts at the overthrow of the state by both communist and right-wing groups culminating in an unsuccessful coup by Adolf Hitler in 1923.

The state of Bavaria included the Palatinate from 1214 until 1945, when it was taken from Bavaria and added to the Rhineland. The present *Bundesland* of Bavaria was formed in 1946. Munich became capital of Bavaria in the reign of Albert IV (1467–1508) and remains capital of the *Bundesland*.

TERRITORY AND POPULATION

Bavaria has an area of 70,550 sq. km. There are seven administrative regions, 25 urban districts, 71 rural districts, 193 unincorporated areas and 2,056 communes, 987 of which are members of 313 administrative associations (as of 31 Dec. 2011). The population (31 Dec. 2011) numbered 12,595,891 (6,199,656 males, 6,396,235 females).

The capital is Munich, with a population (31 Dec. 2011) of 1,378,176.

SOCIAL STATISTICS

Statistics for calendar years:

	Live births	Marriages	Divorces	Deaths
2008	106,298	58,300	27,566	121,109
2009	103,710	58,812	25,427	122,494
2010	105,251	59,092	26,807	123,089
2011	103,668	59,274	27,004	122,955

CONSTITUTION AND GOVERNMENT

The Constituent Assembly, elected on 30 June 1946, passed a constitution on the lines of the democratic constitution of 1919, but with greater emphasis on state rights; this was agreed upon by the Christian Social Union (CSU) and the Social Democrats (SPD). Bavaria has six seats in the Bundesrat. The CSU replaces the Christian Democratic Party in Bavaria.

RECENT ELECTIONS

At the Diet elections on 28 Sept. 2008 the CSU won 92 seats with 43·4% of votes cast (down from 60·7% in 2003), the SPD 39 with 18·6%, the Free Voters 21 with 10·2%, Alliance '90/the Greens 19 with 9·4% and the Free Democratic Party 16 with 8·0%. The Left took 4·3% but won no seats. As a result the CSU lost its absolute majority for the first time since 1962. Turnout was 57·9%.

CURRENT GOVERNMENT

The *Prime Minister* is Horst Seehofer (CSU).

Government Website: http://www.bayern.de

ECONOMY

Performance

Real GDP growth in 2011 was 2·7%, down from 4·2% in 2010.

ENERGY AND NATURAL RESOURCES

Agriculture

Area and yield of the most important products:

	Area (in 1,000 ha.)			Yield (in 1,000 tonnes)		
	2009	2010	2011	2009	2010	2011
Sugar beet	66·8	59·4	65·8	5,145·2	4,253·1	5,563·6
Wheat	543·2	526·7	525·3	3,743·8	3,454·2	3,720·0
Potatoes	45·6	43·4	43·7	1,932·6	1,647·9	2,084·1
Barley	412·7	368·9	365·1	2,420·5	2,006·3	1,962·1
Rye[1]	45·4	40·4	39·3	268·8	176·4	163·6
Oats	34·6	32·3	31·2	164·8	126·3	153·1

[1]From 2010 rye and maslin.

Livestock, Nov. 2012: 3,251,606 cattle (including 1,219,350 dairy cows); 88,300 horses (2010); 3,499,575 pigs; 286,519 sheep; 11,481,300 poultry (2010).

INDUSTRY

On 30 Sept. 2011, 7,048 establishments (with 20 or more employees) employed 1,175,942 persons; of these, 204,046 were employed in the manufacture of machinery and equipment, 172,889 in the manufacture of motor vehicles and 23,422 in the manufacture of textiles and textile products.

Labour

Economically active persons totalled 6,516,000 at the 1% sample survey of the microcensus of 2011. Of the total, 5,656,000 were employees, 781,000 were self-employed, 79,000 were unpaid family workers; 2,030,000 worked in power supply, mining, manufacturing and building; 1,609,000 in commerce, hotels and restaurants, and transport; 155,000 in agriculture and forestry; 2,721,000 in other services.

COMMUNICATIONS

Roads

There were, on 1 Jan. 2012, 41,883 km of 'classified' roads, comprising 2,509 km of Autobahn, 6,535 km of federal roads, 14,026 km of first-class and 18,813 km of second-class highways. Number of motor vehicles on 1 Jan. 2012 was 8,959,539, including 7,110,701 passenger cars, 379,303 lorries, 13,221 buses and 792,637 motorcycles.

Civil Aviation

Munich airport handled 37,593,829 passengers (27,855,522 on international flights) and 303,667 tonnes of freight in 2011. Nuremberg handled 3,933,626 (2,433,432 on international flights) and 7,913 tonnes of freight in 2011. Memmingen handled 755,458 passengers (731,178 on international flights) in 2011.

SOCIAL INSTITUTIONS

Justice

There are a constitutional court (*Verfassungsgerichtshof*), three courts of appeal, 22 regional courts, 73 local courts, two *Bundesland* labour courts, 11 labour courts, a *Bundesland* social court, seven social courts, two finance courts, a higher administrative court

(*Verwaltungsgerichtshof*) and six administrative courts. The supreme *Bundesland* court (*Oberstes Landesgericht*) was abolished in June 2006. Since 1 Jan. 2005 new cases have been transferred to the courts of appeal.

Education

In 2011–12 there were 3,352 primary schools with 43,596 teachers and 645,455 pupils; 352 special schools with 8,273 teachers and 55,175 pupils; 364 intermediate schools with 14,370 teachers and 242,682 pupils; 415 high schools with 25,250 teachers and 355,552 pupils; 227 part-time vocational schools with 8,020 teachers and 277,932 pupils, including 48 special part-time vocational schools with 1,110 teachers and 14,104 pupils; 868 full-time vocational schools with 5,715 teachers and 75,648 pupils including 461 schools for public health occupations with 2,047 teachers and 28,887 pupils; 292 advanced full-time vocational schools with 1,906 teachers and 23,937 pupils; 162 vocational high schools (*Berufsoberschulen, Fachoberschulen*) with 3,523 teachers and 55,909 pupils.

In 2011–12 there were 12 universities with 211,151 students (Augsburg, 17,054; Bamberg, 11,753; Bayreuth, 10,971; Eichstätt, 4,711; Erlangen-Nuremberg, 32,354; Munich, 46,432; Passau, 10,012; Regensburg, 19,547; Würzburg, 23,482; the Technical University of Munich, 30,821; University of the Federal Armed Forces, Munich (*Universität der Bundeswehr*), 3,438; the college of politics, Munich, 576), and three philosophical-theological colleges with 558 students in total (Benediktbeuern, 71; Neuendettelsau, 162; Munich, 325). There were also five colleges of music, two colleges of fine arts and one college of television and film, with 3,457 students in total; 27 vocational colleges (*Fachhochschulen*) with 105,152 students including one for the civil service (*Bayerische Beamtenfachhochschule*) with 3,542 students.

Welfare

In Dec. 2011 there were 42,133 persons receiving benefits of all kinds.

RELIGION

In 2008, 55·6% of the population were Roman Catholics and 20·8% were Protestants.

CULTURE

Tourism

In June 2011 there were 13,337 places of accommodation (with nine beds or more) providing beds for 559,012 people. In 2011 they received 29,837,822 guests of whom 6,732,842 were foreigners. They stayed an average of 2·7 nights each, totalling 80,956,617 nights (14,084,133 nights stayed by foreign visitors).

Festivals

Oktoberfest, Munich's famous beer festival, takes place each year from the penultimate Saturday in Sept. through to the first Sunday in Oct. (extended to 3 Oct. if the last Sunday of the festival falls on 1 or 2 Oct.). There were 6·4m. visitors at the 179th Oktoberfest in 2012.

FURTHER READING

Statistical Information: Bayerisches Landesamt für Statistik und Datenverarbeitung, St.-Martin-Str. 47, 81541 Munich. *President:* Karlheinz Anding. It publishes: *Statistisches Jahrbuch für Bayern.* 1894 ff.—*Bayern in Zahlen.* Monthly (from Jan. 1947).—*Zeitschrift des Bayerischen Statistischen Landesamts.* July 1869–1943; 1948 ff.—*Beiträge zur Statistik Bayerns.* 1850 ff.—*Statistische Berichte.* 1951 ff.— *Kreisdaten.* 1972–2001 (from 2003 incorporated in *Statistisches Jahrbuch für Bayern*).—*Gemeindedaten.* 1973 ff.
State library: Bayerische Staatsbibliothek, Ludwigstr. 16, 80539 Munich. *Director General:* Dr Rolf Griebel.

Berlin

KEY HISTORICAL EVENTS

After the end of World War II, Berlin was divided into four occupied sectors, each with a military governor from one of the victorious Allied Powers (the USA, the Soviet Union, Britain and France). In March 1948 the USSR withdrew from the Allied Control Council and in June blockaded West Berlin until May 1949. In response, the allies flew food and other supplies into the city in what became known as the Berlin Airlift. On 30 Nov. 1948 a separate municipal government was set up in the Soviet sector which led to the political division of the city. In contravention of the special Allied status agreed for the city, East Berlin became 'Capital of the GDR' in 1949 and thus increasingly integrated into the GDR as a whole. In West Berlin, the formal authority of the western allies lasted until 1990.

On 17 June 1953 the protest by workers in East Berlin against political oppression and economic hardship was suppressed by Soviet military forces. To stop refugees, the east German government erected the Berlin Wall to seal off West Berlin's borders on 13 Aug. 1961.

The Berlin Wall was breached on 9 Nov. 1989 as the regime in the GDR bowed to the internal pressure which had been building for months. East and West Berlin were amalgamated on the reunification of Germany in Oct. 1990.

With the move of the national government, the parliament (*Bundestag*), and the federal organ of the *Bundesländer* (*Bundesrat*) in 1999, Berlin once again became a capital city.

TERRITORY AND POPULATION

The area is 891·8 sq. km. Population, 31 Dec. 2011, 3,501,872 (1,784,227 females), including 478,212 foreign nationals; density, 3,927 per sq. km.

SOCIAL STATISTICS

Statistics for calendar years:

	Live births	Marriages	Divorces	Deaths
2008	31,936	11,762	7,716	31,911
2009	32,104	12,557	7,397	31,713
2010	33,393	12,394	8,384	32,234
2011	33,075	12,544	7,930	31,380

CONSTITUTION AND GOVERNMENT

According to the constitutions of Sept. 1950 and Oct. 1995, Berlin is simultaneously a *Bundesland* of the Federal Republic and a city. It is governed by a *House of Representatives* (of at least 130 members); executive power is vested in a *Senate*, consisting of the Governing Mayor, two Mayors and not more than eight senators. Since 1992 adherence to the constitution has been watched over by a Constitutional Court.

Although a proposed merger between Berlin and Brandenburg was rejected in a 1996 referendum, a number of joint institutions have been established on the basis of 20 state treaties.

Berlin has four seats in the Bundesrat.

RECENT ELECTIONS

At the elections of 18 Sept. 2011 turnout was 60·2%. The Social Democratic Party (SPD) won 47 seats with 28·3% of votes cast; the Christian Democratic Union (CDU) 39, with 23·4%; Alliance '90/the Greens 29, with 17·6%; the Left 19, with 11·7%; the Pirate Party 15, with 8·9%.

CURRENT GOVERNMENT

The *Governing Mayor* is Klaus Wowereit (SPD).

Government Website: http://www.berlin.de

ECONOMY

GDP in 2011 was €101,386m. Berlin's real GDP growth in 2011 was 2·3%.

INDUSTRY

In Sept. 2011 there were 737 industrial concerns (20 or more employees) employing 93,118 people. The main industries were: food and animal feed, manufacture of pharmaceutical products, manufacture of electrical and optical equipment, and machine construction.

Labour

Economically active persons totalled 1,618,200 at the 1%-sample survey of the microcensus of 2011. There were on average 218,500 persons registered unemployed in 2011. The unemployment rate in 2011 was 11·9%.

INTERNATIONAL TRADE

Imports and Exports

Total imports (2011): €10,101m.; exports: €12,739m.

COMMUNICATIONS

Roads

On 1 Jan. 2012 there were 5,420·7 km of roads (245·7 km of 'classified' roads, made up of 76·7 km of Autobahn and 169·0 km of federal roads). In Jan. 2012, 1,327,015 motor vehicles were registered, including 1,135,704 passenger cars, 78,367 lorries, 2,133 buses and 97,103 motorcycles. There were 130,010 road accidents in 2011, with 16,933 injured persons.

Civil Aviation

227,186 flights were made from Berlin's two airports—Tegel and Schönefeld—in 2011, carrying a total of 23,991,266 passengers.

SOCIAL INSTITUTIONS

Justice

There are a court of appeal (*Kammergericht*), a regional court, nine local courts, a *Bundesland* Labour court, a labour court, a *Bundesland* social court, a social court, a higher administrative court, an administrative court and a finance court.

Education

In Sept. 2011 there were 321,590 pupils attending schools. There were 424 primary schools with 146,250 pupils, 43 schools for practical education with 5,434 pupils, 89 special schools with 10,883 pupils, 63 secondary modern schools with 10,536 pupils, 116 grammar schools with 80,774 pupils, 60 comprehensive schools with 26,226 pupils and ten *Freie Waldorf* schools with 3,785 pupils. In 2011–12 there were 12 universities, five arts colleges and 23 technical colleges. There were a total of 153,694 (75,195 female) students in higher education.

Health

In 2011 there were 79 hospitals with 19,782 beds, 7,927 doctors and 35,118 medical personnel.

RELIGION

In Dec. 2010 membership and number of places of worship for major religions was as follows:

Religion	Members	Places of Worship
Protestant	660,006[1]	460[2]
Roman Catholic	318,248	108[3]
Jewish	11,441	8[3]
Muslim	249,220	128[3]

[1]2009. [2]2007. [3]2008.

CULTURE

Tourism

In 2011 Berlin had 782 places of accommodation (with nine or more beds) providing 121,056 beds. 9,866,088 visitors (including 3,599,573 foreigners) spent 22,359,470 nights in Berlin (including at campsites).

FURTHER READING

Statistical Information: The Amt für Statistik Berlin-Brandenburg (Behlertstrasse 3a, 14467 Potsdam) was created in Jan. 2007 through the merger of the Statistisches Landesamt Berlin and the Landesbetrieb für Datenverarbeitung und Statistik Land Brandenburg. *President:* Prof. Dr Ulrike Rockmann. It is the main source for the statistics above on Berlin and publishes: *Statistisches Jahrbuch* (from 1867): *Zeitschrift für amtliche Statistik Berlin-Brandenburg* (six a year from 2007).—*100 Jahre Berliner Statistik* (1962).
Website (limited English): http://www.statistik-berlin-brandenburg.de

Kempe, Frederick, *Berlin 1961: Kennedy, Krushchev, and the Most Dangerous Place on Earth.* 2011
Large, David Clay, *Berlin.* 2000
Read, A., and Fisher, D., *Berlin, Biography of a City.* 1994
Richie, Alexandra, *Faust's Metropolis: A History of Berlin.* 1999
Taylor, R., *Berlin and its Culture.* 1997
Till, Karen, E., *The New Berlin: Memory, Politics, Place.* 2005

State library: Zentral- und Landesbibliothek, Blücherplatz 1, 10961 Berlin. *Director General:* Claudia Lux.

Brandenburg

KEY HISTORICAL EVENTS

Brandenburg surrounds the capital city of Germany, Berlin, but the people of the state voted against the recommendations of the Berlin House of Representatives and the Brandenburg State Parliament that the two states should merge. The state capital, Potsdam, is the ancient city of the Emperor Frederic II 'The Great' who transformed the garrison town of his father Frederic I 'The Soldier' into an elegant city.

TERRITORY AND POPULATION

The area is 29,484 sq. km. Population on 31 Dec. 2011 was 2,495,635 (1,258,888 females), including 69,346 foreigners. There are four urban districts, 14 rural districts and 419 communes (31 Dec. 2011).

The capital is Potsdam, with a population (31 Dec. 2011) of 158,902.

SOCIAL STATISTICS

Statistics for calendar years:

	Live births	Marriages	Divorces	Deaths
2008	18,808	11,757	5,060	26,807
2009	18,537	12,066	5,323	27,309
2010	18,954	12,585	5,190	27,894
2011	18,279	12,115	5,344	27,851

CONSTITUTION AND GOVERNMENT

The *Bundesland* was reconstituted on former GDR territory on 14 Oct. 1990. Brandenburg has four seats in the Bundesrat and following the 2009 election 19 in the Bundestag.

At a referendum on 14 June 1992, 93·5% of votes cast were in favour of a new constitution guaranteeing direct democracy and the right to work and housing.

Although a proposed merger between Brandenburg and Berlin was rejected in a 1996 referendum, a number of joint institutions have been established on the basis of 20 state treaties.

RECENT ELECTIONS

At the Diet elections on 27 Sept. 2009 the Social Democrats (SPD) won 31 seats with 33·0% of the vote; the Left 26, with 27·2%; the

Christian Democrats (CDU) 19, with 19·8%; the Free Democratic Party (FDP) 7, with 7·2%; Alliance '90/the Greens 5, with 5·6%. Turnout was 67·0%.

CURRENT GOVERNMENT

The *Prime Minister* is Matthias Platzeck (SPD).

Government Website: http://www.brandenburg.de

ECONOMY

Performance

GDP in 2011 was €55,093m.; real GDP growth was 2·4%.

ENERGY AND NATURAL RESOURCES

Electricity

Power stations in Brandenburg produced 37,626m. kWh in 2011.

Agriculture

Area and yield of the most important crops:

	Area (in 1,000 ha.)			Yield (in 1,000 tonnes)		
	2009	2010	2011	2009	2010	2011
Wheat	143·4	160·4	157·1	987·2	995·7	834·7
Rye	226·2	198·6	193·9	1,101·1	783·5	589·6
Sugar beet	7·2	7·1	8·3	448·6	402·4	519·0
Potatoes	9·6	8·9	9·4	340·6	284·9	357·1
Barley	89·7	76·8	77·1	522·3	447·6	312·0
Rape	131·2	133·5	122·4	538·5	495·0	269·8

Livestock in May 2012: cattle, 557,243 (including 159,165 dairy cows); horses, donkeys and mules (March 2010), 17,892; pigs, 784,610; sheep (Nov. 2012), 79,800; poultry (March 2010), 9,517,705.

INDUSTRY

In 2011, 1,180 establishments (20 or more employees) in the mining and manufacturing industries employed 98,592 persons. The main areas were: the food and animal feed industry; manufacture of chemical products and metal production and treatment. There were 4,702 establishments in the building industry in 2011, employing 34,333 persons.

Labour

In 2011 at the 1%-sample of the microcensus, 1,343,700 persons were economically active. In 2011 there were on average 119,700 unemployed persons (8·9%).

INTERNATIONAL TRADE

Imports and Exports

Total imports (2011): €18,479m.; exports: €13,464m.

COMMUNICATIONS

Roads

On 1 Jan. 2012 there were 1,603,755 registered vehicles including 1,330,774 passenger cars, 113,667 lorries, 2,380 buses and 103,665 motorcycles. There were 80,574 accidents in 2011 with 10,512 injured persons.

SOCIAL INSTITUTIONS

Education

In 2011–12 there were 864 schools providing general education (including special schools) with 222,714 pupils and 62 vocational schools with 47,392 pupils.

In the winter term 2011–12 there were three universities and 11 colleges with 51,676 students.

Health

In 2011 there were 53 hospitals with 15,242 beds, 4,026 doctors and 19,900 medical personnel.

RELIGION

In 2009 there were 435,627 Protestants and 77,830 Roman Catholics.

CULTURE

Tourism

In 2011 there were 1,695 places of accommodation (with nine or more beds), including 451 hotels—providing a total of 82,802 beds—and 183 campsites. 4,053,150 visitors (337,325 foreign) spent a total of 11,056,595 nights (including camping) in Brandenburg in 2011.

FURTHER READING

Statistical office: The Amt für Statistik Berlin-Brandenburg (Dortustrasse 46, 14467 Potsdam) was created in Jan. 2007 through the merger of the Landesbetrieb für Datenverarbeitung und Statistik Land Brandenburg and the Statistisches Landesamt Berlin. *Director:* Prof. Dr Ulrike Rockmann. It is the main source for the statistics above on Brandenburg and publishes *Statistisches Jahrbuch Land Brandenburg* (since 1991).
Website (limited English): http://www.statistik-berlin-brandenburg.de

---

Bremen
Freie Hansestadt Bremen

KEY HISTORICAL EVENTS

The state is dominated by the Free City of Bremen and its port, Bremerhaven. In 1815, when it joined the German Confederation, Bremen was an autonomous city and Hanse port with important Baltic trade. In 1827 the expansion of trade inspired the founding of Bremerhaven on land ceded by Hanover at the confluence of the Geest and Weser rivers. Further expansion followed the founding of the Nord-deutscher Lloyd Shipping Company in 1857. Merchant shipping, associated trade and fishing were dominant until 1940 but there was diversification in the post-war years. In 1939 Bremerhaven was absorbed by the Hanoverian town of Wesermünde. The combined port was returned to the jurisdiction of Bremen in 1947.

TERRITORY AND POPULATION

The area of the *Bundesland*, consisting of the two urban districts and ports of Bremen and Bremerhaven, is 419·2 sq. km. Population, 31 Dec. 2011, 661,301 (322,777 males, 338,524 females).

The capital, Bremen, had a population of 548,319 at 31 Dec. 2011 and Bremerhaven had a population of 112,982.

SOCIAL STATISTICS

Statistics for calendar years:

	Live births	Marriages	Divorces	Deaths
2008	5,569	2,804	1,647	7,353
2009	5,481	2,905	1,590	7,655
2010	5,599	2,978	1,536	7,510
2011	5,388	2,837	1,566	7,411

CONSTITUTION AND GOVERNMENT

Political power is vested in the 83-member House of Burgesses (*Bürgerschaft*) which appoints the executive, called the *Senate*. Bremen has three seats in the Bundesrat.

RECENT ELECTIONS

At the elections of 22 May 2011 the Social Democratic Party won 36 seats with 38·6% of votes cast; Alliance '90/the Greens 21 with 22·5%; the Christian Democratic Union 20 with 20·4%; the Left 5 with 5·6%; and Citizens in Rage 1 with 3·7%. Turnout was 55·5%.

CURRENT GOVERNMENT

The *Burgomaster* is Jens Böhrnsen (SPD).

Government Website: http://www.bremen.de

ENERGY AND NATURAL RESOURCES

Agriculture

Agricultural area comprised (2011) 12,045 ha. Livestock in March 2010: 10,558 cattle (including 3,634 dairy cows); 916 horses; 608 pigs (2007); 160 sheep; 3,736 laying hens.

INDUSTRY

In 2011, 146 establishments (50 or more employees) employed 46,245 persons; of these, 20,356 were employed in the production of cars and car parts and other vehicles; 3,802 in production of metal products; 3,360 in machine construction; 2,026 in fish processing; 1,313 in shipbuilding, including repair and maintenance (except naval engineering); 1,125 in coffee and tea processing.

Labour

Economically active persons totalled 302,000 at the microcensus of 2011. Of the total, 267,000 were employees, 34,000 self-employed; 91,000 in commerce, trade and communications, 60,000 in production industries, 150,000 in other industries and services.

COMMUNICATIONS

Roads

On 1 Jan. 2010 there were 119 km of 'classified' roads, of which 75 km were Autobahn and 44 km federal roads. Registered motor vehicles on 1 Jan. 2012 numbered 311,045, including 269,995 passenger cars, 16,040 lorries, 446 buses, 3,213 tractors and 19,763 motorcycles.

Civil Aviation

Bremen airport handled 2,560,023 passengers in 2011.

Shipping

Vessels entered in 2011, 7,194 of 209,263,000 GT; cleared, 7,028 of 209,644,000 GT. Sea traffic, 2011, incoming 41,197,000 tonnes; outgoing, 39,429,000 tonnes.

SOCIAL INSTITUTIONS

Justice

There are a constitutional court *(Staatsgerichtshof)*, a court of appeal, a regional court, three local courts, a *Bundesland* labour court, two labour courts, a *Bundesland* social court, a finance court, a higher administrative court and an administrative court.

Education

In 2011 there were 294 schools of general education with 5,462 teachers and 66,166 pupils; 39 vocational schools (part-time and full-time) with 23,265 pupils; 23 advanced vocational schools (including institutions for the training of technicians) with 2,725 pupils; six schools for public health occupations with 824 pupils. In 2011 there were 25 special schools with 422 teachers and 1,541 pupils.

In the winter term 2011–12, 18,122 students were enrolled at the University of Bremen and 1,266 at the Jacobs University Bremen. In addition to the universities there were six other colleges in 2011–12 with 13,577 students.

RELIGION

In 2009, 34·7% of the population were Protestants and 12·4% Roman Catholics.

CULTURE

Tourism

In 2011 there were 116 places of accommodation (with nine beds or more) providing 12,701 beds. Of the 1,100,013 visitors 19% were from abroad.

FURTHER READING

Statistical Information: Statistisches Landesamt Bremen (An der Weide 14–16, 28195 Bremen), founded in 1850. *Director:* Jürgen Wayand. Its current publications include: *Statistisches Jahrbuch Bremen* (from 1992). —*Statistische Mitteilungen* (from 1948).—*Statistische Monatsberichte* (1954–2004).—*Statistische Hefte* (2005–09).—*Statistische Berichte* (from 1956).—*Statistisches Handbuch Bremen* (1950–60, 1961; 1960–64, 1967; 1965–69, 1971; 1970–74, 1975; 1975–80, 1982; 1981–85, 1987).—*Bremen im statistischen Zeitvergleich 1950–1976.* 1977.—*Bremen in Zahlen* (from 1975).

Website (German only): http://www.statistik.bremen.de

State and University Library: Bibliotheksstrasse, 28359 Bremen. *Director:* Maria Elisabeth Müller.

Hamburg
Freie und Hansestadt Hamburg

KEY HISTORICAL EVENTS

Hamburg was a free Hanse town owing nominal allegiance to the Holy Roman Emperor until 1806. In 1815 it became part of the German Confederation, sharing a seat in the Federal Diet with Lübeck, Bremen and Frankfurt. During the Empire it retained its autonomy. By 1938 it had become the third largest port in the world and its territory was extended by the cession of land (three urban and 27 rural districts) from Prussia. After World War II, Hamburg became a *Bundesland* of the Federal Republic with its 1938 boundaries.

TERRITORY AND POPULATION

Total area, 755·3 sq. km (2011), including the islands Neuwerk and Scharhörn (7·6 sq. km). Population (31 Dec. 2011), 1,798,836 (880,972 males; 917,864 females). The *Bundesland* forms a single urban district *(Stadtstaat)* with seven administrative subdivisions.

SOCIAL STATISTICS

Statistics for calendar years:

	Live births	Marriages	Divorces	Deaths
2008	16,751	6,615	4,476	17,091
2009	16,779	7,231	3,970	17,188
2010	17,377	7,452	3,659	17,060
2011	17,125	7,022	3,635	17,060

CONSTITUTION AND GOVERNMENT

The constitution of 6 June 1952 vests the supreme power in the House of Burgesses *(Bürgerschaft)* of 121 members. The executive is in the hands of the Senate, whose members are elected by the Bürgerschaft. Hamburg has three seats in the Bundesrat.

RECENT ELECTIONS

The elections of 20 Feb. 2011 had the following results: Social Democrats, 62 seats with 48·3% of votes cast; Christian Democrats, 28 with 21·9%; Alliance '90/the Greens, 14 with 11·2%; the Free Democrats 9 (6·6%); the Left, 8 (6·4%). Turnout was 57·8%.

CURRENT GOVERNMENT

The *First Mayor* is Olaf Scholz (SPD).

Government Website: http://www.hamburg.de

ENERGY AND NATURAL RESOURCES

Agriculture

The agricultural area comprised 14,334 ha. in 2010.

Livestock (2010): cattle, 6,088 (including 1,024 dairy cows); horses, 2,838; pigs (2009), 432; sheep, 1,890; poultry, 3,336.

INDUSTRY

In Sept. 2011, 461 establishments (with 20 or more employees) employed 83,058 persons; of these, 26,524 were employed in manufacturing transport equipment (including motor vehicles, aircraft and ships), 12,600 in manufacturing machinery, 9,499 in manufacturing electrical and optical equipment, 4,555 in manufacturing chemical products and 3,893 in the mineral oil industry.

Labour

Economically active persons totalled 905,000 at the 1%-sample survey of the microcensus of 2011. Of the total, 769,000 were employees and 136,000 were self-employed or unpaid family workers; 286,000 were engaged in commerce and transport, 154,000 in power supply, mining, manufacturing and building, 3,000 in agriculture and forestry, 462,000 in other industries and services.

COMMUNICATIONS

Roads

In 2011 there were 3,928 km of roads, including 83 km of Autobahn and 117 km of federal roads. Number of motor vehicles (1 Jan. 2012), 841,862 of which 731,283 were passenger cars, 48,226 lorries, 1,538 buses, 49,843 motorcycles and 10,972 other motor vehicles.

Civil Aviation

Hamburg airport handled 13,528,395 passengers and 27,585 tonnes of freight in 2011.

Shipping

Hamburg is the largest sea port in Germany.

Vessels		2009	2010	2011
Entered:	Number	10,131	9,843	10,106
	Tonnage (gross)	217,977,077	235,462,280	255,574,100
Cleared:	Number	10,136	9,906	10,161
	Tonnage (gross)	217,632,001	235,371,927	256,534,847

SOCIAL INSTITUTIONS

Justice

There is a constitutional court (Verfassungsgericht), a court of appeal (Oberlandesgericht), a regional court (Landgericht), eight local courts (Amtsgerichte), a Bundesland labour court, a labour court, a Bundesland social court, a social court, a finance court, a higher administrative court and an administrative court.

Education

In 2009–10 there were 414 schools of general education (not including the Internationale Schule) with 180,452 pupils; 45 special schools with 6,777 pupils; 44 part-time vocational schools with 39,604 pupils; 41 schools with 3,990 pupils in manual instruction classes; 45 full-time vocational schools with 9,913 pupils; nine economic secondary schools with 2,298 pupils; two technical Gymnasien with 435 pupils; one pedagogical Gymnasium with 111 pupils; 18 advanced vocational schools with 4,301 pupils; 34 schools for public health occupations with 3,210 pupils; and 17 technical superior schools with 1,311 pupils.

In the winter term 2011–12 there were four universities with 41,011 students; one technical university with 5,916 students; one college of music and one college of fine arts with 1,747 students in total; one university of the Bundeswehr (Helmut Schmidt University) with 2,976 students; 11 professional colleges with a total of 33,029 students.

Health

In 2011 there were 47 hospitals with 12,071 beds, 10,409 doctors and 1,906 dentists.

RELIGION

In 2008, 30·4% of the population went to the Evangelical Church and Free Churches, whilst 10·2% were Roman Catholic.

CULTURE

Tourism

At Dec. 2011 there were 315 places of accommodation with 47,690 beds. Of the 5,083,172 visitors in 2011, 21·4% were foreigners.

FURTHER READING

Statistical Information: Statistisches Amt für Hamburg und Schleswig-Holstein (Standort Hamburg, Steckelhörn 12, 20457 Hamburg). Director: Helmut Eppmann. Publications: Statistische Berichte, Statistisches Jahrbuch, NORD.regional, Statistik informiert spezial. Website (German only): http://www.statistik-nord.de

Hamburger Sparkasse, Hamburg: von Altona bis Zollspieker. 2002
Hamburgische Gesellschaft für Wirtschaftsförderung mbH, Hamburg. 1993
Klessmann, E., Geschichte der Stadt Hamburg. 7th ed. 1994
Kopitzsch, F. and Brietzke, D., Hamburgische Biografie, Personenlexikon. Vol. 1. 2001
Kopitzsch, F. and Tilgner, D., Hamburg Lexikon. 1998
Möller, I., Hamburg. 2nd ed. 1999
Schubert, D. and Harms, H., Wohnen am Hafen. 1993
Schütt, E. C., Die Chronik Hamburgs. 1991

State library: Staats- und Universitätsbibliothek, Carl von Ossietzky, Von-Melle-Park 3, 20146 Hamburg. Director: Prof. Dr Gabriele Beger.

Hessen

KEY HISTORICAL EVENTS

The Bundesland consists of the former states of Hesse-Darmstadt and Hesse-Kassel, and Nassau. Hesse-Darmstadt was ruled by the Landgrave Louis X from 1790. He became grand duke in 1806 with absolute power, having dismissed the parliament in 1803. However, he granted a constitution and bicameral parliament in 1820. Hesse-Darmstadt lost land to Prussia in the Seven Weeks' War of 1866, but retained its independence, both then and as a state of the German Empire after 1871. In 1918 the grand duke abdicated and the territory became a state of the German Republic. In 1945 areas west of the Rhine were incorporated into the new Bundesland of Rhineland-Palatinate, areas east of the Rhine became part of the Bundesland of Greater Hesse.

Hesse-Kassel was ruled by the Landgrave William IX from 1785 until he became Elector in 1805. In 1807 the Electorate was absorbed into the Kingdom of Westphalia (a Napoleonic creation), becoming independent again in 1815 as a state of the German Confederation. In 1831 a constitution and parliament were granted but the Electors remained strongly conservative.

In 1866 the Diet approved alliance with Prussia against Austria; the Elector nevertheless supported Austria. He was defeated by the Prussians and exiled and Hesse-Kassel was annexed to Prussia. In 1867 it was combined with Frankfurt and some areas taken from Nassau and Hesse-Darmstadt to form a Prussian province (Hesse-Nassau). In 1801 Nassau west of the Rhine passed to France; Napoleon also took the northern state in 1806. The remnant of the southern states allied in 1803 and three years later they became a duchy. In 1866 the duke supported Austria against Prussia and the duchy was annexed by Prussia as a result. In 1944 the Prussian province of Hesse-Nassau was split in two: Nassau and Electoral Hesse, also called Kurhessen. The following year these were combined with Hesse-Darmstadt as the Bundesland of Greater Hesse which became known as Hessen.

TERRITORY AND POPULATION

Area, 21,115 sq. km. There are three administrative regions with five urban and 21 rural districts and 426 communes. Population, 31 Dec. 2011, was 6,092,126 (2,993,764 males, 3,098,362 females).

The capital is Wiesbaden, with a population (31 Dec. 2011) of 278,919.

SOCIAL STATISTICS

Statistics for calendar years:

	Live births	Marriages	Divorces	Deaths
2008	51,752	26,685	15,437	60,083
2009	50,744	27,248	14,896	60,676
2010	51,742	27,483	15,088	60,204
2011	51,479	27,468	14,905	60,446

CONSTITUTION AND GOVERNMENT

The constitution was put into force by popular referendum on 1 Dec. 1946. Hessen has five seats in the Bundesrat.

RECENT ELECTIONS

At the Diet elections on 18 Jan. 2009 the Christian Democratic Union (CDU) won 46 of 118 seats with 37·2% of votes cast (up from 36·8% in 2008), the Social Democratic Party (SPD) 29 with 23·7% (down from 36·7% in 2008), the Free Democratic Party (FDP) 20 with 16·2%, Alliance '90/the Greens 17 with 13·7% and the Left 6 with 5·4%.

CURRENT GOVERNMENT

The cabinet is headed by *Prime Minister* Volker Bouffier (CDU).

Government Website (German only): http://www.hessen.de

ECONOMY

Performance

In 2011 the price-adjusted growth of the gross domestic product at market prices (GDP) was 3·2% in comparison with the previous year. The total amount at current prices was €228·5bn. in 2011. The GDP (at current prices) per person engaged in labour productivity was €71,789 in 2011 (€70,355 in 2010).

ENERGY AND NATURAL RESOURCES

Electricity

Electricity production in 2011 was 19,207m. kWh (gross) and 18,081m. kWh (net). Total electricity consumption in 2011 was 34,573m. kWh.

Oil and Gas

Gas consumption in 2011 was 56,810m. kWh. All gas was imported from other parts of Germany.

Agriculture

Area and yield of the most important crops:

	Area (in 1,000 ha.)			Yield (in 1,000 tonnes)		
	2009	2010	2011	2009	2010	2011
Wheat	162·8	166·7	170·5	1,287·8	1,284·6	1,269·3
Sugar beet	15·6	14·2	15·0	1,061·0	965·8	1,167·4
Barley	95·7	87·2	87·7	618·0	560·5	502·8
Rape	66·6	66·8	64·9	296·0	265·2	205·6
Potatoes	4·6	4·3	4·2	189·3	171·5	187·9
Rye	17·2	14·3	14·1	105·7	82·8	72·2
Oats	2·3	11·3	9·8	64·2	57·8	37·5

Livestock, May 2011: cattle, 465,759 (including 149,093 dairy cows); horses, donkeys and mules (March 2010), 32,075; pigs, 649,512; sheep (Nov. 2011), 124,048; poultry (March 2010), 1·68m.

INDUSTRY

In Sept. 2012, 1,403 establishments (with 50 or more employees) employed 358,616 persons; of these, 56,914 were employed in the chemical industry; 48,324 in motor vehicle manufacture; 38,588 in machine construction; 31,163 in production of metal products.

Labour

Economically active persons totalled 3,011,000 at the 1% sample survey of the microcensus in 2011. Of the total, 2,648,000 were employees, 346,000 self-employed, 17,000 unpaid family workers; 821,000 were engaged in commerce, transport, hotels and restaurants, 734,000 in power supply, mining, manufacturing and building, 28,000 in agriculture and forestry and 1,429,000 in other services.

COMMUNICATIONS

Roads

On 1 Jan. 2012 there were 16,668 km of 'classified' roads, comprising 972 km of Autobahn, 3,444 km of federal highways, 7,257 km of first-class highways and 4,996 km of second-class highways. Motor vehicles licensed on 1 Jan. 2012 totalled 4,021,865, including 3,372,935 passenger cars, 5,770 buses, 178,509 lorries, 139,108 tractors and 305,532 motorcycles.

Civil Aviation

Frankfurt/Main airport is one of the most important freight airports in the world. In 2011, 487,162 aeroplanes took off and landed, carrying 56,443,657 passengers, 2,169,304 tonnes of air freight and 82,314 tonnes of air mail.

Shipping

Frankfurt/Main harbour and Hanau harbour are the two most important harbours. In 2011, 7·8m. tonnes of goods were imported into the *Bundesland* and 2·2m. tonnes were exported.

SOCIAL INSTITUTIONS

Justice

There are a constitutional court *(Staatsgerichtshof)*, a court of appeal, nine regional courts, 46 local courts, a *Bundesland* labour court, 12 labour courts, a *Bundesland* social court, seven social courts, a finance court, a higher administrative court *(Verwaltungsgerichtshof)* and five administrative courts.

Education

In 2011 there were 1,238 primary schools with 229,453 pupils (including *Förderstufen*); 158 intermediate schools with 49,685 pupils; 19,142 teachers in the primary and intermediate schools; 234 special schools with 5,656 teachers and 24,469 pupils; 178 high schools with 12,194 teachers and 159,916 pupils; 226 *Gesamtschulen* (comprehensive schools) with 13,526 teachers and 185,214 pupils; 117 part-time vocational schools with 117,771 pupils; 266 full-time vocational schools with 58,955 pupils; 116 advanced vocational schools with 14,355 pupils; 9,445 teachers in the vocational schools.

In the winter term 2011–12 there were four universities (Frankfurt/Main, 40,383 students; Giessen, 25,143; Marburg/Lahn, 22,004; Kassel, 21,242); one technical university in Darmstadt (24,180); two private scientific colleges (2,939); 19 universities of applied sciences (70,664); two Roman Catholic theological colleges and three Protestant theological colleges with a total of 661 students; one college of music and two colleges of fine arts with 1,671 students in total.

RELIGION

In 2010 the churches in Hessen reported 2,390,000 (39·4%) Protestants and 1,485,000 (24·5%) Roman Catholics.

CULTURE

Press

In 2010 there were 76 newspapers published in Hessen with a combined circulation of 1·7m.

Tourism

In 2011, 12·4m. visitors stayed 29·0m. nights in Hessen.

FURTHER READING

Statistical Information: The Hessisches Statistisches Landesamt, Rheinstr. 35–37, 65175 Wiesbaden). *President:* Dr Christel Figgener. Main publications: *Statistisches Jahrbuch für das Land Hessen* (biannual).— *Staat und Wirtschaft in Hessen* (monthly).—*Statistische Berichte.— Hessische Gemeindestatistik* (annual, 1980 ff.).

Website (limited English): http://www.statistik-hessen.de

State library: Hessische Landesbibliothek Wiesbaden, Rheinstr. 55–57, 65185 Wiesbaden. *Director:* Dr Marion Grabka.

Website: http://www.hlb-wiesbaden.de

Lower Saxony

Niedersachsen

KEY HISTORICAL EVENTS

The *Bundesland* consists of the former states of Hanover, Oldenburg, Schaumburg-Lippe and Brunswick. It does not include the cities of Bremen or Bremerhaven. Oldenburg, Danish from 1667, passed to the bishopric of Lübeck in 1773; the Holy Roman Emperor made it a duchy in 1777. As a small state of the Confederation after 1815 it supported Prussia, becoming a member of the Prussian Zollverein (1853) and North German Confederation (1867). The grand duke abdicated in 1918 and was replaced by an elected government.

Schaumburg-Lippe was a small sovereign principality. As such it became a member of the Confederation of the Rhine in 1807 and of the German Confederation in 1815. Surrounded by Prussian territory, it also joined the Prussian-led North German Confederation in 1867. Part of the Empire until 1918, it then became a state of the new republic.

Brunswick, a small duchy, was taken into the Kingdom of Westphalia by Napoleon in 1806 but restored to independence in 1814. In 1830 the duke, Charles II, was forced into exile and replaced in 1831 by his more liberal brother, William. The succession passed to a Hanoverian claimant in 1913 but the duchy ended with the Empire in 1918.

As a state of the republican Germany, Brunswick was greatly reduced under the Third Reich. Its boundaries were restored by the British occupation forces in 1945.

Hanover was an autonomous Electorate of the Holy Roman Empire whose rulers were also kings of Great Britain from 1714 to 1837. From 1762 they ruled almost entirely from England. After Napoleonic invasions Hanover was restored in 1815. A constitution of 1819 made no radical change and had to be followed by more liberal versions in 1833 and 1848. Prussia annexed Hanover in 1866; it remained a Prussian province until 1946. On 1 Nov. 1946 all four states were combined by the British military administration to form the *Bundesland* of Lower Saxony.

TERRITORY AND POPULATION

Lower Saxony has an area of 47,614 sq. km, and is divided into eight urban districts, 38 rural districts and 1,008 communes. Population, on 31 Dec. 2011, was 7,913,502 (3,895,921 males; 4,017,581 females).

The capital is Hanover, with a population (31 Dec. 2011) of 525,875.

SOCIAL STATISTICS

Statistics for calendar years:

	Live births	Marriages	Divorces	Deaths
2008	64,887	39,234	20,368	84,874
2009	62,228	38,116	19,181	85,673
2010	63,130	38,373	18,974	85,794
2011	61,280	37,645	18,953	85,489

CONSTITUTION AND GOVERNMENT

The *Bundesland* Niedersachsen was formed on 1 Nov. 1946 by merging the former Prussian province of Hanover with Brunswick, Oldenburg and Schaumburg-Lippe. Lower Saxony has six seats in the Bundesrat.

RECENT ELECTIONS

At the Diet elections on 20 Jan. 2013 the Christian Democratic Union won 54 of 137 seats, receiving 36·0% of votes cast (down from 42·5% in 2008), the Social Democratic Party 49 with 32·6% (up from 30·3% in 2008), the Alliance '90/The Greens 20 with 13·7% and the Free Democrats 14 with 9·9%.

CURRENT GOVERNMENT

The *Prime Minister* is Stephan Weil (SPD).

Government Website: http://www.niedersachsen.de

ECONOMY

Banking and Finance

185 credit institutions were operating in 2011. Deposits totalled €51,803m.

ENERGY AND NATURAL RESOURCES

Electricity

Electricity production in 2011 was 46,454m. kWh.

Agriculture

Area and yield of the most important crops:

	Area (in 1,000 ha.)			Yield (in 1,000 tonnes)		
	2009	2010	2011	2009	2010	2011
Sugar beet	102	98	102	7,138	6,172	7,594
Potatoes	118	113	113	5,507	4,590	5,251
Wheat	434	434	402	3,667	3,430	3,081
Barley	228	197	182	1,538	1,270	1,036
Rye	150	121	113	960	594	595
Oats	15	11	12	69	42	55

Livestock, March 2010 unless otherwise indicated: cattle (March 2011), 2,518,128 (including 781,801 dairy cows); horses, donkeys and mules, 70,811; pigs, 8,428,731; sheep, 205,569; poultry 56,609,004.

INDUSTRY

In Sept. 2011, 3,660 establishments employed 501,023 persons; of these 54,184 were employed in machine construction.

Labour

Economically active persons totalled 3,778,000 in 2011. Of the total, 3,368,000 were employees, 382,000 self-employed, 26,000 unpaid family workers; 1,016,900 were engaged in power supply, mining, manufacturing and building, 970,500 in commerce and transport, 102,300 in agriculture and forestry, and 1,688,300 in other industries and services.

COMMUNICATIONS

Roads

At 1 Jan. 2012 there were 28,271 km of 'classified' roads, comprising 1,433 km of Autobahn, 4,796 km of federal roads, 8,330 km of first-class and 13,712 km of second-class highways. Number of motor vehicles, 1 Jan. 2012, was 5,142,420 including

4,255,217 passenger cars, 238,668 lorries, 7,481 buses (1 Jan. 2011), 229,985 tractors and 383,048 motorcycles.

Rail
In 2011, 51·4m. tonnes of freight came into the *Bundesland* by rail and 42·5m. tonnes left by rail.

Civil Aviation
68,309 planes landed at Hanover airport in 2011, which saw 2,660,093 passenger arrivals and 2,642,394 departures. 8,441 tonnes of freight (including post) left by air and 8,374 tonnes (including post) came in.

SOCIAL INSTITUTIONS

Justice
There are a constitutional court *(Staatsgerichtshof)*, three courts of appeal, 11 regional courts, 80 local courts, a *Bundesland* labour court, 15 labour courts, a *Bundesland* social court, eight social courts, a finance court, a higher administrative court and seven administrative courts.

Education
In 2011–12 there were 1,796 primary schools with 293,000 pupils; 463 post-primary schools with 69,303 pupils; 336 special schools with 34,416 pupils; 133 secondary schools with 8,236 pupils; 466 secondary modern schools with 161,152 pupils; 258 grammar schools with 233,342 pupils; 37 co-operative comprehensive schools with 42,027 pupils; and 70 integrated comprehensive schools with 45,289 pupils.

In the winter term 2011–12 there were seven universities (Göttingen, 24,573 students; Hanover, 21,621; Hildesheim, 5,715; Lüneburg, 7,138; Oldenburg, 10,786; Osnabrück, 11,034; Vechta, 3,245); two technical universities (Braunschweig, 15,204; Clausthal, 4,004); the medical college of Hanover (3,134); the veterinary college in Hanover (2,484).

Health
In 2011 there were 28,885 doctors and 197 hospitals with 5·3 beds per 1,000 population.

RELIGION
In 2010 there were 49·1% Protestants and 17·4% Roman Catholics.

CULTURE

Tourism
In 2011, 12,452,049 guests spent 39,319,170 nights in Lower Saxony.

FURTHER READING
Statistical Information: Niedersächsischer Landesbetrieb für Statistik und Kommunikationstechnologie, Postfach 910764, 30427 Hanover. *Chairman:* Dr Christoph Lahmann. Main publications are: *Statistische Monatshefte Niedersachsen* (from 1947).—*Statistische Berichte Niedersachsen.— Statistisches Taschenbuch Niedersachsen 2012* (biennial).
State libraries: Niedersächsische Staats- und Universitätsbibliothek, Platz der Göttinger Sieben 1, 37073 Göttingen. *Director:* Prof. Dr Norbert Lossau; Gottfried Wilhelm Leibniz Bibliothek—Niedersächsische Landesbibliothek, Waterloostr. 8, 30169 Hanover. *Director:* Dr Georg Ruppelt.

Mecklenburg-West Pomerania
Mecklenburg-Vorpommern

KEY HISTORICAL EVENTS
Pomerania was at one time under Swedish control while Mecklenburg was an independent part of the German Empire.

The two states were not united until after the Second World War, and after a short period when it was subdivided into three districts under the GDR, it became a state of the Federal Republic of Germany in 1990. The people of the region speak a dialect known as Plattdeutsch (Low German). The four main cities of this state are Hanseatic towns from the period when the area dominated trade with Scandinavia. Rostock on the North Sea coast became the home of the GDR's biggest shipyards.

TERRITORY AND POPULATION
The area is 23,194 sq. km. It is divided into two urban districts, six rural districts and 805 communes. Population on 31 Dec. 2011 was 1,634,734 (825,531 females). It is the most sparsely populated of the German *Bundesländer*, with a population density of 70 per sq. km in 2011.

The capital is Schwerin, with a population (31 Dec. 2011) of 95,300.

SOCIAL STATISTICS
Statistics for calendar years:

	Live births	Marriages	Divorces	Deaths
2008	13,098	10,464	3,195	17,818
2009	13,014	10,493	3,221	18,342
2010	13,337	10,751	3,238	18,738
2011	12,638	10,400	3,407	18,572

CONSTITUTION AND GOVERNMENT
The *Bundesland* was reconstituted on former GDR territory in 1990. It has three seats in the Bundesrat.

RECENT ELECTIONS
At the Diet elections of 4 Sept. 2011 the Social Democrats (SPD) won 27 seats with 35·6% of the vote; the Christian Democrats (CDU), 18 with 23·0%; the Left Party, 14 with 18·4%; the Greens, 7 with 8·7%; the far-right National Democratic Party, 5 with 6·0%; the Free Democrats, no seats with 2·8%. Turnout was 51·5%.

CURRENT GOVERNMENT
The *Prime Minister* is Erwin Sellering (SPD).
Government Website: http://www.mecklenburg-vorpommern.eu

ENERGY AND NATURAL RESOURCES

Agriculture
Area and yield of the most important crops:

	Area (in 1,000 ha.)			Yield (in 1,000 tonnes)		
	2009	2010	2011	2009	2010	2011
Wheat	323·8	350·3	352·3	2,582·7	2,465·1	2,350·2
Sugar beet	22·7	24·6	27·6	1,295·1	1,296·9	1,719·7
Barley	142·9	119·1	118·4	987·8	870·5	644·8
Rape	244·9	252·0	204·9	1,102·1	1,011·7	558·0
Potatoes	14·3	13·9	13·6	564·3	434·1	473·3
Rye	90·1	62·8	70·5	489·0	277·5	301·7

Livestock in 2011: cattle, 533,455 (including 175,242 dairy cows); horses (2010), 13,869; pigs, 814,742; sheep, 67,500; poultry (2010), 8,722,482.

Fisheries
Sea catch, 2011: 19,575 tonnes (5,735 tonnes frozen, 13,840 tonnes fresh). Freshwater catch, 2011: 580 tonnes. Fish farming, 2011: 747 tonnes.

INDUSTRY
In 2011 there were 690 enterprises (with 20 or more employees) employing 57,055 persons.

Labour

784,000 persons (364,100 females) were employed at the 1%-sample survey of the microcensus (2011 average at the place of residence), including 407,400 white-collar workers, 268,400 manual workers and 75,400 self-employed and family assistants. 31,300 persons were employed as officials. Employment by sector (2011 average at the place of work): public and private services, 258,400; trade, guest business, transport and communications, 193,200; financing, leasing and services for enterprises, 115,500; manufacturing, 75,700; construction, 54,400; agriculture, forestry and fisheries, 23,300; mining, energy and water resources, 10,900; total, 731,400.

COMMUNICATIONS

Roads

In 2011 there were 10,013 km of 'classified' roads, comprising 554 km of Autobahn, 1,993 km of federal roads, 3,308 km of first-class and 4,158 km of second-class highways. Number of motor vehicles at 1 Jan. 2012 was 983,970, including 819,575 passenger cars, 69,663 lorries and 55,935 motorcycles.

Shipping

There is a lake district of some 555 lakes greater than 0·1 sq. km. The ports of Rostock, Stralsund and Wismar are important for shipbuilding and repairs. In 2011 the cargo fleet consisted of 201 vessels (including 12 tankers) of 2,950,000 GT. Marine freight traffic in 2011 totalled 26,790,000 tonnes.

SOCIAL INSTITUTIONS

Justice

There is a court of appeal (Oberlandesgericht), four regional courts (Landgerichte), 21 local courts (Amtsgerichte), four labour courts, four social courts, a finance court and two administrative courts.

Education

In 2011 there were 569 schools with 132,677 pupils, including 49,375 pupils in primary schools, 61,525 in secondary schools and 12,712 candidates for the school-leaving examination; and 9,065 pupils in special needs schools.

There are universities at Rostock and Greifswald with (in 2011–12) 27,762 students and 6,353 academic staff, and six institutions of equivalent status with 12,709 students and 1,471 academic staff.

RELIGION

In 2010 the Evangelical Lutheran Church of Mecklenburg had 192,900 adherents, 233 pastors and 268 parishes. The Pomeranian Evangelical Church had 94,000 adherents, 105 pastors and 186 parishes in 2010. Roman Catholics numbered 53,700, with 36 priests and 33 parishes.

CULTURE

Tourism

In July 2011 there were 2,850 places of accommodation (with nine or more beds; excluding camping places) providing a total of 172,907 beds. 5,853,054 guests stayed an average of 3·8 nights each in 2011 (excluding camping places).

FURTHER READING

Statistical office: Statistisches Amt Mecklenburg-Vorpommern, Postfach 120135, 19018 Schwerin.
Main publications are: Statistische Hefte Mecklenburg-Vorpommern (since 1991); Gemeindedaten Mecklenburg-Vorpommern (since 1999; electronic); Statistische Berichte (since 1991; various); Statistisches Jahrbuch Mecklenburg-Vorpommern (since 1991); Statistische Sonderhefte (since 1992; various).
Website (German only): http://www.statistik-mv.de

North Rhine-Westphalia
Nordrhein-Westfalen

KEY HISTORICAL EVENTS

Historical Westphalia consisted of many small political units, most of them absorbed by Prussia and Hanover before 1800. In 1807 Napoleon created a Kingdom of Westphalia for his brother Joseph. This included Hesse-Kassel, but was formed mainly from the Prussian and Hanoverian lands between the rivers Elbe and Weser.

In 1815 the kingdom ended with Napoleon's defeat. Most of the area was given to Prussia, with the small principalities of Lippe and Waldeck surviving as independent states. Both joined the North German Confederation in 1867. Lippe remained autonomous after the end of the Empire in 1918; Waldeck was absorbed into Prussia in 1929.

In 1946 the occupying forces combined Lippe with most of the Prussian province of Westphalia to form the Bundesland of North Rhine-Westphalia. On 1 March 1947 the allied Control Council formally abolished Prussia.

TERRITORY AND POPULATION

The Bundesland comprises 34,098 sq. km. It is divided into five administrative regions, 22 urban districts, 31 rural districts and 396 communes. Population, 31 Dec. 2011, 17,841,956 (9,123,937 females, 8,718,019 males).

The capital is Düsseldorf, with a population (31 Dec. 2011) of 592,393.

SOCIAL STATISTICS

Statistics for calendar years:

	Live births	Marriages	Divorces	Deaths
2008	150,007	81,515	46,098	189,586
2009	145,029	81,861	45,978	190,814
2010	147,333	81,662	45,711	192,137
2011	143,097	80,829	44,501	188,944

CONSTITUTION AND GOVERNMENT

Since Oct. 1990 North Rhine-Westphalia has had six seats in the Bundesrat.

RECENT ELECTIONS

At the Diet elections on 13 May 2012 the Social Democratic Party won 99 of 237 seats with 39·1% of votes cast, the Christian Democratic Union 67 (26·3%), Alliance '90/the Greens 29 (11·3%), the Free Democratic Party 22 (8·6%) and the Pirate Party 20 (7·8%). Turnout was 59·6%. The Christian Democrats suffered their worst ever result in Germany's most populous Bundesland.

CURRENT GOVERNMENT

North Rhine-Westphalia is governed by the Social Democratic Party (SPD) and the Greens.
Prime Minister: Hannelore Kraft (SPD).

Government Website (German only): http://www.nrw.de

ECONOMY

North Rhine-Westphalia has the highest GDP of any German Bundesland (€569·0bn. in 2011). Foreign direct investment is also higher than in any other Bundesland.

Budget

The predicted total revenue for 2012 was €56,010·1m. and the predicted total expenditure was €56,013·2m.

ENERGY AND NATURAL RESOURCES

Agriculture
Area and yield of the most important crops:

	Area (in 1,000 ha.)			Yield (in 1,000 tonnes)		
	2009	2010	2011	2009	2010	2011
Sugar beet	56·6	53·7	60·0	3,992·2	3,705·5	4,587·8
Wheat	295·6	286·8	227·4	2,551·2	2,229·4	2,244·7
Potatoes	30·3	31·1	32·6	1,421·9	1,417·5	1,619·5
Barley	187·4	171·4	156·3	1,402·6	1,138·3	974·4
Rye	19·3	16·1	17·6	127·5	92·7	101·2
Oats	14·7	12·4	11·5	76·7	50·0	60·1

Livestock, May 2011: cattle, 1,404,155 (including 398,027 dairy cows); pigs, 6,428,342; sheep, 131,727; poultry (May 2010), 11,741,044.

INDUSTRY

In Sept. 2011, 9,613 establishments (with 20 or more employees) employed 1,199,644 persons: 272,103 were employed in metal production and manufacture of metal goods; 198,710 in machine construction; 116,381 in manufacture of office machines, computers, electrical and precision engineering and optics; 100,750 in the chemical industry; 96,599 in production of food and tobacco; and 83,570 in motor vehicle manufacture. 70·9% of the workforce is now employed in the services sector. Of the total population, 9·5% were engaged in industry.

Labour
Economically active persons totalled 8,267,700 at the 1%-sample survey of the microcensus of 2011. Of the total, 7,385,570 were employees, 850,375 self-employed and 31,750 unpaid family workers; 2,340,791 were engaged in power supply, mining, manufacturing, water supply and building, 2,090,096 in commerce, hotel trade and transport, 68,300 in agriculture, forestry and fishing, and 3,768,484 in other industries and services.

COMMUNICATIONS

Roads
There were (1 Jan. 2012) 29,582 km of 'classified' roads, comprising 2,207 km of Autobahn, 4,767 km of federal roads, 12,837 km of first-class and 9,771 km of second-class highways. Number of motor vehicles (1 Jan. 2012): 10,729,393, including 9,153,264 passenger cars, 504,453 lorries, 16,118 buses and 781,993 motorcycles.

Civil Aviation
In 2011, 108,257 aircraft landed at Düsseldorf, bringing 10,179,903 incoming passengers; and 59,068 aircraft landed at Cologne-Bonn, bringing 4,791,151 incoming passengers.

SOCIAL INSTITUTIONS

Justice
There are a constitutional court (Verfassungsgerichtshof), three courts of appeal, 19 regional courts, 130 local courts, three Bundesland labour courts, 30 labour courts, one Bundesland social court, eight social courts, three finance courts, a higher administrative court and seven administrative courts.

Education
In 2011 there were 3,695 primary schools with 55,779 teachers and 827,902 pupils; 695 special schools with 19,066 teachers and 94,532 pupils; 564 intermediate schools with 18,199 teachers and 308,860 pupils; 284 Gesamtschulen (comprehensive schools) with 20,421 teachers and 259,409 pupils; 627 high schools with 41,609 teachers and 598,762 pupils; there were 291 part-time vocational schools with 366,206 pupils; 226 vocational preparatory year schools with 22,095 pupils; 317 full-time vocational schools with 104,033 pupils; 209 vocational high schools with 32,065 pupils; 205 full-time vocational schools leading up to vocational colleges with 24,809 pupils; 268 advanced full-time vocational schools with 49,959 pupils; 416 schools for public health occupations with 12,638 teachers and 43,864 pupils.

In the winter term 2011–12 there were 15 universities (Bielefeld, 18,779 students; Bochum, 36,330; Bonn, 28,660; Cologne, 45,568; Dortmund, 26,585; Düsseldorf, 20,560; Duisburg-Essen, 37,264; Münster, 38,069; Paderborn, 17,207; Siegen, 15,707; Witten/Herdecke, 1,315; Wuppertal, 16,437; the Technical University of Aachen, 35,782; Fernuniversität at Hagen, 67,515; German Police University/DHPol, 242); the College for physical education in Cologne, 4,589; three Roman Catholic and two Protestant theological colleges with a total of 428 students. There were also four colleges of music, four colleges of fine arts with 6,155 students in total; 39 Fachhochschulen (vocational colleges) with 163,168 students.

Health
In 2011 there were 401 hospitals in North Rhine-Westphalia with 121,556 beds, which had an average occupancy rate of 75·6%.

RELIGION

In 2009 there were 42·0% Roman Catholics and 27·5% Protestants.

CULTURE

Tourism
In Dec. 2011 there were 5,639 places of accommodation (nine beds or more) providing 310,939 beds altogether. In 2011, 19,509,825 visitors (4,126,367 foreigners) spent 44,245,100 nights in North Rhine-Westphalia.

FURTHER READING

Statistical Information: Information und Technik Nordrhein-Westfalen (IT NRW) (Mauerstr. 51, 40476 Düsseldorf) was founded in 1946 as the Landesamt für Datenverabeitung und Statistik Nordrhein-Westfalen by amalgamating the provincial statistical offices of Rhineland and Westphalia. It was renamed on 1 Jan. 2009. *President:* Hans-Josef Fischer. IT NRW publishes (from 1949): *Statistisches Jahrbuch Nordrhein-Westfalen.* More than 550 other publications yearly.
Website (German only): http://www.it.nrw.de
Bundesland Library: Universitätsbibliothek, Universitässtr. 1, 40225 Düsseldorf. *Director:* Dr Irmgard Siebert.

Rhineland-Palatinate

Rheinland-Pfalz

KEY HISTORICAL EVENTS

The *Bundesland* was formed from the Rhenisch Palatinate and the Rhine valley areas of Prussia, Hesse-Darmstadt, Hesse-Kassel and Bavaria.

From 1214 the Palatinate was ruled by the Bavarian house of Wittelsbach, with its capital as Heidelberg. In 1797 the land west of the Rhine was taken into France, and Napoleon divided the eastern land between Baden and Hesse. In 1815 the territory taken by France was restored to Germany and allotted to Bavaria. The area and its neighbours formed the strategically important Bavarian Circle of the Rhine. The rule of the Wittelsbachs ended in 1918 but the Palatinate remained part of Bavaria until the American occupying forces detached it in 1946. The new *Bundesland*, incorporating the Palatinate and other territory, received its constitution in April 1947.

TERRITORY AND POPULATION

Rhineland-Palatinate has an area of 19,854 sq. km. It comprises 12 urban districts, 24 rural districts and 2,294 other communes. Population (at 31 Dec. 2011), 3,999,117 (2,032,222 females).

The capital is Mainz, with a population (31 Dec. 2011) of 200,957.

SOCIAL STATISTICS

Statistics for calendar years:

	Live births	Marriages	Divorces	Deaths
2008	32,223	20,059	10,273	42,932
2009	30,881	19,867	10,609	43,903
2010	31,574	20,172	10,483	43,465
2011	31,081	20,212	11,041	43,645

CONSTITUTION AND GOVERNMENT

The constitution of the *Bundesland* Rheinland-Pfalz was approved by the Consultative Assembly on 25 April 1947 and by referendum on 18 May 1947, when 579,002 voted for and 514,338 against its acceptance. It has four seats in the Bundesrat.

RECENT ELECTIONS

At the elections of 27 March 2011 the Social Democratic Party won 42 seats of the 101 in the state parliament with 35·7% of votes cast, the Christian Democrats 41 with 35·2% and Alliance '90/the Greens 18 with 15·4%. The Free Democrats received 4·2% of the vote and the Left 3·0%, meaning that neither party won any seats. Turnout was 61·8%.

CURRENT GOVERNMENT

The cabinet is headed by *Prime Minister* Malu Dreyer (SPD). There is currently an SPD–Green Party coalition.

Government Website (limited English): http://www.rlp.de

ENERGY AND NATURAL RESOURCES

Agriculture

Area and yield of the most important products:

	Area (in 1,000 ha.)			Yield (in 1,000 tonnes)		
	2009	2010	2011	2009	2010	2011
Sugar beet	18·9	18·0	19·6	1,305·6	1,292·5	1,429·9
Wheat	111·5	119·9	117·5	830·0	833·4	706·3
Barley	87·7	76·6	79·5	509·7	450·3	370·5
Potatoes	7·9	7·6	7·9	303·4	289·8	326·4
Rye	11·3	11·2	10·9	75·2	68·2	57·0
Oats	7·2	6·2	5·4	33·6	26·2	20·0
Wine	62·6	62·6	62·3	6,088·4[1]	4,606·7[1]	6,162·2[1]

[1]1,000 hectolitres.

Livestock (2011, in 1,000): cattle, 363·1 (including dairy cows, 117·8); horses, donkeys and mules, 19·6; pigs, 242·5; sheep, 70·9; poultry, 1,543·2.

Forestry

Total area covered by forests in Dec. 2011 was 8,335·0 sq. km, or 42·0% of the total area.

INDUSTRY

In 2011, 2,219 establishments (with 20 or more employees) employed 283,362 persons; of these 45,420 were employed in the chemical industry; 37,492 in metal production and manufacture of metal goods; 36,277 in machine construction; 26,522 in motor vehicle manufacture; 18,192 in production of food.

Labour

Economically active persons totalled 1,945,300 in 2011. Of the total, 1,730,700 were employees, 202,000 were self-employed, 12,700 were unpaid family workers; 547,400 were engaged in power supply, mining, manufacturing and building, 490,900 in commerce, transport, hotels and restaurants, 39,900 in agriculture and forestry, and 867,000 in other industries and services.

COMMUNICATIONS

Roads

In 2012 there were 18,413 km of 'classified' roads, comprising 875 km of Autobahn, 2,945 km of federal roads, 7,229 km of first-class and 7,365 km of second-class highways. Number of motor vehicles, 1 Jan. 2012, was 2,796,124, including 2,290,720 passenger cars, 121,861 lorries, 4,825 buses, 140,136 tractors and 224,531 motorcycles.

SOCIAL INSTITUTIONS

Justice

There are a constitutional court (*Verfassungsgerichtshof*), two courts of appeal, eight regional courts, 47 local courts, a *Bundesland* labour court, five labour courts, a *Bundesland* social court, four social courts, a finance court, a higher administrative court and four administrative courts.

Education

In 2011 there were 983 primary schools with 10,888 teachers and 142,147 pupils; 477 secondary schools with 22,307 teachers and 284,860 pupils; 138 special schools with 2,973 teachers and 14,823 pupils; 117 vocational and advanced vocational schools with 5,900 teachers and 127,094 pupils.

In higher education, in the winter term 2012–13 (provisional figures) there were the University of Mainz (36,259 students), the University of Trier (14,962 students), the University of Koblenz-Landau (14,057 students), the University of Kaiserslautern (13,134 students), the *Deutsche Universität für Verwaltungswissenschaften* in Speyer (338 students), the *Wissenschaftliche Hochschule für Unternehmensführung* (Otto Beisheim Graduate School) in Vallendar (937 students), the Roman Catholic Theological College in Trier (390 students) and the Roman Catholic Theological College in Vallendar (213 students). There were also eight *Fachhochschulen* with 37,089 students and three *Verwaltungsfachhochschulen* with 2,478 students.

RELIGION

In 2010 there were 44·7% Roman Catholics and 30·5% Protestants.

CULTURE

Tourism

In 2011, 3,551 places of accommodation provided 154,581 beds for 7,502,593 visitors.

FURTHER READING

Statistical Information: Statistisches Landesamt Rheinland-Pfalz (Mainzer Str., 14–16, 56130 Bad Ems). *President:* Jörg Berres. Its publications include: *Statistisches Jahrbuch Rheinland-Pfalz* (since 1948); *Statistische Monatshefte Rheinland-Pfalz* (since 1958); *Rheinland-Pfalz heute* (since 1973); *Statistik von Rheinland-Pfalz* (from 1946 to 2004), then renamed *Statistische Bände* (since 2004) 400 vols. to date; *Kreisfreie Städte und Landkreise* (since 2004); *Rheinland-Pfalz—ein Ländervergleich in Zahlen* (since 2005); *Die Wirtschaft in Rheinland-Pfalz* (since 2007).

Website (German only): http://www.statistik.rlp.de

Saarland

KEY HISTORICAL EVENTS

Long disputed between Germany and France, the area was occupied by France in 1792. Most of it was allotted to Prussia at the close of the Napoleonic wars in 1815. In 1870 Prussia defeated France and when, in 1871, the German Empire was founded under Prussian leadership, it was able to incorporate Lorraine. This part of France was the Saar territory's western neighbour so the Saar was no longer a vulnerable boundary state. It began to develop industrially, exploiting Lorraine coal and iron.

In 1919 the League of Nations took control of the Saar until a plebiscite of 1935 favoured return to Germany. In 1945 there was a French occupation, and in 1947 the Saar was made an international area, but in economic union with France. In 1954 France and Germany agreed that the Saar should be a separate and autonomous state, under an independent commissioner. This was rejected by referendum and France agreed to return Saarland to Germany; it became a *Bundesland* of the Federal Republic on 1 Jan. 1957.

TERRITORY AND POPULATION

Saarland has an area of 2,569 sq. km (including a mutual area with Luxembourg). It comprises six rural districts and 52 communes. Population, 31 Dec. 2011, 1,013,352 (493,714 males, 519,638 females).

The capital is Saarbrücken, with a population (31 Dec. 2011) of 176,135.

SOCIAL STATISTICS

Statistics for calendar years:

	Live births	Marriages	Divorces	Deaths
2008	7,158	4,936	2,734	12,547
2009	6,927	4,874	2,639	12,588
2010	7,066	4,804	2,718	12,296
2011	7,088	4,866	2,659	12,331

CONSTITUTION AND GOVERNMENT

Saarland has three seats in the Bundesrat.

RECENT ELECTIONS

At the elections to the Saar Diet of 25 March 2012 the Christian Democrats (CDU) won 19 of 51 seats with 35·2% of votes cast; the Social Democrats (SPD) 17, with 30·6%; the Left 9, with 16·1%; the Pirate Party 4, with 7·4%; and the Greens 2, with 5·0%. Turnout was 61·6%.

CURRENT GOVERNMENT

In Oct. 2009 the right-leaning CDU, the right-of-centre FDP and the Greens agreed to form Germany's first ever so-called 'Jamaica coalition' (the traditional colours of the three parties—black, yellow and green—being the same as those of the Jamaican flag). However, it collapsed in Jan. 2012. The *Prime Minister* is Annegret Kramp-Karrenbauer (CDU).

Government Website: http://www.saarland.de

ENERGY AND NATURAL RESOURCES

Electricity
In 2011 electricity production was 7,740m. kWh. End-user consumption totalled 7,718m. kWh in 2011.

Oil and Gas
8,017m. kWh of gas was used in 2011.

Agriculture
The cultivated area (2011) occupied 110,531 ha. or 43·0% of the total area.

Area and yield of the most important crops:

	Area (in 1,000 ha.)			Yield (in 1,000 tonnes)		
	2009	2010	2011	2009	2010	2011
Wheat	9·5	9·7	9·7	66·5	67·7	53·9
Barley	5·1	4·4	4·4	28·2	24·5	21·1
Rye	4·1	3·5	3·2	24·4	20·3	15·6
Oats	2·1	1·9	1·8	9·3	7·9	6·4
Potatoes	0·2	0·1	0·1	5·9	4·8	5·7

Livestock, Nov. 2012: cattle, 49,460 (including 14,639 dairy cows); pigs, 7,044; sheep, 7,335; horses (March 2010), 5,667; poultry (March 2010), 160,374.

Forestry
The forest area (2011: 87,275 ha.) comprises 34·0% of the total (256,978 ha.).

INDUSTRY

In June 2012, 244 establishments (with 50 or more employees) employed 86,572 persons; of these 19,473 were engaged in manufacturing of motor vehicles, parts and accessories, 17,263 in machine construction, 10,165 in iron and steel production, 2,869 in steel construction, 1,981 in coalmining and quarrying of stone, sand and clay and 1,618 in electrical engineering. In 2011 the coalmines produced 1·74m. tonnes of coal. Two blast furnaces and eight steel furnaces produced 4·2m. tonnes of pig iron and 5·2m. tonnes of crude steel in 2011.

Labour
Economically active persons totalled 456,500 at the 1%-sample survey of the microcensus of 2011. Of the total, 418,400 were employees and 38,100 self-employed; 133,600 were engaged in power supply, mining, manufacturing and building, 113,700 in commerce and transport, fewer than 5,000 in agriculture and forestry, and 206,400 in other industries and services.

COMMUNICATIONS

Roads
At 1 Jan. 2012 there were 2,044 km of classified roads, comprising 240 km of Autobahn, 333 km of federal roads, 845 km of first-class and 626 km of second-class highways. Number of registered motor vehicles, 1 Jan. 2012, 702,616, including 594,513 passenger cars, 30,873 lorries, 1,275 buses, 16,536 tractors and 56,310 motorcycles.

Shipping
In 2011, 2,001 ships docked in Saarland ports, bringing 2·4m. tonnes of freight. In the same year 2,001 ships left the ports, carrying 1·1m. tonnes of freight.

SOCIAL INSTITUTIONS

Justice
There is a constitutional court (*Verfassungsgerichtshof*), a regional court of appeal, a regional court, ten local courts, a *Bundesland* labour court, three labour courts, a *Bundesland* social court, a social court, a finance court, a higher administrative court and an administrative court.

Education
In 2011–12 there were 161 primary schools with 30,925 pupils; 38 special schools with 3,738 pupils; 54 *Realschulen* and *Erweiterte Realschulen* with 19,985 pupils; 35 high schools with 26,919 pupils; 18 comprehensive high schools with 12,547 pupils; four *Freie Waldorfschulen* with 1,334 pupils; two evening intermediate schools with 285 pupils; one evening high school with 180 pupils; one Saarland College with 105 pupils; 36 part-time vocational schools with 20,716 pupils; year of commercial basic training: 50 institutions with 1,801 pupils; 13 advanced full-time vocational schools and schools for technicians with 2,444 pupils; 40 full-time vocational schools with 3,496 pupils; 35 *Fachoberschulen* (full-time vocational schools leading up to vocational colleges) with 6,554 pupils; nine business and technical grammar schools with 1,577 pupils; 33 schools for public health occupations with 2,906 pupils. The number of pupils attending the vocational schools amounts to 39,494.

In the winter term 2011–12 there was the University of the Saarland with 17,635 students; one academy of fine art with 367 students; one academy of music with 491 students; one vocational college (economics and technics) with 5,312 students; one vocational college for public administration with 482 students; and one university of applied science (health care and prevention) with 2,577 students.

Health

In 2011 the 23 hospitals in the Saarland contained 6,451 beds and treated 259,106 patients. The average occupancy rate was 86·9%. There were also 19 out-patient and rehabilitation centres that treated 28,562 patients in 2010. On average they were using 73·7% of their capacity.

RELIGION

In 2008, 64·1% of the population were Roman Catholics and 19·5% were Protestants.

CULTURE

Tourism

In 2011, 18,514 beds were available in 289 places of accommodation (of nine or more beds). 802,023 guests spent 2,330,386 nights in the Saarland, staying an average of 2·9 days each.

FURTHER READING

Statistical Information: Landesamt für Zentrale Dienste, Statistisches Amt Saarland (Virchowstrasse 7, 66119 Saarbrücken). *Director:* Michael Sossong. The most important publications are: *Statistisches Jahrbuch Saarland* (annual).—*Saarland in Zahlen* (special issues).—*Einzelschriften zur Statistik des Saarlandes* (special issues).—*Statistik-Journal* (quarterly magazine).

Website (German only): http://www.statistik.saarland.de

Saxony

Freistaat Sachsen

KEY HISTORICAL EVENTS

The former kingdom of Saxony was a member state of the German Empire from 1871 until 1918, when it became the state of Saxony and joined the Weimar Republic. After the Second World War it was one of the five states in the German Democratic Republic until German reunification in 1990. It has been home to much of Germany's cultural history. In the 18th century, the capital of Saxony, Dresden, became the cultural capital of northern Europe earning the title 'Florence of the North', and the other great eastern German city, Leipzig, was a lively commercial city with strong artistic trends. The three cities of Dresden, Chemnitz and Leipzig formed the industrial heartland of Germany which, after World War II, was the manufacturing centre of the GDR.

TERRITORY AND POPULATION

The area is 18,420 sq. km. It is divided into three administrative regions, three urban districts, ten rural districts and 458 communes. Population on 31 Dec. 2011 was 4,137,051 (2,109,025 females, 2,028,026 males); density, 225 per sq. km.

The capital is Dresden, with a population (31 Dec. 2011) of 529,781.

SOCIAL STATISTICS

Statistics for calendar years:

	Live births	Marriages	Divorces	Deaths
2008	34,411	17,397	7,715	48,997
2009	34,093	17,585	7,687	50,365
2010	35,091	18,391	7,285	50,909
2011	34,423	17,580	7,146	50,628

CONSTITUTION AND GOVERNMENT

The *Bundesland* was reconstituted as the Free State of Saxony on former GDR territory in 1990. It has four seats in the Bundesrat.

RECENT ELECTIONS

At the Diet elections of 30 Aug. 2009 the Christian Democratic Union won 58 of 132 seats, with 40·2% of the vote; the Left, 29, with 20·6%; the Social Democratic Party, 14, with 10·4%; the Free Democrats, 14, with 10·0%; Alliance '90/the Greens, 9, with 6·4% and the extreme right-wing National Democratic Party, 8, with 5·6%. Turnout was 52·2%.

CURRENT GOVERNMENT

The *Prime Minister* is Stanislaw Tillich (CDU).

Government Website: http://www.sachsen.de

ENERGY AND NATURAL RESOURCES

Agriculture

Area and yield of the most important crops:

	Area (in 1,000 ha.)			Yield (in 1,000 tonnes)		
	2010	2011	2012	2010	2011	2012
Maize	84·3	96·1	106·2	2,597·9	3,523·9	3,527·3
Fodder	223·5	225·5	226·7	1,455·4	1,662·3	1,574·8
Wheat	198·2	198·2	164·1	1,367·7	1,307·2	1,118·5
Barley	124·8	117·9	131·6	798·2	647·2	825·5
Potatoes	7·0	7·4	6·8	276·9	329·8	295·6
Rye	38·1	34·7	41·1	175·9	140·2	230·9

Livestock in May 2012 (in 1,000): cattle, 500 (including dairy cows, 188); pigs, 636; sheep (Nov. 2011), 81.

INDUSTRY

In Sept. 2012, 1,315 establishments (with 50 or more employees) employed 213,263 persons.

Labour

The unemployment rate was 9·9% in Nov. 2012.

COMMUNICATIONS

Roads

On 1 Jan. 2012 there were 541·7 km of autobahn and 2,525·6 km of main roads. There were 2,468,071 registered motor vehicles, including 2,081,384 passenger cars, 222,574 lorries and tractors, 3,730 buses and 145,827 motorcycles.

Civil Aviation

Leipzig/Halle airport handled 2,263,668 passengers in 2011.

SOCIAL INSTITUTIONS

Education

In 2012–13 there were 831 primary schools (*Grundschulen*) with 124,235 pupils and 8,371 teachers; 336 secondary schools (*Mittelschulen*) with 94,536 pupils and 8,563 teachers; 153 grammar schools (*Gymnasien*) with 88,818 pupils and 7,814 teachers; and 158 high schools (*Förderschulen*) with 18,948 pupils and 3,202 teachers. There were five *Freie Waldorfschulen* (private) with 1,494 pupils and 126 teachers; 282 professional training schools with 105,106 students and 6,226 teachers; and ten adult education colleges with 2,485 students and 190 teachers. In 2011–12 there were seven universities with 78,253 students, 11 polytechnics with 29,628 students, six art schools with 2,708 students and two management colleges with 1,046 students; in 2011–12 there were a total of 111,653 students in higher education institutions.

Health

In 2011 there were 80 hospitals with 26,467 beds. There were 15,569 doctors and 3,861 dentists.

RELIGION

In 2010, 20·3% of the population belonged to the Evangelical Church and 3·6% were Roman Catholic.

CULTURE

Tourism

In 2011 there were 119,998 beds in 2,159 places of accommodation. There were 6,559,975 visitors during the year.

FURTHER READING

Statistical office: Statistisches Landesamt des Freistaates Sachsen, Postfach 1105, 01911 Kamenz. It publishes *Statistisches Jahrbuch des Freistaates Sachsen* (since 1990).

Saxony-Anhalt
Sachsen-Anhalt

KEY HISTORICAL EVENTS

Saxony-Anhalt has a short history as a state in its own right. Made up of a patchwork of older regions ruled by other states, Saxony-Anhalt existed between 1947 and 1952 and then, after reunification in 1990, it was re-established. Geographically, it lies at the very heart of Germany and despite the brevity of its federal status, the region has some of the oldest heartlands of German culture.

TERRITORY AND POPULATION

The area is 20,450 sq. km. It is divided into three county-free cities, 11 rural districts and 219 communes. Population on 31 Dec. 2011 was 2,313,280.

The capital is Magdeburg, with a population (31 Dec. 2009) of 230,500.

SOCIAL STATISTICS

Statistics for calendar years:

	Live births	Marriages	Divorces	Deaths
2008	17,697	10,515	4,994	29,905
2009	17,144	10,346	4,730	30,480
2010	17,300	10,453	4,500	30,729
2011	16,837	10,264	4,808	30,183

CONSTITUTION AND GOVERNMENT

The *Bundesland* was reconstituted on former GDR territory in 1990. It has four seats in the Bundesrat.

RECENT ELECTIONS

At the Diet election on 20 March 2011 the CDU received 34·3% of votes cast giving them 41 of 105 seats, the Left Party (former Party for Democratic Socialism) 24·6% (29 seats), the SPD 21·6% (26 seats) and Alliance '90/the Greens 6·7% (9). Both the National Democratic Party and the Free Democratic Party failed to win any seats, receiving 4·6% and 3·5% of the vote respectively. Turnout was 51·2%.

CURRENT GOVERNMENT

The *Prime Minister* is Reiner Haseloff (CDU).

Government Website: http://www.sachsen-anhalt.de

ENERGY AND NATURAL RESOURCES

Agriculture
Area and yield of the most important crops:

	Area (in 1,000 ha.)			Yield (in 1,000 tonnes)		
	2009	2010	2011	2009	2010	2011
Cereals	599·4	579·4	569·0	4,467·1	4,032·7	3,371·8
Sugar beet	46·3	45·4	48·9	2,828·9	2,681·2	3,358·8
Potatoes	12·8	12·5	13·8	578·4	516·8	678·8
Maize	18·3	17·7	19·4	148·7	140·4	187·0

Livestock in 2011 (in 1,000): cattle, 341·1 (including dairy cows, 123·8); pigs, 1,235·1; sheep, 83·0.

INDUSTRY

In 2011, 1,430 establishments (with 20 or more employees) employed 130,241 persons; of these, 60,665 were employed in basic industry, 37,575 in capital goods industry and 19,890 in food industry. Major sectors are extraction of metal, metalworking, metal articles; the nutrition industry; mechanical engineering; and the chemical industry.

Labour
Economically active persons totalled 1,105,500 in 2011. Of the total, 1,016,600 were employees and 88,900 self-employed; 318,800 were engaged in mining, manufacturing and building, 266,900 in commerce and transport, 24,200 in agriculture and forestry, and 495,600 in other industries and services.

COMMUNICATIONS

Roads
At 1 Jan. 2012 there were 591 km of motorways, 2,331 km of main and 4,067 km of local roads. At 1 Jan. 2012 there were 1,411,183 registered motor vehicles, including 1,191,910 passenger cars, 90,797 lorries, 2,102 buses and 79,284 motorcycles.

SOCIAL INSTITUTIONS

Education
In 2011–12 there were 938 schools with 177,800 pupils. There were 11 universities and institutes of equivalent status with 55,761 students in 2011.

RELIGION

In 2008, 14·8% of the population were Protestants and 3·5% were Roman Catholics.

CULTURE

Tourism
1,064 places of accommodation provided 61,260 beds in Dec. 2011. There were 2,885,685 visitors during the year.

FURTHER READING

Statistical office: Statistisches Landesamt Sachsen-Anhalt, Postfach 20 11 56, 06012 Halle. It publishes *Statistisches Jahrbuch des Landes Sachsen-Anhalt* (since 1991).

Schleswig-Holstein

KEY HISTORICAL EVENTS

The *Bundesland* is formed from two states formerly contested between Germany and Denmark. Schleswig was a Danish dependency ruled since 1474 by the King of Denmark as Duke of Schleswig. He also ruled Holstein, its southern neighbour, as Duke of Holstein, but he did so recognizing that it was a fief of the Holy Roman Empire. As such, Holstein joined the German Confederation in 1815.

Disputes between Denmark and the powerful German states were accompanied by rising national feeling in the duchies, where the population was part-Danish and part-German. There was war in 1848–50 and in 1864, when Denmark surrendered its claims to Prussia and Austria. Following her defeat of Austria in 1866 Prussia annexed both duchies.

North Schleswig (predominantly Danish) was awarded to Denmark in 1920. Prussian Holstein and south Schleswig became the present *Bundesland* in 1946.

TERRITORY AND POPULATION
The area of Schleswig-Holstein is 15,800 sq. km. It is divided into four urban and 11 rural districts and 1,116 communes. The population (estimate, 31 Dec. 2011) numbered 2,837,641 (1,391,708 males, 1,445,933 females).

The capital is Kiel, with a population (31 Dec. 2011) of 242,041.

SOCIAL STATISTICS
Statistics for calendar years:

	Live births	Marriages	Divorces	Deaths
2008	22,678	16,590	7,459	30,719
2009	21,923	16,345	7,286	31,014
2010	22,578	16,456	7,389	31,201
2011	21,331	16,019	7,431	30,981

CONSTITUTION AND GOVERNMENT
The *Bundesland* has four seats in the Bundesrat.

RECENT ELECTIONS
At the elections of 6 May 2012 the Christian Democrats won 22 of the 69 available seats with 30·8% of votes cast, the Social Democrats 22 with 30·4%, Alliance '90/the Greens 10 with 13·2%, the Free Democrats 6 with 8·2%, the Pirate Party 6 with 8·2% and the South Schleswig Voter Federation 3 with 4·6%. Turnout was 60·1%.

CURRENT GOVERNMENT
The *Prime Minister* is Torsten Albig (b. 1965; SPD).

Following the election of May 2012 the SPD formed a coalition government with Alliance '90/the Greens and the South Schleswig Voter Federation, which represents the Danish and Frisian minorities. Its participation in the coalition signifies the first time in German history that a minority party has been part of a *Bundesland* government.

Government Website: http://www.schleswig-holstein.de

ENERGY AND NATURAL RESOURCES
Agriculture
Area and yield of the most important crops:

	Area (in 1,000 ha.)			Yield (in 1,000 tonnes)		
	2009	2010	2011	2009	2010	2011
Wheat	196	208	211	1,861	1,843	1,679
Sugar beet	7	7	9	476	463	645
Barley	75	52	50	613	407	296
Potatoes	5	5	5	222	190	185
Rye	29	20	19	211	121	104
Oats	7	4	7	43	18	35

Livestock, March 2010: 1,137,172 cattle (including 364,240 dairy cows); 43,584 horses; 1,620,161 pigs; 281,728 sheep; 3,075,226 poultry.

Fisheries
In 2011 the yield of small-scale deep-sea and inshore fisheries was 46,923 tonnes. The catch was valued at €52·3m. in 2011.

INDUSTRY
In Sept. 2011, 1,249 mining, quarrying and manufacturing establishments (with 20 or more employees) employed 121,003 persons; of these, 25,686 were employed in machine construction; 21,296 in food and related industries; 6,926 in electrical engineering; 4,508 in shipbuilding (except naval engineering).

Labour
Economically active persons totalled 1,360,000 in 2011. Of the total, 1,194,000 were employees, 166,000 were self-employed or unpaid family workers; 365,000 were engaged in commerce and transport, 297,000 in power supply, mining, manufacturing and building, 36,000 in agriculture and forestry, and 662,000 in other industries and services.

COMMUNICATIONS
Roads
There were (1 Jan. 2012) 9,891 km of 'classified' roads, comprising 533 km of Autobahn, 1,559 km of federal roads, 3,675 km of first-class and 4,124 km of second-class highways. In Jan. 2012 the number of motor vehicles was 1,816,170 including 1,499,358 passenger cars, 95,850 lorries, 2,532 buses, 50,037 tractors and 135,229 motorcycles.

Shipping
The Kiel Canal *(Nord-Ostsee-Kanal)* is 98·7 km long; in 2011, 33,522 vessels of 154m. GT passed through it.

SOCIAL INSTITUTIONS
Justice
There are a court of appeal, four regional courts, 22 local courts, a *Bundesland* labour court, five labour courts, a *Bundesland* social court, four social courts, a finance court, an upper administrative court and an administrative court.

Education
In 2011–12 there were 553 primary schools *(Grundschulen)* with 7,575 teachers and 103,087 pupils; 169 lower secondary schools *(Hauptschulen)* with 747 teachers and 9,897 pupils; 157 intermediate secondary schools *(Realschulen)* with 1,822 teachers and 26,430 pupils; 107 grammar schools *(Gymnasien)* with 6,715 teachers and 88,528 pupils; 186 comprehensive schools *(Gesamtschulen and Gemeinschaftsschulen)* with 5,331 teachers and 60,381 pupils; 341 other schools (including special schools) with 3,189 teachers and 26,804 pupils; 309 vocational schools with 4,915 teachers and 102,430 pupils.

In the winter term of the academic year 2011–12 there were 31,989 students at the three universities (Kiel, Flensburg and Lübeck) and 22,673 students at 11 further education colleges.

RELIGION
In 2008, 53·8% of the population were Protestants and 6·0% Roman Catholics.

CULTURE
Tourism
4,176 places of accommodation provided 177,816 beds in 2011 for 5,357,001 visitors.

FURTHER READING
Statistical Information: Statistisches Amt für Hamburg und Schleswig-Holstein (Fröbelstr. 15–17, 24113 Kiel). *Director:* Helmut Eppmann. Publications: *Statistisches Taschenbuch Schleswig-Holstein,* since 1954.—*Statistisches Jahrbuch Schleswig-Holstein,* since 1951.—*Statistische Monatshefte Schleswig-Holstein,* since 1949.—*Statistische Berichte,* since 1947.—*Beiträge zur historischen Statistik Schleswig-Holstein,* from 1967.—*Lange Reihen,* from 1977.
Website (German only): http://www.statistik-nord.de

Handbuch Schleswig-Holstein. 35th ed. 2010
Ibs, Jürgen, (ed.) *Historischer Atlas Schleswig-Holstein: Vom Mittelalter bis 1867.* 2004
Lange, Ulrich, (ed.) *Historischer Atlas Schleswig-Holstein: seit 1945.* 1999
Momsen, Ingwer, (ed.) *Historischer Atlas Schleswig-Holstein: 1867 bis 1945.* 2001

State library: Schleswig-Holsteinische Landesbibliothek, Kiel, Schloss. *Director:* Dr Jens Ahlers.

Thuringia
Thüringen

KEY HISTORICAL EVENTS

Thuringia with its capital Erfurt is criss-crossed by the rivers Saale, Werra and Weisse Elster and dominated in the south by the mountains of the Thuringian Forest. Martin Luther spent his exile in Eisenach where he translated the New Testament into German while he lived in protective custody in the castle. Weimar became the centre of German intellectual life in the 18th century. In 1919 Weimar was the seat of a briefly liberal Republic. Only ten miles from Weimar lies Buchenwald, the site of a war-time Nazi concentration camp, which is now a national monument to the victims of fascism.

TERRITORY AND POPULATION

The area is 16,172 sq. km. Population on 31 Dec. 2011 was 2,221,222 (1,123,918 females); density, 137 per sq. km. It is divided into six urban districts, 17 rural districts and 913 communes.

The capital is Erfurt, with a population (31 Dec. 2009) of 203,800.

SOCIAL STATISTICS

Statistics for calendar years:

	Live births	Marriages	Divorces	Deaths
2008	17,332	9,810	4,417	26,276
2009	16,854	9,755	4,344	26,774
2010	17,527	10,074	4,113	26,701
2011	17,073	9,750	4,197	26,720

CONSTITUTION AND GOVERNMENT

The *Bundesland* was reconstituted on former GDR territory in 1990. It has four seats in the Bundesrat.

RECENT ELECTIONS

At the Diet elections of 30 Aug. 2009 the Christian Democrats (CDU) won 30 of 88 seats, with 31·2% of the vote; the Left 27, with 27·4%; the Social Democrats (SPD) 18, with 18·5%; the Free Democrats (FDP) 7, with 7·6%; and Alliance '90/the Greens 6, with 6·2%. Turnout was 56·2%.

CURRENT GOVERNMENT

The *Prime Minister* is Christine Lieberknecht (CDU).

Government Website: http://www.thueringen.de

ENERGY AND NATURAL RESOURCES

Agriculture
Area and yield of the most important crops:

	Area (in 1,000 ha.)			Yield (in 1,000 tonnes)		
	2009	2010	2011	2009	2010	2011
Wheat	231·6	239·9	239·8	1,722·2	1,509·4	1,570·5
Sugar beet	8·9	8·0	9·3	592·3	493·3	646·6
Barley	113·7	102·2	102·8	770·0	665·1	557·8
Potatoes	2·3	2·1	2·1	92·4	77·3	87·8
Rye	13·2	11·7	11·2	95·9	61·4	57·9
Oats	5·7	5·1	4·6	30·1	19·8	18·9

Livestock, 2011: 340,981 cattle (including 108,839 dairy cows); 8,606 horses (2010); 850,200 pigs; 146,600 sheep; 2,842,804 poultry (2010).

INDUSTRY

In 2011, 1,826 establishments (with 20 or more employees) employed 165,528 persons; of these, 78,931 were employed by producers of materials and supplies, 51,598 by producers of investment goods, 7,839 by producers of durables and 27,160 by producers of non-durables.

Labour
Economically active persons totalled 1,113,000 in 2011, including 551,000 professional workers, 400,000 manual workers and 114,000 self-employed. 364,000 were engaged in production industries, 246,000 in commerce, transport and communications, 26,000 persons in agriculture and forestry, and 478,000 in other sectors. The unemployment rate in 2011 was 9·8%.

COMMUNICATIONS

Roads
At 1 Jan. 2012 there were 498 km of motorways, 1,611 km of federal roads, 4,562 km of first- and second-class highways and 3,078 km of district highways. Number of registered motor vehicles, Jan. 2012, 1,403,296, including 1,160,958 private cars, 94,241 lorries, 2,278 buses, 50,485 tractors and 86,302 motorcycles.

SOCIAL INSTITUTIONS

Education
In 2011–12 there were 467 primary schools with 65,163 pupils, 236 core curriculum schools with 46,842 pupils, 99 grammar schools with 49,572 pupils and 85 special schools with 8,381 pupils; there were 56,511 pupils in technical and professional education, and 1,929 in professional training for the disabled; there were 13 universities and colleges with 53,668 students enrolled.

Health
In 2011 there were 45 hospitals with 16,193 beds. There were 8,412 doctors (one doctor per 264 population).

Welfare
2011 expenditure on social welfare was €498m.

RELIGION

In 2010, 530,988 persons were Protestant and 173,519 persons were Roman Catholic. In 2011, 850 were Jewish.

CULTURE

Tourism
In July 2011 there were 1,417 places of accommodation (with nine or more beds). There were 3,555,700 visitors who stayed 9,486,700 nights in 2011.

FURTHER READING

Statistical information: Thüringer Landesamt für Statistik (Postfach 900163, 99104 Erfurt; Europaplatz 3, 99091 Erfurt). *President:* Günter Krombholz. Publications: *Statistisches Jahrbuch Thüringen,* since 1993. *Kreiszahlen für Thüringen,* since 1995. *Gemeindezahlen für Thüringen,* since 1998. *Thüringen-Atlas,* since 1999. *Statistische Monatshefte Thüringen,* since 1994. *Statistische Berichte,* since 1991. *Faltblätter,* since 1991.

Website (German only): http://www.statistik.thueringen.de
State library: Thüringer Universitäts- und Landesbibliothek, Jena. *Director:* Dr Sabine Wefers.

GHANA

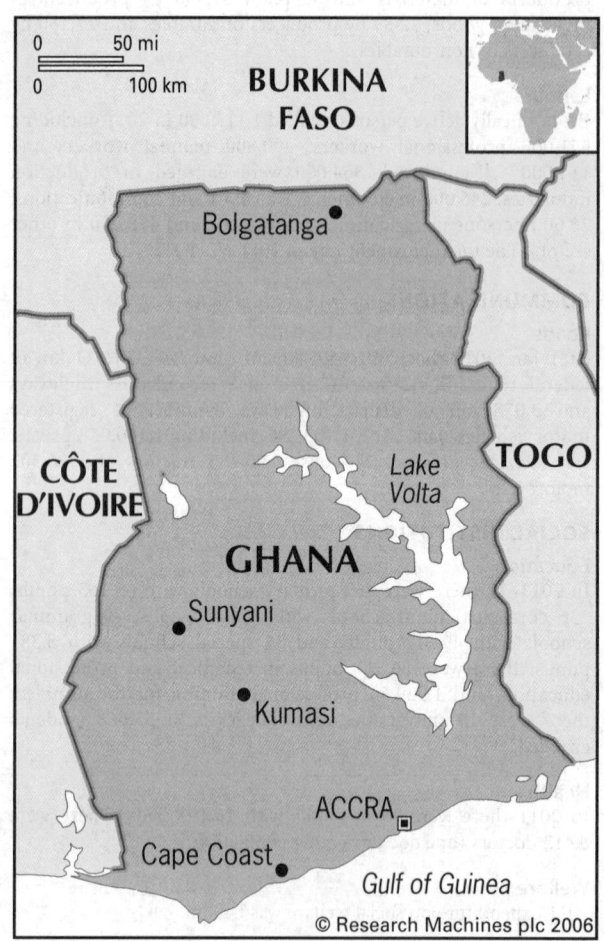

© Research Machines plc 2006

Republic of Ghana

Capital: Accra
Population projection, 2015: 27·32m.
GNI per capita, 2011: (PPP$) 1,584
HDI/world rank: 0·541/135
Internet domain extension: .gh

KEY HISTORICAL EVENTS

Hunter gatherers inhabited Ghana from 8000 BC. There was animal husbandry and agriculture from 1500 BC, with iron technology developing from 100 BC. A centuries-long migration from the north was led by the Guan who moved south from present-day Burkina Faso. The northern Dagomba and Mamprusi kingdoms traded with western Sudanese counterparts from the 12th century, coming under the influence of Islam. In the early 15th century gold and kola nuts from the forested south were exchanged for goods from the savannah regions at towns including Begho.

Portuguese navigators reached Edina (Elmina) on the coast in 1472 and traded with Akan-speaking clans including the Denkyira and Asante. The Portuguese built the São Jorge fort in 1482 and named the region Costa da Mina (Gold Coast). Centred on Tafo and Kumasi, Asante power grew from the 16th century through the supply of gold, ivory, timber and, increasingly, slaves. By the 18th century around 5,000 slaves were shipped annually from European-controlled fortresses on the Gold Coast to plantations in the Caribbean and North America.

After Dutch forces captured Elmina from the Portuguese in 1642, the Dutch West India Company vied with rivals from Britain, Denmark, Sweden and Prussia for control of maritime trade. More than 40 fortresses built along the coast over the next 150 years were frequently attacked.

Britain became the dominant European power on the Gold Coast in the 19th century, focusing trade on timber, ivory and other natural resources. British forces clashed with the expanding Asante kingdom, ruled by Osei Bonsu (1800–24) and then Osei Yaw Akoto (1824–34), until peace was declared in 1831. This gave Britain control over the coastal provinces in return for Asante access to maritime trade. Disputes over control of the gold trade reignited in the 1860s. In 1874 British forces captured Kumasi and proclaimed the Gold Coast Colony, governed from Accra. Following the exile of the Asante king, the British annexed his kingdom and the Northern Territories in Jan. 1902, establishing boundaries with French West Africa.

Mining, agriculture and infrastructure developed with British rule, notably under the governorship of Frederick Guggisberg (1919–27). In the 1930s an educated middle class pushed for inclusion in the colony's administration. While the United Gold Coast Convention sought a gradual shift away from colonial rule after the Second World War, others demanded more rapid change. The Convention People's Party (CPP), formed in 1949 and led by Dr Kwame Nkrumah, campaigned for 'positive action', including strikes, boycotts and civil disobedience.

A new constitution, enacted in 1951, paved the way for legislative elections in which Nkrumah won a seat while serving a three-year jail sentence. Once released, he was a key figure in the parliamentary struggle that led to independence. On 6 March 1957 Ghana became the first independent African nation south of the Sahara, joining the Commonwealth on 1 July 1960, with Nkrumah as president.

In 1966 Nkrumah's regime was overthrown by the military who ruled until 1969 until handing over to a civilian regime under a new constitution. The armed forces seized power again in 1972. In 1979 the Supreme Military Council was toppled in a coup led by Flight-Lieut. J. J. Rawlings. The new government permitted scheduled elections that resulted in victory for Dr Hilla Limann and the People's National Party. However, in Dec. 1981 Rawlings led another coup and established the Provisional National Defence Council to rule.

A pluralist democratic constitution was approved by referendum in April 1992. Rawlings was elected president later that year as a member of the centre-left National Democratic Congress (NDC). He was re-elected in 1996. John Kufuor, of the centre-right New Patriotic Party, became president in Dec. 2000, defeating John Atta Mills of the NDC. During his second term, Kufuor announced the discovery of substantial offshore oil reserves. He was succeeded in Jan. 2009 by John Atta Mills.

TERRITORY AND POPULATION

Ghana is bounded west by Côte d'Ivoire, north by Burkina Faso, east by Togo and south by the Gulf of Guinea. The area is 238,533 sq. km; the 2010 census population was 24,658,823 giving a density of 103·4 persons per sq. km.

The UN gives a projected population for 2015 of 27·32m.

In 2011, 52·2% of the population were urban. An estimated 3m. Ghanaians lived abroad in 2006.

Ghana is divided into ten regions:

Regions	Area (sq. km)	Population, census 2010	Capital
Ashanti	24,389	4,780,380	Kumasi
Brong-Ahafo	39,557	2,310,983	Sunyani
Central	9,826	2,201,863	Cape Coast
Eastern	19,323	2,633,154	Koforidua
Greater Accra	3,245	4,010,054	Accra
Northern	70,384	2,479,461	Tamale
Upper East	8,842	1,046,545	Bolgatanga
Upper West	18,476	702,110	Wa
Volta	20,570	2,118,252	Ho
Western	23,921	2,376,021	Sekondi-Takoradi

In 2010 the capital, Accra, had a population of 1,848,614. Other major cities are Kumasi, Tamale, Sekondi-Takoradi and Ashiaman.

About 42% of the population are Akan. Other tribal groups include Moshi (23%), Ewe (10%) and Ga-Adangme (7%). About 75 languages are spoken; the official language is English.

SOCIAL STATISTICS

2008 estimates: births, 756,000; deaths, 259,000. Rates, 2008 estimates (per 1,000 population): births, 32·4; deaths, 11·1. 2007 life expectancy, 55·6 years for men and 57·4 for women. Infant mortality, 50 per 1,000 live births (2010). Annual population growth rate, 2000–08, 2·2%; fertility rate, 2008, 4·0 births per woman.

CLIMATE

The climate ranges from the equatorial type on the coast to savannah in the north and is typified by the existence of well-marked dry and wet seasons. Temperatures are relatively high throughout the year. The amount, duration and seasonal distribution of rain is very marked, from the south, with over 80" (2,000 mm), to the north, with under 50" (1,250 mm). In the extreme north, the wet season is from March to Aug., but further south it lasts until Oct. Near Kumasi, two wet seasons occur, in May and June and again in Oct., and this is repeated, with greater amounts, along the coast of Ghana. Accra, Jan. 80°F (26·7°C), July 77°F (25°C). Annual rainfall 29" (724 mm). Kumasi, Jan. 77°F (25°C), July 76°F (24·4°C). Annual rainfall 58" (1,402 mm). Sekondi-Takoradi, Jan. 77°F (25°C), July 76°F (24·4°C). Annual rainfall 47" (1,181 mm). Tamale, Jan. 82°F (27·8°C), July 78°F (25·6°C). Annual rainfall 41" (1,026 mm).

CONSTITUTION AND GOVERNMENT

After the coup of 31 Dec. 1981, supreme power was vested in the Provisional National Defence Council (PNDC), chaired by Flight-Lieut. Jerry John Rawlings.

A new constitution was approved by 92·6% of votes cast at a referendum on 28 April 1992. The electorate was 8,255,690; turnout was 43·8%. The constitution sets up a presidential system on the US model, with a multi-party parliament and an independent judiciary. The *President* is elected by universal suffrage for a four-year term renewable once.

The unicameral *Parliament* has 275 members, elected for a four-year term in single-seat constituencies.

National Anthem

'God bless our Homeland, Ghana'; words by the government, tune by P. Gbeho.

GOVERNMENT CHRONOLOGY

Heads of State since 1960. (CPP = Convention People's Party; PNP = People's National Party; NDC = National Democratic Congress; NPP = New Patriotic Party; n/p = non-partisan)

President of the Republic
1960–66   CPP   Kofi Kwame Nkrumah

Chairmen of the National Liberation Council
1966–69   military   Joseph Arthur Ankrah
1969   military   Akwasi Amankwaa Afrifa

Presidential Commission
1969–1970   Akwasi Amankwaa Afrifa (chairman), John Willie Kofi Harlley, Albert Kwesi Ocran

Presidents of the Republic
1970–72   n/p   Edward Akufo-Addo

Chairman of the National Redemption Council
1972–75   military   Ignatius Kutu Acheampong

Chairmen of the Supreme Military Council
1975–78   military   Ignatius Kutu Acheampong
1978–79   military   Frederick Kwasi Akuffo

Chairman of the Armed Forces Revolutionary Council
1979   military   Jerry John Rawlings

President of the Republic
1979–81   PNP   Hilla Limann

Chairman of the Provisional National Defence Council
1981–93   military   Jerry John Rawlings

Presidents of the Republic
1993–2001   NDC   Jerry John Rawlings
2001–09   NPP   John Agyekum Kufuor
2009–12   NDC   John Atta Mills
2012–   NDC   John Dramani Mahama

RECENT ELECTIONS

Presidential elections were held on 7 and 8 Dec. 2012. In the first round the incumbent president John Dramani Mahama of the National Democratic Congress (NDC) won 50·7% of the vote and Nana Akufo-Addo of the New Patriotic Party (NPP) 47·7%. Consequently a second round was not required although the opposition disputed the result. There were six other candidates. Turnout was 79·4%. In parliamentary elections held simultaneously the NDC won 148 of 275 seats, the NPP 123, the People's National Convention 1 and ind. 3. Turnout was 80·0%.

CURRENT GOVERNMENT

President: John Dramani Mahama; b. 1958 (NDC; sworn in 24 July 2012 on the death of President John Atta Mills and for a full term on 7 Jan. 2013).

Vice-President: Kwesi Bekoe Amissah-Arthur.

In March 2013 the government comprised the following:

Minister of Chieftaincy and Traditional Affairs: Henry Seidu Daanaa. *Communications:* Edward Omane Boamah. *Defence:* Mark Woyongo. *Education:* Naana Jane Opoku-Agyemang. *Employment and Labour Relations:* Nii Armah Ashitey. *Energy and Petroleum:* Emmanuel Armah-Kofi Buah. *Environment, Science and Technology:* Joe Oteng-Adjei. *Finance and Economic Planning:* Seth Terkper. *Fisheries and Aquaculture:* Nayon Bilijo. *Food and Agriculture:* Clement Kofi Humado. *Foreign Affairs and Regional Integration:* Hanna Tetteh. *Gender, Children and Social Protection:* Nana Oye Lithur. *Government Business in Parliament:* Benjamin Bewa-Nyog Kunbuor. *Health:* Hanny-Sherry Ayitey. *Information and Media Relations:* Mahama Ayariga. *Interior:* Kwesi Ahwoi. *Justice and Attorney General:* Marietta Brew Appiah-Oppong. *Lands and Natural Resources:* Alhaji Inusah Fuseini. *Local Government and Rural Development:* Akwasi Oppong Ofosu. *Roads and Highways:* Alhaji Amin Amidu Sulemana. *Tourism, Culture and Creative Arts:* Elizabeth Ofosu Agyare. *Trade and Industry:* Haruna Idrissu. *Transport:* Dzifa Aku Attivor. *Water Resources, Works and Housing:* Alhaji Collins Dauda. *Youth and Sports:* Elvis Afriyie Ankrah.

Government Website: http://www.ghana.gov.gh

CURRENT LEADERS

John Dramani Mahama

Position
President

Introduction
John Dramani Mahama became president in July 2012 after the death of John Atta Mills. In line with the constitution, Mahama as vice-president succeeded Mills, also replacing him as leader of the National Democratic Congress (NDC).

Early Life
Mahama was born on 29 Nov. 1958, in Damongo, in the Northern region of Ghana. His father was the first member of parliament for the West Gonja constituency. Mahama graduated in history from the University of Ghana in 1981 and completed a postgraduate degree in communication studies five years later. He then undertook studies in social psychology at the Institute of Social Sciences in Moscow.

Mahama taught history at secondary school level before working in the information, culture and research office of the Japanese embassy in Accra from 1991 until 1995. From 1995–96 he was the international relations, sponsorship, communications and grants manager at the anti-poverty non-governmental organization, Plan International.

Mahama was elected to parliament in 1996 for the Bole/Bamboi constituency. In April 1997 he was appointed deputy minister of communications, becoming minister the following year. He held the position until 2001, when the NDC handed power to the New Patriotic Party. Mahama nonetheless twice retained his parliamentary seat.

From 2003–11 he was a member of the Pan-African Parliament based in Pretoria, South Africa, continuing a keen interest in international affairs. In 2005 he was appointed the minority spokesman for foreign affairs. The NDC regained power in 2008 but in 2009 he gave up his seat in parliament to become vice-president to John Atta Mills.

Career in Office
Following Atta Mills' death, Mahama broadly continued his predecessor's policies. Confronted by a large budget deficit and a tarnished global image, he championed a stimulus package to boost the economy and pledged his commitment to Atta Mills' Better Ghana Agenda, aimed at improving economic health, environmental, educational and employment provision.

Mahama has campaigned to counter the problem of plastic pollution in Africa. He keenly supports the use of information and communication technology to stimulate economic transformation and has sought to improve agricultural productivity and to encourage young people to see farming as a viable business. In the presidential election held on 7–8 Dec. 2012 Mahama secured a further term with 50·7% of the vote. He was sworn in on 7 Jan., although the main opposition New Patriotic Party appealed to the Supreme Court to overturn the result, citing voting irregularities.

DEFENCE

Defence expenditure totalled US$105m. in 2008 (US$4 per capita), representing 0·7% of GDP.

Army
Total strength (2007), 10,000.

Navy
The Navy, based at Sekondi and Tema, numbered 2,000 in 2007.

Air Force
The main air base is at Accra. Personnel strength (2007), 1,500. There were nine combat-capable aircraft although their serviceability was in doubt.

ECONOMY

Agriculture accounted for 31·8% of GDP in 2009, industry 19·0% and services 49·2%.

Overview
One of Africa's biggest borrowers, Ghana is committed to the reform programmes of the IMF and World Bank. Its completion of the Heavily Indebted Poor Countries (HIPC) scheme in 2004 paved the way for debt cancellation by major donors. Nonetheless, the poor infrastructure (particularly power, water and roads) is an obstacle to economic growth and a disincentive to foreign investment.

Steady growth averaging 3–6% since the mid-1980s has resulted from careful macroeconomic management and an increase in exports. More recently growth has been driven by increases in both private and public investment in road-building and agricultural upgrading. A privatization programme was mounted in 1988, with over 100 state-owned enterprises having been sold off to become more than 200 privately-owned companies. Privatization has raised over US$900m., with the Ashanti Goldfields sell-off worth more than US$400m. Only South Africa among sub-Saharan African nations has raised more from privatization.

Ghana's main exports are gold, cocoa and timber, while tourism is increasingly important. Stabilization measures initiated by the new government, as well as continued strong cocoa and gold exports, helped Ghana weather the global economic downturn in 2009. The stabilization plan aims to reduce further the debt-to-(non-oil)-GDP ratio. Oil production, which started in Dec. 2010, led to Ghana being one of the world's fastest-growing economies in 2011, at over 14%.

Currency
The monetary unit is the *cedi* (GHS) of 100 *pesewas*. It was introduced in July 2007 and is equal to 10,000 old cedi (GHC). Inflation was 19·3% in 2009, falling to 10·7% in 2010 and 8·7% in 2011. In July 2011 foreign exchange reserves were US$4,345m., total money supply was ₵15,260m. and gold reserves totalled 281,000 troy oz.

Budget
In 2009 revenues totalled ₵6·0bn. and expenditures ₵7·3bn. Tax revenue accounted for 74·2% of revenues in 2009; current expenditure accounted for 66·9% of expenditures.

Performance
Real GDP growth was 4·0% in 2009 and 8·0% in 2010. In 2011 it increased significantly to 14·4%. Total GDP was US$39·2bn. in 2011.

Banking and Finance
The Bank of Ghana (*Acting Governor*, Kofi Wampah) was established in 1957 as the central bank and bank of issue. At Dec. 2011 its total assets were ₵12,470·6bn. In 2009 there were nine commercial banks, three development banks, three merchant banks, three universal banks and over 100 rural banks.

Foreign investment is actively encouraged with the Ghana Free Zone Scheme offering particular incentives such as full exemption of duties and levies on all imports for production and exports from the zones, full exemption on tax on profits for ten years, and no more than 8% after ten years. It is a condition of the scheme that at least 70% of goods made within the zones must be exported. Within 18 months of the scheme being set up in 1995, 50 projects had been registered.

There is a stock exchange in Accra.

ENERGY AND NATURAL RESOURCES

Environment
Ghana's carbon dioxide emissions from the consumption and flaring of fossil fuels in 2008 were the equivalent of 0·3 tonnes per capita.

Electricity

Installed capacity was 2·1m. kW in 2007. Production (2007) 7·0bn. kWh, approximately 53% from two hydro-electric stations operated by the Volta River Authority, Akosombo (six units) and Kpong (four units). Consumption per capita was 313 kWh in 2007. Since 1998 droughts have caused power cuts owing to the reliance on hydro-electricity. However, production is becoming more dependent on gas—the 678-km West Africa Gas Pipeline bringing gas flows from Nigeria resumed operations in March 2010 following a year-long suspension.

Oil and Gas

Ghana is pursuing the development of its own gas fields and plans to harness gas at the North and South Tano fields located off the western coast. Natural gas reserves, 2005, totalled 24bn. cu. metres. Oil reserves in 2005 were 17m. bbls, although the discovery of an offshore field containing in excess of 300m. bbls of oil was announced in June 2007. Two months later a second offshore field was discovered, potentially with even larger oil resources. Ghana became the world's newest oil producer in Dec. 2010.

Minerals

Gold is one of the mainstays of the economy; Ghana ranks second only to South Africa among African gold producers. Production in 2005 was 63,100 kg. There are also large reserves of bauxite, diamonds and manganese. In 2005 estimated diamond production was 1·0m. carats; manganese (2006), 1·66m. tonnes; bauxite (2006), 886,000 tonnes.

Agriculture

The rural poor earn little and many small farmers have reverted to subsistence farming. The agricultural population in 2002 was 12·97m., of whom 5·74m. were economically active. There were 7·15m. ha. of land under cultivation in 2007 and 11,000 ha. were irrigated. There were 154 registered combine harvesters in 2007. In southern and central Ghana the main food crops are maize, rice, cassava, plantains, groundnuts, yam and taro; and in northern Ghana groundnuts, rice, maize, sorghum, millet and yams. Agriculture presently operates at only 20% of its potential and is an area that is to be a major focus of investment.

Production of main food crops (2007 unless otherwise indicated, in 1,000 tonnes): cassava, 10,218; yams, 4,376; plantains, 3,234; cocoyams, 1,690; maize 1,220; cocoa beans (2006–07), 615; coconuts (2002), 315; oranges (2002), 300; chillies and green peppers (2002), 270; rice (milled), 185; sorghum, 155; millet, 133. Cocoa is the main cash crop. The government estimates that more than 40% of the population relies either directly or indirectly on cocoa as a source of income. It contributes approximately 13% of GDP. Ghana is the second largest cocoa bean producer in the world after Côte d'Ivoire, and the second largest producer of both yams and cocoyams, after Nigeria.

Livestock, 2002: goats, 3·23m.; sheep, 2·92m.; cattle, 1·33m.; pigs, 310,000; chickens, 24m.

Forestry

There were 4·94m. ha. of forest in 2010, or 22% of the total land area. Reserves account for some 30% of the total forest lands. Timber production in 2007 was 35·49m. cu. metres.

Fisheries

In 2007 total catch was 330,486 tonnes, of which 245,730 tonnes came from sea fishing. Imports of fishery commodities were valued at US$129m. in 2008.

INDUSTRY

Ghana's industries include mining, lumbering, light manufacturing and food processing.

Labour

In 2000 the number of economically active persons totalled 8,292,100. Females constituted 50% of the workforce in 2000. Ghana has among the highest proportions of women in the workforce in the world.

The unemployment rate was 8·2% in 2000.

INTERNATIONAL TRADE

Imports and Exports

In 2007 imports (c.i.f.) were valued at US$7,206·7m.; exports (f.o.b.) totalled US$4,186·7m. Principal imported commodities, 2007: machinery and transport equipment, 46·4%; manufactured goods, 17·9%; food and live animals, 12·9%; chemicals and related products, 12·4%. Principal exports, 2007: gold, 34·8%; cocoa, 25·1%; fruit and vegetables, 7·9%; animal and vegetable oils and fats, 7·0%. Main import suppliers in 2007: Germany, 14·6%; China, 11·2%; USA, 7·7%. Main export markets in 2007: South Africa, 31·3%; Netherlands, 16·9%; USA, 10·8%.

COMMUNICATIONS

Roads

In 2005 there were 57,614 km of roads, including 11,177 km of highways, main and national roads. About 14·9% of all roads are paved. A Road Sector Strategy and Programme to develop the road network ran from 1995 to 2000. There were 493,800 passenger cars in use in 2007, 158,400 lorries and vans, and 121,100 buses and coaches. Motorcycles and mopeds numbered 149,100.

Rail

Total length of railways in 2005 was 953 km of 1,067 mm gauge. In 2005 railways carried 1·8m. tonnes of freight and 2·3m. passengers.

Civil Aviation

There is an international airport at Accra (Kotoka). In 2003 scheduled airline traffic of Ghana-based carriers flew 12m. km, carrying 241,000 passengers (all on international flights). Accra handled 623,000 passengers (all on international flights) in 2001.

Shipping

The chief ports are Tema and Takoradi. In 2008, 8·7m. tonnes of cargo were handled at Tema and 4·0m. tonnes at Takoradi. There is inland water transport on Lake Volta. In Jan. 2009 there were 14 ships of 300 GT or over registered, totalling 15,000 GT. The Volta, Ankobra and Tano rivers provide 168 km of navigable waterways for launches and lighters.

Telecommunications

Ghana Telecom was privatized in 1996. There were 277,900 fixed telephone lines in 2010 (11·4 per 1,000 inhabitants). Mobile phone subscribers numbered 17·44m. in 2010. There were 85·5 internet users per 1,000 inhabitants in 2010. Fixed internet subscriptions totalled 92,700 in 2009 (3·9 per 1,000 inhabitants). In June 2012 there were 1·3m. Facebook users.

SOCIAL INSTITUTIONS

Justice

The Courts are constituted as follows:

Supreme Court. The Supreme Court consists of the Chief Justice who is also the President, and not less than four other Justices of the Supreme Court. The Supreme Court is the final court of appeal in Ghana. The final interpretation of the constitution is entrusted to the Supreme Court.

Court of Appeal. The Court of Appeal consists of the Chief Justice with not less than five other Justices of the Appeal court and such other Justices of Superior Courts as the Chief Justice may nominate. The Court of Appeal is duly constituted by three Justices. The Court of Appeal is bound by its own previous decisions and all courts inferior to the Court of Appeal are bound to follow the decisions of the Court of Appeal on questions of law. Divisions of the appeal court may be created, subject to the discretion of the Chief Justice.

High Court of Justice. The Court has jurisdiction in civil and criminal matters as well as those relating to industrial and labour disputes including administrative complaints. The High Court of Justice has supervisory jurisdiction over all inferior Courts and any adjudicating authority and in exercise of its supervisory jurisdiction has power to issue such directions, orders or writs including writs or orders in the nature of habeas corpus, certiorari, mandamus, prohibition and quo warranto. The High Court of Justice has no jurisdiction in cases of treason. The High Court consists of the Chief Justice and not less than 12 other judges and such other Justices of the Superior Court as the Chief Justice may appoint.

Under the Provisional National Defence Council which ruled from 1981 to 2001 public tribunals were established in addition to the traditional courts of justice.

The population in penal institutions in Sept. 2006 was 12,736 (55 per 100,000 of national population).

Education
Schooling is free and compulsory, and consists of six years of primary, three years of junior secondary and three years of senior secondary education. In 2006–07 there were 3·37m. pupils in primary schools with 105,257 teachers; and 1·13m. pupils with 67,005 teachers in junior secondary schools. University education is free. In 2007 there were 140,017 students in tertiary education and 4,011 academic staff. Adult literacy in 2008 was 66%. In 1970 adult literacy was just 31%.

In 2005 public expenditure on education came to 5·5% of GNI.

Health
In 2002 there were 1,842 physicians, 13,102 nurses, 4,094 midwives and 1,433 pharmacists. In 2003 there were an estimated 350,000 people living with HIV, mainly women.

Ghana has been one of the most successful countries in reducing undernourishment in the past 20 years. Between 1990–92 and 2001–03 the proportion of undernourished people declined from 37% of the population to just 12%.

RELIGION

An estimated 30% of the population are Muslim and 24% Christian, with 38% adherents to indigenous beliefs and 8% other religions. In Feb. 2013 the Roman Catholic church had one cardinal.

CULTURE

World Heritage Sites
Ghana has two sites on the UNESCO World Heritage List: Forts and Castles, Volta, Greater Accra, Central and Western Regions (inscribed on the list in 1979), Portuguese trading posts built between 1482 and 1786 along the coast; Asante Traditional Buildings (1980), the remains of the Asante civilization that peaked in the 18th century.

Press
There were 12 paid-for daily newspapers in 2008 with a combined circulation of 210,000 plus 95 paid-for non-dailies.

Tourism
There were 587,000 foreign tourists in 2007, spending US$1,172m.

DIPLOMATIC REPRESENTATIVES

Of Ghana in the United Kingdom (13 Belgrave Sq., London, SW1X 8PN)
High Commissioner: Prof. Kwaku Danso-Boafo.

Of the United Kingdom in Ghana (Osu Link, off Gamel Abdul Nasser Ave., Accra)
High Commissioner: Peter Jones.

Of Ghana in the USA (3512 International Dr., NW, Washington, D.C., 20008)
Ambassador: Daniel Ohene Agyekum.

Of the USA in Ghana (24, 4th Circular Rd, Cantonments, Accra)
Ambassador: Gene Cretz.

Of Ghana to the United Nations
Ambassador: Ken Kanda.

Of Ghana to the European Union
Ambassador: Yaw Konadu-Yiadom.

FURTHER READING

Aryeetey, Ernest and Kanbur, Ravi, (eds.) *The Economy of Ghana: Analytical Perspectives on Stability, Growth and Poverty.* 2008
Boafo-Arthur, Kwame, (ed.) *Ghana: One Decade of the Liberal State.* 2007
Carmichael, J., *Profile of Ghana.* 1992.—*African Eldorado: Ghana from Gold Coast to Independence.* 1993
Gocking, Roger S., *The History of Ghana.* 2005
Herbst, J., *The Politics of Reform in Ghana, 1982–1991.* 1993
Ninsin, Kwame A. (ed.) *Ghana: Transition to Democracy.* 2002
Odotei, Irene K. and Awedoba, Albert K. (eds.) *Chieftaincy in Ghana: Culture, Governance and Development.* 2006
Petchenkine, Y., *Ghana in Search of Stability, 1957–1992.* 1992
Rimmer, D., *Staying Poor: Ghana's Political Economy, 1950–1990.* 1993
Talton, Benjamin, *Politics of Social Change in Ghana: The Konkomba Struggle for Political Equality.* 2010
Tettey, Wisdom, Puplampu, Korbia P. and Berman, Bruce J. (eds.) *Critical Perspectives in Politics and Socio-Economic Development in Ghana.* 2003

National Statistical Office: Ghana Statistical Service, P. O. Box GP 1098, Ministry of Finance and Economic Planning (MoFEP) Head Office Building, Accra.

GREECE

© Research Machines plc 2006

Elliniki Dimokratia
(Hellenic Republic)

Capital: Athens
Population projection, 2015: 11·49m.
GNI per capita, 2011: (PPP$) 23,747
HDI/world rank: 0·861/29
Internet domain extension: .gr

KEY HISTORICAL EVENTS

The land that is now Greece was first inhabited between 2000–1700 BC by tribes from the North. This period was followed by the Mycenaean Civilization which was overthrown by the Dorians at the end of the 12th century BC. Its dominant citadels were at Tiryns and Mycenae. What little is known about this period is from stories such as those by Homer written in the 9th or 8th century BC.

The following period, known as the Greek Dark Ages, ended by the 6th century BC when the *polis*, or city state, was formed. Built mainly on coastal plains, the two principal cities were Sparta and Athens. With government based on consensus of a ruling class, and rich in theatre, art and philosophy, the *polis* was the pinnacle of the Greek Classical Age. It was the era of Euripides, Theusidades and Socrates. With strong trade links, Greece also had territories in Southern Italy, Sicily, Southern France and Asia Minor.

Two Persian invasions in the 5th century were checked at Marathon (490 BC) and Thermopylae (480 BC) where Spartans held off a great force of Persian soldiers. In 431 BC rivalry between the dominant city states erupted into the Peloponnesian War. In 404 BC Sparta defeated Athens, but in the next century Sparta itself fell to Thebes (371 BC).

Led by Philip II of Macedon, the Macedonians defeated the city states in 338 BC. The *poleis* were forced to unify under his rule. With Plato and Aristotle active at this time, the latter serving as a tutor to Philip's son Alexander, this was a period of cultural enrichment. When Philip was assassinated in 336 BC, Alexander, then aged of 20, succeeded him. He spent the next 13 years on a relentless campaign to expand the Macedonian territories. The Greek Empire stretched to the edge of India and encompassed most of the known civilized world.

Following Alexander's death in 323 BC, the empire gradually disintegrated. By the end of the 2nd century AD, the Romans had defeated the Macedonians and Greece was incorporated into the Roman Empire. It remained in Roman hands until it became part of the Byzantine Empire in the 4th century AD. A population of Greek-speaking Christians had its power base in Constantinople.

Over the next six centuries Greece was invaded by Franks, Normans and Arabs but remained part of the Byzantine Empire. Following the Empire's decline in the 11th century, Greece was incorporated into the Ottoman Empire in 1460. Apart from a period under Venetian control between 1686–1715, Greece was part of Turkey until the Greek War of Independence.

Greece broke away from the Ottoman Empire in the 1820s and was declared a kingdom under the protection of Great Britain, France and Russia. Many Greeks were left outside the new state but Greece's area increased by 70%, the population growing from 2·8m. to 4·8m., after the Treaty of Bucharest (1913) recognized Greek sovereignty over Crete.

King Constantine opted for neutrality in the First World War, while Prime Minister Venezelos favoured the Entente powers. This National Schism led to British and French intervention which deposed Constantine on 11 June 1917. When his son Alexander died on 25 Oct. 1920, he returned and reigned until 1922. He was forced to abdicate by a coup after defeat by Turkey and the loss of Smyrna. The Treaty of Lausanne (1923) recognized Smyrna as Turkish with Eastern Thrace and the islands of Imvros and Tenedos, all of which had been ceded to Greece by the 1920 Treaty of Sevres. An exchange of Christian and Muslim populations followed. Resistance to Italian demands brought Greece into the Second World War when Germany had to come to the aid of the hard-pressed Italians. Athens was occupied on 27 April 1941. The occupation lasted until 15 Oct. 1944.

Shortly before the German withdrawal the leading communist resistance movement established a provisional government to supplant the monarchy and the existing government-in-exile. British attempts to oversee a coalition government between the communists and royalist groups collapsed in Dec. 1944. Two months of fierce fighting saw the communists claim most of the country bar Athens and Salonika, before the uprising was suppressed by the British. The communists boycotted the general election of March 1946, which returned a royalist government. When the king was restored to the throne in Sept. 1946 the communists responded with a guerrilla war. The Greek army, heavily backed by the USA, defeated the insurgents and the civil war came to an end in Oct. 1949, with 50,000 dead and around half a million people displaced.

The late 1950s saw the emergence of the Left, capitalizing on the movement for union with Cyprus and unease over NATO membership (1952). A military coup in 1967 led to the authoritarian rule of the 'Colonels' headed by George Papadopoulos. A republic was declared on 29 July 1973.

Papadopoulos was ousted by Brigadier-General Demetrios Ioannidis, head of the military police, who returned some civil powers but kept much power for himself. In 1974 an Athens-supported coup attempt against Cyprus's President Makarios led to a Turkish invasion of the island and the establishment of the 'Turkish Republic of Northern Cyprus'. Ioannidis' government fell and Konstantinos Karamanlis returned from 13 years in exile to head a civilian government of national unity. The monarchy was abolished by a referendum on 8 Dec. 1974 and a new constitution the following year established a parliamentary republic with an executive president. The 1981 election brought Andreas

Papandreou to power at the head of a socialist government. Earlier that year Greece had become the tenth member of the EU. Re-elected in 1985, Papandreou imposed economic austerity to combat inflation and soaring budgets but industrial unrest and evidence of widespread corruption led to his fall and a succession of weak governments. Papandreou returned to power in Oct. 1993 but ill-health forced his resignation two years later. His successor Constantinos Simitis took a more pro-European stance, instituting economic reforms to prepare the way for entry into European Monetary Union (EMU).

Kostas Karamanlis led the Conservative New Democracy party to power in 2004, ending over a decade of rule by Pasok. In the same year Athens hosted the Olympic Games. Karamanlis won a second term in Sept. 2007. In March 2008 his government blocked Macedonia's accession to NATO because of a long-running dispute about Macedonia's name. Greece already had a constituent province called Macedonia.

Crippled by the global financial crisis of 2008–09, Greece threatened to default on its public debt. EU partners agreed to guarantee a potential €110bn. bail-out in return for a swathe of austerity measures that prompted a series of general strikes in 2010 and 2011. By late 2011 the crisis had deepened amid fears that the entire eurozone could be destabilized. In Oct. 2011 eurozone leaders agreed to a 50% write-off of Greek debt in return for additional austerity measures. The Greek prime minister, George Papandreou, resigned after widespread criticism of his plan to put the rescue package to a referendum. A national unity government under Lucas Papademos was established ahead of elections that were held in May 2012. In Feb. 2012 the economy received a further €130bn. bailout from the EU and IMF.

TERRITORY AND POPULATION

Greece is bounded in the north by Albania, the Former Yugoslav Republic of Macedonia (FYROM) and Bulgaria, east by Turkey and the Aegean Sea, south by the Mediterranean and west by the Ionian Sea. The total area is 131,957 sq. km (50,949 sq. miles), of which the islands account for 25,026 sq. km (9,663 sq. miles).

The population was 10,787,690 (provisional) according to the census of March 2011, giving a density of 81·8 per sq. km.

The UN gives a projected population for 2015 of 11·49m.

In 2011, 61·7% of the population lived in urban areas. There were 761,813 resident foreign nationals in 2001. A further 5m. Greeks are estimated to live abroad.

Areas and populations according to the 2001 census:

Geographic Region/ Department[1]	Area (sq. km)	Population[2]	Chief town
Aegean Islands	9,122	508,807	
Chios	904	53,408	Chios
Cyclades	2,572	112,615	Hermoupolis
Dodecanese	2,714	190,071	Rhodes
Lesbos	2,154	109,118	Mytilene
Samos	778	43,595	Samos
Attica	3,808	3,761,810	Athens (Athinai)
Central Greece and Euboea	21,010	829,758	
Aetolia and Acarnaia	5,461	224,429	Messolonghi
Boeotia	2,952	131,085	Levadeia
Euboea	4,167	215,136	Chalcis
Evrytania	1,869	32,053	Karpenissi
Phocis	2,120	48,284	Amphissa
Phthiotis	4,441	178,771	Lamia
Crete	8,336	601,131	
Canea	2,376	150,387	Canea
Heraklion	2,641	292,489	Heraklion
Lassithi	1,823	76,319	Aghios Nikolaos
Rethymnon	1,496	81,936	Rethymnon
Epirus	9,203	353,820	
Arta	1,662	78,134	Arta
Ioannina	4,990	170,239	Ioannina
Preveza	1,036	59,356	Preveza
Thesprotia	1,515	46,091	Hegoumenitsa

Geographic Region/ Department[1]	Area (sq. km)	Population[2]	Chief town
Ionian Islands	2,307	212,984	
Cephalonia	904	39,488	Argostoli
Corfu	641	111,975	Corfu
Leucas	356	22,506	Leucas
Zante	406	39,015	Zante
Macedonia	34,177	2,424,765	
Cavalla	2,111	145,054	Cavalla
Chalcidice	2,918	104,894	Polygyros
Drama	3,468	103,975	Drama
Florina	1,924	54,768	Florina
Grevena	2,291	37,947	Grevena
Imathia	1,701	143,618	Veroia
Kastoria	1,720	53,483	Kastoria
Kilkis	2,519	89,056	Kilkis
Kozani	3,516	155,324	Kozani
Mount Athos	336	2,262	Karyai
Pella	2,506	145,797	Edessa
Pieria	1,516	129,846	Katerini
Serres	3,968	200,916	Serres
Thessaloniki (Salonika)	3,683	1,057,825	Thessaloniki
Peloponnese	21,379	1,155,019	
Achaia	3,271	322,789	Patras
Arcadia	4,419	102,035	Tripolis
Argolis	2,154	105,770	Nauplion
Corinth	2,290	154,624	Corinthos
Elia	2,618	193,288	Pyrgos
Laconia	3,636	99,637	Sparti
Messenia	2,991	176,876	Calamata
Thessaly	14,037	753,888	
Karditsa	2,636	129,541	Karditsa
Larissa	5,381	279,305	Larissa
Magnesia	2,636	206,995	Volos
Trikala	3,384	138,047	Trikala
Thrace	8,578	362,038	
Evros	4,242	149,354	Alexandroupolis
Rhodope	2,543	110,828	Comotini
Xanthi	1,793	101,856	Xanthi

[1]On 1 Jan. 2011 the departments were replaced with seven decentralized administrations plus Mount Athos, which retained its self-governing status. The new administrations are: Aegean; Attica; Crete; Epirus and Western Macedonia; Macedonia and Thrace; Peloponnese, Western Greece and the Ionian Islands; Thessaly and Central Greece. [2]De facto population.

The largest cities (2001 census populations) are Athens (the capital), 745,514; Thessaloniki, 363,987; Piraeus, 175,697; Patras, 160,400; Peristerion, 137,918; Heraklion, 130,914; Larissa, 124,394; Kallithea, 109,609; Volos, 82,439. The department of Attica, composed of Athens, the port of Piraeus and a number of suburbs, contains about one third of the Greek population. It also contains about 50% of the country's industry and is the principal commercial, financial and diplomatic centre. Efforts have, however, been made to decentralize the economy. The second city, Thessaloniki, with its major port, has grown rapidly in population and industrial development.

The Monastic Republic of **Mount Athos** (or Agion Oros, i.e. 'Holy Mountain'), the easternmost of the three prongs of the peninsula of Chalcidice, is a self-governing community composed of 20 monasteries. The peninsula is administered by a Council of four members and an Assembly of 20 members, one deputy from each monastery. The constitution of 1927 gives legal sanction to the Charter of Mount Athos, drawn up by representatives of the 20 monasteries on 20 May 1924, and its status is confirmed by the 1952 and 1975 constitutions. Women are not permitted to enter. Population, 2001, 2,262.

The modern Greek language had two contesting literary standard forms, the archaizing Katharevousa ('purist'), and a version based on the spoken vernacular, 'Demotic'. In 1976 Standard Modern Greek was adopted as the official language, with Demotic as its core.

SOCIAL STATISTICS

2009: 117,933 live births; 108,316 deaths; 59,212 marriages; 13,163 divorces (2008); 505 still births; 7,749 births to unmarried mothers. 2009 rates: birth (per 1,000 population), 10·5; death, 9·6; marriage, 5·3; divorce, 1·2 (2008). Population growth rate, 2005, 0·4%. In 2006 the suicide rate per 100,000 population was 3·5 (men, 5·9; women, 1·2). Expectation of life at birth, 2007, 76·9 years for males and 81·3 years for females. In 2005 the most popular age range for marrying was 25–29 for females and 30–34 for males. Infant mortality, 2009, 3·2 per 1,000 live births; fertility rate, 2005, 1·2 births per woman (one of the lowest rates in the world). In 2007 Greece received 25,113 asylum applications (up from 4,469 in 2004), equivalent to 2·2 per 1,000 inhabitants. In 2010, 90% of all illegal immigrants into the European Union entered through Greece.

CLIMATE

Coastal regions and the islands have typical Mediterranean conditions, with mild, rainy winters and hot, dry, sunny summers. Rainfall comes almost entirely in the winter months, though amounts vary widely according to position and relief. Continental conditions affect the northern mountainous areas, with severe winters, deep snow cover and heavy precipitation, but summers are hot. Athens, Jan. 48°F (8·6°C), July 82·5°F (28·2°C). Annual rainfall 16·6" (414·3 mm).

CONSTITUTION AND GOVERNMENT

Greece is a presidential parliamentary democracy. A new constitution was introduced in June 1975 and was amended in March 1986, April 2001 and May 2008. The 300-member *Chamber of Deputies* is elected for four-year terms by proportional representation. There is a 3% threshold. Extra seats are awarded to the party which leads in an election. The Chamber of Deputies elects the head of state, the *President*, for a five-year term.

National Anthem

'Imnos eis tin Eleftherian' ('Hymn to Freedom'); words by Dionysios Solomos, tune by N. Mantzaros.
    (Same as Cyprus.)

GOVERNMENT CHRONOLOGY

(EEK = National Unionist Party; EK = Center Union; EPEK = National Progressive Center Union; ERE = National Radical Union; ES = Hellenic Union; FDK = Liberal Democratic Center; KF = Liberal Party; LK = People's Party; ND = New Democracy; Pasok = Panhellenic Socialist Movement; n/p = non-partisan)

Presidents since 1973.

1973	n/p	Georgios C. (George) Papadopoulos
1973–74	military	Phaidon D. Gizikis
1974–75	n/p	Michail D. Stasinopoulos
1975–80	ND	Konstantinos D. Tsatsos
1980–85	ND	Konstantinos G. Karamanlis
1985–90	n/p	Christos A. Sartzetakis
1990–95	ND	Konstantinos G. Karamanlis
1995–2005	n/p	Konstantinos (Kostis) Stephanopoulos
2005–	Pasok	Karolos G. Papoulias

Prime Ministers since 1945.

1945	military	Nikolaos Plastiras
1945	military	Petros Voulgaris
1945	EEK	Panagiotis Kanellopoulos
1945–46	KF	Themistoklis P. Sophoulis
1946	n/p	Panagiotis Poulitsas
1946–47	LK	Konstantinos S. Tsaldaris
1947	n/p	Dimitrios E. Maximos
1947	LK	Konstantinos S. Tsaldaris
1947–49	KF	Themistoklis P. Sophoulis
1949–50	n/p	Alexandros N. Diomidis
1950	LK	Ioannis G. Theotokis
1950	KF	Sophoklis E. Venizelos
1950	EPEK	Nikolaos Plastiras
1950–51	KF	Sophoklis E. Venizelos
1951–52	EPEK	Nikolaos Plastiras
1952	n/p	Dimitrios Kiousopoulos
1952–55	ES	Alexandros L. Papagos
1955–58	ES, ERE	Konstantinos G. Karamanlis
1958–61	ERE	Konstantinos G. Karamanlis
1961–63	ERE	Konstantinos G. Karamanlis
1963	ERE	Panagiotis Pipinelis
1963	n/p	Stilianos Mavromichalis
1963	EK	Georgios A. Papandreou, Sr
1963–64	n/p	Ioannis Paraskevopoulos
1964–65	EK	Georgios A. Papandreou, Sr
1965	EK	Georgios T. Athanasiadis-Novas
1965	n/p	Elias I. Tsirimokos
1965–66	FDK	Stephanos C. Stephanopoulos
1966–67	n/p	Ioannis Paraskevopoulos
1967	ERE	Panagiotis Kanellopoulos
1967	n/p	Konstantinos V. Kollias
1967–73	military	Georgios C. (George) Papadopoulos
1973	n/p	Spiros V. Markezinis
1973–74	n/p	Adamantios Androutsopoulos
1974–80	ND	Konstantinos G. Karamanlis
1980–81	ND	Georgios I. Rallis
1981–89	Pasok	Andreas G. Papandreou
1989	ND	Tzannis P. Tzannetakis
1989–90	n/p	Xenophon E. Zolotas
1990–93	ND	Konstantinos K. Mitsotakis
1993–96	Pasok	Andreas G. Papandreou
1996–2004	Pasok	Costandinos G. (Kostas) Simitis
2004–09	ND	Konstantinos A. (Kostas) Karamanlis
2009–11	Pasok	Georgios Papandreou
2011–12	n/p	Lucas Papademos
2012	n/p	Panagiotis Pikrammenos (interim)
2012–	ND	Antonis Samaras

RECENT ELECTIONS

Karolos Papoulias was re-elected president by the 300-member parliament on 3 Feb. 2010, receiving 266 votes. No other candidates stood.

Parliamentary elections were held on 17 June 2012. Turnout was 62·5%. Seats gained (and % of vote): New Democracy, 129 (29·7%); Syriza (Coalition of the Radical Left), 71 (26·9%); Pasok (Panhellenic Socialist Movement), 33 (12·3%); Independent Greeks, 20 (7·5%); Golden Dawn, 18 (6·9%); Democratic Left, 17 (6·3%); Communist Party, 12 (4·5%).

European Parliament

Greece has 22 (24 in 2004) representatives. At the June 2009 elections turnout was 52·6% (63·2% in 2004). Pasok won 8 seats with 36·7% of votes cast (political affiliation in European Parliament: Progressive Alliance of Socialists and Democrats); ND, 8 with 32·3% (European People's Party); the Communist Party, 2 with 8·4% (European United Left/Nordic Green Left); the LAOS, 2 with 7·2% (Europe of Freedom and Democracy); the SIN, 1 with 4·7% (European United Left/Nordic Green Left); the Ecologist Greens, 1 with 3·5% (Greens/European Free Alliance).

CURRENT GOVERNMENT

President: Karolos Papoulias; b. 1929 (Pasok; sworn in 12 March 2005 and re-elected 3 Feb. 2010).

In March 2013 the coalition government comprised:
Prime Minister: Antonis Samaras; b. 1951 (ND; sworn in 20 June 2012).

Minister of Administrative Reform and e-Governance: Antonis Manitakis. *Interior:* Evripidis Stylianidis. *Finance:* Yannis Stournaras. *Foreign Affairs:* Dimitris Avramopoulos. *National*

Defence: Panos Panagiotopoulos. *Development, Competitiveness, Infrastructure, Transport and Networks:* Kostis Hatzidakis. *Environment, Energy and Climate Change:* Evangelos Livieratos. *Education, Religious Affairs, Culture and Sports:* Konstantinos Arvanitopoulos. *Labour, Social Security and Welfare:* Giannis Vroutsis. *Health:* Andreas Lykourentzos. *Rural Development and Food:* Athanasios Tsaftaris. *Justice, Transparency and Human Rights:* Antonis Roupakiotis. *Public Order and Citizens' Protection:* Nikos Dendias. *Tourism:* Olga Kefalogianni. *Shipping and Aegean:* Kostas Mousouroulis. *Macedonia and Thrace:* Theodoros Karaoglou. *Minister of State:* Dimitris Stamatis.

Office of the Prime Minister: http://www.primeminister.gr

CURRENT LEADERS

Karolos Papoulias

Position
President

Introduction
Karolos Papoulias was sworn in as president of Greece on 12 March 2005, having been elected by an unprecedented parliamentary majority of 279 out of the 300 available votes. A founding member of the Panhellenic Socialist Movement (Pasok) and foreign minister throughout the 1980s and 1990s, Papoulias succeeded Kostis Stephanopoulos in this largely ceremonial role and was re-elected in Feb. 2010.

Early Life
Papoulias was born on 4 June 1929 in the northwestern city of Ioannina and went on to study law at the University of Athens and at the University of Milan in Italy, followed by a doctorate in private international law at the University of Cologne in Germany.

In 1967, following a coup that saw the right-wing Greek government replaced by a military dictatorship under Georgios Papadopoulos, Papoulias left Greece for Cologne. There he founded the resistance organization, the Overseas Socialist Democratic Union, which mobilized exiled Greeks against the military regime. From 1967–74 Papoulias broadcast regularly on Deutsche Welle Radio's Greek programme, denouncing the military government. With the fall of the Papadopoulos dictatorship and the establishment of the democratic Third Hellenic Republic in 1974, Papoulias returned to Greece, where, with fellow returned exile Andreas Papandreou, he helped to found Pasok. With its principles of 'National Independence, Popular Sovereignty, Social Emancipation and Democratic Process', Pasok was to dominate Greek political life throughout the 1980s and 1990s.

At the Nov. 1974 elections Pasok won 13·5% of the vote, coming third in the electoral battle behind the Liberal Party and the conservative New Democracy Party. By Nov. 1977, however, Pasok had doubled its percentage of the votes and become the official opposition. In the elections of Oct. 1981 Pasok won a resounding 48% of the vote and, with Papoulias' long-time associate Andreas Papandreou as prime minister, formed the first socialist government in the history of Greece. Papoulias served as secretary of Pasok's International Relations Committee from 1975–85, and from 1976–80 he was also a member of the party's Co-ordinating Council. In 1977 Papoulias entered parliament for the first time, representing Ioannina as a Pasok member. He was to be re-elected eight times, serving a total of 27 years continuously until 2004. In Oct. 1981 Papoulias gave up his law practice to take up a full-time post as deputy foreign minister in the Pasok government. He held his post until 1984, becoming foreign minister from 1985–90, and again from 1993–96.

Under the leadership of Papoulias, Pasok foreign policy in the Balkan states contributed significantly to the stability of at least some parts of this historically volatile area. In 1976 the Greek government initiated an inter-Balkan conference on economic and technical co-operation, attended by representatives of Yugoslavia, Romania, Bulgaria and Turkey. Similar conferences followed in 1979 and 1982, leading in 1984 to talks on the denuclearization of the Balkan region. Despite the two counties having been officially at war since 1940, Greco-Albanian relations improved dramatically during the mid-1980s and, in 1985, the Greco-Albanian border was reopened for the first time in 45 years, with full normalization of relations in 1987.

Following the death of Andreas Papandreou in June 1996, and the general election of Sept. of that year, Papoulias left the cabinet to become the Greek representative at the Organization for Security and Co-operation in Europe (OSCE).

On 12 Dec. 2004 Prime Minister Karamanlis (New Democracy) and leader of the opposition George Papandreou (Pasok) named Papoulias as the only presidential candidate in the Feb. 2005 election. Gaining 279 out of 300 votes Papoulias was elected by a huge majority of MPs representing all the parliamentary parties.

Career in Office
The appointment of Papoulias to the role of president of the Third Hellenic Republic ended months of speculation that Pasok MPs might withhold the votes required for the endorsement of a new president, forcing early elections just one year after the centre-right New Democracy Party came to power. Papoulias, who enjoyed popularity across the political spectrum, spoke of his desire to see a united Cyprus and expressed hope that Turkey's EU membership talks would be a trigger for progress on the issue.

Following defeat in a snap general election in Oct. 2009, Prime Minister Karamanlis resigned and Papoulias asked Georgios Papandreou, the leader of Pasok, to form a new government. Papoulias was subsequently re-elected for a second presidential term unopposed in Feb. 2010. The Pasok government struggled to address Greece's deteriorating economy and burgeoning sovereign debt crisis, and in Nov. 2011 Papandreou submitted his resignation to Papoulias who subsequently swore in Lucas Papademos, a respected economist and banker, as interim prime minister of a national unity administration. Elections in May 2012 failed to produce a working government but following a further election a month later a three-party conservative-led coalition was formed under Antonis Samaras.

Antonis Samaras

Position
Prime Minister

Introduction
Antonis Samaras was sworn in as prime minister on 20 June 2012. His appointment ended a seven-week leadership vacuum in which successive parties had failed to form a workable coalition. He is faced with rebuilding an economy damaged by the global financial crisis.

Early Life
Samaras was born in Athens on 23 May 1951 and educated at the elite Athens College. A champion tennis player in his youth, he was a close friend of his future political rival, the socialist prime minister, Georgios Papandreou.

Samaras graduated in economics from Amherst College, a private liberal arts college in the USA, and completed an MBA at Harvard Business School. He was elected to parliament in 1977 as the New Democracy representative for Messenia and was appointed finance minister in 1989. He was named minister of foreign affairs in the cabinet of Konstantinos Mitsotakis three years later.

Samaras led hard-line opposition to the former Yugoslav Republic of Macedonia using that name as a newly independent state on the grounds that it is also the name of a bordering Greek province. It was a stance that led to his expulsion from the Greek government. He responded by establishing his own political party, Political Spring, creating a schism on the Greek right that prompted the demise of Mitsotakis' government in 1993.

Political Spring remained a fringe party until its dissolution in 2004, when Samaras rejoined New Democracy. He was elected to the European Parliament that year, resigning in 2007 after winnng a seat in the Greek parliament. In 2009 he defeated Dora Bakogiannis, the daughter of Konstantinos Mitsotakis, for the party leadership. She retaliated by forming her own party but returned to the New Democracy fold shortly after the inconclusive general election of May 2012.

Although New Democracy emerged as the largest party, Samaras was unable to mediate a coalition. The stalemate forced a second election in June 2012, with New Democracy claiming 129 of the 300 available seats (an increase of 21 on the May result). Samaras forged a coalition with the radical left-wing Syriza, which gained the second largest share of the vote, and the centre-left Pasok.

Career in Office

Samaras' first task was to secure a bailout from the European Union, IMF and European Central Bank. Efforts to draft an austerity plan meeting the terms of the bailout were met with public hostility. The coalition suffered a number of defections before the plan passed through parliament. While seeking to increase foreign trade and secure Greece's future inside the EU, Samaras is faced with rising domestic extremism and strong anti-immigrant sentiment.

DEFENCE

Prior to 2001 conscription was generally: (Army) 18 months, (Navy) 21 months, (Air Force) 20 months. However, following a gradual shortening of military service, in 2011 conscription was nine months for all three branches of the armed forces.

In 2008 defence expenditure totalled US$10,141m. (US$946 per capita), representing 2·9% of GDP (the highest percentage in the EU).

Army

The Field Army is currently being reorganized. There are three military regions, with one Army, five corps and five divisional headquarters. Total Army strength (2007) 93,500 (around 38% conscripts). There is also a National Guard of 34,500 whose role is internal security.

Navy

The current strength of the Hellenic Navy includes eight diesel submarines, 14 frigates and three corvettes. Main bases are at Salamis, Patras and Soudha Bay (Crete). Personnel in 2007 totalled 20,000 (including 4,000 conscripts).

Air Force

The Hellenic Air Force (HAF) had a strength (2007) of 31,500 (including 11,000 conscripts). There were 357 combat-capable aircraft including A-7s, F-4s, F-16s and Mirage 2000s. The HAF is organized into Tactical, Air Defence, Air Support and Air Training Commands.

ECONOMY

Agriculture accounted for 4% of GDP in 2007, industry 23% and services 73%.

Greece's 'shadow' (black market) economy is estimated to constitute approximately 27% of the country's official GDP.

In 2011 Greece gave US$425m. in international aid. In terms of a percentage of GNI, however, Greece was one of the least generous major industrialized countries, giving just 0·15%.

Overview

Greece's economy was the smallest in the European Union before ten new members joined in 2004. Until the early 1990s the state held up to 70% of all industrial assets but in 1998 the government began a programme of privatization to meet EU membership criteria. Financial aid from the EU brought improvements in road, rail, harbour and airport links.

The economy grew strongly in the 2000s, outstripping both EU and OECD average growth rates, with real GDP growing by nearly 4% per year on average in the period 1995–2007. It benefited substantially from EU aid, equal to 3·3% of annual GDP, and was further boosted by spending on the 2004 Olympic Games. Tourism is also an important revenue earner, with over 16m. foreign tourists in 2011.

However, much of this development was reversed with the global financial crisis. Real GDP contracted by 3·1% in 2009 and the country fell into a prolonged recession. The government deficit was 15·6% of GDP in 2009 while public debt, at 129·3%, was the highest in the eurozone. In early 2009 the government announced a public sector wage freeze and a surcharge on higher income earners. Its credit rating was downgraded in Dec. 2009. The same month, the incumbent Prime Minister, Georgios Papandreou, outlined proposals to reduce public spending further and to counter rampant tax evasion.

In early 2010 the government pledged to reduce the budget deficit to 8·7% by the end of the year. A US$11·2bn. austerity plan was passed in March 2010, including US$6·5bn. in savings through rises in sales tax, lower holiday bonuses to civil servants and a pension freeze. Public anger over the measures resulted in a succession of general strikes while concerns over Greek finances contributed to a plunge in the value of the euro in early 2010.

In March 2010 the 16 eurozone members reached an agreement in conjunction with the IMF to provide loans worth up to €22bn. in the event that Greece was unable to secure market loans. In April 2010 the Athens government requested a bailout, arguing that high interest rates for Greece prohibited further market borrowing. On 2 May 2010 the EU and the IMF agreed to provide up to €110bn. over three years. In return, Greece agreed to yet deeper spending cuts and to improve revenue raising, prompting more domestic protests. Nonetheless, Greece was unable to meet its pledged targets in 2010, recording a budget deficit of 10·3% and with debt standing at 144·9% of GDP.

Following renewed fears that the government was set to default, eurozone leaders together with the IMF and private lenders agreed to lend a further €109bn. in July 2011. As a result, credit rating agency Standard & Poor's downgraded Greece's credit rating from CCC to CC, citing plans to restructure the economy's debt as, effectively, a 'selective default'. In Nov. 2011 an emergency coalition government led by Lucas Papademos was formed with a mandate to complete bailout and debt-cutting talks as a preliminary to a general election.

In Feb. 2012 a €130bn. rescue package was agreed with the EU and the IMF to support Greece's economic restructuring over four years. Following a successful bond swap deal in early March 2012 when private creditors agreed to a sharp devaluing of their holdings, Athens averted the immediate threat of an uncontrolled default. An €11·5bn. package of spending cuts was agreed in March 2012 as part of bailout conditions. The June 2012 election was won by Antonis Samaras' conservative New Democracy party, easing fears of a Greek eurozone exit.

Reforms are focused on stabilizing the financial system, creating a more competitive labour market and improving tax collection. Spending on pensions accounted for 17% of GDP in 2012, five percentage points higher than the eurozone average, but reforms aimed to reduce this to 14% in 2013. In Sept. 2012 unemployment stood at 26%, with youth unemployment at 58%, the highest rates in the EU. In 2013 Greece entered its sixth year of recession. According to the IMF, the debt burden is unsustainable without debt relief or long-term transfers.

Currency

In June 2000 EU leaders approved a recommendation for Greece to join the European single currency, the euro, and on 1 Jan. 2001 the euro (EUR) became the legal currency at the irrevocable conversion rate of 340·750 drachmas to 1 euro. The euro, which consists of 100 cents, has been in circulation since 1 Jan. 2002. On the introduction of the euro there was a 'dual circulation' period before the drachma ceased to be legal tender on 28 Feb. 2002. Euro banknotes in circulation on 1 Jan. 2002 had a total value of €13·4bn.

Inflation rates (based on OECD statistics):

2002	2003	2004	2005	2006	2007	2008	2009	2010	2011
3·9%	3·4%	3·0%	3·5%	3·3%	3·0%	4·2%	1·3%	4·7%	3·1%

Foreign exchange reserves were US$116m. in Sept. 2009 (US$17,726m. in 1999) and gold reserves 3·61m. troy oz. Total money supply in Aug. 2009 was €97,507m.

Budget

Central government revenue in 2009 totalled €86,663m. (€91,587m. in 2008) and expenditure €120,414m. (€111,627m. in 2008). Of the revenue in 2009, €45,429m. came from taxes. Of the expenditure in 2009, €49,002m. went on social benefits.

Greece's budget deficit in 2011 was 9·1% of GDP (2010, 10·3%; 2009, 15·6%). The required target set by the EU is a budget deficit of no more than 3%.

VAT was raised from 21% to 23% on 1 July 2010. There are reduced rates of 13% and 6·5%.

Performance

Real GDP growth rates (based on OECD statistics):

2002	2003	2004	2005	2006	2007	2008	2009	2010	2011
3·4%	5·9%	4·4%	2·3%	5·5%	3·5%	−0·2%	−3·1%	−4·9%	−7·1%

Up until 2007 Greece had recorded economic growth above the EU average every year since 1996. Total GDP in 2011 was US$289·6bn.

Banking and Finance

The central bank and bank of issue is the Bank of Greece. Its *Governor* is George Provopoulos. There were 39 commercial banks in 2002 (17 Greek and 22 foreign). Total assets of all banks were 41,819bn. drachmas (€127,295m.) in 1999. The six leading banks in 2000 accounted for nearly 80% of assets of all Greek banks. Ranked by size of assets the largest banks were National Bank of Greece, Alpha Bank, Agricultural Bank, Emporiki Bank, EFG Eurobank and Piraeus Bank. In June 2012 gross external debt amounted to US$526,151m. Foreign direct investment was US$1,823m. in 2011.

There is a stock exchange in Athens.

ENERGY AND NATURAL RESOURCES

In 2008, 8·0% of energy consumption came from renewables (wind power, solar power, hydro-electric power, tidal power, geothermal energy and biomass), compared to the European Union average of 10·3%. A target of 18% has been set by the EU for 2020.

Environment

Carbon dioxide emissions from the consumption and flaring of fossil fuels in Greece were the equivalent of 10·0 tonnes per capita in 2008.

Electricity

Installed capacity in 2007 was 13·7m. kW. A national grid supplies the mainland, and islands near its coast. Power is produced in remoter islands by local generators. Total production in 2007 was 63·5bn. kWh; consumption per capita in 2007 was 6,062 kWh. 92% of electricity was produced in 2007 by thermal power stations (mainly using lignite) and the rest was from hydro-electric and geothermal generation.

Oil and Gas

Output of crude petroleum, 2007, 74,000 tonnes; proven reserves, 2007, 5m. bbls. The oil sector plays a critical role in the Greek economy, accounting for more than 70% of total energy demand. Supply is mostly imported but oil prospecting is intensifying. Natural gas was introduced in Greece in 1997 through a pipeline from Russia, and an additional source of supply is liquefied natural gas from Algeria. In 2007 natural gas production ran to 27m. cu. metres.

Minerals

Greece produces a variety of ores and minerals, including in 2006 (with production, in tonnes): bauxite (2,162,900); gypsum and anhydrite (865,216 in 2005); pumice (850,000); magnesite (crude) (463,277); aluminium (164,800); caustic magnesia (68,065 in 2005); nickel ore (21,670); zinc (16,414); silver (25,900 kg); marble (250,000 cu. metres). There is little coal, and the lignite is of indifferent quality (66·31m. tonnes in 2007). Salt production (2005), 198,024 tonnes.

Agriculture

In 2002 there were 2·72m. ha. of arable land and 1·13m. ha. of permanent crops.

The Greek economy was traditionally based on agriculture, with small-scale farming predominating, except in a few areas in the north. There were 824,000 farms in 2003. However, there has been a steady shift towards industry and although agriculture still employs nearly 17% of the population, it accounted for only 7% of GDP in 2002. Nevertheless, prior to the accession of the ten new member countries in May 2004 Greece had a higher percentage of its population working in agriculture than any other European Union member country. Agriculture accounts for 33·1% of exports and 17·9% of imports.

Production (2002, in 1,000 tonnes):

Olives	2,863	Oranges	1,193
Sugar beets	2,834	Potatoes	901
Maize	2,219	Grapes	747
Wheat	2,038	Watermelons	696
Tomatoes	1,753	Peaches	687
Cotton	1,306	Olive oil	382
Alfalfa	1,230	Must	368

Livestock (2003, in 1,000): 9,426 sheep, 5,287 goats, 1,082 pigs, 733 cattle, 56 asses, 28 horses, 28 mules, 38,500 poultry. Livestock products, 2002 (in 1,000 tonnes): milk, 2,069; meat, 472; cheese, 170.

Forestry

Area covered by forests in 2010 was 3·90m. ha., or 30% of the total land area. Timber production in 2007 was 1·74m. cu. metres.

Fisheries

Total catch in 2009 was 83,328 tonnes, mainly from sea fishing. In 2009 the fishing fleet consisted of 17,291 vessels totalling 88,400 GT.

INDUSTRY

The leading company by market capitalization in Greece in March 2012 was Coca-Cola Hellenic Bottling Company (US$7·0bn.).

The main products are canned vegetables and fruit, fruit juice, beer, wine, other alcoholic beverages, cigarettes, textiles, yarn, leather, shoes, synthetic timber, paper, plastics, rubber products, chemical acids, pigments, pharmaceutical products, cosmetics, soap, disinfectants, fertilizers, glassware, porcelain sanitary items, wire and power coils and household instruments.

Production in 1,000 tonnes (2002 unless otherwise indicated): cement (2005), 15,166; residual fuel oil (2007), 7,116; distillate fuel oil (2007), 6,562; petrol (2007), 4,318; crude steel (2006), 2,416; jet fuel (2007), 1,719; iron (concrete-reinforcing bars), 1,454; fertilizers, 1,374; alumina, 750; sulphuric acid, 468; packing materials, 318; soap, washing powder and detergents, 239; textile yarns, 130; soft drinks, 573·6m. litres; beer, 465·1m. litres; wine, 167·2m. litres; cigarettes, 30·5bn. units; glass (2001), 382,343 sq. metres.

Although manufacturing accounts for more than 21% of GDP, Greece's performance is hampered by the proliferation of small, traditional, low-tech firms, often run as family businesses. Food, drink and tobacco processing are the most important sectors, but there are also some steel mills and several shipyards. Shipping is of prime importance to the economy. In addition, there are major programmes under way in the fields of power, irrigation and land reclamation.

Labour
Of the total workforce of 4,822,800 in the period April–June 2004, 4,329,700 persons were employed. 748,200 were engaged in wholesale and retail trade; 569,700 in manufacturing; 533,100 in agriculture, animal breeding, hunting and forestry; and 350,000 in construction. Workers in Greece put in among the longest hours of any country in the world. In 2005 the average worker put in 2,053 hours. Retirement age is 65 years for men and 60 for women, although many men retire before the age of 60. Unemployment was 26·2% in Sept. 2012 (up from 17·7% in 2011 as a whole); unemployment among under 25s stood at 57·6% in Sept. 2012.

INTERNATIONAL TRADE
Following the normalization of their relations, Greece lifted its trade embargo (imposed in Feb. 1994) on Macedonia on 13 Oct. 1995. There are quarrels with Turkey over Cyprus, oil rights under the Aegean and ownership of uninhabited islands close to the Turkish coast.

Imports and Exports
In 2010 imports (c.i.f.) were valued at US$63,321m. and exports (f.o.b.) at US$21,560m. In 2007 principal imports were: machinery and transport equipment, 29·7%; mineral fuels, lubricants and related materials (including petroleum and petroleum products), 15·1%; manufactured goods, 14·5%; chemicals and related products, 13·9%; miscellaneous manufactured articles, 12·4%. Principal exports in 2007 were: manufactured goods, 21·8%; food and live animals, 14·7%; chemicals and related products, 13·8%; machinery and transport equipment, 13·6%; mineral fuels, lubricants and related materials (including petroleum and petroleum products), 12·2%.

In 2007 Germany was the principal supplier of imports (12·8% of the total), ahead of Italy (11·7%), Russia (5·6%), France (5·5%) and China (5·0%). Germany was the leading export market (11·5% of the total), followed by Italy (10·7%), Bulgaria (6·5%), Cyprus (6·5%) and the UK (5·4%). Fellow EU member countries accounted for 57·7% of imports in 2007 and 63·8% of exports.

COMMUNICATIONS
Roads
There were 116,631 km of roads in 2005, including 868 km of motorway, 9,299 km of national roads and 30,864 km of secondary roads. Number of motor vehicles in 2005: 4,303,129 passenger cars (388 per 1,000 inhabitants), 1,186,483 trucks and vans, 1,124,172 motorcycles and 26,829 buses. There were 1,612 road deaths in 2007. With 14·4 deaths per 100,000 population in 2007, Greece has among the highest death rates in road accidents of any industrialized country.

Rail
In 2005 the state network, Hellenic Railways (OSE), totalled 2,576 km, of which 1,743 km were of standard 1,435 mm gauge and 833 km were of narrow gauge (1,000 mm, 750 mm and 600 mm). Railways carried 3·0m. tonnes of freight and 10·0m. passengers in 2005. A 52-km long metro opened in Athens in 2000.

Civil Aviation
There are international airports at Athens (Spata 'Eleftherios Venizelos') and Thessaloniki-Makedonia. The airport at Spata opened in 2001. The national carrier, Olympic Airlines, ceased operations in Sept. 2009 and Olympic Air, the new airline formed from its privatization, commenced flights that month. Apart from the international airports there are a further 25 provincial airports. 5·70m. passengers were carried in 2005, of whom 2·90m. were on domestic and 2·80m. on international flights. Olympic Airlines operates routes from Athens to all important cities of the country, Europe, the Middle East and USA. In 2006 Athens airport (Spata) handled 15,079,708 passengers (9,611,095 on international flights).

Shipping
In Jan. 2009 there were 1,127 ships of 300 GT or over registered, totalling 37·14m. GT. Of the 1,127 vessels registered, 412 were oil tankers, 260 bulk carriers, 252 passenger ships, 120 general cargo ships, 45 container ships, 26 chemical tankers and 12 liquid gas tankers. The Greek-controlled fleet comprised 3,094 vessels of 1,000 GT or over in Jan. 2009, of which 2,361 were under foreign flags and only 737 under the Greek flag.

There is a canal (opened 9 Nov. 1893) across the Isthmus of Corinth (about 7 km). The principal port is Piraeus, which handled 10,477,000 tonnes of cargo in 2008 (4,463,000 tonnes loaded and 6,014,000 tonnes discharged). Other seaports are Thessaloniki, Patras, Volos, Igoumenitsa and Heraklion. Greece has 123 seaports with cargo and passenger handling facilities.

Telecommunications
In 2008 there were 5,975,000 main (fixed) telephone lines. In the same year active mobile phone subscribers numbered 13,799,000 (1,239·0 per 1,000 persons). There were 4,845,000 internet users in 2008. There were 19·9 fixed broadband subscribers per 100 inhabitants in Dec. 2010. In March 2012 there were 3·6m. Facebook users.

SOCIAL INSTITUTIONS
Justice
Judges are appointed for life by the President after consultation with the judicial council. Judges enjoy personal and functional independence. There are three divisions of the courts—administrative, civil and criminal—and they must not give decisions which are contrary to the Constitution. Final jurisdiction lies with a Special Supreme Tribunal.

The Office of Ombudsman (Synigoros) was instituted in 1998. The Ombudsman is Kalliopi Spanou.

The population in penal institutions in Nov. 2008 was 12,300 (109 per 100,000 of national population). The death penalty was abolished for all crimes in 2004.

Education
Public education is provided in nursery, primary and secondary schools, starting at 5½–6½ years of age and free at all levels. Estimated adult literacy rate, 2009, 97·2% (male 98·3%; female 96·1%).

In 2005–06 there were 5,715 nursery schools with 11,461 teachers and 143,401 pupils; 5,753 primary schools with 58,376 teachers and 639,685 pupils; 3,308 high schools (lycea) with 62,149 teachers and 569,887 pupils; 660 secondary technical, vocational and ecclesiastic schools with 16,066 teachers and 123,436 students. In 2002–03 there were 68 technical, vocational and ecclesiastic schools in third level education with 11,357 teachers and 146,270 students; and 19 universities with 11,079 academic staff and 175,597 students.

In 2005 public expenditure on education came to 3·5% of GNI and 9·2% of total government spending.

Health

Doctor and hospital treatment within the Greek national health system is free, but patients have to pay 25% of prescription charges. Those living in remote areas can reclaim a proportion of private medical expenses. In 2002 there were 326 hospitals with a total of 51,781 beds, plus 188 health centres. In 2005 there were 55,556 doctors and 13,438 dentists. In 2007 Greece spent 9·6% of its GDP on health. In 2008, 46·3% of Greek adult males and 33·5% of females smoked on a daily basis. Greece has among the highest smoking rates of any country.

Welfare

The majority of employees are covered by the Social Insurance Institute, financed by employer and employee contributions. Benefits include pensions, medical expenses and long-term disability payments. Social insurance expenditure in 2004 totalled €36,038bn.

The basic pension is available to men (aged 65) and women (aged 62) who have at least 4,500 days of contributions. Men aged 63 and six months qualify if they have 10,000 days of contributions and women aged 58 and six months qualify if they have 10,800 days of contributions. Men and women aged 59 can claim if they have 11,000 days. The basic pension is calculated on the length of the insurance period and on pensionable earnings in the last five years. A reduced early pension is available to men aged 60 and women aged 57 with 4,500 days of contributions. The minimum pension for a single person is €487 per month and the maximum is €2,773.

RELIGION

The Christian Eastern (Greek) Orthodox Church is the established religion to which 91% of the population belong. It is under an archbishop and 67 metropolitans, one archbishop and seven metropolitans in Crete, and four metropolitans in the Dodecanese. The primate of the Greek Orthodox Church (also called the Church of Greece) is Archbishop Hieronymos II of Athens and All Greece (b. 1938). Roman Catholics have three archbishops (in Naxos and Corfu and, not recognized by the State, in Athens) and one bishop (for Syra and Santorin). The Exarchs of the Greek Catholics and the Armenians are not recognized by the State. There are 360,000 Muslims.

Complete religious freedom is recognized by the constitution of 1974, but proselytizing from, and interference with, the Greek Orthodox Church is forbidden.

CULTURE

World Heritage Sites

Greece has 17 sites on the UNESCO World Heritage List: the Temple of Apollo Epicurius at Bassae (1986); the archaeological site of Delphi (1987); The Acropolis, Athens (1987); Mount Athos (1988); Meteora (1988); the Paleochristian and Byzantine monuments of Thessaloniki (1988); the Archaeological Site of Epidaurus (1988); the Medieval City of Rhodes (1988); the archaeological site of Olympia (1989); Mystras (1989); Delos (1990); the monasteries of Daphni, Hossios Luckas and Nea Moni of Chios (1990); the Pythagoreion and Heraion of Samos (1992); the archaeological site of Vergina (1996); the archaeological sites of Mycenae and Tiryns (1999); the historical sites on the Island of Patmos (1999); and the old town of Corfu (2007).

Press

There were 45 daily newspapers published in 2008 (41 paid-for and four free) with a combined daily circulation of 1,447,000. The papers with the highest circulation are the free *City Press* and *Metro*.

Tourism

Tourism is Greece's biggest industry, in 2011 accounting for 16·5% of GDP. In 2011 there were 16,427,000 foreign tourists (of which 10,698,000 from citizens of other European Union countries), up from 15,007,000 in 2010 and 14,915,000 in 2009. There were 397,660 hotel rooms and 763,407 hotel beds in 2010 (358,721 rooms and 682,050 beds in 2005).

Festivals

Independence Day is celebrated on 25 March. The Athens and Epidaurus Festival is the leading arts festival and takes place from May to Oct. Thessaloniki International Film Festival occurs in Nov./Dec. Rockwave Festival, held in Athens in June–July, is one of the biggest music festivals.

DIPLOMATIC REPRESENTATIVES

Of Greece in the United Kingdom (1A Holland Park, London, W11 3TP)
Ambassador: Konstantinos Bikas.

Of the United Kingdom in Greece (1 Ploutarchou St., 106 75 Athens)
Ambassador: Dr David Landsman, OBE.

Of Greece in the USA (2221 Massachusetts Ave., NW, Washington, D.C., 20008)
Ambassador: Christos Panagopoulos.

Of the USA in Greece (91 Vasilissis Sophias Blvd, 101 60 Athens)
Ambassador: Daniel B. Smith.

Of Greece to the United Nations
Ambassador: Anastasios Mitsialis.

Of Greece to the European Union
Permanent Representative: Theodoros Sotiropoulos.

FURTHER READING

Clogg, Richard, *A Concise History of Greece.* 2nd ed. 2002
Couloumbis, Theodore A., Kariotis, Theodore and Bellou, Fotini, (eds.) *Greece in the Twentieth Century.* 2003
Dimitrakopoulos, Dionyssis G. and Passas, Argyris G. (eds.) *Greece in the European Union.* 2004
Doumanis, Nicholas, *A History of Greece.* 2009
Jougnatos, G. A., *Development of the Greek Economy, 1950–91: an Historical, Empirical and Econometric Analysis.* 1992
Legg, K. R. and Roberts, J. M., *Modern Greece: A Civilization on the Periphery.* 1997
Mitsopoulos, Michael and Pelagidis, Theodore, *Understanding the Crisis in Greece: From Boom to Bust.* Revised ed. 2012
Pettifer, J., *The Greeks: the Land and the People since the War.* 1994
Veremis, T., *The Military in Greek Politics: From Independence to Democracy.* 1997
Woodhouse, C. M., *Modern Greece: a Short History.* Rev. ed. 1991

National Statistical Office: National Statistical Service; 46 Pireos & Eponiton str., 185 10 Piraeus.
Website: http://www.statistics.gr

GRENADA

Capital: St George's
Population projection, 2015: 107,000
GNI per capita, 2011: (PPP$) 6,982
HDI/world rank: 0·748/67
Internet domain extension: .gd

KEY HISTORICAL EVENTS

Carib Indians inhabited Grenada when it was sighted by Christopher Columbus in 1498. The Caribs prevented European settlement until French forces landed in 1654. The British took control of Grenada in 1783 and established sugar plantations using African slave labour. Eric Gairy led a violent uprising of impoverished plantation workers in 1951 and became the island's dominant political figure in the lead-up to independence on 7 Feb. 1974. He was ousted by a leftist coup on 13 March 1979. The army took control on 19 Oct. 1983 after a power struggle led to the killing of the prime minister, Maurice Bishop. At the request of a group of Caribbean countries, Grenada was invaded by US-led forces on 25–28 Oct. On 1 Nov. a state of emergency was imposed which ended later in the year with the restoration of the 1973 constitution.

TERRITORY AND POPULATION

Grenada is the most southerly island of the Windward Islands with an area of 344 sq. km (133 sq. miles); the state also includes the Southern Grenadine Islands to the north, chiefly Carriacou (58·3 sq. km) and Petite Martinique. The total population at the 2001 census was 102,632; density, 298 per sq. km. 2011 provisional census population: 103,328.

The UN gives a projected population for 2015 of 107,000.

In 2011, 39·7% of the population were urban. The Borough of St George's, the capital, had 35,559 inhabitants in 2001. 52% of the population is Black, 40% of mixed origins, 4% Indian and 1% White.

The official language is English. A French-African patois is also spoken.

SOCIAL STATISTICS

Births, 2008 estimates, 2,000; deaths, 600. Rates per 1,000 population, 2008 estimates: birth, 19·4; death, 6·1. Life expectancy, 2007: 73·7 years for males, 76·7 years for females. Infant mortality, 2010, nine per 1,000 live births. Annual population growth rate, 2000–08, 0·3%; fertility rate, 2008, 2·3 births per woman.

CLIMATE

The tropical climate is very agreeable in the dry season, from Jan. to May, when days are warm and nights quite cool, but in the wet season there is very little difference between day and night temperatures. On the coast, annual rainfall is about 60" (1,500 mm) but it is as high as 150–200" (3,750–5,000 mm) in the mountains. Average temperature, 27°C.

CONSTITUTION AND GOVERNMENT

The head of state is the British sovereign, represented by an appointed *Governor-General.* There is a bicameral legislature, consisting of a 13-member *Senate,* appointed by the Governor-General, and a 15-member *House of Representatives,* elected by universal suffrage.

National Anthem

'Hail Grenada, land of ours'; words by I. M. Baptiste, tune by L. A. Masanto.

RECENT ELECTIONS

At the elections of 19 Feb. 2013 for the House of Representatives the opposition New National Party won all 15 seats, with 60·0% of the votes cast, against 39·3% for the National Democratic Congress, who failed to win any seats. Turnout was 85%.

CURRENT GOVERNMENT

Governor-General: Cecile La Grenade.

In March 2013 the government comprised:

Prime Minister, Minister of Finance, Energy, National Security, Public Administration, Disaster Management, Home Affairs and Implementation: Keith Mitchell; b. 1946 (New National Party; since 20 Feb. 2013, having previously held office from June 1995 to July 2008).

Deputy Prime Minister, Attorney General, Minister for Legal Affairs, Labour, Local Government, and Carriacou and Petite Martinique Affairs: Elvin Nimrod.

Minister of Agriculture, Lands, Forestry, Fisheries, and the Environment: Roland Bhola. *Communication and Works, Physical Development, Public Utilities and Information Communication Technology:* Gregory Bowen. *Economic Development, Trade, Planning and Co-operatives:* Oliver Joseph. *Education and Human Resource Development:* Anthony Boatswain. *Foreign Affairs and International Business:* Nickolas Steele. *Health and Social Security:* Claris Modeste. *Social Development, Housing and Community Development:* Delma Thomas. *Tourism, Civil Aviation and Culture:* Alexandria Otway-Noel. *Youth, Sports and Ecclesiastical Relations:* Emmalin Pierre.

Government Website: http://www.gov.gd

CURRENT LEADERS

Keith Mitchell

Introduction

Keith Mitchell, leader of the New National Party (NNP), was sworn in as prime minister in Feb. 2013. He previously served as premier over three terms from June 1995 to July 2008.

Early Life

Keith Claudius Mitchell was born on 12 Nov. 1946 in St George's, Grenada. He graduated in mathematics and chemistry from the University of the West Indies in Barbados in 1971. After briefly teaching at a boys' college he continued his studies, first in Barbados then in the USA where he gained a PhD in mathematics from the American University in Washington, D. C. From 1977–83 he was professor of mathematics at Howard University in Washington, D.C. and acted as a statistical consultant to assorted US government agencies and private corporations.

In 1984 he returned to Grenada and entered parliament for the ruling NNP. He served as minister of works, communications and public utilities from 1984 until the NNP lost power in 1989. When the NNP returned to power in 1995, Mitchell became prime minister.

Career in Office

In his first term, Mitchell diversified the agriculture-reliant economy by expanding the offshore banking sector. He chaired the Caribbean Community (CARICOM) in 1998 and has been responsible for its science, technology and human resource development since 1995. In 1999 he became the first Grenadian prime minister to win two consecutive terms as premier and to claim all 15 parliamentary seats at an election.

In 2000 Mitchell established a Truth and Reconciliation Commission to confront lingering tensions arising from Grenada's 'Revolutionary Years' of 1976–83. In 2001 the country was placed on a blacklist of international money-laundering and tax havens compiled by the G7-founded Financial Action Task Force (FATF). Mitchell's government duly revoked more than 20 offshore banking licences and strengthened financial regulations. After the 11 Sept. attacks on New York and Washington, he suspended the sale of passports to non-citizens. Grenada was removed from the FATF blacklist in 2003. In the same year, Mitchell secured a new term in office.

In 2004 and 2005 hurricanes Ivan and Emily killed 40 in Grenada and damaged housing and infrastructure. Mitchell responded by calling on the UN to set up a special fund for small countries in crisis. In 2005 he renewed diplomatic relations with Beijing. When the NNP was comprehensively defeated by the National Democratic Congress (NDC) at the general election of June 2008, winning only four of 15 available seats, Mitchell was succeeded as premier by Tillman Thomas.

He remained as head of the NNP and in Feb. 2013 led the party to a repeat of its 1999 electoral performance, winning every parliamentary seat. His priority is to tackle the stagnant economy and high unemployment. On taking office, Mitchell installed a smaller cabinet than his predecessor and pledged to honour public sector salary payments. He aims to deepen Grenada's involvement with CARICOM and has defended his track record of heavy expenditure on sporting facilities, pledging government support for cricket in particular.

DEFENCE

Royal Grenada Police Force

Modelled on the British system, the 730-strong police force includes an 80-member paramilitary unit and a 30-member coastguard.

ECONOMY

Agriculture accounted for 6·7% of GDP in 2006, industry 29·0% and services 64·3%.

Overview

Massive damage caused by hurricanes in 2004 and 2005 led the government to launch a medium-term economic reform package to tackle the large public debt, ensure sustainable growth and alleviate poverty. This was supported by the IMF's Poverty Reduction and Growth Facility arrangement.

Inflation rose sharply in 2008 as a result of increasing world food and fuel prices and the depreciating US dollar, while fiscal performance deteriorated, reflecting high capital expenditure and shortfalls in grants from, among others, the EU. GDP growth was negative in 2009 and 2010 (down from 4% in 2007) owing to declines in tourism and remittances prompted by the global financial crisis.

A fragile recovery began in 2011, with GDP growth of 0·4% stemming principally from improved performance by the tourism and agricultural sectors. Nonetheless, continued high food and fuel prices saw the current account deficit remain at about 25% of GDP. High public sector debt, high current account deficits, and a vulnerable financial sector pose risks to the fragile recovery.

Currency

The unit of currency is the *East Caribbean dollar* (XCD). Foreign exchange reserves were US$100m. in July 2005 and total money supply was EC$380m. Following inflation of 8·0% in 2008, there was deflation of 0·3% in 2009. Inflation was 3·4% in 2010 and 3·0% in 2011.

Budget

In 2009 revenue was EC$431·1m. and expenditure EC$533·7m. Income tax has been abolished. VAT of 15% (reduced rate, 10%) was introduced on 1 Feb. 2010.

Performance

Real GDP contracted by 5·7% in 2009 and 1·3% in 2010 but there was growth of 0·4% in 2011. Total GDP in 2011 was US$0·8bn.

Banking and Finance

Grenada is a member of the Eastern Caribbean Central Bank. The *Governor* is Sir Dwight Venner. In 2002 there were three commercial banks and four foreign banks. The Grenada Agricultural Bank was established in 1965 to encourage agricultural development; in 1975 it became the Grenada Agricultural and Industrial Development Corporation. In 1995 bank deposits were EC$666·8m. (US$249·7m.). Total foreign currency deposits in 1995 amounted to US$11·8m.

External debt amounted to US$525m. in 2007.

Grenada is affiliated to the Eastern Caribbean Securities Exchange in Basseterre, St Kitts and Nevis.

ENERGY AND NATURAL RESOURCES

Environment

Grenada's carbon dioxide emissions from the consumption and flaring of fossil fuels in 2008 were the equivalent of 3·7 tonnes per capita.

Electricity

Installed capacity in 2007 was an estimated 32,000 kW. Production in 2007 was 171m. kWh, with consumption per capita 1,596 kWh.

Agriculture

There were about 2,000 ha. of arable land in 2007 and 10,000 ha. of permanent crops. Principal crop production (2003, in 1,000 tonnes): sugarcane, 7; coconuts, 6; bananas, 4; avocados, 2; grapefruit and pomelos, 2; mangoes, 2. Nutmeg, corn, pigeon peas, citrus, root-crops and vegetables are also grown, in addition to small scattered cultivations of cotton, cloves, cinnamon, pimento, coffee and fruit trees. Grenada is the second largest producer of nutmeg in the world, after Indonesia.

Livestock (2003): sheep, 13,000; goats, 7,000; pigs, 5,000; cattle, 4,000.

Forestry

In 2010 the area under forests was 17,000 ha., or 50% of the total land area.

Fisheries

The catch in 2008 was 2,384 tonnes, entirely from marine waters.

INDUSTRY

Main products are wheat flour, soft drinks, beer, animal feed, rum and cigarettes.

Labour

In 1993 the labour force was estimated at 27,820. Unemployment was 11% in Dec. 2000.

INTERNATIONAL TRADE

Imports and Exports

In 2008 imports (c.i.f.) totalled US$363·3m. and exports (f.o.b.) US$30·5m. Major import commodities in 2005 were: machinery and transport equipment, 23·4%; food and live animals, 14·2%; manufactures of metals, 6·9%. The principal exports were: nutmeg, mace and cardamom, 29·7%; machinery and transport equipment, 15·2%; fish and shellfish, 12·7%.

In 2005 the main import suppliers were the USA (37·5%), Trinidad and Tobago (21·0%), UK (5·8%), Japan (4·0%) and China (3·2%). Main export destinations were the USA (21·4%), Netherlands (14·1%), Trinidad and Tobago (10·1%), St Lucia (9·4%) and Barbados (6·2%).

COMMUNICATIONS

Roads

In 2000 there were 1,127 km of roads, of which 61·0% were hard-surfaced.

Civil Aviation

The main airport is Point Salines International. Union Island and Carriacou have smaller airports. There were direct flights from Point Salines in 2003 to Anguilla, Antigua, Barbados, the British Virgin Islands, Dominica, Frankfurt, London, Montego Bay, New York, Philadelphia, Puerto Rico, St Kitts, St Lucia, St Maarten, St Vincent, Tobago and Trinidad. In 2001 Point Salines handled 344,064 passengers (342,124 on international flights) and 2,747 tonnes of freight.

Shipping

The main port is at St George's; there are eight minor ports. In Jan. 2009 there were four ships of 300 GT or over registered, totalling 2,000 GT.

Telecommunications

There were 28,400 fixed telephone lines in 2010 (271·5 per 1,000 inhabitants). Mobile phone subscribers numbered 121,900 in 2010. There were 334·6 internet users per 1,000 inhabitants in 2010. Fixed internet subscriptions totalled 10,900 in 2009 (104·8 per 1,000 inhabitants).

SOCIAL INSTITUTIONS

Justice

The Grenada Supreme Court, situated in St George's, comprises a High Court of Justice, a Court of Magisterial Appeal (which hears appeals from the lower Magistrates' Courts exercising summary jurisdiction) and an Itinerant Court of Appeal (to hear appeals from the High Court). Grenada was one of ten countries to sign an agreement in Feb. 2001 establishing a Caribbean Court of Justice (CCJ) to replace the British Privy Council as the highest civil and criminal court. In the meantime the number of signatories has risen to 12. The court was inaugurated at Port-of-Spain, Trinidad on 16 April 2005. However, Grenada has not yet accepted the CCJ as its final court of appeal. For police *see* DEFENCE, *above*.

The population in penal institutions in 2007 was 367 (equivalent to 408 per 100,000 of national population).

Education

Adult literacy was 96·0% in 2004. In 2007 there were 13,733 pupils in primary schools (871 teaching staff) and 13,060 pupils (886 teaching staff in 2005) in secondary schools. The Grenada National College was established in 1988. There is also a branch of the University of the West Indies. In 2003 public expenditure on education amounted to 5·9% of GNI and 12·9% of total government spending.

Health

In 2005 there were 41 hospital beds per 10,000 inhabitants. There were 127 physicians in 2003. In 1998 there were 20 dentistry personnel and 326 nursing and midwifery personnel. In 2005 expenditure on health was 9·6% of total government spending.

RELIGION

At the 2001 census 53% of the population were Roman Catholic, 14% Anglican and the remainder other religions.

CULTURE

Press

In 2008 there were five weekly newspapers and several others that were published irregularly.

Tourism

In 2005 there were 98,548 non-resident tourists and 275,080 cruise passenger arrivals. Tourism receipts (excluding passenger transport) totalled US$71m. in 2005.

DIPLOMATIC REPRESENTATIVES

Of Grenada in the United Kingdom (The Chapel, Archel Rd, London, W14 9QH)
High Commissioner: Ruth Elizabeth Rouse.

Of the United Kingdom in Grenada
High Commissioner: Paul Brummell (resides in Bridgetown, Barbados).

Of Grenada in the USA (1701 New Hampshire Ave., NW, Washington, D.C., 20009)
Ambassador: Gillian M. S. Bristol.

Of the USA in Grenada
Ambassador: Larry Palmer (resides in Bridgetown, Barbados).

Of Grenada to the United Nations
Ambassador: Dessima Williams.

Of Grenada to the European Union
Ambassador: Stephen Fletcher.

FURTHER READING

Ferguson, J., *Grenada: Revolution in Reverse.* 1991
Heine, J. (ed.) *A Revolution Aborted: the Lessons of Grenada.* 1990
Steele, Beverley A., *Grenada: A History of its People.* 2003

GUATEMALA

República de Guatemala
(Republic of Guatemala)

Capital: Guatemala City
Population projection, 2015: 16·33m.
GNI per capita, 2011: (PPP$) 4,167
HDI/world rank: 0·574/131
Internet domain extension: .gt

KEY HISTORICAL EVENTS

From 1524 Guatemala was part of a Spanish captaincy-general, comprising the whole of Central America. It became independent in 1821 and formed part of the Confederation of Central America from 1823 to 1839. The overthrow of the right-wing dictator Jorge Ubico in 1944 opened a decade of left-wing activity which alarmed the USA. In 1954 the leftist regime of Jacobo Arbenz Guzmán was overthrown by a CIA-supported coup. A series of right-wing governments failed to produce stability while the toll on human life and the violation of human rights was such as to cause thousands of refugees to flee to Mexico. Elections to a National Constituent Assembly were held on 1 July 1984, and a new constitution was promulgated in May 1985. Amidst violence and assassinations, the presidential election was won by Marco Vinicio Cerezo Arévalo. On 14 Jan. 1986 Cerezo's civilian government was installed—the first for 16 years and only the second since 1954. Violence continued, however, and there were frequent reports of torture and killings by right-wing 'death squads'. The presidential and legislative elections of Nov. 1995 saw the return of open politics for the first time in over 40 years. Meanwhile the Guatemalan Revolutionary Unit (URNG) declared a ceasefire. On 6 May and 19 Sept. 1996 the government agreed reforms to military, internal security, judicial and agrarian institutions. A ceasefire was concluded in Oslo on 4 Dec. 1996 and a final peace treaty was signed on 29 Dec. 1996. In Nov. 1999 the country's first presidential elections took place since the end of the 36-year-long civil war, which had claimed over 200,000 lives.

TERRITORY AND POPULATION

Guatemala is bounded on the north and west by Mexico, south by the Pacific ocean and east by El Salvador, Honduras and Belize, and the area is 108,889 sq. km (42,042 sq. miles). In March 1936

Guatemala, El Salvador and Honduras agreed to accept the peak of Mount Montecristo as the common boundary point.

The population was 11,237,196 at the census of Nov. 2002; density, 103 per sq. km. The estimated population in 2012 was 15,073,400.

The UN gives a projected population for 2015 of 16·33m.

In 2011, 49·9% of the population were urban. In 2000, 33% were Amerindian, of 21 different groups descended from the Maya; 64% Mestizo (mixed Amerindian and Spanish). 51% speak Spanish, the official language of Guatemala, with the remainder speaking one or a combination of the 23 Indian dialects.

Guatemala is administratively divided into 22 departments, each with a governor appointed by the president. Area and population, 2002:

Departments	Area (sq. km)	Population	Departments	Area (sq. km)	Population
Alta Verapaz	8,686	776,246	Petén	35,854	366,735
Baja Verapaz	3,124	215,915	Quezaltenango	1,951	624,716
Chimaltenango	1,979	446,133	Quiché	8,378	655,510
Chiquimula	2,376	302,485	Retalhuleu	1,858	241,411
El Progreso	1,922	139,490	Sacatepéquez	465	248,019
Escuintla	4,384	538,746	San Marcos	3,791	794,951
Guatemala City	2,126	2,541,581	Santa Rosa	2,955	301,370
Huehuetenango	7,403	846,544	Sololá	1,061	307,661
Izabal	9,038	314,306	Suchitepéquez	2,510	403,945
Jalapa	2,063	242,926	Totonicapán	1,061	339,254
Jutiapa	3,219	389,085	Zacapa	2,690	200,167

In 2002 Guatemala City, the capital, had a population of 942,348. Populations of other major towns, 2002 (in 1,000): Mixco, 384; Villa Nueva, 302; Quezaltenango, 120; Petapa, 94; Escuintla, 87.

SOCIAL STATISTICS

Births, 2006, 368,399; deaths, 69,756. 2006 rates per 1,000 population: birth, 28·4; death, 5·4. Life expectancy, 2007: male 66·7 years, female 73·7. Annual population growth rate, 2005–10, 2·5%. Infant mortality, 2010, 25 per 1,000 live births; fertility rate, 2008, 4·1 births per woman.

CLIMATE

A tropical climate, with little variation in temperature and a well marked wet season from May to Oct. Guatemala City, Jan. 63°F (17·2°C), July 69°F (20·6°C). Annual rainfall 53" (1,316 mm).

CONSTITUTION AND GOVERNMENT

A new constitution, drawn up by the Constituent Assembly elected on 1 July 1984, was promulgated in June 1985 and came into force on 14 Jan. 1986. In 1993, 43 amendments were adopted, reducing *inter alia* the President's term of office from five to four years. The President and Vice-President are elected by direct election (with a second round of voting if no candidate secures 50% of the first-round votes) for a non-renewable four-year term. The unicameral *Congreso de la República* comprises 158 members, elected partly from constituencies and partly by proportional representation to serve four-year terms.

National Anthem

'¡Guatemala Feliz!' ('Happy Guatemala'); words by J. J. Palma, tune by R. Alvárez.

GOVERNMENT CHRONOLOGY

Heads of State since 1944. (CAO = Organized Aranista Central; DCG = Guatemalan Christian Democracy; FRG = Guatemalan Republican Front; GANA = Grand National Alliance; MAS =

Solidarity Action Movement; MLN = National Liberation Movement; PAN = National Advancement Party; PAR = Revolutionary Action Party; PID = Democratic Institutional Party; PP = Patriotic Party; PR = Revolutionary Party; PRDN = National Democratic Reconciliation/Redemption Party; UNE = National Union of Hope; n/p = non-partisan

Military Junta

1944–45		Maj. Francisco Javier Arana; Capt. Jacobo Arbenz Guzmán; Jorge Toriello Garrido

Presidents

1945–51	PAR	Juan José Arévalo Bermejo
1951–54	PAR	Jacobo Arbenz Guzmán
1954	military	Carlos Enrique Díaz de León

Military Juntas

1954		Col. Elfego Hernán Monzón Aguirre; Col. José Ángel Sánchez; Col. José Luis Cruz Salazar; Col. Carlos Enrique Díaz de León; Col. Mauricio Dubois
1954		Col. Carlos Castillo Armas; Col. Mauricio Dubois; Maj. Enrique Trinidad Oliva; Col. Elfego Hernán Monzón Aguirre; Col. José Luis Cruz Salazar

Presidents

1954–57	military	Carlos Castillo Armas
1957–58	military	Guillermo Flores Avendaño
1958–63	PRDN	José Ramón Ydígoras Fuentes
1963–66	military	Alfredo Enrique Peralta Azurdia
1966–70	PR	Julio César Méndez Montenegro
1970–74	military, MLN	Carlos Manuel Arana Osorio
1974–78	military, MLN/ PID	Kjell Eugenio Laugerud García
1978–82	military, PID/ PR/ CAO	Fernando Romeo Lucas García
1982–83	military	José Efraín Ríos Montt
1983–86	military	Óscar Humberto Mejía Víctores
1986–91	DCG	Marco Vinicio Cerezo Arévalo
1991–93	MAS	Jorge Antonio Serrano Elías
1993–96	n/p	Ramiro de León Carpio
1996–2000	PAN	Álvaro Enrique Arzú Yrigoyen
2000–04	FRG	Alfonso Antonio Portillo Cabrera
2004–08	GANA	Óscar Rafael Berger Perdomo
2008–12	UNE	Álvaro Colom Caballeros
2012–	PP	Otto Pérez Molina

RECENT ELECTIONS

In the first round of presidential elections on 11 Sept. 2011 Otto Pérez Molina of the Partido Patriota (PP, Patriotic Party) won with 36·0% of the vote. Manuel Baldizón of the Libertad Democrática Renovada (LIDER, Renewed Democratic Party) took 23·2% and Eduardo Suger of Compromiso, Renovación y Orden (CREO, Commitment, Renovation and Order) 16·4%. Seven other candidates received less than 10% of the vote each. Turnout was 69·3%. In the run-off held on 6 Nov. 2011, Pérez won 53·7% of the vote and Baldizón 46·3%.

Congressional elections were held on 11 Sept. 2011. The Partido Patriota won 56 seats with 26·6% of the vote; Unidad Nacional de la Esperanza (National Unity of Hope) together with the Gran Alianza Nacional (Grand National Alliance) 48 with 22·6%; Union del Cambio Nacionalista (Nationalist Change Union) 14 with 9·5%; Libertad Democrática Renovada 14 with 8·9%; and Compromiso, Renovación y Orden 12 with 8·7%. Nine other parties received fewer than ten seats. Turnout was 69·3%.

CURRENT GOVERNMENT

President: Otto Pérez Molina; b. 1950 (PP; sworn in 14 Jan. 2012).

Vice-President: Roxana Baldetti; b. 1962 (PP; took office on 14 Jan. 2012).

In March 2013 the government comprised:

Minister of Agriculture, Livestock and Food: Elmer López Rodríguez. *Communications, Infrastructure and Housing:* Alejandro Sinibaldi. *Culture and Sports:* Carlos Batzín. *Defence:* Col. Ulises Noé Anzueto. *Economy:* Sergio de la Torre. *Education:* Cynthia del Águila. *Energy and Mines:* Erick Archila. *Environment and Natural Resources:* Roxana Sobenes. *External Relations:* Fernando Carrera. *Finance:* Pavel Centeno. *Interior:* Mauricio López Bonilla. *Labour and Social Welfare:* Carlos Contreras. *Public Health and Social Assistance:* Jorge Alejandro Villavicencio Álvarez.

Government Website (Spanish only): http://www.guatemala.gob.gt

CURRENT LEADERS

Otto Pérez Molina

Position
President

Introduction
Otto Pérez Molina took office as president in Jan. 2012. A former director of military intelligence, he has adopted a hardline policy towards tackling crime.

Early Life
Born in Guatemala City on 1 Dec. 1950, Pérez Molina graduated from the country's military academy in 1973. He rose through the army ranks, serving in counterinsurgency campaigns during the 1980s, particularly in the Quiché region. Allegations of atrocities committed under his command have persisted but have never been proven.

In 1983 he backed the coup that removed President Ríos Montt. In the late 1980s he studied at the Army School of the Americas and the Inter-American Defense College, both in the USA, after which he worked in military intelligence, serving as director from 1992–93. After helping to force the resignation of President Serrano in 1993, he served from 1993–95 as chief of staff to President de León Carpio.

Appointed inspector general of the army in 1996, he represented the military in negotiations with guerrilla forces that resulted in a peace treaty ending 36 years of civil war. He was also Guatemala's delegate on the Inter-American Defense Board from 1998–2000. In Feb. 2001 Pérez Molina founded the conservative Patriotic Party (PP), guiding it into the Grand National Alliance (GANA) two years later.

Elected to parliament in Nov. 2003, he subsequently withdrew the PP from GANA while serving as commissioner for defence and security. In 2007 he unsuccessfully contested the presidential election as the PP candidate, but stood again and won in 2011 promising tough policing and increased social spending.

Career in Office
Pérez Molina took office on 14 Jan. 2012. He asked for international help in combating the drugs trade and has introduced a system of checkpoints aimed at curtailing violent crime. His other challenges include reducing poverty and unemployment, as well as managing the conflict between the mining industry and indigenous groups defending their land rights.

DEFENCE

In 2008 defence expenditure totalled US$180m. (US$14 per capita), representing 0·4% of GDP.

Army

The Army numbered (2007) 13,400 and is organized in 15 military zones. It includes a special forces unit. There is a paramilitary national police of 19,000 including 2,500 treasury police.

Navy

The Navy was (2007) 990-strong of whom 650 were marines. Main bases are Santo Tomás de Castilla (on the Atlantic Coast) and Puerto Quetzal (Pacific).

Air Force

There is an Air Force with ten combat-capable aircraft (PC-7s and A-37s). Strength was (2007) 1,070.

ECONOMY

In 2007 agriculture accounted for 11% of GDP, industry 28% and services 62%.

Overview

Guatemala's income per capita is high relative to other Central American countries. However, income inequality is also high and around half the population lives in poverty. Social indicators fall below the average of low-income countries and in 2008 President Colom launched a series of programmes to support the poorest families. The agricultural sector accounts for almost half of all employment and a large part of exports. Remittances are the primary source of foreign income. Since the end of the civil war in 1996 the economy has drawn more foreign capital investment (boosted in 2006 by the implementation of the Central America Free Trade Agreement), contributing to the growth of imports.

Annual growth averaged 4·2% between 2004–07 but fell to 3·3% in 2008. Guatemala only narrowly avoided a recession in 2009 as remittances and investment declined. Natural disasters in 2010 caused damage equalling 2·4% of GDP. Nonetheless, the economy showed signs of bouncing back in late 2010.

Currency

The unit of currency is the *quetzal* (GTQ) of 100 *centavos*, established on 7 May 1925. In July 2005 foreign exchange reserves were US$3,685m., total money supply was Q.30,083m. and gold reserves were 221,000 troy oz. Inflation was 3·9% in 2010 and 6·2% in 2011.

Budget

Budgetary central government revenue in 2010 totalled Q.37,293·7m. (Q.33,968·1m. in 2009) and expenditure was Q.42,143·1m. (Q.38,244·5m. in 2009).

VAT is 12%.

Performance

Real GDP growth was 0·5% in 2009, 2·9% in 2010 and 3·9% in 2011. Total GDP in 2011 was US$46·9bn.

Banking and Finance

The Banco de Guatemala is the central bank and bank of issue (*President*, Edgar Baltazar Barquín Durán). In 2002 there were 27 national banks (four state-owned and 23 private). The international banks and the foreign banks are authorized to operate as commercial banks. Foreign debt was US$14,349m. in 2010, representing 35·9% of GNI.

There are two stock exchanges.

ENERGY AND NATURAL RESOURCES

Environment

Carbon dioxide emissions from the consumption and flaring of fossil fuels in 2008 were the equivalent of 0·9 tonnes per capita.

Electricity

Installed capacity in 2007 was 2·2m. kW. Production, 2007, 8·75bn. kWh. Consumption per capita in 2007 was 647 kWh.

Oil and Gas

There were proven natural gas reserves in 2007 of 3·1bn. cu. metres. In 2007 crude petroleum reserves were 83m. bbls; output in 2007 was 833,000 tonnes.

Minerals

There are deposits of gold, silver and nickel.

Agriculture

There were 1·36m. ha. of arable land in 2001 and 0·55m. ha. of permanent crops. 130,000 ha. were irrigated in 2001. Output, 2003 estimates (in 1,000 tonnes): sugarcane, 17,500; maize, 1,054; bananas, 1,000; melons and watermelons, 314; plantains, 268; potatoes, 248; coffee, 210. Guatemala is one of the largest producers of essential oils (citronella and lemongrass). Livestock (2003 estimates): cattle, 2·54m.; pigs, 780,000; sheep, 260,000; horses, 124,000; goats, 112,000; chickens, 27m.

Forestry

In 2010 the area under forests was 3·66m. ha., or 34% of the total land area. Timber production in 2007 was 17·41m. cu. metres.

Fisheries

In 2010 the total catch was 21,859 tonnes, mainly from sea fishing.

INDUSTRY

Manufacturing contributed 20% of GDP in 2007. The principal industries are food and beverages, tobacco, chemicals, hides and skins, textiles, garments and non-metallic minerals. Cement production in 2007 was an estimated 2,500,000 tonnes; raw sugar production was 2,015,000 tonnes in 2005.

Labour

In 2002 there were 3,463,000 employed persons. The main areas of activity were: agriculture, hunting, forestry and fishing, 1,457,000; wholesale and retail trade and restaurants and hotels, 571,700; manufacturing, 466,000; services, 266,000; construction, 207,900. There is a working week of a maximum of 44 hours.

INTERNATIONAL TRADE

In 2004 Guatemala signed the Central America-Dominican Republic-United States Free Trade Agreement (CAFTA-DR), along with Costa Rica, the Dominican Republic, El Salvador, Honduras, Nicaragua and the USA. The agreement entered into force for Guatemala on 1 July 2006.

Imports and Exports

Values in US$1m. were:

	2003	2004	2005	2006	2007
Imports c.i.f.	6,718·7	7,812·1	10,499·5	9,539·7	12,731·4
Exports f.o.b.	2,634·7	2,931·8	5,380·8	3,198·1	6,900·4

In 2007 the main imports were: machinery and transport equipment, 25·1%; petroleum and petroleum products, 16·1%; chemicals and related products, 14·8%; food and live animals, 9·8%. Principal exports in 2007 were: apparel and clothing accessories, 20·1%; chemicals and related products, 10·8%; coffee and coffee substitutes, 8·5%; cane sugar, 5·2%; bananas, 4·7%. Main import suppliers, 2007: USA, 34·1%; Mexico, 8·8%; China, 5·7%; El Salvador, 4·8%; South Korea, 3·4%. Main export markets, 2007: USA, 42·6%; El Salvador, 12·2%; Honduras, 8·6%; Mexico, 6·7%; Nicaragua, 3·9%.

COMMUNICATIONS

Roads

In 2002 there were 14,891 km of roads, of which 74 km were motorways. 37·6% of all roads were paved in 2002. There is a highway from coast to coast via Guatemala City. There are two highways from the Mexican to the Salvadorean frontier: the

Pacific Highway serving the fertile coastal plain and the Pan-American Highway running through the highlands and Guatemala City. Vehicles in use in 2007 numbered 1,558,100.

Rail
Ferrovías Guatemala (a subsidiary of Railroad Development Corporation, which secured a 50-year concession to upgrade Guatemala's decrepit rail network in 1997) operated 322 km of railway in 2006, with 11 locomotives carrying 93,000 tonnes of freight. However, after a contractual dispute with the government, the company suspended its operations in Sept. 2007 pending the conclusion of legal action. There have been no passenger services since the late 1990s.

Civil Aviation
There are international airports at Guatemala City (La Aurora) and Flores. In 2000 La Aurora handled 1,258,919 passengers and 58,118 tonnes of freight. In 1999 scheduled airline traffic of Guatemalan-based carriers flew 5·3m. km, carrying 506,000 passengers (472,000 on international flights).

Shipping
The chief ports on the Atlantic coast are Puerto Barrios and Santo Tomás de Castilla: on the Pacific coast, Puerto Quetzal and Champerico. Santo Tomás de Castilla, Guatemala's busiest port, handled 4·7m. tonnes of cargo in 2008.

Telecommunications
The government own and operate the telecommunications services. There were 1·50m. fixed telephone lines in 2010 (104·1 per 1,000 inhabitants). Mobile phone subscribers numbered 18·07m. in 2010. There were 105·0 internet users per 1,000 inhabitants in 2010. In Dec. 2011 there were 1·7m. Facebook users.

SOCIAL INSTITUTIONS

Justice
Justice is administered in a Constitution Court, a Supreme Court, six appeal courts and 28 courts of first instance. Supreme Court and appeal court judges are elected by Congress. Judges of first instance are appointed by the Supreme Court.

The death penalty is authorized for murder and kidnapping. There were two executions in 2000, but none since. There were 5,174 homicides in 2012, down from a peak of 6,498 in 2009.

A new National Civil Police force under the authority of the Minister of the Interior was created in 1996. It was 19,000-strong in 2007.

The population in penal institutions in July 2010 was 11,140 (77 per 100,000 of national population).

Education
In 2007 there were 2,448,976 pupils at primary schools and 864,154 pupils at secondary level. The adult literacy rate in 2008 was 74%. There is one state university—the University of San Carlos of Guatemala (Universidad de San Carlos de Guatemala), founded in 1676—as well as several private universities. In 2006 there were 112,215 students and 3,843 academic staff in tertiary education.

In 2007 public expenditure on education came to 3·1% of GNI.

Health
Guatemala had 4,969 public sector doctors and 12,514 nursing staff in 2005. There were 66 hospitals with 8,270 beds (6·4 per 10,000 inhabitants) in 2006.

Welfare
A comprehensive system of social security was outlined in a law of 30 Oct. 1946.

RELIGION
Roman Catholicism is the prevailing faith (8·9m. adherents in 2001) and there is a Roman Catholic archbishopric. The remainder of the population are followers of other religions (mainly Evangelical Protestantism).

CULTURE

World Heritage Sites
There are three UNESCO sites in Guatemala: Tikal National Park (inscribed on the list in 1979); Antigua Guatemala (1979); and the Archaeological Park and Ruins of Quiriguá (1981).

Press
In 2008 there were nine paid-for daily newspapers, the main ones being *Nuestro Diario* and *Prensa Libre*.

Tourism
There were 1,876,000 non-resident visitors in 2010 (up from 1,777,000 in 2009 and 1,715,000 in 2008).

DIPLOMATIC REPRESENTATIVES
Of Guatemala in the United Kingdom (13A Fawcett St., London, SW10 9HN)
Ambassador: Acisclo Valladares Molina.

Of the United Kingdom in Guatemala (Avenida La Reforma 16-00, Zona 10, Edificio Torre Internacional, Nivel 11, Guatemala City)
Ambassador: Sarah Dickson.

Of Guatemala in the USA (2220 R. St., NW, Washington, D.C., 20008)
Ambassador: Francisco Villagrán de León.

Of the USA in Guatemala (7–01 Avenida de la Reforma, Zone 10, Guatemala City)
Ambassador: Arnold A. Chacon.

Of Guatemala to the United Nations
Ambassador: Gert Rosenthal.

Of Guatemala to the European Union
Ambassador: Jorge Skinner-Klée Arenales.

FURTHER READING

Benson, Peter and Fischer, Edward F., *Broccoli and Desire: Global Connections and Maya Struggles in Post-War Guatemala.* 2006
Jonas, Susanne, *Of Centaurs and Doves: Guatemala's Peace Process.* 2001
Reeves, René, *Ladinos with Ladinos, Indians with Indians: Land, Labor, and Regional Ethnic Conflict in the Making of Guatemala.* 2006
Sanford, Victoria, *Buried Secrets: Truth and Human Rights in Guatemala.* 2003

National Statistical Office: Instituto Nacional de Estadística, 8a calle 9–55, Zona 1, Guatemala City.
Website (Spanish only): http://www.ine.gob.gt

GUINEA

© Research Machines plc 2006

République de Guinée
(Republic of Guinea)

Capital: Conakry
Population projection, 2015: 11·32m.
GNI per capita, 2011: (PPP$) 863
HDI/world rank: 0·344/178
Internet domain extension: .gn

KEY HISTORICAL EVENTS

In 1888 Guinea became a French protectorate, in 1893 a colony, and in 1904 a constituent territory of French West Africa. Forced labour and other colonial depredations ensued, although a form of representation was introduced in 1946. The independent Republic of Guinea was proclaimed on 2 Oct. 1958, after the territory of French Guinea had decided to leave the French community. Guinea became a single-party state. In 1980 the armed forces staged a coup and dissolved the National Assembly. Following popular disturbances a multi-party system was introduced in April 1992.

In 2000 fierce fighting broke out between Guinean government troops and rebels, believed to be a mix of Guinean dissidents and mercenaries from Liberia and Sierra Leone. More than 250,000 refugees were caught up in what the United Nations High Commissioner for Refugees described as the world's worst refugee crisis. In 2003 the governments of Guinea, Liberia and Sierra Leone reached a deal on measures to secure mutual borders and to fight insurgency.

In 2008 President Lansana Conté died after 24 years of authoritarian rule. Power fell to the military, who appointed Capt. Moussa Dadis Camara as president. Camara promised presidential elections in Jan. 2010 and parliamentary elections in March 2010. In Sept. 2009 the army opened fire on an opposition rally, killing at least 150 people. The European Union, African Union and USA imposed sanctions the following month, the UN set up an enquiry into the incident and in Feb. 2010 the International Criminal Court condemned the massacre as a crime against humanity. After surviving an attempted assassination in Dec. 2009, Camara agreed to recuperate abroad while rule fell to

his deputy, Gen. Sekouba Konaté. Konaté appointed Jean-Marie Doré as interim prime minister to oversee a return to civilian rule.

TERRITORY AND POPULATION

Guinea is bounded in the northwest by Guinea-Bissau and Senegal, northeast by Mali, southeast by Côte d'Ivoire, south by Liberia and Sierra Leone, and west by the Atlantic Ocean.

The area is 245,860 sq. km (94,930 sq. miles), including 140 sq. km (50 sq. miles) of inland water. In 1996 the census population was 7,156,406 (density 29·1 per sq. km). Estimate, 2010, 9·98m.; density, 41 per sq. km.

The UN gives a projected population for 2015 of 11·32m.

The capital is Conakry. In 2011, 35·9% of the population were urban.

Guinea is divided into seven provinces and a special zone (national capital). These are in turn divided into 34 administrative regions. The major divisions (with their areas in sq. km) are: Boké, 34,231; Conakry (special zone—national capital), 308; Faranah, 38,272; Kankan, 71,085; Kindia, 26,749; Labé, 21,150; Mamou, 13,560; Nzérékoré, 40,502.

The main towns are Conakry (population estimate, 2010, 1,715,000), Kindia, Nzérékoré, Kankan, Guéckédougou and Kissidougou.

The ethnic composition is Fulani (38·6%), Malinké (or Mandingo, 23·2%), Susu (11·0%), Kissi (6·0%) and Kpelle (4·6%) in Guinée-Forestière, and Dialonka, Loma and others (16·6%).

The official language is French.

SOCIAL STATISTICS

2008 estimates: births, 390,000; deaths, 108,000. Rates, 2008 estimates (per 1,000 population): births, 39·6; deaths, 11·0. infant mortality, 2010, 81 per 1,000 live births. Life expectancy, 2007, 55·3 years for males and 59·3 for females. Annual population growth rate, 2000–08, 2·0%; fertility rate, 2008, 5·4 births per woman.

CLIMATE

A tropical climate, with high rainfall near the coast and constant heat, but conditions are a little cooler on the plateau. The wet season on the coast lasts from May to Nov., but only to Oct. inland. Conakry, Jan. 80°F (26·7°C), July 77°F (25°C). Annual rainfall 172" (4,293 mm).

CONSTITUTION AND GOVERNMENT

There is a 114-member *National Assembly* (currently suspended), 38 of whose members are elected on a first-past-the-post system, and the remainder from national lists by proportional representation. It was dissolved following the military coup of Dec. 2008.

On 11 Nov. 2001 a referendum was held in which 98·4% of votes cast were in favour of President Conté remaining in office for a third term, requiring an amendment to the constitution (previously allowing a maximum two presidential terms). The referendum, which also increased the presidential mandate from five to seven years, was boycotted by opposition parties.

National Anthem

'Peuple d'Afrique, le passé historique' ('People of Africa, the historic past'); words anonymous, tune by Fodeba Keita.

RECENT ELECTIONS

The first democratic presidential elections since independence in 1958 were held on 27 June 2010. Former prime minister Cellou Dalein Diallo of the Union of Democratic Forces of Guinea

(UFDG) won 39·7% of the vote against 20·7% for Alpha Condé of the Rally of the Guinean People (RPG) and 15·6% for another former prime minister, Sidya Touré of the Union of Republican Forces (UFR). There were 21 further candidates in the first round. Turnout was 51·6%. A second round between the two leading candidates was initially scheduled for 18 July but was postponed until 7 Nov. Condé took 52·5% of the vote against 47·5% for Diallo. Turnout was 67·9%. A state of emergency was declared owing to outbreaks of violence following the announcement of the result but was lifted just over three weeks later.

Parliamentary elections took place on 30 June 2002. The PUP gained 85 out of 114 seats with 61·6% of votes cast, Union for Progress and Renewal (UPR) 20 seats with 26·6%, Union for the Progress of Guinea (UPG) 3 with 4·1%, Democratic Party of Guinea (PDG) 3 with 3·4%, National Alliance for Progress (ANP) 2 with 2·0% and Party of the Union for Development (PUD) 1 with 0·7%. Turnout was 71·6%. The Rally of the Guinean People, the main opposition party, boycotted the election.

CURRENT GOVERNMENT

President and Minister of National Defence: Alpha Condé; b. 1938 (Rally of the Guinean People/RPG; sworn in 21 Dec. 2010).

In March 2013 the cabinet comprised:

Prime Minister: Mohamed Said Fofana; b. 1952 (ind.; since 24 Dec. 2010).

Minister of State for Economy and Finance: Kerfalla Yansané. *Minister of State for Energy:* Papa Koly Kourouma. *Minister of State for Foreign Affairs and Guineans Abroad:* Louncény Fall. *Minister of State for Justice and Keeper of the Seals:* Christian Sow. *Minister of State for Public Works and Transportation:* Ousmane Bah. *Minister of State at the Presidency:* Nantou Chérif Konaté.

Minister of Agriculture: Emile Yombouno. *Communication:* Togba Césaire Kpoghomou. *Culture, the Arts and Heritage:* Ahmed Tidiane Cissé. *Employment, Technical Education and Vocational Training:* Damantan Albert Camara. *Environment, Water and Forests:* Ibrahima Boiro. *Fisheries and Aquaculture:* Moussa Condé. *Health and Public Sanitation:* Dr Edouard Gnankoye Lama. *Higher Education and Scientific Research:* Téliwel Bailo Diallo. *Hotels, Tourism and Handicrafts:* Hadja Mariama Baldé. *Human Rights and Civil Liberties:* Kalifa Gassama Diaby. *Industry, Small and Medium Enterprises:* Ramatoulaye Bah. *International Co-operation:* Koutoubou Moustapha Sanoh. *Livestock:* Saramady Touré. *Mines and Geology:* Mohamed Lamine Fofana. *Planning:* Sékou Traoré. *Pre-university Education:* Ibrahima Kourouma. *Security, Civil Protection and Reform of Security Services:* Mouramani Cissé. *Social Affairs, and Promotion of Women and Children:* Hadja Diaka Diakité. *Telecommunications, Post and Information Technology:* Oyé Guilavogui. *Territorial Administration and Political Affairs:* Alhassane Condé. *Town Planning, Housing and Construction:* Ibrahima Bangoura. *Trade:* Mohamed Dorval Doumbouya. *Youth, Youth Employment and Sports:* Sanoussy Bantama Sow.

CURRENT LEADERS

Alpha Condé

Position
President

Introduction
Alpha Condé was sworn in as Guinea's first democratically elected president on 21 Dec. 2010. Though his election was marred by violence and allegations of fraud it marked a break with half a century of authoritarian rule.

Early Life
Born in Boke, French Guinea on 4 March 1938, Alpha Condé was educated in Paris at the Institut d'Études Politiques de Paris and the Sorbonne. He earned a PhD in public law before teaching at the Université Paris 1 Panthéon-Sorbonne and the School of Post, Telephone and Telecommunications (PTT).

In the 1950s Condé was heavily involved with the Fédération des Étudiants d'Afrique Noire en France (FEANF), campaigning for Guinea's independence from France. After independence in 1958, Condé remained in France and became a vocal opponent of President Ahmed Sékou Touré and the one party system. Touré had Condé sentenced to death in absentia in 1970 for his opposition. In 1977 Condé co-founded the National Democratic Movement (NDM), a party that evolved into the Rally of the Guinean People (RPG) that Condé currently heads.

Condé returned to Guinea in 1991 and ran for the presidency in 1993 in the nation's first multiparty elections, losing to Conté. In 1998 he again lost out to Conté, who gained 51·7% of the vote. On 16 Dec. 1998, two days after the poll, Condé was arrested along with other opposition leaders, accused of attempting to destabilize the government. Despite international pressure, Condé was imprisoned without trial for 20 months. In Sept. 2000 he was sentenced to five years in prison but was pardoned by Conté in May 2001 on condition that he withdraw from politics. Condé left for France soon after but returned to Guinea in July 2005.

He initially supported the National Council for Democracy and Development (CNDD) and Capt. Moussa Dadis Camara, who seized power in a military coup in Dec. 2008. However, he soon called for a return to civilian rule and free elections. In Feb. 2010 he announced his candidature for the June 2010 presidential elections. Condé was placed second with 20·7% of the vote in the first round but won the delayed run-off with 52·5% of the vote. His victory was confirmed on 3 Dec. 2010 by the Supreme Court, two weeks after the provisional results were challenged by his opponent, Cellou Dalein Diallo.

Career in Office
Nine days after being sworn into office, Condé announced that in an attempt to break the links with the previous regimes and to address the high unemployment rate among young graduates, all civil servants in the offices of the president and prime minister would be replaced. Condé was expected to review the country's mining contracts, reform the military, increase public access to basic amenities and form a Truth, Justice and Reconciliation Commission to investigate human rights abuses and reconcile ethnic divisions. In April 2012 he indefinitely postponed parliamentary elections that had previously been scheduled for late 2011 and then set for July 2012.

DEFENCE

There is selective conscription for two years. Defence expenditure totalled US$51m. in 2008 (US$5 per capita), representing 1·0% of GDP.

Army

The Army strength (2007) was 8,500. There are also three paramilitary forces: People's Militia (7,000), Gendarmerie (1,000) and Republican Guard (1,600) although only 2,600 are active.

Navy

A small force of around 400 (2007) operates from bases at Conakry and Kakanda.

Air Force

Personnel (2007) 800. There were seven combat-capable aircraft including MiG-17s and MiG-21s, although their serviceability was in doubt.

ECONOMY

Agriculture produced 16·9% of GDP in 2009, industry 31·0% and services 52·1%.

Overview

Guinea is rich in natural resources, with around a third of the world's bauxite reserves as well as large gold and diamond deposits along with potential for hydroelectric power. Having achieved relatively strong growth in the 1990s, economic performance deteriorated from 2000, with slowing GDP growth and rising inflation. Aside from the mining sector, the economy is dominated by the processing of agricultural products.

Guinea achieved its 1990s growth on the back of tight financial policies and favourable commodity prices. Inflation was also kept low. Subsequent poor performance stems from a weak policy framework, a fall in commodity export prices (including bauxite, which accounts for 50% of exports) and the knock-on effect of conflicts in neighbouring countries. In the 1990s nearly half a million refugees arrived from surrounding states suffering from civil conflict, notably Liberia and Sierra Leone. Guinea itself experienced a coup in Dec. 2008 following the death of President Lansana Conté, resulting in the suspension of the constitution and the dissolution of parliament. Political instability is a major impediment to growth.

Donor assistance is focused on improving the economic infrastructure, particularly in rural areas. The IMF approved a three-year Poverty Reduction and Growth Facility arrangement for Guinea in 2007 but a review was interrupted by the 2008 coup, which also delayed the completion of the Heavily Indebted Poor Countries initiative.

Weaknesses in governance, particularly corruption, continue to represent a challenge to development. An estimated 47% of the population was living under the poverty line in 2006.

Currency

The monetary unit is the *Guinean franc* (GNF). Inflation was 15·5% in 2010 and 21·4% in 2011. Foreign exchange reserves were US$95m. in Dec. 2005 and total money supply was 1,394·2bn. Guinean francs.

Budget

Revenue for 2008 was 3,854,400m. Guinean francs and expenditure 3,735,600m. Guinean francs.

Tax revenue accounted for 81·9% of total revenue in 2008; current expenditure accounted for 65·2% of total expenditure.

VAT is 18%.

Performance

The economy contracted by 0·3% in 2009 but there was growth of 1·9% in 2010 and 3·9% in 2011. Total GDP in 2011 was US$5·1bn.

Banking and Finance

In 1986 the Central Bank (*Governor*, Lounceny Nabe) and commercial banking were restructured, and commercial banks returned to the private sector. There were seven commercial banks in 2002. There is an Islamic bank.

Foreign debt was US$2,923m. in 2010, representing 69·1% of GNI.

ENERGY AND NATURAL RESOURCES

Environment

Guinea's carbon dioxide emissions from the consumption and flaring of fossil fuels in 2008 were the equivalent of 0·1 tonnes per capita.

Electricity

In 2007 installed capacity was 361,000 kW. Production was 973m. kWh in 2007; consumption per capita was 104 kWh.

Minerals

Mining accounted for 23% of state revenue in 2007. Guinea has the world's largest bauxite reserves, possessing nearly a third of the global total, and is the fifth largest producer. Output: bauxite (2008), 17,682,300 tonnes; alumina (2008), 593,900 tonnes; gold (2007), 15,628 kg. Diamond production in 2008, 445,400 carats.

There are also deposits of chrome, copper, granite, iron ore, lead, manganese, molybdenum, nickel, platinum, uranium and zinc.

Agriculture

Subsistence agriculture supports about 70% of the population. There were around 2·2m. ha. of arable land in 2007 and 670,000 ha. of permanent crops. The chief crops (production, 2008, in 1,000 tonnes) are: rice, 1,534; cassava, 1,122; maize, 952; plantains (2007 estimate), 436; millet (2007 estimate), 323; groundnuts, 316; sugarcane (2007 estimate), 283; sweet potatoes (2007), 200; mangoes and guavas (2007 estimate), 165; bananas (2007 estimate), 160; cocoyams and taro, 105; pineapples, 102.

Livestock (2008): cattle, 4·15m.; goats, 1·53m.; sheep, 1·28m.; pigs, 82,000; chickens (2007 estimate), 18m.

Forestry

The area under forests in 2010 was 6·54m. ha., or 27% of the total land area. In 2007, 12·44m. cu. metres of roundwood were cut.

Fisheries

In 2007 the total catch was 74,823 tonnes, almost entirely from sea fishing.

INDUSTRY

Manufacturing accounted for 5·2% of GDP in 2009. Cement, corrugated and sheet iron, beer, soft drinks and cigarettes are produced.

Labour

In 2010 the labour force was 4,092,000 (54·8% males). The agricultural sector employs 80% of the workforce.

INTERNATIONAL TRADE

Imports and Exports

Imports and exports for calendar years in US$1m.:

	2004	2005	2006	2007	2008
Imports c.i.f.	955·0	1,647·8	1,063·9	1,281·5	1,835·5
Exports f.o.b.	628·7	795·7	770·5	1,059·0	1,430·5

In 2008 main import sources (in US$1m.) were: Netherlands (377·6); France (185·4); UK (144·3). Main export markets in 2008 (in US$1m.) were: France (349·8); Switzerland (278·4); Russia (151·3).

Principal imports in 2008 (in US$1m.) were: mineral fuels, lubricants and related materials (605·0); machinery and transport equipment (538·6); manufactured goods (233·3). Principal exports in 2008 (in US$1m.) were: aluminium ores and concentrates, including alumina (734·5); gold (458·0); printed matter (112·7). Guinea ranks among the world's largest exporters of bauxite.

COMMUNICATIONS

Roads

In 2008 there were 6,758 km of roads, 35·4% of which were asphalted. In 2003 there were an estimated 47,500 passenger cars, 11,500 trucks and vans and 20,900 buses.

Rail

A railway connects Conakry with Kankan (662 km). A line 134 km long linking bauxite deposits at Sangaredi with Port Kamsar was opened in 1973 (carried 24m. tonnes in 2007) and a third line links Conakry and Fria (144 km; carried 1·2m. tonnes in 2004). The Kindia Bauxite Railway (102 km), linking Débéle with Conakry, carried 2·3m. tonnes in 2007. A further railway used by the bauxite industry runs from Tougué to Dabola (130 km).

Civil Aviation

There is an international airport at Conakry (Gbessia). In 2003 there were scheduled flights to Abidjan, Accra, Bamako, Banjul, Bissau, Brussels, Casablanca, Dakar, Freetown, Lagos and Paris.

In 2006 there were 103,200 air arrivals and 153,800 departures plus 9,600 passengers in transit. A total of 8·53m. tonnes of air freight were handled in 2006.

Shipping
There are ports at Conakry and for bauxite exports at Kamsar (opened 1973). Merchant shipping totalled 1,000 GT in 2008.

Telecommunications
The Société des Télécommunications de Guinée, which was privatized in 1995, became 100% state-owned again in 2008 after Telekom Malaysia sold its 60% stake in the company. There were 18,000 fixed telephone lines in 2010 (1·8 per 1,000 inhabitants). Mobile phone subscribers numbered 3·49m. in 2009. There were 9·6 internet users per 1,000 inhabitants in 2010.

SOCIAL INSTITUTIONS
Justice
There are *tribunaux du premier degré* at Conakry and Kankan, and a *juge de paix* at Nzérékoré. The High Court, Court of Appeal and Superior Tribunal of Cassation are at Conakry. The death penalty is in force, but has not been used since 2001.

The population in penal institutions in 2008 was 2,780 (28 per 100,000 of national population).

Education
In 2009 adult literacy was estimated at 39·5%. In 2007 there were 1,317,791 pupils with 29,049 teaching staff in primary schools; and 530,590 pupils with 13,907 teaching staff in secondary schools. In 2006 there were 42,711 students and 1,439 academic staff in tertiary education.

Besides French, there are eight official languages taught in schools: Fulani, Malinké, Susu, Kissi, Kpelle, Loma, Basari and Koniagi.

In 2005 public expenditure on education came to 1·7% of GNI.

Health
In 2006 there were 35 hospitals. There were (2006) 689 doctors, 109 pharmacists, 279 midwives and (2004) 4,061 nursing personnel.

RELIGION
79% of the population are Muslim, 9% Christian. Traditional animist beliefs are still found. In Feb. 2013 there was one cardinal in the Roman Catholic church.

CULTURE
World Heritage Sites
Guinea shares one site with Côte d'Ivoire on the UNESCO World Heritage List: Mount Nimba Strict Nature Reserve (inscribed on the list in 1981 and 1982). The dense forested slopes are home to viviparous toads and chimpanzees among other fauna.

Press
In 2008 there were two daily newspapers (circulation 25,000).

Tourism
In 2007, 30,000 non-resident tourists arrived at Conakry airport.

DIPLOMATIC REPRESENTATIVES
Of Guinea in the United Kingdom (258 Belsize Rd, London, NW6 4BT)
Ambassador: Vacant.
Chargé d'Affaires a.i.: Dondo Sylla.

Of the United Kingdom in Guinea (BP 6729, Conakry)
Ambassador: Graham Styles.

Of Guinea in the USA (2112 Leroy Pl., NW, Washington, D.C., 20008)
Ambassador: Blaise Chérif.

Of the USA in Guinea (Transversale 2, Ratoma, Conakry)
Ambassador: Alexander Laskaris.

Of Guinea to the United Nations
Ambassador: Mamadi Touré.

Of Guinea to the European Union
Ambassador: Ousmane Sylla.

FURTHER READING
Bulletin Statistique et Économique de la Guinée. Monthly.

National Statistical Office: Direction Nationale de la Statistique, BP 221, Conakry.
Website (French only): http://www.stat-guinee.org

GUINEA-BISSAU

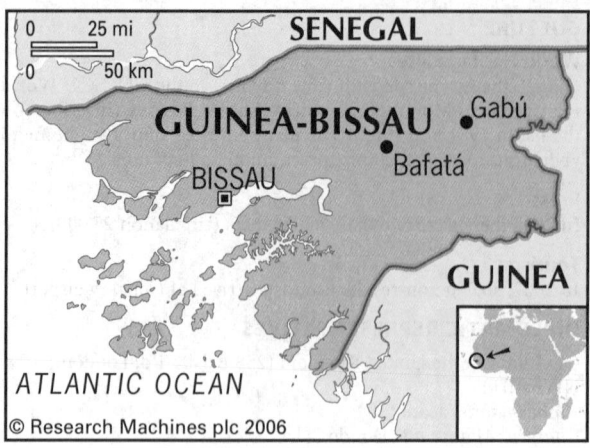

© Research Machines plc 2006

Republica da Guiné-Bissau
(Republic of Guinea-Bissau)

Capital: Bissau
Population projection, 2015: 1·68m.
GNI per capita, 2011: (PPP$) 994
HDI/world rank: 0·353/176
Internet domain extension: .gw

KEY HISTORICAL EVENTS

Portugal was the major power in the area throughout the colonial period. In 1974, after the Portuguese revolution, Portugal abandoned the struggle to keep Guinea-Bissau and independence was formally recognized on 10 Sept. 1974. In 1975 Cape Verde also became independent but the two countries remained separate sovereign states. On 14 Nov. 1980 a coup d'état was in part inspired by resentment in Guinea-Bissau over the privileges enjoyed by Cape Verdians. Guineans obtained a more prominent role under the new government. In May 1984 a new constitution was approved based on Marxist principles but after 1986 there was a return to private enterprise in an attempt to solve critical economic problems and to lift the country out of poverty. A year-long civil war broke out in 1998 between army rebels and the country's long-time ruler. Neighbouring Senegal and Guinea sent troops in to aid the government. In May 1999 President João Bernardo Vieira was ousted in a military coup led by former chief of staff Gen. Ansumane Mané, whom the president had dismissed in 1998. Following the coup Mané briefly headed a military junta before National Assembly speaker Malam Bacaï Sanhá took power as acting president. After presidential elections in Nov. 1999 and Jan. 2000 Kumba Ialá gained the presidency in a landslide victory. Marking a change towards a democratic future in Guinea-Bissau's politics, Ialá rejected a demand made by the outgoing junta for special consultative status following the elections. Kumba Ialá was overthrown in a coup in Sept. 2003 led by army chief of staff Gen. Veríssimo Correia Seabra. Vieira returned from exile to win the 2005 presidential election but was murdered in March 2009 by a group of soldiers following the assassination of his rival Batista Tagme Na Waie, the Army Chief of Staff. Malam Bacaï Sanhá became president in July 2009. In July and Aug. 2011 Prime Minister Carlos Gomes Júnior survived popular calls for his resignation amid spiralling food prices. In Jan. 2012 Sanhá died after a long illness. After the first round of polls two months later to find a successor, the military staged a coup, deposing the

interim president and detaining the front-runner in the election. Manuel Serifo Nhamadjo was installed as head of a transitional government promising to oversee new elections within a year.

TERRITORY AND POPULATION

Guinea-Bissau is bounded by Senegal in the north, the Atlantic Ocean in the west and by Guinea in the east and south. It includes the adjacent archipelago of Bijagós. Area, 36,125 sq. km (13,948 sq. miles). 2009 census population, 1,520,830 (783,196 females); density, 42·1 per sq. km. In 2011, 30·2% of the population were urban.

The UN gives a projected population for 2015 of 1·68m.

The area, population, and chief town of the capital and the eight regions:

Region	Area in sq. km	Population (2009 census)	Chief town
Bissau City	78	387,909	—
Bafatá	5,981	210,007	Bafatá
Biombo	840	97,120	Quinhámel
Bolama	2,624	34,563	Bolama
Cacheu	5,175	192,508	Cacheu
Gabú	9,150	215,530	Gabú
Oio	5,403	224,644	Farim
Quinara	3,138	63,610	Fulacunda
Tombali	3,736	94,939	Catió

The largest ethnic group are the Balanta (nearly a third of the population), Fulani, Manjaco and Mandinga. Portuguese remains the official language, but Crioulo is spoken throughout the country.

SOCIAL STATISTICS

2008 births (estimates), 65,000; deaths, 27,000. Estimated rates per 1,000 population, 2008: births, 41·2; deaths, 17·2. Annual population growth rate, 2000–08, 2·4%. Life expectancy, 2007: male, 46·0 years; female, 49·1. Infant mortality, 2010, 92 per 1,000 live births; fertility rate, 2008, 5·7 births per woman.

CLIMATE

The tropical climate has a wet season from June to Nov., when rains are abundant, but the hot, dry Harmattan wind blows from Dec. to May. Bissau, Jan. 76°F (24·4°C), July 80°F (26·7°C). Annual rainfall 78" (1,950 mm).

CONSTITUTION AND GOVERNMENT

A new constitution was promulgated on 16 May 1984 and has been amended five times since, most recently in 1996. The Revolutionary Council, established following the 1980 coup, was replaced by a 15-member Council of State, while in April 1984 a new National People's Assembly was elected comprising 150 representatives elected by and from the directly-elected regional councils for five-year terms. The sole political movement was the *Partido Africano da Independência da Guiné e Cabo Verde* (PAIGC), but in Dec. 1990 a policy of 'integral multi-partyism' was announced, and in May 1991 the National Assembly voted unanimously to abolish the law making the PAIGC the sole party. The *President* is Head of State and Government and is elected for a five-year term. The *National Assembly* now has a maximum of 102 members.

In the wake of the coup of April 2012 the military junta suspended the constitution and dissolved parliament (although it was resumed in Nov. 2012), and defied international demands for the constitution's restoration.

National Anthem

'Sol, suor, o verde e mar' ('Sun, sweat, the green and the sea'); words and tune by A. Lopes Cabral.

RECENT ELECTIONS

In the first round of presidential elections held on 18 March 2012 former prime minister Carlos Gomes Júnior (African Party for the Independence of Guinea and Cape Verde/PAIGC) took 49·0% of votes cast, ahead of former president Mohamed Ialá Embaló (formerly known as Kumba Ialá) (Party for Social Renewal/PRS) with 23·4%, Manuel Serifo Nhamadjo (ind.) with 15·7% and former interim president Henrique Roas (ind.) with 5·4%. Five other candidates all received less than 4% of the votes cast. Turnout was 55%. A run-off between Carlos Gomes Júnior and Mohamed Ialá Embaló was scheduled for 29 April 2012 but following a coup on 12 April both candidates were arrested by the Military Command. The Economic Community of West African States (ECOWAS) declared that presidential elections must be held within 12 months.

At the parliamentary elections on 16 Nov. 2008 turnout was 82%. The PAIGC won 49·8% of the vote (67 of 100 seats), the PRS 25·3% (28 seats) and the Republican Party for Independence and Development (PRID) 7·5% (3 seats). Two parties took one seat each.

CURRENT GOVERNMENT

Transitional President: Manuel Serifo Nhamadjo; b. 1958 (ind.; since 11 May 2012).

In March 2013 the cabinet comprised:

Transitional Prime Minister: Rui Duarte Barros (since 16 May 2012).

Minister of Agriculture and Fisheries: Malam Mané. *Civil Service, Work and State Reform:* Carlos Joaquim Vamain. *Commerce, Industry and Promotion of Local Products:* Abubacar Baldé. *Defence and War Veterans:* Col. Celestino Carvalho. *Economy and Regional Integration:* Degol Mendes. *Finance:* Abubacar Demba Dahaba. *Foreign Affairs and International Co-operation:* Faustino Fudut Imbali. *Infrastructure:* Fernando Gomes. *Interior:* Antonio Suca Intchama. *Justice:* Mamadu Saico Baldé. *National Education, Youth, Culture and Sport:* Vicente Pungura. *Natural Resources and Energy:* Daniel Gomes. *Public Health and Social Solidarity:* Agostinho Cá. *Territorial Administration and Local Power:* Baptista Té. *President of the Council of Ministers, Social Communication and Relations with Parliament, and Government Spokesperson:* Fernando Vaz.

CURRENT LEADERS

Manuel Serifo Nhamadjo

Position
Interim President

Introduction
Manuel Serifo Nhamadjo was installed as interim president on 11 May 2012. He heads a transitional government formed after a military coup ousted the previous interim president, Raimundo Pereira, and ended an electoral process intended to find a permanent successor to Malam Bacaí Sanhá, who had died in office. Nhamadjo's administration lacks comprehensive international recognition and its leaders are subject to sanctions imposed by the United Nations Security Council.

Early Life
Manuel Serifo Nhamadjo was born in 1958 in what was then Portuguese Guinea. Having previously served as parliamentary speaker, he came third in the presidential elections that followed the death of Sanhá in Jan. 2012. The election was to have been decided by a run-off between a former prime minister, Carlos Gomes Júnior, and a former president, Mohamed Ialá Embaló. However, a coup against Pereira prevented it from being held. Both Carlos Gomes Júnior and Pereira were arrested but were subsequently allowed to leave the country.

A deal between the coup leaders and opposition parties was brokered by the Economic Community of West African States (ECOWAS), allowing power to be ceded to a civilian government charged with the restitution of democracy. Nhamadjo was nominated as president with a one-year mandate. However, international organizations including the United Nations, European Union (EU) and Community of Portuguese-speaking Countries (CPLP) contend that the government remains under military control.

Career in Office
A rebel attack on an army barracks outside Bissau in Oct. 2012 sparked fighting that claimed six lives. The communications minister, Fernando Vaz, announced that 'the government considers Portugal, the CPLP and Carlos Gomes Júnior as the instigators of this attempt at destabilization'. Nhamadjo also criticized the EU for withdrawing aid, arguing that resulting financial pressures made it impossible to organize elections before the expiry of his mandate in May 2013.

DEFENCE

There is selective conscription. Defence expenditure totalled US$18m. in 2008 (US$12 per capita), representing 3·7% of GDP.

Army

Army personnel in 2007 numbered 6,800. There is a paramilitary Gendarmerie 2,000 strong.

Navy

The naval flotilla, based at Bissau, numbered an estimated 350 in 2007.

Air Force

Formation of a small Air Force began in 1978. Personnel (2007) 100 with three combat-capable aircraft (MiG-17s).

ECONOMY

In 2006 agriculture accounted for 61·8% of GDP (the highest percentage of any country), industry 11·5% and services 26·8%.

Overview

Guinea-Bissau is one of the poorest economies in the world. Having gained independence from Portugal in 1974, its development has been hindered by political instability. Civil war from 1998–99 and a series of military coups culminating in the murder of President Vieira and the army chief of staff, Gen. Tagme Na Waie, in March 2009 have undermined the economic and social infrastructure.

90% of exports are from agriculture, in particular cashew nuts, the country's main export, leaving the economy susceptible to external shocks and adverse weather. Economic growth during the global downturn has been robust despite lower prices for cashews and falling remittances. However, the external debt-to-GDP ratio is high, at over 125%. The economy is reliant on donor support and is vulnerable to drug traffickers who use the country as a trans-shipment point for trade between Latin America and Europe.

Currency

In May 1997 Guinea-Bissau joined the French Franc Zone, and the *peso* was replaced by the franc CFA at 65 pesos = one franc CFA. The *franc CFA* (XOF) has a parity rate of 655·957 francs CFA to one euro. Foreign exchange reserves were US$96m. in June 2005 and total money supply was 58,030m. francs CFA. There was inflation of 1·1% in 2010 and 5·0% in 2011.

Budget

Revenue in 2009 was 95,900m. francs CFA; expenditure totalled 88,700m. francs CFA.

Performance

Real GDP growth was 3·5% in 2010 and 5·3% in 2011. Total GDP in 2011 was US$1·0bn.

Banking and Finance
The bank of issue and the central bank is the regional Central Bank of West African States (BCEAO). The *Governor* is Tiémoko Meyliet Koné. There are four other banks (Banco da Africa Occidental; Banco Internacional de Guiné-Bissau; Caixa de Crédito de Guiné; Caixa Económica Postal).

In 2010 external debt totalled US$1,095m., equivalent to 124·8% of GNI.

ENERGY AND NATURAL RESOURCES
Environment
Carbon dioxide emissions from the consumption and flaring of fossil fuels in 2008 were the equivalent of 0·3 tonnes per capita. An *Environmental Performance Index* compiled in 2008 ranked Guinea-Bissau 140th in the world out of 149 countries analysed, with 49·7%. The index examined various factors in six areas—air pollution, biodiversity and habitat, climate change, environmental health, productive natural resources and water resources.

Electricity
Installed capacity in 2007 was estimated at 21,000 kW. Production was about 70m. kWh in 2007; consumption per capita was an estimated 50 kWh.

Minerals
Mineral resources are not exploited. There are estimated to be 200m. tonnes of bauxite and 112m. tonnes of phosphate.

Agriculture
Agriculture employs 80% of the labour force. There were an estimated 300,000 ha. of arable land in 2007 and 250,000 ha. of permanent crops. Main crops (2003 estimates, in 1,000 tonnes): rice, 97; cashew nuts, 80; coconuts, 46; plantains, 38; cassava, 34; millet, 22. Livestock (2003 estimates): cattle, 520,000; pigs, 360,000; goats, 300,000; sheep, 290,000; chickens, 2m.

Forestry
The area covered by forests in 2010 was 2·02m. ha., or 72% of the total land area. In 2007, 592,000 cu. metres of roundwood were cut.

Fisheries
Total catch in 2010 came to an estimated 6,800 tonnes, almost entirely from sea fishing.

INDUSTRY
Manufacturing accounted for 10·1% of GDP in 2001. Output of main products: vegetable oils (3·4m. litres in 2000), sawnwood (16,000 tonnes in 2001), soap (2,500 tonnes in 2000) and animal hides and skins (1,400 tonnes in 2001).

Labour
The labour force in 2010 was 648,000 (52·7% males).

INTERNATIONAL TRADE
Imports and Exports
Imports in 2008 were US$159m. and exports US$98m. Main imports in 2001 were: foodstuffs, 18·7%; transport equipment, 13·2%; equipment and machinery, 7·7%. Exports: cashew nuts, 95·6%; cotton, 2·3%; logs, 1·5%. Guinea-Bissau supplies more than 10% of the world market of cashew nuts. Main import suppliers, 2001: Portugal, 30·9%; Senegal, 28·3%; China, 11·3%. Main export markets, 2001: India, 85·6%; Portugal, 3·8%; Senegal, 2·5%.

COMMUNICATIONS
Roads
In 2002 there were about 4,400 km of roads, of which 2,400 km were national roads. In 2008 there were 42,200 passenger cars in use (27 per 1,000 inhabitants in 2007) and 9,300 lorries and vans.

Civil Aviation
There is an international airport serving Bissau (Osvaldo Vieira). In 2010 there were scheduled flights to Conakry, Dakar, Lisbon and Praia.

Shipping
The main port is Bissau; minor ports are Bolama, Cacheu and Catió.

Telecommunications
There were an estimated 5,000 fixed telephone lines in 2010 (3·3 per 1,000 inhabitants) and 402,000 mobile phone subscriptions in 2011 (or 259·8 per 1,000 inhabitants). There were 24·5 internet users per 1,000 inhabitants in 2010. Fixed internet subscriptions totalled 699 in 2009 (0·5 per 1,000 inhabitants).

SOCIAL INSTITUTIONS
Justice
The death penalty was abolished for all crimes in 1993.

Education
Adult literacy was estimated at 52·2% in 2009 (male, 66·9%; female, 38·0%). In 1999–2000 there were 150,041 pupils at primary schools (3,405 teachers), 25,736 at secondary schools (1,226 teachers) and 463 students in tertiary education. In 1999–2000 total expenditure on education came to 2·3% of GNP.

Health
In 2006 there was one national hospital, seven regional hospitals and 26 prefectorial hospitals. In 2007 there were seven hospital beds per 10,000 inhabitants. There were 188 physicians and 998 nurses in 2006. In 2004 there were 22 dentistry personnel.

RELIGION
In 2001 about 38% of the population were Muslim and about 12% Christian (mainly Roman Catholic). The remainder held traditional animist beliefs.

CULTURE
Press
There are no daily newspapers. In 2008 there were six non-daily papers, which had a combined weekly circulation of 10,000 copies.

Tourism
In 2007, 30,000 non-resident tourists arrived by air.

DIPLOMATIC REPRESENTATIVES
Of Guinea-Bissau in the United Kingdom
Ambassador: Vacant (resides in Paris).

Of the United Kingdom in Guinea-Bissau
Ambassador: John Marshall (resides in Dakar, Senegal).

Of Guinea-Bissau in the USA (PO Box 33813, Washington, D.C., 20033)
Ambassador: Vacant.

Of the USA in Guinea-Bissau
Ambassador: Lewis Lukens (resides in Dakar, Senegal).

Of Guinea-Bissau to the United Nations
Ambassador: João Soares da Gama.

Of Guinea-Bissau to the European Union
Ambassador: Alfredo Lopes Cabral.

FURTHER READING
Barry, Boubacar-Sid, Creppy, Edward G. E., Gacitua-Mario, Estanislao and Wodon, Quentin, *Conflict, Livelihoods, and Poverty in Guinea-Bissau.* 2007
Forrest, J. B., *Lineages of State Fragility: Rural Civil Society in Guinea-Bissau.* 2003

National Statistical Office: Instituto Nacional de Estadística e Censos (INEC), CP 06 Bissau.
Website (Portuguese only): http://www.stat-guinebissau.com

GUYANA

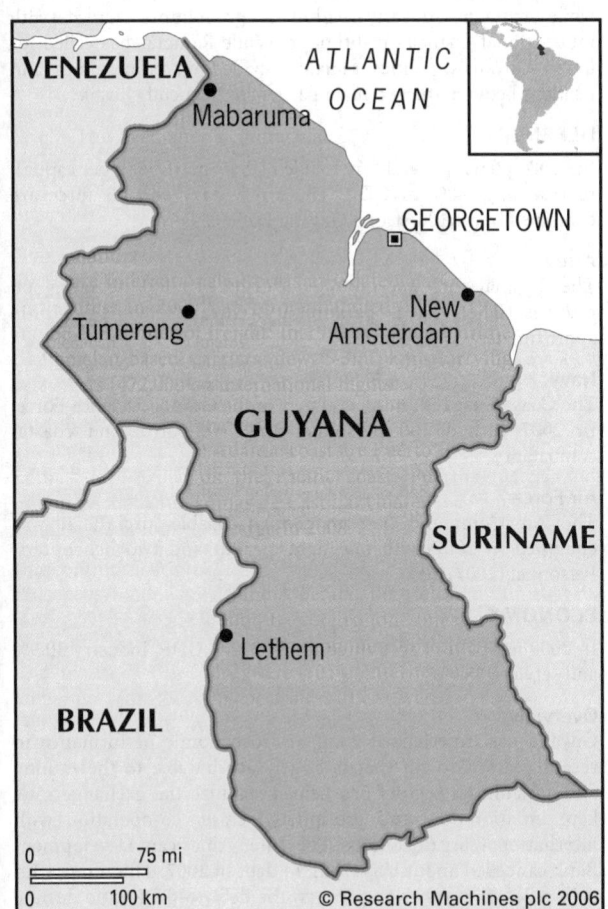

Co-operative Republic of Guyana

Capital: Georgetown
Population projection, 2015: 763,000
GNI per capita, 2011: (PPP$) 3,192
HDI/world rank: 0·633/117
Internet domain extension: .gy

KEY HISTORICAL EVENTS

First settled by the Dutch West Indian Company about 1620, the territory was captured by Britain to which it was ceded in 1814 and named British Guiana. African slaves were transported to Guyana in the 18th century to work the sugar plantations, with East Indian and Chinese indentured labourers following in the 19th century. From 1950 the anti-colonial struggle was spearheaded by the People's Progressive Party (PPP) led by Cheddi Jagan and Forbes Burnham. By the time internal autonomy was granted in 1961 Burnham had split with Jagan to form the more moderate People's National Congress (PNC). Guyana became an independent member of the Commonwealth in 1966 with Burnham as the first prime minister, later president. By the 1980s, desperate economic straits had forced Guyana to seek outside help which came on condition of restoring free elections. Dr Jagan returned to power in 1992. Following his death in March 1997 his wife, Janet Jagan, was sworn in as president.

TERRITORY AND POPULATION

Guyana is situated on the northeast coast of Latin America on the Atlantic Ocean, with Suriname on the east, Venezuela on the west and Brazil on the south and west. Area, 214,999 sq. km (83,013 sq. miles). In 2002 the census population was 751,223; density 3·5 per sq. km.

The UN gives a projected population for 2015 of 763,000.

Guyana has the highest proportion of rural population in South America, with only 28·7% living in urban areas in 2011. Ethnic groups by origin: 49% Indian, 36% African, 7% mixed race, 7% Amerindian and 1% others. The capital is Georgetown (2002 census population, 33,366; urban agglomeration, 134,497); other towns are Linden, New Amsterdam, Anna Regina and Corriverton.

Venezuela demanded the return of the Essequibo region in 1963 (nearly 75% of the area of Guyana). It was finally agreed in March 1983 that the UN Secretary-General should mediate, but the dispute is ongoing. There is also an ongoing unresolved claim by Suriname for the return of a triangle of uninhabited rainforest between the New River and the Courantyne River, near the Brazilian border. In Sept. 2007 the UN settled a long-standing maritime boundary dispute between Guyana and Suriname. The coastal area off both countries is believed to hold significant oil and gas deposits.

The official language is English.

SOCIAL STATISTICS

2009 estimates: births, 14,000; deaths, 5,000. Rates, 2009 estimates (per 1,000 population): birth, 18; death, 6. Life expectancy at birth in 2007: male 63·7 years and female 69·6 years. Annual population growth rate, 2000–05, 0·4%. Infant mortality, 2010, 25 per 1,000 live births; fertility rate, 2008, 2·3 births per woman.

CLIMATE

A tropical climate, with rainy seasons from April to July and Nov. to Jan. Humidity is high all the year but temperatures are moderated by sea-breezes. Rainfall increases from 90" (2,280 mm) on the coast to 140" (3,560 mm) in the forest zone. Georgetown, Jan. 79°F (26·1°C), July 81°F (27·2°C). Annual rainfall 87" (2,175 mm).

CONSTITUTION AND GOVERNMENT

A new constitution was promulgated in Oct. 1980. There is an *Executive Presidency* and a *National Assembly.* The president is elected by simple majority vote as the designated candidate of a party list in parliamentary elections; there are no term limits. The National Assembly has 65 elected members who serve five-year terms, plus not more than four non-voting members and two non-voting parliamentary secretaries appointed by the president. Of the 65 elected members, 25 are elected in multi-seat constituencies and 40 through a party-list proportional representation system.

National Anthem

'Dear land of Guyana'; words by A. L. Luker, tune by R. Potter.

RECENT ELECTIONS

Donald Ramotar and the People's Progressive Party/Civic (PPP/C) won the presidential and parliamentary elections of 28 Nov. 2011. In the presidential election Donald Ramotar received 166,340 votes (48·6% of the vote), with David A. Granger of A Partnership for National Unity receiving 139,678 (40·8%) and Khemraj Ramjattan of Alliance for Change 35,333 (10·3%). The PPP/C won 32 seats in the parliamentary election, followed by A

Partnership for National Unity 26 and Alliance for Change 7. Turnout was 72·9%.

CURRENT GOVERNMENT

President: Donald Ramotar; b. 1950 (PPP/C; sworn in 3 Dec. 2011).

In March 2013 the government comprised:

Prime Minister: Samuel Hinds; b. 1943 (PPP/C; first sworn in 9 Oct. 1992 and now in office for the third time).

Attorney General and Minister of Legal Affairs: Anil Nandlall. *Minister of Agriculture:* Leslie Ramsammy. *Amerindian Affairs:* Pauline Sukhai. *Culture, Youth and Sports:* Frank Anthony. *Education:* Priya Manickchand. *Finance:* Ashni Kumar Singh. *Foreign Affairs:* Carolyn Rodrigues-Birkett. *Health:* Dr Behri Ramsarran. *Home Affairs:* Clement Rohee. *Housing and Water, and Tourism, Industry and Commerce (acting):* Irfaan Ali. *Human Services and Social Security:* Jennifer Webster. *Labour:* Nanda Gopaul. *Local Government and Regional Development:* Ganga Persaud. *Natural Resources and Environment:* Robert Persaud. *Public Service:* Jennifer Westford. *Public Works:* Robeson Benn.

Government Information Agency Website: http://www.gina.gov.gy

CURRENT LEADERS

Donald Ramotar

Position
President

Introduction
Donald Ramotar became president in Dec. 2011, succeeding Bharrat Jagdeo. An economist, Ramotar has spent most of his career working within the socialist People's Progressive Party/ Civic (PPP/C), whose support base is the Indo-Guyanese community. The party has been in power since 1992 but lost its parliamentary majority for the first time at the Nov. 2011 elections.

Early Life
Donald Rabindranauth Ramotar was born on 22 Oct. 1950 in Guyana's Essequibo-West Demerera province. He was at school in Georgetown and graduated in economics from the University of Guyana before taking a masters degree at the Patrice Lumumba Peoples' Friendship University in Moscow.

Returning to Guyana, he worked in the timber industry and, from 1966, the Guyana Import-Export Company, a commercial venture under the opposition PPP/C. Between 1975 and 1983 Ramotar was the manager of Freedom House, the Georgetown headquarters of the PPP/C, before becoming the editor of a journal, *Problems of Peace and Socialism.*

He served as international secretary of the Guyana Agricultural Workers' Union (GAWU) between 1988 and 1993, when he assumed the role of PPP/C executive secretary. The party had been restored to power in Oct. 1992 under Cheddi Jagan. Following President Jagan's death in March 1997, Ramotar replaced him as party general secretary, a post he still holds. Janet Jagan led the PPP/C to victory in the Dec. 1997 elections although she was forced to retire through ill-health in 1999 and was succeeded by Bharrat Jagdeo.

Ramotar has served on the Africa Caribbean Pacific (ACP)-EU Joint Parliamentary Assembly and on several corporate boards. In April 2011 he was selected as the PPP/C's presidential candidate. Later that month he was appointed political adviser to President Jagdeo, a move criticized by the opposition Partnership for National Unity (APNU).

Ramotar emerged victorious in the election on 28 Nov. 2011 and his PPP/C took 32 seats in parliament (compared with 26 for APNU) to form a minority administration, the first since independence in 1966.

Career in Office
Ramotar was sworn in as president on 3 Dec. 2011. He pledged to build on the achievements of the previous administration, developing the country's infrastructure and raising competitiveness while reducing poverty within a programme agreed with international financial institutions. While Ramotar has called for unity, Guyana's politics remains divided on ethnic lines and relations between the two main parties are frequently hostile.

DEFENCE

In 2008 defence expenditure totalled US$67m. (US$89 per capita), representing 5·8% of GDP. The army, navy and air force are combined in a 1,100-strong Guyana Defence Force.

Army
The Guyana Army had (2007) a strength of 1,400 including 500 reserves. There is a paramilitary Guyana People's Militia approximately 1,500 strong.

Navy
The Coast Guard is an integral part of the Guyana Defence Force. In 2007 it had 100 personnel and five patrol and coastal combatants.

Air Force
The Air Command has no combat-capable aircraft. It was equipped in 2007 with one light aircraft and two helicopters. Personnel (2007) 100.

ECONOMY

In 2009 agriculture accounted for 24·0% of GDP, industry 30·9% and services 45·1%.

Overview
Guyana has experienced good macroeconomic performance in recent years. Growth that is largely attributable to the mining and agricultural sectors has helped stabilize the exchange rate, kept inflation low and precipitated closer co-operation with international organizations. The Inter-American Development Bank cancelled about US$470m. of debt in 2007. This along with debt relief from other donors saw the debt-to-GDP ratio decline from 183% in 2006 to 120% in 2007.

Agriculture remains important although its contribution to GDP fell from 30·8% in 2002 to 24·3% in 2010. A shortage of skilled labour is a problem, along with a sub-standard infrastructure. Main exports include sugar, gold, diamonds, bauxite, shrimp, timber and rice. Increasingly, the country benefits from high FDI inflows, especially in the mining sector. Guyana has been a member of CARICOM since 1973 and entered the Caricom Single Market and Economy in 2006.

Protecting the environment, managing sea level rise, improving governance and preventing crime are key for future growth. The country launched a Low Carbon Development Strategy in 2010 to promote economic development while combating climate change.

Currency
The unit of currency is the *Guyana dollar* (GYD) of 100 *cents.* Inflation was 3·7% in 2010 and 5·0% in 2011. Foreign exchange reserves were US$223m. in July 2005 and total money supply was G$35·3bn.

Budget
Revenues in 2008 totalled G$99,513m. (current revenue, 82·9%) and expenditures G$105,838m. (current expenditure, 59·5%).

VAT of 16% was introduced in 2007.

Performance
Real GDP growth was 3·3% in 2009, 4·4% in 2010 and 5·4% in 2011. Total GDP was US$2·6bn. in 2011.

Banking and Finance

The bank of issue is the Bank of Guyana (*Governor*, Lawrence Williams), established 1965. There are five commercial banks and three foreign-owned. At Dec. 2009 the total assets of commercial banks were G$253,760m. Savings deposits were G$130,764m.

In 2010 external debt totalled US$1,354m., equivalent to 52·8% of GNI.

ENERGY AND NATURAL RESOURCES

Environment

Guyana's carbon dioxide emissions from the consumption and flaring of fossil fuels were the equivalent of 2·2 tonnes per capita in 2008.

Electricity

Capacity in 2007 was 0·3m. kW. In 2007 production was 867m. kWh and consumption per capita 1,154 kWh.

Minerals

Placer gold mining commenced in 1884, and was followed by diamond mining in 1887. In 2007 output of bauxite was 2,242,928 tonnes and of gold 7,412 kg. Other minerals include copper, tungsten, iron, nickel, quartz and molybdenum.

Agriculture

In 2007 Guyana had an estimated 420,000 ha. of arable land and 30,000 ha. of permanent crops. Agricultural production, 2007 unless otherwise indicated (in 1,000 tonnes): sugarcane, 3,099; rice (2009), 360; coconuts, 70; cassava, 20; maize (estimate), 8; pumpkins and squash (estimate), 7; oranges (estimate), 6; bananas, 6; green beans, 5; mangoes and guavas, 4.

Livestock (2008 estimate): sheep, 130,000; cattle, 110,000; goats, 79,000; pigs, 13,500; chickens, 20m. Livestock products, 2009: meat, 29,000 tonnes; milk, 33m. litres; eggs, 19m. units.

Forestry

In 2010 the area under forests totalled 15·21m. ha. (77% of the land area). Timber production in 2007 was 426,000 cu. metres.

Fisheries

Fish landings in 2010 came to 45,186 tonnes, almost exclusively from sea fishing.

INDUSTRY

The main industries are agro-processing (particularly sugar and rice) and mining (notably gold and bauxite). Production: sugar (2009), 233,736 tonnes; flour (2008), 29,425 tonnes; rum (2008), 14·2m. litres; beer (2008), 8·2m. litres; soft drinks (2006), 4,050,000 cases; clothes (2008), 1,256,000 items; footwear (2008), 25,901 pairs; margarine (2008), 1,528,121 kg; edible oil (2005), 928,500 litres; paint (2008), 2,488,667 litres; sawnwood (2007), 74,000 cu. metres.

Labour

In 2010 the estimated economically active population was 342,000 (66% males).

INTERNATIONAL TRADE

Imports and Exports

In 2008 imports were valued at US$1,289m. and exports at US$782m. Main commodities imported, 2004: petroleum, petroleum products and related materials, 27·4%; machinery and transport equipment, 23·0%; manufactured goods, 16·3%; food and live animals, 10·9%. Principal commodities exported, 2004: sugar, 20·5%; gold, 18·4%; diamonds, 16·4%; fish, crustaceans and molluscs, 11·4%. Rice, timber and bauxite are also exported. Major import suppliers, 2004: USA, 30%; Trinidad and Tobago, 27%; Netherlands Antilles, 9%; Japan, 6%; UK, 5%. Main export markets in 2004: Canada, 19%; USA, 16%; Belgium, 14%; UK, 12%; Trinidad and Tobago, 7%.

COMMUNICATIONS

Roads

In 2002 there were an estimated 7,970 km of roads, of which 590 km were paved. In 2008 there were 44,700 passenger cars in use, plus 28,100 lorries and vans, and 37,100 motorcycles and mopeds.

Rail

There is a government-owned railway in the North West District, while the Guyana Mining Enterprise operates a standard gauge railway of 133 km from Linden on the Demerara River to Ituni and Coomacka.

Civil Aviation

There is an international airport at Georgetown (Timehri), which handled 465,962 passengers in 2007. In 2003 there were direct flights to Anguilla, Antigua, Barbados, Dominica, Miami, New York, Paramaribo, Port of Spain, St Kitts and the British Virgin Islands.

Shipping

The major port is Georgetown; there are two other ports. In Jan. 2009 there were 28 ships of 300 GT or over registered, totalling 25,000 GT. There are 217 nautical miles of river navigation. There are ferry services across the mouths of the Demerara, Berbice and Essequibo rivers.

Telecommunications

In 2011 there were 152,600 landline telephone subscriptions (equivalent to 201·8 per 1,000 inhabitants) and 518,800 mobile phone subscriptions (or 686·2 per 1,000 inhabitants). In 2010, 29·9% of the population were internet users. In March 2012 there were 124,000 Facebook users.

SOCIAL INSTITUTIONS

Justice

The law, both civil and criminal, is based on the common and statute law of England, save that the principles of the Roman–Dutch law have been retained for the registration, conveyance and mortgaging of land.

The Supreme Court of Judicature consists of a Court of Appeal, a High Court and a number of courts of summary jurisdiction. Guyana was one of ten countries to sign an agreement in Feb. 2001 establishing a Caribbean Court of Justice to replace the British Privy Council as the highest civil and criminal court having ended appeals to the Privy Council when it gained independence in 1970. In the meantime the number of signatories has risen to 12. The court was inaugurated at Port-of-Spain, Trinidad on 16 April 2005 at which point it replaced the Guyana Court of Appeal as the country's final court of appeal.

In 2006 there were 2,756 reported serious crimes, including 173 homicides. The population in penal institutions in Dec. 2006 was 1,955 (260 per 100,000 of national population).

Education

In 2004–05 there were 428 pre-primary schools with 1,958 teachers for 31,730 pupils; 440 primary schools with 4,013 teachers for 113,971 pupils; and 349 secondary schools with 3,392 teachers for 65,638 pupils. In 2004–05 there were 7,689 students at university level.

Adult literacy in 2001 was 98·6% (male, 99·0%; female, 98·2%). The literacy rates are the highest in South America. An OECD report published in 2005 showed that Guyana loses a greater proportion of its graduates (83%) to OECD member countries than any other non-OECD member.

In 2004–05 total expenditure on education came to 7·4% of GNP and 12·4% of total government spending.

Health

In 2006–07 there were 35 hospitals (seven private), 135 health centres and 207 health posts. There were 1,836 hospital beds in

2006–07. There were 5·1 physicians per 10,000 inhabitants in 2007 and 9·9 nurses per 10,000 inhabitants.

RELIGION

In 2002, 28·8% of the population were Hindus, 17·0% Pentecostalists, 8·1% Roman Catholics, 7·3% Muslims and 7·0% Anglicans. There were also significant numbers of other Christians.

CULTURE

Press

In 2008 there were three daily newspapers (the state-owned *Guyana Chronicle* and the privately-owned *Kaieteur News* and *Stabroek News*) with a combined average daily circulation of 32,000.

Tourism

141,000 non-resident tourists arrived at Timehri airport in 2009 (130,000 in 2008).

Festivals

There are a number of Christian, Hindu and Muslim festivals throughout the year.

DIPLOMATIC REPRESENTATIVES

Of Guyana in the United Kingdom (3 Palace Ct, London, W2 4LP)
High Commissioner: Laleshwar K. N. Singh.

Of the United Kingdom in Guyana (44 Main St., Georgetown)
High Commissioner: Andrew Ayre.

Of Guyana in the USA (2490 Tracy Pl., NW, Washington, D.C., 20008)
Ambassador: Bayney R. Karran.

Of the USA in Guyana (99–100 Young and Duke Streets, Kingston, Georgetown)
Ambassador: D. Brent Hardt.

Of Guyana to the United Nations
Ambassador: George Wilfred Talbot.

Of Guyana to the European Union
Ambassador: Patrick Ignatius Gomes.

FURTHER READING

Braveboy-Wagner, J. A., *The Venezuela-Guyana Border Dispute: Britain's Colonial Legacy in Latin America.* 1984
Daly, V. T., *A Short History of the Guyanese People.* 3rd. ed. 1992
Gafar, John, *Guyana: From State Control to Free Markets.* 2003
Seecoomar, Judaman, *Democratic Advance and Conflict Resolution in Post-Colonial Guyana.* 2006

National Statistical Office: Bureau of Statistics, 57 High Street, Kingston, P. O. Box 1070, Georgetown.
Website: http://www.statisticsguyana.gov.gy

HAITI

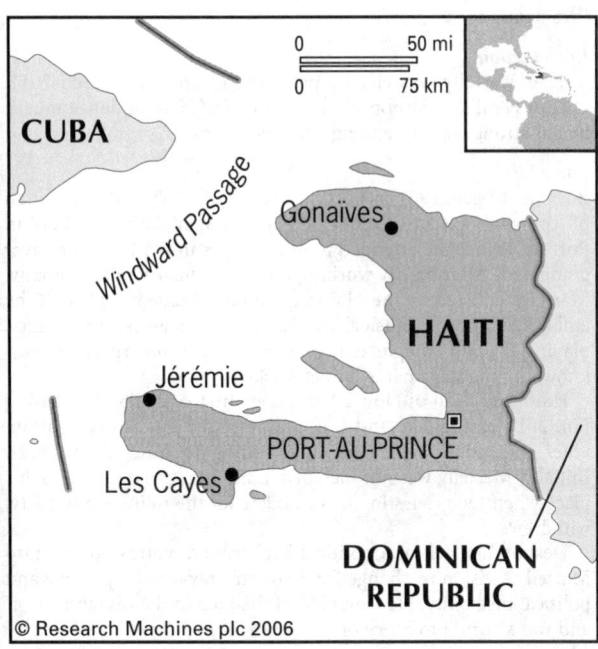

© Research Machines plc 2006

République d'Haïti
(Republic of Haiti)

Capital: Port-au-Prince
Population projection, 2015: 10·65m.
GNI per capita, 2011: (PPP$) 1,123
HDI/world rank: 0·454/158
Internet domain extension: .ht

KEY HISTORICAL EVENTS

In the 16th century, Spain imported large numbers of African slaves whose descendants now populate the country. The colony subsequently fell under French rule. In 1791 a slave uprising led to the 13-year-long Haitian Revolution. In 1801 Toussaint Louverture, one of the leaders of the revolution, succeeded in eradicating slavery. He proclaimed himself governor-general for life over the whole island. He was captured and sent to France, but Jean-Jacques Dessalines, one of his generals, led the final battle that defeated Napoleon's forces. The newly-named Haiti declared its independence on 1 Jan. 1804, becoming the first independent black republic in the world. Ruled by a succession of self-appointed monarchs, Haiti became a republic in the mid-19th century. From 1915 to 1934 Haiti was under United States occupation.

A corrupt regime was dominated by François Duvalier (widely known as Papa Doc) from 1957 to 1964 when he was succeeded by his son, Jean-Claude Duvalier (Baby Doc). He fled the country on 7 Feb. 1986. After a period of military rule, Father Jean-Bertrand Aristide was elected president in Dec. 1990.

On 30 Sept. 1991 President Aristide was deposed by a military junta and went into exile. Under international pressure, parliament again recognized Aristide as president in June 1993. However, despite a UN led naval blockade, the junta showed no sign of stepping down. 20,000 US troops moved into Haiti on 19 Sept. in an uncontested occupation. President Aristide returned to office on 15 Oct. 1994 and on 1 April 1995 a UN peacekeeping force (MANUH) took over from the US military mission. Aristide was succeeded by René Préval who was generally assumed to be a stand-in for his predecessor. Jean-Bertrand Aristide subsequently won the presidential elections held in Nov. 2000. In Dec. 2001 there was an unsuccessful coup led by former police and army officers. After armed rebels took control of the north of the country President Aristide stood down in Feb. 2004 and fled into exile.

After a period of interim government, René Préval was elected president for a second time in 2006. In July 2009 the World Bank and IMF cancelled US$1·2bn. (80%) of the national debt. In Jan. 2010 an earthquake of magnitude 7·0 hit the capital, Port-au-Prince, and its surrounding region, killing at least 217,000 people, displacing over 1m. and seriously undermining Haiti's economic prospects.

TERRITORY AND POPULATION

Haiti is bounded in the east by the Dominican Republic, to the north by the Atlantic and elsewhere by the Caribbean Sea. The area is 27,065 sq. km (10,450 sq. miles). The Île de la Gonâve, some 40 miles long, lies in the gulf of the same name. Among other islands is La Tortue, off the north peninsula. Census population, 2003, 8,373,750; density, 309 per sq. km. The United Nations population estimate for 2003 was 9,075,000. On 1 July 2009 the official population estimate was 9,923,243. In 2011, 53·6% of the population were urban.

The UN gives a projected population for 2015 of 10·65m.

Areas, populations and chief towns of the ten departments:

Department	Area (in sq. km)	2009 estimated population	Chief town
Artibonite	4,887	1,571,020	Gonaïves
Centre	3,487	678,626	Hinche
Grande Anse	1,912	425,878	Jérémie
Nippes	1,268	311,497	Miragoâne
Nord	2,115	970,495	Cap Haïtien
Nord-Est	1,623	358,277	Fort-Liberté
Nord-Ouest	2,103	662,777	Port-de-Paix
Ouest	4,983	3,664,620	Port-au-Prince
Sud	2,654	704,760	Les Cayes
Sud-Est	2,034	575,293	Jacmel

The capital is Port-au-Prince (2009 estimated population, 875,978; urban agglomeration, 2,296,386); the other main cities are Gonaïves (228,725 in 2009) and Cap Haïtien (155,505). Most of the population is of African or mixed origin.

The official languages are French and Créole. Créole is spoken by all Haitians; French by only a small minority.

SOCIAL STATISTICS

2009 estimates: births, 266,000; deaths, 89,000. Rates, 2009 estimates (per 1,000 population): birth, 27; death, 9. Annual population growth rate, 1998–2008, 1·7%. Expectation of life at birth, 2007, 59·1 years for males and 62·9 years for females. Infant mortality, 2010, 70 per 1,000 live births; fertility rate, 2008, 3·5 births per woman.

CLIMATE

A tropical climate, but the central mountains can cause semi-arid conditions in their lee. There are rainy seasons from April to June and Aug. to Nov. Hurricanes and severe thunderstorms can occur. The annual temperature range is small. Port-au-Prince, Jan. 77°F (25°C), July 84°F (28·9°C). Annual rainfall 53" (1,321 mm).

CONSTITUTION AND GOVERNMENT

A new constitution was signed off by President Martelly in June 2012, replacing one promulgated in 1987. The 2012 constitution had received parliamentary backing in May 2011 but was not written into law until after a year of legal wrangling. Among its provisions is the re-legalization of dual citizenship, which had been criminalized under the 1987 constitution. Haitians living abroad, who are responsible for remittances equivalent to 20% of GDP, have the right to own land in Haiti and to stand for political office (with the exceptions of the presidency, premiership, as a senator or a member of the Chamber of Deputies).

The constitution also established a permanent constitutional court to mediate in disputes between parliament and the executive, as well as an electoral council to oversee free and fair elections. Women are required to hold at least 30% of government posts.

There is a bicameral legislature (a 99-member *Chamber of Deputies* and a 30-member *Senate*) and an executive *President*, directly elected for a five-year term.

National Anthem

'La Dessalinienne' ('The Dessalines Song'); words by J. Lhérisson, tune by N. Geffrard.

RECENT ELECTIONS

Presidential elections were held on 28 Nov. 2010. Mirlande Manigat won 31·4% of the vote, Jude Célestin (President Préval's chosen successor) 22·5% and Michel Martelly 21·8%. There were 16 other candidates. There were widespread claims of voting irregularities. In the second round, held on 20 March 2011, Michel Martelly won with 67·6% of the vote against 31·7% for Mirlande Manigat.

Parliamentary elections were also held on 28 Nov. 2010 and 20 March 2011. In the vote for the Chamber of Deputies, Unity won 33 seats, Alternative for Progress and Democracy 14, Ansanm Nou Fò 9, Haiti in Action 8, Lavni Organisation 7 and Rally 4. Independents won 2 seats. The remaining seats went to ten smaller parties. Elections for 11 seats in the Senate were held on the same days. Alternative for Progress and Democracy took 5 seats, Unity 5 and Lavni Organisation 1.

CURRENT GOVERNMENT

President: Michel Martelly; b. 1961 (Farmers' Response Party; sworn in 14 May 2011).

In April 2013 the government comprised:

Prime Minister and Minister of Planning and External Co-operation: Laurent Lamothe; b. 1972 (sworn in 16 May 2012).

Minister of Defence: Jean Rodolphe Joazile. *Youth, Sports and Civil Action:* Magalie Racine. *Environment:* Jean François Thomas. *Haitians Abroad:* Bernice Fidélia. *Justice and Public Security:* Jean Renel Sanon. *Interior and Territorial Collectivities:* David Bazile. *Culture, and Communication (acting):* Josette Darguste. *Women's Affairs and Women's Rights:* Yanick Mézile. *Tourism:* Stéphanie Balmir Villedrouin. *National Education and Professional Training:* Vanneur Pierre. *Economy and Finance, and Trade and Industry (acting):* Wilson Laleau. *Agriculture, Natural Resources and Rural Development:* Thomas Jacques. *Public Health and Population:* Florence Duperval Guillaume. *Social Affairs and Labour:* Charles Jean-Jacques. *Public Works, Transport and Telecommunications:* Jacques Rousseau. *Foreign Affairs and Religion:* Pierre-Richard Casimir. *Relations with Parliament:* Ralph Théano.

Minister Delegate in the Office of the Prime Minister in Charge of Energy Security: René Jean Jumeau. *Minister Delegate in the Office of the Prime Minister in Charge of Human Rights and the Fight Against Extreme Poverty:* Marie Carmelle Rose Anne Auguste. *Minister Delegate in the Office of the Prime Minister in Charge of the Promotion of Peasantry:* Marie Mimose Félix.

Office of the Prime Minister (French only): http://primature.gouv.ht

CURRENT LEADERS

Michel Martelly

Position
President

Introduction
Michel Martelly was elected president of Haiti in March 2011. One of Haiti's most popular performers of Haitian dance music, he has strong support among younger voters.

Early Life
Born on 12 Feb. 1961 in Port-au-Prince, Martelly is the son of an oil company executive. Educated at a Roman Catholic school in Port-au-Prince, he attended junior colleges in the USA but never graduated. After briefly working for a US construction company, Martelly enlisted at the Haitian Military Academy. In 1986 he embarked upon a musical career as a keyboardist and singer, playing Haitian *compas* dance music in Port-au-Prince. He was known by the stage name 'Sweet Micky'.

Haiti's election of Nov. 2010 was its first since the devastating Jan. 2010 earthquake and was characterized by disorganization, voter intimidation and fraud according to foreign observers. Initially, Martelly was not included in the run-off poll but took his place when Jude Célestin, the candidate for the ruling Unity party, withdrew.

Despite his lack of a political background, voters bought into Martelly's vision of change for a country ravaged by poverty and political instability. He won 68% of the vote in the March run-off and was sworn into office on 14 May 2011.

Career in Office
Despite voter turnout of only 25%, his defeated opponent, the conservative former first lady Mirlande Manigat, did not challenge the result. The incumbent prime minister, Jean-Max Bellerive, resigned on 15 May 2011 to allow Martelly to choose his own premier. However, Martelly's first two nominations for premier, entrepreneur Daniel Rouzier and former justice minister Bernard Gousse, were rejected by parliament in June and Aug. 2011 respectively. Bellerive remained in a caretaker capacity until the appointment of Garry Conille as prime minister in Oct. 2011.

Martelly pledged to improve the faltering post-earthquake reconstruction programme. In Aug. 2011 he announced plans to reinstate the military, which had been disbanded by former president Jean-Bertrand Aristide. In Sept. 2011 he created an advisory board aimed at improving the economy and in Dec. 2011 he announced voting reforms.

Martelly has been criticized for his handling of the UN peacekeeping mission in Haiti. From Sept. 2010 there were popular anti-UN protests in the capital after a serious and prolonged outbreak of cholera was blamed on UN staff. The Senate in response passed a resolution for UN withdrawal, although the mission remains in operation.

Martelly's tenure hit further problems when Conille resigned on 24 Feb. 2012 after challenging Martelly's refusal to co-operate with an investigation into whether some government officials held dual nationality in contravention of the constitution. Martelly's nomination as Conille's successor was the foreign minister, Laurent Lamothe, who was confirmed in the post on 4 May 2012. In Sept. and Oct. protesters in Port-au-Prince called for the president's resignation, citing his failure to tackle poverty, the high cost of living and alleged corruption.

DEFENCE

After the restoration of civilian rule in 1994 the armed forces and police were disbanded and an Interim Public Security Force formed, although this was later also dissolved. In 1995 a new police force—Police Nationale d'Haiti (PNH)—was recruited from former military personnel and others not implicated in human rights violations. The PNH currently has about 2,000 members.

A UN peacekeeping force, MINUSTAH, has been in Haiti since 2004. Following the earthquake of Jan. 2010 the UN Security Council passed a resolution recommending an increase in overall force levels to support the immediate recovery, reconstruction and stability efforts in the country. As of Jan. 2013 MINUSTAH consisted of 9,298 uniformed personnel.

In 2003 defence expenditure totalled US$22m. (US$3 per capita), representing 0·8% of GDP.

Army

The Army was disbanded in 1995 but there are now plans to revive it.

Navy

There is a small Coast Guard, which was created in 1996 and is a specialized unit of the PNH.

Air Force

The Air Force was disbanded in 1995.

ECONOMY

Trade and restaurants accounted for 26·8% of GDP in 2006–07, agriculture and forestry 25·3%, and finance and real estate 12·1%.

Overview

Haiti is the poorest economy in the western hemisphere following decades of economic decline and instability. Real per capita GDP has declined on average by 0·7% per year over the past 40 years. The economy has been characterized by macroeconomic instability, rampant inflation and susceptibility to external shocks. Social and environmental indicators are both poor.

Nonetheless, performance in the opening decade of the century suggested some progress, with inflation falling and GDP recording modest growth. However, on 12 Jan. 2010 an earthquake killed 220,000 people, as well as damaging or destroying 13 out of 15 government ministries, 4,200 schools and 60% of hospitals. Nearly US$10bn. of relief funding was pledged for post-earthquake relief, with total damage estimated at 120% of GDP.

Boosted by reconstruction, GDP grew by 5·6% in 2010–11 after a 5·4% contraction in 2009–10. Inflation receded to single-digit levels and the external position has strengthened, reflecting a substantial increase in textile exports to the USA. However, significant challenges remain—the unemployment rate is above 60%, most Haitians live below the poverty line, and political uncertainty and civil unrest continue to impede sustained growth.

Currency

The unit of currency is the *gourde* (HTG) of 100 *centimes*. Inflation was 14·4% in 2008, falling to 3·4% in 2009 and rising again to 7·4% in 2011. In July 2005 foreign exchange reserves were US$74m. and total money supply was 19,263m. gourdes.

Budget

The fiscal year begins on 1 Oct. In 2008–09 revenues were US$1,206m. and expenditures US$1,536m.

Performance

The economy grew by 2·9% in 2008–09. Although the earthquake of Jan. 2010 and the fragile global economic situation led to the economy contracting by 5·4% in 2009–10, it rebounded in 2010–11 with a real GDP growth rate of 5·6%. Total GDP in 2011 was US$7·3bn.

Banking and Finance

The Banque Nationale de la République d'Haïti is the central bank and bank of issue (*Governor*, Charles Castel). In 1999 there were 12 commercial banks (three foreign-owned) and a development bank.

In 2010 foreign debt totalled US$492m., equivalent to 732% of GNI.

ENERGY AND NATURAL RESOURCES

Environment

Carbon dioxide emissions from the consumption and flaring of fossil fuels in 2008 were the equivalent of 0·2 tonnes per capita.

Electricity

Power cuts are common. Installed capacity was 0·2m. kW in 2007. Production in 2007 was 469m. kWh, with consumption per capita 54 kWh.

Minerals

Until the supply was exhausted in the 1970s, a small quantity of bauxite was mined.

Agriculture

There were 780,000 ha. of arable land in 2001 and 320,000 ha. of permanent crops. 65% of the workforce, mainly smallholders, make a living by agriculture carried on in seven large plains, from 0·2m. to 25,000 acres, and in 15 smaller plains down to 2,000 acres. Irrigation is used in some areas and in 2001 covered 75,000 ha. The main crops are (2003 production estimates, in 1,000 tonnes): sugarcane, 1,050; cassava, 340; bananas, 300; plantains, 283; mangoes, 261; yams, 199; maize, 198; sweet potatoes, 175; rice, 105; sorghum, 95. Livestock (2003 estimates): goats, 1·9m.; cattle, 1·5m.; pigs, 1·0m.; horses, 500,000; chickens, 6m.

Forestry

The area under forests in 2010 was 0·10m. ha., or 4% of the total land area. In 2007, 2·26m. cu. metres of roundwood were cut.

Fisheries

The estimated total catch in 2008 was 10,000 tonnes, of which 97% was from marine waters.

INDUSTRY

Manufacturing is largely based on the assembly of imported components: toys, sports equipment, clothing, electronic and electrical equipment. Textiles, steel, soap, chemicals, paint and shoes are also produced. Many jobs were lost to other Central American and Caribbean countries during the 1991–94 trade embargo, after President Aristide was deposed.

Labour

In 2010 the labour force was 4,161,000 (53·0% males). The unemployment rate in 2009 was around 70%.

INTERNATIONAL TRADE

Imports and Exports

In 2006 imports totalled US$1,548·2m. and exports US$494·4m. The leading imports are petroleum products, foodstuffs, textiles, machinery, animal and vegetable oils, chemicals, pharmaceuticals, raw materials for transformation industries and vehicles. The USA is by far the leading trading partner. Main import suppliers in 1999 were the USA, 60%; Dominican Republic, 4%; France, 3%; Japan, 3%. The USA accounted for 90% of exports in 1999.

COMMUNICATIONS

Roads

Total length of roads was estimated at 4,160 km in 2002, of which 1,010 km were surfaced. There were 58,100 passenger cars in 2002 (7·1 per 1,000 inhabitants), plus 39,100 trucks and vans.

Civil Aviation

There is an international airport at Port-au-Prince. Cap Haïtien also has scheduled flights to the Turks and Caicos Islands. In 2003 there were international flights to Aruba, Boston, Cayenne, Curaçao, Fort de France, Kingston, Miami, Montego Bay, Montreal, New York, Panama City, Paramaribo, Pointe-à-Pitre,

Raleigh/Durham, Sint Maarten, Santiago (Cuba), Santiago (Dominican Republic), Santo Domingo and Washington, D.C. In 2001 Port-au-Prince handled 913,022 passengers (771,656 on international flights) and 13,455 tonnes of freight.

Shipping
Port-au-Prince and Cap Haïtien are the principal ports, and there are 12 minor ports. In Jan. 2009 there were three ships of 300 GT or over registered, totalling 2,000 GT.

Telecommunications
The state telecommunications agency is Teleco. There were 108,300 fixed telephone lines in 2009 (11 per 1,000 inhabitants). Mobile phone subscribers numbered 3·65m. in 2009. There were 83·7 internet users per 1,000 inhabitants in 2010. Fixed internet subscriptions totalled 100,000 in 2007 (ten per 1,000 inhabitants). In Dec. 2011 there were 295,000 Facebook users.

SOCIAL INSTITUTIONS
Justice
The Court of Cassation is the highest court in the judicial system. There are four Courts of Appeal and four Civil Courts. Judges are appointed by the President. The legal system is basically French.

The population in penal institutions in Oct. 2007 was 6,370 (71 per 100,000 of national population).

Education
The adult literacy rate in 2006 was an estimated 48·7% (53·4% among males and 44·6% among females). Education is divided into nine years 'education fondamentale', followed by four years to 'Baccalaureate' and university/higher education. The school system is based on the French system and instruction is in French and Créole. About 20% of education is provided by state schools; the remaining 80% by private schools, including Church and Mission schools.

In 1994–95 there were 360 primary schools (221 state, 139 religious), 21 public *lycées*, 123 private secondary schools, 18 vocational training centres and 42 domestic science centres.

There is a state university, several private universities and an Institute of Administration and Management.

In 2000–01 total expenditure on education came to 1·1% of GNP and 10·9% of total government spending.

Health
In 1998 there were 1,949 physicians and 834 nursing and midwifery personnel. There were eight beds per 10,000 population in 2000. Much of the health care infrastructure was destroyed in the earthquake of Jan. 2010. There were 44 health institutions (government and mixed) operating in May 2010.

RELIGION
Since the Concordat of 1860 Roman Catholicism has been given special recognition, under an archbishop with nine bishops. The Episcopal Church has one bishop. 60% of the population are nominally Roman Catholic, while other Christian churches number perhaps 20%. Probably two-thirds of the population to some extent adhere to Voodoo, recognized as an official religion in 2003.

CULTURE
World Heritage Sites
Haiti has one site on the UNESCO World Heritage List: National History Park—Citadel, Sans-Souci, Ramiers (inscribed on the list in 1982), 19th century monuments to independence.

Press
There were two paid-for daily newspapers in 2008 with a combined circulation of 23,000.

Tourism
In 2005 there were 112,267 tourists, spending US$110m. (excluding passenger transport). Cruise passenger arrivals in 2005 numbered 368,021.

DIPLOMATIC REPRESENTATIVES
The Haitian Embassy in London closed on 30 March 1987.

Of the United Kingdom in Haiti
Ambassador: Stephen Fisher (resides in Santo Domingo, Dominican Republic).

Of Haiti in the USA (2311 Massachusetts Ave., NW, Washington, D.C., 20008)
Ambassador: Paul Altidor.

Of the USA in Haiti (Tabarre 41, Blvd 15 Octobre, Port-au-Prince)
Ambassador: Pamela A. White.

Of Haiti to the United Nations
Ambassador: Jean Wesley Cazeau.

Of Haiti to the European Union
Ambassador: Raymond Magloire.

FURTHER READING
Girard, Philippe, *Haiti: The Tumultuous History—From Pearl of the Caribbean to Broken Nation.* 2010
Heinl, Robert & Nancy, revised by Michael Heinl, *Written in Blood.* 1996
Nicholls, D., *From Dessalines to Duvalier: Race, Colour and National Independence in Haiti.* 3rd ed. 1996
Pierre, Hyppolite, *Haiti, Rising Flames from Burning Ashes: Haiti the Phoenix.* 2006
Shamsie, Yasmine and Thompson, Andrew S. (eds.) *Haiti: Hope for a Fragile State.* 2006
Thomson, I., *Bonjour Blanc: a Journey through Haiti.* 1992
Weinstein, B. and Segal, A., *Haiti: the Failure of Politics.* 1992
Wucker, Michele, *Why the Cocks Fight: Dominicans, Haitians, and the Struggle for Hispaniola.* 2000
National library: Bibliothèque Nationale, 193 Rue du Centre, Port-au-Prince.

National Statistical Office: Institut Haïtien de Statistique et d'Informatique (IHSI), 1 Angle rue Joseph Janvier et Blvd Harry Truman, HT6110 Port-au-Prince.
Website (French only): http://www.ihsi.ht

HONDURAS

© Research Machines plc 2006

República de Honduras
(Republic of Honduras)

Capital: Tegucigalpa
Population projection, 2015: 8·39m.
GNI per capita, 2011: (PPP$) 3,443
HDI/world rank: 0·625/121
Internet domain extension: .hn

KEY HISTORICAL EVENTS

Discovered by Columbus in 1502, Honduras was ruled by Spain until independence in 1821. Political instability was endemic throughout the 19th and most of the 20th century. The end of military rule seemed to come in 1981 when a general election gave victory to the more liberal and non-military party, PLH (Partido Liberal de Honduras). However, power remained with the armed forces. Internal unrest continued into the 1990s with politicians and military leaders at loggerheads, particularly over attempts to investigate violations of human rights. In Oct. 1998 Honduras was devastated by Hurricane Mitch, the worst natural disaster to hit the area in modern times. In June 2009 President Manuel Zelaya was deposed in a military coup, leading to international condemnation and the suspension of aid. A presidential poll was held in Nov. 2009, with Porfirio Lobo Sosa of the National Party emerging victorious after Zelaya's refusal to recognize the election. The following month Congress rejected proposals to return Zelaya to power and in Jan. 2010 Porfirio Lobo was sworn in as Zelaya went into exile.

TERRITORY AND POPULATION

Honduras is bounded in the north by the Caribbean, east and southeast by Nicaragua, west by Guatemala, southwest by El Salvador and south by the Pacific Ocean. The area is 112,492 sq. km (43,433 sq. miles). In 2001 the census population was 6,535,344 (3,304,386 females), giving a density of 58·1 per sq. km. Estimate, 2010, 7·60m. In 2011, 52·2% of the population lived in urban areas.

The UN gives a projected population for 2015 of 8·39m.

The chief cities and towns are (2009 estimated populations): Tegucigalpa, the capital (990,600), San Pedro Sula (646,300), Choloma (223,900), La Ceiba (172,900), El Progreso (122,000), Choluteca (91,000), Comayagua (78,300), Puerto Cortés (68,400), La Lima (67,100), Danlí (62,100).

Areas and 2001 populations of the 18 departments:

Department	Area (in sq. km)	Population
Atlántida	4,372	344,099
Choluteca	3,923	390,085
Colón	4,360	246,708
Comayagua	8,249	352,881
Copán	5,124	288,766
Cortés	3,242	1,202,510
El Paraíso	7,489	350,054
Francisco Morazán	8,619	1,180,676
Gracias a Dios	16,997	67,384
Intibucá	3,123	179,862
Islas de la Bahía	236	38,073
La Paz	2,525	156,560
Lempira	4,228	250,067
Ocotepeque	1,630	108,029
Olancho	23,905	419,561
Santa Bárbara	5,024	342,054
Valle	1,665	151,841
Yoro	7,781	465,414

The official language is Spanish. The Spanish-speaking population is of mixed Spanish and Amerindian descent (87%), with 6% Amerindians.

SOCIAL STATISTICS

2009 estimates: births, 201,000; deaths, 37,000. Rates, 2009 estimates (per 1,000 population): birth, 27; death, 5. 2007 life expectancy, 69·6 years for men and 74·4 for women. Annual population growth rate, 1998–2008, 2·0%. Infant mortality, 2010, 20 per 1,000 live births; fertility rate, 2008, 3·3 births per woman. Abortion is illegal.

CLIMATE

The climate is tropical, with a small annual range of temperature but with high rainfall. Upland areas have two wet seasons, from May to July and in Sept. and Oct. The Caribbean Coast has most rain in Dec. and Jan. and temperatures are generally higher than inland. Tegucigalpa, Jan. 66°F (19°C), July 74°F (23·3°C). Annual rainfall 64" (1,621 mm).

CONSTITUTION AND GOVERNMENT

The present Constitution came into force in 1982 and was amended in 1995. The *President* is elected for a single four-year term. Members of the *National Congress* (total 128 seats) and municipal mayors are elected simultaneously on a proportional basis, according to combined votes cast for the Presidential candidate of their party.

In March 2009 the incumbent president, Manuel Zelaya, proposed a referendum to approve an assembly to revise the constitution. His opponents feared that he was seeking revisions to allow him to stand for re-election. A constitutional crisis culminated in a military coup and Zelaya's exile to Costa Rica.

National Anthem

'Tu bandera' ('Thy Banner'); words by A. C. Coello, tune by C. Hartling.

RECENT ELECTIONS

Presidential and parliamentary elections took place on 29 Nov. 2009. In the presidential elections Porfirio Lobo Sosa (National Party, PNH) won 56·5% of votes cast against 38·1% for his chief rival, Elvin Santos (Liberal Party, PLH). There were three other candidates. Turnout was 50·0%. In the elections to the National Congress held on the same day the National Party won 71 of 128 seats, the Liberal Party 45, the Christian Democratic Party 5,

Democratic Unification Party 4 and the Innovation and Unity Party 3.

Presidential and parliamentary elections are scheduled to take place on 10 Nov. 2013.

CURRENT GOVERNMENT

In May 2013 the government consisted of:

President: Porfirio Lobo Sosa; b. 1947 (PNH; sworn in 27 Jan. 2010).

Vice-Presidents: María Antonieta de Bográn *(also Minister of the Presidency)*; Marlon Tábora.

Minister of Agriculture: Jacobo Regalado. *Communications:* Miguel Ángel Bonilla. *Culture, Art and Sport:* Tulio Mariano González. *Defence:* Marlon Pascua. *Education:* Marlon Oniel Escoto Valerio. *Family:* María Elena Zepeda. *Finance:* Wilfredo Cerrato. *Foreign Relations:* Mireya Agüero. *Health:* Roxana Araujo. *Industry and Commerce:* José Adonis Lavaire. *Interior:* África Madrid. *Justice and Human Rights:* Ana Pineda. *Labour:* Felicito Ávila. *National Institute of Women:* María Antonieta Botto. *Natural Resources and Environment:* Rigoberto Cuéllar. *Public Works, Transport and Housing:* Miguel Ángel Gámez. *Security:* Arturo Corrales. *Tourism:* Nelly Jérez.

Office of the President (Spanish only):
http://www.presidencia.gob.hn

CURRENT LEADERS

Porfirio Lobo Sosa

Position
President

Introduction
Porfirio Lobo Sosa was elected president on 29 Nov. 2009, ending months of political turmoil that followed the ousting of President Zelaya in a coup. The right-wing former agronomist has faced the challenge of uniting the country, re-establishing regional alliances and combating rising lawlessness and violence.

Early Life
Porfirio Lobo Sosa was born on 22 Dec. 1947 in Trujillo, Colón district, the son of a wealthy politician who served in Honduras' National Congress in the 1950s. Lobo grew up near Juticalpa, Olancho, attending a local Catholic school and then the San Francisco Institute of Tegucigalpa from 1961–65. He went to the University of Miami in 1966 to study business administration before returning to Honduras in 1970 to work in his family's agricultural business and to teach politics and economics at a college in Juticalpa. In the 1970s Lobo travelled to the Soviet Union and enrolled at Patrice Lumumba University in Moscow. He is reputed to have joined the Communist Party of Honduras on his return before making a political about-turn to join the right-wing National Party (PNH), becoming president of the party's Olancho branch in 1986.

In the general election of Nov. 1989 Lobo secured a seat in the National Congress for the PNH. He worked in the department for agriculture and economics under the new president, Rafael Leonardo Callejas, and headed the corporation for forestry development until 1994. Lobo was elected president of the PNH's central committee in June 1999 and served as president of congress from 2002–06. Selected as the PNH candidate to contest the presidential election of 27 Nov. 2005, he took a hard line on crime, promising the death penalty for convicted gang members. This contrasted with the approach of his rival, José Manuel Zelaya, of the centre-right Liberal Party (PLH) who pledged to introduce re-education programmes for criminals. Lobo was defeated with 46% of the vote to Zelaya's 50%.

Lobo took over as leader of the opposition PNH in Jan. 2006. He criticized Zelaya's lurch to the political left in 2007 and the president's alliance with the Venezuelan leader, Hugo Chávez,

who persuaded Honduras to join regional leftist alliances. Zelaya's popularity was dented by his attempts in 2008 to hold a referendum to change the constitution that barred him from standing for re-election—a path taken by Chávez in Venezuela and President Morales in Bolivia. Zelaya pushed ahead with the referendum, despite opposition from the PNH, national legal bodies and much of the military.

On 28 June 2009, after the Supreme Court had ruled that the bid to change the constitution was illegal, the army launched a coup and forced Zelaya into exile in Costa Rica. A wave of international criticism (and suspension from the Organization of American States—OAS) ushered in five months of sometimes violent turmoil between Zelaya's supporters and backers of the interim president, Roberto Micheletti. In the presidential election of 29 Nov. 2009 (scheduled prior to the coup), Lobo secured 56% of the vote and was sworn in on 27 Jan. 2010.

Career in Office
Lobo promptly granted amnesty to those involved in the political crisis and paved the way for Zelaya to leave for exile in the Dominican Republic. The move was one of the conditions of an accord signed in Oct. 2009 after efforts by the OAS to broker a political settlement. Lobo promised to 're-establish channels of friendship with all nations' and to seek foreign investment to revive the economy. In 2010 Honduras was readmitted to the Central American Integration System and in May 2011 an internationally-brokered agreement to allow Zelaya to return to the country prompted Honduras' resumption of participation in OAS proceedings in June. In Sept. 2011 Lobo replaced some members of his cabinet, including foreign minister Mario Canahuati and interior minister Óscar Álvarez, reportedly over political differences. Drug-related crime has meanwhile continued to increase, and Honduras has been ranked among the most violent countries in the world.

DEFENCE

Conscription was abolished in 1995. In 2008 defence expenditure totalled US$96m. (US$12 per capita), representing 0·7% of GDP.

Army

The Army numbered (2007) 8,300. There is also a paramilitary Public Security Force of 8,000.

Navy

Personnel (2007), 1,400 including 830 marines. Bases are at Puerto Cortés, Puerto Castilla and Amapala.

Air Force

There were 16 combat-capable aircraft in 2007 (A-37B Dragonfly and F-5E/F Tiger II fighters). Total strength was (2007) 2,300 personnel.

ECONOMY

Agriculture accounted for 11·9% of GDP in 2009, industry 26·8% and services 61·3%.

Overview

Honduras is a lower-middle-income country with a diversified economy based on international trading of manufactures and agricultural commodities. One of the poorest countries in Latin America with social indicators among the weakest in the region, the economy is highly susceptible to external shocks and natural disasters.

GDP fell by 0·5% a year for three years after Hurricane Mitch (1998) ruined many small-scale farmers with knock-on damage to banking. Debt relief from the Enhanced Heavily Indebted Poor Countries Initiative, Paris Club creditors, the Multilateral Debt Relief Initiative and Inter-American Development Bank helped reduce external debt from 78% of GDP in 1999 to 16% in 2007. In the five years to 2008 growth was above the Latin American average although living standards for the majority barely improved. Inequality levels are among the highest in the world, with the top 10% of the population consuming about 45% of output while 65% of the population live below the poverty line.

Growth is attributed to increased remittances and strong export performance, particularly by the *maquila* sector (re-export business), and private investment encouraged by the Central American Free Trade Agreement (CAFTA). The US recession of the late 2000s resulted in a decline in remittances, FDI and exports from *maquilas*. In Jan. 2010 Porfirio Lobo Sosa became president following a military-backed coup against President Zelaya in June 2009. As a result of subsequent domestic political instability along with the global economic downturn, GDP growth fell by 2·1% in 2009 and external debt increased to 19% of GDP in 2010.

An 18-month Stand-By Arrangement and a Standby Credit Facility with the IMF totalling US$207m. was approved in Oct. 2010. It was aimed at restoring economic stability, strengthening public finances, rebuilding investor confidence and supporting economic recovery.

Currency

The unit of currency is the *lempira* (HNL) of 100 *centavos*. In July 2005 foreign exchange reserves were US$2,164m., total money supply was 19,865m. lempiras and gold reserves were 21,000 troy oz. Inflation was 4·7% in 2010 and 6·8% in 2011.

Budget

In 2008 revenues were 52,343m. lempiras and expenditures 58,650m. lempiras. Tax revenue accounted for 80·5% of revenues in 2008; current expenditure accounted for 78·7% of expenditures.

There is a sales tax of 12%.

Performance

The economy contracted by 2·1% in 2009 but grew by 2·8% in 2010 and 3·6% in 2011. Total GDP in 2011 was US$17·4bn.

Banking and Finance

The central bank of issue is the Banco Central de Honduras (*President*, María Elena Mondragón). It had total reserves at Dec. 2002 of US$1,531m. There is an agricultural development bank, Banadesa, for small grain producers, a state land bank and a network of rural credit agencies managed by peasant organizations. The Central American Bank for Economic Integration (CABEI) has its head office in Tegucigalpa. In 1999 there were 40 private banks, including four foreign.

In 2010 external debt totalled US$4,168m., representing 28·2% of GNI.

There are stock exchanges in Tegucigalpa and San Pedro Sula.

ENERGY AND NATURAL RESOURCES

Environment

Carbon dioxide emissions from the consumption and flaring of fossil fuels in 2008 were the equivalent of 1·1 tonnes per capita.

Electricity

Installed capacity was 1·6m. kW in 2007 (0·5m. kW hydro-electric). Production in 2007 was 6·32m. kWh (35% hydro-electric); consumption per capita (2007) was 840 kWh.

Minerals

Output in 2006: zinc, 37,646 tonnes; lead, 11,775 tonnes; silver, 55,036 kg. Small quantities of gold are mined, and there are also deposits of tin, iron, copper, coal, antimony and pitchblende.

Agriculture

There were around 1·07m. ha. of arable land in 2007 and 0·36m. ha. of permanent crops. Legislation of 1975 provided for the compulsory redistribution of land, but in 1992 the grounds for this were much reduced, and a 5-ha. minimum area for land titles was abolished. Members of the 2,800 co-operatives set up in 1975

received individual shareholdings which can be broken up into personal units. Since 1992 women may have tenure in their own right. The state monopoly of the foreign grain trade was abolished in 1992. In 1996 the Agricultural Incentive Program was created (Ley de Incentivo Agrícola, LIA) which involves the redistribution of land for agricultural development.

Estimated crop production in 2003 (in 1,000 tonnes): sugarcane, 4,200; bananas, 965; maize, 502; plantains, 260; oranges, 167; coffee, 150; melons and watermelons, 145; palm oil, 112; dry beans, 69; pineapples, 62; sorghum, 52.

Livestock (2003 estimates): cattle, 2·40m.; pigs, 478,000; horses, 181,000; mules, 70,000; chickens, 19m.

Forestry

In 2010 forests covered 5·20m. ha., or 46% of the total land area. In 2007, 9·46m. cu. metres of roundwood were cut.

Fisheries

Shrimp and lobster are important catches. The total catch in 2010 was approximately 11,100 tonnes, almost entirely from sea fishing.

INDUSTRY

Industry is small-scale and local. Output (in 1,000 tonnes): cement (2008), 1,784; raw sugar (2001), 316; wheat flour (2001), 113; fabrics (1995), 11,641 metres; beer (2003), 96·1m. litres; rum (1995), 2·37m. litres.

Labour

The workforce was 2,438,000 in Sept. 2001. Of 2,334,600 persons in employment in Sept. 2001, 766,800 were in agriculture, hunting, forestry and fishing, 559,200 in wholesale and retail trade and restaurants and hotels, 380,300 in community, social and personal services and 356,000 in manufacturing. Unemployment rate, Sept. 2001: 4·2%.

INTERNATIONAL TRADE

In 2004 Honduras signed the Central America-Dominican Republic-United States Free Trade Agreement (CAFTA-DR), along with Costa Rica, the Dominican Republic, El Salvador, Guatemala, Nicaragua and the USA. The agreement entered into force for Honduras on 1 April 2006.

Imports and Exports

Imports in 2008 were valued at US$8,807m. and exports at US$2,558m.

Main imports are machinery and electrical equipment, industrial chemicals, and mineral products and lubricants. Main exports are coffee, bananas, shrimp and lobster, gold, lead and zinc, timber and refrigerated meats. Principal import suppliers, 2004: USA, 34·6%; Guatemala, 7·7%; El Salvador, 5·0%; Costa Rica, 4·9%. Principal export markets, 2004: USA, 41·5%; El Salvador, 10·9%; Guatemala, 7·3%; Germany, 5·9%.

COMMUNICATIONS

Roads

Honduras is connected with Guatemala, El Salvador and Nicaragua by the Pan-American Highway. Out of a total of 13,603 km of roads in 2002, 20·4% were paved. In 2007 there were 487,700 passenger cars in use, 31,500 buses and coaches, 165,200 lorries and vans, and 94,400 motorcycles and mopeds.

Rail

The small government-run railway was built to serve the banana industry and is confined to the northern coastal region and does not reach Tegucigalpa. In 2005 there were 595 km of track in three gauges, which in 1994 carried 1m. passengers and 1·2m. tonnes of freight.

Civil Aviation

There are four international airports: San Pedro Sula (Ramón Villeda) and Tegucigalpa (Toncontín) are the main ones, plus

Roatún and La Ceiba, with over 80 smaller airstrips in various parts of the country. In addition to domestic flights and services to other parts of Central America and the Caribbean, there were flights in 2003 to Barcelona, Dallas/Fort Worth, Houston, Las Vegas, Los Angeles, Madrid, Miami, New Orleans, New York, Oklahoma City, Orange County, Phoenix and San Jose. In 2001 San Pedro Sula handled 496,000 passengers (386,000 on international flights) and 7,500 tonnes of freight, and Tegucigalpa handled 451,000 passengers (327,000 on international flights) and 3,800 tonnes of freight.

Shipping

The largest port is Puerto Cortés on the Atlantic coast. There are also ports at Henecán (on the Pacific) and Puerto Castilla and Tela (northern coast). In Jan. 2009 there were 359 ships of 300 GT or over registered, totalling 489,000 GT. Honduras is a flag of convenience registry.

Telecommunications

In 2011 there were 609,200 landline telephone subscriptions (equivalent to 78·6 per 1,000 inhabitants) and 8,062,200 mobile phone subscriptions (or 1,039·7 per 1,000 inhabitants). There were 110·9 internet users per 1,000 inhabitants in 2010. Fixed internet subscriptions totalled 72,400 in 2009 (9·7 per 1,000 inhabitants). In Dec. 2011 there were 1·1m. Facebook users.

SOCIAL INSTITUTIONS

Justice

Judicial power is vested in the Supreme Court, with nine judges elected by the National Congress for four years; it appoints the judges of the courts of appeal, and justices of the peace.

There were 6,236 homicides in 2010, up from just 2,155 in 2004 and 5,265 in 2009. At 77 per 100,000 persons, Honduras had the highest murder rate of any country in 2010 (largely as a result of the drugs trade).

The population in penal institutions in Dec. 2010 was 11,846 (154 per 100,000 of national population).

Education

Adult literacy in 2007 was 83·6% (male, 83·7%; female, 83·5%). Education is free, compulsory (from 6 to 15 years) and secular. There is a high drop-out rate after the first years in primary education. In 2007 there were 214,051 children in pre-primary schools (8,178 teaching staff in 2006); 1,308,119 children in primary schools (46,308 teaching staff in 2006); 554,297 pupils in secondary schools (16,667 teaching staff in 2004). There were an estimated 148,000 students in tertiary education in 2007–08 and 5,000 academic staff. The leading institution of higher learning is the National Autonomous University of Honduras (Universidad Nacional Autónoma de Honduras), founded in 1847, in Tegucigalpa.

In 1998–99 expenditure on education came to 4·2% of GNP.

Health

In 2000 there were 3,676 physicians, 1,371 dentistry personnel and 8,528 nursing and midwifery personnel. In 2010 there were 28 public hospitals. There were ten hospital beds per 10,000 inhabitants in 2002.

RELIGION

Roman Catholicism is the prevailing religion (5,740,000 followers in 2001), but the constitution guarantees freedom to all creeds, and the State does not contribute to the support of any. In 2001 there were 690,000 Evangelical Protestants with the remainder of the population followers of other faiths. In Feb. 2013 there was one cardinal.

CULTURE

World Heritage Sites

Honduras has two sites on the UNESCO World Heritage List: Maya Site of Copán (inscribed on the list in 1980), a centre of the Mayan civilization abandoned in the early 10th century; and Río Plátano Biosphere Reserve (1982), one of the few remains of the Central American rain forest.

Press

Honduras had six national daily papers in 2008, with a combined circulation of 200,000.

Tourism

In 2009 there were 870,000 non-resident tourists, down from 899,000 in 2008 although up from 831,000 in 2007.

Festivals

There are a number of festivals and religious celebrations held throughout the year in Honduras. The Fiesta de San Isidro is a week-long carnival held in May every year to honour the city's patron saint.

DIPLOMATIC REPRESENTATIVES

Of Honduras in the United Kingdom (115 Gloucester Pl., London, W1U 6JT)
Ambassador: Iván Romero-Martínez.

Of the United Kingdom in Honduras (embassy in Tegucigalpa closed in Dec. 2003)
Ambassador: Sarah Dickson (resides in Guatemala City).

Of Honduras in the USA (3007 Tilden St., NW, Washington, D.C., 20008)
Ambassador: Jorge Ramón Hernández Alcerro.

Of the USA in Honduras (Av. La Paz, Tegucigalpa)
Ambassador: Lisa Kubiske.

Of Honduras to the United Nations
Ambassador: Mary Elizabeth Flores.

Of Honduras to the European Union
Ambassador: Roberto Flores-Bermúdez.

FURTHER READING

Banco Central de Honduras. *Honduras en Cifras 2009–11.* Online only
Euraque, Darío A., *Reinterpreting the Banana Republic: Region and State in Honduras, 1870–1972.* 1997
Loker, William M., *Changing Places: Environment, Development and Social Change in Rural Honduras.* 2004
Meyer, H. K. and Meyer, J. H., *Historical Dictionary of Honduras.* 2nd ed. 1994

National Statistical Office: Instituto Nacional de Estadísticas, Tegucigalpa.
Website (Spanish only): http://www.ine.gob.hn

HUNGARY

Magyar Köztársaság
(Hungarian Republic)

Capital: Budapest
Population projection, 2015: 9·90m.
GNI per capita, 2011: (PPP$) 16,581
HDI/world rank: 0·816/38
Internet domain extension: .hu

KEY HISTORICAL EVENTS

Records date back to 9 BC, when the Romans subdued the Celts to establish Pannonia. From the 5th century both Romans and Celts retreated before attacks from the Huns who were followed by the Avars in the 7th century and the Magyars in the 9th. It was then that the name *On ogur* ('ten arrows') was adopted for the country that was to become Hungary. The founding date of Hungary is put at 896 after which Árpád, leader of one of the Magyar tribes, forged a dynasty which ruled Hungary until 1301. Forays into Italy, Germany, the Balkans and Spain ended after the Magyars were defeated by Holy Roman Emperor Otto I at the battle of Lechfeld in 955, and the Ostmark (Austria) was returned to Germanic control.

In seeking a truce with Otto I, the Árpád leader Géza invited him to send Catholic missionaries into Hungary. He had his son István (Stephen) crowned as King of Hungary and replaced the tribal structure with a system of counties (*megye*), administered by royal officials. A disputed succession led to intervention by the Holy Roman Emperor who established temporary suzerainty over Hungary. By the end of the 11th century, Slovakia, Carpathian Ruthenia and Transylvania were all under the crown of St Stephen. In a struggle for control of the ports on the Adriatic, Venice and Hungary went to war on 21 occasions between 1115–1420.

Andrew III, the last Árpád monarch, could do little to hold the country together against the opposition of feuding nobles. His death in 1301 led to a seven-year interregnum, after which, with two exceptions, Hungary was ruled by foreign kings. Linked to the Árpáds through marriage, Charles Robert of Anjou was elected to the throne. His primary task was to restore royal authority over the nobles. An economic boom coincided with Hungary becoming the leading gold producer in Europe and trade links with European neighbours were fostered.

Ottoman Threat

His successors had to contend with the growing power of the Ottoman Empire. Assaults on Hungary increased after the fall of Constantinople in 1453, but in 1456 János Hunyadi, acting as military regent, broke the siege of Belgrade to keep the Turks at bay for another 70 years.

Rival magnates reacted to Hunyadi's death from the plague in 1456 by trying to wipe out the omnipotent Hunyadi clan, but in 1457 the Diet appointed his 15-year-old son Matthias Corvinus as king. Matthias was an enlightened despot. He built up one of Europe's finest libraries—destroyed a century later by the Ottomans—and encouraged writers and artists, many of whom were Italian, to come and work in Hungary. The heirless Matthias was succeeded in 1490 by Bohemia's ruler Vladislav, or King Ulászló II (1490–1516), known as 'Rex Bene', because 'dobre', or 'good' was his reply to almost everything. He managed to repel a Habsburg invasion of Hungary but indulged the nobles with disproportionate powers and relied heavily on foreign financing. Vladislav II was succeeded in 1516 by his son Louis II, who held both the Hungarian and Bohemian thrones. A ten-year-old, he could do little to discourage the onslaught of the Turks, to whom Belgrade was lost in 1521. The Hungarians were defeated by the Turks under Suleiman II at the battle of Mohács on 29 Aug. 1526. Louis was killed in battle and Hungary lost its independence, not to be regained until 1918.

Hungary was partitioned, the largest section going to the Turks, royal Hungary to the Habsburgs and Transylvania, though theoretically autonomous, becoming a vassal state of the Ottomans. The Transylvanians were at constant war with the Habsburgs, who in turn fought the Ottomans. The economy along with the Magyar language declined and much agricultural land, mainly the Hungarian Plain, went to waste.

The Treaty of Vienna of 1606 was meant to set peaceful boundaries, but was soon violated. A series of costly territorial struggles culminated in the Ottoman siege of Vienna in 1683. Repelled by the Habsburgs, it marked a turning point for the Turks who, by 1699, had ceded most of their Hungarian territory. The Habsburgs became hereditary rulers pursuing a policy of divide and rule which led to anti-Habsburg risings. The second, under Ferenc Rákóczi, the last independent prince of Transylvania, united both nobles and peasants, and lasted from 1703–11. It was concluded by the signing of the Peace of Szatmár, in which the Habsburgs guaranteed political freedom for the three 'nations'—the ethnic Magyar, Saxon and Székelys groups. State education, introduced by Maria Theresa and Joseph II, led to greater Germanization.

Challenge to Habsburg Rule

Power was concentrated on the Magyar nobility, descendents of the Árpád royal line, who owned vast estates and were exempt from land tax. In March 1848 the Hungarian Diet renounced Viennese rule and legislated for a sovereign Magyar state, which was approved by Emperor Ferdinand. However, what began peacefully soon deteriorated as national minorities such as the Croats, the Romanians, Serbs and Slovaks demanded the same rights. In the War of Independence, heavy fighting broke out between the Hungarians and the Austrians, the Hungarians being led by Lajos Kossuth (1802–94).

When Emperor Franz Joseph I took the throne in 1848, the Hungarians refused to recognize him. This provoked an Austrian invasion, which was repelled, and in Feb. 1849 the diet in

Debrecen declared Hungary an independent republic under Kossuth's leadership. Franz Joseph reacted by accepting the assistance of Tsar Nicholas I of Russia in suppressing the revolution. The Magyars chose to surrender to the Russians rather than the Austrians but the aftermath of the war witnessed mass executions and imprisonment of rebel factions. Kossuth escaped into exile. Direct rule was imposed from Vienna.

Dual Monarchy

Having lost territory to Sardinia in 1859 and to Prussia in 1866, Austria recognized the need for a compromise with Hungary. What became known as the 'Ausgleich' created a dual monarchy to preside over the Austro-Hungarian Empire. Hungary gained internal autonomy but while the Ausgleich profited Magyars and Austro-Germans it did little to benefit national minorities.

Bosnia and Herzegovina were annexed in 1908, which outraged Serbia, but Austria tried a number of tricks to prevent retaliation including the Zagreb Treason Trial of 1909, when evidence was produced of a Serb-Croat conspiracy to bring down the Habsburgs. It was the Czech professor and future president Tomáš Masaryk who proved the evidence to be fake.

On 28 June 1914 the heir to the Habsburg throne, Archduke Franz Ferdinand, and his wife were shot in Sarajevo by a Bosnian Serb. Austria-Hungary declared war on Serbia a month later, precipitating the First World War. The Entente of France, Britain and Russia united against the Central Powers of Germany and Austria-Hungary, with other nations soon joining in one or other alliance. By the Treaty of Versailles, the territories of Hungary and Austria were reduced drastically. Hungary became a republic in Nov. 1918, with Mihály Károlyi as president. Transylvania was handed over to Romania. New countries including Czechoslovakia and Yugoslavia were created, all of which gained former Hungarian territory.

On 21 March 1919, Károlyi was replaced by the Bolshevik leader, Béla Kun, who was in power for 133 days. His downfall was brought about by a non-communist revolutionary movement fighting to regain Slovakia and Romania. The Allies persuaded Romania to retreat, and Hungary's borders were finalized by the Treaty of Trianon on 4 June 1920. Two-thirds of Hungary's territory and over half of the population were assigned to neighbouring countries.

In 1919 the Hungarian Kingdom was restored under Count Miklós Horthy, who ruled as regent and appointed a chiefly aristocratic government. Despite Horthy's efforts to amend the Trianon treaty, Hungary's boundaries remained unchanged until the Second World War. Germany and Italy backed the 'Vienna awards' of Nov. 1938 which restored to Hungary southern Slovakia and southern Subcarpathian Ruth, and in Aug. 1940, Transylvanian and Romanian territory. Hitler's support, including favourable trading terms, drew Hungary into fighting with Germany against the Soviet army in 1941, a tactical error which led to enormous losses.

In March 1944 the Germans occupied Hungary. Horthy was forced to abdicate and Hitler appointed a government of Ferenc Szálasi and his fascist Arrow Cross movement. Large-scale deportation of Jews and political dissidents began. Around 400,000 Jews are estimated to have been murdered. With civilian and military losses, almost a million Hungarians died in the war.

Soviet Rule

With the Soviets as the occupying power, post-war Communist rule was established in 1948–49 after a three-year multi-party democracy which the Communists conspired to undermine. Mátyás Rákosi and his Hungarian Workers' Party headed a dictatorship which went unchallenged until 1953, the year of Stalin's death. In July of that year Rákosi was ousted by reformers led by Imre Nagy. Appointed prime minister, Nagy began what he called 'the new stage in building socialism', which entailed industrial and economic reforms and the restoration of human rights. But disagreements within the Soviet leadership gave an advantage to Rákosi who was still general secretary of the Workers' Party. Nagy was forced out of office in April 1955.

On 23 Oct. 1956 a student-led demonstration demanded democratic reforms and Nagy's reinstatement as prime minister. Soviet troops fired into crowds trying to occupy the radio station. The next day Imre Nagy was reappointed prime minister but was unable to quell the riots. Revolutionary committees were set up and there was a general strike to promote the three aims of the revolution: national independence, a democratic political structure and the protection of social benefits. All of this, along with armed rebels in the capital, put pressure on the hardliners in the party to accept reform.

A ceasefire, called by Nagy on 28 Oct., was honoured and Soviet troops retreated from Budapest. A multi-party democracy was announced, and the State Security Authority abolished. Even so, there were continuing demands for a clean sweep of all Stalinist-Rákosist ministers and total Soviet withdrawal. Nagy believed that such a transition should occur gradually and peacefully, but when he voiced the nation's support for neutrality and a withdrawal from the Warsaw pact, it was a step too far for Moscow. János Kádár was encouraged to form a counter-government with Soviet military backing. The Soviet Army marched into Budapest on 4 Nov., crushing all resistance.

Soviet hopes that Nagy would resign after this resounding defeat and support Kádár were disappointed. Kádár returned from Moscow on 7 Nov. after the heaviest fighting was over, to be confronted by a less than compromising nation. The renamed Hungarian Socialist Workers' Party declared all Oct. events as a counter-revolution, and began a series of revenge attacks. Nagy was hanged on 16 June 1958 along with several of his reformist associates. Many opponents of the regime were deported to labour camps in the Soviet Union and over 200,000 people fled the country.

Gradual Reform

János Kádár was party leader from 1956–89, and prime minister in the years 1956–58 and 1961–65. In the early '60s Kádár made a gradual shift towards liberalization. After the wave of executions, a distinction was made between political crime and mere error, and people were no longer required to be active in the party in order to succeed professionally. Trade unions were allowed to play a more active role, as was the press, so long as the government was not openly criticized.

The now-recognized need to loosen state control of the economy gave rise to the New Economic Mechanism (NEM) in 1968, which relaxed price controls, acknowledged the profit motive, improved manufacturing quality and shifted the emphasis from heavy to light industry. Subsidies were reduced and enterprise encouraged. Growing demands for a more open market economy coincided with the first signs of a weakening of the Soviet system. A group of Hungarian dissidents were sufficiently encouraged by the liberal trend in Moscow to form the Hungarian Democratic Forum. Led by their secretary general, Imre Pozsgay, they produced a manifesto 'Turn and Reform' which argued for a total overhaul of the economy.

The subsequent debate reopened divisions between hardliners and reformists, and throughout the country there were demonstrations and strikes. The conservative old school of the Hungarian Socialist Workers' Party was gradually phased out by the reformists. A committee was set up to investigate the events of 1956, which concluded that it had been a popular uprising and not a counter-revolution. This called for the ceremonial reburial of Imre Nagy's remains on 16 June 1989, an event attended by a quarter of a million people who gathered in Heroes' Square, Budapest.

Post Communism

When prime minister Miklós Németh opened the borders with Austria, the flood of refugees from East Germany precipitated the fall of the Berlin wall. Multi-party democracy was enshrined in law in Sept. 1989 and Hungary ceased to be a People's Republic on 23 Oct. A unicameral National Assembly was formed and the first free elections took place on 25 March 1990. Of the 386 members elected to the National Assembly, only 21 had ever served in parliament before, and of the six successful parties, three were entirely new. The Hungarian Democratic Forum (MDF) and Alliance of Free Democrats advocated democracy, political pluralism, a market economy and a 'return' to Europe. The MDF came out ahead but having failed to secure a majority, formed a coalition with the Independent Smallholders' Party and the Christian Democratic People's Party.

A largely inexperienced government set about economic reform while trying to contain trade and budget deficits and high inflation. Social unrest prompted the government to slow down its privatization programme which proved popular with the electorate until they realized that the economy was stalling. In 1993 Iván Szabó became finance minister and adopted much stricter policies, cutting social budgets and devaluing the forint. This again led to domestic hardship. Unemployment, a hitherto unknown phenomenon, grew to over 12%. A nostalgia for a Communist past where jobs, housing and benefits were secure was perceptible in voting patterns at the 1994 elections.

Although the MDF's 'shock tactic' policies were praised by the West, and attracted foreign investment, the electorate opted for an updated version of the Hungarian Socialist Party (MSzP). Former Communist and leader Gyula Horn touted the party as one free of ideological limitations, playing down the traditional left and promising a higher standard of living along with continued economic reform under the guidance of László Bekesy, finance minister of the former Communist government. Horn became prime minister of a coalition led by the Alliance of Free Democrats (SzDSz) and the MSzP. Economic reforms were put back on the agenda but the government moved cautiously in an effort to carry public opinion.

The 1998 elections produced another coalition led by Viktor Orbán of the Federation of Young Democrats (later called Fidesz). He was succeeded in May 2002 by Péter Medgyessy, the Socialists' candidate, who formed a coalition with the SzDSz. In June 2002 revelations that Medgyessy had worked as a counter-intelligence agent for the communist regime highlighted the transitional problems for former Eastern Bloc nations. Hungary became a member of NATO in 1999 and of the EU on 1 May 2004. The economy was hit by the global economic downturn in 2008 and by the overspill from the subsequent euro crisis. A series of austerity measures came into force from 2009 when Hungary accepted emergency funding from the IMF. János Áder became president in May 2012 after his predecessor, Pál Schmitt, resigned following allegations of plagiarism involving a doctoral dissertation he had written 20 years earlier.

TERRITORY AND POPULATION

Hungary is bounded in the north by Slovakia, northeast by Ukraine, east by Romania, south by Croatia and Serbia, southwest by Slovenia and west by Austria. The peace treaty of 10 Feb. 1947 restored the frontiers as of 1 Jan. 1938. The area of Hungary is 93,030 sq. km (35,919 sq. miles), including 690 sq. km (266 sq. miles) of inland waters.

At the census of 1 Oct. 2011 the population (provisional) was 9,982,000 (52·8% females).

The UN gives a projected population for 2015 of 9·90m.

67·7% of the population was urban in Jan. 2008; population density, Oct. 2011, 108·1 per sq. km. Hungary's population has been falling at such a steady rate since 1980 that its 2011 population was the same as that in the early 1960s.

Ethnic minorities, 2001: Roma (Gypsies), 5·3%; Ruthenians, 2·9%; Germans, 2·4%; Romanians, 1·0%; Slovaks, 0·9%. A law of 1993 permits ethnic minorities to set up self-governing councils. There is a worldwide Hungarian diaspora of about 4·7m. (including 1·5m. in the USA and Canada; 1·4m. in Romania; 0·5m. in Slovakia; 0·3m. in Serbia, mainly in Vojvodina; 0·2m. in Israel; 0·2m. in Ukraine; 0·1m. in Brazil; 0·1m. in Germany). In total, 2·5m. Hungarians live in neighbouring countries.

Hungary is divided into 19 counties (*megyék*) and the capital, Budapest, which has county status.

Area (in sq. km) and population (in 1,000) of counties and chief towns:

Counties	Area	2009 population	Chief town	2009 population
Bács-Kiskun	8,445	530	Kecskemét	111
Baranya	4,430	395	Pécs	157
Békés	5,631	371	Békéscsaba	65
Borsod-Abaúj-Zemplén	7,247	701	Miskolc	170
Csongrád	4,263	424	Szeged	169
Fejér	4,359	428	Székesfehérvár	102
Győr-Moson-Sopron	4,089	447	Győr	130
Hajdú-Bihar	6,211	542	Debrecen	206
Heves	3,637	314	Eger	56
Jász-Nagykún-Szolnok	5,582	395	Szolnok	75
Komárom-Esztergom	2,265	314	Tatabánya	70
Nógrád	2,544	208	Salgótarján	38
Pest	6,393[1]	1,213[2]	Budapest	1,712
Somogy	6,036	322	Kaposvár	68
Szabolcs-Szatmár-Bereg	5,936	565	Nyíregyháza	118
Tolna	3,703	236	Szekszárd	34
Vas	3,336	261	Szombathely	80
Veszprém	4,613	360	Veszprém	63
Zala	3,784	290	Zalaegerszeg	62
Budapest	525	1,712	(has county status)	

[1]Excluding area of Budapest. [2]Excluding population of Budapest.

The official language is Hungarian. 98·5% of the population have Hungarian as their mother tongue. Ethnic minorities have the right to education in their own language.

SOCIAL STATISTICS

2011: births, 88,049; deaths, 128,795; marriages, 35,812; divorces, 23,335. In 2000 the number of births rose for the first time in a decade. There were 2,422 suicides in 2011. Rates (per 1,000 population), 2011: birth, 8·8; death, 12·9; marriage, 3·6; divorce, 2·3. Population growth rate, 2009, –0·2%. The suicide rate, at 24·6 per 100,000 population in 2009, is one of the highest in the world (although it has fallen since the mid-1980s when it was over 44 per 100,000). Expectation of life at birth, 2011, 70·9 years for males and 78·2 years for females. Infant mortality, 2010, 5 per 1,000 live births. Fertility rate, 2011, 1·2 births per woman.

CLIMATE

A humid continental climate, with warm summers and cold winters. Precipitation is generally greater in summer, with thunderstorms. Dry, clear weather is likely in autumn, but spring is damp and both seasons are of short duration. Budapest, Jan. 32°F (0°C), July 71°F (21·5°C). Annual rainfall 25" (625 mm). Pécs, Jan. 30°F (–0·7°C), July 71°F (21·5°C). Annual rainfall 26·4" (661 mm).

CONSTITUTION AND GOVERNMENT

On 18 Oct. 1989 the National Assembly approved by an 88% majority a constitution which abolished the People's Republic, and established Hungary as an independent, democratic, law-based state.

In April 2011 parliament passed proposals for a new constitution, known as the 'Easter constitution', by a vote of 263 to 44 (with one abstention). It came into force on 1 Jan. 2012. Two of the three main opposition parties refused to vote in protest at what critics claimed were attacks by the ruling Fidesz party on the rights of various minority groups, including those with mental illness, the gay and lesbian community, and pro-abortion bodies. The constitution's preamble emphasizes Hungary's Christian heritage while other clauses restrict the voting rights of those with 'limited mental ability'. It defines marriage as a union of a man and a woman, and stipulates that the life of a foetus should be protected from conception. The German government subsequently warned that such clauses strained compatibility with EU law. The constitution also states Hungary's 'responsibility for the destiny of Hungarians living outside her borders', limits the jurisdiction of the constitutional court, reduces the number of parliamentary ombudsmen and determines that the national debt should be no more than 50% of the previous year's GDP except in exceptional circumstances. In Jan. 2012 the EU Commission requested clarification from Hungary on several aspects of the constitution, including curbs on the independence of the central bank that could violate EU law.

The head of state is the *President*, who is elected for five-year terms by the National Assembly.

The single-chamber *National Assembly* currently has 386 members, made up of 176 individual constituency winners, 152 allotted by proportional representation from county party lists and 58 from a national list, but in May 2010 the Constitution was amended and the number of members for the next election reduced to 200 (although an additional 13 MPs may be elected for the representation of national and ethnic minorities). It is elected for four-year terms. A *Constitutional Court* was established in Jan. 1990 to review laws under consideration.

National Anthem
'Isten áldd meg a magyart' ('God bless the Hungarians'); words by Ferenc Kölcsey, tune by Ferenc Erkel.

GOVERNMENT CHRONOLOGY
(Fidesz-MPP = Fidesz-Hungarian Civic Party; Fidesz-MPSz = Fidesz-Hungarian Civic Alliance; FKgP = Independent Party of Smallholders, Agrarian Workers and Citizens; MDF = Hungarian Democratic Forum; MDP = Hungarian Workers' Party; MKP = Hungarian Communist Party; MSzMP = Hungarian Socialist Workers' Party; MSzP = Hungarian Socialist Party; SzDSz = Alliance of Free Democrats; n/p = non-partisan)

Presidents since 1946.

1946–48	FKgP	Zoltán Tildy
1948–50	MDP	Árpád Szakasits
1950–52	MDP	Sándor Rónai
1952–67	MSzMP	István Dobi
1967–87	MSzMP	Pál Losonczi
1987–88	MSzMP	Károly Németh
1988–89	MSzMP	Bruno Ferenc Straub
1989–90	MSzP	Mátyás Szűrös
1990–2000	SzDSz	Árpád Göncz
2000–05	n/p	Ferenc Mádl
2005–10	n/p	László Sólyom
2010–12	Fidesz-MPSz	Pál Schmitt
2012	Fidesz-MPSz	László Kövér (acting)
2012–	Fidesz-MPSz	János Áder

Prime Ministers since 1946.

1946–47	FKgP	Ferenc Nagy
1947–48	FKgP	Lajos Dinnyés
1948–52	MDP	István Dobi
1952–53	MDP	Mátyás Rákosi
1953–55	MDP	Imre Nagy
1955–56	MDP	András Hegedüs
1956	MDP	Imre Nagy
1956–58	MSzMP	János Kádár
1958–61	MSzMP	Ferenc Münnich
1961–65	MSzMP	János Kádár
1965–67	MSzMP	Gyula Kállai
1967–75	MSzMP	Jenő Fock
1975–87	MSzMP	György Lázár
1987–88	MSzMP	Károly Grósz
1988–90	MSzP	Miklós Németh
1990–93	MDF	József Antall
1993–94	MDF	Péter Boross
1994–98	MSzP	Gyula Horn
1998–2002	Fidesz-MPP	Viktor Orbán
2002–04	n/p (MSzP)	Péter Medgyessy
2004–09	MSzP	Ferenc Gyurcsány
2009–10	n/p	Gordon Bajnai
2010–	Fidesz-MPSz	Viktor Orbán

Leaders of the Communist Party, 1945–89.

General Secretary of MKP/MDP

1945–	Mátyás Rákosi

First Secretaries of MDP/MSzMP

1953–56	Mátyás Rákosi
1956	Ernő Gerő
1956–88	János Kádár
1988–89	Károly Grósz

Collective Chairmanship of MSzMP

1989	Rezső Nyers; Miklós Németh; Károly Grósz; Imre Pozsgay

RECENT ELECTIONS
János Áder was elected president by the National Assembly on 2 May 2012 by 262 votes to 40, with two abstentions. Áder was the only candidate.

In the Hungarian parliamentary elections on 11 and 25 April 2010 the alliance of Fidesz-Hungarian Civic Alliance and the Christian Democratic People's Party (Fidesz-MPSz and KDNP) won 263 seats in the 386-seat National Assembly (164 in 2006); the Hungarian Socialist Party (MSzP) 59 (186 in 2006); Movement for a Better Hungary (Jobbik) 47 (none in 2006); Politics Can Be Different (LMP) 16 (none in 2006). One seat went to an independent candidate. Turnout in the first round was 64·4% and in the second round 44·2%.

European Parliament
Hungary has 22 (24 in 2004) representatives. At the June 2009 elections turnout was 36·3% (38·5% in 2004). The alliance of Fidesz-Hungarian Civic Alliance and the Christian Democratic People's Party (Fidesz-MPSz and KDNP) won 14 seats with 56·4% of votes cast (political affiliation in European Parliament: European People's Party); the MSzP, 4 with 17·4% (Progressive Alliance of Socialists and Democrats); Jobbik, 3 with 14·8% (non-attached); the Hungarian Democratic Forum (MDF), 1 with 5·3% (European Conservatives and Reformists).

CURRENT GOVERNMENT
President: János Áder; b. 1959 (Fidesz-MPSz; in office since 10 May 2012).

In March 2013 the government comprised:
Prime Minister: Viktor Orbán; b. 1963 (Fidesz-MPSz; sworn in 29 May 2010).

Deputy Prime Minister and Minister of Public Administration and Justice: Tibor Navracsics. *Deputy Prime Minister:* Zsolt Semjén.

Minister of Defence: Csaba Hende. *Foreign Affairs:* János Martonyi. *Human Resources:* Zoltán Balog. *Interior:* Sándor Pintér. *National Development:* Zsuzsanna Németh. *National Economy:* Mihály Varga. *Rural Development:* Sándor Fazekas.

Office of the Prime Minister: http://www.meh.hu

CURRENT LEADERS

János Áder

Position
President

Introduction
János Áder was elected president by parliament on 2 May 2012. He took over from László Kövér, who had held the post in a caretaker capacity after Pál Schmitt resigned on 2 April 2012. Áder was scheduled to serve a five-year term in the largely ceremonial role.

Early Life
János Áder was born on 9 May 1959 in Csorna. He graduated in law from Eötvös Loránd University (ELTE) in Budapest in 1983. From 1986–90 he was a researcher at the Sociological Research Institute of the Hungarian Academy of Sciences. He joined Fidesz (Federation of Young Democrats) at the party's inception in 1988 and took part in the 'round table' talks in 1989 that led to the end of communist single-party rule. Áder managed Fidesz's 1990 election campaign and was himself elected a member of parliament, retaining his seat until 2009.

From 1990–92 he served as deputy leader of the party's parliamentary group before being appointed chair of the national steering committee of Fidesz. In 1994 he led the party's election campaign, a role he repeated in 1998. He served as Fidesz's vice president from 1995–97 until his appointment as deputy parliamentary speaker.

From June 1998 until May 2002 Áder was parliamentary speaker and also held his party's vice presidency from 1999–2000. From 2002–06 he led Fidesz's parliamentary group while it was in opposition. In 2006 he returned as the deputy speaker of parliament, holding the post until 2009, when he became a member of the European Parliament. As an MEP, Áder demonstrated himself a Fidesz-Hungarian Civic Alliance (as it was by then called) party loyalist by voting along party lines on 98% of issues. In 2011 a number of domestic laws that Áder helped draft were ratified, including controversial reforms of the electoral system and the judiciary that critics argued infringed judicial independence.

On 16 April 2012 Áder was nominated by Prime Minister Orbán for the national presidency following Pál Schmitt's resignation over allegations of plagiarism in regard to his doctoral thesis. Áder resigned as an MEP and gave up his Fidesz-Hungarian Civic Alliance party membership in preparation for taking on the post. His election to office on 2 May 2012 was assured by Fidesz-Hungarian Civic Alliance's two-thirds majority in parliament. He formally assumed the presidency on 10 May.

Career in Office
As a strong Orbán ally and a former member of Fidesz-Hungarian Civic Alliance, Áder's appointment was strongly attacked by those who questioned his adherence to the constitutional impartiality of the presidential office.

Viktor Orbán

Position
Prime Minister

Introduction
Viktor Orbán previously served as Hungary's prime minister from 1998 until 2002. A lawyer, in 1988 he co-founded the Federation of Young Democrats–Fidesz. A prominent pro-democracy campaigner during the communist era, as president of Fidesz he led the party to victory at the 1998 general election. He became head of a coalition government and, as prime minister, campaigned for Hungary to join the European Union (EU). He became prime minister for a second time in May 2010 following the landslide victory of Fidesz-Hungarian Civic Alliance in the April general election.

Early Life
Orbán was born on 31 May 1963 in Székesfehérvár. After his schooling he did military service from 1981–82. In 1987 he graduated in law from Hungary's ELTE University and worked for two years as a sociological researcher.

In March 1988 he helped set up Fidesz, serving as its chief spokesman until Oct. 1989. During this period he attended Imre Nagy's reburial in Budapest and made a widely publicized speech calling for free elections and the removal of Soviet troops from Hungarian soil. During the summer of 1989 he was involved with the Opposition Roundtable negotiations that sought to resolve key issues on Hungary's future. Later in the year he took up a political philosophy research scholarship at Oxford University's Pembroke College.

Having won a parliamentary seat at Hungary's first post-communist free elections in early 1990, Orbán led the Fidesz parliamentary group. In May 1993 he became party president and from 1994–98 was the leader of various government committees preparing for European integration. An advocate of close ties with the USA, in 1996 he helped found the Committee for Political Co-operation of the New Atlantic Initiative. In addition in the 1990s he held several prominent posts in the Liberal International.

At the elections of 1998 Fidesz won 28% of the vote, and 14 more seats that the Socialists who gained 32%. Orbán was able to form a coalition in alliance with the Independent Party of Smallholders and the Hungarian Democratic Forum. He was sworn in as prime minister on 6 July 1998.

Gaining entry into the EU dominated Orbán's tenure. Hungary, Poland and the Czech Republic were considered the frontline nations likely to secure earliest entry, but plans for integration in 2003 faltered as the individual nations struggled to satisfy EU criteria. Orbán campaigned vigorously to ensure entry for Hungary was not delayed, arguing the irrelevancy of the state of other East European economies.

Orbán stood down as party president in Jan. 2000 in order to concentrate on the premiership. Later in the year he visited Romania for the first time after an industrial accident in which Romanian waste devastated several Hungarian rivers strained relations between the two countries.

Relations with other East European nations came under scrutiny in 2001 when Orbán commented that the Hungarian economy would require an influx of several million workers. He suggested ethnic Hungarians returning from abroad would be targeted to boost employment, as laid out in the draft Status Law. This was regarded by some observers as an attempt to assure West European nations that there would not be a worker deluge from the East.

Orbán lost the premiership in April 2002 after a narrow defeat at the general elections. Unable to secure a majority, the new government was formed by a Socialist-Alliance of Free Democrats coalition, led by Péter Medgyessy.

Career in Office
Following a comprehensive win for Fidesz-Hungarian Civic Alliance (as Fidesz is now known) in the general election of April 2010 in which it claimed a two-thirds majority, Orbán took office as prime minister on 29 May. Having campaigned on a platform of tax cuts and job creation, he faced major challenges to meet his targets in light of the country's weak financial position. He immediately introduced money-saving measures, reducing the cabinet from 15 members to nine.

In Oct. 2010 Orbán's government declared a state of emergency after a torrent of toxic chemical sludge spilled through a reservoir wall at an aluminium plant at Ajka in the west of the country. Nine people were killed in the industrial accident, which also caused widespread environmental and property damage.

In April 2011 the Fidesz-controlled parliament passed a revised and controversial new constitution, the legitimacy and democratic basis of which were questioned by the main opposition parties who boycotted the vote. The opposition was similarly critical of further legislation approved by parliament in Dec. and deemed to entrench the power of the Fidesz administration. In particular, one law demoted the status of the central bank and handed more control over monetary policy to the government, and another reduced the number of parliamentary members and redrew electoral boundaries.

Having previously decried the financial rescue package negotiated by the preceding government with the International Monetary Fund (IMF) in 2008, Orbán announced in Nov. 2011 that he would reopen talks with the IMF and the European Union (EU) for a precautionary credit line in the light of Hungary's continuing financial and economic woes. EU and IMF opposition to the law affecting the central bank initially stalled negotiations, but they resumed in mid-2012 following fears of an economic default by Hungary and some relaxation by the government of its measures to centralize power. However, in Sept. Orbán rejected stern IMF conditions and said he would pursue an alternative course.

In April 2012 Pál Schmitt resigned as president after being stripped of his university doctorate for alleged plagiarism. The following month parliament elected János Áder, another Orbán loyalist, in his place.

DEFENCE

The President of the Republic is C.-in-C. of the armed forces.

Conscription was abolished in 2004.

In 2008 defence expenditure totalled US$1,869m. (US$188 per capita), representing 1·2% of GDP. In 1985 defence expenditure had represented 6·8% of GDP.

Army
The strength of the Army was (2005) 15,814. There is an additional force of 12,000 border guards.

Air Force
The Air Force had a strength (2005) of 6,545. There were 77 combat aircraft in 2005, including MiG-21s, MiG-23s, MiG-29s and Su-22s, plus 55 in store, and 45 attack helicopters.

INTERNATIONAL RELATIONS

Hungary held a referendum on EU membership on 12 April 2003, in which 83·8% of votes cast were in favour of accession, with 16·2% against, although turnout was only 45·6%. It became a member of the EU on 1 May 2004. In 2000 Hungary introduced a visa requirement for Russians entering the country as one of the conditions for EU membership.

Hungary has had a long-standing dispute with Slovakia over the Gabčíkovo-Nagymaros Project, involving the building of dam structures in both countries for the production of electric power, flood control and improvement of navigation on the Danube as agreed in a treaty signed in 1977 between Hungary and Czechoslovakia. In late 1998 Slovakia and Hungary signed a protocol easing tensions between the two nations and settling differences over the dam.

ECONOMY

Agriculture accounted for 4% of GDP in 2008, industry 29% and services 66%.

Overview
Aided by market-oriented policies in the last two decades of communist rule and structural and stabilization measures implemented in the 1990s, Hungary's transition from communism was among the smoothest of the former Eastern Bloc nations. Since the collapse of communism, services have accounted for an increasing share of GDP. However,

manufacturing, concentrated in low-cost industrial assembly and processing, has been the main engine of growth.

Since the mid-1990s the majority of state assets have been privatized and the private sector now accounts for over 80% of GDP. By 1998 Hungary was attracting nearly half of all foreign direct investment in Central Europe. However, growth slowed from 2005 and was further weakened by a fiscal consolidation package in mid-2006. In the second half of 2008 the Hungarian forint lost nearly 20% of its value against the euro and the dollar, while unemployment rose to 10·1% in 2009 (compared to 7·8% in 2008). High levels of foreign debt increased strain and in Oct. 2008 a US$25bn. rescue package was agreed jointly with the IMF, EU and the World Bank to alleviate the effects of the global financial crisis.

Domestic demand continued to contract in 2010 and 2011 with exports the sole driver of growth, helped by strong links with the German export sector. However, the eurozone crisis has severely disrupted Hungary's external demand. Fiscal policy has been tightened substantially, with VAT and excise tax rates raised. Elements of 2011's 'Szell Kalman Plan'—aimed at improving medium-term growth potential through structural reforms on the expenditure side—have been adopted but investor worries pushed ten-year bond yields up to 9% in 2012 (significantly above the 7% that saw Greece, Ireland and Portugal forced to seek international bailouts). A sharp decline in consumer confidence, high unemployment, weak wage growth and tightening credit continue to hinder domestic private consumption, with the *Economist* describing the country in June 2012 as 'the worst performer in central Europe'.

Currency
A decree of 26 July 1946 instituted a new monetary unit, the *forint* (HUF) of 100 *fillér*. The forint was made fully convertible in Jan. 1991 and moves in a 15% band against the euro either side of a central rate of €1=282·4 forints. Inflation rates (based on OECD statistics):

2002	2003	2004	2005	2006	2007	2008	2009	2010	2011
5·3%	4·7%	6·7%	3·6%	3·9%	8·0%	6·0%	4·2%	4·9%	3·9%

The inflation rate of 3·6% in 2005 was the lowest in more than 25 years. Foreign exchange reserves were US$42,988m. in Sept. 2009 and gold reserves 99,000 troy oz. Total money supply in Aug. 2009 was 5,930·4bn. forints.

Budget
Budgetary central government revenue and expenditure (in 1bn. forints):

	2007	2008	2009
Revenue	6,847·2	7,754·7	7,771·9
Expenditure	8,111·8	8,700·0	8,706·0

Principal sources of revenue in 2009: taxes on goods and services, 3,463·1bn. forints; taxes on income, profits and capital gains, 2,476·3bn. forints; grants, 390·2bn. forints. Main items of expenditure by economic type in 2009: grants, 2,617·6bn. forints; compensation of employees, 1,450·5bn. forints; social benefits, 1,165·6bn. forints. There is a flat income tax rate of 16%.

Hungary registered a budget surplus of 4·3% of GDP in 2011. There were budget deficits of 4·2% in 2010 and 4·6% in 2009. The required target set by the EU is a budget deficit of no more than 3%.

VAT is 27% (reduced rates, 18% and 5%). Hungary has the highest rate of VAT of any country.

Performance
Real GDP growth rates (based on OECD statistics):

2002	2003	2004	2005	2006	2007	2008	2009	2010	2011
4·5%	3·9%	4·8%	4·0%	3·9%	0·1%	0·9%	−6·8%	1·3%	1·6%

Total GDP was US$140·0bn. in 2011.

Banking and Finance

In 1987 a two-tier system was established. The National Bank (*Governor*, György Matolcsy) remained the central state financial institution. It is responsible for the operation of monetary policy and the foreign exchange system. In Dec. 2009 the Hungarian financial system comprised 40 credit institutions, 140 co-operatives, 266 financial enterprises, 25 investment enterprises, 36 investment funds, 37 insurance companies, 20 pension funds and 37 health-related funds. They are all supervised by the Hungarian Financial Supervisory Authority (HFSA).

The largest bank is OTP Bank Rt. (the National Savings Bank plc) with assets in 2009 of 6,566bn. forints. Other leading banks are K+H (Hungarian Commercial and Credit Bank) and MKB (Hungarian Foreign Trade Bank). A law of June 1991 sets capital and reserve requirements, and provides for foreign investment in Hungarian banks. Permission is needed for investments of more than 10%.

In June 2012 gross external debt totalled US$203,265m.

Foreign direct investment was US$2,377m. in 2010. At the end of 2010 foreign direct investment stocks totalled US$91·9bn.

The Hungarian International Trade Bank opened in London in 1973. In 1980 the Central European International Bank was set up in Budapest with seven western banks holding 66% of the shares.

A stock exchange was opened in Budapest in June 1990.

ENERGY AND NATURAL RESOURCES

In 2008, 6·6% of energy consumption came from renewables (wind power, solar power, hydro-electric power, tidal power, geothermal energy and biomass), compared to the European Union average of 10·3%. A target of 13% has been set by the EU for 2020.

Environment

Hungary's carbon dioxide emissions from the consumption and flaring of fossil fuels in 2008 were the equivalent of 5·7 tonnes per capita.

Electricity

Installed capacity in 2007 was 8·5m. kW, about a fifth of which is nuclear. There is a 2,000 MW nuclear power station at Paks with four reactors. In 2007 Hungary produced 40·0bn. kWh of electricity and 14·7bn. kWh were imported. Hungary is a net importer of electricity. Total consumption in 2007 was 43·9bn. kWh. Consumption per capita in 2007 was 4,370 kWh.

Oil and Gas

Oil and natural gas are found in the Szeged basin and Zala county. Oil production in 2007 was 849,000 tonnes. Natural gas production in 2007 was 2·7bn. cu. metres, with proven reserves of 8bn. cu. metres in the same year. However, Hungary relies on Russia for almost all of its oil and much of its gas.

Minerals

Production in 1,000 tonnes: lignite (2007), 9,818; bauxite (2009), 317.

Agriculture

Agricultural land was collectivized in 1950. It was announced in 1990 that land would be restored to its pre-collectivization owners if they wished to cultivate it. A law of April 1994 restricts the area of land that may be bought by individuals to 300 ha., and prohibits the sale of arable land and land in conservation zones to companies and foreign nationals. Today, although 90% of all cultivated land is in private hands, most farms are little more than smallholdings. In 2009 the agricultural area was 5·78m. ha. (equivalent to 64% of the total land area); arable land constituted 4·59m. ha.

Agricultural production has dropped drastically since 1989. Production figures (2011, in 1,000 tonnes): maize, 8,089 (6,747 in 1989); wheat, 4,130 (6,509 in 1989); vegetables (excluding potatoes), 1,600; sunflower seeds, 1,368; barley, 989; sugar beets, 771 (5,277 in 1989); potatoes, 564; fruit, 542.

Livestock has also drastically decreased since 1989 from 7·7m. pigs to 3·2m. by 2010, from 1·6m. cattle to 0·7m., and from 2·1m. sheep to 1·2m. Thus the pig stock, cattle stock and sheep stock have all declined to levels not seen in more than fifty years.

The north shore of Lake Balaton, Villány and the Tokaj area are important wine-producing districts. Wine production in 2007 was 322m. litres.

Forestry

The forest area in 2010 was 2·03m. ha., or 23% of the land area. Timber production in 2007 was 5·64m. cu. metres.

Fisheries

There are fisheries in the rivers Danube and Tisza and Lake Balaton. In 2010 total catch was 6,216 tonnes, exclusively from inland fishing.

INDUSTRY

The leading companies by market capitalization in Hungary in March 2012 were MOL Magyar Olaj-és Gázipari Rt (Hungarian Oil and Gas Plc), US$8·7bn.; and OTP Bank Rt., US$4·8bn.

Manufacturing output grew by an average of 7·7% annually between 1992 and 2000 although it slowed between 2001 and 2007, with annual average growth of 5·2%, and contracted by 1·0% in 2008.

Production, 2009 unless otherwise indicated, in 1,000 tonnes: rolled steel products (2007), 103,347; cement (2007), 3,552; distillate fuel oil, 3,428; crude steel (2007), 2,232; plastics, 1,325; petrol (2007), 1,322; fertilizers, 825; alumina (2006), 270; residual fuel oil (2007), 200; refrigerators and freezers, 2,430,092 units; radio sets, 2,341,000 units; beer, 651·2m. litres.

Labour

In 2009 out of an economically active population of 4,202,600 there were 3,781,900 employed persons, of which 3,309,900 were employees. Among the employed persons in 2009, 64·2% worked in services, 31·2% in industry and construction, and 4·6% in agriculture. Average gross monthly wages of full-time employees in 2009: 199,837 forints. Minimum monthly wage, 2009, 71,500 forints (more than twice the 2000 level). There were a total of 6,474 working days lost to strike action in 2009, down from 25,004 in 2008. The unemployment rate was 10·9% in Nov. 2012. Long-term unemployment is particularly high, with 50·6% of the labour force in 2010 having been out of work for more than a year.

The normal retirement age is 62 but it is set to be increased gradually to 65 for both men and women by 2017.

INTERNATIONAL TRADE

Imports and Exports

Imports and exports for calendar years in US$1m.:

	2003	2004	2005	2006	2007
Imports c.i.f.	47,674·6	60,248·7	65,919·6	76,978·6	94,659·7
Exports f.o.b.	43,003·7	55,468·3	62,271·9	74,055·5	94,590·9

Hungary's foreign trade was expanding at a very fast rate until the global economic downturn of 2008–09, with the value of both its imports and its exports increasing fourfold between 1997 and 2007.

Machinery and transport equipment accounted for 52·5% of imports and 62·4% of exports in 2007, and manufactured goods 32·1% of imports and 26·5% of exports.

79·1% of exports in 2007 went to European Union member countries, the highest share of any of the central and eastern European countries that joined the EU in May 2004. In 2007, 26·8% of imports came from Germany and 28·4% of exports went to Germany. Russia was the second biggest supplier of imports in 2007 (6·9% of the total) and Italy the second biggest market for exports (5·6%). In 2007, 3·1% of exports went to Russia, down from 13·1% in 1992.

COMMUNICATIONS

Roads
In 2007 there were 195,719 km of roads, including 1,157 km of motorways, 6,745 km of main roads and 23,280 km of secondary roads; 37·7% of roads were paved. Passenger cars numbered 3,012,200 in 2007; lorries and vans, 829,800; motorcycles and mopeds, 135,900; and buses and coaches, 17,900. In 2007 there were 20,635 road accidents with 1,232 fatalities.

Rail
In 2008 the rail network was 7,608 km in length; 42·1m. tonnes of freight and 111·7m. passengers were carried. There is a metro in Budapest (30·1 km), and tram/light rail networks in Budapest (332·0 km), Debrecen, Miskolc and Szeged.

Civil Aviation
Budapest airport (Ferihegy) handled 8,095,367 passengers in 2009 (all on international flights) and 54,355 tonnes of freight. Malév, the former national carrier, ceased operations in Feb. 2012. The largest Hungarian airline is now Wizz Air, which started flying in 2004 and is Central and Eastern Europe's largest low-cost carrier.

Shipping
In 2008 there were 1,440 km of navigable waterways. In 2009 the Hungarian river fleet comprised 318 pushed or towed barges, 83 self-propelled barges and 80 other pushed or towed vessels. In 2009, 7·75m. tonnes of cargo and 859,000 passengers were carried. The Hungarian Shipping Company (MAHART) has agencies at Amsterdam, Alexandria, Algiers, Beirut, Rijeka and Trieste. It has 23 ships and runs scheduled services between Budapest and Esztergom.

Telecommunications
In 2008 there were 3,094,000 main (fixed) telephone lines. In the same year mobile phone subscribers numbered 12,224,000 (1,220·9 per 1,000 persons). Matav, the privatized former national telephone company, still has more than 80% of the fixed line market. Internet users numbered 5,873,000 in 2008. There were 19·6 fixed broadband subscribers per 100 inhabitants in Dec. 2010. In March 2012 there were 3·8m. Facebook users.

SOCIAL INSTITUTIONS

Justice
The administration of justice is the responsibility of the Procurator-General, elected by Parliament for six years. There are 111 local courts, 20 labour law courts, 20 county courts, six district courts and a Supreme Court. Criminal proceedings are dealt with by the regional courts through three-member councils and by the county courts and the Supreme Court in five-member councils. A new Civil Code was adopted in 1978 and a new Criminal Code in 1979.

Regional courts act as courts of first instance; county courts as either courts of first instance or of appeal. The Supreme Court acts normally as an appeal court, but may act as a court of first instance in cases submitted to it by the Public Prosecutor. All courts, when acting as courts of first instance, consist of one professional judge and two lay assessors, and, as courts of appeal, of three professional judges. Local government Executive Committees may try petty offences.

Regional and county judges and assessors are elected by the appropriate local councils; members of the Supreme Court by Parliament.

The Office of Ombudsman was instituted in 1993. He or she is elected by parliament for a six-year term, renewable once.

There are also military courts of the first instance. Military cases of the second instance go before the Supreme Court.

The death penalty was abolished in Oct. 1990.

The population in penal institutions in Sept. 2008 was 14,911 (149 per 100,000 of national population). There were 80,618 convictions of adults and 6,283 of juvenile offenders in 2009. Of 394,034 crimes registered in 2009, 23,914 were against the person (including 138 homicides).

Education
Adult literacy rate in 2008 was 99%. Education is free and compulsory from five to 16. Primary schooling ends at 14; thereafter education is continued at secondary, secondary technical or secondary vocational schools, which offer diplomas entitling students to apply for higher education, or at vocational training schools which offer tradesmen's diplomas. Students at the latter may also take the secondary school diploma examinations after two years of evening or correspondence study. Optional religious education was introduced in schools in 1990.

In 2010–11 there were: 4,358 kindergartens with 30,359 teachers and 338,162 children; 3,306 primary schools with 73,565 teachers and 758,566 pupils; and 2,617 secondary schools (including vocational schools) with 48,953 teachers and 662,808 pupils (of which 578,301 were full-time). 361,347 students were enrolled in tertiary education at 69 institutions in 2010–11 (240,727 full-time).

In 2007–08 total expenditure on education came to 5·2% of GNI and 14·4% of total government spending.

Health
In 2009 there were 30,276 doctors, 4,920 dentists, 54,352 nurses and midwives and (2006) 5,364 pharmacists. While there is an excess supply of doctors, there are too few nurses and wages for both groups are exceptionally low. In 2009 there were 175 hospitals with 71,489 beds. Spending on health accounted for 7·3% of GDP in 2009.

Welfare
In 1998 the Hungarian parliament decided to place the financial funds of health and pension insurance under government supervision. The self-governing bodies which had previously been responsible for this were dissolved. Medical treatment is free. Patients bear 15% of the cost of medicines. Sickness benefit is 75% of wages, old age pensions 60–70%. In 2010, 1·9trn. forints was spent on pensions and pension-like benefits for 2·94m. recipients. Family benefits totalled 2·2% of GDP in 2009. On a monthly basis in 2010, 1·2m. families were receiving family allowance and child care allowance was being paid for 178,532 children.

RELIGION
Church-state affairs are regulated by a law of Feb. 1990 which guarantees freedom of conscience and religion and separates church and state by prohibiting state interference in church affairs. Religious matters are the concern of the Department for Church Relations, under the auspices of the Prime Minister's Office.

According to the 2001 census, 51·9% of the population was Roman Catholic (5·3m. people), 15·9% Calvinist (1·6m.), 3·0% Lutheran (0·3m.) and 2·6% Greek Catholic (0·27m.). Adherents to smaller Christian faiths, including Baptists, other Protestant groups, Adventists and a range of Orthodox denominations, numbered around 98,000. About 0·1% of the population was Jewish in 2001.

The Primate of Hungary is Péter Erdő, Archbishop of Esztergom-Budapest, installed in Jan. 2003. There are 11 dioceses, all with bishops or archbishops. There is one Uniate bishopric. In Feb. 2013 the Roman Catholic church had two cardinals.

CULTURE

World Heritage Sites
Sites under Hungarian jurisdiction which appear on UNESCO's World Heritage List are (with year entered on list): Budapest, and specifically the Banks of the Danube and the Buda Castle Quarter

(1987 and 2002); Hollókő (1987), a preserved settlement developed during the 17th and 18th centuries; Millenary Benedictine Monastery of Pannonhalma and its Natural Environment (1996), first settled by Benedictine monks in 996; Hortobágy National Park (1999), a large area of plains and wetlands in eastern Hungary; Pécs (Sopianae) Early Christian Cemetery (2000), a series of decorated tombs dating from the 4th century; Tokaj Wine Region Historic Cultural Landscape (2002), a thousand-year-old wine-producing area.

Hungary also shares two UNESCO sites: the Caves of Aggtelek and Slovak Karst (1995, 2000 and 2008), a complex of 712 temperate-zone karstic caves, is shared with Slovakia; the Cultural Landscape of Fertö/Neusiedlersee (2001), an area that has acted as a meeting place for different cultures for 8,000 years, is shared with Austria.

Press

In 2008 there were 30 daily newspapers with a combined circulation of 1,630,000, at a rate of 176 per 1,000 inhabitants. The most widely read newspapers are the free tabloid *Metropol* and the paid-for tabloid *Blikk*. A total of 13,239 book titles were published in 2007 in 42·63m. copies. In the 2010 *World Press Freedom Index*, compiled by Reporters Without Borders, Hungary was ranked 23rd equal out of 178 countries.

Tourism

In 2011, 3,822,000 non-resident tourists and 4,199,000 domestic tourists stayed in holiday accommodation (3,462,000 and 4,011,000 respectively in 2010). The main countries of origin of non-resident tourists in 2011 were: Germany (542,000), Austria (295,000), the UK (221,000) and Romania (216,000).

Festivals

The Budapest Spring Festival, comprising music, theatre, dance etc., takes place in March. The Balaton Festival is in May, Miskolc Opera Festival in June, the Szeged Open-Air Theatre Festival in July–Aug, and the Sziget music festival in Budapest in mid-Aug. The flower carnival in Debrecen, a five-day celebration of flowers, music and dance, culminates on 20 Aug. (St Stephen's Day).

DIPLOMATIC REPRESENTATIVES

Of Hungary in the United Kingdom (35 Eaton Pl., London, SW1X 8BY)
Ambassador: János Csák.

Of the United Kingdom in Hungary (Harmincad Utca 6, Budapest 1051)
Ambassador: Jonathan Knott.

Of Hungary in the USA (3910 Shoemaker St., NW, Washington, D.C., 20008)
Ambassador: György Szapáry.

Of the USA in Hungary (Szabadság Tér 12, Budapest 1054)
Ambassador: Eleni Tsakopoulos Kounalakis.

Of Hungary to the United Nations
Ambassador: Csaba Kőrösi.

Of Hungary to the European Union
Permanent Representative: Péter Györkös.

FURTHER READING

Central Statistical Office. *Statisztikai Évkönyv*. Annual since 1871.— *Magyar Statisztikai Zsebkönyv*. Annual.—*Statistical Yearbook.— Statistical Handbook of Hungary.—Monthly Bulletin of Statistics.*

Bozóki, A., *et al.*, (eds.) *Post-Communist Transition: Emerging Pluralism in Hungary.* 1992
Burawoy, M. and Lukács, J., *The Radiant Past: Ideology and Reality in Hungary's Road to Capitalism.* 1992
Cartledge, Bryan, *The Will to Survive: A History of Hungary.* 2011
Cox, T. and Furlong, A. (eds.) *Hungary: the Politics of Transition.* 1995
Geró, A., *Modern Hungarian Society in the Making: the Unfinished Experience*; translated from Hungarian. 1995
Halpern, László and Wyplosz, Charles, (eds.) *Hungary: Towards a Market Economy.* 2011
Kontler, László, *A History of Hungary.* 2002
Körösényi, András, *Government and Politics in Hungary.* 1999
Lendvai, Paul, *The Hungarians: A Thousand Years of Victory in Defeat.* 2004
Mitchell, K. D. (ed.) *Political Pluralism in Hungary and Poland: Perspectives on the Reforms.* 1992
Molnár, Miklós, *A Concise History of Hungary.* 2001
Rose-Ackerman, Susan, *From Elections to Democracy: Building Accountable Government in Hungary and Poland.* 2007
Schiemann, John W., *The Politics of Pact-Making: Hungary's Negotiated Transition to Democracy in Comparative Perspective.* 2005
Sugar, Peter F., *A History of Hungary.* 1990
Székely, István and Newbery, David M. G., *Hungary: An Economy in Transition.* 2008
Vardy, Steven B., *Historical Dictionary of Hungary.* 1997

National library: Széchényi Library, Budavári Palota 'F' épület, 1827 Budapest. *Director General:* Andrea Sajó.
National Statistical Office: Központi Statisztikai Hivatal/Central Statistical Office, Keleti Károly u. 5/7, H-1024 Budapest. *President:* Gabriella Vukovich.
Website: http://portal.ksh.hu

ICELAND

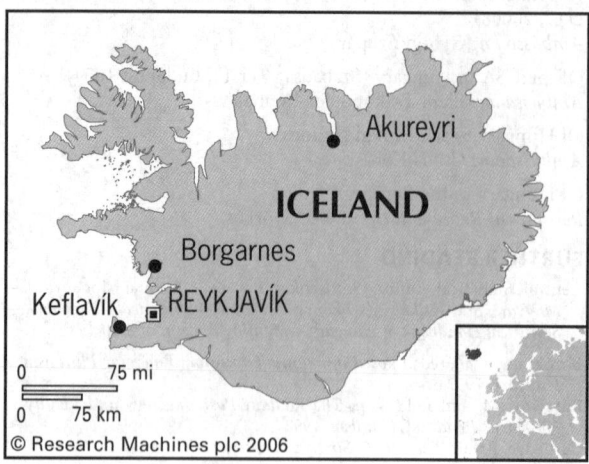

ICELAND

Akureyri

Borgarnes

Keflavík · REYKJAVÍK

0   75 mi

0   75 km

© Research Machines plc 2006

Lyðveldið Ísland
(Republic of Iceland)

Capital: Reykjavík
Population projection, 2015: 339,000
GNI per capita, 2011: (PPP$) 29,354
HDI/world rank: 0·898/14
Internet domain extension: .is

KEY HISTORICAL EVENTS

Scandinavia's North Atlantic outpost was first settled in 874. According to the *Landnámabók* or 'book of settlements', the first to land was Ingólfr Arnarson, who came from Norway to live on the site of present-day Reykjavík. He was followed by some 400 migrants, mainly from Norway but also from other Nordic countries and from Norse settlements in the British Isles.

A ruling class was soon formed by chieftains, known as the *godar*. In 930 they established the first ever democratic national assembly, the *Alþingi* (Althing). Primarily an adjudicating body, it also served as a legislature and as a fair, a marriage mart and as a national celebration in which a large proportion of the Icelandic population participated for two weeks each June. The first notable event in its history occurred in 1000 when, by majority decision, Christianity was adopted as Iceland's official religion. Despite the change, the *godar* remained politically important and some of them were ordained. Bishoprics were established at Skálholt in 1056 and at Hólar in 1106. It was not until the 1800s, after the bishoprics were united, that Reykjavík became the new episcopal see, making it the leading community.

Trade flourished with homespun woollen cloth as the chief export, although certain materials such as grain and timber had to be imported. Iceland's only indigenous wood, birchwood, which grew in abundance yet proved unsuitable for building, later became valuable in making charcoal.

In the mid-13th century there were power struggles between the *godar*. With the 'Old Treaty' of 1262, the *godar* were persuaded to swear allegiance to the king of Norway, bringing Iceland under Norwegian rule but leaving it with internal autonomy. When Norway was joined with Denmark in 1380, Iceland still retained the Althing as well as its own code of law.

In the 14th century, the expansion of fishing to satisfy European demand stimulated agriculture and other basic industries. Iceland's newfound prosperity encouraged trade between the Icelandic

fisherman and traders in Bergen, Norway. English traders in Bergen were keen to bypass Norwegian importers and instead began trading directly with Iceland. The Danish were largely unsuccessful in preventing this and it was not until the 16th century, when the English turned to the North American fishing grounds, that hostilities ceased.

Iceland's economic progress was checked when birchwood became depleted. Coupled with over-grazing, this led to soil erosion and put an end to crop growth. Further troubles came in the 15th century when Iceland fell victim to the Black Death, on two occasions losing around half of the population.

With the advent of Lutheranism in the first half of the 16th century, Iceland resisted Denmark's efforts to impose the Reformation on their North Atlantic possession. The bishoprics of Skálholt and Hólar were eventually overcome in 1550, marking the consolidation of Danish power over Iceland. In 1602 a royal decree gave all foreign trading rights in Iceland exclusively to Danish merchants. This restriction, which lasted until 1787, virtually ended Iceland's contacts with England and Germany, their one-time trading partners. Absolutism in Denmark and Norway under King Frederick III was recognized by Iceland in 1662, further strengthening external rule, and after economic hardship in the 18th century (in the 1780s famine killed one-fifth of the population) Iceland's reduced status was confirmed. When Norway split from Denmark in 1814 there were no similar calls for secession from Iceland.

Home Rule

In the 1830s a Danish consultative assembly was formed in which Iceland was given two seats. Denmark's transition to a system of representative democratic government after Frederick VII relinquished absolute power in 1848 did not extend to Iceland. After failures to reach agreement over the country's status, the Althing decided that 1874, the year that marked a thousand years of settlement, should be chosen as the year when it gained a new constitution. This was to provide the Althing with legislative if not executive control. During this period Iceland's economy continued to fare badly. With soil erosion still a problem, the strains of population growth forced mass emigration to North America. Around 15,000 emigrants left Iceland between 1870 and 1914. In 1904, after several decades of pressure for autonomy and, from 1901 onwards, support from the governing Danish Liberal party, Iceland finally achieved home rule.

Economic progress was led by modernization of the fishing industry and an expanded labour force. In 1916 a national trade unions organization was established and a process of urbanization began as the population moved towards the coastal fishing villages. Educational reform brought the introduction of compulsory education and in 1911, the establishment of the University of Iceland at Reykjavík.

In 1918 Iceland became a separate state under the Danish crown, with only foreign affairs remaining under Danish jurisdiction. The following decades were overshadowed by the influence of the 1930s' depression and the Spanish Civil War in 1936, the latter bringing an end to the lucrative fish trade with Spain.

In 1944 Iceland declared independence since Denmark was then occupied by Nazi Germany. The termination of the union was little more than a formality, the German invasion of Denmark in 1940 having effectively ended that country's responsibility for Iceland's foreign relations.

Iceland was occupied peacefully by Britain in 1940 but US troops took over a year later. They improved roads and docks, built an airport and paid high wages. An American request for a long-term lease on three military bases was reviewed sympathetically. But the continuing presence of American forces somehow gave the lie to the independence so recently celebrated.

The answer was for Iceland to join NATO. Objections to the American-run Keflavík airbase were gradually withdrawn and in 1951 a defence agreement with the United States allowed for an increase in the number of troops brought in 'to defend Iceland and ... to ensure the security of the seas around the country'. Greater integration with Europe came with joining the European Free Trade Association in 1970.

Fish remained central to the economy, accounting for 90% of the export trade. But there were worries about over-dependence on a single product and concern that other nations were taking too large a share of the Icelandic catch. In 1948 the demarcation of new fishing zones was made subject to Icelandic jurisdiction. Two years later one mile was added to the three-mile offshore zone which Iceland had administered since 1901. This was just acceptable to other fishing nations but when, in 1958, the limit was extended to 12 miles, Britain sent naval vessels to protect trawlers from harassment and arrest. This was the first Cod War, a cat-and-mouse game between the British navy and coastguard patrols which continued to 1961. At that point Britain and West Germany, the other fishing nation involved in the dispute, accepted the 12-mile zone on condition that if Iceland intended to widen her jurisdiction still further she had to give six months' notice of her intention and, if challenged, refer her claims to the International Court of Justice at The Hague.

In 1971, however, the government fulfilled its promise to do something about over-fishing by unilaterally extending the offshore zone to 50 miles. Despite a clear contravention of treaty commitment, Iceland gained sympathy as the tiny nation fighting the giants. Also in Iceland's favour was the move by Britain and members of the European Community to extend their jurisdiction over the continental shelf. A law officially expanding the Icelandic fishery limits to 50 miles came into force on 1 Sept. 1972. However, a second Cod War began shortly after as British and German trawlers continued to fish within the new zone. Hostility intensified, with the Icelandic Coast Guard deploying net cutters to prevent the ships securing their catch. An agreement was signed on 8 Nov. 1973 confining British trawlers to specific areas within the 50 mile catch zone, and limiting their annual catch to 130,000 tonnes.

This agreement expired in Nov. 1975, after which Iceland declared the ocean up to 200 miles from its coast to be under Icelandic authority, and a third Cod War began. When the talks reached stalemate in Dec. 1976, British vessels were nevertheless banned from Icelandic waters. In 2006 the country undertook its first commercial whale hunt after a 21-year moratorium. The government took control of the country's three leading banks in late 2008 and requested help from the IMF in a bid to stabilize the near-bankrupt economy amid the global economic crisis. In April 2010 a volcanic explosion produced an ash cloud that paralysed the European air industry for several weeks.

TERRITORY AND POPULATION

Iceland is an island in the North Atlantic, close to the Arctic Circle. Area, 102,819 sq. km (39,698 sq. miles).

There are eight regions:

Region	Inhabited land (sq. km)	Mountain pasture (sq. km)	Waste-land (sq. km)	Total area (sq. km)	Popula-tion (1 Jan. 2010)
Capital area	} 1,266	716	—	1,982	200,907
Southern Peninsula					21,359
West	5,011	3,415	275	8,701	15,370
Western Fjords	4,130	3,698	1,652	9,470	7,362
Northland West	4,867	5,278	2,948	13,093	7,394
Northland East	9,890	6,727	5,751	22,368	28,900
East	} 16,921	17,929	12,555	21,991	12,459
South				25,214	23,879
Iceland	42,085	37,553	23,181	102,819	317,630

Of the population of 317,630 in 2010, 20,428 were domiciled in rural districts and 297,202 (93·6%) in towns and villages (of over 200 inhabitants). Population density (2010), 3·1 per sq. km.

The UN gives a projected population for 2015 of 339,000.

The population is predominantly Icelandic. On 1 Jan. 2010 foreigners numbered 21,701 (9,583 Polish, 1,536 Lithuanian, 1,033 German, 884 Danish, 624 Latvian, 619 Portuguese, 603 Filipinos, 534 UK, 520 Thai, 499 US).

The capital, Reykjavík, had on 1 Jan. 2010 a population of 117,505; other towns were: Akranes, 6,549; Akureyri, 17,295; Bolungarvík, 970; Dalvík, 1,435; Eskifjörður, 1,062; Garðabær, 10,643; Grindavík, 2,837; Hafnarfjörður, 25,913; Húsavík, 2,229; Ísafjörður, 2,677; Kópavogur, 30,357; Neskaupstaður, 1,451; Ólafsfjörður, 852; Reykjanesbær, 14,091; Sauðárkrókur, 2,640; Selfoss, 6,493; Seltjarnarnes, 4,395; Seyðisfjörður, 706; Siglufjörður, 1,214; Vestmannaeyjar, 4,135.

The official language is Icelandic.

SOCIAL STATISTICS

Statistics for calendar years:

	Live births	Still-born	Marriages	Divorces	Deaths	Infant deaths	Net immigration
2006	4,415	13	1,752	516	1,903	6	5,255
2007	4,560	7	1,797	526	1,943	9	5,132
2008	4,835	11	1,704	560	1,987	12	1,144
2009	5,027	12	1,480	550	2,002	12	−4,835

2009 rates per 1,000 population: births, 15·8; deaths, 6·3. 64·4% of births are to unmarried mothers, the highest percentage in Europe. Population growth rate, 2009, −0·5%. In 2009 the most popular age range for marrying was 30–34 for males and 25–29 for females. Life expectancy, 2009: males, 79·7 years; females, 83·3. Infant mortality, 2009, 2·4 per 1,000 live births (one of the lowest rates in the world); fertility rate, 2009, 2·2 births per woman. Iceland legalized same-sex marriage in July 2010.

CLIMATE

The climate is cool temperate oceanic and rather changeable, but mild for its latitude because of the Gulf Stream and prevailing S.W. winds. Precipitation is high in upland areas, mainly in the form of snow. Reykjavík, Jan. 31·1°F (−0·5°C), July 51·1°F (10·6°C). Annual rainfall, 2009: 28·1″ (713 mm).

CONSTITUTION AND GOVERNMENT

The present constitution came into force on 17 June 1944 and has been amended four times since, most recently on 24 June 1999. The President is elected by direct, popular vote for a period of four years (no term limits).

The *Alþingi* (parliament) is elected in accordance with the electoral law of 1999, which provides for an *Alþingi* of 63 members. The country is divided into a minimum of six and a maximum of seven constituencies. There are currently six constituencies: Northwest (10 seats); Northeast (10 seats); South (10); Southwest (11); Reykjavík north (11); and Reykjavík south (11).

National Anthem

'Ó Guð vors lands' ('Oh God of Our Country'); words by M. Jochumsson, tune by S. Sveinbjörnsson.

GOVERNMENT CHRONOLOGY

Presidents since 1944.

1944–52	Sveinn Björnsson
1952–68	Ásgeir Ásgeirsson
1968–80	Kristján Thórarinsson Eldjárn
1980–96	Vigdís Finnbogadóttir
1996–	Ólafur Ragnar Grímsson

Prime Ministers since 1944. (AF = People's Party; FSF = Progressive Party; SF = Social Democratic Alliance; SSF = Independence Party)

1944–47	SSF	Ólafur Thors
1947–49	AF	Stefán Jóhann Stefánsson
1949–50	SSF	Ólafur Thors
1950–53	FSF	Steingrímur Steinthórsson
1953–56	SSF	Ólafur Thors
1956–58	FSF	Hermann Jónasson
1958–59	AF	Emil Jónsson
1959–63	SSF	Ólafur Thors
1963–70	SSF	Bjarni Benediktsson
1970–71	SSF	Jóhann Hafstein
1971–74	FSF	Ólafur Jóhannesson
1974–78	SSF	Geir Hallgrímsson
1978–79	FSF	Ólafur Jóhannesson
1979–80	AF	Benedikt Gröndal
1980–83	SSF	Gunnar Thoroddsen
1983–87	FSF	Steingrímur Hermannsson
1987–88	SSF	Thorsteinn Pálsson
1988–91	FSF	Steingrímur Hermannsson
1991–2004	SSF	Davíð Oddsson
2004–06	FSF	Halldór Ásgrímsson
2006–09	SSF	Geir Haarde
2009–	SF	Jóhanna Sigurðardóttir

RECENT ELECTIONS

In presidential elections held on 30 June 2012, incumbent Ólafur Ragnar Grímsson won 52·8% of the vote, Thóra Arnórsdóttir 32·2% and Ari Trausti Guðmundsson 8·6%. There were three other candidates. Turnout was 69·2%.

In the parliamentary election held on 27 April 2013 the Independence Party (SSF) won 19 of the 63 seats with 26·7% of the votes cast. The Progressive Party (FSF) also won 19 seats but with 24·4% of the vote. The Social Democratic Alliance (SF) won 9 with 12·9%, the Left-Green Movement (VG) 7 with 10·9%, Bright Future 6 with 8·3% and the Pirate Party 3 with 5·1%. Turnout was 81·4%. After the election Sigmundur Davíð Gunnlaugsson of the FSF was invited by the president to form a government. For the latest information *see* www.statesmansyearbook.com.

CURRENT GOVERNMENT

President: Ólafur Ragnar Grímsson; b. 1943 (ind.; sworn in 1 Aug. 1996, and re-elected 2000, 2004, 2008 and 2012).

Prior to the April 2013 election the government was a coalition of the Social Democratic Alliance backed by the Left-Green Movement, in power since the fall of Geir Haarde's government in 2009. In April 2013 it comprised:

Prime Minister: Jóhanna Sigurðardóttir; b. 1942 (SF; sworn in 1 Feb. 2009).

Minister of Education, Science and Culture: Katrín Jakobsdóttir (VG). *Environment and Natural Resources:* Svandís Svavarsdóttir (VG). *Finance and Economic Affairs:* Katrín Júlíusdóttir (SF). *Foreign Affairs:* Össur Skarphéðinsson (SF). *Industries and Innovation:* Steingrímur J. Sigfússon (VG). *Interior:* Ögmundur Jónasson (VG). *Welfare:* Guðbjartur Hannesson (SF).

Government Offices of Iceland Website: http://www.government.is

CURRENT LEADERS

Ólafur Ragnar Grímsson

Position
President

Introduction
Ólafur Ragnar Grímsson was leader of the People's Alliance until becoming president in 1996. Observers feared his background would politicize the presidency, which is traditionally a non-partisan, ceremonial post, but he has enjoyed broad popular support and retained the office in 2000, 2004, 2008 and 2012.

Early Life
Grímsson was born on 14 May 1943 in Ísafjörður. He studied economics and political science at Manchester University in the UK, graduating with a doctorate in 1970. He took up a lecturing post at the University of Iceland and was appointed professor in 1973. From 1966 until 1973 he was on the board of the youth wing of the Progressive Party and between 1971 and 1973 he sat on the party's executive board.

He moved to the People's Alliance and was elected to the *Alþingi* (Parliament) in 1978 as a member for Reykjavík. From 1980 until 1983, when he failed to win re-election to parliament, Grímsson led the People's Alliance in the *Alþingi*. During 1987–96 he was party chairman and between 1988–91 served as the minister of finance. Between 1984–90 he held senior posts with Parliamentarians for Global Action, an international organization with a membership of 1,800 throughout the world. Grímsson also held positions in the Council of Europe during the 1980s and 1990s.

In 1995 he led the People's Alliance to a poor showing at the polls, in which they secured less than 15% of the vote. Shortly afterwards Grímsson announced his candidacy for the presidency at the following year's elections. In June 1996 he was elected with 41% of the vote, defeating three other candidates.

Career in Office
The presidency is a largely ceremonial office and Grímsson's election prompted some observers to fear he would politicize the position. His relationship with the then prime minister, Davíð Oddsson, had been poor ever since the two had clashed as leaders of rival parties. Nevertheless, Grímsson was reappointed as president for a second term (without an election as there were no opposing candidates) and then re-elected by popular vote on 26 June 2004 with nearly 86% of the poll. During his presidency Grímsson has used his international profile to vigorously promote Iceland and its industrial potential, particularly in emerging sectors such as information technology. His reappointment in Aug. 2008 was unopposed.

In Dec. 2009 the *Alþingi* narrowly passed legislation to reimburse the UK and the Netherlands governments for bailing out British and Dutch depositors in Icesave, an Internet operation owned by the failed Icelandic bank Landsbanki. Grímsson, however, refused to sign the law—to popular acclaim—and it was put to a national referendum in March 2010. A large majority rejected the measure and negotiations on a settlement continued. In Dec. 2010 new reimbursement legislation was proposed, including more favourable repayment terms for Iceland, and was passed by the *Alþingi* in Feb. 2011. However, Grímsson again refused to sign the measure, leading to a further referendum and another voter rejection in April that year.

In June 2012 he was elected for a record fifth presidential term.

Jóhanna Sigurðardóttir

Position
Prime Minister

Introduction
Jóhanna Sigurðardóttir became prime minister in Jan. 2009 following economic crisis and the collapse of Geir Haarde's administration. Heading a coalition including her centre-left Social Democratic Alliance (SF) along with the Left-Green Movement (VG), she became the country's first female leader. She announced in Sept. 2012 that she would retire from politics at the end of her term of office in 2013. After the election of April 2013 her tenure continued as opposition parties sought to establish a coalition.

Early Life
Sigurðardóttir was born on 4 Oct. 1942 in Reykjavík. After studying business at Iceland's Commercial College, she undertook a varied career that included flight attendant, trade union

organizer and office administrator. In 1978 she entered *Alþingi* (Parliament) as the Social Democratic Party representative for the Reykjavík constituency. Subsequent boundary changes saw her become the MP for Reykjavík South and then for the North. In 1979 and from 1983–84 she was the speaker of the *Alþingi* and from 1987–94 she was minister of social affairs but resigned and made an unsuccessful run for her party's leadership.

She subsequently left the Social Democrats to form a new party, the National Movement, which won four seats at the 1995 general election. It merged five years later with her old party and two others to form the SF in a bid to end the dominance of the Independence Party (SSF). In 2007 the Alliance joined the Independence Party in a coalition headed by Haarde. Sigurðardóttir was reappointed to the social affairs portfolio. Working on behalf of the elderly, disabled and disadvantaged, she retained high personal approval ratings even in the depths of the national financial crisis in 2008.

When the country's independent banking system collapsed in late 2008, Haarde and his cabinet came under pressure to resign. On 23 Jan. 2009 Haarde called elections for early May but within three days the coalition had fallen apart. Talks between the SF and the SSF to form a new coalition failed and Alliance turned instead to the Left-Green Movement. With SF leader Ingibjorg Gisladóttir suffering ill health, Sigurðardóttir was proposed for prime minister. By the end of the month VG had agreed to form a coalition ahead of a general election scheduled for 25 April.

Career in Office
Sigurðardóttir's appointment received international media coverage as she became not only the first woman premier of Iceland but also the world's first openly gay head of government. She led her party to victory at the April 2009 general election, receiving 29·8% of the vote.

Her major challenge was to restore economic stability. In Feb. 2009 she engineered the removal of the central bank head, former prime minister Davíð Oddsson, who was widely blamed for the banking collapse. In July 2009 her government applied for EU membership, with negotiations beginning a year later despite an ongoing dispute with the UK and the Netherlands over compensation relating to the banking collapse. When a proposal to solve the issue was rejected by referendum in April 2011, the opposition Independence Party tabled a no-confidence motion against Sigurðardóttir and her government, which she narrowly survived. The dispute was subsequently referred to the European Free Trade Association Surveillance Authority.

Meanwhile, in April and May 2010 ash clouds from an Icelandic volcano led to unprecedented bans on flights and airport closures across Europe, causing disruption and heavy financial losses.

In the wake of the 2008 financial collapse, voters in a consultative referendum in Oct. 2012 supported a new draft constitution prepared by a council of 25 ordinary citizens. It advocated more direct democracy and greater control of the country's natural resources, such as fish and geothermal energy.

Sigurðardóttir was set to be succeeded by Sigmundur Davíð Gunnlaugsson of the Progressive Party, who sought to build a coalition after his party tied with the Independence Party on 19 seats each at the election of April 2013.

DEFENCE

Iceland possesses no armed forces. Under the North Atlantic Treaty, US forces were stationed for many years in Iceland as the Iceland Defence Force. In Sept. 2006 an agreement was signed between USA and Iceland, withdrawing all US forces from the island.

Navy
There is a paramilitary coastguard of 120.

ECONOMY

Agriculture, hunting, forestry and fishing contributed 6·4% of GDP in 2008, industry 27·3% and services 66·3%.

According to the anti-corruption organization *Transparency International*, Iceland ranked 11th in a 2012 survey of countries with the least corruption in business and government. It received 82 out of 100 in the annual index.

Overview
The economy experienced strong growth in the mid-1990s as a result of privatization and deregulation, per capita income doubling in the two decades to 2007. The strong housing market, in combination with a tight job market, fuelled domestic demand. Household spending further contributed to the growth of GDP, which expanded by more than 20% between 2003 and 2008.

However, this rapid expansion left the economy with large macroeconomic imbalances and high dependency on foreign financing. Financial sector assets amounted to over 1,000% of GDP and gross external indebtedness to roughly 550% of GDP by the end of 2007. The global credit squeeze hit the domestic financial markets. In Oct. 2008 all three of the country's major banks were nationalized to stabilize the financial system. The IMF approved a US$2·1bn. loan in Nov. 2008, supplemented by US$3bn. in loans from Iceland's neighbours, to support the króna. The key interest rate reached a record high of 18%. In Jan. 2009 the government collapsed following political turmoil prompted by the crisis.

A new government, elected in April 2009, worked with the IMF to rebuild the economy. Among the divisive questions was how to repay the Dutch and British governments for bailing out their depositors in Icesave, owned by the failed Icelandic bank Landsbanki. A law outlining a repayment programme was narrowly passed by parliament in Dec. 2009 but a presidential refusal to ratify the draft law led to a referendum on 6 March 2010 in which 93·2% of voters rejected the legislation. In Dec. 2011 the EFTA Surveillance Authority brought the case to the EFTA Court. In Jan. 2013 the court found that Iceland did not break European free trade laws on deposit guarantee schemes by refusing to compensate foreign depositors.

A rebound in consumption and investment saw growth resume in 2011, while unemployment declined to 7%. Restructuring of the banking sector led to a decline in non-performing loans from 40% in 2010 to 23% in 2012 but vulnerabilities in the sector remain. Inflation continues to be higher than the central bank target of 2·5%, standing at 4·1% in Aug. 2012, while gross external debt constituted 250% of GDP in 2011, although it is expected to decline.

Currency
The unit of currency is the *króna* (ISK) of 100 *aurar* (singular: *eyrir*). Foreign exchange markets were deregulated on 1 Jan. 1992. The króna was devalued 7·5% in June 1993. Inflation rates (based on OECD statistics):

2002	2003	2004	2005	2006	2007	2008	2009	2010	2011
5·2%	2·1%	3·2%	4·0%	6·7%	5·1%	12·7%	12·0%	5·4%	4·0%

Foreign exchange reserves were US$3,263m. and gold reserves 64,000 troy oz in Sept. 2009. Total money supply in April 2008 was 420,423m. kr. Note and coin circulation in 2009 was 28,958m. kr.

Budget
Total central government revenue and expenditure for calendar years (in 1m. kr.):

	2004	2005	2006	2007	2008	2009
Revenue	306,851	363,568	412,839	454,588	476,903	447,448
Expenditure	297,591	317,985	350,866	403,199	668,588	571,089

Central government debt was 1,176,436m. kr. on 31 Dec. 2009. Foreign debt amounted to 75,314m. kr. VAT is 25·5% (reduced rates, 14%, 7% and 0%).

Performance
Real GDP growth rates (based on OECD statistics):

2002	2003	2004	2005	2006	2007	2008	2009	2010	2011
0·1%	2·4%	7·8%	7·2%	4·7%	6·0%	1·2%	−6·6%	−4·0%	2·6%

GDP in 2011 totalled US$14·0bn. In 2009 GDP per capita was US$38,035.

Banking and Finance

The Central Bank of Iceland (founded 1961; *Governor*, Már Guðmundsson) is responsible for note issue and carries out the central banking functions. There were five commercial banks and 12 savings banks operating in 2010. On 31 Dec. 2009 the accounts of the Central Bank balanced at 1,178,082m. kr. Gross external debt amounted to US$106,359m. in June 2012.

There is a stock exchange in Reykjavík.

ENERGY AND NATURAL RESOURCES

Iceland is aiming to become the world's first 'hydrogen economy'; its buses started to convert to fuel cell-powered vehicles in late 2003. Ultimately it aims to run all its transport and even its fishing fleet on hydrogen produced in Iceland.

Environment

Iceland's carbon dioxide emissions from the consumption and flaring of fossil fuels in 2008 were the equivalent of 11·1 tonnes per capita. An *Environmental Performance Index* compiled in 2008 ranked Iceland 11th in the world, with 87·6%. The index examined various factors in six areas—air pollution, biodiversity and habitat, climate change, environmental health, productive natural resources and water resources.

Electricity

The installed capacity of public electrical power plants at the end of 2009 totalled 2,579,300 kWh; installed capacity of hydro-electric plants was 1,882,800 kWh. Electricity production in public-owned plants totalled 16,835m. kWh in 2009. Virtually all of Iceland's electricity is produced from hydro power and geothermal energy. Consumption per capita was estimated in 2009 to be 52,858 kWh (the highest in the world).

Agriculture

Of the total area, about six-sevenths is unproductive, but only about 1·3% is under cultivation, which is largely confined to hay and potatoes. Arable land totalled 15,500 ha. in 2007. In 2009 the total hay crop was 2,105,238 cu. metres; the crop of potatoes, 9,500 tonnes; of tomatoes, 1,481 tonnes; and of cucumbers, 1,452 tonnes. Livestock (2009): sheep, 469,429; horses, 77,158; cattle, 73,498 (dairy cows, 26,489); pigs, 3,818; poultry, 61,095. Livestock products (2009): lamb and mutton, 8,841 tonnes; poultry, 7,146 tonnes; pork, 6,375 tonnes; beef, 3,761 tonnes. Consumption of dairy products (2007): milk, 124,817 tonnes; cheese, 4,903 tonnes; butter and dairy margarines, 1,753 tonnes.

Forestry

In 2010 forests covered 30,000 ha. (equivalent to less than 0·5% of the total land area).

Fisheries

Fishing is of vital importance to the economy. Fishing vessels in 2009 numbered 1,582 with a gross tonnage of 158,253. Total catch in 2006: 1,322,914; 2007: 1,395,716, 2008: 1,283,078; 2009: 1,129,621. Virtually all the fish caught is from marine waters. Iceland has received international praise for its management system, which aims to avoid the over-fishing that has decimated stocks in other parts of the world. Commercial whaling was prohibited in 1989, but recommenced in 2006. In 2009 fisheries accounted for 6·3% of GDP, down from 16·8% in 1996. The per capita consumption of fish and fishery products is the second highest in the world, after that of the Maldives.

INDUSTRY

Production, 2009, in 1,000 tonnes: aluminium, 813·9; ferro-silicon, 135·8. 132,438 tonnes of cement were sold in 2007.

Labour

In 2009 the economically active population was 180,900. The unemployment rate in Dec. 2012 was 5·4% (down from 7·1% in 2011 as a whole). In the period 1996–2005 Iceland averaged 401 working days lost to strikes per 1,000 employees—the highest number in any western European country. In 2009 agriculture and fishing employed 5·2% of the economically active population, industry 19·8% and services 75·0% (including: health services and social work, 16·0%; wholesale, retail trade and repairs, 12·6%).

INTERNATIONAL TRADE

The economy is heavily trade-dependent.

Imports and Exports

Total value of imports (c.i.f.) and exports (f.o.b.) in 1m. kr.:

	2005	2006	2007	2008	2009
Imports	313,855	432,106	429,469	514,739	446,128
Exports	194,355	242,740	305,096	466,860	500,855

Main imports, 2009 (in 1m. kr.): industrial supplies, 140,178 (of which primary, 6,335; processed, 133,844); capital goods (except for transport), 94,279; consumer goods, 68,936; fuels and lubricants, 54,464. Main exports, 2009 (in 1m. kr.): industrial supplies, 225,654; food and beverages, 198,490; transport equipment, 33,449 (of which aeroplanes, 22,583); capital goods (except for transport), 19,159.

Value of trade with principal countries for three years (in 1,000 kr.):

	2007 Imports (c.i.f.)	2007 Exports (f.o.b.)	2008 Imports (c.i.f.)	2008 Exports (f.o.b.)	2009 Imports (c.i.f.)	2009 Exports (f.o.b.)
Belgium	6,725,400	4,997,700	6,809,500	7,214,100	6,871,900	9,208,600
Brazil	526,400	263,300	1,635,600	575,800	18,298,000	391,300
Canada	7,583,100	1,463,800	5,313,800	2,201,400	8,584,800	2,293,600
China	21,601,800	2,375,900	34,110,300	10,276,900	22,161,800	11,728,300
Denmark	31,660,900	10,080,300	37,695,700	14,642,500	32,453,600	13,512,300
France	12,201,000	7,827,300	12,157,300	14,287,300	8,635,700	17,575,100
Germany	51,582,500	40,815,000	52,819,200	52,777,400	36,847,100	56,403,200
Ireland	7,193,000	23,077,400	5,547,900	1,350,300	6,821,100	15,228,400
Italy	14,550,000	1,991,200	14,411,700	3,647,500	12,299,800	5,106,100
Japan	20,161,600	12,775,600	19,130,900	20,453,000	15,317,300	9,304,500
Netherlands	24,012,800	64,885,800	31,234,400	160,477,500	38,506,200	153,981,700
Nigeria	1,300	3,720,800	200	5,442,900	300	10,187,500
Norway	19,682,300	11,565,000	57,644,400	20,344,800	57,865,600	29,066,700
Poland	7,305,500	1,909,400	7,072,600	3,545,200	5,257,600	5,321,200
Portugal	1,012,200	7,560,000	1,154,700	8,791,700	1,088,800	8,804,700
Spain	5,694,200	14,158,800	6,020,800	17,861,000	5,455,900	24,130,500
Sweden	42,848,300	2,214,400	46,277,000	3,912,700	35,937,700	4,000,800
Switzerland	8,555,200	3,856,800	16,314,300	4,271,000	12,241,000	6,609,300
UK	22,875,100	40,333,800	22,467,500	54,189,000	20,298,400	63,970,400
USA	57,558,600	16,049,700	41,353,800	25,720,100	30,972,200	19,402,900

COMMUNICATIONS

Roads

On 1 Jan. 2009 the length of the public roads (including roads in towns) was 12,888 km. Of these 7,829 km were main and secondary roads and 5,059 km were provincial roads. Total length of surfaced roads was 4,566 km. A ring road of 1,400 km runs just inland from much of the coast; about 80% of it is smooth-surfaced. Motor vehicles registered at the end of 2009 numbered 238,149, of which 207,226 were passenger cars (643 per 1,000 inhabitants) and 30,923 lorries and vans; there were also 9,420 motorcycles. There were 15 fatal road accidents in 2009 with 17 persons killed.

Civil Aviation

Icelandair is the national carrier. In 2007 it served 18 destinations in western Europe and six in north America. In 2011 it carried 1·7m. passengers. The main international airport is at Keflavík (Leifsstöd), with Reykjavík for flights to the Faroe Islands, Greenland and domestic services. Keflavík handled 2,112,014 passengers in 2011 (of which 412,440 transit passengers) and 36,628 tonnes of freight.

Shipping

On 1 Jan. 2008 the merchant fleet consisted of 52 vessels totalling 8,515 GT, including 49 passenger ships and ferries of 7,669 GT.

Telecommunications

The number of telephone main lines was 184,851 in 2009; mobile phone subscribers, 329,932 (more than the population of Iceland and equivalent to 1,039 subscriptions per 1,000 population). In 2008, 90·6% of the population (the highest percentage in the world) were internet users. The fixed broadband penetration rate in Dec. 2010 was 33·7 subscribers per 100 inhabitants. In March 2012 there were 210,000 Facebook users.

SOCIAL INSTITUTIONS

Justice

In 1992 jurisdiction in civil and criminal cases was transferred from the provincial magistrates to eight new district courts, separating the judiciary from the prosecution. From the district courts there is an appeal to the Supreme Court in Reykjavík, which has eight judges. The population in penal institutions in Sept. 2008 was 140 (44 per 100,000 of national population).

Education

Primary education is compulsory and free from 6–16 years of age. Optional secondary education from 16 to 19 is also free. In 2009 there were 42,929 pupils in primary schools, 26,364 in secondary schools (22,262 on day courses) and 17,738 tertiary-level students (13,888 on day courses). Some 10·6% of tertiary-level students study abroad.

There are seven universities and five specialized colleges at tertiary level in Iceland. A total of 17,449 students were enrolled in universities in 2007. Universities (with total students, 2007): University of Iceland (founded 1911), Reykjavík, 9,586; Reykjavík University, 2,907; Iceland University of Education, 2,241; University of Akureyri, 1,305; Bifröst University, 744; Iceland Academy of Arts, 380; Agricultural University of Iceland, 286.

In 2009 public sector spending on education was 7·8% of GDP. The adult literacy rate is at least 99%.

Health

In 2002 there were 23 hospitals with 2,228 beds, equivalent to 78 per 10,000 population In 2011 there were 1,121 doctors and surgeons, 2,765 nurses, 357 pharmacists and 283 dentists. There were 3·5 doctors per 1,000 inhabitants in 2011. Iceland has one of the lowest alcohol consumption rates in Europe, at 7·53 litres of alcohol per adult per year (2007). Iceland spent 7·6% of its GDP on health in 2011.

Welfare

The main body of social welfare legislation is consolidated in six acts:

(i) The social security legislation (a) health insurance, including sickness benefits; *(b)* social security pensions, mainly consisting of old age pension, disablement pension and widows' pension, and also children's pension; *(c)* employment injuries insurance.

(ii) The unemployment insurance legislation, where daily allowances are paid to those who have met certain conditions.

(iii) The subsistence legislation. This is controlled by municipal government.

(iv) The tax legislation. Prior to 1988 children's support was included in the tax legislation. Since 1988 family allowances are paid directly to all children age 0–15 years. The amount is increased with the second child in the family, and children under the age of seven get additional benefits. Single parents receive additional allowances.

(v) The rehabilitation legislation

(vi) Child and juvenile guidance

Health insurance covers the entire population. Citizenship is not demanded and there is a six-month waiting period. Most hospitals are both municipally and state run, a few solely state run and all offer free medical help. Medical treatment out of hospitals is partly paid by the patient; the same applies to medicines, except medicines of lifelong necessary use, which are paid in full by the health insurance. Dental care is partly paid by the state for children under 17 years old and also for old age and disabled pensioners. Sickness benefits are paid to those who lose income because of periodical illness.

The pension system is composed of the public social security system and some 90 private pension funds. The social security system pays basic old age and disablement pensions of a fixed amount regardless of past or present income, as well as supplementary pensions to individuals with low present income. The pensions are index-linked, i.e. are changed in line with changes in wage and salary rates in the labour market. In the public social security system, entitlement to old age and disablement pensions at the full rates is subject to the condition that the beneficiary has been resident in Iceland for 40 years at the age period of 16–67. For shorter periods of residence, the benefits are reduced proportionally. Entitled to old age pension are all those who are 67 years old, and have been residents in Iceland for three years of the age period of 16–67. Old age and disablement pension are of equally high amount; in the year 2009 the total sum was 351,528 kr. for an individual. Married pensioners receive double the basic pension. Pensioners with little or no other income are entitled to an income supplement; in 2009 the maximum annual income supplement was 1,155,513 kr.

The employment injuries insurance covers medical care, daily allowances, disablement pension and survivors' pension and is applicable to practically all employees.

RELIGION

The national church, the Evangelical Lutheran, is endowed by the state. There is complete religious liberty. The affairs of the national church are under the superintendence of a bishop. In 2010, 251,487 persons (79·2% of the population) were members of it (93·2% in 1980). 16,497 persons (5·2%) belonged to Lutheran free churches. 39,310 persons (12·4%) belonged to other religious organizations and 10,336 persons (3·3%) did not belong to any religious community.

CULTURE

CULTURE

World Heritage Sites
There are two UNESCO sites in Iceland: Þingvellir National Park (2004), located on an active volcanic site; and Surtsey (2008), an island formed by volcanic eruptions in 1963–67.

Press
In 2008 there were four daily newspapers (two paid-for and two free) and 20 non-daily newspapers. Combined circulation was 336,459 (of which dailies accounted for 278,154 and non-dailies 58,305). Iceland has the highest circulation rates of daily newspapers in the world, at 817 per 1,000 adult inhabitants in 2008.

Iceland publishes more books per person than any other country in the world. In 2008, 1,637 volumes of books and booklets were published.

Tourism
There were 459,252 visitors in 2010; revenue totalled 152,941m. kr. Overnight stays in hotels and guest houses in 2009 numbered 1,939,667 (of which foreign travellers, 1,553,927; Icelanders, 385,740). Tourism accounts for 19·4% of foreign currency earnings.

Festivals
Iceland's national day is celebrated on 17 June. The Reykjavík Arts Festival, an annual programme of international artists and performers, is held every May–June. Iceland Airwaves, an annual music festival in Reykjavík, takes place in Oct. There is also a folk festival in March, a blues festival at Easter and a jazz festival in Aug.

DIPLOMATIC REPRESENTATIVES

Of Iceland in the United Kingdom (2A Hans St., London, SW1X 0JE)
Ambassador: Benedikt Jónsson.

Of the United Kingdom in Iceland (Laufásvegur 31, 101 Reykjavík)
Ambassador: Stuart Gill.

Of Iceland in the USA (2900 K St. NW, Suites 508/509, Washington, D.C., 20007)
Ambassador: Guðmundur Árni Stefánsson.

Of the USA in Iceland (Laufásvegur 21, 101 Reykjavík)
Ambassador: Luis Arreaga.

Of Iceland to the United Nations
Ambassador: Gréta Gunnarsdóttir.

Of Iceland to the European Union
Ambassador: Thórir Ibsen.

FURTHER READING

Statistics Iceland, *Landshagir* (Statistical Yearbook of Iceland).—*Hagtíðindi* (Statistical Series)
Central Bank of Iceland. *Monetary Bulletin* (four a year).—*The Economy of Iceland.*

Boyes, Roger, *Meltdown Iceland.* 2009
Byock, Jesse, *Viking Age Iceland.* 2001
Karlsson, G., *The History of Iceland.* 2000
Smiley, Jane, (ed.) *The Sagas of Icelanders: A Selection.* 2002
Thorhallsson, Baldur, (ed.) *Iceland and European Integration: On the Edge.* 2004

National library: Landsbókasafn Islands—Háskólabókasafn, Arngrímsgata 3, 107 Reykjavík. *Librarian:* Ingibjörg Steinunn Sverrisdóttir.
National Statistical Office: Statistics Iceland, Bogartúni 21a, IS-150 Reykjavík.
Website: http://www.hagstofa.is
Central Bank of Iceland: Kalkofnsvegi 1, 150 Reykjavik.
Website: http://www.sedlabanki.is

INDIA

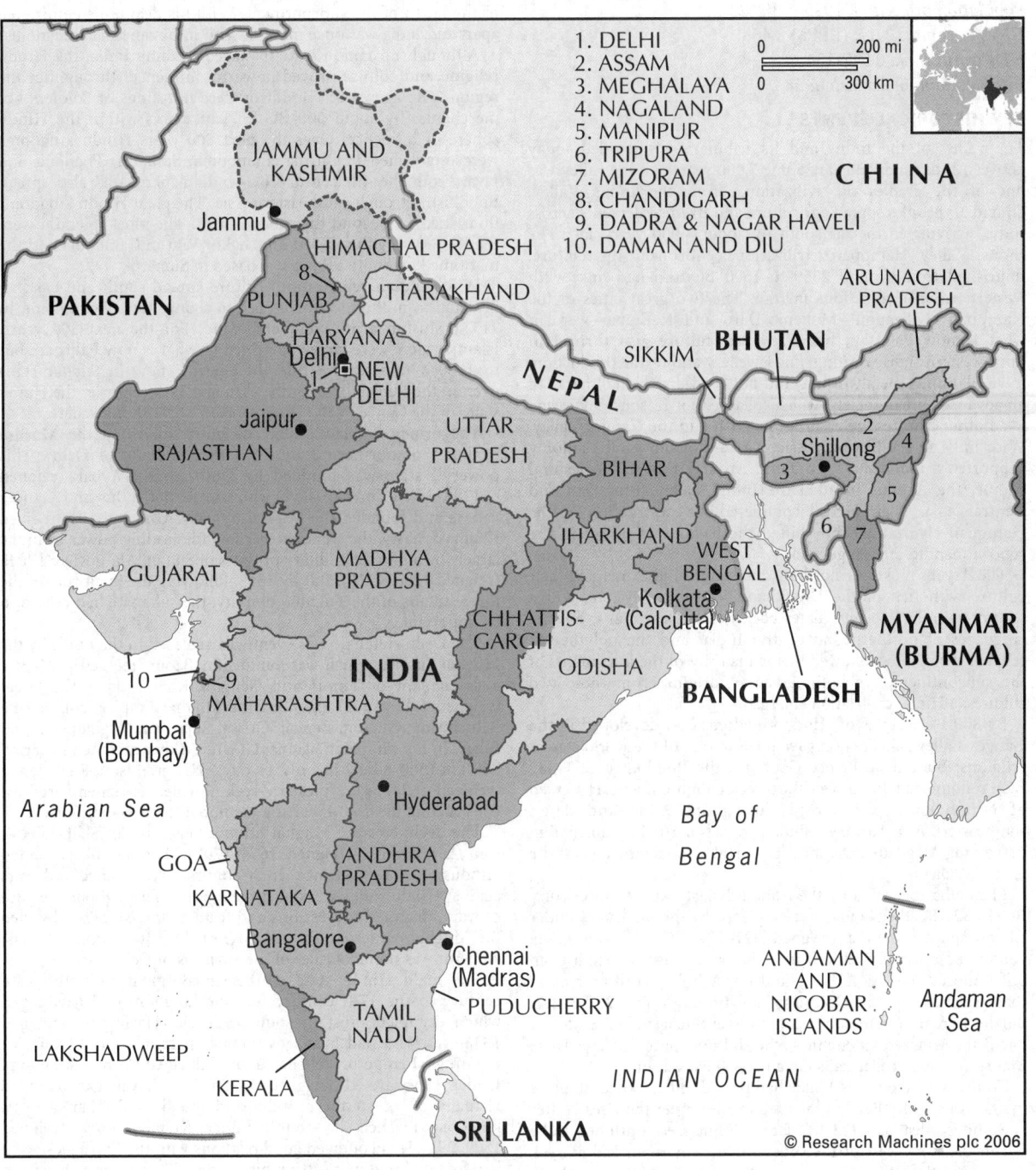

1. DELHI
2. ASSAM
3. MEGHALAYA
4. NAGALAND
5. MANIPUR
6. TRIPURA
7. MIZORAM
8. CHANDIGARH
9. DADRA & NAGAR HAVELI
10. DAMAN AND DIU

0   200 mi

0   300 km

CHINA

JAMMU AND KASHMIR

Jammu

HIMACHAL PRADESH

PAKISTAN

8 PUNJAB

UTTARAKHAND

HARYANA

Delhi

1 NEW DELHI

SIKKIM

NEPAL

BHUTAN

ARUNACHAL PRADESH

Jaipur

RAJASTHAN

UTTAR PRADESH

BIHAR

Shillong

2

4

3

5

6 7

GUJARAT

MADHYA PRADESH

JHARKHAND

WEST BENGAL

MYANMAR (BURMA)

10   9

MAHARASHTRA

CHHATTIS-GARGH

INDIA

Kolkata (Calcutta)

ODISHA

BANGLADESH

Mumbai (Bombay)

Arabian Sea

Hyderabad

Bay of Bengal

GOA

ANDHRA PRADESH

KARNATAKA

Bangalore

Chennai (Madras)

PUDUCHERRY

ANDAMAN AND NICOBAR ISLANDS

Andaman Sea

LAKSHADWEEP

TAMIL NADU

KERALA

INDIAN OCEAN

SRI LANKA

© Research Machines plc 2006

Map. Based upon Survey of India Map with the permission of the Surveyor General of India. The responsibility for the correctness of internal details rests with the publisher. The territorial waters of India extend into the sea to a distance of 12 nautical miles measured from the appropriate base line. The external boundaries and coastlines of India agree with the Record/Master Copy certified by the Survey of India.

Bharat
(Republic of India)

Capital: New Delhi
Population projection, 2015: 1,308·22m.
GNI per capita, 2011: (PPP$) 3,468
HDI/world rank: 0·547/134
Internet domain extension: .in

KEY HISTORICAL EVENTS

The valley of the Indus and its tributaries is divided today between India and Pakistan. Some 7,000 years ago the valley was one of the cradles of civilization. From the Indus Valley, Dravidian peoples spread agriculture and fixed settlements across India, arriving in the far south by about 4,000 years ago. The Indus Valley Harappan civilization, a Bronze Age culture, flourished from around 2300 to 1500 BC and had links with western Asian civilizations in Iran. The two great cities of the Harappan civilization—Mohenjo-Daro and Harappa—were in what is now Pakistan, but Harappan culture also thrived in northwestern India. Writing, fine jewellery and textile production, town planning, metalworking and pottery were the hallmarks of an advanced urban society, which collapsed for reasons unknown.

Another Bronze Age civilization existed in the Ganges Valley. With links to southeastern Asia, a rice-growing rural economy supported a number of city-states. Around 1500 BC a pastoral people, the Aryans, invaded the Indus Valley from Iran and Central Asia. Their arrival completed the destruction of the Harappan civilization and shifted the balance of power in the subcontinent to the Ganges Valley.

The Aryans took over northern and central India, merging their culture with that of the Dravidians. The caste system, still a feature of Indian society, dates back to the Dravidians, but the languages of northern and central India, and the polytheistic religion that is now followed by the majority of the inhabitants of the subcontinent, are both Aryan in origin. From these two cultures, a Hindu civilization emerged.

By 800 BC a series of Hindu kingdoms had developed in the Ganges Valley. This region gave birth to one of the world's great religions: Buddhism. Prince Gautama, the Buddha (*c.* 563–483 BC), renounced a life of wealth to seek enlightenment. His creed of non-violence was spread throughout India and, later, southeastern Asia. However, Buddhism was partly instrumental in destroying Magdalha, the most powerful of ancient states of the Ganges Valley.

Magdalha was ruled by the Nanda dynasty in the 4th century BC. In 321 BC the Nandas were replaced by the Mauryans under Chandragupta Maurya (reigned 321–297 BC). Chandragupta conquered most of northern India before his ascetic death from self-imposed starvation. His grandson, Ashoka, ruled an empire that stretched from the Deccan to Afghanistan from *c.* 272–*c.* 231, but he is mainly remembered for his enthusiasm for Buddhist pacifism. Attacked by enemies who did not share this creed, the Mauryan empire collapsed soon after Ashoka's death.

To the west, the Indus Valley had passed to the Persian Empire by the 5th century BC and then fell to Alexander the Great. After Alexander's death in 323 BC, Greek influences continued to be felt in the northwest of the subcontinent where an Indo-Greek civilization flourished for at least 200 years. This was brought to an end by nomadic invasions from Central Asia between the 1st and 5th centuries AD. By then, India had been divided into many small warring states, most of them short-lived.

Empire Building

However, two strong states emerged briefly to reunite much of India: the Gupta empire and the Harsa empire. The Gupta empire was founded in the Ganges Valley by Chandragupta I (reigned *c.* 320–30). His warrior son, Samudragupta (reigned *c.* 330–80),

won most of north India, but the empire was destroyed by succession disputes and a Hun invasion in the middle of the 5th century. The Harsa empire was the personal creation of a Buddhist convert, Harsa (reigned 606–47), who briefly ruled most of the north of the subcontinent. With his death, his empire fell apart and India was once more divided into many rival kingdoms.

Although no Hindu state managed to unite India, the Hindu religion and culture proved powerful influences throughout the region. The agents of Hinduism were not kings or soldiers but merchants. By about 500 BC Sri Lanka was within the Hindu sphere of influence. Over the next 800 years Hindu kingdoms were established in Burma, Cambodia, Sumatra, Thailand and Java. From the 4th century BC, Indian merchants also spread Buddhism through southeastern Asia. The great Hindu kingdoms flourished far beyond the subcontinent. The most splendid were the Khmer kingdom based on Angkor Wat in Cambodia and the maritime kingdom of Sriwijaya, based in Sumatra.

While Indian religion and culture spread south and east, an invasion from the west threatened to change the subcontinent. In 713 a Muslim army conquered Sind. For the next 300 years, Islamic rulers were largely confined to what is now Pakistan, but in 1000 a raid by the ruler of Ghazni (now in Afghanistan) overran the Punjab. During the 11th and 12th centuries the Hindu states of the Ganges Valley were toppled by Muslim invaders.

The principal Islamic state of India, following the Muslim conquest of northern India, was the sultanate of Delhi. This powerful state was founded by Qutb-ud-Din Aybak (reigned *c.* 1208–10), a former slave, who united the Indus and Ganges valleys and founded the Mu'izzi dynasty. Under the short-lived Khaljis dynasty, the sultanate became the leading power in India, largely owing to the military prowess of Sultan 'Ala-ud-Din Khalji (reigned 1296–1316). But by 1388, following the inept rule of the three sultans of the Tughluq dynasty, the sultanate had ceased to be important.

The Delhi sultanate was eventually replaced in the north by the Mughal Empire, which was founded by Babur (reigned 1526–30), a descendant of Timur and Genghis Khan. Akbar the Great (reigned 1556–1605) extended the Mughal Empire, conquering Baluchistan, Gujarat, Bengal, Orissa, Rajasthan, Afghanistan and Bihar. In his campaign against Gujarat, Akbar marched his army 800 km (500 miles) in only 11 days. His grandson, Shah Jahan (reigned 1628–58), a pleasure-seeking ruler, is remembered for constructing the Taj Mahal as a memorial to his favourite wife.

The decline of the Mughal Empire began under Shah Jahan's son, Aurangzeb I (reigned 1658–1707). Aurangzeb persecuted Hindus with a vengeance. Inter-community violence and wars against Hindu states weakened the empire. Throughout the 18th century, disputed successions and fears of assassination diverted the Mughal emperors. By the close of the 18th century the last emperor was nominal ruler of the environs of Delhi.

The main Hindu state of the subcontinent from the 14th century to the 17th century was the kingdom of Vijayanagar, which occupied most of southern India. Harihara I (reigned 1336–54), who had been governor of part of central India for the Mughal emperor, rebelled and established his own kingdom. Under Devaraya II (reigned 1425–47), Vijayanagar included virtually all of southern India and much of Sri Lanka. This kingdom reached its zenith under Krsnadevaraya (reigned 1509–29). He encouraged good relations with the Portuguese who had founded trading posts on his shores. Vijayanagar collapsed in civil wars (1614–46).

The Bahmani sultanate of the Deccan was an Islamic state, which dominated central India from the mid-14th century until the 16th century. This state was founded by 'Ala-ud-Din Hasan Bahmani Shah (reigned 1347–58), the local governor for the sultan of Delhi who rebelled against Delhi and established his own dynasty. For a time the Bahmani sultanate was the most powerful state in central India, but defeats at the hands of the kingdom of Vijayanagar in the 15th century weakened the Bahmani sultans.

On the death of the last Bahmani sultan in 1518 the kingdom was divided by the provincial governors into small states.

European Influence

By the 16th century European traders were established along India's coasts. The first to arrive were the Portuguese in 1498. In 1510 the Portuguese took Goa, which was to remain the centre of the fragmented possessions of Portuguese India until 1962. The creation of the (English) East India Company in 1600 heralded the beginning of what was to become the British Indian Empire. Forts were established on the coast in 1619 and in 1661 England took possession of Bombay.

Initially, the Europeans were only interested in trade but they soon became involved in local politics, in particular, the disputed successions that bedevilled Indian states. Portugal and England were not alone in attempting to establish outposts in India. The Dutch were active in the 17th century but were effectively eliminated from the competition before 1759, when Britain took Chinsura, the headquarters of Dutch administration in India. Two small Danish colonies lasted from 1618 until 1858. However, the main threat to British rule in India was France. Although the East India Company controlled parts of Bengal and the Ganges Valley, France was supreme in the Deccan where French forces, and Indian rulers allied to France, held sway over an area twice the size of France itself.

In the 1750s Britain and France fought out their European wars overseas. The defeat of French forces, and France's Indian allies, at the battle of Plassey (1757), by British forces led by Robert Clive (1725–74), confirmed British rule in Bengal and Bihar and ejected France from the Deccan. Henceforth, France was restricted to five small coastal possessions.

The Maratha state was the major power in central and southern India in the 17th and 18th centuries. This empire was founded by Sivaji (1627–80), who built the state between 1653 and 1660. The Hindu Sivaji came into conflict with the fanatical Muslim Mughal emperor Aurangzeb, who imprisoned Sivaji. After his famous escape from captivity, concealed in a fruit basket, Sivaji made himself emperor of his Maratha state in 1674. This pious monarch ruled competently, establishing an efficient administration, but by the time of his grandson, Shahu (reigned 1707–27), the power of the Maratha emperors had been eclipsed by that of their hereditary chief minister, the Peshwa. In 1727 the Peshwa Baji Rao I (reigned 1720–40) effectively replaced the emperor and established his own dynasty. Baji Rao made the Maratha state the strongest in India. His descendant, Baji Rao II (reigned 1795–1817), raised a weakened state against the British and was crushed. He was the last important Indian monarch outside British influence.

East India Company

In the first half of the 19th century, wars against Sind (1843) and the Sikhs in Punjab (1849) extended the borders of British India. By the middle of the 19th century about 60% of the subcontinent was controlled by the East India Company. The remaining 40% was divided between about 620 Indian states, which were, in theory, still sovereign and ruled by their own maharajas, sultans, nawabs and other monarchs, each advised by a British resident. The Indian states ranged from large entities the size of European countries (such as Hyderabad, Baroda, Mysore and Indore) to tiny states no bigger than an English parish.

British rule brought land reform in the areas controlled by the East India Company. The traditional patterns of land holdings was broken up and private land ownership was introduced. This had the unintended result of concentrating ownership in the hands of a small number of powerful landlords. As a result, landless peasants and dispossessed princes united in their opposition to British rule. In 1857 a mutiny by soldiers of the East India Company quickly spread into full-scale rebellion. Throughout India those who resented the speed and nature of the changes brought about by British rule made one final attempt to eject the occupiers. The Indian Mutiny took 14 months to put down.

After the Mutiny the British government replaced the East India Company as the ruler of an Indian colonial empire (1858), and the modernization of India began apace. Emphasis was placed on building up an Indian infrastructure, particularly roads and railways. The participation of Indians within the civil administration, the construction of a vast national railway system and the imposition of the English language did much to forge a national identity overriding the divisions of local state and caste. But in deference to British manufacturers industry in India remained backward. In 1877 the Indian Empire was proclaimed and Queen Victoria became Empress of India (Kaiser-i-Hind).

Growing Nationalism

The (Hindu-dominated) Indian National Congress, the forerunner of the Congress Party, first met in 1885, and in 1906 the rival Muslim League was founded. Demands for Home Rule grew in the early years of the 20th century, and nationalist feeling was fuelled when British troops fired on a nationalist protest meeting—the Amritsar Massacre (1919).

Realizing that change was inevitable, the British government reformed the administration in 1919 and 1935. The creation of an Indian federation removed many of the differences between the Crown territories and the Indian states and granted an Indian government limited autonomy. The pace of reform was, however, too slow for popular opinion.

In 1920 the Congress party began a campaign of non-violence and non-cooperation with the British colonial authorities. Congress was led in its struggle by the charismatic figure of Mahatma Gandhi (1869–1948). The British authorities were forced to concede Gandhi's moral influence but he himself was opposed by the traditional rulers of the Indian states, whose own positions were at risk.

By the start of the Second World War (1939–45), relations between the Hindu and Muslim communities in India had broken down, with the Muslims demanding a separate independent Islamic state, later, Pakistan. During the war, Assam and other northeastern areas were faced with the threat of a Japanese invasion. Although many Indians served in the Allied forces during the war, a minority supported Japan as a possible liberator.

In 1945 Britain had neither the will nor the resources to maintain the Indian Empire. But while Britain accepted independence, religious tension made partition inevitable. In 1947 the sub-continent was divided between India, a predominantly Hindu state led by Jawaharlal (Pandit) Nehru (1889–1964) of the Congress Party, and Pakistan, a Muslim state led by Mohammad Ali Jinnah (1876–1948) of the Muslim League. The rulers of the Indian states were entitled to choose their allegiance while British Crown territories were assigned to either India or Pakistan.

Partitions

Partition brought enormous upheaval. More than 70m. Hindus and Muslims became refugees as they trekked across the new boundaries. Many thousands were killed in intercommunal violence. The Muslim ruler of the large, mainly Hindu, southern Indian state of Hyderabad, declared independence and the adherence of his state to India was only achieved through Indian military intervention. The Hindu ruler of mainly Muslim Kashmir opted to join India, against the wishes of his people. Elsewhere the border remained disputed in many places. Tension increased when Gandhi was assassinated by a Hindu fundamentalist (1948). In 1950 India became a republic.

Tension between India and Pakistan erupted into war in 1947–49 when the two countries fought over Kashmir. The region was divided along a ceasefire line, although neither side recognized this as an international border. India and Pakistan went to war again over Kashmir in 1965 and again in 1971 when Bangladesh (formerly East Pakistan) gained its independence as a result of Indian military intervention. Indian forces saw action in 1961 when Indian troops invaded and annexed Portuguese India and in 1962 in a border war with China. France had already ceded its small enclaves to India in 1950 and 1955. In 1975 India annexed the small Himalayan kingdom of Sikkim.

Despite its involvement in several wars, India assumed joint leadership of the non-aligned world. Pandit Nehru, premier from 1947 to 1964, was briefly succeeded by Lal Bahadur Shastri. In 1966 Nehru's daughter Indira Gandhi (1917–84) became premier. Under Mrs Gandhi, India continued to assert itself as a regional power and the rival of Pakistan. Although non-aligned, India developed close relations with the Soviet Union.

In 1971 Mrs Gandhi's government abolished the titles, pensions and privileges guaranteed to the Indian princes at independence as compensation for merging their states into India. India was wracked by local separatism and communal unrest. From 1975 to 1977 Mrs Gandhi imposed a state of emergency. Her actions split the Congress Party, allowing Morarji Desai (1896–1995) of the Janata Party to form India's first non-Congress administration. However, his coalition soon shattered and a wing of Congress, led by Mrs Gandhi, was returned to power in 1980.

Violence in Sikh areas, fanned by demands by militant Sikhs for an independent homeland (called Khalistan) increased tensions. In 1984 Mrs Gandhi ordered that the Golden Temple in Amritsar be stormed after it had been turned into a storehouse for weapons by Sikh extremists. Soon afterwards, Mrs Gandhi was assassinated by her Sikh bodyguards.

Mrs Gandhi was succeeded as premier by her son, Rajiv (1944–91), during whose period of office India became involved in Sri Lanka, supporting the central government against the separatist Tamil Tigers movement. Rajiv Gandhi was assassinated by a Tamil Tiger suicide bomber during the 1991 election campaign.

Recent Politics

By 1989 personality clashes and separatists tendencies had shattered the unity of the once all-powerful Congress Party. Regional parties and Hindu nationalist parties came to the fore and, since 1989, when Rajiv Gandhi left office, coalitions have held office. Seven prime ministers have led India since 1989: the longest periods in office have been enjoyed by Manmohan Singh (premier since 2004), Atal Bihari Vajpayee (premier in 1996 and again from 1998–2004) and P. V. Narasimha Rao (who led a coalition from 1991 to 1996). The right-wing Hindu nationalist Bharatiya Janata Party (BJP) has joined most of these coalitions. Support for the BJP increased following violence between Hindus and Muslims over a campaign, begun in 1990, to build a Hindu temple on the site of a mosque in the holy city of Ayodhya.

Since the fall of the Soviet Union (1991), India has retreated from state ownership and protectionism. Privatization has been accompanied by an economic revolution that has seen the development of high tech industries. At the same time, India has become a nuclear power. Although India exploded its first nuclear device in 1974, tests in 1998 confirmed the nation's capability to deliver these weapons.

There have been 35,000 deaths since the outbreak of the Kashmir insurgency in 1988. Negotiations with Pakistan over the future of the disputed territory began in July 1999. Hopes of avoiding further violence were set back in Dec. 2001, in an attack on the Indian parliament by suicide bombers. 13 people died. Although no group claimed responsibility, Kashmiri separatists were blamed. However, Pakistani President Pervez Musharraf's subsequent crackdown on militants helped to bring the two countries back from the brink of war. Tension between India and Pakistan increased following an attack on an Indian army base in Indian-occupied Kashmir on 14 May 2002. The attack, which killed 31 people, was linked to Islamic terrorists infiltrating the Kashmir valley from Pakistan. It drew widespread criticism of President Musharraf for failing to combat terrorism in the disputed region. In Feb. 2002, 58 Hindu pilgrims returning from

Ayodhya were killed when their train was set on fire following a confrontation with a Muslim crowd at Godhra in Gujarat. The incident led to three months of intermittent communal rioting, during which at least 800 Muslims died in attacks by Hindus. Relations between India and Pakistan cooled again in Aug. 2003 when 50 people were killed by terrorist bombings in Mumbai (Bombay). The two countries then embarked on a new phase of peace negotiations. In May 2004 India elected Manmohan Singh as its first Sikh prime minister. The peace process was set back when unidentified terrorists killed over 200 people and injured 700 more in a series of co-ordinated train bombings in Mumbai on 11 July 2006. In Dec. 2008 talks were temporarily suspended, a month after nearly 200 people died when gunmen launched a series of attacks on buildings in Mumbai's tourism and financial district. India claimed that Pakistani-based militants were responsible.

TERRITORY AND POPULATION

India is bounded in the northwest by Pakistan, north by China (Tibet), Nepal and Bhutan, east by Myanmar, and southeast, south and southwest by the Indian Ocean. The far eastern states and territories are almost separated from the rest by Bangladesh. The area is 3,287,263 sq. km. A Sino-Indian agreement of 7 Sept. 1993 settled frontier disputes dating from the war of 1962. Population, 2011 census (provisional): 1,210,193,422 (51·5% male and 48·5% female), giving a density of 382 persons per sq. km. There are also 20m. Indians and ethnic Indians living abroad, notably in Malaysia, the USA, Saudi Arabia, the UK and South Africa. 68·8% of the population was rural in 2011. Goa is the most urban state, at 62·2% in 2011; and Himachal Pradesh the most rural, at 90·0% in 2011.

The UN gives a projected population for 2015 of 1,308·22m.

By 2050 India is expected to have a population of 1·66bn. It is projected to overtake China as the world's most populous country around 2026.

Area and population of states and union territories (2011 census, provisional):

States	Area in sq. km	Population	Density per sq. km
Andhra Pradesh (AP)	275,045	84,665,533	308
Arunachal Pradesh (AC)	83,743	1,382,611	17
Assam (AS)	78,438	31,169,272	397
Bihar (BI)	94,163	103,804,637	1,102
Chhattisgarh (CG)	135,191	25,540,196	189
Goa (GO)	3,702	1,457,723	394
Gujarat (GU)	196,024	60,383,628	308
Haryana (HA)	44,212	25,353,081	573
Himachal Pradesh (HP)	55,673	6,856,509	123
Jammu and Kashmir (JK)	222,236	12,548,926	56
Jharkhand (JH)	79,714	32,966,238	414
Karnataka (KA)	191,791	61,130,704	319
Kerala (KE)	38,863	33,387,677	859
Madhya Pradesh (MP)	308,245	72,597,565	236
Maharashtra (MH)	307,713	112,372,972	365
Manipur (MN)	22,327	2,721,756	122
Meghalaya (ME)	22,429	2,964,007	132
Mizoram (MZ)	21,081	1,091,014	52
Nagaland (NA)	16,579	1,980,602	119
Odisha (formerly Orissa) (OD)	155,707	41,947,358	269
Punjab (PB)	50,362	27,704,236	550
Rajasthan (RJ)	342,239	68,621,012	201
Sikkim (SI)	7,096	607,688	86
Tamil Nadu (TN)	130,058	72,138,958	555
Tripura (TR)	10,486	3,671,032	350
Uttar Pradesh (UP)	240,928	199,581,477	828
Uttarakhand (formerly Uttaranchal) (UK)	53,483	10,116,752	189
West Bengal (WB)	88,752	91,347,736	1,029
Union Territories			
Andaman and Nicobar Islands (AN)	8,249	379,944	46
Chandigarh (CH)	114	1,054,686	9,252
Dadra and Nagar Haveli (DN)	491	342,853	698

States	Area in sq. km	Population	Density per sq. km
Daman and Diu (DD)	112	242,911	2,169
Delhi (DL)	1,483	16,753,235	11,297
Lakshadweep (LK)	32	64,429	2,013
Puducherry (formerly Pondicherry) (PY)	479	1,244,464	2,598

Urban agglomerations with populations over 2m., together with their core cities at the 2011 census (provisional):

	State/ Union Territory	Urban agglomeration	Core city
Mumbai (Bombay)	Maharashtra	18,414,288	12,478,447
Delhi	Delhi	16,314,838	11,007,835
Kolkata (Calcutta)	West Bengal	14,112,536	4,486,679
Chennai (Madras)	Tamil Nadu	8,696,010	4,681,087
Bangalore	Karnataka	8,499,399	8,425,970
Hyderabad	Andhra Pradesh	7,749,334	6,809,970
Ahmedabad	Gujarat	6,352,254	5,570,585
Pune (Poona)	Maharashtra	5,049,968	3,115,431
Surat	Gujarat	4,585,367	4,462,002
Jaipur	Rajasthan	3,073,350	3,073,350
Kanpur	Uttar Pradesh	2,920,067	2,767,031
Lucknow	Uttar Pradesh	2,901,474	2,815,601
Nagpur	Maharashtra	2,497,777	2,405,421
Ghaziabad	Uttar Pradesh	2,358,525	1,636,068
Indore	Madhya Pradesh	2,167,447	1,960,631
Coimbatore	Tamil Nadu	2,151,466	1,061,447
Kochi	Kerala	2,117,990	601,574
Patna	Bihar	2,046,652	1,683,200
Kozhikode	Kerala	2,030,519	432,097

Smaller urban agglomerations and cities with populations over 250,000 (with provisional 2011 census populations, in 1,000):

Agartala (TR)	400	Darbhanga (BI)	306
Agra (UP)	1,746	Davangere (KA)	435
Ahmadnagar (MH)	380	Dehra Dun (UK)	714
Aizawl (MZ)	292	Dewas (MP)	289
Ajmer (RJ)	551	Dhanbad (JH)	1,195
Akola (MH)	427	Dhule (MH)	376
Aligarh (UP)	910	Dindigul (TN)	292
Allahabad (UP)	1,217	Durgapur (WB)	581
Alwar (RJ)	341	Eluru (AP)	251
Amravati (MH)	647	Erode (TN)	522
Amritsar (PB)	1,184	Etawah (UP)	257
Anand (GU)	287	Faizabad (UP)	259
Anantapur (AP)	342	Faridabad Complex (HA)	1,405
Arrah (BI)	261	Farrukhabad (UP)	291
Asansol (WB)	1,243	Firozabad (UP)	604
Aurangabad (MH)	1,189	Ganganagar (RJ)	250
Baharampur (WB)	306	Gaya (BI)	471
Barddhaman (WB)	347	Gorakhpur (UP)	693
Bareilly (UP)	980	Gulbarga (KA)	542
Bathinda (PB)	286	Guntur (AP)	674
Begusarai (BI)	251	Gurgaon (HA)	902
Belgaum (KA)	610	Guwahati (AS)	969
Bellary (KA)	410	Gwalior (MP)	1,102
Bhagalpur (BI)	410	Habra (WB)	305
Bharatpur (RJ)	253	Hardwar (UK)	311
Bhavnagar (GU)	606	Hisar (HA)	307
Bhilai (CG)	1,064	Hubli-Dharwad (KA)	944
Bhilwara (RJ)	360	Ichalkaranji (MH)	326
Bhiwandi (MH)	737	Imphal (MN)	414
Bhopal (MP)	1,883	Ingraj Bazar (WB)	324
Bhubaneswar (OD)	882	Jabalpur (MP)	1,268
Bihar Sharif (BI)	297	Jalandhar (PB)	874
Bijapur (KA)	326	Jalgaon (MH)	460
Bikaner (RJ)	648	Jalna (MH)	285
Bilaspur (CG)	453	Jammu (JK)	652
Bokaro Steel City (JH)	563	Jamnagar (GU)	600
Brahmapur (OR)	356	Jamshedpur (JH)	1,337
Chandigarh (CH)	1,026	Jhansi (UP)	549
Chandrapur (MH)	321	Jodhpur (RJ)	1,138
Cherthala (KE)	455	Junagadh (GU)	320
Cuttack (OD)	659	Kadapa (AP)	344

Kakinada (AP)	443	Rampur (UP)	349
Kannur (KE)	1,643	Ranchi (JH)	1,127
Karimnagar (AP)	300	Ranippettai (TN)	262
Karnal (HA)	303	Ratlam (MP)	274
Kayamkulam (KE)	427	Rohtak (HA)	373
Khammam (AP)	262	Roorkee (UK)	274
Kharagpur (WB)	294	Rourkela (OD)	553
Kolhapur (MH)	562	Sagar (MP)	370
Kollam (KE)	1,110	Saharanpur (UP)	703
Korba (CG)	365	Salem (TN)	919
Kota (RJ)	1,001	Sambalpur (OD)	270
Kottayam (KE)	358	Sangli-Miraj (MH)	514
Kurnool (AP)	478	Santipur (WB)	289
Latur (MH)	383	Satna (MP)	283
Ludhiana (PB)	1,614	Shahjahanpur (UP)	346
Madurai (TN)	1,462	Shiliguri (WB)	701
Malappuram (KE)	1,699	Shillong (ME)	354
Malegaon (MH)	576	Shimoga (KA)	322
Mangalore (KA)	620	Sholapur (MH)	951
Mathura (UP)	455	Sonipat (HA)	292
Maunath Bhanjan (UP)	279	Srinagar (JK)	1,273
Meerut (UP)	1,425	Surendranagar (GU)	254
Moradabad (UP)	890	Thanjavur (TN)	291
Muzaffarnagar (UP)	495	Thiruvananthapuram	
Muzaffarpur (BI)	394	(KE)	1,687
Mysore (KA)	984	Thrissur (KE)	1,855
Nanded (MH)	551	Tiruchirapalli (TN)	1,022
Nashik (MH)	1,563	Tirunelveli (TN)	499
Navsari (GU)	283	Tirupati (AP)	460
Nellore (AP)	564	Tiruppur (TN)	963
Nizamabad (AP)	310	Tumkur (KA)	306
Noida (UP)	642	Tuticorin (TN)	411
Palakkad (KE)	294	Udaipur (RJ)	475
Panipat (HA)	442	Ujjain (MP)	515
Parbhani (MH)	307	Vadodara (GU)	1,817
Patiala (PB)	445	Varanasi (UP)	1,435
Puducherry (PY)	654	Vasai-Virar (MH)	1,221
Purnea (BI)	311	Vellore (TN)	482
Raipur (CG)	1,123	Vijayawada (AP)	1,491
Rajahmundry (AP)	478	Visakhapatnam (AP)	1,730
Rajkot (GU)	1,391	Warangal (AP)	760
Ramagundam (AP)	252	Yamunanagar (HA)	383

SOCIAL STATISTICS

Many births and deaths go unregistered. The Registrar General's data suggests a birth rate for 2009 of 22·5 per 1,000 population and a death rate of 7·3, which would indicate in a year approximately 27·2m. births and 8·8m. deaths. The growth rate is, however, slowing, and by 2010 had dropped to 1·4%, having been over 2% in 1991. Expectation of life at birth, 2007, 62·0 years for males and 64·9 years for females. In 2010, 50% of the population was aged under 25.

Many marriages and divorces go unregistered. A marriage can be registered under either of the two Marriage Acts: the Hindu Marriage Act, 1955; or the Special Marriage Act, 1954. To be eligible for marriage the minimum age limit is 21 for males and 18 for females. However, a survey carried out in 2005–06 found that 44·5% of women aged 20–24 had been married before the legal age of 18 (which has applied since 1978). Population growth rate, 2001–11, 17·64% (the lowest since 1941–51). Infant mortality, 2009, 50 per 1,000 live births; fertility rate, 2008, 2·7 births per woman. Child deaths (under the age of five) fell from 123 per 1,000 in 1990 to 64 per 1,000 in 2009.

CLIMATE

India has a variety of climatic sub-divisions. In general, there are four seasons. The cool one lasts from Dec. to March, the hot season is in April and May, the rainy season is June to Sept., followed by a further dry season until Nov. Rainfall, however, varies considerably, from 4" (100 mm) in the N.W. desert to over 400" (10,000 mm) in parts of Assam.

Range of temperature and rainfall: New Delhi, Jan. 57°F (13·9°C), July 88°F (31·1°C). Annual rainfall 26" (640 mm).

Chennai, Jan. 76°F (24·4°C), July 87°F (30·6°C). Annual rainfall 51" (1,270 mm). Cherrapunji, Jan. 53°F (11·7°C), July 68°F (20°C). Annual rainfall 432" (10,798 mm). Darjeeling, Jan. 41°F (5°C), July 62°F (16·7°C). Annual rainfall 121" (3,035 mm). Hyderabad, Jan. 72°F (22·2°C), July 80°F (26·7°C). Annual rainfall 30" (752 mm). Kochi, Jan. 80°F (26·7°C), July 79°F (26·1°C). Annual rainfall 117" (2,929 mm). Kolkata, Jan. 67°F (19·4°C), July 84°F (28·9°C). Annual rainfall 64" (1,600 mm). Mumbai, Jan. 75°F (23·9°C), July 81°F (27·2°C). Annual rainfall 72" (1,809 mm). Patna, Jan. 63°F (17·2°C), July 90°F (32·2°C). Annual rainfall 46" (1,150 mm).

On 26 Dec. 2004 an undersea earthquake centred off the Indonesian island of Sumatra caused a huge tsunami that flooded coastal areas in southern India resulting in 16,000 deaths. In total there were more than 225,000 deaths in 14 countries.

CONSTITUTION AND GOVERNMENT

The Constitution was passed by the Constituent Assembly on 26 Nov. 1949 and came into force on 26 Jan. 1950. It has since been amended 96 times.

India is a republic and comprises a Union of 28 States and seven Union Territories. Each State is administered by a Governor appointed by the President for a term of five years while each Union Territory is administered by the President through a Lieut.-Governor or an administrator appointed by him. The head of the Union (head of state) is the *President* in whom all executive power is vested, to be exercised on the advice of ministers responsible to Parliament. The President, who must be an Indian citizen at least 35 years old and eligible for election to the House of the People, is elected by an electoral college of all the elected members of Parliament and of the state legislative assemblies, holds office for five years and is eligible for re-election. There is also a *Vice-President* who is *ex officio* chairman of the Council of States.

There is a *Council of Ministers* to aid and advise the President; this comprises Ministers who are members of the Cabinet and Ministers of State and deputy ministers who are not. A Minister who for any period of six consecutive months is not a member of either House of Parliament ceases to be a Minister at the expiration of that period. The *Prime Minister* is appointed by the President; other Ministers are appointed by the President on the Prime Minister's advice. The salary of each Minister is ₹50,000 per month.

Parliament consists of the President, the *Council of States* (*Rajya Sabha*) and the *House of the People* (*Lok Sabha*). The Council of States, or the Upper House, normally consists of 245 members; in Nov. 2011 there were 232 elected members, ten members nominated by the President and three vacancies. The election to this house is indirect; the representatives of each State are elected by the elected members of the Legislative Assembly of that State. The Council of States is a permanent body not liable to dissolution, but one-third of the members retire every second year. The House of the People, or the Lower House, normally consists of 545 members—543 directly elected on the basis of adult suffrage from territorial constituencies in the States and the Union territories, and two nominated members of the Anglo-Indian Community. In March 2010 the Council of States approved a bill that would reserve a third of seats in the House of the People (and in each state legislature) for women. The bill is yet to go before the lower house, where it requires two-thirds support to become law. In Nov. 2011 in the House of the People there were 542 elected members, two nominated members and one vacancy. Unless sooner dissolved it continues for a period of five years from the date appointed for its first meeting; in emergency, Parliament can extend the term by one year.

State Legislatures

For every State there is a legislature which consists of the Governor, and (a) two Houses, a Legislative Assembly and a Legislative Council, in the States of Bihar, Jammu and Kashmir, Karnataka, Madhya Pradesh (where it is provided for but not in operation), Maharashtra and Uttar Pradesh, and (b) one House, a Legislative Assembly, in the other States. Every Legislative Assembly, unless sooner dissolved, continues for five years from the date appointed for its first meeting. In emergency the term can be extended by one year. Every State Legislative Council is a permanent body and is not subject to dissolution, but one-third of the members retire every second year. Parliament can, however, abolish an existing Legislative Council or create a new one, if the proposal is supported by a resolution of the Legislative Assembly concerned.

Legislation

The various subjects of legislation are enumerated in three lists in the seventh schedule to the constitution. List I, the Union List, consists of 97 subjects (including defence, foreign affairs, communications, currency and coinage, banking and customs) with respect to which the Union Parliament has exclusive power to make laws. The State legislature has exclusive power to make laws with respect to the 66 subjects in list II, the State List; these include police and public order, agriculture and irrigation, education, public health and local government. The powers to make laws with respect to the 47 subjects (including economic and social planning, legal questions and labour and price control) in list III, the Concurrent List, are held by both Union and State governments, though the former prevails. But Parliament may legislate with respect to any subject in the State List in circumstances when the subject assumes national importance or during emergencies.

Fundamental Rights

Two chapters of the constitution deal with fundamental rights and 'Directive Principles of State Policy'. 'Untouchability' is abolished, and its practice in any form is punishable. The fundamental rights can be enforced through the ordinary courts of law and through the Supreme Court of the Union. The directive principles cannot be enforced through the courts of law; they are nevertheless fundamental in the governance of the country.

Citizenship

Under the Constitution, every person who was on the 26 Jan. 1950 domiciled in India and (a) was born in India or (b) either of whose parents was born in India or (c) who has been ordinarily resident in the territory of India for not less than five years immediately preceding that date became a citizen of India. Special provision is made for migrants from Pakistan and for Indians resident abroad. The right to vote is granted to every person who is a citizen of India and who is not less than 18 years of age on a fixed date and is not otherwise disqualified.

Parliament

Parliament and the state legislatures are organized according to the following schedule (figures show distribution of seats in Oct. 2011 for the Lok Sabha, the Rajya Sabha and the State Legislatures):

| | Parliament | | State Legislatures | |
| | House of the People (Lok Sabha) | Council of States (Rajya Sabha) | Legislative Assemblies (Vidhan Sabhas) | Legislative Councils (Vidhan Parishads) |
States:				
Andhra Pradesh	42	18	295[1]	90[2]
Arunachal Pradesh	2	1	60	—
Assam	14	7	126	—
Bihar	40	16	243	75
Chhattisgarh	11	5	90	—
Goa	2	1	40	—
Gujarat	26	11	182	—
Haryana	10	5	90	—
Himachal Pradesh	4	3	68	—
Jammu and Kashmir	6	4	89[3,4]	36[5]
Jharkhand	14	6	81	—
Karnataka	28	12	225[1]	75

States:	Parliament House of the People (Lok Sabha)	Council of States (Rajya Sabha)	State Legislatures Legislative Assemblies (Vidhan Sabhas)	Legislative Councils (Vidhan Parishads)
Kerala	20	9	141[1]	—
Madhya Pradesh	29	11	231[1]	—
Maharashtra	48	19	289[1]	78
Manipur	2	1	60	—
Meghalaya	2	1	60	—
Mizoram	1	1	40	—
Nagaland	1	1	60	—
Odisha	21	10	147	—
Punjab	13	7	117	—
Rajasthan	25	10	200	—
Sikkim	1	1	32	—
Tamil Nadu	39	18	235[1]	—
Tripura	2	1	60	—
Uttar Pradesh	80	31	404[1]	100
Uttarakhand	5	3	70	—
West Bengal	42	16	295[1]	—
Union Territories:				
Andaman and Nicobar Islands	1	—	—	—
Chandigarh	1	—	—	—
Dadra and Nagar Haveli	1	—	—	—
Daman and Diu	1	—	—	—
Delhi	7	3	70	—
Lakshadweep	1	—	—	—
Puducherry	1	1	30	—
Nominated by the President under Article 80 (1) (a) of the Constitution	—	12	—	—
Total	545[6]	245	4,130[7]	454

[1]Includes one nominated member.
[2]Includes 12 nominated members.
[3]Includes two nominated members.
[4]Excludes 24 seats for Pakistan-occupied areas of the State which are in abeyance.
[5]Excludes seats for the Pakistan-occupied areas.
[6]Includes two nominated members to represent Anglo-Indians.
[7]Includes ten nominated members.

The number of seats allotted to the scheduled castes and the scheduled tribes in the House of the People is 84 and 47 respectively. Of the 4,120 elective seats in the state Legislative Assemblies, 570 are reserved for the scheduled castes and 532 for the scheduled tribes.

Language
The Constitution provides that the official language of the Union shall be Hindi in the Devanagari script. Hindi is spoken by over 30% of the population. It was originally provided that English should continue to be used for all official purposes until 1965. But the Official Languages Act 1963 provides that, after the expiry of this period of 15 years from the coming into force of the Constitution, English might continue to be used, in addition to Hindi, for all official purposes of the Union for which it was being used immediately before that day, and for the transaction of business in Parliament. According to the Official Languages (Use for official purposes of the Union) Rules 1976, an employee may record in Hindi or in English without being required to furnish a translation thereof in the other language and no employee possessing a working knowledge of Hindi may ask for an English translation of any document in Hindi except in the case of legal or technical documents.

The 58th amendment to the Constitution (26 Nov. 1987) authorized the preparation of a constitution text in Hindi.

The following 22 languages are included in the Eighth Schedule to the Constitution (with 2003 estimate of speakers): Assamese (16·5m.), Bengali (87·6m.), Bodo (1·5m.), Dogri (2·7m.), Gujarati (51·2m.), Hindi (424·7m.), Kannada (41·2m.), Kashmiri (5·0m.), Konkani (2·2m.), Maithili (9·8m.), Malayalam (38·3m.), Manipuri (1·6m.), Marathi (78·7m.), Nepali (2·6m.), Odia (35·3m.), Punjabi (29·4m.), Sanskrit (fewer than 1m.), Santhali (6·6m.), Sindhi (2·7m.), Tamil (66·7m.), Telugu (83·1m.), Urdu (54·7m.). It is estimated that over 850 different languages are spoken throughout the country.

Mohanty, Biswaranjan, *Constitution, Government and Politics in India.* 2009

National Anthem
'Jana-gana-mana' ('Thou art the ruler of the minds of all people'); words and tune by Rabindranath Tagore.

GOVERNMENT CHRONOLOGY

Prime Ministers since 1947. (BJP = Bharatiya Janata Party; BLD = Indian People's Party/Bharatiya Lok Dal; INC = Indian National Congress (a.k.a. Indian Congress Party); INC(i) = Indian National Congress-Indira Gandhi faction; JD = People's Party/Janata Dal; JD(s) = Janata Dal-Chandra Shekhar faction; JP = People's Party/ Janata Dal; UPA = United Progressive Alliance)

1947–64	INC	Jawaharlal Nehru
1964	INC	Gulzarilal Nanda
1964–66	INC	Lal Bahadur Shastri
1966	INC	Gulzarilal Nanda
1966–77	INC	Indira Gandhi
1977–79	JP	Morarji Desai
1979–80	JP/BLD	Charan Singh
1980–84	INC(i)	Indira Gandhi
1984–89	INC(i)	Rajiv Gandhi
1989–90	JD	Vishwanath Pratap Singh
1990–91	JD(s)	Chandra Shekhar
1991–96	INC(i)	Pamulaparti Venkata Narasimha Rao
1996	BJP	Atal Bihari Vajpayee
1996–97	JD	Haradanahalli Dodde Deve Gowda
1997–98	JD	Inder Kumar Gujral
1998–04	BJP	Atal Bihari Vajpayee
2004–	INC, UPA	Manmohan Singh

Presidents of the Union since 1950.

1950–62	Rajendra Prasad
1962–67	Sarvepalli Radhakrishnan
1967–69	Zakir Husain
1969–74	Varahgiri Venkata Giri
1974–77	Fakhruddin Ali Ahmed
1977–82	Neelam Sanjiva Reddy
1982–87	Zail Singh
1987–92	Ramaswamy Iyer Venkataraman
1992–97	Shankar Dayal Sharma
1997–2002	Kocheril Raman Narayanan
2002–07	Avul Pakir Jainulabdeen Abdul Kalam
2007–12	Pratibha Patil
2012–	Pranab Mukherjee

RECENT ELECTIONS

Presidential elections were held on 19 July 2012. Former finance minister Pranab Mukherjee (backed by the ruling United Progressive Alliance) was elected by federal and state legislators, with 713,763 votes (69·3%), against 315,987 (30·7%) for former Speaker of the *Lok Sabha* P. A. Sangma.

Parliamentary elections were held in five phases between 16 April and 13 May 2009. Turnout was an estimated 60%. The Indian National Congress (INC) and its allies were just ten seats short of a majority after they gained 262 seats (217 seats in 2004), with the INC winning 206 seats; the National Democratic Alliance (NDA) gained 159 seats (185 seats in 2004) with the Bharatiya Janata Party (BJP) winning 116 seats; the Third Front won 79 seats, with the Left Front (LF) winning 24; the Fourth Front won 27 seats, with the Samajwadi Party (SP) winning 23; other parties and independents won 16 seats.

Wallace, Paul and Roy, Ramashray, (eds.) *India's 2009 Elections: Coalition Politics, Party Competition and Congress Continuity.* 2011

CURRENT GOVERNMENT

President: Pranab Mukherjee; b. 1935 (sworn in 25 July 2012).

Vice-President: Hamid Ansari.

After the 2004 elections, despite emotional appeals from her supporters, Congress President Sonia Gandhi declined the premiership on 18 May. Manmohan Singh became India's first Sikh prime minister on 22 May.

In March 2013 the INC-led coalition government (known as the United Progressive Alliance) was composed as follows:

Prime Minister and Minister of Personnel, Public Grievances and Pensions, Planning, Atomic Energy and Space: Manmohan Singh; b. 1932 (INC; since 22 May 2004).

Minister of Railways: Pawan Kumar Bansal (INC). *Finance:* Palaniappan Chidambaram (INC). *Agriculture, and Food Processing Industries:* Sharad Pawar (Nationalist Congress Party). *Defence:* A. K. Antony (INC). *Home Affairs:* Sushilkumar Shinde (INC). *External Affairs:* Salman Khurshid (INC). *Science and Technology, and Earth Sciences:* Jaipal Reddy (INC). *Health and Family Welfare:* Ghulam Nabi Azad (INC). *New and Renewable Energy:* Dr Farooq Abdullah (Jammu and Kashmir National Conference). *Petroleum and Natural Gas:* M. Veerappa Moily (INC). *Overseas Indian Affairs:* Vayalar Ravi (INC). *Civil Aviation:* Ajit Singh (Rashtriya Lok Dal). *Labour and Employment:* Mallikarjun Kharge (INC). *Human Resource Development:* Pallam Raju (INC). *Communications and Information Technology:* Kapil Sibal (INC). *Commerce, Industry and Textiles:* Anand Sharma (INC). *Road Transport and Highways:* C. P. Joshi (INC). *Housing and Urban Poverty Alleviation:* Ajay Maken (INC). *Culture:* Chandresh Kumari Katoch (INC). *Shipping:* G. K. Vasan (INC). *Urban Development, and Parliamentary Affairs:* Kamal Nath (INC). *Water Resources:* Harish Rawat (INC). *Social Justice and Empowerment:* Kumari Selja (INC). *Chemicals and Fertilizers:* M. K. Alagiri (Dravida Progressive Federation). *Heavy Industries and Public Enterprises:* Praful Patel (Nationalist Congress Party). *Coal:* Shriprakash Jaiswal (INC). *Law and Justice:* Ashwani Kumar (INC). *Minority Affairs:* Rahman Khan (INC). *Mines:* Dinsha J. Patel (INC). *Tribal Affairs, and Panchayati Raj:* V. Kishore Chandra Deo (INC). *Steel:* Beni Prasad Verma (INC). *Rural Development:* Jairam Ramesh (INC).

Office of the Prime Minister of India: http://www.pmindia.nic.in

CURRENT LEADERS

Pranab Mukherjee

Position
President

Introduction
A veteran politician with experience in previous administrations, Pranab Mukherjee was elected president in July 2012. He has a reputation as a skilled political broker with a track record in securing cross-party co-operation.

Early Life
Pranab Mukherjee was born on 11 Dec. 1935 in the village of Mirati in the Birbhum district of West Bengal. He studied law at Suri Vidyasagar College, affiliated to the University of Calcutta, where he also attained a masters degree in history and politics. He worked in teaching and journalism until 1969, when, with Indira Gandhi's support, he was elected to the upper house of parliament for the Indian National Congress party.

He rose rapidly, serving in a series of ministerial posts, covering industry (1973–74) and finance, revenue and banking (1975–77). He reached cabinet level in 1980 when he was appointed minister of commerce, steel and mines. Named finance minister in 1982,

he resisted IMF pressure to implement austerity measures, instead pursuing a combination of tax reforms and government investment. His policies were claimed as a success when the economy recovered, allowing India to return part of an IMF loan.

Following Indira Gandhi's assassination in 1984, Mukherjee left government to head the West Bengal Pradesh Congress committee. In 1986 he formed the Rashtriya Samajwadi Congress (RSC) but after a lacklustre performance in the 1987 regional assembly elections, the RSC merged with the Congress Party of Rajeev Gandhi in 1989. Two years later Mukherjee became deputy chairman of the planning commission in Narasimha Rao's government, a post he held until 1996. He re-entered cabinet in 1993 as minister of commerce and served as foreign minister from 1995–96.

Following the Congress' electoral defeat in 1996, Mukherjee supported Sonia Gandhi's successful bid for party presidency. He served as general secretary of the All India Congress Committee (AICC) from 1998–99 and as president of the West Bengal Pradesh Congress committee from 2000. He returned to government in 2004 in the Congress-led United Progressive Alliance (UPA), becoming leader of the lower house. He served as defence minister from 2004–06 and as foreign minister from 2006–09.

In 2009 he was appointed finance minister and faced the task of tightening fiscal discipline while maintaining inclusive growth strategies. Reforms proved difficult to implement, largely owing to political deadlock. While he won praise early in his tenure, he was later criticized for lack of progress on tax reforms, cutting subsidies and opening markets to foreign investment. He resigned in June 2012 to stand as the Congress' presidential nominee.

Career in Office
Mukherjee took office in July 2012, having gained 69% of the electoral college vote to defeat his nearest rival, P. A. Sangma. While the role is largely ceremonial, Mukherjee is viewed as well-placed to broker political agreements. Should the 2014 election prove inconclusive, he will be influential in shaping the next government.

Manmohan Singh

Position
Prime Minister

Introduction
After three decades as a civil servant, the former academic and economist was sworn in as India's first Sikh prime minister on 22 May 2004. His appointment followed the general election victory of the Indian National Congress (INC) over the Bharatiya Janata Party (BJP; Indian People's Party) and Sonia Gandhi's unexpected rejection of the top job. A low-profile technocrat and adviser throughout the 1970s and 1980s, Singh came to the fore in 1991 when he was appointed finance minister in the cabinet of P. V. Narasimha Rao. India was in severe financial crisis and Singh was credited with bringing about a recovery, becoming known as the 'architect of India's economic reform'. His coalition won an emphatic victory at the 2009 elections, coming close to securing an absolute majority in parliament.

Early Life
Manmohan Singh was born in Gah, West Punjab (now in Pakistan), on 26 Sept. 1932, the son of a shopkeeper. He was educated at Punjab University in Chandigarh and also attended the universities of Cambridge and Oxford in England on scholarships, winning Cambridge's prestigious Adam Smith Prize in 1956. Returning to India as an economics lecturer, he remained at Punjab University before being made professor in 1963. Three years later he joined UNCTAD (the UN Conference on Trade and Development) at the United Nations Secretariat in New York, as

economic affairs officer. In 1969 Singh returned to India to the University of Delhi as professor of international trade.

Cutting short his academic career in 1971, Singh joined Indira Gandhi's New Congress Party-led government to serve as an economic adviser to the ministry of foreign trade and, from 1972–76, as chief economic adviser in the finance ministry. Stronger ties with the USSR, which influenced Indian economic policy and brought in new aid agreements, marked this period. In 1976 Singh became director of the Reserve Bank of India, a post he held for four years. From 1982–85 he was its governor and then deputy chairman of the Planning Commission from 1985–87, undertaking various assignments at the International Monetary Fund and the Asian Development Bank. He was first selected for the Rajya Sabha (the upper house of parliament) in 1991, representing the Congress.

In 1991, with India in financial crisis, Singh was appointed finance minister in P. V. Narasimha Rao's cabinet. Foreign exchange reserves were nearly exhausted and the country was close to defaulting on its international debt. Singh brought in an ambitious and unprecedented economic reform programme. He slashed red tape, simplified the tax system and ended the 'license Raj' regulations that forced businesses to get government approval for most decisions. He also devalued the rupee, cut subsidies for domestically produced goods, and privatized some state-run companies. Singh spoke of wanting to 'release the innovative, entrepreneurial spirit which was always there in India in such a manner that our economy would grow at a much faster pace, sooner than most people believed.' The programme worked; industry picked up, inflation was checked, and growth rates remained consistently high through the 1990s (his policies were broadly continued by the BJP-led coalition after they were elected in 1996).

Career in Office

When Singh was sworn in as prime minister on 22 May 2004, he took on a healthy economy. GDP growth was at 7%, foreign exchange reserves were comfortable at US$118bn. and inflation stood at just 4%. However, hundreds of millions of Indians were still living in poverty and Singh faced a tough task in bringing about improvements in living standards, while balancing the demands of leftist and communist parties in the coalition. His first address as prime minister called for 'economic reforms with a human face' stressing the need to achieve friendly relations with neighbouring countries, especially Pakistan. Although Singh has a reputation for honesty and even-handedness, there were some questions about his lack of election-winning political experience— he had failed to win a seat in the Lok Sabha (Lower House) elections for South Delhi in 1999.

While India's economy continued to perform strongly, Singh's early premiership had to contend with the devastating effects of a number of severe natural disasters, including the tsunami across the Indian Ocean in Dec. 2004 which hit coastal communities in the south of the country and the Andaman and Nicobar Islands, floods and landslides in Maharashtra in July 2005 and an earthquake in Kashmir in Oct. 2005.

Terrorism has remained a serious problem for the government. Bomb attacks on commuter trains and railway stations in Mumbai in July 2006 killed about 200 people. This atrocity, blamed by India on Pakistan's intelligence services, threatened to undermine previous improvements in the volatile relations between the two nuclear-armed neighbours. A further terrorist bomb attack in Feb. 2007 on a train travelling from New Delhi to Lahore in Pakistan killed about 70 people. Terrorist incidents continued in 2008, particularly in Nov. when suspected Islamic extremists launched a co-ordinated series of attacks on prominent landmarks in India's commercial capital, Mumbai, killing civilians with grenades and machine guns and taking foreign hostages before being overcome by the security forces. The slaughter, in which some 190 people died, led to the resignation of the home affairs minister. India

blamed militants from Pakistan for the atrocity, leading Singh's government to lodge a formal protest. Bilateral talks between the two countries resumed following Singh's positive meeting with his Pakistani counterpart in April 2010 at a summit meeting of the South Asian Association for Regional Co-operation in Bhutan and, despite further terrorist attacks in Mumbai and Delhi in July and Sept. 2011, the two leaders met again in Nov. pledging to open a fresh chapter in bilateral relations. Singh has nevertheless continued to insist that Pakistan do more to dismantle terror organizations operating from its territory.

Although India is not party to the Nuclear Non-Proliferation Treaty, Singh signed a controversial agreement with the USA in March 2006 giving India access to civilian nuclear energy technology in return for having its nuclear sites inspected. The deal was approved by the US Congress in Dec. 2006 but was stalled by political opposition within the Indian parliament. In July 2008 left-wing parties withdrew their parliamentary support for Singh's government, but it survived a no-confidence vote that cleared the way for it to try and finalize the agreement. In the USA the deal was signed into law in Oct. In the same month, Singh signed a security co-operation agreement with Japan.

Following the victory at the general election in April and May 2009 of the United Progressive Alliance (of which the INC won nearly 80% of the vote), Singh became only the second Indian prime minister after Indira Gandhi to hold office for two consecutive terms. Since then, although the economy has continued to prosper, Singh has faced considerable domestic criticism for his handling of a series of high-profile corruption scandals, for the poor organization that plagued the preparations for the Commonwealth Games staged in New Delhi in Oct. 2010, and for controversial retail sector reform plans (which were approved by parliament in Dec. 2012). There have also been protests in Andhra Pradesh both for and against the government's proposal to create a separate state called Telangana, as well as disquiet over a high court ruling that one of the most bitterly contested religious sites in India at Ayodhya should be divided between Muslims and Hindus.

In 2012 the INC suffered several poor results in state elections and its political relations with some of its coalition allies deteriorated, prompting Singh to conduct an extensive cabinet reshuffle in Oct. A prominent promotion was the appointment of Salman Khurshid as foreign minister, only the third Muslim to hold the position in India's history. Meanwhile Pranab Mukherjee, a senior INC figure and former finance minister, was elected in July as the new state president by the federal and state assemblies.

DEFENCE

The Supreme Command of the Armed Forces is vested in the president. As well as armed forces of 1,325,000 personnel in 2006, there are 1,721,000 active paramilitary forces including 208,000 members of the Border Security Force based mainly in the troubled Jammu and Kashmir region. Military service is voluntary but, under the amended constitution, it is regarded as a fundamental duty of every citizen to perform National Service when called upon. Defence expenditure in 2008 was US$30,030m. (US$25 per capita and 2·5% of GDP), a real-term increase of 5% on the previous year. In 2007 defence spending represented 2·5% of GDP. In the period 2007–11 India's spending on major conventional weapons was the highest of any country, at US$12·7bn. In Sept. 2003 India announced that it would be buying 66 Hawk trainer fighter jets; 24 were received directly from BAE Systems, with the last arriving in Nov. 2009, and 42 are being manufactured in India. A further agreement was signed in July 2010 ensuring India would receive an additional 57 aircraft. In Oct. 2003 agreement was reached for India to purchase Israel's sophisticated US$1bn. Phalcon early-warning radar system.

Nuclear Weapons

India's first nuclear test was in 1974. Its most recent tests were a series of five carried out in May 1998. According to the Stockholm International Peace Research Institute, India's nuclear arsenal was estimated to consist of 80–100 nuclear warheads in Jan. 2012. India, known to have a nuclear weapons programme, has not signed the Comprehensive Nuclear-Test-Ban-Treaty, which is intended to bring about a ban on any nuclear explosions. According to *Deadly Arsenals*, published by the Carnegie Endowment for International Peace, India has chemical weapons and has a biological weapons research programme. In 2006 the USA and India announced a civil nuclear co-operation initiative. Under the terms of the deal, India was exempted from a ban on nuclear energy sales that had previously covered non-signatories of the international non-proliferation treaty, of which India is one. In return India agreed to open up 14 of its 22 nuclear installations to international inspections. The agreement was signed in Oct. 2008.

Army

The Army is organized into six commands covering different areas, which in turn are subdivided into sub-areas, plus a training command.

The strength of the Army in 2006 was 1·1m. There are four 'RAPID' divisions, 18 infantry divisions, ten mountain divisions, three armoured divisions and two artillery divisions. Each division consists of several brigades. Officers are trained at the Indian Military Academy, Dehra Dun (Uttarakhand). Army reserves number 300,000 with a further 500,000 personnel available as a second-line reserve force. There is a volunteer Territorial Army of 40,000. There are numerous paramilitary groups including the Ministry of Defence *Rashtriya Rifles* (numbering 57,000), the Indo-Tibetan Border Police (36,300), the State Armed Police (450,000), the Civil Defence (500,000), the Central Industrial Security Force (94,000) and the Ministry of Home Affairs Assam Rifles (63,900). An Army Aviation Corps was established in 1986.

Navy

The Navy has three commands; Eastern (at Visakhapatnam), Western (at Mumbai) and Southern (at Kochi), the latter a training and support command. The fleet is divided into two elements, Eastern and Western; and well-trained, all-volunteer personnel operate a mix of Soviet and western vessels. In May 2003 India held joint naval exercises with Russia in the Arabian Sea for the first time since the collapse of the Soviet Union.

The principal ship is the light aircraft carrier, *Viraat*, formerly HMS *Hermes*, of 29,000 tonnes, completed in 1959 and transferred to the Indian Navy in 1987 after seeing service in the Falklands War. In 2003 India began construction of another aircraft carrier and began negotiations to purchase a third from the Russian navy. The fleet includes 12 Soviet-built diesel submarines and four new German-designed submarines. There are also 25 destroyers and frigates. The Naval Air force was 7,000-strong in 2006; equipment includes 34 combat aircraft. Main bases are at Mumbai (main dockyard), Goa, Visakhapatnam and Kolkata on the sub-continent and Port Blair in the Andaman Islands.

Naval personnel in 2006 numbered 55,000 including 5,000 Naval Air Arm and 1,200 marines.

Air Force

Units of the IAF are organized into five operational commands— Central at Allahabad, Eastern at Shillong, Southern at Thiruvananthapuram, South-Western at Gandhinagar and Western at Delhi. There is also a training command and a maintenance command. The air force has 170,000 personnel.

Equipment includes more than 850 combat aircraft. Major combat types include Su-30s, MiG-21s, MiG-23s, MiG-27s, MiG-29s, *Jaguars* and Mirage 2000s. Air Force reserves numbered 140,000 in 2006.

ECONOMY

Agriculture accounted for 17·6% of GDP in 2008 (down from 55% in 1950), industry 28·2% (up from 15% in 1950) and services 54·2% (up from 30% in 1950).

Since the late 1990s a divide has become increasingly pronounced between the south and west, where a modern economy is booming in cities such as Bangalore, Hyderabad and Chennai, and the poorer and politically volatile areas in the north and east.

Overview

After independence a broad but lacklustre industrial base was built behind a tariff wall. In 1991 a balance of payments and foreign currency reserve crisis prompted market reforms that have helped India to become one of the world's fastest-growing economies. A 2007 report by Goldman Sachs suggested that if current growth patterns continued, India could be the world's second largest economy by 2050, behind China and ahead of the USA.

Industrial licensing (determining how much entrepreneurs could manufacture) was abolished and trade barriers lowered after India joined the World Trade Organization in 1995. Foreign direct investment rose from almost nothing to US$8bn. in 2005 and over US$31bn. in 2011.

The economy achieved annual growth of over 8% on average between 2004 and 2008, while poverty fell from 37·2% in 2004–05 to 29·8% in 2009–10. In 2008 and 2009 the economy experienced a decline from previous high rates of growth as a result of global economic cooling. At the request of the government, the World Bank approved four loans worth US$4·3bn. in Sept. 2009. The economy began to recover in mid-2009, led by domestic demand and, in particular, investment in infrastructure. Inflation rose to 11% in the first half of 2010 and, although it has since fallen back to single digits, remains a concern.

Liberalization has contributed to increased foreign participation and growth in durable consumer goods, including cars, scooters, electronics, computer systems and white goods. The service sector has become the economy's most dynamic, with India a world leader in telecommunications, IT and pharmaceuticals. It is also an international centre for outsourcing services, particularly for the USA. Half of the nation's workforce is employed in agriculture, which accounts for 16% of total output. Sustained poverty reduction is unlikely until agricultural productivity is raised and the rural infrastructure developed.

Human development indicators are extremely low, especially in rural areas. 37·2% of the population, about 410m. people, live below the poverty line. They account for one-third of the world's poor. Programmes to improve living conditions are under way, using resources generated from recent growth.

The government's fiscal deficit accounted for 9·2% of GDP in 2010, while public debt stands at around 55%. A dip in growth in the final quarter of 2011 to just below 7% led to concerns over India's ability to sustain its rapid growth in the long term. Growth fell to 5·3% in the second quarter of 2012, its lowest rate in seven years. The rupee also experienced the greatest depreciation among major Asian currencies in 2011, partly owing to India's high current account deficit. As well as the global economic slowdown, years of government failure to enact major structural reforms have contributed to the economic cooling, leading to a loss in investor confidence. Reducing the budget deficit, confronting corruption and improving the quality of transport, health and education services are crucial to long-term growth.

Currency

The unit of currency is the *Indian rupee* (INR) of 100 *paise*. Since July 2010 the Indian rupee has been represented by the symbol ₹. Foreign exchange reserves were US$261,247m. in Aug. 2009 and gold reserves 11·50m. troy oz. Inflation rates (based on IMF statistics):

2002	2003	2004	2005	2006	2007	2008	2009	2010	2011
4·5%	3·7%	3·9%	4·0%	6·3%	6·4%	8·3%	10·9%	12·0%	8·9%

India's 2010 inflation rate was the highest since 1998. The official exchange rate was abolished on 1 March 1993; the rupee now has a single market exchange rate and is convertible. The pound sterling is the currency of intervention. Total money supply in July 2009 was ₹12,133·0bn.

Budget

Central government revenues (provisional) in 2009–10 (fiscal year beginning 1 April 2009) totalled ₹7,392,100m.; expenditures (provisional) totalled ₹10,683,400m.

Principal sources of revenue in 2007–08 were: taxes on income, profits and capital gains, ₹3,118,700m.; taxes on goods and services, ₹1,767,500m.; taxes on international trade and transactions, ₹1,041,200m. Main items of expenditure by economic type in 2007–08 were: grants, ₹2,160,300m.; subsidies, ₹1,705,200m.; interest, ₹1,659,500m.

VAT was introduced on 1 April 2005, at 12·5% (reduced rates, 4% and 1%).

Performance

India has one of the fastest-growing economies in Asia. Real GDP growth rates (based on IMF statistics):

2003	2004	2005	2006	2007	2008	2009	2010	2011
6·9%	7·6%	9·0%	9·5%	10·0%	6·9%	5·9%	10·1%	6·8%

Recent years have seen a growing disparity between the performance of India's richest states, mainly in the south and the west, and the poorest states, generally in the east and the north.

Total GDP in 2011 was US$1,848·0bn.

Banking and Finance

The Reserve Bank, the central bank for India, was established in 1934 and started functioning on 1 April 1935 as a shareholder's bank; it became a nationalized institution on 1 Jan. 1949. It has the sole right of issuing currency notes. The *Governor* is Duvvuri Subbarao. The Bank acts as adviser to the government on financial problems and is the banker for central and state governments, commercial banks and some other financial institutions. It manages the rupee public debt of central and state governments and is the custodian of the country's exchange reserve. The Bank has extensive powers of regulation of the banking system, directly under the Banking Regulation Act, 1949, and indirectly by the use of variations in Bank rate, variation in reserve ratios, selective credit controls and open market operations.

Scheduled commercial banks are categorized in five different groups according to their ownership and/or nature of operation: the State Bank of India and its six associates; 19 nationalized banks; regional rural banks; foreign banks; and other scheduled commercial banks (in the private sector). Total deposits in commercial banks, March 2007, stood at ₹26,970,000m. The State Bank of India acts as the agent of the Reserve Bank for transacting government business as well as undertaking commercial functions. In 2011 India received US$31·6bn. worth of foreign direct investment. FDI inflows in 2010 were more than six times the level of 2000.

External debt totalled US$290,282m. in 2010 (US$120,222m. in 2005), equivalent to 16·9% of GNI.

There are stock exchanges in Ahmedabad, Chennai, Delhi, Kolkata, Mumbai and 18 other centres.

ENERGY AND NATURAL RESOURCES

Environment

India's carbon dioxide emissions from the consumption and flaring of fossil fuels in 2008 accounted for 4·9% of the world total

(the fourth highest after China, the USA and Russia). However, this was equivalent to just 1·3 tonnes per capita, well below the global average and the lowest figure for any major industrial country. An *Environmental Performance Index* compiled in 2008 ranked India 120th in the world out of 149 countries analysed, with 60·3%. The index examined various factors in six areas—air pollution, biodiversity and habitat, climate change, environmental health, productive natural resources and water resources.

Electricity

Installed capacity in 2007–08 was 168m. kW. In 2002 nearly 520,000 villages out of 600,000 had electricity. Production of electricity in 2005–06 was 623·8bn. kWh, of which 81·1% came from thermal stations, 16·3% from hydro-electric stations and 2·8% from nuclear stations. In 2010 there were 19 nuclear reactors in use. An additional four reactors were under construction. Electricity consumption per capita in 2009–10 was 779 kWh. Electricity demand exceeds supply, making power surges and cuts frequent. According to the 2011 census 67% of households had electricity, up from 56% in 2001; 93% of urban houdeholds had electricity in 2011 compared to 55% of rural households.

Oil and Gas

Oil and Natural Gas Corporation Ltd and Oil India Ltd are the only producers of crude oil. Production 2008, 36·1m. tonnes. The main fields are in Assam and Gujarat and offshore in the Gulf of Cambay (the Mumbai High field). India imports 70% of its annual oil requirement. There were proven reserves of 5·8bn. bbls in 2008. Oil refinery capacity, 2008, was 3·0m. bbls daily. Natural gas production in 2008 was 30·6bn. cu. metres with 1,090bn. cu. metres of proven reserves.

Wind

India is one of the world's largest wind-power producers, with an installed capacity of 13,065 MW in 2010 (the fifth highest after China, the USA, Germany and Spain).

Water

By 2005–06, 82·62m. ha. of irrigation potential had been created of which 60·19m. ha. was utilized. Irrigation projects have formed an important part of all the Five-Year Plans. The possibilities of diverting rivers into canals being nearly exhausted, the emphasis is now on damming the monsoon surplus flow and diverting that. Ultimate potential of irrigation is assessed at 110m. ha. by 2025, total cultivated land being 185m. ha.

A Ganges water-sharing accord was signed with Bangladesh in 1997, ending a 25-year dispute which had hindered and dominated relations between the two countries.

Minerals

The coal industry was nationalized in 1973. Production, 2007–08, 457m. tonnes; recoverable reserves were estimated at 56bn. tonnes (2005). Production of other minerals (in 1,000 tonnes): iron ore (2005), 140,000; lignite (2007–08), 33,980; salt (2008), 16,000; bauxite (2005), 11,957; chromite (2008), 3,900; manganese ore (2008), 2,400; aluminium (2005), 898; silver (2005), 31,900 kg; gold (2005), 3,200 kg. Other important minerals are lead, zinc, limestone, apatite and phosphorite, dolomite, magnesite and uranium. Value of mineral production (excluding atomic minerals), 2008–09, ₹1,732,485·3m. (2007–08, ₹1,216,850·0m.).

Agriculture

About 60% of the people are dependent on the land for their living. The farming year runs from July to June through three crop seasons: kharif (monsoon), rabi (winter) and summer. In 2002 there were 161,715,000 ha. of arable land and 8,400,000 ha. of permanent cropland. 57,198,000 ha. were irrigated in 2002. There were 1,525,000 tractors and 4,200 harvester-threshers in 2002. The average size of holdings for the whole of India is estimated at 1·4 ha.

Agricultural production, 2003 (in 1,000 tonnes): sugarcane, 289,630; rice, 132,013; wheat, 65,129; potatoes, 23,161; bananas, 16,450; maize, 14,800; millet, 10,700; mangoes, 10,500; coconuts, 9,500; aubergines, 8,200; sorghum, 8,000; groundnuts, 7,500; tomatoes, 7,420; cassava, 7,100; soybeans, 6,800; seed cotton, 6,300; cabbage, 6,100; onions, 5,000; cauliflowers, 4,800; cotton-seed, 4,199; chick-peas, 4,130; rapeseed, 3,842. Jute is grown in West Bengal (70% of total yield), Bihar and Assam: total yield, 1,976,000 tonnes. The coffee industry is growing: the main cash varieties are Arabica and Robusta (main growing areas Karnataka, Kerala and Tamil Nadu). India is the world's leading producer of a number of agricultural crops, including mangoes, millet, bananas and chick-peas.

The tea industry is important, with production concentrated in Assam, West Bengal, Tamil Nadu and Kerala. India is the world's largest tea producer. The 2003 crop was 885,000 tonnes; exports in 2001–02, 180,100 tonnes, valued at US$360m.

Livestock (2003): cattle, 185·2m.; goats, 124·4m.; buffaloes, 97·9m.; sheep, 61·5m.; pigs, 13·5m.; horses and ponies, 751,000; donkeys, 650,000; camels, 632,000; poultry, 489m. There are more cattle and buffaloes in India than in any other country.

Fertilizer use in 2004–05 was 18·4m. tonnes.

Opium
By international agreement the poppy is cultivated under licence, and all raw opium is sold to the central government. Opium, other than for wholly medical use, is available only to registered addicts.

Forestry
The lands under the control of the state forest departments are classified as 'reserved forests' (forests intended to be permanently maintained for the supply of timber, etc., or for the protection of water supply, etc.), 'protected forests' and 'unclassed' forest land. In 2010 the total forest area was 68·43m. ha. (23% of the land area). Main types are teak and sal. About 16% of the area is inaccessible, of which about 45% is potentially productive. In 2007, 330·21m. cu. metres of roundwood were produced, making India the second largest producer after the USA (9·2% of the world total in 2007).

Fisheries
Total catch in 2008–09 was 7·62m. tonnes (2·98m. tonnes marine, 4·64m. tonnes inland). Fishing provides a livelihood for over 14m. people, contributing about 1% of total GDP and 4·6% of the GDP from the agriculture sector. 541,701 tonnes of fish were exported in 2007–08.

INDUSTRY

The leading companies by market capitalization in India in March 2012 were: Reliance Industries, a chemical production company (US$48·2bn.); Oil and Natural Gas Corporation Ltd (ONGC), US$45·1bn.; and Tata Consultancy Services, a software and computer services company (US$44·9bn.).

The information technology industry has become increasingly important; its contribution to GDP rose from 1·2% in 1998–99 to 5·2% by 2007–08. The National Association of Software and Services Companies (NASSCOM) estimated that in 2007–08 the IT industry registered a growth rate of 28% and revenues of US$52bn. (up from US$40bn. in 2006–07).

There is expansion in petrochemicals, based on the oil and associated gas of the Mumbai High field, and gas reserves offshore in the Krishna Godavari basin and the Bassein field, and onshore in Andhra Pradesh, Assam and Gujarat.

In 2006–07 there were an estimated 12·8m. micro and small enterprises, accounting for about 39% of the gross value of output in the manufacturing sector.

Industrial production (2006–07, in 1,000 tonnes): cement, 154,746; distillate fuel oil (2007–08), 59,032; finished steel, 50,196;

sugar, 24,187; nitrogenous and phosphate fertilizers, 16,153; residual fuel oil (2007–08), 15,804; petrol (2005–06), 10,502; jet fuel (2007–08), 9,107; kerosene (2005–06), 9,078; sulphuric acid, 7,156; paper and paperboard, 6,129; pig iron, 4,550; caustic soda, 1,929; jute goods (2003–04), 1,424. Other products (2006–07): 8,436,186 motorcycles, mopeds and scooters; 3,171,000 diesel engines; 1,238,737 cars; 520,000 commercial vehicles; 85·8bn. cigarettes.

Labour
In 2007 the workforce numbered 516·4m. (compared to 402·5m. at the 2001 census) and the unemployment rate was estimated at 7·2%. Workdays lost by industrial disputes through strikes and lockouts, 2005, 23·27m.

Companies
The total number of companies limited by shares at work as on 31 March 2002 was 589,246; estimated paid-up capital was ₹3,870,239m. Of these, 76,279 were public limited companies with an estimated paid-up capital of ₹2,587,149m., and 512,967 private limited companies (₹1,283,090m.).

During 2001–02 there were 21,059 new limited companies registered in the Indian Union under the Companies Act 1956 with a total authorized capital of ₹53,156m.; 14 were government companies (₹5,781m.). There were 479 companies with unlimited liability and 3,007 companies with liability limited by guarantee and association not for profit also registered in 2001–02. During 2001–02, 760 non-government companies with an aggregate paid-up capital of ₹144·7m. went into liquidation or were struck off the register.

On 31 March 2002 there were 1,261 government companies at work with a total paid-up capital of ₹1,099,155m.; 658 were public limited companies and 603 were private limited companies. There were 587,985 non-government companies at work on 31 March 2002. Of these 75,621 were public limited companies and 512,364 were private limited companies.

On 31 March 2002, 1,285 companies incorporated elsewhere were reported to have a place of business in India; 241 were of UK and 286 of US origin.

Co-operative Movement
In 2001–02 there were 390,080 co-operative societies (146,206 credit, 243,874 non-credit) with a total membership of 209·3m. These included Primary Co-operative Marketing Societies, State Co-operative Marketing Federations and the National Agricultural Co-operative Marketing Federation of India. There were also State Co-operative Commodity Marketing Federations and Special Commodities Marketing Federations.

There were, in 2001–02, 368 Central Co-operative Banks, 94,825 Primary Agricultural Credit Societies, 770 Primary Land Development Banks and 1,047 banks and societies which provide long-term credits. There are 31 State Co-operative Banks.

INTERNATIONAL TRADE

Foreign investment is encouraged by a tax holiday on income up to 6% of capital employed for five years. There are special depreciation allowances, and customs and excise concessions, for export industries. Proposals for investment ventures involving up to 51% foreign equity require only the Reserve Bank's approval under new liberalized policy. In Feb. 1991 India resumed trans-frontier trade with China, which had ceased in 1962.

Imports and Exports
The external trade of India (excluding land-borne trade with Tibet and Bhutan) was as follows (in ₹100,000):

	Imports	Exports and Re-exports
2003–04	35,910,766	29,336,675
2004–05	50,106,454	37,533,953
2005–06	66,040,890	45,641,786
2006–07	84,050,631	57,177,929

The main trading partners were as follows in the year ended 31 March 2007 (in ₹100,000):

Countries	Value of Imports
Australia	3,171,090
Belgium	1,874,160
China	7,900,861
France	1,905,933
Germany	3,414,675
Indonesia	1,886,486
Japan	2,079,488
Korea (Republic of)	2,174,700
Malaysia	2,395,876
Saudi Arabia	6,056,150
Singapore	2,483,997
Switzerland	4,128,317
UAE	3,917,494
UK	1,888,930
USA	5,310,541

Countries	Value of Exports
Belgium	1,572,170
China	3,752,978
Germany	1,800,723
Hong Kong	2,117,938
Italy	1,621,243
Japan	1,295,361
Korea (Republic of)	1,137,901
Netherlands	1,208,248
Saudi Arabia	1,171,137
Singapore	2,746,161
South Africa	1,016,528
Sri Lanka	1,020,638
UAE	5,444,497
UK	2,542,129
USA	8,536,849

In 2006–07 the main import suppliers (percentage of total trade) were: China, 9·4%; Saudi Arabia, 7·2%; USA, 6·3%; Switzerland, 4·9%; United Arab Emirates, 4·7%. Main export markets in 2006–07 were: USA, 14·9%; United Arab Emirates, 9·5%; China, 6·6%; Singapore, 4·8%; UK, 4·4%.

The value (in ₹100,000) of the leading articles of merchandise was as follows in the year ended 31 March 2007:

Imports	Value
Artificial resins, plastic materials, etc.	1,169,600
Coal, coke and briquettes, etc.	2,071,000
Electronic goods	7,227,500
Fertilizers (crude and manufactured)	1,373,200
Gold and silver	7,154,000
Iron and steel	2,907,100
Machinery (electrical and non-electrical)	7,154,000
Metalliferous ores and metal scrap	3,776,400
Non-ferrous metal	1,178,700
Organic and inorganic chemicals	3,543,300
Pearls, precious and semi-precious stones	3,388,100
Petroleum, crude oil and related products	25,857,200
Professional instruments, optical equipment, etc.	1,059,300
Transport equipment	4,270,900
Vegetable oils (fixed)	954,000

Exports	Value
Basic chemicals	4,958,800
Cotton yarn, fabrics and finished articles	1,908,900
Electronic goods	1,291,400
Engineering goods	11,987,500
Gems and jewellery	7,229,500
Iron ore	1,765,600
Leather including garments and goods	1,327,800
Man-made yarn, fabrics and finished articles	997,500
Marine products	800,100
Mica, coal and other ores and minerals including processed minerals	1,403,000
Oil meals	550,400
Petroleum products	8,452,000

Exports	Value
Plastics and linoleum	1,471,800
Ready-made garments, including clothing accessories of all textile materials	4,023,700
Rice	703,600

Technology industries have become increasingly important in recent years; the software and services exports sector grew by 29% in 2007–08 to register revenues of US$40·4bn. (up from US$31·4bn. in 2006–07).

COMMUNICATIONS

Roads

In 2006 there were 3·32m. km of roads, including 200 km of motorway. Roads are divided into six main administrative classes, namely: national highways, state highways, other public works department (PWD) roads, *Panchayati Raj* roads, urban roads and project roads. The national highways (198,489 km in 2006) connect capitals of states, major ports and foreign highways. The national highway system is linked with the UN Economic and Social Commission for Asia and the Pacific international highway system. The state highways are the main trunk roads of the states, while the other PWD roads and *Panchayati Raj* roads connect subsidiary areas of production and markets with distribution centres, and form the main link between headquarters and neighbouring districts. A ten-year highway plan, the National Highways Development Project, is currently under way; it aims to link India's main cities, ports and regions through the construction of new highways and the widening of existing ones by 2015.

In 2006 there were 11,526,000 passenger cars, 64,743,000 motorcycles and scooters, 992,000 buses and coaches, and 4,436,000 lorries and vans. In 2007 there were 476,219 road accidents resulting in 114,444 deaths.

Rail

The Indian railway system is government-owned (under the control of the Railway Board). Following reconstruction there are 16 zones, seven of which were created in 2002:

Zone	Headquarters	Year of Creation
Central	Mumbai	1951
Southern	Chennai	1951
Western	Mumbai	1951
Eastern	Kolkata	1952
Northern	Delhi	1952
North Eastern	Gorakhpur	1952
South Eastern	Kolkata	1955
North East Frontier	Guwahati	1958
South Central	Secunderabad	1966
East Central	Hajipur	2002
East Coast	Bhubaneswar	2002
North Central	Allahabad	2002
North Western	Jaipur	2002
South East Central	Bilaspur	2002
South Western	Hubli	2002
West Central	Jabalpur	2002

The total length of the Indian railway network was 64,460 km in March 2011 (19,607 electrified), with the Northern zone having the longest network, at 6,968 km.

The Konkan Railway (760 km of 1,676 mm gauge) linking Roha and Mangalore opened in 1996. It is operated as a separate entity.

Principal gauges are 1,676 mm (55,188 km) and 1 metre (6,809 km), with networks also of 762 mm and 610 mm gauge (2,463 km).

Passenger-km travelled in 2009–10 came to 903·5bn. and revenue earning freight tonne-km to 600·5bn. Revenues (2009–10) ₹871,047m.; expenses, ₹829,154m.

There are metros in Bangalore (6·7 km), Chennai (19·7 km), Delhi (189·6 km) and Kolkata (25·5 km). An extensive metro system is being constructed in Mumbai with the first line expected to open in Sept. 2013.

Civil Aviation

The main international airports are at Chennai, Delhi (Indira Gandhi), Kolkata, Mumbai and Thiruvananthapuram, with some international flights from Ahmedabad, Amritsar, Bangalore, Calicut, Goa and Hyderabad. Air transport was nationalized in 1953 with the formation of two Air Corporations: Air India for long-distance international air services, and Indian Airlines for air services within India and to adjacent countries. Indian (as Indian Airlines became in 2005) merged into Air India in Feb. 2011. Both domestic and international air transport have been opened to private companies, the largest of which is Jet Airways. Two leading budget airlines, IndiGo and SpiceJet, now operate international as well as domestic services. All operational airports handled a total of 116·9m. passengers (87·1m. domestic and 29·8m. international) in the year to 31 March 2008. Total aircraft movements reached 1·31m. and freight volumes increased to over 1·7m. tonnes.

In 2003 Air India operated routes to Africa (Dar es Salaam and Nairobi); to Mauritius; to Europe (Frankfurt, London, Moscow, Paris, Vienna and Zürich); to western Asia (Abu Dhabi, Al Ain, Bahrain, Damman, Doha, Dubai, Jeddah, Kuwait, Muscat and Riyadh); to east Asia (Bangkok, Hong Kong, Jakarta, Kuala Lumpur, Osaka, Seoul, Singapore and Tokyo); and to North America (Chicago and New York). Indian Airlines (subsequently renamed Indian) operated international flights in 2003 to Almaty, Bahrain, Bangkok, Bishkek, Colombo, Dhaka, Doha, Dubai, Fujairah, Kathmandu, Kuala Lumpur, Kuwait, Malé, Muscat, Rangoon (Yangon), Ras-al-Khaimah, Sharjah and Singapore. Flights from Delhi to Lahore were restored in Jan. 2004.

In 2007 Mumbai was the busiest airport, handling 25·2m. passengers, followed by Delhi, with 23·3m. passengers. They were ranked the world's 55th and 61st busiest airports respectively for the year 2006. Both airports were privatized in 2006, with extensive modernization.

Shipping

In Jan. 2009 there were 625 ships of 300 GT or over registered, totalling 8·57m. GT. Of the 625 vessels registered, 301 were general cargo ships, 117 oil tankers, 108 bulk carriers, 51 passenger ships, 22 liquid gas tankers, 16 container ships and ten chemical tankers. The Indian-controlled fleet comprised 403 vessels of 1,000 GT or over in Jan. 2009, of which 347 were under the Indian flag and 56 under foreign flags.

Cargo traffic of major ports for fiscal years, in 1,000 tonnes:

Port	2005–06	2006–07	2007–08
Chennai	47,248	53,414	57,154
Cochin	13,887	15,257	15,810
Ennore	9,168	10,714	11,563
Jawaharlal Nehru	37,836	44,815	55,756
Kandla	45,907	52,982	64,893
Kolkata and Haldia	53,143	55,050	57,282
Mormugao	31,688	34,241	35,128
Mumbai	44,190	52,364	57,039
New Mangalore	34,451	32,042	36,019
Paradip	33,109	38,517	42,438
Tuticorin	17,139	18,001	21,480
Visakhapatnam	55,801	56,385	64,597

The busiest container port is Jawaharlal Nehru, which handled 3·9m. 20-ft equivalent units (TEUs) in 2008.

There are about 3,700 km of major rivers navigable by motorized craft, of which 2,000 km are used. Canals, 4,300 km, of which 900 km are navigable by motorized craft.

Telecommunications

The telephone system is in the hands of the Telecommunications Department, except in Delhi and Mumbai, which are served by a public corporation. In 2008 there were 37·9m. main (fixed) telephone lines. In the same year mobile phone subscribers numbered 346·9m. (293·6 per 1,000 persons). The number of mobile phone subscribers more than doubled between 2006 and 2008, while the number of fixed line subscribers has been gradually falling since 2003. India's largest mobile phone operator is Bharti Airtel, with a 30·3% market share in the third quarter of 2012, ahead of Vodafone and Idea Cellular. There were an estimated 51·8m. internet users in 2008. In March 2012 there were 45·0m. Facebook users.

SOCIAL INSTITUTIONS

Justice

All courts form a single hierarchy, with the Supreme Court at the head, which constitutes the highest court of appeal. Immediately below it are the High Courts and subordinate courts in each state. Every court in this chain administers the whole law of the country, whether made by Parliament or by the state legislatures.

The states of Andhra Pradesh, Assam (in common with Nagaland, Meghalaya, Manipur, Mizoram, Tripura and Arunachal Pradesh), Bihar, Gujarat, Himachal Pradesh, Jammu and Kashmir, Karnataka, Kerala, Madhya Pradesh, Maharashtra (in common with Goa and the Union Territories of Daman and Diu, and Dadra and Nagar Haveli), Odisha, Punjab (in common with the state of Haryana and the Union Territory of Chandigarh), Rajasthan, Tamil Nadu (in common with the Union Territory of Puducherry), Uttar Pradesh, West Bengal and Sikkim each have a High Court. There is a separate High Court for Delhi. For the Andaman and Nicobar Islands the Calcutta High Court, for Puducherry the High Court of Madras and for Lakshadweep the High Court of Kerala are the highest judicial authorities. The Allahabad High Court has a Bench at Lucknow, the Bombay High Court has Benches at Nagpur, Aurangabad and Panaji, the Gauhati High Court has Benches at Kohima, Aizwal, Imphal and Agartala, the Madhya Pradesh High Court has Benches at Gwalior and Indore, the Patna High Court has a Bench at Ranchi and the Rajasthan High Court has a Bench at Jaipur. Judges and Division Courts of the Guwahati High Court also sit in Meghalaya. Similarly, judges and Division Courts of the Calcutta High Court also sit in the Andaman and Nicobar Islands. High Courts have also been established in the new states of Chhattisgarh, Jharkhand and Uttarakhand. Below the High Court each state is divided into a number of districts under the jurisdiction of district judges who preside over civil courts and courts of sessions. There are a number of judicial authorities subordinate to the district civil courts. On the criminal side magistrates of various classes act under the overall supervision of the High Court.

In Oct. 1991 the Supreme Court upheld capital punishment by hanging. In Nov. 2012 India carried out its first execution since 2004. A second execution in the space of three months followed in Feb. 2013.

The population in penal institutions in Dec. 2006 was 373,271 (33 per 100,000 of national population).

Police

The states control their own police forces. The Home Affairs Minister of the central government co-ordinates the work of the states. The Indian Police Service provides senior officers for the state police forces. The Central Bureau of Investigation functions under the control of the Cabinet Secretariat.

The cities of Ahmedabad, Bangalore, Chennai, Delhi, Hyderabad, Kolkata, Mumbai, Nagpur and Pune have separate police commissionerates.

Education

Adult literacy was 62·8% in 2006 (75·2% among males and 50·8% among females). Of the states and territories, Kerala and Mizoram have the highest rates.

Educational Organization. Education is the concurrent responsibility of state and Union governments. In the Union Territories it is the responsibility of the central government. The Union government is also directly responsible for the central universities and all institutions declared by parliament to be of national importance; the promotion of Hindi as the federal language and co-ordinating and maintaining standards in higher education, research, science and technology. Professional education rests with the Ministry or Department concerned. There is a Central Advisory Board of Education to advise the Union and the State governments on any educational question which may be referred to it.

School Education. The school system has four stages: primary, middle, secondary and senior secondary.

Primary education is imparted either at independent primary (or junior basic) schools or primary classes attached to middle or secondary schools. The period of instruction varies from four to five years and the medium of instruction is in most cases the mother tongue of the child or the regional language. Free primary education is available for all children. Legislation for compulsory education has been passed by some state governments and Union Territories but it is not practicable to enforce compulsion when the reasons for non-attendance are socio-economic. There are residential schools for country children. The period for the middle stage varies from two to three years. In 2005, 47·2% of children who enrolled in the first grade went on to finish the eighth grade. In the same year it was estimated that 42m. children aged 6–14 were not attending school. In Aug. 2009 legislation was passed making education free and compulsory for all children between the ages of six and 14.

School statistics for 2004–05:

Type of recognized institution	No. of institutions	No. of students on rolls	No. of teachers[1]
Primary/junior basic schools	767,520	130,800,000	2,161,000
Middle/senior basic schools	274,731	51,200,000	1,589,000
High/higher secondary schools	152,049	37,100,000	2,083,000

[1]Provisional.

Higher Education. Higher education is given in arts, science or professional colleges, universities and all-India educational or research institutions. The majority of universities act as affiliating bodies for colleges; they are responsible for course content, examinations and awarding degrees although the teaching is conducted at the college. In 2005–06 there were 19,753 higher education institutions, comprising: 350 universities, institutions deemed to be universities and institutions of national importance; 12,751 general education colleges (including arts, science and commerce colleges); 5,179 professional education colleges (including engineering, technology, architecture, medical and teacher training colleges); and 1,473 other institutions (including research institutions). Total enrolment at universities, 2004–05, 10,481,000, of which 4,234,000 were women.

Adult Education. The Directorate of Adult Education, established in 1971, is the national resource centre.

There is also a National Literacy Mission.

Expenditure. Total budgeted central expenditure on revenue account of education and other departments for 2004–05 was estimated at ₹96,694m. Total public expenditure on education during the Tenth (2002–07) Plan, ₹438,250m. (₹138,250m. on secondary and higher education, ₹300,000m. on elementary education and literacy). In 2004–05 total expenditure on education came to 3·4% of GDP and 12·1% of total government spending.

Health

Medical services are primarily the responsibility of the states. The Union government has sponsored major schemes for disease prevention and control which are implemented nationally.

Total public expenditure on health and family welfare during the Tenth (2002–07) Plan, ₹373,530m. In 2002 there were 15,741 hospitals and 607,100 doctors. In 2002 there were 15 beds per 10,000 inhabitants.

In the period 2001–03, 20% of the population were undernourished. In 1979, 38% of the population had been undernourished. In 2004–05, 27·5% of the population lived below the poverty line, compared to 38·9% in 1987–88.

Approximately 2·5m. Indians are HIV-infected, a number only exceeded in South Africa and Nigeria.

RELIGION

India is a secular state; any worship is permitted, but the state itself has no religion. The principal religions in 2001 were: Hindus, 828m. (80% of the population); Muslims, 138m. (13%); Christians, 24m.; Sikhs, 19m.; Buddhists, 8m.; Jains, 4m. In addition to having the largest Hindu population of any country, India has the third highest number of Muslims, after Indonesia and Pakistan. In Feb. 2013 the Roman Catholic church had seven cardinals.

CULTURE

World Heritage Sites

There are 29 sites under Indian jurisdiction that appear on the UNESCO World Heritage List. They are (with year entered on list): Ajanta Caves (1983), Ellora Caves (1983), Agra Fort (1983), Taj Mahal (1983), Sun Temple, Konârak (1984), Monuments at Mahabalipuram (1984), Kaziranga National Park (1985), Manas Wildlife Sanctuary (1985), Keoladeo National Park (1985), Churches and Convents of Goa (1986), Monuments at Khajuraho (1986), Monuments at Hampi (1986), Fatehpur Sikri (1986), Monuments at Pattadakal (1987), Elephanta Caves (1987), Great Living Chola Temples (1987 and 2004), Sundarbans National Park (1987), Nanda Devi National Park (1988 and 2005), Buddhist Monuments at Sanchi (1989), Humayan's Tomb, Delhi (1993), Qutb Minar and its Monuments, Delhi (1993), Mountain Railways of India (1999, 2005 and 2008), Mahabodhi Temple Complex at Bodh Gaya (2002), the Rock Shelters of Bhimbetka (2003), Champaner-Pavagadh Archaeological Park (2004), the Chhatrapati Shivaji Terminus—formerly Victoria Terminus—in Mumbai (2004), the Red Fort Complex, Delhi (2007); Jantar Mantar, Jaipur (2010); and the Western Ghats mountain chain (2012).

Press

There were 64,998 registered newspapers in March 2007 (up from 60,413 in March 2005 and 62,483 in March 2006), with a total circulation of 192·1m. In 2007 there were 2,337 dailies with a total circulation of 98·84m. (up from 73·54m. in 2004). India's circulation of paid-for dailies is reported to have overtaken that of China to become the highest of any country in 2008. Hindi papers have the highest number and circulation, followed by Marathi, English and Mayalam. The newspaper with the highest circulation is the *Times of India* (daily average of 3·4m. copies in 2008). The Hindi-language paper with the highest circulation is *Dainik Jagran*, with a daily average of 2·4m. copies in 2008 (although its readership is much larger than that of the *Times of India*).

Tourism

In 2007 there were 5,082,000 non-resident overnight tourists (up from 2,726,000 in 2003). Of these, 1,900,000 were from Europe, 1,050,000 from the Americas and 982,000 from elsewhere in Southern Asia. Tourist receipts amounted to US$10·7bn. in 2007 (excluding passenger transport).

Calendar

The Indian National Calendar, adopted in 1957, is dated from the Saka era (Indian dynasty beginning AD 78). It uses the same year-length as the Gregorian calendar (also used for administrative and informal purposes) but begins on 22 March. Local and religious variations are also used.

Festivals

Independence Day is celebrated on 15 Aug. The Hindu festival of Diwali or Deepavali ('festival of lights') is held over five days between mid-Oct. and mid-Nov. Clay lamps are lit to symbolize the triumph of good over evil and firework displays take place. The biggest rock festival is the Independence Rock Festival (Oct.) in Mumbai. The International Film Festival of India takes place in Nov.–Dec. in Goa.

DIPLOMATIC REPRESENTATIVES

Of India in the United Kingdom (India House, Aldwych, London, WC2B 4NA)
High Commissioner: Jaimini Bhagwati.

Of the United Kingdom in India (Chanakyapuri, New Delhi 110021)
High Commissioner: James Bevan, CMG.

Of India in the USA (2107 Massachusetts Ave., NW, Washington, D.C., 20008)
Ambassador: Nirupama Rao.

Of the USA in India (Shanti Path, Chanakyapuri, New Delhi 110021)
Ambassador: Nancy J. Powell.

Of India to the United Nations
Ambassador: Hardeep Singh Puri.

Of India to the European Union
Ambassador: Dinkar Khullar.

FURTHER READING

Adeney, Katherine and Wyatt, Andrew, *Contemporary India*. 2010
Bhambhri, C. P., *The Political Process in India, 1947–91*. 1991
Bose, S. and Jalal, A. (eds.) *Nationalism, Democracy and Development: State and Politics in India*. 1997
Brown, J., *Modern India: The Origins of an Asian Democracy*. 2nd ed. 1994
Fernandes, Edna, *Holy Warriors: A Journey into the Heart of Indian Fundamentalism*. 2007
Gandhi, Rajmohan, *Gandhi: The Man, His People and the Empire*. 2007
Guha, Ramachandra, *India After Gandhi: The History of the World's Largest Democracy*. 2007

Gupta, S. P., *Globalisation, Economic Reforms and Employment Strategy in India*. 2007
Hardgrave, Robert L. and Kochanek, Stanley A., *India: Government and Politics in a Developing Nation*. 7th ed. 2007
Jaffrelot, C. (ed.) *L'Inde Contemporain de 1950 à nos Jours*. 1996
James, L., *Raj: The Making and Unmaking of British India*. 1997
Joshi, V. and Little, I. M. D., *India's Economic Reforms, 1991–2000*. 1996
Kamdar, Mira, *Planet India: The Turbulent Rise of the World's Largest Democracy*. 2007
Kapur, Akash, *India Becoming: A Portrait of Life in Modern India*. 2012
Keay, John, *India: A History*. 2000
Khan, Yasmin, *The Great Partition: The Making of India and Pakistan*. 2007
Khilnani, S., *The Idea of India*. 1997
King, R., *Nehru and the Language Politics of India*. 1997
Metcalf, Barbara D. and Metcalf, Thomas R., *A Concise History of India*. 2001
Mohan, C. Raja, *Crossing the Rubicon: The Shaping of India's New Foreign Policy*. 2003
New Cambridge History of India. 2nd ed. 5 vols. 1994–96
Nilekani, Nandan, *Imagining India: The Idea of a Renewed Nation*. 2009
Panagariya, Arvind, *India: The Emerging Giant*. 2008
Paul, T. V., *The India-Pakistan Conflict: An Enduring Rivalry*. 2005
Rajadhyaksha, Niranjan, *The Rise of India: Its Transformation from Poverty to Prosperity*. 2006
Robb, Peter, *A History of India*. 2nd ed. 2011
SarDesai, D. R., *India: The Definitive History*. 2008
Tseng, Wanda and Cowen, David, *India's and China's Recent Experience with Reform and Growth*. 2007
Tully, Mark, *India: The Road Ahead*. 2011
Vohra, R., *The Making of India: A Historical Survey*. 1997
Von Tunzelmann, Alex, *Indian Summer: The Secret History of the End of an Empire*. 2007

National library: Belvedere, Kolkata 700027. *Director General:* Prof. Swapan K. Chakravorty.
National Statistical Office: Ministry of Statistics and Programme Implementation, Computer Centre, East Block-10, RK Puram, New Delhi 110066.
Website: http://mospi.nic.in
Census India Website: http://www.censusindia.net

Other more specialized titles are listed under CONSTITUTION AND GOVERNMENT *and* RECENT ELECTIONS *above*.

STATES AND TERRITORIES

GENERAL DETAILS

The Republic of India is composed of the following 28 States and seven centrally administered Union Territories:

States	Capital	States	Capital
Andhra Pradesh	Hyderabad	Maharashtra	Mumbai
Arunachal Pradesh	Itanagar	Manipur	Imphal
Assam	Dispur	Meghalaya	Shillong
Bihar	Patna	Mizoram	Aizawl
Chhattisgarh	Raipur	Nagaland	Kohima
Goa	Panaji	Odisha	Bhubaneswar
Gujarat	Gandhinagar	Punjab	Chandigarh
Haryana	Chandigarh	Rajasthan	Jaipur
Himachal Pradesh	Shimla	Sikkim	Gangtok
Jammu and Kashmir	Srinagar	Tamil Nadu	Chennai
Jharkhand	Ranchi	Tripura	Agartala
Karnataka	Bangalore	Uttar Pradesh	Lucknow
Kerala	Thiruvananthapuram	Uttarakhand	Dehra Dun
Madhya Pradesh	Bhopal	West Bengal	Kolkata

Union Territories. Andaman and Nicobar Islands; Chandigarh; Dadra and Nagar Haveli; Daman and Diu; Delhi; Lakshadweep; Puducherry.

Andhra Pradesh

KEY HISTORICAL EVENTS

Constituted a separate state on 1 Oct. 1953, Andhra Pradesh was the undisputed Telugu-speaking area of Madras. To this region was added, on 1 Nov. 1956, the Telangana area of the former Hyderabad State, comprising the districts of Hyderabad, Medak, Nizamabad, Karimnagar, Warangal, Khammam, Nalgonda and Mahbubnagar, parts of the Adilabad district, some taluks of the Raichur, Gulbarga and Bidar districts and some revenue circles of the Nanded district. On 1 April 1960, 221·4 sq. miles in the Chingleput and Salem districts of Madras were transferred to Andhra Pradesh in exchange for 410 sq. miles from Chittoor district. The district of Prakasam was formed on 2 Feb. 1970. Hyderabad was split into two districts on 15 Aug. 1978 (Ranga Reddy and Hyderabad). A new district, Vizianagaram, was formed in 1979.

TERRITORY AND POPULATION

Andhra Pradesh is in south India and is bounded in the south by Tamil Nadu, west by Karnataka, north and northwest by

Maharashtra, northeast by Chhattisgarh and Odisha and east by the Bay of Bengal. The state has an area of 275,045 sq. km and a population (2011 census, provisional) of 84,665,533; density, 308 per sq. km. The principal language is Telugu. Cities with over 250,000 population, *see* INDIA: Territory and Population. Other large cities (2001): Anantapur, 243,143; Ramagundam, 237,686; Karimnagar, 218,302; Eluru, 215,804; Khammam, 198,620; Vizianagaram, 195,801; Machilipatnam, 179,353; Chirala, 166,294; Adoni, 162,458; Nandyal, 157,120; Ongole, 153,829; Tenali, 153,756; Chittoor, 152,654; Proddutur, 150,309; Bheemavaram, 142,064; Mahbubnagar, 139,662; Adilabad, 129,403; Hindupur, 125,074; Mancherial, 118,195; Srikakulam, 117,320; Guntakal, 117,103; Gudivada, 113,054; Nalgonda, 111,380; Madanapalle, 107,449; Kottagudem, 105,266; Dharmavaram, 103,357; Tadepalligudem, 102,622.

SOCIAL STATISTICS

Growth rate 2001–11, 11·10%.

CONSTITUTION AND GOVERNMENT

Andhra Pradesh has a bicameral legislature; the Legislative Council was abolished in June 1985 but reinstated in April 2007. The *Legislative Assembly* consists of 295 members (one of which is nominated) and the *Legislative Council* of 90 members (12 of which are nominated). For administrative purposes there are 23 districts in the state. The capital is Hyderabad.

RECENT ELECTIONS

At the State Assembly elections held on 16 and 23 April 2009 the Indian National Congress won 156 seats (36·6%), the Telugu Desam Party 92 (28·1%), the Praja Rajyam Party 18 (16·2%), Telangana Rashtra Samithi 10 (4·0%) and All India Majlis-e-Ittehadul Muslimeen 7 (0·8%). Five other parties received a total of 11 seats.

CURRENT GOVERNMENT

Governor: E. S. L. Narasimhan; b. 1945 (since 27 Dec. 2009—acting until 22 Jan. 2010).

Chief Minister: Kiran Kumar Reddy; b. 1960 (since 25 Nov. 2010).

ECONOMY

Budget

Revenue receipts for 2008–09, ₹628,584·5m.; expenditure, ₹618,542·2m. Budget estimates for 2009–10: revenue receipts, ₹789,637·0m.; expenditure, ₹765,571·7m.

ENERGY AND NATURAL RESOURCES

Electricity

There are 13 hydro-electric plants, 11 thermal stations and two gas-based units. Installed capacity, Oct. 2008, 11,911 MW; electricity consumption per capita in 2009–10 was 967 kWh. In Sept. 2008 all 26,613 inhabited villages had electricity.

Oil and Gas

Crude oil is refined at Visakhapatnam in Andhra Pradesh. Oil/gas structures are found in Krishna-Godavari basin which encompasses an area of 20,000 sq. km on land and 21,000 sq. km up to 200 metres isobath off-shore. In 2008–09, 1,524m. cu. metres of natural gas were produced from onshore fields. Reserves of the land basin are estimated at 760 metric tonnes of oil and oil equivalent of gas.

Water

In 2000 more than 120 irrigation projects had created irrigation potential of 6m. ha. The Telugu Ganga joint project with Tamil Nadu, begun in the early 1980s, will eventually irrigate about 233,000 ha., besides supplying drinking water to Chennai city (Tamil Nadu).

Minerals

The state is an important producer of asbestos and barytes. The Cuppadah basin is a major source of uranium and other minerals. Other important minerals are copper ore, coal, iron and limestone, steatite, mica and manganese.

Agriculture

There were (1999) about 10·7m. ha. of cropland, of which 6·8m. ha. were under foodgrains. Irrigated area, 2000, 6m. ha. Production in 1999 (in tonnes): bananas, 13·73m.; pulses, 10·94m.; foodgrains, 10·37m. (rice, 8·51m.); oil seeds, 1·3m.; sugarcane, 0·2m.

Livestock (2003): sheep, 21·38m.; buffaloes, 10·63m.; cattle, 9·30m.; goats, 6·28m.; poultry, 102·28m.

Forestry

In 2007 forests occupied 16·4% of the total area of the state, or 45,102 sq. km; main forest products are teak, eucalyptus, cashew, casuarina, softwoods and bamboo.

Fisheries

Production 2008–09, 1,253,000 tonnes of marine and freshwater fish and crustaceans. The state has a coastline of 974 km. In 2005 there were 289,500 people engaged in fishing and allied activities.

INDUSTRY

The main industries are textile manufacture, sugar-milling, machine tools, pharmaceuticals (Andhra Pradesh commands 40% of India's pharmaceuticals industry), electronic equipment, heavy electrical machinery, aircraft parts and paper-making. There is an oil refinery at Visakhapatnam, where India's major shipbuilding yards are situated. A major steel plant at Visakhapatnam and a railway repair shop at Tirupati are functioning. In 2007–08 there were 16,741 factories employing 1,041,000 people. There are cottage industries and sericulture. District Industries Centres have been set up to promote small-scale industry. Tourism is growing; the main centres are Hyderabad, Nagarjunasagar, Warangal, Arakuvalley, Horsley Hills and Tirupati.

COMMUNICATIONS

Roads

In March 2008 there were 345,012 km of roads (including 4,472 km of national highways and 10,518 km of state highways), of which 189,316 km were surfaced. Andhra Pradesh has the longest road network of any Indian state. There were 7,208,000 motor vehicles registered in March 2008, up from 5,002,000 in March 2003.

Rail

There were 5,185 route-km of railway as at March 2009.

Civil Aviation

There are airports at Hyderabad, Tirupati, Vijayawada and Visakhapatnam, with regular scheduled services to Mumbai, Delhi, Kolkata, Bangalore, Chennai and Bhubaneswar. International flights are operated from Hyderabad to Bangkok, Dubai, Jeddah, Kuala Lumpur, Kuwait, Muscat, Sharjah and Singapore. A new Hyderabad airport (Rajiv Gandhi International Airport, located at Shamshabad) opened in March 2008.

Shipping

The chief port is Visakhapatnam, which handles about 65m. tonnes of cargo annually. There are minor ports at Kakinada, Machilipatnam, Bheemunipatnam, Narsapur, Krishnapatnam, Nizampatnam, Vadarevu and Kalingapatnam.

SOCIAL INSTITUTIONS

Justice

The high court of Judicature at Hyderabad has a Chief Justice and a sanctioned strength of 39 judges.

Education

In 2001, 60·5% of the population were literate (70·3% of men and 50·4% of women). There were, in 1999, 51,836 primary schools (6,237,700 students); 8,713 upper primary (2,440,000); 8,819 high schools (3,732,000). Education is free for children up to 14.

In 1995–96 there were 1,818 junior colleges (676,455 students). In 1996–97 there were 805 degree colleges (427,652 students); 46 oriented colleges and 18 universities: Osmania University, Hyderabad; Andhra University, Waltair; Sri Venkateswara University, Tirupati; Kakatiya University, Warangal; Nagarjuna University, Guntur; Sri Jawaharlal Nehru Technological University, Hyderabad; Hyderabad University, Hyderabad; N. G. Ranga Agricultural University, Hyderabad; Sri Krishnadevaraya University, Anantapur; Smt. Padmavathi Mahila Vishwavidyalayam (University for Women), Tirupati; Dr B. R. Ambedkar Open University, Hyderabad; Patti Sriramulu Telugu University, Hyderabad; N. T. R. University of Health Science, Vijayawada; Moulana Azad National Urdu University, Hyderabad; Dravidian University, Chittoor; Rashtriya Sanskrit Vidyapeeth, Tirupati; Sri Satya Sai Institute of Higher Learning, Prashanti Nilayam; National Academy of Legal Studies and Research University, Hyderabad.

Health

In 2008 there were 16 district hospitals, 566 Ayurvedic hospitals and dispensaries, 202 Unani hospitals and dispensaries and 292 homoeopathy hospitals and dispensaries. In addition there were 1,570 primary health centres with 2,214 doctors, 12,522 sub-centres and 167 community health centres with 20 physicians and 235 specialists in total.

RELIGION

At the 2001 census Hindus numbered 67,836,651; Muslims, 6,986,856; Christians, 1,181,917.

Arunachal Pradesh

KEY HISTORICAL EVENTS

Before independence the North East Frontier Agency of Assam was administered for the viceroy by a political agent working through tribal groups. After independence it became the North East Frontier Tract, administered for the central government by the Governor of Assam. In 1972 the area became the Union Territory of Arunachal Pradesh; statehood was achieved in Dec. 1986.

TERRITORY AND POPULATION

The state is in the extreme northeast of India and is bounded in the north by China, east by Myanmar, west by Bhutan and south by Assam and Nagaland. It has 16 districts and comprises the former frontier divisions of Kameng, Tirap, Subansiri, Siang and Lohit; it has an area of 83,743 sq. km and a population (2011 census, provisional) of 1,382,611; density, 17 per sq. km.

The state is mainly tribal; there are 106 tribes using about 50 tribal dialects. The official languages are English and Hindi.

SOCIAL STATISTICS

Growth rate 2001–11, 25·92%.

CONSTITUTION AND GOVERNMENT

There is a *Legislative Assembly* of 60 members. The capital is Itanagar (population, 2001, 35,022).

RECENT ELECTIONS

At the State Assembly elections held on 13 Oct. 2009 the Indian National Congress won 42 seats; the Nationalist Congress Party, 6; the All Indian Trinamool Congress, 5; the Bharatiya Janata Party, 3; others, 4. Turnout was 72%.

CURRENT GOVERNMENT

Governor: Joginder Jaswant Singh; b. 1945 (since 27 Jan. 2008).
   Chief Minister: Jarbom Gamlin; b. 1961 (since 5 May 2011).

ECONOMY

Budget

2008–09 revenue receipts, ₹38,559·6m.; expenditure, ₹28,716·8m. Budget estimates for 2009–10: revenue receipts, ₹32,569·6m.; expenditure, ₹36,070·8m.

ENERGY AND NATURAL RESOURCES

Electricity

Total installed capacity (Oct. 2008), 180 MW. Electricity consumption per capita in 2009–10 was 470 kWh. In Sept. 2008, 2,195 out of 3,863 inhabited villages had electricity.

Oil and Gas

Onshore production, 2008–09, 102,400 tonnes of crude oil and 30m. cu. metres of gas.

Minerals

Coal reserves are estimated at 90·23m. tonnes; dolomite, 154·13m. tonnes; limestone, 409·35m. tonnes.

Agriculture

Production of foodgrains, 1999, 204,000 tonnes.

Forestry

Area under forest in 2007 was 67,353 sq. km (80·4% of total area).

INDUSTRY

In 2009 there were 17 medium and 2,526 small industries, 76 craft or weaving centres and 25 sericulture demonstration centres. Industries include coal, textiles, jute, iron and steel, chemicals, tea and leather.

COMMUNICATIONS

Roads

In March 2008 there were 16,494 km of roads (including 392 km of national highways), of which 9,755 km were surfaced. There were 22,000 motor vehicles registered in 1997.

SOCIAL INSTITUTIONS

Education

In 2001, 54·3% of the population were literate (63·8% of men and 43·5% of women). There were (1996–97) 1,256 primary schools with 147,676 students, 301 middle schools with 42,197 students, 157 high and higher secondary schools with 24,951 students, six colleges and two technical schools. Arunachal University, established in 1985, had four colleges and 3,240 students in 1994–95.

Health

There were (2004) 14 hospitals, 19 community health centres, 58 primary health centres and 273 sub-centres. In 1996 there were two TB hospitals and 11 leprosy and other hospitals. Total number of beds (2002), 2,641.

RELIGION

At the 2001 census Hindus numbered 379,935; Christians, 205,548; Buddhists, 143,028.

FURTHER READING

Bose, M. L., *History of Arunachal Pradesh.* 1997

Assam

KEY HISTORICAL EVENTS

Assam first became a British Protectorate at the close of the first Burmese War in 1826. In 1832 Cachar was annexed; in 1835 the Jaintia Hills were included in the East India Company's dominions, and in 1839 Assam was annexed to Bengal. In 1874 Assam was detached from Bengal and made a separate chief commissionership. On the partition of Bengal in 1905, it was united to the Eastern Districts of Bengal under a Lieut.-Governor. From 1912 the chief commissionership of Assam was revived, and in 1921 a governorship was created. On the partition of India almost the whole of the predominantly Muslim district of Sylhet was merged with East Bengal (Pakistan). Dewangiri in North Kamrup was ceded to Bhutan in 1951. The Naga Hill district, administered by the Union government since 1957, became part of Nagaland in 1962. The autonomous state of Meghalaya within Assam, comprising the districts of Garo Hills and Khasi and Jaintia Hills, came into existence on 2 April 1970, and achieved full independent statehood in Jan. 1972, when it was also decided to form a Union Territory, Mizoram (now a state), from the Mizo Hills district.

TERRITORY AND POPULATION

Assam is in northeast India, almost separated from central India by Bangladesh. It is bounded in the west by West Bengal, north by Bhutan and Arunachal Pradesh, east by Nagaland, Manipur and Myanmar, south by Meghalaya, Bangladesh, Mizoram and Tripura. The area of the state is now 78,438 sq. km. Population (2011 census, provisional) 31,169,272; density, 397 per sq. km. Cities with over 250,000 population, *see* INDIA: Territory and Population. Other large cities (2001): Silchar, 184,105; Jorhat, 137,814; Dibrugarh, 137,661; Nagaon, 123,265; Tinsukia, 108,123; Tezpur, 105,377. The principal language is Assamese.

The central government is constructing a boundary fence to prevent illegal entry from Bangladesh.

SOCIAL STATISTICS

Growth rate 2001–11, 16·93%.

CONSTITUTION AND GOVERNMENT

Assam has a unicameral legislature of 126 members. The capital is Dispur. The state has 23 districts.

RECENT ELECTIONS

In the elections of 4 and 11 April 2011 the Indian National Congress (INC) took 78 of 126 seats, All India United Democratic Front 18, Bodoland People's Front 12, Asom Gana Parishad 10, Bharatiya Janata Party 5, All India Trinamool Congress 1 and ind. 2.

CURRENT GOVERNMENT

Governor: Janaki Ballabh Patnaik; b. 1927 (took office on 11 Dec. 2009).

Chief Minister: Tarun Gogoi; b. 1936 (took office on 18 May 2001).

ECONOMY

Budget

Revenue receipts for 2008–09, ₹180,770·3m.; expenditure, ₹142,433·1m. Budget estimates for 2009–10: revenue receipts, ₹230,636·5m.; expenditure, ₹292,698·6m.

ENERGY AND NATURAL RESOURCES

Electricity

In Oct. 2008 there was an installed capacity of 980 MW. In Sept. 2008, 19,741 out of 25,124 inhabited villages had electricity.

Oil and Gas

Assam contains important oilfields and produces about 16% of India's crude oil. Production (1999): crude oil, 5·00m. tonnes; gas (1999), 1,333m. cu. metres.

Minerals

Coal production (2002–03), 633,000 tonnes. The state also has limestone, refractory clay, dolomite and corundum.

Agriculture

Over three-quarters of the workforce is engaged in agriculture. Assam produces 50% of India's tea—in 2003 there were 1,196 registered tea estates in the state. Production in 2005 was 487·5m. kg. 82% of the cultivable area is used. Over 72% of the cultivated area is under food crops, of which the most important is rice. Total cereal production, 2006–07, 3·00m. tonnes of which rice, 2·92m. tonnes; wheat, 67,200 tonnes. Pulses (2006–07), 116,700 tonnes. Main cash crops: tea, jute, cotton, oilseeds, sugarcane, fruit and potatoes. Livestock (2003): 8·4m. cattle; 3·0m. goats; 1·5m. pigs; 21·7m. poultry. Livestock products (2006–07): milk, 823m. litres; 535m. eggs; meat, 28,800 tonnes.

Forestry

Area under forest in 2007 was 27,692 sq. km (35·3% of total area).

INDUSTRY

Sericulture and hand-loom weaving, both silk and cotton, are important home industries together with the manufacture of brass, cane and bamboo articles. 17% of workers are employed in the tea industry. The main heavy industry is petrochemicals; there are four oil refineries in the region. Other industries include manufacturing paper, nylon, electronic goods, cement, fertilizers, sugar, jute and plywood products, rice and oil milling.

There were 1,859 factories in 2007–08 employing 134,300 people.

COMMUNICATIONS

Roads

In March 2008 there were 230,334 km of roads (including 2,836 km of national highways and 3,134 km of state highways), of which 26,612 km were surfaced. There were 1,116,000 motor vehicles registered in March 2008, up from 657,000 in March 2003.

Rail

The route-km of railways in 2008–09 was 2,284 km.

Civil Aviation

Daily scheduled flights connect the principal towns with the rest of India. There are airports at Guwahati, Tezpur, Jorhat, North Lakhimpur, Silchar and Dibrugarh.

SOCIAL INSTITUTIONS

Justice

The seat of the High Court is Guwahati. It has a Chief Justice and a sanctioned strength of 19 judges.

Education

In 2001, 63·3% of the population were literate (71·3% of men and 54·6% of women). In 1999–2000 there were 31,888 primary/junior basic schools with 3,293,835 students; 8,019 middle/senior basic schools with 1,406,818 students; 4,514 high/higher secondary schools with 1,465,518 students. There were 247 colleges for general education, six medical colleges, three engineering and one agricultural, 24 teacher training colleges and a fisheries college at Raha. There were five universities: Assam Agricultural University, Jorhat; Dibrugarh University, Dibrugarh with 86 colleges and 55,982 students (1992–93); Gauhati University, Guwahati with 128 colleges and 80,363 students (1992–93); and two central universities, at Silchar and Tezpur.

Health

In 2000 there were 164 hospitals (12,900 beds), 618 primary health centres and 323 dispensaries.

RELIGION

At the 2001 census Hindus numbered 17,296,455; Muslims, 8,240,611; Christians, 986,589; Buddhists, 51,029.

FURTHER READING

Baruah, Sanjib, *India Against Itself: Assam and the Politics of Nationality*. 1999

Bihar

KEY HISTORICAL EVENTS

Bihar was part of Bengal under British rule until 1912 when it was separated together with present-day Odisha. The two were joined until 1936 when Bihar became a separate province. As a state of the Indian Union it was enlarged in 1956 by the addition of land from West Bengal.

The state contains the ethnic areas of North Bihar, Santhal Pargana and Chota Nagpur. In 1956 some areas of Purnea and Manbhum districts were transferred to West Bengal. In 2000 the state of Jharkhand was carved from the mineral-rich southern region of Bihar, substantially reducing the state's revenue-earning power.

TERRITORY AND POPULATION

Bihar is in north India and is bounded north by Nepal, east by West Bengal, south by Jharkhand, southwest and west by Uttar Pradesh. After the formation of Jharkhand the area of Bihar is 94,163 sq. km (previously 173,877 sq. km). Population (2011 census, provisional), 103,804,637, with a density of 1,102 per sq. km. Population of principal towns, *see* INDIA: Territory and Population. Other large towns (2001): Bihar Sharif, 232,071; Arrah, 203,380; Purnea, 197,211; Katihar, 190,873; Munger, 188,050; Chapra, 179,190; Sasaram, 131,172; Saharsa, 125,167; Hajipur, 119,412; Dehri, 119,057; Bettiah, 116,670; Siwan, 109,919; Motihari, 108,428; Begusarai, 107,623.

The state is divided into 38 districts. The capital is Patna.

The official language is Hindi (spoken by 80·9% at the 2001 census), the second, Urdu (9·9%), and the third, Bengali (2·9%).

SOCIAL STATISTICS

Growth rate 2001–11, 25·07%.

CONSTITUTION AND GOVERNMENT

Bihar has a bicameral legislature. The *Legislative Assembly* consists of 243 elected members, and the Council 75.

RECENT ELECTIONS

In elections held on 21, 24 and 28 Oct., and 1, 9 and 20 Nov. 2010 the Janata Dal (United) won 115 of 243 seats; Bharatiya Janata Party, 91; Rashtriya Janata Dal, 22; Indian National Congress, 4; Lok Janshakti Party, 3; Communist Party of India, 1; Jharkhand Mukti Morcha, 1; ind., 6.

CURRENT GOVERNMENT

Governor: Dnyandeo Yashwantrao Patil; b. 1935 (took office on 22 March 2013).

Chief Minister: Nitish Kumar; b. 1951 (took office for a second time on 24 Nov. 2005).

ECONOMY

Budget

Revenue receipts for 2008–09, ₹329,806·8m.; expenditure, ₹285,115·7m. Budget estimates for 2009–10: revenue receipts, ₹418,374·3m.; expenditure, ₹357,150·3m.

ENERGY AND NATURAL RESOURCES

Electricity

Installed capacity (Oct. 2008) 1,970 MW. In Sept. 2008, 20,620 out of 39,015 inhabited villages had electricity.

Minerals

Before the creation of Jharkhand, Bihar was very rich in minerals. The truncated state has only deposits of bauxite, mica, glass sand and salt.

Agriculture

The irrigated area was 4·92m. ha. in 2008–09. Cropped area, 2009–10, 7·30m. ha. out of a total area of 9·42m. ha. Production (2009–10): rice, 3·63m. tonnes; wheat, 4·56m.; total foodgrains, 10·15m. Other major food crops are maize and pulses. Main cash crops are jute, sugarcane, oilseeds, tobacco and potatoes.

Forestry

Forests in 2007 covered 6,804 sq. km. There is one National Park and 12 wildlife sanctuaries.

INDUSTRY

(Including Jharkhand). Iron, steel and aluminium are produced and there is an oil refinery. Other important industries are heavy engineering, machine tools, fertilizers, electrical engineering, manufacturing drugs and fruit processing. There were 1,783 factories in 2007–08 employing 73,700 people.

COMMUNICATIONS

Roads

In March 2008 there were 120,127 km of roads (including 3,642 km of national highways and 3,767 km of state highways), of which 58,136 km were surfaced. There were 1,739,000 motor vehicles registered in March 2008, up from 1,121,000 in March 2003.

Rail

In 2008–09 the state had 3,515 route-km of railway line.

Civil Aviation

There are airports at Patna and Gaya with regular scheduled services to Kolkata and Delhi.

Shipping

(Including Jharkhand). The length of waterways open for navigation is 1,300 km.

SOCIAL INSTITUTIONS

Justice

There is a High Court (constituted in 1916) at Patna with a Chief Justice, 31 puisne judges and four additional judges.

Police

The police force is under a Director General of Police; in 2009 there were 853 police stations.

Education

At the census of 2001, 47·0% of the population were literate (59·7% of males and 33·1% of females). There were, 1996–97, 53,652 primary schools with 9·63m. pupils, 13,834 middle schools with 2·42m. pupils and 4,149 high and higher secondary schools with 1·08m. pupils. Education is free for children aged 6–11.

There are 12 universities: Patna University (founded 1917) with 14,699 students (1994–95); Babasaheb Bhimrao Ambedkar Bihar University, Muzaffarpur (1952) with 95 colleges, and 84,873 students (1989–90); Tilka Manjhi Bhagalpur University (1960) with 140,718 students (1990–91); Kameshwara Singh Darbhanga Sanskrit University (1961); Magadh University, Gaya (1962) with 186 colleges and 122,019 students (1994–95); Rajendra Agricultural University, Samastipur (1970); Lalit Narayan Mithila University (1972), Darbhanga; Nalanda Open University, Nalanda (1987); Jai Prakash University, Chapra (1990); BN Mandal University, Madhepura (1992); Mazrul Haque Arabi-Farsi University, Patna; and Veer Kunwar Singh University, Arrah. Including Jharkhand, there were 742 degree colleges, 11 engineering colleges, 31 medical colleges and 15 teacher training colleges in 1996–97. Ranchi University, Bisra Agricultural College and Sidhu Kanhu University, all formerly in Bihar, are now part of Jharkhand.

Health

(Including Jharkhand). In 2000 there were 1,636 hospitals and dispensaries with 12,123 beds.

RELIGION

At the 2001 census Hindus numbered 69,076,919, Muslims 13,722,048 and Christians 53,137.

Chhattisgarh

KEY HISTORICAL EVENTS

Created from 16 mainly tribal districts of Madhya Pradesh, the state became the twenty-sixth state of India on 1 Nov. 2000. Originally known as South Kosala, archaeological excavations made in recent times indicate that the region was a hive of artistic and cultural experimentation in ancient times. During the Sarabhapuriyas, Nalas, Pandavamsis and Kalchuris dynasties between the 6th and 8th centuries many brick temples were built in the area. The British took control from the Mahrattas in the early 19th century. Despite possessing its own cultural identity Chhattisgarh was constantly swallowed up by other regions and in 1956, as a direct result of the Indian Union of 1949, it was made part of the new region of Madhya Pradesh. Protestors maintained that the revenue generated by their region, from rice and minerals, was insufficiently reinvested in the area. In 2000 the National Democratic Alliance negotiated the passage of a bill through both houses of the Indian parliament which carved out three new Indian states, Chhattisgarh among them.

TERRITORY AND POPULATION

Chhattisgarh is in central eastern India and is bounded by Jharkhand to the east, Odisha to the southeast, Andhra Pradesh to the south and Maharashtra and Madhya Pradesh to the west. Chhattisgarh has an area of 135,191 sq. km. Population (2011 census, provisional) 25,540,196; density, 189 per sq. km. The principal language is Hindi.

Cities with over 250,000 population, *see* INDIA: Territory and Population. Other large cities (2001): Rajnandgaon, 143,770; Raigarh, 115,908; Jagdalpur, 103,123.

SOCIAL STATISTICS

Growth rate 2001–11, 22·59%.

CONSTITUTION AND GOVERNMENT

Chhattisgarh is the twenty-sixth state of India. In creating Chhattisgarh it was decided that the 90 members of the Madhya Pradesh Legislative Assembly from Chhattisgarhi districts would become the members of the new state's *Legislative Assembly*. For administrative purposes the region is divided into 16 districts. The

council of ministers consists of 15 cabinet ministers and eight ministers of state.

The capital and seat of government is at Raipur.

RECENT ELECTIONS

At elections in Nov. 2008 the Bharatiya Janata Party won 50 seats and the Congress (I) Party took 38. Other parties won two seats.

CURRENT GOVERNMENT

Governor: Sekhar Dutt; b. 1945 (took office 23 Jan. 2010).

Chief Minister: Dr Raman Singh; b. 1952 (took office 7 Dec. 2003).

ECONOMY

Budget

Revenue receipts for 2008–09, ₹156,627·6m.; expenditure, ₹137,937·0m. Budget estimates for 2009–10: revenue receipts, ₹188,972·1m.; expenditure, ₹180,910·4m.

ENERGY AND NATURAL RESOURCES

Electricity

In Sept. 2008, 18,877 out of 19,744 inhabited villages had electricity. There was an installed capacity of 3,528 MW in Oct. 2008.

Water

1·21m. ha. of land is under irrigation. 44,750 residential areas have sufficient drinking water supplies while 7,315 residential areas only have partial supplies and 2,751 areas have insufficient supplies. In total there were 102,063 hand pumps in the state and 701 water fulfilment plans in place in 1999.

Minerals

The state has extensive mineral resources including (1999 estimates): over 27,000m. tonnes of tin ore, 2,000m. tonnes of iron ore, 525m. tonnes of dolomite (accounting for 24% of India's entire share) and 73m. tonnes of bauxite. There are also significant deposits of limestone, copper ore, rock phosphate, corundum, tin, coal and manganese ore. Deobogh in the Raipur district contains deposits of diamonds.

Agriculture

Agriculture is the occupation for 1·7m. of the population (around 80%). 5·8m. ha. of land is agricultural and the area provides food grain for over 600 rice mills. The great plains of Chhattisgarh produce 10,000 varieties of rice. Other crops include maize, millet, groundnuts, soybeans and sunflower. More than 25% of the land in Chhattisgarh is double cropped.

COMMUNICATIONS

Roads

In March 2008 there were 74,434 km of roads (including 2,184 km of national highways and 3,419 km of state highways), of which 43,528 km were surfaced. There were 1,935,000 motor vehicles registered in March 2008, up from 1,076,000 in March 2003.

Rail

Raipur is at the centre of the state's railway network, linking Chhattisgarh to the states of Odisha and Madhya Pradesh. There were 1,186 route-km of railway at March 2009.

SOCIAL INSTITUTIONS

Education

In 2001, 64·7% of the population were literate (77·4% of men and 51·9% of women). There are three universities in Chhattisgarh. Ravishankar University (founded 1964), at Raipur, had 89 affiliated colleges (1992–93); Indira Gandhi Krishi Vishwavidyalaya, Raipur, a music and fine arts institution (founded in 1956); and Guru

Ghasidas University, Bilaspur which had 58 colleges and 34,717 students (1992–93).

Health

In 2004 there were 138 hospitals with 5,565 beds and 285 doctors.

RELIGION

At the 2001 census Hindus numbered 19,729,670; Muslims, 409,615; Christians, 401,035; Sikhs, 69,621; Buddhists, 65,267; Jains, 56,103.

Goa

KEY HISTORICAL EVENTS

The coastal area was captured by the Portuguese in 1510 and the inland area was added in the 18th century. In Dec. 1961 Portuguese rule was ended and Goa incorporated into the Indian Union as a Territory together with Daman and Diu. Goa was granted statehood on 30 May 1987. Daman and Diu remained Union Territories.

TERRITORY AND POPULATION

Goa, bounded on the north by Maharashtra and on the east and south by Karnataka, has a coastline of 105 km. The area is 3,702 sq. km. Population (2011 census, provisional) 1,457,723; density, 394 per sq. km. Marmagao is the largest town; population (urban agglomeration, 2001) 104,758. The capital is Panaji; population (urban agglomeration, 2001) 99,677. The state has two districts. There are 183 village Panchayats. The languages spoken are Konkani (official language; 51·5%), Marathi 33·4%, Kannada 4·6%, Hindi and English.

SOCIAL STATISTICS

Growth rate 2001–11, 8·17%.

CONSTITUTION AND GOVERNMENT

The Indian Parliament passed legislation in March 1962 by which Goa became a Union Territory with retrospective effect from 20 Dec. 1961. On 30 May 1987 Goa attained statehood. There is a *Legislative Assembly* of 40 members. In March 2005 the state was put under president's rule following a controversy over a vote of confidence in the Legislative Assembly. It was lifted in June 2005 after Pratapsingh Rane was sworn in as the new chief minister.

RECENT ELECTIONS

Of the 40 seats available at the elections for the State Assembly on 3 March 2012 the Bharatiya Janata Party won 21; Indian National Congress, 9; Maharashtrawadi Gomantak Party, 3; Goa Vikas Party, 2. Five independents were elected.

CURRENT GOVERNMENT

Governor: B. V. Wanchoo; b. 1951 (took office on 4 May 2012).
　Chief Minister: Manohar Parrikar; b. 1955 (took office on 9 March 2012).

ECONOMY

Goa had a per capita annual income of ₹192,652 in 2011–12, the highest of any Indian state.

Budget

2008–09 revenue receipts, ₹35,282·7m.; expenditure, ₹34,254·1m. Budget estimates for 2009–10: revenue receipts, ₹41,367·3m.; expenditure, ₹44,851·4m.

ENERGY AND NATURAL RESOURCES

Electricity

In Oct. 2008 installed capacity was 352 MW, but Goa receives most of its power supply from the states of Maharashtra and Karnataka. In Sept. 2008 all 347 inhabited villages had electricity. Electricity consumption per capita in 2009–10 was 2,264 kWh, compared to 779 kWh for India as a whole.

Minerals

Resources include bauxite, ferro-manganese ore and iron ore, all of which are exported. Iron ore production (2006–07) 28,723,000 tonnes. There are also reserves of limestone and clay.

Agriculture

Agriculture is the main occupation, important crops being rice, pulses, ragi, mango, cashew and coconuts. Area under rice (2007–08) 52,191 ha.; production, 121,670 tonnes. Area under cashew nuts, 55,612 ha.; pulses, 11,477 ha.; sugarcane, 1,034 ha. Total production of foodgrains, 2007–08, 133,101 tonnes.
　Government poultry and dairy farming schemes produced 14m. eggs and 58,000 tonnes of milk in 2007–08. Poultry (2007–08), 531,000; cattle, 72,000; pigs, 59,000.

Forestry

Forests covered 1,424 sq. km in 2007–08.

Fisheries

Fish is the state's staple food. In 2005 there were 5,900 people engaged in fishing and allied activities. In 2007–08 the catch of seafish was 91,185 tonnes. There is a coastline of about 104 km.

INDUSTRY

In 2007–08 there were 522 factories registered with a workforce of 50,847. Production included: automotive components, electronic goods, fertilizers, footwear, nylon fishing nets, pesticides, pharmaceuticals, ready made clothing, shipbuilding and tyres.

COMMUNICATIONS

Roads

In March 2008 there were 10,569 km of roads (including 269 km of national highways and 279 km of state highways), of which 7,664 km were surfaced. There were 624,000 motor vehicles registered in March 2008, up from 397,000 in March 2003.

Rail

There were 69 route-km of railway as at March 2009.

Civil Aviation

An airport at Dabolim is connected with Agatti, Bangalore, Chennai, Delhi, Kochi, Kozhikode, Mumbai and Pune. It also receives international charter flights and scheduled flights from Kuwait and Sharjah.

Shipping

There are seaports at Panaji, Marmagao and Margao.

SOCIAL INSTITUTIONS

Justice

There is a bench of the Bombay High Court at Panaji.

Education

In 2001, 82·0% of the population were literate (88·4% of men and 75·4% of women). In 2001 there were 1,268 primary schools (97,457 students), 440 middle schools (72,726 students) and 445 high and higher secondary schools (85,217 students). In 1996–97 there were also two engineering colleges, four medical colleges, two teacher training colleges, 21 other colleges and six polytechnic institutes. Goa University, Taleigao (1985) had 33 colleges and 16,977 students in 1994–95.

Health

In 2001 there were 120 hospitals (4,865 beds), 201 rural medical dispensaries, health and sub-health centres and 268 family planning units.

RELIGION

At the 2001 census Hindus numbered 886,551; Christians, 359,568; Muslims, 92,210.

FURTHER READING

Hutt, A., *Goa: A Traveller's Historical and Architectural Guide*. 1988

Gujarat

KEY HISTORICAL EVENTS

The Gujarati-speaking areas of India were part of the Moghul empire, coming under Mahratta domination in the late 18th century. In 1818 areas of present Gujarat around the Gulf of Cambay were annexed by the British East India Company. The remainder consisted of a group of small principalities, notably Baroda, Rajkot, Bhavnagar and Nawanagar. British areas became part of the Bombay Presidency.

At independence the area now forming Gujarat became part of Bombay State except for Rajkot and Bhavnagar which formed the state of Saurashtra until incorporated into Bombay in 1956. In 1960 Bombay State was divided and the Gujarati-speaking areas became Gujarat.

In early 2002 at least 800 people, mostly Muslims, were killed in Gujarat in ethnic violence.

TERRITORY AND POPULATION

Gujarat is in western India and is bounded in the north by Pakistan and Rajasthan, east by Madhya Pradesh, southeast by Maharashtra, south and west by the Indian ocean and Arabian sea. The area of the state is 196,024 sq. km and the population (2011 census, provisional) 60,383,628; density, 308 per sq. km. Principal cities, *see* INDIA: Territory and Population. Other important towns (2001 census) are: Navsari (232,411), Surendranagar (219,585), Anand (218,486), Porbandar (197,382), Nadiad (196,793), Gandhinagar (195,985), Morbi (178,055), Bharuch (176,364), Veraval (158,032), Gandhidham (151,693), Valsad (145,592), Mehesana (141,453), Bhuj (136,429), Godhra (131,172), Palanpur (122,300), Patan (113,749), Anklesvar (112,643), Dahod (112,026), Kalol (112,013), Jetpur (104,312), Botad (100,194). Gujarati and Hindi in the Devanagari script are the official languages.

SOCIAL STATISTICS

Growth rate 2001–11, 19·17%.

CLIMATE

Summers are intensely hot: 33–45°C. Winters: 7–13°C. Monsoon season: 22–36°C. Annual rainfall varies from 35 cm to 189 cm.

CONSTITUTION AND GOVERNMENT

Gujarat has a unicameral legislature, the *Legislative Assembly*, which has 182 elected members.

The capital is Gandhinagar. There are 25 districts.

RECENT ELECTIONS

In elections held on 13 and 17 Dec. 2012 the Bharatiya Janata Party retained power with a slightly reduced majority, winning 115 seats against 61 for Congress, with ind. and others winning six seats.

CURRENT GOVERNMENT

Governor: Dr Kamla Beniwal; b. 1927 (took office on 27 Nov. 2009).

Chief Minister: Shri Narendrabhai Modi; b. 1950 (took office on 7 Oct. 2001).

ECONOMY

Budget

Revenue receipts for 2008–09, ₹386,757·1m.; expenditure, ₹387,414·5m. Budget estimates for 2009–10: revenue receipts, ₹418,151·8m.; expenditure, ₹457,283·7m.

ENERGY AND NATURAL RESOURCES

Electricity

In Oct. 2008 total installed capacity was 11,591 MW. In Sept. 2008, 18,014 out of 18,066 inhabited villages had electricity.

Oil and Gas

There are large crude oil and gas reserves. Production, 2008–09: crude oil, 5·9m. tonnes; natural gas, 2,605m. cu. metres.

Water

Water resources are limited. In 2003 irrigation potential was 6·49m. ha.

Minerals

Chief minerals produced in 2006–07 (in tonnes) included limestone (22·5m.), lignite (9·8m.), bauxite (3·5m.), quartz and silica (1·2m.), bentonite (896,000), crude china clay (469,000), dolomite (325,000) and fire clay (232,000). Value of production (2005–06) ₹60,325m.

Agriculture

3·5m. ha. of the cropped area was irrigated in June 2003.

Production of principal crops, 2002–03: foodgrains, 3·6m. tonnes (wheat, 0·86m. tonnes); rice, 0·6m. tonnes from 481,000 ha.; pulses, 327,000 tonnes; cotton, 1·69m. bales of 170 kg. Tobacco and groundnuts are important cash crops.

Livestock (2003): buffaloes, 7·14m.; other cattle, 7·42m.; sheep and goats, 6·60m.; pigs, 351,000; poultry, 8·15m.

Forestry

Forests covered 14,620 sq. km in 2007 (7·5% of total area). The State has four National Parks and 24 wildlife sanctuaries.

Fisheries

There were 158,000 people engaged in fisheries in 2005. In 2006–07 there were 31,370 fishing vessels; the total catch was 754,000 tonnes.

INDUSTRY

Gujarat ranks among India's most industrialized states. In 2007–08 there were 15,107 factories employing 1,045,475 people. There were 171 functioning industrial estates in 2010. Principal industries are textiles, general and electrical engineering, oil-refining, fertilizers, petrochemicals, machine tools, automobiles, heavy chemicals, pharmaceuticals, dyes, sugar, soda ash, cement, man-made fibres, salt, sulphuric acid, paper and paperboard.

In 2002 state production of soda-ash was 1·88m. tonnes, salt production was 13·08m. tonnes and cement production 10·78m. tonnes.

COMMUNICATIONS

Roads

In March 2008 there were 146,630 km of roads—including 3,245 km of national highways and 18,447 km of state highways—of which 132,321 km were surfaced. There were 10,289,000 motor vehicles registered in March 2008 (of which 7,579,000 were two-wheelers), up from 6,508,000 in March 2003.

Rail

The route-km of railways in 2008–09 was 5,328 km.

Civil Aviation

Sardar Vallabhbhai Patel International Airport at Ahmedabad is the main airport. There are some international flights and regular internal services between Ahmedabad and Delhi, Jaipur and Mumbai, and within Gujarat between Ahmedabad and Bhavnagar, Bhuj, Rajkot and Vadodara (Baroda). There are five other airports: Jamnagar (which also has some international flights), Kandla, Keshod, Porbandar and Surat.

Shipping

The largest port is Kandla. There are 40 other ports. Cargo handled at the ports in 2003–04 totalled 130·9m. tonnes (41·5m. tonnes at Kandla).

SOCIAL INSTITUTIONS

Justice

The High Court of Judicature at Ahmedabad has a Chief Justice and 30 puisne judges.

Education

In 2001, 69·1% of the population were literate (79·7% of males and 57·8% of females). Primary and secondary education up to Standard XII are free. Education above Standard XII is free for girls. In 2006–07 there were 39,064 primary schools with 8·28m. students and 7,967 secondary schools with 2·67m. students.

There are 11 universities in the state. Gujarat University, Ahmedabad, founded in 1950, is teaching and affiliating; it has 154 affiliated colleges and 143,692 students (all student figures for 1998–99). The Maharaja Sayajirao University of Vadodara (1949) is residential and teaching; it has 12 colleges and 26,511 students. The Sardar Patel University, Vallabh-Vidyanagar (1955), has 20 constituent and affiliated colleges and 17,913 students. Saurashtra University at Rajkot (1968) has 113 affiliated colleges and 72,234 students. South Gujarat University at Surat (1967) has 58 colleges and 59,600 students. Bhavnagar University (1978) is residential and teaching with 15 affiliated colleges and 11,195 students. North Gujarat University was established at Patan in 1986 and has 73 colleges and 54,720 students. Gujarat Vidyapith at Ahmedabad is deemed a university under the University Grants Commission Act. There are also Gujarat Agricultural University, Banaskantha, Gujarat Ayurved University, Jamnagar and Dr Babasaheb Ambedkar Open University, Ahmedabad.

There were 903 higher education institutions in 2006–07, with a total of 409,000 students enrolled.

Health

At March 2006 there were 273 community health centres, 1,072 primary health centres and 7,274 sub-centres. In 2003–04, 41·3m. patients were treated.

RELIGION

At the 2001 census Hindus numbered 45,143,074; Muslims, 4,592,854; Jains, 525,305; Christians, 284,092.

FURTHER READING

Desai, I. F., *Untouchability in Rural Gujarat.* 1977
Sharma, R. N., *Gujarat Holocaust (Communalism in the Land of Gandhi).* 2002

Haryana

KEY HISTORICAL EVENTS

The state of Haryana, created on 1 Nov. 1966 under the Punjab Reorganization Act, 1966, was formed from the Hindi-speaking parts of the state of Punjab (India). It comprises the districts of Ambala, Bhiwani, Faridabad, Fatehabad, Gurgaon, Hisar, Jhajjar, Jind, Kaithal, Karnal, Kurukshetra, Mahendragarh, Mewat, Palwal, Panchkula, Panipat, Rewari, Rohtak, Sirsa, Sonipat and Yamunanagar.

TERRITORY AND POPULATION

Haryana is in north India and is bounded north by Himachal Pradesh, east by Uttar Pradesh, south and west by Rajasthan and northwest by Punjab. Delhi forms an enclave on its eastern boundary. The state has an area of 44,212 sq. km and a population (2011 census, provisional) of 25,353,081; density, 573 per sq. km. Principal cities, *see* INDIA: Territory and Population. Other large towns (2001) are: Gurgaon (228,820), Sonipat (225,074), Karnal (221,236), Bhiwani (169,531), Ambala Sadar (168,316), Sirsa (160,735), Panchkula Urban Estate (140,925), Ambala (139,279), Jind (135,855), Bahadurgarh (131,925), Thanesar (122,319), Kaithal (117,285), Palwal (100,722), Rewari (100,684). The principal language is Hindi.

SOCIAL STATISTICS

Growth rate 2001–11, 19·90%.

CONSTITUTION AND GOVERNMENT

The state has a unicameral legislature with 90 members. The capital (shared with Punjab) is Chandigarh. Its transfer to Punjab, originally intended for 1986, has been postponed. There are 21 districts.

RECENT ELECTIONS

In the elections of 13 Oct. 2009 the Indian National Congress won 40 seats, the Indian National Lok Dal 32, the Haryana Janhit Congress 6, the Bharatiya Janata Party 4 and others 8. Turnout was 69·4%.

CURRENT GOVERNMENT

Governor: Jagannath Pahadia; b. 1932 (took office on 27 July 2009).

Chief Minister: Bhupinder Singh Hooda; b. 1947 (took office on 5 March 2005).

ECONOMY

Budget

Revenue receipts for 2008–09, ₹184,523·0m.; expenditure, ₹205,347·3m. Budget estimates for 2009–10: revenue receipts, ₹224,370·0m.; expenditure, ₹258,210·6m.

ENERGY AND NATURAL RESOURCES

Electricity

Installed capacity (Oct. 2008) was 4,590 MW. In Sept. 2008 all 6,764 inhabited villages had electricity.

Minerals

Minerals include placer gold, barytes, tin and rare earths. Value of production, 2002–03, ₹1,487m.

Agriculture

Haryana has sandy soil and erratic rainfall, but the state shares the benefit of the Sutlej-Beas scheme. Agriculture employs over 82% of the working population; in 1995–96 there were about 1·7m. holdings (average 2·1 ha.), and the gross irrigated area was 2·05m. ha. in 1993–94. Area under foodgrains, 1995–96, 4·02m. ha. Foodgrain production, 1999–2000, 10·36m. tonnes (rice 2·59m. tonnes in 2000, wheat 7·35m. tonnes in 1995–96); pulses, 416,400 tonnes in 1995–96; cotton, 1·5m. bales of 170 kg in 1995–96; sugar (gur) and oilseeds are important. Haryana produces a surplus of wheat and rice.

Forestry

Forests covered 1,594 sq. km in 2007.

INDUSTRY

Haryana has a large market for consumer goods in neighbouring Delhi. In 2007–08 there were 4,707 factories providing employment to 510,000 persons. The main industries are cotton textiles, agricultural machinery and tractors, woollen textiles, scientific instruments, glass, cement, paper and sugar milling, cars, tyres and tubes, motorcycles, bicycles, steel tubes, engineering goods, electrical and electronic goods. An oil refinery at Panipat was commissioned in 1999 and includes a diesel hydro desulphurization plant.

COMMUNICATIONS

Roads

In March 2008 there were 29,726 km of roads (including 1,512 km of national highways and 2,523 km of state highways), of which 27,703 km were surfaced. There were 3,973,000 motor vehicles registered in March 2008, up from 2,279,000 in March 2003.

Rail

The state is crossed by lines from Delhi to Agra, Ajmer, Ferozepur and Chandigarh. In 2008–09 the state had 1,553 route-km of railway line. The main stations are at Ambala and Kurukshetra.

Civil Aviation

There is no airport within the state but Delhi is on its eastern boundary.

SOCIAL INSTITUTIONS

Justice

Haryana shares the High Court of Punjab and Haryana at Chandigarh.

Education

In 2001, 67·9% of the population were literate (78·5% of men and 55·7% of women). In 1996–97 there were 5,651 primary schools with 1,981,993 students, 3,233 high and higher secondary schools with 511,377 students, 1,631 middle schools with 832,886 students and 129 colleges of arts, science and commerce, nine engineering and technical colleges and ten medical colleges. There are four universities: Haryana Agricultural University, Hisar; Kurukshetra University, Kurukshetra with 70 colleges and 80,000 students (1999); Maharshi Dayanand University, Rohtak; and Guru Jambeshwar University, Hisar.

Health

In 2003 there were 49 hospitals, 64 community health centres, 402 primary health centres and 2,299 health sub-centres. A further 12 primary health centres were under construction.

RELIGION

At the 2001 census Hindus numbered 18,655,925; Muslims, 1,222,916; Sikhs, 1,170,622; Jains, 57,167.

FURTHER READING

Yadav, K. C., *Modern Haryana: History and Culture.* 2002

Himachal Pradesh

KEY HISTORICAL EVENTS

Thirty small hill states were merged to form the Territory of Himachal Pradesh in 1948; the state of Bilaspur was added in 1954 and parts of the Punjab in 1966. The whole territory, a Himalayan region of hill-tribes, rivers and forests, became a state in Jan. 1971. Its main areas are Chamba, a former principality, dominated in turn by Moghuls and Sikhs before coming under British influence in 1848; Bilaspur, an independent Punjab state until it was invaded by Gurkhas in 1814 (the British East India Company forces drove out the Gurkhas in 1815); Simla district around the town built by the Company near Bilaspur on land reclaimed from Gurkha troops (the summer capital of India from 1865 until 1948); Mandi, a principality until 1948 and Kangra and Kullu districts, originally Rajput areas which had become part of the British-ruled Punjab. These were all incorporated into Himachal Pradesh in 1966 when the Punjab was reorganized.

TERRITORY AND POPULATION

Himachal Pradesh is in north India and is bounded north by Kashmir, east by Tibet, southeast by Uttarakhand, south by Haryana, southwest and west by Punjab. The area of the state is 55,673 sq. km and the population (2011 census, provisional) 6,856,509; density, 123 per sq. km. Principal languages are Hindi and Pahari. The capital is Shimla, population (2001 census) of the urban agglomeration, 144,975.

SOCIAL STATISTICS

Growth rate 2001–11, 12·81%.

CONSTITUTION AND GOVERNMENT

Full statehood was attained, as the 18th State of the Union, on 25 Jan. 1971. On 1 Sept. 1972 districts were reorganized and three new districts created, Solan, Hamirpur and Una, making a total of 12.

There is a unicameral *Legislative Assembly* with 68 members.

RECENT ELECTIONS

Elections were held on 4 Nov. 2012. The opposition Indian National Congress won 36 seats; Bharatiya Janata Party, 26; ind., 6.

CURRENT GOVERNMENT

Governor: Urmila Singh; b. 1946 (took office on 25 Jan. 2010).

Chief Minister: Virbhadra Singh; b. 1934 (took office for a fourth time on 25 Dec. 2012).

ECONOMY

Budget

2008–09 revenue receipts, ₹93,079·9m.; expenditure, ₹94,381·3m. Budget estimates for 2009–10: revenue receipts, ₹104,783·3m.; expenditure, ₹102,217·6m.

ENERGY AND NATURAL RESOURCES

Electricity

In March 2011, 17,209 out of 17,495 inhabited villages had electricity. The state has huge hydropower potential—there is an estimated potential of 23,000 MW. Out of this hydropower potential only 29% has been harnessed. In Oct. 2008 there was an installed capacity of 1,896 MW. The Nathpa Jhakri project is India's largest hydroelectric power plant. The plant, which incorporates a 28 km power tunnel, came online in Oct. 2003. Electricity consumption per capita (2009–10), 1,380 kWh.

Water

An artificial confluence of the Sutlej and Beas rivers has been made, directing their united flow into Govind Sagar Lake.

Minerals

The state has rock salt, slate, gypsum, limestone, barytes, dolomite, pyrites, copper, gold and sulphur. However, Himachal Pradesh supplies only 0·2% of the national mineral output.

Agriculture

Agriculture employed 69% of the workforce in 2001. Irrigated area is 19% of the area sown. Main crops are seed potatoes, off season vegetables, wheat, maize, rice, hops and flowers, and fruits such as apples, peaches, apricots, kiwi fruit and strawberries; 1,027,820 tonnes of fruits were produced in 2010–11.

Production (2009–10): foodgrains 1,017,200 tonnes (of which maize 543,190 tonnes, wheat 327,140 tonnes and rice 105,900 tonnes), plus vegetables 1,390,671 tonnes and dry ginger 3,120 tonnes.

Livestock (2007 census): cattle, 2,279,000; goats, 1,241,000; sheep, 901,000; buffaloes, 762,000.

Forestry

Himachal Pradesh forests covered 14,668 sq. km in 2007 (26·4% of the state's area). The forests also ensure the safety of the catchment areas of the Yamuna, Sutlej, Beas, Ravi and Chenab rivers. Commercial felling of green trees has been totally halted and forest working nationalized.

INDUSTRY

The main sources of employment are the forests and their related industries; there are factories making turpentine and rosin. The state also makes fertilizers, cement, electronic items, TV sets, watches, computer parts and electronic toys. Sericulture is a major industry. There is a foundry and a brewery. Other industries include salt production and handicrafts, including weaving. The state has 173 large and medium units, 27,000 small scale units (providing employment for 140,000 people), seven industrial estates and 21 industrial areas. 300 mineral based industries have also been established.

COMMUNICATIONS

Roads

In March 2008 there were 36,298 km of roads (including 1,208 km of national highways and 1,824 km of state highways), of which 21,197 km were surfaced. There were 371,000 motor vehicles registered in March 2008, up from 269,000 in March 2003.

Rail

There is a line from Chandigarh to Shimla, and the Jammu-Delhi line runs through Pathankot. A Nangal-Talwara rail link has been approved by the central government. There are two narrow gauge lines, from Shimla to Kalka (96 km) and Jogindernagar to Pathankot (103 km), and a broad gauge line from Una to Nangal (16 km). Route-km in 2008–09, 285 km.

Civil Aviation

The state has airports at Bhuntar near Kullu, at Jubbarhatti near Shimla and at Gaggal in Kangra district. There are also 12 state-run helipads across the state.

SOCIAL INSTITUTIONS

Justice

The state has its own High Court at Shimla with eight judges.

Education

In 2011, 83·8% of the population were literate (90·8% of men and 76·6% of women). There were (2010–11) 10,767 primary schools with 422,700 students, 2,303 middle schools with 311,700 students, 2,094 high and senior secondary schools with 400,400 students, and 65 government-run colleges with 86,900 students. The Central University of Himachal Pradesh was established in 2009. There are also three state universities (Himachal Pradesh University, Shimla (1970); Chaudhary Sarwan Kumar Himachal Pradesh Krishi Vishvavidyalaya, Palampur (1978); Dr Y. S. Parmar University of Horticulture and Forestry, Solan (1985)) and 11 private universities. In 2010–11 there were three medical colleges, 109 general education colleges, 72 B.Ed (Bachelor of Education) colleges (typically for teaching at secondary level), 29 polytechnic institutions, 27 junior basic training colleges (typically for teaching at primary level), 23 Sanskrit institutions and four dental colleges in the state.

Health

There were (2005) 50 hospitals (8,832 beds), 505 primary and community health centres and 2,068 sub-health centres.

RELIGION

At the 2001 census Hindus numbered 5,800,222, Muslims 119,512, Buddhists 75,859 and Sikhs 72,355.

FURTHER READING

Verma, Vishwashwar, *The Emergence of Himachal Pradesh: A Survey of Constitutional Development.* 1995

Jammu and Kashmir

KEY HISTORICAL EVENTS

The state of Jammu and Kashmir was brought into being in 1846 at the close of the First Sikh War. By the Treaty of Amritsar, Gulab Singh, *de facto* ruler of Jammu and Ladakh, added Kashmir to his existing territories, in return for paying the indemnity imposed by the British on the defeated Sikh empire. Of the state's three parts, Ladakh and Kashmir were ancient polities, Ladakh having been an independent kingdom since the 10th century AD until its conquest by Gulab Singh's armies in 1834–42. Kashmir lost its independence to the Mughal empire in 1586, and was conquered in turn by the Afghans (1756) and the Sikhs (1819). Jammu was a collection of small principalities until consolidated by Gulab Singh and his brothers in the early nineteenth century.

British supremacy was recognized until the Indian Independence Act, 1947, when all states decided on accession to India or Pakistan. Kashmir asked for standstill agreements with both. Pakistan agreed, but India wanted further discussion with the government of Jammu and Kashmir State. Meantime the state was subject to armed attack from Pakistan. The Maharajah acceded to India on 26 Oct. 1947. India approached the UN in Jan. 1948, and the conflict ended by ceasefire in Jan. 1949. The major part of the state remained with India after territory in the north and west went to Pakistan. Hostilities between the two countries broke out in 1965 and again in 1971, but notwithstanding bilateral agreements—the Tashkent Declaration (Jan. 1966) and the Simla Agreement (July 1972)—the issue remains unresolved. With Muslims in the majority, both India and Pakistan regard the state as a touchstone of their divergent political raisons d'être—Pakistan as a Muslim nation, and India a secular one—and hence their position as non-negotiable. Intermittent violence between nationalistic factions has led to further negotiations between India and Pakistan with both sides pledging a peaceful solution. In Dec. 2002 the new provincial government promised to open talks with separatist groups.

TERRITORY AND POPULATION

The state is in the extreme north and is bounded north by China, east by Tibet, south by Himachal Pradesh and Punjab and west by Pakistan. The area is 222,236 sq. km and the population (2011 census, provisional) 12,548,926; density, 56 per sq. km. Srinagar (population, 2001, 988,210) is the summer and Jammu (612,163) the winter capital. The official language is Urdu; other commonly spoken languages are Kashmiri, Hindi, Dogri, Gujri, Pahari, Ladakhi and Punjabi.

SOCIAL STATISTICS

Growth rate 2001–11, 23·71%.

CONSTITUTION AND GOVERNMENT

The Maharajah's son, Yuvraj Karan Singh, took over as Regent in 1950 and, on the ending of hereditary rule (17 Oct. 1952), was sworn in as Sadar-i-Riyasat. On his father's death (26 April 1961) Yuvraj Karan Singh was recognized as Maharajah by the Indian government. The permanent Constitution of the state came into force in part on 17 Nov. 1956 and fully on 26 Jan. 1957. There is a bicameral legislature; the *Legislative Council* has 36 members and the *Legislative Assembly* has 89 (two of which are nominated). Since the 1967 elections the six representatives of Jammu and Kashmir in the central House of the People are directly elected; there are four representatives in the Council of States. There was a period of President's rule in 1977, and since then President's rule has been imposed on four further occasions—in 1986, from 1990–96, in 2002 and in 2008–09.

The state has 14 districts.

RECENT ELECTIONS

Elections to the State Assembly were held in between 17 Nov. and 24 Dec. 2008. The ruling pro-India Jammu and Kashmir National Conference won 28 of the 87 seats (28 in 2002); the People's Democratic Party won 21 (16 in 2002); the Indian National Congress, 17 (20 in 2002); the Bharatiya Janata Party, 11 (1 in 2002); and the Jammu and Kashmir National Panthers Party, 3 (0 in 2002). Despite ongoing violence and calls from Kashmiri separatist groups to boycott the elections, turnout was 61%. Following the elections, a coalition government was formed by the National Conference and the Indian National Congress.

CURRENT GOVERNMENT

Governor: Narinder Nath Vohra; b. 1936 (since 25 June 2008).

Chief Minister: Omar Abdullah; b. 1970 (took office on 5 Jan. 2009).

ECONOMY

Budget

Budget estimates for 2009–10 showed revenue receipts of ₹193,621·1m. and revenue expenditure of ₹147,257·9m.

ENERGY AND NATURAL RESOURCES

Electricity

The state has exploitable hydropower potential of about 15,000 MW. The gas turbine station at Srinagar is an important contributor. Installed capacity (Oct. 2008) 2,009 MW. In Sept. 2008, 6,304 out of 6,417 villages had electricity.

Minerals

Minerals include coal, bauxite and gypsum.

Agriculture

About 80% of the population are supported by agriculture. Rice, wheat and maize are the major cereals. The total area under foodgrains (1998–99) was estimated at 908,000 ha. Total foodgrains produced, 1998–99, 1·45m. tonnes (rice, 0·55m. tonnes; wheat, 0·43m. tonnes); pulses, 17,000 tonnes. Fruit is important: production, 1994–95, 0·9m. tonnes; exports, 0·76m. tonnes.

Irrigated area, 1993–94, 442,000 ha.

Livestock (2003): cattle, 3·08m.; buffaloes, 1·04m.; goats, 2·06m.; sheep, 3·41m.; poultry, 5·57m.

Forestry

Forests cover 22,686 sq. km (2007), forming an important source of revenue, besides providing employment to a large section of the population.

INDUSTRY

There are two central public sector industries and 30 medium-scale. There are 672 factories (2007–08) employing 52,700 people. There are industries based on horticulture; traditional handicrafts are silk spinning, wood-carving, papier mâché and carpet-weaving. 738 tonnes of silk cocoons were produced in 2008–09.

The handicraft sector had a production turnover of ₹16,145m. in 2007–08 of which carpets accounted for ₹7,613m.

COMMUNICATIONS

Roads

In March 2008 there were 22,323 km of roads (including 1,245 km of national highways and 67 km of state highways), of which 10,141 km were surfaced. There were 620,000 motor vehicles registered in March 2008, up from 399,000 in March 2003.

Rail

Kashmir is linked with the Indian railway system by the line between Jammu and Pathankot; route-km of railways in the state, 2008–09, 239 km.

Civil Aviation

Major airports are at Srinagar and Jammu. There is a third airport at Leh. There are services connecting Jammu with Amritsar, Chandigarh, Delhi and Srinagar; and services connecting Srinagar with Ahmedabad, Amritsar, Chandigarh, Delhi, Jammu, Leh and Mumbai.

SOCIAL INSTITUTIONS

Justice

The High Court, at Srinagar and Jammu, has a Chief Justice and four puisne judges.

Education

The proportion of literate people was 55·5% in 2001 (66·6% of men and 43·0% of women). Education is free. There were (1996–97) 1,351 high and higher secondary schools with 227,699 students, 3,104 middle schools with 405,598 students and 10,483 primary schools with 893,005 students. Jammu University (1969) has five constituent and 13 affiliated colleges, with 15,278 students (1992–93); Kashmir University (1948) has 18 colleges (17,000 students, 1992–93); there are two other universities: Sher-e-Kashmir University of Agricultural Sciences and Technology and Hemwati Nandan Bahuguna Garhwal University at Srinagar. There are four medical colleges, two engineering and technology colleges, four polytechnics, eight oriental colleges and an Ayurvedic college, 34 arts, science and commerce colleges and four teacher training colleges.

Health

In 2001 there were 43 hospitals with (2000) 2,076 beds, 337 primary health centres and 1,700 sub-centres, and 53 community health centres. There is a National Institute of Medical Sciences.

RELIGION

The majority of the population, except in Jammu, are Muslims (making it the only Indian state to have a Muslim majority). At the 2001 census Muslims numbered 6,793,240, Hindus 3,005,349, Sikhs 207,154 and Buddhists 113,787.

FURTHER READING

Behera, Navnita Chadha, *Demystifying Kashmir.* 2007
Hewitt, Vernon, *Towards the Future?: Jammu and Kashmir in the 21st Century.* 2001
Lamb, A., *Kashmir: a Disputed Legacy, 1846–1990.* 1991
Wirsing, R. G., *India, Pakistan and the Kashmir Dispute: on Regional Conflict and its Resolution.* 1995

Jharkhand

KEY HISTORICAL EVENTS

The state was carved from Bihar to become the twenty-eighth state of India on 15 Nov. 2000. Located in the plateau regions of eastern India, Jharkhand (literally land of forests) is mentioned in ancient Indian texts as an area inaccessible to the rest of India owing to its unforgiving terrain and the warring forest tribes. The Mughals attacked the region in 1385 and again in 1616, imprisoning the King of Jharkhand while they collected money from local chieftains. In the 17th century Jharkhand was a part of the Mughal empire and spread over areas of present-day Madhya Pradesh and Bihar. The East India Company was granted revenue-collecting power in 1765 and the permanent settlement of 1796 increased the company's grip on the area. In 1858 sovereignty was transferred to the English crown. From 1793 until 1915 there were periodic tribal rebellions throughout Jharkhand. In 1912 Jharkhand was constituted as part of the province of Bihar and present-day Odisha after the former was separated from West Bengal. In 1995 the Jharkhand Party submitted a request to the State Reorganisation Committee for Jharkhand to become a separate state. In 2000 Jharkhand came into being after legislation initiated by the National Democratic Alliance.

TERRITORY AND POPULATION

Jharkhand is in central eastern India and is bounded by Bihar to the north, West Bengal to the east, Odisha to the south and Chhattisgarh to the west. Jharkhand has an area of 79,714 sq. km. Population (2011 census, provisional) 32,966,238; density: 414 per sq. km. Cities with over 250,000 population, see INDIA: Territory and Population. Other large cities (2001): Phusro (174,402), Hazaribag (135,473), Deogar (112,525), Ramgarh (110,496), Chirkunda (106,227), Giridih (105,634). The principal language is Hindi.

SOCIAL STATISTICS

Growth rate 2001–11, 22·34%.

CONSTITUTION AND GOVERNMENT

Jharkhand is the twenty-eighth state of India. After the region was carved from Bihar it was decided that the 81 Members of the *Legislative Assembly* (MLAs) from Jharkhandi districts would become the members of the new state's legislative assembly. For administrative purposes the region is divided into 24 districts.

The capital and seat of government is at Ranchi.

Presidential rule was imposed on 19 Jan. 2009 following the resignation of Chief Minister Shibu Soren but was lifted on 30 Dec. 2009. It was imposed again on 1 June 2010, once more following the resignation of Chief Minister Shibu Soren, and was lifted on 9 Sept. 2010. On 18 Jan. 2013 it was imposed for the third time since Jharkhand came into existence in 2000 following the resignation of Chief Minister Arjun Munda.

RECENT ELECTIONS

In elections held on 27 Nov. and 2, 8, 12 and 18 Dec. 2009 the Bharatiya Janata Party won 18 of 81 seats; Jharkhand Mukti Morcha, 18; the Indian National Congress, 14; Jharkhand Vikas Morcha (Prajatantrik), 11; Rashtriya Janata Dal, 5; Janata Dal (United), 2; ind. and other parties, 13.

CURRENT GOVERNMENT

Governor: Syed Ahmad (b. 1945; took office on 4 Sept. 2011).
Chief Minister: Vacant.

ECONOMY

Budget

Budget estimates for 2009–10 showed revenue receipts of ₹179,356·0m. and revenue expenditure of ₹182,129·5m.

ENERGY AND NATURAL RESOURCES

Electricity

In Sept. 2008, 9,119 out of 29,354 inhabited villages had electricity (equivalent to 31%—the lowest proportion of any state). Installed capacity was 2,153 MW in Oct. 2008.

Minerals

Jharkhand is very rich in minerals, with about 40% of national production, including 90% of the country's cooking coal deposits, 40% of its copper, 37% of known coal reserves and 2% of iron ore. Other important minerals: bauxite, quartz, building stones and ceramics, graphite, limestone, kyanite, manganese, lead and silver. The state has 176 coal mines with an annual production of 78·7m. tonnes. Annually the state mines 8·6m. tonnes of iron ore, 1·2m. tonnes of copper ores, 1·0m. tonnes of bauxite, 50,000 tonnes of fire clays and 18,700 tonnes of manganese.

INDUSTRY

There is a major engineering corporation in Jharkhand as well as India's largest steel plant at Bokaro. Other important industries are aluminium and copper plants, forging, explosives, refractories and glass production. Jharkhand contains large thermal plants at Patratu, Tenughat, Chandrapura and Bokaro.

COMMUNICATIONS

Roads

In March 2008 there were 17,531 km of roads (including 1,805 km of national highways and 1,886 km of state highways), of which 10,037 km were surfaced. There were 1,850,000 motor vehicles registered in March 2008, up from 1,101,000 in March 2003.

Rail

Ranchi and the steel city of Bokaro are at the hub of the state's railway network linking Jharkhand to its neighbouring states as well as to Kolkata; route-km of railways in the state, 2008–09, 1,968 km.

SOCIAL INSTITUTIONS

Education

In 2001, 53·6% of the population were literate (67·3% of men and 38·9% of women). There are five universities: Ranchi University (founded 1960), with 106 colleges and 55,731 students (1994–95); Bisra Agricultural University at Ranchi (1980); Sidhu Kanhu University at Dumka; Binova Bhave University at Hazaribag; B. I. T. Mesra University at Ranchi (formerly Birla Institute of Technology). There are two law colleges, two agricultural colleges, five engineering colleges and ten medical colleges.

Health

There were 83 hospitals in 2003.

RELIGION

At the 2001 census Hindus numbered 18,475,681; Muslims, 3,731,308; Christians, 1,093,382; Sikhs, 83,358.

FURTHER READING

Corbridge, Stuart, Jewitt, Sarah and Kumar, Sanjay, *Jharkhand: Environment, Development, Ethnicity.* 2004

Karnataka

KEY HISTORICAL EVENTS

The state of Karnataka, constituted as Mysore under the States Reorganization Act, 1956, brought together the Kannada-speaking people distributed over five states. It consists of the territories of the old states of Mysore and Coorg, the Bijapur, Kanara and Dharwad districts and the Belgaum district (except one taluk) in former Bombay, the major portions of the Gulbarga, Raichur and Bidar districts in former Hyderabad, the South Kanara district (apart from the Kasaragod taluk) and the Kollegal taluk of the

Coimbatore district in Madras. The state was renamed Karnataka in 1973.

TERRITORY AND POPULATION

The state is in south India and is bounded north by Maharashtra, east by Andhra Pradesh, south by Tamil Nadu and Kerala, west by the Indian ocean and northeast by Goa. The area of the state is 191,791 sq. km, and its population (2011 census, provisional), 61,130,704; density, 319 per sq. km. Principal cities, see INDIA: Territory and Population. The capital is Bangalore. Other large towns (2001) are: Tumkur (248,929), Raichur (207,421), Bidar (174,257), Hospet (164,240), Bhadravati (160,662), Robertson Pet (157,084), Gadag (154,982), Hassan (133,262), Mandya (131,179), Udupi (127,124), Chitradurga (125,170), Kolar (113,907), Gangawati (101,392), Chikmagalur (101,251).

Kannada is the language of administration and is spoken by about 66% of the people. Other languages include Urdu (9%), Telugu (8·2%), Marathi (4·5%), Tamil (3·6%), Tulu and Konkani.

SOCIAL STATISTICS

Growth rate 2001–11, 15·67%.

CONSTITUTION AND GOVERNMENT

Karnataka has a bicameral legislature. The *Legislative Council* has 75 members. The *Legislative Assembly* consists of 225 members (one of which is nominated).

The state has 30 districts grouped in four divisions: Bangalore, Belgaum, Gulbarga and Mysore.

RECENT ELECTIONS

At the state elections on 10, 16 and 22 May 2008 the BJP won 110 seats; the INC 80; the Janata Dal (Secular) 28. Six independents were elected. Turnout was 65·1%.

CURRENT GOVERNMENT

Governor: Hans Raj Bhardwaj; b. 1937 (took office on 29 June 2009).

Chief Minister: Jagadish Shettar; b. 1955 (took office on 12 July 2012).

ECONOMY

Budget

Revenue receipts for 2008–09, ₹432,906·7m.; expenditure, ₹416,592·9m. Budget estimates for 2009–10: revenue receipts, ₹483,890·4m.; expenditure, ₹472,376·6m.

ENERGY AND NATURAL RESOURCES

Electricity

In Oct. 2008 the state's installed capacity was 9,116·7 MW. Electricity generated, 2008–09, 41,799m. kWh. In Sept. 2008, 27,126 out of 27,481 inhabited villages had electricity.

Minerals

Karnataka is an important source of gold and silver. The state produces 84% of India's gold. The estimated reserves of high grade iron ore are 8,798m. tonnes. These reserves are found mainly in the Chitradurga belt. The National Mineral Development Corporation of India has indicated total reserves of nearly 332m. tonnes of magnesite and iron ore (with an iron content ranging from 25 to 40) which have been found in Kudremukh Ganga-Mula region in Chikmagalur District. Value of production (2002–03) ₹10,580m. The estimated reserves of manganese are over 320m. tonnes.

Limestone is found in many regions; production (2006–07) was about 14·7m. tonnes.

Karnataka is the largest producer of chromite. It is one of only two states in India producing magnesite. The other minerals of industrial importance are corundum and garnet. Karnataka produces 63% of India's moulding sand annually and 57% of the country's quartzite, and is the only producer of felsite.

Agriculture

Agriculture forms the main occupation of more than three-quarters of the population. Physically, Karnataka divides into four regions—the coastal region, the southern and northern plains, comprising roughly the districts of Bangalore, Tumkur, Chitradurga, Kolar, Bellary, Mandya and Mysore, and the hill country, comprising the districts of Chikmagalur, Hassan and Shimoga. Rainfall is heavy in the hill country, and there is dense forest. The greater part of the plains are cultivated. Coorg district is essentially agricultural.

The main food crops are rice paddy and jowar, and ragi which is also about 30% of the national crop. Total foodgrains production (2007–08), 11·42m. tonnes (including rice, 3·18m. tonnes); pulses 1·23m. tonnes. Sugar, groundnut, castor-seed, safflower, mulberry silk and cotton are important cash crops. The state grows about 70% of the national coffee crop.

Production, 2008–09: sugarcane, 23·33m. tonnes; cotton, 775,000 bales of 170 kg.

Livestock (2003): cattle, 9·54m.; sheep, 7·26m.; goats, 4·48m.; buffaloes, 3·99m. Livestock products (2008–09): meat, 114,520 tonnes; milk, 4·5m. tonnes; 2·4bn. eggs.

Forestry

Total forest in the state (2007) is 30,720 sq. km, producing sandalwood, bamboo and other timbers.

Fisheries

The catch in 2008–09 totalled 361,854 tonnes. In 2005, 83,400 people were working in fishing and related activities.

INDUSTRY

There were 11,983 factories, 159 industrial estates and 5,787 industrial sheds employing 1,079,700 in March 2009. In 2008–09, 377,725 small industries employed 2,173,100 persons. The Vishveshwaraiah Iron and Steel Works is situated at Bhadravati, while at Bangalore are national undertakings for the manufacture of aircraft, machine tools, telephones, light engineering and electronics goods. The Kudremukh iron ore project is of national importance. An oil refinery is in operation at Mangalore. Other industries include textiles, vehicle manufacture, cement, chemicals, sugar, paper, porcelain and soap. In addition, much of the world's sandalwood is processed, the oil being one of the most valuable products of the state. Sericulture is a more important cottage industry giving employment, directly or indirectly, to about 2·7m. persons; production of silk, 2006–07, 8,205 tonnes or 46% of national output.

COMMUNICATIONS

Roads

In 2009 the state had 228,810 km of roads, including 4,491 km of national highway and 20,905 km of state highway. There were 6,217,000 motor vehicles registered in March 2008, up from 3,738,000 in March 2003.

Rail

In 2008–09 there were 3,007 route-km of railway in the state.

Civil Aviation

There are airports at Bangalore, Hubli, Mysore, Mangalore, Bellary and Belgaum, with regular scheduled services to Chennai, Delhi, Kolkata and Mumbai. Bangalore is being upgraded to an international airport—the present airport already receives international flights from a number of destinations. A new Bangalore international airport has been constructed at Devanahalli, 34 km from the city, and opened in May 2008.

Shipping

Mangalore is a deep-water port for the export of mineral ores. Karwar is being developed as an intermediate port.

SOCIAL INSTITUTIONS

Justice

The seat of the High Court is at Bangalore. It has a Chief Justice and 42 puisne judges.

Education

In 2001, 66·6% of the population were literate (76·1% of men and 56·9% of women). In 2009 the state had 57,520 primary schools with 7,570,000 students, 11,753 secondary schools with 2,522,000 students, 3,530 pre-university colleges, 284 polytechnic colleges, 161 engineering colleges, 110 medical colleges, 40 dentistry colleges and 16 universities including the National Law School of India. Education is free up to pre-university level.

Universities: Mysore (1916); Karnataka (1949) at Dharwad; University of Agricultural Sciences (1964) at Hebbal, Bangalore; Bangalore; Gulbarga; Kannada; Mangalore; University of Agricultural Sciences, Dharwad; Kuvempu University, Shimoga; Karnataka State Open University, Mysore; Rajiv Gandhi University of Health Sciences, Bangalore; and Visveswaraiah Technological University, Nehrunagar. The Indian Institute of Science, in Bangalore, the Manipal Academy of Higher Education and the National Institute of Mental Health and Neuro Sciences, also in Bangalore, all have university status.

Health

There were in 2008–09, 172 hospitals including 103 Ayurvedic hospitals, 324 community health centres and dispensaries, 2,193 primary health centres and 8,143 health subcentres. Total number of beds in 2008–09, 51,800.

RELIGION

At the 2001 census Hindus numbered 44,321,279; Muslims, 6,463,127; Christians, 1,009,164; Buddhists, 393,300.

Kerala

KEY HISTORICAL EVENTS

The state of Kerala was created in 1956, bringing together the Malayalam-speaking areas. It includes most of the former state of Travancore-Cochin and small areas from the state of Madras. Cochin, a safe harbour, was an early site of European trading in India. In 1795 the British took it from the Dutch and British influence remained dominant. Travancore was a Hindu state which became a British protectorate in 1795, having been an ally of the British East India Company for some years. Cochin and Travancore were combined as one state in 1947, and reorganized and renamed Kerala in 1956.

TERRITORY AND POPULATION

Kerala is in south India and is bounded north by Karnataka, east and southeast by Tamil Nadu, southwest and west by the Indian ocean. The state has an area of 38,863 sq. km. The 2011 census showed a population (provisional) of 33,387,677; density, 859 per sq. km. Chief cities, see INDIA: Territory and Population. Other principal towns (2001): Palakkad (197,369), Kottayam (172,878), Malappuram (170,409), Cherthala (141,558), Guruvayur (138,681), Kanhangad (129,367), Vadakara (124,083).

Languages spoken in the state are Malayalam, Tamil and Kannada.

SOCIAL STATISTICS

Growth rate 2001–11, 4·86%.

CONSTITUTION AND GOVERNMENT

The state has a unicameral legislature of 141 members (one of which is nominated) including the Speaker.

The state has 14 districts. The capital is Thiruvananthapuram.

RECENT ELECTIONS

At the elections of 13 April 2011 the United Democratic Front (UDF), led by the Indian National Congress, won 72 out of 140 seats with the remaining 68 seats going to the Left Democratic Front (LDF), led by the Communist Party of India (Marxist).

CURRENT GOVERNMENT

Governor: Nikhil Kumar; b. 1941 (since 23 March 2013).

Chief Minister: Oommen Chandy; b. 1943 (since 18 May 2011).

ECONOMY

Budget

Revenue receipts for 2008–09, ₹245,121·7m.; expenditure, ₹282,238·8m. Budget estimates for 2009–10: revenue receipts, ₹281,538·9m.; expenditure, ₹311,618·4m.

ENERGY AND NATURAL RESOURCES

Electricity

Installed capacity (Oct. 2008), 3,514 MW. Much of the state's electricity is produced by the Idukki hydro-electric plant and the Sabarigiri scheme. The state had a power deficit until the inauguration of the Kayamkulam thermal power plant in 1999. In Sept. 2008 all 1,364 inhabited villages had electricity.

Minerals

The beach sands of Kerala contain monazite, ilmenite, rutile, zircon, sillimanite, etc. There are extensive white clay deposits; other minerals of commercial importance include titanium, copper, magnesite, china clay, limestone, quartz sand and lignite. Iron ore has been found at Kozhikode (Calicut).

Agriculture

Area under irrigation in 2000–01 was 458,000 ha. The chief agricultural products are rice, tapioca, coconut, arecanut, cashew nuts, oilseeds, pepper, sugarcane, rubber, tea, coffee and cardamom. About 98% of Indian black pepper and about 95% of Indian rubber is produced in Kerala. Production of principal crops, 2002–03: total rice, 688,859 tonnes (from 310,521 ha.); tapioca, 2,413,217 tonnes; rubber, 594,917 tonnes; pepper, 67,358 tonnes; cashew nuts, 66,087 tonnes; coffee, 63,322 tonnes; tea, 53,480 tonnes; ginger, 32,412 tonnes; sugarcane, 31,283 tonnes; coconuts, 5,709m. nuts.

Livestock (2003): cattle, 2·12m.; goats, 1·21m.; poultry, 12·22m. In 2001–02 milk production was 2·73m. tonnes; egg production, 2,055m. units.

Forestry

Forest occupied 17,324 sq. km in 2007 (44·6% of the state's total area), including teak, sandalwood, ebony and blackwood and varieties of softwood.

Fisheries

The total catch in 2008–09 was 866,000 tonnes. In 2005 there were 211,300 people working in fishing and allied activities.

INDUSTRY

There are numerous cashew and coir factories. Important industries include rubber, tea, coffee, tiles, automotive tyres, watches, electronics, oil, textiles, ceramics, fertilizers and chemicals, pharmaceuticals, zinc-smelting, sugar, cement, rayon, glass, matches, pencils, monazite, ilmenite, titanium oxide, rare earths, aluminium, electrical goods, paper and shark-liver oil. The state has a refinery and a shipyard at Kochi (Cochin).

There are 5,584 factories (2007–08) employing 356,000 people.

COMMUNICATIONS

Roads
In March 2008 there were 204,757 km of roads (including 1,457 km of national highways and 4,137 km of state highways), of which 116,446 km were surfaced. There were 4,430,000 motor vehicles registered in March 2008, up from 2,552,000 in March 2003.

Rail
There is a coastal line from Mangalore in Karnataka which connects with Tamil Nadu. The state had 1,050 route-km of railway line in 2008–09.

Civil Aviation
There are airports at Kozhikode, Kochi and Thiruvananthapuram with regular scheduled internal services to Chennai, Delhi and Mumbai. In addition Kochi has international flights to a number of destinations in the Gulf states plus Colombo and Singapore, and Kozhikode and Thiruvananthapuram also have flights to the Gulf.

Shipping
Port Kochi, administered by the central government, is one of India's major ports; in 1983 it became the out-port for the Inland Container Depot at Coimbatore in Tamil Nadu. There are 12 other ports and harbours.

SOCIAL INSTITUTIONS

Justice
The High Court at Ernakulam has a Chief Justice and 29 puisne judges.

Education
Kerala is the most literate Indian state, with 25·49m. literate people at the 2001 census (90·9%; 94·2% of men and 87·7% of women). Education is free up to the age of 14.

In 2000 there were 6,726 primary schools with 2·79m. students, 2,968 upper primary schools with 1·84m. students and 3,511 high and higher secondary schools with 1·07m. students. There were 169 junior colleges in 1996–97 with 210,074 pupils.

Kerala University (established 1937) at Thiruvananthapuram is affiliating and teaching; in 1995–96 it had 52 affiliated colleges with 113,569 students. The University of Kochi is federal, and for post-graduate studies only. The University of Calicut (established 1968) is teaching and affiliating and has 95 affiliated colleges with 122,343 students (1995–96). Kerala Agricultural University (established 1971) has seven constituent colleges. Mahatma Gandhi University at Kottayam was established in 1983 and has 64 affiliated colleges with 112,992 students (1995–96). There are two other universities, Sree Sankaracharya University at Ernakulam and Kannur (formerly Malabar) University. There were also (2000) seven medical colleges, 20 pharmacy colleges, three dental colleges, four homeopathy colleges, 32 engineering colleges, 59 technology colleges, three nursing colleges, 19 teacher training colleges and 191 arts and science colleges.

Health
In 2000 there were 1,425 hospitals and health centres, including 113 Ayurvedic hospitals and 30 homeopathic hospitals. There were 41,462 hospital beds plus 2,604 beds in Ayurvedic hospitals and 970 beds in homeopathic clinics and hospitals.

RELIGION

At the 2001 census Hindus numbered 17,883,449; Muslims, 7,863,842; Christians, 6,057,427.

FURTHER READING
Jeffrey, R., *Politics, Women and Well-Being: How Kerala Became a Model.* 1992
Tharamangalam, Joseph, (ed.) *Kerala: The Paradoxes of Public Action and Development.* 2006

Madhya Pradesh

KEY HISTORICAL EVENTS
The state was formed in 1956 to bring together the Hindi-speaking districts including the 17 Hindi districts of the old Madhya Pradesh, most of the former state of Madhya Bharat, the former states of Bhopal and Vindhya Pradesh and a former Rajput enclave, Sironj. This was an area which the Mahrattas took from the Moghuls between 1712 and 1760. The British overcame the Mahrattas in 1818 and established their own Central Provinces. Nagpur became the Province's capital and was also the capital of Madhya Pradesh until, in 1956, boundary changes transferred it to Maharashtra. The present capital, Bhopal, was the centre of a Muslim princely state from 1723. Bhopal, an ally of the British against the Mahrattas, with neighbouring small states, became a British-protected agency in 1818. After independence Bhopal acceded to the Indian Union in 1949. In 1956 the states of Madhya Bharat and Vindhya Pradesh were combined with Bhopal and Sironj and renamed Madhya Pradesh. In 2000, 16 mainly tribal districts were carved from Madhya Pradesh to form Chhattisgarh.

TERRITORY AND POPULATION
The state is in central India and is bounded north by Uttar Pradesh, east by Chhattisgarh, south by Maharashtra, and west by Gujarat and Rajasthan. Since the creation of Chhattisgarh in 2000, Madhya Pradesh is no longer the largest Indian state in size. Its revised area is 308,245 sq. km (previously 443,446 sq. km), making it the second largest state in the country (after Rajasthan). Population (2011 census, provisional), 72,597,565; density, 236 per sq. km.

Cities with over 250,000 population, *see* INDIA: Territory and Population. Other large cities (2001): Ratlam, 234,419; Dewas, 231,672; Satna, 229,307; Burhanpur, 193,725; Murwara, 187,029; Singrauli, 185,190; Rewa, 183,274; Khandwa, 172,242; Bhind, 153,752; Chhindwara, 153,552; Morena, 150,959; Shivpuri, 146,892; Guna, 137,175; Damoh, 127,967; Vidisha, 125,453; Mandsaur, 117,555; Nimach, 112,852; Chhatarpur, 109,078; Itarsi, 107,831; Khargone, 103,448.

Hindi, Marathi, Urdu and Gujarati are spoken. In April 1990 Hindi, which predominates in the state, became the sole official language.

SOCIAL STATISTICS
Growth rate 2001–11, 20·30%.

CONSTITUTION AND GOVERNMENT
Madhya Pradesh is one of the nine states for which the Constitution provides a bicameral legislature, but the Vidhan Parishad or Upper House (to consist of 90 members) has yet to be formed. The Vidhan Sabha or Lower House has 231 members (one of which is nominated).

For administrative purposes the state has been split into nine revenue divisions with a Commissioner at the head of each; the headquarters of these are located at Bhopal, Gwalior, Hoshangabad, Indore, Jabalpur, Morena, Rewa, Sagar and Ujjain. There are 22,029 *gram* (village) panchayats, 313 *janpad* (intermediate) panchayats and 45 *zila* (district) panchayats. There are 50 districts in the state grouped into ten divisions.

The seat of government is at Bhopal.

RECENT ELECTIONS

At the election in Nov. 2008 the Bharatiya Janata Party (BJP) held power with 142 seats (173 in 2003). The Indian National Congress (INC) won 72 seats (38 in 2004). Other parties won 16 seats.

CURRENT GOVERNMENT

Governor: Ram Naresh Yadav; b. 1928 (took office on 8 Sept. 2011).

Chief Minister: Shivraj Singh Chauhan; b. 1959 (took office on 29 Nov. 2005).

ECONOMY

Budget

Revenue receipts for 2008–09, ₹335,772·0m.; expenditure, ₹295,138·8m. Budget estimates for 2009–10: revenue receipts, ₹399,610·3m.; expenditure, ₹382,621·1m.

ENERGY AND NATURAL RESOURCES

Electricity

Madhya Pradesh is rich in low-grade coal suitable for power generation, and also has immense potential for hydro-electric energy. Total installed capacity, 2008–09, 3,051 MW. Power generated, 16,315m. kWh in 2007–08. There are eight hydro-electric power stations of 747·5 MW installed capacity. In Sept. 2008, 50,218 out of 52,117 inhabited villages had electricity.

Water

Major irrigation projects include the Chambal Valley scheme, the Tawa project in Hoshangabad district, the Barna and Hasdeo schemes, the Mahanadi canal system and schemes in the Narmada valley at Bargi and Narmadasagar. Area under irrigation, 2007–08, 5·87m. ha.

Minerals

Much of the state's extensive mineral deposits were in the area that became Chhattisgarh in 2000. In 2004–05 the output of diamonds was 78,315 carats; Madhya Pradesh is India's only diamond producer. Production of other minerals included: 52·68m. tonnes of coal, 24·94m. tonnes of limestone, 2·05m. tonnes of copper ore (making Madhya Pradesh India's largest copper ore producer), 447,000 tonnes of manganese ore, 201,000 tonnes of iron ore and 186,000 tonnes of bauxite. In 2004–05 value of production was ₹36,996m.

Agriculture

The creation in 2000 of Chhattisgarh, previously known as the 'rice bowl' of Madhya Pradesh, had serious implications for the state. Agriculture is the mainstay of the state's economy and 76·8% of the people are rural. 43·7% of the land area is cultivable, of which 30·5% is irrigated. Production of principal crops, 2004–05 (in tonnes): foodgrains, 7·83m.; pulses, 3·35m.; cotton, 0·32m. bales of 170 kg.

Livestock (2003): cattle, 18·91m.; buffaloes, 7·58m.; goats, 8·14m.; sheep, 546,000.

Forestry

The forested area totalled 77,700 sq. km in 2007, or 25·2% of the state. The forests are chiefly of sal, saja and teak species. They are the chief source in India of best-quality teak; they also provide firewood for about 60% of domestic fuel needs, and form valuable watershed protection.

INDUSTRY

The major industries are steel, aluminium, paper, cement, motor vehicles, ordnance, textiles and heavy electrical equipment. Other industries include electronics, telecommunications, sugar, fertilizers, straw board, vegetable oil, refractories, potteries, textile machinery, steel casting and rerolling, industrial gases, synthetic fibres, drugs, biscuit manufacturing, engineering, optical fibres, plastics, tools, rayon and art silk. There are 3,165 factories (2007–08) employing 255,000 people.

There are 23 'growth centres' in operation, and five under development. The Government of India has set up a Special Economic Zone at Indore.

COMMUNICATIONS

Roads

In March 2008 there were 165,740 km of roads (including 4,670 km of national highways and 8,729 km of state highways), of which 82,426 km were surfaced. There were 5,523,000 motor vehicles registered in March 2008, up from 3,459,000 in March 2003.

Rail

The main rail route linking northern and southern India passes through Madhya Pradesh. Bhopal, Bina, Gwalior, Indore, Itarsi, Jabalpur, Katni, Khandwa, Ratlam and Ujjain are important junctions for the central, south, eastern and western networks. Route length (2008–09), 4,949 km.

Civil Aviation

There are domestic airports at Bhopal, Gwalior, Indore and Khajuraho with regular scheduled services to Agra, Delhi, Mumbai, Raipur and Varanasi.

SOCIAL INSTITUTIONS

Justice

The High Court of Judicature at Jabalpur has a Chief Justice and 29 puisne judges. Its benches are located at Gwalior and Indore. A National Institute of Law and a National Judicial Academy have been set up at Bhopal.

Education

In 2001, 63·7% of the population were literate (76·1% of men and 50·3% of women). Education is free for children aged up to 14.

In 2001 there were 97,273 government schools (63,712 pre-primary and primary schools, 25,090 middle schools, 8,471 high and higher secondary schools) and 22,620 private schools (10,473 pre-primary and primary schools, 7,474 middle schools, 4,673 high and higher secondary schools). Total enrolment in 2001: pre-primary and primary schools, 8·44m.; middle schools, 2·69m.; high and higher secondary schools, 1·52m.

Madhya Pradesh has ten universities (with towns): Barkatullah University (Bhopal) has 56 government colleges; Dr Harisingh Gour University (Sagar), 47 government colleges; Jiwaji Univeristy (Gwalior), 46 government colleges; Devi Ahilya University (Indore), 42 government colleges; Awadhesh Pratap Singh University (Rewa), 41 government colleges; Rani Durgavati University (Jabalpur), 40 government colleges; Vikram University (Ujjain), 40 government colleges; Mahatma Gandhi Gramodaya Vishwavidyalaya University (Chitrakoot); National Law Institute University (Bhopal); M. P. Bhoj Open University (Bhopal). There are a total of 312 colleges affiliated to the state's universities: 228 undergraduate and 84 postgraduate. Students studying in Madhya Pradesh totalled 272,296 in 2005–06 (of which arts students, 128,442; commerce, 73,211; science, 57,128).

Health

In 2001–02 there were 45 district hospitals, 57 urban civil hospitals, 1,194 primary health centres, 8,835 sub-health centres, 229 community health centres, seven TB hospitals and two TB sanatoriums.

RELIGION

At the 2001 census Hindus numbered 55,004,675; Muslims, 3,841,449; Jains, 545,446; Buddhists, 209,322; Christians, 170,381; Sikhs, 150,772.

Maharashtra

KEY HISTORICAL EVENTS

The Bombay Presidency region grew in the early 17th century from a collection of British East India Company trading posts. The island of Bombay was a Portuguese possession until it came under the control of Charles II of England on his marriage to Catherine of Braganza in 1661. It was then leased to the British East India Company in 1668 for £10 per annum. The Presidency expanded, overcoming the surrounding Mahratta chiefs until Mahratta power was finally conquered in 1818. After independence Bombay State succeeded to the Presidency; its area was altered in 1956 by adding Kutch and Saurashtra and the Marathi-speaking areas of Hyderabad and Madhya Pradesh, while taking away Kannada-speaking areas (which were added to Mysore). In 1960 the Bombay Reorganization Act divided Bombay State between Gujarati and Marathi areas, the latter becoming Maharashtra.

TERRITORY AND POPULATION

Maharashtra is in central India and is bounded north by Madhya Pradesh, east by Chhattisgarh, south by Andhra Pradesh, Karnataka and Goa, west by the Indian ocean and northwest by Daman and Gujarat. The state has an area of 307,713 sq. km. The population in 2011 (census, provisional) was 112,372,972; density, 365 per sq. km. In 2001 the area of Greater Mumbai was 603 sq. km and its population 16·4m. For other principal cities, *see* INDIA: Territory and Population. Other large towns (2001): Jalna (235,795), Bhusawal (187,564), Nalasopara (184,538), Vasai (174,396), Yavatmal (139,835), Bid (138,196), Kamthi (136,491), Gondia (120,902), Virar (118,928), Wardha (111,118), Satara (108,048), Achalpur (107,316), Barsi (104,785), Panvel (104,058).

The official language is Marathi.

SOCIAL STATISTICS

Growth rate 2001–11, 15·99%.

CONSTITUTION AND GOVERNMENT

Maharashtra has a bicameral legislature. The *Legislative Council* has 78 members. The *Legislative Assembly* has 288 elected members and one member nominated by the Governor to represent the Anglo-Indian community.

The Council of Ministers consists of the Chief Minister, 16 other Ministers and 19 Ministers of State.

The capital is Mumbai (Bombay). The state has 35 districts, which are grouped into six administrative divisions.

RECENT ELECTIONS

At the elections held on 13 Oct. 2009 the Indian National Congress won 82 of the 288 seats, the Nationalist Congress Party 62, the Bharatiya Janata Party 46, Shiv Sena 44, the Maharashtra Navnirman Sena 13, Republican Left Democratic Front 9, other parties 7 and ind. 25. Turnout was approximately 60%.

CURRENT GOVERNMENT

Governor: Kateekal Sankaranarayanan; b. 1932 (took office on 22 Jan. 2010).

Chief Minister: Prithviraj Chavan; b. 1946 (took office on 11 Nov. 2010).

ECONOMY

Budget

Revenue receipts for 2008–09, ₹812,706·8m.; expenditure, ₹756,939·1m. Budget estimates for 2009–10: revenue receipts, ₹890,606·5m.; expenditure, ₹961,840·2m.

ENERGY AND NATURAL RESOURCES

Electricity

Installed capacity, Oct. 2008, 20,036 MW. In Sept. 2008, 36,296 out of 41,095 inhabited villages had electricity; consumption per capita in 2009–10 was 1,028 kWh.

Oil and Gas

Mumbai High (India's largest offshore oil field) produced 12·11m. tonnes of crude oil in 2007–08 and 16·7bn. cu. metres of natural gas in 2008–09. Oil production has declined by one-third since the early 1990s. A recovery plan for the ageing field began in 2000. Phase one of this plan was completed in 2006 and two further phases to increase production are now under way.

Minerals

The state has coal, silica sand, dolomite, kyanite, chromite, limestone, iron ore, manganese and bauxite. Value of mineral production, 2001, ₹21,340m. of which 94% is contributed by coal. Coal production in 2000–01 was 28·8m. tonnes. Manganese is the second most valuable mineral.

Agriculture

3·3m. ha. of the cropped area of 21·4m. ha. are irrigated. In normal seasons the main food crops are rice, wheat, jowar, bajra and pulses. Main cash crops: cotton, sugarcane, groundnuts. Production, 2000–01 (in tonnes): sugarcane, 36·5m.; foodgrains, 11·9m. (rice, 2·4m.; wheat, 1·11m.); pulses, 1·7m.; groundnuts, 0·6m.; cotton, 476,000.

Livestock (2003 census, in 1,000): buffaloes, 6,145; other cattle, 16,303; sheep and goats, 13,778; poultry, 37,968.

Forestry

Forests occupied 50,650 sq. km in 2007 (16·5% of the state's area).

Fisheries

In 2000–01 the marine fish catch was estimated at 403,000 tonnes and the inland fish catch at 123,000 tonnes. In 2005 there were 153,900 people engaged in fishing and allied activities.

INDUSTRY

Industry is concentrated mainly in Mumbai, Nashik, Pune and Thane. The main groups are chemicals and products, textiles, electrical and non-electrical machinery, petroleum and products, aircraft, rubber and plastic products, transport equipment, automobiles, paper, electronic items, engineering goods, pharmaceuticals and food products. In 2007–08 there were 18,304 factories employing 1·4m. people.

COMMUNICATIONS

Roads

In March 2008 Maharashtra had 223,322 km of roads—including 4,176 km of national highways and 33,675 km of state highways—of which 178,045 km were surfaced. There were 13,335,000 motor vehicles registered in March 2008 (the most in any Indian state), up from 8,134,000 in March 2003.

Rail

Maharashtra had 5,602 route-km of railway line in 2008–09. The main junctions and termini are Mumbai, Dadar, Manmad, Akola, Nagpur, Pune and Sholapur.

Civil Aviation

The main airport is Mumbai, which has national and international flights. Nagpur airport is on the route from Mumbai to Kolkata and there are also airports at Pune and Aurangabad.

Shipping

Maharashtra has a coastline of 720 km. Mumbai is the major port, and there are 48 minor ports.

SOCIAL INSTITUTIONS

Justice

The High Court has a Chief Justice and 60 judges. The seat of the High Court is Mumbai, but it has benches at Nagpur, Aurangabad and Panaji (Goa).

Education

The number of literate people, according to the 2001 census, was 63·97m. (76·9%; 86·0% of men and 67·0% of women). In 2001 there were 10,225 high and 3,981 higher secondary schools with (1995) 2,795,567 pupils; 15,070 middle schools with (1995) 4,753,257 pupils; and 66,369 primary schools with (1995) 11,685,598 pupils. There are 111 engineering and technology colleges, 156 medical colleges (including dental and Ayurvedic colleges), 244 teacher training colleges, 152 polytechnics and 820 arts, science and commerce colleges.

The University of Mumbai, founded in 1857, is mainly an affiliating university. It has 276 colleges with a total (1993–94) of 234,469 students. Nagpur University (1923) is both teaching and affiliating. It has 258 colleges with 95,664 students. Pune University, founded in 1948, is teaching and affiliating; it has 167 colleges and 151,990 students. The SNDT Women's University had 33 colleges with a total of 33,343 students. Dr B. R. Ambedkar Marathwada University, Aurangabad was founded in 1958 as a teaching and affiliating body to control colleges in the Marathwada or Marathi-speaking area, previously under Osmania University; it has 190 colleges and 195,806 students. Shivaji University, Kolhapur, was established in 1963 to control affiliated colleges previously under Pune University. It has 205 colleges and 115,553 students. Amravati University has 130 colleges and 74,484 students. Other universities are: Marathwada Krishi Vidyapeeth, Parbhani; Y. Chavan Maharashtra Open University, Nashik; North Maharashtra University, Jalgaon, with 101 colleges and 66,092 students; Mahatma Phule Krishi University, Rahuri; Dr Punjabrao Deshmukh Krishi University, Akola; Konkan Krishi University, Dapoli; Dr Babasaheb Ambedkar Technological University, Lonere; Swami Ramanand Teerth Marathwad University, Nanded; Tilak Maharashtra Vidyapeeth, Pune; Bharati Vidyapeeth, Pune; Gokhale Institute of Politics and Economics, Pune; Deccan College, Pune; Indian Institute of Technology, Mumbai; Indira Gandhi Institute of Developmental Research, Mumbai; International Institute for Population Sciences, Mumbai; Tata Institute of Social Sciences, Mumbai. The Central Institute of Fisheries Education in Mumbai also has university-equivalent status.

Health

There were 3,446 hospitals with 99,062 beds in 2000; there were also 1,768 primary health centres, 9,725 sub-health centres and 351 community health centres in 2001.

RELIGION

At the 2001 census Hindus numbered 77,859,385; Muslims, 10,270,485; Buddhists, 5,838,710; Jains, 1,301,843; Christians, 1,058,313; Sikhs, 215,337.

FURTHER READING

Waite, Louise, *Embodied Working Lives: Work and Life in Maharashtra.* 2005

Manipur

KEY HISTORICAL EVENTS

Formerly a state under the political control of the government of India, Manipur entered into interim arrangements with the Indian Union on 15 Aug. 1947 and the political agency was abolished. Under a merger agreement, the administration was taken over by the government of India on 15 Oct. 1949 to be centrally administered by the government of India through a Chief Commissioner. In 1950–51 an Advisory government was introduced. In 1957 this was replaced by a Territorial Council of 30 elected and two nominated members. Later, in 1963, a Legislative Assembly of 30 elected and three nominated members was established under the government of Union Territories Act 1963. Because of the unstable party position in the Assembly, it had to be dissolved on 16 Oct. 1969 and president's rule introduced. The status of the administrator was raised from Chief Commissioner to Lieut.-Governor from 19 Dec. 1969. On 21 Jan. 1972 Manipur became a state and the status of the administrator was changed from Lieut.-Governor to Governor. In June 2001 Manipur was placed under central rule, but returned to self government after the 2002 elections.

TERRITORY AND POPULATION

The state is in northeast India and is bounded north by Nagaland, east by Myanmar, south by Myanmar and Mizoram, and west by Assam. Manipur has an area of 22,327 sq. km and a population (2011 census, provisional) of 2,721,756; density, 122 per sq. km. The valley, which is about 1,813 sq. km, is 800 metres above sea-level. The largest city is Imphal with a population of 250,234 (2001 census). The hills rise in places to 3,000 metres, but are mostly about 1,500–1,800 metres. The average annual rainfall is 165 cm. The hill areas are inhabited by various hill tribes who constitute about one-third of the total population of the state. There are about 30 tribes and sub-tribes falling into two main groups of Nagas and Kukis. Manipuri and English are the official languages. A large number of dialects are spoken.

SOCIAL STATISTICS

Growth rate 2001–11, 18·65%.

CONSTITUTION AND GOVERNMENT

Manipur has a *Legislative Assembly* of 60 members, of which 19 are from reserved tribal constituencies. There are nine districts. The capital is Imphal.

RECENT ELECTIONS

Elections were held on 28 Jan. 2012. The Indian National Congress party won 42 seats; the All India Trinamool Congress, 7; Manipur State Congress Party, 5; Naga People's Front, 4; Nationalist Congress Party, 1; Lok Janshakti Party, 1.

CURRENT GOVERNMENT

Governor: Gurbachan Jagat; b. 1942 (took office on 23 July 2008).

Chief Minister: Okram Ibobi Singh; b. 1948 (took office on 7 March 2002).

ECONOMY

Budget

2008–09 revenue receipts, ₹38,726·1m.; expenditure, ₹26,222·7m. Budget estimates for 2009–10: revenue receipts, ₹40,046·0m.; expenditure, ₹30,559·1m.

ENERGY AND NATURAL RESOURCES

Electricity

Installed capacity (Oct. 2008) was 158 MW. In Sept. 2008, 1,979 out of 2,315 inhabited villages had electricity.

Water

The main power, irrigation and flood-control schemes are the Loktak Lift Irrigation scheme (irrigation potential, 40,000 ha.); the Singda scheme (potential 4,000 ha., and improved water supply for Imphal); the Thoubal scheme (potential 34,000 ha.); and four other large projects. In 2003–04, 39,980 ha. were irrigated.

Minerals

Chromite is the only significant mineral resource—it is extracted from a single mine.

Agriculture

Rice is the principal crop; with wheat, maize and pulses. Total foodgrains, 2008–09, 421,000 tonnes (rice, 397,000 tonnes).

Around 75% of the cultivable area is under paddy. Fruit and vegetables are important in the valley, including pineapples, oranges, bananas, mangoes, pears, peaches and plums. Soil erosion, produced by shifting cultivation, is being halted by terracing. Fruit production in 2007–08, 0·27m. tonnes.

Forestry

Forests occupied 17,280 sq. km in 2007. The main products are teak, jurjan and pine; there are also large areas of bamboo and cane, especially in the Jiri and Barak river drainage areas, yielding about 0·3m. tonnes annually.

INDUSTRY

Handloom weaving is a cottage industry. Manipur is one of the least industrialized states of India. Location, limited infrastructure and insufficient power hold back industrial development. Larger-scale industries include the manufacture of bicycles and TV sets, sugar, cement, starch, vegetable oil and glucose. Sericulture produces about 45 tonnes of raw silk annually. Estimated non-agricultural workforce, 229,000.

COMMUNICATIONS

Roads

In March 2008 Manipur had 16,502 km of roads (including 959 km of national highways and 1,137 km of state highways), of which 6,682 km were surfaced. There were 147,000 motor vehicles registered in March 2008, up from 97,000 in March 2003.

Rail

A railway link was opened in 1990, linking Karong with the Assamese railway system.

Civil Aviation

There is an airport at Imphal with regular scheduled services to Delhi and Kolkata.

SOCIAL INSTITUTIONS

Education

In 2001, 70·5% of the population were literate (80·3% of men and 60·5% of women). In 1996–97 there were 2,548 primary schools with 230,230 students, 555 middle schools with 106,200 students, 553 high and higher secondary schools with 66,160 students, 50 colleges, one medical college, two teacher training colleges, three polytechnics, Manipur University with 62 colleges and 52,352 students (1997–98) and an agricultural university (Central Agricultural University, Imphal).

Health

In 2001 there were 85 hospitals and public health centres, 28 dispensaries, 16 community health centres, 420 sub-centres and 26 other facilities.

RELIGION

At the 2001 census Hindus numbered 46% of the population; Christians, 34%; Muslims, 9%.

Meghalaya

KEY HISTORICAL EVENTS

The state was created under the Assam Reorganization (Meghalaya) Act 1969 and inaugurated on 2 April 1970. Its status was that of a state within the State of Assam until 21 Jan. 1972 when it became a fully-fledged state of the Union. It consists of the former Garo Hills district and United Khasi and Jaintia Hills district of Assam.

TERRITORY AND POPULATION

Meghalaya is bounded in the north and east by Assam, south and west by Bangladesh. The area is 22,429 sq. km and the population (2011 census, provisional) 2,964,007; density, 132 per sq. km. The people are mainly of the Khasi, Jaintia and Garo tribes. The main languages of the state are Khasi, Garo and English.

SOCIAL STATISTICS

Growth rate 2001–11, 27·82%.

CONSTITUTION AND GOVERNMENT

Meghalaya has a unicameral legislature. The *Legislative Assembly* has 60 seats.

There are seven districts. The capital is Shillong (population, 2001 census, 267,662 in the urban agglomeration).

Presidential rule was imposed on 18 March 2009 despite the state government winning a tightly contested confidence vote a day earlier. It was lifted on 8 May after the Indian National Congress managed to form an alliance with the United Democratic Party.

RECENT ELECTIONS

In elections held on 23 Feb. 2013 the Indian National Congress won 29 seats; United Democratic Party, 8; Hill State People's Democratic Party, 4; Nationalist Congress Party, 2; National People's Party, 2. Independents and others won 15 seats.

CURRENT GOVERNMENT

Governor: Ranjit Shekhar Mooshahary (took office on 1 July 2008).

Chief Minister: Mukul Sangma (took office on 20 April 2010).

ECONOMY

Budget

2008–09 revenue receipts, ₹28,106·5m.; expenditure, ₹26,827·7m. Budget estimates for 2009–10: revenue receipts, ₹38,063·1m.; expenditure, ₹35,882·7m.

ENERGY AND NATURAL RESOURCES

Electricity

Total installed capacity (Oct. 2008) was 288 MW. In Sept. 2008, 3,428 out of 5,782 inhabited villages had electricity.

Minerals

The Khasi Hills, Jaintia Hills and Garo Hills districts produce coal, sillimanite (95% of India's total output), limestone, fire clay, dolomite, feldspar, quartz and glass sand. The state also has deposits of coal (estimated reserves 600m. tonnes), limestone (3,000m.), fire clay (6m.) and sandstone which are so far virtually untapped. Coal production in 2000–01 was 5,149,000 tonnes; limestone production in 2000–01 was 585,000 tonnes.

Agriculture

About 71% of the people depend on agriculture. Principal crops are rice, maize, potatoes, cotton, oranges, ginger, tezpata, areca nuts, jute, mesta, bananas and pineapples. Production 2000–01 (in tonnes) of principal crops: rice, 179,000; potatoes, 144,000; ginger, 45,000; jute, 36,000; citrus fruits, 32,000; maize, 24,000; cotton, 8,000; rape and mustard, 5,000. Poultry and cattle are the principal livestock.

Forestry

Forests covered 17,321 sq. km in 2007. Forest products are one of the state's chief resources.

INDUSTRY

Apart from agriculture the main source of employment is the extraction and processing of minerals; there are also important timber processing mills and cement factories. Other industries include electronics, tantalum capacitors, beverages and watches. The state has five industrial estates, two industrial areas and one growth centre. In 2007–08 there were 90 registered factories employing 5,587 people. In 2000, 17,800 workers were involved in manufacturing and processing. There were also, in 2001–02, 1,812 sericultural villages, six sericultural farms, eight silk units and nine weaving centres. In 2000 there were more than 5,400 *khadi* and village industrial units.

COMMUNICATIONS

Roads

In March 2008 there were 9,839 km of roads (including 810 km of national highways and 1,134 km of state highways), of which 5,472 km were surfaced. There were 128,000 motor vehicles registered in March 2008, up from 73,000 in March 2003.

Civil Aviation

Umroi airport (35 km from Shillong) connects the state with main air services. There are regular flights to Kolkata. Umroi is to be upgraded to receive larger aircraft. However, the main airport serving the state is Borjhar, at Guwahati, 21 km across the state border but only 124 km from Shillong. Guwahati has air links with several major north Indian cities.

SOCIAL INSTITUTIONS

Justice

The Guwahati High Court is common to Assam, Meghalaya, Nagaland, Manipur, Mizoram, Tripura and Arunachal Pradesh—there are 19 judges. There is a bench of the Guwahati High Court at Shillong.

Education

In 2001, 62·6% of the population were literate (65·4% of men and 59·6% of women). In 2000–01 the state had 4,685 primary and middle schools with 445,443 students, and 1,613 senior middle, secondary and higher secondary schools with 181,068 students. There were 35 colleges and other institutions of higher education including ten teacher training schools, one college and one polytechnic, with a total enrolment of 31,975 students. The North Eastern Hill University started functioning at Shillong in 1973; in 1993–94 it had 41 colleges and 54,803 students.

Health

In 2000–01 there were ten government hospitals, 88 primary health centres and 12 additional health centres, 38 government dispensaries and 413 sub-centres. Total beds (hospitals and health centres), 2,377. There were 389 doctors, 384 staff nurses and 915 paramedics.

RELIGION

At the 2001 census Christians numbered 1,628,986; Hindus, 307,822; Muslims, 99,169.

Mizoram

KEY HISTORICAL EVENTS

On 21 Jan. 1972 the former Mizo Hills District of Assam was created a Union Territory. A long dispute between the Mizo National Front (originally Separatist) and the central government was resolved in 1986. Mizoram became a state by the Constitution (53rd Amendment) and the State of Mizoram Acts, July 1986.

TERRITORY AND POPULATION

Mizoram is one of the easternmost Indian states, lying between Bangladesh and Myanmar, and having on its northern boundaries Tripura, Assam and Manipur. There are eight districts. The area is 21,081 sq. km and the population (2011 census, provisional) 1,091,014; density, 52 per sq. km. The main languages spoken are Mizo and English.

SOCIAL STATISTICS

Growth rate 2001–11, 22·78%.

CONSTITUTION AND GOVERNMENT

Mizoram has a unicameral *Legislative Assembly* with 40 seats. The capital is Aizawl (population, 2001, 228,280).

RECENT ELECTIONS

In the elections of Dec. 2008 distribution of seats was: Indian National Congress, 32; Mizo National Front, 3; Mizoram People's Conference, 2; Zoram Nationalist Party, 2; Maraland Democratic Front, 1.

CURRENT GOVERNMENT

Governor: Vakkom Purushothaman; b. 1928 (took office on 2 Sept. 2011).
   Chief Minister: Lal Thanhawla; b. 1942 (took office on 11 Dec. 2008).

ECONOMY

Budget

2008–09 revenue receipts, ₹26,531·3m.; expenditure, ₹23,138·0m. Budget estimates for 2009–10: revenue receipts, ₹30,092·1m.; expenditure, ₹28,316·9m.

ENERGY AND NATURAL RESOURCES

Electricity

Installed capacity (Oct. 2008), 119 MW. In Sept. 2008, 570 out of 707 inhabited villages had electricity.

Agriculture

About 60% of the people are engaged in agriculture, either on terraced holdings or in shifting cultivation. Principal crop production, 2007–08 (in tonnes): bananas, 151,519; turmeric, 83,500; ginger, 57,010; passion fruit, 44,720; oranges, 41,567; squash, 26,418; rice, 15,688.

Forestry

State forest area, 2007–08, 7,406 sq. km.

INDUSTRY

Handloom weaving and other cottage industries are important. There were 205 registered small-scale businesses in 2007–08.

COMMUNICATIONS

Roads

In March 2008 Mizoram had 6,158 km of roads (including 927 km of national highways and 259 km of state highways), of which 5,169 km were surfaced. There were 66,000 motor vehicles registered in March 2008, up from 37,000 in March 2003.

Rail

There is a metre-gauge rail link at Bairabi, 130 km from Aizawl.

Civil Aviation

Lengpui Airport, Aizawl is connected by air with Silchar in Assam and with Kolkata three days a week.

SOCIAL INSTITUTIONS

Education

In 2001, 88·8% of the population were literate (90·7% of men and 86·7% of women). In 2007–08 there were 1,752 primary schools with 134,656 students, 1,090 middle schools with 57,399 students, and 590 high and higher secondary schools with 56,491 students; there was one university, 22 colleges, one teacher training college, two polytechnics and 12 other training institutes.

Health
In 2007–08 there were ten hospitals, 66 health centres and 366 health sub-centres. Total beds, over 1,700. The state pays particular attention to immunization programmes.

RELIGION
At the 2001 census Christians numbered 772,809 and Buddhists 70,494.

Nagaland

KEY HISTORICAL EVENTS
The state was created in 1961, effective 1963. It consisted of the Naga Hills district of Assam and the Tuensang Frontier Agency. The agency was a British-supervised tribal area on the borders of Myanmar. Its supervision passed to the government of India at independence, and in 1957 Tuensang and the Naga Hills became a Centrally Administered Area, governed by the central government through the Governor of Assam.

A number of Naga leaders fought for independence until a settlement was reached with the Indian government at the Shillong Peace Agreement of 1975. However, calls for a greater Naga state, potentially incorporating parts of neighbouring Manipur, Arunachal Pradesh and Assam, continued to be voiced, notably through the National Socialist Council of Nagaland (NSCN), which had been active since 1954. The national government and NSCN met in Delhi in early Jan. 2003 to hold their first joint talks in 37 years, after which the NSCN declared 'the war is over'.

TERRITORY AND POPULATION
The state is in the northeast of India and is bounded in the north by Arunachal Pradesh, west by Assam, east by Myanmar and south by Manipur. Nagaland has an area of 16,579 sq. km and a population (2011 census, provisional) of 1,980,602; density, 119 per sq. km. The major towns are the capital, Kohima (2001 population, 77,030) and Dimapur (98,096). Other towns include Wokha, Mon, Zunheboto, Mokokchung and Tuensang. The chief tribes in numerical order are: Angami, Ao, Sumi, Konyak, Chakhesang, Lotha, Phom, Khiamngan, Chang, Yimchunger, Zeliang-Kuki, Rengma, Sangtam and Pochury. The official language of the state is English; Nagamese, a variant language form of Assamese and Hindi, is the most widely spoken language.

CONSTITUTION AND GOVERNMENT
An Interim Body (*Legislative Assembly*) of 42 members elected by the Naga people and an Executive Council (Council of Ministers) of five members were formed in 1961, and continued until the State Assembly was elected in Jan. 1964. The Assembly has 60 members. The Governor has extraordinary powers, which include special responsibility for law and order. Presidential rule was imposed in Jan. 2008 after the dismissal of the government but revoked in March 2008.

The state has 11 districts (Dimapur, Kiphire, Kohima, Longleng, Mokokchung, Mon, Peren, Phek, Tuensang, Wokha and Zunheboto). The capital is Kohima.

RECENT ELECTIONS
At the elections to the State Assembly on 23 Feb. 2013 the Naga People's Front won 38 seats (26 in 2008); Indian National Congress party won 8 seats (24 in 2008); Nationalist Congress Party, 4; Bharatiya Janata Party, 1; Janata Dal (United), 1; ind., 7. One seat was vacant.

CURRENT GOVERNMENT
Governor: Ashwani Kumar; b. 1950 (took office on 21 March 2013).

Chief Minister: Neiphiu Rio; b. 1950 (since 12 March 2008, having previously been in office March 2003–Jan. 2008).

ECONOMY
Budget
2008–09 revenue receipts, ₹34,008·9m.; expenditure, ₹28,895·3m. Budget estimates for 2009–10: revenue receipts, ₹39,098·9m.; expenditure, ₹31,700·7m.

ENERGY AND NATURAL RESOURCES
Electricity
Installed capacity (Oct. 2008) 103 MW. In Sept. 2008, 823 out of 1,278 inhabited villages had electricity.

Oil and Gas
Oil has been located in three districts. Reserves are estimated at 600m. tonnes.

Minerals
In addition to oil, other minerals include: coal, limestone, marble, chromite, magnesite, nickel, cobalt, chromium, iron ore, copper ore, clay, glass sand and slate.

Agriculture
90% of the people derive their livelihood from agriculture. The Angamis, in Kohima district, practise a fixed agriculture in the shape of terraced slopes, and wet paddy cultivation in the lowlands. In the other two districts a traditional form of shifting cultivation (*jhumming*) still predominates, but some farmers have begun tea and coffee plantations and horticulture. About 61,000 ha. were under terrace cultivation and 74,040 ha. under *jhumming* in 1994–95. Production of rice (1999) was 187,000 tonnes, total foodgrains 227,300 tonnes and pulses 13,000 tonnes.

Forestry
Forests, including open forests, covered 13,464 sq. km in 2007, of which forest area excluding open forest was 6,171 sq. km.

INDUSTRY
There is a forest products factory at Tijit; a paper-mill (100 tonnes daily capacity) at Tuli, a distillery unit and a sugar-mill (1,000 tonnes daily capacity) at Dimapur, and a cement factory (50 tonnes daily capacity) at Wazeho. Bricks and TV sets are also made, and there are 1,850 small units. There is a ceramics plant and sericulture is also important.

COMMUNICATIONS
Roads
In March 2008 there were 22,304 km of roads (including 494 km of national highways and 404 km of state highways), of which 9,540 km were surfaced. There were 226,000 motor vehicles registered in March 2008, up from 162,000 in March 2003.

Rail
Dimapur has a rail-head. Railway route-km in 2008–09, 13 km.

Civil Aviation
There are scheduled services from Dimapur to Guwahati, Imphal and Kolkata.

SOCIAL INSTITUTIONS
Justice
A permanent bench of the Guwahati High Court has been established in Kohima. There are 19 judges.

Education

In 2001, 66·6% of the population were literate (71·2% of men and 61·5% of women). In 1996–97 there were 1,414 primary schools with 271,932 students, 416 middle schools with 63,437 students, 244 high and higher secondary schools with 24,547 students, 36 colleges, two teacher training colleges and two polytechnics. The North Eastern Hill University opened at Kohima in 1978. Nagaland University was established in 1994.

Health

In 2005 there were eight hospitals (1,300 beds), 93 primary and 21 community health centres, 16 dispensaries, 412 sub-centres, ten TB centres and 36 leprosy centres.

RELIGION

At the 2001 census Christians numbered 1,790,349 and Hindus 153,162.

FURTHER READING

Aram, M., *Peace in Nagaland*. 1974

Odisha

KEY HISTORICAL EVENTS

Odisha, known as Orissa until Sept. 2011, was divided between Mahratta and Bengal rulers when conquered by the British East India Company, the Bengal area in 1757 and the Mahratta in 1803. The area which now forms the state then consisted of directly controlled British districts and a large number of small princely states with tributary rulers. The British districts were administered as part of Bengal until 1912 when, together with Bihar, they were separated from Bengal to form a single province. Bihar and Odisha were separated from each other in 1936. In 1948 a new state government took control of the whole state, including the former princely states (except Saraikella and Kharswan which were transferred to Bihar, and Mayurbhanj which was not incorporated until 1949).

In Oct. 1999 Odisha was hit by a devastating cyclone which resulted in more than 10,000 deaths.

TERRITORY AND POPULATION

Odisha is in eastern India and is bounded north by Jharkhand, northeast by West Bengal, east by the Bay of Bengal, south by Andhra Pradesh and west by Chhattisgarh. The area of the state is 155,707 sq. km, and its population (2011 census, provisional) 41,947,358; density 269 per sq. km. Cities with over 250,000 population, *see* INDIA: Territory and Population. Other large cities (2001): Sambalpur, 226,469; Puri, 157,837; Baleshwar, 156,430; Baripada, 100,651. The principal and official language is Odia.

SOCIAL STATISTICS

Growth rate 2001–11, 13·97%.

CONSTITUTION AND GOVERNMENT

The *Legislative Assembly* has 147 members.

There are 30 districts. The capital is Bhubaneswar (18 miles south of Cuttack).

RECENT ELECTIONS

At the state elections of 16 and 23 April 2009 the Biju Janata Dal won 103 seats (with 38·9% of the vote); the INC, 27 (29·1%); the BJP, 6 (15·1%); the Nationalist Congress Party, 4 (1·3%); and the Communist Party of India, 1 (0·5%). Six independents were elected.

CURRENT GOVERNMENT

Governor: Senayangba Chubatoshi Jamir; b. 1931 (took office on 21 March 2013).

Chief Minister: Naveen Patnaik; b. 1946 (took office on 5 March 2000).

ECONOMY

Budget

Revenue receipts for 2008–09, ₹246,100·1m.; expenditure, ₹211,901·2m. Budget estimates for 2009–10: revenue receipts, ₹265,500·9m.; expenditure, ₹289,191·7m.

ENERGY AND NATURAL RESOURCES

Electricity

The Hirakud Dam Project on the river Mahanadi irrigates 628,000 acres. The Upper Indravati Hydro Electric Project has an installed capacity of 600 MW. Hydro-electric power is now serving a large part of the state. Total installed capacity (Oct. 2008) 4,072 MW. In Sept. 2008, 26,535 out of 47,529 inhabited villages had electricity.

Minerals

Odisha is India's leading producer of chromite (97% of national output), graphite (80%), bauxite (71%), dolomite (50%), fire-clay (34%), iron ore (33% of national reserves), manganese ore (32%), limestone (20%), quartz-quartzite (18%) and iron ore (16%). Kaliapani is the centre of chromite mining and processing. Daitari is the major centre for iron production. Production in 2002–03 (1,000 tonnes): coal, 52,229; iron ore, 21,518; bauxite, 4,904; chromite, 3,047; limestone, 2,362; dolomite, 959; manganese ore, 616. Value of production in 2005–06 was ₹81,068m.

Agriculture

The cultivation of rice is the principal occupation of about 80% of the workforce, and only a very small amount of other cereals is grown. Production of foodgrains (1998–99) totalled 6·35m. tonnes from 4·7m. ha. (rice 6·2m. tonnes, wheat 60,000 tonnes); pulses, 0·28m. tonnes; oilseeds, 0·21m. tonnes; sugarcane, 1,114,000 tonnes. Turmeric is cultivated in the uplands of the districts of Ganjam, Phulbani and Koraput, and is exported.

Livestock (2003): buffaloes, 1·39m.; other cattle, 13·90m.; sheep, 1·62m.; goats, 5·80m.; poultry, 17·61m.

Forestry

Forests occupied 48,855 sq. km in 2007 (31·4% of the state). The most important species are sal, teak, kendu, sandal, sisu, bija, kusum, kongada and bamboo.

Fisheries

There were, in 2005, 641 fishing villages. Fish production in 2007–08 was 130,767 tonnes of marine fish and 218,716 tonnes of freshwater fish. Hundreds of fishing boats are engaged in illegal shrimp fishing. The state has four fishing harbours. In 2005 there were 273,800 people working in fishing and allied activities.

INDUSTRY

In 2011 there were 149,247 handicrafts and cottage industries, 59,079 small-scale industries and 334 large and medium industries in operation. The latter ones are mostly based on minerals: steel, pig iron, ferrochrome, ferromanganese, ferrosilicon, aluminium, cement, automotive tyres and synthetic fibres.

Other industries of importance are caustic soda, fertilizers, glass, heavy machine tools, industrial explosives, paper, salt, sugar, a coach-repair factory, a rerolling mill, textile mills and electronics. There is an oil refinery. In the past decade there has been much investment and expansion in biotechnology, electronics, leather and marine-based industries. Also, there were 1,822 factories in 2007–08 employing 184,900 persons. Handloom weaving and the manufacture of baskets, wooden articles, hats and nets, silver filigree work and hand-woven fabrics are particularly significant.

COMMUNICATIONS

Roads
In March 2008 there were 215,404 km of roads (including 3,704 km of national highways and 3,806 km of state highways), of which 30,645 km were surfaced. There were 2,370,000 motor vehicles registered in March 2008, up from 1,359,000 in March 2003.

Rail
In 2008–09 there were 2,385 route-km of railway in the state.

Civil Aviation
There is an airport at Bhubaneswar with regular scheduled services to Bangalore, Delhi, Hyderabad, Kolkata, Mumbai and Raipur.

Shipping
The busiest port is Paradip, which handled 42·4m. tonnes of cargo in 2007–08. There is a smaller port at Gopalpur.

SOCIAL INSTITUTIONS

Justice
The High Court of Judicature at Cuttack has a Chief Justice and 16 puisne judges.

Education
The percentage of literate people in the population in 2001 was 63·1% (males, 75·3%; females, 50·5%).

In 1996–97 there were 42,104 primary schools with 3·95m. students, 12,096 middle schools with 1·3m. students and 6,198 high and higher secondary schools with 945,000 students. There are ten engineering and technology colleges, 20 medical colleges, 13 teacher training colleges, 15 engineering schools/polytechnics, 497 arts, science and commerce colleges and 440 junior colleges.

Utkal University was established in 1943 at Cuttack and moved to Bhubaneswar in 1962; it is both teaching and affiliating. It has 368 affiliated colleges and 14,000 students (1993–94). Berhampur University has 33 affiliated colleges with 33,755 students, and Orissa University of Agriculture and Technology has eight constituent colleges with 641 students. Sambalpur University has 97 affiliated colleges and 43,982 students. Sri Jagannath Sanskrit Viswavidyalaya at Puri was established in 1981 for oriental studies.

Health
There were (1999–2000) 180 hospitals, 150 dispensaries, 1,351 primary health centres and units, and 5,929 health subcentres, with a total of 13,786 beds. There were also 462 homeopathic and 519 Ayurvedic dispensaries.

RELIGION
At the 2001 census Hindus numbered 34,726,129; Christians, 897,861; Muslims, 761,985.

Punjab (India)

KEY HISTORICAL EVENTS

The Punjab was constituted an autonomous province of India in 1937. In 1947 it was partitioned between India and Pakistan as East and West Punjab. The name of East Punjab was changed to Punjab. On 1 Nov. 1956 Punjab and Patiala and East Punjab States Union (PEPSU) were integrated to form the state of Punjab. On 1 Nov. 1966, under the Punjab Reorganization Act,

1966, the state was reconstituted as a Punjabi-speaking state comprising the districts of Gurdaspur (excluding Dalhousie), Amritsar, Kapurthala, Jullundur, Ferozepur, Bhatinda, Patiala and Ludhiana; parts of Sangrur, Hoshiarpur and Ambala districts; and part of Kharar tehsil. The remaining area comprising 47,000 sq. km and an estimated (1967) population of 8·5m. was shared between the new state of Haryana and the Union Territory of Himachal Pradesh. The existing capital of Chandigarh was made joint capital of Punjab and Haryana; its transfer to Punjab alone (originally scheduled for 1986) has been delayed while the two states seek agreement as to which Hindi-speaking districts shall be transferred to Haryana.

TERRITORY AND POPULATION

The Punjab is in north India and is bounded at its northernmost point by Jammu and Kashmir, northeast by Himachal Pradesh, southeast by Haryana, south by Rajasthan, west and northwest by Pakistan. The area of the state is 50,362 sq. km, with a population (2011 census, provisional) of 27,704,236; density, 550 per sq. km. For chief cities, see INDIA: Territory and Population. Other principal towns (2001): Bathinda (217,256); Pathankot (168,485); Hoshiarpur (149,668); Batala (147,872); Moga (135,279); Abohar (124,339); S.A.S. Nagar (123,484); Maler Kotla (107,009); Khanna (103,099); Phagwara (102,253). The official language is Punjabi.

SOCIAL STATISTICS
Growth rate 2001–11, 13·73%.

CONSTITUTION AND GOVERNMENT

Punjab (India) has a unicameral legislature, the *Legislative Assembly*, of 117 members. Presidential rule was imposed in May 1987 after outbreaks of communal violence. In March 1988 the Assembly was officially dissolved. Presidential rule was lifted in Feb. 1992.

There are 17 districts. The capital is Chandigarh.

RECENT ELECTIONS

Legislative Assembly elections were held on 30 Jan. 2012. The Shiromani Akali Dal (SAD) won 56 seats, the Congress Party (INC) 46, the Bharatiya Janata Party (BJP) 12, ind. 3.

CURRENT GOVERNMENT

Governor: Shivraj Patil; b. 1935 (took office on 22 Jan. 2010).

Chief Minister: Parkash Singh Badal; b. 1927 (took office on 2 March 2007 for the fourth time, having previously been chief minister from March 1970–June 1971, June 1977–Feb. 1980 and Feb. 1997–Feb. 2002).

ECONOMY

Budget
Revenue receipts for 2008–09, ₹207,127·9m.; expenditure, ₹245,689·9m. Budget estimates for 2009–10: revenue receipts, ₹240,723·4m.; expenditure, ₹303,062·6m.

ENERGY AND NATURAL RESOURCES

Electricity
Installed capacity, Oct. 2008, was 6,780 MW. In Sept. 2008 all 12,278 inhabited villages had electricity. Electricity consumption per capita in 2009–10 was 1,527 kWh.

Agriculture
About 75% of the population depends on agriculture, which is technically advanced. The irrigated area rose from 2·2m. ha. in 1950–51 to 4·2m. ha. in 1996–97. 95·1% of cropland in Punjab is irrigated. In 2001 wheat production was 15·5m. tonnes; potatoes, 10·0m.; rice, 9·1m.; kinnow, 0·2m.; plus large amounts of chillies, mangoes, grapes, pears, peaches and lemons. Total foodgrains, 24·90m. tonnes; sugarcane, 1·3m. tonnes; oilseeds, 61,000 tonnes. Cotton, 1·91m. bales of 170 kg, representing 12·4% of India's

cotton. Punjab contributes 22·6% of India's wheat. Agriculture in Punjab is more advanced and mechanized than in most other parts of India. Emphasis has recently been on diversification with new crops including hyola seeds, soybeans, sunflower, spring maize and floriculture, and the use of bio-fertilizers.

Livestock (2003 census): buffaloes, 5,995,000; other cattle, 2,039,000; sheep and goats, 498,000; poultry, 10,535,000.

Forestry
In 2007 there were 1,664 sq. km of forest land.

INDUSTRY

In March 2001 the number of registered industrial units was 202,356, employing about 1,184,550 people. In 2001 there were 620 large and medium industries and 201,736 small industrial units, investment ₹43,310m. The chief manufactures are metals, textiles (especially hosiery and fabrics), yarn, sports goods, hand tools, sugar, bicycles, electronic goods, machine tools, hand tools, automobiles and vehicle parts, surgical goods, vegetable oils, tractors, chemicals and pharmaceuticals, fertilizers, food processing, electronics, railway coaches, paper and newsprint, cement, engineering goods and telecommunications items. There is an oil refinery.

COMMUNICATIONS

Roads
In March 2008 there were 45,178 km of roads (including 1,557 km of national highways and 1,393 km of state highways), of which 37,487 km were surfaced. There were 4,573,000 motor vehicles registered in March 2008, up from 3,308,000 in March 2003.

Rail
The Punjab possesses an extensive system of railway communications, served by the Northern Railway. Route-km (2008–09), 2,133 km.

Civil Aviation
There is an airport at Amritsar, and Chandigarh airport is on the northeastern boundary; both have regular scheduled services to Delhi, Jammu, Srinagar and Leh. There are also Vayudoot services to Ludhiana. Amritsar is now an international airport with charter flights from Europe and from several Middle East destinations.

SOCIAL INSTITUTIONS

Justice
The Punjab and Haryana High Court exercises jurisdiction over the states of Punjab and Haryana and the territory of Chandigarh. It is located in Chandigarh. In 2010 it consisted of a Chief Justice and 47 judges.

Education
Compulsory education was introduced in April 1961; at the same time free education was introduced up to 8th class for boys and 9th class for girls as well as fee concessions. The aim is education for all children of 6–11. In 2001, 69·7% of the population were literate (75·2% of men and 63·4% of women).

In 1996–97 there were 12,590 primary schools with 2,081,965 students, 2,545 middle schools with 968,762 students, 2,159 high schools with 490,888 students and 1,134 higher secondary schools with 259,718 students.

Punjab University was established in 1882 at Lahore as an examining, teaching and affiliating body. It divided in 1947 with the Indian part moving to Shimla, and in 1956 moved again to Chandigarh (in 1993–94 it had 94 colleges and 77,868 students). In 1962 Punjabi University was established at Patiala (it had 66 colleges with 40,712 students) and Punjab Agricultural University at Ludhiana. Guru Nanak Dev University was established at Amritsar in 1969 to mark the 500th anniversary celebrations for Guru Nanak Dev, first Guru of the Sikhs (it had 85 colleges and

80,330 students, 1992–93). The Thapar Institute of Engineering and Technology, at Patiala, has university status and there is also the Baba Farid University of Health Science, at Faridkot. Altogether there are 293 affiliated colleges.

Health
There were (2000) 207 hospitals, 12 hospitals/health centres, 55 community health centres, 38 community primary health centres, 446 primary health centres, and 1,470 dispensaries and clinics. There were six Ayurvedic hospitals and 507 Ayurvedic dispensaries, plus one homeopathic hospital and 105 homeopathic dispensaries. There were over 25,000 hospital beds in 2000.

RELIGION

At the 2001 census Sikhs numbered 14,592,387; Hindus, 8,997,942; Muslims, 382,045; Christians, 292,800.

FURTHER READING
Singh, Khushwant, *A History of the Sikhs.* 2 vols. 1999

Rajasthan

KEY HISTORICAL EVENTS

The state is in the largely desert area formerly known as Rajputana. The Rajput princes were tributary to the Moghul emperors when they were conquered by the Mahrattas' leader, Mahadaji Sindhia, in the 1780s. In 1818 Rajputana became a British protectorate and was recognized during British rule as a group of princely states including Jaipur, Jodhpur and Udaipur. After independence the Rajput princes surrendered their powers and in 1950 were replaced by a single state government. In 1956 the state boundaries were altered; small areas of the former Bombay and Madhya Bharat states were added, together with the neighbouring state of Ajmer. Ajmer had been a Moghul power base; it was taken by the Mahrattas in 1770 and annexed by the British in 1818. In 1878 it became Ajmer-Merwara, a British province, and survived as a separate state until 1956.

TERRITORY AND POPULATION

Rajasthan is in northwest India and is bounded north by Punjab, northeast by Haryana and Uttar Pradesh, east by Madhya Pradesh, south by Gujarat and west by Pakistan. Since the area of Madhya Pradesh was reduced by the creation of Chhattisgarh in 2000, Rajasthan has become the largest Indian state in size, with an area of 342,239 sq. km. Population (2011 census, provisional), 68,621,012; density 201 per sq. km. For chief cities, *see* INDIA: Territory and Population. Other major towns (2001): Ganganagar (222,858), Bharatpur (205,235), Pali (187,641), Sikar (185,925), Tonk (135,689), Hunumangarh (129,556), Beawar (125,981), Kishangarh (116,222), Gangapur (105,396), Sawai Madhopur (101,997), Churu (101,874), Jhunjhunun (100,485). The main languages spoken are Rajasthani and Hindi.

SOCIAL STATISTICS
Growth rate 2001–11, 21·44%.

CONSTITUTION AND GOVERNMENT

There is a unicameral legislature, the *Legislative Assembly*, having 200 members. The capital is Jaipur. There are 32 districts.

RECENT ELECTIONS

After the election in Dec. 2008 the Indian National Congress Party came to power. Congress (I) won 96 seats; Bharatiya Janata Party (BJP), 78; Bahujan Samaj Party (BSP), 6; ind. and others, 20.

CURRENT GOVERNMENT

Governor: Margaret Alva; b. 1942 (took office on 12 May 2012).

Chief Minister: Ashok Gehlot; b. 1951 (took office on 13 Dec. 2008, having previously been chief minister from Dec. 1998–Dec. 2003).

ECONOMY

Budget

Revenue receipts for 2008–09, ₹334,688·0m.; expenditure, ₹342,956·0m. Budget estimates for 2009–10: revenue receipts, ₹382,679·6m.; expenditure, ₹396,766·1m.

ENERGY AND NATURAL RESOURCES

Electricity

Installed capacity in Oct. 2008, 6,426 MW. In Sept. 2008, 27,162 out of 39,753 inhabited villages had electricity.

Minerals

There are 64 different minerals mined in the state. It is the sole producer of garnet and jasper in India, and by far the leading producer of zinc, calcite, gypsum and asbestos. Others include silver, tungsten, granite, marble, kaolin (44% of India's production), dolomite, lignite, lead (80% of India's production), fluorite (59% of India's production), emeralds, soapstone, feldspar (70% of India's production), copper, barytes (53% of India's production), limestone and salt. Total revenue from minerals in 2002, ₹3,000m. Four blocs are being explored for mineral oils and gas.

Agriculture

The state has suffered drought and encroaching desert for several years. The cultivable area is (1999) about 25·6m. ha., of which 4·65m. ha. is irrigated. Production of principal crops (in tonnes), 1999: pulses, 2·64m.; total foodgrains, 11·40m. (wheat, 6·7m.; rice, 190,000); cotton, 868,000 tonnes.

The total irrigable area of the state is 13·6m. ha., which is 53% of the cultivable area. The Indira Gandhi Nahar Canal—India's largest irrigation project—is the main canal system, of which 189 km of main canal, 204 km of feeder and more than 3,400 km of distributors have been built. There were 37,560 villages with full or partial drinking water facilities in Jan. 2004, out of 37,889 villages.

Livestock (2003): buffaloes, 10·41m.; other cattle, 10·85m.; sheep, 10·05m.; goats, 16·81m.; camels, 498,000; poultry, 6·19m.

Forestry

Forests covered 16,036 sq. km in 2007 (4·7% of the state's area).

INDUSTRY

In 2001 there were 221,369 small industrial units with an investment of ₹31,160·6m. and employment of 857,000. Of these units 45,705 were agro-based, 26,842 forest-based, 27,397 metal-working and 24,861 textiles. There were 321 industrial areas in 2010. 6,337 factories were registered in 2007–08. Chief manufactures are textiles, dyeing, printing cloth, cement, glass, sugar, sodium, oxygen and acetylene units, pesticides, insecticides, dyes, caustic soda, calcium, carbide, synthetic fibres, fertilizers, shaving equipment, automobiles and automobile components, tyres, watches, nylon tyre cords and refined copper. The state is a major textile centre and is the leading producer of polyester and viscose yarns in India and the second largest producer of suiting material; out of 862 spinning mills in India, 69 are in Rajasthan.

COMMUNICATIONS

Roads

In March 2008 there were 171,479 km of roads (including 5,585 km of national highways and 11,240 km of state highways), of which 123,594 km were surfaced. There were 5,902,000 motor vehicles registered in March 2008, up from 3,487,000 in March 2003.

Rail

Jodhpur, Marwar, Udaipur, Ajmer, Jaipur, Kota, Bikaner and Sawai Madhopur are important junctions of the northwestern network. In 2008–09 there were 5,854 route-km of railway in the state. The major cities of the state are integrated with the national broad-gauge network.

Civil Aviation

There are airports at Jaipur (Sanganer Airport), Jodhpur, Kota and Udaipur with regular scheduled services to Ahmedabad, Delhi and Mumbai. Sanganer has been upgraded and now receives charter international flights as well as scheduled flights from Dubai and other gulf destinations.

SOCIAL INSTITUTIONS

Justice

The seat of the High Court is at Jodhpur. There is a Chief Justice and 32 puisne judges. There is also a bench of High Court judges at Jaipur.

Education

In 2001, 60·4% of the population were literate (75·7% of men and 43·9% of women).

There were 35,015 primary schools with 7,540,000 students in 2001, 16,336 middle schools with 2,327,000 students, 4,124 high schools and 1,923 higher secondary schools with 1,560,000 students between them. Elementary education is free but not compulsory.

In 2001 there were 280 colleges. Rajasthan University, established at Jaipur in 1947, is teaching and affiliating; in 1993–94 it had 135 colleges and 160,000 students. There are 11 other universities: Rajasthan Agricultural University, Bikaner; Mohanlal Sukhadia University, Udaipur; Maharishi Dayanand Saraswati University, Ajmer; Jai Narayan Vyas University, Jodhpur; Kota Open University, Kota; National Law University, Jodhpur; Rajasthan Sanskrit University, Jaipur; Birla Institute of Science and Technology, Pilani; Jain Vishwa Bharti, Ladnu; Rajasthan Vidyapeeth, Udaipur; Vanasthali Vidyapeeth, Vanasthali. There are also 280 colleges: 111 government colleges (including teacher training colleges and 27 polytechnics), 75 government colleges and research institutes, 92 non-aided colleges and two other institutes.

Health

In 2001 there were 113 hospitals with 17,459 beds, 263 community health centres, 1,674 primary health centres and 9,926 sub-centres.

RELIGION

At the 2001 census Hindus numbered 50,151,452; Muslims, 4,788,227; Sikhs, 818,420; Jains, 650,493.

FURTHER READING

Balzani, Marzia, *Modern Indian Kingship: Tradition, Legitimacy and Power in Rajasthan.* 2003

Sharma, S. K. and Sharma, Usha, (eds.) *History and Geography of Rajasthan.* 2000

Sikkim

KEY HISTORICAL EVENTS

A small Himalayan kingdom between Nepal and Bhutan, Sikkim was independent in the 1830s although in continual conflict with larger neighbours. In 1839 the British took the Darjeeling district. British political influence increased in the 19th century, when Sikkim was a buffer between India and Tibet. However, Sikkim remained an independent kingdom ruled by the 14th-century

Namgyal dynasty. In 1950 a treaty was signed with the government of India, declaring Sikkim an Indian Protectorate. Indian influence increased from then on. Political unrest coming to a head in 1973 led to the granting of constitutional reforms in 1974. Agitation continued until Sikkim became a 'state associated with the Indian Union' later that year. In 1975 the king was deposed and Sikkim became an Indian state, a change approved by referendum.

TERRITORY AND POPULATION

Sikkim is in the Eastern Himalayas and is bounded north by Tibet, east by Tibet and Bhutan, south by West Bengal and west by Nepal. Area, 7,096 sq. km. It is inhabited chiefly by the Lepchas, a tribe indigenous to Sikkim, the Bhutias, who originally came from Tibet, and the Nepalis, who entered from Nepal in large numbers in the late 19th and early 20th century. Population (2011 census, provisional), 607,688; density, 86 per sq km. The capital is Gangtok (population of 29,354 at the 2001 census).

SOCIAL STATISTICS

Growth rate 2001–11, 12·36%.

CONSTITUTION AND GOVERNMENT

The Assembly has 32 members.

English is the principal language and the official language of the government. Lepcha, Bhutia, Nepali and Limboo have also been declared official languages.

Sikkim is divided into four districts for administration purposes, Gangtok, Mangan, Namchi and Gyalshing being the headquarters for the Eastern, Northern, Southern and Western districts respectively.

RECENT ELECTIONS

At the State Assembly election of 30 April 2009 the Sikkim Democratic Front won all 32 seats (65·9% of the vote). The INC took 27·6% of the vote but failed to win any seats.

CURRENT GOVERNMENT

Governor: Balmiki Prasad Singh; b. 1942 (took office on 9 July 2008).

Chief Minister: Pawan Kumar Chamling; b. 1950 (took office on 12 Dec. 1994).

ECONOMY

Budget

2008–09 revenue receipts, ₹26,712·4m.; expenditure, ₹22,936·1m. Budget estimates for 2009–10: revenue receipts, ₹29,892·3m.; expenditure, ₹25,675·1m.

ENERGY AND NATURAL RESOURCES

Electricity

Installed capacity (Oct. 2008) 193 MW. There are four hydro-electric power stations. In Sept. 2008, 425 out of 450 inhabited villages had electricity.

Minerals

Copper, zinc and lead are mined.

Agriculture

There are 70,000 ha. of cultivable land. The economy is mainly agricultural; main crops are apples, barley, buckwheat, cardamom, ginger, maize, mandarin oranges, millet, potatoes, rice and wheat. Foodgrain production, 1999, 98,000 tonnes (maize, 56,000; rice, 21,000 tonnes; wheat, 14,000 tonnes); potatoes, 28,000 tonnes; pulses, 6,000 tonnes. Tea is grown. Medicinal herbs are exported. Sericulture produces 179 kg of silk per annum.

Forestry

Forests occupied 3,357 sq. km in 2007 (47·3% of the state's area).

INDUSTRY

Small-scale industries include cigarettes, distilling, tanning, fruit preservation, carpets and watchmaking. Local crafts include carpet weaving, making handmade paper, wood carving and silverwork.

COMMUNICATIONS

Roads

There were 1,873 km of roads in March 2008 (including 62 km of national highways and 179 km of state highways), of which 1,418 km were surfaced. Sikkim has the shortest road network of any Indian state. In March 2008 there were 26,000 motor vehicles registered, up from 15,000 in March 2003.

Rail

The nearest railhead is at Shiliguri (115 km from Gangtok).

Civil Aviation

The nearest airport is at Bagdogra (128 km from Gangtok), linked to Gangtok by helicopter service.

SOCIAL INSTITUTIONS

Education

In 2001, 68·8% of the population were literate (76·0% of men and 60·4% of women). Sikkim had (1999) 739 pre-primary schools with 23,538 students, 335 primary schools with 84,986 students, 122 junior high schools with 23,949 students, 72 high schools with 3,331 students and 27 higher secondary schools with 1,484 students. Education is free up to class XII; text books are free up to class V. There are 500 adult education centres. There is also a training institute for primary teachers, two degree colleges and a teacher training college.

Health

In 2002 there was one state hospital, four community health centres, 24 primary health centres and 147 sub-primary health centres, with a total of 920 beds. Some 28,244 patients were treated in 2000–01.

RELIGION

At the 2001 census Hindus numbered 329,548 and Buddhists 152,042.

Tamil Nadu

KEY HISTORICAL EVENTS

The first trading establishment made by the British in the Madras State was at Peddapali (now Nizampatnam) in 1611 and then at Masulipatnam. In 1639 the British were permitted to create a settlement at the place which is now Chennai, and Fort St George was founded. By 1801 the whole of the country from the Northern Circars to Cape Comorin (with the exception of certain French and Danish settlements) had been brought under British rule.

Under the provisions of the States Reorganization Act, 1956, the Malabar district (excluding the islands of Laccadive and Minicoy) and the Kasaragod district taluk of South Kanara were transferred to the new state of Kerala; the South Kanara district (excluding Kasaragod taluk and the Amindivi Islands) and the Kollegal taluk of the Coimbatore district were transferred to the new state of Mysore. The Laccadive, Amindivi and Minicoy Islands were constituted a separate Territory. Four taluks of the Trivandrum district and the Shencottah taluk of Quilon district were transferred from Travancore-Cochin to the new Madras State. On 1 April 1960, 1,049 sq. km from the Chittoor district of Andhra Pradesh were transferred to Madras in exchange for 844 sq. km from the Chingleput and Salem districts. In Aug. 1968 the state was renamed Tamil Nadu.

TERRITORY AND POPULATION

Tamil Nadu is in south India and is bounded north by Karnataka and Andhra Pradesh, east and south by the Indian Ocean and west by Kerala. Area, 130,058 sq. km. Population (2011 census, provisional), 72,138,958; density 555 per sq. km. Tamil is the principal language and has been adopted as the state language with effect from 14 Jan. 1958. For the principal towns, *see* INDIA: Territory and Population. Other large towns (2001 census): Tuticorin (243,415), Thanjavur (215,314), Nagercoil (208,179), Dindigul (196,955), Kanchipuram (188,733), Kumbakonam (160,767), Cuddalore (158,634), Karur (153,365), Neyveli (138,035), Tiruvannamalai (130,567), Pollachi (128,458), Arcot (126,671), Karaikkudi (125,717), Rajapalaiyam (122,307), Sivakasi (121,358), Pudukkottai (109,217), Bhavani (104,646), Vaniyambadi (103,950), Coonoor (101,490), Gudiyatham (100,115). The capital is Chennai (Madras).

SOCIAL STATISTICS

Growth rate 2001–11, 15·60%.

CONSTITUTION AND GOVERNMENT

There is a unicameral legislature; the *Legislative Assembly* has 235 members (one of which is nominated) including the Speaker. There are 30 districts.

RECENT ELECTIONS

Elections were held on 13 April 2011. The All India Anna Dravida Munnetra Kazhagam (AIADMK) won 203 seats against 31 for the Dravida Munnetra Kazhagam (DMK). Turnout was 77·8%, the highest in any election conducted in Tamil Nadu.

CURRENT GOVERNMENT

Governor: Konijeti Rosaiah; b. 1933 (since 31 Aug. 2011).
   Chief Minister: Jayaram Jayalalitha; b. 1948 (since 16 May 2011).

ECONOMY

Budget
Revenue receipts for 2008–09, ₹550,425·2m.; expenditure, ₹535,902·6m. Budget estimates for 2009–10: revenue receipts, ₹582,711·4m.; expenditure, ₹592,952·8m.

ENERGY AND NATURAL RESOURCES

Electricity
Installed capacity in Jan. 2005 was 10,139 MW, of which 1,995 MW was hydro-electric, 6,424 MW thermal, 1,362 MW wind powered and 358 MW nuclear (the Kalpakkam nuclear power plant became operational in 1983). In Sept. 2008 all 15,400 inhabited villages had electricity.

Minerals
The state has magnesite, lignite, bauxite, limestone, manganese, fireclay and feldspar.

Agriculture
The land is a fertile plain watered by rivers flowing east from the Western Ghats, particularly the Cauvery and the Tambaraparani. Temperature ranges between 6°C and 40°C, rainfall between 442 mm and 934 mm. Of the total land area (13m. ha.), 6,519,000 ha. were cropped and 349,000 ha. of wasteland were cultivable in 1999–2000. Total area under irrigation in 2000–01, 3·49m. ha. The staple food crops grown are paddy, maize, jowar, bajra, pulses and millets. Important commercial crops are sugarcane, oilseeds, cotton, tobacco, coffee, rubber and pepper. In 2003–04, 3·2m. tonnes of paddy, 1·76m. tonnes of sugarcane, 918,000 tonnes of groundnuts, 888,000 tonnes of millets and other cereals, and 201,000 tonnes of pulses were produced.

Livestock (2003): buffaloes, 1·66m.; other cattle, 9·14m.; sheep, 5·59m.; goats, 8·18m.; poultry, 86·59m.

Forestry
Forest area, 2007, 23,338 sq. km. Products include timber, teak, wattle, sandalwood, pulp wood and sapwood.

Fisheries
In 2010–11 marine production totalled 404,000 tonnes and inland production 171,000 tonnes. In 2005 there were 314,400 people engaged in fishing and related activities.

INDUSTRY

In 2002–03 there were 448,905 registered small-scale industrial units, employing 3,142,335 workers. In 2007–08 the number of working factories totalled 21,042, with 1,550,000 workers. The biggest central sector project is Salem steel plant. Textiles constitute one of the major industries; in 2003 Tamil Nadu produced nearly 40% of India's cotton textiles and accounted for 42% of Indian leather exports. Other important industries are automobile ancillaries (constituting 27·5% of exports from India as at mid-2003), chemicals and petrochemicals, agricultural and food processing, biotechnology and computer software (accounting for about 17% of India's software exports).

COMMUNICATIONS

Roads
In March 2008 there were 181,213 km of roads—including 4,462 km of national highways and 9,264 km of state highways—of which 147,346 km were surfaced. There were 11,930,000 motor vehicles registered in March 2008 (of which 9,446,000 were two-wheelers), up from 8,005,000 in March 2003.

Rail
In 2008–09 there were 4,107 route-km of railway in Tamil Nadu. Chennai and Madurai are the main centres.

Civil Aviation
There are airports at Chennai, Coimbatore, Tiruchirapalli and Madurai, with regular scheduled services to Delhi, Kolkata and Mumbai. Chennai is an international airport and the main centre of airline routes in south India. In 2003–04 Chennai handled 2,054,043 international passengers, 2,501,778 domestic passengers and 154,123 tonnes of freight.

Shipping
Chennai, Tuticorin and Ennore are the chief ports. Important minor ports are Cuddalore and Nagapattinam.

SOCIAL INSTITUTIONS

Justice
There is a High Court at Chennai with a Chief Justice and 26 judges.

Police
In 2003–04 the strength of the police force was 95,412, with 1,217 police stations.

Education
At the 2001 census 73·5% of the population were literate (82·4% of men and 64·4% of women).
   Education is free up to pre-university level. In 2003–04 there were 32,242 primary schools with 4·3m. students, 6,825 middle schools with 2·2m. students, 4,859 high schools with 1·9m. students and 4,136 higher secondary schools with 4·5m. students. There were 18 universities in 2003–04: Madras University (founded in 1857); Annamalai University, Annamalainagar (1929); Gandhigram Rural Institute, Gandhigram (1956); Madurai Kamaraj University, Palkalainagar (1966); Tamil Nadu Agricultural University, Coimbatore (1971); Anna University, Chennai (1978); Tamil University, Thanjavur (1981); Bharathidasan University,

Tiruchirapalli (1982); Bharathiyar University, Coimbatore (1982); Mother Teresa Women's University, Kodaikanal (1984); Alagappa University, Karaikkudi (1985); Sri Ramachandra Medical College and Research Institute, Chennai (1985); Tamil Nadu Dr M. G. R. Medical University, Chennai (1987); Avinashilingam Institute for Home Science and Higher Education for Women, Coimbatore (1988); Tamil Nadu Veterinary and Animal Sciences University, Chennai (1989); Manonmaniam Sundaranar University, Tirunelveli (1990); Sri Chandrasekarendra Saraswathi Viswa Mahavidyalaya University, Enathur (1993); Thanthai Periyar University, Salem (1997).

Health

In 2002 there were 408 hospitals and 512 dispensaries, with about 61,000 beds.

RELIGION

At the 2001 census Hindus numbered 54,985,079; Christians, 3,785,060; Muslims, 3,470,647.

CULTURE

Tourism

In 2009, 2·37m. foreign tourists visited Tamil Nadu (making it the most visited Indian state).

FURTHER READING

Statistical Information: The Department of Statistics (Fort St George, Chennai) was established in 1948 and reorganized in 1953. Main publications: *Annual Statistical Abstract; Decennial Statistical Atlas; Season and Crop Report; Quinquennial Wages Census; Quarterly Abstract of Statistics.*

Tripura

KEY HISTORICAL EVENTS

Tripura is a Hindu state of great antiquity having been ruled by the Maharajahs for 1,300 years before its accession to the Indian Union on 15 Oct. 1949. With the reorganization of states on 1 Sept. 1956 Tripura became a Union Territory, and was so declared on 1 Nov. 1957. The Territory was made a State on 21 Jan. 1972.

TERRITORY AND POPULATION

Tripura is bounded by Bangladesh, except in the northeast where it joins Assam and Mizoram. The major portion of the state is hilly and mainly jungle. It has an area of 10,486 sq. km. Population, 3,671,032 (2011 census, provisional); density, 350 per sq. km.

The official languages are Bengali and Kokborok. Manipuri is also spoken.

SOCIAL STATISTICS

Growth rate 2001–11, 14·75%.

CONSTITUTION AND GOVERNMENT

The territory has four districts, namely Dhalai, North Tripura, South Tripura and West Tripura. The capital is Agartala (population, 2001, 189,998).

The *Legislative Assembly* has 60 members.

RECENT ELECTIONS

The Communist Party of India (Marxist) won the Legislative Assembly elections on 14 Feb. 2013 with 49 seats; Indian National Congress took 10; Communist Party of India, 1.

CURRENT GOVERNMENT

Governor: Devanand Konwar (took office on 25 March 2013).

Chief Minister: Manik Sarkar; b. 1949 (took office on 11 March 1998).

ECONOMY

Budget

2008–09 revenue receipts, ₹40,767·8m.; expenditure, ₹31,294·5m. Budget estimates for 2009–10: revenue receipts, ₹46,138·7m.; expenditure, ₹42,668·6m.

ENERGY AND NATURAL RESOURCES

Electricity

Installed capacity in 2005 was 220·46 MW, of which 76·11 MW were hydro-electric and 144·35 MW were thermal. In Sept. 2008, 491 out of 858 inhabited villages had electricity.

Oil and Gas

The state has significant natural gas resources in non-associate form, with established reserves of 31bn. cu. metres.

Agriculture

About 24% of the land area is cultivable. The tribes practise shifting cultivation, but this is being replaced by modern methods. The main crops are rice, wheat, jute, mesta, potatoes, oilseeds and sugarcane. In 2002–03 there were 246,000 ha. under cereal cultivation. In 2001 tea gardens covered 6,700 ha.

Forestry

Forests covered 8,073 sq. km in 2007 (77·0% of the state's area). Commercial rubber plantation is being encouraged.

INDUSTRY

Main small industries: aluminium utensils, rubber, saw-milling, soap, piping, fruit canning, handloom weaving and sericulture. In 2007–08 there were 340 factories that employed 22,896 persons. 384,000 persons were employed in handloom and handicrafts industries in 2003–04.

COMMUNICATIONS

Roads

Tripura had 31,733 km of roads in March 2008 (including 400 km of national highways and 689 km of state highways), of which 12,182 km were surfaced. In March 2008 there were 131,000 motor vehicles registered, up from 66,000 in March 2003.

Rail

There is a railway between Kumarghat and Kalkalighat (Assam). Route-km in 2008–09, 151 km.

Civil Aviation

There is one airport and three airstrips. The airport (Agartala) has regular scheduled services to Kolkata.

SOCIAL INSTITUTIONS

Education

In 2001, 73·2% of the population were literate (81·0% of men and 64·9% of women). In Sept. 2003 there were 1,776 primary schools (451,731 pupils), 1,001 middle schools (186,651), and 652 high and higher secondary schools (118,006). There were 14 colleges of general education, two engineering and technical institutes, and five professional and other colleges. Tripura University, established in 1987, has 20 affiliated colleges.

Health

There were (2002) 27 hospitals, with 2,000 beds. There were 58 primary health centres, 539 sub-centres and 11 community health centres in 2001.

RELIGION

At the 2001 census Hindus numbered 2,739,310; Muslims, 254,442; Christians, 102,489; Buddhists, 98,922.

Uttar Pradesh

KEY HISTORICAL EVENTS

In 1833 the then Bengal Presidency was divided into two parts, one of which became the Presidency of Agra. In 1836 the Agra area was styled the North-West Province and placed under a Lieut.-Governor. In 1877 the two provinces of Agra and Oudh came under one administrator, Lieut.-Governor of the North-West Province and Chief Commissioner of Oudh. In 1902 the name was changed to 'United Provinces of Agra and Oudh', under a Lieut.-Governor, and the Lieut.-Governorship was altered to a Governorship in 1921. In 1935 the name was shortened to 'United Provinces'. On independence, the states of Rampur, Banaras and Tehri-Garwhal were merged with United Provinces. In 1950 the name of the United Provinces was changed to Uttar Pradesh. In 2000 Uttaranchal (officially renamed Uttarakhand on 1 Jan. 2007) was carved from the northern, mainly mountainous, region of Uttar Pradesh.

TERRITORY AND POPULATION

Uttar Pradesh is in north India and is bounded north by Uttarakhand and Nepal, east by Bihar and Jharkhand, south by Madhya Pradesh and Chhattisgarh and west by Rajasthan, Haryana and Delhi. Since the formation of Uttarakhand in 2000 the area of Uttar Pradesh is 240,928 sq. km (previously 294,411 sq. km). Population (2011 census, provisional), 199,581,477; density, 828 per sq. km. Despite the decline in the population caused by the creation of Uttarakhand, Uttar Pradesh still has the highest population of any of the Indian states. If Uttar Pradesh were a separate country it would have the fifth highest population in the world (after China, India, the USA and Indonesia). Cities with more than 250,000 population, see INDIA: Territory and Population. Other important towns (2001 census): Farrukhabad (242,997), Maunath Bhanjan (212,657), Hapur (211,983), Etawah (210,453), Faizabad (208,162), Mirzapur (205,053), Sambhal (182,478), Bulandshahr (176,425), Rae Bareli (169,333), Bahraich (168,323), Amroha (165,129), Jaunpur (160,055), Fatehpur (152,078), Sitapur (151,908), Budaun (148,029), Unnao (144,662), Modinagar (139,929), Banda (139,436), Orai (139,318), Hathras (126,355), Pilibhit (124,245), Lakhimpur (121,486), Loni (120,945), Gonda (120,301), Mughal Sarai (116,308), Hardoi (112,486), Lalitpur (111,892), Basti (107,601), Etah (107,110), Mainpuri (104,851), Deoria (104,227), Chandausi (103,749), Ghazipur (103,298), Ballia (101,465), Sultanpur (100,065). The sole official language has been Hindi since April 1990.

SOCIAL STATISTICS

Growth rate 2001–11, 20·09%.

CONSTITUTION AND GOVERNMENT

Uttar Pradesh has had an autonomous system of government since 1937. There is a bicameral legislature. The *Legislative Council* has 100 members; the *Legislative Assembly* has 404 (one of which is nominated).

There are 18 administrative divisions, each under a Commissioner, and 75 districts.

The capital is Lucknow.

RECENT ELECTIONS

Elections were held in Feb.–March 2012. The Samajwadi Party (SP) won 224 seats; the Bahujan Samaj Party (BSP), 80; the Bharatiya Janata Party (BJP), 47; the Indian National Congress (INC), 28; Rashtriya Lok Dal (RLD), 9; Nationalist Congress Party (NCP), 1; ind., 14.

CURRENT GOVERNMENT

Governor: Banwari Lal Joshi; b. 1936 (took office on 28 July 2009).

Chief Minister: Akhilesh Yadav; b. 1973 (took office on 15 March 2012).

ECONOMY

Budget

Revenue receipts for 2008–09, ₹778,307·3m.; expenditure, ₹759,688·9m. Budget estimates for 2009–10: revenue receipts, ₹944,398·4m.; expenditure, ₹928,666·5m.

ENERGY AND NATURAL RESOURCES

Electricity

Installed capacity in Jan. 2005 was 8,102·6 MW, of which 829·6 MW were hydro-electric, 7,135·0 MW thermal and 138·0 MW nuclear. In Sept. 2008, 86,450 out of 97,942 inhabited villages had electricity.

Minerals

The state's minerals include magnesite, granite, dolomite, coal, marble, limestone, bauxite, uranium and silica sand. In 2003–04, 15·8m. tonnes of coal were produced.

Agriculture

In 2003–04 Uttar Pradesh had almost 19·2m. ha. of land under foodgrain cultivation. It is India's largest producer of foodgrains: production (2002–03), 38·3m. tonnes (wheat, 23·7m. tonnes; rice, 9·6m. tonnes). The state is also one of India's main producers of sugar and potatoes: 2002–03 production of sugarcane, 120·9m. tonnes; and potatoes, 10·2m. tonnes.

Forestry

Forests covered 14,341 sq. km in 2007 (6·0% of the state's area). In 1995 forests had accounted for 51,663 sq. km, but much of this area is now in Uttarakhand.

INDUSTRY

Sugar production is important; other industries include cement, vegetable oils, textiles, cotton yarn, jute and glassware. In 2007–08 there were 10,717 registered factories employing 751,165 workers.

COMMUNICATIONS

Roads

In March 2008 there were 284,673 km of roads—including 5,874 km of national highways and 8,391 km of state highways—of which 202,492 km were surfaced. There were 9,826,000 motor vehicles registered in March 2008 (of which 7,737,000 were two-wheelers), up from 5,928,000 in March 2003.

Rail

Lucknow is the main junction of the northern network; other important junctions are Agra, Kanpur, Allahabad, Mughal Sarai and Varanasi. Route-km in 2008–09, 8,703 km.

Civil Aviation

The main airports are at Lucknow, Kanpur, Varanasi, Allahabad, Agra and Gorakhpur.

SOCIAL INSTITUTIONS

Justice

The High Court of Judicature at Allahabad (with a bench at Lucknow) has a Chief Justice and 63 puisne judges including additional judges. The state is divided into 46 judicial districts.

Education

At the 2001 census 75·72m. people were literate (56·3%; 68·8% of men and 42·2% of women). In 2002–03 there were 98,220 primary schools with 15·60m. students, 23,696 middle schools with 5·54m. students and 11,524 higher secondary schools with 3·87m. students.

Universities: Allahabad University (founded 1887); the Banaras Hindu University, Varanasi (1916); Aligarh Muslim University (1920); Lucknow University (1921); Agra University (1927); Gorakhpur University (1957); Sampurnanand Sanskrit Vishwavidyalaya, Varanasi (1958); Ch. Charan Singh University (1966); Kanpur University (1966); Bundelkhand University, Jhansi (1975); C. S. Azad University of Agriculture and Technology, Kanpur (1975); Dr Ram Manohar Lohia Awadh, Faizabad (1975); Narendra Deva University of Agriculture and Technology, Faizabad (1975); Rohilkhand University, Bareilly (1975); Purvanchal University, Jaunpur (1987).

In 2005 there were also four institutions with university status: Indian Veterinary Research Institute; Central Institute of Higher Tibetan Studies; Sanjai Gandhi Post Graduate Institute of Medical Sciences; and Dayal Bagh Educational Institute. In 2004–05 there were 34 medical colleges, 42 veterinary colleges, 69 engineering colleges, eight agricultural colleges, two law colleges, 21 teacher training colleges and 1,009 arts, science and commerce colleges. There were 491 oriental learning colleges in 2000–01.

Health

In 2007 there were 4,595 allopathic, 2,362 Ayurvedic and Unani and 1,482 homoeopathic hospitals and dispensaries. There were 3,660 primary health centres, 20,521 sub-centres and 386 community health centres in 2006–07. In Dec. 2003 there were 44,927 doctors registered with the state medical council.

RELIGION

At the 2001 census Hindus numbered 133,979,263; Muslims, 30,740,158; Sikhs, 678,059; Buddhists, 302,031; Christians, 212,578; Jains, 207,111.

FURTHER READING

Hasan, Z., *Quest for Power: Oppositional Movements and Post-Congress Politics in Uttar Pradesh.* 1998

Kudaisya, Gyanesh, *Region, Nation, 'Heartland': Uttar Pradesh in India's Body Politic.* 2006

Lieten, G. K. and Srivastava, R., *Unequal Partners: Power Relations, Devolution and Development in Uttar Pradesh.* 1999

Misra, S., *A Narrative of Communal Politics, Uttar Pradesh, 1937–39.* 2001

Uttarakhand

KEY HISTORICAL EVENTS

The state was carved from Uttar Pradesh and became the twenty-seventh state of India on 9 Nov. 2000. It is located in the hilly and mountainous region of the northern border of the Indian subcontinent. The regions of Kumaon and Garhwal contained in the new state were referred to as Uttarakhand in ancient Hindu scriptures. The Chinese suppression of revolt in Tibet in 1959 saw a rapid influx of Tibetan exiles to the region. The Indo-Chinese conflict of 1962 initiated a modernization programme throughout the Indian Himalayas with the development of roads and communication networks in the previously backward region. From the 1970s the hill people agitated for separation from Uttar Pradesh, established in 1950. On 1 Aug. 2000 the Uttar Pradesh Reorganisation Bill was passed, allowing for a separate state, called Uttaranchal (officially renamed Uttarakhand on 1 Jan. 2007), to incorporate 12 hill districts and, controversially, the lowland area of Udham Singh Nagar.

TERRITORY AND POPULATION

Uttarakhand is located in northern India and is bounded in the northeast by China and in the east by Nepal. The state of Uttar Pradesh is to the southwest, Haryana to the west and Himachal Pradesh to the northwest. Uttarakhand has an area of 53,483 sq. km. Population (2011 census, provisional), 10,116,752; density, 189 per sq. km. The principal languages are the Hindi dialects of Garhwali and Kumaoni. Cities with over 250,000 population, *see* INDIA: Territory and Population. Other large cities (2001 census): Hardwar (220,767), Haldwani (158,896), Roorkee (115,278).

SOCIAL STATISTICS

Growth rate 2001–11, 19·17%.

CONSTITUTION AND GOVERNMENT

Uttarakhand is the twenty-seventh state of India. After the region was carved from Uttar Pradesh it was decided that the 22 members of the Legislative Assembly from Uttarakhandi districts would become the members of the new state's *Legislative Assembly*. Subsequently this was increased to 30 when the provisional assembly was established, and to 70 as a result of the elections to the Legislative Assembly of Feb. 2002. For administrative purposes the region is divided into 13 districts.

The interim capital and seat of government is at Dehra Dun.

RECENT ELECTIONS

On the formation of the new state in 2000 the Bharatiya Janata Party (BJP) was the single largest party with 17 seats, enabling them to form a majority administration in the 23-seat assembly with Nityanand Swamy becoming the state's first chief minister.

State assembly elections were held on 30 Jan. 2012. The Indian National Congress Party (INC) won 32 seats; Bharatiya Janata Party (BJP), 31; the Bahujan Samaj Party (BSP), 3; the Uttarakhand Kranti Dal (UKKD), 1; ind. 3.

CURRENT GOVERNMENT

Governor: Aziz Qureshi; b. 1940 (took office on 15 May 2012).

Chief Minister: Vijay Bahuguna; b. 1947 (took office on 13 March 2012).

ECONOMY

Budget

2008–09 revenue receipts, ₹86,348·9m.; expenditure, ₹83,953·6m. Budget estimates for 2009–10: revenue receipts, ₹109,476·7m.; expenditure, ₹111,611·0m.

ENERGY AND NATURAL RESOURCES

Electricity

In Oct. 2008 the state had an installed capacity of 2,383 MW. In Sept. 2008, 15,213 out of 15,761 inhabited villages had electricity.

Water

Uttarakhand suffers from an acute shortage of water for drinking and irrigation. Only 10% of the water potential is currently utilized.

Minerals

There are deposits of limestone, gypsum, iron ore, graphite and copper.

Agriculture

Agriculture is the occupation for approximately 50% of the population. Subsistence farming is the norm, as only 9% of the land in the state is cultivable.

Forestry

In 2007, 24,495 sq. km were covered by forest (45·8% of the state's area).

INDUSTRY

Tourism is by far the most important industry. The state can offer ski resorts, adventure tourism, mountaineering, hiking and several areas of religious interest. Other industries include: horticulture, floriculture, fruit-processing and medicine production. In the Terai region there are around 350 industrial units and 130 in the Doon Valley.

COMMUNICATIONS

Roads

In March 2008 there were 41,041 km of roads (including 1,991 km of national highways and 1,576 km of state highways), of which 20,192 km were surfaced. There were 731,000 motor vehicles registered in March 2008, up from 457,000 in March 2003.

Rail

Four main railway lines in the south of the state link several districts to Uttar Pradesh and Himachal Pradesh. Railways along the foothills connect Dehra Dun, Hardwar, Rishikesh, Roorkee, Kotdwaar, Ram Nagar, Kathgodam and Tanakpur. The rest of the state is not connected to the rail network. Route-km in 2008–09, 345 km.

Civil Aviation

There are airports at Dehra Dun and Udham Singh Nagar.

SOCIAL INSTITUTIONS

Education

In 2001, 71·6% of the population were literate (83·3% of men and 59·6% of women).

RELIGION

At the 2001 census Hindus numbered 7,212,260, Muslims 1,012,141 and Sikhs 212,025.

West Bengal

KEY HISTORICAL EVENTS

Bengal was under the overlordship of the Moghul emperor and ruled by a Moghul governor (*nawab*) who declared himself independent in 1740. The British East India Company based at Calcutta was in conflict with the *nawab* from 1756 until 1757 when British forces defeated him at Plassey and installed their own *nawab* in 1760. The French were also in Bengal; the British captured their trading settlement at Chandernagore in 1757 and in 1794, restoring it to France in 1815.

The area of British Bengal included modern Odisha and Bihar, Bangladesh and (until 1833) Uttar Pradesh. Calcutta was the capital of British India from 1772 until 1912.

The first division into East and West took place from 1905–11 and was not popular. However, at Partition in 1947 the East (Muslim) chose to join what was then East Pakistan (now Bangladesh), leaving West Bengal as an Indian frontier state and promoting a steady flow of non-Muslim Bengali immigrants from the East. In 1950 West Bengal incorporated the former princely state of Cooch Behar and, in 1954, Chandernagore. Small areas were transferred from Bihar in 1956.

TERRITORY AND POPULATION

West Bengal is in northeast India and is bounded north by Sikkim and Bhutan, east by Assam and Bangladesh, south by the Bay of Bengal, southwest by Odisha, west by Jharkhand and Bihar and northwest by Nepal. The total area of West Bengal is 88,752 sq. km. Population (2011 census, provisional), 91,347,736; density, 1,029 per sq. km. The capital is Kolkata (Calcutta). Population of

chief cities, *see* INDIA: Territory and Population. Other major towns (2001): Habra, 239,209; Ingraj Bazar (English Bazar), 224,415; Raiganj, 175,047; Haldia, 170,673; Baharampur, 170,322; Medinipur, 149,769; Krishnanagar, 148,194; Ranaghat, 145,285; Balurghat, 143,321; Santipur, 138,235; Bankura, 128,781; Navadvip, 125,341; Birnagar, 115,127; Alipur Duar, 114,035; Puruliya, 113,806; Basirhat, 113,159; Darjiling (Darjeeling), 108,830; Cooch Behar, 103,008; Bangaon, 102,163; Chakdaha, 101,320; Jalpaiguri, 100,348.

The principal language is Bengali.

SOCIAL STATISTICS

Growth rate 2001–11, 13·93%.

CONSTITUTION AND GOVERNMENT

The state of West Bengal came into existence as a result of the Indian Independence Act, 1947. The territory of Cooch-Behar State was merged with West Bengal on 1 Jan. 1950, and the former French possession of Chandernagore became part of the state on 2 Oct. 1954. Under the States Reorganization Act, 1956, certain portions of Bihar State (an area of 3,157 sq. miles with a population of 1,446,385) were transferred to West Bengal.

The *Legislative Assembly* has 295 seats (294 elected and one nominated).

For administrative purposes there are three divisions (Jalpaiguri, Burdwan and Presidency), under which there are 19 districts, including Kolkata. The Kolkata Metropolitan Development Authority has been set up to co-ordinate development in the metropolitan area (1,350 sq. km). For the purposes of local self-government there are 18 *zilla parishads* (district boards) plus Shiliguri *mahakuma parishad* (which straddles Darjeeling district and Jalpaiguri district), 341 *panchayat samities* (regional boards) and 3,354 *gram* (village) *panchayats*. There are 118 municipalities, six Corporations and three Notified Areas. The Kolkata Municipal Corporation is headed by a mayor in council.

RECENT ELECTIONS

In elections held on 18, 23 and 27 April and 3, 7 and 10 May 2011 the All India Trinamool Congress won 184 of 294 seats, the Indian National Congress 42, the Communist Party of India (Marxist) 40, the All India Forward Bloc 11, the Revolutionary Socialist Party 7 and others 10. After 34 years, the victory for the All India Trinamool Congress signalled the end of the world's longest-running democratically-elected Communist government.

CURRENT GOVERNMENT

Governor: Mayankote Kelath Narayanan; b. 1934 (since 24 Jan. 2010).

Chief Minister: Mamata Banerjee; b. 1955 (since 20 May 2011).

ECONOMY

Budget

Revenue receipts for 2008–09, ₹369,044·0m.; expenditure, ₹516,133·3m. Budget estimates for 2009–10: revenue receipts, ₹423,124·1m.; expenditure, ₹602,525·2m.

ENERGY AND NATURAL RESOURCES

Electricity

Installed capacity as at Jan. 2005 was 6,762 MW, of which 261 MW were hydro-electric, 6,499 MW thermal and 2 MW wind powered. In Sept. 2008, 36,462 out of 37,945 inhabited villages had electricity.

Minerals

Value of production, 2002–03, ₹23,933m. The state has coal (the Raniganj field is one of the three biggest in India) including coking coal. Coal production (2002–03), 20·48m. tonnes.

Agriculture

About 5·84m. ha. were under rice-paddy in 2002–03. Total foodgrain production, 2002–03, 15·52m. tonnes (rice 14·39m. tonnes, wheat 887,000 tonnes, pulses 167,000 tonnes). Other principal crops (2002–03): potatoes, 6·9m. tonnes; sugarcane, 1·28m. tonnes; oilseeds, 476,000 tonnes; jute, 8·5m. bales of 180 kg (76·3% of the national output). The state produces around 200,000 tonnes of tea each year.

Livestock (2003): 18,913,000 cattle; 1,086,000 buffaloes; 1,525,000 sheep; 18,774,000 goats; 60,656,000 poultry.

Forestry

Forests covered 12,994 sq. km in 2007 (14·6% of the state's area).

Fisheries

Fish production, 2007–08, 1,447,000 tonnes, of which inland 1,264,000 tonnes. In 2005 there were 128,500 people working in fishing and allied activities. West Bengal has the largest annual fish catches of any Indian state.

INDUSTRY

In 2008 there were 14,389 factories (provisional) employing a daily average of 927,282 workers. There are 100 coal mines.

There is a large automobile factory at Uttarpara, and an aluminium rolling-mill at Belur. There is a steel plant at Burnpur (Asansol) and a spun pipe factory at Kulti. Durgapur has a large steel plant and other industries under the state sector—a thermal power plant, coke oven plant, fertilizer factory, alloy steel plant and ophthalmic glass plant. There is a locomotive factory at Chittaranjan and a cable factory at Rupnarayanpur. A refinery and fertilizer factory are operating at Haldia. Other industries include chemicals, engineering goods, electronics, textiles, automobile tyres, paper, cigarettes, distillery, aluminium foil, tea, pharmaceuticals, carbon black, graphite, iron foundry, silk and explosives.

Small industries are important; 361,051 units were registered in March 2007, employing some 2m. persons.

COMMUNICATIONS

Roads

In March 2008 there were 211,770 km of roads (including 2,524 km of national highways and 1,682 km of state highways), of which 49,111 km were surfaced. There were 2,762,000 motor vehicles registered in March 2008, up from 2,366,000 in March 2003.

Rail

The route-km of railways within the state was 3,890 km in 2008–09. The main centres are Asansol, Burdwan, Howrah, Kharagpur, New Jalpaiguri and Sealdah. There is a metro in Kolkata (16·5 km).

Civil Aviation

The main airport is Kolkata's Netaji Subhas Chandra Bose International Airport, which has national and international flights. In 2009 it handled 7,691,268 passengers (6,584,741 on domestic flights) and 102,121 tonnes of freight. The second airport is at Bagdogra in the extreme north, which in 2010 had scheduled services to Bangkok, Chennai, Delhi, Guwahati, Kolkata and Paro.

Shipping

Kolkata is the chief port: a barrage has been built at Farakka to control the flow of the Ganges and to provide a rail and road link between North and South Bengal. A second port has been developed at Haldia, between the present port and the sea, which is intended mainly for bulk cargoes. West Bengal has about 800 km of navigable canals.

SOCIAL INSTITUTIONS

Justice

The High Court of Judicature at Kolkata has a Chief Justice and 45 puisne judges. The Andaman and Nicobar Islands come under its jurisdiction.

Education

In 2001, 68·6% of the total population were literate (men, 77·0%; women, 59·6%). There were 52,426 primary schools, 2,883 junior high schools and 9,620 high and higher secondary schools in 2001 with (1998–99) 1,881,226 students. Education is free up to higher secondary stage.

In 2001 there were nine universities: the University of Calcutta (founded 1857); University of Jadavpur, Kolkata (1955); Burdwan University (1960); Kalyani University (1960); University of North Bengal (1962); Rabindra Bharati University (1962); Vidyasagar University, Medinipur (1981); Bengal Engineering College (deemed to have university status from 1992); Netaji Subhas Open University (1998). The enrolment of students in universities for 2001 totalled 37,461. There were 36 government degree colleges and institutes in 2001 and 370 non-government colleges and institutes; enrolment of students, 550,989.

Health

As at 1 Jan. 2002 there were 411 hospitals with 55,279 beds. There were 1,262 primary health centres, 8,126 sub-centres and 99 community health centres in 2001.

RELIGION

At the 2001 census Hindus numbered 58,104,835; Muslims, 20,240,543; Christians, 515,150; Buddhists, 243,364; Sikhs, 66,391; Jains, 55,223.

FURTHER READING

Bagchi, Jasodhara, (ed.) *The Changing Status of Women in West Bengal, 1970–2000: The Challenge Ahead.* 2005
Chatterjee, P., *The Present History of West Bengal: Essays in Political Criticism.* 1997

UNION TERRITORIES

Andaman and Nicobar Islands

GENERAL DETAILS

The Andaman and Nicobar Islands are administered by the President of the Republic of India acting through a Lieut.-Governor. There is a 30-member Pradesh Council, five members of which are selected by the Administrator as advisory counsellors. The seat of administration is at Port Blair, which is connected with Kolkata (1,255 km away) and Chennai (1,190 km) by steamer service which calls about every ten days; there are air services from Kolkata and Chennai. There are 1,111 km of paved roads and 200 km of other roads.

The population (2011 census) was 379,944. The area is 8,249 sq. km and the density 46 per sq. km. Growth rate 2001–11, 6·68%. Port Blair (2001), 99,984.

The climate is tropical, with little variation in temperature. Heavy rain (125" annually) is mainly brought by the southwest monsoon. Humidity is high. The islands were severely affected by the tsunami of 26 Dec. 2004.

Budget figures for 2009–10 showed receipts of ₹1,726·2m. and expenditure of ₹27,208·6m.

There is installed electricity capacity of 38,805 KW. In Sept. 2008, 331 out of 501 inhabited villages had electricity.

In 2001, 26,524 ha. were under cultivation, of which 10,885 ha. were under rice. 48,167 tonnes of rice were grown. There were 79,219 goats, 63,554 cattle, 52,201 pigs and 16,211 buffaloes in 2003.

In 2008–09, 32,490 tonnes of fish were landed. There were 1,966 registered fishing boats in 2002 and 2,721 fishermen.

There were 6,662 sq. km of forests in 2007. In 2002, 4,712 cu. metres of sawn timber were extracted.

In 2003 there were 48 factories and 1,479 small-scale industrial units, employing 5,032 people.

In 2001 there were 207 primary schools with 43,000 students, 56 middle schools with 23,000 students, 45 high schools with 11,000 students and 48 higher secondary schools with 4,000 students. There is a teacher training college, two polytechnics and two colleges. Literacy (2001 census), 81·3% (86·3% of men and 75·2% of women).

In 2003 there were three hospitals, 28 health centres and 107 primary health sub-centres.

At the 2001 census Hindus numbered 246,589; Christians, 77,178.

Lieut.-Governor: Bhopinder Singh; b. 1946 (since 29 Dec. 2006).

The **Andaman Islands** lie in the Bay of Bengal, 193 km from Cape Negrais in Myanmar, 1,255 from Kolkata and 1,190 from Chennai. Five large islands grouped together are called the Great Andamans, and to the south is the island of Little Andaman. There are some 239 islets and a total of 572 islands, islets and rocks, the two principal groups being the Ritchie Archipelago and the Labyrinth Islands. The Great Andaman group is about 467 km long and, at the widest, 51 km broad.

The original inhabitants live in the forests by hunting and fishing. The total population of the Andaman Islands (including about 400–450 aboriginals) was 343,125 in 2011. Main aboriginal tribes: Andamanese, Onges, Jarawas and Sentinelese.

The Great Andaman group, densely wooded (forests covered 83% of the Andaman Islands in 2007), contains hardwood and softwood and supplies the match and plywood industries. Annually the Forest Department export about 25,000 tonnes of timber to the mainland. Coconut, coffee and rubber are cultivated. The islands are slowly being made self-sufficient in paddy and rice, and now grow approximately half their annual requirements. Livestock, 2003: 63,460 goats, 54,645 cattle, 16,144 buffaloes and 8,517 pigs. Fishing is important. There is a sawmill at Port Blair and a coconut-oil mill. Little Andaman has a palm-oil mill.

The islands possess a number of harbours and safe anchorages, notably Port Blair in the south, Port Cornwallis in the north and Elphinstone and Mayabandar in the middle.

The **Nicobar Islands** are situated to the south of the Andamans, 121 km from Little Andaman. The Danes were in possession 1756–1869, and then the British until 1947. There are 19 islands, seven uninhabited; total area, 1,841 sq. km. The islands are usually divided into three sub-groups (southern, central and northern), the chief islands in each being respectively Great Nicobar, Camotra with Nancowrie and Car Nicobar. There is a harbour between the islands of Camotra and Nancowrie, Nancowrie Harbour.

The population numbered 36,819 in 2011. The Nicobarese and Shompen tribes form the majority of its population. The coconut and areca nut are the main items of trade, and coconuts are a major item in the people's diet. Livestock, 2003: 43,684 pigs, 15,759 goats, 8,909 cattle and 67 buffaloes.

FURTHER READING

Dhingra, Kiran, *The Andaman and Nicobar Islands in the Twentieth Century: A Gazetteer.* 2006

Chandigarh

On 1 Nov. 1966 the city of Chandigarh and the area surrounding it was constituted a Union Territory. Population (2011), 1,054,686; density, 9,252 per sq. km; growth rate 2001–11, 17·10%. Area, 114 sq. km. It serves as the joint capital of both Punjab (India) and the state of Haryana, and is the seat of a High Court. The city, which had a population of 808,515 inhabitants at the 2001 census, will ultimately be the capital of just the Punjab; joint status is to last while a new capital is built for Haryana.

Budget figures for 2009–10 showed receipts of ₹15,810·0m. and expenditure of ₹19,319·7m.

In Sept. 2008 all 23 inhabited villages had electricity.

There is some cultivated land and some forest (27·5% of the territory).

In 2007–08 there were 294 factories, employing 12,800 people.

In 1996–97 there were 44 primary schools (60,012 students), 33 middle schools (34,095 students), 50 high schools (18,510 students) and 47 higher secondary schools (16,710 students). There were also two engineering and technology colleges, 12 arts, science and commerce colleges, two polytechnic institutes and a university (Panjab University). Other institutes have university status: Chandigarh College of Architecture; the Chandigarh Government College of Art; Chandigarh Institute of Postgraduate Medicinal Education and Research; Punjab Engineering College.

In 2001, 81·9% of the population were literate (86·1% of men and 76·5% of women).

In 2000 there were 43 dispensaries, 16 general hospitals and 72 private hospitals with a total of 2,530 beds.

At the 2001 census Hindus numbered 707,978 and Sikhs 145,175.

Administrator: Shivraj Patil; b. 1935 (took office as Governor of Punjab on 22 Jan. 2010).

Dadra and Nagar Haveli

GENERAL DETAILS

Formerly Portuguese, the territories of Dadra and Nagar Haveli were occupied in July 1954 by nationalists, and a pro-India administration was formed; this body made a request for incorporation into the Union on 1 June 1961. By the 10th amendment to the constitution the territories became a centrally administered Union Territory with effect from 11 Aug. 1961, forming an enclave at the southernmost point of the border between Gujarat and Maharashtra, approximately 30 km from the west coast. Area 491 sq. km; population (census 2011), 342,853; density 698 per sq. km; growth rate 2001–11, 55·50%. There is an Administrator appointed by the government of India. The day-to-day business is done by various departments, co-ordinated by the Secretaries, Assistant Secretary, Collector and Resident Deputy Collector. The capital is Silvassa, which had a population of 21,893 at the 2001 census. 78·82% of the population is tribal and organized in 140 villages. Languages used are dialects classified under Bhilodi (91·1%), Bhilli, Gujarati, Marathi and Hindi.

CURRENT GOVERNMENT

Administrator: Narendra Kumar.

ECONOMY

Budget

Budget figures for 2009–10 showed receipts of ₹4,835·8m. and expenditure of ₹2,883·2m.

ENERGY AND NATURAL RESOURCES

Electricity

In Sept. 2008 all 70 inhabited villages had electricity. A major sub-station at Kharadpada village has been completed. Installed capacity was 79 MW in Oct. 2008. Electricity consumption per capita in 2009–10 was 11,864 kWh

Minerals

There are few natural mineral resources although there is some ordinary sand and quarry stone.

Agriculture

Farming is the chief occupation and 21,015 ha. were under net crop in 2007–08. Much of the land is terraced and there is a 100% subsidy for soil conservation. The major food crops are rice and ragi; wheat, small millets and pulses are also grown. There is coverage of irrigation over 7,408 ha. During 2007–08 the administration distributed 90 tonnes of high-yielding paddy and wheat seed and 1,310 tonnes of manures and fertilizers.

Forestry

20,359 ha. or 40·8% of the total area is forest, mainly of teak, sadad and khair. In 1985 a moratorium was imposed on commercial felling to preserve the environmental function of the forests and ensure local supplies of firewood, timber and fodder. The tribal peoples have been given exclusive right to collect minor forest produce from the reserved forest area for domestic use. 92 sq. km of reserved forest was declared a wildlife sanctuary in 2000.

INDUSTRY

There is no heavy industry, and the Territory is a 'No Polluting Industry District'. Industrial estates for small and medium scales have been set up at Piparia, Masat and Khadoli. There were 2,270 industrial units permanently registered in 2005–06, of which 1,863 were small scale and 407 medium and large scale units.

The Micro, Small and Medium Enterprises Development Act was implemented in Dadra and Nagar Haveli on 2 Oct. 2006. Micro, small and medium enterprises provided direct and indirect employment to 62,284 people at 31 March 2011.

Labour

The Labour Enforcement Office ensures the application of the Monitoring of Minimum Wages Act (1948), the Industrial Disputes Act (1947), the Contract Labour (Regulation and Abolition) Act (1970) and the Workmen's Compensation Act (1923). During 2010–11, 144 cases under the Industrial Disputes Act were settled. Under the Contract Labour (Regulation and Abolition) Act (1970), 171 certificates of registration and 118 licences were issued to industrial establishments. 66 cases under the Workman's Compensation Act (1923) were settled.

COMMUNICATIONS

Roads

In March 2008 there were 632 km of road of which 610 km were surfaced. There were 57,532 registered motor vehicles as at 31 March 2008. The National Highway no. 8 passes through Vapi, 18 km from Silvassa.

Rail

Although there are no railways in the territory the line from Mumbai to Ahmedabad runs through Vapi, 18 km from Silvassa.

Civil Aviation

The nearest airport is at Mumbai, 180 km from Silvassa.

SOCIAL INSTITUTIONS

Justice

The territory is under the jurisdiction of the Bombay (Maharashtra) High Court. There is a District and Sessions Court and one Junior Division Civil Court at Silvassa.

Education

Literacy was 57·6% of the population at the 2001 census (71·2% of men and 40·2% of women). In 2001–02 there were 195 primary and middle schools (35,637 students) and 17 high and higher secondary schools (8,887 students).

Health

The territory had (2001–02) a civil hospital, 6 primary health centres, 36 sub-centres, three dispensaries and a mobile dispensary. A Community Health Centre has been established at Khanvel, 20 km from Silvassa. The Pulse Polio Immunisation programme was organized in 1999 and 54,128 polio doses were provided to children below five years of age. There has been a sharp fall in the incidence of malaria, especially cerebral malaria, owing to the sustained efforts of the administration. Hepatitis B vaccination of all inmates in the social welfare hostels was completed with the co-operation of voluntary organizations. A blood testing centre has been established for HIV testing.

Welfare

The Social Welfare Department implements the welfare schemes for poor Scheduled castes, Scheduled tribes, women and physically disabled persons, etc.

RELIGION

Numbers of religious followers (2001 census): Hindus, 206,203 (94% of the population), with some Muslims and Christians.

CULTURE

Press

One weekly newspaper and two fortnightly news magazines are published.

Tourism

The territory is a rural area between the industrial centres of Mumbai and Surat-Vapi. The Tourism Department is developing areas of natural beauty to promote eco-friendly tourism. Several gardens and the Madhuban Dam are among the tourist sites. A lion safari park has been set up at Vasona over 20 ha.

Daman and Diu

GENERAL DETAILS

Daman (Damão) on the Gujarat coast, 100 miles (160 km) north of Mumbai, was seized by the Portuguese in 1531 and ceded to them (1539) by the Shar of Gujarat. The island of Diu, captured in 1534, lies off the southeast coast of Kathiawar (Gujarat); there is a small coastal area. Former Portuguese forts on either side of the entrance to the Gulf of Cambay, in Dec. 1961 the territories were occupied by India and incorporated into the Indian Union; they were administered as one unit together with Goa, to which they were attached until 30 May 1987, when Goa was separated from them and became a state.

TERRITORY AND POPULATION

The territory has an area of 112 sq. km and a population of 242,911 at the 2011 census. Density, 2,169 sq. km. Daman has an area of 72 sq. km, population (2001) 113,989; Diu, 40 sq. km, population 44,215. Daman is the capital of the territory. The main language spoken is Gujarati.

The chief towns are (with 2001 populations) Daman (35,770) and Diu (21,578).

Daman and Diu have been governed as parts of a Union Territory since Dec. 1961, becoming the whole of that Territory on 30 May 1987. There are two districts.

SOCIAL STATISTICS

Growth rate 2001–11, 53·54%.

CURRENT GOVERNMENT

Administrator: Narendra Kumar.

ECONOMY

The main activities are tourism, fishing and tapping the toddy palm (preparing palm tree sap for consumption). In Daman there is rice-growing, some wheat and dairying. Diu has fine tourist beaches, grows coconuts and pearl millet, and processes salt.

Budget

Budget figures for 2009–10 showed receipts of ₹3,912·5m. and expenditure of ₹2,777·6m.

ENERGY AND NATURAL RESOURCES

Electricity

In Sept. 2008 all 23 inhabited villages had electricity. There was an installed capacity of 69 MW in Oct. 2008. Electricity consumption per capita in 2009–10 was 7,118 kWh

SOCIAL INSTITUTIONS

Education

In 2001, 78·2% of the population were literate (86·8% of men and 65·6% of women). In 1996–97 there were 53 primary schools with 14,531 students, 20 middle schools with 6,834 students, 20 high schools with 3,220 students and 3 higher secondary schools with 1,202 students. There is a degree college and a polytechnic.

RELIGION

Numbers of religious followers (2001 census): Hindus, 141,901 (90% of the population), with some Muslims and Christians.

Delhi

GENERAL DETAILS

Delhi became a Union Territory on 1 Nov. 1956 and was designated the National Capital Territory in 1995.

TERRITORY AND POPULATION

The territory forms an enclave near the eastern frontier of Haryana and the western frontier of Uttar Pradesh in north India. Delhi has an area of 1,483 sq. km. Its population (2011 census) is 16,753,235 (density per sq. km, 11,297). Growth rate 2001–11, 20·96%.

CONSTITUTION AND GOVERNMENT

The Lieut.-Governor is the Administrator. Under the New Delhi Municipal Act 1994 New Delhi Municipal Council is nominated by central government and replaces the former New Delhi Municipal Committee. There are nine districts.

RECENT ELECTIONS

Elections for the 70-member Legislative Assembly were held on 29 Nov. 2008. The Indian National Congress won 42 seats (47 in 2003); Bharatiya Janata Party, 23 (20 in 2003); Bahujan Samaj Party, 2 (none in 2003); others, 2 (three in 2003). The INC won a by-election held on 13 Dec. 2008, bringing their total to 43 seats.

CURRENT GOVERNMENT

Lieut.-Governor: Tejendra Khanna (took office on 9 April 2007).
  Chief Minister: Sheila Dikshit (took office on 3 Dec. 1998).

ECONOMY

Budget

2008–09 revenue receipts, ₹163,522·1m.; expenditure, ₹117,625·5m. Budget estimates for 2009–10: revenue receipts, ₹190,993·0m.; expenditure, ₹137,033·1m.

ENERGY AND NATURAL RESOURCES

Electricity

In Sept. 2008 all 158 inhabited villages had electricity. The installed capacity was 3,677 MW in Oct. 2008.

Minerals

The Union Territory has deposits of kaolin (china clay), quartzite and fire clay.

Agriculture

The contribution to the economy is not significant. In 2003–04 about 41,500 ha. were cropped (of which 22,700 ha. were irrigated). Animal husbandry is increasing and mixed farms are common. Chief crops are wheat, bajra, paddy, sugarcane, gram, jowar and vegetables. Buffaloes are kept as a source of milk; pigs and goats are kept for meat.

INDUSTRY

The modern city is the largest commercial centre in northern India and an important industrial centre. Since 1947 a large number of industrial units have been established; these include factories for the manufacture of razor blades, sports goods, electronic goods, bicycles and parts, plastic and PVC goods including footwear, textiles, chemicals, fertilizers, medicines, hosiery, leather goods, soft drinks and hand tools. The largest single industry is the manufacture of garments (15·9% of all factories in 2008–09 manufactured wearing apparel). There are also metal forging, casting, galvanizing, electro-plating and printing enterprises. There were 3,026 factories in 2008–09, down from 3,198 in 2007–08. The total output of factories in 2008–09 was ₹274,759m., up from ₹267,383m. in 2007–08.

Some traditional handicrafts, for which Delhi was formerly famous, still flourish; among them are ivory carving, miniature painting, gold and silver jewellery and papier mâché work. The handwoven textiles of Delhi are particularly fine; this craft is being successfully revived.

Delhi is a major market for manufactures, imports and agricultural goods; there are specialist fruit and vegetable, food grain, fodder, cloth, bicycle, hosiery, dry fruit and general markets.

COMMUNICATIONS

Roads

Five national highways pass through the city. As at 31 March 2008 there were 5,899,421 registered motor vehicles including 3,616,417 two-wheelers.

Rail

Delhi is an important rail junction with three main stations: New Delhi, Delhi Junction and Hazrat Nizamuddin. The first of three lines of the Delhi metro system opened in 2002: when complete it will consist of 34·5 km subway, 35·5 km elevated and 111 km surface running. There is a 40-km long electric ring railway, but it has significantly lost its relevance since the opening of Delhi's metro.

Civil Aviation
Indira Gandhi International Airport operates international flights; Palam airport operates internal flights.

SOCIAL INSTITUTIONS
Education
The proportion of literate people to the total population was 81·7% at the 2001 census (87·3% of males and 74·7% of females). In 2003–04 there were 2,126 primary schools with 924,493 students and 22,930 teachers, 681 middle schools with 230,362 students and 9,192 teachers, 1,678 high schools and higher secondary schools with 1,747,884 students and 59,064 teachers. In 1996–97 there were nine engineering and technology colleges, nine medical colleges and 25 polytechnics.

The University of Delhi was founded in 1922; it had 78 affiliated colleges in 2002–03 and 189,332 students in 1994–95. There are also Jawaharlal Nehru University, Indira Gandhi National Open University, the Jamia Millia Islamia University, the Guru Gobind Singh Indraprastha University, Jamia Hamdard University and Shri Lal Bahadur Shastri Rashtriya Sanskrit Vidyapeeth University; the Indian Institute of Technology at Hauz Khas; the Indian Agricultural Research Institute at Pusa; the All India Institute of Medical Science at Ansari Nagar and the Indian Institute of Public Administration are the other important institutions.

Health
In 2001 there were 11 government hospitals plus 71 private hospitals, 167 government dispensaries plus 489 other dispensaries, 73 mobile health clinics and 64 school health clinics.

RELIGION
At the 2001 census Hindus numbered 11,358,049; Muslims, 1,623,520; Sikhs, 555,602; Jains, 155,122; Christians, 130,319.

CULTURE
Press
Delhi publishes major daily newspapers, including the *Times of India, Hindustan Times, The Hindu, Indian Express, National Herald, Patriot, Economic Times, The Pioneer, The Observer of Business and Politics, Financial Express, Statesman, Asian Age* and *Business Standard* (all in English); *Nav Bharat Times, Rashtriya Sahara, Jansatta* and *Hindustan* (all in Hindi); and three Urdu dailies.

Lakshadweep

The territory consists of an archipelago of 36 islands (ten inhabited), about 300 km off the west coast of Kerala. It was constituted a Union Territory in 1956 as the Laccadive, Minicoy and Amindivi Islands, and renamed in Nov. 1973. The total area of the islands is 32 sq. km. The northern portion is called the Amindivis. The remaining islands are called the Laccadives (except Minicoy Island). The inhabited islands are: Androth (the largest), Amini, Agatti, Bitra, Chetlat, Kadmat, Kalpeni, Kavaratti, Kiltan and Minicoy. Androth is 4·8 sq. km, and is nearest to Kerala. An Advisory Committee associated with the Union Home Minister and an Advisory Council to the Administrator assist in the administration of the islands; these are constituted annually. Population (2011 census), 64,429, nearly all Muslims. Density, 2,013 per sq. km; growth rate 2001–11, 6·23%. The language is Malayalam, but the language in Minicoy is Mahl. Budget figures for 2009–10 showed receipts of ₹542·2m. and expenditure of ₹6,791·5m. Installed electric capacity (1998) 8,120 kW. In Sept. 2008 all eight inhabited villages had electricity. A number of solar photovoltaic power plants have been constructed in the islands. Guaranteeing supplies of potable water is problematic in most of the islands. Rain water harvesting schemes have been introduced as well as desalination plants. There are several small factories processing fibre from coconut husks: in 2002 these employed 316 workers. There are two handicraft training centres. The major industry is fishing—in 1999 there were 375 registered fishing boats. The principal catches are tuna and shark. Tuna is canned at a factory at Minicoy. There is an experimental pearl culture scheme at the uninhabited island of Bangarem. In 2001, 86·7% of the population were literate (92·5% of men and 80·5% of women). There were, in 1996–97, nine high schools (2,043 students) and nine nursery schools (1,197 students), 19 junior basic schools (9,015 students), four senior basic schools (4,797 students) and two junior colleges. There are two hospitals and four primary health centres plus 14 health sub-centres. At the 2001 census Muslims numbered 57,903 (95% of the population). The staple products are copra and fish; coconut is the only major crop. Headquarters of administration, Kavaratti, population 10,113 (2001 census), on Kavaratti Island. An airport opened on Agatti Island in1988. The islands are also served by ship from the mainland and have helicopter inter-island services. There are two catamaran-type high-speed inter-island ferries and four barges. The islands have 172 km of roads, all with paved surfaces. The islands have great tourist potential.

Administrator: Amar Nath (took office on 11 July 2011).

Puducherry

GENERAL DETAILS
Formerly the chief French settlement in India, Puducherry (known as Pondicherry until 2006) was founded by the French in 1673, taken by the Dutch in 1693 and restored to the French in 1699. The English took it in 1761, restored it in 1765, retook it in 1778, restored it a second time in 1785, retook it a third time in 1793 and finally restored it to the French in 1816. Administration was transferred to India on 1 Nov. 1954. A Treaty of Cession (together with Karaikal, Mahé and Yanam) was signed on 28 May 1956; instruments of ratification were signed on 16 Aug. 1962 from which date (by the 14th amendment to the Indian Constitution) Pondicherry, comprising the four territories, became a Union Territory.

TERRITORY AND POPULATION
The territory is composed of enclaves on the Coromandel Coast of Tamil Nadu and Andhra Pradesh, with Mahé forming two enclaves on the coast of Kerala. The total area of Puducherry is 479 sq. km, divided into 11 enclaves that are grouped into four Districts. On Tamil Nadu coast: Puducherry (290 sq. km; population, 2001 census, 735,332), Karaikal (161; 170,791). On Kerala coast: Mahé (9; 36,828). On Andhra Pradesh coast (although the enclave lies back from the shore but at no point does its territory touch the coast): Yanam (20; 31,394). Total population (2011 census), 1,244,464; density, 2,598 per sq. km. Puducherry Municipality had (2001) 220,865 inhabitants and the urban agglomeration had 505,959 inhabitants. The principal languages spoken are Tamil, Telugu, Malayalam, French and English.

SOCIAL STATISTICS
Growth rate 2001–11, 27·72%.

CONSTITUTION AND GOVERNMENT

By the government of Union Territories Act 1963 Puducherry is governed by a Lieut.-Governor, appointed by the President, and a Council of Ministers responsible to a Legislative Assembly.

RECENT ELECTIONS

In the elections of 13 April 2011 the All Indian National Congress in alliance with the All India Anna Dravida Munnetra Kazhagam won 20 seats, followed by the Dravida Munnetra Kazhagam with 9 seats. One seat went to an independent candidate.

CURRENT GOVERNMENT

Lieut.-Governor: Iqbal Singh; b. 1945 (took office on 27 July 2009).

Chief Minister: N. Rangaswamy; b. 1950 (took office on 16 May 2011).

ECONOMY

Budget

2008–09 revenue receipts, ₹24,585·0m.; expenditure, ₹25,704·7m. Budget estimates for 2009–10: revenue receipts, ₹30,665·4m.; expenditure, ₹31,936·3m.

ENERGY AND NATURAL RESOURCES

Electricity

Power is bought from neighbouring states. In Sept. 2008 all 92 inhabited villages had electricity. There was an installed capacity of 257 MW in Oct. 2008.

Agriculture

Nearly 45% of the population is engaged in agriculture and allied pursuits; 90% of the cultivated area is irrigated. The main food crop is rice. Foodgrain production, 58,785 tonnes in 2003. Rice production, 2003, 57,514 tonnes from 24,142 ha. Principal cash crops are sugarcane (209,496 tonnes in 2003) and groundnuts; minor food crops include cotton, ragi, bajra and pulses.

Fisheries

In 2007–08 the marine catch was 33,273 tonnes and the inland catch 5,919 tonnes. In 2005 there were 20,400 people working in fishing and related activities.

INDUSTRY

In March 2003 there were 55 large and 139 medium-scale enterprises manufacturing items such as textiles, sugar, cotton yarn, spirits and beer, potassium chlorate, rice bran oil, vehicle parts, soap, amino acids, paper, plastics, steel ingots, washing machines, glass and tin containers and bio polymers. These factories employed 25,095 people. There were also 6,876 small industrial units (2003) engaged in varied manufacturing.

COMMUNICATIONS

Roads

In March 2008 there were 2,696 km of road of which 2,364 km were surfaced. There were 484,008 registered motor vehicles as at 31 March 2008.

Rail

Puducherry is connected to Villupuram Junction. Route-km in 2008–09, 11 km.

Civil Aviation

The nearest main airport is Chennai.

SOCIAL INSTITUTIONS

Education

In 2001, 81·2% of the population were literate (88·6% of men and 73·9% of women). There were, in 2000–01, 223 pre-primary schools (22,462 pupils), 337 primary schools (38,405), 110 middle schools (34,034), 128 high schools (65,451) and 65 higher secondary schools (76,726). There were (2000–01) eight general education colleges, three medical colleges, a law college, five engineering colleges, an agricultural college and a dental college, and five polytechnics. Pondicherry University had around 1,600 students in 2005.

Health

There were 14 hospitals in 2004 (with 3,064 beds); and 39 primary health centres, 75 sub-centres and four community health centres in 2001.

RELIGION

At the 2001 census Hindus numbered 845,449; Christians, 67,688; Muslims, 59,358.

INDONESIA

© Research Machines plc 2006

Republik Indonesia
(Republic of Indonesia)

Capital: Jakarta
Population projection, 2015: 251·88m.
GNI per capita, 2011: (PPP$) 3,716
HDI/world rank: 0·617/124
Internet domain extension: .id

KEY HISTORICAL EVENTS

The Indonesian archipelago was populated from the north from around 3000 BC. Indian scholars described the Dvipantera civilization of Java and Sumatra as early as 200 BC and Indian-influenced Hindu kingdoms began to appear in the west of the archipelago from the 1st century AD. Small maritime trading settlements evolved into Srivijaya, a Buddhist Malay kingdom centred on southeast Sumatra. The Srivijaya empire controlled trade between India and China through the Melaka strait and by the late 7th century it had expanded to encompass much of Sumatra, the Malay peninsula and western Java. Other Indianized kingdoms developed in central Java, including Mataram (8th–10th centuries) and Mahayana (9th century).

Power shifted to east Java, culminating in the Majapahit empire (founded by Wijaya in 1293 and developed by Gaja Mada), which controlled much of present-day Indonesia. Islam was brought to northern Sumatra by Arab merchants from the 7th century but spread only gradually until the rise of the sultanate of Melaka from the early 15th century. The north Javanese coastal kingdoms that converted to Islam competed with the Majapahit empire, contributing to its decline by the early 16th century.

Portuguese forces, led by Afonso de Albuquerque, having conquered the sultanate of Melaka in 1511 attempted to control the spice trade, although local powers including Aceh (north Sumatra) and Makassar (Sulawesi) competed by opening new routes. Dutch mariners arrived in west Java in 1596 and Jan Pietersoon Coen established a base on Java (Batavia, which became Jakarta) in 1619. The capture of Melaka in 1641 heralded 150 years of Dutch control of the spice trade through its United East India Company (VOC).

Following the VOC's collapse at the end of the 18th century and after a short period of British rule under Thomas Stamford Raffles, the Dutch state took control of the archipelago. The Anglo-Dutch Treaty of 1824 delineated the border between British

Malaya and the Dutch East Indies. Between 1825–50 the Dutch suppressed a rebellion on Java, initially led by Prince Diponegoro, and partly fuelled by opposition to the 'Cultivation System' whereby locals had to devote a percentage of their land to cultivating government-approved export crops. The Sarekat Islam (Islamic Union), founded in 1912, was the country's first major nationalist movement. In March 1942 Japanese forces invaded and began to dismantle the Dutch power base. The Indonesian nationalist leaders Sukarno and Mohammad Hatta worked with the Japanese occupiers whilst pushing for independence. On 17 Aug. 1945, two days after Japan's surrender, Sukarno and Hatta proclaimed an independent republic with Sukarno as its president, although the Netherlands did not concede unconditional sovereignty until 27 Dec. 1949.

In 1960 President Sukarno dismissed parliament after a dispute over the government's budget and dissolved political parties. In their place he set up the National Front and the Provisional People's Consultative Assembly. On 11–12 March 1966 the military commanders under the leadership of Lieut.-Gen. Suharto seized executive power, leaving President Sukarno as head of state. The Communist party, which had twice attempted to overthrow the government, was outlawed. On 22 Feb. 1967 Sukarno handed over all his powers to Gen. Suharto. Re-elected president at five-year intervals, on the final occasion on 10 March 1998, Suharto presided over a booming economy but one that was characterized by corruption and cronyism.

These weaknesses became apparent in 1997 when a failure of economic confidence spread from Japan across Asia. By May 1998 food prices had trebled and riots broke out in Jakarta. The risk of society fragmenting along ethnic and religious lines was emphasized by the sufferings of the Chinese community. President Suharto was forced to stand down on 21 May 1998 and was succeeded by his vice-president, Bacharuddin Jusuf Habibie, who promised political and economic reforms. Continuing protest centred on the Suharto family, which still exercised control over large parts of the economy. In Aug. 1999 Timor-Leste, the former Portuguese colony that Indonesia invaded in 1975, voted for independence, a move that was eventually approved by the Indonesian parliament after violent clashes between independence supporters and pro-Indonesian militia groups. It gained independence on 20 May 2002.

Abdurrahman Wahid was elected president by the People's Consultative Assembly in Oct. 1999. He oversaw some reform and economic growth but was forced to step down amid allegations of corruption on 29 Jan. 2001. His vice-president, Megawati Sukarnoputri, took control and assumed the presidency on 23 July 2001. In Oct. 2002 around 200 people died in a car-bomb explosion outside a nightclub in Bali. Jemaah Islamiyah, an extremist group with alleged links to Al-Qaeda, was implicated.

Indonesia's first direct presidential election took place in Sept. 2004 when Susilo Bambang Yudhoyono beat Megawati in a runoff vote. On 26 Dec. 2004 northwest Sumatra was hit by a devastating tsunami. The death toll in Indonesia was put at 166,000, mostly in Aceh. However, the disaster revived the peace process, initiated in late 2002, between the government and the separatist Free Aceh Movement (GAM). Accords signed in Helsinki, Finland in Aug. 2005 created a framework for a military stand-down on both sides.

TERRITORY AND POPULATION

Indonesia, with a land area of 1,910,931 sq. km (737,615 sq. miles), consists of 17,507 islands (6,000 of which are inhabited)

extending about 3,200 miles east to west through three time-zones (East, Central and West Indonesian Standard time) and 1,250 miles north to south. The largest islands are Sumatra, Java, Kalimantan (Indonesian Borneo), Sulawesi (Celebes) and Papua, formerly West Papua (the western part of New Guinea). Most of the smaller islands except Madura and Bali are grouped together. The two largest groups of islands are Maluku (the Moluccas) and Nusa Tenggara (the Lesser Sundas). On the island of Timor, Indonesia is bounded in the east by Timor-Leste.

Population at the 2010 census was 237,641,326; density, 124·4 per sq. km. Indonesia has the fourth largest population in the world, after China, India and the USA. In 2011, 44·6% of the population were urban.

The UN gives a projected population for 2015 of 251·88m.

Area, population and chief towns of the provinces, autonomous regions and major islands:

	Area (in sq. km)	Population (2010 census)	Chief town	Population (2010 census)
Bali	5,780	3,890,757	Denpasar	788,589
Nusa Tenggara Barat	18,572	4,500,212	Mataram	402,843
Nusa Tenggara Timur	48,718	4,683,827	Kupang	336,239
Bali and Nusa Tenggara	73,070	13,074,796		
Banten	9,663	10,632,166	Serang	577,785
DKI Jakarta[1]	664	9,607,787	Jakarta	9,607,787
Jawa Barat	35,378	43,053,732	Bandung	2,394,873
Jawa Tengah	32,801	32,382,657	Semarang	1,555,984
Jawa Timur	47,800	37,476,757	Surabaya	2,765,487
Yogyakarta[1]	3,133	3,457,491	Yogyakarta	388,627
Java	129,439	136,610,590		
Kalimantan Barat	147,307	4,395,983	Pontianak	554,764
Kalimantan Selatan	38,744	3,626,616	Banjarmasin	625,481
Kalimantan Tengah	153,565	2,212,089	Palangkaraya	220,962
Kalimantan Timur	204,534	3,553,143	Samarinda	727,500
Kalimantan Utara[1]			Tanjung Selor	30,486
Kalimantan	544,150	13,787,831		
Maluku	46,914	1,533,506	Ambon	331,254
Maluku Utara	31,983	1,038,087	Ternate	185,705
Papua[2]	319,036	2,833,381	Jayapura	256,705
Papua Barat[2]	97,024	760,422	Manokwari	187,726
Maluku and Papua	494,957	6,165,396		
Gorontalo	11,257	1,040,164	Gorontalo	180,127
Sulawesi Barat	16,787	1,158,651	Mamuju	336,973
Sulawesi Selatan	46,717	8,034,776	Makassar	1,338,663
Sulawesi Tengah	61,841	2,635,009	Palu	336,532
Sulawesi Tenggara	38,068	2,232,586	Kendari	289,966
Sulawesi Utara	13,852	2,270,596	Manado	410,481
Sulawesi	188,522	17,371,782		
Aceh[2]	57,956	4,494,410	Banda Aceh	223,446
Bangka-Belitung	16,424	1,223,296	Pangkalpinang	174,758
Bengkulu	19,919	1,715,518	Bengkulu	308,544
Jambi	50,058	3,092,265	Jambi	531,857
Kepulauan Riau	8,202	1,679,163	Tanjung Pinang	187,359
Lampung	34,624	7,608,405	Bandar Lampung	881,801
Riau	87,024	5,538,367	Pekanbaru	897,767
Sumatera Barat	42,013	4,846,909	Padang	833,562
Sumatera Selatan	91,592	7,450,394	Palembang	1,455,284
Sumatera Utara	72,981	12,982,204	Medan	2,097,610
Sumatra	480,793	50,630,931		

[1]Kalimantan Utara, consisting of four regencies and one city formerly in Kalimantan Timur, was established in Oct. 2012.
[2]Province with special status.

The capital, Jakarta, had an estimated population of 9·61m. in 2010. Other major cities (2010 census population in 1m.): Surabaya, 2·77; Bandung, 2·39; Bekasi, 2·33; Medan, 2·10; Tangerang, 1·80; Depok, 1·74; Semarang, 1·52.

The principal ethnic groups are the Acehnese, Bataks and Minangkabaus in Sumatra, the Javanese and Sundanese in Java, the Madurese in Madura, the Balinese in Bali, the Sasaks in Lombok, the Menadonese, Minahasans, Torajas and Buginese in Sulawesi, the Dayaks in Kalimantan, the Irianese in Papua and the Ambonese in the Moluccas. There were an estimated 6·5m. Chinese resident in 2005.

Bahasa Indonesia (Indonesian) is the official language; Dutch is spoken as a colonial inheritance.

SOCIAL STATISTICS

Estimated births, 2008, 4,222,000; deaths, 1,434,000. 2008 estimated birth rate, 18·6 per 1,000 population; death rate, 6·3. Life expectancy in 2007 was 68·5 years for men and 72·5 for women. Annual population growth rate, 2000–08, 1·3%. Infant mortality, 2010, 27 per 1,000 live births; fertility rate, 2008, 2·2 births per woman.

CLIMATE

Conditions vary greatly over this spread of islands, but generally the climate is tropical monsoon, with a dry season from June to Sept. and a wet one from Oct. to April. Temperatures are high all the year and rainfall varies according to situation on lee or windward shores. Jakarta, Jan. 78°F (25·6°C), July 78°F (25·6°C). Annual rainfall 71" (1,775 mm). Padang, Jan. 79°F (26·1°C), July 79°F (26·1°C). Annual rainfall 177" (4,427 mm). Surabaya, Jan. 79°F (26·1°C), July 78°F (25·6°C). Annual rainfall 51" (1,285 mm).

On 26 Dec. 2004 an undersea earthquake centred off Sumatra caused a huge tsunami that flooded large areas along the coast of northwestern Indonesia resulting in 166,000 deaths. In total there were more than 225,000 deaths in 14 countries.

CONSTITUTION AND GOVERNMENT

The constitution originally dates from Aug. 1945 and was in force until 1949; it was restored on 5 July 1959.

The political system is based on *pancasila*, in which deliberations lead to a consensus. There is a 560-member *Dewan Perwakilan Rakyat* (House of People's Representatives), with members elected for a five-year term by proportional representation in multi-member constituencies. The constitution was changed on 10 Aug. 2002 to allow for direct elections for the president and the vice-president.

There is no limit to the number of presidential terms. Although predominantly a Muslim country, the constitution protects the religious beliefs of non-Muslims.

National Anthem

'Indonesia, tanah airku' ('Indonesia, our native land'); words and tune by W. R. Supratman.

GOVERNMENT CHRONOLOGY

Presidents since 1949. (Golkar = Party of the Functional Groups; PD = Democratic Party; PDIP = Indonesian Democratic Party–Struggle; PKB = National Awakening Party; PNI = Indonesian National Party)

1949–67	PNI	(Ahmed) Sukarno
1967–98	Golkar	(Mohamed) Suharto
1998–99	Golkar	Bacharuddin Jusuf Habibie
1999–2001	PKB	Abdurrahman Wahid
2001–04	PDIP	Megawati Sukarnoputri
2004–	PD	Susilo Bambang Yudhoyono

RECENT ELECTIONS

Elections to the House of People's Representatives were held on 9 April 2009. The Democrat Party (PD) won 20·9% of the vote (148

of 560 seats), the Party of the Functional Groups (Golkar) 14·5% (108), the Indonesian Democratic Party–Struggle (PDIP) 14·0% (93), the Prosperous Justice Party 7·9% (59), the National Mandate Party 6·0% (42), the United Development Party 5·3% (39), the National Awakening Party 4·9% (26), the Great Indonesia Movement Party 4·5% (30) and the People's Conscience Party 3·8% (15).

In the presidential election of 8 July 2009 incumbent Susilo Bambang Yudhoyono (PD) won 60·8% of the vote, former president Megawati Sukarnoputri (PDIP) 26·8% and vice-president Jusuf Kalla (Golkar) 12·4%.

CURRENT GOVERNMENT

President: Susilo Bambang Yudhoyono; b. 1949 (PD; sworn in 20 Oct. 2004).

  Vice-President: Boediono.

In April 2013 the cabinet was composed as follows:

  Co-ordinating Ministers: (*Political, Legal and Security Affairs*) Djoko Suyanto; (*Economic Affairs*) Hatta Rajasa (also *Acting Minister of Finance*); (*People's Welfare*) Agung Laksono.

  Minister of Agriculture: Suswono. *Communication and Information:* Tifatul Sembiring. *Defence:* Purnomo Yusgiantoro. *Energy and Mineral Resources:* Jero Wacik. *Foreign Affairs:* Marty Natalegawa. *Forestry:* Zulkifli Hasan. *Health:* Nafsiah Mboi. *Home Affairs:* Gamawan Fauzi. *Industry:* Mohamad Suleman Hidayat. *Justice and Human Rights:* Amir Syamsuddin. *Manpower and Transmigration:* Abdul Muhaimin Iskandar. *Maritime Affairs and Fisheries:* Sharif Cicip Sutardjo. *National Education:* Mohammad Nuh. *Public Works:* Joko Kirmanto. *Religious Affairs:* Suryadharma Ali. *Social Affairs:* Salim Segaf Al Jufrie. *Tourism and Creative Economy:* Marie Elka Pangestu. *Trade:* Gita Irawan Wirjawan. *Transportation:* Evert Ernest Mangindaan. *State Secretary:* Sudi Silalahi.

Office of the President: http://www.presidenri.go.id

CURRENT LEADERS

Susilo Bambang Yudhoyono

Position
President

Introduction
Retired general Susilo Bambang Yudhoyono, known widely by his acronym SBY, succeeded Megawati Sukarnoputri as president of Indonesia on 20 Oct. 2004. In the country's first direct presidential election he polled 61% of an estimated 125m. votes and he became the first incumbent to be re-elected in July 2009.

Early Life
Susilo Bambang Yudhoyono was born on 9 Sept. 1949 in the small town of Pacitan, in the east of the Indonesian island of Java. His family were observant Muslims and he attended a traditional *pesantren* (Muslim boarding school). He graduated from Indonesia's military academy in 1973 and joined the army, which was then, with Gen. Suharto as president, the country's dominant authority. He served as a senior officer in Indonesia's 1975 invasion of Timor-Leste, then a Portuguese colony. Gen. Suharto's 'New Order' political system was characterized by a strongly anti-communist foreign policy and relatively good relations with the USA. Yudhoyono travelled to the USA in 1976 and 1982, attending military training programmes at Fort Benning, Georgia. He later took a masters degree in business management from Webster University in Missouri and has since described the USA as his 'second home'. Between 1984 and 1987 Yudhoyono returned to Timor-Leste and commanded Battalion 744 in the city of Dili. By the mid-1990s he had risen through the ranks to become chief-of-staff in the Jakarta command. Questions have been asked about his knowledge of a raid by security forces on the Jakarta offices of the Indonesian Democratic Party (PDI) on

27 July 1996 (then chaired by Megawati Sukarnoputri), which left five dead and 23 missing.

In 1996 Yudhoyono served as chief military observer with the United Nations force in Bosnia. Two years later, with Indonesia in turmoil following the ousting of President Suharto in March 1998, he left the army and was appointed the minister for mining and energy in the administration of Abdurrahman Wahid. When the Muslim cleric was succeeded as president in 2001 by Mrs Megawati, daughter of former president Sukarno, Yudhoyono joined her cabinet as chief security minister. He was praised for the way he handled the aftermath of the Oct. 2002 Bali bombing that killed 202 people. He subsequently helped draft Indonesia's first counter-terrorist law and attempted to broker a peace agreement with separatist rebels in the historically troubled province of Aceh in Sumatra in 2003, which collapsed in May of that year. In March 2004 Yudhoyono resigned from Megawati's increasingly unpopular cabinet to establish the Democratic Party (PD). In the first round of elections in April (for choosing the members of parliament and three tiers of local officials) the PD had a strong showing. On 5 July, when Yudhoyono, along with his running mate Jusuf Kalla—a business tycoon with ties to many of the country's Islamic clerics—fought in the country's first direct presidential elections, no candidate won more than 50% of the vote. This forced a run-off election between Yudhoyono and Megawati on 20 Sept. which Yudhoyono won with 60·9% of the vote. He was officially sworn in as president on 20 Oct. 2004.

Career in Office
In interviews with the international media Yudhoyono vowed to fight terrorism and eradicate corruption in his five-year term. He also promised to restore Indonesian institutions and the rule of law and to rebuild the economy. The president set himself the goal of creating jobs for 50m. unemployed Indonesians. He also pledged to repair the often fractious relationship with Australia. In the aftermath of the Indian Ocean tsunami of 26 Dec. 2004, which is estimated to have killed some 166,000 people on the Indonesian island of Sumatra, Yudhoyono was quick to accept aid and expertise from the international community. Handling relief and reconstruction was an opportunity for him to be a more decisive and approachable leader than his predecessor. It was also an opportunity for him to improve relations between Jakarta and Aceh—the region worst-affected by the tsunami—and in Aug. 2005 his government signed a peace agreement with separatist leaders granting greater political autonomy to the province. Elections for a provincial governor and district officials in Aceh took place in Dec. 2006.

Yudhoyono's government was confronted by renewed terrorism, as suicide bombers again targeted the tourist resort of Bali in Oct. 2005 killing 19 people, and by further natural disasters. In Nov. 2008 three Islamic terrorists convicted for their part in the Bali bombing in 2002 were executed by firing squad. However, extremist activity continued, notably the launching of suicide bomb attacks in July 2009 on two luxury hotels in Jakarta which killed nine people and injured at least 50 more. In the first half of 2010 a number of suspected Islamist militants were arrested in a series of anti-terrorist raids by security forces. Meanwhile, an earthquake in May 2006 and another tsunami in July killed around 6,500 people on Java, floods in Jakarta in Feb. 2007 left an estimated 340,000 people homeless, another earthquake in Sumatra the following month killed more than 50 and, in Sept. 2009, an earthquake off the coast of Sumatra left more than 1,000 dead. Further extensive casualties followed another earthquake off the western coast of Sumatra and the volcanic eruption of Mount Merapi in Oct. 2010.

By 2008 Yudhoyono's political popularity was being undermined by continuing unemployment, rising prices and a cut in fuel subsidies, despite increased spending on anti-poverty programmes and significant progress in his anti-corruption drive. Nevertheless, in parliamentary elections in April 2009 his

Democrat Party emerged as the largest party and in July he was returned to office with 60·8% of the vote in the presidential poll.

Following a final report in July 2008 by a joint investigative commission that blamed Indonesia for human rights violations in the run-up to Timor-Leste's independence in 2002, Yudhoyono expressed the Indonesian government's deep regret but did not apologize.

More recently, Yudhoyono has been criticized for a failure to counter incidents of religious intolerance and sectarian violence by hardline Islamic groups. He is ineligible to stand for a third term in the presidential elections scheduled for 2014.

DEFENCE

There is selective conscription for two years. Defence expenditure in 2008 totalled US$5,108m. (US$22 per capita), representing 1·0% of GDP. Real-term military spending was cut by 7% in 2008.

Army
Army strength in 2007 was estimated at 233,000 with a strategic reserve (KOSTRAD) of 40,000 and further potential mobilizeable reserves of 400,000.

There is a paramilitary police some 280,000 strong; and a part-time local auxiliary force, KAMRA (People's Security), which numbers around 40,000.

Navy
The Navy in 2007 numbered about 45,000, including some 20,000 marines and around 1,000 in Naval Aviation. Combatant strength in 2007 included two diesel submarines and 11 frigates.

The Navy's principal command is split between the Western Fleet, at Teluk Ratai (Jakarta), and the Eastern Fleet, at Surabaya.

Air Force
Personnel (2007) 24,000. There were 94 combat-capable aircraft, including A-4s, F-16s, F-5s and British Aerospace *Hawks*.

INTERNATIONAL RELATIONS

Indonesia was in dispute with Malaysia over sovereignty of two islands in the Celebes Sea. Both countries agreed to accept the Judgment of the International Court of Justice which decided in favour of Malaysia in Dec. 2002.

ECONOMY

Agriculture accounted for 15·3% of GDP in 2009, industry 47·7% and services 37·0%.

Overview
The economy has grown dramatically since the early 1990s, primarily through increased labour productivity which, according to McKinsey Global, accounts for 60% of growth. Indonesia was the only G20 member besides China and India to continue growth during the global economic crisis. Its fiscal stimulus package equalled about 1·1% of GDP, half the average stimulus in the G20. Indonesia was also the only G20 member with declining debt, which fell from 61% of GDP in 2003 to 27·5% in 2009.

As a consequence of the crisis, growth slowed in 2009 but was still an encouraging 4·6%, boosted by domestic consumption and sovereign ratings upgrades. Provisional figures suggest that the economy grew by 6% in 2012, driven by vibrant domestic consumption and government spending. The budget made provision for US$21bn. spending on infrastructure projects, including new ports and airport expansion.

Indonesia has a wealth of natural resources, including natural gas and petroleum. In 2011 the industrial sector grew by 4·1%, despite China and India lowering their demand for coal and other industrial products. However, corruption and an inefficient legal system undermine entrepreneurship and job growth, with the country rated 100th out of 183 on Transparency International's Corruption Perceptions Index for 2011. These factors and poor institutional frameworks pose major obstacles to the government's aim to be among the world's top ten economies by 2025.

In agriculture, yields need to improve by 60% to feed the growing population. Around 32m. people live below the poverty line and half of the population was clustered around the line in 2011. It is estimated that 62·2% of the population is employed in the informal sector, with a corresponding impact on tax revenues. The IMF has advised continued efforts to improve public infrastructure and bolster the business climate, and the World Bank has invested in institutional projects to better organize the economy.

Currency
The monetary unit is the *rupiah* (IDR) notionally of 100 *sen*. Inflation rates (based on IMF statistics):

2002	2003	2004	2005	2006	2007	2008	2009	2010	2011
11·8%	6·8%	6·1%	10·5%	13·1%	6·7%	9·8%	4·8%	5·1%	5·4%

In Aug. 2009 foreign exchange reserves were US$55,440m., gold reserves were 2·35m. troy oz and total money supply was 490,111bn. rupiahs.

Budget
The fiscal year used to start 1 April but since 2001 has been the calendar year. In 2010 budgetary central government revenues totalled 970,651bn. rupiah (of which taxes 699,594bn. rupiah) and expenditures 928,087bn. rupiah. Main items of expenditure by economic type in 2010: grants, 325,316bn. rupiah; subsidies, 177,888bn. rupiah; compensation of employees, 147,206bn. rupiah.

The standard rate of VAT is 10%.

Performance
Real GDP growth rates (based on IMF statistics):

2002	2003	2004	2005	2006	2007	2008	2009	2010	2011
4·5%	4·8%	5·0%	5·7%	5·5%	6·3%	6·0%	4·6%	6·2%	6·5%

The Asian economic crisis of 1997 affected Indonesia more than any other country, leading to a recession in 1998 when the economy shrank by 13·1%. In 2011 total GDP was US$846·8bn.

Banking and Finance
The Bank Indonesia, successor to De Javasche Bank established by the Dutch in 1828, was made the central bank of Indonesia on 1 July 1953. Its *Governor* is Darmin Nasution. It had an original capital of 25m. rupiahs, a reserve fund of 18m. rupiahs and a special reserve of 84m. rupiahs. In Jan. 2000 independent auditors declared that the bank was technically bankrupt. In response the IMF stated that future loans would probably depend on recapitalization and an internal reorganization.

In 2003 there were 138 commercial banks, 26 regional government banks, 76 private national banks and 31 foreign banks and joint banks. The leading banks are Bank Mandiri (with assets of US$25·6bn. in June 2005), Bank Central Asia and Bank Rakyat Indonesia. All state banks are authorized to deal in foreign exchange.

The government owns one Savings Bank, Bank Tabungan Negara, and 1,000 Post Office Savings Banks. There are also over 3,500 rural and village savings banks and credit co-operatives. At least 16 banks closed in the wake of the 1997 financial crisis.

Foreign debt was US$179,064m. in 2010, representing 26·1% of GNI.

There is a stock exchange in Jakarta.

ENERGY AND NATURAL RESOURCES

Environment
Indonesia's carbon dioxide emissions from the consumption and flaring of fossil fuels in 2008 were the equivalent of 1·8 tonnes per capita.

Electricity
Installed capacity in 2007 was 27·8m. kW and production in 2007 totalled 142·24bn. kWh (11·29bn. kWh hydro-electric). Consumption per capita was 630 kWh in 2007.

Oil and Gas
The importance of oil in the economy is declining. The 2008 output of crude oil was 49·1m. tonnes, down from 76·5m. tonnes in 1995. Proven reserves in 2008 totalled 3·7bn. bbls. With domestic demand having surpassed production, Indonesia became a net importer of oil in 2005.

Natural gas production, 2008, was 69·7bn. cu. metres with 3,180bn. cu. metres of proven reserves. In 2001 a 640-km gas pipeline linking Indonesia's West Natuna field with Singapore came on stream.

Minerals
The high cost of extraction means that little of the large mineral resources outside Java is exploited; however, there is copper mining in Papua, nickel mining and processing on Sulawesi, and aluminium smelting in northern Sumatra. Open-cast coal mining has been conducted since the 1890s, but since the 1970s coal production has been developed as an alternative to oil. Reserves are estimated at 28,000m. tonnes. Coal production (2007), 178·9m. tonnes. Other minerals: bauxite (2008), 1,152,000 tonnes; copper (2006), 818,000 tonnes (metal content); salt (2005 estimate), 680,000 tonnes; nickel (2006), 150,000 tonnes (metal content); tin (2006), 117,500 tonnes (metal content); silver (2005), 329 tonnes; gold (2005), 167 tonnes.

Agriculture
There were approximately 22·0m. ha. of arable land in 2007 and 15·5m. ha. of permanent crops. 7·89m. ha. were irrigated in 2005. Production (2003, in 1,000 tonnes): rice, 52,079; sugarcane, 25,600; cassava, 18,474; coconuts, 15,630; maize, 10,910; bananas, 4,312; palm kernels, 2,187; sweet potatoes, 1,998; natural rubber, 1,792; cabbage, 1,450; groundnuts, 1,377; copra, 1,272; potatoes, 851; onions, 780; oranges, 733; mangoes, 731; coffee, 702; soybeans, 672; green beans, 620; chillies and green peppers, 553. Annual nutmeg production is 6,000 tonnes, more than two-thirds of the world total. Indonesia is the world's largest producer of coconuts.

Livestock (2003): goats, 13·28m.; cattle, 11·40m.; sheep, 8·13m.; pigs, 6·35m.; buffaloes, 2·46m.; chickens, 1·29bn.; ducks, 48m. Only China has more chickens.

Forestry
In 2010 the area under forests was 94·43m. ha., or 52·1% of the total land area. The annual loss of 685,000 ha. between 2005 and 2010 was exceeded during the same period only in Brazil and Australia. In 2007, 103·42m. cu. metres of roundwood were cut.

Fisheries
In 2010 total catch was 5,380,266 tonnes, of which 5,035,294 tonnes were sea fish. The annual catch increased by nearly a third between 2000 and 2010. About 90% of the country's total fish production is consumed domestically.

INDUSTRY
The largest companies in Indonesia by market capitalization in March 2012 were: Astra International, an automobile company (US$32·7bn.); Bank Central Asia (US$21·4bn.); and Bank Rakyat Indonesia (US$18·6bn.).

There are shipyards at Jakarta Raya, Surabaya, Semarang and Ambon. There are textile factories, large paper factories, match factories, automobile and bicycle assembly works, large construction works, tyre factories, glass factories, a caustic soda and other chemical factories. Production (2007, in 1,000 tonnes): cement, 35,033; palm oil, 16,900; distillate fuel oil, 11,828; residual fuel oil, 9,538; petrol, 8,363; fertilizers, 7,923; paper and paperboard, 7,223; kerosene, 6,894; raw sugar, 2,814; plywood, 4·2m. cu. metres;

464,800 cars and commercial vehicles (2009); 220bn. cigarettes (2005). Indonesia is the world's largest palm oil producer.

Labour

The labour force numbered 111,947,000 in 2008 (69,144,000 men and 42,803,000 women). In 2008, 40·3% of employed persons worked in agriculture, forestry, hunting and fisheries, 16·7% in wholesale and retail trade, repair of motor vehicles, motorcycles and personal and household goods, 12·2% in manufacturing and 6·0% in transport, storage and communications. National monthly average wage, 2008, 976,923 rupiahs. Unemployment in 2008 was 8·4%.

INTERNATIONAL TRADE

Since 1992 foreigners have been permitted to hold 100% of the equity of new companies in Indonesia with more than US$50m. part capital, or situated in remote provinces.

Pressure on Indonesia's currency and stock market led to an appeal to the IMF and World Bank for long-term support funds in Oct. 1997. A bail-out package worth US$38,000m. was eventually agreed on condition that Indonesia tightened financial controls and instituted reforms, including the establishment of an independent privatization board, liberalizing foreign investment, cutting import tariffs and phasing out export levies.

Imports and Exports

Imports and exports in US$1m.:

	2006	2007	2008	2009	2010
Imports c.i.f.	61,066	74,473	129,244	96,829	135,663
Exports f.o.b.	100,799	114,101	137,020	116,510	157,779

Principal import items: machinery and transport equipment, basic manufactures and chemicals. Principal export items: gas and oil, forestry products, manufactured goods, rubber, coffee, fishery products, coal, copper, tin, pepper, palm products and tea. Main import suppliers, 2004: Japan, 14·2%; Singapore, 11·2%; China, 9·5%. Main export markets, 2004: Japan, 15·9%; USA, 13·6%; Singapore, 9·3%.

COMMUNICATIONS

Roads

In 2006 there were 324,150 km of classified roads (27,668 km of highways or main roads), of which 54% was surfaced. Motor vehicles, 2005: passenger cars, 5,494,034; buses and coaches, 1,184,918; trucks and vans, 2,920,828; motorcycles, 28,556,498. There were 11,451 fatalities in road accidents in 2005.

Rail

In 2005 the national railways totalled 6,482 km of 1,067 mm gauge, comprising 3,012 km on Java (of which 565 km electrified), 1,348 km on Sumatra and 2,122 km which was non-operational. Passenger-km travelled in 2008 came to 18·5bn. and freight tonne-km to 5·5bn.

Civil Aviation

Garuda Indonesia is the state-owned national flag carrier. Merpati Nusantara Airlines is their domestic subsidiary. There are international airports at Jakarta (Sukarno-Hatta), Denpasar (on Bali), Medan (Sumatra), Pekanbaru (Sumatra), Ujung Pandang (Sulawesi), Manado (Sulawesi), Solo (Java) and Surabaya Juanda (Java). Jakarta is the busiest airport, in 2001 handling 11,192,000 passengers (6,685,000 on domestic flights) and 280,900 tonnes of freight. Denpasar handled 4,431,000 passengers in 2001 and Surabaya Juanda 2,380,000. In 2003 scheduled airline traffic of Indonesia-based carriers flew 211m. km, carrying 20,358,000 passengers (1,984,000 on international flights).

Shipping

There are 16 ports for ocean-going ships, the largest of which is Tanjung Priok, which serves the Jakarta area and has a container

terminal. In 2007 cargo traffic at Tanjung Priok totalled 42·0m. tonnes. The national shipping company Pelajaran Nasional Indonesia (PELNI) maintains inter-island communications. In Jan. 2009 there were 1,856 ships of 300 GT or over registered, totalling 5·15m. GT. Of the 1,856 vessels registered, 1,095 were general cargo ships, 261 oil tankers, 246 passenger ships, 92 bulk carriers, 85 container ships, 67 chemical tankers and ten liquid gas tankers.

Telecommunications

In 2008 there were 30,378,000 main (fixed) telephone lines; mobile phone subscribers numbered 140,578,000 in 2008 (618·3 per 1,000 persons). Indonesia had 18·0m. internet users in 2008, up from 400,000 in 2000. In March 2012 there were 43·5m. Facebook users.

SOCIAL INSTITUTIONS

Justice

There are around 250 district courts of first instance, 20 high courts of appeal and a Supreme Court of Justice (Mahkamah Agung) for the whole of Indonesia in Jakarta. Religious sharia courts with limited jurisdiction are also in place to handle civil cases between Muslim spouses.

The current legal system is a mixture of 'adat' (customary) law, Dutch colonial law and the national law that was brought in following independence in 1945. As in Dutch civil law, the rule of precedence does not apply.

The present criminal law has been in force since 1915 and is codified and based on European penal law. The death penalty is still in use; an execution in March 2013 was the first in more than four years.

The population in penal institutions in Oct. 2008 was 136,017 (58 per 100,000 of national population).

Education

Adult literacy in 2008 was 92·2%. In 2007 there were 29,796,705 pupils and 1,583,589 teaching staff at primary schools, and 18,716,929 pupils and 1,434,874 teachers at secondary schools. There were 3,755,187 students in higher education in 2007 and 265,527 academic staff. The University of Indonesia in Jakarta, founded in 1849, is the leading institution in the tertiary sector. Other prominent institutions include the Gadjah Mada University (Universitas Gadjah Mada) in Yogyakarta, the Parahyangan Catholic University (Universitas Katolik Parahyangan) in Bandung, the Bandung Institute of Technology (Institut Teknologi Bandung) and the Bogor Agricultural Institute (Institut Pertanian Bogor).

In 2007 public expenditure on education came to 3·6% of GNI and 17·5% of total government spending.

Health

In 2000 there were 34,347 doctors, 92,371 nurses, 11,547 midwives and 2,406 dentists. There were 1,162 hospitals in 2002, with a provision of six beds per 10,000 population.

Welfare

There are currently no unemployment benefits or family allowance programmes. Establishments with at least ten employees (or a monthly payroll of 1m. rupiahs or more) are obliged to contribute towards old-age, sickness and maternity benefits for employees with contracts of more than three months.

RELIGION

Religious liberty is granted to all denominations. In 2001 there were 185·1m. Muslims (making Indonesia the world's biggest Muslim country), 12·8m. Protestants and 7·6m. Roman Catholics. There were also significant numbers of Hindus and Buddhists. In Feb. 2013 there was one cardinal.

CULTURE

World Heritage Sites
There are eight UNESCO World Heritage sites in Indonesia (the first four inscribed in 1991): Borobudur Temple Compounds, Ujung Kulon National Park, Komodo National Park, Prambanan Temple Compounds, Sangiran Early Man Site (1996), Lorentz National Park (1999), the Tropical Rainforest Heritage of Sumatra (2004) and Cultural Landscape of Bali (2012), which includes *subak*, a water management system for rice terraces. *Subak* is also a reflection of the philosophical concept of *Tri Hita Karana*.

Press
In 2008 there were 226 paid-for daily newspapers (total estimated circulation of 5,450,000).

Tourism
In 2010 there were 7,003,000 international tourist arrivals (excluding same-day visitors), up from 6,324,000 in 2009 and 5,002,000 in 2005. The main countries of origin of non-resident tourists in 2010 were: Singapore (1,129,000), Malaysia (1,035,000), Australia (731,000) and China (422,000).

Festivals
Independence from the Dutch is celebrated on 17 Aug. with musical and theatrical performances, carnivals and sporting events. The military parades on Armed Forces Day (5 Oct.) and women are celebrated on Kartini Day (21 April) in memory of Raden Ajeng Kartini, a symbol of female emancipation. In Bali the Hindu new year is marked by a day of silence, Nyepi, followed by a day of feasting. Muslim, Hindu, Buddhist and Christian festivals are marked throughout the country.

DIPLOMATIC REPRESENTATIVES

Of Indonesia in the United Kingdom (38 Grosvenor Sq., London, W1K 2HW)
Ambassador: Teuku Mohammed Hamzah Thayeb.

Of the United Kingdom in Indonesia (Jalan M. H. Thamrin 75, Jakarta 10310)
Ambassador: Mark Canning, CMG.

Of Indonesia in the USA (2020 Massachusetts Ave., NW, Washington, D.C., 20036)
Ambassador: Dino Patti Djalal.

Of the USA in Indonesia (Medan Merdeka Selatan 5, Jakarta)
Ambassador: Scot Marciel.

Of Indonesia to the United Nations
Ambassador: Desra Percaya.

Of Indonesia to the European Union
Ambassador: Arif Havas Oegroseno.

FURTHER READING

Central Bureau of Statistics. *Statistical Yearbook of Indonesia.—Monthly Statistical Bulletin: Economic Indicator.*

Cribb, R., *Historical Dictionary of Indonesia.* 1993.
Cribb, R. and Brown, C., *Modern Indonesia: a History since 1945.* 1995
Day, Tony, (ed.) *Identifying with Freedom: Indonesia after Suharto.* 2007
Elson, R. E., *Suharto; a Political Biography.* 2001
Forrester, Geoff and May, R. J. (eds.) *The Fall of Soeharto.* 1999
Friend, Theodore, *Indonesian Destinies.* 2003
Glassburner, Bruce, (ed.) *The Economy of Indonesia: Selected Readings.* 2007
Holt, Claire, (ed.) *Culture and Politics in Indonesia.* 2007
Kingsbury, Damien, *The Politics of Indonesia.* 3rd ed. 2005
Ricklefs, M. C., *A History of Modern Indonesia since c. 1200.* 4th ed. 2008
Schwarz, Adam, *A Nation in Waiting: Indonesia's Search for Stability.* Revised ed. 1999
Schwarz, Adam and Paris, Jonathan, (eds.) *The Politics of Post-Suharto Indonesia.* 1999
Vatikiotis, M. R. J., *Indonesian Politics under Suharto: Order, Development and Pressure for Change.* 2nd ed. 1994

National Statistical Office: Central Bureau of Statistics, Jl. Dr. Sutomo 6–8, Jakarta, 10710.
Website: http://www.bps.go.id

IRAN

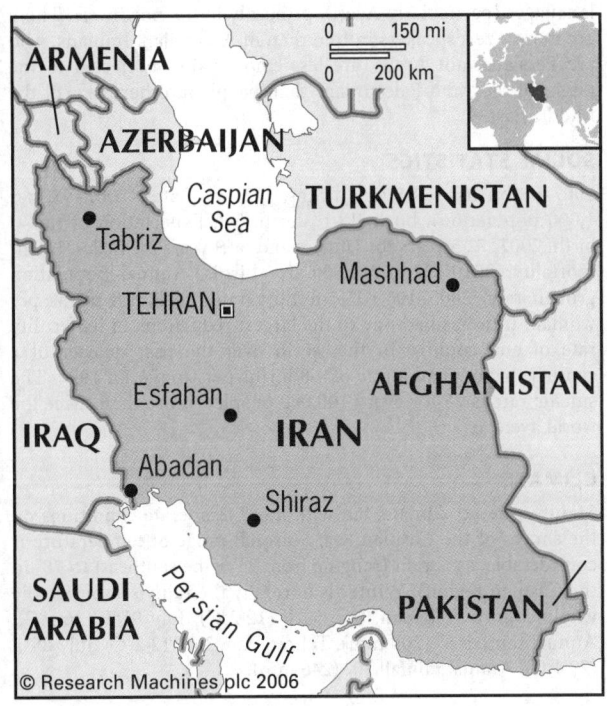

© Research Machines plc 2006

Jomhuri-e-Eslami-e-Iran
(Islamic Republic of Iran)

Capital: Tehran
Population projection, 2015: 77·91m.
GNI per capita, 2011: (PPP$) 10,164
HDI/world rank: 0·707/88
Internet domain extension: .ir

KEY HISTORICAL EVENTS

Neolithic farmers established settlements in the Zagros mountains, in the west of modern Iran, from 6000 BC. From around 2700 BC the southwestern region of Khuzestan was inhabited by Elamite societies, a formative influence on the first Persian empire, established by Cyrus the Great in 550 BC. His Achaemenian dynasty lasted until around 320 BC and was ruled from Persepolis. Persia was subsequently controlled by the Parthian and Sassanian dynasties, during which the Zoroastrian religion took hold. Arabians arriving in the 7th century AD spread the Islamic faith. Their armies defeated the Sassanians at Nahavand in 641, ushering in a period of control by Arab caliphs and, from the 10th century, Seljuk Turks. Persia came under the control of Ghengis Khan's Mongol armies in the 1220s and was then ruled by Timur from 1370.

The Safavid dynasty (1502–1736) was founded by Shah Ismail, who restored internal order and established the Shia sect of Islam as the state religion. The dynasty reached its zenith in the reign of Shah Abbas I (1587–1628), during which Esfahan became the Persian capital. In 1779, following the death of Mohammad Karim Khan Zand, Agha Mohammad Khan, a leader of the Qajars (a Turkmen tribe from modern Azerbaijan), attempted to reunify Persia. He regained Persian control over much of the Caucasus and established his capital at Tehran.

The Qajars fought with an expansionist Russia during the early 1800s and in 1828 Fath Ali Shah was forced to sign the Treaty of Turkmanchai, acknowledging Russian sovereignty over present-day Armenia and Azerbaijan. The reign of Naser o-Din Shah (1848–96) saw a period of modernization and the increasing influence of the Russian and British empires in Persia, which continued after the adoption of the first constitution and the establishment of a national assembly in Aug. 1906. The discovery of oil in Khuzestan province in 1908 led to further British and Russian jockeying for control in the region.

Following a bloodless coup in 1921, Reza Khan began his rise to power. He was crowned Reza Shah Pahlavi on 12 Dec. 1925 and set about a programme of reforms, encouraging the development of industry, education and a modern infrastructure. Responding to Iran's support for Germany in the Second World War, the Allies occupied the country and forced Reza Shah to abdicate in favour of his son, Muhammad Reza Shah, who was sworn in on 17 Sept. 1941.

The British-controlled oil industry was nationalized in March 1951 in line with the policy of the National Front Party, whose leader, Dr Muhammad Mussadeq, became prime minister in April 1951. He was opposed by the Shah, who fled the country until Aug. 1953, when Mussadeq was deposed in an Anglo-American sponsored monarchist coup. The Shah's policy, which included the redistribution of land to small farmers and the enfranchisement of women, was opposed by Shia religious scholars. Despite economic growth, unrest was caused by the Shah's repressive measures and his extensive use of the *Savak* (secret police). The opposition, led by Ayatollah Ruhollah Khomeini, the Shia Muslim spiritual leader who had been exiled in 1965, was increasingly successful. Following intense civil unrest in Tehran, the Shah left Iran with his family on 17 Jan. 1979 (and died in Egypt on 27 July 1980).

The Ayatollah Khomeini returned from exile on 1 Feb. 1979, the Shah's government resigned and parliament dissolved itself on 11 Feb. Following a referendum in March, an Islamic Republic was proclaimed. The Constitution gave supreme authority to a religious leader (*wali faqih*), a position held by Ayatollah Khomeini until his death in 1989. Ayatollah Ali Khamenei then became the nation's supreme leader. In Sept. 1980 border fighting with Iraq escalated into full-scale war. A UN-arranged ceasefire came into effect on 20 Aug. 1988, and in Aug. 1990, following Iraq's invasion of Kuwait, Iraq offered peace terms and began the withdrawal of troops from Iranian soil. 30,000 political opponents of the regime are believed to have been executed shortly after the end of the war.

In 1997 the election of Mohammad Khatami as president signalled a shift away from Islamic extremism. A clampdown on Islamic vigilantes who were waging a violent campaign against Western 'decadence' was evidence of a cautiously liberal integration of the constitution. However, the conservative faction led by Ayatollah Ali Khamenei retained considerable power, including the final say on defence and foreign policy.

In July 1999 riot police fought pitched battles with pro-democracy students in Tehran in the worst unrest since the 1979 revolution. Islamic leaders remain divided on the degree of overlap between politics and religion.

The election of Mahmoud Ahmadinejad as president in June 2005 was seen by some analysts as signalling a return to the extreme conservatism that preceded Khatami's tenure. Ahmadinejad's anti-Israeli rhetoric caused international concern. Under his leadership, Iran recommenced uranium conversion research. Despite Tehran's insistence that the research programme was for peaceful purposes only, increasing international disquiet culminated with the International Atomic Energy Agency reporting Iran to the UN Security Council in Feb. 2006. In April 2006

President Ahmadinejad announced that Iran had successfully enriched uranium.

The country witnessed an upsurge of civil unrest in June 2009 following a disputed presidential election in which incumbent president Mahmoud Ahmadinejad was declared the victor by a large majority.

TERRITORY AND POPULATION

Iran is bounded in the north by Armenia, Azerbaijan, the Caspian Sea and Turkmenistan, east by Afghanistan and Pakistan, south by the Gulf of Oman and the Persian Gulf, and west by Iraq and Turkey. It has an area of 1,648,195 sq. km (636,368 sq. miles) including 116,600 sq. km (45,020 sq. miles) of inland water, but a vast portion is desert. Population (2011 census): 75,149,669. Population density: 46 per sq. km. The population was 71·3% urban in 2011.

The UN gives a projected population for 2015 of 77·91m.

In 2010 Iran had 1·07m. refugees, mainly from Afghanistan. Only Pakistan has more refugees.

The areas, populations and capitals of the provinces (ostan) are:

Province	Area (sq. km)	Census Oct. 2006	Census Oct. 2011	Capital
Ardabil	17,881	1,228,155	1,248,488	Ardabil
Azarbayejan, East	45,481	3,603,456	3,724,620	Tabriz
Azarbayejan, West	37,463	2,873,459	3,080,576	Orumiyeh
Bushehr	23,168	886,267	1,032,949	Bushehr
Chahar Mahal and Bakhtyari	16,201	857,910	895,263	Shahr-e-Kord
Esfahan	107,027	4,559,256	4,879,312	Esfahan
Fars	121,825	4,336,878	4,596,658	Shiraz
Gilan	13,952	2,404,861	2,480,874	Rasht
Golestan	20,893	1,617,087	1,777,014	Gorgan
Hamadan	19,547	1,703,267	1,758,268	Hamadan
Hormozgan	71,193	1,403,674	1,578,183	Bandar-e-Abbas
Ilam	20,150	545,787	557,599	Ilam
Kerman	181,714	2,652,413	2,938,988	Kerman
Kermanshah	24,641	1,879,385	1,945,227	Kermanshah
Khuzestan	63,213	4,274,979	4,531,720	Ahvaz
Kohgiluyeh and Boyer Ahmad	15,563	634,299	658,629	Yasuj
Kordestan	28,817	1,440,156	1,493,645	Sanandaj
Lorestan	28,392	1,716,527	1,754,243	Khorramabad
Markazi	29,406	1,351,257	1,413,959	Arak
Mazandaran	23,833	2,922,432	3,073,943	Sari
North Khorasan	28,434	811,572	867,727	Bojnurd
Qazvin	15,491	1,143,200	1,201,565	Qazvin
Qom	11,237	1,046,737	1,151,672	Qom
Razavi Khorasan	144,681	5,593,079	5,994,402	Mashhad
Semnan	96,816	589,742	631,218	Semnan
Sistan and Baluchestan	178,431	2,405,742	2,534,327	Zahedan
South Khorasan	69,555	636,420	662,534	Birjand
Tehran[1]	19,196	13,422,366	14,595,904	Tehran
Yazd	128,811	990,818	1,074,428	Yazd
Zanjan	21,841	964,601	1,015,734	Zanjan

[1]Including Alborz province, which was created from Tehran province in June 2010.

At the 2006 census the populations of the principal cities were:

	Population		Population
Tehran	7,797,520	Hamadan	479,640
Mashhad	2,427,316	Arak	446,760
Esfahan	1,602,110	Yazd	432,194
Tabriz	1,398,060	Ardabil	418,262
Karaj	1,386,030	Bandar-e-Abbas	379,301
Shiraz	1,227,331	Eslamshahr	357,389
Ahvaz	985,614	Qazvin	355,338
Qom	959,116	Zanjan	349,713
Kermanshah	794,863	Khorramabad	333,945
Orumiyeh	583,255	Sanandaj	316,862
Zahedan	567,449	Gorgan	274,438
Rasht	557,366	Sari	261,293
Kerman	515,991	Kashan	253,509

The official language is Farsi or Persian, spoken by 45·6% of the population in 2003. 28·5% spoke related languages, including Kurdish (9·1%) and Luri in the west, Gilaki and Mazandarami in the north, and Baluchi in the southeast; 22·3% speak Turkic languages (particularly Azeri), primarily in the northwest. There are more Azeri speakers in Iran than in Azerbaijan. Iranians, who are Persians, not Arabs, are less emotionally connected to the plight of the Arab Palestinians than people in other parts of the Middle East.

SOCIAL STATISTICS

2007–08 births, 1,286,716; deaths, 412,735. Rates (2007–08, per 1,000 population): birth, 18·1; death, 5·8. Expectation of life at birth, 2007, 72·5 years for females and 69·9 years for males. Infant mortality, 2010, 22 per 1,000 live births. Annual population growth rate, 2005–10, 1·1%; fertility rate, 2008, 1·8 births per woman. Iran has had one of the largest reductions in its fertility rate of any country in the world over the past quarter of a century, having had a rate of 4·8 births per woman in 1990. The suicide rate is 25 for every 100,000 people—more than twice the world average.

CLIMATE

Mainly a desert climate, but with more temperate conditions on the shores of the Caspian Sea. Seasonal range of temperature is considerable, as is rain (ranging from 2" in the southeast to 78" in the Caspian region). Winter is normally the rainy season for the whole country. Abadan, Jan. 54°F (12·2°C), July 97°F (36·1°C). Annual rainfall 8" (204 mm). Tehran, Jan. 36°F (2·2°C), July 85°F (29·4°C). Annual rainfall 10" (246 mm).

CONSTITUTION AND GOVERNMENT

The Constitution of the Islamic Republic was approved by a national referendum in Dec. 1979. It was revised in 1989 to expand the powers of the presidency and eliminate the position of prime minister. It gives supreme authority to the Spiritual Leader (wali faqih), a position which was held by Ayatollah Khomeini until his death on 3 June 1989. Ayatollah Seyed Ali Khamenei was elected to succeed him on 4 June 1989. Following the death of the previous incumbent, Ayatollah Ali Khamenei was proclaimed the Source of Knowledge (Marja e Taghlid) at the head of all Shia Muslims in Dec. 1994.

The 86-member Assembly of Experts was established in 1982. It is popularly elected every eight years. Its mandate is to interpret the constitution and select the Spiritual Leader. Candidates for election are examined by the Council of Guardians.

The Islamic Consultative Assembly has 290 members, elected for a four-year term in single-seat constituencies. All candidates have to be approved by the 12-member Council of Guardians.

The President of the Republic is popularly elected for not more than two 4-year terms plus a third non-consecutive term and is head of the executive; he appoints Ministers subject to approval by the Islamic Consultative Assembly (Majlis). The president is Iran's second highest-ranking official.

Legislative power is held by the Islamic Consultative Assembly, directly elected on a non-party basis for a four-year term by all citizens aged 17 or over. A new law passed in Oct. 1999 raised the voting age from 16 to 17, thus depriving an estimated 1·5m. young people from voting. Two-thirds of the electorate is under 30. Voting is secret but ballot papers are not printed; electors must write the name of their preferred candidate themselves. Five seats are reserved for religious minorities. All legislation is subject to approval by the Council of Guardians who ensure it is in accordance with the Islamic code and with the Constitution. The Spiritual Leader appoints six members, as does the judiciary.

National Anthem

'Sar zad az ofogh mehr-e khavaran' ('Rose from the horizon the affectionate sun of the East'); words by a group of poets, tune by Dr Riahi.

GOVERNMENT CHRONOLOGY

Spiritual Leaders of the Islamic Republic since 1980.

1980–89	Ayatollah Seyed Ruhollah Mousavi Khomeini
1989–	Ayatollah Seyed Mohammad Ali Hoseyn Khamenei

Heads of State since 1941.

Emperor (Shah)

1941–79	Mohammad Reza Pahlavi

Leader of the Revolution (Rahbar-e Enqelab)

1979–80	Ayatollah Seyed Ruhollah Mousavi Khomeini

President of the Republic

1980–81	Abolhasan Bani-Sadr

Interim Presidential Commission
1981

President of the Republic

1981	Mohammad Ali Rajai

Interim Presidential Commission
1981

Presidents of the Republic

1981–89	Hojatoleslam Seyed Mohammad Ali Hoseyn Khamenei
1989–97	Hojatoleslam Ali Akbar Hashemi Rafsanjani
1997–2005	Hojatoleslam Seyed Mohammad Khatami
2005–	Mahmoud Ahmadinejad

RECENT ELECTIONS

In presidential elections held on 12 June 2009 official results indicated that incumbent president Mahmoud Ahmadinejad took 62·6% of the vote ahead of Mir-Hossein Mousavi with 33·8%, Mohsen Rezaee with 1·7% and Mehdi Karroubi with 0·9%. However, there were widespread accusations against the government of vote rigging and international observers also expressed doubt about the legitimacy of the results. Mir-Hossein Mousavi called for the result to be annulled. There were mass protests and civil unrest in the capital, with several deaths. The official turnout was put at 85%.

Elections to the Islamic Consultative Assembly were held on 2 March and 4 May 2012. In the first round of elections conservative opponents of President Mahmoud Ahmadinejad won a majority of the 290 seats. In the second round, with 65 of the 290 seats being decided, President Mahmoud Ahmadinejad's opponents won 41 seats and his supporters 13, with 11 going to independents.

Elections to the Assembly of Experts were held on 15 Dec. 2006. 68 of the 86 available seats went to representatives of the conservative Combatant Clergy Association.

Presidential elections were scheduled to take place on 14 June 2013.

CURRENT GOVERNMENT

In March 2013 the cabinet was composed as follows:

President: Mahmoud Ahmadinejad; b. 1956 (sworn in 6 Aug. 2005 and re-elected 12 June 2009).

First Vice-President: Mohammad Reza Rahimi.

Head of Presidential Office: Seyed Hassan Mousavi.

Vice-President and Head of Cultural Heritage and Tourism Organization: Mohammad Sharif Malekzadeh. *Vice-President and Head of Environmental Protection Organization:* Mohammad-Javad Mohammadizadeh. *Vice-President and Head of Foundation for Martyrs and Veterans' Affairs:* Masoud Zaribafan. *Vice-President and Head of National Atomic Energy Organization:* Fereydoon Abbasi-Davani. *Vice-President for Executive Affairs:* Hamid Baghaie. *Vice-President for Implementation of the Constitution:* Mohammad Reza Mir Tajedini. *Vice-President for International Affairs:* Ali Saeedlou. *Vice-President for Legal Affairs:* Fatima Bodaghi. *Vice-President for Management, Development and Human Resources:* Gholam-Hossein Elham. *Vice-President for Parliamentary Affairs:* Lotfollah Forouzandeh. *Vice-President for Planning and Strategic Supervision:* Behrouz Moradi. *Vice-President for Science and Technology:* Nasrin Soltankhah. *Vice-President for Social Affairs:* Abdol Reza Sheikholeslami.

Minister of Agricultural Jihad: Sadeq Khalilian. *Communications and Information Technology:* Mohammad Hassan Nami. *Culture and Islamic Guidance:* Mohammad Hosseini. *Defence and Logistics:* Ahmad Vahidi. *Economy and Finance:* Shamseddin Hosseini. *Education:* Hamid Reza Haji Babaie. *Energy:* Majid Namjou. *Foreign Affairs:* Ali Akbar Salehi. *Health and Medical Education (acting):* Mohammad-Hasan Tarighat. *Industry, Mining and Trade:* Mahdi Ghazanfari. *Intelligence:* Heidar Moslehi. *Interior:* Mostafa Mohammad Najjar. *Justice:* Morteza Bakhtiari. *Labour and Social Affairs (acting):* Asadollah Abbasi. *Oil:* Rostam Ghasemi. *Roads and Urban Development:* Ali Nikzad. *Science, Research and Technology:* Kamran Daneshjoo. *Sport and Youth Affairs:* Mohammad Abbasi. *Cabinet Secretary:* Ali Sadoughi.

Speaker of the Islamic Consultative Assembly (Majlis): Ali Larijani.

Presidency Website: http://www.president.ir

CURRENT LEADERS

Ayatollah Seyed Ali Khamenei

Position
Spiritual Leader (wali faqih)

Introduction
Seyed Ali Khamenei succeeded Ayatollah Khomeini as Iran's supreme spiritual leader on the latter's death in June 1989, having previously served from 1981 as the third president of the Islamic Republic.

Early Life
Khamenei was born in Mashhad on 15 July 1939. He attended theological colleges in Qom, where he was a pupil of Ayatollah Khomeini, and Mashhad. From 1963 he was involved with the Islamic opposition to the regime of the Shah, for which he spent three years in prison and a year in exile. Active in the Islamic revolution of 1979, Khamenei was appointed to the Revolutionary Council and became deputy minister of defence. He was also leader of the Friday congregational prayers in Tehran from mid-1980 and, from Aug. 1981, was appointed secretary-general of the Islamic Republican Party (IRP), dissolved in 1987.

Career in Office
On 2 Oct. 1981, as the IRP candidate, Khamenei was the first cleric to be elected as president, with 95% of the popular vote. Ayatollah Khomeini had previously barred the clergy from the office. He succeeded Mohammad Ali Radjai who had been assassinated in Aug. In Aug. 1985 he was re-elected, again overwhelmingly, for a second four-year term. He was injured in a bomb blast in June 1986. On the death of Khomeini, Khamenei was elected to succeed him on 4 June 1989 by an Assembly of Experts. Previously a middle-ranking cleric (Hojatoleslam), he assumed the title of Ayatollah, a constitutional precondition of appointment to the Islamic republic's spiritual leadership. In Dec. 1994 he was proclaimed the Marja e Taghlid (Source of Knowledge) at the head of all Shia Muslims.

The standoff between President Khatami's reformist government and the hard-line conservative Council of Guardians reached a critical point in the run-up to the Feb. 2004 parliamentary elections. The Council's disqualification of over 2,000 reformist candidates provoked threats of resignations in government and boycott in the electorate. Khamenei intervened in Jan. 2004 on state television, calling for review of the Council's decisions and backing the 83 Majlis deputies whose candidacies had been rejected. However, over a third of the Majlis' deputies resigned in protest on 1 Feb. 2004 and in the subsequent election religious conservatives regained parliamentary control.

Khamenei has since presided over a serious decline in relations with the Western powers over Iran's uranium enrichment activities, the confrontational foreign policy stance of the hard-line Islamic state president, Mahmoud Ahmadinejad, and Iran's support for the Assad regime in Syria's civil war. Iran's alleged ambitions to acquire nuclear weapons have incurred damaging international economic sanctions but Khamenei has nevertheless maintained the country's right to nuclear technology for peaceful purposes. Ahmadinejad was first elected as state president in June 2005 and was formally endorsed by Khamenei. Furthermore, Khamenei upheld his disputed re-election in June 2009, dismissing calls for the poll to be rerun and endorsing the ensuing violent security crackdown on opposition to the regime. However, he has distanced himself increasingly from the president during the latter's second and final term of office, criticizing Ahmadinejad's economic stewardship, which has seen a dramatic fall in the value of the currency, and also his inflammatory rhetoric.

In parliamentary elections in March–May 2012 to the Islamic Consultative Assembly, conservative candidates supporting Khamenei and opposing Ahmadinejad won the majority of seats.

Mahmoud Ahmadinejad

Position
President

Introduction
Mahmoud Ahmadinejad won the run-off in Iran's presidential election on 24 June 2005. The ultra-conservative former Revolutionary Guard and mayor of Tehran promised to tackle domestic poverty and corruption, and analysts expected an end to the fragile social reforms made under his predecessor, President Mohammad Khatami. In foreign policy he has hardened Iran's stance towards the West, particularly over its nuclear programme, and engaged in anti-Israeli rhetoric, which has heightened international tensions. Signs of internal dissension during his first term were reflected in his hotly disputed re-election in June 2009, which provoked waves of opposition protests and repressive government retaliation for the rest of the year and into 2010. He has since maintained his uncompromising position on Iran's nuclear development activities and sought to strengthen ties with left-leaning Latin American governments in Venezuela, Nicaragua, Cuba and Ecuador. However, in his second (and final term) of office he has overseen a further decline in the Iranian economy, aggravated by ongoing international sanctions, and has reportedly alienated Iran's Supreme Leader, Ayatollah Khamenei.

Early Life
The son of a blacksmith, Mahmoud Ahmadinejad was born in 1956 in the village of Aradan in northern Iran. The family moved to Tehran a year later. In 1976 he took up a place to study civil engineering at the Iran University of Science and Technology (IUST). As a conservative student, he was supportive of Ayatollah Khomeini's Islamic revolution in 1979. Some of the 52 Americans who were held hostage in the US embassy after the revolution allege that Ahmadinejad was among those who captured them, though he strongly denies the claim. He remained at the IUST until the late 1980s, taking a masters degree in civil engineering, followed by a PhD in traffic and transportation engineering and planning, and then winning a professorship.

Ahmadinejad was drawn into the long-running Iran–Iraq war in 1986, when he joined the Islamic Revolutionary Guards and fought on the Iraqi border near Kirkuk. When the war ended in 1988, he worked as an engineer in the local government offices of Maku and Khvoy in the province of West Azarbayejan. In 1993 he became governor of the northwestern province of Ardabil until he was ousted following the election of the reform-minded President Mohammad Khatami in 1997. Returning to Tehran, Ahmadinejad rejoined the IUST's civil engineering faculty, where he remained until May 2003.

Ahmadinejad was elected mayor of Tehran on 3 May 2003, and pursued conservative policies. He closed down some fast-food restaurants and banned an advertising campaign that featured a Western celebrity. His views were at odds with President Khatami, who barred him from attending cabinet meetings, a privilege normally accorded to mayors of the capital. With the backing of conservative groups, Ahmadinejad contested the June 2005 presidential elections. His campaign was aimed at the poor and disadvantaged, as well as religious hardliners. He emphasized his working-class upbringing and promised to redistribute the country's income from oil. In a run-off against the former president, Ali Akbar Hashemi Rafsanjani, on 24 June 2005, Ahmadinejad emerged victorious, with 63·4% of the vote, although there were complaints of voting irregularities.

Career in Office
On 3 Aug. Ahmadinejad received the formal approval of Ayatollah Khamenei, and he became president on 6 Aug. 2005. In his inaugural address, he called for unity and the building of a model state based on principles of 'modern, advanced, and strong Islamic government'. However, he quickly caused consternation both at home and abroad.

Within Iran, Ahmadinejad instituted a purge of various branches of government, state economic agencies and the diplomatic service, drawing accusations that he was exceeding his constitutional powers. By the end of 2006 there were signs of domestic opposition to his policies. In Dec. his supporters fared badly in elections to local councils and to the powerful Assembly of Experts. Iran's economic malaise under his management incurred parliamentary rebuke in Jan. 2007, reportedly supported by Khamenei, and in June 2007 his government introduced petrol rationing, provoking public protests. There was also increasing evidence in 2007 of an unpopular crackdown on civil liberties. However, in parliamentary elections in March–April 2008 (in which many pro-reform candidates were barred from standing) there was a strong showing by the president's supporters.

The June 2009 presidential election saw Ahmadinejad win 62·6% of the vote, ahead of his nearest rival Mir-Hossein Mousavi with 33·8%. However, the opposition challenged the official results and accused the government of election rigging, resulting in violent clashes between police and demonstrators. Opposition unrest continued over the following months and was met with a repressive backlash by security forces. The government, meanwhile, claimed that outside interference was responsible for fomenting the upheaval. In Dec. 2010 Ahmadinejad dismissed his foreign minister, Manouchehr Mottaki, considered a political rival within the Iranian leadership, and replaced him with Ali Akbar Salehi. However, his reported demand for the intelligence minister, Heidar Moslehi, to resign in April 2011 was overruled by Ayatollah Khamenei, fuelling speculation of a rift between them. In parliamentary elections in March–May 2012 to the Islamic Consultative Assembly, conservative candidates supporting Khamenei and opposing Ahmadinejad won the majority of seats.

On the international stage, Iran's resumption of uranium enrichment from Aug. 2005 intensified Western concerns over nuclear weapons proliferation. Ahmadinejad has maintained a belligerent stance on Iran's refusal to suspend enrichment, which has resulted in a series of punitive sanctions by the United Nations Security Council, the USA and the European Union. International suspicions were heightened in Sept. 2009 by the identification of a secret uranium enrichment plant (Fordo) near the city of Qom and by Iran's test-firing of missiles capable of reaching targets across the Middle East. Subsequent negotiations over the nuclear issue between Iran and the major Western powers have to date come to nothing as reports of an increasing enrichment capability at the Fordo facility have emanated from the International Atomic Energy Agency. Tensions were further exacerbated by US accusations in Oct. 2011 of Iranian complicity in an alleged plot to kill the Saudi ambassador to Washington and by the ransacking of the British embassy in Tehran by an Iranian crowd in Dec. that year, in response to which all Iranian diplomats were expelled from the UK. In Jan. 2012 Ahmadinejad's government threatened to block the shipping of oil exports through the Strait of Hormuz and warned its Arab Gulf neighbours not to back Western attempts to isolate Iran. Nevertheless, further punitive US and EU sanctions, targeting Iranian oil exports in particular, took effect in June and July that year.

Meanwhile, anti-Israeli speeches by Ahmadinejad and his denunciation of the Holocaust as a myth have continued to provoke international condemnation. In Sept. 2010, in his address to the UN General Assembly, he accused the US government of orchestrating the 11 Sept. 2001 suicide hijacking attacks to reverse its declining influence in the Middle East and to prop up Israel, prompting a walkout by Western diplomats. His remarks were labelled 'abhorrent and delusional'. In his UN address in Sept. 2012 he accused Israel of intimidating Iran with threats to attack its nuclear facilities. Ahmadinejad was not eligible to stand again in presidential elections scheduled for June 2013.

DEFENCE

18 months' military service is compulsory (ten months in the case of university graduates). Military expenditure totalled US$9,595m. in 2008 (equivalent to US$146 per capita), representing 2·8% of GDP (compared to 7·7% in 1985).

Iran has on a number of occasions successfully tested Shahab-3 medium-range ballistic missiles, initially with a range of 1,300 km, and most recently in Sept. 2009 an upgraded version that reportedly has a range of 2,000 km. In Nov. 2008, and again in Dec. 2009, it tested a new missile, the Sajil, which also reportedly has a range of 2,000 km. Unlike the Shahab-3, the Sajil is a solid fuel missile.

Nuclear Weapons

Although Iran is a member of the Non-Proliferation Treaty (NPT), there are widespread international concerns that its nuclear programme includes a military dimension. In Nov. 2011 IAEA inspectors reported that Iran was carrying out research 'relevant to the development of a nuclear explosive device'. Tehran maintains that its nuclear programme is solely for civilian use but the UN Security Council has pressed for Iran to end uranium enrichment until its peaceful intentions are confirmed.

According to *Deadly Arsenals*, published by the Carnegie Endowment for International Peace, Iran has a chemical and biological weapons programme.

Army

Strength (2006), 350,000 (about 220,000 conscripts). Reserves are estimated to be around 350,000, made up of ex-service volunteers.

Revolutionary Guard (*Pasdaran Inqilab*)

Numbering some 125,000, the Guard is divided between ground forces (100,000), naval forces (some 20,000) and marines (5,000). It controls the Basij, a volunteer 'popular mobilization army' of about 300,000, which can number 1m. strong in wartime.

Navy

The fleet includes six submarines (including three ex-Soviet *Kilo* class) and three ex-UK frigates. Personnel numbered 18,000 in 2006 including 6,000 in Naval Aviation and 2,600 marines.

The Naval Aviation wing operated 21 aircraft and 30 helicopters in 2006.

The main naval bases are at Bandar-e-Abbas, Bushehr and Chah Bahar.

Air Force

In 2006 there were 281 combat aircraft including US F-14 Tomcat, F-5E Tiger II and F-4D/E Phantom II fighter-bombers, and a number of MiG-29 interceptors and Su-24 strike aircraft. The serviceability of the aircraft varies with only 60–80% operational.

Strength (2006) estimated at 52,000 personnel (about 15,000 air defence).

ECONOMY

Agriculture accounted for 10·4% of GDP in 2006, industry 44·6% and services 45·0%.

Overview

Between 2000 and 2005 GDP grew by an average annual rate of 5·5% as a result of high oil prices and strong private sector growth. Falling oil prices prompted an economic slowdown in 2008 but growth rebounded in 2010 thanks to a recovery in agriculture production as well as oil prices.

The economy remains over-reliant on oil, with the energy sector providing about 80% of export revenue. The government's reform programme has focused on privatization, reducing the role of government and introducing a stronger market mechanism for pricing of energy and agricultural goods. Investment reforms were announced in 2008. Large public commercial banks were privatized in 2008–09, boosting the market share of private banks from 13% in 2007 to 56% in 2011. In Dec. 2010 Iran removed annual product subsidies worth 15% of GDP, prompting increased prices for energy and agricultural products.

In 2010 the USA intensified sanctions against Iran, with particular focus on the petroleum sector. Iran was the world's fourth largest oil producer in 2011, although its oil production has declined every year since 2007. Crude exports declined from around 2·4m. bbls per day in 2011 to about 1m. a day in 2012. The auto industry is the biggest manufacturing sector, with Iran the 13th largest auto maker in the world in 2011. However, car and component production dropped steeply in 2012 as demand dropped in response to higher costs of components.

In 2012 the official inflation rate stood at 25%, though some economists suggested the country was experiencing hyper-inflation. The *rial* depreciated by over 80% in 2012, while prices for staples at least doubled. In Oct. 2012 unemployment was officially 12%, although the real figure has been estimated at three times as much. According to 2009 IMF data, Iran is experiencing a significant brain drain, with more than 150,000 of its educated and most highly skilled citizens leaving annually.

Currency

The unit of currency is the *rial* (IRR) of which 10 = 1 *toman*. Total money supply in April 2008 was 440,095bn. rials. Inflation rates (based on IMF statistics) for fiscal years:

2003	2004	2005	2006	2007	2008	2009	2010	2011
15·6%	15·3%	10·4%	11·9%	18·4%	25·4%	10·8%	12·4%	21·5%

Budget

The financial year runs from 21 March. Revenues in 2008–09 totalled 948,745bn. rials and expenditures 923,015bn. rials. Petroleum and natural gas revenues accounted for 73·5% of all revenues and taxes 19·0%. Current expenditure accounted for 65·8% of all expenditures.

Performance

Real GDP growth rates (based on IMF statistics):

2002	2003	2004	2005	2006	2007	2008	2009	2010	2011
8·2%	8·1%	6·1%	4·7%	6·2%	6·4%	0·6%	3·9%	5·9%	2·0%

Total GDP in 2009 was US$331·0bn.

Banking and Finance

The Central Bank is the note issuing authority and government bank. Its *Governor* is Mahmud Bahmani. All other banks and insurance companies were nationalized in 1979, and reorganized into new state banking corporations. In April 2000 the government announced that it would permit the establishment of private banks for the first time since the revolution in 1979, ending the state monopoly on banking. The first private bank since the revolution came into existence in Aug. 2001 with the creation of Bank Eghtesad Novin (Modern Economic Bank). A further five private banks have opened in the meantime. In 2002 there were 11 commercial banks, two development banks, one housing bank and around 30 foreign banks.

Iran's external debt amounted to US$12,570m. in 2010 (equivalent to 4·1% of GNI), down from US$21,879m. in 2005 (equivalent to 11·6%).

A stock exchange reopened in Tehran in 1992.

ENERGY AND NATURAL RESOURCES

Environment

Iran's carbon dioxide emissions from the consumption and flaring of fossil fuels were the equivalent of 7·8 tonnes per capita in 2008.

Electricity

Total installed capacity in 2007–08 was 50·8m. kW; production (2007–08), 211·78bn. kWh (including 193·65bn. kWh thermal and 17·99bn. kWh hydro-electric). Consumption per capita in 2007–08 was 2,932 kWh. Iran's first nuclear reactor has been built by Russia at Bushehr. It reached full capacity in Aug. 2012.

Oil and Gas

Oil is Iran's chief source of revenue. The main oilfields are in the Zagros Mountains where oil was first discovered in 1908. Oil companies were nationalized in 1979 and operations of crude oil and natural gas exploitation are now run by the National Iranian Oil Company. Iran produced 209·8m. tonnes of oil in 2008 (5·3% of the world total oil output); in 2008 it had reserves amounting to 137·6bn. bbls. In 1999 the most important discovery in more than 30 years was made, with the Azadegan oilfield in the southwest of the country being found to have reserves of approximately 26bn. bbls. In 2009 revenue from oil exports amounted to US$56bn. Iran depends on oil for some 86% of its exports, but domestic consumption has been increasing to such an extent that it is now as high as exports.

Iran has 16% of proven global gas reserves. Natural gas reserves in 2008 were 29,610bn. cu. metres, the second largest behind Russia. Natural gas production was 116·3bn. cu. metres in 2008, the fourth highest in the world. In Dec. 1997 the first natural gas pipeline linking Iran with the Caspian Sea via Turkmenistan was opened. The 200-km line links gas fields in western Turkmenistan to industrial markets in northern Iran.

Minerals

Production (in 1,000 tonnes), 2004: iron ore, 18,205; gypsum, 12,594; decorative stone, 6,450; salt, 1,791; coal, 1,246; bauxite, 366; zinc, 244; aluminium, 213; copper, 190; chromite, 139; manganese, 129. It was announced in Feb. 2003 that uranium deposits had been discovered in central Iran. In Nov. 2003 the International Atomic Energy Agency announced that Iran had admitted to enriching uranium at an electric plant outside Tehran.

Agriculture

There were an estimated 15·02m. ha. of arable land in 2002 and 2·07m. ha. of permanent crops. Around 7·5m. ha. were irrigated in 2002. Crop production (2002, in 1,000 tonnes): wheat, 12,450; sugar beets, 6,098; tomatoes, 4,109; potatoes, 3,756; sugarcane, 3,712; melons and watermelons, 3,388; barley, 3,085; rice (paddy),

2,888; grapes, 2,704. Livestock (2002): 53·9m. sheep; 25·8m. goats; 8·7m. cattle; 1·6m. asses; 270m. chickens.

Forestry

Approximately 7% of Iran was forested (11·08m. ha.) in 2010, much of it in the Caspian region. Timber production in 2007 was 865,000 cu. metres.

Fisheries

In 2010 the total catch was 443,650 tonnes, of which 368,505 tonnes came from sea fishing and 75,145 tonnes from inland fishing.

INDUSTRY

Major industries: petrochemical, automotive, food, beverages and tobacco, textiles, clothing and leather, wood and fibre, paper and cardboard, chemical products, non-metal mining products, basic materials, machinery and equipment, copper, steel and aluminium. The textile industry uses local cotton and silk; carpet manufacture is an important industry. The country's steel industry is the largest in the Middle East.

Production includes: cement (2008 estimate), 44·4m. tonnes; residual fuel oil (2007–08), 25·2m. tonnes; distillate fuel oil (2007–08), 25·1m. tonnes; petrol (2007–08), 12·1m. tonnes; crude steel (2009), 10·9m. tonnes; kerosene (2007–08), 6·6m. tonnes; naphthas (2001–02), 2·6m. tonnes; cottonseed oil (1998), 994,000 tonnes; jet fuel (2007–08), 992,000 tonnes; sugar (2001), 911,000 tonnes; stockings (2000), 18·5m. pairs; building bricks (2000), 10,077m. units.

Labour

The economically active population numbered 20m. in 2002, of which 17·6m. were employed. Approximately 12·2% of the workforce are unemployed and 800,000 Iranians enter the workforce every year.

INTERNATIONAL TRADE

There had been a limit on foreign investment, but legislation of 1995 permits foreign nationals to hold more than 50% of the equity of joint ventures with the consent of the Foreign Investment Board.

Imports and Exports

In 2005–06 imports totalled US$40,969m. and exports US$60,013m. Main imports: machinery and motor vehicles, iron and steel, chemicals, pharmaceuticals, food. Main exports: oil, carpets, pistachios, leather and caviar. Crude oil exports (2003): 2,396,300 bbls a day. Oil exports account for more than 80% of hard currency earnings. Carpet exports are the second largest hard currency earner. Main import suppliers, 1998–99: Germany, 11·6%; Italy, 8·3%; Japan, 7·0%; Belgium, 6·3%; United Arab Emirates, 5·3%; Argentina, 4·4%. Main export markets in 1998–99: UK, 16·8%; Japan, 15·7%; Italy, 8·6%; United Arab Emirates, 6·7%; Greece, 5·0%; South Korea, 5·0%.

COMMUNICATIONS

Roads

In 2006 the total length of roads was 174,301 km, of which 1,429 km were motorways, 27,256 km main roads, 41,129 km secondary regional roads and 104,487 km other local roads. In 2007 there were 920,100 passenger cars; 862,600 motorcycles and mopeds; 179,700 vans and lorries; 4,900 buses and coaches. In 2006 there were 165,130 road accidents resulting in 6,380 deaths.

Rail

The State Railways totalled 7,172 km in 2005, of which 148 km were electrified. The railways carried 26·2m. passengers in 2008 and 28·7m. tonnes of freight in 2005. An isolated 1,676 mm gauge line (94 km) in the southeast provides a link with Pakistan Railways. A rail link to Turkmenistan was opened in May 1996. A link between Sangan in the east of the country and Herat in Afghanistan is currently under construction. Metro systems have been opened in Tehran in 1999 and Mashhad in 2011. A metro system is currently under construction in Esfahan.

Civil Aviation

There are international airports at Tehran (Mehrabad), Shiraz and Bandar-e-Abbas. Tehran is the busiest airport, in 2000 handling 8,474,000 passengers (6,473,000 on domestic flights). The Imam Khomeini International Airport, construction of which began in 1977 before being halted in 1979, was inaugurated in Feb. 2004. The first flight arrived at the airport in May 2004 but it was then shut down by Iran's Revolutionary Guard, citing breaches of security by the foreign operators. The state-owned IranAir is the flag-carrying airline. In 2003 scheduled airline traffic of Iranian-based carriers flew 89m. km, carrying 11,664,000 passengers (2,282,000 on international flights).

Shipping

In Jan. 2009 there were 202 ships of 300 GT or over registered, totalling 959,000 GT. Of the 202 vessels registered, 148 were general cargo ships, 18 bulk carriers, 12 oil tankers, 12 passenger ships, seven container ships, four chemical tankers and there was one liquid gas tanker. The Iranian-controlled fleet comprised 191 vessels of 1,000 GT or over in Jan. 2009, of which 126 were under foreign flags and 65 under the Iranian flag. The principal port is Bandar-e-Abbas, which handled 52,373,000 tonnes of cargo in 2008 (17,185,000 tonnes loaded and 35,161,000 tonnes discharged).

Telecommunications

In 2008 there were 24·8m. main (fixed) telephone lines. In the same year mobile phone subscribers numbered 43·0m. (586·5 per 1,000 persons). There were 23·0m. internet users in 2008.

SOCIAL INSTITUTIONS

Justice

A legal system based on Islamic law (*Sharia*) was introduced by the 1979 constitution. A new criminal code on similar principles was introduced in Nov. 1995. The President of the Supreme Court and the public Prosecutor-General are appointed by the Spiritual Leader. The Supreme Court has 16 branches and 109 offences carry the death penalty. To these were added economic crimes in 1990. The population in penal institutions in Aug. 2007 was 158,351 (222 per 100,000 of national population). Amnesty International reported that there were at least 314 executions in 2012. Executions are frequently held in public.

Police

Women rejoined the police force in 2003 for the first time since the 1979 revolution.

Education

Adult literacy in 2006 was 82%. Most primary and secondary schools are state schools. Elementary education in state schools and university education is free; small fees are charged for state-run secondary schools. In 2007 there were 7,152,492 pupils and 372,859 teaching staff at primary schools; and in 2005, 9,942,201 pupils and 530,190 teaching staff at secondary schools.

In 2007 there were 2,828,528 pupils and 133,484 academic staff at institutions of higher education. The University of Tehran, established in 1851 and with university status since 1934, is the largest and oldest institute of tertiary education in Iran. Other leading universities include Sharif University of Technology, in Tehran, and Esfahan University of Technology.

In 2007 public expenditure on education came to 5·6% of GNI and represented 19·5% of total government expenditure.

Health

There were 717 hospitals in 2001, with 109,152 beds. In 2005 there were 61,870 physicians, 13,210 dentists, 13,900 pharmacists and 111,107 nurses and midwives.

Welfare

The official retirement ages are 60 years (men) or 55 (women) with at least 16 years of contributions; age 50 (men) or 45 (women) with at least 30 years of contributions; and at any age with at least 35 years of contributions or between 20 and 25 years of work in an unhealthy or physically demanding natural environment. The pension is equal to 1/30th of the insured's average earnings during the last 24 months multiplied by the number of years of contributions. The minimum old-age pension is 2,196,000 rials a month (the minimum wage of an unskilled labourer).

RELIGION

The official religion is the Shia branch of Islam. Adherents numbered approximately 85% of the population in 2001; 5% were Sunni Muslims. However, less than 2% of the population now attend Friday prayers.

CULTURE

World Heritage Sites

There are 15 UNESCO World Heritage sites in Iran: Tchogha Zanbil (inscribed on the list in 1979), the ruins of the holy city of the kingdom of Elam founded around 1250 BC; Persepolis (1979), the palace complex founded by Darius I in 518 BC and capital of the Achaemenid empire (the first Persian empire); Meidan Imam (1979), the square built in Esfahan by Abbas I in the early 17th century, which is bordered on all sides by monumental buildings linked by a series of arcades; Takht-e Soleyman (2003), a Sasanian royal residence with important Zoroastrian religious architecture and decoration; Bam and its Cultural Landscape (2004 and 2007), a fortified medieval town where 26,000 people lost their lives in the earthquake of 2003; Pasargadae (2004), the first dynastic capital of the great multicultural Achaemenid Empire in Western Asia; Soltaniyeh (2005), the capital of the Ilkhamid dynasty that stands as a monument to Persian and Islamic architecture; Bisotun (2006), an ancient town on the trade route that linked the Iranian high plateau with Mesopotamia; the Armenian Monastic Ensembles (2008), three monastic buildings dating back to the 7th century; Shushtar Historical Hydraulic System (2009), a homogeneous hydraulic system, designed globally and completed in the 3rd century AD; Sheikh Safi al-din Khanegah and Shrine Ensemble in Ardabil (2010), a spiritual retreat built in the Sufi tradition between the 16th and 18th centuries; Tabriz Historic Bazaar Complex (2010), one of the most complete examples of the traditional commercial and cultural system of Iran; the Persian Garden (2011), a collection of nine gardens selected from various regions of Iran; Gonbad-e Qabus (2012), a 53-metre high tomb built in AD 1006 for Qabus Ibn Voshmgir and the only remaining evidence of Jorjan, a former hub of science and arts in the Muslim world; and Masjed-e Jamé of Esfahan (2012), the 'Friday Mosque'complex, a stunning example of the evolution of mosque architecture over 12 centuries starting in AD 841.

Press

In 2008 there were 183 paid-for daily newspapers and more than 3,300 non-dailies. Approximately 80% of the Iranian press is printed in Farsi; much of the remaining 20% is in English or Arabic. In the 2011–12 *World Press Freedom Index* compiled by Reporters Without Borders, Iran ranked 175th out of 179 countries.

Tourism

There were 2,735,000 non-resident tourists in 2006, spending US$1,760m.

Calendar

The Iranian year is a solar year starting on varying dates between 19 and 22 March. The current solar year is 1392 (21 March 2013 to 20 March 2014). The Islamic *hegira* (AD 622, when Mohammed left Makkah for Madinah) year 1434 corresponds to 14 Nov. 2012–4 Nov. 2013, and is the current lunar year.

Festivals

Iran celebrates Revolution Day on 11 Feb. to mark the anniversary of the overthrow of the Shah in 1979. Nowruz (New Year's Day) falls on 21 March while Constitution Day is on 5 Aug. The feast of Shab-e Yelda, held to mark the longest night of the year, is in Dec.

DIPLOMATIC REPRESENTATIVES

Of Iran in the United Kingdom (embassy in London closed in Nov. 2011). Iran has an Interests Section in the Omani embassy in London.

Of the United Kingdom in Iran (embassy in Tehran closed in Nov. 2011). The United Kingdom has an Interests Section in the Swedish embassy in Tehran.

The USA does not have diplomatic relations with Iran, but Iran has an Interests Section in the Pakistani embassy in Washington, D.C., and the USA has an Interests Section in the Swiss embassy in Tehran.

Of Iran to the United Nations
Ambassador: Mohammad Khazaee.

Of Iran to the European Union
Ambassador: Mahmoud Barimani.

FURTHER READING

Abrahamian, Ervand, *Khomeinism: Essays on the Islamic Republic.* 1993.— *A History of Modern Iran.* 2008

Adib-Moghaddam, Arshin, *Iran in World Politics: The Question of the Islamic Republic.* 2010

Alizadeh, Parvin, (ed.) *The Economy of Iran: The Dilemma of an Islamic State.* 2001

Amuzegar, J., *Iran's Economy Under the Islamic Republic.* 1992

Ansari, Ali M., *Modern Iran Since 1921: The Pahlavis and After.* 2003

Axworthy, Michael, *A History of Iran: Empire of the Mind.* 2010

Buchan, James, *Days of God: The Revolution in Iran and Its Consequences.* 2012

The Cambridge History of Iran. 7 vols. 1968–91

Coughlin, Con, *Khomeini's Ghost: Iran Since 1979.* 2009

Daniel, Elton L., *The History of Iran.* 2008

Daneshvar, P., *Revolution in Iran.* 1996

Ehteshami, A., *After Khomeini: the Iranian Second Republic.* 1994

Ehteshami, A. and Zweiri, M., *Iran and the Rise of its Neoconservatives: The Politics of Tehran's Silent Revolution.* 2007

Fuller, G. E., *Centre of the Universe: Geopolitics of Iran.* 1992

Gheissari, Ali, (ed.) *Contemporary Iran: Economy, Society, Politics.* 2009

Gheissari, Ali and Nasr, Vali, *Democracy in Iran: History and the Quest for Liberty.* 2009

Goodarzi, Jubin, *Syria and Iran: Diplomatic Alliance and Power Politics in the Middle East.* 2006

Hunter, S. T., *Iran after Khomeini.* 1992

Kamrava, M., *Political History of Modern Iran: from Tribalism to Theocracy.* 1993

Kinzer, Stephen, *All the Shah's Men: an American Coup and the Roots of Middle East Terror.* 2003

Martin, Vanessa, *Creating an Islamic State: Khomeini and the Making of a New Iran.* 2000

Mir-Hosseini, Ziba, *Islam and Gender: The Religious Debate in Contemporary Iran.* 1999

Modaddel, M., *Class, Politics and Ideology in the Iranian Revolution.* 1992

Moin, Baqer, *Khomeini: Life of the Ayatollah.* 1999

Omid, H., *Islam and the Post-Revolutionary State in Iran.* 1994

Polk, William R., *Understanding Iran.* 2009

Rahnema, A. and Behdad, S. (eds.) *Iran After the Revolution: the Crisis of an Islamic State.* 1995

Rostami-Povey, Elaheh, *Iran's Influence: A Religious-Political State and Society in its Region.* 2010

Takeyh, Ray, *Hidden Iran: Paradox and Power in the Islamic Republic.* 2006

Wright, Robin, (ed.) *The Iran Primer: Power, Politics, and U.S. Policy.* 2010

National Statistical Office: Statistical Centre of Iran, Dr Fatemi Avenue, Tehran 1414663111, Iran.

IRAQ

© Research Machines plc 2006

Jumhouriya al 'Iraqia
(Republic of Iraq)

Capital: Baghdad
Population projection, 2015: 36·98m.
GNI per capita, 2011: (PPP$) 3,177
HDI/world rank: 0·573/132
Internet domain extension: .iq

KEY HISTORICAL EVENTS

Around 3000 BC the Sumerian culture flourished in Mesopotamia —the part of the Fertile Crescent between and around the Tigris and Euphrates rivers. Incursions from Semitic peoples of the Arabian Peninsula led to Akkadian supremacy after the victory of Sargon the Great (*c.* 2340 BC). The Sumerian cities, such as Ur, reasserted their independence until 1700 BC, when King Hammurabi established the first dynasty of Babylon. Hammurabi and his son, Samsu-iluna, presided over the political and cultural apogee of Babylon; it was a time of great prosperity and relative peace. Babylonia was challenged by the Anatolian Hittites, who sacked Babylon in 1595 BC. A weakened Babylonia fell to the Kassites from the Zagros mountains, who held sway for over 400 years. The power-vacuum in northern Babylonia was filled by the Hurrian kingdom of Mitanni until Assyria's dominance in the 13th century BC. The Semitic Assyrians built an empire that stretched from Tarsus on the Mediterranean to Babylon, which they sacked in 1240 BC.

Elamite invasions in the 12th century BC allowed the establishment of a second Babylonian dynasty—Isin, or Pashe— but the assertiveness of its king, Nebuchadnezzar I, provoked Assyrian retaliation. Assyrian control of Babylonia was regained but tempered by massive immigration of Aramaeans from Upper Mesopotamia and Syria. Nevertheless, the Assyrians achieved considerable imperial expansion under Ashurnasirpal II in the early ninth century BC. Assyrian decline and revival was repeated

in the eighth century. Babylon was recaptured in 729 BC and most of the Fertile Crescent, from the Nile Delta to the Persian Gulf, was subjugated. However, the empire soon crumbled after the death of the great King Ashurbanipal in 627 BC. Revolts in Babylonia were led by the Chaldeans, who had settled in the south from the ninth century. An alliance of old enemies—the Medes and the Scythians—ravaged the Assyrian Empire and in 612 BC, the capital, Nineveh, fell to the Medes.

Babylon, known at this time as Chaldea, assumed control of much of the Fertile Crescent. In 586 BC, Nebuchadnezzar II conquered Phoenicia and Judah, destroying Jerusalem and deporting 15,000 Judaeans as labourers for Babylon. This Babylonian revival withered under his successors, who were defeated by Achaemenid Persia. Cyrus the Great captured Babylon in 539 BC. His rule was strengthened by his self-association with the Babylonian throne and by his religious tolerance; the Babylonian deity Bel-Marduk was restored and the Temple of Jerusalem rebuilt. Xerxes I (485–465 BC) styled himself the Persian Emperor and seized the Bel-Marduk statue, provoking several Babylonian rebellions.

Alexander

The last of the Achaemenids, Darius III, was defeated at the Battle of Gaugamela (near Mosul) in 331 BC by Alexander the Great of Macedon, who established the Hellenistic Age of the Near East. Having assumed the Persian throne, he died at Babylon in 323 BC. His empire was split in four; Seleucus took control of Mesopotamia and Persia and declared himself king in 305 BC. Babylon was soon eclipsed by a new capital at Seleucia on the Tigris and was abandoned during the third century. Parthia, Bactria and Anatolia were lost by Seleucus' successors until Antiochus III (223–187 BC) reasserted his lordship over the lost provinces. However, his foray into Greece was repulsed by Rome, which forced a heavy indemnity on the Seleucid Empire. Rapid territorial losses to Rome, Ptolemaic Egypt and local rebellions led to a Parthian invasion of Babylonia in 129 BC.

The Parthian Empire, with its winter capital at Ctesiphon on the Tigris, reached its territorial zenith under Mithridates II (123–88 BC), who defeated Armenia and repelled the Scythians. Though a looser political unit than the Seleucid state, Mithridates' empire was a conscious inheritor of the great traditions—Persian, Babylonian and Hellenistic—in culture, language and symbolism. Intrigue over its nominal vassal, Armenia, brought Parthia into conflict with Rome. At Carrhae the Parthians inflicted a crushing defeat on a Roman army under Crassus in 53 BC. Several wars followed until Vologases I achieved a settlement with Emperor Nero over the Armenian buffer-state in AD 63. Dynastic disputes bedevilled Parthia and its vassal kingdoms. The invasion of Armenia by Osroes I (AD 109–129) sparked a Roman invasion in AD 113 under Trajan, who annexed Armenia and occupied most of Mesopotamia. Roman control ended after Trajan's death but Vologases IV was forced to cede western Mesopotamia to Rome. However, Mesopotamia remained a battleground, with Emperor Severus invading in AD 195 and looting Ctesiphon, further weakening the Parthian state. In AD 224 Artabanus IV, the last of the Parthian kings, was defeated by Ardashir (Artaxerxes), ruler of Persia and founder of the Sassanid Dynasty.

The Sassanian Persians emulated the Achaemenids and attempted to regain their empire, leading to inevitable conflict with Rome. Ardashir's son, Shapur I, continued his father's expansion in the east and attacked Rome's Levantine provinces. Syria and Armenia were overrun and the Roman Emperor Valerian captured at Edessa in 259. Shapur II (309–379) consolidated Sassanid power, defeating threats from Arabia and

667

Central Asia and wresting control of the Tigris and Armenia from the Romans. The religious policies of the Zoroastrian Sassanids fluctuated from tolerance to persecution. Khosrau I (531–579) revived imperial expansion; his grandson, Khosrau II, was restored to the throne by the Byzantine Emperor Maurice, who was rewarded with Armenia and northeastern Mesopotamia. However, Khosrau retook Mesopotamia after Maurice's murder, beginning a Persian rampage through the Byzantine East. The sack and pillage of Jerusalem provoked Emperor Heraclius, who struck the Persian heartland. In 627 Heraclius entered Ctesiphon and destroyed the palace of Khosrau, who was murdered. Sassanid Persia, exhausted by conflict with Rome, quickly fell to the Arab invasion.

Led by Sa'd ibn Abi Waqqas, the Arab forces of Islam defeated the Sassanians at the Battle of Al-Qadisiyyah (c. 636) on the Euphrates and at Nahavand, western Iran, in 642. By 639 most of Iraq (*Erak*, 'lower Iran'), comprising the centre and south of modern republic, had been conquered; as had Al-Jazirah ('The Island'), the area north of Tikrit. Mass Arab immigration saw the establishment of garrison towns at Kufa (near Babylon) and Basra and later at Mosul. After the first four caliphs, the Caliphate effectively became hereditary under the Ummayads, based at Damascus. However, their rule was disputed, especially in Iraq. The death of Ali's second son, Husayn, at Karbala in 680 left a body of opposition, the Shias, or 'partisans' of Ali. Iraq was controlled by a governor and, from the 690s, Arabic became the language of administration.

Rise of Baghdad
In 743 civil war came to the Caliphate. Having failed to resolve the tensions between rival Arab military groups, the Umayyads succumbed to the rebellion of the Abbasids, who called for a return to strong Islamic leadership. In 750 the last Ummayad caliph, Marwan II, was deposed by Abu al-'Abbas (As-Saffah), supported by Iranian and Iraqi Shias. However, As-Saffah installed himself as caliph, rejecting a Shia imam. In 754 he was succeeded by his brother Al-Mansur, who moved the capital to Baghdad on the Tigris. This move symbolized the end of the hegemony of Syrian and Yemeni Arabs over the Caliphate. Nevertheless, an overburdened Iraq provided numerous threats to Abbasid authority; Al-Mansur had to quell Shia revolts in Iraq in 763. Caliph Al-Mu'tasim moved his capital to Samarra in 836 to remove his Turkic Mamluk soldiers from Baghdad. The suppression of the Zanj Revolt (869–879) of African slaves around Basra prompted the return to Baghdad.

Rapid political fragmentation in the 930s broke the Caliphate. The Shia Buyids took Iraq in 946, depriving the Abbasid caliphs of temporal power. From the 970s, Egypt was ruled by a rival caliphate, the Ismaili Fatimids. Baghdad was taken by the Seljuk Turks under Toghrül, the Sultan of Iran, in 1055. Despite the Seljuk territories fragmenting after Malik Shah I died in 1092, Iraq remained under Seljuk authority until the Mongol invasion. In 1258 Baghdad was sacked by the Mongol Hulagu Khan; the city was ravaged, its people slaughtered and its Grand Library destroyed. The sack ended the Abbasid Caliphate and Baghdad's role as a major cultural centre. Hulagu established the Il-Khanid Dynasty of Iran. Buddhism and Nestorian Christianity flourished under the patronage of Hulagu's successors until the conversion of Khan Ghazan to Sunni Islam in 1292. The Il-Khanate fragmented in the 1330s and Iraq was ruled by the Mongol Jalayirids.

Ottoman Rule
In 1401 Baghdad was sacked by Timur (also known as Tamburlaine), the greatest Central Asian warrior of the period. His death in 1405 allowed the Black Sheep Turkmen (Kara Koyunlu) to overthrow their Jalayirid masters. However, their rapid expansion ended in defeat in 1466 at the hands of the White Sheep Turkmen (Ak Koyunlu). Rivalry with the Ottoman Turks

in Anatolia weakened the White Sheep Turkmen, who were forced to withdraw from Iraq by the Turkic Safavid rulers of Iran. Shah Ismail I took Baghdad in 1509 and made Shi'ism the state religion; all other creeds were banned. However, Iraq soon fell to the Sunni Ottomans, with Sultan Suleyman the Magnificent taking Baghdad in 1534. Shah Abbas reclaimed Iraq for Iran in 1603, brutally suppressing a major Kurdish rebellion in 1610. Ottoman authority was reimposed by Sultan Murad IV, who led his army into Baghdad in 1638.

Centuries of neglect and war had devastated the irrigation systems and agricultural wealth of Iraq. The Ottomans treated Iraq as a buffer state against Iran and allowed Kurdish and Bedouin tribes to dominate. Mamluks asserted their power in Iraq until 1831, when Baghdad was devastated by flooding. Serious administrative reform (*tanzimat*) came in 1869 with the appointment of Midhat Pasha as governor of Baghdad, with great improvements in the army, the law and education. The Young Turks revolution of 1908 gave Iraq limited political representation.

British Mandate
Anglo-German rivalry led to a British invasion of southern Iraq in Nov. 1914. Although Basra fell in 1915, the British suffered a major defeat at Al-Kut in 1916. Nevertheless, Baghdad was taken in March 1917. After the First World War the Allies entrusted Iraq to Britain under the Sykes-Picot Agreement, which protected British oil interests in the region. The State of Iraq became a League of Nations mandate under British Control in Nov. 1920. Rebellions in Kurdish and southern areas were suppressed with bombing campaigns. A Hashemite monarchy was installed, under Amir Faysal ibn Husayn from Mecca, a wartime ally, and an indigenous army created. The monarchy was supported by a plebiscite in 1921. Kurdish-dominated Mosul province—vital for its massive oil reserves—was granted to Iraq by the League of Nations in 1925. Britain's mandate ended in 1932. Rebellions followed in Kurdish areas, led by Mustafa Barzani until he fled to the USSR in 1945. Rejecting the partition of Palestine, Iraq went to war with Israel in 1948, leading to the emigration of 120,000 Iraqi Jews.

The monarchy was overthrown in a military coup on 14 July 1958. King Faisal II and Nuri al Said, the prime minister, were killed. A republic was established, controlled by a military-led Council of Sovereignty under Gen. Abdul Karim Qassim. In 1963 Qassim was overthrown and Gen. Abdul Salam Aref was made president, with a partial return to a civilian government. But on 17 July 1968 a successful coup was mounted by the Pan-Arabist Ba'ath Party. Gen. Ahmed Al Bakr became president, prime minister, and chairman of a newly established nine-member Revolutionary Command Council. In July 1979 Saddam Hussein, the vice-president and a Sunni Muslim, assumed the presidency, having persuaded the ailing Al Bakr to resign. Saddam promptly carried out a purge of the party, resulting in 22 executions.

The 1979 Iranian Revolution was perceived as a threat to the delicate Sunni-Shia balance in Iraq. In Sept. 1980 Iraq invaded Iran, ostensibly over territorial rights in the Shatt-al-Arab waterway. The war claimed over a million lives and saw the use of chemical weapons by the Iraqi army. The al-Anfal campaign (1986–89) countered Kurdish rebellions, killing 182,000 Kurds. Chemical weapons were prominent, most notably at Halabja, where 5,000 died in one day. A UN-arranged ceasefire took place on 20 Aug. 1988 and UN-sponsored peace talks continued in 1989. On 15 Aug. 1990 Iraq accepted the pre-war border and withdrew troops from Iranian soil.

1991 War
On 2 Aug. 1990 Iraqi forces invaded and rapidly overran Kuwait, on the pretext of alleged Kuwaiti 'slant-drilling' across the Iraqi border. The UN Security Council voted to impose economic sanctions on Iraq until it withdrew from Kuwait and the USA sent a large military force to Saudi Arabia. Further Security Council

resolutions included authorization for the use of military force if Iraq did not withdraw by 15 Jan. 1991. On the night of 16–17 Jan. coalition forces (US and over 30 allies) began an air attack on strategic targets in Iraq. A land offensive followed on 24 Feb. The Iraqi army was routed and Kuwait City was liberated on 28 Feb. Iraq agreed to the conditions of a provisional ceasefire, including withdrawal from Kuwait. Subsequent Kurdish and Shia rebellions were brutally suppressed.

In June 1991 UNSCOM, the United Nations Special Commission, conducted its first chemical weapons inspection in Iraq in accordance with UN Resolution 687. In Sept. a UN Security Council resolution permitted Iraq to sell oil worth US$1,600m. to pay for food and medical supplies. In Oct. the Security Council voted unanimously to prohibit Iraq from all nuclear activities. Imports of materials used in the manufacture of nuclear, biological or chemical weapons were banned, and UN weapons inspectors received wide powers to examine and retain data throughout Iraq.

In Aug. 1992 the USA, UK and France began to enforce air exclusion zones over southern and northern Iraq in response to the government's persecution of Shias and Kurds. Following Iraqi violations of this zone and incursions over the Kuwaiti border, US, British and French forces made air and missile attacks on Iraqi military targets in Jan. 1993. On 10 Nov. 1994 Iraq recognized the independence and boundaries of Kuwait. In the first half of 1995 UN weapons inspectors secured information on an extensive biological weapons programme. At the beginning of Sept. 1996 Iraqi troops occupied the town of Irbil in a Kurdish haven in support of the Kurdish Democratic Party faction which was at odds with another Kurdish faction, the Patriotic Union of Kurdistan. On 3 Sept. 1996 US forces fired missiles at targets in southern Iraq and extended the no-fly area northwards to the southern suburbs of Baghdad.

Weapons Inspection

Relations with the USA deteriorated still further in 1997 when Iraq refused co-operation with UN weapons inspectors. The USA and the UK threatened retaliatory action and a renewal of hostilities looked probable until late Feb. 1998 when Kofi Annan, the UN Secretary General, forged an agreement in Baghdad allowing for 'immediate, unconditional and unrestricted access' to all suspected weapons sites. In Aug. 1998 Saddam Hussein engineered another stand-off with the UN arms inspectors, demanding a declaration that Iraq had rid itself of all weapons of mass destruction. This was refused by the UN chief inspector. In Nov. all UN personnel left Iraq as the USA threatened air strikes unless Iraq complied with UN resolutions. Russia and France urged further diplomatic efforts, but on 16 Dec. the USA and Britain launched air and missile attacks aimed at destroying Saddam Hussein's suspected arsenal of nuclear, chemical and biological weapons.

In Feb. 2000 the UN Security Council nominated Sweden's Hans Blix to head the new arms inspectorate to Iraq but he was refused entry into the country. In Feb. 2001 the USA and Britain launched a further series of air attacks on military targets in and around Baghdad. A new UN Security Council resolution was passed in May 2002. Constituting the biggest change since the introduction in 1966 of a UN-administered Oil-for-Food scheme to alleviate the suffering among the civilian population, the new resolution limited import restrictions to a number of specific sensitive goods. In Nov. 2002 the UN Security Council adopted Resolution 1441, holding Iraq in 'material breach' of disarmament obligations. Weapons inspectors, under the leadership of Hans Blix, returned to Iraq four years after their last inspections, but US and British suspicion that the Iraq regime was failing to comply led to increasing tension, resulting in the USA, the UK and Spain reserving the right to disarm Iraq without the need for a further Security Council resolution. Other Security Council members, notably China, France, Germany and Russia, opposed the proposed action.

Fall of Saddam

On 20 March 2003 US forces, supported by the UK, began a war aimed at 'liberating Iraq'. UK troops entered Iraq's second city, Basra, on 6 April. On 9 April 2003 American forces took control of central Baghdad, effectively ending Saddam Hussein's rule. Widespread looting and disorder followed the capital's fall. The bloodless capture of Tikrit, Saddam Hussein's hometown, on 14 April marked the end of formal Iraqi resistance. An interim government was planned until democratic elections could be held. On 22 May the UN Security Council voted to lift economic sanctions against Iraq and to support the US and UK occupation 'until an internationally recognized, representative government is established by the people of Iraq'. Only Syria opposed the resolution by boycotting the session. A 25-man Iraqi-led governing council (IGC) met in Baghdad for the first time in July 2003.

Resistance to the occupying forces increased from late summer. Bomb attacks in Aug. targeted the UN's Baghdad office, killing the UN special representative. Ayatollah Mohammed Baqr al-Hakim, the most senior Shia cleric in Iraq, was assassinated with 100 others in Najaf. Saddam Hussein was captured by American forces at Al-Dawr, near Tikrit, on 13 Dec. 2003. His trial for crimes against humanity, war crimes and genocide began in July 2004. In Feb. 2004 over 100 Kurds were killed in attacks in Irbil and the Shia community suffered 270 deaths in Baghdad and Karbala. In May 2004 accusations surfaced of abuse of Iraqi prisoners by American and British soldiers.

On 30 Jan. 2005 the first democratic elections to a Transitional National Assembly were won by the Shia-dominated United Iraqi Alliance. Jalal Talabani became the country's new president on 6 April 2005, and on 3 May 2005 Iraq's first democratically elected government under Prime Minister Ibrahim al-Jaafari was sworn in. In Oct. 2005 a new federal constitution was approved in a nationwide referendum (although without the support of the Sunni community) and the trial of former dictator Saddam Hussein for mass murder opened in Baghdad. Despite the political advances, insurgent violence has continued against foreign troops, domestic security forces and civilians. In Dec. 2005 a general election for a new parliament was won by the United Iraqi Alliance. In April 2006 after months of deadlock Nouri al-Maliki was appointed the new prime minister. Insurgent violence continues against foreign troops, domestic security forces and civilians. Nobody can be certain how many Iraqis have been killed since the start of the US-led invasion in March 2003. Estimates vary from 118,000 (to April 2013) by Iraq Body Count and the Oxford Research Group on the basis of media reports through to about 600,000 in a John Hopkins University study of 2006 funded by the Massachusetts Institute of Technology based on interviews of households. In Nov. 2006 Saddam Hussein was sentenced to death. His execution on 30 Dec. 2006 drew mixed reaction both in Iraq and abroad.

In Jan. 2007 President Bush announced a 'troop surge' in Iraq, with 21,500 extra troops to be deployed to assist the Iraqi army in fighting insurgents and al-Qaeda forces. 4,000 US troops were stationed in Al-Anbar province, with the remainder sent to Baghdad. Iraqi forces gradually took control of internal security, with responsibility for Basra (Al-Basrah) province from Dec. 2007 and Al-Anbar from Sept. 2008. UK combat operations ended in April 2009 and in June US troops withdrew from Iraqi towns and cities. The last US troop convoys left Iraq in Dec. 2011, in accordance with the US–Iraq Status of Forces Agreement of 2008.

TERRITORY AND POPULATION

Iraq is bounded in the north by Turkey, east by Iran, southeast by the Persian Gulf, south by Kuwait and Saudi Arabia, and west by Jordan and Syria. In April 1992 the UN Boundary Commission redefined Iraq's border with Kuwait, moving it slightly northwards in line with an agreement of 1932. Area, 434,128 sq. km. Population, 1997 census, 22,046,244; density, 50·8 per sq. km. 2009 estimate, 31,664,466, density 72·9 per sq. km. In 2009, 69·0% of the population lived in urban areas. More than 1·3m. Iraqis protected and assisted by the Office of the United Nations High Commissioner for Refugees are displaced within the country. In 2010 there were 1·7m. Iraqi refugees living abroad, mainly in Syria and to a lesser extent Jordan.

The UN gives a projected population for 2015 of 36·98m.

The areas, populations and capitals of the governorates:

Governorate	Area (sq. km)	Population 1997 census	Population 2009 estimate	Capital
Al-Anbar	138,501	1,023,776	1,483,359	Ar-Ramadi
Babil (Babylon)	6,468	1,181,751	1,729,666	Al-Hillah
Baghdad	734	5,423,964	6,702,538	Baghdad
Basra (Al-Basrah)	19,070	1,556,445	2,405,434	Basra
Dahuk	6,553	402,970	1,072,324	Dahuk
Dhi Qar	12,900	1,184,796	1,744,398	An-Nasiriyah
Diyala	19,076	1,135,223	1,371,035	Ba'qubah
Irbil	14,471	1,095,992	1,532,081	Irbil
Karbala	5,034	594,235	1,013,254	Karbala
Kirkuk[1]	10,282	753,171	1,325,853	Kirkuk
Maysan	16,072	637,126	922,890	Al-Amarah
Al-Muthanna	51,740	436,825	683,126	As-Samawah
An-Najaf	28,824	775,042	1,221,228	An-Najaf
Ninawa (Nineveh)	37,323	2,042,852	3,106,948	Mosul
Al-Qadisiyah	8,153	751,331	1,077,614	Ad-Diwaniyah
Salah ad-Din	24,751	904,432	1,337,786	Tikrit
As-Sulaymaniyah	17,023	1,362,739	1,784,853	As-Sulaymaniyah
Wasit	17,153	783,614	1,150,079	Al-Kut

[1]Also known as At-Ta'mim.

The most populous cities are Baghdad (the capital), with an estimated population of 6,150,000 in 2011, Mosul and Basra. Other large cities include Irbil, Karbala, Kirkuk, An-Najaf and As-Sulaymaniyah.

The population is approximately 80% Arab, 17% Kurdish (mainly in the north of the country) and 3% Turkmen, Assyrian, Chaldean or other. Shia Arabs (predominantly in the south of the country) constitute approximately 60% of the total population and Sunni Arabs (principally in the centre) 20%.

The official language is Arabic. Other languages spoken are Kurdish (official in Kurdish regions), Assyrian and Armenian.

SOCIAL STATISTICS

2008 estimates: births, 940,000; deaths, 177,000; marriages, 171,000. Birth and death rates, 2008 estimates (per 1,000 population): births, 31·2; deaths, 5·9. Life expectancy at birth, 2007, was 64·2 years for men and 71·8 years for women. Annual population growth rate, 2000–08, 2·5%. Infant mortality, 2008: 36 per 1,000 live births. Fertility rate, 2008: 4·1 births per woman. Estimated maternal mortality rate per 10,000 live births, 2005: 30.

CLIMATE

The climate is mainly arid, with limited and unreliable rainfall and a large annual range of temperature. Summers are very hot and winters are cold. Baghdad, Jan. 50°F (10°C), July 95°F (35°C). Annual rainfall 6" (140 mm). Basra, Jan. 55°F (12·8°C), July 92°F (33·3°C). Annual rainfall 7" (175 mm). Mosul, Jan. 44°F (6·7°C), July 90°F (32·2°C). Annual rainfall 15" (384 mm).

CONSTITUTION AND GOVERNMENT

Until the fall of Saddam Hussein, the highest state authority was the Revolutionary Command Council (RCC) but some legislative power was given to the 220-member National Assembly. The only legal political grouping was the National Progressive Front (founded 1973) comprising the Arab Socialist Renaissance (Ba'ath) Party and various Kurdish groups, but a law of Aug. 1991 legalized political parties provided they were not based on religion, racism or ethnicity.

In July 2003 a 25-man Iraqi-led governing council met in Baghdad for the first time since the US-led war in an important staging post towards full self-government. The temporary Coalition Provisional Authority was dissolved on 28 June 2004. Power was handed over to the interim Iraqi government which assumed full sovereign powers for governing Iraq. It became a transitional government after elections in Jan. 2005. The 275-member Transitional National Assembly approved a draft new constitution on 29 Aug. 2005, 14 days after the original deadline. It was approved in a nationwide referendum held on 15 Oct., with 78·6% of votes cast in favour. Shias and Kurds generally supported the constitution. Most Sunnis opposed it because of its reference to federalism and the risk that Iraq could ultimately break up, as Iraq's oil resources are in the Shia and Kurdish areas. The constitution states that Iraq is a democratic, federal, representative republic and a multi-ethnic, multi-religious and multi-sect country. Islam is the official religion of the state and a basic source of legislation. Elections were held in Dec. 2005 for the new 275-member Council of Representatives. In Dec. 2009 the number of seats was increased from 275 to 325 ahead of the March 2010 elections.

National Anthem

'Mawtini' ('My Homeland'); words by I. Touqan, tune by M. Fuliefil.

RECENT ELECTIONS

In parliamentary elections to the permanent Iraqi National Assembly held on 7 March 2010 the Iraqi National Movement coalition won 25·9% of the vote, taking 91 of 325 seats, ahead of the State of Law coalition with 25·8% and 89 seats, the National Iraqi Alliance (19·4% and 70), the Kurdistan List (15·3% and 43) and the Movement for Change (4·4% and 8). Four parties took 16 seats with less than 3% of the vote each and eight seats were taken by minority parties.

CURRENT GOVERNMENT

President: Jalal Talabani; b. 1933 (sworn in 7 April 2005 and re-elected 11 Nov. 2010).

Vice President: Khudair al-Khuzaie.

In March 2013 the cabinet consisted of:

Prime Minister: Nouri al-Maliki; b. 1950 (sworn in 20 May 2006).

Deputy Prime Minister for Energy: Hussain al Shahristani. Deputy Prime Ministers: Saleh al-Mutlaq; Dr Rozh Shaways.

Minister of Agriculture: Izz al-Din al-Dawla. Communications: Mohammed Tawfiq Allawi. Culture, and Defence (acting): Saadun al-Dulaimi. Displacement and Migration: Dindar Najman. Education: Mohammed Tamim. Electricity: Abdulkarim Aftan. Environment: Sargon Lazon Sliwah. Finance: Rafi Hiyad al-Issawi. Foreign Affairs: Hoshyar Zebari. Health: Majid Mohammed Amin. Higher Education and Scientific Research: Ali al-Adeeb. Housing and Development: Muhammad al-Darraji. Human Rights: Muhammad Shiya al-Sudani. Industry and Minerals: Ahmad Nassar Dali al-Karbouli. Interior (acting): Adnan al-Asadi. Justice: Hasan al-Shammari. Labour and Social Affairs: Nassar al-Rubayie. Municipalities and Public Works: Adel Mohoder. Oil: Abd al-Karim Luaybi. Planning: Ali Youssif al-Shukri. Science and Technology: Abd al-Karim al-Samarrai. Tourism and Antiquities: Liwaa Semeism. Trade: Khayrullah Babakir. Transportation: Hadi Al-Amiri. Water Resources: Mohaned al-Saadi. Youth and Sport: Jassem Jaafar.

Government Website (limited English): http://www.cabinet.iq

CURRENT LEADERS

Jalal Talabani

Position
President

Introduction
Jalal Talabani, an experienced Iraqi Kurdish politician, was named state president of Iraq on 6 April 2005 by the Iraqi National Assembly. He was elected by parliament to a second term in April 2006 and re-elected in Nov. 2010. He was previously the founder and secretary general of the Patriotic Union of Kurdistan (PUK), and later a prominent member of the Iraqi Governing Council which was established following the US-led invasion of Iraq in 2003.

Early Life
Jalal Talabani was born in Kelkan, Irbil province in Iraqi Kurdistan in 1933. He attended the Law College in Baghdad from 1952–55 before being forced to leave because of his political activities as a young member of the Kurdistan Democratic Party (KDP). Following the Iraqi revolution in 1958 and the overthrow of the monarchy, Talabani rejoined the college, and graduated in 1959. He subsequently served in the Iraqi army before working as a journalist.

When the Kurdish north launched an armed uprising against the Iraqi government in Sept. 1961, Talabani joined the forces led by Mulla Mustafa al-Barzani (the *peshmerga*) and fought in the Kirkuk and As-Sulaymaniyah areas. He also led Kurdish diplomatic delegations to Europe and the Middle East and negotiated with the secular Ba'ath party, whose members dominated Iraq's governing council following a coup led by Abdul Salam Aref in Feb. 1963.

By 1964, when profound disagreements were emerging within the KDP, Talabani established a more secular, urban and left-leaning faction, criticizing al-Barzani for 'conservative and tribal' politics. Factional divisions throughout the late 1960s and early 1970s occasionally erupted into armed confrontations. Although deals that secured some autonomy for the Kurds were struck between the KDP and the ruling Ba'ath party, arguments broke out over access to the region's oil supplies and whether Kurds could maintain an army. When the Kurdish revolt collapsed in 1975 (partly as a result of Iran withdrawing its support), Talabani formed a new party, the Patriotic Union of Kurdistan (PUK).

The PUK opposed the Ba'ath party's enforced resettlement of Kurds to Arab areas of Iraq in the late 1970s, and there were also numerous armed confrontations with the KDP. In the aftermath of Iraqi leader Saddam Hussein's chemical weapons attack that killed around 5,000 Kurds at Halabja in 1988 and the subsequent military action that led to more than 100,000 Kurds fleeing to Turkey, Talabani made efforts to bring unity to Kurdish politics. He improved relations between the PUK and the KDP (then led by Mas'ud al-Barzani) and later formed the Iraqi Kurdistan Front, seeking international support for Kurdish autonomy.

Following elections in the haven created for Kurds by the Western alliance after the first Gulf War, a PUK-KDP joint administration was formed in 1992. However, tensions resurfaced and led to serious confrontations between the two groups in 1994. Both parties signed a peace deal in Washington, D.C. in 1998 and the accord was cemented in Oct. 2002 when the regional parliament reconvened in a session attended by both parties' MPs.

Following the US-led invasion of Iraq and the fall of Saddam in April 2003, Talabani joined the US-appointed Iraqi Governing Council (IGC), distancing himself from the movement for Kurdish independence and pledging to support Iraqi federalism. In the Iraqi elections on 30 Jan. 2005, a Shia alliance won a slim majority in parliament and the Kurdish coalition came second in the polls. For over two months, with the country under sustained attacks from insurgents, both groups argued about the formation of the new government before electing Talabani as the president (a largely ceremonial role) on 6 April 2005.

Career in Office
A presidential council of Talabani and two vice-presidents appointed Ibrahim al-Jaafari, a conservative from the majority Shia community, as prime minister on 7 April 2005. Talabani promised as president to represent all the country's ethnic and religious groups and to reach out to Iraq's Arab and Islamic neighbours.

Against a backdrop of continuing violence in Iraq, many analysts questioned the strength of the Shia-Kurdish alliance, given that the two groups had little previous common ground beyond resistance against Saddam Hussein. In Oct. 2005 a new Iraqi constitution was approved narrowly in a national referendum, heralding fresh parliamentary elections on 15 Dec. 2005. After months of political deadlock, Iraq's parliament convened on 22 April 2006 to fill the top leadership posts and Talabani was elected by parliament to a second presidential term. On the same day he appointed the Shia politician Nouri al-Maliki as prime minister designate after the latter was nominated by his Shia coalition, the United Iraqi Alliance (UIA).

Following the restoration of Iraqi-Syrian diplomatic ties in Nov. 2006, Talabani became the first Iraqi head of state to visit Damascus for 30 years in Jan. 2007.

In Nov. 2010 Talabani was again re-elected as president by the National Assembly on the second ballot. He then tasked Prime Minister Nouri al-Maliki with forming a new coalition government and bringing to an end the political stalemate prevailing since inconclusive parliamentary elections the previous March.

Nouri al-Maliki

Position
Prime Minister

Introduction
Nouri al-Maliki was appointed Iraq's prime minister designate by the president, Jalal Talabani, in April 2006. He succeeded his ally and fellow member of the conservative Shia Muslim al-Dawa group, Ibrahim al-Jaafari, who had been unable to curb the violent insurgency or create alliances with Sunni and Kurdish factions since elections in Dec. 2005. Al-Maliki, who once commanded Shia forces against Saddam Hussein's regime from exile in Syria, promised an inclusive government. Although his first administration struggled to achieve the basis for a lasting political consensus, levels of violence did subside and in Nov. 2008 parliament approved an agreement with the USA that all US troops leave the country by the end of 2011. Al-Maliki's State of Law coalition came second at elections in March 2010, but he remained in office and sought to construct a workable coalition that was eventually established and approved by parliament at the end of that year. Sectarian and ethnic divisions and frequent associated violence have nevertheless continued to undermine political and social stability in Iraq.

Early Life
Nouri Kamel al-Maliki was born in Hindiyah, southern Iraq in 1950. While studying Arabic at Baghdad University in the early 1970s he joined al-Dawa, which was opposed to the secularism of the ruling Ba'ath party. In 1980 he was forced into exile, initially in Iran and from 1990 in Syria.

Following the US-led invasion of Iraq in March 2003, al-Maliki returned home. In July 2003 he was selected as a member of the US-backed Interim Governing Council, serving on a committee formed to purge Saddam's Ba'athist allies from political life. The committee was widely criticized for heavy-handedness and many Sunni Muslims resented what they saw as a Shia plot to deny them a role in post-Saddam Iraq. As a senior member of al-Dawa, al-Maliki worked closely with party leader al-Jaafari to forge a coalition of Shia parties, called the United Iraqi Alliance (UIA), which won a parliamentary majority in the elections in Jan. 2005.

Elected to the transitional National Assembly, al-Maliki was the senior Shia member charged with drafting the new constitution. In protracted negotiations, he resisted efforts by Sunnis to reduce the autonomy given to Kurds in the north and Shias in the south. Attempts by Prime Minister al-Jaafari to form a broad-based coalition government to reflect the results of the elections on 15 Dec. 2005 became deadlocked and he stepped down on 21 April 2006. Al-Maliki emerged as the UIA's premiership candidate and was named prime minister-designate on 22 April 2006.

Career in Office

Calling for an end to sectarian divisions, al-Maliki announced a national reconciliation plan in June 2006, including a conditional amnesty for insurgents and intra-communal dialogue between political leaders, clerics, armed militias and civil society representatives. However, sectarian violence continued and Sunni opinion was further inflamed by the widely criticized conduct of Saddam Hussein's execution for crimes against humanity at the end of Dec. 2006. In Jan. 2007 President Bush announced that he would send 21,000 extra US troops to Iraq to reassert the authority of al-Maliki's government. Meanwhile, al-Maliki remained under pressure to find a political consensus on divisive ethnic and sectarian issues, an impasse aggravated by the withdrawal from the national unity government of radical Shia members, secular-leaning Iraqis and the main Sunni coalition group (who rejoined in July 2008).

In Dec. 2007 the UK military contingent in Iraq handed control of Basra province to Iraqi forces, which in March 2008 launched a crackdown on radical Shia Mahdi Army militia. Also in March, Iranian President Mahmoud Ahmadinejad made an unprecedented two-day visit to Iraq for talks with al-Maliki, who returned the visit in June. The following month he first raised the prospect of a timetable for a US withdrawal, leading in Nov. to the approval by Iraq's parliament of a security pact under which all troops would leave the country by the end of 2011.

On 1 Jan. 2009 the Iraqi government took control of Baghdad's fortified Green Zone from US forces and assumed authority over foreign troops in the country. In the same month al-Maliki's allies did well in provincial elections. British troops formally ended their combat mission in Iraq in April 2009 and in June the Iraqi government declared a holiday to mark National Sovereign Day as US combat troops completed their withdrawal from towns and cities. In Oct. 2011 the US government confirmed that all remaining US troops would be out of Iraq by the end of the year.

In Oct. 2009 al-Maliki announced the formation of a multi-confessional nationalist State of Law grouping to contest the forthcoming general election after a split in the broad Shia coalition that won the 2005 polls. Unexpectedly, his alliance was narrowly defeated at elections in March 2010 by the Iraqi National Movement of former prime minister Iyad Allawi, although al-Maliki challenged the result. He remained in office during the drawn-out process of negotiating a coalition, which was eventually concluded in Dec. that year when, having been reappointed in Nov. as premier, he named a new cabinet which was approved by parliament. Sectarian tensions in the government and parliament nevertheless continued through 2011 and 2012, prompting the arrest on terrorism charges and subsequent flight abroad of the Sunni vice president, Tariq al-Hashemi (who was convicted and sentenced to death *in absentia*), and fuelling further violence across the country. At the same time Iraq was not isolated from the wave of popular disaffection that swept across much of the Arab world from early 2011, and there were also increasing tensions between the central government and the administration of autonomous Iraqi Kurdistan, in part over oil contracts with foreign companies.

DEFENCE

Following the downfall of Saddam Hussein, recruitment began in July 2003 for a new professional army run by the US military. Saddam Hussein's forces numbered 400,000 at their peak. In Nov. 2008 Iraq's parliament approved a plan that saw the last American troops leave the country in Dec. 2011. In 2012 proposals were made for the restoration of compulsory conscription.

Army

A New Iraqi Army is being developed to replace Saddam's army with a professional force. In 2011 personnel numbered an estimated 193,000. In July 2004 the Civil Defense Corps (23,100 personnel in April 2004) was disbanded and converted into a National Guard. It was in turn merged into the Army in Jan. 2005.

Navy

A 1,100-strong (2007) Iraqi Navy (initially called the Coastal Defense Force) has been re-established since the 2003 war. It began operations in Oct. 2004.

Air Force

An Iraqi Air Force (initially called the Army Air Corps) has been reconstructed since 2003. There were a total of 1,200 personnel in Nov. 2007, with 22 aircraft and 37 helicopters.

ECONOMY

The oil sector accounted for 76·1% of GDP in 2001; agriculture accounted for 7·8%, manufacturing 1·6% and services 13·8%.

Iraq featured among the ten most corrupt countries in the world in a 2011 survey of 183 countries carried out by the anti-corruption organization *Transparency International*.

Overview

Iraq continues to face development challenges following decades of dictatorship, international sanctions and war. After the military victory of the US-led coalition in 2003, GDP declined by 41%. However, substantial progress has since been made, with inflation falling to single figures and economic growth resuming. In Nov. 2004 the Paris Club of creditors agreed to write off 80% of Iraq's external debt.

Although reliable data is lacking, the poverty rate is near 23% while unemployment is estimated at 12%. The government has pledged to introduce financial reforms and increase spending in social sectors. Oil reserves are amongst the largest in the world and production has increased to levels last seen in the 1980s. However, despite an improving security situation, sectarian unrest, corruption and crime impede growth. Strengthening public sector governance, improving security and reducing dependence on oil are key to long-term development.

Currency

From 15 Oct. 2003 a new national currency, the new *Iraqi dinar* (NID), was introduced to replace the existing currencies in circulation in the south and north of the country. There was deflation of 2·2% in 2009 but inflation of 2·4% in 2010 and 5·6% in 2011.

Budget

In 2007 revenues were ID58,714bn. and expenditures ID48,153bn. Crude oil export revenue accounted for 80·3% of all revenues in 2007; current expenditure accounted for 79·6% of all expenditures.

Performance

Real GDP growth was 2·9% in 2009, 3·0% in 2010 and 8·9% in 2011. Total GDP in 2011 was US$115·4bn.

Banking and Finance

All banks were nationalized in 1964. Following the Gulf War in 1991 the formation of private banks was approved, although they were prohibited from conducting international transactions. A new post-Saddam banking law in Oct. 2003 authorized private banks to process international payments, remittances and foreign currency letters of credit. The Trade Bank of Iraq has been established as an export credit agency to facilitate trade financing. The independent Central Bank of Iraq is the sole bank of issue; its *Acting Governor* is Abdul Basit Turki. All domestic interest rates were liberalized on 1 March 2004.

The Iraq Stock Exchange opened in Baghdad in June 2004 as the successor to the former Baghdad Stock Exchange.

ENERGY AND NATURAL RESOURCES

Environment
Iraq's carbon dioxide emissions from the consumption and flaring of fossil fuels were the equivalent of 3·6 tonnes per capita in 2008.

Electricity
Installed capacity was 7·3m. kW in 2007. Production in 2007 was 33·18bn. kWh, with consumption per capita 1,164 kWh.

Oil and Gas
Proven oil reserves in 2011 totalled 143·1bn. bbls, the fifth highest in the world. Oil production in 2011 totalled 136·9m. tonnes, the highest total since 1989. The sector provides 95% of government revenue.

In 2008 Iraq had natural gas reserves of 3,170bn. cu. metres.

Minerals
The principal minerals extracted are sulphur (30,000 tonnes in 2006) and salt (an estimated 25,000 tonnes in 2006).

Agriculture
There were around 5·75m. ha. of arable land in 2002 and 0·34m. ha. of permanent crops. An estimated 3·53m. ha. were irrigated in 2002. Production (2003 estimates, in 1,000 tonnes): wheat, 2,553; barley, 1,316; tomatoes, 1,000; dates, 910; potatoes, 625; melons and watermelons, 575; cucumbers and gherkins, 350.

Livestock (2003 estimates): sheep, 6·2m.; goats, 1·6m.; cattle, 1·5m.; asses, 380,000; chickens, 23m.

Forestry
In 2010 forests covered 0·83m. ha., representing 2% of the land area. Timber production in 2007 was 117,000 cu. metres.

Fisheries
Catches in 2010 totalled 25,720 tonnes, of which about 52% from marine waters.

INDUSTRY

Iraq remains under-developed industrially. Production figures (2007, in 1,000 tonnes): residual fuel oil, 5,595; distillate fuel oil, 4,995; cement (estimate), 4,500; petrol, 2,212; kerosene, 739; jet fuel, 409.

Labour
In 1996 the labour force was 5,573,000 (75% males). Unemployment was 33% in Aug. 2005, down from 55% in Aug. 2003.

INTERNATIONAL TRADE

Imports and Exports
In 2007 imports amounted to US$18,289m. and exports to US$39,590m. Manufactures and food are the main import commodities. Crude oil is the main export commodity. Imports and exports have both increased significantly since the Saddam era. In Oct. 2012 oil exports totalled 81·3m. bbls, their highest level in more than 30 years.

COMMUNICATIONS

Roads
In 2002 there were an estimated 44,900 km of roads, of which 84·3% were paved. Vehicles in use in 2006 included 785,000 passenger cars and 1,345,000 lorries and vans. In 2005 there were 1,789 road accident deaths. Considerable post-war road reconstruction since 2003 reflects heavy military use and lack of maintenance.

Rail
In 2005 railways comprised 2,032 km of 1,435 mm gauge route. Passenger-km travelled in 2004 came to 24m. and freight tonne-km to 90m.

Civil Aviation
In 2000 there were international flights for the first time since the 1991 Gulf War, with air links being established between Iraq and Egypt, Jordan and Syria. Since 2003 the two international airports at Baghdad and Basra have undergone post-war reconstruction. Major domestic airports are at Mosul, Kirkuk and Irbil. In May 2010 the government dissolved the state airline, Iraqi Airways, owing to a legal dispute with Kuwait dating back to the Iraqi invasion in 1990.

Shipping
In Jan. 2009 there were 18 ships of 300 GT or over registered, totalling 86,000 GT. A 565-km canal was opened in 1992 between Baghdad and the Persian Gulf for shipping, irrigation, the drainage of saline water and the reclamation of marsh land. Iraq has three oil tanker terminals at Basra, Khor Al-Amaya and Khor Al-Zubair. Its single deep-water port is at Umm Qasr.

Telecommunications
There were 1·11m. fixed telephone lines in 2009 (36·1 per 1,000 inhabitants). Mobile phone subscribers numbered 19·72m. in 2009. Mobile phones were banned during the Saddam Hussein era. There were 56·0 internet users per 1,000 inhabitants in 2010. Fixed internet subscriptions totalled 3,072 in 2008 (0·1 per 1,000 inhabitants). In March 2012 there were 1·6m. Facebook users.

SOCIAL INSTITUTIONS

Justice
The Iraqi Constitution of 2005 provided for a judiciary comprised of a Higher Judicial Council (to oversee federal judicial affairs), a Supreme Court, a Court of Cassation and other federal courts including the Central Criminal Court. There is also a Judiciary Oversight Commission. The judicial system remains in a period of transition.

The death penalty was introduced for serious theft in 1992; amputation of a hand for theft in 1994. It is believed that during the Saddam era there were hundreds of executions annually. The death penalty was suspended in April 2003 after the fall of Saddam, but reinstated in Aug. 2004. There were at least 129 confirmed executions in 2012, up from 68 in 2011.

The population in penal institutions in June 2008 was 27,366 (93 per 100,000 of national population).

Police
A new post-war national police force has been established and numbered about 135,000 in Nov. 2007. The personnel includes both former officers who are being retrained and new recruits.

Education
Primary education became compulsory in 1976. Primary school age is 6–11. Secondary education is for six years, of which the first three are termed intermediate. The medium of instruction is Arabic; Kurdish is used in primary schools in northern districts.

In 2005 there were 92,769 pre-primary school children with 5,981 teaching staff; 4·43m. primary school children with 215,795 teaching staff; and 1·75m. secondary school pupils with 93,219 teaching staff. Adult literacy rate was an estimated 78·1% in 2009 (male, 86·3%; female, 69·9%). Most schools were closed in March–April 2003 when UNICEF estimates that 200 were destroyed and a further 2,750 looted. By Oct. 2003 all 22 universities and 43 technical institutes and colleges were open, as were nearly all primary and secondary schools. There were 424,908 students and 19,231 academic staff in tertiary education in 2005. Expenditure on education in 2003 was an estimated US$384·5m.

Health
According to the World Health Organization, in 2005 there were 19,010 physicians, 3,460 dentists, 3,170 pharmacists and 38,001 nurses and midwifery personnel. There are approximately 240 hospitals and 1,200 primary health care clinics operating in post-war Iraq.

RELIGION
The constitution proclaims Islam the state religion, but also stipulates freedom of religious belief and expression. In 2001 the population was 97% Muslim; there were also 750,000 Christians, although their numbers have declined since then to 500,000. In Feb. 2013 there was one Roman Catholic cardinal. *See also* TERRITORY AND POPULATION *above*.

CULTURE
World Heritage Sites
Iraq has three UNESCO World Heritage sites: Hatra (inscribed in 1985), a large fortified city of the Parthian (Persian) Empire; Ashur (Qal'at Sherqat) (2003), the first capital and the religious centre of the Assyrians from the 14th to the 9th centuries BC; and Samarra Archaeological City (2007), the site of the capital of the former Abbasid Empire.

Press
In 2008 several hundred daily and weekly publications appeared regularly, the most popular of which, *Al-Sabah* ('The Morning'), had an average circulation of 50,000.

Tourism
In 2010 there were 1,518,000 foreign tourists, up from 864,000 in 2008.

DIPLOMATIC REPRESENTATIVES
Of Iraq in the United Kingdom (3 Elvaston Pl., London, SW7 5QH)
Ambassador: Vacant.
Chargé d'Affaires a.i.: Muhieddin Hussien Abdullah Al-Taaie.

Of the United Kingdom in Iraq (International Zone, Baghdad)
Ambassador: Simon Collis.

Of Iraq in the USA (1801 P St., NW, Washington., D.C., 20036)
Ambassador: Jabir Habib Jabir.

Of the USA in Iraq (APO AE 09316, Baghdad)
Ambassador: Robert S. Beecroft.

Of Iraq to the United Nations
Ambassador: Hamid Al-Bayati.

Of Iraq to the European Union
Ambassador: Mohammed Abdullah Al-Humaimidi.

FURTHER READING
Aburish, S. K., *Saddam Hussein: The Politics of Revenge.* 2000
Allawi, Ali A., *The Occupation of Iraq: Winning the War, Losing the Peace.* 2007

Anderson, Liam and Stansfield, Gareth, *The Future of Iraq: Dictatorship, Democracy or Division?* 2004
Blix, Hans, *Disarming Iraq: The Search for Weapons of Mass Destruction.* 2004
Butler, R., *Saddam Defiant: The Threat of Weapons of Mass Destruction and the Crisis of Global Security.* 2000
Herring, Eric and Rangwala, Glen, *Iraq in Fragments: The Occupation and its Legacy.* 2006
Mackey, Sandra, *The Reckoning: Iraq and the Legacy of Saddam Hussein.* 2002
Marr, Phebe, *The Modern History of Iraq.* 2003
Polk, William R., *Understanding Iraq: The Whole Sweep of Iraqi History, from Genghis Khan's Mongols to the Ottoman Turks to the British Mandate to the American Occupation.* 2006
Shahid, Anthony, *Night Draws Near: Iraq's People in the Shadow of America's War.* 2005
Sluglett, Marion Farouk and Sluglett, Peter, *Iraq Since 1958: From Revolution to Dictatorship.* 3rd ed. 2001
Stansfield, Gareth, *Iraq: People, History, Politics.* 2007
Stiglitz, Joseph E. and Bilmes. Linda J., *The Three Trillion Dollar War: The True Cost of the Iraq Conflict.* 2008
Tripp, Charles, *A History of Iraq.* 3rd ed. 2007

National Statistical Office: Central Organization of Statistics & Information Technology, Baghdad.
Website: http://cosit.gov.iq

Kurdistan

The Kurdistan Region of Iraq ('Iraqi Kurdistan') is the only area of Kurdistan to be recognized officially as an autonomous federal entity. After decades of insurgency Iraq granted limited independence in 1970. *De facto* independence was established following the Kurdish uprising at the end of the Gulf War in 1991. This led to the creation of the Kurdistan Regional Government by the Iraqi Kurdistan Front a year later. Self-governance was disrupted in 1994 by civil war between the Kurdistan Democratic Party (KDP) and the Patriotic Union of Kurdistan (PUK). Rival administrations were set up in Irbil and As-Sulaymaniyah. Peace was restored in Sept. 1998 with the signing of the US-mediated Washington Agreement. The region was acknowledged officially in the 2005 Iraqi constitution and power-sharing began in 2006.

Area, 40,643 sq. km (15,692 sq. miles); population (2002 estimate), 3,757,058. The region comprises three governorates, Irbil, As-Sulaymaniyah and Dahuk, and claims territory in other Kurdish areas although borders remain a contentious issue with Iraq. The unified government has been based at the capital, Irbil, since 2006.

In presidential elections held on 25 July 2009 Massoud Barzani was re-elected with 69·6% of the vote against 25·3% for Kamal Mirawdily. In parliamentary elections held on the same day the Kurdistani List won 57·3% of the vote and 59 of 111 seats, the Change List 23·8% (25 seats) and the Service and Reform List 12·8% (13 seats).

President: Massoud Barzani.
Prime Minister: Nechervan Idris Barzani.

Oil and gas are set to become a major source of revenue although exploitation has been hindered over rights' disputes with the Baghdad government. Oil exports began in June 2009 with Baghdad receiving 88% of revenues under current agreements. There are airports at As-Sulaymaniyah and Irbil.

IRELAND

Éire

Capital: Dublin
Population projection, 2015: 4·73m.
GNI per capita, 2011: (PPP$) 29,322
HDI/world rank: 0·908/7
Internet domain extension: .ie

KEY HISTORICAL EVENTS

Ireland was first inhabited around 7500 BC by Mesolithic hunter-gatherers who travelled across the land bridge that connected southwest Scotland with the northern part of Ireland (it was submerged around 6700 BC). Farmers from the Middle-East arrived in Ireland around 3500 BC. Their elaborate graves are also a feature of Neolithic communities in Brittany and the Iberian peninsula. From the sixth century BC, the island was invaded by waves of Celtic tribes from central Europe, including the Gaels, who established pastoral communities within massive stone forts. By AD 200 the Gaels dominated the island, though there was no central control: society was based on a complex structure of hundreds of small kingdoms. The Romans, who dominated much of northern Europe, never reached Ireland. The Gaels traded with other Celtic peoples and sent raiding parties to form settlements in Scotland (Dál Riata) and west Wales.

Christian missionaries reached Ireland during the 3rd century AD. St Patrick, born on the west coast of Britain, was consecrated as a bishop in Gaul and lived and preached in Ireland from *c.* 432 until his death *c.* 465. Monasteries were founded and, in an overwhelmingly agrarian society, they became important centres of learning and the dissemination of the written word. In contrast to much of northern Europe, ravaged by fragmentary forces following the collapse of the Roman Empire, Christianity found a haven in Ireland. Later, Irish missionaries took Celtic Christianity to Britain and continental Europe. By the 5th century AD there were five leading Gaelic kingdoms, which roughly correspond to the latter-day provinces of Ulster, Leinster, Munster and Connacht (the fifth kingdom occupied land in the modern counties of Meath and Westmeath). Each kingdom was dominated by one or two families—the Uí Néill clan was especially powerful in the north and east. The south (Munster) was dominated by the Eóganachta family.

Nordic Invasion

Viking longboats first appeared off the Irish coast in the late seventh century. 795 saw a full-scale Viking invasion, which heralded more than two hundred years of Scandinavian influence. The Vikings were great traders and established the first towns along the east and south coasts—the towns of Wexford, Waterford, Cork and Limerick became prosperous centres of manufacturing and commerce. Dublin, said to be founded in 841 by the Norse king Thurgesius, became a key outpost in a Viking diaspora stretching as far as Sicily and Russia. Gaelic kings made military alliances with the Viking settlers to support their struggles with neighbouring dynasties. In 976 the warrior Brian Boru (Bóruma) became king of Munster following a series of victories against the powerful Eóganachta. Following Boru's defeat of the Leinster groups and their Norse allies at Clontarf in 1014 he seemed destined to be the first high king of all Ireland, but was murdered shortly after his famous victory.

In the mid-12th century the Pope gave his blessing to an expedition of Anglo-Normans to Ireland. They were sent by the English King Henry II, who had been approached for military support by the deposed king of Leinster, Dermot MacMurrough (Díarmait Mac Murchada). Returning to Ireland in 1169 with Norman barons and Welsh mercenaries, MacMurrough recovered part of his former territories and captured Dublin. Richard de Clare (Strongbow), a powerful Norman invader, became MacMurrough's heir after marrying his daughter. During the 13th century various Anglo-Norman adventurers began to establish themselves in Ireland. Dublin Castle was built in 1204 on the site of a Norse fort and the first parliament sat there in 1264. After his decisive defeat of English forces at the Battle of Bannockburn in 1314, Edward Bruce, the brother of Robert Bruce, king of Scotland, dreamed of establishing a Celtic kingdom. In 1315 he landed in Ulster and attempted to overthrow the English. Within a year he controlled most of Ireland north of Dublin, but his troops left a trail of destruction and soon lost support. Bruce was defeated and killed at Dundalk in 1317 by a Norman-Irish army reinforced from England under orders from King Edward II.

The descendants of the Anglo-Norman settlers were gradually identified with the native Irish, whose language, habits, and laws they adopted. To counteract this, the Anglo-Irish Parliament passed the Statute of Kilkenny in 1366, decreeing heavy penalties against all who allied themselves with the Irish. This statute, however, remained inoperative; although Richard II went to Ireland in 1394 and 1399 to reassert royal authority, he failed to achieve any practical result. During the subsequent Wars of the Roses in England the authority of the English crown became limited to the Pale, a coastal district around Dublin.

King Edward IV, of the House of York, came to the English throne in 1461 and appointed Gerald (Gearóid Mór) FitzGerald, 8th earl of Kildare as viceroy of Ireland. The FitzGeralds were wealthy Yorkists, well-connected to a network of Anglo-Norman and Gaelic families. Gearóid Mór wielded considerable power, and managed to hold onto it even after the return of the Lancastrians in 1485. He was eventually replaced in 1494 by Sir Edward Poynings, who, representing English interests, brought in legislation providing for the reduction of the power of the

Anglo-Irish lords. The Poynings Laws removed the legal rights of the Irish parliament to legislate independently.

Henry VII reappointed Gearóid Mór as viceroy in 1496. For the next 38 years the FitzGeralds (Geraldines) ruled Ireland from Maynooth Castle, paying deference to the English crown. Henry VIII was determined to centralize power and reduce the influence of provincial magnates. He introduced the Reformation to Ireland in 1537 and began to dissolve the monasteries with little resistance.

Ulster Rebellion

Elizabeth I, through her deputy in Ireland, Sir Henry Sidley, removed the Irish chiefdoms from their positions of power. An uprising in Munster in the early 1570s was quickly suppressed and only Ulster now provided a stumbling block to Tudor domination. It was from Ulster that Hugh O'Neill and Red Hugh O'Donnell launched an open rebellion. In 1598 O'Neill ambushed and defeated a government force of over 4,000 at the Battle of Yellow Ford near Armagh. Spoken of as 'Prince of Ireland', his ambitions were thwarted by the arrival of 20,000 troops under Lord Mountjoy in 1600. Reinforcements of Spanish soldiers in 1601 were insufficient and O'Neill left for the Continent with his followers in the 1607 'Flight of the Earls'.

The Earls' lands were seized by the English crown and in 1609 Ulster-Scottish and English settlers were invited to colonize. Swathes of land were cleared of farms and woodland, and 23 walled new towns were created, including Belfast. By the early 1620s the Anglo-Scottish population of Ulster was more than 20,000. English politics in the 1630s was dominated by struggles between the crown and parliament (the Puritans) and the Irish in Ulster took advantage, rebelling against the planters in late 1641 in a series of attacks in which thousands of Protestants were killed. The following year Owen Roe O'Neill, who had fled to Spain with his uncle Hugh in 1607, returned to Ireland and led the Confederate forces. A provisional government was established at Kilkenny and by the end of 1642 O'Neill controlled the whole island apart from Dublin and parts of Ulster.

Victory for the English parliamentarians under Oliver Cromwell and the execution of Charles I in 1649 had a profound impact on Ireland. Cromwell was determined to avenge the 1641 massacre of the Ulster planters. With his New Model Army, he stormed Drogheda and murdered its garrison of 2,000. Wexford fell, and by 1652 all of Ireland was in Cromwellian hands. Hundreds of thousands of acres of land were confiscated and given to a new wave of Protestant settlers.

Following the restoration of the English monarchy in 1660, Catholics in Ireland hoped to be rewarded for their former loyalty, but Charles II restored only a small number of Catholic estates. King James II, however, was a declared Catholic and under his viceroy in Ireland, Richard Talbot, earl of Tyrconnel, Catholics were advanced to positions of state and placed in control of the military. Protestant power was on the wane in England and the Protestant aristocracy invited William of Orange (the Dutch husband of James II's daughter Mary) to claim the English crown. James II fled to France, then travelled to Ireland with French soldiers. They moved north, aiming to subjugate Protestant Ulster. In the spring of 1689 only the walled towns of Derry/Londonderry and Enniskillen remained in Protestant hands. Derry/Londonderry was besieged, but it held out for 15 weeks until the arrival of William's forces, which defeated James at the Battle of the Boyne. The Jacobites retreated to Limerick, where they negotiated the Treaty of Limerick of 1691. Catholics were permitted some religious freedom, and the restoration of their lands. However, the treaty was not honoured by the English parliament and 11,000 Irish Jacobites set sail to join the French army.

Religious Divide

The defeat of the Catholic cause was followed by more confiscation of land and the introduction of the Penal Laws which prevented Catholics from buying freehold land, holding public office or bearing arms. During the American revolution, fear of a French invasion led Irish Protestants to form the Protestant Volunteer army. Led by Henry Grattan, they used their military strength to extract concessions from Britain. Trade concessions were granted in 1779 and the Poynings Laws repealed three years later. However, Catholics continued to be denied the right to hold political office.

The principles of the French Revolution found their most powerful expression in Ireland in the Society of United Irishmen, which, led by the protestant lawyer Theobald Wolfe Tone, mounted a rebellion in 1798. Without the expected French assistance, the rebellion was crushed by crown troops led by Gen. Lake. The British prime minister, William Pitt, was convinced that the 'Irish problem' could be solved by the abolition of the Irish parliament, legislative union with Britain and Catholic emancipation. The first two goals were achieved in 1801, but the opposition of George III and British Protestants prevented the enactment of the Catholic Emancipation act until 1829, when it was accomplished largely through the efforts of Daniel O'Connell.

After 1829 the Irish representatives in the British Parliament, led by O'Connell, sought a repeal of the Act of Union. Calls for land reforms were drowned out by a disastrous potato famine. Between 1845–49 a blight wiped out the potato crop, the staple food of the Irish population and resulted in mass starvation. Of a population of 8·5m. almost 1m. died and over 1m. emigrated, mostly to the United States. Irish Catholics in the United States formed the secret Fenian movement, dedicated to achieving full Irish independence.

Home Rule Campaign

Charles Parnell, a Home Rule League MP, came to the fore of the nationalist movement in 1877 as president of the Home Rule Confederation of Great Britain. Parnell led parliamentary obstruction in response to the House of Lords' rejection of limited land reform in Ireland. Parnell's Irish Land League saw limited gains in Gladstone's 1881 Land Act but Parnell, voicing continuing discontent, was imprisoned in Dublin. His release in 1882 and the subsequent Kilmainham Treaty, granting more concessions to tenants, was seen by London as the quickest solution to an increasingly anarchic Ireland.

The Home Rule Party, led by Parnell, brought down the Conservative government at Westminster by voting with the Liberals, allowing William Gladstone to form a government in 1886. Gladstone attempted to resolve the 'Irish problem' by introducing a Home Rule Bill—seen as Parnell's greatest achievement—which would give the Irish Parliament the right to appoint the executive of Ireland. However, Home Rule was greatly opposed in Ulster and England and failed at Westminster in 1886 and 1893. Parnell's domination of Irish politics came to end with the disclosure of his affair with Kitty O'Shea—an English woman of aristocratic background—in 1890. After Gladstone rejected him, he lost control of the Irish parliamentarians and was condemned by the Catholic clergy.

During the 1880s a new pride in traditional Irish culture took root, symbolized by the establishment of the Gaelic Athletic Association in 1884 and the Gaelic League in 1893, which successfully campaigned for the return of the Irish language to the school curriculum. Though not political organizations, they provided a link between the conservative Catholic church and the Fenians (nationalists). In 1905 the Irish political leader and journalist Arthur Griffith founded Sinn Féin ('we ourselves') to promote Irish economic welfare and achieve complete political independence. However, at the time the dominant nationalist group remained the Home Rule party of John Redmond.

A Home Rule Bill was finally passed in 1914 but the act was suspended for the duration of the First World War. Redmond pledged the support of Ireland to the British war effort, which angered some nationalists. The Irish Republican Brotherhood (IRB) plotted a rebellion while Britain was at war, soliciting German support. On Easter Monday 1916 the IRB seized the General Post Office in Dublin and Patrick Pearse read out the proclamation of the Republic of Ireland. Though the Easter Rising was over in under a week, the emotional impact was heightened when the British executed 16 of the rebel leaders. Sinn Féin, linked in the Irish public's mind with the rising, scored a dramatic victory in the parliamentary elections of 1918. Its members refused to take their seats in Westminster, declared the *Dáil Éireann* ('Diet of Ireland') and proclaimed the Irish republic. The British outlawed Sinn Féin and the Dáil, which went underground and associated military groups including the Irish Republican Army (IRA) engaged in guerrilla warfare against the local authorities representing the Union. The British sent police reinforcements (the Black and Tans) who further inflamed the situation.

Civil War

A new Home Rule Bill was passed in 1920, establishing two parliaments, one in Belfast and the other in Dublin. The Unionists of the six counties accepted this scheme, and a Northern Parliament was duly elected in May 1921. Sinn Féin rejected the plan, but in autumn 1921 British Prime Minister Lloyd George negotiated with Griffith and Michael Collins of the Dáil a treaty granting Catholic Ireland dominion status within the British Empire. Collins managed to gain approval in the Dáil by a slim majority. The Republicans in the Dáil, led by Éamon de Valera, rejected the treaty, which had divided Ireland and fell short of full independence. A brutal civil war ensued; Collins, who had assumed command of the army, was assassinated in Aug. 1922 by anti-treaty rebels. The treaty supporters emerged as victors and the Irish Free State was established in Jan. 1922. William Cosgrave became the first prime minister and his Fine Gael party led for ten years. In 1932 de Valera, leader of the Fianna Fáil party, became prime minister (*taoiseach*). Five years later he brought in a new constitution establishing the sovereign nation of Ireland and abolishing the oath of allegiance sworn by Irish parliamentarians to the British crown.

Independence

Ireland remained neutral in the Second World War, though in the harsh economic conditions of the time tens of thousands of people emigrated to Britain for work and many thousands joined the war effort. In 1948 Prime Minister John Costello demanded total independence from Britain and reunification with the six counties of Northern Ireland. Independence came the following year and in 1955 the Republic of Ireland was admitted to the United Nations, but nothing came of the claim to the six Ulster counties under British rule. Economic relations between the Republic and Northern Ireland improved in the 1950s and '60s, though both decades were marked by large-scale emigration from the Republic, chiefly to the United States. Trouble in the North flared up in the late '60s over Catholic demands for civil rights and equality in the allocation of housing. Confrontation between the two religious communities intensified and in 1969 British troops were deployed to keep the peace. The British military soon lost the confidence of the Catholic community, the IRA increased its activity and more violence ensued. In 1972 the Unionist government in Belfast resigned and direct rule from London was imposed.

On 1 Jan. 1973 the Republic of Ireland became a member state of the European Economic Community. Jack Lynch led Fianna Fáil into power in 1977, though over the next decade there were party splits while general elections were held against a backdrop of soaring unemployment. Emigration increased, especially among young people, reaching a peak of 44,000 in 1989 under Charles Haughey's premiership. The 1990s were marked by an economic upturn, buoyed by EU subsidies and foreign investment. The legalization of divorce in 1995 symbolized the Irish Republic's embrace of modern European values. In the north, a ceasefire between the IRA and Protestant militias in 1994 formed the basis for the signing of the Good Friday Agreement in April 1998. On 2 Dec. 1999 the Irish constitution was amended to remove the articles laying claim to Northern Ireland. Prime Minister Bertie Ahern, who came to power in the 1997 general election, took Ireland into the single European currency in Jan. 2002. In Oct. 2002 the Northern Irish Assembly was suspended for the fourth time in its history over allegations of IRA spying at the Northern Ireland Office. Direct rule from London was subsequently reimposed. In May 2007 a devolved Northern Ireland government replaced direct rule from London.

TERRITORY AND POPULATION

The Republic of Ireland lies in the Atlantic Ocean, separated from Great Britain by the Irish Sea to the east, and bounded in the northeast by Northern Ireland (UK). In 2011, 62·3% of the population lived in urban areas. The population at the 2011 census was 4,588,252 (2,315,553 females), giving a density of 67·0 persons per sq. km. The census population in 2011 was the highest figure since 1861 when the census recorded a population of 4·40m.

The UN gives a projected population for 2015 of 4·73m.

The capital is Dublin (Baile Átha Cliath). Town populations, 2011: Greater Dublin, 1,110,627; Cork, 198,582; Limerick, 91,454; Galway, 76,778; Waterford, 51,519.

Counties and Cities[1]	Area in ha[2]	Population, 2011		
		Males	Females	Totals
Province of Leinster				
Carlow	89,655	27,431	27,181	54,612
Dublin City	11,758	257,303	270,309	527,612
Dun Laoghaire-Rathdown	12,638	98,567	107,694	206,261
Fingal	45,467	134,488	139,503	273,991
Kildare	169,540	104,658	105,654	210,312
Kilkenny	207,289	47,788	47,631	95,419
Laois	171,990	40,587	39,972	80,559
Longford	109,116	19,649	19,351	39,000
Louth	82,613	60,763	62,134	122,897
Meath	234,207	91,910	92,225	184,135
Offaly	200,117	38,430	38,257	76,687
South Dublin	22,364	129,544	135,661	265,205
Westmeath	183,965	42,783	43,381	86,164
Wexford	236,685	71,909	73,411	145,320
Wicklow	202,662	67,542	69,098	136,640
Total of Leinster	1,980,066	1,233,352	1,271,462	2,504,814
Province of Munster				
Clare	345,004	58,298	58,898	117,196
Cork City	3,953	58,812	60,418	119,230
Cork	746,042	198,658	201,144	399,802
Kerry	480,689	72,629	72,873	145,502
Limerick City	2,087	27,947	29,159	57,106
Limerick	273,504	67,868	66,835	134,703
Tipperary, N. R.	204,627	35,340	34,982	70,322
Tipperary, S. R.	225,845	44,244	44,188	88,432
Waterford City	4,103	22,921	23,811	46,732
Waterford	181,556	33,543	33,520	67,063
Total of Munster	2,467,410	620,260	625,828	1,246,088
Province of Connacht				
Galway City	5,057	36,514	39,015	75,529
Galway	609,820	88,244	86,880	175,124
Leitrim	159,003	16,144	15,654	31,798
Mayo	558,605	65,420	65,218	130,638
Roscommon	254,819	32,353	31,712	64,065
Sligo	183,752	32,435	32,958	65,393
Total of Connacht	1,771,056	271,110	271,437	542,547

Counties and Cities[1]	Area in ha[2]	Population, 2011		
		Males	Females	Totals
Province of Ulster (part of)				
Cavan	193,177	37,013	36,170	73,183
Donegal	486,091	80,523	80,614	161,137
Monaghan	129,508	30,441	30,042	60,483
Total of Ulster (part of)	808,776	147,977	146,826	294,803
Total	7,027,308	2,272,699	2,315,553	4,588,252

[1]Cities were previously known as County Boroughs.
[2]Area details provided by Ordnance Survey.

The official languages are Irish (the national language) and English; according to the 2006 census, Irish is spoken by 1·66m. persons in the Republic of Ireland (41·9% of the population, down from 42·8% in 2002). It is a compulsory subject at school.

SOCIAL STATISTICS

Statistics for five calendar years:

	Births	Marriages	Deaths
2003	61,529	20,302	29,074
2004	61,972	20,619	28,665
2005	61,372	21,355	28,260
2006	64,237	21,841	27,479
2007	70,620	22,544	28,050

2007 rates (per 1,000 population): birth, 16·3; death, 6·5; marriage, 5·2. Annual population growth rate, 2005–10, 1·5%. Expectation of life at birth, 2006, 76·8 years for males and 81·6 years for females.

In 2009 the suicide rate per 100,000 population was 11·8 (men, 19·0; women, 4·7). Infant mortality in 2010, three per 1,000 live births; fertility rate (2008), 2·0 births per woman.

At a referendum on 24 Nov. 1995 on the legalization of civil divorce the electorate was 1,628,580; 818,852 votes were in favour, 809,728 against.

The number of immigrants in the year to April 2006 was 107,800 while emigrants numbered 36,000 in the same period. Immigration peaked at 109,500 in 2006–07 but by 2009–10 had fallen to a provisional 30,800 (the lowest figure since 1993–94). Provisional figures for 2009–10 suggest that the number of emigrants, at 65,300 (including 27,700 Irish citizens), was the highest total since 1988–89. In 2009 Ireland received 2,689 asylum applications, down from 11,634 in 2002.

A UNICEF report published in 2010 showed that 8·4% of children in Ireland live in relative poverty (living in a household in which disposable income—when adjusted for family size and composition—is less than 50% of the national median income).

CLIMATE

Influenced by the Gulf Stream, there is an equable climate with mild southwest winds, making temperatures almost uniform over the whole country. The coldest months are Jan. and Feb. (39–45°F, 4–7°C) and the warmest July and Aug. (57–61°F, 14–16°C). May and June are the sunniest months, averaging 5·5 to 6·5 hours each day, but over 7 hours in the extreme southeast. Rainfall is lowest along the eastern coastal strip. The central parts vary between 30–44" (750–1,125 mm), and up to 60" (1,500 mm) may be experienced in low-lying areas in the west. Dublin, Jan. 40°F (4°C), July 59°F (15°C). Annual rainfall 30" (750 mm). Cork, Jan. 42°F (5°C), July 61°F (16°C). Annual rainfall 41" (1,025 mm).

CONSTITUTION AND GOVERNMENT

Ireland is a sovereign independent, democratic republic. Its parliament exercises jurisdiction in 26 of the 32 counties of the island of Ireland. The first Constitution of the Irish Free State came into operation on 6 Dec. 1922. Certain provisions which were regarded as contrary to the national sentiments were gradually removed by successive amendments, with the result that at the end of 1936 the text differed considerably from the original document. On 14 June 1937 a new constitution was approved by Parliament and enacted by a plebiscite on 1 July 1937. This constitution came into operation on 29 Dec. 1937. Under it the name Ireland (Éire) was restored. In its original form the Irish Constitution provided that the territory of Ireland comprised the whole island, and thus included that of Northern Ireland. This position was modified by referendum in 1998 following the Good Friday Agreement of that year. The former territorial claim has now been replaced with a statement that, while it is the aspiration of the Irish nation to unite the peoples of the island and the current territory of Ireland is not final, unification shall not take place without the consent of majorities in both jurisdictions.

The head of state is the *President*, whose role is largely ceremonial, but who has the power to refer proposed legislation which might infringe the Constitution to the Supreme Court.

The *Oireachtas* or National Parliament consists of the President, a House of Representatives (*Dáil Éireann*) and a Senate (*Seanad Éireann*). The *Dáil*, consisting of 166 members, is elected by adult suffrage on the Single Transferable Vote system in constituencies of three, four or five members. Of the 60 members of the Senate, 11 are nominated by the *Taoiseach* (Prime Minister), six are elected by the universities and the remaining 43 are elected from five panels of candidates established on a vocational basis, representing the following public services and interests: (1) national language and culture, literature, art, education and such professional interests as may be defined by law for the purpose of this panel; (2) agricultural and allied interests, and fisheries; (3) labour, whether organized or unorganized; (4) industry and commerce, including banking, finance, accountancy, engineering and architecture; (5) public administration and social services, including voluntary social activities. The electing body comprises members of the *Dáil*, Senate, county boroughs and county councils.

A maximum period of 90 days is afforded to the Senate for the consideration or amendment of Bills sent to that House by the *Dáil*, but the Senate has no power to veto legislative proposals.

No amendment of the Constitution can be effected except with the approval of the people given at a referendum.

National Anthem
'Amhrán na bhFiann' ('The Soldier's Song'); words by P. Kearney, tune by P. Heeney and P. Kearney.

GOVERNMENT CHRONOLOGY
(FF = Fianna Fáil; FG = Fine Gael; L = Labour; n/p = non-partisan)

Presidents since 1938.

1938–45	FF	Douglas Hyde
1945–59	FF	Séan Thomas O'Kelly
1959–73	FF	Éamon de Valera
1973–74	FF	Erskine Hamilton Childers
1974–76	FF	Cearbhall O'Dalaigh
1976–90	FF	Patrick John Hillery
1990–97	n/p	Mary Terese Robinson
1997–2011	FF	Mary Patricia McAleese
2011–	L	Michael Daniel Higgins

Prime Ministers since 1932.

1932–48	FF	Éamon de Valera
1948–51	FG	John Aloysius Costello
1951–54	FF	Éamon de Valera
1954–57	FG	John Aloysius Costello
1957–59	FF	Éamon de Valera
1959–66	FF	Séan Francis Lemass

1966–73	FF	John (Jack) Mary Lynch
1973–77	FG	Liam Thomas Cosgrave
1977–79	FF	John (Jack) Mary Lynch
1979–81	FF	Charles James Haughey
1981–82	FG	Garret Michael FitzGerald
1982	FF	Charles James Haughey
1982–87	FG	Garret Michael FitzGerald
1987–92	FF	Charles James Haughey
1992–94	FF	Albert Reynolds
1994–97	FG	John Gerard Bruton
1997–2008	FF	Bartholomew (Bertie) P. Ahern
2008–11	FF	Brian Cowen
2011–	FG	Enda Kenny

RECENT ELECTIONS

A general election was held on 25 Feb. 2011: Fine Gael (FG) gained 76 seats with 36·1% of first preference votes (in 2007, 51 seats); Labour Party (L), 37 with 19·4% (20); Fianna Fáil (FF) 20 with 17·4% (78), Sinn Féin 14 with 9·9% (4); United Left Alliance, 5; New Vision, 1; ind., 13. Turnout was 70·1%.

Following elections to the Senate in April 2011, FG held 19 of the 60 seats, FF had 14, L had 12, Sinn Féin had 3 and independents held 12 seats.

Presidential elections took place on 27 Oct. 2011. Michael D. Higgins (L) won 39·6% of the first preference votes, Seán Gallagher (ind.) 28·5%, Martin McGuinness (Sinn Féin) 13·7%, Gay Mitchell (FG) 6·4%, David Norris (ind.) 6·2%, Dana Rosemary Scallon (ind.) 2·9% and Mary Davis (ind.) 2·7%. Higgins won in the final count, with 61·6% of votes against 38·4% for Gallagher. Turnout was 56·1%.

European Parliament

Ireland has 12 (13 in 2004) representatives. At the June 2009 elections turnout was 58·6% (58·6% in 2004). Fine Gael won 4 seats with 29·1% of votes cast (political affiliation in European Parliament: European People's Party); Fianna Fáil, 3 with 24·1% (Alliance of Liberals and Democrats for Europe); Labour Party, 3 with 13·9% (Progressive Alliance of Socialists and Democrats); Socialist Party, 1 with 2·8% (European United Left/Nordic Green Left). One independent was elected, with 4·6% (Alliance of Liberals and Democrats for Europe).

CURRENT GOVERNMENT

President: Michael D. Higgins (b. 1941; L), elected out of seven candidates on 27 Oct. 2011 and inaugurated 11 Nov. 2011.

In March 2013 the Fine Gael–Labour coalition government was composed as follows:

Taoiseach (Prime Minister): Enda Kenny; b. 1951 (FG; in office since 9 March 2011).

Tánaiste (Deputy Prime Minister) and Minister for Foreign Affairs and Trade: Eamon Gilmore (b. 1955; L). *Finance:* Michael Noonan (b. 1943; FG). *Health:* James Reilly (b. 1955; FG). *Justice, Equality and Defence:* Alan Shatter (b. 1951; FG). *Environment, Community and Local Government:* Phil Hogan (b. 1960; FG). *Enterprise, Jobs and Innovation:* Richard Bruton (b. 1953; FG). *Agriculture, Food and the Marine:* Simon Coveney (b. 1972; FG). *Education and Skills:* Ruairí Quinn (b. 1946; L). *Transport, Tourism and Sport:* Leo Varadkar (b. 1979; FG). *Children:* Frances Fitzgerald (b. 1950; FG). *Social Protection:* Joan Burton (b. 1949; L). *Public Expenditure and Reform:* Brendan Howlin (b. 1956; L). *Arts, Heritage and Gaeltacht Affairs:* Jimmy Deenihan (b. 1952; FG). *Communications, Energy and Natural Resources:* Pat Rabbitte (b. 1949; L).

There are 15 Ministers of State.

Attorney General: Máire Whelan.

Chairman of Dáil Éireann: Seán Barrett.

Government Website: http://www.gov.ie

CURRENT LEADERS

Michael D. Higgins

Position
President

Introduction
Michael Daniel Higgins took office as president on 11 Nov. 2011. Prior to his appointment he was the head of the Labour Party. He has published several volumes of poetry.

Early Life
Known popularly as Michael D., Higgins was born on 18 April 1941 in Limerick and raised in County Clare, in the west of Ireland. After working as a clerk he enrolled as a mature student at Galway University in 1962, where he would later lecture in sociology and politics. He undertook further studies at Manchester University in the UK and at the University of Indiana, USA.

Having twice stood unsuccessfully for election to the Dáil as a Labour Party candidate, Higgins was appointed to the Senate by Taoiseach Liam Cosgrave in 1973. He was elected to the Dáil in 1981 but lost his seat in Nov. 1982 following the collapse of the governing coalition. Known as a radical, he served as mayor of Galway from 1982–83, a position he again held from 1991–92. Re-elected to parliament in 1987, he retained his Galway West seat until opting not to stand at the 2011 election.

Having previously opposed Labour's entry into the ruling coalition, Higgins accepted the post of minister of the arts and culture in 1993. During his four years in the post he repealed the controversial Section 31, which allowed censorship over media coverage of the Troubles in Northern Ireland. He also oversaw the establishment of the Irish-language broadcaster, TG4.

Higgins was a vocal opponent of the 2003 Iraq War and has campaigned for human rights in Somalia, Chile, Nicaragua, Gaza and Cambodia. At the 2011 presidential election he defeated Seán Gallagher, a well-known entrepreneur, and Martin McGuinness, a former commander of the IRA and later deputy leader of the Northern Irish power-sharing assembly.

Career in Office
During campaigning for the largely ceremonial post, Higgins pledged to promote justice and equality as the country emerges from economic crisis.

Enda Kenny

Position
Prime Minister

Introduction
Enda Kenny became prime minister (Taoiseach) on 9 March 2011. He has been leader of the centrist Fine Gael since 2002 and is currently Ireland's longest-serving member of parliament. He is a former minister of state for tourism and trade.

Early Life
Kenny was born in 1951 in Castlebar, County Mayo. He studied at the University of Galway and St Patrick's College of Education in Dublin and then worked as a primary school teacher for a year.

His father was the Fine Gael member of parliament for Mayo West and a parliamentary secretary. Enda regularly assisted with constituency matters and on his father's death in 1975 took over the Mayo West seat after winning a by-election. Over the next decade Kenny served as a backbencher but in 1986 was appointed junior minister for education and labour in the government of Garret FitzGerald.

After Fine Gael lost the 1987 election, Kenny served on the opposition front bench. He became party chief whip in 1993 and a year later, after the collapse of Albert Reynolds' Fianna Fáil government, helped negotiate a 'rainbow coalition' with the Labour Party and the Democratic Left. He served in that government as minister for tourism and trade and chaired the European Union Council of Trade Ministers during Ireland's six-month presidency of the EU in 1996.

Fine Gael left government after the 1997 general election and Kenny ran unsuccessfully for party leader in 2001. In 2002, following the party's worst ever election performance, he was elected its new leader. In 2004 Fine Gael made significant gains at the European elections and out-polled Fianna Fáil, the first time Fianna Fáil had come second in elections across the country since 1927. In 2006 Kenny became vice-president of the European People's Party (EPP), the centre-right grouping in the European Parliament.

Kenny led Fine Gael to a strong showing at the 2007 general election despite coming second to Fianna Fáil. In 2010 he held off a leadership challenge before overseeing victory at the 2011 general election on the back of strong anti-Fianna Fáil sentiment in light of the country's economic collapse. With 76 of a possible 166 seats, Fine Gael was the largest single party for the first time in its history and Kenny formed a coalition government with the Labour Party.

Career in Office
Kenny's primary challenge has been to spur a recovery in the financial system. Prior to the election he pledged to renegotiate the conditions of the €85bn. IMF–EU bail-out package agreed by the previous government, and in July 2011 EU leaders agreed to cut the interest rate on the rescue deal and to extend the repayment period. Ireland's fiscal position has since improved under an ongoing austerity regime of tax rises and spending cuts, but the economy remains vulnerable to debt levels and external pressures.

In July 2011 Kenny courted diplomatic controversy with an attack on the Vatican, which he accused of having tried to cloud the extent of sexual abuse of children by Catholic clergy in Ireland. In response, the Vatican recalled its envoy from Dublin.

In June 2012 Irish voters in a referendum approved an EU fiscal pact agreed by heads of government earlier in the year that imposes strict budgetary discipline on member states.

DEFENCE

The President of Ireland is the supreme commander of the Irish Defence Forces. Military command and administrative powers in relation to the Defence Forces are exercisable by the Government through and by the Minister for Defence. The Defence Acts, as amended, provides the legislative basis for the Defence Forces (Óglaigh na hÉireann). The Act and associated Defence Force Regulations set out the guiding principles under which the Defence Forces operate.

The Defence Forces comprise the Permanent Defence Force (the regular Army, the Air Corps and the Naval Service) and the Reserve Defence Force (comprising a First Line Reserve of members who have served in the Permanent Defence Force, a second-line Army Reserve and a second-line Naval Reserve).

The total strength of the Permanent Defence Force in Dec. 2011 was 9,438 (including 565 women) and the total strength of the Reserve Defence Force was 5,220. In Dec. 2011, 529 Defence Forces personnel were involved in 11 peace-support missions throughout the world.

Defence expenditure in 2011 totalled €926·83m., representing 0·6% of GDP.

Army
The Army strength in Dec. 2011 was 7,650 personnel with 4,995 reservists. The Army currently provides the deployable military capabilities for overseas peace support, crisis management and humanitarian operations augmented by personnel from the Air Corps and Naval Service.

Navy
The Naval Service is based at Haulbowline in Co. Cork. The strength in Dec. 2011 was 997 with 225 reservists. It operates eight offshore patrol vessels.

Air Corps
The Air Corps has its headquarters at Casement Aerodrome, Baldonnel, Co. Dublin. As the air component of the Defence Forces, the Air Corps provides air support capabilities to the other components in carrying out their roles. The Air Corps strength in Dec. 2011 was 791 personnel. The Corps operates 17 fixed-wing aircraft and ten helicopters.

INTERNATIONAL RELATIONS

On 12 June 2008 Ireland became the first European Union member to reject the Treaty of Lisbon when it held a national referendum in which 53·4% of votes cast were against the reform treaty and only 46·6% in favour. Turnout was 53·1%. However, at a second referendum on 2 Oct. 2009, the treaty was approved by 67·1% to 32·9% (turnout, 59·0%). Parliamentary ratification followed on 23 Oct. 2009.

ECONOMY

Agriculture accounted for 1% of GDP in 2009, industry 32% and services 67%.

Overview
Ireland is a high income OECD nation. In the decade to 2008 it experienced rapid growth fuelled by a housing and credit boom. However, the global financial crisis sent Ireland into recession, with GDP contracting by 5·5% in 2009 and 0·8% in 2010. In Sept. 2008 the government moved to stabilize the domestic financial system with a €400bn. plan to protect all deposits, bonds and debts in six banks and building societies for two years. By Feb. 2009 the state had nationalized Anglo Irish Bank and provided emergency funding to Allied Irish Banks and the Bank of Ireland.

In Nov. 2010, following widespread concerns in Europe over the Irish government's ability to service its banking sector's debts, the country accepted an €85bn. bailout from the EU, with the IMF providing €22·5bn. The government undertook a programme of austerity and tax rises to cut the budget deficit, amounting to some €25bn. (or 16% of output) between 2008 and 2013. The government aims to bring down the deficit to 3% by 2015. In Feb. 2013 the government negotiated a deal with the European Central Bank to extend the bailout period of the Anglo Irish Bank from ten to 40 years, so cutting borrowing needs by €20bn. over ten years.

GDP grew by 1·4% in 2011, driven by exports and assisted by a low corporation tax rate of 12·5%. Foreign investment into financial services, pharmaceuticals and information technology has increased and exports grew by 2% in 2012. However, domestic demand remains low owing to high levels of household debt (averaging 209% of disposable income). The export-led recovery is fragile and is particularly susceptible to downturns in the UK and eurozone economies.

Currency
On 1 Jan. 1999 the euro (EUR) became the legal currency in Ireland at the irrevocable conversion rate of 0·787564 Irish pounds to 1 euro. The euro, which consists of 100 cents, has been in circulation since 1 Jan. 2002. On the introduction of the euro there was a 'dual circulation' period before the Irish pound ceased to be legal tender on 9 Feb. 2002. Euro banknotes in circulation on 1 Jan. 2002 had a total value of €6·8bn.

Inflation rates (based on OECD statistics):

2002	2003	2004	2005	2006	2007	2008	2009	2010	2011
4·7%	4·0%	2·3%	2·2%	2·7%	2·9%	3·1%	−1·7%	−1·6%	1·2%

The Central Bank has the sole right of issuing legal tender notes; token coinage is issued by the Minister for Finance through the Bank. Gold reserves were 176,000 troy oz in Sept. 2009 and foreign exchange reserves US$574m. Total money supply was €83,181m. in Aug. 2009.

Budget
Current revenue and expenditure (in €1m.):

Current Revenue	2007	2008
Customs duties	266	248
Excise duties	5,838	5,443
Capital taxes	3,498	1,762
Stamp duties	3,186	1,651
Income tax	13,572	13,177
Corporation tax	6,391	5,066
Value-added tax	14,497	13,430
Levies	3	1
Non-tax revenue	638	847
Total	47,887	41,624
Current expenditure		
Industry and labour	1,499	1,547
Agriculture	1,363	1,446
Fisheries, Forestry, Tourism	214	267
Health	14,281	15,356
Education	7,891	8,465
Social Welfare	15,498	17,807
Security	3,475	3,746
Other	4,386	4,749
Gross current	48,607	53,384
Less: Receipts, e.g. social security	11,647	12,626
Net total (including non-voted central fund)	36,960	40,758

VAT is 23·0% (reduced rates of 13·5%, 9·0% and 4·8%).

Ireland's budget deficit in 2011 was 13·1% of GDP (2010, 31·2%; 2009, 14·0%). The required target set by the EU is a budget deficit of no more than 3%.

Performance
Real GDP growth rates (based on OECD statistics):

2002	2003	2004	2005	2006	2007	2008	2009	2010	2011
5·7%	3·9%	4·4%	5·9%	5·4%	5·4%	−2·1%	−5·5%	−0·8%	1·4%

During the late 1990s Ireland had the fastest-growing economy in the European Union, with real GDP growth of 10·9% in 1997 and growth averaging 9·0% between 1995 and 1999. From a GDP per head of only 69% of the EU average in 1987, it was estimated to have risen to 148% of the EU average by 2007. Total GDP in 2011 was US$217·3bn.

Banking and Finance
In Oct. 2010 a new Central Bank of Ireland was created with a single fully-integrated structure and a unitary board—the Central Bank Commission. This replaced the boards of the Central Bank and Financial Services Authority of Ireland and the Irish Financial Services Regulatory Authority.

The Board of Directors of the Central Bank of Ireland consists of a Governor, appointed for a seven-year term by the President on the advice of the government, three *ex officio* directors (Director-General of Central Bank, Head of Financial Regulation and Secretary-General of Department of Finance) and between six and eight other directors, all appointed by the Minister for Finance. The *Governor* is Patrick Honohan. In 2009 the Bank's profit was €933·81m.; €745·47m. was paid to the Exchequer.

As at Nov. 2010, 420 credit institutions (including branches) were registered with the Central Bank.

At Sept. 2010 total assets of within-the-State offices of all credit institutions amounted to €1,312bn.

Gross external debt totalled US$2,123,300m. in June 2012.

There is a stock exchange in Dublin.

ENERGY AND NATURAL RESOURCES
In 2008, 3·8% of energy consumption came from renewables (wind power, solar power, hydro-electric power, tidal power, geothermal energy and biomass), compared to the European Union average of 10·3%. A target of 16% has been set by the EU for 2020.

Environment
Ireland's carbon dioxide emissions from the consumption and flaring of fossil fuels in 2008 were the equivalent of 10·7 tonnes per capita.

Electricity
The total generating capacity in 2004 was 5,592 MW, as averaged on a daily basis. This included wind generation, small renewable and small Combined Heat and Power (CHP). In 2003 there were approximately 1,804,680 customers connected to the network consuming 22,286 GWh.

Oil and Gas
Oil accounts for 56% of primary energy demand in Ireland while gas makes up 25% of demand. Over 0·6m. sq. km of the Irish continental shelf has been designated an exploration area for oil and gas; at the furthest point the limit of jurisdiction is 520 nautical miles from the coast. In the offshore there is a vast Continental Shelf in which a number of major basins and troughs have been identified. In March 2012 Providence Resources announced the discovery of a large oil field located 50 km off the Cork coast. Providing a flow in excess of 3,500 bbls per day, it is the first commercially viable oil field in Irish waters.

Natural gas reserves in 2007 totalled 10bn. cu. metres. Output in 2006 was 510m. cu. metres and consumption 4·6bn. cu. metres. 91% of natural gas supplies for the Irish Market are imported through the two sub-sea interconnectors connecting Ireland with Scotland and the remaining 9% is supplied from the Kinsale Head gas field, about 50 km off the south coast of Ireland. Deliveries from the Corrib gas field, about 80 km off the west coast, are scheduled to start by 2015.

Natural gas transmission and distribution is currently carried out by Bord Gáis Éireann (Irish Gas Board). The gradual liberalization of the Irish gas market, which started in July 2004, was completed in July 2007 when domestic as well as business users became free to select any licensed natural gas supplier.

Peat
The country has very little indigenous coal, but possesses large reserves of peat, the development of which is handled largely by Bord na Móna (Peat Board). To date, the Board has acquired and developed 85,000 ha. of bog and has 27 locations around the country. In 2006–07 the Board sold 2·4m. tonnes of milled peat for use in three milled peat electricity generating stations. 231,000 tonnes of briquettes were produced for sale to the domestic heating market. Bord na Móna also sold 1·8m. cu. metres of horticultural peat, mainly for export.

Minerals
Ireland has three zinc-lead mines, which in 2009 produced a combined total of 357,000 tonnes of zinc in concentrate and 43,000 tonnes of lead in concentrate, together with some silver (in lead). Ireland is Europe's leading zinc mine producer (36% of European output in 2009) and ranks tenth in the world for total zinc production. Gross value of metal production (lead and zinc) in Ireland in 2009 was €314m. Production of gypsum is significant (400,000 tonnes in 2009). Aggregate production in 2009 was an estimated 65m. tonnes. In June 2010, 37 companies held 514 prospecting licences; a total of €20·2m. was spent on exploration in 2009. The main target is base metals but there is also interest in gold.

Agriculture

The Central Statistics Office's *Quarterly National Household Survey* showed that in the quarter of March–May 2012 there were 87,100 people whose primary source of income was from agriculture, forestry and fisheries. In 2010 a total of 272,016 people worked on farms on a regular basis, working the equivalent of 168,387 full-time jobs. There were 139,860 farm holdings in Ireland, almost all of which were family farms. Average farm size was 32·7 ha. The land area of Ireland is 6·9m. ha., of which 4·6m. ha. were used for agriculture in 2011.

Agriculture, fisheries and forestry represented 2·5% of gross value added in 2010 (provisional). More than 80% of the agricultural area is devoted to pasture, hay and grass silage (3·74m. ha.), 10% to rough grazing (0·45m. ha.) and 8% to crop production (0·37m. ha.). In 2011 beef and milk production accounted for 57% of goods output at producer prices.

In 2011: barley accounted (in ha.) for 180,600; wheat, 94,200; oats, 21,400; oilseed rape, 12,400; potatoes, 10,400. Production figures in 2011 (in 1,000 tonnes) were: barley, 1,412; wheat, 669; potatoes, 420; oats, 148; oilseed rape, 28.

Goods output at producer prices including changes in stock for 2011 was estimated at €6·3bn.; operating surplus (aggregate income) was €2·4bn. Direct income payments, financed or co-financed by the EU, amounted to €1·8bn. It is estimated that net subsidies (subsidies on products plus subsidies on production less taxes on products and taxes on production) represented 76% of aggregate income. Livestock figures in 2010 were: 6,606,585 cattle; 5,078,952 sheep; 1,518,332 pigs; 11,025,441 poultry.

Forestry

Total forest area in 2010 was 0·74m. ha. (11% of total land area). Timber production in 2007 was 2·71m. cu. metres.

Fisheries

In 2012 approximately 11,100 people were engaged full- or part-time in the sea fishing industry; in 2004 the fishing fleet consisted of 1,425 vessels. The quantities and values of fish landed in 2004 were: wetfish, 256,170 tonnes, value €106·2m.; shellfish, 62,331 tonnes, value €80·1m. Total quantity (2004): 318,501 tonnes; total value, €186·4m. The main types of fish caught in 2004 were mackerel (63,000 tonnes), blue whiting (49,000 tonnes), horse mackerel (37,000 tonnes) and herring (29,000 tonnes). More than 98% of fish caught is from sea fishing.

INDUSTRY

The leading companies by market capitalization in Ireland in March 2012 were: Ryanair Holdings (US$8·8bn.); Élan, a pharmaceuticals and biotechnology company (US$8·6bn.); and Kerry Group, a chilled food manufacturer (US$8·1bn.).

The census of industrial production for 2005 gives the following details of the values (in €1m.) of gross and net output for the principal manufacturing industries.

	Gross output	Net output
Mining and quarrying	1,436	742
Manufacture of food products, beverages and tobacco	18,226	10,187
Manufacture of wood and wood products	1,061	417
Manufacture of pulp, paper and paper products; publishing and printing	13,661	12,219
Manufacture of chemicals, chemical products and man-made fibres	28,943	23,914
Manufacture of rubber and plastic products	1,356	634
Manufacture of other non-metallic mineral products	2,027	1,004
Manufacture of basic metals and fabricated metal products	2,050	881
Manufacture of machinery and equipment n.e.c.	1,784	871
Manufacture of electrical and optical equipment	29,371	12,910
Manufacture of transport equipment	1,128	553

	Gross output	Net output
Manufacturing n.e.c.	2,573	665
Electricity, gas and water supply	5,425	2,873
Total (all industries)	109,576	68,160

In 2004 gross output was €103,751m. and net output €66,289m.

Labour

The total labour force in 2008 was 2,239,600, of whom 126,700 were out of work. The unemployment rate in 2008 was 6·0%, down from nearly 16% in 1993. However, it rose sharply as a consequence of the global economic crisis and in Dec. 2012 was 14·7%. Of those at work in 2008, 1,246,400 were employed in the services sector, 492,000 in the industrial sector and 113,800 in the agricultural sector. Employment rose by approximately 40% between 1998 and 2008. In 2001 there were only 69,400 unemployed people, down from 226,000 in 1987, although this figure has risen steadily back up since then and in 2009 exceeded the 1987 total, with 264,600 unemployed. Ireland, along with the UK and Sweden, decided to open its labour market to nationals of the new EU member states in May 2004. Poles in particular went to Ireland following the EU expansion and by 2006 had become the second largest ethnic minority after the British; there were 63,276 Polish citizens in Ireland at the time of the 2006 census. On 1 Feb. 2011 the minimum hourly wage was lowered to €7·65 from €8·65. The normal retirement age is 65 years.

INTERNATIONAL TRADE

Imports and Exports

Value of imports and exports of merchandise for calendar years (in €1m.):

	2004	2005	2006	2007	2008
Imports	51,105	57,465	60,857	63,486	57,585
Exports	84,410	86,732	86,772	89,226	86,394

The values of the chief imports and total exports are shown in the following table (in €1m.):

	Imports		Exports	
	2005	2006	2005	2006
Animal and vegetable oils and waxes	130	163	18	20
Beverages and tobacco	776	805	1,103	1,359
Chemicals	7,419	7,964	40,421	39,696
Live animals and food	3,681	4,086	6,380	7,034
Machinery and transport equipment	25,002	25,571	22,710	23,186
Manufactured articles	7,086	7,360	9,066	8,651
Manufactured goods	4,947	5,569	1,755	1,714
Mineral fuels and lubricants	4,020	4,719	616	562
Raw materials	935	1,069	1,077	1,493

Ireland is one of the most trade-dependent countries in the world. However, export levels have fallen slightly since the peak in 2002, when merchandise exports amounted to €93·7bn., generating a trade surplus of €38·0bn. In 2006 merchandise imports from other European Union countries accounted for 60·4% of total imports while merchandise exports to other EU countries accounted for 63·5% of total exports. Information technology has become increasingly important, and by 1999 Ireland had become the largest exporter of software products in the world.

Import and export totals for Ireland's top ten export markets in 2005 and 2006 (€1m.):

	Imports		Exports	
	2005	2006	2005	2006
Belgium	1,061	1,212	13,540	12,217
China	5,164	5,746	1,766	1,794
France	1,980	2,219	5,713	5,079
Germany	4,512	5,012	6,625	6,970

	Imports		Exports	
	2005	2006	2005	2006
Italy	1,254	1,563	3,713	3,613
Japan	2,105	1,731	2,233	1,980
Netherlands	2,287	2,445	3,610	3,402
Spain	845	990	2,980	3,261
United Kingdom	18,271	19,425	15,352	15,566
United States of America	8,000	6,808	15,475	16,182

In 2008 exports accounted for 81% of GDP.

COMMUNICATIONS

Roads
At 31 Dec. 2012 there were 95,811 km of public roads, consisting of 5,515 km of National Primary Roads (including 1,187 km of motorway), 2,716 km of National Secondary Roads, 11,607 km of Regional Roads and 78,773 km of Local Roads.

Number of licensed motor vehicles at 31 Dec. 2011: private cars, 1,887,810; public service vehicles, 33,405; goods vehicles, 320,966; agricultural and industrial vehicles, 71,677; motorcycles, 36,582; other vehicles, 74,716. In 2011 a total of 186 people were killed in road accidents.

Rail
The total length of railway open for traffic in 2009 was 1,919 km (52 km electrified), all 1,600 mm gauge. A massive investment in public transport infrastructure is taking place in Ireland. The second National Development Plan running from Jan. 2007 to Dec. 2013 allows for €12·9bn. to be invested in public transport, particularly in the Greater Dublin area.

Railway statistics for years ending 31 Dec.	2005	2006
Passengers (journeys)	37,700,000	43,300,000
Receipts (€1)	181,860,000	227,696,000
Expenditure (€1)	446,785,000	439,113,000

A light railway system was launched in Dublin in 2004.

Civil Aviation
Aer Lingus and Ryanair are the two major airlines operating in Ireland.

Aer Lingus was founded in 1936 as a State-owned enterprise. Its principal business is the provision of passenger and cargo services to the UK, Europe and the USA. It was privatized in 2006, with Ryanair now holding a 29·8% stake in the company and the government a 25·1% stake. Ryanair began operations in 1985 and now operates to a range of destinations in the UK and Europe. In 2005 Ryanair carried 33·4m. passengers (all on international flights); passenger-km totalled 31·2bn. The total number of passengers carried by Ryanair on international flights in 2005 was the second highest of any airline, behind Lufthansa. However, it has in the meantime overtaken Lufthansa to become the airline carrying the highest number of international passengers.

In addition to Aer Lingus and Ryanair, there are 16 other independent air transport operators. The main operators in this group are Aer Arann Express and Cityjet.

The principal airports (Dublin, Shannon and Cork) are operated by the Dublin Airport Authority plc. In 2006 Dublin handled 21·2m. passengers (an increase of 14·9% on 2005) and 150,000 tonnes of freight. Shannon handled 3·6m. passengers (increase of 10·2%) and 45,015 tonnes of freight. Cork was the third busiest, with 3·0m. passengers (increase of 10·3%) and 8,300 tonnes of freight.

There are six privately owned regional airports. The government part funds the scheduled services from Dublin to five of these airports and to the City of Derry airport in Northern Ireland to ensure efficient and speedy access to the more isolated regions of the state for both business and tourist travellers.

Shipping
In Jan. 2009 there were 32 ships of 300 GT or over registered, totalling 122,000 GT. Total cargo traffic passing through the country's ports amounted to 41,880,000 tonnes in 2009 (down from 51,081,000 in 2008). Dublin handled 18·6m. tonnes of cargo in 2009 and Cork 8·0m. tonnes.

Inland Waterways
The principal inland waterways open to navigation are the Shannon Navigation (270 km), which includes the Shannon-Erne Waterway (Ballinamore/Ballyconnell Canal), and the Grand Canal and Barrow Navigation (249 km). Merchandise traffic has now ceased and navigation is confined to pleasure craft operated either privately or commercially. The Royal Canal (146 km) from Dublin to Mullingar (53 km) was reopened for navigation in 1995.

Telecommunications
The Minister for Communications, Energy and Natural Resources, a member of the government, has overall policy responsibility for the development of the sector. The core policy objective is to contribute to sustained macro-economic growth and competitiveness, and ensure that Ireland is best placed to avail of the emerging opportunities provided by the information and knowledge society.

The communications sector within the department is divided into five divisions: the business and technology division, the communications development and electronic commerce division, communications policy division, knowledge society division and postal division.

Ireland's telecommunications sector has been fully liberalized with effect from 1 Dec. 1998 when the last remaining elements of Telecom Éireann's (now called eircom) exclusive privilege were removed. All elements of the market are now open to competition from other licensed operators. The largest mobile telephone operators in terms of subscribers are Vodafone Ireland and O2 Ireland.

The Government has also sold the state's entire remaining stake of 50·1% in eircom by way of an initial public offering of shares in the company. The sale took place in July 1999.

eircom plc—Operational Information
The dominant operator in the telecommunications sector is eircom plc (previously Telecom Éireann). Telecom Éireann was a statutory body set up under the Postal and Telecommunications Services Act, 1983. In 1996, 20% of the State's holding was sold to KPN/Telia, a Dutch–Swedish consortium, who had an option of a further 15%, which was taken up in July 1999. In 1998 the government concluded an Employee Share Ownership Scheme under which 14·9% of the company was to be made available to employees and also held an Initial Public Offer (IPO) of shares in the company in July 1999. In Oct. 1999 the newly-privatized Telecom Éireann became eircom plc.

The level of network digitalization is 100%. In 2008 there were 2,204,000 main (fixed) telephone lines. In the same year mobile phone subscribers numbered 5,357,000 (1,207·4 per 1,000 persons). In 2008, 25m. text messages were sent every day. Ireland had 2·78m. internet users in 2008. The fixed broadband penetration rate stood at 21·1 subscribers per 100 inhabitants in Dec. 2010. In March 2012 there were 2·1m. Facebook users.

SOCIAL INSTITUTIONS

Justice
The Constitution provides that justice shall be administered in public in Courts established by law by Judges appointed by the President on the advice of the government. The jurisdiction and organization of the Courts are dealt with in the Courts (Establishment and Constitution) Act, 1961, the Courts (Supplemental Provisions) Acts, 1961–91, and the Courts and Court Officers Acts, 1995–2002. These Courts consist of Courts of First Instance and a Court of Final Appeal, called the Supreme

Court. The Courts of First Instance are the High Court with full original jurisdiction and the Circuit and the District Courts with local and limited jurisdictions. A judge may not be removed from office except for stated misbehaviour or incapacity and then only on resolutions passed by both Houses of the Oireachtas. Judges of the Supreme Court and High Court are appointed from among practising barristers or solicitors of not less than 12 years standing or by the elevation of an existing member of the judiciary. Judges of the Circuit Court are appointed from among practising barristers or solicitors of not less than ten years standing or a County Registrar who has practised as a barrister or solicitor for not less than ten years before being appointed to that post or by the elevation of a District Court Judge. Judges of the District Court are appointed from among practising barristers or solicitors of not less than ten years standing.

The Supreme Court, which consists of the Chief Justice (who is *ex officio* an additional judge of the High Court) and seven ordinary judges, may sit in two Divisions and has appellate jurisdiction from all decisions of the High Court. The President may, after consultation with the Council of State, refer a Bill, which has been passed by both Houses of the Oireachtas (other than a money bill and certain other bills), to the Supreme Court for a decision on the question as to whether such Bill or any provision thereof is repugnant to the Constitution.

The High Court, which consists of a President (who is *ex officio* an additional Judge of the Supreme Court) and 31 ordinary judges (or 32 when a High Court Judge is appointed as a Commissioner of the Law Reform Commission, as is currently the case), has full original jurisdiction in and power to determine all matters and questions, whether of law or fact, civil or criminal. In all cases in which questions arise concerning the validity of any law having regard to the provisions of the Constitution, the High Court alone exercises original jurisdiction. The High Court on Circuit acts as an appeal court from the Circuit Court.

The Court of Criminal Appeal consists of the Chief Justice or an ordinary Judge of the Supreme Court, together with either two ordinary judges of the High Court or the President and one ordinary judge of the High Court. It deals with appeals by persons convicted on indictment where the appellant obtains a certificate from the trial judge that the case is a fit one for appeal, or, in case such certificate is refused, where the court itself, on appeal from such refusal, grants leave to appeal. The decision of the Court of Criminal Appeal is final, unless that court, the Attorney-General or the Director of Public Prosecutions certifies that the decision involves a point of law of exceptional public importance, in which case an appeal is taken to the Supreme Court.

The Offences against the State Act, 1939 provides in Part V for the establishment of Special Criminal Courts. A Special Criminal Court sits without a jury. The rules of evidence that apply in proceedings before a Special Criminal Court are the same as those applicable in trials in the Central Criminal Court. A Special Criminal Court is authorized by the 1939 Act to make rules governing its own practice and procedure. An appeal against conviction or sentence by a Special Criminal Court may be taken to the Court of Criminal Appeal. On 30 May 1972 Orders were made establishing a Special Criminal Court and declaring that offences of a particular class or kind (as set out) were to be scheduled offences for the purposes of Part V of the Act, the effect of which was to give the Special Criminal Court jurisdiction to try persons charged with those offences.

The High Court exercising criminal jurisdiction is known as the Central Criminal Court. It consists of a judge or judges of the High Court, nominated by the President of the High Court. The Court tries criminal cases which are outside the jurisdiction of the Circuit Court.

The Circuit Court consists of a President (who is *ex officio* an additional judge of the High Court) and 33 ordinary judges. The country is divided into eight circuits. The jurisdiction of the court in civil proceedings is subject to a financial ceiling, save by consent of the parties, in which event the jurisdiction is unlimited. In criminal matters it has jurisdiction in all cases except murder, treason, piracy, rape, serious and aggravated sexual assault and allied offences. The Circuit Court acts as an appeal court from the District Court. The Circuit Court also has jurisdiction in the Family Law area such as divorce.

The District Court, which consists of a President and 54 ordinary judges, has summary jurisdiction in a large number of criminal cases where the offence is not of a serious nature. In civil matters the Court has jurisdiction in contract and tort (except slander, libel, seduction, slander of title and false imprisonment) where the claim does not exceed €6,348·69; in proceedings founded on hire-purchase and credit-sale agreements, the jurisdiction is also €6,348·69. The District Court also has jurisdiction in Family Law matters such as maintenance, custody, access and the issuing of barring orders. The District Court also has jurisdiction in a large number of licensing (intoxicating liquor) matters.

All criminal cases, except those of a minor nature, and those tried in the Special Criminal Court, are tried by a judge and a jury of 12. Generally, a verdict need not be unanimous in a case where there are not fewer than 11 jurors if ten of them agree on the verdict.

The Courts Service Act, 1998, provided for the transfer of responsibility for the day to day management of the Courts from the Minister for Justice, Equality and Law Reform to a new body known as the Courts Service. The Board of the Courts Service consists of 17 members including members of the judiciary, the legal profession, staff and trade union representatives, a representative of court users, a person with commercial/financial experience and a Chief Executive Officer. The Courts Service was formally established on 9 Nov. 1999. While the Minister retains political responsibility to the Oireachtas, the courts are now administered independently by the Board and CEO.

In 2009 the police force, the Garda Síochána, had a total staff of 14,547. There were 284,485 crimes recorded in 2009, of which 87 homicide offences. The National Juvenile Office received 23,952 referrals relating to 18,519 individual children during 2009. The population in penal institutions in Oct. 2007 was 3,325 (76 per 100,000 of national population).

Education

Education is compulsory from six to 16 years of age. In 2005 public expenditure on education came to 4·8% of GDP and 13·9% of total government spending. The adult literacy rate is at least 99%.

Elementary. Elementary education is free and was given in about 3,303 national schools (including 128 special schools) in 2008–09. The total number of pupils on rolls in 2008–09 was 498,914, including pupils in special schools and classes; the number of teachers of all classes was about 31,349 in 2008–09, including remedial teachers and teachers of special classes. The total expenditure for first level education during the financial year ended 31 Dec. 2009 was €3,133m. The total salaries for teachers for 2009, including superannuation etc., was €2,520m.

Special. Special provision is made for children with disabilities in special schools which are recognized on the same basis as primary schools, in special classes attached to ordinary schools and in certain voluntary centres where educational services appropriate to the needs of the children are provided. Integration of children with disabilities in ordinary schools and classes is encouraged wherever possible, if necessary with special additional support. There are also part-time teaching facilities in hospitals, child guidance clinics, rehabilitation workshops, special 'Saturday-morning' centres and home teaching schemes. Special schools (2008–09) numbered 128 with approximately 6,653 pupils. There were also some 9,668 pupils enrolled in about 1,131 special classes within ordinary schools. There is a National Education Officer for travelling children.

Secondary. Voluntary secondary schools are under private ownership and are conducted in most cases by religious orders. These schools receive grants from the State and are open to inspection by the Department of Education. The number of recognized secondary schools during the school year 2008–09 was 388, and the number of pupils in attendance was 184,329. There were 13,448 teachers in 2007–08.

Vocational Education Committee schools provide courses of general and technical education. Pupils are prepared for State examinations and for entrance to universities and institutes of further education. The number of vocational schools during the school year 2008–09 was 253, the number of full-time students in attendance was 103,732 and the number of teachers (2007–08) 8,106. These schools are controlled by the local Vocational Education Committees; they are financed mainly by State grants and also by contributions from local rating authorities and Vocational Education Committee receipts. These schools also provide adult education facilities for their own areas.

Comprehensive and Community Schools. Comprehensive schools which are financed by the State combine academic and technical subjects in one broad curriculum so that pupils may be offered educational options suited to their needs, abilities and interests. Pupils are prepared for State examinations and for entrance to universities and institutes of further education. The number of comprehensive and community schools during the school year 2008–09 was 91 and the number of students in attendance was 53,251. These schools also provide adult education facilities for their own areas and make facilities available to voluntary organizations and to the adult community generally.

The total current expenditure from public funds for second level and further education for 2003 was €2,304·8m.

Third-Level Education. The third-level education system funded by the State comprises the university sector, the technological sector and colleges of education—these are autonomous and self-governing. In addition there are a number of independent private colleges offering third-level qualifications.

Ireland has a binary system of higher education, designed to ensure maximum flexibility and responsiveness to the needs of students and to the wide variety of social and economic requirements. However, within each sector, a diversity of institutions offer differing types and levels of programmes. The Universities are essentially concerned with undergraduate and postgraduate programmes, together with basic and applied research. The main work of the Institutes of Technology is in undergraduate programmes, with a smaller number of postgraduate programmes and a growing involvement in research. Numbers in third-level education have expanded dramatically since the mid-1960s, from 21,000 full-time students in 1965 to 148,197 in 2008–09.

The total current expenditure from public funds on third-level education during the financial year ended 31 Dec. 2009 was €1·8bn.

There are seven universities in the Republic of Ireland. These are: NUI, Galway; NUI, Maynooth; University College Cork; University College Dublin; Trinity College, Dublin; University of Limerick; and Dublin City University. Four of the universities (NUI Galway, NUI Maynooth, UCC and UCD) are constituent universities of the National University of Ireland. The National University of Ireland also has a number of recognized colleges, including the Royal College of Surgeons in Ireland, National College of Art and Design and Shannon College of Hotel Management.

The Irish higher education university system offers degree programmes—at bachelor, master's and doctorate level—in the humanities, social sciences, scientific, technological and social sciences, and in the medical area. Typically teaching at undergraduate level is by way of a programme of lectures supplemented by tutorials and, where appropriate, practical demonstration and laboratory work. Master's degrees are usually taken by course work, research work or some combination of both. Doctoral degrees are awarded on the basis of research. Institutions award their own degrees using external examiners to ensure consistency of standards. The institutions also have continuing and some distance education programmes and also engage in research work.

There are 14 Institutes of Technology. They are Dublin Institute of Technology, Athlone IT, IT Blanchardstown, Cork IT, IT Carlow, Dundalk IT, Dun Laoghaire Institute of Art, Design and Technology, Letterkenny IT, Galway-Mayo IT, Limerick IT, IT Sligo, IT Tallaght, IT Tralee and Waterford IT.

HETAC (the Higher Education and Training Awards Council) was established on 11 June 2001, under the Qualifications (Education and Training) Act 1999. It is the successor to the National Council for Educational Awards (NCEA) and is the qualifications awarding body for third-level education and training institutions outside the university sector. The Qualifications (Education and Training) Act 1999 extended autonomy to allow Institutes of Technology to apply to HETAC for delegation of authority to make their own awards. Delegation of authority also allows institutes to validate their own programmes subject to the policies and criteria determined by HETAC and within the parameters of the National Framework of Qualifications.

Initial teacher education is provided by the Colleges of Education (primary teacher education) and the Universities (post-primary teacher education).

Health

Health boards are responsible for administering health services in Ireland. There are currently ten health boards established: three area health boards located in the eastern region under the guidance of the Eastern Regional Health Authority (ERHA) and seven regional health boards covering the rest of the country. Each health board is responsible for the provision of health and social services in its area. The boards provide many of the services directly and they arrange for the provision of other services by health professionals, private health service providers, voluntary hospitals and voluntary/community organizations.

A health service reform programme is currently being implemented which will result in the most significant structural changes in the Irish health services in recent decades. The existing health boards will be replaced by a single Health Service Executive (HSE) with four regional administrative areas. A Health Information and Quality Authority (HIQA) will also be established.

Everybody ordinarily resident in Ireland has either full or limited eligibility for the public health services.

A person who satisfies the criteria of a means test receives a medical card, which confers Category 1 or full eligibility on them and their dependants. This entitles the holder to the full range of public health and hospital services, free of charge, i.e. family doctor, drugs and medicines, hospital and specialist services as well as dental, aural and optical services. Maternity care and infant welfare services are also provided.

The remainder of the population has Category 2 or limited eligibility. Category 2 patients receive public consultant and public hospital services subject to certain charges. Persons in Category 2 are liable for a hospital in-patient charge of €75 per night up to a maximum of €750 in any 12 consecutive months. Persons in Category 2 are liable for a charge of €100 if they attend the Accident and Emergency Department of a hospital or receive out-patient services without a letter from a General Practitioner.

The Long Term Illness Scheme entitles persons to free drugs and medicines, which are prescribed in respect of 15 specific illnesses. The needs of individuals with significant or ongoing medical expenses are met by a range of other schemes, which provide assistance towards the cost of prescribed drugs and medicines. The *Drug Payment Scheme* was introduced on 1 July 1999 and replaced the Drug Cost Subsidisation Scheme (DCSS) and the Drug Refund Scheme (DRS). Under this scheme no individual or family will have to pay more than €144 in any calendar month for approved prescribed drugs, medicines and appliances for use by the person or his/her family in that month.

Services for People with Disabilities: The Department of Health and Children provides, through the health boards and the Eastern Regional Health Authority, a wide range of services for people with disabilities. These include day care, home support (including personal assistance services), therapy services, training, employment, sheltered work and residential respite care. The following allowances and grants for eligible people with disabilities come under the aegis of the Department of Health and Children and are administered by the health boards and the Eastern Regional Health Authority:

Disability Allowance—payable to persons who have an injury, disease, illness or disability that substantially restricts their capacity to work. The full rate of €188·00 per week has, since Jan. 2007, also been made available to long-term residents in institutions.

Blind Welfare Allowance—provides supplementary financial support to unemployed blind persons who are not maintained in an institution and who are in receipt of a Department of Social, Community and Family Affairs payment, such as Disability Allowance, Blind Pension or Old Age Pension.

Rehabilitative Training Bonus—payable to persons who are attending approved rehabilitative training programmes. The payment of €31·80 replaced the Disabled Persons Rehabilitation Allowance (DPRA) from 1 Aug. 2001.

Domiciliary Care Allowance (DCA)—provides home care for severely disabled or mentally handicapped children up to the age of 16. The maximum rate of DCA in Jan. 2013 was €309·50 per month.

Respite Care Grant (RCG)—an annual payment of €1,375 (per person cared for) to help carers obtain respite care.

Health Contributions—A health contribution of 2% of income is payable by those with Category 2 eligibility. Employers meet the levy in respect of those employees who have a medical card.

In 2003 there were 59 publicly funded acute hospitals in operation with an 85% occupancy rate. The average number of in-patient beds available for use over the year was 12,300. There were 96,499 wholetime equivalent numbers employed in health board/regional authority and voluntary/joint board hospitals and homes for the mentally handicapped at 31 Dec. 2003. Of these 6,792 were medical/dental staff, 12,690 were health and social care professionals and 33,766 were nursing staff. In 2008 Ireland spent 8·7% of its GDP on health.

Welfare

The Department of Social Protection is responsible for the day-to-day administration and delivery of social welfare schemes and services through a network of local, regional and decentralized offices. The Department's local delivery of services is structured on a regional basis. There are a total of seven regions, with offices in Waterford, Cork, Galway, Longford, Sligo and two in the Dublin area.

There are, in addition, three statutory agencies under the aegis of the Department:

—the *Pensions Board*, which has the function of promoting the security of occupational pensions, their development and the general issue of pensions coverage.

—the *Citizens Information Board*, which has the function of ensuring that all citizens have easy access to the highest quality of information, advice and advocacy on social services.

—*Office of the Pensions Ombudsman*, which investigates and decides complaints and disputes involving occupational pension schemes and Personal Retirement Savings Accounts (PRSAs). The Ombudsman is independent of the Minister and the Department in the performance of his functions.

A programme of significant reform has taken place in the Department including developments such as: the transfer of the General Register Office to the Department of Social Protection in Jan. 2008; the transfer of the Community Welfare Service from the Health Service Executive to the Department in Oct. 2011; and the integration of FÁS Employment and Community Employment Services into the operations of the Department of Social Protection in Jan. 2012.

The following schemes and services have also transferred to the Department of Social Protection: the Rural Social Scheme; the Community Services Programme; Domiciliary Care Allowance; and the Redundancy and Insolvency Scheme.

In 2010 social welfare expenditure accounted for 13·5% of GDP.

Social Welfare Schemes. The social welfare supports can be divided into three categories:

—*Social Insurance (Contributory)* payments made on the basis of a Pay Related Social Insurance (PRSI) record. Such payments are funded by employers, employees and the self-employed. Any deficit in the fund is met by Exchequer subvention.

—*Social Assistance (Non Contributory)* payments made on the basis of satisfying a means test. These payments are financed entirely by the Exchequer.

—*Universal payments* such as Child Benefit or Free Travel, which do not depend on PRSI or a means test.

State Pension (Contributory). The State Pension (Contributory) is available to those aged 66 who have social insurance coverage beginning before 56 years of age.

People who reached pension age before 6 April 2002 must have 156 qualifying paid contributions (a total of three years although they do not have to be consecutive). This means that to be eligible people must have actually paid full-rate contributions.

People who reached pension age on or after 6 April 2002 but before 6 April 2012 need to have 260 paid contributions (effectively five years contributions although they need not be consecutive). However, anyone who paid voluntary contributions on or before 6 April 1997 only needs to have 156 paid contributions providing they have a yearly average of at least 20 contributions.

People who reach pension age on or after 6 April 2012 will need to have 520 paid contributions (ten years paid contributions). In this case, not more than 260 of the 520 contributions may be voluntary contributions. However, anyone who paid voluntary contributions on or before 6 April 1997 and has a yearly average of ten contributions may also meet the requirements. They will need to have a total of 520 contributions, but only 156 need to be compulsory paid contributions.

There is also a means-tested non-contributory pension available to citizens aged 66 or older with limited means.

The Social Welfare and Pensions Act 2011 made a number of changes to the qualifying age for state pensions. The qualifying age will rise to 66 in 2014, 67 in 2021 and 68 in 2028. For recipients born on or after 1 Jan. 1948 the minimum qualifying state pension age will be 66; for recipients born on or after 1 Jan. 1955 the minimum qualifying state pension age will be 67; for recipients born on or after 1 Jan. 1961 the minimum qualifying state pension age will be 68.

The *Social Welfare Appeals Office (SWAO)* is an independent office responsible for determining appeals against decisions on social welfare entitlements.

RELIGION

According to the census of population taken in 2006 the principal religious professions were as follows:

	Leinster	Munster	Connacht	Ulster (part of)	Total
Roman Catholics	1,949,269	1,047,502	453,086	231,589	3,681,446
Church of Ireland (including Protestants)	73,312	28,788	11,200	12,285	125,585
Other Christian religion n.e.c.	19,117	6,320	2,600	1,169	29,206
Presbyterians	10,310	2,704	1,401	9,131	23,546
Muslims	22,176	6,423	3,109	831	32,539
Methodists	7,008	3,069	1,128	955	12,160
Orthodox	15,208	3,612	1,415	563	20,798
Other stated religions	37,557	13,078	5,080	2,213	57,928
Not stated or no religion	161,166	61,844	25,102	8,528	256,640

Seán Brady (b. 1939) is the Roman Catholic Cardinal of Armagh and Primate of All Ireland. In Feb. 2013 there were two cardinals.

In May 1990 the General Synod of the Church of Ireland voted to ordain women.

CULTURE

World Heritage Sites

There are two UNESCO sites in Ireland: Archaeological Ensemble of the Bend of the Boyne (inscribed in 1993), the three principal sites of the Brúna Bóinne Complex, a major centre of prehistoric megalithic art; Skellig Michael (1996), a monastic complex on a craggy island from the 7th century.

Press

In 2008 there were 11 dailies and nine paid-for Sunday newspapers (all in English) with a combined circulation of 2,129,000. In 2008 a record 1,902 book titles were published in Ireland (1,807 in 2007).

Tourism

Total number of overseas tourists in 2008 was 7,839,000 (a 2·2% fall from 2007). In 2008 earnings from all visits to Ireland, including cross-border visits, amounted to €4,781m. 49% of visits in 2008 were from Great Britain. Irish residents made 7,877,000 visits abroad in 2008 (a 2·1% increase on 2007).

Festivals

Ireland's national holiday, St Patrick's Day (17 March), is celebrated annually. Among the most popular festivals are Clonmel Junction arts festival (July), Galway Arts Festival (July), Ballyshannon Folk and Traditional Festival in Co. Donegal (July–Aug.), Fleadh Cheoil na hEireann (national festival of music) in Co. Cavan (Aug.), Dublin Theatre Festival and Dublin Fringe Festival (Sept.–Oct.), Wexford Opera Festival (Oct.) and Cork Jazz Festival (Oct.).

DIPLOMATIC REPRESENTATIVES

Of Ireland in the United Kingdom (17 Grosvenor Pl., London, SW1X 7HR)
Ambassador: Bobby McDonagh.

Of the United Kingdom in Ireland (29 Merrion Rd, Ballsbridge, Dublin 4)
Ambassador: Dominick Chilcott, CMG.

Of Ireland in the USA (2234 Massachusetts Ave., NW, Washington, D.C., 20008)
Ambassador: Michael Collins.

Of the USA in Ireland (42 Elgin Rd, Ballsbridge, Dublin 4)
Ambassador: Daniel M. Rooney.

Of Ireland to the United Nations
Ambassador: Anne Anderson.

Of Ireland to the European Union
Permanent Representative: Rory Montgomery.

FURTHER READING

Central Statistics Office. *National Income and Expenditure* (annual), *Statistical Abstract* (annual), *Census of Population Reports* (quinquennial), *Census of Industrial Production Reports* (annual), *Trade and Shipping Statistics* (annual and monthly), *Trend of Employment and Unemployment*, *Reports on Vital Statistics* (annual and quarterly), *Statistical Bulletin* (quarterly), *Labour Force Surveys* (annual), *Trade Statistics* (monthly), *Economic Series* (monthly).

Adshead, Maura and Tonge, Jonathan, *Politics in Ireland: Convergence and Divergence in a Two-Polity Island.* 2009
Ardagh, J., *Ireland and the Irish: a Portrait of a Changing Society.* 1994
Bartlett, Thomas, *Ireland: a History.* 2010
Chubb, B., *Government and Politics in Ireland.* 3rd ed. 1992
Cronin, Mike, *A History of Ireland.* 2001
Cronin, Mike, Gibbons, Luke and Kirby, Peadar, (eds.) *Reinventing Ireland: Culture, Society and the Global Economy.* 2002
Delanty, G. and O'Mahony, P., *Rethinking Irish History: Nationalism, Identity and Ideology.* 1997
Foster, R. F., *The Oxford Illustrated History of Ireland.* 1991
Gallagher, Michael and Marsh, Michael, (eds.) *How Ireland Voted 2011: The Full Story of Ireland's Earthquake Election.* 2011
Garvin, T., *1922: The Birth of Irish Democracy.* 1997
Harkness, D., *Ireland in the Twentieth Century: a Divided Island.* 1995
Kirby, Peadar, *The Celtic Tiger in Collapse.* 2nd ed. 2010
Kostick, C., *Revolution in Ireland – Popular Militancy 1917–1923.* 1997
Laffan, Brigid and O'Mahony, Jane, *Ireland and the European Union.* 2008
Lalor, Brian, (ed.) *The Encyclopedia of Ireland.* 2003
Lynch, David J., *When the Luck of the Irish Ran Out: The World's Most Resilient Country and its Struggle to Rise Again.* 2010
O'Beirne Ranelagh, J., *A Short History of Ireland.* 2nd ed. 1999
O'Malley, Eoin, *Contemporary Ireland.* 2011
O'Sullivan, Michael J., *Ireland and the Global Question.* 2006
Patterson, Henry, *Ireland Since 1939: The Persistence of Conflict.* 2006
Vaughan, W. E. (ed.) *A New History of Ireland*, 6 vols. 1996
Wyndham, Andrew Higgins, (ed.) *Re-Imagining Ireland.* 2006

National Statistical Office: Central Statistics Office, Skehard Road, Cork. *Director-General:* Pádraig Dalton.
Website: http://www.cso.ie

ISRAEL

LEBANON
GOLAN HEIGHTS
Haifa
Sea of Galilee
SYRIA
Mediterranean Sea
Netanya
TEL AVIV
WEST BANK
Jerusalem
ISRAEL
GAZA STRIP
Dead Sea
Beersheba
EGYPT
JORDAN

0   25 mi
0   25 km

© Research Machines plc 2006

Medinat Israel
(State of Israel)

Capital: Jerusalem
Population projection, 2015: 8·06m.
GNI per capita, 2011: (PPP$) 25,849
HDI/world rank: 0·888/17
Internet domain extension: .il

KEY HISTORICAL EVENTS

A settled agricultural community by 6000 BC, the oasis of Jericho is possibly the world's oldest continuously inhabited settlement. Canaan—probably derived from 'Land of Purple', from the purple sea snail dye—described the Eastern Mediterranean coast and hinterland from the 3rd millennium BC. As part of the Fertile Crescent, it became an important caravan route between Egypt and Mesopotamia. 'Canaanite' has come to be associated with the Semitic group of languages and peoples of the pre-Classical Levant.

In the reign of Pharaoh Pepi I (*c.* 2313–2279 BC), Canaan was invaded five times by Egyptian forces. Egyptian authority collapsed

in the 17th century, marking the end of the Middle Kingdom. Egyptian control was re-established with the reunification of Egypt in the 16th century. Thutmose III (1479–1425 BC), campaigning against the Mitanni Kingdom in Syria, defeated a Canaanite coalition at the Battle of Megiddo, subjugating Canaan and deporting thousands to Egypt. Egyptian power was challenged by the Hittites of Anatolia until Ramesses II concluded a peace treaty (the first recorded in the world) with the Hittite King Hattusilis III in 1258 BC, setting the border in northern Canaan.

The Israelite (or Hebrew) group occupied the hills of southern Canaan by the late 13th century. Around 1200 BC the Eastern Mediterranean littoral was attacked by the 'Sea Peoples' (probably including the Philistines), who destroyed coastal cities and settled on the coastal plain. The Israelite kingdom was formed from tribes supposedly returned from captivity in Egypt. In the late 11th century, Saul became king but it was his successor, David, who greatly expanded the Israelite state over most of southern Canaan. With Hittite and Egyptian power at low ebb, David conquered the trans-Jordanian states of Ammon, Edom and Moab, subjected Aram (lower Syria) to vassalage and made Jerusalem his capital. After the reign of Solomon (mid-10th century), who built the Temple of Jerusalem, the kingdom split into two: Judah in the south and the more populous Israel, centred on Samaria, in the north.

Having refused to pay tribute, Israel was conquered by Assyria's Sargon II in 722 BC and many of its people were deported; subsequent inhabitants of the Assyrian province of 'Samerina' became known as Samaritans, a mixed race of Israelites and immigrants from Mesopotamia and Persia. Sargon also besieged Jerusalem but was distracted by a Babylonian uprising. The resurgent Babylonians conquered Judah in 586 BC, having destroyed Philistia in 605 to clear access to Egypt. Nebuchadnezzar II had taken Jerusalem the previous year, deporting much of the Judaean (Jewish) nobility to Babylon. Having conquered Babylon in 539 BC, Cyrus II of Persia allowed the return of the Jews to Jerusalem, as Persian vassals, and the rebuilding of the Temple. Persia's defeat by Alexander the Great of Macedon brought the region, by then known as Palestine (derived from Philistia), under Hellenistic control. The Hellenistic period saw an influx of Arab groups, including the Nabataeans, who replaced the Edomites south of the Dead Sea.

Roman Rule

A revolt against the religious intolerance of the Seleucid King Antiochus IV began in 167 BC, led by Judas Maccabaeus, who established the Hasmonean Dynasty in Judaea. Relations with the Samaritans, who also followed the Torah (the first five books of the Hebrew bible), deteriorated when the Hasmonean King John Hyrcanus destroyed the Samaritan Temple at Mount Gerizim in 128 BC. The entire region was conquered for Rome by Pompey in 67 BC; Judaea, including Samaria, was administered as a client kingdom. After the Parthian invasion of Judaea in 40 BC, an Idumaean, Herod, was installed by Rome as king of Judaea. On Herod's death in 4 BC, the kingdom was split amongst three of his sons, who ruled as tetrarchs. Herod's grandson, Herod Agrippa, was granted a reunited Judaea by Emperor Claudius in AD 41, as a reward for supporting Claudius' claim to the imperial throne. However, Herod Agrippa was assassinated in AD 44 and Judaea placed under a Roman procurator. Jewish resentment against loss of autonomy grew until the Great Jewish Revolt (AD 66–73), which was brutally suppressed. Jerusalem was destroyed and hundreds of thousands were massacred or sold into slavery.

Jewish rebellions across the East in 115 (the Kitos War) were quickly suppressed. Emperor Hadrian's attempts to enforce cultural uniformity across the Empire included rebuilding Jerusalem as Aelia Capitolina and forbidding Jewish custom. Simon Bar Kokhba, supported by the Sanhedrin (Jewish sages), led a major revolt in AD 132 and established a Jewish government in Jerusalem. However, Roman armies prevailed in 135, with the death of around half a million Jews. Hadrian reacted to the rebellion by suppressing Judaism, banning Jews from Aelia Capitolina, deporting large numbers as slaves and renaming the province Syria Palaestina; the province was split in three around 390.

Christianity

Under the Christian Byzantine Empire, Palestine became a centre of Christianity (Jerusalem was recognized as a patriarchate in 451), bringing pilgrims and prosperity. It also received lavish imperial patronage, such as Constantine's Church of the Holy Sepulchre (c. 326). The Samaritans made a bid for independence in 529 but were crushed by Justinian I and the Ghassanid Arabs. Persecuted by Christians, Jews and later by Muslims, Samaritan numbers dwindled over the following centuries. Byzantine administration of Palestine ended temporarily during the Persian occupation of 614–28; Jerusalem was sacked, its churches burned and the city turned over to the Jews. A spectacular campaign in 628, led by Emperor Heraclius, forced the Persians to cede Palestine and Syria. However, Byzantine rule ended permanently after the Arabs conquered the region; Jerusalem was taken in 638.

The Arabs retained the existing system of administration in the provinces of Jund Filastine (the south) and Jund Urdunn (the north). Taxes and restrictions on religious practice and office-holding imposed on non-Muslims caused large-scale conversions. The Ummayad caliphs moved the capital to Damascus and built the Dome of the Rock on the site of the Jewish Temple in Jerusalem in the 690s. The Christian and Jewish communities of Palestine were partly administered by their own religious leaders. Under the Ummayads' successors, the Abbasids, the capital moved to Baghdad in 762, drawing Asian trade away from Palestine. Fragmentation of the Caliphate in the 9th century saw Egyptian independence under the Tulunids, who seized Palestine and Syria in 878. Although Palestine was retaken in 906 for the Caliphate, in 935 it again fell to Egypt, this time under the Ikhshid Dynasty and, in 970, to its successors, the Fatimids.

Crusades

1070 saw the arrival of the Seljuk Turks, who rapidly overran the Byzantine East. Seljuk restrictions on Christian pilgrimage led to the European Crusader invasions of Palestine and Syria in the 12th century. Having been wrested from the Seljuks by the Fatimids in 1098, Jerusalem was taken the following year by a Crusader army, which massacred the population. Baldwin, count of Edessa, became king of Jerusalem in 1100. Responding to Crusader threats to Mecca (Makkah), Saladin (Salah ad-Din), the Kurdish sultan of Egypt, recaptured Jerusalem in 1187, bringing Palestine under the Ayyubid Dynasty. A treaty in 1192 with Richard I of England allowed Christian pilgrimage to Jerusalem and secured the rump Crusader states along the coast. A treaty of 1229 gave much of Palestine (the Kingdom of Jerusalem) to the Holy Roman Emperor, Frederick II, though Jerusalem was destroyed by Central Asian Khwarezmians in 1244, on behalf of the Ayyubids. The fall of Acre (Akko) to the Mamluks, rulers of Egypt, in 1291 ended Crusader rule in the Holy Land.

Under Mamluk suzerainty, Palestine was administered by Muslim emirates. Economic decline was exacerbated by the arrival of the Black Death in 1351. Although the Mamluk sultanate successfully held off Mongol invasions, Palestine fell to the Ottomans in 1516, bringing it (as part of the Damascus-Syria province) and most of the Islamic world under the rule of Turkish İstanbul. Suleyman the Magnificent rebuilt Jerusalem's walls in 1537.

Zionism

Palestine was briefly invaded in 1799 by Napoleon Bonaparte of France, who had occupied Egypt. Muhammad Ali, the renegade Ottoman viceroy of Egypt, invaded Palestine and Syria in 1831, defeating the Ottoman army. However, British intervention at Beirut, on behalf of the sultan, forced the viceroy's withdrawal to Egypt. Jews from central and eastern Europe arrived in Palestine from the 1880s as part of a nascent Zionist movement. In 1897 the first Zionist Congress met in Basle, Switzerland. Attempts were made in vain to gain the approval of Sultan Abdul Hamid II for Jewish settlement. However, by 1914, about 85,000 Jews were living in Palestine, many on agricultural collectives (kibbutz), in part funded by Western Europe's Jewry. Tel Aviv (originally called Ahuzat Bayit 'homestead') was founded by Jews in 1909 as a dormitory settlement for workers in Jaffa.

Britain, France and Russia declared war on the Ottoman Empire in Nov. 1914, in retaliation for its co-operation with Germany. Having repelled Ottoman attacks on the Suez Canal, British forces invaded Ottoman Palestine, seizing Rafah in Jan. 1917. Two abortive attacks on Gaza were followed by British-led success at Beersheba in Oct. 1917, leading to the fall of Gaza in Nov. and Jerusalem in Dec. The British won a major victory at Megiddo in Sept. 1918, effectively ending Ottoman rule in Palestine. The British were granted Palestine and Transjordan as mandates under the League of Nations, established in 1919 at the Versailles Peace Conference.

Britain supported a 'national home' for the Jews in Palestine, as laid out in the Balfour Declaration of 1917. Jewish immigration, though limited by the British authorities, increased in the 1920s. Land ownership disputes aggravated Jewish-Arab relations, leading to paramilitary communal attacks. While Transjordan was granted independence in 1928, proposals for an Arab–Jewish partition in Palestine were rejected. In 1936 an Arab strike degenerated into insurrection—the 'Great Uprising' or 'Great Revolt'—under the leadership of Amin al-Husayni, the Grand Mufti of Jerusalem. The revolt was suppressed by the British by 1939, aided informally by the Jewish paramilitary Haganah. Al-Husayni fled to Germany, where he declared jihad on the Allies during the Second World War. The Italian air force bombed Haifa and Tel Aviv in 1940. Some Jewish groups, such as the Lehi, fought the British during the war on account of the British ban on Jewish immigration to Palestine.

Arab Israeli Wars

In 1947 the United Nations intervened, recommending partition of Palestine and an international administration for Jerusalem. The plan was accepted by the Jewish Agency (not representative of all Jewish groups) but rejected by the Palestinian Arab leadership; inter-communal war followed. On 14 May 1948 the British Government terminated its mandate and the Jewish leaders proclaimed the State of Israel. No independent Arab state was established in Palestine. Instead the neighbouring Arab states invaded Israel on 15 May 1948. The Jewish state defended itself successfully, and the ceasefire in Jan. 1949 left Israel with one-third more land than had been originally assigned by the UN.

In 1956 Israel was subject to international criticism for its involvement in the Suez Crisis. When Egypt nationalized the Suez Canal, France and the UK resorted to military action. Under the premiership of David Ben-Gurion, Israel joined forces with the European powers and agreed to lead an initial attack on the Egyptian-controlled Gaza Strip and the Sinai Peninsula. The plan was that the UK and France would offer to intervene in the conflict and reoccupy the areas. Nasser's expected refusal of the offer would be the pretext for an invasion that would reclaim the Canal. The Israeli incursions began in late Oct. 1956 but amid widespread international condemnation, most damagingly from the USA, the three nations were forced into a humiliating withdrawal by Dec. Nasser promoted the affair as a victory for pan-Arabism.

In 1967, following some years of uneasy peace, local clashes on the Israeli–Syrian border were followed by Egyptian mass concentration of forces on the borders of Israel. Israel struck out at Egypt on land and in the air on 5–9 June 1967. Jordan joined in the conflict which spread to the Syrian borders. By 11 June the Israelis had occupied the Gaza Strip and the Sinai peninsula as far as the Suez Canal in Egypt, West Jordan as far as the Jordan valley and the heights east of the Sea of Galilee, including the Syrian city of Quneitra, which was destroyed during the conflict.

A further war broke out on 6 Oct. 1973 when Egyptian and Syrian offensives were launched. Following UN Security Council resolutions a ceasefire came into force on 24 Oct. In Sept. 1978 Egypt and Israel agreed on frameworks for peace in the Middle East. A treaty was signed in Washington on 26 March 1979 whereby Israel withdrew from the Sinai Desert in two phases; part one was achieved on 26 Jan. 1980 and the final withdrawal was completed on 26 April 1982.

In June 1982 Israeli forces invaded the Lebanon. On 16 Feb. 1985 the Israeli forces started a withdrawal, leaving behind an Israeli trained and equipped Christian Lebanese force to act as a buffer against Muslim Shia or Palestinian guerrilla attacks.

Peace Process

In 1993, following declarations by Prime Minister Yitzhak Rabin recognizing the Palestine Liberation Organization (PLO) as representative of the Palestinian people, and by Yasser Arafat, leader of the PLO, renouncing terrorism and recognizing the State of Israel, an agreement was signed in Washington providing for limited Palestinian self-rule in the Gaza Strip and Jericho. The treaty marked the end of six years of violent opposition to the Israeli occupation in what is known as the first *intifada*, during which over 2,000 Palestinians and 150 Israelis were killed. Negotiations on the permanent status of the West Bank and Gaza began in 1996. On 4 Nov. 1995 Yitzhak Rabin was assassinated by a Jewish religious extremist. In the subsequent election, a right-wing coalition led by Binyamin Netanyahu took office. Peace talks with the Palestinians then stalled. In Oct. 1998 Israel accepted partial withdrawal from the West Bank on condition that the Palestinians cracked down on terrorism. The following month, 2% of the West Bank was handed over to Palestinian control. Further moves were put on hold after the collapse of the Netanyahu coalition and the announcement of early elections.

In Sept. 1999 Ehud Barak provided the first evidence that the Middle East peace process was back on track by releasing nearly 200 Palestinian prisoners and by handing over 430 sq. km of land on the West Bank. In May 2000 Israel completed its withdrawal from south Lebanon, 22 years after the first invasion. By Oct. 2000 violence had broken out again between Israelis and Palestinians, fuelled by the conflict over control of Jerusalem, with terrorist acts a daily occurrence, leading to heavy casualties on both sides. This second *intifada* ended in 2005. With peace talks stalled once again, Barak called for a nationwide vote of confidence by putting himself up for re-election as prime minister. Defeated by the right-wing Ariel Sharon in Feb. 2001, he retired from politics. As violence escalated, in Dec. 2001 Israel ended all contact with Yasser Arafat, besieging his compound at Ramallah and putting him under virtual house arrest. Israeli incursions into Palestinian-controlled areas of the West Bank and the Gaza Strip, and suicide attacks by Palestinians, continued unabated in early 2002 with heavy loss of life. In June 2002 Israel began constructing a barrier to cut off the West Bank, with the aim of shielding the country from suicide bombers. Arafat died on 11 Nov. 2004 and was succeeded by Mahmoud Abbas in Jan. 2005. In Feb. 2005 Israeli prime minister Ariel Sharon and Mahmoud Abbas agreed to a 'cessation of hostilities' between the two peoples, a move which encouraged hopes of a resumption of the peace process. In Aug. 2005 Israeli troops and police evicted the 8,500 Jewish settlers from the Gaza Strip in accordance with an agreement between Israel and the Palestinians. This was the

first time Israel had withdrawn from Palestinian land captured in the 1967 war.

In July 2006, after Hizbollah forces in Lebanon had captured two Israeli soldiers, Israel launched a large-scale military campaign against Lebanon with a series of bombing raids, destroying large parts of the civilian infrastructure.

In Dec. 2008 Israel began a military assault on Gaza aimed at destroying Hamas strongholds responsible for rocket and mortar attacks on Israeli targets. Three weeks of air and ground operations resulted in many Palestinian civilian deaths and the destruction of much of Gaza's civilian infrastructure. Israel's action was widely criticized, particularly after the bombardment of the UN's relief and works headquarters. The UN Security Council called for an immediate ceasefire, with all members voting in favour of the motion bar the USA.

TERRITORY AND POPULATION

The area of Israel, including the Golan Heights (1,154 sq. km) and East Jerusalem, is 22,072 sq. km (8,522 sq. miles), of which 21,643 sq. km (8,357 sq. miles) are land. The population in Dec. 2011 was 7·84m. (5·91m. Jews, 1·61m. Arabs and 0·32m. others), including East Jerusalem, the Golan Heights and Israeli settlers in the West Bank but excluding 200,000 foreign workers. Population density, 362 per sq. km.

The UN gives a projected population for 2015 of 8·06m.

In 2011, 91·9% of the population lived in urban areas.

Jewish population by place of origin as of 2011: former USSR, 644,900; Morocco, 151,300; North America and Oceania, 93,900; Romania, 86,200; Ethiopia, 74,000; Iraq, 61,200; Poland, 49,500; Iran, 48,700.

The Jewish Agency, which, in accordance with Article IV of the Palestine Mandate, played a leading role in establishing the State of Israel, continues to organize immigration.

Israel is administratively divided into six districts:

District	Area (sq. km)	Population, 2011	Chief town
Northern	4,473	1,304,600	Nazareth
Haifa	866	926,700	Haifa
Central	1,294	1,894,400	Ramla
Tel Aviv	172	1,295,000	Tel Aviv
Jerusalem[1]	653	968,800	Jerusalem
Southern	14,185	1,121,600	Beersheba

[1]Includes East Jerusalem.

A further 325,500 people lived in Judaea and Samara (roughly corresponding to the West Bank) in 2011.

On 23 Jan. 1950 the Knesset proclaimed Jerusalem the capital of the State and on 14 Dec. 1981 extended Israeli law into the Golan Heights. Population of the main towns (Dec. 2011): Jerusalem, 804,400; Tel Aviv/Jaffa, 404,800; Haifa, 270,300; Rishon le-Ziyyon, 232,400; Ashdod, 212,300; Petach Tikva, 210,400.

The official languages are Hebrew and Arabic.

SOCIAL STATISTICS

2008 births, 156,923; deaths, 39,484; marriages, 50,038; divorces, 13,488. 2008 crude birth rate per 1,000 population of Jewish population, 20·4; Non-Jewish: Muslims, 28·5; Christians, 16·5; Druzes, 21·0. Crude death rate per 1,000 (2008), Jewish, 6·2; Muslims, 2·5; Christians, 4·7; Druzes, 3·0. Infant mortality rate per 1,000 live births (2005–09), 4·0 (Jewish, 2·9; Muslims, 7·5; Christians, 2·3; Druzes, 5·2). Life expectancy, 2007, 78·5 years for males and 82·7 for females. Average annual population growth rate, 2000–05, 1·9%. Fertility rate, 2008, 2·8 births per woman. There were 16,892 immigrants in 2011, up from 13,699 in 2008 but down from 199,516 in 1990 and 176,100 in 1991 following the fall of communism in Eastern Europe and the break-up of the former Soviet Union.

CLIMATE

From April to Oct., the summers are long and hot, and almost rainless. From Nov. to March, the weather is generally mild, though colder in hilly areas, and this is the wet season. Jerusalem, Jan. 12·8°C, July 28·9°C. Annual rainfall, 657 mm. Tel Aviv, Jan. 17·2°C, July 30·2°C. Annual rainfall, 803 mm.

CONSTITUTION AND GOVERNMENT

Israel is an independent sovereign republic, established by proclamation on 14 May 1948.

In 1950 the Knesset (*Parliament*), which in 1949 had passed the Transition Law dealing in general terms with the powers of the Knesset, President and Cabinet, resolved to enact from time to time fundamental laws, which eventually, taken together, would form the Constitution. The eleven fundamental laws that have been passed are: the Knesset (1958), Israel Lands (1960), the President (1964), the State Economy (1975), the Army (1976), Jerusalem, capital of Israel (1980), the Judicature (1984), the State Comptroller (1988), Human Dignity and Liberty (1992), Freedom of Occupation (1994) and the Government (2001).

The *President* (head of state) is elected by the Knesset by secret ballot by a simple majority; his term of office is seven years. He may only serve for one term.

The Knesset, a one-chamber Parliament, consists of 120 members. It is elected for a four-year term by secret ballot and universal direct suffrage. Under the system of election introduced in 1996, electors vote once for a party and once for a candidate for Prime Minister. To be elected Prime Minister, a candidate must gain more than half the votes cast, and be elected to the Knesset. If there are more than two candidates and none gain half the vote, a second round is held 15 days later. The Prime Minister forms a cabinet (no fewer than eight members and no more than 18) with the approval of the Knesset.

National Anthem

'Hatikvah' ('The Hope'); words by N. H. Imber; folk-tune.

GOVERNMENT CHRONOLOGY

Prime Ministers since 1948. (Avoda = Labour Party; Herut = Freedom Movement; Kadima = 'Forward'; Likud = 'Consolidation'; Mapai = Israeli Workers' Party)

1948–53	Mapai	David Ben-Gurion
1953–55	Mapai	Moshe Sharett
1955–63	Mapai	David Ben-Gurion
1963–69	Mapai	Levi Eshkol
1969–74	Avoda	Golda Meir
1974–77	Avoda	Yitzhak Rabin
1977–83	Herut/Likud	Menahem Begin
1983–84	Herut/Likud	Yitzhak Shamir
1984–86	Avoda	Shimon Peres
1986–92	Likud	Yitzhak Shamir
1992–95	Avoda	Yitzhak Rabin
1995–96	Avoda	Shimon Peres
1996–99	Likud	Binyamin Netanyahu
1999–2001	Avoda	Ehud Barak
2001–06	Likud, Kadima	Ariel Sharon
2006–09	Kadima	Ehud Olmert
2009–	Likud	Binyamin Netanyahu

RECENT ELECTIONS

In the parliamentary (Knesset) elections on 22 Jan. 2013, the Likud-Yisrael Beiteinu bloc (an alliance of Likud and Yisrael Beiteinu) won 31 of 120 seats with 23·3% of votes cast, Yesh Atid 19 (14·3%), Labour Party 15 (11·4%), Jewish Home 12 (9·1%), Shas 11 (8·8%), United Torah Judaism 7 (5·2%), Hatnuah 6 (5·0%), Meretz 6 (4·6%), the United Arab List 4 (3·7%), Hadash 4 (3·0%), Balad 3 (2·6%) and Kadima 2 (2·1%). Turnout was 67·8%.

In a parliamentary vote for the presidency on 13 June 2007, Shimon Peres was elected in the second round with 86 votes in favour and 23 against after his two opponents from the first round had withdrawn.

CURRENT GOVERNMENT

President: Shimon Peres; b. 1923 (since 15 July 2007).

The government is formed by a coalition of the Likud-Yisrael Beiteinu bloc, Yesh Atid, Jewish Home and Hatnuah. In March 2013 the cabinet was composed as follows:

Prime Minister, and Minister of Foreign Affairs (acting): Binyamin Netanyahu; b. 1949 (Likud-Yisrael Beiteinu; since 31 March 2009, having previously held office from June 1996–July 1999).

Minister of Agriculture and Rural Development: Yair Shamir (Likud-Yisrael Beiteinu). *Communications and Home Front Defence:* Gilad Erdan (Likud-Yisrael Beiteinu). *Culture and Sport:* Limor Livnat (Likud-Yisrael Beiteinu). *Defence:* Moshe Ya'alon (Likud-Yisrael Beiteinu). *Economy and Trade:* Naftali Bennett (Jewish Home). *Education:* Shai Piron (Yesh Atid). *Environmental Protection:* Amir Peretz (Hatnuah). *Finance:* Yair Lapid (Yesh Atid). *Health:* Yael German (Yesh Atid). *Housing:* Uri Ariel (Jewish Home). *Immigrant Absorption:* Sofa Landver (Likud-Yisrael Beiteinu). *Interior:* Gideon Sa'ar (Likud-Yisrael Beiteinu). *Internal Security:* Yitzhak Aharonovitch (Likud-Yisrael Beiteinu). *International, Intelligence and Strategic Affairs:* Yuval Steinitz (Likud-Yisrael Beiteinu). *Justice:* Tzipi Livni (Hatnuah). *Pensioners Affairs:* Uri Orbach (Jewish Home). *Regional Development, Water and Energy Resources:* Silvan Shalom (Likud-Yisrael Beiteinu). *Science and Technology:* Yaakov Peri (Yesh Atid). *Tourism:* Uzi Landau (Likud-Yisrael Beiteinu). *Transportation:* Yisrael Katz (Likud-Yisrael Beiteinu). *Welfare:* Meir Cohen (Yesh Atid).

Israeli Parliament: http://main.knesset.gov.il

CURRENT LEADERS

Binyamin Netanyahu

Position
Prime Minister

Introduction
Following the general election of 10 Feb. 2009, right-wing Likud party leader Binyamin Netanyahu, who had previously been premier from 1996–99, formed a broad but politically volatile coalition government consisting of centre-right, centre-left and far-right parties. Further elections were held in Jan. 2013 amid a continuing stalemate over the status of the Palestinians, further controversial Israeli settlement building in occupied territories, concerns over the nuclear capability of Iran and civil war in neighbouring Syria. The poll resulted in an alliance of the Likud and Yisrael Beiteinu parties, headed by Netanyahu, becoming the largest parliamentary grouping, although wider support was needed to form a viable administration.

Early Life
Binyamin Netanyahu was born on 21 Oct. 1949 in Tel Aviv. After military service, he studied in the USA at the Massachusetts Institute of Technology and Harvard. Having worked in business consultancy, he then joined the Israeli diplomatic service in Washington, D.C. in 1982 and served as Israel's ambassador to the UN from 1984–88.

In 1988 Netanyahu was elected to the Knesset for Likud and named deputy foreign minister. From 1991–92 he served as deputy minister in the Prime Minister's Office before succeeding Yitzhak Shamir in 1993 as Likud chairman and leader of the opposition. Sceptical of the Sept. 1993 Oslo Accords, Netanyahu played on Israeli fears over security at a time of escalating Palestinian violence and oversaw a slim victory at the elections of May 1996. However, he was defeated in the 1999 general election by Labour's Ehud Barak and subsequently lost the Likud leadership to Ariel Sharon.

Netanyahu returned to politics in 2002 as foreign minister and in 2003 took the finance portfolio in Sharon's cabinet. In Aug. 2005 he resigned over the Gaza disengagement plan, but in Dec. he regained the Likud leadership and became opposition leader following the Kadima party's victory in the 2006 parliamentary elections.

At the Feb. 2009 elections Netanyahu claimed victory, despite winning 27 seats to Kadima's 28, on the basis that his right-wing coalition partners had won the majority of the vote.

Career in Office
Netanyahu's primary challenges on taking office were the economic downturn, the continuing crisis with the Palestinians and concerns over Iran's nuclear ambitions.

In June 2009 he expressed for the first time his acceptance of a two-state solution for Israel and Palestine, provided that the Palestinian state was demilitarized and that the Palestinians recognized Israel as the state of the Jewish people. There appeared to be a broad Israeli consensus in favour of this policy, but substantive progress in the peace process has remained elusive, particularly over the divisive issue of Jewish settlements in the West Bank and East Jerusalem. Having resisted early diplomatic pressure from US President Obama to stop all settlement building on Palestinian land, Netanyahu announced in Nov. 2009 a plan for a ten-month freeze in the West Bank (but not East Jerusalem) in a bid to restart peace negotiations. However, the move was deemed inadequate by the Palestinians and, following its expiry without renewal in Sept. 2010, the opportunity for further meaningful dialogue was effectively abandoned in Dec. that year.

Tensions were heightened in 2011 as Palestinian President Abbas confirmed in Sept. that he was applying for formal United Nations recognition of Palestinian independent statehood in defiance of Israel and the USA. Nevertheless, in Oct. Netanyahu presided over a major prisoner exchange with the Palestinian Hamas faction to secure the release of an Israeli soldier captured in 2006, and in Jan. 2012 Israeli and Palestinian officials met in Jordan for their first, if inconclusive, direct contacts for well over a year. However, in Nov. 2012 the UN General Assembly voted overwhelmingly to recognize Palestine's enhanced status as a non-member observer state, in response to which the Netanyahu government announced that Israel would extend Jewish settlement building in the West Bank despite international criticism.

Addressing the UN General Assembly in Sept. 2009, Netanyahu had stated that Iran posed a threat to world peace and that it must be prevented from acquiring nuclear weapons. He also condemned the Iranian president's denial of the Holocaust. Such antagonism has since fuelled speculation about possible pre-emptive Israeli military action against Iran. In 2012 Netanyahu was quick to pin responsibility for bomb attacks aimed at Israeli envoys in the capitals of India and Georgia in Feb. and against Israeli tourists in Bulgaria in July on Iran. In Sept. he claimed that Iran would have a nuclear bomb capability by the middle of 2013.

Israel's international reputation was damaged in 2010 by its alleged involvement in the killing of a Palestinian militant in Dubai in Feb. by assassins using forged passports, and also by its military assault in May on a convoy of ships bringing aid to the Gaza Strip in defiance of the Israeli blockade. Nine Turkish nationals were killed in the latter incident, which severely undermined Israeli–Turkish diplomatic relations and led in Sept. 2011 to the suspension of all defence links and the expulsion of the Israeli ambassador.

In Feb. 2011, in response to mounting concern over political turmoil in Egypt, Netanyahu said that Israel would review its security arrangements if Egypt reneged on the 1979 bilateral peace treaty. In Sept. that year bilateral relations came under further strain as a violent Egyptian crowd stormed the Israeli embassy in Cairo, prompting an airlift of diplomats and their dependants out of the country. Similarly mindful of the escalation of civil war in Syria and its potential security implications, Israel reportedly

launched a bombing raid on an unspecified target near Damascus in Jan. 2013.

Netanyahu's government has meanwhile been confronted by widespread domestic unrest in protest at rising living costs and other economic grievances.

DEFENCE

Conscription (for Jews and Druze only) is three years (usually four years for officers; 21 months for women). Israel is one of the few countries with female conscription. The Israel Defence Force is a unified force, in which army, navy and air force are subordinate to a single chief-of-staff. The Minister of Defence is de facto C.-in-C.

Defence expenditure in 2008 totalled US$14,772m., representing 7·4% of GDP. Expenditure per capita in 2008 was US$2,077, a figure exceeded only by the United Arab Emirates, Kuwait, the USA and Qatar.

Nuclear Weapons
Israel has an undeclared nuclear weapons capability. Although known to have a nuclear bomb, it pledges not to introduce nuclear testing to the Middle East. According to the Stockholm International Peace Research Institute, the nuclear arsenal was estimated to have about 80 warheads in Jan. 2012. Israel is one of three countries not to have signed the Nuclear Non-Proliferation Treaty (the others being India and Pakistan). Israel has never admitted possessing biological or chemical weapons, but according to Deadly Arsenals, published by the Carnegie Endowment for International Peace, it does have a chemical and biological weapons programme.

Army
Strength (2006) 125,000 (conscripts 105,000). There are also 380,000 reservists available on mobilization. In addition there is a paramilitary border police of about 8,000.

Navy
The Navy, tasked primarily for coastal protection and based at Haifa, Ashdod and Eilat, includes three small diesel submarines and three corvettes.

Naval personnel in 2006 totalled about 8,000 (including a Naval Commando of 300) of whom 2,500 are conscripts. There are also 11,500 naval reservists available on mobilization.

Air Force
The Air Force (including air defence) has a personnel strength (2006) of 35,000, with 402 combat aircraft, all jets, of Israeli and US manufacture including F-15s and F-16s, and 95 armed helicopters. There are 24,500 Air Force reservists.

ECONOMY

Services account for about 82% of GDP, industry 16% and agriculture 2%.

Overview
Israel's economy is diversified relative to its neighbours. Over the past two decades electronics manufacturing has replaced industries such as footwear and clothing. Until the 1990s traditional industries benefited from protectionist policies but have since undergone structural changes.

In 2000 the government opened the telecommunications sector to foreign competition. High-tech industries have benefited from Israel's high standard of education and its investment in military research and development. The 1990s saw strong growth and in 2000 the economy grew by 9·2%. However, in 2001 and 2002 Israel experienced its worst recession in 50 years, a result of high security costs arising from the second intifada, a sharp decline in tourism and difficulties in the high-tech sector.

The economy rebounded and GDP grew by an annual average of over 5% from 2004–07 thanks to improved internal security, strong external demand, sound financial conditions and prudent macroeconomic policies. In 2008 the global financial crisis led to reduced exports and GDP growth, but the economy began to revive in the second quarter of 2009. GDP grew by 4·6% in 2011, driven by private consumption and investment. Recent discoveries of natural gas fields may see Israel become a net energy exporter in coming years. The authorities plan to place natural resource revenues in a sovereign wealth fund although the global downturn is expected to hold back growth for the foreseeable future.

Reducing the high public debt, which stands at around 75% of GDP, is a key challenge, while poverty remains among the highest in OECD countries.

Currency
The unit of currency is the *shekel* (ILS) of 100 *agorot*. Foreign exchange reserves were US$58,426m. in Sept. 2009. Gold reserves have been negligible since 1998. Total money supply in Nov. 2008 was 83,131m. shekels.

Inflation rates (based on OECD statistics):

2002	2003	2004	2005	2006	2007	2008	2009	2010	2011
5·7%	0·7%	−0·4%	1·3%	2·1%	0·5%	4·6%	3·3%	2·7%	3·5%

Budget
In 2008 revenues were 310·9bn. shekels and expenditures 320·9bn. shekels. Tax revenue accounted for 57·9% of revenues in 2008; debt repayment accounted for 21·7% of expenditures, defence 17·4% and education 11·2%.

VAT is 17%.

Performance
Real GDP growth rates (based on OECD statistics):

2002	2003	2004	2005	2006	2007	2008	2009	2010	2011
−0·1%	1·5%	4·9%	4·9%	5·8%	5·9%	4·1%	1·1%	5·0%	4·6%

Total GDP was US$242·9bn. in 2011.

Banking and Finance
The Bank of Israel was established by law in 1954 as Israel's central bank. Its Governor is appointed by the President on the recommendation of the Cabinet for a five-year term. The *Governor* is Prof. Stanley Fischer. Central bank reserves in Dec. 2002 were US$24·1bn. As part of a government scheme several banks were privatized in the years 1993–2006.

In 2001 there were 23 commercial banks headed by Bank Leumi le-Israel, Bank Hapoalim and Israel Discount Bank, two merchant banks, three foreign banks, eight mortgage banks and nine lending institutions specifically set up to aid industry and agriculture.

Gross external debt amounted to US$95,263m. in June 2012.

There is a stock exchange in Tel Aviv.

ENERGY AND NATURAL RESOURCES
Environment
Carbon dioxide emissions from the consumption and flaring of fossil fuels in 2008 were the equivalent of 9·9 tonnes per capita.

Electricity
Installed capacity in 2007 was 11·6m. kW. Electric power production amounted to 55·09bn. kWh in 2007; consumption per capita was 7,383 kWh in 2007.

Oil and Gas
In 2009 large quantities of natural gas were discovered off the coast of Israel that are expected to meet the country's needs for 35 years. In 2010 the reserves totalled 238bn. cu. metres. Crude petroleum reserves in 2007 were 2m. bbls.

Minerals
The most valuable natural resources are the potash, bromine and other salt deposits of the Dead Sea. Production figures in 1,000 tonnes: phosphate rock (2004), 2,947; potash (2004), 2,060; lignite (2004), 439; salt (2004 estimate), 398.

Agriculture
There were about 338,000 ha. of arable land in 2002 and 86,000 ha. of permanent crops. Production, 2002 (in 1,000 tonnes): melons and watermelons, 405; potatoes, 394; tomatoes, 383; grapefruit and pomelos, 256; wheat, 179; oranges, 159; cucumbers and gherkins, 144; apples, 126.

Livestock (2003 estimates): 395,000 sheep; 390,000 cattle; 190,000 pigs; 63,000 goats; 35m. poultry.

Types of rural settlement: (1) the *Kibbutz* and *Kvutza* (communal collective settlement), where all property and earnings are collectively owned and work is collectively organized (117,700 people lived in 267 *Kibbutzim* in 2005). (2) The *Moshav* (workers' co-operative smallholders' settlement) which is founded on the principles of mutual aid and equality of opportunity between the members, all farms being equal in size (213,600 in 402 *Moshavim* in 2005). (3) The *Moshav Shitufi* (co-operative settlement), which is based on collective ownership and economy as in the *Kibbutz,* but with each family having its own house and being responsible for its own domestic services (17,000 in 40 *Moshavim Shitufi'im* in 2005). (4) Other rural settlements in which land and property are privately owned and every resident is responsible for his own well-being. In 2005 there were a total of 240 non-cooperative villages with a population of 159,400.

Forestry
In 2010 forests covered 0·15m. ha. or 7% of the total land area. Timber production was 27,000 cu. metres in 2007.

Fisheries
Catches in 2010 totalled 2,588 tonnes, mainly from marine waters.

INDUSTRY
The leading companies by market capitalization in Israel in April 2012 were: Teva Pharmaceutical Industries Ltd (US$41·0bn.); and Check Point Software, a security software and computer hardware company (US$12·8bn.).

Products include chemicals, metal products, textiles, tyres, diamonds, paper, plastics, leather goods, glass and ceramics, building materials, precision instruments, tobacco, foodstuffs, electrical and electronic equipment.

Labour
The economically active workforce was 2,270,500 in 2001 (1,236,200 males). The principal areas of activity were: manufacturing, mining and quarrying, 394,200; wholesale and retail trade/repair of motor vehicles, motorcycles and personal and household goods, 299,800; education, 283,700; and real estate, renting and business activities, 277,200. Unemployment was 6·9% in Dec. 2012.

INTERNATIONAL TRADE
Imports and Exports
Imports (c.i.f.) in 2007 totalled US$56,619m. and exports (f.o.b.) US$54,091m. Main imports in 2007 were: manufactured goods, 28·6% (including diamonds); machinery and transport equipment, 28·5%; mineral fuels, lubricants and related materials (including petroleum and petroleum products), 15·8%. Diamonds constituted 17·7% of Israel's imports in 2007. Leading import suppliers in 2007: USA, 13·9%; Belgium, 7·9%; Germany, 6·2%. Main exports in 2007 were: manufactured goods, 39·5% (including diamonds); chemicals and related products, 15·1%; machinery and transport equipment, 14·4%. Diamonds constituted 34·0% of exports in 2007. The leading export markets in 2007 were: USA, 35·0%; Belgium, 7·5%; Hong Kong, 5·8%.

COMMUNICATIONS

Roads

There were 17,870 km of paved roads in 2007, including 344 km of motorway. Motor vehicles in use in 2007 totalled 1,805,400 passenger cars, 362,200 lorries and vans, 94,800 motorcycles and mopeds, and 21,300 buses and coaches. There were 398 fatalities as a result of road accidents in 2007.

Rail

There were 909 km of standard gauge line in 2005. 26·8m. passengers and 7·5m. tonnes of freight were carried in 2005. One of the smallest metro systems in the world (1,800 metres) was opened in Haifa in 1959. A tram system in Jerusalem opened in Aug. 2011.

Civil Aviation

There are international airports at Tel Aviv (Ben Gurion), Eilat (J. Hozman), Haifa and Ovda. Tel Aviv is the busiest airport, in 2001 handling 8,305,950 passengers (7,864,200 on international flights) and 296,054 tonnes of freight. El Al is the flag carrier. In 2005 scheduled airline traffic of Israeli-based carriers flew 97·9m. km and carried 4,382,200 passengers. In 2003 services (mainly domestic) were also provided by another Israeli airline, Arkia, and by over 40 international carriers.

Shipping

Israel has three commercial ports—Haifa, Ashdod and Eilat. In Jan. 2009 there were 15 ships of 300 GT or over registered, totalling 428,000 GT.

Telecommunications

In 2008 there were 3,224,000 main (fixed) telephone lines. In the same year mobile phone subscribers numbered 8,982,000 (1,273·8 per 1,000 persons). There were 3·3m. internet users in 2007. In March 2012 there were 3·5m. Facebook users.

SOCIAL INSTITUTIONS

Justice

Law. Under the Law and Administration Ordinance, 5708/1948, the first law passed by the Provisional Council of State, the law of Israel is the law which was obtaining in Palestine on 14 May 1948 in so far as it is not in conflict with that Ordinance or any other law passed by the Israel legislature and with such modifications as result from the establishment of the State and its authorities.

Capital punishment was abolished in 1954, except for support given to the Nazis and for high treason.

The law of Palestine was derived from Ottoman law, English law (Common Law and Equity) and the law enacted by the Palestine legislature, which to a great extent was modelled on English law.

Civil Courts. Municipal courts, established in certain municipal areas, have criminal jurisdiction over offences against municipal regulations and bylaws and certain specified offences committed within a municipal area. Magistrates courts, established in each district and sub-district, have limited jurisdiction in both civil and criminal matters. District courts, sitting at Jerusalem, Tel Aviv and Haifa, have jurisdiction, as courts of first instance, in all civil matters not within the jurisdiction of magistrates courts, and in all criminal matters, and as appellate courts from magistrates courts and municipal courts. The 14-member Supreme Court has jurisdiction as a court of first instance (sitting as a High Court of Justice dealing mainly with administrative matters) and as an appellate court from the district courts (sitting as a Court of Civil or of Criminal Appeal).

In addition, there are various tribunals for special classes of cases. Settlement Officers deal with disputes with regard to the ownership or possession of land in settlement areas constituted under the Land (Settlement of Title) Ordinance.

Religious Courts. The rabbinical courts of the Jewish community have exclusive jurisdiction in matters of marriage and divorce, alimony and confirmation of wills of members of their community and concurrent jurisdiction with the civil courts in all other matters of personal status of all members of their community with the consent of all parties to the action.

The courts of the several recognized Christian communities have a similar jurisdiction over members of their respective communities.

The Muslim religious courts have exclusive jurisdiction in all matters of personal status over Muslims who are not foreigners, and over Muslims who are foreigners, if under the law of their nationality they are subject in such matters to the jurisdiction of Muslim religious courts.

Where any action of personal status involves persons of different religious communities, the President of the Supreme Court will decide which court shall have jurisdiction, and whenever a question arises as to whether or not a case is one of personal status within the exclusive jurisdiction of a religious court, the matter must be referred to a special tribunal composed of two judges of the Supreme Court and the president of the highest court of the religious community concerned in Israel.

In 2001 government expenditure on public security and justice totalled 7,238m. shekels. The population in penal institutions in March 2008 was 22,788 (326 per 100,000 of national population).

Education

The adult literacy rate in 2003 was 96·9% (male, 98·3%; female, 95·6%). There is free and compulsory education from five to 18 years. There is a unified state-controlled elementary school system with a provision for special religious schools. The standard curriculum for all elementary schools is issued by the Ministry of Education with a possibility of adding supplementary subjects comprising not more than 25% of the total syllabus.

In 2004–05 there were 1,614,000 Hebrew pupils and 436,000 Arab pupils in the education system. In primary schools and kindergartens in 2004–05 there were 888,000 Hebrew children and 302,000 Arab children. There were 57,000 Hebrew teachers and 16,000 Arab teachers in primary education in 2004–05. In post-primary education there were 472,000 Hebrew pupils and 132,000 Arab pupils in 2004–05, with 64,000 Hebrew teachers and 11,000 Arab teachers. In special education there were 11,180 pupils in 2004–05. In post-secondary education, such as colleges, universities and vocational institutions, there were 255,000 pupils, of which 253,000 were Hebrew.

The Hebrew University of Jerusalem, founded in 1925, comprises faculties of the humanities, social sciences, law, science, medicine and agriculture. In 2004–05 it had 21,985 students. The Technion–Israel Institute of Technology in Haifa had 12,810 students. The Weizmann Institute of Science in Rehovoth, founded in 1949, had 960 students in 2004–05.

Tel Aviv University had 28,740 students in 2004–05. The religious Bar-Ilan University at Ramat Gan, opened in 1965, had 25,025 students, the Haifa University had 16,270 students and the Ben Gurion University had 18,640 students.

In 2008 public expenditure on education came to 5·9% of GDP and accounted for 13·7% of total government spending.

Health

In 2010 there were 121 hospitals. There were 60 hospital beds per 10,000 inhabitants in 2007. There were 25,138 physicians, 7,726 dentists, 4,958 pharmacists and 42,609 nurses and midwives in 2006. In 2009 health spending represented 7·5% of GDP.

Welfare

The National Insurance Law of 1954 provides for old-age pensions, survivors' insurance, work-injury insurance, maternity insurance, family allowances and unemployment benefits. In 2001 recipients of allocations from the National Insurance Institute included (monthly averages): child allowances, 2,154,735; old age pensions, 571,200; general disability allowances, 142,440; income support benefits, 142,011; maternity grants, 129,089; survivors' pensions, 105,818; unemployment benefits, 104,707.

RELIGION

Religious affairs are under the supervision of a special ministry, with departments for the Christian and Muslim communities. The religious affairs of each community remain under the full control of the ecclesiastical authorities concerned: in the case of the Jews, the Ashkenazi and Sephardi Chief Rabbis, in the case of the Christians, the heads of the various communities, and in the case of the Muslims, the Qadis. The Druze were officially recognized in 1957 as an autonomous religious community.

In 2001 there were: Jews, 4,960,000; Muslims, 930,000; others (mainly Christians and Druze), 360,000.

The Chief Rabbis are Yona Metzger (Ashkenazi) and Shlomo Amar (Sephardi).

CULTURE

World Heritage Sites

There are seven UNESCO sites in Israel. Masada and the old city of Acre were both inscribed in 2001. Masada was built as a palace complex and fortress by Herod the Great. It was the site of the mass suicide of about 1,000 Jewish patriots in the face of a Roman army in the 1st century AD and is a symbol of the ancient kingdom of Israel. The port city of Acre preserves remains of its medieval Crusader buildings beneath the existing Muslim fortified town dating from the 18th and 19th centuries. The White City of Tel Aviv—the Modern Movement (2003) is an example of early 20th century town planning, based on the plan of Sir Patrick Geddes. The Biblical Tels, a series of prehistoric settlement mounds with biblical connections, and the Incense Route, four Nabatean towns along the spice and incense trail, were added to the list in 2005. The Bahá'i Holy Places in Haifa and the Western Galilee (2008) is a complex of buildings including the Shrine of Bahá'u'lláh in Acre and the Shrine of the Báb in Haifa that are visited as part of the Bahá'i pilgrimage. Sites of Human Evolution at Mount Carmel: the Nahal Me'arot/Wadi el-Mughara Caves (2012) is an archaeological site displaying 500,000 years of human evolution with evidence of burials, early stone architecture and the transition from a hunter-gathering lifestyle to agriculture and animal husbandry.

Press

In 2008 there were 18 daily newspapers with a combined circulation of 1·3m. The most widely read paper is *Yedioth Ahronoth*.

Tourism

In 2011 there were 2,820,000 tourist arrivals (excluding same-day visitors), up from 2,803,000 in 2010 and 2,417,000 in 2000. The main countries of origin of non-resident tourists in 2011 were the USA (21%), followed by Russia (13%) and France (10%). 86% of all tourist arrivals in 2011 were by air and 14% were by land border crossings.

Calendar

The Jewish year 5773 corresponds to 17 Sept. 2012–4 Sept. 2013; 5774 corresponds to 5 Sept. 2013–24 Sept. 2014.

DIPLOMATIC REPRESENTATIVES

Of Israel in the United Kingdom (2 Palace Green, Kensington, London, W8 4QB)
Ambassador: Daniel Taub.

Of the United Kingdom in Israel (192 Hayarkon St., Tel Aviv 63405)
Ambassador: Matthew Gould, MBE.

Of Israel in the USA (3514 International Dr., NW, Washington, D.C., 20008)
Ambassador: Michael Oren.

Of the USA in Israel (71 Hayarkon St., Tel Aviv)
Ambassador: Daniel B. Shapiro.

Of Israel to the United Nations
Ambassador: Ron Prosor.

Of Israel to the European Union
Ambassador: David Walzer.

FURTHER READING

Central Bureau of Statistics. *Statistical Abstract of Israel.* (Annual)— *Statistical Bulletin of Israel.* (Monthly)

Beitlin, Y., *Israel: a Concise History.* 1992
Bregman, Ahron, *History of Israel.* 2002
Freedman, R. (ed.) *Israel Under Rabin.* 1995
Garfinkle, A., *Politics and Society in Modern Israel: Myths and Realities.* 1997
Gelvin, James L., *The Israel-Palestine Conflict: One Hundred Years of War.* 2005
Gilbert, Martin, *Israel: A History.* 1998
Kershner, Isabel, *Barrier: The Seam of the Israeli-Palestinian Conflict.* 2005
Sachar, H. M., *A History of Israel: From the Rise of Zionism to Our Time.* 3rd ed. 2007
Segev, T., *1949: The First Israelis.* 1986
Shulman, David, *Dark Hope: Working for Peace in Israel and Palestine.* 2007
Smith, Charles D., *Palestine and the Arab-Israeli Conflict.* 7th ed. 2010
Thomas, Baylis, *How Israel Was Won: A Concise History of the Arab–Israeli Conflict (1900–1999).* 2000
Wasserstein, Bernard, *Israel and Palestine: Why They Fight and Can They Stop?* 2003

Other more specialized titles are entered under PALESTINIAN TERRITORIES.

National library: The Jewish National and University Library, Edmond Safra Campus, Givat Ram, PO Box 39105, Jerusalem 91390.
National Statistical Office: Central Bureau of Statistics, Prime Minister's Office, POB 13015, Jerusalem 91130.
Website: http://www.cbs.gov.il

Palestinian Territories

KEY HISTORICAL EVENTS

After Israel declared independence on 14 May 1948, Arab League troops invaded the former British Mandate for Palestine. The first Arab–Israeli War (known in Israel as the War of Independence) ended with an armistice in July 1949. Under its terms 77% of Palestine came under Israeli control (56% had been allocated by the UN Partition Plan of 1947). Around 700,000 Palestinians were displaced to the West Bank, the Gaza Strip or to neighbouring countries. Up to 150,000 Palestinians remained in Israel. Gaza came under Egyptian control and in April 1950 the West Bank and East Jerusalem were annexed into Jordan.

Border clashes were frequent in the early 1950s. When Egypt nationalized the Suez Canal in July 1956 Israeli troops occupied Gaza and the Sinai Peninsula until the arrival of UN Emergency Forces. Fatah (the Palestine National Liberation Movement) emerged in Gaza in the late 1950s, led by Yasser Arafat and Khalil al-Wazir. It became the leading faction in the Palestine Liberation Organization (PLO), launched by the Arab League in 1964.

Guerrilla attacks by Fatah on Israel began in Jan. 1965. Tensions between Egypt and Israel rose in 1967 when UN forces withdrew. Israel's pre-emptive strike on Egyptian air bases ignited the Six Day War on 5 June, culminating in Israeli control of the West Bank, Gaza Strip and Golan Heights. In Nov. 1967 the UN stated that Israel should withdraw its forces from the territories occupied during the war in return for peace with its Arab neighbours. However, the PLO refused to accept Israel's right to exist. Following the Yom Kippur war in 1973, the possibility of a settlement was lost when radical factions broke from the PLO.

In Oct. 1974 the Arab League recognized the PLO as the 'sole, legitimate representative of the Palestinian people'.

When Israel and Egypt signed the Camp David Accords in Sept. 1978, Israel withdrew from Sinai. Plans for Palestinian autonomy in Gaza and the West Bank excluded the PLO and were rejected by Palestinians, while Israel made clear its intention to maintain a military presence in the Occupied Territories and expand Jewish settlements in the West Bank.

Israel's invasion of southern Lebanon in June 1982 ended the PLO's presence there. The 88-day siege of Beirut forced 10,000 militia into Yemen, Sudan and other Arab countries. Arafat established new headquarters in Tunisia. In Nov. 1988, almost a year after the first Palestinian *intifada* (uprising) against Israel began, the PLO declared a Palestinian state. Arafat also recognized Israel's right to exist and renounced terrorism, paving the way for the 1993 Oslo Accords.

After the Israeli prime minister, Yitzhak Rabin, recognized the PLO an agreement was signed in Washington providing for limited Palestinian self-rule (through the Palestinian National Authority) in the Gaza Strip and part of the West Bank. The six-year *intifada* ended, during which over 2,000 Palestinians and 150 Israelis were killed.

In Nov. 1995 Rabin was assassinated by a Jewish extremist. Under an agreement of 1995 the Israeli army withdrew from six of the seven largest Palestinian towns in the West Bank and from 460 smaller towns and villages. The rest of the West Bank stayed under Israeli control with further withdrawals at six-month intervals. Negotiations on the permanent status of the West Bank and Gaza began in May 1996.

After elections in 1996, a right-wing coalition under Binyamin Netanyahu approved plans for an expansion of Jewish settlement in the West Bank. In Oct. 1998 Israel accepted partial withdrawal from the West Bank on condition that the Palestinians cracked down on terrorism. The following month, 2% of the West Bank was handed over.

Netanyahu's defeat by Ehud Barak in elections in 1999 improved relations with the PLO. In March 2000 Arafat accepted plans for expanded self-rule in the West Bank, involving the transfer of another 6% of the area to the Palestinian Authority. However, by Oct. 2000 a second *intifada* had begun, fuelled by conflict over control of Jerusalem. When talks stalled, Barak went to the polls in Feb. 2001 but was defeated by Ariel Sharon. Amid escalating violence, in Dec. 2001 Israel broke off contact with Arafat, putting him under virtual house arrest.

Violence on both sides continued in 2002 with heavy loss of life. In March 2002 the UN endorsed a Palestinian state for the first time. In June 2002 Israel began constructing a barrier cutting off the West Bank to shield against suicide bombers. Arafat died in Nov. 2004 and was succeeded by Mahmoud Abbas. In Feb. 2005 Sharon and Abbas agreed to cease hostilities. In Aug. 2005 Israel evicted 8,500 Jewish settlers from Gaza.

In Jan. 2006 Palestinian legislative elections were won by Change and Reform (Hamas), which does not recognize Israel and has called for its destruction. Western aid was thus suspended. In Dec. 2006, after deadlock between Fatah and Hamas over forming a national unity government, Abbas called new elections. Tensions between Fatah and Hamas peaked in June 2007, with Hamas seizing control of the Gaza Strip while the West Bank remained under Fatah. Abbas established a new government, recognized by Fatah and Israel but not by Hamas, under the premiership of Salam Fayyad.

In Dec. 2008 Israel began an assault in Gaza aimed at destroying Hamas strongholds responsible for attacks on Israeli targets. The international community called for a ceasefire amid concerns over the high civilian death toll and infrastructural damage. On 13 Jan. 2009 the UN headquarters in Gaza was bombed by Israeli forces. Israel apologized, claiming its forces were attacked by militants taking refuge there. On 18 Jan. Israel and Hamas announced unilateral ceasefires. An estimated 1,300

Palestinians and 13 Israelis were killed during the three-week offensive, and 50,000 displaced.

In May 2011 Mahmoud Abbas and Khaled Meshaal, leaders of Fatah and Hamas respectively, agreed that a joint caretaker administration should hold power until Palestinian elections— initially scheduled for May 2012 but subsequently postponed.

TERRITORY AND POPULATION

The 2007 census population of the Palestinian territory was 3,767,126 (2,895,683 in 1997). In 2011, 74·4% of the population were urban. Life expectancy at birth, 2007, was 74·9 years for females and 71·7 years for males. The UN gives a projected population for 2015 of 4·65m.

The West Bank (preferred Palestinian term, Northern District) has an area of 5,655 sq. km; the 2007 census population was 2,350,583, in addition to 275,000 Jewish settlers and 10,000 troops deployed there. 99·8% of the population in 1997 were Palestinians. In 2001 there were 1,860,000 Muslims, 230,000 Jews and 200,000 Christians and others. By 2009 the number of Jewish settlers had risen above 300,000. In 2006 there was a Palestinian diaspora of 5·0m. The birth rate in 2004 was estimated at 39·6 per 1,000 population and the death rate 4·8 per 1,000. In 1995–99 the infant mortality rate was 24·4 per 1,000 live births. The fertility rate in 1999 was 5·5 births per woman. In 2003 there were 31,646 private cars and 14,521 commercial vehicles and trucks registered. There were (2003–04) 542,520 pupils in basic stage education and 59,909 in secondary stage. In 1998–99 there were 36,224 students in institutions of higher education. In 2003 there were 54 hospitals.

The Gaza Strip (preferred Palestinian term, Gaza District) has an area of 365 sq. km; the 2007 census population was 1,416,543. The population doubled between 1975 and 1995. Crude birth rate in 2004 was 43·7 per 1,000 population. The death rate was estimated at 3·9 per 1,000 population. The fertility rate in 1999 was 6·8 births per woman. Infant mortality, 1995–99, 27·3 per 1,000 live births. Agricultural production, 2002 estimates, in 1,000 tonnes: oranges, 105; tomatoes, 48; potatoes, 35; cucumbers and gherkins, 18; grapefruit and pomelos, 10. Total fish catch in 2005 for the Palestinian-Administered Territories was 1,805 tonnes. In 2003–04 there were 374,713 students in basic stage education, 41,185 in secondary stage and 30,058 students in higher education (1998–99). In 2003 there were 17 hospitals.

The chief town is Gaza itself. Over 98% of the population are Arabic-speaking Muslims. In 1995 an estimated 94·2% of the population lived in urban areas. In 2003 there were 38,677 private cars and 9,392 commercial vehicles and trucks registered. Gaza International Airport, at the southern edge of the Gaza Strip, opened in Nov. 1998. Telecommunications development has been rapid, the number of fixed line telephone subscribers more than trebling between 1997 and 2000. In 2003 there were 243,494 subscribers. In 2003 life expectancy at birth was 71·7 years.

CONSTITUTION AND GOVERNMENT

In April 1996 the Palestinian Council removed from its Charter all clauses contrary to its recognition by Israel, including references to armed struggle as the only means of liberating Palestine, and the elimination of Zionism from Palestine. The *President* is directly elected and heads the executive organ, the Palestinian National Authority, one fifth of whose members he appoints, while four fifths are elected by the *Legislative Council*. The latter comprises 132 members (88 until 2005), of which 66 members are chosen by district voting and the other 66 by proportional representation. The 2007 Election Law passed by President Mahmoud Abbas during a period of emergency rule introduced proportional representation for all seats. Hamas claimed the reforms were illegal as they were not ratified by the Legislative Council. The Palestinian Authority was created by agreement of the PLO and Israel as an interim instrument of self-rule for Palestinians living on the West Bank and Gaza Strip. The failure of the PLO and Israel to strike a permanent status agreement has resulted in the Authority retaining its powers. It is entitled to establish ministries and subordinate bodies, as required to fulfil its obligations and responsibilities. It possesses legislative and executive powers within the functional areas transferred to it in the 1995 Interim Agreement. Its territorial jurisdiction is restricted to Areas A and B in the West Bank and approximately two-thirds of the Gaza Strip.

Following an Israeli-Palestinian agreement on customs duties and VAT in Aug. 1994 the Palestinians set up their own customs and immigration points into Gaza and Jericho. Israel collects customs dues on Palestinian imports through Israeli entry points and transfers these to the Palestinian treasury.

A special committee is working on drafting a new Palestinian constitution. In March 2003 parliament approved the creation of the position of prime minister. Yasser Arafat nominated Mahmoud Abbas, the PLO Secretary General, to be the first premier. The president may dismiss the prime minister but parliament has to approve any new government.

There is a Palestinian *Council for Reconstruction and Development*.

RECENT ELECTIONS

Legislative Council elections were held on 25 Jan. 2006. Change and Reform (Hamas) won 74 seats; Fatah Movement, 45; Popular Front for the Liberation of Palestine, 3; the Alternative, 2; Independent Palestine, 2; Third Way, 2; ind. and others, 4. Turnout was 74·6%.

Presidential elections were held on 9 Jan. 2005. Mahmoud Abbas was elected president by 67·4% of votes cast, ahead of Mustafa Barghouti with 21·0%. There were five other candidates.

CURRENT GOVERNMENT

President of the Palestinian Authority: Mahmoud Abbas (Fatah); b. 1935.

   Prime Minister: Vacant.

INTERNATIONAL RELATIONS

The Palestinian National Council unilaterally declared Palestine an independent state in 1988. In 2011 President Mahmoud Abbas of the Palestinian National Authority submitted an application for membership of the United Nations. Membership requires the backing of two-thirds of member states, including all five of the permanent members of the Security Council. As of March 2013, 131 member states of the United Nations—67·9% of the total—recognized Palestine, but only two of the Security Council members did (China and Russia). Palestine was granted 'non-member observer state' status in the UN in Nov. 2012.

ECONOMY

Overview

Following the outbreak of the second *intifada* in Sept. 2000, conditions in the Palestinian-Administered Territories deteriorated, with real GDP in 2007 estimated to be around 18% lower than its peak in 1999.

   Revival followed the appointment of a new caretaker government in the West Bank in 2007, with growth estimated to have reached 5% in 2008 and accelerating further in 2009. This expansion has been driven by donor assistance, alongside increased 'tunnel activity' (shadow economic relations with neighbouring Egypt) and a relaxation of the blockade in Gaza. However, growth reflects recovery from the very low base reached during the second intifada and the economic situation remains poor.

   Unemployment has been amongst the highest in the world during the past decade. In 2009 just over one-fifth of the population lived in poverty, a 4% decline compared to 2004.

Currency

Israeli currency is in use.

Performance

The total GDP of the West Bank and the Gaza Strip was US$4·0bn. in 2007.

Banking and Finance

Banking is regulated by the Palestinian Monetary Authority. Palestine's leading bank is Arab Bank. A securities exchange, the Palestine Securities Exchange, opened in Nablus in Feb. 1997.

ENERGY AND NATURAL RESOURCES

Forestry

In 2010 forests covered 9,000 ha. or 2% of the total land area.

COMMUNICATIONS

Telecommunications

There were 368,200 landline telephone subscriptions in 2009 (equivalent to 93·7 per 1,000 inhabitants) and 1,314,400 mobile phone subscriptions (or 343·5 per 1,000 inhabitants) in 2008. In 2009, 32·2% of the population were internet users. In March 2012 there were 915,000 Facebook users.

SOCIAL INSTITUTIONS

Justice

The Palestinian police consists of some 15,000; they are not empowered to arrest Israelis, but may detain them and hand them over to the Israeli authorities. There were six executions in 2012 (three in 2011). All of the executions in 2011 and 2012 were in the Hamas-controlled Gaza Strip.

Education

Adult literacy was 94·6% in 2009 (97·4% among males and 91·7% among females).

CULTURE

World Heritage Sites

The Birthplace of Jesus: Church of the Nativity and the Pilgrimage Route, Bethlehem was the first Palestinian site added to the UNESCO World Heritage List (2012). The Church of the Nativity is considered to be the oldest continuously operating Christian church in the world. The site also includes numerous convents and churches, bell towers, terraced gardens and a pilgrimage route.

Tourism

In 2008, 387,000 non-resident tourists stayed in hotels and similar accommodation (up from 264,000 in 2007 and 123,000 in 2006).

FURTHER READING

Chehab, Zaki, *Inside Hamas: The Untold Story of the Militant Islamic Movement.* 2007
Gelvin, James L., *The Israel-Palestine Conflict: One Hundred Years of War.* 2005
Hilal, Jamil, *Where Now for Palestine?: The Demise of the Two-State Solution.* 2007
Kershner, Isabel, *Barrier: The Seam of the Israeli-Palestinian Conflict.* 2005
Kimmerling, B. and Migdal J. S., *Palestinians: the Making of a People.* 1994.—*The Palestinian People: A History.* 2003
Mishal, Shaul and Sela, Avraham, *The Palestinian Hamas: Vision, Violence, and Coexistence.* 2006
Pappe, Ilan, *A History of Modern Palestine: One Land, Two Peoples.* 2003.—*The Forgotten Palestinians: A History of the Palestinians in Israel.* 2011
Peleg, Ilan and Waxman, Dov, *Israel's Palestinians: The Conflict Within.* 2011
Rubin, B., *Revolution Until Victory? The Politics and History of the PLO.* 1994
Segev, T., *One Palestine, Complete.* 2000
Smith, Charles D., *Palestine and the Arab-Israeli Conflict.* 7th ed. 2010
Stendel, O., *The Arabs in Israel.* 1996
Wasserstein, Bernard, *Israel and Palestine: Why They Fight and Can They Stop?* 2003

Statistical office: Palestinian Central Bureau of Statistics.
Website: http://www.pcbs.gov.ps

ITALY

SWITZ.   AUST.
                SLOV.
Milan   Venice
Turin Genoa SAN
             MARINO   CROATIA
FRANCE   Florence
         ITALY   Adriatic Sea
            ROME
Corsica
(France)
         VATICAN
         CITY   Naples
         Sardinia
       Tyrrhenian Sea
Mediterranean
     Sea   Palermo
0   100 mi   Ionian
0   150 km   Sea
                    Sicily
© Research Machines plc 2006

Repubblica Italiana
(Italian Republic)

Capital: Rome
Population projection, 2015: 61·24m.
GNI per capita, 2011: (PPP$) 26,484
HDI/world rank: 0·874/24
Internet domain extension: .it

KEY HISTORICAL EVENTS

Excavations at Isernia have uncovered remains of Palaeolithic Neanderthal man that date back 70,000 years. New Stone Age settlements have been found across the Italian peninsula and at the beginning of the Bronze Age there were several Italic tribes, including the Ligurians, Veneti, Apulians, Siculi and the Sardi. The Etruscans were established in Italy by around 1200 BC. Their highly civilized society flourished between the Arno and Tiber valleys, with other important settlements in Campania, Lazio and the Po valley. The Etruscans were primarily navigators and travellers competing for the valuable trading routes and markets with the Phoenicians and Greeks. During the 8th century BC the Greeks had begun to settle in southern Italy and presented a challenge to Etruscan domination of sea trade routes. Greek settlements were established along the southern coast, on the island of Ischia in the Bay of Naples and in Sicily where the Corinthians founded the city of Syracuse. These colonies were known as *Magna Graecia* and flourished for six centuries. Magna Graecia eventually succumbed to the growing power of Rome where the impact of the Hellenic culture had already been felt.

According to legend, Rome was founded on 21 April 753 BC by Romulus (a descendant of Aeneas, a Trojan) who, after killing his twin brother, Remus, declared himself the first king of Rome.

The Etruscan dynasty of Tarquins gained control in 616 BC and expanded Roman agriculture and trade to rival the Greeks. The Romans overthrew the Tarquins in 510 BC and the first Roman Republic was born.

With the Republic came the establishment of the 'Roman Code', a collection of principles of political philosophy that enshrined the sovereign rights of Roman citizens. The early Roman Senate was dominated by a few patrician families, who held a monopoly on public office with the *equites* (the highest class of non-noble rich).

With the exception of the Greek city-states, Italy was unified by the Romans, who then set their sights on the Mediterranean, controlled by Carthage. Between 264–146 BC Carthage and Rome fought three wars (the Punic Wars) for supremacy of the Mediterranean trade routes. At the start Carthage was the more powerful, with a colonial empire that stretched as far as Morocco and included Sicily, Corsica, Sardinia and parts of Spain. Rome was also inexperienced in maritime war. In 218 BC the second Punic War started when Hannibal crossed the Alps and marched south, defeating the Romans in a series of battles in Italy. Without taking Rome itself, he crossed over to Zama in North Africa where he was finally defeated by Scipio in 202 BC. But by the end of the third Punic War in 146 BC the destruction of Carthage was total and Macedonian Greece was added to Rome's provinces. Rome incorporated Spain into her colonies and became the dominant power in the Mediterranean.

This dominance of trade routes led to great riches for Rome and the ensuing corruption among the upper ruling classes gave rise to social unrest. Sulla, a patrician general, marched on Rome in 82 BC, took the city in a bloody coup and instituted a new constitution. Nine years later Spartacus, an escaped slave, led 70,000 of his fellow slaves in a rampage throughout the peninsula. Out of the ensuing chaos, Julius Caesar emerged as leader. He had already conquered Gaul and declared southern Britain a part of Rome in 54 BC. His disregard for the Senate led to his legions being disbanded but he remained popular and returned to Rome a hero. His strength and charisma led to his assassination by members of the Senate on the Ides of March 44 BC. After his death, various rival successors fought to gain control, including Mark Anthony (Marcus Antonius), Marcus Junius Brutus and Gaius Cassius. But it was Caesar's nephew Octavian, having defeated Mark Anthony in 31 BC, who was crowned the first emperor of Rome in 27 BC, assuming the title Augustus.

Roman Domination

Augustus reigned for 45 years. With the aid of a professional army and an imperial bureaucracy he established the *Pax Romana* while extending the empire and disseminating its laws and civic culture. The arts thrived with writers, dramatists and philosophers such as Cicero, Plautus, Terence, Virgil, Horace and Ovid developing Latin into an expressive and poetic language. In 100 BC Rome itself had more than 1·5m. inhabitants and the Roman Empire was a unified diversity of many races and creeds. It had more than 100,000 km of paved roads, a complex of sophisticated aqueducts, and an efficient army and administrative system.

In AD 14 Augustus was succeeded by his stepson, Tiberius, who ruled in an era that saw the rise of Christianity. Successive emperors tried to suppress the new religion which spread quickly throughout the empire. The deranged Emperor Nero, who came to power in AD 54, intensified persecution of the Christians and was accused of setting Rome on fire. His death in AD 68 brought the Julio-Claudian dynasty to a close and, after a period of instability, Vespasian, the son of a provincial civil servant, took the throne and began some of the most ambitious building projects the Empire had seen. He started the Colosseum (completed by his son Titus) and the Arco di Tito (where the Via Sacra joins the Forum).

In AD 98 the Senate elected Trajan as emperor. Beginning a century of successful rule by the Antonine dynasty, he expanded the empire with the conquests of Dacia (Romania), Mesopotamia, Persia, Syria and Armenia. By the end of his reign the Roman Empire stretched from the Persian Gulf to Britain, from the Caspian Sea to Morocco and from the Sahara to the Danube. Trajan was responsible for several great architectural projects. A huge column depicting his Dacian campaigns served as his tomb in Rome. Trajan's successor, Hadrian, continued this programme of huge constructions, including Hadrian's Wall in Britain. After his death in 138, his tomb was converted into the fortress of Castel Sant'Angelo on the banks of the Tiber.

Under pressure from Teutonic tribes along the Danube and as a result of the increasingly strong influence of the Eastern religions, Rome began to lose control of its empire at the start of the 3rd century. In 306 Constantine became emperor. After he converted to Christianity in 313 his Edict of Milan established Rome as the headquarters of the Christian religion. A new building programme of Christian cathedrals and churches began throughout Italy. At the same time, Constantine cultivated the wealthy eastern regions of the Empire and, in 324, he moved his capital to Constantinople (now İstanbul). The decline of the Roman Empire continued when, after the death of Constantine, two brothers, Valens and Valentian, divided the Empire. The west and east gradually became alienated, separated by invaders, language and religious interpretation. 'Rome' endured in the east as the Byzantine Empire, the most powerful medieval state in the Mediterranean.

Fall of Rome
The western half of the Roman Empire, having embraced Christianity as the state religion, came under repeated attacks from Central European ('Barbarian') tribes. The Germanic Vandals had cut off Rome's corn supplies from North Africa, and the Visigoths, a Teutonic tribe, controlled the northern Mediterranean coast and northern Italy. In 452 Attila the Hun, from the steppes of Central Asia, invaded and forced the people of northeastern Italy onto a lagoon haven that became Venice. Rome was captured and sacked in 455 by the Vandals and in 476 a Germanic mercenary captain, Odovacar, deposed Romulus Augustus, the last of the Western Roman Emperors. This date is generally accepted as the end of the Roman Empire in the West.

In 493 Odovacar was succeeded by Theodoric, an Ostrogoth who had acquired a taste for Roman culture. Theodoric ruled from Ravenna and by the time he died in 527 he had managed to restore peace to Italy. On his death, Italy was reconquered by an emperor of the Eastern Roman Empire, Justinian, who together with his wife Theodora laid the foundations of the Byzantine period. Although the Lombards drove back the Justinian conquest, Byzantine emperors managed to retain control of parts of southern Italy until the 11th century.

In the mid-5th century Attila the Hun had been persuaded not to attack Rome by Pope Leo I ('The Great'). This and a document known as the 'Donation of Constantine' secured the Western Roman Empire for the Catholic Church. In 590 Gregory I became pope and set about an extensive programme of reforms, including improved conditions for slaves and the distribution of free bread in Rome. He oversaw the Christianization of Britain, repaired Italy's network of aqueducts and created the foundations for Catholic services and rituals and church administration.

The invasion of Italy by the Lombards began before Gregory became pope and, although they eventually penetrated as far south as Spoleto and Benevento, they were unable to take Rome. They settled around Milan, Pavia and Brescia and abandoned their own language and customs in favour of the local culture. However, they were sufficiently threatening to cause the pope to invite the Franks under King Pepin to invade. In 756 the Franks overthrew the Lombards and established the Papal States (which survived until 1870). Pepin issued his 'Donation of Pepin', which gave the land still controlled by the Byzantine Empire to Pope Stephen II, proclaiming him and future popes the heirs of the Roman emperors. Pepin's son, Charlemagne, succeeded him and was crowned emperor on Christmas Day 800 by Pope Leo III in St Peter's Basilica in Rome. The installation of a 'Roman' emperor in the West—what was to become the Holy Roman Empire— endorsed the separation between Rome and Byzantium and moved the seat of European political power north of the Alps.

After Charlemagne's death it proved impossible to keep the enormous Carolingian Empire intact. In the period of anarchy that followed, many small independent rival states were established while in Rome the aristocratic families fought over the Papacy. Meanwhile, southern Italy was prospering under Muslim rule. By 831 Muslim Arabs had invaded Sicily and made Palermo their capital. Syracuse fell to them in 878. They created a Greek style civilization with Muslim philosophers, physicians, astronomers, mathematicians and geographers. Cotton, sugarcane and citrus fruits appeared for the first time in Italy and taxes were lowered. Hundreds of mosques were built and all over the region centres of academic and medical learning sprang up. Southern Italy lived harmoniously under Arab influence for more than 200 years while the north remained unsettled. After the collapse of the Carolingian Empire, warfare broke out between local rulers, forcing many people to take refuge in fortified hill towns. In 962 Otto I, a Saxon, was crowned Holy Roman Emperor, the first of a succession of Germanic emperors that was to continue until 1806.

At the beginning of the 11th century the Normans began to enter southern Italy in great numbers, where they had originally been recruited to fight the Arabs. Establishing themselves in Apulia and Calabria, they assimilated much of the eastern culture, co-existing peacefully with the Arabs. The architecture of churches and cathedrals built during this period shows the merging of the two cultural and religious influences. Roger II of Sicily (reigned 1112–54), nephew of the adventurer Robert Guiscard, extended Norman Hauteville power over southern Italy and his navy was dominant in the Mediterranean. He presided over a famous court of scholars and artists, many from the Muslim world, making Palermo a model of tolerance and learning.

North South Divide
Meanwhile, the delicate relationship between the Holy Roman Empire based in the north of Europe and the Papacy in the south was maintained by a common desire to recapture the Holy Land from the Muslims. Crusades were launched, mostly from the northern states, but achieved little. Germanic claims to the southern territories grew and after Frederick I (known as Barbarossa) was crowned Holy Roman Emperor in 1155, he married off his son Henry to the heir to the Norman throne in Sicily. Frederick II, Barbarossa's grandson, came to the throne of Sicily as a child in 1197 and was crowned Holy Roman Emperor in 1220. An enlightened and tolerant ruler, he was known as 'Stupor Mundi' ('Wonder of the World'). An accomplished warrior, he valued scholarship and the Arab culture and allowed Muslims and Jews freedom to follow their own religions. He founded the University of Naples in 1224 with the intention of producing a generation of administrators for his kingdom and moved the court of the Holy Roman Empire to the newly built octagonal masterpiece, Castel del Monte, in Apulia.

During this period a new middle class emerged; with the seat of government so far south, some of the northern cities began to free themselves from feudal control and set themselves up as autonomous states under the protection of either the pope or the emperor. Milan, Cremona, Bologna, Florence, Pavia, Modena, Parma and Lodi were the most important of these new states, each dominated by a powerful family, exercising governmental power in the form of *signorie*. These states functioned autonomously within larger regional areas: Veneto, Lombardy,

Tuscany, the Papal States and the Southern Kingdom. In 1265 Charles of Anjou (a Frenchman who had beheaded Frederick II's grandson) was crowned king of Sicily. Greatly increased taxes, especially on rich landowners, made him unpopular despite his programme of road building, reform of the monetary system, improvement of the ports and the opening of silver mines. In 1282 an uprising known as the Sicilian Vespers was sparked off by a French soldier assaulting a Sicilian woman. As a consequence of the opposition to the French in southern Italy, Palermo declared itself an independent republic while supporting the Spaniard Peter of Aragon as king. By 1302 the Anjou dynasty had established itself in Naples.

Plague

The Black Death (La Peste), the deadly plague that swept throughout Europe towards the end of the 13th century, ravaged the populations of the major cities, which were already struggling with famine after years of war. Despite this, the strength of the northern and central Italian city-states was increasing. The rival maritime republics of Venice and Genoa had their own fleets. Venice had added the ports of Dalmatia, the Peloponnese and Cyprus to its possessions and Genoa's influence stretched as far as the Black Sea. Meanwhile, the pope and the Church turned their crusading zeal from the East towards European heretics. Pope Boniface, elected in 1294, came from Italian nobility and was determined to safeguard the interests of his own family. He claimed papal supremacy in worldly and spiritual affairs with his Papal Bull (Unam Sanctam) in 1302.

Meanwhile, a rival Papacy had appeared in Avignon, where John XXII was based. Rome had lost most of her former glory and had become little more than a battleground for the power struggles between the Orsini and Colonna families. The Papal claim to be temporal rulers of Rome was under threat and the Papal States began to fall apart. The period 1305–77, when seven successive popes ruled in Avignon, became known as the 'Babylonian Captivity'. In 1377 Pope Gregory XI returned to Rome after Cardinal Egidio d'Albornoz managed to restore the Papal States with his Egidian Constitutions. Rome was in such a ruined state that Gregory was obliged to set up his court in the Vatican, which was fortified and protected by the proximity of the Castel Sant'Angelo. Gregory died a year later and the Roman cardinals elected one of their own, Urban VI, as his successor. Urban's unpopularity was such that the French cardinals rebelled, electing their own pope, Clement VII, who set up his rival claim in Avignon. Yet another rival pope set himself in Pisa and thus began the Great Schism that would separate the papacy from Rome for nearly half a century.

Renaissance

In 1418 the Great Schism was brought to an end by the Council of Constance and Rome began to recapture her previous glory. Italy was at the forefront of the Renaissance, a flowering of artistic and intellectual humanist expression in the city-states. After the Peace of Lodi in 1454, the powerful ruling families—among others the Medici in Florence, the Gonzaga in Mantua and the d'Este in Ferrara—were at leisure to sponsor the Renaissance and Rome became again the centre of Italian political, cultural and intellectual life. In Florence the Signoria was taken over by a wealthy merchant, Cosimo de Medici. His nephew, Lorenzo Il Magnifico, became one of the great patrons of the arts. Feudal lords like Lorenzo de Medici frequently switched allegiance between the popes and the emperors, becoming wealthy bankers and captains of adventure in the process. Having defeated its arch-rival, Genoa, in 1381, Venice grew enormously, transforming its commercial maritime empire into a territorial empire that stretched almost to Milan.

The peace was shattered in 1494 by the invasion of Charles VIII, king of France. Encouraged to pursue his claim to the crown of Naples by Ludovico Sforza, duke of Milan, Charles shocked the

Italian cities into an alliance to expel his army. As cities competed to become the richest and most cultured, a Dominican monk, Girolamo Savonarola, preached against humanism in Florence. He persuaded Charles VIII to overthrow the Medici family and declare a Florentine republic. Although he was eventually excommunicated, hanged and burned at the stake, Savonarola exerted a lasting influence on Florentine politics. The Venetian expansion, through diplomatic and military guile, had alienated Venice's neighbours, who formed in 1508 the League of Cambrai, which came close to eradicating the Venetian Republic.

The appearance of Spanish power in Naples began the Habsburg-Valois wars that used Italy as a battlefield until the Peace of Cateau-Cambrésis in 1559. These Italian Wars radically altered the political landscape of the peninsula, leaving Spain dominant in Italy. Florence's time as a republic was brief. The Emperor Charles V, who had sacked Rome in 1527, reinstated the Medici, who went on to rule Florence for the next 210 years.

By the second half of the 16th century, the Church of Rome was obliged to respond to the rise of the Protestant movement (the Reformation), inspired in Germany by Martin Luther. During the Counter-Reformation, the Inquisition, backed by Catholic Spain, was used to suppress heresy. Spain succeeded in dominating Italy during the second half of the 16th century but when Charles II (the last of the Spanish Habsburgs) died in 1700, the War of the Spanish Succession saw Italy become a prize for the dominant European powers. Italy was divided amongst the Austrian Habsburgs, the Spanish Bourbons, Savoy and the independent states. The papacy became less influential, the Jesuits were expelled from Portugal, France and Spain and, thanks to intermarriage between many of the ruling houses of Europe and new trading laws, many national barriers were broken down. The 18th century Age of Enlightenment gave Italy some of its greatest thinkers and writers as well as liberal legal reforms.

Unification

In 1796 Napoleon Bonaparte invaded Italy and declared an Italian Republic under his personal rule. In creating a single political entity, he laid the basis for modern Italy. The Congress of Vienna, which met after the defeat of Napoleon in 1815, reinstated Italy's former rulers. Secret societies, made up of disillusioned middle class intellectuals, fought for a new constitution to reunify the country. One such was founded in 1830 by a Genoan, Giuseppe Mazzini. His 'Young Italy' was committed to liberating the country from foreign dominance and to establishing a unified state under a republican government, a campaign that came to be known as Il Risorgimento. During the 1830s and 1840s Mazzini instigated a series of unsuccessful uprisings until he was exiled. By 1848 revolutionary uprisings were taking place all over Europe and the Italian Nationalist movement was gaining ground. Two supporters of the Nationalist cause, Cesare Balbo and Count Camillo Benso di Cavour, advocated an Italian constitution and a bicameral legislature.

As nationalist feeling increased, Giuseppe Garibaldi, whose terrorist activities for Young Italy had obliged him to flee to South America, returned to Italy and allied himself with the Italian National Society. Cavour, the prime minister of Sardinia-Piedmont, attempted to remove Austria from Italy with French help but it was not until Garibaldi and 1,000 volunteers (the Red Shirts) took Sicily and Naples from the Bourbons in 1860 that unification became a real possibility. Garibaldi handed over these kingdoms to Victor Emmanuel II, king of Sardinia-Piedmont. This was to the relief of Cavour, who had feared that Garibaldi might institute a rival republican government in the south. Although Italy was declared a kingdom in 1861 under Victor Emmanuel II, the country was still not unified. Venice was in the hands of the Austrians while France held Rome. In 1866 the Italians took the Veneto from the Prussians and in 1870 Rome was recaptured from the French. Only the Papal troops resisted the advance of the Italian army in 1870 and Pope Pius IX refused to recognize the Kingdom of Italy. In retaliation, the government stripped the pope of his temporal powers. Thus Italy was fully unified.

Twentieth Century

The turn of the 20th century saw popular support fluctuate between left-wing socialist and right-wing imperialist political parties. When the First World War broke out in 1914, Italy remained neutral although the State was associated with the British, French and Russian allies while the Papacy declared for Catholic Austria. In 1919 Benito Mussolini founded the Italian Fascist Party, whose black shirts and Roman salutes were to become the symbols of aggressive nationalism in Italy for the next two decades. In the elections of 1921 the Fascist Party won 35 of the 135 seats in the Italian parliament. A year later, Mussolini raised a militia of 40,000 'Black Shirts' and marched on Rome to 'liberate' it from the socialists. In 1922 the king asked Mussolini to form a government. His Fascist party won the elections of 1924 and Mussolini assumed the title *Il Duce*. By the end of 1925 Mussolini had expelled all opposition parties from parliament and gained control of the trade unions. Four years later, he signed a pact with Pope Pius XI declaring Catholicism the sole religion of Italy and recognizing the Vatican as an independent state. In return, the pope finally recognized the United Kingdom of Italy.

Mussolini's aggressive foreign policy resulted in disputes with Greece over Corfu and military campaigns in the Italian colony of Libya. In 1935 Italy invaded Abyssinia (now Ethiopia) and captured Addis Ababa. The newly formed League of Nations condemned this action and imposed sanctions. In the face of international isolation, Mussolini formed an alliance with the German dictator, Adolf Hitler, and in 1936 the Rome-Berlin Axis was formed. Having annexed Albania in April 1939, Italy entered the Second World War in June 1940. Mussolini's armies invaded Greece from Albania in Oct. 1940 but were repelled, forcing Hitler to invade Yugoslavia and Greece in April 1941. This diversion of German troops has been seen as a critical factor in the ultimate failure of the invasion of the USSR, delayed from May to June 1941. The Italian colonies of East Africa were lost in 1941 and Italian forces in North Africa surrendered in May 1943. The Allied armies landed in Sicily in July 1943 and, in the face of diminishing popular support for fascism and Hitler's refusal to assign more troops to the defence of Italy, the king led a coup against Mussolini and had him arrested. In the 45 days that followed, Italy exploded in a series of uprisings against the war. The king signed an armistice with the Allies and declared war on Germany but Nazi troops had already overrun northern Italy. The Germans rescued Mussolini from prison and installed him as a puppet ruler. In 1945 after trying to flee the country, Mussolini was recaptured by Italian partisans and shot. After the Italian Resistance suffered huge losses against the Germans, the allies liberated northern Italy in May 1945.

Post-War Period

In the years following the end of the Second World War, Italy's political forces attempted to regroup. The Marshall Plan, America's post-war aid programme, exerted considerable political and economic influence. The constitutional monarchy was abolished in 1946 by referendum and a republic was formed with a president (elected for a seven-year term by an electoral college), a two-chamber parliament and a separate judiciary. Initially the newly formed Christian Democrats under Alcide De Gasperi were in power with both the Communist Party and the Socialist Party, participating in a series of coalition governments until they were both excluded by De Gasperi in 1947. More than 300 separate political factions have struggled for power throughout the post-war era and no government has lasted longer than four years. Despite this instability, the war-damaged Italian economy began to pick up in the early 1950s. The industrialized northern regions thrived while the less industrialized south remained underdeveloped. The Cassa per il Mezzogiorno (a state fund for the South) was founded to try to redress the balance but with limited success.

In 1957 Italy became a founder member of the European Economic Community (EEC). The rapid growth of the motor industry, most notably Fiat in Turin, saw huge migrations of peasants from the south to work in the factories. By the mid-1960s the Communist Party, which had been gradually increasing its share of the poll at each election, had more card carrying members than the Christian Democrats but without participating in government. Social unrest was commonplace and in 1969 a series of strikes, demonstrations and riots followed on the heels of unrest elsewhere in Europe. Various terrorist groups were active including the extreme left-wing socialist group, the Red Brigade, founded in 1970. Right-wing neo-fascist terrorists were also in action, and in the less developed south, the Mafia, a loose coalition of crime 'families', flourished. Most of Italy's social, economic and political structures were manipulated by these unofficial organizations.

In 1963 Aldo Moro, a Christian Democrat, was appointed prime minister (a post he held until 1968) and invited the Socialists into his government. By the 1970s he was working towards a compromise to allow the Communists to enter government when he was captured, held hostage and finally murdered by the Red Brigade. This national outrage prompted the government to appoint Carabinieri Gen. Carlo Alberto dalla Chiesa to wipe out the terrorist groups. He instituted a system of *pentiti* (informants) who, in return for collaboration, would receive greatly reduced prison sentences. In 1980 he was asked to expand his area of operations to include the Mafia but was assassinated in Palermo a few months later. Throughout the 1970s Italy experienced radical social and political change. The country was divided into regional administrative areas with their own elected governments. Divorce became legal, women's rights were expanded (Italian women only achieved full suffrage after the Second World War) and abortion was legalized. In 1983 the minority Christian Democratic government handed the premiership to the Socialists under Bettino Craxi.

Italy was well on its way to becoming one of the world's leading economic powers but the 1990s brought fresh crises in the economic and political arenas. Unemployment and inflation rose sharply which, combined with a huge national debt and unstable lira, led to instability. On the political front, the Communist Party split with the hard-liners forming the Rifondazione Communista, led by Fausto Bernotti, while the more moderate members set up the Democratic Party of the Left. In early 1992 the arrest of a Socialist Party worker on charges of accepting bribes in exchange for public works contracts sparked off Italy's largest ever political corruption scandal. Investigations into 'Tangentopoli' ('kick-back city') implicated thousands of politicians, public officials and businessmen. Former Prime Minister Bettino Craxi was forced to resign as party secretary after he came under investigation for bribery. Allied to Italy's humiliating exit from Europe's Exchange Rate Mechanism (ERM), the old political establishment was driven out of office. In the April 1992 elections, the Christian Democrat share of the vote dropped by 5% while the Lega Nord (the Northern League), under Umberto Bossi, took 9% of the vote on an anti-corruption, federalist platform. Oscar Luigi Scalfaro was elected president on a promise to set about reforming electoral laws and clearing up the Tangentopoli scandal. Investigations into corruption continued, despite reprisals from the Mafia. Craxi was convicted *in absentia* while Giulio Andreotti, who was prime minister three times between 1972 and 1992, was brought to trial in 1995 on charges of dealing with the Sicilian Mafia.

In the 1994 elections a right-wing coalition was elected. The Freedom Alliance, including the neo-fascist National Alliance and the federalist Northern League, was led by Silvio Berlusconi, a multi-millionaire media tycoon. Berlusconi lost his majority when the Northern League withdrew after nine months. Under mounting criticism for his failure to disassociate himself from his business interests and after receiving a vote of no confidence, Berlusconi resigned. After leaving the Freedom Alliance, the Northern League became more fanatical, advocating a 'Northern Republic of Padania', a separation of the rich northern regions from the poorer southern ones. The 1996 elections brought to power the centre-left 'Olive Tree' alliance with Romano Prodi as prime minister. Prodi aimed to balance the budget and create a stable political environment. He gained his first objective with a succession of economic measures that prepared the way for Italy's entry into EMU.

Prodi was succeeded by Massimo D'Alema in 1998 who, in turn, was replaced by Giuliano Amato in 2000. By the time of the 2001 elections Berlusconi's popularity had revived and he formed a new centre-right coalition. He introduced the first major constitutional reforms in 55 years, allowing the nation's 20 regions increased responsibility for their own tax, education and environmental programmes.

Berlusconi's tenure was dogged by questions over his private business interests. In Oct. 2002 parliament passed new criminal reform legislation that critics claimed was partly designed to allow Berlusconi to escape charges of corruption. He nonetheless stood trial in May 2003 on corruption charges related to his business dealings in the 1980s but the trial was halted the following month when the new law granted the prime minister immunity from prosecution. The legislation was declared void by the constitutional court in Jan. 2004 and his trial resumed three months later, culminating in his acquittal in Dec. 2004.

The proposed EU constitution was approved by parliament in April 2005, shortly before Berlusconi's government fell after a poor showing in regional elections. He was then asked by the president to form a new government but was beaten by Prodi in the general election of April 2006. In May 2006 Giorgio Napolitano became president. Prodi resigned in Feb. 2007 when the Senate refused to back his foreign policy but resumed his premiership after winning confidence votes in both the upper and lower houses.

In early 2008 Prodi's coalition split when a minor partner withdrew its support. Despite surviving a vote of no confidence in the lower house, Prodi lost a similar vote in the Senate. Parliament was dissolved in Feb. 2008 and Prodi was asked to remain as caretaker prime minister ahead of a general election in April 2008, in which Silvio Berlusconi was returned to power.

The economy was in a parlous state as the global financial crisis deepened. Berlusconi responded by imposing austerity measures but faced several votes of confidence as the economy continued to falter and revelations emerged about his private life. With the IMF warning that the country needed to reduce its public debt, Berlusconi resigned in Nov. 2011 following an impasse in parliament over a new austerity package. He was replaced by Mario Monti, a technocrat charged with restoring economic stability. Monti introduced an austerity programme before his government collapsed in Dec. 2012. Parliamentary elections in Feb. 2013 resulted in deadlock until April when Giorgio Napolitano was re-elected president and appointed Enrico Letta as premier.

TERRITORY AND POPULATION

Italy is bounded in the north by Switzerland and Austria, east by Slovenia and the Adriatic Sea, southeast by the Ionian Sea, south by the Mediterranean Sea, southwest by the Tyrrhenian Sea and Ligurian Sea and west by France.

The area is 301,338 sq. km. Populations at successive censuses (in 1,000) were as follows:

10 Feb. 1901	33,778	15 Oct. 1961	50,624
10 June 1911	36,921	24 Oct. 1971	54,137
1 Dec. 1921	37,856	25 Oct. 1981	56,557
21 April 1931	41,043	20 Oct. 1991	56,778
21 April 1936	42,399	21 Oct. 2001	56,996
4 Nov. 1951	47,516	9 Oct. 2011	59,434

Population in 2011, 59,433,744 (30,688,237 females). Density: 197 per sq. km.

The UN gives a projected population for 2015 of 61·24m.

In 2011, 68·6% of the population lived in urban areas.

The following table gives area and population of the Autonomous Regions (censuses 2001 and 2011):

Regions	Area in sq. km	Resident pop. census, 2001	Resident pop. census, 2011	Density per sq. km, 2011
Piedmont (Piemonte)	25,399	4,214,677	4,363,916	172
Valle d'Aosta[1]	3,263	119,548	126,806	39
Lombardy (Lombardia)	23,861	9,032,554	9,704,151	407
Trentino-Alto Adige[1]	13,607	940,016	1,029,475	76
Bolzano (Bozen)	7,400	462,999	504,643	68
Trento	6,207	477,017	524,832	85
Veneto	18,392	4,527,694	4,857,210	264
Friuli-Venezia Giulia[1]	7,855	1,183,764	1,218,985	155
Liguria	5,421	1,571,783	1,570,694	290
Emilia Romagna	22,124	3,983,346	4,342,135	196
Tuscany (Toscana)	22,997	3,497,806	3,672,202	160
Umbria	8,456	825,826	884,268	105
Marche	9,694	1,470,581	1,541,319	159
Lazio	17,207	5,112,413	5,502,886	320
Abruzzi	10,798	1,262,392	1,307,309	121
Molise	4,438	320,601	313,660	71
Campania	13,595	5,701,931	5,766,810	424
Puglia	19,362	4,020,707	4,052,566	209
Basilicata	9,992	597,768	578,036	58
Calabria	15,080	2,011,466	1,959,050	130
Sicily (Sicilia)[1]	25,708	4,968,991	5,002,904	195
Sardinia (Sardegna)[1]	24,090	1,631,880	1,639,362	68

[1]With special statute.

Communes of more than 100,000 inhabitants, with population resident at the census of 9 Oct. 2011:

Rome (Roma)	2,617,175	Ravenna	153,740
Milan (Milano)	1,242,123	Cagliari	149,883
Naples (Napoli)	962,003	Foggia	147,036
Turin (Torino)	872,367	Rimini	139,601
Palermo	657,561	Salerno	132,608
Genoa (Genova)	586,180	Ferrara	132,545
Bologna	371,337	Sassari	123,782
Florence (Firenze)	358,079	Monza	119,856
Bari	315,933	Siracusa (Syracuse)	118,385
Catania	293,902	Latina	117,892
Venice (Venezia)	261,362	Pescara	117,166
Verona	252,520	Forlì	116,434
Messina	243,262	Bergamo	115,349
Padua (Padova)	206,192	Trento	114,198
Trieste	202,123	Vicenza	111,500
Taranto	200,154	Terni	109,193
Brescia	189,902	Giugliano in	
Prato	185,456	Campania	108,793
Reggio di Calabria	180,817	Bolzano (Bozen)	102,575
Modena	179,149	Novara	101,952
Parma	175,895	Ancona	100,497
Perugia	162,449	Piacenza	100,311
Reggio nell'Emilia	162,082	Andria	100,052
Livorno	157,052		

The official language is Italian, spoken by 92·8% of the population in 2003. Romanians, numbering 997,000 in Dec. 2010, make up the largest ethnic minority group. In Jan. 2010, 7·0% of the population was foreign-born

In addition to Sicily and Sardinia, there are a number of other Italian islands, the largest being Elba (363 sq. km), and the most distant Lampedusa, which is 205 km from Sicily but only 113 km from Tunisia.

SOCIAL STATISTICS

Vital statistics (and rates per 1,000 population), 2008: births, 576,659 (9·6); deaths, 585,126 (9·8). Marriages in 2007, 250,360 (4·2); divorces in 2006, 49,534 (0·8). Infant mortality rate, 2010 (up to one year of age): three per 1,000 live births. Expectation of life, 2007: females, 84·0 years; males, 78·1. In 2010, 20·3% of the population was over 65—one of the highest percentages in the world.

Annual population growth rate, 2000–05, 0·6%; fertility rate, 2008, 1·4 births per woman. With only 17·7% of births being to unmarried mothers in 2007 (albeit up from 8·1% in 1995), Italy has one of the lowest rates of births out of marriage in Europe.

In 2006 there were 3,701 suicides; 76·8% were men.

At 1 Jan. 2007 there were 2,938,922 foreigners living in Italy, up from 2,670,514 a year earlier. In 2005, 53,931 people emigrated from Italy and there were 304,960 immigrants into the country (compared to 440,301 immigrants in 2003). Since 1992 Italy has experienced net immigration every year, peaking in 2003, although for most of the 20th century up until the mid-1960s it saw mass emigration. Italy received 30,324 asylum applications in 2008, up from 14,053 in 2007. New legislation was introduced in 2002 to tighten up immigration rules.

CLIMATE

The climate varies considerably with latitude. In the south, it is warm temperate, with little rain in the summer months, but the north is cool temperate with rainfall more evenly distributed over the year. Florence, Jan. 47·7°F (8·7°C), July 79·5°F (26·4°C). Annual rainfall 33" (842 mm). Milan, Jan. 38·7°F (3·7°C), July 73·4°F (23·0°C). Annual rainfall 38" (984 mm). Naples, Jan. 50·2°F (10·1°C), July 77·4°F (25·2°C). Annual rainfall 36" (935 mm). Palermo, Jan. 52·5°F (11·4°C), July 78·4°F (25·8°C). Annual rainfall 35" (897 mm). Rome, Jan. 53·4°F (11·9°C), July 76·3°F (24·6°C). Annual rainfall 31" (793 mm). Venice, Jan. 43·3°F (6·3°C), July 70·9°F (21·6°C). Annual rainfall 32" (830 mm).

CONSTITUTION AND GOVERNMENT

The Constitution dates from 1948. Italy is 'a democratic republic founded on work'. Parliament consists of the *Chamber of Deputies* and the *Senate*. The Chamber is elected for five years by universal and direct suffrage and consists of 630 deputies. The Senate is elected for five years on a regional basis by electors over the age of 25, each Region having at least seven senators. The total number of senators is 319, of which 315 are directly elected. The Valle d'Aosta is represented by one senator only and Molise by two. The President of the Republic can nominate five senators for life from eminent persons in the social, scientific, artistic and literary spheres. The President may become a senator for life. The *President* is elected in a joint session of Chamber and Senate, to which are added three delegates from each Regional Council (one from the Valle d'Aosta). A two-thirds majority is required for the election, but after a third indecisive scrutiny the absolute majority of votes is sufficient. The President must be 50 years or over; term of office, seven years. The Speaker of the Senate acts as the deputy President. The President can dissolve the chambers of parliament, except during the last six months of the presidential term. An attempt to create a new constitution, which had been under consideration for 18 months, collapsed in June 1998.

There is a *Constitutional Court* that consists of 15 appointed judges, five each by the President, Parliament (in joint session) and the highest law and administrative courts. The Court can decide on the constitutionality of laws and decrees, define the powers of the State and Regions, judge conflicts between the State and Regions and between the Regions, and try the President and Ministers.

The revival of the Fascist Party is forbidden. Direct male descendants of King Victor Emmanuel are excluded from all public offices and have no right to vote or to be elected; their estates are forfeit to the State. For 56 years they were also banned from Italian territory until the constitution was changed in 2002 to allow them to return from exile. Titles of nobility are no longer recognized, but those existing before 28 Oct. 1922 are retained as part of the name.

A referendum was held in June 1991 to decide whether the system of preferential voting by indicating four candidates by their listed number should be changed to a simpler system, less open to abuse, of indicating a single candidate by name. The electorate was 46m. Turnout was 62·5% (there was a 50% quorum). 95·6% of votes cast were in favour of the change. As a result, an electoral reform of 1993 provides for the replacement of proportional representation by a system in which 475 seats in the Chamber of Deputies are elected by a first-past-the-post single-round vote and 155 seats by proportional representation in a separate single-round vote on the same day. There are 27 electoral regions. There is a 4% threshold for entry to the Chamber of Deputies.

At a further referendum in April 1993, turnout was 77%. Voters favoured the eight reforms proposed, including a new system of election to the Senate and the abolition of some ministries. 75% of the Senate is now elected by a first-past-the-post system, the remainder by proportional representation; no party may present more than one candidate in each constituency. Subsequent proposed constitutional reforms included a directly elected president with responsibility for defence and foreign policy, a reduction in the number of seats in the Senate and in the lower house and the creation of a third chamber to speak on behalf of the regions. Greater powers over tourism, transport and welfare policy have been devolved to the regions.

National Anthem

'Fratelli d'Italia' ('Brothers of Italy'); words by G. Mameli, tune by M. Novaro, 1847.

GOVERNMENT CHRONOLOGY

Presidents of the Council of Ministers (Prime Ministers) since 1944. (DC = Christian Democrats; DP = Democratic Party; DS = Democrats of the Left-Party of the European Socialism; FI = Forza Italia; PA = Action Party; PdL = People of Freedom; PRI = Italian Republican Party; PSI = Italian Socialist Party; Ulivo = Olive Tree; n/p = non-partisan)

1944–45	n/p	Ivanoe Bonomi
1945	PA	Ferruccio Parri
1945–53	DC	Alcide De Gasperi
1953–54	DC	Giuseppe Pella
1954	DC	Amintore Fanfani
1954–55	DC	Mario Scelba
1955–57	DC	Antonio Segni
1957–58	DC	Adone Zoli
1958–59	DC	Amintore Fanfani
1959–60	DC	Antonio Segni
1960	DC	Fernando Tambroni
1960–63	DC	Amintore Fanfani
1963	DC	Giovanni Leone
1963–68	DC	Aldo Moro
1968	DC	Giovanni Leone
1968–70	DC	Mariano Rumor
1970–72	DC	Emilio Colombo
1972–73	DC	Giulio Andreotti
1973–74	DC	Mariano Rumor
1974–76	DC	Aldo Moro
1976–79	DC	Giulio Andreotti
1979–80	DC	Francesco Cossiga
1980–81	DC	Arnaldo Forlani
1981–82	PRI	Giovanni Spadolini
1982–83	DC	Amintore Fanfani
1983–87	PSI	Benedettino Craxi
1987	DC	Amintore Fanfani
1987–88	DC	Giovanni Giuseppe Goria
1988–89	DC	Ciriaco De Mita
1989–92	DC	Giulio Andreotti
1992–93	PSI	Giuliano Amato
1993–94	n/p	Carlo Azeglio Ciampi

1994–95	FI	Silvio Berlusconi
1995–96	n/p	Lamberto Dini
1996–98	n/p	Romano Prodi
1998–2000	DS	Massimo D'Alema
2000–01	n/p	Giuliano Amato
2001–06	FI	Silvio Berlusconi
2006–08	Ulivo	Romano Prodi
2008–11	PdL	Silvio Berlusconi
2011–13	n/p	Mario Monti
2013–	DP	Enrico Letta

RECENT ELECTIONS

Parliamentary elections were held on 24–25 Feb. 2013. Pier Luigi Bersani's centre-left coalition won 345 of 630 seats in the Chamber of Deputies (including the Democratic Party with 297 seats), Silvio Berlusconi's centre-right coalition 125 (including People of Freedom, 98), Beppe Grillo's Five Star Movement 109 and Prime Minister Mario Monti's coalition 47, with smaller parties accounting for the remainder. In the Senate, Bersani's coalition won 123 of 315 seats (including the Democratic Party, 111), Berlusconi's coalition 117 (including People of Freedom, 98), the Five Star Movement 54 and Monti 19, again with smaller parties accounting for the remainder.

Giorgio Napolitano was elected re-president by an assembly of lawmakers and regional representatives on 20 April 2013, winning 738 votes to Stefano Rodotà's 217 in the sixth round of voting after five earlier rounds had failed to produce a clear result.

European Parliament

Italy has 73 (78 in 2004) representatives. At the June 2009 elections turnout was 65·1% (71·7% in 2004). People of Freedom won 29 seats with 35·3% of votes cast (political affiliation in European Parliament: European People's Party); Democratic Party, 21 with 26·1% (Progressive Alliance of Socialists and Democrats); the Northern League, 9 with 10·2% (Europe of Freedom and Democracy); Italy of Values, 7 with 8·0% (Alliance of Liberals and Democrats for Europe); the Union of Christian and Centre Democrats, 5 with 6·5% (European People's Party); South Tyrolean People's Party, 1 with 0·5% (European People's Party).

CURRENT GOVERNMENT

President: Giorgio Napolitano; b. 1925 (sworn in 15 May 2006 and re-elected 20 April 2013).

In April 2013 the coalition government comprised:

President of the Council of Ministers (Prime Minister): Enrico Letta; b. 1966 (PD; sworn in 28 April 2013).

Undersecretary to the Presidency of the Council: Filippo Patroni Griffi.

Deputy Prime Minister and Minister of Interior: Angelino Alfano.

Minister of Agriculture, Food and Forestry: Nunzia De Girolamo. *Cultural Heritage and Tourism:* Massimo Brai. *Defence:* Mario Mauro. *Economic Development:* Flavio Zanonato. *Economy and Finance:* Fabrizio Saccomanni. *Education, Universities and Research:* Maria Chiara Carrozza. *Environment:* Andrea Orlando. *Foreign Affairs:* Emma Bonino. *Health:* Beatrice Lorenzin. *Infrastructure and Transport:* Maurizio Lupi. *Justice:* Annamaria Cancellieri. *Labour and Social Policy:* Enrico Giovannini. *Ministers without Portfolio:* Gaetano Quagliariello (for *Constitutional Reforms*); Josefa Idem (for *Equal Opportunities, Sport and Youth Policies*); Enzo Moavero Milanesi (for *European Affairs*); Cécile Kyenge (for *Integration*); Giampiero D'Alia (for *Public Administration and Simplification*); Graziano Delrio (for *Regional Affairs and Autonomy*); Dario Franceschini (for *Relations with Parliament*); Carlo Trigilia (for *Territorial Cohesion*).

Government Website (Italian only): http://www.governo.it

CURRENT LEADERS

Enrico Letta

Position
Prime Minister

Introduction
Enrico Letta became prime minister in April 2013 at the head of a grand coalition that united his centre-left Democratic Party (PD) with the centre-right People of Freedom party (PdL) of Silvio Berlusconi. Letta's appointment was the culmination of two months of post-electoral negotiation. He argues the need for policies to boost economic growth alongside existing austerity measures.

Early Life
Born on 20 Aug. 1966 in Pisa, Enrico Letta studied international law at the University of Pisa and completed a PhD in European Community law at the city's Sant'Anna School of Advanced Studies. Having joined Christian Democracy (DC), from 1991–95 he was president of the umbrella organization Youth of the European People's Party. In this period he also worked for the Agency of Research and Legislation (AREL) think tank, becoming its secretary general in 1993.

After the demise of the DC in 1994, Letta joined the successor Italian People's Party (PPI), serving as its deputy leader from 1997–98. In Nov. 1998 he was appointed minister of European affairs in the government of Massimo D'Alema before taking on the industry portfolio from 1999 until 2001. He was elected to parliament in 2001 representing Democracy is Freedom–The Daisy Party (DL), into which the PPI had merged. From 2001–04 he was the DL's shadow minister for economic policy.

From 2004 until 2006 he sat as a member of the European Parliament, serving on the committee for economic and monetary affairs. In 2006 he returned to national politics, joining Romano Prodi's centre-left coalition government as secretary of the council of ministers. When the DL became part of the newly formed Democratic Party (PD) in 2007, Letta challenged for the leadership but secured only 11% of the vote in a poll won by Walter Veltroni. In 2008 Letta won a parliamentary seat as a PD member and proposed a bill to reform MPs' pay. In 2009 he became the PD's deputy leader under Pier Luigi Bersani. During the 2011–13 tenure of Prime Minister Mario Monti, Letta broadly supported the premier's programme of fiscal austerity.

In the inconclusive general election of Feb. 2013, when a quarter of the vote went to the anti-establishment Five Star Movement, the PD emerged as the largest party in the Chamber of Deputies. After two months of political deadlock and the rejection of successive presidential nominees, Bersani and the rest of the PD leadership resigned. In April 2013 the incumbent president, Giorgio Napolitano, was persuaded to stand for a second term. On securing re-election, he nominated Letta as prime minister at the head of a grand coalition of the PD and the PdL.

Career in Office
Letta was sworn in on 28 April 2013, having assembled a cabinet composed of figures from the left and right of politics. He urged that policies to boost growth should run alongside existing fiscal austerity measures, making visits to European neighbours to press the case. As well as reviving the economy, he faces the challenge of restoring public confidence in government while balancing the demands of his own party against those of the PdL. His task is further complicated by Italy's reliance on economic support from the EU and other international financial institutions.

DEFENCE

Head of the armed forces is the Defence Chief of Staff. Conscription was abolished at the end of 2004 with the military becoming all-professional from 2005. In Aug. 1998 the government voted to allow women into the armed forces.

In 2008 defence expenditure totalled US$40·6bn. (US$689 per capita). In 2007 defence spending represented 1·8% of GDP.

Army
Strength (2006) 110,000 (about 2,000 conscripts). Equipment includes 120 *Leopard,* 300 *Centauro* and 200 *Ariete* tanks. There are 35,500 Army reserves.

The paramilitary Carabinieri (police force with military status) number 111,400.

Navy
The principal ships of the Navy are the aircraft carriers *Giuseppe Garibaldi* and *Cavour,* commissioned in 1985 and 2008 respectively. The combatant forces also include six diesel submarines, five destroyers and 12 frigates. The Naval Air Arm, 2,000 strong, operates 15 combat aircraft and 63 armed helicopters.

Main naval bases are at La Spezia, Brindisi, Taranto and Augusta. The personnel of the Navy numbered 33,100 in 2006. There were 21,000 naval reservists.

Air Force
Control is exercised through two regional headquarters near Taranto and Milan.

Air Force strength in 2006 was about 44,000 (1,200 conscripts). There were 234 combat aircraft in operation in 2006 including Typhoons and Tornados.

ECONOMY
Agriculture accounted for 2% of GDP, industry 28% and services 70% in 2007.

Italy's 'shadow' (black market) economy is estimated to constitute approximately 27% of the country's official GDP.

In 2011 Italy gave US$4·3bn. in international aid. In terms of a percentage of GNI, however, Italy was one of the least generous major industrialized countries, giving just 0·20%.

According to the anti-corruption organization *Transparency International,* Italy ranked equal 72nd in the world in a 2012 survey of the countries with the least corruption in business and government, down from 29th in 2001.

Overview
Italy is the eurozone's third largest economy, with a diversified industrial base and average income levels on a par with those of other leading economies. The economic structure is in line with the OECD average, with a small and diminishing primary sector and a large gross value added contribution from the service sector. Manufacturing accounted for 16·7% of GDP in 2010.

Since 1988 economic performance has trailed that of other developed countries. Growth averaged only 1·5% per year in the decade leading up to the global financial crisis, with Italy losing significant market share in world trade since the mid-1990s. This has resulted from specialization in slow-growing sectors of world demand, comparatively weak FDI and low investment in research and development.

Small and medium-sized family-owned companies are the strongest component of the economy, producing high-quality consumer goods such as clothing, furniture and white goods. These companies have resisted public funding but face pressure from global economic integration and competition and are vulnerable to foreign acquisition. Italy's most dynamic region is the northeast, where most value added production is concentrated. In 1951 income per capita in Italy's southern regions was about 50% of that of the north and despite progress in the Mezzogiorno (the south and Sicily and Sardinia), the economic gap between north and south remains wide and has been magnified by the economic crisis.

Economic growth was 2·3% in 2006 following years of stagnant performance. With the global financial crisis came negative growth in both 2008 and 2009, leading to the worst recession in Italy since the Second World War. The economy contracted by 5·5% in 2009, with weak levels of fixed investment, net exports and private consumption. Exports fell 25% in a year, the first decline since 2000. Public debt, which decreased from over 120% in the mid-1990s to just over 115% prior to the financial crisis, once again broke through 120% in 2011. The financial sector, however, proved resilient and growth returned in the third quarter of 2009.

In July 2011 the government announced a €40bn. austerity package with the aim of balancing the budget by the end of 2013 but the measures failed to reassure the financial markets and there was a sharp rise in the price of Italian bond yields. This prompted the European Central Bank to intervene, purchasing Italian bonds to prevent spiralling borrowing costs. A second emergency budget was passed in Sept. 2011, worth €45·5bn. Measures included a 1% VAT rise to 21% and an increase in the retirement age for women in the private sector from 60 in 2014 to 65 in 2026.

The economy fell back into recession in the third quarter of 2011 as a result of sharp falls in consumption and investment. In Nov. 2011 unsustainable levels of borrowing prompted fears of an Italian default. A new government under Mario Monti initiated an austerity plan the following month with fiscal adjustments of US$40bn. over three years, including tax increases and pension reforms. The plan saw Italy's ten-year borrowing rate reduce from 7%, where it had stood for most of 2011, to close to 6% in early 2012. Nonetheless, Standard & Poor's downgraded Italy's credit rating in Jan. 2012 to BBB+ and downgraded 34 Italian banks a month later, predicting weak profitability in the banking sector over the coming years.

In March 2012 parliament passed a comprehensive liberalization package including labour market reforms and adjustments aimed at lowering the cost of doing business. In April 2012 a rule concerning a structural balanced budget was added to the constitution. In the fourth quarter of 2012 GDP decreased by 0·9% compared to the third quarter. In Jan. 2013 unemployment reached 11·7%. Youth unemployment stood at 38·7%, the third highest in the EU.

With low birth rates and a high and rising ratio of people over 65, Italy faces one of the greatest challenges from population ageing of any OECD country. In 2007 it spent the largest proportion of national income on pensions in the OECD. The OECD country with the second oldest population after Japan in 2010, pension spending in 2011 amounted to 11·5% of GDP. The IMF predicts that population ageing will increase pension spending by approximately 2% of GDP over the coming years and health spending by 3%. Along with the continuing trouble in the eurozone, low productivity, an ageing society and loss of competitiveness are expected to constrain Italy's future growth prospects.

For further developments *see* www.statesmansyearbook.com.

Currency
On 1 Jan. 1999 the euro (EUR) became the legal currency in Italy at the irrevocable conversion rate of 1,936·27 lire to 1 euro. The euro, which consists of 100 cents, has been in circulation since 1 Jan. 2002. On the introduction of the euro there was a 'dual circulation' period before the lira ceased to be legal tender on 28 Feb. 2002. Euro banknotes in circulation on 1 Jan. 2002 had a total value of €97·4bn.

Inflation rates (based on OECD statistics):

2002	2003	2004	2005	2006	2007	2008	2009	2010	2011
2·6%	2·8%	2·3%	2·2%	2·2%	2·0%	3·5%	0·8%	1·6%	2·9%

In Sept. 2009 gold reserves were 78·83m. troy oz and foreign exchange reserves US$35,474m. Total money supply in Aug. 2009 was €711,993m.

Budget

In 2009 central government revenues totalled €594·01bn. and expenditures €668·64bn. Principal sources of revenue in 2009: social security contributions, €210·79bn.; taxes on income, profits and capital gains, €191·44bn.; taxes on goods and services, €118·29bn. Main items of expenditure by economic type in 2009: social benefits, €289·25bn.; grants, €139·05bn.; compensation of employees, €99·86bn.

Italy's budget deficit in 2011 was 3·9% of GDP (2010, 4·6%; 2009, 5·4%). The required target set by the EU is a budget deficit of no more than 3%.

VAT was increased from 20% to 21% in Sept. 2011. The reduced rates were kept at 10% and 4%. The standard rate was scheduled to rise to 22% from 1 July 2013.

The public debt at 31 Dec. 2006 totalled €1,256,946m.

Performance

Real GDP growth rates (based on OECD statistics):

2002	2003	2004	2005	2006	2007	2008	2009	2010	2011
0·4%	0·0%	1·6%	1·1%	2·3%	1·5%	−1·2%	−5·5%	1·8%	0·6%

According to the National Institute of Statistics, the real GDP growth rate was −2·4% (provisional) in 2012. Italy's average economic growth rate since 1998 has been the slowest in the EU. Total GDP was US$2,194·0bn. in 2011.

Banking and Finance

The bank of issue is the Bank of Italy (founded 1893). It is owned by public-sector banks. Its *Governor* (Ignazio Visco) is nominated by the government for a six-year term, renewable once. In 1991 it received increased responsibility for the supervision of banking and stock exchange affairs, and in 1993 greater independence from the government.

The number of banks has gradually been declining in recent years, from 1,176 in 1990 to 807 (32,818 branches) in 2007. Of these, 439 were mutual banks and 39 were co-operative banks. Italy's largest bank in terms of assets is UniCredit (until May 2008 known as UniCredito Italiano). In June 2005 UniCredito Italiano finalized an agreement to acquire Germany's HypoVereinsbank in Europe's biggest cross-border banking takeover. In 2006 it had Italian assets of €282bn. (€752bn. including assets from its German and eastern European operations). In Aug. 2006 Italy's second and third largest banks, Banca Intesa and Sanpaolo IMI, agreed to merge. The merger was approved in Dec. 2006, creating the largest Italian bank, Intesa Sanpaolo, with assets of €541bn.

The 'Amato' law of July 1990 gave public sector banks the right to become joint stock companies and permitted the placing of up to 49% of their equity with private shareholders. In 1999 the last state-controlled bank was sold off.

On 31 Dec. 2005 banks had total deposits of €690,746m.

Gross external debt totalled US$2,356,371m. in June 2012.

Italy attracted US$29·1bn. worth of foreign direct investment in 2011, up from US$9·2bn. in 2010 but down from the record US$43·8bn. of 2007.

Legislation reforming stock markets came into effect in Dec. 1990. In 1996 local stock exchanges, relics of pre-unification Italy, were closed, and stock exchange activities concentrated in Milan.

ENERGY AND NATURAL RESOURCES

In 2008, 6·8% of energy consumption came from renewables (wind power, solar power, hydro-electric power, tidal power, geothermal energy and biomass), compared to the European Union average of 10·3%. A target of 17% has been set by the EU for 2020.

Environment

Italy's carbon dioxide emissions from the consumption and flaring of fossil fuels in 2008 were the equivalent of 7·8 tonnes per capita.

Electricity

In 2005 installed capacity was 88,345 MW and the total power generated was 309·0bn. kWh (13·9% hydro-electric). Consumption in 2005 was 309·8bn. kWh, of which: industry, 153·7bn. kWh; services, 83·8bn. kWh; domestic use, 66·9bn. kWh; agriculture, 5·4bn. kWh. Consumption per capita was 5,273 kWh in 2005. Italy has four nuclear reactors in permanent shutdown, the last having closed in 1990.

Oil and Gas

Oil production, 2008, 5·2m. tonnes. Proven oil reserves in 2008 were 0·8bn. bbls. In 2008 natural gas production was 8·4bn. cu. metres with proven reserves of 120bn. cu. metres.

Minerals

Fuel and mineral resources fail to meet needs. Only sulphur and mercury yield a substantial surplus for exports.

Production of metals and minerals (in tonnes) was as follows:

	1998	1999	2000	2001	2002
Sulphur	3,413,522	3,338,162	3,339,761	—	—
Feldspar	2,503,541	2,493,846	2,851,289	3,240,457	3,159,569
Bentonite	580,209	562,674	636,589	579,029	463,231
Lead	10,102	9,734	5,961	4,016	4,709
Zinc	5,242	—	—	—	—

Agriculture

In 2000, 1,120,000 persons were employed in agriculture, of whom 451,000 were dependent (148,000 female); independently employed were 669,000 (203,000 female). At the fifth agricultural census, held on 22 Oct. 2000, there were 13,212,652 sq. km of agricultural and forest lands, distributed as follows (in 1,000 ha.): woods, 4,711; cereals, 4,052; forage and pasture, 3,414; olive trees, 1,081; vines, 676; leguminous plants, 66. In 2002 there were 8·29m. ha. of arable land and 2·78m. ha. of permanent crops. In 2003 organic crops were grown in an area covering 1·17m. ha. (the third largest area after Australia and Argentina), representing 8·0% of all farmland.

At the 2000 census agricultural holdings numbered 2,593,090 and covered 19,607,094 ha. 2,457,960 owners (95·7%) farmed directly 13,868,478 ha. (70·3%); 132,935 owners (3·9%) worked with hired labour on 5,706,993 ha. (29·1%); the remaining 2,195 holdings (0·4%) of 31,623 ha. (0·6%) were operated in other ways. 97,307 share-croppers tilled 1,445,826 ha. Only 13,212,652 sq. km was in active agricultural use.

Agriculture and fishing accounted for 1·5% of exports and 3·4% of imports in 2001.

In 2005, 3,069,599 tractors were in use and 27,445 harvester-threshers.

Output of principal crops (in 1,000 tonnes) in 2005: sugar beets, 14,156; maize, 10,428; wheat, 7,717; grapes, 6,892; tomatoes, 6,640; olives, 3,775; oranges, 2,261; apples, 2,192; potatoes, 1,756; peaches and nectarines, 1,693; rice, 1,445; barley, 1,214.

Wine production in 2008 totalled 51,500,000 hectolitres (18·1% of the world total) making Italy the world's largest wine producer. Wine consumption has declined considerably in recent times, from more than 110 litres per person in 1966 to 50·1 litres per person in 2008.

Livestock, 2005: cattle, 6,251,925; sheep and goats, 8,900,062; pigs, 9,200,270; horses, 278,471; buffaloes, 205,093; chickens, 90,387,988. Livestock products, 2003 (in 1,000 tonnes): cow's milk, 11,000; sheep's milk, 790; buffalo's milk, 140; goat's milk, 112; pork, bacon and ham, 1,587; poultry meat, 1,156; beef and veal, 1,125; cheese, 1,099; butter, 130; eggs, 672. Italy is the second largest producer of sheep's milk, after China.

Forestry

In 2010 forests covered 9·15m. ha. or 31% of the total land area. Timber production was 8·12m. cu. metres in 2007.

Fisheries

In 2008 the fishing fleet comprised 13,683 vessels of 196,313 GT. The catch in 2010 was 234,101 tonnes, of which more than 98% were from marine waters. Imports of fishery commodities were valued at US$5,453m. in 2008 and exports at US$793m.

INDUSTRY

The leading companies by market capitalization in Italy in March 2012 were: Eni, an integrated oil company (US$93·8bn.); Enel, an electricity and gas company (US$34·0bn.); and Intesa Sanpaolo, a banking group (US$29·2bn.).

In 2007 industry accounted for 28% of GDP, with manufacturing contributing 18%.

Industrial products (in tonnes): cement (2008), 43·0m.; distillate fuel oil (including San Marino; 2007), 41·1m.; crude steel (2007), 32·0m.; petrol (including San Marino; 2007), 21·4m.; residual fuel oil (including San Marino; 2007), 17·4m.; pig iron (2006), 11·5m.; lime (2005), 5·9m.; polyethylene (2004), 1·4m.

Motor vehicle production in 2009 totalled 843,239 units (661,100 cars). In 2007, 1·34m. fridge-freezers and 9·83m. washing and drying machines were produced. Italy was the leading mineral water producer in 2004, with 13,117·4m. litres. In 2004, 3,766·3m. litres of soft drinks and 1,369·2m. litres of beer were produced.

Labour

In 2003 the workforce was 24,150,000 (69·3% males and 42·7% females) of whom 21,829,000 were employed. 2,096,000 were unemployed and looking for work. Until the summer of 2007 the unemployment rate had been declining steadily for some years; it was 7·7% in 2005, 6·8% in 2006 and 6·1% in 2007 as a whole, down from 10·8% in 2000. However, it rose to 6·8% in 2008 and climbed steadily throughout 2009 to reach 8·5% in Dec. Unemployment stood at 11·2% in Dec. 2012. In 2002, 63·2% of the workforce were in services, 31·8% in industry and 5·0% in agriculture. There are strong indications of labour markets having become less rigid, especially in the north. In the northeast unemployment was 3·3% in 2003, in the northwest 4·4% and in the centre 6·6%; in the south it was 18·3%. In 2006 the difference in the unemployment rates in the north and in the south was more than 8%, compared to a difference of just 2% in the 1960s. Long-term unemployment is particularly high, with 48·5% of the labour force in 2010 having been out of work for more than a year. Pensionable retirement age rose from 57 to 58 in 2008, and is set to rise to 66 by 2018 and 67 in 2022.

In 1997 parliament approved the so-called 'Treu Package', which involves a large number of institutional changes regarding working hours and apprenticeships, mainly for young people from the south, and the introduction of employment agencies. As a consequence, the share of temporary workers over total employees had grown from 6·2% in 1993 to 9·9% in 2002.

INTERNATIONAL TRADE

Imports and Exports

The following table shows the value of Italy's foreign trade (in US$1bn.):

	2006	2007	2008	2009	2010
Imports c.i.f.	442·6	511·8	561·0	414·8	486·6
Exports f.o.b.	417·2	500·2	541·8	406·5	447·5

Percentage of trade with other EU countries in 2006: imports, 55·2%; exports, 58·0%. Principal import suppliers, 2007 (% of total trade): Germany, 16·7%; France, 9·0%; China, 5·9%; Netherlands, 5·2%. Principal export markets: Germany, 12·8%; France, 11·4%; Spain, 7·3%; USA, 6·8%.

Imports/exports by category, 2007 (in US$1bn.):

	Imports	Exports
Animal and vegetable oils and fats	3·2	2·0
Beverages and tobacco	4·7	6·7
Chemicals and related products	63·4	49·7
Food and live animals	32·2	22·1
Inedible crude materials, excluding fuels	21·3	5·4
Machinery and transport equipment	143·1	186·1
Manufactured goods and articles	139·5	188·0
Mineral fuels and lubricants	60·4	18·8
Other products	36·9	13·3

COMMUNICATIONS

Roads

Roads totalled 175,430 km in 2005, of which 6,542 km were motorways, 21,524 km were highways and main roads, and 147,364 km were regional and provincial roads. In 2005 there were 47,104,048 motor vehicles, including: passenger cars, 34,882,476 (594 per 1,000 inhabitants); buses and coaches, 96,477; vans and trucks, 3,982,001. There were 5,426 fatalities in road accidents in 2005.

Rail

The length of state-run railway (Ferrovie dello Stato) in 2005 was 16,225 km (11,364 km electrified). Italy's first section of high-speed railway opened in 1981; by 2009 the total length had reached 923 km. In 2005 the railways carried 759·9m. passengers and 89·8m. tonnes of freight. There are metros in Milan (76·0 km), Rome (38·0 km), Naples (29·8 km), Turin (9·6 km), Genoa (5·3 km) and Catania (3·8 km), and tram/light rail networks in Bergamo, Cagliari, Florence, Genoa, Messina, Milan, Naples, Padua, Perugia, Rome, Sassari, Trieste, Turin and Venice. A driverless automated metro system opened in Brescia in March 2013.

Civil Aviation

There are major international airports at Bologna (G. Marconi), Genoa (Cristoforo Colombo), Milan (Linate and Malpensa), Naples (Capodichino), Pisa (Galileo Galilei), Rome (Leonardo da Vinci/Fiumicino), Turin (Caselle) and Venice (Marco Polo). A number of other airports have a small selection of international flights. Alitalia commenced operations in Jan. 2009 as a privately-owned company (25%-owned by Air France-KLM), having taken over the name, landing rights and significant assets of the former national carrier (also Alitalia, which went bankrupt in 2008) and having merged with rival airline Air One. There are a number of other Italian airlines, notably Meridiana. In 2003 scheduled airline traffic of Italian-based carriers flew 398m. km, carrying 36,077,000 passengers (13,613,000 on international flights). The busiest airport for passenger traffic is Rome (Fiumicino), which in 2001 handled 24,331,558 passengers (12,244,136 on international flights), plus 381,956 passengers in transit and 169,648 tonnes of freight. Milan Malpensa was the second busiest for passengers, handling 18,457,037 (14,169,573 on international flights), plus 109,652 passengers in transit, but the busiest for freight, with 289,382 tonnes. Linate, which handled 7,131,604 passengers in 2001 (4,995,000 on domestic flights), plus 738 passengers in transit, had been the principal Milan airport and for many years Italy's second busiest for passenger traffic, but in 1998 a new terminal was opened at Malpensa with many foreign operators subsequently using it instead of Linate.

Shipping

In Jan. 2009 there were 779 ships of 300 GT or over registered, totalling 13·32m. GT. Of the 779 vessels registered, 280 were passenger ships, 212 oil tankers, 131 general cargo ships, 62 bulk carriers, 45 chemical tankers, 26 liquid gas tankers and 23 container ships. The Italian-controlled fleet comprised 745 vessels of 1,000 GT or over in Jan. 2009, of which 527 were under the Italian flag and 218 under foreign flags. The chief ports are Genoa (which handled 54,218,000 tonnes of cargo in 2008), Trieste (48,279,000 tonnes in 2008) and Taranto (43,271,000 tonnes in 2008). Gioia Tauro, the busiest container port, handled 3·5m. 20-ft equivalent units (TEUs) in 2008.

Telecommunications

In 2008 there were 21,246,000 main (fixed) telephone lines. In May 1999 Olivetti bought a controlling stake in the telephone operator Telecom Italia, and in July 2001 Pirelli, backed by the Benetton clothing empire, in turn paid €7bn. (US$6·1bn.) to take over control of Telecom Italia. In 2008 mobile phone subscribers numbered 90,341,000, equivalent to 1,515·7 per 1,000 persons (among the highest penetration rates in the world). TIM (Telecom Italia Mobile) is the largest operator, with a 34% share of the market, just ahead of Vodafone Italia, which has a 33% share. There were 25·0m. internet users in 2008. There were 22·1 fixed broadband subscribers per 100 inhabitants in Dec. 2010. In March 2012 there were 20·9m. Facebook users.

SOCIAL INSTITUTIONS

Justice

Italy has one court of cassation, in Rome, and is divided for the administration of justice into 29 appeal court districts, subdivided into 164 tribunal *circondari* (districts). There are also 93 first degree assize courts and 29 assize courts of appeal. For civil business, besides the magistracy above mentioned, *Giudici di pace* have jurisdiction in petty plaints.

2,709,888 crimes were reported in 2008. In June 2008 there were 55,057 persons in prison (including persons imprisoned by San Marino and the Vatican City). In 1947 the re-established democracy rewrote the Legislative Order; the constitution of the Italian Republic abolished the death penalty sanctioned in 1930 by Codice Penale, commonly known as Codice Rocco. Although the death penalty was abolished for ordinary crimes in 1947, it was not until 1994 that it was abolished for all crimes.

Education

Five years of primary and five years of secondary education are compulsory from the age of six. In 2005–06 there were 24,845 pre-school institutions with 1,662,139 children and 140,687 teachers (state and non-state schools); 18,218 primary schools with 2,790,254 pupils and 293,091 teachers (state and non-state schools); 7,886 compulsory secondary schools (*scuole secondarie primo grado*) with 1,764,230 pupils and 211,093 teachers (state and non-state schools); and 6,565 higher secondary schools with 2,691,713 pupils and 305,383 teachers (state and non-state schools).

Higher secondary education is subdivided into classical (*ginnasio* and classical *liceo*), scientific (scientific *liceo*), language lyceum, professional institutes and technical education: agricultural, industrial, commercial, technical, nautical institutes, institutes for surveyors, institutes for girls (five-year course) and teacher-training institutes (five-year course).

In 2005–06 there were 98 universities (79 state and 19 non-state), of which two are universities of Italian studies for foreigners, three specialized universities (commerce; education; Roman Catholic), three polytechnical university institutes; seven specialized university institutes (architecture; bio-medicine; modern languages; naval studies; oriental studies; social studies; teacher training). In 2005–06 there were 1,823,886 university students and 61,097 academic staff.

Estimated adult literacy rate, 2009, 98·9% (male 99·2%; female 98·6%).

In 2008 public expenditure on education came to 4·6% of GDP and 9·4% of total government spending.

Health

In 2003 the National Health Service included 1,367 hospitals of which 746 were public with 184,796 beds and 621 private hospitals with 55,059 beds. There were 246,834 physicians, 28,566 dentists and 59,580 pharmacists in 2008; and 379,213 nursing and midwifery personnel in 2009. A survey published by the World Health Organization in June 2000 to measure health systems in all of the sovereign countries and find which country has the best

overall health care ranked Italy in second place, behind France. In 2009 Italy spent 9·5% of its GDP on health.

Welfare

Social expenditure is made up of transfers which the central public departments, local departments and social security departments make to families. Payment is principally for pensions, family allowances and health services. Expenditure on subsidies, public assistance to various classes of people and people injured by political events or national disasters are also included.

In Jan. 2008 the minimum retirement age was raised from 57 to 58 (or after 36 years of work, whichever comes first). It is rising gradually and is set to be 66 by 2018. The age restriction will not apply in the case of workers who have made 40 years of pension contributions. Those in 'arduous' professions, which encompasses 6% of the population, will be able to retire at 57. The average age for Italian men to stop work is 60 years 8 months, among the youngest in the EU. Pensions account for around 15% of GDP.

Citizens who have entered the workforce since 1996 will not qualify for this seniority pension, but will receive an old-age pension instead. There are three categories of pension. The first is for people who have joined the workforce since 1996, and is available to claimants aged 57 or above with at least five years of contributions. The second category is for men aged at least 65 or women aged at least 60, with under 18 years of contributions as of 1995. The third category is for men aged at least 65 or women aged at least 60, with at least 18 years of contributions as of 1995. It is paid at a rate of 0·9%–2% of earnings multiplied by number of years of contributions, up to a maximum of 40.

Public pensions are indexed to prices; 23,257,480 pensions were paid in 2005, with payments totalling €214,881·3m. (including 16,875,341 private sector, with payments totalling €152,483·5m.). The average annual pension in 2005 was €9,239. Social contributions in 2005 totalled €184,642m.

RELIGION

The treaty between the Holy See and Italy of 11 Feb. 1929, confirmed by article 7 of the Constitution of the republic, lays down that the Catholic Apostolic Roman Religion is the only religion of the State. Other creeds are permitted, provided they do not profess principles, or follow rites, contrary to public order or moral behaviour.

The appointment of archbishops and of bishops is made by the Holy See; but the Holy See submits to the Italian government the name of the person to be appointed in order to obtain an assurance that the latter will not raise objections of a political nature. In Feb. 2013 there were 49 cardinals.

Catholic religious teaching is given in elementary and intermediate schools. Marriages celebrated before a Catholic priest are automatically transferred to the civil register. Marriages celebrated by clergy of other denominations must be made valid before a registrar.

There were 46,260,000 Roman Catholics in 2001, 680,000 Muslims, 1,350,000 adherents of other religions and 9,600,000 non-religious and atheists.

CULTURE

Milan is scheduled to host Expo 2015 under the theme 'Feeding the Planet, Energy for Life'.

World Heritage Sites

There are 47 UNESCO sites in Italy (the most of any country): the Rock Drawings in Valcamonica near Brescia (inscribed in 1979); Santa Maria delle Grazie with 'The Last Supper' by Leonardo da Vinci (1980); San Paolo Fuori le Mura Historic Centre of Florence (1982); Venice and its Lagoon (1987); Piazza del Duomo, Pisa (1987 and 2007); Historic Centre of San Gimignano (1990); I Sassi di Matera (1993); Vicenza, the City of Palladio and the Villas of the Veneto (1994 and 1996); Historic Centre of Siena (1995);

Historic Centre of Naples (1995); Ferrara and its Po Delta (1995 and 1999); Crespi d'Adda (1995); Castel del Monte (1996); Trulli of Alberobello (1996); Early Christian Monuments and Mosaics of Ravenna (1996); Historic Centre of the City of Pienza (1996); The 18th-Century Royal Palace at Caserta with the Park, the Aqueduct of Vanvitelli and the San Leucio Complex (1997); Residences of the Royal House of Savoy (1997); Botanical Garden (Orto Botanico), Padua (1997); Cathedral, Torre Civica and Piazza Grande, Modena (1997); Archaeological Areas of Pompeii, Ercolano and Torre Annunziata (1997); Villa Romana del Casale (1997); Su Nuraxi di Barumini (1997); Portovenere, Cinque Terre and the Islands (Palmaria, Tino and Tinetto) (1997); The Costiera Amalfitana (1997); Archaeological Area of Agrigento (1997); Cilento and Vallo di Diano National Park (1998); Historic Centre of Urbino (1998); Archaeological Area and the Patriarchal Basilica of Aquileia (1998); Villa Adriana (1999); Aeolian Islands (2000); Assisi (2000); the City of Verona (2000); Villa d'Este, Tivoli (2001); the Late Baroque Towns of the Val di Noto (2002); Monte San Giorgio (2003); Sacri Monti of Piedmont and Lombardy (2003); Val d'Orcia (2004), part of the agricultural hinterland of Siena; the Etruscan Necropolises of Cerveteri and Tarquinia (2004); Syracuse and the Rocky Necropolis of Pantalica (2005); Genoa (2006), featuring the Strade Nuove and the system of the Palazzi dei Rolli; Mantua and Sabbioneta (2008); the Dolomites (2009); and the Longobards in Italy, Places of Power, 568–774 AD, comprising seven groups of important buildings throughout the country (2011).

The Historic Centre of Rome, the properties of the Holy See in that city enjoying extraterritorial rights (1980 and 1990), is shared with Vatican City State. The Rhaetian Railway in the Albula/Bernina Landscapes (2008) is shared with Switzerland and the Prehistoric Pile dwellings around the Alps (2011) are shared with Austria, France, Germany, Slovenia and Switzerland.

Press
There were, in 2008, 84 paid-for dailies with a combined circulation of 5·3m. copies and nine free dailies with a combined circulation of 4·4m. copies. Several of the papers are owned or supported by political parties. The church and various economic groups exert strong right of centre influence on editorial opinion. Most newspapers are regional but *Corriere della Sera* (which has the highest circulation of any Italian newspaper), *La Repubblica*, *Il Sole 24 Ore* and *La Stampa* are the most important of those papers that are nationally circulated. In 2008 a total of 58,829 book titles were published in 213m. copies.

Tourism
In 2010, 43·6m. international tourists visited Italy (43·2m. in 2009); receipts from tourism in 2010 were US$38·8bn. (US$40·2bn. in 2009). Only France, the USA, China and Spain received more foreign tourists in 2010.

Festivals
One of the most traditional festivals in Italy is the Carnival di Ivrea which lasts for a week in late Feb. or early March. Among the famous arts festivals is the Venice Film Festival in Sept. Venice also plays host, in the ten days before Ash Wednesday, to a large carnival. Major music festivals are the Maggio Musicale Fiorentino in Florence (May–June), the Ravenna Festival (June–July), the Spoleto Festival (June–July), the Rossini Opera Festival at Pesaro (Aug.) and the Verona Arena Opera Festival (June–Aug.). The biggest rock festival is the Italia Wave Love Festival in Livorno (known as the Arezzo Wave Festival until 2007), which is held in July, while the largest jazz festival is the Umbria Jazz Festival, also in July. Until 2010 Europe's largest reggae festival, Rototom Sunsplash, was held at Osoppo, in Tuscany. However, it was forced to relocate to Spain after the Italian authorities accused the festival and its organizers of facilitating the use of drugs.

DIPLOMATIC REPRESENTATIVES
Of Italy in the United Kingdom (14 Three Kings Yard, Davies St., London, W1K 4EH)
Ambassador: Alain Economides.

Of the United Kingdom in Italy (Via XX Settembre 80A, 00187, Rome)
Ambassador: Christopher Prentice, CMG.

Of Italy in the USA (3000 Whitehaven St., NW, Washington, D.C., 20008)
Ambassador: Claudio Bisogniero.

Of the USA in Italy (Via Vittorio Veneto 119/A, Rome)
Ambassador: David H. Thorne.

Of Italy to the United Nations
Ambassador: Cesare Maria Ragaglini.

Of Italy to the European Union
Permanent Representative: Ferdinando Nelli Feroci.

FURTHER READING
Istituto Nazionale di Statistica. *Annuario Statistico Italiano.—Compendio Statistico Italiano* (Annual).—*Italian Statistical Abstract* (Annual).—*Bollettino Mensile di Statistica* (Monthly).

Baldoli, Claudia, *A History of Italy.* 2009
Bufacchi, Vittorio and Burgess, Simon, *Italy since 1989.* 1999
Burnett, Stanton H. and Mantovani, Luca, *The Italian Guillotine: Operation 'Clean Hands' and the Overthrow of Italy's First Republic.* 1999
Cotta, Maurizio and Verzichelli, Luca, *Political Institutions of Italy.* 2007
Di Scala, S. M., *Italy from Revolution to Republic: 1700 to the Present.* 1995
Doumanis, Nicholas, *Italy: Inventing the Nation.* 2001
Duggan, Christopher, *A Concise History of Italy.* 1994.—*The Force of Destiny: A History of Italy Since 1796.* 2007
Emmott, Bill, *Good Italy, Bad Italy: Why Italy Must Conquer Its Demons to Face the Future.* 2012
Foot, John, *Modern Italy.* 2003
Frei, M., *Italy: the Unfinished Revolution.* 1996
Furlong, P., *Modern Italy: Representation and Reform.* 1994
Gilbert, M., *Italian Revolution: the Ignominious End of Politics, Italian Style.* 1995
Gilmour, David, *The Pursuit of Italy: A History of a Land, its Regions and their Peoples.* 2011
Ginsborg, Paul, *Italy and its Discontents, 1980–2001.* 2002.—*A History of Contemporary Italy: Society and Politics, 1943–1988.* 2003
Gundie, S. and Parker, S. (eds.) *The New Italian Republic: from the Fall of the Berlin Wall to Berlusconi.* 1995
Plant, Margaret, *Venice: Fragile City 1797–1997.* 2002
Putnam, R., *et al.*, *Making Democracy Work: Civic Traditions in Modern Italy.* 1993
Richards, C., *The New Italians.* 1994
Smith, D. M., *Modern Italy: A Political History.* 1997
Volcanasek, Mary L., *Constitutional Politics in Italy.* 1999

National library: Biblioteca Nazionale Centrale, Vittorio Emanuele II, Viale Castro Pretorio, Rome.
National Statistical Office: Istituto Nazionale di Statistica (ISTAT), 16 Via Cesare Balbo, 00184 Rome.
Website (limited English): http://www.istat.it

JAMAICA

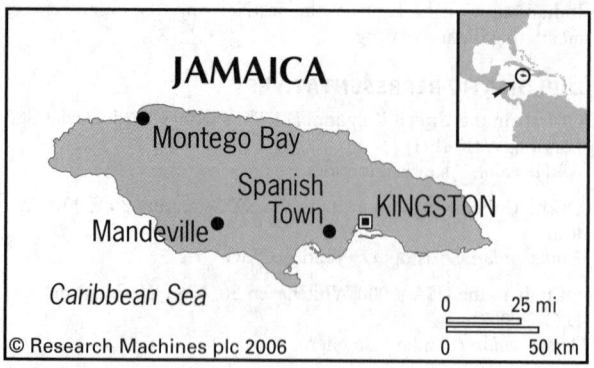

Capital: Kingston
Population projection, 2015: 2·79m.
GNI per capita, 2011: (PPP$) 6,487
HDI/world rank: 0·727/79
Internet domain extension: .jm

KEY HISTORICAL EVENTS

Jamaica was discovered by Columbus in 1494 and was occupied by the Spaniards from 1509 until 1655 when the island was captured by the English. In 1661 a representative constitution was established consisting of a governor, privy council, legislative council and legislative assembly. The slavery introduced by the Spanish was augmented as sugar production increased in value and extent in the 18th century. The plantation economy collapsed with the abolition of the slave trade in the late 1830s. The 1866 Crown Colony government was introduced with a legislative council. In 1884 a partially elective legislative council was instituted. Women were enfranchised in 1919. By the late 1930s, demands for self-government increased. The constitution of Nov. 1944 introduced a freely-elected house of representatives of 32 members, a legislative council (the upper house) of 15 members, and an executive council. In 1958 Jamaica joined with Trinidad, Barbados, the Leeward Islands and the Windward Islands to create the West Indies Federation. In 1959 internal self-government was achieved. Jamaica withdrew from the West Indies Federation in 1961 and became an independent state within the British Commonwealth in 1962.

Power alternated between the Jamaica Labour Party and the People's National Party. The latter held power from 1989 until 2007, with Portia Simpson-Miller becoming the country's first female premier in 2006. In May 2010 a state of emergency was declared after the death of dozens of people in an operation to arrest one of the country's leading drug lords, Christopher 'Dudus' Coke. He was subsequently captured and extradited to the USA.

TERRITORY AND POPULATION

Jamaica is an island in the Caribbean Sea about 150 km south of Cuba. The area is 10,991 sq. km (4,244 sq. miles). The population at the census of April 2011 was 2,697,983, distributed on the basis of the 13 parishes of the island as follows: Kingston and St Andrew, 662,426; St Catherine, 516,218; Clarendon, 245,103; Manchester, 189,797; St James, 183,811; St Ann, 172,362; St Elizabeth, 150,205; Westmoreland, 144,103; St Mary, 113,615; St Thomas, 93,902; Portland, 81,744; Trelawny, 75,164; Hanover, 69,533. 2011 density: 245 per sq. km. There is a worldwide Jamaican diaspora of more than 2m.

The UN gives a projected population for 2015 of 2·79m.

Chief towns (in 1,000), 2011: Kingston (metropolitan area), 585; Portmore, 182; Spanish Town, 147; Montego Bay, 110; May Pen, 62.

In 2011, 52·1% of the population were urban. The population is about 92% of African ethnic origin. The official language is English. Patois, a combination of English and African languages, is widely spoken.

SOCIAL STATISTICS

Vital statistics (2006): births, 46,277 (17·4 per 1,000 population); deaths, 16,317 (6·1); marriages, 23,181 (8·7); divorces, 1,768 (0·7). There were 17,100 emigrants in 2006, mainly to the USA. Expectation of life at birth, 2007, 68·3 years for males and 75·1 years for females. Annual population growth rate, 2008–10, 0·3%; infant mortality, 2010, 20 per 1,000 live births; fertility rate, 2008, 2·4 births per woman.

CLIMATE

A tropical climate but with considerable variation. High temperatures on the coast are usually mitigated by sea breezes, while upland areas enjoy cooler and less humid conditions. Rainfall is plentiful over most of Jamaica, being heaviest in May and from Aug. to Nov. The island lies in the hurricane zone. Kingston, Jan. 76°F (24·4°C), July 81°F (27·2°C). Annual rainfall 32" (800 mm).

CONSTITUTION AND GOVERNMENT

Under the constitution of Aug. 1962 the Crown is represented by a Governor-General appointed by the Crown on the advice of the Prime Minister. The Governor-General is assisted by a Privy Council of six appointed members. The Legislature comprises the *House of Representatives* and the *Senate*. The Senate consists of 21 senators appointed by the Governor-General, 13 on the advice of the Prime Minister, eight on the advice of the Leader of the Opposition. The House of Representatives (increased from 60 to 63 members for the 2011 election) is elected by universal adult suffrage for a period not exceeding five years. Electors and elected must be Jamaican or Commonwealth citizens resident in Jamaica for at least 12 months before registration.

National Anthem

'Eternal Father, bless our land'; words by H. Sherlock, tune by R. Lightbourne.

RECENT ELECTIONS

In parliamentary elections held on 29 Dec. 2011 the opposition People's National Party (PNP) took 42 of the 63 seats with 53·3% of votes cast (up from 28 in 2007) and the Jamaica Labour Party (JLP) 21 with 46·6% (down from 32 in 2007).

CURRENT GOVERNMENT

Governor-General: Patrick Allen.

In March 2013 the cabinet comprised:

Prime Minister and Minister of Defence, Development, Information and Sports: Portia Simpson-Miller, b. 1945 (PNP; sworn in 5 Jan. 2012, having previously been in office from March 2006–Sept. 2007).

Minister of Agriculture and Fisheries: Roger Clarke. *Education:* Ronald Thwaites. *Finance, Planning and the Public Service:* Peter Phillips. *Foreign Affairs and Foreign Trade:* Arnold Nicholson. *Health:* Dr Fenton Ferguson. *Industry, Investment and Commerce:* Anthony Hylton. *Justice:* Mark Golding. *Labour and Social Security:* Derrick Kellier. *Local Government and Community Development:* Noel Arscott. *National Security:* Peter Bunting.

Science, Technology, Energy and Mining: Phillip Paulwell. *Tourism and Entertainment:* Wykeham McNeil. *Transport, Works and Housing:* Omar Davies. *Water, Land, Environment and Climate Change:* Robert Pickersgill. *Youth and Culture:* Lisa Hanna. *Ministers without Portfolio:* Horace Dalley; Sandrea Falconer; Morais Guy; Natalie Neita-Headley.

Cabinet Website: http://www.cabinet.gov.jm

CURRENT LEADERS

Portia Simpson-Miller

Position
Prime Minister

Introduction
Portia Simpson-Miller was sworn in as prime minister for a second time on 5 Jan. 2012 after the People's National Party defeated the Jamaica Labour Party in elections held on 29 Dec. 2011. She previously served in the post from 30 March 2006 to 11 Sept. 2007. A member of the Council of Women World Leaders, Simpson-Miller was her country's first democratically elected female leader on taking office in 2006.

Early Life
Portia Simpson-Miller was born on 12 Dec. 1945 in Wood Hall, St Catherine's Parish, Jamaica. She was educated at St Martin's High School and the Union Institute, Miami, USA, where she graduated with a degree in public administration. She entered politics as a councillor for the left-leaning PNP in 1974, winning the inner-city seat of Trench Town West in the Kingston & St Andrew Corporation.

In 1977 she was appointed parliamentary secretary in the ministry of local government and the following year was elected vice-president of the PNP, a post she held for the next 27 years. Despite the PNP's crushing defeat in the 1980 general election, she retained her councillor's seat and from 1983–89 served as the party spokesperson on women's affairs and pensions, social security and consumer affairs.

In the 1989 general election, when Michael Manley returned the PNP to power, Simpson-Miller was elected MP for South West St Andrew and was appointed minister of labour, social security and sport. Following Manley's resignation in 1992 she unsuccessfully challenged P. J. Patterson for the premiership. Promoted to minister for tourism and sport in 2000, Simpson-Miller won plaudits for her work to rebuild the tourism sector in the wake of the 11 Sept. 2001 attacks on the USA.

Following the general election of Oct. 2002 she regained the local government portfolio in an expanded ministry of local government, community development and sport. On 25 Feb. 2006 she narrowly defeated Peter Phillips to succeed Patterson as head of the PNP and become prime minister-elect. She was sworn into office on 30 March 2006. During her 18 months as premier Simpson-Miller sought to tackle poverty and crime. However, with unemployment running at 9% and crime still endemic, Simpson-Miller's PNP was narrowly defeated by the Labour Party in the parliamentary elections of Sept. 2007, ending 18 years of PNP rule. The following year, Simpson-Miller's leadership of the PNP was challenged by Peter Phillips but she defeated her rival and continued to lead the party in opposition.

Simpson-Miller performed strongly on the 2011 election campaign trail, with her charismatic populism widely credited with swinging public opinion in favour of the PNP. In what had been predicted to be a tight election, she led her party to a landslide win, taking 42 of 63 available seats and ensuring the two-thirds majority necessary to alter the constitution.

Career in Office
Simpson-Miller's priorities were to slash the rising rate of unemployment and tackle the massive debt-to-GDP ratio. In a televised election debate, she said that she would consider reviewing Jamaica's controversial criminalization of homosexuality. An avowed republican, she has stated her intention to move towards the severance of ties with the UK, though any such decision must be ratified by popular referendum. Simpson-Miller assumed the defence portfolio upon taking office as prime minister.

DEFENCE

In 2008 defence expenditure totalled US$96m. (US$34 per capita), representing 0·7% of GDP.

Army
The Jamaica Defence Force consists of a Regular and a Reserve Force. Total strength (Army, 2009): 3,377, including 877 reservists.

Navy
The Coast Guard, numbering 250 in 2009 including 60 reservists, operates nine patrol craft based at Port Royal and Pedro Cays.

Air Force
The Air Wing of the Jamaica Defence Force was formed in July 1963 and has since been expanded and trained successively by the British Army Air Corps and Canadian Air Force personnel. There are no combat aircraft. Personnel (2009), 156 (including 16 reservists).

ECONOMY

In 2009 agriculture accounted for 6·2% of GDP, industry 22·1% and services 71·7%.

Overview
Jamaica has a small, open economy dominated by the service sector. Tourism is the biggest source of employment. Worker remittances attract revenue roughly equal to that of the tourist sector. The 2008–09 global economic crisis had a severe impact, with remittances declining sharply. Tourism, however, proved more resilient, with arrivals increasing in 2009.

Leading exports are alumina, bauxite, bananas and sugar. After experiencing spiky but strong growth in the early part of the 1990s the economy fell into recession in 1997. Growth from 2001–08 averaged 1·5–2·0%, but there were three years of negative growth in 2008, 2009 and 2010. The poverty rate declined from 30·5% to 9·9% between 1989 and 2007.

Unemployment remains high, especially among the young. The national debt burden is over 130% of GDP.

Currency
The unit of currency is the *Jamaican dollar* (JMD) of 100 *cents*. The Jamaican dollar was floated in Sept. 1990. Inflation was 22·0% in 2008 but has since fallen, and was 12·6% in 2010 and 7·5% in 2011. Foreign exchange reserves were US$2,423m. in July 2005 and total money supply was J$68,141m.

Budget
Budgetary central government revenue and expenditure for fiscal years ending 31 March (in J$1m.):

	2008	2009	2010[1]
Revenue	249,512	272,460	241,766
Expenditure	275,223	325,731	412,992
		[1]Provisional.	

The chief items of current revenue are income taxes, consumption taxes and customs duties. The chief items of current expenditure are public debt, education and health.

There is a General Consumption Tax of 16·5%.

Performance
After suffering major economic difficulties with negative growth in 1996, 1997 and 1998, Jamaica's economy recovered slightly, with growth of 2·9% in 2006 and 1·4% in 2007. However, there

was again negative growth in 2008, of –0·8%, in 2009, of –3·5%, and in 2010, of –1·5%. In 2011 there was a slight recovery with the economy growing by 1·3%; total GDP was US$14·4bn.

Banking and Finance
The central bank and bank of issue is the Bank of Jamaica. The *Governor* is Brian Wynter.

In 2002 there were five commercial banks, three development banks and two other banks (National Export-Import Bank of Jamaica and the National Investment Bank of Jamaica). Total assets of commercial banks at March 2006 were J$385,759·5m.; deposits were J$255,315·4m.

Foreign debt was US$13,865m. in 2010, representing 104·2% of GNI.

There is a stock exchange in Kingston, which participates in the regional Caribbean exchange.

ENERGY AND NATURAL RESOURCES
Environment
In 2008 carbon dioxide emissions from the consumption and flaring of fossil fuels were the equivalent of 4·7 tonnes per capita.

Electricity
The Jamaica Public Service Co. is the public supplier. Total installed capacity was 1·2m. kW in 2007. Production in 2007 totalled 7·78bn. kWh; consumption per capita in 2007 was 2,886 kWh.

Oil and Gas
There is an oil refinery in Kingston.

Minerals
Jamaica is ranks among the world's largest producers of bauxite. Ceramic clays, marble, silica sand and gypsum are also commercially viable. Production in 2005 (in tonnes): bauxite ore, 14·1m.; limestone, 2·6m.; sand and gravel, 2·4m.; gypsum, 302,066.

Agriculture
In 2007 there were an estimated 174,000 ha. of arable land and 110,000 ha. of permanent crops.

2003 production (in 1,000 tonnes): sugarcane, 2,400; coconuts, 170; yams, 151; oranges, 140; bananas, 130; grapefruit and pomelos, 42; pumpkins and squash, 36; cabbage, 29; plantains, 29.

Livestock (2003 estimates): goats, 440,000; cattle, 430,000; pigs, 180,000; chickens, 11m. Livestock products, 2003 estimates (in 1,000 tonnes): beef and veal, 14; pork, bacon and ham, 5; poultry meat, 81.

Forestry
Forests covered 0·34m. ha. in 2010, or 31% of the total land area. Timber production was 834,000 cu. metres in 2007.

Fisheries
Catches in 2010 totalled 15,440 tonnes, almost exclusively from sea fishing.

INDUSTRY
Alumina production, 2005, 4·1m. tonnes. Output of other products (2007 unless otherwise indicated, in tonnes): cement (2005), 844,840; residual fuel oil, 473,000; sugar (2002), 174,949; distillate fuel oil, 190,000; petrol, 104,000; molasses (2003), 72,631; wheat flour (2002), 32,000; fertilizer (2002), 22,400; cigarettes (2003), 889m. units; rum (2003), 25·5m. litres. In 2005 industry accounted for 33·1% of GDP, with manufacturing contributing 13·6%.

Labour
Total labour force (2003), 1·10m., of whom 957,300 were employed. In 2003, 257,000 were employed in community, social and personal services; 210,000 in wholesale and retail trade, restaurants and hotels; 188,000 in agriculture, hunting, forestry and fishing; and 90,000 in construction. In 2005 the unemployment rate was 10·0%.

INTERNATIONAL TRADE
Imports and Exports
Value of imports and domestic exports for calendar years (in US$1m.):

	2002	2003	2004	2005	2006
Imports f.o.b.	3,179·6	3,328·2	3,546·1	4,245·5	5,077·1
Exports f.o.b.	1,309·1	1,385·6	1,601·6	1,664·3	2,133·6

Principal imports in 2006 (% of total): petroleum and petroleum products 23·6%, machinery and transport equipment 22·6%, manufactured goods 13·9%, food and live animals 12·5% and chemicals and related products 11·3%.

Principal domestic exports in 2006 (% of total): crude materials (excluding fuels) 63·3%, petroleum oils 13·5%, cane sugar 4·5%, alcoholic beverages 4·2% and chemicals 3·7%.

Main import suppliers, 2006: USA, 36·8%; Trinidad and Tobago, 11·5%; Venezuela, 10·7%; Japan, 4·2%; China, 4·1%. Main export markets, 2006: USA, 30·4%; Canada, 15·6%; China, 15·1%; UK, 10·3%; Netherlands, 7·0%.

COMMUNICATIONS
Roads
In 2007 the island had 22,121 km of roads, including 44 km of motorway and 4,922 km of main roads. In 2006 there were 373,700 passenger cars in use and 29,100 motorcycles and mopeds. There were 350 fatalities in traffic accidents in 2007.

Rail
Passenger traffic ceased in 1992 but there are plans to revive passenger services. Freight transport continues on a limited basis, mainly for carrying bauxite to docks.

Civil Aviation
International airlines operate through the Norman Manley and Sangster airports at Palisadoes and Montego Bay. Sangster International is the busiest for passenger traffic, handling 3,378,000 passengers in 2006–07. Norman Manley airport is busier for freight, handling 16,136 tonnes of freight in 2006 but only 1,715,078 passengers. Air Jamaica, originally set up in conjunction with BOAC and BWIA in 1966, became a new company, Air Jamaica (1968) Ltd. In 1969 it began operations as Jamaica's national airline. It was acquired in May 2010 by Caribbean Airlines, the national airline of Trinidad and Tobago, but the name Air Jamaica has been retained. In 2006 scheduled airline traffic of Jamaica-based carriers flew 52m. km and carried 1,527,000 passengers.

Shipping
In Jan. 2009 there were 21 ships of 300 GT or over registered, totalling 229,000 GT. Kingston handled 16·3m. tonnes of cargo in 2008 (down from 17·7m. tonnes in 2007).

Telecommunications
In 2011 there were 272,100 landline telephone subscriptions (equivalent to 98·9 per 1,000 inhabitants) and 2,974,700 mobile phone subscriptions (or 1,081·2 per 1,000 inhabitants). There were 261·0 internet users per 1,000 inhabitants in 2010. Fixed internet subscriptions totalled 114,600 in 2009 (42·0 per 1,000 inhabitants). In Dec. 2011 there were 684,000 Facebook users.

SOCIAL INSTITUTIONS
Justice
The Judicature comprises a Supreme Court, a court of appeal, resident magistrates' courts, petty sessional courts, coroners' courts, a traffic court and a family court which was instituted in 1975. The Chief Justice is head of the judiciary. Jamaica was one of ten countries to sign an agreement in Feb. 2001 establishing a Caribbean Court of Justice to replace the British Privy Council as the highest civil and criminal court. In the meantime the number

of signatories has risen to 12. The court was inaugurated at Port-of-Spain, Trinidad on 16 April 2005 but Jamaica has yet to accept it as its final court of appeal.

In 2010 there were 1,428 murders, down from the record high of 1,682 in 2009. The rate of 52 per 100,000 persons (61 per 100,000 in 2009) is more than ten times that of the USA and ranks among the highest in the world. The death penalty is permitted but has not been used since 1988.

The population in penal institutions in Oct. 2007 was 4,709 (174 per 100,000 of national population).

Police
The Constabulary Force in 2005 stood at approximately 8,011 officers, sub-officers and constables (men and women).

Education
Adult literacy was an estimated 86·4% in 2009 (91·1% among females but only 81·2% among males).

Education is free in government-operated schools. Enrolment in 2007 in primary institutions was 310,000; in secondary institutions, 257,000; and in tertiary institutions (2008), 61,000. Numbers of teaching staff: primary (2005), 11,793; secondary (2007), 13,006; tertiary (2003), 2,006.

The University of the West Indies, which was founded in 1948 and has its main campus at Kingston, is the oldest, fully regional institution of higher learning in the Commonwealth Caribbean. In 2006 there were two public universities and two private universities as well as six teacher training colleges, five community colleges, and several technical/vocational training institutes and specialist colleges. Large numbers of educated Jamaicans have left the island over the past 30 years, but in the early part of the 21st century there are signs that young professionals are increasingly returning to Jamaica. However, 72% of Jamaican graduates live in OECD member countries.

In 2005 public expenditure on education came to 5·6% of GNI and 8·8% of total government spending.

Health
In 2001 there were 27 hospitals with 4,606 beds. There were 2,253 physicians, 4,374 nurses and midwives, and 212 dentists in 2003.

Welfare
The official retirement age is 65 years (men) or 60 years (women). The old-age pension is made up of a basic benefit of J$900 a week (reduced to J$675 a week with annual average contributions of between 26 and 38 weeks; J$450 with 13 weeks to 25 weeks), plus an earnings-related benefit of J$0·06 a week for every J$13 of employer-employee contributions paid during the working lifetime.

Jamaica's social welfare projects also cover disability and survivor benefits, sickness and maternity, and work injury. Jamaica has no unemployment programmes.

RELIGION
Freedom of worship is guaranteed under the Constitution. The main Christian denominations are Anglican, Baptist, Roman Catholic, Methodist, Church of God, United Church of Jamaica and Grand Cayman (Presbyterian-Congregational-Disciples of Christ), Moravian, Seventh-Day Adventist, Pentecostal, Salvation Army and Quaker. Pocomania is a mixture of Christianity and African survivals. Non-Christians include Hindus, Jews, Muslims, Bahai followers and Rastafarians.

CULTURE
Press
In 2008 there were three daily newspapers with a combined circulation of 115,000.

Tourism
In 2011 there were a record 1,951,752 non-resident overnight tourists and 1,125,481 cruise passenger arrivals (down from a peak of 1,336,994 in 2006).

DIPLOMATIC REPRESENTATIVES
Of Jamaica in the United Kingdom (1–2 Prince Consort Rd, London, SW7 2BZ)
High Commissioner: Aloun Ndombet-Assamba.

Of the United Kingdom in Jamaica (28 Trafalgar Rd, Kingston 10)
High Commissioner: Howard Drake, OBE.

Of Jamaica in the USA (1520 New Hampshire Ave., NW, Washington, D.C., 20036)
Ambassador: Stephen Vasciannie.

Of the USA in Jamaica (142 Old Hope Rd, Kingston 6)
Ambassador: Pamela Bridgewater.

Of Jamaica to the United Nations
Ambassador: Raymond Wolfe.

Of Jamaica to the European Union
Ambassador: Vilma McNish.

FURTHER READING
Planning Institute of Jamaica. *Economic and Social Survey, Jamaica.* Annual.—*Survey of Living Conditions.* Annual
Statistical Institute of Jamaica. *Statistical Abstract.* Annual.—*Demographic Statistics.* Annual.—*Production Statistics.* Annual

Hart, R., *Towards Decolonisation: Political, Labour and Economic Developments in Jamaica 1938–1945.* 1999
Henke, H. W. and Mills, D., *Between Self-Determination and Dependency: Jamaica's Foreign Relations 1972–1989.* 2000

National library: National Library of Jamaica, 12 East Street, Kingston.
National Statistical Office: Statistical Institute of Jamaica (STATIN), 7 Cecelio Ave., Kingston 10. *Director General:* Sonia Jackson.
Website: http://www.statinja.com

JAPAN

© Research Machines plc 2006

Nihon (or Nippon[1]) Koku
(Land of the Rising Sun)

Capital: Tokyo
Population projection, 2015: 126·07m.
GNI per capita, 2011: (PPP$) 32,295
HDI/world rank: 0·901/12
Internet domain extension: .jp

KEY HISTORICAL EVENTS

When the last ice sheets covered much of Asia, the sea level fell low enough for a land bridge to appear between Japan and the Asian mainland. This route was taken by hunter-gatherers from Asia who crossed into previously uninhabited Japan. By 10,000 BC the first pottery was produced in Japan and there was some cultivation. Rice was introduced, probably from Korea, by about 400 BC, and the use of metals around a century later, but agriculture and fixed settlements were confined to the south for a long period. During this time waves of migrants came from mainland Asia, bringing with them skills and technologies, including the Chinese characters for writing.

Religion, too, came from China: both Buddhism and Confucianism entered Japan, the former gaining a large following. In time traditional beliefs consolidated into Shintoism, which became the national religion. But, until the first millennium AD, there was no Japanese nation, although the legends of Japan tell us otherwise. According to myth, the first Japanese emperor was Jimmu around 600 BC, said to be a descendant of the sun goddess, Amaterasu.

In the first century BC, another wave of migrants entered Japan from Korea. The first Japanese state appeared in the central region of Honshu in the 7th century. This state soon controlled most of the west and centre of the island. In 710 the first permanent Japanese capital was established in Nara by Empress Genmei. In 794 the seat of power moved to Heian-kyo (present-day Kyoto).

Following the court and government tradition of China, Japan cut itself off from the outside world. As the imperial office became increasingly religious the day-to-day power passed into the hands of powerful nobles, such as the Fujiwara clan. Fujiwara Yoshifusa (804–872) was a powerful regent of Japan from 857 until his death, and by the 11th century the Fujiwaras were unchallenged rulers of the country. In the 12th century, however, Japan entered into a period of anarchy. The country passed under the control of barons, the *daimyo,* who exercised power through the warrior class known as the *samurai.*

Shogun

The anarchy ended when Taira Kiyamori seized power and made himself dictator. A civil war, the Gempei War, followed (1180–85). When Taira was defeated, power passed to Minamoto Yoritomo (1147–99), a distant descendant of the imperial family. Yoritomo established a new office, the *shogun.* For the next 700 years Japan was ruled by a military dictator, the shogun, while the emperor lived reclusively as a religious and national symbol.

At first the shogunate was seated in Kamakura, near modern Tokyo. Nine shoguns ruled during the Kamakura epoch (1185–1333) although latterly the Kamakura shogun was a puppet of the Hojo clan. In 1274 and 1281 Mongol attempts to invade Japan were unsuccessful: in 1281 the invasion was thwarted by a sudden typhoon that became known as the 'divine wind' (kami-kaze).

In 1334 a brief restoration of power to the emperor was ended by Ashikaga Takauji (1305–58), who established a strong military government. Subsequent members of the Ashikaga family ruled as shoguns based in Kyoto. Eventually this system, too, collapsed into anarchy, the victim of the ambitions of rival warlords. From 1467 to 1603 Japan suffered the Fighting Principalities (*Sengokujidai*). It was when the country was at its weakest that another powerful outside influence began to exert itself.

From 1543 Portuguese traders and missionaries arrived on the southern and western coasts. At first, trade was welcomed. Christianity, too, made converts after the Spanish Jesuit missionary St Francis Xavier landed in Japan in 1549. Along with western ideas and religion, the Portuguese, and later the Dutch, brought firearms. Three warlords in turn used western weapons to seize power and reunite the country. The last of this trio was Tokugawa Ieyasu (1542–1616), who held power from 1600. Ieyasu ordered the nobles to destroy their fortifications, except their principal residences, and encouraged the arts and learning as a preferred alternative to warfare.

Isolation

As the true rulers of Japan until 1869, the Tokugawa shogunate established itself at Edo (present-day Tokyo). They ruled harshly, subduing the warring lords by holding members of their families hostages. The Tokugawa perceived foreign influences as unsettling and a danger to their supremacy. For this reason, they decreed that Japan should become a closed society. In 1636 Japanese were forbidden to emigrate. Europeans were expelled, except for a single Dutch trading post in Nagasaki, which after 1639 became Japan's only contact with the outside world. Christianity was suppressed and the ownership of firearms, except by the central authorities, was made illegal. Japan entered 220 years of self-imposed isolation.

Cut off from outside influences, Japan gained stability and a strong sense of national identity. Yet this isolation came at a price. In 1853 a US fleet led by Commodore Matthew C. Perry appeared

---

[1]Both forms are valid, and derive from different pronunciations of a Chinese character.

off the Japanese coast. Japan was forced to open up to international trade through the threat of invasion. Other western nations followed the American example. Japan was thrust into a modern world for which it was ill suited. The voices for reform grew and the Tokugawa shogunate, humiliated by Perry's mission, collapsed. Reformers seized Kyoto and parts of the west, but they needed a national symbol to legitimize their rule. In 1869 the shadowy figure of the emperor was called out of his cloistered life. His city, Edo, had by then been renamed Tokyo, meaning 'eastern capital'. The emperor surprised the country by his zeal for modernization which led to a period of rapid reform and transformed Japan into a modern nation.

But while a constitution was introduced, the resemblance to a western democracy was skin deep. Though the peasants were freed from serfdom, power remained in the hands of the nobility. Priority was given to developing industry and modern technology. Japan's rise as an industrial state began.

Rise of the Military
Much emphasis was given to modernizing the armed forces. A revitalized Japan defeated China in the First Sino-Japanese War in 1894–95 and gained Taiwan. In 1900 Japan intervened alongside the western powers against the Boxer Rebellion in China. An even greater shock was Japan's victory against Russia in 1903–04 in a war over Korea and Manchuria. Having contained Russian land forces in Manchuria, the Japanese decisively defeated the Russian Baltic Fleet in the Tsushima Strait. Russia's influence in the region faded and Japan received half of Sakhalin and the Kurile Islands. Later, with Russia removed from the scene, Japan annexed Korea (1910) and took control of parts of Manchuria.

In 1902 Japan made an alliance with Britain. To emphasize Japan's western credentials, Tokyo entered the First World War against Germany in 1914. Japanese forces took the German island colonies in the north and central Pacific and received these archipelagoes as a League of Nations Trust Territory in 1919. But greater rewards for their efforts in the war had been expected and Tokyo's disillusion with the west began. The collapse of world trade at the end of the 1920s brought hardship and helped the rise of political extremism and nationalism.

Japan began a phase of aggressive expansionism. In 1931 Japan invaded Manchuria and, two years later, installed the deposed last emperor of China as puppet emperor of Manchukuo. From 1932 Japanese forces entered various coastal and border areas of China, and in 1937 there was a full-scale war with China. Japanese forces took Shanghai in 1937, Guangzhou in 1938 and Nanjing in 1940. By the end of 1940 Japan had occupied French Indochina and formed a triple alliance (or Axis) with Nazi Germany and Fascist Italy.

Pearl Harbor
Under premier Gen. Tojo Hideki (1884–1948), Japan attacked the US fleet in Pearl Harbor, Hawaii in Dec. 1941. This action brought the United States into the Second World War (1939–45) and ranged Japan against forces that were superior in size and technology. Nevertheless, the war was initially in Japan's favour. Japanese forces swept through the Pacific and into Malaya and the Dutch East Indies (now Indonesia). The speed of Japan's ruthless advance overwhelmed the Allied powers as British and American positions were surrendered. The tide turned with the American victory at Midway in late 1942, but by the time Germany surrendered in May 1945 Japanese forces were still in control of large areas of the Pacific and Southeast Asia. In Aug. 1945 US planes dropped atomic bombs on the Japanese cities of Hiroshima and Nagasaki, devastating the two cities and causing more than 200,000 deaths. The emperor Hirohito (1926–89) surrendered.

The war had cost Japan dearly. Not only had two cities suffered the horror of atomic warfare, but many more Japanese had died in combat. Nearly 2m. Japanese were abandoned in China, most of whom were shipped to Siberia as prisoners. Japan was to be

reformed by the occupying US forces under Gen. MacArthur. In 1945 Shintoism, which had become associated with aggressive nationalism, ceased to be the state religion. In the following year, the emperor renounced his divinity.

A new liberal constitution was introduced in 1946. Japan signed a peace treaty in 1951 at San Francisco and the American occupation of Japan ended in April 1952 when the country regained its independence. A separate peace treaty was concluded later between Japan and China. There was, however, no agreement with the Soviet Union, which, at American behest, had declared war against Japan in the closing days of the Second World War. Soviet forces occupied Sakhalin and the Kurile islands to which Japan still lays claim.

The new Japan remained a monarchy, albeit one in which the emperor was a figurehead. Japan renounced war and the threat or use of force, but retained 'Self Defence Forces'. Japanese cities and industry were rebuilt. An astonishing economic recovery was led by an aggressive export policy. Huge investment in new technology gave the country a dominant position in many industries including motor vehicles, shipbuilding, electrical goods, electronics and computers. Japan grew to be the world's second biggest economy. This success is owed, in part, to the protection of domestic markets.

The power of Japanese industry was reflected in the political power of a small number of major corporations. From 1955 until 1993 the political scene was dominated by the centre-right pro-business Liberal Democrats (LDP). However, a series of major financial scandals broke the party's monopoly and coalition governments followed. By 2000 the LDP had resumed its dominant role.

In recent years, Japan has shown more confidence in international relations. The country is a major aid donor to developing countries. In 1992 the Diet (parliament) approved the involvement of Japanese military personnel and equipment in UN peacekeeping missions and in 2002 Japan contributed naval support vessels to the US-led intervention in Afghanistan. However, the country faces severe economic problems. A heavy international debt and domestic deflation left the Japanese economy in the doldrums at the turn of the century.

In March 2011 a huge offshore earthquake and a resulting tsunami devastated large areas of the northeast of Japan, causing between US$120bn. and US$235bn. of damage and killing over 10,000. The disaster also caused extensive damage to the Fukushima nuclear plant.

TERRITORY AND POPULATION

Japan consists of four major islands, Honshu, Hokkaido, Kyushu and Shikoku, and many small islands, with an area of 377,950 sq. km. Census population of 1 Oct. 2010 (2005 census in brackets), 128,057,352 (127,767,994); of which males, 62,327,737 (62,348,977), females, 65,729,615 (65,419,017); population density (land area only), 351 per sq. km (351 per sq. km).

The UN gives a projected population for 2015 of 126·07m. The population started to decline in 2009; the UN projects that by 2050 it will only be 109m.

In 2011, 67·0% of the population lived in urban areas. Foreigners registered on 31 Dec. 2007 were 2,152,973: including 606,889 Chinese, 593,489 Koreans, 316,967 Brazilians, 202,592 Filipinos, 59,696 Peruvians, 51,851 Americans, 41,384 Thais, 36,860 Vietnamese, 25,620 Indonesians, 20,589 Indians, 17,328 British, 11,459 Canadians, 11,255 Bangladeshis, 11,033 Australians, 9,332 Pakistanis and 1,573 stateless persons. In 2008 Japan accepted 57 asylum seekers (54 of whom were from Myanmar).

Japanese overseas, Oct. 2007, 1,085,671; of these 374,732 lived in the USA, 127,905 in China, 63,526 in the UK, 63,459 in Australia, 61,527 in Brazil, 47,376 in Canada, 42,736 in Thailand, 32,755 in Germany, 29,279 in France and 25,969 in Singapore.

The official language is Japanese.

A law of May 1997 'on the promotion of Ainu culture' marked the first official recognition of the existence of an ethnic minority in Japan. The Ainu were recognized as a people in their own right through a resolution passed by parliament in June 2008.

Japan is divided into 43 prefectures, one metropolis (Tokyo), one territory (Hokkaido) and two urban prefectures (Kyoto and Osaka). The populations and chief cities are:

Prefecture	Census pop. 2010	Chief city
Aichi	7,410,719	Nagoya
Akita	1,085,997	Akita
Aomori	1,373,339	Aomori
Chiba	6,216,289	Chiba
Ehime	1,431,493	Matsuyama
Fukui	806,314	Fukui
Fukuoka	5,071,968	Fukuoka
Fukushima	2,029,064	Fukushima
Gifu	2,080,773	Gifu
Gumma	2,008,068	Maebashi
Hiroshima	2,860,750	Hiroshima
Hokkaido	5,506,419	Sapporo
Hyogo	5,588,133	Kobe
Ibaraki	2,969,770	Mito
Ishikawa	1,169,788	Kanazawa
Iwate	1,330,147	Morioka
Kagawa	995,842	Takamatsu
Kagoshima	1,706,242	Kagoshima
Kanagawa	9,048,331	Yokohama
Kochi	764,456	Kochi
Kumamoto	1,817,426	Kumamoto
Kyoto	2,636,092	Kyoto
Mie	1,854,724	Tsu
Miyagi	2,348,165	Sendai
Miyazaki	1,135,233	Miyazaki
Nagano	2,152,449	Nagano
Nagasaki	1,426,779	Nagasaki
Nara	1,400,728	Nara
Niigata	2,374,450	Niigata
Oita	1,196,529	Oita
Okayama	1,945,276	Okayama
Okinawa	1,392,818	Naha
Osaka	8,865,245	Osaka
Saga	849,788	Saga
Saitama	7,194,556	Saitama
Shiga	1,410,777	Otsu
Shimane	717,397	Matsue
Shizuoka	3,765,007	Shizuoka
Tochigi	2,007,683	Utsunomiya
Tokushima	785,491	Tokushima
Tokyo	13,159,388	Tokyo
Tottori	588,667	Tottori
Toyama	1,093,247	Toyama
Wakayama	1,002,198	Wakayama
Yamagata	1,168,924	Yamagata
Yamaguchi	1,451,338	Yamaguchi
Yamanashi	863,075	Kofu

The leading cities, with population in 2010 (in 1,000), are:

Akashi	291	Higashiosaka	510
Akita	324	Himeji	536
Amagasaki	454	Hirakata	408
Aomori	300	Hiratsuka	261
Asahikawa	347	Hiroshima	1,174
Chiba	962	Ibaraki	275
Fuchu	256	Ichihara	280
Fuji	254	Ichikawa	474
Fujisawa	410	Ichinomiya	379
Fukui	267	Iwaki	342
Fukuoka	1,464	Kagoshima	606
Fukushima	293	Kakogawa	267
Fukuyama	461	Kanazawa	462
Funabashi	609	Kashiwa	404
Gifu	413	Kasugai	306
Hachioji	580	Kawagoe	343
Hakodate	279	Kawaguchi	501
Hamamatsu	801	Kawasaki	1,426

Kitakyushu	977	Otsu	338
Kobe	1,544	Sagamihara	718
Kochi	343	Saitama	1,222
Koriyama	339	Sakai	842
Koshigaya	326	Sapporo	1,914
Kumamoto	734	Sasebo	261
Kurashiki	476	Sendai	1,046
Kurume	302	Shimonoseki	281
Kyoto	1,474	Shizuoka	716
Machida	427	Suita	356
Maebashi	340	Takamatsu	419
Matsudo	484	Takasaki	371
Matsuyama	517	Takatsuki	357
Mito	269	Tokorozawa	342
Miyazaki	401	Tokushima	265
Morioka	298	Tokyo	8,946
Nagano	382	Toyama	422
Nagaoka	283	Toyohashi	377
Nagasaki	444	Toyonaka	389
Nagoya	2,264	Toyota	421
Naha	316	Tsu	286
Nara	367	Utsunomiya	512
Niigata	812	Wakayama	370
Nishinomiya	483	Yamagata	254
Oita	474	Yao	271
Okayama	710	Yokkaichi	308
Okazaki	372	Yokohama	3,689
Osaka	2,665	Yokosuka	418

The Tokyo conurbation, with a population in 2010 of 36.9m., is the largest in the world, having overtaken New York around 1970.

SOCIAL STATISTICS

Statistics (in 1,000) for calendar years:

	2001	2002	2003	2004	2005	2006	2007
Births	1,171	1,154	1,124	1,111	1,063	1,093	1,090
Deaths	970	982	1,015	1,029	1,084	1,084	1,108

Birth rate of Japanese nationals in present area in 2007, 8·6 per 1,000 population (1947: 34·3); death rate, 8·8. Marriage rate in 2007 (per 1,000 persons), 5·7; divorce rate, 2·0. In 2007 the mean age at first marriage was 30·1 for males and 28·3 for females. The infant mortality rate per 1,000 live births, 2 (2010), is one of the lowest in the world. In 2007 only 2·0% of births were outside marriage. Life expectancy at birth was 86·0 years for women and 79·0 years for men in 2007. Japan's life expectancy is the highest of any sovereign country. The World Health Organization's *World Health Statistics 2009* put the Japanese in first place in a 'healthy life expectancy' list, with an expected 76 years of healthy life for babies born in 2007. Japan has a very quickly ageing population, stemming from a sharply declined fertility rate and one of the highest life expectancies in the world. In 2007 the total fertility rate was 1·34 births per woman (compared to 1·36 in 2000, 1·91 in 1975 and 3·65 in 1950). The percentage of the population over 65 rose from 10% in 1985 to more than 21% in 2007. Japan had an estimated 40,400 centenarians (35,000 women) in Sept. 2009—an increase of over 4,000 from Sept. 2008, and the 39th consecutive year the number has risen. In 2008 the population fell by 0·1%.

There was a total of 33,093 suicides in 2007, a rate of 35·8 males per 100,000 and 13·7 females per 100,000. The rate among women is one of the highest in the world.

A UNICEF report published in 2010 showed that 14·9% of children in Japan live in relative poverty (living in a household in which disposable income—when adjusted for family size and composition—is less than 50% of the national median income), compared to just 4·7% in Iceland.

CLIMATE

The islands of Japan lie in the temperate zone, northeast of the main monsoon region of southeast Asia. The climate is temperate with warm, humid summers and relatively mild winters except in the island of Hokkaido and northern parts of Honshu facing the Sea of Japan. There is a month's rainy season in June–July, but the best seasons are spring and autumn, although Sept. may bring typhoons. Tokyo, Jan. 5·8°C, July 25·4°C. Annual rainfall 1,467 mm. Hiroshima, Jan. 5·3°C, July 26·9°C. Annual rainfall 1,541 mm. Nagasaki, Jan. 6·8°C, July 26·6°C. Annual rainfall 1,960 mm. Osaka, Jan. 5·8°C, July 27·2°C. Annual rainfall 1,306 mm. Sapporo, Jan. −4·1°C, July 20·5°C. Annual rainfall 1,128 mm.

CONSTITUTION AND GOVERNMENT

The Emperor is Akihito (b. 23 Dec. 1933), who succeeded his father, Hirohito on 7 Jan. 1989 (enthroned, 12 Nov. 1990); married 10 April 1959, to Michiko Shoda (b. 20 Oct. 1934). *Offspring:* Crown Prince Naruhito (Hironomiya; b. 23 Feb. 1960); Prince Akishino (Akishinomiya; b. 30 Nov. 1965); Princess Sayako (Norinomiya; b. 18 April 1969). Prince Naruhito married Masako Owada (b. 9 Dec. 1963) 9 June 1993. *Offspring:* Princess Aiko (b. 1 Dec. 2001). Prince Fumihito (henceforth to adopt his new title Prince Akishino) married Kawashima Kiko (b. 11 Sept. 1966) 29 June 1990. *Offspring:* Princess Mako (23 Oct. 1991); Princess Kako (29 Dec. 1994); Prince Hisahito (6 Sept. 2006). Princess Sayako married Yoshiki Kuroda (b. 17 April 1965) 15 Nov. 2005 and gave up her imperial title in doing so as required by law. The succession to the throne is fixed upon the male descendants. Prince Hisahito was the first male born into the imperial family since 1965. The 1947 constitution supersedes the Meiji constitution of 1889. In it the Japanese people pledge themselves to uphold the ideas of democracy and peace. The Emperor is the symbol of the unity of the people. Sovereign power rests with the people. The Emperor has no powers related to government. Fundamental human rights are guaranteed.

Legislative power rests with the *Diet*, which consists of the *House of Deputies* (Shugi-in), elected by men and women over 20 years of age for a four-year term, and an upper house, the *House of Councillors* (Sangi-in) of 242 members (96 elected by party list system with proportional representation according to the d'Hondt method and 146 from prefectural districts), one-half of its members being elected every three years. The number of members has been reduced in recent years. There had been 252 members until 2001 and 247 members from 2001 until elections of July 2004.

The number of members in the House of Deputies was reduced from 500 to 480 for the election of June 2000, of whom 300 were to be elected from single-seat constituencies, and 180 by proportional representation on a base of 11 regions. There is a 2% threshold to gain one of the latter seats. Donations to individual politicians are to be supplanted over five years by state subsidies to parties.

A new electoral law passed in Oct. 2000 gives voters a choice between individual candidates and parties when casting ballots for the proportional representation seats in the *House of Councillors*.

On becoming prime minister in April 2001 Junichiro Koizumi established a panel to consider introducing the direct election of prime ministers by popular vote.

National Anthem

'Kimigayo' ('The Reign of Our Emperor'); words 9th century, tune by Hayashi Hiromori. On 9 Aug. 1999 a law on the national flag and the national anthem was enacted. The law designates the Hinomaru and 'Kimigayo' as the national flag and national anthem of Japan. The 'Kimi' in 'Kimigayo' indicates the Emperor who is the symbol of the State and of the unity of the people, deriving his position from the will of the people with whom resides sovereign power; 'Kimigayo' depicts the state of being of the country as a whole.

GOVERNMENT CHRONOLOGY

Prime ministers since 1945. (DPJ = Democratic Party of Japan; JNP = Japan New Party; JSP = Japan Socialist Party; Jt = Liberal Party; LDP = Liberal Democratic Party; Mt = Democratic Party; SDP = Social Democratic Party; SSt = Renewal Party; n/p = non party)

1945	military	Kantaro Suzuki
1945	military	Naruhito Kigashi-Kuni
1945–46	n/p	Kijuro Shidehara
1946–47	Jt	Shigeru Yoshida
1947–48	JSP	Tetsu Katayama
1948	Mt	Hitoshi Ashida
1948–54	Jt	Shigeru Yoshida
1954–56	LDP	Ichiro Hatoyama
1956–57	LDP	Tanzan Ishibashi
1957–60	LDP	Nobusuke Kishi
1960–64	LDP	Hayato Ikeda
1964–72	LDP	Eisaku Sato
1972–74	LDP	Kakuei Tanaka
1974–76	LDP	Takeo Miki
1976–78	LDP	Takeo Fukuda
1978–80	LDP	Masayoshi Ohira
1980	LDP	Masayoshi Ito
1980–82	LDP	Zenko Suzuki
1982–87	LDP	Yasuhiro Nakasone
1987–89	LDP	Noboru Takeshita
1989	LDP	Sosuke Uno
1989–91	LDP	Toshiki Kaifu
1991–93	LDP	Kiichi Miyazawa
1993–94	JNP	Morihiro Hosokawa
1994	SSt	Tsutomu Hata
1994–96	SDP	Tomiichi Murayama
1996–98	LDP	Ryutaro Hashimoto
1998–2000	LDP	Keizo Obuchi
2000	LDP	Michio Aoki
2000–01	LDP	Yoshiro Mori
2001–06	LDP	Junichiro Koizumi
2006–07	LDP	Shinzo Abe
2007–08	LDP	Yasuo Fukuda
2008–09	LDP	Taro Aso
2009–10	DPJ	Yukio Hatoyama
2010–11	DPJ	Naoto Kan
2011–12	DPJ	Yoshihiko Noda
2012–	LDP	Shinzo Abe

RECENT ELECTIONS

Elections to the House of Deputies were held on 16 Dec. 2012. The opposition Liberal Democratic Party (LDP; Jiminto) won 294 seats (with 43·0% of the single-seat constituency vote), up from 119 in the 2009 elections; the ruling Democratic Party of Japan (DPJ, Minshuto) 57 seats (22·8%), down from 308 in 2009; Restoration Party (JRP, Ishin no Kai), 54 (11·6%); New Komeito Party (NKP, Komeito), 31 (1·5%); Your Party (YP, Minna no To), 18 (4·7%); Tomorrow Party (TPJ, Mirai no To), 9 (5·0%); Communist Party of Japan (JCP, Nihon Kyosanto), 8 (7·9%); Social Democratic Party (SDP; Shakai Minshuto), 2 (0·8%); New Party Daichi (NPD, Shinto Daichi), 1 (0·5%); People's New Party (Kokumin Shinto), 1 (0·2%). Independents took five seats. Turnout was 59·3%, the lowest since the Second World War.

Elections to 121 seats of the House of Councillors were held on 11 July 2010. The LDP gained 51 seats, DPJ 44, Your Party 10, New Komeito 9, Communist Party of Japan 3, Social Democratic Party 2, New Renaissance Party 1 and Sunrise Party of Japan 1. As a result the DPJ held 106 seats, LDP 84, New Komeito Party 19, Your Party 11, Communist Party of Japan 6, Social Democratic Party 4, People's New Party 3, Sunrise Party of Japan 3, New Renaissance Party 2, New Nippon Party 1, Happiness Realization Party 1 and ind. 2. The DPJ–New Nippon Party coalition held 110 of the 242 seats and the LDP–New Komeito Party–New Renaissance Party coalition 105.

CURRENT GOVERNMENT

Prime Minister: Shinzo Abe; b. 1954 (LDP; sworn in 26 Dec. 2012, having previously held office from Sept. 2006–Sept. 2007).

Following a resounding election victory for the Liberal Democratic Party in the 2012 parliamentary election, in March 2013 the coalition government consisting of the Liberal Democratic Party and the New Komeito Party comprised:

Deputy Prime Minister: Taro Aso (also *Minister of Finance, Minister of State for Financial Services, and Minister in Charge of Overcoming Deflation and Counting Yen Appreciation*).

Minister of Internal Affairs and Communications, Minister of State for Decentralization Reform, and Minister in Charge of Regional Revitalization, and Regional Government: Yoshitaka Shindo. *Justice:* Sadakazu Tanigaki. *Foreign Affairs:* Fumio Kishida. *Education, Culture, Sports, Science and Technology and Minister in Charge of Rebuilding Education:* Hakubun Shimomura. *Health, Labour and Welfare:* Norihisa Tamura. *Agriculture, Forestry and Fisheries:* Yoshimasa Hayashi. *Economy, Trade and Industry, Minister of State for the Corporation in Support of Compensation for Nuclear Damage, and Minister in Charge of Nuclear Incident Economic Countermeasures, and Industrial Competitiveness:* Toshimitsu Motegi. *Land, Infrastructure, Transport and Tourism:* Akihiro Ota. *Environment and Minister of State for the Nuclear Emergency Preparedness:* Nobuteru Ishihara. *Defence:* Itsunori Onodera. *Chief Cabinet Secretary and Minister in Charge of Strengthening National Security:* Yoshihide Suga. *Reconstruction and Minister in Charge of Comprehensive Policy Co-ordination for Revival from the Nuclear Accident at Fukushima:* Takumi Nemoto. *Chairman of the National Public Safety Commission, Minister of State for Disaster Management, and Minister in Charge of the Abduction Issue, and the Nation's Infrastructure Resilience:* Keiji Furuya. *Minister of State for Okinawa and Northern Territories Affairs, Science and Technology Policy, and Space Policy, and Minister in Charge of Information Technology Policy, and Ocean Policy and Territorial Issues:* Ichita Yamamoto. *Minister of State for Consumer Affairs and Food Safety, Measures for Declining Birthrate, and Gender Equality, and Minister in Charge of Support for Women's Empowerment and Child-Rearing:* Masako Mori. *Minister of State for Economic and Fiscal Policy, and Minister in Charge of Total Reform of Social Security and Tax, and Economic Revitalization:* Akira Amari. *Minister of State for Regulatory Reform, and Minister in Charge of Administrative Reform, Civil Service Reform, 'Cool Japan' Strategy, and 'Challenge Again' Initiative:* Tomomi Inada.

Office of the Prime Minister: http://www.kantei.go.jp

CURRENT LEADERS

Shinzo Abe

Position
Prime Minister

Introduction
Shinzo Abe, who first served as prime minister from 2006–07, became president of the Liberal Democratic Party (LDP) for the second time in Sept. 2012. He was sworn in as premier for a second spell after leading the party to victory at the Dec. 2012 election. A conservative from a prominent political family, Abe has consistently argued for Japan to take a more assertive role on the world stage. During his first term he oversaw an improvement in relations with North Korea and China, although domestic disagreements over Japanese logistical support for American troops in Afghanistan clouded his tenure. He started his second tenure promising to boost the economy and to defend Japanese interests while seeking improved relations with China.

Early Life
Abe was born on 21 Sept. 1954 in Nagato, Yamaguchi Prefecture. He is the son of Shintaro Abe, a former secretary-general of the centre-right LDP. Shinzo Abe graduated in political science from Seikei University in 1977 before undertaking further study at the University of Southern California. Returning to Japan in 1979 he took up employment at Kobe Steel Ltd and entered the political scene in 1982 as executive assistant to his father, who was minister for foreign affairs in the government of Yasuhiro Nakasone.

In elections for the house of representatives in July 1993 Abe won the seat in Yamaguchi Prefecture that had been held by his father (who died in 1991), although the ruling LDP lost its overall majority for the first time since 1955 and was replaced by an eight-party alliance headed by Morihiro Hosokawa. Appointed to the house of representatives' committee on foreign affairs, Abe also served as director of the LDP's social affairs division, where he focused on pensions and social security against the backdrop of deep recession. The economy began to revive in 1999 after the government, led by the LDP's Keizo Obuchi, spent more than US$1trn. on public works and other programmes to stimulate growth.

In 2000 Abe was appointed deputy chief cabinet secretary in the second cabinet of Prime Minister Yoshiro Mori. In the wake of the 11 Sept. attacks in America, Abe led the parliamentary campaign for co-operation with the US-led 'war on terror', securing widespread support after initial opposition. In 2002 he accompanied Junichiro Koizumi, leader of the LDP and prime minister since April 2001, to Pyongyang to attend landmark summit talks with the North Korean leader, Kim Jong Il. Abe won plaudits for his tough stance with North Korea over the repatriation of Japanese nationals who had been abducted by North Koreans in the 1970s and 1980s.

He became secretary-general of the LDP in 2003. Following the party's landslide victory in the Sept. 2005 general election, he was appointed chief cabinet secretary and was central to the government's reform programme. During Koizumi's final parliamentary session the LDP passed 82 of its 91 proposed bills, including the controversial privatization of the postal service, set for 2017. On 20 Sept. 2006 Abe was elected to succeed Koizumi, who had stepped down as leader of the LDP in accordance with party rules. Six days later Abe was elected prime minister with 339 of 475 votes in the lower house and a majority in the upper house.

Career in Office
Abe pledged to continue Koizumi's economic reforms and to work to improve relations with China. He announced a 30% cut in his pay as a 'good model' for cutting government spending. His visits to China and South Korea within weeks of taking office demonstrated an assertive approach in international affairs. He also made it clear that he intended to pursue his predecessor's goals of revising Japan's pacifist constitution and securing a permanent seat on the UN Security Council.

However, his premiership went into meltdown in May 2007 when his minister of agriculture, Toshikatsu Matsuoka, committed suicide shortly before facing questions over a financial scandal. The LDP fared badly at elections to the upper house in July 2007 and Abe's position was further weakened when the new agriculture minister, Norihiko Akagi, resigned. He appointed a new cabinet in Aug. but another agriculture minister, Takehiko Endo, resigned after being linked to a separate financial scandal.

Abe faced stern parliamentary opposition to his anti-terrorism bill, which included provision for Japanese logistical support to the USA in Afghanistan. His failure to win passage for the law was widely believed to have influenced his decision to resign on 12 Sept. 2007. Health concerns were also cited, and he subsequently received treatment for ulcerative colitis.

He was replaced as prime minister and LDP president by Yasuo Fukuda but retained his seat in the House of Representatives.

After the LDP went into opposition in 2009, he argued for measures to strengthen the existing social order and ease restrictions on defence institutions. He also called for increased government investment in the economy, especially the technology sector.

In Sept. 2012 Abe was reappointed as LDP president and in Nov. 2012, as opposition leader, secured a promise of early elections in return for supporting a government finance bill. In the subsequent Dec. 2012 general election, the LDP secured an outright majority. Abe took office as premier for the second time on 26 Dec. 2012. As in his first term, Japan's faltering economy and continuing regional tensions, notably with China, are likely to dominate his tenure.

DEFENCE

Japan has renounced war as a sovereign right and the threat or the use of force as a means of settling disputes with other nations. Its troops had not previously been able to serve abroad, but in 1992 the House of Representatives voted to allow up to 2,000 troops to take part in UN peacekeeping missions. A law of Nov. 1994 authorizes the Self-Defence Force to send aircraft abroad in rescue operations where Japanese citizens are involved. Following the attacks on New York and Washington of 11 Sept. 2001, legislation was passed allowing Japan's armed forces to take part in operations in the form of logistical support assisting the US-led war on terror. The legislation permits troops to take part in limited overseas operations but not to engage in combat. In May 2003 parliament passed a series of measures in response to North Korea's nuclear programme. Central government won increased control over the military which now has greater freedom to requisition civilian property in the event of attack.

In Jan. 1991 Japan and the USA signed a renewal agreement under which Japan pays 40% of the costs of stationing US forces and 100% of the associated labour costs. US forces in Japan totalled 38,660 in 2006 (mostly marines and air force personnel).

Total armed forces in 2006 numbered 260,250.

Defence expenditure in 2008 totalled US$46,296m. (US$361 per capita). In 2007 defence spending represented 0·9% of GDP.

Army

The 'Ground Self-Defence Force' is organized in five regional commands and in 2006 had a strength of 148,200. Equipment includes 980 main battle tanks.

Navy

The 'Maritime Self-Defence Force' is tasked with coastal protection and defence of the sea lanes to 1,000 nautical miles range from Japan. The main elements of the fleet are organized into four escort flotillas based at Yokosuka, Kure, Sasebo, Maizuru and Ominato. The submarines are based at Yokosuka and Kure.

Personnel in 2006 numbered 44,400. The combatant fleet, all home-built, includes 18 diesel submarines, 45 destroyers and nine frigates. The Air Arm operated 80 combat aircraft in 2006. Air Arm personnel was estimated at 9,800 in 2006.

Air Force

An 'Air Self-Defence Force' was inaugurated on 1 July 1954. Its equipment includes (2006) F-15 *Eagles*, F-4E *Phantoms* and Mitsubishi F-1 fighters.

Strength (2006) 45,600 operating 300 combat aircraft.

ECONOMY

In 2007 services accounted for 68·4% of GDP, industry 30·2% and agriculture, forestry and fisheries 1·4%.

According to the anti-corruption organization *Transparency International*, Japan ranked equal 17th in the world in a 2012 survey of the countries with the least corruption in business and government. It received 74 out of 100 in the annual index.

In terms of total aid given, Japan was the fifth most generous country in the world in 2011 after the USA, Germany, the UK and France, donating US$10·8bn. in international aid in the course of the year. However, this represented only 0·18% of its GNI (compared to the UN target of 0·7%).

Overview

Following China's recent growth, Japan is now the world's third largest economy at market exchange rates, having been the second largest up to 2010. It is a major donor of global aid, capital and credit.

From 1973 until the asset market bubble burst in 1991, Japan's economy grew at an average of more than 4% a year. Lax monetary policies, however, caused a bubble economy and in its aftermath there was deflation, static wages and declining investment. In the decade 1991–2001 growth slowed to a yearly rate of 1·1%. This improved to 1·8% over the period 2002–07. Exports were the initial driving force (its share of GDP rising from 10·6% to 16·5%) before momentum shifted to strong domestic demand and private investment.

The IMF attributed the strengthening of the economy to reforms in labour and product markets, bank restructuring and corporate efforts to eliminate excess capacity and debt. The traditional system of lifetime employment, limiting labour market flexibility, is slowly being replaced by more flexible contracts. Japan's labour productivity in 2010 was 20th among 34 OECD countries, up from 22nd in 2009.

Nonetheless, global financial turmoil saw the economy experience four consecutive quarters of contraction in 2008 and 2009, with a collapse in overseas demand compounded by a strong yen. Difficulties were accentuated by a reliance on manufactured exports, notably cars and consumer technology. The economy emerged from recession in the second quarter of 2009 on a rebound in exports and fiscal stimulus but contracted in 2011 as a result of a magnitude 9·0 earthquake and subsequent tsunami that caused damage prompting a reconstruction programme costing US$245bn. There was a trade deficit in 2011 for the first time in 31 years but the Japanese economy began to recover in 2012, led by reconstruction activity and private consumption. Unemployment stood at 4·2% in Jan. 2013.

The system of *keiretsu* (closely knit production chains linking manufacturers, suppliers and distributors) is gradually eroding. Foreign direct investment nearly tripled between 1998 and 2002, reaching record levels in 2008, before turning negative in 2010 and 2011. Japan ranks third among major industrialized countries for expenditure on science and technology, with R&D spending equalling 3·6% of GDP in 2011.

The Long-Term Trade Agreement (LTTA) led to Japan becoming China's most important trading partner in the late 1970s, with Japan relying on China for its oil. Since the adoption of market policies in China, domestic demand for energy has increased rapidly, constraining the export market and causing friction between the two countries. Expanding international ties has been a priority of the Chinese government, leading to Japan's relative decline as a major trading partner. China currently accounts for 19·7% of Japanese exports and 21·5% of imports.

The banking system has been strengthened by tighter regulation introduced in the Program for Financial Revival (PFR). Bank profitability remains low by OECD standards. Nonetheless, the banking system has remained relatively well insulated from the global financial crisis as a result of its low exposure to toxic securities and bad debts. Commercial banks have increased lending to firms as credit has dried up, helped by government guarantees for small and medium-sized businesses.

In 2012 public debt stood at an estimated 234·5% of GDP, the highest level in the world. Government revenues stood at 27·6% of GDP in 2010 and are among the lowest in the OECD. A Sustainability Report by the IMF in 2011 noted that Japan is experiencing 'moderate' to 'large' fiscal and private saving imbalances, which have grown to unsustainable levels over the last two decades.

Japan has the world's most aged and fastest ageing population, with the elderly share of the population increasing by 14% between 1980 and 2010. It is one of the few OECD countries where the working-age population started declining in the early 2000s. The ratio of over-65s to working-age population is the world's highest and is expected to rise from 38% in 2010 to 57% in 2030. Between 1990 and 2010 spending on social security increased by 60%, putting further pressure on public finances.

For further developments see www.statesmansyearbook.com.

Currency

The unit of currency is the *yen* (JPY). Inflation rates (based on OECD statistics):

2002	2003	2004	2005	2006	2007	2008	2009	2010	2011
−0·9%	−0·3%	0·0%	−0·6%	0·2%	0·1%	1·4%	−1·4%	−0·7%	−0·3%

Japan's foreign exchange reserves totalled US$996,552bn. in Dec. 2009 (US$203·2bn. in 1998)—second only to those of China. Gold reserves in Sept. 2009 were 24·60m. troy oz. In Dec. 2009 the currency in circulation consisted of 80,954,000m. yen Bank of Japan notes and 4,556,000m. yen subsidiary coins.

Budget

Ordinary revenue and expenditure for fiscal year ending 31 March 2009 balanced at 88,548,000m. yen.

Of the proposed revenue (in yen) in 2009, 46,103,000m. was to come from taxes and stamps, 33,294,000m. from public bonds. Main items of expenditure (in yen): social security, 24,834,000m.; local government, 16,111,000m.; public works, 7,070,000m.; education, 5,310,000m.; defence, 4,774,000m.

The outstanding national debt incurred by public bonds was 680,448,000m. yen in March 2008.

The estimated 2009 budgets of the prefectures and other local authorities forecast a total revenue of 82,556,000m. yen, to be made up partly by local taxes and partly by government grants and local loans.

VAT is 5%.

Performance

Real GDP growth rates (based on OECD statistics):

2002	2003	2004	2005	2006	2007	2008	2009	2010	2011
0·3%	1·7%	2·4%	1·3%	1·7%	2·2%	−1·0%	−5·5%	4·5%	−0·7%

Provisional figures suggest that the economy grew by 2·0% in 2012 according to the Cabinet Office. In 2011 Japan's total GDP was US$5,687·2bn., the third highest in the world after the USA and China, which replaced Japan as the second largest economy in 2010.

Banking and Finance

The Nippon Ginko (Bank of Japan), founded 1882, finances the government and the banks, its function being similar to that of a central bank in other countries. The Bank undertakes the management of Treasury funds and foreign exchange control. Its *Governor* is Haruhiko Kuroda (appointed March 2013). Its gold bullion and cash holdings at 31 Dec. 2002 stood at 638,000m. yen.

There were in Feb. 2004, six city banks, 64 regional banks, 27 trust banks, two long-term credit banks, 50 member banks of the second association of regional banks, 309 Shinkin banks (credit associations), 185 credit co-operatives, 72 foreign banks and six others. There is also a public corporation Japan Post handling postal savings which amounted to 229,938,100m. yen in Sept. 2003. Total savings by individuals, including insurance and securities, stood at 1,209,453,100m. yen on 30 Sept. 2003, and about 61% of these savings were deposited in banks and the post office. In 1999 a number of important mergers were announced in the banking sector, most notably the proposed merger of the Industrial Bank of Japan, Dai-Ichi Kangyo and Fuji Bank, which

in Sept. 2000 created Mizuho Financial Group, at the time the world's biggest bank in terms of assets, at over 135,000bn. yen (US$1·3trn.). The second and fourth biggest banks, the Mitsubishi Tokyo Financial Group and UFJ Holdings, announced in Aug. 2004 that they had reached a basic agreement to merge to create Japan's largest bank. The new bank, named Bank of Tokyo–Mitsubishi UFJ, came into existence in Jan. 2006 with assets of 190,000bn. yen (US$1·6trn.). In Oct. 2007 the newly-created Japan Post Bank became the world's largest bank by assets, at US$3·1trn. (although in April 2012 it was only the fourth largest, with assets of US$2·5trn.).

Japan's banks are in a situation where many of them would be insolvent if they admitted the market value of the loans, shares and property they hold. At 31 March 2003 it was estimated that the banking system's bad loans amounted to 21,441bn. yen.

Gross external debt amounted to US$3,092,027m. in June 2012.

Foreign direct investment was US$11·9bn. in 2009, down from a record US$24·4bn. in 2008.

There are five stock exchanges, the largest being in Tokyo.

ENERGY AND NATURAL RESOURCES

Environment

Japan's carbon dioxide emissions from the consumption and flaring of fossil fuels in 2008 accounted for 4·0% of the world total and were equivalent to 9·5 tonnes per capita. An *Environmental Performance Index* compiled in 2008 ranked Japan 21st in the world, with 86·3%. The index examined various factors in six areas—air pollution, biodiversity and habitat, climate change, environmental health, productive natural resources and water resources.

Electricity

Japan is poor in energy resources, and nuclear power generation is important in reducing dependence on foreign supplies. However, following the March 2011 earthquake and Fukushima disaster all of the 54 reactors in operation before the earthquake—with a capacity of 48,960 MW—were shut for safety checks. The last one was shut down in May 2012, but two months later one of three reactors at Ohi became the first reactor to be restarted since the disaster. When the last of the reactors was taken offline for maintenance in May 2012 it left Japan without nuclear power for the first time since 1970. In 2009 nuclear reactors produced approximately 29% of electricity. Total installed generating capacity was 274·5m. kW in 2006. Electricity produced in 2005–06 was 1,157,911m. kWh. In 2006–07, ten regional publicly-held supply companies produced 70·7% of output. Consumption per capita in 2007 was an estimated 8,990 kWh.

Oil and Gas

Output of crude petroleum, 2008, was 985,680 kilolitres, almost entirely from oilfields on the island of Honshu, but 234·4m. kilolitres of crude oil had to be imported. Output of natural gas, 2008, 3,735m. cu. metres; with reserves (2005) of 40bn. cu. metres.

Minerals

Production in tonnes: zinc (2006), 7,169; copper (2002), 1,519; lead (2006), 777; iron (2005), 736; silver (2006), 11,463 kg; gold (2008), 6,868 kg. Output of other minerals (2008 unless otherwise indicated, in 1,000 tonnes): limestone, 156,813; quartzite, 10,682; gypsum (2004), 5,865; silica sand, 3,664; dolomite, 3,370; coal (2004), 1,339; salt (2004), 1,225.

Agriculture

The agricultural population was 4·08m. in 2004 (of whom 2·17m. were economically active), compared to 12·98m. in 1980. Land under cultivation in 2009 was 4·6m. ha., down from 6·1m. ha. in 1961. In 2002 Japan had 0·34m. ha. of permanent crops. Average farm size was 1·6 ha. in 2002. In 2002 there were 2,028,000 tractors and 1,042,000 harvester-threshers.

Rice is the staple food, but its consumption is declining. Rice cultivation accounted for 1,627,000 ha. in 2008. Output of rice (in 1,000 tonnes) was 10,748 in 1995, 9,490 in 2000 and 8,823 in 2008.

Production in 2008 (in 1,000 tonnes) of sugar beets was 4,248; potatoes, 2,743; sugarcane, 1,598; cabbage, 1,389; onions, 1,271; wheat, 881; tomatoes, 733; carrots, 657; cucumbers, 627; lettuce, 544; aubergines, 366; spinach, 293; soybeans, 262; pumpkins and squash, 243; barley, 217; yams, 181. Sweet potatoes, which in the past mitigated the effects of rice famines, have, in view of rice over-production, decreased from 4,955,000 tonnes in 1965 to 1,011,000 tonnes in 2008. Domestic sugar production accounted for only 36·7% of consumption in 2006. 1·31m. tonnes were imported, of which 42·0% from Australia, 39·9% from Thailand and 13·8% from South Africa.

Fruit production, 2008 (in 1,000 tonnes): apples, 911; oranges (2006), 910; melons and watermelons, 611; pears, 362; persimmons, 267; grapes, 201.

Livestock (2008): 9·75m. pigs, 4·42m. cattle (including about 1·53m. dairy cows), 285m. chickens; and (2003) 34,000 goats, 20,000 horses and 11,000 sheep. Livestock products, 2008 (in 1,000 tonnes): milk, 7,982; meat (2003), 2,991; eggs, 2,554.

Forestry

Forests covered 24·98m. ha. in 2010, or 69% of the land area. Of the total area under forests 19% was primary forest in 2010, 40% other naturally regenerated forest and 41% planted forest. Timber production was 17·75m. cu. metres in 2007.

Fisheries

The catch in 2008 was 4,302,264 tonnes. More than 99% of fish caught are from marine waters. Japan is the second largest importer of fishery commodities (after the USA), with imports in 2009 totalling US$13·26bn.

INDUSTRY

The leading companies by market capitalization in Japan in March 2012 were: Toyota Motor Corporation (US$149·6bn.); NTT DoCoMo, a mobile telecommunications company (US$72·9bn.); and Mitsubishi UFJ Financial Group (US$70·9bn.).

The industrial structure is dominated by corporate groups (*keiretsu*) either linking companies in different branches or linking individual companies with their suppliers and distributors.

Japan's industrial capacity, 2004, numbered 246,603 plants of all sizes, employing 8·11m. production workers.

Output in 2003 included: watches, 523·5m.; personal computers (2007), 8·35m.; television sets (2005), 11·07m.; refrigerators, 2·86m.; radio sets (2005), 1·77m. The chemical industry ranks fourth in shipment value after machinery, metals and food products. Production, 2008, included (in tonnes): sulphuric acid, 7·23m.; caustic soda (2005), 3·89m.; ammonium sulphate, 1·41m.; compound fertilizers, 1·23m. A total of 11,564,000 motor vehicles were manufactured in Japan in 2008, making it the world's largest vehicle producer. It is also the largest producer of passenger cars (9,916,000 in 2008).

Output, in 1,000 tonnes, 2009: crude steel, 87,534; pig iron, 66,943; cement (2008), 62,810; ordinary rolled steel, 62,024.

2007 production (in 1,000 tonnes): distillate fuel oil, 55,300; petrol, 42,801; residual fuel oil, 29,545; kerosene, 18,783.

In 2008 paper production was 18·83m. tonnes; paperboard, 11·80m. tonnes.

Output of woven fabrics, 2008, 1,554m. sq. metres. Output of cotton yarn, 2007, 71,669 tonnes; and of cotton woven fabrics (2008), 327m. sq. metres. Output, 2007, 3,870 tonnes of woollen yarns and (2008) 61m. sq. metres of wool fabrics. Output, 2008, of synthetic woven fabrics, 1,008m. sq. metres; rayon woven fabrics, 36m. sq. metres; silk fabrics, 14m. sq. metres.

3,213m. litres of beer were produced in 2008–09; 13,649m. litres of soft drinks and mineral water in 2003; 878,000 tonnes of sugar in 2008–09.

Shipbuilding orders in 2005 totalled 8,698,000 GRT. In 2007, 17,240,220 GRT were launched of which 5,284,893 GRT were tankers.

Labour

Total labour force, 2009, was 66·17m., of which 10·73m. were in manufacturing, 10·55m. in wholesale and retail trade, 6·21m. in health and welfare, 5·17m. in construction, 4·63m. in services, 3·80m. in hotels and restaurants, 3·48m. in transport and postal activities, 2·87m. in education and 2·42m. in agriculture and forestry. The declining population means that the United Nations expects the working-age population in 2050 to be lower than it was in the 1950s. Retirement age is being raised progressively from 60 years to reach 65 by 2025. However, in 2007 the average actual retirement age was 69 for men and 66 for women, one of the highest in the OECD.

In July 2009 unemployment stood at 5·7%, the highest rate on record (up from 3·9% in 2007 and 4·0% in 2008 as a whole). The rate has since fallen, and was 4·2% in Jan. 2013. In 2005, 5,629 working days were lost in industrial stoppages (down from 9,755 in 2004). Between 1996 and 2005 strikes cost Japan an average of just one day per 1,000 employees a year—one of the lowest rates in the industrialized world. In 2006 the average working week was 38·38 hours.

INTERNATIONAL TRADE

Imports and Exports

Trade (in US$1m.):

	2007	2008	2009	2010
Imports	622,243	762,534	551,985	694,059
Exports	714,327	781,412	580,719	769,774

In 2011 merchandise imports exceeded exports for the first time in nearly half a century. In 2008 Japanese imports accounted for 4·6% of the world total imports, and exports 4·9% of the world total exports.

Distribution of trade by countries (customs clearance basis) (US$1m.):

	Imports		Exports	
	2005	2006	2005	2006
Australia	22,963	27,291	11,626	12,209
China	101,621	115,818	74,991	90,695
Germany	16,701	18,032	17,464	19,964
Hong Kong	1,468	1,487	33,681	35,618
Korea, Republic of	22,870	26,703	43,669	49,146
Saudi Arabia	26,909	36,340	3,912	4,537
Taiwan	16,921	19,872	40,809	43,113
Thailand	14,579	16,502	21,028	22,392
UAE	23,719	30,854	4,540	5,915
USA	60,030	66,471	125,636	142,286

China has in the meantime overtaken the USA as Japan's leading trading partner. In 2005 machinery and transport equipment accounted for 26·6% of Japan's imports and 65·6% of exports; chemicals, manufactured goods classified chiefly by material and miscellaneous manufactured articles 32·0% of imports and 32·1% of exports; mineral fuels, lubricants and related materials 25·5% of imports and 0·7% of exports; food, live animals, beverages and tobacco 9·8% of imports and 0·5% of exports; inedible crude materials (except fuels), and animal and vegetable oil and fats 6·1% of imports and 1·1% of exports.

The importation of rice was prohibited until the emergency importation of 1m. tonnes from Australia, China, Thailand and the USA in 1993–94 to offset a poor domestic harvest. The prohibition was lifted in line with WTO agreements. Until 2000 rice imports had limited access; the market is now fully open.

COMMUNICATIONS

Roads

The total length of roads (including urban and other local roads) was 1,196,217 km at 1 April 2008. There were 54,736 km of national roads of which 49,756 km were paved. In 2006, 79·2% of all roads were paved. Motor vehicles, at 31 March 2010, numbered 78,693,000, including 40,419,000 passenger cars and 6,362,000 trucks. In 2007 there were 5,353,648 new vehicle registrations. In 2009 there were 4,914 road deaths (10,679 in 1995).

The world's longest undersea road tunnel, spanning Tokyo Bay, was opened in Dec. 1997. The Tokyo Bay Aqualine, built at a cost of 1·44trn. yen (US$11·3bn.), consists of a 4·4 km (2·7 mile) bridge and a 9·4 km tunnel that allows commuters to cross the bay in about 15 minutes.

Rail

The first railway was completed in 1872, between Tokyo and Yokohama (29 km). Most railways are of 1,067 mm gauge, but the high-speed 'Shinkansen' lines are standard 1,435 mm gauge. In April 1987 the Japanese National Railways was reorganized into seven private companies, the Japanese Railways (JR) Group—six passenger companies and one freight company. Total length of railways in 2008–09 was 27,343 km, of which the JR had 19,987 km and other private railways 7,356 km. In 2008–09 the JR carried 8,984m. passengers (other private, 13,992m.) and 33m. tonnes of freight (other private, 13m.). An undersea tunnel linking Honshu with Hokkaido was opened to rail services in 1988.

There are metros in Tokyo (two metro systems, total 304·1 km in 2008), Fukuoka (29·8 km), Hiroshima (18·4 km), Kobe (30·6 km), Kyoto (28·8 km), Nagoya (89·1 km), Osaka (137·8 km), Sapporo (48·0 km), Sendai (14·8 km) and Yokohama (40·4 km). There are over 40 electric light railways and tram networks in 14 cities.

Japan was ranked second only to Switzerland for quality of rail infrastructure in the World Economic Forum's *Global Competitiveness Report 2009–2010*.

Civil Aviation

The main international airports are at Fukuoka, Hiroshima, Kagoshima, Nagoya, Naha, Niigata, Osaka (Kansai International), Sapporo, Sendai and two serving Tokyo—at Narita (New Tokyo International) and Haneda (Tokyo International). The principal airlines are Japan Airlines International (JAL), formed when Japan Airlines and Japan Air System merged in 2001, and All Nippon Airways. In Jan. 2010 JAL filed for bankruptcy protection after making a single-quarter loss of nearly 100bn. yen. The move prompted a restructuring of the company and the expected loss of 15,600 jobs. In the financial year 2008 Japanese companies carried 92·89m. passengers on domestic services and 16·43m. passengers on international services. JAL flew 361·8m. km in 2002 and carried 33,525,752 passengers, All Nippon Airways flew 259·1m. km and carried 43,680,438 passengers, and Japan Air System flew 106·9m. km and carried 21,426,817 passengers.

In 2007 Narita handled 35,478,146 passengers (mainly on international flights) and 2,254,421 tonnes of freight (making it the 7th busiest airport in the world for freight). Although Tokyo Haneda is mainly used for domestic flights, the opening of a dedicated international terminal in 2010 in conjunction with the completion of a fourth runway has allowed the number of international flights serving the airport to increase considerably; it handled 66,823,414 passengers in 2007 (making it the 4th busiest airport in the world for overall traffic volume).

Shipping

In Jan. 2009 there were 2,524 ships of 300 GT or over registered, totalling 12·29m. GT. Of the 2,524 vessels registered, 966 were general cargo ships, 499 oil tankers, 370 bulk carriers, 306 passenger ships, 249 chemical tankers, 162 liquid gas tankers and

22 container ships. The Japanese-controlled fleet is the largest in the world, comprising 3,674 vessels of 1,000 GT or over in Jan. 2009. Only 646 of the 3,474 vessels in Jan. 2009 were flying the Japanese flag. The busiest ports are Nagoya (218,130,000 freight tons handled in 2008), Chiba, Yokohama, Kitakyushu and Osaka.

Coastguard

The 'Japan Coast Guard' consists of one main headquarters, 11 regional headquarters, 66 offices, one maritime guard and rescue office, 53 stations, six info-communication management centres, seven traffic advisory service centres, 14 air stations, one transnational organized crime strike force station, one special security station, one special rescue station, one national strike team station, five district communications centres, four hydrographic observatories, one Loran navigation system centre and 39 aids-to-navigation offices (with 5,604 aids-to-navigation facilities); and controlled 52 large patrol vessels, 44 medium patrol vessels, 23 small patrol vessels, 233 patrol craft, 13 hydrographic service vessels, five large firefighting boats, four medium firefighting boats, 87 special guard and rescue boats, one aids-to-navigation evaluation vessel, four buoy tenders and 50 aids-to-navigation tenders in the financial year 2003. Personnel numbered 12,258. The 'Japan Coast Guard' aviation service includes 29 fixed-wing aircraft and 46 helicopters.

Telecommunications

Telephone services have been operated by private companies (NTT and others) since 1985. In 2008 there were 48,427,000 main (fixed) telephone lines. In the same year mobile phone subscribers numbered 110,395,000 (867·3 per 1,000 persons). In 2008 there were an estimated 96·0m. internet users—a figure exceeded only in China and the USA. Approximately 70% of internet users are men. Internet commerce, or e-commerce, amounted to 1·97trn. yen (US$17·07bn.) in 2007. The fixed broadband penetration rate in Dec. 2010 was 26·7 subscribers per 100 inhabitants. In March 2012 there were 7·7m. Facebook users (only 6% of the population).

SOCIAL INSTITUTIONS

Justice

The Supreme Court is composed of the Chief Justice and 14 other judges. The Chief Justice is appointed by the Emperor, the other judges by the Cabinet. Every ten years a justice must submit himself to the electorate. All justices and judges of the lower courts serve until they are 70 years of age.

Below the Supreme Court are eight regional higher courts, district courts in each prefecture (four in Hokkaido) and the local courts.

The Supreme Court is authorized to declare unconstitutional any act of the Legislature or the Executive which violates the Constitution.

Jury trials were reintroduced in Aug. 2009 for the first time since the Second World War.

In 2008, 2,533,351 penal code offences were reported, including 1,297 homicides. The death penalty is authorized; there were no executions in 2011 but seven in 2012, including one woman. The average daily population in penal institutions in 2008 was 76,881 (60 per 100,000 population).

Education

Education is compulsory and free between the ages of six and 15. Almost all national and municipal institutions are co-educational. In May 2009 there were 13,516 kindergartens with 110,692 teachers and 1,630,000 pupils; 21,970 elementary schools with 419,518 teachers and 7,064,000 pupils; 10,785 lower secondary schools with 250,771 teachers and 3,600,000 pupils; 5,074 upper secondary schools with 307,914 teachers and 3,347,000 pupils; 406 junior colleges with 33,040 teachers and 161,000 pupils; and 64 technical colleges with 6,525 teachers and 59,386 pupils. There were also 935 special schools for children with physical disabilities (74,141 teachers, 117,035 pupils).

Japan has seven main state universities: Tokyo University (1877); Kyoto University (1897); Tohoku University, Sendai (1907); Kyushu University, Fukuoka (1910); Hokkaido University, Sapporo (1918); Osaka University (1931); and Nagoya University (1939). In addition, there are various other state and municipal as well as private universities. There are 773 colleges and universities altogether with (May 2009) 2,846,000 students and 352,514 teaching staff (172,039 full-time).

In 2007 expenditure on education came to 3·5% of GDP and 9·4% of total government spending.

The adult literacy rate is at least 99%.

Health
Hospitals on 1 Oct. 2008 numbered 8,794 with 1,609,403 beds. The hospital bed provision of 126 per 10,000 population was one of the highest in the world. Physicians in 2008 numbered 286,699; dentists, 99,426. In 2007 Japan spent 8·1% of its GDP on health, with public spending accounting for 81·9% of total expenditure on health and private spending 18·1%.

Welfare
There are various types of social security schemes in force, such as health insurance, unemployment insurance and age pensions. The old age pension system in Japan is made up of a two-tiered public benefit. The first tier of the public pension is the basic pension which is payable from age 65 with 25 years' contributions. To receive the full benefit amount, 40 years' contributions to the system are necessary. There is an earnings floor for contributions at approximately 28% of average earnings. The monthly premium of the National Pension is uniformly fixed (13,300 yen in fiscal year 2004). The full basic pension was a flat amount of 804,200 yen per annum in fiscal year 2002, paid in two-monthly instalments. A reduced early pension is available between the ages 60–64, while pensions may also be deferred up to age 69.

14 weeks maternity leave is statutory.

Social security expenditure in 2006–07 was 89,109·8bn. yen, including 47,325·3bn. yen on pensions and 28,102·7bn. on medical care. In 2006, 18,166,704 persons and 12,909,835 households received some form of regular public assistance. A proposed reform of the pension system involves the public making higher payments for lower benefits.

RELIGION
State subsidies have ceased for all religions, and all religious teachings are forbidden in public schools. In Dec. 2007 Shintoism claimed 105·83m. adherents, Buddhism 89·54m.; these figures overlap. Christians numbered 2·14m.

CULTURE

World Heritage Sites
Japan has 16 sites on the UNESCO World Heritage List (date of inscription on the list in brackets): the Buddhist Monuments in the Horyu-ji Area (1993); Himeji-jo (1993); Yakushima (1993); Shirakami-Sanchi (1993); the Historic Monuments of Ancient Kyoto (Kyoto, Uji and Otsu Cities) (1994), including 13 of Kyoto's Buddhist temples, three Shinto shrines and one castle—temples include Byōdo-in, Daigo-ji, Enryaku-ji, Ginkaku-ji, Kinkaku-ji, Kiyomizu-dera, Kōzan-ji, Ninna-ji, Nishi Hongan-ji, Ryōan-ji, Saihō-ji, Tenryū-ji and Tō-ji; the Historic Villages of Shirakawa-go and Gokayama (1995); Hiroshima Peace Memorial (Genbaku Dome) (1996); Itsukushima Shinto Shrine (1996); the Historic Monuments of Ancient Nara (1998), including five Buddhist temples—Tōdai-ji, Kōfuku-ji, Gango-ji, Yakushi-ji and Tōshōdai-ji—and three listed shrines—Kasuga Taisha, Kasuga Yama Primeval Forest and the remains of Heijō-kyō Palace; Shrines and Temples of Nikko (1999); Gusuku Sites and Related Properties of the Kingdom of Ryukyu (2000); Sacred sites and pilgrimage routes in the Kii mountain range (2004); marine and land ecosystems at Shiretoko (2005); the Iwami Ginzan Silver Mine and its cultural landscape (2007); Hiraizumi—temples, gardens and archaeological sites representing the Buddhist Pure Land (2011); and Ogasawara Islands (2011).

Press
In 2009 daily newspapers numbered 121 with aggregate circulation of 65·08m. including four major English-language newspapers. The newspapers with the highest circulation are *Yomiuri Shimbun* (daily average of 10·0m. copies in 2008) and *Asahi Shimbun* (daily average of 8·0m. copies in 2008). They are also the two most widely read newspapers in the world. Japan has one of the highest circulation rates of daily newspapers in any country.

In 2008, 78,013 book titles were published.

Tourism
In 2007, 9,152,186 foreigners visited Japan, 2,845,556 of whom came from South Korea, 1,428,873 from Taiwan and 1,140,419 from mainland China. Japanese travelling abroad totalled 17,294,935. Tourism receipts in 2007 totalled US$12·4bn.

Festivals
Japan has a huge number of annual festivals, among the largest of which are the Sapporo Snow Festival (Feb.); Hakata Dontaku, Fukuoka City (May); the Sanja Festival of Asakusa Shrine, Tokyo (May); the Tanabata Festival in Hiratsuka City (July) and Sendai City (Aug.); the Nebuta Festival in Aomori City (Aug.); and Jidai Matsuri, Kyoto (Oct.).

DIPLOMATIC REPRESENTATIVES
Of Japan in the United Kingdom (101–104 Piccadilly, London, W1J 7JT)
Ambassador: Keiichi Hayashi.

Of the United Kingdom in Japan (1 Ichiban-cho, Chiyoda-ku, Tokyo 102-8381)
Ambassador: Timothy Hitchens, CMG, LVO.

Of Japan in the USA (2520 Massachusetts Ave., NW, Washington, D.C., 20008)
Ambassador: Kenichiro Sasae.

Of the USA in Japan (10–5, Akasaka 1-chome, Minato-ku, Tokyo)
Ambassador: John V. Roos.

Of Japan to the United Nations
Ambassador: Tsuneo Nishida.

Of Japan to the European Union
Ambassador: Kojiro Shiojiri.

FURTHER READING

Statistics Bureau of the Prime Minister's Office (up to 2000) and Statistics Bureau of the Ministry of Internal Affairs and Communications (from 2001): *Statistical Yearbook* (from 1949).—*Statistical Handbook* (from 1958).—*Monthly Statistics of Japan* (from 1947–2006; online only since 2006 as *Japan Monthly Statistics*).—*Historical Statistics* (from 1868–2002)
Economic Planning Agency (up to 2000) and Economic and Social Research Institute (from 2001) of the Cabinet Office: *Economic Survey* (annual), *Economic Statistics* (monthly), *Economic Indicators* (monthly)
Ministry of International Trade and Industry (up to 2000) and the Ministry of Economy, Trade and Industry (from 2001): *Foreign Trade of Japan* (annual)

Allinson, G. D., *Japan's Postwar History*. 1997
Argy, V. and Stein, L., *The Japanese Economy*. 1996
Bailey, P. J., *Post-war Japan: 1945 to the Present*. 1996
Beasley, W. G., *The Rise of Modern Japan: Political, Economic and Social Change Since 1850*. 3rd ed. 2000
Buruma, Ian, *Inventing Japan: 1853–1964*. 2003
The Cambridge Encyclopedia of Japan. 1993
Cambridge History of Japan. Vols. 1–5. 1990–93
Campbell, A. (ed.) *Japan: an Illustrated Encyclopedia*. 1994
Clesse, A., *et al.*, (eds.) *The Vitality of Japan: Sources of National Strength and Weakness*. 1997

Henshall, K. G., *A History of Japan: From Stone Age to Superpower*. 3rd ed. 2012

Kingston, Jeff, *Contemporary Japan: History, Politics and Social Change Since the 1980s.* 2010

McCargo, Duncan, *Contemporary Japan.* 3rd ed. 2012

McClain, James, *Japan: A Modern History.* 2001

Morton, W. Scott and Olenik, J. Kenneth, *Japan: Its History and Culture.* 2004

Perren, R., *Japanese Studies From Pre-History to 1990.* 1992

Schirokauer, C., *Brief History of Japanese Civilization.* 1993

Stockwin, J., *Dictionary of the Modern Politics of Japan.* 2003

Takao, Yasuo, *Reinventing Japan: From Merchant Nation to Civic Nation.* 2008

Woronoff, J., *The Japanese Economic Crisis.* 2nd ed. 1996

Yoda, Tomiko, *Japan After Japan: Social and Cultural Life from the Recessionary 1990s to the Present.* 2006

National library: The National Diet Library, 1-10-1 Nagata-cho, Chiyoda-ku, Tokyo 100-8924.

National Statistical Office: Statistics Bureau, Ministry of Internal Affairs and Communications, 19-1 Wakamatsu-cho, Shinjuku-ku, Tokyo 162-8668.

Website: http://www.stat.go.jp

JORDAN

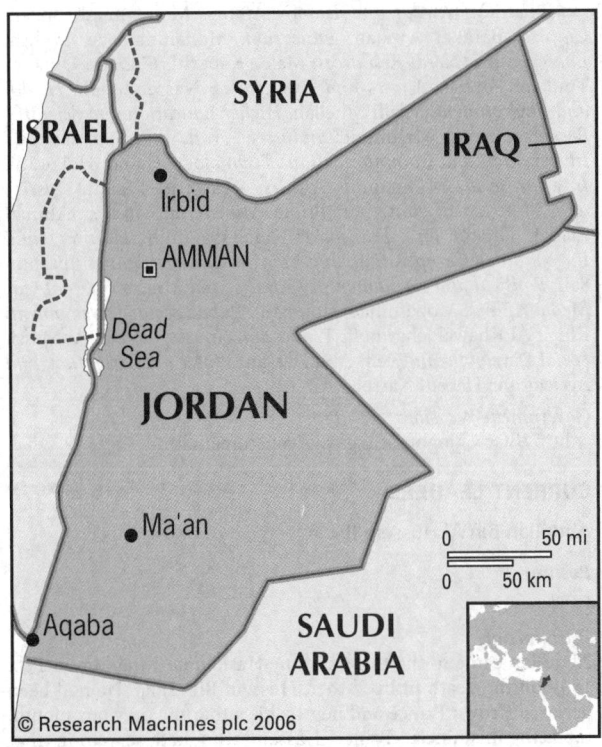

© Research Machines plc 2006

Al-Mamlaka Al-Urduniya Al-Hashemiyah
(Hashemite[1] Kingdom of Jordan)

Capital: Amman
Population projection, 2015: 6·80m.
GNI per capita, 2011: (PPP$) 5,300
HDI/world rank: 0·698/95
Internet domain extension: .jo

KEY HISTORICAL EVENTS

Egyptian control was established over Semitic Amorite tribes in the Jordan valley in the 16th century BC. However, Egypt's conflict with the Hittite Empire allowed the development of autonomous kingdoms such as Edom, Moab, Gilead and Ammon (centred on modern Amman). The Israelites settled on the east bank of the Jordan in the 13th century and crossed into Canaan. David subjugated Moab, Edom and Ammon in the 10th century but the Assyrians wrested control in the 9th century, remaining until 612 BC. Nabataea expanded in the south during the Babylonian and Persian periods until conquered for Rome by Pompey in the 1st century BC. After Trajan's campaign of AD 106, the Jordan area was absorbed as Arabia Petraea.

Rome (later Byzantium) and Sassanid Persia clashed over the area but a Muslim army under Khalid ibn al-Walid defeated Byzantium in 636 at the Yarmuk River. After the fall of the Umayyad Caliphate in 750, the centre of power moved from Damascus to Baghdad. The principality of Oultre Jourdain, established by the Christian crusader kingdom of Jerusalem in the early 12th century, was destroyed by Saladin in 1187. The Mamluk Empire held power until the advent of the Turkish Ottoman Empire in the 16th century.

The Arabs of the Ottoman Damascus province rebelled with British support in 1916. The Hashemite Prince Faisal ibn Husayn took Aqaba in 1917 and the British took Amman and Damascus in 1918. The First World War victors decreed two mandates—British Palestine and French Syria. Britain created the Transjordan Emirate in 1922, ruled semi-autonomously by Faisal's brother, Abdullah. Full independence was achieved on 25 May 1946 as the Hashemite Kingdom of Transjordan (Jordan from 1949).

Transjordan declared war on the Israeli state in May 1948, taking the West Bank and East Jerusalem, an occupation supported only by Britain. Palestinian resistance to the annexation culminated in King Abdullah's assassination in 1951. Talal, his son and successor, was deemed mentally unfit in 1952 and Hussein Bin Talal was installed in 1953. After an attempted coup in 1957, instigated by West Bank Palestinians, King Hussein banned political parties and ended Palestinian representation. A brief union with Iraq, ruled by his cousin, ended after an Iraqi republican coup in 1958. Hussein turned to Britain and the USA for military and financial support.

Fatah and the Palestine Liberation Organization (PLO) maintained terrorist attacks on Israel from Jordan, provoking Israeli retaliation in the West Bank. Despite secret co-operation with Israel over containing the Palestinians, Hussein allied with Syria and Egypt in the war of June 1967. Israel repelled Jordanian forces from the West Bank, moving the *de facto* border to the River Jordan. This devastated the Jordanian economy but removed Palestinian opposition to the Hashemite regime. However, Jordan's relations with the Palestinians deteriorated; in Sept. 1970 four airliners were destroyed by Palestinian extremists in the Jordanian desert. Jordan, with US and British assistance, repelled a Syrian invasion and evicted the PLO. Relations with Israel also worsened from 1977 with the Jewish settlement programme in the West Bank.

On 31 July 1988 Hussein dissolved Jordan's legal and administrative ties with the West Bank in reaction to the *intifada*, which he saw as a threat to his regime. Elections in 1989 led the way to the suspension of martial law—in place from 1967–91. Hussein, constrained by Jordan's economic and political ties with Iraq, refused to abandon Saddam Hussein in the 1991 Gulf War, creating a rift with Jordan's Western partners. Multi-party elections were held in 1993, giving Hussein parliamentary support. He signed a peace treaty with Israel in 1994. In Jan. 1999 Hussein replaced as crown prince his brother, Hassan, with his son, Abdullah, who succeeded on his father's death a month later.

TERRITORY AND POPULATION

Jordan is bounded in the north by Syria, east by Iraq, southeast and south by Saudi Arabia and west by Israel. It has an outlet to an arm of the Red Sea at Aqaba. Its area is 89,342 sq. km (including 540 sq. km inland water). The 2004 census population was 5,103,639; Dec. 2010 estimate, 6,113,000, giving a density of 68·8 per sq. km. The United Nations population estimate for 2004 was 5,400,000.

The UN gives a projected population for 2015 of 6·80m.

In 2011, 78·6% of the population lived in urban areas. Populations of the 12 governorates:

Governorate	Census 2004	Governorate	Census 2004
Ajloun	118,725	Karak	204,185
Amman	1,942,066	Ma'an	94,253
Aqaba	102,097	Madaba	129,960
Balqa	346,354	Mafraq	244,188
Irbid	928,292	Tafilah	75,267
Jerash	153,602	Zarqa	764,650

---

[1]'Hashemite' denotes a descendant of the prophet Mohammed.

The largest towns, with 2004 census population, are: Amman, the capital, 1,036,330; Zarqa, 395,227; Irbid, 250,645.

Jordan's population includes an estimated 2·0m. Palestinian refugees. About 750,000 Iraqi refugees have entered Jordan since the start of the war in 2003.

The official language is Arabic.

SOCIAL STATISTICS

Births (est.), 2008, 180,000; deaths, 20,000. Rates, 2008 per 1,000 population: birth (est.), 31; death (est.), 4. Annual population growth rate, 2008–10, 2·2%. Life expectancy at birth in 2007; 70·7 years for men, 74·3 for women. Infant mortality, 2010, 18 per 1,000 live births; fertility rate, 2008, 3·1 births per woman.

CLIMATE

Predominantly a Mediterranean climate, with hot dry summers and cool wet winters, but in hilly parts summers are cooler and winters colder. Those areas below sea-level are very hot in summer and warm in winter. Eastern parts have a desert climate. Amman, Jan. 46°F (7·5°C), July 77°F (24·9°C). Annual rainfall 13·4" (340·6 mm). Aqaba, Jan. 61°F (16°C), July 89°F (31·5°C). Annual rainfall 1·4" (36·7 mm).

CONSTITUTION AND GOVERNMENT

The Kingdom is a constitutional monarchy headed by H. M. King **Abdullah Bin Al Hussein** II, born 30 Jan. 1962, married H. M. Queen Rania (Rania Al-Yassin, b. 31 Aug. 1970) on 10 June 1993. He succeeded on the death of his father, H. M. King Hussein, on 7 Feb. 1999. *Sons:* Hussein, b. 28 June 1994; Hashem, b. 30 Jan. 2005; *daughters:* Iman, b. 27 Sept. 1996; Salma, b. 26 Sept. 2000.

The Constitution ratified on 8 Dec. 1952 provides that the Cabinet is responsible to Parliament. It was amended in 1954, 1958, 1960, 1973, 1974, 1976, 1984 and 2011. The legislature consists of a *Senate* of 60 members appointed by the King and a *House of Representatives* of 150 members (15 are reserved for women) elected by universal suffrage. Nine seats are reserved for Christians, and three for Circassians or Chechens. A law of 1993 restricts each elector to a single vote.

The lower house was dissolved in 1976 and elections postponed because no elections could be held in the West Bank under Israeli occupation. Parliament was reconvened on 9 Jan. 1984. By-elections were held in March 1984 and six members were nominated for the West Bank, bringing Parliament to 60 members. Women voted for the first time in 1984. On 9 June 1991 the King and the main political movements endorsed a national charter which legalized political parties in return for the acceptance of the constitution and monarchy. Movements linked to, or financed by, non-Jordanian bodies are not allowed.

National Anthem

'Asha al Malik' ('Long Live the King'); words by A. Al Rifai, tune by A. Al Tanir.

GOVERNMENT CHRONOLOGY

Kings since 1946.

1946–51	Abdullah Bin Al Hussein Al Hashimi I
1951–52	Talal Bin Abdullah Al Hashimi
1953–99	Hussein Bin Talal Al Hashimi
1999–	Abdullah Bin Al Hussein Al Hashimi II

RECENT ELECTIONS

Early elections to the House of Representatives were held on 23 Jan. 2013 after the King dissolved the parliament in Oct. 2012. Independents, mostly loyal to King Abdullah, won 123 of the 150 seats. The Muslim Brotherhood and four smaller parties boycotted the election in protest against the legislation governing the process, and afterwards claimed that there had been electoral fraud and vote-buying. The official turnout was 56·7%.

CURRENT GOVERNMENT

In March 2013 the government consisted of:

Prime Minister and Minister of Defence: Abdullah Ensour; b. 1939 (ind., took office on 11 Oct. 2012).

Minister of Awqaf and Islamic Affairs: Mohammad Qudah. *Culture:* Barakat Awajan. *Education:* Mohammad Al Wahsh. *Energy and Mineral Resources:* Malek Kabariti. *Finance:* Umayya Toukan. *Foreign Affairs and Expatriates:* Nasser Judeh. *Health and Environment:* Mjalli Mheilan. *Higher Education and Scientific Research:* Amin Mahmoud. *Industry, Trade and Supplies, and Information and Communications Technology:* Hatem Halawani. *Interior and Municipial Affairs:* Hussein Majali. *Justice and Minister of State for Prime Ministerial Affairs:* Ahmad Ziadat. *Labour and Transport:* Nidal Qatamin. *Planning and International Co-operation, and Tourism and Antiquities:* Ibrahim Saif. *Political and Parliamentary Affairs, and Minister of State for Media Affairs:* Mohammad Moumani. *Public Sector Development:* Khleif Al Khawaldeh. *Public Works and Housing:* Walid Al Masri. *Social Development:* Reem Abu Hassan. *Water and Irrigation, and Agriculture:* Hazem Nasser.

Government Website:
http://www.kinghussein.gov.jo/government.html

CURRENT LEADERS

Abdullah Bin Al Hussein II

Position
King

Introduction
Abdullah came to the throne of the Hashemite Kingdom in Feb. 1999 on the death of his father, Hussein Bin Talal. He had been declared Crown Prince and heir by his father the previous month, replacing his uncle, Prince Hassan, who had served in that capacity since 1965. Abdullah has maintained the moderate regional policies of his late father. He has aimed to reconcile the domestically unpopular 1994 peace agreement with Israel and friendly relations with the USA with the need to appease Jordan's more militant Arab neighbours and its own large Palestinian population. Following the wave of discontent that swept across the Arab world in 2011, he faced popular demands for democratic reform.

Early Life
Born in Amman on 30 Jan. 1962, Abdullah was educated at St Edmund's School in Surrey, England, then Eaglebrook School in Massachusetts and Deerfield Academy in the USA. With his uncle holding office as Crown Prince, Abdullah focused on the military, enrolling in the Royal Military Academy Sandhurst, England in 1980. Having then attended Oxford University and Georgetown University in Washington, D.C., for studies in international relations, he moved up through the ranks of Jordan's armed forces to become Major-Gen. in May 1998.

Career in Office
Abdullah became Crown Prince on 25 Jan. 1999 after King Hussein had rescinded the 1965 constitutional amendment in favour of his younger brother Hassan. Two weeks later, on 7 Feb. 1999, Hussein died and Abdullah assumed the throne. Consistent with the policy of his father, Abdullah has deterred Islamic militancy (particularly the activities of the radical Palestinian Hamas group), while extending economic liberalization. He revived the privatization programme, oversaw Jordan's admission to the World Trade Organization and concluded a free trade accord with the USA.

Abdullah has supported the wider Arab-Israeli peace process and maintains a close affinity with the USA. He also backs Palestinian statehood in the West Bank, a policy that takes account of Jordan's large Palestinian population. He made early overtures towards Jordan's moderate Arab neighbours, visiting Egypt, Saudi Arabia, Oman and the United Arab Emirates in the first few months of his reign, and also tried to forge closer relations with Syria, a traditional antagonist. Relations with Iran have wavered since Abdullah's accession. In 2003 Abdullah backed the US intervention in Iraq, a decision not wholly popular with Jordanian citizens. The subsequent insurgency in Iraq spilled over into Jordan in Nov. 2005 when nearly 70 people were killed in the capital, Amman, in co-ordinated suicide bomb attacks apparently perpetrated by an Iraqi wing of the al-Qaeda terrorist network. Nevertheless, in Aug. 2008 Abdullah became the first leader of an Arab state to visit Iraq since the 2003 US invasion in a move signifying a rapprochement across the confessional Sunni-Shia divide and also growing international confidence in the Iraqi government.

Although Abdullah retains the power to rule by decree, there is an elected Chamber of Deputies to which a large majority of non-partisan candidates loyal to the King were returned in polling in Nov. 2007. Abdullah subsequently appointed Nader Dahabi as the new prime minister, replacing Marouf al-Bakhit who had been in office since Nov. 2005. In Nov. 2009 the King unexpectedly dissolved the Chamber only halfway through its four-year term and called for early elections. No official reason was given, but the assembly had reportedly been accused of inaction and inept handling of legislation. Abdullah appointed Samir Zaid al-Rifai as the new prime minister in Dec. and a new electoral law was introduced in May 2010. Fresh parliamentary elections were held in Nov. 2010, despite a boycott by the opposition Islamic Action Front, and the poll returned mainly pro-government candidates to the 120-member Chamber, provoking street protests.

Jordan was not immune from the anti-government disaffection across much of the Arab world from early 2011. Although on a lesser scale than in some neighbouring countries, the protests prompted Abdullah to appoint new prime ministers in Feb. and Oct. 2011 and in April and Oct. 2012 charged with balancing political reform with establishment concerns over the increasing influence of the Islamist opposition. Following changes to the electoral law in mid-2012 that increased the number of seats for political parties, in Oct. Abdullah announced that parliamentary elections would be held in Jan. 2013. The Islamic Action Front, dissatisfied with the scope of the changes, nevertheless boycotted the poll.

DEFENCE

Defence expenditure in 2008 totalled US$2,127m. (US$347 per capita), representing 10·6% of GDP (the highest percentage of any country).

Army

Total strength (2007) 88,000. In addition there were 60,000 army reservists, a paramilitary Public Security Directorate of approximately 10,000 and a civil militia 'People's Army' of approximately 35,000.

Navy

The Royal Naval Force numbered an estimated 500 in 2007 and operates 13 patrol and coastal combatants, all based at Aqaba.

Air Force

Strength (2007) 12,000 personnel, 100 combat-capable aircraft (including F-5Es, F-16As and Mirage F1s) and some 20 attack helicopters.

INTERNATIONAL RELATIONS

A 46-year-old formal state of hostilities with Israel was brought to an end by a peace agreement on 26 Oct. 1994.

ECONOMY

Services accounted for 65·5% of GDP in 2009, industry 31·6% and agriculture 2·9%.

Overview

Classified by the World Bank as a small, lower middle-income country, Jordan has a shortage of water and, unlike many of its neighbours, limited energy supplies. Jordan imported most of its oil from Iraq prior to 2003 but in recent years has depended on other Gulf states. Aside from potash and phosphate there are few natural resources. Development relies on human capital, as well as remittances from abroad and one of the world's highest levels of unilateral financial transfers (accounting for about 20–25% of GDP). Services account for more than 75% of jobs. The country is highly urbanized, with about 78% of the population living in cities, and has one of the youngest populations among middle-income countries, with 35% under the age of 14.

The country compares favourably to other middle-income economies in human development indicators, with more than 25% of GDP spent on human development initiatives. Authorities have focused on opening up the economy to the private sector and foreign trade and much work has been done to build modern regulatory institutions. As a result of reforms, Jordan's investment climate ranking improved between 2009 and 2010.

Since 2000 the economy has grown 7% per year on average, per capita GDP has more than doubled, the sizeable foreign debt has been reduced and the current account balance narrowed. Growth has been led by manufacturing, construction, real estate and services. Exports have grown significantly since the signing of a free trade accord with the USA in 2000 and the development of Jordanian Qualifying Industrial Zones, where exports are allowed duty free to the USA.

Growth slowed with the global economic crisis. Jordan remains vulnerable to fluctuation in world oil and food prices. High unemployment, dependency on remittances and increasing pressure on scarce natural resources, particularly water, need to be addressed.

Currency

The unit of currency is the *Jordan dinar* (JOD), usually written as JD, of 1,000 *fils*, pegged to the US dollar since 1995 at a rate of one dinar = US$1·41. There was deflation of 0·7% in 2009 but inflation of 5·0% in 2010 and 4·4% in 2011. Foreign exchange controls were abolished in July 1997. Foreign exchange reserves were US$5,601m. and gold reserves 411,000 troy oz in July 2005. Total money supply in May 2005 was JD 3,487m.

Budget

In 2007 revenues totalled JD 3,971·5m. and expenditures JD 4,540·1m. Tax revenue constituted 75·4% of revenues in 2007; social protection accounted for 28·0% of expenditures, defence 16·7% and education 13·9%.

There is a sales tax of 16% (reduced rates, 4% and 0%).

Performance

Total GDP was US$28·8bn. in 2011. Real GDP growth was 2·3% in 2010 and 2·6% in 2011.

Banking and Finance

The Central Bank of Jordan was established in 1964 (*Governor*, Ziad Fariz). In 2002 there were nine national banks, seven foreign banks and 11 specialized credit institutions. Assets and liabilities of the banking system (including the Central Bank, commercial banks, the Housing Bank and investment banks) totalled JD 8,430·4m. in 1995.

Foreign debt was US$7,822m. in 2010, representing 27·9% of GNI.

There is a stock exchange in Amman (Amman Financial Market).

ENERGY AND NATURAL RESOURCES

Environment
Carbon dioxide emissions in 2008 were the equivalent of 3·6 tonnes per capita.

Electricity
Installed capacity was 2·2m. kW in 2007. Production (2007) 12·84bn. kWh; consumption per capita was 2,249 kWh. Nuclear power is seen as the solution to meeting the country's growing energy needs. The construction of its first reactor is expected to begin in the course of 2013.

Oil and Gas
Natural gas reserves in 2007 totalled 6bn. cu. metres, with production (2007) 180m. cu. metres. Jordan relies heavily on oil from other Arab countries, importing around 96% of its energy needs.

Water
99% of the total population and 100% of the urban population has access to safe drinking water.

Minerals
Phosphate ore production in 2009 was 5·28m. tonnes; potash, 1·20m. tonnes.

Agriculture
The country east of the Hejaz Railway line is largely desert; northwestern Jordan is potentially of agricultural value and an integrated Jordan Valley project began in 1973. In 2007 there were 81,090 ha. of irrigated land. The agricultural cropping pattern for irrigated vegetable cultivation was introduced in 1984 to regulate production and diversify the crops being cultivated. In 1986 the government began to lease state-owned land in the semi-arid southern regions for agricultural development by private investors, mostly for wheat and barley. In 2007 there were 142,958 ha. of arable land and 99,484 ha. of permanent crops. There were 5,357 tractors in 2007 and 76 harvester-threshers.

Production in 2007 (in 1,000 tonnes): tomatoes, 617; cucumbers and gherkins, 154; olives, 125; watermelons, 116; potatoes, 97; pumpkins and squash, 55; cauliflowers, 44; cabbage, 36; bananas, 34; apples, 32; lemons and limes, 22.

Livestock (2007): 2·5m. sheep; 559,600 goats; 88,200 cattle; 9,600 asses; 8,000 camels; 36m. chickens. Total meat production was 154,300 tonnes in 2007; milk, 344,980 tonnes.

Forestry
Forests covered 0·10m. ha. in 2010, or 1% of the land area. In 2007, 281,000 cu. metres of roundwood were cut.

Fisheries
Fish landings in 2010 totalled 486 tonnes, mainly from inland waters.

INDUSTRY
The largest company by market capitalization in Jordan in April 2012 was Arab Bank (US$6·0bn.).

The number of industrial establishments in 2006 was 20,214, employing 174,368 persons. The principal industrial concerns are the production or processing of phosphates, potash, fertilizers, cement and oil.

Production, 2007 (in 1,000 tonnes) includes: cement, 4,051; distillate fuel oil, 1,292; residual fuel oil, 1,215; fertilizers, 831; phosphoric acid, 480.

Labour
The workforce in 2004 was 2,000,900. In 2002, 692,070 persons worked in social and public administration, 150,922 in commerce, 122,741 in mining and manufacturing, and 31,095 in transport and communications. In 2003 approximately 10% of the total labour force worked in agriculture. Unemployment was officially 12% in Oct. 2000 but was estimated by economists to be more than 20%. In 2000 Jordan had more than 600,000 foreign workers, many of them Iraqis.

INTERNATIONAL TRADE
Legislation of 1995 eases restrictions on foreign investment and makes some reductions in taxes and customs duties.

Imports and Exports
Imports (c.i.f.) in 2008 totalled US$16,764m. and exports (f.o.b.) US$7,788m. Major exports are phosphate, potash, fertilizers, foodstuffs, pharmaceuticals, clothes, cement, fruit and vegetables, textiles and plastics.

Principal imports in 2004 were from: Saudi Arabia, 19·9%; China, 8·4%; Germany, 6·8%; and USA, 6·7%. Main exports in 2004 were to: USA, 26·2%; Iraq, 18·8%; Free zones, 7·6%; and India, 6·5%. In 2000 Jordan became the first Arab country to sign a free trade agreement with the USA.

COMMUNICATIONS

Roads
Total length of roads, 2007, 7,768 km, of which 3,206 km were main roads. In 2007 there were 536,700 passenger cars (94 per 1,000 inhabitants), 2,800 motorcycles and mopeds, 17,200 coaches and buses, and 230,800 lorries and vans. There were 992 deaths in road accidents in 2007 (388 in 1992).

Rail
The 1,050 mm gauge Hedjaz Jordan Railway (HJR) runs from the Syrian border to Amman. HJR controls 496 km of track but much of it is out of use. The Aqaba Railway Corporation (ARC) leases a section of track (169 km) south of Menzil from HJR for freight traffic and owns a 115 km stretch from Batn el Ghul to the port town of Aqaba. Passenger-km travelled in 2008 came to 0·4m. Freight tonne-km travelled amounted to 353m. on ARC in 2009 and 0·4m. on HJR in 2008.

Civil Aviation
The Queen Alia International airport is at Zizya, 30 km south of Amman. There are also international flights from Amman's second airport. Queen Alia International handled 2,231,806 passengers in 2001 (2,209,168 on international flights) and 87,679 tonnes of freight. Royal Jordanian is the national carrier. The government owns 26% of its shares; around 70% of all shares belong to Jordanians. In 2003 scheduled airline traffic of Jordanian-based carriers flew 36m. km, carrying 1,353,000 passengers (all on international flights).

Shipping
In Jan. 2009 there were 21 ships of 300 GT or over registered, totalling 315,000 GT. The main port is Aqaba, which handled 17·3m. tonnes of foreign cargo in 2008.

Telecommunications
In 2011 there were 465,400 landline telephone subscriptions (equivalent to 73·5 per 1,000 inhabitants) and 7,482,600 mobile phone subscriptions (or 1,182·0 per 1,000 inhabitants). In 2000 the government sold a 40% stake in Jordan Telecommunications Company (Jordan Telecom) to France Télécom. In 2006 France Télécom became the majority shareholder when it purchased a further 11% of Jordan Telecom from the government. Jordan Telecom's monopoly on fixed-line services ended on 1 Jan. 2005. In 2011, 34·9% of the population were internet users. In March 2012 there were 2·2m. Facebook users.

SOCIAL INSTITUTIONS

Justice
The legal system is based on Islamic law (Sharia) and civil law, and administers justice in cases of civil, criminal or administrative disputes. The constitution guarantees the independence of the

judiciary. Courts are divided into three tiers: regular courts (courts of first instance, magistrate courts, courts of appeal, Court of Cassation/High Court of Justice); religious courts (Sharia courts and Council of Religious Communities); special courts (e.g. police court, military councils, customs court, state security court).

The death penalty is authorized, and was last used in 2006. The murder rate in 2007 stood at 1·7 per 100,000 population. The population in penal institutions in April 2008 was approximately 7,500 (123 per 100,000 of national population).

Education

Adult literacy in 2007 was 91·1% (male, 95·7%; female, 88·4%). Basic primary and secondary education is free and compulsory. In 2006–07 there were 1,267 kindergartens (1,259 private) with 4,834 teachers and 93,236 pupils; 2,996 basic schools (706 private) with 66,075 teachers and 1,294,075 pupils; 1,268 secondary schools (160 private) with 17,771 teachers and 184,663 pupils; and 27 vocational schools with 3,581 teachers and 31,432 pupils. In 2006–07 there were ten state and 15 private universities. 16,753 Jordanians were studying abroad in 2007.

In 1999–2000 total expenditure on education came to 5·0% of GNP and 20·6% of total government spending.

Health

There were 13,460 physicians, 18,439 nurses and midwives, 4,330 dentists and 7,360 pharmacists in 2005. In 2007 there were a total of 11,029 hospital beds in 103 hospitals.

Welfare

There are numerous government organizations involved in social welfare projects. The General Union of Voluntary Societies finances and supports the Governorate Unions, voluntary societies, and needy individuals through financial and in-kind aid. There are also 240 day care centres run by non-governmental organizations.

RELIGION

About 94% of the population are Sunni Muslims.

CULTURE

World Heritage Sites

There are four sites on the World Heritage List: the rose-red rock-carved city of Petra and Quseir Amra (both entered on the list in 1985); Um er-Rasas (2004); and the Wadi Rum protected area (2011). Petra is over 2,000 years old and contains more than 800 monuments, some built but most carved out of the natural rock. Quseir Amra is the best preserved of Jordan's 'desert castles' and is noted for its frescoes. Um er-Rasas (Kastron Mefa'a) is an archaeological site largely unexcavated, containing remains from the Roman, Byzantine and Early Muslim periods. The Wadi Rum protected area features a varied desert landscape with rock carvings and inscriptions that trace the evolution of human thought and the early development of the alphabet.

Press

In 2007 there were seven paid-for daily newspapers with a combined circulation of 257,000 and 23 paid-for non-dailies.

Tourism

In 2009 there were 3,789,000 non-resident tourists (excluding same-day visitors), up from 3,729,000 in 2008 and 3,431,000 in 2007.

DIPLOMATIC REPRESENTATIVES

Of Jordan in the United Kingdom (6 Upper Phillimore Gdns, Kensington, London, W8 7HA)
Ambassador: Mazen Kemal Homoud.

Of the United Kingdom in Jordan (PO Box 87, Abdoun, Amman 11118)
Ambassador: Peter Millett.

Of Jordan in the USA (3504 International Dr., NW, Washington, D.C., 20008)
Ambassador: Alia Hatough Bouran.

Of the USA in Jordan (Al-Omawyeen, Abdoun, Amman 11118)
Ambassador: Stuart Jones.

Of Jordan to the United Nations
Ambassador: Zeid Ra'ad Zeid Al-Hussein.

Of Jordan to the European Union
Ambassador: Montaser Oklah Al-Zou'bi.

FURTHER READING

Department of Statistics. *Statistical Yearbook*
Central Bank of Jordan. *Monthly Statistical Bulletin*

Dallas, R., *King Hussein, The Great Survivor*. 1998
George, Alan, *Jordan: Living in the Crossfire*. 2006
Lucas, Russell E., *Institutions and the Politics of Survival in Jordan: Domestic Responses to External Challenges, 1988–2001*. 2006
Rogan, E. and Tell, T. (eds.) *Village, Steppe and State: the Social Origins of Modern Jordan*. 1994
Salibi, Kamal, *The Modern History of Jordan*. 1998

National Statistical Office: Department of Statistics, P. O. Box 2015, Amman.
Website: http://www.dos.gov.jo

KAZAKHSTAN

Qazaqstan Respūblīkasy
(Republic of Kazakhstan)

Capital: Astana
Population projection, 2015: 16·89m.
GNI per capita, 2011: (PPP$) 10,585
HDI/world rank: 0·745/68
Internet domain extension: .kz

KEY HISTORICAL EVENTS

Turkestan (part of the territory now known as Kazakhstan) was conquered by the Russians in the 1860s. In 1866 Tashkent was occupied, followed in 1868 by Samarkand. Subsequently further territory was conquered and united with Russian Turkestan. In the 1870s Bokhara was subjugated, with the amir, by an agreement of 1873, recognizing Russian suzerainty. In the same year Khiva became a vassal state to Russia. Until 1917 Russian Central Asia was divided politically into the Khanate of Khiva, the Emirate of Bokhara and the Governor-Generalship of Turkestan. In the summer of 1919 the authority of the Soviet Government extended to these regions. The Khan of Khiva was deposed in Feb. 1920, and a People's Soviet Republic was set up, the medieval name of Khorezm being revived. In Aug. 1920 the Amir of Bokhara suffered the same fate and a similar regime was set up in Bokhara. The former Governor-Generalship of Turkestan was constituted an Autonomous Soviet Socialist Republic within the RSFSR on 11 April 1921.

In the autumn of 1924 the Soviets of the Turkestan, Bokhara and Khiva Republics decided to redistribute their territories on a nationality basis; at the same time Bokhara and Khiva became Socialist Republics. The redistribution was completed in May 1925, when the new states of Uzbekistan, Turkmenistan and Tajikistan were accepted into the USSR as Union Republics. The remaining districts of Turkestan populated by Kazakhs were united with Kazakhstan which was established as an Autonomous Soviet Republic in 1925 and became a constituent republic in 1936. Independence was declared on 16 Dec. 1991 when Kazakhstan joined the CIS. Nursultan Nazarbayev became president, and legislation has been introduced to award him privileges for life. Over a million of the country's ethnic Russians and Germans have returned to their homelands in the last ten years. Kazakhstan has been focusing on border disputes with

China and Uzbekistan and fighting fundamentalism along with other Central Asian governments.

TERRITORY AND POPULATION

Kazakhstan is bounded in the west by the Caspian Sea and Russia, in the north by Russia, in the east by China and in the south by Uzbekistan, Kyrgyzstan and Turkmenistan. The area is 2,724,900 sq. km (1,052,090 sq. miles). The population at the census of Feb. 2009 was 16,009,597 (density of 5·9 per sq. km), of whom Kazakhs accounted for 63·1% and Russians 23·7%. There are also Uzbeks, Ukrainians, Uigurs, Tatars, Germans and smaller minorities. In 2009 the population was 51·7% female; it was 58·8% urban in 2011. During the 1990s some 1·5m. people left Kazakhstan—mostly Russians and Germans returning to their homelands. In 2008 around 4m. ethnic Kazakhs were living abroad.

The UN gives a projected population for 2015 of 16·89m.

Kazakhstan's administrative divisions consist of 14 provinces and three cities as follows, with area and population:

	Area (sq. km)	Population (2009 census)
Almaty[1]	224,000	1,807,894
Almaty City	300	1,365,632
Aqmola[2]	146,200	737,495
Aqtöbe	300,600	757,768
Astana City	700	613,006
Atyraū[3]	118,600	510,377
Batys Qazaqstan	151,300	598,880
Bayqonyr (city)	(6,700)	—[4]
Mangghystaū	165,600	485,392
Ongtüstik Qazaqstan	117,300	2,469,357
Pavlodar	124,800	742,475
Qaraghandy	428,000	1,341,700
Qostanay	196,000	885,570
Qyzylorda	226,000	678,794
Shyghys Qazaqstan	283,200	1,396,593
Soltüstik Qazaqstan	98,000	596,535
Zhambyl[5]	144,300	1,022,129

[1]Formerly Alma-Ata. [2]Formerly Tselinograd and then Akmola. [3]Formerly Gurev. [4]As the space base of Bayqonyr is under Russian administration, its 6,700 sq. km and estimated 70,000 inhabitants are excluded from overall Kazakhstan figures. The lease was extended until 2050 in 2004. [5]Formerly Dzhambul.

In Dec. 1997 the capital was moved from Almaty to Aqmola, which was renamed Astana in May 1998 (the name of the province remained as Aqmola). Astana has a population of 613,006 (Feb. 2009 census). Other major cities, with Feb. 2009 populations: Almaty (1,365,632); Shymkent (603,499); Qaraghandy (459,778).

The official languages are Kazakh and Russian; Russian is more widely spoken.

SOCIAL STATISTICS

2007: births, 321,963; deaths, 158,297; marriages, 146,379; divorces, 36,107. Rates, 2007 (per 1,000 population): birth, 20·8; death, 10·2; marriage, 9·5; divorce, 2·3. Suicides in 2007 numbered 4,168 (rate of 26·9 per 100,000 population). Annual population growth rate, 2000–05, 0·4%. Expectation of life at birth, 2007, 59·1 years for males and 71·2 years for females. Infant mortality, 2010, 29 per 1,000 live births; fertility rate, 2008, 2·3 births per woman.

CLIMATE

The climate is generally fairly dry. Winters are cold but spring comes earlier in the south than in the far north. Almaty, Jan. −4°C, July 24°C. Annual rainfall 598 mm.

CONSTITUTION AND GOVERNMENT

Relying on a judgement of the Constitutional Court that the 1994 parliamentary elections were invalid, President Nazarbayev dissolved parliament on 11 March 1995 and began to rule by decree. A referendum on the adoption of a new constitution was held on 30 Aug. 1995. The electorate was 8·8m.; turnout was 80%. 89% of votes cast were in favour. The Constitution thus adopted allowed the President to rule by decree and to dissolve parliament if it holds a no-confidence vote or twice rejects his nominee for Prime Minister. It established a parliament consisting of a 39-member Senate (two selected by each of the elected assemblies of Kazakhstan's 16 principal administrative divisions plus seven appointed by the president); and a lower house (*Majlis*) of 77 (67 popularly elected by single mandate districts, with ten members elected by party-list vote). The constitution was amended in Oct. 1998 to provide for a seven-year presidential term. It was amended again in May 2007 to lift the term-limit clause on the president, reduce the presidential term to five years with effect from 2012, oblige the president to consult with parliament when choosing a prime minister and adopt proportional representation for the lower house. The amendment also raised from seven to 15 the number of senators appointed by the president (increasing the total number of senators to 47) and from 77 to 107 the number of lower house deputies (with 98 elected by proportional representation from party lists and nine elected by the Assembly of the People of Kazakhstan—a body comprising the various ethnic groups in the country, which itself is appointed by the president). In June 2010 parliament approved an amendment to the constitution giving President Nazarbayev the title 'Leader of the Nation' and with it a wide range of privileges after the end of his presidential term.

A Constitutional Court was set up in Dec. 1991 and a new constitution adopted on 28 Jan. 1993, but President Nazarbayev abolished the Constitutional Court in 1995. In June 2000 a bill to provide President Nazarbayev with life-long powers and privileges was passed into law.

National Anthem

'Mening Qazaqstan' ('My Kazakhstan'); words by Z. Nazhimedenov and N. Nazarbayev, tune by S. Kaldayakov.

GOVERNMENT CHRONOLOGY

Presidents since 1991.
1991– Nursultan Abishuly Nazarbayev

RECENT ELECTIONS

At the presidential elections of 3 April 2011 Nursultan Nazarbayev was re-elected with 95·6% of votes cast against three other candidates. Turnout was 90·0%.

National Assembly elections were held on 15 Jan. 2012. President Nursultan Nazarbayev's Nur Otan (Light of the Fatherland) Party took 83 of the 98 seats with 81·0% of votes cast, the Democratic Party of Kazakhstan Ak Zhol (Bright Path) 8 with 7·5% and the Communist People's Party of Kazakhstan 7 with 7·2%. Turnout was 75·4%. There were widespread allegations that the elections were fraudulent and failed to meet international standards.

CURRENT GOVERNMENT

President: Nursultan Nazarbayev; b. 1940 (elected in 1991 and re-elected in 1999, 2005 and 2011).

In March 2013 the government comprised:
First Deputy Minister and Minister of Regional Development: Bakytzhan Sagintayev.
Prime Minister: Serik Akhmetov; b. 1958 (since 24 Sept. 2012).
Deputy Prime Ministers: Krymbek Kusherbayev; Kairat Kelimbetov; Yerbol Orynbayev; Asset Isekeshev (also *Minister of Industry and New Technologies*).

Minister of Agriculture: Assylzhan Mamytbekov. *Culture and Information:* Mukhtar Kul-Mukhammed. *Defence:* Adilbek Dzhaksybekov. *Economic Integration Affairs:* Zhanar Aitzhanova. *Economy and Budget Planning:* Yerbolat Dossayev. *Education and Science:* Bahytzhan Zhumagulov. *Emergency Situations:* Vladimir Bozhko. *Environmental Protection:* Nurlan Kapparov. *Finance:* Bolat Zhamishev. *Foreign Affairs:* Yerlan Idrisov. *Health:* Salidat Kairbekova. *Internal Affairs:* Kalmukhanbet Kassymov. *Justice:* Berik Imashev. *Labour and Social Protection:* Serik Abdenov. *Oil and Gas:* Sauat Mynbayev. *Transport and Communications:* Askar Zhumagaliyev.

Chairman, Senate (Upper House): Kairat Mami.
Chairman, Majlis (Lower House): Ural Mukhamedzhanov.

Government Website: http://www.government.kz

CURRENT LEADERS

Nursultan Abishuly Nazarbayev

Position
President

Introduction
Nursultan Nazarbayev, leader of the Nur Otan (Light of the Fatherland) Party, was first elected president of Kazakhstan in 1991, leading the country to independence after the collapse of the USSR. He has sought to exploit the nation's rich mineral resources although much of the population remains poor. He has also sought close ties with southern regional neighbours as well as Russia, China and the West. His autocratic regime, however, has been widely accused of corruption and human rights abuses.

Early Life
Nursultan Nazarbayev was born on 6 July 1940 in Chemolgan in the Almaty region. He was employed by the Karagandy metallurgical works in 1960 and graduated in engineering from a higher technical college in 1967. Having joined the Soviet Communist Party in 1962, he became secretary of the party's regional committee in 1977 and rose through the ranks to the central committee. In 1984 he was appointed chairman of the Republic's council of ministers and became a full member of the Politburo five years later. In the same year Nazarbayev was named first secretary of the Kazakh Communist party. In April 1990 he was chosen by the Supreme Soviet as president of the Republic of Kazakhstan.

Career in Office
Nazarbayev spoke out in support of Soviet leader Mikhail Gorbachev during a coup attempt in Moscow in Aug. 1991. Nevertheless, Kazakhstan seceded from the USSR in Dec. 1991 to join the Commonwealth of Independent States (CIS). In the same month Nazarbayev's position as head of state was consolidated in presidential elections. Earlier in the year he had closed the nuclear test ground at Semipalatinsk.

In 1992 Nazarbayev secured Kazakhstan's membership of the UN and of the Conference on Security and Co-operation in Europe (the precursor of the OSCE). Despite parliamentary opposition, he implemented a series of economic reforms, including a programme of privatization. He sought close co-operation with his CIS partners and signed up Kazakhstan to the treaties on strategic arms reduction and nuclear non-proliferation. Two years later he signed an agreement on economic and military co-operation with Russia. In the same year his term of office was extended by referendum to 2000 amid accusations that he was becoming increasingly autocratic.

In 1997 Nazarbayev announced the transfer of the national capital from Almata to Aqmola (renamed Astana in 1998) to take advantage of Aqmola's central location and its seismatically less sensitive position. He won presidential elections brought forward to Jan. 1999 with 79·8% of the vote, but earned international criticism for the disqualification of the leading opposition figure, Akezhan Kazhegeldin, from the polls. Kazhegeldin was accused of corruption, went into exile and was sentenced *in absentia* to ten years imprisonment. Parliamentary elections held later in the year were criticized by the OSCE.

Nazarbayev implemented heightened security measures against Islamist militants following increased activity in the region in 2000. That year the government passed constitutional amendments granting him wide-ranging influence once he has retired from office. In June 2001 Kazakhstan joined the Shanghai Co-operation Organization (along with China, Russia, Kyrgyzstan, Uzbekistan and Tajikistan) to bolster regional co-operation in economics and against ethnic and religious activism. In the aftermath of the 11 Sept. attacks on Washington and New York in 2001, Nazarbayev met US President George W. Bush to consolidate relations between the two countries. In 2000 the government passed constitutional amendments granting Nazarbayev wide-ranging influence once he has retired from office. In Nov. 2001 he purged his government of founding members of Democratic Choice, a group seeking to reduce presidential powers. Leading Democratic Choice figures were subsequently imprisoned on disputed charges, as were journalists critical of his regime.

In June 2003 the then prime minister, Imangali Tasmagambetov, resigned in protest at land reforms allowing private ownership for the first time in the nation's history. Nazarbayev's Otan Party (the predecessor of the Nur Otan Party) won a majority of National Assembly seats in parliamentary elections in 2004, and on 4 Dec. 2005 he was re-elected president with 91% of the votes cast. There was some movement towards further democratization, including the establishment in March 2006 of a new state commission chaired by the president to oversee a widening of the powers of legislative bodies, strengthening of the judiciary and law enforcement agencies, and constitutional development. Kazakhstan also ratified two international covenants on civil, political, economic and cultural rights. However, the political opposition remained sceptical in the light of parliament's vote in May 2007 to allow Nazarbayev to stay in office for an unlimited number of terms and then the legislative elections later in the year which returned all the seats to his Nur Otan Party. Moreover, in mid-2010 parliament approved legislation giving Nazarbayev the title of 'Leader of the Nation' and granting him additional powers including immunity from prosecution. In Jan. 2010 Kazakhstan became the first former Soviet state to chair the OSCE, despite widespread reservations about Nazarbayev's commitment to democracy.

After almost a decade of average annual growth of around 10%, Kazakhstan's economy started to slow markedly in 2008 in the wake of the global credit crisis. This prompted Nazarbayev in Oct. to announce a US$10bn. injection of reserves from the National Fund (established in 2000 to accumulate revenues from the expanding oil and gas sector) into the economy, with a further US$5bn. in support for struggling local banks.

In April 2009, responding to an earlier initiative by the International Atomic Energy Agency, Nazarbayev announced his country's readiness to host an international nuclear fuel bank to ensure other countries do not need to develop their own sources and to curtail nuclear proliferation.

In July 2010 a customs union established between Kazakhstan, Belarus and Russia came into force.

Nazarbayev secured a further term in the presidential elections of April 2011, taking 96% of the vote. In Nov. he brought forward the date of parliamentary elections to Jan. 2012, at which his Nur Otan Party was again dominant. In Sept. 2012 Karim Massimov resigned as prime minister to join the presidential office as chief of staff and Nazarbayev appointed Serik Akhmetov in his place at the head of a largely unchanged cabinet. The following month Vladimir Kozlov, a prominent opposition figure, was jailed on charges of orchestrating unrest among oil workers.

DEFENCE

Defence expenditure in 2008 totalled US$1,608m. (US$105 per capita), representing 1·2% of GDP. There is conscription for two years.

Nuclear Weapons

When Kazakhstan gained independence from the USSR in 1991 it became the world's fourth largest nuclear power. However, all the weapons systems were returned to Russia unilaterally and Kazakhstan's nuclear infrastructure was dismantled.

Army

Personnel, 2007, 30,000. Paramilitary units: Presidential Guard (2,000), Government Guard (500), Internal Security Troops (approximately 20,000), State Border Protection Forces (approximately 9,000).

Navy

A 3,000-strong Navy was established in 2003. In 2009 it was equipped with 14 inshore patrol craft.

Air Force

In 2007 there were 12,000 personnel (including Air Defence) with 163 combat-capable aircraft, including MiG-29, MiG-31 and Su-27 interceptors and Su-24 and Su-25 strike aircraft.

INTERNATIONAL RELATIONS

In Jan. 1995 agreements were reached for closer integration with Russia, including the combining of military forces, currency convertibility and a customs union.

In 1998 President Nazarbayev signed major treaties with Russia and China, Kazakhstan's neighbours to the north, west and east, in the hope of improving relations with both countries.

ECONOMY

Agriculture accounted for 6·4% of GDP in 2009, industry 40·3% and services 53·3%.

Overview

Kazakhstan's economy shrunk dramatically after the break-up of the Soviet Union until economic reform and privatization schemes implemented during the late 1990s prompted a strong recovery in the 2000s. From 2002–07 real GDP growth averaged nearly 10% per year, driven by expansion in construction and financial services. Youth unemployment fell dramatically.

Kazakhstan was hit by the global economic crisis but a rebound in commodity prices helped spur recovery. However, rising global food and fuel prices raised inflation above 7%.

The country is a net energy exporter and industry is heavily geared towards the exploitation of its vast natural resources. The reliance on extractive industries has encouraged the government to diversify the economic base. Efforts to create a more business-friendly environment earned Kazakhstan top reformer status in the *Doing Business* 2011 report, improving its overall ranking from 74th in 2010 to 59th in 2011.

Currency

The unit of currency is the *tenge* (KZT) of 100 *tiyn*, which was introduced on 15 Nov. 1993 at 1 tenge = 500 roubles. It became the sole legal tender on 25 Nov. 1993. Inflation was running at nearly 1,880% in 1994, but dropped dramatically and remained relatively stable for several years before rising again to 17·1% in 2008. In 2010 inflation fell to 7·1% only to rise again to 8·3% in 2011. In July 2005 foreign exchange reserves were US$6,953m. and gold reserves amounted to 1·88m. troy oz. Total money supply was 876,981m. tenge in June 2005.

Budget

In 2007 revenues were 2,896bn. tenge and expenditures 2,678bn. tenge. Tax revenue accounted for 81·4% of revenues in 2007;

social security accounted for 18·8% of expenditures, education 17·0% and health 11·2%.

Performance
The break-up of the Soviet Union triggered an economic collapse as orders from Russian factories for Kazakhstan's metals and phosphates, two mainstays of the economy, dried up. Real GDP growth was –1·9% in 1998 but there was a slight recovery in 1999, with growth of 2·7%. Growth was an impressive 9·8% in 2000 and an even more spectacular 13·5% in 2001. Driven by increased oil production, the economy continued to expand, with growth of 8·9% in 2007 and 3·2% in 2008. Even in 2009 Kazakhstan managed to avoid a recession, achieving real GDP growth of 1·2%. The economy then grew by 7·3% in 2010 and 7·5% in 2011. Total GDP in 2011 was US$188·1bn.

Banking and Finance
The central bank and bank of issue is the National Bank (*Governor*, Grigorii Marchenko). In 2001 there were 44 domestic banks, with assets totalling US$5·3bn. The largest bank is Kazkommertsbank (KKB), with assets of US$1·8bn. in Dec. 2002. The other major banks are Bank TuranAlem and Halyk Bank. In 2001 there were also 12 branches or representative offices of foreign banks. External debt was US$118,723m. in 2010, representing 94·3% of GNI. Foreign direct investment amounted to US$10·0bn. in 2010, down from US$13·8bn. in 2009.

ENERGY AND NATURAL RESOURCES
Environment
Carbon dioxide emissions from the consumption and flaring of fossil fuels in 2008 were the equivalent of 13·0 tonnes per capita.

Electricity
Installed capacity was an estimated 18·7m. kW in 2007. Output in 2007 was 76·6bn. kWh. Consumption per capita was 4,926 kWh in 2007.

Oil and Gas
Proven oil reserves in 2008 were 39·8bn. bbls. The onshore Tengiz field has estimated oil reserves between 6bn. and 9bn. bbls; the onshore Karachaganak field has oil reserves of 2bn. bbls, and gas reserves of 600bn. cu. metres. Output in 2008 of oil, 72·0m. tonnes; natural gas, 30·2bn. cu. metres with proven reserves of 1,820bn. cu. metres. The first major pipeline for the export of oil from the Tengiz field was opened in March 2001, linking the Caspian port of Atyraŭ with the Russian Black Sea port of Novorossiisk. In Sept. 1997 Kazakhstan signed oil agreements with China worth US$9·5bn.; although only a small portion of that investment ever materialized. However, a 962-km oil pipeline linking Atasu in Kazakhstan and Alashankou in China opened in Dec. 2005. Oil and gas investment by foreign companies is now driving the economy. In 1997 oil production sharing deals were concluded with two international consortia to explore the North Caspian basin and to develop the Karachaganak gas field. A huge new offshore oilfield in the far north of the Caspian Sea, known as East Kashagan, was discovered in early 2000. The field could prove to be the largest find in the last 30 years, and estimates suggest that it may contain 50bn. bbls of oil. Commercial production was expected to begin in June 2013.

It is believed that there may be as much as 14bn. tonnes of oil and gas reserves under Kazakhstan's portion of the Caspian Sea.

A state-owned national company, KazMunaiGaz, was created in 2002 to manage the oil and natural gas industries.

Minerals
Kazakhstan is extremely rich in mineral resources, including coal, bauxite, cobalt, vanadium, iron ores, chromium, phosphates, borates and other salts, copper, lead, manganese, molybdenum, nickel, tin, gold, silver, tungsten and zinc. Production figures (2003 unless otherwise indicated), in tonnes: coal (2004), 86·00m.; iron ore, 19·28m.; bauxite, 4·74m.; lignite (2004), 3·95m.; copper, 485,000; zinc, 394,000; uranium (2010), 17,803; silver, 827·4; gold, 19·3. Kazakhstan overtook Canada as the world's largest producer of uranium in 2009.

Agriculture
Kazakh agriculture has changed from primarily nomad cattle breeding to production of grain, cotton and other industrial crops. In 2006 agriculture accounted for 6% of GDP. There were 21·54m. ha. of arable land and 0·14m. ha. of permanent crops in 2001. 2·35m. ha. were irrigated in 2001. In 1993, 181·3m. ha. were under cultivation, of which private subsidiary agriculture accounted for 0·3m. ha. and commercial farming 6·3m. ha. in 16,300 farms. Around 60,000 private farms have emerged since independence.

Tobacco, rubber plants and mustard are also cultivated. Kazakhstan has rich orchards and vineyards. Kazakhstan is noted for its livestock, particularly its sheep, from which excellent quality wool is obtained. Livestock (2003): 4·56m. cattle (down from 9·57m. in 1993), 9·79m. sheep (down from 33·63m. in 1993), 1·49m. goats, 1·23m. pigs, 1·02m. horses and 23·79m. chickens.

Output of main agricultural products (in 1,000 tonnes) in 2003: wheat, 11,537; potatoes, 2,308; barley, 2,154; watermelons, 604; tomatoes, 448; maize, 438; sugar beets, 424; cabbage, 328; onions, 320; sunflower seeds, 293; rice, 273. Livestock products, 2002 (in 1,000 tonnes): cow's milk, 4,110; meat, 676; eggs, 117. Kazakhstan is a major exporter of grain to Russia, but in recent years there has been a significant reduction in the quantity exported as a result of low crop yields coupled with the need to meet domestic demand.

Forestry
Forests covered 3·31m. ha. in 2010, or 1% of the land area. In 2007, 852,000 cu. metres of timber were cut.

Fisheries
Catches in 2008 totalled 55,581 tonnes, exclusively freshwater fish.

INDUSTRY
Kazakhstan was heavily industrialized in the Soviet period, with non-ferrous metallurgy, heavy engineering and the chemical industries prominent. Output was valued at 2,000bn. tenge in current prices in 2001, up from 1,798bn. tenge in 2000. Production, 2007 unless otherwise indicated (in 1,000 tonnes) includes: cement (2008), 5,837; crude steel, 4,784; distillate fuel oil, 4,295; pig iron, 3,240; petrol, 2,633; residual fuel oil, 2,584; wheat flour (2003), 2,123; ferroalloys, 1,713; cotton woven fabrics (2002), 14m. sq. metres; leather footwear (2002), 250,000 pairs; TV sets (2001), 347,000 units.

Labour
In 2002 the economically active labour force numbered 6,708,900, with the main areas of activity as follows: agriculture, hunting and forestry, 2,366,700; trade, restaurants and hotels, 1,007,200; industry, 824,000; education, 589,000; transport, storage and communications, 503,700. In 2003 the unemployment rate was 8·8% (down from 13·5% in 1999).

INTERNATIONAL TRADE
In Jan. 1994 an agreement to create a single economic zone was signed with Kyrgyzstan and Uzbekistan. Since Jan. 1992 individuals and enterprises have been able to engage in foreign trade without needing government permission, except for goods 'of national interest' (fuel, minerals, mineral fertilizers, grain, cotton, wool, caviar and pharmaceutical products) which may be exported only by state organizations.

In Jan. 2010 Kazakhstan joined the newly established Customs Union, together with Russia and Belarus.

Imports and Exports

In 2007 imports (c.i.f.) were valued at US$32,756·4m. (compared to US$8,408·3m. in 2003) and exports (f.o.b.) at US$47,755·3m. (US$12,926·6m. in 2003). In 2007, 35·5% of imports came from Russia, 10·7% from China, 7·9% from Germany and 5·0% from the USA. Main export markets in 2007 were Italy, 16·3%; Switzerland, 15·7%; China, 11·8%; Russia, 9·8%. Main imports: machinery and transport equipment; manufactured goods (including oil and steel); mineral fuels, lubricants and related materials (including petroleum and petroleum products); and chemicals and related products. Main exports: mineral fuels, lubricants and related materials (including petroleum and petroleum products); manufactured goods (including iron and steel, and copper); and inedible crude material (except fuels). Oil and gas account for 65% of exports.

COMMUNICATIONS

Roads

In 2007 there were 93,123 km of roads, of which 23,507 were highways, main or national roads. Passenger cars in use in 2007 numbered 2,183,100, and there were also 359,200 lorries and vans, 83,400 buses and coaches, and 45,200 motorcycles and mopeds. There were 4,365 fatalities as a result of road accidents in 2007. With 28·2 deaths per 100,000 population in 2007, Kazakhstan has among the highest death rates in road accidents of any country.

Rail

In 2005 there were 14,195 km of 1,520 mm gauge railways. Passenger-km travelled in 2008 came to 14·5bn. and freight tonne-km to 214·9bn. The first section of a metro in Almaty, covering 8·6 km, opened in 2011. Eventually it is expected to reach 45 km in length.

Civil Aviation

The national carrier is Air Astana. There is an international airport at Almaty. In 2005 scheduled airline traffic of Kazakhstan-based carriers flew 93·3m. km, carrying 583,600 passengers.

Shipping

There is one large port, Aktau. In Jan. 2009 there were nine ships of 300 GT or over registered, totalling 33,000 GT.

Telecommunications

There were 4·01m. fixed telephone lines in 2010 (250·3 per 1,000 inhabitants). Mobile phone subscribers numbered 19·77m. in 2010. There were 182·0 internet users per 1,000 inhabitants in 2009. Fixed internet subscriptions totalled 846,900 in 2010 (52·8 per 1,000 inhabitants). In March 2012 there were 452,000 Facebook users.

SOCIAL INSTITUTIONS

Justice

Jury trials for serious offences were introduced in Jan. 2007. In 2009, 121,667 crimes were reported, of which 1,638 homicides or attempted homicides. The population in penal institutions in Jan. 2008 was 56,012 (378 per 100,000 of national population).

Education

In 2007, 330,897 children were attending pre-school institutions, there were 947,807 pupils at primary schools, 1,874,213 pupils at secondary schools and 772,600 students in tertiary education. Adult literacy rate is more than 99%.

In 2007 public government expenditure on education came to 3·2% of GNI.

Health

In 2002 there were 894 hospitals with a provision of 81 beds per 10,000 inhabitants (2007). There were 57,514 physicians, 5,612 dentists, 113,098 nurses and midwives and 14,048 pharmacists in 2006.

Welfare

In 1998 the former social insurance system was replaced by mandatory individual accounts. Pension contributions are 10% of employees' monthly income. The basic state old-age pension is 40% of the monthly minimum wage. There were 1·7m. old-age pensioners in 2006.

RELIGION

There were some 4,000 mosques in 1996 (63 in 1990). An Islamic Institute opened in 1991 to train imams. A Roman Catholic diocese was established in 1991. In 2001 there were 6,988,000 Muslims, 1,216,000 Russian Orthodox and 318,000 Protestants. The remainder of the population followed other religions or were non-religious.

CULTURE

World Heritage Sites

Kazakhstan has three sites on the UNESCO World Heritage List: the Mausoleum of Khoja Ahmed Yasawi (inscribed on the list in 2003), an excellent and well preserved example of late 14th century Timurid architecture; Petroglyphs within the Archaeological Landscape of Tamgaly (2004), a concentration of some 5,000 rock carvings; and Saryarka-Steppe and the Lakes of Northern Kazakhstan (2008), 450,344 ha. of wetlands that contain the Naurzum State Nature Reserve and the Korgalzhyn State Nature Reserve and are of outstanding importance for migratory water birds.

Press

There were 1,900 newspapers and magazines in 2008. The leading newspapers are the Kazakh-language *Egemen Kazakhstan* and the Russian-language *Kazakhstanskaya Pravda*. In the 2011–12 *World Press Freedom Index* compiled by Reporters Without Borders, Kazakhstan ranked 154th out of 179 countries

Tourism

In 2010 there were 3,393,000 non-resident tourists, up from 3,118,000 in 2009. There were 1,460 hotels in 2010.

DIPLOMATIC REPRESENTATIVES

Of Kazakhstan in the United Kingdom (33 Thurloe Sq., London, SW7 2SD)
Ambassador: Kairat Abusseitov.

Of the United Kingdom in Kazakhstan (62 Kosmonavtov St., Renco Building, 6th Floor, Astana 010000)
Ambassador: Carolyn Browne.

Of Kazakhstan in the USA (1401 16th St., NW, Washington, D.C., 20036)
Ambassador: Kairat Umarov.

Of the USA in Kazakhstan (22–23 Str., No. 3, Ak Bulak 4, Astana)
Ambassador: Kenneth J. Fairfax.

Of Kazakhstan to the United Nations
Ambassador: Byrganym Aitimova.

Of Kazakhstan to the European Union
Ambassador: Almaz Khamzayev.

FURTHER READING

Alexandrov, M., *Uneasy Alliance: Relations Between Russia and Kazakhstan in the Post-Soviet Era, 1992–1997.* 1999
Cummings, Sally, *Kazakhstan: Power and the Elite.* 2005
Nazpary, J., *Post-Soviet Chaos: Violence and Dispossession in Kazakhstan.* 2001
Olcott, Marta Brill, *The Kazakhs.* 1987.—*Kazakhstan: Unfilled Promise.* 2001

National Statistical Office: Agency of Kazakhstan on Statistics, House of Ministries, 4th entry, Astana 010000, Kazakhstan.
Website: http://www.stat.kz

KENYA

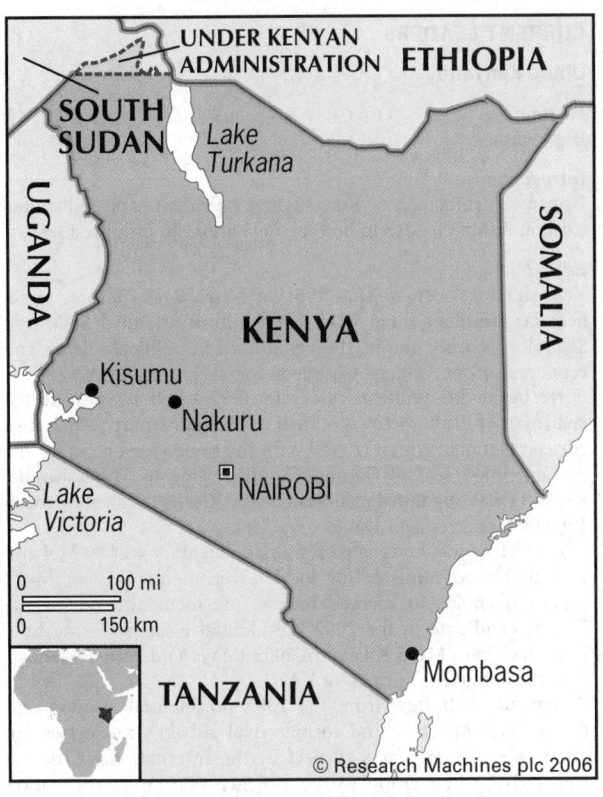

UNDER KENYAN ADMINISTRATION ETHIOPIA
SOUTH SUDAN
Lake Turkana
UGANDA
SOMALIA
KENYA
Kisumu
Nakuru
Lake Victoria
NAIROBI
0   100 mi
0   150 km
TANZANIA
Mombasa
© Research Machines plc 2006

Jamhuri ya Kenya
(Republic of Kenya)

Capital: Nairobi
Population projection, 2015: 46·33m.
GNI per capita, 2011: (PPP$) 1,492
HDI/world rank: 0·509/143
Internet domain extension: .ke

KEY HISTORICAL EVENTS

Prior to colonialism, the area comprised African farming communities, notably the Kikuyu and the Masai. From the 16th century through to the 19th, they were loosely controlled by the Arabic rulers of Oman. In 1895 the British declared part of the region the East Africa Protectorate, which from 1920 was known as the Colony of Kenya. The influx of European settlers was resented by Africans not only for the whites' land holdings but also for their exclusive political representation in the colonial Legislative Council. A state of emergency existed between Oct. 1952 and Jan. 1960 during the period of the Mau Mau uprising. Over 13,000 Africans and 100 Europeans were killed. The Kenya African Union was banned and its president, Jomo Kenyatta, imprisoned. The state of emergency ended in 1960. Full internal self-government was achieved in 1962 and in Dec. 1963 Kenya became an independent member of the Commonwealth. In 1982 Kenya became a one-party state and in 1986 party preliminary elections were instituted to reduce the number of parliamentary candidates at general elections. Only those candidates obtaining over 30% of the preliminary vote were eligible to stand. On the death of Kenyatta in Aug. 1978 Daniel T. arap Moi, the vice-president, became acting president and was elected in 1979, and then re-elected in 1983, 1988, 1992 and 1997. An attempted coup in 1982 was unsuccessful. A multi-party election was permitted in 1992 and again in 1997, the first genuinely competitive elections since 1963. In the 2002 elections the opposition united behind Mwai Kibaki, who won a landslide victory against Moi's successor Uhuru Kenyatta. Kibaki became the first non-Kenya African National Union president of independent Kenya. In Nov. 2005 a new draft constitution was rejected amid criticism that it gave too much power to the president. In Dec. 2007 Kibaki claimed victory in the presidential election, although the opposition and international observers alleged irregularities, prompting a wave of civil unrest that claimed around 1,500 lives. In Feb. 2008 Kibaki and the opposition leader, Raila Odinga, agreed to a power-sharing deal. In March 2013 Uhuru Kenyatta won elections for the presidency. An appeal against the result by Odinga, his chief rival, was rejected by the Supreme Court.

TERRITORY AND POPULATION

Kenya is bounded by South Sudan and Ethiopia in the north, Uganda in the west, Tanzania in the south and Somalia and the Indian Ocean in the east. The total area is 581,313 sq. km. The 2009 census gave a population of 38,610,097 (19,417,639 females); density, 66 per sq. km. In 2009, 70·2% of the population were rural. In 2006 more than 30,000 Somali refugees entered Kenya to escape the fighting that escalated in Somalia the course of the year.

The UN gives a projected population for 2015 of 46·33m.

Kenya is divided into seven provinces and one national capital area (Nairobi). The land areas, populations and capitals are:

Province	Sq. km	Census 2009	Capital	Census 2009
Rift Valley	173,868	10,006,805	Nakuru	286,411
Eastern	159,891	5,668,123	Embu	35,736
Nyanza	16,162	5,442,711	Kisumu	259,258
Central	13,176	4,383,743	Nyeri	63,626
Western	8,360	4,334,282	Kakamega	69,502
Coast	83,603	3,325,307	Mombasa	915,101
Nairobi	684	3,138,369		
North-Eastern	126,902	2,310,757	Garissa	110,383

Other large towns (2009): Eldoret (252,061), Ruiru (236,961), Kikuyu (190,208), Thika (136,576).

Most of Kenya's 38·61m. people belong to 13 tribes, the main ones including Kikuyu (about 22% of the population), Luhya (14%), Luo (13%), Kalenjin (12%), Kamba (11%), Gusii (6%), Meru (5%) and Mijikenda (5%). There is a large Somali minority.

Swahili and English are both official languages, but people belonging to the different tribes have their own language as their mother tongue.

SOCIAL STATISTICS

2008 births (estimates), 1,503,000; deaths, 451,000. Estimated birth rate in 2008 was 38·8 per 1,000 population; estimated death rate, 11·6. Annual population growth rate, 2000–08, 2·6%. Expectation of life at birth in 2007 was 53·2 years for males and 54·0 years for females. Infant mortality, 2010, 55 per 1,000 live births. Fertility rate, 2008, 4·9 births per woman. In 2005, 46% of Kenyans lived below the poverty line (down from 52% in 1997).

CLIMATE

The climate is tropical, with wet and dry seasons, but considerable differences in altitude make for varied conditions between the hot, coastal lowlands and the plateau, where temperatures are very much cooler. Heaviest rains occur in April and May, but in some parts there is a second wet season in Nov. and Dec. Nairobi, Jan. 65°F (18·3°C), July 60°F (15·6°C). Annual rainfall 39" (958 mm). Mombasa, Jan. 81°F (27·2°C), July 76°F (24·4°C). Annual rainfall 47" (1,201 mm).

CONSTITUTION AND GOVERNMENT

A new constitution was approved in a referendum on 4 Aug. 2010 with 66·9% of votes cast in favour. Under its terms, the *President* and *Parliament* will have five-year fixed terms. The president may not serve more than two terms. To be elected president, a candidate must secure at least 50% of votes cast, with at least a quarter coming from more than half of the county constituencies. The old 46 local government districts were restructured into 47 counties, with each county having a governor and a senator. Senators sit in a newly-created 68-member upper house, providing the country with a bicameral legislature following the elections of March 2013. The *National Assembly*, the lower house, is made up of 350 members following the election with 290 directly elected, 47 women, 12 nominated plus the Speaker (up from 224 previously, with 210 directly elected, 12 appointed plus the Speaker and the Attorney General). The *Senate* consists of 47 elected senators, 20 nominated senators plus the Speaker. Each county assembly must return at least one female MP. Parliament has the power to vet key appointments previously appointed by order of the president. The constitution also provides for a supreme court (the highest court in the land) backed by a court of appeals. Judges are subject to review by a judicial appointments panel.

National Anthem

'Ee Mungu nguvu yetu' ('Oh God of all creation'); words by a collective, tune traditional.

GOVERNMENT CHRONOLOGY

President since 1964. (DP = Democratic Party; KANU = Kenya African National Union; NARC = National Rainbow Coalition; PNU = Party of National Unity; TNA = The National Alliance)

1964–78	KANU	Jomo Kenyatta
1978–2002	KANU	Daniel arap Moi
2002–13	NARC/DP, PNU	Mwai Kibaki
2013–	TNA	Uhuru Kenyatta

RECENT ELECTIONS

In presidential elections held on 4 March 2013 Uhuru Kenyatta of The National Alliance (TNA) received 50·1% of vote casts, Raila Odinga of the Orange Democratic Movement (ODM) 43·3% and Musalia Mudavadi of the United Democratic Forum (UDF) 3·9%. There were five other candidates. Turnout was 85·9%.

National Assembly elections also held on 4 March 2013 were the first after the passing of the new constitution in 2010. Uhuru Kenyatta's Jubilee alliance won 167 seats (of which TNA won 89); Raila Odinga's Coalition for Reforms and Democracy (CORD) alliance, 141 (of which ODM won 96); Musalia Mudavadi's Amani coalition, 24 (of which UDF won 12); and the Eagle coalition, 2. Unaffiliated parties and independent candidates obtained 15 seats. In elections to the Senate (which was established under the new constitution) the same day the Jubilee coalition obtained 30 of 67 seats; CORD coalition, 28; Amani coalition, 6; and the Alliance Party of Kenya (APK), 3.

CURRENT GOVERNMENT

President: Uhuru Kenyatta; b. 1961 (TNA; sworn in 9 April 2013).

Kenyatta nominated a scaled-down cabinet on 25 April, his electoral victory having been upheld by the Supreme Court after a legal challenge by his defeated opponent and the former prime minister, Raila Odinga. Kenyatta presented a list comprised largely of technocrats. However, before the cabinet could be sworn in, each nominee was to be vetted by a parliamentary-appointed committee, a process that had yet to be completed at the time of

going to print. Members of the public were also invited to submit complaints about individual nominations. For the latest information *see* www.statesmansyearbook.com.

Cabinet Office: http://www.cabinetoffice.go.ke

CURRENT LEADERS

Uhuru Kenyatta

Position
President

Introduction
Uhuru Kenyatta, son of Kenya's first president, won a disputed election in March 2013 to become the nation's fourth president.

Early Life
Kenyatta was born in Oct. 1961 in Nairobi. His father, Jomo, became president three years later. Uhuru attended St Mary's School in Nairobi and in 1985 graduated in political science and economics from Amherst College in Massachusetts, USA.

He began his political career in 1997 when he was elected chairman of his hometown branch of the ruling party, the Kenya African National Union (KANU). In the same year he ran for the parliamentary seat of Gatundu South, losing to Moses Muhia. Kenyatta was appointed chairman of the Kenya Tourism Board in 1999 by then President Daniel arap Moi.

In 2001 he was nominated to a parliamentary seat by Moi and was appointed minister for local government. He was Moi's preferred choice to succeed him as president and ran as the KANU candidate in the 2002 presidential election, losing by a large margin to Mwai Kibaki. In 2005 Kenyatta defeated Nicholas Biwott to become chairman of KANU.

Kenyatta withdrew from the 2007 presidential election and threw his support behind former rival Kibaki's re-election. In 2010 he was named as a suspect by the International Criminal Court (ICC) in relation to the violence that swept the nation in the aftermath of the election, which Kibaki won and the opposition disputed. Specifically, Kenyatta was accused of organizing attacks against members of groups who supported Kibaki's rival, Raila Odinga.

Kenyatta was appointed minister of local government by Kibaki in 2008 and subsequently held the posts of deputy prime minister and minister of trade. In 2009 he became minister of finance. In 2012 he took over the National Alliance (TNA), which formed a coalition with the United Republican Party of William Ruto to oppose Raila Odinga's Coalition for Reforms and Democracy in the 2013 presidential election.

Career in Office
Five days after polling on 4 March 2013, Kenyatta was declared president. Odinga unsuccessfully lodged a formal challenge with the Supreme Court, claiming the election was fraudulent.

Kenyatta faced the prospect of carrying out his duties while on trial at the ICC accused of crimes against humanity for his part in the violence that left over 1,200 dead following the 2007 election. His trial was scheduled to begin in July 2013. Kenyatta maintains his innocence and contests the nature of the evidence to be used against him. Nonetheless, the allegations leave relations with the international community strained and weaken his authority domestically.

DEFENCE

In 2008 defence expenditure totalled US$735m. (US$19 per capita), representing 2·1% of GDP.

Army

Total strength (2007) 20,000. In addition there is a paramilitary Police General Service Unit of 5,000.

Navy

The Navy, based in Mombasa, consisted in 2007 of 1,620 personnel (including 120 marines).

Air Force

An air force, formed on 1 June 1964, was built up with RAF assistance. Personnel (2007) 2,500, with 29 combat-capable aircraft and 11 attack helicopters although their serviceability is in doubt.

INTERNATIONAL RELATIONS

In Nov. 1999 a treaty was signed between Kenya, Tanzania and Uganda to create a new East African Community as a means of developing East African trade, tourism and industry and laying the foundations for a future common market and political federation.

ECONOMY

Agriculture contributed 22·6% of GDP in 2009, industry 15·3% and services 62·1%.

Overview

Following independence in 1963 Kenya was among the leading East African economies, with average annual GDP growth of 6·5%. However, from 1974 economic performance declined. The effect of years of mismanagement and corruption was made worse in 2000 by one of the longest droughts in living memory. Up to US$1bn. in international aid was frozen during the tenure of Daniel arap Moi (1978–2002) after Kenya failed to pass anti-corruption legislation. From 2003, under the presidency of Mwai Kibaki, GDP increased, driven by the agriculture and service sectors. Flowers, fruits and vegetables lead Kenya's exports to Europe (its largest export market) along with coffee, tea and petroleum goods. In the service sectors, telecommunications and tourism have taken off while economic growth stimulated the domestic construction industry.

Mass unrest following the contentious re-election of President Kibaki in Dec. 2007 resulted in the displacement of over 200,000 people. A subsequent spike in international commodity prices, poor rainfall from Oct.–Nov. 2008 and spillover effects from the global financial crisis brought growth down to 1·3% in 2008 from 7·0% the previous year. As a result Kenya sought IMF assistance under the Rapid Access Component of the Exogenous Shocks Facility in May 2009. Fiscal stimulus aimed at infrastructure and agriculture, along with favourable weather conditions, helped the economy bounce back in 2009, with growth broad-based. However, inflationary pressures driven by global oil and food prices, along with further weather disturbances, have severely tested the recovery.

Currency

The monetary unit is the *Kenya shilling* (KES) of 100 *cents*. The currency became convertible in May 1994. The shilling was devalued by 23% in April 1993. The annual rate of inflation was 15·1% in 2008, falling to 4·1% in 2010 but then rising to 14·0% in 2011. Foreign exchange reserves were US$1,653m. in July 2005, total money supply was K Sh 222,558m. and gold reserves were 1,000 troy oz.

Budget

In 2008–09 revenues totalled K Sh 511,355m. and expenditures K Sh 621,909m. Tax revenue accounted for 85·5% of revenues; current expenditure accounted for 74·3% of expenditures. The fiscal year ends on 30 June.

Performance

Real GDP growth was 2·7% in 2009, 5·8% in 2010 and 4·4% in 2011. Total GDP in 2011 was US$33·6bn.

Banking and Finance

The central bank and bank of issue is the Central Bank of Kenya (*Governor*, Njuguna Ndung'u). There are 43 banks, two non-banking financial institutions and a couple of building societies. In Dec. 2003 their combined assets totalled K Sh 567,600m. In 1998 the government offloaded 25% of its stake in the Kenya Commercial Bank, which lowered its shareholding to 35%. In 2004 it further lowered its shareholding, to 26·2%. It was lowered again to 23·1% in 2008 and a third rights issue exercise in 2010 further reduced its shareholding to 17·7%.

Foreign debt was US$8,400m. in 2010, representing 26·9% of GNI.

There is a stock exchange in Nairobi.

ENERGY AND NATURAL RESOURCES

Environment

Kenya's carbon dioxide emissions from the consumption and flaring of fossil fuels in 2008 were the equivalent of 0·3 tonnes per capita.

Electricity

Installed generating capacity was 1·22m. kW in 2007, mostly provided by hydropower from power stations on the Tana river with some from oil-fired power stations and by geothermal power. Production in 2007 was 6·77bn. kWh, with consumption per capita 181 kWh. In 1999 it was decided to encourage the private sector to take part in electricity generation alongside the state-owned Kenya Electricity Generating Company as a means of bringing to an end the shortage of power and the frequent blackouts. In June 2000 a rationing scheme was introduced in much of the country restricting the power supply to 12 hours a day, and sometimes less.

Oil and Gas

Oil was discovered in the northwest of the country in March 2012, although the commercial viability of the find is still to be established.

Minerals

Production, 2005 (in 1,000 tonnes): lime and limestone (estimate), 1,085; soda ash, 360; fluorite, 97. Other minerals include gold (616 kg exported in 2005), raw soda, diatomite, garnets, salt and vermiculite.

Agriculture

As agriculture is possible from sea-level to altitudes of over 2,500 metres, tropical, sub-tropical and temperate crops can be grown and mixed farming is pursued. In 2006 there were around 5·31m. ha. of arable land and 444,000 ha. of permanent crop land. 14,000 ha. were irrigated in 2006. There were 26 tractors and two harvester-threshers per 10,000 ha. of arable land in 2006. Four-fifths of the country is range-land which produces mainly livestock products and the wild game which is a major tourist attraction.

Tea, coffee and horticultural products, particularly flowers, are all major foreign exchange earners.

Kenya has about 131,450 ha. under tea production, and is the world's fourth largest producer and largest exporter of tea. The production is high quality tea, raised in near-perfect agronomic conditions. In 2003 production was 294,000 tonnes; exports were worth US$434m.

Coffee output in 2003 was 55,000 tonnes; 170,000 ha. is under coffee production. However, the annual crop is now a quarter of that at the time of independence in 1963. Some 75% of the total hectarage under coffee is cultivated by smallholders, although their production has been in decline in recent years. Other major agricultural products (2003 estimates, in 1,000 tonnes): sugarcane, 4,500; maize, 2,300; potatoes, 900; plantains, 830; cassava, 600; pineapples, 600; sweet potatoes, 520; cabbage, 270; tomatoes, 260; dry beans, 255.

Livestock (2003): cattle, 12·8m.; goats, 11·5m.; sheep, 9·5m.; camels, 863,000; pigs, 337,000; chickens, 28·6m.

More than half the agricultural labour force is employed in the livestock sector, accounting for 10% of GDP.

Forestry

Forests covered 3·47m. ha. in 2010 (6% of the land area). There are coniferous, broad-leaved, hardwood and bamboo forests. Timber production was 27·65m. cu. metres in 2007.

Fisheries

Catches in 2010 totalled 143,111 tonnes, of which 134,847 tonnes were freshwater fish (mostly from Lake Victoria). Marine fishing has not reached its full potential, despite a coastline of 640 km. Fish landed from the sea totals less than 9,000 tonnes annually, but there is an estimated potential of 150,000 tonnes in tuna and similar species.

INDUSTRY

In 2006 industry accounted for 18·8% of GDP, with manufacturing contributing 11·5%. In 2003 there were 579 manufacturing firms employing more than 50 persons. The main products are textiles, chemicals, vehicle assembly and transport equipment, leather and footwear, printing and publishing, food and tobacco processing and oil refining. Production (2003 unless otherwise indicated) included (in tonnes): cement (2008), 2,829,000; residual fuel oil (2005), 549,000; sugar, 448,489; distillate fuel oil (2005), 374,000; petrol (2007), 207,000; wheat flour, 179,866; maize meal, 120,942; cattle feed, 99,616; kerosene (2007), 116,000.

Labour

The labour force in 1998–99 was 12,326,000. In 1998–99 the unemployment level was estimated to be 1·8m.

INTERNATIONAL TRADE

Imports and Exports

Imports (c.i.f.) in 2009 totalled US$10,202·0m. and exports (f.o.b.) US$4,463·4m. The gap between imports and exports has widened considerably since 2005, when imports (c.i.f.) totalled only US$5,846·2m. but exports (f.o.b.) were US$3,419·9m. Principal imports in 2007: machinery and transport equipment, 30·5%; mineral fuels, lubricants and related materials (including petroleum and petroleum products), 21·3%; manufactured goods, 15·9%. Exports: food and live animals (including tea), 37·2%; inedible crude materials (except fuels), 13·9%; manufactured goods, 13·1%. Tea constituted 17·1% of Kenya's exports in 2007. Main import suppliers, 2007: United Arab Emirates, 14·8%; India, 9·4%; China, 7·6%. Main export markets, 2007: Uganda, 12·2%; UK, 10·5%; Tanzania, 8·1%.

COMMUNICATIONS

Roads

In 2004 there were 63,265 km of roads (6,527 km of highways, national and main roads). There were, in 2007, 562,400 passenger cars in use, 210,900 vans and lorries, 180,800 motorcycles and mopeds, and 20,100 buses and coaches. There were 2,893 fatalities as a result of road accidents in 2007.

Rail

In 2006 there were 2,064 km of railways (metre gauge). Most of the network (1,918 km, including non-operational sections) is managed by Rift Valley Railways (Kenya) Ltd. In 2008–09, 4·4m. passengers and 1·6m. tonnes of freight were carried. The Magadi Railway Co. Ltd manages a 146 km stretch of line from Manzi to Konza to carry soda ash for export through Mombasa.

Civil Aviation

There are international airports at Nairobi (Jomo Kenyatta International) and Mombasa (Moi International). The national carrier is the now privatized Kenya Airways. KLM has a 26·7% share of Kenya Airways. In 2006 scheduled airline traffic of Kenyan-based carriers flew 51m. km and carried 2,548,000 passengers. In 2010 Jomo Kenyatta International handled 5,484,771 passengers and Moi International 1,271,078.

Shipping

The main port is Mombasa, which handled 16·4m. tonnes of cargo in 2008; container traffic totalled 616,000 TEUs (twenty-foot equivalent units) in 2008. In Jan. 2009 there were six ships of 300 GT or over registered, totalling 6,000 GT.

Telecommunications

Kenya had 283,500 landline telephone subscribers in 2011, or 6·8 per 1,000 persons. Since 1999 the government has been introducing measures to liberalize the telecommunications sector that have led to massive price reductions and improved services. In 2011 mobile phone subscribers numbered 26,980,800. The main mobile providers are Safaricom and Airtel Kenya. There were 209·8 internet users per 1,000 inhabitants in 2010. Fixed internet subscriptions totalled 8,300 in 2009 (0·2 per 1,000 inhabitants). In June 2012 there were 1·4m. Facebook users.

SOCIAL INSTITUTIONS

Justice

The courts of Justice comprise the Supreme Court, Court of Appeal and the High Court, beneath which are specialized courts, magistrates courts and Kadhis (Islamic) courts. The Chief Justice is the president of the Supreme Court, which also comprises the Deputy Chief Justice and five other judges. Only the Supreme Court can hear and determine any case challenging the election of the president. The Court of Appeal, comprising at least 12 judges, is based in Nairobi and in the course of its Appellate duties visits Mombasa, Kisumu, Nakuru, Nyeri and Eldoret. The High Court, with full jurisdiction in both civil and criminal matters, comprises the Chief Justice and not fewer than 11 but not more than 50 puisne judges. There are 15 High Court stations in the country. Kadhis courts are established in areas of concentrated Muslim populations: Mombasa, Nairobi, Malindi, Lamu, Garissa, Kisumu and Marsabit.

There were 63,476 recorded crimes in 2008; the prison population was 47,036 in Sept. 2006 (130 per 100,000 of national population).

Education

The adult literacy rate in 2008 was 87%. Free primary education was introduced in 2003. In 2007 there were 1,691,093 children in pre-primary schools with 76,323 teaching staff, 6,687,510 pupils were in primary schools with 146,796 teaching staff and 2,729,040 pupils in secondary schools with 102,449 teaching staff. There were 139,524 students in higher education in 2007.

In 2005 public expenditure on education came to 7·3% of GNI and 17·9% of total government spending.

Health

In 2003 there were 4,813 physicians, 772 dentists, 40,081 nurses and 1,881 pharmacists. There were 526 hospitals (with 63,407 beds), 649 health centres and 3,382 sub-centres and dispensaries in 2003. Free medical service for all children and adult out-patients was launched in 1965.

RELIGION

In 2001 there were 6·78m. Roman Catholics, 6·40m. African Christians, 6·17m. Protestants, 2·90m. Anglicans and 2·24m. Muslims. Traditional beliefs persist. In Feb. 2013 there was one Roman Catholic cardinal.

CULTURE

World Heritage Sites

Kenya has six sites on the UNESCO World Heritage List: Mount Kenya National Park/Natural Forest (1997), including the second highest peak in Africa; Lake Turkana National Parks (1997 and 2001), a breeding ground for Nile crocodiles and hippopotami; Lamu Old Town (2001), the oldest and best-preserved Swahili settlement in East Africa; the Sacred Mijikenda Kaya Forests (2008), 11 separate forest sites containing the remains of numerous fortified villages; Fort Jesus, Mombasa (2011), a 16th century Portuguese fort; and the Kenya Lake System in the Great Rift Valley (2011), consisting of three lakes that are home to numerous threatened bird species.

Press

In 2008 there were eight paid-for daily papers with a total average daily circulation of 320,000 plus 15 paid-for non-dailies. The most widely read paper is the English-language *Daily Nation*.

Tourism

In 2005 there were 1,536,000 foreign visitors (1,199,000 in 2004). In 2005 receipts from tourism amounted to US$579m., up from US$486m. in 2004. Tourism is the country's leading source of hard currency.

DIPLOMATIC REPRESENTATIVES

Of Kenya in the United Kingdom (45 Portland Pl., London, W1B 1AS)
High Commissioner: Ephraim Waweru Ngare.

Of the United Kingdom in Kenya (Upper Hill Rd, Nairobi)
High Commissioner: Dr Christian Turner.

Of Kenya in the USA (2249 R. St., NW, Washington, D.C., 20008)
Ambassador: Elkanah Odembo Absalom.

Of the USA in Kenya (United Nations Ave., Gigiri, Nairobi)
Ambassador: Robert Godec.

Of Kenya to the United Nations
Ambassador: Macharia Kamau.

Of Kenya to the European Union
Ambassador: James Kembi-Gitura.

FURTHER READING

Anderson, David, *Histories of the Hanged: The Dirty War in Kenya and the End of Empire.* 2005

Elkins, Caroline, *Britain's Gulag.* 2005; US title: *Imperial Reckoning: The Untold Story of the End of Empire in Kenya.* 2005

Haugerud, A., *The Culture of Politics in Modern Kenya.* 1995

Kyle, Keith, *The Politics of the Independence of Kenya.* 1999

Miller, N. N., *Kenya: the Quest for Prosperity.* 2nd ed. 1994

Murunga, Godwin R. and Nasong'o, Shadrack W., *Kenya: The Struggle for Democracy.* 2007

Ogot, B. A. and Ochieng, W. R. (eds.) *Decolonization and Independence in Kenya, 1940–93.* 1995

Throup, David and Hornsby, Charles, *Multi-Party Politics in Kenya.* 1999

National Statistical Office: Kenya National Bureau of Statistics, PO Box 30266—00100 GPO, Nairobi.

Website: http://www.knbs.or.ke

KIRIBATI

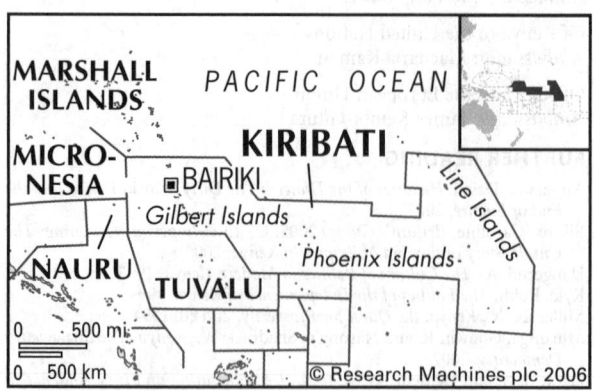

Abaiang	5,502	North Tabiteuea	3,689
Tarawa	56,284	South Tabiteuea	1,290
North Tarawa	6,102	Beru	2,099
South Tarawa	34,427	Nikunau	1,907
Betio	15,755	Onotoa	1,519
Maiana	2,027	Tamana	951
Abemama	3,213	Arorae	1,279
Kuria	980	Kanton	31
Aranuka	1,057	Teraina	1,690
Nonouti	2,683	Tabuaeran	1,960
Tabiteuea	4,979	Kiritimati	5,586

The remaining 12 atolls have no permanent population; the seven Phoenix Islands comprise Birnie, Rawaki (formerly Phoenix), Enderbury, Manra (formerly Sydney), Orona (formerly Hull), McKean and Nikumaroro (formerly Gardner), while the others are Malden and Starbuck in the Central Line Islands, and Millennium Island (formerly Caroline), Flint and Vostok in the Southern Line Islands. The population is almost entirely Micronesian.

English is the official language; I-Kiribati (Gilbertese) is also spoken.

SOCIAL STATISTICS

2005 estimates: births, 2,460; deaths, 810. Rates, 2005 estimates (per 1,000 population): births, 26·6; deaths, 8·7. Infant mortality rate (2010), 39 per 1,000 live births; life expectancy (2005), 61·0 years. Annual population growth rate, 2000–05, 1·8%; fertility rate, 2008, 3·1 births per woman.

CLIMATE

The Line Islands, Phoenix Islands and Banaba have a maritime equatorial climate, but the islands further north and south are tropical. Annual and daily ranges of temperature are small; mean annual rainfall ranges from 50" (1,250 mm) near the equator to 120" (3,000 mm) in the north. Typhoons are prevalent (Nov.–March) and there are occasional tornadoes. Tarawa, Jan. 83°F (28·3°C), July 82°F (27·8°C). Annual rainfall 79" (1,977 mm).

CONSTITUTION AND GOVERNMENT

Under the constitution founded on 12 July 1979 the republic has a unicameral legislature, the *House of Assembly* (Maneaba ni Maungatabu), comprising 46 members, 44 of whom are elected by popular vote, and two (the Attorney-General *ex officio* and a representative from the Banaban community) appointed for a four-year term. The *President* is both Head of State and government. Presidential candidates are initially selected by members of parliament before facing a popular vote.

National Anthem

'Teirake kain Kiribati' ('Stand up, Kiribatians'); words and tune by U. Ioteba.

RECENT ELECTIONS

The last House of Assembly elections were held on 21 and 28 Oct. 2011. Boutokaan Te Koaua (BK; 'Pillars of Truth') won 15 seats, Karikirakean Tei-Kiribati (KTK; 'United Coalition Party') 10 seats and the Maurin Kiribati party (MKP) 3, with remaining seats going to independents.

On 12 Jan. 2012 Anote Tong (BK) was re-elected president with 42·2% of the vote, defeating Tetaua Taitai who took 35·0% and Rimeta Beniamina with 22·8%.

CURRENT GOVERNMENT

President and Minister of Foreign Affairs and Immigration: Anote Tong (elected 4 July 2003; re-elected in Oct. 2007 and Jan. 2012).

Ribaberikin Kiribati
(Republic of Kiribati)

Capital: Bairiki (Tarawa)
Population, 2010: 103,000
GNI per capita, 2011: (PPP$) 3,140
HDI/world rank: 0·624/122
Internet domain extension: .ki

KEY HISTORICAL EVENTS

The islands that now constitute Kiribati were first settled by early Austronesian-speaking peoples long before the 1st century AD. Fijians and Tongans arrived about the 14th century and subsequently merged with the older groups to form the traditional I-Kiribati Micronesian society and culture. The Gilbert and Ellice Islands were proclaimed a British protectorate in 1892 and annexed at the request of the native governments as the Gilbert and Ellice Islands Colony on 10 Nov. 1915. On 1 Oct. 1975 the Ellice Islands severed constitutional links with the Gilbert Islands and took on a new name, Tuvalu. The Gilberts achieved full independence as Kiribati in 1979. Internal self-government was obtained on 1 Nov. 1976 and independence on 12 July 1979 as the Republic of Kiribati.

TERRITORY AND POPULATION

Kiribati (pronounced Kiribahss) consists of three groups of coral atolls and one isolated volcanic island, spread over a large expanse of the Central Pacific with a total land area of 811 sq. km (313 sq. miles). It comprises **Banaba** or Ocean Island (6 sq. km), the 16 **Gilbert Islands** (280 sq. km), the eight **Phoenix Islands** (29 sq. km) and eight of the 11 **Line Islands** (496 sq. km), the other three Line Islands (Jarvis, Palmyra Atoll and Kingman Reef) being uninhabited dependencies of the USA. The capital is the island of Bairiki in Tarawa. The gradual rise in sea levels in recent years is slowly reducing the area of the islands. Most of the land is less than 3 metres above sea level.

Population, 2010 census, 103,058 (52,262 females); density, 127 per sq. km.

In 2011, 44·0% of the population lived in urban areas.

Banaba, all 16 Gilbert Islands, Kanton (or Abariringa) in the Phoenix Islands and three atolls in the Line Islands (Teraina, Tabuaeran and Kiritimati—formerly Washington, Fanning and Christmas Islands respectively) are inhabited; their populations in 2010 (census) were as follows:

Banaba (Ocean Is.)	295	Butaritari	4,346
Makin	1,798	Marakei	2,872

In March 2013 the government comprised:

Vice President and Minister of Internal Affairs: Teima Onorio.

Minister of Commerce and Co-operatives: Pinto Katia. *Communications, Transport and Tourism Development:* Taberannang Timeon. *Education:* Maere Tekanene. *Environment and Agriculture:* Tiarite Kwong. *Finance:* Tom Murdoch. *Fisheries and Natural Resources:* Tinian Reiher. *Health and Medical Services:* Dr Kautu Tenaua. *Labour and Human Resources Development:* Boutu Bateriki. *Line and Phoenix Islands:* Tawita Temoku. *Works and Energy:* Kirabuke Teiaua. *Attorney General:* Tiitabu Taabane.

Parliament Website: http://www.parliament.gov.ki

CURRENT LEADERS

Anote Tong

Position
President

Introduction
Anote Tong became president of the Pacific Ocean republic in July 2003 and was re-elected in Oct. 2007 and Jan. 2012. The president is chief of state and head of government.

Early Life
Anote Tong was born in the British Gilbert and Ellice Islands colony in 1952, the son of a Chinese father and a Gilbertian mother. He was an undergraduate at the University of Canterbury, Christchurch, New Zealand and graduated in 1988 with an MSc from the London School of Economics.

He entered politics in 1976 as an assistant secretary in the ministry of education, then served in the ministry of communications and works during the 1980s. Between 1994–96 he was minister of environment and natural resources in the government of President Teburoro Tito, after which he represented Maiana Island in parliament. When Tito lost a parliamentary confidence motion shortly after being elected to serve for a third term, the speaker, Taomati Iuta, led an interim government until presidential elections in July 2003. Anote Tong stood for the opposition Boutokanto Koaava party (Pillars of Truth) against his older brother, the government candidate Dr Harry Tong. Anote won by 13,500 to 12,500 votes.

Career in Office
Following Tong's election his brother mounted a court challenge alleging electoral fraud but in Oct. 2003 an Australian judge ruled in President Tong's favour. During his campaign the president had promised to review the lease on a satellite-tracking base used by China. Within six months of taking office the president established relations with Taiwan, which had paid a large sum for fishing rights in Kiribati's territorial waters. China severed relations with Kiribati and abandoned the satellite-tracking base.

During 2004 the political scene was dominated by a dispute between the president and his brother over the extent of Taiwanese influence in Kiribati affairs. The government was also faced by high unemployment and rising sea levels caused by climate change, a pressing threat to the 33-island archipelago, whose land rises only a few metres above sea level. In March 2006 the government created a large marine reserve, banning fishing around the Phoenix Islands.

Following parliamentary elections in Aug. 2007, Tong was returned in Oct. for another presidential term. He has continued to publicize the threat posed to Kiribati by rising sea levels and the prospect that the population may eventually need to be resettled elsewhere. He won a third successive term in elections held in Jan. 2012.

ECONOMY

Agriculture accounted for 26% of GDP in 2009, industry 8% and services 66%.

Overview
Following independence in 1979, the economy experienced moderate growth from the mid-1980s until 2003. Growth was driven by fishing licence fees from its Exclusive Economic Zone, drawings from the Revenue Equalization Reserve Fund (RERF, constituting pre-1979 revenues accrued from phosphate mining), passport fees and foreign aid (including remittances from Kiribati seamen). The economy has a narrow production base with exports limited to copra, seaweed and fish (from which around 80% of households make a living). The public sector dominates the economy, accounting for nearly 80% of employment.

Limited resources and geographic isolation make transport and communications costly. The country is also one of the most vulnerable to climate change and sea-level rise. Kiribati was badly hit by the global economic crisis through a fall in remittances and decline in value of the RERF and Kiribati Provident Fund. Following two years of contraction, the economy began recovery in the second half of 2010, thanks in part to copra subsidies and increased civil servant wages.

Currency
The currency in use is the Australian *dollar*. Inflation was 8·8% in 2009 but there was deflation of 2·8% in 2010. In 2011 inflation stood at 2·8%.

Budget
Foreign financial aid, mainly from the UK and Japan, has amounted to 25–50% of GDP in recent years. Revenues in 2008 totalled $A161·7m. and expenditures $A183·0m.

Performance
The economy contracted by 2·3% in 2009 but grew by 1·4% in 2010 and 1·8% in 2011. Total GDP in 2011 was US$167m.

Banking and Finance
ANZ Bank (Kiribati) Ltd is 25% government-owned and 75% owned by ANZ Bank. In 1999 it had total assets of $A46·3m. There is also a Development Bank of Kiribati and a network of village lending banks and credit institutions.

ENERGY AND NATURAL RESOURCES

Environment
Carbon dioxide emissions from the consumption and flaring of fossil fuels were the equivalent of 0·4 tonnes per capita in 2008.

Electricity
Installed capacity (2007 estimate), 6,000 kW; production (2007), 24m. kWh.

Agriculture
In 2007 there were about 2,000 ha. of arable land and 35,000 ha. of permanent crops. Copra and fish represent the bulk of production and exports. The principal tree is the coconut; other food-bearing trees are the pandanus palm and the breadfruit. The only vegetable which grows in any quantity is a coarse calladium (alocasia) with the local name 'bwabwai', which is cultivated in pits; taro and sweet potatoes are also grown. Coconut production (2003), 99,000 tonnes; copra, 7,000 tonnes; bananas, 5,000 tonnes; taro, 2,000 tonnes. Principal livestock: pigs (12,000 in 2003).

Forestry
There were 12,000 ha. of forest in 2010, or 15% of the total land area.

Fisheries
Fishing license fees provide key sources of income. Total catches in 2010 amounted to 44,559 tonnes, exclusively from sea fishing.

INDUSTRY

Industry is concentrated on fishing and handicrafts.

Labour

The economically active population classified as cash workers (not including village workers engaged in subsistence activities) totalled 13,133 in 2005. In 2005, 52·9% of cash workers were employed in public administration, 11·2% in transport and communication, 9·0% in retail trade, and 7·1% in agriculture and fishing. 6·1% of the labour force were unemployed in 2005; the unemployment rate in 2005 including village workers was 64·5%.

INTERNATIONAL TRADE

Imports and Exports

Total imports (2005), US$74·0m.; exports, US$3·6m. Main import sources in 2005: Australia, 35·5%; Fiji Islands, 20·9%; Japan, 17·0%; New Zealand, 5·4%. Main export markets in 2005: Free zones, 33·3%; Australia, 22·2%; Fiji Islands, 16·7%; Hong Kong, 8·3%. Principal exports: copra, seaweed, fish; imports: foodstuffs, machinery and equipment, manufactured goods and fuel.

COMMUNICATIONS

Roads

In 2002 there were 670 km of roads. There were 9,600 cars, 4,320 trucks and vans and 2,080 motorcycles in 2008.

Civil Aviation

The national airline is the state-owned Air Kiribati. In 2010 there were scheduled services from Tarawa (Bonriki) to the Fiji Islands as well as domestic flights linking the main islands of Kiribati.

Shipping

The main port is at Betio (Tarawa). Other ports of entry are Banaba, English Harbor and Kanton. There is also a small network of canals in the Line Islands. In Jan. 2009 there were 58 ships of 300 GT or over registered, totalling 245,000 GT.

Telecommunications

In 2008 there were 4,000 main (fixed) telephone lines. There were 90 internet users per 1,000 inhabitants in 2010.

SOCIAL INSTITUTIONS

Justice

Kiribati's police force is under the command of a Commissioner of Police who is also responsible for prisons, immigration, fire service (both domestic and airport) and firearms licensing. There is a Court of Appeal and High Court, with judges at all levels appointed by the President.

The population in penal institutions in May 2007 was 88 (equivalent to 82 per 100,000 of national population).

Education

In 2005 there were 16,133 pupils and 654 teachers at primary schools and 7,487 pupils in general secondary education with 665 teachers. 51·6% of males and 49·5% of females had received secondary or higher education in 2005 compared to 27·1% of males and 20·6% of females in 1995. There is a regional campus of the University of the South Pacific on Tarawa. Other post-secondary institutions include a teacher training college, a marine training centre, a fisheries training centre, a school of nursing and and a technical institute.

Health

The government maintains free medical and other services. In 2004 there were 20 physicians, three dentistry personnel and 260 nursing and midwifery personnel. There is a national referral hospital in South Tarawa and three other hospitals. In 2005 there were 15 hospital beds per 10,000 inhabitants.

RELIGION

In 2005, 55% of the population were Roman Catholic and 36% Kiribati Protestant; there are also small numbers of Seventh-Day Adventists, Latter-day Saints (Mormons), Bahais and Church of God.

CULTURE

World Heritage Sites

Kiribati has one site on the UNESCO World Heritage List: Phoenix Islands Protected Area (inscribed on the list in 2010).

Press

In 2008 there were three newspapers with a combined circulation of 4,000.

Tourism

There were 5,687 visitors to Kiribati in 2007.

DIPLOMATIC REPRESENTATIVES

Of Kiribati in the United Kingdom
Acting High Commissioner: Makurita Baaro (resides in Kiribati).
Honorary Consul: Michael Walsh (The Great House, Llanddewi Rhydderch, Monmouthshire, NP7 9UY).

Of the United Kingdom in Kiribati
Acting High Commissioner: Martin Fidler (resides in Suva, Fiji Islands).

Of the USA in Kiribati
Ambassador: Frankie A. Reed (resides in Suva, Fiji Islands).

FURTHER READING

Tearo, T., *Coming of Age.* 1989

National Statistical Office: Kiribati Statistics Office, PO Box 67, Bairiki.
Website: http://www.spc.int/prism/Country/ki/Stats

KOREA, NORTH

Chosun Minchu-chui Inmin Konghwa-guk
(Democratic People's Republic of Korea)

Capital: Pyongyang
Population projection, 2015: 24·85m.
GNI per capita, 2010: US$503
Internet domain extension: .kp

KEY HISTORICAL EVENTS

The Korean peninsula was first settled by tribal peoples from Manchuria and Siberia who provided the basis for the modern Korean language. By 3000 BC agriculture-based communities had emerged. The earliest known colony in the region was established at Pyongyang in the 12th century BC. Among the most prominent agricultural communities was Old Choson, which by 194 BC had evolved into a league of tribes ruled by Wiman or 'Wei Man'. His realm was taken over by the Han empire of China in 108 BC and replaced by four Chinese colonies.

The rest of the peninsula evolved into tribal states; Puyo in the north and Chin south of the Han River. Chin was itself split into three tribal states (Mahan, Chinhan and Pyonhan); these states then evolved into three rival kingdoms (Koguryo, Paekche and Silla). Three powerful figures, King T'aejo (AD 53–146) of Koguryo, King Koi (AD 234–86) of Paekche and King Naemul (AD 356–402) of Silla, established hereditary monarchies while powerful aristocracies developed from tribal chiefdoms.

With China's support Silla conquered the other two kingdoms; Paekche in 660 and Koguryo in 668. In 676 Silla drove out the Chinese and gained complete control of the peninsula. Survivors from Koguryo established Parhae, under the leadership of Tae Cho-yong, in the northern region. After a period of conflict with Silla, Parhae grew into a prosperous state in its own right before being taken over by northern nomadic peoples. In Silla an absolute monarchy replaced the council of nobles (its former decision making body) with a central administrative body called

the chancellery (*Chipsabu*), thus undermining aristocratic power. Meanwhile, the capital Kumsong (now Kyongju in South Korea) was developed. The state was divided into administrative units by province (*chu*), prefecture (*kun*), and county (*hyon*), and five provincial capitals prospered as cultural centres. Avatamsaka Buddhism was the dominant religion.

Divisions within the aristocracy in the 8th century led to the restoration of the Council of Nobles and the overthrow of the monarchy. Forced to pay taxes to powerful provincial families and central government, the peasants rebelled. Two provincial leaders, Kyonhwon and Kungye, established the Later Paekche (892) and Later Koguryo (901) as rivals to Silla.

National Unity
The powerful leader Wang Kon founded Koryo (now Kaesong, North Korea) in 918, and established a unified kingdom in the Korean peninsula in 936. Three chancelleries and the royal secretariat formed the supreme council of state and governed the kingdom. Koryo's leaders were then largely aristocratic, and the political system greatly favoured those in the top five tiers of the nine hierarchical levels. That the military was not eligible for any hierarchical position above the second level and received little land, led to a military coup in 1170. Gen. Ch'oe Ch'ung-hon established a military regime which held power for the next sixty years. Zen Buddhism and the allied ideology of Confucianism had grown but were suppressed under the Ch'oe regime. Many monks fled to the mountains, where they formed what became Korean Buddhism, the *Chogye*.

In 1231 the Mongols invaded Koryo but were resisted by the Ch'oe leaders for nearly three decades, until a peasant uprising saw the Ch'oe overthrown. A power-sharing agreement between the rebels and the Mongols came into force in 1258. Despite some interference from the Mongols, Koryo retained its identity as a unified state. Conflict between the aristocracy and the bureaucratic class led to a rebellion. With the help of the Ming dynasty in China and supported by government officials, Gen. Yi Song-gye seized power in 1392. A new system of land distribution ended the Koryo dynasty.

Gen. Yi named the state Choson, designating Hanyang (now Seoul, South Korea) as the capital. Buddhism was dropped in favour of a new Chinese-influenced Confucian ethical system and the state was governed by a hereditary aristocracy (the *yangban*), who controlled all aspects of Korean society. In 1420 the Hall of Worthies (*Chiphyonjon*) was established for scholars, and after 1443 the Korean phonetic alphabet (*hangul*) was adopted. Later in the period, a centralized yangban government was formed and the country divided into eight administrative regions, with standardized laws and a central decision-making and judicial body.

In 1592 Japan, newly unified under the command of Toyotomi Hideyoshi, sent an army to Korea supposedly as part of an invasion of China. Korea's naval forces, under Admiral Yi Sun-shin, were able to repel the invaders. Swelling anti-Japanese sentiment prompted Koreans from all hierarchical divisions to fight in the war alongside troops dispatched from Ming China. However, Japanese forces did not withdraw completely until Toyotomi's death in 1598, leaving Korea in ruins.

Despite joint efforts by China and Korea to stem the advances of the nomadic Manchu in the early 17th century, Seoul was captured in 1636. The Manchu established the Ch'ing dynasty several years later and demanded tribute from Korea.

During the 17th and 18th centuries, advances in irrigation increased the output of rice, tobacco and ginseng. By the late 18th century many Korean scholars had turned to Roman Catholicism, leading to government suppression of Christianity in a bid to preserve the dominance of Confucianism. However, European priests maintained strong links in the country.

Japanese Influence

In the 19th century, a succession of monarchs yet to attain the age of majority undermined national stability. In 1864 Taewon'gun, the father of the child-king Kojong, took power and pursued a programme of controversial political reform that increasingly isolated Korea from the outside world. When Taewon'gun was eventually forced to step down, Korea came under pressure from Japan to open up its ports. Nervous of growing Japanese influence, China placed troops in Korea following a failed coup attempt by pro-Taewon'gun forces. There followed a trade agreement which greatly benefited Chinese commercial interests. Further treaties with France, Germany, Russia, the UK and the USA followed in the 1880s. As foreign influence increased, Korea's ruling elite divided between moderates and radicals. The radicals carried out a coup in 1884 but were quickly defeated by Chinese troops. An agreement to maintain a balance of power in the region was signed by Japan and China the following year.

As modernization gathered pace, government spending increased, adding to the burden of reparations payments to Japan. The peasants turned to Tonghak ('Eastern Learning'), a new religion established by an old yangban scholar and based on traditional beliefs. A Tonghak rebellion in 1894 caused China to send in troops. Japan responded by sending its own forces and war broke out. By the following year Japan had secured control of the peninsula.

Korea declared neutrality at the outbreak of war between Japan and Russia in 1904 but was pressured by Japan into allowing use of Korean territory. Japan achieved victory in 1905 and made Korea a protectorate. An unsuccessful appeal to the international peace conference at The Hague further undermined relations between Japan and Korea. Anti-Japanese guerrilla fighters in the southern provinces were active during 1908–09 but were crushed the following year when Korea was annexed by Japan.

Japan established a government in Korea and implemented a programme designed to supplant the Korean identity. There were restrictions on freedom of speech, press and assembly while the language and history of Japan were taught in schools. Many Koreans were dispossessed as Japan built new transport and communications infrastructures. After the Japanese suppressed a mass demonstration in 1919, independence leaders set up a provisional government in Shanghai and named Syngman Rhee as president. Hoping to calm dissent, Japan lifted certain press restrictions and replaced the gendarmerie with an ordinary police force, but uncompromising colonial rule remained in place.

Korea became a market for Japanese goods and attracted much capital investment but at the expense of agriculture, leading to a long-term shortage of rice. Tokyo reimposed military rule in 1931 when war broke out between Japan and China. Magazines, newspapers and academic organizations operating in the Korean language were banned and hundred of thousands of Koreans were made to fight in the Japanese army or work in Japanese mines and factories to support Japan in the Second World War. The provisional government, having moved from Shanghai to Chungking in southwest China, declared war on Japan in Dec. 1941. An army of Korean resistance fighters joined the Allied forces in China and fought with them until the Japanese surrender in 1945.

Korea Divided

Korea was promised independence by China, Britain and the USA at the Cairo conference of 1943 but at the end of the war, after Japan's collapse, Korea was divided along the 38th Parallel. A planned four-way power share in Korea, involving Britain and the Republic of China, was abandoned after US troops took control of the south of the country while the USSR occupied the north. The Soviet forces helped to establish a Communist-led provisional government under Kim Il-sung. As relations between the USA and the Soviet Union worsened, trade ceased between the two zones.

General elections in Korea were proposed by the United Nations in Sept. 1947. However, a commission to oversee voting was denied entry by Soviet troops. Rhee was elected in the South while the North appointed Kim Il-sung as leader. In 1948 the southern Republic of Korea (with Seoul as the capital) and the northern Democratic People's Republic of Korea (with Pyongyang as capital) were formally recognized.

Soviet and US troops left the peninsula in 1949 and war broke out between the North and South in June 1950. A US-led UN force under Gen. Douglas MacArthur entered South Korea and pushed back the North Korean forces. The UN pressed on into North Korea and established a commission for the reunification. China, which at that point had no representation in the United Nations, entered the war adding 1·2m. troops to the North Korean side. Peace negotiations began in 1951 and a new international boundary and demilitarized zone were declared in 1953. The USA offered South Korea financial support and signed a mutual security pact with Rhee, who had been reluctant to accept the division of the country. The issue of prisoner returns, particularly of North Koreans unwilling to return to the communist state, remained a point of contention. The war left 4m. people dead or injured.

In the aftermath, Kim set about tightening his grip on his country. He established a dictatorship based on a personality cult and introduced his philosophy of Juche, by which the country was to develop without any help from outside. Industrialization and military spending gathered pace in the later 1950s and the 1960s despite North Korea's international isolation. However, by the late 1970s North Korea had fallen far behind its southern neighbour and a period of stagnation began.

Kim maintained close relations with China and the Soviet Union, although his allegiance wavered between the two as Sino-Soviet relations deteriorated in the 1960s. When the Soviet Union collapsed in 1990–91, North Korea went into an economic crisis that included widespread famine. Attempts to improve relations with South Korea in the early 1990s faltered over the North's alleged nuclear capacity although in 1994 North Korea agreed to shut down controversial reactors in return for aid and oil. Kim died in 1994 and power passed to his son, Kim Jong-il.

Under the younger Kim the economy has collapsed to subsistence level although spending on the military remains high. In 1997 the UN World Food Programme estimated that 2m. North Koreans faced starvation. More than 5% of the population starved to death during the 1990s. In 2000 Kim received South Korean President Kim Dae-jung as relations between the North and South appeared to be thawing. The two leaders agreed that reunification was the eventual aim of both Koreas but relations again deteriorated in 2002 after a naval battle in the Yellow Sea between North and South forces killed four South Korean and around 30 North Korean sailors. Kim Jong-il blamed the USA and South Korea for the attack. South Korean president Kim Dae-jung suspended rice shipments to the north and demanded an apology.

Relations with the USA worsened during 2002 and 2003 after the USA claimed that North Korea had a secret nuclear programme. US president George W. Bush accused North Korea of forming part of what he called the 'Axis of Evil' along with Iraq and Iran. North Korea subsequently reactivated a nuclear plant and demanded the withdrawal of inspectors from the UN International Atomic Energy Agency. Pyongyang claimed it had been forced to reopen the reactor in response to US plans for a pre-emptive nuclear strike. North Korea then announced its withdrawal from the nuclear non-proliferation treaty, although it denied any intention to produce nuclear weapons. Many observers suggested Kim carried out these manoeuvres to pressurize the USA into direct talks with a view to signing a mutual non-aggression pact. North Korea's nuclear programme has in turn unsettled relations with regional neighbours including

South Korea and Japan. In Feb. 2005 North Korea publicly admitted for the first time that it possessed nuclear weapons. On 9 Oct. 2006 it conducted its first nuclear test, carrying out an underground explosion. In Feb. 2007 at talks between North Korea, South Korea, Japan, Russia, China and the USA, Pyongyang agreed to close its chief nuclear reactor in return for fuel aid. In Oct. 2007 it agreed to close a further three installations and was scheduled to surrender its nuclear stockpile in 2008. However, Pyongyang missed a Dec. 2007 deadline to disclose full details of all of its nuclear facilities.

In Oct. 2007 North and South Korea issued a joint declaration calling for a permanent peace on the peninsula to replace the armistice in place since the end of the Korean War. However, relations soon deteriorated and at the end of 2008 Pyongyang accused Seoul of fostering hostility. The South accused the North of testing long-range missile technology in April 2009 and a month later North Korea walked out of international talks aimed at ending its nuclear programme. In May 2010 an international report concluded that North Korea had been responsible for the sinking of a South Korean ship two months earlier. A collapse in trading and diplomatic relations followed.

In Dec. 2011 it was announced that Kim Jong-il had died of a heart attack. State-run television called on the public to support his son, Kim Jong-un, as 'the great successor'.

TERRITORY AND POPULATION

North Korea is bounded in the north by China, east by the Sea of Japan (East Sea of Korea), west by the Yellow Sea and south by South Korea, from which it is separated by a demilitarized zone of 1,262 sq. km. Its area is 122,762 sq. km.

The census population in 2008 was 24,052,231; density 195·9 per sq. km. In 2011 60·3% of the population were urban.

The UN gives a projected population for 2015 of 24·85m.

The area, 2008 census population (in 1,000) of the provinces and Pyongyang (directly governed city):

	Area in sq. km	Population	Chief Town
Chagang	16,968	1,300	Kanggye
North Hamgyong[1]	17,570	2,327	Chongjin
South Hamgyong	18,970	3,066	Hamhung
North Hwanghae[2]	9,262	2,114	Sariwon
South Hwanghae	8,002	2,310	Haeju
Kangwon[3]	11,152	1,478	Wonsan
North Pyongan[4]	12,191	2,729	Sinuiju
South Pyongan	12,330	4,052	Pyongsong
Pyongyang (directly governed city)	2,000	3,255	
Yanggang	14,317	719	Hyesan

[1]Area and population include Rason directly governed city.
[2]Area and population include Kaesong industrial region.
[3]Area and population include Kumgangsan tourist region.
[4]Area and population include Sinuiju special administrative region.

Pyongyang, the capital, had a 2008 census population of 2,581,076. Other large towns (census, 2008): Hamhung (703,610); Chongjin (614,892); Sinuiju (334,031).

The official language is Korean.

SOCIAL STATISTICS

2008 estimated births, 327,000; deaths, 238,000. 2008 estimated birth rate, 13·7 per 1,000 population; death rate, 10·0. Annual population growth rate, 2000–08, 0·5%. Marriage is discouraged before the age of 32 for men and 29 for women. Life expectancy at birth, 2007, was 64·9 years for men and 69·1 years for women. Infant mortality, 2010, 26 per 1,000 live births; fertility rate, 2008, 1·9 births per woman.

CLIMATE

There is a warm temperate climate, though winters can be very cold in the north. Rainfall is concentrated in the summer months. Pyongyang, Jan. 18°F (–7·8°C), July 75°F (23·9°C). Annual rainfall 37" (916 mm).

CONSTITUTION AND GOVERNMENT

North Korea adopted a new constitution in April 2009 that formalized *songun* or 'military first' politics as a guiding principle of state but dropped the word 'communism'. The Constitution provides for a 687-seat *Supreme People's Assembly* elected every five years by universal suffrage. Citizens of 17 years and over can vote and be elected. The government consists of the *Administration Council* directed by the Central People's Committee.

In 1998, four years after the death of Kim Il-sung, the title of president was abolished. On the death of Kim Jong-il on 19 Dec. 2011 his son and designated successor, Kim Jong-un (b. 1983), assumed the role of 'Supreme Leader'.

About 3m. people are affiliated with the ruling party, the Workers' Party of Korea. There are also the puppet religious Chongu and Korean Social Democratic Parties and various organizations combined in a Fatherland Front.

National Anthem

'A chi mun bin na ra i gang san' ('Shine bright, o dawn, on this land so fair'); words by Pak Se Yong, tune by Kim Won Gyun.

RECENT ELECTIONS

Elections to the Supreme People's Assembly were held on 8 March 2009. Only the list of the Democratic Front for the Reunification of the Fatherland (led by the Korean Workers' Party) was allowed to participate. 687 deputies were elected unopposed.

CURRENT GOVERNMENT

Supreme Leader, Commander of the Korean People's Army, First Secretary of the Workers' Party of Korea and First Chairman of the National Defence Commission: Kim Jong-un.

In April 2013 the government comprised:
Prime Minister: Pak Pong-ju; b. 1939 (appointed 1 April 2013).
Vice Prime Ministers: Han Kwang-bok (also *Minister of Electronic Industry*), Pak Su-gil (also *Minister of Finance*), Jo Pyong-ju (also *Minister of Machine-Building Industry*), Jon Ha-chol, Kang Nung-su, Kang Sok-ju, Kim Rak-hui, Ri Mu-yong, Ro Tu-chol.

Minister of Agriculture: Kim Chang-sik. *Capital City Construction:* Kim Ung-gwan. *Chemical Industry:* Yi Mu-yong. *Coal Industry:* Kim Hyong-sik. *Commerce:* Kim Pong-chol. *Construction and Building Materials Industry:* Tong Chong-ho. *Culture:* An Tong-chun. *Education:* Kim Yong-chin. *Extractive Industries:* Kang Min-chol. *Fisheries:* Pak Thae-won. *Food Procurement and Administration:* Mun Ung-jo. *Foodstuffs and Daily Necessities Industry:* Jo Yong-chol. *Foreign Affairs:* Pak Ui-chun. *Foreign Trade:* Ri Ryong-nam. *Forestry:* Kim Kwang-yong. *Labour:* Jong Yong-su. *Land and Environment Protection:* Kim Chang-ryong. *Land and Marine Transport:* Ra Tong-hui. *Light Industry:* An Jong-su. *Metal Industry:* Kim Tae-bong. *Oil:* Kim Hui-yong. *People's Security:* Ri Myong-su. *Physical Culture and Sports:* Pak Myong-chol. *Post and Telecommunications:* Ryu Yong-sop. *Power Industry:* Ho Taek. *Public Health:* Choe Chang-sik. *Railways:* Chon Kil-su. *State Construction Control:* Pae Tal-chun. *State Inspection:* Kim Ui-sun. *Urban Management:* Hwang Hak-won.

President, Supreme People's Assembly Presidium: Kim Yong-nam. *Vice Presidents:* Yang Hyong-sop, Kim Yong-dae.

In addition there is one minister who is not in the cabinet.
Minister of the People's Armed Forces: Kim Kyok-sik.

In practice the country is ruled by the Korean Workers' (i.e. Communist) Party, which elects a Central Committee that in turn appoints a Politburo.

Government Website: http://www.korea-dpr.com

CURRENT LEADERS

Kim Jong-un

Position
Supreme Leader

Introduction
Kim Jong-un was named Supreme Commander of North Korea's military in Dec. 2011 after the death of his father, Kim Jong-il. Officially, Kim is a member of a triumvirate—with Premier Choe Yong-rim and President of the Supreme People's Assembly Presidium Kim Yong-nam—that heads the executive branch of government. However, it is widely understood that Kim, like his late father, yields absolute power over the state, party and army.

Early Life
Kim Jong-un was born in 1983 or 1984. He was educated in Switzerland before attending the Kim Il-sung Military University. As the youngest son, Kim was not in line to succeed to the leadership until his elder brother, Kim Jong-nam, reportedly fell out of favour with their father in 2001.

He became clear favourite to take over from his father when he was appointed a four-star *Daejang* (General) of the People's Army on 27 Sept. 2010, despite having no previous military experience. A day later he was given the post of vice-chairman of the Central Military Commission of the Workers' Party. In the months before Kim Jong-il's death, Kim accompanied him to public events apparently confirming reports that he was being groomed for the leadership. Following Kim Jong-il's sudden death, Kim was proclaimed as 'the great successor' by the North Korean media before his appointment as Supreme Commander of the Army on 30 Dec. 2011.

Career in Office
Despite the government's insistence that the change in leadership would not signal major policy shifts, there was initial optimism within the international community that the new administration might be open to reform. However, a moratorium on missile tests, brokered with the USA in Feb. 2012 in return for food aid, was broken when North Korea unsuccessfully tested a long-range rocket in April. Then, in Oct., Pyongyang claimed that the country held missiles capable of reaching the US mainland. Two months later a long-range rocket was successfully test-launched, prompting condemnation from the international community. Meanwhile, Kim was named First Secretary of the ruling Workers' Party and First Chairman of the National Defence Commission in April and in July he assumed the highest military rank of marshal.

In March 2013 the UN imposed new sanctions in response to a nuclear test the previous month. Pyongyang reacted by threatening pre-emptive nuclear strikes against South Korea and the USA.

DEFENCE

The Supreme Commander of the Armed Forces is Kim Jong-un. Military service is compulsory at the age of 16 for periods of 5–12 years in the Army, 5–10 years in the Navy and 3–4 years in the Air Force, followed by obligatory part-time service in the Pacification Corps to age 40. Total armed forces troops were estimated to number 1,106,000 in 2007, up from 840,000 in 1986 although down from 1,160,000 in 1997. Around 70% of the troops are located along or near the Demilitarized Zone between North and South Korea.

Defence expenditure in 2003 totalled US$5,500m. (US$243 per capita), and represented 25·0% of GDP.

In 1998 North Korea tested a medium-range nuclear-capable Taepo Dong-1 missile. It has also developed a shorter-range No-Dong ballistic missile in addition to Scud B and Scud C missiles, and is developing a longer-range inter-continental ballistic missile, the two- or three-stage Taepo Dong-2, which experts believe could reach Alaska and the westernmost Hawaiian islands. A first

unsuccessful test was carried out in July 2006. A second unsuccessful test, of the three-stage Taepo Dong-2, followed in April 2009 and a third unsuccessful launch came in April 2012.

Nuclear Weapons
North Korea was for many years suspected of having a secret nuclear-weapons programme, and perhaps enough material to build two warheads. In Oct. 2002 it revealed that it had developed a nuclear bomb in violation of an arms control pact agreed with the USA in 1994. North Korea has not signed the Comprehensive Nuclear-Test-Ban-Treaty, which is intended to bring about a ban on any nuclear explosions. It ratified the Nuclear Non-Proliferation Treaty in 1985 but withdrew in 2003. In Feb. 2005 it declared that it had manufactured nuclear weapons and stated that it would not re-enter multilateral negotiations on its disarmament. It carried out its first test of a nuclear weapon in Oct. 2006. In July 2007 Pyongyang closed the Yongbyon nuclear complex in return for international aid and in Oct. pledged to disable a further three facilities, ahead of surrendering its nuclear stockpile in 2008. However, Pyongyang missed a Dec. 2007 deadline to give a full account of all of its nuclear facilities although it eventually did so in June 2008. The following day North Korea destroyed Yongbyon's cooling tower. Further dismantling of its nuclear facilities was postponed after the USA refused to remove North Korea from its list of state sponsors of terrorism until Pyongyang produced verification of its nuclear downgrading. North Korea was removed from the list in Oct. 2008 and pledged to resume dismantling the Yongbyon reactor.

In April 2009 North Korea was accused by South Korea and the UN of testing long-range nuclear missile technology. North Korea responded by walking out of international talks to wind up its nuclear programme. The following month Pyongyang claimed it had successfully completed underground nuclear tests. In June the UN imposed new sanctions, with Pyongyang stating its intent to weaponize plutonium supplies. A 'miniaturized' nuclear device was tested underground at the Punggye-ri test site in Feb. 2013, prompting renewed international condemnation.

Army
One of the world's biggest, the Army was estimated at 950,000 personnel in 2007 with 600,000 reserves. There is also a para-military worker-peasant Red Guard of some 3·5m. and a ministry of public security force of 189,000 including border guards.

Equipment includes some 3,500 T-34, T-54/55, T-62 and Type-59 main battle tanks.

Navy
The Navy, principally tasked to coastal patrol and defence, comprises 63 diesel submarines, three small frigates and five corvettes. Personnel in 2007 totalled about 46,000 with 65,000 reserves.

Air Force
The Air Force had a total of 590 combat-capable aircraft and 110,000 personnel in 2007. Combat-capable aircraft include J-5/6/7s (Chinese built versions of MiG-17/19/23s), MiG-23s, MiG-29s, Su-7s and Su-25s.

ECONOMY

Agriculture is estimated to account for approximately 25% of GDP, industry 60% and services 15%.

In 2012 North Korea received approximately US$126m. in foreign aid.

North Korea was rated the joint most corrupt country in the world in a 2012 survey of 176 countries carried out by the anti-corruption organization *Transparency International*.

Overview
Noted for its rigid state-control, economic progress in North Korea has been impeded by an all-powerful bureaucracy and a reluctance to depart from the Marxist-Stalinist line. Aid agencies

estimate that food shortages caused by natural disasters and economic mismanagement have resulted in up to 2m. deaths since the mid-1990s. The economy is dependent on international aid. Food shortages became critical in 2003 but US food assistance resumed in June 2008.

International sanctions over North Korea's nuclear programme have led to greater reliance on China, which now accounts for four-fifths of its trade. In June 2011 it was announced that two new 'special economic zones', at Rason and Hwanggumpyong, would be developed in co-operation with China, an idea long promoted by Pyongyang.

Currency

The monetary unit is the *won* (KPW) of 100 *chon*. Banknotes were replaced by a new issue in July 1992. Exchanges of new for old notes were limited to 500 won. In Nov. 2009 the government readjusted the value of the won, with 100 old won worth one new won. Officially the won trades at 135 per US dollar but unofficially it can sometimes trade at up to 3,000 per dollar. North Korea has periodically suffered from high inflation rates since 2000. Following a failed currency reform in Nov. 2009 the inflation rate soared, peaking at an estimated monthly rate of 496% in March 2010.

Budget

Estimated revenue, 1999, 19,801m. won; expenditure, 20,018m. won.

Performance

The real GDP growth rate was 6·2% in 1999 following a decade of negative growth. This was followed in 2000 by growth of 1·3%, rising in 2001 to 3·7%. In both 2002 and 2003 there was growth of 0·8%. GDP per head was put at US$494 in 2003, or less than a twentieth of that of South Korea.

Banking and Finance

The bank of issue is the Central Bank of Korea. In 2002 there were seven state banks, seven joint venture banks and two foreign investment banks.

ENERGY AND NATURAL RESOURCES

Environment

Carbon dioxide emissions from the consumption and flaring of fossil fuels in 2008 were the equivalent of 3·1 tonnes per capita.

Electricity

Installed capacity was an estimated 9·5m. kW in 2007 (approximately 53% hydro-electric). Production in 2007 was 21·52bn. kWh. Consumption per capita was 905 kWh in 2007. In Feb. 2003 North Korea reactivated its nuclear reactor at Yongbyon that had been dormant since 1994. It was again shut down in July 2007 in exchange for economic aid and political concessions following negotiations with China, Japan, Russia, South Korea and the USA. In April 2009 it was reactivated again following United Nations condemnation of a controversial long-range rocket launch.

Oil and Gas

Oil wells went into production in 1957. An oil pipeline from China came on stream in 1976. China's supplies account for 70% of North Korea's oil consumption. Refinery distillation output amounted to 2·5m. tonnes in 1998.

Minerals

North Korea is rich in minerals. Estimated reserves in tonnes: coal, 11,990m.; manganese, 6,500m.; iron ore, 3,300m.; uranium, 26m.; zinc, 12m.; lead, 6m.; copper, 2·15m. 23·9m. tonnes of coal were mined in 2007, 6·5m. tonnes of lignite in 2007, 5m. tonnes of iron ore in 2006, 500,000 tonnes of salt in 2006 and 12,000 tonnes of copper in 2006. 2006 production of silver was 20 tonnes; gold, 2,000 kg.

Agriculture

In 2007 there were approximately 2·8m. ha. of arable land and 200,000 ha. of permanent crop land. An estimated 3·12m. persons were economically active in agriculture in 2007.

Collectivization took place between 1954 and 1958. 90% of the cultivated land is farmed by co-operatives. Land belongs either to the State or to co-operatives, and it is intended gradually to transform the latter into the former, but small individually-tended plots producing for 'farmers' markets' are tolerated as a 'transition measure'.

There is a large-scale tideland reclamation project. In 2002 around 1·46m. ha. were under irrigation, making possible two rice harvests a year. There were an estimated 64,000 tractors in 2002. The technical revolution in agriculture (nearly 95% of ploughing, etc., is mechanized) has considerably increased the yield of wheat (sown on 58,000 ha.). Areas harvested of other major crops, 2003: rice, 593,000 ha.; maize, 495,000 ha.; potatoes, 188,000 ha. Production (2003, in 1,000 tonnes): rice, 2,284; potatoes, 2,023; maize, 1,725; cabbage, 680; apples, 660.

Livestock, 2003: pigs, 3·18m.; goats, 2·72m.; cattle, 576,000; sheep, 171,000; 20m. chickens.

A chronic food shortage has led to repeated efforts by UN agencies to stave off famine. In Jan. 1998 the UN launched an appeal for US$378m. for food for North Korea, the largest ever relief effort mounted by its World Fund Programme.

Forestry

Forest area in 2010 was 5·67m. ha. (47% of the land area). Timber production was 7·37m. cu. metres in 2007.

Fisheries

In 2010 total catch was an estimated 205,000 tonnes, of which 98% were sea fish.

INDUSTRY

Industries were intensively developed by the Japanese occupiers, notably cotton spinning, hydro-electric power, cotton, silk and rayon weaving, and chemical fertilizers. Production: cement (2007), 6·1m. tonnes; crude steel (2007), 1·23m. tonnes; pig iron (2007), 900,000 tonnes; textile fabrics (1994), 350m. metres; TV sets (1995), 240,000 units; cars (2000), 6,600 units; ships (1995), 50,000 GRT. Industrial production is estimated to have halved between 1990 and 2000.

Labour

The labour force totalled 11,881,000 (55% males) in 1996. Nearly 29% of the economically active population in 2002 were engaged in agriculture.

INTERNATIONAL TRADE

Joint ventures with foreign firms have been permitted since 1984. A law of Oct. 1992 revised the 1984 rules: foreign investors may now set up wholly-owned facilities in special economic zones, repatriate part of profits and enjoy tax concessions. In Jan. 2012 foreign debt was estimated at US$20bn. The USA imposed sanctions in Jan. 1988 for alleged terrorist activities. Since June 1995 South Korean businesses and individuals have been permitted to make investments and set up branch offices in North Korea.

Imports and Exports

Imports in 2001 were US$1,847m.; exports, US$826m. In 2001 China was the biggest import supplier (31%), followed by Japan (13%) and South Korea (12%); Japan was the main export destination (27%), ahead of South Korea (21%) and China (20%). The chief imports are machinery and petroleum products, the chief exports metal ores and products.

COMMUNICATIONS

Roads
There were 25,554 km of road in 2006. There were 262,000 passenger cars in 2000. The first of two planned cross-border roads between the two Koreas opened in Feb. 2003.

Rail
Rail transport is provided by Korean State Railways. There is an extensive network of standard gauge lines totalling over 6,000 km and a network of 762 mm narrow gauge lines covering some 350 km. Main lines cover around 2,500 km. In June 2000 it was agreed to start consultations to restore the railway from Sinuiju, on the North Korean/Chinese border, to Seoul by rebuilding an 8 km long stretch from Pongdong-ni to Changdan, on the North Korean/South Korean border, and a 12 km long stretch in South Korea. Two passenger trains crossed the border between North and South Korea on 17 May 2007 (one northbound and one southbound), completing the first cross-border journey in more than 50 years. Regular freight services between the two Koreas were resumed in Dec. 2007 but usage is very limited.

There is a metro and two tramways in Pyongyang.

Civil Aviation
There is an international airport at Pyongyang (Sunan). There were flights in 2010 to Bangkok, Beijing, Shenyang and Vladivostok. The national carrier is Air Koryo.

Shipping
The leading ports are Chongjin, Wonsan and Hungnam. Pyongyang is connected to the port of Nampo by railway and river. In Jan. 2009 there were 223 ships of 300 GT or over registered, totalling 884,000 GT.

The biggest navigable river is the Yalu, 698 km up to the Hyesan district.

Telecommunications
There were 1,180,000 main (fixed) telephone lines in 2008. A mobile phone service was introduced in Dec. 2008 four years after a previous service had been shut down without explanation. In March 2009 there were 20,000 subscribers. North Korea is one of only two countries in the world where there are more landline than mobile phone subscriptions.

SOCIAL INSTITUTIONS

Justice
The judiciary consists of the Supreme Court, whose judges are elected by the Assembly for three years; provincial courts; and city or county people's courts. The procurator-general, appointed by the Assembly, has supervisory powers over the judiciary and the administration; the Supreme Court controls the judicial administration.

In May 2011 some 150,000–200,000 political prisoners were being held at a number of detention camps in the country. North Korea does not divulge figures on its use of the death penalty; however, Amnesty International reported that there were at least six executions in 2012.

Education
Free compulsory universal technical education lasts 11 years: one pre-school year, four years primary education starting at the age of six, followed by six years secondary. In 2000 there were 1·5m. children in 27,017 nursery schools, 748,416 children in 14,167 kindergartens, 1·6m. pupils in 4,886 primary schools and 2·1m. pupils in 4,772 secondary schools. Nearly 1·9m. students attended more than 300 colleges and universities.

The adult literacy rate in 2008 was 100%.

Health
Medical treatment is free. In 2003 there were 74,597 doctors (one per 312·5 population), 87,330 nurses and 6,084 midwives. In 2002 there were 214,647 hospital beds (92 per 10,000 population). 6·3% of GDP was spent on health care in 2004.

North Korea has been one of the least successful countries in the battle against undernourishment in the past 20 years. The proportion of undernourished people rose from 18% of the population in the period 1990–92 to 35% in 2001–03.

RELIGION
The Constitution provides for 'freedom of religion as well as the freedom of anti-religious propaganda'. In 2001 there were 3·0m. Chondoists. Another 3·4m. followed traditional beliefs. There were also significant numbers of Christians and Buddhists.

CULTURE

World Heritage Sites
There is one UNESCO site in North Korea: the Complex of Koguryo Tombs (2004).

Press
There were three national daily newspapers and 12 regional dailies in 2008 with a combined circulation of 4·5m. The party newspaper is *Nodong* (or *Rodong*) *Sinmun* (Workers' Daily News). In the 2010 *World Press Freedom Index* compiled by Reporters Without Borders, North Korea ranked 177th out of 178 countries.

Tourism
A 40-year ban on non-Communist tourists was lifted in 1986. In 2002 there were 400,000 foreign tourists. On 19 Nov. 1998 North Korea received its first tourists from South Korea, on a cruise and tour organized by the South Korean firm Hyundai.

Calendar
A new yearly calendar was announced on 9 July 1997 based on Kim Il-sung's birthday on 15 April 1912. Thus 1912 became *Juche* year 1; 2013 is *Juche* 102.

DIPLOMATIC REPRESENTATIVES
Of North Korea in the United Kingdom (73 Gunnersbury Ave., London, W5 4LP)
Ambassador: Hyon Hak Bong.

Of the United Kingdom in North Korea (Munsu Dong Diplomatic Compound, Pyongyang)
Ambassador: Michael Gifford.

Of North Korea to the United Nations
Ambassador: Sin Son Ho.

Of North Korea to the European Union
Ambassador: Vacant.

FURTHER READING

Becker, Jasper, *Rogue Regime: Kim Jong Il and the Looming Threat of North Korea.* 2005

Cha, Victor D, *The Impossible State: North Korea, Past and Future.* 2012

Cha, Victor D. and Kang, David C., *Nuclear North Korea: A Debate on Engagement Strategies.* 2003

Cumings, Bruce, *North Korea: Another Country.* 2004

Harrison, S., *Korean Endgame: A Strategy for Reunification and US Disengagement.* 2002

Hassig, Ralph and Oh, Kongdan, *The Hidden People of North Korea: Everyday Life in the Hermit Kingdom.* 2009

Hunter, H., *Kim Il-Song's North Korea.* 1999

Kleiner, J., *Korea: a Century of Change.* 2001

Myers, B. R., *The Cleanest Race: How North Koreans See Themselves and Why It Matters.* 2010

Oh, K. and Hassig, R. C., *North Korea Through the Looking Glass.* 2000

O'Hanlon, Michael E. and Mochizuki, Mike, *Crisis on the Korean Peninsula: How to Deal with a Nuclear North Korea.* 2003

Sigal, L. V., *Disarming Strangers: Nuclear Diplomacy with North Korea.* 1999

Smith, H., *et al.*, (eds.) *North Korea in the New World Order.* 1996

National Statistical Office: Central Statistics Bureau, Pyongyang.

KOREA, SOUTH

NORTH KOREA

Sea of Japan (East Sea)

SEOUL

Incheon

SOUTH KOREA

Daejeon

Yellow Sea

Daegu

Gwangju

Busan

Jeju

JAPAN

0   75 mi
0   100 km

© Research Machines plc 2006

Daehan Minguk
(Republic of Korea)

Capital: Seoul
Population projection, 2015: 49·12m.
GNI per capita, 2011: (PPP$) 28,230
HDI/world rank: 0·897/15
Internet domain extension: .kr

KEY HISTORICAL EVENTS

The Korean peninsula was first settled by tribal peoples from Manchuria and Siberia who provided the basis for the modern Korean language. By 3000 BC agriculture-based communities had emerged. The earliest known colony in the region was established at Pyongyang in the 12th century BC. Among the most prominent agricultural communities was Old Choson, which by 194 BC had evolved into a league of tribes ruled by Wiman or 'Wei Man', a leader widely held to have defected from China, although he may have been a native of the Choson region. His realm was taken over by the Han empire of China in 108 BC and replaced by four Chinese colonies.

The rest of the peninsula developed into tribal states; Puyo in the north and Chin south of the Han River. Chin was itself split into three tribal states (Mahan, Chinhan and Pyonhan); these states then evolved into three rival kingdoms, Koguryo, Paekche, and Silla. Three powerful figures, King T'aejo (AD 53–146) of Koguryo, King Koi (AD 234–86) of Paekche and King Naemul (AD 356–402) of Silla, established hereditary monarchies while powerful aristocracies developed from tribal chiefdoms.

With China's support Silla conquered the other two kingdoms; Paekche in 660 and Koguryo in 668. In 676 Silla drove out the Chinese and gained complete control of the peninsula. Survivors from Koguryo established Parhae, under the leadership of Tae Cho-yong, in the northern region. After a period of conflict with Silla, Parhae grew into a prosperous state in its own right before being taken over by northern nomadic peoples. In Silla an absolute monarchy replaced the council of nobles (its former decision making body) with a central administrative body called the chancellery (*Chipsabu*), thus undermining aristocratic power. Meanwhile, the capital Kumsong (now Kyongju in South Korea) was developed. The state was divided into administrative units by province (*chu*), prefecture (*kun*), and county (*hyon*), and five provincial capitals prospered as cultural centres. Avatamsaka Buddhism was the dominant religion.

Divisions within the aristocracy in the 8th century led to the restoration of the Council of Nobles and the overthrow of the monarchy. Forced to pay taxes to powerful provincial families and central government, the peasants rebelled. Two provincial leaders, Kyonhwon and Kungye, established the Later Paekche (892) and Later Koguryo (901) as rivals to Silla.

National Unity

The powerful leader Wang Kon founded Koryo (now Kaesong, North Korea) in 918, and established a unified kingdom in the Korean peninsula in 936. Three chancelleries and the royal secretariat formed the supreme council of state and governed the kingdom. Koryo's leaders were then largely aristocratic, and the political system greatly favoured those in the top five tiers of the nine hierarchical levels. That the military was not eligible for any hierarchical position above the second level and received little land, led to a military coup in 1170. Gen. Ch'oe Ch'ung-hon established a military regime which held power for the next 60 years. Zen Buddhism and the allied ideology of Confucianism had grown popular but were suppressed under the Ch'oe regime. Many monks fled to the mountains, where they formed what became Korean Buddhism, the *Chogye*.

In 1231 the Mongols invaded Koryo but were resisted by the Ch'oe leaders for nearly three decades, until a peasant uprising saw the Ch'oe overthrown. A power-sharing agreement between the rebels and the Mongols came into force in 1258. Despite some interference from the Mongols, Koryo retained its identity as a unified state. The aristocracy established seats of power throughout the country, encouraging peasants to seek protection as serfs. This, however, led to reduced tax revenues and when the government did not have sufficient resources to reward its bureaucratic class, a rebellion ensued. Led by General Yi Song-gye, and with the help of the Ming dynasty in China, government officials seized power in 1392 and established a new system of land distribution, thus ending the Koryo dynasty.

Gen. Yi named the state Choson, designating Hanyang (now Seoul, South Korea) as the capital. Buddhism was dropped in favour of a new Chinese-influenced Confucian ethical system and the state was governed by a hereditary aristocracy (the *yangban*), who controlled all aspects of Korean society. In 1420 the Hall of Worthies (*Chiphyonjon*) was established for scholars, and after 1443 the Korean phonetic alphabet (*hangul*) developed. Later in the period, a centralized yangban government was formed and the country divided into eight administrative regions, with standardized laws and a central decision-making and judicial body.

In 1592 Japan, newly unified under the command of Toyotomi Hideyoshi, sent an army to Korea supposedly as part of an invasion of China. Korea's naval forces, under Admiral Yi Sun-shin, were able to repel the invaders. Swelling anti-Japanese sentiment prompted Koreans from all hierarchical divisions to fight in the war alongside troops dispatched from Ming China. However, Japanese forces did not withdraw completely until Toyotomi's death in 1598, leaving Korea in ruins.

Despite joint efforts by China and Korea to stem the advances of the nomadic Manchu in the early 17th century, Seoul was captured in 1636. The Manchu established the Ch'ing dynasty several years later and demanded tribute from Korea.

During the 17th and 18th centuries, Korea's agriculture developed as irrigation improved and rice, tobacco and ginseng became increasingly important crops. By the late 18th century many Korean scholars had turned to Roman Catholicism, leading to government suppression of Christianity in a bid to preserve the dominance of Confucianism. However, European priests maintained strong links in the country.

Japanese Influence

In the 19th century, a succession of monarchs yet to attain the age of majority undermined national stability. In 1864 Taewon'gun, the father of the child-king Kojong, took power and pursued a programme of controversial political reform that increasingly isolated Korea from the outside world. When Taewon'gun was eventually forced to step down, Korea came under pressure from Japan to open up its ports. Nervous of growing Japanese influence, China placed troops in Korea following a failed coup attempt by pro-Taewon'gun forces. There followed a trade agreement which greatly benefited Chinese commercial interests. Further treaties with France, Germany, Russia, the UK and the USA followed in the 1880s. As foreign influence increased, Korea's ruling elite divided between moderates and radicals. The radicals carried out a coup in 1884 but were quickly defeated by Chinese troops. An agreement to maintain a balance of power in the region was signed by Japan and China the following year.

As modernization gathered pace, government spending increased, adding to the burden of reparations payments to Japan. The peasants turned to *Tonghak* ('Eastern Learning'), a new religion established by an old yangban scholar and based on traditional beliefs. A Tonghak rebellion in 1894 caused China to send in troops. Japan responded by sending its own forces and war broke out. By the following year Japan had secured control of the peninsula.

Korea declared neutrality at the outbreak of war between Japan and Russia in 1904 but was pressured by Japan into allowing use of Korean territory. Japan achieved victory in 1905 and made Korea a protectorate. An unsuccessful appeal to the international peace conference at The Hague further undermined relations between Japan and Korea. Anti-Japanese guerrilla fighters in the southern provinces were active during 1908–09 but were crushed the following year when Korea was annexed by Japan.

Japan established a government in Korea and implemented a programme designed to supplant the Korean identity. There were restrictions on freedom of speech, press and assembly and the language and history of Japan was taught in schools at the expense of those of Korea. Many Koreans were dispossessed of their land as Japan built new transport and communications infrastructures. When the Japanese brutally suppressed a 2m.-strong demonstration in 1919, independence leaders established a provisional government in Shanghai and named Syngman Rhee as president. Hoping to calm dissent, Japan lifted certain press restrictions and replaced the gendarmerie with an ordinary police force, but uncompromising colonial rule remained in place.

Korea became a market for Japanese goods and attracted much capital investment but at the expense of agriculture, leading to a long-term shortage of rice. Tokyo reimposed military rule in 1931 when war broke out between Japan and China and attempted to quash all manifestations of a separate Korean identity over the following decade. Magazines, newspapers and academic organisations operating in the Korean language were banned.

Hundreds of thousands of Koreans were made to fight in the Japanese army, or work in Japanese mines and factories in order to support Japan's military efforts during the Second World War. The provisional government, having moved from Shanghai to Chungking in southwest China, declared war on Japan in Dec. 1941. An army of Korean resistance fighters joined the Allied forces in China and fought with them until the Japanese surrender in 1945.

Korea Divided

Korea was promised independence by China, Britain and the USA at the Cairo conference of 1943 but at the end of the war, after Japan's collapse, Korea was divided in two along the 38° Parallel. Initially the USA and the USSR had agreed informally to a four-way power share in Korea, involving Britain and the Republic of China. However, in order to hasten a Japanese surrender, US troops controlled the south of the country while the USSR took command of the north. The Soviet forces helped to establish a Communist-led provisional government under Kim Il Sung. As relations between the USA and the Soviet Union worsened, trade ceased between the two zones, causing economic hardship because industry was concentrated in the north and agriculture in the south.

In Sept. 1947 the United Nations urged elections in both sectors. However, a commission to oversee voting in the North was denied entry by Soviet troops. Rhee was elected in the South while the North appointed Kim Il Sung as leader. In 1948 the southern Republic of Korea (with Seoul as the capital) and the northern Democratic People's Republic of Korea (with Pyongyang as capital) came into being.

Soviet and US troops left the peninsula in 1949 and war broke out between the North and South in June 1950. A US-led UN force under Gen. Douglas MacArthur entered South Korea and pushed back the North Korean forces. The UN pressed on into North Korea and established a commission for the reunification and rehabilitation of Korea. China, which at that point had no representation in the United Nations, entered the war and contributed 1·2m. troops to the North Korean side. Peace negotiations began in 1951 and a new international boundary and demilitarized zone were declared in 1953. The USA offered South Korea financial support and signed a mutual security pact with Rhee, who had been reluctant to accept the division of the country. The issue of prisoner returns, particularly of North Koreans unwilling to return to the communist state, remained a point of contention. The war left 4m. people dead or injured.

Traditionally an agricultural region, South Korea faced severe economic problems after partition. Limited resources, war damage and a flood of refugees from North Korea all pointed towards economic disaster, and the country became dependent on foreign aid, particularly from the USA.

The authoritarian rule of President Rhee, marred by corruption, received widespread condemnation. The elections of 1960 were blighted by violence and fraud and when the police shot 125 students during a demonstration, the government was forced to step down. Rhee was exiled. Subsequent leaders failed to solve the country's problems and in May 1961 Gen. Park Chung-hee led a military coup. As leader of the Democratic Republican Party, he was elected president in 1963, 1967 and, following a constitutional amendment to allow a third term in office, 1971.

Park's government was powerful and efficient, reviving the economy through the development of manufacturing for export and attracting increased foreign investment, especially from America. In 1972 Park proclaimed martial law and abolished the national assembly. In 1979 he was assassinated and the country collapsed into chaos. Chun Doo-hwan became leader in 1980 in another military coup and, as leader of the Democratic Justice Party (DJP), revived the national assembly.

In 1983 a South Korean passenger jet strayed into Soviet airspace and was shot down with the loss of all 269 lives on board. The incident increased tensions between Moscow and Washington.

Popular dissatisfaction with the South Korean government grew throughout the decade and a new constitution in 1987 stipulated that the president be elected by popular vote and his term of office reduced to five years.

Roh Tae-woo was elected president in 1988 as leader of the DJP and later of the Democratic Liberal Party. Fighting rising inflation, he established diplomatic relations with China and the Soviet Union and developed a better relationship with opposition parties in his own country. North and South Korea met several times during the 1980s in a bid to improve relations. In 1991 the two countries signed a treaty of non-aggression, with each country promising not to interfere in the internal affairs of the other.

In 1992 Kim Young-sam, the former opposition leader who had merged his party with Roh's, became the first civilian to be elected president since the Korean War. He launched an anti-corruption campaign and continued to pursue closer relations with North Korea. During the financial crisis that affected East Asia in 1997 South Korea was forced to ask the International Monetary Fund for help, though it largely avoided long-term economic damage and remains one of Asia's most affluent countries.

In Dec. 1997 Kim Dae-jung, a pro-democracy dissident during the years of military dictatorship, was elected president. Kim forged a 'sunshine policy' aimed at closer ties with the North and received the Nobel peace prize for his efforts. The two Koreas subsequently undertook a series of joint commercial and infrastructural projects. Constitutionally disqualified from standing for the presidency again in 2002, Kim was replaced by Roh Moo-hyun who continued the 'sunshine policy', despite North Korea's deteriorating relationship with the USA.

In Oct. 2007 South and North Korea jointly called for a permanent peace on the peninsula to replace the armistice in place since the end of the Korean War. However, relations soon deteriorated and at the end of 2008 Pyongyang accused Seoul of fostering hostility. The South accused the North of testing long-range missile technology in April 2009 and a month later North Korea walked out of international talks aimed at ending its nuclear programme. In May 2010 an international report concluded that the sinking of a South Korean ship two months earlier had been the responsibility of North Korea, prompting a collapse in trading and diplomatic relations between the two countries.

TERRITORY AND POPULATION

South Korea is bounded in the north by the demilitarized zone (separating it from North Korea), east by the Sea of Japan (East Sea), south by the Korea Strait (separating it from Japan) and west by the Yellow Sea. The area is 99,461 sq. km. The population at the census of 1 Nov. 2010 was 48,580,293; density, 488·4 per sq. km (one of the highest in the world). In 2011 the urban population was 83·3%.

The UN gives a projected population for 2015 of 49·12m.

The official language is Korean. In July 2000 the Korean government introduced a new Romanization System for the Korean Language to romanize Korean words into English.

There are nine provinces (do) and seven metropolitan cities with provincial status. Area and population in 2010:

Province	Area (in sq. km)	Population (in 1,000)
Gyeonggi	10,135	11,379
Gyeongsangnam	10,516	3,160
Gyeongsangbuk	19,024	2,600
Chungcheongnam	8,586	2,082
Jeollabuk	8,050	1,777
Jeollanam	11,987	1,741
Chungcheongbuk	7,432	1,512
Gangwon	16,572	1,472
Jeju	1,846	532
Seoul (city)	606	9,794

Province	Area (in sq. km)	Population (in 1,000)
Busan (city)	760	3,415
Incheon (city)	965	2,663
Daegu (city)	886	2,446
Daejeon (city)	540	1,502
Gwangju (city)	501	1,476
Ulsan (city)	1,056	1,083

Cities with over 500,000 inhabitants (census 2010):

Seoul	9,794,304	Suwon	1,071,913	Cheongju	666,924
Busan	3,414,950	Changwon	1,058,021	Jeonju	649,728
Incheon	2,662,509	Seongnam	949,964	Anyang	602,122
Daegu	2,446,418	Goyang	905,076	Cheonan	574,623
Daejeon	1,501,859	Yongin	856,765	Namyangju	529,898
Gwangju	1,475,745	Bucheon	853,039	Pohang	511,390
Ulsan	1,082,567	Ansan	728,775		

SOCIAL STATISTICS

2008: births, 465,900; deaths, 246,100; marriages, 327,700; divorces, 116,500. Rates per 1,000 population in 2008: birth, 9·7; death, 5·1; marriage, 6·8; divorce, 2·4. In 2006 only 1·5% of births were outside marriage, one of the lowest rates in the world. Suicides numbered 10,688 in 2006 (21·9 per 100,000 people). Expectation of life at birth, 2007, 82·4 years for females and 75·8 for males. Life expectancy had been 47 in 1955 and 62 in 1971. Infant mortality, 2010, four per 1,000 live births; fertility rate, 2008, 1·2 births per woman (the joint lowest rate in the world). Annual population growth rate, 2005–10, 0·5%. In 2009 the average age of first marriage was 31·6 for men and 28·7 for women.

South Korea has one of the most rapidly ageing populations in the world, partly owing to an ever-decreasing birth rate. In 2009, 10·7% of the population were over 65, up from 2·9% in 1960. There were 16·92m. households in 2009, with on average 2·8 members per household. According to the UN Human Development Report 2009, South Korea has an emigration rate of 3·1%; North America is the main destination, with 50·3% of South Korean migrants living there. Within South Korea, there are 551,200 foreign migrants, representing 1·2% of the total population.

CLIMATE

The country experiences continental temperate conditions. Rainfall is concentrated in the period April to Sept. and ranges from 40" (1,020 mm) to 60" (1,520 mm). Busan, Jan. 36°F (2·2°C), July 76°F (24·4°C). Annual rainfall 56" (1,407 mm). Seoul, Jan. 23°F (–5°C), July 77°F (25°C). Annual rainfall 50" (1,250 mm).

CONSTITUTION AND GOVERNMENT

The 1988 constitution provides for a *President*, directly elected for a single five-year term, who appoints the members of the *State Council* and heads it, and for a *National Assembly* (*Gukhoe*), currently of 299 members, directly elected for four years (243 from constituencies and 56 from party lists in proportion to the overall vote). The current constitution created the Sixth Republic. The minimum voting age is 20.

National Anthem

'Aegukga' ('A Song of Love for the Country'); words anonymous, tune by Ahn Eaktay.

GOVERNMENT CHRONOLOGY

Heads of State of South Korea since 1948. (DJP = Democratic Justice Party; DLP = Democratic Liberal Party; DP = Democratic Party; DRP = Democratic Republican Party; GNP = Grand National Party; LP = Liberal Party; MDP = Millennium Democratic Party; NCNP = National Congress for New Politics; NDP = New Democratic Party; NFP = Saenuri Party; NKP = New Korea Party; UD = Uri Party)

Presidents

1948–60	LP	Syngman Rhee
1960–62	DP, NDP	Yun Po-sun

Chairman of the Supreme Council for National Reconstruction

1962–63	military	Park Chung-hee

Presidents

1963–79	DRP	Park Chung-hee
1979–80	DRP	Choi Kyu-hah
1980–88	military, DJP	Chun Doo-hwan
1988–93	DJP, DLP	Roh Tae-woo
1993–98	DLP, NKP	Kim Young-sam
1998–2003	NCNP, MDP	Kim Dae-jung
2003–08	MDP, UD	Roh Moo-hyun
2008–13	GNP, NFP	Lee Myung-bak
2013–	NFP	Park Geun-hye

RECENT ELECTIONS

Presidential elections were held on 19 Dec. 2012. Park Geun-hye of the Saenuri Party (New Frontier Party/NFP) won with 51·6% of votes cast, ahead of Moon Jae-in of the Democratic United Party (DUP) with 48·0% and four other candidates. Turnout was 75·8%.

Elections to the National Assembly were held on 11 April 2012. Turnout was 54·3%. The Saenuri Party, formerly the Grand National Party, won 152 out of 300 seats with 42·8% of votes cast; the DUP 127 with 36·5%; the Unified Progressive Party (UPP) 13 with 10·3%; and the Liberty Forward Party (LFP) 5 with 3·2%. Three seats went to independents.

CURRENT GOVERNMENT

President: Park Geun-hye; b. 1952 (Saenuri Party; sworn in 25 Feb. 2013).

In March 2013 the cabinet comprised:

Prime Minister: Chung Hong-won; b. 1944 (Saenuri Party; since 26 Feb. 2013).

Minister of Culture, Sport and Tourism: Yoo Jin-ryong. *Education, Science and Technology:* Seo Nam-soo. *Employment and Labour:* Phang Ha-nam. *Environment:* Yoon Seong-kyu. *Food, Agriculture, Forestry and Fisheries:* Lee Dong-phil. *Foreign Affairs and Trade:* Yun Byung-se. *Gender Equality and Family:* Cho Yoon-sun. *Health and Welfare:* Chin Young. *Justice:* Hwang Kyo-ahn. *Knowledge Economy:* Yoon Sang-jick. *Land, Transport and Maritime Affairs:* Suh Seung-hwan. *National Defence:* Kim Kwan-jin. *Public Administration and Security:* Yoo Jeong-bok. *Strategy and Finance:* Bahk Jae-wan. *Unification:* Yoo Kil-jae.

National Assembly Speaker: Kang Chang-hee.

Government Website: http://www.korea.net

CURRENT LEADERS

Park Geun-hye

Position
President

Introduction
Park Geun-hye became the country's first female president on 25 Feb. 2013 at the head of one of the world's most male-dominated governments. She must contend with the legacy of her late father, Park Chung-hee, a former president both revered as the driving force behind South Korea's economic miracle and condemned for his suppression of opposition.

Early Life
Park Geun-hye was born on 2 Feb. 1952 in Jung-gu, Daegu. She graduated in electronic engineering from Seoul's Sogang University in 1974. She then studied at the University of Grenoble, France, before returning to Seoul following the death of her mother on 15 Aug. 1974 in a botched assassination attempt on Park Chung-hee. Park stepped into the role of first lady until

26 Oct. 1979 when her father was assassinated by his intelligence chief.

Following her father's death, Park retreated from politics and served on the boards of various charities and educational institutions. In 1997 she joined the Grand National Party (GNP) and won a seat in the National Assembly after winning a by-election for Dalseong, Daegu, in 1998. In the run-up to the 2004 general election the GNP, beleaguered by scandals, appointed Park as leader. Despite its lacklustre performance at the election, under Park's leadership the GNP recovered in the following years by winning all the by-elections it contested.

In 2006 Park stepped down from the party leadership in a bid to become the GNP candidate at the following year's presidential election. She was narrowly defeated at the party primaries by Lee Myung-bak, who went on to become president. In Dec. 2011 Park was chosen to chair the GNP's emergency committee, overseeing the party's name change to Saenuri (New Frontier) to signal a fresh start in the face of growing voter dissatisfaction. Stepping back from the party in May 2012 to run for the presidency, she won the election on 19 Dec. 2012 with 51·6% of the vote.

Career in Office
Park's centrist agenda aims to continue economic growth, close the wealth gap, increase spending on social welfare and reduce unemployment, especially among the young. She has pledged to improve relations with North Korea on condition that the North abandons its nuclear programme.

DEFENCE

Peacetime operational control, which had been transferred to the United Nations Command (UNC) under a US general in July 1950 after the outbreak of the Korean War, was restored to South Korea on 1 Dec. 1994. In the event of a new crisis, operational control over the Korean armed forces will revert to the Combined Forces Command (CFC). However, in June 2010 it was agreed that in 2015 South Korea would resume wartime command of its military. Conscription is 21 months in the Army, 23 months in the Navy and 24 months in the Air Force. In Sept. 2007 it was announced that the length of conscription will be gradually reduced and that conscientious objectors will be allowed to choose community service in place of military service. In 2004 the USA and South Korea agreed to the redeployment of 12,500 US personnel in three phases that would continue until 2008. In April 2008 the number of troops had been reduced to 28,000 (mainly army and air force personnel) from 37,000 in 2002.

Defence expenditure in 2008 totalled US$24,172m. (US$501 per capita). In 2007 defence spending represented 2·7% of GDP. In the period 2007–11 South Korea's spending on major conventional weapons, at US$7·1bn., was the second highest behind India.

Army

Strength (2007) 560,000 (140,000 conscripts). Paramilitary Civilian Defence Corps, 3·5m. The armed forces reserves numbered 4·5m.

Navy

In 2007 the Navy had a substantial force of 63,000 (around 19,000 conscripts), including 28,000 marine corps troops. In 2007 the fleet included 12 submarines, seven destroyers, nine frigates, 28 corvettes and around 75 patrol and coastal combatants. Naval Aviation operated eight combat-capable aircraft. The fleet head-quarters is at Jinhae.

Air Force

In 2007 the Air Force had a strength of 64,000 men and 555 combat-capable aircraft (F-4s, F-5s, F-15s and F-16s).

INTERNATIONAL RELATIONS

Defections to South Korea from North Korea totalled a record 2,927 in 2009 (1,387 in 2005, 312 in 2000 and 9 in 1990).

The aim of Korea's foreign policy is to secure international support for peace and stability in Northeast Asia, including a means to reunify the Korean Peninsula without confrontation.

ECONOMY

Agriculture accounted for 3% of GDP in 2008, industry 37% and services 60%.

In 2011 South Korea gave US$1·3bn. in international aid. In terms of a percentage of GNI, however, it was one of the least generous major industrialized countries, giving just 0·12%.

Overview

South Korea has seen rapid economic development since the 1950s, when it was ranked amongst the poorest nations. It is now among the top 15 richest. Economic convergence with the developed world came only in the late 1980s. With few natural resources, imports were focused on raw materials and technology at the expense of consumer goods. Furthermore, domestic savings and investment were encouraged over consumption, resulting in an extremely high fixed-investment expenditure share of GDP.

Korea's development model was based on export-oriented industrialization. Export production was dominated by *chaebol* (conglomerates such as Samsung, Hyundai and LG), controlled by founding families with close government ties. Subsidized businesses (and the careers of the associated bureaucrats) were contingent on export success. Textile manufacturing was the first internationally competitive sector but the country is now competitive in electronics, automobiles, shipbuilding, chemicals and steel. According to the World Bank, Korea ranked eighth out of 183 economies for the ease of doing business in 2012, up from 15th in 2011.

The 1997 Asian financial crisis exposed weaknesses in the development model. High debt-to-equity ratios, heavy foreign borrowing and an undisciplined financial sector with a significant amount of non-performing loans weighed heavily on the economy. Eleven of the 30 largest *chaebol* that existed in June 1997 collapsed between July 1997 and June 1999. Daewoo, a once-major *chaebol*, was dismantled by the government in 1999 following a massive bankruptcy. Since 1997 government reforms and large-scale bankruptcies have changed the role of the *chaebol*. IMF-designed reforms to strengthen competition and the financial sector helped growth rebound in 1999. Strong domestic demand, spurred by a credit card boom, helped weather the global slowdown in 2001–02 but a credit card crisis hit in 2003, with 3·7m. people defaulting by the end of the year. Although household consumption accelerated in 2005, many households remained heavily indebted.

Reliance on oil makes the economy vulnerable to fluctuations in the oil price. China, a major export and FDI destination, poses an increasingly competitive threat to many of Korea's leading industries. Nonetheless, real GDP performance was positive before the global financial crisis, with growth in 2006 (at over 5%) at its highest level for four years as a result of robust exports and strong domestic demand.

In 2008 exports slumped and capital left the country at a higher rate than during the Asian crisis. GDP growth declined to 2·3% that year and 0·3% in 2009 (the slowest rate since 1998), before recovering strongly in 2010—one of the earliest bounce-backs in the OECD. The economy expanded by more than 6% in 2010 and the unemployment rate reverted to pre-crisis levels in 2011.

Since 2011 growth has moderated in line with global economic developments, including the euro area crisis. Household debt at 125% of household disposable income remains a drag on private consumption. Responding to a weaker-than-expected outlook and low headline inflation, a modest stimulus package was adopted in June 2012 and the Bank of Korea cut its base rate by 0·25% in July 2012 after putting it on hold for a year.

Korea was the world's seventh largest holder of international reserves in 2011 and its banks are well-capitalized, with low levels of non-performing loans. The United States–Korea Free Trade Agreement (KORUS FTA) was signed in 2007 and approved by the US Congress in Oct. 2011 and the Korean National Assembly the following month. It came into force in March 2012.

The OECD has highlighted service sector reform as essential for further growth, especially since the country will soon be faced with a shrinking labour force. Population ageing poses risks to future growth as the old age dependency ratio is expected to increase to 65% by 2050, potentially a major strain on fiscal resources. The government has committed to medium-term fiscal consolidation in a bid to address the challenges of an ageing population.

Currency

The unit of currency is the *won* (KRW). Inflation rates (based on OECD statistics):

2002	2003	2004	2005	2006	2007	2008	2009	2010	2011
2·8%	3·5%	3·6%	2·8%	2·2%	2·5%	4·7%	2·8%	2·9%	4·0%

Foreign exchange reserves were US$240,915m. in Aug. 2009 (US$51,963m. in 1998) and gold reserves 463,000 troy oz. Total money supply in June 2009 was 103,242bn. won.

Budget

In 2007 central government revenue was 236,006bn. won and expenditure 196,369bn. won. Principal sources of revenue in 2007: taxes on income, profits and capital gains, 74,273bn. won; taxes on goods and services, 59,835bn. won. Main items of expenditure by economic type in 2007: grants, 89,203bn. won; social benefits, 30,801bn. won.

VAT is 10%.

Performance

Real GDP growth rates (based on OECD statistics):

2002	2003	2004	2005	2006	2007	2008	2009	2010	2011
7·2%	2·8%	4·6%	4·0%	5·2%	5·1%	2·3%	0·3%	6·3%	3·6%

Total GDP in 2011 was US$1,116·2bn.

Banking and Finance

The central bank and bank of issue is the Bank of Korea (*Governor*, Kim Choong-soo). There are four major financial groups—Woori Finance Holdings (assets in April 2010 of US$230·52bn.), KB Financial Group (US$212·40bn.), Shinhan Financial Group (US$208·99bn.) and Hana Financial Group (US$125·57bn.). Specialized banks include the National Agricultural Co-operative Federation (NACF), Korea Development Bank and Industrial Bank of Korea. Foreign-owned banks operating in South Korea include ABN AMRO, Bank of America, Bank of Tokyo-Mitsubishi UFJ, Deutsche Bank and Hongkong and Shanghai Banking Corporation. The use of real names in financial dealings has been required since 1994.

Gross external debt amounted to US$418,607m. in June 2012, up from US$399,249m. in June 2011.

South Korea has started to open up once protected industries to foreign ownership, and in 2011 attracted US$4·7bn. in foreign direct investment.

There is a stock exchange in Seoul.

ENERGY AND NATURAL RESOURCES

Environment

South Korea's carbon dioxide emissions from the consumption and flaring of fossil fuels in 2008 were the equivalent of 11·2 tonnes per capita.

Electricity

Installed capacity in 2007 was 73·4m. kW. Electricity generated (2007) was 427,317m. kWh (thermal, 278,891m. kWh; nuclear, 142,937m. kWh; hydro-electric, 5,042m. kWh; geothermal,

447m. kWh). There were 20 nuclear reactors in use in 2008. Consumption per capita in 2007 was 8,776 kWh.

Oil and Gas

South Korea is the world's fifth largest importer of crude oil and second largest importer of liquefied natural gas (LNG). Domestic oil and gas production is negligible. In 2009 total gross petroleum imports amounted to 3·1m. bbls per day; daily consumption was 2·2m. bbls. Three of the world's ten largest oil refineries are in South Korea. In 2009 consumption of natural gas totalled 34·09bn. cu. metres. South Korea has no international gas pipeline connections and all imports are delivered by LNG tankers.

Minerals

Output, 2008, included (in tonnes): limestone, 82·25m.; anthracite coal, 2·77m.; salt, 0·38m.; iron ore, 0·37m.; silver, 1,462; gold, 38.

Agriculture

Cultivated land was 1·74m. ha. in 2009, of which 1·01m. ha. were rice paddies; the area of dry field was 727,000 ha. As at Dec. 2009 the farming population was 3·12m. and there were 1·20m. farm households. 1·76m. people were employed in agriculture, forestry and fishing in Sept. 2010. There were 243,662 tractors in 2007. Food and live animals accounted for 0·7% of exports and 3·7% of imports in 2007.

Production (2008, in 1,000 tonnes): rice, 6,919; cabbage, 2,902; onions, 1,540; watermelons, 857; tangerines, mandarins and clementines, 636; potatoes, 605; apples, 471; pears, 471; persimmons, 431; tomatoes, 408; chillies and green peppers, 386; cucumbers and gherkins, 384; garlic, 375; grapes, 334.

Livestock in 2008 (in 1,000): pigs, 9,153; cattle, 2,894; goats, 266; chickens, 120,000.

Livestock products in 2008 (in 1,000 tonnes): pork 1,056; beef, 246; poultry meat, 542; milk, 2,204; eggs, 595.

Forestry

Forest area was 6·22m. ha. in 2010 (63% of the land area). Of the total area under forests 48% was primary forest in 2010, 23% other naturally regenerated forest and 29% planted forest. Timber production was 5·15m. cu. metres in 2007.

Fisheries

The fish catch totalled 3·18m. tonnes in 2009 (coastal aquaculture, 1·31m. tonnes; inshore-offshore fisheries, 1·23m. tonnes; deep sea, 604,880 tonnes; inland waters, 30,060 tonnes), worth 6,910·6bn. won. Imports of fishery commodities were valued at US$2,928m. in 2008 and exports at US$1,287m.

INDUSTRY

The leading companies by market capitalization in South Korea in March 2012 were: Samsung Electronics Company Ltd (US$181·8bn.); Hyundai Motor (US$45·3bn.); and Pohang Iron and Steel Company (POSCO), US$29·2bn.

Manufacturing industry is concentrated primarily on oil, petrochemicals, chemical fibres, construction, iron and steel, mobile phones, cement, machinery, chips, shipbuilding, automobiles and electronics. Tobacco manufacture is a semi-government monopoly. Industry is dominated by giant conglomerates (*chaebol*). There were 3·19m. businesses in 2004, of which 263,128 were incorporated. 878,294 businesses were in wholesale and retail trades, 643,773 in hotels and restaurants, 331,458 in transport and communications and 328,338 in manufacturing. The leading *chaebol* are Samsung (with assets of US$252·5bn. and revenue of US$173·4bn. in 2008), LG, Hyundai and SK Group.

Production in 2005: petroleum products, 922·9m. bbls; mobile phones, 201·2m. units; refrigerators, 6·92m. units; TV sets, 5·85m. units; cars, 3·70m. units; cigarettes, 107·2bn. units. Production (2005) in 1,000 tonnes: cement, 51,391; crude steel, 47,770;

distillate fuel oil (2007), 34,314; residual fuel oil (2007), 27,363; pig iron (2009), 27,475; beer (2004), 2,016m. litres.

Shipbuilding orders totalled 21·96m. GT in 2005.

Labour

In Sept. 2010 the population of working age was 40·68m.; the economically active population was 24·91m. (14·51m. males and 10·40m. females) including 16·44m. persons employed in services, 5·85m. in construction, manufacturing and mining, and 1·76m. in agriculture, fisheries and forestry. 5·61m. persons were self-employed in Sept. 2010. Unemployment was 3·0% in Dec. 2012—one of the lowest rates in the industrialized world. Long-term unemployment is particularly low, with only 0·3% of the labour force in 2010 having been out of work for more than a year. An annual legal minimum wage is set by the *Minimum Wage Act* (enforced from 1988), which applies to all industries. In Jan. 2009 it was increased to 4,000 won per hour and 32,000 won per eight-hour day.

In 2008 the average monthly wage was 2·72m. won and the working week averaged 39·4 hours. In July 2004 the working week in the civil service, for financial and insurance firms, and for employers with 1,000 or more staff was reduced from 44 to 40 hours. A five-day working week for smaller companies has being gradually phased in; small businesses with between five and 20 employees did not have to introduce the shorter working week until July 2011. Workers in South Korea put in among the longest hours in the industrialized world. In 2008 full- and part-time workers put in an average of 2,256 hours (although this has been declining every year since 2000 when the average was 2,520 hours).

INTERNATIONAL TRADE

Total external foreign debt was US$381,100m. in 2008. In May 1998 the government removed restrictions on foreign investment in the Korean stock market. It also began to allow foreign businesses to engage in mergers and acquisitions. From July 1998 foreigners were allowed to buy plots of land for both business and non-business purposes. Since Aug. 1990 South Korean businesses and individuals have been permitted to make investments and set up branch offices in North Korea, on an approval basis.

Imports and Exports

Imports (c.i.f.) and exports (f.o.b.) for calendar years in US$1m.:

	2003	2004	2005	2006	2007
Imports	178,825·9	224,460·9	261,235·6	309,379·5	356,841·0
Exports	193,817·3	253,844·6	284,418·2	325,457·2	371,477·1

Both imports and exports almost doubled between 2003 and 2007.

Leading import sources in 2007 were: China, 17·7%; Japan, 15·8%; USA, 10·5%; Saudi Arabia, 5·9%. The principal export markets in 2007 were: China, 22·1%; USA, 12·4%; Japan, 7·1%; Hong Kong, 5·0%.

In 2004 machinery and transport equipment accounted for 33·6% of imports and 63·0% of exports; chemicals, manufactured goods classified chiefly by material and miscellaneous manufactured articles 33·4% of imports and 30·7% of exports; mineral fuels, lubricants and related materials 22·4% of imports and 4·1% of exports; inedible crude materials (except fuels), and animal and vegetable oil and fats 6·3% of imports and 1·0% of exports; and food, live animals, beverages and tobacco 4·4% of imports and 1·2% of exports.

Rice imports were prohibited until 1994, but following the GATT Uruguay Round the rice market opened to foreign imports in 1995.

COMMUNICATIONS

Roads

In 2007 there were 102,061 km of roads, comprising 3,103 km of motorways, 14,225 km of highways and main roads and 84,733 km of secondary roads; 77·6% of roads (79,189 km) were paved. In 2006, 97,854m. passenger-km were travelled by road and 12,545m. tonne-km of freight were moved.

In 2007 motor vehicles in use included 12,020,700 passenger cars, 4,189,000 vans and lorries, 182,100 buses and coaches, and 1,821,300 motorcycles and mopeds. In 2007 there were 6,166 fatalities as a result of road accidents (9,353 in 2000).

The first of two planned cross-border roads between the two Koreas opened in Feb. 2003.

Rail

In 2009 Korail's system totalled 3,380 km of 1,435 mm gauge (including 240 km of high speed railways). In 2009 passenger-km travelled came to 31·3bn. and freight tonne-km to 9·3bn. In June 2000 it was agreed to start consultations to restore the railway from Seoul to Sinuiju, on the North Korean/Chinese border, by rebuilding a 12 km long stretch from Munsan, in South Korea, to Jangdan, on the South Korean/North Korean border, and an 8 km long stretch in North Korea.

Two passenger trains crossed the border between North and South Korea on 17 May 2007 (one northbound and one southbound), completing the first cross-border journey in more than 50 years. Freight services between the two Koreas were resumed in Dec. 2007 but usage is very limited.

There is an extensive metro system in Seoul and smaller ones in Busan, Daegu, Daejeon, Gwangju and Incheon.

Civil Aviation

There are six international airports in South Korea: at Seoul (Incheon), Busan (Gimhae), Daegu, Jeju, Yangyang and Cheongju. Incheon airport, 50 km to the west of Seoul and built on reclaimed land made up of four small islands, opened in March 2001 and is the largest airport in Asia. It replaced Gimpo Airport as Seoul's International Airport; Gimpo remains open for domestic flights and is the second busiest airport with 14·3m. passengers in 2008. Incheon handled 30·0m. passengers in 2008, while Jeju handled 12·4m. and Busan 7·2m.

The national carrier is Korean Air, which in June 2009 operated flights to 101 cities in 39 countries. In 2005 Korean Air carried 22·0m. passengers (11·0m. on international flights); passenger-km totalled 50·3bn. The other main Korean carrier is Asiana Airlines, which was established in 1988.

Shipping

In 2005 there were 52 ports (28 for international trade), including Busan, Incheon, Gunsan, Mokpo, Yeosu, Pohang, Donghae, Jeju, Masan, Ulsan, Daesan and Kwangyang. In Jan. 2009 there were 1,128 ships of 300 GT or over registered, totalling 13·41m. GT. Of the 1,128 vessels registered, 350 were general cargo ships, 260 oil tankers, 242 bulk carriers, 81 container ships, 80 passenger ships, 67 chemical tankers and 48 liquid gas tankers. The busiest port is Busan, which was visited by 48,343 vessels of 354,350,000 GRT in 2005. Cargo handled in 2005 totalled 217,217,000 tonnes (112,103,000 tonnes loaded and 105,114,000 tonnes discharged).

In 2005, 11,099,554 domestic passengers and 2,104,939 international passengers took ferries and other ocean-going vessels. There were 7,119 registered vessels accounting for a tonnage of 10,068,379.

Telecommunications

In 2008 there were 21,325,000 main (fixed) telephone lines. In the same year mobile phone subscribers numbered 45,607,000 (947·1 per 1,000 persons). The largest operator, SK Telecom, has 46% of the market share, ahead of KT, with 29%.

There were 36·8m. internet users in 2008. In Dec. 2010 there were 89·8 wireless broadband subscribers per 100 inhabitants and 34·0 fixed broadband subscribers per 100. In March 2012 there were 6·4m. Facebook users.

SOCIAL INSTITUTIONS

Justice

Judicial power is vested in the Supreme Court, High Courts, District Courts and Family Court, as well as the Administrative Court and Patent Court. The single six-year term Chief Justice is appointed by the President with the consent of the National Assembly. The other 13 Justices of the Supreme Court are appointed by the President with the consent of the National Assembly, upon the recommendation of the Chief Justice, for renewable six-year terms; the Chief Justice appoints other judges. There has been an unofficial moratorium on executions since 1998—the death penalty was last used in Dec. 1997. In Dec. 2009 there were 2,468 judges, 1,699 prosecutors and 11,016 registered private attorneys.

In 2008 there were 2·19m. recorded offences; the most serious offences (including murder, rape, theft and assault) numbered 439,000. The population in the 50 penal institutions in Jan. 2010 was 47,514 (98 per 100,000 of national population); 5·4% were females and 3·2% foreigners.

Education

The Korean education system consists of a six-year elementary school, a three-year middle school, a three-year high school and college and university (two to four years). Elementary education for 6–11 year olds and middle school education are compulsory. Mandatory middle school education began in 2002.

The total number of schools has increased sixfold from 3,000 in 1945 to 19,586 in 2005, with 11,934,863 enrolled students. In 2005 there were 8,275 kindergartens with 541,603 pupils and 31,033 teachers; 5,646 elementary schools with 4,023,806 pupils and 160,143 teachers; 2,010,704 pupils and 103,835 teachers at 2,935 middle schools; and 2,095 high schools with 1,762,896 pupils and 116,411 teachers. In 2005 there were 158 colleges and universities with 1,859,639 students and 49,200 teachers; 11 universities of education with 25,141 students and 798 teachers; 1,051 graduate schools with 282,225 students; and 18 industrial universities with 188,753 students and 2,658 teachers. In 2005, 5·6% of the population was enrolled in tertiary education, up from just 0·6% in 1970. Around 214,000 South Koreans were studying abroad in 2005.

In 2005 public expenditure on education came to 4·3% of GDP and 15·3% of total government spending. Total expenditure on tertiary education in 2005 was among the highest in the world at 2·4% of GDP. Private spending on education in 2005 came to 2·9% of GDP, the highest share of any industrialized country. The adult literacy rate is at least 99%.

According to the OECD's 2009 PISA (Programme for International Student Assessment) study, 15-year-olds in Korea rank first among OECD countries in reading and mathematics. The three-yearly study compares educational achievement of pupils in the major industrialized countries.

Health

In 2004 there were 282 general hospitals (with 117,323 beds), 25,346 other hospitals and clinics (189,044 beds), 9,303 oriental medical hospitals and clinics (9,585 beds) and 15,406 dental hospitals and clinics. In 2004 there were 81,998 physicians (587 people per doctor), 15,406 oriental medical doctors, 21,344 dentists, 8,626 midwives, 202,012 nurses and 53,492 pharmacists.

In 2008 South Korea spent 6·5% of its GDP on health, with public spending accounting for 55·3% of total expenditure on health and private spending 44·7%. According to a 2006 survey, 55·4% of men aged 25–64 smoked (down from 69·9% in 1999) but only 3·2% of women.

Welfare

In 2006, 12·8m. persons were covered by the National Pension System introduced in 1988. Employers and employees should make contributions amounting to 4·5% of the standard monthly income respectively, based on the employees' earned income. The Scheme covers old age pensions, disability pensions and survivors' pensions.

Under a system of unemployment insurance introduced in July 1995, workers laid off after working at least six months for a member employer are entitled to benefits averaging 50% of their previous wage for a period of 90 up to 240 days.

RELIGION

Traditionally, Koreans have lived under the influence of shamanism, Buddhism (introduced AD 372) and Confucianism, which was the official faith from 1392 to 1910. Catholic converts from China introduced Christianity in the 18th century, but a ban on Roman Catholicism was not lifted until 1882. The Anglican Church was introduced in 1890 and became an independent jurisdiction in 1993 under the Archbishop of Korea. Religious affiliations of the population at the census in 2005: Buddhism, 22·8%; Protestantism, 18·3%; Roman Catholicism, 10·9%; Confucianism, 0·2%; others, 1·3%; no religion, 46·5%. In Feb. 2013 there was one Roman Catholic cardinal.

CULTURE

World Heritage Sites

There are ten sites in South Korea that appear on the UNESCO World Heritage List. They are (with year entered on list): the Sokkuram Grotto and Pulguksa Temple (1995), the Temple of Haeinsa (1995), Chongmyo Shrine (1995), Changdeokgung Palace, Seoul (1997), Hwasong Fortress, Suwon (1997), the dolmens of Gochang, Hwasun and Ganghwa (2000), Gyeongju historic area (2000), Jeju volcanic island and lava tubes (2007), the Royal Tombs of the Joseon Dynasty (2009) and the Historic Villages of Hahoe and Yangdong (2010).

Press

There were 288 daily newspapers in 2008. With 13m. paid daily newspaper subscriptions in 2008, South Korea has the fifth highest newspaper circulation among developed countries. 37% of the population read a newspaper regularly in 2008. The most widely read dailies are *Chosun Ilbo* (average daily circulation of 2·3m. per issue in 2008), *JoongAng Ilbo* (2·2m. copies) and *Dong-A Ilbo* (2·1m. copies). In the 2010 *World Press Freedom Index*, compiled by Reporters Without Borders, South Korea was ranked 42nd equal out of 178 countries.

Tourism

A record 9,795,000 foreign nationals visited South Korea in 2011 (up from 8,798,000 in 2010 and 6,023,000 in 2005). The leading countries of origin of non-resident tourists in 2011 were: Japan (3,289,000), mainland China (2,220,000), the USA (662,000), Taiwan (428,000) and the Philippines (337,000). 12,694,000 South Koreans travelled abroad in 2011 (up from 12,488,000 in 2010

and 10,080,000 in 2005). In Nov. 1998 the first South Korean tourists to visit North Korea went on a cruise and tour organized by the South Korean firm Hyundai.

Festivals

Korean New Year or Seollal is celebrated on the first day of the lunar calendar and generally falls on the same date as the Chinese New Year. Other traditional festivals include the festival for the first full moon of the year known as Daeboreum; Dano or Suritnal (spring festival) which is held on the fifth day of the fifth month of the lunar calendar; and Chuseok (harvest festival) on the 15th day of the eighth month. The biggest music festivals are the Tongyeong International Music Festival (March), specializing in Western and Asian classical music, and the Pentaport Rock Festival in Incheon (July).

DIPLOMATIC REPRESENTATIVES

Of the Republic of Korea in the United Kingdom (60 Buckingham Gate, London, SW1E 6AJ)
Ambassador: Park Suk-hwan.

Of the United Kingdom in the Republic of Korea (Taepyeongno 40, 4 Jeong-dong, Jung-gu 100-120, Seoul)
Ambassador: Scott Wightman, CMG.

Of the Republic of Korea in the USA (2450 Massachusetts Ave., NW, Washington, D.C., 20008)
Ambassador: Choi Young-jin.

Of the USA in the Republic of Korea (32 Sejongno, Jongno-gu 110-710, Seoul)
Ambassador: Sung Y. Kim.

Of the Republic of Korea to the United Nations
Ambassador: Kim Sook.

Of the Republic of Korea to the European Union
Ambassador: Kim Chang-beom.

FURTHER READING

National Bureau of Statistics. *Korea Statistical Yearbook*
Bank of Korea. *Economic Statistics Yearbook*

Castley, R., *Korea's Economic Miracle.* 1997
Cumings, B., *Korea's Place in the Sun: A Modern History.* 1997
Kang, M.-H., *The Korean Business Conglomerate: Chaebol Then and Now.* 1996
Kim, D.-H. and Tat, Y.-K. (eds.) *The Korean Peninsula in Transition.* 1997
Kim, Myung Oak and Jaffe, Sam, *The New Korea: An Inside Look at South Korea's Economic Rise.* 2010
Lie, John, *Han Unbound: The Political Economy of South Korea.* 2000
Simons, G., *Korea: the Search for Sovereignty.* 1995
Smith, H., *Industry Policy in Taiwan and Korea in the 1980s.* 2000
Song, P.-N., *The Rise of the Korean Economy.* 3rd ed. 2003
Tennant, R., *A History of Korea.* 1996

National Statistical Office: Statistics Korea, Government Complex Daejeon, 139 Seonsaro, Seo-gu, Daejeon 302-701.
Website: http://kostat.go.kr

KUWAIT

© Research Machines plc 2006

Dowlat al Kuwait
(State of Kuwait)

Capital: Kuwait
Population projection, 2015: 3·09m.
GNI per capita, 2011: (PPP$) 47,926
HDI/world rank: 0·760/63
Internet domain extension: .kw

KEY HISTORICAL EVENTS

The ruling dynasty was founded by Sheikh Sabah al-Awwal, who ruled from 1756 to 1772. In 1899 Sheikh Mubarak concluded a treaty with Great Britain wherein, in return for the assurance of British protection, he undertook to support British interests. In 1914 the British Government recognized Kuwait as an independent government under British protection. On 19 June 1961 an agreement reaffirmed the independence and sovereignty of Kuwait and recognized the Government of Kuwait's responsibility for the conduct of internal and external affairs. On 2 Aug. 1990 Iraqi forces invaded the country. Following the expiry of the date set by the UN for the withdrawal of Iraqi forces, an air offensive was launched by coalition forces, followed by a land attack on 24 Feb. 1991. Iraqi forces were routed and Kuwait City was liberated on 26 Feb. On 10 Nov. 1994 Iraq recognized the independence and boundaries of Kuwait. In 2006 Sheikh Jaber, who had been Amir since 1977, died and was replaced by Sheikh Sabah.

TERRITORY AND POPULATION

Kuwait is bounded in the east by the Persian Gulf, north and west by Iraq and south and southwest by Saudi Arabia, with an area of 17,818 sq. km. In 1992–93 the UN Boundary Commission redefined Kuwait's border with Iraq, moving it slightly northwards in conformity with an agreement of 1932. The population at the 2005 census was 2,193,651; density, 123 per sq. km. At the 2005 census 60·8% were non-Kuwaitis. In 2011, 98·4% of the population were urban.

The UN gives a projected population for 2015 of 3·09m.

The country is divided into six governorates: the capital (comprising Kuwait City, Kuwait's nine islands and territorial and shared territorial waters) (2005 census population, 254,503); Farwaniya (620,935); Hawalli (482,127); Ahmadi (390,927); Jahra (269,915); Mubarak al-Kabir (175,244). The capital city is Kuwait, with a population in 2005 of 31,574. Other major cities are (2005 populations): Qalib ash-Shuyukh (179,425), as-Salimiya (145,314), Hawalli (104,901), Hitan-al-Janubiyah (92,475).

The Neutral Zone (Kuwait's share, 2,590 sq. km), jointly owned and administered by Kuwait and Saudi Arabia from 1922 to 1966, was partitioned between the two countries in May 1966, but the exploitation of the oil and other natural resources continues to be shared.

Over 78% speak Arabic, the official language. English is also used as a second language.

SOCIAL STATISTICS

Births, 2008, 54,571; deaths, 5,701. The birth rate in 2009 was 21·9 per 1,000 population and death rate 2·3 per 1,000 population (one of the lowest in the world). Expectation of life at birth, 2007, was 76·0 years for males and 79·8 years for females. Infant mortality, 2010, ten per 1,000 live births. Annual population growth rate, 2000–05, 2·8%.

Fertility rate, 2008, 2·2 births per woman. Kuwait has had one of the largest reductions in its fertility rate of any country in the world over the past 30 years, having had a rate of 7·2 births per woman in 1975.

CLIMATE

Kuwait has a dry, desert climate which is cool in winter but very hot and humid in summer. Rainfall is extremely light. Kuwait, Jan. 56°F (13·5°C), July 99°F (36·6°C). Annual rainfall 5" (125 mm).

CONSTITUTION AND GOVERNMENT

The ruler is HH Sheikh Sabah al-Ahmed al-Jaber al-Sabah, the 15th Amir of Kuwait, who succeeded on 29 Jan. 2006. *Crown Prince:* Sheikh Nawwaf al-Ahmed al-Sabah (b. 1937). The present constitution was approved and promulgated on 11 Nov. 1962.

In 1990 the *National Council* was established, consisting of 50 elected members and 25 appointed by the Amir. It was replaced by a *National Assembly* or *Majlis al-Umma* in 1992, consisting at the time of 50 elected members. It now has 65 members, of whom 50 are elected. The franchise extends to Kuwaiti citizens who are 21 or older, with the exception of those who are serving in the armed forces and citizens who have been naturalized for fewer than 30 years. In May 1999 the cabinet approved a draft law giving women the right to vote and run for parliament. However, in Dec. 1999 parliament rejected the bill allowing women to vote by a margin of 32 to 20. Women were granted the right to vote and run for office in May 2005 when parliament voted in favour of amending the election law by a margin of 35 votes to 23. Women were eligible to stand for election and to vote in a council by-election held in April 2006 and in the full parliamentary election held in June 2006.

Executive authority is vested in the *Council of Ministers*.

National Anthem

'Watanil Kuwait salemta lilmajdi, wa ala jabeenoka tali ossaadi,' ('Kuwait, my fatherland! May you be safe and glorious! May you always enjoy good fortune!'); words by Moshari al-Adwani, tune by Ibrahim Nassar al-Soula.

GOVERNMENT CHRONOLOGY

Amirs since 1950.

1950–65	Sheikh Abdullah al-Salem al-Sabah
1965–77	Sheikh Sabah al-Salem al-Sabah
1977–2006	Sheikh Jaber al-Ahmed al-Jaber al-Sabah
2006–	Sheikh Sabah al-Ahmed al-Jaber al-Sabah

RECENT ELECTIONS

In Oct. 2012 Sheikh Sabah dissolved parliament, leading to new elections on 1 Dec. 2012. The parliament is dominated by non-attached pro-regime members. The Shia minority won 17 seats and the Sunni Islamists four seats. Three women were elected. Turnout was only 40% as the opposition boycotted the elections. Prior to the elections the Amir passed a law cutting the number of votes per citizen from four to one.

CURRENT GOVERNMENT

In March 2013 the government comprised:

Prime Minister: Sheikh Jaber Mubarak al-Hamad al-Sabah; b. 1942 (sworn in 4 Dec. 2011).

Deputy Prime Ministers: Sheikh Ahmad al-Humoud al-Sabah (also *Minister of Interior*); Sheikh Sabah Khaled al-Sabah (also *Minister of Foreign Affairs*); Sheikh Ahmed Khaled al-Sabah (also *Minister of Defence*); Mustafa al-Shamali (also *Minister of Finance*).

Minister of Commerce and Industry: Anas Khalid al-Saleh. *Communications:* Salem al-Othaina. *Education and Higher Education:* Nayef Falah al-Hajraf. *Electricity and Water, and Public Works:* Abdulaziz Abdullatif al-Ibrahim. *Health:* Mohammad Barrak al-Haifi. *Information:* Sheikh Salman Sabah Salem Al-Sabah. *Justice, and Awqaf and Islamic Affairs:* Sharida Abdullah al-Mousharij. *Oil:* Hani Abdulaziz al-Hussein. *Social Affairs and Labour:* Thikra Ayed al-Rashidi.

Government Website: http://www.da.gov.kw

CURRENT LEADERS

Sheikh Sabah al-Ahmed al-Jaber al-Sabah

Position
Amir

Introduction
Sheikh Sabah became Amir of Kuwait in Jan. 2006, ending a brief constitutional crisis in the wake of the death of Sheikh Jaber. As foreign minister for over 40 years, Sheikh Sabah oversaw the positioning of Kuwait as a key Western ally in the Gulf, allowing the USA to use the country as a launch pad for its invasion of Iraq in 2003. He has faced increasing calls for greater democratization.

Early Life
Sabah IV al-Ahmed al-Jaber al-Sabah was born in 1929 in Kuwait, then a British protectorate. He is the fourth son of Sheikh Ahmed al-Jaber al-Sabah, the founder of modern Kuwait and its leader from 1921–50. Educated at al Mubarakya School and by tutors, Sheikh Sabah became a member of the central committee municipality council in 1954. He also served as a member of the building and construction council at a time when the Amir, Sheikh Abdullah al-Salem al-Sabah, was pumping much of the state's new oil wealth into an ambitious public works programme.

From 1956–62 Sheikh Sabah chaired the printing and publishing authority and was then appointed minister of information in the first post-independence cabinet. He was promoted to foreign minister in 1963 and headed Kuwait's inaugural delegation to the UN later that year. He presided over a generally low-profile, neutralist foreign policy. Palestinian rights received strong support; Fatah was founded in Kuwait.

On 16 Feb. 1978 Sheikh Sabah was appointed deputy prime minister while keeping the foreign affairs portfolio. A broadly pro-Iraqi orientation was adopted during the early stages of the Iran–Iraq War and Kuwait became a member of the Gulf Co-operation Council. After Iraq's invasion of Kuwait in 1990, Sheikh Sabah joined other government members in exile in Saudi Arabia. The al-Sabahs' flight caused resentment among those left behind in Kuwait but post-liberation elections and the creation of a National Assembly in 1992 were well received. In 1996 Sheikh Sabah joined the Supreme Council of Planning.

The increasingly frail Sheikh Jaber issued a decree separating the posts of crown prince and prime minister on 13 July 2003 and appointed Sheikh Sabah as premier. Endowed with considerable executive powers, he continued the reforms begun by the Amir, appointing the first woman minister and promoting religious tolerance in schools. After the death of Sheikh Jaber in Jan. 2006, the 76-year old crown prince, Sheikh Saad al-Abdullah al Salim al-Sabah, was deemed too ill to take the Amir's oath of office. He was voted out of office by parliament on 24 Jan. 2006 and the cabinet nominated Sheikh Sabah to take over. He was sworn in on 29 Jan. 2006.

Career in Office
Sheikh Sabah promised to speed up reforms promoting greater economic transparency, full political rights for women and democratic elections. Women had their first opportunity to vote in April 2006 in a municipal election, and were also allowed to stand as candidates in the parliamentary elections in June although none were successful. However, two women ministers were included in a new cabinet appointed in March 2007. In March 2008 Sheikh Sabah dissolved parliament and called fresh elections in May, in which Islamists won 30 of the 50 National Assembly seats. Women candidates were again unsuccessful. A year later he dissolved parliament again when the cabinet resigned on 16 March 2009 to prevent questioning of the Amir's nephew, Prime Minister Sheikh Nasser Muhammad al-Ahmad al-Sabah, on charges of misuse of public funds. Sheikh Nasser was reappointed prime minister following elections held on 16 May 2009 in which four women candidates were returned as parliamentary members for the first time.

Reflecting the political disaffection that spread across much of the Arab world from early 2011, there were demonstrations calling for reforms in Kuwait in March. The cabinet again resigned, but Sheikh Nasser was reinstated once more and in May formed a new administration with no changes to key personnel. However, after a crowd stormed parliament in Nov. in protest at alleged high-level corruption, Sheikh Nasser stood down again. The Amir accepted his resignation and appointed Sheikh Jaber Mubarak al-Hamad al-Sabah as the new prime minister. In Feb. 2012 Sheikh Jaber also resigned, but his administration was promptly reinstated with no changes to key ministries other than defence.

Islamists won a majority of seats in the Feb. 2012 elections but the Constitutional Court annulled the results in June. The Amir called for new elections in Dec. while issuing a decree in Oct. amending the electoral law. The opposition claimed that the law favoured government candidates, prompting violent street protests. The Dec. poll was boycotted by the opposition and was further undermined by a low voter turnout.

DEFENCE

In Sept. 1991 the USA signed a ten-year agreement with Kuwait to store equipment, use ports and carry out joint training exercises. In March 2013, 15,000 US troops were stationed in Kuwait.

Defence expenditure in 2008 totalled US$6,812m. (US$2,623 per capita), representing 4·4% of GDP. The expenditure per capita in 2008 was second highest in the world after that of the United Arab Emirates.

Army

Strength (2007) 11,000. In addition there is a National Guard of around 6,600.

Navy

Personnel in 2007 numbered an estimated 2,000, including 500 Coast Guard personnel.

Air Force

From a small initial combat force the Air Force has grown rapidly, although it suffered heavy losses after the Iraqi invasion of 1990–91. Equipment includes F/A-18 *Hornet* strike aircraft and BAe *Hawks*. Personnel strength was estimated (2007) at 2,500, with 50 combat-capable aircraft and 16 attack helicopters.

ECONOMY

Oil and natural gas accounted for 59·3% of GDP in 2008; finance and real estate, 15·0%; public administration, defence and services, 12·2%.

Overview

After the liberation that followed invasion by Iraq in 1991, the economy achieved relative stability. The economy is dominated by oil, which accounts for 80% of government revenue. Kuwaiti nationals make up only a third of the population, with the private sector dependent on expatriate labour.

High oil prices from 2003 produced large trade surpluses. Growth was strong up to 2008, until the economy felt the impact of the global financial crisis. Real GDP contracted by 7·8% in 2009 following a drop in domestic and global oil demand and prices. Activity recovered in 2010, driven primarily by the government's expansionary fiscal stance. In early 2010 the government launched a four-year development plan, emphasizing investment in health, education and infrastructure to transform the economy into a regional trade and financial centre, as well as expanding the role of the private sector.

Currency

The unit of currency is the *Kuwaiti dinar* (KWD), usually written as KD, of 1,000 *fils*. Inflation was 4·0% in 2010 and 4·7% in 2011. Foreign exchange reserves were US$7,544m. in July 2005, monetary gold reserves were 2·54m. troy oz and total money supply was KD 3,581m.

In 2001 the six Gulf Arab states—Kuwait, along with Bahrain, Oman, Qatar, Saudi Arabia and the United Arab Emirates— signed an agreement to establish a single currency by 2010. In June 2009 it was agreed to postpone the implementation of the new currency, the *khaleeji*, until 2013. It has since been put back further still. Both Oman and the United Arab Emirates have now withdrawn from the scheme, in 2007 and 2009 respectively.

Budget

The fiscal year begins on 1 April. Budgetary central government revenue in 2006–07 totalled KD 15,304m. and expenditure KD 11,192m. Oil accounts for 94% of government revenues. Expenditure by function in 2006–07 (in KD 1m.): social protection, 2,964; economic affairs, 1,405; defence, 1,060; general public services, 1,030.

Performance

The economy contracted by 7·8% in 2009 but grew by 2·5% in 2010 and 8·2% in 2011. Total GDP in 2011 was US$176·6bn.

Banking and Finance

The *Governor* of the Central Bank is Dr Mohammad Al-Hashel. There is also the Kuwait Finance House. In 2002 there were 11 national banks and one Islamic banking firm. The combined assets of banks operating in Kuwait totalled KD 18,818m. in Dec. 2003. Foreign banks were not permitted until 2005.

There is a stock exchange, linked with those of Bahrain and Oman.

ENERGY AND NATURAL RESOURCES

Environment

Kuwait's carbon dioxide emissions from the consumption and flaring of fossil fuels were the equivalent of 31·6 tonnes per capita in 2008.

Electricity

Installed capacity was 10·9m. kW in 2007. Production in 2007 was 48·8bn. kWh; consumption per capita was 20,223 kWh.

Oil and Gas

Oil production in 2008, 137·3m. tonnes. Kuwait had reserves in 2008 amounting to 101·5bn. bbls. Most of the oil is in the Great Burgan area (reserves of approximately 70bn. bbls), comprising the Burgan, Maqwa and Ahmadi fields located south of Kuwait City. Natural gas production was 12·8bn. cu. metres in 2008, with 1,780bn. cu metres of proven reserves.

Water

The country depends upon desalination plants. Fresh mineral water is pumped and bottled at Rawdhatain. Underground brackish water is used for irrigation, street cleaning and livestock. Production, 2003, 127,185m. gallons (95,174m. gallons fresh, 32,011m. gallons brackish). Consumption, 2003, 119,521m. gallons (94,987m. gallons fresh, 24,534m. gallons brackish).

Agriculture

There were 10,400 ha. of arable land in 2003 and 2,100 ha. of permanent crops. Production of main crops, 2003 (in 1,000 tonnes): tomatoes, 64; cucumbers and gherkins, 35; potatoes, 21; dates, 16; aubergines, 15; chillies and green peppers, 8; pumpkins and squash, 8; cauliflowers, 7.

Livestock (2003): sheep, 481,000; goats, 194,000; cattle, 27,000; poultry, 30m. Milk production (2003), 43,000 tonnes.

Forestry

Forests covered 6,000 ha. in 2010, or 0·3% of the land area.

Fisheries

The total catch in 2008 was 3,979 tonnes, exclusively from sea fishing. In the space of a month in 2001 more than 2,000 tonnes of dead fish were washed ashore. Some experts claimed the cause was the alleged pumping of raw sewage into the Gulf while others attributed it to waste from the oil industry. Shrimp fishing was important, but has declined since the 1990–91 war through oil pollution of coastal waters. Before the discovery of oil, pearls were at the centre of Kuwait's economy, but today pearl fishing is only on a small scale.

INDUSTRY

The leading companies by market capitalization in Kuwait in March 2012 were: National Bank of Kuwait (US$16·6bn.); Zain Group, a mobile telecommunications company (US$12·2bn.); and Kuwait Finance House, a banking group (US$8·0bn.).

Industries, apart from oil, include boat building, fishing, food production, petrochemicals, gases and construction. Production figures in 2007 (in 1,000 tonnes): residual fuel oil, 11,559; distillate fuel oil, 11,475; kerosene, 6,081; liquefied petroleum gas, 3,447; petrol, 2,852; jet fuel, 2,827; cement (2008 estimate), 2,200.

Labour

In June 2004 the labour force totalled 1,551,342 (81·3% non-Kuwaitis). Of the total labour force, 52·1% worked in social, community and personal services, 15·2% in trade, hotels and restaurants, 7·2% in construction and 5·8% in manufacturing. Registered unemployment in June 2004 was 1·7%. Approximately 95% of nationals work for the government, with around 95% of private jobs being filled by expatriates.

INTERNATIONAL TRADE

Kuwait, along with Bahrain, Oman, Qatar, Saudi Arabia and the United Arab Emirates entered into a customs union in Jan. 2003.

Imports and Exports

Imports (c.i.f.) were valued at US$24,840m. in 2008 and exports (f.o.b.) at US$87,457m. Oil accounts for 92% of revenue from exports. In 2008 the leading import suppliers were China, the USA and Japan; the main export markets were Japan, South Korea and India.

COMMUNICATIONS

Roads

There were 5,749 km of roads in 2004, 85% of which were paved. There were 750,600 passenger cars in use in 2007 (282 per 1,000 inhabitants), 573,200 lorries and vans, and 27,300 buses and coaches. There were 45,376 road accidents in 2003 involving injury with 1,704 fatalities.

Civil Aviation

There is an international airport (Kuwait International). The national carrier is the state-owned Kuwait Airways. Kuwait's first low-cost airline, Jazeera Airways, began operations in Oct. 2005. In 2005 scheduled airline traffic of Kuwait-based carriers flew 50·7m. km and carried 1,944,200 passengers. Kuwait International airport handled 4,260,136 passengers in 2003 and 144,727 tonnes of freight.

Shipping

The port of Kuwait formerly served mainly as an entrepôt, but this function is declining in importance with the development of the oil industry. The largest oil terminal is at Mina Ahmadi. Three small oil ports lie to the south of Mina Ahmadi: Mina Shuaiba, Mina Abdullah and Mina al-Zor. In Jan. 2009 there were 52 ships of 300 GT or over registered (including 22 oil tankers), totalling 2·33m. GT.

Telecommunications

In 2010 Kuwait had an estimated 566,000 landline telephone subscriptions (equivalent to 207 per 1,000 inhabitants) and 4·4m. mobile phone subscriptions (or 1,608 per 1,000 inhabitants). In 2011, 74·2% of the population were internet users. In March 2012 there were 899,000 Facebook users.

SOCIAL INSTITUTIONS

Justice

In 1960 Kuwait adopted a unified judicial system covering all levels of courts. These are: Courts of Summary Justice, Courts of the First Instance, Supreme Court of Appeal, Court of Cassation and a Constitutional Court. Islamic Sharia is a major source of legislation. The death penalty is still in use. There was one confirmed execution in 2007 but none since.

The population in penal institutions in 2010 was 4,179 (137 per 100,000 of national population).

Education

Education is free and compulsory from six to 14 years. In 2007 there were 211,576 pupils in primary schools with 22,016 teaching staff, and 247,233 pupils in secondary schools with 26,050 teaching staff. In 2006 there were 37,521 students in tertiary education and 1,986 academic staff. There are two state-supported higher education institutions, Kuwait University and the Public

Authority for Applied Education and Training. There were approximately 28,000 students at Kuwait University in 2009. There are also a number of private universities and colleges, a teacher training college, a music academy and several Quranic schools. The Arab Open University which opened in Nov. 2002 is based in Kuwait and has branches in several other Middle Eastern countries. There were around 17,000 enrolments in 2003–04. Adult literacy rate in 2007 was 94%. In 2005 public expenditure on education came to 4·7% of GDP and accounted for 12·7% of total government spending.

Health

Medical services are free to all residents. In 2003 there were 15 hospitals and sanatoria, with a provision of 4,712 beds (19 per 10,000 population). There were 4,840 doctors (18 per 10,000 population), 810 dentists, 9,940 nurses and midwives and 1,340 pharmacists in 2005. There were 74 clinics and other health centres and 1,569,549 people were admitted to public hospitals in 2003.

RELIGION

In 2001, 1,020,000 people were Sunni Muslims, 680,000 Shia Muslims, 230,000 other Muslims and 340,000 other (mostly Christian and Hindu).

CULTURE

Press

In 2008 there were 17 daily newspapers, with a combined circulation of 630,000. Formal press censorship was lifted in Jan. 1992.

Tourism

There were 4,482,000 non-resident visitors in 2007 (up from 2,072,000 in 2001), bringing revenue of US$512m.

DIPLOMATIC REPRESENTATIVES

Of Kuwait in the United Kingdom (2 Albert Gate, London, SW1X 7JU)
Ambassador: Khaled al-Duwaisan, GCVO.

Of the United Kingdom in Kuwait (Arabian Gulf St., Dasman, Kuwait)
Ambassador: Frank Baker, OBE.

Of Kuwait in the USA (2940 Tilden St., NW, Washington, D.C., 20008)
Ambassador: Salem Abdulla al-Jaber al-Sabah.

Of the USA in Kuwait (Al-Masjed al-Aqsa St., Bayan, Kuwait)
Ambassador: Matthew H. Tueller.

Of Kuwait to the United Nations
Ambassador: Mansour Ayyad al-Otaibi.

Of Kuwait to the European Union
Ambassador: Nabeela Abdulla al-Mulla.

FURTHER READING

Al-Yahya, M.A., *Kuwait: Fall and Rebirth.* 1993
Boghardt, Lori Plotkin, *Kuwait Amid War, Peace and Revolution: 1979–1991 and New Challenges.* 2007

National Statistical Office: Kuwait Central Statistics Bureau, Al Sharq, Arabian Gulf Street, Kuwait City.
Website (limited English): http://www.csb.gov.kw

KYRGYZSTAN

Kyrgyz Respublikasy
(Kyrgyz Republic)

Capital: Bishkek
Population projection, 2015: 5·63m.
GNI per capita, 2011: (PPP$) 2,036
HDI/world rank: 0·615/126
Internet domain extension: .kg

KEY HISTORICAL EVENTS

Kyrgyzstan became part of Soviet Turkestan, which itself became a Soviet Socialist Republic within the Russian Soviet Federal Socialist Republic (RSFSR) in April 1921. In 1924, when Central Asia was reorganized territorially on a national basis, Kyrgyzstan was separated from Turkestan. In Dec. 1936 Kyrgyzstan was proclaimed one of the constituent Soviet Socialist Republics of the USSR. With the collapse of the Soviet Empire, the republic asserted its claim to sovereignty in 1990 and declared independence in Sept. 1991. Askar Akayev became president in 1990 and subsequently expanded presidential powers. Kyrgyzstan became a member of the CIS in Dec. 1991.

Incursions into Kyrgyz territory by Islamic rebels and border skirmishes in the Fergana Valley are a cause for concern for all Central Asian governments. Kyrgyzstan tripled its defence budget for 2001 to combat terrorism. Allegations of widespread government corruption and disputed parliamentary elections in Feb. 2005 led to popular protests. The Supreme Court declared the elections void and Kurmanbek Bakiyev was appointed prime minister and acting president. Akayev, in exile in Russia, resigned as president in April 2005. Bakiyev was confirmed as president by winning the elections held in July 2005. His tenure was marked by elections that fell below accepted international standards, concerns over civil liberties and growing popular resentment at his failure to address corruption. The doubling of household utility costs in Jan. 2010 sparked a wave of protests that saw Bakiyev step down in April 2010. Roza Otunbayeva headed up the government until Almazbek Atambayev took office as president in Dec. 2011.

TERRITORY AND POPULATION

Kyrgyzstan is situated on the Tien-Shan mountains and bordered in the east by China, west by Kazakhstan and Uzbekistan, north by Kazakhstan and south by Tajikistan. Area, 199,945 sq. km (77,199 sq. miles). Population (census 2009), 5,362,800 (51·3% females); density, 27 per sq. km. In 2009, 65·9% of the population lived in rural areas.

The UN gives a projected population for 2015 of 5·63m.

The republic comprises seven provinces (Batken, Djalal-Abad, Issyk-Kul, Naryn, Osh, Talas and Chu) plus the city of Bishkek, the capital (formerly Frunze; 2009 census population, 822,000). Other large towns are Osh (233,800 in 2009), Djalal-Abad (89,000), Karakol (formerly Przhevalsk, 63,400), Tokmak (53,200), Uzgen (49,400), Balykchy (42,400) and Karabalta (37,800).

The Kyrgyz are of Turkic origin and formed 69·2% of the population in 2008; the rest included Uzbeks (14·5%), Russians (8·7%) and Dungans (1·2%).

The official languages are Kyrgyz and Russian. After the break-up of the Soviet Union, Russian was only the official language in provinces where Russians are in a majority. However, in May 2000 parliament voted to make it an official language nationwide, mainly in an attempt to stem the ever-increasing exodus of skilled ethnic Russians. The Cyrillic alphabet is still used although the reintroduction of the Roman alphabet (in use 1928–40) remains a source of political debate.

SOCIAL STATISTICS

2009 births, 135,494; deaths, 35,898; marriages (2006), 43,760. Rates, 2009 (per 1,000 population): birth, 26·4; death, 7·0; infant mortality (per 1,000 live births, 2010), 33. Life expectancy, 2007, 63·9 years for males and 71·4 for females. In 2003 the most popular age for marrying was 20–24 for females and 25–29 for males. Annual population growth rate, 2000–05, 0·9%; fertility rate, 2008, 2·5 births per woman.

CLIMATE

The climate varies from dry continental to polar in the high Tien-Shan, to sub-tropical in the southwest (Fergana Valley) and temperate in the northern foothills. Bishkek, Jan. 9°F (–13°C), July 70°F (21°C). Annual rainfall 14·8" (375 mm).

CONSTITUTION AND GOVERNMENT

A new constitution was adopted in June 2010 after it won overwhelming support in a referendum following the ousting of the incumbent president, Kurmanbek Bakiyev, in April 2010. The referendum was held two weeks after ethnic violence between Kyrgyz and Uzbek groups killed up to 2,000 and left 400,000 displaced. Nonetheless, international observers deemed the vote fair. Turnout was put at 72·2%, with 90·5% of votes cast in favour.

Under the terms of the constitution, greater power is invested in parliament at the expense of the presidency. The president is allowed to serve a maximum of one six-year term and cannot seek re-election, although the office does retain its power of veto and has the authority to appoint heads of various state institutions. The unicameral parliament (*Jogorku Kenesh*) is comprised of 120 seats, with no single party allowed to hold more than 65. Political parties cannot be constituted on religious or ethnic grounds and members of the armed forces, the judiciary and the police are banned from party membership.

National Anthem

'Ak möngülüü aska yoolor, talaalar' ('High mountains, valleys and fields'); words by D. Sadykov and E. Kuluev, tune by N. Davlyesov and K. Moldovasanov.

RECENT ELECTIONS

Presidential elections were held on 30 Oct. 2011. Former prime minister Almazbek Atambayev won with 63·2% of the vote, ahead of Adakhan Madumarov with 14·7% and Kamchybek Tashiyev with 14·3%. There were 13 other candidates. Reported turnout was over 57%. Madumarov and Tashiyev warned in advance of potential attempts to rig the voting and concerns were raised over the transparency of the elections.

Parliamentary elections were held on 10 Oct. 2010 in which 28 of 120 seats were won by Ata-Zhurt (Fatherland), 26 by the Social Democratic Party of Kyrgyzstan, 25 by Ar-Namys, 23 by Respublika and 18 by Ata-Meken. Turnout was 55·9%.

CURRENT GOVERNMENT

In March 2013 the government comprised:

President: Almazbek Atambayev; b. 1956 (Social Democratic Party of Kyrgyzstan; since 1 Dec. 2011).

Prime Minister: Zhantoro Satybaldiyev; b. 1956 (ind.; since 5 Sept. 2012).

First Deputy Prime Minister: Dzhoomart Otorbayev. *Deputy Prime Minister in Charge of Security, Law and Order, and Borders Issues:* Shamil Atakhanov. *Deputy Prime Ministers:* Taiyrbek Sarpashev; Kamila Talieva.

Minister of Agriculture and Land Reclamation: Chyngysbek Uzakbayev. *Culture, Information and Tourism:* Sultan Rayev. *Defence:* Taalaibek Omuraliev. *Economy and Anti-Monopoly Policy:* Temir Sariev. *Education and Science:* Kanat Sadykov. *Emergency Situations:* Kubatbek Boronov. *Energy and Industry:* Avtandil Kalmambetov. *Finance:* Olga Lavrova. *Foreign Affairs:* Erlan Abdyldayev. *Health:* Dinara Sagimbayeva. *Interior:* Abdylda Suranchiyev. *Justice:* Almambet Shykmamatov. *Labour, Migration and Youth:* Aliyasbek Alymkulov. *Social Development:* Kylychbek Sultanov. *Transport and Communications:* Kalykbek Sultanov.

CURRENT LEADERS

Almazbek Atambayev

Position
President

Introduction
Almazbek Atambayev was sworn in as president on 1 Dec. 2011. His election represented the first peaceful handover of power since the country gained independence during the dissolution of the Soviet Union. He succeeded Roza Otunbayeva, who assumed the post on a temporary basis after the previous incumbent, Kurmanbek Bakiyev, was ousted in a violent uprising in April 2010. A successful entrepreneur, Atambayev favours closer ties with Russia.

Early Life
Atambayev was born on 17 Sept. 1956 in Chui, a northern region of what is now Kyrgyzstan. He studied at the Moscow Institute of Management, from where he graduated with a degree in economics. From 1983 until 1987, when the Kyrgyz Republic was still a constituent part of the Soviet Union, he served on the Supreme Council of the Republic.

In 1993, two years after the country had declared independence, Atambayev was one of the founders of the Social Democratic Party of Kyrgyzstan. Having established himself as a prosperous businessman in the post-Soviet era, he became chairman of the party in 1999. In 2000 he won 6% of the vote in a presidential election. In Dec. 2005 he was appointed minister of industry, trade and tourism, but resigned five months later. In Nov. 2006 he was among the leaders of anti-government protests in the capital, Bishkek. He served as prime minister from March to Nov. 2007, and ran again for president in April 2009 but withdrew on the day of the ballot, citing electoral fraud.

Kurmanbek Bakiyev was overthrown in April 2010, prompting a wave of ethnic violence between Kyrgyz and Uzbek communities in the south of the country. A national referendum approved a new constitution that weakened the authority of the president and shifted power towards the legislature. Parliamentary elections followed and a coalition government was established for the first time in the country's history. Atambayev was appointed

prime minister once more, a position he held until the presidential election in Oct. 2011.

Atambayev gained a comfortable victory at that election which, though criticized by some observers, was regarded as a significant democratic achievement.

Career in Office
Atambayev came to power in a country riven by ethnic division. He has pledged to foster national unity and has described Russia as the country's 'main strategic partner'. Soon after coming to power he said that he would look to close a US military base in the country when its lease expires in 2014. In Sept. 2012 Atambayev appointed Zhantoro Satybaldiyev, a technocrat, as the new prime minister. This followed the collapse of the coalition government in the wake of corruption allegations, a poor economic record and the resignation of former premier Omurbek Babanov.

DEFENCE

Conscription is for 12 months. Defence expenditure in 2008 totalled US$47m. (US$9 per capita), representing 0·9% of GDP. The USA opened a military base in Kyrgyzstan in 2001 to aid the war in Afghanistan against the Taliban. The base was scheduled to close by the end of Aug. 2009 after an eviction notice was served on 20 Feb. 2009 giving the US military 180 days to vacate the site. However, on 23 June 2009 the Kyrgyz and US governments agreed a new deal that allowed a one-year extension of the lease. The current lease expires in mid-2014 but President Atambayev, who took office in Dec. 2011, has stated that the base must close then. In Sept. 2003 Kyrgyzstan agreed to allow Russia to open an air force base in the country. For the time being Kyrgyzstan is the only country in the world to host both a US and a Russian military base.

Army

Personnel, 2007, 8,500. In addition there are 5,000 border guards, 3,500 interior troops and a National Guard of 1,000.

Air Force

Personnel, 2007, 2,400, with 52 combat-capable aircraft (mainly MiG-21 fighters) and nine attack helicopters.

ECONOMY

Agriculture accounted for 21·0% of GDP in 2009, industry 27% and services 52%.

Overview

Following independence from the Soviet Union in 1991, the loss of trade preferences and Soviet subsidies caused a major downturn in the economy, characterized by declining GDP, hyperinflation and widespread poverty. However, economic performance and macroeconomic stability improved after the 1998 Russian financial crisis, with GDP growth averaging 5% per year. Traditionally, agriculture and mining are the leading sectors and international development aid has focused on agriculture since the early 1990s. In recent years, however, growth has been driven by the construction, power, transportation, trade and communications sectors.

Despite a decline in the poverty rate (particularly for those in extreme poverty), 22% of the population lives below the international poverty line of US$1·25 (PPP) per day according to the 2010 Human Development Report. Inflation threatens to become entrenched. Following a period in which it remained in the 3–5% range, a global rise in food and energy prices led to an inflation surge in late 2007.

When the political crisis in 2010 forced the resignation and exile of President Kurmanbek Bakiyev, there were knock-on impacts for the economy. Thousands of people were displaced and around 3,500 houses, 700 commercial establishments and 79 public buildings damaged or destroyed. The total damage was put at US$490m., equivalent to 13% of GDP. At a meeting of high level donors in July 2010, US$1·1bn. was pledged for 30 months. Finance for the recovery and reconstruction programme over the same period was estimated at US$1·0bn. Reduced private sector confidence is expected to affect investment, including foreign direct investment.

The political situation remains fragile and rising inflation is a concern. The IMF points to the need to reduce vulnerabilities in the banking sector and to improve the business environment.

Currency

On 10 May 1993 Kyrgyzstan introduced its own currency unit, the *som* (KGS), of 100 *tiyin*, at a rate of 1 som = 200 roubles. Inflation was 24·5% in 2008, falling to 7·8% in 2010 before increasing to 16·6% in 2011. Gold reserves totalled 83,000 troy oz in July 2005, foreign exchange reserves US$509m. and total money supply 13,884m. soms.

Budget

Budgetary central government revenue totalled 32,670·1m. soms in 2007 and expenditure 25,666·4m. soms. Tax revenues in 2007 were 23,266·0m. soms. Main items of expenditure by economic type in 2007 were: compensation of employees (6,845·1m. soms) and use of goods and services (6,370·6m. soms).

VAT is 20%.

Performance

After growth of 3·1% in 2006, 8·5% in 2007, 7·6% in 2008 and 2·9% in 2009, the economy contracted by 0·5% in 2010 but grew again by 5·7% in 2011. Total GDP in 2011 was US$5·9bn.

Banking and Finance

The central bank and bank of issue is the National Bank (*Chair*, Zina Asankojoeva). There were 22 commercial banks, including three foreign banks, in 2002.

Foreign debt was US$3,984m. in 2010, representing 89·2% of GNI.

There is a stock exchange in Bishkek.

ENERGY AND NATURAL RESOURCES

Environment

Kyrgyzstan's carbon dioxide emissions from the consumption and flaring of fossil fuels were the equivalent of 1·1 tonnes per capita in 2008.

Electricity

Installed capacity was an estimated 3·7m. kW in 2007. Production in 2007 was 16·24bn. kWh, of which 85·9% hydro-electric; consumption per capita was 2,624 kWh.

Oil and Gas

Output of oil, 2007, 69,000 tonnes; natural gas, 2007, 15m. cu. metres.

Minerals

In 2009 lignite production totalled 535,000 tonnes and hard coal production 67,000 tonnes. Gold is also mined. Output in 2009 (gold content) was 16,950 kg. Gold is by far the country's leading mineral commodity in terms of value.

Agriculture

Kyrgyzstan is famed for its livestock breeding, in particular the small Kyrgyz horse. In 2004 there were 2,882,000 sheep, 1,003,000 cattle, 795,000 goats, 340,000 horses and 2m. chickens. Yaks are bred as meat and dairy cattle, and graze on high altitudes unsuitable for other cattle. Crossed with domestic cattle, hybrids give twice the yield of milk.

There were 1·34m. ha. of arable land in 2003 and 67,000 ha. of permanent crops. Number of peasant farms (2003), 255,822.

Principal crops include wheat, barley, corn and vegetables. Fodder crops for livestock are grown, particularly lucerne; also

sugar beets, cotton, tobacco and medicinal herbs. Sericulture, fruit, grapes and vegetables are major branches.

Output of main agricultural products (in 1,000 tonnes) in 2003: potatoes, 1,308; wheat, 1,014; sugar beets, 812; corn for grain, 399; barley, 198; tomatoes, 144; carrots, 126; raw cotton, 106; cabbage, 104; onions, 104; cucumbers, 50. Livestock products, 2003, in 1,000 tonnes: beef and veal, 94; mutton and goat meat, 44; milk, 1,192; eggs, 268m. units.

Forestry

In 2010 forests covered 954,000 ha., or 5% of the land area. Timber production in 2007 was 27,000 cu. metres.

Fisheries

The catch in 2010 was 27 tonnes, entirely from freshwater fishing.

INDUSTRY

Industrial enterprises include food, timber, textile, engineering, metallurgical, oil and mining. There are also sugar refineries, tanneries, cotton and wool-cleansing works, flour-mills and a tobacco factory. In 2006 industry accounted for 20·1% of GDP, with manufacturing contributing 12·9%. In 2003 output was valued at 48,940·1m. soms at current prices.

Production, 2003: cement, 757,300 tonnes; carpets, 13·4m. sq. metres; cotton woven fabrics, 1m. sq. metres; footwear, 238,000 pairs.

Labour

Out of 1,837,000 people in employment in 2003, 951,200 were engaged in agriculture, hunting and forestry; 205,800 in wholesale and retail trade/repair of motor vehicles, motorcycles and personal and household goods; 151,900 in education; and 113,700 in manufacturing. In 2004 the unemployment rate was 2·9%.

INTERNATIONAL TRADE

In Jan. 1994 an agreement to create a single economic zone was signed with Kazakhstan and Uzbekistan. In March 1996 Kyrgyzstan joined a customs union with Russia, Kazakhstan and Belarus.

Imports and Exports

Imports (c.i.f.) were valued at US$4,072m. in 2008 and exports (f.o.b.) at US$1,618m.

In 2008 main import sources (in US$1m.) were: Russia (1,492); China (728); Kazakhstan (377). Main export markets in 2008 (in US$1m.) were: Switzerland (445); Russia (186); Uzbekistan (167).

Principal imports in 2008 (in US$1m.) were: machinery and transport equipment (646); manufactured goods (514); food (401). Principal exports in 2008 (in US$1m.) were: gold (464); articles of apparel and clothing accessories (99); non-metallic mineral manufactures (90).

COMMUNICATIONS

Roads

There were 34,000 km of roads in 2007. Passenger cars in use in 2007 numbered 229,700 (44 per 1,000 inhabitants). There were 1,252 road accident fatalities in 2007.

Rail

In the north a railway runs from Lugovaya through Bishkek to Rybachi on Lake Issyk-Kul. Towns in the southern valleys are linked by short lines with the Ursatyevskaya–Andizhan railway in Uzbekistan. Total length of railway, 2005, 424 km. Passenger-km travelled in 2008 came to 90m. and freight tonne-km to 946m.

Civil Aviation

There is an international airport at Bishkek (Manas). The national carrier, Kyrgyzstan Airlines, ceased operations in 2005. In 2003 Bishkek handled 217,576 passengers (112,487 on international flights) and 1,978 tonnes of freight. In 2003 scheduled airline traffic of Kyrgyzstan-based carriers flew 6m. km, carrying 206,000 passengers (103,000 on international flights).

Telecommunications

In 2010 there were 489,100 landline telephone subscriptions (equivalent to 91·7 per 1,000 inhabitants) and 5,275,500 mobile phone subscriptions (or 989·0 per 1,000 inhabitants). Fixed internet subscriptions totalled 68,900 in 2010 (12·9 per 1,000 inhabitants).

SOCIAL INSTITUTIONS

Justice

In 2004, 32,616 crimes were reported, including 419 murders and attempted murders. The population in penal institutions in March 2008 was 8,427 (156 per 100,000 of national population). The new constitution that came into force in Jan. 2007 abolished the death penalty.

Education

In 2008–09 there were 72,000 children in pre-primary education (3,000 teaching staff), 392,000 pupils in primary education (16,000 teaching staff), an estimated 679,000 pupils in secondary education (51,000 teaching staff) and 17,000 university level lecturers for 294,000 students. There were 49 higher educational institutions and 75 secondary professional education establishments in 2004–05. Kyrgyz University had 20,855 students in 2004–05. Adult literacy was 99·2% in 2009.

In 2005 public expenditure on education came to 4·9% of GDP.

Health

In 2006 there were 12,710 physicians, 1,017 dentists and 30,824 nurses and midwives. In 2003 there were 151 hospitals.

Welfare

In Dec. 2007 there were 529,000 pensioners.

RELIGION

In 2001, 75% of the population was Sunni Muslim. There were 1,784 mosques, 359 Christian congregations, one synagogue and one Buddhist temple in 2008.

CULTURE

World Heritage Sites

Sulaiman-Too Sacred Mountain in the Fergana Valley was inscribed on the UNESCO World Heritage List in 2009. Situated at the crossroads of important routes on the Central Asian Silk Roads system, Sulaiman-Too is the site of numerous ancient places of worship and caves with petroglyphs.

Press

There were three national daily newspapers in 2008, with a combined circulation of 40,000.

Tourism

In 2010 there were 1,316,000 non-resident tourists, down from 2,147,000 in 2009. This was as a consequence of the political upheaval in April 2010 and the ethnic conflict that ensued.

DIPLOMATIC REPRESENTATIVES

Of Kyrgyzstan in the United Kingdom (Ascot House, 119 Crawford St., London, W1U 6BJ)
Ambassador: Vacant.
Chargé d'Affaires a.i.: Aibek Tilebaliev.

Of the United Kingdom in Kyrgyzstan (215 Manaschy Sagynbaya St., Bishkek 720010)
Ambassador: Judith Farnworth.

Of Kyrgyzstan in the USA (2360 Massachusetts Ave., NW, Washington, D.C., 20008)
Ambassador: Muktar Djumaliev.

Of the USA in Kyrgyzstan (171 Prospekt Mira, Bishkek 720016)
Ambassador: Pamela L. Spratlen.

Of Kyrgyzstan to the United Nations
Ambassador: Talaibek Kydyrov.

Of Kyrgyzstan to the European Union
Ambassador: Jyrgalbek Kumarovich Azylov.

FURTHER READING

Abazov, Rafis, *Historical Dictionary of Kyrgyzstan.* 2004
Anderson, J., *Kyrgyzstan: Central Asia's Island of Democracy?* 1999
Marat, Erica, *The Tulip Revolution: Kyrgyzstan One Year After.* 2006

National Statistical Office: National Statistical Committee of the Kyrgyz Republic, 374 Frunze Street, Bishkek City 720033.

LAOS

After decades of isolationism, the authorities responded to the collapse of the Soviet Union by cautiously embracing economic reform and forging international ties. In 1995 the USA ended its aid embargo and two years later Laos joined the Association of Southeast Asian Nations. In 2005 the USA normalized its trade relations with the country. A stock market was opened in the capital, Vientiane, in 2011 as the authorities sought to engage with capitalism. Having applied to join the World Trade Organization in 1997, Laos became a member in Feb. 2013.

TERRITORY AND POPULATION

Laos is a landlocked country of 236,800 sq. km (91,428 sq. miles) bordered on the north by China, the east by Vietnam, the south by Cambodia and the west by Thailand and Myanmar. Apart from the Mekong River plains along the border of Thailand, the country is mountainous, particularly in the north, and in places densely forested.

The population (2005 census) was 5,621,982 (2,821,431 females); density, 24 per sq. km. 2009 estimate: 6,128,000. In 2011, 34·3% of the population lived in urban areas.

The UN gives a projected population for 2015 of 6·63m.

There are 16 provinces and one prefecture divided into 141 districts and one special region (*khetphiset*). Area, population and administrative centres in 2005:

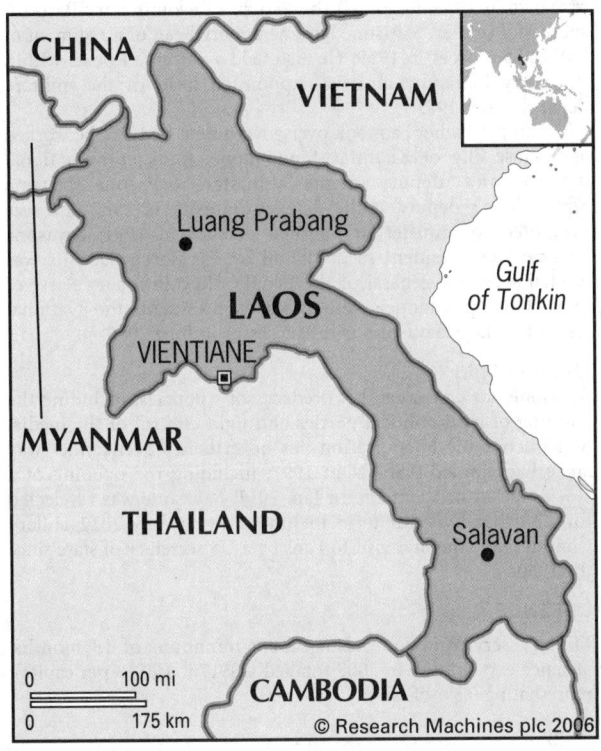

Province	Sq. km	Population (in 1,000)	Administrative centre
Attopeu	10,320	112·1	Samakhi Xai
Bokeo	6,196	145·3	Ban Houei Xai
Bolikhamxai	14,863	225·3	Paksan
Champassak	15,415	607·4	Pakse
Houa Phan	16,500	280·9	Xam Neua
Khammouane	16,315	337·4	Thakhek
Luang Namtha	9,325	145·3	Luang Namtha
Luang Prabang	16,875	407·0	Luang Prabang
Oudomxai	15,370	265·2	Muang Xai
Phongsali	16,270	165·9	Phongsali
Salavan	10,691	324·3	Salavan
Savannakhet	21,774	825·9	Shanthabouli
Sayabouri	16,389	338·7	Sayabouri
Sekong	7,665	85·0	Sekong
Vientiane	18,526	388·9	Phonghong
Vientiane[1]	3,920	698·3	Vientiane
Xaisomboun	4,506	39·4	Ban Muang Cha
Xieng Khouang	15,880	229·6	Phonsavanh

[1]Prefecture.

The capital and largest town is Vientiane, with a population of (2005 estimate) 570,000. Other important towns are Savannakhet, Pakse, Xam Neua and Luang Prabang.

The population is divided into three groups: about 67% Lao-Lum (Valley-Lao); 17% Lao-Theung (Lao of the mountain sides); and 7·4% Lao-Sung (Lao of the mountain tops), who comprise the Hmong and Yao (or Mien). Lao is the official language. French and English are spoken.

SOCIAL STATISTICS

2009 estimates: births, 141,000; deaths, 37,000. Rates, 2009 estimates (per 1,000 population): birth, 23; death, 6. Infant mortality (per 1,000 live births, 2010), 42. Life expectancy, 2007: 63·2 years for men and 65·9 for women. Annual population growth rate, 2000–05, 1·7%. Fertility rate, 2008, 3·5 births per woman.

CLIMATE

A tropical monsoon climate, with high temperatures throughout the year and very heavy rains from May to Oct. Vientiane, Jan. 70°F (21·1°C), July 81°F (27·2°C). Annual rainfall 69" (1,715 mm).

Sathalanalath Pasathipatai Pasasonlao
(Lao People's Democratic Republic)

Capital: Vientiane
Population projection, 2015: 6·63m.
GNI per capita, 2011: (PPP$) 2,242
HDI/world rank: 0·524/138
Internet domain extension: .la

KEY HISTORICAL EVENTS

The Kingdom of Laos, once called Lanxang (the Land of a Million Elephants), was founded in the 14th century. In 1893 Laos became a French protectorate and in 1907 acquired its present frontiers. In 1945, after French authority had been suppressed by the Japanese, an independence movement known as Lao Issara (Free Laos) set up a government which collapsed with the return of the French in 1946. Under a new constitution of 1947 Laos became a constitutional monarchy under the Luang Prabang dynasty and in 1949 became an independent sovereign state within the French Union. An almost continuous state of war began in 1953 between the Royal Lao Government, supported by American bombing and Thai mercenaries, and the Patriotic Front Pathet Lao, supported by North Vietnamese troops. Peace talks resulted in an agreement on 21 Feb. 1973 providing for the formation of a provisional government of national union and the withdrawal of foreign troops. A provisional coalition government was duly formed in 1974. However, after the Communist victories in neighbouring Vietnam and Cambodia in April 1975, the Pathet Lao took over the running of the whole country, maintaining only a façade of a coalition. On 29 Nov. 1975 HM King Savang Vatthana abdicated and the People's Congress proclaimed a People's Democratic Republic of Laos on 2 Dec. 1975.

CONSTITUTION AND GOVERNMENT

In Aug. 1991 the National Assembly adopted a new constitution. The head of state is the President, elected by the National Assembly, which consists of 132 members (115 prior to the elections of April 2011).

Under the constitution the People's Revolutionary Party of Laos (PPPL) remains the 'central nucleus' of the 'people's democracy'; other parties are not permitted. The PPPL's Politburo comprises 11 members, including Choummaly Sayasone (PPPL, *President*).

National Anthem

'Xatlao tangtae dayma lao thookthuana xeutxoo sootchay' ('For the whole of time the Lao people have glorified their Fatherland'); words by Sisana Sisane, tune by Thongdy Sounthonevichit.

RECENT ELECTIONS

The Sixth Legislature of the National Assembly elected Choummaly Sayasone as president on 8 June 2006. Bouasone Bouphavanh was elected prime minister on the same day. Sayasone and incumbent prime minister Thongsing Thammavong were re-elected by the Seventh Legislature on 15 June 2011.

There were parliamentary elections on 30 April 2011 in which the People's Revolutionary Party of Laos (PPPL) won 128 of 132 seats. Only four (approved) non-partisan candidates won seats. Turnout was 99·6%.

CURRENT GOVERNMENT

President: Lieut.-Gen. Choummaly Sayasone; b. 1936 (PPPL; elected on 8 June 2006 and re-elected 15 June 2011).

Vice President: Boungnang Volachit.

In March 2013 the government consisted of:

Prime Minister: Thongsing Thammavong; b. 1944 (PPPL; elected on 23 Dec. 2010).

Deputy Prime Ministers: Maj. Gen. Asang Laoli (also *Chairman of State Control Commission*); Thongloun Sisoulit (also *Minister of Foreign Affairs*); Lieut.-Gen. Douangchai Phichit (also *Minister of Defence*); Somsavat Lengsavad.

Minister of Agriculture and Forestry: Vilayvanh Phomkhe. *Education and Sports:* Phankham Viphavanh. *Energy and Mining:* Soulivong Dalavong. *Finance:* Phouphet Khamphounvong. *Industry and Commerce:* Nam Viyaket. *Information, Culture and Tourism:* Bosengkham Vongdara. *Interior:* Khampane Philavong. *Justice:* Chaleun Yiabaoher. *Labour and Social Welfare:* Onchanh Thammavong. *Natural Resources and Environment:* Noulin Sinhbandith. *Planning and Investment:* Somdy Duangdy. *Post, Telecommunication and Communication:* Hiem Phommachanh. *Public Health:* Eksavang Vongvichit. *Public Security:* Thongbanh Sengaphone. *Public Works and Transport:* Sommad Pholsena. *Science and Technology:* Boviengkham Vongdara. *Ministers in the Prime Minister's Office:* Bounheuang Duangphachanh; Sinlavong Khoutphaythoune; Bounpheng Mouphosay; Bountiem Phitsamay; Khempheng Pholsena; Duangsavad Souphanouvong.

National Assembly Website: http://www.na.gov.la

CURRENT LEADERS

Choummaly Sayasone

Position
President

Introduction
Choummaly Sayasone was elected president in June 2006, succeeding his long-time mentor, Khamtay Siphandone. Part of the Lao ruling elite for decades, Sayasone had previously served as vice president and minister of national defence. He was re-elected by the National Assembly in June 2011.

Early Life
Born on 6 March 1936 into a farming family in Vat Neua village, Attopeu province, Sayasone took up arms with the revolutionary Pathet Lao guerrilla forces in 1954. While fighting in Houaphan province in 1955, he joined the People's Revolutionary Party. A successful soldier, Sayasone became deputy head of a regiment of Pathet Lao forces in 1959. He also held a variety of posts within the party hierarchy and was appointed head of the military department in 1972.

When the Pathet Lao took over government in 1975, he worked as a close ally of Khamtay Siphandone, minister of national defence and deputy prime minister. Sayasone became Siphandone's deputy at the defence ministry in 1982 and was promoted to minister of national defence in 1991. Sayasone became vice president in 2001 and on 21 March 2006 he was elected secretary general of the People's Revolutionary Party of Laos. On Siphandone's retirement as president, the National Assembly chose Sayasone as his successor in June 2006.

Career in Office
Sayasone has continued his predecessor's policies, including the banning of rival political parties and tight control of the media. Some economic liberalization has nevertheless been under way since Laos joined ASEAN in 1997, including the opening of a stock market in Vientiane in Jan. 2011. Sayasone was re-elected for a further five-year term in June 2011. In July 2012 Hillary Clinton made the first visit to Laos by a US secretary of state since the 1950s.

DEFENCE

Military service is compulsory for a minimum of 18 months. Defence expenditure in 2008 totalled US$17m. (US$3 per capita), representing 0·3% of GDP.

Army

There are four military regions. Strength (2007) about 25,000. In addition there are local defence forces totalling over 100,000.

Navy

There is an Army Marine Section of about 600 personnel (2007).

Air Force

The Air Force has 22 combat-capable aircraft, including MiG-21 fighters, although serviceability is in doubt. Personnel strength, 3,500 in 2007.

ECONOMY

In 2009 agriculture accounted for 35·2% of GDP, industry 25·5% and services 39·3%.

Overview

In 1997 Laos economy came under severe strain when the currency collapsed as a result of the Asian economic crisis. However, the economy has achieved impressive and steady growth in recent years, in large part owing to the enhanced role of export-oriented mining and hydropower projects. Growth in the natural resources sector stems particularly from the Nam Theun hydropower project (NT2), which began operation in 2010. With the majority of its electricity exported to Thailand, it is estimated it will earn Laos US$80m. a year in its first 25 years of operation.

Previously high levels of FDI for the production of gold, copper and export-oriented hydropower declined during the global financial crisis but rebounded in 2010. The number of households categorized as poor fell from 45% in 1992–93 to 31% in 2005. The National Socio-Economic Development Plan (NSEDP), which has received World Bank backing, aims to sustain economic growth and reduce poverty while promoting modernization and safeguarding environmental resources. The seventh five-year plan (2011–15) aims to remove Laos from the UN list of least developed countries by 2020.

Currency

The unit of currency is the *kip* (LAK). Inflation was 128·4% in 1999 but has since fallen steeply reaching 0·0% in 2009. The following year it increased to 6·0% and in 2011 further to 7·6%. Foreign exchange reserves were US$199m. in March 2005 and gold reserves were 117,000 troy oz. Total money supply was 1,364·1bn. kip in Feb. 2005.

Budget

The fiscal year begins on 1 Oct. Revenues in 2008–09 were 8,065bn. kip (tax revenue, 78·6%) and expenditures 9,783bn. kip (current expenditure, 58·3%).

VAT is 10%.

Performance

Real GDP growth was 7·5% in 2009, 8·1% in 2010 and 8·0% in 2011. Total GDP in 2011 was US$8·3bn.

Banking and Finance

The central bank and bank of issue is the State Bank (*Governor*, Sompao Phaysith). There were 17 commercial banks in 2002 (seven foreign; branches only permitted). Total savings and time deposits in 1991 amounted to 4,075m. kip. External debt was US$5,559m. in 2010, representing 79·0% of GNI.

A stock exchange opened in Vientiane in Jan. 2011.

ENERGY AND NATURAL RESOURCES

Environment

In 2008 carbon dioxide emissions from the consumption and flaring of fossil fuels were the equivalent of 0·2 tonnes per capita.

Electricity

Total installed capacity in 2007 was an estimated 742,000 kW, of which around 516,000 kW was hydro-electric. In 2007 production was about 3,667m. kWh, almost exclusively hydro-electric. Consumption per capita was an estimated 153 kWh; approximately 2,970 kWh were exported.

Minerals

Estimated production in 2009 (in tonnes): gypsum, 775,000; lignite (2007), 682,000; coal (2007), 305,000; refined copper, 68,000; salt, 35,000.

Agriculture

In 2002, 76·1% of the economically active population were engaged in agriculture. There were an estimated 920,000 ha. of arable land in 2002 and 81,000 ha. of permanent crop land. The chief products (2002 estimates in 1,000 tonnes) are: rice, 2,416; sugarcane, 222; sweet potatoes, 194; maize, 124; melons and watermelons, 116; cassava, 83; bananas, 53; pineapples, 36; potatoes, 35; coffee, 32; oranges, 29; tobacco, 27.

Livestock (2003 estimates): pigs, 1·65m.; cattle, 1·20m.; buffaloes, 1·08m.; chickens, 20m.

Livestock products (2003 estimates, in 1,000 tonnes): pork, bacon and ham, 36; beef and veal, 22; poultry, 18; milk, 6; eggs, 13.

Forestry

Forests covered 15·75m. ha. in 2010, or 68% of the land area. They produce valuable woods such as teak. Timber production, 2007, 6·14m. cu. metres.

Fisheries

The catch in 2007 was 28,410 tonnes, entirely from inland waters.

INDUSTRY

Production in 2002: cement, 201,000 tonnes; iron bars, 13,000 tonnes; detergent, 650 tonnes; nails, 650 tonnes; corrugated iron, 2·8m. sheets; plywood, 2·1m. sheets; mineral water, 235m. litres; beer, 60·49m. litres; soft drinks, 13·15m. litres; oxygen, 21,500 cylinders; cigarettes, 38·3m. packets; lumber, 155,000 cu. metres.

Labour

The working age is 16–55 for females and 16–60 for males. In 2003 the labour force totalled 2,672,900. 82·2% of the employed population were engaged in agriculture in 2003.

INTERNATIONAL TRADE

Since 1988 foreign companies have been permitted to participate in Lao enterprises.

Imports and Exports

Imports were estimated at US$2,816m. in 2008 and exports at US$1,6390m. The main imports in 2000 were: consumption goods, 50·6%; mineral fuels, 13·9%; materials for garment assembly, 10·6%. Main exports: electricity, 32·0%; garments, 26·1%; wood products, 24·8%. Main import suppliers, 2001: Thailand, 52·0%; Vietnam, 26·5%; China, 5·7%; Singapore, 3·3%. Main export markets, 2001: Vietnam, 41·5%; Thailand, 14·8%; France, 6·1%; Germany, 4·6%.

COMMUNICATIONS

Roads

In 2006 there were 29,811 km of roads, of which 13·5% were paved. In 2007 there were 12,800 passenger cars (two per 1,000 inhabitants), 109,000 lorries and vans, 6,400 buses and coaches, and 506,500 motorcycles and mopeds. There were 5,198 traffic accidents with 608 fatalities in 2006. A bridge over the River Mekong, providing an important north-south link, was opened in 1994.

Rail

A 3·5-km stretch of railway from Nongkhai, on the Thai bank of the Mekong River, across the Thai–Lao Friendship Bridge to Thanaleng in Laos was opened in 2009.

Civil Aviation

There are three international airports at Vientiane (Wattay), Pakse and Luang Prabang. The national carrier is Lao Airlines, which in 2005 operated domestic services and international flights to Bangkok, Chiang Mai, Hanoi, Ho Chi Minh City, Kunming, Phnom Penh and Siem Reap (Cambodia). In 2006 scheduled airline traffic of Laos-based carriers flew 4m. km, carrying 327,000 passengers (81,000 on international flights).

Shipping

The River Mekong and its tributaries are an important means of transport. In Jan. 2008 there were two ships of 300 GT or over registered, totalling 3,000 GT.

Telecommunications

In 2011 there were 107,600 landline telephone subscriptions (equivalent to 17·1 per 1,000 inhabitants) and 5,480,900 mobile phone subscriptions (or 871·6 per 1,000 inhabitants). In 2011, 9·0% of the population were internet users. In March 2012 there were 156,000 Facebook users.

SOCIAL INSTITUTIONS

Justice

Criminal legislation of 1990 established a system of courts and a prosecutor's office. Polygamy became an offence.

Education

In 2007 there were 891,807 pupils in primary schools with 29,604 teaching staff, and 403,833 pupils and 17,110 teaching staff at secondary level.

There are eight teacher training institutes (four teacher training colleges and four teacher training schools) and one college of Pali. In June 1995 the National University of Laos (NUOL) was established by merging nine existing higher education institutes and a centre of agriculture. NUOL comprises faculties in agriculture, pedagogy, political science, economics and management, forestry, engineering and architecture, medical science, humanities and social science, science, and literature. In 2007 there were 75,003 students in higher education and 3,030 academic staff.

Adult literacy in 2005 was 73%. Laos has only a small educated elite.

In 2007 public expenditure on education came to 3·6% of GNI and 15·8% of total government spending.

Health
In 2003 there were 24 hospitals (with 2,711 beds), 125 district-level hospitals and 662 primary health care centres. In 2003 there were 1,283 physicians, 83 dentists and 5,291 nurses. Only 37% of the population had access to safe drinking water in 2000.

RELIGION
In 2001 some 2·75m. were Buddhists (Hinayana), but about 40% of the population follow tribal religions.

CULTURE
World Heritage Sites
Laos has two sites on the UNESCO World Heritage List: the Town of Luang Prabang (inscribed on the list in 1995), a unique blend of Lao and European colonial architecture; and Vat Phou and Associated Ancient Settlements within the Champasak Cultural Landscape (2001), including a Khmer era Hindu temple complex.

Press
In 2008 there were six paid-for national dailies with a combined circulation of 25,000. In the 2010 *World Press Freedom Index* compiled by Reporters Without Borders, Laos ranked 168th out of 178 countries.

Tourism
There were 1,624,000 non-resident visitors in 2007 (including 1,404,000 from elsewhere in Asia and Oceania, 152,000 from Europe and 61,000 from the Americas); revenue from tourism (excluding passenger transport) amounted to US$233m.

Festivals
The national day is celebrated on 2 Dec. The Boun Bang Fai (Rocket Festival) takes place at the beginning of the rainy season in May and is thought to have originated in the pre-Buddhist fertility rites and ceremonies to induce the coming of the rains. It is celebrated with parades, songs and dances and homemade bottle rockets and also coincides with the Visakha Puja festival that commemorates the life of Buddha.

DIPLOMATIC REPRESENTATIVES
Of Laos in the United Kingdom
Ambassador: Khouanta Phalivong (resides in Paris).

Of the United Kingdom in Laos (Rue J. Nehru, Phonexay, Saysettha District, Vientiane)
Ambassador: Philip Malone.

Of Laos in the USA (2222 S. St., NW, Washington, D.C., 20008)
Ambassador: Seng Soukhathivong.

Of the USA in Laos (19 Rue Bartholonie, That Dam, Vientiane)
Ambassador: Karen B. Stewart.

Of Laos to the United Nations
Ambassador: Saleumxay Kommasith.

Of Laos to the European Union
Ambassador: Southam Sakonhninhom.

FURTHER READING
National Statistical Centre. *Basic Statistics about the Socio-Economic Development in the Lao P.D.R.* Annual.

Evans, Grant, *A Short History of Laos: The Land in Between.* 2002
Stuart-Fox, M., *Laos: Politics, Economics and Society.* 1986—*History of Laos.* 1997

National Statistical Office: National Statistical Centre, Committee for Planning and Investment, Luang Prabang Road, Vientiane.
Website: http://www.nsc.gov.la

LATVIA

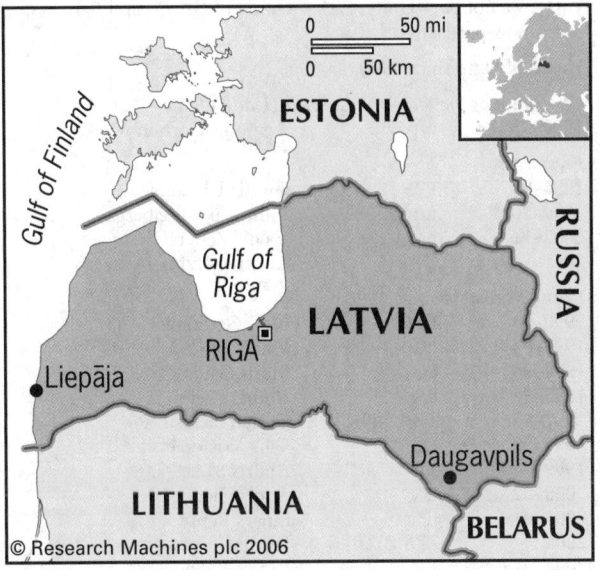

Latvijas Republika
(Republic of Latvia)

Capital: Riga
Population projection, 2015: 2·21m.
GNI per capita, 2011: (PPP$) 14,293
HDI/world rank: 0·805/43
Internet domain extension: .lv

KEY HISTORICAL EVENTS

The name Latvia derives from *latvis*, a 'forest clearer'. Human inhabitation dates from around 9000 BC and the Balts (or proto-Balts) probably arrived around 2000 BC. In addition to the Finnic predecessors of the Estonians and Livs, four Baltic tribal groups emerged during the Iron Age: the Couronians (Kurši), Selonians (Sēļi), Semigallians (Zemgaļi) and Latgallians (Latgaļi). Linguistic evidence points to the habitation of central Latvia by Latgallians and Lithuanians while coastal areas were populated by Couronians, Semigallians, Selonians and Prussians. The north of the country, occupied mainly by Finnic Livs, was separated by sparsely inhabited areas, which accounts for the lack of cultural mixing between the ethnic groupings.

Scandinavian settlements were established after AD 650, disappearing abruptly around 850. In the 10th century the Baltic tribes came under attack from Varangians (Swedish Vikings), attracted in part by amber which was traded up the Daugava River to Russia and south to Byzantium. Slavic incursions from the east were hampered by primeval forest and marshes.

The region was transformed by German colonization in the 13th century. The Archbishop of Bremen ordered the Christian conquest of the Eastern Baltic sending his nephew Albert, who founded Riga in 1201 and became its bishop. He also created the Sword Brothers (or Livonian Order), a small military order that carved out a feudal state ruled mainly by German aristocracy. It subdued Livonia (southern Estonia and northeastern Latvia), Courland and Zemgale, forming the Livonian Confederation. In 1237 the weakened Sword Brothers were incorporated as an autonomous order into the Order of the Teutonic Knights, which completed the conquest of modern-day Latvia. Riga joined the Hanseatic League in 1282.

The Teutonic Knights were defeated in 1410 at the Battle of Grünwald (Tannenberg) by a Polish–Lithuanian army. Russian Tsar Ivan IV invaded in 1558 in an attempt to gain access to the Baltic Sea. The Livonian War ended with the disbandment of the Livonian Order and the partition of Livonia in 1561. The north went to Lithuania, while Courland became a Lithuanian fief. During the early 17th century Latvia was a theatre of war between Sweden and the Commonwealth of Lithuania–Poland. In 1605 the small hussar army of Jan Karol Chodkiewicz destroyed a much larger Swedish force at the Battle of Kircholm (Salaspils), near Riga. However, Sweden had taken control of Livonia by 1621 and its empire in the Baltic did not end until the signing of the Treaty of Nystad in 1721. The Livonian territory of Vidzeme passed to Russia which acquired Latgale (Latgallia) in 1772 from Poland, and Courland in 1795.

The German landowners remained within the Russian Empire, bolstered by unification with other Germans in the Baltic. The abolition of serfdom in 1817 created new tensions between the Latvian and German communities, though there were German elements who supported land reform which duly arrived in 1847. Nascent Latvian nationalism in the 1860s prompted the Russian authorities to centralize power away from the German aristocracy. Tensions between the communities erupted during the 1905 Russian Revolution when attacks on hundreds of German settlements precipitated a wave of German emigration.

After the German occupation of 1915 demands for the creation of a Latvian state grew until the 1917 Russian Revolution. Soviet rule was proclaimed in Dec. 1917 but was overthrown when the Germans reinvaded in Feb. 1918, with Russia ceding its claims on Latvia by the Treaty of Brest-Litovsk. After the armistice, Latvia declared independence but the Soviets reasserted power following the German withdrawal in Dec. 1918. The Soviets were again overthrown between May–Dec. 1919 by combined British naval and German military forces and a democratic government was established.

In the wake of economic depression, the democratic regime fell in a May 1934 coup when Prime Minister Kārlis Ulmanis dissolved parliament and, in 1936, merged his office with that of the president. When the secret protocol of the Soviet–German agreement of 23 Aug. 1939 assigned Latvia to the Soviet sphere of interest, most of the ancient Baltic German community emigrated. Formal annexation came in 1940 and over 15,000 Latvians were deported to Siberia. The German occupation of 1941–45 caused the deaths of over 90,000 Latvians, mostly Jews. Violent resistance to Soviet reoccupation led to the deportation of another 43,000 Latvians to Siberia and substantial Soviet immigration followed.

The Latvian Supreme Soviet declared sovereignty in July 1989 and on 4 May 1990, having declared the 1940 Soviet occupation illegal, re-established the 1922 constitution. This was annulled by Soviet President Mikhail Gorbachev, sparking violent protest and police crackdowns. Following a referendum, independence was declared on 21 Aug. 1991 and recognized in Sept. by the USSR. With independence, issues of ethnic identity came to the fore, with the large Russian minority initially disadvantaged by new citizenship and language laws (later repealed). President Vaira Vīķe-Freiberga was elected as the former Communist bloc's first female president in 1999. Latvia became a member of NATO and the European Union in 2004.

TERRITORY AND POPULATION

Latvia is situated in northeastern Europe. It is bordered by Estonia on the north and by Lithuania on the southwest, while on the east there is a frontier with the Russian Federation and to the southeast with Belarus. Territory, 64,559 sq. km (larger than Denmark, the Netherlands, Belgium and Switzerland), including 2,402 sq. km of inland waters. Population (2011 census), 2,070,371; density, 32·1 per sq. km.

The UN gives a projected population for 2015 of 2·21m.

In 2006, 68·0% of the population were urban. Major ethnic groups in 2006: Latvians 59·0%, Russians 28·5%, Belarusians 3·8%, Ukrainians 2·5%, Poles 2·4%, Lithuanians 1·4%, Jews 0·4%, Roma 0·4%, Germans 0·2%, Estonians 0·1%.

There are 110 municipalities (*novadi*) and nine republican cities (*republikas pilsētas*). The capital is Riga (658,640, or nearly a third of the country's total population, at the 2011 census); other principal towns, with 2011 populations, are Daugavpils (93,312), Liepāja (76,731), Jelgava (59,511), Jurmala (50,840) and Ventspils (38,750).

The official language is Latvian. Latgalian is also spoken.

SOCIAL STATISTICS

2010: births, 19,219 (rate of 8·6 per 1,000 population); deaths, 30,040 (13·4 per 1,000 population); marriages, 9,290 (4·1 per 1,000 population); divorces, 4,930 (2·2 per 1,000 population); infant mortality, 5·7 per 1,000 live births (2010). In 2007 life expectancy was 67·1 years for males but 77·1 years for females. In 2005 the most popular age range for marrying was 25–29 for males and 20–24 for females. The annual population growth rate in the period 2000–05 was –0·6%. Fertility rate, 2011, 1·2 births per woman (the joint lowest rate in the world). The suicide rate was 22·9 per 100,000 population in 2009 (rate among males, 40·0). In 2005 there were 1,886 immigrants and 2,450 emigrants.

CLIMATE

Owing to the influence of maritime factors, the climate is relatively temperate but changeable. Average temperatures in Jan. range from –2·8°C in the western coastal town of Liepāja to –6·6°C in the inland town of Daugavpils. The average summer temperature is 20°C.

CONSTITUTION AND GOVERNMENT

The Declaration of the Renewal of the Independence of the Republic of Latvia dated 4 May 1990, and the 21 Aug. 1991 declaration re-establishing *de facto* independence, proclaimed the authority of the Constitution (*Satversme*). The Constitution was fully re-instituted as of 6 July 1993, when the fifth Parliament (*Saeima*) was elected.

The head of state in Latvia is the *President*, elected by parliament for a period of four years and for a maximum of two terms.

The highest legislative body is the one-chamber parliament comprised of 100 deputies and elected in direct, proportional elections by citizens 18 years of age and over. Deputies serve for four years and parties must receive at least 5% of the national vote to gain seats in parliament.

In a referendum on 3 Oct. 1998, 53% of votes cast were in favour of liberalizing laws on citizenship, which would simplify the naturalization of the Russian-speakers who make up nearly a third of the total population and who were not granted automatic citizenship when Latvia regained its independence from the former Soviet Union in 1991. Around half of the 650,000 ethnic Russians in Latvia have not taken out Latvian citizenship. Ethnic Russians who are not Latvian citizens do not have the right to vote. A seven-member *Constitutional Court* was established in 1996 with powers to invalidate legislation not in conformity with the constitution. Its members are appointed by parliament for ten-year terms.

Executive power is held by the *Cabinet of Ministers*.

National Anthem

'Dievs, svēti Latviju' ('God bless Latvia'); words and tune by Kārlis Baumanis.

GOVERNMENT CHRONOLOGY

(JL = New Era; LC = Latvian Way; LTF = Latvian Popular Front; LZP = Latvian Green Party; LZS = Latvian Farmers' Union; TB/LNNK = For Fatherland and Freedom/LNNK; TP = People's Party; Vienotība = Unity, ZZS = Union of Greens and Farmers; n/p = non-partisan)

Heads of State since 1990.

Chairman of the Supreme Council/Head of State
1990–93	n/p, LC	Anatolijs Gorbunovs

Presidents
1993–99	LZS	Guntis Ulmanis
1999–2007	n/p	Vaira Vīķe-Freiberga
2007–11	n/p	Valdis Zatlers
2011–	ZZS	Andris Bērziņš

Prime Ministers since 1990.
1990–93	LTF	Ivars Godmanis
1993–94	LC	Valdis Birkavs
1994–95	LC	Māris Gailis
1995–97	n/p	Andris Šķēle
1997–98	TB/LNNK	Guntars Krasts
1998–99	LC	Vilis Krištopāns
1999–2000	TP	Andris Šķēle
2000–02	LC	Andris Bērziņš
2002–04	JL	Einars Repše
2004	ZZS (LZP)	Indulis Emsis
2004–07	TP	Aigars Kalvītis
2007–09	LC	Ivars Godmanis
2009–	JL, Vienotība	Valdis Dombrovskis

RECENT ELECTIONS

In presidential elections held on 2 June 2011 the first round of voting proved inconclusive with Andris Bērziņš receiving 50 votes and incumbent Valdis Zatlers 43. In a second round held later on the same day Bērziņš was elected with 53 votes against 41 for Zatlers.

Parliamentary elections were held on 17 Sept. 2011. Harmony Centre (Saskanas Centrs/SC; comprises Social Democratic Party 'Harmony' and Socialist Party of Latvia) won 31 seats with 28·4% of the votes cast; Zatlers' Reform Party 22 with 20·8%; Unity (Vienotība) 20 seats with 18·8%; National Alliance (Nacionālā Apvienība) 14 seats with 13·9%; and Union of Greens and Farmers (Zaļo un Zemnieku savienība/ZZS; comprises Latvian Farmers' Union and Latvian Green Party), 13 with 12·2%. Turnout was 59·5%.

European Parliament
Latvia has nine representatives. At the June 2009 elections turnout was 53·7% (41·3% in 2004). The Civic Union (Pilsoniskā savienība) won 2 seats with 24·3% of votes cast (political affiliation in European Parliament: European People's Party); SC, 2 with 19·6% (one with Progressive Alliance of Socialists and Democrats and one with European United Left/Nordic Green Left); For Human Rights in a United Latvia (Par cilvēka tiesībām vienotā Latvijā), 1 with 9·7% (Greens/European Free Alliance); Latvia's First Party/Latvian Way Party (Latvijas Pirmā Partija/ Savienība 'Latvijas ceļš'), 1 with 7·5% (Alliance of Liberals and Democrats for Europe); For Fatherland and Freedom/ LNNK ('Tēvzemei un Brīvībai'/LNNK), 1 with 7·5% (European Conservatives and Reformists); New Era (Jaunais laiks), 1 with 6·7% (European People's Party).

CURRENT GOVERNMENT

President: Andris Bērziņš; b. 1944 (ZZS; sworn in 8 July 2011).

Prime Minister and Minister for Regional Development and Local Government: Valdis Dombrovskis; b. 1971 (Unity; took office on 12 March 2009). In March 2013 the coalition of Unity, Zatlers' Reform Party (ZRP) and the National Alliance comprised:

Minister for Defence: Artis Pabriks (Unity). *Foreign Affairs:* Edgars Rinkēvičš (ZRP). *Economics:* Daniels Pavļuts (ZRP). *Finance:* Andris Vilks (Unity). *Education and Science:* Robert Ķīlis (ZRP). *Culture:* Žaneta Jaunzeme-Grende (National Alliance). *Welfare:* Ilze Viņķele (Unity). *Environmental Protection and Regional Development:* Edmunds Sprūdžs (ZRP). *Transport:* Aivis Ronis (ind.). *Justice:* Jānis Bordāns (National Alliance). *Interior:* Rihards Kozlovskis (ZRP). *Health:* Ingrīda Circene (Unity). *Agriculture:* Laimdota Straujuma (Unity).

Office of the President: http://www.president.lv

CURRENT LEADERS

Andris Bērziņš

Position
President

Introduction
Andris Bērziņš became president in June 2011 after an election dominated by rows over parliamentary links to business. A former banker, Bērziņš leads the Union of Greens and Farmers and advocates economic reforms and the consolidation of ties with the European Union (EU).

Early Life
Andris Bērziņš was born on 10 Dec. 1944 in Nītaure. After schooling in Nitaure and Sigulda, he studied radio engineering at the Riga Polytechnic Institute from 1966–71. He worked for R/A Elektrons from 1971–88, eventually as managing director. From 1987–88 he studied industrial planning at the University of Latvia, and from 1988–89 served as deputy minister of municipal services of the Latvian SSR. He was elected to the district council of the city of Valmiera in 1989, serving as its chairman until 1993. From 1990–93 he represented Valmiera on the Supreme Council of the Republic of Latvia, where he joined the Popular Front grouping. On 4 May 1990 he voted for Latvian independence from the USSR.

Between 1990 and 1992 Bērziņš served on the supervisory council of the Bank of Latvia, as the country began its conversion to a market economy. In 1993 he was appointed president of the newly founded commercial bank, Latvijas Unibanka, overseeing its rapid growth until leaving the post in 2004. During this period he extended his own business interests into property.

In 2005 Bērziņš unsuccessfully stood for mayor of Riga, representing the Union of Greens and Farmers. He was president of the Latvian Chamber of Commerce and Industry from 2006 until 2010, and from 2007–09 was chairman of the supervisory council of the state-owned Latvenergo electricity company. In 2010 he entered parliament as a member of the Union of Greens and Farmers, which formed a coalition government with the Unity party.

From 2010 he served a year as chairman of the economic, agricultural, environmental and regional policy committee. In June 2011 he was elected president by parliament, beating the incumbent Valdis Zatlers in a second round of voting after Zatlers had accused MPs of being lax on corruption.

Career in Office
Bērziņš took office on 8 July 2011, promising to work independently of Latvia's oligarchs. After an inconclusive general election in Sept. 2011, he reappointed the incumbent, Valdis Dombrovskis, as prime minister, and saw his own party left out of the coalition government. Bērziņš has prioritized economic growth, consolidating Latvia's position in the EU and restoring trust in the nation's democratic institutions.

Valdis Dombrovskis

Position
Prime Minister

Introduction
Valdis Dombrovskis became prime minister in March 2009 following the resignation of Ivars Godmanis. The former finance minister's tenure has been mainly focused on reversing the country's economic crisis.

Early Life
Dombrovskis was born on 5 Aug. 1971 in Riga, while Latvia was part of the USSR. He studied physics and economics at the University of Latvia in Riga and the Riga Technological University. He was employed at Germany's University of Mainz and at the University of Latvia's Institute of Solid-State Physics, and in 1998 became a research assistant at Maryland University in the USA.

Later that year he began working for the Bank of Latvia, leaving in 2002 after a year as chief economist. A member of the centre-right New Era party, he joined its governing board in 2002 and was also elected to parliament, serving until 2004 as finance minister in Einars Repše's government. From 2003–04 he was Latvia's observer at the Council of the European Union and became a member of the European Parliament (MEP) in 2004.

Against the backdrop of global economic turmoil, Latvia's economy crashed in 2008. The economy shrank by over 4%. Worse was to follow, with a contraction of 18% in 2009. In Dec. 2008, having nationalized the country's second biggest bank, Prime Minister Ivars Godmanis turned to the IMF, World Bank and European Union for a US$9·5bn. bail-out package. In return he was forced to accept public spending cuts and tax increases. After an anti-government riot in Riga in Jan. 2009, his coalition fell the following month. President Valdis Zatlers nominated Dombrovskis to form a new administration.

Career in Office
Having resigned as an MEP, Dombrovskis formed a six-party coalition and won parliamentary approval on 12 March 2009. In Aug. 2009 his government reached agreement with unions and employers on deep spending cuts and tax rises aimed at staving off bankruptcy and persuading the IMF and EU to release further tranches of loans. Subsequently, as a condition for these disbursements, the government committed to further cuts in fiscal expenditures to contain the budget deficit. Having lost his parliamentary majority in March 2010, Dombrovskis called an election in Oct. at which he was returned to power heading a Unity and Union of Greens and Farmers coalition. In further legislative elections in Sept. 2011, following a controversial dissolution of the *Saeima*, the pro-Russian Harmony Centre emerged as the largest party, but Dombrovskis retained the premiership at the head of a new coalition comprising Unity, Zatlers' Reform Party and the National Alliance, which together controlled 56 of the 100 parliamentary seats. In Feb. 2012 voters in a referendum rejected a proposal to make Russian an official language of Latvia.

DEFENCE

The National Armed Forces (NAF) were created in 1994 and comprise the Land Forces, which are based on an infantry brigade and the National Guard, the Naval Forces, the Air Forces, the Logistic Command, the Training Doctrine Command and the National Defence Academy. Compulsory military service was abolished in Jan. 2007.

In 2008 military expenditure totalled US$542m. (US$241 per capita), representing 1·6% of GDP.

Army

The Land Forces were 1,526 strong in 2007. There is a National Guard reserve numbering 10,483 in 2007.

Navy

The Naval Forces, based at Riga and Liepāja, numbered 603 in 2007. Latvia, Estonia and Lithuania have established a joint naval unit 'BALTRON' (Baltic Naval Squadron), with bases at Liepāja, Riga and Ventspils in Latvia, Tallinn in Estonia and Klaipėda in Lithuania.

Air Force

Personnel numbered 480 in 2007. There are no combat-capable aircraft.

INTERNATIONAL RELATIONS

Latvia held a referendum on EU membership on 20 Sept. 2003, in which 67·4% of votes cast were in favour of accession, with 32·6% against. It became a member of NATO on 29 March 2004 and the EU on 1 May 2004.

ECONOMY

Services accounted for 76·1% of GDP in 2009, industry 20·6% and agriculture 3·3%.

Overview

With Latvia's transition from communism in the early 1990s, the economy suffered from raw material and energy shortages, the loss of Soviet export markets and weak international competitiveness. After independence, market reforms were introduced including privatization, price liberalization, land reforms and the establishment of a local currency and an independent central bank.

After joining the EU in May 2004, growth accelerated to double-digit figures driven by private sector capital inflows and EU funds. However, by mid-2006 the economy showed signs of overheating, inflation accelerated and the current account deficit peaked at over 20% of GDP. In 2008 the global financial crisis led to a slowdown in lending driven by the withdrawal of funds from foreign banks. In Dec. 2008 Latvia accepted a US$9·5bn. IMF-led bailout, causing widespread social turmoil and the resignation of the ruling coalition in Feb. 2009.

A new coalition government agreed a series of spending cuts, including significant reductions in public sector pay and pensions to meet IMF targets to reduce the budget deficit. The economy contracted by nearly 18% in 2009 and unemployment was the highest in the EU at nearly 21%. However, the economy emerged from recession in the first quarter of 2010 and fiscal adjustment measures brought the fiscal deficit down from 9·7% in 2009 to 4% of GDP in 2011. International reserves have risen above pre-crisis levels, while wage and price cuts, along with productivity growth, have improved competitiveness and reduced external imbalances. Nonetheless, unemployment remains at over 16%, poverty rates are among the highest in Europe, and the overspill effects of the eurozone crisis have left the economy vulnerable to further downturns.

Currency

The unit of currency is the *lats* (LVL) of 100 *santims*. The lats is pegged to the euro at a rate of one euro to 0·7028 lats. Latvia hopes to adopt the euro as its currency from 1 Jan. 2014. Inflation, which reached a high of 109·1% in 1993, was 15·3% in 2008 but plummeted and there was deflation of 1·2% in 2010. In 2011 inflation stood at 4·2%. Gold reserves were 249,000 troy oz in July 2005, foreign exchange reserves US$2,076m. and total money supply 1,736m. lats.

Budget

The financial year is the calendar year. Central government revenue totalled 3,783·3m. lats in 2010 (3,916·4m. lats in 2009)

and expenditure was 4,537·2m. lats (4,603·3m. lats in 2009). Tax revenue constituted 43·0% of revenues in 2010 and social contributions 28·9%; social benefits accounted for 36·5% of expenditures and subsidies 24·2%. There is a flat income tax rate of 24%.

Latvia's budget deficit in 2011 was 3·5% of GDP (2010, 8·2%; 2009, 9·8%). The required target set by the EU is a budget deficit of no more than 3%.

The standard rate of VAT is 21·0% (reduced rate, 12·0%).

Performance

In 2007 real GDP growth was 9·6%. However, having contracted by 3·3% in 2008 the economy shrank by 17·7% in 2009 (the worst rate in the world) and a further 0·3% in 2010. It then began to recover, growing by 5·5% in 2011. Latvia has the lowest GDP per capita of any of the ten countries that joined the EU in May 2004. Total GDP was US$28·3bn. in 2011.

Banking and Finance

The Bank of Latvia both legally and practically is a completely independent institution. Governor of the Bank and Council members are appointed by Parliament for office for six years (present *Governor*, Ilmārs Rimšēvičs). In 2010 there were 29 banks in Latvia, including nine foreign banks' branches. Latvia's banks have 11 branches abroad. Latvian banks' assets to GDP exceeded 170% at the end of 2010. The banking sector assets constitute almost 90% of Latvia's financial sector assets. Foreign direct investment net inflows in 2010 totalled US$369m. The accumulated FDI at the end of 2010 reached US$10·75bn. Foreign debt was US$39,555m. in 2010, equivalent to 164·3% of GNI.

NASDAQ OMX Riga (formerly Riga Stock Exchange) is the only regulated secondary securities market in Latvia.

ENERGY AND NATURAL RESOURCES

In 2008, 29·9% of energy consumption came from renewables (wind power, solar power, hydro-electric power, tidal power, geothermal energy and biomass), compared to the European Union average of 10·3%. A target of 40% has been set by the EU for 2020.

Environment

Latvia's carbon dioxide emissions from the consumption and flaring of fossil fuels in 2008 were the equivalent of 4·4 tonnes per capita. An *Environmental Performance Index* compiled in 2008 ranked Latvia eighth in the world, with 88·8%. The index examined various factors in six areas—air pollution, biodiversity and habitat, climate change, environmental health, productive natural resources and water resources. Latvia's greenhouse gas emissions fell by 55·6% between 1990 and 2008, mainly owing to the decline of polluting industries from the Soviet era.

Electricity

Electricity production in 2007 totalled 4·77bn. kWh. Consumption per capita in 2007 was 3,414 kWh. 57% of electrical power produced in Latvia is generated in hydro-electric power stations. Installed capacity was about 2·1m. kW in 2007.

Oil and Gas

Latvia produces virtually no oil and is dependent on imports, although the Latvian Development Agency estimates that there are 733m. bbls of offshore reserves in the Latvian areas of the Baltic Sea. Consumption of oil products was 1,488,000 tonnes in 2010. All Latvia's natural gas supplies are imported from Russia. Consumption in 2010 totalled 620m. cu. metres.

Minerals

Peat deposits extend over 645,000 ha. or about 10% of the total area, and it is estimated that total deposits in Jan. 2012 were 168m. tonnes. Peat output in 2011 (provisional) totalled 1,387,689 tonnes.

Production of other minerals in 2011 (provisional, in 1,000 tonnes): gravel, pebbles, shingle and flint, 5,642; silica and construction sand, 2,338; crushed stone for concrete aggregate, 2,051. Crude dolomite, clays, gypsum and limestone are also produced.

Agriculture
In 2010 there were 1·17m. ha. of arable land and 6,800 ha. of permanent crops (excluding strawberries). Field crops and dairy farming are the chief agricultural occupations. In 2010 there were 83,300 economically active agricultural holdings. 34% of holdings have fewer than 5 ha of utilized agricultural land. There were 50,000 tractors, 6,500 grain combine harvesters and 1,500 potato combine harvesters in 2010. In 2010, 8·8% of the economically active population were employed in agriculture, forestry and fishing.

Output of crops (in 1,000 tonnes), 2010: wheat, 989; potatoes, 484; barley, 229; rapeseeds, 226; oats, 101; rye, 70; cabbage, 61; carrots, 34; apples, 10. Livestock, 2010: pigs, 390,000; cattle, 380,000; sheep, 77,000; poultry, 4·9m. Livestock products (2010, in 1,000 tonnes): meat, 80; milk, 835; eggs, 45.

Forestry
In 2010 Latvia's total forest area was 3·35m. ha., or 54% of the land area. Growing stock was 633m. cu. metres in 2010 (335m. cu. metres coniferous and 298m. cu. metres broadleaved). Timber production in 2007 was 12·17m. cu. metres.

The share of the forest sector in gross industrial output is between 13 and 15%. Timber and timber products exports account for 37–38% of Latvia's total exports.

To provide the protection of forests there are three forest categories: commercial forests, 70·4%; restricted management forests, 18·6%; protected forests, 11·0%.

Fisheries
In 2010 the total catch was 164,819 tonnes, almost exclusively marine fish. The Latvian fishing fleet numbered 758 vessels of 9,100 gross tonnes in 2011.

INDUSTRY
Industry accounted for 20·6% of GDP in 2009, with manufacturing contributing 9·9%.

Industrial output in 1,000 tonnes: cement (2009 estimate), 650; crude steel (2009 estimate), 550; steel products (2000), 549; sawnwood (2009), 2·23m. cu. metres; wood-based panels (2009), 492,323 cu. metres; plywood (2009), 212,612 cu. metres; beer (2009), 129·2m. litres.

Labour
The total labour force (persons aged 15–74) in 2011 numbered 1,028,200. In 2011 there were 861,600 persons in employment in Latvia. The leading areas of activity were: wholesale and retail trade/repair of motor vehicles and motorcycles, 136,200; manufacturing, 114,400; education, 88,800. In 2011 women constituted 52% of the workforce. In 2011 there was a monthly minimum wage of 200 lats. Average gross monthly salary was 464 lats in 2011. The average gross monthly salary in the public sector in 2011 was 492 lats.

The unemployment rate (persons aged 15–74) in the second quarter of 2012 was 16·1%, one of the highest rates in the EU.

INTERNATIONAL TRADE
Imports and Exports
Imports (f.o.b.) were valued at US$11,143m. in 2010 and exports (f.o.b.) at US$8,851m. The leading imports are machinery and mechanical appliances (21·2%), products of chemical and allied industries (10·5%), mineral products (9·7%), metals and products thereof (8·4%). The main exports are wood and wood products (33·6%), base metals and articles of base metals (13·2%), textiles and textile articles (12·8%). Main import suppliers (2006):

Germany, 15·5%; Lithuania, 13·0%; Russia, 7·8%; Estonia, 7·7%; Poland, 7·2%. Main export markets (2006): Lithuania, 14·7%; Estonia, 12·7%; Germany, 10·1%; Russia, 8·9%; UK, 7·8%.

COMMUNICATIONS
Roads
In 2005 there were 66,319 km of roads, including 20,182 km of national roads. Public road transport totalled 2,869m. passenger-km in 2005 and freight 8,547m. tonne-km. There were 442 fatalities in traffic accidents in 2005. With 19·2 deaths per 100,000 population in 2005 Latvia has one of the highest death rates in road accidents of any industrialized country. Passenger cars in 2005 numbered 742,447 (324 per 1,000 inhabitants), in addition to which there were 113,113 trucks and vans, 25,193 motorcycles, 7,284 mopeds and 10,644 buses and coaches.

Rail
In 2005 there were 2,270 km of 1,520 mm gauge route (257 km electrified). In 2008, 56·1m. tonnes of cargo and 20·4m. passengers were carried by rail. The main groups of freight transported are oil and oil products, mineral fertilizers, ferrous metals and ferrous alloys.

Civil Aviation
There is an international airport at Riga. A new national carrier, airBaltic, assumed control of Latavio and Baltic International Airlines in Aug. 1995 and began flying in Oct. 1995. It went on to become eastern Europe's first low-cost airline; in 2012 it carried 3·08m. passengers and operated scheduled services to 55 destinations. It is 99·8% state-owned, with Transaero owning the remaining 0·2%. In 2010 Riga handled 4,663,692 passengers and 12,247 tonnes of freight.

Shipping
There are two major ports. Riga handled 28·6m. tonnes of cargo in 2008 and Ventspils 27·4m. tonnes. There is a smaller port at Liepāja. A total of 54·1m. tonnes were loaded at the three ports in 2008 and 6·0m. tonnes unloaded. In Jan. 2009 there were 28 ships of 300 GT or over registered, totalling 240,000 GT.

Telecommunications
Telecommunications are conducted by companies in which the government has a 51% stake, under the aegis of the state-controlled Lattelecom. There were 516,300 landline telephone subscriptions in 2011 (equivalent to 230·2 per 1,000 inhabitants) and 2,303,600 mobile phone subscriptions in 2009 (or 1,018·7 per 1,000 inhabitants). There were 684·2 internet users per 1,000 inhabitants in 2010. Fixed internet subscriptions totalled 320,700 in 2007 (140·5 per 1,000 inhabitants). In March 2012 there were 319,000 Facebook users.

SOCIAL INSTITUTIONS
Justice
A new criminal code came into force in 1998. Judges are appointed for life. There is a Supreme Court, regional and district courts and administrative courts. The death penalty was abolished for all crimes in Jan. 2012, having previously been abolished for peacetime offences in 1999. In 2009, 56,748 crimes were reported; 10,855 people were convicted for offences (79 for intentional homicide). In Jan. 2008 there were 6,548 people in penal institutions, giving a prison population rate of 288 per 100,000 population.

Education
Adult literacy rate in 2009 was estimated at 99·8% (99·8% among both males and females). The Soviet education system has been restructured on the UNESCO model. Education begins with two years of compulsory attendance at pre-primary education institutions. From the age of six or seven education is compulsory for nine years in comprehensive schools. This may be followed by

three years in secondary school or one to six years in art, technical or vocational schools. In 2011–12 there were 839 schools with 218,442 pupils.

State-financed education is available in Latvian and four national minority languages (Russian, Polish, Ukrainian and Belarusian), although the use of Latvian in the classroom is being increased. A bilingual curriculum had to be implemented by all minority primary schools from the start of the 2002–03 school year. Secondary schools started to implement minority education curricula with an increased Latvian-language component (60% of all teaching) from Sept. 2004. In 2011–12, 158,828 pupils were taught solely in Latvian, 58,087 received instruction in Russian and 1,527 in other minority languages.

In 2011–12 there were 34 state-recognized higher education institutions. Courses at state-financed universities are conducted in Latvian. A number of private educational institutions have languages of instruction other than Latvian.

Public expenditure on education in 2009 came to 5·6% of GDP and accounted for 12·1% of total government expenditure.

Health

At the end of 2010 there were 7,951 physicians and dentists, 10,024 nurses and 408 midwives. There were 67 hospitals in 2010 with a provision of 53 beds per 10,000 persons.

Welfare

The official retirement age is age 62 years for men and women. In 2008 the minimum pension was just over the state social security allowance of 45 lats a month at 49·50 lats. The minimum pension is increased by 1·1% for an insurance period of at least 20 years, by 1·3% for an insurance period of 20 to 30 years and by 1·5% for an insurance period of more than 30 years. In Dec. 2007 there were 567,400 pension recipients.

The government runs an unemployment benefit scheme in which the amount awarded is determined by the number of insurance contributions and the length of previous employment.

RELIGION

In order to practise in public, religious organizations must be licensed by the Department of Religious Affairs attached to the Ministry of Justice. New sects are required to demonstrate loyalty to the state and its traditional religions over a three-year period. Traditionally Catholics and Lutherans constitute the largest churches, with about 500,000 and 400,000 members respectively in 2002. Congregations in Feb. 2003: Lutherans, 307; Roman Catholics, 252; Russian Orthodox, 117; Baptists, 90; Old Believers, 67; Adventists, 47; Jews, 13; others, 47. In Feb. 2013 the Roman Catholic church had one cardinal.

CULTURE

Riga will be one of two European Capitals of Culture for 2014. The title attracts large European Union grants.

World Heritage Sites

Latvia has two sites on the UNESCO World Heritage List: the Historic Centre of Riga (inscribed on the list in 1997), a late-medieval Hanseatic centre; and the Struve Geodetic Arc (2005). The Arc is a chain of survey triangulations spanning from Norway to the Black Sea that helped establish the exact shape and size of the earth and is shared with nine other countries.

Press

Latvia had 19 daily newspapers in 2008 (17 paid-for and two free) with a combined circulation of 370,000. The leading newspapers in terms of readership in 2008 were *Diena* and *Latvijas Avīze*, both of which are in Latvian, and the Russian-language *Vesti Segodnya*.

Tourism

In 2010 there were 1,373,000 overnight non-resident tourists (1,323,000 in 2009). The main countries of origin of non-resident tourists in 2010 were Russia (189,000), Lithuania (182,000), Sweden (157,000) and Estonia (130,000).

Festivals

There is an annual Riga Opera Festival in June and the city also hosts the Lielais Kristaps National Film Festival in Sept.–Oct. The biggest music festival is Positivus Festival which takes place in July at Salacgrīva. The National Song Festival (held every five years) will next be held in 2018.

DIPLOMATIC REPRESENTATIVES

Of Latvia in the United Kingdom (45 Nottingham Place, London, W1U 5LY)
Ambassador: Eduards Stiprais.

Of the United Kingdom in Latvia (5 Alunana ielā, Riga, LV 1010)
Ambassador: Andrew Soper.

Of Latvia in the USA (2306 Massachusetts Ave., NW, Washington, D.C., 20008)
Ambassador: Andris Razāns.

Of the USA in Latvia (Samnera Velsa iela 1, Riga, LV 1510)
Ambassador: Mark A. Pekala.

Of Latvia to the United Nations
Ambassador: Normans Penke.

Of Latvia to the European Union
Permanent Representative: Ilze Juhansone.

FURTHER READING

Central Statistical Bureau. *Statistical Yearbook of Latvia.—Latvia in Figures.* Annual.

Dreifeld, J., *Latvia in Transition.* 1997
Hood, N., *et al.,* (eds.) *Transition in the Baltic States.* 1997
Kasekamp, Andres, *A History of the Baltic States.* 2010
Lieven, A., *The Baltic Revolution: Estonia, Latvia, Lithuania and the Path to Independence.* 2nd ed. 1994
Misiunas, R. J. and Taagepera, R., *The Baltic States: the Years of Dependence, 1940–90.* 2nd ed. 1993
O'Connor, Kevin, *The History of the Baltic States.* 2003
Plakans, Andrejs, *A Concise History of the Baltic States.* 2011
Smith, David J., Purs, Aldis, Pabriks, Artis and Lane, Thomas, (eds.) *The Baltic States: Estonia, Latvia and Lithuania.* 2002

National Statistical Office: Central Statistical Bureau, Lācplēša ielā 1, 1301 Riga.
Website: http://www.csb.lv

LEBANON

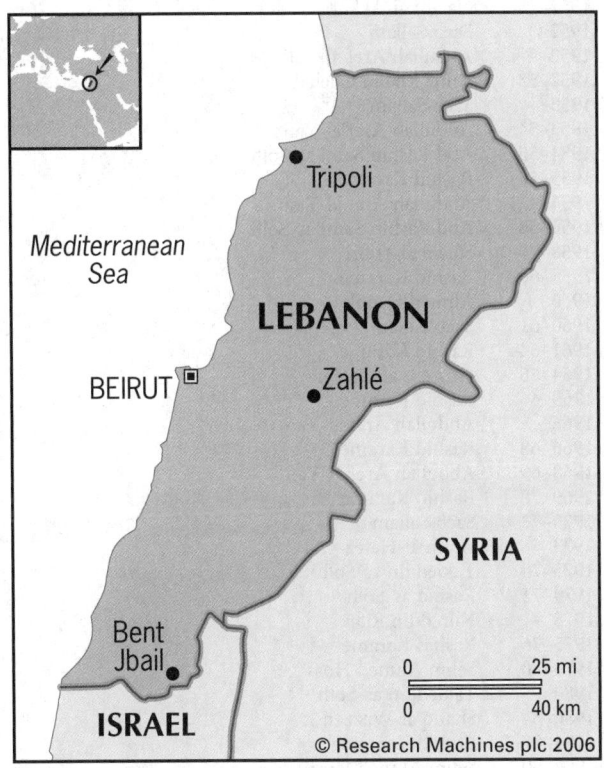

© Research Machines plc 2006

Jumhouriya al-Lubnaniya
(Republic of Lebanon)

Capital: Beirut
Population projection, 2015: 4·39m.
GNI per capita, 2011: (PPP$) 13,076
HDI/world rank: 0·739/71
Internet domain extension: .lb

KEY HISTORICAL EVENTS

The Ottomans invaded Lebanon, then part of Syria, in 1516–17 and held nominal control until 1918. After 20 years' of French mandatory regime, Lebanon was proclaimed independent on 26 Nov. 1941. In early May 1958 the Muslim opposition to President Chamoun rose in insurrection and for five months the Muslim quarters of Beirut, Tripoli, Sidon and the northern Bekaa were in insurgent hands. On 15 July the US Government landed army and marines who re-established Government authority. Internal problems were exacerbated by the Palestinian problem. An attempt to regulate the activities of Palestinian fighters through the secret Cairo agreement of 1969 was frustrated both by the inability of the Government to enforce its provisions and by an influx of battle-hardened fighters expelled from Jordan in Sept. 1970. From March 1975 Lebanon was beset by civil disorder bringing the economy to a virtual standstill.

By Nov. 1976 large-scale fighting had been brought to an end by the intervention of the Syrian-dominated Arab Deterrent Force. Large areas of the country, however, remained outside governmental control, including West Beirut, which was the scene of frequent conflict between opposing militia groups. In March 1978 there was an Israeli invasion following a Palestinian attack

inside Israel. Israeli troops eventually withdrew in June, but instead of handing over all their positions to UN Peacekeeping Forces, they installed Israeli-controlled Christian Lebanese militia forces in border areas. In June 1982 Israeli forces once again invaded, this time in massive strength, and swept through the country, eventually laying siege to and bombing Beirut. In Sept. Palestinian forces, together with the PLO leadership, evacuated Beirut. Israeli forces started a withdrawal on 16 Feb. 1985 but it was not until the end of 1990 that the various militias which had held sway in Beirut withdrew. A new Government of National Reconciliation was announced on 24 Dec. 1990. The dissolution of all militias was decreed by the National Assembly in April 1991, but the Shia Muslim militia Hizbollah was allowed to remain active. Following a 17-day Israeli bombardment of Hizbollah positions in April 1996, a US-brokered unsigned 'understanding' of 26 April 1996 guaranteed that Hizbollah guerrillas and Palestinian radical groups would cease attacks on civilians in northern Israel and granted Israel the right to self-defence. Hizbollah maintained the right to resist Israel's occupation of Lebanese soil. In May 2000 Israel completed its withdrawal from south Lebanon, 22 years after the first invasion.

On 14 Feb. 2005 former Prime Minister Rafiq al-Hariri was assassinated in a bomb attack, sparking international condemnation of the murder and mass public protests at the continued presence of Syrian soldiers in the country. Soon afterwards Syria began withdrawing and redeploying its 14,000 troops and intelligence agents from Beirut. By the end of April 2005 all Syrian troops had been withdrawn from Lebanon. In July 2006, after Hizbollah forces in Lebanon had captured two Israeli soldiers, Israel launched a large-scale military campaign against Lebanon with a series of bombing raids, destroying large parts of the civilian infrastructure. More than 1,200 Lebanese, the majority of them civilians, were killed during the conflict. Following the resignation of five Shia Muslim cabinet ministers and the assassination of industry minister Pierre Gemayel in Nov. 2006 there were anti-government demonstrations in Beirut in which over 800,000 protesters—nearly a quarter of the population of Lebanon—demanded the resignation of Prime Minister Fouad Siniora.

In Nov. 2007 Siniora's cabinet assumed presidential powers in the wake of Emile Lahoud's resignation until parliament chose Michel Suleiman to fill the office in May 2008. Suleiman reappointed Siniora as premier. Diplomatic relations were established with Syria in Oct. 2008, the first time since both countries became independent. Following a general election in June 2009, Saad al-Hariri of the pro-Western 14 March Alliance formed a coalition government in Nov. In Jan. 2011 the coalition collapsed, although he agreed to stay on as premier in a caretaker capacity until a new administration could be formed.

TERRITORY AND POPULATION

Lebanon is mountainous, bounded on the north and east by Syria, on the west by the Mediterranean and on the south by Israel. The area is 10,201 sq. km (3,939 sq. miles). Population (2007 estimate), 3·76m.; density, 369 per sq. km. The last census was in 1932. In 2011, 87·4% of the population were urban.

The UN gives a projected population for 2015 of 4·39m.

The principal towns, with estimated population (1998), are: Beirut (the capital), 1·5m.; Tripoli, 160,000; Zahlé, 45,000; Saida (Sidon), 38,000.

The official language is Arabic. French and, increasingly, English are widely spoken in official and commercial circles. Armenian is spoken by a minority group.

SOCIAL STATISTICS

2008 estimates: births, 66,000; deaths, 29,000. Estimated rates, 2008 (per 1,000 population): births, 15·7; deaths, 7·0. Infant mortality was 19 per 1,000 live births in 2010; expectation of life (2007), 69·8 years for males and 74·1 for females. Annual population growth rate, 2000–08, 1·3%; fertility rate, 2008, 1·8 births per woman.

CLIMATE

A Mediterranean climate with short, warm winters and long, hot and rainless summers, with high humidity in coastal areas. Rainfall is largely confined to the winter months and can be torrential, with snow on high ground. Beirut, Jan. 55°F (13°C), July 81°F (27°C). Annual rainfall 35·7" (893 mm).

CONSTITUTION AND GOVERNMENT

The first Constitution was established under the French Mandate on 23 May 1926. It has since been amended in 1927, 1929, 1943 (twice), 1947 and 1990. It is based on a separation of powers, with a President, a single-chamber *National Assembly* elected by universal suffrage at age 21 in 12 electoral constituencies, and an independent judiciary. The President serves a six-year term, although in both 1995 and 2004 the terms of office of the then Presidents were extended from six to nine years through 'exceptional' constitutional amendments. The executive consists of the President and a Prime Minister and Cabinet appointed after consultation between the President and the National Assembly. The system is adapted to the communal balance on which Lebanese political life depends by an electoral law which allocates deputies according to the religious distribution of the population, and by a series of constitutional conventions whereby, e.g. the President is always a Maronite Christian, the Prime Minister a Sunni Muslim and the Speaker of the Assembly a Shia Muslim. There is no party system. In Aug. 1990, and again in July 1992, the National Assembly voted to increase its membership, and now has 128 deputies with equal numbers of Christians and Muslims (although Muslims make up a clear majority of the population).

On 21 Sept. 1990 President Haraoui established the Second Republic by signing constitutional amendments which had been negotiated at Taif (Saudi Arabia) in Oct. 1989. These institute an executive collegium between the President, Prime Minister and Speaker, and remove from the President the right to recall the Prime Minister, dissolve the Assembly and vote in the Council of Ministers.

National Anthem

'Kulluna lil watan lil 'ula lil 'alam' ('All of us for our country, flag and glory'); words by Rashid Nakhlé, tune by W. Sabra.

GOVERNMENT CHRONOLOGY

Presidents since 1943.

1943–52	Béchara Khalil El-Khoury
1952–58	Camille Nemr Chamoun
1958–64	Fouad Abdallah Chehab
1964–70	Charles Alexandre Hélou
1970–76	Soleiman Kabalan Franjieh
1976–82	Elias Sarkis
1982–88	Amine Pierre Gemayel
1989	René Anis Moawad
1989–98	Elias Khalil Haraoui
1998–2007	Emile Geamil Lahoud
2007–08	Fouad Siniora (acting)
2008–	Gen. Michel Suleiman

Prime Ministers since 1943.

1943–45	Riyad as-Solh
1945	Abdulhamid Karame
1945–46	Abd' Rashin Sami as-Solh
1946	Saadi al-Munla
1946–51	Riyad as-Solh
1951	Hussein al-Oweini
1951–52	Abdullah Aref al-Yafi
1952	Abd Rashin Sami as-Solh
1952	Nazim al-Akkari
1952	Saeb Sallam
1952	Abdullah Aref al-Yafi
1952–53	Amir Khalid Chehab
1953	Saeb Sallam
1953–54	Abdullah Aref al-Yafi
1954–55	Abd Rashin Sami as-Solh
1955–56	Rashid Karame
1956	Abdullah Aref al-Yafi
1956–58	Abd Rashin Sami as-Solh
1958	Khalil al-Hibri
1958–60	Rashid Karame
1960	Ahmed Daouk
1960–61	Saeb Sallam
1961–64	Rashid Karame
1964–65	Hussein al-Oweini
1965–66	Rashid Karame
1966	Abdullah Aref al-Yafi
1966–68	Rashid Karame
1968–69	Abdullah Aref al-Yafi
1969–70	Rashid Karame
1970–73	Saeb Sallam
1973	Amin al-Hafez
1973–74	Takieddin as-Solh
1974–75	Rashid as-Solh
1975	Nureddin Rifai
1975–76	Rashid Karame
1976–80	Sélim Ahmed Hoss
1980	Takieddin as-Solh
1980–84	Shafiq al-Wazzan
1984–87	Rashid Karame
1987–90	Sélim Ahmed Hoss
1990–92	Omar Karame
1992	Rashid as-Solh
1992–98	Rafiq al-Hariri
1998–2000	Sélim Ahmed Hoss
2000–04	Rafiq al-Hariri
2004–05	Omar Karame
2005	Najib Mikati
2005–09	Fouad Siniora
2009–11	Saad al-Hariri
2011–	Najib Mikati

RECENT ELECTIONS

Elections were held on 7 June 2009. The ruling anti-Syrian 14 March Alliance—including the Movement of the Future, the Progressive Socialist Party, the Lebanese Forces, the Kataeb Party and their allies—won 71 of 128 seats; Hizbollah, Amal and their allies, 29 seats; and the Free Patriotic Movement and its allies, 28.

The presidential election, originally scheduled for 25 Sept. 2007, was put back repeatedly amid continued deadlock between rival political leaders. At the 20th attempt on 25 May 2008 Gen. Michel Suleiman was elected president, receiving 118 of 127 votes in parliament.

CURRENT GOVERNMENT

President: Gen. Michel Suleiman; b. 1948 (took office 25 May 2008).

In March 2013 the caretaker government comprised:

Prime Minister: Najib Mikati; b. 1955 (took office 13 June 2011), having previously been in office from April–July 2005).

Deputy Prime Minister: Samir Mouqbel.

Minister of Agriculture: Hussein Hajj Hassan. *Culture:* Gaby Layyoun. *Defence:* Fayez Ghosn. *Displaced Persons:* Alaaeddine Terro. *Economy and Trade:* Nicholas Nahhas. *Education:* Hassan

Diab. *Energy and Water:* Gebran Bassil. *Environment:* Nazim Khoury. *Finance:* Mohammed Safadi. *Foreign Affairs:* Adnan Mansour. *Health:* Ali Hassan Khalil. *Industry:* Vrej Sabounjian. *Information:* Walid Daouk. *Interior and Municipalities:* Marwan Charbel. *Justice:* Shakib Qortbawi. *Labour:* Salim Jreissati. *Public Works and Transport:* Ghazi Aridi. *Social Affairs:* Wael Abou Faour. *Sports and Youth:* Faisal Karami. *Telecommunications:* Nicolas Sehnawi. *Tourism:* Fadi Abboud.

President's Website: http://www.presidency.gov.lb

CURRENT LEADERS

Michel Suleiman

Position
President

Introduction
Gen. Michel Suleiman was sworn into office on 25 May 2008, filling a vacuum created when Emile Lahoud's term ended in Nov. 2007. Suleiman was appointed as a 'compromise candidate' after negotiations between the Siniora Western-backed government at that time and the Hizbollah-led opposition.

Early Life
Michel Suleiman was born in Amsheet on 21 Nov. 1948 to a prominent Maronite Christian family. He joined the armed forces in 1967 and graduated from Lebanon's Military Academy in 1970. He went on to complete a degree in political and administrative sciences at the Lebanese University.

Suleiman rose rapidly through the ranks of the armed forces at a time when Syria played a dominant role in Lebanon's military. On 21 Dec. 1998 he was appointed commander of the armed forces when Emile Lahoud left the post to take over the presidency. During his tenure Suleiman managed to maintain the military's non-partisan status and built good relationships with all sides. Following the Feb. 2005 assassination of Rafiq al-Hariri, Suleiman refused to crack down on anti-Syrian demonstrations or sanction military intervention. His stance was credited with paving the way to Syria's withdrawal from Lebanon. Suleiman also ensured that the military stood back when Hizbollah and Israel fought a 34-day war in 2006 and oversaw a successful operation against Fatah al-Islam militants at the Nahr el-Bared refugee camp in 2007.

On 25 May 2008 Suleiman won the presidency with 118 of 127 parliamentary votes after Qatari-brokered talks on Lebanon's political future. His appointment ended six months of political deadlock, with the government and opposition agreeing to a power-sharing deal.

Career in Office
Suleiman reappointed pro-Western Fouad Siniora as prime minister on 28 May 2008 and invited him to form a national unity cabinet. In Oct. 2008 Lebanon and Syria established diplomatic ties. However, Suleiman still faced formidable challenges, including the implementation of the UN Security Council resolution calling for all militias in Lebanon to be disarmed. Following general elections in June 2009 in which the pro-Western 14 March Alliance won a majority of seats in the National Assembly, Suleiman designated as the new prime minister Alliance leader Saad al-Hariri, who eventually succeeded in forming a national unity government in Nov. that year. Al-Hariri's government then collapsed in Jan. 2011 after Hizbollah withdrew support and Najib Mikati was appointed prime minister-designate by Suleiman. However, only after five months was Mikati able to successfully negotiate the composition of a new administration, with Hizbollah's backing, which Suleiman swore into office in June 2011.

Najib Mikati

Position
Prime Minister

Introduction
Najib Mikati was sworn into office on 13 June 2011 having formed a government at the invitation of President Michel Suleiman following the collapse of Saad al-Hariri's government on 12 Jan. 2011. Mikati's centrist politics have appealed to the country's major political groups, although some critics labelled him pro-Syrian after he secured Hizbollah backing.

Early Life
Najib Azmi Mikati was born in Tripoli, Lebanon, on 24 Nov. 1955 to a Sunni family. He was educated at the American University of Beirut, earning a bachelor's degree and then an MBA. He also studied at the INSEAD business school in France and at Harvard University in the USA. While in education, Mikati co-founded a construction company, M1 Group, with his brother. In 1982, at the height of the civil war, the brothers saw the potential of the emerging telecommunications market and set up Investcom.

In 1998 Mikati was made minister of public works and transport, retaining both portfolios until 2004. He was elected to the National Assembly in 2000 but decided not to run against Rafiq al-Hariri for the premiership. Following al-Hariri's death in Feb. 2005, Mikati was invited to form a government. His assurances that he would remove the country's security commanders and oversee new elections won him the support of the anti-Syrian opposition. He subsequently stood down as prime minister after the June 2005 polls.

In June 2006 Investcom was sold to South Africa's MTN Group, although Mikati's M1 Group remained the second largest shareholder of MTN. Mikati was re-elected to parliament in June 2009 but did not join the government of Saad al-Hariri. After al-Hariri's coalition collapsed in Jan. 2011, Mikati was appointed prime minister designate after 68 parliamentarians approved his Hizbollah-backed nomination.

Career in Office
Mikati was sworn into office in June 2011. After months of negotiating, his government (with a pro-Hizbollah majority) won parliamentary approval in July. He was faced with forging a workable policy towards the disputed Special Tribunal for Lebanon investigating the 2005 assassination of Rafiq al-Hariri.

In 2012 the conflict in neighbouring Syria threatened to destabilize Lebanon's fragile political and sectarian balance, particularly following the assassination in Beirut in Oct. of the head of police intelligence, a critic of Syria's President Assad. Amid rising tensions, Mikati resigned on 22 March 2013 but remained in office while his potential successor, Tammam Salam, attempted to form a new government. At the time of going to print this process was ongoing.

DEFENCE

There were 14,000 Syrian troops in the country in early 2005, but in March 2005 Lebanon and Syria agreed that the troops would be redeployed to the Bekaa Valley in the east of the country. They were subsequently all withdrawn from Lebanon. The United Nations Interim Force in Lebanon (UNIFIL), created in 1978, had a strength of 1,990 in June 2006. Following the conflict between Israel and Lebanon of July–Aug. 2006 the Security Council established UNIFIL II, a more powerful peacekeeping force deployed to maintain the ceasefire, support the Lebanese armed forces and aid humanitarian efforts. In March 2013 UNIFIL II comprised 11,028 peacekeepers from 38 countries.

Conscription was reduced from 12 months to six in 2005, and was finally abolished in Feb. 2007.

Defence expenditure in 2008 totalled US$751m. (US$189 per capita), representing 2·7% of GDP.

Army

The strength of the Army was 53,900 in 2007 and includes a Presidential Guard and five special forces regiments. There is an internal security force, run by the Ministry of the Interior, some 20,000 strong.

Navy

A force of 1,100 personnel (2007) operate 21 patrol and coastal combatants. An additional seven inshore patrol craft are operated by customs.

Air Force

The Air Force had (2007) 1,000 personnel. There are no combat-capable aircraft although eight attack helicopters were in operation in 2007.

INTERNATIONAL RELATIONS

A Treaty of Brotherhood, Co-operation and Co-ordination with Syria of May 1991 provides for close relations in the fields of foreign policy, the economy, military affairs and security. The treaty stipulates that Lebanese government decisions are subject to review by six joint Syrian–Lebanese bodies.

ECONOMY

Agriculture accounted for 5·9% of GDP in 2009, industry 23·4% and services 70·7%.

Overview

Prior to civil war the country was a model for economic development in the Middle East, displaying impressive growth, strong investment and high social indicators. After the civil war Lebanon found itself with heavy reconstruction costs and little foreign aid, leaving the country with US$35bn. of debt and the second highest debt-to-GDP ratio in the world in 1991. However, the economy recovered and growth rates were strong until the country fell back into recession in 1999. Austerity measures helped the economy recover strongly from 2001–04.

In 2005 the economy again stalled after the assassination of former prime minister Rafiq al-Hariri. From July–Aug. 2006 Israeli bombing damaged the economy, with the government putting the cost at US$6·5bn. Political troubles hampered economic activity until an agreement was reached in May 2008 and a national unity government was formed in 2009. This relative stability boosted investment and tourism and growth slowed only very slightly in 2009 in the wake of the global financial crisis. The government debt-to-GDP ratio declined from 157% of GDP in 2008 to 148% in 2009. However, government debt is still among the world's highest. The national unity government's collapse in early 2011 coupled with regional turmoil led to declines in growth, FDI, tourism and remittances.

Currency

The unit of currency is the *Lebanese pound* (LBP) of 100 *piastres*. Inflation was 4·5% in 2010 and 5·0 in 2011. In July 2005 foreign exchange reserves totalled US$10,627m., gold reserves were 9·22m. troy oz and total money supply was £Leb.3,005·7bn. The Lebanese pound has been pegged to the US dollar since Sept. 1999 at £Leb.1,507·5 = 1 US$.

Budget

The fiscal year is the calendar year.

In 2007 budgetary central government revenue totalled £Leb.8,390bn. and expenditure £Leb.11,816bn. Tax revenues in 2007 were £Leb.5,593bn. Main items of expenditure by economic type in 2007 were interest (£Leb.4,695bn.) and compensation of employees (£Leb.3,198bn.).

VAT of 10% was introduced in 2002.

Performance

Total GDP was US$40·1bn. in 2011. Real GDP growth was 8·5% in 2009, 7·0% in 2010 and 1·5% in 2011.

Banking and Finance

The Bank of Lebanon (*Governor*, Riad Salameh) is the bank of issue. In Jan. 2008 there were 66 functioning banks (45 domestic and 21 foreign) of which 51 were commercial, 11 were investment and/or private and four were Islamic. Commercial bank deposits in May 2011 totalled US$110·5bn.

There is a stock exchange in Beirut (closed 1983–95).

ENERGY AND NATURAL RESOURCES

Environment

Lebanon's carbon dioxide emissions from the consumption and flaring of fossil fuels in 2008 were the equivalent of 3·6 tonnes per capita.

Electricity

Installed capacity in 2007 was approximately 2·3m. kW. Production in 2007 was 11·35bn. kWh and consumption per capita 3,277 kWh.

Minerals

There are no commercially viable deposits.

Agriculture

In 2002 there were around 170,000 ha. of arable land and 143,000 ha. of permanent crop land. Crop production (in 1,000 tonnes), 2002: potatoes, 397; tomatoes, 270; olives, 184; oranges, 155; apples, 150; cucumbers and gherkins, 132; wheat, 119; grapes, 102; melons and watermelons, 100.

Livestock (2002): goats, 409,000; sheep, 298,000; cattle, 88,000; pigs, 21,000; chickens, 33m.

Forestry

The forests of the past have been denuded by exploitation and in 2010 covered 0·14m. ha., or 13% of the total land area. Timber production was 87,000 cu. metres in 2007.

Fisheries

The catch in 2006 was 3,811 tonnes, of which 3,541 tonnes were sea fish.

INDUSTRY

In 2001 industry accounted for 21·9% of GDP, with manufacturing contributing 10·3%. Industrial production, 2001 (in 1,000 tonnes): cement, 2,890; flour, 420; sulphuric acid, 357; mineral water, 276·8m. litres.

Labour

The economically active population in 2007 was 1,228,800 (921,600 males and 307,100 females), of whom 1,118,400 (842,400 males and 276,000 females) were in employment.

INTERNATIONAL TRADE

Imports and Exports

Trade, 2010, in US$1m.: imports (c.i.f.), 17,970; exports (f.o.b.), 4,254. Main imports, 2010 (in US$1m.), were: machinery and transport equipment (4,100); mineral fuels, lubricants and related materials (3,674); manufactured goods (2,736). Main exports, 2010 (in US$1m.), were: gold (830); arms and ammunition (333); non-metallic mineral manufactures (273). In 2010 major import sources (in US$1m.) were: USA (1,919); China (1,639); Italy (1,394). Major export markets in 2010 (in US$1m.) were: Switzerland (503); UAE (419); France (349).

COMMUNICATIONS

Roads

There were 6,970 km of roads in 2005, including 170 km of motorway. Passenger cars in 2002 numbered 1,253,700, and there were also 97,200 trucks and vans; in 1997 there were 61,470 motorcycles and mopeds and 6,830 buses and coaches. In 2007 there were 4,281 road accidents resulting in 487 deaths.

Rail

Railways are state-owned. There is 222 km of standard gauge track.

Civil Aviation

Beirut International Airport was served in 2003 by nearly 30 airlines. It handled 2,373,056 passengers (all on international flights) in 2001 and 62,789 tonnes of freight. The national airline is the state-owned Middle East Airlines. In 2003 scheduled airline traffic of Lebanese-based carriers flew 20m. km, carrying 935,000 passengers (all on international flights).

Shipping

Beirut is the largest port, followed by Tripoli, Jounieh and Saida (Sidon). In Jan. 2009 there were 46 ships of 300 GT or over registered, totalling 140,000 GT.

Telecommunications

In 2010 there were 887,800 landline telephone subscriptions (equivalent to 210·0 per 1,000 inhabitants) and 2,874,800 mobile phone subscriptions (or 680·0 per 1,000 inhabitants). In 2011, 52·0% of the population were internet users. In March 2012 there were 1·4m. Facebook users.

SOCIAL INSTITUTIONS

Justice

The population in penal institutions in 2007 was 5,870 (159 per 100,000 of national population). The death penalty is still in force. It was last used in 2004 when three people were executed.

Education

There are state and private primary and secondary schools. In 2007 there were 450,566 pupils with 32,412 teaching staff at primary schools; and 368,359 pupils with 40,919 teaching staff in secondary education. The Lebanese University, founded in 1951, is the only public university. Private universities include the American University of Beirut, founded in 1866 as the Syrian Protestant College; the Lebanese American University, established in 1924; and the Lebanese International University, established in 2001. In 2007 there were 187,055 students in tertiary education and 21,778 academic staff. Adult literacy was 90% in 2007. In 2007 public expenditure on education came to 2·7% of GNI and 9·6% of total government spending.

There is a Lebanese Academy of Fine Arts, which is part of the University of Balamand.

Health

There were 162 hospitals in 2008. The provision of hospital beds was 36 per 10,000 inhabitants in 2005. There were 9,876 doctors in 2007, and 4,316 dentists, 4,990 pharmacists and 1,699 nurses in 2008.

RELIGION

In 2001 it was estimated that the population was 56·6% Muslim (34·8% Shia and 21·8% Sunni), 36·0% Christian (mainly Maronite) and 7·4% Druze. In 1996 there were 119 Roman Catholic bishops. In Feb. 2013 there were two cardinals.

CULTURE

World Heritage Sites

There are five UNESCO sites in Lebanon. Four were entered on the list in 1984: the ruins of Anjar, a city founded by the Muslim Arab caliph Walid I at the beginning of the 8th century; Baalbek, the most impressive ancient site in Lebanon and one of the most important Roman ruins in the Middle East; Byblos, the site of multi-layered ruins of one of the most ancient cities of Lebanon, dating back to Neolithic times; and Tyre, which has important archaeological remains, principally from Roman times. The Qadisha Valley and Bcharre district, inscribed in 1998, has been the site of monastic communities since the earliest years of Christianity. Its cedar trees, among the most highly prized building materials of the ancient world, are survivors of a sacred forest.

Press

In 2009 there were 14 paid-for daily newspapers with a combined circulation of 244,000 and two free dailies. The newspapers with the highest circulation are *An-Nahar* and *As-Safir*.

Tourism

In 2009 there were 1,844,000 non-resident tourists (excluding Syrians, Palestinians, students and same-day visitors), up from 1,333,000 in 2008.

Festivals

Major annual cultural events are the Al Bustan Festival of music, dance and theatre in Feb.–March; Hamra Festival in June; Baalbek International Festival, which reopened in 1997 after an absence of 23 years, in June–Aug.; Beiteddine Festival in July–Aug.; Tyre Festival; Byblos Festival; and the Beirut Film Festival.

DIPLOMATIC REPRESENTATIVES

Of Lebanon in the United Kingdom (21 Palace Garden Mews, London, W8 4RB)
Ambassador: Inaam Osseiran.

Of the United Kingdom in Lebanon (Embassies Complex, Army St., Zkak Al-Blat, Serail Hill, PO Box 11–471, Beirut)
Ambassador: Tom Fletcher, CMG.

Of Lebanon in the USA (2560 28th St., NW, Washington, D.C., 20008)
Ambassador: Antoine Chedid.

Of the USA in Lebanon (PO Box 70-840, Antelias, Beirut)
Ambassador: Maura Connelly.

Of Lebanon to the United Nations
Ambassador: Nawaf Salam.

Of Lebanon to the European Union
Ambassador: Adnan Mansour.

FURTHER READING

Choueiri, Y. M., *State and Society in Syria and Lebanon.* 1994
Fisk, R., *Pity the Nation: Lebanon at War.* 3rd ed. 2001
Gemayel, A., *Rebuilding Lebanon.* 1992
Harris, William, *The New Face of Lebanon: History's Revenge.* 2005
Hiro, D., *Lebanon Fire and Embers: a History of the Lebanese Civil War.* 1993
Hirst, David, *Beware of Small States: Lebanon, Battleground of the Middle East.* 2010
Young, Michael, *The Ghosts of Martyrs Square: An Eyewitness Account of Lebanon's Life Struggle.* 2010

National library: Dar el Kutub, Parliament Sq., Beirut.
National Statistical Office: Service de Statistique Générale, Beirut.
Website: http://www.cas.gov.lb

LESOTHO

SOUTH AFRICA

LESOTHO

Hlotse

MASERU · Bokong

Moyeni · SOUTH AFRICA

0   25 mi
0   40 km

© Research Machines plc 2006

Muso oa Lesotho
(Kingdom of Lesotho)

Capital: Maseru
Population projection, 2015: 2·29m.
GNI per capita, 2011: (PPP$) 1,664
HDI/world rank: 0·450/160
Internet domain extension: .ls

KEY HISTORICAL EVENTS

The Basotho nation was constituted in the 19th century under the leadership of Moshoeshoe I, bringing together refugees from disparate tribes scattered by Zulu expansionism in southern Africa. After war with land-hungry Boer settlers in 1856 (and again in 1886), Moshoeshoe appealed for British protection. This was granted in 1868, and in 1871 the territory was annexed to the Cape Colony (now Republic of South Africa), but in 1883 it was restored to the direct control of the British government through the High Commissioner for South Africa. In 1965 full internal self-government was achieved under King Moshoeshoe II. On 4 Oct. 1966 Basutoland became an independent and sovereign member of the British Commonwealth as the Kingdom of Lesotho. Chief Leabua Jonathan, leader of the Basotho National Party and prime minister from 1965, suspended the constitution when the elections of 1970 were declared invalid. On 20 Jan. 1986, after a border blockade by the Republic of South Africa, Chief Jonathan was deposed in a bloodless military coup led by Maj.-Gen. Justin Lekhanya who granted significant powers to the king. King Moshoeshoe II was deposed in Nov. 1990 and replaced by King Letsie III. Lekhanya was deposed in May 1991. A democratic constitution was promulgated in April 1993. The elections in May 1998 were won by the ruling Lesotho Congress for Democracy. In Sept. 1998 an army mutiny prompted intervention from South Africa to support the government.

TERRITORY AND POPULATION

Lesotho is an enclave within South Africa. The area is 30,355 sq. km (11,720 sq. miles).

The census in 2006 showed a total population of 1,876,633 (963,835 females); density, 61·8 per sq. km. In 2006 the population was 77·2% rural.

The UN gives a projected population for 2015 of 2·29m.

There are ten districts, all named after their chief towns, except Berea (chief town, Teyateyaneng). Area and population:

Region	Area (in sq. km.)	2006 census population
Berea	2,222	250,006
Butha-Buthe	1,767	110,320
Leribe	2,828	293,369
Mafeteng	2,119	192,621
Maseru	4,279	431,998
Mohale's Hoek	3,530	176,928
Mokhotlong	4,075	97,713
Qacha's Nek	2,349	69,749
Quthing	2,916	124,048
Thaba-Tseka	4,270	129,881

In 2006 the capital, Maseru, had a population of 197,907. Other major towns (with 2006 census population) are: Teyateyaneng, 61,475; Mafeteng, 32,148; Maputsoe, 30,800 (estimate); Mohale's Hoek, 28,310.

The official languages are Sesotho and English.

The population is more than 98% Basotho. The rest is made up of Xhosas, approximately 3,000 expatriate Europeans and several hundred Asians.

SOCIAL STATISTICS

2008 estimated births, 59,000; deaths, 35,000. Rates, 2008 estimates: birth (per 1,000 population), 28·9; death, 16·9. Annual population growth rate, 2000–08, 1·0%. Life expectancy at birth in 2007 was 43·9 years for males and 45·5 years for females. Life expectancy has declined dramatically over the last 15 years, largely owing to the huge number of people in the country with HIV. In 2007, 23·2% of all adults between 15 and 49 were infected with HIV. Infant mortality, 2010, 65 per 1,000 live births; fertility rate, 2008, 3·3 births per woman.

CLIMATE

A healthy and pleasant climate, with variable rainfall, but averaging 29" (725 mm) a year over most of the country. The rain falls mainly in the summer months of Oct. to April, while the winters are dry and may produce heavy frosts in lowland areas and frequent snow in the highlands. Temperatures in the lowlands range from a maximum of 90°F (32·2°C) in summer to a minimum of 20°F (–6·7°C) in winter.

CONSTITUTION AND GOVERNMENT

Lesotho is a constitutional monarchy with the King as Head of State. Following the death of his father, Moshoeshoe II, **Letsie III** succeeded to the throne in Jan. 1996.

The 1993 constitution provided for a *National Assembly* comprising an elected 80-member lower house and a *Senate* of 22 principal chiefs and 11 members nominated by the King. For the elections of May 2002 a new voting system was introduced, increasing the number of seats in the National Assembly to 120, elected for a five-year term as before, but with 80 members in single-seat constituencies and 40 elected by proportional representation.

National Anthem

'Lesotho fatsela bontat'a rona' ('Lesotho, land of our fathers'); words by F. Coillard, tune by L. Laur.

RECENT ELECTIONS

Following the elections of May 1998 the King swore allegiance to a new constitution and the Military Council was dissolved.

Parliamentary elections were held on 26 May 2012. The Democratic Congress won 48 of 120 seats, All Basotho Convention 30, Lesotho Congress for Democracy 26, Basotho National Party 5, Popular Front for Democracy 3, National Independent Party 2. Six other parties won a single seat each.

CURRENT GOVERNMENT

In March 2013 the Council of Ministers comprised:

Prime Minister and Minister of Defence, Police and National Security: Tom Motsoahae Thabane; b. 1939 (All Basotho Convention; sworn in 8 June 2012).

Deputy Prime Minister, and Minister of Local Government and Chieftainship Affairs: Mothetjoa Metsing.

Minister of Agriculture and Food Security: Lits'oane Lits'oane. *Communications, Science and Technology:* Ts'eliso Mokhosi. *Development Planning:* Moeketsi Majoro. *Education and Training:* 'Makabelo Mosothoane. *Employment and Labour:* Lebesa Maloi. *Energy, Meteorology and Water Affairs:* Timothy Thahane. *Finance:* Leketekete Ketso. *Foreign Affairs and International Affairs:* Mohlabi Tsekoa. *Forestry and Land Reclamation:* Khotso Matla. *Gender, Youth, Sport and Recreation:* Thesele 'Maseribane. *Health:* Dr Pinkie Manamolela. *Home Affairs:* Joang Molapo. *Human Rights, Law and Constitutional Affairs:* Haae Phoofolo. *Justice and Correctional Services:* Mophato Monyake. *Mining:* Tlali Khasu. *Public Service:* Motloheloa Phooko. *Public Works and Transport:* Keketso Rant'so. *Social Development:* 'Matebatso Doti. *Tourism, Environment and Culture:* 'Mamahele Radebe. *Trade and Industry, Co-operatives and Marketing:* Temeki Ts'olo. *Minister in the Prime Minister's Office:* Molobeli Soulo.

The *College of Chiefs* settles the recognition and succession of Chiefs and adjudicates cases of inefficiency, criminality and absenteeism among them.

Government Website: http://www.gov.ls

CURRENT LEADERS

Tom Motsoahae Thabane

Position
Prime Minister

Introduction
Tom Motsoahae Thabane became prime minister in June 2012 as head of a coalition government, ending 14 years of Democratic Congress rule. His priorities include addressing corruption and economic development.

Early Life
Born in 1939, Tom Motsoahae Thabane worked in the ministry of health and education from 1970–72 and at the ministry of justice from 1972–76. In 1976 he became principal secretary for the interior and from 1978–83 was principal secretary for health in the government of Leabua Jonathan. He then took over the foreign affairs portfolio until 1985 before serving a further year as principal secretary for the interior.

After the government was ousted in Jan. 1986, Thabane served in the military administration that ruled for seven years. As secretary to the military council, he negotiated the return of political exiles and smoothed relations between the military government and civilian political groups. He was minister of foreign affairs, information and broadcasting in 1990 and 1991 and chaired the committee responsible for creating the constituent assembly that in turn paved the way for democratic multi-party elections.

From 1991–94 Thabane also worked as a development consultant, basing himself in South Africa. In 1995 he began a three-year stint as special political adviser to the prime minister,

Ntsu Mokhehle. From 1998–2006 in the government of Pakalitha Mosisili he served variously as foreign minister, minister of home affairs and public safety, and minister of communications, science and technology. However, in 2006 he broke with the ruling Lesotho Congress for Democracy (LCD) to form the All Basotho Congress (ABC), which relied heavily on urban support.

At the 2007 election the LCD retained power and the ABC came third. With strained relations between Thabane and Mosisili, Mosisili's newly-formed Democratic Congress won the most seats at the 2012 election but was unable to form a coalition. The ABC, the second largest party in parliament, agreed to enter government with the LCD and two smaller parties. Thabane was duly sworn in as premier on 8 June 2012.

Career in Office
Thabane has pledged to tackle poverty, create jobs, improve welfare provision and invest in national infrastructure. In Aug. 2012 he declared an emergency when agricultural production dropped by more than 70% as a result of adverse weather conditions.

In Jan. 2013 Thabane urged the police to arrest any serving and former ministers suspected of corruption. As he sought to build relations between the coalition partners, there were encouraging economic signs when the IMF's mission chief in the country reported robust growth.

DEFENCE

South African and Batswanan troops intervened after a mutiny by Lesotho's armed forces in Sept. 1998. The foreign forces were withdrawn in May 1999.

The Royal Lesotho Defence Force has about 2,000 personnel. Defence expenditure totalled US$36m. in 2008 (US$17 per capita), representing 2·1% of GDP.

ECONOMY

In 2009 agriculture accounted for 7·7% of GDP, industry 32·9% and services 59·4%.

Overview

Lesotho's economy is heavily integrated with that of South Africa. Its only significant natural resource is water, exported to South Africa via the Lesotho Highlands Water Project. Subsistence agriculture dominates the economy and the garment sector also has a significant role in generating employment and exports. The economy is heavily dependent on Southern African Customs Union (SACU) receipts and workers' remittances. Lesotho is a member of the Common Monetary Area together with South Africa, Namibia and Swaziland.

Following several sluggish years, economic growth exceeded 4% in three consecutive years from 2006 to 2008 before slowing to 3·8% in 2009 as a result of the global economic slowdown. The crisis resulted in a loss of textile and diamond exports and a drop in SACU revenues and remittances. Growth rose to 5·2% in 2010 but the continuing decline in SACU revenues raises medium-term uncertainties. The economy needs sustained and broad-based growth and the acceleration of structural reforms to help reduce poverty. According to the 2010 Human Development Report, 43% of the population lives below the international poverty line of US$1·25 (PPP) per day.

Currency

The unit of currency is the *loti* (plural *maloti*) (LSL) of 100 *lisente*, at par with the South African rand, which is legal tender. Total money supply in July 2005 was 1,659m. maloti and foreign exchange reserves were US$539m. Inflation was 10·7% in 2008, falling to 3·6% in 2010 but then increasing to 5·6% in 2011.

Budget

The fiscal year is 1 April–31 March. Revenues in 2008–09 were 8,818m. maloti and expenditures 6,462m. maloti. Tax revenue accounted for 88·0% of revenues in 2008–09; wages and salaries accounted for 36·0% of expenditures and grants 11·9%.

The standard rate of VAT is 14%.

Performance

Real GDP growth was 3·8% in 2009, 5·2% in 2010 and 4·9% in 2011. Total GDP in 2011 was US$2·4bn. Lesotho ranks among the countries most reliant on remittances from abroad, which accounted for 29% of total GDP in 2010. In particular remittances from South Africa contributed 21% of Lesotho's GDP.

Banking and Finance

The Central Bank of Lesotho (*Governor*, Rets'elisitsoe Adelaide Matlanyane) is the bank of issue, founded in 1982 to succeed the Lesotho Monetary Authority. There are four commercial banks (First National Bank of Lesotho, Lesotho Post Bank, Nedbank Lesotho and Standard Lesotho Bank). Gross national savings were estimated at 12·2% of GNI in 2010, compared to 25·8% of GNI in 2009. The decline in savings mainly reflected lower private sector savings as household disposable income fell during the year. Foreign debt was US$726m. in 2010, representing 28·4% of GNI.

ENERGY AND NATURAL RESOURCES

Environment

Lesotho's carbon dioxide emissions from the consumption and flaring of fossil fuels in 2008 were the equivalent of 0·1 tonnes per capita.

Electricity

Capacity (2007) 76,00 kW. Production in 2007 was 200m. kWh (all hydro-electric). Consumption in 2007 was 223m. kWh.

Minerals

Diamonds are the main product; 2007 output was 454,014 carats (up from 721 carats in 2002). The huge increase was owing to the start of commercial operations at Letseng Diamonds mine in 2004. Gravel and crushed rock production (2007), 300,000 cu. metres.

Agriculture

Agriculture employs two-thirds of the workforce. The chief crops were (2002 production in 1,000 tonnes): maize, 108; potatoes, 90; wheat, 45; sorghum, 38; dry beans, 7. Soil conservation and the improvement of crops and pasture are matters of vital importance. In 2002 there were an estimated 330,000 ha. of arable land and 4,000 ha. of permanent crop land. There were 2,000 tractors in 2002.

Livestock (2003 estimates): sheep, 850,000; cattle, 540,000; goats, 650,000; asses, 154,000; horses, 100,000; chickens, 2m.

Forestry

In 2010 Lesotho's total forest area was 44,000 ha., or 1% of the land area. Timber production was 2·07m. cu. metres in 2007.

Fisheries

The catch in 2010 was 45 tonnes, exclusively from inland waters.

INDUSTRY

Important industries are food products, beverages, textiles and chemical products. In 2009 industry accounted for 33% of GDP, with manufacturing contributing 16%.

Labour

The labour force in 1996 was 847,000 (63% males). In 1998, 76,100 were working in mines in South Africa.

INTERNATIONAL TRADE

Lesotho is a member of the Southern African Customs Union (SACU) with Botswana, Namibia, South Africa and Swaziland.

Imports and Exports

Trade, 2008, in US$1m.: imports (c.i.f.), 1,066; exports (f.o.b.), 245.

The principal imports in 2008 (in US$1m.) were: machinery and transport equipment (208); food (177); miscellaneous manufactured articles (152). The principal exports in 2008 (in US$1m.) were: machinery and transport equipment (100); miscellaneous manufactured articles (65); food (33).

South Africa accounted for 95% of imports in 2008 and 83% of exports. Other significant trading partners are Japan and Germany for imports and the USA for exports.

COMMUNICATIONS

Roads

The road network in 2002 totalled 7,091 km, of which 19·8% were paved. In 2002 there were 4,800 passenger cars (2·1 per 1,000 inhabitants) plus 13,000 trucks and vans. There were 402 deaths in 2007 as a result of road accidents.

Rail

A branch line built by the South African Railways, one mile long, connects Maseru with the Bloemfontein–Natal line at Marseille for transport of cargo.

Civil Aviation

There are direct flights from Maseru to Johannesburg. In 2000 Maseru handled 28,613 passengers (28,503 on international flights).

Telecommunications

In 2010 there were 38,600 landline telephone subscriptions (equivalent to 17·8 per 1,000 inhabitants) and 987,400 mobile phone subscriptions (or 454·8 per 1,000 inhabitants). There were 38·6 internet users per 1,000 inhabitants in 2010.

SOCIAL INSTITUTIONS

Justice

The legal system is based on Roman-Dutch law. The Lesotho High Court and the Court of Appeal are situated in Maseru, and there are Magistrates' Courts in the districts. A total of 47,971 crimes were recorded in 2005, of which 15,847 were classed as 'serious' (including 734 homicides).

The population in penal institutions in July 2007 was 2,701 (144 per 100,000 of national population).

Education

Education levels: pre-school, 3 to 5 years; first level (elementary), 6 to 12; second level (secondary or teacher training or technical training), 7 to 13; third level (university or teacher training college). Free primary education was introduced in 2000. Lesotho has the highest proportion of female pupils at secondary schools in Africa, with 56% in 2007. In 2006 there were 424,855 pupils in primary schools with 10,513 teaching staff, 93,996 pupils in secondary schools with 3,725 teaching staff and 8,500 students in higher education with 638 academic staff. The National University of Lesotho was established in 1975 at Roma; enrolment in 2006–07, 8,566 students. There are eight government-supported technical and vocational training institutions as well as a teacher training college, the Lesotho College of Education. The adult literacy rate in 2008 was 90% (among the highest in Africa).

In 2005 public expenditure on education came to 13·8% of GDP and 29·8% of total government spending (one of the highest percentages of any country).

Health

In 2003 there were 89 physicians, 16 dentistry personnel and 1,123 nursing and midwifery personnel. There were 13 hospital beds per 10,000 inhabitants in 2006.

RELIGION

In 2001 there were 0·82m. Roman Catholics, 0·28m. Protestants, 0·26m. African Christians and the remainder followed other religions.

CULTURE

Press

There were 14 non-daily newspapers and periodicals in 2008, but no dailies.

Tourism

In 2010 there were a record 425,870 non-resident visitors, up from 343,743 in 2009 and 293,073 in 2008.

Festivals

The Morija Arts & Cultural Festival is held annually at Morija, where missionaries first arrived in Lesotho in 1833.

DIPLOMATIC REPRESENTATIVES

Of Lesotho in the United Kingdom (7 Chesham Pl., Belgravia, London, SW1 8HN)
Acting High Commissioner: Maana Mapetja.

Of the United Kingdom in Lesotho (High Commission in Maseru closed in 2005)
High Commissioner: Dr Nicola Brewer (resides in Pretoria, South Africa).

Of Lesotho in the USA (2511 Massachusetts Ave., NW, Washington, D.C., 20008)
Ambassador: Eliachim Molapi Sebatane.

Of the USA in Lesotho (254 Kingsway Ave., Maseru 100)
Ambassador: Michele T. Bond.

Of Lesotho to the United Nations
Ambassador: Motlatsi Ramafole.

Of Lesotho to the European Union
Ambassador: Mamoruti Tiheli.

FURTHER READING

Bureau of Statistics. *Statistical Reports.* [Various years]

Haliburton, G. M., *A Historical Dictionary of Lesotho.* 1977
Machobane, L. B. B. J., *Government and Change in Lesotho, 1880–1966: A Study of Political Institutions.* 1990
Rosenberg, Scott, Weisfelder, Richard F. and Frisbie-Fulton, Michelle, (eds.) *Historical Dictionary of Lesotho.* 2003

National Statistical Office: Bureau of Statistics, PO Box 455, Maseru 100.
Website: http://www.bos.gov.ls

LIBERIA

© Research Machines plc 2006

Republic of Liberia

Capital: Monrovia
Population projection, 2015: 4·56m.
GNI per capita, 2011: (PPP$) 265
HDI/world rank: 0·329/182
Internet domain extension: .lr

KEY HISTORICAL EVENTS

The Republic of Liberia was created on the Grain Coast for freed American slaves. In 1822 a settlement was formed near the spot where Monrovia now stands. On 26 July 1847 the state was constituted as the Free and Independent Republic of Liberia.

On 12 April 1980 President Tolbert was assassinated and his government overthrown in a coup led by Master-Sergeant Samuel Doe. In 1990 rebel forces entered Liberia from the north and fought their way southwards to confront President Doe's forces in Monrovia. The rebels comprised the National Patriotic Front of Liberia (NPFL) led by Charles Taylor, and the hostile breakaway Independent National Patriotic Front led by Prince Johnson. A peacekeeping force dispatched by the Economic Community of West African States (ECOWAS) disembarked at Monrovia on 25 Aug. 1990. On 9 Sept. Doe was assassinated by Johnson's rebels. ECOWAS installed a provisional government led by Amos Sawyer. Charles Taylor declared himself president, as did the former vice-president, Harry Moniba. A succession of ceasefires was negotiated and broken. An ECOWAS-sponsored peace agreement was signed on 17 Aug. 1996 in Abuja, providing for the disarmament of all factions by the end of Jan. 1997 and the election of a president on 31 May 1997. By the end of Jan. 1997, 20,000 out of approximately 60,000 insurgents had surrendered their arms. It is estimated that up to 200,000 people died in the civil war and up to 1m. were made homeless. Charles Taylor was elected president in July 1997. In Feb. 2002 Taylor declared a state of emergency after an attack by rebels on the town of Kley, where thousands of refugees from Sierra Leone were encamped.

In Aug. 2003 the UN called for the immediate deployment of an ECOWAS peacekeeping force, to be replaced by a full UN force on 1 Oct. Nigerian peacekeepers arrived on 4 Aug. 2003. Taylor relinquished power to his vice-president, Moses Blah, and to a transitional government on 11 Aug. In Nov. 2005 Ellen Johnson-Sirleaf won presidential elections to become Africa's first elected female head of state. In May 2012 Charles Taylor received a 50-year prison sentence for war crimes from the UN-sponsored Special Court for Sierra Leone.

TERRITORY AND POPULATION

Liberia is bounded in the northwest by Sierra Leone, north by Guinea, east by Côte d'Ivoire and southwest by the Atlantic ocean. The total area is 97,036 sq. km. At the last census, in 2008, the population was 3,476,608; density, 36 per sq. km.

The UN gives a projected population for 2015 of 4·56m. Liberia's population has doubled since the mid-1990s.

In 2007, 59·5% of the population lived in urban areas. English is the official language spoken by 20% of the population. The rest belong in the main to three linguistic groups: Mande, West Atlantic and the Kwa. These are in turn subdivided into 16 ethnic groups: Bassa, Bella, Gbandi, Mende, Gio, Dey, Mano, Gola, Kpelle, Kissi, Krahn, Kru, Lorma, Mandingo, Vai and Grebo.

The population of Monrovia (the capital) was 970,824 in 2008 including its suburbs.

There are 15 counties, whose areas, populations and capitals are as follows:

County	Sq. km	2008 population	Chief town
Bomi	1,942	84,119	Tubmanburg
Bong	8,769	333,481	Gbarnga
Gbarpolu	9,685	83,388	Bepolu
Grand Bassa	7,932	221,693	Buchanan
Grand Cape Mount	5,160	127,076	Robertsport
Grand Gedeh	10,480	125,258	Zwedru
Grand Kru	3,894	57,913	Barclayville
Lofa	9,978	276,863	Voinjama
Margibi	2,615	209,923	Kakata
Maryland	2,296	135,938	Harper
Montserrado	1,908	1,118,241	Bensonville
Nimba	11,546	462,026	Saniquillie
River Cess	5,592	71,509	Cesstos City
River Gee	5,110	66,789	Fish Town
Sinoe	10,133	102,391	Greenville

SOCIAL STATISTICS

2008 births, estimate, 145,000; deaths, 40,000. 2008 rates (per 1,000 population), estimate: birth, 38·3; death, 10·5. Annual population growth rate, 2000–08, 3·7%. Life expectancy at birth (2007): 56·5 years for men and 59·3 years for women. Infant mortality in 2010 was at 74 per 1,000 live births. Fertility rate, 2008, 5·9 births per woman.

CLIMATE

An equatorial climate, with constant high temperatures and plentiful rainfall, although Jan. to May is drier than the rest of the year. Monrovia, Jan. 79°F (26·1°C), July 76°F (24·4°C). Annual rainfall 206" (5,138 mm).

CONSTITUTION AND GOVERNMENT

A constitution was approved by referendum in July 1984 and came into force on 6 Jan. 1986. Under it the *National Assembly* consisted of a 26-member *Senate* and a 64-member *House of Representatives*. For the elections of 2005 the number of seats in the Senate was increased to 30 and in 2010 a further nine seats were added to the House of Representatives, bringing the total to 73. The executive power of the state is vested in the *President*, who may serve up to two six-year terms.

National Anthem

'All hail, Liberia, hail!'; words by President Daniel Warner, tune by O. Luca.

RECENT ELECTIONS

In the first round of presidential elections held on 11 Oct. 2011 Ellen Johnson-Sirleaf of the Unity Party (UP) received 43·9% of the vote, followed by Winston Tubman of the Congress for Democratic Change with 32·7% and Prince Yormie Johnson of the National Union for Democratic Progress with 11·6%. There were 13 other candidates. Turnout was 71·6%. With no candidate receiving an absolute majority a run-off was held on 8 Nov. 2011. Tubman claimed that the first round had been rigged and called on his supporters to boycott the run-off, in which turnout was consequently only 38·6%. Sirleaf gained 90·7% of the vote.

In the elections to the House of Representatives also on 11 Oct. 2011 the UP won 24 seats; the Congress for Democratic Change, 11; the Liberal Party, 7; the National Union for Democratic Progress, 6; the National Democratic Coalition, 5; the Alliance for Peace and Democracy, 3; the National Patriotic Party, 3; and the Movement for Progressive Change, 2. Three parties won one seat each and independents nine. Elections for half of the seats in the Senate were also held on 11 Oct. 2011. As a result the UP held 10 of the 30 seats; the National Patriotic Party, 6; the Congress for Democratic Change, 3; the Alliance for Peace and Democracy, 2; and the National Union for Democratic Progress, 2. Four parties held one seat each and independents three.

CURRENT GOVERNMENT

President: Ellen Johnson-Sirleaf; b. 1938 (Unity Party; sworn in 16 Jan. 2006 and re-elected in Nov. 2011).

In March 2013 the government comprised:

Vice President: Joseph Boakai.

Minister of Agriculture: Florence Chenoweth. *Commerce and Industry*: Miatta Beysolow. *Defence*: Brownie J. Samukai. *Education*: Etmonia David Tarpeh. *Finance, and Planning and Economic Affairs*: Amara Konneh. *Foreign Affairs*: Augustine Kpehe Ngafuan. *Gender Development*: Julia Duncan-Cassell. *Health and Social Welfare*: Dr Walter Gwenigale. *Information, Culture and Tourism*: Lewis Brown. *Internal Affairs*: Blamon Nelson. *Justice and Attorney General*: Christiana Tah. *Labour*: Varbah Gayflor. *Land, Mines and Energy*: Patrick Sendolo. *National Security*: Samuel Brisbane. *Posts and Telecommunications*: Frederick Norkeh. *Public Works*: Samuel Kofi Woods. *Rural Development*: Ernest C. B. Jones. *Transport*: Eugene Nagbe. *Youth and Sports*: S. Tornolah Varpilah. *Minister of State for Presidential Affairs*: Edward McClain, Jr.

Government Website: http://www.emansion.gov.lr

CURRENT LEADERS

Ellen Johnson-Sirleaf

Position
President

Introduction
Ellen Johnson-Sirleaf became Africa's first elected female president in Jan. 2006, having defeated the former footballer, George Weah, in a run-off. A US-educated economist, she returned from exile to attempt to resurrect Liberia's shattered economy after 14 years of civil war. She was re-elected in presidential polling held in Oct.–Nov. 2011, pledging to continue her reform and anti-corruption agenda.

Early Life
Ellen Johnson-Sirleaf was born in Monrovia, Liberia on 29 Oct. 1938. She was educated at the College of West Africa in Monrovia from 1948–55, before graduating in accountancy in 1964 from the University of Wisconsin in the USA. From 1967 she served as special assistant to the secretary of the treasury in Liberia before undertaking an MA in public administration at America's Harvard University from 1969–71. Returning to Liberia, Sirleaf became assistant minister of finance in the administration of William R. Tolbert, Jr. Following public criticisms of Tolbert's presidency she resigned and left the country, taking up a post as a loan officer for several Latin American countries at the World Bank. In 1977 she was invited to return home to become deputy minister of finance for fiscal and banking affairs. In Aug. 1979 she replaced James T. Philips as minister of finance.

Shortly after a coup d'état and Tolbert's assassination on 12 April 1980, the new military leader, Sgt Samuel Doe, appointed Sirleaf president of the Liberia Bank for Development and Investment. However, she resigned in Dec. 1980 and returned to the World Bank, before becoming vice president of Citibank in Nairobi, Kenya in mid-1981. She stood in Liberia's general elections in Oct. 1985, at which Doe was controversially elected president. Sirleaf was elected senator but was sentenced to ten years in jail as part of Doe's crackdown on 'opponents' following a failed coup in Nov. 1985. Pardoned and released in June 1986, she again left Liberia for the USA, where she worked for the Equator Bank in Washington, D.C., followed by the UN Development Programme (UNDP) in New York.

While in the USA, Sirleaf joined other Liberian exiles in criticizing Doe and helped raise funds for a fellow exile, Charles Taylor, to lead the National Patriotic Front of Liberia (NPFL) into Liberia from the Côte d'Ivoire in 1989. It triggered a devastating civil war that led to the deaths of over 200,000 people by the time a ceasefire was declared in Aug. 1996. Disillusioned with Taylor, Sirleaf resigned as director of the UNDP's Bureau for Africa (a post she held from July 1992) and stood against him on behalf of the Unity Party in presidential elections in 1997. She received only 10% of the vote (against 75% for Taylor) and was later charged with treason by him. Forced into exile again, she became active in various humanitarian projects, including investigations into the 1994 Rwandan genocide for the Organization for African Unity and serving on the board of the International Crisis Group and the Nelson Mandela Foundation. Liberia again descended into civil war but Sirleaf returned after Taylor was forced into exile in Aug. 2003 (to be later imprisoned in 2012 by the International Criminal Court for abetting war crimes in neighbouring Sierra Leone). She headed the governance reform commission until resigning in March 2005 to enter the presidential race.

During her campaign, Sirleaf criticized the transitional government's inability to fight corruption. She went through to a run-off against George Weah, a former World Footballer of the Year who was representing the Congress for Democratic Change, and on 11 Nov. the national elections commission declared Sirleaf the winner. Although Weah accused her of fraud, her victory was confirmed on 23 Nov. Independent observers declared the vote to be free, fair and transparent and her inauguration took place on 16 Jan. 2006.

Career in Office
In her inaugural speech, Sirleaf vowed to wage a war on corruption, promising that leading civil servants and ministers would have to declare their assets. She also pledged to work towards reconciliation by bringing former opponents into a government of national unity, and spoke of establishing peaceful relations with neighbouring West African states. She appointed a number of women to ministerial positions and controversially nominated a Nigerian soldier to head Liberia's army. While rebuilding the country's shattered economy—with a road network in ruins, no national telephone network, no national electricity

grid and no piped water—has remained a major challenge, the World Bank and other international bodies have praised her government's efforts in office. A Truth and Reconciliation Commission was inaugurated with a mandate to investigate human rights abuses during the long civil war, and she has made progress in confronting poor governance and corrupt officialdom.

In Nov. 2010 Sirleaf appointed a 22-member acting cabinet after placing their predecessors on 'mass administrative leave'. Several ministers were subsequently reappointed to their posts. She retained the presidency in Nov. 2011 when she was re-elected in a second round of voting, but there was a low turnout after rival candidate William Tubman boycotted the process because of alleged electoral fraud.

In Oct. 2011 Sirleaf was jointly awarded the Nobel Peace Prize.

DEFENCE

In June 2003 UN Secretary-General Kofi Annan called for an international peacekeeping force to restore peace after fighting broke out between government forces and Liberians United for Reconciliation and Democracy (LURD). An ECOWAS peacekeeping force of over 3,000 troops was deployed initially, but this has been replaced by the UN Peacekeeping Mission in Liberia (UNMIL), totalling 8,119 uniformed personnel in Jan. 2013.

The Armed Forces of Liberia were created in 2007 to replace the Liberian Army, which was demobilized in 1999. In 2009 there were approximately 2,100 troops although they are not expected to be fully operational until 2014.

Defence expenditure totalled US$7m. in 2009, representing 0·8% of GDP.

ECONOMY

Agriculture accounted for 54% of GDP in 2007 (one of the highest proportions of any country), industry 19% and services 27%.

Overview

After years of civil war destroyed much of the economy's human and physical capital, post-war reconstruction has accelerated. On the basis of its Poverty Reduction Strategy Paper, the government has made progress in repaying long-standing arrears to the World Bank and African Development Bank, strengthening public revenues and public finance management and improving institutional governance. The international community has provided financial support, estimated by the IMF at US$85–100 per capita.

Real GDP growth rose to 13·2% in 2007 (up from 9·0% in 2006), supported by continued recovery in agriculture as well as mining and other services. However, the country remains one of the world's poorest, with per capita GDP estimated at US$216 in 2009.

Gross international reserves increased in 2011. FDI is strong in the iron ore and palm oil sectors, while prices in the rubber sector are high, and the construction sector is growing rapidly. The agriculture and forestry sectors, however, are hampered by low productivity and poor infrastructure.

In the near term, the authorities plan to focus on job creation and infrastructure rehabilitation. Slower global demand for commodities and weakened FDI inflows pose challenges to the mid- to long-term economic outlook.

Currency

US currency is legal tender. There is a *Liberian dollar* (LRD), in theory at parity with the US dollar. Between 1993 and March 2000 different notes were in use in government-held Monrovia and the rebel-held country areas, but on 27 March 2000 a set of new notes went into circulation to end the years of trading in dual banknotes. Inflation was 17·5% in 2008, falling to 7·3% in 2010 but then increasing slightly to 8·5% in 2011. Total money supply

was L$4,316m. in July 2005 and foreign exchange reserves were US$21m.

Budget

Revenue in 2007 was L$10,222m.; expenditure was L$9,498m. Customs and excise duties accounted for 44·3% of revenues in 2007; general administration accounted for 41·5% of expenditures.

Performance

Real GDP growth was 5·3% in 2009, 6·1% in 2010 and 8·2% in 2011. Total GDP was US$1·5bn. in 2011.

Banking and Finance

The National Bank of Liberia opened on 22 July 1974 to act as a central bank. The *Governor* of the bank is Joseph Mills Jones. There were only three banks in operation in Jan. 2004. External debt was US$228m. in 2010, representing 28·3% of GNI.

ENERGY AND NATURAL RESOURCES

Environment

Liberia's carbon dioxide emissions from the consumption and flaring of fossil fuels were the equivalent of 0·2 tonnes per capita in 2008.

Electricity

Installed capacity in 2007 was estimated at 198,000 kW. Production in 2007 was approximately 353m. kWh. Consumption per capita in 2007 was about 93 kWh.

Minerals

2010 estimates: gold production 666 kg and diamond production 26,591 carats.

Agriculture

In 2007 the agricultural population was approximately 2·31m., of which about 853,000 were economically active. There were an estimated 385,000 ha. of arable land in 2007 and 215,000 ha. of permanent crops. Principal crops (2003 estimates) in 1,000 tonnes: cassava, 480; sugarcane, 255; bananas, 110; rice, 110; rubber, 108; palm oil, 42; plantains, 40; taro, 26; yams, 20. Coffee, cocoa and palm kernels are produced mainly by the traditional agricultural sector.

Livestock (2003 estimates): cattle, 36,000; pigs, 130,000; sheep, 210,000; goats, 220,000; chickens, 6m.

Livestock products (2003 estimates) in tonnes: meat, 23,000; milk, 1,000; eggs, 4,000.

Forestry

Forest area was 4·33m. ha. (45% of the land area) in 2010. In 2007, 6·62m. cu. metres of roundwood were cut. There are rubber plantations.

Fisheries

Fish landings in 2008 were 7,890 tonnes, of which approximately 90% from sea fishing.

INDUSTRY

There are a number of small factories. Production of cement, cigarettes, soft drinks, palm oil and beer are the main industries.

Labour

In 2010 the labour force was 1,374,000 (52·3% males).

INTERNATIONAL TRADE

Imports and Exports

Imports in 2008 were US$865m. and exports US$262m. Main import sources in 2004 were South Korea, 38·1%; Japan, 21·9%; Singapore, 12·6%. Major export destinations in 2004 were the USA, 61·4%; Belgium, 29·5%; China, 5·3%.

Main imports are food and live animals, petroleum and petroleum products, and machinery and transport equipment. Main exports are rubber, and logs and timber.

COMMUNICATIONS

Roads
There were about 10,600 km of roads in 2002 (only 6·2% of which were paved). In 2007 there were 7,400 passenger cars in use and 2,800 lorries and vans.

Rail
There is a total of 490 km single track. A 148-km freight line connects iron mines to Monrovia. There is a line from Bong to Monrovia (78 km). The railways were out of use for many years because of the civil wars but there is now some traffic, both freight and passenger. However, large sections of track have been dismantled.

Civil Aviation
There are two international airports (Roberts International and Sprigg Payne), both near Monrovia. In 2003 there were services to Abidjan, Accra, Brussels, Freetown and Lagos.

Shipping
There are ports at Buchanan, Greenville, Harper and Monrovia. Over 2,000 vessels enter Monrovia each year. The Liberian government requires only a modest registration fee and an almost nominal annual charge and maintains no control over the operation of ships flying the Liberian flag. In Jan. 2009 there were 2,203 ships of 300 GT or over registered, totalling 80·15m. GT (a figure only exceeded by Panama's fleet). Of the 2,203 vessels registered, 741 were container ships, 669 oil tankers, 404 bulk carriers, 262 general cargo ships, 83 liquid gas tankers, 40 chemical tankers and four passenger ships.

Telecommunications
In 2009 Liberia had just 2,200 main (fixed) telephone lines, but there were 1,058,000 mobile phone subscribers. No other country had such a high ratio of mobile phone subscriptions to fixed telephone lines in 2009. There were 20,000 internet users in 2007.

SOCIAL INSTITUTIONS

Justice
Liberia is governed by a dual system of statutory law. The modern sector is regulated by Anglo-American common law with the indigenous sector following customary law based on unwritten tribal practices. Following a 2003 proposal by UNMIL (United Nations Mission in Liberia), a scheme to rebuild the post civil war justice system was introduced. However, reforms are progressing at a slow rate and the judiciary remains severely dysfunctional.

Education
Schools are classified as: (1) Public schools, maintained and run by the government; (2) Mission schools, supported by foreign Missions and subsidized by the government, and operated by qualified Missionaries and Liberian teachers; (3) Private schools, maintained by endowments and sometimes subsidized by the government.

Adult literacy in 2009 was estimated at 59·1%.

Health
In 2004 there were 103 physicians (approximately one for every 33,300 inhabitants), 13 dentists, 613 nurses, 422 midwives and 35 pharmacists. The John F. Kennedy Memorial Hospital in Monrovia is the country's leading health care institution.

RELIGION
There were (2001) about 1·27m. Christians and 520,000 Sunni Muslims, plus 1·39m. followers of traditional beliefs.

CULTURE

Press
There were seven paid-for daily newspapers in 2008 with a combined circulation of 55,000, plus 24 paid-for non-dailies.

DIPLOMATIC REPRESENTATIVES
Of Liberia in the United Kingdom (23 Fitzroy Sq., London, W1T 6EW)
Ambassador: Wesley Momo Johnson.

Of the United Kingdom in Liberia
Ambassador: Ian Hughes (resides in Freetown, Sierra Leone).

Of Liberia in the USA (5201 16th St., NW, Washington, D.C., 20011)
Ambassador: Jeremiah C. Sulunteh.

Of the USA in Liberia (PO Box 98, 502 Benson St., Monrovia)
Ambassador: Deborah R. Malec.

Of Liberia to the United Nations
Ambassador: Marjon V. Kamara.

Of Liberia to the European Union
Ambassador: R. Francis Tuan Karpeh.

FURTHER READING
Daniels, A., *Monrovia Mon Amour: a Visit to Liberia.* 1992
Ellis, Stephen, *The Mask of Anarchy: The Destruction of Liberia and the Religious Dimension of an African Civil War.* 1999
Sawyer, A., *The Emergence of Autocracy in Liberia: Tragedy and Challenge.* 1992

National Statistical Office: The Liberia Institute of Statistics and Geo-Information Services (LISGIS), 9th Street, Sinkor, Monrovia.
Website: http://www.lisgis.org

Discover even more in the Archive at www.statesmansyearbook.com

LIBYA

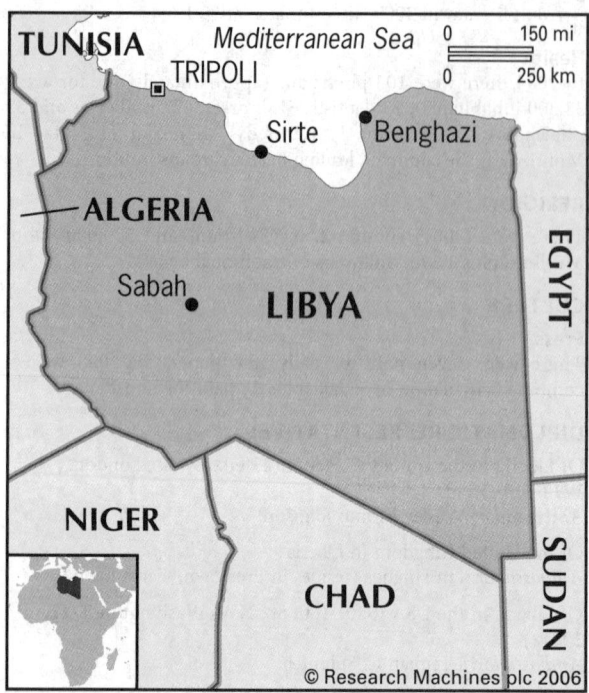

Capital: Tripoli
Population projection, 2015: 6·61m.
GNI per capita, 2011: (PPP$) 12,637
HDI/world rank: 0·760/64
Internet domain extension: .ly

KEY HISTORICAL EVENTS

Libya's earliest inhabitants were the semi-nomadic Berbers, whose descendants still live throughout North Africa's Atlas Mountains. Phoenician merchants from the Levant began to settle in what is today Libya from around 1000 BC and founded Carthage in what is now Tunisia in 814 BC. Carthage became the leading port in the western Mediterranean, extending its influence over Libya and most of coastal North Africa for the next five centuries. The port of Oea (now Tripoli) was founded by Phoenicians around 500 BC. Greek traders settled in the Cyrenaica region (eastern Libya) in the seventh century BC, founding Cyrene in 630 BC. Roman settlements appeared in Libya in the third century BC, growing in importance following Rome's sacking of Carthage in 146 BC. The port of Leptis Magna was founded in Tripolitania in the first century AD, becoming a major centre of commerce until it fell to the Vandals early in the fifth century.

The first of several waves of Arab conquerors arrived in 630, spreading the Islamic faith, initially along the coastal fringes. Libya became part of the powerful Umayyad and Abbasid Caliphates, the latter centred on Baghdad where it reached its apotheosis under Harun al Rashid (786–809). From 971 until 1045 Libya, along with present-day Algeria and Tunisia, were ruled by the Zirid amirs who were loyal to the Fatimid Caliphs of the Nile Valley. Bedouin Arabs from the Nile Valley, known as the Banu Hilal, entered the region in around 1000 spreading Islam and the Arabic language throughout Libya over the next three centuries.

By the beginning of the 15th century the Libyan, or Barbary, coast had become infamous as a haven for pirates. Habsburg Spain occupied Tripoli in 1510 but it fell to Ottoman corsairs in 1551. Direct control from Istanbul was superseded by rule through local Turkish governors (known as beys). In 1711 Ahmed Karamanli, an Ottoman cavalry officer, seized power and declared Libyan independence. Direct Ottoman rule was re-established in 1835 when Sultan Mahmud II made Libya a province of the Sublime Ports.

On 29 Sept. 1911 Italy declared war on the ailing Ottoman Empire and occupied Tripoli four days later. Italian control was confirmed following the signing of the Treaty of Ouchy in 1912 but local resistance, particularly by the Sanusi Order in Cyrenaica, confined Italian power to the coastal cities of Tripoli, Benghazi, Tobruk and Derna.

During the Second World War the British army expelled the Italians and their German allies, placing Tripolitania and Cyrenaica under British military administration and Fezzan under French control. This continued under a UN directive until 1951 when Libya gained independence. The former Amir of Cyrenaica, Muhammad Idris al Senussi, was crowned king.

The discovery of oil in 1959 brought rapid economic growth but resentment grew as wealth remained in the hands of the elite. Idris was deposed in Sept. 1969 by a group of army officers, 12 of whom formed the Revolutionary Command Council which, chaired by Col. Muammar Gaddafi, proclaimed the Libyan Arab Republic. In 1977 the Revolutionary Command Council was superseded by a more democratic People's Congress, though Gaddafi remained head of state. Throughout the 1980s Libya found itself at odds with its neighbours while deteriorating relations with the USA and other Western countries culminated in the US bombing of Tripoli in April 1986, a punishment for Gaddafi's alleged support of international terrorism. A US trade embargo was also enforced that year.

In 1992 the UN imposed sanctions after Libya refused to surrender suspects in the 1988 bombing of a Pan Am flight over Lockerbie in Scotland. In April 1999 Libya handed over two suspects to face trial under Scottish law in the Netherlands. In Jan. 2001 Abdelbaset Ali Mohmed Al-Megrahi was found guilty of murder and sentenced to life imprisonment. The UN suspended sanctions in 1999 but they were not lifted formally until Sept. 2003. The USA lifted its remaining sanctions in Sept. 2004 after Col. Gaddafi pledged to end his weapons of mass destruction programme. Libya's auction of oil and gas exploration licences in Jan. 2005 led to the return of US energy companies and full diplomatic ties between the two countries were resumed in May 2006.

In Aug. 2009 Al-Megrahi was released from prison on compassionate grounds following a ruling by the Scottish courts. He was said to be suffering terminal cancer. His subsequent enthusiastic reception in Libya raised tensions with the USA.

In Feb. 2011 Gaddafi faced popular demands, originating in Benghazi and spreading to other cities, for him to step down. In March he began a military campaign to seize back the east of the country from insurgents. In response, the UN, with the backing of the Arab League, authorized a no-fly zone over the country and air strikes as required in an attempt to protect civilians. In May the International Criminal Court issued an arrest warrant for Gaddafi, citing his 'widespread and systematic attacks' on civilians. In Aug. rebel forces seized control of Tripoli and a National Transitional Council (NTC) became the de facto government as Gaddafi went into hiding. The following month, rebels launched a fresh assault on Gaddafi's home town of Sirte, one of the last remaining pockets of loyalist support. On 20 Oct.

2011 Gaddafi was found hiding in the city by National Transitional Council troops and killed. One of the Gaddafi's sons, Mutassim Gaddafi, a senior army officer and security adviser, was also killed. The death of the Muammar Gaddafi brought an end to 42 years of violent rule and on 23 Oct. 2011 in Benghazi the NTC formally announced the liberation of the country.

TERRITORY AND POPULATION

Libya is bounded in the north by the Mediterranean Sea, east by Egypt and Sudan, south by Chad and Niger and west by Algeria and Tunisia. The area is 1,759,540 sq. km. The population at the 2006 census was 5,657,692; density, 3·2 per sq. km. In 2011, 78·1% of the population lived in urban areas. Ethnic composition, 2000: Libyan Arab and Berber, 64%; other (mainly Egyptians, Sudanese and Chadians), 36%.

The UN gives a projected population for 2015 of 6·61m.

Libya is divided into 22 districts (sha'biyat). Capitals and populations (2006 census) were as follows:

	Capital	Population
Al Butnan	Tubruq	157,747
Al Jabal al Akhdar	Al Bayda	206,180
Al Jabal al Gharbi	Gharyan	302,705
Al Jifarah	Al' Aziziyah	451,175
Al Jufrah	Hun	52,092
Al Kufrah	Al Jawf	48,328
Al Marj	Al Marj	184,531
Al Marqab	Al Hums	427,886
Al Wahah	Ajdabiya	179,155
An Nuqat al Khams	Zuwarah	287,359
Ash Shati'	Birat	78,563
Az Zawiyah	Az Zawiyah	290,637
Benghazi	Benghazi	674,951
Darnah	Darnah	162,857
Ghat	Ghat	23,199
Misratah	Misratah	543,129
Murzuq	Murzuq	78,772
Nalut	Nalut	93,896
Sabha	Sabha	133,206
Surt	Surt	141,495
Tarabulus (Tripoli)	Tarabulus (Tripoli)	1,063,571
Wadi al Hayat	Awbari	76,258

The two largest cities are Tripoli, the capital (population of 1,063,571 in 2006), and Benghazi (912,000 in 2000).

The official language is Arabic.

SOCIAL STATISTICS

Estimates, 2008: births, 147,000; deaths, 26,000. Estimated rates, 2008 (per 1,000 population): births, 23·3; deaths, 4·1. Life expectancy (2007), 71·6 years for men and 76·8 for women. Annual population growth rate, 2000–08, 2·0%; infant mortality, 2010, 13 per 1,000 live births; fertility rate, 2008, 2·7 births per woman.

CLIMATE

The coastal region has a warm temperate climate, with mild wet winters and hot dry summers, although most of the country suffers from aridity. Tripoli, Jan. 52°F (11·1°C), July 81°F (27·2°C). Annual rainfall 16" (400 mm). Benghazi, Jan. 56°F (13·3°C), July 77°F (25°C). Annual rainfall 11" (267 mm).

CONSTITUTION AND GOVERNMENT

Following the uprising in 2011 that culminated in the capture and killing of Libya's incumbent leader, Col. Gaddafi, the National Transitional Council (NTC)—formed in Feb. 2011—formally announced the country's 'liberation' in Oct. and appointed an executive committee to serve as the de facto interim government.

In accordance with its own guidelines set out in Aug. 2011, the NTC scheduled elections for a 200-seat General National Congress for July 2012 (with 120 individual candidates elected in constituencies and 80 members chosen from party lists using proportional representation). The General National Congress has appointed a prime minister and is expected to form a constituent assembly to draft a new constitution. Once the constitution has won approval by referendum, parliamentary elections are required to be held within six months.

National Anthem

'Libya, Libya, Libya'; words by Al Bashir Al Arebi, tune by Mohammed Abdel Wahab. Originally the Kingdom of Libya's national anthem from 1951–69, 'Libya, Libya, Libya' became the country's new national anthem following the death of Muammar Gaddafi in 2011.

GOVERNMENT CHRONOLOGY

Leaders since 1951.

1951–69	(King) Muhammad Idris I al Senussi
1969–2011	Col. Muammar Abu Minyar Gaddafi (Chairman of the Revolutionary Command Council until 1977; General Secretary of the General People's Congress until 1979; de facto leader 1979–2011)

Chairman of the National Transitional Council
2011–12   Mustafa Abdul Jalil

Interim President
2012–   Mohamed Magariaf

Heads of Government since 2011.

Prime Ministers
2011	Mahmoud Jibril (interim)
2011	Ali Tarhouni (interim)
2011–12	Abdurrahim al-Keib (interim)
2012–	Ali Zeidan

RECENT ELECTIONS

Elections to the General National Congress were held on 7 July 2012. The National Forces Alliance won 39 seats, the Justice and Construction Party (the Muslim Brotherhood's political party in Libya) 17, the National Front Party 3, National Centrist Party 2, Union for Homeland 2 and Wadi Al-Hayah Gathering 2. 15 other parties won one seat each and 120 seats went to independents.

CURRENT GOVERNMENT

Interim President: Mohamed Magariaf; b. 1940 (National Front Party; sworn in 10 Aug. 2012).

In March 2013 the government comprised:

Prime Minister: Ali Zeidan; b. 1950 (ind.; in office since 31 Oct. 2012).

Deputy Prime Ministers: Sadiq Abdulkarim Abdulrahman; Awad al-Barasi; Abdussalam al-Qadi.

Minister of Agriculture: Ahmed Ali al-Urfi. *Communications:* Osama Abdurauf Siala. *Culture:* Habib Mohammed al-Amin. *Defence:* Mohammed Mahmoud al-Bargati. *Economy:* Mustafa Mohammed Abufunas. *Education:* Mohammed Hassan Abubaker. *Electricity:* Mohammed Muhairiq. *Finance:* Alkilani al-Jazi. *Foreign Affairs and International Co-operation:* Mohamed Abdulaziz. *Health:* Nurideen Abdulhamid Dagman. *Higher Education:* Abdulasalm Bashir Sharif. *Housing:* Ali Hussein al-Sharif. *Industry:* Suleiman Ali al-Taif al-Fituri. *Interior:* Ashour Shwayel. *Justice:* Salah Bashir Margani. *Labour and Retraining:* Mohamed Fitouri Sualim. *Local Government:* Abubaker al-Hadi Mohammed. *Martyrs and Missing Persons:* Sami Mustafa al-Saadi. *Oil:* Abdulbari al-Arusi. *Planning:* Mahdi Ataher Genia. *Religious Affairs:* Abdulsalam Mohammed Abusaad. *Social Affairs:* Kamila Khamis al-Mazini. *Sports and Youth:* Abdulsalam Abdullah Guaila. *Tourism:* Ikram Abdulsalam Imam. *Transport:* Mohamed al-Ayib. *Water Resources:* Alhadi Suleiman Hinshir. *Minister of State for General National Congress Affairs:* Muaz Fathi al-Kujah. *Minister of State for Wounded People:* Ramadan Ali Mansour Zarmuh.

CURRENT LEADERS

Mohamed Magariaf

Position
President of the General National Congress

Introduction
Mohamed Magariaf was elected leader of the General National Congress (GNC) in Aug. 2012, so becoming *de facto* head of state and interim president. Magariaf will hold office until a new constitution is drafted, with elections scheduled to take place during 2013.

Early Life
Mohamed Magariaf was born in 1940 in Benghazi. He graduated in economics from Garyounis University, Benghazi, in 1962. Working as an economics professor at his *alma mater*, he was known for his criticism of Muammar Gaddafi's regime.

In 1972 Magariaf was appointed auditor general (with the rank of minister). Following continued criticism of the ruling regime, he was posted to India as ambassador in 1978. In July 1980 he resigned his post and cut all ties with the Gaddafi government, claiming it did not represent the Libyan people. Defecting first to Morocco and then Egypt, he settled in Georgia in the USA. Meanwhile, a Libyan military court sentenced him to death *in absentia*.

In Oct. 1981 Magariaf co-founded the National Front for the Salvation of Libya (NFSL), the first major opposition group to call for democratic reforms and the removal of Gaddafi. In May 1982 Magariaf became secretary general of the NFSL, a position he held until 2001.

Following the overthrow of the Gaddafi regime in 2011, Magariaf returned to Libya to lead the National Front Party (the successor of the NFSL which was dissolved in May 2012). On 9 Aug. 2012 he became leader of the GNC after winning 113 votes against 85 for his opponent, Ali Zeidan, in a second round of voting. On his election, Magariaf stepped down as head of the National Front Party.

Career in Office
Regarded as a moderate Islamist despite his ties to the NFSL, known in the 1980s as a militant group, Magariaf has spoken in support of secularism. Although partially retracting his comments, he continues to argue that Libya is ready for neither a theocratic or entirely secular government.

His chief challenge is to improve security in the face of ongoing agitation by armed militias responsible for events such as the Sept. 2012 attack on the US consulate in Benghazi. In the longer term he must aim to revive the economy and reduce dependency on oil.

Ali Zeidan

Position
Prime Minister

Introduction
Ali Zeidan became prime minister in Oct. 2012. Having spent 30 years in exile opposing Muammar Gaddafi, Zeidan was a key figure in securing Western support for the rebellion that overthrew the former dictator in 2011.

Early Life
Ali Zeidan was born on 15 Dec. 1950 and grew up in the town of Waddan in central Libya. In the 1970s he served in the Libyan embassy in India under Mohamed Magariaf, now president of the General National Congress and *de facto* interim president. Zeidan defected from Libya in 1980 and the following year, along with Magariaf, joined the opposition National Front for the Salvation of Libya. Zeidan spent the next three decades in exile, working as a lawyer in Geneva and campaigning for human rights in his homeland.

During the 2011 Libyan uprising, Zeidan served as European envoy for the National Transitional Council (the rebel movement's chief political arm). He was widely credited with convincing Western leaders, including France's President Nicolas Sarkozy, to support the insurgency.

In Libya's first ever democratic elections in July 2012, Zeidan was elected as an independent for Al Jufrah. He stood for the premiership after a no-confidence vote was carried against the interim leader, Abdurrahim al-Keib. In a vote in the 200-seat General National Congress on 14 Oct. 2012, he secured the support of a liberal coalition led by the National Forces Alliance, gaining 93 votes against 85 for the only other candidate, local government minister Mohammed al-Hrari (who had been the choice of the Justice and Construction Party founded by the Muslim Brotherhood).

Career in Office
Parliament approved Zeidan's cabinet on 31 Oct. 2012. His primary task was to unite a country riven by political and regional affiliations. At his swearing-in, he pledged that 'this government will give its utmost best to the nation based on the rule of law, human rights, democracy, rights, and the belief in God, his Prophet and a state based on Islam'. He has prioritized the recruitment and training of a professional army and police force, who must work in an environment awash with arms and militias in the wake of the revolution.

DEFENCE

Defence expenditure in 2008 totalled US$800m. (US$129 per capita), representing 1·2% of GDP. Building a professional national army is a top priority in the post-Gaddafi Libya, although little progress has been made. The inability to disarm militias continues to be the principal obstacle.

Nuclear Weapons
In Dec. 2003 then leader Col. Muammar Gaddafi agreed to dismantle his weapons of mass destruction programmes. He also agreed unconditionally to allow inspectors from international organizations to enter Libya.

Army
Strength (2007) 50,000 (including an estimated 25,000 conscripts). In addition there is a People's Militia of some 40,000 that acts as a reserve force.

Navy
The fleet, a mixture of Soviet and West European-built ships, includes two diesel submarines, two frigates and one corvette although serviceability is in doubt. There is a small Naval Aviation wing operating seven helicopters.

Personnel in 2007 totalled 8,000, including coastguard. The main naval bases are at Tripoli, Benghazi, Tubruq and Al Hums.

Air Force
The Air Force has 374 combat-capable aircraft, including MiG-21s, MiG-23s, MiG-25s and Mirage F1s, but many are in storage. Personnel total (2007) 18,000.

ECONOMY

Petroleum and natural gas accounted for 71·6% of GDP in 2007; public administration, defence and services 6·9%; finance, insurance and real estate 6·2%; and construction 4·3%.

Overview
Libya enjoyed solid growth performance in the first decade of the century. In 2003 UN sanctions were lifted after Libya admitted complicity in the 1988 Lockerbie aircraft bombing. In 2004 the USA lifted almost all of its unilateral sanctions after Libya agreed to give up its nuclear weapons programmes.

The production of hydrocarbons accounts for most of the government's revenues and the country's export earnings. Oil accounts for 98% of GDP and the country is the fourth largest oil producer in Africa. Construction and services dominate the non-oil economy.

Libya suffered a recession in 2009 following the global economic downturn. As a result, there was a greater contribution from non-oil industries following increased government spending and liberalization of the trade, tourism and service sectors (including the abolition of most of its import monopolies).

In the wake of revolution, GDP is estimated to have fallen by 60% in 2011. The Energy Investment Authority put the decline in production of light and sweet crude oil at 80%.

Libya is dependent on imports for food and medical supplies. Civil unrest has resulted in a humanitarian crisis. Social development indicators are high relative to regional variations but educational standards are low while unemployment is at 30%.

Currency

The unit of currency is the *Libyan dinar* (LYD) of 1,000 *millemes*. The dinar was devalued 15% in Nov. 1994, and alongside the official exchange rate a new rate was applied to private sector imports. Foreign exchange reserves were US$29,315m. in June 2005. Total money supply in May 2005 was 11,552m. dinars. There was inflation of 2·5% in 2010, increasing to 15·9% in 2011.

Budget

In 2008 revenues totalled 72,741m. dinars and expenditures 44,115m. dinars. Oil accounts for 88·6% of government revenues.

Performance

The economy contracted by 2·3% in 2009 but grew by 4·2% in 2010. Total GDP in 2009 was US$62·4bn. The Libyan armed conflict led to the economy contracting by 59·7% in 2011—the highest percentage decrease of any country that year.

Banking and Finance

A National Bank of Libya was established in 1955; it was renamed the Central Bank of Libya in 1972. The current *Governor* is Saddek Omar Elkaber. All foreign banks were nationalized by Dec. 1970. In 1972 the government set up the Libyan Arab Foreign Bank. The Agricultural Bank was set up to give loans and subsidies to farmers and to assist them in marketing their crops. Following the popular uprising against Col. Gaddafi in 2011, international sanctions were imposed on the central bank and its subsidiaries. In March 2011 rebels assumed control of the Central Bank of Libya, setting up a temporary headquarters in Benghazi. Following the overthrow of Gaddafi in Aug. 2011, authority passed to the National Transitional Council.

A stock exchange was opened in Tripoli in March 2007.

ENERGY AND NATURAL RESOURCES

Environment

Libya's carbon dioxide emissions from the consumption and flaring of fossil fuels in 2008 were the equivalent of 9·3 tonnes per capita.

Electricity

Installed capacity in 2007 was an estimated 5·5m. kW. Production was 25·69bn. kWh in 2007 and consumption per capita 4,161 kWh.

Oil and Gas

Oil accounts for 30% of Libya's GDP. Oil production in 2008 totalled 86·2m. tonnes. Proven reserves (2008) 43·7bn. bbls. Some analysts believe total reserves may be as high as 100bn. bbls. The National Oil Corporation (NOC) is the state's organization for the exploitation of oil resources. Libya's first oilfields were discovered in 1959, but the offshore sector remains relatively unexplored although the decision to abandon programmes for developing weapons of mass destruction in 2003 led to greatly increased interest among foreign oil companies. Oil export revenues more than doubled between 1998 and 2003. Production increased by a third between 2002 and 2008.

Proven natural gas reserves totalled 1,540bn. cu. metres in 2008. Production (2008), 15·9bn. cu. metres.

Water

Since 1984 a US$20bn. project has been under way to bring water from aquifers underlying the Sahara to the inhabited coastal areas of Libya. This scheme, called the 'Great Man-Made River', is intended, on completion, to bring 6,000 cu. metres of water a day along some 4,000 km of pipes. Phase I was completed in Aug. 1991; Phase II of the project (covering the west of Libya) was completed in Sept. 1996. Work on Phase III began in Feb. 2008. The river is providing Libya's main centres of population with clean water as well as making possible the improvement and expansion of agriculture. The whole project is more than three-quarters complete.

Minerals

Iron ore deposits have been found in the south.

Agriculture

Only the coastal zone, which covers an area of about 17,000 sq. miles, is really suitable for agriculture. Of some 25m. acres of productive land, nearly 20m. are used for grazing and about 1m. for static farming. Agriculture employs around 17% of the workforce. The sub-desert zone produces the alfalfa plant. The desert zone and the Fezzan contain some fertile oases. In 2002 there were around 1·82m. ha. of arable land and 0·34m. ha. of permanent crops. 470,000 ha. were irrigated in 2002. There were about 40,000 tractors in 2002 and 3,400 harvester-threshers.

Cyrenaica has about 10m. acres of potentially productive land and is suitable for grazing. Certain areas are suitable for dry farming; in addition, grapes, olives and dates are grown. About 143,000 acres are used for settled farming; about 272,000 acres are covered by natural forests. The Agricultural Development Authority plans to reclaim 6,000 ha. each year for agriculture. In the Fezzan there are about 6,700 acres of irrigated gardens and about 297,000 acres are planted with date palms.

Production (2003 estimates, in 1,000 tonnes): watermelons, 218; potatoes, 195; onions, 180; tomatoes, 160; olives, 150; dates, 140; wheat, 125; barley, 80; oranges, 42.

Livestock (2003 estimates): 4·1m. sheep, 1·3m. goats, 130,000 cattle, 47,000 camels, 25m. chickens.

Forestry

Forest area in 2010 was 0·22m. ha. (0·1% of the land area). In 2007, 1·03m. cu. metres of roundwood were cut.

Fisheries

The catch in 2009 was 52,110 tonnes, entirely from marine waters.

INDUSTRY

Industry employs nearly 30% of the workforce. Small-scale private sector industrialization in the form of partnerships is permitted. Output (in 1,000 tonnes): cement (2006 estimate), 5,300; distillate fuel oil (2007), 4,542; residual fuel oil (2007), 4,310; petrol (2007), 1,295.

Labour

The labour force in 2010 was 2,379,000 (72·0% males).

INTERNATIONAL TRADE

In 1986 the USA applied a trade embargo on the grounds of Libya's alleged complicity in terrorism. Many of the economic sanctions were suspended in April 2004, and in June 2006 Libya was removed from Washington's list of state sponsors of terror. In 1992 UN sanctions were imposed for Libya's refusal to deliver suspected terrorists for trial in the UK or USA, but these were formally lifted in 2003.

In Feb. 1989 Libya signed a treaty of economic co-operation with the four other Maghreb countries: Algeria, Mauritania, Morocco and Tunisia.

Libya applied for WTO membership in Dec. 2001 and accession negotiations commenced in 2004.

Imports and Exports

In 2006 imports were valued at US$13·2bn. and exports at US$37·5bn. Some 80% of GDP derives from trade. Oil accounts for over 95% of exports. Main import suppliers in 2000 were Italy (24%), Germany (12%), Tunisia (9%) and the UK (7%); main export markets were Italy (33%), Germany (24%), Spain (10%) and France (5%).

COMMUNICATIONS

Roads

There were 100,024 km of roads in 2002 (57·2% paved). In 2007 there were 1,388,200 passenger cars in use (225 per 1,000 inhabitants), plus 310,500 lorries and vans. There were 1,080 deaths as a result of road accidents in 1996.

Rail

Although there have not been any operational railways since 1965, some routes were under construction at the outbreak of the civil conflict in Feb. 2011. The projects were then abandoned but may be resurrected as Libya begins its post-conflict recovery.

Civil Aviation

The UN ban on air traffic to and from Libya enforced since April 1992 was lifted in April 1999 following the handing over for trial of two suspected Lockerbie bombers. The national flag carrier, Libyan Airlines, was grounded in March 2011 as a result of the Libyan revolution but has now resumed operations. In 2006 scheduled airline traffic of Libya-based carriers flew 17m. km, carrying 1,152,000 passengers.

Shipping

In Jan. 2009 there were 28 ships of 300 GT or over registered, totalling 239,000 GT.

Telecommunications

There were 1·23m. fixed telephone lines in 2010 (193·3 per 1,000 inhabitants). Mobile phone subscribers numbered 9·53m. in 2009. There were 108·0 internet users per 1,000 inhabitants in 2009. Fixed internet subscriptions totalled 772,500 in 2009 (123·3 per 1,000 inhabitants). In June 2012 there were 560,000 Facebook users.

SOCIAL INSTITUTIONS

Justice

The Civil, Commercial and Criminal codes are based mainly on the Egyptian model. Matters of personal status of family or succession matters affecting Muslims are dealt with in special courts according to the Muslim law. All other matters, civil, commercial and criminal, are tried in the ordinary courts, which have jurisdiction over everyone.

There are civil and penal courts in Tripoli and Benghazi, with subsidiary courts at Misratah and Darnah; courts of assize in Tripoli and Benghazi, and courts of appeal also in Tripoli and Benghazi.

The population in penal institutions in June 2007 was 12,748 (209 per 100,000 of national population). The death penalty is in force; there were at least four executions in 2009 and 18 in 2010. There were no judicial executions in 2011 or 2012.

Education

In 2006 there were 755,338 primary school pupils and 732,614 secondary level pupils. In 2009 the government launched a five-year US$9bn. plan to reform higher education and scientific research through strengthening international ties and improving the information technology network as well as creating a National Authority for Scientific Research. There are 12 state universities in Libya (of which Al Fateh University in Tripoli and Garyounis University in Benghazi are the largest), eight other higher education institutions and eight oil/bank training centres/ petroleum training and qualifying institutes. In 2003 there were 375,028 tertiary level students and 15,711 academic staff. Adult literacy in 2009 was estimated at 88·9%.

Health

There were 7,070 physicians, 850 dentistry personnel and 27,160 nursing and midwifery personnel in 2004. Provision of hospital beds in 2006 was 37 per 10,000 population.

RELIGION

Islam is declared the State religion, but the right of others to practise their religion is provided for. In 2001, 92% were Sunni Muslims.

CULTURE

World Heritage Sites

Libya has five sites on the UNESCO World Heritage List: the Archaeological Site of Leptis Magna (inscribed on the list in 1982); the Archaeological Site of Sabratha (1982); the Archaeological Site of Cyrene (1982); the Rock-art Sites of Tadrart Acacus (1985); and the Old Town of Ghadamès (1986).

Press

In 2008 there were six daily newspapers with a combined circulation of 100,000.

Tourism

In 2007 there were 106,000 non-resident visitors, down from 125,000 in 2006.

DIPLOMATIC REPRESENTATIVES

Of Libya in the United Kingdom (15 Knightsbridge, London, SW1X 7LY)
Ambassador: Mahmud Mohammed Nacua.

Of the United Kingdom in Libya (24th Floor, Tripoli Towers, Tripoli)
Ambassador: Sir Dominic Asquith, KCMG.

Of Libya in the USA (2600 Virginia Ave., NW, Suite 705, Washington, D.C., 20037)
Ambassador: Ali Aujali.

Of the USA in Libya (Sidi Slim Area/Walie Al-Ahed Rd, Tripoli)
Ambassador: Vacant.
Chargé d'Affaires a.i.: William Roebuck.

Of Libya to the United Nations
Ambassador: Abdul Rahman Mohammad Shalgam.

Of Libya to the European Union
Ambassador: Alhadi Ahmed Hadeiba.

FURTHER READING

Simons, G., *Libya: the Struggle for Survival.* 1993.—*Libya and the West: From Independence to Lockerbie.* 2004
St John, Ronald Bruce, *Libya: From Colony to Revolution.* 2011
Vandewalle, D. (ed.) *Qadhafi's Libya, 1969–1994.* 1995.—*A History of Modern Libya.* 2006.—*Libya Since 1969: Qadhafi's Revolution Revisited.* 2008

LIECHTENSTEIN

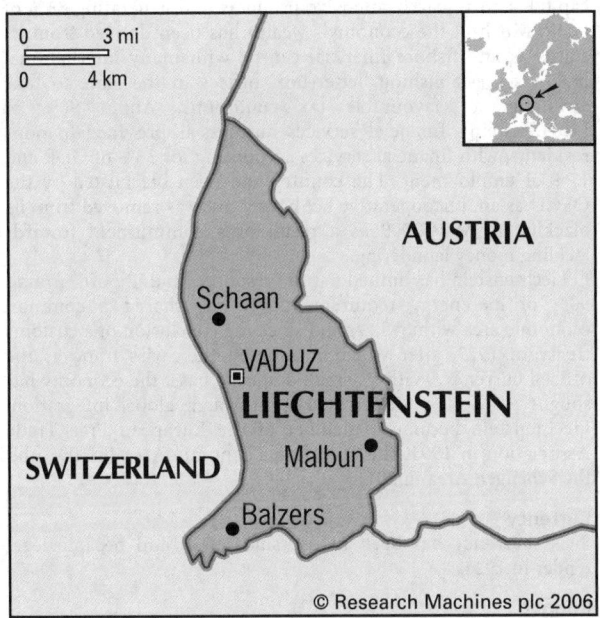

Fürstentum Liechtenstein
(Principality of Liechtenstein)

Capital: Vaduz
Population, 2011: 36,000
GNI per capita, 2010: US$144,207
HDI/world rank: 0·905/8
Internet domain extension: .li

KEY HISTORICAL EVENTS

Liechtenstein is a sovereign state with a history dating back to 1342 when Count Hartmann III became ruler of the county of Vaduz. Additions were later made to the count's domains and by 1434 the territory reached its present boundaries. On 23 Jan. 1719 Emperor Charles VI constituted the two counties as the Principality of Liechtenstein. In 1862 the constitution established an elected diet. After the First World War, Liechtenstein was represented abroad by Switzerland. Swiss currency was adopted in 1921. On 5 Oct. 1921 a new constitution based on that of Switzerland extended democratic rights, but in March 2003 the people of Liechtenstein voted in a referendum to give their prince the power to govern without reference to elected representatives.

TERRITORY AND POPULATION

Liechtenstein is bounded on the east by Austria and the west by Switzerland. Total area 160 sq. km (61·8 sq. miles). The population (Dec. 2011) was 36,475 (18,433 females), including 12,144 resident foreigners, giving a density of 228 per sq. km.

The population of Liechtenstein is predominantly rural. Population of Schaan (2011), 5,853; Vaduz (2011), 5,236.

The official language is German.

SOCIAL STATISTICS

In 2008 there were 350 births and 205 deaths (rates of 9·9 per 1,000 population and 5·8 respectively). The annual population growth rate was 1·2% over the period 2000–05.

CLIMATE

There is a distinct difference in climate between the higher mountains and the valleys. In summer the peaks can often be foggy while the valleys remain sunny and warm, while in winter the valleys can often be foggy and cold whilst the peaks remain sunny and comparatively warm. Vaduz, Jan. 0°C, July 20°C. Annual rainfall 1,090 mm.

CONSTITUTION AND GOVERNMENT

Liechtenstein is a constitutional monarchy ruled by the princes of the House of Liechtenstein.

The reigning Prince is **Hans-Adam II**, b. 14 Feb. 1945; he succeeded his father Prince Francis Joseph, 13 Nov. 1989 (he exercised the prerogatives to which the Sovereign is entitled from 26 Aug. 1984); married on 30 July 1967 to Countess Marie Kinsky von Wchinitz und Tettau. *Offspring:* Hereditary Prince Alois (b. 11 June 1968), married Duchess Sophie of Bavaria on 3 July 1993 (*offspring:* Prince Joseph Wenzel, b. 24 May 1995; Marie Caroline, b. 17 Oct. 1996; Georg Antonius, b. 20 April 1999; Nikolaus Sebastian, b. 6 Dec. 2000); Prince Maximilian (b. 16 May 1969), married Angela Brown on 29 Jan. 2000 (*offspring:* Alfons, b. 18 May 2001); Prince Constantin (b. 15 March 1972), married Countess Marie Kálnoky de Köröspatak on 17 July 1999 (*offspring:* Moritz, b. 27 May 2003; Georgina, b. 23 July 2005; Benedikt, b. 18 May 2008); Princess Tatjana (b. 10 April 1973), married Philipp von Lattorff on 5 June 1999 (*offspring:* Lukas, b. 13 May 2000; Elisabeth, b. 25 Jan. 2002; Marie, b. 18 Jan. 2004; Camilla, b. 4 Nov. 2005; Anna, b. 3 Aug. 2007; Sophie, b. 30 Oct. 2009; Maximilian, b. 17 Dec. 2011). The monarchy is hereditary in the male line.

The present constitution of 5 Oct. 1921 provided for a unicameral parliament (*Landtag*) of 15 members elected for four years, but this was amended to 25 members in 1988. Election is on the basis of proportional representation. The prince can call and dismiss the parliament, and following a referendum held on 16 March 2003, dismiss the government and veto bills. On parliamentary recommendation, he appoints the ministers. According to the constitution, the Government is a collegial body consisting of five ministers including the prime minister. Each minister has an Alternate who takes part in the meetings of the collegial Government if the minister is unavailable. Any group of 1,000 persons or any three communes may propose legislation (initiative). Bills passed by the parliament may be submitted to popular referendum. A law is valid when it receives a majority approval by the parliament and the prince's signed concurrence. The capital is Vaduz.

National Anthem

'Oben am jungen Rhein' ('Up above the young Rhine'); words by H. H. Jauch; tune, 'God save the Queen'.

RECENT ELECTIONS

At the elections on 1 and 3 Feb. 2013 the opposition Progressive Citizens' Party (FBP) gained 10 seats (40·0% of votes cast); the Patriotic Union (VU), 8 (33·5% of votes); the Independents, 4 (15·3%); the Free List (FL), 3 (11·1% of votes). Turnout was 79·8%.

CURRENT GOVERNMENT

Head of Government, and Minister for Finance and General Government Affairs: Adrian Hasler; b. 1964 (FBP; sworn in 27 March 2013).

In March 2013 the cabinet comprised:

Deputy Head of Government and Minister for Economic Affairs, Interior and Justice: Thomas Zwiefelhofer. *Environmental Affairs, Infrastructure and Sports:* Marlies Amann-Marxer. *Social Affairs:* Mauro Pedrazzini. *Cultural Affairs, Education and Foreign Affairs:* Aurelia Frick.

Princely House Website: http://www.fuerstenhaus.li

CURRENT LEADERS

Hans-Adam II

Position
Prince

Introduction
Hans-Adam II succeeded his father, Francis Joseph II, as Prince of Liechtenstein in 1989. A successful banker with a large personal fortune, in March 2003 he was granted extensive legal rights which effectively made him Europe's only absolute monarch. He handed day-to-day responsibility for running the country to his son, Crown Prince Alois, in 2004.

Early Life
Hans-Adam, whose full name is Johannes Adam Pius Ferdinand Alois Josef Maria Marko d'Aviano von und zu Liechtenstein, was born on 14 Feb. 1945. The eldest son of the ruling Prince Francis Joseph II, he was brought up with his three brothers and one sister in Vaduz castle. He was schooled in Austria and Vienna, worked for a short while at a London bank and studied at the St Gallen School of Economics and Social Sciences, graduating with a masters degree in 1969.

In 1970 he was named head of the Prince of Liechtenstein Foundation, a position he retained until 1984. In 1972 his father put him in charge of running the royal estate, during which time he won a reputation for sound management and an interest in the wider economic sphere. In Aug. 1984 Franz Joseph transferred much of his executive power to Hans-Adam, who formally acceded to the throne on his father's death in Nov. 1989.

Career in Office
Hans-Adam has striven to maintain Liechtenstein's strong economy, consolidating its position as an offshore financial centre. In 1990 he successfully concluded membership talks with the United Nations. A year later, despite having previously declared his support for European unity, he ruled out a bid for membership of the European Union.

In March 2003 Hans-Adam called a national referendum on constitutional amendments which would award him the right to dissolve the government, appoint judges and unilaterally veto legislation. In return he proposed that his right to rule by emergency decree would be reduced to six months, his entitlement to nominate government officials be terminated and that the future of the monarchy be subject to referendum. He threatened to leave for Vienna if the proposals were rejected, a move many Liechtensteiners feared would severely diminish the country's economic standing. Despite the presence of a strong pro-democracy group within Liechtenstein and the threat that the nation might lose its membership of the Council of Europe if the motion was passed, the reforms won 64·3% backing.

In Aug. 2004 Hans-Adam formally transferred responsibility for day to day affairs to his son, Alois. However, he reiterated he had no intention of abdicating the throne. In July 2012 voters rejected by a large margin a proposal to end the ruling prince's political power to veto the results of national referendums.

ECONOMY

Liechtenstein is one of the world's richest countries with a well diversified economy. Low taxes and bank secrecy laws have made Liechtenstein a successful financial centre.

Overview

Liechtenstein is a diversified and highly industrialized economy, with one of the world's highest GDP per capita ratios. Industry and manufacturing comprises the largest sector, producing mainly capital- and research-intensive products and generating 36% of GDP. Much of the economy's wealth has been derived from its status as an offshore financial centre, with many international businesses establishing 'letter-box' offices in the state to take advantage of favourable tax conditions. About 90% of Liechtenstein's financial services business is provided to non-residents, with financial services accounting for 33% of GDP and 17% of employment. The country had been blacklisted by the OECD as an 'uncooperative tax haven' but was removed from its blacklist in May 2009 as a result of a commitment towards tackling money laundering.

Liechtenstein has limited natural resources and imports around 90% of its energy requirements. It has shared a common economic area with Switzerland since the conclusion of a customs treaty in 1923, after which it adopted the Swiss franc as the official currency. With its small domestic base, the economy has sought to capture foreign markets through global integration. Liechtenstein became a member of the European Free Trade Association in 1991, the European Economic Area in 1995 and the Schengen Area in 2011.

Currency

Swiss currency has been in use since 1920 and became legal tender in 1924.

Budget

Budget (in 1,000 Swiss francs), 2008: revenue, 1,103,000; expenditure, 1,229,600. There is no public debt.

Performance

Real GDP growth was –5·0% in 2009. Total GDP in 2009 was US$4·8bn.

Banking and Finance

There were 16 banks in 2010. Combined total assets were 52,466·4m. Swiss francs in 2010.

ENERGY AND NATURAL RESOURCES

Electricity

In 2008 the consumption of electricity was 386,290 MWh (imported 314,897 MWh; produced in Liechtenstein 71,393 MWh).

Agriculture

In 2007 there were 3,743 ha. of agricultural land. In 2009, 1,005 ha. (26·9% of all agricultural land—the highest proportion of any sovereign country) was set aside for organic farming. The rearing of cattle on the Alpine pastures is highly developed. In 2009 there were 6,078 cattle (including 2,993 dairy cows), 3,963 sheep, 1,811 pigs, 495 horses and 452 goats. Total production of dairy produce in 2009 was 13,308 tonnes.

Forestry

In 2010 there were 7,000 ha. of forest (43% of the land area). Timber production in 2007 was 22,000 cu. metres.

INDUSTRY

Liechtenstein has a broadly diversified economic structure with a significant emphasis on industrial production. The most important branches of the heavily export-oriented industry are mechanical engineering, plant construction, manufacturing of precision instruments, dental technology and the food-processing industry.

Labour

The workforce was 32,435 in 2007, including employees commuting from abroad (16,242 in 2007). The farming population went down from 70% in 1930 to 1·1% in 2007. The rapid change-over has led to the immigration of foreign workers (Austrians, Germans, Italians, Swiss).

INTERNATIONAL TRADE

Liechtenstein has been in a customs union with Switzerland since 1923.

Imports and Exports

Value of the imports and exports of goods, excluding trade with and through Switzerland: imports, 2008, 2,461m. Swiss francs; exports, 2008; 4,245m. Swiss francs. Machinery, electronic goods and fabricated metals are both the leading imports and leading exports. Most trade is with other European countries.

COMMUNICATIONS

Roads

There are 400 km of roads. Postal buses are the chief means of public transportation within the country and to Austria and Switzerland. There were 24,293 cars in 2006. There were 420 road accidents in 2007 (none fatal).

Rail

The 10 km of main railway passing through the country is operated by Austrian Federal Railways.

Telecommunications

In 2007 there were 19,518 main telephone lines. There were 32,013 mobile phone subscribers in 2007 and 23,500 internet users in 2008.

SOCIAL INSTITUTIONS

Justice

The principality has its own civil and penal codes. The lowest court is the county court, *Landgericht*, presided over by one judge, which decides minor civil cases and summary criminal offences. The criminal court, *Kriminalgericht*, with a bench of five judges is for major crimes. Another court of mixed jurisdiction is the court of assizes (with three judges) for misdemeanours. Juvenile cases are treated in the Juvenile Court (with a bench of three judges). The superior court, *Obergericht*, and Supreme Court, *Oberster Gerichtshof*, are courts of appeal for civil and criminal cases (both with benches of five judges). An administrative court of appeal from government actions and the State Court determines the constitutionality of laws.

The death penalty was abolished in 1989. Some persons convicted by Liechtenstein are held in Austrian prisons.

Police

The principality has no army. In 2009 there were 88 police officers (excluding civilian staff, auxiliary police officers, village policemen and Swiss customs officers).

Education

In 2009 there were 16 primary, three upper, seven secondary and two grammar schools, with approximately 5,000 pupils and 640 teachers. There is a university, a music school and an art school.

Health

There is an obligatory sickness insurance scheme. In 2008 there was one hospital, but Liechtenstein has an agreement with the Swiss cantons of St Gallen and Graubünden and the Austrian Federal State of Vorarlberg that her citizens may use certain hospitals. In 2008 there were 87 physicians, 27 dentists and two pharmacies.

RELIGION

In 2003, 80·4% of the population was Roman Catholic and 7·1% Protestant; 12·5% belonged to other religions.

CULTURE

Press

In 2008 there were two daily newspapers (*Liechtensteiner Vaterland* and *Liechtensteiner Volksbatt*) with an estimated total circulation of 20,000.

Tourism

In 2008, 77,957 overnight tourists visited Liechtenstein.

DIPLOMATIC REPRESENTATIVES

In 1919 Switzerland agreed to represent the interests of Liechtenstein in countries where it has diplomatic missions and where Liechtenstein is not represented in its own right. In so doing Switzerland always acts only on the basis of mandates of a general or specific nature, which it may either accept or refuse, while Liechtenstein is free to enter into direct relations with foreign states or to set up its own additional diplomatic missions.

Of the United Kingdom in Liechtenstein
Ambassador: Sarah Gillett, CMG, MVO (resides in Berne).

Of Liechtenstein to the USA (2900 K. St. NW, Washington, D.C., 20007)
Ambassador: Claudia Fritsche.

Of the USA in Liechtenstein
Ambassador: Donald S. Beyer, Jr (resides in Berne).

Of Liechtenstein to the United Nations
Ambassador: Christian Wenaweser.

Of Liechtenstein to the European Union
Ambassador: Kurt Jäger.

FURTHER READING

Amt für Volkswirtschaft. *Statistisches Jahrbuch.* Vaduz
Rechenschaftsbericht der Fürstlichen Regierung. Vaduz. Annual, from 1922
Jahrbuch des Historischen Vereins. Vaduz. Annual since 1901
National library: Landesbibliothek, Vaduz

Beattie, David, *Liechtenstein: A Modern History.* 2004

National Statistical Office: Amt für Volkswirtschaft, Gerberweg 5, 9490 Vaduz.
Website (limited English): http://www.as.llv.li

The world in focus at
www.statesmansyearbook.com

LITHUANIA

© Research Machines plc 2006

Lietuvos Respublika
(Republic of Lithuania)

Capital: Vilnius
Population projection, 2015: 3·25m.
GNI per capita, 2011: (PPP$) 16,234
HDI/world rank: 0·810/40
Internet domain extension: .lt

KEY HISTORICAL EVENTS

Lithuania has been inhabited since the 10th millennium BC, with agriculture developing in the 3rd millennium BC. Baltic tribes settled in the area around 2000 BC. In the 13th century AD their lands came under threat from two German religious orders, the Teutonic Knights and the Livonian Brothers of the Sword, prompting several of the tribes to establish a defensive union. The union defeated the Livonians in 1236 and in 1250 its leader, Mindaugas, signed a peace treaty with the Teutonic Order.

In 1253 Lithuania was proclaimed a state, with Mindaugas its crowned head. In the second half of the 13th century the Grand Duchy of Lithuania suffered internal unrest and was repeatedly raided by the Turks and Mongols of the Golden Horde. In the 14th century, led by Grand Duke Gediminas, Lithuania repulsed the threat from the Golden Horde and expanded eastwards. Gediminas established relations with the Christian church while retaining pagan beliefs.

In 1386 Lithuania's ruler Jogaila married Jadwiga, queen of Poland, and became king of Poland. During the 14th century Lithuania was Christianized. Lithuania continued its expansion and by 1430 extended from the Baltic to the Black Sea. Lithuania and Poland were allied from 1447 and from 1501 they shared the same leader. In 1569 the Lublin Union legislated for a Lithuanian-Polish commonwealth. Polish culture was increasingly influential and Polish became the official state language in 1696. The constitution promulgated on 3 May 1791 is generally regarded as Europe's first national constitution (and the world's second after the USA). However, its democratic and egalitarian tone provoked Prussia and Russia.

During partitions of the Polish-Lithuanian Commonwealth by Russia, Prussia and Austria in 1772, 1793 and 1795, Lithuania was divided between Russia and Prussia. From 1795 Russia ruled most of Lithuania, including its capital, Vilnius. Uprisings were quelled in 1831 and 1863. In the second half of the 19th century, a cultural awakening interacted with an independence movement. On 5 Dec. 1905 at the Great Seimas (Congress) of Vilnius, Lithuanian representatives demanded political autonomy within the Russian Empire. The demand failed but in its aftermath pro-independence political parties were formed.

During the First World War Lithuania was occupied by Germany. Following the Russian revolution, heavy fighting occurred between Soviet Russian, German, Polish and Lithuanian forces in Feb. 1918. In Nov. 1918, after Germany's surrender, Lithuania declared full independence. In April 1919 the Soviets withdrew and the reformed Lithuanian government established a democratic republic. Lithuanian independence was recognized by the Treaty of Versailles later that year. Territorial disputes led to Lithuania supporting an uprising in the Memelland (under French jurisdiction from 1920) in Jan. 1923. In May 1924 Memelland became an autonomous part of Lithuania. There was also continued conflict with Poland over ownership of Vilnius. Following the establishment of a Polish-controlled mini-state around the city in 1922, Lithuania maintained a formal state of war with Poland.

In Dec. 1926 an internal coup deposed Lithuania's elected government and a non-democratic regime was installed. In 1938, under Polish pressure, diplomatic relations with Poland were restored. The secret protocol of the Soviet–German non-aggression pact of 1939 assigned the greater part of Lithuania to the Soviet sphere of influence. Soviet troops occupied Lithuania in June 1940 and it became part of the USSR on 3 Aug. 1940. Following the German invasion of the USSR in 1941, Lithuania was occupied by Germany. Lithuanian armed groups fought against or with German troops, according to regional and ideological loyalties. Pogroms took place against the Jewish population, which had risen to 250,000 following influxes of refugees from Poland. In 1944 the USSR reclaimed Lithuania as a Soviet Republic, with the agreement of the USA and Britain. An estimated 350,000 Lithuanians were deported to Siberia. In 1949 the Soviets closed most Lithuanian churches. More deportations occurred in 1956, when Poles and Russians were encouraged to move to Vilnius.

In 1988 the Lithuanian Movement for Reconstruction (Sajudis) drew up a programme of democratic and national rights. In the same year the ruling communists relaxed anti-nationalist measures and legalized a multi-party system. On 11 March 1990 the newly-elected Lithuanian Supreme Soviet declared independence, a move rejected by the USSR. Initially despatched to Vilnius to enforce conscription, Soviet army units occupied key buildings in the face of mounting popular unrest. On 13 Jan. 1991 the army fired on demonstrators. A referendum held in Feb. 1991 produced a 90·5% vote in favour of independence. The USSR recognized Lithuania's independence on 6 Sept. 1991, with all Russian troops withdrawn by Aug. 1993. The 1992 Lithuanian constitution provided for a presidency and Algirdas Brazauskas won the first presidential elections the following year. Lithuanian became the official language in Jan. 1995, prompting protests from some Polish and Russian speakers. Lithuania joined NATO on 29 March 2004 and became a member of the EU on 1 May 2004.

TERRITORY AND POPULATION

Lithuania is bounded in the north by Latvia, east and south by Belarus, and west by Poland, the Russian enclave of Kaliningrad and the Baltic Sea. The total area is 65,300 sq. km (25,212 sq. miles), including 2,265 sq. km (875 sq. miles) of inland waters, and the population (2011 census) 3,043,429 (1,640,825 females);

density, 48·3 per sq. km. The United Nations population estimate for 2011 was 3,307,000.

The UN gives a projected population for 2015 of 3·25m.

In 2011, 67·1% of the population lived in urban areas. Of the 2011 census population, Lithuanians accounted for 84·2%, Poles 6·6%, Russians 5·8% (9·4% in 1989), Belarusians 1·2% and Ukrainians 0·5%.

There are ten counties (with capitals of the same name): Alytus; Kaunas; Klaipėda; Marijampolė; Panevėžys; Šiauliai; Tauragė; Telšiai; Utena; Vilnius.

The capital is Vilnius (2011 census population, 535,631). Other large towns are Kaunas (315,933 in 2011), Klaipėda (162,360), Šiauliai (109,328) and Panevėžys (99,690).

The official language is Lithuanian, but ethnic minorities have the right to official use of their language where they form a substantial part of the population. All residents who applied by 3 Nov. 1991 received Lithuanian citizenship, requirements for which are ten years' residence and competence in Lithuanian.

SOCIAL STATISTICS

2009: births, 36,682; deaths, 42,032; marriages, 20,542; divorces, 9,270; infant deaths, 181. Rates (per 1,000 population): birth, 11·0; death, 12·6; marriage, 6·2; divorce, 2·8. The population started to decline in 1993, a trend which is set to continue. Annual population growth rate, 2000–05, –0·5%. In 2002, 8,386 births were registered to unmarried mothers and there were 18,907 legally induced abortions. Life expectancy at birth in 2007 was 65·9 years for males and 77·7 years for females. In 2006 the most popular age range for marrying was 25–29 for males and 20–24 for females. Infant mortality, 2010, five per 1,000 live births; fertility rate, 2008, 1·3 births per woman (one of the lowest rates in the world). In 2002 there were 7,086 emigrants and 5,110 immigrants.

Lithuania has one of the world's highest suicide rates, at 30·4 per 100,000 inhabitants in 2007 (a rate of 53·9 among males but only 9·8 among women).

CLIMATE

Vilnius, Jan. –2·8°C, July 20·5°C. Annual rainfall 520 mm. Klaipėda, Jan. –0·6°C, July 19·4°C. Annual rainfall 770 mm.

CONSTITUTION AND GOVERNMENT

A referendum to approve a new constitution was held on 25 Oct. 1992. Parliament is the 141-member *Seimas*. Under a new electoral law passed in July 2000, 71 of the parliament's 141 members will defeat rivals for their seats if they receive the most votes in a single round of balloting. Previously they had to win 50% of the votes or face a run-off against the nearest competitor. The parliament's 70 other seats are distributed according to the proportional popularity of the political parties at the ballot box.

The *Constitutional Court* is empowered to rule on whether proposed laws conflict with the constitution or existing legislation. It comprises nine judges who serve nine-year terms, one third rotating every three years.

National Anthem

'Lietuva, tėvyne mūsų' ('Lithuania, our fatherland'); words and tune by V. Kurdirka.

GOVERNMENT CHRONOLOGY

(LDDP = Democratic Labour Party of Lithuania; LDP = Liberal Democratic Party; LKP = Communist Party of Lithuania; LLS = Lithuanian Liberal Union; LSDP = Social Democratic Party of Lithuania; Sąjūdis = 'Unity'/Reform Movement of Lithuania; TS(LK) = Homeland Union (Conservatives of Lithuania); TS-LKD = Homeland Union-Lithuanian Christian Democrats; n/p = non-partisan)

Heads of State since 1990.

Chairman of the Supreme Council
1990–92	Sąjūdis	Vytautas Landsbergis

Chairman of the Seimas (Parliament)
1992–93	LDDP	Algirdas Brazauskas

Presidents of the Republic
1993–98	LDDP	Algirdas Brazauskas
1998–2003	n/p	Valdas Adamkus
2003–04	LDP	Rolandas Paksas
2004–09	n/p	Valdas Adamkus
2009–	n/p	Dalia Grybauskaitė

Prime Ministers since 1990.
1990–91	LKP/LDDP	Kazimiera Prunskienė
1991	Sąjūdis	Albertas Simenas
1991–92	Sąjūdis	Gediminas Vagnorius
1992	n/p	Aleksandras Abišala
1992–93	LDDP	Bronislovas Lubys
1993–96	LDDP	Adolfas Šleževičius
1996	LDDP	Mindaugas Stankevičius
1996–99	TS(LK)	Gediminas Vagnorius
1999	TS(LK)	Rolandas Paksas
1999–2000	TS(LK)	Andrius Kubilius
2000–01	LLS	Rolandas Paksas
2001–06	LSDP	Algirdas Brazauskas
2006–08	LSDP	Gediminas Kirkilas
2008–12	TS-LKD	Andrius Kubilius
2012–	LSDP	Algirdas Butkevičius

RECENT ELECTIONS

Presidential elections were held on 17 May 2009. Dalia Grybauskaitė, Lithuania's European Union commissioner, won 69·1% of the vote, ahead of Algirdas Butkevičius with 11·8%, Valentinas Mazuronis 6·2%, Valdemar Tomaševski 4·7%, former prime minister Kazimiera Prunskienė 3·9%, Loreta Graužinienė 3·6% and Česlovas Jezerskas 0·6%. Turnout was 51·7%.

Parliamentary elections were held in two rounds on 14 and 28 Oct. 2012. The opposition Social Democratic Party of Lithuania (LSDP) won 38 of the 141 seats (with 18·4% of the proportional representation vote), of which 23 were in single-member constituencies and 15 through proportional representation; Homeland Union-Lithuanian Christian Democrats 33 (15·1% of the proportional representation vote) of which 20 were in single-member constituencies and 13 through proportional representation; Labour Party 29 (19·8%) of which 12 were in single-member constituencies and 17 through proportional representation; Order and Justice 11 (7·3%); Liberal Movement 10 (8·6%); Electoral Action of Poles in Lithuania 8 (5·8%); the Way of Courage 7 (8·0%); Peasant and Greens Union 1 (3·9%). Three seats went to independents and one was vacant. Turnout was 52·9%.

European Parliament

Lithuania has 12 (13 in 2004) representatives. At the June 2009 elections turnout was 21·0% (48·4% in 2004). The Homeland Union-Lithuanian Christian Democrats won 4 seats with 26·9% of votes cast (political affiliation in European Parliament: European People's Party); Social Democratic Party of Lithuania, 3 with 18·6% (Progressive Alliance of Socialists and Democrats); 'Order and Justice' coalition, 2 with 12·2% (Europe of Freedom and Democracy); Labour Party, 1 with 8·8% (Alliance of Liberals and Democrats for Europe); Lithuanian Poles' Electoral Action, 1 with 8·4% (European Conservatives and Reformists); Liberals' Movement of the Republic of Lithuania, 1 with 7·4% (Alliance of Liberals and Democrats for Europe).

CURRENT GOVERNMENT

President: Dalia Grybauskaitė; b. 1956 (took office on 12 July 2009).

Prime Minister: Algirdas Butkevičius; b. 1958 (LSDP; in office since 13 Dec. 2012).

In March 2013 the coalition government comprised:

Minister of Agriculture: Vigilijus Jukna. *Culture:* Šarūnas Birutis. *Economy:* Birutė Vėsaitė. *Education and Science:* Dainius Pavalkis. *Energy:* Jaroslav Neverovič. *Environment:* Valentinas Mazuronis. *Finance:* Rimantas Šadžius. *Foreign Affairs:* Linas Antanas Linkevičius. *Health:* Vytenis Povilas Andriukaitis. *Interior:* Dailis Alfonsas Barakauskas. *Justice:* Juozas Bernatonis. *National Defence:* Juozas Olekas. *Social Security and Labour:* Algimanta Pabedinskienė. *Transport and Communications:* Rimantas Sinkevičius.

Seimas Speaker: Vydas Gedvilas.

Government of the Republic of Lithuania: http://www.lrvk.lt

CURRENT LEADERS

Dalia Grybauskaitė

Position
President

Introduction
Dalia Grybauskaitė was sworn in as the first female president of Lithuania on 12 July 2009 after a landslide election victory. Grybauskaitė left her job as EU commissioner for financial programming and budget to stand as an independent candidate, backed by the incumbent centre-right government. Renowned as a tough negotiator and skilled economist, she took office during the worst economic crisis since the dissolution of the Soviet Union.

Early Life
Grybauskaitė was born on 1 March 1956 in Vilnius, the then capital of the Lithuanian Soviet Socialist Republic, and went on to study political and economic science at Zhdanov University (now St Petersburg State University).

After graduating, Grybauskaitė returned to Lithuania and embarked on a career as a lecturer at the department of political economy at Vilnius Higher Party School, in which role she worked for seven years. She received her doctorate in economic sciences from the Moscow Academy of Social Sciences in 1988.

Following the dissolution of the USSR and Lithuanian independence, Grybauskaitė completed a special course for leaders at Georgetown University in Washington in 1991. She subsequently entered government, going on to head departments in the ministries of international economic relations and foreign affairs between 1991 and 1994. She was Lithuania's representative when it entered into the European Union Free Trade Agreement in 1993.

She continued her involvement with the EU in 1994, when she was appointed envoy extraordinary and minister plenipotentiary at the Lithuanian mission to the EU. She continued her diplomatic work when she moved to the Lithuanian embassy in the USA.

In 1999 Grybauskaitė was appointed vice-minister of finance and foreign affairs in the cabinet of Prime Minister Andrius Kubilius. In this role she was responsible for conducting negotiations with international institutions including the World Bank and the IMF. In 2001 she was made minister of finance in the government of Algirdas Brazauskas.

Lithuania acceded to the EU on 1 May 2004, with Grybauskaitė appointed EU commissioner responsible for managing the EU budget and embarking on an ambitious programme of reform. In 2005 she criticized the UK's presidency of the EU, stating that the 'main obstacle' to reaching agreement on budgetary reform was Britain's insistence that it retain its rebate.

On 26 Feb. 2009 Grybauskaitė ended months of speculation by announcing her intention to stand in the forthcoming presidential elections. A popular figure across party lines, her entry into the race rendered the election a foregone conclusion, with opposition candidates either withdrawing or remaining in the race only to develop support ahead of elections to the European parliament.

In the event, Grybauskaitė gained 69·1% of the vote, against 11·8% for her nearest challenger, Algirdas Butkevičius.

Career in Office
Grybauskaitė's victory was broadly popular, with several media commentators comparing the optimism that her election inspired with that seen in the United States on the inauguration of President Obama. In her inaugural address she promised to invest in local government and to narrow the gap between rich and poor in Lithuanian society, although she supported austerity measures to reduce the long-term impact of the prevailing economic crisis.

Despite her close association with the EU, Grybauskaitė promised that greater integration into Europe would not come at the price of deteriorating bilateral relations with Russia. She promised a more considered, less confrontational approach to issues of international relations. She sparked a parliamentary inquiry in Nov. 2009 after conceding that she harboured 'indirect suspicions' that the CIA had used a Lithuanian riding school as a secret jail for the detention of suspected Al-Qaeda militants captured in Afghanistan.

She described the reinforcement of Lithuania's energy security as among her top priorities and has sought to secure the provision of extra funds from the EU to develop energy links with the West, thus reducing dependence on Russia.

In Dec. 2012, following parliamentary elections in Oct., Grybauskaitė approved Algirdas Butkevičius, the leader of the opposition Social Democrats and a former rival for the presidency, as prime minister at the head of a new coalition government.

Algirdas Butkevičius

Position
Prime Minister

Introduction
Leader of the Social Democratic Party of Lithuania (LSDP), Algirdas Butkevičius became prime minister in Dec. 2012 at the head of a coalition government. While espousing fiscal responsibility, he is expected to moderate the austerity programme of the previous government.

Early Life
Born on 19 Nov. 1958 in Paežeriai in Radviliškis district, Butkevičius graduated in engineering economics from the Vilnius Civil Engineering Institute in 1984 and worked in the Vilkaviškis district for the industrial association Žemūktechnika from 1982–85. From 1985–90 he was an architect and inspector for the state construction and architecture unit. In 1991 he gained a diploma in technical management from the Lithuanian Management Academy and in 1995 was awarded a masters degree in management from the Kaunas University of Technology.

After serving as deputy governor of Vilkaviškis district from 1991–95, Butkevičius spent a year as director of research and marketing for the private company, AB Vilkauta. He joined the LSDP in 1992 and was chairman of the party's Vilkaviškis branch from 1995–97. He sat on the Vilkaviškis municipal council from 1990–97 and again from 2000–02. Elected to parliament in 1996, he chaired its budget and finance committee from 2001–04. In 2004 he was appointed finance minister, serving for five months in the administration of Prime Minister Brazauskas. He resigned in April 2005 in protest at government proposals to introduce a corporate turnover tax. From 2006–08 he was minister for transport and communications under Prime Minister Kirkilas, overseeing Lithuania's participation in the 'east-west corridor' project to develop a transport network from Sweden to Vilnius.

After the LSDP lost power in 2008, Butkevičius was elected party chairman in March 2009 and unsuccessfully contested the presidency later that year. After four years of stringent austerity by the government of Andrius Kubilius, Butkevičius campaigned ahead of the 2012 parliamentary election on a platform of observing fiscal discipline and keeping the budget deficit under control while relaxing wage restraints to boost the economy. In Oct. 2012 the LSDP became the biggest party in parliament and entered negotiations to form a coalition government.

Career in Office

Butkevičius took office on 13 Dec. 2012 at the head of a four-party coalition after prolonged disagreement with President Grybauskaitė, who opposed the inclusion in the coalition of Labour Party members facing allegations of misconduct. Butkevičius has to combat high unemployment and low economic growth, while keeping to agreed borrowing targets. He is expected to encourage closer integration with the EU, with the aim of joining the single currency within the next few years. He has also signalled his intention to build better relations with Russia.

DEFENCE

Conscription ended on 1 July 2009. In 2008 military expenditure totalled US$547m. (US$153 per capita), representing 1·2% of GDP.

Army

The Land Forces numbered 7,800 in 2007 and included one motorized infantry brigade ('Iron Wolf'). Reserves numbered 6,700 in 2007 and there were also 4,700 active reservists in the National Defence Voluntary Forces. There was a paramilitary riflemen union numbering 9,600 and a state border guard service of 5,000 operates under the ministry of internal affairs.

A joint Polish/Lithuanian battalion (LITPOLBAT), which was a component of the EU's rapid reaction forces, was disbanded in 2007. However, plans are under way to replace it with a Lithuanian/Polish/Ukrainian Brigade (LITPOLUKRBRIG), which is expected to become operational by 2016.

Navy

In 2007 the Navy numbered 450 personnel (150 conscripts). It operates several vessels including one frigate. Lithuania, Estonia and Latvia have established a joint naval unit 'BALTRON' (Baltic Naval Squadron), with bases at Klaipėda in Lithuania, Tallinn in Estonia, and Liepāja, Riga and Ventspils in Latvia.

In addition there is a 540-strong Coast Guard.

Air Force

The Air Force consisted of 900 personnel (100 conscripts) in 2007. There are no combat-capable aircraft.

The joint Baltic Regional Air Surveillance Network (BALTNET), established in co-operation between the air forces of Estonia, Latvia and Lithuania, has its co-ordination centre in Karmėlava in Lithuania.

INTERNATIONAL RELATIONS

Lithuania held a referendum on EU membership on 10–11 May 2003, in which 91·0% of votes cast were in favour of accession, with 9·0% against. It became a member of NATO on 29 March 2004 and the EU on 1 May 2004.

ECONOMY

Agriculture accounted for 3·4% of GDP in 2009, industry 26·9% and services 69·7%.

Overview

Recession in the early years of Lithuania's transition from a command economy to a market economy was more pronounced than in many of its Baltic neighbours, partly because its export sector had closer ties to the Russian market. From 1995 to 1998 the economy grew at a strong and stable rate but the country fell into recession in 1999 following the 1998 Russian financial crisis. In 2000 Lithuania rebounded and enjoyed one of the highest annual growth rates of any transition economy.

Most enterprises have been privatized and the government has maintained a stable macroeconomic policy while encouraging economic relations with other countries. It is a member of the World Trade Organization and joined the EU in May 2004. It has attracted significant foreign investment from EU partners, particularly the Nordic countries, and the World Bank ranks Lithuania 27th of 183 economies for the ease of doing business.

Core exports are mineral products (25·5% of total exports in 2011), and machinery and equipment (10·4%). In 2011, 61% of exports went to the EU. Russia's share of Lithuanian exports fell from 30% in 1997 to 17% in 2011. Nonetheless, 32% of imports (mostly of energy) still came from Russia. Unemployment fell from 17·4% in 2001 to a record low of 4·3% in 2007. Thriving exports and domestic consumer demand ensured average growth of 8% per year in the four years before the 2008 financial crisis. Between 1998 and 2008 real per capita incomes rose from about two-fifths to two-thirds of the EU average, principally driven by strong exports.

The downturn prompted by the global financial crisis has been among the most severe in the region. A contraction in domestic demand was compounded by Lithuania's main export partners also falling into recession. The economy shrank by 14·8% in 2009 while unemployment surged to 18·3% in June 2010.

In the boom years a large structural deficit was accumulated, putting government finances under strain during the crisis. With the currency fixed against the euro and lending from the international bond markets unforthcoming, the government implemented a package of austerity measures in Dec. 2009. Public spending was cut by 30%, public sector wages fell by between 20–30%, pensions were cut by up to 11%, VAT rose from 18% to 21%, corporate tax went from 15% to 20% and there were significant tax rises on alcohol and pharmaceuticals. The measures resulted in savings equivalent to 9% of GDP without calling on the assistance of the IMF.

Lithuania experienced a strong recovery in 2011, recording 6% growth largely thanks to a resurgent exports. However, growth was predicted to slow and the unemployment rate remained high, at 13·3% in the second quarter of 2012. The ongoing eurozone crisis is the main threat to growth prospects.

Currency

The unit of currency is the *litas* (plural *litai*) (LTL) of 100 *cents*, which was introduced on 25 June 1993 and became the sole legal tender on 1 Aug. The litas was pegged to the US dollar on 1 April 1994 at US$1 = four litai, but since 2 Feb. 2002 it has been pegged to the euro at 3·4528 litai = one euro. Inflation, which reached a high of 1,161% in the early 1990s, was 11·1% in 2008 but fell to 1·2% in 2010 and stood at 4·1% in 2011.

Total money supply was 13,884m. litai in July 2005, foreign exchange reserves were US$3,411m. and gold reserves 186,000 troy oz.

Budget

Budgetary central government revenue and expenditure (in 1m. litai):

	2007	2008	2009[1]
Revenue	19,543	21,497	17,485
Expenditure	17,825	21,808	21,698

[1]Provisional.

Principal sources of revenue in 2008: taxes on goods and services, 12,322m. litai; taxes on income, profits and capital gains, 4,780m. litai; grants, 2,680m. litai. Main items of expenditure by economic type in 2008: compensation of employees, 5,902m. litai; grants, 4,936m. litai; use of goods and services, 4,164m. litai. There is a flat income tax rate of 15%.

Lithuania's budget deficit in 2011 was 5·5% of GDP (2010, 7·2%; 2009, 9·4%). The required target set by the EU is a budget deficit of no more than 3%.

VAT is 21%.

Performance

Between 2000 and 2006 growth averaged 7·2%. In 2007 and 2008 the economy continued to grow, by 9·8% and 2·9% respectively, although Lithuania was then one of the countries most affected by the economic crisis in 2009, with the GDP contracting by 14·8%. There was a slight recovery in 2010 when the economy grew by 1·4%, continuing in 2011 when it grew by 5·9%. Total GDP in 2011 was US$42·7bn.

Banking and Finance

The central bank and bank of issue is the Bank of Lithuania (*Governor*, Vitas Vasiliauskas). A programme to restructure and privatize the state banks was started in 1996. In 2007 there were 11 commercial banks and foreign bank branches, the central credit union of Lithuania and 66 credit unions in operation. The largest private bank in Lithuania is SEB Vilniaus bankas, which controls approximately 36% of the total banking assets in the country. In 2006 it was estimated that total assets of domestic commercial banks amounted to 59bn. litai.

Foreign debt amounted to US$29,602m. in 2010, up from US$11,433m. in 2005 and US$4,723m. in 2000. Lithuania's foreign debt in 2010 was equivalent to 83·0% of GNI (up from 42·0% in 2000).

Lithuania attracted US$1,217m. worth of foreign direct investment in 2011, up from US$753m. in 2010 but down from the record US$2,015m. of 2007.

A stock exchange opened in Vilnius in 1993. In Nov. 2007 its capitalization was €6·8bn. The trading turnover in 2006 was €1·6bn.

ENERGY AND NATURAL RESOURCES

In 2008, 15·3% of energy consumption came from renewables (wind power, solar power, hydro-electric power, tidal power, geothermal energy and biomass), compared to the European Union average of 10·3%. A target of 23% has been set by the EU for 2020.

Environment

According to Lithuania's Ministry of Environment, carbon dioxide emissions were the equivalent of 5·1 tonnes per capita in 2008. Lithuania's greenhouse gas emissions fell by 51·1% between 1990 and 2008, mainly owing to the decline of polluting industries from the Soviet era.

Electricity

Installed capacity was 4·60m. kW in 2007; production was 14·01bn. kWh in 2007. A nuclear power station (with two reactors) in Ignalina was responsible for 70·2% of total output in 2007, and there are also two large hydro-electric, five public and five autoproducer thermal plants. At the time no other country had such a high percentage of its electricity generated through nuclear power. However, as a condition of entry into the European Union the government agreed to close down Ignalina. The process to close the first reactor began on 31 Dec. 2004. The whole facility was shut down on 31 Dec. 2009. Plans to build a successor near the old site are being reconsidered after 64·8% opposed the idea in a consultative referendum held in Oct. 2012. Prior to the closure of Ignalina, Lithuania was one of the countries most reliant on nuclear energy. Electricity consumption per capita in 2007 was 3,743 kWh.

Oil and Gas

Oil production started from a small field at Kretinga in 1990. Reserves were 12m. bbls in 2007. Oil production in 2007 was 154,000 tonnes. Lithuania relies on Russia for almost all of its oil and gas.

Minerals

Output of minerals in 2006 (1,000 cu. metres): dolomite, 1,600; limestone, 900; peat, 400. Quarrying of gravel, clay and sand totalled 8·7m. cu. metres in 2006.

Agriculture

In 2002 agriculture employed about 17·2% of the workforce. As of 1 Jan. 2003 the average farm size was 15·2 ha., one of the lowest in eastern Europe; the agricultural land area was 3,956,200 ha. In 2002 there were 2·93m. ha. of arable land and 59,000 ha. of permanent crops. In 2002, 242,000 persons were employed in agriculture and forestry.

Output of main agricultural products (in 1,000 tonnes) in 2002: potatoes, 1,531; wheat, 1,218; sugar beets, 1,052; barley, 871; rye, 170; rapeseed, 105; cabbage, 98; oats, 97. Value of agricultural production, 2002 (in 1m. litai), was 4,303·3, of which from individual farm holdings, 3,396·2; and from agricultural partnerships and enterprises, 907·1.

Livestock, Jan. 2003 (in 1,000): cattle, 779·1 (of which dairy cows, 443·3); pigs, 1,061·0; sheep and goats, 35·6; horses, 60·7; poultry, 6,848·1. There were 103,000 tractors in use in 2002. Animal products, 2002 (in 1,000 tonnes): meat, 173·6; milk, 770·9; eggs, 779m. units.

Forestry

In 2010 forests covered 2·2m. ha., or 34% of the land area, and consist of conifers, mostly pine. Of the total area under forests 1% was primary forest in 2010, 75% other naturally regenerated forest and 24% planted forest. Timber production in 2007, 5·86m. cu. metres.

Fisheries

In 2011 the fishing fleet comprised 171 vessels of a combined 45,960 gross tonnes. Total catch in 2010 amounted to 149,851 tonnes (mainly from sea fishing), down from 187,513 tonnes in 2007 although up from 44,002 tonnes in 1997. Imports of fishery commodities were valued at US$283m. in 2008 and exports at US$288m.

INDUSTRY

Industrial output in 2006 included (in 1,000 tonnes): petrol, 2,303; cement, 1,100; sulphuric acid, 730; sugar, 97; woollen fabrics, 22·5m. sq. metres; cotton fabrics, 20·4m. cu. metres; linen, 12·8m. sq. metres; television picture tubes, 1,240,000 units; TV sets, 711,300 units; bicycles, 330,000 units.

Labour

In 2002 the workforce was 1·6m. (69·9% in private enterprises and 30·1% in the public sector). Employed population by activity (as a percentage): manufacturing, 18·6; wholesale and retail trade, 15·0; education, 9·9; health and social work, 6·7; construction, 6·6; transport and communications, 6·2; real estate, 3·9. Employment skills, 33·2% with tertiary education, 52·5% with upper secondary education, 11·8% with lower secondary. There were a total of 31,601 working days lost to strike action in 2008 (9,559 in 2007). In 2002 the average monthly wage was 1,013·9 litai; legal minimum wage was 450 litai in 2003.

In 2002 the old age pension for men started at 62 years and for women at 58. The average number of persons entitled to pensions in 2001 was 636,900. The unemployment rate in the period Oct.–Dec. 2010 was 17·4%, one of the highest rates in the EU.

INTERNATIONAL TRADE

Foreign investors may purchase up to 100% of the equity companies in Lithuania.

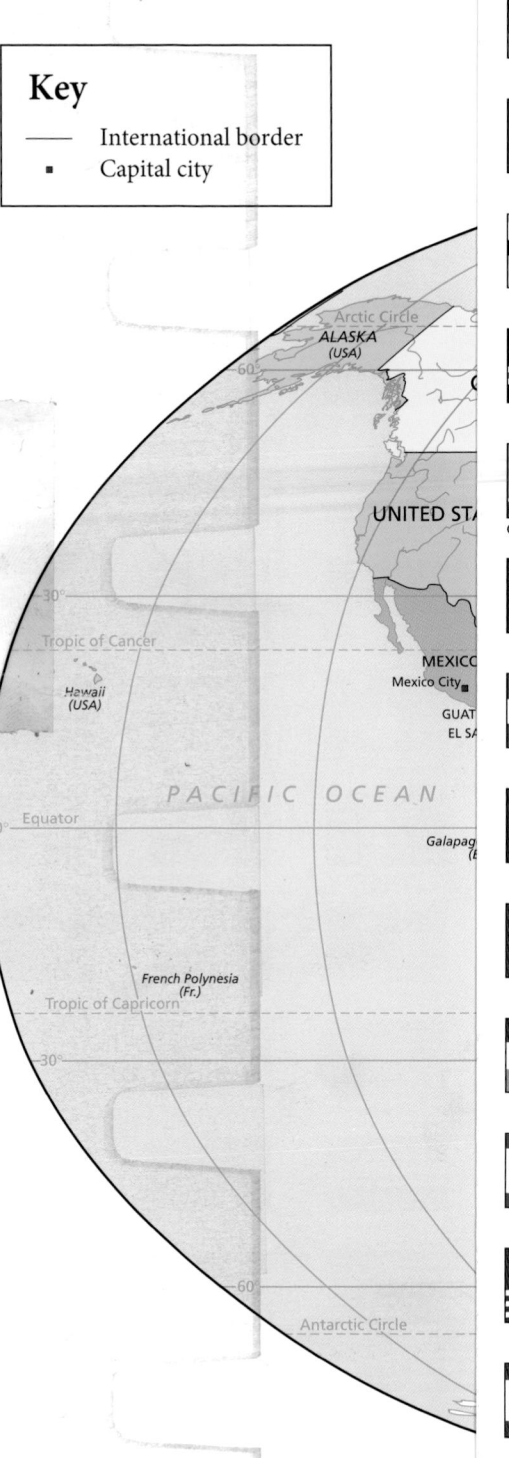

Key

— International border
▪ Capital city

Andorra

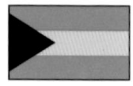
Angola

Antigua and Barbuda

Argentina

Azerbaijan

Bahamas

Bahrain

Bangladesh

Belize

Benin

Bhutan

Bolivia

Brunei

Bulgaria

Burkina Faso

Burundi

Cape Verde

Central African Republic

Chad

Chile

Congo, Democratic Republic of

Congo, Republic of

Costa Rica

Côte d'Ivoire

Czech Republic

Denmark

Djibouti

Dominica

El Salvador

Equatorial Guinea

Eritrea

Estonia

France

Gabon

The Gambia

Georgia

Grenada

Guatemala

Guinea

Guinea-Bissau

Hungary

Iceland

India

Indonesia

Israel

Italy

Jamaica

Japan

Kiribati

Korea, North

Korea, South

Kuwait

Lebanon

Lesotho

Liberia

Libya

Macedonia

Madagascar

Malawi

Malaysia

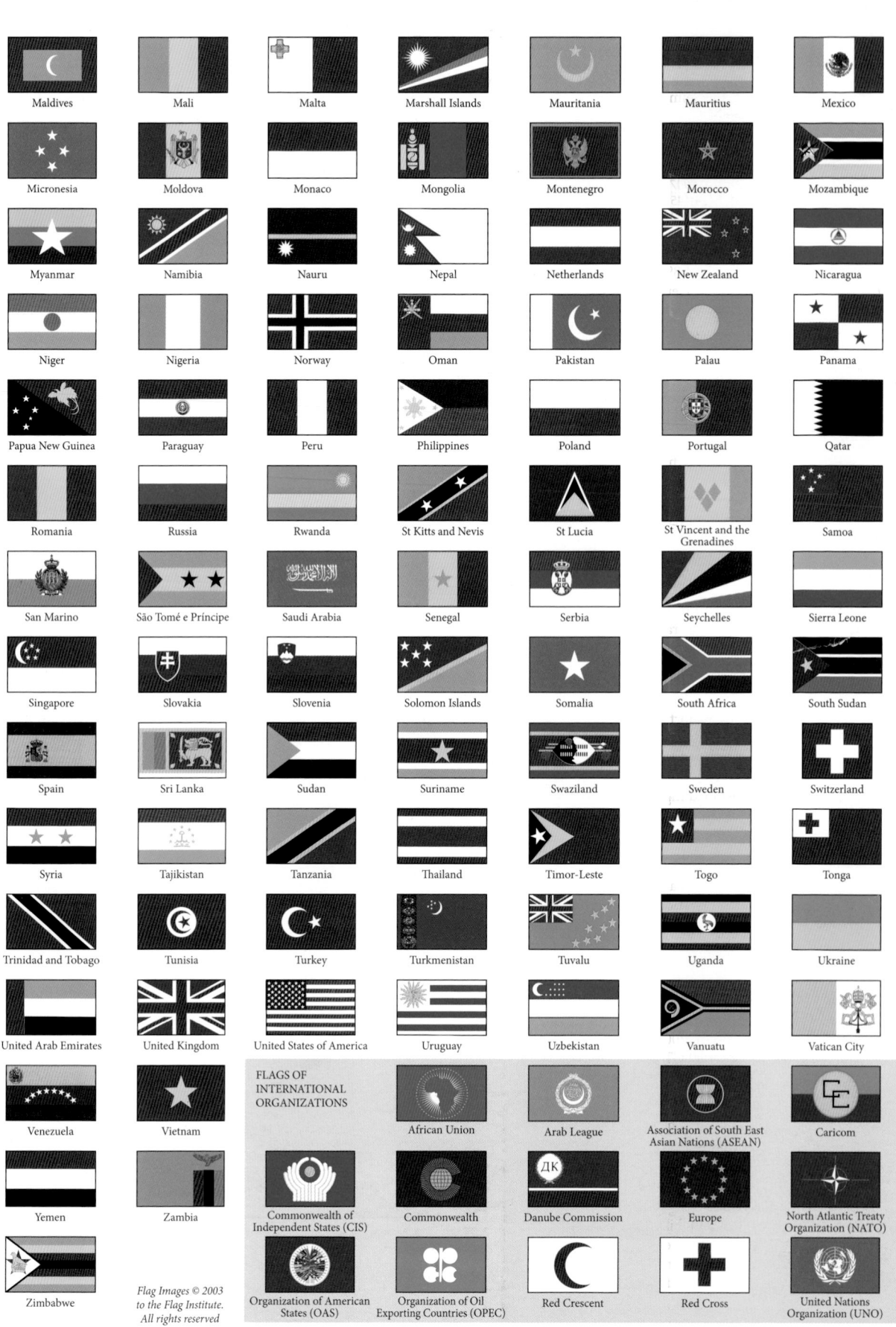

Maldives | Mali | Malta | Marshall Islands | Mauritania | Mauritius | Mexico
Micronesia | Moldova | Monaco | Mongolia | Montenegro | Morocco | Mozambique
Myanmar | Namibia | Nauru | Nepal | Netherlands | New Zealand | Nicaragua
Niger | Nigeria | Norway | Oman | Pakistan | Palau | Panama
Papua New Guinea | Paraguay | Peru | Philippines | Poland | Portugal | Qatar
Romania | Russia | Rwanda | St Kitts and Nevis | St Lucia | St Vincent and the Grenadines | Samoa
San Marino | São Tomé e Príncipe | Saudi Arabia | Senegal | Serbia | Seychelles | Sierra Leone
Singapore | Slovakia | Slovenia | Solomon Islands | Somalia | South Africa | South Sudan
Spain | Sri Lanka | Sudan | Suriname | Swaziland | Sweden | Switzerland
Syria | Tajikistan | Tanzania | Thailand | Timor-Leste | Togo | Tonga
Trinidad and Tobago | Tunisia | Turkey | Turkmenistan | Tuvalu | Uganda | Ukraine
United Arab Emirates | United Kingdom | United States of America | Uruguay | Uzbekistan | Vanuatu | Vatican City
Venezuela | Vietnam | FLAGS OF INTERNATIONAL ORGANIZATIONS | African Union | Arab League | Association of South East Asian Nations (ASEAN) | Caricom
Yemen | Zambia | Commonwealth of Independent States (CIS) | Commonwealth | Danube Commission | Europe | North Atlantic Treaty Organization (NATO)
Zimbabwe | | Organization of American States (OAS) | Organization of Oil Exporting Countries (OPEC) | Red Crescent | Red Cross | United Nations Organization (UNO)

Individual laws on three free economic zones (namely the laws on Šiauliai, Klaipėda and Kaunas) have been cleared by Lithuania's Parliament, the Seimas.

Imports and Exports

Imports and exports for calendar years in US$1m.:

	2006	2007	2008	2009	2010
Imports c.i.f.	19,388·4	24,445·1	31,294·7	18,340·6	23,378·0
Exports f.o.b.	14,135·2	17,162·4	23,769·9	16,496·3	20,813·9

Leading import suppliers, 2006: Russia, 24·2%; Germany, 14·9%; Poland, 9·5%; Latvia, 4·8%. Principal export markets, 2006: Russia, 12·8%; Latvia, 11·1%; Germany, 8·6%; Estonia, 6·5%.

Main imports are machinery and apparatus, crude petroleum, road vehicles, and chemicals and chemical products. Main exports are mineral products, electrical equipment, textiles and textile articles, transport equipment, TV sets, chemical products and prepared foodstuffs.

COMMUNICATIONS

Roads

In 2007 there were 80,715 km of roads (including 309 km of motorways), of which 28·6% were paved. There were 1,587,900 passenger cars in use in 2007 (470 per 1,000 inhabitants), plus 14,000 buses and coaches, 14,500 lorries and vans, and 35,300 motorcycles and mopeds. There were 6,448 traffic accidents in 2007, with 740 fatalities.

Rail

In 2005 there were 1,817 km of railway track in operation in Lithuania. The majority of rail traffic was diesel propelled, although 122 km of track was electrified. In 2008, 5·1m. passengers and 55·0m. tonnes of freight were carried.

Civil Aviation

The main international airport is based in the capital, Vilnius. Other international airports are at Kaunas and Palanga. FlyLAL–Lithuanian Airlines, formerly Lithuania's largest airline, ceased operations in Jan. 2009. In 2008 a number of international airlines ran regular scheduled flights to Lithuania. Vilnius handled 1,308,065 passengers in 2009 and 4,336 tonnes of freight. Kaunas handled 456,698 passengers in 2009 and Palanga 105,195.

Shipping

The ice-free port of Klaipėda plays a dominant role in the national economy and Baltic maritime traffic. It handled 29,880,000 tonnes of cargo in 2008 (22,218,000 tonnes loaded and 7,662,000 tonnes discharged); container traffic totalled 373,000 TEUs (twenty-foot equivalent units) in 2008. A 412 ha. site at the port is dedicated a Free Economic Zone, which offers attractive conditions to foreign investors.

In Jan. 2009 there were 52 ships of 300 GT or over registered, totalling 348,000 GT. In Jan. 2009 the Lithuanian-controlled fleet comprised 64 vessels of 1,000 GT or over, of which 37 were under the Lithuanian flag and 27 under foreign flags.

Telecommunications

A majority stake in Lithuanian Telecom (the only fixed telephone service provider) was sold to the Finnish and Swedish consortium SONERA in 1998 and by Jan. 2003 the telecommunications market was fully liberalized. In 2010 there were 733,700 landline telephone subscriptions (equivalent to 220·8 per 1,000 inhabitants) and 4,891,000 mobile phone subscriptions (or 1,471·6 per 1,000 inhabitants). There were 621·2 internet users per 1,000 inhabitants in 2010. Fixed internet subscriptions totalled 636,000 in 2009 (190·3 per 1,000 inhabitants). In March 2012 there were 983,000 Facebook users.

SOCIAL INSTITUTIONS

Justice

The general jurisdiction court system consists of the Supreme Court, the Court of Appeal, five county courts and 54 district courts. Specialized administrative courts were established in 1999. In 2006 there were 732 judges: 469 in district courts, 144 in county courts, 27 in the Court of Appeal, 34 in the Supreme Court, 43 in the administrative county courts and 15 in the High Administrative Court.

75,474 crimes were reported in 2006, of which 43% were solved. In 2006 there were 294 murders. 14,717 persons were convicted of offences in 2006. There were 8,079 prisoners in 2006, 7,082 of whom had been convicted. The death penalty was abolished for all crimes in 1998.

Education

Education is compulsory from seven to 16. In 2002–03 there were 686 pre-school establishments with 90,860 pupils and 2,172 general schools with 49,286 teachers and 594,313 pupils, in the following categories:

Type of School	No. of Schools	No. of Pupils
Nursery	148	12,219
Primary	683	35,819
Junior	25	2,326
Basic	645	118,415
Special	67	7,212
Secondary	574	400,566
Adult	27	17,318

119,548 students (70,777 females) attended 19 institutions of higher education and 22,367 (13,735 females) attended vocational colleges in 2002–03. The adult literacy rate in 2009 was estimated at 99·7% (99·7% for both males and females).

In 2006 public expenditure on education represented 5·0% of GNI and 14·4% of total government expenditure.

Health

In 2006 there were 13,510 physicians, 2,249 dentists, 26,140 nurses and midwives and 2,184 pharmacists. There were 196 hospitals with 31,031 beds in 2002.

Welfare

The social security system is financed by the State Social Insurance Fund. In 2002, 625,000 persons were eligible for retirement pensions, 188,000 for disability provisions and 219,000 for widow's/widower's pensions. In 2002 the average state social insurance old age pension was 323 litai (monthly).

RELIGION

Under the Constitution, the state recognizes traditional Lutheran churches and religious organizations, as well as other churches and religious organizations if their teaching and rituals do not contradict the law. In 2001, 79% of the population was Roman Catholic. As of 2006 there were 677 Roman Catholic churches with 710 priests, and 50 Orthodox parishes with 47 priests. In 2007 the Lutheran Church had 54 parishes and 20 pastors headed by a bishop. In Feb. 2013 there was one cardinal.

CULTURE

World Heritage Sites

Lithuania has four sites (two shared) on the UNESCO World Heritage List: Vilnius Historic Centre (inscribed on the list in 1994) and Kernavė Archaeological Site (2004).

Lithuania shares the Curonian Spit (2000) with the Russian Federation as a UNESCO site. A sand-dune spit between Zelenogradsk, Kaliningrad Region, and Klaipėda, Lithuania, the Spit was subject to massive protective engineering in the 19th century. Lithuania also shares the Struve Geodetic Arc (2005). The Arc is a chain of survey triangulations spanning from

Norway to the Black Sea that helped establish the exact shape and size of the earth and is shared with nine other countries.

Press

In 2008 there were 327 newspapers (24 paid-for dailies, one free daily and 302 paid-for non-dailies). The papers with the highest circulation are the free *15 minučiu* and the paid-for *Vakaro žinios* and *Lietuvos rytas*. In the 2010 *World Press Freedom Index*, compiled by Reporters Without Borders, Lithuania was ranked 11th equal out of 178 countries.

Tourism

In 2010 accommodation establishments received 1,552,900 guests (up from 1,325,600 in 2005), of whom 840,400 were foreigners (681,500 in 2005). The leading countries of origin of non-resident overnight visitors in 2010 were: Poland (135,900), Russia (105,900), Germany (105,800) and Belarus (71,400). Lithuania had 908 accommodation establishments in 2010 with 50,087 beds, including 342 hotels with 23,137 beds.

Festivals

The Lithuanian Song and Dance Celebration is held every four years, focusing international attention on the country's culture. There is also a Kaunas Jazz Festival and the Pažaislis Classical Music Festival (also in Kaunas). The main rock festivals are the summer festivals, Be2gether at Norviliškės near the Belarusian border and Rock Nights at Zarasai. Major film festivals include Cinema Spring in Vilnius and the Kaunas International Film Festival. The annual Klaipėda Sea Festival attracts nearly half a million visitors each year.

DIPLOMATIC REPRESENTATIVES

Of Lithuania in the United Kingdom (Lithuania House, 2 Bessborough Gdns, London SW1V 2JE)
Ambassador: Asta Skaisgirytė-Liauškienė.

Of the United Kingdom in Lithuania (Antakalnio str. 2, 10308 Vilnius)
Ambassador: David Hunt.

Of Lithuania in the USA (2622 16th St., NW, Washington, D.C., 20009)
Ambassador: Žygimantas Pavilionis.

Of the USA in Lithuania (Akmenu 6, 03106 Vilnius)
Ambassador: Vacant.
Chargé d'Affaires a.i.: Anne Hall.

Of Lithuania to the United Nations
Ambassador: Raimonda Murmokaitė.

Of Lithuania to the European Union
Permanent Representative: Raimundas Karoblis.

FURTHER READING

Department of Statistics to the Government. *Statistical Yearbook of Lithuania.* Annual. *Economic and Social Development in Lithuania.* Monthly.

Hood, N., *et al.,* (eds.) *Transition in the Baltic States.* 1997
Kasekamp, Andres, *A History of the Baltic States.* 2010
Lane, Thomas, *Lithuania: Stepping Westward.* 2001
Lieven, A., *The Baltic Revolution: Estonia, Latvia, Lithuania and the Path to Independence.* 2nd ed. 1994
Misiunas, R. J. and Taagepera, R., *The Baltic States: the Years of Dependence, 1940–90.* 2nd ed. 1993
O'Connor, Kevin, *The History of the Baltic States.* 2003
Plakans, Andrejs, *A Concise History of the Baltic States.* 2011
Smith, David J., Purs, Aldis, Pabriks, Artis and Lane, Thomas, (eds.) *The Baltic States: Estonia, Latvia and Lithuania.* 2002
Vardys, V. S. and Sedaitis, J. B., *Lithuania: the Rebel Nation.* 1997

National Statistical Office: Department of Statistics to the Government, Gedimino Pr. 29, LT 01 500 Vilnius. *Director General:* Vilija Lapėnienė.
Website: http://www.stat.gov.lt

LUXEMBOURG

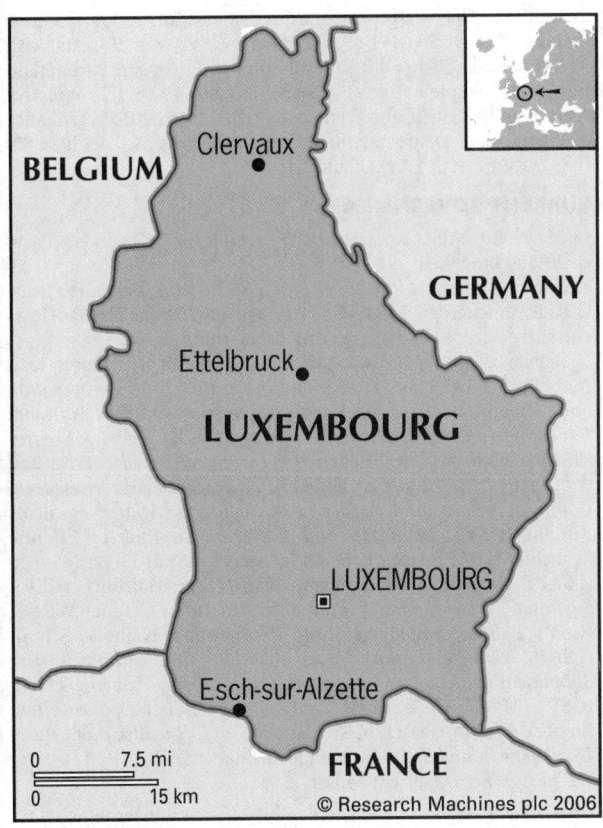

BELGIUM
Clervaux
GERMANY
Ettelbruck
LUXEMBOURG
LUXEMBOURG
Esch-sur-Alzette
FRANCE

0   7.5 mi
0   15 km
© Research Machines plc 2006

Grand-Duché de Luxembourg
(Grand Duchy of Luxembourg)

Capital: Luxembourg
Population projection, 2015: 543,000
GNI per capita, 2011: (PPP$) 50,557
HDI/world rank: 0·867/25
Internet domain extension: .lu

KEY HISTORICAL EVENTS

Celtic tribes, with origins in the Danube basin, settled in the Ardennes hills and surrounding plains from at least 1000 BC. The Romans advanced north into the region (then part of Gaul) from around 50 BC and controlled much of it over the next five centuries from garrisons such as that at Trier. Frankish clans from the middle Rhine valley spread across present-day Luxembourg from around the 5th century AD and intermarried with the Gallo-Romans. Anglo-Saxon missionaries were active in the region during the 7th century, and a Benedictine monastery was founded at Echternach in 698. From the early 9th century the region was controlled by Charlemagne, the Roman Emperor in the West, who ruled from Hungary to the Atlantic Ocean. The empire's division, following the signing of the Treaty of Verdun in 843, enabled the rise of several feudal states, one of which became Luxembourg; the castle of Lutzilinburhurch was founded by Count Sigefroi in 963 on an outcrop overlooking the river Alzette.

Subsequent Counts of Luxembourg were influential in the wider European arena. From 1353, under Wenzel I, Luxembourg expanded to its greatest extent, covering four times the area of the

present state. The House of Luxembourg subsequently went into decline and from 1443 was ruled by the Burgundians from their capital, Brussels. In 1477 Luxembourg became one of the 17 provinces of the Netherlands to come under Habsburg rule, initially under Maximilian of Austria and later under Charles of Ghent, who became the King of Spain in 1516 and Holy Roman Emperor in 1520. Luxembourg was ruled as part of the Spanish Netherlands for much of the next 200 years, although there were brief periods of French dominance under Louis XIV in the 1680s and 1690s. The transfer of the Spanish Netherlands to Austrian rule in 1715 heralded a period of relative tranquility for Luxembourg, which lasted until the territory was occupied by French revolutionary forces in the 1790s.

The 1815 Congress of Vienna made Luxembourg a grand duchy, which subsequently came under the jurisdiction of the house of Orange-Nassau, the ruling house of the Netherlands. At the same time it became part of the German Confederation. In 1839 the Walloon-speaking area was joined to Belgium, which had achieved independence in 1831, giving the Lëtzebuergesch-speaking provinces autonomy. At the London Conference of 1867 the European powers declared Luxembourg a neutral territory and in 1890 the union with the Netherlands was ended. Full independence was confirmed in the same year when Adolf of Nassau-Weilburg became grand duke, founding the present line of rulers.

Luxembourg was invaded and occupied by Germany in both the First and Second World Wars. In June 1942 the Grand Duchy staged a general strike against occupation. In 1948 a Benelux customs union formed by Belgium, the Netherlands and Luxembourg allowed for standardization of prices, taxes and wages and the free movement of labour among the three countries. Luxembourg joined NATO in 1949 and was one of six founding countries of the European Economic Community in 1957. On 24 Dec. 1999 Prime Minister Jean-Claude Juncker announced Grand Duke Jean's decision to abdicate the throne on 7 Oct. 2000. He was succeeded by Prince Henri who assumed the title of Grand Duke.

TERRITORY AND POPULATION

Luxembourg has an area of 2,586 sq. km (999 sq. miles) and is bounded on the west by Belgium, south by France and east by Germany. A census took place on 1 Feb. 2011; the population was 512,353 (including 220,522 foreigners); density, 198 per sq. km. The percentage of foreigners living in Luxembourg has increased dramatically in recent years, from 26% in 1986 to 43% in 2011 (the highest percentage in the EU). The main countries of origin of foreigners living in Luxembourg are Portugal (82,400 in Feb. 2011), France (31,500) and Italy (18,100).

In 2011, 85·4% of the population were urban. The capital, Luxembourg, has (Feb. 2011 census) 95,058 inhabitants; Esch-sur-Alzette, the centre of the mining district, 30,125; Differdange, 21,935; Dudelange, 18,781; Pétange, 16,085; Sanem, 14,470; Hésperange, 13,335.

The UN gives a projected population for 2015 of 543,000.

Lëtzebuergesch is spoken by most of the population, and since 1984 has been an official language with French and German.

SOCIAL STATISTICS

Statistics (figures in parentheses indicate births and deaths of resident foreigners):

	Births	Deaths	Marriages	Divorces
2007	5,477 (2,959)	3,886 (721)	1,969	1,106
2008	5,596 (3,126)	3,595 (612)	1,917	977
2009	5,638 (2,952)	3,655 (659)	1,739	1,052
2010	5,874 (2,845)	3,760 (711)	1,749	1,083

2010 rates per 1,000 population; birth, 11·6; death, 7·4; marriage, 3·5; divorce, 2·1. Nearly half of annual births are to foreigners. In 2008 the most popular age range for marrying was 30–34 for males and 25–29 for females. Life expectancy at birth in 2007 was 76·5 years for males and 82·0 years for females. Annual population growth rate, 2000–05, 0·9%. Infant mortality, 2010, two per 1,000 live births (one of the lowest rates in the world); fertility rate, 2008, 1·7 births per woman. In 2009 Luxembourg received 477 asylum applications.

CLIMATE

In general the country resembles Belgium in its climate, with rain evenly distributed throughout the year. Average temperatures are Jan. 0·8°C, July 17·5°C. Annual rainfall 30·8" (782·2 mm).

CONSTITUTION AND GOVERNMENT

The Grand Duchy of Luxembourg is a constitutional monarchy.

The reigning Grand Duke is **Henri**, b. 16 April 1955, son of the former Grand Duke Jean and Princess Joséphine-Charlotte of Belgium; succeeded 7 Oct. 2000 on the abdication of his father; married Maria Teresa Mestre 14 Feb. 1981. *Offspring*: Prince Guillaume, b. 11 Nov. 1981 (married Countess Stéphanie Marie Claudine Christine de Lannoy, b. 18 Feb. 1984, on 19 Oct. 2012); Prince Félix, b. 3 June 1984; Prince Louis, b. 3 Aug. 1986 (married Tessy Antony, b. 28 Oct. 1985, on 29 Sept. 2006; *offspring*, Gabriel, b. 12 March 2006; Noah, b. 21 Sept. 2007); Princess Alexandra, b. 16 Feb. 1991; Prince Sébastien, b. 16 April 1992.

The constitution of 17 Oct. 1868 was revised in 1919, 1948, 1956, 1972, 1979, 1983, 1988 and more than 20 occasions since then.

The separation of powers between the legislature and the executive is not very strong, resulting in much interaction between the two bodies. Only the judiciary is completely independent.

The 12 cantons are divided into four electoral districts: the South, the East, the Centre and the North. Voters choose between party lists of candidates in multi-member constituencies. The parliament is the *Chamber of Deputies*, which consists of a maximum of 60 members elected for five years. Voting is compulsory and there is universal suffrage. Seats are allocated according to the rules of proportional representation and the principle of the smallest electoral quote. There is a *Council of State* of 21 members appointed by the Sovereign. Membership is for a maximum period of 15 years, with retirement compulsory at the age of 72. It advises on proposed laws and any other question referred to it.

The head of state takes part in the legislative power, exercises executive power and has a part in the judicial power. The constitution leaves to the sovereign the right to organize the government, which consists of a Minister of State, who is Prime Minister, and of at least three Ministers. Direct consultation by referendum is provided for in the Constitution.

National Anthem
'Ons Hemecht' ('Our Homeland'); words by M. Lentz, tune by J. A. Zinnen.

GOVERNMENT CHRONOLOGY

Prime Ministers since 1937. (CSV = Christian Social Party; DP = Democratic Party)

1937–53	CSV	Pierre Dupong
1953–58	CSV	Joseph Bech
1958–59	CSV	Pierre Frieden
1959–74	CSV	Pierre Werner
1974–79	DP	Gaston Thorn
1979–84	CSV	Pierre Werner
1984–95	CSV	Jacques Santer
1995–	CSV	Jean-Claude Juncker

RECENT ELECTIONS

Elections took place on 7 June 2009. The Christian Social Party (CSV) won 26 seats (with 38·0% of the vote), the Socialist Workers' Party (LSAP) 13 (21·6%), the Democratic Party (DP) 9 (15·0%), the Greens (Déi Gréng) 7 (11·7%), the Alternative Democratic Reform Party (ADR) 4 (8·1%) and the Left 1 (3·3%). Turnout was 85·2%.

European Parliament
Luxembourg has six representatives. At the June 2009 elections turnout was 90·8% (91·4% in 2004). CSV won 3 seats with 31·3% of votes cast (political affiliation in European Parliament: European People's Party); LSAP, 1 with 19·4% (Progressive Alliance of Socialists and Democrats); DP, 1 with 18·7% (Alliance of Liberals and Democrats for Europe); the Greens, 1 with 16·8% (Greens/European Free Alliance).

CURRENT GOVERNMENT

In March 2013 the Christian Social Party–Socialist Workers' Party coalition comprised:

Prime Minister, Minister of State, and of the Treasury: Jean-Claude Juncker; b. 1954 (CSV; sworn in 20 Jan. 1995). He is currently Europe's longest-serving prime minister.

Deputy Prime Minister, Minister of Foreign Affairs: Jean Asselborn (LSAP). *The Family and Integration, and Co-operation and Humanitarian Affairs*: Marie-Josée Jacobs (CSV). *National Education and Professional Training*: Mady Delvaux-Stehres (LSAP). *Finance*: Luc Frieden (CSV). *Justice, Civil Service and Administrative Reform, Higher Education and Research, Communications and Media, and Religious Affairs*: François Biltgen (CSV). *Economy and External Commerce*: Etienne Schneider (LSAP). *Health and Social Security*: Mars Di Bartolomeo (LSAP). *Interior and Defence*: Jean-Marie Halsdorf (CSV). *Sustainable Development and Infrastructure*: Claude Wiseler (CSV). *Labour, Employment and Immigration*: Nicholas Schmit (LSAP). *Culture, Parliamentary Relations, and Administrative Simplification*: Octavie Modert (CSV). *Housing*: Marco Schank (CSV). *Middle Classes and Tourism, and Equal Opportunities*: Françoise Hetto-Gaasch (CSV). *Agriculture, Viticulture and Rural Development, and Sport*: Romain Schneider (LSAP).

The *Speaker* is Laurent Mosar.

Government Website (French only): http://www.gouvernement.lu

CURRENT LEADERS

Jean-Claude Juncker

Position
Prime Minister

Introduction
Jean-Claude Juncker was appointed prime minister in Jan. 1995, replacing Jacques Santer who became president of the European Commission. He is the leader of the Christian Social Party (CSV). Having been re-elected as prime minister in 1999, 2004 and 2009, he is currently Europe's longest-serving head of government. He is committed to European integration, and played an important role in the decisions leading up to the creation of the European Union's single currency (euro).

Early Life
Juncker was born in Redange-sur-Attert on 9 Dec. 1954. He obtained his primary and secondary education in Luxembourg and Belgium. Having studied law at the University of Strasbourg, he was admitted to the Bar of Luxembourg in Feb. 1980. He was an active member of the CSV and chaired its youth organization from 1979–84. Juncker was appointed state secretary for employment and social affairs in 1982. In 1984 he was elected to Parliament for the first time as minister of labour, minister of social security and minister in charge of the budget. When Luxembourg held the presidency of the European Community in 1985, Juncker chaired the council of ministers for social affairs and the budget. In 1990 he was elected party leader of the CSV. As president of the EC Economic and Finance Council in 1991, Juncker was among the core co-authors of the Treaty of

Maastricht. He was a governor of the World Bank from 1989–95, and since 1995 has been the country's governor of the European Investment Bank and the International Monetary Fund.

Career in Office

Juncker concurrently holds the position of prime minister, minister of state and of the Treasury. In Oct. 2000 his government oversaw the abdication of the King, Grand Duke Jean, in favour of his son Prince Henri. In Feb. 2002 Juncker was awarded the Légion d'Honneur by French President Jacques Chirac. Following his re-election in mid-2004, he formed a new CSV coalition government with the Socialist Workers' Party. From Jan.–June 2005 he led Luxembourg's six-month presidency of the European Union. In Dec. 2008 Luxembourg's parliament voted to amend the constitution so that bills no longer need the approval of Grand Duke Henri before passing into law following a controversy over proposed euthanasia legislation. In the June 2009 elections the CSV increased its vote share and its representation in the Chamber of Deputies and Juncker began his fourth term as prime minister at the head of the CSV–LSAP coalition.

DEFENCE

There is a volunteer light infantry battalion of (2009) 900, of which only the career officers are professionals. In recent years Luxembourg soldiers and officers have been actively participating in peacekeeping missions, mainly in the former Yugoslavia. There is also a Gendarmerie of 600. In 2000 the Gendarmerie and the police force merged to form the Police grand-ducale. NATO maintains a squadron of E-3A *Sentries.*

In 2008 military expenditure totalled US$232m. (US$478 per capita), representing 0·4% of GDP.

ECONOMY

Services accounted for 85% of GDP in 2007 and industry 14%.

According to the anti-corruption organization *Transparency International*, Luxembourg ranked 12th in the world in a 2012 survey of the countries with the least corruption in business and government. It received 80 out of 100 in the annual index.

Luxembourg gave US$409m. in international aid in 2011, which at 0·97% of GNI made it the third most generous developed country as a percentage of its gross national income, behind Sweden and Norway. Luxembourg was one of only five countries to exceed the UN target of 0·7%.

Overview

Luxembourg's post-Second World War economic growth was based primarily on its steel industry. The small industrial sector has since diversified to include chemical, rubber and other manufactured products. The main engine of recent growth has been the financial services sector, which contributed 26% of GDP in 2009. Luxembourg hosts Europe's largest investment fund industry and second largest money market industry. Other dynamic sectors of the economy include telecommunications, audio-visual and multimedia, industrial plastics and air transport.

Given the large financial sector, Luxembourg was badly damaged by the global financial crisis. Gross investment decreased by 23% in 2009, two major banks were bailed out, three smaller banks failed and unemployment increased. This led to the economy's worst performance for 30 years. A fiscal stimulus package was put in place and exports, particularly of financial services and metal products, started to bounce back during the second quarter of 2010. Authorities target a balanced budget by 2014.

Addressing vulnerabilities in the financial sector, increasing employment, an ageing population and volatility within the eurozone are key challenges.

Currency

On 1 Jan. 1999 the euro (EUR) became the legal currency in Luxembourg at the irrevocable conversion rate of 40·3399

Luxembourg francs to 1 euro. The euro, which consists of 100 cents, has been in circulation since 1 Jan. 2002. On the introduction of the euro there was a 'dual circulation' period before the Luxembourg franc ceased to be legal tender on 28 Feb. 2002. Euro banknotes in circulation on 1 Jan. 2002 had a total value of €5·6bn.

Inflation rates (based on OECD statistics):

2002	2003	2004	2005	2006	2007	2008	2009	2010	2011
2·1%	2·5%	3·2%	3·8%	3·0%	2·7%	4·1%	0·0%	2·8%	3·7%

Foreign exchange reserves were US$285m. in Sept. 2009 (none in 2002) and gold reserves 73,000 troy oz. Total money supply was €84,140m. in March 2009.

Budget

Revenue and expenditure for calendar years in €1m.:

	2005	2006	2007
Revenue	8,499	9,213	10,015
Expenditure	8,883	9,455	9,727

Public debt in 2007 was €532·0m.

Luxembourg's budget deficit in 2011 was 0·6% of GDP (2010, 0·9%; 2009, 0·8%). In 2008 it had a surplus of 3·0%. The required target set by the EU is a budget deficit of no more than 3%.

VAT is 15%, with reduced rates of 12%, 6% and 3%. Income taxes and business taxes have been reduced to preserve competitiveness in the international environment. The normal tax rate for companies at 1 Jan. 2006 was 29·63%, compared with 40·3% in 1996.

Performance

In terms of GDP per head, Luxembourg ranks among the richest countries in the world with a purchasing power parity (PPP) per capita GDP of 79,258 current international dollars in 2009.

Real GDP growth rates (based on OECD statistics):

2002	2003	2004	2005	2006	2007	2008	2009	2010	2011
4·1%	1·7%	4·4%	5·3%	4·9%	6·6%	−0·8%	−4·1%	2·9%	1·7%

Total GDP in 2011 was US$59·2bn.

Banking and Finance

Luxembourg's Central Bank (formerly the Monetary Institute) was established in July 1998 (*Director-General*, Gaston Reinesch). In Dec. 2004 there were 162 banks. German banks make up nearly a third of all the banks. Total deposits in 2004 were €560·7bn.; net assets in unit trusts, €504·0bn.; net assets in investment companies, €600·3bn. There is a stock exchange.

In 2004 the financial sector accounted for 18·1% of gross added value at basic prices and the banks showed a net profit of €2·9bn. The total number of approved insurance companies in 2004 was 95, with reinsurance companies numbering 271; the amount of premiums due was €8,737·5m.

Luxembourg's gross external debt amounted to US$2,197,354m. in June 2012.

ENERGY AND NATURAL RESOURCES

In 2008, 2·1% of energy consumption came from renewables (wind power, solar power, hydro-electric power, tidal power, geothermal energy and biomass), compared to the European Union average of 10·3%. A target of 11% has been set by the EU for 2020.

Environment

Carbon dioxide emissions from the consumption and flaring of fossil fuels in 2008 were the equivalent of 23·9 tonnes per capita (compared to the European average of 7·8 tonnes per capita).

Electricity

Apart from hydro-electricity and electricity generated from fossil fuels, Luxembourg has no national energy resources. Installed capacity in 2007 was 1·7m. kW. Production was 4,002m. kWh in 2007 and consumption per capita 16,588 kWh.

Agriculture

The contribution of agriculture, viticulture and forestry to the economy has been gradually declining over the years, accounting for only 0·5% of gross added value at basic prices in 2004. However, the actual output of this sector has nearly tripled during the past 30 years, a trend common to many EU countries. There were 4,975 workers engaged in agricultural work (including wine-growing and forestry) in 2004. In 2007 there were 2,303 farms with an average area of 63·4 ha.; 130,884 ha. were under cultivation in 2007.

Production, 2007 (in tonnes) of main crops: grassland and pasturage, 749,101; maize, 197,508; forage crops, 129,096; bread crops, 77,435; potatoes, 19,968; colza (rape), 18,302. Production, 2007 (in 1,000 tonnes) of meat, 26·8; milk, 274·2. In 2007–08, 142,000 hectolitres of wine were produced. Total tractors and other agriculture vehicles, 2008: 15,238.

Livestock (15 May 2007): 4,334 horses, 191,928 cattle, 83,255 pigs, 9,339 sheep.

Forestry

In 2010 there were 87,000 ha. of forests (33% of the total land area). In 2011, 261,000 cu. metres of roundwood were cut.

INDUSTRY

The largest company by market capitalization in Luxembourg in April 2012 was ArcelorMittal (US$32·3bn.).

In 2004 there were 3,038 industrial enterprises, of which 1,972 were in the building industry. Production, 2007 (in tonnes): rolled steel products, 2,933,000; steel, 2,858,000. The world's largest steel producer, ArcelorMittal, has its headquarters in Luxembourg. Created in June 2006 through the merger of Arcelor and Mittal Steel, it produces in excess of 100m. tonnes of steel annually and accounts for approximately 10% of world steel output. The steel industry mainly relies on imported ore.

Labour

In 2004 the estimated total workforce was 301,000. The government fixes a legal minimum wage. Retirement is at 65. Employment creation was 3·2% in 2004–05. In Dec. 2012 the unemployment rate was 5·3%. The minimum wage in Jan. 2005 was €8·48 an hour.

Between 1996 and 2005 strikes cost Luxembourg an average of just one days per 1,000 employees a year.

There was a 2·6% increase in employment in 2004. Of the new jobs created, around two-thirds went to so-called *frontaliers*, workers living in surrounding countries who commute into Luxembourg to work. More than 100,000 people cross into Luxembourg every day from neighbouring France, Germany and Belgium to work, principally in the financial services industry.

INTERNATIONAL TRADE

Imports and Exports

Imports in 2007 (provisional figures) totalled €16,262·2m. and exports €11,823·0m. In 2005 exports reached 158% of GDP. In 2007, 90·5% of imports were from other EU member countries and 85·9% of exports went to other EU member countries.

Principal imports and exports by standard international trade classification (provisional figures) in €1m.:

	Imports 2007	Exports 2007
Food and live animals	1,120·7	539·2
Beverages and tobacco	425·3	195·0
Crude materials, oils, fats and waxes	1,297·9	251·3

	Imports 2007	Exports 2007
Mineral fuels and lubricants	2,076·4	91·9
Chemicals and related products	1,568·2	798·7
Manufactured goods in metals	1,250·8	1,821·1
Other manufactured goods classified chiefly by material	1,789·4	3,928·5
Machinery	2,547·1	2,136·0
Transport equipment	2,602·4	885·7
Other manufactured goods	1,583·9	1,175·6
Total	16,262·2	11,823·0

Trade with selected countries (provisional figures) in €1m.:

	Imports 2007	Exports 2007
Austria	137·8	243·1
Belgium	5,496·3	1,481·2
France	1,899·8	1,993·0
Germany	4,800·0	3,124·1
Italy	356·6	641·7
Netherlands	1,002·6	644·9
Poland	109·0	196·1
Spain	168·9	413·5
Sweden	112·9	172·7
UK	282·1	553·7
(Total EU)	14,722·7	10,154·4)
China	95·1	194·4
Switzerland	152·2	144·7
USA	623·2	298·7
Total (including others)	16,262·2	11,823·0

Trade Fairs

The *Foires Internationales de Luxembourg* occur twice a year, and there are a growing number of specialized fairs.

COMMUNICATIONS

Roads

On 1 Jan. 2008 there were 2,894 km of roads of which 147 km were motorways. Motor vehicles registered at 1 Jan. 2008 numbered 394,917 including 321,520 passenger cars, 27,043 trucks, 1,455 coaches and 14,946 motorcycles. In 2009 there were 47 fatalities in road accidents.

Rail

In 2009 there were 275 km of railway (standard gauge) of which 262 km were electrified; passenger-km totalled 333m. in 2008.

Civil Aviation

Findel is the airport for Luxembourg. 1,643,000 passengers and 856,450 tonnes of freight were handled in 2007. The national carrier is Luxair, in which the state has a 39·04% stake directly along with a further 21·81% indirectly through the Banque et Caisse d'Epargne de l'Etat (State and Savings Bank). Cargolux has developed into one of the major international freight carriers. In 2006 scheduled airline traffic of Luxembourg-based carriers flew 93m. km, carrying 928,000 passengers (all on international flights).

Shipping

A shipping register was set up in 1990; 143 vessels were registered in Dec. 2006.

Telecommunications

In 2006 there were 362,722 main (fixed) telephone lines. In 2008 active mobile phone subscribers numbered 707,000 (1,471·1 per 1,000 persons). There were an estimated 387,000 internet users in 2008. The fixed broadband penetration rate in Dec. 2010 was 33·5 subscribers per 100 inhabitants. In March 2012 there were 190,000 Facebook users.

SOCIAL INSTITUTIONS

Justice

The Constitution makes the Courts of Law independent in performing their functions, restricting their sphere of activity, defining their limit of jurisdiction and providing a number of procedural guarantees. The Constitution has additionally laid down a number of provisions designed to ensure judges remain independent of persons under their jurisdiction, and to ensure no interference from the executive and legislative organs. All judges are appointed by Grand-Ducal order and are irremovable.

The judicial organization comprises three Justices of the Peace (conciliation and police courts). The country is, in addition, divided into two judicial districts—Luxembourg and Diekirch. District courts deal with matters such as civic and commercial cases. Offences which are punishable under the Penal Code or by specific laws with imprisonment or hard labour fall within the jurisdiction of the criminal chambers of District Courts, as the Assize Court was repealed by law in 1987. The High Court of Justice consists of a Supreme Court of Appeal and a Court of Appeal.

The judicial organization of the Grand-Duchy does not include the jury system. A division of votes between the judges on the issue of guilt/innocence may lead to acquittal. Society before the Courts of Law is represented by the Public Prosecutor Department, composed of members of the judiciary directly answerable to the government.

In 1999 a new Administrative Tribunal, Administrative Court and Constitutional Court were established.

The population in penal institutions in Sept. 2007 was 745.

Education

The adult literacy rate in 2004 was 100%. Education is compulsory for all children between the ages of four and 15 (including two years of pre-primary school attendance). In 2006–07 there were 13,672 children in pre-primary school (pre-nursery education, 3,671; nursery education, 10,001) with 1,227 teachers; 33,136 pupils in primary schools; 34,970 pupils in secondary schools. In higher education (2003–04) the Higher Institute of Technology (IST) had 358 students and there were 401 students in teacher training. In 2006–07 the University Centre of Luxembourg had 3,180 students. Many students go abroad, predominantly to France, Germany and Belgium. In 2006–07, 7,222 students pursued university studies abroad.

In 2005 public expenditure on education came to 3·8% of GDP.

Health

In 2004 there were 1,591 doctors (411 GPs and 840 specialists) and 340 dentists. There were 17 hospitals and 3,045 hospital beds in 2004. In 2007 Luxembourg spent 7·1% of its GDP on health.

Welfare

The official retirement age is 65 years for both men and women. To be eligible, a pensioner must have paid 120 months contributions. The maximum old-age pension is €7,011·23 per month. The minimum pension varies from €757·22 to €1,154·43 per month. A minimum pension is paid with at least 20 years of coverage.

Unemployment benefit is 80% (85% if the insured has a dependant child) of the basis salary during the previous three months, up to 2·5 times the social minimum wage. Recent graduates receive 70% of the social minimum wage whereas self-employed persons receive 80% of the social minimum wage.

RELIGION

The population was 91% Roman Catholic in 2001. There are small Protestant, Jewish, Greek Orthodox, Russian Orthodox and Muslim communities as well.

CULTURE

World Heritage Sites

Luxembourg has one site on the UNESCO World Heritage List: the City of Luxembourg—its Old Quarters and Fortifications (inscribed on the list in 1994).

Press

There were eight paid-for daily newspapers in 2008 with an average circulation of 117,000 and two free dailies with an average circulation of 127,000; there were also 15 non-dailies. The German-language *Luxemburger Wort* has the highest circulation, with an average of 72,000 copies in 2008.

Tourism

In 2010 there were 907,000 overnight tourists and 2,256,000 overnight stays; there were 7,751 hotel rooms in 2010. Tourists spent US$3,620m. in 2006 (excluding passenger transport). Camping is widespread; there were 739,000 overnight stays at campsites in 2010.

Festivals

The Festival International Echternach (May–June) and the Festival of Wiltz (June–July) are annual events. Both feature a variety of classical music, jazz, theatre and recitals.

DIPLOMATIC REPRESENTATIVES

Of Luxembourg in the United Kingdom (27 Wilton Cres., London, SWIX 8SD)
Ambassador: Vacant.
Chargé d'Affaires a.i.: Béatrice Kirsch.

Of the United Kingdom in Luxembourg (5 Boulevard Joseph II, L-1840 Luxembourg)
Ambassador: Hon. Alice Walpole.

Of Luxembourg in the USA (2200 Massachusetts Ave., NW, Washington, D.C., 20008)
Ambassador: Jean-Louis Wolzfeld.

Of the USA in Luxembourg (22 Blvd. Emmanuel Servais, L-2535 Luxembourg)
Ambassador: Robert A. Mandell.

Of Luxembourg to the United Nations
Ambassador: Sylvie Lucas.

Of Luxembourg to the European Union
Permanent Representative: Christian Braun.

FURTHER READING

STATEC. *Annuaire Statistique 2011.—Le Luxembourg en chiffres 2012*

Arblaster, Paul, *A History of the Low Countries.* 2005
Newcomer, J., *The Grand Duchy of Luxembourg: The Evolution of Nationhood, 963 AD to 1983.* 2nd ed. 1995

National library: 37 Boulevard Roosevelt, Luxembourg City, L-2450 Luxembourg.
National Statistical Office: Service Central de la Statistique et des Études Économiques (STATEC), CP 304, Luxembourg City, L-2013 Luxembourg. *Director:* Serge Allegrezza.
Website: http://www.statec.public.lu

MACEDONIA

Republika Makedonija
(The Republic of Macedonia)
(Former Yugoslav Republic of Macedonia)

Capital: Skopje
Population projection, 2015: 2·07m.
GNI per capita, 2011: (PPP$) 8,804
HDI/world rank: 0·728/78
Internet domain extension: .mk

KEY HISTORICAL EVENTS

The history of Macedonia can be traced to the reign of King Karan (808–778 BC), but the country was at its most powerful at the time of Philip II (359–336 BC) and Alexander the Great (336–323 BC). At the end of the 6th century AD Slavs began to settle in Macedonia. There followed a long period of internal fighting but the spread of Christianity led to consolidation and the creation of the first Macedonian Slav state, the Kingdom of Samuel, 976–1018. In the 14th century it fell to Serbia, and in 1355 to the Turks. After the Balkan wars of 1912–13 Turkey was ousted and Serbia received part of the territory, the rest going to Bulgaria and Greece. In 1918 Yugoslav Macedonia was incorporated into Serbia as South Serbia, becoming a republic in the Socialist Federal Republic of Yugoslavia. Claims to the historical Macedonian territory have long been a source of contention with Bulgaria and Greece. Macedonia declared its independence on 18 Sept. 1991. In April 1999 the Kosovo crisis which led to NATO air attacks on Yugoslavian military targets set off a flood of refugees into Macedonia, though subsequently most returned home.

In March 2001 there were a series of clashes between government forces and ethnic Albanian separatists near the border between Macedonia and Kosovo. As violence escalated Macedonia found itself on the brink of civil war. In May 2001 the new national unity government gave ethnic Albanian rebels a 'final warning' to end their uprising. As the crisis worsened, a stand-off within the government between the Macedonian and the ethnic Albanian parties was only resolved after the intervention of Javier Solana, the EU's foreign and security policy chief. A number of Macedonian soldiers were killed in clashes with the rebels, and following reverses in the military campaign the commander of the Macedonian army, Jovan Andrevski, resigned in June 2001. In Aug. 2001 a peace accord was negotiated.

TERRITORY AND POPULATION

Macedonia (referred to within the United Nations as the Former Yugoslav Republic of Macedonia) is bounded in the north by Serbia, in the east by Bulgaria, in the south by Greece and in the west by Albania. Its area is 25,710 sq. km, including 490 sq. km of inland water. According to the 2002 census final results, the population on 1 Nov. 2002 was 2,022,547. The main ethnic groups in 2002 were Macedonians (1,297,981), Albanians (509,083), Turks (77,959), Romas (53,879), Serbs (35,939) and Vlachs (9,695). Estimate, 31 Dec. 2010, 2,057,284; density, 82 per sq. km. Ethnic Albanians predominate on the western side of Macedonia. Minorities are represented in the Council for Inter-Ethnic Relations. In 2011, 59·4% of the population lived in urban areas.

The UN gives a projected population for 2015 of 2·07m.

Macedonia is divided into 84 municipalities. The major cities (with 2002 census population) are: Skopje, the capital, 506,926; Kumanovo, 76,275; Bitola, 74,550; Prilep, 69,704; Tetovo, 52,915.

The official language is Macedonian, which uses the Cyrillic alphabet. Around 25% of the population speak Albanian.

SOCIAL STATISTICS

In 2011: live births, 22,770; deaths, 19,465; marriages, 14,736; divorces, 1,753; infant deaths, 172. Rates (per 1,000 population): live births, 11·1; deaths, 9·5; marriages, 7·2; divorces, 0·9. Infant mortality, 2011 (per 1,000 live births), 7·6. Expectation of life at birth in 2007 was 71·7 years for males and 76·5 years for females. Annual population growth rate, 2005–10, 0·2%. In 2004 the most popular age range for marrying was 25–29 for males and 20–24 for females. Fertility rate, 2011, 1·6 births per woman.

Migration within the Republic of Macedonia, 2004: 9,326. International (external) migration: emigrated persons, 669; immigrated persons 1,381. Net migration in 2004 was 712.

CLIMATE

Macedonia has a mixed Mediterranean-continental type climate, with cold moist winters and hot dry summers. Skopje, Jan. –0·4°C, July 23·1°C.

CONSTITUTION AND GOVERNMENT

At a referendum held on 8 Sept. 1991 turnout was 74%; 99% of votes cast were in favour of a sovereign Macedonia. On 17 Nov. 1991 parliament promulgated a new constitution which officially proclaimed Macedonia's independence. This was replaced by a constitution adopted on 16 Nov. 2001 which for the first time included the recognition of Albanian as an official language. It also increased access for ethnic Albanians to public-sector jobs.

The *President* is directly elected for five-year terms. Candidates must be citizens aged at least 40 years. The parliament is a 123-member single-chamber *Assembly* (*Sobranie*), elected by universal suffrage for four-year terms. There is a *Constitutional Court* whose members are elected by the assembly for non-renewable eight-year terms, and a *National Security Council* chaired by the President. Laws passed by the Assembly must be countersigned by the President, who may return them for reconsideration, but cannot veto them if they gain a two-thirds majority.

Political Parties

The Law on Political Parties makes a distinction between a political party and an association of citizens. The signatures of 500 citizens with the right to vote must be produced for a party to be legally registered. As of 2002 the country had 89 legally registered parties.

National Anthem

'Denes nad Makedonija se radja novo sonce na slobodata' ('Today a new sun of liberty appears over Macedonia'); words by V. Maleski, tune by T. Skalovski.

RECENT ELECTIONS

Parliamentary elections were held on 5 June 2011. The ruling Internal Macedonian Revolutionary Organization-Democratic Party for Macedonian National Unity (VMRO-DPMNE) won 56 seats with 39·2% of votes cast. The Social Democratic Union of Macedonia (SDSM) took 42 seats with 32·8% of the vote, the Democratic Union for Integration (DUI) 15 with 10·3%, the Democratic Party of Albanians (DPA) 8 with 5·9% and the National Democratic Revival (NDP) 2 with 2·7%. Turnout was 63·5%.

Presidential elections were held on 22 March 2009. Gjorgje Ivanov (VMRO-DPMNE), took 35·1% of the vote, Ljubomir Frčkoski (SDSM) 20·5%, Imer Selmani (New Democracy) 15·0%, Ljube Boškoski (ind.) 14·9%, Agron Buxhaku (DUI) 7·5%, Nano Ružin (Liberal Democratic Party) 4·1%, and Mirushe Hoxha (DPA) 3·1%. In the run-off on 5 April Ivanov won with 63·1% against Frčkoski with 36·9%. Turnout was an estimated 56·4% in the first round and 42·6% in the second.

CURRENT GOVERNMENT

President: Gjorgje Ivanov; b. 1960 (VMRO-DPMNE; sworn in 12 May 2009).

Prime Minister: Nikola Gruevski; b. 1970 (VMRO-DPMNE; in office since 27 Aug. 2006).

In March 2013 the multi-party coalition government was composed as follows:

Deputy Prime Minister and Minister of Finance: Zoran Stavrevski. *Deputy Prime Minister in Charge of Economic Affairs:* Vladimir Pesevski. *Deputy Prime Minister in Charge of European Affairs:* Fatmir Besimi. *Deputy Prime Minister in Charge of Implementing the Ohrid Agreement:* Musa Xhaferri.

Minister of Agriculture, Forestry and Water Supply: Ljupcho Dimovski. *Culture:* Elizabeta Kanceska Milevska. *Defence:* Talat Xhaferi. *Economy:* Valon Saracini. *Education and Science:* Pance Kralev. *Environment:* Abdulakim Ademi. *Foreign Affairs:* Nikola Poposki. *Health:* Nikola Todorov. *Information Society:* Ivo Ivanovski. *Interior:* Gordana Jankulovska. *Justice:* Blerim Bexheti. *Labour and Social Policy:* Spiro Ristovski. *Local Self-Government:* Tahir Hani. *Transport and Communications:* Mile Janakieski. *Ministers without Portfolio:* Nezdet Mustafa; Hadi Neziri; Bill Pavleski; Vele Samak.

Government Website: http://www.vlada.mk

CURRENT LEADERS

Gjorgje Ivanov

Position
President

Introduction
Gjorgje Ivanov was sworn into office on 12 May 2009 after winning the second round of the presidential election on 5 April. Although his duties during his five-year term are largely ceremonial, Ivanov is supreme commander of the army and has decision-making powers on foreign policy and the judiciary.

Early Life
Gjorgje Ivanov was born on 2 May 1960 in Valandovo. He graduated in law from Ss. Cyril and Methodius University of Skopje in 1982 and was appointed assistant law professor in 1995.

In 1998 he became an associate professor in political theory and political philosophy at the same time as completing his doctorate.

In 1999 Ivanov was made a professor of post-graduate studies at the University of Athens before joining the political science departments of the University of Bologna and the University of Sarajevo the following year. He returned to his alma mater in 2001 to take up the post of director of political studies. In 2008 Ivanov received his political science professorship and was appointed as chair of the Macedonian Higher Education Accreditation Council.

On 25 Jan. 2009 the ruling conservative VMRO-DPMNE announced Ivanov as its presidential candidate for the 2009 election. In the first round of voting on 22 March 2009 he took 35% of the vote and then won with 63% in a second round.

Career in Office
Ivanov pledged to continue Macedonia's campaign for membership of the European Union and NATO. In Dec. 2011 the International Court of Justice ruled against Greece's obstruction of Macedonia's bid to join the military alliance in their dispute over the country's name (*see* GREECE: Key Historical Events on page 550). He also aimed to diffuse ongoing tensions with Macedonia's ethnic Albanian minority. However, in March 2012 there were violent clashes between youths from both communities in Skopje.

In elections in June 2011 Ivanov's VMRO-DPMNE was returned to office under Prime Minister Nikola Gruevski but without a parliamentary majority.

DEFENCE

The President is the C.-in-C. of the armed forces. Compulsory national military service was abolished in 2006.

Defence expenditure in 2008 totalled US$192m. (US$93 per capita), representing 2·1% of GDP.

The European Union's first ever peacekeeping force (EUFOR) officially started work in Macedonia on 1 April 2003, replacing the NATO-led force that had been in the country since 2001. EUFOR left the country in Dec. 2003.

Army

Army strength was 9,760 in 2007. There is a paramilitary police force of 7,600.

Navy

In 2007 the Marine Wing operated four river patrol craft.

Air Force

The Army Air Force numbered 1,130 in 2007, and had four combat-capable aircraft (in storage) and ten attack helicopters.

INTERNATIONAL RELATIONS

On 13 Sept. 1995 under the auspices of the UN, Macedonia and Greece agreed to normalize their relations.

ECONOMY

Agriculture accounted for 11·3% of GDP in 2010, industry 27·8% and services 60·9%.

Overview

Following fighting between the government and ethnic Albanian rebels in 2001, the economy has recovered its pre-conflict levels, aided by structural reforms, prudent macroeconomic policies and improvements in the business environment. Foreign trade accounts for 130% of GDP, leaving the economy prone to external shocks. Candidacy for accession to the EU has been a major influence on policy since 2005.

As a result of the global economic crisis, external demand collapsed and foreign direct investment was in decline by the end of 2008, although the situation had improved by mid-2009. Supporting growth while reducing external imbalances (including a large current account deficit) remains a key challenge.

Currency

The national currency of Macedonia is the *denar* (MKD), of 100 *deni*. Foreign exchange reserves were US$910m. in July 2005, gold reserves 197,000 troy oz and total money supply was 29,745m. denars. Inflation was 8·4% in 2008 but there was deflation of 0·8% in 2009. Inflation was 1·5% in 2010 and 3·9% in 2011.

Budget

In 2009 revenues totalled 128,498m. denars and expenditures 139,393m. denars. Tax revenue accounted for 85·5% of revenues in 2009; current expenditure accounted for 90·4% of expenditures.

Performance

In 2001 the political turmoil in the country resulted in the economy contracting by 4·5%. There was then a slight recovery in 2002, with a growth rate of 0·9%. The economy contracted by 0·9% in 2009 but there was growth of 2·9% in 2010 and 3·1% in 2011. Total GDP in 2011 was US$10·2bn.

Banking and Finance

The central bank and bank of issue is the National Bank of Macedonia. Its *Governor* is Dimitar Bogov (since May 2011). Privatization of the banking sector was completed in 2000. In 2001 there were 20 commercial banks, six of which were majority foreign-owned. As of 31 Dec. 1998 commercial banks' total non-government deposits were 23,136m. denars, and non-government savings deposits were 15,095m. denars. The largest banks are Stopanska Banka, followed by Komercijalna Banka; between them they control more than half the total assets of all banks in Macedonia.

External debt was US$5,804m. in 2010, representing 65·1% of GNI.

A stock exchange opened in Skopje in 1996.

ENERGY AND NATURAL RESOURCES

Environment

Macedonia's carbon dioxide emissions from the consumption and flaring of fossil fuels were the equivalent of 3·6 tonnes per capita in 2008.

Electricity

Installed capacity in 2007 was 1·6m. kW. Output in 2007: 6·73bn. kWh, of which 1·01bn. kWh were from hydro-electric plants. Consumption per capita was 4,512 kWh in 2007.

Oil and Gas

A 230-km long pipeline bringing crude oil to Macedonia from Thessaloniki in Greece opened in July 2002. Built at a cost of over US$130m., it has the capacity to provide Macedonia with 2·5m. tonnes of crude oil annually.

Minerals

Macedonia is relatively rich in minerals, including lead, zinc, copper, iron, chromium, nickel, antimony, manganese, silver and gold. Output in 2009, unless otherwise indicated (in tonnes): lignite, 7,454,000; copper ore, 3,767,000; gypsum, 155,000; lead concentrate, 52,000; copper concentrate, 35,430; zinc concentrate, 32,000.

Agriculture

In 2002 the agricultural population numbered 833,000 persons, of whom 109,000 were economically active. In 2004 there were 560,264 ha. of arable land, 703,830 ha. of pasture and 44,000 ha. of permanent crops. In 2004, 101,004 ha. of arable land were owned by agricultural organizations and 459,260 ha. by individual farmers. There were 65,338 tractors in use in 2004.

Crop production, 2004 (in 1,000 tonnes): wheat, 356; grapes, 194; potatoes (2003), 175; barley, 149; wine, 142; maize (2003), 141; watermelons, 115; chillies and green peppers (2003), 111; tomatoes (2002), 109; lucerne, 98; apples, 82; cabbage (2002), 71;

sugar beets, 47; onions (2003), 31; cucumbers and gherkins, 27; plums, 26. In 2004, 119,000 tonnes of wine were produced.

Livestock, 2004 (in 1,000): sheep, 1,432; cattle, 255; pigs, 158; horses, 40; chickens, 2,725. Livestock products, 2004 (in 1,000 tonnes): beef, 9; pork, bacon and ham, 9; mutton, 7; poultry, 3; cow's milk, 213m. litres; sheep's milk, 49m. litres; eggs (total), 340m.

Forestry

Forests covered 1·00m. ha. in 2010 (39% of the land area), chiefly oak and beech. 752,000 cu. metres of timber were cut in 2007.

Fisheries

Total catch in 2008 was 122 tonnes, entirely from inland waters.

INDUSTRY

In 1999 there were 94,404 enterprises (90,426 private, 1,112 public, 1,257 co-operative, 1,577 mixed and 32 state-owned). Production, 2004 (in tonnes): cement, 585,000; distillate fuel oil, 359,000; residual fuel oil, 282,000; petrol, 146,377; sulphuric acid (2001), 101,058; ferroalloys, 72,082; detergents, 14,507.

Labour

In April 2004 there were 522,995 employed persons, including: 116,300 in manufacturing; 87,608 in agriculture, hunting and forestry; 74,218 in wholesale and retail trade/repair of motor vehicles, motorcycles and personal and household goods; and 33,635 in education. The number of unemployed persons in 2004 was 309,286, giving an unemployment rate of 37·2%.

INTERNATIONAL TRADE

Imports and Exports

In 2009 imports (c.i.f.) were valued at US$5,403m. and exports (f.o.b.) at US$2,692m.

Leading imports in 2010 (in US$1m.) were: manufactured goods (1,202); machinery and transport equipment (1,038); chemicals and related products (566). Leading exports in 2010 (in US$1m.) were: articles of apparel and clothing accessories (584); iron and steel (169); fruit and vegetables (143).

In 2010 major import sources (in US$1m.) were: Germany (518); Russia (495); Greece (439). Major export markets in 2010 (in US$1m.) were: Serbia (652); Germany (450); Greece (290).

COMMUNICATIONS

Roads

In 2007 there were 221 km of motorways, 690 km of other main roads, 3,774 km of regional roads and 9,155 km of local roads. There were 248,800 passenger cars in use in 2007, plus 2,300 buses and coaches, and 26,600 lorries and vans. In the same year there were 4,037 road accidents with 173 fatalities.

Rail

In 2009 there were 699 km of railways (234 km electrified). 1·5m. passengers and 2·9m. tonnes of freight were transported in 2009. The former Macedonian Railways was reorganized in 2007 with two new entities being created—Macedonian Railways Infrastructure (PE Makedonski Železnici Infrastructure, or MŽ-I), which is responsible for the maintenance and operation of the infrastructure, and Macedonian Railways Transport (MŽ Transport AD, or MŽ-T), which is responsible for the operation of passenger and freight services.

Civil Aviation

There are international airports at Skopje and Ohrid. A new Macedonia-based carrier, Aeromak, has been established to replace MAT Macedonian Airlines, the former flag carrier which ceased operations in 2009. In 2009 Skopje handled 602,298 passengers (658,366 in 2008) and 2,326 tonnes of freight. The much smaller airport at Ohrid handled 36,652 passengers in 2009 (44,413 in 2008).

Telecommunications

In 2011 there were 413,500 landline telephone subscriptions (equivalent to 200·3 per 1,000 inhabitants) and 2,257,100 mobile phone subscriptions (or 1,093·6 per 1,000 inhabitants). In 2011, 56·7% of the population were internet users. In 2002 the Hungarian firm Matav acquired a 51% stake in MakTel, the state monopoly telecommunications provider, in the most significant economic development in the country's history. The deal was worth €618·2m. (US$568·4m.) over two years. In March 2012 there were 880,000 Facebook users.

SOCIAL INSTITUTIONS

Justice

Courts are autonomous and independent. Judges are tenured and elected for life on the proposal of the *Judicial Council*, whose members are themselves elected for renewable six-year terms. The highest court is the Supreme Court. There are 27 courts of first instance and three higher courts.

The population in penal institutions in March 2008 was 2,200 (107 per 100,000 of national population).

Education

The literacy rate was 96·1% in 2003 (98·2% among males and 94·1% among females). Education is free and compulsory for nine years. In 2004, 36,392 children attended 51 pre-school institutions and 486 infant schools of elementary education. In 2004–05 there were 227,254 pupils enrolled in 1,012 primary schools, 95,268 in 96 secondary schools and (2001–02) 343,587 students in higher education. There are universities at Skopje (Cyril and Methodius, founded in 1949; 36,509 students and 1,314 academic staff in 2004–05) and Bitola (founded 1979; 10,043 students and 233 academic staff in 2004–05). There are two private universities at Skopje (1,075 students and 41 academic staff in 2004–05) and Tetovo (1,737 students and 86 academic staff in 2004–05).

In 2002 public expenditure on education came to 3·5% of GDP.

Health

In 2004 there were 4,490 doctors, 1,134 dentists, 322 pharmacologists and 63 hospitals with 9,699 beds.

Welfare

In 2004 social assistance was paid to 67,260 households. Child care and special supplements went to 46,203 children, and 16,970 underage and 95,053 adults received social benefits. There were 260,075 pensioners in 2004.

RELIGION

Macedonia is traditionally Orthodox but the church is not established and there is freedom of religion. In 2001 there were 1·21m. Serbian (Macedonian) Orthodox and 580,000 Sunni Muslims. In 1967 an autocephalous Orthodox church split off from the Serbian. Its head is the Archbishop of Ohrid and Macedonia whose seat is at Skopje. It has five bishoprics in Macedonia and representatives in USA, Canada and Australia. It has some 300 priests.

The Muslim Religious Union has a superiorate at Skopje. The Roman Catholic Church has a seat at Skopje.

CULTURE

World Heritage Sites

The Former Yugoslav Republic of Macedonia has one site on the UNESCO World Heritage List: Ohrid Region with its Cultural and Historic Aspect and its Natural Environment (inscribed on the list in 1979 and 1980), a rich repository of Byzantine art and architecture.

Press

There were 12 daily newspapers in 2008 with a circulation of 295,000 copies. *Dnevnik* is the most popular with a daily circulation of 50,000 copies in 2008.

There are four news agencies in Macedonia: the Macedonian Information Agency (national); and Macedonian Information Centre, Makfax and Net Press (privately owned). Net Press is exclusively an internet news agency.

Tourism

There were 261,696 foreign tourists in 2010, the highest total since 1991. The main countries of origin of non-resident tourists in 2010 were: Serbia (13·7%), Greece (10·3%), Turkey (7·7%) and Albania (6·5%).

Festivals

The main festivals are Days of Macedonian Music in Skopje (March), the Balkan Festival of Folk Songs and Dances in Ohrid (July), the Ohrid Summer Festival focusing on music and drama (July–Aug.) and the Skopje Jazz Festival (Oct.).

DIPLOMATIC REPRESENTATIVES

Of Macedonia in the United Kingdom (Suites 2·1 and 2·2, Buckingham Court, 75–83 Buckingham Gate, London, SW1E 6PE)
Ambassador: Vacant.
Chargé d'Affaires a.i.: Mile Prangoski.

Of the United Kingdom in Macedonia (Salvador Aljende 73, 1000 Skopje)
Ambassador: Christopher Yvon.

Of Macedonia in the USA (2129 Wyoming Ave., NW, Washington, D.C., 20008)
Ambassador: Zoran Jolevski.

Of the USA in Macedonia (ul. Samoilova 21, 1000 Skopje)
Ambassador: Paul Wohlers.

Of Macedonia to the United Nations
Ambassador: Pajo Avirovikj.

Of Macedonia to the European Union
Ambassador: Andrej Lepavcov.

FURTHER READING

Danforth, L. M., *The Macedonian Conflict: Ethnic Nationalism in a Transnational World.* 1996
Phillips, John, *Macedonia: Warlords and Rebels in the Balkans.* 2004
Poulton, H., *Who Are the Macedonians?* 1996

National Statistical Office: State Statistical Office, Dame Gruev 4, Skopje.
  Director: Blagica Novkovska.
Website: http://www.stat.gov.mk

MADAGASCAR

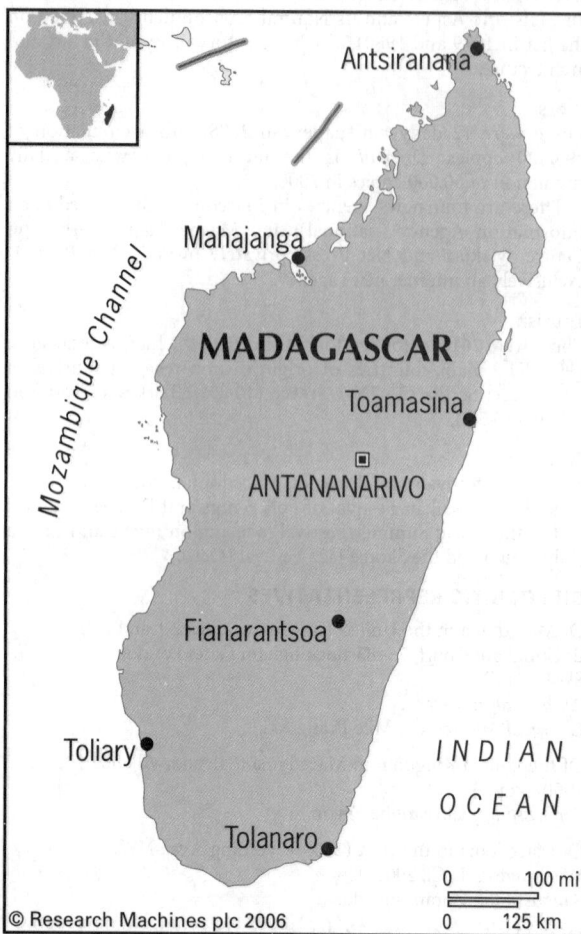

© Research Machines plc 2006

Constitutional Court called for a run-off. On 22 Feb. 2002 Ravalomanana declared himself president and imposed a state of emergency. However, incumbent Didier Ratsiraka and his government set up a rival capital in Toamasina. In April 2002 both men agreed to a recount of votes to solve the dispute. Ravalomanana was declared president.

Re-elected in 2006, the suppression of political opponents led to an army mutiny against Ravalomanana in early 2009. As civil unrest increased, he resigned the presidency and handed power to the military which installed Andry Rajoelina, Ravalomanana's chief political rival, as president.

TERRITORY AND POPULATION

Madagascar is situated 400 km (250 miles) off the southeast coast of Africa, from which it is separated by the Mozambique channel. Its area is 587,041 sq. km (226,658 sq. miles), including 5,500 sq. km (2,120 sq. miles) of inland water. At the 1993 census the population was 12,092,157 (50·45% female); density, 20·6 per sq. km. The estimate for 2011 was 20,696,100; density, 35·6 per sq. km. 69·8% of the population lived in rural areas in 2010.

The UN gives a projected population for 2015 of 23·85m.

Regions	Area in sq. km.	Population (2011 estimate)	Chief town
Alaotra Mangoro	31,948	973,200	Ambatondrazaka
Amoron'i Mania	16,141	677,500	Ambositra
Analamanga	16,911	3,173,100	Antananarivo
Analanjirofo	21,930	980,800	Fenoarivobe
Androy	19,317	695,400	Ambovombe
Anosy	25,731	636,600	Taolagnaro
Atsimo Andrefana	66,236	1,247,700	Toliary
Atsimo Atsinanana	18,863	851,500	Farafangana
Atsinanana	21,934	1,204,000	Toamasina
Betsiboka	30,025	278,100	Maevatanana
Boeny	31,046	757,700	Mahajanga
Bongolava	16,688	433,400	Tsiroanomandidy
Diana	19,266	663,300	Antsiranana
Haute Matsiatra	21,080	1,136,300	Fianarantsoa
Ihorombe	26,391	295,900	Ihosy
Itasy	6,993	694,400	Miarinarivo
Melaky	38,852	274,400	Maintirano
Menabe	46,121	561,000	Morondava
Sava	25,518	929,300	Sambava
Sofia	50,100	1,181,600	Antsohihy
Vakinankaratra	16,599	1,708,700	Antsirabe
Vatovavy Fitovinany	19,605	1,342,100	Manakara

The indigenous population is of Malayo-Polynesian stock, divided into 18 ethnic groups of which the principal are Merina (24%) of the central plateau, the Betsimisaraka (13%) of the east coast and the Betsileo (11%) of the southern plateau. Foreign communities include Europeans (mainly French), Indians, Chinese, Comorians and Arabs.

Malagasy, French and (since 2007) English are all official languages.

SOCIAL STATISTICS

2008 estimates: births, 686,000; deaths, 176,000. Rates, 2008 estimates (per 1,000 population): births, 35·9; deaths, 9·2. Infant mortality, 2010 (per 1,000 live births), 43. Expectation of life in 2007 was 58·3 years for males and 61·5 for females. Annual population growth rate, 2000–08, 2·8%. Fertility rate, 2008, 4·7 births per woman.

CLIMATE

A tropical climate, but the mountains cause big variations in rainfall, which is very heavy in the east and very light in the west.

Repoblikan'i Madagasikara
(Republic of Madagascar)

Capital: Antananarivo
Population projection, 2015: 23·85m.
GNI per capita, 2011: (PPP$) 824
HDI/world rank: 0·480/151
Internet domain extension: .mg

KEY HISTORICAL EVENTS

The island was settled by people of African and Indonesian origin when it was visited by the Portuguese explorer, Diego Diaz, in 1500. The island was unified under the Imérina monarchy between 1797 and 1861. A French protectorate was established in 1895. Madagascar became a French colony on 6 Aug. 1896 and achieved independence on 26 June 1960.

In Feb. 1975 Col. Richard Ratsimandrava, head of state, was assassinated. The 1975 constitution instituted a 'Democratic Republic' allowing for only one political party. After six months of anti-government unrest an 18-month transitional administration was agreed. A new constitution instituted the Third Republic in Sept. 1992.

Following the presidential election of Dec. 2001 the opposition candidate Marc Ravalomanana claimed victory, although the High

Antananarivo, Jan. 70°F (21·1°C), July 59°F (15°C). Annual rainfall 54" (1,350 mm). Toamasina, Jan. 80°F (26·7°C), July 70°F (21·1°C). Annual rainfall 128" (3,256 mm).

CONSTITUTION AND GOVERNMENT

A new constitution was promulgated on 10 Dec. 2010, having won 74·2% support at a referendum held in Nov. 2010 following a coup in March 2009. However, the referendum was boycotted by the three main opposition parties and turnout was 52·6%, with registered voters accounting for only about a third of the population. On the day of the vote there was an unsuccessful army mutiny against the rule of President Rajoelina, who had come to power during the 2009 coup.

The new constitution reduces the minimum age requirement for the presidency from 40 years to 35, qualifying the 36-year old Rajoelina to stand in the next presidential election. It also demands that presidential candidates should be resident in the country in the six months leading up to an election, effectively ruling out Rajoelina's exiled predecessor, Marc Ravalomanana, from standing against him. Rajoelina's critics argued the constitution was designed to bolster the interim president's personal standing.

National Anthem

'Ry tanindrazanay malala ô!' ('O our beloved Fatherland'); words by Pastor Rahajason, tune by N. Raharisoa.

RECENT ELECTIONS

Presidential elections were held on 3 Dec. 2006. Incumbent president Marc Ravalomanana won with 54·8% of the vote, ahead of Jean Lahiniriko with 11·7%, former president Roland Ratsiraka with 10·1% and Herizo Razafimahaleo with 9·0%. There were ten other candidates who each took less than 5% of the vote. Turnout was 61·5%.

In parliamentary elections held on 23 Sept. 2007 President Marc Ravalomanana's I Love Madagascar party won 105 of the 127 seats, independents 11 and Fanjava Velogno 2 with nine smaller parties each winning one seat.

Presidential elections were scheduled to take place on 24 July 2013 with a second round if needed on 25 Sept. 2013, the same day as parliamentary elections.

CURRENT GOVERNMENT

Interim President: Andry Rajoelina; b. 1974 (Young Malagasies Determined/TGV, assumed office 17 March 2009 and sworn in 21 March 2009).

In March 2013 the transitional government was composed as follows:

Prime Minister, and Minister of Environment and Forests (acting): Omer Beriziky; b. 1950 (LEADER-Fanilo; sworn in 2 Nov. 2011).

Deputy Prime Minister and Minister of Development and Land Management: Andrianiainarivelo Hajo Herivelona. *Deputy Prime Minister and Minister of Economy and Industry:* Pierrot Botozaza.

Minister of Agriculture: Rolland Ravatomanga. *Armed Forces:* Gen. André-Lucien Rakotoarimasy. *Commerce:* Olga Ramalason. *Communications and Government Spokesman:* Harry-Laurent Rahajason. *Culture and Heritage:* Elia Ravelomanantsoa. *Decentralization:* Ruffine Tsiranana. *Energy:* Nestor Razafindroariaka. *Finance and Budget:* Hery Rajaonarimampianina. *Fisheries and Fishing Resources:* Sylvain Manoriky. *Foreign Affairs:* Pierrot Rajaonarivelo. *Higher Education:* Étienne-Hilaire Razafindehibe. *Hydrocarbons:* Marcel Bernard. *Interior:* Florent Rakotoarisoa. *Internal Security:* Arsène Rakotondrazaka. *Justice:* Christine Harijaona Razanamahasoa. *Livestock:* Ihanta Randriamandranto. *Mines:* Daniella Randrianfeno Tolotrandry Rajo. *National Education:* Régis Manoro. *Population:* Olga Vaomalala. *Posts, Telecommunications and New Technologies:* Ny Hasina Andriamanjato. *Promotion of Handicrafts:* Elisa Zafitombo Alibena. *Public Health:* Johanita Ndahimananjara. *Public Service:* Tabera Randriamanantsoa. *Public Works and Meteorology:* Col. Botomanovatsara. *Relations with Institutions:* Victor Manantsoa. *Sports:* Gérard Botralahy. *Technical Education and Vocational Training:* Jean-André Ndremanjary. *Tourism:* Jean-Max Rakotomamonjy. *Transport:* Benjamina Ramanantsoa. *Water:* Julien Reboza. *Youth and Leisure:* Jacques-Ulrich Randriantiana.

CURRENT LEADERS

Andry Rajoelina

Position
President

Introduction
Andry Rajoelina was sworn in as president on 21 March 2009. He was installed by the military after the former president, Marc Ravalomanana, ceded power following three months of political turmoil. On assuming office Rajoelina suspended parliament and set up a transitional authority to run the country. His mandate is not universally recognized by the international community.

Early Life
Rajoelina was born on 30 May 1974 into the wealthy family of a colonel in the Malagasy army. He rose to prominence as a disc jockey in Antananarivo, before setting up a TV and radio station and running an advertising company. Rajoelina's brash personality earned him the nickname TGV, after the French high-speed train. The initials went on to serve as the acronym for his political movement, Tanora malaGasy Vonona (Young Malagasies Determined). He harnessed his public profile to win the Antananarivo mayoral election in Dec. 2007.

In Dec. 2008 and Jan. 2009 Rajoelina's radio and TV networks were shut down by the government, which accused them of 'inciting civil disobedience'. Rajoelina called a general strike, resulting in widespread disorder. On 17 March Ravalomanana stepped down under pressure from military chiefs who immediately installed Rajoelina as his successor. The African Union (AU) denounced the change of government as a coup and suspended Madagascar's membership. Rajoelina's ascent to power was also condemned by the European Union and the USA.

Career in Office
Aged 34, Rajoelina was the youngest president in Madagascar's history, although the prevailing constitution stipulated that presidential candidates must be at least 40 years of age. At that time he promised a new constitution and elections within two years.

On 17 April 2009 Rajoelina issued a warrant for the arrest of Ravalomanana, who was then tried in absentia (having fled to South Africa) and sentenced in June to four years in prison for abuse of office. In Aug. 2009 a power-sharing agreement, sponsored by international mediators, was signed between the rival Rajoelina and Ravalomanana political camps with the aim of establishing a transitional unity government. However, continued disputes prevented its effective implementation and in Dec. 2009 Rajoelina announced that he was abandoning the agreement. This prompted the AU to impose targeted sanctions against his administration in March 2010. Then, in Aug., a court sentenced Ravalomanana in absentia to life imprisonment for conspiracy to commit murder.

A referendum on a new constitution took place in Nov. 2010. The referendum was boycotted by the main opposition parties, which regarded the revision as an illegal attempt to consolidate Rajoelina's hold on power by lowering the age requirement for the presidency from 40 to 35. 74% of participants voted in favour. At the same time, an attempted coup against Rajoelina failed as loyal troops arrested a group of dissident army officers.

In Nov. 2011 a transitional cross-party administration took office pending fresh elections. Two-stage presidential polling was later scheduled for July–Sept. 2013.

DEFENCE

There is conscription (including civilian labour service) for 18 months. Defence expenditure totalled US$103m. in 2008 (US$5 per capita), representing 1·1% of GDP.

Army

Strength (2007) approximately 12,500 and gendarmerie 8,100.

Navy

In 2007 the Navy had a strength of 500 (including some 100 marines).

Air Force

Personnel (2007) 500. There are no combat-capable aircraft.

ECONOMY

In 2009 agriculture contributed 29·1% of GDP, industry 16·0% and services 54·9%.

Overview

Economic decline over several decades saw Madagascar's per capita income fall from US$473 in 1970 to US$410 in 2008. More than two-thirds of the population lived in poverty in 2005. Key commodities such as vanilla and oil are highly susceptible to price- and weather-related shocks. Nonetheless, the economy grew at an average of over 6% per year from 2003–08 and poverty declined from its 2002 peak, driven by strong growth in the tourism and mining sectors, public investment and improved rice productivity.

In Feb. 2009 a political crisis prompted a decline in economic growth and the postponement of international aid. The economy's problems were exacerbated by the global financial crisis, with exports declining by 50% between 2008 and 2010. However, some sectors experienced growth in 2009, notably agriculture and mining.

Currency

In July 2003 then President Marc Ravalomanana announced that the *Ariary* (MGA) would become the official currency, replacing the *Malagasy franc* (MGFr). The Ariary became legal tender on 1 Aug. 2003 at a rate of 1 *Ariary* = 5 *Malagasy francs*. The Ariary is subdivided into five *Iraimbilanja*.

In July 2005 foreign exchange reserves were US$435m. and total money supply was 1,324·0bn. ariarys. Inflation was 9·3% in 2010 and 10·0% in 2011.

Budget

Revenues totalled 2,685·4bn. ariarys in 2008 and expenditures 2,998·7bn. ariarys. Tax revenue accounted for 77·7% of revenues in 2008; current expenditure accounted for 58·5% of expenditures.
VAT is 20%.

Performance

Total GDP in 2011 was US$9·9bn. There was a recession in 2002 with the economy contracting by 12·4% as a result of the six-month long political crisis, but a recovery followed in 2003 and 2004 with real GDP growth of 9·8% and 5·3% respectively. The recovery continued until 2009 when the economy contracted by 4·1%. However, there was growth of 0·4% in 2010 and 1·8% in 2011.

Banking and Finance

A Central Bank, the Banque Centrale de Madagascar, was formed in 1973, replacing the former Institut d'Émission Malgache as the central bank of issue. The *Acting Governor* is Guy Richard Ratovondrahona. All commercial banking and insurance was nationalized in 1975 and privatized in 1988. Of the six other

banks, the largest are the Bankin'ny Tantsaha Mpamokatra and the BNI—Crédit Lyonnais de Madagascar.

External debt was US$2,295m. in 2010, representing 26·6% of GNI.

ENERGY AND NATURAL RESOURCES

Environment

Madagascar's carbon dioxide emissions from the consumption and flaring of fossil fuels in 2008 were the equivalent of 0·1 tonnes per capita.

Electricity

Installed capacity was around 0·2m. kW in 2007. Production in 2007 was 935m. kWh, with consumption per capita 50 kWh.

Oil and Gas

Several oil blocks both on land and offshore were discovered in 2005.

Minerals

Mining production in 2009 (estimates) included: ilmenite concentrate, 160,000; salt, 75,000 tonnes; chromite, 60,000 tonnes; graphite (2008 estimate), 5,000 tonnes. There have also been discoveries of precious and semi-precious stones in various parts of the country, in particular sapphires, topaz and garnets. Madagascar is believed to have the world's largest reserves of sapphires.

Agriculture

75–80% of the workforce is employed in agriculture. There were an estimated 2·95m. ha. of arable land in 2007 and 0·6m. ha. of permanent crops. 890,000 ha. were irrigated in 2007. The principal agricultural products in 2003 were (in 1,000 tonnes): rice, 2,800; cassava, 2,367; sugarcane, 2,236; sweet potatoes, 509; potatoes, 298; bananas, 290; mangoes, 210; tomatoes, 210; taro, 200; maize, 181; coconuts, 84; oranges, 83; dry beans, 70. Rice is produced on some 40% of cultivated land. Madagascar is the world's largest producer of vanilla.

Cattle breeding and agriculture are the chief occupations. There were, in 2003, 10·5m. cattle, 1·6m. pigs, 1·2m. goats, 650,000 sheep and 24m. chickens.

Forestry

In 2010 the area under forests was 12·55m. ha., or 22% of the total land area. The forests contain many valuable woods, while gum, resins and plants for tanning, dyeing and medicinal purposes abound. Timber production was 13·35m. cu. metres in 2007.

Fisheries

The catch of fish in 2010 was 128,836 tonnes (mainly from marine waters).

INDUSTRY

Industry, hitherto confined mainly to the processing of agricultural products, is now extending to cover other fields.

Labour

In 2003 the economically active population totalled 7,573,900 (50·5% males). In 2002 approximately 82·2% of the economically active population were engaged in agriculture, hunting and forestry.

INTERNATIONAL TRADE

Imports and Exports

Trade, 2010, in US$1m.: imports (c.i.f.), 2,546; exports (f.o.b.), 1,082. Leading imports in 2010 (in US$1m.) were: machinery and transport equipment (752); manufactured goods (634); mineral fuels, lubricants and related materials (386). Leading exports in 2010 (in US$1m.) were: articles of apparel and clothing accessories (303); food and live animals (248); machinery and transport equipment (120).

In 2010 main import sources (in US$1m.) were: France (368); China (310); South Africa (197). Main export markets in 2010 (in US$1m.) were: France (359); Germany (79); China (57).

COMMUNICATIONS

Roads
In 2002 there were about 65,663 km of roads, 11·6% of which were paved. There were 146,300 passenger cars, 280,800 buses and coaches and 83,800 lorries and vans in 2008. 550 people died in road accidents in 2006.

Rail
In 2005 there were 854 km of railways, all metre gauge. In 2005, 100,000 passengers and 300,000 tonnes of freight were transported.

Civil Aviation
There are international airports at Antananarivo (Ivato) and Mahajanga (Amborovy). The national carrier is Air Madagascar, which is 90·6% state-owned. In 2003 scheduled airline traffic of Madagascar-based carriers flew 9m. km, carrying 452,000 passengers (140,000 on international flights). In 2001 Antananarivo handled 699,074 passengers (348,238 on domestic flights) and 15,499 tonnes of freight.

Shipping
The main ports are Toamasina, Mahajanga, Antsiranana and Toliary. In Jan. 2009 there were 19 ships of 300 GT or over registered, totalling 18,000 GT.

Telecommunications
In 2011 there were 130,100 landline telephone subscriptions (equivalent to 6·5 per 1,000 inhabitants) and 8,159,600 mobile phone subscriptions (or 382·8 per 1,000 inhabitants). There were 17·0 internet users per 1,000 inhabitants in 2010. Fixed internet subscriptions totalled 8,300 in 2009 (0·4 per 1,000 inhabitants). In June 2012 there were 233,000 Facebook users.

SOCIAL INSTITUTIONS

Justice
The Supreme Court and the Court of Appeal are in Antananarivo. In most towns there are Courts of First Instance for civil and commercial cases. For criminal cases there are ordinary criminal courts in most towns.

The population in penal institutions in Dec. 2006 was 17,495 (91 per 100,000 of national population).

Education
Education is compulsory from six to 14 years of age. In 2007 there were 78,743 teaching staff for 3,837,343 pupils in primary schools, 835,539 pupils at secondary level with 34,320 teaching staff and 58,313 students at tertiary level with 3,032 academic staff. Adult literacy rate in 2008 was 71%. In 2007 public expenditure on education came to 3·4% of GNI and 16·4% of total government spending.

Health
There were three hospital beds per 10,000 population in 2005. In 2004 there were 5,201 physicians, 410 dentistry personnel and 5,661 nursing and midwifery personnel. In 2005 government expenditure on health came to 9·6% of total government spending.

Welfare
In 2006 social security accounted for 10·4% of government expenditure.

RELIGION
About 48% of the population practise the traditional religion, 43% are Christians (of whom approximately half are Roman Catholic and half are Protestant, mainly belonging to the Fiangonan'i Jesosy Kristy eto Madagasikara) and 9% are followers of other religions (predominantly Islam).

CULTURE

World Heritage Sites
Tsingy de Bemaraha Strict Nature Reserve joined the UNESCO World Heritage List in 1990. The undisturbed forests, lakes and mangrove swamps are the habitat for rare and endangered lemurs and birds. The Royal Hill of Ambohimanga was added in 2001, a royal city and burial site and a symbol of Malagasy identity. The rainforests of Atsinanana, comprising six national parks in the eastern part of the island, were inscribed on the list in 2007.

Press
In 2008 there were 13 daily newspapers with a total circulation of 115,000.

Tourism
In 2011, 225,005 non-resident tourists arrived by air (excluding same-day visitors), up from 196,052 in 2010 although down from the peak of 375,010 in 2008.

DIPLOMATIC REPRESENTATIVES
Of Madagascar in the United Kingdom (embassy in London closed in 2011)
Ambassador: Vacant (resides in Paris).

Of the United Kingdom in Madagascar (Tour Zital Ankorondrado, Ravoninahitriniarivo St., Antananarivo 101)
Ambassador: Timothy Smart.

Of Madagascar in the USA (2374 Massachusetts Ave., NW, Washington, D.C., 20008)
Ambassador: Vacant.
Chargé d'Affaires a.i.: Velotiana Rakotoanosy Raobelina.

Of the USA in Madagascar (Lot 207A, Point Liberty, Andranoro, Antehiroka, 105 Antananarivo)
Ambassador: Vacant.
Chargé d'Affaires a.i.: Eric M. Wong.

Of Madagascar to the United Nations
Ambassador: Zina Andrianarivelo.

Of Madagascar to the European Union
Ambassador: Jeannot Rakotomalala.

FURTHER READING
Banque des Données de l'État. *Bulletin Mensuel de Statistique*
Allen, P. M., *Madagascar*. 1995

National Statistical Office: Institut National de la Statistique (INSTAT), BP 485 Anosy, Antananarivo 101.
Website (French only): http://www.instat.mg

MALAŴI

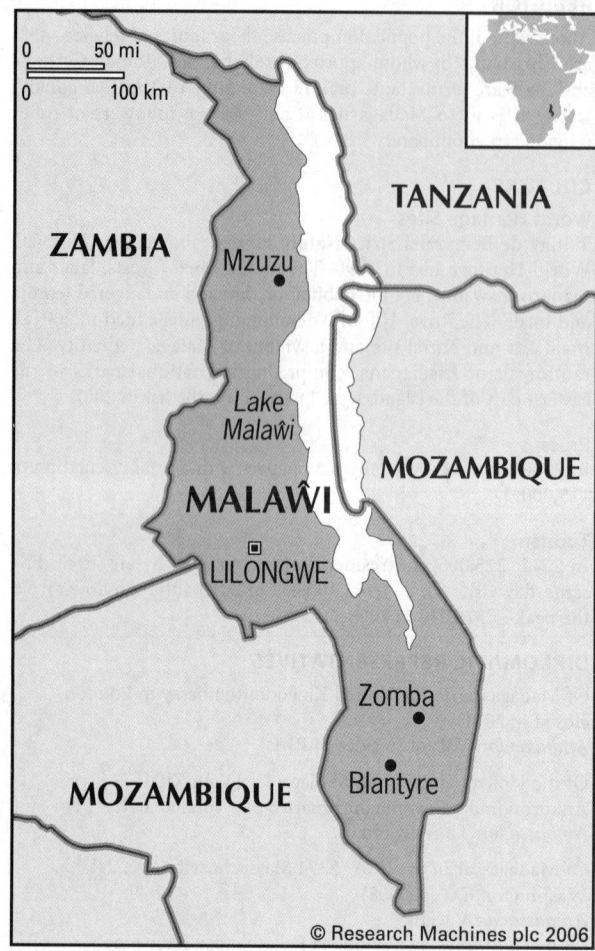

Dziko la Malaŵi
(Republic of Malaŵi)

Capital: Lilongwe
Population projection, 2015: 17·52m.
GNI per capita, 2011: (PPP$) 753
HDI/world rank: 0·400/171
Internet domain extension: .mw

KEY HISTORICAL EVENTS

The area was dominated by the Twa and Fulani tribes until the 1st century AD when Bantu-speaking tribes made inroads. The explorer David Livingstone reached Lake Nyasa, now Lake Malaŵi, in 1859 and it was the land along the lake's western shore that became, in 1891, the British Protectorate of Nyasaland. In 1884 the British South Africa Company applied for a charter to trade. Pressure on land, the colour bar and other grievances generated Malaŵian resistance. In 1893 it was renamed the British Central African Protectorate. This became Nyasaland in 1907. By the mid-1940s a nationalist movement had emerged, spearheaded by the Nyasaland African Congress. In 1953 Nyasaland was joined with Southern Rhodesia (Zimbabwe) and Northern Rhodesia (Zambia) to form the Federation of Rhodesia and Nyasaland,

under British control. This union was dissolved in 1963. Nyasaland was self-governing until on 6 July 1964 it became independent, adopting the name Malaŵi. In 1966 Malaŵi was declared a republic and Dr Hastings Banda became the first president, establishing a one party dictatorship which lasted for 30 years. In 1994 Malaŵi returned to multi-party democracy.

TERRITORY AND POPULATION

Malaŵi lies along the southern and western shores of Lake Malaŵi (the third largest lake in Africa), and is otherwise bounded in the north by Tanzania, south by Mozambique and west by Zambia. Area (including the inland water areas of Lake Malombe, Chilwa, Chiuta and the Malaŵi portion of Lake Malaŵi, which total 24,208 sq. km), 118,484 sq. km (45,747 sq. miles).

Census population (2008), 13,077,160 (6,718,227 females); density, 138·7 per sq. km. The United Nations population estimate for 2008 was 14,005,000. In 2011, 20·3% of the population were urban.

The UN gives a projected population for 2015 of 17·52m.

Population of main towns (2008): Lilongwe, 674,448; Blantyre, 661,256; Mzuzu, 133,968; Zomba, 88,314. Population of the regions (2008): Northern, 1,708,930; Central, 5,510,195; Southern, 5,858,035.

The official languages are Chichewa, spoken by over 58% of the population, and English.

SOCIAL STATISTICS

2008 estimates: births, 597,000; deaths, 182,000. Estimated rates, 2008 (per 1,000 population): births, 40·2; deaths, 12·3. Annual population growth rate, 2000–08, 2·8%. Expectation of life at birth in 2007 was 51·3 years for males and 53·4 for females. Infant mortality, 2010, 58 per 1,000 live births; fertility rate, 2008, 5·5 births per woman.

CLIMATE

The tropical climate is marked by a dry season from May to Oct. and a wet season for the remaining months. Rainfall amounts are variable, within the range of 29–100" (725–2,500 mm), and maximum temperatures average 75–89°F (24–32°C), and minimum temperatures 58–67°F (14·4–19·4°C). Lilongwe, Jan. 73°F (22·8°C), July 60°F (15·6°C). Annual rainfall 36" (900 mm). Blantyre, Jan. 75°F (23·9°C), July 63°F (17·2°C). Annual rainfall 45" (1,125 mm). Zomba, Jan. 73°F (22·8°C), July 63°F (17·2°C). Annual rainfall 54" (1,344 mm).

CONSTITUTION AND GOVERNMENT

The *President* is also head of government. Malaŵi was a one-party state, but following a referendum on 14 June 1993, in which 63% of votes cast were in favour of reform, a new constitution was adopted on 17 May 1994 which ended Hastings Banda's life presidency and provided for the holding of multi-party elections. At these Bakili Muluzi was elected president with 47·16% of votes cast, beating President Banda and two other opponents. There is a *National Assembly* of 193 members, elected for five-year terms in single-seat constituencies.

National Anthem

'O God Bless our Land of Malaŵi'; words and tune by M.-F. Sauka.

RECENT ELECTIONS

At parliamentary elections of 19 May 2009 the Democratic Progressive Party (DPP) won 114 seats, the Malaŵi Congress Party (MCP—formerly the only legal party) 27 and the United

Democratic Front (UDF) 17. Independents took 32 seats and others three.

At the concurrent presidential elections incumbent Bingu wa Mutharika (DPP) won with 66·0% of the vote, ahead of John Tembo (MCP) with 30·7%. There were five other candidates who all received less than 1% of the vote.

CURRENT GOVERNMENT

President and C.-in-C. of the Malaŵi Defence Force and Malaŵi Police Service: Joyce Banda; b. 1950 (People's Party; sworn in 7 April 2012).

Vice President: Khumbo Hastings Kachali.

The government consisted of the following in March 2013:

Minister of Agriculture and Food Security: Peter Mwanza. *Defence:* Ken Kandodo. *Disability and Elderly Affairs:* Reen Kachere. *Economic Planning and Development:* Ralph Jooma. *Education, Science and Technology:* Eunice Kazembe. *Energy:* Ibrahim Matola. *Environment and Climate Change Management:* Jennifer Deborah Chilunga. *Finance:* Ken Lipenga. *Foreign Affairs and International Co-operation:* Ephraim Mganda Chiume. *Gender, Children and Social Welfare:* Anita Kalinde. *Health:* Catherine Gotani Hara. *Home Affairs:* Uladi Mussa. *Industry and Trade:* Sosten Gwenge. *Information:* Moses Kunkuyu. *Justice:* Ralph Kasambara. *Labour:* Eunice Makangala. *Lands and Housing:* Henry Dama Phoya. *Local Government and Rural Development:* Grace Zinenani Maseko. *Mining:* John Bande. *Tourism and Culture:* Rachel Zulu Mazombwe. *Transport and Public Infrastructure:* Sidik Muhamed Mia. *Water Development and Irrigation:* Richie Bizwick Muheya. *Youth and Sports:* Enoch Chihana. *Attorney General:* Anthony Kamanga.

Government Website: http://www.malawi.gov.mw

CURRENT LEADERS

Joyce Banda

Position
President

Introduction
Joyce Banda was sworn in as president in April 2012 following the death of President Bingu wa Mutharika. A lifelong campaigner for women's rights, Banda is the first female president of Malaŵi and only the second female president of any African state. She is the leader of the People's Party, which she formed in 2011 after a dispute with Mutharika that saw her expelled from the Democratic Progressive Party (DPP).

Early Life
Joyce Banda was born in Malaŵi's Zomba District on 12 April 1950. In 1975 she left her abusive husband, who worked at Malaŵi's embassy in Nairobi, Kenya. Her decision, she said later, was inspired by a Kenyan women's movement.

Banda secured her financial independence by establishing a clothing manufacturing business, the first of several enterprises she set up. In 1990 she founded the National Association of Business Women, a non-governmental group aimed at improving the prospects of small, female-led businesses. In 1997 she set up the Joyce Banda Foundation, focusing on education.

She entered politics in 1999 as the member of parliament for the Zomba-Malosa constituency, serving under President Bakili Muluzi as minister of gender and community services. She was re-elected to parliament in 2004 but left the defeated United Democratic Front to join the DPP, the party of new president Bingu wa Mutharika. Banda was named foreign minister on 1 June 2006. During her time in the role she severed ties with Taiwan in favour of closer relations with the government in Beijing, paving the way for communist Chinese investment in Malaŵi's infrastructure.

In 2009 Mutharika selected Banda as his running mate in his bid for re-election, seeking to take advantage of her popularity in rural areas. However, once elected, Mutharika moved to isolate Banda, who had refused to support the planned succession to the presidency of Mutharika's brother. Having been ousted from the DPP, Banda established her own People's Party in 2011. She refused to relinquish the vice-presidency, and Mutharika lacked the constitutional powers to remove her.

Mutharika's sudden death from a heart attack on 5 April 2012 threatened a prolonged power struggle but after two days of political jockeying, constitutional procedure prevailed and Banda was sworn into office on 7 April.

Career in Office
Within days of becoming president Banda dismissed the chief of police, Peter Mukhito, who had been accused of mishandling anti-government riots in 2011 in which at least 19 people were killed. She replaced him with Lot Dzonzi, a human rights supporter. She also acted quickly to remove the minister of information, Patricia Kaliati, who had attempted to block Banda's accession after the death of Mutharika. In May 2012 she oversaw a currency devaluation to comply with International Monetary Fund conditions for a restoration of external funding, but the move triggered steep increases in the price of basic goods.

DEFENCE

All services form part of the Army. Defence expenditure totalled US$43m. in 2008 (US$3 per capita), representing 1·5% of GDP.

Army
Personnel (2007) 5,300. In addition there is a paramilitary mobile police force totalling 1,500.

Navy
The Maritime Wing, based at Monkey Bay on Lake Malaŵi, numbered 220 personnel in 2007.

Air Wing
The Air Wing acts as infantry support and numbered 200 in 2007 with no combat-capable aircraft.

ECONOMY

Agriculture accounted for 34·2% of GDP in 2006, industry 19·7% and services 46·1%.

Overview
Malaŵi has enjoyed solid growth since the mid- to late 2000s. Having qualified for the Heavily Indebted Poor Countries (HIPC) initiative in Aug. 2006, government finances improved and growth peaked at 9·7% in 2008, compared to around 2% before 2005. Growth was supported by good weather, strong tobacco harvests and a government programme of fertilizer subsidies.

Heavy dependence on agriculture (which accounts for 80% of export earnings and supports 85% of the population) makes the economy vulnerable to weather shocks and natural disasters including droughts and heavy rainfall. Tobacco, tea, cotton, coffee and sugar are the primary exports, with tobacco accounting for more than half the total. Social indicators are poor. According to the 2010 Human Development Report, 74% of the population lives below the international poverty line of US$1·25 (PPP) per day. Malaŵi relies on assistance from the World Bank, IMF and other international donors. However, gains have seen HIV/AIDS prevalence down from 14% to 12%, while health care, education and environmental conditions are showing improvements.

Challenges are numerous—persistent energy constraints mean only 7% of the population have access to electricity services, a weak balance-of-payments threatens exchange rate and price stability, and prospects are uncertain for future international tobacco consumption. Medium-term growth is expected to be supported by expansion at the Kayelekera uranium mine.

Currency

The unit of currency is the *kwacha* (MWK) of 100 *tambala*. Foreign exchange reserves were US$119m. in June 2005, gold reserves 13,000 troy oz and total money supply was K.29,579m. In May 2012 Malaẁi ended its currency peg to the dollar and adopted a floating exchange rate regime. Inflation was 7·4% in 2010 and 7·6% in 2011, having been as high as 83·1% in 1995.

Budget

The fiscal year runs from 1 July–30 June. In 2008–09 revenues were K.187·40bn. and expenditures K.223·50bn. Tax revenue accounted for 62·4% of revenues in 2008–09; current expenditure accounted for 82·0% of expenditures.

VAT is 16·5%.

Performance

Since shrinking by 4·1% during a recession in 2001 the economy has recovered, with real GDP growth rates of 9·0% in 2009, 6·5% in 2010 and 4·3% in 2011. Total GDP was US$5·6bn. in 2011.

Banking and Finance

The central bank and bank of issue is the Reserve Bank of Malaẁi (founded 1964). The *Governor* is Charles Chuka. In 2002 there were four commercial banks, one development bank, three merchant banks and a savings bank.

Foreign debt was US$922m. in 2010, representing 18·5% of GNI.

There is a stock exchange in Blantyre.

ENERGY AND NATURAL RESOURCES

Environment

Carbon dioxide emissions from the consumption and flaring of fossil fuels in 2008 were the equivalent of 0·1 tonnes per capita.

Electricity

The Electricity Supply Commission of Malaẁi is the sole supplier. Installed capacity was 0·5m. kW in 2007. Production was 1,637m. kWh in 2007; consumption per capita was 123 kWh. Only 6% of the population has access to electricity.

Oil and Gas

Malaẁi does not produce any oil or natural gas and imports all its fuel products. There are plans for an oil pipeline to be constructed linking Nsanje in the south of the country with the port of Beira in neighbouring Mozambique.

Minerals

Mining operations have been limited to small-scale production of coal, limestone, rubies and sapphires, but companies are now moving in to start exploration programmes. Bauxite reserves are estimated at 29m. tonnes and there are proven reserves of clays, diamonds, glass and silica sands, graphite, limestone, mercurate, phosphates, tanzanite, titanium and uranium. Output in 2007: crushed stone, 226,351 tonnes; coal, 58,550 tonnes; limestone, 42,088 tonnes; gemstones, 3,710 kg.

Agriculture

Malaẁi is predominantly an agricultural country. Agricultural produce contributes 90% of export earnings. There were an estimated 3·0m. ha. of arable land in 2007 and 120,000 ha. of permanent crops. Maize is the main subsistence crop and is grown by over 95% of all smallholders. Tobacco is the chief cash crop, employing 12% of the workforce and generating a quarter of tax earnings. Also important are groundnuts, cassava, millet and rice. There are large plantations which produce sugar, tea and coffee. Production (2003 estimates, in 1,000 tonnes): maize, 1,901; sugarcane, 1,900; cassava, 1,774; potatoes, 1,100; plantains, 200; groundnuts, 158; dry beans, 94; bananas, 93; rice, 87; tobacco, 70; sorghum, 45; tea, 45.

Livestock in 2003: goats, 1·7m.; cattle, 750,000; pigs, 456,000; sheep, 115,000; chickens, 15m.

Forestry

In 2010 the area under forests was 3·24m. ha., or 34% of the total land area. Timber production in 2007 was 5·76m. cu. metres.

Fisheries

Landings in 2010 were 98,299 tonnes, entirely from inland waters. The annual catch has doubled since 2002, when it was 41,329 tonnes.

INDUSTRY

Index of industrial production in 2001 (1984 = 100): total general industrial production, 101·9; of this goods for the domestic market were at 73·2 and export goods were at 101·5. Electricity and water were at 231·7.

Labour

The labour force in 2010 was 6,708,000 (51·5% female). Approximately 80% of the economically active population in 2010 were engaged in agriculture.

INTERNATIONAL TRADE

Imports and Exports

In 2010 imports (c.i.f.) amounted to US$2,173m. and exports (f.o.b.) US$1,066m.

Principal imports in 2010 (in US$1m.) were: chemicals and related products (532); machinery and transport equipment (515); manufactured goods (347). Principal exports in 2010 (in US$1m.) were: tobacco and tobacco products (585); uranium ores and concentrates (114); tea (81).

In 2010 major import sources (in US$1m.) were: South Africa (654); China (198); India (165). Major export markets in 2010 (in US$1m.) were: Belgium (133); Canada (118); Egypt (98).

Trade Fairs

The annual Malaẁi International Trade Fair takes place in Blantyre, the commercial capital.

COMMUNICATIONS

Roads

The road network consisted of 15,451 km in 2003, of which 45·0% were paved. There were 53,300 passenger cars and 59,800 vans and trucks in 2007.

Rail

In 2005 Malaẁi Railways operated 797 km on 1,067 mm gauge, providing links to the Mozambican ports of Beira and Nacala. In 2009 passenger-km travelled came to 44m. and freight tonne-km to 47m.

Civil Aviation

The national carrier is Air Malaẁi. It flies to a number of regional centres in Ethiopia, Kenya, South Africa, Zambia and Zimbabwe. In 2003 scheduled airline traffic of Malaẁi-based carriers flew 4m. km, carrying 109,000 passengers (68,000 on international flights). There are international airports at Lilongwe (Lilongwe International Airport) and Blantyre (Chileka). In 2000 Lilongwe handled 175,915 passengers (120,575 on international flights) and 4,182 tonnes of freight, and Blantyre had 101,809 passengers (53,426 on international flights) and 680 tonnes of freight.

Telecommunications

In 2011 there were 173,500 landline telephone subscriptions (equivalent to 11·3 per 1,000 inhabitants) and 3,855,800 mobile phone subscriptions (or 250·7 per 1,000 inhabitants). In 2011, 3·3% of the population were internet users. In June 2012 there were 140,000 Facebook users.

SOCIAL INSTITUTIONS

Justice

Justice is administered in the High Court and in the magistrates' courts. Traditional courts were abolished in 1994. Appeals from magistrates' courts lie to the High Court, and appeals from the High Court to Malaŵi's Supreme Court of Appeal.

The population in penal institutions in Dec. 2007 was 10,830 (78 per 100,000 of national population).

Education

The adult literacy rate in 2009 was estimated at 73·7% (80·6% among males and 67·0% among females). Fees for primary education were abolished in 1994. In 2007 the number of pupils in primary schools was 2,943,248 (44,048 teaching staff). The primary school course is of eight years' duration, followed by a four-year secondary course. In 2007 there were 574,003 pupils in secondary schools. English is taught from the 1st year and becomes the general medium of instruction from the 4th year.

The University of Malaŵi (consisting of four colleges and one polytechnic) had 6,257 students and 676 academic staff in 2007. A university at Mzuzu opened in 1998 and provides courses for secondary school teachers. In 2007 there were 6,458 students in higher education and 861 academic staff.

In 2009 public expenditure on education came to 2·8% of GDP.

Health

In 2004 there were 266 physicians and 7,264 nursing and midwifery personnel. In 2007 there were 11 hospital beds per 10,000 inhabitants.

RELIGION

2001 estimates: 2,600,000 Roman Catholic; 2,070,000 Protestant (mostly Presbyterian); 1,770,000 African Christian; 1,560,000 Muslim; 820,000 traditional beliefs. The remainder follow other religions.

CULTURE

World Heritage Sites

Malaŵi has two sites on the UNESCO World Heritage List: Lake Malaŵi National Park (inscribed on the list in 1984); and the Chongoni Rock-Art area (2006).

Press

There were two paid-for dailies and nine paid-for non-dailies in 2008. The two dailies are *The Nation* (average circulation of 15,000 copies daily in 2008); and *The Daily Times* (7,000 copies daily in 2008).

Tourism

There were 755,031 non-resident tourists in 2009 (excluding same-day visitors), up from 742,457 in 2008.

DIPLOMATIC REPRESENTATIVES

Of Malaŵi in the United Kingdom (36 John St., London, WC1N 2AT)
High Commissioner: Bernard H. Sande.

Of the United Kingdom in Malaŵi (PO Box 30042, Lilongwe 3)
High Commissioner: Michael Nevin.

Of Malaŵi in the USA (1156 15th St., NW, Suite 320, Washington, D.C., 20005)
Ambassador: Steve Matenje.

Of the USA in Malaŵi (Area 40, Plot 24, Kenyatta Rd, Lilongwe 3)
Ambassador: Jeanine Jackson.

Of Malaŵi to the United Nations
Ambassador: Charles P. Msosa.

Of Malaŵi to the European Union
Ambassador: Brave Rona Ndisale.

FURTHER READING

National Statistical Office. *Monthly Statistical Bulletin*
Ministry of Economic Planning and Development. *Economic Report.* Annual

Kalinga, Owen J. M., *Historical Dictionary of Malawi.* 4th ed. 2011
Sindima, Harvey J., *Malawi's First Republic: An Economic and Political Analysis.* 2002

National Statistical Office: National Statistical Office, POB 333, Zomba.

MALAYSIA

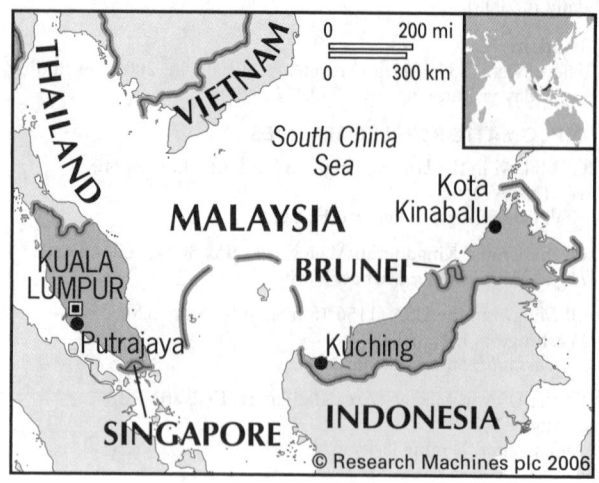

Persekutuan Tanah Malaysia (Federation of Malaysia)

Capitals: Putrajaya (Administrative),
Kuala Lumpur (Legislative and Financial)
Population projection, 2015: 30·71m.
GNI per capita, 2011: (PPP$) 13,685
HDI/world rank: 0·761/61
Internet domain extension: .my

KEY HISTORICAL EVENTS

Excavations at Niah in Sarawak, East Malaysia have uncovered evidence of human settlement from 38,000 BC (the oldest relic of *homo sapiens* in southeast Asia). There are numerous sites in the north of Peninsular Malaysia where evidence of hunter-gatherers has been dated to around 10,000 BC. These Hoabinhians were spread across the region from present-day Myanmar to southern China between 12,000 and 3,000 BC. After 3,000 BC Mon-Khmer speaking immigrants moved south into Peninsular Malaysia and introduced a more advanced Neolithic culture, engaging in simple farming. The indigenous people known as Orang Asli, who still live in the remoter, mountainous areas of the northern Malay Peninsula, are considered to be descendents of the Neolithic farmers. Indian traders first visited the Malay Peninsula in the 1st century BC and introduced political ideas, art forms and the Sanskrit language. Hinduism and Buddhism gained a foothold and were practised alongside traditional animist beliefs.

Various Hinduized city-states were established, one of which was located in Kedah. In the 7th century AD Kedah came under the control of the Hinduized Srivijaya empire, centred on Palembang in Sumatra. Srivijaya rule ended in the late 13th century when Sumatra fell to a Javan invasion, after which the king of Sukothai sent forces south into the Malay Peninsula. The Sumatran kingdom of Melayu next ruled over the southern part of the Peninsula, followed by the Madjapahit, the last Hindu empire of Java. In the mid-15th century Melaka emerged as the key trading port in the region—it was host to indigenous Malays, Sumatrans, Javans, Gujaratis, Arabs, Persians, Filipinos and Chinese—and grew rapidly in prosperity. A pattern of government was established in Melaka that became the basis of Malay identity and it was emulated by subsequent Malay kingdoms. Gujarati sailors introduced Islam to the region through

Melaka in the 15th century. In 1511 the port was captured by the Portuguese navigator Alfonso de Albuquerque (who had seized Goa in western India the previous year), and who sought to dominate the route by which precious spices were shipped to Europe.

The sultan of Melaka fled to Johor and some of the Muslim mercantile elite relocated to Brunei in northwest Borneo. Sultanates also emerged in Pahang and Perak, which subsequently received large numbers of immigrants from Indonesian islands, notably Acehnese, Bugis and Minangkabau settlers, who displaced the Orang Asli from their coastal communities and drove them to the Malay Peninsula's interior. Conflict arose between the sultanates of Johor and Aceh and the Portuguese as they vied for control over the Straits of Melaka. In the late 16th century the northern Peninsular states of Kedah, Kelantan and Terengganu came under the control of the Thai state of Phetburi. The early 17th century saw the arrival of Dutch traders in the strait of Melaka. As part of the United Netherlands East India Company (Vereenigde Oostindische Compagnie, VOC) they made an alliance with Johor to besiege Melaka, capturing it in 1641. The Dutch brokered a peace deal between Aceh and Johor in the same year, ushering in an era of relative peace and prosperity for Johor under Laksamana Tun Abdul Jamil.

In the late 17th century the Malay Peninsula came under the influence of Bugis merchants from the Indonesian island of Sulawesi, who began settling in Selangor to trade in tin. The Bugis were formidable warriors, renowned for their navigational and commercial skills. By the 1740s they controlled many of the key shipping routes across the Indonesian archipelago and influenced all areas of government in Johor and the Riau archipelago, although Sultan Suliaman was permitted to remain as a figurehead.

British Influence

In the mid-18th century Johor and Melaka became entrepôts for the trade in tea between China and Europe. Ships owned by the British East India Company (EIC) began plying the Melaka straits in greater numbers. The British foothold in India allowed them to expand eastwards, and their control of India's poppy fields enabled them to dominate the lucrative opium trade. In 1786 Francis Light of the EIC leased the island of Penang from the Sultan Abdullah of Kedah who hoped the British would provide protection against attacks from Siam or Burma. Penang grew swiftly, luring trade away from Melaka (which remained in Dutch control) and Johor-Riau. The British sought to increase their control over the maritime route to China and Sir Thomas Stamford Raffles was ordered to establish an entrepôt in the southern reaches of the Melaka Straits. In 1819 he signed a treaty with Sultan Husein Syah of Johor and founded Singapore. Five years later the British formally acquired Melaka from the Dutch. From 1826 Penang, Singapore and Melaka were ruled by the British authorities in India under a joint administration known as the Straits Settlements. By 1831 the population of Singapore had reached 18,000 (a large proportion were Chinese immigrants) and the following year the port replaced Penang as the capital of the Straits Settlements. Meanwhile, the northern provinces of Kedah and Perak came under the influence of the Siamese Chakri dynasty.

The discovery of tin deposits at Larut (western Malay Peninsula) in the 1850s led to large-scale immigration by Chinese miners and labourers. They organized themselves into *hui* (brotherhoods), which eventually became powerful political and economic organizations. Vast profits could be made from tin and there were clashes between rival developers. At the same time, piracy was on the increase in the Melaka straits, and merchants asked the British to intervene and restore order. A series of agreements in 1874 introduced the British Residential system to Perak, Selangor and Sungei Ujung. In each region, a British Resident functioned as an adviser to the Malay Sultan on all aspects of administration apart from matters relating to the Islamic faith and Malay tradition.

Colonial Rule

In 1896 the three states and Pahang were grouped together as the Federated Malay States, presided over by a British Resident-General at Kuala Lumpur in the heart of the tin-mining district. By the end of the century, a British colonial infrastructure was taking shape in the form of public buildings, municipal services, rubber plantations and road and rail construction, which required a stream of cheap workers. Tamils from south India and Sri Lanka arrived as indentured (and later as licensed) labourers. Negotiations between the British and Siamese in the early years of the 20th century led to British control over the northern states of Kedah, Perlis, Kelantan and Terengganu. A rubber boom followed the expansion of the motor car industry in Europe and North America. Rubber plants, originally from the forests of Brazil and introduced to Malaysia in the 1880s, were planted in every state in Malaysia by 1908 and by 1913 rubber had eclipsed tin as the country's chief export.

Sabah, Sarawak and Brunei, which had come under the control of the North Borneo Chartered Company and was granted protectorate status in 1888, experienced slower economic development than British Malaya but gold, antimony and coal were mined and oil was discovered at Miri in 1910. The colony was hit hard by the global depression of 1929–31 and widespread unemployment in the mines and plantations caused the repatriation of Chinese and Indian workers. Rubber, tin and oil made Malaya a focus for Imperial Japan from early in the Second World War. When Pearl Harbor and Hong Kong came under attack from Japanese forces in Dec. 1941, other Japanese divisions came ashore at Kota Bharu and Miri. British forces retreated south to Singapore, but the 'impregnable' island capitulated within a few weeks, on 15 Feb. 1942. Japanese troops quickly took over from British colonial officers and controlled Malaya from Singapore (Shonan), meting out harsh treatment to the Chinese population. Thailand allied itself with Japan and was granted control of the northern Malay states in 1943.

Post-War Period

Returning in 1946, the British reorganized the colony into the Malayan Union. The Malay elite, fearing an end to their privileges as a consequence of equal rights for Chinese and Indian subjects, campaigned via the United Malays National Organization (UMNO, led by Datuk Onn) to demand the continuation of the sultanates. The British were forced to compromise and established the Federation of Malaya in Feb. 1948, consisting of the nine Malay states, Melaka and Penang and administered by a High Commissioner in Kuala Lumpur. Within months the Federation was under attack by the Chinese-dominated Malayan Communist Party (MCP) which had grown during the Japanese occupation, when it controlled various anti-Japanese National Salvation Organizations. In 1950 the High Commissioner, Sir Henry Gurney, responded to guerrilla attacks by the MCP by putting Malaya on a war-footing, including tightening security, recruiting soldiers from other colonies and dispersing the Chinese squatter settlements that harboured the MCP. More than 500,000 Chinese were resettled by the mid-1950s. The Communist insurrection, known by the British as 'the Emergency', hastened the transition to Malayan independence and local elections were held in Penang in late 1951. Four years later the first federal-level election was won convincingly by the Alliance Party, a loose coalition of Malay, Chinese and Indian parties, led by Tuanku (Prince) Abdul Rahman. On 31 Aug. 1957 the Federation of Malaya became an independent state with Tuanku Abdul Rahman as its first premier.

The concept of Malaysia as a broader federation including Sabah, Sarawak, Singapore and the British protectorate of Brunei, was first suggested by Abdul Rahman in 1961. It was opposed by neighbouring Indonesia and the Philippines, but public support in Sabah and Sarawak led to Malaysia's formation in Sept. 1963, although Brunei declined to join. The new nation faced continuing hostility from Indonesia, led by Ahmed Sukarno, over the sovereignty of Borneo. Disagreements with Singapore's Prime Minister Lee Kuan Yew ended with Singapore declaring independence in 1965. Tension arose between the Chinese and Malay communities over the use of the Malay language and Malay fears about Chinese economic dominance. The 1969 elections were fought on the highly emotional issues of education and language. When the Alliance party failed to obtain a majority, rioting and serious inter-ethnic violence followed and an emergency government was set up, led by Deputy Prime Minister Abdul Razak. Parliamentary rule was restored in 1971 when Razak launched the New Economic Policy—a series of five-year plans to eradicate poverty and restructure society to improve ethnic relations, specifically by encouraging ethnic Malays, the *bumiputera*, to shift from subsistence agriculture into the mainstream economy.

Mahathir Mohamad was the first non-royal or non-aristocrat to become prime minister of Malaysia, winning the 1981 elections for the UMNO and leading the National Front coalition to further victories in 1986, 1990, 1995 and 1999. Mahathir shifted the economy away from dependence on commodities and towards manufacturing, services and tourism, aided by substantial Japanese and east Asian investment in manufacturing. The prolonged spell of economic growth and stability was broken by the 1997–98 recession but Mahathir refused to accept financial aid from the International Monetary Fund. In Sept. 1998 Mahathir dismissed Anwar Ibrahim, his finance minister, deputy prime minister and heir apparent. Anwar was found guilty of corruption charges in 1999 and sentenced to prison for six years. In 2002 Mahathir announced that he would resign from the presidency of UMNO and he stepped down as prime minister on 31 Oct. 2003, to be succeeded by Abdullah Ahmad Badawi. Badawi won a landslide victory in the March 2004 general elections for the National Front. In Sept. 2004 Anwar was unexpectedly released after being acquitted by the Federal Court. Najib Razak succeeded Badawi in 2009, winning re-election in 2013.

TERRITORY AND POPULATION

The federal state of Malaysia comprises the 13 states and three federal territories of Peninsular Malaysia, bounded in the north by Thailand, and with the island of Singapore as an enclave on its southern tip; and, on the island of Borneo to the east, the state of Sabah (which includes the federal territory of the island of Labuan), and the state of Sarawak, with Brunei as an enclave, both bounded in the south by Indonesia and in the northwest and northeast by the South China and Sulu Seas.

The area of Malaysia is 330,803 sq. km (127,724 sq. miles), and the 2010 census population 28,334,135; density, 85·7 per sq. km. Malaysia's national waters cover 515,256 sq. km. In 2011, 73·0% of the population lived in urban areas.

The UN gives a projected population for 2015 of 30·71m.

The growth of the population has been:

Year	Peninsular Malaysia	Sarawak	Sabah/Labuan	Total Malaysia
1980	10,944,844	1,235,553	955,712	13,136,109
1991	14,131,723	1,642,771	1,788,926	17,563,420
2000	17,649,266	2,009,893	2,539,117	22,198,276
2010	22,569,345	2,471,140	3,293,650	28,334,135

The areas, populations and chief towns of the states and federal territories are:

Peninsular states	Area (in sq. km)	Population (2010 census)	Chief town	Population (2000 census)
Johor	19,210	3,348,283	Johor Bharu	642,944
Kedah	9,500	1,947,651	Alor Star	186,433
Kelantan	15,099	1,539,601	Kota Bharu	251,801
Kuala Lumpur[1]	243	1,674,621	Kuala Lumpur	1,305,792[1]
Melaka	1,664	821,110	Melaka	151,082
Negeri Sembilan	6,686	1,021,064	Seremban	290,709
Pahang	36,137	1,500,817	Kuantan	288,727
Perak	21,035	2,352,743	Ipoh	536,832
Perlis	821	231,541	Kangar	54,282
Pulau Pinang (Penang)	1,048	1,561,383	Penang (Georgetown)	181,380
Putrajaya[1]	46	72,413	Putrajaya	—
Selangor	8,108	5,462,141	Shah Alam	314,440
Terengganu	13,035	1,035,977	Kuala Terengganu	255,518
Other states				
Labuan[1]	91	86,908	Victoria	—
Sabah	73,631	3,206,742	Kota Kinabalu	306,920
Sarawak	124,450	2,471,140	Kuching	422,240

[1]Federal territory.

Other large cities (2000 census population): Klang (626,699), Ampang Jaya (478,613), Subang Jaya (437,121), Petaling Jaya (432,619), Sandakan (276,791) and Tawau (213,745).

Putrajaya, a planned new city described as an 'intelligent garden city', became the administrative capital of Malaysia in 1999 and was created a federal territory on 1 Feb. 2001.

Bahasa Malaysia (Malay) is the official language of the country —50% of the population are Malays. The government promotes the use of the national language to foster national unity. However, the people are free to use their mother tongue and other languages. English as the second language is widely used in business. In Peninsular Malaysia Chinese dialects and Tamil are also spoken. In Sabah there are numerous tribal dialects and Chinese (Mandarin and Hakka dialects predominate). In Sarawak Mandarin and numerous tribal languages are spoken. In addition to Malays, 22% of the population are Chinese, 12% other indigenous ethnic groups, 7% Indians and 9% others.

SOCIAL STATISTICS

2007 estimated births, 481,000; deaths, 120,000. 2007 rates (per 1,000 population): birth, 18·1; death, 4·5. Life expectancy, 2007: males, 71·9 years; females, 76·6 years. Annual population growth rate, 2000–08, 1·9%. Infant mortality, 2010, five per 1,000 live births; fertility rate, 2007, 2·3 births per woman. Today only 6% of Malaysians live below the poverty line, compared to 50% in the early 1970s.

CLIMATE

Malaysia lies near the equator between latitudes 1° and 7° North and longitudes 100° and 119° East. Malaysia is subject to maritime influence and the interplay of wind systems which originate in the Indian Ocean and the South China Sea. The year is generally divided into the South-East and the North-East Monsoon seasons. The average daily temperature throughout Malaysia varies from 21°C to 32°C. Humidity is high.

CONSTITUTION AND GOVERNMENT

The Constitution of Malaysia is based on the Constitution of the former Federation of Malaya, but includes safeguards for the special interests of Sabah and Sarawak. It was amended in 1983. The Constitution provides for one of the Rulers of the Malay States to be elected from among themselves to be the *Yang di-Pertuan Agong* (Supreme Head of the Federation). He holds office for a period of five years. The Rulers also elect from among

themselves a Deputy Supreme Head of State, also for a period of five years. In Feb. 1993 the Rulers accepted constitutional amendments abolishing their legal immunity.

Supreme Head of State (Yang di-Pertuan Agong). Tuanku Abdul Halim Muadzam Shah ibni Al-Marhum Sultan Badlishah, b. 1927, acceded 13 Dec. 2011.

Sultan of Johor Tuanku Ibrahim Ismail ibni Al-Marhum Sultan Iskandar, b. 1958, acceded 23 Jan. 2010.

Sultan of Kedah Tuanku Abdul Halim Muadzam Shah ibni Al-Marhum Sultan Badlishah, b. 1927, acceded 15 July 1958.

Sultan of Kelantan. Sultan Muhammad V (Tuanku Muhammad Faris Petra Ibni Sultan Ismail Petra), b. 1979, acceded 13 Sept. 2010 (but disputed by Tuanku Ismail Petra Sultan Yahya Petra— the previous sultan and his ailing father).

Yang Di-Pertuan Besar Negeri Sembilan. Tuanku Muhriz ibni Al-Marhum Tuanku Munawir, b. 1948, acceded 29 Dec. 2008.

Sultan of Pahang. Sultan Haji Ahmad Shah Al-Musta'in Billah ibni Al-Marhum Sultan Abu Bakar Ri'Ayatuddin Al-Mu'Adzam Shah, b. 1930, acceded 7 May 1974.

Sultan of Perak. Sultan Azlan Muhibbuddin Shah ibni Al-Marhum Sultan Yusuff Izzuddin Shah Ghafarullah, b. 1928, acceded 3 Feb. 1984.

Raja of Perlis. Tuanku Syed Sirajuddin ibni Al-Marhum Syed Putra Jamalullail, b. 1943, acceded 17 April 2000.

Sultan of Selangor. Sultan Sharafuddin Idris Shah Alhaj ibni Al-Marhum Sultan Salahuddin Abdul Aziz Shah Alhaj, b. 1945, acceded 22 Nov. 2001.

Sultan of Terengganu. Tuanku Mizan Zainal Abidin ibni al-Mahrum Sultan Mahmud Al-Muktafi Billah Shah, b. 1962, acceded 15 May 1998.

Yang di Pertua Negeri Melaka. Tun Mohd Khalil bin Yaakob, b. 1937, appointed 4 June 2004.

Yang di-Pertua Negeri Pulau Pinang (Penang). Tun Haji Abdul Rahman bin Haji Abbas, b. 1938, appointed 1 May 2001.

Yang di-Pertua Negeri Sabah. Tun Haji Juhar bin Haji Mahiruddin, b. 1953, appointed 1 Jan. 2011.

Yang di-Pertua Negeri Sarawak. Tun Datuk Patinggi Abang Haji Muhammad Salahuddin, b. 1921, appointed 4 Dec. 2000.

The federal parliament consists of the *Yang di-Pertuan Agong* and two *Majlis* (Houses of Parliament) known as the *Dewan Negara* (Senate) of 70 members (26 elected, two by each state legislature; and 44 appointed by the *Yang di-Pertuan Agong*) and the *Dewan Rakyat* (House of Representatives) of 222 members. Appointment to the Senate is for three years. The maximum life of the House of Representatives is five years, subject to its dissolution at any time by the *Yang di-Pertuan Agong* on the advice of his Ministers.

National Anthem

'Negaraku' ('My Country'); words collective, tune by Pierre de Béranger.

GOVERNMENT CHRONOLOGY

Supreme Heads of State since 1957.

1957–60	Tuanku Abdul Rahman ibni al-Marhum
1960	Tuanku Hisamuddin Alam Shah ibni al-Marhum
1960–65	Syed Harun Petra ibni al-Marhum
1965–70	Tuanku Ismail Nasiruddin Shah ibni al-Marhum
1970–75	Tuanku Abdul Halim Muadzam Shah ibni al-Marhum
1975–79	Tuanku Yahaya Petra ibni al-Marhum
1979–84	Tuanku Ahmad Shah al-Mustain Billah ibni al-Marhum
1984–89	Tuanku Mahmud Iskandar ibni al-Marhum
1989–94	Tuanku Azlan Muhibuddin Shah ibni al-Marhum

1994–99	Tuanku Jaafar ibni al-Marhum
1999–2001	Tuanku Salehuddin Abdul Aziz Shah ibni al-Marhum
2001–06	Tuanku Syed Sirajuddin ibni al-Marhum
2006–11	Tuanku Mizan Zainal Abidin ibni al-Marhum Sultan Mahmud
2011–	Tuanku Abdul Halim Muadzam Shah ibni al-Marhum Sultan Badlishah

Prime Ministers since 1957. (UMNO = United Malays National Organization)

1957–59	UMNO	Tunku Abdul Rahman Putra
1959	UMNO	Tun Abdul Razak bin Hussein (acting)
1959–70	UMNO	Tunku Abdul Rahman Putra
1970–76	UMNO	Tun Abdul Razak bin Hussein
1976–81	UMNO	Hussein bin Onn
1981–2003	UMNO	Mahathir bin Mohamad
2003–09	UMNO	Abdullah bin Haji Ahmad Badawi
2009–	UMNO	Najib Tun Razak

RECENT ELECTIONS

Elections to the *Dewan Rakyat* and 12 state assemblies (except Sarawak) were held on 5 May 2013. The 13-party National Front coalition (BN; Barisan Nasional) gained 133 seats, obtaining 46·7% of the votes cast (the predominant partner, the United Malays National Organization/UMNO, winning 88 seats). The opposition People's Front coalition (PR; Pakatan Rakyat) won 89 seats with 50·1% of the vote (with the Democratic Action Party winning 38 seats and the People's Justice Party of leader of the opposition Anwar Ibrahim winning 30). The National Front coalition gained a majority in nine of the 12 state assemblies.

CURRENT GOVERNMENT

In March 2013 the government comprised:

Prime Minister and Minister of Finance: Dato' Sri Haji Mohd Najib bin Tun Haji Abdul Razak; b. 1953 (UMNO; took office on 3 April 2009 and re-elected in May 2013).

Deputy Prime Minister and Minister of Education: Tan Sri Dato' Haji Muhiyiddin bin Mohamed Yassin. *Minister of Transport:* Datuk Seri Kong Cho Ha. *Plantation Industries and Commodities:* Tan Sri Bernard Giluk Dompok. *Home Affairs:* Dato' Seri Hishammuddin bin Tun Hussein. *Information, Communications, Arts and Culture:* Dato' Seri Utama Dr Rais Yatim. *Energy, Green Technology and Water:* Datuk Peter Chin Fah Kui. *Rural and Regional Development:* Dato' Seri Haji Mohd Shafie bin Haji Apdal. *Higher Education:* Dato' Seri Mohamed Khaled bin Nordin. *International Trade and Industry:* Dato' Mustapa bin Mohamed. *Science, Technology and Innovation:* Datuk Dr Maximus Johnity Ongkili. *Natural Resources and the Environment:* Datuk Douglas Uggah Embas. *Tourism:* Dato' Sri Dr Ng Yen Yen. *Agriculture and Agro-Based Industry:* Dato' Haji Noh bin Omar. *Defence:* Dato' Seri Dr Ahmad Zahid bin Hamidi. *Works:* Dato' Shaziman bin Abu Mansor. *Health:* Dato' Sri Liow Tiong Lai. *Youth and Sports:* Dato' Ahmad Shabery Cheek. *Human Resources:* Datuk Dr S. Subramaniam. *Domestic Trade and Consumer Affairs:* Dato' Sri Ismail Sabri bin Yaakob. *Women, Family and Community Development:* Dato' Seri Shahrizat Abdul Jalil. *Foreign Affairs:* Datuk Anifah bin Haji Aman. *Federal Territories:* Dato' Raja Nong Chik bin Dato' Raja Zainal Abidin. *Housing and Local Government:* Datuk Wira Chor Chee Heung. *Second Minister of Finance:* Dato' Haji Ahmad Husni bin Mohamad Hanadzlah. *Ministers in Prime Minister's Department:* Tan Sri Dr Koh Tsu Koon; Dato' Seri Mohamad Nazri bin Abdul Aziz; Tan Sri Nor Mohamed bin Yakcop; Dato' Maj.-Gen. Jamil Khir bin Baharom; Dato' Sri Idris Jala; Datuk G. Palanivel.

Office of the Prime Minister: http://www.pmo.gov.my

CURRENT LEADERS

Tuanku Abdul Halim Muadzam Shah ibni Sultan Badlishah

Position
King (Yang di-Pertuan Agong)

Introduction
Tuanku (His Majesty) Abdul Halim was installed as the 14th Yang di-Pertuan Agong on 13 Dec. 2011 following his election to the five-year post by the Conference of Rulers in Oct. 2011. He had previously held the largely ceremonial role from 1970–75.

Early Life
Tuanku Abdul Halim Abdul Halim Muadzam Shah was born on 28 Nov. 1927 in Anak Bukit, Kedah. He received his early education at Alor Merah and Titi Gajah Malay schools and from 1946–48 studied at the Sultan Abdul Hamid College, Alor Star. On 6 Aug. 1949 he was appointed Raja Muda (heir apparent) to his father, the Sultan of Kedah.

In 1952 Tuanku Abdul Halim attended Wadham College, Oxford, where he obtained a diploma in social science and public administration. On his return to Malaya, he joined the Kedah Administrative Service, serving with the Alor Star district office and the state treasury.

On his father's death on 15 July 1958, Tuanku Abdul Halim became the 28th Sultan of Kedah. He was crowned on 20 Feb. 1959. On 21 Sept. 1965 he was selected as deputy Yang di-Pertuan Agong by the Conference of Rulers and on 21 Sept. 1970 was appointed to the throne as the fifth Yang di-Pertuan Agong. He completed his term on 20 Sept. 1975. During his tenure he presided over the first transition of civilian governments when the prime minister, Tunku Abdul Rahman, resigned in favour of his deputy, Tun Abdul Razak.

On 2 Nov. 2006 Tuanku Abdul Halim was elected deputy Yang di-Pertuan Agong for the second time. At a special meeting of the Conference of Rulers on 14 Oct. 2011 he was selected to serve a second term as the Yang di-Pertuan Agong. He became the first of the nine hereditary state rulers of the Conference of Rulers to serve two terms.

Career in Office
The role of Yang di-Pertuan Agong is mainly ceremonial although royal assent is necessary for all laws passed in parliament and the appointment of members of the cabinet and the judiciary. Tuanku Abdul Halim also serves as the nominal commander-in-chief and as the Islamic head of a large population of Malay Muslims. He took on the role at a time when race and religious relations in Malaysia were fragile.

Dato' Sri Haji Mohd Najib bin Tun Haji Abdul Razak

Position
Prime Minister

Introduction
Dato' Sri Haji Mohd Najib bin Tun Haji Abdul Razak became prime minister on 3 April 2009, replacing Abdullah bin Haji Ahmad Badawi, who left office following the poor showing of the Barisan Nasional (National Front Coalition) at the general election of March 2008. Najib had replaced Badawi as head of the United Malays National Organization (UMNO), the senior party in the coalition, in March 2009. Najib won a new term after electoral victory in May 2013.

Early Life
Najib was born on 23 July 1953 in Kuala Lipis, Pahang, into a political family. His father was independent Malaysia's second prime minister and his uncle was its third. Najib was educated at St John's Institution, Kuala Lumpur, and at Malvern College in

England before graduating from the University of Nottingham in 1974 with a bachelor's degree in industrial economics.

In 1976 he became Malaysia's youngest member of parliament when he stood uncontested for his late father's seat of Pekan. In his first year as an MP, Najib was appointed deputy minister of energy, telecommunications and posts. He was later appointed deputy minister of education and deputy minister of finance. In 1981 he joined UMNO's Supreme Council and the following year became the Menteri Besar (Chief Executive) of Pahang state.

Najib became vice president of UMNO Youth in 1982, a post he also held from 1987–93. Having lost his Pekan parliamentary seat, he regained it at the elections of 1986 and was appointed minister of culture, youth and sports. He went on to hold several other cabinet portfolios including defence and education. On 7 Jan. 2004 Najib was selected as Badawi's deputy and given the defence portfolio. In July 2004 he stood unopposed for the vice presidency of UMNO.

Although Barisan Nasional won the election of 2008, it was with a much reduced majority. Badawi named Najib as his likely successor and on 17 Sept. 2008 Najib was handed the finance portfolio as part of a gradual power transfer. On 26 March 2009 he stood unopposed for the UMNO presidency, ensuring him the post of prime minister when Badawi resigned on 2 April 2009.

Career in Office
Najib took office promising reform and change but faced the immediate challenges of a severe economic downturn, a divided UMNO and the increasing unpopularity of the ruling coalition that opponents accused of corruption and complacency. He slimmed the cabinet from 32 ministers to 28, but a financial stimulus of RM60bn. was met with a lukewarm response. He has since faced controversies relating to allegations of an ongoing political conspiracy against former deputy prime minister Anwar Ibrahim, the murder of a Mongolian woman in which several associates (including his wife) were implicated, and problems with the national service programme that he devised and in which several conscripts have died.

In March 2010 Najib unveiled details of a New Economic Model, intended to more than double per capita income by 2020. The plan aimed to promote greater private-sector investment and to revise a controversial ethnically-based affirmative action policy as the nation strives to become a high-income economy.

In response to anti-government demonstrations in Kuala Lumpur in July 2011, which met with a heavy police response, Najib announced in Aug. that year the establishment of a parliamentary committee to consider the protesters' demands for electoral reform. The following month he promised to repeal longstanding internal security legislation and to relax press censorship.

At the parliamentary election of 5 May 2013, Najib led Barisan Nasional to victory, winning 133 seats to the 89 of the People's Front coalition. He was sworn in for a new term as prime minister the following day.

DEFENCE
The Constitution provides for the Head of State to be the Supreme Commander of the Armed Forces who exercises his powers in accordance with the advice of the Cabinet. Under their authority, the Armed Forces Council is responsible for all matters relating to the Armed Forces other than those relating to their operational use. The Ministry of Defence has established bilateral defence relations with countries within as well as outside the region. Malaysia is a member of the Five Powers Defence Arrangement with Australia, New Zealand, Singapore and the UK.

The Malaysian Armed Forces has participated in 25 UN peacekeeping missions in Africa, the Middle East, Indo-China and Europe.

Since 2004 a lottery system has been in place to choose conscripts to serve three months of national service. In 2008 defence expenditure totalled US$4,370m. (US$173 per capita), representing 2·0% of GDP.

Army
Strength (2007) about 80,000. There is a paramilitary Police General Operations Force of 18,000 and a People's Volunteer Corps of 240,000 of which some 17,500 are armed.

Navy
The Royal Malaysian Navy is commanded by the Chief of the Navy from the integrated Ministry of Defence in Kuala Lumpur. The main base is at Lumut, with other bases at Kuantan, Labuan, Sandakan, Semporna, Sepanggar and Tanjung Pengelih. Further bases are under construction at Langkawi and Sejingkat. The peacetime tasks include fishery protection and anti-piracy patrols. The fleet includes two submarines, three frigates and eight corvettes. The two Scorpene-class submarines were jointly built by French and Spanish firms and were delivered in Sept. 2009 and July 2010 respectively. A Naval aviation squadron operates six helicopters although serviceability is in doubt.

Navy personnel in 2007 totalled 14,000 including 160 Naval Air personnel. There were 1,000 naval reserves.

In addition, there is a maritime enforcement agency some 4,500 strong and 2,100 marine police.

Air Force
Formed on 1 June 1958, the Royal Malaysian Air Force is equipped primarily to provide air defence and air support for the Army, Navy and Police. Its secondary role is to render assistance to government departments and civilian organizations.

Personnel (2007) totalled 15,000, with 68 combat-capable aircraft including F-5Es, MiG-29s and British Aerospace *Hawks*. There were 600 Air Force reserves.

INTERNATIONAL RELATIONS
Malaysia was in dispute with Indonesia over sovereignty of two islands in the Celebes Sea. Both countries agreed to accept the Judgment of the International Court of Justice which decided in favour of Malaysia in Dec. 2002.

ECONOMY
In 2009 agriculture accounted for 9·5% of GDP, industry 43·8% and services 46·7%.

Overview
Malaysia has transformed itself from an economy dependent on mineral production and agriculture into one based on industrialization and manufacturing. The electrical and electronics industry accounted for 34·1% of the country's total exports in 2011, making Malaysia vulnerable to fluctuations in global electronics demand. It also exports oil and gas and has benefited from high energy prices. The oil and gas reserves of the state oil company, Petronas, rank 28th in the world. The company is the country's largest source of revenue and accounts for as much as 45% of the government budget, although efforts are underway to reduce this dependency.

In the decade up to the 1997 Asian financial crisis, the economy grew at an annual rate of 7–10%. In 1998 it was hit hard by twin currency and banking crises, shrinking by 7·4%. In response, the government sought to boost growth via spending on large infrastructure projects.

When global recession cut export demand in 2001, the country managed to avoid recession by introducing a US$1·9bn. fiscal stimulus package. Despite continued weak global demand and the outbreak of SARS, the economy recovered in 2002 with growth of 5·4%. A rebound in the global economy and electronics demand buoyed the economy in 2004 and 2005. The depegging of the ringgit from the dollar in July 2005 was seen as an important step toward broad-based growth.

The economy was hit hard by the global financial crisis in 2008 and experienced a sharp downturn in 2009. However, a fiscal stimulus package and revivals in external demand, private consumption and investment helped drive a strong recovery in 2010, with FDI inflows returning to pre-crisis levels. The recovery has since moderated, with export growth and GDP growth both slowing. The financial system has remained sound but gross household debt increased by 12·3% to 76% of GDP between 2008 and 2011. Owing to its openness, the economy is at risk from continued low growth in major advanced economies and from exposure to the European banking crisis.

The IMF recommends reforms to improve the business climate, enhance competition and upgrade workers' skills. The government aims to bring Malaysia to advanced country status by 2020.

Currency
The unit of currency is the Malaysian *ringgit* (RM) of 100 *sen*. For seven years it was pegged to the US dollar at 3·8 ringgit = 1 US$, but since the revaluation of the Chinese yuan on 21 July 2005 it has been allowed to operate in a managed float. Foreign exchange reserves were US$92,217m. and gold reserves 1·17m. troy oz in Sept. 2009. Inflation rates (based on IMF statistics):

2002	2003	2004	2005	2006	2007	2008	2009	2010	2011
1·8%	1·1%	1·4%	3·0%	3·6%	2·0%	5·4%	0·6%	1·7%	3·2%

Total money supply in April 2009 was RM179,274m.

Budget
Budgetary central government revenue and expenditure (in RM1m.):

	2008	2009	2010
Revenue	159,793	158,639	159,653
Expenditure	150,664	154,485	149,763

Taxes accounted for 68·6% of revenues in 2010; compensation of employees accounted for 31·2% of expenditures, grants 18·6% and use of goods and services 15·9%.

There is a sales tax of 6–10%.

Performance
Malaysia was badly affected by the Asian financial crisis, with the economy contracting by 7·4% in 1998. The country narrowly avoided a recession in 2001 but failed to do so in the global financial crisis of 2009. Real GDP growth rates (based on IMF statistics):

2003	2004	2005	2006	2007	2008	2009	2010	2011
5·8%	6·8%	5·0%	5·6%	6·3%	4·8%	−1·5%	7·2%	5·1%

Total GDP in 2010 was US$237·8bn.

Banking and Finance
The central bank and bank of issue is the Bank Negara Malaysia (*Governor*, Dr Zeti Akhtar Aziz). In 2002 there were 47 domestic commercial banks, merchant banks and finance companies. Total deposits of commercial banks, finance companies and merchant banks at 31 Dec. 2005 were RM140·6bn. The largest commercial bank is Malayan Banking Berhad (Maybank), with assets in 2009 of RM238·3bn. The Islamic Bank of Malaysia began operations in July 1983. In Jan. 2006 there were 54 banks licensed by the Labuan Offshore Financial Services Authority (LOFSA).

External debt amounted to US$81,497m. in 2010, up from US$51,855m. in 2005 and US$41,765m. in 2000—this was equivalent to 35·4% of GNI, compared to 48·5% in 2000.

Malaysia attracted a record US$11,966m. worth of foreign direct investment in 2011, up from US$9,103m. in 2010 and US$1,453m. of 2009.

There is a stock exchange at Kuala Lumpur, known as Bursa Malaysia.

ENERGY AND NATURAL RESOURCES

Environment
Malaysia's carbon dioxide emissions from the consumption and flaring of fossil fuels in 2008 were the equivalent of 6·4 tonnes per capita.

Electricity
Installed capacity in 2007, 23·0m. kW. In 2007, 101,325m. kWh were generated (94,840m. kWh thermal and 6,485m. kWh hydro-electric). Consumption per capita in 2007 was 3,574 kWh.

Oil and Gas
Oil reserves, 2008, 5·5bn. bbls. Oil production, 2008, was 34·3m. tonnes. In 2007 exports of oil stood at 16·4m. tonnes. Natural gas reserves, 2008, 2,390bn. cu. metres. Production of natural gas in the same year was 62·5bn. cu. metres. In April 1998 Malaysia and Thailand agreed to share equally the natural gas jointly produced in an offshore area (the Malaysian-Thailand Joint Development Area) which both countries claim as their own territory.

Minerals
In 2004 mining contributed 7·0% of GDP. Production in 2004 (in tonnes): aggregate, 51,236,000; limestone, 19,968,000; sand and gravel, 18,371,000; iron ore, 663,732; silica sand, 631,402; coal, 389,176; kaolin, 326,928; ilmenite concentrate, 61,471; gold, 4,221 kg.

Agriculture
In 2002 agriculture contributed 9·2% of GDP. There were an estimated 1·80m. ha. of arable land in 2002 and 5·79m. ha. of permanent crops. In 2002 approximately 365,000 ha. were irrigated. Production in 2003 (in 1,000 tonnes): palm kernels, 3,550; rice, 2,145; sugarcane, 1,600; coconuts, 740; rubber, 589; bananas, 500; cassava, 370; pineapples, 255. Livestock (2002): pigs, 1·82m.; cattle, 748,000; goats, 248,000; sheep, 118,000; buffaloes, 154,000; chickens, 161m. Oil palms account for 75% of Malaysia's agricultural area. Malaysia's output of palm kernels is the highest of any country.

Forestry
In 2010 there were 20·46m. ha. of forests, or 62% of the total land area, down from 22·38m. ha. in 1990 and 21·59m. ha. in 2000. Of the total area under forests 19% was primary forest in 2010, 72% other naturally regenerated forest and 9% planted forest. Timber production in 2007 was 25·15m. cu. metres.

Fisheries
Total catch in 2010 amounted to 1,433,427 tonnes, almost entirely from sea fishing. There were 31,503 licensed motorized fishing vessels in 2010, up 44% from 21,871 in 2005; non-motorized vessels fell from 170 to 89 in the same period. Imports of fishery commodities were valued at US$582m. in 2008 and exports at US$777m.

INDUSTRY

The leading companies by market capitalization in Malaysia in March 2012 were: Malayan Banking (US$22·1bn.); Sime Darby, a diversified conglomerate (US$19·1bn.); and CIMB Group Holdings, a financial services provider (US$18·7bn.).

In 2001 industry accounted for 48·3% of GDP, with manufacturing contributing 30·5%. Production figures for 2001 (in 1,000 tonnes): cement, 13,820; palm oil (2002), 11,909; distillate fuel oil (2007), 8,891; petrol (2007), 5,029; residual fuel oil (2007), 2,006; refined sugar, 1,210; wheat flour, 664; plywood

(2002), 4,341,000 cu. metres; cigarettes, 25·6bn. units; radio sets, 28·8m. units; pneumatic tyres, 13·1m. units; TV sets, 9·5m. units.

Labour
In 2001 the workforce was 9,892,000 (46·7% female in 2000), of whom 9,535,000 were employed (22·6% in manufacturing, 14·2% in agriculture, forestry and fishing, 10·5% in government services and 8·9% in construction). Unemployment was 3·8% in 2002. It is estimated that Malaysia has some 500,000 illegal workers.

INTERNATIONAL TRADE
Privatization policy permits foreign investment of 25–30% generally; total foreign ownership is permitted of export-oriented projects.

Imports and Exports
In 2006 imports totalled US$124,144m. and exports US$160,842m. The trade surplus in 2006 was US$36·7bn., up from US$18·1bn. in 2002.

Main imports, 2006: electrical machinery, apparatus and appliances, 31·9%; manufactured goods, 11·6%; petroleum and petroleum products, 8·3%; chemicals and related products, 7·8%. Chief exports, 2006: electrical machinery, apparatus and appliances, 21·3%; office machines and computers, etc., 17·4%; telecommunications, sound recording and reproducing equipment, 9·0%; petroleum and petroleum products, 8·9%.

The principal import sources in 2006 were: Japan (13·2%), USA (12·5%), China (12·1%), Singapore (11·7%). The leading export markets were: USA (18·8%), Singapore (15·4%), Japan (8·9%), China (7·2%).

COMMUNICATIONS

Roads
Total road length in 2004 was 109,333 km, of which 82·8% were paved. In 2006 there were 7,024,000 passenger cars in use, 60,000 buses and coaches, 836,600 lorries and vans, and 7,458,100 motorcycles and mopeds. There were 6,287 deaths as a result of road accidents in 2006, which at 24·1 per 100,000 people ranks among the highest rates in the world.

Rail
Length of route in 2004, 1,667 km, of which 150 km were electrified. The Malayan Railway carried 34·6m. passengers and 4·1m. tonnes of freight in 2005; the Sabah State Railway carried 347,000 passengers and 500,000 tonnes of freight. A railway from Kuala Lumpur to the international airport opened in 2002 and carried 10m. passengers in 2003. There are two metro systems in Kuala Lumpur with a combined length of 56 km.

Civil Aviation
There are a total of 19 airports of which five are international airports and 14 are domestic airports at which regular public air transport is operated. *International airports:* Kuala Lumpur, Penang, Kota Kinabalu, Kuching and Langkawi. *Domestic airports:* Johor Bharu, Alor Star, Ipoh, Kota Bharu, Kuala Terengganu, Kuantan, Melaka, Sandakan, Lahad Datu, Tawau, Labuan, Bintulu, Sibu and Miri. There are 39 Malaysian airstrips of which ten are in Sabah, 15 in Sarawak and 14 in peninsular Malaysia.

In 2003, 40 international airlines operated through Kuala Lumpur (KLIA-Sepang). Malaysia Airlines, the national airline, is 52% state-owned, and operates domestic flights within Malaysia and international flights to nearly 40 different countries. A low-cost airline, AirAsia, began operations in Nov. 1996; its budget sister long-haul carrier, Air Asia X, started flying in Nov. 2007. Air Asia and its subsidiaries (Air Asia X, Indonesia AirAsia, AirAsia Japan, AirAsia Philippines and Thai AirAsia) now rank among Asia's ten largest airlines. In 2005 scheduled airline traffic of Malaysian-based carriers flew 282·7m. km, carrying 23,026,000 passengers. In 2001 Kuala Lumpur handled 14,208,055 passengers (10,044,013 on international flights) and 423,712 tonnes of freight.

Kota Kinabalu handled 2,912,802 passengers in 2001 and Kuching 2,544,502.

Shipping
The major ports are Port Kelang, Pulau Pinang, Johor Pasir Gudang, Tanjung Beruas, Miri, Rajang, Pelabuhan Sabah, Port Dickson, Kemaman, Teluk Ewa, Kuantan, Kuching and Bintulu. Port Kelang, the busiest port, handled 152,348,000 freight tons of cargo in 2008; container throughput in 2008 was 7,974,000 TEUs (twenty-foot equivalent units), making it Malaysia busiest container port. In Jan. 2009 there were 457 ships of 300 GT or over registered, totalling 6·43m. GT. Of the 457 vessels registered, 189 were general cargo ships, 130 oil tankers, 45 container ships, 34 liquid gas tankers, 26 passenger ships, 20 chemical tankers and 13 bulk carriers. The Malaysian-controlled fleet comprised 307 vessels of 1,000 GT or over in Jan. 2009, of which 242 were under the Malaysian flag and 65 under foreign flags.

Telecommunications
In 2008 there were 4,292,000 main (fixed) telephone lines. In the same year mobile phone subscribers numbered 27,713,000 (1,025·9 per 1,000 persons). There were 15,074,000 internet users in 2008. In March 2012 there were 12·4m. Facebook users.

SOCIAL INSTITUTIONS

Justice
The highest judicial authority and final court of appeal is the Federal Court. There are also two High Courts, one for Peninsular Malaysia and one for the States of Sabah and Sarawak, as well as a system of subordinate courts (comprising Magistrate Courts and Sessions Courts).

The Federal Court comprises the Chief Justice—the head of the Malaysian judiciary—the President of the Court of Appeal, the Chief Judges of the two High Courts and four other judges. It has jurisdiction to determine the validity of any law made by Parliament or by a State legislature and disputes between States or between the Federation and any State. It also has jurisdiction to hear and determine appeals from the High Courts.

The death penalty is authorized, and was reportedly used in 2011. The population in penal institutions in 2007 was 50,305 (192 per 100,000 of national population).

Education
School education is free; tertiary education is provided at a nominal fee. There are six years of primary schooling starting at age seven, three years of universal lower secondary, two years of selective upper secondary and two years of pre-university education. During the Seventh Plan period (1996–2000), a number of major changes were introduced to the education and training system with a view to strengthening and improving the system. These efforts were aimed at improving the quality of output to meet the manpower needs of the nation, particularly in the fields of science and technology. In addition, continued emphasis will be given to expand educational opportunities for those in the rural and remote areas. Under the Seventh Plan, the Education Ministry allocated RM8,437,200 on this education programme and RM1,661,600 for training purposes.

In 2005 there were 3,133,399 pupils at primary schools with 194,872 teaching staff, 2,489,117 pupils with 146,503 teaching staff at secondary schools and (2006) 749,165 students and 39,809 academic staff at higher education institutions.

Adult literacy was 92% in 2008.

In 2006 public expenditure on education came to 4·7% of GNI.

Health
In 2001 there were 15,619 doctors, 2,144 dentists, 31,129 nurses and 2,333 pharmacists. In 2001 the Ministry of Health ran a total of 855 health clinics and 1,744 dental clinics. In 2002 there were 323 hospitals (provision of 16 beds per 10,000 inhabitants).

Welfare

The Employment Injury Insurance Scheme (SOCSO) provides medical and cash benefits and the Invalidity Pension Scheme provides protection to employees against invalidity as a result of disease or injury from any cause. Other supplementary measures are the Employees' Provident Fund, the pension scheme for all government employees, free medical benefits for all who are unable to pay and the provision of medical benefits particularly for workers under the Labour Code. In 1998 there were 49 welfare service institutions with capacity for 7,170.

RELIGION

Malaysia has a multi-racial population divided between Islam, Buddhism, Taoism, Hinduism and Christianity. Under the Federal constitution, Islam is the official religion of Malaysia but there is freedom of worship. In 2001 there were an estimated 10·77m. Muslims, 5·45m. adherents of Chinese traditional religions, 1·88m. Christians, 1·66m. Hindus and 1·50m. Buddhists.

CULTURE

World Heritage Sites

There are four sites in Malaysia that appear on the UNESCO World Heritage List: the Gunung Mulu National Park (inscribed on the list in 2000) with its limestone caves; Kinabalu Park/Mount Kinabalu (2000); the historic cities of Melaka and Georgetown (2008) in the Straits of Malacca; and the Archaeological Heritage of the Lenggong Valley (2012), one of the longest records of early man in a single locality and the oldest outside the African continent.

Press

The Malaysian Media Agencies are comprised of the press, magazine and press agencies/local media, which are further divided into home and foreign news. In 2008 there were 50 daily newspapers (49 paid-for and one free) with a combined circulation of 4,750,000. The dailies with the highest circulation are the Malay-language *Mingguan Malaysia* and the Chinese-language *Sin Chew Daily*.

Tourism

In 2009, 23,646,000 international tourists visited Malaysia (up from 22,052,000 in 2008), making it the ninth most popular tourist destination; receipts from tourism in 2009 totalled US$15,772m.

Festivals

National Day (31 Aug.) is celebrated in Kuala Lumpur at the Dataran Merdeka and marks Malaysia's independence. The Rainforest World Music Festival has been held annually in July in Kuching, Borneo since 1998.

DIPLOMATIC REPRESENTATIVES

Of Malaysia in the United Kingdom (45 Belgrave Sq., London, SW1X 8QT)
High Commissioner: Datuk Zakaria Sulong.

Of the United Kingdom in Malaysia (185 Jalan Ampang, 50450 Kuala Lumpur)
High Commissioner: Simon Featherstone.

Of Malaysia in the USA (3516 International Court, NW, Washington, D.C., 20008)
Ambassador: Othman bin Hashim.

Of the USA in Malaysia (376 Jalan Tun Razak, Kuala Lumpur)
Ambassador: Paul W. Jones.

Of Malaysia to the United Nations
Ambassador: Datuk Haniff Hussein.

Of Malaysia to the European Union
Ambassador: Dato' Zainuddin bin Yahya.

FURTHER READING

Department of Statistics: Kuala Lumpur. *Yearbook of Statistics, Malaysia* (2011); *Yearbook of Statistics, Sabah* (2011); *Yearbook of Statistics, Sarawak* (2011); *Vital Statistics, Malaysia* (2011).
Prime Minister's Department: Economic Planning Unit. *Malaysian Economy in Figures.* Annual, 2012

Andaya, B. W. and Andaya, L. Y., *A History of Malaysia.* 2nd ed. 2001
BNM: Kuala Lumpur. *Bank Negara Malaysia, Annual Report.* 2012
Drabble, J., *An Economic History of Malaysia, c. 1800–1990.* 2001
Gomez, Edmund Terence, *Politics in Malaysia.* 2009
Kahn, J. S. and Wah, F. L. K., *Fragmented Vision: Culture and Politics in Contemporary Malaysia.* 1992
Stockwell, A. J., *The Making of Malaysia.* 2005
Swee-Hock, Saw, *Malaysia: Recent Trends and Challenges.* 2006

National Statistical Office: Department of Statistics, Block C6, Parcel C, Federal Government Administrative Centre, 62514 Putrajaya.
Website: http://www.statistics.gov.my

MALDIVES

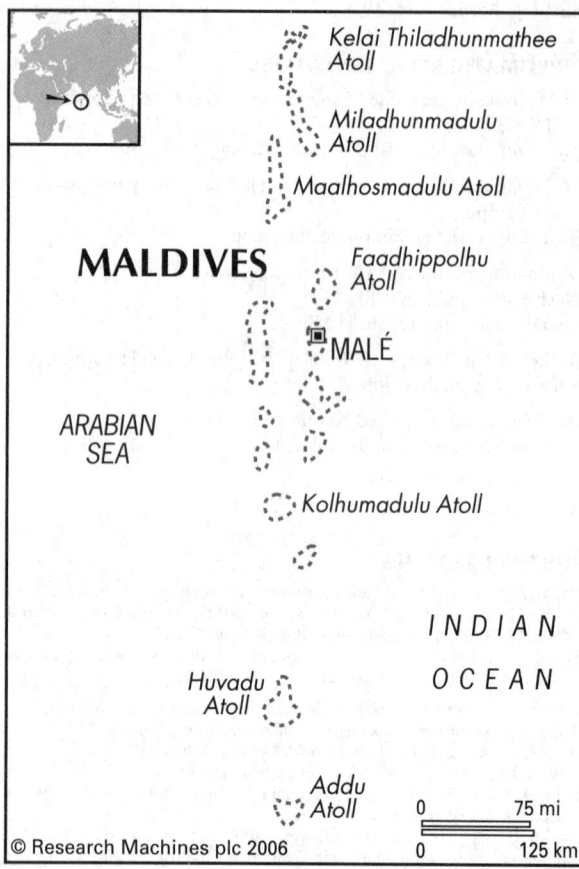

Kelai Thiladhunmathee Atoll

Miladhunmadulu Atoll

Maalhosmadulu Atoll

MALDIVES

Faadhippolhu Atoll

⊡MALÉ

ARABIAN SEA

Kolhumadulu Atoll

INDIAN OCEAN

Huvadu Atoll

Addu Atoll

0   75 mi
0   125 km

© Research Machines plc 2006

Divehi Raajjeyge Jumhooriyyaa
(Republic of the Maldives)

Capital: Malé
Population projection, 2015: 337,000
GNI per capita, 2011: (PPP$) 5,276
HDI/world rank: 0·661/109
Internet domain extension: .mv

KEY HISTORICAL EVENTS

Divehi-speaking people (a language related to Sinhalese) have lived on the Maldives since at least AD 400. Visited by Middle Eastern merchants from around AD 1000, the archipelago became an Islamic sultanate in 1153. Portuguese explorers occupied the island of Malé (the modern capital) from 1558 until they were expelled by Muhammad Thakurufaanu Al-Azam in 1573. The Dutch, who replaced the Portuguese as the dominant power in Ceylon in the mid-1600s, controlled Maldivian affairs until 1796, although the sultanate held sway over local administration. Thereafter the Maldives came under British protection (formalized in an agreement in 1887) until complete independence was achieved on 26 July 1965. A republic was declared on 11 Nov. 1968.

TERRITORY AND POPULATION

The republic, some 650 km to the southwest of Sri Lanka, consists of 1,192 low-lying (the highest point is 2·4 metres above sea-level) coral islands, grouped into 19 atolls and two cities. 199 are

inhabited. Area 298 sq. km (115 sq. miles). At the 2006 census the population was 298,968; density, 1,003·2 per sq. km.

The UN gives a projected population for 2015 of 337,000.

In 2011, 41·3% of the population lived in urban areas. Capital, Malé (2006 population, 92,555).

The official and spoken language is Divehi.

SOCIAL STATISTICS

2006 births, 5,827; deaths, 1,084. Birth rate, 2006, per 1,000 population, 19·5; death rate, 3·6. Annual population growth rate, 20050–10, 1·4%. Life expectancy at birth in 2007 was 69·7 years for males and 72·7 years for females. Infant mortality, 2010, 14 per 1,000 live births; fertility rate, 2008, 2·0 births per woman. The Maldives has had the largest reduction in its fertility rate of any country in the world over the past quarter of a century, having had a rate of 6·1 births per woman in 1990.

CLIMATE

The islands are hot and humid, and affected by monsoons. Malé: average temperature 81°F (27°C), annual rainfall 59" (1,500 mm).

CONSTITUTION AND GOVERNMENT

The present constitution came into effect on 1 Jan. 1998. There is a Citizens' *Majlis* (Parliament) which consists of 77 members all of whom are directly elected for a term of five years. Political parties were not permitted until the introduction of a multiparty system in June 2005. In a referendum held on 18 Aug. 2007 voters supported the retention of a presidential system, with 62·0% of votes cast in favour and 38·0% for a switch to a parliamentary system. The President of the Republic is elected by the Citizens' Majlis.

National Anthem

'Gavmii mi ekuverikan matii tibegen kuriime salaam' ('In national unity we salute our nation'); words by M. J. Didi, tune by W. Amaradeva.

RECENT ELECTIONS

The Maldives' first multi-party presidential elections were held on 8 and 29 Oct. 2008. Turnout in the first round was 85·4%. Incumbent president Maumoon Abdul Gayoom (Dhivehi Rayyithunge Party; DRP) won 40·3% of the vote, Mohamed Nasheed ('Anni') (Maldivian Democratic Party; MDP) 24·9%, Hassan Saeed (ind.) 16·7%, Qasim Ibrahim (Jumhooree Party) 15·2%, Umar Naseer (Islamic Democratic Party) 1·4% and Ibrahim Ismail (Social Liberal Party) 0·8%. A second round run-off was held on 29 Oct. 2008 which Mohamed Nasheed won with 54·2% of the vote against 45·8% for Maumoon Abdul Gayoom. Turnout was 86·6%.

Legislation passed in 2005 allowed political parties to stand for the first time in elections to the Majlis held on 9 May 2009. The DRP won 28 seats, the MDP 26, the People's Alliance 7, the Dhivehi Qaumee Party 2 and the Justice Party 1. 13 seats went to independents.

Presidential elections are scheduled to take place on 7 Sept. 2013.

CURRENT GOVERNMENT

In March 2013 the government consisted of:

President: Dr Mohamed Waheed Hassan Manik; b. 1953 (National Unity Party; in office since 7 Feb. 2012).

Vice President: Mohamed Waheed Deen.

Minister of Defence and National Security, and Transport and Communications (acting): Mohamed Nazim. *Economic Development,* and Finance and Treasury (acting): Ahmed Mohamed. *Education:* Asim Ahmed. *Environment and Energy:* Mariyam Shakeela. *Fisheries and Agriculture:* Ahmed Shafeeu. *Foreign Affairs:* Abdul Samad Abdulla. *Health:* Dr Ahmed Jamsheed Mohamed. *Home Affairs:* Mohamed Jameel Ahmed. *Housing and Infrastructure:* Mohamed Muizzu. *Human Resources, Youth and Sport:* Mohamed Hussain Shareef. *Islamic Affairs:* Sheikh Mohamed Shaheem Ali Saeed. *Tourism, Arts and Culture:* Ahmed Adheeb Abdul Ghafoor. *Attorney General, and Gender, Family and Human Rights (acting):* Uza Aishath Azima Shakooru.

Speaker of Citizens' Majlis: Abdulla Shahid.

Office of the President: http://www.presidencymaldives.gov.mv

CURRENT LEADERS

Mohammed Waheed Hassan

Position
President

Introduction
Mohammed Waheed Hassan took office as president on 7 Feb. 2012 after incumbent Mohamed Nasheed's resignation following public protests over the arrest of a senior judge. Waheed, formerly Nasheed's deputy, pledged to uphold the rule of law and is expected to remain in office until presidential elections scheduled for Sept. 2013.

Early Life
Mohammed Waheed Hassan was born on 3 Jan. 1953 in the capital, Malé. He graduated in English Language and gained a diploma in teaching from the American University of Beirut in 1976, before returning to Malé to teach. In 1978 he became the first anchor on Maldivian television with TVM (Television Maldives).

Waheed won a scholarship to Stanford University, California, where he completed a masters degree in education planning in 1979. He joined the Maldivian ministry of education and in 1980 was appointed by President Gayoom as an adviser to a special session of the *Majlis* (parliament) tasked with reviewing the constitution.

In 1982 Waheed returned to Stanford to study political science. In 1987 he was awarded his PhD in international development education. In 1988 he was appointed director of educational services at the ministry of education and in 1989 won the seat of Malé in the general election. Having resigned over the poor human rights record of Gayoom's government, he left the Maldives in 1991 to work as a freelance consultant with the UN on educational development programmes.

In 2008 he set up the Gaumee Itthihaad party, which formed the 'MDP Itthihaad' with the Maldivian Democratic Party (MDP) to contest the 2008 elections. The coalition emerged victorious and on 11 Nov. 2008 Nasheed—a co-founder of the MDP in 2003—and Waheed were sworn into office as president and vice-president respectively. On 7 Feb. 2012 Nasheed resigned amid growing public unrest following his arrest of the chief justice, Abdullah Mohamed. Waheed was sworn in as president later that day, leading an administration consisting largely of technocrats and Gayoom appointees.

Career in Office
Waheed's first task was to rebut Nasheed's claim that his successor helped organize his removal from office. Waheed denied the charge but Nasheed's supporters have continued to protest and demand early elections. Waheed appealed to the international community for support and in Aug. 2012 a Commonwealth-backed commission dismissed Nasheed's assertion that he had been ousted in a coup and confirmed the constitutional legitimacy of the transfer of power. In Oct. Nasheed was arrested for defying a summons to stand trial for the arrest of the chief justice, which had triggered the crisis in Feb.

DEFENCE

In 2008 military expenditure totalled US$43m. (US$111 per capita), representing 3·4% of GDP.

ECONOMY

Fisheries accounts for approximately 7% of GDP, industry 15% and services 78%.

Overview
Tourism accounts for approximately 28% of total GDP. Fisheries is the next biggest sector, with fish the largest export commodity.

The Maldives was badly affected by the tsunami of Dec. 2004 that left many dead and thousands displaced. GDP growth has been variable in recent years. In 2005 it stood at –8·7% (the result of the tsunami) but jumped to nearly 20% in 2006 prompted by a revival in tourism and the construction of new resorts. Government expenditure nearly doubled as a share of GDP from 2004–08, reaching 63% of GDP in 2008.

The global financial crisis had a severe impact, with sharp declines in tourism, construction, exports and investment. The banking sector was battered by an increase in non-performing loans. This combined with the high public spending of previous years led to a balance of payments crisis. A stand-by arrangement was approved by the IMF in Dec. 2009 and economic results since then have been better than anticipated, with tourism enjoying a particularly strong rebound. In Jan. 2011 the Maldives became only the third country to graduate from the UN's list of Least Developed Countries (LDCs) to 'developing country' status.

Currency
The unit of currency is the *rufiyaa* (MVR) of 100 *laari*. Inflation was 12·3% in 2008, falling to 4·7% in 2010 but increasing to 14·1% in 2011. Gold reserves were 2,000 troy oz in July 2005, foreign exchange reserves were US$218m. and total money supply was 3,163m. rufiyaa.

Budget
In 2008 budgetary central government revenue totalled 7,414·2m. rufiyaa (including taxes, 3,366·8m. rufiyaa); expenditure totalled 7,463·2m. rufiyaa.

Performance
After shrinking by 8·7% in 2005 as a consequence of the devastation wreaked by the tsunami, the economy rebounded spectacularly in 2006 with a real GDP growth rate of 19·6%, driven by the return of tourists, reconstruction efforts and new development. Although the economy contracted by 4·7% in 2009 there was growth of 5·7% in 2010 and 5·8% in 2011. Total GDP in 2011 was US$2·1bn.

Banking and Finance
The Maldives Monetary Authority (*Governor*, Fazeel Najeeb), established in 1981, is endowed with the regular powers of a central bank and bank of issue. There is one domestic commercial bank (Bank of Maldives) and branches of four foreign banks.

Total foreign debt amounted to US$562m. in 2007.

There is a stock exchange in Malé.

ENERGY AND NATURAL RESOURCES

Environment
Carbon dioxide emissions from the consumption and flaring of fossil fuels were the equivalent of 2·3 tonnes per capita in 2008.

Electricity
Installed capacity was 61,000 kW in 2007. Production in 2007 was 245m. kWh; consumption per capita in 2007 was 804 kWh.

Minerals

Inshore coral mining has been banned as a measure against the encroachment of the sea.

Agriculture

There were approximately 4,000 ha. of arable land in 2007 and 8,000 ha. of permanent crops. Principal crops in 2009 (estimates, in 1,000 tonnes): coconuts, 5; bananas, 4. Various other types of fruit, vegetables, roots and tubers, and nuts are also produced.

Fisheries

The total catch in 2010 was 94,953 tonnes. The Maldives has the highest per capita consumption of fish and fishery products of any country in the world. In the period 2005–07 the average person consumed 142 kg (313 lb) a year, or more than eight times the average for the world as a whole.

INDUSTRY

The main industries are fishing, tourism, shipping, lacquerwork and garment manufacturing.

Labour

In 2005 the economically active workforce totalled 99,000 of whom 96,000 were employed. More than two-thirds of the working population are engaged in tourism.

INTERNATIONAL TRADE

Imports and Exports

In 2008 imports (c.i.f.) were valued at US$1,387·5m and exports (f.o.b.) at US$126·4m. Tuna is the main export commodity. It is exported principally to Thailand, Sri Lanka and some European markets. Main import suppliers in 2008 were Singapore (21·3%), UAE (18·0%), India (10·4%), Malaysia (7·7%), Sri Lanka (5·9%). Leading export destinations were Thailand (49·0%), Sri Lanka (9·5%), France (8·8%), Italy (8·3%), UK (7·6%).

COMMUNICATIONS

Roads

In 2007 there were 3,060 passenger cars in use (10 per 1,000 inhabitants), 26,780 motorcycles and mopeds, 2,870 lorries and vans, and 74 buses and coaches.

Civil Aviation

The former national carrier Air Maldives collapsed in April 2000 with final losses in excess of US$50m. In 2003 there were 1,833,620 passenger arrivals, 21m. pieces of cargo and 100,352 pieces of mail handled at Malé's international airport. There are four domestic airports. In 2001 scheduled airline traffic of Maldives-based carriers flew 7m. km, carrying 367,000 passengers (226,000 on international flights).

Shipping

In Jan. 2009 there were 58 ships of 300 GT or over registered, totalling 130,000 GT.

Telecommunications

There were 48,000 fixed telephone lines in 2010 (152·0 per 1,000 inhabitants). Mobile phone subscribers numbered 494,400 in 2010. There were 283·0 internet users per 1,000 inhabitants in 2010. Fixed internet subscriptions totalled 20,100 in 2009 (64·4 per 1,000 inhabitants). In March 2012 there were 120,000 Facebook users.

SOCIAL INSTITUTIONS

Justice

Justice is based on the Islamic Sharia.

Education

Adult literacy in 2006 was 98%. Education is not compulsory. In 2004 there were 81 government schools (57,139 pupils), 176 community schools (38,043 pupils) and 337 private schools (104,214 pupils) with a total of 5,239 teachers. In 2005 public expenditure on education came to 7·8% of GDP and 15·0% of total government spending.

Health

In 2003 there were 236 beds at the Indira Gandhi Memorial Hospital in Malé, six regional hospitals (226 beds) and 27 health centres. In 2003 there were 315 doctors and 785 nurses, 251 pharmacists and 409 midwives.

RELIGION

The State religion is Islam.

CULTURE

Press

In 2008 there were six paid-for daily newspapers and around 200 independent newspapers and periodicals in total.

Tourism

Tourism is the major foreign currency earner. There were a record 791,917 tourist arrivals in 2010, spending US$714m.

DIPLOMATIC REPRESENTATIVES

Of the Maldives in the United Kingdom (22 Nottingham Pl., London, W1U 5NJ)
Acting High Commissioner: Ahmed Shiaan.

Of the United Kingdom in the Maldives
High Commissioner: John Rankin (resides in Colombo, Sri Lanka).

Of the Maldives in the USA and to the United Nations (800 2nd Ave., Suite 400E, New York, NY 10017)
Ambassador: Ahmed Sareer.

Of the USA in the Maldives
Ambassador: Michele J. Sison (resides in Colombo, Sri Lanka).

Of the Maldives to the European Union
Ambassador: Ali Husain Didi.

FURTHER READING

Gayoom, M. A., *The Maldives: A Nation in Peril.* 1998

National Statistical Office: Statistics Section, Ministry of Planning and National Development.
Website: http://www.planning.gov.mv

MALI

République du Mali
(Republic of Mali)

Capital: Bamako
Population projection, 2015: 17·82m.
GNI per capita, 2011: (PPP$) 1,123
HDI/world rank: 0·359/175
Internet domain extension: .ml

KEY HISTORICAL EVENTS

Mali's power reached its peak between the 11th and 13th centuries when its gold-based empire controlled much of the surrounding area. The country was annexed by France in 1904. As French Sudan it was part of French West Africa. The country became an autonomous state within the French Community on 24 Nov. 1958, and on 4 April 1959 joined with Senegal to form the Federation of Mali. The Federation achieved independence on 20 June 1960, but Senegal seceded on 22 Aug. and Mali proclaimed itself an independent republic on 22 Sept. There was an army coup on 19 Nov. 1968, which brought Moussa Traoré to power. Ruling the country for over 22 years, he wrecked the economy. A further coup followed in March 1991.

In Jan. 1991 a ceasefire was signed with Tuareg insurgents in the north and in April 1992 a national pact was concluded providing for a special administration for the Tuareg north.

Under President Alpha Oumar Konaré, two elections for the National Assembly were held. The first (April 1997) was cancelled by the constitutional court and the second, in July 1997, was boycotted by opposition parties. Amadou Toumani Touré, a former military ruler, won presidential elections held in April and May 2002. In July 2005 severe food shortages led to more than 1m. people facing starvation. In March 2012 Touré was ousted in a military coup. A civilian interim government was nominally established the following month, with Dioncounda Traoré installed as president. However, after allegations of a counter-coup by supporters of Touré, the military reasserted control in May. In Aug. a government of national unity was sworn in to oversee the transition to civilian government.

In May 2012 Tuareg and Islamic militant groups declared the northern Azawad region an independent Islamic state. In Sept. 2012 the government in Bamako agreed to host 3,000 ECOWAS troops in a bid to displace the rebels. The following month Mali was readmitted to the African Union, having been suspended following the March coup.

In Jan. 2013 President Traoré appealed to France for assistance after an escalation in militant operations. French forces quickly seized the cities of Gao and Tombouctou (Timbuktu), dislodging the rebels from their principal strongholds.

TERRITORY AND POPULATION

Mali is bounded in the west by Senegal, northwest by Mauritania, northeast by Algeria, east by Niger and south by Burkina Faso, Côte d'Ivoire and Guinea. Its area is 1,248,574 sq. km (482,077 sq. miles) and it had a population of 14,528,662 at the 2009 census; density, 11·6 per sq. km. In 2011, 36·6% of the population were urban.

The UN gives a projected population for 2015 of 17·82m.

The areas, populations and chief towns of the regions are:

Region	Sq. km	2009 census population	Chief town
Gao	170,572	542,304	Gao
Kayes	119,743	1,993,615	Kayes
Kidal	151,430	67,739	Kidal
Koulikoro	95,848	2,422,108	Koulikoro
Mopti	79,017	2,036,209	Mopti
Ségou	64,821	2,338,349	Ségou
Sikasso	70,280	2,643,179	Sikasso
Tombouctou	496,611	674,793	Tombouctou
Capital District	252	1,810,366	Bamako

In 2009 the capital, Bamako, had a population of 1,809,000. The second largest town, Sikasso, had a population of 226,000 in 2009.

The Bambara, Khassonké, Malinké and Soninké, all of which belong to the broader Mandé group, make up 50% of the population; the other leading groups are the Fula (17%), Voltaic (12%), Songhai (6%), and Tuareg and Moor (10%). The official language is French; Bambara is spoken by about 68% of the population.

SOCIAL STATISTICS

2008 estimates: births, 541,000; deaths, 200,000. Rates, 2008 estimates (per 1,000 population): births, 42·6; deaths, 15·7. Infant mortality, 2010 (per 1,000 live births), 99. Expectation of life in 2007 was 47·4 years for males and 48·8 for females. Annual population growth rate, 2000–08, 2·4%; fertility rate, 2008, 6·5 children per woman.

CLIMATE

A tropical climate, with adequate rain in the south and west, but conditions become increasingly arid towards the north and east. Bamako, Jan. 76°F (24·4°C), July 80°F (26·7°C). Annual rainfall 45" (1,120 mm). Kayes, Jan. 76°F (24·4°C), July 93°F (33·9°C). Annual rainfall 29" (725 mm). Tombouctou, Jan. 71°F (21·7°C), July 90°F (32·2°C). Annual rainfall 9" (231 mm).

CONSTITUTION AND GOVERNMENT

A constitution was approved by a national referendum in 1974; it was amended by the National Assembly on 2 Sept. 1981. The sole legal party was the *Union démocratique du peuple malien* (UDPM).

A national conference of 1,800 delegates agreed a draft constitution enshrining multi-party democracy in Aug. 1991, and this was approved by 99·76% of votes cast at a referendum in Jan. 1992. Turnout was 43%.

The *President* is elected for not more than two terms of five years.

There is a *National Assembly*, consisting of 147 deputies (formerly 116) plus 13 Malinese living abroad.

A *Constitutional Court* was established in 1994.

In May 2012 the rebel National Movement for the Liberation of Azawad (MNLA) and Ansar Dine, an Islamist militant group, declared Azawad—a region in the north covering over half of Mali's total land area—a breakaway Islamic state. The unilateral declaration went unrecognized by Bamako and the international community. Relations between the MNLA and Ansar Dine soon strained. By the end of June 2012 Ansar Dine had militarily defeated the MNLA to claim control of the region.

National Anthem

'A ton appel, Mali' ('At your call, Mali'); words by S. Kouyate, tune by B. Sissoko.

RECENT ELECTIONS

Presidential elections were held on 29 April 2007. Amadou Toumani Touré was re-elected with 71·2% of votes cast, against Ibrahim Boubacar Keita (19·2%) and five other candidates. Turnout was 36·2%. The Front for Democracy and the Republic, the coalition backing Keita, refused to accept the result, alleging widespread fraud. However, foreign observers declared it mostly fair.

Parliamentary elections were held in two rounds on 1 and 22 July 2007. The Alliance for Democracy in Mali (ADEMA) won 51 seats (67 in 2002), Union for the Republic and Democracy 34, Rally for Mali 11, Patriotic Movement for Renewal 8, National Congress for Democratic Initiative 7, Party for National Rebirth 4, African Solidarity for Democracy and Independence 4, Union of Democrats for Citizenship and Development 3, Movement for the Independence, Renaissance and Integration of Africa 2, Popular Party for Progress 2 and Alternation Bloc for Renewal, Integration and African Co-operation 2. Four other parties won a single seat each and 15 seats went to independents. The Alliance for Democracy and Progress, a coalition that supports President Touré and includes the Alliance for Democracy in Mali and the Union for the Republic and Democracy, won 113 of the 147 seats. The opposition coalition, the Front for Democracy and the Republic (which includes Rally for Mali), won 15 seats. Turnout was about 33% in the first round and about 11% in the second.

Presidential elections were scheduled to take place on 7 July 2013 with a second round if needed on 21 July 2013, the same day as parliamentary elections.

CURRENT GOVERNMENT

President: Dioncounda Traoré; b. 1942 (in office since 12 April 2012).

Interim Prime Minister: Diango Cissoko; b. 1949 (in office since 11 Dec. 2012).

In March 2013 the transitional government comprised:

Minister of Economy, Finance and Budget: Tiéna Coulibaly. *Defence and Veterans Affairs:* Gen. Yamoussa Camara. *Foreign Affairs and International Co-operation:* Tieman Coulibaly. *Territorial Administration, Decentralization and Regional Development:* Col. Moussa Sinko Coulibaly. *Mines:* Amadou Baba Sy. *Education, Literacy and the Promotion of National Languages:* Bocar Moussa Diarra. *Higher Education and Scientific Research:* Messaoud Ould Lahbib. *Labour, Public Service and Relations with Institutions:* Mamadou Namory Traoré. *Malians Abroad and African Integration:* Demba Traoré. *Internal Security and Civil Protection:* Gen. Tiéfing Konaté. *Agriculture:* Baba Berthé. *Justice and Attorney General:* Malick Coulibaly. *Equipment and*

Transport: Abdoulaye Koumaré. *Health:* Soumana Makadji. *Commerce and Industry:* Abdel Karim Konaté. *Crafts and Tourism:* Yéhia Ag Mohamed Ali. *Housing, Urban Development and Land Affairs:* David Sagara. *Employment and Vocational Training:* Diallo Dédia Mahamane Kattra. *Posts and New Technologies:* Bréhima Tolo. *Family and the Promotion of Women and Children:* Alwata Ichata Sahi. *Energy and Water:* Makan Tounkara. *Environmental Sanitation:* Ousmane Ag Rhissa. *Youth and Sports:* Hameye Founé Mahalmadane. *Humanitarian Action, Solidarity and Seniors:* Mamadou Sidibé. *Livestock and Fisheries:* Diane Mariame Koné. *Religious Affairs and Cults:* Yacouba Traoré. *Communications and Government Spokesman:* Manga Dembelé. *Culture:* Bruno Maiga.

CURRENT LEADERS

Dioncounda Traoré

Position
President

Introduction
Dioncounda Traoré was sworn in as interim president on 8 April 2012, ending a brief period of military rule after the removal of the former president, Amadou Toumani Touré, in a coup on 21 March. As parliamentary speaker, Traoré was next in the constitutional succession after Touré's enforced resignation.

Early Life
Dioncounda Traoré was born in Kati on 23 Feb. 1942, the son of army colonel Sékou Traoré. He read mathematics in Moscow from 1963 to 1965 before continuing his studies at the University of Algiers. He was later awarded a doctorate from the University of Nice. He started work as a teacher at the Ecole Normale Supérieure in Bamako in 1977 but was jailed in 1980 for his trade union activities.

A militant trade unionist and pro-democracy activist throughout the 1980s, Traoré was a founding member of the Alliance for Democracy in Mali (ADEMA), a coalition of opponents to the dictatorship of Moussa Traoré. After Moussa Traoré was overthrown in 1991, ADEMA evolved into ADEMA-PASJ (African Party for Solidarity and Justice), with Dioncounda Traoré elected the party's second vice-president.

After Mali's first democratic elections, Traoré was appointed to the council of ministers by President Alpha Oumar Konaré in June 1992. He variously held the portfolios for public works, defence and foreign affairs before resigning from the council in 1997 to take his place in the National Assembly as the representative for Nara.

In 2000 Ibrahim Boubacar Kéïta resigned as prime minister and leader of ADEMA-PASJ, with Traoré elected to succeed him in the party post. In 2007 Traoré became president of the National Assembly but fled Mali during the 2012 coup. He returned when the junta, led by Capt. Amadou Haya Sanogo, agreed to an Economic Community of West African States (ECOWAS)-brokered handover to an interim administration. The junta nevertheless retained considerable power and influence.

Career in Office
In the period between the military coup and Traoré's installation as interim president, Tuareg and militant Islamist insurgents made significant gains in the country's north, culminating in a declaration of independence in April 2012. While the claim went unrecognized internationally, Traoré entered office under pressure to prevent the spread of insurgency to neighbouring regions. Preparations for military action against the rebel groups in the north, with the aid of ECOWAS forces and with United Nations and African Union backing, proceeded slowly over the following months. However, tensions between the civilian government and junta, which led in Dec. to Prime Minister Cheick Modibo Diarra's forced resignation by the junta and the appointment of

Diango Cissoko as premier by Traoré, threatened to undermine plans for intervention and prompted international demands for a restoration of constitutional rule.

DEFENCE

There is selective conscription for two years. In 2008 military expenditure totalled US$157m. (US$12 per capita), representing 2·1% of GDP.

Army

Strength (2007) 7,350. There are also paramilitary forces of 4,800.

Navy

There is a Navy of around 50 operating three patrol craft although their serviceability is in doubt.

Air Force

Personnel (2007) total 400. There were around 16 combat-capable aircraft.

ECONOMY

Agriculture accounted for 36·9% of GDP in 2006, industry 24·0% and services 39·1%.

Overview

Mali is ranked 175th of 187 nations on the UN Human Development Index. The economy is heavily dependent on agriculture, which accounts for 37% of GDP and provides income to 80% of the population.

In the 15 years from 1996, GDP growth averaged over 5% per year and macroeconomic stability was maintained during the global crisis in 2008. Per capita income rose from US$240 in 1994 to US$500 in 2007. However, over 70% of the population lives below the poverty line. Exports are concentrated in gold, cotton and livestock, although reliance on the last two makes the economy vulnerable to drought. As a landlocked country, Mali is dependent on ports in neighbouring countries.

Mali aims to establish itself as a tourism destination, receiving 200,000 visitors in 2011 spending on average at least US$100 a day to make the sector the economy's third biggest revenue raiser. However, increasing militant activity in the aftermath of the March 2012 coup resulted in only 10,000 visitors entering the country in 2012.

GDP grew by 2·7% in 2011, the result of poor harvests and regional instability. Civil conflict then resulted in an estimated contraction of over 4% in 2012.

Currency

The unit of currency is the *franc CFA* (XOF), which replaced the Mali franc in 1984. It has a parity rate of 655·957 francs CFA to one euro. Total money supply in June 2005 was 573,692m. francs CFA and foreign exchange reserves were US$827m. Inflation was 1·3% in 2010 and 3·1% in 2011.

Budget

Budgetary central government revenue and expenditure in 1bn. francs CFA:

	2007	2008	2009
Revenue	778·6	752·4	909·5
Expenditure	509·4	525·7	620·7

Principal sources of revenue in 2009 were: taxes on goods and services, 265·7bn. francs CFA; grants, 184·6bn. francs CFA. Main items of expenditure by economic type in 2009 were: compensation of employees, 213·5bn. francs CFA; use of goods and services, 190·2bn. francs CFA.

VAT is 18%.

Performance

Real GDP growth was 4·5% in 2009, 5·8% in 2010 and 2·7% in 2011. Total GDP in 2011 was US$10·6bn.

Banking and Finance

The bank of issue and the central bank is the regional Central Bank of West African States (BCEAO). The *Governor* is Tiémoko Meyliet Koné. In 2002 there were eight commercial and two development banks.

External debt was US$2,326m. in 2010, representing 26·1% of GNI.

There is a stock exchange in Bamako.

ENERGY AND NATURAL RESOURCES

Environment

Carbon dioxide emissions from the consumption and flaring of fossil fuels were the equivalent of 0·1 tonnes per capita in 2008. An *Environmental Performance Index* compiled in 2008 ranked Mali 145th in the world out of 149 countries analysed, with 44·3%. The index examined various factors in six areas—air pollution, biodiversity and habitat, climate change, environmental health, productive natural resources and water resources.

Electricity

Installed capacity in 2007 was 0·2m. kW. Production in 2007 totalled about 495m. kWh, approximately 55% of it hydro-electric. Consumption per capita was an estimated 40 kWh in 2007.

Minerals

Gold (51,957 kg in 2006) is the principal mineral produced. There are also deposits of iron ore, uranium, diamonds, bauxite, manganese, copper, salt, limestone, phosphate, gypsum and lithium.

Agriculture

About 80% of the population depends on agriculture, mainly carried on by small peasant holdings. Mali is second only to Egypt among African cotton producers. In 2002 there were an estimated 4·66m. ha. of arable land and 40,000 ha. of permanent cropland. There were around 2,600 tractors in 2002 and 650 harvester-threshers. Production in 2003 included (estimates, in 1,000 tonnes): millet, 815; rice, 693; sorghum, 650; seed cotton, 464; maize, 365; sugarcane, 300; cotton lint, 250; cottonseed, 185; groundnuts, 170; sweet potatoes, 74, tomatoes, 50.

Livestock, 2003 estimates: goats, 11·46m.; sheep, 7·97m.; cattle, 7·31m.; asses, 700,000; camels, 470,000; horses, 170,000; chickens, 29m.

Livestock products, 2003 estimates (in 1,000 tonnes): beef and veal, 113; goat meat, 46; lamb and mutton, 36; poultry meat, 34; goat's milk, 227; cow's milk, 179; sheep's milk, 117; eggs, 10.

Approximately 138,000 ha. were irrigated in 2002.

Forestry

In 2010 forests covered 12·50m. ha., or 10% of the total land area. Timber production in 2007 was 5·56m. cu. metres.

Fisheries

In 2010 an estimated 100,000 tonnes of fish were caught, exclusively from inland waters.

INDUSTRY

The main industries are food processing, followed by cotton processing, textiles and clothes. Cement and pharmaceuticals are also produced.

Labour

There were 2,491,800 employed persons (58·4% males) in 2004. In 2003 approximately 79·0% of the total labour force were engaged in agriculture. Large numbers of Malians emigrate temporarily to work abroad, principally in Côte d'Ivoire.

INTERNATIONAL TRADE

Imports and Exports

The following table shows the value of Mali's foreign trade (in US$1m.):

	2006	2007	2008
Imports c.i.f.	1,819·8	2,184·8	3,338·9
Exports f.o.b.	1,526·1	1,440·6	1,918·3

Principal import commodities are machinery and equipment, foodstuffs, construction materials, petroleum and textiles. Principal export commodities are cotton and livestock (between them accounting for three-quarters of Mali's annual exports) and gold.

The main import suppliers are France and its former colonies (in particular Côte d'Ivoire), western Europe and China. Main export markets are also France and its former colonies, western Europe and China.

COMMUNICATIONS

Roads

There were 18,912 km of roads in 2005, of which 19·0% were paved. In 2007 there were 87,000 passenger cars (seven per 1,000 inhabitants), 26,800 lorries and vans, and 10,000 motorcycles and mopeds.

Rail

Mali has a railway from Kayes to Koulikoro by way of Bamako, a continuation of the currently non-operational Dakar–Kayes line in Senegal; total length, 2005, 643 km (metre-gauge). In 2005, 179,000 passengers and 1·7m. tonnes of freight were transported.

Civil Aviation

There is an international airport at Bamako (Senou), which handled 312,000 passengers (305,000 on international flights) and 4,400 tonnes of freight in 2001. In 2003 Trans African Airlines operated services to Abidjan, Brazzaville, Cotonou, Dakar, Lomé and Pointe-Noire. There were also international flights to Accra, Addis Ababa, Banjul, Bobo-Dioulasso, Casablanca, Conakry, Douala, Kano, Lagos, Libreville, N'Djaména, Niamey, Nouakchott, Ouagadougou, Paris and Tripoli. In 2001 scheduled airline traffic of Mali-based carriers flew 1m. km, carrying 46,000 passengers (all on international flights).

Shipping

For about seven months in the year small steamboats operate a service from Koulikoro to Tombouctou and Gao, and from Bamako to Kouroussa.

Telecommunications

In 2008 there were 81,100 main (fixed) telephone lines; mobile phone subscribers numbered 3,439,000 in the same year (27·1 per 100 persons). There were 200,000 internet users in 2008. In June 2012 there were 141,000 Facebook users.

SOCIAL INSTITUTIONS

Justice

The Supreme Court was established at Bamako in 1969 with both judicial and administrative powers. The Court of Appeal is also at Bamako, at the apex of a system of regional tribunals and local *juges de paix*.

The population in penal institutions in mid-2009 was 6,700 (52 per 100,000 of national population).

Education

The adult literacy rate in 2006 was 26%. In 2007 there were 1,510 teaching staff for 54,591 children in pre-primary schools, 33,230 teaching staff for 1,716,956 pupils in primary schools, 15,013 teaching staff for 533,849 secondary school pupils and 50,787 students in tertiary education with 976 academic staff. During the period 1990–95 only 19% of females of primary school age were enrolled in school but by 2007 this had risen to 56%.

In 2005 public expenditure on education came to 4·1% of GDP.

Health

In 2001 there were 17 hospitals. In 2000 there were 529 physicians, 1,501 nurses and 284 midwives.

RELIGION

The state is secular, but predominantly Sunni Muslim. About 15% of the population follow traditional animist beliefs and there is a small Christian minority.

CULTURE

World Heritage Sites

Mali has four sites on the UNESCO World Heritage List: Old Towns of Djenné (inscribed on the list in 1988), a market centre established in 250 BC and an important Islamic centre in the 16th century—its buildings are all mudbrick, plastered annually with adobe; Tomboctou (1988), an important Islamic centre containing the Koranic Sankore University and the famous Djingareyber Mosque; the Cliff of Bandiagara (Land of the Dogons) (1989), for its natural and architectural wonders; and the Tomb of Askia (2004).

Press

In 2008 there were 12 daily newspapers with an estimated combined circulation of 40,000.

Tourism

There were 164,000 non-resident tourists staying at hotels and similar establishments in 2007 (including 104,000 from Europe and 29,000 from other African countries); tourist revenue totalled US$175m. in 2006.

DIPLOMATIC REPRESENTATIVES

Of Mali in the United Kingdom
Ambassador: Ibrahim Bocar Ba (resides in Brussels).

Of the United Kingdom in Mali (Immeuble Semega, Koulikoro Rd, Hippodrome, PO Box 2069, Bamako). Consular assistance is provided by the embassy in Senegal.
Ambassador: Vacant.

Of Mali in the USA (2130 R. St., NW, Washington, D.C., 20008)
Ambassador: Al-Maamoun Baba Lamine Keita.

Of the USA in Mali (ACI 2000, Rue 243, Porte 297, Bamako)
Ambassador: Mary Beth Leonard.

Of Mali to the United Nations
Ambassador: Oumar Daou.

Of Mali to the European Union
Ambassador: Ibrahim Bocar Ba.

FURTHER READING

Bingen, R. James, *Democracy and Development in Mali.* 2000

National Statistical Office: Direction National de la Statistique et de l'Informatique, BP 12 rue Archinard, Porte 233, Bamako.

MALTA

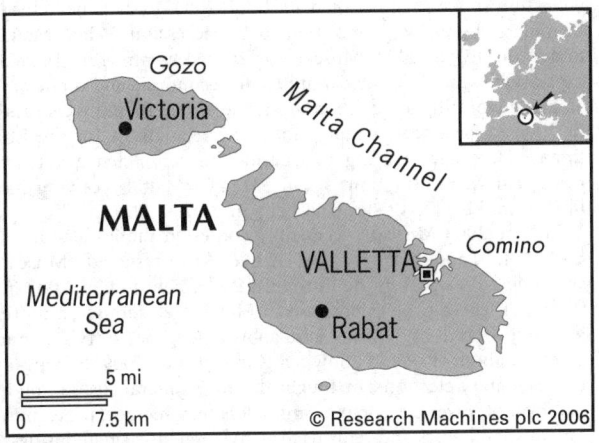

Repubblika ta' Malta
(Republic of Malta)

Capital: Valletta
Population projection, 2015: 423,000
GNI per capita, 2011: (PPP$) 21,460
HDI/world rank: 0·832/36
Internet domain extension: .mt

KEY HISTORICAL EVENTS

Malta was held in turn by Phoenicians, Carthaginians and Romans, and was conquered by Arabs in 870. From 1090 it was subject to the same rulers as Sicily until 1530, when it was handed over to the Knights of St John, who ruled until dispersed by Napoleon in 1798. The Maltese rose in rebellion against the French and the island was blockaded by the British, aided by the Maltese from 1798 to 1800. The Maltese people freely requested the protection of the British Crown in 1802 on condition that their rights and privileges be preserved. The island was finally annexed to the British Crown by the Treaty of Paris in 1814. Malta became independent on 21 Sept. 1964 and a republic within the Commonwealth on 13 Dec. 1974. On 1 May 2004 Malta joined the European Union.

TERRITORY AND POPULATION

The three Maltese islands and minor islets lie in the Mediterranean 93 km (at the nearest point) south of Sicily and 288 km east of Tunisia. The area of Malta is 246 sq. km (94·9 sq. miles); Gozo, 67 sq. km (25·9 sq. miles) and the virtually uninhabited Comino, 3 sq. km (1·1 sq. miles); total area, 316 sq. km (121·9 sq. miles). The census population (provisional) in 2011 was 416,055 (Malta island, 384,912; Gozo and Comino, 31,143); density, 1,317 per sq. km.

The UN gives a projected population for 2015 of 423,000.

In 2011, 94·8% of the population were urban. Chief town and port, Valletta, population 5,784 (2011 census, provisional). Other towns: Birkirkara, 21,533; Mosta, 19,650; St Paul's Bay, 16,478; Qormi, 16,312; Zabbar, 14,823.

The constitution provides that the national language and language of the courts is Maltese, but both Maltese and English are official languages. Italian is also spoken.

SOCIAL STATISTICS

2009: births, 3,713; deaths, 3,221; marriages, 2,353; emigrants, 9,708 (1,771 Maltese); immigrants, 8,147 (1,190 returning Maltese). 2009 rates per 1,000 population: birth, 9·4; death, 7·8; marriage, 5·7. Abortion is illegal, as was divorce until Oct. 2011. Parliament voted in July 2011 to legalize divorce following a vote in favour of the reform by 52·7% to 47·3% in a referendum held in May. Until then Malta had been one of only three countries still to outlaw divorce (the others being the Philippines and the Vatican). In 2008 the most popular age range for marrying was 25–29 for both males and females. Life expectancy at birth in 2007: 77·7 years for males and 81·3 years for females. Annual population growth rate, 2000–05, 1·1%. Infant mortality in 2010: five per 1,000 live births; fertility rate, 2008, 1·3 births per woman.

CLIMATE

The climate is Mediterranean, with hot, dry and sunny conditions in summer and very little rain from May to Aug. Rainfall is not excessive and falls mainly between Oct. and March. Average daily sunshine in winter is six hours and in summer over ten hours. Valletta, Jan. 12·8°C (55°F), July 25·6°C (78°F). Annual rainfall 578 mm (23").

CONSTITUTION AND GOVERNMENT

Malta is a parliamentary democracy. The constitution of 1964 provides for a *President*, a *House of Representatives* of members elected by universal suffrage and a Cabinet consisting of the Prime Minister and such number of Ministers as may be appointed. The Constitution makes provision for the protection of fundamental rights and freedom of the individual, and for freedom of conscience and religious worship, and guarantees the separation of executive, judicial and legislative powers. The House of Representatives may have between 65 and 69 members (currently 69). Malta uses the single transferable vote system.

National Anthem

'Lil din l'art helwa, l'omm li tatna isimha' ('Guard her, O Lord, as ever Thou hast guarded'); words by Dun Karm Psaila, tune by Dr Robert Samut.

RECENT ELECTIONS

At the elections of 9 March 2013 the electorate was 315,357; turnout was 93·3%. The opposition Labour Party (MLP) gained 39 seats with 54·8% of votes cast; the Nationalist Party (NP), 30 with 43·3%.

European Parliament
Malta has six representatives. At the June 2009 elections turnout was 78·8% (82·4% in 2004). The MLP won 3 seats with 54·8% of votes cast (political affiliation in European Parliament: Progressive Alliance of Socialists and Democrats); and the NP 2 with 40·5% (European People's Party).

CURRENT GOVERNMENT

President: George Abela; b. 1948 (MLP; sworn in 4 April 2009).

In March 2013 the government comprised:

Prime Minister: Joseph Muscat; b. 1974 (MLP; sworn in 11 March 2013).

Deputy Prime Minister and Minister of European Affairs and Implementation of the Electoral Manifesto: Louis Grech.

Minister of Education and Employment: Evarist Bartolo. *Energy and Conservation of Water:* Konrad Mizzi. *Finance:* Edward Scicluna. *Economy, Investment and Small Business:* Christian Cardona. *Family and Social Solidarity:* Marie-Louise Coleiro Preca. *Foreign Affairs:* Dr George Vella. *Gozo:* Anton Refalo. *Home Affairs and National Security:* Emanuel Mallia. *Health:* Dr Godfrey Farrugia. *Social Dialogue, Consumer Affairs and Civil Liberties:* Helena Dalli. *Sustainable Development, the Environment and Climate Change:* Leo Brincat. *Tourism:* Karmenu Vella. *Transport and Infrastructure:* Joe Mizzi.

   Speaker: Angelo Farrugia.

Government Website: http://www.gov.mt

CURRENT LEADERS

George Abela

Position
President

Introduction
George Abela was sworn in as president on 4 April 2009. He was nominated by then Prime Minister Lawrence Gonzi, the first time a president has been appointed from the opposition (having been a member of the MLP before renouncing political affiliation to take up the office). He was also the first president to be elected with the unanimous approval of parliament.

Early Life
Abela was born in Qormi, a small town in central Malta, in April 1948. The son of a postal worker, he was educated at the University of Malta, gaining a degree in English, Maltese and History before graduating as a lawyer in 1975. He subsequently took up private legal practice specializing in civil, commercial and industrial law. For 25 years he was the legal consultant for Malta's General Workers' Union, representing its membership in negotiations including the 2002 Air Malta rescue plan.

   In 1982, having previously served as treasurer and president of his hometown football team, Abela was appointed president of the Maltese Football Association, a position he held for ten years. He also sat as an arbitrator at the Court of Arbitration for Sport in Lausanne. In 1992 he was elected deputy leader of the MLP, resigning in 1998 in protest at then leader Alfred Sant's call for an early election. Abela served as the Labour Party representative on the Malta–EU Steering and Action Committee (MEUSAC), having been involved in pre-accession talks regarding Malta's bid for European Union membership. He has also served as director of the Central Bank of Malta. In June 2008 he was defeated in a bid for the MLP leadership by Joseph Muscat.

Career in Office
Abela has aimed to harness the political unity evidenced by his unanimous election to the presidency, a largely ceremonial role. A devout Catholic, he has emphasized the importance of family values. Nevertheless, in July 2011 parliament passed legislation legalizing divorce after the proposal won support in a referendum the previous May.

Joseph Muscat

Position
Prime Minister

Introduction
Joseph Muscat became prime minister in March 2013. A former member of the European Parliament (MEP), Muscat is viewed as a modernizer and centrist.

Early Life
Joseph Muscat was born on 22 Jan. 1974 in Pietà, Malta. In 1997 he graduated from the University of Malta with a master's degree in European studies, management and public policy. From 1997–99 he worked as a market intelligence manager, and from 2000–04 as an investment adviser. He was a journalist on the Malta Labour Party radio station from 1992–96 and assistant head of news on its television channel from 1996–97.

   He served as financial secretary of the Labour Youth Forum from 1994–97 and was its acting chairperson in 1997. He also served on the national executive of the Malta Labour Party from

1994 until 2001 and, under Alfred Sant's Labour government, was a member of the national commission for fiscal morality from 1997–98.

   After Labour returned to opposition, Muscat was the party's education spokesman between 2001 and 2003. In 2003 he helped formulate the party's opposition to EU accession. When Malta joined the EU in 2004, Muscat was elected as an MEP. In this capacity he served on the committee for economic and monetary affairs, advocating a reduction in mobile phone roaming charges. In 2006 he produced a report into new regulations for the EU financial services. During this period he completed academic research into multinationals and SMEs in Malta, receiving his PhD from the UK's Bristol University in 2007.

   In June 2008 Muscat successfully contested the leadership of the Malta Labour Party after Alfred Sant resigned. Muscat surrendered his seat in the European Parliament in Oct. 2008 in favour of leading the opposition in Malta. He espoused a more inclusive brand of politics and abbreviated the party's name to the Labour Party. Following gains at the 2009 European Parliamentary elections, he fought the 2013 general election on a platform of lower electricity charges, less bureaucracy, protection for whistle-blowers and reforms to reinvigorate the small business sector. He emerged from the polls on 9 March 2013 with a majority of nine seats.

Career in Office
Muscat took office on 11 March 2013. He faces the challenge of building economic confidence at a time of continued financial uncertainty among Malta's European neighbours.

DEFENCE

The Armed Forces of Malta (AFM) are made up of a Headquarters and three Regiments. In 2007 they had a strength of 1,609 personnel. An Emergency Volunteer Reserve Force was introduced in 1998 on a small scale (40 in 2007). There were also 50 individuals reserves in 2007. In addition to infantry and light air defence artillery weapons, the AFM are equipped with helicopters, light fixed wing and trainer aircraft. There is no conscription.

   Apart from normal military duties, AFM are also responsible for Search and Rescue, airport security, surveillance of Malta's territorial and fishing zones, harbour traffic control and anti-pollution duties.

   In 2008 military expenditure totalled US$49m. (US$122 per capita), representing 0·6% of GDP.

Navy

There is a maritime squadron that operated nine patrol and coastal combatants in 2007.

Air Force

The Air Wing had four combat-capable aircraft in 2007 although they were in storage.

INTERNATIONAL RELATIONS

Malta held a referendum on EU membership on 9 March 2003, in which 53·6% of votes cast were in favour of accession, with 46·4% against. It became a member of the EU on 1 May 2004.

ECONOMY

Services accounted for 65% of GDP in 2009, industry 32% and agriculture 3%.

Overview

Malta is the smallest economy in the eurozone. After joining the EU in 2004, GDP growth was stimulated by a public investment boom largely financed by EU grants. Fiscal policy was directed towards reducing the budget deficit, which had expanded to 56% of GDP by the late 1990s. By 2006 the government had brought the general deficit down to 2·5% of GDP, allowing Malta to join the eurozone on 1 Jan. 2008.

A reliance on FDI inflows and EU funding leaves the economy vulnerable to external shocks. The global financial crisis led to a downturn from late 2008, with manufacturing and tourism the worst hit sectors. However, the economy strongly rebounded in 2010 and has weathered the eurozone debt crisis better than many of its partners, supported by strong private consumption and growth in tourism and services exports. Unemployment fell to 6·4% from a crisis peak of 6·9%, while fiscal adjustments saw the general government deficit stabilize around 3% of GDP. Though spillovers from the eurozone crisis have been contained, Malta has a large financial sector and highly open economy that puts it at risk of contagion. Banking sector assets are worth eight times GDP.

Currency

On 1 Jan. 2008 the euro (EUR) replaced the Maltese lira (MTL) as the legal currency of Malta at the irrevocable conversion rate of Lm0·4293 to one euro. Inflation was 2·0% in 2010 and 2·5% in 2011. Foreign exchange reserves stood at US$2,390m. in March 2005, gold reserves were 4,000 troy oz in May 2005 and total money supply was Lm1,413m. in July 2005.

Budget

In 2008 revenues were €2,132·2m. and expenditures €2,365·3m. Income tax accounted for 34·5% of revenues in 2008 and VAT 21·4%; recurrent expenditure accounted for 90·1% of expenditures.

Malta's budget deficit in 2011 was 2·7% of GDP (2010, 3·7%; 2009, 3·8%). The required target set by the EU is a budget deficit of no more than 3%.

The standard rate of VAT is 18·0% (reduced rates, 7% and 5%).

Performance

The economy contracted by 2·6% in 2009 but grew by 2·5% in 2010 and 2·1% in 2011. Total GDP in 2011 was US$8·9bn.

Banking and Finance

The Central Bank of Malta (*Governor*, Josef Bonnici) was founded in 1968. In Jan. 2004 there were 16 licensed credit institutions carrying out domestic and international banking activities. In addition 13 local financial institutions licensed in terms of the Financial Institutions Act 1994 also provide services that range from exchange bureau related business to merchant banking.

Gross external debt amounted to US$44,388m. in June 2012.

There is a stock exchange in Valletta.

ENERGY AND NATURAL RESOURCES

In 2008, 0·2% of energy consumption came from renewables (wind power, solar power, hydro-electric power, tidal power, geothermal energy and biomass), compared to the European Union average of 10·3%. A target of 10% has been set by the EU for 2020.

Environment

Malta's carbon dioxide emissions from the consumption and flaring of fossil fuels in 2008 were the equivalent of 7·9 tonnes per capita.

Electricity

Electricity is generated at two interconnected thermal power stations located at Marsa (267 MW) and Delimara (304 MW). Marsa is scheduled to close in Oct. 2013. The primary transmission voltages are 132,000, 33,000 and 11,000 volts while the low-voltage system is 400/230V, 50Hz with neutral point earthed. Installed capacity was 571,000 kW in 2006. Production in 2007 was 2·3bn. kWh; consumption per capita was 5,630 kWh.

Oil and Gas

Malta's large offshore area, made up of geological extensions of southeast Sicily, east Tunisia and northwest Libya, contains significant hydrocarbon reserves. Active exploration is at present being carried out by TGS-Nopec, Pancontinental Oil & Gas and TM Services Ltd in offshore areas. Discussions are also under way with oil companies with a view to awarding new licences. The policy of Malta in the oil and gas sector is to intensify exploration by offering oil companies competitive terms and returns that are commensurate with the risk undertaken.

Water

The demand for water in 2003 was 34m. cu. metres. Seawater desalination (Reverse Osmosis Plants) provides 54% of the total potable water requirements.

Agriculture

Malta is self-sufficient in fresh vegetables, pig meat, poultry, eggs and fresh milk. The main crops are potatoes (the spring crop being the country's primary agricultural export), vegetables and fruits, with some items such as tomatoes serving as the main input in the local canning industry. In 2001 there were about 1,524 full-time farmers and 12,589 part-time. There were around 11,959 agricultural holdings and 943 intensive livestock farm units. In 2001 there were 9,000 ha. of arable land and 1,000 ha. of permanent crops.

Agriculture contributes around 2·6% of GDP annually. 2001 production figures (in 1,000 tonnes): potatoes, 25; tomatoes, 18; melons, 12; wheat, 10; onions, 7; cauliflowers, 6.

Livestock in 2001: cattle, 18,417; pigs, 80,481; sheep, 10,376; chickens, 1·9m.

Livestock produce accounted for 60·7% of the total value of agricultural production during 2001.

Fisheries

In 2008 the fishing fleet comprised 1,152 vessels of 10,961 GT. Total catch in 2008 was 1,279 tonnes.

INDUSTRY

Besides manufacturing (food, clothing, chemicals, electrical machinery parts and electronic components and products), the mainstays of the economy are ship repair and shipbuilding, agriculture, small crafts units, tourism and the provision of other services such as the freeport facilities. The majority of state-aided manufacturing enterprises operating in Malta are foreign-owned or with foreign interests. The Malta Development Corporation is the government agency responsible for promoting investment, while the Malta Export Trade Corporation serves as a catalyst to the export of local products.

Labour

The labour supply in Dec. 2002 was 144,016 (females, 40,185), including 35,571 in private direct production (agriculture and fisheries, 2,203; manufacturing, 28,970; oil drilling, construction and quarrying, 6,398), 50,059 in private market services, 47,992 in the public sector (including government departments, armed forces, revenue security corps, independent statutory bodies and companies with government majority shareholding) and 1,206 in temporary employment. There were 7,188 registered unemployed (5·0% of labour supply).

INTERNATIONAL TRADE

Imports are being liberalized. Marsaxlokk is an all-weather freeport zone for transhipment activities. The Malta Export Trade Corporation promotes local exports.

Imports and Exports

In 2010 imports (c.i.f.) amounted to US$4,245·8m. and exports (f.o.b.) US$3,357·5m.

Principal imports in 2010 (in US$1m.) were: machinery and transport equipment (1,923·1); food (480·6); chemicals and related products (465·1). Principal exports in 2010 (in US$1m.) were: machinery and transport equipment (2,029·7); miscellaneous manufactured articles (388·8); food and live animals (271·9).

Main import suppliers (in US$1m.) in 2010 were: Italy (985·0); UK (352·3); Germany (348·0); France (337·4). Main export markets in 2010 (in US$1m.) were: Singapore (495·2); USA (370·4); Hong Kong (344·4); Germany (311·2).

Trade Fairs

The Malta Trade Fairs Corporation organizes the International Fair of Malta (early July).

COMMUNICATIONS

Roads

In 2004 there were 3,096 km of roads, including 185 km of highways. 87·5% of roads are paved. Malta has one of the densest road networks in the world. Motor vehicles in use in 2007 included 203,900 passenger cars, 23,600 vans and lorries, 10,600 motorcycles and mopeds, and 690 buses and coaches. There were 1,209 casualties in traffic accidents in 2007, including 14 fatalities (equivalent to 3·4 fatalities per 100,000 population, giving Malta the lowest death rate in road accidents of any industrialized country).

Civil Aviation

The national carrier is Air Malta, which is 98% state-owned. There were scheduled services in 2003 to around 30 different countries. In 2001 there were 32,652 commercial aircraft movements at Malta International Airport. 2,806,013 passengers and 12,925 tonnes of freight/mail were handled. In 2003 scheduled airline traffic of Maltese-based carriers flew 22m. km and carried 1,309,000 passengers (all on international flights).

Shipping

There is a car ferry between Malta and Gozo. In Jan. 2009 there were 1,487 ships of 300 GT or over registered, totalling 31·65m. GT; Malta's fleet was the sixth largest in terms of the number of ships and the eighth largest on the basis of gross tonnage. Of the 1,487 vessels registered, 498 were general cargo ships, 464 bulk carriers, 346 oil tankers, 78 container ships, 56 passenger ships, 24 liquid gas tankers and 21 chemical tankers.

The Malta Freeport plays an important role in the economy as it is effectively positioned to act as a distribution centre in the Mediterranean.

Telecommunications

In 2008 there were 241,100 main (fixed) telephone lines; mobile phone subscribers numbered 385,600 in 2008 (94·6 per 100 persons). There were 198,800 internet users in the same year. In March 2012 there were 192,000 Facebook users.

SOCIAL INSTITUTIONS

Justice

The number of persons arrested between 1 Jan. 2001 and 31 Oct. 2001 was 5,451; those found guilty numbered 2,180. 184 persons were committed to prison.

In Jan. 2003 total police strength was 1,841 including 107 officers (92 males and 15 females) and 1,734 other ranks (250 females).

Malta abolished the death penalty for all crimes in 2000.

Education

Adult literacy rate, 2005, 92·4% (male, 91·2%; female, 93·5%).

Education is compulsory between the ages of 5 and 16 and free in government schools from kindergarten to university. Kindergarten education is provided for three- and four-year old children. The primary school course lasts six years. In 2003 there were 19,300 children enrolled in 77 state primary schools. There

are education centres for children with special needs, but they are taught in ordinary schools if possible.

Secondary schools, trade schools and junior lyceums provide secondary education in the state sector. At the end of their primary education, pupils sit the 11+ examination to start a secondary education course. Pupils who qualify are admitted in the junior lyceum, while the others attend secondary schools. In 2003–04, 11 junior lyceums had a total of 9,700 students (5,600 girls and 4,100 boys). About 8,200 pupils attend secondary schools. Five centres providing secondary education for under-performing students have a registered student population of about 900, of which 500 are boys. Secondary schools and junior lyceums offer a five-year course leading to the Secondary Education Certificate and the General Certificate of Education, Ordinary Level.

At the end of the five-year secondary course, students may opt to follow a higher academic or technical or vocational course of one to four years. The academic courses generally lead to Intermediate and Advanced Level examinations set by the British universities. The Matriculation Certificate, which qualifies students for admission to university, is a broad-based holistic qualification covering—among others—the humanities and the sciences, together with systems of knowledge.

About 35% of the student population attend non-state schools, from kindergarten to higher secondary level. In Oct. 2003 there were about 25,700 pupils attending non-state schools, 800 of whom were at post-compulsory secondary level, 17,100 were in schools run by the Roman Catholic Church, while 8,500 students were attending private schools. Under an agreement between the government and the Church, the government subsidizes Church schools and students attending these schools do not pay any fees. During 2001 the government introduced tax rebates for parents whose children attended independent schools.

More than 9,800 students (including 750 from overseas) were following courses at the University in 2003. University students receive a stipend.

A post-compulsory vocational college, the Malta College of Arts, Science and Technology, provides vocational and technical courses up to degree level. In Oct. 2003 about 9,000 students (47% of which were females) were following post-compulsory education in state colleges and institutes.

In 2008 public expenditure on education came to 5·9% of GDP.

Health

In 2006 there were 1,564 doctors, 190 dentists, 790 pharmacists and 2,411 nurses and midwives. There were eight hospitals (three private) with 2,122 beds in 2003. There are also nine health centres.

Welfare

Legislation provides a national contributory insurance scheme and also for the payment of non-contributory allowances, assistances and pensions. It covers the payment of marriage grants, maternity benefits, child allowances, parental allowances, disabled child allowance, family bonus, sickness benefit, injury benefits, disablement benefits, unemployment benefit, contributory pensions in respect of retirement, invalidity and widowhood, and non-contributory medical assistance, free medical aids, social assistance, a carers' pension and pensions for the visually impaired, disabled or severely disabled persons and the aged.

From Jan. 2007 the statutory retirement age has been gradually increased from 61 years for men and 60 years for women to reach 65 for both sexes by 2015.

RELIGION

98% of the population belong to the Roman Catholic Church, which is established by law as the religion of the country, although full liberty of conscience and freedom of worship are guaranteed. In Feb. 2013 there was one cardinal.

CULTURE

World Heritage Sites

Malta has three sites on the UNESCO World Heritage List (all inscribed on the list in 1980): Hal Saflieni Hypogeum, a prehistoric underground necropolis; the City of Valletta, a highly concentrated centre marked by the influences of Romans, Byzantines and Arabs and the Knights of St John; and the Megalithic Temples of Malta (reinscribed in 1992), seven temples on Malta and Gozo containing Bronze Age structures.

Press

In 2008 there were two English paid-for dailies (*The Times* and *The Malta Independent*) and two Maltese dailies (*In-Nazzjon* and *L-Orizzont*). There were seven paid-for non-dailies and six Sunday newspapers (three in English and three in Maltese).

Tourism

Tourism is a major component of the Maltese economy. In 2010 there were 1,336,000 staying foreign tourists, spending US\$1,130m.; 31% of tourists in 2010 were from the UK and 16% from Italy. Cruise passenger visits totalled 491,201 in 2010 (more than double the 2000 total of 170,782).

Festivals

Major festivals include the Malta Song Festival; Carnival Festivals at Valletta (Feb.); History and Elegance Festival at Valletta (April); National Folk Singing; Malta International Arts Festival; Malta Jazz Festival; International Food and Beer Festival (June/July); Festa Season (June–Sept.); Malta International Choir Festival (Nov.).

DIPLOMATIC REPRESENTATIVES

Of Malta in the United Kingdom (36–38 Piccadilly, London, W1J 0LE)
High Commissioner: Joseph Zammit Tabona.

Of the United Kingdom in Malta (Whitehall Mansions, Ta'Xbiex Seafront, Ta'Xbiex, XBX 1026)
High Commissioner: Rob Luke.

Of Malta in the USA (2017 Connecticut Ave., NW, Washington, D.C., 20008)
Ambassador: Joseph Cole.

Of the USA in Malta (Ta' Qali National Park, Attard, ATD 4000)
Ambassador: Gina Abercrombie-Winstanley.

Of Malta to the United Nations
Ambassador: Christopher Grima.

Of Malta to the European Union
Permanent Representative: Marlene Bonnici.

FURTHER READING

National Statistics Office (Lascaris, Valletta). *Abstract of Statistics*, a quarterly digest of statistics, quarterly and annual trade returns, annual vital statistics and annual publications on shipping and aviation, education, agriculture, industry, National Accounts and Balance of Payments. *Malta in Figures 2012.*

Department of Information (3 Castille Place, Valletta). *The Malta Government Gazette, Malta Information, Economic Survey [year], Reports on the Working of Government Departments, Business Opportunities on Malta, Acts of Parliament and Subsidiary Legislation, Laws of Malta, Constitution of Malta 1992.*

Central Bank of Malta. *Annual Reports.*

Chamber of Commerce (annual). *Trade Directory.*

Berg, W. G., *Historical Dictionary of Malta.* 1995

Pace, Roderick, *The European Union's Mediterranean Enlargement: Cyprus and Malta.* 2006

The Malta Year Book. Annual

National Statistical Office: National Statistics Office, Lascaris, Valletta CMR 02.
Website: http://www.nso.gov.mt

MARSHALL ISLANDS

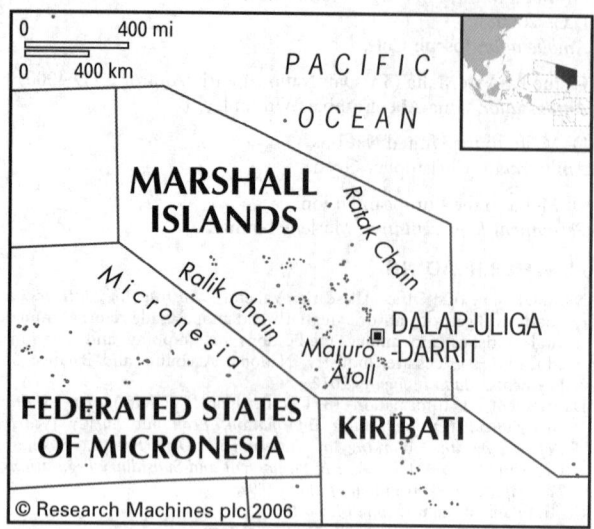

Republic of the Marshall Islands

Capital: Dalap-Uliga-Darrit
Population, 2011: 53,000
GNI per capita, 2010: US$3,756
Internet domain extension: .mh

KEY HISTORICAL EVENTS

The Pacific archipelago was populated by emigrants from southeast Asia from around 2000 BC and first documented by Portuguese mariners in 1528. The islands owe their name to the English seafarer, John Marshall, who visited in 1788. They became part of the protectorate of German New Guinea in 1886 and administrative affairs were managed by private German and Australian interests. Japan seized control in 1914 and received a League of Nations mandate over the islands in 1919. The Marshall Islands were occupied by Allied forces in 1944 and became part of the UN Trust Territory of the Pacific Islands on 18 July 1947 (administered by the USA). On 21 Oct. 1986 the islands gained independence. A Compact of Free Association with the USA that came into force at the time was extended by 20 years in May 2004.

TERRITORY AND POPULATION

The Marshall Islands lie in the North Pacific Ocean north of Kiribati and east of Micronesia, and consist of an archipelago of 31 coral atolls, five single islands and 1,152 islets strung out in two chains, eastern and western. Of these, 25 atolls and islands are inhabited. The land area is 181 sq. km (70 sq. miles). At the 2011 census the population was 53,158 (27,243 males); density, 294 per sq. km. The capital is Dalap-Uliga-Darrit (1999 population, 15,486) on Majuro Atoll in the eastern chain. The largest atoll in the western chain is Kwajalein, containing the only other town, Ebeye (1999 population, 9,345). The two archipelagic island chains of Bikini and Enewetak are former US nuclear test sites; Kwajalein is now used as a US missile test range. The islands lay claim to the US territory of Wake Island. In 2011 the population was 72·1% urban. About 88% of the population are Marshallese, a Micronesian people.

English is universally spoken and is the official language. Two major Marshallese dialects from the Malayo-Polynesian family and Japanese are also spoken.

SOCIAL STATISTICS

2006 births, estimate, 1,576; deaths, 318. 2006 rates per 1,000 population, estimates: birth, 30·3; death, 6·1. Infant mortality rate, 2010, 22 per 1,000 live births. Life expectancy, 2008: male, 68·9 years; female, 73·0. Annual population growth rate, 1998–2008, 1·6%; fertility rate, 2008, 3·7 births per woman.

CLIMATE

Hot and humid, with wet season from May to Nov. The islands border the typhoon belt. Jaluit, Jan. 81°F (27·2°C), July 82°F (27·8°C). Annual rainfall 161" (4,034 mm).

CONSTITUTION AND GOVERNMENT

Under the constitution which came into force on 1 May 1979, the Marshall Islands form a republic with a *President* as head of state and government, who is elected for four-year terms by the parliament. The parliament consists of a 33-member *House of Assembly* (Nitijela), directly elected by popular vote for four-year terms. There is also a 12-member appointed *Council of Chiefs* (Iroij) which has a consultative and advisory capacity on matters affecting customary law and practice.

National Anthem
'Forever Marshall Islands'; words and tune by Amata Kabua.

RECENT ELECTIONS

At the House of Assembly elections on 21 Nov. 2011, 33 members were elected. The Aelon Kein Ad party (AKA), joined by independent candidates, won a majority of seats. An indirect presidential election was held on 3 Jan. 2012. Christopher Loeak won with 21 votes against 11 for incumbent Jurelang Zedkaia.

CURRENT GOVERNMENT

President: Christopher Loeak; b. 1952 (AKA; took office on 10 Jan. 2012).

In March 2013 the government comprised:

Minister in Assistance to the President: Tony A. de Brum. *Education:* Hilda C. Heine. *Finance:* Dennis P. Momotaro. *Foreign Affairs:* Phillip H. Muller. *Health:* David Kabua. *Internal Affairs:* Wilbur Heine. *Justice:* Thomas Heine. *Natural Resources and Development:* Michael Konelios. *Public Works:* Hiroshi V. Yamamura. *Transportation and Communications:* Rien Morris.

Office of the President: http://www.rmigovernment.org/index.jsp

CURRENT LEADERS

Christopher Loeak

Position
President

Introduction
Christopher Loeak was sworn into office on 10 Jan. 2012 following his appointment by the 33-member *Nitijela* (parliament) on 3 Jan. 2012. He is scheduled to serve a four-year term.

Early Life
Christopher Loeak was born on 11 Nov. 1952 on Ailinglaplap Atoll to a family of tribal leaders. He was educated at Marshall Island High School before attending the Hawaii Pacific College. He completed his legal training at Gonzaga University, Washington in 1982.

Loeak then joined the Kwajalein Atoll Corporation as a lobbyist based in Hawaii. In 1983, whilst still in Hawaii, he unsuccessfully stood for the Ailinglaplap parliamentary seat. Two years later he contested it again and won, serving in the cabinet of Amata

Kabua as minister of justice from 1988–92. In 1992 he became minister of social services and in 1996 took on the education portfolio in Kunio Lemari's government, a post he retained after Imata Kabua (the younger cousin of Amata Kabua) took over as president.

Loeak was appointed minister for the Ralik Chain of islands in 1998 and a year later was also named minister in assistance to the president. Loeak was re-elected to the *Nitijela* in 2007 and in 2008 was reappointed as minister in assistance to President Litokwa Tomeing. He successfully defended his seat in the Nov. 2011 elections and on 2 Jan. 2012 was elected president when he won parliamentary backing by 21 votes to 11 against the incumbent president, Jurelang Zedkaia.

Career in Office
Loeak's main challenge is to raise levels of education and to gain a measure of financial independence from the USA, on which the Marshall Islands are heavily reliant for financial aid.

DEFENCE
The Compact of Free Association gives the USA responsibility for defence in return for US assistance. In 2003 the US lease of Kwajalein Atoll, a missile testing site, was extended by 50 years.

ECONOMY
Agriculture accounts for approximately 15% of GDP, industry 13% and services 72%.

Overview
The 1986 Compact of Free Association with the USA was renegotiated from 1999–2003 to allow for an increase in aid. This led to seven years of continuous growth. However, growth slowed in 2005, and the economy contracted in 2008 and 2009. Growth returned in 2010, driven by the fisheries and export sectors. Unemployment, particularly among young people, remains high, contributing to increased emigration to the USA, made easier under the Compact agreement.

The economy is highly vulnerable to external shocks owing to its geographical isolation, narrow production base and reliance on foreign aid. The planned steady decline in Compact grants means fiscal consolidation and structural reforms will be needed to ensure long-term fiscal sustainability, although progress towards this goal has been held back by the global financial crisis. Structural reforms aim to stimulate the private sector so that it can replace the government as the primary engine of growth.

Migration and climate change constitute long-term challenges.

Currency
US currency is used. Inflation was 17·5% in 2008 although it dropped to 9·6% in 2009.

Budget
Revenue in 2007 was US$98·9m.; expenditure was US$99·9m. Under the terms of the Compact of Free Association, the USA provides approximately US$65m. a year in aid. The fiscal year begins on 1 Oct.

Performance
Total GDP in 2011 was US$174m.; GDP per capita in 2006 was US$2,770. Real GDP growth was 5·2% in 2009–10.

Banking and Finance
There are three banks: the Bank of Marshall Islands, the Marshall Islands Development Bank and the Bank of Guam.

ENERGY AND NATURAL RESOURCES

Electricity
Total installed capacity (2007 estimate), 17,200 kW. Production (2007 estimate), 108m. kWh.

Minerals
High-grade phosphate deposits are mined on Ailinglaplap Atoll. Deep-seabed minerals are an important natural resource.

Agriculture
A small amount of agricultural produce is exported: coconuts, tomatoes, melons and breadfruit. Other important crops include copra, taro, cassava and sweet potatoes. Pigs and chickens constitute the main livestock. In 2007 there were approximately 2,000 ha. of arable land and 8,000 ha. of permanent crop land.

Forestry
There were 13,000 ha. of forest in 2010, or 70% of the total land area.

Fisheries
Total catch in 2010 amounted to 59,730 tonnes. There is a commercial tuna-fishing industry with a processing plant on Majuro. Seaweed is cultivated. Fisheries offer some of the best opportunities for economic growth.

INDUSTRY
The main industries are copra, fish, tourism, handicrafts (items made from shell, wood and pearl), mining, manufacturing, construction and power.

Labour
In 2004 the labour force was estimated at 17,342. Approximately 34% were unemployed in 2004. In 2007, 37% of employed people worked in the private sector. In the same year 35% of workers were employed in public administration, 18% in wholesale and retail trade, 12% in extra-territorial organizations and bodies and 8% in construction. Agriculture, hunting, forestry and fishing accounted for just 3%.

INTERNATIONAL TRADE
The Compact of Free Association with the USA is the major source of income for the Marshall Islands, and accounts for about 70% of total GDP.

Imports and Exports
Imports (mainly oil) were US$54·7m. in 2000; exports, US$9·1m. Main import suppliers in 2000: USA, 56·7%; Australia, 10·0%; Japan, 9·3%; Hong Kong, 5·9%. The USA accounted for approximately 71·0% of exports in 2000. Main exports: coconut oil, copra cake, chilled and frozen fish, pet fish, shells and handicrafts.

COMMUNICATIONS

Roads
There are paved roads on major islands (Majuro, Kwajalein); roads are otherwise stone-, coral- or laterite-surfaced. In 2004 there were 1,555 passenger cars and 159 trucks and buses.

Civil Aviation
There were two international airports and 30 airfields on 24 atolls and islands in 2004. The main airport is Majuro International. In 2003 there were flights to Guam, Honolulu, Johnston Island, Kiribati and Micronesia as well as domestic services. The national carrier is Air Marshall Islands.

Shipping
Majuro is the main port. In Jan. 2009 there were 1,125 ships of 300 GT or over registered, totalling 41·58m. GT (a figure exceeded only by the fleets of Panama, Liberia and the Bahamas). Of the 1,125 vessels registered, 434 were oil tankers, 311 bulk carriers, 192 container ships, 110 general cargo ships, 56 liquid gas tankers, 15 chemical tankers and seven passenger ships. The ship's register of the Marshall Islands is a flag of convenience register.

Telecommunications

In 2010 there were 4,400 main (fixed) telephone lines. There is a US satellite communications system on Kwajalein and two Intelsat satellite earth stations (Pacific Ocean). The National Telecommunications Authority provides domestic and international services. Mobile phone subscribers numbered 1,000 in 2008 (1·7 per 100 persons) and there were 2,200 internet users (3·6 per 100 persons).

SOCIAL INSTITUTIONS

Justice

The Supreme Court is situated on Majuro. There is also a High Court, a District Court and 23 Community Courts. A Traditional Court deals with disputes involving land properties and customs.

Education

In 2008–09 there were 8,000 pupils enrolled in primary schools and 5,000 pupils in secondary schools. There is a College of the Marshall Islands and a subsidiary of the University of the South Pacific on Majuro. In 2004 public expenditure on education came to 11·8% of GDP.

Health

There were two hospitals in 2003, with a total of 140 beds. There were 31 doctors, 189 nurses and four dentists in 2003; and two pharmacists in 2000.

RELIGION

The population is mainly Protestant, with Roman Catholics next. Other Churches and denominations include Latter-day Saints (Mormons), Jehovah's Witnesses, Baptists, Bahais, Seventh Day Adventists and Assembly of God.

CULTURE

World Heritage Sites

The Marshall Islands has one site on the UNESCO World Heritage List: Bikini Atoll (inscribed on the list in 2010), the site used by the US military to carry out 67 nuclear tests between 1946 and 1958.

Press

There is a publication called *Micronitor* (The Marshall Islands Journal).

Tourism

In 2007 there were 6,959 foreign tourists. Tourism offers one of the best opportunities for economic growth.

Festivals

Custom Day and the Annual Canoe Race are the main festivals.

DIPLOMATIC REPRESENTATIVES

Of the United Kingdom in the Marshall Islands
Ambassador: Stephen Lillie (resides in Manila, Philippines).

Of the Marshall Islands in the USA (2433 Massachusetts Ave., NW, Washington, D.C., 20008)
Ambassador: Charles Paul.

Of the USA in the Marshall Islands (PO Box 1379, Majuro, MH 96960)
Ambassador: Tom Armbuster.

Of the Marshall Islands to the United Nations
Ambassador: Amatlain Elizabeth Kabua.

Of the Marshall Islands to the European Union
Ambassador: Vacant.

FURTHER READING

Barker, Holly, *Bravo for the Marshallese: Regaining Control in a Post-Nuclear, Post-Colonial World.* 2003

National Statistical Office: Economic Policy, Planning and Statistics Office (EPPSO), Office of the President, PO Box 7, Majuro, MH 96960.
Website: http://www.spc.int/prism/country/mh/stats

Explore the world at
www.statesmansyearbook.com

MAURITANIA

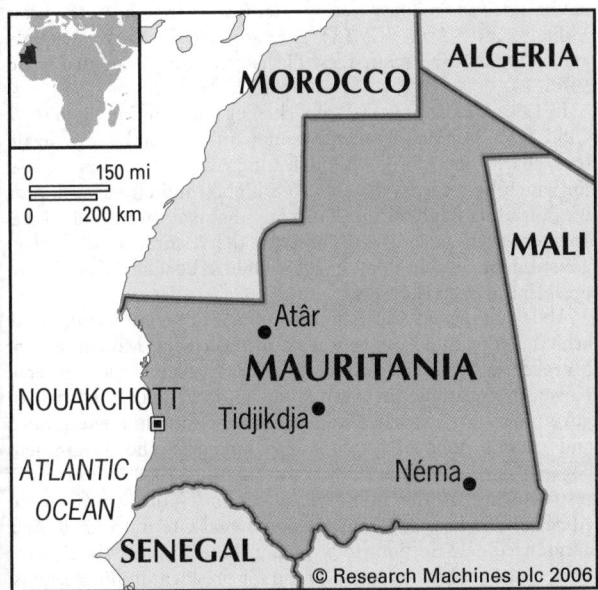

Region	Area	Population	Chief town
Açâba	36,600	242,265	Kiffa
Adrar	215,300	69,542	Atâr
Brakna	33,800	247,006	Aleg
Dakhlet Nouâdhibou	22,300	79,516	Nouâdhibou
Gorgol	13,600	242,711	Kaédi
Guidimaka	10,300	177,707	Sélibaby
Hodh ech-Chargui	182,700	281,600	Néma
Hodh el-Gharbi	53,400	212,156	Aïoun el Atrouss
Inchiri	46,800	11,500	Akjoujt
Nouakchott District	1,000	558,195	Nouakchott
Tagant	95,200	76,620	Tidjikdja
Tiris Zemmour	252,900	41,121	Zouérate
Trarza	67,800	268,220	Rosso

Principal town (2000 census population): 664,759 including the suburbs of Nouâdhibou and Kaédi.

In 2000 there were also 0·23m. nomads.

The major ethnic groups are (with numbers in 1993): Moors (of mixed Arab, Berber and African origin), 1,513,400; Wolof, 147,000; Tukulor, 114,600; Soninke, 60,000.

Arabic is the official language. French no longer has official status. Pulaar, Soninke and Wolof are national languages.

Al-Jumhuriyah al-Islamiyah al-Muritaniyah (Islamic Republic of Mauritania)

Capital: Nouakchott
Population projection, 2015: 3·87m.
GNI per capita, 2011: (PPP$) 1,859
HDI/world rank: 0·453/159
Internet domain extension: .mr

KEY HISTORICAL EVENTS

Mauritania became a French protectorate in 1903 and a colony in 1920. It achieved full independence on 28 Nov. 1960. Mauritania was made a one-party state in 1964.

The 1980s were marked by territorial disputes with Morocco and Senegal. Seizing power in 1984, Lieut.-Col. Maaouya Ould Sid'Ahmed Taya prepared the way for a new constitution allowing for a multi-party political system, which also gave extensive powers to the president. A coup attempt against Ould Taya failed in June 2003. But in Aug. 2005, while out of the country, he was overthrown in a bloodless coup by a group of army officers who set up a Military Council for Justice and Democracy. Under the leadership of Col. Ely Ould Mohammed Vall, the Council pledged to hold democratic elections within two years. In June 2006 a new constitution was approved, limiting the president to two 5-year terms. Sidi Mohamed Ould Cheikh Abdallahi won presidential elections in March 2007 but was ousted in a coup in Aug. 2008.

TERRITORY AND POPULATION

Mauritania is bounded west by the Atlantic Ocean, north by Western Sahara, northeast by Algeria, east and southeast by Mali, and south by Senegal. The total area is 1,030,700 sq. km (398,000 sq. miles) of which 47% is desert, and the population at the census of 2000 was 2,508,159; density, 2·4 per sq. km. In 2011, 41·7% of the population lived in urban areas.

The UN gives a projected population for 2015 of 3·87m.

Area (in sq. km), population (at the 2000 census) and chief towns of the Nouakchott Capital District and 12 regions:

SOCIAL STATISTICS

2008 estimates: births, 108,000; deaths, 33,000. 2008 rates, estimate (per 1,000 population): births, 33·6; deaths, 10·3. Expectation of life at birth in 2007 was 54·7 years for males and 58·5 for females. Annual population growth rate, 2000–08, 2·6%. Infant mortality, 2010, 75 per 1,000 live births; fertility rate, 2008, 4·5 births per woman.

CLIMATE

A tropical climate, but conditions are generally arid, even near the coast, where the only appreciable rains come in July to Sept. Nouakchott, Jan. 71°F (21·7°C), July 82°F (27·8°C). Annual rainfall 6" (158 mm).

CONSTITUTION AND GOVERNMENT

A referendum was held on 25 June 2006 to approve a new constitution. Turnout was 76·5%; 96·99% of votes cast were in favour.

The constitution imposes a limit of two five-year terms for a president, to be elected by popular vote. It also sets a maximum age of 75 for a president. There is a 56-member *Senate* (53 elected and three appointed) and a 95-member *National Assembly*.

Following the coup d'état in Aug. 2008 a transitional government took power, headed by an 11-member High Council of State (all of whom came from the military). Nonetheless, the junta has retained the constitution and vowed to protect the country's democratic institutions. In April 2009 Gen. Mohamed Ould Abdel Aziz stood down as head of government to run in the presidential elections of July 2009.

National Anthem

'Kun lil-ilahi nasiran' ('Be a helper for God'); words by Baba Ould Cheikh, tune by Tolia Nikiprowetzky.

RECENT ELECTIONS

Presidential elections originally scheduled for 6 June 2009 were held on 18 July 2009. Mohamed Ould Abdel Aziz, the leader of the 2008 coup, received 52·6% of votes cast, Messaoud Ould Boulkheir 16·3%, Ahmed Ould Daddah 13·7%, Mohamed Jemil Ould Mansour 4·8%, Ibrahima Moctar Sarr 4·6% and Ely Ould Mohamed Vall 3·8%. There were four other candidates. Turnout was 65·1%. The opposition contested the results but the appeal was rejected in the Constitutional Court.

Elections for the National Assembly—the first since the coup in Aug. 2005 that ended 20 years of authoritarian rule—were held on 19 Nov. and 3 Dec. 2006. An 11-member Coalition of Forces for Democratic Change, comprising groups opposed to former President Taya, won 41 of the 95 seats; the parties supporting the former president, 13 (comprising the Republican Party for Renewal and Democracy 7, the Rally for Democracy and Unity 3 and the Union for Democracy and Progress 3); Alternative (Mauritania), 1; the Union of the Democratic Centre, 1; the National Rally for Liberty, Democracy and Equality, 1; ind., 38. In the Senate elections of 21 Jan. and 4 Feb. 2007 Coalition of Forces for Democratic Change won 15 seats; Republican Party for Democracy and Renewal, 3; Union for Democracy and Progress, 1; ind., 34.

CURRENT GOVERNMENT

President: Gen. Mohamed Ould Abdel Aziz; b. 1956 (since 5 Aug. 2009).

Prime Minister: Moulaye Ould Mohamed Laghdaf; b. 1957 (ind.; in office since 14 Aug. 2008).

In March 2013 the government comprised:

Minister of State for National Education, Higher Education and Scientific Research: Ahmed Ould Bahia.

Minister of Justice: Abidine Ould El Khaire. *Foreign Affairs and Co-operation:* Hamadi Ould Hamadi. *National Defence:* Ahmedou Ould Dey Ould Mohamed Radhi. *Interior and Decentralization:* Mohamed Ould Boilil. *Economic Affairs and Development:* Sidi Ould Tah. *Finance:* Thiam Dioumbar. *Islamic Affairs and Basic Education:* Ahmed Ould Neini. *Civil Service and Administrative Modernization:* Maty Mint Hamadi. *Health:* Ahmedou Ould Hademine Ould Jervoune. *Oil, Energy and Mines:* Taleb Ould Abdi Vall. *Fisheries and Maritime Economy:* Ghdafna Ould Eyih. *Trade, Handicrafts, Tourism and Industry:* Bomba Ould Daramane. *Housing, Town Planning and Land Management:* Yahya Ba. *Rural Development:* Brahim Ould M'Bareck Ould Mohamed El Moctar. *Equipment and Transport:* Yahya Ould Hademine. *Water Supply and Sanitation:* Mohamed Lemine Ould Aboye. *Culture, Youth and Sports:* Cissé Mint Cheikh Ould Boyde. *Communications and Relations with Parliament:* Hamdy Ould Mahjoub. *Social Affairs, Children and Family:* Moulaty Mint El Moctar.

Government Website (French and Arabic only):
   http://www.mauritania.mr

CURRENT LEADERS

Gen. Mohamed Ould Abdel Aziz

Position
President

Introduction
Gen. Mohamed Ould Abdel Aziz was elected president of Mauritania in July 2009, having called elections after seizing power in a coup against his predecessor Sidi Mohamed Ould Cheikh Abdallahi in Aug. 2008. Aziz led the country at the head of a governing council until April 2009.

Early Life
Ould Abdel Aziz was born on 20 Dec. 1956 in Akjoujt, Mauritania, half way between the coastal capital and the Sahara desert. He was born into the Oulad Bou Sbaa Berber-Arab tribe, from which a number of powerful Mauritanian figures have emerged. In 1977 he attended officer training at the Royal Military Academy in Meknès, Morocco. After a stint in the army, he returned to the Military Academy in 1980 to receive training in logistics.

From 1978 Mauritania had a string of *de facto* governments following the overthrow of the civilian president Ould Daddah. Aziz became a staff officer in 1982 and in 1984 was appointed aide-de-camp to Col. Maaouya Ould Sid'Ahmed Taya, the nation's *de facto* military leader. After further training at the École Militaire Inter-Arme (EMIA), Aziz attained the rank of captain and, at Ould Taya's request, established an elite presidential guard (BASEP).

In 1992 Aziz became the army chief of staff, continuing to serve Ould Taya. In 1998, by then a lieutenant colonel, he once again took charge of BASEP. After defeating coups in 2003 and 2004, for which he received the country's highest military award, Aziz was a leading figure in the 2005 coup that overthrew Ould Taya. Aziz emerged as a driving force for the restoration of civilian government and in March 2007 Ould Abdallahi was elected president in open elections.

Abdallahi named Aziz (still head of BASEP) commander of the armed forces, and even sent Aziz to meet King Mohammed of Morocco in an official state visit. In 2008 Aziz became a general. However, relations between the military and the civilian government grew strained as the economic situation deteriorated and Islamic political forces gained strength. The government began a curb of military power and in Aug. 2008 announced a restructuring of the military. In response, the army ousted Abdallahi. As head of a military-dominated interim council, Aziz announced new elections to be held in June 2009. In March 2009 Aziz announced his intention to run for president in the elections, rescheduled for July.

Career in Office
His assumption of office in Aug. 2009 was attacked by France (the former colonial power), the African Union (AU), the USA and Algeria. However, he was received warmly in neighbouring Morocco and in Libya.

The AU continued to impose sanctions in 2010, including a travel ban in AU countries for military personnel who had supported the coup and a seizure of their assets in AU banks. Aziz's electoral pledges included re-establishing civilian governance, improving national unity and consolidating republican institutions. In July 2010 his government adopted new anti-terrorism legislation to give security forces greater powers.

There were opposition demonstrations in May 2012 calling on Aziz to resign, and in Oct. he was wounded in a shooting incident, apparently by troops at a military checkpoint. Described officially as an accident, other sources claimed it was a coup attempt.

DEFENCE

Conscription is authorized for two years. Defence expenditure in 2008 totalled US$20m. (US$7 per capita), representing 0·7% of GDP.

Army

There are six military regions. Army strength was 15,000 in 2007. In addition there was a Gendarmerie of about 3,000 and a National Guard of 2,000.

Navy

The Navy, some 620 strong in 2007, has bases at Nouâdhibou and Nouakchott.

Air Force

Personnel (2007), 250 with 18 aircraft (none combat-capable).

ECONOMY

In 2006 agriculture accounted for 13·1% of GDP, industry 47·8% and services 39·1%.

Overview
Mauritania is a developing economy with limited agrarian resources but extensive mineral deposits and rich fishing grounds. Mauritania has a long history of mining and the sector represents nearly 75% of exports but less than 3% of employment. Substantial gold and copper deposits have been exploited since 2006. However, over-dependence on mining and fisheries for export earnings leaves the economy vulnerable to external shocks.

Oil was discovered in 2001. Production in the main oil field started in 2006 but quickly ran into technical difficulties, with output falling from 75,000 bbls per day in early 2006 to 23,000 by the end of that year. Production has steadily declined ever since, with levels at 5,900 bbls per day in 2010. Deposits are expected to be depleted around 2015.

Mauritania's economy was hit hard in 2008 by a military coup, which led to political crisis and the suspension of donor assistance, coupled with the global recession and a fuel and food crisis. Economic growth, however, has rebounded in recent years. Nonetheless, Mauritania faces key challenges, including diversifying its economy and strengthening the business climate (it was 159th of 183 countries in the World Bank's 2012 *Doing Business* report). Poverty levels remain high. According to the 2010 Human Development Report, 21% of the population live below the international poverty line of US$1·25 (PPP) per day.

Currency
The monetary unit is the *ouguiya* (MRO) which is divided into five *khoums*. The ouguiya was devalued in Oct. 1992 by 28% and again in July 1998 by 18%. Inflation was 6·3% in 2010 and 5·7% in 2011. Total money supply in March 2004 was 34,318m. ouguiya, foreign exchange reserves were US$390m. and gold reserves 12,000 troy oz.

Budget
Revenues were 188·5bn. ouguiya in 2009 and expenditures 242·9bn. ouguiya.

VAT is 18%.

Performance
The economy contracted by 1·2% in 2009 but grew by 5·1% in 2010 and 4·0% in 2011. Mauritania's total GDP in 2011 was US$4·1bn.

Banking and Finance
The Central Bank (created 1973) is the bank of issue (*Governor*, Sid'Ahmed Ould Raiss). In 2010 there were ten commercial banks. Bank deposits totalled 35·2bn. ouguiya in 2009.

In 2010 external debt totalled US$2,461m., equivalent to 67·0% of GNI.

ENERGY AND NATURAL RESOURCES

Environment
In 2008 carbon dioxide emissions from the consumption and flaring of fossil fuels were the equivalent of 0·9 tonnes per capita. An *Environmental Performance Index* compiled in 2008 ranked Mauritania 146th in the world out of 149 countries analysed, with 44·2%. The index examined various factors in six areas—air pollution, biodiversity and habitat, climate change, environmental health, productive natural resources and water resources.

Electricity
Installed capacity was 156,000 kW in 2007. Production in 2007 was 587m. kWh; consumption per capita was 227 kWh.

Oil and Gas
Oil was discovered off the coast of Mauritania in 2001. Production began in Feb. 2006, initially with 75,000 bbls a day.

Minerals
There are reserves of copper, gold, phosphate, gypsum, platinum and diamonds. Iron ore, 11·2m. tonnes of which were mined in 2006, accounts for about 69% of exports. Prospecting licences have also been issued for diamonds.

Agriculture
Only 1% of the country receives enough rain to grow crops, so agriculture is mainly confined to the south, in the Senegal river valley. In 2007 the agricultural population numbered 1,598,000 of whom 674,000 were economically active. There were an estimated 450,000 ha. of arable land in 2007 and 12,000 ha. of permanent crops. Production (2006, in 1,000 tonnes): sorghum, 84; rice, 70; dates, 20; maize, 17; dry beans, 10 (estimate); dry peas, 10 (estimate); yams, 3 (estimate); millet, 2; potatoes, 2 (estimate).

Herding is the main occupation of the rural population and accounted for 11% of GDP in 2005. In 2006–07 there were 17·15m. sheep and goats; 1·37m. cattle; 1·35m. camels; 4·3m. chickens (estimate).

Forestry
There were 0·24m. ha. of forests in 2010, chiefly in the southern regions, where wild acacias yield the main product, gum arabic. In 2007, 1·71m. cu. metres of roundwood were cut.

Fisheries
Total catch in 2010 was 276,238 tonnes, of which 95% came from marine waters. Mauritania's coastal waters are among the world's most abundant fishing areas, earning it significant amounts of hard currency through licensing agreements. The fishing sector accounts for an estimated 20% of Mauritania's national budget.

INDUSTRY
Output, 2002 (in tonnes): residual fuel oil, 364,000; petrol, 235,000; distillate fuel oil, 155,000; frozen and chilled fish (2001), 27,000; hides and skins (2001), 5,400.

Labour
In 2008 the economically active population was estimated at 1,353,000 (58% males). Of those in employment, 24·6% worked in commerce, 15·2% in administration, 14·9% in services and 10·5% in agriculture. The unemployment rate in 2008 was 31·2%.

INTERNATIONAL TRADE
In Feb. 1989 Mauritania signed a treaty of economic co-operation with the four other Maghreb countries—Algeria, Libya, Morocco and Tunisia.

Imports and Exports
In 2007 imports were valued at US$1,376·6m. and exports at US$1,302·4m. Main imports in 2007 were petroleum products (30·6% of total imports), foodstuffs and capital goods. Main exports in 2007 were iron ore (41·3% of total exports), oil and fish products. Main import suppliers in 2007 were France (15·9%), Brazil (6·0%), China (5·8%) and Belgium (4·5%). Principal export markets in 2006 were China (26·3%), Italy (11·8%), France (10·2%) and Belgium (6·8%).

COMMUNICATIONS

Roads
There were about 11,066 km of roads in 2006, of which 26·8% were paved. In 2002 there were 7,100 passenger cars and 5,700 commercial vehicles.

Rail
A 704-km railway links Zouérate with the port of Point-Central, 10 km south of Nouâdhibou, and is used primarily for iron ore exports. In 2005 it carried 10·8m. tonnes of freight.

Civil Aviation
There are international airports at Nouakchott, Nouâdhibou and Néma. In 2003 scheduled airline traffic of Mauritania-based carriers flew 1m. km, carrying 116,000 passengers (14,000 on international flights).

Shipping

The major ports are at Point-Central (for mineral exports), Nouakchott and Nouâdhibou.

Telecommunications

In 2008 there were 76,400 main (fixed) telephone lines; mobile phone subscribers numbered 2,092,000 in 2008 (65·1 per 100 persons). There were 60,000 internet users in 2008.

SOCIAL INSTITUTIONS

Justice

There are courts of first instance at Nouakchott, Atâr, Kaédi, Aïoun el Atrouss and Kiffa. The Appeal Court and Supreme Court are situated in Nouakchott. Islamic jurisprudence was adopted in 1980.

The population in penal institutions in 2010 was 1,700 (50 per 100,000 of national population).

Education

Basic education is compulsory for all children between the ages of six and 14. In 2007 there were 483,776 pupils and 11,379 teaching staff in primary schools, 102,130 secondary level pupils with 3,843 teaching staff and 11,794 tertiary level students with (2006) 353 academic staff. The University of Nouakchott, founded in 1981, is the leading tertiary education institution. Adult literacy rate in 2008 was 57%.

Public expenditure on education came to 2·8% of GNI in 2006.

Health

In 2006 there were four hospital beds per 10,000 persons. There were 477 physicians, 73 dentists, 1,636 nurses and 328 midwives in 2005.

In 2006, 60% of the population had access to improved drinking water sources.

RELIGION

Over 99% of Mauritanians are Sunni Muslim, mainly of the Qadiriyah sect.

CULTURE

World Heritage Sites

Mauritania has two sites on the UNESCO World Heritage List: Banc d'Arguin National Park (inscribed on the list in 1989), a coastal park of dunes and swamps; and the Ancient Ksour of Ouadane, Chinguetti, Tichitt and Oualata (1996), Islamic trading and religious centres in the Sahara.

Press

In 2008 there were four daily newspapers with a circulation of 9,000.

Tourism

There were 30,000 foreign tourists in 2000; spending by tourists totalled US$25m.

DIPLOMATIC REPRESENTATIVES

Of Mauritania in the United Kingdom
Ambassador: Vacant (resides in Paris).
Chargé d'Affaires a.i.: Mohamed Yahya Sidi Haiba.

Of the United Kingdom in Mauritania
Ambassador: Clive Alderton (resides in Rabat, Morocco).

Of Mauritania in the USA (2129 Leroy Pl., NW, Washington, D.C., 20008)
Ambassador: Mohamed Lemine El Haycen.

Of the USA in Mauritania (Rue Abdallaye, Nouakchott)
Ambassador: Jo Ellen Powell.

Of Mauritania to the United Nations
Ambassador: Ahmed Ould Teguedi.

Of Mauritania to the European Union
Ambassador: Mohamed Mahmoud Ould Brahim Khlil.

FURTHER READING

Belvaud, C., *La Mauritanie.* 1992

National Statistical Office: Office National de la Statistique, BP240, Nouakchott.
Website (French only): http://www.ons.mr

MAURITIUS

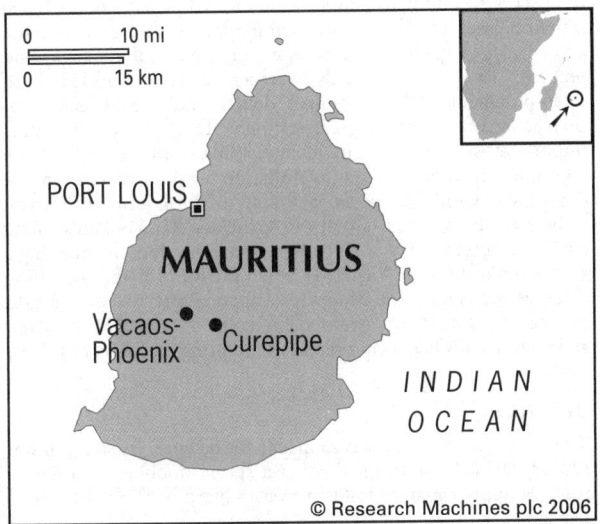

© Research Machines plc 2006

Republic of Mauritius

Capital: Port Louis
Population projection, 2015: 1·33m.
GNI per capita, 2011: (PPP$) 12,918
HDI/world rank: 0·728/77
Internet domain extension: .mu

KEY HISTORICAL EVENTS

Mauritius was visited by Middle Eastern and Malay merchants from around AD 1000 and documented by Portuguese seafarers between 1507 and 1512. In 1598 the Dutch admiral, Van Warwyck, established a settlement and named the island after Prince Maurice of Nassau, the stadtholder of Holland and Zeeland. French forces settled the island in 1722, renamed it Isle de France and brought African slaves to cultivate sugarcane. The British occupied the island in 1810 and it was formally ceded to Great Britain by the Treaty of Paris in 1814. Following the abolition of slavery in 1835, indentured labourers were transported from India. Independence was attained within the Commonwealth on 12 March 1968. Mauritius became a republic on 12 March 1992.

TERRITORY AND POPULATION

Mauritius, the main island, lies 800 km (500 miles) east of Madagascar. Rodrigues is 560 km (350 miles) east. The outer islands are Agalega and the St Brandon Group. Area and population:

Island	Area in sq. km	2011 census population
Mauritius	1,865	1,196,383
Rodrigues	104	40,434
Outer Islands	71	274
Total	2,040	1,237,091

Port Louis is the capital (128,851 inhabitants in 2010). Other towns: Beau Bassin-Rose Hill, 110,687; Vacaos-Phoenix, 106,404; Curepipe, 84,487; Quatre Bornes, 77,495. In 2010, 58·3% of the population were rural.

The UN gives a projected population for 2015 of 1·33m.

Ethnic composition, 2000: Indo-Mauritian, 66·0%; General Population (Creole and Franco-Mauritian), 32·0%; Sino-Mauritian, 2·0%. Mauritius has no indigenous population.

The official language is English, although French is widely used. Creole and Bhojpuri are vernacular languages.

SOCIAL STATISTICS

2007: births, 17,034 (rate of 13·5 per 1,000 population); deaths, 8,498 (6·7 per 1,000); marriages, 11,547 (9·2 per 1,000); divorces, 1,302 (1·0 per 1,000). In 2007 the suicide rate was 15·7 per 100,000 population among men and 4·7 per 100,000 among women. Population growth rate in 2007 was 0·65%. In 2007 the most popular age range for marrying was 25–29 for males and 20–24 for females. Life expectancy at birth in 2007 was 69·1 years for males and 75·8 for females. Infant mortality, 2010, 13 per 1,000 live births; fertility rate, 2007, 2·0 births per woman.

CLIMATE

The sub-tropical climate is humid. Most rain falls in the summer. Rainfall varies between 40" (1,000 mm) on the coast to 200" (5,000 mm) on the central plateau, though the west coast only has 35" (875 mm). Mauritius lies in the cyclone belt, whose season runs from Nov. to April, but is seldom affected by intense storms. Port Louis, Jan. 73°F (22·8°C), July 81°F (27·2°C). Annual rainfall 40" (1,000 mm).

CONSTITUTION AND GOVERNMENT

The present constitution came into effect on 12 March 1968 and was amended on 12 March 1992. The head of state is the *President*, elected by a simple majority of members of the National Assembly. The role of *President* is largely a ceremonial one.

The 69-seat *National Assembly* consists of 62 elected members (three each for the 20 constituencies of Mauritius and two for Rodrigues) and seven additional seats in order to ensure a fair and adequate representation of each community within the Assembly. The government is headed by the *Prime Minister* and a Council of Ministers. Elections are held every five years on the basis of universal adult suffrage.

National Anthem

'Glory to thee, Motherland'; words by J. G. Prosper, tune by P. Gentille.

RECENT ELECTIONS

Parliamentary elections were held on 5 May 2010. The Alliance of the Future (consisting of the Mauritius Labour Party, Mauritian Social Democrat Party and Militant Socialist Movement) won 45 seats with 49·7% of votes cast, followed by the Alliance of the Heart (Mauritian Militant Movement, National Union and Mauritian Socialist Democrat Movement) with 20 seats (42·0% of votes cast). The Rodrigues Movement won two seats and the Mauritian Solidarity Front and the Organization of the People of Rodrigues won one seat each. Turnout was 77·8%.

CURRENT GOVERNMENT

President: Rajkeswur Purryag; b. 1947 (Mauritius Labour Party; since 21 July 2012).

In March 2013 the cabinet was composed as follows:

Prime Minister, Minister of Defence, Home Affairs and External Communications, and Minister for Rodrigues: Navin Ramgoolam; b. 1947 (Social Alliance; took office 5 July 2005, having previously been prime minister from Dec. 1995 to Sept. 2000).

First Deputy Prime Minister and Minister of Energy and Public Utilities: Ahmed Rashid Beebeejaun. *Deputy Prime Minister and Minister of Finance and Economic Development:* Xavier-Luc Duval. *Deputy Prime Minister and Minister of Public Infrastructure, National Development Unit, Land Transport and Shipping:* Anil Kumar Bachoo.

Minister of Agro-Industry and Food Security: Satya Veyash Faugoo. *Arts and Culture:* Mookhesswur Choonee. *Business, Enterprise and Co-operatives:* Jangbahadoorsing Iswurdeo Mola Roopchand Seetaram. *Civil Service and Administrative Reform:* Sutyadeo Moutia. *Education and Human Resources:* Vasant Kumer Bunwaree. *Environment and Sustainable Development:* Devanand Virahsawmy. *Fisheries:* Louis Joseph Von-Mally. *Foreign Affairs, Regional Integration and International Trade:* Arvin Boolell. *Gender Equality, Child Development and Family Welfare:* Maria Francesca Mireille Martin. *Health and Quality of Life:* Lormus Bundhoo. *Housing and Lands:* Abu Kasenally. *Industry, Trade and Consumer Protection:* Sayyad Abd-Al-Cader Sayed-Hossen. *Information and Communication Technology:* Tassarajen Pillay Chedumbrum. *Labour, Industrial Relations and Employment:* Shakeel Ahmed Yousuf Abdul Razack Mohamed. *Local Government and Outer Islands:* Hervé Aimée. *Social Integration and Economic Development:* Surendra Dayal. *Social Security, National Solidarity and Institutional Reform:* Sheilabai Bappoo. *Tertiary Education, Science, Research and Technology:* Rajeshwar Jeetah. *Tourism and Leisure:* John Michael Tzoun Sao Yeung Sik Yuen. *Youth and Sports:* Satyaprakash Ritoo. *Attorney General:* Yatindra Nath Varma.

Government Website: http://www.gov.mu

CURRENT LEADERS

Navin Ramgoolam

Position
Prime Minister

Introduction
Navin Ramgoolam was returned as prime minister at the elections of July 2005, defeating Paul Bérenger, leader of the Mauritian Militant Movement. Ramgoolam had previously held the post from Dec. 1995 to Sept. 2000, when he lost to former Mauritian president Sir Anerood Jugnauth.

Early Life
Navin Ramgoolam was born in Mauritius on 14 July 1947, the son of Seewoosagur Ramgoolam, the country's first president following independence in 1968. The younger Ramgoolam studied sciences at the Royal College at Curepipe in Mauritius before moving to Dublin, Ireland to train as a doctor at the Royal College of Surgeons in 1968. He gained full registration with the UK General Medical Council in 1977. Over the next ten years he worked as a senior medical officer and as a general practitioner in Mauritius, also holding the post of resident medical officer at the Yorkshire Clinic in the UK.

In 1987 Ramgoolam abandoned medicine to study for a masters degree in law at the London School of Economics. However, he subsequently abandoned a legal career in favour of politics, becoming leader of the Mauritius Labour Party in 1991. He went on to succeed Sir Anerood Jugnauth as prime minister in 1995.

Career in Office
In 2000, towards the end of Ramgoolam's first period in office, Mauritius secured a temporary seat on the United Nations Security Council. Having lost the premiership to his predecessor, Jugnauth, at elections later that year, he then formed the Social Alliance, a coalition led by the Mauritian Labour Party and including the Mauritian Party of Xavier-Luc Duval, the Mauritian

Social Democrat Party, the Greens, the Republican Movement and the Militant Socialist Movement (MSM).

At the election of July 2005 the Social Alliance won 42 of a possible 70 seats, giving Ramgoolam a further term as prime minister. On coming to power he announced plans to tackle rising inflation and high levels of unemployment, and sought trade agreements to protect Mauritian exports, particularly sugar and textiles. By 2008 he had overseen a reduction in both unemployment and the budget deficit and the attraction of increasing levels of foreign investment. However, as Mauritius imports most of its food and energy, rising world prices for these commodities then pushed up inflation while recession in the developed world has posed a threat to the country's tourism industry and export potential. Nevertheless, Ramgoolam's ruling coalition won the May 2010 parliamentary election, maintaining a secure majority until Aug. 2011 when the MSM withdrew, citing disagreements with the Mauritius Labour Party. In March 2012 the state president, Sir Anerood Jugnauth of the MSM, similarly resigned from office, being replaced following elections in July by Rajkeswur Purryag.

DEFENCE

The Police Department is responsible for defence. Its strength was (2008) 8,000. In addition there is a special mobile paramilitary force of approximately 1,400, a Coast Guard of about 700 and a helicopter unit of about 100.

Defence expenditure totalled US$36m. in 2008 (US$28 per capita), representing 0·4% of GDP.

ECONOMY

Agriculture accounted for 4·3% of GDP in 2009, industry 29·1% and services 66·6%.

Overview

At independence in 1968 Mauritius had a low-income, agro-based economy reliant on sugar. It has subsequently developed into an upper middle income economy with the second highest GDP per capita in Africa and is considered a development success story by the World Bank. In the World Bank's 'Doing Business 2010' report, Mauritius ranked 20th of 183 countries and first in Africa.

The pillars of the economy are sugar, tourism, textiles and financial services. Early growth was largely derived from preferential trade status for textiles and sugar with the USA and the EU. However, the phasing out of such agreements for sugar exports to the EU, the expiration of textile quotas at the end of 2004 and rising oil prices in 2006 prompted the government to embark on structural reforms.

Attempts to diversify the economy into new sectors including seafood and technology have shown early signs of success. GDP growth averaged 4·8% annually between 2006 and 2008, against an average of 3·3% between 2002 and 2005. Foreign direct investment increased, the unemployment rate dropped and public debt declined from almost 70% of GDP in 2006 to 53% in 2008.

Currency

The unit of currency is the *Mauritius rupee* (MUR) of 100 *cents*. There are Bank of Mauritius notes, cupro-nickel coins, nickel-plated steel coins and copper-plated steel coins. Inflation was 2·9% in 2010 and 6·5% in 2011. In July 2005 foreign exchange reserves were US$1,365m. and gold reserves totalled 62,000 troy oz. Total money supply was Rs 22,646m. in June 2005.

Budget

Central government revenue in 2007–08 was Rs 58,125m. (Rs 46,222m. in 2006–07); expenditure was Rs 51,265m. (Rs 46,726m. in 2006–07). Principal sources of revenue, 2007–08: taxes on goods and services, Rs 26,604m.; taxes on income, profits and capital gains, Rs 10,558m.; taxes on international trade and transactions, Rs 6,646m.; taxes on property Rs 4,003m. Main items of expenditure in 2007–08 were: compensation of employees,

Rs 17,341m.; social benefits, Rs 12,584m.; interest, Rs 7,971m.; use of goods and services, Rs 5,749m. Since 1 Jan. 2010 the fiscal year has been the calendar year.

VAT is 15%.

Performance

Real GDP growth was 3·0% in 2009, 4·2% in 2010 and 4·1% in 2011. Total GDP in 2011 was US$11·3bn. Thanks to tourism, financial services and the traditional industries of sugar and textiles, Mauritius is now one of Africa's richest and most developed countries.

Banking and Finance

The Bank of Mauritius (founded 1967) is the central bank. The *Governor* is Rundheersing Bheenick. In 2007 there were 19 commercial banks and one development bank. Since 2005 there has been no distinction between onshore and offshore banks. Non-bank financial intermediaries are the Post Office Savings Bank, the State Investment Corporation Ltd, the Mauritius Leasing Company, the National Mutual Fund, the National Investment Trust and the National Pension Fund. Other financial institutions are the Mauritius Housing Company and the Development Bank of Mauritius. External debt was US$857m. at June 2007. There is also a stock exchange in Port Louis.

ENERGY AND NATURAL RESOURCES

Environment

Carbon dioxide emissions were the equivalent of 3·6 tonnes per capita in 2008.

Electricity

Installed capacity was 0·75m. kW in 2007. Production (2007) was 2·46bn. kWh. Consumption per capita in 2007 was 1,956 kWh.

Agriculture

68,523 ha. were planted with sugarcane in 2007; yield in 2007 was 4,235,849 tonnes. Main secondary crops (2007, in 1,000 tonnes): potatoes, 15; tomatoes, 11; bananas, 9; tea, 8; pumpkins and squash, 7; cucumbers, 6; onions, 6; pineapples, 6. In 2003 there were 100,000 ha. of arable land and 6,000 ha. of permanent cropland; 21,619 ha. were irrigated.

Livestock, 2007: cattle, 7,000; goats and sheep, 26,000; pigs, 17,000.

Livestock products (2003) in tonnes: beef and veal, 2,580; pork, bacon and ham, 1,040; milk, 4,000; eggs, 12,500.

Forestry

The total forest area was 35,000 ha. in 2010 (17% of the land area). In 2007 timber production totalled 15,000 cu. metres.

Fisheries

The catch in 2010 totalled 7,786 tonnes, exclusively sea fish.

INDUSTRY

Manufacturing includes: sugar, textile products, footwear and other leather products, diamond cutting, jewellery, furniture, watches and watchstraps, sunglasses, plastic ware, chemical products, electronic products, pharmaceutical products, electrical appliances, ship models and canned food. There were six sugar mills in 2010 producing 452,473 tonnes of sugar. Production figures for other leading commodities in 2010: beer and stout, 36·8m. litres; animal feeds, 175,250 tonnes; molasses, 145,752 tonnes.

Labour

In 2007 the labour force was estimated at 548,900. Manufacturing employed the largest proportion, with 30·8% of total employment; agriculture, forestry and fishing, 7·2%; wholesale and retail trade, 6·3%. In 2007 the unemployment rate was 8·5%.

INTERNATIONAL TRADE

Imports and Exports

Imports in 2010 were valued at Rs 135,394m. and exports at Rs 69,556m. In 2010 imports valued at Rs 30,239m. came from India, Rs 18,027m. from China, Rs 11,393m. from South Africa and Rs 10,992m. from France. Exports valued at Rs 13,542m. went to the UK in 2010, Rs 10,376m. to France, Rs 6,229m. to the USA and Rs 4,052m. to Spain.

Major imports in 2010 included machinery and transport equipment, Rs 27,451m.; mineral fuels, lubricants and related products, Rs 25,929m.; manufactured goods (paper, textiles, iron and steel), Rs 25,901m.; food and live animals, Rs 24,006m. Major exports (2010) included articles of apparel and clothing, Rs 23,004m.; fish and fish preparations, Rs 7,782m.; sugar, Rs 7,740m.; textile yarns, fabrics and finished articles, Rs 1,899m.

COMMUNICATIONS

Roads

In 2007 there were 75 km of motorway, 962 km of main roads and 991 km of secondary and other roads. In 2007 there were 144,400 passenger cars, 142,600 motorcycles and mopeds, 40,900 lorries and vans, and 4,000 buses and coaches. In 2007 there were 140 deaths as a result of road accidents.

Civil Aviation

In 2007, 2,412,200 passengers were handled at Sir Seewoosagur Ramgoolam International Airport. The national carrier is Air Mauritius, which is partly state-owned. In 2006 scheduled airline traffic of Mauritius-based carriers flew 47m. km, carrying 1,150,000 passengers (1,067,000 on international flights).

Shipping

A free port was established at Port Louis in Sept. 1991. In 2008–09 Port Louis handled 6,295,000 tonnes of cargo. In Jan. 2009 there were three ships of 300 GT or over registered, totalling 13,000 GT.

Telecommunications

In 2008 there were 364,500 main (fixed) telephone lines; mobile phone subscribers numbered 1,033,300 in 2008 (80·7 per 100 persons). Communication with other parts of the world is by satellite and microwave links. There were 282,000 internet users in 2008. In June 2012 there were 324,000 Facebook users.

SOCIAL INSTITUTIONS

Justice

There is an Ombudsman. The death penalty was abolished for all crimes in 1995.

The population in penal institutions in May 2008 was 2,223 (171 per 100,000 of national population).

Education

The adult literacy rate in 2009 was estimated at 87·9% (90·6% among males and 85·3% among females). Primary and secondary education is free, primary education being compulsory. Almost all children aged 5–11 years attend schools. In 2008 there were 114,007 pupils in 286 primary schools and 112,995 pupils in 175 secondary schools in the island of Mauritius, and 5,015 pupils in 13 primary schools and 3,508 in six secondary schools in Rodrigues. In 2007, 3,945 teachers were enrolled for training at the Mauritius Institute of Education.

In 2007–08 there were 7,794 students and 487 academic staff at the University of Mauritius.

In 2007–08 total expenditure on education came to 3·2% of GDP and 12·7% of total government spending.

Health

In 2007 there were 1,444 physicians, 228 dentists, 3,300 nurses and midwives, and 327 pharmacists. There were 12 hospitals in 2006 with a provision of 28 beds per 10,000 inhabitants.

RELIGION

In 2001 there were 610,000 Hindus, 330,000 Roman Catholics and 190,000 Muslims.

CULTURE

World Heritage Sites

Mauritius has two sites on the UNESCO World Heritage List: Aapravasi Ghat (inscribed on the list in 2006), the site where the modern indentured labour diaspora began; and Le Morne cultural landscape (2008), a rugged mountain jutting into the Pacific Ocean that was used by runaway slaves (maroons) as a shelter in the 18th and 19th centuries.

Press

In 2008 there were four daily papers with a combined circulation of 110,000, plus 16 non-dailies.

Tourism

In 2010 there were 934,827 visitors (including 605,401 from Europe and 226,207 from other African countries), bringing in US$1,227m. in tourist revenue.

Festivals

Independence Day is marked by an official celebration at the Champ de Mars racecourse on 12 March. The Hindu festival of Cavadee is celebrated by the Tamil community at the beginning of the year; the major three-day Hindu festival of Maha Shivarati takes place around Feb./March. Other Hindu festivals include Diwali and Ganesh Chaturhi, which is celebrated around Aug./Sept. The Spring Festival is celebrated on the eve of the Chinese New Year; Ougadi, the Telegu new year, is celebrated in March; the Tamil new year, Varusha Pirappu, takes place in April.

Muslim festivals include Eid al-Fitr and Eid al-Adha. On 9 Sept. pilgrims visit the grave of the 19th century missionary Père Laval who is regarded as a national saint.

DIPLOMATIC REPRESENTATIVES

Of Mauritius in the United Kingdom (32–33 Elvaston Pl., London, SW7 5NW)
High Commissioner: Abhimanu Mahendra Kundasamy.

Of the United Kingdom in Mauritius (Les Cascades Bldg, Edith Cavell St., Port Louis)
High Commissioner: Nick Leake.

Of Mauritius in the USA (1709 N. St., NW, Washington, D.C., 20036)
Ambassador: Somduth Soborun.

Of the USA in Mauritius (Rogers House, John Kennedy St., Port Louis)
Ambassador: Shari Villarosa.

Of Mauritius to the United Nations
Ambassador: Milan Jaya Nyamrajsingh Meetarbhan.

Of Mauritius to the European Union
Ambassador: Jagdish Dharamchand Koonjul.

FURTHER READING

Central Statistical Information Office. *Bi-annual Digest of Statistics.*

Bowman, L. W., *Mauritius: Democracy and Development in the Indian Ocean.* 1991

National Statistical Office: Central Statistics Office, LIC Building, President John Kennedy St., Port Louis.
Website: http://www.gov.mu/portal/site/cso

The world in focus at
www.statesmansyearbook.com

MEXICO

Estados Unidos Mexicanos
(United States of Mexico)

Capital: Mexico City
Population projection, 2015: 120·06m.
GNI per capita, 2011: (PPP$) 13,245
HDI/world rank: 0·770/57
Internet domain extension: .mx

KEY HISTORICAL EVENTS

The first settlers of the New World arrived in Alaska from Asia about 15,000 years ago. From about 2000 BC the people of Ancient Mexico began to settle in villages and to cultivate maize and other crops. From about 1000 BC the chief tribes were the Olmec on the Gulf Coast, the Maya in the Yucatán peninsula and modern day Chiapas, the Zapotecs and Mixtecs in Oaxaca, the Tarascans in Michoacán and the Toltecs in central Mexico. One of the largest and most powerful cities in ancient Mexico was Teotihuacán, which in the 6th century AD was one of the six largest cities in the world. By the time the Spanish *conquistadores* arrived in 1519, the dominant people were the Mexica, more commonly known as the Aztecs, whose capital Tenochtitlán became Mexico City after the conquest.

Hernán Cortés landed on the Gulf Coast in 1519 and by 1521 his small band of Spaniards, assisted by an army of indigenous peoples, had destroyed the Aztec state. The land conquered by Cortés was named New Spain, and was ruled by the Spanish Crown for three centuries. The new colony was the personal property of the King, whose representative, the Viceroy, was charged with extracting the maximum income for the Crown. The mainstays of the colonial economy were silver and land. Rich silver mines were discovered and large estates (*haciendas*) were formed. Spain controlled trade with the colonies and discouraged manufacturing to maximize profits for the King. Acapulco became Spain's sole port for trade with Asia.

One early result of the Conquest was a collapse of the indigenous population caused by social dislocation and European diseases. In 1520 the native population was probably 20m. By 1540 it had fallen to 6·5m. and by 1650 the figure was just over 1m.

The beginning of the end of Spanish rule came on 16 Sept. 1810 when the parish priest of Dolores, Miguel Hidalgo y Costilla, called for independence (the 'grito de Dolores') and led a popular army against the Spaniards. Hidalgo's revolution failed as did that of the insurrectionary José María Morelos y Pavón. Independence from Spain was declared in the Plan of Iguala on 24 Feb. 1821 when Agustín de Iturbide proclaimed himself Emperor of Mexico. He ruled for two years.

There followed half a century of coups and counter coups. Spain invaded Tampico in 1829. Texas declared secession in 1836. The Mexican dictator Antonio de Santa Anna marched north but was defeated by the Texans. France invaded Veracruz in 1838 (the 'Pastry War'). In 1846 the USA declared war on Mexico. The war was ended in 1848 by the Treaty of Guadalupe which forced Mexico to cede a huge swathe of its territory to the USA. Liberals and conservatives fought the War of the Reform from 1858–61. The liberal government of Benito Juárez abolished the *fueros* (clerical and military privileges) and hereditary titles, confiscated the church's lands and attempted far-reaching land reform. This was followed by the French Intervention (1862–67), which installed the Habsburg Archduke Maximilian of Austria as Emperor of Mexico. The French were resisted stubbornly by President Juárez but the republicans were forced into the resource-poor and sparsely populated north. Napoleon III withdrew his troops from Mexico in 1867 despite a pledge to support Maximilian, allowing the republicans to take back the country virtually unopposed. Asserting Mexico's independence, Juárez ordered the execution of Maximilian.

From 1876–1910, a period known as the *porfiriato*, Mexico was ruled (with one interlude from 1880–84) by Gen. Porfirio Díaz. Díaz imposed a degree of stability and order. He encouraged foreign investment, which funded a rapid expansion of the railways and an export-led economic boom. The economy faltered in the first decade of the 20th century. Díaz was deposed in 1911 by Francisco Madero, whose Plan of San Luís Potosí launched the Mexican Revolution.

Madero was deposed and assassinated in 1913. There followed a civil war fought by the armies of Venustiano Carranza, Pancho Villa and Emiliano Zapata. A new constitution was written in 1917. Zapata was ambushed and killed in 1919 and Carranza was assassinated in 1920. Villa retired the same year but was assassinated in 1923.

In the 1920s Mexico was ruled by Alvaro Obregón and Plutarco Elías Calles. Obregón's assassination in 1928 led to the formation of the Natural Revolutionary Party (PRN), later the Institutional Revolutionary Party (PRI), which ruled Mexico for the rest of the century. Lázaro Cárdenas was president from 1934–40. He nationalized the oil industry and accelerated the distribution of land to the peasantry. The election of Miguel Alemán in 1946 was opposed unsuccessfully by the last military rebellion in Mexico's history. Alemán's pro-business administration began a long period of relative economic prosperity, the 'Mexican Miracle'.

However, by the late 1960s the Mexican economic and political system was under increasing strain. An uprising led by students ended in a bloody massacre in the Tlatelolco district of Mexico City in 1968. Successive PRI presidents made gestures towards democratization and effective opposition gradually developed. Financial and economic problems in the 1980s increased the pressure on the political system. The crisis came in 1988 when the PRI candidate, Carlos Salinas de Gortari, defeated Cuauhtémoc Cárdenas, son of the former president and candidate of the Democratic Revolutionary Party (PRD), in a rigged election. Salinas took Mexico into the North American Free Trade Agreement (NAFTA) with the USA and Canada in 1992. Salinas' choice as the PRI's presidential candidate, Luís Donaldo Colosio, was assassinated in Tijuana on 23 March 1994. He was replaced by Ernesto Zedillo. In the same year the Zapatista National Liberation Army (EZLN) led an uprising in Chiapas, which is ongoing.

In 2000 Vicente Fox Quesada of the National Action Party (PAN) was elected to the presidency. Fox attempted to address two key issues: Mexico's economic and financial weakness and illegal migration to the USA. However, the PRI majority in Congress blocked Fox's fiscal reforms and the Bush administration was unwilling to support Fox's proposal to liberalize immigration. In the 2006 presidential elections the conservative Felipe Calderón and the socialist Andrés Manuel López Obrador both claimed victory. Calderón was finally declared the winner more than two months after the election, but in Nov. López Obrador proclaimed himself the 'legitimate' president. Nevertheless, Calderón was sworn in as scheduled in Dec. 2006.

TERRITORY AND POPULATION

Mexico is bounded in the north by the USA, west and south by the Pacific Ocean, southeast by Guatemala, Belize and the Caribbean Sea, and northeast by the Gulf of Mexico. It comprises 1,964,375 sq. km (758,464 sq. miles), including uninhabited islands (5,127 sq. km) offshore.

Population at recent censuses: 1970, 48,225,238; 1980, 66,846,833; 1990, 81,249,645; 2000, 97,483,412; 2005, 103,263,388; 2010, 112,336,538 (57,481,307 females). Population density, 57·2 per sq. km (2010). 78·1% of the population were urban in 2011.

The UN gives a projected population for 2015 of 120·06m.

Area, population and capitals of the Federal District and 31 states:

	Area (Sq. km)	Population (2010 census)	Capital
Federal District	1,486	8,851,080	Mexico City
Aguascalientes	5,618	1,184,996	Aguascalientes
Baja California	71,446	3,155,070	Mexicali
Baja California Sur	73,922	637,026	La Paz
Campeche	57,924	822,441	Campeche
Chiapas	73,289	4,796,580	Tuxtla Gutiérrez
Chihuahua	247,455	3,406,465	Chihuahua
Coahuila de Zaragoza	151,563	2,748,391	Saltillo
Colima	5,625	650,555	Colima
Durango	123,451	1,632,934	Victoria de Durango
Guanajuato	30,608	5,486,372	Guanajuato
Guerrero	63,621	3,388,768	Chilpancingo de los Bravo
Hidalgo	20,847	2,665,018	Pachuca de Soto
Jalisco	78,599	7,350,682	Guadalajara
México	22,357	15,175,862	Toluca de Lerdo
Michoacán de Ocampo	58,643	4,351,037	Morelia
Morelos	4,893	1,777,227	Cuernavaca
Nayarit	27,815	1,084,979	Tepic
Nuevo Léon	64,220	4,653,458	Monterrey
Oaxaca	93,793	3,801,962	Oaxaca de Juárez
Puebla	34,290	5,779,829	Heroica Puebla de Zaragoza
Querétaro Arteaga	11,684	1,827,937	Santiago de Querétaro
Quintana Roo	42,361	1,325,578	Chetumal
San Luis Potosí	60,983	2,585,518	San Luis Potosí
Sinaloa	57,377	2,767,761	Culiacán Rosales
Sonora	179,503	2,662,480	Hermosillo
Tabasco	24,738	2,238,603	Villahermosa
Tamaulipas	80,175	3,268,554	Ciudad Victoria
Tlaxcala	3,991	1,169,936	Tlaxcala de Xicohténcatl
Veracruz-Llave	71,820	7,643,194	Xalapa-Enríquez
Yucatán	39,612	1,955,577	Mérida
Zacatecas	75,539	1,490,668	Zacatecas
Total	1,959,248	112,336,538	

The *de facto* official language is Spanish, the mother tongue of over 93% of the population (2005), but there are some indigenous language groups (of which Náhuatl, Maya, Zapotec, Otomi and Mixtec are the most important) spoken by 6,011,202 persons over five years of age (census 2005).

The populations (2010 census) of the largest cities (250,000 and more) were:

Mexico City	8,555,272	Ciudad López Mateos	489,160
Ecatepec de Morelos	1,655,015	Cuautitlán Izcalli	484,573
Guadalajara	1,495,182	Apodaca	467,157
Heroica Puebla de Zaragoza	1,434,062	Heroica Matamoros	449,815
Juárez	1,321,004	San Nicolás de los Garza	443,273
Tijuana	1,300,983	Veracruz	428,323
León de los Aldama	1,238,962	Xalapa-Enríquez	424,755
Zapopan	1,142,483	Tonala	408,759
Monterrey	1,135,512	Mazatlán	381,583
Ciudad Nezahualcoyotl	1,104,585	Irapuato	380,941
Chihuahua	809,232	Nuevo Laredo	373,725
Naucalpan de Juárez	792,211	Xico	356,352
Mérida	777,615	Villahermosa	353,577
San Luis Potosí	722,772	Escobedo	352,444
Aguascalientes	722,250	Celaya	340,387
Hermosillo	715,061	Cuernavaca	338,650
Saltillo	709,671	Tepic	332,863
Mexicali	689,775	Ixtapaluca	322,271
Culiacán Rosales	675,773	Ciudad Victoria	305,155
Guadalupe	673,616	Ciudad Obregón	298,625
Acapulco de Juárez	673,479	Tampico	297,284
Tlalnepantla	653,410	Villa Nicolás Romero	281,799
Cancún	628,306	Ensenada	279,765
Santiago de Querétaro	626,495	San Francisco Coacalco	277,959
Chimalhuacan	612,383	Ciudad Santa Catarina	268,347
Torreón	608,836	Uruapan	264,439
Morelia	597,511	Gómez Palacio	257,352
Reynosa	589,466	Los Mochis	256,613
Tlaquepaque	575,942	Pachuca de Soto	256,584
Tuxtla Gutiérrez	537,102	Oaxaca de Juárez	255,029
Durango	518,709	Soledad Díez Gutiérrez	255,015
Toluca de Lerdo	489,333		

SOCIAL STATISTICS

Statistics for calendar years:

	Births	Deaths	Marriages	Divorces
2005	2,567,906	495,240	595,713	70,184
2006	2,505,939	494,471	586,978	72,396
2007	2,655,083	514,420	595,209	77,255
2008	2,636,110	539,530	589,352	81,851
2009	2,577,214	564,673	558,913	84,302

Rates per 1,000 population, 2009: births, 18·0; deaths, 4·9. In 2006 the most popular age range for marrying was 20–24 for both males and females. Infant mortality was 14·7 per 1,000 live births in 2009. Life expectancy at birth in 2007 was 73·6 years for males and 78·5 years for females. Annual population growth rate, 2005–10, 1·8%. Fertility rate, 2008, 2·2 births per woman (less than half the number in the late 1970s). Much of the population still lives in poverty, with the gap between the modern north and the backward south constantly growing.

CLIMATE

Latitude and relief produce a variety of climates. Arid and semi-arid conditions are found in the north, with extreme temperatures, whereas in the south there is a humid tropical climate, with temperatures varying with altitude. Conditions on the shores of the Gulf of Mexico are very warm and humid. In general, the rainy season lasts from May to Nov. Mexico City, Jan. 55°F (12·9°C), July 61°F (16·2°C). Annual rainfall 31" (787·6 mm). Guadalajara, Jan. 63°F (17·0°C), July 72°F (22·1°C). Annual rainfall 39" (987·6 mm). La Paz, Jan. 62°F (16·8°C), July 86°F (29·9°C). Annual rainfall 7" (178·3 mm). Mazatlán, Jan. 68°F (20·0°C), July 84°F (29·0°C). Annual rainfall 32" (822·1 mm). Mérida, Jan. 73°F (23·0°C), July 81°F (27·4°C). Annual rainfall 39" (990·0 mm). Monterrey, Jan. 58°F (14·3°C), July 83°F (28·1°C).

Annual rainfall 23" (585·4 mm). Puebla de Zaragoza, Jan. 52°F (11·4°C), July 62°F (16·9°C). Annual rainfall 36" (900·8 mm).

CONSTITUTION AND GOVERNMENT

A new constitution was promulgated on 5 Feb. 1917 and has occasionally been amended. Mexico is a representative, democratic and federal republic, comprising 31 states and a federal district, each state being free and sovereign in all internal affairs, but united in a federation established according to the principles of the Fundamental Law. The head of state and supreme executive authority is the *President*, directly elected for a non-renewable six-year term. The constitution was amended in April 2001, granting autonomy to 10m. indigenous peoples. The amendment was opposed both by the National Congress of Indigenous Peoples and Zapatista rebels who claimed it would leave many indigenous people worse off.

There is complete separation of legislative, executive and judicial powers (Art. 49). Legislative power is vested in a General Congress of two chambers, a *Chamber of Deputies* and a *Senate*. The Chamber of Deputies consists of 500 members directly elected for three years, 300 of them from single-member constituencies and 200 chosen under a system of proportional representation. In 1990 Congress voted a new Electoral Code. This established a body to organize elections (IFE), an electoral court (TFE) to resolve disputes, new electoral rolls and introduce a voter's registration card. Priests were enfranchised in 1991.

The Senate comprises 128 members. In each of the 31 states and the Federal District the party coming first wins two seats and the party coming second wins one seat, making 96 in total. An additional 32 seats are filled through proportional representation from national party lists. Members of both chambers are not immediately re-eligible for election. Congress sits from 1 Sept. to 31 Dec. each year; during the recess there is a permanent committee of 15 deputies and 14 senators appointed by the respective chambers.

National Anthem

'Mexicanos, al grito de guerra' ('Mexicans, at the war-cry'); words by F. González Bocanegra, tune by Jaime Nunó.

GOVERNMENT CHRONOLOGY

Presidents since 1940. (PRI = Institutional Revolutionary Party; PAN = National Action Party)

1940–46	PRI	Manuel Ávila Camacho
1946–52	PRI	Miguel Alemán Valdés
1952–58	PRI	Adolfo Ruiz Cortines
1958–64	PRI	Adolfo López Mateos
1964–70	PRI	Gustavo Díaz Ordaz Bolaños
1970–76	PRI	Luis Echeverría Álvarez
1976–82	PRI	José López Portillo y Pacheco
1982–88	PRI	Miguel de la Madrid Hurtado
1988–94	PRI	Carlos Salinas de Gortari
1994–2000	PRI	Ernesto Zedillo Ponce de León
2000–06	PAN	Vicente Fox Quesada
2006–12	PAN	Felipe de Jesús Calderón Hinojosa
2012–	PRI	Enrique Peña Nieto

RECENT ELECTIONS

In the presidential elections of 1 July 2012 Enrique Peña Nieto of the Partido Revolucionario Institucional (Institutional Revolutionary Party/PRI) won 39·1% of the vote, Andrés Manuel López Obrador of the Partido de la Revolución Democrática (Party of the Democratic Revolution/PRD) 32·4%, Josefina Vázquez Mota of the ruling Partido Acción Nacional (National Action Party/PAN) 26·0% and Gabriel Quadri de la Torre of the Partido Nueva Alianza (New Alliance Party/PNA) 2·4%. Turnout was 63·1%. As in 2006 López Obrador claimed electoral irregularities and there were widespread protests denouncing the result.

In elections to the Chamber of Deputies held on 5 July 2009 the Institutional Revolutionary Party (PRI) won 241 seats (36·7% of the vote), the National Action Party (PAN) 147 (28·0%), the Party of the Democratic Revolution (PRD) 72 (12·2%), the Ecologist Green Party of Mexico (PVEM) 17 (6·7%), the Labour Party (PT) 9 (3·7%), the New Alliance Party (PNA) 8 (3·4%) and Convergence 6 (2·5%).

CURRENT GOVERNMENT

President: Enrique Peña Nieto; b. 1966 (Institutional Revolutionary Party; sworn in 1 Dec. 2012).

In March 2013 the government comprised:

Secretary of the Interior: Miguel Ángel Osorio Chong. *Foreign Affairs:* José Antonio Meade. *Defence:* Gen. Salvador Cienfuegos Zepeda. *Naval Affairs:* Adm. Vidal Soberón. *Finance and Public Credit:* Luis Videgaray. *Social Development:* Rosario Robles. *Environment and Natural Resources:* Juan José Guerra Abud. *Energy:* Pedro Joaquín Coldwell. *Economy:* Ildefonso Guajardo. *Agriculture, Livestock, Rural Development, Fisheries and Food:* Enrique Martínez y Martínez. *Communication and Transport:* Gerardo Ruiz Esparza. *Public Education:* Emilio Chuayffet. *Health:* Mercedes Juan López. *Labour and Social Welfare:* Alfonso Navarrete. *Agrarian Reform:* Jorge Carlos Ramírez Marín. *Tourism:* Claudia Ruiz Massieu. *Attorney General:* Jesús Murillo Karam. *Undersecretary of Constituent Services:* Julián Olivas. *Undersecretary of Public Security and Institutional Planning:* Manuel Mondragón. *Head of the Presidential Office:* Aurelio Nuño.

Presidency Website: http://www.presidencia.gob.mx

CURRENT LEADERS

Enrique Peña Nieto

Position
President

Introduction
Enrique Nieto of the Institutional Revolutionary Party (PRI) became president in Dec. 2012. His election brought to an end the 12-year presidency of the National Action Party (PAN).

Early Life
Nieto was born in July 1966 in the city of Atlacomulco to an electrical engineer father and a teacher mother. He graduated in law from the Universidad Panamericana, Mexico City, and received a masters degree in business from the Monterrey Institute of Technology.

Nieto began his political career as secretary of the Citizen Movement of Zone 1, a group affiliated to the PRI-aligned National Confederation of Popular Organisations (CNOP). He worked in several PRI-linked organizations, serving as a delegate to the National Front for Organisations and Citizens and, from 1993–98, as private secretary to the state of México's secretary for economic development. In 1999 Nieto was financial co-ordinator for the successful state gubernatorial campaign of Arturo Montiel Rojas. In 2001 Rojas appointed Nieto under-secretary of the interior and he went on to hold senior roles overseeing family, health and social security programmes.

In 2003 he was elected deputy of the XIII Local District in Atlacomulco. Two years later he successfully campaigned as the PRI candidate for the state of México governorship, overseeing a reduction in the state's debt during his eight years in office. He ran for national president in 2012, winning with 39% of the vote.

Career in Office
Nieto inherited a stagnant economy and high unemployment. His campaign pledges included the privatization of Mexico's state-owned oil company, PEMEX. However, with a lack of congressional support he will struggle to fulfil this commitment.

Nieto also faces a continuing war on drugs after promising to reduce the murder rate in Mexico City by 50% by the end of his six-year term. He has pledged to stamp down on corruption.

DEFENCE

Conscription is for 12 months. In 2008 defence expenditure totalled US$4,346m. (US$40 per capita), representing 0·4% of GDP.

Army

Enlistment into the regular army is voluntary, but there is also one year of conscription (four hours per week) by lottery. Strength of the regular army (2007) 178,000. There are reserve forces numbering 39,899 in 2007. In addition there is a rural defence militia of 18,000.

Navy

The Navy is primarily equipped and organized for offshore and coastal patrol duties. It includes one destroyer and six frigates. Naval Aviation, 1,250 strong, operates eight combat-capable aircraft.

Naval personnel in 2007 totalled 46,400, including Naval Aviation. In addition there were 12,600 marines.

Air Force

The Air Force had (2007) a strength of 11,700 with 78 combat-capable aircraft, including PC-7s and F-5Es.

ECONOMY

Agriculture accounted for 4·0% of GDP in 2009, industry 33·7% and services 62·3%.

Overview

Mexico is the second largest economy in Latin America, with some commentators predicting it will overtake Brazil as the region's largest by 2022. However, Mexico's growth has been volatile. In 2009 GDP contracted by 6·0% but grew by 5·6% in 2010, followed by 3·9% in 2011.

Mexico is the world's seventh largest producer of oil, which provides a third of government revenues. The government plans to attract major private investment to state-owned energy firms to maximize potential growth.

The economy benefits from its proximity to the USA, providing cheap labour for US firms. Mexico's manufacturing sector grew by 10% in 2011 alone as foreign firms relocated their production facilities in the country. With rising labour and production costs in China, Mexico is looking to make up ground on its global competitor as a major industrial exporter.

The transfer to manufactured exports started with the economic liberalization of the 1980s and was accelerated by membership of the North American Free-Trade Agreement in 1994. Nearly half of the country's total exports are produced in *maquiladoras* (in-bond assembly plants for re-export). However, reliance on the US economy for 80% of exports leaves the economy susceptible to any downturn in US economic fortunes.

An innovative financial sector has created the first multi-peril CAT bond which seeks to manage financial risks arising from interest rate volatility. The central bank has built its foreign reserves and negotiated a flexible credit line with the IMF that offers access to US$73bn. The 2011 budget introduced a programme of fiscal consolidation through tax increases and lower public expenditure, with the aim of returning to a balanced budget in the short- to mid-term.

Tackling poverty is a significant challenge, with 42·6% living below the poverty line in 2010. There is a high degree of income inequality, with the World Bank estimating that the richest 10% of the population earns over 40% of total income while the poorest 10% accounts for only 1·1%. Corruption and violence hamper growth prospects, while educational reform is a major priority to ensure a skilled and competitive workforce. Drug-related violence has cost around 60,000 lives since 2006.

Currency

The unit of currency is the *Mexican peso* (MXN) of 100 *centavos*. A new peso was introduced on 1 Jan. 1993: 1 new peso = 1,000 old pesos. The peso was devalued by 13·94% in Dec. 1994. Foreign exchange reserves were US$82,023m. and gold reserves 288,000 troy oz in Sept. 2009. Inflation rates (based on OECD statistics):

2002	2003	2004	2005	2006	2007	2008	2009	2010	2011
5·0%	4·5%	4·7%	4·0%	3·6%	4·0%	5·1%	5·3%	4·2%	3·4%

Total money supply in Aug. 2009 was 1,391·5bn. new pesos.

Budget

In 2008 revenues were 2,857·1bn. new pesos and expenditures 2,865·3bn. new pesos.

VAT is 16% (11% in the frontier region).

Performance

Real GDP growth rates (based on OECD statistics):

2002	2003	2004	2005	2006	2007	2008	2009	2010	2011
0·1%	1·4%	4·0%	3·2%	5·1%	3·2%	1·2%	−6·0%	5·6%	3·9%

In 2011 total GDP was US$1,153·3bn.

Banking and Finance

The Bank of Mexico, established 1 Sept. 1925, is the central bank of issue (*Governor*, Agustín Carstens Carstens). It gained autonomy over monetary policy in 1993. Exchange rate policy is determined jointly by the bank and the Finance Ministry. Banks were nationalized in 1982, but in May 1990 the government approved their reprivatization. The state continues to have a majority holding in foreign trade and rural development banks. In 1999 Congress approved the removal of regulations limiting foreign holdings to 49%.

In 2007 there were 38 commercial banks (including seven development banks) and 81 representative offices of foreign banks. Mexico's largest banks are BBVA Bancomer with assets of US$85·0bn. in June 2009, followed by Banamex with assets of US$74·9bn. and Santander with assets of US$48·7bn. Most of Mexico's leading banks are now foreign-owned.

Foreign debt was US$200,081m. in 2010, representing 19·5% of GNI.

Mexico received US$19·6bn. worth of foreign direct investment in 2011, down from a record US$31·5bn. in 2007.

There is a stock exchange in Mexico City.

ENERGY AND NATURAL RESOURCES

Environment

Mexico's carbon dioxide emissions from the consumption and flaring of fossil fuels in 2008 were the equivalent of 4·0 tonnes per capita.

Electricity

Installed capacity, 2007, 50·8m. kW. Output in 2007 was 257·45bn. kWh and consumption per capita 2,384 kWh. In 2010 there were two nuclear reactors in operation.

Oil and Gas

Oil production was 157·4m. tonnes in 2008. Mexico produced 4·0% of the world total oil output in 2008, and had reserves amounting to 11·9bn. bbls. Revenues from oil exports provide about a third of all government revenues. Natural gas production was 54·9bn. cu. metres in 2008 with 500bn. cu. metres in proven reserves.

Minerals

Output (in 1,000 tonnes): iron ore (2006), 14,568; lignite (2007), 10,456; salt (2005), 9,508; gypsum and anhydrite (2006), 6,076; silica (2005), 2,121; coal (2007), 2,058; sulphur (2005), 1,590; fluorite (2005), 876; aluminium (2005), 574; zinc (2005), 476; copper (2005), 429; feldspar (2005), 373; barite (2005), 269; lead (2005), 134; manganese (2005), 133; silver (2005), 2·9; gold (2005), 30,356 kg. Mexico is the biggest producer of silver in the world.

Agriculture

In 2007 Mexico had an estimated 24·5m. ha. of arable land and around 2·4m. ha. of permanent cropland. There were 5·4m. ha. of irrigated land in 2006. There were 238,830 tractors and some 22,500 harvester-threshers in 2007. In 2007 agriculture, fishing and forestry contributed 3·6% of GDP. In 1992 the Mexican constitution was amended to permit the voluntary privatization of *ejidos*, communal land in which each member farms an independent plot, to combat the low productivity resulting from the fragmentation of farming units, some 58% of which were less than 5 ha. in 1991.

Sown areas, 2003 (in 1,000 ha.) included: maize, 7,781; beans, 1,948; sorghum, 1,879; coffee beans, 744; wheat, 627; sugarcane, 639; barley, 384; chick-peas, 150; chillies and green peppers, 141; safflower seeds, 85; rice, 50. Production in 2003 (in 1,000 tonnes): sugarcane, 45,126; maize, 19,652; sorghum, 6,462; oranges, 3,970; wheat, 3,000; tomatoes, 2,148; bananas, 2,027; chillies and green peppers, 1,854; lemons and limes, 1,825; potatoes, 1,735; mangoes, 1,503; beans, 1,400; barley, 1,109; avocados, 1,040.

Livestock (2003): cattle, 30·80m.; sheep, 6·56m.; pigs, 18·10m.; goats, 9·50m.; horses, 6·26m.; mules, 3·28m.; asses, 3·26m.; chickens, 540m. Production, 2003 (in 1,000 tonnes): beef and veal, 1,496; pork, bacon and ham, 1,043; horse, 79; goat meat, 42; lamb and mutton, 40; poultry meat, 2,204; cow's milk, 9,842; goat's milk, 148; eggs, 1,882; cheese, 130; honey, 56.

Forestry

Forests extended over 64·80m. ha. in 2010, representing 33% of the land area, containing pine, spruce, cedar, mahogany, logwood and rosewood. There are 14 forest reserves (nearly 0·8m. ha.) and 47 national park forests of 0·75m. ha. Timber production was 44·91m. cu. metres in 2007.

Fisheries

The total catch in 2010 was 1,523,889 tonnes, of which 1,407,479 tonnes came from sea fishing.

INDUSTRY

The leading companies by market capitalization in Mexico in March 2012 were: América Móvil, S.A.B. de C.V. (a mobile phone company), US$66·0bn.; Wal-Mart de México, S.A.B. de C.V. (general retailers, formerly Cifra), US$59·6bn.; and Femsa (a beverages company), US$26·7bn.

In 2001 the manufacturing industry provided 19·6% of GDP. Output (in 1,000 tonnes): cement (2008), 37,139; petrol (2007), 21,563; distillate fuel oil (2007), 18,011; residual fuel oil (2007), 17,252; crude steel (2009), 13,957; sugar (2002), 5,073; pig iron (2009), 3,925; wheat flour (2001), 2,611; cigarettes (2001), 56·1bn. units; soft drinks (2001), 13,005·0m. litres; beer (2007), 8,051·0m. litres. Car production has increased from 857,000 in 1994 to 1,098,000 in 2006.

Labour

In the period March–June 2001 the employed population totalled 39,004,300. The principal areas of activity were (in 1,000): wholesale and retail trade/repair of motor vehicles, motorcycles and personal and household goods, 8,839·2; manufacturing, 7,373·0; agriculture, hunting and forestry, 6,920·7; construction, 2,396·9; hotels and restaurants, 1,982·2; education, 1,971·6.

Unemployment rate, Dec. 2012, 4·9%. The daily minimum wage at Jan. 2009 ranged from 51·95 new pesos to 54·80 new pesos.

INTERNATIONAL TRADE

In Sept. 1991 Mexico signed the free trade Treaty of Santiago with Chile, envisaging an annual 10% tariffs reduction from Jan. 1992. The North American Free Trade Agreement (NAFTA), between Canada, Mexico and the USA, was signed on 7 Oct. 1992 and came into effect on 1 Jan. 1994. A free trade agreement was signed with Costa Rica in March 1994. Some 8,300 products were free from tariffs, with others to follow over ten years. The Group of Three (G3) free trade pact with Colombia and Venezuela came into effect on 1 Jan. 1995. A free trade agreement was signed with the European Union in 1999.

Imports and Exports

Trade for calendar years in US$1m.:

	2003	2004	2005	2006	2007
Imports c.i.f.	170,546	196,809	221,819	256,086	281,927
Exports f.o.b.	164,907	187,980	214,207	249,961	271,821

Of imports in 2007, 49·6% came from the USA, 10·6% from China, 5·8% from Japan, 4·5% from South Korea and 3·8% from Germany. Of exports in 2007, 82·2% went to the USA, 2·4% to Canada, 1·5% to Germany, 1·4% to Spain and 1·1% to Colombia.

The in-bond (*maquiladora*) assembly plants generate the largest flow of foreign exchange. Although originally located along the US border when the programme was introduced in the 1960s, they are now to be found in almost every state. In 2009 there were over 5,200 'foreign to Mexico' manufacturing companies, employing more than 1·6m. people. Manufactured goods account for 90% of trade revenues.

COMMUNICATIONS

Roads

The total road length in 2007 was 360,075 km, of which 6,565 km were motorways, 40,631 km other main roads, 73,874 km secondary roads and 239,005 km other roads. In 2005 there were 14,074,669 passenger cars, 7,111,172 trucks and vans and 264,726 buses and coaches. There were 5,398 fatalities as a result of road accidents in 2007.

Rail

The National Railway, *Ferrocarriles Nacionales de México*, was split into four companies in 1995 as a preliminary to privatization. It ceased operations in 1999. The rail network comprises 26,677 km of 1,435 mm gauge. In 2007, 99·9m. tonnes of freight were transported. Passenger traffic declined dramatically between 1997 and 2008, when a suburban rail network was opened between Mexico City and Cuautitlán in the state of México. There is a 202 km metro in Mexico City with 11 lines. There are light rail lines in Guadalajara (24 km) and Monterrey (32 km).

Civil Aviation

There is an international airport at Mexico City (Benito Juárez) and 55 other international and 29 national airports. Each of the larger states has a local airline which links it with main airports. The national carrier is Aeroméxico, which was privatized in 1988. Two former flag carriers, Aviacsa and Mexicana, both ceased operations in 2010 (in April and Aug. respectively). In 2005 scheduled airline traffic of Mexican-based carriers flew 366·7m. km. and carried 20,218,800 passengers. In 2001 Mexico City handled 20,599,064 passengers (13,711,141 on domestic flights). Cancún was the second busiest airport for passengers in 2001, with 7,640,007 (5,905,813 on international flights). Guadalajara handled 5,020,631 passengers (3,337,339 on domestic flights).

Shipping

Mexico had 114 ports and terminals in 2007 (66 ocean navigation), of which the most important are Altamira, Progreso, Tampico, Tuxpan and Veracruz on the Gulf coast and Manzanillo on the Pacific coast. Mexico's busiest port is Manzanillo, which handled 21·21m. tonnes of cargo in 2007 (7·53m. tonnes loaded and 13·68m. tonnes discharged). A law to privatize port operations was passed in 1993.

In 2007 the merchant marine numbered 2,387 vessels with a total tonnage of 1,727,000 GRT. In 2005 vessels totalling 60,527,000 NRT entered ports and vessels totalling 144,652,000 NRT cleared.

Telecommunications

Telmex (Teléfonos de México), a former state-run company privatized in 1991, is the leading provider of fixed-line telephone services and broadband, with around 80% of the market. In 2008 there were 20,668,000 fixed telephone lines and 75,305,000 mobile phone subscribers (693·7 per 1,000 persons). The leading mobile phone operator is Telcel (part of América Móvil), which has about 70% of the market. There were 23·6m. internet users in 2008. There were 10·4 fixed broadband subscribers per 100 inhabitants in Dec. 2010. In Dec. 2011 there were 31·0m. Facebook users.

SOCIAL INSTITUTIONS

Justice

Magistrates of the Supreme Court are appointed for six years by the President and confirmed by the Senate; they can be removed only on impeachment. The courts include the Supreme Court with 21 magistrates, 12 collegiate circuit courts with three judges each and nine unitary circuit courts with one judge each, and 68 district courts with one judge each.

The penal code of 1 Jan. 1930 abolished the death penalty, except for the armed forces. Mexico abolished the death penalty for all crimes in Dec. 2005—the last execution had been in 1961.

There were 24,374 murders in 2010 (a rate of 22 per 100,000 population), up from 8,867 in 2007, 14,006 in 2008 and 19,803 in 2009. The population in penal institutions in Aug. 2006 was 214,450 (196 per 100,000 of national population). Following the collapse of the main Colombian drug cartels in the 1990s, Mexican cartels are estimated to control 70% of illicit foreign drugs entering the US market. In 2009, 9,616 people died in drug-related killings in Mexico.

Education

Adult literacy was 93·4% in 2009 (male, 94·9%; female, 92·1%). Primary and secondary education is free and compulsory, and secular, although religious instruction is permitted in private schools.

In 2002–03 there were:

	Establishments	Teachers	Students (in 1,000)
Pre-school	74,758	163,282	3,636
Primary	99,463	557,278	14,857
Secondary	29,749	325,233	5,660
Baccalaureate	9,668	202,161	2,936
Vocational training	1,659	31,683	359
Medium/Professional	664	17,280	167
Higher education	2,539	192,593	1,932
Postgraduate education	1,283	21,685	138

In 2006 public expenditure on education came to 4·8% of GDP and 22·0% of total government spending (the highest share in the OECD).

Health

There were 3,039 hospitals in 2003, with a total provision of 73,446 beds. In 2001 there were 172,266 physicians, 9,669 dentists and 222,389 nurses. In 2008 Mexico spent 5·9% of its GDP on health. In 2006, 70% of Mexicans were considered overweight (one of the highest rates in the world) and 30% obese (having a body mass index over 30).

Welfare

As of 1 July 1997 all workers had to join the private insurance system, while the social insurance system was being phased out. At retirement, employees covered by the social insurance system before 1997 can choose to receive benefits from either the social insurance system or the private insurance system. The official retirement age is 65 years but to be eligible, a pensioner must have paid 1,250 weeks of contributions. The guaranteed minimum pension in 2011 was 2,095·99 new pesos.

Unemployment benefit exists under a labour law which requires employers to pay a dismissed employee a lump sum equal to three months' pay plus 20 days' pay for each year of service. Social security pays an unemployment benefit of between 75% and 95% of the old-age pension for unemployed persons aged 60 to 64.

RELIGION

In 2005 an estimated 90% of the population was Roman Catholic, down from 98% in 1950. In Feb. 2013 there were four cardinals. The Church is separated from the State, and the constitution of 1917 provided strict regulation of this and all other religions. In Nov. 1991 Congress approved an amendment to the 1917 constitution permitting the recognition of churches by the state, the possession of property by churches and the enfranchisement of priests. Church buildings remain state property. In 2001 there were estimated to be 3·82m. Protestants, plus followers of various other religions. There were 1,044,000 Latter-day Saints (Mormons) in 2006.

CULTURE

World Heritage Sites

Mexico has 31 UNESCO World Heritage sites. They are (with year entered on list): the Sian Ka'an nature reserve; the historic centre of Mexico City and the canal and island network of Xochimilco; Puebla's historic centre; the pre-Hispanic city of Teotihuacán, now a major archaeological site; the Historic Centre of Oaxaca and archaeological site of Monte Alban; and Palenque —lying in the foothills of the Altos de Chiapas, the Maya ruins of Palenque are surrounded by waterfalls, rainforest and fauna (all 1987); the historic town of Guanajuato and adjacent disused silver mines; and the pre-Hispanic city of Chichen-Itza, Yucatan (both 1988); the historic centre of Morelia, on the southern Pacific coast (1991); the pre-Hispanic city of El Tajin, Veracruz (1992); the El Vizcaino whale sanctuary; the historic centre of Zacatecas, once a major silver mining centre; and the Sierra de San Francisco rock paintings (all 1993); the 14 early 16th-century monasteries on Popocatépetl, to the southeast of Mexico City (1994); the Maya town of Uxmal, Yucatán, with its preserved pyramids and sculptures; and the historic monuments zone of Querétaro (both 1996); the Hospicio (Hospice) Cabañas, Guadalajara (1997); the historic monuments zone of Tlacotalpan; and the archaeological zone of Paquimé, Casas Grandes in Chihuahua (both 1998); the historic fortified town of Campeche; and Xochicalco's archaeological monuments zone, Morelos state (both 1999); the Ancient Maya City of Calakmul, Campeche (2002); the Franciscan Missions in the Sierra Gorda of Querétaro (2003); Luis Barragán House and Studio in Mexico City (2004); the islands and protected areas of the Gulf of California (2005 and 2007); the Agave landscape and ancient industrial facilities of Tequila (2006); the Central University City Campus of the Universidad Nacional Autónoma de México (2007); the Monarch Butterfly Biosphere Reserve (2008); the fortified town of San Miguel and Sanctuary of Jesús Nazareno de Atotonilco (2008); Camino Real de Tierra Adentro, also known as the Silver Route, between Mexico City and the USA; and the Prehistoric Caves of Yagul and Mitla in the Central Valley of Oaxaca (both 2010).

Press

In 2008 there were 462 daily newspapers with a circulation of 4,590,000. The three leading dailies are *Esto* (average daily circulation of 320,000 in 2008), *La Prensa* (315,000) and *El Universal Gráfico* (300,000).

Tourism

There were 21·45m. non-resident tourists in 2009 (excluding same-day visitors), making Mexico the tenth most popular tourist destination; spending amounted to US$11,275m. in 2009.

DIPLOMATIC REPRESENTATIVES

Of Mexico in the United Kingdom (16 St George St., London, W1S 1FD)
Ambassador: Eduardo Medina Mora Icaza.

Of the United Kingdom in Mexico (Rio Lerma 71, Col. Cuauhtémoc, 06500 México, D.F.)
Ambassador: Judith Macgregor.

Of Mexico in the USA (1911 Pennsylvania Ave., NW, Washington, D.C., 20006)
Ambassador: Eduardo Medina-Mora Icaza.

Of the USA in Mexico (Paseo de la Reforma 305, 06500 México, D.F.)
Ambassador: Earl Anthony Wayne.

Of Mexico to the United Nations
Ambassador: Luis Alfonso de Alba.

Of Mexico to the European Union
Ambassador: Sandra Fuentes-Beráin.

FURTHER READING

Instituto Nacional de Estadística, Geografía e Informática. *Anuario Estadístico de los Estados Unidos Mexicanos. Mexican Bulletin of Statistical Information.* Quarterly.

Bethell, L. (ed.) *Mexico since Independence.* 1992
Castañeda, Jorge G., *Mañana Forever?: Mexico and the Mexicans.* 2011
Hamnett, Brian R., *A Concise History of Mexico.* 1999
Krauze, E., *Mexico, Biography of Power: A History of Modern Mexico, 1810–1996.* 1997
Levy, Daniel C., *Mexico: The Struggle for Democratic Development.* 2006
Mentinis, Mihalis, *Zapatistas: The Chiapas Revolt and What it Means for Radical Politics.* 2006
Philip, G. (ed.) *The Presidency in Mexican Politics.* 1991
Randall, Laura, *Changing Structure of Mexico: Political, Social and Economic Prospects.* 2005
Ruíz, R. E., *Triumphs and Tragedy: a History of the Mexican People.* 1992
Snyder, Richard, *Politics After Neoliberalism: Reregulation in Mexico.* 2006
Whiting, V. R., *The Political Economy of Foreign Investment in Mexico: Nationalism, Liberalism, Constraints on Choice.* 1992

National Statistical Office: Instituto Nacional de Estadística, Geografía e Informática (INEGI), Av. Héroe de Nacozari Sur 2301, Fracc. Jardines del Parque, CP 20276 Aguascalientes.
Website (Spanish only): http://www.inegi.org.mx

MICRONESIA

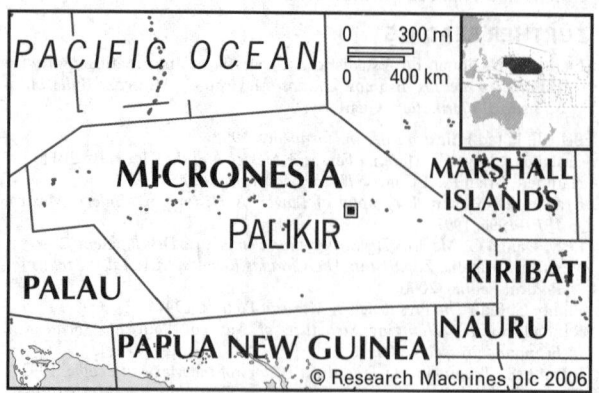

Federated States of Micronesia

Capital: Palikir
Population projection, 2015: 114,000
GNI per capita, 2011: (PPP$) 2,935
HDI/world rank: 0·636/116
Internet domain extension: .fm

KEY HISTORICAL EVENTS

Spain acquired sovereignty over the Caroline Islands in 1886 but sold the archipelago to Germany in 1899. Japan occupied the Islands at the beginning of the First World War and in 1921 they were mandated to Japan by the League of Nations. Captured by Allied Forces in the Second World War, the Islands became part of the UN Trust Territory of the Pacific Islands created on 18 July 1947 and administered by the USA. The Federated States of Micronesia came into being on 10 May 1979. American trusteeship was terminated on 3 Nov. 1986 by the UN Security Council and on the same day Micronesia entered into a 15-year Free Association with the USA. An amended 20-year Compact of Free Association was signed into law on 17 Dec. 2003, guaranteeing US$1·8bn. to Micronesia in grants for a government trust fund.

TERRITORY AND POPULATION

The Federated States lie in the North Pacific Ocean between 137° and 163° E, comprising 607 islands with a total land area of 701 sq. km (271 sq. miles). The 2010 census population was 102,843; density, 147 per sq. km.

The UN gives a projected population for 2015 of 114,000.

In 2011, 22·8% of the population lived in urban areas.

The areas and populations of the four major groups of island states (east to west) are as follows:

State	Area (sq. km)	Population (2010 census)	Headquarters
Kosrae	110	6,616	Tofol
Pohnpei	345	36,196	Kolonia
Chuuk	127	48,654	Weno
Yap	119	11,377	Colonia

Kosrae consists of a single island. Its main town is Lelu (2,160 inhabitants in 2010). Pohnpei comprises a single island (covering 334 sq. km) and eight scattered coral atolls. Kolonia (6,074 inhabitants in 2010) was the national capital until 1989. The new capital, Palikir (6,647 inhabitants in 2010), lies approximately 10 km southwest in the Palikir valley. Chuuk consists of 542 islets in

a 7,190 sq. km reef-fringed lagoon (40,465 inhabitants in 2000); the state also includes coral atolls (13,130 inhabitants in 2000), the most important being the Mortlock Islands. The chief town is Weno (13,856 inhabitants in 2010). Yap comprises a main group of four islands (covering 100 sq. km with 7,391 inhabitants in 2000) and 15 outer islands (3,850 inhabitants), the main ones being Ulithi and Woleai. Colonia is its chief town (3,126 inhabitants in 2010).

English is used in schools and is the official language. Trukese, Pohnpeian, Yapese and Kosrean are also spoken.

SOCIAL STATISTICS

2009 estimates: births, 2,800; deaths, 700. Rates, 2009 estimates (per 1,000 population): birth, 25; death, 6. Infant mortality rate (2010), 34 per 1,000 live births. 2007 life expectancy, 67·6 years for men and 69·2 years for women. Population growth rate, 2006, 0·1%; fertility rate, 2008, 3·6 births per woman.

CLIMATE

Tropical, with heavy year-round rainfall, especially in the eastern islands, and occasional typhoons (June–Dec.). Kolonia, Jan. 80°F (26·7°C), July 79°F (26·1°C). Annual rainfall 194" (4,859 mm).

CONSTITUTION AND GOVERNMENT

Under the Constitution founded on 10 May 1979, there is an executive presidency and a 14-member *National Congress*, comprising ten members elected for two-year terms from single-member constituencies of similar electorates, and four members elected one from each State for a four-year term by proportional representation. The *Federal President* and *Vice-President* first run for the Congress before they are elected by members of Congress for a four-year term.

National Anthem

'Patriots of Micronesia'; words anonymous, tune adapted from J. Brahms' 'Academic Festival Overture'.

RECENT ELECTIONS

The last election for Congress was held on 5 March 2013. Only non-partisans were elected. Immanuel Mori was re-elected president by Congress unopposed on 11 May 2011.

CURRENT GOVERNMENT

President: Immanuel 'Manny' Mori; b. 1948 (took office 11 May 2007 and re-elected in May 2011).

Vice-President: Alik L. Alik.

In March 2013 the government comprised:

Minister of Education: Casiano Shoniber. *Finance and Administration:* Finley Perman. *Foreign Affairs:* Lorin Robert. *Health and Social Affairs:* Dr Vita Skilling. *Justice:* Maketo Robert. *Resources and Development:* Marion Henry. *Transportation, Communications and Infrastructure:* Francis Itimai. *Chief Public Defender:* Julius Joey Sapelalut.

Speaker of the Congress: Isaac V. Figir.

Government Website: http://www.fsmgov.org

CURRENT LEADERS

Immanuel Mori

Position
President

Introduction
Immanuel 'Manny' Mori became the seventh president of the Federated States of Micronesia (FSM) on 11 May 2007. He is the

second president to come from the state of Chuuk. The president, who is both head of state and of the unicameral government, serves for a four-year term.

Early Life

Mori was born on 25 Dec. 1948 in Sapore on the island of Fefan, Chuuk. He spent his childhood in Fefan and attended the Xavier High School in Chuuk's capital, Weno. From 1969–73 he studied at the University of Guam, graduating in business management. He joined Citycorp Credit as a management intern, becoming assistant manager at the Saipan branch.

Mori returned to the FSM in 1976 and joined the Trust Territory social security office. In 1979 he became responsible for Chuuk's tax and revenue office. From 1981–83 he served as comptroller of the Development Bank of the FSM before being appointed its president and CEO. He held the post until Feb. 1997, when the bank's board ousted him in an attempt to encourage reform. He was subsequently named vice-president.

Mori's political career began in July 1999 when he was elected a Chuuk congressman. He held several positions including chair of the ways and means committee and vice-chair of the committee on judiciary and government operation. He also sat on the committees for resource and development and for health and social affairs. In 2003 he was elected senator-at-large for Chuuk for a four-year term. In this capacity Mori served on the task force for national government restructuring and on the planning council of the College of Micronesia. He also served as a CEO of the Chuuk Public Utility Corporation and on the board of the Pacific Island Development Bank.

In May 2007 parliament selected Mori to replace Joseph Urusemal as president. Alik L. Alik was chosen as vice-president.

Career in Office

Under a Compact of Free Association Micronesia is guaranteed US aid until 2023. Chief among Mori's challenges has been to prepare for economic self-reliance and to reinvigorate the stagnant private sector. He was re-elected for a second term in May 2011.

ECONOMY

Overview

Following independence in 1986, the country signed a Compact of Free Association with the USA, providing it with grants upon which it relies heavily. From 1987 to 2001 these grants made up half of GDP, although a renegotiated Compact Agreement effective since 2004 steadily lowers aid payments until 2023. It also includes the creation of a trust fund in a bid to make Micronesia self-reliant, although the global financial crisis has led to lower returns than anticipated.

The economy experienced low or negative GDP growth from 2001–09. Compact-related spending cuts reduced public sector employment, stimulating migration to the USA where citizens can work without a visa. Higher fuel and food import costs put further pressure on state finances.

The economy returned to growth in 2008–09 and grew by 2·5% in 2009–10, driven mainly by the construction, fisheries and public administration sectors. The private sector's share of the economy has remained unchanged at about 25% since the early 1990s and contributes little to GDP growth.

The decline in Compact grants, slow private sector growth and the global slowdown continue to pose challenges in the medium term. Migration and climate change constitute long-term challenges.

Currency

US currency is used. Foreign exchange reserves were US$50m. and total money supply was US$23m. in June 2005.

Budget

US compact funds are an annual US$100m. Revenue (2006–07), US$145·2m. (of which US$92·5m. grants); expenditure, US$151·6m. The financial year runs from 1 Oct.–30 Sept.

Performance

In 2011 total GDP was US$310m.; real GDP growth was 1·0% in 2008–09 and 2·5% in 2009–10.

Banking and Finance

There are three commercial banks: Bank of Guam, Bank of Hawaii and Bank of the Federated States of Micronesia. There is also a Federated States of Micronesia Development Bank and a regulatory Banking Board.

ENERGY AND NATURAL RESOURCES

Electricity

Electricity production in 2007 was 67m. kWh.

Minerals

The islands have few mineral deposits except for high-grade phosphates.

Agriculture

Agriculture consists mainly of subsistence farming: coconuts, breadfruit, bananas, sweet potatoes and cassava. A small amount of crops are produced for export, including copra, tropical fruits, peppers and taro. Production (2008 estimates, in 1,000 tonnes): coconuts, 41; cassava, 12; sweet potatoes, 3; bananas, 2. Livestock (2008 estimates): pigs, 33,000; cattle, 14,000; goats, 4,000. In 2007 there were approximately 2,500 ha. of arable land and 18,000 ha. of permanent crops.

Forestry

The total forest area was 64,000 ha. in 2010 (92% of the land area)

Fisheries

In 2009 the catch amounted to 27,706 tonnes, almost entirely from marine waters. In 2007, 303 fleet vessels were licensed to fish in Micronesia. Fishing licence fees were an estimated US$12m. in 2007 and are a primary revenue source, making up around 7% of GDP.

INDUSTRY

The chief industries are construction, fish processing, tourism and handicrafts (items from shell, wood and pearl).

Labour

In 2007 just over half the labour force were government employees. In 2007, 41·7% of employees worked in public administration, 21·1% in wholesale and retail trade and repairs and 7·0% in transport, storage and communications. Agriculture, hunting, forestry and fishing accounted for 1·7% of employees. The unemployment rate was 22·0% in 2000.

INTERNATIONAL TRADE

Imports and Exports

Total imports (c.i.f.) in 2007, US$142·7m.; exports (f.o.b.), US$16·2m. Main import suppliers, 2007: US mainland, 41·2%; Guam, 14·4%; Singapore, 8·7%. Main export markets, 2007: Guam, 22·5%; US mainland, 17·2%; Northern Mariana Islands, 4·3%. The main imports are mineral products, foodstuffs and beverages, machinery, mechanical and electrical appliances, animals and animal products, and vegetable products. Main exports: marine products, kava, handicrafts and souvenirs, citrus fruits and garments.

COMMUNICATIONS

Roads

In 2000 there were 240 km of roads (42 km paved).

Civil Aviation

There are international airports on Pohnpei, Chuuk, Yap and Kosrae. Services are provided by Continental Airlines. In 2003 there were international flights to Guam, Honolulu, Manila, the Marshall Islands and Palau in addition to domestic services. There were five airports in 1996 (four paved).

Shipping

The main ports are Kolonia (Pohnpei), Colonia (Yap), Lepukos (Chuuk), Okat and Lelu (Kosrae). In Jan. 2009 there were 11 ships of 300 GT or over registered, totalling 9,000 GT.

Telecommunications

Micronesia had 36,100 telephone subscribers in total in 2007, or 325·2 per 1,000 population. Mobile phone subscribers numbered 27,400 in 2007. The islands are interconnected by shortwave radiotelephone. There are four earth stations linked to the Intelsat satellite system. There were 15,000 internet users in 2007.

SOCIAL INSTITUTIONS

Justice

There is a Supreme Court headed by the Chief Justice with two other judges, and a State Court in each of the four states with 13 judges in total.

Education

In 2007 there were 18,512 pupils in primary schools with 1,113 teaching staff; and 14,742 pupils in secondary schools. The College of Micronesia in Pohnpei, initially founded as the Micronesian Teacher Education Center in 1963, is the only institute of higher education and now has campuses on all four major island groups, including the Fisheries and Maritime Institute on the Yap Islands.

In 2001–02 total expenditure on education came to 6·7% of GNP.

Health

In 2006 there were 33 hospital beds per 10,000 population. There were 72 physicians, 13 dentists and 305 nursing staff in 2007.

RELIGION

The population is predominantly Christian. Yap is mainly Roman Catholic; Protestantism is prevalent elsewhere.

CULTURE

Tourism

In 2007 there were 21,146 visitors, up from 19,136 in 2006 and 18,958 in 2005.

DIPLOMATIC REPRESENTATIVES

Of the United Kingdom in Micronesia
Ambassador: Stephen Lillie (resides in Manila, Philippines).

Of Micronesia in the USA (1725 N St., NW, Washington, D.C., 20036)
Ambassador: Asterio Takesy.

Of the USA in Micronesia (POB 1286, Kolonia, Pohnpei)
Ambassador: Doria Rosen.

Of Micronesia to the United Nations
Ambassador: Jane J. Chigiyal.

FURTHER READING

Wuerch, W. L. and Ballendorf, D. A., *Historical Dictionary of Guam and Micronesia.* 1995

National Statistical Office: FSM Statistics Division, P. O. Box PS 253, Palikir, Pohnpei, FM 96941.
Website: http://www.spc.int/prism/country/fm/stats

MOLDOVA

© Research Machines plc 2006

Republica Moldova
(Republic of Moldova)

Capital: Chişinău
Population projection, 2015: 3·45m.
GNI per capita, 2011: (PPP$) 3,058
HDI/world rank: 0·649/111
Internet domain extension: .md

KEY HISTORICAL EVENTS

In antiquity the territories that make up Moldova were inhabited by the Dacians. In 1359 the region was subsumed into the Principality of Moldavia, founded by Dragoş of Bedeu. From the 16th to 19th centuries Russia and the Ottoman Empire wrestled for influence. In 1812 the Treaty of Bucharest gave Russia control of eastern Moldavia, or Bessarabia (the area between the River Prut and the Dniester, which corresponds to much of present-day Moldova). The Ottomans ruled western Moldavia. In 1918 Romania absorbed Bessarabia, while the Soviet Union controlled the territory east of the Dniester from 1924. Bessarabia reverted from Romanian rule to become part of the Moldavian Socialist Republic within the USSR in 1940.

In Dec. 1991 Moldova became a member of the Commonwealth of Independent States, a decision ratified by parliament in April 1994. Fighting took place in 1992 between government forces and separatists in the (largely Russian and Ukrainian) area east of the River Nistru (Transnistria). An agreement signed by the presidents of Moldova and Russia on 21 July 1992 brought to an end the armed conflict and established a 'security zone' controlled by peacekeeping forces from Russia, Moldova and Transnistria. On 21 Oct. 1994 a Moldo-Russian agreement obliged Russian troops to withdraw from the territory

of Moldova over three years but the agreement was not ratified by the Russian Duma. On 8 May 1997 an agreement between Transnistria and the Moldovan government to end the separatist conflict stipulated that Transnistria would remain part of Moldova as it was territorially constituted in Jan. 1990. In 1997 some 7,000 Russian troops were stationed in Transnistria. In the autumn of 1999 Ion Sturza's centre-right coalition collapsed, along with privatization plans for the wine and tobacco industries. Communist President Vladimir Voronin, who was elected in 2001, has proposed giving the Russian language official status and joining the Russia–Belarus union.

In Aug. 2009 the Communist government fell to a coalition of four opposition parties after earlier disputed elections and a failure to elect a new president.

TERRITORY AND POPULATION

Moldova is bounded in the east and south by Ukraine and on the west by Romania. The area is 33,848 sq. km (13,067 sq. miles). At the last census, in 2004, the population was 3,938,679 (52·2% female). Population estimate, Jan. 2012: 4,077,000; density, 120 per sq. km.

The UN gives a projected population for 2015 of 3·45m. (excluding Transnistria).

In 2011, 47·7% of the population lived in urban areas. Ethnicity (2004): Moldovans accounted for 69·6%, Ukrainians 11·3%, Russians 9·3%, Gagauz 3·9%, Bulgarians 2·0%, Roma (Gypsy) 1·9% and others 2·0%.

Apart from Chişinău, the capital (population estimate of 667,600 in 2012), major towns are Tiraspol (147,800 in 2012), Bălţi (144,300 in 2012) and Tighina (93,300 in 2012). The official Moldovan language (i.e. Romanian) was written in Cyrillic prior to the restoration of the Roman alphabet in 1989. It is spoken by 62% of the population; the use of other languages (Russian, Gagauz) is safeguarded by the Constitution.

SOCIAL STATISTICS

2007: births, 37,973; deaths, 43,050. Rates, 2007 (per 1,000 population): births, 10·6; deaths, 12·0. In 2006 the most popular age at first marriage was 20–24 for both males and females. Life expectancy at birth in 2007 was 65·0 years for males and 72·6 years for females. Annual population growth rate, 2000–05, –0·2%. Infant mortality, 2010, 16 per 1,000 live births; fertility rate, 2008, 1·3 births per woman (one of the lowest rates in the world). In the period 2000–06, 48·5% of the population were classified as living below the national poverty line.

CLIMATE

The climate is temperate, with warm summers, crisp, sunny autumns and cold winters with snow. Chişinău, Jan. –7°C, July 20°C. Annual rainfall 677 mm.

CONSTITUTION AND GOVERNMENT

A declaration of republican sovereignty was adopted in June 1990 and in Aug. 1991 the republic declared itself independent. A new constitution came into effect on 27 Aug. 1994, which defines Moldova as an 'independent, democratic and unitary state'. At a referendum on 6 March 1994 turnout was 75·1%; 95·4% of votes cast favoured 'an independent Moldova within its 1990 borders'. The referendum (and the Feb. parliamentary elections) were not held by the authorities in Transnistria. In a further referendum on 4 June 1999, on whether to switch from a parliamentary system to a presidential one, turnout was 58% with the majority of the votes cast being in favour of the change.

Parliament (*Parlamentul*) has 101 seats and is elected for four-year terms. There is a 4% threshold for election; votes falling below this are redistributed to successful parties. The *President* is now elected for four-year terms by parliament, after the constitution had been amended to abolish direct presidential elections.

The 1994 constitution makes provision for the autonomy of Transnistria and the Gagauz (Gagauzi Yeri) region. Work began in July 2003 on the drafting of a new constitution to resolve the conflict between Moldova and Transnistria.

Transnistria. In the predominantly Russian-speaking areas of Transnistria a self-styled 'Dniester Republic' was established in Sept. 1991, and approved by a local referendum in Dec. 1991. A Russo-Moldovan agreement of 21 July 1992 provided for a special statute for Transnistria and a guarantee of self-determination should Moldova unite with Romania. The population at the 2004 census was 555,347. Romanian here is still written in the Cyrillic alphabet. At a referendum on 24 Dec. 1995, 81% of votes cast were in favour of adopting a new constitution proclaiming independence.

On 17 June 1996 the Moldovan government granted Transnistria a special status as 'a state-territorial formation in the form of a republic within Moldova's internationally recognized border'.

At elections to the Supreme Council held on 12 Dec. 2010, which were not internationally recognized as legitimate, Renewal won 25 of 43 seats, Republic Party 16, Breakthrough 1 and Pridnestrovie Communist Party also 1. In a referendum held on 17 Sept. 2006, 97·2% of votes cast were in favour of independence and possible future union with Russia, although no international organization or foreign country recognized the referendum. Elections for president were held on 11 Dec. 2011. Turnout was 58·9%. Yevgeny Shevchuk won 38·6% of the vote, Anatoly Kaminsky 26·3% and incumbent Igor Smirnov 24·7%. In the run-off held on 25 Dec. 2011, Shevchuk won 73·9% and Kaminsky 19·7%. Turnout was 52·5%.

Gagauz Yeri. This was created an autonomous territorial unit by Moldovan legislation of 13 Jan. 1995. In 2004 the census population was 155,646. There is a 35-member *Popular Assembly* directly elected for four-year terms and headed by a *Governor*, who is a member of the Moldovan cabinet. At the elections of 9 and 23 Sept. 2012 independents took 25 seats, the Party of Communists of the Republic of Moldova 7, the Liberal Democrat Party of Moldova 2 and the Party of Socialists of the Republic of Moldova 1.

Governor: Mihail Formuzal; b. 1959.

National Anthem

The Romanian anthem was replaced in 1994 by a traditional tune, 'Lîmbă noastră' ('Our Tongue'); words by Alexei Mateevici, tune by Alexandru Cristi.

RECENT ELECTIONS

On 16 Dec. 2011 the only presidential candidate, Marian Lupu failed to obtain the requisite 61 votes from parliament to be elected after the communists boycotted the vote. On 12 Jan. 2012 the Constitutional Court ruled the elections invalid owing to violations of secret voting procedures. Nicolae Timofti was then elected president on 16 March 2012, with 62 votes in the 101-seat parliament, bringing an end to nearly three years of political deadlock.

Parliamentary elections were held on 28 Nov. 2010. The ruling Party of Communists of the Republic of Moldova won 42 seats with 39·3% of the vote, the Liberal Democratic Party of Moldova 32 with 29·4%, the Democratic Party of Moldova 15 with 12·7% and the Liberal Party 12 with 10·0%. Turnout was 59·1%.

CURRENT GOVERNMENT

In April 2013 the government comprised:

President: Nicolae Timofti; b. 1948 (ind.; since 23 March 2012).

Acting Prime Minister, and Minister of Foreign Affairs and European Integration: Iurie Leancă; b. 1963 (Liberal Democratic Party of Moldova; since 25 April 2013).

Deputy Prime Ministers: Valeriu Lazăr (also *Minister of the Economy*); Mihai Moldovanu; Eugen Carpov.

Minister of Agriculture and Food Industries: Vasile Bumacov. *Culture:* Boris Focşa. *Defence:* Vitalie Marinuta. *Education:* Maia Sandu. *Environment:* Gheorghe Şalaru. *Finance:* Veaceslav Negruţă. *Health:* Andrei Usatii. *Information Technologies and Communications:* Pavel Filip. *Internal Affairs:* Dorin Recean. *Justice:* Oleg Efrim. *Labour, Social Protection and Family:* Valentina Buliga. *Regional Development and Construction:* Marcel Răducan. *Transport and Road Infrastructure:* Anatol Şalaru. *Youth and Sports:* Octavian Ticu.

Government Website: http://www.gov.md

CURRENT LEADERS

Nicolae Timofti

Position
President

Introduction
Nicolae Timofti took office on 23 March 2012 after securing 62 votes in the 101-seat parliament—one vote more than the minimum required—ending three years of political deadlock.

Early Life
Nicolae Timofti was born on 22 Dec. 1948 in Ciutuleşti, north-eastern Moldova. Graduating in law from Moldova State University, Chişinău, in 1972, he completed two years national service in the Soviet Army. In 1974 he became a consultant for the ministry of justice and in 1976 joined the Chişinău district court. In 1980 he was appointed to the supreme court and a decade later was named its vice-president.

In 1996 he joined the court of appeal and later became its president. In 2003 he was appointed to the Chişinău court of appeal and was made a member of the supreme council of magistrates. Two years later he was posted to the supreme court of justice and in 2011 became president of the supreme council of magistrates.

Timofti was nominated for the presidency in early 2012 by the ruling Alliance for European Integration coalition. On 16 March 2012 he was elected into office by parliament, despite a Communist Party boycott. He was the first permanently-appointed president since the end of Vladimir Voronin's tenure in 2009.

Career in Office
Timofti, widely considered a neutral figure in Moldovan politics, has pledged to fight corruption and support democratic and free-market reforms. However, critics point to his failure to overhaul the notoriously corrupt judiciary while a senior judicial official.

Timofti supports Moldovan EU membership and is keen to maintain close ties with the USA and Russia. He also aims to solve the issue of the breakaway region of Transnistria.

In March 2013 the prime minister, Vladimir Filat, resigned after his government lost a confidence vote. Timofti asked him to form an interim administration but the following month the constitutional court banned Filat from holding the office. The foreign minister, Iurie Leancă, was sworn in as premier on 25 April 2013.

DEFENCE

Conscription is for 12 months (three months for higher education graduates). In 2008 military expenditure totalled US$22m. (US$5 per capita), representing 0·4% of GDP.

Russian troops remained in Transnistria after Moldova gained independence, but in Nov. 1999 the Organization for Security and Co-operation in Europe (OSCE) passed a resolution at its summit requiring Russia to withdraw its troops to Russia by Dec. 2002, unconditionally and under international observation. This deadline was extended to Dec. 2003 but around 400 Russian troops still remained in the region in 2013 as part of a joint peacekeeping force.

Army
Personnel, 2007, 5,150 (3,479 conscripts). In 2007 there was also a paramilitary Interior Ministry force of 2,379, riot police numbering 900 and combined forces reserves of 66,000.

Air Force
Personnel (including air defence), 2007, 850.

ECONOMY
Agriculture accounted for 10·1% of GDP in 2009, industry 13·1% and services 76·8%.

Overview
After poor performance through most of the 1990s, the economy began to recover in 2000, with real GDP growth exceeding 6% every year between 2001 and 2005. In spite of economic shocks (such as a 200% price increase for imported natural gas and a Russian ban on wine imports in 2006), growth remained robust for several years before the economy shrank by 6% in 2009 in the wake of the global downturn.

Growth rebounded to 7·1% in 2010 and the economy was among the fastest growing in Central and Eastern Europe in 2011, driven by private domestic demand and strong exports. However, high unemployment, political instability, volatile commodity prices and the fragile global economy pose a long-term risk to recovery.

Currency
A new unit of currency, the *leu* (MDL), replaced the *rouble* in Nov. 1993. Inflation was 12·7% in 2008 and 0·0% in 2009, down from a peak of 2,198% in the early 1990s. It increased to 7·4% in 2010 and 7·6% in 2011. Foreign exchange reserves were US$506m. in July 2005. Total money supply in June 2005 was 6,523m. lei.

Budget
Budgetary central government revenue and expenditure in 1m. lei (years ending 31 Dec.):

	2005	2006	2007
Revenue	7,941	9,823	14,059
Expenditure	7,031	9,121	12,164

Principal sources of revenue in 2007 were: taxes 10,900m. lei; grants, 967m. lei. Main items of expenditure by economic type in 2007: grants, 4,597m. lei; compensation of employees, 2,634m. lei; subsidies, 1,262m. lei.

VAT is 20% (reduced rates, 8% and 6%).

Performance
Moldova's economy shrank by 6·5% in 1998 and 3·4% in 1999. Between 2002 and 2008 it grew by an average 6·4% including growth of 7·8% in 2008. With the global economic downturn the economy contracted by 6·0% in 2009 but there was growth of 7·1% in 2010 and 6·4% in 2011.

In 2002 the level of GDP was estimated to be only 38% of that in 1989. Total GDP was US$7·0bn. in 2011 (excluding Transnistria). The private sector accounts for over 50% of official GDP. Moldova ranks among the countries most reliant on remittances from abroad, which accounted for 23% of total GDP in 2009.

Banking and Finance
The central bank and bank of issue is the National Bank (*Governor*, Dorin Drăguţanu). At June 2002 there were 21 commercial banks and one savings bank. Foreign debt was US$4,615m. in 2010, representing 73·5% of GNI. There is a stock exchange in Chişinău.

ENERGY AND NATURAL RESOURCES

Environment
Moldova's carbon dioxide emissions from the consumption and flaring of fossil fuels in 2008 were the equivalent of 1·7 tonnes per capita.

Electricity
Installed capacity in 2007 was 504,000 kW. Production was 1·10bn. kWh in 2007; consumption per capita in 2007 was 1,129 kWh.

Minerals
There are deposits of lignite, phosphorites, gypsum and building materials.

Agriculture
There were 387,400 people employed in agriculture in 2008. Land under cultivation in 2008 was 2·5m. ha., of which 0·3m. ha. was accounted for by private subsidiary agriculture and 668,600 ha. by farms. In 2008 there were 1·82m. ha. of arable land and 303,000 ha. of permanent crops. The agricultural and food sector accounts for 38% of Moldova's total exports.

Output of main agricultural products (in 1,000 tonnes) in 2008: maize, 1,479; wheat, 1,286; sugar beets, 961; grapes, 636; sunflower seeds, 372; barley, 353; potatoes, 271.

Livestock (2008): 853,000 sheep and goats, 299,000 pigs, 232,000 cattle, 17m. chickens.

Livestock products, 2008 (in 1,000 tonnes): milk, 542; meat, 78; eggs, 541m. units.

Forestry
In 2010 forests covered 0·39m. ha., or 12% of the total land area. Timber production in 2007 was 188,000 cu. metres.

Fisheries
The catch in 2010 (exclusively freshwater fish) was 1,633 tonnes.

INDUSTRY
There are canning plants, wine-making plants, woodworking and metallurgical factories, a factory of ferro-concrete building materials, footwear, dairy products and textile plants. Manufacturing accounted for 14·8% of GDP in 2007. Production (in tonnes): crude steel (2007), 995,000; cement (2006), 837,000; flour (2007), 113,300; canned fruit and vegetables (2007), 94,000; granulated sugar (2007), 74,000; footwear (2007), 3·8m. pairs; 5·0bn. cigars and cigarettes (2007); wine (2008), 397·9m. litres.

Labour
In 2007 the labour force totalled 1,314,000. A total of 1,247,000 persons were in employment in 2007, including 409,000 engaged in agriculture, hunting, forestry and fisheries, 250,000 in public administration, education, heath and social work, 198,000 in wholesale and retail trade/hotels and restaurants and 128,000 in manufacturing. In 2007 the unemployment rate was 5·1%.

INTERNATIONAL TRADE

Imports and Exports
Imports and exports for calendar years in US$1m.:

	2002	2003	2004	2005	2006
Imports f.o.b.	1,037·5	1,428·1	1,748·2	2,296·1	2,644·4
Exports f.o.b.	659·7	805·1	994·1	1,104·6	1,053·0

Chief import sources in 2006 were: Ukraine, 19·2%; Russia, 15·5%; Romania, 12·8%; Germany, 7·9%. Main export markets in

2006 were: Russia, 17·3%; Romania, 14·8%; Ukraine, 12·2%; Italy, 11·1%.

Moldova's leading imports are mineral products and fuel, machinery and equipment, chemicals and textiles. The main export commodity is wine, ahead of tobacco. Fruit and vegetables, textiles and footwear, and machinery are also significant exports.

COMMUNICATIONS

Roads

There were 9,343 km of public roads in 2009 (94·3% hard surfaced). Registered passenger cars (including taxis) in 2008 numbered 366,351, there were 115,967 goods vehicles and 21,491 buses and minibuses. In 2005 there were 2,289 road accidents resulting in 391 deaths.

Rail

Total length in 2005 was 1,026 km of 1,520 mm gauge. Passenger-km travelled in 2005 came to 355m. and freight tonne-km to 2,980m.

Civil Aviation

The main Moldovan-based airline is Air Moldova, which had flights in 2003 to Amman, Amsterdam, Athens, Bucharest, İstanbul, Larnaca, Moscow, Paris, Prague, Rome and Vienna. In 2000 the airport at Chişinău handled 254,234 passengers (all on international flights) and 2,159 tonnes of freight. In 2003 scheduled airline traffic of Moldovan-based carriers flew 5m. km, carrying 179,000 passengers (all on international flights).

Shipping

In 2008, 0·11m. passengers and 0·20m. tonnes of freight were carried on inland waterways. In Jan. 2009 there were 63 ships of 300 GT or over registered, totalling 170,000 GT.

Telecommunications

There were 2,962,700 telephone subscribers in total in 2007 (781·0 per 1,000 persons). There were 1,882,800 mobile phone subscribers in 2007—up from 338,200 in 2002—and 700,000 internet users. In March 2012 there were 221,000 Facebook users.

SOCIAL INSTITUTIONS

Justice

A total of 24,362 crimes were recorded in 2007. The population in penal institutions in Sept. 2007 was 8,130 (227 per 100,000 of national population). The death penalty was abolished for all crimes in 1995.

Education

In 2007 there were 103,811 children and 10,517 teaching staff in pre-schools; 160,528 pupils and 9,876 teaching staff in primary schools; and 367,636 pupils and 30,376 teaching staff in secondary schools. There were 148,449 students (8,570 academic staff) in tertiary education in 2007. In 2006 there were 16 public and 15 private higher education institutions. Adult literacy rate in 2008 was 98%.

In 2007 public expenditure on education came to 7·3% of GNI and represented 19·8% of total government expenditure.

Health

In 2007 there were 83 hospitals with 21,892 beds, a provision of 61 per 10,000 inhabitants. In 2007 there were 12,733 physicians (11% private sector), 1,566 dentists, 20,868 nurses and 2,834 pharmacists (2006).

Welfare

There were 469,600 age pensioners and 169,800 other pensioners in 2008.

RELIGION

Religious affiliation in 2001: Romanian Orthodox, 1·26m.; Russian (Moldovan) Orthodox, 342,000.

CULTURE

World Heritage Sites

Moldova has one site on the UNESCO World Heritage List: the Struve Geodetic Arc (inscribed in 2005). The Arc is a chain of survey triangulations spanning from Norway to the Black Sea that helped establish the exact shape and size of the earth and is shared as a UNESCO site with nine other countries.

Press

In 2008 there were seven paid-for daily newspapers and 240 non-dailies. The dailies had a combined circulation of 303,000, with the most widely read being the Russian-language *Komsomolskaya Pravda v Moldove*.

Tourism

In 2010, 64,000 non-resident tourists stayed in holiday accommodation; 24% of tourists in 2010 were from Romania, 10% from Russia and 10% from Ukraine.

DIPLOMATIC REPRESENTATIVES

Of Moldova in the United Kingdom (5 Dolphin Sq., Edensor Rd, Chiswick, London, W4 2ST)
Ambassador: Iulian Fruntaşu.

Of the United Kingdom in Moldova (18 Nicolae Iorga St., Chişinău MD-2012)
Ambassador: Philip Batson.

Of Moldova in the USA (2101 S St., NW, Washington, D.C., 20008)
Ambassador: Igor Munteanu.

Of the USA in Moldova (103 Strada Alexei Matveevici, Chişinău)
Ambassador: William. H. Moser.

Of Moldova to the United Nations
Ambassador: Vladimir Lupan.

Of Moldova to the European Union
Ambassador: Eugen Caras.

FURTHER READING

Gribincea, M., *Agricultural Collectivization in Moldavia.* 1996
King, C., *Post-Soviet Moldova: A Borderland in Transition.* 1997.—*The Moldovans: Romania, Russia, and the Politics of Culture.* 2000
Kolsto, Pal, *National Integration and Violent Conflict in Post-Soviet Societies: The Cases of Estonia and Moldova.* 2002
Mitrasca, M., *Moldova: A Romanian Province Under Russian Rule: Diplomatic History from the Archives of the Great Powers.* 2002

National Statistical Office: National Bureau of Statistics of Moldova, MD-2019, Chişinău mun., 106 Grenoble St.
Website: http://www.statistica.md

MONACO

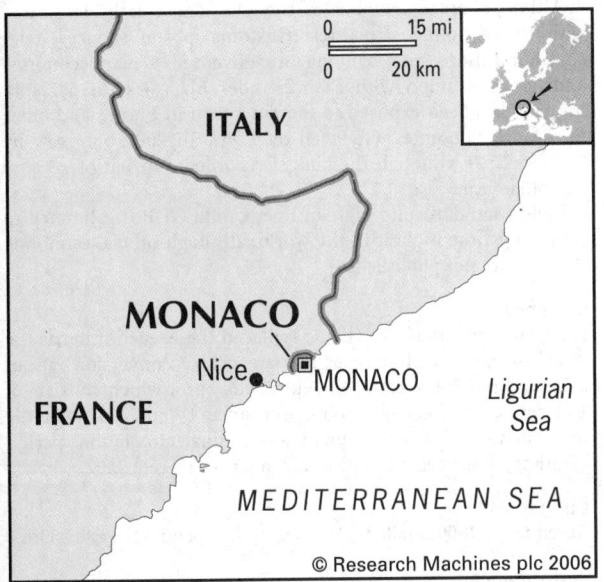

© Research Machines plc 2006

**Principauté de Monaco
(Principality of Monaco)**

Capital: Monaco
Population, 2011: 36,000
GNI per capita, 2010: US$153,177
Internet domain extension: .mc

KEY HISTORICAL EVENTS

Monaco's natural harbour was settled by Phoenicians, Greeks and Ligurians and later by Saracens. A fortress, built where the palace now stands, was captured by the Grimaldi family of Genoa in 1297. It was passed on through the male line until 1731, when control of Monaco passed to Louise Hippolyte, daughter of Antoine I and wife of Jacques de Goyon Matignon, who took the name of Grimaldi. The Principality was placed under the protection of the Kingdom of Sardinia by the Treaty of Vienna in 1815, and under that of France in 1861. Prince Rainier III succeeded his grandfather, Louis II, in 1949 and ruled the Principality until his death on 6 April 2005 when his son, Prince Albert II, inherited the throne.

TERRITORY AND POPULATION

Monaco is bounded in the south by the Mediterranean and elsewhere by France (Department of Alpes Maritimes). The area is 1·97 sq. km (0·8 sq. miles), making it the second smallest sovereign country—only the Vatican City is smaller. The Principality is divided into four districts: Monaco-Ville, la Condamine, Monte-Carlo and Fontvieille. Population (2008 census), 31,109; there were 6,687 Monegasques (22%), 8,785 French (28%) and 5,778 Italian (19%). In Dec. 2011 the population was estimated at 36,371. The population is 100% urban.

The official language is French.

SOCIAL STATISTICS

2008: births, 970; deaths, 545. 2005 marriages, 161; divorces, 69. Rates per 1,000 population, 2008: birth, 31·2; death, 17·5; marriage (2005), 5·0; divorce (2005), 2·1. Annual population growth rate,

1998–2008, 0·4%; fertility rate, 2008, 1·5 births per woman. Infant mortality per 1,000 live births (2010), 3.

CLIMATE

A Mediterranean climate, with mild moist winters and hot dry summers. Monaco, Jan. 50°F (10°C), July 74°F (23·3°C). Annual rainfall 30" (758 mm).

CONSTITUTION AND GOVERNMENT

On 17 Dec. 1962 a new constitution was promulgated which maintains the hereditary monarchy.

The reigning Prince is **Albert II**, b. 14 March 1958, son of Prince Rainier III, 1923–2005, and Grace Kelly, 1929–1982; married Charlene Wittstock on 1 and 2 July 2011. Prince Albert succeeded his father Rainier III, who died on 6 April 2005.

Sisters of the Prince. Princess Caroline Louise Marguerite, b. 23 Jan. 1957; married Philippe Junot on 28 June 1978, divorced 9 Oct. 1980; married Stefano Casiraghi on 29 Dec. 1983 (died 3 Oct. 1990); married Prince Ernst of Hanover on 23 Jan. 1999. *Offspring:* Andrea, b. 8 June 1984; Charlotte, b. 3 Aug. 1986; Pierre, b. 5 Sept. 1987; Alexandra, b. 20 July 1999. Princess Stéphanie Marie Elisabeth, b. 1 Feb. 1965, married Daniel Ducruet on 1 July 1995, divorced 4 Oct. 1996; married Adans López Peres on 12 Sept. 2003; divorced 24 Nov. 2004. *Offspring:* Louis, b. 26 Nov. 1992; Pauline, b. 4 May 1994; Camille, b. 15 July 1998.

Prince Rainier III renounced the principle of divine right. Executive power is exercised jointly by the Prince and a five-member *Council of government*, headed by a Minister of State (a French citizen). A 24-member *National Council* is elected for five-year terms.

The constitution can be modified only with the approval of the National Council. Laws of 1992, 2003 and 2005 permit Monegasque women to give their nationality to their children.

National Anthem

'Principauté Monaco ma patrie' ('Principality of Monaco my fatherland'); words by T. Bellando de Castro, tune by C. Albrecht.

RECENT ELECTIONS

In parliamentary elections held on 10 Feb. 2013 Horizon Monaco won 20 of 24 seats against 3 for Union Monégasque; Renaissance won one seat. Turnout was 74·5%.

CURRENT GOVERNMENT

Chief of State: Prince Albert II.

In March 2013 the cabinet comprised:
Minister of State: Michel Roger; b. 1949 (sworn in 29 March 2010).

Minister of External Relations: José Badia. *Facilities, Environmental Affairs and Town Planning:* Marie-Pierre Gramaglia. *Finance and Economics:* Marco Piccinini. *Interior:* Paul Masseron. *Social Affairs and Health:* Stéphane Valeri.

President of the National Council: Jean-François Robillon.

Government Website: http://www.monaco.gouv.mc

CURRENT LEADERS

Albert II

Position
Prince

Introduction
Albert II became ruler of Monaco on 6 April 2005 following the death of his father, Prince Rainier III, who had ruled the principality for 56 years. Albert II has maintained the status quo,

upholding the low-tax regime that has made Monaco a haven for the super-rich.

Early Life
Albert Alexandre Louis Pierre Grimaldi was born in Monaco on 14 March 1958, the second child and only son of Prince Rainier III and Grace Kelly, a US cinema actress. He attended the Lycée Albert Ier, then the principality's sole secondary school, where he developed a passion for sport. Having received his baccalaureate diploma in 1976, Albert enrolled the following year at Amherst College in Massachusetts, USA, and graduated with a degree in political science in 1981. From Sept. 1981–April 1982 he served in the French Navy as a sub-lieutenant on the aircraft carrier *Jeanne d'Arc*. On 14 Sept. 1982 his mother was killed in a car crash in the mountains near Monaco. Subsequently, he became vice-president of the Princess Grace-USA Foundation, which grants scholarships to talented young musicians, actors and dancers. In the same year he also became president of the Monaco Red Cross.

During the mid-1980s Albert undertook work experience at an investment bank and an international law firm in New York, as well as the French luxury goods group, Moet-Hennessy, in Paris. Back in Monaco he chaired the principality's prestigious Yacht Club, the International Television Festival and Monaco's Athletic Federation. Albert also became increasingly involved in the Olympic movement, both as an administrator (as a member of the International Olympic Committee in 1985 and president of Monaco's Olympic Committee in 1993) and competitor (in the principality's bobsleigh team at four Winter Olympics between 1988 and 2002).

During the 1990s he began to increase his involvement in the day-to-day administration of Monaco. In 1997 he organized the 700th anniversary celebrations of the Grimaldi family's control over the principality. He also assisted his father and the government in preparing reports that strongly denied allegations by French parliamentarians in 2000 that Monaco's lax policies had facilitated money laundering.

Long described in the press as the world's 'most eligible bachelor', Albert's unmarried status became a matter of political concern, casting doubt on the succession of the Grimaldi family and the independence of the principality. A change to the constitution was formulated in April 2002, however, allowing the throne to continue through the female line. On 31 March 2005 the Palace of Monaco announced that Albert would take over the duties of his father as Regent, after Prince Rainier III, who had been admitted to hospital, was no longer able to rule. Following the death of his father on 6 April 2005, he became Sovereign Prince of Monaco, and was enthroned on 12 July.

Career in Office
In his first public statement as Prince Albert II, he said that the death of his father, who had governed for 56 years, had left the people of the principality feeling orphaned and united in a profound sense of loss. He did not make reference to the future direction of policy, but analysts expected him to retain the famously low-tax regime and continue to develop tourism, as well as nurturing Monaco's precision engineering, fish canning, banking and pharmaceutical industries. He was also expected to rule in a more consensual style than his father. In July 2011 he married Charlene Wittstock, a former Olympic swimmer for South Africa.

ECONOMY

Overview
A tiny economy with sparse natural resources, Monaco is primarily geared towards tourism and finance, which together account for 95% of GDP. Tourism provides 15% of revenues, focused around Monte Carlo's casino and the Formula 1 Grand Prix hosted every May. Many foreign companies are drawn by low corporate taxes. Having been identified as a tax haven by the IMF

in 2003, it was subsequently placed on the OECD's blacklist of uncooperative tax havens. However, in 2009 it was removed following its commitment to implement standards of transparency and effective exchange of information.

Although not a member of the EU, it is closely associated with its economic structures. Customs, postal services, telecommunications and banking are governed via an economic and customs union with France under EU rules. As a result, Monaco has been exposed to the downturn in France and other European economies. GDP fell by 11·5% in 2009 but grew by 2·5% in 2010, although the budget recorded a deficit of 1·9% of GDP that year.

Living standards are high and per capita GDP is estimated to be amongst the highest in the world, although official economic statistics are not published.

Currency
On 1 Jan. 1999 the euro (EUR) replaced the French franc as the legal currency in Monaco at the irrevocable conversion rate of 6·55957 French francs to one euro. The euro, which consists of 100 cents, has been in circulation since 1 Jan. 2002. On the introduction of the euro there was a 'dual circulation' period before the franc ceased to be legal tender on 17 Feb. 2002.

Budget
Revenues in 2009 totalled €744·21m. and expenditures €805·53m.

Performance
In 2009 total GDP was US$6·1bn.

Banking and Finance
There were 43 banks in 2004 of which 19 were Monegasque banks.

ENERGY AND NATURAL RESOURCES

Electricity
Electricity is imported from France. 503 GWh were supplied to 24,178 customers in 2004. In 2001 output capacity was 83 MW.

Oil and Gas
In 2004, 61 GWh of gas were supplied to 3,935 customers; output capacity was 21 MW.

Water
Total consumption (2004), 5·38m. cu. metres.

INDUSTRY
The main industry is tourism. There is some production of cosmetics, pharmaceuticals, glassware, electrical goods and precision instruments.

Labour
There were 42,637 persons employed in Jan. 2004. 38,773 worked in the private sector; 3,864 in the public sector. 26,017 French citizens worked in Monaco in 2004.

INTERNATIONAL TRADE

Imports and Exports
There is a customs union with France. Imports for 2004 totalled €512m.; exports, €528m. Main imports: pharmaceuticals, perfume, clothing, paper, synthetic and non-metallic products, and building materials.

COMMUNICATIONS

Roads
There were 77 km of roads in 2007. In 2004 there were 33,275 vehicles. Monaco has the densest network of roads of any country in the world. In 2004, 5,141,964 people travelled by bus.

Rail

The 1·7 km of main line passing through the country are operated by the French National Railways (SNCF). In 2004, 3,953,859 people arrived at or departed from Monaco railway station.

Civil Aviation

There are helicopter flights to Nice with Heli Air Monaco and Heli Inter. Helicopter movements (2004) at the Heliport of Monaco (Fontvieille), 37,521; the number of passengers carried was 112,379. The nearest airport is at Nice in France.

Shipping

In 2004 there were 3,829 vessels registered, of which 12 were over 100 tonnes. 2,636 yachts put in to the port of Monaco and 1,193 at Fontvieille in 2004. 178 liners put in to port in Monaco; 10,581 people embarked, 10,195 disembarked and 104,202 were in transit.

Telecommunications

There were 34,100 fixed telephone lines in 2010 (964·0 per 1,000 inhabitants). Mobile phone subscribers numbered 26,300 in 2010. Monaco is one of only two countries in the world where there are more landline than mobile phone subscriptions. Internet users numbered 22,000 in 2008.

SOCIAL INSTITUTIONS

Justice

There are the following courts: *Tribunal Suprème, Cour de Révision, Cour d'Appel*, a Correctional Tribunal, a Work Tribunal, a Tribunal of the First Instance, two Arbitration Commissions for Rents (one commercial, one domestic), courts for Work-related Accidents and Supervision, a *Juge de Paix*, and a Police Tribunal. There is no death penalty.

Police

In 2004 the police force (Sûreté Publique) comprised 516 personnel. Monaco has one of the highest number of police per head of population of any country in the world.

Education

In 2004, in the public sector, there were six pre-school institutions (*écoles maternelles*) with 750 pupils; four elementary schools with 1,367 pupils; two secondary schools with 2,359 pupils. There were 277 primary teachers and 150 secondary school teachers in total in 2004. In the private sector there were two pre-schools and three primary schools with 179 and 481 pupils respectively; and one secondary school with 700 pupils. In 2005 the government allocated 5·2% of its total budget to education.

The International University of Monaco had 300 students in 2006.

Health

In 2005 the government allocated 6·4% of its total budget to public health. There were 191 doctors and 19 dentists in 2004 and 19 childcare nurses in 2002. There were 503 hospital beds in 2002. Monaco has the highest provision of hospital beds of any country: in 2002 there were 162 per 10,000 population.

RELIGION

90% of the resident population are Roman Catholic. There is a Roman Catholic archbishop.

CULTURE

Press

Monaco has no domestically-published daily newspaper. In 2008 there were two state weeklies: *Journal de Monaco* (published by the government) and *Monaco Hebdo*.

Tourism

In 2009, 264,540 foreign visitors (212,966 leisure and 51,574 business) spent a total of 778,451 nights in Monaco; the main visitors were French, followed by Italians and British. There were also 235,904 cruise ship passengers in 2009. There are three casinos run by the state, including the one at Monte Carlo.

DIPLOMATIC REPRESENTATIVES

Of Monaco in the United Kingdom (7 Upper Grosvenor St., London, W1K 2LX)
Ambassador: Evelyne Genta.

British Consul-General (resident in France): Vacant.
British Honorary Consul: Eric G. F. Blair.

Of Monaco in the USA (Suite 2K-100, 3400 International Drive, NW, Washington, D.C., 20008)
Ambassador: Gilles Noghès.

Of the USA in Monaco
Ambassador: Charles Rivkin (resides in Paris).

Of Monaco to the United Nations
Ambassador: Isabelle Picco.

Of Monaco to the European Union
Ambassador: Gilles Tonelli.

FURTHER READING

Journal de Monaco. Bulletin Officiel. 1858 ff.

National Statistical Office: Institut Monégasque de la Statistique et des Etudes Economiques, 9 rue du Gabian, MC 98000 Monaco.
Website (French only): http://www.gouv.mc/Action-Gouvernementale/ L-Economie/Analyses-et-Statistiques

MONGOLIA

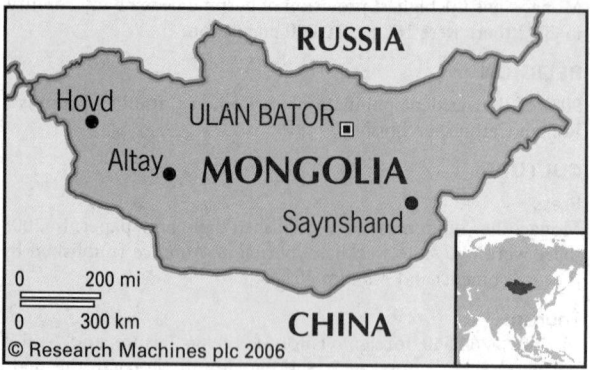

Mongol Uls

Capital: Ulan Bator
Population projection, 2015: 2·98m.
GNI per capita, 2011: (PPP$) 3,391
HDI/world rank: 0·653/110
Internet domain extension: .mn

KEY HISTORICAL EVENTS

Temujin became khan of Hamag Mongolia in 1190. Having united by conquest various Tatar and Mongolian tribes he was confirmed as 'Universal' ('Genghis', 'Chingiz') khan in 1206. The expansionist impulse of his nomadic empire (Beijing captured in 1215; Samarkand in 1220) continued after his death in 1227. Tamurlaine (died 1405) was the last of the conquering khans. In 1368 the Chinese drove the Mongols from Beijing, and for the next two centuries Sino-Mongolian relations alternated between war and trade. In 1691 Outer Mongolia accepted Manchu rule. The head of the Lamaist faith became the symbol of national identity, and his seat ('Urga', now Ulan Bator) was made the Mongolian capital. When the Manchu dynasty was overthrown in 1911 Outer Mongolia declared its independence under its spiritual ruler and turned to Russia for support against China. Soviet and Mongolian revolutionary forces set up a provisional government in March 1921. On the death of the spiritual ruler a people's republic and new constitution were proclaimed in May 1924. With Soviet help Japanese invaders were fended off during the Second World War. The Mongols then took part in the successful Soviet campaign against Inner Mongolia and Manchuria. On 5 Jan. 1946 China recognized the independence of Outer Mongolia. Until 1990 sole power was in the hands of the (Communist) Mongolian People's Revolutionary Party (MPRP), but an opposition Mongolian Democratic Party, founded in Dec. 1989, achieved tacit recognition and held its first congress in Feb. 1990. Following demonstrations and hunger-strikes, on 12 March the entire MPRP Politburo resigned and political opposition was legalized.

TERRITORY AND POPULATION

Mongolia is bounded in the north by the Russian Federation, and in the east and south and west by China. Area, 1,564,100 sq. km (603,900 sq. miles). Population (2010 census), 2,647,545. Density in 2010 was 1·7 per sq. km, making Mongolia the most sparsely populated country in the world. In 2011, 62·5% of the population were urban. More Mongols live in China than in Mongolia (5·8m. according to China's 2010 census).

The UN gives a projected population for 2015 of 2·98m.

The population is predominantly made up of Mongolian peoples (78·8% Khalkh). There is a Turkic Kazakh minority (3·7% of the population) and 21 Mongol minorities. The official language is Khalkh Mongol, which uses a modified Cyrillic alphabet.

The republic is administratively divided into 21 provinces (*aimag*) and the capital, Ulan Bator. The provinces are sub-divided into 334 districts or counties (*suums*).

SOCIAL STATISTICS

Births, 2005, 45,326; deaths, 16,480. 2001 rates: birth, 17·8 per 1,000 population; death, 6·5 per 1,000; marriage, 5·9 per 1,000; divorce, 0·6 per 1,000. Annual population growth rate, 2000–05, 1·3%. Infant mortality rate, 2010, 26 per 1,000 live births. Expectation of life in 2007 was 63·0 years for males and 69·6 for females. Fertility rate, 2008, 2·0 births per woman. Mongolia has had one of the most impressive reductions in its fertility rate of any country in the world over the past quarter of a century, having had a rate of 4·2 births per woman in 1990.

CLIMATE

A very extreme climate, with six months of mean temperatures below freezing, but much higher temperatures occur for a month or two in summer. Rainfall is very low and limited to the months from mid-May to mid-Sept. Ulan Bator, Jan. –14°F (–25·6°C), July 61°F (16·1°C). Annual rainfall 8" (208 mm).

CONSTITUTION AND GOVERNMENT

The constitution of 12 Feb. 1992 abolished the 'People's Democracy', introduced democratic institutions and a market economy and guarantees freedom of speech.

The *President* is directly elected for renewable four-year terms.

Since June 1992 the legislature has consisted of a single-chamber 76-seat parliament, the *Great Hural (Ulsyn Ich-Chural)*, which elects the Prime Minister.

National Anthem

'Darkhan manai khuvsgalt uls' ('Our sacred revolutionary country'); words by Tsendiyn Damdinsüren, tune by Bilegin Damdinsüren and Luvsanjamts Murjorj.

RECENT ELECTIONS

At the parliamentary elections of 28 June 2012 the Democratic Party won 31 of the 76 available seats with 35·3% of the vote, the Mongolian People's Party (formerly the communist Mongolian People's Revolutionary Party; MPRP), 25 with 31·3%, the Justice Coalition (comprising a faction of the former MPRP and the Mongolian National Democratic Party) 11 with 22·3% and the Civic Will–Green Party 2 with 5·5%. Independent won three seats and five remained vacant. Turnout was 65·2%.

In presidential elections on 24 May 2009 former prime minister Tsakhiagiin Elbegdorj (Democratic Party) won with 51·2% of the vote against incumbent Nambaryn Enkhbayar (Mongolian People's Revolutionary Party) with 47·4%. Turnout was 73·5%.

CURRENT GOVERNMENT

President: Tsakhiagiin Elbegdorj; b. 1963 (Democratic Party; sworn in 18 June 2009).

In March 2013 the Democratic Party-Justice Coalition government comprised:

Prime Minister: Norov Altankhuyag; b. 1958 (Democratic Party; sworn in 10 Aug. 2012).

Deputy Prime Minister: Dendev Terbishdagva.

Minister of Construction and Urban Development: Tsevelmaa Bayarsaikhan. *Culture, Sports and Tourism:* Tsedevdamba Oyungerel. *Defence:* Dashdemberel Bat-Erdene. *Economic Development:* Nyamjar Batbayar. *Education and Science:* Luvsannyam Gantumur. *Energy:* Mishig Sonompil. *Finance:* Chultem Ulaan. *Foreign Affairs:* Luvsanvandan Bold. *Health:* Natsag Udval. *Industry and Agriculture:* Khaltmaa Battulga. *Justice:* Khishigdemberel Temuujin. *Labour:* Yadamsuren Sanjmyatav. *Mining:* Davaajav Gankhuyag. *Nature, Environment and Green Development:* Sanjaasuren Oyun. *Population Development and Social Welfare:* Sodnomzundui Erdene. *Roads and Transportation:* Amarjargal Gansukh. *Minister of the Cabinet Office:* Chimed Saikhanbileg.

Office of the President: http://www.president.mn

CURRENT LEADERS

Tsakhiagiin Elbegdorj

Position
President

Introduction
Tsakhiagiin Elbegdorj took office on 18 June 2009, the first president not to have been a member of the Mongolian People's Revolutionary Party (MPRP). Campaigning on a platform of change and anti-corruption that appealed to urban voters, he won the election with just over 51% of the vote on 24 May 2009, beating MPRP candidate and incumbent president Nambaryn Enkhbayar.

Early Life
Elbegdorj was born on 30 March 1963 in the district of Zereg. In 1988 he graduated from the Military Political Institute in Ukraine (then part of the USSR) and went on to attain a masters degree in public administration from Harvard University in the USA.

Elbegdorj was active in the fight against communism and in 1989 helped establish a recognized pro-democracy movement. The following year he founded Mongolia's first independent newspaper, *Democracy.* A few months later he was elected to government and was a key figure in drafting a new constitution, adopted in Jan. 1992. In 1996, as leader of the Democratic Party (DP), Elbegdorj led the Democratic Union Coalition to victory in parliamentary elections, paving the way for the country's first peaceful power transition.

Elbegdorj was elected prime minister in April 1998 but a banking scandal forced his resignation shortly afterwards. After the 2004 elections he was again appointed to the premiership, heading a fragile MPRP-DP coalition. During his tenure, he announced measures to boost tourism, reorganized domestic political structures, loosened control of the news media and oversaw the replacement of Russian by English as Mongolia's second language. In Jan. 2006 he was again forced to resign when the MPRP withdrew from the coalition, sparking public demonstrations in Ulan Bator.

At elections in May 2009 Elbegdorj won by only a narrow margin over the incumbent president. Nevertheless, the MPRP accepted the result, producing a peaceful outcome after fears of a repeat of the violence that followed the DP's claims of fraud at the 2008 parliamentary polling.

Career in Office
On taking office Elbegdorj faced the challenges of working with a parliament dominated at the time by the MPRP and tackling high levels of corruption and unemployment. He also aimed to spread the profits of Mongolia's mineral wealth more widely. On the foreign stage, he was expected to cultivate Western ties to counterbalance the influence of Russia and China.

Following parliamentary elections in June 2012, Elbegdorj's DP was returned as the largest party but without an overall majority.

This heralded the formation in Aug. of a DP-led government including the Justice Coalition, with DP chairman Norov Altankhuyag as prime minister.

Norov Altankhuyag

Position
Prime Minister

Introduction
Norov Altankhuyag became prime minister in Aug. 2012 after securing 72% support from parliament. His appointment ended weeks of political uncertainty after the Democratic Party failed to win the required seats at the June 2012 elections to form their own government.

Early Life
Norov Altankhuyag was born on 20 Jan. 1958 in Ulaangom, Uvs Province. After graduating in physics from the National University of Mongolia, he took a teaching position in 1981 at the mathematics and physics department of his *alma mater.*

In 1990 he was involved in anti-Communist protests and helped to establish the Social Democratic Party, holding key posts including general secretary. He was elected to the *Great Hural* (the national parliament) in 1996, firstly for the Social Democratic Party and then the Democratic Party. He served as minister of agriculture and industry from 1998–99 and as minister of finance from 2004–06.

Altankhuyag was elected Democratic Party leader on 30 Aug. 2008 after Tsakhiagiin Elbegdorj stood down in the aftermath of disputed elections. Altankhuyag also became first deputy prime minister in a coalition government headed by the Mongolian People's Revolutionary Party (MPRP). He served as acting prime minister for one day on 28 Oct. 2009 after the incumbent, Sanj Bayar, resigned as a result of ill health. Sukhbaataryn Batbold was sworn in as premier the following day.

At a parliamentary election in June 2012 the Democratic Party won 31 seats. After weeks of negotiations, it formed a coalition with the Justice Coalition, which had come third in the election. On 10 Aug. 2012 Altankhuyag was confirmed as premier at a parliamentary session.

Career in Office
Altankhuyag's principal challenge is to manage the country's vast natural resources and secure the confidence of foreign investors while maintaining his fragile coalition between its pro-market Democratic members and the nationalist MPRP wing. He must also tackle widespread corruption and address the disparity in wealth between urban and rural communities.

DEFENCE

Conscription is for one year for males aged 18–25 years. Defence expenditure in 2008 totalled US$52m. (US$17 per capita), representing 1·1% of GDP.

Army
Strength (2007) 7,500 (3,300 conscripts). There is a border guard of 6,000, 1,200 internal security troops and 300 Construction Troops.

Air Force
The Air Force had a strength of 800 in 2007 with 11 attack helicopters and nine aircraft.

ECONOMY

In 2009 agriculture accounted for 19·6% of GDP, industry 33·0% and services 47·4%.

Overview
Traditionally a nomadic pastoral economy, substantial progress has been made in the past decade towards a private sector-led

open economy. The private sector now accounts for over 80% of GDP. The country is rich in mineral resources, especially copper and gold.

Assistance to Mongolia has been dominated by loans to the energy and transport sectors, as a result of the high cost of constructing and maintaining infrastructure in the country. However, since the recent opening-up of one of the world's largest copper mines, development agencies have focused on using mining revenues to fund development projects.

The economy was hit hard by the global economic crisis owing to its dependence on mineral exports. An 18-month IMF Stand-by Arrangement in April 2009 helped to stabilize the economy and growth rebounded to 6·4% in 2010. Mineral exports are projected to increase substantially over the next few years thanks to new mining projects. However, double digit inflation, resulting from heavy government spending and higher prices for imported oil and food, threaten short-term growth and macroeconomic stability. According to the 2010 Human Development Report, 22% of the population lives below the international poverty line of US$1·25 (PPP) per day.

Currency
The unit of currency is the *tugrik* (MNT) of 100 *möngö*. The tugrik was made convertible in 1993. In March 2005 foreign exchange reserves were US$276m. and gold reserves totalled 112,000 troy oz. Total money supply was 222,085m. tugriks in Dec. 2004. Inflation, which reached a high of 268% in 1993, was just 0·9% in 2002. It rose to 26·8% in 2008, fell to 6·3% in 2009 but went up again to 10·2% in 2010. In 2011 inflation stood at 7·7%.

Budget
In 2010 revenues were 2,451·1bn. tugriks and expenditures 2,376·2bn. tugriks; taxes accounted for 78·8% of revenue and wages and salaries 24·9% of expenditure.

VAT is 10%.

Performance
The economy contracted by 1·3% in 2009 but grew by 6·4% in 2010 and an even more spectacular 17·5% in 2011. Total GDP in 2011 was US$8·8bn.

Banking and Finance
The Mongolian Bank (established 1924) is the bank of issue, being also a commercial, savings and development bank: the *Governor* is Lhanaasuren Purevdorj. It has 21 main branches. There were 25 other banks in 2002. The largest bank is the state-owned Trade and Development Bank.

External debt was US$2,444m. in 2010, representing 44·3% of GNI.

A stock exchange opened in Ulan Bator in 1992.

ENERGY AND NATURAL RESOURCES
Environment
In 2008 carbon dioxide emissions from the consumption and flaring of fossil fuels in Mongolia were the equivalent of 2·4 tonnes per capita.

Electricity
Installed capacity was 0·8m. kW in 2007. Production, 2007, 3·70bn. kWh; consumption per capita in 2007 was 1,486 kWh. Mongolia imports electricity from Russia to meet increasing demand.

Minerals
There are large deposits of copper, gold, nickel, zinc, molybdenum, phosphorites, tin, wolfram and fluorspar; production of the latter in 2005, 367,000 tonnes. There are major coalmines near Ulan Bator and Darhan. A huge gold and copper mine at Oyu Tolgoi was scheduled to begin production in June 2013. In 2007 lignite production was 7·73m. tonnes; coal production, 1·51m. tonnes. Copper production, 2008, 126,796 tonnes; gold production, 2008, 15,184 kg.

Agriculture
The prevailing Mongolian style of life is pastoral nomadism. Livestock production comprises 79·5% of the total agricultural production. In 2003 there were 10·8m. sheep, 2·0m. horses, 1·8m. cattle and 257,000 camels. The number of goats rose from 5·5m. to 10·7m. between 1992 and 2003 as production of cashmere has increased along with the market economy. The winter of 2009–10 was extremely harsh, resulting in the loss of some 7·8m. animals, representing 17% of Mongolia's livestock. In spite of this, livestock outnumber humans by 14 to 1.

In 2007 there were around 851,000 ha. of arable land and 2,000 ha. of permanent crop land. In 2003 output of major crops was 165,000 tonnes of wheat (down from 607,000 in the period 1989–91) and 79,000 tonnes of potatoes (down from 128,000 in 1989–91). Vegetables produced in 2003 included (in tonnes): turnips, 25,000; cabbage, 15,000; carrots, 10,000. Livestock products, 2003 (in 1,000 tonnes): milk, 292; meat, 153. There were 48 tractors and seven harvester-threshers per 10,000 ha. of arable land in 2006.

Forestry
Forests, chiefly larch, cedar, fir and birch, occupied 10·90m. ha. in 2010 (7% of the land area). Timber production was 791,000 cu. metres in 2007.

Fisheries
The catch in 2009 was 90 tonnes, entirely from inland waters.

INDUSTRY
Industry is still small in scale and local in character. The food industry accounts for 25% of industrial production. The main industrial centre is Ulan Bator; others are at Darhan and Erdenet. Production figures: cement (2009), 235,000 tonnes; lime (2009), 43,000 tonnes; bread (2003), 22,000 tonnes; carpets (2003), 663,000 sq. metres; sawnwood (2002), 300,000 cu. metres.

Labour
Out of 870,800 people in employment in Dec. 2002, 381,400 were engaged in agriculture, hunting, fishing and forestry; 104,500 in wholesale and retail trade/repair of motor vehicles, motorcycles and personal and household goods; 59,300 in education; and 55,600 in manufacturing. In July 2003 there were 37,300 registered unemployed persons.

INTERNATIONAL TRADE
Mongolia is dependent on foreign aid. Total foreign assistance to Mongolia exceeds US$300m. annually. Japan is the largest bilateral donor.

Imports and Exports
In 2006 imports (f.o.b.) were valued at US$1,356·7m. and exports (f.o.b.) at US$1,545·2m. Main exports, 2006: copper concentrate, 42·7%; gold, 18·1%; refined copper, 7·2%. Minerals account for over 80% of Mongolia's exports. Principal import suppliers in 2006: Russia, 36·9%; China, 27·2%; Japan, 6·6%; South Korea, 5·6%. Main export markets, 2006: China, 67·8%; Canada, 11·1%; USA, 7·7%; Russia, 2·9%.

COMMUNICATIONS
Roads
The total road network covered 49,250 km in 2002, including 11,121 km of highway. There are 1,185 km of surfaced roads running around Ulan Bator, from Ulan Bator to Darhan, at points on the frontier with the Russian Federation and towards the south. Truck services run where there are no surfaced roads. Vehicles in use in 2007 included 110,200 passenger cars and 37,300 lorries and vans. In 2008 passenger transport totalled

1,215m. passenger-km and freight 782m. tonne-km. In 2007 there were 562 fatalities as a result of road accidents.

Rail
The Trans-Mongolian Railway (1,815 km of 1,520 mm gauge in 2005) connects Ulan Bator with the Russian Federation and China. There are spur lines to Erdenet and to the coalmines at Nalayh and Sharyn Gol. A separate line connects Choybalsan in the east with Borzaya on the Trans-Siberian Railway. Passenger-km travelled in 2007 came to 1,401m. and freight tonne-km to 8,373m.

Civil Aviation
MIAT-Mongolian Airlines operates internal services, and in 2003 flew from Ulan Bator to Beijing, Berlin, Frankfurt, Hohhot, Irkutsk, Moscow, Seoul and Tokyo. In 2003 scheduled airline traffic of Mongolian-based carriers flew 9m. km, carrying 289,000 passengers (139,000 on international flights). In 2001 Ulan Bator handled 285,399 passengers and 2,660 tonnes of freight.

Shipping
There is a steamer service on the Selenge River and a tug and barge service on Hövsgöl Lake.

Telecommunications
In 2008 there were 200,500 main (fixed) telephone lines; mobile phone subscribers numbered 1,763,200 in 2008 (66·8 per 100 persons). There were 330,000 internet users in 2008. In March 2012 there were 459,000 Facebook users.

SOCIAL INSTITUTIONS

Justice
The Procurator-General is appointed, and the Supreme Court elected, by parliament for five years. There are also courts at province, town and district level. Lay assessors sit with professional judges. The death penalty is retained but has been subject to a moratorium since Jan. 2010; it was last used in 2008.

The population in penal institutions in 2006 was 6,593 (244 per 100,000 of national population).

Education
Adult literacy was 97·8% in 2003 (male, 98·0%; female, 97·5%). Schooling begins at the age of seven. In 2007 there were 94,702 children in pre-primary education and 3,262 teaching staff; 239,262 pupils and 7,572 teaching staff in primary education; and 328,009 pupils and 16,605 teaching staff in secondary schools. There were 142,411 students in tertiary education in 2007 and 8,754 academic staff. In 2003 there were eight state and three private universities, and 38 public and 134 private institutes and colleges.

In 2007 public expenditure on education came to 5·2% of GNI.

Health
In 2003 there were 385 state hospitals with the equivalent of 73 beds per 10,000 inhabitants. There were 6,823 physicians, 469 dentists, 612 midwives and 7,802 nurses in 2002.

RELIGION
Tibetan Buddhist Lamaism is the prevalent religion; the Dalai Lama is its spiritual head. In 2009 there were 457 registered placed of worship; 239 of these were Buddhist, 161 Christian and 44 Muslim.

CULTURE

World Heritage Sites
Mongolia has three sites (one shared) on the UNESCO World Heritage List: the Orkhon Valley Cultural Landscape (inscribed on the list in 2004), an extensive area on both banks of the Orkhon River including the archaeological remains of Kharkhorum, the 13th and 14th century capital of Genghis Khan's vast Empire; and the Petroglyphic Complexes of the Mongolian Altai (2011), numerous rock carvings and funerary monuments that date back to 11,000 BC. The third site falls under joint Mongolian and Russian jurisdiction: Uvs Nuur Basin (2003), an important saline lake system supporting a rich wildlife, especially the snow leopard and Asiatic ibex.

Press
In 2008 there were 15 paid-for daily newspapers with a combined circulation of 61,000 and 115 paid-for non-dailies with a circulation of 197,000. The leading paid-for dailies are *Udriin Sonin* (Daily News) and *Onoodor*.

Tourism
In 2011 there were 458,000 non-resident visitors to Mongolia; visitor numbers have doubled since 2003 and trebled since 2000.

DIPLOMATIC REPRESENTATIVES

Of Mongolia in the United Kingdom (7 Kensington Ct, London, W8 5DL)
Ambassador: Bulgaa Altangerel.

Of the United Kingdom in Mongolia (30 Enkh Taivny Gudamzh, Bayanzurkh District, Ulan Bator 13381)
Ambassador: Christopher Stuart.

Of Mongolia in the USA (2833 M St., NW, Washington, D.C., 20007)
Ambassador: Khasbazaryn Behkbat.

Of the USA in Mongolia (Micro Region 11, Big Ring Rd, Ulan Bator)
Ambassador: Piper Campbell.

Of Mongolia to the United Nations
Ambassador: Od Och.

Of Mongolia to the European Union
Ambassador: Avirmed Battur.

FURTHER READING

State Statistical Office: *Mongolian Economy and Society in [year]: Statistical Yearbook.—National Economy of the MPR, 1924–1984: Anniversary Statistical Collection.* Ulan Bator, 1984

Akiner, S. (ed.) *Mongolia Today.* 1992
Becker, J., *The Lost Country.* 1992
Bruun, O. and Odgaard, O. (eds.) *Mongolia in Transition.* 1996
Griffin, K. (ed.) *Poverty and the Transition to a Market Economy in Mongolia.* 1995
Hanson, Jennifer L., *Mongolia.* 2003
Rossabi, Morris, *Modern Mongolia: From Khans to Commissars to Capitalists.* 2005

National Statistical Office: Government Building-III, Bagatoiruu-44, Ulan Bator-11.
Website: http://www.nso.mn/v3

MONTENEGRO

0 ____ 40 mi
0 ____ 50 km

BOSNIA AND
HERZEGOVINA

SERBIA

MONTENEGRO
Nikšić ●   Ivangrad ●

PODGORICA
□

Adriatic
Sea

ALBANIA

Ulcinj ●

© Research Machines plc 2006

Republika Crna Gora
(Republic of Montenegro)

Capital: Podgorica
Population projection, 2015: 634,000
GNI per capita, 2010: US$6,570
HDI/world rank: 0·771/54
Internet domain extension: .me

KEY HISTORICAL EVENTS

Montenegro emerged as a separate entity on the break-up of the Serbian Empire in 1355. Owing to its mountainous terrain, it was never effectively subdued by Turkey. It was ruled by Bishop Princes until 1851, when a royal house was founded. The Treaty of Berlin (1828) recognized the independence of Montenegro and doubled the size of the territory.

The assassination of Archduke Franz Ferdinand of Austria in Sarajevo on 28 June 1914 precipitated the First World War. In the winter of 1915–16 the Serbian army was forced to retreat to Corfu, where the government aimed at a centralized, Serb-run state. But exiles from Croatia and Slovenia wanted a South Slav federation. This was accepted by the victorious Allies as the basis for the new state. The Croats were forced by the pressure of events to join Serbia and Montenegro on 1 Dec. 1918. From 1918–29 the country was known as the Kingdom of the Serbs, Croats and Slovenes. The remains of King Nicholas I, who was deposed in 1918, were returned to Montenegro for reburial in Oct. 1989.

A constitution of 1921 established an assembly but the trappings of parliamentarianism could not bridge the gulf between Serbs and Croats. The Croat peasant leader Radić was assassinated in 1928; his successor, Vlatko Maček, set up a separatist assembly in Zagreb. On 6 Jan. 1929 the king suspended the constitution and established a royal dictatorship, redrawing provincial

boundaries without regard for ethnicity. In Oct. 1934 he was murdered by a Croat extremist while on an official visit to France.

During the regency of Prince Paul, the government pursued a pro-fascist line. On 25 March 1941 Paul was persuaded to adhere to the Axis Tripartite Pact. On 27 March he was overthrown by military officers in favour of the boy king Peter. Germany invaded on 6 April. Within ten days Yugoslavia surrendered; king and government fled to London. Resistance was led by a royalist group and the communist-dominated partisans of Josip Broz, nicknamed Tito. Having succeeded in liberating Yugoslavia, Tito set up a Soviet-type constitution but he was too independent for Stalin, who sought to topple him. However, Tito made a *rapprochement* with the west and it was the Soviet Union under Khrushchev that had to extend the olive branch in 1956. Yugoslavia was permitted to evolve its 'own road to socialism'. Collectivization of agriculture was abandoned; and Yugoslavia became a champion of international 'non-alignment'. A collective presidency came into being with the death of Tito in 1980.

Dissensions in Kosovo between Albanians and Serbs, and in parts of Croatia between Serbs and Croats, reached crisis point after 1988. On 25 June 1991 Croatia and Slovenia declared independence. Fighting began in Croatia between Croatian forces and Serb irregulars from Serb-majority areas of Croatia. On 25 Sept. the UN Security Council imposed a mandatory arms embargo on Yugoslavia. A three-month moratorium agreed at EU peace talks on 30 June having expired, both Slovenia and Croatia declared their independence from the Yugoslav federation on 8 Oct. After 13 ceasefires had failed, a fourteenth was signed on 23 Nov. under UN auspices. A Security Council resolution of 27 Nov. proposed the deployment of a UN peacekeeping force if the ceasefire was kept. Fighting, however, continued. On 15 Jan. 1992 the EU recognized Croatia and Slovenia as independent states. Bosnia and Herzegovina was recognized on 7 April 1992 and Macedonia on 8 April 1993. A UN delegation began monitoring the ceasefire on 17 Jan. and the UN Security Council on 21 Feb. voted to send a 14,000-strong peacekeeping force to Croatia and Yugoslavia. On 27 April 1992 Serbia and Montenegro created a new Federal Republic of Yugoslavia.

On 30 May, responding to further Serbian military activities in Bosnia and Croatia, the UN Security Council voted to impose sanctions. In mid-1992 NATO committed air, sea and eventually land forces to enforce sanctions and protect humanitarian relief operations in Bosnia. At a joint UN-EC peace conference on Yugoslavia held in London on 26–27 Aug. some 30 countries and all the former republics of Yugoslavia endorsed a plan to end the fighting in Croatia and Bosnia, install UN supervision of heavy weapons, recognize the borders of Bosnia and Herzegovina and return refugees. At a further conference at Geneva on 30 Sept. the Croatian and Yugoslav presidents agreed to make efforts to bring about a peaceful solution in Bosnia, but fighting continued. Following the Bosnian-Croatian-Yugoslav (Dayton) agreement all UN sanctions were lifted in Nov. 1995.

In July 1997 Slobodan Milošević switched his power base to become president of federal Yugoslavia. The former Yugoslav foreign minister, Milan Milutinović, succeeded Milošević as Serbian President. Meanwhile, in Montenegro, the pro-western Milo Đukanović succeeded a pro-Milošević president.

Following the break-up of Yugoslavia, in March 1998 a coalition government was formed between the Socialist Party of Slobodan Milošević and the ultra-nationalist Serb Radical Party.

Kosovo

In 1998 unrest in Kosovo, with its largely Albanian population, led to a bid for outright independence. Violence flared resulting in what a US official described as 'horrendous human rights violations', including massive shelling of civilians and destruction of villages. A US-mediated agreement to allow negotiations to proceed during an interim period of autonomy allowed for food and medicine to be delivered to refugees; American support for a degree of autonomy (short of independence), accepted in principle by President Milošević, lifted the immediate threat of NATO air strikes. Further outbreaks of violence in early 1999 were followed by the departure of the 800-strong team of international 'verifiers' of the fragile peace. Peace talks in Paris broke down without a settlement though subsequently Albanian freedom fighters accepted terms allowing them broad autonomy. The sticking point on the Serbian side was the international insistence on having 28,000 NATO-led peacemakers in Kosovo to keep apart the warring factions. Meanwhile, the scale of Serbian repression in Kosovo persuaded the NATO allies to take direct action. On the night of 24 March 1999 NATO aircraft began a bombing campaign against Yugoslavian military targets. Further Serbian provocation in Kosovo caused hundreds of thousands of ethnic Albanians to seek refuge in neighbouring countries. On 9 June after 78 days of air attacks NATO and Yugoslavia signed an accord on the Serb withdrawal from Kosovo, and on 11 June NATO's peacekeeping force, KFOR, entered Kosovo.

When the general election held on 24 Sept. 2000 resulted in a victory for the opposition democratic leader Vojislav Koštunica, President Milošević demanded a second round of voting. A strike by miners at the Kolubara coal mine on 29 Sept. was followed by a mass demonstration in Belgrade on 5 Oct. when the parliament building was set on fire and on 6 Oct. Slobodan Milošević accepted defeat. He was arrested on 1 April 2001 after a 30-hour confrontation with the authorities. On 28 June he was handed over to the United Nations War Crimes Tribunal in the Hague to face charges of crimes against humanity. Prime Minister Zoran Žižić resigned the next day. On 12 Feb. 2002 the trial of Slobodan Milošević, on charges of genocide and war crimes in the Balkans over a period of nearly ten years, began at the International Criminal Tribunal in The Hague. Milošević defended himself and questioned the legitimacy of the court but died in March 2006 before his trial had ended.

On 14 March 2002 Montenegro and Serbia agreed to remain part of a single entity called Serbia and Montenegro, thus relegating the name Yugoslavia to history. The agreement was ratified in principle by the federal parliament and the republican parliaments of Serbia and Montenegro on 9 April 2002. The new union came into force on 4 Feb. 2003. Most powers in this loose confederation were divided between the two republics. After 4 Feb. 2006 Serbia and Montenegro had the right to vote for independence. Following a referendum held on 21 May 2006, Montenegro declared independence on 3 June and was recognized as such by Serbia on 15 June. Montenegro became the 192nd member state of the United Nations on 28 June 2006.

TERRITORY AND POPULATION

Montenegro is a mountainous country which opens to the Adriatic in the southwest. It is bounded in the west by Croatia, northwest by Bosnia and Herzegovina, in the northeast by Serbia and in the southeast by Albania. The capital is Podgorica (2011 census population, 150,977), although some capital functions have been transferred to Cetinje, the historic capital of the former kingdom of Montenegro. Its area is 13,812 sq. km. Population at the 2011 census was 620,029; population density per sq. km, 44·9. The main ethnic groups in 2011 were: Montenegrins (44·98%); Serbs (28·73%); Bosniaks (8·65%); Albanians (4·91%). 61·5% of the population lived in urban areas in 2011.

The UN gives a projected population for 2015 of 634,000.

The official language is the Serbian language of the Iekavian dialect. The Roman and Cyrillic alphabets have equal status.

SOCIAL STATISTICS

Statistics for calendar years:

	Live births	Deaths	Marriages	Divorces
2005	7,352	5,839	3,291	499
2006	7,531	5,968	3,462	470
2007	7,834	5,979	4,005	453
2008	8,258	5,708	3,445	460

Life expectancy, 2007, 71·6 years for men and 76·5 years for women. Infant mortality per 1,000 births (2010), 7.

CLIMATE

Mostly a central European type of climate, with cold winters and hot summers. Podgorica, Jan. 2·8°C, July 26·5°C. Annual rainfall 1,499 mm.

CONSTITUTION AND GOVERNMENT

The *President* is elected by direct vote to serve a five-year term. There is an 81-member single-chamber *National Assembly*, elected through a party list proportional representation system to serve four-year terms. The *Prime Minister* is nominated by the President and has to be approved by the National Assembly.

A referendum was held on 29 Feb.–1 March 1992 to determine whether Montenegro should remain within a common state, Yugoslavia, as a sovereign republic. The electorate was 412,000, of whom 66% were in favour. The then president, Milo Đukanović, had pledged a referendum on independence in May 2002, but this was postponed with the announcement of the creation of the new entity of Serbia and Montenegro, which came into being on 4 Feb. 2003.

Montenegro held a referendum on 21 May 2006 in which 55·5% voted for independence. The margin required for victory was 55·0%. Turnout was 86·6%.

RECENT ELECTIONS

In presidential elections held on 7 April 2013 incumbent president Filip Vujanović was re-elected for a third term with 51·2% of the vote against Miodrag Lekić with 48·8%. Turnout was 63·9%.

Parliamentary elections were held on 14 Oct. 2012. The ruling Coalition for European Montenegro (consisting of the Democratic Party of Socialists of Montenegro/DPS CG, the Social Democratic Party of Montenegro and the Liberal Party of Montenegro) won 39 of 81 seats (with 45·6% of votes cast), the Democratic Front (consisting of the Movement for Changes and New Serb Democracy) 20 (22·8%), the Socialist People's Party 9 (11·1%), Positive Montenegro 7 (8·2%), Bosniak Party 3 (4·2%), FORCA for Unity 1 (1·5%), Albanian Coalition 1 (1·1%) and Croatian Civic Initiative 1 (0·4%). Turnout was 72·8%.

CURRENT GOVERNMENT

President: Filip Vujanović; b. 1954 (DPS CG; sworn in 22 May 2003, and re-elected in April 2008 and April 2013).

In March 2013 the cabinet comprised:

Prime Minister: Milo Đukanović; b. 1962 (DPS CG; sworn in 4 Dec. 2012, having previously held office from Feb. 1991 to Feb. 1998, Jan. 2003 to Nov. 2006 and Feb. 2008 to Dec. 2010).

Deputy Prime Ministers: Igor Lukšić (also *Minister of Foreign Affairs and European Integration*); Vujica Lazović (also *Minister of Information Society and Telecommunications*); Duško Marković (also *Minister of Justice*); Rafet Husović.

Minister of Agriculture and Rural Development: Petar Ivanović. *Culture:* Branislav Mićunović. *Defence:* Milica Pejanović-Đurišić. *Economy:* Vladimir Kavarić. *Education and Sports:* Slavoljub Stijepović. *Finance:* Radoje Žugić. *Health:* Miodrag Radunović. *Human and Minority Rights:* Suad Numanović. *Interior:* Raško Konjević. *Labour and Social Welfare:* Predrag Bošković. *Science:*

Sanja Vlahović. *Tourism and Sustainable Development:* Branimir Gvozdenović. *Transport and Maritime Affairs:* Ivan Brajović. *Minister without Portfolio:* Marija Vučinović.

Government Website: http://www.gov.me

CURRENT LEADERS

Filip Vujanović

Position
President

Introduction
Filip Vujanović became Montenegro's president following independence in June 2006. He had served as president of Montenegro within the confederation of Serbia and Montenegro for the previous three years. Economic and structural reforms have included a programme of privatization, while closer links have been pursued with the European Union including an official application for EU membership.

Early Life
Born on 1 Sept. 1954 in Belgrade, then the capital of Yugoslavia, Filip Vujanović was educated in Nikšić and studied law at the University of Belgrade. After graduating in 1978 he worked at a municipal court and then as an official at the Belgrade District Court. In 1981 he moved south to Podgorica.

Following a period as secretary to the Podgorica district court, Vujanović worked as an attorney from 1981–93. Following the break-up of the Yugoslav Federation, Serbia and Montenegro formed the Federal Republic of Yugoslavia in 1992. In March 1993 the Montenegrin Prime Minister Đukanović appointed Vujanović as minister for justice. He served for two years, becoming a close ally of Đukanović, who adopted a pro-independence and pro-European stance. In May 1995 Vujanović took over as minister of the interior. When the ruling Democratic Party of Socialists (DPS CG) split into two factions in 1996, Vujanović backed Đukanović against his rival Momir Bulatović, a former ally of Serbian leader Slobodan Milošević.

In 1997 Đukanović was elected president of Montenegro and in Feb. 1998 Vujanović took up the premiership. While supporting Montenegro's independence campaign, he continued to maintain good relations with Serbia. He offered his resignation in April 2002 during a constitutional crisis over ratification of a looser federation between Serbia and Montenegro.

Following legislative elections in Oct. 2002 Vujanović was appointed parliamentary speaker and stood as the DPS CG candidate for Montenegro's presidency. He won the first round with 86% of the vote in Dec. 2002 and a second round with 81% in Feb. 2003 but the result was declared invalid as less than 50% of the electorate voted. Parliament subsequently abolished the minimum turnout rule and Vujanović became president in March 2003 with 63% of the vote.

Career in Office
Vujanović identified Montenegro's accession to the EU and entry into NATO's Partnership for Peace programme as priorities. To these ends, he oversaw the privatization of key parts of the economy, including the banking sector and Kombinat Aluminijuma Podgorica, the country's largest industrial company. He also introduced social and judicial reforms and made a start in combating organized crime. A long-time advocate of cross-border co-operation, he opened talks on joint commercial ventures with neighbouring states including Croatia, Slovenia and Serbia.

On 3 June 2006, following a referendum, Montenegro declared independence and on 28 June 2006 became a member of the United Nations. The country has since joined the World Bank and International Monetary Fund, and in Dec. 2008 submitted a formal application to join the EU (achieving candidate status in Dec. 2010). Its decision in Jan. 2010 to establish diplomatic relations with Kosovo, which had unilaterally declared itself independent from Serbia in 2008, has aggravated tensions with the Serbian government.

Vujanović was re-elected for a further five-year term in the presidential election of April 2008. In Dec. 2010 he nominated Igor Lukšić as prime minister following the resignation of Đukanović, but following parliamentary elections in Oct. 2012 Đukanović resumed the premiership at the head of the ruling centre-left coalition. Vujanović secured a further term by winning the presidential election held in April 2013, defeating Miodrag Lekić of the Democratic Front.

DEFENCE

The all-professional Military of Montenegro was formed from part of the Armed Forces of Serbia and Montenegro when the two countries became independent in 2006.

Defence expenditure in 2008 totalled US$71m. (US$105 per capita), representing 2·3% of GDP.

Army

Strength (2007) 2,500. There are paramilitary forces numbering around 10,100.

Navy

Strength (2007) 3,300 including 900 marines.

ECONOMY

In 2009 agriculture accounted for 10·0% of GDP, industry 20·1% and services 69·9%.

Overview

From its independence in 2006 up to the global crisis of 2008, Montenegro's economy experienced strong growth and expansion. Real GDP growth exceeded 10% in 2007, with tourism bringing large inflows of FDI and generating strong construction activity. The contribution of tourism to GDP rose from 15% in 2004 to 21% in 2007. In Dec. 2008 Montenegro applied to join the European Union.

The global economic crisis prompted a decline in tourism and a contraction in GDP, although tourism picked up in 2010, while metal production also bounced back. After contracting for almost two years, GDP began to grow in the second half of 2010 but was still much below pre-crisis levels by the end of the year. The construction sector remains depressed. Structural reforms, including improving labour participation and business environment, remain a top policy priority.

Currency

On 2 Nov. 1999 the pro-Western government decided to make the Deutsche Mark legal tender alongside the dinar. Subsequently it was made the sole official currency, and consequently the euro (EUR) became the currency of Montenegro on 1 Jan. 2002. Inflation was 9·0% in 2008, falling to 0·7% in 2010 before rising to 3·1% in 2011.

Budget

In 2006 total revenue was €582,258m. and total expenditure was €579,780m.

VAT is 17% (reduced rate, 7%).

Performance

Total GDP in 2011 was US$4·5bn. The economy contracted by 5·7% in 2009 but there was real GDP growth of 2·5% in 2010 and 2·4% in 2011.

Banking and Finance

The Central Bank of Montenegro (*Governor*, Milojica Dakić) was established in Nov. 2000. Montenegro has 11 commercial banks. Foreign debt was US$1,554m. in 2010, representing 39·1% of GNI.

ENERGY AND NATURAL RESOURCES

Electricity
Electricity production in 2006 was 2·95bn. kWh.

Minerals
Lignite production in 2006 totalled 1,502,000 tonnes; bauxite production was 659,370 tonnes.

Agriculture
In 2011 the cultivated area was 189,100 ha. Yields (2011, in 1,000 tonnes): potatoes, 180·1; grapes, 32·8; plums, 12·1; maize, 11·7; oranges and tangerines, 7·9; wheat, 2·4. Livestock (1 Dec. 2011, 1,000 head): poultry, 470; sheep, 209; cattle, 87; pigs, 21.

Forestry
Forest area in 2010 was 0·54m. ha., covering 40% of the total land area. Timber cut in 2007: 457,000 cu. metres.

Fisheries
In 2010 the catch amounted to 1,144 tonnes, of which 53% came from marine waters.

INDUSTRY
Production (2006, in 1,000 tonnes): alumina, 237; steel bars, 137; bread, 24; steel ingots, 20; wheat flour, 17; cigarettes, 433m. units; beer, 51·7m. litres.

Labour
In 2011 there were 163,082 people employed, including 37,820 in wholesale and retail trade, repair of vehicles, personal and household goods; 19,195 in public administration and defence, and compulsory social security; 14,368 in manufacturing; 12,429 in accommodation and food service activities; 12,223 in education; 10,565 in health and social work; 9,188 in transport, storage and communications. Average gross monthly wages for 2011 were €722 and average net wages €484. Unemployment rate in 2011 was 19·7%.

COMMUNICATIONS

Roads
In 2004 there were 7,314 km of roads. Passenger-km in 2004 were 100·6m.; tonne-km of freight carried, 64·5m.

Rail
In 2009 there were 249 km of railway. 1·1m. passengers and 2m. tonnes of freight were carried in 2009.

Civil Aviation
The national carrier is Montenegro Airlines, which has flights to a number of cities throughout Europe. There are airports at Podgorica and Tivat, which handled 450,504 and 532,148 passengers in 2009 respectively.

Shipping
In Jan. 2009 there were seven ships of 300 GT or over registered, totalling 13,000 GT.

Telecommunications
There were 174,000 landline telephone subscriptions in 2008 (276·6 per 1,000 inhabitants) and 1,158,000 mobile phone subscriptions (1,840·5 per 1,000 inhabitants). In 2008, 17·5% of households had internet access. In March 2012 there were 293,000 Facebook users.

SOCIAL INSTITUTIONS

Justice
There is a Supreme Court, two High Courts, 15 Municipal Courts, two Commercial Courts, one Appellate Court and one Administrative Court. In 2011 the Supreme Court had 18 judges, the High Courts 54, the Municipal Courts 147, the Commercial Courts 22, the Appellate Court 12 and the Administrative Court 10.

Education
In 2011–12 there were: 108 pre-schools with 14,155 pupils and 841 teachers; 163 central primary schools with 69,461 pupils and 269 regional primary schools with 5,048 teachers; and 49 secondary schools with 31,914 pupils and 2,396 teachers. There were three universities in 2011 (one public and two private). A total of 25,313 students were enrolled at the universities and other higher education institutions at the beginning of the 2011–12 academic year, including 3,086 in postgraduate and doctoral studies.

Health
In 2011 there were eight general hospitals with 1,706 beds. A total of 66,451 patients were admitted in 2011.

RELIGION
The Serbian Orthodox Church is the official church in Montenegro. The Montenegrin church was banned in 1922, but in Oct. 1993 a breakaway Montenegrin church was set up under its own patriarch.

CULTURE

World Heritage Sites
There are two sites on the UNESCO World Heritage List: the Natural and Culturo-Historical Region of Kotor (inscribed on the list in 1979); and Durmitor National Park (1980 and 2005).

Press
In 2008 there were four daily newspapers with a combined circulation of 46,000.

Tourism
In 2010 there were 1,087,794 non-resident overnight tourist arrivals. The main countries of origin were: Serbia (314,836); Russia (150,194); Bosnia and Herzegovina (103,025); France (42,099).

DIPLOMATIC REPRESENTATIVES

Of Montenegro in the United Kingdom (11 Callcott Pl., London, W8 7SU)
Ambassador: Ljubiša Stanković.

Of the United Kingdom in Montenegro (Ulcinjska 8, Gorica C, 81000 Podgorica)
Ambassador: Catherine Knight-Sands.

Of Montenegro in the USA (1610 New Hampshire Ave., Washington, D.C., 20009)
Ambassador: Srđan Darmanović.

Of the USA in Montenegro (Dzona Djeksona 2, 81000 Podgorica)
Ambassador: Sue K. Brown.

Of Montenegro to the United Nations
Ambassador: Milorad Šćepanović.

Of Montenegro to the European Union
Ambassador: Ivan D. Leković.

FURTHER READING

Bieber, Florian, *Montenegro in Transition: Problems of Identity and Statehood*. 2003
Fleming, Thomas, *Montenegro: The Divided Land*. 2002
Roberts, Elizabeth, *Realm of the Black Mountain: A History of Montenegro*. 2007
Stevenson, Francis Seymour, *A History of Montenegro*. 2002
Treadway, J. D., *The Falcon and the Eagle: Montenegro and Austria-Hungary, 1908–1914*. 1998

National Statistical Office: Statistical Office of the Republic of Montenegro, IV Proleterske No. 2, 81000 Podgorica.
Website: http://www.monstat.org

MOROCCO

0   200 mi
0   175 km

© Research Machines plc 2006

**Mamlaka al-Maghrebia
(Kingdom of Morocco)**

Capital: Rabat
Population projection, 2015: 33·57m.
GNI per capita, 2011: (PPP$) 4,196
HDI/world rank: 0·582/130
Internet domain extension: .ma

KEY HISTORICAL EVENTS

Neolithic settlements in western North Africa date from around 6000 BC. Semi-nomadic Berber clans established a foothold in the region at the end of the second millennium BC. Phoenician and Carthaginian merchants founded coastal settlements in northern Morocco from around 500 BC. In the following centuries Roman traders established bases, including Tangier and Volubilis. Roman influence increased after the fall of Carthage in 146 BC and endured until the 5th century AD, when the region was attacked by the Vandals in AD 429 and Byzantines in AD 533.

An Arab invasion of Morocco in 682 made Islam the dominant religion although several Jewish colonies remained. The first Arab rulers established an independent Kingdom in 788 under Idris I and held sway until around 900, when Morocco fragmented into Arab and Berber tribal states. Conflict ensued with the Fatimids of Tunisia and the Umayidds of Andalusia until 1062, when the Almoravids of Marrakesh established a kingdom that stretched from Spain to Senegal.

Portuguese forces captured the port of Ceuta in 1415 and subsequently took most other Moroccan coastal towns, except Melilla and Larache which fell to Spain. The threat from Christian Portugal and Spain spurred resistance among Islamic families, culminating in the rise of Ahmed I al-Mansur, whose Sharifian dynasty unified the country between 1579 and 1603. Moors and Jews expelled from Spain settled in Morocco during this time and the country flourished.

The second Sharifian dynasty emerged in 1660 and by 1700 Moroccan forces had regained control of most ports. During the 18th and early 19th centuries the North African (Barbary) coast was notorious for piracy. The European powers vied to exploit Morocco's strategic position and resources as the century progressed, with French forces defeating Sultan Abd ar Rahman in 1844 and Spain invading in 1860.

As part of the Entente Cordiale, Britain recognized Morocco as a French sphere of influence and in 1904 Morocco was divided between France and Spain. In 1912 a Franco–Spanish agreement divided Morocco into four administrative areas: a French protectorate, the largest share of the territory, centred on Rabat; a Spanish protectorate with its capital at Tétouan; a Southern Protectorate of Morocco, administered as part of the Spanish Sahara; and the international zone of Tangier.

A revolt led by Abd al-Krim in the northern region of Rif in 1921 was crushed after five years. A nationalist movement gradually gained ground and an independence party, the Istiqlal, was formed during the Second World War. Facing growing disorder and nationalist revolts in its North African colonies, France relinquished control over Morocco in March 1956 and on 29 Oct. 1956 the international status of the Tangier Zone was abolished. Morocco became an independent kingdom on 18 Aug. 1957, with the Sultan taking the title Mohammed V.

Succeeding his father on 3 March 1961, King Hassan II tried to establish an elected House of Representatives but, following political unrest, he seized legislative and executive powers in June 1965. The country returned to partial democracy under a controversially amended constitution. Morocco annexed Western Sahara in 1975, which sparked a guerrilla war with Algerian-backed pro-independence forces (Polisario) that continued until 1991. The territory's status remains unresolved. Hassan II received plaudits for his work promoting peace in the Middle East but criticism for human rights abuses. Following his death in 1999 his son, Mohammed VI, introduced a more liberal economic and social regime but retains sweeping powers.

TERRITORY AND POPULATION

Morocco is bounded by Algeria to the east and southeast, Mauritania to the south, the Atlantic Ocean to the northwest and the Mediterranean to the north. Excluding the Western Saharan territory claimed and retrieved since 1976 by Morocco, the area is 458,730 sq. km. The population at the 2004 census (including Western Sahara) was 29,891,708; density (including Western Sahara), 42·1 per sq. km. At the 2004 census Western Sahara had an area of 252,120 sq. km and a population of about 356,000. The Moroccan superficie is 710,850 sq. km. The population was 58·8% urban in 2011.

The UN gives a projected population for 2015 of 33·57m.

Morocco has 16 states (*wilaya'at*) divided further into 71 prefectures and provincial units. Areas of the states and census populations in 2004:

State	Area in sq. km	Population
Chaouia-Ouardigha	16,760	1,655,660
Doukkala-Abda	13,285	1,984,039
Fès-Boulemane	19,795	1,573,055
Gharb-Chrarda-Béni Hssen	8,805	1,859,540
Grand Casablanca	1,615	3,631,061
Guelmin-Es Semara	71,970	462,410
Laâyoune-Boujdour-Sakia El Hamra[1]	—	256,152
Marrakesh-Tensift-Al Haouz	31,160	3,102,652
Meknès-Tafilalet	79,210	2,141,527
Oriental	82,820	1,918,094
Oued Eddahab-Lagouira[1]	—	99,367
Rabat-Salé-Zemmour-Zaer	9,580	2,366,494
Souss Massa-Draâ	70,880	3,113,653

State	Area in sq. km	Population
Tadla-Azilal	17,125	1,450,519
Tangier-Tétouan	11,570	2,470,372
Taza-Al Hoceima-Taounate	24,155	1,807,113

[1]Laâyoune-Boujdour-Sakia El Hamra and Oued Eddahab-Lagouira correspond roughly to Western Sahara.

The chief cities (with populations in 1,000, 2004) are as follows:

Casablanca	2,934	Tangier	670	Safi	285
Rabat	1,623	Meknès	536	Mohammedia	189
Fez	947	Oujda	401	Khouribga	166
Marrakesh	823	Kénitra	359	Béni Mellal	163
Agadir	679	Tétouan	321		

The official language are Arabic, spoken by 65% of the population, and Berber (since July 2011). Berber languages, including Tachelhit (or Soussi), Tamazight and Tarafit (or Rifia), are spoken by about half the population. French (widely used for business), Spanish (in the north) and English are also spoken.

SOCIAL STATISTICS

2008 estimates: births, 645,000; deaths, 184,000. Estimated rates, 2008 (per 1,000 population): birth, 20·4; death, 5·8. Annual population growth rate, 2000–08, 1·2%. Life expectancy at birth in 2007 was 68·8 years for males and 73·3 years for females. Infant mortality, 2010, 30 per 1,000 live births; fertility rate, 2008, 2·4 births per woman.

CLIMATE

Morocco is dominated by the Mediterranean climate which is made temperate by the influence of the Atlantic Ocean in the northern and southern parts of the country. Central Morocco is continental while the south is desert. Rabat, Jan. 55°F (12·9°C), July 72°F (22·2°C). Annual rainfall 23" (564 mm). Agadir, Jan. 57°F (13·9°C), July 72°F (22·2°C). Annual rainfall 9" (224 mm). Casablanca, Jan. 54°F (12·2°C), July 72°F (22·2°C). Annual rainfall 16" (404 mm). Marrakesh, Jan. 52°F (11·1°C), July 84°F (28·9°C). Annual rainfall 10" (239 mm). Tangier, Jan. 53°F (11·7°C), July 72°F (22·2°C). Annual rainfall 36" (897 mm).

CONSTITUTION AND GOVERNMENT

The ruling King is **Mohammed VI**, born on 21 Aug. 1963, married to Salma Bennani on 21 March 2002; succeeded on 23 July 1999, on the death of his father Hassan II, who reigned 1961–99. *Offspring:* Hassan, b. 8 May 2003; Khadija, b. 28 Feb. 2007. The King holds supreme civil and religious authority, the latter in his capacity of Emir-el-Muminin or Commander of the Faithful. He resides usually at Rabat, but occasionally in one of the other traditional capitals, Fez (founded in 808), Marrakesh (founded in 1062), or at Skhirat.

In Feb. and March 2011, Morocco experienced popular protests echoing those occurring in other North African states. In response King Mohammed established a commission to bring about 'comprehensive constitutional reform'. It recommended that: the king select the prime minister from the party with the greatest parliamentary representation; the prime minister, not the king, be head of government and have the power to dissolve parliament; parliament have greater influence on civil rights, nationality issues and electoral law; women be guaranteed 'civic and social' equality with men; the Berber language become an official state language; and a reference to the king as 'sacred' be removed from the constitution. It also provides for the independence of the judiciary. On 1 July 2011 the reforms won 98·5% support in a referendum (turnout: 73·5%), although some opponents claimed electoral irregularities.

The new constitution came into effect on 29 July 2011, ahead of the parliamentary elections four months later. It replaced an earlier constitution that was approved in March 1972 and amended several times. The Kingdom of Morocco is a constitutional monarchy. Parliament consists of a Chamber of Representatives composed of 395 deputies (up from 325 for the 2007 elections) directly elected for five-year terms. A referendum on 13 Sept. 1996 established a second Chamber of Counsellors, composed of 270 members serving nine-year terms, of whom 162 are elected by local councils, 81 by chambers of commerce and 27 by trade unions. One third are renewed every three years. The Chamber of Counsellors has power to initiate legislation, issue warnings of censure to the government and ultimately to force the government's resignation by a two-thirds majority vote.

An electoral code of March 1997 fixed voting at 20 and made enrolment on the electoral roll compulsory. In Dec. 2002 King Mohammed VI announced that the voting age was to be lowered from 20 to 18.

National Anthem

'Manbit al Ahrah, mashriq al anwar' ('Fountain of freedom, source of light'); words by Ali Squalli Houssaini, tune by Leo Morgan.

GOVERNMENT CHRONOLOGY

Kings since 1955.

1955–61	Mohammed V ibn Yusuf (sultan from 1955–57)
1961–99	Hassan II ibn Mohammed
1999–	Mohammed VI ibn al-Hasan

RECENT ELECTIONS

Elections to the Chamber of Representatives took place on 25 Nov. 2011. The Islamist Parti de la Justice et du Développement (PJD/Party of Justice and Development) gained 107 seats of 395 seats; Parti de l'Indépendance/Istiqlal (PI/Independence Party) 60; Rassemblement National des Indépendants (RNI/National Rally of Independents) 52; Parti Authenticité et Modernité (PAM/Authenticity and Modernity Party) 47; Union Socialiste des Forces Populaires (USFP/Socialist Union of Popular Forces) 39; Mouvement Populaire (MP/Popular Movement) 32; Union Constitutionnelle (UC/Constitutional Union) 23; Parti du Progrès et du Socialisme (PPS/Party of Progress and Socialism) 18; Parti Travailliste (PT/Labour Party) 4. A further nine parties obtained fewer than two seats each. Turnout was 45·4%.

In indirect elections to the Chamber of Counsellors on 3 Oct. 2009 PAM gained 22 of 90 available seats, Istiqlal 17, MP 11, USFP 10 and RNI 9 with the remaining seats going to 11 other parties.

CURRENT GOVERNMENT

In March 2013 the government comprised:
Prime Minister: Abdelilah Benkirane; b. 1954 (Party of Justice and Development; in office since 29 Nov. 2011).
Minister of State: Abdellah Baha.
Minister for Interior: Mohand Laenser. *Foreign Affairs and Co-operation:* Saad-Eddine El Othmani. *Justice and Freedoms:* Mustafa Ramid. *Endowments and Islamic Affairs:* Ahmed Toufiq. *Economy and Finance:* Nizar Baraka. *Housing, Town Planning and Urban Policy:* Nabil Benabdellah. *Agriculture and Marine Fisheries:* Aziz Akhenouch. *Education:* Mohamed El Ouafa. *Higher Education, Scientific Research and Executive Training:* Lahcen Daoudi. *Youth and Sports:* Mohamed Ouzzine. *Equipment and Transportation:* Aziz Rabbah. *Health:* El Hossein El Ouardi. *Communication and Government Spokesperson:* Mustapha El Khalfi. *Energy, Mines, Water and Environment:* Fouad Douiri. *Employment and Vocational Training:* Abdelouahed Souhail. *Industry, Trade and New Technologies:* Abdelkader Aâmara. *Tourism:* Lahcen Haddad. *Solidarity, Women, Family and Social Development:* Bassima Hakkaoui. *Culture:* Mohamed Amine Sbihi.

Crafts: Abdessamad Qaiouh. *Relations with Parliament and Civil Society:* Lahbib Choubani.

Office of the Prime Minister (French and Arabic only): http://www.pm.gov.ma

CURRENT LEADERS

Mohammed VI ibn al-Hasan

Position
King

Introduction
Mohammed VI ibn al-Hasan was crowned King in July 1999 after the death of his father, King Hassan II. Less austere than his father, he pledged to improve Morocco's democratic institutions and encourage private investment in key economic sectors.

Early Life
King Mohammed VI was born on 21 Aug. 1963 in Rabat, Morocco. In 1985 he graduated from the College of Law in the Rabat Mohammed V University. In 1987 he took a degree in political science and in 1993 was awarded a law doctorate from the French University of Nice-Sophia Antipolis.

Mohammed undertook his first official royal duty aged 11 and by the time he was 20 he had led a Moroccan delegation to the Franco-African conference and negotiated with the Organization of African Unity (now the African Union) over the Western Sahara conflict.

Appointed head of the general staff of the Royal Armed Forces in 1985, he succeeded to the throne in 1999.

Career in Office
Mohammed voiced support for developing a market economy and urged increased private sector investment. However, economic liberalization did not significantly alleviate Morocco's widespread poverty or reduce unemployment. Political power remained concentrated in the monarchy until protests in early 2011, reflecting disaffection across much of the Arab world, prompted Mohammed to cede constitutional reforms in July that transferred some of the monarchy's powers to parliament and the prime minister. In line with the new constitution, Mohammed was obliged to choose Abdelilah Benkirane as his new premier following the Islamist Party of Justice and Development's victory in parliamentary elections in Nov. 2011.

In foreign policy, Mohammed has co-operated with the USA in its anti-terror initiatives since the Sept. 2001 attacks. Morocco has itself been targeted by terrorist violence, including co-ordinated suicide bombings in Casablanca in May 2003 and further incidents in 2007 and April 2011. Elsewhere, Mohammed has expressed support for Palestinian claims to their own independent state.

In 2002 the King came into conflict with Spain when Moroccan forces landed on the uninhabited island of Perejil (a Spanish possession since 1668) off the Moroccan coast. Spain launched a bloodless counter-assault before withdrawing its troops on the understanding that neither country would occupy the island. In Jan. 2006 Prime Minister Rodríguez Zapatero became the first Spanish leader in 25 years to make an official visit to the enclaves of Melilla and Ceuta. However, a subsequent visit to the territories by Spain's King Juan Carlos in Nov. 2007 was criticized by Mohammed, and tension arose again in Aug. 2010 following incidents near the Melilla border.

DEFENCE

Compulsory national military service was abolished in 2006. Defence expenditure in 2008 totalled US$2,977m. (US$96 per capita), representing 3·5% of GDP.

Army
The Army is deployed in two commands: Northern Zone and Southern Zone. There is also a Royal Guard of 1,500. Strength (2007), 175,000 (100,000 conscripts). There is also a Royal Gendarmerie of 20,000, an Auxiliary Force of 30,000 and reserves of 150,000.

Navy
The Navy includes three frigates, 27 patrol and coastal combatants and four amphibious craft.

Personnel in 2007 numbered 7,800, including 1,500 marines. Bases are located at Casablanca, Agadir, Al Hoceima, Dakhla and Tangier.

Air Force
Personnel strength (2007) about 13,000, with 89 combat-capable aircraft, including F-5s and Mirage F-1s.

ECONOMY

Agriculture accounted for 16·4% of GDP in 2009, industry 28·6% and services 55·0%.

Overview
Morocco has achieved strong macroeconomic growth since 2001, thanks to diversification of the non-agricultural sector. The implementation of broad-based structural reforms has seen an increase in foreign direct investment, containment of inflation and a reduction in financial sector vulnerabilities. As a result, unemployment fell and the poverty rate shrank by 6·5% from 2000–08. Both the unemployment and poverty rates, however, remain around 9%.

Trade is focused on the EU, which in 2010 accounted for 59·3% of exports and 57·2% of imports. In March 2012 an agreement between the EU and Morocco came into force abolishing tariffs on a range of industrial goods. In the previous month the European parliament approved an agreement to cut tariffs on specified agricultural and fishery produce over a ten-year transitional period. However, instability within the eurozone in 2011, along with rising international food and oil prices, poses long-term challenges. In 2004 Morocco signed free trade agreements with the USA, Turkey, Egypt, Jordan and Tunisia.

Currency
The unit of currency is the *dirham* (MAD) of 100 *centimes*, introduced in 1959. Foreign exchange reserves were US$14,710m. in July 2005, gold reserves 708,000 troy oz and total money supply was DH353,598m. Inflation was 1·0% in 2010 and 0·9% in 2011.

Budget
Revenues in 2007 totalled DH167,904m. and expenditures DH168,959m. The main revenue items were VAT (29·6%), corporate taxes (18·1%) and income tax (16·5%). Current expenditure accounted for 78·5% of total expenditures.

VAT is 20% (reduced rates, 14%, 10% and 7%).

Performance
Real GDP growth was 4·9% in 2009, 3·7% in 2010 and 4·9% in 2011. Total GDP in 2011 was US$100·2bn.

Banking and Finance
The central bank is the Bank Al Maghrib (*Governor*, Abdellatif Jouahri) which had assets of DH226,667m. in Dec. 2011. The largest bank is Attijariwafa Bank, with assets of US$40·0bn. in Dec. 2011. Other leading banks are Banque Populaire and BMCE Bank. Morocco's external debt amounted to US$25,403m. in 2010 (equivalent to 28·1% of GNI), up from US$16,174m. in 2005.

There is a stock exchange in Casablanca.

ENERGY AND NATURAL RESOURCES

Environment
Carbon dioxide emissions from the consumption and flaring of fossil fuels in 2008 were the equivalent of 1·3 tonnes per capita.

Electricity

Installed capacity was 5·8m. kW in 2007. Production was 22·86bn. kWh (approximately 93% thermal) in 2007 and consumption per capita 843 kWh.

Oil and Gas

Natural gas reserves in 2007 were 1·6bn. cu. metres; output (2007), 65m. cu. metres.

Minerals

The principal mineral exploited is phosphate (Morocco has the largest reserves in the world), the output of which was 27·24m. tonnes in 2006. Other minerals (in tonnes, 2006) are: barytine, 628,400; salt, 506,700; zinc, 148,700; lead, 59,100; iron, 35,500; copper, 17,800; manganese, 4,815; silver, 246.

Agriculture

Agricultural production is subject to drought; about 1·35m. ha. were irrigated in 2002. 85% of farmland is individually owned. Only 1% of farms are over 50 ha.; most are under 3 ha. There were 8·40m. ha. of arable land in 2002 and 887,000 ha. of permanent crops. Main land usage, 2003 (in 1,000 ha.): wheat, 2,989; barley, 2,267; maize, 247. There were 43,226 tractors in 2001 and 3,763 harvester-threshers. Production in 2003 (in 1,000 tonnes): wheat, 5,147; sugar beets, 3,428; barley, 2,620; potatoes, 1,435; tomatoes, 1,004. Livestock, 2002: sheep, 16·34m.; goats, 5·09m.; cattle, 2·67m.; asses, 982,000; chickens, 137m. Livestock products in 2003 included (in 1,000 tonnes): milk, 1,311; meat, 598.

Forestry

Forests covered 5·13m. ha. in 2010, or 11% of the total land area. Timber production was 1·04m. cu. metres in 2007.

Fisheries

Total catch in 2010 was 1,136,240 tonnes (sea fish, 1,129,014 tonnes). Morocco's annual catch is the highest of any African country.

INDUSTRY

The leading companies by market capitalization in Morocco in March 2012 were Maroc Telecom, US$14·3bn.; and Attijariwafa Bank, US$8·3bn.

In 2006 industry contributed 27·8% of GDP, with manufacturing accounting for 16·5%. Production in 1,000 tonnes (2004 unless otherwise indicated): cement, 9,828; residual fuel oil, 2,264; distillate fuel oil, 2,254; sugar (2005), 1,059; petrol, 257; paper and paperboard (2005 estimate), 129; olive oil (2004–05), 50.

Labour

Of 9,927,728 persons in employment in 2006, 43·3% were engaged in agriculture, fishing and forestry, 12·4% in commerce, 12·3% in industry (including handicrafts), 8·0% in construction and public works, 5·4% in general administration and public services, 4·0% in transport and communication and 14·5% in other services. The unemployment rate in 2006 was 9·7%. In Nov. 2006 the minimum hourly wage for non-agricultural workers was DH9·66. The minimum wage for agricultural workers is set at DH50 per day.

INTERNATIONAL TRADE

In 1989 Morocco signed a treaty of economic co-operation with the four other Maghreb countries: Algeria, Libya, Mauritania and Tunisia. Morocco is an active participant in the Euromed process, which aims to create a Euro-Mediterranean Free Trade Area.

Imports and Exports

Imports (c.i.f.) in 2010 were US$35,379m. and exports (f.o.b.) US$17,765m. Imports in 2006 included: machinery and transport equipment, 27·7%; mineral fuels, 21·6%; chemicals and related products, 9·7%; food and live animals, 7·0%. Exports included: apparel and clothing accessories, 25·8%; machinery and transport equipment, 17·2%; chemicals and related products, 13·4%; fish and seafood, 9·3%. Main import suppliers, 2006: France, 16·5%; Spain, 11·6%; Saudi Arabia, 6·8%; Italy, 6·4%. Main export markets, 2006: France, 28·4%; Spain, 20·8%; UK, 6·0%; Italy, 4·9%.

COMMUNICATIONS

Roads

In 2007 there were 57,799 km of classified roads, including 813 km of motorways and 11,251 km of main roads. By 2010 the motorway network had been extended to 1,042 km. In 2007 freight transport totalled 697m. tonne-km. In 2007 there were 1,644,500 passenger cars in use, 525,300 lorries and vans and 22,800 motorcycles and mopeds. There were 58,924 road accidents in 2007 (3,838 fatalities).

Rail

In 2005 there were 1,907 km of railways, of which 1,003 km were electrified. Passenger-km travelled in 2009 came to 4·19bn. and freight tonne-km to 4·11bn. In 2003 the construction of two 38 km-long rail tunnels under the Straits of Gibraltar was agreed with Spain although there are ongoing talks as to the project's feasibility.

Civil Aviation

The national carrier is Royal Air Maroc. The major international airport is Mohammed V at Casablanca; there are eight other airports. Casablanca handled 3,457,209 passengers in 2001 (2,612,998 on international flights) and 41,140 tonnes of freight; Marrakesh (Menara) handled 1,371,851 passengers and 2,471 tonnes of freight and Agadir (Al Massira) 1,052,181 passengers and 2,351 tonnes of freight. In July 1997 Morocco launched its first private air company, Regional Air Lines, to serve the major regions of the kingdom, in addition to southern Spain and the Canary Islands. In 2005 scheduled airline traffic of Moroccan-based carriers flew 70·3m. km, carrying 4,423,100 passengers.

Shipping

The busiest ports are Casablanca (which handled 26,572,000 tonnes of foreign cargo in 2008), Mohammedia, Nador, Tanger Med and Tangier. In Jan. 2009 there were 37 ships of 300 GT or over registered, totalling 341,000 GT.

Telecommunications

In 2008 there were 2,991,200 main (fixed) telephone lines; mobile phone subscribers numbered 22,815,700 in 2008 (72·2 per 100 persons). The main telecommunication company is Maroc Telecom, which was privatized in 2001. Maroc Telecom's principal competitor is Méditel. There were 10,442,500 internet users in 2008. In June 2012 there were 4·6m. Facebook users.

SOCIAL INSTITUTIONS

Justice

The legal system is based on French and Islamic law codes. There are a Supreme Court, 21 courts of appeal, 65 courts of first instance, 196 centres with resident judges and 706 communal jurisdictions for petty offences.

The population in penal institutions in Dec. 2006 was 53,580 (167 per 100,000 of national population). On ascending to the throne in July 1999, King Mohammed VI pardoned and ordered the release of 7,988 prisoners and reduced the terms of 38,224 others.

Education

The adult literacy rate in 2009 was 56·1% (68·9% among males and 43·9% among females). Education in Berber languages has been permitted since 1994; Berber languages were officially added to the syllabus in 2003. Education is compulsory from the age of six to 14 but is expected to be extended to 15 by 2015. In 2008–09 pre-primary schools had an estimated 33,000 teachers for 722,000 children and there were 145,000 teachers at primary schools for

3,851,000 pupils. In 2006–07 there were 2,173,000 pupils in secondary schools. There were 20,000 teaching staff at universities for 419,000 students in 2008–09. There is an English-language university at Ifrane.

In 2008 public expenditure on education came to 5·6% of GDP.

Health
In 2006 there were 18,269 physicians, 3,473 dentists, 27,658 paramedical personnel and 8,002 pharmacists. There were 133 public hospitals in 2006 with about 27,000 beds, a provision of 87 beds per 100,000 inhabitants.

RELIGION
Islam is the established state religion. 98% of the population are Sunni Muslims of the Malekite school and 0·2% are Christians, mainly Roman Catholic, and there is a small Jewish community.

CULTURE
World Heritage Sites
Morocco has nine sites on the UNESCO World Heritage List: the Medina of Fez (inscribed on the list in 1981); the Medina of Marrakesh (1985); the Ksar of Ait-Ben-Haddou (1987); the Historic City of Meknès (1996); the Archaeological Site of Volubilis (1997 and 2008); the Medina of Tétouan (1997); the Medina of Essaouira/Magador (2001); the Portuguese City of Mazagan, now part of the city of El Jadida (2004); and Rabat, Modern Capital and Historic City (2012).

Press
In 2008 there were 33 paid-for daily newspapers. The leading dailies are the Arabic-language *Al-Massae, Assabah* and *Al-Ahdath al-Maghrebia* and the French-language *Le Matin du Sahara et du Maghreb.*

Tourism
In 2010 there were a record 9,288,000 non-resident tourists (excluding same-day visitors), up from 8,341,000 in 2009 and 8,209,000 in 2008.

DIPLOMATIC REPRESENTATIVES
Of Morocco in the United Kingdom (49 Queen's Gate Gdns, London, SW7 5NE)
Ambassador: HH Princess Lalla Joumala Alaoui.

Of the United Kingdom in Morocco (28 avenue S.A.R. Sidi Mohammed, Souissi, Rabat)
Ambassador: Clive Alderton.

Of Morocco in the USA (1601 21st St., NW, Washington, D.C., 20009)
Ambassador: Mohamed Rachad Bouhlal.

Of the USA in Morocco (2 Ave. de Mohamed el Fassi, Rabat)
Ambassador: Samuel L. Kaplan.

Of Morocco to the United Nations
Ambassador: Mohammed Loulichki.

Of Morocco to the European Union
Ambassador: Menouar Alem.

FURTHER READING
Direction de la Statistique. *Annuaire Statistique du Maroc.—Conjoncture Économique.* Quarterly.—*Bulletin Officiel.* Weekly.

Bourqia, Rahma and Gilson Miller, Susan, (eds.) *In the Shadow of the Sultan: Culture, Power and Politics in Morocco.* 2000
Pennell, C. R., *Morocco: From Empire to Independence.* 2003

National library: Bibliothèque Générale et Archives, 5 Avenue Ibn Batouta, BP 1003, Rabat.
National Statistical Office: Direction de la Statistique, Haut-Commissariat au Plan, BP 178, Rabat.
Website (French only): http://www.hcp.ma

Western Sahara

GENERAL DETAILS
The Western Sahara was designated by the United Nations in 1975, its borders having been marked as a result of agreements made between France, Spain and Morocco in 1900, 1904 and 1912. Sovereignty of the territory is in dispute between Morocco and the Polisario Front (Popular Front for the Liberation of the Saguia el Hamra and Rio de Oro), which formally proclaimed a government-in-exile of the Sahrawi Arab Democratic Republic (SADR) in Feb. 1976. According to a UN Security Council resolution adopted in July 2003, Western Sahara should be a semi-autonomous region of Morocco for five years. There would then be a referendum to decide whether it should remain part of Morocco or becomes a separate state. However, the Moroccan government rejected the plan.

Area 252,120 sq. km (97,346 sq. miles). Around 356,000 inhabitants (2004 estimate) are within Moroccan jurisdiction. Another estimated 196,000 Sahrawis live in refugee camps around Tindouf in southwest Algeria. The main towns are El-Aaiún (Laâyoune), the capital (183,691 inhabitants in 2004), Dakhla and Es-Semara.

Life expectancy at birth (1997 est.) male, 46·7 years; female, 50·0 years. Birth rate (1997 est.) per 1,000 population: 46·1; death rate: 17·5. The UN gives a projected population for 2015 of 624,000.

The population is Arabic-speaking, and almost entirely Sunni Muslim.

President: Mohammed Abdelaziz.
Prime Minister: Abdelkader Taleb Oumar.

Rich phosphate deposits were discovered in 1963 at Bu Craa. Morocco holds 100% of the shares of the former Spanish state-controlled company Phosboucraa. Production reached 5·6m. tonnes in 1975, but exploitation has been severely reduced by guerrilla activity. After a nearly complete collapse, production and transportation of phosphate resumed in 1978, ceased again, and then resumed in 1982. Installed electrical capacity was an estimated 58,000 kW in 2007, with production in 2007 of approximately 90m. kWh. Carbon dioxide emissions from the consumption and flaring of fossil fuels in 2008 were the equivalent of 0·8 tonnes per capita. There are about 6,100 km of motorable tracks, but only about 500 km of paved roads. There are airports at El-Aaiún and Dakhla. As most of the land is desert, less than 19% is in agricultural use, with about 2,000 tonnes of grain produced annually. Forests covered 0·71m. ha. in 2010, or 3% of the total land area. In 1994 there were 100 physicians, equivalent to one per 2,504 inhabitants.

FURTHER READING
Sheley, Toby, *Endgame in the Western Sahara: What Future for Africa's Last Colony?* 2004

MOZAMBIQUE

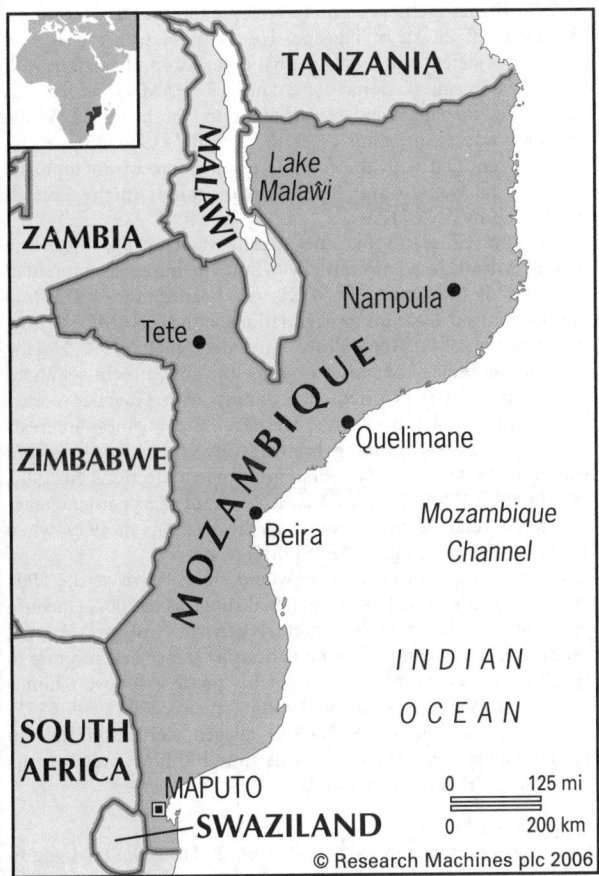

© Research Machines plc 2006

República de Moçambique
(Republic of Mozambique)

Capital: Maputo
Population projection, 2015: 26·16m.
GNI per capita, 2011: (PPP$) 898
HDI/world rank: 0·322/184
Internet domain extension: .mz

KEY HISTORICAL EVENTS

Mozambique was at first ruled as part of Portuguese India but a separate administration was created in 1752. Following a decade of guerrilla activity, independence was achieved on 25 June 1975. A one-party state dominated by the Liberation Front of Mozambique (FRELIMO) was set up but armed insurgency led by the Mozambican National Resistance (RENAMO) continued until 4 Oct. 1992. The peace treaty provided for all weapons to be handed over to the UN and all armed groups to be disbanded within six months. In 1994 the country held its first multi-party elections. In early 2000 some 700 people died in the floods that made thousands homeless.

TERRITORY AND POPULATION

Mozambique is bounded east by the Indian ocean, south by South Africa, southwest by Swaziland, west by South Africa and Zimbabwe and north by Zambia, Malawi and Tanzania. It has an area of 799,380 sq. km (308,642 sq. miles) and a population, according to the 2007 census, of 20,252,223 (10,505,533 females), giving a density of 25·3 per sq. km.

The UN gives a projected population for 2015 of 26·16m.

In 2011, 39·2% of the population were urban. The areas, populations and capitals of the provinces are:

Province	Sq. km	2007 census	Capital
Cabo Delgado	82,625	1,606,568	Pemba
Gaza	75,709	1,228,514	Xai-Xai
Inhambane	68,615	1,271,818	Inhambane
Manica	61,661	1,412,248	Chimoio
City of Maputo	300	1,094,628	—
Province of Maputo	26,058	1,205,709	Maputo
Nampula	81,606	3,985,613	Nampula
Niassa	129,056	1,170,783	Lichinga
Sofala	68,018	1,642,920	Beira
Tete	100,724	1,783,967	Tete
Zambézia	105,008	3,849,455	Quelimane

The capital is Maputo (2007 population, 1,094,628). Other large cities (with 2007 populations) are Matola (671,556), Nampula (471,717) and Beira (431,583).

The main ethnolinguistic groups are the Makua/Lomwe (52% of the population), the Tsonga/Ronga (24%), the Nyanja/Sena (12%) and Shona (6%).

Portuguese remains the official language, but vernaculars are widely spoken throughout the country. English is also widely spoken.

SOCIAL STATISTICS

2008 estimates: births, 877,000; deaths, 357,000. Estimated rates per 1,000 population, 2008: births, 39·2; deaths, 15·9. Infant mortality per 1,000 live births, 2010, 92. Life expectancy at birth, 2007, was 46·9 years for males and 48·7 years for females. Annual population growth rate, 2000–08, 2·6%; fertility rate, 2008, 5·1 births per woman.

CLIMATE

A humid tropical climate, with a dry season from June to Sept. In general, temperatures and rainfall decrease from north to south. Maputo, Jan. 78°F (25·6°C), July 65°F (18·3°C). Annual rainfall 30" (760 mm). Beira, Jan. 82°F (27·8°C), July 69°F (20·6°C). Annual rainfall 60" (1,522 mm).

CONSTITUTION AND GOVERNMENT

On 2 Nov. 1990 the People's Assembly unanimously voted in favour of a new constitution, which came into force on 30 Nov. This changed the name of the state to 'Republic of Mozambique', legalized opposition parties, provided for universal secret elections and introduced a bill of rights including the right to strike, press freedoms and *habeas corpus*. The head of state is the *President*, directly elected for a five-year term. Parliament is a 250-member *Assembly of the Republic*, elected for a five-year term by proportional representation.

National Anthem

'Patria Amada' ('Beloved Motherland'); words and tune by J. Sigaulane Chemane.

RECENT ELECTIONS

In the parliamentary elections of 28 Oct. 2009 the Liberation Front of Mozambique (FRELIMO) won 191 of the 250 seats with 74·7% of the vote, the Mozambican National Resistance (RENAMO) 51 with 17·7% and the Democratic Movement of Mozambique 8 with 3·9%.

In the presidential election, also held on 28 Oct. 2009, incumbent Armando Guebuza of FRELIMO took 75·0% of the vote against 16·4% for RENAMO's Afonso Marceta Macacho Dhlakama and 8·6% for Daviz Simango of the Democratic Movement of Mozambique. Turnout was 44·6%.

CURRENT GOVERNMENT

President: Armando Guebuza; b. 1943 (FRELIMO; sworn in 2 Feb. 2005 and re-elected in Oct. 2009).

In March 2013 the government comprised:

Prime Minister: Alberto Vaquina (FRELIMO; sworn in 9 Oct. 2012).

Minister of Agriculture: José Condungua António Pacheco. *Culture:* Armando Artur João. *Defence:* Filipe Jacinto Nhussi. *Development and Planning:* Aiuba Cuereneia. *Education:* Augusto Jone Luís. *Energy:* Salvador Namburete. *Environmental Action Coordinator:* Alcinda Abreu. *Finance:* Manuel Chang. *Fisheries:* Victor Borges. *Foreign Affairs and Co-operation:* Oldemiro Balói. *Health:* Alexandre Lourenço Jaime Manguele. *Industry and Commerce:* Armando Inroga. *Interior:* Alberto Ricardo Mondlane. *Justice:* Maria Benvinda Levi. *Labour:* Helena Taípo. *Mineral Resources:* Esperança Bias. *Public Service:* Vitória Dias Diogo. *Public Works and Housing:* Cadmiel Muthemba. *Science and Technology:* Louis Augusto Mutomene Pelembe. *State Administration:* Caremelita Namashalua. *Tourism:* Carvalho Muária. *Transport and Communications:* Paulo Zucula. *Veterans' Affairs:* Mateus Óscar Kida. *Women's and Social Affairs:* Iolanda Cintura. *Youth and Sport:* Fernando Sumbana Júnior. *Minister of Civilian Affairs (President's Office):* António Correia Sumbana. *Minister of Parliamentary, Municipal and Provincial Affairs (President's Office):* Adelaide Amurane. *Minister of Social Affairs (President's Office):* Feliciano Salomão Gundana.

Government Website (Portuguese only):
   http://www.portaldogoverno.gov.mz

CURRENT LEADERS

Armando Guebuza

Position
President

Introduction
Armando Guebuza, a veteran of Mozambique's fight for independence and one of the nation's wealthiest businessmen, was chosen as the ruling party's candidate for the 2004 presidential elections. Having won a large majority, he took office in Feb. 2005 and was re-elected for a second term in Oct. 2009.

Early Life
Armando Emílio Guebuza was born on 20 Jan. 1943 in Murrupula, in the northern province of Nampula. Politically active from an early age, he was elected in 1963 as president of the Mozambican Centre of African Students, a group created by Eduardo Mondlane, then the leader of Mozambique's fight for independence from Portugal. Later that year Guebuza joined the Liberation Front of Mozambique (FRELIMO) and in 1965 was elected to the organization's central and executive committees. Having undergone military training in Tanzania, Guebuza was involved in guerrilla fighting against the Portuguese administration in northern Mozambique. Following Mondlane's assassination in 1969, FRELIMO was led by Uria Simango and then Samora Machel. Under Machel it grew to include over 7,000 guerrillas and by the early 1970s had control over much of northern and central Mozambique. Guebuza became a general and was also an inspector of the schools run by FRELIMO.

When Marcello Caetano was overthrown in a military coup in Portugal on 25 April 1974, independence was assured for Mozambique. Following the signing of the Lusaka Agreements later in 1974, Guebuza was appointed to the transitional government that led the country to full independence in June 1975. He then served as minister of the interior in the single-party Marxist government led by President Machel. Guebuza was responsible for implementing the notorious '20–24' decree, which gave Portuguese settlers 24 hours to leave the country, carrying a maximum of 20 kg of luggage. He went on to serve as vice minister of defence in 1980, against a backdrop of warfare with the Mozambican National Resistance (RENAMO), which was backed by the apartheid government in South Africa. While Guebuza was again minister of the interior (1983–85) he was heavily identified with the forcible resettlement of unemployed residents of Maputo and Beira to work-camps in the isolated northern province of Niassa.

Joaquim Chissano became president in 1986, following Machel's death in an aircraft crash, and Guebuza was appointed minister of transport. In 1990 he headed the FRELIMO government's delegation to negotiations with RENAMO, leading to the signing of the Rome Peace Agreement in Oct. 1992. Having formally renounced Marxism in 1989, the government set about developing a market-oriented economy with Guebuza spearheading many of the reforms. He developed business interests in many sectors, including brewing, investment banking and shipping. In the country's first multi-party elections in 1994, won by FRELIMO, Guebuza was elected head of its parliamentary group. He retained that position in the elections of 1999, when Joaquim Chissano again led FRELIMO to victory.

Chissano announced that he would stand down at the 2004 elections. During FRELIMO's national congress in 2002, Guebuza was elected the party's secretary-general and presidential candidate. His uncompromising nationalist stance and promise to continue the economic reforms of his predecessor won him a large majority in the presidential polling in Dec. 2004 (with 63·7% of the vote), although RENAMO alleged electoral fraud. In parliamentary elections at the same time FRELIMO retained its majority in the National Assembly.

Career in Office
Guebuza was sworn in as president on 2 Feb. 2005, pledging to fight poverty, tackle corruption and seek further foreign investment to build infrastructure. In mid-2005 a trade and investment agreement was signed with the USA, whose officials cited Mozambique as 'a positive model because of its impressive track record on democracy, political stability, economic growth, openness to foreign direct investment and expanding exports'. In July 2006 the World Bank cancelled most of the country's debt under a scheme backed by the major industrialized nations.

Guebuza and FRELIMO increased their respective vote shares in the presidential and parliamentary elections in Oct. 2009, although RENAMO again disputed the results. In Sept. 2010 there were riots in Maputo and other cities over food price rises and several people were killed as police fired on protesters.

In Sept. 2012 Guebuza was re-elected as head of FRELIMO. In a surprise cabinet reshuffle a month later, he dismissed Aires Ali as prime minister after only nine months in the post and replaced him with Alberto Vaquina, previously a provincial governor.

DEFENCE

The President of the Republic is C.-in-C. of the armed forces. Defence expenditure totalled US$76m. in 2008 (US$4 per capita), representing 0·8% of GDP. Conscription for both men and women is for two years.

Army

Personnel numbered around 9–10,000 in 2007.

Navy

Naval personnel in 2007 were believed to total 200.

Air Force

Personnel (2007) 1,000 (including air defence units). There were four attack helicopters although their serviceability was in doubt but no combat-capable aircraft.

ECONOMY

Agriculture accounted for 31·5% of GDP in 2009, industry 23·6% and services 44·9%.

Overview

Following the end of civil war in 1992, Mozambique achieved average annual growth of 8% between 1994 and 2007. Although heavy flooding in 2000 devastated the economy and infrastructure, growth fully recovered by 2001. The IMF's Poverty Reduction and Growth Facility (PRGF) programme, including a low-interest lending facility, has helped maintain macroeconomic stability, reduce inflation and provide a sustainable fiscal and external position.

Despite the global economic crisis, growth fell less than expected owing to strong performances in the construction, energy and financial sectors. Reducing poverty, diversifying the economic base (especially through expansion of the transport and electricity infrastructure) and fostering private sector activity remain long-term challenges.

Currency

The unit of currency is the *new metical* (MZN) of 100 *centavos*, which replaced the *metical* (MZM) in July 2006. The currency was revalued at a rate of 1 new metical = 1,000 meticais. Inflation was 3·3% in 2009, increasing to 12·7% in 2010 before falling to 10·4% in 2011. Foreign exchange reserves were US$979m. in July 2005 and total money supply was 18,258·3bn. meticais.

Budget

In 2008 revenues were 69,107m. meticais and expenditures 83,220m. meticais.

Performance

GDP growth has averaged 8·0% since 2001, making Mozambique one of Africa's fastest-expanding economies. There was real GDP growth of 6·3% in 2009, 7·1% in 2010 and 7·3% in 2011. Total GDP in 2011 was US$12·8bn.

Banking and Finance

Most banks had been nationalized by 1979. The central bank and bank of issue is the Bank of Mozambique (*Governor*, Ernesto Gouveia Gove) which hived off its commercial functions in 1992 to the newly-founded Commercial Bank of Mozambique. It in turn merged in 2001 with Banco Internacional de Moçambique, which now trades as Millennium bim and is the country's largest bank. In July 2010 the Bank of Mozambique had external assets of US$2,105·0m. In 2008 there were 14 commercial banks, one microbank and six credit co-operatives. The Mozambique Stock Exchange opened in Maputo in Oct. 1999. By the late 1990s financial services had become one of the fastest-growing areas of the economy. Foreign debt was US$4,124m. in 2010, representing 43·8% of GNI.

ENERGY AND NATURAL RESOURCES

Environment

Carbon dioxide emissions from the consumption and flaring of fossil fuels in 2008 were the equivalent of 0·1 tonnes per capita.

Electricity

Installed capacity was 2·5m. kW in 2007. Production in 2007 was 16·08bn. kWh; consumption per capita was 615 kWh.

Oil and Gas

Natural gas finds, particularly offshore, have been explored for potential exploitation, and both onshore and offshore foreign companies are prospecting for oil. In 2007 natural gas reserves were 127bn. cu. metres; output, 2·8bn. cu. metres. Some river basins, especially the Rovuma, Zambezi and Limpopo, are of interest to oil prospectors.

Water

Although the country is rich in water resources, the provision of drinking water to rural areas remains a major concern.

Minerals

There are deposits of pegamite, tantalite, graphite, apatite, tin, iron ore and bauxite. Other known reserves are: nepheline, syenite, magnetite, copper, garnet, kaolin, asbestos, bentonite, limestone, gold, titanium and tin. Mozambique also has extensive —largely unexploited—coal reserves. Output in 2005 (in 1,000 tonnes): aluminium, 554; sea salt (estimate), 80; bauxite, 10; coal, 3.

Agriculture

All land is owned by the state but concessions are given. There were an estimated 4·2m. ha. of arable land in 2002 and 0·24m. ha. of permanent crops. Around 107,000 ha. were irrigated in 2002. There were about 5,750 tractors in 2002. Production in 1,000 tonnes (2003): cassava, 6,150; maize, 1,248; sugarcane, 400; sorghum, 314; coconuts, 265; rice, 200; groundnuts, 110. Livestock, 2003 estimates: 1·32m. cattle, 392,000 goats, 180,000 pigs, 125,000 sheep, 28m. chickens.

Forestry

In 2010 there were 39·02m. ha. of forests, or 50% of the land area, including eucalyptus, pine and rare hardwoods. In 2007 timber production was 18·03m. cu. metres.

Fisheries

The catch in 2010 was 150,634 tonnes, of which 118,893 tonnes were from sea fishing.

INDUSTRY

Although the country is overwhelmingly rural, there is some substantial industry in and around Maputo (steel, engineering, textiles, processing, docks and railways). A huge aluminium smelter, Mozal, was constructed in two phases—the last phase was completed in 2003. Production exceeds its theoretical annual capacity of 506,000 tonnes and is a focal point in the country's strategy of attracting foreign investment.

Labour

The economically active population in 2010 totalled an estimated 11,261,000 (52% females). In 2007, 75% of the employed workforce were engaged in agriculture, forestry and fisheries. The leading occupations in non-agricultural sectors were commerce and services.

INTERNATIONAL TRADE

Imports and Exports

Imports (c.i.f.) totalled US$2,869m. in 2006 (US$2,408m. in 2005). Exports (f.o.b.) totalled US$2,381m. in 2006 (US$1,783m. in 2005). Principal imports in 2003: mineral fuels, 16·5%; machinery and apparatus, 16·2%; foodstuffs, 12·3%; transport equipment, 9·0%. Principal exports in 2003: aluminium, 54·4%; electricity, 10·9%; prawns, 7·3%; cotton, 3·1%. Main import suppliers in 2003: South Africa, 37·3%; Australia, 12·1%; USA, 5·9%; India, 4·2%. Main export markets in 2003: Belgium, 43·5%; South Africa, 16·2%; Spain, 6·7%; Portugal, 3·7%.

COMMUNICATIONS

Roads

In 2008 there were 29,323 km of roads, of which 17·9% were paved. There were 290,600 vehicles in 2008. There were 5,438 road accidents in 2008, with 1,529 fatalities. The flooding of early 2000 washed away at least one fifth of the country's main road linking the north and the south.

Rail

The railway system consists of three separate networks, with principal routes on 1,067 mm gauge radiating from the ports of Maputo, Beira and Nacala. Total length in 2009 was 3,116 km, mainly on 1,067 mm gauge with some 762 mm gauge lines, but only 1,929 km was operational. In 2009 passenger-km travelled on the Mozambique Ports and Railways network came to 164m. and freight tonne-km to 2,078m.

Civil Aviation

There are international airports at Maputo and Beira. The national carrier is the state-owned Linhas Aéreas de Moçambique (LAM). It provides domestic services and in 2003 operated international routes to Comoros, Dar es Salaam, Durban, Harare, Johannesburg and Lisbon. In 2001 Maputo handled 394,671 passengers (213,612 on international flights) and Beira 106,586 (98,590 on domestic flights). In 2003 scheduled airline traffic of Mozambique-based carriers flew 6m. km, carrying 281,000 passengers (103,000 on international flights).

Shipping

The principal ports are Maputo, Beira, Nacala and Quelimane. In Jan. 2009 there were six ships of 300 GT or over registered, totalling 5,000 GT.

Telecommunications

Telephone subscribers numbered 2,406,300 in 2006 (119·4 per 1,000 persons). There were 2,339,300 mobile phone subscribers in 2006, up from just 254,800 in 2002, and 200,000 internet users in 2007. In June 2012 there were 248,000 Facebook users.

SOCIAL INSTITUTIONS

Justice

The 1990 constitution provides for an independent judiciary, *habeas corpus*, and an entitlement to legal advice on arrest. The death penalty was abolished in Nov. 1990.

The population in penal institutions in 2007 was approximately 15,000 (53 per 100,000 of national population).

Education

The adult literacy rate in 2008 was 54%.

In 2007 there were 4,563,633 pupils with 70,389 teaching staff in primary schools; and 444,926 pupils with 12,064 teaching staff at secondary schools. Private schools and universities were permitted to function in 1990. There were 28,298 students and 3,009 academic staff in higher education in 2005. The largest higher education institution is the Eduardo Mondlane University, founded in 1962 and granted university status in 1968.

In 2006 public expenditure on education came to 5·8% of GNI and 21·0% of total government spending.

Health

There were (2004) 46 hospitals, 722 health centres and 479 medical posts. There were two psychiatric hospitals. In 2000 there were 435 doctors, 1,414 midwives, 3,664 nursing personnel, 136 dentists and 419 pharmacists. Private health care was introduced alongside the national health service in 1992.

RELIGION

About 55% of the population follow traditional animist religions. In 2001 there were 6·18m. Christians (mainly Roman Catholic) and 2·04m. Muslims. In Feb. 2013 there was one cardinal.

CULTURE

World Heritage Sites

Mozambique has one site on the UNESCO World Heritage List: the Island of Mozambique (inscribed on the list in 1991), a Portuguese trading post with a style of architecture unchanged since the 16th century.

Press

There were two well-established daily newspapers in 2008 (*Notícias* and *Diário* in Maputo and Beira respectively) with a combined circulation of 13,000.

Tourism

Tourism is a potential growth area for the country. There were 3,110,000 non-resident visitors in 2009 (2,617,000 in 2008).

DIPLOMATIC REPRESENTATIVES

Of Mozambique in the United Kingdom (21 Fitzroy Sq., London, W1T 6EL)
High Commissioner: Carlos dos Santos.

Of the United Kingdom in Mozambique (Ave. Vladimir I. Lenine 310, Maputo)
High Commissioner: Shaun Cleary.

Of Mozambique in the USA (1990 M. St., NW, Washington, D.C., 20036)
Ambassador: Amélia Matos Sumbana.

Of the USA in Mozambique (Ave. Kenneth Kaunda 193, Maputo)
Ambassador: Douglas Griffiths.

Of Mozambique to the United Nations
Ambassador: Antonio Gumende.

Of Mozambique to the European Union
Ambassador: Ana Nemba Uaiene.

FURTHER READING

Alden, Chris, *Mozambique and the Construction of the New African State: From Negotiations to Nation Building.* 2001
Cabrita, João M., *Mozambique: The Tortuous Road to Democracy.* 2001
Finnegan, W., *A Complicated War: the Harrowing of Mozambique.* 1992
Manning, Carrie L., *The Politics of Peace in Mozambique: Post-Conflict Democratization, 1992–2000.* 2002
Newitt, M., *A History of Mozambique.* 1996
Pitcher, M. Anne, *Transforming Mozambique: The Politics of Privatization, 1975–2000.* 2002

National Statistical Office: Instituto Nacional de Estatística, Av. Ahmed Sekou Touré, No. 21.
Website: http://www.ine.gov.mz

MYANMAR

© Research Machines plc 2006

Pyidaunzu Thanmăda Myăma Nainngandaw (Republic of the Union of Myanmar)

Capitals: Naypyidaw/Pyinmana (Administrative and Legislative), Yangon/Rangoon (Commercial)
Population projection, 2015: 49·19m.
GNI per capita, 2011: (PPP$) 1,535
HDI/world rank: 0·483/149
Internet domain extension: .mm

KEY HISTORICAL EVENTS

After Burma's invasion of the kingdom of Assam, the British East India Company retaliated in defence of its Indian interests and in 1826 drove the Burmese out of India. Territory was annexed in south Burma but the kingdom of Upper Burma, ruled from Mandalay, remained independent. A second war with Britain in 1852 ended with the British annexation of the Irrawaddy Delta. In 1885 the British invaded and occupied Upper Burma. In 1886 all Burma became a province of the Indian empire. There were violent uprisings in the 1930s and in 1937 Burma was separated from India and permitted some degree of self-government.

Independence was achieved in 1948. In 1958 there was an army coup, and another in 1962 led by Gen. Ne Win, who installed a Revolutionary Council and dissolved parliament.

The Council lasted until March 1974 when the country became a one-party socialist republic. On 18 Sept. 1988 the Armed Forces seized power and set up the State Law and Order Restoration Council (SLORC). Since then civil unrest has cost more than 10,000 lives. On 19 June 1989 the government changed the name of the country in English to the Union of Myanmar. Aung San Suu Kyi, leader of the National League for Democracy, was put under house arrest in July 1989. In spite of her continuing detention, her party won the 1990 election by a landslide, but the military junta refused to accept the results. She was eventually freed in July 1995. She was detained and later released on several occasions over the years, most recently being released in Nov. 2010.

In Aug. 2007 the government implemented fuel price hikes that prompted public protests, led chiefly by students and political activists. Despite the authorities taking a hard line, the demonstrations were given renewed force when several thousand monks came out in support in mid-Sept. The government crackdown, during which thousands of monks were reportedly rounded up and several protesters killed, brought condemnation from the international community.

In May 2008 the Irrawaddy delta suffered a cyclone that caused massive damage, claimed an estimated 145,000 lives and left 1m. people displaced. Nonetheless, the following week the government proceeded with a referendum on a new constitution, which it claimed secured 92% support. In Nov. 2010 the pro-government Union Solidarity and Development Party won the first elections for 20 years. Thein Sein became president in a nominal transition to civilian rule.

TERRITORY AND POPULATION

Myanmar is bounded in the east by China, Laos and Thailand, and west by the Indian Ocean, Bangladesh and India. Three parallel mountain ranges run from north to south; the Western Yama or Rakhine Yama, the Bagu Yama and the Shaun Plateau. The total area of the Union is 676,590 sq. km (261,230 sq. miles), including 23,070 sq km (8,910 sq. miles) of inland water. At the last census, in 1983, the population was 35,307,913. Estimate, 2010, 47·96m.; density, 73 per sq. km. In 2011, 34·3% of the population lived in urban areas.

The UN gives a projected population for 2015 of 49·19m.

The administrative capital is Naypyidaw (Pyinmana); its population was an estimated 1,026,000 in 2010. The largest city is Yangon (Rangoon), with an estimated population of 4,356,000 in 2010. Other leading towns are Mandalay (2010 population estimate of 1,035,000), Moulmein, Bago (Pegu), Bassein, Sittwe (Akyab), Taunggye and Monywa. In Nov. 2005 the government began relocating from Yangon to the new administrative and legislative capital, Pyinmana, subsequently renamed Naypyidaw. The move was completed in Feb. 2006.

Myanmar's constitution of 2008 established a new administrative order dividing the country into one union territory (Naypyidaw), one self-administered division (Wa), five self-administered zones (Danu, Kokang, Naga, Pa Laung and Pa-O), seven states (Chin, Kachin, Kayah, Kayin (Karen), Mon, Rakhine (Arakan) and Shan) and seven regions (Ayeyarwady (Arawaddy), Bago (Pegu), Magway, Mandalay, Sagaing, Tanintharyi and Yangon (Rangoon)).

Myanmar is inhabited by many ethnic nationalities. There are 135 national groups with the Bamars, comprising about 68% of the population, forming the largest group. The Shan and the Karen account for 9% and 7% of the population respectively.

The official language is Burmese; English is also in use.

SOCIAL STATISTICS

2008 estimates: births, 1,020,000; deaths, 496,000. Estimated birth rate in 2008 was 21 per 1,000 population; estimated death rate, 10. Annual population growth rate, 2000–08, 0·8%. Life expectancy at birth, 2007, was 59·0 years for males and 63·4 years for females. Infant mortality, 2010, 50 per 1,000 live births; fertility rate, 2008, 2·3 births per woman.

CLIMATE

The climate is equatorial in coastal areas, changing to tropical monsoon over most of the interior, but humid temperate in the extreme north, where there is a more significant range of temperature and a dry season lasting from Nov. to April. In coastal parts, the dry season is shorter. Very heavy rains occur in the monsoon months May to Sept. Yangon, Jan. 77°F (25°C), July 80°F (26·7°C). Annual rainfall 104" (2,616 mm). Sittwe, Jan. 70°F (21·1°C), July 81°F (27·2°C). Annual rainfall 206" (5,154 mm). Mandalay, Jan. 68°F (20°C), July 85°F (29·4°C). Annual rainfall 33" (828 mm).

CONSTITUTION AND GOVERNMENT

In Nov. 1997 the country's ruling generals changed the name of the government to the *State Peace and Development Council* (SPDC). It nominally ceded power to an elected president in Feb. 2011 and was abolished a month later.

In May 2008 an army-drafted constitution won 92·5% support in a referendum. The constitution specified that multi-party elections should be scheduled for 2010; 25% of parliamentary seats were automatically allocated to the military. It called for the creation of a National Defence and Security Council, dominated by military appointments, with the power to suspend the constitution under certain circumstances. It also laid out rules that would ban opposition leader Aung San Suu Kyi from holding public office. The constitution was formally adopted on 30 May 2008. The previous constitution, dating from 3 Jan. 1974, had been suspended since 1988. Amendments to the Political Party Registration law in Oct. 2011 now allow Aung San Suu Kyi to hold public office.

The 440-member lower chamber, the House of Representatives (*Pythu Hluttaw*), has 330 elected seats with 110 appointed and the 224-member upper chamber, the House of Nationalities (*Amyotha Hluttaw*), has 168 elected seats with 56 appointed. Parliament convened in Jan. 2011 for the first time since 1988.

National Anthem

'Gba majay Bma' ('We shall love Burma for ever'); words and tune by Saya Tin.

RECENT ELECTIONS

In elections in May 1990 the opposition National League for Democracy (NLD), led by Aung San Suu Kyi (b. 1945), won 392 of the 485 People's Assembly seats contested with some 60% of the valid vote. Turnout was 72%, but 12·4% of ballots cast were declared invalid. The military ignored the result and refused to hand over power.

The next elections were not held until 7 Nov. 2010 and were boycotted by the NLD who claimed that they would not be free and fair. This was confirmed when the government Union Solidarity and Development Party (USDP) won large majorities in both houses and pro-democracy candidates complained of ballot-rigging. Of the 330 elected seats in the House of Representatives, the Union Solidarity and Development Party won 259. In the House of Nationalities they won 129 of 168 elected seats.

By-elections to fill vacant parliamentary seats took place on 1 April 2012. The opposition NLD won all 37 seats contested in the House of Representatives and four of the six seats contested in the House of Nationalities. Opposition leader Aung San Suu Kyi won a House of Representatives seat in Yangon.

On 4 Feb. 2011 parliament elected prime minister Thein Sein as president, winning 408 of 659 votes. Tin Aung Myint Oo (171 votes) and Sai Mauk Kham (75) became vice-presidents.

CURRENT GOVERNMENT

In March 2013 the government comprised:

President: Thein Sein; b. 1945 (since 30 March 2011).

Vice Presidents: Nyan Tun; Sai Mauk Kham.

Minister of Defence: Lieut.-Gen. Wai Lwin. *Home Affairs:* Lieut.-Gen. Ko Ko. *Border Affairs:* Lieut.-Gen. Thein Htay. *Foreign Affairs:* U Wunna Maung Lwin. *Information:* U Aung Kyi. *Agriculture and Irrigation:* U Myint Hlaing. *Environmental Conservation and Forestry:* U Win Tun. *Finance and Revenue:* U Win Shein. *Construction:* U Kyaw Lwin. *Livestock and Fisheries:* U Ohn Myint. *Commerce:* U Win Myint. *Communications, Posts and Telegraphs:* U Thein Tun. *Labour:* U Maung Myint. *Mines:* Myint Aung. *Co-operatives:* U Kyaw San. *Transport:* U Nyan Tun Aung. *Sports:* Tint San. *National Planning and Economic Development:* Kan Zaw. *Industry:* U Aye Myint. *Rail Transportation:* Maj.-Gen. Zayar Aung. *Energy:* U Than Htay. *Electric Power:* U Khin Maung Soe. *Hotels and Tourism:* U Htay Aung. *Education:* Mya Aye. *Health:* Dr Pe Thet Khin. *Religious Affairs:* Thura U Myint Maung. *Science and Technology:* Ko Ko Oo. *Immigration and Population:* U Khin Yi. *Social Welfare, Relief and Resettlement:* Myat Myat Ohn Khin. *Culture:* U Aye Myint Kyu. *Ministers in the President's Office:* U Thein Nyunt; U Soe Maung; U Aung Min; U Soe Thein; U Hla Htun; Tin Naing Thein.

CURRENT LEADERS

Thein Sein

Position
President

Introduction
Thein Sein was sworn into office on 30 March 2011 after his party, the Union Solidarity and Development Party (USDP), dominated the Nov. 2010 polls. He is the country's first civilian president in nearly 50 years, succeeding the hardline military regime of Senior Gen. Than Shwe. Although Thein Sein has maintained his close links to the armed forces, prompting international scepticism over the validity of the 2010 elections, he has nevertheless embraced political reform and has sought to improve Myanmar's image as an aspiring democracy.

Early Life
Thein Sein was born on 20 April 1945 in Bassein, Irrawaddy Division. Graduating from the Defence Services Academy in 1968, he rose through the ranks to become a major in 1988. In 1989 he graduated from the Command and General Staff College and was appointed commander of the 89th Infantry Battalion in Sagaing Division.

From 1992–95 he served at the war office under Than Shwe, then commander-in-chief of the armed forces. He took charge of a military command in Yangon before leading a newly-formed regional military command in Kengtung, Shan State, in 1996. Promoted to lieutenant-general in 2003, he was appointed first secretary of the State Peace and Development Council in 2004 and then, until 2007, chaired the national convention commission that oversaw the drafting of a new constitution. Following his appointment as prime minister in 2007, Thein Sein was promoted to the rank of general and represented Myanmar in regional negotiations. In the aftermath of Cyclone Nargis in May 2008 he was criticized when the government obstructed aid efforts.

On 29 April 2010 Thein Sein retired from the military to set up and lead the USDP. He successfully contested a constituency in

Naypyidaw Union Territory in the election of Nov. 2010 and in Feb. 2011 he was elected president by parliament.

Career in Office

Since taking office Thein Sein has pursued a dialogue with pro-democracy leader Aung San Suu Kyi, freed a number of political prisoners, eased media restrictions, agreed ceasefires with Shan, Kachin and Karen rebels, and passed laws allowing for trade unions and peaceful protest.

Myanmar has been granted the 2014 ASEAN chair and hosted visits from US Secretary of State Hillary Clinton (in Dec. 2011) and President Barack Obama (in Nov. 2012), who have welcomed its democratic progress. The European Union suspended all non-military sanctions for a year from April 2012. The opposition National League for Democracy (NLD) has also cautiously backed the changes, re-registering ahead of a series of by-elections in April 2012 in which it won landslide victories and which saw Aung San Suu Kyi returned to the federal parliament.

DEFENCE

Military expenditure in 2006 totalled US$6,920m. (US$147 per capita), representing 18·7% of GDP. A law making conscription obligatory was introduced in 2010 but has not been implemented.

Army

The strength of the Army was reported to be about 375,000 in 2007. The Army is organized into 12 regional commands. There are three paramilitary units: People's Police Force (72,000), People's Militia (35,000) and People's Pearl and Fishery Ministry (approximately 250).

Navy

Personnel in 2007 totalled about 16,000 including 800 naval infantry.

Air Force

The Air Force is intended primarily for internal security duties. Personnel (2007) approximately 15,000 operating 125 combat-capable aircraft, including F-7s.

ECONOMY

In 2009 agriculture accounted for 38·1% of GDP, industry 24·5% and services 37·4%.

Myanmar featured among the ten most corrupt countries in the world in a 2012 survey of 176 countries carried out by the anti-corruption organization *Transparency International*.

Overview

Myanmar has one of the least developed economies in the world, having suffered decades of stagnation, economic mismanagement and isolation under military rule. The country has significant natural resources, including offshore oil and gas deposits, and is a major source of gems. However, GDP is mainly derived from agriculture, forestry, livestock and fisheries. Myanmar is highly dependent on China and Thailand as trading partners.

Since the end of 2010 political reform has seen the country gradually open to the world economy. GDP grew by 5·3% in 2010–11, aided by fiscal spending programmes and buoyant exports, and both the EU and USA eased long-term sanctions after elections in April 2012. The government has embarked on economic reforms, including steps to unify multiple informal market exchange rates, reduce the fiscal deficit, increase social spending and modernize the financial system.

Currency

The unit of currency is the *kyat* (MMK) of 100 *pyas*. Total money supply was K.1,742,810m. in June 2005. Foreign exchange reserves were US$691m. and gold reserves 231,000 troy oz in May 2005. Myanmar adopted a managed float for the kyat in April 2012, having had a fixed exchange rate for the previous 35 years. Inflation was 22·5% in 2008, falling to 8·2% in 2010 and further to 4·0% in 2011. Since 1 June 1996 import duties have been calculated at a rate US$1 = K.100.

Budget

In 2005–06 revenues were K.819,534m. and expenditures K.1,008,785m. Tax revenue accounted for 58·2% of revenues in 2005–06; general public services accounted for 35·2% of expenditures and economic affairs 34·3%. The fiscal year begins on 1 April.

Performance

Real GDP growth was 5·1% in 2009, 5·3% in 2010 and 5·5% in 2011.

Banking and Finance

The Central Bank of Myanmar was established in 1990. Its *Governor* is U Than Nyein. In 2010 there were four state banks (Myanma Economic Bank, Myanma Foreign Trade Bank, Myanma Agricultural and Rural Development Bank, and Myanma Investment and Commercial Bank) and 19 private banks. Since 1996 foreign banks with representative offices have been permitted to set up joint ventures with Myanmese banks. The foreign partner must provide at least 35% of the capital. The state insurance company is the Myanmar Insurance Corporation. Deposits in state and private banks were K.903,722m. in 2006.

Before being delisted in Oct. 2006, Myanmar was the only country named in a report in June 2006 as failing to co-operate in the fight against international money laundering. The Financial Action Task Force on Money Laundering was set up by the G7 group of major industrialized nations.

Foreign debt was US$6,352m. in 2010.

A stock exchange opened in Yangon in 1996.

ENERGY AND NATURAL RESOURCES

Environment

Myanmar's carbon dioxide emissions from the consumption and flaring of fossil fuels in 2008 were the equivalent of 0·3 tonnes per capita.

Electricity

Total electricity generated, 2007–08, 6·50bn. kWh; consumption per capita in 2007–08 was 113 kWh. Installed capacity was approximately 1·8m. kW in 2007–08.

Oil and Gas

Production (2007–08) of crude oil was 1·0m. tonnes; natural gas (2008), 12·4bn. cu. metres. There were proven natural gas reserves of 490bn. cu. metres in 2008.

Minerals

Myanmar's mineral resources include antimony, coal, copper, lead, limestone, marble, precious stones, tin, tungsten and zinc. Production (in tonnes unless otherwise indicated): hard coal (2007–08), 1,075,000; lignite (2007–08), 414,000; gypsum (2004), 71,155; copper (2004), 31,756; jade (2004), 12,408, ruby, sapphire and spinel (2004), 6,198,915 carats. 90% of the world's rubies are mined in Myanmar.

Agriculture

In 2007 there were 10·6m. ha. of arable land and 1·1m. ha. of permanent crops. 2·2m. ha. were irrigated in 2006–07. Production (2006–07, in 1,000 tonnes): rice, 30,435; sugarcane, 8,039; pulses, 4,198; maize, 1,114; groundnuts, 1,088; onions, 918; plantains (2008 estimate), 630; sesame seeds, 614; potatoes, 508. Opium output was 1,097 tonnes in 2001, falling to 312 tonnes in 2005. In 2009 it was 330 tonnes. Myanmar's opium production is second only to that of Afghanistan.

Livestock (2005–06): cattle, 12·15m.; pigs, 5·79m.; buffaloes, 2·71m.; sheep and goats, 2·44m.; chickens, 85m. In 2006–07 there were 12,000 tractors.

Forestry

Forest area in 2010 was 31·77m. ha., covering 48% of the total land area. Teak resources cover about 6m. ha. (15m. acres). In 2007, 42·55m. cu. metres of roundwood were cut.

Fisheries

In 2010 the total catch was 3,063,210 tonnes (2,060,780 tonnes from sea fishing), the eighth highest in the world. The annual catch has doubled since 2003. Aquaculture production totalled 604,660 tonnes in 2007.

INDUSTRY

Production in 1,000 tonnes (unless otherwise specified): cement (2007), 618; raw sugar (2007), 160; nitrogenous fertilizers (2007), 115; paper and paperboard (2007), 45; sawnwood (2007), 1·5m. cu. metres; cigarettes (2007), 2,755m. units; clay bricks (2004, government production only), 77m. units; bicycles (2007), 53,880 units.

Labour

The estimated economically active population in 2010 was 27,337,000. In 1998 the leading areas of activity (in 1,000) were: agriculture, hunting, forestry and fishing, 11,507; wholesale and retail trade/repair of motor vehicles, motorcycles and personal and household goods, 1,781; manufacturing, 1,666. In 2001 there were 398,300 persons aged 18 years and over registered as unemployed.

INTERNATIONAL TRADE

In Aug. 1991 the USA imposed trade sanctions in response to alleged civil rights violations. A law of 1989 permitted joint ventures, with foreign companies or individuals able to hold 100% of the shares.

Imports and Exports

Since 1990, in line with market-oriented measures, firms have been able to participate directly in trade.

Imports in 2006–07 totalled K.16,835·0m. and exports K.30,026·1m. Main imports, 2006–07: mineral fuels, lubricants and related materials, 24·1%; machinery and transport equipment, 20·3%; manufactured goods, 19·6%; chemicals, 10·7%. Leading import suppliers in 2006–07 were Singapore, 35·2%; China, 24·9%; Thailand, 10·4%; Japan, 5·3%. Main exports in 2006–07: gas, 38·9%; pulses, 11·6%; timber, 9·8%; precious stones and pearls, 7·4%. Main export markets, 2006–07: Thailand, 45·1%; India, 14·0%; China, 11·8%; Hong Kong, 7·7%.

COMMUNICATIONS

Roads

There were 27,000 km of roads in 2005, of which 11·9% were surfaced. In 2005 there were 194,411 passenger cars, 54,482 vans and lorries, 17,985 buses and coaches, and 640,313 motorcycles and mopeds. There were 1,638 deaths as a result of road accidents in 2007.

Rail

In 2005 there were 4,809 km of route on metre gauge. Passenger-km travelled in 2006–07 came to 5,307m. and freight tonne-km to 887m.

Civil Aviation

Myanmar Airways International operates domestic services and in 2003 had international flights to Bangkok, Kuala Lumpur and Singapore. In 2003 scheduled airline traffic of Myanmar-based carriers flew 16m. km, carrying 1,117,000 passengers (691,000 on international flights).

Shipping

There are nearly 100 km of navigable canals. The Irrawaddy is navigable up to Myitkyina, 1,450 km from the sea, and its tributary, the Chindwin, is navigable for 630 km. The Irrawaddy delta has approximately 3,000 km of navigable water. The Salween, the Attaran and the G'yne provide about 400 km of navigable waters around Moulmein. In Jan. 2009 merchant shipping totalled 140,000 GT (vessels of 300 GT and over).

In 2006–07, 26·33m. passengers and 4·28m. tonnes of freight were carried on inland waterways. The ocean-going fleet of the state-owned Myanma Five Star Line in 2006–07 comprised 26 vessels; in addition there were eight chartered vessels. In 2006–07, 51,373 passengers and 10,954,800 tonnes of seaborne cargo were transported coastally and overseas. Myanmar's main port is Yangon, which handles about 90% of the country's imports and exports.

Telecommunications

In 2011 there were 521,100 landline telephone subscriptions (equivalent to 10·8 per 1,000 inhabitants) and 1,243,600 mobile phone subscriptions (or 25·7 per 1,000 inhabitants). In 2011, 1·0% of the population were internet users.

SOCIAL INSTITUTIONS

Justice

The highest judicial authority is the Chief Judge, appointed by the government. In mid-2007 there were 65,063 people (126 per 100,000 of national population) held in prisons. Amnesty International reported in 2007 that before the protests against the government resulting in hundreds more arrests there were around 1,150 political prisoners in the country's jails. In Oct. 2011 President Thein Sein granted amnesty to more than 6,300 prisoners, although it was unclear how many of Myanmar's 2,100 political prisoners would be freed. In Jan. 2012, 651 prisoners were released, including political activists, student leaders and army dissidents.

Education

Education is free in primary, middle and vocational schools; fees are charged in senior secondary schools and universities. In 2007 there were 5,013,582 pupils at primary schools with 172,209 teaching staff; and 2,686,198 pupils at secondary schools with 81,943 teaching staff. In 2006–07 there were 1,055 monastic primary schools (permitted since 1992) with 152,548 pupils, 256 monastic middle schools with 22,992 pupils and two monastic high schools with 3,887 pupils. There were 507,660 students and 10,669 academic staff in tertiary education in 2007.

In higher education in 2005–06 there were 521,702 students enrolled at 35 universities of arts and sciences, including 413,902 at the University of Distance Education. There were also 59,503 students enrolled at vocational universities in 2005–06, with 16,216 studying computers, computer science and technology, 14,516 medicine, 12,079 economics, 6,286 education, 1,738 dentistry, 1,711 technology, 1,387 nursing, 1,287 agriculture, 1,231 pharmacy and 1,077 paramedical science.

The adult literacy rate was an estimated 92·0% in 2009.

In the period 1998–2007, 13% of central government expenditure was allocated to education.

Health

In 2006–07 there were 826 government hospitals with 43,128 beds. In 2006–07 there were 20,501 doctors (65% private), 21,075 nurses, 17,703 midwives and 1,732 dentists. Spending on health in 2005 amounted to 2·2% of GDP although only 10·6% came from the state. General government expenditure on health in 2005 totalled 1·1% of total government spending.

Welfare

In 2006–07 contributions to social security totalled (K.1m.) 3,697·9 (from employers, 2,311·2; from employees, 1,386·7). Benefits paid totalled 935·7, and included: medical care, 764·7; sickness, 77·5; maternity, 41·0; death, 36·0.

RELIGION

About 89·3% of the population—mainly Bamars, Shans, Mons, Rakhines and some Kayins—are Buddhists, while the rest are Christians, Muslims, Hindus and Animists. The Christian population is composed mainly of Kayins, Kachins and Chins. There are about 400,000 monks. Islam and Hinduism are practised mainly by people of Indian origin.

CULTURE

Press

There were eight daily newspapers in 2006, with a combined circulation of 550,000. The three largest newspapers were all government-owned. In the 2010 *World Press Freedom Index* compiled by Reporters Without Borders, Myanmar ranked 174th out of 178 countries.

Tourism

In 2007 there were 264,000 non-resident tourists; spending by tourists totalled US$59m. in 2006.

DIPLOMATIC REPRESENTATIVES

Of Myanmar in the United Kingdom (19A Charles St., London, W1J 5DX)
Ambassador: Kyaw Myo Htut.

Of the United Kingdom in Myanmar (80 Strand Rd, Yangon)
Ambassador: Andrew Heyn.

Of Myanmar in the USA (2300 S. St., NW, Washington, D.C., 20008)
Ambassador: Than Swe.

Of the USA in Myanmar (110 University Ave., Yangon)
Ambassador: Derek J. Mitchell.

Of Myanmar to the United Nations
Ambassador: Tin Kyaw.

Of Myanmar to the European Union
Ambassador: Thant Kyaw.

FURTHER READING

Aung San Suu Kyi, *Freedom from Fear and Other Writings.* 1991
Carey, P. (ed.) *Burma: The Challenge of Change in a Divided Society.* 1997
Hiaing, Kyaw Yin, *Myanmar: Beyond Politics to Social Imperatives.* 2005
Metraux, Daniel A., *Burma's Modern Tragedy.* 2005
Myint, S., *Burma File: A Question of Democracy.* 2004
Rogers, Benedict, *Burma: A Nation at the Crossroads.* 2012
Seekins, Donald M., *Historical Dictionary of Burma.* 2006
Skidmore, Monique, *Burma at the Turn of the Twenty-First Century.* 2005
Smith, Martin, *Burma: Insurgency and the Politics of Ethnicity.* 1999
Steinberg, David I., *Burma: The State of Myanmar.* 2002
Thant Myint-U, *The Making of Modern Burma.* 2001.—*The River of Lost Footsteps: Histories of Burma.* 2007.—*Where China Meets India: Burma and the New Crossroads of Asia.* 2011
Tucker, Shelby, *Burma: The Curse of Independence.* 2001

National Statistical Office: Ministry of National Planning and Economic Development, Yangon.

NAMIBIA

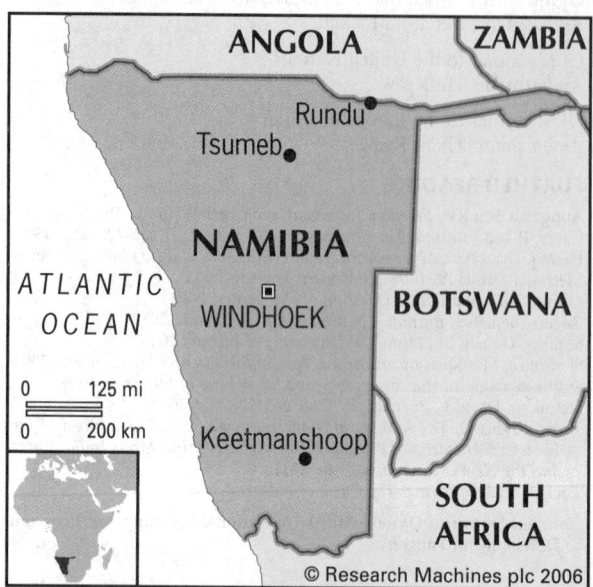

ANGOLA ZAMBIA
Rundu
Tsumeb
NAMIBIA
ATLANTIC OCEAN
WINDHOEK
BOTSWANA
0 125 mi
0 200 km
Keetmanshoop
SOUTH AFRICA
© Research Machines plc 2006

Republic of Namibia

Capital: Windhoek
Population projection, 2015: 2·48m.
GNI per capita, 2011: (PPP$) 6,206
HDI/world rank: 0·625/120
Internet domain extension: .na

KEY HISTORICAL EVENTS

Namibia was first settled by people from the Khoisan language group. The earliest, the nomadic San people, were followed about 2000 years ago by the pastoral Nama, who became dominant in the south. In the 9th century AD the Damara settled the central grasslands (known as Damaraland). Other clans followed and by the 19th century three Bantu peoples were established: the Herero in northeastern and central Namibia (Kaokoland); the Ovambo around the Kunene River in the north; and the Kavango people in the east. In the far east, the Barotse expanded from Zambia to settle the Caprivi Strip while the Tswana (from Botswana) settled the edges of the Kalahari desert.

European traders and settlers arrived in the late 18th century. Walvis Bay came under Dutch (1793) then British (1797) control, and European settlement began on the coast. In the 1830s the Oorlans, from South Africa, expanded into Nama and Damara territory, becoming dominant under the leadership of Jonker Afrikaner.

In 1884 the area then known as South West Africa became a German protectorate. From 1904–08 conflict between German troops and the Herero and Nama peoples saw the deaths of 80% of the Herero population and 50% of the Nama. In the aftermath, Germany introduced racial segregation and used forced labour for diamond mines. In 1915 the Union of South Africa occupied German South West Africa and on 17 Dec. 1920 the League of Nations entrusted the territory as a Mandate to the Union of South Africa. After the Second World War South Africa applied unsuccessfully to annex the territory, continuing to administer it in defiance of the UN. Indigenous opposition to South African rule intensified in the 1950s: the Ovamboland Peoples'

Organization was founded in 1958 (known as the South West Africa Peoples' Organization, or SWAPO, from 1960) and the South West Africa National Union (SWANU) in 1959.

In 1968 the UN changed the territory's name to Namibia. Following widespread strikes in 1971–72, negotiations took place between South Africa and the UN and in 1973 a multi-racial advisory council was appointed in preparation for independence. However, attempts at organizing free elections failed. In 1988, after military defeat in Angola, South Africa withdrew. UN-supervised elections took place in Nov. 1989, delivering a victory for SWAPO. After independence on 21 March 1990, Namibia joined the Commonwealth.

In April 1990 Namibia joined the UN and in June 1990 the Organization of African Unity, forerunner of the African Union. In 2004 the country suffered major flooding. It continues to face a serious AIDS epidemic, with around 20% of adults infected, though in 2007 the rate fell for the first time.

TERRITORY AND POPULATION

Namibia is bounded in the north by Angola and Zambia, west by the Atlantic Ocean, south and southeast by South Africa and east by Botswana. The Caprivi Strip (Caprivi Region), about 300 km long, extends eastwards up to the Zambezi river, projecting into Zambia and Botswana and touching Zimbabwe. The area, including the Caprivi Strip and Walvis Bay, is 825,615 sq. km. South Africa transferred Walvis Bay to Namibian jurisdiction on 1 March 1994. 2011 census population (provisional), 2,104,900 (1,021,600 males); density 2·5 per sq. km. In 2011, 38·6% of the population were urban.

The UN gives a projected population for 2015 of 2·48m.

The largest ethnic group is the Ovambo (about half the population), followed by the Kavango, Damara and Herero. Namibia is administratively divided into 13 regions. Area, population and chief towns in 2011:

Region	Area (in sq. km)	Population (provisional)	Chief town
Caprivi (Liambezi)	14,785	90,100	Katima Mulilo
Erongo	63,539	150,400	Swakopmund
Hardap	109,781	79,000	Mariental
Karas	161,514	76,000	Keetmanshoop
Khomas	36,964	340,900	Windhoek
Kunene	115,260	88,300	Opuwo
Ohangwena	10,706	245,100	Eenhana
Okavango	48,742	222,500	Rundu
Omaheke	84,981	70,800	Gobabis
Omusati	26,551	242,900	Outapi
Oshana	8,647	174,900	Oshakati
Oshikoto	38,685	181,600	Tsumeb
Otjozondjupa	105,460	142,400	Grootfontein

Towns with populations over 10,000 (2011 provisional): Windhoek, 322,500; Rundu, 61,900; Walvis Bay, 61,300; Swakopmund, 44,700; Oshakati, 35,600; Rehoboth, 28,800; Katima Mulilo, 28,200; Otjiwarongo, 28,000; Okahandja, 22,500; Ondangwa, 21,100; Ongwediva, 19,300; Helao Nafidi, 19,200; Tsumeb, 19,200; Gobabis, 19,000; Keetmanshoop, 18,900; Grootfontein, 16,400; Lüderitz, 12,500; Mariental, 12,300.

English is the official language. Afrikaans and German are also spoken.

SOCIAL STATISTICS

Estimates, 2008: births, 59,000; deaths, 18,000. Estimated birth rate in 2008 was 27·6 per 1,000 population; estimated death rate, 8·6. Expectation of life, 2007: males, 59·3 years; females, 61·2. Annual population growth rate, 2000–08, 1·9%; infant mortality,

2010, 29 per 1,000 live births. The fertility rate dropped from 5·5 births per woman in 1994 to 3·4 births per woman in 2008.

CLIMATE

The rainfall increases steadily from less than 50 mm in the west and southwest up to 600 mm in the Caprivi Strip. The main rainy season is from Jan. to March, with lesser showers from Sept. to Dec. Namibia is the driest African country south of the Sahara.

CONSTITUTION AND GOVERNMENT

On 9 Feb. 1990 with a unanimous vote the Constituent Assembly approved the Constitution which stipulated a multi-party republic, an independent judiciary and an executive *President* who may serve a maximum of two five-year terms. The constitution became effective on 12 March 1990 and was amended in 1999 to allow President Sam Nujoma to stand for a third term in office. The bicameral legislature consists of a 78-seat *National Assembly*, 72 members of which are elected for five-year terms by proportional representation and up to six appointed by the president by virtue of position or special expertise, and a 26-seat *National Council* consisting of two members from each Regional Council elected for six-year terms.

National Anthem

'Namibia, land of the brave'; words and tune by Axali Doeseb.

RECENT ELECTIONS

Presidential and parliamentary elections were held on 27–28 Nov. 2009. Hifikepunye Pohamba (South West Africa People's Organization/SWAPO) was elected president with 75·3% of votes cast followed by Hidipo Hamutenya (Rally for Democracy and Progress/RDP) with 10·9%, Katuutire Kaura (Democratic Turnhalle Alliance/DTA) with 3·0%, Kuaima Riruako (National Unity Democratic Organization/NUDO) with 2·9% and Chief Justus Garoëb (United Democratic Front/UDF) with 2·4%. There were seven other candidates. Turnout was an estimated 75%. In the parliamentary elections SWAPO won 54 of the available 72 seats with 74·3% of the vote; the RDP, 8 with 11·2%; DTA, 2 with 3·1%; NUDO, 2 with 3·0%; UDF, 2 with 2·4%; All People's Party, 1 with 1·3%; Republican Party, 1 with 0·8%; Congress of Democrats, 1 with 0·7%; South West Africa National Union, 1 with 0·6%.

CURRENT GOVERNMENT

President: Hifikepunye Pohamba; b. 1935 (SWAPO; sworn in 21 March 2005 and re-elected in Nov. 2009).
    In March 2013 the government comprised:
    Prime Minister: Hage Geingob; b. 1941 (SWAPO; sworn in 4 Dec. 2012, having previously held office from March 1990 to Aug. 2002).
    Deputy Prime Minister: Marco Hausiku.
    Minister of Home Affairs: Pendukeni Iivula-Iithana. *Presidential Affairs and Attorney General:* Albert Kawana. *Foreign Affairs:* Netumbo Nandi-Ndaitwah. *Defence:* Nahas Angula. *Finance:* Saara Kuugongelwa-Amadhila. *Education:* David Namwandi. *Health:* Richard Kamwi. *Mines and Energy:* Isak Katali. *Justice:* Untoni Nujoma. *Regional and Local Government, Housing and Rural Development:* Charles Namoloh. *Agriculture, Water and Forests:* John Mutorwa. *Trade and Industry:* Calle Schlettwein. *Environment and Tourism:* Uahekua Herunga. *Works and Transport:* Erkki Nghimtina. *Lands and Rehabilitation:* Alpheus Naruseb. *Fisheries and Marine Resources:* Bernard Esau. *Safety and Security:* Immanuel Ngatjizeko. *Youth, National Service, Sport and Culture:* Jerry Ekandjo. *Gender Equality and Child Welfare:* Rosalia Nghidinwa. *Labour and Social Welfare:* Doreen Sioka. *Information and Information Technology:* Joel Kaapanda. *Veterans' Affairs:* Nickey Iyambo.

Office of the Prime Minister: http://www.opm.gov.na

CURRENT LEADERS

Hifikepunye Pohamba

Position
President

Introduction
Lucas Hifikepunye Pohamba, representing the ruling South West Africa People's Organization (SWAPO), won a landslide victory at presidential elections in Nov. 2004 and took office in March 2005. He succeeded Namibia's 'founding father' and former president, Sam Nujoma, and has continued with the same broad political programme. He was re-elected in Nov. 2009.

Early Life
Pohamba was born on 18 Aug. 1935 at Okanghudi in South West Africa (modern Namibia) and educated at the Holy Cross Mission School at Onamunama. He worked in the Tsumeb copper mines and joined SWAPO in April 1959. He joined Nujoma in exile in Dar es Salaam (Tanzania) and became a leading figure in SWAPO, representing it in Zambia and Algeria and raising funds. In 1969 he was appointed to SWAPO's central committee and in 1975 became secretary for finance and administration. From 1979 until the late 1980s he was based in Luanda, Angola.
    Following Nujoma's victory in the country's first presidential elections on 7 Nov. 1989, Pohamba was appointed as minister of home affairs. In 1995 he became minister for fisheries and marine resources until 2001, when he took responsibility for lands, resettlement and rehabilitation. As such, he pushed ahead with Namibia's controversial 'land reform' scheme, involving the compulsory purchase of land owned by white farmers for distribution to black citizens.

Career in Office
Since his inauguration on 21 March 2005 Pohamba has pursued established policies, including development of education, the rural water supply and the infrastructure network. He has also continued the controversial compulsory land purchases scheme. In elections in Nov. 2009, Pohamba was returned to the presidency with about 75% of the vote and SWAPO retained its majority of parliamentary seats with a similar vote share. A legal challenge launched in 2010 by opposition parties against the election results was dismissed by the High Court in Feb. 2011 for lack of evidence.
    In Dec. 2012 Pohamba carried out a cabinet reshuffle and appointed SWAPO vice-president and former trade minister, Hage Geingob, as prime minister.

DEFENCE

In 2008 defence expenditure totalled US$287m. (US$137 per capita), representing 3·3% of GDP.

Army

Personnel (2007), 9,000. There is also a 6,000-strong paramilitary police force.

Navy

A force of around 200 (2007) is based at Walvis Bay.

Air Force

The Army has a small air wing that operates two combat-capable aircraft.

ECONOMY

Agriculture accounted for 10·9% of GDP in 2006, industry 30·6% and services 58·5%.
    The Namibian economy is heavily dependent on mining and fisheries.

Overview

A small open economy closely linked to South Africa, Namibia is classified as an upper middle-income country. The economy is stable and has been characterized by high rates of investment and real GDP growth (averaging about 5·1% during the period 2003–07). Exports account for almost 60% of GDP, with diamonds and other minerals accounting for close to 55% of total exports.

Namibia's small market size, openness and heavy dependence on international trade makes it vulnerable to external shocks such as changes in the terms of trade, external demand and climatic variations. The global economic crisis in 2008 impacted through lower demand for commodity exports (especially diamonds).

Poverty is widespread, running at 28%, and unemployment in 2008 was 36·7%. The inequality rate is the highest in the world (Namibia is ranked last in the world in terms of the Gini coefficient, a measure of income equality) and HIV remains a serious threat to development.

However, the Third National Development Plan aims to address poverty, inequality and unemployment. The government is seeking to diversify the economy away from minerals and into tourism and financial services. The discovery of a water source on the border with Angola could provide the north of the country—historically blighted by a shortage of water—with supplies for 400 years at current rates of consumption. Namibia is a member of the Southern African Customs Union.

Currency

The unit of currency is the *Namibia dollar* (NAD) of 100 *cents*, introduced on 14 Sept. 1993 and pegged to the South African rand. The rand is also legal tender at parity. Inflation was 4·5% in 2010 and 5·8% in 2011. In May 2005 foreign exchange reserves were US$333m. Total money supply in Dec. 2003 was N$7,851m.

Budget

The financial year runs from 1 April. In 2008–09 revenues were N$21,973m. and expenditures N$22,469m. Tax revenue constituted 91·9% of revenues in 2008–09; current expenditure accounted for 76·9% of expenditures.

Performance

The economy contracted by 0·4% in 2009 but there was growth of 6·6% in 2010 and 4·9% in 2011; total GDP in 2011 was US$12·3bn.

Banking and Finance

The Bank of Namibia is the central bank. Its *Governor* is Ipumbu Wendelinus Shiimi. Assets in June 2010 were N$12,717·4m. Total assets of other depository corporations (First National Bank of Namibia, Standard Bank of Namibia, Nedbank Namibia, Bank Windhoek, Agribank of Namibia, National Housing Enterprise and the Namibia Post Office Savings Bank) were N$50,990·5m. in June 2010.

A stock exchange (NSE) is in operation in Windhoek.

ENERGY AND NATURAL RESOURCES

Environment

Carbon dioxide emissions from the consumption and flaring of fossil fuels were the equivalent of 1·5 tonnes per capita in 2008.

Electricity

In 2007 electricity production was 1·7bn. kWh. Namibia also imports electricity to meet demand (2·1bn. kWh in 2007, mainly from South Africa). Consumption per capita in 2007 was 1,824 kWh.

Oil and Gas

Natural gas reserves in 2007 totalled 62bn. cu. metres.

Minerals

There are diamond deposits both inshore and off the coast, with production equally divided between the two. Some 1·5bn. carats of diamonds are believed to be lying in waters off Namibia's Atlantic coast. Namibia produced 2·0m. carats in 2004, exclusively of gem quality. Output in 2005 (in tonnes): salt, 573,248; zinc (metal content), 69,368; lead (metal content), 14,320; copper (metal content), 10,157; uranium (metal content), 2,855; silver (metal content), 34; gold (metal content), 2,649 kg.

Agriculture

Namibia is essentially a stock-raising country, the scarcity of water and poor rainfall rendering crop-farming, except in the northern and northeastern parts, almost impossible. There were an estimated 800,000 ha. of arable land in 2007 and 5,000 ha. of permanent crops. There were 39 tractors per 10,000 ha. of arable land in 2006. Generally speaking, the southern half is suited for the raising of small stock, while the central and northern parts are more suited for cattle. Guano is harvested from the coast, converted into fertilizer in South Africa and most of it exported to Europe. In 2007 the agricultural population was an estimated 905,000, of which some 252,000 were economically active.

Principal crops (2003, in tonnes): millet, 51,000; maize, 33,000; wheat, 8,000; grapes, 6,000; seed cotton, 5,000; sorghum, 5,000. Livestock (2003 estimates): 2·51m. cattle, 2·37m. sheep, 1·78m. goats, 3m. chickens. In 2003 an estimated 105,000 tonnes of milk and 83,000 tonnes of meat were produced.

Forestry

Forests covered 7·29m. ha. in 2010, or 9% of the land area.

Fisheries

Namibia has one of the most productive fishing grounds in the world. The catch in 2009 was 370,348 tonnes, of which more than 99% came from marine waters.

INDUSTRY

Of the estimated total of 400 undertakings, the most important branches are food production (accounting for 29·3% of total output), metals (12·7%) and wooden products (7%). The supply of specialized equipment to the mining industry, the assembly of goods from predominantly imported materials and the manufacture of metal products and construction material play an important part. Small industries (including home industries, textile mills, leather and steel goods) have expanded. Products manufactured locally include chocolates, beer, cement, leather shoes, delicatessen meats and game meat products.

Labour

Of 431,800 people in employment in 2000, 126,500 were engaged in agriculture, hunting and forestry; 46,300 in community, social and personal service activities; 39,300 in real estate, renting and business activities; and 38,900 in wholesale and retail trade/repair of motor vehicles, motorcycles and personal and household goods. In 2000 the unemployment rate was 33·8%.

INTERNATIONAL TRADE

Total foreign debt was US$2·0bn. in Dec. 2009. Export Processing Zones were established in 1995 to grant companies with EPZ status some tax exemptions and other incentives. The Offshore Development Company (ODC) is the flagship of the Export Processing Zone regime. The EPZ regime does not restrict; any investor (local or foreign) enjoys the same or equal advantages in engaging themselves in any choice of business (allowed by law).

Imports and Exports

In 2007 imports (c.i.f.) were valued at US$4,026·0m. and exports (f.o.b.) at US$4,040·3m. Exports in 2007 (in US$1m.) included diamonds (705), zinc (624), fish (432), uranium ores and concentrates (350), machinery and transport equipment (280), meat products (153). The largest import supplier in 2006 was South Africa with 82·4%; largest export markets: UK, 25·6%; South Africa, 24·6%.

COMMUNICATIONS

Roads
In 2002 the total road network covered 42,237 km, including 4,550 km of national roads. In 2008 there were 107,800 passenger cars in use and 117,400 lorries and vans. There were 368 deaths as a result of road accidents in 2007.

Rail
The Namibia system connects with the main system of the South African railways at Ariamsvlei. The total length of the line inside Namibia was 2,628 km of 1,065 mm gauge in 2005. In 2002–03 railways carried 150,000 passengers and 1·9m. tonnes of freight.

Civil Aviation
The national carrier is the state-owned Air Namibia. In 2001 the major airport, Windhoek's Hosea Kutako International, handled 379,000 passengers (363,000 on international flights). Eros is used mainly for domestic flights. In 2003 scheduled airline traffic of Namibian-based carriers flew 11m. km, carrying 266,000 passengers (222,000 on international flights).

Shipping
Walvis Bay, the busiest port, handled 4,960,000 tonnes of cargo in 2007–08. There is a harbour at Lüderitz which handles mainly fishing vessels. Merchant shipping totalled 3,000 GT in Jan. 2009.

Telecommunications
Telecom Namibia is the responsible corporation. In 2008 there were 140,000 main (fixed) telephone lines and 1,052,000 mobile phone subscribers (49·4 per 100 persons). There were 113,500 internet users in 2008. In June 2012 there were 172,000 Facebook users.

SOCIAL INSTITUTIONS

Justice
There is a Supreme Court, a High Court and a number of magistrates' and lower courts. An Ombudsman is appointed. Judges are appointed by the president on the recommendation of the Judicial Service Commission.

The population in penal institutions at 31 Dec. 2007 was 4,064 (194 per 100,000 of national population).

Education
Literacy was an estimated 88·5% in 2009 (male, 88·9%; female, 88·1%). Primary education is free and compulsory. In 2007 there were 409,508 pupils at primary schools, 158,162 at secondary schools and (2006) 13,185 students at institutions of higher education.

In 2010 public expenditure on education came to 8·1% of GDP.

Health
There were 45 hospitals (11 private) and 296 health centres and clinics in 2005. In 2006 there were 33 hospital beds per 10,000 population. There were 598 physicians, 113 dentistry personnel and 6,145 nursing and midwifery personnel in 2004.

RELIGION
About 75% of the population is Christian (mainly Protestant).

CULTURE

World Heritage Sites
Namibia has one site on the UNESCO World Heritage List: a large collection of rock engravings at Twyfelfontein (added to the list in 2007).

Press
There were four daily newspapers in 2006 with a combined circulation of 28,000.

Tourism
In 2005 there were 778,000 visitors who spent US$348m.

DIPLOMATIC REPRESENTATIVES

Of Namibia in the United Kingdom (6 Chandos St., London, W1G 9LU)
High Commissioner: George Mbanga Liswaniso.

Of the United Kingdom in Namibia (116 Robert Mugabe Ave., 9000 Windhoek)
High Commissioner: Marianne Young.

Of Namibia in the USA (1605 New Hampshire Ave., NW, Washington, D.C., 20009)
Ambassador: Martin Andjaba.

Of the USA in Namibia (14 Lossen St., Private Bag 12029, Windhoek)
Ambassador: Wanda Nesbitt.

Of Namibia to the United Nations
Ambassador: Wilfried Inotira Emvula.

Of Namibia to the European Union
Ambassador: Hanno Burkhard Rumpf.

FURTHER READING
Kaela, L. C. W., *The Question of Namibia.* 1996
Melber, Henning, *Re-examining Liberation in Namibia: Political Cultures Since Independence.* 2003
Sparks, D. L. and Green, D., *Namibia: the Nation after Independence.* 1992

National Statistical Office: National Planning Commission, Government Office Park, Block D2, Luther Street, Windhoek.
Website: http://www.npc.gov.na/cbs/index.htm

NAURU

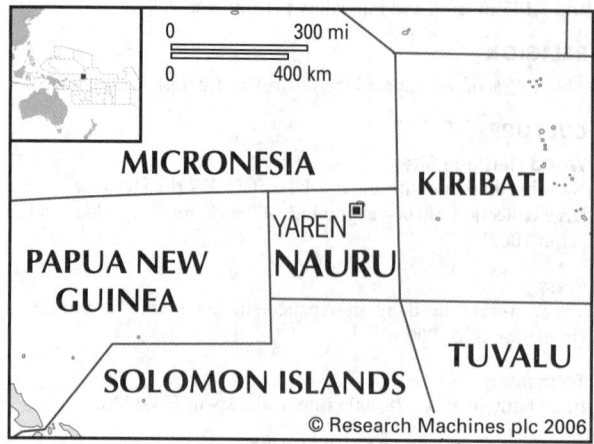

Ripublik Naoero
(Republic of Nauru)
Population, 2011: 10,000
GNI per capita, 2010: US$6,084
Internet domain extension: .nr

KEY HISTORICAL EVENTS

Nauru was originally settled by Melanesians and Polynesians. Tradition holds that among the earliest settlers were castaways from another island, probably Kiribati. The name 'Nauru' is a European corruption of 'A-nao-ero', which means 'I am going to the beach to lay my bones'. The island has had little contact with its neighbours, enabling its distinctive language to survive. By the 18th century the society was organized into 12 matrilineal tribes, each headed by a different chief.

The island was discovered by a British captain, John Fearn, in 1798 but was avoided by most ships in subsequent decades because of the region's notoriety for piracy. In the 1830s European whaling ships began using Nauru as a supply port and European settlers arrived. Though small in number, they had a profound impact by introducing alcohol, firearms and disease. An escalation in tribal conflict resulted in a war from 1878–88, killing 500 people or one third of the population. Germany annexed the island in 1888 to protect its trading interests and in the early 1900s agreed a deal to allow the British Pacific Phosphate Company to mine newly discovered phosphates.

Nauru was surrendered to Australian forces in 1914. In 1920 its administration was formally passed to the UK under a League of Nations mandate but in practice Australia continued to run the island. Australia, Britain and New Zealand set up and jointly ran the British Phosphate Commission, which controlled the phosphate mining industry. During the Second World War Japanese forces occupied from 1942–45, deporting 1,200 Nauruans to Truk (now Chuuk, in present-day Micronesia) as forced labour. Only 737 of the deportees returned. In 1947 Nauru became a UN trust, with Australia, New Zealand and the UK as trustees.

On 31 Jan. 1968 the country gained independence. The government of President Hammer DeRoburt took over the phosphate industry, continuing to run it as a communal trust. Phosphate prices rose and the country enjoyed a boom throughout the 1970s, achieving one of the world's highest rates of GDP per capita. However, mismanagement of the revenues, poor investment decisions and a lack of political accountability led to problems when prices fell in 1988. In 1993 Australia and the

UK paid US$73m. compensation for environmental damage during mining. Nauru developed an offshore banking industry in the 1990s but was accused of money-laundering by the international community. In the mid-1990s the economic crisis deepened with the collapse of the Bank of Nauru and in 2000 the OECD's financial watchdog (Financial Action Task Force on Money Laundering, or FATF) blacklisted the country. In 2001 Nauru agreed to hold detained asylum seekers on behalf of Australia and has subsequently relied heavily on Australian aid.

The 21st century has seen much political turmoil. During 2003 there were six changes of president and in Sept. 2004 President Ludwig Scotty declared a state of emergency. Following regulatory tightening of the offshore banking industry, Nauru was removed from the FATF blacklist in 2005. The island continues to face severe challenges, including a health crisis arising from the world's highest obesity rate.

TERRITORY AND POPULATION

Nauru is a coral island surrounded by a reef situated 0° 32' S. lat. and 166° 56' E. long. Area, 21·2 sq. km (8·2 sq. miles). At the 2011 census the population (provisional) totalled 10,084 (5,105 males). Population density, 476 per sq. km. In 2002, 75% of the population (7,572 out of 10,065) were indigenous Nauruans. The *de facto* capital is Yaren.

Nauruan is the official language, although English is widely used for government purposes.

SOCIAL STATISTICS

2011 births, 370; deaths, 75. Infant deaths (2007), 12. Annual population growth rate, 1998–2008, 0·1%; fertility rate, 2008, 2·9 births per woman.

CLIMATE

A tropical climate, tempered by sea breezes, but with a high and irregular rainfall, averaging 82" (2,060 mm). Average temperature, Jan. 81°F (27·2°C), July 82°F (27·8°C). Annual rainfall 75" (1,862 mm).

CONSTITUTION AND GOVERNMENT

A Legislative Council was inaugurated on 31 Jan. 1966. The constitution was promulgated on 29 Jan. 1968 and was amended on 17 May 1968. An 18-member Parliament is elected on a three-yearly basis.

National Anthem

'Nauru bwiema, ngabena ma auwe' ('Nauru our homeland, the country we love'); words by M. Hendrie, tune by L. H. Hicks.

RECENT ELECTIONS

At the parliamentary elections of 19 June 2010, President Marcus Stephen's supporters won 9 of the 18 seats. On 15 Nov. 2011 Sprent Dabwido was elected president by parliament by 9 votes to 8.

CURRENT GOVERNMENT

In April 2013 the government comprised:
President, Cabinet Chairman and Minister for Public Service, Police and Emergency Services, Home Affairs and Climate Change: Sprent Dabwido; b. 1972 (sworn in 15 Nov. 2011).

Minister for Commerce, Industry and Environment, Transport and Telecommunications, Nauru Utilities Corporation and Nauru Air Corporation: Ridell Akua. *Education:* Aloysius Amwano. *Finance and Sustainable Development, Justice, RONPHOS, Nauru Rehabilitation Corporation, Sports, Fisheries, and Foreign Affairs and Trade:* Roland Kun. *Health:* Shadlog Bernicke.

CURRENT LEADERS

Sprent Dabwido

Position
President

Introduction
Sprent Dabwido became Nauru's third president in less than a week on 15 Nov. 2011, following the resignation of Marcus Stephen amid corruption allegations and the removal of Freddy Pitcher by a parliamentary vote of no confidence. In accordance with the constitution Dabwido became both head of government and head of state.

Early Life
Born on 16 Sept. 1972 Sprent Dabwido was elected to parliament at the 2004 general election when he won the seat of Meneng. He was re-elected in 2007 and 2008, and in 2009 he was appointed minister of telecommunications in Stephen's government. In this role Dabwido oversaw the introduction of a mobile telephone system to Nauru and retained his Meneng seat at the June 2010 general election.

On 15 Nov. 2011 Dabwido cut his ties with the ruling faction in parliament and joined the opposition in support of a motion of no confidence that brought down Pitcher. Dabwido was subsequently elected president by 9 votes to 8 in a parliamentary vote.

Career in Office
Dabwido represented the Pacific Small Island Developing States at the UN Climate Change Conference in Durban in Dec. 2011. He highlighted the problem of rising sea levels for Pacific island nations and called for a legally binding protocol to complement the existing Kyoto Protocol and Bali Action Plan.

In June 2012 he appointed a new cabinet, including opposition members, following an impasse over proposed constitutional reforms.

ECONOMY

Overview
The exhaustion of phosphate deposits has seen GDP per capita recede since peaking in the 1970s and 1980s. The economy is now dependent on external aid and imports owing to its narrow resource base. Attempts to diversify the economy into offshore banking during the 1990s proved unsuccessful and the country became a haven for money laundering, resulting in its blacklisting by international bodies (including the Financial Action Task Force) until 2005. In 2001 Nauru signed an agreement with Australia to house asylum seekers on the island, generating millions of dollars in revenue. However, this agreement ended in 2008.

New phosphate contracts came into force in 2008 resulting in the resumption of growth-driving exports. The economy grew by 4% in the year to 30 June 2011 as phosphate exports increased by 20%. Improved phosphate-loading facilities are expected to further increase exports. The long-term economic outlook depends on the conversion of secondary reserves into alternative sources of income. Further challenges include reducing public debt and promoting sound public and private investment.

Currency
The Australian dollar is in use.

Budget
The fiscal year is 1 July–30 June. Revenues in 2005–06 were $A27·0m. and expenditures $A26·4m.

Performance
Real GDP growth was −27·3% in 2007 as a result of low construction spending on phosphate facilities and the suspension of phosphate mining owing to storm damage to port infrastructure, combined with a contracted public sector.

Banking and Finance
The Bank of Nauru is a state bank and there is a commercial bank, Hampshire Bank and Trust Inc.

Nauru was one of three countries named in a report in June 2005 as failing to co-operate in the fight against international money laundering. In Oct. 2005 Nauru was delisted and its formal monitoring was ended a year later. The Financial Action Task Force on Money Laundering was set up by the G7 group of major industrialized nations.

ENERGY AND NATURAL RESOURCES

Environment
Carbon dioxide emissions from the consumption and flaring of fossil fuels were the equivalent of 14·7 tonnes per capita in 2008.

Electricity
Installed capacity in 2007 was an estimated 11,000 kW; production was estimated at 35m. kWh in 2007.

Minerals
A central plateau contained high-grade phosphate deposits. The interests in the phosphate deposits were purchased in 1919 from the Pacific Phosphate Company by the UK, Australia and New Zealand. In 1967 the British Phosphate Corporation agreed to hand over the phosphate industry to Nauru for approximately $20m. over three years. Nauru took over the industry in July 1969, and the profits from the mining meant that Nauru had, for a brief period, one of the highest rates of GDP per capita in the world. However, production declined (from 1·67m. tonnes in 1985–86 to 162,000 tonnes in 2001–02) and the primary reserves were exhausted in 2003. Mining of a deeper layer of secondary phosphate began in 2006, and it is hoped that this development might resuscitate Nauru's ailing economy. Production was an estimated 84,000 tonnes in 2006. In May 1989 Nauru filed a claim against Australia for environmental damage caused by the mining. In Aug. 1993 Australia agreed to pay compensation of $A73m. In March 1994 New Zealand and the UK each agreed to pay compensation of $A12m.

Agriculture
In 2007 about 1,000 people were economically active in agriculture. In 2009 the crop of coconuts was an estimated 2,200 tonnes. Livestock (2009 estimates): pigs, 3,000.

Fisheries
The catch in 2010 was an estimated 200 tonnes.

INTERNATIONAL TRADE

Imports and Exports
Imports are food, building construction materials, machinery for the phosphate industry and medical supplies. Phosphates—formed from fossilized bird droppings and used for fertilizer—have traditionally accounted for virtually all of Nauru's exports, but the reserves are now largely depleted.

Imports, 2005, A$33·7m.; exports, A$5·0m. The leading import sources in 2005 were Korea, Australia and the USA; the main export markets in 2005 were Korea, Canada and the UK.

COMMUNICATIONS

Roads
In 2002 there were 30 km of roads, 24 km of which were paved.

Civil Aviation
There is an airfield on the island capable of accepting medium size jet aircraft. The national carrier, Our Airline (formerly Air Nauru), is a wholly-owned government subsidiary. It has one aircraft. In 2006 it flew to Brisbane, Honiara, Tarawa and Majuro.

In 2003 Our Airline flew 3m. km, carrying 156,000 passengers (all on international flights).

Shipping
Deep offshore moorings can accommodate medium-size vessels. Shipping coming to the island consists of vessels under charter to the phosphate industry or general purpose vessels bringing cargo by way of imports.

Telecommunications
There were 1,800 main telephone lines in operation in 2008.

SOCIAL INSTITUTIONS

Justice
The highest Court is the Supreme Court of Nauru. It is the Superior Court of record and has the jurisdiction to deal with constitutional matters in addition to its other jurisdiction. There is also a District Court which is presided over by the Resident Magistrate who is also the Chairman of the Family Court and the Registrar of Supreme Court. The laws applicable in Nauru are its own Acts of Parliament. A large number of British statutes and much common law has been adopted insofar as is compatible with Nauruan custom.

Education
Attendance at school is compulsory between the age of six and 16. In 2007–08 there were 1,000 children in pre-primary schools with 40 teachers, 1,000 pupils in primary schools with 100 teachers and 1,000 pupils in secondary education with (2006–07) 30 teachers. There is a technical school and also a mission school. Scholarships are available for Nauruan children to receive secondary and higher education and vocational training in Australia and New Zealand.

In 2007 public expenditure on education came to an estimated 7·5% of total government spending.

Health
In 2004 there were ten physicians and 63 nursing and midwifery personnel.

Nauru has the highest percentage of overweight people of any country, with 94% overweight or obese according to a 2006 World Health Organization survey.

RELIGION
The population is mainly Roman Catholic or Protestant.

DIPLOMATIC REPRESENTATIVES
Of Nauru in the United Kingdom
Honorary Consul: Martin W. L. Weston (Romshed Courtyard, Underriver, Nr Sevenoaks, Kent, TN15 0SD).

Of the United Kingdom in Nauru
Acting High Commissioner: Martin Fidler (resides in Suva, Fiji Islands).

Of Nauru in the USA and to the United Nations (800 Second Ave., New York, NY 10017)
Ambassador: Marlene Inemwin Moses.

Of the USA in Nauru
Ambassador: Frankie A. Reed (resides in Suva, Fiji Islands).

FURTHER READING
McDaniel, Carl N., *Paradise for Sale: Back to Sustainability.* 2000

National Statistical Office: Nauru Bureau of Statistics, Ministry of Finance, Government Offices, Yaren District.
Website: http://www.spc.int/prism/country/nr/stats

NEPAL

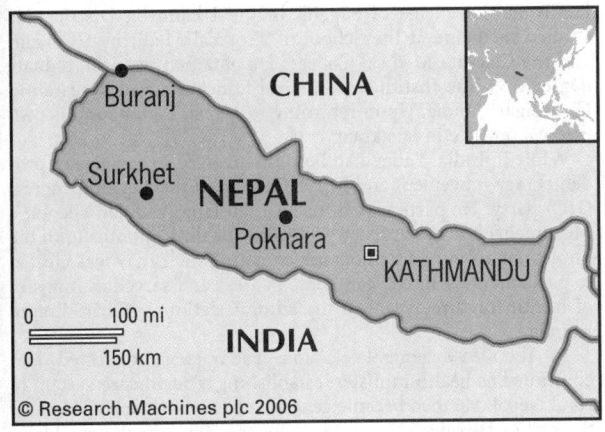

© Research Machines plc 2006

Sanghiya Loktantrik Ganatantra Nepal
(Federal Democratic Republic of Nepal)

Capital: Kathmandu
Population projection, 2015: 32·58m.
GNI per capita, 2011: (PPP$) 1,160
HDI/world rank: 0·458/157
Internet domain extension: .np

KEY HISTORICAL EVENTS

Nepal is an independent Himalayan republic located between India and the Tibetan region of China. From the 8th to the 11th centuries many Buddhists fled to Nepal from India, which had been invaded by Muslims. In the 18th century Nepal was a collection of small principalities (many of Rajput origin) and the three kingdoms of the Malla dynasty: Kathmandu, Patan and Bhadgaon. In central Nepal lay the principality of Gurkha (or Gorkha); its ruler after 1742 was Prithvi Narayan Shah, who conquered the small neighbouring states. Fearing his ambitions, in 1767 the Mallas requested armed support from the British East India Company. In 1769 these forces were withdrawn and Gurkha was then able to conquer the Malla kingdoms and unite Nepal as one state with its capital at Kathmandu. In 1846 the Rana family became the effective rulers of Nepal, establishing the office of prime minister as hereditary. In 1860 Nepal reached agreement with the British in India whereby Nepali independence was preserved and the recruitment of Gurkhas to the British army was sanctioned.

In 1950 the Shah royal family allied itself with Nepalis abroad to end the power of the Ranas. The last Rana prime minister resigned in Nov. 1951, the king having proclaimed a constitutional monarchy in Feb. 1951. A new constitution, approved in 1959, led to confrontation between the king and his ministers; it was replaced by one less liberal in 1962. In Nov. 1990 the king relinquished his absolute power. The Maoists abandoned parliament in 1996 and launched a 'people's war' in the aim of turning the kingdom into a republic. This resulted in over 13,000 deaths over a ten-year period.

In June 2001 the king and queen, along with six other members of the royal family, were shot dead by their son and heir to the throne, Crown Prince Dipendra, allegedly following a dispute over his choice of bride. Prince Dipendra then shot himself. The former monarch's younger brother, Gyanendra, was crowned king. In Nov. 2001 King Gyanendra declared a state of emergency and ordered troops to contain a fresh outbreak of Maoist violence.

The government lifted the state of emergency in Aug. 2002. In Jan. 2003 the government and Maoist rebels reached a ceasefire agreement, seen as a first step towards bringing to an end the rebels' seven-year insurgency. In Feb. 2005 King Gyanendra dismissed his government and once more declared a state of emergency, taking control of the country and suspending democracy for three years. He lifted the state of emergency on 29 April 2005. In April 2006 he agreed to a return to parliamentary democracy after more than two weeks of unrest. On 21 Nov. 2006 a peace agreement was signed between the government and the country's Maoist rebels, bringing a formal end to the decade-long insurgency. In Dec. 2007 an agreement was made to abolish the monarchy and establish Nepal as a republic. On 28 May 2008 the newly-elected Constituent Assembly officially inaugurated the Federal Democratic Republic of Nepal and began the process of creating a new constitution.

TERRITORY AND POPULATION

Nepal is bounded in the north by China (Tibet) and the east, south and west by India. Area 147,181 sq. km; 2011 census population, 26,494,504 (13,645,463 females); density 180·0 per sq. km. In 2011, 19·2% of the population were urban.

The UN gives a projected population for 2015 of 32·58m.

The country is divided into five developmental regions and 75 administrative districts. Area, population and administrative centres are:

Region	Sq. km	Population (2011)	Administrative centre
Central Region	27,410	9,656,985	Kathmandu
East Region	28,456	5,811,555	Dhankuta
West Region	29,398	4,926,765	Pokhara
Mid-West Region	42,378	3,546,682	Surkhet
Far West Region	19,539	2,552,517	Dipayal

Capital, Kathmandu; population (2011) 1,003,285. Other towns include (2011 census population): Pokhara, 264,991; Lalitpur, 226,728; Biratnagar, 204,949; Bharatpur, 147,777.

The indigenous people are of Tibetan origin with a considerable Hindu admixture. The Gurkha clan became predominant in 1559 and has given its name to men from all parts of Nepal. There are 18 ethnic groups, the largest being: Newars, Indians, Tibetans, Gurungs, Mogars, Tamangs, Bhotias, Rais, Limbus and Sherpas. The official language is Nepalese but there are 20 new languages divided into numerous dialects.

SOCIAL STATISTICS

2008 estimates: births, 731,000; deaths, 185,000. Estimated rates per 1,000 population, 2008: births, 25·4; deaths, 6·4. Annual population growth rate, 2000–08, 2·1%. Expectation of life was 65·6 years for males and 66·9 years for females in 2007. Infant mortality, 2010, 41 per 1,000 live births; fertility rate, 2008, 2·9 births per woman.

CLIMATE

Varies from cool summers and severe winters in the north to sub-tropical summers and mild winters in the south. The rainfall is high, with maximum amounts from June to Sept., but conditions are very dry from Nov. to Jan. Kathmandu, Jan. 10°C, July, 25°C. Average annual rainfall, 1,424 mm.

CONSTITUTION AND GOVERNMENT

Following years of political turbulence an interim constitution was approved in Dec. 2006, effectively removing King Gyanendra as the head of the state.

On 23 Dec. 2007 the interim government declared the establishment of the Federal Democratic Republic of Nepal, with the abolition of the monarchy approved by parliament five days later. This change entered into force on 28 May 2008 at the first meeting of a 601-member *Constituent Assembly* (with 240 seats filled on a first-past-the-post system, 335 filled through proportional representation and 26 nominated by the cabinet). The Constituent Assembly was charged with drafting a new constitution but failed to do so even though its deadline was extended several times. Elections to a new Constituent Assembly are tentatively set to take place in the second half of 2013.

National Anthem

'Sayaun thunga phoolka hami eutai mala Nepali' ('From hundreds of flowers, we are one garland Nepali'); words by Byakul Maila, tune by Ambar Gurung.

RECENT ELECTIONS

In elections to the Constituent Assembly held on 10 April 2008 the Communist Party of Nepal (Maoist) (CPN-M) won 220 of the 575 elected seats (26 seats are reserved for nominated members). The Nepali Congress won 110 seats, the Communist Party of Nepal (Unified Marxist-Leninist) 103, the Madhesi Jana Adhikar Forum/Madhesi People's Rights Forum 52 and the Tarai-Madhesh Loktantrik Party 20. The remaining 70 seats were shared among fringe parties and independents.

Following the resignation of Jhalanath Khanal, Baburam Bhattarai of the Unified Communist Party of Nepal (Maoist) was elected prime minister by parliament on 28 Aug. 2011 with 340 votes. His only rival, Ram Chandra Poudel of the Nepali Congress, received 235 votes.

CURRENT GOVERNMENT

President: Ram Baran Yadav; b. 1948 (Nepali Congress; sworn in 23 July 2008).

Vice President: Parmanand Jha.

In March 2013 the interim government comprised:

Prime Minister, and Minister of Defence, Cooperatives and Poverty Alleviation: Khil Raj Regmi; b. 1949 (ind.; since 14 March 2013).

Minister of Agriculture Development, and Forest and Soil Conservation: Tek Bahadur Thapa. *Energy, Science, Technology and Environment, and Irrigation:* Uma Kanta Jha. *Federal Affairs and Local Development, and Health and Population:* Biddhyadhar Mallik. *Finance, Industry, and Commerce and Supplies:* Shankar Koirala. *General Administration, Information and Communications, and Education:* Madhav Paudel. *Home Affairs and Foreign Affairs:* Madhav Prasad Ghimire. *Law, Justice, Constituent Assembly and Parliamentary Affairs, and Labour and Employment:* Hari Prasad Neupane. *Physical Infrastructure and Transport, and Urban Development:* Chhabi Raj Panta. *Women, Children and Social Welfare, and Land Reform and Management:* Riddhi Baba Pradhan. *Youth and Sports, Peace and Reconstruction, and Culture, Tourism and Civil Aviation:* Ram Kumar Shrestha.

Office of the Prime Minister and Council of Ministers:
http://www.opmcm.gov.np

CURRENT LEADERS

Ram Baran Yadav

Position
President

Introduction
Ram Baran Yadav became president in July 2008 after winning a comfortable majority in the first presidential elections since the country became a federal democratic republic. Yadav's role is largely ceremonial, including the performance of traditional Hindu duties previously carried out by the now abolished monarchy.

Early Life
Ram Baran Yadav was born on 4 Feb. 1948 to an ethnic Madhesi family in the village of Sapahi, in the Dhanusha District. He studied medicine at the School of Tropical Medicine (Kolkata) and the Calcutta Medical College. He obtained his postgraduate degree from the Institute of Medical Education and Research in Chandigarh, India. Upon returning to Nepal, Yadav ran his own medical practice in Janakpur.

While in India, Yadav had become involved in the Nepali pro-democracy movement and in 1980 he joined the Nepali Congress (NC) party. He participated in the 1990 Jana Andolan (People's Movement) but was arrested and jailed for three months until the implementation of a democratic constitution. Yadav was elected to parliament as an NC candidate in 1991 and served as minister of health for three years in the administration of Girija Prasad Koirala.

At the 1999 general election Yadav was re-elected and continued as health minister, establishing a healthcare system in rural Nepal. He then became general secretary of the NC. He won the seat of Dhanusa-5 at elections for the constituent assembly in April 2008.

Yadav contested the presidential election of July 2008 on an NC ticket but fell four votes short of the 298 needed to win a simple majority. In a second round of voting he won 308 of 590 votes cast. Yadav's victory came as an upset to the Communist Party of Nepal (Maoist) (CPN-M) who were seeking to form the government after their win in the April 2008 Constituent Assembly election. Yadav was sworn into office on 23 July 2008.

Career in Office
Yadav prioritized the drafting of a new democratic constitution in a bid to end Nepal's long track record of political violence, and has been keen to maintain friendly ties with both India and China. However, he faced opposition from the CPN-M whose leader, Pushpa Kamal Dahal (Prachanda), led a coalition government from Aug. 2008 until May 2009 when he resigned in protest at Yadav's blocking of his controversial attempt to dismiss the country's army head. Madhav Kumar Nepal of the Communist Party (Unified Marxist-Leninist) was sworn in as the prime minister of a new coalition, excluding the CPN-M, later in the month. The CPN-M demanded a return to power, claiming that Yadav had acted unconstitutionally, and kept up prolonged political pressure on Madhav Kumar Nepal until he resigned as prime minister in June 2010. The previous month the coalition government and Maoist opposition had agreed to extend a deadline for drafting a new constitution until May 2011. Repeated efforts to elect a new premier were unsuccessful until Feb. 2011 when, following the withdrawal of Pushpa Kamal Dahal and then Ram Chandra Poudel of the NC as candidates, Jhalanath Khanal of the Communist Party of Nepal (Unified Marxist-Leninist) was approved by parliament and sworn in by Yadav. In a further premiership change in Aug. 2011 Khanal resigned after the government and opposition had failed to meet the May deadline for agreeing a new constitution and was replaced by Baburam Bhattarai of the Unified Communist Party of Nepal (Maoist). Bhattarai similarly missed a constitutional deadline of May 2012 and also failed to hold elections promised for Nov., prolonging the political stalemate and generating increasing friction with Yadav, who called the Maoist government's legitimacy into question.

Khil Raj Regmi

Position
Prime Minister

Introduction
Khil Raj Regmi, a chief justice of the Supreme Court of Nepal, was sworn in as prime minister on 14 March 2013. He heads an

interim government charged with supervising elections initially scheduled for June 2013 but now expected to take place later in the year. He succeeded Maoist leader Baburam Bhattarai, who stepped down amid political stalemate that had left the nascent republic without a functioning parliament since May 2012.

Early Life

Khil Raj Regmi was born on 31 May 1949 in Pokharathok in the Lumbini Zone of southern Nepal. He studied arts at Tribhuvan University in Kathmandu before completing a masters degree in law at the same institution. He became a district judge in 1974, rising through the judicial ranks to become chief justice of the Supreme Court on 6 May 2011.

Regmi's appointment as prime minister in March 2013 was the result of a compromise deal between the four main political parties (the United Democratic Madhesi Front, the Nepali Congress, the Communist Party of Nepal [Unified Marxist-Leninist] and the United Communist Party of Nepal [Maoists]). They had been unable to agree on a suitable candidate since May 2012 when a special assembly was dissolved having failed to draft a new constitution.

Regmi was regarded as a political independent with the bureaucratic skills to lead an interim administration. Though his appointment was well received by the international community, domestic opponents criticized it for blurring the separation between executive and judicial power.

Career in Office

Regmi announced an 11-member cabinet on 19 March 2013, made up entirely of ex-civil servants in accordance with the cross-party agreement that brought him to power. He was charged with overseeing elections to select a constituent assembly to finish drafting the country's first constitution since the monarchy was abolished in 2008. He faces difficult economic conditions as the nation, dependent on foreign aid and tourism, struggles to rebuild after the ten-year civil war that ended in 2006.

DEFENCE

The then King was formerly commander-in-chief of the armed forces, but he was stripped of the position in May 2006. The cabinet now has the power to appoint the army chief.

Defence expenditure in 2008 totalled US$176m. (US$6 per capita), representing 3·5% of GDP.

As at 31 Jan. 2013, 4,462 personnel (including 3,677 troops) were deployed in UN peacekeeping operations.

Army

Strength (2007) 69,000, and there is also a 62,000-strong paramilitary police force (15,000 armed).

Air Force

The Army's air wing has no combat-capable aircraft. Personnel, 2007, 320.

ECONOMY

Agriculture accounted for 35·1% of GDP in 2006, industry 17·4% and services 47·5%.

Overview

In the early 2000s the economy underperformed as a result of political conflict and internal unrest. Tourism, an important source of earnings, also suffered with visitor numbers falling below the peak of the late 1990s. Since conflict ended in 2006 annual growth has averaged about 4% but political tensions and poor security hold back the economy.

Agriculture is the principal area of activity, employing around 80% of the workforce and accounting for a third of GDP. Apart from agricultural land and forests, exploitable natural resources include mica, hydropower and tourism. Remittances accounted for nearly 22% of GDP in 2009. As a result of the global financial crisis, remittances and exports have declined while imports have increased.

Nepal is the poorest country in South Asia, with high unemployment and the highest inequality level. It is ranked 157th of 187 countries on the UN Human Development Index, although poverty levels fell from 42% in 1995 to 24·7% in 2008. An inadequate infrastructure, frequent power shortages, a poor business climate and troublesome labour relations are impediments to growth.

Currency

The unit of currency is the *Nepalese rupee* (NPR) of 100 *paisas*. 50 *paisas* = 1 *mohur*. Inflation was 9·5% in 2010 and 9·6% in 2011. Foreign exchange reserves were US$1,478m. in July 2005 and gold reserves totalled 129,000 troy oz. Total money supply in March 2009 was NRs 177,682m.

Budget

Budgetary central government revenue totalled NRs 102,188m. in 2006–07 (NRs 84,100m. in 2005–06) and expenditure NRs 116,606m. (NRs 94,881m. in 2005–06). Main sources of revenue, 2006–07 (NRs 1m.): taxes, 71,127 (including: taxes on goods and services, 36,434; taxes on international trade and transactions, 16,708); grants, 15,801. Major items of expenditure, 2006–07 (NRs 1m.): general public services, 27,151; economic affairs, 26,571; education, 21,388; defence, 10,958.

VAT is 13%.

Performance

Real GDP growth was 4·5% in 2009, 4·8% in 2010 and 3·9% in 2011. Nepal's total GDP in 2011 was US$18·9bn.

Banking and Finance

The Central Bank is the bank of issue (*Governor*, Dr Yuvraj Khatiwada). In 2002 there were four domestic commercial banks (Kumari Bank; Nepal Bank; Nepal Industrial and Commercial Bank; Rastriya Banijya Bank), ten joint-venture banks and four development finance organizations (Agricultural Development Bank; Nepal Development Bank; Nepal Housing Development Finance Corporation; Nepal Industrial Development Corporation).

External debt totalled US$3,702m. in 2010. As a proportion of GNI, external debt fell from 52·0% in 2000 to 23·4% in 2010.

There is a stock exchange in Kathmandu.

ENERGY AND NATURAL RESOURCES

Environment

Nepal's carbon dioxide emissions from the consumption and flaring of fossil fuels in 2008 were the equivalent of 0·1 tonnes per capita.

Electricity

Installed capacity was approximately 0·6m. kW in 2006–07. Production in 2006–07 was an estimated 2·81bn. kWh (over 99% hydro-electric), with consumption per capita 109 kWh.

Minerals

Production (in tonnes), 2005: limestone, 263,701; red clay, 35,484; agricultural lime (2003), 13,025; coal, 9,289; talc, 5,832; salt, 2,000.

Agriculture

Agriculture is the mainstay of the economy, providing a livelihood for around 80% of the population and accounting for 35% of GDP. In 2002 there were about 3·2m. ha. of arable land and 94,000 ha. of permanent crops. Cultivated land accounts for 26·5% of land use; forest and woodland 42·4%. Crop production (2003, in 1,000 tonnes): rice, 4,155; sugarcane, 2,250; potatoes, 1,480; maize, 1,441; wheat, 1,344; millet, 288.

Livestock (2002); cattle, 6·98m.; goats, 6·61m.; buffaloes, 3·70m.; pigs, 934,000; sheep, 840,000; chickens, 21m.

Livestock products (2003 estimates, in 1,000 tonnes): buffalo meat, 130; beef and veal, 48; goat meat, 39; poultry meat, 15; buffalo's milk, 816; cow's milk, 353; goat's milk, 64; eggs, 27.

Forestry

In 2010 the area under forests was 3·64m. ha., or 25% of the total land area, down from 4·82m. ha. in 1990 and 3·90m. ha. in 2000. There are eight national parks, covering 1m. ha., five wildlife reserves (170,490 ha.) and two conservation areas (349,000 ha.). Timber production was 13·88m. cu. metres in 2007, mainly for use as fuelwood and charcoal. Expansion of agricultural land has led to widespread deforestation.

Fisheries

The catch in 2010 was an estimated 21,500 tonnes, exclusively from inland waters.

INDUSTRY

In 2008–09 manufacturing accounted for 6·8% of GDP. In 2006–07 there were 158 manufacturing establishments with 200 or more staff employing 71,409 people. Production (2007 unless otherwise stated): cement, 1,060,490 tonnes; raw sugar, 140,000 tonnes; soap (2005–06), 55,000 tonnes; jute fibres, 17,000 tonnes; paper and paperboard, 13,000 tonnes; tea, 12,200 tonnes; sawnwood, 630,000 cu. metres; plywood, 30,000 cu. metres; spirits and liqueurs (2007–08), 10·5m. litres; beer (2005–06), 26·0m. litres; cigarettes, 6,081m. units; clay bricks, 12·7m. units; shoes, 10m. pairs.

Labour

The estimated labour force in 2009 totalled 131,315,000 (55% males).

INTERNATIONAL TRADE

Imports and Exports

Imports (c.i.f.) in 2010 totalled US$5,127·5m. and exports (f.o.b.) US$834·0m. The main import suppliers in 2004 were: India (43·0%), China (10·0%), UAE (10·0%), Saudi Arabia (4·4%). The leading export markets in 2004 were: India (48·8%), USA (22·3%), Germany (8·5%), UK (2·8%).

Principal import commodities are petroleum products, transport equipment and parts, chemical fertilizer and raw wool. Principal export commodities are carpets, clothing, leather goods, pulses, raw jute and jute goods, and handicrafts. Hand-knotted woollen carpets are the largest overseas export item constituting almost 32% of foreign exchange earnings.

COMMUNICATIONS

Roads

In 2006 there were 16,834 km of roads, of which 17% were paved.

Rail

59 km (762 mm gauge) connect Jayanagar on the North Eastern Indian Railway with Janakpur. In 2003 the railways carried 1·28m. passengers.

There are plans for a metro system in Kathmandu.

Civil Aviation

There is an international airport (Tribhuvan) at Kathmandu. The national carrier is the state-owned Nepal Airlines (formerly known as Royal Nepal Airlines). It operates domestic services and in 2003 flew to Bangalore, Bangkok, Delhi, Dubai, Hong Kong, Kuala Lumpur, Mumbai, Osaka, Shanghai and Singapore. In 2004 Kathmandu handled 2,066,950 passengers (1,140,660 on international flights) and 15·3m. tonnes of freight. In 2003 scheduled airline traffic of Nepali-based carriers flew 8m. km, carrying 356,000 passengers (279,000 on international flights).

Telecommunications

In 2008 there were 805,100 main (fixed) telephone lines in Nepal and mobile phone subscribers numbered 4,200,000 (14·6 per 100 persons). There were 499,000 internet users in 2008. In March 2012 there were 1·4m. Facebook users.

SOCIAL INSTITUTIONS

Justice

The Supreme Court Act established a uniform judicial system, culminating in a supreme court of a Chief Justice and no more than six judges. Special courts to deal with minor offences may be established at the discretion of the government. The King previously had the power to appoint the Chief Justice, but this power passed to the prime minister under the temporary constitution signed in Dec. 2006.

The death penalty was abolished in 1997. The population in penal institutions in Jan. 2008 was approximately 6,700 (24 per 100,000 of national population).

Education

The adult literacy rate in 2009 was estimated 59·1% (72·0% among males but only 46·9% among females).

In 2007 there were 4,515,059 pupils and 112,827 teaching staff in primary schools and 1,998,990 pupils in secondary schools with (in 2008) 56,294 teaching staff. There were 320,844 students in tertiary education in 2007 with 9,932 academic staff. The oldest and largest university in Nepal is Tribhuvan University, which was established in 1959.

In 2009 public expenditure on education came to 4·7% of GDP and 19·5% of total government spending.

Health

There were 1,259 physicians and 6,216 nurses in 2001. In 2000 there were 133 hospitals, 180 primary health care centres and 711 health posts.

RELIGION

Nepal is a Hindu state. Hinduism was the religion of 82·8% of the people in 2001. Buddhists comprise 8·9% and Muslims 4·2%. Christian missions are permitted, but conversion is forbidden.

CULTURE

World Heritage Sites

Nepal has four sites on the UNESCO World Heritage List: Sagarmatha National Park (inscribed on the list in 1979); Kathmandu Valley (1979 and 2006); Royal Chitwan National Park (1984); and Lumbini, the Birthplace of the Lord Buddha (1997).

Press

In 2008 there were 298 daily newspapers, including the official English-language *Rising Nepal*, 25 bi-weeklies, 1,442 weeklies and 273 fortnightlies. Press censorship was relaxed in 1991, but following the imposition of a state of emergency in 2005 the press was subjected to total censorship.

Tourism

Foreign tourists visiting Nepal numbered 361,200 in 2001, down from 463,600 in 2000, largely as a consequence of the massacre of the royal family and an upsurge in Maoist rebel violence. They have since returned to and surpassed former levels with 526,700 tourists in 2007, an increase of 37·2% on 2006. Gross foreign exchange earnings came to US$230·6m. in 2007, an increase of 41·7% on the previous year. In 2007, 24,700 hotel beds were available. Tourism accounts for approximately 4% of GDP.

Festivals

Hindu, Buddhist and traditional festivals crowd the Nepali lunar calendar. Dasain (Sept./Oct.) is the longest and most widely observed festival in Nepal. The 15 days of celebration include Dashami, when family elders are honoured. Tihar (Oct./Nov.) celebrates the Hindu goddess Laxmi. During the first three days crows, dogs and cows are worshipped, followed by the spirit, or self. It concludes with Bhai Tika ('Brother's Day'). Buddha Jayanti

(May/June) remembers the birth, enlightenment and death of the Buddha. Sherpas gather at Tengboche Monastery near Mount Everest in May to observe Mani Rimdu with meditation, mask dances and Buddhist ceremonies.

DIPLOMATIC REPRESENTATIVES

Of Nepal in the United Kingdom (12A Kensington Palace Gdns, London, W8 4QU)
Ambassador: Dr Suresh Chandra Chalise.

Of the United Kingdom in Nepal (Lainchaur, Kathmandu, POB 106)
Ambassador: John Tucknott, MBE.

Of Nepal in the USA (2131 Leroy Pl., NW, Washington, D.C., 20008)
Ambassador: Shankar Prasad Sharma.

Of the USA in Nepal (Maharajguni, Kathmandu)
Ambassador: Peter W. Bodde.

Of Nepal to the United Nations
Ambassador: Gyan Chandra Acharya.

Of Nepal to the European Union
Ambassador: Ram Mani Pokharel.

FURTHER READING

Central Bureau of Statistics. *Statistical Pocket Book.* [Various years]

Hutt, Michael, (ed.) *Himalayan 'People's War' Nepal's Maoist Rebellion.* 2004

Lawoti, Mahendra, *Towards a Democratic Nepal: Inclusive Political Institutions for a Multicultural Society.* 2005

Sanwal, D. B., *Social and Political History of Nepal.* 1993

Thapa, Deepak, *A Kingdom Under Siege: Nepal's Maoist Insurgency, 1996 to 2004.* 2005

Whelpton, John, *A History of Nepal.* 2005

National Statistical Office: Central Bureau of Statistics, National Planning Commission Secretariat, Kathmandu.
Website: http://www.cbs.gov.np

THE NETHERLANDS

© Research Machines plc 2006

**Koninkrijk der Nederlanden
(Kingdom of the Netherlands)**

Capital: Amsterdam
Seat of government: The Hague
Population projection, 2015: 16·85m.
GNI per capita, 2011: (PPP$) 36,402
HDI/world rank: 0·910/3
Internet domain extension: .nl

KEY HISTORICAL EVENTS

Flint tools found in the Maastricht area have been estimated to be 250,000 years old. The first definable culture (*c.* 3000 BC) was the Late Stone Age 'Funnel-neck Beaker' culture, named after the objects made by a people known for their monolithic burial monuments. The environment of the 'Low Countries' affected the behaviour of its earliest inhabitants, as demonstrated by the *terpen*—islands of earth and clay—built by the autochthonous Frisians (Frisii) *c.* 500 BC as protection from the sea.

The Romans encountered Celtic tribes to the west and south of the Rhine and Germanic tribes, such as the Frisii, to the north and east. In the 1st century BC Julius Caesar attested to the resistance of the Celtic Eburones and Aduatuci. Roman power beyond the Rhine was limited to isolated forts and client kingdoms.

In the 3rd century AD the stagnant Roman borders began to crumble as military posts were abandoned. Among the most prominent of the encroaching Germanic tribes were the Franks, who settled at first in Toxandria (modern Brabant). Like many 'barbarian' tribes, the Franks entered into agreements with Rome, settling and guarding the border region and assimilating Roman culture. The Frisians became important traders, holding strategic territory between the German (North) Sea and the Meuse and Rhine rivers. With the collapse of Roman government in Gaul and the Rhine in the 5th century, the Franks extended their

power, centred on Austrasia (the central Rhine region). The spread of Christianity in the 7th century, first from the bishoprics of Arras, Tournai and Cambrai, assisted Frankish expansion into the northern Low Countries, where the missionary bishopric of Utrecht was established.

Viking raids on the North Sea coast devastated the flourishing Frisian economy. The Frisian trading centre of Dorestad was destroyed four times between 834–37 by raiders seeking Carolingian silver. Frisia came under Frankish domination during the reign of Pippin the Short, the founder of the Carolingian Empire.

The High Middle Ages saw the development of independent and semi-autonomous principalities, both secular and ecclesiastical. Great landlords established the large counties (Flanders, Hainault, Namur and Holland and Zeeland) and duchies (Brabant, Limburg and Guelders), increasing their authority and size through dynastic alliances and inheritance. The majority fell broadly under the authority of the German king, heirs to the Eastern Frankish realm, though the feudal relationship allowed the growth of a tradition of independence that became a defining characteristic of Dutch politics. The growth of population and its pressure on the land increased the need for land reclamation. Dykes were built from Friesland to Flanders to drain the bogs and marshes for pasturage and, later, agrarian use. The development of urban centres outside the feudal structure was encouraged by the strength of trade and the merchant classes.

The Burgundian era in the Low Countries was born of a series of dynastic matches, most importantly that of Duke Philip II (the Bold) of Burgundy and Margaret, Countess of Flanders and Artois in 1369. Their son, Philip III (the Good), brought most of the northern Low Countries under one lord by inheriting Brabant and Hainault-Holland in the 1430s as well as Luxembourg in 1443. Although the dukes attempted to rule through new centralized bodies, the Burgundian Low Countries were held in a personal union and did not constitute a state. The duke appointed *stadhouders* (stadtholders) and governors to represent him in each of his territories. The summoning of the Estates in 1464 in Brugge (Bruges) represented the first parliamentary assembly in the Low Countries and the importance of the *Nederlands* in the Burgundian realm.

Burgundian Rule

The reign of Charles the Bold, or Rash (1467–77), saw the brief land connection of the realm (by the acquisition of Lorraine) and the first explicit attempt to create a unitary kingdom—an echo of the Middle Frankish Kingdom, Lotharingia. Charles failed in his bid to make himself regent of this kingdom in 1473 and his death at the Battle of Nancy left his domains to his daughter, Mary. The duchess was soon stripped of the Duchy of Burgundy by the French king and was forced to concede privileges to the provinces. Her marriage to Maximilian of Habsburg, the future Holy Roman Emperor, brought the Low Countries into personal union with Austria and, later, Spain. Mary's son, Philip the Handsome, inherited the Spanish throne through his wife, Juana the Mad, forging a massive and disparate empire of kingdoms, principalities and lordships. Philip's son, Charles V, though born in Ghent, spent little time in the Low Countries after succeeding to the Spanish throne. They were administered by governors-general, normally taken from the ruler's family. Centralization, though consistently opposed, continued to be pressed on the inhabitants of the Low Countries. The 17 provinces were brought together formally in 1548 as the 'Burgundian *Kreis*' and the sovereign succession regulated by Pragmatic Sanction the following year.

Brussels became the centre of government, being the location of the court and most organs of government.

Philip (II of Spain) imposed a new ecclesiastical hierarchy, sanctioned by papal bull in 1559, in an attempt to use the church as a centralizing force. The traditional resistance of the towns and provinces was given added fervour by the religious controversies attributable to the Reformation. Erasmus, a leading Dutch humanist, openly attacked the abuses and corruptions of the Church but rejected the theology of the reformers such as Martin Luther. However, the works of the radical Jean Calvin arrived in Antwerp in 1545, spreading throughout the region rapidly after their translation in 1560. Calvinism appealed to the intellectual middle classes, as well as the artisans, whose work ethic it extolled. The government focused its repressive efforts on the Anabaptists, whose refusal to swear allegiance to the prince was an affront to temporal and spiritual authority. The iconoclastic purges of 1566 provoked Philip to send the duke of Alba to restore his authority, thereby sparking full-scale revolt and the Eighty Years War (1568–1648).

The causes of the Dutch Revolt were numerous; religious tensions, resentment towards 'Spanish' authority, the heavy burden of taxes and absolutist government and the perceived desecration of traditional privileges were combined with years of hardship caused by climatic conditions and wars with France. However, in the earlier years of the revolt, the 'legitimate' *casus belli* claimed by the Dutch was the influence of 'evil advisers' around the prince—few openly rejected Philip's sovereignty. The *Geuzen*, an army of beggars, pillaging and pirating in the name of William of Orange, took the port of Brielle in 1572. This began the expulsion of Spanish authority from the northern provinces, a process completed by 1574.

The conversion of William (the Silent) to Calvinism in 1572, in response to his selection as stadtholder of Holland and Zeeland, was a political move to gain support for a united Netherlands of Catholics and Protestants. The 1576 Pacification of Ghent brought together predominantly Catholic and Protestant provinces in the face of bloody repression meted out by Alba's Council of Troubles (Council of Blood) and the notorious 'Spanish Fury' massacre in Antwerp. The mainly Catholic southern provinces were largely regained for Philip by the brilliant Alessandro Farnese, duke of Parma in 1578, forcing a 'closer union'—the Union of Utrecht—in the north in 1579, committed to resisting the Spanish. This marked the birth of the United Provinces of the Netherlands, or the 'Dutch Republic', with power concentrated in the hands of the stadtholders, nominally representing the hereditary prince. Philip's refusal to compromise led to his 'forfeiture of sovereignty' in the States-General Act of Abjuration in 1581, on the grounds of persistent tyranny.

The constitutional position of the Republic was unclear. The House of Orange was recognized as the traditional stadtholders of each province, though the lordship of the territories was tendered to both France and England in the 1580s. Maurice of Nassau, the son of William of Orange, was named stadtholder of Holland and Zeeland in 1587. Maurice's victories over Farnese came to be called the 'closing of the garden', giving the United Provinces the approximate borders it has maintained to the modern day. With recognition from England and France, the government negotiated the Twelve Year Truce in 1609 with Spain, which recognized the independence of the United Provinces.

The Calvinist church divided between the followers of two prominent clerics, Jacobus Arminius (the Remonstrants) and Franciscus Gomarus (the Contra-Remonstrants). The Arminians, championed by the elite of Holland and the towns, objected to the repressive orthodoxy of the Gomarists and demanded an inclusive reformed church to protect trade and foreign relations. The execution of the Remonstrant Johan van Oldenbarnvelt, the Advocate of Holland, signified the triumph of Maurice's Contra-Remonstrants and made permanent peace with Spain impossible. After initial Spanish success at Breda in 1621, Maurice's successor,

Frederick Henry, turned the tide, taking Maastricht in the far south. Ending the persecution of the Remonstrants, Frederick Henry augmented the authority of his princely house, even earning an honorific royal title from the French King. Lasting peace with Spain was finally won at the 1648 Treaty of Münster, which formally recognized the Dutch Republic.

Independence and the Golden Age
The 17th century has traditionally been called the Golden Age of the Dutch. From the Twelve Year Truce, the Dutch economy expanded massively, principally through trade in the Baltic and with France, Iberia and the colonies of the West and East Indies. The United East Indies Company, chartered in 1602, held quasi-sovereign authority over its colonies in Sri Lanka, India and Indonesia. Dutch banking financed the northern European markets, chiefly through foreign government bonds. The increase of wealth stimulated the arts. Prosperous life in Dutch towns was painted by Jan Vermeer and Amsterdam's burghers by Rembrandt. Although Calvinism had been officially adopted, Catholics were left unmolested but public worship was prohibited.

After the death of Frederick Henry's bellicose son, William II, in 1650, the republic experienced its first 'stadtholderless' period when the prosperous province of Holland dominated the Netherlands. Relations with Republican England deteriorated because of the execution of Charles I, who was closely related to the House of Orange. More importantly, competition for trade and shipping between the two great maritime powers caused skirmishes in America and Europe and a series of Anglo-Dutch Wars, conducted at sea. The destruction of the English fleet at Chatham in 1667 destroyed relations with Charles II, who had supported Orangist interests in the Netherlands.

The House of Orange reassumed the leadership of the Netherlands when William III took the stadtholdership of Holland in 1672 and defeated the French and the English in naval encounters. The Dutch supported William in his invasion of England—the Glorious Revolution—in 1688, claiming the throne with his wife, Mary Stuart. His death without issue in 1702 heralded the second stadtholderless period, when the councillor pensionaries of Holland asserted the province's leadership. However, the oligarchic nature of government attracted little support, especially during Dutch humiliations at the hands of the French in the War of the Austrian Succession (1740–48). William IV of Orange was elected to all provinces in 1747, the House of Orange being seen as the natural leaders of the Dutch people.

Both William IV and William V resisted calls for a more relaxed rule. The Patriot Movement took advantage of the Dutch defeat in the Fourth Anglo-Dutch War of the 1780s to depose William V. However, Prussia's intervention restored the stadtholder and many Patriots fled to France, then on the brink of revolution.

Revolutionary France's invasion of Belgium (the Spanish Netherlands) in 1794 was soon extended to the United Provinces. William V fled to England and the Patriots, supported by the French, assumed control of government. The new 'Batavian Republic', styled after the supposedly original inhabitants, was in reality a protectorate of France. This truly republican period enabled political modernization, much of which has lasted to the modern day. An elected national assembly was instituted (though the franchise was retained by property owners only), with new electoral constituencies to replace the old provinces. Religious toleration was adopted, with all denominations awarded equal treatment. However, the economy declined, partly because of the seizure of the Dutch colonies in the name of William V by Great Britain, which had declared war on France.

Napoleon
The republic was ended in 1806 when Napoleon incorporated the Netherlands into his empire. He installed his brother, Louis, as king of Holland. Louis adopted the cause of his new subjects, frequently defying his brother's orders in favour of Dutch

interests. Napoleon ended his brother's reign in 1810 and brought his kingdom under French rule. Gijsbert Karel van Hogendorp, who drew up the new constitution after the French withdrawal in 1813, led the opposition to France. The new constitution provided for a constitutional monarchy, with William V's son proclaimed king (William I), as demanded by the Congress of Vienna. The northern provinces were united with Belgium and Luxembourg under the Kingdom of the Netherlands.

William I saw the revival of the economy as the first priority. Using his personal resources as well as the treasury, he invested heavily in the re-establishment of Dutch shipping, especially to the restored colonies. Domestically, William was not so successful. In 1830 Belgium proclaimed its independence, rejecting a common identity with the predominantly Protestant north—the declaration of Dutch as the sole official language had alienated the French-speaking Walloons in Brussels. Though defeated by the Dutch army, the Belgians gained their independence in 1839 thanks to French and British intervention in 1832.

In response to the European revolutions of 1848, the king granted a liberal constitution. Support for the king was bolstered by the patriotic reaction in the northern provinces to the Belgian secession. The reintroduction of the Catholic hierarchy in 1853 won over a community which made up over a third of the population.

Dutch imperialism was consolidated in the second half of the 19th century. Having lost numerous colonies in the Americas, southern Africa and India, attention focused on the Indonesian archipelago. War with Aceh in northern Sumatra, famous for its piracy, was long and bloody but secured the archipelago for the Netherlands. The division of New Guinea was settled with Germany and Great Britain in 1875. Personal union with Luxembourg came to an end on the accession of Wilhelmina in 1890, barred by Salic Law from inheriting the Grand Duchy.

European War

In 1917 universal male suffrage was granted in return for the secular parties' acceptance of funding for religious schools, thus concluding the 30-year School Conflict. Female suffrage followed in 1922. Wilhelmina, though less active in government than her father, William III, strongly advocated neutrality in the conflicts of the early 20th century, keeping the Netherlands out of the First World War. The German Kaiser, Wilhelm II, was granted asylum in the Netherlands.

The inter-war years were a period of social and political continuity. The *zuilen* system expanded, cementing what has been described as a bourgeois consensus, though worldwide depression hit the Netherlands hard in the 1930s. In 1932 the IJsselmeer dam was completed, transforming the Zuider Zee, an inlet of the North Sea, into a freshwater lake, the IJsselmeer.

The neutrality of the Netherlands was not respected by Germany in the Second World War, despite assurances from Hitler after the invasion of Poland. Control of the Netherlands and Belgium was seen as essential to protect the industrial centres of the Ruhr and to gain broader access to the North Sea. The Dutch armed forces were overwhelmed within a week in May 1940. The queen and government went into exile in London. Persecution of the Jews began in Oct. 1941. The first transports left in July 1942, mostly to Auschwitz. 107,000 Dutch Jews died. Dutch resistance took the form of civilian sabotage and the hiding of Jews and *onkerduikers* ('underdivers')—underground military operatives.

The Netherlands saw some of the bitterest fighting near the close of the war when Allied troops made airborne incursions— Arnhem Bridge in Sept. 1944—to speed victory over Germany. By the end of the war the Dutch were on the brink of famine. The destruction of the economy and much of the infrastructure caused large-scale emigration. In 1947 the Netherlands accepted US$1bn. for reconstruction from the Marshall Plan and entered the Benelux Economic Union with Belgium and Luxembourg (fully

established in 1958). The Netherlands abandoned its neutrality when it joined NATO in 1949, the year it granted Indonesia independence. Further changes to Dutch overseas possessions took place in 1954 under the *Statute for the Kingdom*, which gave the territories in the West Indies equal status. Dutch New Guinea (Irian Jaya) was ceded to Indonesia in 1963 and Suriname was given its independence in 1974.

Dutch politics saw several important changes in the post-war years, such as the introduction of proportional representation in elections. From the end of the war until 1958, a coalition of Catholic and labour parties held power, taking the Netherlands into the Korean War in 1950. The Netherlands was a founder member of the European Coal and Steel Community (ECSC) in 1951, which later merged with the European Economic Community (EEC).

The economy grew rapidly in the late 1950s when the welfare state was greatly expanded. Social unrest in the 1960s was led by youth and labour groups. Social changes in the '70s included the demise of the traditional *zuilen* and the creation of new political parties across religious divides; most notable of these was the Christian Democratic Appeal (CDA). Newspapers, the voice of the *zuilen*, disassociated themselves from religious denominations, becoming independent commercial enterprises. The decriminalization of personal cannabis use in the 1970s indicated a policy towards drug use and abuse that focused on rehabilitation (for hard drug users) as opposed to punishment. Vocal youth action was seen most clearly in the confrontations between the police and the *krakers*—squatters demanding affordable housing.

Opposition to nuclear weapons grew in the 1980s, sparked by the support given by Prime Minister Andreas van Agt to placing US cruise missiles on Dutch soil. In 1986 the pressures of the Netherlands' population density led to creation of the 12th province, Flevoland, from four polders reclaimed from the IJsselmeer.

The Netherlands joined the coalition forces in the 1991 Gulf War, providing two naval frigates. Serious flooding in Gelderland and the threat of worse to come led to the evacuation of 240,000 people from the province in 1995.

The Netherlands became the first country to legalize homosexual marriage and adoption, in 2001, and euthanasia, in 2002. In April 2002, Prime Minister Wim Kok's government resigned in the wake of a report that criticized Dutch inaction in preventing the massacre at Srebrenica in 1995. During the subsequent election campaign, the right-wing politician Pim Fortuyn was assassinated by an animal-rights activist who opposed Fortuyn's anti-immigration policies.

The coalition government led by Jan Peter Balkenende, formed in July, collapsed in Oct., necessitating fresh elections. Balkenende formed a new government in May 2003. In 2005 the electorate voted in a referendum against adopting a proposed EU constitution. In June 2006 Balkenende's coalition again collapsed, although he remained head of a minority government until elections in Nov. In Feb. 2007, after months of negotiations, he was sworn in for a new tenure as premier, leading a centrist coalition of three parties. It collapsed in Feb. 2010 amid tensions over troop deployment in Afghanistan. The People's Party for Freedom and Democracy (VVD) emerged from elections in June 2010 with the most seats and formed a coalition, with Mark Rutte as prime minister, with the support of the far-right Freedom Party. In April 2012 Rutte offered to resign after the Freedom Party refused to back his austerity measures. After elections in Sept. 2012 he headed a new coalition of the VVD and the Labour Party.

TERRITORY AND POPULATION

The Netherlands is bounded in the north and west by the North Sea, south by Belgium and east by Germany. The area is 41,543 sq. km, of which 33,756 sq. km is land. Projects of sea-flood control and land reclamation (polders) by the construction of dams and drainage schemes have continued since 1920. More than a quarter of the country is below sea level.

The population was 13,060,115 at the census of 1971 and 16,730,348 on 1 Jan. 2012. Population growth in 2011, 0·5%.

The UN gives a projected population for 2015 of 16·85m.

Ongoing 'rolling' censuses have replaced the former decennial counts.

Area, population and density, and chief towns of the 12 provinces on 1 Jan. 2011:

	Area 2008 (in sq. km)	Population 2011	Density 2011 per sq. km land area	Provincial capital
Groningen	2,967·90	579,036	248	Groningen
Friesland	5,740·87	647,282	193	Leeuwarden
Drenthe	2,680·37	491,411	186	Assen
Overijssel	3,420·86	1,134,465	341	Zwolle
Flevoland	2,412·30	391,967	276	Lelystad
Gelderland	5,136·51	2,004,671	403	Arnhem
Utrecht	1,449·12	1,228,794	887	Utrecht
Noord-Holland	4,091·76	2,691,477	1,008	Haarlem
Zuid-Holland[1]	3,418·50	3,528,324	1,252	The Hague
Zeeland	2,933·89	381,530	213	Middelburg
Noord-Brabant	5,081·76	2,454,215	499	's-Hertogenbosch
Limburg	2,209·22	1,122,627	521	Maastricht
Total	41,543·06	16,555,799	490	

[1]Since 29 Sept. 1994 includes inhabitants of the municipality of The Hague formerly registered in the abolished Central Population Register.

In 2011, 83·3% of the population lived in urban areas.

Population of municipalities with over 50,000 inhabitants on 1 Jan. 2011:

Alkmaar	93,936	Kampen	50,403
Almelo	72,599	Katwijk	62,044
Almere	190,655	Lansingerland	54,090
Alphen a/d Rijn	72,680	Leeuwarden	94,838
Amersfoort	146,592	Leiden	117,915
Amstelveen	81,796	Leidschendam-	
Amsterdam	779,808	Voorburg	72,068
Apeldoorn	156,199	Lelystad	75,111
Arnhem	148,070	Maastricht	119,664
Assen	67,177	Nieuwegein	60,947
Barneveld	53,026	Nijmegen	164,223
Bergen op Zoom	66,074	Oosterhout	54,072
Breda	174,599	Oss	84,201
Capelle a/d Ijssel	66,104	Purmerend	79,193
Delft	97,690	Roermond	55,595
Deventer	98,737	Roosendaal	77,541
Doetinchem	56,037	Rotterdam	610,386
Dordrecht	118,810	Schiedam	75,718
Ede	108,285	Sittard-Geleen	94,814
Eindhoven	216,036	Smallingerland	55,436
Emmen	109,259	Spijkenisse	72,244
Enschede	157,838	Stichtse Vecht	63,050
Gouda	71,047	Súdwest Fryslân	82,445
Groningen	189,991	Terneuzen	54,823
Haarlem	150,670	Tilburg	206,240
Haarlemmermeer	143,374	Utrecht	311,367
The Hague (Den Haag)	495,083	Veenendaal	62,267
Hardenberg	59,283	Velsen	67,347
Heerlen	89,212	Venlo	99,793
Den Helder	57,207	Vlaardingen	71,269
Helmond	88,560	Westland	99,776
Hengelo	80,747	Zaanstad	146,940
's-Hertogenbosch	140,786	Zeist	60,824
Hilversum	84,984	Zoetermeer	121,911
Hoogeveen	54,844	Zwolle	120,355
Hoorn	70,697		

Urban agglomerations as at 1 Jan. 2010: Amsterdam, 1,053,413; Rotterdam, 996,183; The Hague, 633,201; Utrecht, 441,866; Eindhoven, 327,245; Leiden, 251,436; Dordrecht, 236,285; Tilburg, 227,614; Groningen, 205,814; Heerlen, 204,825; Haarlem, 197,660;

Amersfoort, 173,674; Breda, 173,299; 's-Hertogenbosch, 165,007; Nijmegen, 162,963; Enschede, 157,052; Apeldoorn, 155,726; Arnhem, 148,513; Sittard-Geleen, 137,495; Zwolle, 119,030; Maastricht, 118,533; Leeuwarden, 94,073.

Dutch is the official language. Frisian, spoken as a first language by 2·2% of the population, is also recognized as an official language in the northern province of Friesland.

SOCIAL STATISTICS

Vital statistics for calendar years:

	Live births		Marriages	Divorces	Deaths
	Total	Outside marriage			
2002	202,083	58,525	85,808	33,179	142,355
2003	200,297	61,439	80,427	31,479	141,936
2004	194,007	63,029	73,441	31,098	136,553
2005	187,910	65,563	72,263	31,905	136,402
2006	185,057	68,575	72,369	31,734	135,372
2007	181,336	71,559	72,485	31,983	133,022

2007 rates per 1,000 population: birth, 11·1; death, 8·1. Annual population growth rate, 2000–05, 0·5%. In 2009 the suicide rate per 100,000 population was 9·3 (men, 13·1; women, 5·5). In 2006 the most popular age range for marrying (excluding same-sex marriages) was 30–34 for males and 25–29 for females. Expectation of life, 2007, was 77·6 years for males and 81·9 for females. Infant mortality, 2008, 3·8 per 1,000 live births; fertility rate, 2008, 1·7 births per woman. Percentage of population by age in 2007: 0–24 years, 30·1%; 25–64, 55·5%; 65 and over, 14·5%. In 2009 the Netherlands received 14,905 asylum applications, up from 13,399 in 2008. In 2001 the Netherlands became the first country to legalize same-sex marriage.

A UNICEF report published in 2010 showed that 6·1% of children in the Netherlands live in relative poverty (living in a household in which disposable income—when adjusted for family size and composition—is less than 50% of the national median income), the joint third lowest of any country.

CLIMATE

A cool temperate maritime climate, marked by mild winters and cool summers, but with occasional continental influences. Coastal temperatures vary from 37°F (3°C) in winter to 61°F (16°C) in summer, but inland the winters are slightly colder and the summers slightly warmer. Rainfall is least in the months Feb. to May, but inland there is a well-defined summer maximum in July and Aug.

The Hague, Jan. 37°F (2·7°C), July 61°F (16·3°C). Annual rainfall 32·8" (820 mm). Amsterdam, Jan. 36°F (2·3°C), July 62°F (16·5°C). Annual rainfall 34" (850 mm). Rotterdam, Jan. 36·5°F (2·6°C), July 62°F (16·6°C). Annual rainfall 32" (800 mm).

CONSTITUTION AND GOVERNMENT

According to the Constitution (promulgated 1815; last revision, 2005), the Kingdom consists of the Netherlands and its overseas countries and territories. Their relations are regulated by the 'Statute' for the Kingdom, which came into force on 29 Dec. 1954 and was revised on 10 Oct. 2010 in recognition of the dissolution of the Netherlands Antilles, by which Curaçao and Sint Maarten became (along with Aruba) independent countries within the Kingdom and Bonaire, Saba and Sint Eustatius became autonomous special municipalities. Each part enjoys full autonomy; they are united, on a footing of equality, for mutual assistance and the protection of their common interests.

The Netherlands is a constitutional and hereditary monarchy. The royal succession is in the direct female or male line in order of birth. The reigning King is **Willem-Alexander**, born 27 April 1967, son of Princess Beatrix Wilhelmina Armgard and Claus von Amsberg; married to Máxima Zorreguieta on 2 Feb. 2002 (born 17 May 1971); succeeded to the crown on 30 April 2013, on the abdication of his mother. *Offspring*: Catharina-Amalia, born 7 Dec. 2003; Alexia, born 26 June 2005; Ariane, born 10 April 2007.

The monarch receives an allowance from the civil list. Prior to her abdication, Queen Beatrix was to receive €5,233,000 under the terms of the 2013 budget; as Crown Prince, Willem-Alexander was scheduled to receive €1,410,000 and Princess Máxima, €634,000.

Other Members of the Royal House. Princess Beatrix, born 31 Jan. 1938; *Prince Constantijn* (the King's brother), born 11 Oct. 1969, married to Laurentien Brinkhorst (*Princess Laurentien*) on 19 May 2001 (*offspring*: Eloise, born 8 June 2002; Claus-Casimir, born 21 March 2004; Leonore, born 3 June 2006); *Princess Margriet Francisca* (sister of Princess Beatrix), born in Ottawa, 19 Jan. 1943, married to *Pieter van Vollenhoven* on 10 Jan. 1967 (*sons*: Prince Maurits, born 17 April 1968; Prince Bernhard, born 25 Dec. 1969; Prince Pieter-Christiaan, born 22 March 1972; Prince Floris, born 10 April 1975). Names in italics represent members of the Royal House.

The central executive power of the State rests with the Crown, while the central legislative power is vested in the Crown and Parliament (the *States-General*), consisting of two Chambers. The upper *First Chamber* is composed of 75 members, elected by the members of the Provincial States. The 150-member *Second Chamber* is directly elected by proportional representation for four-year terms. Members of the States-General must be Netherlands subjects of 18 years of age or over. The Hague is the seat of the Court, government and Parliament; Amsterdam is the capital.

The *Council of State*, appointed by the Crown, is composed of a vice-president and not more than 28 members. The monarch is president, but the day-to-day running of the Council is in the hands of the vice-president. The Council has to be consulted on all legislative matters. The Sovereign has the power to dissolve either Chambers, subject to the condition that new elections take place within 40 days, and the new Chamber be convoked within three months. Both the government and the Second Chamber may propose Bills; the First Chamber can only approve or reject them without inserting amendments. The meetings of both Chambers are public, although each of them may by a majority vote decide on a secret session. A Minister or Secretary of State cannot be a member of Parliament at the same time.

The Constitution can be revised only by a Bill declaring that there is reason for introducing such revision and containing the proposed alterations. The passing of this Bill is followed by a dissolution of both Chambers and a second confirmation by the new States-General by two-thirds of the votes. Unless it is expressly stated, all laws concern only the realm in Europe, and not the overseas parts of the kingdom.

National Anthem

'Wilhelmus van Nassaue' ('William of Nassau'); words by Philip Marnix van St Aldegonde, tune anonymous.

GOVERNMENT CHRONOLOGY

Prime Ministers since 1940. (ARP = Anti-Revolutionary Party; CDA = Christian Democratic Appeal; KVP = Catholic People's Party; PvdA = Labour Party; VDB = Liberal Democratic League; VVD = People's Party for Freedom and Democracy)

1940–45	ARP	Pieter Sjoerds Gerbrandy
1945–46	VDB/PvdA	Willem Schermerhorn
1946–48	KVP	Louis Jozef Maria Beel
1948–58	PvdA	Willem Drees
1958–59	KVP	Louis Jozef Maria Beel
1959–63	KVP	Jan Eduard de Quay
1963–65	KVP	Victor Gérard Marie Marijnen
1965–66	KVP	Joseph Maria Laurens Theo (Jo) Cals
1966–67	ARP	Jelle Zijlstra
1967–71	KVP	Petrus Josephus Sietse (Piet) de Jong
1971–73	ARP	Barend Willem Biesheuvel
1973–77	PvdA	Johannes Marten (Joop) den Uyl
1977–82	CDA	Andreas Maria (Andries) van Agt
1982–94	CDA	Rudolphus Frans Marie (Ruud) Lubber
1994–2002	PvdA	Willem (Wim) Kok
2002–10	CDA	Jan Peter Balkenende
2010–	VVD	Mark Rutte

RECENT ELECTIONS

Party affiliation in the First Chamber as elected on 23 May 2011: People's Party for Freedom and Democracy (VVD), 16 seats; Labour Party (PvdA), 14; Christian Democratic Appeal (CDA), 11; Party for Freedom (PVV), 10; Socialist Party (SP), 8; Democrats '66 (D66), 5; Green Left (GL), 5; Christian Union (CU), 2; Reformed Political Party (SGP), 1; 50Plus (50+), 1; Party for the Animals (PvdD), 1; Independent Group in the Senate, 1.

Elections to the Second Chamber were held on 12 Sept. 2012, after Prime Minister Mark Rutte handed in his government's resignation to then Queen Beatrix on 23 April. The VVD won 26·6% of the vote (41 of 150 seats), PvdA 24·8% (38 seats), PVV 10·1% (15), SP 9·7% (15), CDA 8·5% (13), D66 8·0% (12), CU 3·1% (5), GL 2·3% (4), SGP 2·1% (3), PvdD 1·9% (2) and 50+ 1·9% (2). Turnout was 74·3%.

European Parliament
The Netherlands has 26 (27 in 2004) representatives. At the June 2009 elections turnout was 36·8% (39·3% in 2004). The CDA won 5 seats with 20·1% of votes cast (political affiliation in European Parliament: European People's Party); PVV, 4 with 17·0% (non-attached); PvdA 3 with 12·1% (Progressive Alliance of Socialists and Democrats); VVD, 3 with 11·4% (Alliance of Liberals and Democrats for Europe); D66, 3 with 11·3% (Alliance of Liberals and Democrats for Europe); Green Left, 3 with 8·9% (Greens/European Free Alliance); SP, 2 with 7·1% (European United Left/Nordic Green Left); Christian Union-Reformed Political Party, 2 with 6·8% (one with European Conservatives and Reformists and one with Europe of Freedom and Democracy).

CURRENT GOVERNMENT

Following parliamentary elections in Sept. 2012 a coalition government of VVD and PvdA was sworn in on 5 Nov. 2012. In March 2013 the government comprised:

Prime Minister and Minister of General Affairs: Mark Rutte; b. 1967 (VVD).

Deputy Prime Minister and Minister of Social Affairs and Employment: Lodewijk Asscher (PvdA).

Minister of Defence: Jeanine Hennis-Plasschaert (VVD). *Economic Affairs:* Henk Kamp (VVD). *Education, Culture and Science:* Jet Bussemaker (PvdA). *Finance:* Jeroen Dijsselbloem (PvdA). *Foreign Affairs:* Frans Timmermans (PvdA). *Foreign Trade and Development Co-operation:* Lilianne Ploumen (PvdA). *Health, Welfare and Sport:* Edith Schippers (VVD). *Housing and Central Government Sector:* Stef Blok (VVD). *Infrastructure and Environment:* Melanie Schultz van Haegen (VVD). *Interior and Kingdom Relations:* Ronald Plasterk (PvdA). *Security and Justice:* Ivo Opstelten (VVD).

Government Website: http://www.government.nl

CURRENT LEADERS

Mark Rutte

Position
Prime Minister

Introduction
Mark Rutte became prime minister in Oct. 2010, heading a coalition between his People's Party for Freedom and Democracy (VVD) and the Christian Democratic Appeal (CDA) with outside support of the Party for Freedom (PVV). His premiership ended an era of alternating Labour Party and Christian Democratic

governments uninterrupted since 1918. Following elections in Sept. 2012, he continued as prime minister as the VVD formed a new coalition with the Labour Party (PvdA).

Early Life
Rutte was born in The Hague on 14 Feb. 1967 and read history at Leiden University. He served as national president of the VVD youth party from 1988–91. After graduating in 1992, he was employed by the Unilever group as a human resources manager.

Rutte became a member of the VVD's national board in 1993, resigning in 1997 to focus on his business career. He was a personnel manager at Unilever subsidiary Van den Bergh Nederland until 2000, when he was promoted to Unilever's corporate human resources group. In 2002 he became director of human resources for Unilever subsidiary IgloMora Group.

Rutte worked on the VVD candidate committee at the 2002 general election. Later that year he became state secretary for social affairs and employment in the coalition government. In 2003 he entered the House of Representatives and in 2004 was appointed state secretary for higher education and science.

Despite managing a disappointing VVD municipal election campaign, in May 2006 the party elected him 'lijstrekker' (effective leader). In June 2006 he resigned from Jan Peter Balkenende's cabinet. He won notable victories in televised debates ahead of the June 2010 general election, attracting praise for his stance on economic affairs. The VVD won the election and after months of coalition talks Rutte was appointed premier.

Career in Office
Rutte and the VVD negotiated a coalition agreement with the CDA and the right-leaning PVV. Rutte shared executive posts evenly between VVD and CDA politicians, giving the CDA the ministries of defence, the interior and finance. The PVV, whose popularity sprang from its stance on immigration, was given a stake in policy-making, although controversial leader Geert Wilders subsequently faced trial (but was acquitted in June 2011) for inciting hatred against Muslims.

Rutte pledged to cut the budget deficit from 6% to 0·9% of GDP by 2015, raise the retirement age from 65 to 66 and support nuclear power. He imposed funding cuts on universities, while immigration policy was also tightened, with a clamp-down on repeat asylum applications and low-skilled immigrants, and restrictions on the wearing of the Islamic burqa in public.

In Jan. 2011 Rutte won parliamentary approval to send personnel to Afghanistan (the Netherlands having previously withdrawn from involvement there) to train the Afghan police. In the same month he held talks with the UK prime minister, David Cameron, on re-activating the EU's Service Directive to encourage free-market reforms in the European business community.

In April 2012 Rutte and his government resigned, having failed to secure parliamentary endorsement of austerity measures following the PVV's withdrawal of support for the coalition. He stayed in office in a caretaker capacity until elections in Sept. The VVD remained the largest party, with increased representation, and entered into a new coalition with the PvdA in Nov. 2012.

DEFENCE

Conscription ended on 30 Aug. 1996.

The total strength of the armed forces in 2007 was 45,608. Reserves, 32,200. In 2006 defence expenditure totalled US$9,904m. (US$601 per capita), representing 1·5% of GDP.

Army
The core fighting element of the Royal Netherlands Army is divided into two mechanized brigades and one airborne brigade. The 1st Netherlands Army Corps merged with a German corps to become 1 German/Netherlands Corps in 1995. It is based in Münster, Germany and is a certified NATO Response Force.

Personnel in 2007 numbered 18,266. The core fighting element of the Army consists of a single element divided into two mechanized brigades and one airborne brigade. Some units in the Netherlands may be assigned to the UN as peacekeeping forces. The army is responsible for the training of these units.

There is a paramilitary Royal Military Constabulary, 6,800 strong. In addition there are 22,200 army reservists.

Navy
The principal headquarters and main base of the Royal Netherlands Navy is at Den Helder, with a minor base at Curaçao. Command and control in home waters is exercised jointly with the Belgian Naval Component (submarines excepted).

The combatant fleet includes four diesel submarines, four destroyers and four frigates. In 2007 personnel totalled 10,401 including 3,100 in the Royal Netherlands Marine Corps.

Air Force
The Royal Netherlands Air Force (RNLAF) had 10,141 personnel in 2007. It had 105 combat-capable aircraft in 2007 (F-16s) and 24 attack helicopters. All squadrons are operated by Tactical Air Command.

INTERNATIONAL RELATIONS

On 1 June 2005 the Netherlands became the second European Union member after France to reject the proposed EU constitution, with 61·54% of votes cast in a referendum against the constitution and only 38·46% in favour.

The Hague is the seat of several international organizations, including the International Court of Justice.

ECONOMY

Services accounted for 74% of GDP in 2007, industry 24% and agriculture 2%.

According to the anti-corruption organization *Transparency International*, the Netherlands ranked equal ninth in the world in a 2012 survey of the countries with the least corruption in business and government. It received 84 out of 100 in the annual index.

The Netherlands gave US$6·3bn. in international aid in 2011, which at 0·75% of GNI made it the fifth most generous developed country as a percentage of its gross national income. The Netherlands was one of only five countries to exceed the UN target of 0·7%.

Overview
The Dutch economy boasts one of the world's highest levels of average income and relatively low income inequality. Given its small domestic market, a location at the heart of northwest Europe's economy and favourable harbour facilities, the economy is one of the most open and outward-looking. The Netherlands is among the most competitive destinations for global foreign direct investment (FDI). A favourable tax environment for multinationals has attracted many foreign companies and significant FDI inflows. One of the leading donors of international aid, it has committed itself to poverty reduction in Africa.

Exports accounted for 76·7% of GDP in 2008. Rotterdam is Europe's largest port and generates annual added value equal to 3·7% of total GDP. Trade dependency is even stronger because of the scarcity of industrial raw materials while industry is geared towards processing. Relative to the European Big Four, the manufacturing sector share of GDP is small compared to its agricultural and service sectors. The Netherlands is a leader in horticulture and is a competitive meat and dairy product exporter.

In 2006 the government introduced tax cuts and increased expenditure by 6·2% on the previous year, resulting in greater economic activity, rising employment rates, modest wage rises and favourable fiscal conditions. Private consumption also rose. Annual budget deficits between 2000 and 2008 averaged roughly 1% compared to 3·7% over the previous half decade, despite weaker growth.

However, the global economic crisis of 2008 saw the economy fall into its worst recession for several decades with negative growth of 3·7% in 2009. Exports declined by 25%. Along with a fall in domestic demand, manufacturing suffered. State support was needed to save several financial institutions, with four of the five largest subject to restructuring programmes. Public debt rose to 62·9% of GDP in 2010.

Nonetheless, the Netherlands weathered the crisis better than many of its European partners and, stimulated by strong exports, growth resumed in 2010. The Netherlands also has one of the highest levels of labour productivity in the world. However, the recovery remains fragile, with more than 77% of Dutch exports going to other European countries (less than 3% of total exports go to Asia).

In Oct. 2012 the government announced austerity measures totalling €16bn. (US$20·7bn.). Low consumer confidence was cited as a major factor in the weak performance of the economy in the second half of 2012, with spending impacted by falling real wages, falling house prices and a 2% increase in VAT to 21%. Unemployment rose to a 15-year high while the budget deficit was expected to exceed the EU's 3% limit in 2013. The eurozone debt crisis remains a major threat to the economic outlook. Household debt has grown substantially and is now among the highest of any advanced economy.

The welfare system and labour market institutions follow the German model, with extensive welfare provisions and worker influence at the corporate level. The state's pension liabilities are substantial. An ageing population is a significant challenge and the chief reason for the aggressive strengthening of public finances in recent years.

Currency

On 1 Jan. 1999 the euro (EUR) became the legal currency in the Netherlands at the irrevocable conversion rate of 2·20371 guilders to 1 euro. The euro, which consists of 100 cents, has been in circulation since 1 Jan. 2002. On the introduction of the euro there was a 'dual circulation' period before the guilder ceased to be legal tender on 28 Jan. 2002. Euro banknotes in circulation on 1 Jan. 2002 had a total value of €29·7bn.

Inflation rates (based on OECD statistics):

2002	2003	2004	2005	2006	2007	2008	2009	2010	2011
3·9%	2·2%	1·4%	1·5%	1·7%	1·6%	2·2%	1·0%	0·9%	2·5%

Gold reserves were 19·69m. troy oz in Sept. 2009 and foreign exchange reserves US$10,102m. Total money supply was €217,793m. in Aug. 2009.

Budget

In 2009 central government revenues totalled €234,498m. (€249,637m. in 2008) and expenditures €260,136m. (€241,951m. in 2008). Principal sources of revenue in 2009: social security contributions, €78,959m.; taxes on goods and services, €62,606m.; taxes on income, profits and capital gains, €61,512m. Main items of expenditure by economic type in 2009: social benefits, €118,723m.; grants, €74,672m.; use of goods and services, €19,765m.

The Netherlands' budget deficit in 2011 was 4·7% of GDP (2010, 5·1%; 2009, 5·6%). In 2008 it had a surplus of 0·5%. The required target set by the EU is a budget deficit of no more than 3%.

VAT is 21·0% (reduced rate, 6·0%).

Performance

Real GDP growth rates (based on OECD statistics):

2002	2003	2004	2005	2006	2007	2008	2009	2010	2011
0·1%	0·3%	2·0%	2·2%	3·5%	3·9%	1·8%	−3·7%	1·6%	1·1%

In 2011 total GDP was US$836·1bn.

Banking and Finance

The central bank and bank of issue is the Netherlands Bank (*President*, Klaas Knot), founded in 1814 and nationalized in 1948. Its Governor is appointed by the government for seven-year terms. In 2011 the capital amounted to €500m. There were 82 registered commercial banks in 2011. The largest banks in 2007 were ABN Amro Holding NV (assets in 2007 of €1,025·2bn.) and ING Bank NV (assets in 2007 of €994·1bn.). In Oct. 2007 ABN Amro and a consortium led by the UK's Royal Bank of Scotland (and including Santander from Spain and the Belgian-Dutch company Fortis) agreed a merger worth US$98·5bn., representing Europe's largest banking takeover. The Dutch part of Fortis was nationalized in Oct. 2008 and its shares in ABN Amro were also transferred to the Dutch government. There is a stock exchange in Amsterdam; it is a component of Euronext, which was created in Sept. 2000 through the merger of the Amsterdam, Brussels and Paris bourses.

Gross external debt amounted to US$2,416,352m. in June 2012.

In Dec. 2011, 66·3% of internet users in the Netherlands were using e-banking—the highest proportion of any country.

ENERGY AND NATURAL RESOURCES

In 2008, 3·2% of energy consumption came from renewables (wind power, solar power, hydro-electric power, tidal power, geothermal energy and biomass), compared to the European Union average of 10·3%. A target of 14% has been set by the EU for 2020.

Environment

Carbon dioxide emissions from the consumption and flaring of fossil fuels in 2008 were the equivalent of 15·9 tonnes per capita.

The Netherlands is one of the world leaders in recycling. In 2008 an estimated 59% of municipal waste was recycled or composted, with only 1% going to landfill.

Electricity

Installed capacity was 23·9m. kW in 2007. Production of electrical energy in 2007 was 103·24bn. kWh (approximately 4% nuclear); consumption per capita was 7,375 kWh. There was one nuclear reactor in operation in 2010.

Oil and Gas

Production of natural gas in 2008, 67·5bn. cu. metres. Reserves in 2008 were 1,390bn. cu. metres. The Groningen gas field in the north of the country is the largest in continental Europe. In 2007 crude oil production was 2·1m. tonnes; reserves were 100m. bbls in 2007.

Minerals

In 2009, 6·0m. tonnes of salt were produced. Aluminium production in 2009 totalled 300,000 tonnes.

Agriculture

The Netherlands is one of the world's largest exporters of agricultural produce. There were 76,740 agricultural holdings in 2007. Food and live animals accounted for 12·1% of exports and 8·4% of imports in 2010. The agricultural sector (including forestry and fisheries) employs 2·5% of the workforce. In 2009 there were 1,054,700 ha. of arable land and 35,500 ha. of permanent crops. The total area of cultivated land in 2011 was 1,858,000 ha.: grassland and green fodder crops, 1,225,000 ha.; arable crops, 535,000 ha.; open ground horticulture, 89,000 ha.; glasshouse horticulture, 10,000 ha. In 2010, 212,000 people were employed in agriculture (of which family workers, 148,000; non-family workers, 64,000).

The yield of the more important arable crops, in 1,000 tonnes, was as follows:

Crop	2010	2011
Potatoes	6,843	7,333
Sugar beets	5,280	5,858
Wheat	1,370	1,175
Sown onions	1,252	1,582
Barley	204	205

Other major fruit and vegetable production in 2010 included (in 1,000 tonnes): tomatoes, 815; cucumbers, 435; carrots, 350; sweet peppers, 345; apples, 338; mushrooms, 240; pears, 195.

Cultivated areas of main flowers (2011) in 1,000 ha.: tulips, 11·9; lilies, 5·1. Total area of bulbs, 24,100 ha.

Livestock, 2011 (in 1,000) included: 12,429 pigs; 3,885 cattle; 1,088 sheep; 380 goats; 96,919 chickens.

Animal products in 2010 (in 1,000 tonnes) included: pork, bacon and ham, 1,288; beef and veal, 388; poultry, 751; milk, 11,626; cheese, 753; butter, 133; hens' eggs, 631.

Forestry
Forests covered 0·37m. ha. in 2010, or 11% of the land area. In 2007, 1·02m. cu. metres of roundwood were cut.

Fisheries
Total catch in 2010 was 389,357 tonnes (down from 495,774 tonnes in 2000), of which 387,306 tonnes were from marine waters. In 2008 the fishing fleet comprised 825 vessels of 146,925 GT.

INDUSTRY
In March 2012 the leading companies by market capitalization were: Unilever (Dutch/British), a consumer goods firm (US$94·8bn.); Heineken, a Beverages company (US$32·0bn.); and ING, a finance company (US$31·9bn.)

In 2011 there were 51,170 companies in the manufacturing industry (48% in textiles, paper, wood, furniture and miscellaneous industries; 20% basic metals and metal products; 11% electrical engineering and machinery; 9% food industry; 8% oil, chemicals, rubber and synthetics; 4% transport equipment). Total production value in 2010 was €270·4bn.

The largest industrial sectors by production value in 2010 were oil, chemicals, rubber and synthetics (37%); food industry (22%); textiles, paper, wood, furniture and miscellaneous industries (13%); and electrical engineering and machinery (13%). There were 809,000 employees (full-time equivalent) in manufacturing in 2008: textiles, paper, wood, furniture and other industry, 274,000; electrical engineering and machinery, 138,000; oil, chemicals, rubber and synthetics, 128,000; food industry, 117,000; basic metals and metal products, 112,000; transport equipment, 40,000.

Labour
The total labour force (15–65 years) in 2011 was 7,811,000 persons (3,492,000 women) of whom 419,000 (195,000 women) unemployed. Of the 7,392,000 employed persons, 5,709,000 were in permanent employment, 606,000 were in flexible employment and 1,077,000 were self-employed. Nearly a third of all 15–65-year-old women were working between 20 and 35 hours per week. By education level, the 2011 employed labour force included (in 1,000): primary education, 361; junior secondary education, 1,309; senior secondary education, 3,130; university education, 2,524 (bachelor, 1,639; masters or PhD, 885).

The unemployment rate was 5·8% in Dec. 2012, one of the lowest in the EU. Youth unemployment was only 7·8% in Jan. 2011. Although the Netherlands has a very low unemployment rate, for every 100 people below the age of 65 who are active in the labour market, 35 are not. In 2010, 48·9% of the labour force was in part-time employment with 76·5% of all women employed working part-time. In 2008 the average age for retirement was 63.

In 2010 the average weekly working hours of employees were 34·4. In 2011 employees' average annual working hours totalled 1,379. Workers in the Netherlands put in among the shortest hours of any industrialized country.

Average annual earnings of employees in 2009 totalled €30,700 with average hourly earnings of €20·01. By type of employment hourly earnings in 2009 ranged from €12·47 in hotels and restaurants up to €33·57 in mineral extraction.

INTERNATIONAL TRADE
On 5 Sept. 1944 and 14 March 1947 the Netherlands signed agreements with Belgium and Luxembourg for the establishment of a customs union. On 1 Jan. 1948 this union came into force and the existing customs tariffs of the Belgium–Luxembourg Economic Union and of the Netherlands were superseded by the joint Benelux Customs Union Tariff. It applied to imports into the three countries from outside sources, and exempted from customs duties all imports into each of the three countries from the other two.

Imports and Exports
In 2011 imports totalled €364,000m. (provisional), up from €332,000m. in 2010; exports, €405,000m. (provisional), up from €372,000m. in 2010.

Value of trade with major partners (in €1bn.):

Region/Country	Imports 2010	Exports 2010	Imports 2011 (provisional)	Exports 2011 (provisional)
Europe	205	298	228	327
Belgium	32	41	36	49
France	14	32	17	36
Germany	59	90	61	97
Italy	7	19	8	20
Russia	14	6	17	6
UK	22	30	25	32
Africa	11	11	12	12
Americas	40	26	42	27
USA	25	17	24	17
Asia	74	32	79	34
China	31	5	31	7
Japan	9	3	10	3
Australia and Oceania	1	5	2	4

The main imports in 2011 (provisional) were (in €1bn.): machines and transport equipment, 103 (100 in 2010); mineral fuels, 79 (60); chemical products, 47 (51); manufactured goods, 39 (34); food and live animals, 32 (28); inedible raw materials except fuel, 16 (13). Main exports in 2011 (provisional) were (in €1bn.): machines and transport equipment, 112 (106 in 2010); chemical products, 71 (71); mineral fuels, 65 (51); food and live animals, 48 (45); manufactured goods, 37 (33); inedible raw materials except fuels, 21 (19).

COMMUNICATIONS

Roads
In 2008 the total length of the Netherlands road network was 136,135 km (including 2,637 km of motorways). Number of vehicles (2008): private cars, 7·39m.; trucks and vans, 1·07m.; motorcycles and mopeds, 1·37m. There were 750 fatalities as a result of road accidents in 2008, equivalent to 4·6 fatalities per 100,000 population (one of the lowest death rates in road accidents of any industrialized country).

Rail
All railways are run by the mixed company 'N.V. Nederlandse Spoorwegen'. Route length in 2011 was 3,013 km. Passenger-km travelled in 2009 came to 16·32bn. Goods transported in 2010 totalled 36m. tonnes. There is a metro (44 km) and tram/light rail network (154 km) in Amsterdam and in Rotterdam (76 km and 67 km). Tram/light rail networks operate in The Hague (128 km) and Utrecht (22 km). A tram link between Maastricht and Hasselt in Belgium is currently under construction; it is expected to open in 2017.

Civil Aviation
There are international airports at Amsterdam (Schiphol), Rotterdam, Maastricht and Eindhoven. The Royal Dutch Airlines (KLM) was founded on 7 Oct. 1919. In Oct. 2003 it merged with Air France to form Air France-KLM, in which the French state owns a 15·7% stake.

Airport passenger traffic reached 53·9m. in 2011: Amsterdam handled 49·8m. passengers, Eindhoven 2·6m. and Rotterdam 1·1m. Amsterdam was the fourth busiest airport in Europe in 2011 on the basis of passenger numbers and the 14th busiest in the world. In 2011, 1·6m. tonnes of freight were transported via Dutch airports.

Shipping

In Jan. 2009 there were 810 ships of 300 GT or over registered, totalling 7,428,000 GT. Of the 810 vessels registered, 593 were general cargo ships, 83 container ships, 47 passenger ships, 43 oil tankers, 20 chemical tankers, 18 liquid gas tankers and six bulk carriers. The Dutch-controlled fleet comprised 578 vessels of 1,000 GT or over in Jan. 2009, of which 426 were under the Dutch flag and 152 under foreign flags.

Total throughput at Rotterdam, the busiest port in the Netherlands and Europe and the third busiest in the world, was 386,957,000 tonnes in 2009 (down from a record 421,136,000 tonnes in 2008). Of the total cargo handled in 2009, 273,292,000 tonnes were incoming and 113,665,000 tonnes outgoing. Rotterdam is also the busiest container port in both the Netherlands and Europe; in 2008 it was the ninth busiest container port in the world, handling 10,784,000 TEUs (twenty-foot equivalent units). The Amsterdam ports were the second busiest in the Netherlands in 2009, handling 86,677,000 tonnes of cargo (down from 94,833,000 tonnes in 2008).

There were 6,215 km of inland waterways in 2008, including 2,686 km of canals and 823 km of canalized rivers; 271·5m. tonnes of freight were carried on inland waterways in 2009.

The Netherlands was ranked third in the World Economic Forum's *Global Competitiveness Report 2009–2010* for the quality of its port facilities.

Telecommunications

In 2008 there were 7,317,000 main (fixed) telephone lines. In the same year mobile phone subscribers numbered 20,627,000 (1,248·0 per 1,000 persons). There were 14·3m. internet users in 2008. The Netherlands has one of the highest fixed broadband penetration rates, at 38·1 subscribers per 100 inhabitants in Dec. 2010. In March 2012 there were 5·8m. Facebook users.

SOCIAL INSTITUTIONS

Justice

Justice is administered by the High Court (Court of Cassation), by five courts of justice (Courts of Appeal), by 19 district courts and by 61 cantonal courts. The Cantonal Court, which deals with minor offences, comprises a single judge; more serious cases are tried by the district courts, comprising as a rule three judges (in some cases one judge is sufficient); the courts of appeal are constituted of three and the High Court of five judges. All judges are appointed for life by the Sovereign (the judges of the High Court from a list prepared by the Second Chamber of the States-General). They can be removed only by a decision of the High Court.

At the district court the juvenile judge is specially appointed to try children's civil cases and at the same time charged with administration of justice for criminal actions committed by young persons between 12 and 18 years old, unless imprisonment of more than six months ought to be inflicted; such cases are tried by three judges.

The population in penal institutions in Aug. 2008 was 16,416. Owing to a declining prison population, since Feb. 2010 some Belgian prison inmates have been accommodated at Tilburg prison in the Netherlands. 1,215,000 crimes were recorded by the police in 2007 (1,402,000 in 2002).

Police

The police force is divided into 25 regions. There is also a National Police Service which includes the Central Criminal Investigation Office, which deals with serious crimes throughout the country, and the International Criminal Investigation Office, which informs foreign countries of international crimes.

Education

Statistics for the academic year 2009–10:

	Schools/ institutions	Full-time pupils/students (in 1,000) Total
Primary education	6,895	1,548
Special primary education	311	43
Special schools	323	68
Secondary education	657	935
Senior secondary vocational education	71	522
Higher professional education	43	403
University education	13	233

University student enrolment by subject in 2009–10 (with total students): behaviour and society (49,600); economics (39,600); language and culture (32,300); health (31,500); engineering and technology (30,600); law (28,100); science (20,100); cross-sector programmes (3,200); university teacher-training courses (1,700).

In 2007 there were 29,104 Open University students and 752 staff. There are 12 study centres in the Netherlands and three support centres, plus six study centres in Belgium.

In 2006 public expenditure on education came to 5·4% of GNI and 12·0% of total government spending. The adult literacy rate is at least 99%.

Health

There were 8,673 general practitioners on 1 Jan. 2007; 2,825 pharmacists and 2,197 midwives on 1 Jan. 2006; and 16,346 specialists and 13,355 physiotherapists on 1 Jan. 2005. There were 7,994 dentists in 2005. At 1 Jan. 2006 there were 117 hospitals and 50,209 licensed hospital beds (excluding mental hospitals). The 1919 Opium Act (amended in 1928 and 1976) regulates the production and consumption of 'psychoactive' drugs. Personal use of cannabis is effectively decriminalized and the sale of soft drugs through 'coffee shops' is not prosecuted provided certain conditions are met. Euthanasia became legal when the First Chamber (the Senate) gave its formal approval on 10 April 2001 by 46 votes to 28. The Second Chamber had voted to make it legal by 104 votes to 40 in Nov. 2000. The law came into effect on 1 April 2002. In 2007 euthanasia organizations recorded 2,120 instances of doctors helping patients to die. The Netherlands was the first country to legalize euthanasia. In 2007 the Netherlands spent 8·9% of its GDP on health.

Welfare

The General Old Age Pensions Act (AOW) entitles everyone to draw an old age pension from the age of 65, although this is to rise gradually to 66 by 2019 and 67 by 2024. At 31 Dec. 2010 there were 2,881,000 persons entitled to receive an old age pension, and 98,000 a pension under the General Surviving Relatives Act; 1,932,000 parents were receiving benefits under the General Child Benefit Act. In 2011 there were 825,000 persons claiming incapacity benefits and 270,000 persons claiming benefits under the Unemployment Benefits Act.

RELIGION

Entire liberty of conscience is granted to the members of all denominations. The royal family belong to the Protestant Church in the Netherlands.

Population aged 12 years and over in 2009 was: Roman Catholics, 27%; Protestant Church in the Netherlands, 9%; Calvinist, 3%; other creeds, 10%; no religion, 44%. The Dutch Reformed Church merged with the Reformed Churches in the Netherlands and the Evangelical Lutheran Church in the Kingdom of the Netherlands in May 2004 to form the Protestant Church in the Netherlands—now the second largest church body in the country. The Roman Catholic Church had, Jan. 1992, one archbishop (of Utrecht), six bishops, four assistant bishops and about 1,750 parishes and rectorships. In Feb. 2013 there were two Roman Catholic cardinals. The Old Catholic Church of the Netherlands has one Archbishop (of Utrecht), one Bishop (of Haarlem) and 26 parishes. The Jews had, in 2008, 30 communities. In 2006 there were 825,000 Muslims (5·1% of the population). There were 99,000 Hindus (0·6% of the population) in 2004.

Source: Statistics Netherlands

CULTURE

World Heritage Sites

The Kingdom of the Netherlands has nine sites on the UNESCO World Heritage List: Schokland and its surroundings (inscribed on the list in 1995); the defence line at Amsterdam (1996); the mill network at Kinderdijk-Elshout (1997); the historic area of Willemstad, the inner city and harbour in Curaçao (1997); the D. F. Wouda steam pumping station (1998); Droogmakerij de Beemster (Beemster Polder) (1999); the Rietveld Schröder house (2000); and the Seventeenth-century canal ring area of Amsterdam inside the Singelgracht (2010).

The Netherlands shares the Wadden Sea (2009) with Germany.

Press

In 2006 there were 32 daily newspapers with a combined circulation of 4,769,000. The most widely read daily is *De Telegraaf*, with an average daily circulation of 723,000 copies in 2006.

Tourism

Tourism is a major sector of the economy. In 2011 international tourist spending totalled €10,400m. A total of 11,299,000 non-resident tourists stayed in holiday accommodation in 2011 (up from 10,883,000 in 2010 and 9,921,000 in 2009).

Festivals

Floriade, a world-famous horticultural show, takes place every ten years and is the largest Dutch attraction, being attended by 2·3m. people in 2002. The Maastricht Carnival in April attracts many visitors. The Flower Parade from Noordwijk to Haarlem occurs in late April. Koningsdag on 27 April is a nationwide celebration of King Willem-Alexander's birthday. The Oosterparkfestival, a cultural celebration of that district of Amsterdam, runs for three days in the first week of May. Liberation Day is celebrated every five years on 5 May, with the next occurrence being in 2015. An international music festival, the Holland Festival, is held in Amsterdam throughout June each year and the Early Music Festival is held in Utrecht. The North Sea Jazz Festival, the largest in Europe, takes place in The Hague. Each year the most important Dutch and Flemish theatre productions of the previous season are performed at the Theatre Festival in Amsterdam and Antwerp (Belgium). The Holland Dance Festival is held every other year in The Hague and the Springdance Festival in Utrecht annually. Film festivals include the Rotterdam Film Festival in Feb., the World Wide Video Festival in April, the Dutch Film Festival in Sept. and the International Documentary Film Festival of Amsterdam in Dec.

DIPLOMATIC REPRESENTATIVES

Of the Netherlands in the United Kingdom (38 Hyde Park Gate, London, SW7 5DP)
Ambassador: Laetitia van den Assum.

Of the United Kingdom in the Netherlands (Lange Voorhout 10, 2514 ED The Hague)
Ambassador: Paul Arkwright.

Of the Netherlands in the USA (4200 Linnean Ave., NW, Washington, D.C., 20008)
Ambassador: Rudolf Bekink.

Of the USA in the Netherlands (Lange Voorhout 102, The Hague)
Ambassador: Vacant.
Chargé d'Affaires a.i.: Edwin R. Nolan.

Of the Netherlands to the United Nations
Ambassador: Herman Schaper.

Of the Netherlands to the European Union
Permanent Representative: Pieter de Gooijer.

FURTHER READING

Centraal Bureau voor de Statistiek. *Statistical Yearbook of the Netherlands.* From 1923/24.—*Statistisch Jaarboek.* From 1899/1924.—*CBS Select (Statistical Essays).* From 1980.—*Statistisch Bulletin.* From 1945; weekly. —*Maandschrift.* From 1944; monthly bulletin.—*90 Jaren Statistiek in Tijdreeksen* (historical series of the Netherlands 1899–1989)

Nationale Rekeningen (National Accounts). From 1948–50.—*Statistische onderzoekingen.* From 1977.—*Regionaal Statistisch Zakboek* (Regional Pocket Yearbook). From 1972

Staatsalmanak voor het Koninkrijk der Nederlanden. Annual from 1814

Staatsblad van het Koninkrijk der Nederlanden. From 1814

Staatscourant (State Gazette). From 1813

Andeweg, Rudy B. and Irwin, Galen A., *Governance and Politics of the Netherlands.* 3rd ed. 2009

Arblaster, Paul, *A History of the Low Countries.* 2005

Blom, J. C. H. and Lamberts, E. (eds.) *History of the Low Countries.* Revised ed. 2006

Cox, R. H., *The Development of the Dutch Welfare State: from Workers' Insurance to Universal Entitlement.* 1994

Gladdish, K., *Governing from the Centre: Politics and Policy-Making in the Netherlands.* 1991

National library: De Koninklijke Bibliotheek, Prinz Willem Alexanderhof 5, The Hague.

National Statistical Office: Centraal Bureau voor de Statistiek, Netherlands Central Bureau of Statistics, POB 4000, 2270 JM Voorburg.

Statistics Netherlands Website: http://www.cbs.nl

OVERSEAS COUNTRIES AND TERRITORIES

Landen en gebieden overzee

These fall into two categories: *Autonomous Countries within the Kingdom of the Netherlands* (Aruba, Curaçao and Sint Maarten) and *Autonomous Special Municipalities of the Netherlands* (Bonaire, Saba and Sint Eustatius). Following the dissolution of the Netherlands Antilles on 10 Oct. 2010, Curaçao and Sint Maarten became autonomous countries within the Kingdom of the Netherlands (a status held by Aruba since 1986) while Bonaire, Saba and Sint Eustatius were granted a status comparable to that of the municipalities within the Netherlands itself.

AUTONOMOUS COUNTRIES WITHIN THE KINGDOM OF THE NETHERLANDS

Zelfstandige landen binnen het Koninkrijk der Nederlanden

Aruba

KEY HISTORICAL EVENTS

Discovered by Alonzo de Ojeda in 1499, Aruba was claimed for Spain but not settled. It was acquired by the Dutch in 1634, but apart from garrisons, was left to the indigenous Caiquetious (Arawak) Indians until the 19th century. From 1828 it formed part of the Dutch West Indies and, from 1845, part of the Netherlands Antilles with which, on 29 Dec. 1954, it achieved internal self government. Following a referendum in March 1977 the Dutch government announced on 28 Oct. 1981 that Aruba would proceed to independence separately from the other islands. Aruba was constitutionally separated from the Netherlands Antilles from 1 Jan. 1986. An agreement with the Netherlands government in June 1990 deleted references to eventual independence at Aruba's request.

TERRITORY AND POPULATION

The island, which lies in the southern Caribbean 32 km north of the Venezuelan coast and 68 km west of Curaçao, has an area of 193 sq. km (75 sq. miles) and a population at the last census in Sept. 2010 of 101,484; density 526 inhabitants per sq. km. The UN gives a projected population for 2015 of 109,000. The chief towns are Oranjestad, the capital (2010 census population, 28,294) and San Nicolas. Dutch is the official language, but the language usually spoken is Papiamento, a creole language. Over half the population is of Indian stock, with the balance of Dutch, Spanish and mestizo origin.

SOCIAL STATISTICS

Population growth rate, 2009, 1·0%. Life expectancy at birth in 2010 was 75 years. Birth rate per 1,000 population (2008), 11·6; death rate, 4·9; infant mortality, 1·6.

CLIMATE

Aruba has a tropical marine climate, with a brief rainy season from Oct. to Dec.

CONSTITUTION AND GOVERNMENT

Under the separate constitution inaugurated on 1 Jan. 1986, Aruba is an autonomous part of the Kingdom of the Netherlands with its own legislature, government, judiciary, civil service and police force. The Netherlands is represented by a Governor appointed by the monarch. The unicameral legislature (*Staten*) consists of 21 members elected for a four-year term of office.

RECENT ELECTIONS

Elections were held on 25 Sept. 2009. The Aruban People's Party (AVP) won with 12 out of 21 seats (48·1% of the vote), against 8 seats (35·9%) for the People's Electoral Movement (MEP) and 1 seat (5·7%) for the Real Democracy Party (PDR). Turnout was 86·3%.

CURRENT GOVERNMENT

Governor: Fredis Refunjol; b. 1950 (took office on 11 May 2004).

   Prime Minister: Mike Eman; b. 1961 (sworn in on 30 Oct. 2009).

Government Website: http://www.gobierno.aw

ECONOMY

Currency

The currency is the *Aruban florin* (AWG). Inflation was 3·8% in 2005. Total money supply in July 2005 was 1,030m. Aruban

florins, foreign exchange reserves were US$297m. and gold reserves were 100,000 troy oz.

Budget

In 2007 revenues totalled 1,632·6m. Aruban florins and expenditures 1,685·6m. Aruban florins.

Performance

Real GDP growth was 0·6% in 2006 and 0·4% in 2007.

Banking and Finance

The *President* of the Central Bank of Aruba is Jane Semeleer.

ENERGY AND NATURAL RESOURCES

Environment

Carbon dioxide emissions from the consumption and flaring of fossil fuels were the equivalent of 10·9 tonnes per capita in 2008.

Electricity

In 2007 consumption of electricity was 936,000 MWh.

Fisheries

In 2008 the catch totalled 151 tonnes.

INDUSTRY

The quantity of oil refined in 2001 was 64m. bbls.

Labour

The economically active population in 2000 numbered 44,384 persons of which 41,286 were employed and 3,098 unemployed.

EXTERNAL ECONOMIC RELATIONS

There are two Free Zones at Oranjestad.

Imports and Exports

2008: imports, US$1,113m.; exports, US$101m. Leading import suppliers are the USA, Netherlands, UK and Colombia. Leading export destinations are Panama, Colombia, Venezuela and the USA.

COMMUNICATIONS

Roads

In 2000 there were 39,995 passenger cars and 5,443 commercial vehicles. There were 439 passenger cars per 1,000 inhabitants.

Civil Aviation

There is an international airport (Aeropuerto Internacional Reina Beatrix).

Shipping

Oranjestad has a container terminal and cruise ship port. The port at Barcadera services the offshore and energy sector and a deep-water port at San Nicolas services the oil refinery.

Telecommunications

Aruba had 128,000 mobile phone subscribers in 2009 and 38,300 fixed telephone lines. There were 24,000 internet users in 2007.

SOCIAL INSTITUTIONS

Justice

There is a Joint Court of Justice of Aruba, Curaçao, Sint Maarten and the Caribbean part of the Netherlands. Final Appeal is to the Supreme Court in the Netherlands. The population in penal institutions in Jan. 2005 was 231 (equivalent to 324 per 100,000 population).

Education

In 2007 there were 2,713 pupils in pre-primary schools, 9,511 in primary schools, 2,950 in junior high schools, 2,372 in high schools and 2,192 in technical or vocational schools. Literacy rate

(2008), 98%. In 2007 public spending on education amounted to 4·0% of GDP and 11·4% of total government expenditure.

Health
In 2000 there was one hospital with 305 beds.

RELIGION
In 2000, 86·2% of the population were Roman Catholic.

CULTURE
Press
In 2006 there were four daily newspapers with a combined circulation of 54,000 (667 per 1,000 adult inhabitants).

Tourism
In 2008 there were 826,774 tourists staying 6,264,689 nights and 556,090 cruise passenger visitors.

DIPLOMATIC REPRESENTATIVES
US Consul-General: Valerie Belon (J. B. Gorsiraweg 1, Willemstad, Curaçao).

FURTHER READING
Central Bureau of Statistics Website: http://www.cbs.aw

Curaçao

KEY HISTORICAL EVENTS
The Arawak Amerindians were the first people to settle the island. Spanish sailors sighted it in 1499 but it was not until 1634 that the Dutch established a presence. The Dutch West India Company built the capital, Willemstad, on a natural harbour. The island became an important centre for regional commerce, in particular the slave trade. In 1828 it was integrated into the Dutch West Indies and from 1845 was part of the Netherlands Antilles.

Economic decline took hold after slavery was outlawed by the Dutch in 1863. When oil was discovered in 1914, Royal Dutch Shell and the Dutch government combined to build a refinery. Internal self-government was granted in 1954 and in Oct. 2010 Curaçao became autonomous within the Kingdom of the Netherlands. The Netherlands Antilles was formally dissolved.

TERRITORY AND POPULATION
Situated around 100 km north of the Venezuelan coast, Curaçao covers 444 sq. km and at the 2011 census had a population of 150,563. Willemstad is the capital with a population (2001) of 125,000.

Dutch, Papiamento (a Creole language with elements of Dutch, English, Portuguese and Spanish) and English are all official languages.

CLIMATE
Willemstad, Feb. 27·2°C, Aug. 29·4°C. Annual rainfall 553 mm.

CONSTITUTION AND GOVERNMENT
At a non-binding referendum on 8 April 2005, 68% of votes cast favoured Curaçao seceding from the Netherlands Antilles and becoming a territory of the Netherlands in its own right. In Oct. 2010 the Netherlands Antilles was dissolved and Curaçao became an autonomous state within the Kingdom of the Netherlands, independent in all respects except defence and foreign affairs.

The head of state is the monarch of the Netherlands, represented by a *Governor.* The *Prime Minister* heads the Executive Council. The *Staten* (parliament) is comprised of 21 members elected by popular vote every four years.

RECENT ELECTIONS
Parliamentary elections were held on 19 Oct. 2012. Pueblo Soberano (PS) won 5 seats (22·6% of the vote), Movementu Futuro Korsou (MFK) 5 (21·2%), Partido Antiá Restrukturá (PAR) 4 (19·7%), Partido pa Adelanto i Inovashon Soshal (PAIS) 4 (17·7%), Movishon Antia Nobo (MAN) 2 (9·5%) and Partido Nashonal di Pueblo (PNP) 1 (5·9%). Turnout was a record 74·5%.

CURRENT GOVERNMENT
Acting Governor: Adeel van der Pluijm-Vrede.
Prime Minister: Daniel Hodge, b. 1959 (in office since 31 Dec. 2012).

Government Website (Dutch and Papiamento only):
   http://www.curacao-gov.an

ECONOMY
The economy is reliant on oil refining, tourism and offshore finance.

Currency
It is expected that the *Caribbean guilder* will replace the *Netherlands Antillean guilder* in the course of 2013. Inflation was 1·8% in 2008 (2·2% in 2007).

Performance
Real GDP growth was 2·1% in 2008 (1·6% in 2007).

Banking and Finance
After dissolution, the Bank of the Netherlands Antilles was renamed the Central Bank of Curaçao and Sint Maarten.

ENERGY AND NATURAL RESOURCES
Oil and Gas
The economy was formerly based on oil refining at the Shell refinery but following an announcement that closure was imminent, it was sold to the government in Sept. 1985 and leased to Petróleos de Venezuela to operate on a reduced scale.

INDUSTRY
Labour
Unemployment stood at 9·7% in 2009.

COMMUNICATIONS
Roads
In 2004 there were 75,248 registered vehicles (60,590 passenger cars).

Civil Aviation
There is an international airport (Curaçao International Airport or Hato International Airport), which handled 1,452,015 passengers in 2009.

Shipping
In 2006, 2,889 ships entered Curaçao's port.

RELIGION
At the 2001 census 80·1% of the population was Roman Catholic.

CULTURE
The Historic Area of Willemstad, Inner City and Harbour was inscribed on the UNESCO World Heritage List in 1997.

Tourism
In 2006 Curaçao handled 324,345 cruise ship passengers.

DIPLOMATIC REPRESENTATIVES
US Consul-General: Valerie Belon (J. B. Gorsiraweg 1, Willemstad, Curaçao).

FURTHER READING
Statistical office: Central Bureau of Statistics Curaçao, Fort Amsterdam Z/N, Curaçao.
Website: http://www.cbs.cw

Sint Maarten

KEY HISTORICAL EVENTS

Sighted by Christopher Columbus on 11 Nov. 1493, the feast day for St Martin of Tours, the island was eventually settled by the Dutch in 1631. A centre of salt mining, the British, French and Spanish competed for control with the Dutch. In 1648 the island was divided in two by the Treaty of Concordia, with the French administering the north (called St Martin) and the Dutch the south. Slaves in the south worked on cotton, sugar and tobacco plantations until the Dutch abolished slavery in 1863. Economic decline set in until Sint Maarten became a free port at the outbreak of the Second World War.

In 1954 it was incorporated into the Netherlands Antilles. In 2000 a referendum supported self-government, a status achieved in Oct. 2010 when Sint Maarten became autonomous within the Kingdom of the Netherlands.

TERRITORY AND POPULATION

Situated around 950 km northeast of the Venezuelan coast, Sint Maarten covers 34 sq. km and at 1 Jan. 2009 had a population of 40,917. Philipsburg is the capital. The northern half of the island is the French possession of St Martin.

The official languages are Dutch and English. Creole, Papiamento and Spanish are also spoken.

CLIMATE

Feb. 25·0°C, Aug. 28·5°C. Annual rainfall averages between 931 mm and 955 mm.

CONSTITUTION AND GOVERNMENT

In a referendum on 22 June 2000, 70% of votes cast were in favour of Sint Maarten becoming a self-governing country within the Kingdom of the Netherlands. This status was achieved in all respects except defence and foreign affairs in Oct. 2010 when the Netherlands Antilles was dissolved.

The head of state is the monarch of the Netherlands, represented by a *Governor*. The *Prime Minister* heads a *Council of Ministers*. The *Staten* (parliament) is comprised of 15 members elected by proportional representation every four years.

RECENT ELECTIONS

Parliamentary elections were held on 17 Sept. 2010. The National Alliance (NA) won 7 seats (45·9% of the vote), United People (UP) 6 (36·1%) and Democratic Party (DP) 2 (17·1%). With no party achieving a clear majority, UP and DP agreed to form the government.

CURRENT GOVERNMENT

Governor: Eugene Holiday (in office since 10 Oct. 2010).

Prime Minister: Sarah Wescot-Williams (DP; in office since 10 Oct. 2010).

Government Website: http://www.sintmaartengov.org

ECONOMY

Tourism is the mainstay of the economy, providing 80% of jobs. Virtually all food, energy and manufactured goods are imported.

Currency

It is expected that the *Caribbean guilder* will replace the *Netherlands Antillean guilder* in the course of 2013. Inflation was 1·8% in 2008 (2·2% in 2007).

Performance

The economy grew by 1·6% in 2008, down from 4·5% in 2007. GDP at purchasing power parity was estimated at US$794·7m. in 2008. Unemployment stood at 10·2% in the same year.

Banking and Finance

After dissolution, the Bank of the Netherlands Antilles was renamed the Central Bank of Curaçao and Sint Maarten.

INDUSTRY

Labour

The unemployment rate in June 2009 was 12·2%.

COMMUNICATIONS

Civil Aviation

There is an international airport (Princess Juliana Airport), which handled 1,625,885 passengers in 2009.

Shipping

In 2008 Sint Maarten handled 1,345,812 cruise ship passengers.

RELIGION

At the 2001 census 39% of the population was Roman Catholic, 27% Protestant and 12% Pentecostal.

CULTURE

Tourism

There were 475,410 stay-over tourist arrivals in 2008.

DIPLOMATIC REPRESENTATIVES

US Consul-General: Valerie Belon (J. B. Gorsiraweg 1, Willemstad, Curaçao).

AUTONOMOUS SPECIAL MUNICIPALITIES OF THE NETHERLANDS

Bijzondere gemeenten van Nederland

Bonaire

Inhabited by Arawak Caquetios Indians by AD 1000, the island was sighted by Spanish sailors in 1499 and settled by the Dutch from 1623. The two nations exchanged control of the island until the Dutch established dominance in 1636, with the Dutch West India Company running the island. The British seized control from 1800–03 and also from 1807–16. Slavery was outlawed in 1862. After Nazi Germany invaded the Netherlands in 1940 Bonaire was used as an internment camp. In the post-war years tourism became the staple of the economy. When the Netherlands Antilles was dissolved on 10 Oct. 2010, Bonaire became an Autonomous Special Municipality of the Netherlands.

Bonaire is situated some 100 km north of the Venezuelan coast. It has an area of 288 sq. km and, in Jan. 2009, a population of

12,877. The capital is Kralendijk. Dutch, English and Papiamento are all official languages.

The US dollar replaced the Netherlands Antillean guilder as the legal currency in Jan. 2011. Real GDP grew by 2·7% in 2005 and inflation was 1·3%. The mainstays of the economy are tourism (particularly diving) and the salt industry. Flamingo International Airport handled 573,792 passengers in 2006 (of whom 277,505 were in transit). In 2004 there were 6,766 registered vehicles (4,139 passenger cars). In 2006 there were 63,552 tourist arrivals. Bonaire handled 40,077 cruise ship passengers in 2005.

Lieut.-Governor: Lydia Emerencia (took office on 1 March 2012).

Saba

Archaeological evidence suggests the island was originally inhabited by Arawak or Carib Indans. Sighted by Columbus in 1493, there was no serious attempt at Western colonization until the Dutch West India Company sent a party of settlers in 1640. An English buccaneer, Thomas Morgan, seized the island in 1664 and it served as a centre of piracy in the 17th century. The French, English and Dutch long battled for dominance, with rum and sugar important revenue earners. The Dutch reclaimed Saba in 1816 and have maintained control ever since. After the dissolution of the Netherlands Antilles on 10 Oct. 2010, Saba became an Autonomous Special Municipality of the Netherlands.

The island has an area of 13 sq. km and a population, in Jan. 2009, of 1,601. The capital is The Bottom. Dutch, English and Papiamento are all official languages. Roman Catholicism is the dominant religion. There is a single main road plus the Juancho E. Yrausquin Airport. A ferry service operates regularly to and from Sint Maarten. Tourism is increasingly important to the economy, with some 25,000 visitors a year. Fishing is also a major revenue earner. The US dollar replaced the Netherlands Antillean guilder as the legal currency in Jan. 2011. The Saba University School of Medicine has been in operation since 1986.

Lieut.-Governor: Jonathan G. A. Johnson (took office on 2 July 2008).

Sint Eustatius

Sighted, but not landed, by Christopher Columbus in 1493, the island was settled by a delegation from Zeeland, a province of the Netherlands, in 1636. From 1678 the Dutch West India Company was directly responsible for government. A free port from 1756, it became a focus of smuggling as well as a centre of sugar production. On 16 Nov. 1766 the guns of Fort Oranje were fired to return the salute of a visiting American ship, an act generally considered as the first international recognition of the newly independent USA. Against the backdrop of the Fourth Anglo–Dutch War, the British seized control of the island but yielded power to the French within a year, who transferred sovereignty back to the Dutch in 1784. A referendum in April 2005 came out in favour of remaining within the Netherlands Antilles, but it was the only constituent island in favour of its continuation. Following the dissolution of the Netherlands Antilles on 10 Oct. 2010, Sint Eustatius became an Autonomous Special Municipality of the Netherlands.

The island covers an area of 21 sq. km and in Jan. 2009 had a population of 2,768. Its capital is Oranjestad. Dutch, English and Papiamento are all official languages. The US dollar replaced the Netherlands Antillean guilder as the legal currency in Jan. 2011. The government is the principal employer. F. D. Roosevelt Airport has six scheduled flights from St Maarten every day. The University of Sint Eustatius School of Medicine received its charter in 1999.

Lieut.-Governor: Gerald Berkel (took office on 1 April 2010).

NEW ZEALAND

© Research Machines plc 2006

Aotearoa

Capital: Wellington
Population projection, 2015: 4·60m.
GNI per capita, 2011: (PPP$) 23,737
HDI/world rank: 0·908/5
Internet domain extension: .nz

KEY HISTORICAL EVENTS

The earliest settlers of New Zealand are thought to have come from eastern Polynesia, around the turn of the first millennium. Maori oral traditions point to discovery of the country by Kupe, who gave New Zealand its first name, Aotearoa, or 'Land of the Long White Cloud'. Oral tradition also refers to seven waka leaving a homeland known as Hawaiiki in a Great Fleet. The waka are still remembered in the names of significant tribal groupings and descent lines: *Aotea, Kurahaupo, Mataatua, Tainui, Takitimu, Te Arawa,* and *Tokomaru*.

By Capt. James Cook's arrival in 1769, substantial settlements existed throughout the North Island, with smaller settlements in the South Island. Sporadic warfare was common as tribes, or 'iwi', fought for resources and status, or 'mana'.

The first recorded European contact was with Dutch explorer Abel Tasman in 1642. Believing the South Island to be the beginning of a mythical continent connected to Southern Africa, he bequeathed the name 'Staten Land'. A Dutch cartographer corrected Tasman, giving the name New Zealand to compliment the larger New Holland, as Australia was known at the time.

Earlier on the same voyage Tasman had landed on an island off Australia which he named Van Diemen's Land, later called Tasmania.

Contact between Maori and Europeans, or 'Pakeha', followed Cook's journeys to New Zealand and mapping of the coastline, opening the way for sealing and whaling stations. Coastal trade grew throughout the first decades of the 19th century and trade routes were established between Maori and the new colony of New South Wales as early as the 1820s. The Maori adapted quickly to both a market economy—selling provisions, timber and flax—and to new technologies (notably the musket). Pakeha settlement in the decades following Cook's arrival was often on Maori terms and was used by Maori in the traditional pursuit of mana in the eyes of rivals and neighbours. Mission stations soon appeared: the Church Missionary Society established three stations in the Bay of Islands between 1814 and 1823, and were joined by a Wesleyan Missionary Society station in the Hokianga in the 1820s.

British Ascendancy
With greater contact both Maori and Pakeha saw the need to regulate Pakeha settlement. The Colonial Office in London appointed a Resident, James Busby, in 1833. In 1835, prompted by Busby, thirty-five chiefs signed a Declaration of Independence naming themselves the heads of state of a 'United Tribes of New Zealand'. Relations between Maori and Pakeha were formalized by the signing of the Treaty of Waitangi in 1840. In principle—or at least in the Maori text—this treaty guaranteed Maori chieftainship, or 'rangatiratanga', while granting governorship, or 'kawanatanga', to Queen Victoria. Until 1860 Maori outnumbered Pakeha but in practice—and in the English text—sovereignty was transferred, allowing greater British settlement and control.

Established in 1840, Auckland was chosen as the colony's capital by its first governor, Capt. William Hobson. Immigration was encouraged by the New Zealand Company with settlements at Wellington and Wanganui (1840), New Plymouth (1841), and Nelson (1842). Scottish immigrants founded Dunedin (1848); and Edward Gibbon Wakefield made plans for a model English settlement at Christchurch (1851). In 1852 representative government was established with a constitution providing for a House of Representatives and Legislative Council, as well as six provincial councils. The governor at the time, Sir George Grey, retained the right of veto and was responsible for 'Native' policy. At first, each provincial council exercised extensive powers. Their abolition in 1876 marked the beginnings of central government. The Legislative Council was disbanded in 1950 leaving New Zealand with a single-tier parliament.

Initially, voting—based on individual land ownership—excluded Maori who traditionally owned land collectively. Participation was extended to Maori in 1867 with four Maori seats. James Carroll, Apirana Ngata, Maui Pomare, and Peter Buck (Te Rangi Hiroa) were all prominent: Carroll was the first Maori to enter government. He was minister of native affairs and later acting prime minister

Settlement was not always peaceful: war broke out in the 1840s and 1860s in the central and western North Island. British troops fought alongside local militia and friendly Maori, facing some of the earliest forms of trench and guerrilla warfare. The Native Land Court was set up in 1865 to determine the ownership of Maori land according to Pakeha law. Where Maori land and user rights existed communally, the Court sought to define parcels of land owned individually, thereby facilitating land sales.

Economic Boom

Land speculation fuelled an agricultural boom in the 1840s and 1850s providing the colony's first sustainable export commodity. New Zealand provided 8·6% of Britain's wool imports in 1861 and had 8·5m. sheep by 1867. The development of refrigerated shipping in the 1880s bolstered the pastoral economy through meat exports. Gold rushes in the 1860s and 1870s in Otago, the west coast of the South Island, and in Coromandel also contributed to the economy. Gold exports totalled £46m. by 1890. Wealth brought progress; the 1870s administrations of Julius Vogel and Harry Atkinson borrowed heavily to fund work schemes to encourage immigration and settlement. 1,100 miles of rail track were laid by 1879, and telegraphs linked all the main towns. The population doubled to 500,000 by 1881.

The 1880s saw the beginnings of party politics. A Liberal agenda, with policies of 'one man, one vote' and the compulsory purchase of large estates, was popular and succeeded in extending suffrage to all men. Robert Stout and John Ballance's leasehold land policies in the mid-1880s were also popular. A Liberal Party was formed in 1889 and, backed by unions and the landless, won the 1890 election with Ballance as its leader. The Conservatives formed the first genuine opposition. Richard John Seddon took over the Liberal leadership in 1892 and remained premier until his death in 1906. Among Liberal achievements were the Land and Income Tax Act 1891 and the Advance to Settlers Act 1894, which assisted 17,000 people by 1912. In 1893, New Zealand became the first country to extend the suffrage to women. Other reforms included William Pember Reeves' Industrial Conciliation and Arbitration Act of 1894 and one of the world's first national pension schemes.

In 1901 New Zealand declined the offer to join the Commonwealth of Australia and remained a British colony until 1907 when it gained Dominion status. Parliament remained subordinate to the British parliament until the adoption of the *Statute of Westminster 1931* in 1947, when New Zealand became fully sovereign with the British monarch as head of state. New Zealand annexed the Cook Islands in 1901, and was granted administration of Western Samoa at the Treaty of Versailles in 1918. It administered Samoa until the 1960s. New Zealand contributed around 100,000 soldiers in the First World War from a population of little more than a million; nearly 17,000 did not return. Nearly 9,000 New Zealanders died in the influenza epidemic spread by returning soldiers, with a Maori mortality rate six times that of Pakeha. In the Second World War around 200,000 New Zealanders joined Allied forces from a population of 1·6m.

Twentieth Century

Class-based political divisions intensified in the early twentieth century. Amid industrial unrest in 1912, the Liberal government fell to a vote of no confidence. Reform took power, introducing anti-union legislation. Strikes in Waihi, Wellington and Huntly were quelled. Reform governed until 1928, assisted at first by a wartime coalition with the Liberals, and then with tacit Liberal support. Policies broadly followed the dictates of farmers, creating a national Meat Board (1922) and Dairy Board (1923).

A United–Reform coalition (1931–35) coped with world recession. Employment reached 12%; the national income fell from an estimated £150m. to £90m., and the value of exports fell by 40%. To balance the budget, cuts were made to pensions, education, health and public works. In the absence of an unemployment benefit, men were sent to rural relief camps to work on low-capital, high-labour tasks. Measures such as a Reserve Bank and currency devaluation in 1933 helped farmers but did not address the broader social distress.

Welfare State

Michael Joseph (Micky) Savage's first Labour government (1935–49) reclaimed for New Zealand its title of social laboratory of the world. It introduced one of the world's most comprehensive social welfare systems—incorporating pensions, health, education and family benefits—and increased state housing; introduced state guaranteed prices for farm produce to protect farmers from international price fluctuations; and nationalized the Reserve Bank.

After 1936 rural support rallied around a National Party formed from remnants of the United–Reform coalition. Labour retained power with support of the four Maori seats, all held by the Ratana Party. The National Party won the 1949 election, promising to increase spending power and curb union power and economic controls. National retained power for most of the post-war boom years; brief Labour administrations under Walter Nash (1957–60) and Norman Kirk (1972–75) coincided with unfavourable economic conditions. Keith Holyoake's National government (1960–72) was concerned that Britain's anticipated entry to the EEC would damage New Zealand's exports to the UK. The Equal Pay Act (1972) challenged gender-based pay discrimination. A state-funded workplace injury compensation was set up.

Maori demands for recognition of the Treaty of Waitangi grew in the 1970s. The 1975 Land March saw tens of thousands march on parliament and the occupation of Bastion Point in 1977–78 centred on land compulsorily acquired by the government in 1951. Labour established the Waitangi Tribunal in 1975 to hear Maori claims of Treaty breaches. It lacked authority until 1985 when a Labour government made its powers retrospective to 1840. Tribunal recommendations have formed the basis for negotiations between the Crown and tribal authorities.

Britain's entry into the EEC in 1973 was a set-back for an economy dependent on exports to Britain. Robert Muldoon's National government (1975–84) introduced tariff protection, wage and price freezes, and increased borrowing for 'Think Big' public works. Muldoon won a narrow victory in the 1981 election following civil unrest during the 'Springbok' rugby tour. Riot police faced massive demonstrations as many New Zealanders opposed sporting links with the South African apartheid regime. The country found a new direction in the free market policies of David Lange's Labour government, which came to power in 1984. The economic direction of Roger Douglas, 'Rogernomics', radically altered the socio-economic landscape, reducing trade barriers and selling state assets to fund debt recovery.

In international affairs the Labour government introduced the New Zealand Nuclear Free Zone, Disarmament, and Arms Control Act 1987, declaring the country nuclear free. The legislation—supported by all political parties—led to the end of New Zealand's involvement in the ANZUS military agreement with Australia and the USA. In 1973 Australia and New Zealand tried to halt French nuclear testing in the Pacific through the International Court of Justice. New Zealand sent two frigates to Mururoa Atoll in protest. The 1985 bombing of the *Rainbow Warrior* in Auckland harbour by French secret service agents reopened the issue.

Wrangling over economic direction in the late 1980s led to Lange's resignation. He was replaced by Geoffrey Palmer in 1989, who in turn resigned shortly before the 1990 election. He was succeeded by Mike Moore. Labour lost the 1990 election to a National Party led by Jim Bolger who was determined on free market reforms. Social welfare reform, cuts in tertiary education funding and reform of accident compensation legislation cut back state intervention. The Employment Contracts Act (1991) outlawed compulsory union membership and introduced individual contracts, weakening union power. Jenny Shipley led a leadership coup in 1997 to become the country's first female prime minister, though not the first elected female prime minister. That landmark was achieved by Helen Clark who led the Labour Party to victory in the 1999 election. In 2008 John Key led National back to power.

Electoral reform in the 1990s saw New Zealand move from a first-past-the-post system to proportional representation under the mixed-member-proportional system (MMP). Despite the debacle of the first MMP election in 1996 when a minor party (New Zealand First, formed by disgruntled National supporters) played National off against Labour for two months before forming a coalition with National, the system has provided greater representation for minorities.

TERRITORY AND POPULATION

New Zealand lies southeast of Australia in the south Pacific, Wellington being 1,983 km from Sydney. There are two principal islands, the North and South Islands, besides Stewart Island, Chatham Islands and small outlying islands, as well as the territories overseas.

New Zealand (i.e. North, South and Stewart Islands) extends over 1,750 km from north to south. Area, excluding territories overseas, 267,707 sq. km. The main islands are: North Island, 114,154 sq. km; South Island, 150,416 sq. km; Stewart Island, 1,681 sq. km; Chatham Islands, 963 sq. km. The minor islands included within the geographical boundaries of New Zealand (but not within any local government area) are: Antipodes Islands, Auckland Islands, Bounty Islands, Campbell Island, Kermadec Islands, Snares Islands, Solander Island and Three Kings Islands. With the exception of meteorological station staff on Raoul Island in the Kermadec Group and Campbell Island there are no inhabitants.

The Kermadec Islands were annexed to New Zealand in 1887, have no separate administration and all New Zealand laws apply to them. Situation, 29° 10' to 31° 30' S. lat., 177° 45' to 179° W. long., 1,600 km NNE of New Zealand. The largest of the group is Raoul or Sunday Island, 29 sq. km, smaller islands being Macauley and Curtis, while Macauley Island is 5 km in circuit.

Growth in census population, exclusive of territories overseas:

	Total population	Average annual increase (%)		Total population	Average annual increase (%)
1858	115,461	—	1945[1,2]	1,702,329	0.83
1874	344,985	—	1951[1]	1,939,473	2.37
1878	458,007	7.33	1956[1]	2,174,061	2.31
1881	534,030	5.10	1961[1]	2,414,985	2.12
1886	620,451	3.06	1966[1]	2,676,918	2.11
1891	668,652	1.50	1971[1]	2,862,630	1.35
1896	743,214	2.13	1976[1]	3,129,384	1.80
1901[1]	815,862	1.90	1981[1]	3,175,737	0.29
1906	936,309	2.75	1986[1]	3,307,083	0.82
1911	1,058,313	2.52	1991[1]	3,434,949	0.76
1916[1]	1,149,225	1.50	1996[1]	3,681,546	1.40
1921	1,271,667	2.27	2001[1]	3,820,749	0.74
1926	1,408,140	2.06	2006[1,2]	4,143,279	1.63
1936[2]	1,573,812	1.13			

[1]Excluding members of the Armed Forces overseas.
[2]The census of New Zealand is quinquennial, but the census falling in 1931 was abandoned as an act of national economy, and owing to war conditions the census due in 1941 was not taken until 25 Sept. 1945.

The latest census took place on 7 March 2006. Of the 4,143,279 people counted, 4,027,947 were usually resident in the country and 115,332 were overseas visitors.

The usually-resident populations of the 11 regional councils, five unitary authorities and one special territorial authority (all data conforms with boundaries redrawn after the 1989 reorganization of local government) in 2001 and 2006:

Local Government Region	Total population 2001 census	2006 census	Percentage change 2001–06 (%)
Northland	140,133	148,470	5.9
Auckland[1]	1,158,891	1,303,068	12.4
Waikato	357,726	382,713	7.0
Bay of Plenty	239,412	257,379	7.5
Gisborne[1]	43,974	44,499	1.2

Local Government Region	Total population 2001 census	2006 census	Percentage change 2001–06 (%)
Hawke's Bay	142,947	147,783	3.4
Taranaki	102,858	104,124	1.2
Manawatu-Wanganui	220,089	222,423	1.1
Wellington	423,765	448,959	5.9
Total North Island	2,829,798	3,059,418	8.1
Tasman[1]	41,352	44,625	7.9
Nelson[1]	41,568	42,891	3.2
Marlborough[1]	39,558	42,558	7.6
West Coast	30,303	31,326	3.4
Canterbury	481,431	521,832	8.4
Otago	181,542	193,800	6.8
Southland	91,005	90,876	–0.1
Total South Island	906,753	967,908	6.7
Area outside region[2]	726	618	–14.9
Total New Zealand	3,737,277	4,027,947	7.8

[1]Unitary Authorities. [2]Special Territorial Authority—Chatham Islands.

The UN gives a projected population for 2015 of 4.60m.

In 2011, 86.2% of the population lived in urban areas. Density, 15.3 per sq. km (2006).

Resident populations of main urban areas at the 2006 census were as follows:

North Island			
Auckland	1,208,091	Wanganui	38,988
Gisborne	32,529	Wellington	397,974
Hamilton	184,838	Whangarei	49,080
Kapiti	37,344		
Napier	118,404	*South Island*	
New Plymouth	49,281	Christchurch	360,768
Palmerston North	76,032	Dunedin	110,997
Rotorua	53,766	Invercargill	46,773
Tauranga	108,882	Nelson	56,364

Between 1996 and 2006 the number of people who identified themselves as being of European ethnicity dropped from 83.1% to 77.6%. Pacific Island people made up 6.9% of the population in 2006 (5.8% in 1996); Asian ethnic groups went from 5.0% in 1996 to 9.2% in 2006. Permanent and long-term arrivals in 2009–10 totalled 82,106, including 15,691 from Australia, 15,340 from the UK, 6,924 from India, 5,951 from the People's Republic of China and 3,476 from the USA. Permanent and long-term departures in 2009–10 totalled 67,599, including 33,027 to Australia, 8,809 to the UK, 2,646 to the USA and 2,398 to the People's Republic of China.

Maori population: 1896, 42,113; 1936, 82,326; 1945, 98,744; 1951, 115,676; 1961, 171,553; 1971, 227,414; 1981, 279,255; 1986, 294,201; 1991, 324,000; 1996, 523,374; 2001, 526,281; 2006, 565,329 (14.6% of the total population compared with 15.1% in 1996). In the 2006 census, 157,110 New Zealand residents (4.1%) said they could hold a conversation about everyday matters in Maori. In 2006, 23.7% of people of Maori ethnicity could hold a conversation about everyday matters in Maori.

From the 1970s organizations were formed to pursue Maori grievances over loss of land and resources. The Waitangi Tribunal was set up in 1975 as a forum for complaints about breaches of the Treaty of Waitangi, and in 1984 empowered to hear claims against Crown actions since 1840. Direct negotiations with the Crown have been offered to claimants and a range of proposals to resolve historical grievances launched for public discussion in Dec. 1994. These proposals specify that all claims are to be met over ten years with treaty rights being converted to economic assets. There have been four recent major treaty settlements: NZ$170m. each for Tainui and Ngai Tahu, the NZ$150m. Sealord fishing agreement and NZ$40m. for Whakatohea in the Bay of Plenty. The Maori Land Court has jurisdiction over Maori freehold land and some general land owned by Maoris under the Te Ture Whenue Maori Act 1993.

English and Maori are the official languages.

SOCIAL STATISTICS

Statistics for calendar years:

	Live births	Deaths	Marriages	Divorces
2002	54,021	28,065	20,690	10,292
2003	56,134	28,010	21,419	10,491
2004	58,073	28,419	21,006	10,609
2005	57,745	27,034	20,470	9,972
2006	59,193	28,245	21,423	10,065
2007	64,044	28,522	21,494	9,650

Birth rate, 2007, 15·1 per 1,000 population; death rate, 6·7 per 1,000 population; infant mortality, 2010, five per 1,000 live births. Annual population growth rate, 2000–05, 1·2%. In 2005 there were 514 suicides (382 males). Expectation of life, 2006: males, 78·0 years; females, 82·2. Fertility rate, 2007, 2·2 births per woman. New Zealand legalized same-sex marriage in April 2013.

In 2007 there were 82,572 permanent and long-term immigrants (78,963 in 2005) and 77,081 permanent and long-term emigrants (71,992 in 2005).

CLIMATE

Lying in the cool temperate zone, New Zealand enjoys very mild winters for its latitude owing to its oceanic situation, and only the extreme south has cold winters. The situation of the mountain chain produces much sharper climatic contrasts between east and west than in a north-south direction. Mean daily maximum temperatures and rainfall figures:

	Annual rainfall (mm) in 2004	Jan (°C)	July (°C)
Auckland	23·3	14·5	1,331
Christchurch	22·5	11·3	643
Dunedin	18·9	9·8	765
Wellington	20·3	11·4	1,447

The highest extreme temperature recorded in 2004 was 38·4°C, recorded at Darfield on 1 Jan., and the lowest –12·0°C, at Fairlie on 16 Aug.

CONSTITUTION AND GOVERNMENT

Definition was given to the status of New Zealand by the (Imperial) Statute of Westminster of Dec. 1931, which had received the antecedent approval of the New Zealand Parliament in July 1931. The Governor-General's assent was given to the Statute of Westminster Adoption Bill on 25 Nov. 1947.

The powers, duties and responsibilities of the *Governor-General* and the *Executive Council* are set out in Royal Letters Patent and Instructions thereunder of 11 May 1917. In the execution of the powers vested in him the Governor-General must be guided by the advice of the Executive Council.

At a referendum on 6 Nov. 1993 a change from a first-past-the-post to a proportional representation electoral system was favoured by 53·9% of votes cast.

Parliament is the *House of Representatives*, consisting of 121 members (for the 2011 election 63 were general seats, 51 party list seats and seven Maori seats), elected by universal adult suffrage on the mixed-member-proportional system (MMP) for three-year terms. The seven Maori electoral districts cover the whole country. Maori and people of Maori descent are entitled to register either for a general or a Maori electoral district. As at Oct. 2008 there were 229,666 persons on the Maori electoral roll.

Angelo, Anthony H., *Constitutional Law in New Zealand*. 2011
Joseph, P. A. (ed.) *Essays on the Constitution*. 1995
McGee, D. G., *Parliamentary Practice in New Zealand*. 2nd ed. 1994
Palmer, Geoffrey and Palmer, Matthew, *Bridled Power: New Zealand's Constitution and Government*. 2004
Ringer, J. B., *An Introduction to New Zealand Government*. 1992

National Anthem

'God Defend New Zealand'; words by T. Bracken, tune by J. J. Woods. There is a Maori version, 'Aotearoa', words by T. H. Smith. The UK national anthem has equal status.

GOVERNMENT CHRONOLOGY

Prime Ministers since 1940. (Lab = Labour; Nat = National)

1940–49	Lab	Peter Fraser
1949–57	Nat	Sidney Holland
1957	Nat	Keith Jacka Holyoake
1957–60	Lab	Walter Nash
1960–72	Nat	Keith Jacka Holyoake
1972	Nat	John Ross Marshall
1972–74	Lab	Norman Eric Kirk
1974	Lab	Hugh Watt (acting)
1974–75	Lab	Wallace Edward Rowling
1975–84	Nat	Robert David Muldoon
1984–89	Lab	David Lange
1989–90	Lab	Geoffrey Palmer
1990	Lab	Mike Moore
1990–97	Nat	Jim Bolger
1997–99	Nat	Jenny Shipley
1999–2008	Lab	Helen Clark
2008–	Nat	John Key

RECENT ELECTIONS

At parliamentary elections on 26 Nov. 2011 turnout was 73·2%. The ruling National Party won 60 seats with 48·0%; the Labour Party 34 with 27·1%; the Green Party 13 with 10·6%; New Zealand First 8 with 6·8%; the Maori Party 3 with 1·4%; ACT New Zealand 1 with 1·1%; Mana 1 with 1·0%; and UnitedFuture 1 with 0·6%.

CURRENT GOVERNMENT

Governor-General: Lieut.-Gen. Sir Jeremiah Mateparae, GNZM, QSO (b. 1954; sworn in 31 Aug. 2011).

In March 2013 the cabinet consisted of:

Prime Minister and Minister of Tourism: John Key; b. 1961 (National Party; in office since 19 Nov. 2008).

Deputy Prime Minister and Minister of Finance: Bill English.

Minister of Canterbury Earthquake Recovery, and Transport: Gerry Brownlee. *Commerce, Consumer Affairs, and Broadcasting:* Craig Foss. *Conservation, and Housing:* Nick Smith. *Defence, and State Services:* Jonathan Coleman. *Economic Development, Science and Innovation, and Tertiary Education, Skills and Employment:* Steven Joyce. *Education, and Pacific Island Affairs:* Hekia Parata. *Energy and Resources, and Labour:* Simon Bridges. *Environment, and Communications and Information Technology:* Amy Adams. *Food Safety, Civil Defence, and Youth Affairs:* Nikki Kaye. *Foreign Affairs, and Sport and Recreation:* Murray McCully. *Health, and State-Owned Enterprises:* Tony Ryall. *Internal Affairs, and Local Government:* Chris Tremain. *Justice, the ACC (Accident Compensation Corporation), and Ethnic Affairs:* Judith Collins. *Police, and Corrections:* Anne Tolley. *Primary Industries, and Racing:* Nathan Guy. *Social Development:* Paula Bennett. *Trade, and Climate Change Issues:* Tim Groser. *Attorney General, and Minister for Treaty of Waitangi Negotiations, and Arts, Culture and Heritage:* Christopher Finlayson.

In addition there are eight ministers who are not in the cabinet: *Minister of Building and Construction, Customs, Land Information, and Statistics:* Maurice Williamson (National Party). *Community and the Voluntary Sector, Senior Citizens, and Women's Affairs:* Jo Goodhew (National Party). *Courts:* Chester Borrows (National Party). *Immigration and Veterans' Affairs:* Michael Woodhouse (National Party). *Maori Affairs:* Dr Pita Sharples (Maori Party). *Regulatory Reform, and Small Business:* John Banks (ACT New

Zealand). *Revenue:* Peter Dunne (UnitedFuture). *Whanau Ora, and Disability Issues:* Tariana Turia (Maori Party).

Office of the Prime Minister: http://newzealand.govt.nz

CURRENT LEADERS

John Key

Position
Prime Minister

Introduction
John Key, a former currency trader, was elected prime minister in Nov. 2008. Leading the National Party to electoral victory, he ended nine years of Labour government under Helen Clark. He won a further term at elections in Nov. 2011.

Early Life
John Phillip Key was born on 9 Aug. 1961 in Auckland, where his British father and Austrian–Jewish mother ran a restaurant. When his father died in 1967 the family were left with large debts and lived in state housing in a suburb of Christchurch. After finishing at Burnside School, he graduated in accounting from the University of Canterbury. He then studied management at Harvard University in the USA.

In 1982 Key began working as an auditor, subsequently joining a clothing manufacturer as a project manager. In 1985 he started his career in foreign exchange (forex) trading just as the NZ dollar was floated on currency markets. In 1988 he was recruited by the Bankers Trust in Auckland as head of their forex dealing team, remaining there until 1995.

Key then joined Merrill Lynch as managing director of the Asia forex group in Singapore, subsequently becoming head of Merrill's global forex group in London. In 2001 he moved to Sydney as head of the institution's debt markets. In 2008 Key was listed in the National Business Review's rich list with an estimated wealth of NZ$50m.

In 2001 he joined the National Party, winning the seat for Helensville, a newly-created constituency in northwest Auckland. He won re-election in 2005, a year after joining the opposition front benches as finance spokesman, and again in 2008. In 2006 Key became party leader following the resignation of Don Brash. He led the party to victory at the general election of 8 Nov. 2008 and was sworn in as prime minister 11 days later.

Career in Office
Following the election, the National Party signed deals with ACT New Zealand, the Maori Party and United Future, offering ministerial positions outside the cabinet. Key guaranteed the continuation of a number of Maori-specific parliamentary seats. He has a reputation as a pragmatic centrist and favours privatization, but his first year in office was largely overshadowed by the longest economic recession in the country's history. In Feb. 2009 he launched an NZ$480m. strategy to help small businesses, including a 90-day probation period for workers, the lowering of provisional tax and the relaxing of tax penalties for businesses with incorrect tax returns. The plan also provided for short-term export credit and the fast-tracking of several government building projects.

In foreign affairs, there was a diplomatic rift over alleged interference by New Zealand in Fijian affairs that saw the mutual expulsion of high commissioners in 2009 before relations were restored in July 2012.

In Nov. 2010 Key declared a national state of mourning in response to a mining accident on the South Island in which 29 miners were killed. In Feb. 2011 Christchurch was struck by a 6·3-magnitude earthquake that killed at least 166 and caused damage put at US$12bn.

In parliamentary elections in Nov. 2011 the National Party increased its share of the vote and Key secured a second term in office, heading a new coalition with ACT New Zealand and UnitedFuture.

DEFENCE

The control and co-ordination of defence activities is obtained through the Ministry of Defence. New Zealand forces serve abroad in Australia, Iraq and Singapore, and with UN peace-keeping missions.

Defence expenditure in 2008 totalled US$1,754m. (US$420 per capita), representing 1·4% of GDP.

Army
Personnel total in 2007: 4,580, plus reserves numbering 1,762.

Navy
The Navy includes three frigates. The main base and Fleet headquarters is at Auckland.

The Royal New Zealand Navy personnel totalled 2,034 uniformed plus 291 reserve personnel in 2007.

Air Force
Squadrons are based at RNZAF Base Auckland and RNZAF Base Ohakea. Flying training is conducted at Ohakea and Auckland. Ground training is carried out at RNZAF Base Woodbourne.

The uniform strength in 2007 was 2,437 with 190 reserves. There were six combat-capable aircraft.

ECONOMY

Agriculture accounted for 8% of GDP in 2001, industry 23% and services 69%.

According to the anti-corruption organization *Transparency International*, New Zealand ranked equal first in a 2012 survey of the countries with the least corruption in business and government. It received 90 out of 100 in the annual index.

New Zealand gave US$424m. in international aid in 2011, equivalent to 0·28% of GNI (compared to the UN target of 0·7%).

Overview
Tourism and dairy exports are important sources of revenue, with Australia and China the country's largest export partners. Prior to the 1980s the economy was one of the most regulated and protected in the developed world. Liberalization in the 1990s stalled towards the end of the century, with public ownership of commercial enterprises increasing in the following decade.

The economy grew every year from 1991 until 2007 despite the 1997 Asian crisis, periods of drought and the global slowdown in the wake of the 11 Sept. attacks in the USA. Following an economic downturn in late 2005, domestic demand regained momentum, aided by a recovery in the housing market and improved business and consumer confidence.

However, the economy entered recession in early 2008. A severe drought and the end of a housing boom initiated a slowdown in domestic demand, exacerbated by the global economic crisis. Unemployment rose from 3·7% in 2007 to 6·5% in 2010. Thanks to the absence of a banking crisis, the implementation of a government stimulus package and ties to the fast-growing Asian markets as well as the robust Australian economy, the economy soon began to recover. The recovery stalled following two earthquakes in 2010 and 2011. Reconstruction, estimated to cost 8% of GDP, was expected to provide boosts to several sectors, albeit temporarily.

Weak business investment, low household saving, an overvalued exchange rate and high external debt challenge the long-term outlook.

Currency
The monetary unit is the *New Zealand dollar* (NZD), of 100 *cents*. The total value of notes and coins on issue from the Reserve Bank

in Sept. 2010 was NZ$3,979m. Inflation rates (based on OECD statistics):

2002	2003	2004	2005	2006	2007	2008	2009	2010	2011
2·7%	1·8%	2·3%	3·0%	3·4%	2·4%	4·0%	2·1%	2·3%	4·0%

In Aug. 2009 foreign exchange reserves were US$11,783m. and total money supply was NZ$33,640m. Gold reserves are negligible.

Budget
The government fiscal year begins 1 July; the company and personal financial year begins on 1 April. Total central government revenue for 2009–10 was NZ$74,725m. (NZ$79,506m. in 2008–09). Central government expenditure in 2009–10 was NZ$81,040m. (NZ$83,399m. in 2008–09).

In 2009–10 tax revenue was NZ$50,744m. (NZ$54,681m. in 2008–09). Social security and welfare was the leading item of expenditure in 2009–10, at NZ$21,185m. (NZ$19,382m. in 2008–09).

The gross public debt at June 2010 was NZ$69,733m. (NZ$61,953m. at June 2009).

There is a Goods and Services Tax (GST) of 15%.

Performance
Real GDP growth rates (based on OECD statistics):

2002	2003	2004	2005	2006	2007	2008	2009	2010	2011
4·6%	4·3%	4·3%	2·6%	2·1%	3·4%	−0·6%	−0·2%	0·9%	0·5%

Total GDP was US$159·7bn. in 2011.

Banking and Finance
The central bank and bank of issue is the Reserve Bank (*Governor*, Graeme Wheeler).

The financial system comprises a central bank (the Reserve Bank of New Zealand), registered banks and other financial institutions. Registered banks include banks from abroad, which have to satisfy capital adequacy and managerial quality requirements. Other financial institutions include the regional trustee banks, now grouped under Trust Bank, building societies, finance companies, merchant banks and stock and station agents. The number of registered banks in 2007 was 17, of which only two (TSB Bank Ltd and Kiwibank Ltd) were not wholly overseas-owned. Around 99% of the assets of the New Zealand banking system were under the ownership of a foreign bank parent.

The primary functions of the Reserve Bank are the formulation and implementation of monetary policy to achieve the economic objectives set by the government, and the promotion of the efficiency and soundness of the financial system, through the registration of banks, and supervision of financial institutions. Since 1996 supervision has been conducted on a basis of public disclosure by banks of their activities every quarter.

On 30 June 2007 the assets of the Reserve Bank were NZ$21,095m. (including government securities totalling NZ$4,342m. and marketable securities totalling NZ$12,526m.).

Total overseas debt was NZ$246,462m. in June 2010.

The stock exchange in Wellington conducts on-screen trading, unifying the three former trading floors in Auckland, Christchurch and Wellington. There is also a stock exchange in Dunedin.

ENERGY AND NATURAL RESOURCES

Environment
New Zealand's carbon dioxide emissions from the consumption and flaring of fossil fuels were the equivalent of 9·4 tonnes per capita in 2008. An *Environmental Performance Index* compiled in 2008 ranked New Zealand seventh in the world, with 88·9%. The index examined various factors in six areas—air pollution, biodiversity and habitat, climate change, environmental health, productive natural resources and water resources.

Electricity
On 1 April 1987 the former Electricity Division of the Ministry of Energy became a state-owned enterprise, the Electricity Corporation of N.Z. Ltd. In 1994 Transpower separated out from the company to operate the national grid. The remainder of ECNZ was subsequently divided in stages into four state-owned enterprises (Contact Energy, Genesis Power Limited, Meridian Energy Limited and Mighty River Power Limited), causing a competitive wholesale electricity market to be established. Around 70% of the country's electricity is generated by renewable sources. Hydro-electric plants, mainly based in the South Island, account for some 55% with geothermal power, generated in the North Island, accounting for around 8%. The rest comes from natural gas (22%), coal, wind and landfill gas. Electricity generating capacity, 2007, 9·1m. kW. Consumption per capita was 9,646 kWh in 2006.

Electricity consumption statistics (in GWh) for years ended 31 March are:

	Residential	Commercial	Industrial	Total consumption
2003	11,723	7,734	15,431	34,889
2004	12,255	7,389	16,151	35,795
2005	12,161	7,975	16,190	36,326
2006	12,231	8,383	16,780	37,394

New Zealand also has 12 wind farms.

Oil and Gas
Crude oil production was 1·9m. tonnes in 2007–08. New Zealand's annual crude petroleum imports are more than three times as much as its production. Proven reserves were estimated at 53m. bbls in 2007.

In 2008 gasfields produced 3·8bn. cu. metres. Gas reserves are estimated to last until about 2014. In 2007 proven natural gas reserves were estimated at 25·0bn. cu. metres.

Minerals
Coal production in 2009 was 4·56m. tonnes. Of 26 mines operating in 2006, 22 were opencast and four underground, responsible for 80·9% and 19·1% of total coal production respectively. 47% of coal produced in 2006 was exported, with India and Japan being the main markets.

While New Zealand's best known non-fuel mineral is gold (producing about 13·44 tonnes in 2009 worth NZ$662m.) there is also production of silver, ironsand, aggregate, limestone, clay, aluminium, dolomite, pumice, salt, serpentinite, zeolite and bentonite. In addition, there are resources or potential for deposits of titanium (ilmenite beach sands), platinum, sulphur, phosphate, silica and mercury.

Agriculture
Two-thirds of the land area is suitable for agriculture and grazing. The total area of farmland in use in 2005 was 15,305,478 ha. There were 11,967,000 ha. of grazing, arable, fodder and fallow land, 110,000 ha. of land for horticulture and 1,879,000 ha. of plantations of exotic timber in 2002. In 2001 there were 1·5m. ha. of arable land and 1·87m. ha. of permanent crops.

The largest freehold estates are held in the South Island. The number of occupied holdings as at 30 June 2007 were as follows:

Regional Council	No. of farms	Total area of farms (2002, in 1,000 ha.)
Auckland	4,044	302
Bay of Plenty	5,364	600
Gisborne	1,329	653
Hawke's Bay	3,405	962
Manawatu-Wanganui	5,847	1,545
Northland	5,175	836
Taranaki	3,426	496
Waikato	10,680	1,730
Wellington	2,157	504
Total North Island	41,424	7,627

Regional Council	No. of farms	Total area of farms (2002, in 1,000 ha.)
Canterbury	9,522	3,151
Marlborough	1,788	723
Nelson	117	21
Otago	3,942	2,368
Southland	3,948	1,198
Tasman	1,731	277
West Coast	822	225
Chatham Islands	42	—
Total South Island	21,915	8,013
Total New Zealand	63,339	15,640

Production of main crops (2000, in 1,000 tonnes): potatoes, 500; apples, 482; wheat (2006–07), 344; barley (2006–07), 336; maize (2006–07), 186; pumpkins and squash, 155; tomatoes, 85; carrots, 80; grapes, 80; cauliflower, 63.

Livestock, 2010: sheep, 32·56m.; diary cattle, 5·92m.; beef cattle, 3·95m.; deer, 1·12m.; pigs (2007), 367,000; goats (2007), 112,000; chickens (2000), 13m. Total meat produced in 2006–07 was 1·44m. tonnes (including 624,000 tonnes of beef and veal, and 573,000 tonnes of lamb and mutton). Meat industry products are New Zealand's second largest export income earner, accounting for about 12% of merchandise exports. New Zealand's main meat exports are beef, lamb and mutton. About 73% of lamb, 58% of mutton and 57% of beef produced in New Zealand in 2006–07 was exported overseas. The domestic market absorbs over 99% of the pigmeat and poultry produced in New Zealand. 54% of the world's exported sheepmeat comes from New Zealand.

Production of wool for the year 2006–07 was 216,300 tonnes. Milk production for 2010–11 totalled a record 17,339m. litres. In 1999–2000 butter production totalled 254,639 tonnes and cheese production 296,745 tonnes.

Forestry

Forests covered 8·27m. ha. in 2010 (31% of New Zealand's land area), up from 7·67m. ha. in 1990. In 2007 about 6·2m. ha. was indigenous forest and 1·8m. ha. planted productive forest. New planting was 5,000 ha. in 2006, the lowest amount since 1959. Introduced pines form the bulk of the large exotic forest estate and among these radiata pine is the best multi-purpose tree, reaching log size in 25–30 years. Other species planted are Douglas fir and Eucalyptus species. Total roundwood production in 2006–07 was 20·04m. cu. metres. The table below shows production of rough sawn timber in 1,000 cu. metres for years ending 31 March:

	Indigenous			Exotic			All Species
	Rimu and Miro	Beech	Total (including others)	Radiata Pine	Douglas Fir	Total (including others)	Total
2004	5	8	16	3,992	178	4,206	4,222
2005	5	7	13	4,178	167	4,379	4,392
2006	0	0	12	4,038	156	4,222	4,234
2007	3	6	9	4,102	157	4,292	4,301

In 2006–07 forest industries consisted of approximately 264 sawmills, seven plywood and 11 veneer plants, and six fibreboard mills.

Production of wood pulp in the year ending 31 March 2007 amounted to 1,528,991 tonnes and of paper (including newsprint paper and paperboard) to 871,946 tonnes.

Fisheries

In 2007 the total catch was 494,492 tonnes, almost entirely from sea fishing. The total value of New Zealand fisheries exports in 2006–07 was NZ$1,275m., of which frozen fish exports constituted NZ$345m.

INDUSTRY

Statistics of manufacturing industries (in NZ$1m.):

Production year	Salaries and wages paid	Closing stocks of raw materials	Closing stocks of finished goods	Operating income	Purchases operating and other expenses
2001–02	8,961	2,618	4,945	63,396	47,163
2002–03	9,523	2,595	6,954	65,146	47,770

The following is a statement of the value of the products (including repairs) of the principal industries for the year 2002–03 (in NZ$1m.):

Industry group	Salaries and wages paid	Closing stocks of raw materials	Closing stocks of finished goods	Operating income	Purchases and other operating expenses
Dairy and meat products	1,481	261	3,015	16,057	13,777
Other food	869	213	651	6,939	5,063
Beverage, malt and tobacco	292	154	423	2,926	2,060
Textile and apparel	579	205	296	3,069	2,092
Wood products	674	107	381	4,245	3,227
Paper and paper products	407	110	206	2,870	2,002
Printing, publishing and recorded media	796	84	84	3,392	2,001
Petroleum and industrial chemical	267	195	192	3,202	2,154
Rubber, plastic and other chemical products	728	218	513	4,144	2,840
Non-metallic mineral products	295	52	137	2,041	1,364
Basic metal	314	108	161	2,041	1,524
Structural, sheet and fabricated metal products	780	195	214	4,112	2,849
Transport equipment manufacturing	499	212	145	2,244	1,488
Machinery and equipment	1,138	357	429	5,876	4,010
Furniture and other manufacturing	405	124	106	1,984	1,317

According to the World Bank's *Doing Business 2012* New Zealand is the easiest country in which to start a business.

Labour

There were 2,142,500 persons employed in the year ending Sept. 2007 (1,664,000 full-time and 478,500 part-time). The largest number of employed people worked in the education, health and community, and other services area (27·7%); followed by wholesale and retail trade, restaurants and hotels (22·4%); and finance and insurance, property and business services (14·7%). Average unemployment total for the year ending Sept. 2007 was 81,400. The unemployment rate in 2011 was 6·5%. Long-term unemployment is low, with only 9·0% of the labour force in 2010 having been out of work for more than a year.

The weekly average earnings in the year ending March 2007 was NZ$963 for men, NZ$760 for women. A minimum wage is set by the government annually. As of 1 April 2010 it was NZ$12·75 an hour; a new rate for those entering the labour market for the first time was introduced in April 2008—it was NZ$10·20 an hour in April 2010. In 2006 there were 42 work stoppages with 27,983 person-days of work lost.

INTERNATIONAL TRADE

In 1990 New Zealand and Australia completed the Closer Economic Relations Agreement (initiated in 1983), which provides for mutual free trade in goods.

Imports and Exports

Trade in NZ$1m. for recent years ending 30 June:

	Imports (c.i.f.)	Exports, including re-exports (f.o.b.)	Balance of merchandise trade
2002	31,811	32,332	521
2003	32,161	29,291	−2,870
2004	33,378	29,864	−3,514
2005	35,793	30,618	−5,175

The principal imports for the 12 months ended 30 June 2003 were:

Commodity	Value (NZ$1m. v.f.d.)
Vehicles, parts and accessories	4,985
Mechanical machinery and equipment	4,333
Mineral fuels	3,152
Electrical machinery and equipment	2,699
Plastics and plastic articles	1,279
Optical, medical and measuring equipment	967
Paper, paperboard and paper articles	924
Aircraft and parts	804
Pharmaceutical products	747
Iron or steel articles	491

The principal exports for the 12 months ended 30 June 2003 were:

Commodity	Value (NZ$1m. f.o.b.)
Dairy produce, eggs and honey	4,714
Meat and edible offal	4,111
Wood and articles of wood	2,386
Machinery and mechanical appliances	1,356
Fish, crustaceans and molluscs	1,215
Albuminoidal substances; modified starches; glues; enzymes	1,148
Fruits and nuts (edible)	1,032
Aluminium and aluminium articles	980
Wool, fine or coarse animal hair	943
Electrical machinery, equipment and parts	938

The principal import suppliers in 2002–03 (imports v.f.d., in NZ$1m.) were: Australia, 7,278; USA, 4,067; Japan, 3,876; China, 2,687; Germany, 1,713; UK, 1,120; Malaysia, 864. The leading export destinations in 2002–03 (exports and re-exports f.o.b., in NZ$1m.) were: Australia, 6,050; USA, 4,366; Japan, 3,354; China, 1,457; UK, 1,361; South Korea, 1,178; Germany, 855.

COMMUNICATIONS

Roads

Total length of roads in 2007 was 93,748 km (65·4% paved), including 172 km of motorways. There were 10,893 km of highways, main or national roads. At 30 June 2008 motor vehicles licensed numbered 4,125,932, of which 2,788,938 were passenger cars and vans. In addition there were 577,684 trailers and caravans, 519,992 commercial vehicles, and 130,213 motorcycles and mopeds. In 2007 there were 422 deaths in road accidents.

In 2008 there were 34,590 persons employed in road transport. Total expenditure on roads (including infrastructure) by the central government and local authorities combined amounted to NZ$1,751m. in 2008.

Rail

The national rail operator is Kiwi Rail. In 2008–09 KiwiRail rolling stock included 231 diesel, electric and shunting locomotives, 4,215 freight wagons, 50 passenger carriages and 16 non-passenger coaches.

In 1994 a 24-hour freight link was introduced between Auckland and Christchurch. There were, in 2002, 3,898 km of 1,067 mm gauge railway open for traffic (506 km electrified).

In 2008–09 KiwiRail carried 4·0m. tonnes of freight and 12·4m. passengers. Total income in the financial year 2008–09 was NZ$636·6m. and total expense NZ$573·3m.

Civil Aviation

There are international airports at Wellington, Auckland and Christchurch, with Auckland International being the main airport. The national carrier is Air New Zealand, which was privatized in 1989 but then renationalized in 2001. Trans-Tasman air travel is subject to agreement between Air New Zealand and Qantas.

New Zealand has one of the highest ratios of aircraft to population in the world with 3,530 aircraft in the year to March 2003. In 2005 scheduled airline traffic of New Zealand-based carriers flew 243·0m. km, carrying 11,402,400 passengers. In 2002 there were 113 airports, of which 46 had paved runways.

Shipping

In Jan. 2009 there were 29 ships of 300 GT or over registered, totalling 150,000 GT. The busiest port is Tauranga, which handled a record 13,748,000 tonnes of cargo in 2009–10 (up from 13,458,000 tonnes in 2008–09).

Telecommunications

The predominant telecommunications service provider is the Telecom Corporation of New Zealand (Telecom New Zealand), formed in 1987 and privatized in 1990. The largest mobile phone operators are Vodafone New Zealand and Telecom Mobile. In 2008 there were 1,750,000 main (fixed) telephone lines. In the same year mobile phone subscribers numbered 4,620,000 (1,092·2 per 1,000 persons). There were 3·0m. internet users in 2008. The fixed broadband penetration rate was 24·9 subscribers per 100 inhabitants in Dec. 2010. In Dec. 2011 there were 2·1m. Facebook users.

SOCIAL INSTITUTIONS

Justice

The judiciary consists of the Supreme Court, the Court of Appeal, the High Court and District Courts. All exercise both civil and criminal jurisdiction. The Supreme Court replaced the Privy Council in London as the court of final appeal in 2004. Special courts include the Maori Land Court, the Maori Appellate Court, Family Courts, the Youth Court, Environment Court and the Employment Court. In March 2011 there were 8,755 sentenced inmates of whom 556 were women. Of inmates in 2011, 51·2% (some 4,483) identified themselves as Maori only compared to 33·7% who identified themselves as European only. There were 170,999 convictions, including 14,537 for violent offences, in 2002. The death penalty for murder was replaced by life imprisonment in 1961.

The Criminal Injuries Compensation Act, 1963, which came into force on 1 Jan. 1964, provided for compensation of persons injured by certain criminal acts and the dependants of persons killed by such acts. However, this has now been phased out in favour of the Accident Compensation Act, 1982, except in the residual area of property damage caused by escapees. The Offenders Legal Aid Act 1954 provides that any person charged or convicted of any offence may apply for legal aid which may be granted depending on the person's means and the gravity of the offence etc. Since 1970 legal aid in civil proceedings (except divorce) has been available for persons of small or moderate means. The Legal Services Act 1991 now brings together in one statute the civil and criminal legal aid schemes.

Police

The police are a national body maintained by the central government. In June 2003 there were 7,257 full-time equivalent sworn officers (16% female).

Ombudsmen

The office of Ombudsman was created in 1962. From 1975 additional Ombudsmen have been authorized. There are currently two. Ombudsmen's functions are to investigate complaints under the Ombudsman Act, the Official Information Act and the Local Government Official Information and Meetings Act from members of the public relating to administrative decisions of central, regional and local government. During the year ended 30 June 2003 a total of 4,418 complaints were received. A total of 27 complaints were sustained during the year and 729 were still under investigation.

Education

Education is compulsory between the ages of six and 16. Early childhood services are available for education and care for children from birth to six years of age. In 2009 there were 626 kindergartens and 485 play centres (24 licence exempt). In 2009 there were 39,346 and 15,498 children on the rolls respectively. There were also 464 *te kohanga reo* (providing early childhood education in the Maori language) with 9,288 children, and a number of other providers of early childhood care and education.

In 2009 there were 2,027 primary schools (including full primary, contributing primary and intermediate schools), with 434,810 pupils; the number of teachers was 27,640. A correspondence school for children in remote areas (*Te Kura*) and those otherwise unable to attend school had 6,076 pupils and 270 teachers.

In 2009 there were 336 secondary schools with 20,439 teachers and 273,409 pupils. There were also 149 composite schools with 2,500 teachers and 43,149 pupils.

There were 469,107 (including 43,457 international) enrolments in tertiary institutions in 2009. Of the international students in 2009, 30,946 were from Asia and 4,068 from Europe. The most popular subject areas studied in 2009 by domestic students were society and culture, management and commerce, health, and engineering related technologies. New Zealand has eight state-funded universities—the University of Auckland, Auckland University of Technology, University of Waikato (at Hamilton), Victoria University of Wellington, Massey University (at Palmerston North), the University of Canterbury (at Christchurch), the University of Otago (at Dunedin) and Lincoln University (near Christchurch). The number of students attending universities in 2009 was 154,866. 180,709 students were enrolled in institutes of technology and polytechnics in 2009.

In 2008–09 public expenditure on education came to 6·4% of GDP, representing 17·9% of total government spending. The universities are autonomous bodies. All state-funded primary and secondary schools are controlled by boards of trustees. Education in state schools is free for children under 19 years of age. All educational institutions are reviewed every three years by teams of educational reviewers.

The adult literacy rate is at least 99%.

Health

In 2003 there were 10,355 practising doctors. In 2002 there were 85 public hospitals with 12,484 beds and 360 private hospitals with 11,341 beds. In 2008 New Zealand spent 9·8% of its GDP on health. Total budgeted expenditure on health in 2003–04 was NZ$9·6bn.

Welfare

Non-contributory old-age pensions were introduced in 1898. Large reductions in welfare expenditure were introduced by the government in Dec. 1990.

From 1 Oct. 1998 anyone receiving unemployment benefit, sickness benefit, a training benefit, a 55 plus benefit, or a young job seekers allowance has received a benefit called the Community Wage. In return for receiving the Community Wage, recipients are expected to search for work, meet with Work and Income New Zealand when asked, take a suitable work offer and take part in activities that would improve their chances of finding a job.

On 1 April 1992 the Guaranteed Retirement Income Scheme (GRI) was replaced by the national superannuation scheme which is income-tested. Eligibility has been gradually increased to 65 years. Universal eligibility is available at 70 years. At 1 April 2008 a married couple received NZ$439·80 per week, a single person living alone NZ$285·87 per week.

Social Welfare Benefits

Benefits	Number in force at 30 June 2003	Total expenditure 2003 (NZ$1,000)
Community Wage—Job Seeker	111,906	1,287,730
Community Wage—Training	4,291	37,942
Community Wage—Sickness	39,902	460,209
Invalids' Benefit	68,507	926,515
Domestic Purposes' Benefit	109,295	1,634,477
Orphans' Benefit/Unsupported Child's Benefit	6,789	47,081
Widows' Benefit	8,659	90,265
Transitional Retirement Benefit	2,110	42,013
New Zealand Superannuation	457,278	5,798,873
Veterans' Pension	7,872	87,625
War Pension	22,271	108,862
Total Income Support	838,880	10,521,592

Reciprocity with Other Countries. New Zealand has overseas social security agreements with the United Kingdom, the Netherlands, Greece, Ireland, Australia, Jersey and Guernsey, Denmark and Canada. The main purpose of these agreements is to encourage free movement of labour and to ensure that when a person has lived or worked in more than one country, each of those countries takes a fair share of the responsibility for meeting the costs of that person's social security coverage. New Zealand also pays people eligible for New Zealand Superannuation or veterans' pensions who live in the Cook Islands, Niue or Tokelau.

RELIGION

No direct state aid is given to any form of religion. For the Church of England the country is divided into seven dioceses, with a separate bishopric (Aotearoa) for the Maori. The Presbyterian Church is divided into 23 presbyteries and the Maori Synod. The Moderator is elected annually. The Methodist Church is divided into ten districts; the President is elected annually. The Roman Catholic Church is divided into four dioceses, with the Archbishop of Wellington as Metropolitan Archbishop. In Feb. 2013 there was one cardinal.

Adherents of leading religions at the 2001 census were as follows:

Religious denomination	Adherents
Anglican	584,793
Catholic	486,012
Presbyterian	417,453
Methodist	120,708
Baptist	51,426
Ratana	48,975
Buddhist	41,664
Latter-day Saints (Mormons)	39,915
Hindu	39,876
Pentecostal	30,222
Islam/Muslim	23,637
Brethren	20,406
Jehovah's Witnesses	17,826
Assemblies of God	16,023

Religious denomination	Adherents
Salvation Army	12,618
Seventh-day Adventist	12,600
All other religious affiliations	398,847
No religion	1,028,052
Object to state	239,241
Not specified	211,638
Total	3,841,932[1]

[1]Where a person reported more than one religious affiliation, they have been counted in each applicable group.

CULTURE

World Heritage Sites

There are three UNESCO World Heritage sites under New Zealand jurisdiction. Te Wahipounamu on South Island was listed in 1990; Tongariro National Park, on North Island, was listed in 1990 and 1993; the Sub-Antarctic Islands, consisting of the Auckland Islands, Antipodes Islands, Bounty Islands, Campbell Island and the Snares, were inscribed on the list in 1998.

Press

In 2008 there were 22 paid-for daily newspapers with a combined circulation of 653,000. The *New Zealand Herald,* published in Auckland, had the largest daily circulation in 2008, with an average of 187,000 copies. Other major dailies are *The Dominion Post* and *The Press,* with circulations of 94,000 and 87,000 copies respectively. In 2007 there were also three Sunday newspapers, four paid-for non-dailies and 109 free non-dailies.

Tourism

There were a record 2,617,930 tourists in the year to March 2012 of whom 1,168,316 were from Australia, 222,152 were from the UK, 184,056 were from the USA and 160,268 were from China. Tourism receipts totalled NZ$22,848m. in 2010–11. Employment in tourism in 2010–11 totalled 188,100 (full-time equivalents), of whom 120,700 were directly employed in tourism and 67,400 indirectly.

Festivals

The biennial New Zealand Festival takes place in Wellington in Feb./March in even-numbered years. The biennial Christchurch Arts Festival takes place in July/Aug. in odd-numbered years.

DIPLOMATIC REPRESENTATIVES

Of New Zealand in the United Kingdom (New Zealand House, Haymarket, London, SW1Y 4TQ)
High Commissioner: Derek Leask.

Of the United Kingdom in New Zealand (44 Hill St., Wellington, 6011)
High Commissioner: Vicki Treadell.

Of New Zealand in the USA (37 Observatory Cir., NW, Washington, D.C., 20008)
Ambassador: Mike Moore.

Of the USA in New Zealand (29 Fitzherbert Terr., Wellington)
Ambassador: David Huebner.

Of New Zealand to the United Nations
Ambassador: Jim McLay.

Of New Zealand to the European Union
Ambassador: Vangelis Vitalis.

FURTHER READING

Statistics New Zealand. *New Zealand Official Yearbook.—Key Statistics: a Monthly Abstract of Statistics.—Profile of New Zealand.*

Belich, James, *Making Peoples: a History of the New Zealanders from Polynesian Settlement to the End of the Nineteenth century.* 1997.—*Paradise Reforged: A History of New Zealanders from the 1880s to the Year 2000.* 2002

Harris, P. and Levine, S. (eds.) *The New Zealand Politics Source Book.* 3rd ed. 1999

Massey, P., *New Zealand: Market Liberalization in a Developed Economy.* 1995

Mein Smith, Philippa, *A Concise History of New Zealand.* 2005

Miller, Raymond, *Political Leadership in New Zealand.* 2006

Miller, Raymond, (ed.) *New Zealand Government & Politics.* 5th ed. 2009

Mulgan, Richard, *Politics in New Zealand.* Revised ed. 2004

Rowe, James E., *Economic Development in New Zealand.* 2005

Sinclair, Keith, revised by Raewyn Dalziel, *A History of New Zealand.* 2001

Sinclair, K. (ed.) *The Oxford Illustrated History of New Zealand.* 2nd ed. 1994

Wood, G. A. and Rudd, Chris, *The Politics and Government of New Zealand: Robust, Innovative and Challenged.* 2004

For other more specialized titles see under CONSTITUTION AND GOVERNMENT above.

National Statistical Office: Statistics New Zealand, Statistics House, The Boulevard, Harbour Quays, PO Box 2922, Wellington 6140.
Website: http://www.stats.govt.nz

TERRITORIES OVERSEAS

Territories Overseas coming within the jurisdiction of New Zealand consist of Tokelau and the Ross Dependency.

Tokelau

Tokelau is situated some 500 km to the north of Samoa and comprises three dispersed atolls—Atafu, Fakaofo and Nukunonu. The land area is 10·1 sq. km and the population at the 2011 census was 1,411, giving a density of 140 per sq. km.

The British government transferred administrative control of Tokelau to New Zealand in 1925. Formal sovereignty was transferred to New Zealand in 1948 by act of the New Zealand Parliament. New Zealand statute law, however, does not apply to

Tokelau unless it is expressly extended to Tokelau. In practice New Zealand legislation is extended to Tokelau only with its consent.

Under a programme agreed in 1992, the role of Tokelau's political institutions is being better defined and expanded. The process under way enables the base of Tokelau government to be located within Tokelau's national level institutions rather than as before, within a public service located largely in Samoa. In 1994 the Administrator's powers were delegated to the *General Fono* (the national representative body), and when the *General Fono* is not in session, to the *Council of Faipule.* The Tokelau Amendment Act 1996 conferred on the *General Fono* a power to make rules for Tokelau, including the power to impose taxes. There are no parties—in elections of 19–21 Jan. 2011, 20 independents were elected to the *General Fono*.

Administrator: Jonathan Kings.

Head of Government: Salesio Lui.

Coconuts (the source of copra) are the only cash crop. Pulaka, breadfruit, papayas, the screw-pine and bananas are cultivated as food crops. Livestock comprises pigs, poultry and goats.

Tokelau affirmed to the United Nations in 1994 that it had under active consideration both the Constitution of a self-governing Tokelau and an act of self-determination. It also expressed a strong preference for a future status of free association with New Zealand. A referendum on self-determination took place in Feb. 2006 with 60% voting in favour of the proposal, short of the two-thirds majority needed for the referendum to succeed. Another referendum in Oct. 2007 showed 64% in favour, still short of the required two-thirds majority. In May 2008 Ban Ki-moon, secretary general of the UN, reiterated his desire that the process of 'decolonizing' all of the 16 non-self-governing regions around the world be accelerated.

Ross Dependency

By Imperial Order in Council, dated 30 July 1923, the territories between 160° E. long. and 150° W. long. and south of 60° S. lat. were brought within the jurisdiction of the New Zealand government. The region was named the Ross Dependency. From time to time laws for the Dependency have been made by regulations promulgated by the Governor-General of New Zealand.

The mainland area is estimated at 400,000–450,000 sq. km and is mostly ice-covered. In Jan. 1957 a New Zealand expedition under Sir Edmund Hillary established a base in the Dependency. In Jan. 1958 Sir Edmund Hillary and four other New Zealanders reached the South Pole.

The main base—Scott Base, at Pram Point, Ross Island—is manned throughout the year, about 12 people being present during winter. The annual activities of 200–300 scientists and support staff are managed by a crown agency, Antarctica New Zealand, based in Christchurch.

SELF-GOVERNING TERRITORIES OVERSEAS

The Cook Islands

KEY HISTORICAL EVENTS

The Cook Islands, which lie between 8° and 23° S. lat., and 156° and 167° W. long., were made a British protectorate in 1888, and on 11 June 1901 were annexed as part of New Zealand. In 1965 the Cook Islands became a self-governing territory in 'free association' with New Zealand.

TERRITORY AND POPULATION

The islands fall roughly into two groups—the scattered islands towards the north (Northern group) and the islands towards the south (Southern group). The islands with their populations (provisional) at the census of 2011:

Southern Group—	Area sq. km	Population
Aitutaki	18·3	2,035
Atiu	26·9	481
Mangaia	51·8	573
Manuae and Te au-o-tu	6·2	—
Mauke (Parry Is.)	18·4	307
Mitiaro	22·3	189
Rarotonga	67·1	13,097
Northern Group—	Area sq. km	Population
Manihiki (Humphrey)	5·4	243
Nassau	1·3	73
Palmerston (Avarua)	2·1	60
Penrhyn (Tongareva)	9·8	203
Pukapuka (Danger)	1·3	453
Rakahanga (Reirson)	4·1	77
Suwarrow (Anchorage)	0·4	—
Total	235·4	17,791

Population density in 2011 was 76 per sq. km. In 2007 an estimated 73% of the population lived in urban areas. Between 2001 and 2006 the population had increased by 7·3%. However, the 2011 census saw an 8·0% decline from 2006.

SOCIAL STATISTICS

2005: births, 278; deaths, 86. Birth rate (2005, per 1,000 population), 22·4; death rate, 6·9. Life expectancy was estimated in 2009 at: males, 72 years; females 80. Fertility rate, 2008, 2·6 births per woman.

CLIMATE

Oceanic climate where rainfall is moderate to heavy throughout the year, with Nov. to March being particularly wet. Weather can be changeable from day to day and can end in rainfall after an otherwise sunny day. Rarotonga, Jan. 26°C, July 20°C. Annual rainfall 2,060 mm.

CONSTITUTION AND GOVERNMENT

The Cook Islands Constitution of 1965 provides for internal self-government but linked to New Zealand by a common Head of State and a common citizenship, that of New Zealand. It provides for a ministerial system of government with a Cabinet consisting of a Prime Minister and not more than eight nor fewer than six other Ministers. There is also an advisory council composed of hereditary chiefs, the 15-member House of Ariki, without legislative powers. The New Zealand government is represented by a New Zealand Representative and the Queen, as head of state, by the Queen's Representative. The capital is Avarua on Rarotonga.

The bicameral *Parliament* comprises a Legislative Assembly (or lower house) of 25 members (including one representing overseas voters) elected for a term of five years and a House of Ariki (or upper house) that has an advisory role only and is made up of chiefs.

RECENT ELECTIONS

At the elections of 17 Nov. 2010 the Cook Islands Party won 16 of the 24 seats and the Democratic Party 8 seats.

CURRENT GOVERNMENT

High Commissioner: John Carter.
  Prime Minister: Henry Puna.

Government Website: http://www.cook-islands.gov.ck

ECONOMY

Currency

The Cook Island *dollar* was at par with the New Zealand *dollar*, but was replaced in 1995 by New Zealand currency.

Budget

Revenue, 2008–09 (provisional), NZ$116·7m.; expenditure, NZ$124·4m. Domestic taxes accounted for 29·8% of revenues in 2008–09, taxes on income 27·9% and grants 18·1%.

Performance

The economy contracted by 1·2% in 2008 and by 0·1% in 2009.

Banking and Finance

There are four banks in the Cook Islands. The Cook Islands Savings Bank is state-owned and has deposit services throughout the islands. The Cook Islands Development Bank is a state-owned corporation funded in part by loans from the Asian Development Bank. The two remaining banks are subsidiaries of the Australia and New Zealand Banking Group Limited and the Westpac Bank, which are both Australian-owned and major banks in Australasia.

ENERGY AND NATURAL RESOURCES

Environment

Carbon dioxide emissions from the consumption and flaring of fossil fuels in 2008 were the equivalent of 7·4 tonnes per capita.

Electricity

Production in 2007 was 34m. kWh. Installed capacity was an estimated 8,000 kW in 2007.

Minerals

The islands of the Cook group have no significant mineral resources. However, the seabed, which forms part of the exclusive economic zone, has some of the highest concentrations of manganese nodules in the world. Manganese nodules are rich in cobalt and nickel.

Agriculture

In 2002 there were approximately 4,000 ha. of arable land and 2,000 ha. of permanent crops. Production estimates (2002, in 1,000 tonnes): coconuts, 5; cassava, 3; mangoes, 3. Livestock (2002): 40,000 pigs, 2,000 goats.

Forestry

In 2010 the area under forests was 16,000. ha., or 65% of the total land area. Timber production was 5,000 cu. metres in 2007.

Fisheries

In 2010 the total catch was 10,019 (entirely from sea fishing), up from just 2,695 tonnes in 2009.

INDUSTRY

Labour

In 2001 there were 5,928 employed persons in the Cook Islands and 892 unemployed. Of those employed, 3,386 were men and 2,542 were women.

INTERNATIONAL TRADE

Imports and Exports

Exports (f.o.b.) were valued at NZ$7·1m. in 2007. Main items exported are fish, pearls, and fruit and vegetables. Imports (c.i.f.) in 2007 totalled NZ$144·7m. Leading imports are petroleum and related materials, food, and machinery and transport equipment.

Around two-thirds of imports come from New Zealand. Australia and Singapore are other significant import sources. Japan, China and New Zealand are the main export markets.

COMMUNICATIONS

Roads

In 2003 there were 320 km of roads. There were 706 newly-registered vehicles in 2011 (the lowest total since 1998).

Civil Aviation

New Zealand has financed the construction of an international airport at Rarotonga which became operational for jet services in 1973. There are nine useable airports. Domestic services are provided by Air Rarotonga, and in 2003 there were also services to Auckland, Honolulu, Los Angeles, the Fiji Islands, French Polynesia and Vancouver.

Shipping

Two international shipping services connect Rarotonga with Auckland, Samoa, Tonga and Niue. In Jan. 2009 there were 44 ships of 300 GT or over registered, totalling 143,000 GT.

Telecommunications

In 2008 there were 6,700 main telephone lines in service. There were 6,700 mobile phone subscribers in 2008 and 5,000 internet users.

SOCIAL INSTITUTIONS

Justice

There is a High Court and a Court of Appeal, from which further appeal is to the Privy Council in the UK.

The population in penal institutions in April 2005 was 27 (equivalent to 126 per 100,000 population).

Education

In 2010 there were 24 early childhood education centres, 13 primary schools, four secondary schools and 12 combined primary and secondary schools. Pupils in primary education in 2010 totalled 1,841 with 123 teachers; there were 1,893 pupils in secondary education with 121 teachers.

In 2009–10 the education budget was an estimated NZ$9,928,000.

Health

A user pay scheme was introduced in July 1996 where all Cook Islanders pay a fee of NZ$5·00 for any medical or surgical treatment including consultation. Those under the age of 16 years or over the age of 60 years are exempted from payment of this charge. The dental department is privatized except for the school dental health provision. This service continues to be free to all schools.

The Rarotonga Hospital, which is the referral hospital for the outer islands, consists of 80 beds. The hospital has eight doctors, 33 registered nurses and 11 hospital aides.

RELIGION

From the census of 2001, 55% of the population belong to the Cook Islands Christian Church; about 17% are Roman Catholics, and the rest are Seventh-Day Adventists and Latter-day Saints and other religions.

CULTURE

Press

The *Cook Islands News* (circulation 2,000 in 2006) is the sole daily newspaper. The *Cook Islands Star*, which is published fortnightly, is sold in the Cook Islands and in New Zealand.

Tourism

In 2008 there were 94,776 tourist arrivals.

FURTHER READING

Local statistical office: Ministry of Finance and Economic Management, P. O. Box 41, Rarotonga, Cook Islands.
Statistical office: Cook Islands Statistics Office, PO Box 41, Avarua, Rarotonga.
Website: http://www.stats.gov.ck

Niue

KEY HISTORICAL EVENTS

Capt. James Cook sighted Niue in 1774 and called it 'Savage Island'. Christian missionaries arrived in 1846. Niue became a British Protectorate in 1900 and was annexed to New Zealand in 1901. Internal self-government was achieved in free association with New Zealand on 19 Oct. 1974, with New Zealand taking responsibility for external affairs and defence. Niue is a member of the South Pacific Forum. In Jan. 2004 Cyclone Heta destroyed the capital, Alofi, and a state of emergency was declared, although it was lifted a month later.

TERRITORY AND POPULATION

Niue is the largest uplifted coral island in the world. Distance from Auckland, New Zealand, 2,161 km; from Rarotonga, 933 km. Area, 261 sq. km; height above sea level, 67 metres. The population has been declining steadily, from around 6,000 in the 1960s to 1,625 recorded in the 2006 census, giving a population density of 6 per sq. km. Migration to New Zealand is the main factor in population change. The capital is Alofi.

SOCIAL STATISTICS

Annual growth rate, 1998–2008, –2·7%. 2009: births, 31; deaths, 12. Infant mortality, 2010, 19 per 1,000 live births.

CLIMATE

Oceanic, warm and humid, tempered by trade winds. May to Oct. are cooler months. Temperatures range from 20°C to 28°C.

CONSTITUTION AND GOVERNMENT

There is a Legislative Assembly (*Fono*) of 20 members, 14 elected from 14 constituencies and six elected by all constituencies.

RECENT ELECTIONS

Parliamentary elections were held on 7 May 2011. 20 non-partisan members were elected. On 16 May Toke Talagi was re-elected as premier, receiving 12 votes.

CURRENT GOVERNMENT

High Commissioner: Mark Blumsky.
  Prime Minister: Toke Talagi.

Government Website: http://www.gov.nu

ECONOMY

Budget
Financial aid from New Zealand, 2006–07, totalled NZ$9·1m.

ENERGY AND NATURAL RESOURCES

Electricity
Production in 2007 was about 3m. kWh; installed capacity was estimated at 1,000 kW in 2007.

Agriculture
In 2002 there were approximately 4,000 ha. of arable land and 3,000 ha. of permanent crops. The main commercial crops of the island are coconuts, taro and yams.
    In 2002 there were 2,000 pigs.

Forestry
In 2010 the area under forests was 19,000 ha., or 72% of the total land area.

Fisheries
In 2010 the total catch was 113 tonnes, exclusively from marine waters.

INTERNATIONAL TRADE

Imports and Exports
Imports, 2002, NZ$3·25m.; exports, NZ$0·14m.

COMMUNICATIONS

Civil Aviation
Weekly commercial air services link Niue with New Zealand, Sydney and Samoa.

Telecommunications
There is a wireless station at Alofi, the port of the island. Main telephone lines (2008) 1,000. There were 1,000 internet users in 2008. Niue launched its own mobile phone service in 2011.

SOCIAL INSTITUTIONS

Justice
There is a High Court under a Chief Justice, with a right of appeal to the New Zealand Supreme Court.

Education
In 2002 there was one primary school with 17 teachers and 251 pupils, and one secondary school with 29 teachers and 240 pupils. There is also the University of the South Pacific.

Health
In 2003 there were four doctors, two dentists, three midwives and 14 nursing personnel. The 24-bed hospital at Alofi was destroyed by Cyclone Heta in Jan. 2004.

RELIGION

At the 2006 census, 956 people—62% of the population—belonged to the Congregational Christian Church of Niue (Ekalesia Niue); Roman Catholics (138); Latter-day Saints (127); Jehovah's Witness (28); Seventh Day Adventists (6); other (132); no religion or not stated (151).

CULTURE

Press
A weekly newspaper is published in English and Niuean; circulation about 400.

Tourism
In 2007 there were 3,463 visitor arrivals.

FURTHER READING

Statistical Office: Statistics Niue, Economic Planning Development & Statistics, Premier's Department, Utuko, Alofi.
Website: http://www.spc.int/prism/country/nu/stats

NICARAGUA

0   50 mi
0   75 km

HONDURAS

NICARAGUA

● Matagalpa

León   MANAGUA
              ⊡
              Lake
Granada   Nicaragua

PACIFIC
OCEAN   Caribbean
                              Sea

© Research Machines plc 2006 COSTA RICA

República de Nicaragua
(Republic of Nicaragua)

Capital: Managua
Population projection, 2015: 6·21m.
GNI per capita, 2011: (PPP$) 2,430
HDI/world rank: 0·589/129
Internet domain extension: .ni

KEY HISTORICAL EVENTS

There is evidence of settlement by Paleo-Indians in the region around 4000 BC. Spanish explorers, led by Gil González de Ávila, arrived in the west of present-day Nicaragua in 1523. They made contact with the Niquirano and the Chorotegano tribes, thought to have been linked to the Aztec civilization in Mexico, and the Chontal, who shared cultural traits with the Honduran Maya people. Government up to this point was through tribal monarchies and each grouping had distinct customs. In 1524 Francisco Hernández de Córdoba established Granada on Lake Nicaragua and León on Lake Managua. Many indigenous Indians were killed, died of introduced diseases or were enslaved. Estimates suggest the population fell from 1m. to less than 100,000.

In 1538 the Vice Royalty of New Spain was established, spanning Mexico and most of Central America. In 1570 present-day Nicaragua came under the authority of the Captaincy General of Guatemala. A conservative landholding elite developed around Granada, whilst León was associated with the colonial government. British pirates and adventurers took control of parts of the Mosquito Coast during the 17th century. After it declared independence from Spain in 1821, Nicaragua briefly came under the influence of the Mexican Empire, ruled by Agustín de Iturbide. From 1825–38 it was part of the Central American Federation, which unsuccessfully attempted to build a democratic union along US lines.

The 1840s and '50s saw escalating conflict between the elites of Conservative Granada and Liberal León. In 1855 the American mercenary, William Walker, arrived with 57 men to support the Liberal cause. Having defeated the national Nicaraguan army, he took Granada and proclaimed himself the country's ruler. However, his rule was short-lived and he was driven out of office by Honduran forces, with British assistance, two years later. There followed 30 years of relative stability under Conservative rule.

José Santos Zelaya's defeat of the Conservatives in 1893 led to a Liberal dictatorship, which became synonymous with economic decline. Civil war erupted in 1912, prompting the arrival of US forces. The Bryan–Chamarro Treaty of 1914 entitled the USA to a permanent option for a canal route through Nicaragua, a 99-year option for a naval base in the Bay of Fonseca on the Pacific coast and occupation of the Corn Islands on the Atlantic coast. The treaty was not abrogated until 1970 when the Corn Islands returned to Nicaragua.

The Somoza family dominated Nicaragua from 1933 to 1979. Imposing a brutal dictatorship, they plundered a large share of the national wealth. In 1962 the Sandinista National Liberation Front (FLSN; named after the murdered Liberal general, Augusto César Sandino, who fought against the US military presence in the 1920s) was formed to overthrow the Somozas. After 17 years of civil war the Sandinistas triumphed. On 17 July 1979 President Somoza fled into exile. The USA made efforts to unseat the revolutionary government by supporting the Contras (counter-revolutionary forces). It was not until 1988 that the state of emergency was lifted as part of the Central American peace process. Rebel anti-Sandinista activities had ceased by 1990 and the last organized insurgent group negotiated an agreement with the government in April 1994. In Oct. 1998 Hurricane Mitch devastated the country, killing 3,800. In Nov. 2006 presidential elections were won by Daniel Ortega of the FLSN, returning him to power after 16 years in opposition.

TERRITORY AND POPULATION

Nicaragua is bounded in the north by Honduras, east by the Caribbean, south by Costa Rica and west by the Pacific. Area, 131,812 sq. km (121,428 sq. km dry land). The coastline runs 450 km on the Atlantic and 305 km on the Pacific. The census population in May 2005 was 5,142,098 (density, 39·0 per sq. km). Estimate, June 2012: 6,071,000. 57·6% of the population were urban in 2011.

The UN gives a projected population for 2015 of 6·21m.

15 administrative departments and two autonomous regions are grouped in three zones. Areas (in sq. km), populations at the 2005 census and chief towns:

	Area	Population	Chief town
Pacific Zone	18,429	2,778,257	
Carazo	1,050	166,073	Jinotepe
Chinandega	4,926	378,970	Chinandega
Granada	929	168,186	Granada
León	5,107	355,779	León
Managua	3,672	1,262,978	Managua
Masaya	590	289,988	Masaya
Rivas	2,155	156,283	Rivas
Central-North Zone	35,960	1,647,605	
Boaco	4,244	150,636	Boaco
Chontales	6,378	153,932	Juigalpa
Estelí	2,335	201,548	Estelí
Jinotega	9,755	331,335	Jinotega
Madriz	1,602	132,459	Somoto
Matagalpa	8,523	469,172	Matagalpa
Nueva Segovia	3,123	208,523	Ocotal

	Area	Population	Chief town
Atlantic Zone	67,039	716,236	
Atlántico Norte[1]	32,159	314,130	Puerto Cabezas
Atlántico Sur[1]	27,407	306,510	Bluefields
Río San Juan	7,473	95,596	San Carlos

[1]Autonomous region.

The capital is Managua with (2005 census population) 908,892 inhabitants. Other cities (2005 populations): León, 139,433; Chinandega, 95,614; Masaya, 92,598; Estelí, 90,294; Tipitapa, 85,948; Matagalpa, 80,228; Granada, 79,418; Ciudad Sandino, 72,501; Juigalpa, 42,763.

The population is of Spanish and Amerindian origins with an admixture of Afro-Americans on the Caribbean coast. Ethnic groups in 2000: Mestizo (mixed Amerindian and White), 63%; White, 14%; Black, 8%; Amerindian, 5%. The official language is Spanish.

SOCIAL STATISTICS

2008 estimates: births, 140,000; deaths, 27,000. Estimated rates (per 1,000 population), 2008: births, 24·6; deaths, 4·7. Annual population growth rate, 2000–08, 1·3%. 2007 life expectancy: male 69·8 years, female 75·9. Infant mortality, 2010, 23 per 1,000 live births; fertility rate, 2008, 2·7 births per woman. A law prohibiting abortion was passed in Nov. 2006.

CLIMATE

The climate is tropical, with a wet season from May to Jan. Temperatures vary with altitude. Managua, Jan. 81°F (27°C), July 81°F (27°C). Annual rainfall 38" (976 mm).

CONSTITUTION AND GOVERNMENT

A new constitution was promulgated on 9 Jan. 1987 and underwent reforms in 1995 and 2000. It provides for a unicameral 92-seat *National Assembly* comprising 90 members directly elected by proportional representation for a five-year term, together with one seat for the previous president and one seat for the runner-up in the previous presidential election. Citizens are entitled to vote at the age of 16.

The *President* and *Vice-President* are directly elected for a five-year term commencing on the 10 Jan. following their date of election. At present the President may stand for a second term, but not consecutively. However, in Oct. 2009 the Supreme Court ruled in favour of a petition brought by President Daniel Ortega to remove the barrier against consecutive terms.

National Anthem

'Salve a ti Nicaragua' ('Hail to thee, Nicaragua'); words by S. Ibarra Mayorga, tune by L. A. Delgadillo.

RECENT ELECTIONS

Presidential and parliamentary elections took place on 6 Nov. 2011. In the presidential elections incumbent José Daniel Ortega Saavedra of the Sandinista National Liberation Front (FSLN) was elected with 62·5% of votes cast, defeating Fabio Gadea Mantilla (31·0%) and Arnoldo Alemán (5·9%). Despite accusations of electoral fraud by Gadea, international observers deemed the irregular voting patterns not strong enough to change the results. At the parliamentary elections the Sandinista National Liberation Front won 63 seats, the Independent Liberal Party 27 and the Constitutional Liberal Party 2.

CURRENT GOVERNMENT

President: José Daniel Ortega Saavedra; b. 1945 (FSLN; in office since 10 Jan. 2007, having previously been president from Jan. 1985–April 1990).

Vice President: Omar Halleslevens.

In March 2013 the government comprised:

Minister of Agriculture and Forestry: Ariel Bucardo. *Defence:* Ruth Esperanza Tapia Roa. *Education:* Miriam Ráudez.

Environment and Natural Resources: Juana Argeñal. *Family:* Marcia Ramírez Mercado. *Finance:* Alberto Guevara. *Foreign Affairs:* Samuel Santos López. *Health:* Dr Sonia Castro González. *Industry, Commerce and Development:* Orlando Solórzano Delgadillo. *Interior:* Ana Isabel Morales Mazún. *Labour:* Jeaneth Chávez Gómez. *Tourism:* Mario Salinas. *Transportation and Infrastructure:* Pablo Martínez.

Office of the President (Spanish only):
  http://www.presidencia.gob.ni

CURRENT LEADERS

Daniel Ortega

Position
President

Introduction
An iconic figure of the Sandinista movement since the late 1970s, Daniel Ortega first served as president from 1985–90 and was elected to further five-year terms in Nov. 2006 and Nov. 2011.

Early Life
Ortega was born in Nov. 1945 in La Libertad, Chontales to a middle-class family who opposed the Somoza family's dictatorship. After briefly attending the University of Central America in Managua, Ortega joined the Sandinista National Liberation Front (FSLN) in 1963. In 1967 he was convicted of staging a bank robbery to raise money for arms. He was released in 1974 in exchange for FSLN-held hostages and spent a brief spell in Cuba.

He led the FSLN in Somoza Debayle's overthrow in 1979 and joined the governing junta of national reconstruction. The Sandinistas dominated the junta and established a national constitution and democratic elections, which Ortega won in 1984. His presidency was marked by conflict with the US-backed Contras, resulting in tens of thousands of deaths. US sanctions and high inflation weakened the economy.

At the end of his term Ortega enacted the 'Piñata' laws by which large estates were appropriated and handed to FSLN supporters. In the 1990 elections Ortega suffered a surprise defeat to a US-supported coalition of anti-Sandinista groups led by Violeta de Chamorro. Ortega retained the leadership of the FSLN but was again defeated at the 1996 elections. He survived several damaging scandals, including allegations, in 1998, by his step-daughter of sexual abuse, followed in 1999 by a furore over an immunity pact with then president Miguel Aleman, who was facing charges of corruption. Ortega again made an unsuccessful bid for office in 2001.

Ortega publicly apologized for excesses during his first presidency. His politics moderated and he increasingly embraced Catholicism. The USA openly opposed his candidacy for the 2006 presidential election but he emerged victorious from the polls in Nov. 2006.

Career in Office
Ortega campaigned on a platform of 'unity and reconciliation', symbolized by his appointment of former Contra leader Jaime Morales as vice president. He announced plans to address poverty, affecting 80% of the population, as well as soaring inflation, crippling interest payments on public debt, a national energy shortage and endemic corruption. In Nov. 2006 the FSLN came out in support of a strict anti-abortion law.

On the international stage Ortega sought to reassure foreign investors and the private sector, many of whom feared further deterioration in US–Nicaraguan relations. His tenure received an early boost when the Inter-American Development Bank confirmed its commitment to an assistance loan of US$100m. promised to the previous government.

Ortega committed Nicaragua to the Bolivarian Alternative for the Americas (ALBA) while guaranteeing active membership of a free trade agreement with the USA (CAFTA). He pledged to maintain good relations with the USA while simultaneously strengthening links with Iran, Cuba and China. He also signed off a major natural gas deal with Bolivia. Venezuela pledged US$60m. in debt relief and donations for social development and helped establish a 40% Nicaraguan-owned and 60% Venezuelan-owned oil company, Albanic. In Oct. 2007 the Nicaraguan and Honduran governments accepted an International Court of Justice (ICJ) ruling settling a long-running territorial dispute between their two countries. However, in Nov. 2010 a dispute with neighbouring Costa Rica over a river border area led to the deployment of security forces by both countries. In March 2011 the ICJ ruled that all troops should be withdrawn from the area in a judgment viewed as favouring Costa Rica. In Nov. 2012, in another territorial arbitration the ICJ controversially ruled in Nicaragua's favour to redraw its maritime border in the Caribbean with Colombia.

In 2008 doubts about Ortega's democratic credentials resurfaced and in municipal elections in Nov. opposition parties denounced the Sandinista victory (in 94 of 146 mayorships, including the capital Managua) as rigged. He also circumvented a constitutional barrier to his standing for re-election through a favourable Supreme Court ruling in Oct. 2010. In the Nov. 2011 presidential polls Ortega retained power with 63% of the vote, although international observers expressed concerns over the official results.

DEFENCE

In 2008 defence expenditure totalled US$42m. (US$7 per capita), representing 0·7% of GDP.

Army
There are six regional commands. Strength (2007) around 12,000.

Navy
The Nicaraguan Navy was some 800 strong in 2007.

Air Force
The Air Force has been semi-independent since 1947. Personnel (2007) 1,200, with no combat-capable aircraft.

ECONOMY

In 2006 agriculture accounted for 19·7% of GDP, industry 29·5% and services 50·8%.

Overview
Nicaragua, one of Latin America's poorest countries, suffers from low productivity, high unemployment and a severe external debt. However, in the first decade of the century it experienced sound economic growth, averaging near 4% from 2004–08.

Coffee and meat account for around 40% of exports, and tourism is enjoying a revival. Leading trading partners include the USA, the members of the Central American Common Market and the EU. 30% of exports are directed to the USA while workers' remittances, mainly from the USA and Costa Rica, constitute 13% of GDP. The global financial crisis, Hurricane Felix and a spike in oil and food prices contributed to a 1·5% contraction in growth in 2009. The government's National Human Development Plan (NHDP) sets out plans to accelerate growth and reduce poverty.

Currency
The monetary unit is the *córdoba* (NIO), of 100 *centavos*, which replaced the córdoba oro in 1991 at par. Inflation was 16·8% in 2008, falling to 3·0% in 2010 but increasing to 7·4% in 2011. In July 2005 Nicaragua had foreign exchange reserves of US$628m. In June 2005 total money supply was 4,746m. córdobas.

Budget
In 2008 budgetary central government revenues were 27,042m. córdobas and expenditures 25,032m. córdobas. Tax revenue accounted for 80·4% of revenues in 2008; compensation of employees accounted for 36·2% of expenditures.

The standard rate of VAT is 15%.

Performance
The economy shrank by 1·5% in 2009 but there was growth of 4·5% in 2010 and 4·7% in 2011. Total GDP in 2011 was US$9·3bn.

Banking and Finance
The Central Bank of Nicaragua came into operation on 1 Jan. 1961 as an autonomous bank of issue, absorbing the issue department of the National Bank. The *President* is Alberto Guevara. There were seven private commercial banks in 2000.

In 2010 external debt totalled US$4,786m., equivalent to 76·9% of GNI.

There is a stock exchange in Managua.

ENERGY AND NATURAL RESOURCES

Environment
Nicaragua's carbon dioxide emissions from the consumption and flaring of fossil fuels in 2008 were the equivalent of 0·9 tonnes per capita.

Electricity
Installed capacity in 2007 was 0·8m. kW. In 2007, 3·21bn. kWh were produced; consumption per capita in 2007 was 585 kWh.

Minerals
Production in 2005: gold, 3,674 kg; silver, 2,999 kg. Calcium carbonate, gypsum and limestone are also mined.

Agriculture
In 2002 there were an estimated 1·93m. ha. arable land and 236,000 ha. permanent cropland. Approximately 94,000 ha. were irrigated in 2002. Production (in 1,000 tonnes) in 2003: sugarcane, 3,603; maize, 524; rice, 291; dry beans, 203; sorghum, 99; groundnuts, 82; oranges, 70; coffee, 60; bananas, 59; cassava, 52; pineapples, 48; plantains, 40.

In 2003 there were 3·5m. cattle, 440,000 pigs, 260,000 horses and 16m. chickens. Livestock products (in 1,000 tonnes), 2003: beef and veal, 66; poultry meat, 62; milk, 281; eggs, 22.

Forestry
The forest area in 2010 was 3·11m. ha., or 26% of the land area. Timber production was 6·10m. cu. metres in 2007.

Fisheries
In 2010 the catch was 37,423 tonnes (36,480 tonnes from sea fishing).

INDUSTRY

Industry contributed 26·0% of GDP in 2001, with manufacturing accounting for 14·4%. Important industries include chemicals, textiles, metal products, oil refining and food processing. Production (in 1,000 tonnes): cement (2009 estimate), 530; raw sugar (2000), 398; residual fuel oil (2007), 395; distillate fuel oil (2007), 198; petrol (2007), 87; wheat flour (2001), 60; vegetable oil (2001), 21; rum (1998), 7·7m. litres; sawnwood (2002), 45,000 cu. metres.

Labour
The workforce in 2001 was 1,900,400 (1,315,000 males). In 2001, 1,701,700 persons were in employment, of whom 739,000 were engaged in agriculture, hunting, forestry and fishing; 294,300 in community, social and personal services; 279,800 in wholesale and retail trade, and restaurants and hotels; and 131,600 in manufacturing. There were 159,500 unemployed in 2005, a rate of 7·2%.

INTERNATIONAL TRADE

In 2004 Nicaragua signed the Central America-Dominican Republic-United States Free Trade Agreement (CAFTA-DR), along with Costa Rica, the Dominican Republic, El Salvador, Guatemala, Honduras and the USA. The agreement entered into force for Nicaragua on 1 April 2006.

Imports and Exports
Imports and exports in US$1m.:

	2003	2004	2005	2006	2007
Imports c.i.f.	1,904·6	2,249·7	2,535·6	2,740·7	3,538·0
Exports f.o.b.	605·2	759·8	866·0	758·6	1,194·5

Main imports in 2006 were: petroleum and petroleum products, 24·3%; machinery and transport equipment, 22·9%; chemicals and related products, 16·7%; food and live animals, 9·7%. Principal exports were: coffee, 26·5%; beef, 10·3%; seafood, 9·5%; gold, 7·7%; cane sugar, 6·6%.

Main import suppliers, 2006: USA, 22·8%; Mexico, 14·8%; China, 7·6%; Venezuela, 6·8%; Costa Rica, 5·4%. Main export markets, 2006: USA, 46·5%; Mexico, 6·9%; Canada, 6·0%; Spain, 4·5%; Honduras, 4·4%.

COMMUNICATIONS

Roads
Road length in 2007 was 20,333 km, of which 1,081 km were main roads. In 2007 there were 101,900 passenger cars (18 per 1,000 inhabitants), 7,700 buses and coaches, 179,900 lorries and vans and 61,200 motorcycles and mopeds. 522 fatalities were caused by road accidents in 2007.

Civil Aviation
In 1999 scheduled airline traffic of Nicaragua-based carriers flew 0·8m. km, carrying 59,000 passengers (all on international flights). The Augusto Sandino international airport at Managua handled 754,000 passengers in 2001 (608,000 on international flights) and 20,000 tonnes of freight.

Shipping
In Jan. 2009 there were two ships of 300 GT or over registered, totalling 1,000 GT. The Pacific ports are Corinto (the largest), San Juan del Sur and Puerto Sandino through which pass most of the external trade. The chief eastern ports are El Bluff (for Bluefields) and Puerto Cabezas.

Telecommunications
In 2008 there were 312,000 main (fixed) telephone lines; mobile phone subscribers numbered 3,108,000 in 2008 (54·8 per 100 persons). There were 185,000 internet users in 2008. In Dec. 2011 there were 664,000 Facebook users.

SOCIAL INSTITUTIONS

Justice
The judicial power is vested in a Supreme Court of Justice at Managua, five chambers of second instance and 153 judges of lower courts.

The population in penal institutions in Dec. 2006 was 6,060 (107 per 100,000 of national population).

Education
Adult literacy rate in 2005 was 78·0% (male, 78·1%; female, 77·9%). In 2007 there were 952,964 primary school pupils (31,188 academic staff), 470,520 secondary school pupils (15,126 academic staff) and (2003) 103,577 students at tertiary level with 6,757 academic staff. The largest university is the National Autonomous University of Nicaragua, established in 1812.

In 2003 public expenditure on education came to 3·3% of GNI.

Health
In 2003 there were 32 hospitals, with a provision of nine beds per 10,000 population. There were 8,986 physicians, 1,585 dentists and 5,862 nurses in 2003.

RELIGION
The prevailing form of religion is Roman Catholicism (3·59m. adherents in 2001), but religious liberty is guaranteed by the Constitution. There were also 810,000 Protestants in 2001. There is one arch-bishopric, seven bishoprics and one cardinal.

CULTURE

World Heritage Sites
Nicaragua has two sites on the UNESCO World Heritage List: the Ruins of León Viejo (inscribed on the list in 2000), a 16th century Spanish settlement; and León Cathedral (2011), a Baroque and Neoclassical monument built between 1747 and the early 19th century.

Press
In 2005 there were six daily newspapers in Managua, with a total circulation of 180,000.

Tourism
In 2005 there were 712,000 non-resident tourists, spending US$211m.

DIPLOMATIC REPRESENTATIVES
Of Nicaragua in the United Kingdom (Suite 31, Vicarage House, 58–60 Kensington Church St., London, W8 4DP)
Ambassador: Carlos Argüello Gómez.

Of the United Kingdom in Nicaragua (embassy in Managua closed in March 2004)
Ambassador: Chris Campbell (resides in San José, Costa Rica).

Of Nicaragua in the USA (1627 New Hampshire Ave., NW, Washington, D.C., 20009)
Ambassador: Francisco Obadiah Campbell Hooker.

Of the USA in Nicaragua (Km. 5½ Carretera Sur, Managua)
Ambassador: Phyllis M. Powers.

Of Nicaragua to the United Nations
Ambassador: María Rubiales de Chamorro.

Of Nicaragua to the European Union
Ambassador: Mauricio Lautaro Sandino Montes.

FURTHER READING
Baracco, Luciano, *Nicaragua: The Imagining of a Nation—From Nineteenth-Century Liberals to Twentieth-Century Sandinistas.* 2005
Cruz, Consuelo, *Political Culture and Institutional Development in Costa Rica and Nicaragua: World-making in the Tropics.* 2005
Dijkstra, G., *Industrialization in Sandinista Nicaragua: Policy and Party in a Mixed Economy.* 1992
Horton, Lynn, *Peasants in Arms: War and Peace in the Mountains of Nicaragua, 1979–94.* 1998
Jones, Adam, *Beyond the Barricades: Nicaragua and the Struggle for the Sandinista Press, 1979–1998.* 2002

National Statistical Office: Dirección General de Estadística y Censos, Managua.
Website (Spanish only): http://www.inide.gob.ni

NIGER

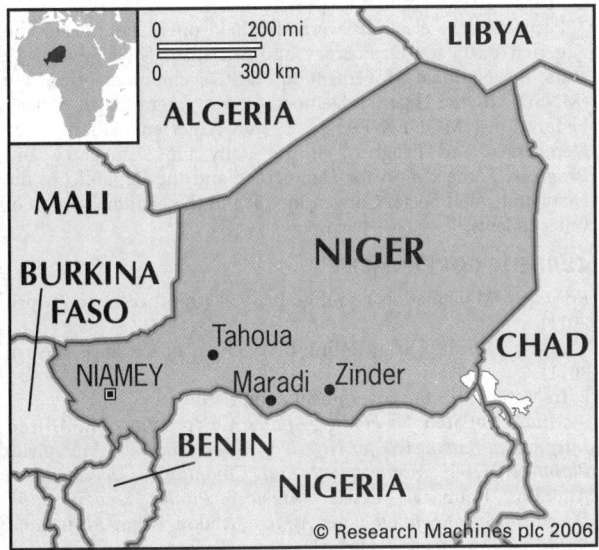

© Research Machines plc 2006

République du Niger
(Republic of Niger)

Capital: Niamey
Population projection, 2015: 18·50m.
GNI per capita, 2011: (PPP$) 641
HDI/world rank: 0·295/186
Internet domain extension: .ne

KEY HISTORICAL EVENTS

Niger has been settled for at least 6,000 years. Early cattle-herding and agricultural economies developed in the Sahara and by the 14th century the Hausa people had established city-states in the south of the region. Meanwhile Berbers dominated trade routes in the north, which came under the control of the Songhai Empire around 1515. After the fall of the Songhai Empire in the late 16th century the Bornu Empire expanded into the centre and east of the region, while the Hausa people retained the south and the Tuareg were prominent in the north. The Djerma people later became established in the southwest.

In the 19th century Fulani Muslims vied for power with the Hausas and Bornus. French forces took control of the region at the end of the century, creating the military district of Niger as part of the larger territory of Haut-Sénégal et Niger. Tuareg resistance led to a major uprising during the First World War, which was put down by combined French and British forces. France formally designated Niger as a colony in 1922.

Gradual decentralization of power began in 1946. Niger became an autonomous state in 1958 and on 3 Aug. 1960 it obtained independence, with Hamani Diori as its first president. In 1968 severe drought culminated in civil unrest. Diori was deposed in a 1974 military coup and Lieut. (later Col.) Seyni Kountché took power. Kountché gradually replaced the military government with civilians, surviving several coup attempts in the process. He was succeeded in 1987 by Ali Seibou, whose National Movement for the Development Society (MNSD), became Niger's only legal political party. A new constitution was approved in 1989 and in the early 1990s Seibou responded to pro-reform demonstrations by legalizing opposition parties.

In 1990 conflict broke out in the north between government forces and the Tuaregs, who demanded autonomy. Relations with Libya deteriorated as the Niger government suspected its neighbour of encouraging the insurgency. In 1995 a ceasefire established land rights for the Tuaregs and formalized their relationship with government. The agreement has largely held despite sporadic conflict.

In July 1991 a constitutional conference removed Seibou's powers and established a transitional government under André Salifou. Multi-party elections were held in 1993 and, following victory for the Alliance of the Forces of Change, Mahamane Ousmane was elected president. However, the MNSD won back control of the national assembly in 1995 and tensions mounted between President Ousmane and the government. In Jan. 1996 Ousmane was deposed in a military coup and replaced by Col. Ibrahim Baré Mainassara (known as Baré).

Strict military rule was eventually relaxed and in July 1996 Baré claimed decisive victory in a disputed election. In April 1999 Baré was killed by his bodyguards and Major Daoude Wanké took power with military backing. Under international pressure, multi-party elections were held in Oct. 1999 and were declared by independent observers to be largely fair. Mamadou Tandja of the MNSD was elected president and his party won control of the national assembly. Niger held its first local elections in July 2004, with most seats being won by parties supporting the president. The new national assembly established amnesties for those involved in the 1996 and 1999 coups. In Dec. 2004 Mamadou Tandja won a second term as president, with Hama Amadou as prime minister.

Niger continues to suffer severe economic problems. In 2005 the government increased tax on basic goods (including food), leading to widespread protests and strikes until the government granted exemptions on some staples. Drought and locust infestations the previous year devastated harvests. After UN World Food Programme warnings that millions faced severe malnutrition, an international relief effort staved off the worst of the disaster but thousands still died of diseases associated with hunger and poverty. In June 2006 unions organized a national strike to protest against rising prices. In response to allegations of corruption by aid donors, several ministers were dismissed from the government.

In 2006 the International Court of Justice settled a land dispute with Benin by awarding Niger most of the river islands along the shared border. In April 2009 the government reached a peace agreement with the rebel Movement of Niger People for Justice after several years of troubles. The following month President Mamadou Tandja dismissed parliament and suspended the constitution after a constitutional court ruling against his attempt to hold a referendum on his running for a third presidential term. In Aug. 2009 the referendum, criticized for lacking legitimacy, supported Tandja's candidacy but elections in Oct. were boycotted by opponents. Tandja claimed a landslide victory but was removed from office in a coup in Feb. 2010, with power falling to a military junta. The presidential elections in Jan.–March 2011 signalled a peaceful return to democracy with outgoing President Salou Djibo heralding them as an example for Africa.

TERRITORY AND POPULATION

Niger is bounded in the north by Algeria and Libya, east by Chad, south by Nigeria, southwest by Benin and Burkina Faso, and west by Mali. Area, 1,186,408 sq. km, with a population at the 2001 census of 11,060,291; density, 9·3 per sq. km. Estimate, 2010, 15,204,000. In 2011, 17·2% of the population were urban.

The UN gives a projected population for 2015 of 18·50m. Niger's population has doubled since the early 1990s.

The country is divided into the capital, Niamey, an autonomous district, and seven departments. Area, population and chief towns at the 2001 census:

Department	Sq. km	Population	Chief town	Population
Agadez	634,209	321,639	Agadez	78,289
Diffa	140,216	346,595	Diffa	23,409
Dosso	31,002	1,505,864	Dosso	43,561
Maradi	38,581	2,235,748	Maradi	148,017
Niamey	670	707,951	Niamey	707,951
Tahoua	106,677	1,972,729	Tahoua	73,002
Tillabéri	89,623	1,889,515	Tillabéri	16,683
Zinder	145,430	2,080,250	Zinder	170,575

The population is composed chiefly of Hausa (53%), Djerma-Songhai (21%), Fulani (10%), Tuareg (10%) and Kanuri-Manga (4%). The official language is French. Hausa, Djerma and Fulani are national languages.

SOCIAL STATISTICS

Estimates, 2008: births, 787,000; deaths, 219,000. Estimated birth rate in 2008 was 53·5 per 1,000 population (the highest in the world); estimated death rate, 14·9. Niger has one of the youngest populations of any country, with 73% of the population under the age of 30 and 49% under 15. Infant mortality, 2010, 73 per 1,000 live births. Annual population growth rate, 2000–08, 3·6%. Expectation of life at birth, 2007, 50·0 years for males and 51·7 for females. Fertility rate, 2008, 7·1 children per woman (the highest anywhere in the world).

CLIMATE

Precipitation determines the geographical division into a southern zone of agriculture, a central zone of pasturage and a desert-like northern zone. The country lacks water, with the exception of the southwestern districts, which are watered by the Niger and its tributaries, and the southern zone, where there are a number of wells. Niamey, 95°F (35°C). Annual rainfall varies from 22" (560 mm) in the south to 7" (180 mm) in the Sahara zone. The rainy season lasts from May until Sept., but there are periodic droughts.

CONSTITUTION AND GOVERNMENT

In May 2009 former president Mamadou Tandja dissolved the 113-member *National Assembly*, allowing a constitutional referendum to be pushed through. At the referendum on 4 Aug. 2009, 92·5% of votes cast were in favour of abolishing the two-term limit that had existed under the constitution of 1999, although both the opposition and the international community condemned the poll. A coup followed in Feb. 2010 with the military junta suspending the constitution and dissolving the cabinet. In March 2010 the military leadership announced it had formed a transitional government of 20 ministers and promised to return Niger to democracy, although no date was set for fresh elections. In Oct. 2010 a new constitution received 90·18% support in a referendum; turnout was 52·65%. Establishing Niger as a secular state, it reimposes a two-term limit (each of five years) on the presidency, prohibits members of the military from running for office and guarantees that the government will release data on national oil and mining revenues.

National Anthem

'Auprès du grand Niger puissant' ('By the banks of the mighty great Niger'); words by M. Thiriet, tune by R. Jacquet and N. Frionnet.

RECENT ELECTIONS

In the first round of presidential elections held on 31 Jan. 2011 former prime minister Mamadou Issoufou won 36·1% of the votes, Seyni Oumarou (a second former prime minister) 23·2%,

Hama Amadou (a third former prime minister) 19·8% and Mahamane Ousmane (a former president) 8·4%. There were six other candidates. Turnout was 52·8%. In the run-off on 12 March 2011 Mahamadou Issoufou won 57·9% of the votes and Seyni Oumarou 42·1%. Turnout was 48·2%

Parliamentary elections were also held on 31 Jan. 2011. The Nigerien Party for Democracy and Socialism (PNDS) won 39 seats; the National Movement for the Development of Society (MNSD), 26; the Nigerien Democratic Movement for an African Federation (MODEN/FA), 23; the Nigerien Alliance for Democracy and Progress, 8; the Rally for Democracy and Progress, 7; the Union for Democracy and the Republic, 6; the Democratic and Social Convention, 3; and the National Union of Independents, 1.

CURRENT GOVERNMENT

President: Mahamadou Issoufou; b. 1952 (in office since 7 April 2011).

Prime Minister: Brigi Rafini; b. 1953 (in office since 7 April 2011).

In March 2013 the government comprised:

Minister of State for Foreign Affairs, Co-operation and African Integration, Responsible for Nigeriens Abroad: Bazoum Mohamed. *Planning, Land Management and Community Development:* Amadou Boubacar Cissé. *Interior, Public Security and Decentralization of Religious Affairs:* Abdou Labo. *Mines and Industrial Development:* Omar Hamidou Tchiana.

Minister of Agriculture: Oua Saidou. *Civil Service and Labour:* Sabo Fatouma Zara Boubacar. *Commerce and Promotion of the Private Sector:* Saley Saidou. *Communication, New Information Technologies and Relations with Institutions:* Salifou Labo Bouché. *Energy and Oil:* Foumakoye Gado. *Equipment:* Saddi Soumaila. *Finance:* Jules Baillet. *Higher Education and Scientific Research:* Mamadou Youba Diallo. *Industrial Development, Handicrafts and Tourism:* Yahaya Baré Haoua Abdou. *Justice and Government Spokesperson:* Marou Amadou. *Livestock:* Mahaman El Ousmane. *National Defence:* Karidjo Mahamadou. *National Education, Literacy and Promotion of National Languages:* Ali Mariama El Hadj Ibrahim. *Parliamentary Affairs:* Elhadj Laouali Chaibou. *Population, the Promotion of Women and Child Protection:* Maitchibi Kadidjatou Dan Dobi. *Professional Training and Employment:* N'Gadé Nana Hadiza Noma Kaka. *Public Health:* Soumana Sanda. *Transport:* Ibrahim Yacouba. *Urban Affairs, Housing and Sanitation:* Moussa Bako Abdoulkarim. *Water and Environment:* Issoufou Issaka. *Youth, Sports and Culture:* Kounou Hassan.

Government Website (French only): http://www.gouv.ne

CURRENT LEADERS

Mahamadou Issoufou

Position
President

Introduction
Mahamadou Issoufou was sworn in as president on 7 April 2011. The veteran opposition leader defeated rival candidate Seyni Oumarou, a former ally of deposed leader Mamadou Tandja, in a run-off on 12 March 2011. The election fulfilled a pledge by the military that it would reinstate civilian rule following the ousting of Tandja in a coup in 2010. Heading the Nigerien Party for Democracy and Socialism (PNDS), Issoufou has previously served as prime minister and as president of the National Assembly.

Early Life
Issoufou was born in 1952 in the town of Dan Daji, in the Tahoua department of central Niger. He is an ethnic Hausa, the country's most populous group. Trained as a mining engineer in France, Issoufou returned to Niger to work as a technical director for

French company Areva. In 1980 he was appointed national director of mines, a post he held for five years before becoming secretary-general of the Mining Company of Niger.

In Feb. 1993, at the country's first open elections, Issoufou was elected to parliament as a representative of the PNDS. Later that month he came third in presidential elections, marking the first of four failed attempts to gain the presidency. He was appointed prime minister by the successful candidate, Mahamane Ousmane, in April 1993 but resigned in Sept. 1994 in protest at Ousmane's decree diminishing prime ministerial influence.

The following year, after parliamentary elections had led to the establishment of a new coalition, he was appointed president of the National Assembly. Following a military coup in Jan. 1996 and subsequent flawed presidential elections, Issoufou spent intermittent periods under house arrest. He made two further unsuccessful bids for the presidency in 1999 and 2004.

In 2009 he was charged with misappropriating funds after the opposition called for a general strike to protest Mamadou Tandja's attempt at changing the constitution to secure a third term. Issoufou briefly fled the country, claiming that the charge against him was politically motivated. However, the ousting of Tandja and the establishment of a transitional junta allowed him to return to national politics. He gained 58% of the vote in the second round of presidential elections in March 2011, defeating Seyni Oumarou.

Career in Office
Issoufou pledged to secure a more equal distribution of the wealth from Niger's substantial uranium reserves. Among his first challenges was to devise a plan to cope with the influx of refugees fleeing conflict in Libya and Côte d'Ivoire. An alleged assassination plot against him was foiled in July 2011.

DEFENCE

Selective conscription for two years operates. Defence expenditure totalled US$58m. in 2008 (US$4 per capita), representing 1·1% of GDP.

Army
There are three military districts. Strength (2007) 5,200. There are additional paramilitary forces of 5,400.

Air Force
In 2007 the Air Force had 100 personnel. There are no combat-capable aircraft.

ECONOMY

Agriculture, forestry and fishing accounted for 43·0% of GDP in 2006; trade and hotels 14·7%; services 8·8%; and finance and real estate 7·6%.

Overview
Niger is among the poorest countries in the world, ranking 186th out of 187 countries in the UN Human Development Index in 2011. Long-term GDP growth has been low and volatile, averaging 1·7% per year between 1970 and 2005 while failing to keep pace with population growth. The economy is dominated by agriculture, uranium mining and development assistance. About 45% of the national budget is financed by development assistance. Over 60% of the population live on less than a dollar a day.

Niger received enhanced debt relief from Dec. 2000 under the IMF programme for Heavily Indebted Poor Countries (HIPC), completing the programme in April 2004. It continues to qualify for the Multilateral Debt Relief Initiative (MDRI), which covers large fiscal and external current account deficits.

Currency
The unit of currency is the *franc CFA* (XOF) with a parity of 655·957 francs CFA to one euro. In June 2005 total money supply was 170,798m. francs CFA and foreign exchange reserves were

US$177m. Inflation was 10·5% in 2008, falling to 0·9% in 2010. In 2011 it stood at 2·9%.

Budget
In 2008 revenue totalled 584·1bn. francs CFA and expenditure 546·0bn. francs CFA.

Performance
The economy contracted by 0·9% in 2009 but there was growth of 8·0% in 2010 and 2·3% in 2011; total GDP in 2011 was US$6·0bn.

Banking and Finance
The regional Central Bank of West African States (BCEAO)— *Governor*, Tiémoko Meyliet Koné—functions as the bank of issue. There were six commercial banks in 2002, three development banks and a savings bank.

In 2010 external debt totalled US$1,127m., equivalent to 20·5% of GNI.

There is a stock exchange in Niamey.

ENERGY AND NATURAL RESOURCES

Environment
In 2008 Niger's carbon dioxide emissions from the consumption and flaring of fossil fuels were the equivalent of 0·1 tonnes per capita. An *Environmental Performance Index* compiled in 2008 ranked Niger 149th in the world out of 149 countries analysed, with 39·1%. The index examined various factors in six areas—air pollution, biodiversity and habitat, climate change, environmental health, productive natural resources and water resources.

Electricity
Installed capacity was approximately 0·1m. kW in 2007. Production in 2007 amounted to about 197m. kWh, with consumption per capita an estimated 42 kWh.

Minerals
Large uranium deposits are mined at Arlit and Akouta. Uranium production (2010), 4,198 tonnes. Niger's uranium production is the fifth largest in the world. Phosphates are mined in the Niger valley, and coal reserves are being exploited by open-cast mining (production of hard coal in 2007 was an estimated 185,000 tonnes). Salt production in 2006 was an estimated 1,300 tonnes.

Agriculture
There were an estimated 4·49m. ha. of arable land in 2002 and 13,000 ha. of permanent crops. About 66,000 ha. were irrigated in 2002. There were about 130 tractors in 2002. Production estimates in 2003 (in 1,000 tonnes): millet, 2,500; sorghum, 797; onions, 270; sugarcane, 220; cabbage, 120; cassava, 105; groundnuts, 100; tomatoes, 100. Livestock (2003 estimates): goats, 6·9m.; sheep, 4·5m.; cattle, 2·3m.; asses, 580,000; camels, 420,000; chickens, 25m. Livestock products (in 1,000 tonnes), 2003 estimates: milk, 305; meat, 130; cheese, 15; eggs, 11.

Forestry
There were 1·20m. ha. of forests in 2010 (1% of the land area). Timber production in 2007 was 9·63m. cu. metres, mainly for fuel.

Fisheries
There are fisheries on the River Niger and along the shores of Lake Chad. In 2009 the catch was 29,884 tonnes, exclusively from inland waters.

INDUSTRY

Some small manufacturing industries, mainly in Niamey, produce textiles, food products, furniture and chemicals. Output of cement in 2007 (estimate), 42,000 tonnes.

Labour
The estimated economically active population in 2009 totalled 4,803,000 (68% males). Agriculture, fisheries and forestry remains the largest sector of employment.

INTERNATIONAL TRADE

Imports and Exports

In 2008 imports (c.i.f.) were valued at US$1,247·5m. and exports at US$503·1m. Main imports in 2008: machinery and transport goods, 20·2%; petroleum and petroleum products, 15·2%; cereal and cereal preparations, 11·4%; medicinal and pharmaceutical products, 9·5%. Main exports in 2008: uranium ores and concentrates, 57·5%; livestock, 9·2%; textile fibres, 6·1%; cotton fabrics, 4·5%. Main import suppliers in 2008 (as % of total): France, 13·2; China, 12·6; USA, 7·6; Netherlands, 7·2. Main export destinations in 2008: France, 33·3; USA, 17·6; Nigeria, 11·8; Japan, 9·9.

COMMUNICATIONS

Roads

In 2007 there were 18,949 km of roads including 3,912 km of paved roads. Niamey and Zinder are the termini of two trans-Sahara motor routes; the Hoggar–Aïr–Zinder road extends to Kano and the Tanezrouft–Gao–Niamey road to Benin. A 648-km 'uranium road' runs from Arlit to Tahoua. There were, in 2005, 57,732 passenger cars, 11,261 vans, 2,613 buses and 1,035 lorries. In 2007 there were 676 road accidents resulting in 265 fatalities.

Civil Aviation

There is an international airport at Niamey (Diori Hamani Airport), which handled 108,000 passengers in 2006 and 2,300 tonnes of freight. In 2003 there were international flights to Abidjan, Bamako, Casablanca, Dakar, Khartoum, Libreville, Ouagadougou, Paris and Tripoli. In 1999 scheduled airline traffic of Niger-based carriers flew 3·0m. km, carrying 84,000 passengers (all on international flights).

Shipping

Sea-going vessels can reach Niamey (300 km inside the country) between Sept. and March.

Telecommunications

There were 83,600 landline telephone subscriptions in 2010 (equivalent to 5·4 per 1,000 inhabitants) and 4,339,900 mobile phone subscriptions in 2011 (or 270·1 per 1,000 inhabitants). In 2011, 1·3% of the population were internet users.

SOCIAL INSTITUTIONS

Justice

There are Magistrates' and Assize Courts at Niamey, Zinder and Maradi, and justices of the peace in smaller centres. The Court of Appeal is at Niamey.

The population in penal institutions in May 2006 was 5,709 (46 per 100,000 of national population).

Education

In 2007 there were 31,131 teaching staff for 1,235,065 primary school pupils and 7,852 teaching staff for 213,991 secondary school pupils. There were 11,208 students in tertiary education (1,095 academic staff) in 2006. There is a university and an Islamic university.

Adult literacy in 2005 was 29%.

In 2006 public expenditure on education came to 3·3% of GNI and 17·6% of total government spending.

Health

In 1998 there were 1·2 hospital beds per 10,000 inhabitants. There were 427 physicians, 28 dentists, 1,988 nurses, 500 midwives and 22 pharmacists in 2008.

RELIGION

In 2001 there were 9·39m. Sunni Muslims. There are some Roman Catholics, and traditional animist beliefs are widespread.

CULTURE

World Heritage Sites

Niger has two sites on the UNESCO World Heritage List: the Aïr and Ténéré Natural Reserves (inscribed on the list in 1991), part of the largest protected area in Africa (7·7m. ha.); and the 'W' National Park of Niger (1996), a savannah and forested area of biodiversity.

Press

In 2005 there was one government-owned daily newspaper and around 15 weekly and monthly newspapers.

Tourism

In 2007 there were 41,000 non-resident tourists; spending by tourists totalled US$41m. in 2007.

DIPLOMATIC REPRESENTATIVES

Of Niger in the United Kingdom
Ambassador: Adamou Seydou (resides in Paris).

Of the United Kingdom in Niger
Ambassador: Vacant (resides in Bamako, Mali). Consular assistance is provided by the embassy in Senegal.

Of Niger in the USA (2204 R. St., NW, Washington, D.C., 20008)
Ambassador: Maman Sambo Sidikou.

Of the USA in Niger (BP 11201, Rue des Ambassades, Niamey)
Ambassador: Bisa Williams.

Of Niger to the United Nations
Ambassador: Baboucar Boureima.

Of Niger to the European Union
Ambassador: Abdou Agbarry.

FURTHER READING

Miles, W. F. S., *Hausaland Divided: Colonialism and Independence in Nigeria and Niger.* 1994

National Statistical Office: Institut National de la Statistique, 182 rue de la Sirba, BP 13416, Niamey.
Website (French only): http://www.stat-niger.org

NIGERIA

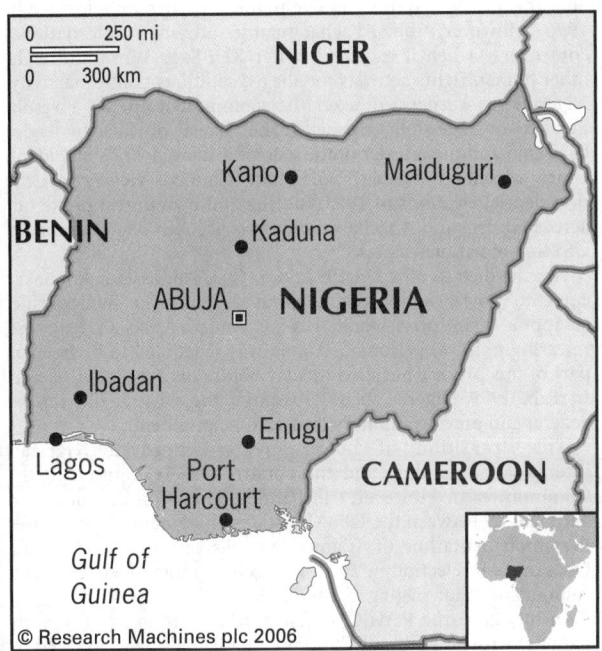

Federal Republic of Nigeria

Capital: Abuja
Population projection, 2015: 179·79m.
GNI per capita, 2011: (PPP$) 2,069
HDI/world rank: 0·459/156
Internet domain extension: .ng

KEY HISTORICAL EVENTS

The earliest evidence of human settlement in Nigeria dates from 9000 BC and by 2000 BC its inhabitants were cultivating crops and domestic animals. However, the first organized society was of the Nok people, from around 800 BC to AD 200. Traces of Nok influence are visible in Nigerian art today, particularly in areas such as Igbo, Ukwe, Esie and Benin City. By AD 1000, Nok had given way to the Kanem, thanks to the trans-Saharan trade route that ran from West Africa to the Mediterranean.

In the 11th century northern Nigeria split into seven independent Hausa city-states, Biram, Daura, Gobir, Kano, Katsina, Rano and Zaria. By the 14th century, two states had developed in the south, Oyo and Benin, with the Igbo people of the southeast living in small village communities. South of the Hausa states and west of the Niger, the Ife flourished between the 11th and 15th centuries. The importance of the Ife civilization is evident today; all Yoruba states claim that their leaders are descended from the Ife as a way of establishing legitimacy, and its ritual is imitated in their modern public ceremonies.

Most of the north was held by the Songhai empire by the early 16th century, only to be taken later in the century by Kanem-Bornu, allowing the Hausa states to retain their autonomy. At the end of the 18th century, Fulani religious groups waged war in the north, merging states to create the single Islamic state of the Sokoto Caliphate.

In the late 15th century Portuguese navigators, following the demise of the spice trade, began to purchase slaves from middlemen in the region. They were followed by British, French and Dutch traders. Wealthy traders established towns such as Bonny, Owome and Okrika. Slave trading had a profound effect on Nigeria. From the 1650s until the 1860s, it caused a forced migration of around 3·5m. people. Within Nigeria itself, the defensive measures adopted to avoid enslavement led to the reinforcement of ethnic distinctions and of the north-south divide.

After the abolition of the slave trade in Britain in 1807, attempts to find a lucrative alternative and to discourage the predominance of slavery in Nigeria (other countries continued to trade in slaves until 1875) led to a large-scale campaign to encourage the production of palm oil for export. This itself caused the development of an internal slave trade, involving slaves in the collection and manufacture of palm fruits, as well as the transportation of the oil. The British also took over the mines at Jos at the expense of the livelihoods of independent tin producers. When heavy reliance on mining exports resulted in the neglect of agricultural work, Nigeria experienced its first food shortage.

Religious missions were active at this time, with Presbyterians, Methodists, Baptists and the Church Missionary Society (CMS) operating in Lagos, Abeokuta, Ibadan, Oyo and Ogbomosho. The CMS pioneered trade on the Niger by encouraging merchants to run steamboats, partially as a means of travel for the missionaries but also to ship goods.

In 1804 Usuman dan Fodio began a 'Holy war' to reform the practice of Islam in the north, conquering the Hausa city-states, though Kanem-Bornu retained its independence. However, by the late 19th century Kanem-Bornu's power was in decline. Usuman's son, Muhammed Bello, established a state centred at Sokoto, controlling most of northern Nigeria for the rest of the century. In the south, the Oyo region was troubled by civil wars, only brought to a close when the British intervened and the Oyo Empire collapsed. Britain took Lagos as its colony in 1861.

In 1879 Sir George Goldie gained all British firms trading on the Niger, and in the 1880s took over French companies trading there, signing treaties with African leaders and enabling Britain's domination of southern Nigeria in 1884–85. In 1887, Jaja, an African trader based in the Niger Delta, was deported following his fierce opposition to European competition. Goldie's firm received a British Royal Charter as the Royal Niger Company to administer the Niger River and north Nigeria, and this monopoly of trade on the river angered Africans and Europeans alike. The Royal Niger Company also lacked sufficient power to control north Nigeria. In 1900 its charter was revoked and British forces moved in, taking Sokoto in 1903. By 1906 Britain controlled Nigeria as the Colony (Lagos), the Protectorate of Southern Nigeria and the Protectorate of Northern Nigeria, amalgamating the regions in 1914 to establish the Colony and Protectorate of Nigeria. The administration was based on existing leadership systems. Yet the appointment of African officials failed to gain wide acceptance from Nigeria's people. The British governor made all major decisions, and the traditional authority of African rulers was weakened irreparably.

British dominance met with major resistance from the Nigerian people. In the south, the tribal Yoruba group, the Ijebu, fought against colonial rule in 1892, as did the Aro in the east and the Aniocha (both Igbo groups) in the west. There were also rebellions in the north. The British forces responded with brutality, destroying the homes of many Nigerians in order to secure their capitulation.

British colonial rule brought development in the transportation and communications systems and a shift towards cash crops. Western and Christian influences prevailed, including widespread use of the English language. This influence spread far more rapidly in the south, where British control had been secure over a longer period, and this added to the growing disparity between north and south. Nigerian forces helped defeat the German army in Cameroon during the First World War, involved in an arduous campaign until 1916.

Growing unrest and widespread anticolonialism became focused in the 1920s as demands for African representation increased. In 1923 Herbert Macaulay, grandson of the first Nigerian to be ordained, established the first Nigerian political party, the Nigerian National Democratic Party. In 1944 he united the party with several others to form the National Council of Nigeria and the Cameroons (NCNC). In response to this activity, the British attempted to quell demands for an end to colonial rule by granting some political reforms. In 1947 they announced a new constitution that they claimed would give traditional authorities a stronger voice. This met with resistance, and in 1951 the British agreed to form a new constitution that would provide for elected representation on a regional basis.

Three political parties developed, the National Council of Nigeria and the Cameroons (later the National Convention of Nigerian Citizens), largely supported by the Igbo, the Action Group, with mostly Yoruba membership, and the Northern People's Congress (NPC). When the constitution failed in 1952, a new one divided Nigeria into three regions, Eastern, Western and Northern, plus the federal territory of Lagos. In 1956 the Western and Eastern regions became self-governing, as did the Northern region in 1959.

In 1960 Nigeria declared independence. Elections failed to elect any one party by a majority, and the NPC and the NCNC formed a coalition government, with Abukar Tafawa Balewa (NPC) as prime minister. Nnamdi Azikiwe, who had helped Herbert Macaulay to establish the NCNC in 1944, was governor-general. When Nigeria became a republic in 1963 Azikiwe became president.

Continuing conflict between north and south undermined the new republic. In 1966 fighting culminated in a military coup that installed Maj.-Gen. Aguiyi-Ironsi, an Igbo, as head of a military government. Another coup later in the year placed Lieut.-Col. Yakubu Gowon in power and saw many northern Igbo massacred. In May 1967 the Igbo people of the south declared their region independent from the rest of the country, naming the breakaway republic Biafra. Civil war raged for three years until federal Nigeria triumphed at the price of 1m. dead and widespread famine and destruction.

This was followed by a period of relative prosperity as oil prices rose. Foreign interest and investment flourished but government overspending and high levels of corruption and crime led to social chaos. In the 1980s recession sent oil prices down, and Nigeria found itself struggling with major debt, rising inflation and mass unemployment.

Gowon's regime was overthrown in 1975 by Gen. Murtala Muhammed whose plans for a new capital to be built at Abuja drained the economy. He was assassinated in 1976, to be succeeded by Gen. Olusegun Obasanjo who oversaw the transition to civilian rule, while juggling the need for Western aid with his support for African nationalist movements.

In 1979 elections brought Alhaji Shehu Shagari to power. Shagari's government came under popular attack for alleged corruption but he was re-elected in 1983, amidst rumours of voting irregularities. Under Shagari relations with the USA improved, heralded by a visit from President Jimmy Carter. However, dogged by worsening economic problems, Shagari was ousted in a military coup in 1983 and replaced by Gen. Muhammadu Buhari. Buhari's regime quickly fell out of favour with the public when it arrested not only the politicians blamed for the country's social and economic problems, but also journalists and academics.

A bloodless coup in 1985 brought to power Maj.-Gen. Ibrahim Babangida, who promulgated a new constitution with the aim of returning to civilian government. Babangida, however, clung to power and refused to accept electoral defeat in 1990, 1992 and 1993. Unrest eventually forced his resignation, but after just three

months of rule by an interim leader, one of Babangida's long-term allies, Gen. Sani Abacha, became president and closed down all unions and political institutions. He extended his military rule for a further three years in 1995, proposing a return to civilian rule after this period. To this end five political parties were formed in 1996. However, the Abacha regime attracted international controversy when it executed writer Ken Saro-Wiwa and eight other human rights activists for alleged seditious political activity. Nigeria was suspended from the Commonwealth as a result. Further outrage followed with the arrest of former leader Obasanjo and the murder of the wife of leading political dissident, Chief Moshood Abiola, who had claimed victory at the presidential elections of 1993. Rioting and civil unrest broke out across Nigeria and Abacha's family was accused of siphoning off US$4bn. of national assets.

Abacha died in office in 1998. Maj.-Gen. Abdusalam Abubakar came to power and brought about a return to civilian rule, scrapping plans that would have extended Abacha's rule and releasing political prisoners. Abiola was scheduled to be freed as part of this process but died the day before his scheduled release. In Feb. 1999 Nigeria chose Obasanjo, the 62-year-old retired general and previous military leader, to be president.

This transition of power greatly improved Nigeria's international standing and the country was readmitted to the Commonwealth. However, tribal and religious conflict continued and fighting between the Igbo Christians and Hausa Muslims over the implementation of Islamic law has left thousands dead. Obasanjo's re-election in 2003 was accompanied by violence and rumours of ballot-rigging and bribery.

In July 2005 the Paris Club (an informal grouping of wealthy nations) wrote off around US$20bn. of Nigeria's debt. Assisted by high oil prices, Nigeria had paid off the remaining US$10bn. by April 2006, the first African nation to fully service its Paris Club debt. In April 2007 Umaru Yar'Adua won disputed presidential elections. In March 2008 Nigeria and Cameroon reached agreement on the long-running dispute over sovereignty of the Bakassi Peninsula.

TERRITORY AND POPULATION

Nigeria is bounded in the north by Niger, east by Chad and Cameroon, south by the Gulf of Guinea and west by Benin. It has an area of 923,768 sq. km (356,667 sq. miles). For sovereignty over the Bakassi Peninsula see CAMEROON: Territory and Population. Census population, 2006, 140,431,790 (69,086,302 females); population density, 152·0 per sq. km. Nigeria is Africa's most populous country. In 2011, 50·5% of the population were urban.

The UN gives a projected population for 2015 of 179·79m.

There were 36 states and a Federal Capital Territory (Abuja) in 2006.

Area, population and capitals of these states:

State	Area (in sq. km)	Population (2006 census)	Capital
Adamawa	36,917	3,178,950	Yola
Bauchi	45,837	4,653,066	Bauchi
Benue	34,059	4,253,641	Makurdi
Borno	70,898	4,171,104	Maiduguri
Gombe	18,768	2,365,040	Gombe
Jigawa	23,154	4,361,002	Dutse
Kaduna	46,053	6,113,503	Kaduna
Kano	20,131	9,401,288	Kano
Katsina	24,192	5,801,584	Katsina
Kebbi	36,800	3,256,541	Birnin-Kebbi
Kogi	29,833	3,314,043	Lokoja
Kwara	36,825	2,365,353	Ilorin
Nassarawa	27,117	1,869,377	Lafia
Niger	76,363	3,954,772	Minna
Plateau	30,913	3,206,531	Jos
Sokoto	25,973	3,702,676	Sokoto
Taraba	54,473	2,294,800	Jalingo

State	Area (in sq. km)	Population (2006 census)	Capital
Yobe	45,502	2,321,339	Damaturu
Zamfara	39,762	3,278,873	Gusau
Federal Capital Territory	7,315	1,406,239	Abuja
Total North	*730,885*	*75,269,722*	
Abia	6,320	2,845,380	Umuahia
Akwa Ibom	7,081	3,902,051	Uyo
Anambra	4,844	4,177,828	Awka
Bayelsa	10,773	1,704,515	Yenagoa
Cross River	20,156	2,892,988	Calabar
Delta	17,698	4,112,445	Asaba
Ebonyi	5,670	2,176,947	Abakaliki
Edo	17,802	3,233,366	Benin City
Ekiti	6,353	2,398,957	Ado Ekiti
Enugu	7,161	3,267,837	Enugu
Imo	5,530	3,927,563	Owerri
Lagos	3,345	9,113,605	Ikeja
Ogun	16,762	3,751,140	Abeokuta
Ondo	14,606	3,460,877	Akure
Osun	9,251	3,416,959	Oshogbo
Oyo	28,454	5,580,894	Ibadan
Rivers	11,077	5,198,716	Port Harcourt
Total South	*192,883*	*65,162,068*	

Abuja replaced Lagos as the federal capital and seat of government in Dec. 1991.

Estimated population of the largest cities, 1995:

Lagos	1,484,000[1]	Ikorodu	180,300
Ibadan	1,365,000	Ilawe-Ekiti	179,900
Ogbomosho	711,900	Owo	178,900
Kano	657,300	Ikirun	177,000
Oshogbo	465,000	Calabar	170,000
Ilorin	464,000	Shaki	169,700
Abeokuta	416,800	Ondo	165,400
Port Harcourt	399,700	Akure	158,200
Zaria	369,800	Gusau	154,000
Ilesha	369,000	Ijebu-Ode	152,500
Onitsha	362,700	Effon-Alaiye	149,300
Iwo	353,000	Kumo	144,400
Ado-Ekiti	350,500	Shomolu	144,100
Abuja (capital)	339,100	Oka	139,600
Kaduna	333,600	Ikare	137,300
Mushin	324,900	Sapele	135,800
Maiduguri	312,100	Deba Habe	135,400
Enugu	308,200	Minna	133,600
Ede	299,500	Warri	122,900
Aba	291,600	Bida	122,500
Ife	289,500	Ikire	120,200
Ila	257,400	Makurdi	120,100
Oyo	250,100	Lafia	119,500
Ikerre	238,500	Inisa	116,800
Benin City	223,900	Shagamu	114,300
Iseyin	211,800	Awka	108,400
Katsina	201,500	Gombe	105,200
Jos	201,200	Ejigbo	103,300
Sokoto	199,900	Igboho	103,300
Ilobu	194,400	Agege	100,300
Offa	192,300	Ugep	100,000

[1]Greater Lagos had a population of 12,763,000 in 1999.

There are about 250 ethnic groups. The largest linguistic groups are the Yoruba (17·5% of the total) and the Hausa (17·2%), followed by Igbo (13·3%), Fulani (10·7%), Ibibio (4·1%), Kanuri (3·6%), Egba (2·9%), Tiv (2·6%), Bura (1·1%), Edo (1·0%) and Nupe (1·0%). The official language is English, but 50% of the population speak Hausa as a *lingua franca*.

SOCIAL STATISTICS

2008 estimates: births, 6,050,000; deaths, 2,420,000. Rates, 2008 estimates (per 1,000 population): births, 40; deaths, 16. Infant mortality, 2010, 88 (per 1,000 live births). Annual population growth rate, 2000–05, 3·0%. Life expectancy at birth, 2007, was 47·2 years for males and 48·2 years for females. Fertility rate, 2008, 5·3 children per woman.

CLIMATE

Lying wholly within the tropics, temperatures everywhere are high. Rainfall varies greatly, but decreases from the coast to the interior. The main rains occur from April to Oct. Lagos, Jan. 81°F (27·2°C), July 78°F (25·6°C). Annual rainfall 72" (1,836 mm). Ibadan, Jan. 80°F (26·7°C), July 76°F (24·4°C). Annual rainfall 45" (1,120 mm). Kano, Jan. 70°F (21·1°C), July 79°F (26·1°C). Annual rainfall 35" (869 mm). Port Harcourt, Jan. 79°F (26·1°C), July 77°F (25°C). Annual rainfall 100" (2,497 mm).

CONSTITUTION AND GOVERNMENT

The constitution was promulgated on 5 May 1999, and entered into force on 29 May. Nigeria is a federation, comprising 36 states and a federal capital territory. The constitution includes provisions for the creation of new states and for boundary adjustments of existing states. The legislative powers are vested in a *National Assembly*, comprising a *Senate* and a *House of Representatives*. The 109-member Senate consists of three senators from each state and one from the federal capital territory, who are elected for a term of four years. The House of Representatives comprises 360 members, representing constituencies of nearly equal population as far as possible, who are elected for a four-year term. The *President* is elected for a term of four years and must receive not less than one-quarter of the votes cast at the federal capital territory. A president may not serve more than two consecutive four-year terms. In 2006 Olusegun Obasanjo sought to alter the constitution to allow him to run for a third term, but he failed to win backing for the amendment.

National Anthem

'Arise, O compatriots, Nigeria's call obey'; words by a collective, tune by B. Odiase.

GOVERNMENT CHRONOLOGY

(NCNC = National Council of Nigeria and the Cameroons; NPN = National Party of Nigeria; PDP = People's Democratic Party; n/p = non-partisan)

Heads of State since 1963.

President of the Republic
1963–66	NCNC	Benjamin Nnamdi Azikiwe

Heads of the Military Government
1966	military	Johnson Aguiyi-Ironsi
1966–75	military	Yakubu Gowon
1975–76	military	Murtala Ramat Muhammed
1976–79	military	Olusegun Obasanjo

President of the Republic
1979–83	NPN	Shehu Shagari

Head of the Federal Military Government
1983–85	military	Muhammadu Buhari

Chairman of the Armed Forces Ruling Council, then Chairman of the National Defence and Security Council
1985–93	military	Ibrahim Babangida

Head of the Interim National Government
1993	n/p	Ernest Shonekan

Chairmen of the Provisional Ruling Council
1993–98	military	Sani Abacha
1998–99	military	Abdulsalam Abubakar

Presidents of the Republic
1999–2007	PDP	Olusegun Obasanjo
2007–10	PDP	Umaru Yar'Adua
2010–	PDP	Goodluck Jonathan

RECENT ELECTIONS

Presidential elections were held on 16 April 2011. Goodluck Jonathan, the candidate for the ruling People's Democratic Party (PDP), won against 19 opponents with 58·9% of the votes cast. His main opponent, Muhammadu Buhari of the Congress for Progressive Change (CPC), received 32·0%, and Nuhu Ribadou of Action Congress of Nigeria (ACN) 5·4%.

Elections to the House of Representatives were held on 9 April 2011, but with voting in several constituencies postponed owing to logistical problems. The PDP won 123 seats, ACN 47, CPC 30, the All Nigeria's People Party (ANPP) 25 and others 9.

In Senate elections held on the same day 45 seats went to the PDP, 13 to the ACN, 7 to the ANPP, 5 to CPC and 4 to others. Again, voting was postponed in some constituencies.

CURRENT GOVERNMENT

President and Minister of Power: Goodluck Jonathan; b. 1957 (PDP; since 9 Feb. 2010—acting until 6 May 2010).

Vice President: Namadi Sambo.

In March 2013 the government comprised:

Minister of Aviation: Stella Oduah-Ogiemwonyi. *Defence:* Kabiru Tanimu Turaki. *Education:* Ruqayyatu A. Rufa'i. *Federal Capital Territory:* Bala Muhammed. *Finance:* Ngozi Okonjo-Iweala. *Foreign Affairs:* Olugbenga Ashiru. *Health:* Prof. Christian Onyebuchi Chukwu. *Information and Communications:* Labaran Maku. *Interior:* Abba Moro. *Justice and Attorney General:* Mohammed Bello Adoke. *Labour:* Chukwuemeka Ngozichineke Wogu. *Mines and Steel Development:* Musa Mohammed Sada. *National Planning:* Shamsudeen Usman. *Niger Delta:* Peter Godsday Orubebe. *Petroleum Resources:* Dieziani Alison-Madueke. *Police Affairs:* Caleb Olubolade. *Power:* Chinedu Osita Nebo. *Science and Technology:* Prof. Ita Okon Bassey Ewa. *Sports:* Yusuf Suleiman. *Trade and Investment:* Olusegun O. Aganga. *Transport:* Idris A. Umar. *Women's Affairs:* Zainab Maina. *Works:* Mike Onolememen. *Youth Development:* Bolaji Abdullahi.

Nigerian Parliament: http://www.nassnig.org

CURRENT LEADERS

Goodluck Ebele Jonathan

Position
President

Introduction
The National Assembly appointed Goodluck Jonathan as acting president in Feb. 2010, replacing President Umaru Yar'Adua, who had been undergoing medical treatment since Nov. 2009. An academic and former governor of Bayelsa State, Jonathan had been vice-president since Nov. 2007. He became president in May 2010 on Yar'Adua's death.

Early Life
Goodluck Ebele Jonathan was born on 20 Nov. 1957 in Otueke, Bayelsa State, in the oil-rich Niger Delta. A member of the indigenous Ijaw people, he was raised as a Pentecostal Christian by his father, a canoe builder and fisherman. After attending local primary schools Jonathan went to Mater Dei High School in Imiringi. He graduated in zoology from Port Harcourt University in 1981.

After completing military service Jonathan pursued an academic career, gaining a masters degree in hydrobiology and fisheries in 1985 and a PhD in zoology ten years later, both from Port Harcourt University. He worked as a lecturer, education inspector and environmental protection officer before entering politics in 1998 with the People's Democratic Party (PDP). He was appointed deputy governor of Bayelsa State in 1999 and again in 2003. When state governor Diepriye Alamieyeseigha was charged with corruption in the UK in 2005, Jonathan replaced him.

In Dec. 2006, months after his wife Patience Faka was accused of, but not charged with, money laundering, Jonathan was elected vice-presidential running mate to Yar'Adua for the 2007 elections. The pair won, although opponents questioned the legitimacy of the vote and Jonathan's house in Bayelsa was bombed shortly after.

Jonathan's knowledge of the Niger Delta region helped Yar'Adua secure a ceasefire and disarmament from Delta rebels, generally considered the biggest achievement of his time in office. In Nov. 2009 Yar'Adua left Nigeria for medical treatment in Saudi Arabia without designating an interim presidential replacement. Jonathan was granted presidential powers after much wrangling by a parliamentary resolution of 9 Feb. 2010, though its constitutional validity was questioned. Yar'Adua returned to Nigeria in Feb. 2010 but remained out of public view, fuelling speculation about his condition.

Career in Office
On becoming acting president, Jonathan made moves to secure his tenuous position. In a government reshuffle he replaced two-thirds of Yar'Adua's appointments, including the justice minister and the national security adviser. He selected a London-based Goldman Sachs banker as finance minister and named the first female oil minister.

Jonathan vowed to calm militancy in the Delta region and address electricity shortages. He met with major oil corporations in Feb. 2010 after one of the main rebel groups renewed its campaign against the oil infrastructure in Dec. 2009. In April 2010 he dismissed the head of the state-run Nigerian National Petroleum Corporation. After Yar'Adua died on 5 May 2010, Jonathan was sworn in as his successor the following day.

In Sept. 2010 Jonathan announced that he would contest presidential elections to be staged in 2011. The following month Nigeria marked 50 years of independence from Britain, although the celebrations were marred by a car bomb incident in Abuja in which at least 12 people were killed. The presidential poll took place in April 2011 and Jonathan was returned with almost 60% of the vote, though voting was divided along religious and ethnic lines. In 2011 and 2012 Jonathan was confronted with serious outbursts of violence between the Christian and Muslim communities. He imposed a state of emergency in response to a campaign of bombings and shootings by Boko Haram, an extremist Islamist sect.

DEFENCE

In 2008 defence expenditure totalled US$1,339m., equivalent to US$9 per capita and representing 0·7% of GDP.

Nigeria's armed forces have over 5,000 personnel in peace-keeping missions in other countries, notably Liberia and Sierra Leone.

Army
Strength (2007) 62,000.

Navy
The Navy includes one frigate with a helicopter and one corvette. There is a small aviation element. Naval personnel in 2007 totalled 8,000, including Coastguard. The main bases are at Apapa (Lagos) and Calabar.

Air Force
Personnel (2007) total 10,000 although the force has a very limited operational capacity. There were 75 combat-capable aircraft in 2007 including MiG-21s and Aero L-39s, but the serviceability of much of the equipment is in doubt. In addition there were five attack helicopters in 2007.

ECONOMY

Agriculture accounted for 32·0% of GDP in 2006, industry 41·9% and services 26·1%.

Overview

Since independence in 1960, weak and corrupt government has damaged the economy. GDP per capita is low (at approximately US$1,452 in 2011 according to World Bank figures), while poverty is widespread and increasing (60·9% in 2010, compared to 43·0% in 1985). Unemployment is also high, at 24% in 2011.

Nonetheless, Nigeria is well-positioned to build a prosperous economy. It is Africa's largest oil exporter, with the continent's greatest reserves of natural gas. Oil and gas account for over 75% of total revenue. It was among the pioneers in adopting and implementing the Extractive Industries Transparency Initiative (EITI), which aims to improve governance in the hydrocarbon sector.

A ten-year rift with the IMF ended with an agreement in Jan. 1999 on a Fund-monitored reform programme, including provisions for abolishing the dual exchange rate, ending subsidies on local fuel and expanding privatization. Between 2004 and 2007, there were wide-ranging reforms, particularly in the management of public finance, governance and the banking sector. External debt was also reduced following negotiations with the Paris and London Clubs to eradicate major arrears.

Since the beginning of the century Nigeria has had one of the highest GDP growth rates in sub-Saharan Africa. The economy weathered the global financial crisis and a 2009 domestic banking crisis, in part resulting from countercyclical policies. A three-part sovereign wealth fund (comprising a stabilization fund, an infrastructure fund and an inter-generational saving fund) was launched in Oct. 2011. However, inflation remains in double digits and the economy is vulnerable to a fall in oil prices. With oil revenue projected to decline over the medium term, the government aims to diversify the economy. Efforts to increase non-oil revenues are currently focused on improving tax administration.

Currency

The unit of currency is the *naira* (NGN) of 100 *kobo*. Foreign exchange reserves were US$44,786m. in May 2009 (US$7,100m. in 1998) and gold reserves were 687,000 troy oz. Inflation rates (based on IMF statistics):

2002	2003	2004	2005	2006	2007	2008	2009	2010	2011
12·9%	14·0%	15·0%	17·9%	8·2%	5·4%	11·6%	12·5%	13·7%	10·8%

In June 2009 total money supply was ₦4,247·6bn.

Budget

The financial year is the calendar year. 2008 budgetary central government revenue, ₦2,386·9m. (tax revenue 2·9%; non-tax revenue 97·1%); expenditure, ₦1,771·3m. (including social benefits, 27·4%; grants, 25·5%).

VAT was raised from 5% to 10% in May 2007, but as a result of a general strike in protest against the increase was lowered back to 5% a month later.

Performance

Real GDP growth rates (based on IMF statistics):

2003	2004	2005	2006	2007	2008	2009	2010	2011
10·3%	10·6%	5·4%	6·2%	7·0%	6·0%	7·0%	8·0%	7·4%

Before the discovery of oil in the early 1970s Nigeria's GDP per head was around US$200. By the early 1980s it had reached around US$800, but has now declined to some US$300. Total GDP in 2011 was US$244·0bn.

Banking and Finance

The Central Bank of Nigeria (CBN) is the bank of issue (*Governor*, Lamido Sanusi).

In 2004 a major banking reform was announced, consolidating the existing institutions at the time (nearly 100) into 25 large universal banks. However, despite attempts to strengthen the banking sector, the financial condition of the banks deteriorated. Subsequently the CBN requested technical assistance from the IMF to strengthen the banking system. Another banking crisis in 2008 prompted a ₦600bn. bailout from the Central Bank to five of the leading banks: Intercontinental Bank, Oceanic International Bank (now Ecobank), Union Bank of Nigeria, Bank PHB (now Keystone Bank) and Afribank (now Mainstreet Bank). More ambitious reforms took effect in Oct. 2010. In 2007 bank reserves at the CBN totalled ₦1,195bn. and CBN foreign assets amounted to ₦6,570bn.

Nigeria was one of three countries and territories named in a report in June 2005 as failing to co-operate in the fight against international money laundering. In June 2006 Nigeria was delisted and its formal monitoring was ended a year later. The Financial Action Task Force on Money Laundering was set up by the G7 group of major industrialized nations.

Foreign debt amounted to US$7,883m. in 2010, down from US$22,060m. in 2005 and US$31,355m. in 2000—this was equivalent to just 4·5% of GNI (down from 22·3% in 2005 and 77·9% in 2000).

Nigeria attracted a record US$8,915m. worth of foreign direct investment in 2011 (the most of any African country), up from US$6,099m. in 2010.

The Nigerian Stock Exchange is in Lagos.

ENERGY AND NATURAL RESOURCES

Environment

Nigeria's carbon dioxide emissions from the consumption and flaring of fossil fuels were the equivalent of 0·7 tonnes per capita in 2008.

Electricity

Installed capacity, 2007 estimate, 5·9m. kW. Production, 2007, 22·98bn. kWh (28% kWh hydro-electric); consumption per capita was 156 kWh in 2007. Power cuts are frequent, with both businesses and homes having to rely on generators as demand for electricity far outweighs supply.

Oil and Gas

Nigeria's oil production, though declining, is the largest of any African country and amounted to 105·3m. tonnes in 2008. Reserves in 2008 totalled 36·2bn. bbls. There are four refineries. For many years oil accounted for over 90% of Nigeria's exports, but in 2009 this declined sharply. In the first quarter of 2009 oil revenue was around US$4·9bn., down 50% from the fourth quarter of 2008 with militant activity blamed for the decline. Most of Nigeria's oil wealth comes from onshore wells, but there are also large untapped offshore deposits.

Natural gas reserves, 2008, were 5,220bn. cu. metres; production, 35·0bn. cu. metres. In March 2007 the 678-km West Africa Gas pipeline was completed to supply natural gas to Benin, Ghana and Togo. After a series of delays resulting from vandalism and fuel quality problems, it was restarted in March 2010 and should help to reduce Nigeria's dependence on oil for government revenue.

The Petroleum Industry Bill was introduced in the Nigerian Senate in Jan. 2009 but has yet to be passed into law. It proposes the break-up of the National Petroleum Corporation into seven independent regulatory agencies and a new commercial national oil company, the National Petroleum Company of Nigeria.

Minerals

Production, 2005 estimates (in tonnes): limestone, 2·10m.; kaolin, 200,000; marble, 149,000. There are large deposits of iron ore, coal (reserves estimate 245m. tonnes), lead and zinc. There are small quantities of gold and uranium. Lead production was an estimated 5,000 tonnes in 2009. Tin is also mined.

Agriculture

Of the total land mass, 75% is suitable for agriculture, including arable farming, forestry, livestock husbandry and fisheries. In 2001, 28·5m. ha. were arable and 2·7m. ha. permanent cropland. 0·23m. ha. were irrigated in 2001. 90% of production was by smallholders with less than 3 ha. in 2000, and less than 1% of farmers had access to mechanized tractors. Main food crops are millet and sorghum in the north, plantains and oil palms in the south, and maize, yams, cassava and rice in much of the country. The north is, however, the main food producing area. Cocoa is the crop that contributes most to foreign exchange earnings. Output, 2000 (in 1,000 tonnes): cassava, 32,697; yams, 25,873; sorghum, 7,520; millet, 5,960; maize, 5,476; taro, 3,835; rice, 3,277; groundnuts, 2,783; plantains, 1,902; sweet potatoes, 1,662; palm oil, 896; pineapples, 881; tomatoes, 879. Nigeria is the biggest producer of yams, accounting for more than two-thirds of the annual world output. It is also the leading cassava and taro producer and the second largest millet producer.

Livestock, 2000: cattle, 19·83m.; sheep, 20·50m.; goats, 24·30m.; pigs, 4·86m.; chickens, 126m. Products (in 1,000 tonnes), 2000: beef and veal, 298; goat meat, 154; mutton and lamb, 91; pork, bacon and ham, 78; poultry meat, 172; milk, 386; eggs, 435.

Forestry

There were 9·04m. ha. of forests in 2010, or 10% of the land area, down from 17·23m. ha. in 1990 and 13·14m. ha. in 2000. Timber production in 2007 was 71·42m. cu. metres.

Fisheries

The total catch in 2010 was 616,981 tonnes, of which 323,599 tonnes came from marine waters and 293,382 tonnes from inland waters.

INDUSTRY

The largest company by market capitalization in Nigeria in April 2012 was Dangote Cement, a diversified conglomerate (US$10·9bn.).

In 2009 manufacturing accounted for 4·2% of GDP. Production, in 1,000 tonnes: cement (2008 estimate), 5,000; palm oil (2006), 1,287; residual fuel oil (2007), 1,002; distillate fuel oil (2007), 623; petrol (2007), 287; jet fuel (2007), 232; kerosene (2007), 98; paper and paperboard (2007 estimate), 19; cigarettes (2005), 1,813m. units; motorcycles (2005), 6,900 units. Also plywood (2007 estimate), 55,000 cu. metres.

Labour

The labour force in 2004 totalled an estimated 55·67m. There were 33 work stoppages in 2003–04 with 407,000 working days lost (233·5m. working days lost in 1994–95).

INTERNATIONAL TRADE

Imports and Exports

Imports (c.i.f.) in 2006 totalled US$22,903m.; exports (f.o.b.) US$59,215m. Principal imports in 2003 were: machinery and transport equipment, 38·0%; manufactured goods, 16·0%; food and livestock, 14·0%; petroleum and petroleum products, 12·3%. In 2003 crude oil amounted to 96·4% of exports by value. Other exports included ships and boats, and natural gas.

In 2006 the main import suppliers were: USA, 15·7%; China, 13·8%; UK, 11·8%; Germany, 5·6%. Leading export destinations in 2006 were: USA, 45·0%; India, 9·3%; Spain, 8·0%; France, 5·7%.

COMMUNICATIONS

Roads

The road network covered 193,200 km in 2004, including 15,688 km of main roads. In 2007 there were 4,560,000 passenger cars in use and 3,040,000 motorcycles and mopeds. There were 17,797 road accidents with 9,390 fatalities in 2007.

Rail

In 2005 there were 3,505 route-km of track (1,067 mm gauge). There are plans to convert the entire network to 1,435 mm gauge. Passenger-km travelled in 2008 came to 773m. and freight tonne-km to 41m.

Civil Aviation

Lagos (Murtala Muhammed) is the major airport, and there are also international airports at Port Harcourt and Kano (Mallam Aminu Kano Airport). The main carrier is Air Nigeria, established in 2004 as Virgin Nigeria Airways and rebranded as Nigerian Eagle Airlines in Sept. 2009 and Air Nigeria in June 2010. It operated direct flights in Sept. 2009 to Abidjan, Accra, Banjul, Cotonou, Dakar, Douala, Libreville and Monrovia as well as providing domestic services. In 2008 Murtala Muhammed International Airport handled 2,688,595 passengers and 130,163 tonnes of freight.

Shipping

In Jan. 2009 there were 107 ships of 300 GT or over registered, totalling 427,000 GT. The principal ports are Lagos and Port Harcourt. There is an extensive network of inland waterways.

Telecommunications

In 2010 there were 1,050,000 main (fixed) telephone lines. In the same year mobile phone subscribers numbered 87,298,000 (551·0 per 1,000 persons), up from 18,587,000 in 2005. Nigeria has now surpassed South Africa as the continent's largest mobile phone market. The largest mobile phone company is MTN Nigeria Communications. There were 24·0m. internet users in 2008, up from 5·0m. users in 2005. In June 2012 there were 5·1m. Facebook users.

SOCIAL INSTITUTIONS

Justice

The highest court is the Federal Supreme Court, which consists of the Chief Justice of the Republic, and up to 15 Justices appointed by the government. It has original jurisdiction in any dispute between the Federal Republic and any State or between States; and to hear and determine appeals from the Federal Court of Appeal, which acts as an intermediate appellate Court to consider appeals from the High Court.

High Courts, presided over by a Chief Justice, are established in each state. All judges are appointed by the government. Magistrates' courts are established throughout the Republic, and customary law courts in southern Nigeria. In each of the northern States of Nigeria there are the Sharia Court of Appeal and the Court of Resolution. Muslim Law has been codified in a Penal Code and is applied through Alkali courts. The northern province of Zamfara introduced *sharia*, or Islamic law, in Oct. 1999, as have the other 11 predominantly Muslim northern provinces in the meantime. The death penalty is in force and was used in 2006, although there have been no reported executions since.

The population in penal institutions in Jan. 2007 was 39,438 (28 per 100,000 of national population).

Education

The adult literacy rate was an estimated 60·8% in 2009. Free, compulsory education for nine years is provided for all children from the age of six. In 2006 there were 22·86m. pupils and 565,646 teaching staff in primary schools; 6·44m. pupils and 202,082 teaching staff in secondary schools; and (2005) 1·39m. students in tertiary education with 37,031 academic staff.

Among the leading institutions of higher education are Ahmadu Bello University in Zaria—Nigeria's largest university—with around 35,000 students, and the University of Ibadan—the country's oldest university—founded in 1948.

Health

Health personnel, 2000: 30,885 doctors, 2,180 dentists and 8,642 pharmacists.

Nigeria has made significant progress in the reduction of undernourishment in the past 25 years. In the period 2001–03 only 9% of the population was undernourished, one of the lowest rates in sub-Saharan Africa.

An estimated 2·9m. people in Nigeria are living with HIV/AIDS, a total exceeded only in South Africa.

RELIGION

Muslims and Christians both constitute about 45% of the population; traditional animist beliefs are also widespread. Northern Nigeria is mainly Muslim; southern Nigeria is predominantly Christian and western Nigeria is evenly divided between Christians, Muslims and animists. Far more Nigerians consider their religion to be of prime importance rather than their nationality. In Feb. 2013 the Roman Catholic church had three cardinals.

CULTURE

World Heritage Sites
The Sukur Cultural Landscape, a hilly area in Adamawa State (northeastern Nigeria), was entered on the UNESCO World Heritage list in 1999. Osun Sacred Grove is one of the last remnants of primary high forest in southern Nigeria and was inscribed in 2005.

Press
In 2005 there were 26 daily newspapers with a combined circulation of 820,000.

Tourism
In 2006 there were 3,056,000 non-resident visitors (including 2,108,000 from other African countries and 506,000 from Europe); spending by tourists totalled US$51m.

DIPLOMATIC REPRESENTATIVES

Of Nigeria in the United Kingdom (Nigeria House, 9 Northumberland Ave., London, WC2N 5BX)
High Commissioner: Dr Dahaltu S. Tafida.

Of the United Kingdom in Nigeria (19 Torrens Close, Maitami, PMB 4808 (Garki), Abuja)
High Commissioner: Dr Andrew Pocock, CMG.

Of Nigeria in the USA (3519 International Court, NW, Washington, D.C., 20008)
Ambassador: Adebowale Ibidapo Adefuye.

Of the USA in Nigeria (Plot 1075, Diplomatic Drive, Central District Area, Abuja)
Ambassador: Terence McCulley.

Of Nigeria to the United Nations
Ambassador: U. Joy Ogwu.

Of Nigeria to the European Union
Ambassador: Usman Alhaji Baraya.

FURTHER READING

Forrest, T., *Politics and Economic Development in Nigeria.* 1993
Maier, K., *This House Has Fallen: Midnight in Nigeria.* 2000
Miles, W. F. S., *Hausaland Divided: Colonialism and Independence in Nigeria and Niger.* 1994
Okafor, Victor Oguejiofor, *A Roadmap for Understanding African Politics.* 2006

National Statistical Office: National Bureau of Statistics, Plot 762, Independence Ave., Central Business District, Garki, P.M.B. 127, Abuja.
Website: http://www.nigerianstat.gov.ng

NORWAY

0 150 mi
0 200 km

Norwegian Sea

Tromsø

SWEDEN

Trondheim

FINLAND

NORWAY

Bergen   Hamar

OSLO

Stavanger

Baltic
Sea   ESTONIA

© Research Machines plc 2006

Kongeriket Norge
(Kingdom of Norway)

Capital: Oslo
Population projection, 2015: 5·05m.
GNI per capita, 2011: (PPP$) 47,557
HDI/world rank: 0·943/1
Internet domain extension: .no

KEY HISTORICAL EVENTS

The first settlers arrived at the end of the Ice Age, as the glaciers retreated north. Archaeological remains in Finnmark in the north and in Rogaland in the southwest of Norway date from between 9500 to 8000 BC and suggest coastal, hunting-fishing communities. By 2500 BC a new influx of settlers brought cattle and crop farming and gradually replaced the earlier hunting-fishing communities. Although there is little evidence of the impact of the bronze and iron ages on Norway as its people had not yet found ways to exploit their natural resources for trade, links with Roman-occupied Gaul in the first four centuries AD were strong. By the time of the collapse of the Roman Empire, tribal groups had started to develop and by AD 800 had each established their own legislative and adjudicatory assemblies, known as *things*.

In the ninth century communities from the Vik, an area between the south coasts of Norway and Sweden, gave their name to the people collectively known as Vikings. The Norwegian Vikings sailed to the Atlantic islands, England, France, Scotland and Ireland, and also colonized Iceland. One of the many whose exploits were faithfully recorded by the saga writers was Eric the Red, who discovered Greenland. His son, Leif Erikson, voyaged across the Davis Strait, to America, becoming possibly the first European to do so.

The first steps towards centralized rule were taken by Harold Fairhair who extended his rule along the coastal region of Norway. Battles with rival chieftains culminated in about 900 when Harold was proclaimed king of the Norwegians. His successors were less assertive and by the mid-tenth century the country was effectively under the suzerainty of Harold Bluetooth, king of Denmark and Skåne. Bluetooth's grandson, Canute the Great, fought successfully to incorporate England into his North Sea Empire before setting his sights on Sweden. But the limitations of royal authority were shown on the death of Canute when the English, unchallenged, simply chose their own king while the Danish and Norwegian nobles decided that whichever of their own monarchs lived longest should take power in both countries, an agreement which for a time resulted in a Norwegian ruler for Denmark.

Viking Strength

The Viking's territorial expansion came to an end with the Norwegian King Harald Hardrada's defeat at the battle of Stamford Bridge in England in 1066. Supported by the English church, the Norwegian monarchy gained strength. By the 12th century the balance of power between the church and monarchy had become a source of civil conflict which was only resolved when Håkon IV became King in 1217. Thus began Norway's 'Golden Age' in which the unity of the kingdom was solidly established. Blood feuds were prohibited, a royal council was created, and primogeniture was introduced to secure the continuity of the monarchic line. Under Håkon's rule, both Greenland and Iceland ceded control to Norway. It was Håkon's son, Magnus VI, known as the Lawmender, who oversaw the codification of a national law system between 1274–76, elements of which have survived to this day. Under Erik II, Magnus' son, much of the royal power was divested to wealthy magnates. His succession by his brother Håkon V in 1299 marked a renewed effort to strengthen the monarchy and also a movement of political power to Oslo.

Union with Sweden came in 1319 with the coronation of Magnus VII, the son of Håkon's daughter and Duke Erik of Sweden. This was to last until 1355, when the Swedish crown passed to Magnus' son. Between 1349–50 Norway fell victim to the Black Death which killed around two-thirds of its population. The effects of this were to dramatically reduce the strength of the nobility and to undermine the cohesion of the government, as many official positions were taken up by Danes and Swedes. Newly vulnerable to the threat of encroachment by the Germans, the incentive for all three Scandinavian kingdoms to unite was strong. When the Danish king died in 1375 his widow, Margaret, claimed the throne on behalf of her five-year-old son, Olav. Acting for her son, Margaret became regent of Denmark and, on the death of Håkon, regent of Norway. Confirmed as regent of Denmark and Norway, Margaret defeated Albrecht, the German claimant to the Swedish throne, thus clearing the way to a Nordic union. With the death of her son in 1387 and unable to take the triple crown for herself, she nominated her five-year-old nephew, Erik of Pomerania, as king of all three countries. His election was formalized at Kalmar in 1397.

From 1450 the Norwegian government was based in Copenhagen and many administrative positions were taken by Germans and Danes. An attempt by the Norwegian council to gain independence in 1523 led to civil war between 1534–36 and the council's subsequent abolition. Norway was then to remain a province of Denmark, with limited control over internal affairs, until the 19th century.

In the Napoleonic Wars, Denmark and Norway were allied with Napoleon I. Napoleon's defeat at the battle of Leipzig in 1813 was followed by a successful attack on Denmark from Sweden which resulted in the Treaty of Kiel (Jan. 1814). With the signing of the treaty Norway was conceded to the Swedish throne and, despite Denmark's continued resistance, its newly written constitution came into force in Nov. 1814. Although the arrangement meant the regency and foreign policy were to be shared with Sweden, the new constitution gave Norway control over internal affairs, with a newly established political base at Christiania.

The economic damage of the Napoleonic Wars was remedied by the rapid expansion of the fishing industry and, from the 1850s onwards, agriculture. In the latter half of the century, the merchant navy grew to become the third largest in the world after the United States and Great Britain.

Independence
From the 1880s, successive steps towards self-government within the union culminated in a referendum in which the overwhelming majority of Norwegians voted for separation. In Oct. 1905 Oscar II renounced his title to the western provinces and a month later a Danish prince was confirmed as Håkon VII of free Norway. Reigning for 52 years, he was succeeded by his son.

At the outset of the First World War Norway declared its neutrality. This did not prevent the loss of almost half of its merchant navy and damage to the economy as a result of trade embargos. But despite the hardships of the 1930s' depression, industrial expansion continued.

From 1940 to 1944, during the Second World War, Norway was occupied by the Germans who set up a pro-German government under Vidkun Quisling. Apart from this wartime episode, the Labour Party held office, and the majority in the *Storting* (parliament), from 1935 to 1965. Norway's first post-war prime minister, Einar Gerhardsen, had spent four years in a concentration camp. He had been vice-chairman of Oslo city council until he became leader of the underground anti-Nazi movement in the early days of the occupation. As recently elected social democrat leader he was the natural choice to head the 1945 caretaker government.

The action needed to restore Norway's prosperity was self evident: to make good the heavy losses in the merchant fleet; to increase the output of hydro-electricity; and to develop new industries. The chief worry for the social democrats was the likely impact of communists who had gained credit for leading the resistance. Talks on a possible merger of the parties were as unproductive as parallel negotiations in Denmark, but the electoral results of each party's going its own way were markedly different in the two countries. More confident of their purpose, the Norwegian social democrats took the electorate by storm, increasing their share of the popular vote in the 1945 election by close on 10%. Their advance gave them the one prize that eluded their colleagues everywhere else in Scandinavia—an absolute majority and the freedom to govern without always looking over their shoulder.

The government was supported wholeheartedly by the trade unions. In return for price controls and food subsidies which stabilized the cost of living for almost five years, and the guarantee of full employment, the unions accepted compulsory arbitration for all wage disputes. Returned in 1949 with an increased majority, the social democrats were able to point to a rise in productivity and living standards well beyond that achieved by most other Western countries. But the general increase in world prices triggered by the Korean War meant that Norway had to pay much more for essential imports. The use of subsidies to counteract price increases reached its limit when they became the largest item in the national budget. In 1950 food prices were allowed to get closer to their market level and the cost of living

started on an upward curve, leading to a 30% increase over three years. Industrial investment suffered a sharp cutback.

The social democrats held on to power until the mid-sixties when a centre right coalition took over led by Per Borten. By 1969 the social democrats had recovered much of their lost ground. The centre right coalition government struggled on with a majority of two until the EEC issue broke through the normally placid surface of Norwegian politics.

Although Norway's application for membership in the EEC in 1969 was successful, a referendum held in 1972 found more than 53% of voters opposed to joining. Norway had been a member of EFTA since that organization's foundation, and continued to sign up to a series of bilateral free-trade treaties with members of the EEC, but opposition to joining the EEC remained strong. On the inception of the EU in 1992, Norway, like its Scandinavian neighbours, applied for membership. But, again, a referendum was won by the anti-European lobby.

Norway's continued reluctance to join the EU hinges on its dependence on the export of petroleum and natural gas. Since the 1960s and the discovery of vast off-shore deposits, the oil and gas export industry has contributed to making Norway one of the world's richest economies.

In 2010 relations with China came under pressure when the Norwegian Nobel committee awarded its Peace Prize to Liu Xiaobo, a jailed dissident. In July 2011 an anti-immigration extremist set off a bomb at the prime minister's office in Oslo before carrying out a gun attack at a summer camp on the island of Utøya in Tyrifjorden, organized by the youth wing of the Labour Party. A total of 77 people were killed in the two attacks.

TERRITORY AND POPULATION

Norway is bounded in the north by the Arctic Ocean, east by Russia, Finland and Sweden, south by the Skagerrak Straits and west by the Norwegian Sea. The total area of mainland Norway is 323,787 sq. km, including 19,539 sq. km of fresh water. Total coastline, including fjords, 25,148 km. There are more than 50,000 islands along the coastline. Exposed mountain (either bare rock or thin vegetation) makes up over 70% of the country. 25% of the land area is woodland and 4% tilled land.

Population (2001 census) was 4,520,947 (2,240,281 males; 2,280,666 females); population density per sq. km, 14·8. Estimated population, 1 Jan. 2011, 4,920,305; population density, 16·2. With the exception of Iceland, Norway is the most sparsely populated country in Europe.

The UN gives a projected population for 2015 of 5·05m.

There are 19 counties (*fylke*). Land area, population and densities:

	Land area (sq. km)	Population (2001 census)	Population (2011 estimate)	Density per sq. km 2011
Østfold	3,922	252,520	274,827	70
Akershus	4,620	476,440	545,653	118
Oslo (City)	427	512,093	599,230	1,404
Hedmark	26,244	187,878	191,562	7
Oppland	23,878	183,302	186,087	8
Buskerud	13,870	239,591	261,110	19
Vestfold	2,157	216,333	233,705	108
Telemark	13,894	165,732	169,185	12
Aust-Agder	8,353	102,848	110,048	13
Vest-Agder	6,706	157,697	172,408	26
Rogaland	8,605	377,579	436,087	51
Hordaland	14,554	411,100	484,240	33
Sogn og Fjordane	17,709	107,261	107,742	6
Møre og Romsdal	14,614	243,888	253,904	17
Sør-Trøndelag	17,909	266,098	294,066	16
Nord-Trøndelag	20,881	127,444	132,140	6
Nordland	36,194	237,561	237,280	7
Troms Romsa	24,950	151,646	157,554	6
Finnmark	45,984	73,936	73,417	2
Mainland total	305,470[1]	4,520,947	4,920,305	16

[1]117,943 sq. miles.

The Arctic territories of Svalbard and Jan Mayen have an area of 61,397 sq. km. Persons staying on Svalbard and Jan Mayen are registered as residents of their home Norwegian municipality.

At Jan. 2011, 79·2% of the population lived in urban areas.

Population of the principal urban settlements on 1 Jan. 2011:

Oslo	906,681	Ålesund	47,772
Bergen	235,046	Haugesund	43,913
Stavanger/Sandnes	197,852	Moss	42,781
Trondheim	164,953	Sandefjord	41,811
Fredrikstad/Sarpsborg	104,382	Bodø	37,834
Drammen	100,303	Arendal	33,303
Porsgrunn/Skien	88,335	Hamar	30,565
Kristiansand	69,380	Larvik	24,252
Tromsø	56,466	Halden	23,711
Tønsberg	48,350		

The official language is Norwegian, which has two versions: Bokmål (or Riksmål) and Nynorsk (or Landsmål).

The Sami, the indigenous people of the far north, number some 40,000 and form a distinct ethnic minority with their own culture and language.

SOCIAL STATISTICS

Statistics for calendar years:

	Births	Still-born	Ourside marriage	Deaths	Marriages	Divorces
2006	58,545	201	31,056	41,253	21,721	10,598
2007	58,459	241	31,849	41,954	23,471	10,280
2008	60,497	221	33,302	41,712	25,125	10,158
2009	61,807	215	34,038	41,449	24,582	10,235
2010	61,442	190	33,655	41,499	23,577	10,264

Rates per 1,000 population, 2010, birth, 12·6; death, 8·5; marriage, 4·8; divorce, 2·1. Average annual population growth rate, 2000–10, 0·86% (2010, 1·28%). In 2009 there were 573 suicides, giving a rate of 11·9 per 100,000 population (men, 17·3 per 100,000; women, 6·5).

Expectation of life at birth, 2010, was 78·9 years for males and 83·2 years for females. Infant mortality, 2010, 2·8 per 1,000 live births; fertility rate, 2010, 1·95 births per woman. 55% of births are to unmarried mothers. In 2009 the average age at marriage was 37·3 years for males and 33·8 years for females (33·8 and 31·0 years respectively for first marriages). Norway legalized same-sex marriage in Jan. 2009.

At 1 Jan. 2011 the immigrant population totalled 600,922, including 60,610 from Poland, 34,108 from Sweden, 31,884 from Pakistan and 27,827 from Iraq. In 2010 Norway received 10,064 asylum applications. Most were from Eritrea (1,711), Somalia (1,397), Afghanistan (979) and Russia (628).

A UNICEF report published in 2010 showed that 6·1% of children in Norway live in relative poverty (living in a household in which disposable income—when adjusted for family size and composition—is less than 50% of the national median income), the joint third lowest of any country.

In the Human Development Index, or HDI (measuring progress in countries in longevity, knowledge and standard of living), Norway was ranked first in the 2011 rankings published in the annual Human Development Report.

CLIMATE

There is considerable variation in the climate because of the extent of latitude, the topography and the varying effectiveness of prevailing westerly winds and the Gulf Stream. Winters along the whole west coast are exceptionally mild but precipitation is considerable. Oslo, Jan. 24·3°F (−4·3°C), July 61·5°F (16·4°C). Annual rainfall 30·0" (763 mm). Bergen, Jan. 34·3°F (1·3°C), July 57·7°F (14·3°C). Annual rainfall 88·6" (2,250 mm). Trondheim, Jan. 26°F (−3·5°C), July 57°F (14°C). Annual rainfall 32·1" (870 mm). Bergen has one of the highest rainfall figures of any European city. The sun never fully sets in the northern area of the country in the summer and even in the south the sun rises at around 3 a.m. and sets at around 11 p.m.

CONSTITUTION AND GOVERNMENT

Norway is a constitutional and hereditary monarchy.

The reigning King is **Harald V**, born 21 Feb. 1937, married on 29 Aug. 1968 to Sonja Haraldsen. He succeeded on the death of his father, King Olav V, on 21 Jan. 1991. *Offspring*: Princess Märtha Louise, born 22 Sept. 1971 (married Ari Behn, b. 30 Sept. 1972, on 24 May 2002; *offspring*, Maud Angelica, b. 29 April 2003; Leah Isadora, b. 8 April 2005; Emma Tallulah, b. 29 Sept. 2008); Crown Prince Haakon Magnus, born 20 July 1973 (married Mette-Marit Tjessem Høiby, b. 19 Aug. 1973, on 25 Aug. 2001; *offspring*, Ingrid Alexandra, b. 21 Jan. 2004; Sverre Magnus, b. 3 Dec. 2005; *offspring* of Crown Princess Mette-Marit from previous relationship, Marius, b. 13 Jan. 1997). The king and queen together receive an annual personal allowance of 9·6m. kroner from the civil list, and the Crown Prince and Crown Princess together 8·0m. kroner. Princess Märtha Louise relinquished her allowance in 2002. Women have been eligible to succeed to the throne since 1990. There is no coronation ceremony. The royal succession is in direct male line in the order of primogeniture. In default of male heirs the King may propose a successor to the *Storting*, but this assembly has the right to nominate another, if it does not agree with the proposal.

The Constitution, voted by a constituent assembly on 17 May 1814 and modified at various times, vests the legislative power of the realm in the *Storting* (Parliament). The royal veto may be exercised; but if the same Bill passes two Stortings formed by separate and subsequent elections it becomes the law of the land without the assent of the sovereign. The King has the command of the land, sea and air forces, and makes all appointments.

The 169-member Storting (increased from 165 for the 2005 election) is directly elected by proportional representation. The country is divided into 19 districts, each electing from 4 to 15 representatives.

The Storting, when assembled, divides itself by election into the *Lagting* and the *Odelsting*. The former is composed of one-fourth of the members of the Storting, and the other of the remaining three-fourths. Each Ting (the Storting, the Odelsting and the Lagting) nominates its own president. Most questions are decided by the Storting, but questions relating to legislation must be considered and decided by the Odelsting and the Lagting separately. In the event of the Odelsting and the Lagting disagreeing the Bill is considered by the Storting in plenary sitting, with a majority of two-thirds of the votes required for a new law to be passed. The same majority is required for alterations of the Constitution, which can only be decided by the Storting in plenary sitting. The Storting elects five delegates, whose duty it is to revise the public accounts. The Lagting and the ordinary members of the Supreme Court of Justice (the *Høyesterett*) form a High Court of the Realm (the *Riksrett*) for the trial of ministers, members of the *Høyesterett* and members of the Storting. The impeachment before the *Riksrett* can only be decided by the Odelsting.

The executive is represented by the King, who exercises his authority through the Cabinet. Cabinet ministers are entitled to be present in the Storting and to take part in the discussions, but without a vote.

National Anthem

'Ja, vi elsker dette landet' ('Yes, we love this land'); words by B. Bjørnson, tune by R. Nordraak.

GOVERNMENT CHRONOLOGY

Prime Ministers since 1945. (AP = Labour Party; DNA = Norwegian Labour Party; H = Conservative Party; KrF = Christian People's Party; Sp = Center Party)

1945–51	DNA	Einar Henry Gerhardsen
1951–55	DNA	Oscar Fredrik Torp

1955–63	DNA	Einar Henry Gerhardsen
1963	H	John Fyrstenberg Lyng
1963–65	DNA	Einar Henry Gerhardsen
1965–71	Sp	Per Borten
1971–72	DNA	Trygve Martin Bratteli
1972–73	KrF	Lars Korvald
1973–76	DNA	Trygve Martin Bratteli
1976–81	DNA	Odvar Nordli
1981	DNA	Gro Harlem Brundtland
1981–86	H	Kåre Isaachsen Willoch
1986–89	DNA	Gro Harlem Brundtland
1989–90	H	Jan Peder Syse
1990–96	DNA	Gro Harlem Brundtland
1996–97	DNA	Thorbjørn Jagland
1997–2000	KrF	Kjell Magne Bondevik
2000–01	DNA	Jens Stoltenberg
2001–05	KrF	Kjell Magne Bondevik
2005–	DNA, AP	Jens Stoltenberg

RECENT ELECTIONS

At the elections for the Storting held on 14 Sept. 2009 the following parties were elected: the ruling Norwegian Labour Party (DNA), winning 64 out of 169 seats (with 35·4% of the vote, up from 61 and 32·7% in 2005); Progress Party (FrP), 41 (22·9%); Conservative Party (H), 30 (17·2%); Socialist Left Party (SV), 11 (6·2%); Centre Party (Sp), 11 (6·2%); Christian People's Party (KrF), 10 (5·5%); Liberal Party (V), 2 (3·9%). Turnout was 75·7%.

Parliamentary elections are scheduled to take place on 9 Sept. 2013.

CURRENT GOVERNMENT

In March 2013 the three-party coalition government comprised:

Prime Minister: Jens Stoltenberg; b. 1959 (Labour Party/AP—formerly DNA; sworn in 17 Oct. 2005 and re-elected in Sept. 2009, having previously held office from March 2000 to Oct. 2001).

Minister of Agriculture and Food: Trygve Slagsvold Vedum (Sp). *Children, Equality and Social Inclusion:* Inga Marte Thorskilden (SV). *Culture:* Hadia Tajik (AP). *Defence:* Anne-Grete Strøm-Erichsen (AP). *Education and Research:* Kristin Halvorsen (SV). *Environment:* Bård Vegar Solhjell (SV). *Finance:* Sigbjørn Johnsen (AP). *Fisheries and Coastal Affairs:* Lisbeth Berg-Hansen (AP). *Foreign Affairs:* Espen Barth Eide (AP). *Government Administration, Reform and Church Affairs, and Nordic Co-operation:* Rigmor Aasrud (AP). *Health and Care Services:* Jonas Gahr Støre (AP). *International Development:* Heikki Holmås (SV). *Justice, Police and Immigration:* Grete Faremo (AP). *Labour:* Anniken Huitfeldt (AP). *Local Government and Regional Development:* Liv Signe Navarsete (Sp). *Petroleum and Energy:* Ola Borten Moe (Sp). *Trade and Industry:* Trond Giske (AP). *Transport and Communication:* Marit Arnstad (Sp). *Minister at the Office of the Prime Minister:* Karl Eirik Schjøtt-Pedersen (AP).

Office of the Prime Minister: http://www.regjeringen.no

CURRENT LEADERS

Jens Stoltenberg

Position
Prime Minister

Introduction
Jens Stoltenberg became prime minister of Norway for a second time on 17 Oct. 2005, following the victory of his centre-left coalition in parliamentary elections, having previously held the office from 2000–01. He was returned to power again in elections in Sept. 2009.

Early Life
Jens Stoltenberg was born in Oslo on 16 March 1959, the son of politicians. He studied economics at Oslo University, where he joined the Norwegian Labour Party (Det Norske Arbeiderpartiet, DNA). In 1985 he was appointed leader of the Labour Youth League and from 1985–89 was vice president of the International Union of Socialist Youth. He also worked briefly at the National Statistics Office and was an economics lecturer at Oslo University before serving for two years as leader of the Oslo Labour Party (1990–92). He was also a state secretary at the department of the environment at this time.

Elected a member of the Storting (parliament) for Oslo in the Sept. 1993 general election, Stoltenberg served as minister of trade and energy from 1993–96 and oversaw Norway's accession to the European Economic Area in 1994. In Oct. 1996 he was made minister of finance, a post he held for a year until the DNA lost power to the conservative Christian People's Party, led by Kjell Magne Bondevik. Bondevik, who attempted to govern with a coalition which held a slim majority, resigned in March 2000 and Stoltenberg (by now deputy leader of the DNA) was asked to form a government as the youngest prime minister in Norway's history.

Career in Office
Stoltenberg controversially ushered in reforms to the welfare state that included the part-privatization of several state-owned services. In the parliamentary elections of Sept. 2001 the party suffered a heavy defeat and Bondevik returned as prime minister of a centre-right coalition. A DNA party leadership battle between Stoltenberg and Jagland (leader since 1992) ensued with Stoltenberg emerging victorious.

Thanks to burgeoning oil and gas exports and high international prices, the economy prospered under Bondevik but the DNA's campaign in the run-up to the Sept. 2005 parliamentary elections centred on increased funding for education, health and care of the elderly. In partnership with the Socialist Left Party and the Centre Party, the DNA took 87 of 169 seats. Stoltenberg was sworn in to office on 17 Oct. 2005.

Stoltenberg vowed to reform the welfare system while creating conditions for Norway to develop as a knowledge-based economy. He also pledged sustainable management of the country's fish and energy resources. In 2006 his administration approved the expansion of oil exploration in the Barents Sea and also the merger of Norway's two largest energy companies, Statoil and Norsk Hydro (with the government having a controlling stake in the combined group). Stoltenberg withdrew the small contingent of Norwegian troops from Iraq, but promised to increase the country's participation in United Nations peacekeeping missions elsewhere in the world.

In Sept. 2009 his centre-left coalition was returned to power in parliamentary elections, with the DNA (renamed the Labour Party/AP in April 2011) marginally increasing its share of the vote.

In July 2011 the government was confronted with an unprecedented act of violence in Norway as a right-wing extremist set off a bomb in central Oslo, killing eight people, before going on a shooting rampage at a political youth camp run by Stoltenberg's AP near the capital and murdering another 69 victims. The perpetrator was sentenced to the maximum allowable prison term in Aug. 2012.

In Sept. 2012 Stoltenberg carried out an extensive cabinet reshuffle as he entered the final year of his term before elections scheduled for Sept. 2013. The changes included the appointment to the culture portfolio of Hadia Tajik, who became Norway's youngest-ever cabinet member and first Muslim minister.

DEFENCE

Conscription is for 12 months, with four to five refresher training periods.

In 2008 defence spending totalled US$5,869m. (US$1,264 per capita), representing 1·3% of GDP. Expenditure per capita was the highest of any European country in 2008.

Army

Strength (2007) 6,700 (including 3,500 conscripts). The Army fast mobilization reserve numbers 83,000.

Navy

The Royal Norwegian Navy has three components: the Navy, Coast Guard and Coastal Artillery. Main naval combatants include six German-built Ula class submarines and five frigates.

The personnel of the Navy totalled 4,100 in 2007, of whom 2,000 were conscripts. 721 (400 conscripts) served in the Coast Guard. The main naval base is at Bergen (Håkonsvern), with a subsidiary base at Ramsund.

Air Force

The Air Force consists of seven air stations, two control and reporting centres, ten squadrons with aircraft and helicopters, and two surface-to-air battalions. Total strength (2007) is 5,000 personnel, including 3,200 conscripts. There were 52 combat-capable aircraft in operation including F-16A/Bs.

Home Guard

The Home Guard is organized in small units equipped and trained for special tasks. Service after basic training is one week a year. The Home Guard consists of the Land Home Guard (strength, 2007, 46,000 reservists in mobilization), Naval Home Guard (1,800) and Anti-Air Home Guard (2,500).

INTERNATIONAL RELATIONS

In a referendum on 27–28 Nov. 1994, 52·2% of votes cast were against joining the EU. The electorate was 3,266,182; turnout was 88·9%.

ECONOMY

Services accounted for 56% of GDP in 2007, industry 43% and agriculture 1%.

Transparency International, the anti-corruption organization, ranked Norway equal seventh in the world in a survey of the countries with the least corruption in business and government in 2012. It received 85 out of 100 in the annual index.

Norway gave US$4·93bn. in international aid in 2011, which at 1·00% of GNI made it the second most generous developed country as a percentage of its gross national income (behind Sweden) and one of only five countries to exceed the UN target of 0·7%.

Overview

Norway has one of the world's highest levels of GDP per capita and one of the lowest levels of income inequality. In 2011 it was the 14th largest oil producer and eighth largest oil exporter in the world. Oil and gas account for about a quarter of GDP. The country is well endowed with other natural resources including hydropower, fish, forests and minerals.

Since emerging as a major oil and gas exporter in the mid-1970s, Norway has enjoyed solid growth linked to global oil prices, allowing the government to run large fiscal surpluses. Fiscal guidelines effective since the 2002 budget hold the central government non-oil deficit to 4% of the assets of the Government Pension Fund—Global. Revenue from oil production is transferred to this fund. In 2011 Statoil (in which the government is the largest shareholder) made two large discoveries at the Skrugard oil field and the Aldous/Avaldsnes oil field, the latter representing one of the ten largest oil finds ever on the Norwegian continental shelf.

The economy combines free market capitalism and an advanced welfare state. GDP growth slowed in 2008 as a result of lower oil prices and the global financial crisis. Norway went into recession in the first half of 2009. However, a recovery started in the second half of the year and by 2011 output had surpassed its pre-recession levels. Unemployment is low, standing at 3·5% in Nov. 2012.

A significant policy issue is how to finance pensions for an ageing population. A flexible retirement act has been implemented gradually since 2010 as part of wider pension reforms. House prices, one of the highest ratios of household debt in the OECD and eurozone turmoil are further causes for concern. The cost of increased security and upgrading of data systems following the twin attacks of July 2011 accounted for 1·45bn. kroner in the 2011 and 2012 budgets.

Currency

The unit of currency is the *Norwegian krone* (NOK) of 100 *øre*. After Oct. 1990 the krone was fixed to the ecu in the EMS of the EU in the narrow band of 2·25%, but it was freed in Dec. 1992. Inflation rates (based on OECD statistics):

2002	2003	2004	2005	2006	2007	2008	2009	2010	2011
1·3%	2·5%	0·5%	1·5%	2·3%	0·7%	3·8%	2·2%	2·4%	1·3%

Foreign exchange reserves were US$42,214m. in Aug. 2009. Gold reserves are negligible. On 30 Sept. 2011 the nominal value of notes and coins in circulation was 49,609m. kroner.

Budget

Central government current revenue and expenditure (in 1m. kroner) for years ending 31 Dec.:

	2006	2007	2008	2009	2010
Revenue	1,100,076	1,146,478	1,310,399	1,150,070	1,217,295
Expenditure	699,677	734,785	798,146	873,414	906,804

The standard rate of VAT is 25·0% (reduced rates of 15% and 8%).

Performance

Real GDP growth rates (based on OECD statistics):

2002	2003	2004	2005	2006	2007	2008	2009	2010	2011
1·5%	1·0%	4·0%	2·6%	2·5%	2·7%	0·0%	−1·7%	0·7%	1·4%

Major oil discoveries on the Norwegian continental shelf coincided with the 1974 and 1979 oil shocks, resulting in a pronounced upswing in the mainland economy which lasted until the 1986 oil price collapse. Norway only began to recover from the subsequent slump in the economy in 1993. The strong performance of the Norwegian economy in 1993–98 lifted mainland GDP by 20%, but there was a significant slowdown in 1998 when the oil price collapsed at a time when the labour market was overheated. Norway's total GDP in 2011 was US$485·8bn.

Banking and Finance

Norges Bank is the central bank and bank of issue. Supreme authority is vested in the Executive Board consisting of seven members appointed by the King and the Supervisory Council consisting of 15 members elected by the Storting. The *Governor* is Øystein Olsen. Total assets and liabilities at 30 Sept. 2011 were 3,375,882m. kroner.

Norway's largest commercial bank is DNB bank (with assets at 31 Dec. 2010 of 1,862bn. kroner); the second largest bank is Nordea Bank Norge. There were 16 commercial banks in 2007 (with total assets of 1,799bn. kroner) and 121 savings banks (with total assets of 1,993bn. kroner).

In June 2012 gross external debt amounted to US$618,646m. There is a stock exchange in Oslo.

ENERGY AND NATURAL RESOURCES

Environment
Norway's carbon dioxide emissions from the consumption and flaring of fossil fuels in 2008 were the equivalent of 8·7 tonnes per capita. An *Environmental Performance Index* compiled in 2010 ranked Norway fifth in the world behind Iceland, Switzerland, Costa Rica and Sweden, with 81·1%. The index examined various factors in six areas—air pollution, biodiversity and habitat, climate change, environmental health, productive natural resources and water resources.

In 2010 there were 33 national parks (total area, 2,996,032 ha.), 2,010 nature reserves (495,771 ha.), 195 landscape protected areas (1,628,844 ha.) and 474 other areas with protected flora and fauna (42,749 ha.).

Norway is one of the world leaders in recycling. In 2008, 52% of all household waste was sent for recovery.

Electricity
Norway is the sixth largest producer of hydropower in the world and the largest in Europe. The potential total hydro-electric power was estimated at 205,937m. kWh in 2010. Installed electrical capacity in 2007 was 30·5m. kW, 95% of it hydro-electric. Production, 2007, was 137,164m. kWh (98% hydro-electric). Consumption per capita in 2008, at 27,023 kWh, was one of the highest in the world. In 1991 Norway became the first country in Europe to deregulate its energy market. Norway is a net importer of electricity.

Oil and Gas
There are enormous oil reserves in the Norwegian continental shelf. In 1966 the first exploration well was drilled. Production of crude oil, 2008, 114·2m. tonnes. Norway ranks among the world's biggest oil exporters, with net oil exports of around 1·9m. bbls a day in 2010. It had proven reserves of 7·5bn. bbls in 2008. In March 1998 Norway announced that it would reduce its output for the year by 100,000 bbls per day as part of a plan to cut global crude production. In June 2001 the Norwegian government sold a 17·5% stake in Statoil, the last major state-owned oil company in western Europe. In Oct. 2007 Statoil and the oil and gas division of Norsk Hydro (a Norwegian energy and metals company) merged to form StatoilHydro, as a result creating the world's largest offshore oil and natural gas producer.

Output of natural gas, 2010, 106·4bn. cu. metres with proven reserves in 2008 of 2,910bn. cu. metres.

Minerals
Production in 2008 unless otherwise indicated (in tonnes): coal (2009), 2,640,000; aluminium, 1,368,000; ilmenite concentrate, 915,000; iron ore, 477,000; zinc, 145,469; nickel, 88,741; refined copper (2007), 34,212.

Agriculture
Norway is barren and mountainous. The arable area is in strips in valleys and around fjords and lakes.

In 2010 the agricultural area was 1,003,010 ha., of which 651,176 ha. were meadow and pasture, 145,818 ha. were sown to barley, 75,810 ha. to oats, 71,945 ha. to wheat and 13,207 ha. to potatoes. Production in 2010 (in 1,000 tonnes): hay, 2,659; barley, 541; wheat, 331; potatoes, 321; oats, 299.

Livestock, 2010 (provisional), 875,169 cattle (308,399 dairy cows), 919,046 sheep (one year and over), 97,318 pigs for breeding, 36,935 dairy goats, 3,891,109 hens, 150,000 silver fox, 105,000 blue and silver blue fox, 650,000 mink and 251,400 tame reindeer.

Forestry
In 2010 the total area under forests was 10·07m. ha., or 33% of the total land area. Productive forest area, 2010, approximately 6·77m. ha. About 80% of the productive area consists of conifers and 20% of broadleaves. In 2008, 8·07m. cu. metres of roundwood were cut.

Fisheries
The total number of fishermen in 2009 was 12,656, of whom 2,528 had another chief occupation. In 2009 the number of registered fishing vessels (all with motor) was 6,506.

The catch in 2009 totalled 2,537,134 tonnes, almost entirely from sea fishing. The catch of herring in 2009 totalled 1,077,250 tonnes, cod 243,659 tonnes and capelin 233,005 tonnes. 4,652 harp seals were caught in 2010. The catch of hooded seals was prohibited in 2007. Commercial whaling was prohibited in 1988, but recommenced in 1993: 484 whales were caught in 2008. Norway is the second largest exporter of fishery commodities, after China. In 2009 exports were valued at US$7·07bn.

INDUSTRY
The leading companies by market capitalization in Norway in March 2012 were: Statoil, US$86·4bn.; Telenor Group, a telecommunications company (US$29·8bn.); and DNB, a financial services group (US$20·9bn.).

Industry is chiefly based on raw materials. Paper and paper products, industrial chemicals and basic metals are important export manufactures. In the following table figures are given for industrial establishments in 2009. The values are given in 1m. kroner.

Industries	Establish-ments	Number of employees	Gross value of production	Value added (in market prices)
Mining and quarrying	745	4,680	11,463	4,032
Food products	2,269	44,207	131,352	23,700
Beverages and tobacco	106	4,487	16,405	9,932
Textiles, clothing and leather	1,338	4,772	6,101	2,183
Wood and wood products	1,930	14,248	21,557	6,429
Paper and paper products	93	5,475	14,470	2,415
Printing and reproduction of recorded media	1,414	7,461	10,851	4,079
Refined petroleum products, chemicals and pharmaceuticals	316	13,443	108,196	23,464
Rubber, plastic and mineral products	1,389	15,887	33,734	10,608
Basic metals	163	10,515	47,753	3,638
Fabricated metal products	2,609	25,225	44,124	15,314
Computer and electrical equipment	822	16,841	38,587	13,563
Machinery and equipment	1,501	20,872	78,404	23,620
Ships, boats, and oil platforms	534	22,761	65,130	16,353
Other transport equipment	182	4,085	6,482	2,157
Furniture and other manufacturing industries	2,056	11,335	13,553	5,371
Repair and installation of machinery and equipment	2,117	17,307	33,856	11,765
Total	19,584	243,601	682,019	178,623

Labour
Norway has a tradition of centralized wage bargaining. Since the early 1960s the contract period has been for two years with intermediate bargaining after 12 months, to take into consideration such changes as the rate of inflation.

The labour force averaged 2,602,000 in 2010 (1,224,000 females). The total number of employed persons in 2010 averaged 2,508,000 (1,187,000 females), of whom 1,835,000 were in full-time employment, 667,000 in part-time employment and 6,000 working unspecified hours. Distribution of employed persons by occupation in 2007 showed 492,700 in health and social work; 364,000 in trade; 300,200 in business services; 286,200 in manufacturing; 185,500 in education; 184,200 in construction; 167,300 in transport; 156,600 in public administration and defence; 78,800 in hotels and restaurants; 60,100 in agriculture.

The unemployment rate in Nov. 2012 was 3·5% (3·3% in 2011 as a whole).

There were 12 work stoppages in 2010 (two in 2009): 500,009 working days were lost (180 in 2009).

INTERNATIONAL TRADE

Imports and Exports

Total imports and exports in calendar years (in 1m. kroner):

	2006	2007	2008	2009	2010
Imports	411,755	468,918	504,481	430,363	466,810
Exports	782,943	795,366	953,154	717,965	792,575

Norway's trade surplus was 325,765m. kroner in 2010. Major import suppliers in 2010 (value in 1m. kroner): Sweden, 65,603·1; Germany, 57,461·5; China, 39,614·9; Denmark, 29,018·6; UK, 27,460·5; USA, 25,324·6. Imports from economic areas: EU, 295,667·6; Nordic countries, 108,068·6; OECD, 360,693·1.

Major export markets in 2010 (value in 1m. kroner): UK, 213,981·8; Netherlands, 94,720·9; Germany, 89,725·8; Sweden, 55,315·6; France, 50,532·8; USA, 39,368·0. Exports to economic areas: EU, 639,798·7; Nordic countries, 93,203·0; OECD, 721,729·5.

Principal imports in 2010 (in 1m. kroner): motor vehicles, 43,800·5 (including passenger cars and station wagons, 26,205·3); transport equipment excluding motor vehicles, 32,708·0; petroleum, petroleum products and related materials, 23,151·6; electrical machinery, apparatus and appliances, 22,784·0; metalliferous ores and metal scrap, 22,288·8; general industrial machinery and equipment, 21,925·9.

Principal exports in 2010 (in 1m. kroner): petroleum, petroleum products and related materials, 315,259·3 (including crude petroleum, 283,337·2); natural and manufactured gas, 188,010·4 (including natural gas, 163,970·3); fish, crustaceans and molluscs, and preparations thereof, 52,307·6; non-ferrous metals, 44,083·6 (including aluminium, 26,493·6); general industrial machinery and equipment, 18,423·9; chemical materials and products, 17,992·3.

COMMUNICATIONS

Roads

In Jan. 2011 the length of public roads (including roads in towns) totalled 93,509 km. Total road length in Jan. 2011 included: national roads, 10,496 km; provincial roads, 44,281 km; local roads, 38,732 km. Number of registered motor vehicles, 2010, included: 2,308,548 passenger cars (including station wagons and ambulances), 397,279 vans, 254,674 tractors and special purpose vehicles, 168,904 mopeds, 146,592 motorcycles, 81,330 goods vehicles (including lorries), 48,432 combined vehicles and 20,348 buses. In 2010, 9,130 injuries were sustained in road accidents, with 208 fatalities. Norway has one of the lowest death rates in road accidents of any industrialized country, at 4·3 deaths per 100,000 people in 2010.

Rail

The length of state railways in 2010 was 4,169 km (2,566 km electrified). In 2009 passenger-km travelled came to 2,669m. and freight tonne-km to 2,804m. Sales and other operating income totalled 11,179m. kroner in 2010.

There is a metro (104 km) and a tram network (146 km) in Oslo.

Civil Aviation

The main international airports are at Oslo (Gardermoen), Bergen (Flesland), Stavanger (Sola), Sandefjord (Torp) and Moss (Rygge). Norway's largest airline is SAS Norge, a wholly-owned subsidiary of the Scandinavian Airlines System (SAS) Group. It was established in 2004 as SAS Braathens through the merger of the Norwegian part of SAS and Braathens, and was renamed SAS Norge in 2007. SAS Norge carries around 10m. passengers a year to 55 destinations. The second largest airline is the low-cost carrier Norwegian.

In 2010 Oslo (Gardermoen) handled 19,140,384 passengers (10,123,605 on international flights). Bergen is the second busiest airport for passenger traffic, with 5,189,714 passengers in 2010 (3,604,882 on domestic flights).

Shipping

The Norwegian International Ship Register was set up in 1987. In 2010, 525 ships were registered (400 Norwegian) totalling 13,792,000 GT. 218 tankers accounted for 6,948,000 GT. There were also 882 vessels totalling 1,917,000 GT on the Norwegian Ordinary Register. These figures do not include fishing boats, tugs, salvage vessels, icebreakers and similar special types of vessels. In Jan. 2011 Norway's merchant fleet represented 3·4% of total world tonnage. In 2009, 41m. passengers travelled on internal ferries. The warm Gulf Stream ensures ice-free harbours throughout the year.

Telecommunications

At 31 Dec. 2010 there were 1,648,927 main (fixed) telephone lines and 5,648,673 mobile phone subscribers (1,148·0 per 1,000 persons). There were 3·9m. internet users in 2008. In March 2012 there were 2·6m. Facebook users. Since 2000 the government has been reducing its interest in Telenor, the country's largest telecommunications operator, and in March 2004 lowered its stake to 54·0%.

SOCIAL INSTITUTIONS

Justice

The judicature is common to civil and criminal cases; the same professional judges preside over both. These judges are state officials. The participation of lay judges and jurors, both summoned for the individual case, varies according to the kind of court and kind of case.

The 96 city or district courts of first instance are in criminal cases composed of one professional judge and two lay judges, chosen by ballot from a panel elected by the local authority. In civil cases two lay judges may participate. These courts are competent in all cases except criminal cases where the maximum penalty exceeds six years imprisonment. In every community there is a Conciliation Board composed of three lay persons elected by the district council. A civil lawsuit usually begins with mediation by the Board which can pronounce judgement in certain cases.

The five high courts, or courts of second instance, are composed of three professional judges. Additionally, in civil cases two or four lay judges may be summoned. In serious criminal cases, which are brought before high courts in the first instance, a jury of ten lay persons is summoned to determine whether the defendant is guilty according to the charge. In less serious criminal cases the court is composed of two professional and three lay judges. In civil cases, the court of second instance is an ordinary court of appeal. In criminal cases in which the lower court does not have judicial authority, it is itself the court of first instance. In other criminal cases it is an appeal court as far as the appeal is based on an attack against the lower court's assessment of the facts when determining the guilt of the defendant. An appeal based on any other alleged mistakes is brought directly before the Supreme Court.

The Supreme Court (Høyesterett) is the court of last resort. There are 18 Supreme Court judges. Each individual case is heard by five judges. Some major cases are determined in plenary session. The Supreme Court may in general examine every aspect of the case and the handling of it by the lower courts. However, in criminal cases the Court may not overrule the lower court's assessment of the facts as far as the guilt of the defendant is concerned.

The Court of Impeachment (*Riksretten*) is composed of five judges of the Supreme Court and ten members of Parliament.

All serious offences are prosecuted by the State. The Public Prosecution Authority consists of the Attorney General, 18 district attorneys and legally qualified officers of the ordinary police force. Counsel for the defence is in general provided for by the State.

The population in penal institutions in Aug. 2008 was 3,276 (69 per 100,000 of national population).

Education

Free compulsory schooling in primary and lower secondary schools was extended to 10 years from 9, and the starting age lowered to 6 from 7, in July 1997. All young people between the ages of 16 and 19 have the statutory right to three years of upper secondary education. In 2010 there were 6,579 kindergartens (children up to six years old) with 277,139 children and 87,401 staff. There were 614,020 pupils at primary and lower secondary schools in 2010; 228,170 pupils at upper secondary schools; and 20,658 students at folk high schools and vocational schools.

There are eight universities: Oslo (founded 1811), with 27,341 students in 2007; Bergen (1948), with 14,057 students; Tromsø (1972), with 5,424 students; the Norwegian University of Science and Technology (1996, formerly the University of Trondheim and the Norwegian Institute of Technology), with 19,351 students; Stavanger (2005, formerly Stavanger University College), with 8,050 students; Norwegian University of Life Sciences (1859, university since 2005—formerly the Agricultural University of Norway), with 2,817 students; Agder (2007, formerly Agder University College), with 7,801 students; and Nordland (2011, formerly Bodø University College), with 5,700 students in 2011. There are also nine specialized university institutions, 21 state university colleges and a number of private colleges. In 2010 the universities had 93,768 students and the state university colleges 89,572 students. The University of Tromsø is responsible for Sami language and studies.

In 2006 public expenditure on education came to 6·6% of GNI and 16·2% of total government spending. The adult literacy rate is at least 99%.

Health

The health care system, which is predominantly publicly financed (mainly by a national insurance tax), is run on both county and municipal levels. Persons who fall ill are guaranteed medical treatment, and health services are distributed according to need. In Aug. 2008 there were 20,035 active physicians in Norway. On 31 Dec. 2009 there were 15,205 hospital beds (excluding those in psychiatric institutions). In 2007 Norway spent 8·9% of its GDP on health. In 2010, 19% of men and 19% of women aged 16–74 smoked on a daily basis.

Welfare

Expenditure on social assistance in 2009 totalled 4,642m. kroner. On 31 Dec. 2010 there were 663,799 old age pensioners (377,170 women) and 301,088 disability pensioners (170,959 women). Maternity leave is either for 44 weeks on 100% of previous salary or 54 weeks on 80% of previous salary; unused portions may pass to the father. In Dec. 2010, 33,433 children aged one to three received cash benefit (27% of all children between one and three years of age).

RELIGION

There is freedom of religion, the Church of Norway (Evangelical Lutheran), however, being the national church, endowed by the State. Its clergy are nominated by the King. Ecclesiastically Norway is divided into 11 dioceses, 100 deaneries and 1,298 parishes. About 80% of Norwegians belong to the Church of Norway (which had 3,848,841 members in 2009) and

approximately 68% of infants were baptised in the Church in 2009. There were 431,287 members of registered and unregistered religious and philosophical communities outside the Church of Norway in 2009, subsidized by central government and local authorities, including 234,772 Christians and 92,744 Muslims. The Roman Catholics are under a Bishop at Oslo, and Prelates at Tromsø and Trondheim.

CULTURE

World Heritage Sites

Norway's UNESCO heritage sites (with year listed) are: the 12–13th century wooden church in Sogn og Fjordane on the west coast, the Urnes Stave Church (1979), a testimony to the city's key role in the Hanseatic League trading route between the 14–16th centuries; the 58 wooden buildings in Bergen's wharf of Bryggen (1979); the wooden houses of the copper mining village of Røros (1980), active between the 17–20th centuries; the pre-historic Rock Drawings of Alta (1995) in the Alta Fjord, near the Arctic Circle, dating from 4200 to 500 BC; Vegaøyan—the Vega Archipelago (2004), a cluster of dozens of islands centred on Vega, just south of the Arctic Circle; the West Norwegian Fjords—Geirangerfjord and Naerøyfjord (2005); and the Struve Geodetic Arc (2005). The Arc is a chain of survey triangulations spanning from Norway to the Black Sea that helped establish the exact shape and size of the earth and is shared with nine other countries.

Press

There were 74 paid-for daily newspapers with a combined average net circulation of 2·19m. in 2008, and in 2007 there were 151 non-dailies with a circulation of 623,000. Norway has among the highest circulation rates of daily newspapers in the world, at 580 per 1,000 adult inhabitants in 2007. In 2007 a total of 7,074 book titles were published.

Tourism

In 2007 there were 3,260,000 foreign holiday and leisure visitors (excluding same-day visitors) who stayed an average of 7·4 nights each, totalling 24,252,000 nights. The main countries of origin were Sweden (761,000), Germany (548,000), Denmark (431,000) and the UK (246,000). In 2010 there were 1,128 hotels and 782 camping sites. Spending by foreign tourists totalled 30·8bn. kroner in 2007.

Festivals

The Bergen International Festival, Norway's oldest festival, is held annually in May/June and includes music, dance and theatre. The biennial Ibsen Festival (theatre) is held in Oslo in Aug./Sept. in even-numbered years. CODA (the Oslo International Dance Festival) runs for three weeks every Sept./Oct.

DIPLOMATIC REPRESENTATIVES

Of Norway in the United Kingdom (25 Belgrave Sq., London, SW1X 8QD)
Ambassador: Kim Traavik.

Of the United Kingdom in Norway (Thomas Heftyesgate 8, 0264 Oslo)
Ambassador: Jane Owen.

Of Norway in the USA (2720 34th St., NW, Washington, D.C., 20008)
Ambassador: Wegger Christian Strommen.

Of the USA in Norway (Henrik Ibsens Gate 48, 0244 Oslo)
Ambassador: Barry B. White.

Of Norway to the United Nations
Ambassador: Geir O. Pedersen.

Of Norway to the European Union
Ambassador: Atle Leikvoll.

FURTHER READING

Statistics Norway (formerly Central Bureau of Statistics). *Statistisk Årbok/ Statistical Yearbook of Norway.—Economic survey* (annual, from 1935; with English summary from 1952, now published in *Økonomiske Analyser*, annual).—*Historisk Statistikk; Historical Statistics.—Statistisk Månedshefte* (with English index)

Norges Statskalender. From 1816; annual from 1877

Archer, Clive, *Norway and an Integrating Europe.* 2004

Danielsen, R., *et al.*, *Norway: a History From the Vikings to Our Own Times.* 1994

Petersson, O., *The Government and Politics of the Nordic Countries.* 1994

Sejersted, Francis, *The Age of Social Democracy: Norway and Sweden in the Twentieth Century.* 2011

National library: The National Library of Norway, Henrik Ibsens gate 110, 0255 Oslo; Finsetveien 2, 8624 Mo i Rana.

National Statistical Office: Statistics Norway, PB 8131 Dep., 0033 Oslo.

Website: http://www.ssb.no

Svalbard

An archipelago situated between 10° and 35° E. long. and between 74° and 81° N. lat. Total area, 61,022 sq. km (23,561 sq. miles). The main islands are Spitsbergen, Nordaustlandet, Edgeøya, Barentsøya, Prins Karls Forland, Bjørnøya, Hopen, Kong Karls Land and Kvitøya. The Arctic climate is tempered by mild winds from the Atlantic.

The archipelago was probably discovered by Norsemen in 1194 and rediscovered by the Dutch navigator Barents in 1596. In the 17th century whale-hunting gave rise to rival Dutch, British and Danish-Norwegian claims to sovereignty; but when in the 18th century the whale-hunting ended, the question of the sovereignty of Svalbard lost its significance. It was again raised in the 20th century, owing to the discovery and exploitation of coalfields. By a treaty, signed on 9 Feb. 1920 in Paris, Norway's sovereignty over the archipelago was recognized. On 14 Aug. 1925 the archipelago was officially incorporated in Norway.

Total population on 1 Jan. 2011 was 2,394, of whom 2,017 lived in the two Norwegian settlements, 370 in the Russian settlement and seven in the Polish settlement. Coal is the principal product. There are two Norwegian and two Russian mining camps. In 2009, 2·6m. tonnes of coal were produced from Norwegian mines.

In 2008, 69% of households on Svalbard had one or more snow scooters but only 49% of households had a car. There are research and radio stations, and an airport near Longyearbyen (Svalbard Lufthavn) opened in 1975.

Greve, T., *Svalbard: Norway in the Arctic.* 1975

Hisdal, V., *Geography of Svalbard.* Rev. ed., 1984

Jan Mayen

This bleak, desolate and mountainous island of volcanic origin and partly covered by glaciers is situated at 71° N. lat. and 8° 30' W. long., 300 miles north-north-east of Iceland. The total area is 377 sq. km (146 sq. miles). Beerenberg, its highest peak, reaches a height of 2,277 metres. Volcanic activity, which had been dormant, reactivated in Sept. 1970.

There exist several unverified and inconclusive reports of the island's discovery. Its present name derives from the Dutch whaling captain Jan Jacobsz May, who mapped the island in 1614. Jan Mayen was subsequently established as a whaling base for the Dutch Noordsche Compagnie. The island was abandoned in 1638 owing to the near extinction of the local whale population, and remained uninhabited, though occasionally visited by seal hunters and trappers, until 1921 when Norway established a radio and meteorological station. On 8 May 1929 Jan Mayen was officially proclaimed as incorporated into the Kingdom of Norway. Its relation to Norway was finally settled by law of 27 Feb. 1930. A LORAN station (1959) and a CONSOL station (1968) have been established.

Bouvet Island
Bouvetøya

This uninhabited volcanic island, mostly covered by glaciers and situated at 54° 25' S. lat. and 3° 21' E. long., was discovered in 1739 by a French naval officer, Jean Baptiste Loziert Bouvet, but no flag was hoisted until, in 1825, Capt. Norris raised the Union Jack. In 1928 Great Britain waived its claim to the island in favour of Norway, which in Dec. 1927 had occupied it. A law of 27 Feb. 1930 declared Bouvetøya a Norwegian dependency. The area is 49 sq. km (19 sq. miles). Since 1977 Norway has had an automatic meteorological station on the island.

Peter I Island
Peter I Øy

This uninhabited island, situated at 68° 48' S. lat. and 90° 35' W. long., was sighted in 1821 by the Russian explorer, Admiral von Bellingshausen. The first landing was made in 1929 by a Norwegian expedition which hoisted the Norwegian flag. On 1 May 1931 Peter I Island was placed under Norwegian sovereignty, and on 24 March 1933 it was incorporated as a dependency. The area is 156 sq. km (60 sq. miles).

Queen Maud Land
Dronning Maud Land

On 14 Jan. 1939 the Norwegian Cabinet placed that part of the Antarctic Continent from the border of Falkland Islands dependencies in the west to the border of the Australian Antarctic Dependency in the east (between 20° W. and 45° E.) under Norwegian sovereignty. The territory had been explored only by Norwegians and hitherto been ownerless. In 1957 it was given the status of a dependency.

OMAN

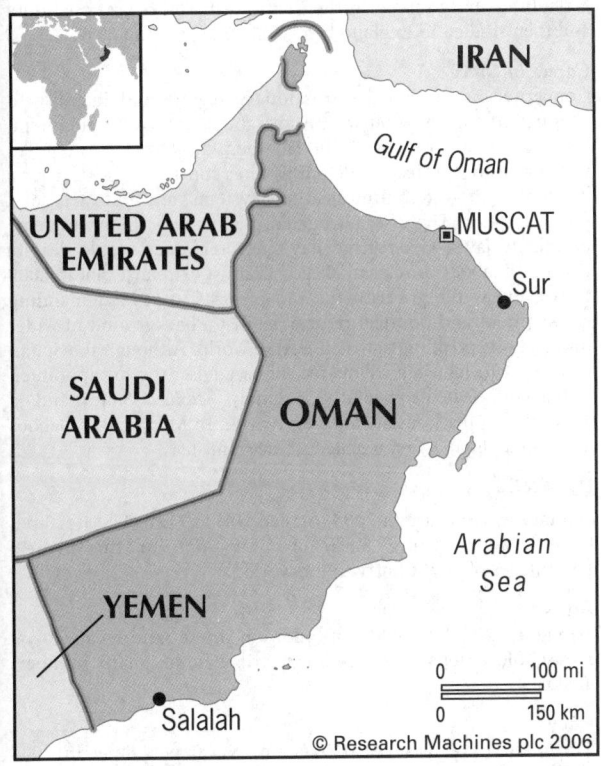

© Research Machines plc 2006

Saltanat 'Uman
(Sultanate of Oman)

Capital: Muscat
Population projection, 2015: 3·06m.
GNI per capita, 2011: (PPP$) 22,841
HDI/world rank: 0·705/89
Internet domain extension: .om

KEY HISTORICAL EVENTS

The ancestors of present day Oman are believed to have arrived in two waves of migration, the first from Yemen and the second from northern Arabia. In the 9th century maritime trade flourished and Sohar became the greatest sea port in the Islamic world. In the early 16th century the Portuguese occupied Muscat. The Ya'aruba dynasty introduced a period of renaissance in Omani fortunes both at home and abroad, uniting the country and bringing prosperity; but, on the death in 1718 of Sultan bin Saif II, civil war broke out over the election of his successor. Persian troops occupied Muttrah and Muscat but failed to take Sohar which was defended by Ahmad bin Said, who expelled the Persians from Oman after the civil war had ended. In 1744 the Al bu Said family assumed power and has ruled to the present day. Oman remained largely isolated from the rest of the world until 1970 when Said bin Taimur was deposed by his son Qaboos in a bloodless coup.

TERRITORY AND POPULATION

Situated at the southeast corner of the Arabian peninsula, Oman is bounded in the northeast by the Gulf of Oman and southeast by the Arabian Sea, southwest by Yemen and northwest by Saudi Arabia and the United Arab Emirates. There is an enclave at the northern tip of the Musandam Peninsula. An agreement of April 1992 completed the demarcation of the border with Yemen, and an agreement of March 1990 finalized the border with Saudi Arabia.

With a coastline of 1,700 sq. km from the Strait of Hormuz in the north to the borders of the Republic of Yemen, the Sultanate is strategically located overlooking ancient maritime trade routes linking the Far East and Africa with the Mediterranean.

The Sultanate of Oman occupies a total area of 309,500 sq. km and includes different terrains that vary from plain to highlands and mountains. The coastal plain overlooking the Gulf of Oman and the Arabian Sea forms the most important and fertile plain in Oman.

The **Kuria Muria** islands were ceded to the UK in 1854 by the Sultan of Muscat and Oman. On 30 Nov. 1967 the islands were retroceded to the Sultan of Muscat and Oman, in accordance with the wishes of the population. They are now known as the **Halaniyat Islands**.

Population at the census of Dec. 2010, 2,773,479 (1,612,411 males); density 9·0 per sq. km. The population comprised 1,957,336 Omanis and 816,143 expatriates.

The UN gives a projected population for 2015 of 3·06m.

In 2011, 73·3% of the population lived in urban areas. The census population of the capital, Muscat, in 2010 was 775,878.

The official language is Arabic; English is in commercial use.

SOCIAL STATISTICS

2008 estimates: births, 61,000; deaths, 8,000. Estimated rates, 2008 (per 1,000 population): births, 22·0; deaths, 2·7. Expectation of life at birth, 2007, was 74·1 years for males and 77·3 years for females. Average annual population growth rate, 2000–08, 1·8%. Fertility rate, 2008, 3·0 births per woman, down from 7·8 in 1988.

Oman has achieved some of the most rapid advances ever recorded. Infant mortality declined from 200 per 1,000 live births in 1960 to eight per 1,000 live births in 2010, and as recently as 1970 life expectancy was just 40.

CLIMATE

Oman has a desert climate, with exceptionally hot and humid months from April to Oct., when temperatures may reach 47°C. Light monsoon rains fall in the south from June to Sept., with highest amounts in the western highland region. Muscat, Jan. 28°C, July 46°C. Annual rainfall 101 mm. Salalah, Jan. 29°C, July 32°C. Annual rainfall 98 mm.

CONSTITUTION AND GOVERNMENT

Oman is a hereditary absolute monarchy. The Sultan legislates by decree and appoints a Cabinet to assist him. The Basic Statute of the State was promulgated on 6 Nov. 1996.

The present Sultan is **Qaboos bin Said Al Said** (b. Nov. 1940).

In 1991 a new consultative assembly, the *Majlis al-Shura*, replaced the former State Consultative Chamber. The Majlis consists of 84 elected members. It debates domestic issues, but has no legislative or veto powers. There is also an upper house, the *Majlis al-Dawla*, which consists of 83 appointed members; it too has advisory powers only.

In Dec. 2002 the Sultan of Oman extended voting rights to all citizens over the age of 21.

National Anthem

'Ya Rabbana elifidh lana jalalat al Saltan' ('O Lord, protect for us his majesty the Sultan'); words by Rashid bin Aziz, tune by Rodney Bashford.

GOVERNMENT CHRONOLOGY

Sultans since 1932.

1932–70   Said bin Taimur Al Said
1970–   Qaboos bin Said Al Said

RECENT ELECTIONS

The last elections to the *Majlis al-Shura* were on 15 Oct. 2011. No parties are allowed. 84 legislators were chosen for four-year terms from over 1,300 candidates, including 76 women (although only one women was elected). Three activists who took part in the Feb. 2011 protests were elected.

CURRENT GOVERNMENT

The Sultan is nominally Prime Minister and Minister of Foreign Affairs, Defence and Finance.

In March 2013 the other Ministers were:

Special Representative of the Sultan: Thuwayni bin Shihab Al Said.

Deputy Prime Minister for Cabinet Affairs: Fahd bin Mahmud Al Said.

Minister Responsible for Defence Affairs: Sayyid Badr bin Saud bin Harib Al Busaidi. *Minister Responsible for Foreign Affairs:* Yusuf bin Alawi bin Abdallah. *Minister Responsible for Financial Affairs:* Darwish bin Ismail bin Ali Al Balushi.

Minister of Agriculture and Fisheries: Fuad bin Jaffer bin Mohammed Al Sajwani. *Awqaf and Religious Affairs:* Abdallah bin Muhammad bin Abdallah Al Salimi. *Civil Service:* Shaikh Khalid bin Omar bin Saeed Al Marhoon. *Commerce and Industry:* Ali bin Masoud bin Ali Al Sunaidi. *Education:* Madeeha bint Ahmed bin Nassir Al Shibaniyah. *Environment and Climate Change:* Mohammed Salem Al Tobi. *Health:* Ahmed bin Mohammed bin Obaid Al Sa'eedi. *Higher Education:* Rawya bint Saud Al Busaidi. *Housing:* Sheikh Saif bin Mohammed bin Saif Al Shabibi. *Information:* Abdulmunem Al Hasni. *Interior:* Sayyid Hamoud bin Faisal Al Busaidi. *Justice:* Shaikh Abdul Malik bin Abdullah Al Khalili. *Legal Affairs:* Abdullah bin Mohammed bin Saeed Al Saeedi. *Manpower:* Abdullah bin Nasser Al Bakri. *National Heritage and Culture:* Sayyid Haitham bin Tariq Al Said. *Oil and Gas:* Muhammad bin Hamad bin Seif Al Rumhi. *Regional Municipalities and Water Resources:* Ahmed bin Abdullah bin Mohammed Al Sahi. *Social Development:* Shaikh Mohammed bin Saeed Al Kalbani. *Sport:* Shaikh Saad bin Mohammed Al Saadi. *Tourism:* Ahmed bin Nasser Al Mahrazi. *Transportation and Communications:* Ahmed bin Mohammed bin Salem Al Futaisi. *Diwan of the Royal Court:* Sayyid Khalid bin Hilal bin Saud Al Busaidi. *Royal Office:* Lieut.-Gen. Sultan bin Mohammed Al Nu'amani. *Minister of State and Governor of the Capital:* Sayyid Soud bin Hilal Al Busaidi. *Minister of State and Governor of Dhofar:* Shaikh Mohammed bin Marhoon bin Ali Al Maamari. *Secretary General of the Council of Ministers:* Shaikh Fadhjil bin Mohammed Al Harthi.

CURRENT LEADERS

Qaboos bin Said

Position
Sultan

Introduction
Qaboos has been the Sultan since 23 July 1970 when he deposed his father, Said bin Taimur. He has carried out an ambitious social and economic modernization programme, opening Oman to the outside world through accession to the League of Arab States, Gulf Co-operation Council and United Nations and pursuing a moderate regional foreign policy while preserving a longstanding political and military relationship with the United Kingdom. He is nominally prime minister, minister of defence, minister of foreign affairs, minister of finance and chairman of the central bank.

Early Life
Born in Salalah on 18 Nov. 1940, Qaboos attended a private school in England from the age of 16. In 1960 he went to the British Royal Military Academy at Sandhurst as an officer cadet. He subsequently served in the British army on operational duty and then studied in England before returning to Oman.

Career in Office
Concerned at his father's reactionary regime and inability to channel Oman's new oil wealth into the country's development, Qaboos led a coup in 1970. He then undertook a range of infrastructure projects, including the construction of roads, hospitals, schools, communications systems, and industrial and port facilities. He also abrogated his father's more extreme moralistic laws. His regime has since remained stable despite periods of labour unrest, an alleged Islamist extremist plot in 2005 to overthrow the government and protests in 2011 demanding employment and political reform against a background of wider disaffection spreading across the Arab world. Although the Sultan continues to legislate by decree, he is advised by an appointed Cabinet, an elected consultative assembly (*Majlis al-Shura*) and an appointed upper house (*Majlis al-Dawla*). In March 2004 Qaboos appointed Oman's first female Cabinet minister.

DEFENCE

Military expenditure in 2008 totalled US$4,671m. (US$1,410 per capita), representing 8·5% of GDP—one of the highest percentages of any country in the world.

Army

Strength (2007) 25,000. In addition there are 6,400 Royal Household troops. A paramilitary tribal home guard numbers 4,000.

Navy

The main naval base is at Wudam. Naval personnel in 2007 totalled 4,200.

The wholly separate Royal Yacht Squadron consists of a yacht, a support ship with helicopter and troop-carrying capability, and a dhow.

Air Force

The Air Force, formed in 1959, has 64 combat-capable aircraft including F-16s, Jaguars and Hawks.

Personnel (2007) 5,000.

ECONOMY

Oil and natural gas accounted for 48% of GDP in 2006; trade, restaurants and hotels 12%; manufacturing 10%; and services 8%.

Overview

Crude oil dominates the economy, accounting for 37% of exports in 2009, with China, Japan and India taking 32%, 17% and 11% respectively. Government attempts to diversify the economy have focused on tourism, shipping and investment in infrastructure. There are also plans to increase natural gas production as a share of gross domestic product to 10% by 2020. Oman holds 0·5% of the world's liquefied natural gas supply.

Growth from 2005 to 2009 averaged 7·1%, supported by high oil prices and accelerated growth in non-hydrocarbon sectors including trade, transport and communications. Higher global commodity prices, domestic demand growth (prompted by fiscal stimuli) and strong private sector credit growth raised inflation to over 12% in 2008 although it has fallen since then.

Public debt was 5·6% of GDP in 2012, while unemployment stood at 15%. The government's eighth Five Year Plan (for 2011 until 2015) aims for GDP growth at a minimum of 3% per year, with RO 12bn. earmarked for investment in the natural gas sector. It is hoped that development of gas-based and non-hydrocarbon industries will reduce unemployment.

Currency

The unit of currency is the *Rial Omani* (OMR). It is divided into 1,000 *baiza*. The rial is pegged to the US dollar. In July 2005 foreign exchange reserves were US$4,511m. and gold reserves totalled 1,000 troy oz (291,000 troy oz in April 2002). Total money supply was RO 1,067m. in May 2005. Inflation was 12·6% in 2008, 3·5% in 2009, 3·3% in 2010 and 4·0% in 2011.

In 2001 the six Gulf Arab states—Oman, along with Bahrain, Kuwait, Qatar, Saudi Arabia and the United Arab Emirates—signed an agreement to establish a single currency by 2010. However, Oman withdrew from the scheme in 2007.

Budget

In 2008 revenues were RO 7,829·4m. and expenditures RO 7,556·7m. Oil revenue accounted for 67·5% of revenues in 2008; current expenditure accounted for 58·5% of expenditures.

Performance

Real GDP growth was 3·9% in 2009, 5·0% in 2010 and 5·4% in 2011. Total GDP in 2009 was US$71·8bn.

Banking and Finance

The bank of issue is the Central Bank of Oman, which commenced operations in 1975 (*President*, Hamood Sangour Al Zadjali). All banks must comply with BIS capital adequacy ratios and have a minimum capital of RO 20m. (minimum capital requirement for foreign banks established in Oman is RO 3m.). In 2002 there were 15 commercial banks (of which nine were foreign) and three specialized banks. The largest bank is BankMuscat SAOG, with assets of RO 1·3bn.

Total foreign debt was US$3,472m. in 2005.

There is a stock exchange in Muscat, which is linked with those in Bahrain and Kuwait.

ENERGY AND NATURAL RESOURCES

Environment

Oman's carbon dioxide emissions from the consumption and flaring of fossil fuels in 2008 were the equivalent of 13·2 tonnes per capita.

Electricity

Installed capacity was 3·4m. kW in 2007. Production in 2007 was 14·44bn. kWh, with consumption per capita 5,264 kWh.

Oil and Gas

The economy is dominated by the oil industry. Oil in commercial quantities was discovered in 1964 and production began in 1967. Production in 2008 was 36·0m. tonnes. In 2007 exports of oil stood at 30·3m. tonnes. Total proven reserves in 2008 were 5·6bn. bbls. It was announced in Aug. 2000 that two new oilfields in the south of the country had been discovered, with a potential combined daily production capacity of 12,200 bbls. Earlier in 2000 oil began to be pumped from two further recently-discovered oilfields.

Gas is likely to become the second major source of income for the country. Oman's proven natural gas reserves were 980bn. cu. metres in 2008 (570bn. cu. metres in 1998). Natural gas production was 24·1bn. cu. metres in 2008 (5·2bn. cu. metres in 1998).

Water

Oman relies on a combination of aquifers and desalination plants for its water, augmented by a construction programme of some 60 recharge dams. Desalination plants at Ghubriah and Wadi Adai provide most of the water needs of the capital area. In 2005 water production was 32,951m. gallons.

Minerals

Production in 2009 (in 1,000 tonnes): sand and gravel, 77,114; limestone, 7,948; chromite, 636 (up from 19 in 2004); marble, 588; gypsum, 333; quartz, 198. The mountains of Oman are rich in mineral deposits; these include chromite, coal, asbestos, manganese, gypsum, limestone and marble.

Agriculture

Agriculture and fisheries are the traditional occupations of Omanis and remain important to the people and economy of Oman to this day. The country now produces a wide variety of fresh fruit, vegetables and field crops. The country is rapidly moving towards its goal of self-sufficiency in agriculture with the total area under cultivation standing at over 70,000 ha. and total output more than 1m. tonnes. This has not been achieved without effort. In a country where water is a scarce commodity it has meant educating farmers on efficient methods of irrigation and building recharge dams to make the most of infrequent rainfall. In 2002 there were an estimated 38,000 ha. of arable land and 43,000 ha. of permanent crops. Approximately 62,000 ha. were irrigated in 2002. In 2002, 35·2% of the economically active population were engaged in agriculture.

The coastal plain (Batinah) northwest of Muscat is fertile, as are the Dhofar highlands in the south. In the valleys of the interior, as well as on the Batinah coastal plain, date cultivation has reached a high level, and there are possibilities of agricultural development. Agricultural products, 2004 estimates (in 1,000 tonnes): dates, 239; tomatoes, 43; bananas, 33; watermelons, 27. Vegetable and fruit production are also important, and livestock are raised in the south where there are monsoon rains. Camels (125,000 in 2004) are bred by the inland tribes. Other livestock, 2004: sheep, 355,000; cattle, 315,000; goats, 1m.; chickens, 3·4m.

Fisheries

The catch was 163,927 tonnes in 2010, exclusively sea fish.

INDUSTRY

Apart from oil production, copper smelting and cement production, there are light industries, mainly food processing and chemical products. The government gives priority to import substitute industries.

Labour

Males constituted 83·9% of the economically active population in 2000. In 2003 there were 482,632 employees in the private sector and 123,045 persons in government service. The employment of foreign labour is being discouraged following 'Omanization' regulations of 1994. More than a quarter of the total population are foreign workers. Following the unrest of early 2011 the private sector monthly minimum wage for national workers was increased in Feb. 2011 from RO 140 to RO 200. There is no minimum wage for foreign workers.

INTERNATIONAL TRADE

Oman, along with Bahrain, Kuwait, Qatar, Saudi Arabia and the United Arab Emirates entered into a customs union in Jan. 2003.

Imports and Exports

Imports and exports in US$1m.:

	2003	2004	2005	2006	2007
Imports c.i.f.	6,801	8,796	8,970	11,038	16,025
Exports f.o.b.	12,196	13,381	18,692	21,586	24,692

Main import suppliers, 2006: United Arab Emirates, 25·5%; Japan, 17·1%; India, 5·2%; USA, 5·2%. In 2004 crude oil exports made up approximately 68% of total exports. Main export markets in 2006 were: China, 26·5%; South Korea, 16·0%; Thailand, 12·4%; United Arab Emirates, 9·5%.

COMMUNICATIONS

Roads

A network of adequate graded roads links all the main sectors of population, and only a few mountain villages are not accessible by motor vehicles. In 2005 there were about 42,300 km of roads (16,500 km paved) including 953 km of dual carriageway. In 2007 there were 453,400 passenger cars in use (174 per 1,000 inhabitants), 113,300 vans and lorries, and 26,400 buses and coaches. In 2007 there were 8,816 road accidents and 798 deaths. With 30·7 deaths per 100,000 population in 2007, Oman has among the highest death rates in road accidents of any country.

Civil Aviation
The national airline is Oman Air, which in 2007 had 15 aircraft and served 26 destinations. Oman formerly had a 50% share in Gulf Air with Bahrain, but withdrew in May 2007. In 2009 Seeb International Airport (Muscat) handled 4,556,502 passengers (3,983,413 international) and 64,418 tonnes of freight.

There are plans to expand Oman's two major airports, Seeb International and Salalah (mainly domestic flights).

Shipping
In Mutrah a deep-water port (named Mina Qaboos) was completed in 1974. In 2008 it handled 6·3m. tonnes of foreign cargo. In Jan. 2009 there were seven ships of 300 GT or over registered, totalling 17,000 GT.

Telecommunications
In 2008 there were 274,200 main (fixed) telephone lines in Oman; mobile phone subscribers numbered 3,219,300 in 2008 (115·6 per 100 persons). There were an estimated 557,000 internet users in 2008. In March 2012 there were 422,000 Facebook users.

SOCIAL INSTITUTIONS

Justice
The population in penal institutions in 2002 was 1,403 (61 per 100,000 of national population). The death penalty is in force, but has not been used since 2001.

Education
Adult literacy was 87% in 2008. In 2003–04 there were 1,022 schools, up from just three when Qaboos bin Said became Sultan in 1970. The total number of pupils in state education in 2003–04 was 576,472 (139,082 in basic education and 437,390 in general education) with 32,345 teachers (13,939 in basic education and 18,406 in general education). Oman's first university, the Sultan Qaboos University, opened in 1986 and in 2003–04 there were 12,437 students. It remains the only public university although a number of private universities and colleges have opened in the meantime.

In 2006 public expenditure on education came to 4·2% of GNI and 31·1% of total government spending (the highest percentage of any country).

Health
In 2003 there were 49 hospitals with 4,501 beds. There were also 129 health centres. In 2002 there were 3,478 doctors, 297 dentists, 594 pharmacists and 8,004 nursing staff.

RELIGION

In 2001, 83·5% of the population were Muslim. There were also Hindu and Christian minorities.

CULTURE

World Heritage Sites
The four sites on the UNESCO World Heritage List under Omani jurisdiction are (with the year entered on the list): Bahla Fort (1987); the archaeological sites of Bat, Al-Khutm and Al-Ayn, a collection of settlements and necropolises of the 3rd millennium BC (1988); the Frankincense Trail, a group of archaeological sites representing the production and distribution of frankincense (2000); and the five ancient Aflaj irrigation systems (2006). A fifth site, the Arabian Oryx Sanctuary, was removed from the list after the Omani government decided to reduce the size of the protected area by 90%.

Press
In 2006 there were six daily newspapers with a combined circulation of 165,000.

Tourism
Non-resident tourists staying at hotels and similar establishments numbered 1,276,000 in 2009 (down from 1,378,000 in 2008 although up from 1,182,000 in 2007).

Festivals
National Day (18 Nov.); Spring Festival in Salalah (July–Aug.); Ramadan (July–Aug. in 2013 and June–July in 2014).

DIPLOMATIC REPRESENTATIVES

Of Oman in the United Kingdom (167 Queen's Gate, London, SW7 5HE)
Ambassador: Sheikh Abdulaziz bin Abdullah bin Zahir al-Hinai.

Of the United Kingdom in Oman (PO Box 185, Mina Al Fahal, Postal Code 116, Muscat)
Ambassador: Jamie Bowden.

Of Oman in the USA (2535 Belmont Rd, NW, Washington, D.C., 20008)
Ambassador: Hunaina Sultan Ahmed Al-Mughairi.

Of the USA in Oman (PO Box 202, Medinat Qaboos, Muscat)
Ambassador: Greta C. Holtz.

Of Oman to the United Nations
Ambassador: Lyutha Al-Mughairy.

Of Oman to the European Union
Ambassador: Sheikh Ghazi Bin Said Al-Bahar Al-Rawas.

FURTHER READING

Ghubash, Hussein, *Oman: The Islamic Democratic Tradition.* 2005
Manea, Elham, *Regional Politics in the Gulf: Saudi Arabia, Oman and Yemen.* 2005
Oman. A Country Study. 2004
Owtram, Francis, *A Modern History of Oman: Formation of the State since 1920.* 2002

National Statistical Office: Ministry of National Economy, Information and Documentation Centre, POB 881, Muscat 113.
Website: http://www.mone.gov.om

PAKISTAN

Islami Jamhuriya e Pakistan
(Islamic Republic of Pakistan)

Capital: Islamabad
Population projection, 2015: 189·65m.
GNI per capita, 2011: (PPP$) 2,550
HDI/world rank: 0·504/145
Internet domain extension: .pk

KEY HISTORICAL EVENTS

The Neolithic settlement of Mehrgarh in Balochistan, western Pakistan dates from around 7000 BC. Continuously occupied for over 4,000 years, it was a precursor to the Indus Valley civilization, which flourished between 3300 and 1700 BC. Indus Valley settlements spread across much of present-day Pakistan and northwest India, from the Arabian Sea to the foothills of the Himalayas, centring on the cities of Mohenjo-Daro and Harappa. The civilization's decline coincided with the arrival of Indo-European-speaking tribes, including the Aryans, from central Asia. Taxila in northern Pakistan became a centre for the development of Vedic/Hindu culture from the 6th century BC.

Under the influence of the Persian Achaemenid Empire from around 550 BC, most of present-day Pakistan was ruled by Darius the Great from Persepolis after 515 BC. Alexander the Great, conqueror of the Persian Empire, invaded northern Pakistan in 326 BC but was superseded by the Maurya dynasty from around 300 BC. Its emperor, Ashoka the Great, ruled over central Asia and much of the Indian sub-continent between 273 and 232 BC. The Indus Valley came under Greco-Bactrian control from around 180 BC, bringing about a fusion of classical Greek culture and Buddhism. Invasions by Scythians and Parthians were followed by the arrival of the Yuechi from the steppes of western China. They established the Buddhist Kushan dynasty in the 1st century AD. Centred on Peshawar, it linked the Silk Road with the Arabian Sea and the Ganges Valley.

Arab settlers introduced Islam early in the 8th century. In 712 Muhammad ibn Qasim conquered Sindh province and incorporated it into the Umayyad Caliphate, ruled from Baghdad. Over the next three centuries the southern provinces of Multan and Balochistan were also absorbed. In 1005 Peshawar was conquered by the Turkic-Afghan warlord, Sultan Mahmud of Ghazni, who went on to take Punjab, Kashmir and Balochistan. The Ghaznavid Dynasty made Lahore one of its key cities—the easternmost outpost of Islam—though it was destroyed during Genghis Khan's Mongol invasion of 1219.

In the late 13th century northern and eastern Pakistan came under the influence of Islamic sultanates. Centred on Delhi, they gradually gained control of most of the Indian subcontinent. Timur-i Lang seized Persia, Afghanistan and western Pakistan in the 1380s, absorbing them into a vast central Asian empire. Babur, the founder of the Moghul dynasty in the early 16th century, initially made Kabul his capital, before power transferred to Lahore, Delhi and Agra. While most of the Indian sub-continent remained part of a united Moghul empire between the 16th and 19th centuries, the northwestern fringe was attacked by Persians in the 1730s and Sindh and Punjab were incorporated into Ahmad Shah Durrani's Afghan state from 1747.

The British, who established a protectorate in Bengal in 1757 and subsequently controlled much of the sub-continent, attempted to secure the anarchic northwest against Russian expansion in the first Afghan War (1838–42). Although Sindh and the Punjab were absorbed into British India in the 1840s, Balochistan and Afghanistan remained independent. Following Britain's failure to win these mountainous regions in the second Afghan War (1878–80), the North-West Frontier Province was created in 1901 as a semi-autonomous region.

Hindu–Muslim tensions escalated and in the 1930s the poet Muhammad Iqbal proposed a separate Muslim nation, an idea taken up by Muhammad Ali Jinnah, leader of the Muslim League. Following the League's victories in most of the majority-Muslim constituencies in the 1946 elections, the British agreed to the formation of East and West Pakistan under the 1947 Independence of India Act. Jinnah became West Pakistan's governor-general in the new capital, Karachi.

The partition of India saw 14m. people leaving their homes, with violence claiming 500,000 lives on both sides of the border. The signing over to India of Kashmir in Oct. 1947 was disputed by Pakistan. War broke out until a UN-brokered ceasefire and temporary border was agreed in 1949. In 1951 Pakistan's first prime minister, Liaquat Ali Kahn, was assassinated. The popularity of the Muslim League declined and Pakistan became increasingly unstable.

1958 saw the first of several periods of martial law, followed by the rule of Field Marshal Mohammad Ayub Khan (until 1969) and Gen. Agha Mohammad Yahya Khan (until 1971). Discontent in East Pakistan with the federal government led to calls by the Awami League for full autonomy. Following East Pakistan's declaration of independence as Bangladesh in March 1971, West Pakistan sent in troops, sparking civil war. Hundreds of thousands died and 10m. refugees fled to India. The surrender of Pakistan's forces to the Indian army in Dhaka on 16 Dec. 1971 cleared the way for an independent Bangladesh.

A new constitution in 1973 provided for a federal parliamentary government with a president and prime minister. Zulfiquar Ali Bhutto, representing the Pakistan People's Party (PPP), became premier. Considered by traditionalists to be insufficiently Islamic, in July 1977 Gen. Mohammad Zia ul-Haq led an army coup against Ali Bhutto, who was hanged for conspiracy to murder. His daughter, Benazir, led the PPP to victory in the elections in 1988 but was dismissed on charges of corruption and incompetence in 1990. Reinstated as prime minister following elections in 1993, President Farooq Leghari overthrew her administration in 1996.

Indo–Pakistani relations have foundered over Kashmir. In May 1998 Pakistan carried out five nuclear tests in the deserts of Balochistan in response to India's tests earlier in the month. US President Bill Clinton invoked sanctions but Pakistan carried out a sixth test. On 11 June, following India's example, Pakistan announced a unilateral moratorium on nuclear tests. On 12 Oct. 1999 Gen. Pervez Musharraf seized power in a coup, overthrowing the democratically-elected government of Nawaz Sharif after Sharif tried to dismiss Musharraf as army chief of staff. Sharif was convicted of corruption and sentenced to life imprisonment.

Negotiations with India over Kashmir began in July 1999. In May 2001 India ended its six-month long ceasefire but invited Pakistan for further talks, prompting hopes of avoiding more violence. Following the attacks on New York and Washington of 11 Sept. 2001 Pakistan found itself central to the war against terrorism. Neighbouring Afghanistan was believed to be sheltering Osama bin Laden and the USA persuaded Musharraf to allow its forces to use Pakistani air bases. In return the USA lifted its remaining sanctions. In Dec. 2001 suicide bombers attacked the Indian parliament. Although no-one claimed responsibility, the Indian authorities suspected Kashmiri separatists, heightening Indo-Pakistani tensions. Musharraf's subsequent crackdown on militants helped to reduce tension between the two countries.

An attack on an Indian army base in Indian-occupied Kashmir in May 2002, killing 31, was linked to terrorists infiltrating from Pakistan. Musharraf drew widespread criticism for failing to combat terrorism in the region. In Nov. 2003 Pakistan and India agreed to another ceasefire along Kashmir's Line of Control. Relations gradually improved and in April 2005 bus services resumed across the divided territory for the first time in 60 years.

In Oct. 2005 Pakistan-administered Kashmir was struck by an earthquake which killed 73,300 people and left 3m. homeless. In 2006 there was an upsurge in violence by tribal groups in Balochistan demanding greater autonomy. Bomb blasts rocked Karachi and Islamabad in early 2007 and 68 passengers were killed in an explosion on a train travelling between Lahore and New Delhi. Musharraf's decision to suspend the chief justice of the Supreme Court, Iftikhar Mohammad Chaudhry, for alleged abuse of power in March 2007 sparked riots in Karachi.

In July 2007 security forces stormed Islamabad's Red Mosque, considered to have close links with militant groups. At the end of a week-long siege, over 100 people were killed and 250 injured. In Oct. 2007 Musharraf won the presidential election, though the Supreme Court refused to confirm the result until it had ruled on Musharraf's eligibility to stand while still head of the army. The court dismissed challenges to the result the following month. Meanwhile, fighting intensified in Waziristan, a stronghold of Islamic militant groups.

The assassination of Benazir Bhutto in Dec. 2007 ahead of elections in Jan. 2008 threw the already fragile political climate into turmoil. The election was rescheduled for Feb. when the leading opposition parties, including Bhutto's, won a resounding victory against pro-Musharraf parties and formed a coalition government. Musharraf's presidency was severely weakened as a result and he eventually resigned in Aug. 2008. In Sept. 2008 Asif Ali Zardari was elected his successor. His term in office has seen continued fighting between government forces and Islamic militant groups linked with several high-profile attacks including the Mumbai bombings of Nov. 2008 and an assault on the visiting Sri Lankan cricket team in March 2009. In Oct. 2009 the military launched an offensive against Taliban forces in the South Waziristan region. Since then more than 500 civilians have been killed in terrorist attacks throughout the country.

TERRITORY AND POPULATION

Pakistan is bounded in the west by Iran, northwest by Afghanistan, north by China, east by India and south by the Arabian Sea. The area (excluding the disputed area of Kashmir) is 796,100 sq. km (307,380 sq. miles), including 25,220 sq. km (9,740 sq. miles) of inland water. 2011 provisional census population (excluding three districts of Balochistan, the agency of South Waziristan in the Federally Administered Tribal Areas, and the autonomous states Azad-Kashmir and Gilgit-Baltistan),

192,288,944. In 2011, 36·2% lived in urban areas. There were 1·7m. refugees in 2011, mostly from Afghanistan, the highest number in any country and 17% of the global total.

The UN gives a projected population for 2015 of 189·65m.

The population of the principal cities is as follows:

		1998 census			
Karachi	9,339,023	Multan	1,197,384	Peshawar	982,816
Lahore	5,143,495	Hyderabad	1,166,894	Quetta	565,137
Faisalabad	2,008,861	Gujranwala	1,132,509	Islamabad	529,180
Rawalpindi	1,409,768				

Population of the four provinces and two territories (census of 1998):

	Area (sq. km)	1998 census population (in 1,000)				Density per sq. km
		Total	Male	Female	Urban	
Khyber Pakhtunkhwa[1]	74,521	17,744	9,089	8,655	2,994	238
Federally Administered Tribal Areas[2]	27,220	3,176	1,652	1,524	85	117
Federal Capital Territory Islamabad[2]	906	805	434	371	529	889
Punjab[3]	205,345	73,621	38,094	35,527	23,019	359
Sindh[3]	140,914	30,440	16,098	14,342	14,840	216
Balochistan[3]	347,190	6,566	3,507	3,059	1,569	19

[1]Province—formerly known as North-West Frontier Province. [2]Territory. [3]Province.

English, the official language, is used in business, higher education and in central government; Urdu is the national language and the *lingua franca*, although only spoken as a first language by about 8% of the population. Around 48% of the population speak Punjabi.

SOCIAL STATISTICS

Estimates, 2008: births, 5,324,000; deaths, 1,224,000. Estimated birth rate in 2008 was 30·1 per 1,000 population; estimated death rate, 6·9. Infant mortality (per 1,000 live births), 70 (2010). Formal registration of marriages and divorces has not been required since 1992. Expectation of life in 2007 was 65·9 years for men and 66·5 years for women. Annual population growth rate, 2000–08, 2·2%. Fertility rate, 2008, 4·0 births per woman.

CLIMATE

A weak form of tropical monsoon climate occurs over much of the country, with arid conditions in the north and west, where the wet season is only from Dec. to March. Elsewhere, rain comes mainly in the summer. Summer temperatures are high everywhere, but winters can be cold in the mountainous north. Islamabad, Jan. 50°F (10°C), July 90°F (32·2°C). Annual rainfall 36" (900 mm). Karachi, Jan. 61°F (16·1°C), July 86°F (30°C). Annual rainfall 8" (196 mm). Lahore, Jan. 53°F (11·7°C), July 89°F (31·7°C). Annual rainfall 18" (452 mm). Multan, Jan. 51°F (10·6°C), July 93°F (33·9°C). Annual rainfall 7" (170 mm). Quetta, Jan. 38°F (3·3°C), July 80°F (26·7°C). Annual rainfall 10" (239 mm).

CONSTITUTION AND GOVERNMENT

Under the 1973 constitution, the *President* was elected for a five-year term by a college of parliamentary deputies, senators and members of the Provincial Assemblies. Parliament is bicameral, comprising a *Senate* of 104 members and a *National Assembly* of 342. In the *Senate*, each of the four provinces is allocated 14 seats, while the federally administered tribal areas and the federal capital are assigned eight and two seats respectively. In addition, each province is conferred four seats for technocrats and four for women. Two seats, one for technocrats and another for women, are reserved for the federal capital. The *National Assembly* is directly elected for five-year terms. 272 members are elected in single-seat constituencies, there are ten seats for non-Muslim minorities and 60 seats for women.

Following the 1999 coup Gen. Musharraf announced that the Constitution was to be held 'in abeyance' and issued a 'Provisional Constitution Order No. 1' in its place. In Aug. 2002 he unilaterally amended the constitution to grant himself the right to dissolve parliament.

During the period of martial law from 1977–85 the Constitution was also in abeyance, but not abrogated. In 1985 it was amended to extend the powers of the President, including those of appointing and dismissing ministers and vetoing new legislation until 1990. Legislation of 1 April 1997 abolished the President's right to dissolve parliament, appoint provincial governors and nominate the heads of the armed services.

Gen. Pervez Musharraf, Chief of the Army Staff, assumed the responsibilities of the chief executive of the country following the removal of Prime Minister Nawaz Sharif on 12 Oct. 1999. He formed a National Security Council consisting of six members belonging to the armed forces and a number of civilians with expertise in various fields. A Federal Cabinet of Ministers was also installed working under the guidance of the National Security Council. Also formed was the National Reconstruction Bureau, a think tank providing institutional advice and input on economic, social and institutional matters. The administration declared that it intended to first restore economic order before holding general elections to install a civilian government. The Supreme Court of Pakistan allowed the administration a three-year period, which expired on 12 Oct. 2002, to accomplish this task. Elections were held on 10 Oct. 2002. On 30 April 2002 a referendum was held in which 97·7% voted in favour of extending Musharraf's rule by a further five years. Turnout was around 50%. He amended the constitution in Aug. 2002 to formally extend his mandate by five years. The constitution was further amended in Dec. 2003 to enhance Musharraf's power and allow a vote of confidence in his presidency.

In March 2007 Musharraf announced his intention to stand for a five-year presidential term. The Supreme Court rejected his candidacy while opposition parties challenged his constitutional right to hold the presidency and head the military. In Sept. 2007 the Supreme Court's earlier judgement was overturned and the following month Musharraf was granted a further term of office by the national parliament and provincial assemblies. However, the Supreme Court refused to sanction the appointment until the surrounding legal questions had been resolved. In Nov. 2007 Musharraf suspended the constitution and imposed martial rule. Several prominent members of the judiciary and opposition leaders, including Benazir Bhutto and Imran Khan, were arrested or jailed. The Constitution was reinstated in Dec. but judicial freedom was severely compromised by the dismissal of senior court officials. Amendments were introduced to extend the military's power to try citizens and facilitate the arrest of political opponents. Following Musharraf's resignation in Aug. 2008, ousted Chief Justice Iftikhar Chaudry was reinstated in March 2009. The 2007 amendments to the constitution were subsequently revoked in July 2009 and in Aug. the Supreme Court ruled that Musharraf's actions had been illegal.

National Anthem

'Pak sarzamin shadbad' ('Blessed be the sacred land'); words by Abul Asr Hafeez Jaulandhari, tune by Ahmad G. Chaagla.

GOVERNMENT CHRONOLOGY

Heads of State since 1947. (ML = Muslim League; n/p = non partisan; PML-N = Pakistan Muslim League-Nawaz Sharif; PML-Q = Pakistan Muslim League (Quaid-e-Azam); PPP = Pakistan People's Party; RP = Republican Party)

Governors-General

1947–48	ML	Mohammad Ali Jinnah
1948–51	ML	Khwaja Nazimaddin

1951–55	ML	Ghulam Mohammad
1955–56	military	Iskander Ali Mirza

Presidents of the Republic

1956–58	RP	Iskander Ali Mirza
1958–69	military	Mohammad Ayub Khan
1969–71	military	Agha Mohammad Yahya Khan
1971–73	PPP	Zulfiqar Ali Bhutto
1973–78	PPP	Fazal Elahi Chaudhry
1978–88	military	Mohammad Zia ul-Haq
1988–93	n/p	Ghulam Ishaq Khan
1993–97	PPP	Farooq Ahmed Khan Leghari
1998–2001	PML-N	Mohammad Rafiq Tarar
2001–08	military	Pervez Musharraf
2008	PML-Q	Mohammadmian Soomro (acting)
2008–	PPP	Asif Ali Zardari

RECENT ELECTIONS

Parliamentary elections were held on 18 Feb. 2008 having been delayed following the assassination of Benazir Bhutto, the leader of the Pakistan People's Party (PPP), in Dec. 2007. Turnout was 44·5%. The PPP gained 87 of the National Assembly's 272 elected seats and received 30·6% of votes cast; the Pakistan Muslim League-N (PML-N, led by ex-prime minister Nawaz Sharif) won 67 seats and received 19·6% of the vote; the pro-Musharraf Pakistan Muslim League (Quaid-e-Azam) (PML-Q) gained 42 seats and 23·0% of the vote; and the United National Movement (MQM) 19 seats with 7·4% of the vote. The remaining seats went to smaller parties and non-partisans. After the election, allocation of seats to women and minority representatives was carried out in accordance with the constitution resulting in the PPP having 121 seats, the PML-N 91 seats, the PML-Q 54 and the MQM 25. On 21 Feb. 2008 the PPP, the PML-N and the Awami National Party agreed to form a coalition government; the PPP's Yousaf Raza Gilani was elected prime minister on 24 March 2008.

In indirect elections held on 6 Sept. 2008 Asif Ali Zardari (PPP) was elected president by federal and provincial lawmakers, winning 481 votes against 153 for Saeeduz Zaman Siddique (PML-N) and 44 for Mushahid Hussain Syed (PML-Q).

Parliamentary elections were scheduled to take place on 11 May 2013.

CURRENT GOVERNMENT

President: Asif Ali Zardari; b. 1955 (Pakistan People's Party; since 9 Sept. 2008).

In April 2013 the caretaker government comprised:

Prime Minister, and Minister of Defence, Finance, and Foreign Affairs: Mir Hazar Khan Khoso; b. 1929 (ind., since 25 March 2013).

Minister of Commerce, and Textile Industries: Maqbool H. H. Rahimtoola. *Communication, Ports and Shipping:* Asadullah Khan Mandokhel. *Education and Training, Science and Technology, and Information Technology:* Sania Nishtar. *Housing, and Works:* Younis Soomro. *Human Rights, Law and Justice, and Parliamentary Affairs:* Ahmer Bilal Soofi. *Industries, and Production:* Shahzada Ahsan Ashraf Sheikh. *Information and Broadcasting, and Postal Services:* Arif Nizami. *Inter-provincial Co-ordination, and National Food Security and Research:* Mir Hassan Domki. *Interior, and Narcotics Control:* Malik Mohammad Habib Khan. *National Harmony, National Regulations and Services, National Heritage and Integration, and Religious Affairs:* Shahzada Jamal Nazir. *Overseas Pakistanis:* Feroze Jamal Shah Kakakhel. *Petroleum and Natural Resources:* Sohail Wajahat H. Siddiqui. *Railways:* Abdul Malik Kasi. *Water and Power:* Musadik Malik.

Office of the President: http://www.president.gov.pk

CURRENT LEADERS

Asif Ali Zardari

Position
President

Introduction
Asif Ali Zardari became Pakistan's president in Sept. 2008 following the resignation of Pervez Musharraf. He has been co-chairman of the Pakistan's People's Party (PPP), the largest grouping in the National Assembly, since the assassination of his wife, Benazir Bhutto, in Dec. 2007. A controversial figure, he has struggled to maintain a stable coalition government.

Early Life
Born on 26 July 1955 in Nawabshah, Zardari was brought up and educated in Karachi. He established a career in business and property and in 1983 unsuccessfully contested local elections in Nawabshah. In 1987 Zardari married Benazir Bhutto, then leader-in-exile of the PPP in London. The following year Bhutto became prime minister when the PPP won the general election.

During Bhutto's first term of office Zardari prospered amid rumours of corruption. In 1990 Bhutto was ousted by President Ghulam Ishaq Khan, and Zardari was arrested and jailed on corruption and blackmail charges although never brought to trial. He claimed the action was politically motivated. While in prison Zardari was elected to the National Assembly and in 1993, when Bhutto won a second term as prime minister, the charges against him were dropped. He then served as minister for the environment from 1993–96 and as minister for investment from 1995–96.

In 1996 Bhutto was again removed from office and Zardari was arrested on charges of murder and corruption. He was detained for eight years, during which time he and Bhutto appealed against a Swiss court conviction for money laundering. With Bhutto in exile in Dubai, Zardari continued to deny all charges and claimed political persecution. In 2004 he was freed and the charges against him in Pakistan were dropped. From 2004–07 he lived primarily in the USA, where he received medical treatment for several conditions.

In Nov. 2007 President Musharraf introduced a measure to cancel criminal charges against National Assembly members, clearing the way for Zardari and Bhutto to return to Pakistan. Following Bhutto's assassination during a PPP rally on 27 Dec. 2007, Zardari assumed joint chairmanship of the party with his 19-year-old son, Bilawal. In Feb. 2008 the PPP won the general election in coalition with the Pakistan Muslim League-Nawaz Sharif (PML-N). Zardari and Sharif sought to establish a coalition government and made preparations to impeach Musharraf, who was refusing to relinquish power. Musharraf finally resigned on 18 Aug. 2008 and four days later the PPP nominated Zardari as their presidential candidate. However, on 25 Aug. Sharif took his party out of the coalition, protesting at the concentration of executive power in Zardari's hands and claiming that he had reneged on a promise to reinstate 60 Supreme Court judges. Zardari was elected president on 6 Sept. 2008.

Career in Office
Zardari's tenure has been marked by political tensions, both domestic and with neighbouring states. In March 2009 he defused mounting friction with the Pakistani judiciary and the PML-N by announcing the reinstatement of former chief justice Iftikhar Chaudhry—sacked by Musharraf in 2007 and since championed by Sharif—in response to widespread popular protests. Zardari also pledged to reduce some of the powers invested in the presidency (under constitutional changes which were later approved by parliament in April 2010), and in Nov. 2009 relinquished control of the country's nuclear weapons to the prime minister in an apparent effort to deflect growing

opposition. This gesture, however, was overshadowed the following month as the Supreme Court quashed an earlier legal amnesty protecting Zardari and several political allies from corruption charges. Zardari's opponents renewed calls for his resignation. Prime Minister Gilani's failure to pursue further investigations into Zardari prompted the Court to charge the premier with contempt in Feb. 2012. In June it disqualified Gilani from holding office and he was succeeded by Raja Pervez Ashraf.

Zardari's attempts to improve relations with India have been hampered by allegations of Pakistani involvement in the major terrorist attack in Mumbai in 2008 and further attacks in Mumbai and Delhi in July and Sept. 2011. Furthermore, longstanding tensions over the disputed territory of Kashmir led to border clashes in Jan. 2013. Zardari has also faced an increasing domestic security threat from Islamic extremists, especially in the volatile northwest region of the country. In response to a series of attacks, government forces launched major military offensives against Taliban militants from April 2009 in the Swat valley and from Oct. 2009 in the tribal area of South Waziristan. However, suicide bombings by militants continued unabated in many of Pakistan's major cities, further undermining Zardari's authority.

In July–Aug. 2010 Zardari was criticized for his government's response to a humanitarian crisis following flooding that devastated large areas of the country. More than 1,600 people were killed and 20m. displaced by the deluge.

Relations with the USA became increasingly strained in 2010 and 2011. Cross-border US air strikes in anti-Taliban operations led Pakistan to temporarily suspend NATO's supply routes into Afghanistan in Sept. 2010 and again in Nov. 2011. Meanwhile, Osama bin Laden, founder of the al-Qaeda militant network, was killed by US special forces in Abbottabad in Pakistan in May 2011. The Pakistani security establishment's apparent ignorance about Bin Laden's presence in the city, and the US government's failure to notify the Pakistani authorities about its military intentions, caused further friction between the two uneasy allies.

Concurrently, there were increasing tensions between the Zardari government and Pakistan's powerful military leadership. In Dec. 2011 Zardari underwent hospital treatment in Dubai, triggering rumours of his resignation under coercion from the army, and in Jan. 2012 a scandal over a leaked memo alleging that senior government officials had sought US aid against a possible military coup further soured the political atmosphere.

In March 2013 parliament dissolved itself ahead of elections scheduled for 11 May. In so doing, it became the first parliament in Pakistan's history to complete a full five-year term.

DEFENCE

A *Council for Defence and National Security* was set up in Jan. 1997, comprising the President, the Prime Minister, the Ministers of Defence, Foreign Affairs, Interior, Finance and the military chiefs of staff. The Council advised the government on the determination of national strategy and security priorities, but was disbanded in Feb. 1997. The Council was revived in Oct. 1999 following the change of government but was to have a wider scope and not restrict itself to defence matters.

Defence expenditure in 2008 totalled US$4,442m. (US$26 per capita), representing 3·0% of GDP. However, following the return of civilian government in March 2008, the 2008–09 defence budget was reduced. Expenditure is also increasingly targeted at developing counter-insurgency capabilities as a result of the threat from Taliban forces along the border with Afghanistan.

As at 31 Jan. 2013 Pakistan had 8,216 personnel serving in UN peacekeeping operations (the second largest contingent of any country, after Bangladesh).

Nuclear Weapons
Pakistan began a secret weapons programme in 1972 to reach parity with India, but was restricted for some years by US sanctions. The Stockholm International Peace Research Institute

estimates that Pakistan possesses 90–110 nuclear warheads. In May 1998 Pakistan carried out six nuclear tests in response to India's tests earlier in the month. Pakistan, known to have a nuclear weapons programme, has not signed the Comprehensive Nuclear-Test-Ban-Treaty, which is intended to bring about a ban on any nuclear explosions. According to *Deadly Arsenals*, published by the Carnegie Endowment for International Peace, Pakistan has both chemical and biological weapon research programmes.

Army

Strength (2007) 550,000. There were also about 304,000 personnel in paramilitary units: National Guard, Frontier Corps and Pakistan Rangers.

Most armoured equipment is of Chinese origin including over 2,450 main battle tanks. There is an air wing with fixed-wing aircraft and 26 attack helicopters.

Navy

The combatant fleet includes five French-built diesel submarines, three midget submarines for swimmer delivery and six ex-British frigates. The Naval Air wing operates 16 combat-capable aircraft.

The principal naval base and dockyard are at Karachi. There are secondary bases at Gwadar and Ormara. Naval personnel in 2007 totalled 24,000 (including an estimated 1,400 marines and some 2,000 Marine Security Agency personnel).

Air Force

The Pakistan Air Force came into being on 14 Aug. 1947. It has its headquarters at Peshawar and is organized within three air defence sectors, in the northern, central and southern areas of the country. There is an Air Force Academy at Risalpur, which includes a College of Aeronautical Engineering.

Total strength in 2007 was 360 combat-capable aircraft and 45,000 personnel. Equipment included Mirage IIIs, Mirage 5s, F-16s, Q-5s and J-7s.

INTERNATIONAL RELATIONS

Following Gen. Musharraf's coup in Oct. 1999, Pakistan was suspended from the Commonwealth's councils although the suspension was ended in May 2004. Pakistan was again suspended in Nov. 2007 after then President Musharraf declared emergency rule but it was readmitted in May 2008.

ECONOMY

Agriculture accounted for 21·2% of GDP in 2010, industry 25·4% and services 53·4%.

Overview

Though growth in the agricultural sector has been falling since the 1980s, it contributes over 21% of GDP and employs 45% of the labour force. Pakistan is one of the world's largest producers of raw cotton but historically textile exports have added little value. Cotton accounted for 8·6% of the value added in agriculture and about 1·8% of GDP in 2009. Chiefly grown in Punjab province, cotton is crucial to the success of the yarn-spinning industry concentrated around Karachi.

From 2004–07 the economy grew by over 5% per year, driven by the industrial and service sectors. Broad-based growth was supported by sound macroeconomic management and progress in implementing structural reforms and privatization. The Oct. 2005 earthquake in northern Pakistan had little impact on Pakistan's total production. However, the cost of reconstruction put major strain on the public finances and the current account balance.

In 2007–08 international oil and food prices rose sharply, with negative effects. A stabilization programme supported by the IMF was initiated in Nov. 2008 to avoid default on debt payments. Severe flooding in July 2010 affected more than 20m. people and disrupted 500,000 ha. of cropped land. A US$451m. emergency assistance package was provided by the IMF. The World Bank estimated the damage at US$9·7bn. and committed US$1bn. towards recovery. In 2011 more floods affected over 6m. people.

Public debt was reduced from 75·1% of GDP in 2001 to 60·1% in 2011. Poverty levels fell from 34·5% in 2002 to 22·3% in 2006. Inflation is persistently high, averaging 10·6% between 2003 and 2012. Pakistan requires an annual average growth rate of 7% to absorb new entrants into the labour force, more than double the growth achieved in 2011.

Political instability, structural problems in the energy sector, poor infrastructure and a volatile security situation pose additional challenges.

Currency

The monetary unit is the *Pakistan rupee* (PKR) of 100 *paisas*. Gold reserves in Sept. 2009 were 2·10m. troy oz; foreign exchange reserves, US$10,418m. Inflation rates (based on IMF statistics):

2002	2003	2004	2005	2006	2007	2008	2009	2010	2011
2·4%	3·2%	4·0%	9·3%	8·0%	7·8%	10·8%	17·6%	10·1%	13·6%

The rupee was devalued by 3·65% in Sept. 1996, 8·5% in Oct. 1996 and 8·7% in Oct. 1997, and by 4·2% in June 1998 in response to the financial problems in Asia. In June 2008 total money supply was Rs3,359·3bn.

Budget

The financial year ends on 30 June. Budgetary central government revenues totalled Rs1,803·2bn. in 2008–09; expenditures totalled Rs2,137·4bn. in 2008–09. Taxes accounted for 65·5% of revenues in 2008–09; interest accounted for 35·2% of expenditures, and goods and services 21·8%.

There is a general sales tax of 16%.

Performance

Real GDP growth rates (based on IMF statistics):

2002	2003	2004	2005	2006	2007	2008	2009	2010	2011
3·1%	4·7%	7·5%	9·0%	5·8%	6·8%	3·7%	1·7%	3·1%	3·0%

Pakistan's total GDP in 2011 was US$210·2bn.

Banking and Finance

The State Bank of Pakistan is the central bank (*Governor*, Yaseen Anwar); it came into operation as the Central Bank on 1 July 1948 and was nationalized in 1974 with other banks. Private commercial bank licences were reintroduced in 1991.

The State Bank of Pakistan is the issuing authority of domestic currency, custodian of foreign exchange reserves and bankers for the federal and provincial governments and for scheduled banks. It also manages the rupee public debt of the federal and provincial governments. The National Bank of Pakistan acts as an agent of the State Bank where the State Bank has no offices of its own.

In Feb. 1994 the State Bank of Pakistan was granted more autonomy to regulate the monetary sector of the economy.

In Dec. 1999 the Supreme Court ruled that Islamic banking methods, whereby interest is not permitted, had to be used from 1 July 2001. However, the decision was rescinded in June 2002. The State Bank offered three options for the implementation of Islamic banking practices: i) banks to establish an independent Islamic bank; ii) the opening of subsidiaries of existing commercial banks; iii) the establishment of new branches to execute Islamic banking procedures.

In Sept. 2003 total assets of public sector commercial banks amounted to Rs980,300m., total assets of local private banks to Rs1,122,400m., total assets of foreign banks to Rs276,900m. and total assets of all commercial banks amounted to Rs2,379,600m. In Dec. 2005 total deposits of scheduled banks (stocks) equalled Rs2,661,697m. and net foreign assets amounted to Rs523,044m.

There were 37 commercial banks in Sept. 2003 (five state-owned and 15 foreign) with assets Rs2,380bn. In 2002 there were 45 leasing banks, operating in accordance with Sharia demands. There is a Federal Bank for Co-operatives.

In 2010 external debt totalled US$1,127m., equivalent to 31·3% of GNI.

Foreign direct investment was US$56,773m. in 2010, down from US$2,338m. in 2009 and US$5,438m. in 2008. The total stock of FDI at the end of 2010 was US$21·49bn.

There are stock exchanges in Islamabad, Karachi and Lahore.

ENERGY AND NATURAL RESOURCES

Environment
Pakistan's carbon dioxide emissions from the consumption and flaring of fossil fuels were the equivalent of 0·9 tonnes per capita in 2008.

Electricity
Installed capacity in 2006–07 was 19·42m. kW, of which 12·48m. kW was thermal, 6·48m. kW was hydro-electric and 0·46m. kW was nuclear. In 2010 there were two nuclear reactors in use. Production in 2006–07 was 95·66bn. kWh, of which 68% was thermal and 14% was hydro-electric. Power shortages are frequent. Consumption per capita in 2006–07 was 601 kWh.

Oil and Gas
Crude petroleum production in 2006–07 was 3·4m. tonnes. Reserves in 2007 were 289m. bbls. Exploitation is mainly through government incentives and concessions to foreign private sector companies. Natural gas production in 2008 was 37·5bn. cu. metres with 850bn. cu. metres of proven reserves. The French oil company Total agreed a US$3bn. deal with the government in July 2003 for exploration in the Arabian Sea.

Water
Pakistan's Indus Basin irrigation system is the largest and oldest in the world. It includes a network of 43 independent canal systems and two storage reservoirs. Total length of main canals is 58,000 km which serve 35m. acres of cultivatable land.

Currently three major surface water projects are under way, as are flood control schemes and programmes to check the problems of waterlogging and salinity.

Minerals
Production (tonnes, 2005): limestone, 14·86m.; coal, 3·37m.; rock salt, 1·65m.; gypsum, 552,496; fire clay, 253,501; dolomite, 199,653; chromite, 46,359; barytes, 42,087; china clay, 37,732; fuller's earth, 17,001; bauxite, 6,504. Other minerals of which useful deposits have been found are copper, magnesite, sulphur, marble, antimony ore, bentonite, celestite, fluorite, phosphate rock, silica sand and soapstone.

Agriculture
The north and west are covered by mountain ranges. The rest of the country consists of a fertile plain watered by five big rivers and their tributaries. Agriculture is dependent almost entirely on the irrigation system based on these rivers. Area irrigated, 2002, about 17·80m. ha. Agriculture employs around half of the workforce. In 2002 there were an estimated 21·45m. ha. of arable land and 672,000 ha. of permanent crops.

Pakistan is self-sufficient in wheat, rice and sugar. Areas harvested, 2003: wheat, 8·07m. ha.; seed cotton, 3·00m. ha.; rice, 2·21m. ha.; chick-peas, 1·68m. ha.; sugarcane, 1·09m. ha.; maize, 0·88m. ha. Production, 2003 (1,000 tonnes): sugarcane, 52,056; wheat, 19,210; rice, 6,751; seed cotton, 5,071; cottonseed, 3,337; potatoes, 1,946; cotton lint, 1,690; onions, 1,400; oranges, 1,400; maize, 1,275; mangoes, 1,036; chick-peas, 672; dates, 650.

A Land Reforms Act of 1977 reduced the upper limit of land holding to 100 irrigated or 200 non-irrigated acres. A new agricultural income tax was introduced in 1995, from which

holders of up to 25 irrigated or 50 unirrigated acres are exempt. Of about 5m. farms, 12% are of less than 10 ha.

Livestock, 2003 (in 1m.): goats, 52·8; buffaloes, 24·8; sheep, 24·6; cattle, 23·3; asses, 4·1; camels, 0·8; chickens, 155·0.

Livestock products, 2003 (in 1,000 tonnes): beef and veal, 445; poultry meat, 375; goat meat, 373; mutton and lamb, 174; buffalo's milk, 18,520; cow's milk, 8,620; goat's milk, 640; eggs, 354; wool, 40.

Forestry
The area under forests in 2010 was 1·69m. ha., some 2% of the total land area. Timber production in 2007 totalled 29·22m. cu. metres.

Fisheries
In 2010 the catch totalled 453,264 tonnes, mainly from marine waters. Exports of fishery commodities were valued at US$193m. in 2008.

INDUSTRY
The leading company by market capitalization in March 2012 was Oil & Gas Development (US$8·0bn.).

Industry is based largely on agricultural processing, with engineering and electronics. Government policy is to encourage private industry, particularly small businesses. The public sector, however, is still dominant in large industries. Steel, cement, fertilizer and vegetable ghee are the most valuable public sector industries.

Production in 1,000 tonnes (in 2006–07 unless otherwise stated): cement (2007 estimate), 21,000; raw sugar (2007), 4,355; distillate fuel oil, 3,697; residual fuel oil, 3,324; cotton yarn (2005–06), 2,547; petrol, 1,337; paper and paperboard (2007), 1,010; pig iron (2005–06), 768; clays, 358; coke, 326; soda ash (2005), 260; caustic soda, 242; woven cotton fabrics, 965m. sq. metres; bicycles, 486,350 items; tractors (2007), 54,610 items.

Labour
Out of 45·29m. economically active people in 2005, 37·81m. were males. The rate of unemployment in 2005 was 6·8%. In 2005 a total of 17·18m. persons were engaged in agriculture, forestry and fishing, 6·67m. in manufacturing, 6·50m. in community, social and personal services and 6·29m. in wholesale and retail trade, restaurants and hotels.

In 2001 there were four industrial disputes and 7,078 working days were lost.

INTERNATIONAL TRADE
Most foreign exchange controls were removed in Feb. 1991. Tax exemptions are available for companies set up before 30 June 1995.

Imports and Exports
Trade in US$1m.:

	2005	2006	2007	2008	2009
Imports c.i.f.	25,097	29,826	32,594	42,327	31,584
Exports f.o.b.	16,050	16,933	17,838	20,279	17,555

Major imports in 2009 (as % of total value): petroleum and petroleum products, 26·2; machinery and transport equipment, 22·9; chemicals and related products, 16·9; manufactured goods, 10·6. Major exports in 2009: textile yarn, fabrics and finished articles, 37·1; clothing and apparel, 19·1; rice, 10·1; petroleum and petroleum products, 4·1.

Major import suppliers in 2009 (as % of total): China, 12·0; Saudi Arabia, 11·1; UAE, 10·6; Kuwait 5·7; USA, 5·7. Major export markets in 2009: USA, 18·3; UAE, 8·8; Afghanistan; 7·8; China, 5·7; UK, 5·4.

COMMUNICATIONS

Roads

In 2006 there were 260,420 km of roads, of which 65·4% were paved. There are ten motorways providing links between Pakistan's major cities. These include the M-1 from Islamabad to Peshawar, the M-2 from Islamabad to Lahore, the M-4 from Faisalabad to Multan and the M-9 from Karachi to Hyderabad. In 2007 there were 1,440,100 passenger cars in use, 187,100 vans and lorries, 170,400 buses and coaches and 2,684,300 motorcycles. There were 10,466 road accidents involving injury in 2007, with 5,465 fatalities.

All traffic in Pakistan drives on the left. All cars must be insured and registered. Minimum age for driving: 18 years.

Rail

In 2005 Pakistan Railways had a route length of 7,791 km (of which 305 km electrified) mainly on 1,676 mm gauge, with some metre gauge line. Passenger-km travelled in 2004–05 came to 24·2bn. and freight tonne-km to 5·0bn.

Civil Aviation

There are international airports at Karachi, Islamabad, Lahore, Peshawar and Quetta.

The national carrier is the state-owned Pakistan International Airlines, or PIA. It operates scheduled services to 46 international and 24 domestic destinations. In 2006, 88,302,000 revenue-km were flown. The revenue passengers carried totalled 5·73m. in 2006 and revenue tonne-km came to 1,801m. Operating revenues of the corporation stood at Rs70,587m. in 2006 and operating expenditure at Rs79,164m.

Shipping

In Jan. 2009 there were 16 ships of 300 GT or over registered, totalling 384,000 GT. The busiest port is Karachi. In 2008–09 cargo traffic totalled a record 38,732,000 tonnes (13,364,000 tonnes loaded and 25,368,000 tonnes discharged). In 2008–09, 2,386 vessels were handled at the port of Karachi. There are also ports at Port Qasim, which handled 25,023,000 tonnes in 2008–09, and Gwadar.

Telecommunications

The telephone system is government-owned. In 2008 there were 4,416,000 main (fixed) telephone lines. In the same year mobile phone subscribers numbered 88,020,000 (497·4 per 1,000 persons). There were 18·5m. internet users in 2008. In March 2012 there were 6·4m. Facebook users.

SOCIAL INSTITUTIONS

Justice

The Federal Judiciary consists of the Supreme Court of Pakistan, which is a court of record and has three-fold jurisdiction; original, appellate and advisory. There are four High Courts in Lahore, Peshawar, Quetta and Karachi. Under the Constitution, each has power to issue directions of writs of *Habeas Corpus, Mandamus, Certiorari* and others. Under them are district and sessions courts of first instance in each district; they have also some appellate jurisdiction. Below these are subordinate courts and village courts for civil matters and magistrates for criminal matters.

The Constitution provides for an independent judiciary, as the greatest safeguard of citizens' rights. There is an Attorney-General, appointed by the President, who has right of audience in all courts and the Parliament, and a Federal Ombudsman.

A Federal Sharia Court at the High Court level has been established to decide whether any law is wholly or partially un-Islamic. In Aug. 1990 a presidential ordinance decreed that the criminal code must conform to Islamic law (Sharia), and in May 1991 parliament passed a law incorporating it into the legal system.

538,048 crimes were reported in 2007 (399,558 in 2002). Execution of the death penalty for murder, in abeyance since 1986, was resumed in 1992. The first judicial execution in four years was carried out in Nov. 2012, although there have been frequent reports of extrajudicial executions since 2008. There were 10,556 murders in 2007 (9,396 in 2002). The population in penal institutions in 2007 was 90,000 (55 per 100,000 of national population).

Education

In 1998 the landmark National Education Policy (1998–2010) was launched with the aim of eradicating illiteracy and spreading basic education. The policy stressed vocational and technical education, disseminating a common culture based on Islamic ideology. The follow-up National Education Policy 2009 reinforced the position that education and training 'should enable the citizens of Pakistan to lead their lives according to the teachings of Islam'. In 2010 the constitution was amended by the addition of an article declaring: 'The state shall provide free and compulsory education to all children of the age of five to sixteen in such a manner as may be determined by law.' The adult literacy rate in 2008 was 54%. Adult literacy programmes are being strengthened.

In 2007 there were 17·98m. primary schools pupils (450,027 teaching staff), 9·15m. secondary school pupils (197,082 teaching staff in 2004) and 955,000 students in tertiary education (52,245 academic staff). There are 70 public and 58 private universities of which Quaid-i-Azam University in Islamabad and the University of the Punjab in Lahore are considered to be the most prestigious.

Public expenditure on education came to 2·8% of GNI and 11·2% of total government spending in 2007.

Health

In 2002 there were 906 hospitals and 4,590 dispensaries (with a total of 98,264 beds) and 862 maternity and child welfare centres. There were 126,350 doctors, 70,698 nurses and midwives and 15,790 dentists in 2005.

Welfare

The official retirement age is 60 (men), 55 (women) or 50 (miners). To qualify for a pension, 15 years of contributions are needed. The minimum old age and survivor pension is Rs3,000 per month (as of 2010).

Medical services, provided mainly through social security facilities, cover cash and medical benefits such as general medical care, specialist care, medicines, hospitalization, maternity care and transportation.

RELIGION

Pakistan was created as a Muslim state. The Muslims are mainly Sunni, with an admixture of 15–20% Shia. Religious groups: Muslims, 93%; Christians, 2%; Hindus, Parsees, Buddhists, Qadianis and others. Pakistan has the second highest number of Muslims, after Indonesia. There is a Minorities Wing at the Religious Affairs Ministry to safeguard the constitutional rights of religious minorities.

CULTURE

There is a Pakistan National Council of the Arts, a cultural organization to promote art and culture in Pakistan and abroad.

World Heritage Sites

There are six sites under Pakistani jurisdiction which appear on the UNESCO World Heritage List. They are (with year entered on list): the archaeological ruins at Moenjodaro (1980), Taxila (1980), the Buddhist ruins at Tahkt-i-Bahi and the neighbouring city remains at Sahr-i-Bahlol (1980), Thatta (1981), the Fort and Shalamar Gardens in Lahore (1981) and Rohtas Fort (1997).

Press

In 2007 there were 400 paid-for dailies and 1,200 paid-for non-daily periodicals. Average combined circulation of all dailies in 2007 was 9,935,000. The most popular daily papers in 2008 were

Jang, with a circulation of 450,000, and *Express,* with a circulation of 375,000. The most widely read English-language paper is *Dawn,* with an average daily circulation of 225,000 copies in 2008.

Tourism

In 2010 there were 906,800 non-resident tourists including 288,200 from the UK, 120,400 from the USA, 110,900 from Afghanistan, 46,200 from Canada and 43,700 from India. 54% of tourists in 2010 visited Punjab and 29% Sindh.

Festivals

Pakistan is rich in culture. Famous festivals include the Eid Festival, Eid-e-Milad un Nabi (Birthday of Prophet Muhammad), the Basnat Festival, Shab-e-Baraat Festival and the Independence Day Festival.

DIPLOMATIC REPRESENTATIVES

Of Pakistan in the United Kingdom (35–36 Lowndes Sq., London, SW1X 9JN)
High Commissioner: Wajid Shamsul Hasan.

Of the United Kingdom in Pakistan (Diplomatic Enclave, Ramna 5, Islamabad)
High Commissioner: Adam Thomson, CMG.

Of Pakistan in the USA (3517 International Court, NW, Washington, D.C., 20008)
Ambassador: Sherry Rehman.

Of the USA in Pakistan (Diplomatic Enclave, Ramna, 5, Islamabad)
Ambassador: Richard G. Olson.

Of Pakistan to the United Nations
Ambassador: Masood Khan.

Of Pakistan to the European Union
Ambassador: Munawar Saeed Bhatti.

FURTHER READING

Federal Bureau of Statistics.—*Pakistan Statistical Yearbook.*—*Statistical Pocket Book of Pakistan.* (Annual)

Ahmed, A. S., *Jinnah, Pakistan and Islamic Identity: The Search for Saladin.* 1997

Ahsan, A., *The Indus Saga and the Making of Pakistan.* 1997

Akhtar, R., *Pakistan Year Book.*

Cohen, Stephen Philip, *The Idea of Pakistan.* 2006

Jaffrelot, Christophe, (ed.) *A History of Pakistan and Its Origins.* Revised ed. 2004

Jetly, Rajshree, *Pakistan in Regional and Global Politics.* 2009

Joshi, V. T., *Pakistan: Zia to Benazir.* 1995

Khan, Hamid, *Constitutional and Political History of Pakistan.* 2005

Lieven, Anatol, *Pakistan: A Hard Country.* 2011

Lodhi, Maleeha, *Pakistan: Beyond 'The Crisis State'.* 2011

Malik, Iftikhar H., *State and Civil Society in Pakistan: the Politics of Authority, Ideology and Ethnicity.* 1996.—*The History of Pakistan.* 2008

Paul, T. V., *The India-Pakistan Conflict: An Enduring Rivalry.* 2005

Rashid, Ahmed, *Pakistan on the Brink: The Future of Pakistan, Afghanistan and the West.* 2012

Riedel, Bruce, *Deadly Embrace: Pakistan, America, and the Future of Global Jihad.* 2011

Talbot, Ian, *Pakistan: A Modern History.* 1999

Wynbrandt, James, *A Brief History of Pakistan.* 2009

Zaidi, S. Akbar, *Issues in Pakistan's Economy.* 2005

Ziring, Lawrence, *Pakistan in the 20th Century: A Political History.* 2004

National library: National Library of Pakistan, Constitution Avenue, Islamabad.

National Statistical Office: Federal Bureau of Statistics, 5-SLIC Bldg., F-6/4, Blue Area, Islamabad.

Website: http://www.pbs.gov.pk

PALAU

Beluu er a Belau
(Republic of Palau)

Capital: Melekeok
Population, 2006: 22,000
GNI per capita, 2010: US$10,073
HDI/world rank: 0·782/49
Internet domain extension: .pw

KEY HISTORICAL EVENTS

Spain acquired sovereignty over the Palau Islands in 1886 but sold the archipelago to Germany in 1899. Japan occupied the islands in 1914 and in 1921 they were mandated to Japan by the League of Nations. Captured by Allied Forces in 1944, the islands became part of the UN Trust Territory of the Pacific Islands created on 18 July 1947 and administered by the USA. Following a referendum in July 1978 in which Palauans voted against joining the new Federated States of Micronesia, the islands became autonomous from 1 Jan. 1981. A referendum in Nov. 1993 favoured a Compact of Free Association with the USA. Palau became an independent republic on 1 Oct. 1994.

TERRITORY AND POPULATION

The archipelago lies in the western Pacific and has a total land area of 488 sq. km (188 sq. miles). It comprises 26 islands and over 300 islets. Only nine of the islands are inhabited, the largest being Babelthuap (396 sq. km), but most inhabitants live on the small island of Koror (18 sq. km) to the south. In Oct. 2006 the capital moved from Koror to Melekeok, a newly-built town in eastern Babelthuap. The total population of Palau at the time of the 2005 census was 19,907 (10,699 males and 9,208 females), giving a density of 40·8 per sq. km. 2006 estimate: 21,669. Koror's population according to the 2005 census was 12,676. In 2000 approximately 70% of the population were Palauans.

In 2011, 84·3% of the population lived in urban areas. Some 6,000 Palauans live abroad. The local language is Palauan; both Palauan and English are official languages.

SOCIAL STATISTICS

2006 births, 259; deaths, 144. Rates, 2006 (per 1,000 population): births, 12·0; deaths, 6·6; infant mortality (2010), 15 per 1,000 live births. Annual population growth rate, 1998–2008, 1·0%. Expectation of life: males, 69 years; females, 73. Fertility rate, 2008, 1·9 births per woman.

CLIMATE

Palau has a pleasantly warm climate throughout the year with temperatures averaging 81°F (27°C). The heaviest rainfall is between July and Oct.

CONSTITUTION AND GOVERNMENT

The Constitution was adopted on 2 April 1979 and took effect from 1 Jan. 1981. The Republic has a bicameral legislature, the *Olbiil Era Kelulau* (National Congress), comprising a 13-member *Senate* and a 16-member *House of Delegates* (one from each of the Republic's 16 states), both elected for a term of four years as are the *President* and *Vice-President*. Customary social roles and land and sea rights are allocated by a matriarchal 16-clan system.

National Anthem

'Belau loba klisiich er a kelulul' ('Palau is coming forth with strength and power'); words anonymous, tune Y. O. Ezekiel.

RECENT ELECTIONS

At the elections on 6 Nov. 2012 Tommy Remengesau, Jr was elected president with 58·0% of votes cast against 42·0% for incumbent president Johnson Toribiong. At the legislative elections that were also held on 6 Nov. 2012 only non-partisans were elected.

CURRENT GOVERNMENT

President: Tommy Remengesau, Jr; b. 1956 (in office since 17 Jan. 2013, having previously been president from Jan. 2001–Jan. 2009).

  Vice-President and Minister of Finance: Kerai Mariur.

  In March 2013 the cabinet consisted of:

  Minister of Community and Cultural Affairs: Faustina Rehuher-Marugg. *Education:* Masa-Aki Emesiochl. *Health:* Dr Stevenson Kuartei. *Justice:* Johnny Gibbons. *Natural Resources, Environment and Tourism:* Harry Fritz. *Public Infrastructure, Industries and Commerce:* Jackson Ngiraingas. *Minister of State for Foreign Affairs:* Dr Victor Yano.

Government Website: http://www.palaugov.net

CURRENT LEADERS

Tommy Remengesau, Jr

Position
President

Introduction
Tommy Remengesau, Jr first became president in Jan. 2001 and was re-elected in Nov. 2004. Out of office from 2009, he won a third term in Nov. 2012. The US-educated career politician has prioritized improvements in basic services and is expected to seek higher levels of foreign investment.

Early Life
Tommy Esang Remengesau, Jr was born on 28 Feb. 1956 on the island of Koror, Palau, in the US-administered Trust Territory of the Pacific Islands (TTPI). The eldest son of Thomas O. Remengesau, Sr (who served as president from 1988–89), he graduated in criminal justice from Grand Valley State College in Michigan, USA in 1978. On his return to Palau, Remengesau

began work at the *Olbiil Era Kelulau* (OEK), Palau's National Congress. In 1984 he became the youngest Palauan to be elected a senator in the OEK. Re-elected in 1988, he served on the committee on ways and means, playing a key role in reducing Palau's budget deficit and securing financial stability.

In 1992 Remengesau was elected vice-president and the following year Palauan voters approved a Compact of Free Association with the USA. With responsibility for the finance portfolio, Remengesau was credited with reforming the financial system and preparing the new sovereign state (established in 1994) for membership of the IMF and the World Bank Group. He won the 2000 presidential elections, claiming 52% of the vote against Peter Sugiyama.

Career in Office
Remengesau set out to reduce his country's dependence on the USA and announced plans to increase revenue from tourism. He pledged to achieve these aims while protecting the country's natural environment, particularly the coral reef, and preserving the island's resources. In 2006 he launched the Micronesia Challenge, which encouraged neighbouring countries to follow Palau's conservation commitments.

Remengesau began a new four-year term following the election of Nov. 2004, in which he claimed 66·5% of the vote against Polycarp Basilius. In Oct. 2006 the institutions of government began relocating to a new capital, Melekeok, in eastern Babelthuap. In 2008 Remengesau announced he would seek a seat in the Senate, having held the presidency for the maximum two consecutive terms allowed by the constitution. However, his campaign was unsuccessful and he came 11th in the election. He was succeeded as president by Johnson Toribiong in Jan. 2009.

In April 2009 an investigation found that Remengesau was guilty of incorrectly filing the financial details of properties in which he had an interest. He was charged with 19 counts of violating Palau's code of ethics in 2002 and 2003, accusations he described as 'selective persecution'. In April 2010 Associate Justice Kathleen Salii fined Remengesau US$156,400, equivalent to an eighth of the amount initially recommended by prosecutor Michael Copeland.

Remengesau was elected to a further four-year term as president in Nov. 2012, having defeated Toribiong on a platform of encouraging foreign investment and improving services for grassroots communities. He took office on 17 Jan. 2013.

ECONOMY

Overview
Following independence in 1994, Palau has been highly dependent on foreign aid. The USA provides assistance under a Compact of Free Association, giving Washington responsibility for Palau's defence and security matters. Grant assistance under the Compact, guaranteed until 2024, increased in 2010 ahead of a gradual decline.

Tourism, which accounts for nearly half of GDP, has been hit by the global downturn and, specifically, financial difficulties in the Taiwanese air industry (with Taipei being a key hub for incoming tourists). In the medium term the government looks to reduce dependence on foreign grants and to fund current fiscal spending from domestic revenues. Such adjustment will require comprehensive tax and fiscal reforms, while authorities are looking to generate new sources of revenue through liberalization of the FDI regime and offshore exploration of oil and gas.

Currency
US currency is used.

Budget
The fiscal year begins on 1 Oct. Revenues for 2008–09 were US$81·3m. and expenditures US$95·0m. Grants accounted for 53·5% of revenues in 2008–09 and tax revenue 36·5%; current expenditure accounted for 77·6% of expenditures in 2008–09.

Performance
Total GDP in 2011 was US$166m. The economy contracted by 4·9% in 2008 and by 2·1% in 2009.

Banking and Finance
The National Development Bank of Palau is situated in Koror. Other banks include the Bank of Guam, the Bank of Hawaii, Bank Pacific, Melekeok Government Bank and the Pacific Savings Bank.

ENERGY AND NATURAL RESOURCES

Electricity
Electricity production was approximately 154m. kWh in 2007; installed capacity was about 52,000 kW in 2007.

Agriculture
Subsistence farming is one of the major economic activities. In 2007 approximately 2,000 people were economically active in agriculture. The main agricultural products are bananas, coconuts, copra, cassava and sweet potatoes. There were about 1,000 ha. of arable land in 2007 and 2,000 ha. of permanent crop land.

Forestry
Forests covered 0·04m. ha. in 2010, or 88% of the land area.

Fisheries
In 2008 the catch totalled 1,007 tonnes, mainly tuna.

INDUSTRY
There is little industry, but the principal activities are food-processing and boat-building.

Labour
In 2005 the total labour force numbered 10,203 (6,214 males and 3,989 females), of whom 9,777 were employed (5,982 males and 3,795 females).

INTERNATIONAL TRADE

Imports and Exports
Imports (2006–07) US$91·3m.; exports (2006–07) US$10·1m. The main trading partner is Japan for exports and the USA for imports.

COMMUNICATIONS

Roads
There were 146 km of roads in 2007 including the 85-km US-funded two-lane highway around Babelthuap, providing a link between the old capital of Koror and the new capital of Melekeok.

Civil Aviation
The main airport is on Koror (Roman Tmetuchl International Airport, near Airai). In 2010 there were scheduled flights to Guam, Manila, Seoul, Taipei and Yap (Micronesia). A new Palau-based carrier, Palau Airways, was founded in 2011 and launched scheduled passenger services between Koror and Taipei in May 2012.

Shipping
There is a port at Malakal.

Telecommunications
In 2008 there were 7,400 main (fixed) telephone lines and 12,200 mobile phone subscribers.

SOCIAL INSTITUTIONS

Justice
There is a Supreme Court and various subsidiary courts. The population in penal institutions in 2010 was 79 (378 per 100,000 national population).

Education
In 2007 there were 1,544 pupils at primary schools and 2,448 at secondary schools. In 2006 there were 21 primary schools (19

public and two private) and six secondary schools (one public and five private). There were 727 students at Palau Community College in 2002–03. The adult literacy rate is 92%.

In 2002 government expenditure on education came to an estimated 10·3% of GDP.

Health
There were 26 doctors, 117 nurses and one midwife in 2006 and five dentists and one pharmacist in 2007.

RELIGION
The majority of the population is Roman Catholic.

CULTURE

World Heritage Sites
Rock Islands Southern Lagoon was added to the UNESCO World Heritage List in 2012. It covers hundreds of large and small forested limestone islands, channels, tunnels, caves, arches and coves scattered within a marine lagoon protected by a barrier reef. The site is of exceptional aesthetic beauty and home to diverse and abundant marine life.

Press
There are no daily newspapers published in Palau. In 2008 there were three weekly newspapers: *Tia Belau* and *Palau Horizon*, published in English, and *Roureur Belau*, published in Palauan.

Tourism
Tourism is a major industry, particularly marine-based. There were 83,795 visitor arrivals in 2009 (down from a record 94,895 in 2004). Of the visitor arrivals in 2009, 68,329 were for tourist purposes. Visitors to Palau in 2009 included: 27,180 from Japan; 16,571 from the Republic of China; 13,193 from the Republic of Korea.

DIPLOMATIC REPRESENTATIVES

Of the United Kingdom in Palau
Ambassador: Stephen Lillie (resides in Manila, Philippines).

Of Palau in the USA (1800 K St., NW, Suite 400, Washington, D.C., 20006)
Ambassador: Hersey Kyota.

Of the USA in Palau (PO Box 6028, PW 96940, Koror)
Ambassador: Helen Reed-Rowe.

Of Palau to the United Nations
Ambassador: Stuart Beck.

Of Palau to the European Union
Ambassador: Carlos Hiroshi Salii.

FURTHER READING
National Statistical Office: Bureau of Planning and Budget, P. O. Box 6011, Melekeok PW 96940.
Website: http://www.palaugov.net/stats

PANAMA

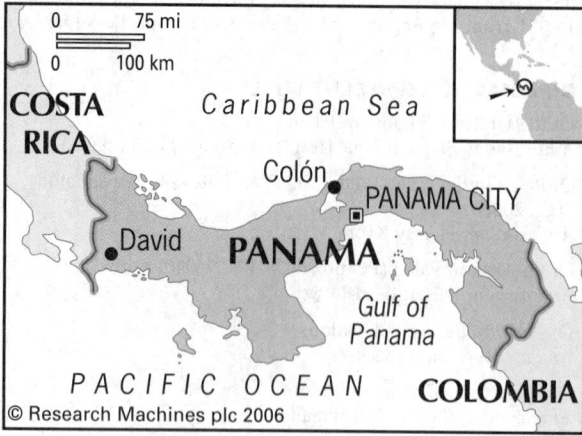

República de Panamá
(Republic of Panama)

Capital: Panama City
Population projection, 2015: 3·78m.
GNI per capita, 2011: (PPP$) 12,335
HDI/world rank: 0·768/58
Internet domain extension: .pa

KEY HISTORICAL EVENTS

Ceramic objects unearthed on Panama's Pacific coast suggest the area was settled from 2,500 BC. Maize cultivation began in southern Panama around 1,500 BC and by 300 BC it supported numerous large, permanent villages. Public architecture was in evidence by AD 500, along with pottery (Gran Coclé), gold ornaments and elaborate burial chambers. From AD 750 present-day Panama was widely settled by Chibchan, Chocoan and Cueva peoples, and there was trade with other groups across central America.

A Spanish adventurer, Rodrigo de Bastidas, explored the isthmus of Panama in 1501, followed by Christopher Columbus the following year. Nuestra Señora de la Asunción de Panamá (present-day Panama City) was founded in 1519. The first European settlement on the Pacific Ocean, it was a staging post for later Spanish conquests across central and south America. A part of the New Kingdom of Granada, from 1549 Panama was ruled by the *Audiencia* of Bogotá. It was a key port for shipping gold, silver and other treasures from the western fringe of South America to Spain.

In the late 1690s the isthmus saw an ill-fated attempt by the Kingdom of Scotland to establish a colony called New Caledonia. Poor planning, insufficient resources, a lack of opportunities for trade and crippling diseases led to the scheme's demise at the hands of Spanish forces in 1700. Panama was formally incorporated into the Viceroyalty of New Granada in 1739, which encompassed present-day Colombia, Ecuador and Venezuela.

An independence movement for New Granada was precipitated by the Napoleonic invasion of Spain in 1808. Antonio Nariño, who took part in an uprising at Bogotá in 1810, and Simón Bolívar became leading revolutionary figures. Bolívar's victory at Boyacá in 1819 was followed by the Spanish surrender of Cartagena in 1821. Panama was incorporated into the new independent Republic of Colombia, which included Venezuela and (from 1822) Ecuador under Bolívar's leadership. Rebellions against rule from Bogotá took place in 1830 and 1840, before the

Bidlack-Mallarino Treaty of 1846 granted the USA rights to both build a railway across Panama (completed in 1855) and intervene militarily in the event of disturbances.

While Colombia was embroiled in the Thousand Day War (1899–1902) Panama again sought independence, with separatists such as José Agustín Arango favouring direct negotiation with the USA over plans for a Panama Canal. The USA chose to back the revolution and Panama declared independence on 3 Nov. 1903. A subsequent treaty granted the USA the use, occupation and control of a Canal Zone, with full sovereign rights in perpetuity. In return the USA guaranteed the independence of the republic. The Canal was opened on 15 Aug. 1914.

US domination of Panama provoked frequent anti-American protests. In 1968 Col. Omar Torrijos Herrera took power in a coup and attempted to negotiate a more advantageous treaty with the USA. Two new treaties between Panama and the USA were agreed in Aug. 1977 and signed in Sept. of the same year. One dealt with the operation and defence of the Canal until the end of 1999 and the other guaranteed permanent neutrality.

Torrijos vacated his position as chief of government in 1978 but remained head of the National Guard until his death in an air crash in 1981. Subsequently, Gen. Manuel Noriega, Torrijos' successor as head of the National Guard, became the strong man of the regime. His position was threatened by internal political opposition and economic pressure applied by the USA but in Oct. 1989 a US-backed coup attempt failed. On 15 Dec. Gen. Noriega declared a 'state of war' with the USA. On 20 Dec. the USA invaded. Gen. Noriega surrendered on 3 Jan. 1990. Accused of drug dealing, he was convicted by a court in Miami. All remaining US troops left the country when the Panama Canal was handed back to Panama at the end of 1999.

In a referendum in 2006, voters backed a US$5·2bn. plan to widen the Panama Canal, potentially doubling its capacity. The conservative retail tycoon Ricardo Martinelli won a landslide victory in the April 2009 presidential election, promising to take a tough stance on crime and rising prices.

TERRITORY AND POPULATION

Panama is bounded in the north by the Caribbean Sea, east by Colombia, south by the Pacific Ocean and west by Costa Rica. The area is 75,001 sq. km. Population at the census of 2010 was 3,405,813 (1,693,229 females); density, 44·9 per sq. km. The population was 75·5% urban in 2011.

The UN gives a projected population for 2015 of 3·78m.

The largest towns (2010) are Panama City, the capital, on the Pacific coast (430,299) and its suburb San Miguelito (315,019). Other large towns are Las Cumbres, Tocumen, David, Arraiján and Colón.

The areas and populations of the nine provinces and the three province-level indigenous districts, plus their capitals, are:

Province	Sq. km	Census 2010	Capital
Bocas del Toro	4,601	125,461	Bocas del Toro
Chiriquí	6,477	416,873	David
Coclé	4,927	233,708	Penonomé
Colón[1]	4,891	241,928	Colón
Darién[1]	11,866	48,378	La Palma
Emberá[2]	4,398	10,001	Cirilo Guainora
Herrera	2,341	109,955	Chitré
Kuna Yala[2]	2,393	33,109	El Porvenir
Los Santos	3,805	89,592	Las Tablas
Ngöbe-Buglé[2]	6,673	156,747	Chichica
Panamá[3]	11,952	1,713,070	Panama City
Veraguas	10,677	226,991	Santiago

[1]Includes the indigenous district of Kuna de Wargandí. [2]Indigenous district. [3]Includes the indigenous district of Kuna de Madungandí.

The population is a mix of African, American, Arab, Chinese, European and Indian immigrants. The official language is Spanish.

SOCIAL STATISTICS

2006 births, 65,764; deaths, 14,358; marriages, 10,747; divorces, 2,866. Birth rate, 2006 (per 1,000 population), 20·0; death rate, 4·4. Annual population growth rate, 2000–05, 2·5%. Expectation of life at birth, 2007, was 73·0 years for males and 78·2 years for females. In 2006 the most popular age range for marrying was 25–29 for both males and females. Infant mortality, 2010, 17 per 1,000 live births; fertility rate, 2006, 2·4 births per woman.

CLIMATE

Panama has a tropical climate, unvaryingly with high temperatures and only a short dry season from Jan. to April. Rainfall amounts are much higher on the north side of the isthmus. Panama City, Jan. 79°F (26·1°C), July 81°F (27·2°C). Annual rainfall 70" (1,770 mm). Colón, Jan. 80°F (26·7°C), July 80°F (26·7°C). Annual rainfall 127" (3,175 mm). Balboa Heights, Jan. 80°F (26·7°C), July 81°F (27·2°C). Annual rainfall 70" (1,759 mm). Cristóbal, Jan. 80°F (26·7°C), July 81°F (27·2°C). Annual rainfall 130" (3,255 mm).

CONSTITUTION AND GOVERNMENT

The 1972 constitution, as amended in 1978, 1983, 1994 and 2004, provides for a *President*, elected for five years, two *Vice-Presidents* and a 72-seat *Legislative Assembly* (since reduced to 71 seats) to be elected for five-year terms by a direct vote. As a result of the amendment of 2004 there has only been one *Vice-President* since the election of May 2009. To remain registered, parties must have attained at least 50,000 votes at the last election. A referendum held on 15 Nov. 1992 rejected constitutional reforms by 64% of votes cast. Turnout was 40%. In a referendum on 30 Aug. 1998 voters rejected proposed changes to the constitution which would allow for a President to serve a second consecutive term.

National Anthem

'Alcanzamos por fin la victoria' ('We achieve victory in the end'); words by J. de la Ossa, tune by Santos Jorge.

GOVERNMENT CHRONOLOGY

(CD = Democratic Change; CNP = National Patriotic Coalition; PA = Arnulfist Party; PL = Liberal Party; PLN = National Liberal Party; PP = Panameñista Party; PR = Republican Party; PRA = Authentic Revolutionary Party; PRD = Revolutionary Democratic Party; n/p = non-partisan)

Heads of State since 1941.

Presidents of the Republic

1941–45	n/p	Ricardo Adolfo de la Guardia Arango
1945–48	PL	Enrique Adolfo Jiménez Brin
1948–49	PL	Domingo Díaz Arosemena
1949	PL	Daniel Chanis Pinzón
1949–51	PRA	Arnulfo Arias Madrid
1951–52	PRA	Alcibíades Arosemena Quinzada
1952–55	CNP	José Antonio Remón Cantera
1955–56	CNP	Ricardo Manuel Arias Espinosa
1956–60	CNP	Ernesto de la Guardia Navarro
1960–64	PLN	Roberto Francisco Chiari Remón
1964–68	PLN	Marco Aurelio Robles Méndez
1968	PP	Arnulfo Arias Madrid

Chairmen of the Provisional Junta of Government

1968–69	military	José María Pinilla Fábrega
1969–72	n/p	Demetrio Basilio Lakas Bahas

Presidents of the Republic

1972–78	n/p	Demetrio Basilio Lakas Bahas
1978–82	n/p	Arístides Royo Sánchez
1982–84	n/p	Ricardo de la Espriella Toral
1984	n/p	Jorge Enrique Illueca Sibauste
1984–85	PRD	Nicolás Ardito Barletta Vallarino
1985–88	PR	Eric Arturo Delvalle Cohen-Henríquez
1989–94	PA	Guillermo David Endara Galimany
1994–99	PRD	Ernesto Pérez Balladares González
1999–2004	PA	Mireya Elisa Moscoso de Arias
2004–09	PRD	Martín Erasto Torrijos Espino
2009–	CD	Ricardo Martinelli Berrocal

De facto rulers from 1968–89.

1968–81	military	Omar Efraín Torrijos Herrera
1982–83	military	Rubén Darío Paredes del Río
1983–89	military	Manuel Antonio Noriega Moreno

RECENT ELECTIONS

In the presidential election on 3 May 2009 Ricardo Martinelli of Democratic Change (CD) won 60·3% of the vote, against 37·3% for Balbina Herrera of the ruling Revolutionary Democratic Party (PRD) and 2·4% for Guillermo Endara (Moral Vanguard of the Fatherland).

At the parliamentary elections, also held on 3 May 2009, the PRD won 26 seats; the Panameñista Party won 21 seats; CD, 15; Patriotic Union Party, 4; Nationalist Republican Liberal Movement (Molirena), 2; independents, 2; People's Party, 1. Turnout was 70·1%.

CURRENT GOVERNMENT

President: Ricardo Martinelli; b. 1952 (Democratic Change; sworn in 1 July 2009).

Vice-President: Juan Carlos Varela Rodríguez.

In March 2013 the government comprised:

Minister of Public Security: José Raúl Mulino. *Government:* Jorge Ricardo Fábrega. *Education:* Lucinda Molinar. *Public Works:* Jaime Ford. *Health:* Javier Díaz. *Labour and Work Development:* Alma Lorena Cortés Aguilar. *Foreign Affairs:* Fernando Núñez Fábrega. *Commerce and Industry:* Ricardo Quijano. *Housing:* Jasmina Pimetel. *Agricultural Development:* Oscar Armando Osorio. *Canal Affairs:* Roberto Roy. *Social Development:* Guillermo Antonio Ferrufino Benítez. *Economy and Finance:* Frank de Lima. *Tourism:* Salomón Shamah Zuchin. *Minister of the Presidency:* Roberto Henríquez.

Office of the President (Spanish only):
  http://www.presidencia.gob.pa

CURRENT LEADERS

Ricardo Martinelli Berrocal

Position
President

Introduction
Businessman Ricardo Martinelli became president in May 2009. He led the Democratic Change party to a landslide victory, ending 40 years of bipartisan government.

Early Life
Ricardo Martinelli was born in March 1952 in Panama City to parents of Italian and Spanish descent. He was schooled in Panama City before attending Staunton Military Academy in Virginia, USA. In 1973 he graduated in business administration from Arkansas University and subsequently obtained a masters degree in the same subject from the Central American Institute of Business Administration in Costa Rica.

He returned to Panama to embark on a business career, first at Citibank before joining the retail company Almacen 99 in 1981. By 1985 he had set up the Super 99 supermarket chain and variously headed the Panamanian Chamber of Commerce, the Italian-Panamanian Chamber of Commerce and the governing association of retail companies. In 1991 he set up the Ricardo

Martinelli Foundation, which grants scholarships to several thousand poor students each year. From 1993–96 Martinelli was involved with the Revolutionary Democratic Party (PRD) and was briefly head of the National Social Security Institute (CSS).

After two years out of the political spotlight, Martinelli established his own party in 1998, called Democratic Change (CD). In 1999 he allied himself with Mireya Moscoso in a conservative coalition against the PRD. When Moscoso went on to win the presidency, CD was rewarded with the ministry for the Panama Canal, just as management of the Canal was being handed over to Panama by the USA. Martinelli was a strong advocate for expansion of the Canal but progress was slow and in 2003 he resigned to stand for the presidency in 2004. Martinelli started with a strong base in local and international commerce, insurance, banking, food, the agricultural and chemical industries, and the media.

Although his 2004 campaign was unsuccessful, it positioned him in the public eye ahead of his second run in 2009. He stood as the candidate for the CD-led Alliance for Change coalition campaigning on a platform of change and fighting corruption. He won with over 60% of the vote.

Career in Office

Martinelli's election bucked the left-leaning trend in neighbouring El Salvador and Nicaragua. Among the policies he advocated were Panama's exit from the Central American Parliament (Parlacen), increased police wages, a monthly stipend of 100 balboas for the unpensioned elderly and the eradication of corruption. In 2009 he appointed the ex-military Gustavo Pérez as head of police and pledged to tackle the 'wild capitalism' of Panama. Martinelli also planned to reform the tax system and to overhaul Panama City's transport system with the construction of a metro network, which is scheduled to open in 2014.

Panama has maintained an impressive annual rate of economic growth during Martinelli's premiership. However, much of the country's wealth continues to stem from the nationally-administered Canal, ports and Colón free trade zone, while extreme poverty, although reduced, remains a problem. In Oct. 2011 Martinelli welcomed the ratification of a free trade agreement with the USA, which had been stalled since 2007.

Martinelli has been criticized for political bullying and remains at loggerheads with his vice-president, who has levelled accusations of corruption against him.

DEFENCE

The armed forces were disbanded in 1990 and constitutionally abolished in 1994. Divided between both coasts, the National Maritime Service, a coast guard rather than a navy, numbered around 600 personnel in 2007. In addition there is a paramilitary police force of 11,000 and a paramilitary national air service of 400 with no combat capable aircraft. In 2008 defence expenditure totalled US$226m. (US$68 per capita), representing 1·0% of GDP. For Police *see* JUSTICE *below*.

ECONOMY

Agriculture accounted for 5·3% of GDP in 2010, industry 16·8% and services 77·9%.

Overview

Panama boasts one of the fastest-growing GDP rates in Latin America, averaging over 9·5% between 2006–08 before slowing to 3·9% in 2009. In 2011 real GDP growth rebounded to 10·6%, driven by buoyant transportation, commerce and tourism sectors, as well as a developed financial sector. However, income inequality is high. The poverty rate was 29% in 2011.

As the link between North and South America, Panama is strategically important for shipping and trade. Much of the country's economic activity is centred on its Canal and the Colón Free Zone, at the Canal's Atlantic entrance. 80% of GDP is derived from related sectors including transportation, packaging, shipping, importing and exporting. It is hoped that a widening and deepening of the Canal, begun in 2007 and scheduled for completion in 2014, will further boost the economy and reduce poverty by 30%.

The banking system is highly integrated with international markets. The national currency is pegged to the US dollar, making it vulnerable to global trends. Panama and the USA entered a Free Trade Agreement in June 2007. This was approved by the Panamanian National Assembly the following month and by the US Congress in Oct. 2011.

Currency

The monetary unit is the *balboa* (PAB) of 100 *centésimos*, at parity with the US dollar. The only paper currency used is that of the USA. US coinage is also legal tender. Inflation was 3·5% in 2010 and 5·9% in 2011. In July 2005 foreign exchange reserves were US$1,018m. and total money supply was 1,587m. balboas.

Budget

Revenues in 2007 were 4,433m. balboas (tax revenue, 48·1%) and expenditures 4,432m. balboas (current expenditure, 78·1%).

VAT is 7%.

Performance

Real GDP growth was 10·1% in 2008, 3·9% in 2009, 7·6% in 2010 and 10·6% in 2011. Total GDP in 2011 was US$26·8bn.

Banking and Finance

There is no statutory central bank. Banking is supervised and promoted by the Superintendency of Banks (formerly the National Banking Commission); the *Superintendent* is Alberto Diamond. Government accounts are handled through the state-owned Banco Nacional de Panama. In 2007 there were two other state banks, 39 banks operating under general licence, 34 under international licence and 11 as representative offices. In Aug. 2007 the combined assets of those banks operating under general and international licences totalled US$50,500m.; their combined deposits were US$36,500m.

Foreign debt was US$11,412m. in 2010, representing 45·8% of GNI.

There is a stock exchange in Panama City.

ENERGY AND NATURAL RESOURCES

Environment

Panama's carbon dioxide emissions from the consumption and flaring of fossil fuels in 2008 were the equivalent of 4·6 tonnes per capita.

Electricity

In 2007 capacity was 1·5m. kW. Production was 6·5bn. kWh in 2007 (57% hydro-electric and 43% thermal), with consumption per capita 1,904 kWh.

Minerals

Limestone, clay and salt are produced. There are known to be copper deposits.

Agriculture

In 2002, 19·3% of the economically active population were engaged in agriculture. In 2002 there were approximately 548,000 ha. of arable land and 147,000 ha. of permanent crops. Production in 2002 (in 1,000 tonnes): sugarcane, 1,441; bananas, 500; rice, 245; plantains, 105; maize, 76; melons and watermelons, 63; oranges, 47; potatoes, 26; yams, 26; pineapples, 23; tomatoes, 21; cassava, 20.

Livestock (2003 estimates): 1,550,000 cattle, 305,000 pigs, 175,000 horses and 14m. chickens.

Livestock products (2002, in 1,000 tonnes): beef and veal, 54; pork, bacon and ham, 18; poultry meat, 89; milk, 178; eggs, 26.

Forestry

Forests covered 3·25m. ha. in 2010 (44% of the land area). There are great timber resources, notably mahogany. Production in 2007 totalled 1·35m. cu. metres.

Fisheries

In 2010 the catch totalled 163,488 tonnes (mainly shrimp), almost entirely from sea fishing. Exports of fishery commodities were valued at US$442m. in 2008.

INDUSTRY

The main industry is agricultural produce processing. Other areas include chemicals and paper-making. Cement production (2008 estimate), 1,843,000 tonnes; sugar (2005), 157,280 tonnes.

Labour

In Aug. 2003 a total of 1,145,982 persons were in employment, with principal areas of activity as follows: agriculture, hunting and forestry, 228,305; wholesale and retail trade/repair of motor vehicles, motorcycles and personal and household goods, 196,418; manufacturing, 105,830; transport, storage and communications, 85,883. In Aug. 2003 the unemployment rate was 13·1%.

INTERNATIONAL TRADE

The Colón Free Zone, the largest free zone in the Americas, is an autonomous institution set up in 1953. More than 2,500 companies were operating there in 2007.

Imports and Exports

Imports and exports in US$1m.:

	2002	2003	2004	2005	2006
Imports f.o.b.	3,035·3	3,122·3	3,592·2	4,152·8	4,817·7
Exports f.o.b.	759·6	805·0	891·1	963·2	1,021·8

Main imports: machinery and apparatus, mineral fuels, chemicals and chemical products. Main exports: marine products, bananas, melons.

Chief import suppliers, 2006: USA, 27%; Curaçao, 10%; Costa Rica, 5%; Japan, 5%. Principal export markets, 2006: USA, 39%; Spain, 8%; Netherlands, 7%; Sweden, 6%.

COMMUNICATIONS

Roads

In 2006 there were 13,365 km of roads, of which 34·1% were paved. The road from Panama City westward to the cities of David and Concepción and to the Costa Rican frontier, with several branches, is part of the Pan-American Highway. The Trans-Isthmian Highway connects Panama City and Colón. In 2007 there were 436,200 passenger cars, 174,500 lorries and vans and 20,100 buses and coaches. There were 425 road accident fatalities in 2007.

Rail

The 1,435 mm gauge Ferrocarril de Panama, which connects Ancón on the Pacific with Cristóbal on the Atlantic along the bank of the Panama Canal, is the principal railway. Traffic in 2004 amounted to 77,000 passengers and 700,000 tonnes of freight. The United Brands Company runs 376 km of railway, and the Chiriquí National Railroad 171 km.

Civil Aviation

There is an international airport at Panama City (Tocumén International). The national carrier is COPA, which flew to 15 different countries in 2003. In 2003 scheduled airline traffic of Panama-based carriers flew 43m. km and carried 1,313,000 passengers (all on international flights). In 2005 Tocumén International handled 2,710,857 passengers and 100,063 tonnes of freight.

Shipping

Panama, a nation with a transcendental maritime career and a strategic geographic position, is the shipping world's preferred flag for ship registry. The Ship Registry System equally accepts vessels of local or international ownership, as long as they comply with all legal parameters. Ship owners also favour Panamanian registry because fees are low. The Panamanian merchant fleet is the largest in the world. In Jan. 2009 there were 6,842 ships of 300 GT or over registered, totalling 180·87m. GT (representing 22·9% of the world total). Of the 6,842 vessels registered, 2,198 were bulk carriers, 2,174 general cargo ships, 1,078 oil tankers, 798 container ships, 221 chemical tankers, 201 liquid gas tankers and 172 passenger ships.

All the international maritime traffic for Colón and Panama runs through the Canal ports of Cristóbal, Balboa and Manzanillo International.

Panama Canal

The Panama Canal Commission is concerned primarily with the operation of the Canal. In Oct. 2002 a new toll structure was adopted based on ship size and type.

Although most of the world's shipping fleet can use the Canal the percentage that is able to is gradually declining as many new ships are too wide for the Canal. A referendum was held in Oct. 2006 on whether to expand the canal and double its capacity. The US$5·25bn. plan was approved by 78% of voters. The expansion work began in Sept. 2007 and is expected to be completed in 2014.

Administrator of the Panama Canal Authority: Jorge Quijano.

Particulars of the ocean-going commercial traffic through the Canal are given as follows:

Fiscal year ending 30 Sept.	No. of vessels transiting	Cargo in long tons	Tolls revenue (in US$1)
2008	13,138	209,705,000	1,316,031,000
2009	12,849	197,896,000	1,436,810,000
2010	12,582	204,816,000	1,480,554,000

Most numerous transits by flag (fiscal year 2010): Panama, 2,636; Liberia, 1,989; Bahamas, 849; Marshall Islands, 703; Hong Kong, 646.

Statistical Information: The Panama Canal Authority Corporate Communications Division

Annual Reports on the Panama Canal, by the Administrator of the Panama Canal

Rules and Regulations Governing Navigation of the Panama Canal. The Panama Canal Authority

Major, J., *Prize Possession: the United States and the Panama Canal, 1903–1979.* 1994

Telecommunications

Panama had 3,505,900 telephone subscribers in 2007, or 1,048·6 per 1,000 persons, including 3,010,600 mobile phone subscribers. There were 745,300 internet users in 2007, including 143,900 broadband subscribers. In Dec. 2011 there were 896,000 Facebook users.

SOCIAL INSTITUTIONS

Justice

The Supreme Court consists of nine justices appointed by the executive. There is no death penalty. The police force numbered 11,000 in 2007, and includes a Presidential Guard.

The population in penal institutions in Nov. 2008 was 10,036 (295 per 100,000 of national population).

Education

Adult literacy was 94% in 2008. Elementary education is compulsory for all children from six to 14 years of age. In 2007 there were 446,176 pupils with 18,183 teaching staff at primary schools and 260,694 pupils with 16,847 teaching staff at secondary schools. In 2006 there were 130,838 students and 11,528 academic staff in tertiary education. The University of Panama (Universidad de Panamá), founded in 1935 in Panama City, is the leading higher education institution.

In 2008 public expenditure on education came to 3·8% of GDP.

Health

In 2002 there were 61 hospitals with a provision of 25 beds per 10,000 persons. There were 4,203 physicians, 897 dentists, 3,451 nurses and 612 pharmacists.

RELIGION

80% of the population is Roman Catholic, 14% Protestant. The remainder of the population follow other religions (notably Islam). There is freedom of religious worship and separation of Church and State. Clergymen may teach in schools but may not hold public office.

CULTURE

World Heritage Sites

Panama has five sites on the UNESCO World Heritage List: the Fortifications on the Caribbean side of Panama: Portobelo-San Lorenzo (inscribed on the list in 1980); Darien National Park (1981); the Archaeological Site of Panamá Viejo and the Historic District of Panamá (1997 and 2003); and the Coiba National Park (2005).

Panama shares a UNESCO site with Costa Rica: the Talamanca Range-La Amistad Reserves (1983 and 1990), an important cross-breeding site for North and South American flora and fauna.

Press

In 2008 there were seven dailies with a combined circulation of 233,000.

Tourism

In 2006 there were 843,000 non-resident tourists (702,000 in 2005); spending by tourists totalled US$1,450m. in 2006 (US$1,108m. in 2005).

Festivals

Leading festivals include: the National Folkloric Festival, held annually (Sept.) in Guararé since 1949; the Panama Jazz Festival (Jan.); the International Fair of the Sea, held in Bocas del Toro province in Sept.; and the International Fair of Changuinola, also in Bocas del Toro in Sept.

DIPLOMATIC REPRESENTATIVES

Of Panama in the United Kingdom (40 Hertford St., London, W1J 7SH)
Ambassador: Ana Irene Delgado.

Of the United Kingdom in Panama (MMG Tower, Calle 53, Apartado/POB 0816-07946, Panama City)
Ambassador: Michael John Holloway, OBE.

Of Panama in the USA (2862 McGill Terr., NW, Washington, D.C., 20008)
Ambassador: Mario E. Jaramillo.

Of the USA in Panama (Edificio 783, Avenida Demetrio Basilio Lakas, Panama City 5)
Ambassador: Jonathan D. Farrar.

Of Panama to the United Nations
Ambassador: Pablo Antonio Thalassinos.

Of Panama to the European Union
Ambassador: Carlos Constantino Arosemena Ramos.

FURTHER READING

Statistical Information: The Controller-General of the Republic (Contraloria General de la República, Calle 35 y Avenida 6, Panama City) publishes an annual report and other statistical publications.

Harding, Robert C., *The History of Panama.* 2006
Lindsay-Poland, John, *Emperors in the Jungle.* 2003
McCullough, D. G., *The Path Between the Seas: The Creation of the Panama Canal, 1870–1914.* 1999
Sahota, G. S., *Poverty Theory and Policy: a Study of Panama.* 1990

Other titles are listed under Panama Canal, *above.*

National library: Biblioteca Nacional, Departamento de Información, Av. Balboa y Federico Boyd, Ciudad de Panama.
Website (Spanish only): http://www.contraloria.gob.pa

PAPUA NEW GUINEA

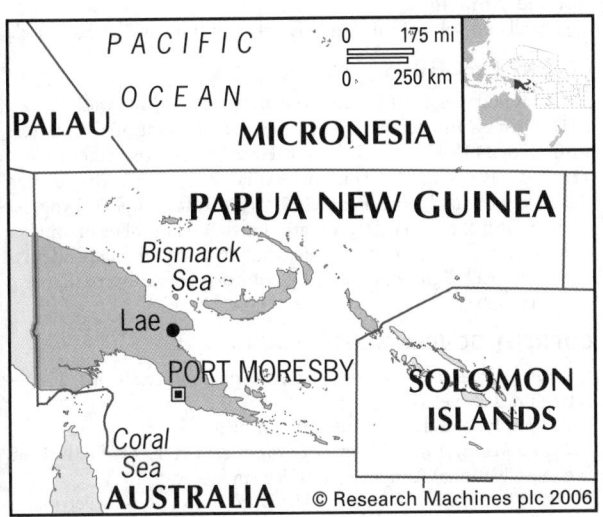

Capital: Port Moresby
Population projection, 2015: 7·65m.
GNI per capita, 2011: (PPP$) 2,271
HDI/world rank: 0·466/153
Internet domain extension: .pg

KEY HISTORICAL EVENTS

The region was settled by Asian peoples around 50,000 years ago, when New Guinea was still part of the main Australian landmass. 10,000 years ago, at the end of the last ice age, the waters of the Torres Straight cut off New Guinea from Australia, giving rise to separate development of the indigenous peoples. Plant-based agriculture developed about 9,000 years ago in the New Guinea highlands. 2,500 years ago a large migration of Austronesian-speaking peoples settled the coastal areas. These communities developed animal husbandry, pottery and fishing and there is evidence of trade with the southeast Asian mainland. The rugged topography meant that contact between communities was limited and they retained their separate languages and customs. This heterogeneity continued down the centuries, so that today Papua New Guinea has an estimated 1,000 different cultural groups. Spanish and Portuguese explorers reached the island in the early 16th century and the Portuguese are thought to have introduced the kaukau (sweet potato), which became a staple crop. Spain laid claim to the western half of the island in 1545, naming it 'New Guinea' because of the supposed resemblance of its inhabitants to the people of Africa's Guinea coast.

From the late 18th century the British and Dutch competed for control of the island. In 1828 the Dutch claimed the western half as part of the Dutch East Indies, while the eastern half came under British influence. Following an abortive annexation attempt by Australian colonists, Britain declared a protectorate over the southern coast and islands adjacent to the eastern half of New Guinea on 6 Nov. 1884. Germany colonized the northern portion of the eastern half, along with New Britain, New Ireland and Bougainville, an arrangement formalized by the 1885 Anglo–German Agreement.

In 1888 Britain formally annexed the area under its control as British New Guinea and passed its administration to Australia in 1902. On 1 Sept. 1906 the Australian administration renamed it the Territory of Papua, Papua being a Malay word used to describe the Melanesian inhabitants' curly hair. The northeastern section of the island remained a German colony until the outbreak of the First World War in 1914, when Australian armed forces occupied it. For the next seven years it was under their administration and remained so after 1921, when it became a League of Nations mandated territory. In the Second World War, Allied and Japanese forces fought a prolonged military campaign on the island. It began with a Japanese victory at the Battle of Rabaul, New Britain in Feb. 1942 and continued throughout the war, with Japanese troops in occupation for much of the time until the Allies' victory in Aug. 1945.

In 1947 the northeastern part of the island became the UN Trust Territory of New Guinea, before merging in 1949 with the Territory of Papua to become the Territory of Papua New Guinea. A legislative council was established in 1953, succeeded by a house of assembly in 1964. On 1 Dec. 1973 Australia granted Papua New Guinea self-government. After a vigorous debate over questions of land-reform and citizenship, a constitution was passed and on 16 Sept. 1975 Papua New Guinea became fully independent. It also became a member of the Commonwealth, recognizing the British monarch as its own. In the same year the new state ignored a unilateral declaration of independence by the island of Bougainville and suppressed its campaign of civil disobedience. In 1988 an armed campaign by tribes claiming traditional land rights against the Australian owner of the massive Panguna copper field escalated into a civil war for Bougainville's secession. Fighting lasted nine years and cost an estimated 20,000 lives before a permanent truce was signed in April 1998. Bougainville gained autonomy in 2005 and is expected to hold a referendum on full independence in the future.

Political volatility caused frequent changes of government in Papua New Guinea during its first three decades of independence. In 2002, following violence in the general election, the electoral system was changed to encourage stable government. The deployment of Australian police on the island in 2003 caused resentment and from 2005, under a new agreement between the two governments, they operated in smaller numbers and with reduced powers. Despite its considerable mineral resources, the country's economy has stagnated in recent years; the difficulty of establishing land ownership has contributed to this, as has the need to protect the environment in what remains a largely subsistence culture. In 2006 the United Nations downgraded Papua New Guinea's status to least developed nation.

In 2011–12 Papua New Guinea experienced a leadership crisis between Sir Michael Somare and Peter O'Neill. In Aug. 2011, with Prime Minister Somare having been out of the country for five months on medical leave, the government declared the post vacant and O'Neill was elected by parliament, replacing acting prime minister Sam Abal. However, in Dec. the Supreme Court ruled the declaration and the election of O'Neill illegal. It also ordered the reinstatement of Somare. Although Somare and a new cabinet were sworn in, O'Neill refused to step down and the power struggle triggered a failed mutiny by rebel soldiers in Jan. 2012. Having the support of the civil service, the police, the defence force and most MPs, O'Neill remained effectively in control of the country. In May 2012 the Supreme Court once again ruled that Somare should be recognized as the legitimate prime minister. O'Neill's government subsequently ordered the arrest of the chief justice on charges of sedition. The prime minister's office was again declared vacant owing to Somare's continued absence. At the end of the month O'Neill was re-elected unopposed and sworn in as prime minister for the third time.

TERRITORY AND POPULATION

Papua New Guinea extends from the equator to Cape Baganowa in the Louisiade Archipelago to 11° 40' S. lat. and from the border of West Irian to 160° E. long. with a total area of 462,840 sq. km. According to the 2011 census the provisional population was 7,059,653 (3,663,249 males); density, 15·3 per sq. km.

The UN gives a projected population for 2015 of 7·65m.

In 2011, 12·6% of the population lived in urban areas (the second lowest percentage in the world). In 2000 population of Port Moresby (National Capital District) was 254,158. Population of other main towns (2000 census): Lae, 123,398; Madang, 32,318; Mount Hagen, 27,877; Wewak, 27,031; Goroka, 19,523; Kimbe, 14,184. The areas, populations and capitals of the provinces are:

Provinces	Sq. km	Census 2011 (provisional)	Capital
Bougainville	9,300	234,280	Arawa
Central	29,500	237,016	Port Moresby
Chimbu	6,100	403,772	Kundiawa
East New Britain	15,500	271,250	Rabaul
East Sepik	42,800	433,481	Wewak
Eastern Highlands	11,200	582,159	Goroka
Enga	12,800	452,596	Wabag
Gulf	34,500	121,128	Kerema
Madang	29,000	487,460	Madang
Manus	2,100	50,321	Lorengau
Milne Bay	14,000	269,954	Alotau
Morobe	34,500	646,876	Lae
National Capital District	240	318,128	—
New Ireland	9,600	161,165	Kavieng
Oro	22,800	176,206	Popondetta
Sandaun	36,300	227,657	Vanimo
Southern Highlands[1]	23,800	868,209	Mendi
West New Britain	21,000	242,676	Kimbe
Western	99,300	180,455	Daru
Western Highlands[2]	8,500	694,862	Mount Hagen

[1]Includes the province of Hela, established in 2012. [2]Includes the province of Jiwaka, established in 2012.

Tok Pisin (or Pidgin, a creole of English), Hiri Motu and English are all official languages.

SOCIAL STATISTICS

Estimates, 2008: births, 207,000; deaths, 52,000. Rates, 2008 estimates (per 1,000 population): births, 31·4; deaths, 7·9. Expectation of life at birth in 2007 was 58·7 years for males and 63·0 years for females. Annual population growth rate, 2000–08, 2·5%. Infant mortality, 2010, 47 per 1,000 live births; fertility rate, 2008, 4·1 births per woman.

CLIMATE

There is a monsoon climate, with high temperatures and humidity the year round. Port Moresby is in a rain shadow and is not typical of the rest of Papua New Guinea. Jan. 82°F (27·8°C), July 78°F (25·6°C). Annual rainfall 40" (1,011 mm).

CONSTITUTION AND GOVERNMENT

The constitution took effect on 16 Sept. 1975. The head of state is the British sovereign, who is represented by a *Governor-General*, nominated by parliament for six-year terms. A single legislative house, known as the *National Parliament*, is made up of 111 members: 89 district representatives and 22 provincial representatives (MPs). The members are elected by universal suffrage; elections are held every five years. All citizens over the age of 18 are eligible to vote and stand for election. Voting is by secret ballot and follows the limited preferential system. The *Prime Minister*, nominated by parliament and appointed by the Governor-General, selects ministers for the National Executive Council. The government cannot be subjected to a vote of no confidence in the first 18 months of office. The 20 provincial assemblies, comprising elected national MPs, appointed members and elected local government representatives, are headed by a Governor, normally the provincial representative in the National Parliament.

National Anthem

'Arise, all you sons of this land'; words and tune by T. Shacklady.

RECENT ELECTIONS

Parliamentary elections were held between 23 June and 17 July 2012. Prime Minister Peter O'Neill's People's National Congress won 27 of 111 seats; the Triumph Heritage Empowerment Party 12; PNG Party 8; the National Alliance Party 7; the United Resources Party 7; the People's Party 6; the People's Progress Party 6; and the Social Democratic Party 3. A number of smaller parties each took one or two seats, with 16 going to independents.

Sir Michael Ogio was elected governor-general by parliament on 14 Jan. 2011.

CURRENT GOVERNMENT

Governor-General: Sir Michael Ogio; b. 1942 (since 20 Dec. 2010—acting until 25 Feb. 2011).

In March 2013 the government comprised:

Prime Minister and Minister of Finance: Peter O'Neill; b. 1965 (People's National Congress; in office since 3 Aug. 2011).

Deputy Prime Minister, Minister for Inter-Government Relations: Leo Dion.

Minister for Agriculture and Livestock: Tommy Tomscoll. *Bougainville Affairs:* Steven Kamma. *Civil Aviation:* Davis Steven. *Communication and Information Technology:* Jim Miringtoro. *Correctional Services:* Jim Simitab. *Defence:* Fabian Pok. *Environment and Conservation:* John Pundari. *Finance, and Education (acting):* James Marape. *Fisheries and Marine Resources:* Mao Zeming. *Foreign Affairs and Immigration:* Rimbink Pato. *Forestry and Climate Change:* Patrick Pruaitch. *Health and HIV/AIDS:* Michael Malabag. *Housing and Urban Development:* Paul Isikeli. *Justice and Attorney General:* Kerenga Kua. *Labour and Industrial Relations:* Mark Maipakai. *Lands and Physical Planning:* Benny Allan. *Mining:* Byron Chan. *National Planning:* Charles Abel. *Petroleum and Energy:* William Duma. *Police:* Nixon Duban. *Public Enterprises and State Investment:* Ben Micah. *Public Service:* Sir Puka Temu. *Religion, Youth and Community Development:* Loujaya Toni. *Sports and Pacific Games:* Justin Tkatchencko. *Tourism, Arts and Culture:* Boka Kondra. *Trade, Commerce and Industry:* Richard Maru. *Transport:* Ano Pala. *Treasury, and Higher Education, Research Science and Technology (acting):* Don Polye. *Works and Implementation:* Francis Awesa.

CURRENT LEADERS

Peter O'Neill

Position
Prime Minister

Introduction
Peter O'Neill was elected by MPs to succeed Sir Michael Somare in Aug. 2011. After Somare refused to acknowledge the result, there were two rival administrations in late 2011. Having been confirmed in office, O'Neill, a businessman and former treasury minister, pledged to fight corruption and develop the nation's infrastructure and services.

Early Life
Born on 13 Feb. 1965 in Pangia District in the Southern Highlands Province, O'Neill graduated in accountancy and commerce from the University of Papua New Guinea in 1988. He practised as a chartered accountant before going into the real estate business. From 1993–97 he was executive chairman of Pangia Enterprises and from 1997–99 served as executive chairman of Pacific Finance, which managed state-owned

enterprises including the PNG Banking Corporation and the National Provident Fund.

O'Neill came in for criticism when several of these public enterprises failed, though his supporters praised his financial management and efforts to improve transparency. In 1999 he became executive chairman of Remington Technologies and in 2002 was elected to parliament for the People's National Congress (PNC), representing the constituency of Ialibu-Pangia. From 2002–03 he served as minister for labour and industrial relations and from 2003–04 was minister for public service and leader of government business.

As head of the PNC, he led the opposition from 2004–07. In 2007 he was appointed minister for public service in Somare's National Alliance Party-led coalition government and from 2010–11 also took on the finance and treasury portfolios. In office, he supported the liquefied national gas projects undertaken in partnership with foreign companies. On 2 Aug. 2011, during Somare's prolonged absence from government for medical treatment, MPs from across the political spectrum elected O'Neill prime minister by 70 votes to 24.

Career in Office

O'Neill's early months were dominated by political instability as Somare challenged the legality of his election. In Dec. 2011 Somare formed a rival administration after the Supreme Court found in favour of the ex-incumbent. Somare's government was initially recognized by Governor-General Sir Michael Ogio but he later reversed his decision and declared O'Neill's premiership to be legitimate. By early 2012 O'Neill had gained widespread recognition, although Somare pursued his legal challenge. In May 2012 the Supreme Court ruled that Somare was still the country's legitimate leader. The chief justice was charged with sedition and a group of police blockaded parliament. However, given that Somare had missed three consecutive sessions of parliament, the prime minister's office was again declared vacant. Having been elected unopposed by 56 of the 109 members of parliament, O'Neill was sworn in as the country's prime minister for the third time on 31 May 2012.

In parliamentary elections held from June–July 2012, the PNC was returned as the largest party, effectively ending the year-long feud with Somare and his supporters. O'Neill promised to act against corruption and to manage a natural gas and minerals boom in the national interest. He faced urgent demands to develop the education and health sectors and to improve the national infrastructure. In Nov. he proposed an amendment to the constitution to extend a ban on votes of no confidence from 18 to 30 months after a government is elected.

DEFENCE

The Papua New Guinea Defence Force had a total estimated strength of 3,100 in 2007 consisting of land, maritime and air elements. The Navy is based at Port Moresby and Manus. Personnel numbered around 400 in 2007. There is an air force, 200 strong in 2007, but it does not possess any combat capable aircraft.

Defence expenditure in 2008 totalled US$35m. (US$6 per capita), representing 0·6% of GDP.

ECONOMY

Agriculture accounted for 19·1% of GDP in 2006, industry 45·2% and services 35·7%.

Overview

Papua New Guinea has experienced sound growth and macroeconomic stability since 2002. The country has large deposits of oil, gas, gold, copper, timber and silver, as well as abundant fisheries. The formal economy is focused on large-scale export of natural resources, while the informal sector consists of subsistence activities by a large proportion of the rural population.

Growth has remained strong through the global economic crisis, particularly in the extractive, construction and telecommunications industries. Prudent macroeconomic management in previous years provided fiscal space to adjust to the external shock. A significant driver in the economy is the Liquefied Natural Gas NG Project, approved in Dec. 2009 with production scheduled to start in 2014. Developing infrastructure remains crucial for further private sector growth.

An estimated 40% of people live on less than US$1 per day.

Currency

The unit of currency is the *kina* (PGK) of 100 *toea*. The kina was floated in Oct. 1994. Foreign exchange reserves were US$543m. in July 2005, gold reserves 63,000 troy oz and total money supply was K2,765m. Inflation was 6·0% in 2010 and 8·4% in 2011.

Budget

In 2009 revenues totalled K6,651·0m. (tax revenue, 74·8%) and expenditures K6,688·0m. (current expenditure, 62·3%).

VAT is 10%.

Performance

Real GDP growth was 6·1% in 2009, 7·6% in 2010 and 8·9% in 2011. Total GDP in 2011 was US$12·9bn.

Banking and Finance

The Bank of Papua New Guinea (*Governor*, Loi Martin Bakani) assumed the central banking functions formerly undertaken by the Reserve Bank of Australia on 1 Nov. 1973. A national banking institution, the Papua New Guinea Banking Corporation, has been established. This bank has assumed the Papua New Guinea business of the Commonwealth Trading Bank of Australia.

In 2002 there were seven commercial banks (Australia and New Zealand Banking Group; Bank of Hawaii; Bank of South Pacific; Maybank; MBf Finance; Papua New Guinea Banking Corporation; Westpac Bank) and a Rural Development Bank.

Total commercial bank deposits, 1992, K1,318·2m. Total savings account deposits, 1992, K226·8m. In addition, the Agriculture Bank of Papua New Guinea had assets of K82·6m. in 1992, and finance companies and merchant banks had total assets of K198·4m.

Foreign debt was US$5,882m. in 2010, representing 62·9% of GNI.

There is a stock exchange in Port Moresby.

ENERGY AND NATURAL RESOURCES

Environment

Carbon dioxide emissions from the consumption and flaring of fossil fuels in 2008 were the equivalent of 0·8 tonnes per capita.

Electricity

Installed capacity was an estimated 0·7m. kW in 2007. Production in 2007 was estimated at 3·05bn. kWh, around 29% of it hydro-electric. Consumption per capita was an estimated 258 kWh.

Oil and Gas

Natural gas reserves in 2008 were 440bn. cu. metres; output in 2006 was 283m. cu. metres. Crude oil production (2007), 2·1m. tonnes. Oil predominantly comes from the Iagifu field in the Southern Highlands. There were 240m. bbls of proven oil reserves in 2007.

Minerals

In 2004 the mineral sector produced 20·8% of GDP. Copper is the main mineral product. Gold, copper and silver are the only minerals produced in quantity. The Misima open-pit gold mine was opened in 1989 but its resources were depleted by the end of 2001. The Porgera gold mine opened in 1990 with an expected life of 20 years. Major copper deposits in Bougainville have proven reserves of about 800m. tonnes; mining was halted by secessionist rebel activity. Copper and gold deposits in the Star Mountains of the Western Province are being developed by Ok Tedi Mining Ltd at the Mt Fubilan mine. Production of gold commenced in 1984 and of copper concentrates in 1987. In 2005 Ok Tedi Mining Ltd produced 192,978 tonnes of copper and 16 tonnes of gold. Gold mining also began at Lihir in 1997. In 2005 total gold production was 67 tonnes; silver production in 2002 was 64 tonnes.

Agriculture
In 2002 agriculture employed 73% of the economically active population. In 2002 there were approximately 220,000 ha. of arable land and 650,000 ha. of permanent cropland. Minor commercial crops include pyrethrum, tea, peanuts and spices. Locally consumed food crops include sweet potatoes, maize, taro, bananas, rice and sago. Tropical fruits grow abundantly. There is extensive grassland. The sugar industry has made the country self-sufficient in this commodity while a beef-cattle industry is being developed.

Production (2002, in 1,000 tonnes): bananas, 860; coconuts, 513; sweet potatoes, 490; sugarcane, 370; palm oil, 316; yams, 280; taro, 250; cassava, 130.

Livestock (2003 estimates): pigs, 1·8m.; cattle, 90,000; chickens, 4m.

Forestry
The forest area totalled 28·73m. ha. in 2010 (63% of the land area). 91% of forest area is primary forest. Timber production is important for both local consumption and export. Timber production was 7·24m. cu. metres in 2007.

Fisheries
Tuna is the major resource. In 2010 the fish catch was 224,507 tonnes (94% sea fish).

INDUSTRY
Secondary and service industries are expanding for the local market. The main industries are food processing, beverages, tobacco, timber products, wood and fabricated metal products. Industry accounted for 45·2% of GDP in 2006, with manufacturing contributing 6·1%. Production (2002): palm oil, 370,000 tonnes; copra, 110,000 tonnes; wood-based panels, 79,000 cu. metres; sawnwood, 70,000 cu. metres.

Labour
In 2000 there were 2,344,734 persons in employment (51·3% males). The rate of unemployment was 2·8%.

INTERNATIONAL TRADE
Australian aid amounts to an annual $A300m. The 'Pactra II' agreement of 1991 established a free trade zone with Australia and protects Australian investments.

Imports and Exports
Imports in 2008 were US$3,550m. (US$2,945m. in 2007); exports were US$5,805m. (US$4,683m. in 2007).

The main imports in terms of value are machinery and transport equipment, manufactured goods, and food and live animals; and the main exports crude petroleum, gold and logs.

Of imports in 2001, Australia furnished 60·0%; Japan, 8·5%; Singapore, 7·8%; USA, 7·1%. Of exports in 2001, Singapore took 27·5%; Japan, 13·2%; Australia 10·6%; China, 4·4%.

COMMUNICATIONS
Roads
In 2002 there were 19,600 km of roads, only about 690 km of which were paved. There were 38,200 passenger cars in use in 2007 and 11,300 lorries and vans.

Civil Aviation
Jacksons International Airport is at Port Moresby. The state-owned national carrier is Air Niugini. In 2003 there were scheduled international flights to Brisbane, Cairns, Honiara,

Manila, Singapore, Sydney and Tokyo. There are a total of 177 airports and airstrips with scheduled services.

Shipping
There are 12 entry and four other main ports served by five major shipping lines; the Papua New Guinea Shipping Corporation is state-owned. In Jan. 2009 there were 55 ships of 300 GT or over registered, totalling 44,000 GT.

Telecommunications
In 2007 there were 360,000 telephone subscribers, or 56·9 for every 1,000 inhabitants. There were 300,000 mobile phone subscribers in 2007 and 110,000 internet users in 2006. In Dec. 2004 the government rejected a bid by a South African joint venture to acquire a 51% stake in the state-owned telecommunications company Telikom PNG.

SOCIAL INSTITUTIONS
Justice
The judicial system consists of a Supreme Court, a National Court, and district and local courts. The Supreme Court sittings are usually held with three or five judges. In 2004 there were 64,709 court cases registered of which 16,459 were civil cases. The death penalty for wilful murder was abolished in 1970 but reintroduced in 1991, although there have not been any executions since 1954.

The population in penal institutions in Dec. 2010 was 4,268 (61 per 100,000 of national population).

Education
The education system has three levels: primary, secondary and tertiary. However, there are no legal provisions regarding either free or compulsory education. In 2001 there were 3,055 elementary and primary schools with 395,129 pupils and 11,307 teachers, 77,451 pupils in secondary schools (2,187 teachers) and 14,333 students in institutes of higher education. There are six universities: the University of Papua New Guinea (UPNG), Port Moresby; the Papua New Guinea University of Technology, Lae; Divine Word University, Madang; Pacific Adventist University, Boroko; the University of Goroka; and the University of Vudal, Rabaul. UPNG, founded in 1965, has two campuses in the capital, five provincial open campuses and 13 study centres. In 2002 there were also ten colleges, eight nursing schools and three academic institutes.

Adult literacy rate was an estimated 60·1% in 2009 (63·6% among males and 56·5% among females).

In 2000–01 total expenditure on education came to 2·4% of GNP and 17·5% of total government spending.

Health
In 2000 there were 275 physicians, 90 dentists and 2,841 nurses. Provision of hospital beds in 1993 was 34 per 10,000 persons.

RELIGION
At the 2000 census there were 4·93m. Christians: Roman Catholics made up 27·0%; Lutherans, 19·5%; United Church, 11·5%; Anglicans, 3·2%. In 1998 the Catholic Church had four archdioceses (Madang, Mount Hagen, Port Moresby and Rabaul), 14 dioceses, 340 parishes and 540 priests.

CULTURE
World Heritage Sites
There is one UNESCO site in Papua New Guinea: Kuk Early Agricultural Site (inscribed on the list in 2008).

Press
In 2008 there were two daily newspapers (the *Papua New Guinea Post-Courier* and the *National*) and a number of weeklies and monthlies. The *Papua New Guinea Post-Courier* is the oldest

(1969) and most widely read, with a daily circulation of 30,000 (2007).

Tourism

In 2008 there were 114,000 non-resident tourists (excluding same-day visitors), up from 104,000 in 2007 and 78,000 in 2006.

Festivals

Alongside the major Christian festivals several cultural shows are held, in Enga (late July), at Mount Hagen (Western Highlands; late Aug.) and at Goroka (Eastern Highlands; mid-Sept.). The Tumbuan Mask Festival takes place at Rabaul (July) and the Hiri Moale Festival in Port Moresby (Sept.) was established to preserve the trading expeditions between the Motu-Koitabu people and the Erema (Kerema). Independence Day is celebrated on 16 Sept.

DIPLOMATIC REPRESENTATIVES

Of Papua New Guinea in the United Kingdom (3rd Floor, 14 Waterloo Pl., London, SW1Y 4AR)
High Commissioner: Winnie Anna Kiap.

Of the United Kingdom in Papua New Guinea (Sec 411 Lot 1 and 2, Kiroki St., Waigani, Port Moresby)
High Commissioner: Jackie Barson.

Of Papua New Guinea in the USA (1779 Massachusetts Ave., NW, Washington, D.C., 20036)
Ambassador: Vacant.
Chargé d'Affaires a.i.: Elias Rahuromo Wohengu.

Of the USA in Papua New Guinea (Douglas St., Port Moresby)
Ambassador: Teddy B. Taylor.

Of Papua New Guinea to the United Nations
Ambassador: Robert Aisi.

Of Papua New Guinea to the European Union
Ambassador: Peter Pulkiye Maginde.

FURTHER READING

National Statistical Office. *Summary of Statistics.* Annual.—*Abstract of Statistics.* Quarterly.
Bank of Papua New Guinea. *Quarterly Economic Bulletin.*

Connell, John, *Papua New Guinea: The Struggle for Development.* 1997
Turner, A., *Historical Dictionary of Papua New Guinea.* 1995

Waiko, John Dademo, *Short History of Papua New Guinea.* 1993.—*Papua New Guinea: A History of Our Times.* 2003

National Statistical Office: National Statistical Office, PO Box 337, Waigani, National Capital District, Port Moresby.
Website: http://www.nso.gov.pg

Bougainville

The region of Bougainville, as part of New Guinea, became a United Nations Trust Territory in 1947 under the jurisdiction of Australia. Bougainville declared independence in 1975 but the uprising was suppressed and the islands became part of the North Solomons Province of the newly independent Papua New Guinea. Following tensions between landowners and Rio Tinto, the owners of the Panguna copper mine, conflict between separatists and Papua New Guinea began in 1989, lasting nearly a decade. A ceasefire was agreed in 1998, overseen by the United Nations. The Bougainville Peace Agreement was signed in Aug. 2001, allowing for the creation of an autonomous government and a referendum on full independence 10–15 years later. After the promulgation of a constitution, elections were held for the autonomous government in May–June 2005. A referendum on independence is now expected to take place between 2015 and 2020.

Area, 9,300 sq. km (3,600 sq. miles); population (2000), 175,160. The autonomous region consists of the main Bougainville island, Buka island and several smaller island groups. The government is currently based at Buka but the capital city will eventually revert to the former capital, Arawa.

In presidential elections held in May 2010 John Momis was elected with 52·4% of votes cast, defeating incumbent president James Tanis who took 20·9%. There were five other candidates. Momis was sworn in on 10 June.

Agriculture is the mainstay of Bougainville's economy. Major crops include cocoa (production in 2006, 16,000 tonnes) and copra (12,472 tonnes). However, the region is still reliant on grants and donors, raising only 20% of its 2007 budget internally. New exploration for minerals is banned but the government is considering reopening Panguna. There is an airport at Buka.

PARAGUAY

BOLIVIA

Bahía Negra

BRAZIL

PARAGUAY

Pedro Juan Caballero

ARGENTINA

ASUNCIÓN

Ciudad del Este

0 ___ 100 mi
0 ___ 200 km

© Research Machines plc 2006

República del Paraguay
(Republic of Paraguay)

Capital: Asunción
Population projection, 2015: 7·03m.
GNI per capita, 2011: (PPP$) 4,727
HDI/world rank: 0·665/107
Internet domain extension: .py

KEY HISTORICAL EVENTS

Paraguay was occupied by the Spanish in 1537 and became a Spanish colony as part of the viceroyalty of Peru. The area gained its independence, as the Republic of Paraguay, on 14 May 1811. Paraguay was then ruled by a succession of dictators. A devastating war fought from 1865 to 1870 between Paraguay and a coalition of Argentina, Brazil and Uruguay reduced Paraguay's population from about 600,000 to 233,000. Further severe losses were incurred during the war with Bolivia (1932–35) over territorial claims in the Chaco inspired by the unfounded belief that minerals existed in the territory. A peace treaty by which Paraguay obtained most of the area her troops had conquered was signed in July 1938.

A new constitution took effect in Feb. 1968 under which executive power is discharged by an executive president. Gen. Alfredo Stroessner Mattiauda was re-elected seven times between 1958 and 1988. Since then, Paraguay has had more or less democratic government. On 23 March 1999 Paraguay's vice-president Luis Maria Argaña was assassinated. The following day, Congress voted to impeach President Raúl Cubas who was said to be implicated in the murder. He then resigned. The victory of Fernando Lugo of the Patriotic Alliance for Change in the April 2008 presidential election brought to an end the 61-year rule of the Colorado Party, at the time the world's longest-ruling party.

TERRITORY AND POPULATION

Paraguay is bounded in the northwest by Bolivia, northeast and east by Brazil and southeast, south and southwest by Argentina. The area is 406,752 sq. km (157,042 sq. miles).

The 2002 census population was 5,163,198 (2,603,242 males), giving a density of 12·7 per sq. km. Estimate, July 2010: 6,451,100. In 2011, 62·1% lived in urban areas.

The UN gives a projected population for 2015 of 7·03m.

In 2002 the capital, Asunción, had a population of 512,112. Other major cities (2002 census populations) are: Ciudad del Este, 222,274; San Lorenzo, 204,356; Luque, 185,127.

There are 17 departments and the capital city. Area and population (2010 estimates):

Department	Area in sq. km	Population
Asunción (city)[1]	117	518,200
Central	2,465	2,068,100
Alto Paraná	14,895	753,700
Itapúa	16,525	535,500
Caaguazú	11,474	480,800
San Pedro	20,002	357,300
Cordillera	4,948	276,900
Paraguari	8,705	239,600
Concepción	18,051	190,500
Guairá	3,846	197,000
Canendiyú	14,667	183,700
Caazapá	9,496	151,300
Amambay	12,933	125,300
Misiones	9,556	117,000
Neembucú	12,147	83,800
Oriental	*159,827*	*6,278,700*
Presidente Hayes	72,907	103,400
Boquerón[2]	91,669	57,800
Alto Paraguay[3]	82,349	11,300
Occidental	*246,925*	*172,500*

[1]Capital district. [2]Incorporates former department of Nueva Asunción. [3]Incorporates former department of Chaco.

The population is mixed Spanish and Guaraní Indian. There are 89,000 unassimilated Indians of other tribal origin, in the Chaco and the forests of eastern Paraguay. The official languages are Spanish and Guaraní: 24·8% of the population speak only Guaraní; 51·5% are bilingual (Spanish/Guaraní); and 7·6% speak only Spanish.

Mennonites, who arrived in three groups (1927, 1930 and 1947), are settled in the Chaco and eastern Paraguay. There are also Korean and Japanese settlers.

SOCIAL STATISTICS

2006 births, 112,659; deaths, 19,298. Rates, 2006 (per 1,000 population): birth, 18·7; death, 3·2. Annual population growth rate, 2000–05, 2·0%. Expectation of life, 2007: 69·6 years for males and 73·8 for females. Infant mortality, 2010, 21 per 1,000 live births; fertility rate, 2008, 3·0 births per woman.

CLIMATE

A tropical climate, with abundant rainfall and only a short dry season from July to Sept., when temperatures are lowest. Asunción, Jan. 81°F (27°C), July 64°F (17·8°C). Annual rainfall 53" (1,316 mm).

CONSTITUTION AND GOVERNMENT

On 18 June 1992 a Constituent Assembly approved a new constitution. The head of state is the *President,* elected for a non-renewable five-year term. Parliament consists of an 80-member *Chamber of Deputies,* elected from departmental constituencies, and a 45-member *Senate,* elected from a single national constituency.

National Anthem

'Paraguayos, república o muerte!' ('Paraguayans, republic or death!'); words by F. Acuña de Figueroa, tune by F. Dupuy.

RECENT ELECTIONS

Parliamentary and presidential elections were held on 21 April 2013. Horacio Cartes of the Colorado Party (ANR) was elected president with 45·8% of votes cast. Efraín Alegre of the Authentic Radical Liberal Party (PLRA) won 36·9% of the vote, Mario Ferreiro of Forward Country 5·9% and Aníbal Carrillo of the Guasú Front 3·3%. There were seven other candidates. Turnout was 68·6%.

In the parliamentary elections the National Republican Association–Colorado Party (ANR-PC) won 44 seats in the Chamber of Deputies, the Authentic Radical Liberal Party (PLRA) 27 seats, the National Union of Ethical Citizens (UNACE) 2, Forward Country 2 and the National Encounter Party (PEN) 2. Three smaller parties won one seat each. In the Senate the ANR-PC won 19 seats, the PLRA 12, the Guasú Front 5, the Democratic Progressive Party 3, Forward Country 2, the UNACE 2, the PEN 1 and the Beloved Fatherland Party 1.

CURRENT GOVERNMENT

President: Federico Franco; b. 1962 (PLRA; sworn in 22 June 2012).

In March 2013 the cabinet comprised:

Minister of Agriculture: Enzo Cardozo. *Defence:* María Liz Arnold. *Education:* Horacio Galeano. *Finance:* Manuel Ferreira. *Foreign Affairs:* José Félix Fernández Estigarribia. *Health:* Dr Antonio Arbo. *Industry and Trade:* Francisco Rivas. *Interior:* Carmelo Caballero. *Justice and Labour:* María Lorena Segovia Azucas. *Public Works:* Enrique Salyn Buzarquis.

General Secretary of the Presidency: Martín Burt.

Office of the President (Spanish only):
http://www.presidencia.gov.py

CURRENT LEADERS

Federico Franco

Position
President

Introduction
Federico Franco became president in June 2012 after the impeachment of his predecessor, Fernando Lugo. Franco had been vice-president since 2003.

Early Life
Luis Federico Franco Gómez was born in Asunción on 24 July 1962. The son of a physician active in the Liberal Party, he attended the San Jose Apostolic College before graduating in medicine from the National University of Asunción in 1986.

After postgraduate study in Cuba, he qualified as a surgeon and established a practice in Fernando de la Mora, a suburb of Asunción. A supporter of the centrist Authentic Radical Liberal Party (PLRA) since his student days, Franco was elected to represent them in the municipality of Fernando de la Mora in 1991 and became mayor in 1996.

In 2000 Franco's brother, Julio César, was elected PLRA party vice president. In the general election of April 2003, Franco was narrowly elected governor of the Central department but his brother lost the presidential race to Nicane Duarte Frutos of the Colorado Party. As governor of Central, Paraguay's most populous and prosperous department, Franco reformed its health services.

In Jan. 2008 Franco was elected leader of the PLRA, taking it into a new 12-party coalition, the Patriotic Alliance for Change (APC). The APC leader, Fernando Lugo, won the presidency on a platform of social reform, bringing an end to 61 years of Colorado Party rule. Franco became vice-president but policy differences with Lugo soon emerged, notably over land reform. Franco accused the president of failing to respect the law in his drive to push through measures to give land to poorer farmers.

Amid growing hostility from Congress and the armed forces, Lugo's authority was further undermined by paternity claims made against him by three women in 2009. In mid-2012 there were violent clashes between police and farmers occupying land that they claimed had been taken from them during the Stroessner dictatorship. Seven police and 11 farmers were killed. Parliament launched an impeachment of the president over his handling of the crisis and a vote on 21 June ended his mandate. In accordance with the constitution, Franco took over as president until Aug. 2013.

Career in Office
Progress on Franco's promised land, tax and energy reforms has been undermined by weak support in Congress and a lack of recognition of his administration by regional neighbours. The governments of Venezuela and Ecuador were particularly critical, arguing that Lugo was not given sufficient time to present the case against his impeachment and that his removal was tantamount to a coup.

Following the presidential election of April 2013, Franco was scheduled to be succeeded as president by Horacio Cartes on 15 Aug. 2013.

DEFENCE

The army, navy and air forces are separate services under a single command. The President of the Republic is the active C.-in-C. Conscription is for 12 months (two years in the navy).

In 2008 defence expenditure totalled US$132m. (US$19 per capita), representing 0·8% of GDP.

Army

Strength (2007) 7,600 (1,500 conscripts). In addition there is a paramilitary Special Police Force numbering 14,800 (4,000 conscripts).

Navy

Personnel in 2007 totalled 1,950 (or which 850 conscripts) including 900 marines (of which 200 conscripts) and 100 naval aviation.

Air Force

The air force had a strength of 1,100 in 2007 (200 conscripts). There are ten combat-capable aircraft including Lockheed T-33s.

ECONOMY

In 2006 agriculture accounted for 21·0% of GDP, industry 18·3% and services 60·7%.

Overview

Paraguay has few mineral resources and the economy is focused on agriculture. The country is the world's fourth largest exporter of soybean and the largest exporter of organic sugar, as well as a major exporter of hydro-electric power.

In the 1990s growth was slow but steady until 1998 when the economy stagnated. In 2002 performance slumped again, partly as a knock-on from the recession in Argentina, resulting in a request for IMF help. There was a return to growth from 2003–08, with restored confidence in public institutions, structural reform and debt reduction leading to fiscal account stability. Economic growth was at its highest level since the 1970s, boosted by strong exports, before a contraction in 2009. However, the economy rebounded the following year, achieving record growth of 13·1%.

Foreign direct investment increased by 46% between 2010 and 2011, but a severe drought in the first quarter of 2012 stifled the agricultural sector and held back further growth.

There is a large informal component of the economy and most enterprises are small. The lack of infrastructure to serve basic needs is a growing concern. 34·7% of the population live in poverty and only 35% of the rural population has access to drinking water. Unemployment was close to 6% in 2011, while underemployment stood at over 20%. Other problems include endemic corruption and Paraguay's reputation as a centre for smuggling, organized crime and money laundering.

Currency
The unit of currency is the *guaraní* (PYG), notionally divided into 100 *céntimos*. In July 2005 total money supply was 4,733·6bn. guaranís and foreign exchange reserves were US$1,109m. Inflation was 4·7% in 2010 and 6·6% in 2011.

Budget
Budgetary central government revenue and expenditure in 1bn. guaranís:

	2005	2006	2007
Revenue	8,430·0	9,556·5	10,812·4
Expenditure	7,157·4	8,175·6	8,975·2

Main sources of revenue in 2007: taxes on goods and services, 4,835·4bn. guaranís; taxes on income, profits and capital gains, 1,228·5bn. guaranís; taxes on international trade and transactions, 837·6bn. guaranís. Leading items of expenditure by economic type in 2007: compensation of employees, 4,456·1bn. guaranís; social benefits, 1,442·6bn. guaranís; grants, 1,218·2bn. guaranís.

VAT is 10% (reduced rate, 5%).

Performance
The economy contracted by 4·0% in 2009 but there was growth of 13·1% in 2010, among the highest rates in the world. In 2011 the economy grew by 4·3%; total GDP was US$23·8bn.

Banking and Finance
The Central Bank is a state-owned autonomous agency with the sole right of note issue, control over foreign exchange and the supervision of commercial banks (*Governor*, Jorge Raúl Corvalán Mendoza). There is a Superintendencia de Bancos under Edgar Virgilio Paredes Álvarez. In 2002 there were five commercial banks and 11 foreign banks.

Foreign debt was US$4,938m. in 2010, representing 25·3% of GNI.

There is a stock exchange in Asunción.

ENERGY AND NATURAL RESOURCES

Environment
Paraguay's carbon dioxide emissions from the consumption and flaring of fossil fuels were the equivalent of 0·6 tonnes per capita in 2008.

Electricity
Installed capacity was 8·1m. kW in 2007. Output (2007), 53·71bn. kWh (almost exclusively hydro-electric); consumption per capita in 2007 was 1,402 kWh. Paraguay is the third largest exporter of electricity (after France and Germany), with 45·1bn. kWh in 2007.

Minerals
The country is poor in minerals. Limestone, gypsum, kaolin and salt are extracted. 2006 estimated output: kaolin, 66,000 tonnes; limestone, 16,000 tonnes.

Agriculture
In 2002, 33·4% of the economically active population were engaged in agriculture. In 2002 there were approximately 3·02m. ha. of arable land and 95,000 ha. of permanent crops.

Output (in 1,000 tonnes), 2002: cassava, 4,430; soybeans, 3,300; sugarcane, 3,210; maize, 867; wheat, 359; oranges, 207; seed cotton, 124; sweet potatoes, 124; watermelons, 115; rice, 105. *Yerba maté*, or strongly flavoured Paraguayan tea, continues to be produced but is declining in importance.

Livestock (2003 estimates): 8·81m. cattle, 3·25m. pigs, 410,000 sheep, 360,000 horses and 16m. chickens.

Forestry
The area under forests in 2010 was 17·58m. ha., or 44% of the total land area. Timber production was 10·30m. cu. metres in 2007.

Fisheries
In 2008 the catch totalled 1,708 tonnes, exclusively from inland waters.

INDUSTRY
Paraguay is one of the least industrialized countries in Latin America. Industries include meat packing, sugar processing, cement, textiles, brewing, wood products and consumer goods. In 2006 industry accounted for 18·3% of GDP, with manufacturing contributing 11·8%.

Labour
The labour force in 2002 totalled 1,980,492 (67·9% males). In 2002, 27% of the economically active population were engaged in agriculture, fisheries, hunting and forestry.

INTERNATIONAL TRADE

Imports and Exports
Trade in US$1m.:

	2003	2004	2005	2006	2007
Imports c.i.f.	2,227·5	3,097·4	3,714·9	5,878·8	5,844·7
Exports f.o.b.	1,241·5	1,625·7	1,687·8	1,906·4	2,784·7

Main imports in 2006: machinery and transport equipment, 46·9%; chemicals and chemical products, 12·2%; petroleum oils, 11·8%; metal tools, 2·6%. Main exports, 2006: soybeans, 23·0%; beef, 21·5%; maize, 8·7%; oilcake and other solid residues, 7·2%. Main import suppliers in 2006: China, 25·1%; Brazil, 19·0%; Argentina, 13·0%; USA, 6·0%. Main export markets, 2006: Uruguay, 22·0%; Brazil, 17·2%; Russia, 11·9%; Cayman Islands, 9·5%.

COMMUNICATIONS

Roads
In 2002 there were around 29,500 km of roads, of which 53·9% were paved. Passenger cars numbered 240,700 in 2007, there were 248,100 lorries and vans, 12,800 buses and coaches, and 134,900 motorcycles and mopeds. There were 845 fatalities as a result of road accidents in 2007.

Rail
The President Carlos Antonio López (formerly Paraguay Central) Railway runs from Asunción to Encarnación, on the Río Alto Paraná, with a length of 441 km (1,435 mm gauge), and connects with Argentine Railways over the Encarnación-Posadas bridge opened in 1989. In 2003 freight carried came to 210,000 tonnes.

Civil Aviation
There is an international airport at Asunción (Silvio Pettirossi). The main Paraguay-based carrier is TAM Airlines (formerly TAM Mercosur). In 2003 scheduled airline traffic of Paraguay-based carriers flew 6m. km, carrying 299,000 passengers (288,000 on international flights). In 2000 Asunción handled 466,000 passengers (422,000 on international flights) and 6,600 tonnes of freight.

Shipping

Asunción, the chief port, is 1,500 km from the sea. In Jan. 2009 there were 32 ships of 300 GT or over registered, totalling 44,000 GT.

Telecommunications

In 2008 there were 491,000 main (fixed) telephone lines; mobile phone subscribers numbered 5,954,400 in 2008 (95·5 per 100 persons). There were 894,200 internet users in 2008. In March 2012 there were 1·0m. Facebook users.

SOCIAL INSTITUTIONS

Justice

The 1992 constitution confers a large measure of judicial autonomy. The highest court is the Supreme Court with nine members. Nominations for membership must be backed by six of the eight members of the Magistracy Council, which appoints all judges, magistrates and the electoral tribunal. The Council comprises elected representatives of the Presidency, Congress and the bar. There are special Chambers of Appeal for civil and commercial cases, and criminal cases. Judges of first instance deal with civil, commercial and criminal cases in six departments. Minor cases are dealt with by Justices of the Peace.

The Attorney-General represents the State in all jurisdictions, with representatives in each judicial department and in every jurisdiction.

The population in penal institutions in Nov. 2008 was 6,031 (97 per 100,000 of national population). The death penalty was abolished for all crimes in 1992.

Education

Adult literacy was 95% in 2007. Education is free and nominally compulsory. In 2005 there were 933,995 pupils at primary schools (with 33,434 teaching staff in 2004) and 529,309 at secondary level (with 44,440 teaching staff in 2004). There were 156,167 students in tertiary education in 2005. Paraguay's leading institute of higher education is the National University of Asunción (Universidad Nacional de Asunción), the country's oldest university, founded in 1889.

In 2007 public expenditure on education came to 4·0% of GDP and 11·9% of total government spending.

Health

In 2003 there were 1,117 health establishments (including 84 hospitals) with 7,167 beds. There were 6,400 physicians, 1,947 dentists and 1,089 nurses in 2000.

RELIGION

Religious liberty was guaranteed by the 1967 constitution. Article 6 recognized Roman Catholicism as the official religion of the country. It had 3·5m. adherents in 2002. There are Mennonite, Anglican and other communities as well. In 2002 followers of other religions (mostly Protestants) totalled 322,000.

CULTURE

World Heritage Sites

Paraguay has one site on the UNESCO World Heritage List: the Jesuit Missions of La Santísima Trinidad de Paraná and Jesús de Tavarangue (inscribed on the list in 1993).

Press

In 2008 there were eight daily newspapers with a combined circulation of 135,000.

Tourism

In 2005 there were 341,000 foreign tourists, spending US$96m.

DIPLOMATIC REPRESENTATIVES

Of Paraguay in the United Kingdom (3rd Floor, 344 High St. Kensington, London, W14 8NS)
Ambassador: Miguel Solano López.

Of the United Kingdom in Paraguay (embassy in Asunción closed in April 2005)
Ambassador: John Freeman (resides in Buenos Aires, Argentina).

Of Paraguay in the USA (2400 Massachusetts Ave., NW, Washington, D.C., 20008)
Ambassador: Rigoberto Gauto.

Of the USA in Paraguay (1776 Mariscal López Ave., Asunción)
Ambassador: James H. Thessin.

Of Paraguay to the United Nations
Ambassador: José Antonio dos Santos.

Of Paraguay to the European Union
Ambassador: Mario Francisco Sandoval Fernández.

FURTHER READING

Gaceta Official, published by Imprenta Nacional, Estrella y Estero Bellaco, Asunción
Anuario Daumas. Annual
Anuario Estadístico de la República del Paraguay. Annual

Nickson, R. A. and Lambert, P. (eds.) *The Transition to Democracy in Paraguay*. 1997

National library: Biblioteca Nacional, Calle de la Residenta, 820 c/ Perú, Asunción.
National Statistical Office: Dirección General de Estadísticas, Encuestas y Censos.
Website (Spanish only): http://www.dgeec.gov.py

PERU

República del Perú
(Republic of Peru)

Capital: Lima
Population projection, 2015: 30·76m.
GNI per capita, 2011: (PPP$) 8,389
HDI/world rank: 0·725/80
Internet domain extension: .pe

KEY HISTORICAL EVENTS

Hunter-gatherers lived in Peru from at least 9000 BC. Irrigation canals discovered recently in the Andean foothills of northern Peru date farming from around 3400 BC. The Chavin culture in central and northern Peru between 900 BC and 200 BC left monumental temples and intricate artwork across a wide area. The Paracas culture emerged on the southern coast in around 300 BC and evolved into the Nazca culture, famed for its exquisite textiles. Further north, the coastal Moche culture flourished between around 100 BC and AD 700, producing distinctive metalwork and pottery. The following centuries saw the rise of inland, Andean cultures and powerful city states such as Chancay, Sipan and Cajamarca.

The Inca civilization is thought to have its origins in a Quechua-speaking tribe that settled in the Cusco Valley from about 1200. By the mid-15th century, under Emperor Tupac Yupanqui (1471–93) and subsequently under Huayna Capac, the empire stretched along the Andes from Ecuador to Chile. The Inca's road network was an engineering masterpiece, as were the terraced fields and cities such as the famed Machu Picchu.

The Spanish adventurer, Francisco Pizarro, landed on the Ecuadorian coast in 1532 and moved south. Relations between the Spanish and the Incas quickly soured and Emperor Atahualpa was captured following the battle of Cajamarca. He was executed in 1533 and a year later Pizarro conquered the city of Cusco. Lima was founded in 1535 and in 1542 it became the seat of the Viceroyalty of Peru, which for a time had jurisdiction over all the Spanish colonies in South America. The Spanish conquistadors amassed vast wealth and power by controlling the trade in Andean gold and silver, while the Incas became increasingly marginalized. Revolts against Spanish rule occurred in the 1780s but concerted demands for independence came only after the French Revolution and Napoleon's conquest of Spain in 1808.

José de San Martin of Argentina (who had ended Spanish rule in Chile in 1818) and Simón Bolívar of Venezuela proclaimed Peruvian independence on 28 July 1821 but it was only confirmed in Dec. 1824, when Antonio José de Sucre defeated Spanish troops at Ayacucho. After independence Peru and its neighbours engaged in various territorial disputes. Chile's victory over Peru and Bolivia in the War of the Pacific (1879–83) resulted in Peru ceding the department of Tarapaca and the provinces of Tacna and Arica to Chile. Gen. Andrés Avelino Cáceres became president in 1885 and tried to breathe life into the crippled economy by encouraging foreign management of the railways and guano (fertilizer) exports.

A businessman, Augusto Leguía, who became president for four years in 1908, pushed through economic reforms. He became increasingly authoritarian during his second term of office (1919–30), and in 1924 Dr Victor Raúl Haya de la Torre founded the Alianza Popular Revolucionaria Americana (APRA), which called for radical reform. The party was banned by Leguia and then outlawed by his successor, Sanchez Cerro, during the 1930s. Peru sided with the Allies during the Second World War and in 1945 Luis Bustamante y Rivero was elected president, with APRA backing. Splits soon emerged and Manuel Odría led a military coup three years later. An inconclusive election in 1962 enabled Gen. Ricardo Pérez Godoy to seize power, although he was deposed in a coup led by Gen. Nicolás Lindley López a year later. There followed a period of civilian rule but the military staged yet another coup in 1968. In 1978–79 a constituent assembly drew up a new constitution, after which a civilian government was installed.

Peru was plagued by political violence between the early 1980s and the late 1990s with 69,000 people killed by Maoist Shining Path insurgents, the smaller Tupac Amaru Revolutionary Movement and government forces. On 5 April 1992 President Alberto Fujimori suspended the constitution, dissolved parliament and implemented drastic economic reforms to tackle rampant inflation. A new constitution was promulgated on 29 Dec. 1993. But while Peru enjoyed stability and economic growth, continuing autocratic rule put some politicians above the law. Embroiled in a bribery and corruption scandal, President Fujimori's discredited administration came to an end in Nov. 2000 when he resigned while out of the country.

A caretaker government under Valentín Paniagua presided over new presidential and congressional elections in April 2001. A new government led by President Alejandro Toledo took office in July 2001. The Toledo government consolidated Peru's return to democracy and presided over a period of strong economic growth, although great inequalities persisted. Alan García, who had been president for five years in the second half of the 1980s, narrowly won a run-off election in April 2006 and was sworn in as president for a second term in July 2006.

TERRITORY AND POPULATION

Peru is bounded in the north by Ecuador and Colombia, east by Brazil and Bolivia, south by Chile and west by the Pacific Ocean. Area, 1,285,216 sq. km (including the area of the Peruvian part of Lake Titicaca).

For an account of the border dispute with Ecuador, *see* ECUADOR: Territory and Population.

Census population, 2007, 27,412,157; density, 21·3 per sq. km. In 2011 the population was 77·3% urban.

The UN gives a projected population for 2015 of 30·76m.

The country is administratively divided into 25 regions and an autonomous province of Lima (with capitals): Amazonas (Chachapoyas), Ancash (Huaráz), Apurímac (Abancay), Arequipa (Arequipa), Ayacucho (Ayacucho), Cajamarca (Cajamarca), Callao (Callao), Cusco (Cusco), Huancavelica (Huancavelica), Huánuco (Huánuco), Ica (Ica), Junín (Huancayo), La Libertad (Trujillo), Lambayeque (Chiclayo), Lima (Huacho), Lima province (Lima), Loreto (Iquitos), Madre de Dios (Puerto Maldonado), Moquegua (Moquegua), Pasco (Cerro de Pasco), Piura (Piura), Puno (Puno), San Martín (Moyobamba), Tacna (Tacna), Tumbes (Tumbes), Ucayali (Pucallpa).

The largest cities (with 2007 census populations) are: Lima, 8,472,935; Arequipa, 784,651; Trujillo, 682,834; Chiclayo, 524,442; Piura, 377,496; Iquitos, 370,962.

In 1991 there were some 100,000 Peruvians of Japanese origin. Indigenous peoples account for 47% of the population.

The official languages are Spanish (spoken by 83·9% of the population in 2007), Quechua (13·2%) and Aymara (1·8%).

SOCIAL STATISTICS

2009 births (estimate), 604,000; 2009 deaths (estimate), 144,000. Rates per 1,000 population (2009): birth, 21; death, 5. Annual population growth rate, 2000–05, 1·5%; infant mortality, 2010, 15 per 1,000 live births. Life expectancy, 2007: males, 70·4 years; females, 75·8. Fertility rate, 2008, 2·6 births per woman.

CLIMATE

There is a very wide variety of climates, ranging from tropical in the east to desert in the west, with perpetual snow in the Andes. In coastal areas, temperatures vary very little, either daily or annually, though humidity and cloudiness show considerable variation, with highest humidity from May to Sept. Little rain is experienced in that period. In the Sierra, temperatures remain fairly constant over the year, but the daily range is considerable. There the dry season is from April to Nov. Desert conditions occur in the extreme south, where the climate is uniformly dry, with a few heavy showers falling between Jan. and March. Lima, Jan. 74°F (23·3°C), July 62°F (16·7°C). Annual rainfall 2" (48 mm). Cusco, Jan. 56°F (13·3°C), July 50°F (10°C). Annual rainfall 32" (804 mm). El Niño is the annual warm Pacific current that develops along the coasts of Peru and Ecuador. El Niño in 1982–83 resulted in agricultural production down by 8·5% and fishing output down by 40%. El Niño in 1991–94 was unusually long. El Niño in 1997–98 resulted in a sudden rise in the surface temperature of the Pacific by 9°F (5°C) and caused widespread damage and loss of life.

CONSTITUTION AND GOVERNMENT

The 1980 constitution provided for a legislative *Congress* consisting of a *Senate* and a *Chamber of Deputies*, and an Executive formed of the President and a Council of Ministers appointed by him. Elections were to be every five years with the President and Congress elected, at the same time, by separate ballots.

On 5 April 1992 President Fujimori suspended the 1980 constitution and dissolved Congress.

A referendum was held on 31 Oct. 1993 to approve the twelfth constitution, including a provision for the president to serve a

consecutive second term. 52·24% of votes cast were in favour. The constitution was promulgated on 29 Dec. 1993. In Aug. 1996 Congress voted for the eligibility of the President to serve a third consecutive term of office.

Congress has 130 members, elected for a five-year term by proportional representation. All citizens over the age of 18 are eligible to vote. Voting is compulsory.

National Anthem

'Somos libres, seámoslo siempre' ('We are free, let us always be so'); words by J. De La Torre Ugarte, tune by J. B. Alcedo.

GOVERNMENT CHRONOLOGY

Heads of State since 1945. (AP = Popular Action; APRA = American Popular Revolutionary Alliance; FDN = National Democratic Front; MDP = Pradista Democratic Movement/ Peruvian Democratic Movement; NM-C90 = New Majority/ Change 90; PAP = Peruvian Aprista Party; PNP = Peruvian Nationalist Party; PP = Peru Possible; PR = Restorer Party)

President of the Republic
1945–48	FDN	José Luis Bustamante y Rivero

Chairmen of the Military Junta of Government
1948–50	military	Manuel Apolinario Odría Amoretti
1950	military	Zenón Noriega Agüero

Presidents of the Republic
1950–56	PR	Manuel Apolinario Odría Amoretti
1956–62	MDP	Manuel Prado y Ugarteche

Junta of Government/Joint Command of the Armed Forces
1962–63	military	Ricardo Pío Pérez Godoy, Nicolás Lindley López, Juan Francisco Torres Matos, Pedro Vargas Prada Peirano

Presidents of the Republic
1963–68	AP	Fernando Belaúnde Terry
1968–75	military	Juan Francisco Velasco Alvarado
1975–80	military	Francisco Morales Bermúdez
1980–85	AP	Fernando Belaúnde Terry
1985–90	APRA	Alan Gabriel Ludwig García Pérez
1990–2000	NM-C90	Alberto Keinya Fujimori Fujimori
2000–01	AP	Valentín Paniagua Corazao
2001–06	PP	Alejandro Celestino Toledo Manrique
2006–11	PAP	Alan Gabriel Ludwig García Pérez
2011–	PNP	Ollanta Moisés Humala Tasso

RECENT ELECTIONS

The first round of presidential elections were held on 10 April 2011. Ollanta Humala Tasso of the Gana Perú alliance (Peru Wins) won 31·7% of the vote, followed by Keiko Fujimori of Fuerza 2011 (Force 2011) with 23·5%, Pedro Pablo Kuczynski of Alianza por el Gran Cambio (Alliance for the Great Change) with 18·5% and former president Alejandro Celestino Toledo Manrique of Perú Posible (Peru Possible) 15·6%. There were six other candidates. In the presidential run-off held on 5 June 2011 Ollanta Humala Tasso won with 51·5% of the vote against 48·5% for Keiko Fujimori.

In the congressional elections held on the same day Gana Perú gained 47 seats with 25·3% of votes cast. Fuerza 2011 came second with 37 seats (23·0%), ahead of the Alianza Electoral Perú Posible (Possible Peru Alliance) with 21 (14·8%), the Alianza por el Gran Cambio 12 (14·4%), Alianza Solidaridad Nacional (National Solidarity Alliance) 9 (10·2%) and Partido Aprista Peruano (American Popular Revolutionary Alliance) 4 (6·4%). Other parties received 5·9% of the vote between them but won no seats.

CURRENT GOVERNMENT

President: Ollanta Moisés Humala Tasso; b. 1962 (Peruvian Nationalist Party; sworn in 28 July 2011).

First Vice-President: Marisol Espinoza Cruz. *Second Vice-President:* Vacant.

In March 2013 the government comprised:

President of the Council of Ministers (Prime Minister): Juan Jiménez Mayor; b. 1964 (ind.; sworn in 23 July 2012).

Minister of Foreign Affairs: Rafael Roncagliolo Orbegozo. *Defence:* Pedro Cateriano. *Economy and Finance:* Luis Miguel Castilla. *Interior:* Wilfredo Pedraza. *Justice and Himan Rights:* Eda Rivas Franchini. *Education:* Patricia Salas O'Brien. *Health:* Midori de Habich. *Agriculture:* Milton von Hesse. *Labour and Employment Promotion:* Teresa Laos Cáceres. *Production:* Gladys Triveño. *Foreign Trade and Tourism:* José Luis Silva Martinot. *Energy and Mines:* Jorge Humberto Merino. *Transport and Communications:* Carlos Paredes Rodríguez. *Housing, Construction and Sanitation:* René Cornejo Díaz. *Women's Affairs and Vulnerable Populations:* Ana Jara. *Environment:* Manuel Pulgar-Vidal. *Culture:* Luis Peirano. *Development and Social Inclusion:* Carolina Trivelli.

President of the Council of Ministers (Spanish only):
    http://www.pcm.gob.pe

CURRENT LEADERS

Ollanta Moisés Humala Tasso

Position
President

Introduction
Ollanta Humala became president in July 2011. Leader of the Peruvian Nationalist Party, he was elected after two rounds of voting, defeating Keiko Fujimori, daughter of former president Alberto Fujimori. Humala was backed by the Gana Perú coalition and succeeded Alan García.

Early Life
Born in Lima in June 1962, Humala is the son of Isaac Humala who developed the ethno-nationalist ideology called 'ethno-cacerism'. Ollanta Humala studied in Lima before joining the military. He trained initially as a paratrooper and amphibious combatant at the School of the Americas in Panama. In 1992, as head of a detachment in Tingo María, he was involved in operations against the rebel group Shining Path, for which he was later accused of human rights abuses. By 2000 he had been promoted to lieutenant-colonel.

Also in 2000, he led an uprising (along with his brother) in a bid to force out President Alberto Fujimori and leading military commanders. Following Fujimori's resignation the interim government pardoned Humala and awarded him the Peru Cross for Military Merit.

From 2000–04 Humala worked for the ministry of defence and in 2001 completed a masters degree in political sciences from the Pontifical Catholic University of Peru. He became the military attaché to the Peruvian embassy in Paris in 2003 and studied for a doctorate in international law at the Sorbonne. In 2004 he was posted to South Korea. While there his name was linked to an uprising in Peru against President Alejandro Toledo.

In 2005 Humala founded the Peruvian Nationalist Party (PNP). With support from the Union for Peru (UPP), he ran for the presidency in 2006 but lost support over questions about his past and his left-wing sympathies. From 2006–11 he fought off several legal cases related to previous coups. In 2011, backed by the Gana Perú coalition, he ran for the presidency again, defeating Keiko Fujimori.

Career in Office
In 2006 Humala had defined himself as 'anti-neoliberal' and 'anti-global capitalist' but in 2011 he adopted a more moderate profile. He distanced himself from associations with Venezuela's Hugo

Chávez and stood on a platform of energy self-sufficiency, a more equitable distribution of wealth and increased taxes on mining. On coming to power, he chose a moderate cabinet, quelling worries in business circles and cutting short an initial shock in the Lima stock market. He increased taxes on mining, and announced social programmes including a non-contributory basic pension for the elderly, a public childcare programme and more scholarships to promote university education. The minimum wage was also increased.

Large-scale mining projects, such as the Conga copper and gold mine, offer potentially lucrative future revenue sources. However, popular opposition to mining schemes is widespread in the country and protests led in Dec. 2011 to the resignation of Prime Minister Salomon Lerner, his replacement by Óscar Valdés and a major cabinet reshuffle. There were further violent anti-mining clashes in the southern Cusco region in May 2012 and in northern provinces in July, prompting the government to declare temporary states of emergency. Also in July, following the resignation of the unpopular Óscar Valdés and his administration in response to the unrest, Humala appointed Juan Jiménez Mayor as the new premier.

DEFENCE

Conscription was abolished in 1999. In 2008 defence expenditure totalled US$1,424m. (US$49 per capita), representing 1·1% of GDP.

Army

There are four military regions. In 2007 the Army comprised 74,000 personnel and 188,000 reserves. In addition there is a paramilitary national police force of 77,000 personnel.

Navy

The principal ship of the Navy is the former Netherlands cruiser *Almirante Grau*, built in 1953. Other combatants include six diesel submarines (two in refit) and eight Italian-built frigates.

Callao is the main base, where the dockyard is located and most training takes place. Smaller bases exist at Iquitos, Paita, Puerto Maldonado, Puno, San Lorenzo Island and Talara.

Naval personnel in 2007 totalled 23,000 including 1,000 Coast Guards, about 800 Naval Air Arm and 4,000 Marines.

Air Force

The operational force is divided into five regions—North, Lima, South, Central and Amazon.

In 2007 there were some 17,000 personnel and 60 combat-capable aircraft (including Su-25s, Mirage 2000s and MiG-29s) and 16 attack helicopters.

ECONOMY

Agriculture produced 6·8% of GDP in 2006, industry 37·5% and services 55·7%.

Overview

From 1990 the economy underwent a transformation, with the ending of a debt crisis and hyperinflation, the implementation of tax and pension reforms, and a programme of liberalization and privatization.

Recession between 1997 and 2001 was caused by a spillover from emerging market crises and commodity price weaknesses. The economy then responded well to a policy of fiscal responsibility, job creation programmes and rising commodity prices. The government oversaw reforms to open the economy, reduce labour costs in the formal sector, and attract private investment.

Following the 1997–2001 dip, the economy grew at its most sustained pace since the 1950s spurred by elevated commodity prices (particularly for minerals and metals) as well as growth in the agriculture, textiles, manufacturing and construction sectors. Per capita income almost tripled between 2001 and 2011. With an expanded middle class, income inequality decreased and the poverty rate fell from 44·5% in 2006 to 30·8% in 2010. The government aims to further reduce the rate to 20% by 2016.

The economy recovered quickly from the global economic crisis. In 2010 real GDP expanded by nearly 9%, driven by private domestic demand, while the annual inflation rate was the lowest in Latin America.

Private investment has been concentrated in the natural resources sector while employment creation has been predominantly in the capital, Lima (home to roughly a third of the population), and the mining-intensive north coast region. The impoverished rural highlands have benefited little and remain without proper access to basic infrastructure and running water. 54% of Peruvians in rural areas were living in poverty in 2010.

Key challenges include improving the investment climate and achieving more inclusive growth. The economy remains vulnerable to fluctuations in world commodity prices and over-reliance on non-renewable natural resources.

Currency
The monetary unit is the *nuevo sol* (PEN), of 100 *céntimos*, which replaced the *inti* in 1991 at a rate of 1m. intis = 1 nuevo sol. Inflation, which had been over 7,000% in 1990, was 1·5% in 2010 and 3·4% in 2011. Foreign exchange reserves were US$14,773m. in July 2005, gold reserves totalled 1·12m. troy oz and total money supply was 24,565m. nuevos soles.

Budget
Central government revenue and expenditure (in 1m. sols), year ending 31 Dec.:

	2007	2008	2009
Revenue	67,324	74,356	67,932
Expenditure	57,165	62,424	67,245

Principal sources of revenue in 2009 were: taxes on goods and services, 26,325m. sols; taxes on income, profits and capital gains, 20,352m. sols; social security contributions, 6,659m. sols. Main items of expenditure by economic type in 2009 were: grants, 24,157m. sols; use of goods and services, 13,727m. sols; compensation of employees, 12,233m. sols.

In Dec. 2005 the World Bank approved a US$150m. loan to assist with the government decentralization process and enhance competitiveness.

VAT is 18%.

Performance
Real GDP growth was 0·9% in 2009, rising to 8·8% in 2010. Peru's real GDP growth was above average for Latin America every year between 2002 and 2009 apart from 2004. In 2011 the economy grew by 6·9%; total GDP was US$176·9bn.

Banking and Finance
The bank of issue is the Banco Central de Reserva (*President*, Julio Velarde Flores), which was established in 1922. The government's fiscal agent is the Banco de la Nación. In 2002 there were three other government banks (Banco Central Hipotecario del Perú, Banco de la Nación and Corporación Financiera de Desarrollo), ten commercial banks, one regional bank and three foreign banks. Legislation of April 1991 permitted financial institutions to fix their own interest rates and reopened the country to foreign banks. The Central Reserve Bank sets the upper limit.

External debt in 2010 totalled US$36,271m., equivalent to 24·6% of GNI.

In 2010 Peru received US$7,328m. of foreign direct investment, up from US$5,576m. in 2009. The total stock of FDI at the end of 2010 was US$41·8bn.

There are stock exchanges in Lima and Arequipa.

ENERGY AND NATURAL RESOURCES
Peru lays claim to 84 of the world's 114 ecosystems; 28 of its climate types; 19% of all bird species; 20% of all plant species; and 25 conservation areas (seven national parks, eight national reserves, seven national sanctuaries and three historic sanctuaries).

Environment
Peru's carbon dioxide emissions from the consumption and flaring of fossil fuels in 2008 were the equivalent of 1·2 tonnes per capita.

Electricity
In 2007 output was 29·93bn. kWh (19·55bn. kWh hydro-electric and 10·38bn. kWh thermal). Total generating capacity was 7·0m. kW in 2007. Consumption per capita in 2007 was 1,092 kWh.

Oil and Gas
Proven oil reserves at the end of 2008 amounted to 1·1bn. bbls. Output, 2008, 5·3m. tonnes. Natural gas reserves in 2008 were 330bn. cu. metres; output in 2007 was 2,831m. cu. metres. Commercial development of the huge Camisea gas field began in late 2004. In June 2010 Peru became a net gas exporter.

Minerals
The mining and fuel sectors accounted for some 8·1% of GDP in 2005. Lead, copper, iron, silver, zinc and petroleum are the chief minerals exploited. Mineral production, 2004 (in 1,000 tonnes): iron, 4,315; zinc, 1,209; copper, 1,036; lead, 306; silver, 3·1; gold, 0·17. 16,000 tonnes of coal were produced in 2004. Early in 1998 Southern Peru Copper, the country's largest mining company, estimated that 3,000 tonnes of copper production had been lost as a result of flooding caused by El Niño.

Agriculture
There are four natural zones: the Coast strip, with an average width of 80 km; the Sierra or Uplands, formed by the coast range of mountains and the Andes proper; the Montaña or high wooded region which lies on the eastern slopes of the Andes; and the jungle in the Amazon Basin, known as the Selva. Legislation of 1991 permits the unrestricted sale of agricultural land. Workers in co-operatives may elect to form limited liability companies and become shareholders.

Production in 2003 (in 1,000 tonnes): sugarcane, 9,550; potatoes, 3,300; rice, 2,139; bananas and plantains, 1,600; maize, 1,471; cassava, 890; onions, 450; oranges, 295; lemons and limes, 255; sweet potatoes, 225.

Livestock, 2003 estimates: sheep, 14·1m.; cattle, 5·0m.; pigs, 2·9m.; alpacas, 2·5m.; goats, 2·0m.; poultry, 95m. Livestock products (in 1,000 tonnes), 2003 estimates: poultry meat, 635; beef and veal, 146; pork, bacon and ham, 85; mutton and lamb, 32; milk, 1,220.

In 2002 there were approximately 3·70m. ha. of arable land and 0·61m. ha. of permanent crops. About 1·2m. ha. were irrigated in 2002.

Coca was cultivated in 2005 on approximately 38,000 ha., down from 115,000 ha. in 1995.

Forestry
In 2010 the area covered by forests was 67·99m. ha., or 53% of the total land area. The forests contain valuable hardwoods; oak and cedar account for about 40%. In 2007 timber production was 9·45m. cu. metres.

Fisheries
Sardines and anchovies are caught offshore to be processed into fishmeal, of which Peru is the world's largest producer (with a 2009 output of 1,346,900 tonnes or 28% of world production). Total catch in 2010 was 4,261,091 tonnes, of which 99% was from sea fishing. In 2009 total exports of fishery commodities came to a value of US$2·21bn.

INDUSTRY

The leading companies by market capitalization in Peru in April 2012 were: Credicorp, a financial holding company (US$12·0bn.); and Buenaventura, a mining company (US$10·7bn.).

About 70% of industries are located in the Lima/Callao metropolitan area. Industry accounted for 29·7% of GDP in 2001, with manufacturing contributing 15·3%. Production (2007 unless otherwise indicated, in 1,000 tonnes): cement (2002), 4,120; distillate fuel oil, 3,063; residual fuel oil, 2,693; petrol, 2,356; prepared animal feeds (2001), 1,508; sugar (2001), 755; soft drinks (2002), 1,179·7m. litres; beer (2005), 797·0m. litres; cigarettes (2002), 3·8bn. units.

Labour

The labour force in 1996 totalled 8,652,000 (71% males). In 1993, 1,852,800 people worked in agriculture, 1,167,000 in commerce, 783,900 in manufacturing, 599,700 in services, 347,500 in transport, 255,000 in building and 72,200 in mining. In 2002 an estimated 8·4% of the workforce was unemployed, up from 5·9% in 1991. On taking office in July 2011 the new president, Ollanta Humala, announced an immediate rise in the monthly minimum wage from 600 to 675 nuevos soles, with a further increase to 750 nuevos soles from Jan. 2012 (although this was subsequently postponed but did take effect from June 2012).

INTERNATIONAL TRADE

An agreement of 1992 gives Bolivia duty-free transit for imports and exports through a corridor leading to the Peruvian Pacific port of Ilo from the Bolivian frontier town of Desaguadero, in return for Peruvian access to the Atlantic via Bolivia's roads and railways. In April 2006 Peru and the USA signed a free trade agreement that eliminates tariffs on each other's goods; it took effect in Jan. 2009.

Imports and Exports

The following table shows the value of Peru's foreign trade (in US$1m.):

	2007	2008	2009	2010
Imports c.i.f.	20,368·3	29,952·8	21,869·7	29,879·5
Exports f.o.b.	28,084·6	31,288·2	26,738·3	35,073·2

The principal import suppliers in 2009 were: USA, 19·8%; China, 14·9%; Brazil, 7·7%; Ecuador, 4·7%. The principal export markets in 2009 were: USA, 17·2%; China, 15·3%; Switzerland, 14·8%; Canada, 8·6%.

The leading imports in 2009 were machinery and transport equipment (34·8%), manufactured goods (15·9%), chemicals and related products (15·0%), and petroleum and petroleum products (13·6%). Leading exports in 2009 were gold (25·2%), copper, copper ores and concentrates (22·9%), petroleum and petroleum products (7·2%), and animal feeds (5·7%).

COMMUNICATIONS

Roads

In 2006 there were 78,986 km of roads, of which 13·9% were paved. In 2007 there were 917,100 passenger cars, 480,900 lorries and vans and 44,400 buses and coaches. There were 67,155 road accidents involving injury in 2006 with 3,481 fatalities.

Rail

Total length (2002), 2,121 km on 1,435- and 914-mm gauges. Passenger-km travelled in 2005 came to 126m. and freight tonne-km to 1,101m. A mass transit system opened in Lima in 2003. Peru's first metro, also in Lima, opened in Jan. 2012.

Civil Aviation

There is an international airport at Lima (Jorge Chávez International). The main airline is the Chilean-owned Lan Perú. In 2003 services were also provided by the domestic airlines Aero Cóndor, AVIANDINA and Transportes Aéreos Nacionales de Selva, and by more than 20 international carriers. In 2003 scheduled airline traffic of Peruvian-based carriers flew 44m. km, carrying 2,226,000 passengers (547,000 on international flights). In 2001 Jorge Chávez International handled 4,089,914 passengers (2,128,872 on international flights) and 112,709 tonnes of freight.

Shipping

In 2004 there were 46 sea-going vessels and 651 lake and river craft. In Jan. 2009 there were nine ships of 300 GT or over registered, totalling 87,000 GT. Callao is the busiest port, handling 18,191,000 tonnes of cargo in 2008. There are also ports at Chimbote, Paita and Talara.

Telecommunications

In 2008 there were 2,878,200 main (fixed) telephone lines; mobile phone subscribers numbered 20,951,800 in 2008 (72·7 per 100 persons). In 2011, 36·5% of the population were internet users. In March 2012 there were 8·2m. Facebook users.

SOCIAL INSTITUTIONS

Justice

The judicial system is a pyramid at the base of which are the justices of the peace who decide minor criminal cases and civil cases involving small sums of money. The apex is the Supreme Court with a president and 12 members; in between are the judges of first instance, who usually sit in the provincial capitals, and the superior courts.

The police had 95,789 personnel in 2008. The population in penal institutions in Oct. 2004 was 32,129 (114 per 100,000 of national population).

Education

Adult literacy was 90% in 2007. Elementary education is compulsory and free between the ages of six and 16; secondary education is also free. In 2007 there were 1,204,022 children in pre-school education with 58,177 teaching staff, 3,993,965 pupils in primary schools with 179,743 teaching staff and 2,861,313 pupils in secondary schools with 158,890 teaching staff. There were 952,437 students in tertiary education in 2006. The leading higher education institute is the National University of San Marcos (Universidad Nacional Mayor de San Marcos), founded in 1551, making it the oldest both in the country and in South America.

In 2007 public expenditure on education came to 2·7% of GNI and 16·4% of total government spending.

Health

There were 483 hospitals with 43,074 beds (provision of 16 beds per 10,000 inhabitants) in 2002. There were 29,138 physicians, 3,190 dentists and 21,351 nurses in 2002.

Peru has been one of the most successful countries in reducing undernourishment in the past 20 years. Between 1990–92 and 2001–03 the proportion of undernourished people declined from 42% of the population to 12%.

Welfare

An option to transfer from state social security (IPSS) to privately-managed funds was introduced in 1993.

RELIGION

Religious liberty exists, but the Roman Catholic religion is protected by the State, and since 1929 only Roman Catholic religious instruction is permitted in schools, state or private. There were 17·0m. Catholics in 2007 as well as 2·6m. Protestants and 1·3m. with other beliefs (including non-religious). In Feb. 2013 there was one cardinal.

CULTURE

World Heritage Sites

There are 11 sites under Peruvian jurisdiction that appear on the UNESCO World Heritage List. They are (with the year entered on the list): the City of Cusco (1983), the Historic Sanctuary of Machu Picchu (1983), Chavin Archaeological Site (1985), Huascarán National Park (1985), Chan Chan Archaeological Zone (1986), Manú National Park (1987), Historic Centre of Lima (1988 and 1991), Río Abiseo National Park (1990 and 1992), Lines and Geoglyphs of Nasca and Pampas de Jumana (1994), the Historical Centre of the City of Arequipa (2000) and the Sacred City of Caral-Supe (2009).

Press

In 2008 there were 89 paid-for daily newspapers, of which 23 were national and 66 regional and local. The leading dailies are *Líbero* (with an average daily circulation in 2008 of 214,000), *Trome* (average daily circulation in 2008 of 213,000) and *El Comercio* (average daily circulation in 2008 of 199,000).

Tourism

There were 1,916,000 non-resident tourists in 2007, up from 1,064,000 in 2002; tourist spending in 2007 totalled US$2,222m., compared to US$836m. in 2002.

Festivals

The Lord of Tremors festival (the second part of March and the first week of April) renders homage to the image of Taitacha Temblores, the Lord of the Earthquakes, and demonstrates the fusion of Andean religions and Christianity; Fiesta de las Cruces (3 May) is celebrated across Peru as well as Spain and other parts of Hispanic America, and includes processions, folk music and dance; Inti Raymi (24 June), one of the biggest celebrations in Peru, celebrates the winter solstice and the Inca sun god; Corpus Christi (June) includes a traditional and colourful procession of saints and virgins; Virgin of the Carmen (second week of July) takes place in the small town of Paucartambo and includes a series of processions in honour of Mamacha Carmen, the patron saint of the mestizo population; All Saints Day (1–2 Nov.) is dedicated to the memory of the dead.

DIPLOMATIC REPRESENTATIVES

Of Peru in the United Kingdom (52 Sloane St., London, SW1X 9SP)
Ambassador: Julio Muñoz-Deacon.

Of the United Kingdom in Peru (Torre Parque Mar, Piso 22, Avenida Jose Larco 1301, Miraflores, Lima)
Ambassador: James Dauris.

Of Peru in the USA (1700 Massachusetts Ave., NW, Washington, D.C., 20036)
Ambassador: Harold Forsyth.

Of the USA in Peru (Avenida La Encalada Cdra 17-Monterrico, Lima)
Ambassador: Rose Likins.

Of Peru to the United Nations
Ambassador: Enrique Armando Román-Moray.

Of Peru to the European Union
Ambassador: Cristina Ronquillo.

FURTHER READING

Instituto Nacional de Estadística e Informática.—*Anuario Estadistico del Perú.—Perú: Compendio Estadístico*. Annual.—*Boletin de Estadistica Peruana.* Quarterly
Banco Central de Reserva. Monthly Bulletin.—*Renta Nacional del Perú.* Annual

Cameron, M. A., *Democracy and Authoritarianism in Peru: Political Coalitions and Social Change.* 1995
Carrion, Julio F., *The Fujimori Legacy: The Rise of Electoral Authoritarianism in Peru.* 2006
Daeschner, J., *The War of the End of Democracy: Mario Vargas Llosa vs. Alberto Fujimori.* 1993
Gorriti, Gustavo, (trans. Robin Kirk) *The Shining Path: A History of the Millenarian War in Peru.* 1999
Starn, Orin, *The Peru Reader: History, Culture, Politics.* 2005

National Statistical Office: Instituto Nacional de Estadística e Informática, Av. Gral. Garzón 654–658, Jesús María, Lima.
Website (limited English): http://www.inei.gob.pe

PHILIPPINES

PHILIPPINES

PACIFIC OCEAN

South China Sea

MANILA

Iloilo
Bacolod
Cebu

Sulu Sea

Zamboanga

Cagayan de Oro

Davao

General Santos

MALAYSIA

Celebes Sea

© Research Machines plc 2006

Republika ng Pilipinas
(Republic of the Philippines)

Capital: Manila
Population projection, 2015: 101·42m.
GNI per capita, 2011: (PPP$) 3,478
HDI/world rank: 0·644/112
Internet domain extension: .ph

KEY HISTORICAL EVENTS

Pottery was being made on the Philippine archipelago from at least 3000 BC, probably by people of Malay origin, and metals were being worked by the first millennium BC. Merchants from south China reached the islands during the 10th century AD (T'ang Dynasty), heralding centuries of Chinese trade with the region. Arab traders brought Islam from the Malay peninsula via Borneo and the Sulu archipelago in the late 13th century, and by the 15th century Islamic influence had spread as far north as Luzon. Most islanders lived in barangays, communities of 30–100 households based largely on kinship.

The Portuguese explorer, Ferdinand Magellan, landed at Samar on 16 April 1521 during his Spanish-financed expedition round the world. Subsequent expeditions consolidated Spanish control over the islands, which were named after Philip II of Spain in 1542. Manila was established by Miguel Lopez de Legaspi in 1571 on the site of an existing Moro (Muslim Filipinos) settlement. By the end of the 16th century the Philippines had become a major trading centre with India, China and the East Indies. The islands came under Dutch control from around 1600.

The waning of the Spanish Empire during the 18th century saw a rise in the power base of the Jesuit orders, which caused resentment and stoked demands for independence. In 1896 revolution in the province of Cavite, led by Emilio Aguinaldo among others, spread through the major islands. However, in Dec. 1898, following the Spanish-American War, the Philippines were ceded to the United States. Aguinaldo fought a guerrilla campaign but was captured in 1901. The US granted the Philippines partial autonomy in 1916 and the Hare-Hawes-Cutting Act of 1932 set a timetable for full independence after a ten-year period of self-governance as a Commonwealth of the USA. Manuel Quezon was elected the first president in Sept. 1935.

In Dec. 1941 the islands were invaded by Japanese troops, who went on to take complete control in 1942. Quezon escaped to the USA and established a government-in-exile in Washington, D.C. Sergio Osmena succeeded Quezon in 1944 and returned to the Philippines with a US-backed liberation force in Oct. 1944. Manuel Roxas defeated Osmena in the election of April 1946, becoming president of the Republic of the Philippines when independence was achieved on 4 July 1946. The USA continued to play a key role in its former colony, particularly in economic policy. In return for assistance in rebuilding the country's war-torn infrastructure, the USA secured 99-year leases over several air and naval bases. Relations with neighbouring countries improved and in 1954 the Philippines joined the Southeast Asia Treaty Organization, precursor to the Association of South East Asian Nations (ASEAN).

Ferdinand Marcos was elected president in 1965 and re-elected four years later, although his rule became increasingly unpopular. In Sept. 1972 Marcos declared martial law and thousands of political opponents were arrested. In May 1980 Benigno Aquino, Jr, the leading opponent of Marcos, was released from prison to go to the USA for medical treatment. His assassination, on his return to the Philippines in 1983, led to growing US pressure for Marcos to restore democracy. In late 1985 he announced a snap presidential election. He was challenged by Aquino's widow, Corazón, who eventually emerged victorious from a controversial poll and was installed as president on 25 Feb. 1986. Marcos fled the country and a new constitution limiting the president to a single, six-year term in office was ratified in Feb. 1987.

More than twenty years of insurgency by the Moro National Liberation Front were ended by a peace agreement of 2 Sept. 1996, providing for a Muslim autonomous region in an area of Mindanao island. The rebellion left more than 120,000 people dead. In Oct. 2000 impeachment proceedings began against President Estrada who was alleged to have received more than US$10·8m. from gambling kickbacks. His impeachment trial collapsed in Jan. 2001 when he was forced from office by mass protests. Subsequently Estrada's supporters tried to overthrow his successor, Gloria Macapagal-Arroyo.

In Nov. 2001 the fragile peace between the government and Islamic militants was shattered. Since then violence has frequently erupted, notably in early 2005, when fighting on the southern island of Jolo left 90 dead and caused 12,000 people to flee. On 14 Feb. 2005 three bombs were detonated killing nine and injuring 130. In Feb. 2006 President Arroyo declared a week-long state of emergency after the military declared it had discovered a coup plot.

TERRITORY AND POPULATION

The Philippines is situated between 21° 25' and 4° 23' N. lat. and between 116° and 127° E. long. It is composed of 7,100 islands and islets, 3,144 of which are named. Approximate land area, 300,076 sq. km (115,859 sq. miles). The largest islands (in sq. km) are Luzon (104,688), Mindanao (94,630), Samar (13,080), Negros (12,710), Palawan (11,785), Panay (11,515), Mindoro (9,735), Leyte (7,214), Cebu (4,422), Bohol (3,865) and Masbate (3,269). The census population in May 2010 was 92,337,852; density, 307·7 per sq. km. In 2011, 49·1% of the population lived in urban areas.

The UN gives a projected population for 2015 of 101·42m.

The area and population of the 17 regions (from north to south):

Region	Sq. km	2010
Ilocos	12,840	4,748,372
Cordillera[1]	18,294	1,616,867
Cagayan Valley	26,838	3,229,163
Central Luzon	21,470	10,137,737
National Capital	636	11,855,975
Calabarzon	16,229	12,609,803
Mimaropa	27,456	2,744,671
Bicol	17,632	5,420,411
Western Visayas	20,223	7,102,438
Central Visayas	14,951	6,800,180
Eastern Visayas	21,432	4,101,322
Northern Mindanao	17,125	4,297,323
Davao	19,672	4,468,563
Soccskargen	18,433	4,109,571
Zamboanga Peninsula	14,811	3,407,353
Muslim Mindanao[2]	12,695	3,256,140
Caraga	18,847	2,429,224

[1]Administrative region. [2]Autonomous region.

City populations (2010 census, in 1,000) are as follows; all on Luzon unless indicated in parenthesis:

Quezon City[1]	2,761	Bacolod (Negros)	512
Manila (the capital)[1]	1,652	Muntinlupa[1]	460
Caloocan[1]	1,489	San Jose del Monte	455
Davao (Mindanao)	1,177[3]	General Santos	
Cebu (Cebu)	866	(Mindanao)	444
Antipolo	678	Iloilo	425
Pasig[1]	670	Marikina[2]	424
Taguig[2]	644	Pasay[1]	393
Zamboanga (Mindanao)	644[3]	Calamba	389
Cagayan de Oro		Malabon[2]	353
(Mindanao)	602	Lapu-Lapu (Cebu)	350
Parañaque[2]	588	Mandaue (Cebu)	331
Dasmariñas	576	Mandaluyong[1]	329
Valenzuela[2]	575	Angeles	326
Las Piñas[2]	553	Baguio[1]	319
Makati[1]	529	Cainta	312
Bacoor	520	Imus	302

[1]City within Metropolitan Manila. Population of Metro Manila in 2007, 11,553,427. [2]Municipality within Metropolitan Manila. [3]Estimate.

Filipino (based on Tagalog) is spoken as a mother tongue by only 29·3%; among the 76 other indigenous languages spoken, Cebuano is spoken as a mother tongue by 23·3% and Ilocano by 9·3%. English, which along with Filipino is one of the official languages, is widely spoken.

In 2000 some 5·5m. Filipinos were living and working abroad, including 2m. in the USA, 850,000 in Saudi Arabia and 620,000 in Malaysia.

SOCIAL STATISTICS

Births, 2007, 1,749,878; deaths, 2007, 441,956. Divorce is illegal. Birth rate per 1,000 population (2007), 19·7; death rate (2007), 5·1. Expectation of life at birth, 2007, was 69·4 years for males and 73·9 years for females. Annual population growth rate, 2000–05,

2·2%. Infant mortality, 2010, 23 per 1,000 live births; fertility rate, 2008, 3·1 births per woman. Abortion is illegal.

CLIMATE

Some areas have an equatorial climate while others experience tropical monsoon conditions, with a wet season extending from June to Nov. Mean temperatures are high all year, with very little variation. Manila, Jan. 77°F (25°C), July 82°F (27·8°C). Annual rainfall 83·3" (2,115·9 mm).

CONSTITUTION AND GOVERNMENT

A new constitution was ratified by referendum in Feb. 1987 with the approval of 78·5% of voters. The head of state is the *President*, directly elected for a non-renewable six-year term.

Congress consists of a 24-member upper house, the *Senate* (elected for a six-year term from 'at large' seats covering the country as a whole, half of them renewed every three years), and a *House of Representatives* of 287 members. In the *House of Representatives* 230 members are directly elected for a three-year term and the rest are chosen from party and minority-group lists.

A campaign led by the president at the time, Fidel Ramos, to amend the constitution to allow him to stand for a second term was voted down by the Senate by 23 to one in Dec. 1996.

National Anthem

'Land of the Morning', lyric in English by M. A. Sane and C. Osias, tune by Julian Felipe; 'Lupang Hinirang', Tagalog lyric by the Institute of National Language.

GOVERNMENT CHRONOLOGY

Presidents since 1946. (KBL = New Society Movement; Lakas-CMD = Lakas-Christian Muslim Democrats; LE-NUCD = People's Power-National Union of Christian Democrats; LMP = Struggle of the Philippine Masses; PL = Liberal Party; PN = Nacionalista Party; UNIDO = Nationalist Democratic Organization)

1946–48	PL	Manuel Roxas y Acuña
1948–53	PL	Elpidio Quirino y Rivera
1953–57	PN	Ramon Magsaysay y del Fierro
1957–61	PN	Carlos Polestico García
1961–65	PL	Diosdado Pañgan Macapagal
1965–86	PN, KBL	Ferdinand Emmanuel Edralin Marcos
1986–92	UNIDO	Corazón Cojuangco Aquino
1992–98	LE-NUCD	Fidel Valdez Ramos
1998–2001	LMP	Joseph Marcelo Ejercito Estrada
2001–10	Lakas-CMD	Gloria Macapagal-Arroyo
2010–	PL	Benigno Aquino III

RECENT ELECTIONS

The presidential elections of 10 May 2010 were won by Benigno 'Noynoy' Aquino III (Liberal Party, PL) with 42·1% of votes cast, ahead of Joseph Estrada (Force of the Filipino Masses, PMP) with 26·3% of the vote, Manny Villar (Nacionalista Party, PN) with 15·4% and Gilberto Teodoro (Lakas-Kampi-Christian Muslim Democrats) with 11·3%. There were five other candidates.

Elections to the House of Representatives were also held on 10 May 2010. 119 seats went to PL, 45 to Lakas-Kampi-Christian Muslim Democrats, 22 to PN and five to PMP. The remaining seats went to independents and smaller parties.

Senate elections were likewise most recently held on 10 May 2010, following which Lakas-Kampi-Christian Muslim Democrats had 4 seats, PL 4, PN 4, Nationalist People's Coalition 2, PMP 2, People's Reform Party 1, Struggle of Democratic Filipinos 1 and ind. 5.

Elections to the House of Representatives and the Senate were scheduled to take place on 13 May 2013.

CURRENT GOVERNMENT

President: Benigno 'Noynoy' Aquino III; b. 1960 (PL; sworn in 30 June 2010). His mother, Corazón Aquino, had been president from 1986–92.

Vice-President: Jejomar Binay (sworn in 30 June 2010).

In March 2013 the government comprised:

Minister of Agrarian Reform: Virgilio de los Reyes. *Agriculture:* Proceso Alcala. *Budget and Management:* Florencio 'Butch' Abad. *Education:* Armin Luistro. *Energy:* Carlos Jericho Petilla. *Environment and Natural Resources:* Ramon Paje. *Finance:* Cesar Purisima. *Foreign Affairs:* Albert del Rosario. *Health:* Enrique Ona. *Interior and Local Government:* Manuel 'Mar' Roxas II. *Justice:* Leila de Lima. *Labour and Employment:* Rosalinda Baldoz. *National Defence:* Voltaire Gazmin. *Public Works and Highways:* Rogelio Singson. *Science and Technology:* Mario Montejo. *Social Welfare and Development:* Corazon 'Dinky' Soliman. *Socioeconomic Planning:* Cayetano Paderanga. *Tourism:* Alberto Lim. *Trade and Industry:* Gregorio Domingo. *Transport and Communications:* Joseph Emilio Abaya.

Executive Secretary: Paquito Ochoa, Jr.

Speaker of the House of Representatives: Feliciano Belmonte, Jr.

Government Website: http://www.gov.ph

CURRENT LEADERS

Benigno Aquino III

Position
President

Introduction
Benigno Simeon Cojuangco Aquino III was sworn in as president in June 2010 after a landslide victory at the election in May. He campaigned on a platform of tackling corruption and eradicating poverty.

Early Life
Aquino, popularly known as Noynoy, was born on 8 Feb. 1960 in Manila into the influential Aquino political family. His father, Benigno Simeon 'Ninoy' Aquino, Jr, was a prominent opposition leader who was assassinated by Ferdinand Marcos' regime. His mother, Corazón Aquino, led a non-violent people's revolution following her husband's death that brought about an end to the Marcos dictatorship and secured her the presidency. Benigno Aquino III graduated in economics from Ateneo de Manila University in 1981 before joining his family in exile in the USA. On his father's death in 1983, he returned to the Philippines.

From 1983–98 Aquino worked in business for companies including Nike, Intra-Strata Assurance Corp., Best Security Agency Corporation and Central Azucarera Tarlac, an organization owned by the Cojuangco clan. In 1989 in an attempted coup led by Gregorio Honasan against his mother's administration Aquino narrowly escaped assassination.

Joining the Liberal Party in 1998, he was elected to the House of Representatives to represent the 2nd District of Tarlac, a seat he held until 2007. He served as deputy speaker of the House from Nov. 2004–Feb. 2006, resigning when he joined other Liberal leaders in calling for the resignation of President Macapagal-Arroyo at the height of the 'Hello Garci' vote-rigging scandal. On 17 March 2006 Aquino was appointed vice chairman of the Liberal Party. Constitutionally unable to seek a fourth term in the House of Representatives, he won a Senate seat in the May 2007 elections.

After the death of his mother in Aug. 2009, there was a swell of public support for Aquino to stand for the presidency. Announcing his candidacy on 9 Sept. 2009 his subsequent campaign used the colour yellow to evoke his mother's 'people power' movement. The front runner leading up to the election, Aquino won with 42% of the vote.

Career in Office
On assuming office, Aquino announced the creation of an independent commission to investigate the alleged corruption of the Arroyo administration (which led to the former president's arrest in Oct. 2012) and appointed a new justice secretary, Leila de Lima, the former head of the commission on human rights. In education, he pushed for the extension of the basic education system from a ten-year programme to 12 years. He also came out in support of a controversial family planning bill which, despite strong opposition from the Catholic Church, was eventually approved by parliament in Dec. 2012. Critics have nevertheless questioned Aquino's ability to introduce reforms that act against the interests of the Philippines' powerful clans, of which he himself is a member.

In Oct. 2012 Aquino signed a framework agreement on peace and autonomy between the government and the Moro Islamic Liberation Front, the Muslim separatist movement operating in the Mindanao region, with the aim of ending years of armed conflict.

At the international level, there were territorial tensions with China in mid-2012 over a disputed area of the South China Sea around the Scarborough Shoal.

DEFENCE

An extension of the 1947 agreement granting the USA the use of several Army, Navy and Air Force bases was rejected by the Senate in Sept. 1991. An agreement of Dec. 1994 authorizes US naval vessels to be repaired in Philippine ports. The Philippines is a signatory of the South-East Asia Collective Defence Treaty.

Defence expenditure in 2008 totalled US$1,427m. (US$15 per capita), representing 0·9% of GDP.

Army
The Army is organized into five area joint-service commands.

Strength (2007) 66,000, with reserves totalling 100,000. The paramilitary Philippines National Police numbered 40,500 in 2007 with 40,000 reservists.

Navy
The Navy consists principally of ex-US ships completed in 1944 and 1945, and serviceability and spares are a problem. The modernization programme in progress has been revised and delayed, but the first 30 inshore patrol craft of US and Korean design have been delivered. The present fleet includes one ex-US frigate.

Navy personnel in 2007 was estimated at 24,000 including 7,500 marines.

Air Force
The Air Force had an estimated strength of 16,000 in 2007, with 30 combat-capable aircraft. There was one fighter squadron of Agusta S-211s.

ECONOMY

Agriculture accounted for 14·2% of GDP in 2006, industry 31·6% and services 54·2%.

Overview
Market-oriented reforms have been implemented over the last two decades. Foreign investment and trade barriers have been dismantled and many industries deregulated. Most state industrial assets were privatized between 1992 and 1995 and monopolies were ended in the telecommunications, oil, civil aviation, shipping, water and power industries.

In 2004 President Arroyo was re-elected to office on a platform of public debt reduction, pledging to address low government revenues (14·5% of GDP in 2004). Chronic public deficits have been reduced, while tax collection has become more aggressive and VAT reform was implemented in early 2006.

Recent growth has been robust, averaging around 5% per year, and the economy has held strong against a number of external shocks. Despite growth plunging as a result of the global economic crisis, the economy avoided a technical recession in 2009, with a combination of remittances and fiscal stimuli supporting growth. 2010 saw a continuation of the recovery with broad-based growth across private consumption, investment and exports, despite the impact of the El Niño drought during the first half of the year. The call centre industry brought in nearly US$10bn. in 2010. There are now more call centre employees in the Philippines than in India. The handover of power following elections in May 2010 was smooth, while in Nov. 2010 Standard & Poor's upgraded the Philippines' credit rating to BB in recognition of its stronger financial position.

However, robust growth has not translated into poverty reduction, with just over a quarter of the population living below the poverty line.

Currency

The unit of currency is the *peso* (PHP) of 100 *centavos*. Inflation rates (based on IMF statistics):

2002	2003	2004	2005	2006	2007	2008	2009	2010	2011
2·7%	2·3%	4·8%	6·6%	5·5%	2·9%	8·2%	4·2%	3·8%	4·7%

Foreign exchange reserves were US$35,493m. in Aug. 2009 and gold reserves 5·08m. troy oz. Total money supply in Feb. 2008 was 836,709m. pesos.

Budget

Budgetary central government revenue and expenditure (in 1bn. pesos):

	2008	2009	2010
Revenue	1,172·8	1,123·1	1,208·0
Expenditure	1,265·6	1,425·3	1,519·6

Taxes accounted for 90·5% of revenues in 2010; compensation of employees accounted for 30·9% of expenditures, use of goods and services 26·7% and interest 19·9%.

VAT was introduced in 1988. The standard rate was raised from 10·0% to 12·0% in 2006.

Performance

Real GDP growth rates (based on IMF statistics):

2002	2003	2004	2005	2006	2007	2008	2009	2010	2011
3·6%	5·0%	6·7%	4·8%	5·2%	6·6%	4·2%	1·1%	7·6%	3·9%

The 2010 growth rate was the highest since democracy was restored to the Philippines in 1986 following the Marcos era. Total GDP in 2011 was US$224·8bn.

Banking and Finance

The Central Bank (*Governor*, Amando Tetangco, Jr) issues the currency, manages foreign exchange reserves and supervises the banking system. At 30 June 2003 there were 42 commercial banks (24 regular commercial banks and 18 universal banks), 93 thrift banks and 771 rural and co-operative banks. In June 2003 the total number of banking institutions was 6,414, with total assets of 3,529,128m. pesos.

External debt amounted to US$72,337m. in 2010, up from US$58,304m. in 2000—this was equivalent to 36·2% of GNI (compared to 72·0% in 2000).

Foreign direct investment was US$1,262m. in 2011, down from a record US$2,921m. in 2006.

The financial crisis that struck southeast Asia in 1997 led to the floating of the peso in July of that year. It subsequently lost 36% of its value against the dollar.

There is a stock exchange in Manila.

ENERGY AND NATURAL RESOURCES

Environment

Carbon dioxide emissions from the consumption and flaring of fossil fuels in 2008 were the equivalent of 0·8 tonnes per capita.

Electricity

Total installed capacity was 16·0m. kW in 2007. Production was 59·65bn. kWh in 2007 (40·79bn. kWh thermal and 8·58bn. kWh hydro-electric). Consumption per capita was 672 kWh in 2007.

Oil and Gas

The largest natural gas field is the Camago-Malampaya gas field, discovered off the island of Palawan in 1992, with reserves initially put at 76bn. cu. metres but now increased to 85bn. cu. metres. The Philippines' total natural gas reserves in 2007 were 99bn. cu. metres.

Crude petroleum reserves were 139m. bbls in 2007.

Water

Total freshwater withdrawal in 2006 was 78·9bn. cu. metres, equivalent to 843 cu. metres per capita. Breakdown of water use in 2006: agricultural, 83%; industrial, 9%; domestic, 7%.

Minerals

Mineral production in 2003, unless otherwise indicated (in tonnes): coal (2004), 2,482,000; salt, 429,160; silica sand, 372,200; copper, 80,920; chromite refractory ore (chromium content), 13,220; nickel bearing ore (2002), 26,532 (nickel content); gold, 37,840 kg; silver, 9,530 kg. Other minerals include rock asphalt, sand and gravel. Total value of mineral production, 2003, 139,597m. pesos.

Agriculture

Agriculture is a mainstay of the economy, contributing up to 30% of national output. In 2001 there were 5·65m. ha. of arable land and 5·0m. ha. of permanent crops. In 2001, 37·4% of the working population was employed in agriculture. In 2002 agricultural production grew by 2·6% (7·4% in 2001).

Output (in 1,000 tonnes) in 2002: sugarcane, 21,417; coconuts, 13,683; rice, 13,271; bananas, 5,275; maize, 4,319; copra, 2,010; pineapples, 1,639; cassava, 1,626. The Philippines is the second largest producer of both coconuts and pineapples. Minor crops are fruits, nuts, vegetables, coffee, cacao, peanuts, ramie, rubber, maguey, kapok, abaca and tobacco.

Livestock, 2003: pigs, 12·36m.; goats, 3·27m.; buffaloes, 3·18m.; cattle, 2·56m.; chickens, 128·51m.; ducks, 9·81m.

Forestry

Forests covered 7·67m. ha. (26% of the land area) in 2010, up from 6·57m. ha. in 1990 and 7·12m. ha. in 2000. Of the total area under forests 11% was primary forest in 2010, 84% other naturally regenerated forest and 5% planted forest. Timber production was 15·79m. cu. metres in 2007.

Fisheries

The catch in 2010 was 2,611,720 tonnes (2,426,314 tonnes from marine waters). Imports of fishery commodities were valued at US$147m. in 2008 and exports at US$645m.

INDUSTRY

The leading companies by market capitalization in the Philippines in March 2012 were: Philippine Long Distance Telephone (US$13·6bn.); San Miguel Brewery (US$10·8bn.); and SM Investments, a shopping mall developer and operator (US$9·4bn.).

Leading sectors are foodstuffs, oil refining and chemicals. Production, 2002 (in 1,000 tonnes): cement, 11,396; residual fuel oil, 3,537; distillate fuel oil, 3,004; sugar, 1,988; petrol, 1,501; paper and paperboard, 1,056; plywood, 409,000 cu. metres.

Labour

In 2003 the total workforce was 34,635,000, of whom 30,418,000 were employed (19,263,000 in non-agricultural work). Employees by sector, 2003: 14·4m. in services; 11·2m. in agriculture, hunting, forestry and fisheries; 4·9m. in industry. 3·9m. persons were registered unemployed in 2004. 868,000 persons worked overseas in 2003 (652,000 land-based).

The unemployment rate in Oct. 2001 was 9·8%.

INTERNATIONAL TRADE

A law of June 1991 gave foreign nationals the right to full ownership of export and other firms considered strategic for the economy.

Imports and Exports

Imports (c.i.f.) in 2007 totalled US$57,995·7m. (US$54,078·0m. in 2006) and exports (f.o.b.) US$50,465·7m. (US$47,410·1m. in 2006).

Main imports: electronics and components, mineral fuels, lubricants and related materials, industrial machinery and equipment, telecommunications equipment and transport equipment. Principal exports: electronics, garments, machinery, transport equipment and apparatus, and processed foods. In 2001 electronics exports were worth US$21·4bn. and constituted 67% of all exports although this had shrunk to US$16·3bn. and 32% by 2007. In 1992 they had been worth just US$3bn.

Main sources of imports in 2007: USA, 14·0%; Japan, 12·4%; Singapore, 11·1%; China, 7·3%. Main export markets, 2007: USA, 17·0%; Japan, 14·5%; Hong Kong, 11·5%; China, 11·4%.

COMMUNICATIONS

Roads

In 2003 roads totalled 200,037 km; of these, 28,266 km were national roads and 49,782 km were regional roads. In 2007 there were 937,600 passenger cars in use, 55,200 buses and coaches, 1,875,300 vans and lorries, and 2,647,500 motorcycles and mopeds. There were 6,240 road accidents involving injury in 2006 with 961 fatalities.

Rail

In 2005 the National Railways totalled 419 km (1,067 mm gauge). In 2004 passenger-km totalled 83m. There is a light metro railway in Manila.

Civil Aviation

There are international airports at Manila (Ninoy Aquino) and Cebu (Mactan International). In Sept. 1998 the Asian economic crisis that had started more than a year earlier forced the closure of the national carrier, Philippine Airlines, after it had suffered huge losses. However, it has since resumed its operations both internally and externally. In 2005 scheduled airline traffic of Philippine-based carriers flew 28·4m. km, carrying 6,610,400 passengers. In 2001 Manila handled 12,545,000 passengers (7,144,000 on international flights) and 356,700 tonnes of freight.

Shipping

The main ports are Cagayan de Oro, Cebu, Davao, Iloilo, Manila and Zamboanga. Manila, the leading port, handled 45,230,000 tonnes of cargo in 2008. In Jan. 2009 there were 838 ships of 300 GT or over registered, totalling 4,771,000 GT. Of the 838 vessels registered, 415 were general cargo ships, 168 passenger ships, 128 oil tankers, 83 bulk carriers, 21 chemical tankers, 15 liquid gas tankers and eight container ships.

Telecommunications

In 2008 there were 4,076,000 main (fixed) telephone lines. In the same year mobile phone subscribers numbered 68,117,000 (753·9 per 1,000 persons). There were 5·6m. internet users in 2008. In March 2012 there were 27·7m. Facebook users.

SOCIAL INSTITUTIONS

Justice

There is a Supreme Court which is composed of a chief justice and 14 associate justices; it can declare a law or treaty unconstitutional by the concurrent votes of the majority sitting. There is a Court of Appeals, which consists of a presiding justice and 50 associate justices. There are 15 regional trial courts, one for each judicial region, with a presiding regional trial judge in each of its 720 branches. Municipal trial courts and municipal circuit trial courts are found in the municipalities of the Philippines. If the court covers one municipality it is a municipal trial court; if it covers two or more municipalities it is a municipal circuit trial court. In Metropolitan Manila the equivalents are metropolitan trial courts, and in the cities outside Metropolitan Manila the courts are known as municipal trial courts in cities.

The Supreme Court may designate certain branches of the regional trial courts to handle exclusively criminal cases, juvenile and domestic relations cases, agrarian cases, urban land reform cases which do not fall under the jurisdiction of quasijudicial bodies and agencies and/or such other special cases as the Supreme Court may determine. The death penalty, abolished in 1987, was officially restored in Dec. 1993 as punishment for 'heinous crimes'. In Feb. 1999 a rapist was executed, the first incident of capital punishment in the Philippines since 1976. The death penalty was abolished again for all crimes in 2006.

In Oct. 2011 there were 140,000 police officers. Local police forces are supplemented by the Philippine Constabulary, which is part of the armed forces.

In 2009 the prison population was 102,267 (111 per 100,000 of national population).

Constabulary

Since 1990 public order has been maintained completely by the Philippine National Police. Qualified Philippine Constabulary personnel were absorbed by the PNP or were transferred to branches or services of the Armed Forces of the Philippines.

Education

Public elementary education is free and schools are established in virtually all parts of the country. The majority of secondary and post-secondary schools are private. Formal education consists of an optional one to two years of pre-school education; six years of elementary education; four years of secondary education; and four to five years of tertiary or college education leading to academic degrees. Three-year post-secondary non-degree technical/vocational education is also considered formal education. In 2007 there were 961,397 children in pre-school institutions with (2006) 27,742 teaching staff; 13,145,210 pupils in primary schools with 390,432 teaching staff; and 6,365,985 pupils in secondary schools with 181,193 teaching staff. In 2005 there were 2,402,649 students in tertiary education with 112,941 academic staff.

Non-formal education consists of adult literacy classes, agricultural and farming training programmes, occupation skills training, youth clubs, and community programmes of instructions in health, nutrition, family planning and co-operatives.

In 2004–05 there were 176 public higher education institutions including 111 state and 50 local universities and colleges, and 1,443 private higher education institutions including 340 religious institutions. The adult literacy rate in 2008 was 94%.

Public expenditure on education in 2005 came to 2·3% of GNI and was equivalent to 15·2% of total government spending.

Health

In 2003 there were 1,723 hospitals (1,061 private) with 85,040 beds (1·1 beds per 1,000 inhabitants). In 2002 there were 91,408 physicians, 44,129 dentists, 347,349 nurses, 140,675 midwives and 47,463 pharmacists.

Welfare

The Social Security System (SSS) is a contributory scheme for employees. Disbursements in 2001 (in 1m. pesos): social security, 37,813 (1,775,996 recipients); employees' compensation, 1,201 (90,356 recipients).

RELIGION

82% of the population are Roman Catholics, 5% Protestants, 5% Muslims and 7% Buddhists or other religions. There were 181,500 Latter-day Saints (Mormons) in 2000.

The Roman Catholic Church has three cardinals, 23 arch-bishoprics, 91 bishoprics, 79 dioceses, 2,328 parishes and some 20,873 chapels or missions.

CULTURE

World Heritage Sites

The Philippines has five sites on the UNESCO World Heritage List: Tubbataha Reefs Natural Park (inscribed on the list in 1993 and 2009); the Baroque Churches of the Philippines (1993); the Rice Terraces of the Philippine Cordilleras (1995); the Historic Town of Vigan (1999); and Puerto-Princesa Subterranean River National Park (1999).

Press

There were 28 daily newspapers in 2008, with a combined circulation of 3,870,000. The leading daily is *Remate*, with an average daily circulation of 620,000 in 2008.

Tourism

In 2007, 3,092,000 foreign tourists brought revenue of US$5,518m.

DIPLOMATIC REPRESENTATIVES

Of the Philippines in the United Kingdom (8 Suffolk St., London, SW1Y 4HH)
Ambassador: Enrique A. Manalo.

Of the United Kingdom in the Philippines (120 Upper McKinley Rd, McKinley Hill, Taguig City 1634, Manila)
Ambassador: Stephen Lillie.

Of the Philippines in the USA (1600 Massachusetts Ave., NW, Washington, D.C., 20036)
Ambassador: Jose L. Cuisia, Jr.

Of the USA in the Philippines (1201 Roxas Blvd, Manila)
Ambassador: Harry K. Thomas, Jr.

Of the Philippines to the United Nations
Ambassador: Libran N. Cabactulan.

Of the Philippines to the European Union
Ambassador: Victoria Sisante Bataclan.

FURTHER READING

National Statistics Office. *Philippine Statistical Yearbook.*

Abinales, Patricio N., *State and Society in the Philippines.* 2005
Balisacan, Arsenio M. and Hill, Hal, (eds.) *The Philippine Economy: Development, Policies, and Challenges.* 2003
Francia, Luis H., *A History of the Philippines: From Indios Bravos to Filipinos.* 2010
Hamilton-Paterson, J., *America's Boy: The Marcoses and the Philippines.* 1998
Hedman, Eva-Lotta, *In the Name of Civil Society: From Free Election Movements to People Power in the Philippines.* 2005
Hedman, Eva-Lotta and Sidel, John, (eds.) *Philippine Politics and Society in the Twentieth Century: Colonial Legacies.* 2000
Larkin, J. A., *Sugar and the Origins of Modern Philippine Society.* 1993

National Statistical Office: National Statistics Office, Solicarel Bldg., 1 Ramon Magsaysay Blvd., Sta Mesa, Manila 1008.
Website: http://www.census.gov.ph

POLAND

© Research Machines plc 2006

Rzeczpospolita Polska
(Polish Republic)

Capital: Warsaw
Population projection, 2015: 38·36m.
GNI per capita, 2011: (PPP$) 17,451
HDI/world rank: 0·813/39
Internet domain extension: .pl

KEY HISTORICAL EVENTS

In the 7th and 8th centuries Slavic peoples first settled on the forest covered plains between the Odra and Vistula rivers. Poland takes its name from the Polanie ('plain dwellers'), whose ruler Mieszko I, first in line of the Piast dynasty, founded the Polish state in 966. Christianity came via Bohemia and Moravia to the Kraków region, and in 991 Mieszko I placed Poland under the Holy Roman See. His son and heir, Bolesław I the Brave (ruled 992–1025) continued his father's territorial expansionism until Poland's boundaries were much as they are today. He established an independent Polish Catholic Church in the year 1000 and was officially crowned the first king of Poland in 1024 with the support of Holy Roman Emperor Otto III. The growing power of the church stimulated economic activity ranging from the manufacture of parchment and glass to building and painting.

In the twelfth century, under the rule of Bolesław III, German infiltration and internecine struggles led to Bolesław's 1138 Testament which divided the kingdom between his three sons. Around this time, many Jewish immigrants from Western Europe were attracted by the offer of asylum. The General Charter of Jewish Liberties was published in 1264 by Bolesław V, the Duke of Kraków.

A series of Mongol invasions in 1241–42 laid waste much of Poland, and in 1308 the crusades of the Teutonic Knights captured Gdańsk, cutting off Poland's access to the sea. In 1320 Władysław I Łokietek (the Short) of Kraków reunited the majority of the Polish lands that had been divided in 1138 and was crowned king of a united Poland. His son Casimir III the Great (Kasimierz, ruled 1333–70) continued this work, and his reign brought prosperity and administrative efficiency. He negotiated a truce with the Teutonic Knights and fostered closer diplomatic relations with the neighbouring kingdoms of Bohemia and Hungary.

Casimir III was the last monarch in the Piast line, and when he died his nephew Louis of Anjou, simultaneously King Lajos I of Hungary, donned the Polish crown. His death led to a disjointed succession. After a brief civil war his eleven-year-old daughter Jadwiga married Jagiełło, the pagan Grand Duke of Lithuania, who converted to Catholicism. Their marital union in 1386 signalled the beginning of the Jagiełłonian dynasty which ruled over Lithuania and Poland, at the time the largest state in Europe. The Jagiełłonian period to 1572 is regarded as an economic and cultural 'golden age'. This joint, multi-ethnic power managed to quell opposition on its eastern and western fronts. Poland–Lithuania crushed the Tatars and in 1410 defeated an army of 27,000 Teutonic Knights at the Battle of Tannenberg. In 1454 the Polish–Teutonic war broke out. King Casimir IV (1427–92) led a successful campaign, taking control of Western Prussia. At the Peace of Toruń in 1466 Gdańsk was returned to the Polish crown. The city, granted autonomy in exchange for its efforts in the war, thrived on shipping trade with the Netherlands, Spain and England among others while the population outgrew that of Warsaw.

The link between Poland and Lithuania was further strengthened by the Union of Lublin in 1569, which was primarily signed to protect both parties from expansionist threats on the Eastern front from Russia's Tsar Ivan IV (the Terrible). Warsaw became the capital of the two kingdoms which were henceforth known as the Commonwealth of Poland–Lithuania.

The last Jagiełłonian, Zygmunt II, died in 1572, after which the nobility introduced an elective monarchy with powers limited by the Acta Henriciana, so called because the first elected king to whom it applied was Henri III de Valois. He was obliged to swear his allegiance to maintaining the elective monarchy, which consulted the nobles on tax and warfare, respected religious tolerance and held a bi-annual meeting of the Sejm, the bicameral assembly dating from 1493. In contrast to many other countries in Europe, the Commonwealth was sufficiently broadminded on religious issues to abide by the Statute of Toleration (1573), although Catholicism was still the official religion.

Polish Wars

During this period, many foreign leaders were elected, partly to neutralize external interests. In 1573 Catherine de Médicis of France organized the election of her third son, Henry, duke of Anjou, to the Polish crown. When he returned to France as king on his brother's death, he was succeeded by a Transylvanian, Prince István Bathory. He increased Poland–Lithuania's military strength—a necessity given Ivan the Terrible's bellicose claims. In campaigns throughout 1578–81 the latter was beaten with a huge loss of Russian lives, and the territories he had encroached upon were restored.

1587 marked the beginning of Vasa rule, with Swedish-born Sigismund III taking the throne. But his succession led to territorial claims from his native land. Disapproving of Sigismund's Catholic persuasion, Calvinist Sweden occupied Livonia and Pomerania. In alliance with Russia, King Karl X of Sweden mounted a full invasion of Poland–Lithuania, devastating Warsaw and Kraków. During the ensuing Polish–Swedish war of 1655–60, support for Poland–Lithuania came from the Netherlands and Denmark. The Poles fought back against the invaders, winning a major battle at Częstochowa, but were eventually defeated. King Jan Kazimiercz (John Casimir), the last in line of the Vasa dynasty, abdicated in 1668.

Hopes of salvation for the Commonwealth came with Jan III Sobieski's election to the throne in 1674. He fought off the Ottomans who were advancing onto Polish territory. But further invasions and wars weakened Poland. The Great Northern War of 1700–21 had Poland as the battleground for fierce fighting between Russia, Denmark–Norway and Saxony–Poland (also Prussia from 1715) on one side against Sweden on the other. Each of the warring factions occupied parts of Poland, which was also subject to internecine fighting. Russia played the dominant role in Polish affairs until Frederick II, king of an increasingly powerful Prussia, proposed the division of Poland between Russia, Prussia and Austria. The outcome was the first Partition of Poland, in 1772. Austria was awarded the Kingdom of Galicia–Lodomeria, with 2·5m. inhabitants. Russia took over an area with a population of over 1m. Prussia contented itself with 0·5m. new citizens, and the long-desired connection between Western Pomerania and East Prussia.

In 1791 Stanisław II, the last king of the remaining Poland–Lithuania, introduced a constitution which amounted to a bid for independence. The three surrounding superpowers nonetheless engaged in a second partition in 1793. A peasant uprising against Russian rule, led by Tadeusz Kościuszko, was crushed, along with Poland itself which lost control of all its territory to Austria, Prussia and Russia in the third partition (1795).

The territory remained a battleground, particularly during the Napoleonic wars. Napoleon established the Grand Duchy of Warsaw in 1807, which had a French-style constitution, but came under Saxon, and later Russian, administration. Polish legions, which fought on the French side against Prussia, incurred heavy losses. In 1815, when the victorious Allies redistributed the territory Napoleon had won, the 'Congress' Kingdom of Poland reappeared, this time under Russian rule, with the Tsar as its hereditary king.

Thereafter the Poles suffered by their colonizers' attempts to assimilate their culture. A series of uprisings against the Russians took place throughout the century. In the November Revolution of 1830 inexperienced military cadets were suppressed by Tsar Nicholas, who led a campaign of bloody reprisals. Around 8,000 Poles emigrated after this defeat—many of them intellectuals, and most headed for France. During the peasants' revolt in Galicia in 1846 up to 2,000 nobles were murdered and their land ravaged. There was a strong insurgent movement among the peasants, but in 1848 they failed once more to topple their oppressors, this time the Prussians.

The January Uprising against the Russians which began in 1863–64 and ended in the spring of 1865 again led to defeat. Wide-scale Russianization followed, though the abolition of serfdom marked a significant concession. As part of Bismarck's 'Kulturkampf'—the Germanization of the Prussian zones— German was introduced as the official language and Polish began to be taught in schools as a foreign language. Anti-Semitism became rife, and pogroms were not unusual. Many Jewish and Gentile Poles fled.

The Habsburg-dominated part of Poland, Galicia, was more tolerant of Polish nationalism which centred on Kraków. At one point the Austrian prime minister, finance minister and foreign minister were all Polish. Newly-formed parties began to gain ground, with the National Democrats under Roman Dmowski campaigning for autonomy and Józef Piłsudski's Socialists engaging in an underground struggle for independence. Piłsudski led an anti-Russian uprising in 1905, and was to take up arms against Russia in the First World War when Poland's territory again bore the brunt of much of the fighting between its three partitioners.

In 1917 a Polish National Committee, formed by Roman Dmowski in Paris, was recognized by the Allies. One of its members and US representative was the pianist Ignacy Jan Paderewski, who urged the Americans to support the cause for Polish independence. President Woodrow Wilson's 'Fourteen Points' for peace addressed the Polish issue, guaranteeing independence and access to the sea under point thirteen. A Polish army was formed in France in 1918. In Poland, Piłsudski set up the Polish legions and a rival government. Poland regained its independence under Piłsudski's leadership on 11 Nov. 1918.

But while the Paris Peace Conference recognized the republic, the question of its borders was highly contentious. Poland challenged Lithuania over Vilnius, the city changing hands more than once before the Second World War. Fighting also took place against Ukraine over the issue of Galicia. A war with Russia followed over the next two years, which Poland narrowly managed to win before signing the Soviet–Polish Peace Treaty in Riga in 1921. The Treaty established the borders between Russia, Ukraine and Belarus, the last two being swallowed up by the USSR the following year. Gdańsk was awarded the status of a free city, and the Polish Corridor was formed between German West and East Prussia and the rest of Germany.

Between the wars there were 16 palatinates, all centrally governed from Warsaw. The new republic was first headed by President Narutowicz, the representative of the left and centre parties, who served for only days before being assassinated by a right-wing fanatic, and replaced in 1922 by Stanisław Wojciechowski. A series of intra-party disputes and factionalisms led the way for Józef Piłsudski to mount a coup in May 1926, seizing the power he maintained under a dictatorship until his death in 1935.

Second World War
In foreign affairs Poland managed to maintain a balance between its two most intimidating neighbours, Germany and the USSR, signing a non-aggression pact with Germany in 1934. However, the Molotov–Ribbentrop non-aggression pact of Aug. 1939 secretly agreed to partition Poland between Germany and the Soviet Union in the event of war. British and French guarantees of Polish independence that had been agreed in April of the same year obliged them to declare war on Nazi Germany two days after Hitler's troops marched into Poland on 1 Sept. 1939.

The response of Britain and France signalled the start of the Second World War. The German army invaded Poland along the entire front from the Baltic Sea to Slovakia, annexing over half of the country within three weeks. Stalin's troops marched into Poland from the Eastern Front on 17 Sept., leaving the country occupied for most of the duration of the war. The Nazis undertook a policy of liquidation—not only of Jews and ethnic 'undesirables' but also of the intelligentsia, so as to avoid any possibility of a Polish leadership class. Many Polish children seen as racially pure were taken away from their parents to be brought up as Germans, while others were deported. A total of over 6m. Polish nationals, or 17% of the population, were killed in the war, half of them Jewish. Not all of the murders were attributable to the Nazis, however. In 1989 Soviet authorities finally admitted to having murdered 15,000 Polish officers who went missing in May 1940. The Soviet secret service had been equally keen to obliterate potential opposition leaders.

Polish forces regrouped on Allied soil under a government-in-exile headed by Gen. Władysław Sikorski, first in Paris and then, after 1940, in London. In Poland an underground national army, the AK, was formed under Gen. Komorowski to fight against the occupiers and to organize resistance. After Germany's invasion of the USSR in 1941, Poland was occupied solely by Nazi forces. Many of the largest concentration camps were built on Polish soil, including Auschwitz near Kraków.

In 1943 the exiled prime minister Gen. Sikorski was killed in a plane crash. He was replaced by Stanisław Mikołajczyk of the Polish Peasants' Party. The same year saw a Jewish uprising in the Warsaw ghetto, and in 1944 there was a second rebellion against the Nazi occupation which lasted for two months. The Red Army was on the threshold of Warsaw throughout the two month revolt, but did not intervene. 150,000 civilians and 18,000 members of the AK lost their lives with virtually the whole of the remaining urban population deported and 85% of the city destroyed. By the time of Warsaw's liberation in Jan. 1945, the Jewish population numbered 200. The decimated underground movement was forced to seek assistance from Moscow, and after a number of compromises the Soviets recognized the Polish Committee of National Liberation, or the 'Lublin Committee', which proclaimed itself the sole legal government when Lublin was liberated in July 1944.

Poland's post-war fate was decided by the Allies at the Yalta and Potsdam conferences. At Yalta, Stalin agreed that the Lublin government should be extended to include non-Communists from the exile government, a promise that he failed to keep. Stanisław Mikołajczyk and three other members joined the provisional cabinet in July 1945. Nonetheless, many Polish politicians left the country. The Potsdam conference set Poland's Western border along the Oder–Neisse line, with all former German territories east of these rivers handed to Poland. As a result, Poles and Germans had to be resettled.

The first post-war elections were held in Jan. 1947. The Stalinist Polish Workers' Party (PPR) managed to crush both official and underground opposition. A Communist-dominated coalition under the leadership of Władysław Gomułka, the 'Democratic Bloc', won over three-quarters of the votes. Bolesław Bierut, leader of the USSR-backed Polish Communist Party, was named president. Defeated, Stanisław Mikołajczyk fled the country. An independently minded politician, Gomułka entered into conflict with Stalin by opposing agricultural collectivization and by speaking out against the formation of Cominform (Communist Information Bureau) in 1947. As a result he was removed as Secretary General of the PPR in Sept. 1948. Expelled from the party in late 1949, he was put under house arrest in July 1951. In 1948 the Polish United Workers' Party (PZPR) was formed, with Bierut as first party secretary. The nationalization of industry, land expropriation and the restructuring of the economy to favour heavy industry, including arms production, were accompanied in 1952 by a Soviet-style constitution and the renaming of the country as the People's Republic of Poland. This 'Stalinization' also included political and religious suppression and persecution, which targeted the Catholic church in particular.

Post-War Reform

In 1955 Poland joined other Eastern bloc countries in signing the Warsaw Pact military treaty. Meanwhile, the planned economy was failing, leading to widespread public unrest as food prices spiralled. Workers' strikes and riots in Poznań in 1956 were brutally suppressed by the authorities resulting in the death of 53 people. At this time Gomułka, the opponent of Stalinism, gained popularity. Readmitted to the party in 1956, Gomułka was reinstated as first secretary of the Party.

He attempted to introduce reforms, winning public support for his pledges of a 'Polish way' to socialism. Gomułka cut the power of the secret police, halted agricultural collectivization and brought an end to attacks on the Catholic Church. However, the suppression of freedom of expression continued and the economy did not improve. Gomułka's popular appeal began to falter. Student riots sprang up throughout the 1960s. In the 'March events' of 1968, the campaign for intellectual freedom led to widespread student riots and a reactive Party campaign against intellectuals and Jews, many of whom were forced to flee abroad.

Unrest and dissatisfaction with the party remained. Increased food prices in Dec. 1970 resulted in riots and strikes in the shipyards of Gdańsk, Szczecin and Gdynia. These were met with armed opposition, the authorities firing into the masses and killing several demonstrators. Gomułka and other leaders subsequently resigned, although Gomułka at least had the satisfaction of procuring West Germany's recognition of the Oder–Neisse line as the official Western border of Poland in Dec. 1970.

Solidarity

Edward Gierek succeeded Gomułka as first secretary, and in the following years launched a reform programme which was chiefly financed by loans from Western banks. He was hoping for a Polish economic miracle, but lacked the will to push through the necessary reforms. Short term rewards were not enough to overcome the problems of a failing infrastructure, economic mismanagement of successive governments and a faltering world economy following the 1973–74 world oil crisis. Further demonstrations took place in several cities in 1976 to protest at more food price increases, and in Radom a Workers' Defence Committee was founded. While the government expressed disapproval, it did not act against the Committee.

In 1978 the election of Karol Wojtyła, Cardinal of Kraków, as Pope John Paul II boosted Poland's national self-esteem, celebrated in his trip to his native country the following year. Nonetheless, increased meat prices in July 1980 led to more waves of strikes, rippling out from the Ursus tractor plant near Warsaw across the country, and culminating in the Lenin shipyards in Gdańsk, where the Solidarity movement was born. The first independent trade union to be established in a communist country soon boasted a membership of 10m. Its leader, Lech Wałęsa, a shipyard electrician, set up a strike committee, the first of a succession across the country, and drew up a 21-point accord, demanding the right to strike and to form independent trade unions, the abolition of censorship, freedom of expression, the release of political prisoners and access to the media. Soviet and Polish communist efforts to curb Solidarity's popularity failed and the group was officially recognized after some government resistance.

The social unrest, coupled with failing health, led to Gierek's resignation in Sept. 1980. In Feb. 1981 Gen. Wojciech Jaruzelski, the defence minister, became prime minister. This brought the military into the political front line and in Dec. 1981 Jaruzelski imposed martial law. A Military Council of National Salvation was established and Solidarity was proscribed. Wałęsa was among the thousands of members who were arrested and imprisoned. Demonstrations and strikes provoked the government into even stricter controls with the banning of all independent trade unions, although martial law was dropped a year later.

New hope was given to Poland in 1983, by the Pope's second visit, and by the award of the Nobel Peace Prize to Lech Wałęsa. Economic difficulties continued throughout the 1980s, and when the government proposed unpopular economic reforms in 1987, support for Solidarity led to nationwide strikes during 1988. Jaruzelski was forced to embark on negotiations with Wałęsa and the Catholic Church. Agreement was reached in April 1989 and Solidarity was given legal status and freedom to fight the upcoming elections, whilst the previously ceremonial post of Presidency was vested with new legislative powers. In return, Solidarity agreed to compete for only 35% of the seats in the Sejm.

At the July 1989 elections Solidarity won virtually all the seats they contested but because of the 35% rule Jaruzelski was voted in as president. However, Solidarity refused to join the communists in a grand coalition and Jaruzelski had to appoint Tadeusz Mazowiecki, an official of Solidarity, to be Poland's first non-communist premier in over 40 years. Jaruzelski subsequently resigned.

The first round of presidential elections in Nov. 1990 pitted Lech Wałęsa against Tadeusz Mazowiecki. Wałęsa won 43% of the votes in the first round and 74% in the second round in Dec., when he was inaugurated. Mazowiecki resigned his premiership and was replaced by Jan Bielcki, whose government held office until Aug. 1991. Genuinely free parliamentary elections did not take place until Oct. 1991, when there was a surprisingly low electoral turnout. In the absence of a clear-cut winner, several parties combined to form a centre-right coalition headed by Jan Olszewski. Owing to disputes both within the party and with President Wałęsa, however, the government lasted only seven months. This factionalism and inability to make compromises was typical of Poland's early post-Communist years. The government formed under Hanna Suchocka, Poland's first female prime minister, fared no better.

As in other post-Communist states, the economic measures necessary for the transition to a profitable market economy were highly unpopular with the electorate, not least when industrial modernization led to unemployment. In 1990 the finance minister, Leszek Balcerowicz, had introduced a range of tight austerity measures, including price rises and currency devaluation in an attempt to stabilize the economy before opening it to market forces. The Polish economy prospered but Wałęsa's popular appeal diminished as his tenure progressed. His skills as Solidarity's leader revolved around his ability to speak for the common people, but in government his tone was often regarded as aggressive and his style of leadership autocratic. Solidarity's loyalties as a trade union were often incompatible with its responsibilities as a political party. The elections of 1993 saw the return of the left under Waldemar Pawlak of the Polish Peasants' Party. After a series of intra-party quarrels and accusations of corruption, Pawlak's premiership ended in Feb. 1995.

Pawlak was replaced by the Communist Józef Oleksy of the Democratic Left Alliance. The left gained further political clout when Wałęsa was ousted in the presidential elections of 1995 by Aleksander Kwaśniewski. Redundancies in the Gdańsk shipyards in 1997 saw a renewed outbreak of nationwide strikes. Revising its political agenda, Solidarity forged a coalition of 25 centre-right parties to create Solidarity Electoral Action. This party emerged as the strongest in the 1997 general election when Jerzy Buzek, a member of Solidarity since its inception, became prime minister. A new constitution came into effect, reducing the powers of the president and committing the country to a social market economy.

Kwaśniewski's communist heritage caused concern among many Western leaders, but he confirmed his intention to press for EU and NATO membership. Market reforms and privatization continued apace. In 1999, at a joint ceremony with Czech president Vaclav Havel, Kwaśniewski signed Poland into NATO. The following year Kwaśniewski secured a second term and in 2001 Buzek was succeeded by Leszek Miller. A former communist turned social democrat, Miller's key aim was to prepare Poland for entry into the EU. Facing a deteriorating economy, he cut the national debt by increases in taxation and spending cuts. On 1 May 2004 Poland became a member of the EU.

Lech Kaczyński succeeded Kwaśniewski in 2005, with his twin brother Jarosław Kaczyński, serving as prime minister in 2006–07. President Kaczyński was killed in April 2010 when his plane crashed over Russia on its way to a war memorial service. The head of the national bank, the entire command of the armed services and several senior government figures, MPs, clergy and academics also perished in the accident.

TERRITORY AND POPULATION

Poland is bounded in the north by the Baltic Sea and Russia, east by Lithuania, Belarus and Ukraine, south by the Czech Republic and Slovakia and west by Germany. Poland comprises an area of 312,685 sq. km (120,728 sq. miles).

At the census of 31 March 2011 the population was 38,511,824, giving a density of 123·2 per sq. km. In 2009, 61·0% of the population lived in urban areas.

The UN gives a projected population for 2015 of 38·36m.

The country is divided into 16 regions or voivodships (*wojewodztwo*), created from the previous 49 on 1 Jan. 1999 following administrative reform. Area (in sq. km) and census population (in 1,000) in 2011 (density per sq. km in brackets).

Voivodship	Area	Population	Density
Dolnośląskie	19,948	2,915	(146)
Kujawsko-Pomorskie	17,970	2,098	(117)
Lubelskie	25,122	2,176	(87)
Lubuskie	13,989	1,023	(73)
Łódzkie	18,219	2,539	(139)
Małopolskie	15,190	3,337	(220)
Mazowieckie	35,560	5,269	(148)
Opolskie	9,412	1,016	(108)
Podkarpackie	17,844	2,127	(119)
Podlaskie	20,186	1,202	(60)
Pomorskie	18,293	2,276	(124)
Śląskie	12,331	4,630	(376)
Świętokrzyskie	11,708	1,281	(109)
Warmińsko-Mazurskie	24,192	1,452	(60)
Wielkopolskie	29,826	3,447	(116)
Zachodniopomorskie	22,897	1,723	(75)

Population (in 1,000) of the largest towns and cities (2011 census):

Warsaw (Warszawa)	1,700·6	Częstochowa	236·8
Cracow (Kraków)	757·6	Radom	221·3
Łódź	728·9	Sosnowiec	216·4
Wrocław	630·1	Toruń	205·0
Poznań	554·7	Kielce	202·2
Gdańsk	460·3	Gliwice	187·5
Szczecin	410·1	Zabrze	181·1
Bydgoszcz	363·9	Rzeszów	179·4
Lublin	349·1	Bytom	176·9
Katowice	310·8	Olsztyn	174·6
Białystok	294·0	Bielsko-Biała	174·5
Gdynia	249·1		

The population is 96·7% Polish. Minorities at the 2002 census included 173,153 Silesians, 152,987 Germans, 48,737 Belarusians and 30,957 Ukrainians. There are an estimated 300,000 people in Poland of Kashubian ethnicity (direct descendants of an early Slavic tribe of Pomeranians). They generally declare Polish nationality and consider themselves both Poles and Kashubians.

A movement for Silesian autonomy has attracted sufficient support to suggest that further moves towards decentralization may soon be considered. A Council of National Minorities was set up in March 1991. There is a large Polish diaspora, some 53% in the USA.

The official language is Polish.

SOCIAL STATISTICS

2010 (in 1,000): births, 415·0; deaths, 378·5; marriages, 228·3; divorces, 61·3; infant deaths, 2·1. Rates (per 1,000 population): birth, 10·8; death, 9·9; marriage, 6·0; divorce, 1·6; infant mortality (per 1,000 live births), 5·0. A law prohibiting abortion was passed in 1993, but an amendment of Aug. 1996 permits it in cases of hardship or difficult personal situation. The most popular age range for marrying in 2010 was 25–29 for both males and females. Expectation of life at birth, 2007, was 71·3 years for males and 79·7 years for females. In 2010 there were 17,360 emigrants (including 6,818 to Germany) and 15,246 immigrants. 70% of Polish emigrants between 2000 and 2005 settled in Germany. Number of suicides, 2008, 5,681; the suicide rate per 100,000 population was 26·4 among males and 4·1 among females in 2008. Population growth rate, 2010, 0·1%; fertility rate, 2008, 1·3 births per woman (one of the lowest rates in the world).

CLIMATE

Climate is continental, marked by long and severe winters. Rainfall amounts are moderate, with a marked summer maximum. Warsaw, Jan. 24°F (−4·3°C), July 64°F (17·9°C). Annual rainfall 18·3" (465 mm). Gdańsk, Jan. 29°F (−1·7°C), July 63°F (17·2°C). Annual rainfall 22·0" (559 mm). Kraków, Jan. 27°F (−2·8°C), July 67°F (19·4°C). Annual rainfall 28·7" (729 mm). Poznań, Jan. 26°F (−3·3°C), July 64°F (17·9°C). Annual rainfall 21·0" (534 mm). Szczecin, Jan. 27°F (−3·0°C), July 64°F (17·7°C). Annual rainfall 18·4" (467 mm). Wrocław, Jan. 24°F (−4·3°C), July 64°F (17·9°C). Annual rainfall 20·7" (525 mm).

CONSTITUTION AND GOVERNMENT

The present Constitution was passed by national referendum on 25 May 1997 and became effective on 17 Oct. 1997. The head of state is the *President*, who is directly elected for a five-year term (renewable once). The President may appoint, but may not dismiss, cabinets.

The authority of the republic is vested in the *Sejm* (Parliament of 460 members), elected by proportional representation for four years by all citizens over 18. There is a 5% threshold for parties and 8% for coalitions, but seats are reserved for representatives of ethnic minorities even if their vote falls below 5%. 69 of the Sejm seats are awarded from the national lists of parties polling more than 7% of the vote. The Sejm elects a *Council of State* and a *Council of Ministers.* There is also an elected 100-member upper house, the *Senate.* The President and the Senate each has a power of veto which only a two-thirds majority of the Sejm can override. The President does not, however, have a veto over the annual budget. The *Prime Minister* is chosen by the President with the approval of the Sejm.

A *Political Council* consultative to the presidency consisting of representatives of all the major political tendencies was set up in Jan. 1991.

National Anthem

'Jeszcze Polska nie zginęła' ('Poland has not yet perished'); words by J. Wybicki, tune by M. Ogiński.

GOVERNMENT CHRONOLOGY

First Secretaries of the Polish United Workers' Party (1943–90) and Presidents of the Republic (since 1990). (PiS = Law and Justice Party; PO = Civic Platform; PZPR = Polish United Workers' Party; SdRP = Social Democracy of the Republic of Poland; SLD = Democratic Left Alliance; n/p = non-partisan)

First Secretaries of PZPR

1943–48	Władysław Gomułka
1948–52	Bolesław Bierut
1952–54	Hilary Minc
1954–56	Bolesław Bierut
1956	Edward Ochab
1956–70	Władysław Gomułka
1970–80	Edward Gierek
1980–81	Stanisław Kania
1981–89	Wojciech Jaruzelski (military)
1989–90	Mieczysław F. Rakowski

Presidents

1989–90	n/p	Wojciech Jaruzelski
1990–95	Solidarność	Lech Wałęsa
1995–2005	SdRP/SLD	Aleksander Kwaśniewski
2005–10	PiS	Lech Kaczyński
2010–	PO	Bronisław Komorowski

Prime Ministers since 1945. (AWS = Solidarity Electoral Action; KLD = Liberal Democratic Congress; PC = Centre Alliance; PiS = Law and Justice Party; PO = Civic Platform; PPR = Polish Workers' Party; PPS = Polish Socialist Party; PSL = Polish Peasants' Party; PZPR = Polish United Workers' Party; RS AWS = Social Movement-Solidarity Electoral Action; SdRP = Social Democracy of the Republic of Poland; SLD = Democratic Left Alliance; UD = Democratic Union)

1945–47	PPS	Edward Osóbka-Morawski
1947–52	PPR, PZPR	Józef A. Z. Cyrankiewicz
1952–54	PZPR	Bolesław Bierut
1954–70	PZPR	Józef A. Z. Cyrankiewicz
1970–80	PZPR	Piotr Jaroszewicz
1980	PZPR	Edward Babiuch
1980–81	PZPR	Józef Pińkowski
1981–85	PZPR/military	Wojciech Jaruzelski
1985–88	PZPR	Zbigniew Messner
1988–89	PZPR	Mieczysław F. Rakowski
1989	PZPR	Czesław Kiszczak
1989–91	Solidarność, UD	Tadeusz Mazowiecki
1991	KLD	Jan Krzysztof Bielecki
1991–92	PC	Jan Olszewski
1992	PSL	Waldemar Pawlak
1992–93	UD	Hanna Suchocka
1993–95	PSL	Waldemar Pawlak
1995–96	SdRP/SLD	Józef Oleksy
1996–97	SdRP/SLD	Włodzimierz Cimoszewicz
1997–2001	RS AWS/AWS	Jerzy Buzek
2001–04	SLD	Leszek Miller
2004–05	SLD	Marek Belka
2005–06	PiS	Kazimierz Marcinkiewicz
2006–07	PiS	Jarosław Kaczyński
2007–	PO	Donald Tusk

RECENT ELECTIONS

Parliamentary elections were held on 9 Oct. 2011. The ruling Civic Platform (PO) won 207 of 460 seats with 39·2% of the votes, ahead of the Law and Justice Party (PiS), with 157 seats and 29·9%; the Palikot Movement won 40 seats with 10·0%; Polish Peasants' Party (PSL) won 28 with 8·4%; the Democratic Left Alliance won 27 with 8·2%; German Minority (MN) won one seat with 0·2%. In the Senate elections on the same day, the Civic Platform won 63 of 100 seats, the Law and Justice Party 31 and the Polish Peasants' Party 2; four seats went to independents. Turnout was 48·9%.

Presidential elections were held in two rounds on 20 June and 4 July 2010. In the first round ten candidates stood; turnout was 54·9%. Bronisław Komorowski of the PO gained 41·5% of votes cast, Jarosław Kaczyński of the PiS 36·5%, Grzegorz Napieralski of the Democratic Left Alliance 13·7% and Janusz Korwin-Mikke of the Liberty and Rule Party 2·5%. Other candidates obtained less than 2%. In the second round run-off Bronisław Komorowski was elected president with 53·0% of the vote against 47·0% for Jarosław Kaczyński.

European Parliament

Poland has 51 (54 in 2004) representatives. At the June 2009 elections turnout was 24·5% (20·9% in 2004). The PO won 25 seats with 44·4% of votes cast (political affiliation in European Parliament: European People's Party); the PiS, 15 with 27·4% (European Conservatives and Reformists); the Democratic Left Alliance–Union of Labour, 7 with 12·3% (Progressive Alliance of Socialists and Democrats); the PSL, 3 with 7·0% (European People's Party).

CURRENT GOVERNMENT

President: Bronisław Komorowski; b. 1952 (PO; since 10 April 2010—acting until 6 Aug. 2010).

In March 2013 the coalition government consisted of:

Prime Minister: Donald Tusk; b. 1957 (PO; sworn in 16 Nov. 2007).

Deputy Prime Ministers and Minister of Economy: Janusz Piechociński (PSL). *Deputy Prime Minister and Minister of Finance:* Jan-Vincent Rostowski (ind.).

Minister of Administration and Digitalization: Michał Boni (ind.). *Agriculture and Rural Development:* Stanisław Kalemba (PSL). *Culture and National Heritage:* Bogdan Zdrojewski (PO). *Environment:* Marcin Korolec (ind.). *Foreign Affairs:* Radosław Sikorski (PO). *Health:* Bartosz Arłukowicz (PO). *Interior:* Bartłomiej Sienkiewicz (ind.). *Justice:* Jarosław Gowin (PO). *Labour and Social Policy:* Władysław Kosiniak-Kamysz (PSL). *National Defence:* Tomasz Siemoniak (PO). *National Education:* Krystyna Szumilas (PO). *Regional Development:* Elżbieta Bieńkowska (ind.). *Science and Higher Education:* Barbara Kudrycka (PO). *Sport and Tourism:* Joanna Mucha (PO). *State Treasury:* Mikołaj Budzanowski (ind.). *Transport, Construction and Marine Economy:* Sławomir Nowak (PO). *Head of the Chancellery of the Prime Minister:* Jacek Cichocki (ind.).

Speaker of the Sejm: Ewa Kopacz (PO).

Office of the Prime Minister: http://www.kprm.gov.pl

CURRENT LEADERS

Bronisław Komorowski

Position
President

Introduction
Bronisław Komorowski was elected president in June 2010, having served as acting president following the death of incumbent Lech Kaczyński in an air crash in April 2010. A member of the governing Civic Platform (PO) party, he supports pro-market economic reforms and favours a more active role in the European Union.

Early Life
Bronisław Komorowski was born on 4 June 1952 in Oborniki Śląskie to an aristocratic family whose lands were confiscated by the Communist government. In 1977 he graduated in history from the University of Warsaw, where he had become politically active. He completed his masters degree the same year and from 1977–80 was editor of a Catholic journal. From 1980–81 he worked at the centre for social research of the trade union Solidarność (Solidarity), campaigning for democracy and against the existing communist regime.

After being interned for four months in 1981 when Poland came under martial law, he spent the next eight years working as a teacher in Niepokalanów. Following the collapse of communism and the election of a Solidarity-led government in 1989, he was deputy minister for defence from 1990–93. At the 1991 general election he was elected as an MP for the Freedom Union party (UW), serving as its general secretary from 1993–95. In 1997 he co-founded a new party which allied itself with the Conservative People's Party to become part of the Solidarity Electoral Action (AWS) grouping.

Following the formation of an AWS–UW government in 1997, Komorowski was appointed head of the parliamentary national defence committee. As a member of Jerzy Buzek's government, he helped oversee Poland's entry into EU and NATO, and from 2000–01 served as defence minister. In 2001 Komorowski joined the new, reformist PO, led by Donald Tusk, and was elected to parliament at the 2001 general election, in which the AWS-UW lost power. Subsequently he served as deputy chairman of the parliamentary national defence committee and as a member of the parliamentary committee for foreign affairs.

In Oct. 2005 he was elected deputy speaker of the Sejm, becoming speaker in Nov. 2007. On 10 April 2010 President Lech Kaczyński died in an air crash. Under the constitution, Komorowski became acting president until presidential elections took place on 20 June 2010. Initially campaigning as a loyal supporter of Prime Minister Tusk's programme of economic reforms and public spending cuts, he promised not to use the presidential veto to block progress in these areas. However, during a closely fought contest he moderated his stance, softening his line on the need for reform of farmers' pensions and abandoning plans to raise the retirement age. Voting went to two rounds and in the run-off on 4 July 2010 he narrowly defeated the late president's twin brother, Jarosław Kaczyński, with 53·0% of the vote.

Career in Office
Komorowski has sought to raise Poland's profile within the EU and to increase economic, defensive and diplomatic ties with key international partners, including the USA. However, in Aug. 2012 he accused US President Barack Obama of betraying Poland's national security over the 2009 cancellation of a controversial anti-ballistic missile system, and he called for Poland to build its own missile shield to ensure the country's defence.

Donald Tusk

Position
Prime Minister

Introduction
Donald Tusk became prime minister on 9 Nov. 2007 following his party's resounding victory in parliamentary elections. The former Solidarity activist has taken a pro-business stance and has been keen to establish closer relations with EU neighbours. He began a second term following elections in Oct. 2011.

Early Life
Donald Franciszek Tusk was born on 22 April 1957 in Gdańsk. His family is part of the city's long-established minority Kashubian community. Following his secondary education he attended the University of Gdańsk where he studied history. A long-time critic of the communist administration, Tusk helped to establish the student committee of the Solidarity movement, which grew out of the nationwide industrial unrest centred on the Gdańsk shipyard during the summer of 1980. He subsequently co-founded the Independent Polish Students' Association (NZS). Following the authorities' crackdown on Solidarity in 1981, Tusk and other activists were forced into the shadows. He earned a living as a builder, an experience subsequently presented as evidence of his empathy with 'ordinary people'.

In the late 1980s Tusk left Solidarity to join the nascent liberal movement and in 1991 joined the Liberal Democratic Congress (KLD), which contested the first multi-party elections in Oct. 1991 on a free-market platform calling for privatization, freedom of movement and accession to the EU. Tusk took one of the KLD's 37 seats in the Sejm. Although he was re-elected as a deputy in the 1993 elections, the KLD fared poorly and in March 1994 merged with the Democratic Union to form a new centre-right party, Freedom Union (UW). The party secured 13·4% of the vote in the 1997 elections, becoming the junior partner in Jerzy Buzek's coalition government. Tusk was elected to the Senate, where from 1998–2001 he served as vice-speaker.

Having failed to win the chairmanship of the UW in 2000, Tusk resigned from the party. He joined Andrzej Olechowski (who had performed creditably in the 2000 presidential contest) and Maciej Płażyński in establishing the secular, liberal Civic Platform (PO) in early 2001, with Płażyński at the helm. The PO performed strongly in the 2001 elections, taking 65 seats in the Sejm and becoming the largest opposition party to Leszek Miller's government. In June 2003 Tusk became the PO's chairman. He was a vocal critic of the left-leaning SLD government, particularly its economic policies. His standing improved as the SLD became mired in corruption scandals but he failed in his 2005 bid for the presidency, losing to Lech Kaczyński of the socially conservative, nationalist Law and Justice Party. Later in 2005 the PO suffered further electoral defeat to Law and Justice, led by Jarosław Kaczyński (Lech's twin brother), who became prime minister.

Tusk remained leader of the PO and took a more aggressive approach in the run-up to the early election called for Oct. 2007. The election followed the collapse of the Law and Justice-led coalition amid allegations of corruption. Tusk accused Kaczyński of incompetence on international relations—notably deteriorating relations with Germany—and of failing to prevent the mass movement of Poles to Britain and Ireland in search of work. Tusk campaigned on a platform to speed up privatization, lower taxes and reduce business bureaucracy to encourage investors.

In parliamentary elections on 21 Oct. 2007 the PO emerged victorious, taking around 41% of the vote against 32% for Law and Justice. Tusk took office as prime minister on 16 Nov. and his cabinet won a confidence vote in the Sejm on 24 Nov. 2007.

Career in Office

Tusk pledged to create jobs and promote economic development by cutting bureaucracy and regulation. However, the global financial downturn in 2008 undermined Poland's growth prospects, prompting the government to launch an economic stimulus programme in Dec. that year and to negotiate a one-year US$20·6bn. credit line with the International Monetary Fund which was approved in May 2009. In July 2010 Tusk's political position was strengthened by the election of the PO's candidate, Bronisław Komorowski, as state president in place of Lech Kaczyński, who had been killed in an air crash the previous April.

On the international stage, Tusk oversaw the withdrawal in Oct. 2008 of Poland's last troops stationed in Iraq, fulfilling a key electoral pledge. However, plans agreed in 2008 for Poland to host a controversial missile defence shield for the USA were effectively abandoned in Sept. 2009 when the US president announced the scrapping of key elements of the system. Despite the Polish government's disappointment at the decision, Tusk insisted that the USA and Poland would remain close allies. In July 2011 Poland assumed the EU's six-month rotating presidency for the first time since the country's accession to the organization in 2004.

In Oct. 2011 Tusk's PO was returned to power as the largest party in national parliamentary elections. However, having introduced unpopular fiscal measures in May 2012—including increasing the retirement age—his administration lost ground in opinion polls to the opposition Law and Justice Party. In Oct. the government narrowly won a parliamentary vote of confidence after pledging US$95bn. in infrastructure and other investments to boost economic growth.

DEFENCE

Poland is divided into two military districts: Pomeranian (North) and Silesian (South). In 2008 military expenditure totalled US$10,176m. (US$264 per capita), representing 1·9% of GDP.

Conscription ended on 1 Jan. 2010.

Army

Strength (2007) 79,000 (including 39,000 conscripts). In accordance with a programme of modernization of the armed forces, the strength has been gradually declining, from 230,000 in the socialist era in 1988 to 186,000 in 1995 and further to the current figure of under 80,000. In addition there were 188,000 Army reservists in 2007 and 14,100 border guards.

Navy

The fleet comprises four ex-Soviet and one ex-Norwegian diesel submarines, three frigates and five corvettes. There is a small Naval Aviation force.

Personnel in 2007 totalled 11,600 including 600 conscripts and 1,900 in Naval Aviation. Bases are at Gdynia, Hel, Świnoujście and Kolobrzeg.

Air Force

The Air Force had a strength (2007) of 28,466 (466 conscripts). There are two air defence corps (North and South) with 103 combat-capable aircraft (including MiG-29s and Su-22s).

INTERNATIONAL RELATIONS

A treaty of friendship with Germany signed on 17 June 1991 renounced the use of force, recognized Poland's western border as laid down at the Potsdam conference of 1945 (the 'Oder–Neisse line') and guaranteed minority rights in both countries.

A referendum held on 8 June 2003 approved accession to the EU, with 77·4% of votes cast for membership and 22·6% against. Poland became a member of the EU on 1 May 2004.

Poland's Senate approved the European Union's Treaty of Lisbon on 2 April 2008, the day after the *Sejm* had done so. However, then President Lech Kaczyński did not ratify it until 12 Oct. 2009, following acceptance of the Treaty by Ireland in a referendum ten days earlier.

ECONOMY

Agriculture accounted for 4·4% of GDP in 2006, industry 31·7% and services 63·9%.

Overview

Under communism the economy was skewed towards heavy industry to the neglect of services. Since the collapse of the old regime in 1990, the service sector has gained, contributing 63% of GDP in 2011 compared to 33% by the industrial sector. Most of the banking sector has been privatized, as have many large industries. The private sector now accounts for around three-quarters of GDP and 63·4% of employment. In 1990 and 1991 the economy shrank by 11·5% and 7·0% respectively but subsequent economic restructuring helped to achieve productivity gains.

The Economist Intelligence Unit estimates that from 1994–2003 total factor productivity grew at an annual average of over 3%. This resulted from sound macroeconomic management combined with a raft of transition policies, including price liberalization and lower import barriers. After accession to the EU in May 2004 growth was buoyant following half a decade of below-par expansion, with exporters benefiting from EU market integration. An unspectacular yet solid growth performance since the initial transition shock has helped raise living standards, with per capita income levels increasing by 250% between 1993 and 2010. The economic environment is friendly with transparent investment rules and equality for domestic and foreign firms.

Unemployment declined steadily (from 19·8% in 2002 to 7·1% in 2008) until the effects of the global financial crisis began to be felt, with unemployment reaching 10·6% in Dec. 2012. Unemployment is highest where state farms were once the rule (primarily in the northeast), while former industrial areas have proved more dynamic. Though accounting for only 3·6% of GDP in 2011, agriculture makes up a large share of total employment and is politically powerful. In 2012 the highest growth in average gross monthly wages (14·9%) was observed in agriculture, forestry and fishing.

Despite suffering during the financial crisis, Poland fared better than many of its neighbours. Real GDP growth fell to 5·0% in 2008, down from 6·8% in 2007. In response the central bank cut the seven-day reference interest rate from 5% to 4·25%. Output in the industrial sector fell by 4·4% year-on-year in Dec. 2008 and manufacturing output, which makes the largest contribution to industrial output, fell by 3·9%, mirroring a slowdown in the export market. Exports contracted by 30% year-on-year in the first quarter of 2009 and FDI fell by 50%.

In April 2009 the Polish government approached the IMF for US$20·5bn. from its flexible credit line to shore up borrowing needs in the face of the crisis. This helped Poland become the only EU country to avoid recession in 2009, with real GDP growth of 1·7% aided by the economy's large domestic market, stable private consumption and limited reliance on exports. Tax cuts enacted in 2006 and 2007 combined with government counter-cyclical fiscal stimulus during the crisis led to a rise in the government deficit from 2% of GDP in 2007 to 7% in 2009.

Eurozone countries accounted for 54·2% of exports and 46·6% of imports in 2011. About 66% of assets in the Polish banking sector belong to foreign financial institutions, mainly from the EU. Reduced export demand and lower public investment as a result of the eurozone crisis has caused growth to moderate. As a result, in Sept. 2012 the government stated it would not meet its deficit target of 3% of GDP in 2012.

The IMF has stressed the need for continuing privatization and structural reforms to strengthen the labour market and improve the business climate. Another priority is to reduce public debt, which stood at 56·3% of GDP in 2011.

Currency
The currency unit is the *złoty* (PLN) of 100 *groszy*. A new złoty was introduced on 1 Jan. 1995 at 1 new złoty = 10,000 old złotys. Inflation rates (based on OECD statistics):

2002	2003	2004	2005	2006	2007	2008	2009	2010	2011
1·9%	0·7%	3·4%	2·2%	1·3%	2·4%	4·2%	3·8%	2·6%	4·2%

Inflation, in single figures since 1999, had been nearly 250% in 1990. The złoty became convertible on 1 Jan. 1990. In 1995 the złoty was subject to a creeping devaluation of 1·2% per month; it was allowed to float in a 14% (+/–7%) band from 16 May 1995. In April 2000 Poland introduced a floating exchange rate. Foreign exchange reserves were US$72,280m. and gold reserves 3·31m. troy oz in Sept. 2009. In July 2009 total money supply was 363,655m. złotys.

Budget
Central government revenue and expenditure (in 1bn. złotys):

	2007	2008	2009
Revenue	389·4	414·6	412·8
Expenditure	402·8	449·3	480·2

Poland's budget deficit in 2011 was 5·1% of GDP (2010, 7·8%; 2009, 7·4%). The required target set by the EU is a budget deficit of no more than 3%.

VAT is 23·0% (reduced rates, 8% and 5%). Taxes accounted for 53·2% of revenues in 2009; social benefits accounted for 44·5% of expenditures.

Performance
Real GDP growth rates (based on OECD statistics):

2002	2003	2004	2005	2006	2007	2008	2009	2010	2011
1·5%	3·9%	5·2%	3·6%	6·2%	6·8%	5·0%	1·7%	3·9%	4·3%

Poland was the only EU member country not to experience a recession in 2009 and the only one with positive growth. Total GDP in 2011 was US$514·5bn. The private sector accounts for more than 70% of GDP.

Banking and Finance
The National Bank of Poland (established 1945) is the central bank and bank of issue (*President*, Marek Belka). There were 73 banks operating at the end of 2000, of which only seven were controlled—directly or indirectly—by the Polish government through its state treasury. Poland's leading banks are PKO Bank Polski with assets of 156·5bn. złotys in Dec. 2009 and Bank Pekao (assets of 130·6bn. złotys in Dec. 2009).

Gross external debt amounted to US$331,790m. in June 2012.

In 2010 Poland received US$9,681m. of foreign direct investment, down from US$13,698m. in 2009 and US$23,561m. in 2007. It receives the most foreign direct investment of any of the former socialist countries of central and eastern Europe. The total stock of FDI at the end of 2010 was US$193·1bn.

There is a stock exchange in Warsaw.

ENERGY AND NATURAL RESOURCES
In 2008, 7·9% of energy consumption came from renewables (wind power, solar power, hydro-electric power, tidal power, geothermal energy and biomass), compared to the European Union average of 10·3%. A target of 15% has been set by the EU for 2020.

Environment
Poland's carbon dioxide emissions from the consumption and flaring of fossil fuels in 2008 were the equivalent of 7·8 tonnes per capita.

Electricity
Installed capacity was 32·5m. kW in 2007. Production (2007) 159·35bn. kWh; consumption per capita was 4,040 kWh in 2007. Poland is a net exporter of electricity.

Oil and Gas
Total oil reserves (2007) amount to some 96m. bbls; natural gas reserves (2008), 110bn. cu. metres. Crude oil production was 721,000 tonnes in 2007; natural gas (2008), 4·1m. cu. metres. The largest oil distributor is Polski Koncern Naftowy ORLEN SA, created by the merger of Petrochemia Płock and Centrala Produktów Naftowych.

Minerals
Poland is a major producer of coal (reserves of some 7,500m. tonnes), copper (56m. tonnes) and sulphur. Production (in tonnes): coal (2007), 88·3m.; brown coal (2007), 57·5m.; salt (2006), 4·0m.; copper (2006), 497,000; silver (2006), 1,265.

Agriculture
In 2007, 15·2% of the economically active population were engaged in agriculture. In 2007 there were 11·87m. ha. of arable land. There were 2·6m. farms in 2007; private farms accounted for 89·1% of the total area of agricultural land and state-owned farms for 10·9%. In 2007 agriculture, hunting and forestry contributed 3·8% of GDP.

Output in 2007 (in 1,000 tonnes): sugar beets, 12,682; potatoes, 11,791; wheat, 8,317; barley, 4,008; rye, 3,126; rapeseed, 2,130; maize, 1,722; oats, 1,462; cabbage, 1,325; apples, 1,040; carrots, 938; onions, 752. Poland is the third largest producer of rye, after Russia and Germany.

Livestock, 2007: pigs, 18·13m.; cattle, 5·70m. (including cows, 2·79m.); sheep, 332,000; horses, 329,000; chickens, 124m.

Livestock products, 2007: pork, bacon and ham, 2,150,700 tonnes; beef and veal, 379,500 tonnes; poultry meat, 1,139,600 tonnes; milk, 12,096,000 tonnes; cheese (including soft cheese), 670,000 tonnes; eggs, 547,000 tonnes.

In 2007 there were 1,553,400 tractors in use.

Forestry
In 2010 forest area was 9·34m. ha. (predominantly coniferous), or 30% of the land area. Timber production in 2007 was 35·93m. cu. metres.

Fisheries
The catch was 185,179 tonnes in 2007, of which 72% were sea fish. In 2007 there were 4,500 people employed in the fishing industry. Imports of fishery commodities were valued at US$1,255m. in 2008 and exports at US$1,174m.

INDUSTRY
The leading companies by market capitalization in Poland in March 2012 were: PKO Bank (US$13·5bn.); Bank Pekao (US$13·0bn.); and Polska Grupa Energetyczna (US$11·6bn.).

In 2007 there were 572 state firms, 216,887 limited liability companies, 324,246 other companies and 18,128 co-operatives. Production in 2007 unless otherwise indicated (in 1,000 tonnes): cement, 17,000; crude steel, 10,631; distillate fuel oil, 8,787; pig iron, 5,804; petrol, 3,867; paper and paperboard, 2,992; fertilizers, 2,835; residual fuel oil, 2,831; plastics in primary forms, 2,778; ammonia, 2,417; nitric acid, 2,270; sulphuric acid, 2,010; sugar, 1,857; soda ash, 1,215; paints and lacquers, 1,117; sulphur, 834; beer, 3,690m. litres; mineral water, 2,708m. litres; fruit and vegetable juice, 664m. litres; vodka, 93m. litres; cigarettes, 124bn. units; bricks, 839m. units; television receivers, 14,929,000 units; refrigerators and freezers, 2,305,000 units; washing machines, 1,938,000 units; telephone sets, 797,000 units; cars, 698,000 units; tractors, 7,400 units; public transport vehicles, 3,600 units.

Output of light industry in 2007: cotton woven fabrics, 142·9m. sq. metres; silk fabrics, 18·1m. sq. metres; woollen woven fabrics, 5·6m. sq. metres; shoes, 43·6m. pairs.

Between 1990 and 2006 employment in the Polish mining industry fell from 388,000 to 119,000 and more than 30 mines closed. For some time the Polish government had been trying to reduce employment further, but in 2006 a mining strategy suggested that the industry may face serious labour shortages by 2015. More than 125,000 jobs have been lost in the steel industry since the early 1990s; by 2009 it employed just 19,400 people. In 2002 the four largest state-owned steel enterprises were regrouped into one company, Polskie Huty Stali SA, which was privatized in 2003. It was bought by Mittal Steel (now ArcelorMittal) in 2004.

Labour
In 2008 a total of 14,037,000 persons were in employment. In Dec. 2008, 3,103,000 persons worked in industry, 2,269,000 in trade and repairs, 1,133,000 in property, renting and business activities, 1,039,000 in education, 840,000 in construction, 809,000 in transport, storage and communications, and 748,000 in health and social services. There were a total of 176,300 working days lost to strike action in 2008. The unemployment rate increased steadily for several years peaking at 20·0% in 2002, compared to the EU average of 7·7%. It has declined considerably since then, and in Dec. 2012 stood at 10·6%. Unemployment among the under 25s was 26·0% in the third quarter of 2011. Workers made redundant are entitled to one month's wages after one year's service, two months after two years' service and three months after three or more years' service. A five-day working week was introduced in May 2001. The number of hours worked was reduced to 40 in 2003. Despite strong public opposition an amendment to the Act on Pensions was passed in 2012 increasing the retirement age to 67 for both men and women (having previously been 65 for men and 60 for women). Starting from Jan. 2013 the retirement age is increasing by three months every year and is set to reach 67 for men in 2020 and for women in 2040.

INTERNATIONAL TRADE
Imports and Exports
In 2007 imports (c.i.f.) totalled US$161·94bn. (US$124·73bn. in 2006); exports (f.o.b.), US$137·83bn. (US$109·33bn. in 2006). The main imports in 2005 were electrical equipment (14·4%); chemicals and chemical products (13·3%); transportation equipment (12·4%); mineral fuels (11·3%); machinery and apparatus (11·2%). Leading exports were transportation equipment (21·1%); base and fabricated metals (12·4%); electrical equipment (11·0%); food (8·4%); machinery and apparatus (8·4%).

Main import suppliers, 2006: Germany, 24·0%; Russia, 9·7%; Italy, 6·8%; China, 6·1%; France, 5·5%. Main export markets, 2006: Germany, 27·2%; Italy, 6·5%; France, 6·2%; United Kingdom, 5·7%; Czech Republic, 5·5%. In 2006 trade with the European Union accounted for 63·2% of Polish imports and 77·4% of Polish exports.

COMMUNICATIONS
Roads
In 2007 there were 258,910 km of roads, including 663 km of motorways; 67·6% of roads were paved. In 2006 there were 13,384,000 passenger cars, 2,393,000 lorries and vans, 84,000 buses and 784,000 motorcycles and mopeds. In 2006 public transport totalled 28,148m. passenger-km and freight 136,490m. tonne-km. There were 5,583 road accident fatalities in 2007. With 14·7 deaths per 100,000 population in 2007, Poland has among the highest death rates in road accidents of any industrialized country.

Rail
In 2007 there were 20,107 km of railways in use managed by Polish State Railways (11,898 km electrified). Over 98% is standard 1,435 mm gauge with the rest broad gauge (1,520 mm). All narrow gauge lines (511 km in 2001) have been closed or sold to private or local authorities. In 2007 railways carried 279·7m. passengers and 245·3m. tonnes of freight. Passenger-km travelled in 2007 came to 19·9bn. and freight tonne-km to 54·3bn. Some regional railways are operated by local authorities. An 11 km metro opened in Warsaw in 1995, extended by 2007 to 18 km, and there are 14 tram/light rail networks with a total length of 930 km.

Civil Aviation
The main international airport is at Warsaw (Frederic Chopin), with some international flights from Kraków (John Paul II Balice International), Bydgoszcz, Gdańsk, Katowice, Łódź, Poznań, Rzeszów, Szczecin and Wrocław. The national carrier is LOT-Polish Airlines (68·0% state-owned). It flew 107·7m. km in 2011, carrying 6,491,199 passengers (5,377,869 on international flights). In 2011 Warsaw handled 9,324,635 passengers (8,253,153 on international flights) and 60,625 tonnes of freight.

Shipping
The principal ports are Gdańsk, Szczecin, Świnoujście and Gdynia. 59·48 tonnes of cargo were handled at Polish ports in 2005. In 2005, 9·36m. tonnes of freight and 714,133 passengers were carried by the maritime transport fleet. In 2005 the merchant marine totalled 1,862,265 GRT. Ships with a capacity of 739,686 GRT were built in 2005. Vessels totalling 52,004,843 NRT entered ports in 2005 and vessels totalling 52,199,635 NRT cleared. In 2005 there were 3,638 km of navigable inland waterways; 3·4m. tonnes of freight were carried on inland waterways in 2009.

Telecommunications
In 2008 there were 9,711,000 main (fixed) telephone lines. In the same year mobile phone subscribers numbered 43,926,000 (1,152·8 per 1,000 persons). The privatization of Telekomunikacja Polska (TP SA), the former state telecom operator, was completed in 2001. In April 2012 it was rebranded as Orange. France Télécom, the biggest foreign investor in Poland, now owns a 49·8% stake in the company. There were 18·7m. internet users in 2008. There were 14·2 fixed broadband subscribers per 100 inhabitants in Dec. 2010. In March 2012 there were 7·5m. Facebook users.

SOCIAL INSTITUTIONS
Justice
The penal code was adopted in 1969. Espionage and treason carry the severest penalties. For minor crimes there is provision for probation sentences and fines. In 1995 the death penalty was suspended for five years; it had not been applied since 1988. A new penal code abolishing the death penalty was adopted in June 1997.

In 2009 there were the following courts: one Supreme Court, one Supreme Administrative Court, 16 administrative courts of first instance, 11 appeal courts, 45 regional courts, 320 district courts, 68 family consultative centres and 35 juvenile institutions.

Judges are appointed by the President of the Republic from candidatures proposed by the National Council of the Judiciary and have life tenure. An ombudsman's office was established in 1987.

Family consultative centres were established in 1977 for cases involving divorce and domestic relations, but in 1990 divorce suits were transferred to ordinary courts. In 2008, 420,729 criminal sentences were passed. There were 763 ascertained homicides in 2009. The population in penal institutions in Dec. 2009 was 84,003 (220 per 100,000 of national population).

Education

Education from six to 18 is free and compulsory, although from 16 to 18 it may be part-time. Secondary education is then optional in general or vocational schools. In the 2009–10 school year there were: pre-primary schools, 17,444 with 994,138 pupils and 53,103 full-time equivalent teachers; primary schools, 13,972 with 2,235,018 pupils and 224,351 full-time equivalent teachers; lower secondary schools, 7,810 with 1,346,112 pupils and 104,755 full-time equivalent teachers; upper secondary schools, 10,709 with 1,770,013 pupils and 134,573 full-time equivalent teachers; post-secondary schools, 3,108 with 266,057 pupils and 8,210 full-time equivalent teachers; tertiary institutions, 563 with 1,918,793 students and 105,309 academic staff (excluding postgraduate and doctoral courses). In 2009 institutions of higher education included 19 universities, 23 technical universities, seven agricultural schools, 80 schools of economics, 15 theological schools, nine medical schools and 102 teacher training colleges. Since the early 1990s there has been a boom in private higher education—in 2009–10 more than 30% of all students in higher education were at private colleges.

The adult literacy rate in 2008 was over 99%.

Religious (Catholic) instruction was introduced in all schools in 1990; for children of dissenting parents there are classes in ethics.

In 2009 total expenditure on education came to 5·8% of GDP and 11·5% of total government spending.

Health

Medical treatment is free and funded from public sources. Medical care is also available in private clinics. In 2009 there were 795 general hospitals with a total of 193,400 beds (including beds and incubators for newborn babies). 82,700 physicians, 12,100 dentists, 193,100 nurses, 24,200 pharmacists and 22,400 thousand midwives worked directly with patients within hospital and outpatient care in 2009. In Jan. 1999 reform of the health care system was inaugurated. All citizens can now choose their own doctor, who is paid by the public sector under the social security system. The share of health insurance paid mainly by employees, equalling 7·5% of the amount earned by them, is assigned for the financing of the health care system. In 2008 Poland spent 7·0% of its GDP on health, with 27·8% of health expenditure coming from private sources.

Welfare

All social security benefits are administered by the State Insurance Office and funded 45% by a payroll tax and 55% from the state budget. Pensions, disability payments, child allowances, survivor benefits, maternity benefits, funeral subsidies, sickness compensation and alimony supplements are provided. In 2010 social benefits totalling 198,479·6m. złotys were paid (including 170,879·5m. złotys for retirement pay and pensions, 7,830·1m. złotys for family benefits and supplements to the family benefits, nursing benefits and allowances, 7,213·9m. złotys for sick benefits, 3,038·2m. złotys for maternity benefits, 2,622·9m. złotys for unemployment benefits and 2,435·5m. złotys for funeral benefits). There were a total of 9,243,427 pensioners in 2010. Unemployment benefits are paid from a fund financed by a 3% payroll tax. It is indexed in various categories to the average wage and payable for 12 months.

A retirement pension is available to an insured person who fulfils the following conditions: has attained the retirement age dependent on gender, specific employment conditions or functions, and an appropriate contributory and non-contributory period. The basic retirement pay for people born before 1949 is defined by job seniority and wages in chosen work periods. For people who were born after 31 Dec. 1948 and who joined the Open Pension Fund (OPF), or who are not members of the OPF and who did not fulfil the requirements needed to obtain retirement pay under the previous rules, basic retirement pay depends on the capital that was built up in individual insurance accounts.

On 1 Jan. 2009 a law on bridging pensions came into force entitling men and women who had not attained the age of 60 and 65 respectively to obtain a retirement pension. The retirement age is being raised gradually from 60 for men and 65 for women to 67 for both men and women, by 2020 and 2040 respectively.

RELIGION

State relations are regulated by laws of 1989 which guarantee religious freedom, grant the Church radio and TV programmes and permit it to run schools, hospitals and old age homes. The Church has a university (Lublin) and seminaries. On 28 July 1993 the government signed a Concordat with the Vatican regulating mutual relations. The religious capital is Gniezno. Its archbishop, Henryk Muszyński (b. 1933) is the primate of Poland. Kazimierz Nycz was appointed archbishop of Warsaw on 1 April 2007. In Oct. 1978 Cardinal Karol Wojtyła, archbishop of Kraków, was elected Pope as John Paul II. In Feb. 2013 there were seven cardinals.

Statistics of major churches as at Dec. 2009:

Church	Congregations	Places of Worship	Clergy	Adherents
Roman Catholic	10,157	13,662[1]	30,142	33,695,233
Uniate	134	98	74	55,000
Old Catholics	142	148	128	47,043
Polish Orthodox	225	415	390	504,150
Protestant (45 churches)	1,225	962	2,423	156,201
Muslim	19	16	33	2,667
Jewish	10	19	6	1,410
Jehovah's Witnesses	1,816	878	—	128,292

[1]1999.

CULTURE

World Heritage Sites

There are 13 UNESCO World Heritage sites in Poland. They are: Kraków's Historic Centre (inscribed on the list in 1978), Poland's former capital; Wieliczka Salt Mine (1978 and 2008), a mine since the 13th century; Auschwitz Concentration Camp (1979), the German concentration camp and nearby Birkenau death camp; Historic Centre of Warsaw (1980), celebrating the 20th century reconstruction of the city's 18th century heart decimated during World War II; Old City of Zamość (1992), a 16th century town; Medieval Town of Toruń (1997); Castle of the Teutonic Order in Malbork (1997), a medieval brick castle; Kalwaria Zebrzydowska: the Mannerist Architectural and Park Landscape Complex and Pilgrimage Park (1999); Churches of Peace in Jawor and Świdnica (2001), Europe's biggest timber-framed religious buildings; Wooden Churches of Southern Little Poland (2003); the Centennial Hall in Wrocław (2006).

Poland and Belarus are jointly responsible for Belovezhskaya Pushcha/Białowieża Forest (1979 and 1992), in the Baltic/Black Sea region; and Poland and Germany are jointly responsible for Muskauer Park/Park Mużakowski (2004), a landscaped park astride the Neisse river.

Press

In 2011 there were 32 daily newspapers with a combined daily circulation of 3,108,400 (81 per 1,000 inhabitants). The most popular newspapers are *Fakt*, *Gazeta Wyborcza*, *Super Express* and *Rzeczpospolita*. 7,713 magazine titles were published in 2011 with a combined total of 1,437m. copies. In 2011, 31,515 book titles were published.

Tourism
In 2011 there were 13,350,000 tourist arrivals, up from 12,470,000 in 2010 and 11,890,000 in 2009. The main countries of origin of non-resident tourists in 2011 were Germany (4,590,000), Ukraine (1,580,000), Belarus (1,220,000) and Lithuania (630,000).

Festivals
The International Chopin Festival at Duszniki Zdrój is held in Aug. and the Warsaw Autumn Festival takes place in Sept. Kraków Film Festival runs from May–June. Popular music festivals include Metalmania in Katowice (March), Open'er Festival in Gdynia (July), Przystanek Woodstock at Kostrzyn nad Odrą (July–Aug.), Off Festival at Mysłowice (Aug.) and Zaduszki Jazzowe (All Soul's Day Jazz Festival) in Kraków (Nov.).

DIPLOMATIC REPRESENTATIVES
Of Poland in the United Kingdom (47 Portland Pl., London, W1B 1JH)
Ambassador: Witold Sobków.

Of the United Kingdom in Poland (ul. Kawalerii 12, 00-468, Warsaw)
Ambassador: Robin Barnett, CMG.

Of Poland in the USA (2640 16th St., NW, Washington, D.C., 20009)
Ambassador: Ryszard Schnepf.

Of the USA in Poland (Aleje Ujazdowskie 29/31, 00-540 Warsaw)
Ambassador: Stephen Mull.

Of Poland to the United Nations
Ambassador: Ryszard Sarkowicz.

Of Poland to the European Union
Permanent Representative: Marek Prawda.

FURTHER READING
Central Statistical Office, *Rocznik Statystyczny*. Annual—*Concise Statistical Yearbook of Poland—Statistical Bulletin*. Monthly

Biskupski, M. B., *The History of Poland*. 2000
Chodakiewicz, Marek Jan, *Poland's Transformation: A Work in Progress*. 2006
Kochanski, Halik, *The Eagle Unbowed: Poland and the Poles in the Second World War*. 2012
Lukowski, Jerzy and Zawadzki, Hubert, *A Concise History of Poland*. 2001
Prazmowska, Anita J., *History of Poland*. 2nd ed. 2011.—*Poland: A Modern History*. 2010
Sikorski, R., *The Polish House: An Intimate History of Poland*. 1997; US title: *Full Circle*. 1997
Slay, B., *The Polish Economy: Crisis, Reform and Transformation*. 1994
Staar, R. F. (ed.) *Transition to Democracy in Poland*. 1993
Wedel, J., *The Unplanned Society: Poland During and After Communism*. 1992
Zamoyski, Adam, *Poland: A History*. 2009

National library: Biblioteka Narodowa, al. Niepodległości 213, 02-086 Warsaw.
National Statistical Office: Central Statistical Office, Aleje Niepodległości 208, 00-925 Warsaw.
Website: http://www.stat.gov.pl

PORTUGAL

Madeira
Funchal

Braga

Oporto

ATLANTIC
OCEAN

Coimbra

SPAIN

PORTUGAL

Amadora
LISBON
Setubal

0   50 mi

0   75 km

Faro

© Research Machines plc 2006

República Portuguesa
(Republic of Portugal)

Capital: Lisbon
Population projection, 2015: 10·70m.
GNI per capita, 2011: (PPP$) 20,573
HDI/world rank: 0·809/41
Internet domain extension: .pt

KEY HISTORICAL EVENTS

The western fringe of the Iberian peninsula was inhabited from 8000 BC by Neolithic peoples known as Iberians. Celtic tribes settled in the north and west of the peninsula in the first millennium BC with Phoenician settlements in the southwest around Cádiz from around 800 BC. From 241 BC the Iberian peninsula came under the influence of Carthage, and then Rome after 206 BC. The Romans made their way north to what is now central Portugal and clashed with a Celtic federation, the Lusitanians. They resisted the Roman advance under their leader Viriathus until he was killed in 140 BC, after which the Romans were able to move north across the Douro river. In 25 BC Augustus founded Augustus Emirita (now Mérida) as the capital of Lusitania.

From AD 409, with the Roman Empire in decline, the Iberian Peninsula was invaded by Germanic tribes from central Europe, including the Suevi and Visigoths, who established Christian kingdoms. Southern Galicia was settled by the Suevi, who were converted to Christianity by St Martin of Braga in around AD 550. Following the arrival of Muslim armies in Iberia in 711, the southern part of what is now Portugal became part of the Muslim dominion of al-Andalus. The northern and western fringes of Iberia remained largely agrarian, poor and Christian.

From 850 the Christians began to push southward: the region between the rivers Minho and Douro became known as Territorium Portugualense and was ruled by Mumadona Dias after 931. Fernando I, King of Castile, drove the Muslims from the city of Viseu in 1058 and reconquered Coimbra in 1064. Fernando's successor, Afonso VI of Leon, set up his power base in the town of Braga. His daughter, Teresa, who was married to Henry of Burgundy, then governed Portugal as regent for their son, Afonso Henriques. Teresa eventually lost the support of many of the powerful local barons, who united behind Afonso and made him the first king of Portugal in 1139.

Muslim chroniclers refer to Afonso I as 'the cursed of Allah'. He crusaded southwards through the Muslim strongholds, capturing Lisbon in 1147. However, the Portuguese reconquest was not completed for 150 years, when Afonso III finally took Algarve in the far south. Afonso III established the first Cortes (government) at Leiria in 1254. Large swathes of the newly conquered lands were given over to the army and monastic orders to ensure their protection. Afonso's son, Dinis (1279–1325), became one of the most celebrated of the Burgundian dynasty. He established trade links with other European powers and in 1317 worked with a Genoese admiral to establish a navy. Dinis made the vernacular, rather than Latin, the official language and founded the first university in Lisbon in 1290.

The later kings of the House of Burgundy were entangled in various marriage alliances with neighbouring Castile. Fernando I (1367–83) inherited the Portuguese crown as a battle raged in Castile between King Pedro 'the Cruel' and his half-brother Enrique de Trastámara. Both sides attempted to garner support from outside the kingdom, with the English supporting Pedro (Peter) and his heirs and the French backing Enrique (Henry) and his supporters. Enrique eventually prevailed, being crowned Enrique II of Castile in 1369. The new king offered his support to Fernando, who accepted, and Castilian rule was duly established in Portugal.

After Ferenando's heiress, Beatriz, married Juan I of Castile the nobility broadly supported Castilian rule but Portugal's coastal towns wanted independence and rebelled. Their choice for ruler was João of Avis, half brother of Fernando, who was declared King João I in 1384. A year later, with English support, the Portuguese defeated Juan I and his Castilian army at the battle of Aljbarrota. The Anglo–Portuguese alliance was cemented by King João's marriage in 1387 to Philippa of Lancaster, sister of England's future King Henry IV.

Empire Building

Having made peace with Spain, João turned his attention overseas. The capture of the town of Ceuta on the north African coast in 1415 was the beginning of an era of discovery by Portuguese mariners, spearheaded by João's third son Henry who became known as Henry the Navigator. He founded a school of navigation at Sagres and organized numerous expeditions along the west coast of Africa. Madeira and the Azores were also discovered and settled during this period.

Relations with Castile deteriorated in the reign of Afonso V (1438–81). Afonso married Juana, daughter of Enrique IV of Castile, and laid claim to the Castilian throne. The marriage of Fernando II and Isabella I of Castile and the merging of the powerful kingdoms of Aragon and Castile, weakened Afonso's claim. There were lengthy battles in the Zamora and Toro regions, which Afonso lost in 1476. Peace was made three years later by Afonso's heir, João II. Overseas explorations were resumed, and Portugal became a haven for tens of thousands of Jews fleeing persecution in Spain.

In 1487 Bartolomeu Dias rounded the southern cape of Africa, but the Portuguese crown rejected Christopher Columbus' proposal for finding a new westward route to the Indies. Backed instead by Fernando II and Isabella I of Castile, he reached the New World in 1492. The 1493 Treaty of Tordesillas gave the newly-unified Spain all lands west of a vertical line drawn 370 degrees west of the Cape Verde Islands. Land to the east, including Brazil, went to Portugal. In 1497, with backing from King Manuel I, Vasco da Gama set out to map a sea route to India. Indian spices were used for preserving food, in the preparation of medicines and in glues, perfumes, dyes and varnishes. The Portuguese built an administrative capital at Goa and by 1550 it was considered Portugal's second city.

Fortified trading posts were later established along the coast of East Africa and India, and commercial centres set up further east. The profits generated by the spice trade made Manuel I 'the Fortunate' one of the wealthiest rulers in Europe. No longer dependent on taxes, Manuel was able to rule independently. The Cortes did not meet between 1502–25 and the expulsion of the Jews in 1496 (a condition of Manuel's marriage to Princess Isabella of Castile, daughter of Isabella I) dealt a heavy blow to the economy.

In 1568 King Sebastião launched a disastrous crusade to eradicate Islam in the Maghreb. More than 10,000 Portuguese troops, including Sebastião, were killed by superior Moroccan forces. After the elderly and childless Cardinal Henrique, the line of succession passed to the cardinal's nephew, Felipe II of Spain.

Spanish Rule

Felipe II annexed Portugal in 1580. The Spanish empire was at its height and Portuguese merchants saw commercial advantages in forming an alliance with Spain. Portugal was granted autonomy but this was gradually eroded. The Inquisition was established in Portugal and the ports of Lisbon and Porto (Oporto) were closed to English and Dutch ships, which then made their own way to the east and snatched control of the spice trade. The 'Spanish domination' of Portugal lasted for 60 years, though after 1621 Spain was considerably weakened by the cost of defending its empire against France and England. A rebellion in Catalonia spurred the Portuguese to stage their own revolution. They rallied round the duke of Bragança, who was crowned King João IV in 1640.

João IV was anxious to formalize new alliances with the other European powers. Spain's peace with France, set out in the 1659 Treaty of the Pyrenees, made João's successor, Afonso VI, anxious to strengthen Portugal's alliance with England. Thus, Catherine of Bragança was married to King Charles II in 1662. Her dowry included the right to trade with the Portuguese colonies and the cession of Bombay and Tangier. In return, England agreed to defend Portugal and its colonies. In the late 1600s gold and then diamonds were discovered in Brazil. Money poured into the Portuguese court, but the crown's wealth did little to enrich the nation. João V (1706–50) aped Louis XIV of France, lavishing money on ambitious building projects, including a gigantic convent-palace at Mafra.

Portuguese political development lagged behind that of many European states and it remained comparatively untouched by the Enlightenment until the Marquess of Pombal was made chief minister shortly after the great Lisbon earthquake in 1755. His methods were harsh and he made enemies quickly. His principal victims were the conservative Jesuits with their vast array of privileges. Pombal is also credited with reforming the education system but ultimately his legacy was limited. When Maria I came to the throne in 1777, she immediately banished Pombal to his estates and rescinded many of his reforms.

After Louis XVI of France was guillotined in 1793, the new French Republic invaded Catalonia and the Basque provinces. Following these defeats, King Charles IV of Spain formed an alliance with Napoleon. In 1801 France and Spain demanded that Portugal abandon its alliance with Britain, open its ports to French and Spanish shipping and hand over some of its colonies to Spain. Portugal refused and a French army marched into Lisbon in 1807. King João VI and his family escaped to Brazil with the help of the British navy and the defence of Portugal was left in the hands of British Generals Wellesley and Beresford. The French were finally driven out of Portugal in 1811 and the British, as a reward, were granted free access to the Brazilian ports, a concession which damaged the Portuguese economy and stoked-up popular resentment.

After an army-backed revolution in Porto in 1820, an unofficial Cortes was set up to devise a new liberal constitution. The Cortes was to be a single chamber parliament elected by universal male suffrage, and feudal and clerical rights would be abolished. Returning from Brazil the following year, João VI accepted the new constitution. Brazil declared its independence, with Pedro IV (João's elder son) as emperor.

Following João's death in 1826, Pedro became king of Portugal in defiance of the Brazilian constitution. Forced to abdicate in favour of his daughter, Maria II, then aged seven, he chose his brother Miguel as her steward. Miguel seized the throne in 1828, defeating the liberals and repealing Pedro's constitution. Pedro IV returned to Portugal in 1832 to lead the liberals in the Miguelist Wars. Maria was eventually restored to the throne. The late 1850s saw improvements in the nation's infrastructure but by the start of the reign of Carlos I in 1889, Portugal's economy was in a poor way. Explorations in Africa strengthened Portugal's hold on Angola and Mozambique but the British refused to give up territory which would have linked the two colonies. Carlos attempted to end inefficiency and corruption, establishing a dictatorship in 1906 under the conservative João Franco. Amidst growing public discontent, Carlos and his eldest son Prince Luís Filipe were assassinated in 1908. Manuel II succeeded to the throne but in 1910 a republican revolution forced his abdication and flight to Britain.

In the First World War Portugal was at first neutral, then joined the Allies in 1916. The economy deteriorated further and in 1926 there was a military coup. Gen. Carmona became president. António de Oliveira Salazar was made finance minister in 1928 with a brief to reform the economy.

Dictatorship to Democracy

Salazar became prime minister in 1932. His *Estado Novo* (New State) had a strongly nationalist and dictatorial flavour. Political parties, unions and strikes were abolished and dissent was crushed by the notorious PIDE (Polícia Internacional e de Defesa do Estado) secret police force. Portugal was neutral in the Second World War but allowed the Allies to establish naval and air bases. After Goa was seized by India in 1961, Salazar was determined to cling on to the African territories. By 1968, over 100,000 Portuguese were fighting independence movements in Angola, Guinea-Bissau and Mozambique.

In 1968 Salazar suffered a stroke and was replaced by Marcello Caetano. Under Caetano repression was eased but the unpopular wars in Africa continued. In 1974, amid mounting public discontent, a group of officers formed the Movement of Armed Forces and toppled the government in a bloodless coup known as 'the Revolution of the Carnations'. Gen. António de Spínola was appointed head of the ruling military junta. The secret police force was abolished. All political prisoners were released; full civil liberties, including freedom of the press and of all political parties, were restored and, in 1975, Angola, Mozambique, São Tomé e Príncipe and Cape Verde were granted independence. Timor-Leste was forcibly taken by Indonesia. Following an attempted revolt in late 1975, the military junta was dissolved and a Supreme Revolutionary Council ruled until a new constitutional government resumed the following year. During the late 1970s several moderate, Socialist-dominated governments tried unsuccessfully to stabilize the country. In 1982 a centre-right coalition revised the constitution, reducing presidential power and the right of the military to intervene in politics. From 1983 to 1985 a coalition government under Socialist leader Mário Soares made progress in reducing the chaos and poverty that were the legacy of Salazar's long dictatorship.

In 1985 the centrist Social Democratic party under Aníbal Cavaco Silva won an undisputed majority in parliament. In 1986 Soares was elected to the presidency, and Portugal was admitted to the European Community. Political stability and economic reforms created a favourable business climate, especially for renewed foreign investment, and Portugal became one of the fastest-growing economies in Europe. The Socialists returned to power as a minority government after the 1995 parliamentary elections. Macao, Portugal's colony on the south coast of China, was handed back to China in 1999. Portugal joined the single European currency in 2001. In Jan. 2006 the centre-right candidate Aníbal Cavaco Silva was elected president, beginning a period of political 'cohabitation' alongside the Socialist prime minister, Jóse Sócrates. In March 2011, following an appeal to the EU for assistance to avoid a debt default, the government resigned.

TERRITORY AND POPULATION

Mainland Portugal is bounded in the north and east by Spain and south and west by the Atlantic Ocean. The Atlantic archipelagoes of the Azores and of Madeira form autonomous but integral parts of the republic, which has a total area of 92,207 sq. km. Population (2011 census), 10,562,178 (5,515,578 females).

Mainland Portugal is divided into five regions. At the 2011 census the regions, with their populations, were: North (3,689,682); Central (2,327,755); Lisbon (2,821,876); Alentejo (757,302); Algarve (451,006). Population of the Azores, 246,772; Madeira, 267,785. Density (2011), 114·5 per sq. km (North, 173; Central, 83; Lisbon, 940; Alentejo, 24; Algarve, 90; Azores, 106; Madeira, 334).

The UN gives a projected population for 2015 of 10·70m.

In 2011, 61·3% of the population lived in urban areas. The populations of the districts and Autonomous Regions (2011 census):

Areas	Population	Areas	Population
North	3,689,682	Médio Tejo	220,661
Alto Trás-os-Montes	204,381	Oeste	362,540
Ave	511,737	Pinhal Interior Norte	131,468
Cávado	410,169	Pinhal Interior Sul	40,705
Douro	205,902	Pinhal Litoral	260,942
Entre Douro e Vouga	274,859	Serra da Estrela	43,737
Grande Porto	1,287,282	Lisbon	2,821,876
Minho-Lima	244,836	Grande Lisboa	2,042,477
Tâmega	550,516	Península de Setúbal	779,399
Central	2,327,755	Alentejo	757,302
Baixo Mondego	332,326	Alentejo Central	166,822
Baixo Vouga	390,822	Alentejo Litoral	97,925
Beira Interior Norte	104,417	Alto Alentejo	118,410
Beira Interior Sul	75,028	Baixo Alentejo	126,692
Cova da Beira	87,869	Lezíria do Tejo	247,453
Dão-Lafões	277,240	Algarve	451,006

In 2010 there were 443,055 foreign citizens with legal residency status, with the leading nationalities as follows: Brazil, 119,195;

Ukraine, 49,487, Cape Verde, 43,510; Romania, 36,830; Angola, 23,233; Guinea-Bissau, 19,304. Large numbers of immigrants have come to Portugal from eastern Europe since 1999, mainly from Ukraine, Romania and Moldova.

The capital is Lisbon (Lisboa), with a population of 547,733 in 2011 (metropolitan area population, 2,821,876 in 2011). Other major cities are Porto, 237,591 in 2011 (metropolitan area population, 1,672,670 in 2011), Almada, Amadora, Braga, Funchal (in Madeira) and Vila Nova de Gaia.

The official language is Portuguese.

The Azores islands lie in the mid-Atlantic Ocean, between 1,200 and 1,600 km west of Lisbon. They are divided into three widely separated groups with clear channels between, São Miguel (759 sq. km) together with Santa Maria (97 sq. km) being the most easterly; about 160 km northwest of them lies the central cluster of Terceira (382 sq. km), Graciosa (62 sq. km), São Jorge (246 sq. km), Pico (446 sq. km) and Faial (173 sq. km); still another 240 km to the northwest are Flores (143 sq. km) and Corvo (17 sq. km), the latter being the most isolated and undeveloped of the islands. São Miguel contains over half the total population of the archipelago.

Madeira comprises the island of Madeira (745 sq. km), containing the capital, Funchal; the smaller island of Porto Santo (40 sq. km), lying 46 km to the northeast of Madeira; and two groups of uninhabited islets, Ilhas Desertas (15 sq. km), being 20 km southeast of Funchal, and Ilhas Selvagens (4 sq. km), near the Canaries.

SOCIAL STATISTICS

2006: births, 105,449; deaths, 101,990; marriages, 47,857; divorces, 22,881. Rates per 1,000 population in 2006: birth, 9·9; death, 9·6; marriage, 4·5; divorce, 2·2. Annual population growth rate, 2000–05, 0·6%. Expectation of life at birth, 2007, was 75·3 years for males and 81·8 years for females. Infant mortality in 2010 was three per 1,000 live births, down from 77 per 1,000 live births in 1960, representing the greatest reduction in infant mortality rates in Europe over the past half century. Fertility rate, 2008, 1·4 births per woman. Around one in five babies are born outside marriage, up from one in 14 in 1970. In 2005 the most popular age range for marrying was 25–29 for both males and females. Portugal legalized same-sex marriage in June 2010.

On 11 Feb. 2007 a national referendum was held on whether to decriminalize abortion up until the 10th week of pregnancy. Turnout was low at 44%; 59·3% of voters approved the motion. Despite results only having to be legally binding if the turnout exceeded 50%, parliament voted overwhelmingly in favour on 9 March 2007. In 1998 the same question had been put in another referendum. 51% voted against the motion; turnout was 32%.

In 2010 Portugal received 160 asylum applications.

CLIMATE

Because of westerly winds and the effect of the Gulf Stream, the climate ranges from the cool, damp Atlantic type in the north to a warmer and drier Mediterranean type in the south. July and Aug. are virtually rainless everywhere. Inland areas in the north have greater temperature variation, with continental winds blowing from the interior. Lisbon, Jan. 52°F (11°C), July 72°F (22°C). Annual rainfall 27·4" (686 mm). Porto, Jan. 48°F (8·9°C), July 67°F (19·4°C). Annual rainfall 46" (1,151 mm).

CONSTITUTION AND GOVERNMENT

Portugal is governed under the constitution of April 1976, amended in 1982, 1989, 1992, 1997, 2001, 2004 and 2005. The 1982 revision abolished the (military) Council of the Revolution and reduced the role of the President under it. Portugal is a sovereign, unitary republic. Executive power is vested in the President, directly elected for a five-year term (for a maximum of two consecutive terms). Political parties may support a candidate in presidential elections but not actually field a candidate.

The President appoints a Prime Minister and, upon the latter's nomination, other members of the Council of Ministers.

The 230-member *National Assembly* is a unicameral legislature elected for four-year terms by universal adult suffrage under a system of proportional representation. Women did not have the vote until 1976.

National Anthem
'Herois do mar, nobre povo' ('Heroes of the sea, noble breed'); words by Lopes de Mendonça, tune by Alfredo Keil.

GOVERNMENT CHRONOLOGY

(PS = Socialist Party; PSD = Social Democratic Party; UN = National Union; n/p = non-partisan)

Presidents since 1926.

1926–51	UN/military	António (Óscar de) Fragoso Carmona
1951–58	UN/military	Francisco (Higino de) Craveiro Lopes
1958–74	UN/military	Américo (de Deus Rodrigues) Thomaz
1974		National Salvation Junta (all military)
1974	military	António (Sebastião Ribeiro) de Spínola
1974–76	military	Francisco da Costa Gomes
1976–86	military, n/p	(António dos Santos) Ramalho Eanes
1986–96	PS	Mário (Alberto Nobre Lopes) Soares
1996–2006	PS	Jorge (Fernando Branco de) Sampaio
2006–	PSD	Aníbal (António) Cavaco Silva

Prime Ministers since 1932.

1932–68	UN	António de Oliveira Salazar
1968–74	UN	Marcello (das Neves Alves) Caetano
1974	n/p	Adelino da Palma Carlos
1974–75	military	Vasco (dos Santos) Gonçalves
1975–76	military	José (Batista) Pinheiro de Azevedo
1976–78	PS	Mário (Alberto Nobre Lopes) Soares
1978	n/p	Alfredo (Jorge) Nobre da Costa
1978–79	n/p	Carlos (Alberto) da Mota Pinto
1980	PSD	Francisco (Manuel Lumbrales de) Sá Carneiro
1981–83	PSD	Francisco (José Pereira) Pinto Balsemão
1983–85	PS	Mário (Alberto Nobre Lopes) Soares
1985–95	PSD	Aníbal (António) Cavaco Silva
1995–2002	PS	António (Manuel de Oliveira) Guterres
2002–04	PSD	José Manuel Durão Barroso
2004–05	PSD	Pedro (Miguel de) Santana Lopes
2005–11	PS	José Sócrates (Carvalho Pinto de Sousa)
2011–	PSD	Pedro (Manuel Mamede) Passos Coelho

RECENT ELECTIONS

At the presidential elections of 23 Jan. 2011, incumbent Aníbal Cavaco Silva won 52·9% of the vote, Manuel Alegre (PS) 19·8%, Fernando Nobre (ind.) 14·1%, Francisco Lopes (Communist Party) 7·1%, José Manuel Coelho (New Democracy Party) 4·5% and Defensor Moura (ind.) 1·6%. Turnout was 46·6%.

At the parliamentary elections of 5 June 2011 the Social Democratic Party (PSD) won 108 seats (38·7% of votes cast); the ruling Socialist Party (PS), 74 (28·0%); the Democratic and Social Centre–People's Party, 24 (11·7%); the Democratic Unity Coalition, 16 (7·9%); and the Left Bloc, 8 (5·2%). Turnout was 58·0%.

European Parliament
Portugal has 22 (24 in 2004) representatives. At the June 2009 elections turnout was 36·8% (38·6% in 2004). The PSD won 8 seats with 31·7% of votes cast (political affiliation in European

Parliament: European People's Party); the PS 7 with 26·5% (Progressive Alliance of Socialists and Democrats); the BE, 3 with 10·7% (European United Left/Nordic Green Left); the UDC, 2 with 10·6% (European United Left/Nordic Green Left); CDS-PP, 2 with 8·4% (European People's Party).

CURRENT GOVERNMENT

President: Aníbal Cavaco Silva; b. 1939 (ind.; sworn in 9 March 2006 and re-elected 23 Jan. 2011).

In March 2013 the government comprised:
Prime Minister: Pedro Passos Coelho; b. 1964 (PSD; sworn in 21 June 2011).
Ministers of State: Vítor Gaspar (also *Minister of Finance*); Paulo Portas (also *Minister of Foreign Affairs*).
Minister of Agriculture, Sea, Environment and Territorial Administration: Assunção Cristas. *Economy and Labour:* Álvaro Santos Pereira. *Education and Science:* Nuno Crato. *Health:* Paulo Macedo. *Internal Affairs:* Miguel Macedo. *Justice:* Paula Teixeira da Cruz. *National Defence:* José Pedro Aguiar Branco. *Parliamentary Affairs:* Miguel Relvas. *Social Solidarity and Social Security:* Pedro Mota Soares.

Government Website (Portuguese only):
http://www.portugal.gov.pt

CURRENT LEADERS

Aníbal Cavaco Silva

Position
President

Introduction
Aníbal Cavaco Silva was elected president on 22 Jan. 2006, the first centre-right politician to fill the largely ceremonial post since the 1974 revolution. The free-market economist played a key role in preparing Portugal's entry in 1986 to the European Economic Community (EEC; later the European Union) and served as prime minister from 1985–95. He was re-elected president in Jan. 2011.

Early Life
Aníbal António Cavaco Silva was born in Boliqueime, Algarve on 15 July 1939. Educated in Faro and Lisbon, he graduated in finance in 1964. He worked as a researcher for the Calouste Gulbenkian Foundation in Lisbon from 1967 to 1971 before studying for a PhD in economics at the University of York in the UK.

Returning to Portugal in 1974, the year when the socialist Armed Forces Movement toppled the Caetano dictatorship, Cavaco Silva taught economics at the Catholic University of Portugal. He joined the newly formed centre-right Popular Democratic Party, which became the Social Democratic Party (PSD) in 1976. From 1977 he worked as director of the research and statistics department of the Bank of Portugal. Elected to parliament for the PSD in Oct. 1980, he served as minister of finance and planning, initially under PSD leader and prime minister, Francisco Sá Carneiro, and then under Francisco Balsemão.

An advocate of free-market economics, Cavaco Silva's reforms, combined with a constitutional reduction in presidential power, paved the way for Portugal's entry into the EEC in 1986. Elected head of the PSD in June 1985, he led the party to victory in elections the following Oct. He retained the position for ten years, the longest tenure of any democratically elected prime minister in Portuguese history. The PSD won a clear majority of seats in legislative elections in both 1987 and 1991, with analysts attributing Cavaco Silva's success to economic liberalization, tax cuts and the flow of funds from the EEC.

Cavaco Silva stepped down as leader of the PSD prior to the 1995 elections, which were won by the Socialist Party (PS). He contested the 1996 presidential election but, after losing to the

Socialist candidate, retired from politics to advise at the Bank of Portugal and teach economics at the Catholic University of Portugal. In Oct. 2005 he returned to the political fray and announced his candidacy for the forthcoming presidential election. He received 50·5% of the votes cast on 22 Jan. 2006 and was sworn in on 9 March 2006.

Career in Office
Cavaco Silva's victory over the two Socialist candidates was a setback for the Socialist prime minister, Jóse Sócrates, who had presided over a period of economic stagnation. However, the result ushered in a new era of 'cohabitation' in Portuguese politics, with analysts predicting that the two leaders would find common ground to implement economic reform. Although a Roman Catholic, Cavaco Silva endorsed legislation in April 2007 liberalizing abortion and aligning Portuguese law with that of most other EU countries. In Oct. 2009, following parliamentary elections the previous month, he invited Sócrates to form a new PS government.

In Jan. 2011 Cavaco Silva's re-election as president heralded a turbulent economic period. High debt levels and flat growth forced the negotiation in May of a financial rescue package with stringent deficit reduction conditions from the European Union and International Monetary Fund. Parliamentary elections in June 2011 saw the PS government replaced by a centre-right PSD/Democratic and Social Centre–People's Party coalition. Continuing austerity prompted social and labour unrest in 2012.

Pedro Passos Coelho

Position
Prime Minister

Introduction
Pedro Passos Coelho became prime minister in June 2011 when Portugal was suffering a severe sovereign debt crisis. A centre-right politician, he has advocated cutting government expenditure and privatizing many of Portugal's state-owned businesses to improve efficiency and to restore the confidence of international investors.

Early Life
Born in Coimbra on 24 July 1964, Pedro Passos Coelho was educated in Silva Porto, Portugal and Luanda, Angola, where his family lived from 1969–74. He joined the Social Democratic Party (PSD) and served on the national council of its junior wing from 1980–82.

In 1982 he began studying mathematics at Lisbon University but left before graduating to work in private enterprise and to pursue his political career. Leader of the junior PSD from 1990–95, he was elected to parliament in 1991, representing a Lisbon constituency. He served as vice-chairman of the PSD parliamentary group from 1996–99 and, after standing unsuccessfully for mayor of Amadora in 1997, served as a municipal councillor from 1997–2001.

He left parliament in 1999 following the PSD general election defeat and studied economics at Lusíada University, Lisbon, graduating in 2001. After working in consultancy firms, he joined the Fomentinvest investment holding company in 2004 and became finance director under chairman Ângelo Correia, a founder of the PSD. Coelho also served as vice president of the PSD from 2005–06, gaining a business-friendly and pragmatic reputation.

In 2008 he unsuccessfully contested the party leadership, and also established a think tank to develop policies based on economic liberalization and partial privatization. In May 2010 he again contested the leadership and won with 61% of the vote.

As leader of the opposition, Coelho helped vote down the government budget in March 2011, arguing it relied too heavily on tax increases rather than spending cuts. The government

collapsed and a caretaker administration then agreed a €78bn. IMF–EU bailout, with stringent austerity conditions attached.

Coelho campaigned at the June 2011 parliamentary election on a platform of economic reform. The PSD won the most votes and formed a coalition with the centre-right Democratic and Social Centre–People's Party.

Career in Office
In its early months Coelho's government increased taxes, cut state spending and announced a privatization programme. However, after discovering further debt, he announced the transfer of €6bn. of banks' pension funds to the state in order to reduce the short-term deficit and boost liquidity. To comply with the terms of the financial bailout, the government was required to implement labour law reform. Its reform proposals and the austerity measures, however, provoked strike action and street protests in 2012. Having been granted an additional year by the IMF and EU to meet deficit targets, Coelho's government presented a draft budget in Oct. providing for further retrenchment, which was passed by parliament the following month.

DEFENCE

Conscription was abolished in Nov. 2004. Portugal now has a purely professional army.

In 2008 defence expenditure totalled US$3,729m. (US$349 per capita), representing 1·5% of GDP.

Army
Strength (2007) 26,700. There are Army reserves totalling 210,000. Paramilitary forces include the National Republican Guard (26,100) and the Public Security Police (21,600).

Navy
In 2007 the combatant fleet comprised one French-built diesel submarine, 12 frigates and seven corvettes. Naval personnel in 2007 totalled 9,110 including 335 recalled reservists and 1,725 marines. There were 900 naval reserves.

Air Force
The Air Force in 2007 had a strength of 7,100. There were 25 combat-capable aircraft (mainly F-16s).

ECONOMY

Services accounted for about 73% of GDP in 2007, industry 24% and agriculture 3%.

Overview
The economy has suffered badly as a result of the global financial downturn and the eurozone crisis. The economy grew by only 1% between 2003 and 2012 and entered its eighth quarter of recession at the start of 2013. Unemployment in 2010 stood at 15·5%.

Since the 1980s the economy has become increasingly dependent on services, which accounted for 74·5% of GDP in 2011. Low productivity, weak competitiveness and high debt have hindered growth and led to high fiscal imbalances. The deficit was set to reach 120% of GDP by the end of 2013.

In April 2011 the government applied to the European Commission for financial assistance to avoid a debt default. The Extended Fund Facility for Portugal was agreed in May 2011, providing €78bn. (US$116bn.) over three years, funded by the European Union, the European Central Bank and the IMF. Income tax rose from 24·5% to 28·5% with effect from Jan. 2013 to meet the terms of the bailout, with a resulting squeeze on living standards and reduced consumer spending. Poor employment prospects and lower living standards have resulted in increased emigration rates—including to the oil-rich former Portuguese colony of Angola—prompting fears of a brain drain.

Required by the bailout to sell €5bn. worth of state assets, the government has undertaken a 'fire sale' of publicly owned companies to foreign investors. This resulted, for instance, in 21%

of the utility company Energias de Portugal being bought by the Chinese Three Gorges firm and 25% of the state electric grid being sold to the Chinese State Grid.

The sovereign debt crisis in the eurozone threatens recovery and with Spain accounting for a fifth of Portuguese exports, instability in the Spanish economy has knock-on effects in Portugal. However, Portugal re-entered the bond market in Jan. 2013 (owing to lower borrowing costs and the ten-year bond yield falling back to pre-recession levels of 7%) with the hope of increasing confidence in the wider economy.

Currency
On 1 Jan. 1999 the euro (EUR) became the legal currency in Portugal at the irrevocable conversion rate of 200·482 escudos to 1 euro. The euro, which consists of 100 cents, has been in circulation since 1 Jan. 2002. On the introduction of the euro there was a 'dual circulation' period before the escudo ceased to be legal tender on 28 Feb. 2002. Euro banknotes in circulation on 1 Jan. 2002 had a total value of €10·6bn.

Inflation rates (based on OECD statistics):

2002	2003	2004	2005	2006	2007	2008	2009	2010	2011
3·7%	3·3%	2·5%	2·1%	3·0%	2·4%	2·7%	−0·9%	1·4%	3·6%

Gold reserves were 12·30m. troy oz in Sept. 2009 and foreign exchange reserves US$774m. Total money supply was €52,659m. in Aug. 2009.

Budget
In 2008 central government revenues totalled €62,321m. and expenditures €68,413m. Taxes accounted for 59·4% of revenues in 2008 and social contributions 31·1%; social benefits accounted for 46·3% of expenditures and compensation of employees 24·6%.

Portugal's budget deficit in 2011 was 4·2% of GDP (2010, 9·8%; 2009, 10·2%). The required target set by the EU is a budget deficit of no more than 3%.

The standard rate of VAT is 23·0% (reduced rates, 13% and 6%).

Performance
Real GDP growth rates (based on OECD statistics):

2002	2003	2004	2005	2006	2007	2008	2009	2010	2011
0·8%	−0·9%	1·6%	0·8%	1·4%	2·4%	0·0%	−2·9%	1·4%	−1·7%

In the years since Portugal joined the European Union its GDP per head has risen from being 53% of the EU average to 80% in 2010. Portugal's total GDP in 2011 was US$237·4bn.

Banking and Finance
The central bank and bank of issue is the Bank of Portugal, founded in 1846 and nationalized in 1974. Its *Governor* is Carlos Costa.

In 2006 there were 5,039 branches of banks and savings banks and 676 branches of agricultural credit co-operatives. Deposits in all monetary establishments totalled €146·7bn. in 2006. The largest Portuguese bank is the state-owned Caixa Geral de Depósitos, with assets of €111·1bn. in 2008. Other major banks are Banco Comercial Português, Banco Espírito Santo and Banco Português de Investimento.

Gross external debt amounted to US$485,816m. in June 2012.

There are stock exchanges in Lisbon and Porto.

ENERGY AND NATURAL RESOURCES

In 2008, 23·2% of energy consumption came from renewables (wind power, solar power, hydro-electric power, tidal power, geothermal energy and biomass), compared to the European Union average of 10·3%. A target of 31% has been set by the EU for 2020.

Environment
Portugal's carbon dioxide emissions from the consumption and flaring of fossil fuels in 2008 were the equivalent of 5·4 tonnes per capita.

Electricity
Installed capacity was 15m. kW in 2007. Production in 2007 was 47·25bn. kWh; consumption per capita was 5,160 kWh.

Minerals
Portugal possesses considerable mineral wealth. Production in tonnes (2005): limestone, marl and calcite, 51,025,000; granite (2006), 27,489,000; marble, 752,000; salt, 597,945; kaolin, 164,072; copper, 89,541; tungsten, 816.

Agriculture
There were 274,563 agricultural holdings in 2007. The agricultural sector employs 11·6% of the workforce. In 2007 there were 1·08m. ha. of arable land and 796,000 ha. of permanent crops.

The following figures show the production (in 1,000 tonnes) of the chief crops:

Crop	2005	2006	2007	Crop	2005	2006	2007
Cabbage, broccoli, etc.[1]	200	145	150	Maize	511	535	605
				Olive oil[2]	318	518	353
Carrots and turnips[1]	150	160	170	Olives	212	373	375[1]
Fruits				Onions[1]	118	118	121
oranges	218	234	211	Potatoes	570	611	657
apples	252	258	247	Rice	120	149	156
grapes	989	1,029	822	Sugar beets	605	320	254
pears	130	175	141	Tomatoes	1,085	983	1,236
Lettuce and chicory[1]	95	100	100	Wheat	82	250	102
				Wine[2]	7,064	7,338	5,842

[1]Estimates. [2]In 1,000 hectolitres.

Livestock (1,000 head):

	2005	2006	2007
Cattle	1,441	1,407	1,443
Goats	551	547	509
Pigs	2,344	2,295	2,374
Sheep	3,583	3,549	3,356
Poultry[1]	42,000	43,200	44,500

[1]Estimates.

Animal products in 2007 (1,000 tonnes): meat, 844·8; eggs, 121·6; cheese, 79·5; milk, 2,029m. litres.

Forestry
Forests covered 3·46m. ha. (38% of the land area) in 2010. Portugal is the world's largest producer of cork, averaging 157,000 tonnes annually. Timber production was 10·80m. cu. metres in 2007.

Fisheries
The fishing industry is important, although much less so than in the past, and the Portuguese eat more fish per person than in any other European Union member country (more than twice the EU average). In 2006 there were 8,754 registered fishing vessels (7,153 with motors) and 17,261 registered fishermen. The catch was 141,683 tonnes in 2006 (almost exclusively from marine waters).

The 2006 fishing catch consisted of:

Species	Tonnes	Value (in €1m.)
Sardine	48,096	26,334
Mackerel	30,486	23,235
Shellfish	17,501	62,394
Other	45,600	132,337
Total	141,683	244,300

INDUSTRY

The leading companies by market capitalization in Portugal in March 2012 were: Jerónimo Martins, a food distribution group (US$12·8bn.); Galp Energia, SGPS, SA, an oil and gas company (US$12·7bn.); and EDP—Energias de Portugal (US$8·5bn.).

Output of major industrial products (in tonnes unless otherwise specified):

Product	2001	2002
Ready-mix concrete	25,658,038	25,567,852
Portland cement	10,162,310	9,760,964
Refined sugar	381,626	399,621
Preparation of animal food feeds	3,933,649	3,905,501
Beer (hectolitres)	6,829,719	7,124,710
Woven fabrics of synthetic staple fibres[1]	60,624	56,251
Footwear with leather uppers (1,000 pairs)	72,373	68,757
Wood pulp	1,784,347	1,806,403
Paper and cardboard	1,341,576	1,453,105
Petrol	2,619,805	2,484,639
Glass bottles (1,000)	3,663,323	3,890,782

[1]In 1,000 sq. metres.

Labour

The maximum working week was reduced from 44 hours to 40 in 1997. A minimum wage is fixed by the government. In 2011 the minimum wage was €485 a month. Retirement is at 65 years for men and 62 for women. In 2003, out of a working population of 5,460,300 (2,947,900 male), 5,118,000 (2,787,100 male) were employed. In Dec. 2012 the unemployment rate was 16·5% (up from 12·9% in 2011 as a whole). Employment (in 1,000) by sector, 2003 (males in parentheses): services, 2,823·1 (1,283·6); industry, construction, energy and water, 1,652·8 (1,174·7); agriculture, forestry and fishing, 642·1 (328·7). The immigrant population makes up 10% of the labour force.

INTERNATIONAL TRADE

Imports and Exports

In 2010 imports (c.i.f.) totalled US$75,573m. (US$69,985m. in 2009); exports (f.o.b.), US$48,744m. (US$43,397m. in 2009).

In 2010 main import sources (in US$1m.) were: Spain (23,557); Germany (10,460); France (5,466); Italy (4,292); Netherlands (3,866). Main export markets in 2010 (in US$1m.) were: Spain (12,917); Germany (6,262); France (5,729); UK (2,656); Angola (2,533).

Principal imports in 2010 (in US$1m.) were: machinery and transport equipment (22,828); manufactured goods (11,294); mineral fuels, lubricants and related materials (11,026); chemicals and chemical products (9,481); food (8,392). Principal exports in 2010 (in US$1m.) were: machinery and transport equipment (13,225); manufactured goods (10,871); miscellaneous manufactured articles including articles of apparel, clothing accessories and footwear (8,037); chemicals and chemical products (4,030); food (3,411).

COMMUNICATIONS

Roads

In 2005 there were 2,613 km of motorways, 5,883 km of national roads, 4,406 km of secondary roads and 63,900 km of other roads. In 2006 the number of vehicles registered included 5,234,500 passenger cars, 535,300 motorcycles and mopeds, 119,000 lorries and vans and 29,700 buses and coaches. In 2007 there were 854 deaths in road accidents.

Rail

In 2008 total railway length was 2,842 km. Passenger-km travelled in 2008 came to 3·81bn. and freight tonne-km to 2·55bn. There is a metro (19 km) and tramway (94 km) in Lisbon. New light rail systems were opened in Porto in 2002 and Almada in 2007.

Civil Aviation

There are international airports at Portela (Lisbon), Pedras Rubras (Porto), Faro (Algarve) and Funchal (Madeira). The national carrier is the state-owned TAP-Air Portugal, with some domestic and international flights being provided by Portugália. In 2006 scheduled airline traffic of Portuguese-based carriers flew 171m. km, carrying 9,449,000 passengers (6,449,000 on international flights). In 2007 Lisbon handled 13,393,000 passengers (11,249,000 on international flights) and 82,645 tonnes of freight. Faro was the second busiest in terms of passenger traffic, with 5,471,000 passengers, and Porto was the second busiest for freight, with 31,991 tonnes.

Shipping

In 2007, 15,226 vessels of 151·82m. tonnes entered all Portuguese ports; 367,391 passengers embarked and 368,095 disembarked during 2007. 21·17m. tonnes of cargo were loaded in 2007 and 47·05m. tonnes unloaded. In Jan. 2009 there were 154 ships of 300 GT or over registered, totalling 981,000 GT.

Telecommunications

Portugal Telecom (PT) was formed from a merger of three state-owned utilities in 1994. It is now fully privatized. In 2008 there were 4,111,000 main (fixed) telephone lines. In the same year mobile phone subscribers numbered 14,910,000 (1,396·4 per 1,000 persons). There were 4·5m. internet users in 2008. There were 19·8 broadband subscribers per 100 inhabitants in Dec. 2010. In March 2012 there were 4·2m. Facebook users.

SOCIAL INSTITUTIONS

Justice

There are four judicial districts (Lisbon, Porto, Coimbra and Évora) divided into 58 circuits. In 2007 there were 335 courts, including 329 common courts of first instance. There are also six higher courts (five courts of appeal and a Supreme Court in Lisbon).

Capital punishment was abolished completely in the constitution of 1976.

In 2006 there were 54 prisons with an inmate capacity of 12,115. The population in penal institutions in Nov. 2008 was 11,017 (104 per 100,000 of national population).

Education

Adult literacy rate was an estimated 94·9% in 2009. Compulsory education has been in force since 1911.

In 2007 there were 263,887 children in pre-school establishments with 16,599 teaching staff, 753,646 pupils in primary schools with 64,274 teaching staff and 680,338 pupils in secondary schools with 92,965 teaching staff.

In 2006 public tertiary education institutions included 14 universities and a non-integrated university institution; 15 polytechnics and a number of polytechnic schools integrated in universities; 9 non-integrated nursing schools; 4 university-level military schools; and 5 polytechnic military schools. In the private sector there were 34 university level institutions and 66 polytechnics as well as a Catholic university. Portugal's oldest university is the University of Coimbra (Universidade de Coimbra), initially established in Lisbon in 1290; its largest is the University of Porto (Universidade do Porto), with 27,184 students in 2007–08. In 2007 there were 366,729 students in higher education with 36,069 academic staff.

Public expenditure on education came to 5·5% of GNI in 2006 (11·3% of total government expenditure).

Health

There were 200 hospitals in 2006 with 36,563 beds, and 378 clinics. In 2007 there were 37,904 doctors, 5,629 dentists, 10,117 pharmacists and 54,079 nurses. In 2007 Portugal spent 10·0% of its GDP on health.

Welfare

In 2001, €25,817m. were paid in social security benefits. Cash payments in euros (and types) were: 9,984m. (old age); 8,070m. (sickness); 3,186m. (disability); 1,846m. (survivors); 1,458m. (family); 940m. (unemployment); 328m. (social exclusion); 6m. (housing).

Pensions are available to men and women aged at least 65 with 15 years of contributions. Pensions are available at a younger age to workers in specified industries including mining, dancing, and the maritime and aviation sectors. The pension value is 2% of the average lifetime salary for each year of contributions, up to 40 years. Until 2017 pensions may also be calculated using an older system (2% of average earnings for the best ten of the last 15 years, multiplied by the number of years of contributions).

RELIGION

There is freedom of worship, both in public and private, with the exception of creeds incompatible with morals and the life and physical integrity of the people. There were 9·52m. Roman Catholics in 2001. In Feb. 2013 there were three cardinals.

CULTURE

The Community of Portuguese-Speaking Countries (CPLP, comprising Angola, Brazil, Cape Verde, Guinea-Bissau, Mozambique, Portugal and São Tomé e Príncipe) was founded in July 1996 with headquarters in Lisbon, primarily as a cultural and linguistic organization.

World Heritage Sites

(With year entered on list). In the Central Zone of the Town of Angra do Heroísmo in the Azores (1983) are the fortresses of San Sebastião and San Filipe, the latter built around 1590 on the orders of King Phillip II of Spain. The Monastery of the Hieronymites was built at the turn of the 16th century in Belém, Lisbon, while the capital's Tower of Belém was constructed as a monument to Vasco da Gama's explorations (1983 and 2008). The Monastery of Batalha (1983) near Leiria was built from 1388. The Convent of Christ in Tomar (1983) was originally built in 1160 as the centre of the Templar order. It was taken over by the Order of Christ in 1360 of which Henry the Navigator was made governor in 1418, and was greatly enriched in the 16th century. Other sites are the medieval walled Historic Centre of Évora (1988), the Gothic Cistercian 12th century Monastery of Alcobaça, north of Lisbon (1989), the Cultural Landscape of Sintra (1995), the Historic Centre of Porto (1996) and the Laurisilva of Madeira (1999), an area of biodiverse laurel forest. In 2001 two more sites were added: the Alto Douro Wine Region, famous for its port wine since the 18th century, and the Historic Centre of Guimarães, a town closely associated with the formation of Portuguese identity. The Landscape of the Pico Island Vineyard Culture followed in 2004. The Garrison Border Town of Elvas and its Fortifications (2012) houses the remains of an enormous war fortress, extensively fortified from the 17th to 19th centuries. It is the largest bulwarked dry ditch system in the world.

The Prehistoric Rock Art Sites in the Côa Valley and Siega Verde (1998) is shared with Spain.

Press

In 2006 there were 35 daily newspapers (morning and evening editions) including seven in the Azores and two in Madeira, with a combined annual circulation of 671,329,640. In addition there were 2,019 periodicals in 2006 with a combined circulation of 223,765,806.

Tourism

In 2010, 6,831,600 non-resident tourists stayed in holiday accommodation (6,478,700 in 2009) including: 1,375,800 from Spain; 1,111,200 from the UK; 728,800 from Germany; 574,800 from France. There were 2,011 hotel establishments with 279,506 beds in 2010.

Festivals

Popular rock and pop music festivals include Rock in Rio in Lisbon (May), Optimus Alive at Oeiras (June), Super Bock Super Rock at Meco (July), Paredes de Coura Festival (Aug.), Festival Sudoeste at Zambujeria do Mar (Aug.) and the Jazz in August festival in Lisbon. Other music events are the Festival Músicas do Mundo that takes place in Sines (July), the Estoril Music Festival (July) and Sintra Music Festival (July). The Almada Theatre Festival takes place in Lisbon in July and the Festival Internacional de Cinema de Setúbal in June. Porto's most important festival, Festa de São João, takes place on 24 June.

DIPLOMATIC REPRESENTATIVES

Of Portugal in the United Kingdom (11 Belgrave Sq., London, SW1X 8PP)
Ambassador: João de Vallera.

Of the United Kingdom in Portugal (Rua de São Bernardo 33, 1249-082 Lisbon)
Ambassador: Jill Gallard.

Of Portugal in the USA (2125 Kalorama Rd, NW, Washington, D.C., 20008)
Ambassador: Nuno Filipe Alves Salvador e Brito.

Of the USA in Portugal (Ave. das Forças Armadas, 1600 Lisbon)
Ambassador: Allan J. Katz.

Of Portugal to the United Nations
Ambassador: José Filipe Moraes Cabral.

Of Portugal to the European Union
Permanent Representative: Domingos Fezas Vital.

FURTHER READING

Instituto Nacional de Estatística. *Anuário Estatístico de Portugal/Statistics Year-Book.—Estatísticas do Comércio Externo.* 2 vols. Annual from 1967

Birmingham, David, *A Concise History of Portugal.* 1993
Maxwell, K., *The Making of Portuguese Democracy.* 1995
Page, Martin, *The First Global Village: How Portugal Changed the World.* 2002
Saraiva, J. H., *Portugal: A Companion History.* 1997
Wheeler, D. L., *Historical Dictionary of Portugal.* 1994

National library: Biblioteca Nacional de Lisboa, Campo Grande 83, 1749-081 Lisbon.
National Statistical Office: Instituto Nacional de Estatística (INE), Avenida António José de Almeida, 1000-043 Lisbon.
Website: http://www.ine.pt

QATAR

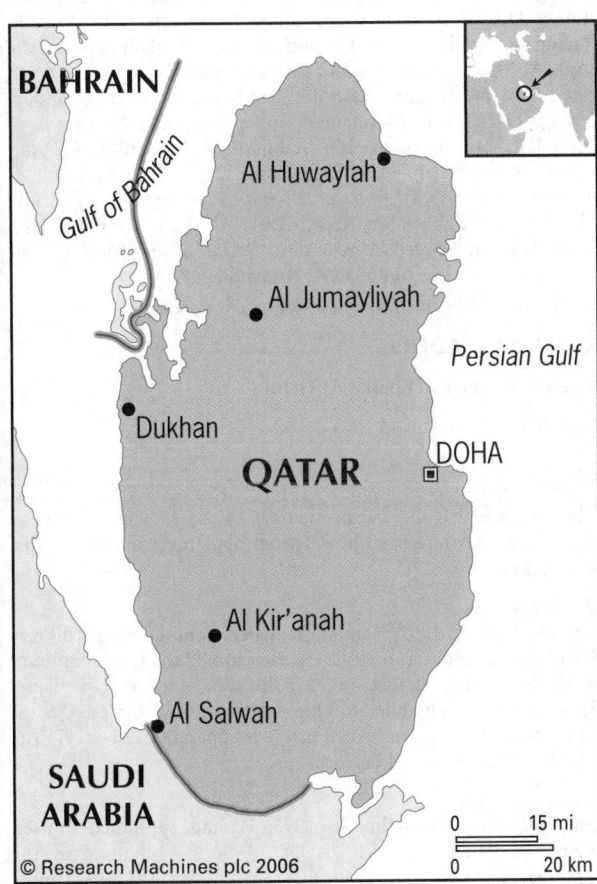

BAHRAIN
Gulf of Bahrain
Al Huwaylah
Al Jumayliyah
Persian Gulf
Dukhan
QATAR
DOHA
Al Kir'anah
Al Salwah
SAUDI ARABIA

0   15 mi
0   20 km

© Research Machines plc 2006

Dawlat Qatar
(State of Qatar)

Capital: Doha
Population projection, 2015: 2·03m.
GNI per capita, 2010: US$71,117
HDI/world rank: 0·831/37
Internet domain extension: .qa

KEY HISTORICAL EVENTS

Qatar has rock carvings, inscriptions and fragments of pottery dating from 4000 BC. The early population was swelled by seasonal migration of Arab tribes and the peninsula became a centre for fishing and pearls. Commercial activity declined in the Roman era, when trade was concentrated in the Red Sea, but recovered in the 3rd century AD.

Islam was established in Qatar in the mid-7th century AD. During the Abbasid period (750–1258) Qatar enjoyed strong relations with the Caliphs in Baghdad. After briefly coming under Portuguese influence in the early 16th century, Qatar fell under Ottoman sovereignty. For the next four centuries it was nominally part of the Ottoman Empire, though considerable power remained with local tribal sheikhs.

Pearling and trading settlements were established along the coast in the 18th century, which also saw the rise of the Al Thani family who were originally from Saudi Arabia. In the mid-19th century Sheikh Mohammed bin Thani moved the family to the growing coastal town of Doha and established control of the surrounding region. Territorial disputes with the Al Khalifa family in neighbouring Bahrain led to a war in 1867 in which Doha was almost destroyed. The British intervened to recognize the Al Thani family as rulers of Qatar and in 1878 Sheikh Mohammed was succeeded by his son, Sheikh Qassim, who became the first Amir. With the collapse of the Ottoman Empire during the First World War, Qatar came under British rule. Under the treaties of 1916 and 1934 Qatar ceded Britain control over its external affairs in return for British military protection.

Oil was discovered in 1939 and, after a delay caused by the Second World War, exports began in 1949. In Dec. 1961 Qatar joined the Organization of the Petroleum Exporting Countries (OPEC). In 1968 British troops left Qatar and in 1970 Qatar adopted a constitution confirming the emirate as an absolute monarchy. It led negotiations to establish a union of Arab emirates but terms could not be agreed. On 3 Sept. 1971 Qatar assumed full independence under the rule of Sheikh Ahmad and joined the Arab League and the United Nations. In 1972 Sheikh Ahmad was ousted in a coup and chief minister Sheikh Khalifa bin Hamad Al Thani assumed power.

The discovery in 1971 of a large offshore oil field gave further impetus to the economy and Qatar rapidly developed a modern infrastructure, building up its health and education services. In 1981 it was a founder member of the Gulf Co-operation Council (GCC) and in 1988 established diplomatic relations with the USSR and the People's Republic of China. Qatar allied itself with Saudi Arabia on many regional and international issues and in 1991 joined the US-led international alliance against Iraq following the invasion of Kuwait.

In Jan. 1992 pressure for political reform culminated in demands from 50 prominent Qataris for a consultative assembly. On 27 June 1995 Amir Sheikh Khalifa was ousted by his son, Sheikh Hamad, who announced plans to introduce democratic reforms. In 1996 he survived an assassination attempt, part of an abortive attempt to restore his father to power. In 1996 the Arabic-language news agency Al-Jazeera was established in Doha, with the Amir's personal support and state financial backing. The station has won widespread respect and influence, despite pressure from some powers to tone down what is seen as an anti-West bias.

In 2001 a long-standing territorial dispute between Qatar and Bahrain was settled by the International Court of Justice, with Qatar recognizing Bahrain's sovereignty over the Hamar Islands in return for Bahrain renouncing claims on parts of mainland Qatar. In 2003 Qatar supported the UN-backed, American- and British-led invasion of Iraq, with Doha hosting the coalition headquarters.

In recent years Qatar's economy has grown rapidly. In 2003 a referendum was held to approve the country's constitution, which provides for a parliament with 30 elected and 15 appointed members. In the same year Qatar's first female minister, Sheikha Ahmad Al Mahmoud, was appointed.

TERRITORY AND POPULATION

Qatar is a peninsula running north into the Persian Gulf. It is bounded in the south by Saudi Arabia. The territory includes a number of islands in the coastal waters of the peninsula, the most important of which is Halul, the storage and export terminal for the offshore oilfields. The area of Qatar is 11,571 sq. km. Population at the census of April 2010, 1,699,435; density 146·9 per sq. km. In 2011, 95·9% of the population lived in urban areas.

The UN gives a projected population for 2015 of 2·03m.

Areas and populations of the seven municipalities:

Municipalities	Sq. km	2010 census population
Doha	234	796,947
Al Rayyan	5,818	445,623
Al Khour	1,551	193,983
Al Wakra	2,520	141,222
Umm Salal	310	60,509
Al Daayen	236	43,176
Al Shamal	902	7,975

The capital is Doha, which is the main port, and had a census population in 2010 of 521,283. Other towns are Dukhan (the centre of oil production), Umm Said (the oil terminal of Qatar), Ruwais, Wakra, Al Khour, Umm Salal Mohammad and Umm Bab.

About 40% of the population are Arabs, 18% Indian, 18% Pakistani and 10% Iranian. Other nationalities make up the remaining 14%. Only about 20% of the population are Qatari citizens.

The official language is Arabic.

SOCIAL STATISTICS

Births, 2008, 17,210; deaths, 1,942; marriages, 3,235; divorces, 939. 2008 rates per 1,000 population: births, 11·9; deaths, 1·3. Qatar's 2008 death rate was among the lowest in the world. Infant mortality, 2010 (per 1,000 live births), 7. Expectation of life in 2007 was 74·8 years for males and 76·8 for females. Annual population growth rate, 2000–08, 9·1% (the highest in the world). Fertility rate, 2008, 2·4 births per woman.

CLIMATE

The climate is hot and humid. Doha, Jan. 62°F (16·7°C), July 98°F (36·7°C). Annual rainfall 2·5" (62 mm).

CONSTITUTION AND GOVERNMENT

Qatar is ruled by an *Amir*. HH Sheikh Hamad bin Khalifa Al Thani, KCMG (b. 1952) assumed power after deposing his father on 27 June 1995. The heir apparent was Sheikh Hamad's third son, Sheikh Jasim bin Hamad Al Thani (b. 1978), but in Aug. 2003 he named his fourth son, Sheikh Tamim bin Hamad Al Thani (b. 1979), as heir apparent instead.

Qatar's first written constitution was approved in June 2004 and came into force on 9 June 2005. It allows for a 45-member *Consultative Assembly* or *Majlis al-Shura*, with 30 members directly elected and 15 appointed by the Amir.

A *Council of Ministers* is assisted by a 35-member nominated Advisory Council.

National Anthem

'As-Salam Al-Amiri' ('Peace for the Amir'); words by Sheikh Mubarak bin Saïf al-Thani, tune by Abdul Aziz Nasser Obaidan.

GOVERNMENT CHRONOLOGY

Amirs since 1971.
1971–72   Sheikh Ahmad bin Ali Al Thani
1972–95   Sheikh Khalifa bin Hamad Al Thani
1995–   Sheikh Hamad bin Khalifa Al Thani

RECENT ELECTIONS

30 of the 45 members of the Consultative Assembly may be elected for a four-year term for the first time during 2013.

CURRENT GOVERNMENT

In March 2013 the government comprised:

Amir, Minister of Defence and C.-in-C. of the Armed Forces: HH Sheikh Hamad bin Khalifa Al Thani; b. 1952.

Prime Minister and Minister of Foreign Affairs: Sheikh Hamad bin Jasim bin Jabir Al Thani; b. 1959 (in office since 3 April 2007).

Deputy Prime Ministers: Abdallah bin Hamad Al Attiyah; Ahmed bin Abdullah Al Mahmud (also *Minister of State for Cabinet Affairs*).

Minister of Awqaf and Islamic Affairs: Ghaith bin Mubarak bin Imran Al Kuwari. *Business and Trade:* Jassim bin Abdul Aziz bin Jassim Al Thani. *Culture, Arts and Heritage:* Dr Hamad bin Abdul Aziz Al Kawari. *Economy and Finance:* Yusif Husayn Al Kamal. *Education and Higher Education:* Saad bin Ibrahim Al Mahmoud. *Energy and Industry:* Mohamed bin Saleh al-Sada. *Environment:* Abdullah bin Mubarak bin Aaboud Al Midhadhi. *Interior:* Abdullah bin Khalid Al Thani. *Justice:* Hasan bin Abdallah Al Ghanim. *Municipal Affairs and Urban Planning:* Sheikh Abdul Rahman bin Khalifa bin Abdul Azziz Al Thani. *Public Health:* Abdallah bin Khaled Al Qahtani. *Social Affairs, and Labour (acting):* Nasser bin Abdullah Al Humaidi.

Government Website: http://www.gov.qa/wps/portal

CURRENT LEADERS

Sheikh Hamad bin Khalifa Al Thani

Position
Amir

Introduction
Sheikh Hamad is the eighth member of the Al Thani family to rule Qatar, having seized power from his father, Sheikh Khalifa, on 27 June 1995.

Early Life
Born in Doha in 1952, Hamad graduated from the Royal Military Academy, Sandhurst in 1971. He then joined the Qatari military with the rank of major. In 1975 he was promoted to major-general and commander-in-chief of the armed forces. On his appointment as Crown Prince and heir apparent in May 1977, he also became minister of defence.

Career in Office
Having ousted his father in 1995, Hamad appointed himself prime minister. However, in Oct. 1996 he relinquished the premiership to his younger brother, Sheikh Abdallah. The Amir is credited with initiating plans for an elected consultative council (through the new constitution which he approved in 2004 and which took effect in June 2005), giving women the right to vote in municipal elections (from 1999) and ending official media censorship. He has also encouraged foreign investment in Qatar's oil and natural gas industries.

In Aug. 2003 he named his fourth son as his heir apparent, and in April 2007 appointed his foreign minister and cousin, Sheikh Hamad bin Jasim bin Jabir Al Thani, prime minister. To pre-empt pressure for political liberalization in the wake of the discontent that spread across much of the Arab world from early 2011, Hamad announced in Nov. that year that elections to Qatar's Advisory Council would be held in 2013 although by April 2013 no fixed date had been set.

In foreign relations he has overseen the resolution of long-standing border disputes with Bahrain and Saudi Arabia. In 2011 Qatar was involved in international military operations against the Gaddafi regime in Libya, a fellow Arab state. The country's profile on the world stage was raised in Dec. 2010 when it successfully bid to host the football World Cup finals in 2022. In Sept. 2012 Hamad called on Arab nations to form a coalition to intervene in the Syrian conflict, and in Oct. he made the first visit to Gaza by a head of state since the Hamas government took control of the Palestinian territory.

DEFENCE

Defence expenditure in 2008 totalled US$1,756m. (US$2,129 per capita), representing 1·8% of GDP. The expenditure per capita in

2008 was the fourth highest in the world after that of the United Arab Emirates, Kuwait and the USA.

Army
Personnel (2007) 8,500.

Navy
Personnel in 2007 totalled 1,800 including Marine Police; the base is at Doha.

Air Force
The Air Force operates 18 combat-capable aircraft (including Mirage 2000 fighters), 11 attack helicopters and 14 other helicopters. Personnel (2007) 1,500.

INTERNATIONAL RELATIONS
In March 2001 the International Court of Justice ruled on a long-standing dispute between Bahrain and Qatar over the boundary between the two countries and ownership of certain islands. Both countries accepted the decision.

ECONOMY
Oil, natural gas and other mining accounted for 62% of GDP in 2006, public administration and defence 9%, finance and real estate 8% and manufacturing 7%.

Overview
Qatar is an oil and natural gas exporter; with 13% of world supply and an oil production capacity of 800,000 bbls a day. Qatar has welcomed foreign investment in the development of its gas fields over the last decade and became the world's leading liquefied natural gas exporter in 2007. Oil and gas contribute 50% of GDP, 85% of export earnings and 70% of government revenues.

Considered to be one of the most competitive economies in the world, Qatar is highly rated by credit agencies. The economy is open, barriers to trade are low and no income taxes are levied. There has been double-digit growth every year since 2006. Japan is the main export market accounting for 25·7%, followed by South Korea (17·7%) and India (9·6%). Manufacturing, construction, commercial ship repairs, the air industry and financial services are leading non-energy sector industries.

Tourism has also thrived, with the country aiming to rival Dubai as a regional destination. Construction has boomed on the back of the expansion of Doha International Airport, while preparations for the 2022 football World Cup include infrastructural spending on a Metro system and the Qatar–Bahrain Causeway.

Qatar also aims to compete with Bahrain and Dubai in establishing itself as a regional financial centre, and is home to branches of leading universities including Georgetown, Carnegie Mellon and Weill Cornell Medical College.

Currency
The unit of currency is the *Qatari riyal* (QAR) of 100 *dirhams*, introduced in 1973. Foreign exchange reserves were US$4,370m. in July 2005, gold reserves were 19,000 troy oz and total money supply was 19,159m. riyals. There was deflation of 2·4% in 2010 but inflation of 1·9% in 2011.

In 2001 the six Gulf Arab states—Qatar, along with Bahrain, Kuwait, Oman, Saudi Arabia and the United Arab Emirates—signed an agreement to establish a single currency by 2010. In June 2009 it was agreed to postpone the implementation of the new currency, the *khaleeji*, until 2013. It has since been put back further still. Both Oman and the United Arab Emirates have now withdrawn from the scheme, in 2007 and 2009 respectively.

Budget
The fiscal year is 1 April–31 March. Revenue (2009–10) 154,674m. riyals; expenditure, 108,085m. riyals. Oil and natural gas account for about 45% of revenues.

Performance
In terms of GDP per head, Qatar ranks among the richest countries in the world with a purchasing power parity (PPP) per capita GDP of 77,568 current international dollars in 2009. Qatar has been experiencing a period of rapid economic expansion, driven by rising oil prices and the increased exploitation of its natural gas reserves, the third largest in the world. Real GDP growth was 12·0% in 2009 and 16·7% in 2010—the highest rate of any country. Total GDP was US$173·0bn. in 2011, when the economy grew by 14·1%.

Banking and Finance
The Qatar Monetary Agency, which functioned as a bank of issue, became the Central Bank in 1995 (*Governor*, Abdullah bin Saud Al-Thani). In 2003 there were eight commercial domestic banks and seven foreign banks. The largest bank is the Qatar National Bank, with assets in 2008 of US$41·8bn.

A stock exchange was established in Doha by the Amir's decree in 1995.

ENERGY AND NATURAL RESOURCES

Environment
Qatar's carbon dioxide emissions from the consumption and flaring of fossil fuels in 2008 were the equivalent of 74·1 tonnes per capita, the highest of any sovereign country.

Electricity
Installed capacity was an estimated 3·2m. kW in 2007. Production was 16·08bn. kWh in 2007; consumption per capita was 13,113 kWh.

Oil and Gas
Proven reserves of oil (2008) 27·3bn. bbls. Output, 2008, 60·8m. tonnes. Production rose by 13·2% in 2008 compared to 2007—the largest increase of any oil producing country.

The North Field, the world's biggest single reservoir of gas and containing 12% of the known world gas reserves, is half the size of Qatar itself. Development cost is estimated at US$25bn. In 2008 natural gas reserves were 25,460bn. cu. metres (the third largest after Russia and Iran); output in 2008 was 76·6bn. cu. metres.

Agriculture
In 2002, 1·2% of the economically active population were engaged in agriculture. An estimated 13,000 ha. were irrigated in 2002. There were approximately 18,000 ha. of arable land in 2002 and 3,000 ha. of permanent crops. Production (2002) in 1,000 tonnes: dates, 16; tomatoes, 11; pumpkins and squash, 8; aubergines, 5; barley, 5; cucumbers and gherkins, 5; melons and watermelons, 5; onions, 4.

Livestock (2003 estimates): sheep, 200,000; goats, 180,000; camels, 51,000; cattle, 15,000; chickens, 4m. Livestock products, 2003 estimates (in 1,000 tonnes): meat, 15; milk, 22; eggs, 4.

Fisheries
The catch in 2010 totalled 13,760 tonnes, entirely from sea fishing —down from 17,688 tonnes in 2008 although up from 7,140 tonnes in 2000.

INDUSTRY
The leading companies by market capitalization in Qatar in March 2012 were Qatar National Bank (US$26·3bn.); and Industries Qatar, a producer of petrochemicals, fertilizer and steel (US$21·4bn.).

Production (2005, in 1,000 tonnes): urea, 2,979; ammonia, 2,134; petrol, 1,656; butane, 1,075; cement, 1,049; distillate fuel oil, 926; jet fuel, 905; steel bars, 791; ethylene, 544; residual fuel oil, 418; polyethylene, 415. There is an industrial zone at Umm Said.

Labour
In 2011 the economically active population totalled 1,271,100. Males constituted 88% of the labour force in 2011; foreigners

make up 94% of the workforce. Qatar has the lowest percentages of females in the workforce of any country.

INTERNATIONAL TRADE

Qatar, along with Bahrain, Kuwait, Oman, Saudi Arabia and the United Arab Emirates entered into a customs union in Jan. 2003.

Imports and Exports

Total imports and exports in calendar years (in US$1m.):

	2003	2004	2005	2006
Imports	4,897·4	6,004·6	10,060·9	16,440·1
Exports	13,382·6	18,685·1	25,762·5	34,051·3

The main imports are machinery and equipment, consumer goods, food and chemicals. Main exports are petroleum and petroleum products (52%) and liquefied natural gas (35%). Qatar is by far the world's largest exporter of liquefied natural gas. Principal import suppliers in 2004: France, 26·7%; USA, 9·6%; Saudi Arabia, 9·5%. Leading export markets, 2004: Japan, 41·6%; South Korea, 15·7%; Singapore, 9·1%.

COMMUNICATIONS

Roads

In 2007 there were about 7,790 km of roads. Vehicles in use in 2007 totalled 605,700. In 2007 there were 199 fatalities as a result of road accidents.

Civil Aviation

Gulf Air was formed as a partnership between Qatar, Bahrain, Oman and Abu Dhabi, but Qatar withdrew in 2002 (and Abu Dhabi and Oman have withdrawn in the meantime). In 2003 it operated services from Doha International to Abu Dhabi and Bahrain. A Qatari airline, Qatar Airways, operates on the same routes, and in 2003 additionally flew to Amman, Bangkok, Beirut, Cairo, Casablanca, Colombo, Damascus, Damman, Dhaka, Dubai, Frankfurt, Hyderabad, Islamabad, Jakarta, Jeddah, Karachi, Kathmandu, Khartoum, Kochi, Kuala Lumpur, Kuwait, Lahore, London, Malé, Manchester, Manila, Milan, Mumbai, Munich, Muscat, Paris, Peshawar, Riyadh, Salalah, Sana'a, Sharjah and Thiruvananthapuram. In June 2003 Qatar Airways commissioned 32 aircraft worth US$5·1bn. from Airbus SAS. Doha handled 2,759,000 passengers (all on international flights) and 64,000 tonnes of freight in 2001.

Telecommunications

In 2008 there were 263,400 main (fixed) telephone lines; mobile phone subscribers numbered 1,683,000 in 2008 (131·4 per 100 persons). There were 436,000 internet users in 2008. In March 2012 there were 481,000 Facebook users.

SOCIAL INSTITUTIONS

Justice

The Judiciary System is administered by the Ministry of Justice which comprises three main departments: legal affairs, courts of justice and land and real estate register. In 2004 a High Judicial Council was established to oversee the court system, which as a result of a new Judicial Authority Law that took effect at the same time comprises the Court of Cassation, the Court of Appeal and the Court of First Instance. The courts proclaim sentences in the name of HH the Amir. The death penalty is in force. There was one execution in 2003, but none since. The population in penal institutions in Dec. 2008 was 551 (43 per 100,000 of national population).

All issues related to personal affairs of Muslims under Islamic Law embodied in the Holy Koran and Sunna are decided by Sharia Courts.

Education

Adult literacy rate was 94·7% in 2009. There were, in 2007, 75,451 pupils at primary schools (with 6,639 teaching staff in 2006), 61,226 pupils at secondary schools (6,200 teaching staff in 2006) and 8,881 students with 1,153 academic staff in higher education. There were 265 Arab and foreign private schools with 56,183 pupils and 4,092 teachers in 2002–03. The University of Qatar had 8,801 students in 2008 and 637 full-time academic staff in 2007–08. Education City, a campus built in the outskirts of Doha, houses branches of eight prestigious international universities along with the Qatar Faculty of Islamic Studies.

Students abroad (2003–04) numbered 374. In 2002–03, 2,009 men and 940 women attended night schools and literacy centres.

In 2005 public expenditure on education accounted for 19·6% of total government spending.

Health

There were three government and two private hospitals in 2002. In 2005 there were 2,150 physicians, 1,100 pharmacists, 690 dentists and 4,880 nursing and midwifery personnel.

RELIGION

The population is predominantly Muslim, although there is a small Christian minority among expatriates.

CULTURE

Press

There are three Arabic language daily newspapers—*Al-Rayah*, *Al-Sharq* and *Al-Watan*. *The Gulf Times* and *Al-Jazeera* (The Peninsula) are English dailies. In 2006 the combined circulation was 81,000.

Tourism

In 2008, 1,405,000 non-resident tourists stayed in hotels (964,000 in 2007).

DIPLOMATIC REPRESENTATIVES

Of Qatar in the United Kingdom (1 South Audley St., London, WIK 1NB)
Ambassador: Khalid Rashid Salem Al-Homoudi Al-Mansouri.

Of the United Kingdom in Qatar (PO Box 3, Doha, Qatar)
Ambassador: Michael O'Neill.

Of Qatar in the USA (2555 M St., NW, Washington, D.C., 20037)
Ambassador: Mohammed bin Abdullah Al-Rumaihi.

Of the USA in Qatar (22 February St., Doha)
Ambassador: Susan Ziadeh.

Of Qatar to the United Nations
Ambassador: Sheikh Meshal bin Hamad Al-Thani.

Of Qatar to the European Union
Ambassador: Sheikh Ali bin Jassim Al-Thani.

FURTHER READING

Central Statistical Organization. *Annual Statistical Abstract.*

El-Nawawy, Mohammed and Iskandar, Adel, *Al-Jazeera: How the Free Arab News Network Scooped the World and Changed the Middle East.* 2002

National Statistical Office: Central Statistical Organization, Presidency of the Council of Ministers, Doha.
Website: http://www.qsa.gov.qa

ROMANIA

© Research Machines plc 2006

România

Capital: Bucharest
Population projection, 2015: 21·24m.
GNI per capita, 2011: (PPP$) 11,046
HDI/world rank: 0·781/50
Internet domain extension: .ro

KEY HISTORICAL EVENTS

Neolithic peoples with links to Anatolia settled from 5000 BC at Lake Golovita, close to Romania's Black Sea coast. The subsequent Boian culture spread across the lower Danube valley by 3500 BC. Later, Indo European peoples, collectively known as the Thracians, entered the Carpathian-Balkan region. The Dacii (or Getae to the Greeks, who had established colonies on the western Black Sea coast by the 7th century BC) occupied much of present-day Romania. Dacian power grew under King Burebista (82–44 BC), attracting Roman attention. In AD 106 the Roman emperor Trajan succeeded in making the kingdom a frontier province.

Under Barbarian attack, in 271 Emperor Aurelian withdrew the Roman Army and administration south of the Danube to Dobrogea (Constanța), although much of Dacia remained in the Roman-Byzantine sphere through trade along the Danube. As Rome's influenced waned, Dacia was populated between the 3rd and 9th centuries by tribes from the Steppes to the east, including Goths, Huns, Avars, Bulgars and Magyars. Magyar power was centred on Transylvania and for most of the 10th–13th centuries it was a Hungarian dependency. Mongol Invasion in 1241 was followed by the foundation of the feudal 'Danubian Principalities' of Wallachia and Moldavia in the late 13th and early 14th centuries. In 1415, under Prince Mircea, Wallachia became a vassal of the Ottoman Empire, although retaining some independence. Prince Vlad Țepeș (known as Dracula) ruled Wallachia from 1456–62 and Stephen the Great reigned in Moldavia from 1457–1504.

Following the Magyar defeat at the battle of Mohács in 1526, Transylvania came under Ottoman vassalage. Moldavia succumbed to Ottoman domination in 1538. Michael the Brave, ruler of Wallachia from 1593, defeated an Ottoman invasion in 1595 and conquered Transylvania four years later. In 1600 he briefly unified the three Romanian provinces. Several independent princes then ruled Transylvania in the 17th century until the Habsburgs expanded eastward in the 1680s. In 1686, through the Treaty of Vienna, Transylvania accepted Habsburg protection under Leopold I. In Wallachia, the late 17th century saw prolonged struggle against Ottoman rule, with Princes Cantacuzino and Brâncoveanu even negotiating with the Austrian and Russian Empires. The Ottomans responded by backing the suzerainty of the İstanbul-based Phanariots, whose rule included phases of corruption and enlightened reform. Prince Mavrocordat abolished serfdom in 1746. The treaty of Kuchuk Kainarji (1774) boosted Russian influence in Wallachia and Moldavia.

Phanariot rule ended in 1821, following a revolt led by Tudor Vladimirescu and the outbreak of the Greek War of Independence. Local princes were reinstated in Wallachia and Moldavia. The 1829 Treaty of Adrianople that concluded the Russo–Turkish war prompted Russian repression in the principalities. Transylvania remained under Habsburg influence until revolution in all three provinces in 1848, sparked by opposition to foreign rule and the power of local landowners (*boyars*). That year the Diet of Cluj voted for union between Transylvania and the other three provinces of Hungary. Support grew for the union of Moldavia and Wallachia. The election of Alexander Cuza as prince of both territories in 1859 paved the way for union as Romania in 1862, with Bucharest as its capital. In 1866 Carol of Hohenzollern was crowned and adopted a constitution based on that of Belgium. Romania was declared independent by the Treaty of Berlin of 1878.

Despite economic expansion, the peasantry remained downtrodden and rebelled in 1907. Romania joined the First World War on the allied side in 1916. Victory brought Transylvania (with large Hungarian and German populations), Bessarabia, Bucovina and Dobrudja into the union with the 'Old Kingdom'. The world recession then drew Romania into Germany's economic orbit. The fascist Iron Guard assassinated Liberal leader Ion G. Duca in 1933 and King Carol II became increasingly totalitarian. Following Nazi and Soviet annexations of Romanian territory in 1940, he abdicated in favour of his son Mihai. The fascist government of Ion Antonescu declared war on the USSR on 22 June 1941 but on 23 Aug. 1944 Mihai, with the backing of opposition parties, deposed Antonescu and switched sides.

The armistice of Sept. 1944 gave the Soviet army control of Romania. This, and the 'spheres of influence' diplomacy of the Allies, ensured Romania's communist future. Transylvania was restored to Romania (although it lost Bessarabia and Northern Bucovina), and large estates were broken up for the benefit of the peasantry. Following elections in Nov. 1946, marred by intimidation and fraud, Mihai was forced to abdicate and a People's Republic proclaimed. The communist leader, Gheorghe Gheorghiu-Dej, purged his fellow leaders in the early 1950s. Under Nicolae Ceaușescu, who became the centre of power in 1965, Romania took a relatively independent stand in foreign affairs while domestic repression and impoverishment increased.

In Dec. 1989 there were mass demonstrations, backed by the army, against the government. Ceaușescu fled the capital and the National Salvation Front (FSN) proclaimed itself the provisional government. Ceaușescu and his family were captured, secretly tried and, on 25 Dec., executed. A day later the FSN's Ion Iliescu became president, overseeing a reformist administration inhibited by its communist origins.

With the economy stalled, a four-party coalition led by Emil Constantinescu came to power in 1996. Iliescu returned as president in 2000 and in 2004 was replaced by Traian Băsescu. On 1 Jan. 2007 Romania joined the EU. An IMF-led bail-out followed the global economic crisis of late 2008. In Dec. 2009 Băsescu was re-elected president and Emil Boc formed a new coalition government. However, Boc resigned in Feb. 2012 amid widespread protests against austerity measures. Mihai-Răzvan Ungureanu replaced him as premier but left office in May 2012 after losing a confidence vote. Victor Ponta of the Social Democratic Party succeeded him as prime minister.

TERRITORY AND POPULATION

Romania is bounded in the north by Ukraine, in the east by Moldova, Ukraine and the Black Sea, south by Bulgaria, southwest by Serbia and northwest by Hungary. The area is 238,391 sq. km. Population (2011 census, provisional), 19,042,936; density, 79·9 per sq. km. The United Nations estimate for 2011 was 21,436,000. In 2011, 58·0% of the population lived in urban areas. Romania's population has been falling at such a steady rate since 1990 that its population at the time of the 2002 census was the same as that in the late 1970s.

The UN gives a projected population for 2015 of 21·24m.

Romania is divided into 41 counties (judeţ) and the municipality of Bucharest (Bucureşti).

County	Area in sq. km	Population (2011 census)	Capital	Population (in 1,000) (2011)
Bucharest (Bucureşti)[1]	228	1,677,985	—	—
Alba	6,242	327,224	Alba Iulia	59
Arad	7,754	409,072	Arad	148
Argeş	6,826	591,353	Piteşti	148
Bacău	6,621	583,588	Bacău	133
Bihor	7,544	549,752	Oradea	183
Bistriţa-Năsăud	5,355	277,861	Bistriţa	70
Botoşani	4,986	398,938	Botoşani	101
Brăila	4,766	304,925	Brăila	168
Braşov	5,363	505,442	Braşov	228
Buzău	6,103	432,054	Buzău	108
Călăraşi	5,088	285,050	Călăraşi	57
Caraş-Severin	8,520	274,277	Reşiţa	66
Cluj	6,674	659,370	Cluj-Napoca	309
Constanţa	7,071	630,679	Constanţa	255
Covasna	3,710	206,261	Sf. Gheorghe	54
Dâmboviţa	4,054	501,996	Tîrgovişte	74
Dolj	7,414	618,335	Craiova	244
Galaţi	4,466	507,402	Galaţi	231
Giurgiu	3,526	265,494	Giurgiu	55
Gorj	5,602	334,238	Tîrgu Jiu	79
Harghita	6,639	304,969	Miercurea-Ciuc	38
Hunedoara	7,063	396,253	Deva	57
Ialomiţa	4,453	258,669	Slobozia	43
Iaşi	5,476	723,553	Iaşi	263
Ilfov[1]	1,593	364,241	—	—
Maramureş	6,304	461,290	Baia Mare	115
Mehedinţi	4,933	254,570	Drobeta-Turnu Severin	86
Mureş	6,714	531,380	Tîrgu Mureş	128
Neamţ	5,896	452,900	Piatra-Neamţ	77
Olt	5,498	415,530	Slatina	64
Prahova	4,716	735,903	Ploieşti	198
Sălaj	3,864	217,895	Zalău	53
Satu Mare	4,418	329,079	Satu Mare	95
Sibiu	5,432	375,992	Sibiu	137
Suceava	8,553	614,451	Suceava	86
Teleorman	5,790	360,178	Alexandria	42
Timiş	8,697	649,777	Timişoara	304
Tulcea	8,499	201,462	Tulcea	67
Vâlcea	5,765	355,320	Râmnicu Vâlcea	93
Vaslui	5,318	375,148	Vaslui	51
Vrancea	4,857	323,080	Focşani	74

[1]Bucharest municipality and surrounding localities of Ilfov cover 1,821 sq. km.

At the 2002 census the following ethnic minorities numbered over 50,000: Hungarians, 1,431,807 (mainly in Transylvania); Roma (Gypsies), 535,140; Ukrainians, 61,098; Germans, 59,764. A *Council of National Minorities* made up of representatives of the government and ethnic groups was set up in 1993. The actual number of Roma is estimated to be nearer 2m. Romania has the largest Roma population of any country.

The official language is Romanian.

SOCIAL STATISTICS

2010 (in 1,000): births, 212·2; deaths, 259·7; marriages, 115·8; divorces, 32·6. Rates, 2010 (per 1,000 population): live births, 9·9; deaths, 12·1; marriages, 5·4; divorces, 1·5. Infant mortality, 2010 (per 1,000 live births), 9·8. Expectation of life at birth, 2007, was 69·0 years for males and 76·1 years for females. In 2005 the most popular age range for marrying was 25–29 for males and 20–24 for females. Measures designed to raise the birth rate were abolished in 1990, and abortion and contraception legalized. The annual abortion rate, at approximately 41 per 1,000 women, ranks among the highest in the world. Population growth rate, 2010, –0·2%; fertility rate, 2008, 1·3 births per woman.

CLIMATE

A continental climate with an annual average temperature varying between 8°C in the north and 11°C in the south. Bucharest, Jan. 27°F (–2·7°C), July 74°F (23·5°C). Annual rainfall 23·1" (579 mm). Constanţa, Jan. 31°F (–0·6°C), July 71°F (21·7°C). Annual rainfall 15" (371 mm).

CONSTITUTION AND GOVERNMENT

A new constitution was approved by a referendum on 18–19 Oct. 2003. Turnout was 55·7%, and 89·7% of votes cast were in favour. The Constitution, which replaces the previous one from 1991, defines Romania as a republic where the rule of law prevails in a social and democratic state. Private property rights and a market economy are guaranteed. The new pro-European constitution was aimed at helping Romania achieve EU membership.

The head of state is the *President*, elected by direct vote for a maximum of two five-year terms. The president is not allowed to be affiliated with any political party while in office. The President appoints the *Prime Minister*, who then has to be approved by a vote in parliament. The President is empowered to veto legislation unless it is upheld by a two-thirds parliamentary majority. The National Assembly consists of a 412-member *Chamber of Deputies* and a 176-member *Senate*; both are elected for four-year terms from 43 constituencies through a proportional mixed member system. 18 seats in the Chamber of Deputies are reserved for ethnic minorities. There is a 3% threshold for admission to either house. Votes for parties not reaching this threshold are redistributed.

There is a *Constitutional Court*.

National Anthem

'Deşteaptăte, Române, din somnul cel de moarte' ('Wake up, Romanians, from your deadly slumber'); words by A. Muresianu, tune by A. Pann.

GOVERNMENT CHRONOLOGY

(FDSN = Democratic National Salvation Front; FSN = National Salvation Front; PCR = Romanian Communist Party; PD = Democratic Party; PDSR = Party of Social Democracy in Romania; PD-L = Democratic Liberal Party; PNL = National Liberal Party; PNTCD = National Peasant Party Christian Democratic; PSD = Social Democratic Party; n/p = non-partisan)

Heads of State since 1940.

King
1940–47   Mihai I

Presidents of the Presidium of the Grand National Assembly

1947–52	PCR	Constantin Ion Parhon
1958	PCR	Anton Moisescu
1958–61	PCR	Ion Gheorghe Maurer

Chairmen of the Council of State

1961–65	PCR	Gheorghe Gheorghiu-Dej
1965–67	PCR	Chivu Stoica
1967–74	PCR	Nicolae Ceauşescu

Presidents

1974–89	PCR	Nicolae Ceauşescu
1989–96	PCR, n/p, FSN, FDSN, PDSR	Ion Iliescu
1996–2000	PNTCD	Emil Constantinescu
2000–04	PDSR, PSD	Ion Iliescu
2004–07	PD	Traian Băsescu
2007	PSD	Nicolae Văcăroiu (acting for Traian Băsescu)
2007–12	PD, PD-L	Traian Băsescu
2012	PNL	Crin Antonescu (acting for Traian Băsescu)
2012–	PD-L	Traian Băsescu

Heads of Government since 1945.

Chairmen of the Council of Ministers

1945–52	PCR	Petru Groza
1952–55	PCR	Gheorghe Gheorghiu-Dej
1955–61	PCR	Chivu Stoica
1961–74	PCR	Ion Gheorghe Maurer
1974–79	PCR	Manea Mănescu
1979–82	PCR	Ilie Verdeţ
1982–89	PCR	Constantin Dăscalescu

Prime Ministers

1989–91	FSN	Petre Roman
1991–92	n/p	Teodor Stolojan
1992–96	n/p, PDSR	Nicolae Văcăroiu
1996–98	PNTCD	Victor Ciorbea
1998–99	PNTCD	Radu Vasile
1999–2000	n/p	Mugur Isărescu
2000–04	PDSR/PSD	Adrian Năstase
2004–08	PNL	Călin Popescu-Tăriceanu
2008–12	PD-L	Emil Boc
2012	n/p	Mihai-Răzvan Ungureanu
2012–	PSD	Victor Ponta

RECENT ELECTIONS

Presidential elections were held in two rounds on 22 Nov. and 6 Dec. 2009. In the first round President Traian Băsescu of the Democratic Liberal Party (PD-L) received 32·4% of votes cast, Mircea Geoană of the Social Democratic Party (in alliance with the Conservative Party) (PSD-PC) 31·2%, Crin Antonescu of the National Liberal Party (PNL) 20·0% and Corneliu Vadim Tudor of the Greater Romania Party 5·6%. There were eight other candidates. In the second round run-off Băsescu retained the presidency with 50·3% of the vote against 49·7% for Geoană. However, the opposition contested the results and accused Băsescu of ballot-rigging.

In parliamentary elections held on 9 Dec. 2012 the Social Liberal Union (USL) alliance—including the Social Democratic Party and the National Liberal Party—took 273 seats (58·6% of the vote) in the lower house and 122 (60·1% of the vote) in the Senate, the Right Romania Alliance—including the Democratic Liberal Party—56 seats (16·5%) and 24 (16·7%), the People's Party–Dan Diaconescu 47 seats (14·0%) and 21 (14·6%), and the Democratic Union of Hungarians in Romania (UDMR) 18 seats (5·2%) and 9 (5·3%). Turnout was 41·7%.

European Parliament

Romania has 33 (35 in 2007) representatives. In June 2009 elections, turnout was 27·7% (29·5% in 2007). The PSD-PC won 11 seats with 31·1% of the vote (political affiliation in European Parliament: Progressive Alliance of Socialists and Democrats), the PD-L 10 seats with 29·7% (European People's Party), the PNL 5 seats with 14·5% (Alliance of Liberals and Democrats for Europe), the UDMR 3 seats with 8·9% (European People's Party), Greater Romania Party 3 seats with 8·7% (non-attached) and independents 1 seat with 4·2% (European People's Party).

CURRENT GOVERNMENT

President: Traian Băsescu; b. 1951 (Democratic Liberal Party; since 20 Dec. 2004, but suspended from 20 April–23 May 2007; re-elected 6 Dec. 2009 and suspended again from 10 July–28 Aug. 2012).

In March 2013 the government comprised:

Prime Minister: Victor Ponta; b. 1972 (Social Democratic Party; sworn in 7 May 2012).

Deputy Prime Minister and Minister of Regional Development and Public Administration: Liviu Nicolae Dragnea. *Deputy Prime Minister and Minister of Public Finance:* Daniel Chiţoiu. *Deputy Prime Minister:* Gabriel Oprea.

Minister of Agriculture and Rural Development: Daniel Constantin. *Culture and Cults:* Daniel Constantin Barbu. *Economy:* Varujan Vosganian. *Education:* Remus Pricopie. *Environment and Climate Change:* Rovana Plumb. *European Funds:* Eugen Teodorovici. *Foreign Affairs:* Titus Corlăţean. *Health:* Eugen Gheorghe Nicolăescu. *Information Society:* Dan Nica. *Internal Affairs:* Radu Stroe. *Justice:* Mona-Maria Pivniceru. *Labour, Family, Social Protection and the Elderly:* Mariana Câmpeanu. *Transport:* Relu Fenechiu. *National Defence:* Mircea Duşa. *Water, Forests and Fisheries:* Lucia Ana Varga. *Youth and Sports:* Nicolae Bănicioiu.

Government Website: http://www.gov.ro

CURRENT LEADERS

Traian Băsescu

Position
President

Introduction
Traian Băsescu, a former ship's captain and mayor of Bucharest, successfully fought the country's 2004 presidential elections on an anti-corruption platform. He took over from Ion Iliescu, who had served as president for much of the post-communist period. Băsescu was suspended from office in April 2007 but resumed his role a month later after a national referendum supported his reinstatement. He was narrowly re-elected in Dec. 2009.

Early Life
Băsescu was born in the village of Basarabi near the Romanian port of Constanţa on 4 Nov. 1951. He studied at the Marine Institute in Constanţa, graduating in 1976. He then joined the merchant navy, controlled in Romania's communist era by NAVROM, and worked his way through the ranks, becoming a captain in 1981. He went on to captain some of the country's largest merchant ships and was promoted to Admiral of Romania's merchant fleet in the mid-1980s. In 1987 Băsescu travelled to Antwerp, Belgium, to work as head of the NAVROM Agency. Two years later he returned to Bucharest and entered the political scene as general director of the State Inspectorate of Civic Navigation in the ministry of transport, in what turned out to be the final months of Nicolae Ceauşescu's 24-year grip on power. After the dramatic collapse of Ceauşescu's regime in Dec. 1989, Băsescu was promoted to deputy minister of transport. He became minister of transport in 1991 in the government dominated by the National Salvation Front (FSN), which had

received mass support in the first post-communist elections on 20 May 1990.

Following a split in the FSN in 1992, Băsescu joined Petre Roman in the newly-established centre-left Democratic Party and, in 1996, he co-ordinated Roman's unsuccessful presidential campaign. Băsescu was re-elected as a Democratic Party MP in 1996 and served as minister of transport until 2000, when he stood as the Democratic Party candidate in the Bucharest mayoral election. He won and began co-ordinating the regeneration of large areas of the city, gaining praise for his direct approach to addressing problems—from cracking down on the notorious packs of stray dogs to improving traffic flow and municipal central heating systems.

Following disagreements with Petre Roman, Băsescu replaced him as leader of the (opposition) Democratic Party in 2001. Two years later, in Sept. 2003, Băsescu became a co-chairman of the centre-right Justice and Truth Alliance (DA), forged between his Democratic Party and the National Liberal Party (PNL). Băsescu's energetic rule as mayor of Bucharest proved popular and he was re-elected to the post in June 2004. Three months later he decided to contest the 2004 presidential election as the DA candidate. Campaigning on an anti-corruption and pro-Western platform, he defeated the PSD candidate in the second round run-off in Dec.

Career in Office

Băsescu's first task as president was the formation of a new DA government, which became possible when the small Humanist Party (since renamed the Conservative Party) pledged its support, in addition to the backing of the ethnic Hungarian Democratic Federation of Romania. Băsescu appointed the PNL leader and former minister of the economy Călin Popescu-Tăriceanu to the post of prime minister. In his inaugural address Băsescu said that fighting corruption would remain his priority and that he intended to steer Romania towards membership of the European Union. This was achieved on 1 Jan. 2007 as Romania and Bulgaria became the EU's 26th and 27th member states. He also stressed the need to strengthen strategic partnerships with the USA and the UK, as well as to improve relations with Russia and the former Soviet states.

Although allied within the ruling DA, Băsescu's relations with his then prime minister, Călin Popescu-Tăriceanu, deteriorated to the point where, in April 2007, Popescu-Tăriceanu won a vote in parliament to suspend Băsescu for 'grave infringements of the constitution' and the Democratic Party was excluded from the government. Former prime minister Nicolae Văcăroiu was appointed interim president. However, Băsescu returned to office on 23 May 2007 after he was backed by 74·5% of voters in a referendum on his leadership. He then promised to campaign for electoral reform to increase MPs' accountability to the electorate and for a law that could be used to remove former senior officials of the Ceaușescu regime from their offices. In Dec. 2007 he called on Popescu-Tăriceanu to dismiss his justice minister because of corruption allegations. Also in Dec. 2007 the Democratic Party (PD) merged with the Liberal Democratic Party (PLD) to form the Democratic Liberal Party (PD-L). Concerns over high-level corruption also prompted the EU in Feb. 2008 to threaten sanctions against Romania if serious failings were not corrected. In April 2008 Băsescu hosted a summit meeting of NATO leaders in Bucharest.

Following inconclusive parliamentary elections in Nov. 2008, Băsescu nominated Theodor Stolojan, an economist and former prime minister, to head a new government. Stolojan soon withdrew his acceptance and Băsescu then asked Emil Boc, the leader of the Democratic Liberal Party, to form the new administration, which was approved by parliament in Dec. 2008. However, in Oct. 2009 this government lost a confidence vote, although Boc carried on in a caretaker capacity until the pending presidential elections as Băsescu's subsequent nominees for prime minister-designate were not acceptable to parliament.

In 2009 Băsescu stood for presidential re-election. Following a close first round of voting, a run-off between Băsescu and Geoană of the PSD-PC alliance was held on 6 Dec. Băsescu's narrow win with 50·3% of the vote was challenged by his opponent who claimed electoral fraud. Băsescu then turned again to Boc to form a new government, which was approved by parliament but which laboured with economic recession and austerity. With his government having become increasingly unpopular, Boc resigned in Feb. 2012. Mihai-Răzvan Ungureanu succeeded him as prime minister but his government fell after only two months when it received a vote of no confidence in parliament. Victor Ponta of the Social Liberal Union then established a new administration to govern until elections later in the year.

In July 2012 Băsescu faced a further challenge to his authority, this time from his new prime minister, and was suspended from office following a parliamentary vote pending efforts to impeach him. However, he survived an impeachment referendum as the majority vote to remove him failed to meet the required minimum turnout of the electorate and he resumed office in Aug. In the Dec. 2012 parliamentary elections Ponta's government won a clear majority against parties backed by Băsescu, after which the president and prime minister signed an agreement guaranteeing their institutional co-operation and respect for the constitution.

Victor Ponta

Position
Prime Minister

Introduction
Victor Ponta became prime minister in May 2012 following the collapse of the short-lived administration led by Mihai-Răzvan Ungureanu. Ponta's Social Liberal Union administration (USL, including the Social Democratic Party and the National Liberal Party) secured a parliamentary mandate following elections in Dec. 2012.

Early Life
Victor Ponta was born on 20 Sept. 1972 in Bucharest. He graduated in law from the University of Bucharest in 1995 and worked as a prosecutor specializing in corruption cases. Employed in the Supreme Court of Justice from 1998–2001, he was also an assistant professor in public law at the Romanian-American University (RAU) in Bucharest. He received his masters degree from Italy's University of Catania in 2000 and graduated from the National Defence College in 2002. That same year he began lecturing at the RAU, receiving his PhD from the University of Bucharest in 2003 and becoming an associate professor at RAU in 2007.

A prominent member of the youth wing of the Social Democratic Party (PSD), in 2002 Ponta joined the party's national council and was appointed chairman of Social Democratic Youth, a post he held until 2006. In 2004 he entered the chamber of deputies, securing re-election four years later. In Dec. 2006 he was elected PSD vice-president and became party president in 2010.

In 2008 Ponta was named minister responsible for liaison with parliament in the coalition of Emil Boc of the Democratic Liberal Party. Ponta was instrumental in controversial legislation that widened the criminal definition of self-defence and banned therapeutic abortions after the 24th week of pregnancy. He and his PSD colleagues resigned from the cabinet in Oct. 2009 in protest at the sacking of the interior minister, Dan Nica.

The government collapsed, but Boc formed a new coalition that brought in austerity measures to meet IMF conditions. In 2011 Ponta led the PSD into an alliance with the National Liberal Party, forming the USL. Amid growing popular discontent, Boc resigned from office in Feb. 2012 and Mihai-Răzvan Ungureanu replaced him. However, that government fell two months later when it lost a confidence vote. President Băsescu then invited Ponta to form a new administration.

Career in Office

On 1 May 2012 Ponta announced his cabinet, with the USL's constituent parties sharing portfolios equally. His initial mandate lasted only until elections scheduled for late 2012. Ponta pledged to create jobs and 'right the social injustices' caused by his predecessors' austerity measures. As his relationship with Băsescu quickly deteriorated, Ponta launched an unsuccessful attempt to impeach the president in July. Their mutual hostility persisted until Ponta's USL won a comfortable victory in parliamentary elections in Dec., after which they signed an agreement to set aside their differences to end the political turmoil.

DEFENCE

Compulsory national military service was abolished in 2006.

In 2008 military expenditure totalled US$3,005m. (US$135 per capita), representing 1·5% of GDP.

Army

Strength (2007) 42,200. There is a joint reserve of 45,000. The Ministry of the Interior operates a paramilitary Border Guard (22,900 strong in 2007) and a Gendarmerie (around 57,000).

Navy

The fleet includes three frigates and four corvettes. There is also a naval infantry force.

The headquarters of the Navy is at Mangalia with the main base at Constanţa. The Danube flotilla is based at Brăila. Personnel in 2007 totalled 8,067.

Air Force

The Air Force numbered some 10,500 in 2007, with 74 combat-capable aircraft (MiG-21s).

INTERNATIONAL RELATIONS

At the European Union's Helsinki Summit in Dec. 1999 Romania, along with five other countries, was invited to begin full negotiations for membership in Feb. 2000. Romania joined the EU on 1 Jan. 2007. Romania became a member of NATO on 29 March 2004.

ECONOMY

Agriculture accounted for 10·5% of GDP in 2006, industry 37·9% and services 51·5%. The percentage of Romania's GDP coming from agriculture is the highest of any EU member country.

Overview

Economic growth averaged 6·2% annually in the period 2002–06. Having signed an accession treaty in 2005, Romania joined the EU in Jan. 2007, with the aim of adopting the euro by 2014 (although the target is now 2015).

The economy was on the brink of collapse in the late 1980s as the Soviet Union disintegrated. When Romania made the transition to a market economy in 1990 it failed to introduce the structural reforms necessary for macroeconomic stability. Negative economic growth followed and poverty doubled. In 2000 a reform programmed was belatedly initiated, including tight monetary and fiscal policies leading to greater stability and a fall in inflation.

After seven years of strong growth, economic activity declined in the last quarter of 2008 and continued to fall in 2009, with real GDP contracting by 7·6% in the first half of the year. In March 2009 Romania secured a €20bn. (US$27bn.) loan from the IMF, European Union and other lenders to stabilize the economy in the wake of the global financial crisis. Growth returned in 2011 and the trade balance improved thanks to strong exports, especially in the machinery and automotive sectors.

However, the recovery remains fragile. The economy is overly dependent on the export and banking sectors, the latter of which is largely foreign-owned. Reform of state-owned enterprises remains an important goal in securing long-term growth.

Currency

The monetary unit has since 1 July 2005 been the *new leu*, pl. *new lei* (RON) notionally of 100 *bani*, which replaced the *leu* (ROL) at a rate of one new leu = 10,000 lei. Foreign exchange reserves were US$41,571m. and gold reserves 3·34m. troy oz in Sept. 2009. Inflation rates (based on IMF statistics):

2002	2003	2004	2005	2006	2007	2008	2009	2010	2011
22·5%	15·4%	11·9%	9·0%	6·6%	4·8%	7·8%	5·6%	6·1%	5·8%

Total money supply was 64,201m. new lei in Aug. 2009.

Budget

Central government revenue and expenditure (in 1m. new lei) for calendar years:

	2006	2007	2008
Revenue	86,225	106,905	158,032
Expenditure	87,189	108,621	107,195

Romania's budget deficit in 2011 was 5·2% of GDP (2010, 6·8%; 2009, 9·0%). The required target set by the EU is a budget deficit of no more than 3%.

VAT is 24% (reduced rates, 9% and 5%). There is a flat income tax rate of 16%.

Performance

Real GDP growth rates (based on IMF statistics):

2004	2005	2006	2007	2008	2009	2010	2011
8·5%	4·2%	7·9%	6·3%	7·3%	−6·6%	−1·6%	2·5%

Total GDP in 2011 was US$179·8bn.

Banking and Finance

The National Bank of Romania (founded 1880; nationalized 1946) is the central bank and bank of issue under the Minister of Finance. Its *Governor* is Dr Mugur Isărescu. In 2002 there were 31 banks, plus eight branches of foreign banks. Only one bank remains state-owned. The largest bank is Romanian Commercial Bank (Banca Comercială Română), with a market share of 20% and assets in 2008 of US$24·7bn.; Austria's Erste Bank AG had a 93·6% stake in Feb. 2013. The size of the government's share in the banking sector fell from over 80% in the mid-1990s to just over 40% in 2002.

Foreign debt was US$121,505m. in 2010, representing 76·4% of GNI.

A stock exchange reopened in Bucharest in 1995.

ENERGY AND NATURAL RESOURCES

In 2008, 20·4% of energy consumption came from renewables (wind power, solar power, hydro-electric power, tidal power, geothermal energy and biomass), compared to the European Union average of 10·3%. A target of 24% has been set by the EU for 2020.

Environment

Romania's carbon dioxide emissions from the consumption and flaring of fossil fuels were the equivalent of 4·6 tonnes per capita in 2008.

Electricity

In 2007 installed capacity was approximately 19·9m. kW; output in 2007 was 61·67bn. kWh (26% hydro-electric and 12% nuclear). Consumption per capita in 2007 was 2,766 kWh. A nuclear power plant at Cernavodă began working in 1996. A second reactor became operational there in 2007.

Oil and Gas

Oil production in 2008 was 4·7m. tonnes, but with annual consumption of more than twice as much a large amount has to be imported. There were 0·5bn. bbls of proven oil reserves in 2008.

Romania was the first country to start oil exploration, and in the late 1850s was the world's leading oil producer, with an output of 200 tonnes a year. Natural gas production in 2008 totalled 11·5bn. cu. metres with 630bn. cu. metres in proven reserves.

The oil company Petrom, Romania's largest company, was privatized in 2004 when the government sold a 51% stake to the Austrian oil and gas group ÖMV.

Minerals
The principal minerals are oil and natural gas, salt, lignite, iron and copper ores, bauxite, chromium, manganese and uranium. Output, 2005 (in 1,000 tonnes): lignite, 31,070; salt, 2,420; iron ore, 265; zinc, 14.

Agriculture
Romania has the biggest agricultural area in eastern Europe after Poland. In 2002, 13·7% of the economically active population were engaged in agriculture. There were 13·94m. ha. of agricultural land in 2002 including 8·96m. ha. of arable land and 4·63m. ha. of permanent pasture. There were 3,081,000 ha. of irrigated land in 2001. There were 164,221 tractors and 27,051 harvester-threshers in 2001.

Production (2003, in 1,000 tonnes): maize, 9,577; potatoes, 3,947; wheat, 2,479; sunflower seeds, 1,506; grapes, 1,078; cabbage, 1,019; melons and watermelons, 1,000; tomatoes, 819; sugar beets, 764; wine, 546; barley, 541.

Livestock, 2002 (in 1,000): pigs, 8,229; sheep, 7,221; cattle, 2,865; horses, 909; goats, 737; poultry, 82,000.

A law of Feb. 1991 provided for the restitution of collectivized land to its former owners or their heirs up to a limit of 10 ha. Land could be resold, but there was a limit of 100 ha. on total holdings. In 2000 a law was passed allowing the restitution of state farm land for the first time (up to 50 ha. of farmland and 10 ha. of forest land per family).

Forestry
Total forest area was 6·57m. ha. in 2010 (29% of the land area). Timber production in 2007 was 15·34m. cu. metres.

Fisheries
The catch in 2010 totalled 2,688 tonnes (6,053 tonnes in 2005 and 216,938 tonnes in 1988), of which 2,457 tonnes were from inland waters.

INDUSTRY
In 2001 industry accounted for 37·0% of GDP. Industrial output grew by 7·5% in 2001.

Output of main products (in 1,000 tonnes): cement (2006 estimate), 8,253; rolled steel (2007), 5,589; crude steel (2008), 5,035; distillate fuel oil (2007), 4,660; petrol (2007), 3,799; pig iron (2008), 2,945; lime (2006 estimate), 1,942; fertilizers (2000), 1,931; wheat flour (2001), 1,597; residual fuel oil (2007), 1,186; ammonia (2001), 1,155; steel tubes (2001), 665; caustic soda (2001), 661; soda ash (2001), 451; paper and paperboard (2002), 370.

Labour
The labour force in 2006 totalled 10·04m.; the employed population was 9·31m. In the civilian labour force 29·7% worked in agriculture and 26·7% in manufacturing and construction. In 2006, 46% of the employed workforce were women. The standard retirement age is 65 years for men and 60 for women. A minimum monthly wage was set in 1993; it is 670 new lei for full-time adult employees from 1 Jan. 2011. The average gross monthly wage was 1,845 new lei in 2009. Unemployment was 7·2% in Jan. 2012 (7·3% in Jan. 2011).

INTERNATIONAL TRADE
Imports and Exports
Imports in 2007 were valued at US$69,946m. (US$24,003m. in 2003) and exports US$40,247m. (US$17,618m. in 2003). Principal imports are mineral fuels, machinery and transport equipment, and textiles; main export commodities are textiles, mineral products and chemicals.

Romania's main import sources in 2007 were: Germany (17·2%); Italy (12·8%); Hungary (6·9%); Russia (6·3%); France (6·3%). In 2007 Romania's main export markets were: Italy (17·2%); Germany (17·0%); France (7·7%); Turkey (7·0%).

COMMUNICATIONS
Roads
There were 198,817 km of roads in 2004, including 228 km of motorways, 14,809 km of highways, main and national roads and 36,010 km of secondary roads. Passenger cars in 2005 numbered 3,363,800 (156 per 1,000 inhabitants). In 2007 there were 2,712 fatalities as a result of road accidents.

Rail
Length of standard-gauge route in 2005 was 10,882 km, of which 3,929 km were electrified; there were 425 km of narrow-gauge lines and 57 km of 1,524 mm gauge. Freight carried in 2005, 55·3m. tonnes; passengers, 98·6m. There is a metro (62·4 km) and tram/light rail network (338 km) in Bucharest, and tramways in 13 other cities.

Civil Aviation
Tarom (*Transporturi Aeriene Române*) is the 95·0% state-owned airline. In 2002 it provided domestic services and international flights to Amman, Amsterdam, Ancona, Athens, Beijing, Beirut, Berlin, Bologna, Brussels, Budapest, Cairo, Chişinău, Copenhagen, Damascus, Dubai, Düsseldorf, Frankfurt, İstanbul, Larnaca, London, Luxembourg, Madrid, Milan, Moscow, Munich, New York, Paris, Prague, Rome, Sofia, Stuttgart, Tel Aviv, Thessaloniki, Treviso, Verona, Vienna, Warsaw and Zürich. Other Romanian airlines which operated international flights in 2007 were Blue Air, Carpatair, Chris Air, Ion Ţiriac Air, Jet Tran Air and Romavia. In 2003 scheduled airline traffic of Romanian-based carriers flew 26m. km, carrying 1,255,000 passengers (1,034,000 on international flights).

Bucharest's airports are at Băneasa (mainly domestic flights) and Otopeni (international flights). Constanţa, Cluj-Napoca, Oradea, Arad, Sibiu and Timişoara also have some international flights. Otopeni handled 1,981,508 passengers in 2001 and 11,410 tonnes of freight; Timişoara handled 161,000 passengers in 2001 and Băneasa 74,000.

Shipping
In Jan. 2009 there were 35 ships of 300 GT or over registered, totalling 150,000 GT. The Romanian-controlled fleet comprised 57 vessels of 1,000 GT or over in Jan. 2009, of which 11 were under the Romanian flag and 46 under foreign flags. The main ports are Constanţa and Constanţa South Agigea on the Black Sea and Galaţi, Brăila and Tulcea on the Danube. In 2009 the length of navigable inland waterways was around 1,730 km including 1,075 km on the Danube River.

Telecommunications
In 2008 there were 5,036,000 main (fixed) telephone lines. In the same year active mobile phone subscribers numbered 24,467,000 (1,145·4 per 1,000 persons). The telecommunications sector was fully liberalized on 1 Jan. 2003, ending the monopoly of the Greek-controlled operator Romtelecom. OTE, the major shareholder, increased its stake in Romtelecom to 54% in Jan. 2003, with the government retaining 46% of shares. There were 6·2m. internet users in 2008. In March 2012 there were 4·2m. Facebook users.

SOCIAL INSTITUTIONS
Justice
The legal system is based on the Napoleonic code and the judiciary is constitutionally independent. The High Court of

Cassation and Justice is the highest judicial authority, with its judges appointed by the president after consultation with the Superior Council of the Magistracy (an elected professional body). There is also a Constitutional Court.

Day-to-day hearings are administered through a system of local courts and 40 county courts (and the Bucharest Municipal Court), whose judgements may be challenged in any one of 15 courts of appeal. In 2006 there were 3,799 judges.

As a condition of EU accession and World Bank financing, Romania has been subject to a Judicial Reform Project aimed at increasing the efficiency of the court system, improving transparency and reducing corruption. This has included the implementation of four revised codes (the criminal code, civil code, criminal procedures code and civil procedures code).

The death penalty was abolished in Jan. 1990 and is forbidden by the 1991 constitution. The population in penal institutions in Sept. 2006 was 35,429 (164 per 100,000 of national population).

Education

Education is free and compulsory from the age of six. There is compulsory school attendance for ten years. Primary education comprises four years of study, secondary education comprises lower secondary education (organized in two cycles: grades 5th–8th in elementary schools and grades 9th–10th in high schools or vocational schools) and upper secondary education includes further education in high schools. Further secondary education is also available at *lycées*, professional schools or advanced technical schools.

In 2007 there were 648,862 children and 36,555 teaching staff in pre-primary schools; 917,829 pupils and 55,487 teaching staff in primary schools; and 1,954,077 pupils and 153,805 teaching staff in secondary schools. In 2002–03 primary and secondary education in Hungarian was given to 106,515 pupils, in German to 10,019 pupils and in other national minority languages to 1,536 pupils.

In 2007 there were 928,175 students in tertiary education and 30,583 academic staff. In 2002–03 there were 125 higher education institutions with 742 faculties, 30,000 teaching staff and 596,297 students (545,405 for long-term studies and 50,892 for short-term studies). The distribution of pupils and subjects studied was as follows: pedagogy, 30·3%; economics, 26·5%; technical subjects, 25·6%; law, 10·6%; medicine and pharmacy, 5·4%; arts, 1·5%.

Adult literacy rate in 2009 was estimated at 97·7% (male 98·3%; female 97·0%).

In 2005 public expenditure on education came to 3·6% of GNI and 14·3% of total government spending.

Health

In 2006 there were 419 public hospitals with 141,225 beds; there were 45,815 physicians and 4,360 dentists in the public sector in 2006.

Welfare

In Dec. 2004 pensioners comprised 3,050,500 old age and retirement, 1,441,800 retired farmers, 798,200 disability, 639,500 survivor allowance and 418,000 social assistance. These drew average monthly pensions ranging from 792,698 lei to 3,504,205 lei. The social security spending in 2002 was 10·4% of GDP.

RELIGION

The government officially recognizes 17 religions (which receive various forms of state support); the predominant one is the Romanian Orthodox Church. It is autocephalous, but retains dogmatic unity with the Eastern Orthodox Church. Its *Patriarch* is Daniel (enthroned 30 Sept. 2007). It is made up of five metropolitan sees, with ten archdioceses and 13 dioceses, 158 deaneries and 10,987 parishes. There were 12,320 priests and deacons in 2003. In Feb. 2013 there was one cardinal.

Religious affiliation at the 2002 census included: Romanian Orthodox, 18,817,975 (about 87% of the population); Roman Catholic, 1,026,429; Protestant Reformed Church, 701,077; Pentecostal, 324,462; Greek Catholics or Uniates, 191,556; Baptist, 126,639; Seventh Day Adventist, 93,670; Muslim, 67,257.

CULTURE

World Heritage Sites

Romania has seven sites on the UNESCO World Heritage List: the Danube Delta (inscribed on the list in 1991); the Villages with Fortified Churches in Transylvania (1993 and 1999); the Monastery of Horezu (1993); the Churches of Moldavia (1993); the Historic Centre of Sighişoara (1999); the Dacian Fortresses of the Orastie Mountains (1999); and the Wooden Churches of Maramureş (1999).

Press

In 2008 there were 80 daily papers (75 paid-for and five free)with a combined circulation of 1,634,000. In 2007 there were 2,320 periodicals, including 119 periodicals in minority languages. 15,566 book titles were published in 2007.

Tourism

In 2009, 1,275,600 non-resident tourists stayed in holiday accommodation (down from 1,465,900 in 2008) including: 181,100 from Germany; 141,600 from Italy; 100,300 from France; 76,900 from Hungary.

Festivals

The George Enescu Festival of classical music is held every two years in Aug.–Sept. with concerts across Bucharest, Iaşi and Sibiu. The Peninsula/Félsziget rock festival takes place annually in July or Aug. in Târgu Mureş in Transylvania. Other notable events are the EUROPAfest of music in Bucharest (May), the International Romani Art Festival in Timişoara (July), Garana Jazz Festival at Poiana Lupului (July) and the Romanian National Theatre Festival in Bucharest (Oct.–Nov.).

DIPLOMATIC REPRESENTATIVES

Of Romania in the United Kingdom (Arundel House, 4 Palace Green, London, W8 4QD)
Ambassador: Dr Ion Jinga.

Of the United Kingdom in Romania (24 Strada Jules Michelet, 010463 Bucharest)
Ambassador: Martin Harris, OBE.

Of Romania in the USA (1607 23rd St., NW, Washington, D.C., 20008)
Ambassador: Adrian Cosmin Vieriţa.

Of the USA in Romania (7–9 Strada Tudor Arghezi, Bucharest)
Ambassador: Vacant.
Chargé d'Affaires a.i.: Duane C. Butcher.

Of Romania to the United Nations
Ambassador: Simona-Mirela Miculescu.

Of Romania to the European Union
Permanent Representative: Mihnea Motoc.

FURTHER READING

Comisia Nationala pentru Statistica. *Anuarul Statistic al României/ Romanian Statistical Yearbook.* Annual.—*Revista de Statistica.* Monthly

Carey, Henry F., *Romania since 1989: Politics, Economics and Society.* 2004
Gallagher, T., *Romania after Ceauşescu; the Politics of Intolerance.* 1995
Phinnemore, David, (ed.) *The EU and Romania: Accession and Beyond.* 2006

National Statistical Office: Comisia Nationala pentru Statistica, 16 Libertatii Ave., sector 5, Bucharest.
Website: http://www.insse.ro

RUSSIA

1. KARACHAI-CHERKESSIA	5. CHUVASHIA	10. INGUSHETIA
2. ADYGEYA	6. TATARSTAN	11. NORTH OSSETIA
3. KALMYKIA	7. UDMURTIA	12. KABARDINO-BALKARIA
4. MORDOVIA	8. MARI-EL	13. DAGESTAN
	9. CHECHNYA	

A. ESTONIA
B. LATVIA
C. BELARUS
D. TURKEY

Rossiiskaya Federatsiya
(Russian Federation)

Capital: Moscow
Population projection, 2015: 142·23m.
GNI per capita, 2011: (PPP$) 14,561
HDI/world rank: 0·755/66
Internet domain extension: .ru

KEY HISTORICAL EVENTS

Archaeological evidence points to the influence of Arabic and Turkic cultures prior to the 4th century AD. Avar, Goth, Hun and Magyar invasions punctuated the development of the East Slavs over the next five centuries, while trade with Germanic, Scandinavian and Middle Eastern regions began in the 8th century.

In 882 the Varangian prince Oleg of Novgorod took Kyiv and made it the capital of Kievan Rus, the first unified state of the East Slavs, uniting Finnish and Slavic tribes. During the 10th century, trade was extended between the Baltic and Black Seas, forming Kyiv's main economy. The Varangians, led by Rurik of Jutland, led attacks on Baghdad and Constantinople, subsequently establishing a trade link with the latter.

During the 13th century the area was invaded from the west by Teutonic Knights, Lithuanians and Swedes, and from the south by Mongol and Tatar tribes. In 1223 Genghis Khan's grandson, Batu Khan, conquered Kievan Rus. Despite the ruthless reign of the Mongols, trade flourished during the period and many cities were reinvigorated. The Mongols and Tatars created an ascendency known as the 'Golden Horde' around most of Western Russia and

Central Asia and made Itil (near modern Astrakhan) the capital. Its dominance lasted until the 15th century when internal struggles finally forced the break-up of the empire.

Co-operation between Moscow's leader Ivan and the Mongol Öz Beg (ruled 1312–41), in addition to geographical advantages and natural resources, allowed Moscow to develop and prosper. The city was first consolidated under the Muscovite Grand Duke Ivan III (ruled 1462–1505), who adopted the Roman title of tsar and Byzantine ritual after marrying into Byzantine royalty. Ivan annexed the East Slavic regions, as well as Belarus and the Ukraine, conquered Novgorod in 1478 and opened up contacts with Western Europe.

The empire was strengthened and further expanded by his son Vasily III and reformed by Vasily's successor Ivan IV, a sickly and volatile ruler known as Ivan the Terrible (or 'Awesome', *Grozny*) who came to the throne at the age of 16 in 1547. Ivan's divisive and suppressive administration, *oprichnina*, led a reign of terror from 1566–72 in which thousands were executed (although it is believed that initially the Russian nobles, or boyars, had strong control over the throne and its direction, including local government reforms, a new law code and restrictions of hereditary rights). Ivan bolstered the military and led campaigns against the khanates of Kazan (1552), Astrakhan (1556) and the Crimea, extending Russia's territory towards Siberia and down to the Caspian Sea. But the costly war with Livonia (1558–82) drained Russia's resources. Ivan murdered his son in 1581 leaving a hereditary gap and a struggle for succession.

Russia was ruled nominally by Ivan's mentally subnormal brother Fyodor I—in actuality by Fyodor's brother-in-law Boris Godunov, who succeeded Fyodor in 1598. But in 1601 False

Dmitri claimed to be Ivan IV's son (Dmitri had died in 1591) and challenged Boris for the throne. With the backing of the boyars, the Cossacks and the Polish nobility, Dmitri succeeded Boris as tsar on the latter's death in 1605. There followed a chaotic period of instability as differing sides fought for control of the realm. The following year Dmitri was assassinated and the boyars crowned the rebel leader Vasily Shuysky in return for privileges. But soon a subgroup of boyars led by the Romanovs gave support to a second False Dmitri in 1608, establishing a shadow government just outside Moscow. Shuysky turned to Sweden for help, bargaining away territory and triggering Poland's invasion of Muscovy and the siege of Smolensk (1609). Both governments collapsed and a coalition government was formed. A peace treaty signed with Sweden in 1617 lost Russia Novgorod in exchange for Baltic control, while an armistice with Poland began the following year.

Romanovs

With the Polish occupiers ejected from Moscow, Mikhail Fyodorovich Romanov, the first of a dynasty that would rule until 1917, became tsar of a country ruined by war and with regions occupied by Swedish, Polish or rebel forces. But by avoiding involvement in the Thirty Years' War, in which Sweden and Poland were embroiled, he managed to restore some stability and to strengthen Russia's holdings in the southern regions. His son Aleksey inherited the throne as a child. Unpopular measures implemented by Aleksey's adviser, Boris Ivanovich Morozov, including a crippling salt tax, led to a riot in 1648 and rebellion in Novgorod and Pskov. Eastern Ukraine was annexed, while the support of a Cossack rebellion against Polish rule in the Ukraine degenerated into a costly war with Sweden and Poland over Ukrainian, Baltic and Belorussian territory. Russia consequently lost the Baltic coast to Sweden in 1661 and later Belarus and parts of the Ukraine to Poland. Sophia succeeded Aleksey to the disputed throne in 1682, followed seven years later by her half brother Peter the Great.

The reign of Peter I (1689–1725) signalled a new era for Russia that broke so far with Muscovy tradition as to be seen as the birth of modern Russia. The empire was expanded and strengthened and there was increased trade with Western Europe. His modest upbringing and travels to the West gave Peter a novel pro-European stance. The capital was transferred from Moscow to the newly built St Petersburg (1712), as part of a Europeanization programme. Peter introduced radical structural changes to the Russian body politic, converting it into the Western European mould. The tsardom of Muscovy became the Empire of All Russias and Peter became head of state as opposed to ruling patriarch. Administrative reforms divided Russia into eight main provinces, put the church under state control and introduced compulsory secular education for the nobility, although the rights of the peasantry were abolished and they were forced into serfdom. Peter expanded industry, created the navy, introduced army conscription and strengthened the southern border against the Crimean Tatars. He formed an alliance with Denmark, Poland and Saxony against Sweden, resulting in the Great Northern War (1700–21), which ended with Russia claiming Livonia in the Treaty of Nystad (1721). The expanded empire made Russia the leading Baltic power.

Catherine the Great

Despite Peter's rejection of hereditary rule in favour of appointing a successor, his choice was never named before his sudden death. The rest of the 18th century was marked by disputed succession. After Peter's death, his widow Catherine I was declared empress, though Peter's collaborator Prince Menshikov ruled in her name. A supreme privy council was established to distribute power; Peter's grandson, Peter II, ruled briefly before dying of smallpox. He was succeeded by Peter I's niece, Anna, the duchess of Courland (1730), then by her niece Anna Leopoldovna, before Peter I's daughter, Elizabeth, came to power in 1741 in a bloodless

coup. During her 21-year reign, her father's reforms were consolidated and Western culture and literature flourished. She founded the University of Moscow and established the St Petersburg Academy of Arts. At the end of her reign Russia was involved in the Seven Years' War, occupying Berlin for a short time before Elizabeth's death in 1762. Russia's subsequent withdrawal saved Frederick the Great's Prussia from destruction.

Elizabeth's nephew, Peter III, proved an unpopular ruler. Childless, his politically ambitious wife, Catherine the Great, plotted to depose him, claiming the throne for herself soon after. Influenced by the Enlightenment, she attempted to implement legislative, educative and administrative reforms. But many of these, as well as the emancipation of the serfs, were blocked by the nobility. The imposed Russification of the Ukrainian, Polish and Baltic regions proved unpopular, while civil unrest led to the Pugachev Revolt (1773–75), in which peasants, Cossacks and workers rebelled against the aristocracy. Catherine's foreign policy was an aggressive expansion plan to the south and east to make Russia the leading European power at the expense of the Turks and Tatars. She forged a path through to the Mediterranean Sea to maximize maritime trade routes and developed close relations with Prussia and Austria with whom Poland was shared. But despite two wars with Turkey she failed to take Constantinople, as much an emotional as a political prize.

After Catherine's death in 1796, her son Paul's tyrannical rule led to his murder in 1801. His son Alexander I adopted more liberal policies in administration, science and education. War with France in 1805 led to a crushing defeat at Austerlitz, but when Napoleon invaded Russia in 1812 his army fell victim to the Russian winter. Alexander's death in 1825 provoked instability and uprisings which were quashed by military force. Russia was defeated by Britain, France and Turkey in the Crimean War (1853–56). Alexander II (ruled 1855–81), who followed Nicholas I (ruled 1825–55), implemented reforms, the most important of which was the partial emancipation of the serfs in 1861. Major judicial reform followed three years later and universal military service in 1874. But towards the end of his reign Alexander's increasingly conservative measures exacerbated the revolutionary mood of socialist-influenced university students and the peasantry. He was assassinated in 1881 and was succeeded by Alexander III (1881–94). Labour reforms introduced by Alexander III were harsh and restrictive and the peasants' lot failed to improve. The government's neglect of agricultural policy resulted in crop failure and widespread famine in 1891.

The Russian empire had expanded to the far reaches of Asia, to Afghanistan and into Central Europe. By the end of Alexander III's reign, only half the population spoke Russian or were members of the Orthodox Church.

Revolution

Nicholas II's reign (1894–1917) marked the end of Tsarist Russia. Like his father, he did little to improve social conditions for the masses, concentrating instead on military power. Industrial growth produced an unskilled urban working class whose living conditions fuelled revolutionary feeling. Socialism and Liberalism were also taking hold of the educated middle classes—doctors, teachers and engineers—as well as disaffected civil servants. In 1904 Nicholas embarked on and lost an unpopular war with Japan. The middle classes campaigned for a legislative assembly. In Jan. 1905 the priest, Georgy Gapon, led a protest of factory workers to St Petersburg's Winter Palace. Tsarist troops opened fire on the crowds killing over 100 people. Public outrage to 'Bloody Sunday' soon spread throughout the country. A general strike, paralysing most of Russia, led to violence between monarchists and insurgents well into 1907, while factions in the armed forces rebelled. Yielding to the pressure of the 1905 revolution, the tsar permitted the establishment of the first *Duma* (parliament), which convened in St Petersburg in 1906. But it lasted only 70 days. Violence continued into 1907.

In 1912 the two strands of the Social Democratic Workers' Party—the Bolsheviks (or majority) led by Vladimir Ilyich Ulianov (Lenin), and the Mensheviks (or minority)—split, the Bolsheviks pursuing revolution, the Mensheviks evolutionary change. The outbreak of the First World War in 1914 temporarily unified Russians in the war effort. The tsar took command of the armed forces in 1915, leaving an authoritarian vacuum that allowed the tsarina and the influential adviser Grigori Rasputin to implement various unpopular ministerial changes. Rasputin was assassinated by disgruntled nobles in 1916. Depleting military resources and social unrest caused by hardship forced the end of the tsar's reign. A succession of anti-tsar demonstrations culminated in a mass protest in St Petersburg. Soldiers deserted, allying themselves with the workers, a pattern repeated throughout the country. A provisional government comprising Menshevik and Bolshevik elements was established and Tsar Nicholas abdicated on 2 March 1917. The Royal Family was executed in July 1918.

Tension between moderate Mensheviks and radical Bolsheviks intensified and in Oct. 1917 the Bolsheviks led by Lenin, newly returned from exile, seized control. The new government headed by Lenin, the Council of People's Commissars, created the Soviet constitution the following year. Russia was declared the Soviet Republic of Workers, Soldiers and Peasants and the capital was moved back to Moscow. Russia eventually withdrew from the First World War in 1918 but its forced acceptance of the unfavourable Brest-Litovsk treaty led many to abandon the government. Between 1918–21 a civil war raged between the Bolshevik Red Army, led by Lenin's ally Leon Trotsky, and the White Army, formed by former imperial officers, Cossacks, anti-communists and anarchists. The government imposed 'war communism'—forced labour and expropriation of business and food supplies—to support its cause, and eventually overcame the White Army. Lenin instituted the New Economic Policy (NEP) in 1921 to replace War Communism, reintroducing a monetary system and private ownership of small-scale industry and agriculture. In 1922 the Union of Soviet Socialist Republics was established comprising Russia, the Ukraine, Belarus and Transcaucasia. The Turkmen and Uzbek republics were added two years later, and the Tadzhik republic joined in 1929.

Stalin
On Lenin's death in 1924, Joseph Stalin (Ioseb Dzhugashvili) became general secretary of the Communist Party. Stalin rejected the 'state capitalism' of the NEP, which had failed to provide enough food for the urban workforce. From 1928 Stalin pursued a programme of industrialization and from 1933 agricultural collectivization, which cost the lives of 10m. peasants through famine or persecution. Constructing a personality cult for Lenin and himself, Stalin reasserted his absolute authority in massive purges; in 1934 and 1937 the NKVD (political police) eliminated millions of political dissidents.

Despite a non-aggression pact signed with Germany in Aug. 1939, the USSR was forced into the Second World War (termed the Great Patriotic War) in 1941 when the Nazi's Plan Barbarossa targeted Kyiv, Moscow and Leningrad for invasion. Up to 20m. Soviet lives were lost, almost 1m. in the battle of Stalingrad alone (1942–43). Expansion before and during the war created 15 aligned republics. Transcaucasia was divided into Armenia, Georgia and Azerbaijan, Kazakh and Kirghiz Soviet Socialist Republics were formed, and, along with Latvia, Lithuania, Estonia and Moldavia, were incorporated into the USSR. Following the war, Stalin managed to gain Western acceptance of a Soviet sphere of influence in Eastern Europe. The Baltic States and large tracts of land from neighbouring countries were annexed, while puppet regimes established Poland, Czechoslovakia, East Germany, Hungary, Bulgaria and Romania as satellites of Moscow.

The blockade of West Berlin (1948–49) and the Soviet detonation of an atomic bomb in Aug. 1949 were major factors in the escalation of the Cold War, waged indirectly in the Korean War (1950–53). On Stalin's death, Nikita Khrushchev reversed many of Stalin's policies and condemned his predecessor. In reaction to the famine in his native Ukraine, he developed the vast wheatfields in Kazakhstan. Relaxing control in the Eastern Bloc allowed for some liberalization although the Hungarian Uprising and the Poznań Riots in Poland (both 1956) were brutally suppressed and the Berlin Wall built in 1961. Relations with the Soviet Union's great ideological ally, China, collapsed over differences in interpretation of Marxist doctrine and Chinese opposition to Khrushchev's attempts at détente with the West (which came to be known as 'peaceful co-existence'). The Cuban Missile Crisis of 1962 intensified hostilities with the West and led the world to the brink of nuclear war. Khrushchev's perceived failure in the crisis, coupled with food shortages, led to widespread discontent. He was forced out of office in a 1964 coup led by Leonid Brezhnev, who ruled until 1982.

Soviet Reform
By the 1970s, Russia's international status had reached its zenith. Along with the USA, it was perceived as one of two global superpowers, despite relative economic stagnation. But Brezhnev kept a tight grip on the Eastern Bloc, introducing his 'Brezhnev Doctrine' which permitted the Soviet Union to intervene in the Eastern Bloc countries if Communist rule was ever threatened. In Aug. 1968 the USSR invaded Czechoslovakia to suppress an increasingly liberal regime. Relations with the West were further strained when the Soviets invaded Afghanistan in 1979. By the end of his tenure Brezhnev's failing health mirrored the country's economic decline. The domestic price of Brezhnev's obsessive pursuit of prominence in the space race was the failure of the agricultural and consumer-goods sectors and the decline of living standards. From his death in 1982, the country was led by his aides Yuri Andropov, a short-lived reformer, then Konstantin Chernenko.

When the latter died in 1985, Mikhail Gorbachev became general secretary of the Communist Party. He launched *perestroika*, a policy of economic and structural reform. *Glasnost* ('openness') extended civil liberties, including freedom of the press, and led to official rejection of Stalinist-style totalitarianism. The political system was overhauled, with electoral processes made more democratic and some free-market principles introduced. Gorbachev sought warmer relations with both Communist and Western governments and withdrew troops from Afghanistan in 1989. In a rejection of the 'Brezhnev Doctrine', throughout 1989 and 1990 Gorbachev refused to intervene as one Communist regime after another fell in the Eastern Bloc. Within the USSR, the republics demanded independence. Initially rejected, ethnic tensions arose between and within the republics, with heavy fighting in the Caucasus. Suppressed for so long, the newfound freedom also brought chaos and Gorbachev was blamed. Nonetheless, he won the first USSR presidential elections. Opposition parties were legalized soon after, although he was reluctant to open up the economy to privatization. An attempted coup in Aug. 1991 led to Gorbachev's house arrest for three days and though the coup failed, largely owing to Russian president Boris Yeltsin's intervention (elected June 1991), Gorbachev's leadership was existing on borrowed time. In quick succession Gorbachev resigned his party membership, dismantled the central committee and took KGB and military control away from the Communists. On Christmas Day 1991 Gorbachev resigned as Soviet president and the Soviet Union was dissolved.

Yeltsin
A period of confrontation in 1992–93 between President Yeltsin and parliament climaxed when thousands of armed anti-Yeltsin demonstrators assembled on 3 Oct. 1993 and were urged to seize

the Kremlin and television centre. On 4 Oct. troops took the parliament building by storm after a ten-hour assault in which 140 people died. Vice-President Rutskoi and Speaker Khasbulatov were arrested.

Boris Yeltsin was re-elected president in 1996. Many took this as a signal of confidence in the new, democratic Russia. But the reality was a state in which democratic institutions were weakened to the point of impotence by racketeering and bureaucratic dead-weight. Russia defaulted on its debt, the rouble halved in value, imports fell by 45% and oil revenues slumped. On 17 Aug. 1998 the government freed the rouble, in effect devaluing it, imposed currency controls and froze the domestic debt market.

In Aug. 1999 Boris Yeltsin appointed as prime minister Vladimir Putin, a former KGB colonel and director of the KGB's successor organization, the FSB. On 31 Dec. 1999, Yeltsin resigned the presidency, nominating Putin as his interim successor, a job he retained after a clear-cut victory in the presidential election of March 2000. Under Putin, Russia continued the war with separatist Chechnya that began in Dec. 1994. One of his primary aims has been to reduce the power of the business oligarchs and to fight corruption. Tax cuts have been introduced and in 2000 a programme of regional reform divided Russia's 89 regions into seven new districts run by Kremlin representatives.

Following the attacks on the USA in Sept. 2001, Putin made clear his support for the war on terrorism. In Oct. 2002 a group of Chechen rebels took control of a Moscow theatre and held hostage 800 people for three days before Russian troops stormed the building. An anaesthetic gas, used to combat the rebels, killed many of the hostages. The rebels had been demanding that Russia end the war in Chechnya. The new relationship with the USA faltered as a result of the war with Iraq, which Russia opposed. Russia's vulnerability to terrorism was highlighted in Sept. 2004 when hostage takers seized a school in Beslan, in the Russian republic of North Ossetia. A three-day standoff ended with more than 350 people killed, nearly half of them children. Chechen rebels claimed responsibility for the siege.

In 2006 Russia temporarily cut off oil and gas links to Ukraine, Georgia and Belarus ostensibly over pricing disputes. As well as incurring the wrath of those governments immediately involved, the knock-on effects to pipeline flows throughout Europe prompted official protests. The future of Chechnya continues to be a thorn in Moscow's side. Relations with Georgia took a downturn in 2006 over Russia's implicit support for the break-away Georgian region of South Ossetia. They reached a new low in Aug. 2008 when Russian and Georgian troops fought each other for a week following the Georgian government's military attack on separatist forces in the region.

Putin's human and civil rights records have come under increasing international scrutiny. In the aftermath of the Beslan siege, Putin assumed responsibility for nominating regional governors who had previously been directly elected. Restrictions on press freedom have been criticized as has the absence of a credible opposition and an independent judiciary. Particularly controversial was the imprisonment of Mikhail Khodorkovsky, a multibillion dollar businessman and Putin critic who was sentenced to nine years in 2005 amid claims that the prosecution was politically motivated. In 2006 there was widespread concern at Putin's new powers to monitor non-governmental organizations and potentially expel foreign-based groups, some of which Putin accused of being 'led by puppeteers from abroad'.

In March 2008 Dmitry Medvedev won the presidential election, from which Putin was constitutionally barred. Medvedev named Putin his prime minister in May 2008. At parliamentary elections in Dec. 2011 international observers recorded widespread abuses at the polls and Putin faced popular protests in several cities across the country. In March 2012 he regained the presidency for an extended six-year term and appointed Medvedev as premier.

TERRITORY AND POPULATION

Russia is bounded in the north by various seas (Barents, Kara, Laptev, East Siberian) which join the Arctic Ocean, and in which is a fringe of islands, some of them large. In the east Russia is separated from the USA (Alaska) by the Bering Strait; the Kamchatka peninsula separates the coastal Bering and Okhotsk Seas. Sakhalin Island, north of Japan, is Russian territory. Russia is bounded in the south by North Korea, China, Mongolia, Kazakhstan, the Caspian Sea, Azerbaijan, Georgia, the Black Sea and Ukraine, and in the west by Belarus, Latvia, Estonia, the Baltic Sea and Finland. Kaliningrad (the former East Prussia) is an exclave on the Baltic Sea between Lithuania and Poland in the west. Russia's area is 17,075,400 sq. km and it has nine time zones (11 until March 2010). In 2007 Russia claimed control over 1·2m. sq. km of the Arctic Ocean bed, known to be rich in energy sources. Immediately disputed by the international community, the claim would require validation by the UN Commission on the Limits of the Continental Shelf. The 2010 census population was 142,856,536 density, 8·4 per sq. km. Ethnicity in 2002 showed 79·8% were Russians, 3·8% Tatars, 2·0% Ukrainians, 1·1% Bashkir and 1·1% Chuvash. There are also small numbers of Armenians, Avars, Belarusians, Chechens, Germans, Jews, Kazakhs, Mari, Mordovians and Udmurts.

In 2011, 73·2% of the population lived in urban areas.

The UN gives a projected population for 2015 of 142·23m.

Russia's population has been declining since the break-up of the Soviet Union and will continue to do so in the future. By 2050 its population is projected to be the same as it was in the early 1950s.

The two principal cities are Moscow (Moskva), the capital, with a 2010 census population of 11·50m. and St Petersburg (formerly Leningrad), with 4·88m. Other major cities (with 2010 populations) are: Novosibirsk (1·47m.), Ekaterinburg (1·35m.), Nizhny Novgorod (1·25m.), Samara (1·16m.) and Omsk (1·15m.). In May 2000 President Putin signed a decree dividing Russia into seven federal districts (*okrug*), in the process creating a layer above the various federal subjects (*see* CONSTITUTION AND GOVERNMENT *below*). These, with their administrative centres and 2010 populations in brackets, are: Central (Moscow, 38·43m.), North-Western (St Petersburg, 13·62m.), Southern (Rostov-on-Don, 13·85m.), Volga (Nizhny Novgorod, 29·90m.), Ural (Ekaterinburg, 12·08m.), Siberian (Novosibirsk, 19·26m.) and Far-Eastern (Khaborovsk, 6·29m.). In Jan. 2010 President Medvedev created a new North Caucasus federal district (making eight now in total) by splitting the Southern federal district in two. The 2010 population in the area that constitutes the North Caucasus federal district was 9·43m. Its administrative centre is Pyatigorsk.

The official federal language is Russian, although there are several other officially-recognized languages within individual administrative units.

SOCIAL STATISTICS

2008 births, 1,717,500; deaths, 2,081,000; marriages, 1,178,700; divorces, 703,400. Rates, 2008 (per 1,000 population): birth, 12·1; death, 14·7; marriage, 8·3; divorce, 5·0. At the beginning of the 1970s the death rate had been just 9·4 per 1,000 population. Infant mortality, 2010 (per 1,000 live births), 9. There were 1,582,400 legal abortions in 2006. The annual abortion rate, at approximately 52 per 1,000 women, ranks among the highest in the world. The divorce rate is also among the highest in the world. The most popular age range for marrying in 2008 was 25–34 for males and 18–24 for females. Expectation of life at birth, 2007, was 59·9 years for males and 72·9 years for females. With a difference of 13·0 years, no other country has a life expectancy for females so high compared to that for males. The low life expectancy (down from 64·6 years for males and 74·0 years for females in the USSR as a whole in 1989) and the low birth rate (down from 17·6 per 1,000 population in the USSR in 1989) is causing a demographic crisis. Disease, pollution, poor health care and alcoholism are all contributing to a steady decline in the population. More than 35,000 Russians died of alcohol poisoning in 2005. In 2005, 16% of Russians were living below the subsistence level (down from 27% in 2001). Annual population growth rate, 2000–05, –0·5%; fertility rate, 2008, 1·4 births per woman. The suicide rate, at 30·1 per 100,000 population in 2006, is one of the highest in the world. Among males it was 53·9 per 100,000 population in 2006.

CLIMATE

Moscow, Jan. –9·4°C, July 18·3°C. Annual rainfall 630 mm. Arkhangelsk, Jan. –15°C, July 13·9°C. Annual rainfall 503 mm. St Petersburg, Jan. –8·3°C, July 17·8°C. Annual rainfall 488 mm. Vladivostok, Jan. –14·4°C, July 18·3°C. Annual rainfall 599 mm.

CONSTITUTION AND GOVERNMENT

The Russian Soviet Federative Socialist Republic (RSFSR) adopted a declaration of republican sovereignty by 544 votes to 271 in June 1990. It became a founding member of the Commonwealth of Independent States (CIS) in Dec. 1991, and adopted the name 'Russian Federation'. A law of Nov. 1991 extended citizenship to all who lived in Russia at the time of its adoption and to those in other Soviet republics who requested it.

According to the 1993 constitution the Russian Federation is a 'democratic federal legally-based state with a republican form of government'. The Federation consists of 83 federal subjects (administrative units), of which 21 are republics, one autonomous region, four autonomous districts, nine territories, 46 regions and two federal cities. The state is secular. Individuals have freedom of movement within or across the boundaries of the Federation; there is freedom of assembly and association, and freedom to engage in any entrepreneurial activity not forbidden by law. The state itself is based upon a separation of powers and upon federal principles, including a Constitutional Court. The most important matters of state are reserved for the federal government, including socio-economic policy, the budget, taxation, energy, foreign affairs and defence. Other matters, including the use of land and water, education and culture, health and social security, are for the joint management of the federal and local governments, which also have the right to legislate within their spheres of competence. A central role is accorded to the *President*, who defines the 'basic directions of domestic and foreign policy' and represents the state internationally. The President is directly elected for a six-year term (since Dec. 2008—previously a four-year term), and for not more than two consecutive terms; he must be at least 35 years old, a Russian citizen, and a resident in Russia for the previous ten years. 2m. signatures are needed to validate a presidential candidate not affiliated to a party represented in the State Duma, no more than 2·5% of which may come from any one region or republic. The President has the right to appoint the prime minister, and (on his nomination) to appoint and dismiss deputy prime ministers and ministers, and may dismiss the government as a whole. In the event of the death or incapacity of the President, the Prime Minister becomes head of state.

Parliament is known as the *Federal Assembly* (Federalnoe Sobranie). The 'representative and legislative organ of the Russian Federation', it consists of two chambers: the *Federation Council* (Sovet Federatsii) and the *State Duma* (Gosudarstvennaya Duma). The Federation Council, or upper house, consists of 166 deputies. The State Duma, or lower house, consists of 450 deputies elected for a four-year term. Starting with the elections in Dec. 2007 all deputies to the State Duma are elected from party lists by proportional representation. There is a 7% threshold for the party-list seats. To qualify for candidacy an individual must be nominated by a registered political party. Incumbent parties are automatically included in the ballot; others must obtain a minimum of 200,000 supporting signatures of which no more than 5% may come from any one region. Alternatively, non-incumbent parties may put forward a deposit of 60m. roubles, which is returned if the party manages to win at least 4% of the popular vote. Parties which gain at least 35 seats may register as a faction, which gives them the right to join the Duma Council and chair committees. Any citizen aged over 21 may be elected to the State Duma, but may not at the same time be a member of the upper house or of other representative bodies. The Federation Council considers all matters that apply to the Federation as a whole, including state boundaries, martial law, and the deployment of Russian forces elsewhere. The Duma approves nominations for Prime Minister, and adopts federal laws (they are also considered by the Federation Council but any objection may be overridden by a two-thirds majority; objections on the part of the President may be overridden by both houses on the same basis). The Duma can reject nominations for Prime Minister but after the third rejection it is automatically dissolved. It is also dissolved if it twice votes a lack of confidence in the government, or if it refuses to express confidence in the government when the matter is raised by the Prime Minister.

A law was approved in June 2001 to reduce the proliferation of political parties (numbering some 200 in 2001). It took effect in July 2003. The new law introduced stricter registration criteria and obliging existing parties to reregister within two years. In order to register, political parties are required to have at least 50,000 members and (since Jan. 2006) more than 45 regional branches with a minimum membership of 500 each. Multiple party membership is banned.

There is a 19-member *Constitutional Court*, whose functions under the 1993 constitution include making decisions on the constitutionality of federal laws, presidential and government decrees, and the constitutions and laws of the subjects of the Federation. It is governed by a Law on the Constitutional Court, adopted in July 1994. Judges are elected for non-renewable 12-year terms.

National Anthem

In Dec. 2000 the Russian parliament, on President Putin's initiative, decided that the tune of the anthem of the former Soviet Union should be reintroduced as the Russian national anthem. Written by Alexander Alexandrov in 1943, the anthem was composed for Stalin. New words were written by Sergei Mikhalkov, who had written the original words for the Soviet anthem in 1943. The new anthem is 'Rossiya—svyashennaya nasha derzhava, Rossiya—lyubimaya nasha strana' ('Russia—our holy country, Russia—our beloved country'). Boris Yeltsin had introduced a new anthem during his presidency—'Patriotic Song', from an opera by Mikhail Glinka and arranged by Andrei Petrov.

GOVERNMENT CHRONOLOGY

General/First Secretaries of the Central Committee of the USSR (1922–91) and Presidents of Russia (1991–)

1922–53	Joseph Stalin
1953–64	Nikita Sergeyevich Khrushchev
1964–82	Leonid Ilyich Brezhnev
1982–84	Yuri Vladimirovich Andropov
1984–85	Konstantin Ustinovich Chernenko
1985–91	Mikhail Sergeyevich Gorbachev
1991–99	Boris Nikolayevich Yeltsin
1999–2008	Vladimir Vladimirovich Putin
2008–12	Dmitry Anatolyevich Medvedev
2012–	Vladimir Vladimirovich Putin

RECENT ELECTIONS

In the presidential elections held on 4 March 2012 prime minister Vladimir Putin (United Russia) won 63·6% of the votes cast. Gennady Zyuganov (Communist Party of the Russian Federation; KPRF) took 17·2% of the vote; Mikhail Prokhorov (ind.) 7·9%; Vladimir Zhirinovsky (Liberal Democratic Party; LDPR) 6·2%;

and Sergey Mironov (A Just Russia; SR) 3·9%. Turnout was 65·3%. The OSCE reported 'procedural irregularities' and the Communist Party of the Russian Federation refused to recognize the result.

Elections for the State Duma were held on 4 Dec. 2011. United Russia won 238 seats with 49·5% of the vote (down from 315 seats and 64·3% at the 2007 elections); the KPRF 92 seats with 19·2% (up from 57 and 11·6% in 2007); A Just Russia 64 with 13·2% (up from 38 and 7·7% in 2007); and the LDPR 56 and 11·7% (up from 40 and 8·1% in 2007). Turnout was 60·2%. There were reports of election fraud, leading to major protests.

CURRENT GOVERNMENT

President: Vladimir Putin; b. 1952 (sworn in 7 May 2012 for a second time, having previously been in office from Dec. 1999–May 2008).

Prime Minister: Dmitry Medvedev; b. 1965 (sworn in 8 May 2012).

In May 2013 the government comprised:

First Deputy Prime Minister: Igor Shuvalov.

Deputy Prime Ministers: Arkady Dvorkovich; Olga Golodets; Alexander Khloponin; Dmitry Kozak; Dmitry Rogozin.

Minister of Agriculture: Nikolay Fyodorov. *Communications and Mass Media:* Nikolai Nikiforov. *Culture:* Vladimir Medinksy. *Defence:* Sergei Shoigu. *Economic Development:* Andrey Belousov. *Education and Science:* Dmitry Livanov. *Emergency Situations:* Vladimir Puchkov. *Energy:* Alexander Novak. *Far East Development:* Viktor Ishayev. *Finance:* Anton Siluanov. *Foreign Affairs:* Sergei Lavrov. *Health Care:* Veronika Skvortsova. *Industry and Trade:* Denis Manturov. *Interior (MVD):* Vladimir Kolokoltsev. *Justice:* Alexander Konovalov. *Labour and Social Security:* Maksim Topilin. *Natural Resources and Environmental Protection:* Sergey Donskoy. *Open Government:* Mikhail Abyzov. *Regional Development:* Igor Slyunyayev. *Sport:* Vitaliy Mutko. *Transportation:* Maksim Sokolov.

Chairman of the State Duma: Sergei Naryshkin.

President's Website: http://kremlin.ru

CURRENT LEADERS

Vladimir Vladimirovich Putin

Position
President

Introduction
Vladimir Putin became president again in March 2012, having previously served two presidential terms from 1999–2008 and one term as prime minister since 2008. His initial appointment as president at the end of 1999 had been the culmination of a rapid political rise in the post-Communist era. Little known internationally, his KGB past aroused early concerns but he quickly gained respect within Russia as a modernizer and efficient administrator. He also made a determined effort to re-establish a more influential world role for his country. However, his handling of the Chechen war led to international criticism, while his opposition to the US war in Iraq from 2003 and to US plans to expand its missile defence systems into Eastern Europe strained relations with Washington. He nevertheless maintained a high approval rating among most Russian voters, although his style of government was perceived as authoritarian and imperial. Precluded from serving a third term under the constitution, he oversaw the election of his chosen successor, Dmitry Medvedev, to the presidency in March 2008 and he assumed the post of prime minister in May. In Sept. 2011 Putin was confirmed as United Russia's candidate for the presidential election scheduled for March 2012. Despite a significant fall in support for his United Russia party in parliamentary polls at the end of 2011, he regained the presidency and in May 2012 replaced Medvedev who in turn took over the premiership.

Early Life
Vladimir Putin was born in St Petersburg (then called Leningrad) on 7 Oct. 1952, the son of a war veteran who, with his wife, had survived the siege of Leningrad. Baptized into Russian Orthodoxy, he was an accomplished athlete. After graduating from law school in 1975, he began a 15-year career with the KGB's foreign intelligence arm, stationed in Leningrad and East Germany. When collapse threatened the Soviet Union, he retired as a colonel and embarked on a political career.

In the early 1990s Putin worked in local government in St Petersburg, becoming deputy mayor in 1994. In 1996 President Yeltsin brought him to Moscow and appointed him deputy chief Kremlin administrator. He took charge of relations with Russia's diverse regions and in 1998 became head of the Federal Security Service (successor of the KGB) and secretary of the presidential Security Council. In Aug. 1999 he was named acting prime minister.

Career in Office
Putin became the acting president of Russia after Yeltsin's resignation on 31 Dec. 1999 and was officially elected president on 26 March 2000, taking 53% of the vote. His election programme prioritized a 'dictatorship of law' to combat high crime rates, as well as pledging to tackle poverty and promote family values, patriotism and fair business conditions.

He quickly set about exercising firm control over local government and the economy, with a stated aim of reducing corruption. In a bid to centralize power in Moscow, he restructured 89 legislative regions into seven districts, each with a government-approved leader (the majority of whom had military or security backgrounds). He reversed tax concessions that Yeltsin had brought in to assist the regions and reserved the right to dismiss any democratically-elected politician found to have broken the law.

Putin also removed several high-profile business and media figures from official positions. Yeltsin's daughter was dismissed as a Kremlin adviser but immunity was granted to Yeltsin himself, one of the more controversial moves of the then acting president. Putin appointed former finance minister and Yeltsin ally Mikhail Kasyanov as prime minister while placing other supporters in key Kremlin positions. Putin's economic policy was influenced by his allegiance to Anatoly Chubais, who led the wave of privatization in Russia in the early 1990s.

In April 2001 Putin's Unity party merged with the opposition Fatherland bloc, led by the mayor of Moscow, Yury Luzhkov. To pass legislation the president needed a simple majority of 226, with the merger giving him at least 132. In the summer of 2001 new laws on land, labour and pensions were proposed. These were opposed by the Communists whose leader Gennady Zyuganov called for a national demonstration against the reforms.

After Oct. 2001 Russians were free to buy residential and commercial land for the first time since the Bolsheviks took power in 1917. Farm land, making up 98% of the total, was not covered by the law. Critics feared that a privileged few would buy up the land much as they bought privatized businesses in the 1990s. Supporters maintained that it would attract foreign investment, speed up economic reform and stop the illegal sale of land.

Putin's image was dented by the sinking of the Kursk nuclear submarine in Aug. 2000 when all 118 Russian sailors on board died. He was widely condemned for inaction, refusing to return from his holiday and turning down offers of help from Norway and the UK. He was subsequently dogged by allegations of an official cover-up.

Putin meanwhile used the war with Chechnya to establish his 'strong man' credentials, although heavy Russian losses cost him some support and alleged human rights abuses led to a suspension of Russia's voting rights in the Council of Europe. In Oct. 2002 Chechen rebels took 800 people hostage inside a Moscow theatre, demanding the immediate withdrawal of Russian troops from

Chechnya. The siege lasted three days before the Russian military stormed the building using an anaesthetic gas which killed the rebels but also over 100 hostages. In March 2003 Putin promised greater autonomy for Chechnya. This followed a referendum in the republic supporting a new constitution that would keep Chechnya within Russia but provide for a president and parliament. Moscow claimed 96% support for the proposals although no international observers were present and the referendum was opposed by separatist groups. In May 2003, following two suicide bomb attacks, Putin reaffirmed his determination to defeat Chechnya's rebel forces. He offered an amnesty for rebels who handed over their weapons by 1 Aug. 2003 and for Russian troops accused of human rights violations.

Following the 11 Sept. attacks on New York and Washington, D.C. there was a rapprochement between Russia and the USA. In 2001 Putin gave unprecedented support for UN military action in Afghanistan. His offers of military assistance to the Afghan Northern Alliance, the use of Russian airspace for humanitarian aid and his role in persuading Tajikistan and Uzbekistan to support the campaign were well received in the West where leaders were quick to downplay Russia's role in Chechnya. In May 2002 Putin and US President George W. Bush signed an anti-nuclear deal agreeing to reduce their respective strategic nuclear warheads by two-thirds over the next ten years.

However, tension between the two countries increased over the question of Iraq in late 2002. While President Bush attempted to garner support for military action in Iraq—with Russia holding a power of veto within the UN Security Council—he warned Putin at the same time that he would not support Russian military incursions into Georgia, which Russia claimed was tolerating hostile Chechen activity. Putin's political dealings with 'rogue' nations, including North Korea and Cuba, were also criticized, as was Russia's trading of nuclear fuel and weapons with India, Iran, Iraq and Syria.

When US-led forces began attacking Iraq in March 2003 Russia condemned the action and delayed ratifying the US-Russian strategic arms control treaty (*see above*) until the war was over. Putin refuted US accusations that Russia had breached UN sanctions by selling armaments, including anti-tank missiles and jamming equipment, to Iraq. In May 2003 Russia nevertheless voted to accept a UN resolution on Iraq's future jointly proposed by the USA, UK and Spain. In return for the immediate ending of sanctions, the UN was to co-operate with the occupying forces to form a new government. In addition Russia would be able to complete long-standing contracts with Iraq. Despite tension in Russia's relationship with the UK as a result of differences over Iraq, Putin made an official state visit to Britain, the first by a Russian leader for over a century.

Having previously avoided party politics, Putin publicly endorsed a pro-government party—United Russia—in the Dec. 2003 parliamentary elections. In the March 2004 presidential elections, he was criticized by the international press and by OSCE monitors for manipulating the Russian media to influence voting in his favour. He was re-elected in the poll, claiming 71% of the vote and leaving his closest rival, the Communist Nikolai Kharitonov, with less than 14%. In May 2004 Putin set out his goals for his second term—modernizing Russia and raising living standards while aiming for a more stable democracy able to pursue strategic interests abroad.

Chechen violence had continued in 2003. Following a referendum in which Chechens agreed to a Moscow-approved constitution, Putin announced that a presidential vote would go ahead. In Oct. 2003, with a turnout of 85%, the pro-Moscow leader Akhmad Kadyrov was elected. However, he was then killed in a bomb attack in Chechnya's capital in May 2004 and Putin pledged to send extra troops in response. Alu Alkhanov was elected president of Chechnya on 29 Aug. 2004, and in Chechen parliamentary elections in Nov. 2005 the pro-Moscow United Russia party claimed about 60% of the popular vote.

In the aftermath of the bloodbath that ended the Beslan school siege by Chechen militants in Sept. 2004 in which more than 350 people were killed, Putin controversially took control of the appointment of regional governors who had been directly elected for the previous decade. Critics saw the move as undermining democracy.

In May 2005 Mikhail Khodorkovsky, a billionaire former head of oil-exporting company Yukos, was sentenced to nine years' imprisonment for tax evasion and fraud. His conviction, following the effective renationalization of Yukos in 2004, was widely viewed as politically motivated, owing to his criticism of Putin's regime.

High oil and gas prices continued to underpin Russia's strong economic growth during Putin's presidency. There was increasing state influence over the energy industry, exercised through the giant gas monopoly Gazprom and through Rosneft (another state-run company that acquired the prime assets of Yukos in 2004). However, Russia's reputation as a reliable international energy trader was tarnished. Several cuts in Russian oil and gas supplies to neighbouring countries—particularly Ukraine and Belarus—from 2006, purportedly over pricing disputes but with suspected political overtones, in turn disrupted pipeline flows to EU countries and prompted high-level protests.

There was meanwhile growing concern for civil and political freedoms in Putin's Russia, with foreign commentators citing increasing authoritarianism, the lack of a genuine opposition or independent judiciary and more general lawlessness, including violent xenophobia. The deaths in late 2006 of two prominent government critics—campaigning journalist Anna Politkovskaya and former intelligence officer Alexander Litvinenko—fuelled widespread suspicions of official involvement. Litvinenko's poisoning in London in particular led to a sharp deterioration in Russian-UK relations. Putin's refusal to extradite a prime suspect in the affair to Britain resulted in the mutual imposition of diplomatic sanctions in July 2007.

Plans by the USA to expand its missile defences into the Eastern European member countries of NATO (which have since been scaled back) further soured Russian-US relations in Putin's second term of office. In Nov. 2007 he approved a law suspending Russia's participation in the 1990 Conventional Armed Forces in Europe (CFE) Treaty limiting military deployments.

In Dec. 2007 the United Russia party won a landslide victory in parliamentary elections. With his second term set to expire in 2008, Putin announced that he would nevertheless continue in politics as the next prime minister and that Dmitry Medvedev (the chair of Gazprom) would be his preferred successor as president after elections scheduled for March. Russian voters duly endorsed this political continuity at the polls and Putin and Medvedev were sworn into their respective posts in May 2008. Putin therefore remained a pivotal figure in the Russian leadership, having an increasingly centralized grip on power and policy with President Medvedev.

In Oct. 2009 Putin's ruling United Russia party won a sweeping victory in nationwide regional and local elections. Opposition parties walked out of parliament, claiming that the vote had been rigged, and they threatened to demonstrate in protest. Also in Oct., Putin visited China to conclude trade deals worth US$3·5bn.

Significant issues confronting the Putin administration in mid-2010 included an espionage controversy, as a number of alleged undercover Russian agents posing as ordinary citizens were arrested in the USA and subsequently deported under a prisoner exchange deal, and the environmental consequences of unprecedented wildfires across central and western Russia amid the hottest temperatures in over a century. There were also further acts of terrorism in 2010 and early 2011, including suicide bombings on the Moscow Metro and at Moscow's Domodedovo airport and an attack on the regional parliament building in Chechnya.

In Dec. 2011 Putin's dominant United Russia won the parliamentary elections, although with a reduced majority of seats and slightly less than 50% of the vote. Reports of electoral fraud prompted unprecedented opposition protests on the streets of Moscow. Despite the apparent decline in his personal support, Putin stood successfully as United Russia's candidate for the presidential election in March 2012 and once more took over the office that he had vacated in 2008. He stepped down as leader of United Russia ahead of being sworn in as president in May 2012.

In Aug. 2012 the Russian government was condemned by human rights groups and large parts of the international community for imposing prison sentences on three female members of a punk band that staged an anti-Putin protest in Moscow's cathedral. In foreign affairs Putin maintained his opposition to outside intervention in the civil war in Syria, Russia's long-standing ally in the Middle East.

Dmitry Anatolyevich Medvedev

Position
Prime Minister

Introduction
Dmitry Medvedev's political career has been closely associated with Vladimir Putin's since the mid-1990s and he has been viewed as a fellow modernizer. In March 2008 Medvedev was elected by a landslide vote to the presidency as Putin's preferred successor after the latter had completed the two consecutive presidential terms allowed under the constitution. In 2012, after four years in office, Medvedev stood down as Putin ran successfully again for the presidency, and he was subsequently appointed by Putin as prime minister in May.

Early Life
Medvedev was born in St Petersburg (then called Leningrad) on 14 Sept. 1965. Both his parents were university teachers. He studied law at Leningrad State University, graduating in 1987. After obtaining a PhD in private law in 1990, he worked as an assistant professor at the university until 1999. In 1990 he also worked at the Leningrad Soviet of People's Deputies. From 1991–95 he served as legal adviser to the chairman of Leningrad city council and to the committee for external relations of the St Petersburg mayor's office, headed by Vladimir Putin. In Nov. 1993 he became legal affairs director of Ilim Pulp Enterprise and in 1998 he was elected to the board of governors of Bratskiy LPK paper mill.

Medvedev was appointed deputy head of the presidential administration by Putin in Dec. 1999. In 2000 he was promoted to the rank of first deputy chief of staff and during the 2000 election was in charge of the presidential campaign headquarters. In the same year Medvedev became chairman of the board of directors of Gazprom, Russia's largest company. From 2001 he served as deputy chairman until, in June 2002, he became chairman again. Under Medvedev, Gazprom acquired other energy companies and expanded overseas, increasing its significance both within the Russian economy and as an international supplier.

In 2003 Medvedev was appointed chief of staff of the presidential executive office and in Nov. 2005 Putin made him first deputy prime minister, giving him control of four national infrastructure projects: health care, education, housing and agriculture. Medvedev's initiatives included supporting foster families and developing pre-school education; he also attempted to reshape relationships between the Kremlin and Russia's billionaire oligarchs. In Dec. 2007, having served the constitutional maximum of two terms, Putin named Medvedev as his chosen successor. This was interpreted by many as an arrangement by which the two could continue to govern in partnership. Medvedev was elected president with 70% of the vote on 2 March 2008 and was sworn in on 7 May 2008.

Career in Office
As expected, Medvedev appointed Putin prime minister. Medvedev pledged to continue Putin's policies, indicating that he would encourage a diversification of the economy to reduce reliance on gas revenues. He also signalled his support for a free press and free judiciary and spoke of limiting the influence of the security services.

Early challenges internationally included a deterioration in relations with NATO and Russia's western neighbours. In Aug. 2008 he sent troops to South Ossetia and Abkhazia in Georgia, in response to Tbilisi's military attacks on separatist forces within the breakaway regions. A week of fierce fighting ended with a French-brokered peace deal. Russia's intervention received widespread international criticism, as did Medvedev's announcement of Russia's unilateral recognition of the independence of the two territories.

In his first state-of-the-nation address in Nov. 2008, Medvedev proposed a constitutional change to extend the presidential term of office from four to six years and that of parliament from four to five years, arguing that it was necessary to guarantee effective government. However, critics considered the move undemocratic and designed to perpetuate authoritarian rule from the Kremlin. In the same speech Medvedev threatened to deploy short-range missiles in Kaliningrad to counter the USA's proposed missile shield in central Europe (although a more conciliatory approach followed Barack Obama's assumption of the US presidency in Jan. 2009). He meanwhile claimed that the USA bore responsibility for the global financial crisis that had undermined the Russian banking system and destabilized markets. Also in Nov. 2008 Medvedev undertook an overseas tour of Latin America and Cuba.

Russia's relations with the USA improved from 2009 following the installation of a new US administration under President Obama. Medvedev met Obama in July as the latter made his first official visit to Moscow and they agreed to negotiate cuts in their countries' nuclear weapons arsenals in a new initiative to supersede the 1991 Strategic Arms Reduction Treaty. (A 30% cut was subsequently concluded under an agreement signed in April 2010.) In Sept. 2009 Medvedev welcomed the US decision not to site missile defence bases in Poland and the Czech Republic, the Russian military having earlier confirmed that plans to deploy short-range missiles in the Kaliningrad enclave were being shelved. Nevertheless, in Nov. 2010 he warned of the dangers of a new arms race unless NATO and Russia reached a co-operative accord within a decade on a joint missile-defence mechanism.

Medvedev had meanwhile called in Nov. 2009 for reform of the economy and re-emphasized the need for Russia to end its dependency on gas and oil exports. He also said that Russia's survival depended on rapid modernization, based on democratic institutions and an end to corruption.

In Nov. 2010 Medvedev became the first Russian leader to visit the Russian-occupied Kurile Islands, which have been the subject of a territorial dispute with Japan since the end of World War II.

In Sept. 2011 Vladimir Putin was confirmed as United Russia's candidate for the presidential election scheduled for March 2012. Following his expected victory, he succeeded Medvedev in May as the latter assumed the premiership at the head of the United Russia-dominated government.

In Nov. 2012 the government introduced legislation broadening the definition of treason in what was widely viewed as an attempt to stop Russians working with Western non-governmental organizations and to stifle domestic dissent.

DEFENCE

The President of the Republic is C.-in-C. of the armed forces. Conscription was reduced to 18 months for those drafted in 2007 and was further reduced to one year for those drafted from 1 Jan. 2008.

A presidential decree of Feb. 1997 ordered a cut in the armed forces of 200,000 men, reducing them to an authorized strength of 1,004,000 in 1999. In 2007 armed forces totalled 1,027,000, plus 418,000 personnel in paramilitary forces. There were estimated to be around 20,000,000 reserves (all armed forces) in 2006 of whom 2,000,000 had seen service within the previous five years.

Military expenditure totalled an estimated US$58,600m. in 2008 (less than a tenth of that of the USA), equivalent to US$413 per capita. This made Russia the world's fifth biggest military spender. In the period 1999–2008 military expenditure rose by 173% in real terms. In 2007 defence spending represented 3·5% of GDP.

Nuclear Weapons

Russia's strategic warhead count is now shrinking and stood at an estimated 1,800 in Jan. 2012 according to the Stockholm International Peace Research Institute. There are about a further 8,200 warheads held in reserve or scheduled to be dismantled, giving a total stockpile of around 10,000 warheads. Shortfalls in planned investments to replace current systems as they reach the end of their service lives means the number of strategic warheads will continue to decline.

At the height of the Cold War each side possessed over 10,000 nuclear warheads. The START I arms-control treaty, signed in 1991, limited to approximately 6,000 the number of nuclear warheads that Russia and the USA may each deploy on long-range, or 'strategic', land-based missiles, submarine-launched missiles and bombers. The 1993 START II treaty would have obliged both sides to reduce their stocks of strategic weapons to 3,500 nuclear warheads. However, instruments of ratification were never exchanged and the treaty lapsed. Russia retracted acceptance in June 2002 after the USA withdrew from the Anti-Ballistic Missile treaty. START I expired on 5 Dec. 2009. On 24 May 2002 the USA and Russia signed the Strategic Offensive Reductions Treaty (or Moscow Treaty) to reduce the number of US and Russian warheads to between 1,700 and 2,200 each. The treaty was in force from June 2003 until Feb. 2011. It was then superseded by the Measures to Further Reduction and Limitation of Strategic Offensive Arms (known as New START), signed by the Russian and US presidents in April 2010 and ratified by the US Senate in Dec. 2010 and the Russian Federal Assembly in Jan. 2011. Under its terms, both Russia and the USA were limited to 1,550 warheads, a 30% drop on previous levels, to be implemented within seven years.

Chemical and Biological Weapons

Russia has converted or destroyed its former chemical weapon production facilities and is currently working to complete destruction of its chemical weapon stockpile in accordance with the provisions of the 1993 Chemical Weapons Convention. Russia has the largest declared stockpile of chemical weapons, originally totalling 40,000 tonnes.

The Soviet Union had perhaps the world's largest biological weapon programme until at least the 1980s. Russia has reiterated its commitment to the 1972 Biological and Toxin Weapons Convention.

Arms Trade

Russia was the world's second largest exporter after the USA in 2008, with sales worth US$5,400m. or 17·0% of the world total.

Army

In 2006 Army personnel numbered around 395,000 (including about 190,000 conscripts). There were around 17,000 Russian troops stationed outside Russia in 2006, the majority in various states of the former USSR (including 7,800 in Tajikistan and 3,000 in the Caucasus).

The Army is deployed in six military districts and one Operational Strategic Group. Equipment includes some 23,000 main battle tanks (including T-55s, T-62s, T-64A/-Bs, T-72L/-Ms, T-80/-U/UD/UMs and T-90s) plus 150 light tanks (PT-76s).

Strategic Nuclear Ground Forces

In 2008 there were three rocket armies, which will fall to two by 2016. Each rocket army is divided into launcher groups with ten silos and one control centre. Inter-continental ballistic missiles numbered 506. Personnel, 40,000.

Navy

The Russian Navy continues to reduce steadily and levels of sea-going activity remain very low with activity concentrated on a few operational units in each fleet. The safe deployment and protection of the reduced force of ballistic missile submarines remains its first priority; and the defence of the Russian homeland its second. The strategic missile submarine force operates under command of the Strategic Nuclear Force commander whilst the remainder come under the Main Naval Staff in Moscow, through the Commanders of the fleets.

The Northern and Pacific fleets count the entirety of the ballistic missile submarine force, all nuclear-powered submarines, the sole operational aircraft carrier (the *Admiral Kuznetsov*) and most major surface warships. The Baltic Fleet organization is based in the St Petersburg area and in the Kaliningrad exclave. The Black Sea Fleet is based at facilities in the Crimea, Ukraine. Ukraine had stated that the lease on these three harbours for warships and two airfields would not be extended and that the fleet must leave the main base of Sevastopol by 2017. However, since being elected in Feb. 2010 Ukraine's president Viktor Yanukovych has suggested that he would allow Russia's fleet to remain in his country beyond 2017. There is a small Caspian Sea flotilla. In Nov. 2008 Russia held joint exercises with the Venezuelan Navy in the Caribbean Sea in what was widely perceived as an attempt to provoke the USA.

The material state of all the fleets is suffering from continued inactivity and lack of spares and fuel. The nuclear submarine refitting and refuelling operations in the Northern and Pacific Fleets remain in disarray, given the large numbers of nuclear submarines awaiting defuelling and disposal. The strength of the submarine force has now essentially stabilized, but there are still large numbers of decommissioned vessels awaiting their turn for scrapping in a steadily deteriorating state. In Jan. 2003 it was announced that up to a fifth of the fleet was to be scrapped.

In Jan. 2011 there were 11 operational nuclear-fuelled ballistic-missile submarines (four of Delta-III class, six Delta-IV and one unarmed Typhoon used as a test platform). There was a total of 276 sea-launched SS-N-9, -12, -19, -21 and -22s in Jan. 2009. After a series of test launches, a new submarine-launched ballistic missile 'Bulava' (SS-NX-30) was officially approved for service in Dec. 2011. It is set to be deployed on the new 'Borei'-class nuclear submarines.

The attack submarine fleet comprises a wide range of classes, from the enormous 16,250 tonne 'Oscar' nuclear-powered missile submarine to diesel boats of around 2,000 tonnes. The inventory of tactical nuclear-fuelled submarines comprises five 'Oscar'-class, nine 'Akula'-class, three 'Sierra'-class and four 'Victor III'-class submarines.

The diesel-powered 'Kilo' class, of which the Navy operates 15, is still building at a reduced rate mostly for export.

Cruisers are divided into two categories; those optimized for anti-submarine warfare (ASW) are classified as 'Large Anti-Submarine Ships' and those primarily configured for anti-surface ship operations are classified 'Rocket Cruisers'. The principal surface ships of the Russian Navy include the following classes:

Aircraft Carrier. The *Admiral Kuznetsov* of 67,500 tonnes was completed in 1989. It is capable of embarking 20 aircraft and 15–17 helicopters. All other aircraft carriers have been decommissioned or scrapped.

Cruisers. The ships of this classification are headed by the one ship of the Kirov-class, the largest combatant warships, apart from aircraft carriers, to be built since the Second World War. There are, in addition, three Slava-class and one of the Nikolaev ('Kara') class in operation.

Destroyers. There are seven Udaloy I-class, the first of which entered service in 1981, one Udaloy II-class and five Sovremenny-class guided missile destroyers in operation. In addition there is a single remaining 'modified Kashin'-class ship in operation.

Frigates. There are six frigates in operation including the first of a new class, the 'Gepard', two Neustrashimy-class, two Krivak I-class and one Krivak II-class ships.

The Russian Naval Air Force operates some 266 combat aircraft including 58 Tu-22M bombers and 58 Su-24, 10 Su-25 and 49 Su-27 fighters. In 2006 there were an additional 161 armed helicopters in operation.

Total Naval personnel in 2006 numbered 142,000. Some 11,000 serve in the strategic submarine force, 35,000 in naval aviation and 9,500 naval infantry/coastal defence troops.

Air Force

Air Force personnel is estimated at 160,600 and equipment includes some 1,850 combat aircraft and about 2,000 helicopters.

The Air Force was reorganized in 2009 and now consists of: Operational Strategic Command for Air-Space Defence; First Air Force and Air Defence Command; Second Air Force and Air Defence Command; Third Air Force and Air Defence Command; Fourth Air Force and Air Defence Command; Military Transport Aviation Command; and Long Range Aviation Command. An air force base opened in Kyrgyzstan in Oct. 2003.

Prior to the restructuring equipment of the 37th Air Army included 124 Tu-22Ms plus Tu-95s and Tu-160s.

Tactical Aviation comprised in 2006 (numbers in brackets) Su-24 (400) and Su-25 (275) fighter-bombers and MiG-29 (314), MiG-31 (279) and Su-27 (390) fighters. In addition MiG-25 and Su-24s are used for reconnaissance missions.

Military Transport Aviation Command comprised nine regiments in 2006 and has some 293 aircraft. Funding shortages have reduced serviceability drastically.

ECONOMY

Agriculture accounted for 5·1% of GDP in 2006, industry 38·0% and services 57·0%.

In Oct. 1991 a programme was launched to create a 'healthy mixed economy with a powerful private sector'. The prices of most commodities were freed on 2 Jan. 1992.

Privatization, which is overseen by the State Committee on the Management of State Property, began with small and medium-sized enterprises. A state programme of privatization of state and municipal enterprises was approved by parliament in June 1992, and vouchers worth 10,000 roubles each began to be distributed to all citizens in Oct. 1992. These could be sold or exchanged for shares. Employees had the right to purchase 51% of the equity of their enterprises. 25 categories of industry (including raw materials and arms) remained in state ownership. The voucher phase of privatization ended on 30 June 1994. A post-voucher stage authorized by presidential decree of 22 July 1994 provides for firms to be auctioned for cash following the completion of the sale of up to 70% of manufacturing industry for vouchers. The Ministry of Property Relations was established in 2000 with the mandate of overall federal policies on property issues and the management of state property, and in Dec. 2001 a new Federal Law on Privatization of State and Municipal Property was adopted. By that time a total of 129,811 enterprises had been sold. In 2010 only 30·9% of total employment was still in the public sector, down from 69·1% in 1992 and 37·8% in 2000.

Overview

The economy is based around the energy industry, with natural gas and oil accounting for 70% of revenues. With Russia the largest energy exporter in the world, economic growth averaged

7% per year from 2000 to 2008, largely driven by high energy prices. Although exact figures are unavailable, reserves of oil are estimated to be sufficient to last 20 years at present production levels.

With the economy reliant on energy prices, the global economic crisis saw the Russian economy contract by nearly 8% in 2009. It then recovered well, expanding by over 4% in both 2010 and 2011. Growth has also benefited from low unemployment and high domestic consumption. The government has recognised the need for a more diversified economy and provides 75% of research and development funding, with a particular focus on high-tech products in the automobile and aviation industries.

Obstacles to growth include a shortage of skilled workers, lack of competiveness and an underdeveloped infrastructure. Foreign and domestic private investment is restricted by corruption, organized crime, problems with the judicial system and a perceived excess of government intervention. The European Bank for Reconstruction and Development and the World Bank rated Russia 120th out of 183 countries for the ease of doing business in 2012.

Russia became a member of the World Trade Organization in 2012 after 18 years of negotiations. GNI per capita is $10,730, but there is a significant wealth gap. In 2008 the richest 10% of the population accounted for 33·5% of wealth while the poorest 10% accounted for 2·6%. In 2006 the poverty rate stood at 11·6%. An ageing population, shrinking workforce and lack of immigration present further long-term challenges.

Currency

The unit of currency is the *rouble* (RUB), of 100 *kopeks*. In Jan. 1998 the rouble was redenominated by a factor of a thousand. Foreign exchange reserves were US$383,664m. in Sept. 2009 and gold reserves 19·00m. troy oz. In Feb. 2005 Russia abandoned its *de facto* dollar peg and switched to a euro-dollar basket. Inflation rates (based on IMF statistics):

2002	2003	2004	2005	2006	2007	2008	2009	2010	2011
15·8%	13·7%	10·9%	12·7%	9·7%	9·0%	14·1%	11·7%	6·9%	8·4%

Inflation had been 2,510% in 1992. Total money supply in Dec. 2008 was 7,419·7bn. roubles. In Nov. 2000 then President Putin and President Lukashenka of Belarus agreed the introduction of a single currency, but plans to introduce the Russian rouble to Belarus have since been postponed indefinitely.

Budget

Central government revenue totalled 14,151·8bn. roubles in 2009 (14,127·3bn. roubles in 2008) and expenditure 12,063·3bn. roubles (8,887·8bn. roubles in 2008). Principal sources of revenue in 2009: taxes on international trade and transactions, 2,599·2bn. roubles; social security contributions, 2,354·7bn. roubles; taxes on goods and services, 2,288·9bn. roubles. Main items of expenditure by economic type in 2009: social benefits, 4,146·3bn. roubles; compensation of employees, 1,982·8bn. roubles; grants, 1,693·9bn. roubles.

VAT is 18% (reduced rate, 10%).

Performance

Real GDP growth rates (based on IMF statistics):

2002	2003	2004	2005	2006	2007	2008	2009	2010	2011
4·7%	7·3%	7·2%	6·4%	8·2%	8·5%	5·2%	−7·8%	4·3%	4·3%

GDP growth in 2012 was 3·4% according to the Federal Statistics Service. Total GDP was US$1,857·8bn. in 2011. In June 2002 Russia was acknowledged as a market economy under United States trade law, symbolically underscoring the country's transformation from a state-planned economy.

Banking and Finance

The central bank and bank of issue is the State Bank of Russia (*Chairman*, Sergey Mikhailovich Ignatiev). The Russian Bank for Reconstruction and Development and the State Investment Company were created in 1993 to channel foreign and domestic investment. Foreign bank branches have been operating since Nov. 1992.

By 1995 the number of registered commercial banks had increased to around 5,000 but following the Aug. 1997 liquidity crisis, owing to the ensuing bankruptcies, mergers and the Central Bank's revoking of licences, the number fell to 2,500. This has since fallen to 1,300. Approximately 80% of the commercial banks were state-owned through ministries or state enterprises. Sberbank is the leading bank with assets of US$335·0bn. in Dec. 2011, followed by VTB Bank (formerly Vneshtorgbank) with assets of US$127·3bn. and Gazprombank with assets of US$75·5bn. In Sept. 2011 there were 991 credit institutions with 2,825 branches.

In the wake of one of the worst financial crises that Russia's market economy had experienced, the central bank tripled interest rates to 150% in May 1998 in an effort to restore stability to the financial system. In 2002 the banking sector in Russia was healthier than at any time since the collapse of the former Soviet Union.

External debt was US$384,740m. in 2010, representing 26·9% of GNI.

In 2010 Russia received US$41·2bn. worth of foreign direct investment, up from US$36·5bn. in 2009 but down from the record US$75·0bn. of 2008.

There are stock exchanges in Moscow, Novosibirsk, St Petersburg and Vladivostok.

ENERGY AND NATURAL RESOURCES

Environment

Russia's carbon dioxide emissions from the consumption and flaring of fossil fuels in 2008 accounted for 5·7% of the world total (the third highest after China and the USA), and were equivalent to 12·3 tonnes per capita. An *Environmental Performance Index* compiled in 2008 ranked Russia 28th in the world, with 86·3%. The index examined various factors in six areas—air pollution, biodiversity and habitat, climate change, environmental health, productive natural resources and water resources.

Electricity

In 2007 installed capacity was an estimated 224m. kW and electricity production 1,015·33bn. kWh (675·82bn. kWh thermal, 178·98bn. kWh hydro-electric and 160·04bn. kWh nuclear). Consumption per capita was 7,054 kWh in 2007. There were 31 nuclear reactors in use in 2009.

Oil and Gas

Russia was the largest oil producer in 2009 (overtaking Saudi Arabia's output for the first time) and the second largest exporter in 2008 (after Saudi Arabia). Russia consumes less than a third of the oil that it produces. Oil and gas account for 50% of Russia's export revenues. In 2008 there were proven crude petroleum reserves of 79·0bn. bbls, but they are expected to be exhausted by 2030. 2009 production of oil was 494·2m. tonnes (12·9% of the world total). Oil production rose every year between 1998 and 2007 but fell slightly in 2008 before increasing again in 2009. There is an extensive domestic oil pipeline system. The main export pipeline to Europe is the Druzhba pipeline (crossing Belarus before splitting into northern and southern routes). The main export terminal is at Novorossiisk on the Black Sea. A 964-km pipeline from Skovorodino in Russia to Daqing in China was inaugurated in Jan. 2011, allowing Russia—as the world's largest oil producer—to increase significantly its exports to China. Other export pipeline developments include the Baltic Pipeline System I and II (the first system became operational in Dec. 2001 and the

second in March 2012) and the Caspian Pipeline Consortium's pipeline from Tengiz (Kazakhstan) to Novorossiisk, which was commissioned in March 2001. Construction of a second stage is expected to be completed in the course of 2013.

Output of natural gas in 2011 was a record 607·0bn. cu. metres. Russia is the world's second largest producer but was surpassed as the largest producer by the USA in 2010. It also has the largest reserves of natural gas—in 2008 it had proven reserves of 43,300bn. cu. metres (23% of the world total). There is a comprehensive domestic distribution system (run by state-owned Gazprom, in which the government has a 50·002% stake), as well as gas pipelines linking Russia with former Soviet republics. In Russia's biggest-ever takeover Gazprom agreed in Sept. 2005 to buy a 72·7% stake in Sibneft, a leading oil company. The main export pipelines run from western Siberia through Ukraine and Belarus to European markets. Russia is seeking to diversify its gas export routes and a number of pipeline projects are under development. Russia is also looking to export its natural gas to Asian markets.

The oil and gas boom has meant that the sector's share in Russia's total GDP has risen from 12·7% in 1999 to 31·6% in 2007.

Minerals

Russia contains great mineral resources: coal (17% of the world's total reserves), iron ore, gold, platinum, copper, zinc, lead, tin and rare metals. Output in 2006, unless otherwise indicated (in tonnes): coal (2007), 217·9m.; iron ore (2005), 96·8m.; lignite (2007), 71·1m.; potash (2008), 6·7m.; bauxite, 6·4m.; aluminium, 3·7m.; alumina, 3·3m.; salt, 2·8m.; chrome ore, 966,065; copper (estimate), 739,000; nickel (estimate), 327,000; zinc, 178,000; molybdenum (estimate), 3,100; tin (estimate), 3,000; gold (2007), 157. Diamond production, 2005 estimate: 38·0m. carats. Only Australia produces more diamonds. Annual uranium production is more than 3,000 tonnes.

Agriculture

A presidential decree of Dec. 1991 authorized the private ownership of land on a general basis, but excluded farmland. Nevertheless, large state and collective farms, inherited from the Soviet era, were forced officially to reorganize, with most becoming joint-stock companies. Farm workers could branch off as private farmers by obtaining a grant of land from their parent farm, although they lacked full ownership rights. In 2002 over 90% of Russia's 400m. ha. of farmland remained under the control of the state or former collectives. In Jan. 2003 a new law came into force regulating the possession, use and disposal of land plots designated as agricultural land. The law provides that: the authorities may confiscate farmland if its owners are using it for non-agricultural purposes; regional authorities will have the first option to purchase farmland from its owners; and farmland can only be sold to third parties if authorities refuse their option to buy. The law also deprives foreigners of the right to own agricultural land, although they may lease it for up to 49 years. In 2007 there were 121·57m. ha. of arable land and 1·79m. ha. of permanent crops. There were 4·4m. ha. of irrigated land in 2007.

Output in 2009 (in 1,000 tonnes) included: wheat, 61,740; potatoes, 31,134; sugar beets, 24,892; barley, 17,881; sunflower seeds, 6,454; oats, 5,401; rye, 4,333; maize, 3,963; cabbage, broccoli, etc., 3,312; tomatoes, 2,170; onions, 1,602; apples, 1,596; carrots and turnips, 1,519. Russia was the world's largest producer of barley, oats, rye and sunflower seeds in 2009.

Livestock, 2009: cattle, 21·04m.; sheep, 19·60m.; pigs, 16·16m.; chickens, 366m.

Livestock products in 2009 (in tonnes): poultry meat, 2·3m.; pork, bacon and ham, 2·2m.; beef and veal, 1·7m.; cow's milk, 32·3m.; goat's milk, 0·2m.; eggs, 2·2m.; cheese, 0·6m.

Forestry

Russia has the largest area covered by forests of any country in the world, with 809·1m. ha. in 2010 (49% of the land area). In 2007 timber production was 207·0m. cu. metres. Russia was the world's largest exporter of roundwood in 2007, with 36·2% of the world total.

Fisheries

Total catch in 2010 was 4,069,624 tonnes (down from 8,211,516 tonnes in 1989). Approximately 94% of the fish caught are from marine waters.

INDUSTRY

As a result of Soviet central planning, Russian industry remains dominated by heavy industries, such as energy and metals. As of 1 Jan. 2010 there were 418,600 manufacturing enterprises and organizations (of which 383,300 were private) and 433,700 construction enterprises and organizations (413,400 private). Manufacturing accounted for 13·1% of GDP in 2009 and construction 4·8%.

The leading companies by market capitalization in Russia in March 2012 were: Gazprom, a gas company (US$145·8bn.); Rosneft, an oil and gas company (US$75·7bn.); and Sberbank of Russia (US$72·2bn.)

Output in 2007 (unless otherwise indicated) includes: (in tonnes) crude steel, 72·4m.; residual fuel oil, 67·7m.; distillate fuel oil, 66·3m.; cement, 59·9m.; rolled steel, 59·7m.; pig iron, 51·5m.; petrol, 35·1m.; coke, 32·3m.; jet fuel, 10·7m.; sulphuric acid (2006), 9·5m.; steel pipe, 8·7m.; bread, 7·9m.; paper and paperboard, 7·6m.; raw sugar, 6·1m.; clays, 2·2m.; caustic soda, 1·3m.; (in cu. metres) sawnwood, 22·1m.; plywood, 2·8m.; (in sq. metres) cotton fabrics, 2,108m.; woollen fabrics, 28·7m.; (in units) cigarettes, 397bn.; bricks, 13,090m.; televisions, 6·8m.; refrigerators and freezers, 3·5m.; washing and drying machines, 2·7m.; microwaves, 2·5m.; bicycles, 1·5m.; passenger cars, 1·3m.; trucks, 285,030; combine harvesters, 8,060; tractors, 7,500; (in litres) beer, 11,472m.; soft drinks, 5,982m.; mineral water, 3,632m.; spirits and liqueurs, 1,315m.

Labour

In 2010 the economically active population numbered 75·45m. (38·58m. males and 36·87m. females). Of those in employment in 2010, 18·1% worked in wholesale and retail trade/repair of motor vehicles, motorcycles and personal and household goods, 15·4% in manufacturing, 9·6% in agriculture and forestry, 8·8% in education, 7·9% in transport, 7·8% in construction and 7·0% in health and social work. The unemployment rate was 9·9% in May 2009—with 7·5m. people unemployed using ILO methodology— down from 10·2% in April 2009 although up from 6·1% in Oct. 2008. Average monthly wages were 20,952·2 roubles in 2010 (compared to 8,554·9 roubles in 2005); the monthly minimum wage in 2010 was 4,330 roubles (up from 132 roubles in 2000). In 2010, 18·5m. people, or 13·1% of the population, had an average per capita money income lower than the subsistence minimum (down from 29·0% of the population in 2000). The state Federal Employment Service was set up in 1992. Unemployment benefits are paid by the Service for 12 months, payable at: 75% of the average monthly wage during the last two months preceding unemployment for the first three months; 60% for the next four months; and 45% for the last five months. Annual paid leave is 24 working days. In 2005, 85,900 working days were lost through strikes (6,000,500 in 1996). Retirement age is 55 years for women, 60 for men.

INTERNATIONAL TRADE

In Jan. 2010 Russia joined the newly established Customs Union, together with Belarus and Kazakhstan.

Imports and Exports

The following table shows the value of Russia's foreign trade (in US$1bn.):

	2006	2007	2008	2009	2010
Imports c.i.f.	137·8	199·8	267·1	167·3	229·1
Exports f.o.b.	301·2	351·9	467·6	301·7	396·6

Of imports in 2010, 44·4% by value was machinery and transport equipment, 16·1% chemicals products and rubber, 15·9% foodstuffs and agricultural raw materials, and 7·4% metals and precious stones. Of exports, 68·4% by value was mineral products, 12·8% metals and precious stones, 6·2% chemical products and rubber, and 5·4% machinery and transport equipment. China provided 13·4% of imports in 2009, Germany 12·4%, USA 5·4%, Ukraine 5·3% and France 4·9%. In 2009 the Netherlands accounted for 12·0% of exports, Italy 8·3%, Germany 6·2%, Belarus 5·5% and China 5·5%.

COMMUNICATIONS

Roads

There were 933,000 km of roads in 2006, of which 80·9% were hard surfaced. In 2007, 78bn. passenger-km were travelled by road. There were 29,249,000 passenger cars in use in 2007 plus 4,730,000 lorries and vans and 861,000 buses and coaches. In 2007 there were 33,300 road deaths.

Rail

Length of railways in 2010 was 86,000 km, of which about half is electrified. In 2008, 1,295·6m. passengers and 1,304·7m. tonnes of freight were carried by rail; passenger-km travelled came to 176bn. and freight tonne-km to 2,116bn. There are metro services in Moscow (309 km), St Petersburg (105 km), Nizhny Novgorod (15 km), Novosibirsk (14 km), Samara (10 km), Ekaterinburg (9 km) and Kazan (7 km). Kazan's metro opened in 2005, making it the first metro to be opened in Russia since the breakup of the Soviet Union.

Civil Aviation

The main international airports are at Moscow (Domodedovo, Sheremetyevo and Vnukovo) and St Petersburg (Pulkovo). The national carrier is Aeroflot International Russian Airlines (51·2% state-owned), which carried 11·3m. scheduled passengers in 2010. Rossiya, S7 Airlines, Transaero and UTair also operate internationally.

In 2009 scheduled airline traffic of Russian-based carriers flew 836m. km, carrying 34,403,000 passengers (11,992,000 on international flights). The three busiest airports all serve Moscow. Domodedovo is Russia's busiest airport in terms of passenger traffic (22,255,000 in 2010, a 19% increase on 2009). Sheremetyevo handled 19,329,000 passengers in 2010 (up 31% on 2009), and Vnukovo handled 9,460,000 in 2010 (up 22% on 2009). The fourth busiest airport is Pulkovo, which serves St Petersburg and handled 8,444,000 passengers in 2010 (up 25% on 2009).

Shipping

In Jan. 2009 there were 1,272 ships of 300 GT or over registered, totalling 4·90m. GT. Of the 1,272 vessels registered, 857 were general cargo ships, 296 oil tankers, 58 bulk carriers, 24 chemical tankers, 24 passenger ships, 12 container ships and one liquid gas tanker. The Russian-controlled fleet comprised 1,418 vessels of 1,000 GT or over in Jan. 2009, of which 945 were under the Russian flag and 473 under foreign flags. In 2010, 16m. passengers and 102m. tonnes of freight were carried on 101,000 km of inland waterways. The busiest ports are Novorossiisk (which handled 81,633,000 tonnes in 2008), Primorsk (75,582,000 tonnes in 2008) and St Petersburg (59,945,000 tonnes in 2008).

Telecommunications

In 2008 there were 44,897,000 main (fixed) telephone lines. In the same year mobile phone subscribers numbered 199,522,000

(1,411·1 per 1,000 persons). There were 61·5m. internet users in 2011. In March 2012 there were 5·2m. Facebook users.

SOCIAL INSTITUTIONS

Justice

The Supreme Court is the highest judicial body on civil, criminal and administrative law. The Supreme Arbitration Court deals with economic cases. The KGB, and the Federal Security Bureau which succeeded it, were replaced in Dec. 1992 by the Federal Counter-Intelligence Service. The legal system is, however, crippled by corruption.

A new civil code was introduced in 1993 to replace the former Soviet code. It guarantees the inviolability of private property and includes provisions for the freedom of movement of capital and goods.

12-member juries were introduced in a number of courts after Nov. 1993, but in the years that followed jury trials were not widely used. However, on 1 Jan. 2003 jury trials began to be phased in nationwide. A new criminal code came into force on 1 Jan. 1997, based on respect for the rights and freedoms of the individual and the sanctity of private property. A further new code that entered force on 1 July 2002 introduced new levels of protection for defendants and restrictions on law enforcement officials. The death penalty is retained for five crimes against the person. It is not applied to minors, women or men over 65.

In 2010, 2,629,000 crimes were recorded, including 15,600 murders and attempted murders (down from 31,800 in 2000), 1,108,000 thefts and 4,900 rapes and attempted rapes. In 1996 there were 140 executions (86 in 1995; 1 in 1992). President Yeltsin placed a moratorium on capital punishment in 1996 when Russia joined the Council of Europe, but parliament has refused to abolish the death penalty. The last execution was in Aug. 1996. The prison population in May 2011 was 806,100 (568 per 100,000 population—the second highest rate in the world after the USA). In 2010 there were 1,052 prison establishments and institutions.

Education

Adult literacy rate in 2008 was over 99%. In 2007 there were 5·01m. pupils in primary schools, 10·80m. in secondary schools and 9·37m. students in tertiary education (56·8% of whom were females). There were 5·23m. children in 45,300 pre-school establishments in 2009.

In 2006 public expenditure on education came to 4·0% of GNI.

Russia's largest university is the M. V. Lomonosov Moscow State University. Founded in 1755, it now has 30 faculties and 15 research centres. It has an annual enrolment of 31,000 students. The Russian Academy of Sciences, founded in 1724 and reorganized in 1925 as the Academy of Sciences of the Union of Soviet Socialist Republics, was restored under its present name in 1991. It is the highest scientific self-governing institution in Russia and has 18 divisions on particular areas of science. The Academy also has three regional branches: the Urals Branch, the Siberian Branch and the Far East Branch.

Health

Doctors in 2006 numbered 614,183. In 2006 Russia had 43 doctors per 10,000 people and 97 hospital beds per 10,000 persons. There were 45,628 dentists, 1,214,292 nurses and midwives and 11,521 pharmacists in 2006. Expenditure on health in 2009 was 5·4% of GDP. Russia has among the highest rates of growth of HIV cases in the world; by 31 Dec. 2007 there were 416,113 registered cases. In 2005 there were 150 cases of tuberculosis per 100,000 people. In 2009, 39·1% of Russian adults smoked (males, 60·2%; females, 21·7%).

Welfare

Russia is in the process of implementing a reform of its pensions system, the focus of which is to move away from a distributive system to an accumulating (funded) scheme. Instead of citizens paying 28% of their monthly salary into the state pension fund, since 2004 it has been possible to pay between 2% and 6% to private asset managers.

State welfare provision includes: old age, disability and survivor pensions; sickness and maternity benefits; work injury payments; unemployment benefits; and family allowances. The basic pension is a flat amount provided to all those reaching retirement age (60 for males and 55 for females) with a minimum contribution record of five years. The old-age pension is calculated as the sum of three components. In 2010 the basic monthly flat-rate amount for pensioners under the age of 80 varied from 2,622 roubles to 5,124 roubles according to the number of dependents; the basic monthly flat-rate amount for a pensioner age 80 or older varied from 5,124 roubles to 7,686 roubles according to the number of dependents.

RELIGION

The Russian Orthodox Church is the largest religious association in the country. In early 2003 it had 128 dioceses (compared with 67 in 1989), over 19,000 parishes (6,893 in 1988) and about 480 monasteries (18 in 1980). There were also five theological academies, 26 seminaries, 29 pre-seminaries, two Orthodox universities, a theological institute, a women's pre-seminary and 28 icon-painting schools. In 2001 there were 23·6m. adherents. The total number of theological students was around 6,000. There are still many Old Believers, whose schism from the Orthodox Church dates from the 17th century. The Russian Church is headed by the Patriarch of Moscow and All Russia (Patriarch Kirill I—Metropolitan Kirill of Smolensk and Kaliningrad, b. 1946; elected Jan. 2009), assisted by the Holy Synod, which has seven members—the Patriarch himself and the Metropolitans of Krutitsy and Kolomna (Moscow), St Petersburg and Kyiv ex officio, and three bishops alternating for six months in order of seniority from the three regions forming the Moscow Patriarchate. The Patriarchate of Moscow maintains jurisdiction over 119 eparchies, of which 59 are in Russia; there are parishes of Russian Orthodox abroad, in Belarus, Ukraine, Kazakhstan, Moldova, Uzbekistan, the Baltic states, and in Damascus, Geneva, Prague, New York and Japan. There is a spiritual mission in Jerusalem, and a monastery at Mt Athos in Greece. A Russian Orthodox church was consecrated in Dublin in Ireland in Feb. 2003. Muslims represent the second largest religious community in Russia, numbering 19m. In Feb. 2010 the Supreme Co-ordinating Council of Russian Muslims was established to be co-chaired by the heads of the three major organizations—Talgat Tajuddin of the Central Spiritual Board of Muslims, Ravil Gainutdin of the Council of Muftis of Russia and Ismail Berdiyev of the Co-ordinating Muslim Council of the North Caucasus. There are 1,150 Protestant and 42 Jewish communities.

CULTURE

World Heritage Sites

Russia's World Heritage sites as classified by UNESCO (with year entered on list) are: the Historic Centre of St Petersburg (1990); the Kremlin and Red Square in Moscow (1990); Khizi Pogost (1990); the Historic Monuments of Novgorod and surroundings (1992); Cultural and Historic Ensemble of the Solovetsky Islands (1992); the White Monuments of Vladimir and Suzdal (1992); Architectural Ensemble of the Trinity Sergius Lavra in Sergiev Posad (1993); the Church of the Ascension, Kolomenskoye (1994); Virgin Komi Forests (1995); Lake Baikal (1996); Volcanoes of Kamchatka (1996, 2001); Golden Mountains of Altai (1998); Western Caucasus (1999); the Ensemble of Ferapontov Monastery (2000); Historic and Architectural Complex of the Kazan Kremlin (2000); Central Sikhote-Alin (2001); the Citadel, Ancient City and Fortress Buildings of Derbent (2003); Ensemble of the Novodevichy Convent in southwest Moscow (2004); Natural System of Wrangel Island Reserve (2004); the historical centre of the city of Yaroslavl (2005); the Putorana Plateau (2010); and the Lena Pillars Nature Park (2012).

The Russian Federation also shares three UNESCO sites: the Curonian Spit with Lithuania (2000); Uvs Nuur Basin with Mongolia (2003); and the Struve Geodetic Arc (2005), a chain of survey triangulations spanning from Norway to the Black Sea that helped establish the exact shape and size of the earth, with nine other countries.

Press
In 2008 there were 533 daily newspapers. There were 27,510 non-daily newspapers in 2008. The most popular daily newspaper in 2008 was *Moskovsky Komsomolets*, with an average daily circulation of 750,000, followed by *Komsomolskaya Pravda*, with a circulation of 716,000. A presidential decree of 22 Dec. 1993 brought the press agencies ITAR-TASS and RIA-Novosti under state control. In 2008, 123,336 new or revised books were published, a figure exceeded only by China, the UK and the USA.

Tourism
In 2011 arrivals of non-resident visitors—including Russians living abroad—totalled 24,932,000 (22,281,000 in 2010), of which 2,336,000 were tourists (2,134,000 in 2010). There were 7,866 hotels and similar establishments in 2011, with 537,000 beds.

Festivals
One of the most popular festivals is the White Nights Festival, which is held from May to July in St Petersburg and combines ballet, opera and music performances. The Annual Summer Ballet Festival runs from June to Sept. in Moscow while the Moscow International Film Festival takes place in June. The biggest rock and pop festival is Nashestvie, which is held annually near Moscow in Aug.

DIPLOMATIC REPRESENTATIVES

Of Russia in the United Kingdom (13 Kensington Palace Gdns, London, W8 4QX)
Ambassador: Alexander Yakovenko.

Of the United Kingdom in Russia (Smolenskaya Naberezhnaya 10, 121099 Moscow)
Ambassador: Tim Barrow, CMG, LVO, MBE.

Of Russia in the USA (2650 Wisconsin Ave., NW, Washington, D.C., 20007)
Ambassador: Sergei Kislyak.

Of the USA in Russia (8 Bolshoy Devyatinskiy Pereuulok, 121099 Moscow)
Ambassador: Michael McFaul.

Of Russia to the United Nations
Ambassador: Vitaly Churkin.

Of Russia to the European Union
Ambassador: Vladimir A. Chizhov.

FURTHER READING
Rossiiskii Statisticheskii Ezhegodnik. Annual (title varies)

Acton, E., et al., *Critical Companion to the Russian Revolution.* 1997
Aslund, Anders, (ed.) *Economic Transformation in Russia.* 1994.—*Building Capitalism: the Transformation of the Former Soviet Bloc.* 2002
Bacon, Edwin, *Securitising Russia: The Domestic Politics of Putin.* 2006.—*Contemporary Russia.* 2nd ed. 2010
Brady, Rose, *Kapitalizm: Russia's Struggle to Free its Economy.* 2000
Cambridge Encyclopedia of Russia and the Former Soviet Union. 1995
Evans, Alfred B., *Russian Civil Society: A Critical Assessment.* 2005
Fowkes, B. (ed.) *Russia and Chechnia: The Permanent Crisis, Essays on Russo-Chechen Relations.* 1998
Freeze, G. (ed.) *Russia: A History.* 1997
Gall, C. and de Waal, T., *Chechnya: Calamity in the Caucasus.* 1998
Gessen, Masha, *The Man Without a Face: The Unlikely Rise of Vladimir Putin.* 2012
Granville, Brigitte and Oppenheimer, Peter, (eds.) *Russia's Post-Community Economy.* 2001
Gustafson, Thane, *Capitalism Russian-Style.* 2000
Hollander, Paul, *Political Will and Personal Belief: The Decline and Fall of Soviet Communism.* 2000
Holmes, Stephen, *The State After Communism: Governance in the New Russia.* 2006
Hosking, Geoffrey, *Russia and the Russians, A History from Rus to the Russian Federation.* 2001
Kanet, Roger E. (ed.) *Russia: Re-Emerging Great Power.* 2007
Kochan, L. and Keep, J., *The Making of Modern Russia.* 3rd ed. 1997
Kotkin, Stephen, *Armageddon Averted: the Soviet Collapse 1970–2000.* 2001
Lieven, A., *Chechnya: Tombstone of Russian Power.* 1998
Lloyd, J., *Rebirth of a Nation.* 1998
Marks, Steven, *How Russia Shaped the Modern World: From Art to Anti-Semitism, Ballet to Bolshevism.* 2002
Mendras, Marie, *Russian Politics: The Paradox of a Weak State.* 2012
Paxton, J., *Encyclopedia of Russian History.* 1993.—*Leaders of Russia and the Soviet Union.* 2004
Ponsard, Lionel, *Russia, NATO and Cooperative Security.* 2006
Putin, Vladimir, *First Person*; interviews, translated from Russian. 2000
Remington, Thomas F., *Politics in Russia.* 2005
Remnick, D., *Resurrection: The Struggle for a New Russia.* 1998
Riasanovsky, N. V., *A History of Russia.* 8th ed. 2010
Sakwa, R., *Russian Politics and Society.* 4th ed. 2008.—*Communism in Russia.* 2010
Service, Robert, *A History of Twentieth-Century Russia.* 1997.—*Lenin: A Biography.* 2000.—*Russia: Experiment with a People.* 2002
Shevtsova, Lilia, *Putin's Russia.* 2003
Shiraev, Eric, *Russian Government and Politics.* 2nd ed. 2013
Shriver, G. (ed. and transl.) *Post-Soviet Russia, A Journey Through the Yeltsin Era.* 2000
Tsygankov, Andrei P., *Russia's Foreign Policy: Change and Continuity in National Identity.* 2006
Webber, Stephen L., *Military and Society in Post-Soviet Russia.* 2006
Westwood, J. N., *Endurance and Endeavour: Russian History, 1812–1992.* 4th ed. 1993
White, Stephen, Sakwa, Richard and Hale, Henry E. (eds.) *Developments in Russian Politics 7.* 2009

National Statistical Office: Federal State Statistics Service, 39 Miasnitskaya St., 103450 Moscow.
Website: http://www.gks.ru

THE REPUBLICS

Status
The 21 republics that with Russia itself constitute the Russian Federation were part of the RSFSR in the Soviet period. On 31 March 1992 the federal government concluded treaties with the then 20 republics, except Checheno-Ingushetia and Tatarstan, defining their mutual responsibilities. The *Council of the Heads of the Republics* is chaired by the Russian President and includes the Russian Prime Minister. Its function is to provide an interaction between the federal government and the republican authorities.

Adygeya

Part of Krasnodar Territory. Area, 7,600 sq. km (2,950 sq. miles); population (2002 census), 477,109. Estimated population, 1 Jan.

2010, 443,200. Capital, Maikop (2002 census, 156,931). Established 27 July 1922; granted republican status in 1991.

President: Aslan Tkhakushinov, b. 1947 (in office since 13 Jan. 2007).

Prime Minister: Murat Kumpilov, b. 1973 (in office since 12 May 2008—acting until 28 May 2008).

Chief industries are timber, woodworking and food processing; there is some engineering and gas production. Agriculture consists primarily of crops (beets, wheat, maize), on partly irrigated land. Industry accounted for 15·2% of gross regional product in 2003 and agriculture 13·6%.

In 2004 there were 12,400 pupils in 126 pre-school institutions and 50,700 pupils in 175 primary and secondary day schools. There were 20,000 students at the two institutions of higher education, Adygeya State University and Maikop State Technological Institute.

In 2004 the rates of doctors and hospital beds per 10,000 population were 37·2 and 110 respectively.

Altai

Part of Altai Territory. Area, 92,600 sq. km (35,750 sq. miles); population (2002 census), 202,947. Estimated population, 1 Jan. 2010, 210,700. Capital, Gorno-Altaisk (2002 census, 53,538). Established 1 June 1922 as Oirot Autonomous Region; renamed 7 Jan. 1948; granted republican status in 1991 and renamed in 1992.

Chairman of the Government: Aleksandr Berdnikov (since 20 Jan. 2006).

Cattle breeding predominates. Chief industries are clothing and footwear, foodstuffs, gold mining, timber, chemicals and dairying. Industrial output was valued at 838m. roubles in 2004 and agricultural output at 2,972m. roubles. In 2000, 91,200 people were economically active, of whom 72,000 were in employment.

In 2004 there were 6,600 pupils in 130 pre-school institutions and 32,300 pupils in 202 primary and secondary day schools. There were 6,000 students at Gorno-Altaysk State University.

The rates of doctors and hospital beds per 10,000 population in 2004 were 38·0 and 123 respectively.

Bashkortostan

Area 143,600 sq. km (55,450 sq. miles), population (2002 census), 4,104,336. Estimated population, 1 Jan. 2010, 4,066,000. Capital, Ufa (2002 census population, 1,042,437). Bashkiria was annexed to Russia in 1557. It was constituted as an Autonomous Soviet Republic on 23 March 1919. A declaration of republican sovereignty was adopted in 1990, and a declaration of independence on 28 March 1992. A treaty of Aug. 1994 with Russia preserves the common legislative framework of the Russian Federation while defining mutual areas of competence. The main ethnic groups are Russians, Tatars and Bashkirs. There are also Chuvash and Mari minorities.

A constitution was adopted on 24 Dec. 1993. It states that Bashkiria conducts its own domestic and foreign policy, that its laws take precedence in Bashkiria, and that it forms part of the Russian Federation on a voluntary and equal basis.

President: Rustem Khamitov (in office since 15 July 2010—acting until 19 July 2010).

Industrial production was valued at 354,000m. roubles in 2004 and agricultural output at 57,160m. roubles. The most important industries are oil and oil products; there are also engineering, glass and building materials enterprises. Agriculture specializes in wheat, barley, oats and livestock.

In 2004 there were 144,600 pupils in 1,904 pre-school institutions and 583,300 pupils in 3,187 primary and secondary day schools. There were 150,200 students in 17 institutions of higher education. There is a state university and a branch of the Academy of Sciences with eight learned institutions.

In 2004 the rates of doctors and hospital beds per 10,000 population were 41·9 and 104 respectively.

Buryatia

Area is 351,300 sq. km (135,650 sq. miles). The Buryat Republic, situated to the south of Sakha, adopted the Soviet system on 1 March 1920. This area was penetrated by the Russians in the 17th century and finally annexed from China by the treaties of Nerchinsk (1689) and Kyakhta (1727). Population (2002 census), 981,238. Estimated population, 1 Jan. 2010, 963,500. Capital, Ulan-Ude (2002 census population, 359,391). The main ethnic groups are Russians, followed by Buryats. There are also Ukrainian, Tatar and Belarusian minorities.

There is a 65-member parliament, the *People's Hural.*

Head of the Republic: Vyacheslav Nagovitsyn (since 12 May 2012; previously president from 10 July 2007–12 May 2012).

The main industries are engineering, brown coal and graphite, timber, building materials, sheep and cattle farming. Industrial production was valued at 32,161m. roubles in 2004 and agricultural output at 8,352m. roubles.

In 2004 there were 30,100 pupils in 423 pre-school institutions and 139,000 pupils in 572 primary and secondary day schools. There were 31,900 pupils in five institutions of higher education.

In 2004 the rates of doctors and hospital beds per 10,000 population were 38·3 and 109 respectively. The level of poverty in Buryatia was 38·2% in 2003.

Chechnya

GENERAL DETAILS

The area of the Republic of Chechnya is 15,000 sq. km (5,800 sq. miles). The population at the 2002 census was 1,103,686. The estimated population at 1 Jan. 2010 was 1,268,000. Capital, Dzhohar (since March 1998; previously known as Grozny; 2002 census population, 210,720). The Chechens and Ingushes were conquered by Russia in the late 1850s. In 1920 each nationality were constituted areas within the Soviet Mountain Republic and the Chechens became an Autonomous Region on 30 Nov. 1922. In Jan. 1934 the two regions were united, and on 5 Dec. 1936 constituted as the Checheno-Ingush Autonomous Republic. This was dissolved in 1944 and the population was deported en masse, allegedly for collaboration with the German occupation forces. It was reconstituted on 9 Jan. 1957: 232,000 Chechens and Ingushes returned to their homes in the next two years.

In 1991 rebel leader Jokhar Dudayev seized control of Chechnya and won elections. In Nov. he declared an independent Chechen Republic. Ingush desire to separate from Chechnya led to fighting along the Chechen-Ingush border and a deployment of Russian troops. An agreement to withdraw was reached between Russia and Chechnya on 15 Nov. 1992. The separation of Chechnya and Ingushetia was formalized in Dec. 1992. In April 1993 President Dudayev dissolved parliament. Hostilities continued throughout 1994 between the government and forces loosely grouped under the 'Provisional Chechen Council'. The Russian government, which had never recognized the Chechen declaration of independence of Nov. 1991, moved troops and armour into Chechnya on 11 Dec. 1994. Grozny was bombed and attacked by Russian ground forces at the end of Dec. 1994 and the presidential palace was captured on 19 Jan. 1995, but fighting continued. On 30 July 1995 the Russian and Chechen authorities signed a ceasefire. However, hostilities, raids and hostage-taking continued; Dudayev was killed in April 1996 and a ceasefire was agreed on 30 Aug. 1996.

Fighting broke out again, however, in Sept. 1999 as Russian forces launched attacks on 'rebel bases'. Fighting intensified and more than 200,000 civilians were forced to flee, mostly to neighbouring Ingushetia. By Feb. 2000 much of Grozny had been destroyed and was closed by the Russians. In June 2000 Vladimir Putin declared direct rule. The conflict continues, with estimates of the number of deaths varying from 6,500 to 15,000. Over 4,000 Russian soldiers have been killed. However, on 18 Nov. 2001 the first official meeting between negotiators for the Russian government and Chechen separatists took place. In Oct. 2002 a group of Chechen rebels took control of a Moscow theatre and held hostage 800 people for three days, before Russian troops stormed the building. An anaesthetic gas, used to combat the rebels, also killed many of the hostages.

On 23 March 2003 a referendum was held on a new constitution that would keep Chechnya within Russia but give it greater autonomy, and provide a new president and parliament for the republic. Although 96% of votes cast were in favour of the new constitution there was criticism of the conduct of the referendum. Presidential elections held on 5 Oct. 2003 were won by the Kremlin-backed candidate Akhmad Kadyrov, with 80·8% of the vote, but there was widespread condemnation of the electoral process. President Kadyrov was assassinated on 9 May 2004. Presidential elections held on 29 Aug. 2004, widely seen as rigged, were won by the Kremlin-backed Alu Alkhanov with 73·5% of the vote, against 5·9% for Movsur Khamidov, head of the Chechen department of the Federal Security Service. There were five other candidates. Turnout was 85·2%. On 27 Nov. 2005 the first parliamentary elections took place since Russian troops restored Moscow's control over Chechnya in 1999. In the elections to the People's Assembly (lower chamber) the United Russia party won 19 of 38 seats with 60·7% of the vote, the Communist Party 3 with 12·2%, the Union of Rightist Forces 1 with 12·4%, the Eurasian Union 1 with 3·9%; independents won 14 seats. In the Council of the Republic (upper chamber), United Russia won 14 of 20 seats, the Communist Party 3 and the Union of Rightist Forces 3.

Separatist President Aslan Maskhadov was killed by Russian troops on 8 March 2005, as was his successor Abdul-Khalim Sadulayev on 17 June 2006.

Moscow-backed Head of the Republic: Ramzan Kadyrov; b. 1976.
Prime Minister: Abubakar Edelgeriyev; b. 1974.
Amir of the Caucasus Emirate (self-proclaimed): Doku Umarov; b. 1964.

Checheno-Ingushetia had a major oilfield, and a number of engineering works, chemical factories, building materials works and food canneries. There was a timber, woodworking and furniture industry. In 2003 oil production was 1·8m. tonnes. Chechnya's oil reserves are estimated at some 220m. bbls. Industrial output in the two republics was valued at 213,000m. roubles in 1993, agricultural output at 79,000m. roubles.

In 2004 there were 212,300 pupils in 460 primary and secondary day schools. There were 23,500 students in three institutions of higher education. In 1995 the rates of doctors and hospital beds per 10,000 population were 21·1 and 91 respectively.

FURTHER READING

Jagielski, Wojciech, *Towers of Stone: The Battle of Wills in Chechnya.* 2009
Lieven, A. and Bradner, H., *Chechnya: Tombstone of Russian Power.* 1999

Chuvashia

Area, 18,300 sq. km (7,050 sq. miles); population (2002 census), 1,313,754. Estimated population, 1 Jan. 2010, 1,278,400. Capital, Cheboksary (2002 census population, 440,621). The territory was annexed by Russia in the middle of the 16th century. On 24 June 1920 it was constituted as an Autonomous Region, and on 21 April 1925 as an Autonomous Republic. The main ethnic groups are Chuvash, followed by Russians. There are also Tatar and Mordovian minorities. Republican sovereignty was declared in Sept. 1990.

President: Mikhail Ignatyev (in office since 29 Aug. 2010).
Prime Minister: Ivan Motorin (in office since 23 Dec. 2011).

The timber industry antedates the Soviet period. Other industries include railway repair works, electrical and other engineering industries, building materials, chemicals, textiles and food industries. Grain crops account for nearly two-thirds of all sowings and fodder crops for nearly a quarter. Chuvashia is Russia's main producer of hops and the republic has a significant brewing industry. Industrial output was valued at 49,826m. roubles in 2004 and agricultural output at 12,722m. roubles.

In 2004 there were 47,100 pupils at 433 pre-school institutions and 166,200 pupils in 619 primary and secondary day schools. There were 63,900 students in seven higher educational establishments.

In 2004 the rates of doctors and hospital beds per 10,000 population were 46·8 and 116 respectively.

Dagestan

Area, 50,300 sq. km (19,400 sq. miles); population (2002 census), 2,576,531. Estimated population, 1 Jan. 2010, 2,737,300. Capital, Makhachkala (2002 census population, 462,412). Over 30 nationalities inhabit this republic apart from Russians; the most numerous are Dagestanis and there are also Azerbaijani, Chechen and Jewish minorities. Annexed from Persia in 1723, Dagestan was constituted an Autonomous Republic on 20 Jan. 1921. In 1991 the Supreme Soviet declared the area of republican, rather than autonomous republican, status. Many of the nationalities who live in Dagestan have organized armed militias, and in May 1998 rebels stormed the government building in Makhachkala. In Aug. 1999 Dagestan faced attacks from Islamic militants who invaded from Chechnya. Although Russian troops tried to restore order and discipline, the guerrilla campaign has continued with a series of bombings targeting Russian military personnel.

Acting President: Ramazan Abdulatipov (since 28 Jan. 2013).
Prime Minister: Mukhtar Medzhidov (in office since 31 Jan. 2013).

There are engineering, oil, chemical, woodworking, textile, food and other light industries. Agriculture is varied, ranging from wheat to grapes, with sheep farming and cattle breeding. Industrial output was valued at 6,568m. roubles in 2001 and agricultural output at 13,162m. roubles.

In 2004 there were 55,900 pupils in 593 pre-schools and 443,300 pupils in 1,683 primary and secondary day schools. There were 105,500 students in 15 institutions of higher education. There is a branch of the Russian Academy of Sciences with ten learned institutions.

In 2004 the rates of doctors and hospital beds per 10,000 population were 38·2 and 70 respectively.

Ingushetia

The history of Ingushetia is interwoven with that of Chechnya (*see above*). Ingush desire to separate from Chechnya led to fighting along the Chechen-Ingush border and a deployment of Russian troops. The separation of Ingushetia from Chechnya was formalized by an amendment of Dec. 1992 to the Russian Constitution. On 15 May 1993 an extraordinary congress of the peoples of Ingushetia adopted a declaration of state sovereignty within the Russian Federation. Skirmishes between Ingush refugees and local police broke out in Aug. 1999 and tensions remained high with the danger of further outbreaks of fighting. The Russian attacks on neighbouring Chechnya in Sept. 1999 led to thousands of Chechen refugees fleeing to Ingushetia. In April 2004 President Murat Zyazikov survived an assassination attempt, as did Prime Minister Ibragim Malsagov in Aug. 2005.

The capital is Magas (since 1999; formerly Nazran; 2002 census population, 125,066).

Area, 4,300 sq. km (1,700 sq. miles); population (2002 census), 467,294. Estimated population, 1 Jan. 2010, 516,700.

There is a 27-member parliament. On 27 Feb. 1994 presidential elections and a constitutional referendum were held. Turnout was 70%. At the referendum 97% of votes cast approved a new constitution stating that Ingushetia is a democratic law-based secular republic forming part of the Russian Federation on a treaty basis.

President: Yunus-bek Yevkurov.

Prime Minister: Musa Chiliev.

Industry accounted for 14·6% of gross regional product in 2003 and agriculture 14·1%. A special economic zone for Russian residents was set up in 1994, and an 'offshore' banking tax haven in 1996.

In 2004 there were 2,500 pupils in 20 pre-school institutions and 64,600 pupils in 113 primary and secondary day schools. There were 9,400 students in five institutions of higher education.

In 2004 the rates of doctors and hospital beds per 10,000 population were 22·8 and 41 respectively.

Kabardino-Balkaria

Area, 12,500 sq. km (4,850 sq. miles); population (2002 census), 901,494. Estimated population, 1 Jan. 2010, 893,800. Capital, Nalchik (2002 census population, 274,974). Kabarda was annexed to Russia in 1557. The republic was constituted on 5 Dec. 1936. The main ethnic groups are Kabardinians, followed by Russians and Balkars. There are also Ukrainian, Ossetian and German minorities.

A treaty with Russia of 1 July 1994 defines their mutual areas of competence within the legislative framework of the Russian Federation. The recent history of Kabardino-Balkaria has been marked by the instability that has plagued the whole of the north Caucasus. In Oct. 2005 militants staged a large-scale assault on government buildings in Nalchik, an act for which Chechen rebel leader Shamil Besayev claimed responsibility. All mosques in the capital have been closed.

President: Arsen Kanokov (since 28 Sept. 2005).

Prime Minister: Ruslan Khasanov (since 1 Nov. 2012).

Main industries are ore-mining, timber, engineering, coal, food processing, timber and light industries, building materials. Grain,

livestock breeding, dairy farming and wine-growing are the principal branches of agriculture. Agriculture accounted for 31·9% of gross regional product in 2003 and industry 14·2%.

In 2004 there were 26,100 pupils in 102 pre-school institutions and 123,400 pupils in 371 primary and secondary day schools. There were 28,500 students in four institutions of higher education. There is a branch of the Academy of Sciences with five learned institutions.

In 2004 the rates of doctors and hospital beds per 10,000 population were 41·3 and 101 respectively.

Kalmykia

Area, 76,100 sq. km (29,400 sq. miles); population (2002 census), 292,410. Estimated population, 1 Jan. 2010, 283,200. Capital, Elista (2002 census population, 104,254). The population is mainly Kalmyk and Russian, with small Chechen, Kazakh and German minorities.

The Kalmyks migrated from western China to Russia (Nogai Steppe) in the early 17th century. The territory was constituted an Autonomous Region on 4 Nov. 1920, and an Autonomous Republic on 22 Oct. 1935; this was dissolved in 1943. On 9 Jan. 1957 it was reconstituted as an Autonomous Region and on 29 July 1958 as an Autonomous Republic once more. In Oct. 1990 the republic was renamed the Kalmyk Soviet Socialist Republic; it was given its present name in Feb. 1992.

President: Aleksey Orlov (in office since 24 Oct. 2010).

Prime Minister: Lyudmila Ivanova (in office since 8 Feb. 2011—acting until 15 Feb. 2011).

In April 1993 the Supreme Soviet was dissolved and replaced by a professional parliament consisting of 25 of the former deputies. On 5 April 1994 a specially-constituted 300-member constituent assembly adopted a 'Steppe Code' as Kalmykia's basic law. This is not a constitution and renounces the declaration of republican sovereignty of 18 Oct. 1990. It provides for a *President* elected for five-year terms with the power to dissolve parliament, and a 27-member parliament, the *People's Hural*, elected every four years. It stipulates that Kalmykia is an equal member and integral part of the Russian Federation, functioning in accordance with the Russian constitution.

Main industries are oil and gas production, canning and building materials. Cattle breeding and irrigated farming (mainly fodder crops) are the principal branches of agriculture. Overgrazing during the Soviet period has led to the desertification of Kalmykia's pastures and agricultural output has declined substantially in recent years. Agriculture accounted for 10·4% of gross regional product in 2003 and industry 6·4%.

In 2004 there were 9,500 pupils in 119 pre-school institutions and 44,900 pupils in 213 primary and secondary day schools. There were 10,000 students in two institutions of higher education. In 2004 the rates of doctors and hospital beds per 10,000 population were 51·0 and 136 respectively. The main religion is Buddhism.

Karachai-Cherkessia

Area, 14,300 sq. km (5,500 sq. miles); population (2002 census), 439,470. Estimated population, 1 Jan. 2010, 427,000. Capital, Cherkessk (2002 census population, 116,244). A Karachai Autonomous Region was established on 26 April 1926 (out of a previously united Karachaevo-Cherkess Autonomous Region created in 1922), and dissolved in 1943. A Cherkess Autonomous

Region was established on 30 April 1928. The present Autonomous Region was re-established on 9 Jan. 1957. The Region declared itself a Soviet Socialist Republic in Dec. 1990. Tension between the two ethnic groups increased after the first free presidential election in April 1999 was won by Vladimir Semyonov, an ethnic Karchayev. Despite numerous allegations of fraud the result was upheld by the Supreme Court. There were subsequently fears that the ethnic Cherkess opposition would attempt to set up breakaway government bodies.

President: Rashid Temrezov (in office since 26 Feb. 2011—acting until 1 March 2011).

Prime Minister: Indris Kyabishev (in office since 11 March 2011).

There are ore-mining, engineering, chemical and woodworking industries. The Kuban-Kalaussi irrigation scheme irrigates 200,000 ha. Livestock breeding and grain growing predominate in agriculture. Conflict in the north Caucasus has had a serious impact on the economy and the agricultural sector is supported by central government. Agriculture accounted for 19·4% of gross regional product in 2003 and industry 18·7%.

In 2004 there were 10,900 pupils in 99 pre-school institutions and 60,000 pupils in 190 primary and secondary day schools. There were 16,200 students in two institutions of higher education.

In 2004 the rates of doctors and hospital beds per 10,000 population were 33·9 and 101 respectively.

Karelia

The Karelian Republic, capital Petrozavodsk (2002 census population, 266,160), covers an area of 172,400 sq. km, with a 2002 census population of 716,281. Estimated population, 1 Jan. 2010, 684,000. Russians constitute the majority of the population, with some Karelians, Belarusians and Ukrainians.

Karelia (formerly Olonets Province) became part of the RSFSR after 1917. In June 1920 a Karelian Labour Commune was formed and in July 1923 this was transformed into the Karelian Autonomous Soviet Socialist Republic (one of the autonomous republics of the RSFSR). On 31 March 1940, after the Soviet-Finnish war, practically all the territory (with the exception of a small section in the neighbourhood of the Leningrad area) which had been ceded by Finland to the USSR was added to Karelia, and the Karelian Autonomous Republic was transformed into the Karelo-Finnish Soviet Socialist Republic as the 12th republic of the USSR. In 1946, however, the southern part of the republic, including its whole seaboard and the towns of Viipuri (Vyborg) and Keksholm, was attached to the RSFSR, reverting in 1956 to autonomous republican status within the RSFSR. In Nov. 1991 it declared itself the 'Republic of Karelia'.

Head of the Republic: Aleksandr Khudilainen (in office since 22 May 2012—acting until 24 May 2012).

Karelia has a wealth of timber, some 70% of its territory being forest land. It is also rich in other natural resources, having large deposits of mica, diabase, spar, quartz, marble, granite, zinc, lead, silver, copper, molybdenum, tin, baryta and iron ore. Its lakes and rivers are rich in fish.

There are timber mills, paper-cellulose works, mica, chemical plants, power stations and furniture factories. Industrial output was valued at 512,000m. roubles in 2004. Over half of Karelia's production output is exported annually, principally to EU countries. Exports totalled US$842m. in 2004.

In 2004 there were 28,800 pupils in 496 pre-schools and 81,100 pupils in 293 primary and secondary day schools. There were 21,900 students in three institutions of higher education. There is

a branch of the Russian Academy of Sciences with seven learned institutions.

In 2004 the rates of doctors and hospital beds per 10,000 population were 49·4 and 123 respectively.

Khakassia

Area, 61,900 sq. km (23,900 sq. miles); population (2002 census), 546,072. Estimated population, 1 Jan. 2010, 539,200. Capital, Abakan (2002 census population, 165,197). Established 20 Oct. 1930; granted republican status in 1991.

Acting Head of the Republic: Viktor Zimin.

There are coal- and ore-mining, timber and woodworking industries. The region is linked by rail with the Trans-Siberian line. Industrial output was valued at 25,651m. roubles in 2004 and agricultural output at 3,807m. roubles.

In 2004 there were 16,500 pupils in 157 pre-school institutions and 69,200 pupils in 289 primary and secondary day schools. There were 20,400 students in three higher education institutions.

In 2004 the rates of doctors and hospital beds per 10,000 population were 37·3 and 110 respectively.

Komi

Area, 415,900 sq. km (160,550 sq. miles); population (2002 census), 1,018,674. Estimated population, 1 Jan. 2010, 951,200. Capital, Syktyvkar (2002 census population, 230,011). Annexed by the princes of Moscow in the 14th century, the territory was constituted as an Autonomous Region on 22 Aug. 1921 and as an Autonomous Republic on 5 Dec. 1936. The largest ethnic group are Russians, followed by Komis, with Ukrainian and Belarusian minorities.

A declaration of sovereignty was adopted by the republican parliament in Sept. 1990, and the designation 'Autonomous' dropped from the republic's official name.

Head of the Republic: Vyacheslav Gayzer (since 15 Jan. 2010).

There are coal, oil, timber, gas, asphalt and building materials industries, and light industry is expanding. Livestock breeding (including dairy farming) is the main branch of agriculture. Industrial output was valued at 87·5bn. roubles in 2004 and agricultural output at 4·3bn. roubles.

In 2004 there were 47,700 pupils in 412 pre-schools and 128,300 pupils in 531 primary and secondary day schools. There were 35,400 students in seven institutions of higher education.

In 2004 the rates of doctors and hospital beds per 10,000 population were 43·9 and 116 respectively.

Mari-El

Area, 23,200 sq. km (8,950 sq. miles); population (2002 census), 727,979. Estimated population, 1 Jan. 2010, 698,200. Capital, Yoshkar-Ola (2002 census population, 256,719). The Mari people were annexed to Russia, with other peoples of the Kazan Tatar Khanate, when the latter was overthrown in 1552. On 4 Nov. 1920 the territory was constituted as an Autonomous Region, and on 5 Dec. 1936 as an Autonomous Republic. The republic renamed itself the Mari Soviet Socialist Republic in Oct. 1990, and adopted a new constitution in June 1995. In Dec. 1991 Vladislav Zotin was elected the first president. The main ethnic groups are Russians, followed by Maris, with some Tatars.

President: Leonid Markelov (since 14 Jan. 2001).

Coal is mined. The main industries are metalworking, timber, paper, woodworking and food processing. Crops include grain, flax, potatoes, fruit and vegetables. Industry accounted for 24·4% of gross regional product in 2003 and agriculture 16·5%.

In 2004 there were 25,700 pupils in 265 pre-school institutions and 88,600 pupils in 382 primary and secondary day schools. There were 28,500 students in five institutions of higher education.

In 2004 the rates of doctors and hospital beds per 10,000 population were 35·2 and 125 respectively.

Mordovia

Area, 26,200 sq. km (10,100 sq. miles); population (2002 census), 888,766. Estimated population, 1 Jan. 2010, 826,500. Capital, Saransk (2002 census population, 304,866). By the 13th century the Mordovian tribes had been subjugated by Russian princes. In 1928 the territory was constituted as a Mordovian Area within the Middle-Volga Territory, on 10 Jan. 1930 as an Autonomous Region and on 20 Dec. 1934 as an Autonomous Republic. The main ethnic groups are Russians, followed by Mordovians, with some Tatars.

Head of the Republic: Vladimir Volkov (since May 2012).

Prime Minister: Vladimir Sushkov (since May 2012).

Industries include wood-processing and the production of building materials, furniture, textiles and leather goods. Agriculture is devoted chiefly to grain, sugar beet, sheep and dairy farming. Industrial output was valued at 29,117m. roubles in 2002.

In 2004 there were 23,900 pupils in 241 pre-school institutions and 97,300 students in 713 primary and secondary day schools. There were 42,900 students in four institutions of higher education.

In 2004 the rates of doctors and hospital beds per 10,000 population were 51·4 and 134 respectively.

North Ossetia (Alania)

Area, 8,000 sq. km (3,100 sq. miles); population (2002 census), 710,275. Estimated population, 1 Jan. 2010, 700,900. Capital, Vladikavkaz (2002 census population, 315,608). North Ossetia was annexed by Russia from Turkey and named the Terek region in 1861. On 4 March 1918 it was proclaimed an Autonomous Soviet Republic, and on 20 Jan. 1921 set up with others as the Mountain Autonomous Republic, with North Ossetia as the Ossetian (Vladikavkaz) Area within it. On 7 July 1924 the latter was constituted as an Autonomous Region and on 5 Dec. 1936 as an Autonomous Republic. In the early 1990s there was a conflict with neighbouring Ingushetia to the east, and to the south the decision of the Georgian government to disband the republic of South Ossetia led to ethnic war, with North Ossetia supporting the South Ossetians. Pressure for Ossetian reunification continues. In Sept. 2004 hostage takers seized a school in the town of Beslan. A three-day standoff ended with more than 350 people killed, nearly half of them children. Chechen rebels claimed responsibility for the siege.

A new constitution was adopted on 12 Nov. 1994 under which the republic reverted to its former name, Alania. Ossetians are the largest ethnic group, followed by Russians, with some Chechens, Armenians and Ukrainians.

Head of the Republic: Taimuraz Mamsurov.

Prime Minister: Sergey Takoyev.

The main industries are non-ferrous metals (mining and metallurgy), maize processing, timber and woodworking, textiles, building materials, distilleries and food processing. There is also a varied agriculture. Agriculture accounted for 16·8% of gross regional product in 2003 and industry 13·0%.

In 2004 there were 22,000 pupils in 216 pre-school institutions and 95,200 pupils in 218 primary and secondary day schools. There were 32,200 students in nine institutions of higher education.

In 2004 the rates of doctors and hospital beds per 10,000 population were 68·0 and 115 respectively.

Sakha

The area is 3,103,200 sq. km (1,197,750 sq. miles), making Sakha the largest republic in the Russian Federation; population (2002 census), 949,280. Estimated population, 1 Jan. 2010, 949,300. Capital, Yakutsk (2002 census population, 210,642). The Yakuts were subjugated by the Russians in the 17th century. The territory was constituted an Autonomous Republic on 27 April 1922. The largest ethnic group are Russians, followed by Yakuts, with Ukrainian and Tatar minorities.

President: Yegor Borisov (in office since 31 May 2010—acting until 17 June 2010).

Prime Minister: Galina Danchikova (in office since 18 June 2010).

The principal industries are mining (gold, tin, mica, coal) and livestock-breeding. Silver- and lead-bearing ores and coal are worked. Large diamond fields have been opened up; Sakha produces most of the Russian Federation's output. Timber and food industries are developing. Trapping and breeding of fur-bearing animals (sable, squirrel, silver fox) are an important source of income. Industry accounted for 41·1% of gross regional product in 2003 and agriculture 3·6%.

In 2004 there were 51,900 pupils in 697 pre-school institutions and 168,400 pupils in 692 primary and secondary day schools. There were 43,700 students in eight institutions of higher education.

In 2004 the rates of doctors and hospital beds per 10,000 population were 49·5 and 147 respectively.

Tatarstan

Area, 68,000 sq. km (26,250 sq. miles); population (2002 census), 3,779,265. Estimated population, 1 Jan. 2010, 3,778,500. Capital, Kazan (2002 census population, 1,105,289). From the 10th to the 13th centuries this was the territory of the Volga-Kama Bulgar State; conquered by the Mongols, it became the seat of the Kazan (Tatar) Khans when the Mongol Empire broke up in the 15th century, and in 1552 was conquered again by Russia. On 27 May 1920 it was constituted as an Autonomous Republic. The main ethnic groups are Tatars and Russians, with Chuvash, Ukrainian and Mordovian minorities.

In Oct. 1991 the Supreme Soviet adopted a declaration of independence. At a referendum in March 1992, 61·4% of votes cast were in favour of increased autonomy. A constitution was adopted in April 1992, which proclaims Tatarstan a sovereign state which conducts its relations with the Russian Federation on an equal basis. On 15 Feb. 1994 the Russian and Tatar presidents signed a treaty defining Tatarstan as a state united with Russia on the basis of the constitutions of both, but the Russian parliament has not ratified it.

President: Rustam Minnikhanov (since March 2010).

Prime Minister: Ildar Khalikov (since April 2010).

The republic has engineering, oil and chemical, timber, building materials, textiles, clothing and food industries. Industrial production was valued at 252,037m. roubles in 2003 and agricultural output at 43,639m. roubles. Tatarstan is one of the fastest-growing Russian republics.

In 2004 there were 149,900 pupils in 1,982 pre-school institutions and 472,900 pupils in 2,448 primary and secondary day schools. There were 207,100 students in 35 institutions of higher education. There is a branch of the Russian Academy of Sciences with four learned institutions. In 2004 the rates of doctors and hospital beds per 10,000 population were 44·9 and 109 respectively.

Tuva

Area, 170,500 sq. km (65,800 sq. miles); population (2002 census), 305,510. Estimated population, 1 Jan. 2010, 317,100. Capital, Kyzyl (2002 census population, 104,105). Tuva was incorporated in the USSR as an autonomous region on 11 Oct. 1944 and elevated to an Autonomous Republic on 10 Oct. 1961. The largest ethnic group are Tuvans, followed by Russians. Tuva renamed itself the 'Republic of Tuva' in Oct. 1991.

A new constitution was promulgated on 22 Oct. 1993 which adopts the name 'Tyva' for the republic. This constitution provides for a 32-member parliament (*Supreme Hural*), and a *Grand Hural* alone empowered to change the constitution, asserts the precedence of Tuvan law and adopts powers to conduct foreign policy. It was approved by 62·2% of votes cast at a referendum on 12 Dec. 1993.

Chairman of the Government: Sholban Kara-ool.

Tuva is well-watered and hydro-electric resources are important. The Tuvans are mainly herdsmen and cattle farmers and there is much good pastoral land. There are deposits of gold, cobalt and asbestos. The main exports are hair, hides and wool. There are mining, woodworking, garment, leather, food and other industries. Industrial production was valued at 1,700m. roubles in 2003 and agricultural output at 1,994m. roubles.

In 2004 there were 14,700 pupils in 217 pre-school institutions and 64,400 pupils in 176 primary and secondary day schools. There were 5,800 students at Tuva State University.

In 2004 the rates of doctors and hospital beds per 10,000 population were 42·7 and 178 respectively.

Udmurtia

Area, 42,100 sq. km (16,250 sq. miles); population (2002 census), 1,570,316. Estimated population, 1 Jan. 2010, 1,526,300. Capital, Izhevsk (2002 census population, 632,140). The Udmurts (formerly known as 'Votyaks') were annexed by the Russians in the 15th and 16th centuries. On 4 Nov. 1920 the Votyak Autonomous Region was constituted (the name was changed to Udmurt in 1932), and on 28 Dec. 1934 was raised to the status of an Autonomous Republic. The main ethnic group are Russians, followed by Udmurts, with Tatar, Ukrainian and Mari minorities. A declaration of sovereignty and the present state title were adopted in Sept. 1990.

A new parliament was established in Dec. 1993 consisting of a 50-member upper house, the *Council of Representatives*, and a full-time 35-member lower house.

President: Alexander Alexandrovich Volkov (since April 1995).

Prime Minister: Yury Pitkevich (since Oct. 2000).

Heavy industry includes the manufacture of locomotives, machine tools and other engineering products, most of them for the defence industries, as well as timber and building materials. There are also light industries: clothing, leather, furniture and food. Industrial production was valued at 82,405m. roubles in 2004 and agricultural output at 16,662m. roubles.

In 2004 there were 73,900 pupils in 806 pre-school institutions and 185,400 pupils in 849 primary and secondary day schools. There were 73,300 students in eight institutions of higher education.

In 2004 the rates of doctors and hospital beds per 10,000 population were 56·4 and 131 respectively.

Autonomous Districts and Provinces

Chukot
Situated in Magadan region (Far East); area, 737,700 sq. km, population (2002 census), 53,824. Estimated population, 1 Jan. 2010, 48,600. Capital, Anadyr. Formed 1930. Population chiefly Russian, also Chukchi, Koryak, Yakut, Even. Minerals are extracted in the north, including gold, tin, mercury and tungsten.

Khanty-Mansi
Situated in Tyumen region (western Siberia); area, 523,100 sq. km, population (2002 census), 1,432,817. Estimated population, 1 Jan. 2005, 1,469,000, chiefly Russians but also Khants and Mansi. Capital, Khanty-Mansiisk. Formed 1930.

Nenets
Situated in Archangel region (Northern Russia); area, 176,700 sq. km, population (2002 census), 41,546. Estimated population, 1 Jan. 2005, 42,000. Capital, Naryan-Mar. Formed 1929.

Yamalo-Nenets
Situated in Tyumen region (western Siberia); area, 750,300 sq. km, population (2002 census), 507,006. Estimated population, 1 Jan. 2005, 523,400. Capital, Salekhard. Formed 1930.

Yevreyskaya (Jewish) Autonomous Oblast (Province)
Part of Khabarovsk Territory. Area, 36,000 sq. km (13,895 sq. miles); population (2002 census), 190,915. Estimated population, 1 Jan. 2010, 185,000, chiefly Russians, but also Ukrainians and Jews. Capital, Birobijan (2002 census population, 77,250). Established as Jewish National District in 1928. There is a Yiddish national theatre, newspaper and broadcasting service.

RWANDA

© Research Machines plc 2006

Republika y'u Rwanda
(Republic of Rwanda)

Capital: Kigali
Population projection, 2015: 12·30m.
GNI per capita, 2011: (PPP$) 1,133
HDI/world rank: 0·429/166
Internet domain extension: .rw

KEY HISTORICAL EVENTS

The Twa—hunter-gatherer pygmies—were the first people to inhabit Rwanda. They now comprise 1% of the population. The Hutu were the next group to settle in Rwanda. They arrived at some point between AD 500 and 1100. They were small-scale agriculturalists, led by a king who ruled over clan groups. The final group to migrate to Rwanda was the Tutsi around 1400. Their ownership of cattle and their combat skills allowed them to gain economic and political control of the country. A feudalistic system developed where the Tutsi lent cows to the Hutu in return for labour and military service. At the apex was the Tutsi king, the *mwami* (pl., *abami*), who was believed to be of divine origin. The *abami* consolidated their power by centralizing the monarchy and reducing the power of neighbouring chiefs. Mwami Kigeri IV (reigned 1853–95) established the borders of Rwanda in the 19th century.

The Conference of Berlin in 1885 placed Rwanda under German control. However, no German actually reached the area until 1894 when Count von Götzen became the governor of German East Africa. The Belgians and British had ambitions in the area owing to its strategic position at the juncture of their separate empires. However, by 1910 German rule was accepted.

A consequence of German influence was the arrival of the Catholic Church through the mission of the White Fathers, who established schools and missions from 1899. Germany did not change the political structure of the country but made use of the mwami, Yuhi V (reigned 1896–1931), who accepted German overlordship. The victors of the First World War disrupted this relationship by stripping Germany of all her colonies. Rwanda was occupied by Belgian forces in 1916 and was declared a Belgian mandate in Aug. 1923 by the League of Nations. The Belgians ruled more directly than the Germans, curtailing the mwami's power and favouring the Tutsi minority on more explicitly racial grounds. From 1952 the UN ordered Belgium to integrate Rwandans into the political system. The Belgians continued their policy of favouring the fairer skinned Tutsi and placed them in a position of domination over the Hutu majority. Increasing civil unrest erupted into a civil war by 1959. A state of Ruanda-Urundi was established in 1960, under Belgian trusteeship, following an election. In 1961, while abroad, Mwami Kigeli V was exiled by the Belgians, who refused to allow him to return despite pressure from the UN. On 27 June 1962 the parliament voted to terminate the trusteeship and on 1 July 1962 Rwanda became independent.

Independence

The independent state of Rwanda was first governed by the Parmehutu party (a Hutu party representing the 85% Hutu population), led by Grégoire Kayibanda, but this was not accepted by some Tutsis. An attempted invasion in 1963 by Tutsis who had fled to Uganda and Burundi was repelled. In retaliation over 12,000 Tutsis in Rwanda were massacred by the Hutu. The next massacre in 1972–73 was partly in response to the persecution of Hutus in neighbouring Tutsi-dominated Burundi. An attempt by Kayibanda to revive his waning popularity, the violence instead spurred Maj.-Gen. Juvénal Habyarimana, a senior army commander, to launch a bloodless coup and take over government. In 1975 Habyarimana formed *le Mouvement Révolutionaire National pour le Développement* (MRND), and turned Rwanda into a one-party and tightly controlled police state, discriminating against the Tutsi in favour of the Hutu.

In 1990 the Rwandan Patriotic Front (RPF), of between 5,000 and 10,000 Tutsis, invaded Rwanda from Uganda, starting a civil war. A ceasefire was agreed on 29 March 1991 and on 14 July 1992 the Arusha Accords were signed. These allowed other political parties to stand for election and share power.

Many Hutus opposed Arusha. Multi-partyism led to the rise of far-right Hutu power groups who believed that the only solution to Hutu-Tutsi problems was the extermination of the Tutsi. The assassination of the first legitimately elected Hutu president of Burundi (21 Oct. 1993) by Tutsi army officers and the massacre of over 150,000 Hutus in Burundi served to further destabilize Rwanda. The assassination of Habyarimana in a plane crash on 6 April 1994, probably shot down by Hutu extremists, was the first step in a carefully premeditated genocide which killed around 1m. Rwandans in three months and forced over 2m. to flee to neighbouring countries.

Among the victims was the moderate Hutu prime minister, Agathe Uwilingiyimana. Gangs of *interahamwe* (civilian death squads) roamed the capital, Kigali, killing, looting and raping Tutsis and politically-moderate Hutus. When the RPF, led by Paul Kagame, reached Kigali the killings spread to other parts of the country. The UN sent a peacekeeping force (the United Nations Assistance Mission for Rwanda; UNAMIR) but a lack of resources, an unclear mandate and international apathy made it impotent as the genocide took hold. Eventually France dispatched 2,000 troops on a humanitarian mission on 22 June 1994 to maintain a 'safe zone'. Owing to France's affinity with the French-speaking Hutu former government, this zone served as an escape route for Hutu extremists to flee to Zaïre (now the Democratic Republic of the Congo).

The RPF declared the war over on 17 July 1994 and was quickly recognized as the new government. Genocide trials began in Arusha, Tanzania in Dec. 1996. In Sept. 1998 Jean Kambanda, the former prime minister (April–July 1994), was sentenced to life imprisonment.

Rwanda was destabilized by the presence of Hutu refugee camps on the Zaïre borders. Amongst the 1·1m. refugees were *interahamwe* who used the camps as bases for attacks on Rwanda. These were broken up by Laurent Kabila in May 1997, before he assumed power in Zaïre (later renamed the Democratic Republic of the Congo).

In April 2000 Paul Kagame (the Tutsi vice-president and defence minister) was elected president by parliament, replacing Pasteur Bizimungu, a Hutu who had been appointed by the RPF, in July 1994. Kagame was re-elected president in Aug. 2003 in Rwanda's first democratic elections since the atrocities.

TERRITORY AND POPULATION

Rwanda is bounded south by Burundi, west by the Democratic Republic of the Congo, north by Uganda and east by Tanzania. A mountainous state of 25,314 sq. km (9,774 sq. miles), its western third drains to Lake Kivu on the border with the Democratic Republic of the Congo and thence to the Congo river, while the rest is drained by the Kagera river into the Nile system.

The population was 7,164,994 at the 1991 census, of whom over 90% were Hutu, 9% Tutsi and 1% Twa (pygmy). Following the genocide of 1994 ethnicity was not enumerated at the 2002 census, when the population was 8,128,553. Provisional population at the 2012 census, 10,537,222; density, 416·3 per sq. km.

The UN gives a projected population for 2015 of 12·30m.

In 2011 the population was 19·2% urban.

At the time of the 2002 census there were 12 administrative divisions (11 provinces and Kigali City). Areas and populations were:

Province	Area (in sq. km)	Population (2002 census)
Butare	1,872	725,914
Byumba	1,694	707,786
Cyangugu	1,894	607,495
Gikongoro	1,974	489,729
Gisenyi	2,047	864,377
Gitarama	2,141	856,488
Kibungo	2,964	702,248
Kibuye	1,748	469,016
Kigali City	313	603,049
Kigali-Ngali	2,780	789,330
Ruhengeri	1,657	891,498
Umutara	4,230	421,623

Since Jan. 2006 Rwanda has been reorganized into five provinces (*intara*) as follows (with 2012 provisional census populations): Eastern (2,600,814), Northern 1,729,927), Southern (2,594,110), Western (2,476,943) and Kigali City (1,135,428). Among the reasons given for the change were the reduction of ethnic divisions and the suppression of reminders of the 1994 genocide.

Kigali, the capital, had a provisional population of 1,135,428 in 2012 (603,049 in 2002); other towns are Butare, Gisenyi, Gitarama and Ruhengeri.

Kinyarwanda, the language of the entire population, and English are the official languages (French having ceased to be an official language in Oct. 2008). Swahili is spoken in the commercial centres.

SOCIAL STATISTICS

2008 estimates: births, 400,000; deaths, 141,000. Estimated birth rate in 2008 was 41·1 per 1,000 population; estimated death rate, 14·5. Annual population growth rate, 2000–08, 2·5%. Life expectancy at birth in 2007 was 51·4 years for females and 47·9 for males, up from 23·1 years for females and 22·1 years for males during the period 1990–95 (at the height of the civil war). Infant mortality, 2010, 44 per 1,000 live births; fertility rate, 2008, 5·4 births per woman.

CLIMATE

Despite the equatorial situation, there is a highland tropical climate. The wet seasons are from Oct. to Dec. and March to May. Highest rainfall occurs in the west, at around 70" (1,770 mm), decreasing to 40–55" (1,020–1,400 mm) in the central uplands and to 30" (760 mm) in the north and east. Kigali, Jan. 67°F (19·4°C), July 70°F (21·1°C). Annual rainfall 40" (1,000 mm).

CONSTITUTION AND GOVERNMENT

Under the 1978 constitution the MRND was the sole political organization.

A new constitution was promulgated in June 1991 permitting multi-party democracy.

The Arusha Agreement of Aug. 1994 provided for a transitional 70-member National Assembly, which began functioning in Nov. 1994. The seats won by the MRNDD (formerly MRND) were taken over by other parties on the grounds that the MRNDD was culpable of genocide.

A referendum was held on 26 May 2003 which approved a draft constitution by 93·4% (turnout was 87%). The new constitution, subsequently approved by the Supreme Court, provides for an 80-member *Chamber of Deputies* and a 26-member *Senate*, with the provision that no party may hold more than half of cabinet positions. 53 members of the Chamber of Deputies are directly elected, 24 women are elected by provincial councils, two members are elected by the National Youth Council and one is elected by a disabilities organization.

National Anthem

'Rwanda Nziza' ('Beautiful Rwanda'); words by F. Murigo, tune by Capt. J.-B. Hashakaimana.

RECENT ELECTIONS

In a popular election on 9 Aug. 2010 Paul Kagame was re-elected president for a seven-year term with 93·1% of the vote. Jean Damascene Ntawukuriryayo won 5·1%, Prosper Higiro 1·4% and Alvera Mukabaramba 0·4%. Turnout was 97·5%.

In parliamentary elections held on 15 Sept. 2008, President Kagame's Rwandan Patriotic Front (RPF) and its coalition won 78·8% of the vote. The RPF took 42 of the 53 directly elected seats, the Social Democratic Party 7 and the Liberal Party 4. Following the Sept. 2008 election, of the 80 Members of Parliament there were 45 women (56·25%) and 35 men (43·75%), making Rwanda the first country in the world—and currently the only one—to have a female majority in parliament (in part as a consequence of the adoption of quotas, which stipulate that at least 24 of the 80 seats must be filled by women). In Feb. 2013 there were still 45 female MPs.

CURRENT GOVERNMENT

President: Paul Kagame; b. 1957 (RPF; sworn in 22 April 2000 having been acting president since 24 March 2000 and re-elected in Aug. 2003 and Aug. 2010).

In March 2013 the government comprised:

Prime Minister: Pierre Damien Habumuremyi; b. 1961 (RPF; sworn in 6 Oct. 2011).

Minister of Agriculture and Animal Resources: Agnes Kalibata. *Cabinet Affairs:* Protais Musoni. *Civil Service and Labour:* Anastase Murekezi. *Defence:* Gen. James Kabarebe. *East African Community:* Monique Mukaruliza. *Education:* Vincent Biruta. *Finance and Planning:* Claver Gatete. *Foreign Affairs and Regional Co-operation:* Louise Mushikiwabo. *Health:* Agnès Binagwaho. *Infrastructure:* Silas Lwakabamba. *Internal Security:* Musa Fazil Harerimana. *Justice:* Tharcisse Karugarama. *Local Government:* James Musoni. *Natural Resources:* Stanislas Kamanzi. *Refugees and Disaster Preparedness:* Serafina Mukantabana. *Sports and Culture:* Protais Mitali. *Trade and Industry:* François Kanimba. *Youth, and Information and Communications Technology:* Jean Philbert Nsengimana. *Minister in the President's Office:* Venantia

Tugireyezu. *Minister in the Prime Minister's Office:* Oda Gasinzigwa (*in Charge of Gender and Family Promotion*).

Ministers of State: Emma Françoise Isumbingabo (*in Charge of Energy and Water*); Mathias Harebamungu (*in Charge of Primary and Secondary Education*); Alexis Nzahabwanimana (*in Charge of Transport*); Albert Nsengiyumba (*in the Ministry of Education in Charge of Technical and Vocational Education and Training*); Eugène-Richard Gasana (*in the Ministry of Foreign Affairs and Co-operation*); Anita Assimwe (*in the Ministry of Health in Charge of Public Health and Primary Healthcare*); Alvera Mukabaramba (*in the Ministry of Local Government in Charge of Social Affairs*); Evode Imena (*in the Ministry of Natural Resources in Charge of Mining*).

Government Website: http://www.gov.rw

CURRENT LEADERS

Paul Kagame

Position
President

Introduction
Paul Kagame was elected president by the transitional National Assembly in April 2000 and democratic elections in 2003 cemented his mandate. He is leader of the ruling Rwandan Patriotic Front (RPF) and the first member of the Tutsi minority to be president since Rwanda gained independence in 1962. Although a leading force in the country, he is a low-key public figure. He has openly criticized the United Nations, arguing it could have done more to avoid the genocide of 1994 when around 1m. Rwandans were killed, and also accused France of complicity. He was re-elected president for another seven-year term by an overwhelming margin in Aug. 2010.

Early Life
Kagame was born in Oct. 1957 in the Gitarama prefecture. Following ethnic violence in Rwanda, his family fled to Uganda in 1960 where he grew up in a refugee camp and then studied at Makerere University in Kampala. In 1979 Kagame joined the National Salvation Front (FRONASA), led by Yoweri Museveni, which took part in the Tanzanian removal of Idi Amin's regime in 1979. In 1980 he became a founding member of Museveni's National Resistance Army (NRA) and fought against the dictatorship of Milton Obote in Uganda. He became head of intelligence of the NRA in 1986 and the following year established the Rwandan Patriotic Front (RPF), with support from Museveni, which launched a guerrilla war against President Juvénal Habyarimana and his Hutu government. In 1993 a peace agreement was signed, but the death of Habyarimana in 1994 triggered the Hutu massacres of the Tutsi. The RPF resumed the civil war and soon gained control over the country. The new government of national unity, formed in July 1994, was led by Pasteur Bizimungu, a Hutu, with Kagame as vice president and defence minister.

Career in Office
Rwanda's involvement in the Democratic Republic of the Congo (DRC; known as Zaïre until 1997) began covertly in 1996 with an agreement with Uganda to oust the Zaïrean president Mobutu Sese Seko. Kagame and Museveni sent troops back into the east of the DRC in 1998 to assist rebel groups against President Laurent Kabila and to eliminate the *interahamwe* (death squads from the 1994 genocide in Rwanda). In Nov. 1998 Kagame publicly admitted that Rwandan forces were active in the DRC for reasons of national security. However, divisions began to appear between Kagame and Museveni, as the enlargement and reorganization of the Rwandan forces was interpreted as a direct threat by Museveni.

In March 2000 President Bizimungu resigned, claiming that he and the prime minister, Pierre-Celéstin Rwigema, were hounded from office for being Hutus. Elected president by the transitional National Assembly in April 2000, Kagame relinquished the defence ministry to Col. Emmanuel Habyarimana, a Hutu, and promoted Nyamuasa Kayumba, the army chief, to Major-General, a rank only Kagame himself had held previously. Kagame attempted to portray himself as a civilian and neutral president, calling for ethnic peace and reconciliation.

Following the assassination of Laurent Kabila (president of the DRC) in Jan. 2001, his son and successor, Joseph Kabila, met Kagame who accused the DRC of harbouring Hutu militias. Kabila blamed Rwanda for killing more than 3·5m. inhabitants of the DRC and refused to co-operate until Rwandan troops were withdrawn. However, on 30 July 2002 the two nations signed a peace deal in South Africa under which the DRC agreed to disarm and arrest Hutu rebels and Rwanda withdrew its troops from the DRC in Oct. 2002.

Kagame was elected president by a popular democratic vote on 25 Aug. 2003. His win—with 95% of the vote—prompted allegations of irregularities from his main rival, Faustin Twagiramungu, although he accepted the win in Sept. Museveni's attendance at Kagame's inauguration ceremony demonstrated the easing of tensions between the two men. In 2007 Kagame authorized the release of several thousand prisoners accused of genocide. He also awarded a presidential pardon to Pasteur Bizimungu, who had served three years of a 15-year prison sentence for attempting to form a militia and for embezzlement. Parliamentary elections in Sept. 2008 returned the ruling coalition led by Kagame's RPF to power with almost 79% of the vote.

In Nov. 2006 a French investigative judge had accused Kagame of ordering the assassination of President Habyarimana in 1994. Kagame responded by cutting diplomatic links with France. Then, in Aug. 2008, a Rwandan report commissioned by Kagame made a counter-accusation of complicity by French politicians (including former president François Mitterrand) and army officers in the genocide. However, diplomatic ties were eventually restored in Nov. 2009 and French President Nicolas Sarkozy paid an official visit to Rwanda in Feb. 2010. The two leaders met again in Paris in Sept. 2011 in a further bid to restore trust.

From Aug. 2008 fighting in the DRC intensified between the army and the mainly Tutsi insurgents loyal to Laurent Nkunda. Kabila's government had for some time claimed that Kagame was giving Rwandan support to Nkunda. However, in an apparent reversal of Rwandan policy, troops from both countries launched a joint offensive in Jan. 2009 against Nkunda's headquarters in the DRC and the rebel leader was arrested when he fled into Rwanda. Following the operation, relations between Rwanda and the DRC stabilized, but in 2012 there were further allegations of Rwandan support for rebel forces opposed to the Kabila regime, prompting the USA, UK and the Netherlands to suspend aid contributions.

In Nov. 2009 Rwanda was admitted to the Commonwealth despite (like Mozambique) not having had colonial or constitutional ties to the United Kingdom. In Aug. 2010 Kagame was re-elected as president with another landslide share of the vote, although the opposition claimed that the election was not free and fair.

Pierre Damien Habumuremyi

Position
Prime Minister

Introduction
Pierre Damien Habumuremyi was sworn in as prime minister in Oct. 2011, succeeding Bernard Makuza. Makuza had served for 11 years, the longest serving head of government in Rwanda's post-genocide era, and was credited with bringing a degree of stability

to the country. A technocrat, Habumuremyi was previously the minister of education.

Early Life

Habumuremyi was born in 1961 in Ruhondo, in Musanze District. He graduated in sociology from Lubumbashi University in the Democratic Republic of the Congo, before gaining post-graduate qualifications in education and political science.

He served from 2000–08 on the National Electoral Commission, including five years as executive secretary. In 2008 he was elected to represent the country at the East African Community legislative assembly.

In 2011 Habumuremyi was appointed Rwanda's minister of education, serving five months before being appointed prime minister. His promotion to the premiership surprised many commentators, given his relatively low profile and lack of government experience.

Career in Office

Habumuremyi pledged to rejuvenate the economy and develop its information and communications base. He also promised to fight corruption and has refuted suggestions that he is a puppet of the president.

DEFENCE

In 2008 defence expenditure totalled US$71m. (US$7 per capita), representing 1·8% of GDP.

Army

Strength (2007) about 32,000. There were local defence forces of some 2,000 in 2007.

ECONOMY

Agriculture accounted for 41·3% of GDP in 2006, industry 13·3% and services 45·4%.

Overview

Since the 1994 genocide and civil war, Rwanda has registered strong economic growth and macroeconomic stability. The economy is dominated by subsistence agriculture, which accounts for 36% of GDP, 80% of employment and 45% of exports. Real GDP growth peaked at 11·2% in 2008 before the effects of the global financial crisis slowed growth to 4·1% in 2009. However, a recovery was under way in 2010, driven by growth in agriculture and the resurgent services and construction sectors.

According to the World Bank, 50% of Rwanda's budget is financed by international aid, encompassing around 12% of GDP. Poverty remains high, with nearly 57% of the population living below the national poverty line according to a 2006 household survey. With Rwanda aiming to become a lower-middle income economy by 2020, the Economic Development and Poverty Reduction Strategy (EDPRS) is set to stimulate broad-based economic growth and employment through better economic infrastructure, higher agricultural productivity and stronger political and economic governance. In the World Bank's 2010 'Doing Business' survey Rwanda was ranked as the world's number one business reformer for 2008–09.

Currency

The unit of currency is the *Rwanda franc* (RWF) notionally of 100 *centimes*. On 3 Jan. 1995, 500-, 1,000- and 5,000-Rwanda franc notes were replaced by new issues, demonetarizing the currency taken abroad by exiles. The currency is not convertible. Foreign exchange reserves were US$312m. in July 2005. Total money supply in Dec. 2003 was 82,305m. Rwanda francs. Inflation was 15·4% in 2008, falling to 2·3% in 2010. It stood at 5·7% in 2011.

Budget

In 2007 revenues were 472·3bn. Rwanda francs and expenditures 491·4bn. Rwanda francs.

VAT is 18%.

Performance

Real GDP growth was 24·5% in 1995, following two years of negative growth including a rate of –41·9% in 1994 at the height of the civil war. From 2000–08 growth averaged 8·1%. Real GDP growth was 4·1% in 2009, 7·2% in 2010 and 8·6% in 2011. Total GDP in 2011 was US$6·4bn.

Banking and Finance

The central bank is the National Bank of Rwanda (founded 1960; *Governor*, John Rwangombwa), the bank of issue since 1964. There are seven commercial banks (Banque de Kigali, Banque de Commerce et de Développement Industriel, Banque Continentale Africaine au Rwanda, Banque à la Confiance d'Or, Banque Commerciale du Rwanda, Caisse Hypothécaire du Rwanda and Compagnie Générale de Banque), one development bank (Rwandan Development Bank) and one credit union system (Rwandan Union of Popular Banks).

Foreign debt totalled US$795m. in 2010, representing 14·2% of GNI.

A stock exchange opened in Kigali in Jan. 2011.

ENERGY AND NATURAL RESOURCES

Environment

Carbon dioxide emissions from the consumption and flaring of fossil fuels in 2008 were the equivalent of 0·1 tonnes per capita.

Electricity

Installed capacity was 57,000 kW in 2007. Production was estimated at 169m. kWh in 2007 and consumption per capita an estimated 27 kWh.

Oil and Gas

In 2007 proven natural gas reserves were 57bn. cu. metres.

Minerals

Production (2002): cassiterite, 197 tonnes; wolfram, 153 tonnes.

Agriculture

There were an estimated 1·2m. ha. of arable land in 2007 and 275,000 ha. of permanent crops. Production (2003, in 1,000 tonnes): bananas and plantains, 2,408; potatoes, 1,110; cassava, 1,003; sweet potatoes, 868; dry beans, 239; pumpkins and squash, 210; sorghum, 172; taro, 139; maize, 79; sugarcane, 40; rice, 28; dry peas, 18; tea, 15; wheat, 15.

In 2006 there were 1,330,000 goats, 997,000 cattle, 473,000 sheep, 327,000 pigs and 2m. chickens.

Forestry

Forests covered 0·44m. ha. (18% of the land area) in 2010. Timber production in 2007 was 10·0m. cu. metres.

Fisheries

The catch in 2007 totalled 9,050 tonnes, entirely from inland waters.

INDUSTRY

There are about 100 small-sized modern manufacturing enterprises in the country. Food manufacturing is the dominant industrial activity (64%) followed by construction (15·3%) and mining (9%). There is a large modern brewery.

Labour

In 2005–06 there were 4,377,000 employed persons, with 79% of the economically active population engaged in agriculture, fisheries and forestry.

INTERNATIONAL TRADE

Rwanda, Burundi and the Democratic Republic of the Congo make up the Economic Community of the Great Lakes. Foreign debt was US$1,518m. in 2005.

Imports and Exports

In 2006 imports (f.o.b.) amounted to US$488m. (US$355m. in 2005); exports (f.o.b.) US$145m. (US$128m. in 2005). Leading imports are capital goods, food and energy products; major exports are coffee, tea and tin. Main import suppliers, 2006: Kenya, 26·1%; Uganda, 13·3%; Belgium, 6·9%; UAE, 6·9%. Main export markets, 2006: UK, 21·4%; Kenya, 21·3%; Belgium, 16·2%; Hong Kong, 10·7%.

COMMUNICATIONS

Roads

There were 14,008 km of roads in 2004, of which 19·0% were paved. There are road links with Burundi, Uganda, Tanzania and the Democratic Republic of the Congo. In 2006 there were 4,130 motorcycles, 1,813 cars and jeeps, and 1,270 trucks and pick-ups. There were 308 road deaths in 2007.

Civil Aviation

There is an international airport at Kigali (Gregoire Kayibanda), which handled 103,000 passengers (100,000 on international flights) in 2001. In 2003 there were scheduled flights to Addis Ababa, Brussels, Bujumbura, Douala, Entebbe, Johannesburg and Nairobi. A national carrier, Rwandair Express (since renamed RwandAir), began operations in 2003, flying to Entebbe and Johannesburg.

Telecommunications

Rwanda had 27,000 fixed telephone lines in 2006 and 268,000 mobile phone subscribers. Internet users numbered 100,000 in 2006. In June 2012 there were 144,000 Facebook users.

SOCIAL INSTITUTIONS

Justice

A system of Courts of First Instance and provincial courts refer appeals to Courts of Appeal and a Court of Cassation situated in Kigali. The death penalty was last used in 1998 and was abolished in 2007. A number of people were executed for genocide in the civil war in 1994, including 22 at five different locations throughout the country on 24 April 1998.

The population in penal institutions in Dec. 2007 was 58,598, of which around 39,000 were being held on suspicion of genocide.

Education

In 2008–09 there were 33,000 primary school teachers for 2·3m. pupils; 347,000 secondary pupils with 15,000 teachers; and 55,000 students at university level with 3,000 academic staff. Adult literacy rate in 2008 was 70%.

In 2007 public expenditure on education came to 4·9% of GDP and accounted for 19·0% of total government spending.

Health

In 2005 there were 221 public sector physicians and dentists (148 in 2000), and 4,063 nurses (1,167 in 2000). There were 42 hospitals and 389 health centres in 2007.

There were 1·07m. reported cases of malaria in 2006.

RELIGION

In 2001 approximately 47% of the population were Roman Catholics, 19% Protestants and 7% Muslims. Some of the population follow traditional animist religions. Before the civil war there were nine Roman Catholic bishops and 370 priests. By the end of 1994, three bishops had been killed and three reached retiring age; 106 priests had been killed and 130 had sought refuge abroad.

CULTURE

Press

The English-language *New Times* is published six days a week, with its sister publication the *Sunday Times* appearing on Sundays.

Tourism

In 2007 there were 826,374 tourist arrivals.

DIPLOMATIC REPRESENTATIVES

Of Rwanda in the United Kingdom (120–122 Seymour Place, London, W1H 1NR)
High Commissioner: Ernest Rwamucyo.

Of the United Kingdom in Rwanda (Parcelle No. 1131, Blvd de l'Umuganda, Kacyira-Sud, PB 576, Kigali)
High Commissioner: Benedict Llewellyn Jones, OBE.

Of Rwanda in the USA (1714 New Hampshire Ave., NW, Washington, D.C., 20009)
Ambassador: James Kimonyo.

Of the USA in Rwanda (2657 Ave. de la Gendarmerie, Kigali)
Ambassador: Donald W. Koran.

Of Rwanda to the United Nations
Ambassador: Eugène-Richard Gasana.

Of Rwanda to the European Union
Ambassador: Robert Masozera.

FURTHER READING

Barnett, Michael, *Eyewitness to a Genocide: The United Nations and Rwanda.* 2003
Dallaire, Romeo, *Shake Hands with the Devil: The Failure of Humanity in Rwanda.* 2005
Dorsey, L., *Historical Dictionary of Rwanda.* 1995
Melson, Robert, *Genocide and Crisis in Central Africa: Conflict Roots, Mass Violence and Regional War.* 2001
Melvern, Linda, *A People Betrayed: The Role of the West in Rwanda's Genocide.* 2000
Prunier, G., *The Rwanda Crisis: History of a Genocide.* 1995
Waugh, Colin M., *Paul Kagame and Rwanda: Power, Genocide and the Rwandan Patriotic Front.* 2004

National Statistical Office: National Institute of Statistics of Rwanda, B. P. 46, Kigali.
Website: http://www.statistics.gov.rw

ST KITTS AND NEVIS

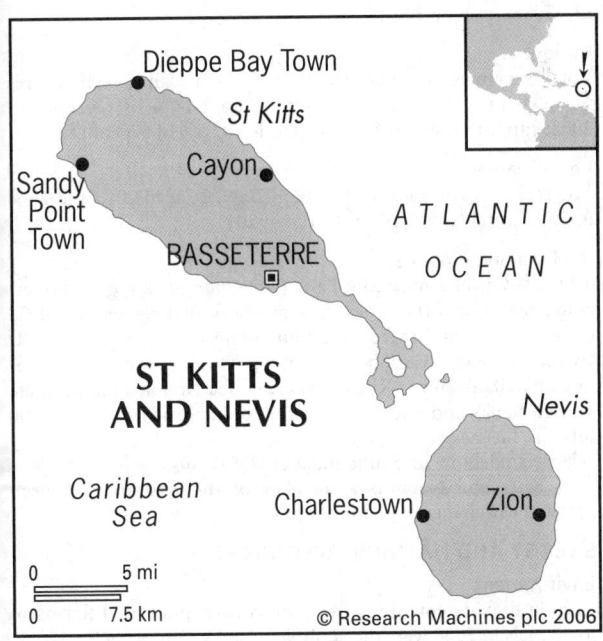

Dieppe Bay Town

St Kitts

Cayon

Sandy Point Town

BASSETERRE

ATLANTIC OCEAN

ST KITTS AND NEVIS

Nevis

Caribbean Sea

Charlestown

Zion

| 0 | 5 mi |
| 0 | 7.5 km |

© Research Machines plc 2006

Federation of St Kitts and Nevis

Capital: Basseterre
Population, 2001: 46,000
GNI per capita, 2011: (PPP$) 11,897
HDI/world rank: 0·735/72
Internet domain extension: .kn

KEY HISTORICAL EVENTS

The islands of St Kitts (formerly St Christopher) and Nevis were discovered and named by Columbus in 1493. They were settled by Britain in 1623 and 1628, but ownership was disputed with France until 1783. In Feb. 1967 colonial status was replaced by an 'association' with Britain, giving the islands full internal self-government. St Kitts and Nevis became fully independent on 19 Sept. 1983. In Oct. 1997 the five-person Nevis legislature voted to end the federation with St Kitts. However, in a referendum held on 10 Aug. 1998 voters rejected independence, only 62% voting for secession when a two-thirds vote in favour was needed. In Sept. 1998 Hurricane Georges caused devastation, leaving 25,000 people homeless, with some 80% of the houses in the islands damaged.

TERRITORY AND POPULATION

The two islands of St Kitts and Nevis are situated at the northern end of the Leeward Islands in the eastern Caribbean. Nevis lies 3 km to the southeast of St Kitts. Population, 2001 census, 46,325. In 2011, 32·6% of the population were urban.

	Sq. km	Census 1991	Census 2001	Chief town	Census 2001
St Kitts	168·0	31,824	35,217	Basseterre	13,251
Nevis	93·3	8,789	11,108	Charlestown	1,790
	261·3	40,613	46,325		

In 2001, 92·4% of the population were of African origin. English is the official and spoken language.

SOCIAL STATISTICS

Births, 2008, 709; deaths, 357. Rates, 2008 (per 1,000 population): births, 13·8; deaths, 7·0. Infant mortality, 2010 (per 1,000 live births), 7. Life expectancy in 2006 was 73 years. Annual population growth rate, 2000–08, 1·3%; fertility rate, 2008, 1·8 births per woman.

CLIMATE

Temperature varies between 21·4–30·7°C, with a sea breeze throughout the year and low humidity. Average annual rainfall is between 1,270 mm and 1,905 mm.

CONSTITUTION AND GOVERNMENT

The British sovereign is the head of state, represented by a Governor-General. The 1983 constitution described the country as 'a sovereign democratic federal state'. It allowed for a unicameral Parliament consisting of 11 elected Members (eight from St Kitts and three from Nevis), three appointed Senators and one *ex officio* member. Nevis was given its own Island Assembly and the right to secession from St Kitts.

National Anthem

'O Land of beauty! Our country where peace abounds'; words and tune by K. A. Georges.

RECENT ELECTIONS

At the National Assembly elections on 25 Jan. 2010 the Labour Party gained 6 seats, the People's Action Movement 2, the Concerned Citizens' Movement 2 and the Nevis Reformation Party 1. Turnout was 79·8%.

CURRENT GOVERNMENT

Governor-General: Sir Edmund Wickham Lawrence, KCMG, OBE, CSM, JP (since 2 Jan. 2013).
  In March 2013 the government comprised:
  Prime Minister, Minister of Finance, Sustainable Development, National Security and Constituency Empowerment: Dr Denzil L. Douglas; b. 1953 (Labour Party; sworn in 7 July 1995 and re-elected in 2000, 2004 and 2010).
  Deputy Prime Minister and Minister of Public Works, Utilities, Housing and Energy: Dr Earl Asim Martin.
  Minister of Agriculture, Marine Resources, Education and Information: Nigel Carty. *Health, Social Services, Community Development, Culture and Gender Affairs:* Marcella Liburd. *International Trade, Industry, Commerce, Consumer Affairs, Tourism and International Transport:* Richard Skerritt. *Youth Empowerment, Sports, Information Technology, Telecommunications and Post:* Glenn Phillip. *Justice and Legal Affairs, Homeland Security, Foreign Affairs and Labour:* Patrice Nisbett. *Attorney General:* Jason Hamilton.
  The *Nevis Island* legislature comprises an Assembly of three nominated members and five elected members, and an Administration consisting of the Premier and two other persons appointed by the Deputy Governor-General.
  The Premier of *Nevis* is Vance Amory.

Government Website: http://www.gov.kn

CURRENT LEADERS

Dr Denzil L. Douglas

Position
Prime Minister

Introduction
Denzil Douglas has been prime minister of St Kitts and Nevis since 1995. In Jan. 2010 he was re-elected for his fourth

consecutive term after his Labour Party secured six out of a possible 11 seats.

Early Life

Born in 1953, Denzil Llewellyn Douglas is a graduate of the University of the West Indies. He worked as a family physician before embarking on a career in politics. A Labour party activist from an early age, he became leader of the St Kitts and Nevis Labour Party in 1989. The 1995 election was called following drug smuggling allegations brought against the prime minister at the time, Kennedy Simmonds.

Career in Office

On becoming prime minister, Douglas sought to mend relations between the islands of St Kitts and Nevis. Nevertheless, in 1998 a referendum on secession was held after occupants of Nevis complained that the federal government in St Kitts was ignoring their needs. Although they failed to gain the two-thirds vote needed to break away, unrest remained. Douglas has since tried to tackle the differences through constitutional reform.

In Nov. 2004 Douglas chaired a CARICOM–UK conference aimed at reducing stigma and discrimination against those living with HIV and AIDS in the Caribbean. In May 2005 he chaired the first Caribbean Forum of Development, aimed at transforming the regional economy, and achieving international competitiveness and sustainable growth. He has also called for the cancellation of international debt, which has been crippling many Caribbean Islands.

While he has been widely praised for promoting poverty reduction and combating crime, he has been criticized for not reviving the islands' ailing sugar industry. In March 2005 the government effectively closed the 300 year-old industry after another loss-making harvest. Although the closure has had an adverse impact on government debt, economic activity was buoyed in 2006 and early 2007 by an expansion in tourism and construction spending related to the 2007 Cricket World Cup.

Re-elected for the third time in Jan. 2010, Douglas holds a number of ministerial posts in addition to the premiership. In July 2012 he announced the completion of a major public debt restructuring programme agreed with the Paris Club of creditor nations.

ECONOMY

Agriculture accounted for 1·5% of GDP in 2009, industry 20·0% and services 78·5%.

Overview

The economy is in transition following the government's closure of the sugar industry in 2005, a move encouraged by the EU and WTO after years of heavy losses. The most important sources of revenue are service industries, offshore finance and tourism but the country is vulnerable to external shocks, such as natural disasters and shifts in the tourist market.

The economy contracted in 2009, 2010 and 2011 following a fall in tourism receipts and FDI-related construction activity, but a slow recovery is under way driven by stronger performance in the USA, the Islands' primary export market. An increasing fiscal deficit led to the implementation of revenue reforms, including the introduction of VAT and excise tax reforms in 2010, as well as expenditure cuts including a freeze on public wages. Inflationary pressures intensified as a result of the VAT introduction and rising oil and food prices, while the public debt-to-GDP ratio increased to 200% in 2010. The Islands have requested financial assistance from the IMF to reduce public debt.

Currency

The *East Caribbean dollar* (XCD) (of 100 *cents*) is in use. Inflation was 0·6% in 2010 but increased to 7·1% in 2011. In July 2005 foreign exchange reserves were US$75m. Total money supply was EC$200m. in June 2005.

Budget

Total revenues in 2008 were EC$641·2m. and expenditures EC$634·4m. Tax revenue accounted for 64·9% of revenues in 2008; current expenditure accounted for 87·8% of expenditures.

Performance

The economy contracted by 5·6% in 2009, 2·7% in 2010 and 2·0% in 2011. Total GDP was US$0·7bn. in 2011.

Banking and Finance

The East Caribbean Central Bank (*Governor*, Sir Dwight Venner) is located in St Kitts. It is a regional bank that serves the OECS countries. In 2002 there were four domestic commercial banks (Bank of Nevis, Caribbean Banking Corporation, Nevis Co-operative Bank and St Kitts-Nevis-Anguilla National Bank), three foreign banks and one development bank. Nevis has some 9,000 offshore businesses registered.

Foreign debt in 2007 amounted to US$274m.

St Kitts and Nevis is a member of the Eastern Caribbean Securities Exchange, based in Basseterre.

ENERGY AND NATURAL RESOURCES

Environment

Carbon dioxide emissions from the consumption and flaring of fossil fuels in 2008 were the equivalent of 4·9 tonnes per capita.

Electricity

Installed capacity was an estimated 22,000 kW in 2007. Production in 2007 was an estimated 137m. kWh.

Minerals

In 2007 an estimated 223,000 tonnes of sand and gravel and 131,000 tonnes of crushed stone were produced.

Agriculture

Main crops are coconuts, cotton, bananas, yams and molasses, and until 2005 sugarcane. The sugar industry was closed down following the 2005 harvest after years of losses at the state-run sugar company. In 2007 there were an estimated 4,000 ha. of arable land. Most of the farms are small-holdings and there are a number of coconut estates amounting to some 400 ha. under public and private ownership. Production, 2003 estimates (in 1,000 tonnes): sugarcane, 193; coconuts, 1.

Livestock (2003 estimates): goats, 14,000; sheep, 14,000; cattle, 4,000; pigs, 4,000.

Forestry

The area under forests in 2010 was 11,000 ha., or 42% of the total land area.

Fisheries

The estimated catch in 2008 was 450 tonnes.

INDUSTRY

There are three industrial estates on St Kitts and one on Nevis. Export products include electronics and data processing equipment, and garments for the US market. Other small enterprises include food and drink processing, and construction.

Labour

Of 24,778 persons on St Kitts aged 15 or over in 2001, 17,044 were economically active of whom 16,171 were employed and 873 were unemployed. The country has a phenomenon of the working poor—the unemployment rate among the poor in 2001 was very low at 5·3% on St Kitts and 5·0% on Nevis.

INTERNATIONAL TRADE

Imports and Exports

Imports (c.i.f.), 2008, US$324·8m.; exports (f.o.b.), US$51·2m. The USA is by far the biggest trading partner. In 2008, 60·1% of imports were from the USA and 84·6% of exports went to the USA. Trinidad and Tobago is the second largest import supplier and the United Kingdom the second largest export destination. Main imports include machinery, manufactures, food and precious jewellery. Major exports are machinery, transport equipment, printed matter and beer.

COMMUNICATIONS

Roads

In 2002 there were about 383 km of roads, of which 42·5% were paved; and 6,900 passenger cars and 2,500 commercial vehicles.

Rail

In 2005 there were 50 km of railway, formerly operated by the sugar industry but now used for tourist purposes.

Civil Aviation

The main airport is the Robert Llewelyn Bradshaw International Airport (just over 3 km from Basseterre). In 2003 there were flights to Anguilla, Antigua, Barbados, British Virgin Islands, Dominica, Grenada, Jamaica, Netherlands Antilles, Nevis (Newcastle), Philadelphia, Puerto Rico, St Lucia, St Vincent, Trinidad and the US Virgin Islands.

Shipping

There is a deep-water port at Bird Rock (Basseterre). In Jan. 2009 there were 200 ships of 300 GT or over registered, totalling 892,000 GT. Among the 200 vessels registered were 132 general cargo ships and 47 oil tankers. The government maintains a commercial motor boat service between the islands.

Telecommunications

In 2008 there were 20,400 main (fixed) telephone lines; mobile phone subscribers numbered 80,000 in 2008 (156·7 per 100 persons). There were 16,000 internet users in 2008 (31·3 per 100 persons).

SOCIAL INSTITUTIONS

Justice

Justice is administered by the Supreme Court and by Magistrates' Courts. They have both civil and criminal jurisdiction. St Kitts and Nevis was one of ten countries to sign an agreement in Feb. 2001 establishing a Caribbean Court of Justice to replace the British Privy Council as the highest civil and criminal court. In the meantime the number of signatories has risen to 12. The court was inaugurated at Port-of-Spain, Trinidad on 16 April 2005 but has not yet been accepted as St Kitts and Nevis' final court of appeal.

The death penalty is in force and was used in 2008 for the first time since 1998.

The population in penal institutions in Nov. 2007 was 232 (equivalent to 588 per 100,000 of national population, the fourth highest rate in the world).

Education

Adult literacy was 98% in 1998–99. Education is compulsory between the ages of 5 and 16. In 2007 there were 2,360 pupils in pre-primary schools with 357 teaching staff, 6,172 pupils with 372 teaching staff in primary schools and 4,522 pupils with 447 teaching staff in secondary schools.

The main post-secondary institution is the Clarence Fitzroy Bryant College (formerly the St Kitts and Nevis College of Further Education). There are six divisions: Adult and Continuing Education; Arts, Sciences and General Studies; Health Science; Hospitality Studies; Teacher Education; and Technical and Vocational Education and Management Studies.

In 2005 public expenditure on education came to 10·9% of GNI.

Health

In 2005 there were 68 physicians; there were 37·9 registered nurses per 10,000 inhabitants in 2005. In 2008 there were four hospitals and 17 health centres. In 2005 there were 55 hospital beds per 10,000 population.

RELIGION

In 2001, 25·6% of the population were Anglican, 25·6% Methodist and 17·9% Pentecostal. There are also followers of other beliefs, including Roman Catholics, Baptists and Church of God.

CULTURE

World Heritage Sites

There is one site on the UNESCO World Heritage List: Brimstone Hill Fortress National Park (inscribed on the list in 1999), a well-preserved example of 17th and 18th century British military architecture.

Press

In 2006 there was one daily newspaper with a circulation of 3,000. There were also four non-dailies.

Tourism

In 2008 there were 533,353 visitors in total including 400,916 cruise ship passengers and 127,705 staying visitors. Receipts from tourism in 2007 (excluding passenger transport) totalled US$106m.

DIPLOMATIC REPRESENTATIVES

Of St Kitts and Nevis in the United Kingdom (2nd Floor, 10 Kensington Ct, London, W8 5DL)
High Commissioner: Kevin Monroe Isaac.

Of the United Kingdom in St Kitts and Nevis
High Commissioner: Paul Brummell (resides in Bridgetown, Barbados).

Of St Kitts and Nevis in the USA (OECS Building, 3216 New Mexico Ave., NW, 3rd Floor, Washington, D.C., 20016)
Ambassador: Jacinth Henry-Martin.

Of the USA in St Kitts and Nevis
Ambassador: Larry Palmer (resides in Bridgetown, Barbados).

Of St Kitts and Nevis to the United Nations
Ambassador: Delano Frank Bart.

Of St Kitts and Nevis to the European Union
Ambassador: Shirley Skerritt-Andrew.

FURTHER READING

Statistics Division. *National Accounts.* Annual.—*St Kitts and Nevis Quarterly.*

Dyde, Brian, *St Kitts: Cradle of the Caribbean.* 1999
Hubbard, Vince, *A History of St Kitts.* 2002

National library: Public Library, Burdon St., Basseterre.
National Statistical Office: Statistics Division, Ministry of Finance, Planning and Development, Church St., Basseterre.

ST LUCIA

© Research Machines plc 2006

Districts	Sq. km	Population
Dennery	70	12,599
Gros Inlet	101	25,210
Laborie	38	6,701
Micoud	78	16,284
Soufrière	51	8,472
Vieux Fort	44	16,284
Central Forest Reserve	78	—

The UN gives a projected population for 2015 of 183,000.

The official language is English, but 80% of the population speak a French Creole.

In 2000, 50% of the population was Black, 44% were of mixed race and 3% of south Asian ethnic origin.

The capital is Castries (population, 1999, 57,000).

SOCIAL STATISTICS

2005: births, 2,298; deaths, 1,107. Rates, 2005 (per 1,000 population): births, 14·0; deaths, 6·7. Infant mortality, 2010 (per 1,000 live births), 14. Expectation of life in 2007 was 71·7 years for males and 75·5 for females. Annual population growth rate, 2000–05, 1·1%; fertility rate, 2008, 2·0 births per woman.

CLIMATE

The climate is tropical, with a dry season from Jan. to April. Most rain falls in Nov.–Dec.; annual amount varies from 60" (1,500 mm) to 138" (3,450 mm). The average annual temperature is about 80°F (26·7°C).

CONSTITUTION AND GOVERNMENT

The head of state is the British sovereign, represented by an appointed Governor-General. There is a 18-seat *House of Assembly* (17 members elected for five years plus the speaker) and an 11-seat *Senate* appointed by the Governor-General.

National Anthem

'Sons and daughters of St Lucia'; words by C. Jesse, tune by L. F. Thomas.

RECENT ELECTIONS

At the elections of 28 Nov. 2011 the opposition St Lucia Labour Party won 11 seats with 51·0% of votes cast against six for the United Workers' Party (47·0%).

CURRENT GOVERNMENT

Governor-General: Dame Pearlette Louisy; b. 1946 (appointed 17 Sept. 1997).

In March 2013 the government comprised:

Prime Minister and Minister of Finance and Economic Affairs, Planning, and Social Security: Kenny Anthony; b. 1951 (St Lucia Labour Party; sworn in 30 Nov. 2011, having previously been prime minister from May 1997 to Dec. 2006).

Minister for Infrastructure, Port Services and Transport: Philip J. Pierre. *Foreign Affairs, International Trade and Civil Aviation:* Alva Baptiste. *Education, Human Resource Development and Labour:* Robert Lewis. *Health, Wellness, Human Services and Gender Relations:* Alvina Reynolds. *Agriculture, Food Production, Fisheries and Rural Development:* Moses Jn Baptiste. *Social Transformation, Local Government and Community Empowerment:* Harold Dalson. *Commerce, Business Development and Consumer Affairs:* Emma Hippolyte. *Legal Affairs, Home Affairs and National Security:* Victor Phillip La Corbiniere. *Tourism, Heritage and Creative Industries:* Lorne Theophilus. *Youth Development and Sports:* Shawn Edward. *Public Service, Sustainable Development, Energy, Science and Technology:* James

Capital: Castries
Population projection, 2015: 183,000
GNI per capita, 2011: (PPP$) 8,273
HDI/world rank: 0·723/82
Internet domain extension: .lc

KEY HISTORICAL EVENTS

The island was probably discovered by Columbus in 1502. An unsuccessful attempt to colonize by the British took place in 1605 and again in 1638 when settlers were soon murdered by the Caribs who inhabited the island. France claimed the right of sovereignty and ceded it to the French West India Company in 1642. St Lucia regularly and constantly changed hands between Britain and France, until it was finally ceded to Britain in 1814 by the Treaty of Paris. Since 1924 the island has had representative government. In March 1967 St Lucia gained full control of its internal affairs while Britain remained responsible for foreign affairs and defence. On 22 Feb. 1979 St Lucia achieved independence, opting to remain in the British Commonwealth.

TERRITORY AND POPULATION

St Lucia is an island of the Lesser Antilles in the eastern Caribbean between Martinique and St Vincent, with an area of 617 sq. km (238 sq. miles). Population (2010 census, provisional) 165,595; density, 268·4 per sq. km. In 2011 the population was 28·1% urban.

Areas and populations of the districts at the 2010 census (provisional) were:

Districts	Sq. km	Population
Anse-la-Raye	31	6,247
Canaries	16	2,044
Castries	79	65,656
Choiseul	31	6,098

Fletcher. *Physical Development, Housing and Urban Renewal:* Stanley Felix.

Government Website: http://www.stlucia.gov.lc

CURRENT LEADERS

Dr Kenny Anthony

Position
Prime Minister

Introduction
Dr Kenny Anthony returned to office as prime minister on 30 Nov. 2011, having previously served two consecutive terms from 1997 to 2006. After electoral defeat in Dec. 2006, he remained leader of the Saint Lucia Labour Party (SLP) in opposition before returning to power.

Early Life
Kenny Davis Anthony was born on 8 Jan. 1951. He studied law at the University of the West Indies before studying for a PhD at Birmingham University in the United Kingdom.

He joined the SLP on his return to St Lucia, serving as minister of education from 1980–81. A former consultant to the United Nations on its development programmes, Anthony was also a member of the Caribbean Community secretariat from 1995–97 and served as chairman of the Organisation of Eastern Caribbean States. He became SLP leader in 1996 and was sworn in as prime minister for the first time in May 1997.

In his first term Anthony was at the centre of a dispute with the USA and World Trade Organization over alleged preferential treatment for former colonial banana suppliers to the European Union. In Sept. 2002 tropical storm Lili hit the island, devastating the banana crop and exacerbating economic problems. The government sought to diversify the economy, with a focus on increasing investment in tourism. Anthony also oversaw extensive domestic infrastructure development, while education spending rose significantly.

However, he was criticized for St Lucia's high levels of borrowing, with the debt-to-GDP ratio standing at around 63% in 2005. Meanwhile, the crime rate remained stubbornly high, despite the passage of a new criminal code. In July 2003 parliament amended the constitution to replace the oath of allegiance to the British monarch with a pledge of loyalty to the St Lucian people (although Queen Elizabeth II remained head of state).

At the general election of 11 Dec. 2006, the SLP won only six seats against 11 for the United Workers Party, headed by Sir John Compton. Anthony nonetheless retained his own constituency and continued as leader of the SLP. Compton died in Sept. 2007 and was succeeded as prime minister by Stephenson King. During his time in opposition, Anthony visited Cuba, where he thanked the Cuban administration for its aid to St Lucia and other Caribbean nations. In the general election held in Nov. 2011 the SLP were returned to power, taking 11 of the available 17 seats.

Career in Office
With the economy in need of urgent attention, Anthony warned the island's population that 'difficult times lie ahead'. He also faced a strained relationship with Taiwan following St Lucia's establishment of diplomatic ties with China during his previous tenure.

ECONOMY

In 2007 agriculture, forestry, hunting and fishing contributed 6·4% of GDP, industry 19·6% and services 76·8%.

Overview
High levels of public debt and unemployment are ongoing problems. The principal source of revenue and employment is tourism. The government is trying to boost banana production, which accounts for 41% of commodities exports. The economy is vulnerable to natural disasters, fluctuating tourist arrivals and oil prices.

Following the 2008 economic crisis, the financial system experienced knock-on effects from the collapse of the Trinidad and Tobago-based CL-Financial Group, which had operations in the country. A sharp decline in tourist arrivals and construction projects resulted in GDP contracting in 2009.

St Lucia, Antigua and Barbuda, Dominica, Grenada, St Kitts and Nevis, and St Vincent and the Grenadines comprise the Organization of Eastern Caribbean States (OECS), which aims at economic and political integration. The Treaty establishing the OECS Economic Union was signed on 18 June 2010. Tourism makes up more than 23% of GDP for the OECS region as a whole and will be an important factor in regional economic recovery.

Currency
The *East Caribbean dollar* (XCD) (of 100 *cents*) is in use. US dollars are also normally accepted. There was deflation of 0·2% in 2009 but inflation of 3·3% in 2010 and 2·8% in 2011. Foreign exchange reserves were US$109m. in July 2005 and total money supply was EC$542m.

Budget
The fiscal year ends on 31 March. Revenues were EC$816·0m. in the fiscal year 2008–09 and expenditures EC$959·1m.

Performance
The economy grew by 0·4% in 2010 and 1·3% in 2011. Total GDP in 2011 was US$1·3bn.

Banking and Finance
The East Caribbean Central Bank based in St Kitts and Nevis functions as a central bank. The *Governor* is Sir Dwight Venner. There are three domestic banks (Caribbean Banking Corporation, St Lucia Co-operative Bank, East Caribbean Financial Holding Company) and three foreign banks.

Foreign debt in 2007 amounted to US$441m.

St Lucia is a member of the Eastern Caribbean Securities Exchange, based in Basseterre.

ENERGY AND NATURAL RESOURCES

Environment
Carbon dioxide emissions from the consumption and flaring of fossil fuels in 2008 were the equivalent of 2·6 tonnes per capita.

Electricity
Installed capacity in 2007 was 76,000 kW. Production in 2007 was 346m. kWh; consumption per capita in 2007 was 2,046 kWh.

Agriculture
In 2007 St Lucia had approximately 3,000 ha. of arable land and 7,000 ha. of permanent crops. Bananas, cocoa, breadfruit and mango are the principal crops, but changes in the world's trading rules and changes in taste are combining to depress the banana trade. Farmers are experimenting with okra, tomatoes and avocados to help make up for the loss. Production, 2003 estimates (in 1,000 tonnes): bananas, 120; mangoes, 28; coconuts, 14; yams, 4; grapefruit and pomelos, 3.

Livestock (2003 estimates): pigs, 15,000; cattle, 12,000; sheep, 12,000; goats, 10,000.

Forestry
In 2010 the area under forests was 47,000 ha. (77% of the total land area).

Fisheries
In 2010 the total catch was 1,844 tonnes.

INDUSTRY

The main areas of activity are clothing, assembly of electronic components, beverages, corrugated cardboard boxes, tourism, lime processing and coconut processing.

Labour

In the period April–June 2004 the labour force totalled 78,210. The unemployment rate was 21·0% in 2004.

INTERNATIONAL TRADE

Imports and Exports

Imports and exports for calendar years in US$1m.:

	2003	2004	2005	2006
Imports c.i.f.	407·7	421·6	485·8	592·4
Exports f.o.b.	62·5	79·8	64·2	93·7

Main imports in 2006: machinery and transport equipment, 25·8%; food and live animals, 15·9%; petroleum and petroleum products, 12·1%; chemicals and related products, 6·9%. Main exports, 2006: petroleum and petroleum products, 21·7%; bananas, 19·0%; machinery and transport equipment, 18·2%; beer, 14·0%. Main import suppliers, 2006: USA, 39·2%; Trinidad and Tobago, 16·9%; UK, 6·9%; Japan, 6·3%. Main export markets, 2006: Trinidad and Tobago, 30·1%; UK, 20·7%; USA, 20·6%; Barbados, 6·6%.

COMMUNICATIONS

Roads

The island had about 1,210 km of roads in 2002, of which 150 km were main roads and a further 150 km secondary roads. Passenger cars numbered 13,100 in 2002.

Civil Aviation

There are two international airports: Hewanorra International (near Vieux-Fort) and George F. L. Charles (near Castries). In 2009 Hewanorra handled 513,959 passengers (483,632 in 2008) and George F. L. Charles—which handles inter-Caribbean flights—309,132 passengers (358,313 in 2008).

Shipping

There are two ports, Castries and Vieux Fort.

Telecommunications

Fixed telephone lines numbered 40,900 in 2008 (240·2 per 1,000 persons) and there were 216,500 mobile phone subscriptions in 2011 (or 1,230·0 per 1,000 inhabitants). Fixed internet subscriptions totalled 20,800 in 2010 (119·4 per 1,000 inhabitants).

SOCIAL INSTITUTIONS

Justice

The island is divided into two judicial districts, and there are nine magistrates' courts. Appeals lie to the Eastern Caribbean Supreme Court of Appeal. St Lucia was one of ten countries to sign an agreement in Feb. 2001 establishing a Caribbean Court of Justice to replace the British Privy Council as the highest civil and criminal court. In the meantime the number of signatories has risen to 12. The court was inaugurated at Port-of-Spain, Trinidad on 16 April 2005. It has not yet replaced the Privy Council as St Lucia's final court of appeal.

The population in penal institutions in Dec. 2010 was 551 (315 per 100,000 of national population).

Education

Primary education is free and compulsory. In 2008–09 there were 1,000 teachers for 20,000 pupils at primary school level; and 16,000 pupils and an estimated 1,000 teachers at secondary level. There is a community college. The adult literacy rate was 90·1% in 2003 (89·5% among males and 90·6% among females).

In 2006 public expenditure on education came to 6·9% of GNI and 19·1% of total government spending.

Health

In 2002 there were five hospitals (with 305 beds) and 35 health centres. There were 70 physicians, seven dentists and 302 nurses.

RELIGION

In 2001, 79% of the population was Roman Catholic.

CULTURE

World Heritage Sites

There is one UNESCO World Heritage site in St Lucia: Pitons Management Area (inscribed on the list in 2004). The site near the town of Soufrière includes the Pitons, two volcanic spires rising side by side from the sea, linked by the Piton Mitan ridge.

Press

There are no daily newspapers. In 2008 there were six paid-for non-daily newspapers: the thrice-weekly *The Voice* and *The Star*; and the weekly *The Mirror*, *The Crusader*, *The Vanguard* and *One Caribbean*.

Tourism

The number of foreign tourists in 2005 was 318,000. Receipts in 2005 totalled US$345m. (excluding passenger transport).

DIPLOMATIC REPRESENTATIVES

Of St Lucia in the United Kingdom (1 Collingham Gdns, Earls Court, London, SW5 0HW)
High Commissioner: Dr Ernest Hilaire.

Of the United Kingdom in St Lucia
High Commissioner: Paul Brummell (resides in Bridgetown, Barbados).

Of St Lucia in the USA (3216 New Mexico Ave., NW, Washington, D.C., 20016)
Ambassador: Sonia Merlyn Johnny.

Of the USA in St Lucia
Ambassador: Larry Palmer (resides in Bridgetown, Barbados).

Of St Lucia to the United Nations
Ambassador: Menissa Rambally.

Of St Lucia to the European Union
Ambassador: Shirley Skerritt-Andrew.

FURTHER READING

National Statistical Office: Central Statistical Office, Chreiki Building, Micoud Street, Castries.
Website: http://www.stats.gov.lc

ST VINCENT AND THE GRENADINES

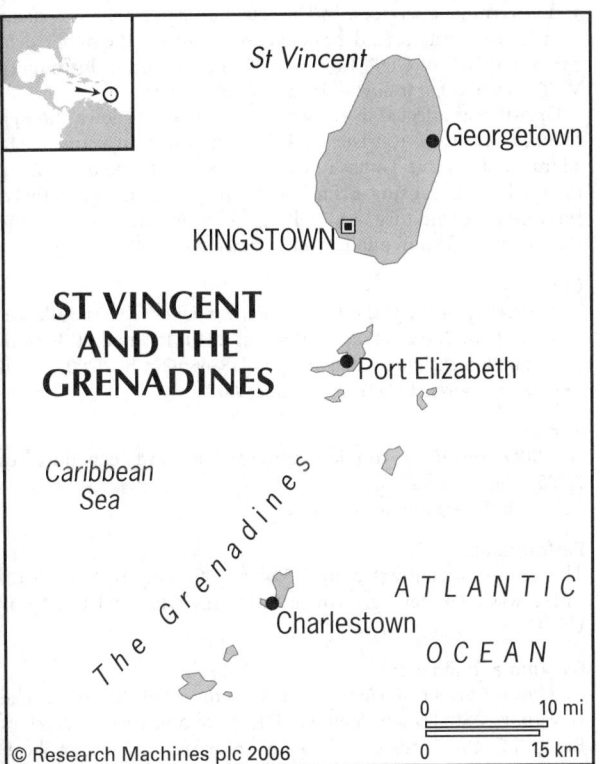

Capital: Kingstown
Population projection, 2015: 109,000
GNI per capita, 2011: (PPP$) 8,013
HDI/world rank: 0·717/85
Internet domain extension: .vc

KEY HISTORICAL EVENTS

St Vincent was discovered by Columbus on 22 Jan. (St Vincent's Day) 1498. British and French settlers occupied parts of the islands after 1627. In 1773 the Caribs recognized British sovereignty and agreed to a division of territory between themselves and the British. Resentful of British rule, the Caribs rebelled in 1795, aided by the French, but the revolt was subdued within a year. On 27 Oct. 1969 St Vincent became an Associated State with the UK responsible for foreign policy and defence. On 27 Oct. 1979 the colony gained full independence as St Vincent and the Grenadines.

TERRITORY AND POPULATION

St Vincent is an island of the Lesser Antilles, situated in the eastern Caribbean between St Lucia and Grenada, from which latter it is separated by a chain of small islands known as the Grenadines. The total area of 389 sq. km (150 sq. miles) comprises the island of St Vincent itself (345 sq. km) and those of the Grenadines attached to it, of which the largest are Bequia, Mustique, Canouan, Mayreau and Union.

The population at the 2001 census was 106,253, of whom 97,638 lived on St Vincent; density, 273 per sq. km. In 2011, 49·8% of the population lived in urban areas.

The UN gives a projected population for 2015 of 109,000.

The capital, Kingstown, had 25,307 inhabitants in 2004 (including suburbs). The population is mainly of Black (65·5%) and mixed (23·5%) origin, with small White, Asian and American minorities.

English is the official language, although French patois is widely spoken.

SOCIAL STATISTICS

Births, 2008 estimate, 1,900; deaths, 800. 2008 estimated rates (per 1,000 population): births, 17·6; deaths, 7·5. Infant mortality, 2010, 19 per 1,000 live births. Life expectancy, 2007, was 69·4 years for males and 73·6 years for females. Annual population growth rate, 2000–08, 0·1%; fertility rate, 2008, 2·1 births per woman.

CLIMATE

The climate is tropical marine, with northeast trades predominating and rainfall ranging from 150" (3,750 mm) a year in the mountains to 60" (1,500 mm) on the southeast coast. The rainy season is from June to Dec., and temperatures are equable throughout the year.

CONSTITUTION AND GOVERNMENT

The head of state is Queen Elizabeth II, represented by a Governor-General. Parliament is unicameral with a 23-member *House of Assembly* consisting of 15 members directly elected for a five-year term from single-member constituencies, six senators appointed by the Governor-General (four on the advice of the Prime Minister and two on the advice of the Leader of the Opposition) and two *ex officio* members.

National Anthem

'St Vincent, land so beautiful'; words by Phyllis Punnett, tune by J. B. Miguel.

RECENT ELECTIONS

At the elections to the House of Assembly on 13 Dec. 2010 the ruling Unity Labour Party (ULP, social democratic) won 8 of the 15 elected seats with 51·1% of the vote, against 7 (48·7%) for the opposition New Democratic Party (NDP, conservative).

CURRENT GOVERNMENT

Governor-General: Sir Frederick Ballantyne (since 2 Sept. 2002).

In March 2013 the government comprised:

Prime Minister and Minister of Finance, National Security, Grenadine Affairs and Telecommunications: Dr Ralph E. Gonsalves; b. 1946 (ULP; sworn in 29 March 2001 and re-elected in Dec. 2005 and Dec. 2010).

Deputy Prime Minister and Minister of Education: Girlyn Miguel.

Minister of National Mobilization, Social Development, Family, Gender Affairs, Persons with Disabilities and Youth: Frederick Stevenson. *National Reconciliation, Public Service, Labour, Information and Ecclesiastical Affairs:* Maxwell Charles. *Health, Wellness and the Environment:* Clayton Burgin. *Foreign Affairs, Foreign Trade and Consumer Affairs:* Douglas Slater. *Transportation and Works, and Urban Development and Local Government:* Julian Francis. *Agriculture, Industry, Forestry, Fisheries and Rural Transformation:* Saboto Caesar. *Tourism, Sports and Culture:* Cecil McKie. *Housing, Informal Human Settlements, Lands and Surveys, and Physical Planning:* Montgomery Daniel.

Government Website: http://www.gov.vc

CURRENT LEADERS

Dr Ralph E. Gonsalves

Position
Prime Minister

Introduction
Dr Ralph E. Gonsalves became prime minister in 2001 after his Unity Labour Party (ULP) won an election brought forward from 2003 following anti-government protests in 2000. He won further terms in 2005 and 2010.

Early Life
'Comrade Ralph' was born in 1946. He studied at the University of the West Indies in Jamaica, gaining a PhD in political science. He later graduated in law from the University of the West Indies in Barbados, before returning home. In 1982 he founded the left-wing Movement for National Unity, which in 1994 amalgamated with the St Vincent Labour Party to form the ULP. Gonsalves succeeded the ULP's first leader, Vincent Beache, in 1998 and led the party to electoral victory in 2001, ending 15 years of New Democratic Party rule.

Career in Office
Gonsalves has sought to tackle the problems of money laundering, gun violence and drug-related crime. In June 2003 St Vincent and Grenadines was removed from a list of uncooperative countries in the fight against money laundering throughout the Caribbean islands.

The country's economy is reliant on the banana trade and the government continues to pay large subsidies to farmers. However, a decline in banana prices and demand has led to some diversification. In June 2004 Gonsalves announced the launch of National Investments Promotions Incorporated to attract investment and boost exports, and the tourism sector has seen significant growth.

Gonsalves has maintained close ties with Taiwan and Cuba. In May 2005 Taiwan made a gift of computer equipment to his government to enhance efficiency. It also helped to build facilities for the 2007 Cricket World Cup. In Jan. 2005 Gonsalves met with Fidel Castro to announce that new diplomatic missions would be established in their respective countries. Castro promised Cuban assistance to Gonsalves' campaign to increase national literacy rates.

Feb. 2005 saw the launch of the World Bank-assisted HIV/AIDS prevention and control project, with the St Vincent government expected to invest at least US$1·7m. over five years. In April 2005 Gonsalves signed agreements with St Lucia and Dominica to introduce measures to combat climate change in the coastal areas of the three Windward Islands (with substantial funding from the USA and Japan). In May 2005 his government ratified the Kyoto protocol.

In parliamentary elections in Dec. 2005 the ULP retained their 12 seats. Gonsalves was sworn in for a new term and took over the national security portfolio. In May 2007 his government introduced VAT to increase public revenue (with a reduced rate for hotel accommodation to help protect the tourist industry). Plans for a new constitution and the establishment of a republic, for which Gonsalves campaigned for acceptance, were rejected by 55% of voters in a referendum in Nov. 2009. The ULP only narrowly retained its parliamentary majority in elections in Dec. 2010, taking eight of the 15 seats.

In 2008 diplomatic relations were established with Iran.

ECONOMY

Agriculture accounted for 6·9% of GDP in 2009, industry 18·9% and services 74·2%.

Overview

With the banana crop accounting for around a third of export earnings, the economy enjoyed vigorous growth in 2006 and 2007 at around 7% per annum, boosted by healthy construction activity and government services. In Oct. 2007 an agreement was made with Italy to write off a debt obligation, reducing the public debt by around 10% of GDP. In May 2007 the authorities introduced VAT, contributing to higher inflation for the year.

Growth was affected by the global economic slowdown through reduced levels of tourism and foreign direct investment. In addition, Hurricane Tomas in Oct. 2010 was followed by flooding in April 2011, causing a further decline in economic activity. Increased expenditure and falling GDP in the wake of the slowdown resulted in public debt increasing to 66·8%.

Currency

The currency in use is the *East Caribbean dollar* (XCD). Inflation was 10·1% in 2008, falling to 0·8% in 2010. It stood at 3·2% in 2011. Foreign exchange reserves were US$66m. in July 2005. Total money supply was EC$319m. in June 2005.

Budget

In 2009 revenues totalled EC$522·5m. and expenditures EC$571·5m.

VAT is 15% (reduced rate, 10%).

Performance

The economy contracted by 2·3% in 2009 and 1·8% in 2010. There was then zero growth in 2011. In 2011 total GDP was US$0·7bn.

Banking and Finance

The East Caribbean Central Bank is the bank of issue. The *Governor* is Sir Dwight Venner. There are branches of Barclays Bank PLC, the Caribbean Banking Corporation, FirstCaribbean International, the Canadian Imperial Bank of Commerce and the Bank of Nova Scotia. Locally-owned banks: First St Vincent Bank, Owens Bank, New Bank, the National Commercial Bank and St Vincent Co-operative Bank. The 'offshore' sector numbered over 11,000 organizations in 2001.

Foreign debt was US$253m. in 2007.

St Vincent and the Grenadines is a member of the Eastern Caribbean Securities Exchange, based in Basseterre.

ENERGY AND NATURAL RESOURCES

Environment

Carbon dioxide emissions from the consumption and flaring of fossil fuels were the equivalent of 2·1 tonnes per capita in 2008.

Electricity

Installed capacity was approximately 34,000 kW in 2007. Production in 2007 was an estimated 135m. kWh; consumption per capita in 2007 was around 1,309 kWh.

Agriculture

In 2007 the agricultural population was an estimated 23,000, of which 11,000 were economically active. There were an estimated 7,000 ha. of arable land and 5,000 ha. of permanent crops in 2007. The sugar industry was closed down in 1985 although some sugarcane is grown for rum production. Production (2003, in 1,000 tonnes): bananas, 50; coconuts, 3; copra, 2; maize, 2; oranges, 2; plantains, 2; sugarcane, 2; sweet potatoes, 2; yams, 2.

Livestock (2003, in 1,000): sheep, 12; pigs, 9; goats, 7; cattle, 5.

Forestry

Forests covered 27,000 ha. in 2010, or 68% of the land area.

Fisheries

Total catch, 2009, 4,058 tonnes (all from sea fishing).

INDUSTRY

Industries include assembly of electronic equipment, manufacture of garments, electrical products, animal feeds and flour, corrugated galvanized sheets, exhaust systems, industrial gases, concrete blocks, plastics, soft drinks, beer and rum, wood products and furniture, and processing of milk, fruit juices and food items.

Labour

The Department of Labour is charged with looking after the interest and welfare of all categories of workers, including providing advice and guidance to employers and employees and their organizations and enforcing the labour laws. In 2001 the total labour force was 43,779, of whom 34,521 (21,274 males and 13,247 females) were employed.

INTERNATIONAL TRADE

Imports and Exports

Imports (c.i.f.) in 2010 totalled US$379·5m. and exports (f.o.b.) US$41·5m. Principal imports are basic manufactures, machinery and transport equipment, and food products. Principal exports are bananas, packaged flour and packaged rice. Main import suppliers, 2005: USA, 33·3%; Trinidad and Tobago, 23·6%; UK, 9·4%. Main export markets, 2005: UK, 26·8%; Barbados, 12·8%; Trinidad and Tobago, 12·3%.

COMMUNICATIONS

Roads

In 2002 there were 829 km of roads, of which 70% were paved. Vehicles in use (2008): 9,250 passenger cars, 12,900 vans and lorries, and 1,220 motorcycles and mopeds.

Civil Aviation

There is an airport (E. T. Joshua) on mainland St Vincent at Arnos Vale. An airport on Union also has regular scheduled services. In 2003 E. T. Joshua handled 273,000 passengers; 29,000 aircraft passed through or landed. A new airport is under construction at Argyle and is scheduled to open in late 2013.

Shipping

In Jan. 2009 there were 580 ships of 300 GT or over registered, totalling 5·10m. GT. Among the 580 vessels registered were 413 general cargo ships, 80 bulk carriers and 31 passenger ships.

Telecommunications

There were 133,400 telephone subscribers in 2007, equivalent to 1,108·1 for every 1,000 inhabitants, including 110,500 mobile phone subscribers. The telephone network has almost 100% geographical coverage. In 2007 there were 57,000 internet users.

SOCIAL INSTITUTIONS

Justice

Law is based on UK common law as exercised by the Eastern Caribbean Supreme Court on St Lucia. Final appeal lies to the UK Privy Council. St Vincent and the Grenadines was one of 12 countries to sign an agreement establishing a Caribbean Court of Justice to replace the British Privy Council as the highest civil and criminal court. The court was inaugurated at Port-of-Spain, Trinidad on 16 April 2005 but has yet to be accepted as the final court of appeal for St Vincent and the Grenadines. Strength of police force (2009), 750.

The population in penal institutions in 2007 was 376 (317 per 100,000 of national population).

Education

In 2007 there were 3,894 children in pre-primary schools with 340 teaching staff, 15,928 pupils in primary schools with 933 teaching staff and (2005) 9,780 pupils in secondary schools. There is a community college. Adult literacy in 2004 was 88·1%.

In 2005 public expenditure on education came to 8·6% of GNI and 16·1% of total government spending.

Health

In 2007 there were 35 hospital beds per 10,000 persons. In 2009 there were 59 physicians, nine dentists, 299 nurses and 27 pharmacists.

RELIGION

In 2001 there were estimated to be 20,000 Anglicans, 17,000 Pentecostalists, 12,000 Methodists, 12,000 Roman Catholics and 52,000 followers of other religions.

CULTURE

Press

In 2006 there was one daily newspaper, *The Herald*. There were also six weekly papers.

Tourism

In 2005 there were 95,505 staying visitors, 69,391 cruise passenger arrivals and 84,610 recorded yacht visitors. Tourism receipts (excluding passenger transport) in 2005 totalled US$105m.

DIPLOMATIC REPRESENTATIVES

Of St Vincent and the Grenadines in the United Kingdom (10 Kensington Ct, London, W8 5DL)
High Commissioner: Cenio Elwin Lewis.

Of the United Kingdom in St Vincent and the Grenadines (Francis Compton Building, 2nd Floor, Waterfront, Castries)
High Commissioner: Paul Brummell (resides in Bridgetown, Barbados).

Of St Vincent and the Grenadines in the USA (3216 New Mexico Ave., NW, Washington, D.C., 20016)
Ambassador: La Celia Aritha Prince.

Of the USA in St Vincent and the Grenadines
Ambassador: Larry Palmer (resides in Bridgetown, Barbados).

Of St Vincent and the Grenadines to the United Nations
Ambassador: Camillo Gonsalves.

Of St Vincent and the Grenadines to the European Union
Ambassador: Shirley Skerritt-Andrew.

FURTHER READING

Sutty, L., *St Vincent and the Grenadines.* 1993

National Statistical Office: Statistical Office, Central Planning Division, Ministry of Finance & Economic Planning, Kingstown.
Website: http://stats.gov.vc

SAMOA

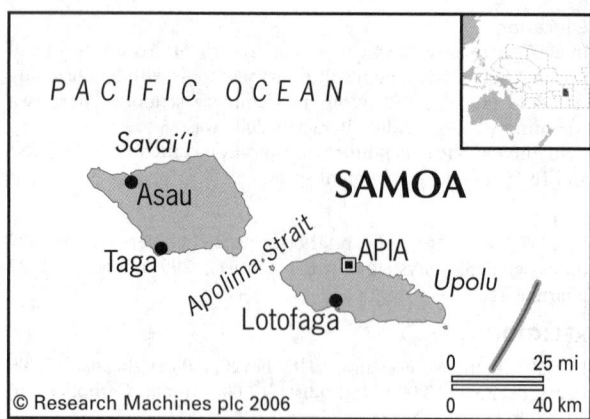

© Research Machines plc 2006

O le Malo Tutoatasi o Samoa
(Independent State of Samoa)

Capital: Apia
Population projection, 2015: 188,000
GNI per capita, 2010: US$3,232
HDI/world rank: 0·688/99
Internet domain extension: .ws

KEY HISTORICAL EVENTS

Polynesians settled in the Samoan group of islands in the southern Pacific from about 1000 BC. Although probably sighted by the Dutch in 1722, the first European visitor was French in 1768. Treaties were signed between the Chiefs and European nations in 1838–39. Continuing strife among the chiefs was compounded by British, German and US rivalry for influence. In the Treaty of Berlin 1889 the three powers agreed to Western Samoa's independence and neutrality. When unrest continued, the treaty was annulled and Western Samoa became a German protectorate until in 1914 it was occupied by a New Zealand expeditionary force. The island was administered by New Zealand from 1920 to 1961. On 1 Jan. 1962 Western Samoa gained independence. In July 1997 the country renamed itself the Independent State of Samoa.

TERRITORY AND POPULATION

Samoa lies between 13° and 15° S. lat. and 171° and 173° W. long. It comprises the two large islands of Savai'i and Upolu, the small islands of Manono and Apolima, and several uninhabited islets lying off the coast. The total land area is 2,785 sq. km (1,075 sq. miles), of which 1,694 sq. km (654 sq. miles) are in Savai'i and 1,091 sq. km (421 sq. miles) in Upolu (including Manono and Apolima). The islands are of volcanic origin, and the coasts are surrounded by coral reefs. Rugged mountain ranges form the core of both main islands. The large area laid waste by lava-flows in Savai'i is a primary cause of that island supporting less than one-third of the population of the islands despite its greater size than Upolu.

The population was 187,820 at the 2011 census; density, 67 per sq. km. The population at the 2011 census was 143,418 in Upolu (including Manono and Apolima) and 44,402 in Savai'i. The capital and chief port is Apia in Upolu (population 36,735 in 2011). In 2011, 20·1% of the population lived in urban areas.

The UN gives a projected population for 2015 of 188,000.

The official languages are Samoan and English.

SOCIAL STATISTICS

2006: births, 4,935; deaths, 728. Rates, 2006 (per 1,000 population): births, 27·3; deaths, 4·0. Expectation of life in 2007 was 68·4 years for males and 74·7 for females. Annual population growth rate, 2001–06, was 0·5%. Infant mortality, 2010, 17 per 1,000 live births; fertility rate, 2006, 4·2 births per woman.

CLIMATE

A tropical marine climate, with cooler conditions from May to Nov. and a rainy season from Dec. to April. The rainfall is unevenly distributed, with south and east coasts having the greater quantities. Average annual rainfall is about 100" (2,500 mm) in the drier areas. Apia, Jan. 80°F (26·7°C), July 78°F (25·6°C). Annual rainfall 112" (2,800 mm).

CONSTITUTION AND GOVERNMENT

HH Malietoa Tanumafili II, who was Head of State for life, died on 11 May 2007. The Head of State is henceforth elected by the Legislative Assembly and holds office for five-year terms.

The executive power is vested in the *Head of State*, who swears in the *Prime Minister* (who is elected by the Legislative Assembly) and, on the Prime Minister's advice, the Ministers to form the Cabinet. The Constitution also provides for a *Council of Deputies* of three members, of whom the chairman is the Deputy Head of State.

The *Legislative Assembly* contains 49 members serving five-year terms. 47 are elected exclusively by *matai* (customary family heads) and the other two by non-Samoans on separate electoral rolls.

National Anthem

'Samoa, tula'i ma sisi ia laufu'a ('Samoa, Arise and Raise your Banner'); words and tune by S. I. Kuresa.

RECENT ELECTIONS

In elections on 4 March 2011 the Human Rights Protection Party (HRPP) won 36 seats, against 13 for the Tautua Samoa Party. Turnout was 87·1%.

Tuiatua Tupua Tamasese Efi was elected head of state unanimously by the Legislative Assembly on 16 June 2007. He was re-elected unopposed for a second term in July 2012.

CURRENT GOVERNMENT

Head of State: Tuiatua Tupua Tamasese Efi; b. 1938 (in office since 20 June 2007).

In March 2013 the cabinet was composed as follows:

Prime Minister and Minister of Foreign Affairs and Trade: Tuila'epa Sailele Malielegaoi; b. 1945 (Human Rights Protection Party; sworn in 23 Nov. 1998, and re-elected in March 2001, March 2006 and March 2011).

Deputy Prime Minister and Minister of Commerce, Industry and Labour: Fontoe Pierre Lauofo.

Minister of Agriculture and Fisheries: Lemamea Ropati. *Communications and Information Technology:* Tuisugaletaua Sofara Aveau. *Education, Sports and Culture:* Magele Mauiliu Magele. *Finance:* Faumuina Liuga. *Health:* Tuitama Leao Tuitama. *Justice and Courts Administration:* Fiame Naomi Mataafa. *Natural Resources and Environment:* Faamoetauloa Faale. *Police and Prisons:* Sala Fata Pinati. *Revenue:* Tuiloma Pule Lameko. *Women, Community and Social Development:* Tolofuaivalelei Falemoe Leiataua. *Works, Transportation and Infrastructure:* Manualesegalala Enokati Posala.

Government Website: http://www.govt.ws

CURRENT LEADERS

Tuila'epa Sailele Malielegaoi

Position
Prime Minister

Introduction
Tuila'epa Sailele Malielegaoi became prime minister in Nov. 1998 and won further terms in March 2001, March 2006 and March 2011. He is leader of the Human Rights Protection Party (HRPP), the traditional ruling party of Samoa since 1982.

Early Life
Tuila'epa Sailele Malielegaoi was born on 14 April 1945 in Lepa, Samoa. He was educated in Samoa and at New Zealand's Auckland University, graduating with a masters degree (the first Samoan to do so) in commerce in 1969.

In 1978 he moved to Brussels to work for the European Economic Community. He entered Samoa's parliament two years later while working as a partner in the accounting firm Coopers and Lybrand. He was elected to the premiership after former prime minister Tofilau Eti Alesana retired in 1998.

Career in Office
Malielegaoi has aimed to diversify an economy dependent on fishing and agriculture and susceptible to natural disasters, focusing particularly on the tourism industry. In Aug. 2004 his government introduced internet access to assist economic development.

Malielegaoi has also been keen to promote education in Samoa. There is a scholarship scheme offering study opportunities in New Zealand, Australia and the Fiji Islands, and in Jan. 2003 a new inter-denominational Christian secondary school was opened. The police and health sectors have also received increased funding. In Jan. 2004 parliament voted to abolish the death penalty and in the same year Australia provided $A7m. to fund training of Samoan security forces.

In foreign relations, Malielegaoi has pursued close relations with China. China agreed to help build an aquatic centre in Samoa for the 2007 South Pacific Games and has been aiding the construction of new buildings for the Samoan parliament and Justice Department. Japan has also invested in Samoa, providing funding for education and vocational training and for land redevelopment. In 2004 Samoa hosted the 35th Pacific Forum, which concentrated on regional economic and political co-operation and the Pacific-wide campaign to tackle HIV and AIDS.

Malielegaoi was re-elected for a third term when his Human Rights Protection Party won the March 2006 election. In May 2007 King Malietoa Tanumafili II died after 45 years on the throne, having been appointed king for life at independence in 1962. Under the constitution, his successor, Tuiatua Tupua Tamasese Efi, was appointed for a five-year term by the Legislative Assembly in June.

Opposition to controversial road traffic legislation by Malielegaoi (changing the driving side from right to left), which took effect in Sept. 2009, had earlier led to defections from the HRPP and to the formation in 2008 of two new political parties— the Tautua Samoa Party and the People's Party. In elections in March 2011 the HRPP won a landslide victory, heralding a fourth term for the premier.

At the end of 2011 Samoa crossed westward over the international date line for trading reasons (to align with Australia and New Zealand) and effectively erased a day from its calendar. In May 2012 Samoa acceded to the World Trade Organization as its 155th member, and in June the country celebrated 50 years of independence from New Zealand.

ECONOMY

Agriculture accounted for 11·8% of GDP in 2006, industry 27·1% and services 61·2%.

Overview

Two-thirds of the population are employed in agriculture, which accounts for 90% of exports (principally coconuts and its by-products, plus coffee). Tourism contributes 25% of GDP and its importance is increasing, with aviation and infrastructural reforms and redevelopment following a tsunami in 2009 helping boost visitor numbers.

Deregulation has spurred growth in the financial sector, while the government deficit fell to 6% of GDP in 2011 as a result of improved fiscal discipline.

Currency

The unit of currency is the *tala* (WST) of 100 *sene*. Inflation was 14·6% in 2009 but there was deflation of 0·2% in 2010. There was then inflation of 2·9% in 2011. Foreign exchange reserves were US$94m. in July 2005. Total money supply was 133m. tala in June 2005.

Budget

The fiscal year begins on 1 July. For 2008–09 revenue was SA$492·0m. (tax revenue, 66·0%); expenditure, SA$551·0m. (current expenditure, 64·7%).

VAT is 12·5%.

Performance

The economy contracted by 5·1% in 2009 but grew by 0·4% in 2010 and 2·0% in 2011. Total GDP in 2011 was US$0·6bn.

Banking and Finance

The Central Bank of Samoa (founded 1984) is the bank of issue. The *Governor* is Atalina Enari. There is one development bank. Commercial banks include: ANZ, Industrial Bank, International Business Bank Corporation, National Bank of Samoa, Samoa Commercial Bank and Westpac Bank Samoa. Total external debt was US$1,140m. in 2007.

ENERGY AND NATURAL RESOURCES

Environment

Samoa's carbon dioxide emissions from the consumption and flaring of fossil fuels in 2008 were the equivalent of 0·8 tonnes per capita.

Electricity

Installed capacity in 2007 was 42,000 kW. Production was 118m. kWh. in 2007 and consumption per capita an estimated 632 kWh.

Agriculture

In 2002 there were 60,000 ha. of arable land and 69,000 ha. of permanent cropland. The main products (2002, in 1,000 tonnes) are coconuts (140), bananas (22), taro (17), copra (11) and pineapples (5).

Livestock (2003 estimates): pigs, 201,000; cattle, 29,000; asses, 7,000.

Forestry

Forests covered 0·17m. ha. (60% of the land area) in 2010. Timber production was 131,000 cu. metres in 2007.

Fisheries

Fish landings in 2007 totalled 4,606 tonnes.

INDUSTRY

Some industrial activity is being developed associated with agricultural products and forestry.

Labour

In 2001 the total labour force numbered 52,945 (36,739 males).

INTERNATIONAL TRADE

Imports and Exports

In 2005 imports were valued at 647m. tala and exports at 236m. tala. Main imports are machinery and transport equipment, foodstuffs and basic manufactures. Principal exports are coconuts, palm oil, taro and taamu, coffee and beer. New Zealand is the principal import source, accounting for 39·2% of imports in 2004. Australia was the largest export market in 2004, accounting for 71·5% of exports. Australia is the second biggest supplier of imports and the USA the second biggest export market.

COMMUNICATIONS

Roads

In 2002 the road network covered 790 km, of which 235 km were main roads. In 2005 there were 5,920 passenger cars plus 4,600 lorries and vans in use.

Civil Aviation

There is an international airport at Apia (Faleolo), which handled 321,973 passengers and 1,175 tonnes of freight in 2009. The national carrier is Virgin Samoa, known until 2011 as Polynesian Blue. In 2007 it operated domestic services and international flights to Auckland, Brisbane and Sydney.

Shipping

In Jan. 2009 there were four ships of 300 GT or over registered, totalling 9,000 GT.

Telecommunications

There are three radio communication stations at Apia. Radio telephone service connects Samoa with American Samoa, the Fiji Islands, New Zealand, Australia, Canada, USA and UK. In 2008 there were 28,800 main (fixed) telephone lines; mobile phone subscribers numbered 124,000 in 2008 (69·3 per 100 persons). There were 9,000 internet users in 2008.

SOCIAL INSTITUTIONS

Justice

The population in penal institutions in Sept. 2007 was 186 (99 per 100,000 of national population). The death penalty, not used in more than 50 years, was abolished in 2004.

Education

There were 30,000 pupils at primary schools in 2008–09 with 1,000 teaching staff and 25,000 pupils at secondary schools with 1,000 teaching staff. The University of the South Pacific has a School of Agriculture in Samoa, at Apia. A National University was established in 1984. In 2006 it had 2,298 students and 151 academic staff. There is also a Polytechnic Institute that provides mainly vocational and training courses.

The adult literacy in 2009 was estimated at 98·8%.

In 2008 public expenditure on education came to 5·7% of GDP and 13·4% of total government spending.

Health

In 2002 there were 33 general hospitals (with 320 beds), one private hospital, 11 district hospitals and 12 primary health care centres. In 2002 there were 43 physicians, six dentists, 333 nurses and 13 midwives.

RELIGION

In 2001 there were 46,200 Latter-day Saints (Mormons), 44,000 Congregationalists, 38,100 Roman Catholics and 21,800 Methodists. The remainder of the population follow other beliefs.

CULTURE

Press

There are two dailies, plus a weekly, a fortnightly and a monthly. The most widely read newspaper is the independent *Samoa Observer*.

Tourism

In 2008 there were 122,163 foreign tourists.

DIPLOMATIC REPRESENTATIVES

Of Samoa in the United Kingdom and to the European Union
High Commissioner: Fatumanava Dr Pa'olelei Luteru (resides in Brussels).
Honorary Consul: Prunella Scarlett, LVO (Church Cottage, Pedlinge, Nr Hythe, Kent, CT12 5JL).

Of the United Kingdom in Samoa
High Commissioner: Vicki Treadell (resides in Wellington).

Of Samoa in the USA and to the United Nations (800 Second Ave., Suite 400J, New York, NY, 10017)
Ambassador: Ali'ioaiga Feturi Elisaia.

Of the USA in Samoa
Ambassador: David Huebner (resides in Wellington).

FURTHER READING

National Statistical Office: Samoa Bureau of Statistics (SBS), Ministry of Finance, Level 1, Government Building (MFMII), P. O. Box 1151, Apia.
Website: http://www.sbs.gov.ws

SAN MARINO

Repubblica di San Marino
(Republic of San Marino)

Capital: San Marino
Population, 2012: 32,000
GNI per capita, 2010: US$41,256
Internet domain extension: .sm

KEY HISTORICAL EVENTS

San Marino is a small republic situated on the Adriatic side of central Italy. According to tradition, St Marinus and a group of Christians settled there to escape persecution. By the 12th century San Marino had developed into a commune ruled by its own statutes and consul. Unsuccessful attempts were made to annex the republic to the papal states in the 18th century and when Napoleon invaded Italy in 1797 he respected the rights of the republic and even offered to extend its territories. In 1815 the Congress of Vienna recognized the independence of the republic. On 22 March 1862 San Marino concluded a treaty of friendship and co-operation, including a *de facto* customs union, with Italy, thus preserving its independence although it is completely surrounded by Italian territory.

TERRITORY AND POPULATION

San Marino is a land-locked state in central Italy, 20 km from the Adriatic. Area is 61·19 sq. km (23·6 sq. miles) and the population (June 2012), 32,368; population density, 529·0 per sq. km. At July 2010, 12,722 citizens lived abroad.

In 2010, 94·1% of the population were urban. The capital, San Marino, has 4,236 inhabitants (June 2012); the largest town is Serravalle (10,540 in June 2012), an industrial centre in the north. The official language is Italian.

SOCIAL STATISTICS

Births registered in 2009, 306; deaths, 233; marriages, 238; divorces, 63. Birth rate, 2009 (per 1,000 population), 9·3; death rate, 6·9. Annual population growth rate, 2000–05, 2·7%; fertility

rate, 2008, 1·5 births per woman; infant mortality rate, 2010, two per 1,000 live births (one of the lowest rates in the world). The World Health Organization's *World Health Statistics 2009* put citizens of San Marino in equal second place in a 'healthy life expectancy' list (level with Switzerland and only behind Japan), with an expected 75 years of healthy life for babies born in 2007.

CLIMATE

Temperate climate with cold, dry winters and warm summers.

CONSTITUTION AND GOVERNMENT

The legislative power is vested in the *Great and General Council* of 60 members elected every five years by popular vote, two of whom are appointed every six months to act as *Captains Regent*, who are the heads of state.

Executive power is exercised by the ten-member *Congress of State*, presided over by the Captains Regent. The *Council of Twelve*, also presided over by the Captains Regent, is appointed by the Great and General Council to perform administrative functions.

National Anthem

No words, tune monastic, transcribed by F. Consolo.

RECENT ELECTIONS

In parliamentary elections on 11 Nov. 2012 the San Marino Common Good coalition won 35 of 60 seats with 50·7% of the vote (Sammarinese Christian Democratic Party–We Sammarinese 29·5% and 21 seats, Party of Socialists and Democrats 14·3% and 10, Popular Alliance 6·7% and 4); the Agreement for the Country coalition won 12 seats with 22·3% (Socialist Party 12·1% and 7, Union for the Republic 8·3% and 5); Active Citizenry won 9 seats with 16·1% (United Left 9·1% and 5, Civic10 6·7% and 4); Civic Movement R.E.T.E. won 4 seats with 6·3%. Turnout was 63·9%.

CURRENT GOVERNMENT

Captains Regent: Antonella Mularoni (since 1 April 2013); Denis Amici (since 1 April 2013).

In March 2013 the Congress of State comprised:

Minister of Foreign and Political Affairs, and Tourism: Pasquale Valentini. *Internal Affairs, Public Administration and Justice:* Gian Carlo Venturini. *Finance, Budget and Post:* Claudio Felici. *Education, Culture and the University:* Giuseppe Maria Morganti. *Health, Social Security, Family, Welfare and Economic Planning:* Francesco Mussoni. *Territory, Environment, Agriculture, Telecommunications, Youth, Sport and Civil Defence:* Matteo Fiorini. *Labour, Co-operation and Information:* Iro Belluzzi. *Industry, Crafts, Trade, Transport and Technological Research:* Marco Arzilli.

Parliament Website (Italian only):
  http://www.consigliograndeegenerale.sm

CURRENT LEADERS

Antonella Mularoni

Position
Captain Regent

Introduction
Antonella Mularoni was sworn in for her first term as Captain Regent on 1 April 2013.

Early Life
Antonella Mularoni was born in San Marino on 27 Sept. 1961 and graduated in law from Bologna University. Between 1991 and

2001 she worked as a lawyer in San Marino and from 2001–07 she served as San Marino's judge at the European Court of Human Rights in Strasbourg, France.

Career in Office
A founder member of the centrist San Marino Popular Alliance of Democrats (linked to the Northern League in Italy), Mularoni was first elected to the Great and General Council (GGC) in 1993. Between then and 2001 she served on several committees and was secretary for finance, budget and economic planning.

Re-elected to the GGC in 2008 as a representative of the centrist Sammarinese Christian Democratic Party, she was appointed secretary of state for foreign and political affairs. In a speech to the United Nations in 2010 she called for the Security Council to be reformed along more democratic, transparent and efficient lines.

Denis Amici

Position
Captain Regent

Introduction
Denis Amici was sworn in for his first term as Captain Regent on 1 April 2013.

Early life
Denis Amici was born on 10 June 1972 in San Marino. A graduate in accountancy and commerce, he took over management of his father's construction company in 1999.

Career in Office
Amici was elected to the Great and General Council (GGC) in 2008, representing Arengo and Freedom (AL), a new party with links to the People of Freedom, a centre-right Italian party. AL formed part of the right-leaning Pact for San Marino, a coalition that won 54·2% of the national vote that year.

Amici served on the permanent committees for constitutional affairs and institutional relations and on the Council of 12 (Consiglio dei XII) that deals with judicial appeals. Following a political crisis in 2011 AL was disbanded and Amici successfully contested the 2012 general election as an independent candidate within the list of the centre-right Sammarinese Christian Democratic Party (PDCS). He again served on the Council of 12 and represented San Marino at the Inter-Parliamentary Union. In Feb. 2013 he joined the PDCS ahead of his inauguration as Captain Regent.

DEFENCE
Military service is not obligatory, but all citizens between the ages of 16 and 55 can be called upon to defend the State. They may also serve as volunteers in the Military Corps. There is a military Gendarmerie.

ECONOMY
Overview
The economy is integrated with that of Italy via a monetary and customs union, close trade links and labour mobility. The economy relies principally on manufacturing and financial services, while tourism draws about 2m. visitors a year.

The global financial crisis and pressure to improve banking transparency led to an economic contraction of nearly 13% in 2009. An Italian tax amnesty in 2009–10 resulted in an outflow of bank deposits. Real GDP declined by an estimated 1% in 2010. Manufacturing and commercial activity improved moderately in early 2011 but the economy remains weak. Consumption has declined and unemployment is rising.

In 2010 San Marino signed Tax Information Exchange Agreements with most major countries to promote international co-operation. Restructuring the financial sector and strengthening economic and financial relations with Italy are key for future growth.

Currency
Since 1 Jan. 2002 San Marino has been using the euro (EUR). Italy has agreed that San Marino may mint a small part of the total Italian euro coin contingent with their own motifs. Inflation was 2·6% in 2010 and 2·0% in 2011. Total money supply in June 2005 was €906m.

Budget
Revenues totalled €547·0m. in 2006 and expenditures €457·3m. VAT accounted for 23·6% of revenue in 2005, social contributions 21·3% and income tax 20·2%; wages and salaries accounted for 35·4% of expenditure in 2005 and social contributions 30·5%.

Performance
The economy contracted by 12·8% in 2009, by 5·2% in 2010 and by 2·6% in 2011. Total GDP was US$1·9bn. in 2008.

Banking and Finance
The Banca Centrale della Repubblica di San Marino (*President*, Renato Clarizia) was established in 2005 as an amalgamation of the Istituto di Credito Sammarinese and the Ispettorato per il Credito e le Valute (the Inspectorate for Credit and Currencies). Many of its functions have since been taken over by the European Central Bank and it has taken on a more supervisory role. Commercial banks include: Banca di San Marino, Credito Industriale Sammarinese, Cassa di Risparmio della Repubblica di San Marino and the Banca Agricola Commerciale della Repubblica di San Marino.

ENERGY AND NATURAL RESOURCES
Electricity
Electricity is supplied by Italy.

Agriculture
There were 1,000 ha. of arable land in 2006. Wheat, barley, maize and vines are grown.

INDUSTRY
Labour
Out of 20,530 people in employment in 2006, 6,247 worked in manufacturing and 2,901 in wholesale and retail trade. In 2006 there were 473 registered unemployed persons.

INTERNATIONAL TRADE
Imports and Exports
Import commodities are a wide range of consumer manufactures and foodstuffs. Export commodities are building stone, lime, wine, baked goods, textiles, varnishes and ceramics. San Marino maintains a customs union with the European Union.

COMMUNICATIONS
Roads
A bus service connects San Marino with Rimini. There are 252 km of public roads and 40 km of private roads, and (2006) 32,263 passenger cars and 5,907 commercial vehicles.

Civil Aviation
The nearest airport is Rimini, 10 km to the east in Italy, which had scheduled flights in 2003 to Berlin, Düsseldorf, Frankfurt, Hamburg, Helsinki, Munich, Naples and Rome.

Telecommunications
San Marino had 21,300 main telephone lines in 2008 and 24,000 mobile phone subscribers. Internet users numbered 17,000 in 2008.

SOCIAL INSTITUTIONS

Justice
Judges are appointed permanently by the Great and General Council; they may not be San Marino citizens. Petty civil cases are dealt with by a justice of the peace; legal commissioners deal with more serious civil cases, and all criminal cases and appeals lie to them from the justice of the peace. Appeals against the legal commissioners lie to two appeals judges as a court of third instance.

Education
Education is compulsory up to 16 years of age. In 2005 there were 15 nursery schools with 1,054 pupils and 141 teachers, 14 elementary schools with 1,497 pupils and 245 teachers, three junior high schools with 805 pupils and 144 teachers, and one high school with 1,289 pupils and 77 teachers. The University of San Marino began operating in 1988.

Health
In 2003 there were 139 hospital beds and 135 doctors. A survey published by the World Health Organization in June 2000 to measure health systems in all of the sovereign countries and find which country has the best overall health care ranked San Marino in third place.

RELIGION
The great majority of the population are Roman Catholic.

CULTURE

World Heritage Sites
There is one UNESCO World Heritage site in San Marino: San Marino Historic Centre and Mount Titano (inscribed on the list in 2008).

Press
San Marino had three daily newspapers in 2006 with a combined daily circulation of 2,000.

Tourism
In 2007, 2·16m. tourists visited San Marino (1·47m. Italians and 696,000 other foreigners).

DIPLOMATIC REPRESENTATIVES
Of San Marino in the United Kingdom
Ambassador: Federica Bigi (resides in Rome).

Of the United Kingdom in San Marino
Ambassador: Christopher Prentice, CMG (resides in Rome).

Of San Marino in the USA (1899 L St., NW, Suite 500, Washington D.C., 20036).
Ambassador: Paolo Rondelli.

Of the USA in San Marino
Ambassador: David H. Thorne (resides in Rome).

Of San Marino to the United Nations
Ambassador: Daniele Bodini.

Of San Marino to the European Union
Ambassador: Gian Nicola Filippi Balestra.

FURTHER READING
National Statistical Office: Ufficio Programmazione Economica e Centro Elaborazione Dati e Statistica, Via 28 Luglio, 192–47893 Borgo Maggiore.
Website: http://www.statistica.sm/on-line/Home.html

Discover even more in the Archive at
www.statesmansyearbook.com

SÃO TOMÉ E PRÍNCIPE

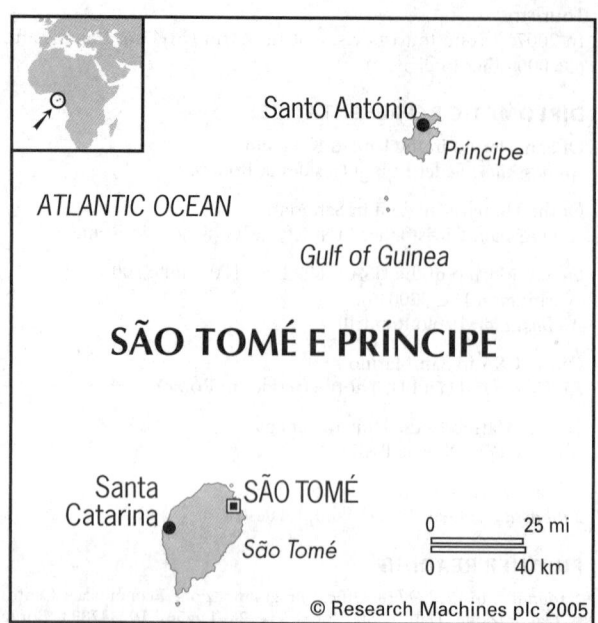

República Democrática de São Tomé e Príncipe
(Democratic Republic of São Tomé e Príncipe)

Capital: São Tomé
Population projection, 2015: 182,000
GNI per capita, 2011: (PPP$) 1,792
HDI/world rank: 0·509/144
Internet domain extension: .st

KEY HISTORICAL EVENTS

The islands of São Tomé and Príncipe off the west coast of Africa were colonized by Portugal in the fifteenth century. There may have been a few African inhabitants earlier but most of the population arrived during the centuries when the islands served as a slave-trading depot for South America. In the 19th century the islands became the first parts of Africa to grow cocoa. In 1876 Portugal officially abolished slavery but in practice it continued with many Angolans, Mozambicans and Cape Verdians brought in to work on the cocoa plantations. Because the slave-descended population was cut off from African culture, São Tomé had a higher proportion than other Portuguese colonies of *assimilados* (Africans acquiring full Portuguese culture and some rights). São Tomé saw serious riots against Portuguese rule in 1953. From 1960 a Movement for the Liberation of São Tomé e Príncipe operated from neighbouring African territories. In 1970 Portugal formed a 16-member legislative council and a provincial consultative council. Following the Portuguese revolution of 1974 a transitional government was formed. Independence came on 12 July 1975. Independent São Tomé e Príncipe officially proclaimed Marxist-Leninist policies but maintained a non-aligned foreign policy and has received aid from Portugal.

The government was overthrown by a coup on 16 July 2003 while President Fradique de Menezes and his foreign minister were abroad. The coup leader, Major Fernando Pereira, installed a junta but accepted a general amnesty from parliament on 24 July after agreeing to allow the ousted president to form a government of national unity.

TERRITORY AND POPULATION

The republic, which lies about 200 km off the west coast of Gabon, in the Gulf of Guinea, comprises the main islands of São Tomé (845 sq. km) and Príncipe and several smaller islets including Pedras Tinhosas and Rolas. It has a total area of 1,001 sq. km (387 sq. miles). Population (census, 2012) 187,356; density, 187 per sq. km. In 2011, 63·0% of the population were urban.

The UN gives a projected population for 2015 of 182,000.

Areas and populations of the two provinces:

Province	Sq. km	Census 2012	Chief town	Census 2001
São Tomé	859	179,814	São Tomé	51,886
Príncipe	142	7,542	São António	1,040

The official language is Portuguese. Lungwa São Tomé, a Portuguese Creole, and Fang, a Bantu language, are the spoken languages.

SOCIAL STATISTICS

2006: births, 5,072; deaths, 1,111. Rates, 2006 (per 1,000 population): birth, 33·4; death, 7·3; infant mortality (2010), 53 per 1,000 live births. Expectation of life, 2006, 63·5 years for males and 68·5 years for females. Annual population growth rate, 2000–05, 1·9%; fertility rate, 2008, 3·8 births per woman.

CLIMATE

The tropical climate is modified by altitude and the effect of the cool Benguela current. The wet season is generally from Oct. to May, but rainfall varies considerably, from 40" (1,000 mm) in the hot and humid northeast to 150–200" (3,800–5,000 mm) on the plateau. São Tomé, Jan. 79°F (26·1°C), July 75°F (23·9°C). Annual rainfall 38" (951 mm).

CONSTITUTION AND GOVERNMENT

The 1990 constitution was approved by 72% of votes at a referendum of March 1990 and became effective in Sept. 1990. It abolished the monopoly of the Movement for the Liberation of São Tomé e Príncipe (MLSTP). The *President* must be over 34 years old, and is elected by universal suffrage for one or two (maximum) five-year terms. He or she is also head of government and appoints a Council of Ministers. The 55-member *National Assembly* is elected for four years.

Since April 1995 **Príncipe** has enjoyed internal self-government, with a five-member regional government and an elected assembly.

National Anthem
'Independência total, glorioso canto do povo' ('Total independence, glorious song of the people'); words by A. N. do Espírito Santo, tune by M. de Sousa e Almeida.

RECENT ELECTIONS

At the presidential election on 17 July 2011 former president Manuel Pinto da Costa (ind.) won 35·8% of the votes cast against former prime ministers Evaristo Carvalho (Independent Democratic Action/ADI) with 21·8%, Maria das Neves (ind.) with 14·0% and Delfim Neves (Democratic Convergence Party-Reflection Group/Force for Change Democratic Movement-Liberal Party) with 13·9%. Six other candidates received less than 5% of votes cast. Turnout was 68·4%. In the run-off held on 7 Aug. 2011, Manuel Pinto da Costa won 52·9% of the vote and Evaristo Carvalho 47·1%. Turnout was 74·0%.

At the National Assembly elections on 1 Aug. 2010 ADI won 26 seats, the Liberation Movement of São Tomé e Príncipe-Social

Democratic Party (MLSTP-PSD) 21, the Democratic Convergence Party-Reflection Group 7 and the Force for Change Democratic Movement-Liberal Party 1. Turnout was 89·0%.

CURRENT GOVERNMENT

President, C.-in-C: Manuel Pinto da Costa; b. 1937 (ind.; sworn in 3 Sept. 2011, having previously held office from 12 July 1975 to 4 March 2001).

In March 2013 the government comprised:

Prime Minister: Gabriel Arcanjo da Costa; b. 1954 (MLSTP-PSD; sworn in 12 Dec. 2012, having previously been prime minister from March–Oct. 2002).

Minister for Agriculture, Fisheries and Rural Development: António Álvaro da Graça Dias. *Defence and Internal Affairs:* Óscar Aguiar Sacramento e Sousa. *Education, Culture and Training:* Jorge Lopes Bom Jesus. *Foreign Affairs, Co-operation and Communities:* Natália Pedro da Costa Umbelina Neto. *Health and Social Affairs:* Leonel Pinto d'Assunção Pontes. *Industry, Trade and Tourism:* Demóstenes Pires dos Santos. *Justice, Public Administration and State Administrative Affairs:* Edite Ramos da Costa Tenjua. *Planning and Finances:* Hélio Silva Vaz d'Almeida. *Public Works, Infrastructures, Environment and Natural Resources:* Osvaldo Cravid Viegas d'Abreu. *Youth and Sport:* Albertino Francisco Boa Morte.

Office of the President (Portuguese only): http://www.presidencia.st

CURRENT LEADERS

Manuel Pinto da Costa

Position
President

Introduction
Manuel Pinto da Costa, a Marxist, came out of retirement to win the presidency in Aug. 2011. He promised to tackle corruption and poverty but opponents warned of a return to authoritarian rule.

Early Life
Manuel Pinto da Costa was born in São Tomé on 5 Aug. 1937. The son of a plantation official, he was educated in São Tomé, Angola and Portugal. He graduated from the University of Lisbon and was a founder of the Committee for the Liberation of São Tomé e Príncipe in 1960. Da Costa spent time in France and Cuba in the early part of the decade before completing a doctorate in economics at Humboldt University in East Berlin.

Elected secretary-general of the new Movement for the Liberation of São Tomé e Príncipe (MLSTP) in 1972, da Costa developed links with other independence groups in Angola, Guinea-Bissau and Mozambique. When Portugal's fascist government was overthrown in April 1974, he spearheaded negotiations that led to independence in July 1975.

As leader of the MLSTP, the sole legal party, da Costa was blamed for the economic hardships that accompanied the nationalization of plantations and the departure of skilled workers to Portugal. After an attempted coup by foreign-based mercenaries in 1978 and amid rising political tensions the following year, da Costa called upon Angolan troops to maintain the peace in a deeply unpopular move.

During the early 1980s he became increasingly authoritarian, but a new constitution ended one-party rule in 1990 and he announced that he would not contest the 1991 multi-party elections. Miguel Trovoada became president, elected unopposed as an independent candidate, and da Costa moved to Angola.

He nonetheless returned to contest the 1996 presidential election, narrowly losing to Trovoada. Re-elected leader of the MLSTP in 1998, he also contested the 2001 presidential election, this time losing to Fradique de Menezes. He remained party leader until retiring in Feb. 2004. However, he returned again to run as an independent candidate for the presidency in July 2011, promising to overcome instability, poverty and corruption. He received sufficient support to force a run-off against the parliamentary speaker, Evaristo Carvalho, and emerged victorious with 52·9% of the vote. He was sworn in on 3 Sept. 2011.

Career in Office
Pinto da Costa's priority was to revitalize the struggling economy by developing tourism, the nascent oil industry and coffee exports, while working in conjunction with Patrice Trovoada, whose Independent Democratic Action party won legislative elections in 2010. In Nov. 2012, however, Trovoada lost a no-confidence vote in the National Assembly and da Costa dismissed his government the following month, appointing Gabriel Arcanjo da Costa as premier.

ECONOMY

In 2005 agriculture accounted for 17% of GDP, industry 21% and services 63%.

Overview

São Tomé e Príncipe is one of the poorest countries in the world. It is heavily reliant on cocoa exports and increasingly dependent on foreign aid. It receives debt relief under the IMF Heavily Indebted Poor Countries Initiative (HIPC) and Multilateral Debt Relief initiatives. However, it remains extremely vulnerable to external shocks because of the narrow export base and reliance on external financing.

Growth averaged around 5% between 2001 and 2007 but slowed in 2009 with the global financial crisis. There were sharp declines in FDI, remittances, tourist arrivals and donor assistance. However, the potential exploitation of large oil reserves in territorial waters could transform the medium- to long-term outlook.

Currency

The unit of currency is the *dobra* (STD) of 100 *centimos*. Inflation was 13·3% in 2010 and 14·3% in 2011. In Dec. 2006 foreign exchange reserves were US$34m. Total money supply in Dec. 2004 was 172,817m. dobras (up from 23,683m. dobras in Dec. 1996).

Budget

In 2008 revenues totalled 1,216bn. dobras and expenditures 841bn. dobras.

Performance

Real GDP growth was 4·5% in 2010 and 4·9% in 2011. In 2011 total GDP was US$248m.

Banking and Finance

In 1991 the Banco Central de São Tomé e Príncipe (*Governor*, Maria do Carmo Trovoada) replaced the Banco Nacional as the central bank and bank of issue. A private commercial bank, the Banco Internacional de São Tomé e Príncipe, began operations in 1993. Foreign debt amounted to US$170m. in 2010, representing 85·3% of GNI.

ENERGY AND NATURAL RESOURCES

Environment

In 2008 carbon dioxide emissions from the consumption and flaring of fossil fuels were the equivalent of 0·6 tonnes per capita.

Electricity

Installed capacity, 2007 estimate, 14,000 kW. Production was about 43m. kWh in 2007, with consumption per capita an estimated 272 kWh.

Oil and Gas

There are large oil reserves around São Tomé e Príncipe that could greatly add to the country's wealth; the Joint Development

Zone was set up with Nigeria to administer the exploitation because the reserves are located in shared waters. The first license to begin exploration was granted in April 2004.

Agriculture

After independence all landholdings over 200 ha. were nationalized into 15 state farms. These were partially privatized in 1985 by granting management contracts to foreign companies, and distributing some state land as small private plots. There were an estimated 9,000 ha. of arable land in 2007 and 47,000 ha. of permanent crops. Production (2003 estimates in 1,000 tonnes): bananas, 27; coconuts, 27; cassava, 6; palm kernels, 4; cocoa beans, 3; maize, 2; palm oil, 2. There were an estimated 5,000 goats, 4,000 cattle, 3,000 sheep and 2,000 pigs in 2003.

Forestry

In 2010 forests covered 27,000 ha., or 28% of the land area. In 2007, 9,000 cu. metres of timber were cut.

Fisheries

There are rich tuna shoals. The total catch in 2007 amounted to an estimated 4,250 tonnes.

INDUSTRY

Manufacturing contributed 6% of GDP in 2005. There are a few small factories in agricultural processing (including beer and palm oil production), timber processing, bricks, ceramics, printing, textiles and soap-making.

Labour

In 2001 the economically active population was 52,150. The unemployment rate was 15·7% in 2001.

INTERNATIONAL TRADE

Imports and Exports

Trade figures for 2006: imports, US$70·9m.; exports, US$3·8m. Cocoa accounts for two-thirds of all exports.

In 2006 the main import suppliers were Portugal (63·6%), Angola (18·3%) and Belgium (4·6%); main export markets were Portugal (33·3%), the Netherlands (27·1%) and Belgium (14·3%).

COMMUNICATIONS

Roads

There were 500 km of roads in 2009, 375 km of which were paved. Approximately 4,500 passenger cars, 2,183 motorcycles and over 1,800 trucks and vans were in use in 2008.

Civil Aviation

São Tomé airport had flights in 2010 to Cape Verde, Libreville, Lisbon, Luanda and Port-Gentil. In 2007 São Tomé handled 50,625 passengers. There is a light aircraft service to Príncipe.

Shipping

São Tomé is the main port, but it lacks a deep water harbour. Neves handles oil imports and is the main fishing port. In Jan. 2009 there were 14 ships of 300 GT or over registered, totalling 19,000 GT.

Telecommunications

In 2008 there were 7,700 main (fixed) telephone lines; mobile phone subscribers numbered 49,000 in 2008 (30·6 per 100 persons). In 2010, 18·8% of the population were internet users.

SOCIAL INSTITUTIONS

Justice

Members of the Supreme Court are appointed by the National Assembly. There is no death penalty. The population in penal institutions in July 2006 was 160 (83 per 100,000 of national population).

Education

Adult literacy was 88% in 2008. Education is free and compulsory. In 2007–08 there were 90 primary schools and 32,616 pupils, and ten secondary schools and 8,380 pupils; 96% of primary age children were attending school in 2006. There are two institutions of higher education.

Health

In 2004 there were 81 physicians, 11 dentists, and 308 nurses and midwives.

RELIGION

In 2001, 81% of the population were Roman Catholic. There is a small Protestant church and a Seventh Day Adventist school.

CULTURE

Press

In 2006 there was one daily newspaper. Two government-owned and six independent papers were also published irregularly.

Tourism

In 2005 there were 11,000 non-resident tourists.

DIPLOMATIC REPRESENTATIVES

Of São Tomé e Príncipe in the United Kingdom
Ambassador: Vacant (resides in Brussels).
Chargé d'Affaires a.i.: Armindo de Brito Fernandes.
Honorary Consul: Natalie Galland-Burkl (Flat 8, Marsham Court, 58 Victoria Drive, London, SW19 6BB).

Of the United Kingdom in São Tomé e Príncipe
Ambassador: Richard Wildash, LVO (resides in Luanda, Angola).

Of São Tomé e Príncipe in the USA and to the United Nations
Ambassador: Carlos Filomeno Agostinho das Neves.

Of the USA in São Tomé e Príncipe
Ambassador: Eric Benjaminson (resides in Libreville, Gabon).

Of São Tomé e Príncipe to the European Union
Ambassador: Carlos Gustavo dos Anjos.

FURTHER READING

National Statistical Office: Instituto Nacional de Estatística, Largo das Alfândegas, Cx. Postal 256, São Tomé.
Website (Portuguese only): http://www.ine.st

The world in focus at
www.statesmansyearbook.com

SAUDI ARABIA

Al-Mamlaka al-Arabiya as-Saudiya
(Kingdom of Saudi Arabia)

Capital: Riyadh
Population projection, 2015: 30·54m.
GNI per capita, 2011: (PPP$) 23,274
HDI/world rank: 0·770/56
Internet domain extension: .sa

KEY HISTORICAL EVENTS

Nomadic tribes have existed across the Arabian peninsula for thousands of years. The pre-Islamic period saw the development of civilizations based on trade in frankincense and spices, notably, from about the 12th century BC, the Minaeans in the southwest of what is now Saudi Arabia and Yemen. The Sabaean and Himyarite kingdoms flourished from around 650 BC and 115 BC respectively, their loose federations of city states lasting until the 6th century AD. Although increased trade brought these civilizations into contact with the Roman and Persian empires—the two great regional powers before the advent of Islam—they remained, for the most part, politically independent. The Nabataeans, an Aramaic people whose capital was at Petra, modern-day Jordan, spread into northern Arabia over a period covering the 1st century BC and the 1st AD before annexation of their territory by Rome. Persian influence was prevalent along Arabia's eastern coast, centred on Dilmun which covered parts of the mainland and the island of Bahrain.

By the 6th century AD the Hejaz region in northwestern Arabia was becoming increasingly powerful and an important link in the overland trade route from Egypt and the Byzantine Empire to the wider East. One of the principal cities of Hejaz was Makkah (Mecca), a staging post on the camel train routes and site of pilgrimage to numerous pre-Islamic religious shrines. The leading tribe in the city was the Quraysh, into which the Prophet Muhammad was born in 570. Muhammad and his followers (known as Muslims) took control of Makkah in 630. He had earlier declared himself a prophetic reformer, destroying the city's pagan idols and declaring it a centre of Muslim pilgrimage dedicated to the worship of Allah (God) alone. Muhammad died in AD 632, by then commanding the loyalty of almost all of Arabia.

The leaders who succeeded Muhammad, known as caliphs, spread the Islamic faith throughout and beyond the Arab world. However, Arabia itself began to fragment and by the latter part of the 7th century it had become a province of the Islamic realm, although the holy cities of Makkah and Madinah retained their spiritual focus. Meanwhile, increasingly remote from the main centres of Islamic authority under the Umayyad and Abbasid caliphate dynasties, Arabia became an arena for sectarian divisions—Shia, Sunni and Kharijite—which developed within the Islamic faith.

After 1269 most of the Hejaz region came under the suzerainty of the Egyptian Mameluks. The Ottoman Turks conquered Egypt in 1517 and, to counter the influence of the Christian Portuguese presence in the Gulf region, extended their nominal control over the whole Arabian Peninsula. Portuguese traders were followed by British, Dutch and French merchants during the 17th and 18th centuries, the British gradually securing political and commercial supremacy in the Gulf and southern Arabia through a system of protectorates and local treaties.

Saudi Arabia's origins as a political entity lay in the rise of the puritanical Wahhabi movement of the 18th century, which called for a return to the original principles of Islam and gained the allegiance of the powerful Al-Saud dynasty (founded in the 15th century) in the Nejd region of central Arabia. By 1811 the Al-Saud/Wahhabi armies controlled most of the peninsula and were seen as a threat to the Ottoman Turkish overlord. The Sultan called on his viceroy in Egypt, Mehmet Ali, to suppress the Wahhabis, who were defeated between 1811 and 1818. Nevertheless, the house of Al-Saud continued to hold sway over the interior of Arabia until 1891 when, after a long period of tribal warfare, the rival Al-Rashid family, with Ottoman support, seized control of the city of Riyadh.

The Al-Saud family was exiled to Kuwait but Abdulaziz Ibn Abdul Rahman (known to Europeans as Ibn Saud) restored Wahhabi fortunes, recapturing Riyadh in 1902 and reasserting Al-Saud control over Nejd by 1906. On the eve of the First World War Abdulaziz gained the al Hasa region east of Nejd on the Gulf from the Ottoman Turks. In 1920 he captured the Asir region and in 1921 added the Jebel Shammar territory (northwest of Nejd) of the Al-Rashid family. In 1925 Abdulaziz completed his conquest of Hejaz, overthrowing Hussein, Sharif of Makkah and a member of the Hashimi family. Abdulaziz became both Sultan of Nejd and King of the Hejaz. Britain recognized Abdulaziz as an independent ruler by the Treaty of Jeddah on 20 May 1927, and in 1932 Nejd and Hejaz were unified as the Kingdom of Saudi Arabia, ruled as an absolute monarchy under Islamic law.

Abdulaziz (died 9 Nov. 1953) concentrated on the political consolidation and modernization of the country. Oil was discovered in 1938 and its exploitation was developed with the support of the USA after the Second World War. Crown Prince Saud succeeded his father and ruled until Nov. 1964, when he was effectively deposed by his brother Faisal. During his reign Saudi relations with the pan-Arabist Nasser regime in Egypt deteriorated, most notably over the 1962 revolution in Yemen.

As king and prime minister, Faisal used oil production revenues to build up the country's economic base. In 1970 came the first of the five-year economic development programmes. Meanwhile, financial support was given to other Arab states in their conflict with Israel. Leading on from the Oct. 1973 Arab-Israeli war Arab producers, including Saudi Arabia, cut supplies to the USA and other Western countries, causing a fourfold increase in oil prices. However, Faisal subsequently adopted a more conciliatory stance than the more radical members of the Organization of Petroleum Exporting Countries (OPEC, founded in 1960) and the close Saudi economic relationship with the USA was reinforced with a co-operation agreement in 1974. In March 1975, when Faisal was assassinated by a nephew, believed to be mentally unstable, his half-brother Khalid became king.

Khalid continued Faisal's policies promoting Islamic solidarity and Arab unity in the wake of hostilities with Israel. In practice his moderate stance was in marked contrast to the militancy of many other Arab states, particularly over oil pricing by OPEC and opposition to Egypt's 1978 peace treaty with Israel. Khalid was also involved in early efforts to stop the civil war in Lebanon and, in 1981, he inaugurated the Gulf Co-operation Council (GCC). Domestically, he maintained his family's absolute political control and the conservative Islamic character of the country. However, opposition to his regime was demonstrated in Nov. 1979 when Sunni Muslim fundamentalists occupied the Grand Mosque at Makkah. A two-week siege ended with over 200 deaths. The second and third five-year development plans (1975–79 and 1980–84), both launched by Khalid, created much of the country's current economic infrastructure. Owing to Khalid's poor health throughout his reign much of his executive responsibility was assumed by his younger half-brother, Crown Prince Fahd.

Fahd succeeded to the throne on 13 June 1982. Like his predecessors, he maintained absolute power but broadened the process of political consultation and decision-making by setting up the Consultative Council (*Majlis Al-Shura*) of royal appointees from 1993. In 1986 he assumed the title of 'Custodian of the Two Holy Mosques' but the Saudi role in protecting religious pilgrims incurred international criticism in 1987 when 400 Iranian worshippers were killed in clashes in Makkah with security forces and again in 1994 when 270 pilgrims died in a stampede. Internationally, Fahd adopted a moderate policy on regional problems and closely allied the kingdom with the USA. Fahd was a key participant in diplomatic efforts to end the Iran-Iraq war in 1988 and, in 1989, he participated in the Taif reconciliation accord ending the 14-year Lebanese civil war. His pro-Western stance and co-operation in the 1990–91 Gulf crisis were crucial to the deployment and successful military operations of the US-led multinational force raised against Iraq following its invasion of Kuwait.

However, anti-Western disaffection among Saudi nationals has become more overt in recent years. In 1996 a bomb exploded at a US military complex at Dhahran, killing 19 and wounding over 300. In 2000 a series of bomb blasts, blamed by Saudi officials on British nationals engaged in criminal activity, was widely believed abroad to be the work of Saudi dissidents. Up to 15 Saudi nationals were involved in the attacks on New York and Washington, D.C. on 11 Sept. 2001, co-ordinated by Saudi dissident Osama bin Laden. In Nov. 2002 the Saudi government refused permission for the US to use its military facilities to attack Iraq, even if sanctioned by the United Nations. In May 2003 suicide bombers killed ten US citizens and many others at housing compounds for Western expatriate workers in Riyadh. In April 2003 the US agreed to pull out most of its troops from the kingdom, while stressing that the two countries would remain allies.

As King Fahd's health declined, his half brother, Crown Prince Abdullah Ibn Abdulaziz Al-Saud, assumed responsibility for government in 1996. When King Fahd died on 1 Aug. 2005, Crown Prince Abdullah was appointed his successor.

TERRITORY AND POPULATION

Saudi Arabia, which occupies nearly 80% of the Arabian peninsula, is bounded in the west by the Red Sea, east by the Persian Gulf, Qatar and the United Arab Emirates, north by Jordan, Iraq and Kuwait and south by Yemen and Oman. For the border dispute with Yemen *see* YEMEN: Territory and Population. The total area is 2,149,690 sq. km (829,995 sq. miles). Riyadh is the political, and Makkah (Mecca) the religious, capital.

Population at the census of April 2010 (provisional), 27,136,977; density, 12·6 per sq. km. Approximately 32% of the population are foreigners. In 2011, 82·3% of the population lived in urban areas.

The UN gives a projected population for 2015 of 30·54m.

Principal cities with 2004 population estimates (in 1m.): Riyadh, 4·09; Jeddah, 2·80; Makkah, 1·29; Madinah, 0·92; Dammam, 0·74; Taif, 0·52.

The Neutral Zone (5,700 sq. km, 3,560 sq. miles), jointly owned and administered by Kuwait and Saudi Arabia from 1922 to 1966, was partitioned between the two countries in 1966, but the exploitation of the oil and other natural resources continues to be shared.

The official language is Arabic.

SOCIAL STATISTICS

2008 estimates: births, 590,000; deaths, 92,000. Birth rate (2008 estimate) was 23·4 per 1,000 population; death rate, 3·6. 75% of the population is under the age of 30. Expectation of life at birth, 2007, was 70·8 years for males and 75·1 years for females. Annual population growth rate, 2000–08, 2·4%. Infant mortality, 2010, was 15 per 1,000 live births, down from 58 in the years 1980–85. Fertility rate, 2008, 3·1 births per woman.

CLIMATE

A desert climate, with very little rain and none at all from June to Dec. The months May to Sept. are very hot and humid, but winter temperatures are quite pleasant. Riyadh, Jan. 58°F (14·4°C), July 108°F (42°C). Annual rainfall 4" (100 mm). Jeddah, Jan. 73°F (22·8°C), July 87°F (30·6°C). Annual rainfall 3" (81 mm).

CONSTITUTION AND GOVERNMENT

The reigning King, **Abdullah Ibn Abdulaziz Al-Saud** (b. 1924), Custodian of the two Holy Mosques, succeeded in Aug. 2005, after King Fahd's death. *Crown Prince:* Prince Salman bin Abdulaziz Al-Saud (b. 1935). The Saudi royal family is around 8,000-strong.

Constitutional practice derives from Sharia law. There is no formal constitution, but three royal decrees of 1 March 1992 established a Basic Law which defines the systems of central and municipal government, and set up a 60-man Consultative Council (*Majlis Al-Shura*) of royal nominees in Aug. 1993. The *Chairman* is Salih bin Abdullah bin Humaid. In July 1997 the King decreed an increase of the Consultative Council to a chairman plus 90 members, selected from men of science and experience; in 2001 it was increased again to a chairman plus 120 members and in 2005 further to a chairman plus 150 members. The Council does not have legislative powers.

Saudi Arabia is an absolute monarchy; executive power is discharged through a *Council of Ministers*, consisting of the King, Deputy Prime Minister, Second Deputy Prime Minister and Cabinet Ministers.

The King has the post of *Prime Minister* and can veto any decision of the Council of Ministers within 30 days.

In Oct. 2003 the government announced that municipal elections would be held in 2004 for the first time (although they were subsequently postponed until 2005), followed by city elections and partial elections to the *Majlis Al-Shura* in the following years. In March 2011 the government announced that the second municipal elections, previously scheduled for 2009, would be held on 22 Sept. 2011 (later delayed to 29 Sept.). Women were not eligible to vote, but they will be at the next elections set for 2015.

National Anthem

'Sarei lil majd walaya' ('Onward towards the glory and the heights'); words by Ibrahim Khafaji, tune by Abdul Rahman al Katib.

GOVERNMENT CHRONOLOGY

Kings since 1932.

1932–53	Abdulaziz bin Abdul Rahman Al-Saud
1953–64	Saud bin Abdulaziz Al-Saud
1964–75	Faisal bin Abdulaziz Al-Saud
1975–82	Khalid bin Abdulaziz Al-Saud
1982–2005	Fahd bin Abdulaziz Al-Saud
2005–	Abdullah bin Abdulaziz Al-Saud

RECENT ELECTIONS

Municipal elections were held on 29 Sept. 2011 for 1,056 seats on the councils of 285 municipalities. Women were not permitted to stand for election or to vote, but four days before the 2011 elections King Abdullah announced that they would be able to run as candidates and vote at the municipal elections scheduled for 2015. There are no political parties.

CURRENT GOVERNMENT

In March 2013 the Council of Ministers comprised:

Prime Minister: King Abdullah bin Abdulaziz Al-Saud; b. 1924.

Deputy Prime Minister and Minister of Defence: Prince Salman bin Abdulaziz Al-Saud.

Second Deputy Prime Minister: Prince Muqrin bin Abdulaziz Al-Saud.

Minister of Interior: Prince Mohammed bin Nayef. *Municipal and Rural Affairs:* Prince Meta'ab bin Abdulaziz Al-Saud. *Foreign Affairs:* Prince Saud Al-Faisal bin Abdulaziz Al-Saud. *Agriculture:* Dr Fahd bin Abdulrahman Balghanaim. *Water and Electricity:* Abdul Rahman Al-Hussayen. *Civil Service:* Abdullah Al-Barrak. *Education:* Prince Faisal bin Abdullah bin Muhammad Al-Saud. *Finance:* Dr Ibrahim bin Abdulaziz Al-Assaf. *Health:* Dr Abdullah bin Abdulaziz Al-Rabeah. *Higher Education:* Dr Khalid bin Mohammed Al-Angary. *Commerce and Industry:* Tawfeeq Al-Rabeeah. *Culture and Information:* Abdulaziz bin Mohieddin Khoja. *Islamic Affairs, Endowments, Call and Guidance:* Sheikh Saleh bin Abdulaziz Al-Ashaikh. *Justice:* Sheikh Dr Mohammed bin Abdulkarim bin Abdulaziz Al-Issa. *Labour:* Adel bin Muhammad Fakieh. *Social Affairs:* Yusuf bin Ahmed Al-Othaimeen. *Petroleum and Mineral Resources:* Ali bin Ibrahim Al-Naimi. *Pilgrimage:* Bandar Al-Hajjar. *Economy and Planning:* Mohammed Al-Jasser. *Communications and Information Technology:* Muhammad bin Jameel Mulla. *Transport:* Dr Jubarah bin Eid Al-Suraiseri. *Housing:* Shwaish bin Saud Al-Duwaihi Al-Mutairi.

Majlis Website: http://www.shura.gov.sa

CURRENT LEADERS

King Abdullah bin Abdulaziz Al-Saud

Position

King

Introduction

King Abdullah administered Saudi Arabia on behalf of his half-brother, King Fahd bin Abdulaziz, between 1996 and Fahd's death on 1 Aug. 2005, following which he was named as successor. Abdullah has maintained the strict Islamic code of governance associated with the Wahhabi Saudis while attempting to rein in the excesses of the princely class. He has gained respect internationally for his efforts in the Middle East peace process but at times relations with the USA administration have been strained.

Early Life

Prince Abdullah bin Abdulaziz was born in Riyadh in 1924, the only son of Fahda bint Asi bin Shurayim Shammar, the eighth wife of Abdulaziz bin Abdul Rahman Al-Saud, then Sultan of Nejd, who founded the Kingdom of Saudi Arabia in 1932. Abdulaziz, known as bin Saud by Europeans, reared his massive family in the Bedouin tradition, educating his sons at court and instilling them with Islamic and Arab virtues.

Abdullah's career began in 1952 when he was given the command of the Saudi National Guard by his half-brother, King Saud, the first son to succeed Abdulaziz. The National Guard comprised descendants of Abdulaziz's Bedouin warriors who took part in the expansion of Saudi power. On his accession in 1975, King Khalid bin Abdulaziz appointed Abdullah second deputy prime minister. In this post Abdullah became involved in foreign policy, visiting the USA in 1976 to meet President Gerald Ford. On the accession of King Fahd in 1982, Abdullah was designated crown prince and first deputy prime minister.

Career in Office

The succession in Saudi Arabia is decided by the Saudi princes, who number over 4,000 (some sources claim an estimated 8,000). A crown prince is traditionally selected by seniority and ability. Since the death of the kingdom's founder, Abdulaziz, in 1953, only his sons have been considered suitable for the succession. Abdullah, who has no full brothers, lacks a fraternal support base and relies on alliances forged with other factions within the family, most notably the sons of King Faisal, Prince Saud (foreign minister since 1975) and Prince Turki (head of Saudi intelligence).

Known as a devout Muslim, Abdullah has 14 sons and 20 daughters by six wives. His reputation for piety has earned him support from religious leaders. Having assumed the position of regent in 1996, he was soon considered the *de facto* ruler of Saudi Arabia on account of King Fahd's recurrent illnesses and absences from the country. It was likely that Abdullah consulted and, to some extent, ruled with Fahd's Sudairi brothers, Sultan and Salman, the second deputy prime minister and the governor of Riyadh respectively.

Abdullah's foreign policy has concentrated on improving relations within the Arab world and on encouraging the peace process in the Middle East (notably putting forward settlement proposals in 2002). He has nevertheless shown support for militant Islamic groups such as Hizbollah and condemned Israeli military action in Lebanon and against the Palestinians. In March 2011 he sanctioned the deployment of Saudi troops to Bahrain in support of its fellow monarchy in response to Shia opposition demands for political reforms in the island kingdom.

Abdullah has been less overtly pro-Western than King Fahd. The attacks on New York, USA, in Sept. 2001, although condemned, created serious tensions owing to the high proportion of Saudi nationals among the perpetrators. He had declined a visit to the USA on two occasions before the attacks, complaining that then President George W. Bush was 'uninformed' about the Middle East and the plight of the Palestinians. Nevertheless, he maintained that the USA would remain a firm ally, and Saudi Arabia has itself since become the target of several suspected al-Qaeda attacks.

In Dec. 2006 the UK government controversially suspended a fraud investigation into the 1980s al-Yamamah defence contract with Saudi Arabia, stating that diplomatic co-operation between the two countries was being put at risk (a decision subsequently confirmed as lawful by the British House of Lords in July 2008). In Nov. 2007 Pope Benedict XVI greeted Abdullah at the Vatican in the first such meeting between the head of the Roman Catholic Church and a Saudi monarch.

Internal liberalization has been slow under Abdullah. He has resisted Western calls for the abolition of Sharia law and the emancipation of women, stating that it is 'absurd to impose on an individual or a society rights that are alien to its beliefs or principles'. However, some key issues have been addressed. Female education, previously the preserve of the *ulema* (religious leaders), was placed under the jurisdiction of the ministry of education in 2002 and Abdullah has supported the increase of female employment. The country's first ever elections in 2005, although with a limited franchise and only for local councils, were a partial response to pressure for political change. A law to reform the judicial system was enacted in Oct. 2007 and provided for new specialized courts and the introduction of a supreme court as a final court of appeal. Plans were also announced to curb the powers of the religious police, which had come under increasing criticism over deaths in custody.

In Feb. 2009 Abdullah made extensive changes in government affecting top positions in the courts, the armed forces, the central bank, the health, education and information ministries, the religious police and the Consultative Council. He also appointed the country's first woman minister, Nora Al-Fayez, as deputy minister responsible for girls' education. In Sept. 2011 he announced that women would be eligible to stand and vote in local elections from 2015, and in Jan. 2013 he granted women seats on the Consultative Council for the first time. In 2011 he responded to political tensions generated by popular discontent across much of the Arab world by announcing a new welfare spending programme for the kingdom.

Abdullah's age and state of health have generated speculation about the issue of succession and internecine royal politics. In late 2006 he announced that a committee of senior princes would be formed to select the future crown prince and reduce the likelihood of family conflicts. The new committee would in theory have the power to remove a king if he was judged to be permanently incapacitated and would lead the country in a caretaker capacity until a successor was chosen. In Oct. 2011 the deputy prime minister and minister of interior, Prince Nayef bin Abdulaziz Al-Saud, was named as heir to the throne. However, he died in June 2012 and was succeeded by defence minister Prince Salman bin Abdulaziz Al-Saud.

DEFENCE

Defence expenditure (including expenditure on public order and safety) in 2008 totalled US$38,223m. (US$1,511 per capita). In 2007 defence spending represented 9·3% of GDP.

5,000 US troops were stationed in Saudi Arabia after the 1991 Gulf War and were joined by a further 20,000 during the 2003 conflict. However, virtually all US troops have now been withdrawn. In March 2011 the Gulf Co-operation Council's Peninsula Shield Force, which is based in Saudi Arabia and whose mission is to protect the security of member states from any external aggression, consisted of approximately 40,000 troops.

Army

Strength (2007) was 75,000. There is a paramilitary Border Guard (10,500) and a National Guard (*see below*).

Navy

The Royal Saudi Naval Forces fleet includes seven frigates and four corvettes. Naval Aviation forces operate 15 armed helicopters.

The main naval bases are at Riyadh (HQ Naval Forces), Jeddah (Western Fleet) and Jubail (Eastern Fleet). Naval personnel in 2007 totalled 15,500, including 3,000 marines.

Air Force

Current combat units include F-15s, F-5Bs, F-5Fs, Tornado strike aircraft and Tornado interceptors. The Air Force operated 278 combat-capable aircraft and numbered 20,000 personnel in 2007.

Air Defence Force

This separate Command was formerly part of the Army. In 2007 it operated surface-to-air missile batteries and had a strength of 4,000.

National Guard

The total strength of the National Guard amounted to 100,000 (75,000 active, 25,000 tribal levies) in 2007. The National Guard's primary role is the protection of the Royal Family and vital points in the Kingdom. It is directly under royal command. The UK

provides small advisory teams to the National Guard in the fields of general training and communications.

Industrial Security Force

This force was established in 2007 to protect state oil facilities in response to attacks in 2006 on Abqaiq oil processing plant. Initial strength, 9,000.

INTERNATIONAL RELATIONS

In April 2001 Saudi Arabia and Iran signed a security pact to fight drug trafficking and terrorism, 13 years after the two countries had broken off relations.

ECONOMY

Agriculture accounted for 2% of GDP in 2008, industry 70% and services 27%.

Overview

The economy is dominated by the oil sector, accounting for 85% of total export revenue. It is the world's second largest oil producer and the largest exporter, with one-fifth of the world's proven oil reserves and over 60% of spare capacity in global oil supply. When Libyan oil output was disrupted in 2011 Saudi Arabia was able to increase production. The industrial sector is based on hydrocarbon resources. The country also has deposits of iron ore, phosphates, bauxite and copper.

In 1998 American and European oil companies were allowed to invest in the energy sector for the first time. Structural reforms were introduced in 1999 to attract foreign investment. The stock market was opened to foreign investors and tax and customs administrations were reformed. Improvements in the investment climate paved the way for full membership of the WTO in 2005. A tourism authority was also established.

In the 1980s the economy posted negative annual growth rates over five years. Nor was its 3·1% average annual growth rate in the 1990s particularly impressive relative to other developing countries, particularly in the Asia Pacific region. In 2001 and 2002 the Saudi economy barely grew at all. However, rising oil prices lifted the economy in 2003 and annual growth accelerated to 7·7%. Per capita GDP grew more robustly after the 2003 rise in oil prices, with the IMF putting the level at nearly US$15,500 in 2007.

Inflation remains in single figures as a result of an open and flexible labour market and an open trade system. Over half of the 2004 fiscal surplus was used to reduce central government debt by 16% to 66% of GDP, while the rest was put into a fund to finance investment in priority areas over a five-year period. In 2008 high oil prices helped ensure record surpluses, some of which was used to repay debt that then stood at 13·5% of GDP.

The global financial crisis caused the stock market to decline by 46% in the last quarter of 2008, while a reduction in oil production contributed to a significant downturn in 2009. However, higher oil prices and recovery in global demand helped the economy to rebound in 2010 and continue to strengthen in 2011.

According to the World Bank, Saudi Arabia ranks 12th out of 183 economies for the ease of doing business. High unemployment, at 10·8% in 2011, is an ongoing concern and the government is focused on improving social services. Dependency on oil must also be addressed in the mid- to long-term.

Currency

The unit of currency is the *rial* (SAR) of 100 *halalah*. Foreign exchange reserves totalled US$27,637m. in Sept. 2009 and gold reserves were 4·60m. troy oz. Total money supply in June 2009 was SAR474,307m. Inflation rates (based on IMF statistics):

2002	2003	2004	2005	2006	2007	2008	2009	2010	2011
0·2%	0·6%	0·4%	0·6%	2·3%	4·1%	9·9%	5·1%	5·4%	5·0%

In 2001 the six Gulf Arab states—Saudi Arabia, along with Bahrain, Kuwait, Oman, Qatar and the United Arab

Emirates—signed an agreement to establish a single currency by 2010. In June 2009 it was agreed to postpone the implementation of the new currency, the *khaleeji*, until 2013. It has since been put back further still. Both Oman and the United Arab Emirates have now withdrawn from the scheme, in 2007 and 2009 respectively.

Budget
In 1986 the financial year became the calendar year. 2007 budget: revenue, SAR642·8bn.; expenditure, SAR466·2bn. Oil revenues accounted for 87·5% of revenue in 2007; current expenditures accounted for 74·5% of expenditure.

Performance
Real GDP growth rates (based on IMF statistics):

2002	2003	2004	2005	2006	2007	2008	2009	2010	2011
0·1%	7·7%	5·3%	5·6%	3·2%	2·0%	4·2%	0·1%	5·1%	7·1%

Total GDP in 2011 was US$567·8bn. Per capita GDP is now around half the level of 1980.

Banking and Finance
The Saudi Arabian Monetary Agency (*Governor*, Fahed Al-Mubarak), established in 1953, functions as the central bank and the government's fiscal agent. In 2002 there were three national banks (the National Commercial Bank, the Al-Rajhi Banking and Investment Corporation and the Riyad Bank), five specialist banks, eight foreign banks and three government specialized credit institutions. The leading banks are National Commercial Bank (assets in 2008 of US$59·1bn.), Samba Financial Group (US$47·7bn. in 2008) and Al-Rajhi Bank (US$44·0bn. in 2008). Sharia (the religious law of Islam) forbids the charging of interest; Islamic banking is based on sharing clients' profits and losses and imposing service charges. In 2005 total assets of commercial banks were US$202·4bn.

A number of industry sectors are closed to foreign investors, including petroleum exploration, defence-related activities and financial services.

There is a stock exchange in Riyadh.

ENERGY AND NATURAL RESOURCES

Environment
Saudi Arabia's carbon dioxide emissions from the consumption and flaring of fossil fuels in 2008 were the equivalent of 16·6 tonnes per capita.

Electricity
Installed capacity was 36·6m. kW in 2007. All electricity is thermally generated. Production was 189·08bn. kWh in 2007; consumption per capita in 2007 was 7,661 kWh.

Oil and Gas
Proven oil reserves in 2011 were 264·5bn. bbls, the second highest after those of Venezuela and around 16% of world resources. However, Saudi Arabia has far more reserves that are readily accessible than Venezuela. Oil production began in 1938 by Aramco, which is now 100% state-owned and accounts for about 99% of total crude oil production. Output in 2009 totalled 459·5m. tonnes (the lowest total since 2002) and accounted for 12·0% of the world total oil output. Saudi Arabia lost its status as the world's largest oil producer to Russia in 2009. In 2009 oil export revenues were US$157bn., the highest of any country.

Production comes from 14 major oilfields, mostly in the Eastern Province and offshore, and including production from the Neutral Zone. The Ghawar oilfield, located between Riyadh and the Persian gulf, is the largest in the world, with estimated reserves of 70bn. bbls. Oil reserves are expected to run out in approximately 2075.

In 2008 natural gas reserves were 7,570bn. cu. metres; output in 2008 was 78·1bn. cu. metres. The gas sector has been opened up to foreign investment.

Water
Efforts are under way to provide adequate supplies of water for urban, industrial, rural and agricultural use. Most investment has gone into sea-water desalination. In 2006 desalination plants produced 2·8m. cu. metres of water a day. Irrigation for agriculture consumes the largest amount, from fossil reserves (the country's principal water source) and from surface water collected during seasonal floods. In 2006 there were 230 dams with a holding capacity of 850·3m. cu. metres. Treated urban waste water is an increasing resource for both agricultural and industrial purposes.

Minerals
Production began in 1988 at Mahd Al-Dahab gold mine, the largest in the country. In 2008 total gold production was 4,527 kg. Deposits of iron, phosphate, bauxite, uranium, silver, tin, tungsten, nickel, chrome, zinc, lead, potassium ore and copper have also been found.

Agriculture
Land ownership is under the jurisdiction of the Ministry of Municipal and Rural Affairs.

Since 1970 the government has spent substantially on desert reclamation, irrigation schemes, drainage and control of surface water and of moving sands. Undeveloped land has been distributed to farmers and there are research and extension programmes. Large scale private investment has concentrated on wheat, poultry and dairy production.

In 2002 there were an estimated 3·60m. ha. of arable land and 194,000 ha. of permanent cropland. Approximately 1·62m. ha. were irrigated in 2002. In 2002, 8·5% of the economically active population were engaged in agriculture (19·1% in 1990).

Production of leading crops, 2002 (in 1,000 tonnes): wheat, 2,431; dates, 829; melons and watermelons, 450; tomatoes, 403; potatoes, 313; sorghum, 239; cucumbers and gherkins, 168; barley, 136; grapes, 117; pumpkins and squash, 101; onions, 80.

Livestock (2002): 8·17m. sheep, 2·50m. goats, 323,000 cattle, 260,000 camels and 130m. chickens. Livestock products (2003, in 1,000 tonnes): milk, 948; meat, 642; eggs, 130.

Forestry
The area under forests was 0·98m. ha. (less than 0·5% of the land area) in 2010.

Fisheries
In 2010 the total catch was 65,142 tonnes, entirely from sea fishing.

INDUSTRY

The largest companies in Saudi Arabia by market capitalization in March 2012 were SABIC (Saudi Basic Industries), at US$87·8bn.; Al-Rajhi Banking (US$32·8bn.); and Saudi Telecom (US$22·0bn.).

In 2005 manufacturing accounted for 9·5% of GDP and construction 4·7%. The government encourages the establishment of manufacturing industries. Its policy focuses on establishing industries that use petroleum products, petrochemicals and minerals. Petrochemical and oil-based industries have been concentrated at eight new industrial cities, with the two principal cities at Jubail and Yanbu. Products include chemicals, plastics, industrial gases, steel and other metals. In 2004 there were 3,657 factories employing 340,000 workers.

Labour
The labour force in 2002 totalled 6,242,000. In 2002 females constituted 14% of the labour force—one of the lowest percentages of females in the workforce of any country. In 2001, 35·7% of the economically active population were engaged in

wholesale and retail trade, 18·7% in manufacturing, 15·7% in construction, 6·7% in research, consultancy and recruitment. There are 6m. foreign workers, including over 1m. Egyptians and over 1m. Indians. Unemployment, which was less than 8% in 1999, reached 12% in 2002. Young people in particular are affected, with nearly a third unemployed.

INTERNATIONAL TRADE

Saudi Arabia, along with Bahrain, Kuwait, Oman, Qatar and the United Arab Emirates entered into a customs union in Jan. 2003.

Imports and Exports

Trade in US$1m.:

	2006	2007	2008	2009	2010
Imports c.i.f.	69,800	90,214	115,134	95,552	106,863
Exports f.o.b.	211,306	234,951	313,462	192,314	251,143

The principal export is crude oil; refined oil, petrochemicals, fertilizers, plastic products and wheat are other major exports. Saudi Arabia is the world's largest exporter of oil (8·7m. bbls per day), accounting for 85·4% of all the country's exports in 2006. Major import suppliers, 2006: USA, 14·5%; China, 8·6%; Germany, 8·1%; Japan, 8·1%. Main export destinations, 2006: Japan, 16·5%; USA, 15·1%; South Korea, 9·2%; China, 6·3%.

COMMUNICATIONS

Roads

In 2005 there was a total road network of 221,372 km (21·5% paved), including 3,891 km of motorway. A causeway links Saudi Arabia with Bahrain. Passenger cars in use in 2005 numbered 3,206,000 (415 per 1,000 inhabitants in 2004) and there were 1,127,900 lorries and vans. Women are not allowed to drive. In 2004–05 there were 293,281 road accidents resulting in 5,168 deaths.

Rail

In 2005, 1,394 km of 1,435 mm gauge lines linked Riyadh and Dammam with stops at Hofuf and Abqaiq. The network is being extended, consisting of links to Jeddah, the Jordanian border, and Makkah and Madinah. The line from Makkah to Madinah via Jeddah will be Saudi Arabia's first high-speed rail link. In 2008 railways carried 1·1m. passengers and 4·6m. tonnes of freight. The first line of a metro system in Makkah opened in 2010, covering 18·1 km.

Civil Aviation

The national carrier is the part-privatized Saudi Arabian Airlines, which in 2006 owned 139 aircraft and served 76 destinations. In 2005 scheduled airline traffic of Saudi-based carriers flew 117·1m. km and carried 11,126,300 passengers. There are four major international airports, at Jeddah (King Abdulaziz), Dhahran, Riyadh (King Khaled), and the newly constructed King Fahd International Airport at Dammam. There are also 22 domestic airports. In 2001 Jeddah handled 10,237,161 passengers (5,413,841 on international flights) and 188,386 tonnes of freight. Riyadh was the second busiest airport in 2001, handling 8,702,697 passengers (5,428,429 on domestic flights) and 155,245 tonnes of freight.

Shipping

The ports of Dammam and Jubail are on the Persian Gulf and Jeddah, Yanbu and Jizan on the Red Sea. There is a deepwater oil terminal at Ras Tanura. In 2009 the major ports handled 142·3m. tonnes of cargo (84·1m. tonnes loaded and 58·2m. tonnes discharged). In Jan. 2009 there were 83 ships of 300 GT or over registered (including 38 oil tankers, 18 general cargo ships and 14 passenger ships), totalling 1·25m. GT.

Telecommunications

In 2008 there were 4·1m. main (fixed) telephone lines. In the same year mobile phone subscribers numbered 36·0m. (1,428·5 per

1,000 persons). The government sold a 30% stake in Saudi Telecom Company (STC) in Dec. 2002. STC lost its monopoly in the mobile phone market in 2005 and in landline services in 2007. The number of internet users in 2008 was 7·8m. In March 2012 there were 5·1m. Facebook users.

SOCIAL INSTITUTIONS

Justice

The religious law of Islam (Sharia) is the common law of the land, and is administered by religious courts, at the head of which is a chief judge, who is responsible for the Department of Sharia Affairs. Sharia courts are concerned primarily with family inheritance and property matters. However, following judicial reforms of Oct. 2007 a newly-established Supreme Court replaced the Supreme Judiciary Council as the highest judicial authority. Specialized courts are also to be established to operate alongside the Sharia courts and there are plans to codify Sharia and introduce the principle of precedent into court practice. The Committee for the Settlement of Commercial Disputes is the commercial court. Other specialized courts or committees include one dealing exclusively with labour and employment matters; the Negotiable Instruments Committee, which deals with cases relating to cheques, bills of exchange and promissory notes; and the Board of Grievances, whose preserve is disputes with the government or its agencies and which also has jurisdiction in trademark-infringement cases and is the authority for enforcing foreign court judgments.

The death penalty is in force for murder, rape, sodomy, armed robbery, sabotage, drug trafficking, adultery and apostasy; executions may be held in public. There were 79 confirmed executions in 2012 (82 confirmed executions in 2011). The population in penal institutions in Jan. 2009 was 44,600 (178 per 100,000 of national population).

Education

The educational system provides students with free education, books and health services. General education consists of kindergarten, six years of primary school and three years each of intermediate and high school. In 2005–06 there were: 1,449 pre-primary schools with 10,150 teachers and 97,137 pupils; 13,163 primary schools with 213,355 teachers and 2,417,811 pupils; 7,086 intermediate schools with 104,675 teachers and 1,071,747 pupils; 4,215 secondary schools with 79,754 teachers and 954,141 pupils. Students can attend either high schools offering programmes in arts and sciences, or vocational schools. Girls' education has traditionally been administered separately, but in Sept. 2009 the country's first mixed-gender university was opened and in Oct. 2009 a trial of mixed-gender education in 15 private elementary schools was launched. In 2005 there were 903 institutions for special needs pupils with 18,958 students. The adult literacy rate in 2009 was an estimated 86·1% (90·0% among males and 81·1% among females). Although Saudi girls were not even allowed to attend school until 1964 women now make up nearly 60% of Saudi Arabia's higher education students.

In 2005 there were 3,775 adult education centres. In 2005–06 there were 11 universities (including two Islamic universities and one university of petroleum and minerals); there were 603,767 students in total in higher education and 26,827 teachers.

Health

In 2005 there were 1,848 health care centres, 1,043 private dispensaries and 364 hospitals with 51,130 beds. Health personnel, 2005: 42,975 physicians, 78,587 nurses and 49,167 technical staff. At Jeddah there is a quarantine centre for pilgrims.

Welfare

The retirement age is 60 (men) or 55 (women), with eligibility based on 120 months of contributions. The minimum monthly old-age pension is SAR1,500, calculated as 2·5% of the average monthly wage during the previous two years multiplied by the number of years of contributions. A 1969 law requires employers with more than 20 employees to pay 100% of wages for the first 30 days of sick leave and 75% of wages for the next 60 days. Unemployment benefits of SAR2,000 (US$535) a month were introduced in 2011 for applicants who proved they were looking for work or undergoing training.

Workers' medical benefits include medical, dental and diagnostic treatment, hospitalization, medicines, appliances, transportation and rehabilitation.

RELIGION

In 2001, 90% of the total population were Sunni Muslims, 4% Shias, 4% Christians and 1% Hindus. The *Grand Mufti*, Sheikh Abdul Aziz bin Abdullah bin Mohammed Al-Sheikh, has cabinet rank. A special police force, the Mutaween, exists to enforce religious norms.

The annual *Hajj*, the pilgrimage to Makkah, takes place from the 8th to the 13th day of Dhu al Hijjah, the last month of the Islamic year. It attracts more than 1·8m. pilgrims annually. In the current Islamic year, 1434, the *Hajj* will begin on 13 Oct. 2013 in the Gregorian calendar.

CULTURE

World Heritage Sites

Saudi Arabia has two sites on the UNESCO World Heritage List: the archaeological site of Madain Salih (Al-Hijr) was inscribed in 2008; and the At-Turaif District in Ad-Dir'iyah, founded in the 15th century and the first capital of the Saudi dynasty, was inscribed in 2010.

Press

In 2006 there were 13 daily newspapers with a combined circulation of 1,397,000. The most widely read newspaper is *Asharq Al-Awsat* ('Middle East'), with an average daily circulation of 272,000 in 2006.

Tourism

There were 8,037,000 foreign tourists in 2005; spending by tourists in 2005 totalled US$5·2bn.

Calendar

Saudi Arabia follows the Islamic *hegira* (AD 622, when Mohammed left Makkah for Madinah), which is based upon the lunar year of 354 days. The Islamic year 1434 corresponds to 14 Nov. 2012–4 Nov. 2013, and is the current lunar year.

DIPLOMATIC REPRESENTATIVES

Of Saudi Arabia in the United Kingdom (30 Charles St., London, W1J 5DZ)
Ambassador: Prince Mohammed Bin Nawaf Bin Abdulaziz Al-Saud.

Of the United Kingdom in Saudi Arabia (PO Box 94351, Riyadh 11693)
Ambassador: Sir John Jenkins, KCMG, LVO.

Of Saudi Arabia in the USA (601 New Hampshire Ave., NW, Washington, D.C., 20037)
Ambassador: Adel bin Ahmed Al-Jubeir.

Of the USA in Saudi Arabia (PO Box 94309, Riyadh)
Ambassador: James B. Smith.

Of Saudi Arabia to the United Nations
Ambassador: Abdallah Yahya Al-Mouallimi.

Of Saudi Arabia to the European Union
Ambassador: Faisal Hassan Trad.

FURTHER READING

Aarts, Paul, *Saudi Arabia in the Balance: Political Economy, Society, Foreign Affairs.* 2006
Al-Rasheed, Madawi, *A History of Saudi Arabia.* 2002
Al-Rasheed, Madawi and Vitalis, Robert, (eds.) *Counter-Narratives: History, Contemporary Society, and Politics in Saudi Arabia and Yemen.* 2004
Bradley, John R., *Saudi Arabia Exposed.* 2005
Hegghammer, Thomas, *Jihad in Saudi Arabia: Violence and Pan-Islamism since 1979.* 2010
Kostiner, J., *The Making of Saudi Arabia: from Chieftaincy to Monarchical State.* 1994
Lacey, Robert, *Inside the Kingdom: Kings, Clerics, Modernists, Terrorists and the Struggle for Saudi Arabia.* 2009
Mackey, Sandra, *The Saudis: Inside the Desert Kingdom.* Revised ed. 2003
Manea, Elham, *Regional Politics in the Gulf: Saudi Arabia, Oman and Yemen.* 2005
Murphy, Caryle, *A Kingdom's Future: Saudi Arabia through the Eyes of its Twentysomethings.* 2013
Peterson, J. E., *Historical Dictionary of Saudi Arabia.* 1994
Vassiliev, Alexei, *King Faisal of Saudi Arabia: Personality, Faith and Times.* 2012
Wright, J. W. (ed.) *Business and Economic Development in Saudi Arabia: Essays with Saudi Scholars.* 1996

National Statistical Office: Ministry of Economy and Planning, Central Department of Statistics and Information, Riyadh.
Website: http://www.cdsi.gov.sa

SENEGAL

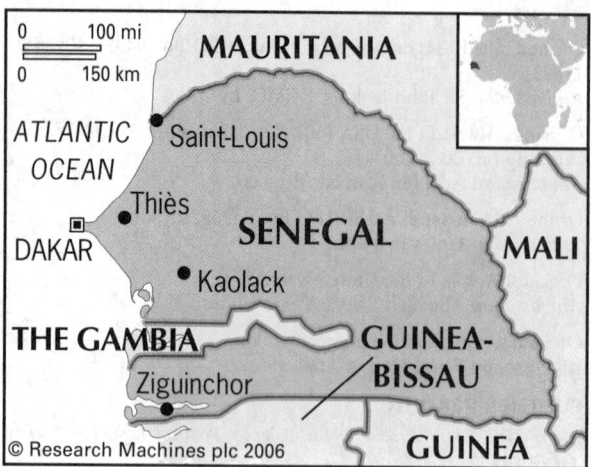

0   100 mi
0   150 km

MAURITANIA

ATLANTIC OCEAN
Saint-Louis
Thiès
SENEGAL
DAKAR
Kaolack
MALI
THE GAMBIA
GUINEA-BISSAU
Ziguinchor
GUINEA
© Research Machines plc 2006

République du Sénégal
(Republic of Senegal)

Capital: Dakar
Population projection, 2015: 14·16m.
GNI per capita, 2011: (PPP$) 1,708
HDI/world rank: 0·459/155
Internet domain extension: .sn

KEY HISTORICAL EVENTS

For much of the 1st millennium AD Senegal was under the influence of the gold-rich Ghana Empire of the Soninke people. In western Senegal the Takrur state was established in the 9th century. Islam was introduced in the 11th century by the Zenega Berbers of southern Mauritania, who gave their name to the region. The power of the Malinke (Madingo) in present-day Mali expanded in the 13th and 14th centuries, especially under Mansa Musa, who subjugated Takrur and the Tukulor in Senegal. The west was dominated by the Jolof empire, which fragmented into four kingdoms in the 16th century.

Portuguese trading colonies were established on Gorée Island and at Rufisque in around 1444, encouraging the growth of the slave trade. The Dutch took control of Senegalese trade in the 17th century, only to be evicted in 1677 by the French, based at Saint-Louis at the mouth of the Sénégal River. Inland, the Tukulor created a Muslim theocracy in Fouta Toro, usurping the Denianké Dynasty in 1776. Tukulor power grew in the 1850s under al-Hajj Umar Tal, whose *jihad* was contained by treaty with the French in 1857. Britain accepted French hegemony in the region in 1814 after half a century of colonial rivalry, while retaining the Gambia River. Railway construction in 1879 cemented French control over western Senegal and Dakar became the capital of French West Africa in 1904. Casamance and eastern Senegal were conquered in the 1890s.

Senegalese service in the French army in the First World War secured representation in Paris and French citizenship for Africans in certain communes. The colonial administration followed a moderate liberalization programme, including the right to form political parties and trade unions. However, the decline in the groundnut trade in the 1930s increased poverty in Senegal. The expansion of the vote after the Second World War gave support to the Democratic Bloc (BDS), which joined the Socialist Party to become the Progressive Union (UPS), dominating the

1959 elections in the newly-autonomous Senegal. Membership of the French Community lasted until independence on 20 June 1960 as part of the Federation of Mali with French Soudan (Mali); the Federation was dissolved on 20 Aug. 1960.

Leopold Sédar Senghor, the BDS founder and leader of the UPS, was elected president on 5 Sept. 1960. Relations with his prime minister, Mamadou Dia, deteriorated and Senghor had him arrested in Dec. 1962 after an attempted coup. Presidential power was augmented by referendum in 1963, allowing Senghor to ban all other parties in 1966. Senghor appointed Abdou Diouf prime minister in 1973 and began relaxing political restrictions. Abdoulaye Wade founded the Democratic Party (PDS) and a Marxist-Leninist party was formed. Recession and political agitation forced Senghor's resignation in Dec. 1980; Diouf succeeded and was confirmed by elections in 1983, 1988 and 1993.

Diouf pursued a vigorous foreign policy via the Organization of African Unity and the Economic Community of West African States. He reinstated the Gambian president, Sir Dawda Jawara, in 1981, creating the Senegambian confederation, which lasted until 1989. Unrest in the southern Casamance region escalated into secessionist civil war in the early 1990s. A skirmish on the Mauritanian border in 1989 resulted in the death of Senegalese and Mauritanians expatriates and the closing of the border, a dispute not resolved until 1994. The deterioration of the economy and the Casamance crisis led to electoral defeat in 2000. He conceded peacefully, handing power to his long-term rival, PDS leader Abdoulaye Wade. Wade was re-elected in 2007 and received legal dispensation to contest a third term in 2012, despite the constitution of 2001 limiting presidents to two terms of office. The decision prompted popular protests, leading to Macky Sall defeating Wade to become the new president in April 2012.

TERRITORY AND POPULATION

Senegal is bounded by Mauritania to the north and northeast, Mali to the east, Guinea and Guinea-Bissau to the south and the Atlantic to the west with The Gambia forming an enclave along that shore. A short section of the boundary with The Gambia is undefined. Area, 196,720 sq. km, including 4,190 sq. km of inland water. Population (2002 census), 9,858,482 (5,005,718 females). Estimate, 2010, 12·43m.; density, 65 per sq. km. In 2011 the population was 42·7% urban.

The UN gives a projected population for 2015 of 14·16m.

About 2m. Senegalese live abroad, particularly in France, Italy, Spain and the USA.

The areas, populations and capitals of the regions at the time of the 2002 census:

Region	Area (in sq. km)	Population 2002 census	Capital
Dakar	550	2,168,314	Dakar
Diourbel	4,359	1,051,941	Diourbel
Fatick	7,935	609,789	Fatick
Kaffrine[1]	—	—	Kaffrine
Kaolack[1]	16,010	1,070,203	Kaolack
Kédougou[2]	—	—	Kédougou
Kolda[3]	21,011	817,438	Kolda
Louga	29,188	677,264	Louga
Matam	25,083	423,967	Matam
Saint-Louis	19,044	694,652	Saint-Louis
Sédhiou[3]	—	—	Sédhiou
Tambacounda[2]	59,602	612,855	Tambacounda
Thiès	6,601	1,322,579	Thiès
Ziguinchor	7,339	409,480	Ziguinchor

[1]Kaffrine (11,853 sq. km), formerly part of Kaolack, was created in 2008. [2]Kédougou (16,896 sq. km), formerly part of Tambacounda, was created in 2008. [3]Sédhiou (7,293 sq. km), formerly part of Kolda, was created in 2008.

Dakar, the capital, had a provisional census population in 2002 of 1,983,093. Other large cities (with 2002 provisional census population) are: Thiès (237,849), Rufisque (179,797), Kaolack (172,305), Saint-Louis (154,555), Mbour (153,503) and Ziguinchor (153,269).

Ethnic groups are the Wolof (36% of the population), Fulani (16%), Serer (16%), Diola (9%), Tukulor (9%), Bambara (6%), Malinké (6%) and Sarakole (2%).

The official language is French; Wolof is widely spoken.

SOCIAL STATISTICS

2005 estimates: births, 430,000; deaths, 132,000. Rates, 2005 estimates (per 1,000 population): births, 39·4; deaths, 12·1. Annual population growth rate, 2000–05, 2·8%; infant mortality, 2010, 50 per 1,000 live births. Life expectancy in 2007 was 53·9 years for men and 56·9 for women. Fertility rate, 2008, 5·0 births per woman. 51% of the population were living in poverty in 2005.

CLIMATE

A tropical climate with wet and dry seasons. The rains fall almost exclusively in the hot season, from June to Oct., with high humidity. Dakar, Jan. 72°F (22·2°C), July 82°F (27·8°C). Annual rainfall 22" (541 mm).

CONSTITUTION AND GOVERNMENT

A new constitution was approved by a referendum held on 7 Jan. 2001. The head of state is the *President*, elected by universal suffrage for not more than two five-year terms (previously two seven-year terms). However, in Jan. 2012 the incumbent, Abdoulaye Wade won a court ruling allowing him to run for a third term on the grounds that the constitutional provision was not enacted until a year after his first term of office had begun. The *President* has the power to dissolve the National Assembly, without the agreement, as had been the case, of a two-thirds majority. The new constitution also abolished the upper house (the Senate), confirmed the status of the prime minister and for the first time gave women the right to own land. Senegal has a bicameral legislature. For the 150-member *National Assembly*, 90 members are elected by simple majority vote in single or multi-member constituencies for five years with 60 elected by a system of party-list proportional representation. The Senate was re-established in Jan. 2007 six years after being dissolved. In Sept. 2012 parliament voted to abolish it after severe floods hit Senegal, with the money that would be saved going towards improving the country's flood defences and aid for flood victims.

National Anthem

'Pincez tous vos koras, frappez les balafos' ('All pluck the koras, strike the balafos'); words by Léopold Sédar Senghor, tune by Herbert Pepper.

RECENT ELECTIONS

Presidential elections took place on 26 Feb. 2012. Incumbent Abdoulaye Wade won 34·8% of the vote, ahead of former prime minister Macky Sall with 26·6%, Moustapha Niasse with 13·2% and Ousmane Tanor Dieng with 11·3%. There were ten other candidates. Turnout was 51·6%. In the run-off held on 25 March, Sall won 65·8% of the vote and Wade 34·2%. Turnout was 55·0%.

Parliamentary elections were held on 1 July 2012. Turnout was 36·8%. The United in Hope coalition backing President Macky Sall took 119 of the 150 seats in the National Assembly and the Senegalese Democratic Party 12 with smaller parties taking four seats or fewer. Of the 150 members of parliament there were 64 women (42·7%), among the highest percentages of women MPs in the world.

In indirect elections to the Senate held on 19 Aug. 2007 the Senegalese Democratic Party won 34 of the available 35 seats, with one going to And-Jëf/African Party for Democracy and Socialism. Elections were then scheduled for Sept. 2012, but shortly

beforehand the president proposed to abolish the Senate and return to a unicameral chamber. Parliament approved the abolition.

CURRENT GOVERNMENT

President: Macky Sall; b. 1961 (Alliance for the Republic; sworn in 2 April 2012).

In March 2013 the government was composed as follows:

Prime Minister: Abdoul Mbaye; b. 1953 (ind.; sworn in 3 April 2012).

Minister of Agriculture and Rural Equipment: Abdoulaye Baldé. *Armed Forces:* Augustin Tine. *Commerce, Industry and the Informal Sector:* Alioune Sarr. *Communication, Telecommunications and the Digital Economy:* Cheikh Mamadou Abiboulaye Dièye. *Culture:* Abdoul Aziz Mbaye. *Economy and Finance:* Amadou Kane. *Energy and Mines:* Aly Ngouille Ndiaye. *Environment and Sustainable Development:* Haydar El Ali. *Fisheries and Maritime Affairs:* Pape Diouf. *Foreign Affairs and Senegalese Abroad:* Mankeur Ndiaye. *Health and Social Action:* Eva Marie Coll Seck. *Higher Education and Research:* Mary Teuw Niane. *Infrastructure and Transport:* Thierno Alassane Sall. *Interior:* Pathé Seck. *Justice:* Aminata Touré. *Livestock:* Aminata Mbengue Ndiaye. *National Education:* Serigne Mbaye Thiam. *Professional Training, Learning and Crafts:* Mamadou Talla. *Promotion of Good Governance and Government Spokesperson:* Abdou Latif Coulibaly. *Public Service, Labour and Relations with Institutions:* Mansour Sy. *Rebuilding and Redevelopment of Flooded Areas:* Khadim Diop. *Regional Planning and Local Government:* Arame Ndoye. *Sports:* Mbagnick Ndiaye. *Tourism and Recreation:* Youssou Ndour. *Urban Development and Housing:* Khoudia Mbaye. *Water and Sanitation:* Oumar Guèye. *Women, Children and Female Entrepreneurs:* Mariama Sarr. *Youth, Employment and Promotion of Civic Values:* Benoît Sambou.

Government Website (French only): http://www.gouv.sn

CURRENT LEADERS

Macky Sall

Position
President

Introduction
Macky Sall, a former prime minister under President Wade and founder of the liberal Alliance for the Republic (APR), was elected president in March 2012. Having broken with Wade's administration over issues of transparency and public spending, Sall pledged to cut the size of government and tackle Senegal's long-standing problems of unemployment, poor infrastructure and food insecurity.

Early Life
Macky Sall was born on 11 Dec. 1961 in Fatick, western Senegal. After graduating in geological engineering from Cheikh Anta Diop University, Dakar, he completed further studies in geophysics at the French Petroleum Institute in Paris from 1992–93. He worked for the state-owned oil company PETROSEN from 1994–2000.

A member of Wade's Senegalese Democratic Party (PDS), Sall became general secretary of its Fatick regional convention in 1998. Following the PDS victory at elections in 2000, he served as director general of PETROSEN from Dec. 2000 until July 2001. He was special presidential adviser on energy and mines from April 2000 to May 2001, and from 2001–03 served as minister for mines, energy and hydraulics.

Sall became mayor of Fatick in 2002 and from 2003–04 was minister of the interior and government spokesperson. In April 2004 he was appointed prime minister by President Wade. He served until June 2007, when he was elected president of the

National Assembly. In Nov. 2007, after Sall initiated an enquiry into spending by an agency headed by the president's son, the term of the National Assembly president was reduced to one year, prompting his resignation.

In Dec. 2008 Sall founded the APR and in 2009 was again elected mayor of Fatick. Against a background of rising public discontent with the government and anger at Wade's decision to circumvent constitutional limits and stand for a third term as president, Sall ran a vigorous 2012 presidential campaign. Arguing for cuts in public spending and increased transparency, he proposed halving the size of government and reducing Senegal's diplomatic representation abroad, using the savings to reduce food prices. He also promised to attract foreign investment to boost the economy and develop infrastructure.

After claiming second place in the first round of voting, Sall won the support of 12 eliminated opposition candidates and on 25 March 2012 took almost 66% of the vote in a run-off. Wade conceded defeat and Sall was sworn in on 2 April 2012.

Career in Office
Sall took office promising to tackle corruption and deliver growth, while dealing with severe food crises in the drought-stricken Sahel region. He was also expected to enter negotiations with neighbouring Gambia to end three decades of conflict in the southern Casamance region.

DEFENCE
There is selective conscription for two years. Defence expenditure totalled US$218m. in 2008 (US$16 per capita), representing 1·7% of GDP.

Army
There are four military zones. The Army had a strength of 11,900 (including conscripts) in 2007. There is also a paramilitary force of gendarmerie and customs of 5,000.

Navy
Personnel (2007) totalled 950, and bases are at Dakar and Casamance.

Air Force
The Air Force, formed with French assistance, has eight combat-capable aircraft. Personnel (2007) 770.

ECONOMY
Agriculture accounted for 13·4% of GDP in 2007, industry 23·6% and services 63·0%.

Overview
In contrast to many of its West African neighbours, Senegal enjoys social and political stability and is open to the outside world. Prompted by the devaluation of the CFA franc in 1994 and aggressive structural reforms, Senegal experienced steady per capita GDP growth between 1995 and 2005. Nonetheless, 34% of the population lives on less than $1·25 a day (PPP) and the unemployment rate is estimated at 48%.

The service sector accounted for 61·0% of GDP in 2010, having grown rapidly during the 1990s (especially transportation and telecommunications). Agriculture and construction have also contributed to recent growth, although the agricultural sector is in long-term decline. In 2000 Senegal became eligible for the Heavily Indebted Poor Countries (HIPC) initiative, reaching the Completion Point of the initiative in 2004 and so earning debt relief totalling US$850m.

In 2006–07 agricultural output decreased as a result of late but heavy rains. The increase in international food and oil prices in 2007–08 caused further deterioration in the economy, while the global financial crisis in 2008 and drought in the Sahel in 2011 were additional challenges. At best, a moderate recovery is anticipated.

The port of Dakar serves as an important centre of trade in West Africa. A 20-ha. dry port project was completed in 2009 while Jafza International of Dubai is developing a Special Economic Zone adjacent to Dakar International Airport. Oil exploration in the Casamance region offers potential for future revenue streams, although work has been hampered by weak transportation links and civil unrest.

Foreign capital enters the country via Senegalese citizens living abroad, and in 2010 remittances made up 10% of GDP.

Currency
The unit of currency is the *franc CFA* (XOF) with a parity of 655·957 francs CFA to one euro. In June 2005 total money supply was 930,246m. francs CFA and foreign exchange reserves totalled US$1,380m. Following deflation of 1·7% in 2009, there was inflation of 1·2% in 2010 and 3·4% in 2011.

Budget
Revenues in 2008 totalled 1,350·9bn. francs CFA (tax revenue, 86·0%) and expenditures 1,678·6bn. francs CFA (current expenditure, 67·1%).

VAT is 18%.

Performance
Real GDP growth was 2·1% in 2009, 4·1% in 2010 and 2·6% in 2011. Senegal's total GDP in 2011 was US$14·3bn.

Banking and Finance
The Banque Centrale des États de l'Afrique de l'Ouest is the bank of issue of the franc CFA for all the countries of the West African Economic and Monetary Union (Benin, Burkina Faso, Côte d'Ivoire, Mali, Niger, Senegal and Togo) but has had its headquarters in Dakar, the Senegalese capital, since 1973. Its *Governor* is Tiémoko Meyliet Koné. There are eight commercial banks, the largest including Banque Internationale pour le Commerce et l'Industrie and Banque de l'Habitat. There are also four development banks and an Islamic bank. Only about 5% of the population have bank accounts.

External debt amounted to US$3,677m. in 2010. As a share of GNI, external debt fell from 78·7% in 2000 to 28·5% in 2010.

Senegal is affiliated to the regional BRVM stock exchange (serving the member states of the West African Economic and Monetary Union), based in Abidjan, Côte d'Ivoire.

ENERGY AND NATURAL RESOURCES
Environment
Senegal's carbon dioxide emissions from the consumption and flaring of fossil fuels in 2008 were the equivalent of 0·5 tonnes per capita.

Electricity
In 2007 installed capacity was an estimated 0·5m. kW. Production in 2007 was 2·12bn. kWh and consumption per capita 194 kWh. Power cuts are common.

Minerals
Output, 2008, included (in tonnes): sand, 6·4m.; calcium phosphate, 645,000; salt, 241,000. Annual gold production was approximately 600 kg for many years but increased to over 5,000 tonnes in 2009 following the opening of a new mine in 2008.

Agriculture
Because of erratic rainfall 25% of agricultural land needs irrigation. Most land is owned under customary rights and holdings tend to be small. In 2007 the economically active population engaged in agriculture was an estimated 3,750,000. In 2006 approximately 2·99m. ha. were used as arable land and 52,000 ha. for permanent crops. An estimated 63,000 ha. were irrigated in 2006. There were about 700 tractors in use in 2006 and 155 harvester-threshers. Production, 2005–06 (in 1,000 tonnes): sugarcane, 829; groundnuts, 703; millet, 609; maize, 400;

rice, 289; cassava, 281; watermelons, 241; tomatoes, 161; sorghum, 144; onions, 76; mangoes, 62.

Livestock (2006): 5·00m. sheep; 4·26m. goats; 3·14m. cattle; 518,000 horses; 415,000 asses; 318,000 pigs.

Meat production (2006, in 1,000 tonnes): beef and veal, 69; lamb and mutton, 17; goat meat, 12; pork, bacon and ham, and camel, 5. Milk production (2006, in 1,000 litres): cow's milk, 135; goat's milk, 46; sheep's milk, 31.

Forestry

Forests covered 8·47m. ha. in 2010 (44% of the land area), down from 9·35m. ha. in 1990 and 8·90m. ha. in 2000. Roundwood production in 2007 amounted to 6·13m. cu. metres.

Fisheries

In 2007 the total catch was 412,835 tonnes, of which 347,104 tonnes came from sea fishing and 38,731 tonnes from inland fishing. Exports of fishery commodities were valued at US$313·5m. in 2007.

INDUSTRY

Predominantly agricultural and fish processing, phosphate mining, petroleum refining and construction materials.

Labour

In 2002 the economically active population numbered 3,699,859, of whom 208,135 were unemployed (5·6%).

INTERNATIONAL TRADE

Imports and Exports

In 2007 imports (f.o.b.) totalled US$4,871·4m. and exports (f.o.b.) US$1,546·3m. Chief imports: petroleum and petroleum products, food and live animals, and machinery and transport equipment. Chief exports: fish, petroleum and petroleum products, and chemicals and related products. Main import suppliers, 2007: France, 22·9%; Nigeria, 8·4%; Netherlands, 7·2%; China, 5·7%; Thailand, 5·3%. Main export markets, 2007: Mali, 24·0%; France, 9·5%; India, 6·7%; Bunkers and ships' stores, 5·5%; Gambia, 5·4%.

COMMUNICATIONS

Roads

The length of roads in 2006 was 14,805 km, of which 29·3% were paved. In 2008 there were 205,704 passenger cars, 56,795 trucks and vans and 15,982 coaches. There were 320 deaths as a result of road accidents in 2007.

Rail

There were previously four railway lines but the total length of the track fell from 1,034 km (metre gauge) in 1986 to 645 km in 2005. Only the Dakar–Kidira line (continuing in Mali) is still theoretically in service although traffic was suspended in 2003 and again since 2009 owing to the poor state of the track. There is also a suburban rail service linking Dakar and Rufisque, which carried 4.9m. passengers in 2009. In 2009, 364,000 tonnes of freight were carried.

Civil Aviation

The international airport is at Dakar/Yoff (Léopold Sédar Senghor), which handled 1,882,242 passengers and 21,816 tonnes of freight in 2008. Air Sénégal International was 49% state-owned and 51% owned by Royal Air Maroc (RAM). After an attempted takeover bid by the government to gain 64% of the shares, RAM pulled out of the airline and flights were suspended on 24 April 2009 owing to financial difficulties. Sénégal Airlines was launched as a replacement national carrier in Oct. 2009 although it is 64% privately-controlled. Flights, initially only within Africa, commenced in Jan. 2011.

Shipping

In Jan. 2009 there were three ships of 300 GT or over registered, totalling 5,000 GT. 10·6m. tonnes of freight were handled in the port of Dakar in 2008. There is a river service on the Senegal from Saint-Louis to Podor (363 km) open throughout the year, and to Kayes (924 km) open from July to Oct. The Senegal River is closed to foreign flags. The Saloum River is navigable as far as Kaolack, the Casamance River as far as Ziguinchor.

Telecommunications

In 2008 there were 237,800 main (fixed) telephone lines; mobile phone subscribers numbered 5,389,100 in 2008 (44·1 per 100 persons). There were 1,020,000 internet users in 2008. In June 2012 there were 666,000 Facebook users.

SOCIAL INSTITUTIONS

Justice

There are *juges de paix* in each *département* and a court of first instance in each region. Assize courts are situated in Dakar, Kaolack, Saint-Louis and Ziguinchor, while the Court of Appeal resides in Dakar. The death penalty, last used in 1967, was abolished in Dec. 2004.

The population in penal institutions in Dec. 2007 was 6,487 (55 per 100,00 of national population).

Education

The adult literacy rate in 2006 was 42%. In 2007 there were 1,572,178 pupils (45,957 teaching staff) in primary schools; 505,097 pupils (20,007 teaching staff) in secondary schools; and 76,949 students in tertiary education. There are two public universities (Cheikh Anta Diop and Gaston Berger) and three private universities (Dakar Bourguiba, Sahel and Suffolk).

In 2006 public expenditure on education came to 4·9% of GNI and 26·3% of total government spending.

Health

In 2006 there were 22 hospitals, 68 health centres and 949 health posts. In 2004–05 medical personnel in government service included: 765 doctors, 64 dentists, 54 pharmacists, 546 midwives and 874 nurses. In 2003 there were 551 doctors and 567 midwives in the private sector. Senegal has been one of the most successful countries in Africa in the prevention of AIDS. Levels of infection have remained low, with the anti-AIDS programme having started as far back as 1986. The infection rate has been kept below 2%.

RELIGION

The population was 93% Sunni Muslim in 2001, the remainder being Christian (mainly Roman Catholic) or animist. There was one Roman Catholic cardinal in Feb. 2013.

CULTURE

World Heritage Sites

Gorée Island, off the coast of Senegal, was added to the UNESCO World Heritage List in 1978. It was formerly the largest slave trading centre on the African coast. The Djoudj Sanctuary in the Senegal River delta (added in 1981), protects 1·5m. birds, including the white pelican, the purple heron, the African spoonbill, the great egret and the cormorant. Niokolo-Koba National Park, along the banks of the Gambia River (added in 1981), is home to the Derby eland (largest of the antelopes), chimpanzees, lions, leopards and a large population of elephants as well as many birds, reptiles and amphibians. The Island of Saint-Louis joined the UNESCO list in 2000 (reinscribed in 2007), as a reminder of its status as the capital between 1872 and 1957. The Saloum Delta (2011) comprises brackish channels encompassing over 200 islands and islets and is marked by numerous shellfish mounds produced by its human inhabitants over 2,000 years. The Bassari Country, comprising the Bassari-Salémata area, the Bedik–Bandafassi area and the Fula–Dindéfello area, was added to the UNESCO list in 2012. The site houses original and still vibrant cultures of the Bassari, Fula and Bedik peoples, featuring their original traits of agro-pastoral, social, ritual and spiritual practices.

Senegal shares a UNESCO site with The Gambia: the Stone Circles of Senegambia (added in 2006) are a collection of 93 stone circles, tumuli and burial mounds from between the 3rd century BC and the 16th century AD.

Press

In 2006 there were 26 daily newspapers with a total average circulation of 120,000 copies and 30 non-dailies.

Tourism

In 2006, 406,000 foreign tourists visited Senegal; revenue in 2006 amounted to US$329m.

Festivals

The 3rd World Festival of Black Arts was held in Dakar in Dec. 2010.

DIPLOMATIC REPRESENTATIVES

Of Senegal in the United Kingdom (39 Marloes Rd, London, W8 6LA)
Ambassador: Abdou Sourang.

Of the United Kingdom in Senegal (20 Rue du Docteur Guillet, Dakar)
Ambassador: John Marshall.

Of Senegal in the USA (2112 Wyoming Ave., NW, Washington, D.C., 20008)
Ambassador: Cheikh Niang.

Of the USA in Senegal (Ave. Jean XXIII, Dakar)
Ambassador: Lewis Lukens.

Of Senegal to the United Nations
Ambassador: Abdou Salam Diallo.

Of Senegal to the European Union
Ambassador: Mame Balla Sy.

FURTHER READING

Centre Français du Commerce Extérieur. *Sénégal: un Marché.* 1993

Adams, A. and So, J., *A Claim in Senegal, 1720–1994.* 1996
Gellar, Sheldon, *Democracy in Senegal: Tocquevillian Analytics in Africa.* 2005
Phillips, L. C., *Historical Dictionary of Senegal.* 2nd ed, revised by A. F. Clark. 1995

National Statistical Office: Direction de la Prévision et de la Statistique, BP 116, Dakar.
Website (French only): http://www.ansd.sn

Explore the world at
www.statesmansyearbook.com

SERBIA

Republika Srbija
(Republic of Serbia)

Capital: Belgrade
Population projection, 2015: 9·81m.
GNI per capita, 2011: (PPP$) 10,236
HDI/world rank: 0·766/59
Internet domain extension: .rs

KEY HISTORICAL EVENTS

The Serbs were converted to Orthodox Christianity by the Byzantines in 891, before becoming a prosperous independent state under Stevan Nemanja (1167–96). A Serbian Patriarchate was established at Peć during the reign of Stevan Dušan (1331–55). Dušan's attempted conquest of Constantinople failed and after he died many Serbian nobles accepted Turkish vassalage. The reduced Serbian state under Prince Lazar received the coup de grace at Kosovo on St Vitus' Day, 1389. However, Turkish preoccupations with a Mongol invasion and wars with Hungary delayed the incorporation of Serbia into the Ottoman Empire until 1459.

The Turks tolerated the Orthodox church though the Patriarchate was abolished in 1776. The native aristocracy was eliminated and replaced by a system of fiefdoms held in return for military or civil service. Local self-government based on rural extended family units (*zadruga*) continued. In its heyday the Ottoman system was no harder on the peasantry than the Christian feudalism it had replaced, but with the gradual decline of Ottoman power, corruption, oppression and reprisals led to economic deterioration and social unrest.

In 1804, murders carried out by mutinous Turkish infantry provoked a Serbian rising under Djordje Karadjordje. By the Treaty of Bucharest (1812) Russia agreed that Serbia, known as Servia until 1918, should remain Turkish. The Turks reoccupied Serbia with ferocious reprisals. A rebellion in 1815 was led by Miloš Obrenović who, with Russian support, won autonomy for Serbia within the Ottoman empire. Obrenović had Karadjordje murdered in 1817. After he was forced to grant a constitution establishing a state council he abdicated in 1839. In 1842 a coup overthrew the Obrenovićs and Alexander Karadjordjević was elected ruler. He was deposed in 1858.

During the reign of the western-educated Michael Obrenović (1860 until his assassination in 1868) the foundations of a modern centralized and militarized state were laid, and the idea of a 'Great Serbia', first enunciated in Prime Minister Garašanin's *Draft Programme* of 1844, took root. Milan Obrenović, adopting the title of king, proclaimed formal independence in 1882. He suffered defeats against Turkey (1876) and Bulgaria (1885) and abdicated in 1889. After Alexander Obrenović was assassinated in 1903 Peter Karadjordjević brought in a period of stable constitutional rule.

Serbia's aim to secure an outlet to the sea was thwarted by Austria. Annexing Bosnia in 1908, Austria forced the Serbs to withdraw from the Adriatic after the first Balkan war (1912).

The assassination of Archduke Franz Ferdinand of Austria in Sarajevo on 28 June 1914 precipitated the First World War. In the winter of 1915–16 the Serbian army was forced to retreat to Corfu, where the government aimed at a centralized, Serb-run state. But exiles from Croatia and Slovenia wanted a South Slav federation. This was accepted by the victorious Allies as the basis for the new state. The Croats were forced by the pressure of events to join Serbia and Montenegro on 1 Dec. 1918. From 1918–29 the country was known as the Kingdom of the Serbs, Croats and Slovenes.

A constitution of 1921 established an assembly but the trappings of parliamentary rule could not bridge the gulf between Serbs and Croats. The Croat peasant leader, Radić, was assassinated in 1928; his successor, Vlatko Maček, set up a separatist assembly in Zagreb. Faced with the threat of his kingdom's dissolution, on 6 Jan. 1929 the king suspended the constitution and established a royal dictatorship, redrawing provincial boundaries without regard for ethnicity and renaming the country Yugoslavia. In Oct. 1934 he was murdered by a Croat extremist while on an official visit to France.

During the regency of Prince Paul, the government pursued a pro-fascist line. On 25 March 1941 Paul was persuaded to adhere to the Axis Tripartite Pact. On 27 March he was overthrown by military officers in favour of the boy king Peter. Germany invaded on 6 April. Within ten days Yugoslavia surrendered; king and government fled to London. Resistance was led by a royalist group and the communist-dominated partisans of Josip Broz, nicknamed Tito. Having succeeded in liberating Yugoslavia, Tito set up a Soviet-type constitution but he was too independent for Stalin who sought to topple him. However, Tito made a *rapprochement* with the west and it was the Soviet Union under Khrushchev that had to extend the olive branch in 1956. Yugoslavia was permitted to evolve its 'own road to socialism'. Collectivization of agriculture was abandoned and Yugoslavia became a champion of international 'non-alignment'. A collective presidency came into being with the death of Tito in 1980.

Dissensions in Kosovo between Albanians and Serbs, and in parts of Croatia between Serbs and Croats, reached crisis point after 1988. In 1988 the Serbian presidency fell to the nationalist Slobodan Milošević, who aimed at the creation of an enlarged Serbian state. On 25 June 1991 Croatia and Slovenia declared independence. Fighting began in Croatia between Croatian forces and Serb irregulars from Serb-majority areas of Croatia. On 25 Sept. the UN Security Council imposed a mandatory arms embargo on Yugoslavia. A three-month moratorium agreed at EU peace talks on 30 June having expired, both Slovenia and Croatia declared their independence from the Yugoslav federation on 8 Oct. After 13 ceasefires had failed, a fourteenth was signed on 23 Nov. under UN auspices. A Security Council resolution of 27 Nov. proposed the deployment of a UN peacekeeping force if the ceasefire was kept. Fighting, however, continued. On 15 Jan. 1992 the EU recognized Croatia and Slovenia as independent states. Bosnia and Herzegovina was recognized on 7 April 1992 and Macedonia on 8 April 1993. A UN delegation began monitoring the ceasefire on 17 Jan. and the UN Security Council on 21 Feb. voted to send a 14,000-strong peacekeeping force to Croatia and Yugoslavia. On 27 April 1992 Serbia and Montenegro created a new federal republic of Yugoslavia.

On 30 May, responding to further Serbian military activities in Bosnia and Croatia, the UN Security Council voted to impose sanctions. In mid-1992 NATO committed air, sea and eventually land forces to enforce sanctions and protect humanitarian relief operations in Bosnia. At a joint UN-EC peace conference on Yugoslavia held in London on 26–27 Aug. some 30 countries and all the former republics of Yugoslavia endorsed a plan to end the fighting in Croatia and Bosnia, install UN supervision of heavy weapons, recognize the borders of Bosnia and Herzegovina and return refugees. At a further conference at Geneva on 30 Sept. the Croatian and Yugoslav presidents agreed to make efforts to bring about a peaceful solution in Bosnia, but fighting continued. Following the Bosnian-Croatian-Yugoslav (Dayton) agreement all UN sanctions were lifted in Nov. 1995.

In July 1997 Slobodan Milošević switched his power base to become president of federal Yugoslavia. The former Yugoslav foreign minister, Milan Milutinović, succeeded Milošević as Serbian President. Meanwhile, in Montenegro, the pro-western Milo Đukanović succeeded a pro-Milošević president.

Following the break-up of Yugoslavia, in March 1998 a coalition government was formed between the Socialist Party of Slobodan Milošević and the ultra-nationalist Serb Radical Party.

Kosovo

In 1998 unrest in Kosovo, with its largely Albanian population, led to a bid for outright independence. Violence flared resulting in what a US official described as 'horrendous human rights violations', including massive shelling of civilians and destruction of villages. A US-mediated agreement for negotiations to proceed during an interim period of autonomy allowed for food and medicine to be delivered to refugees. American support for a degree of autonomy (short of independence), accepted in principle by President Milošević, lifted the immediate threat of NATO air strikes. Further outbreaks of violence in early 1999 were followed by the departure of the 800-strong team of international 'verifiers' of the fragile peace. Peace talks in Paris broke down without a settlement though subsequently Albanian freedom fighters accepted terms allowing them broad autonomy. The sticking point on the Serbian side was the international insistence on having 28,000 NATO-led peacemakers in Kosovo to keep apart the warring factions. Meanwhile, the scale of Serbian repression in Kosovo persuaded the NATO allies to take direct action. On the night of 24 March 1999 NATO aircraft began a bombing campaign against Yugoslavian military targets. Further Serbian provocation in Kosovo caused hundreds of thousands of ethnic Albanians to seek refuge in neighbouring countries. On 9 June after 78 days of air attacks NATO and Yugoslavia signed an accord on the Serb withdrawal from Kosovo, and on 11 June NATO's peacekeeping force, KFOR, entered Kosovo.

When the general election held on 24 Sept. 2000 resulted in a victory for the opposition democratic leader Vojislav Koštunica, President Milošević demanded a second round of voting. A strike by miners at the Kolubara coal mine on 29 Sept. was followed by a mass demonstration in Belgrade on 5 Oct. when the parliament building was set on fire. On 6 Oct. Slobodan Milošević accepted defeat. He was arrested on 1 April 2001 after a 30-hour confrontation with the authorities. On 28 June he was handed over to the United Nations War Crimes Tribunal in The Hague to face charges of crimes against humanity. Prime Minister Zoran Žižić resigned the next day. On 12 Feb. 2002 the trial of Slobodan Milošević, on charges of genocide and war crimes in the Balkans over a period of nearly ten years, began at the International Criminal Tribunal in The Hague. However, Milošević's death in March 2006 from heart failure while in the custody of the Tribunal brought proceedings to a close.

On 14 March 2002 Serbia and Montenegro agreed to remain part of a single entity called Serbia and Montenegro, thus relegating the name Yugoslavia to history. The agreement was ratified in principle by the federal parliament and the republican parliaments of Serbia and Montenegro on 9 April 2002. The new union came into force on 4 Feb. 2003. The country's fragile political structures came under the spotlight on 12 March 2003 when the Serbian prime minister Zoran Đinđić, a key figure in the toppling of the Milošević regime, was shot dead on the stairway of Serbia's chief government building.

After 4 Feb. 2006 Serbia and Montenegro had the right to vote for independence. Following a referendum on 21 May 2006, Montenegro declared independence and was recognized as such by Serbia on 15 June. As a result of Montenegro's vote, Serbia's parliament formally proclaimed Serbia to be independent for the first time since 1918 on 5 June 2006.

TERRITORY AND POPULATION

Serbia is bounded in the northwest by Croatia, in the north by Hungary, in the northeast by Romania, in the east by Bulgaria, in the south by Macedonia and in the west by Albania, Montenegro and Bosnia and Herzegovina. According to the constitution it includes the two provinces of Kosovo and Metohija in the south and Vojvodina in the north. With these Serbia's area is 88,361 sq. km; without, 55,968 sq. km. Population at the 2011 census was (with Vojvodina but without Kosovo and Metohija) 7,186,862; population density per sq. km, 92·8. Population at the 2011 census without both Vojvodina and Kosovo and Metohija was 5,255,053. The population was 56·4% urban in 2011.

The UN gives a projected population for 2015 of 9·81m.

The capital is Belgrade (2011 census population, 1,166,763). Populations (2011 census) of principal towns:

Belgrade	1,166,763	Subotica	97,910
Novi Sad	231,798	Zrenjanin	76,511
Priština	200,000[1]	Pančevo	76,203
Niš	183,164	Čačak	73,331
Kragujevac	150,835	Novi Pazar	66,527

[1]2011 estimate.

The official language is Serbian.

SOCIAL STATISTICS

In 2008 there were a total of 69,083 live births in Serbia (without Kosovo and Metohija), a rate of 9·4 per 1,000 inhabitants. There were 102,711 deaths (14·0 per 1,000) and 38,285 marriages (5·2 per 1,000). Population growth rate, 2007, −0·4%. Life expectancy in 2007 was 71·6 years for men and 76·3 for women. Infant mortality was 6 per 1,000 live births in 2010.

CLIMATE

Most parts have a central European type of climate, with cold winters and hot summers. 2000, Belgrade, Jan. –1·0°C, July 23·5°C. Annual rainfall 367·7 mm.

CONSTITUTION AND GOVERNMENT

A new constitution was approved in a referendum held on 28–29 Oct. 2006, with 53·0% of the electorate (and 96·6% of those voting) supporting the proposed constitution. It declares the province of Kosovo and Metohija an integral part of Serbia and grants Vojvodina financial autonomy. Kosovo Albanians were not able to vote. Turnout was 54·9%. The *President* is elected by universal suffrage for not more than two five-year terms. There is a 250-member single-chamber *National Assembly*.

National Anthem

'Bože pravde' ('God of Justice'); words by Jovan Đorđević, tune by Davorin Jenko.

RECENT ELECTIONS

Parliamentary elections were held on 6 May 2012. The 'Let's Get Serbia Moving' coalition, led by the Serbian Progressive Party (SNS), won 73 seats (24·0% of the vote); the 'Choice for a Better Life' coalition, led by the Democratic Party (DS), 67 (22·1%); the Socialist Party of Serbia–Party of United Pensioners of Serbia–United Serbia (SPS–PUPS–JS) coalition, 44 (14·5%); the Democratic Party of Serbia (DSS), 21 (7·0%); the 'U-Turn' coalition, led by the Liberal Democratic Party (LDP), 19 (6·5%); the 'United Regions of Serbia' coalition, led by G17+, 16 (5·5%). The remaining seats went to smaller parties and independents. Turnout was 57·8%.

First round of presidential elections were also held on 6 May 2012. Boris Tadić (DS) took 25·3% of the vote, followed by Tomislav Nikolić (SNS) with 25·0%, Ivica Dačić (SPS) with 14·2% and Vojislav Koštunica (DSS) with 7·4%. There were eight other candidates. In the run-off held on 20 May 2012, Nikolić won with 49·5% of the vote against 47·3% for Tadić. Turnout was 57·8% in the first round and 46·3% in the second.

CURRENT GOVERNMENT

President: Tomislav Nikolić; b. 1952 (ind.; sworn in 31 May 2012).

In March 2013 the multi-party coalition government comprised:

Prime Minister and Minister of the Interior: Ivica Dačić; b. 1966 (SPS; took office on 27 July 2012).

First Deputy Prime Minister in Charge of Defence, Security, Fight Against Corruption and Crime, and Minister of Defence: Aleksandar Vučić. *Deputy Prime Minister and Minister of Foreign and Domestic Trade, and Telecommunications:* Rasim Ljajić. *Deputy Prime Minister and Minister of Labour, Employment and Social Policy:* Jovan Krkobabić. *Deputy Prime Minister and Minister of Regional Development and Local Self-Government:* Verica Kalanović. *Deputy Prime Minister for European Integration:* Suzana Grubješić.

Minister of Finance and Economy: Mlađan Dinkić. *Foreign Affairs:* Ivan Mrkić. *Transport:* Milutin Mrkonjić. *Construction and Urban Planning:* Velimir Ilić. *Justice and Public Administration:* Nikola Selaković. *Agriculture, Forestry and Water Management:* Goran Knežević. *Education, Science and Technological Development:* Žarko Obradović. *Health:* Slavica Đukić-Dejanović. *Energy, Development and Environmental Protection:* Zorana Mihajlović. *Culture and Information:* Bratislav Petković. *Natural Resources, Mining and Spatial Planning:* Milan Bačević. *Youth and Sport:* Alisa Marić. *Minister without Portfolio:* Sulejman Ugljanin.

Government Website: http://www.srbija.gov.rs

CURRENT LEADERS

Tomislav Nikolić

Position
President

Introduction
Tomislav Nikolić was elected president in May 2012. Once an ultra-nationalist, he reinvented himself as a moderate supporting Serbian EU membership while maintaining Serbian claims over Kosovo.

Early Life
Born on 15 Dec. 1952 in Kragujevac, Tomislav Nikolić was educated in Kragujevac and Novi Sad. He worked in building construction from 1971–78 before taking up an investment and management role with a company in Kragujevac.

A member of the National Radical Party, in Feb. 1991 he oversaw its merger with the Serb National Renewal Party to form the ultra-nationalist Serb Radical Party (SRS). In the same year he became party vice-president under the leadership of Vojislav Šešelj and was elected to the national assembly.

The SRS formed periodic alliances with the governments of Slobodan Milošević in the 1990s, endorsing military action in Bosnia and Kosovo and advocating the formation of a 'Greater Serbia'. Nikolić served as deputy prime minister of Serbia from March 1998–Nov. 1999 and deputy premier of the Federal Republic of Yugoslavia from 1999–2000. In 2000, 2003 and 2004 he launched unsuccessful bids for the national presidency, each time unsuccessfully.

In Feb. 2003 Nikolić took over the leadership of the SRS after Šešelj was indicted for war crimes by the International Criminal Tribunal for the Former Yugoslavia (ICTY). In 2007 he was briefly speaker of the national assembly and in 2008 again stood for the presidency, losing narrowly to the incumbent Boris Tadić. In Sept. 2008 Nikolić broke with the SRS and formed the Serbian Progressive Party (SNS) following Šešelj's accusations that he had betrayed SRS ideology by expressing pro-Western sentiment (specifically support for EU accession).

Grooming the SNS as a more moderate party than the SRS, Nikolić showed willingness to accept EU conditions for membership and sought to build ties with Western nations, rather than looking towards Russia as Šešelj advocated. Nonetheless, Nikolić did not abandon the nationalist agenda, arguing that Kosovo remained an integral part of Serbia and denying Kosovan sovereignty after its declaration of independence in Feb. 2008.

In April 2011 Nikolić led protests, including an aborted hunger strike, in an unsuccessful bid to bring about early elections amid growing public frustration at rampant corruption and the country's grave economic situation. He contested his fifth presidential election in 2012, winning despite controversy over his political and educational track records. He defeated Tadić in a second round of voting on 20 May 2012 and resigned the leadership of the SNS four days later.

Career in Office
Nikolić was sworn into office on 31 May 2012. Ivica Dačić, leader of the Socialist Party of Serbia, became prime minister at the head of a coalition with the Serbian Progressive Party (SNS).

At home Nikolić must contend with the struggling economy and high unemployment. On the international stage he faces opposition to Serbia's EU membership application, notably over the refusal to recognize Kosovo's independence. Relations with a number of neighbouring countries remain strained, a situation not eased by his accusations of ICTY bias against Serbia.

DEFENCE

Conscription was abolished with effect from 1 Jan. 2011. In 2008 military expenditure totalled US$1,034m. (US$139 per capita), representing 2·1% of GDP.

Army

Strength (2007) 11,180 including 1,724 conscripts. The head-quarters of the Army are at Niš. Equipment includes 224 main battle tanks.

Air Force

The Air Force and Air Defence is based at Zemun. Strength (2007) 4,155.

ECONOMY

Agriculture accounted for 13% of GDP in 2007, industry 28% and services 59%.

Overview

Since democratic reforms in 2000, Serbia has enjoyed strong growth and positive structural changes. GDP grew on average by 5·5% per year in the first five years after transition and remained strong prior to the global downturn that began in 2008. Inflation has fallen significantly since the 1990s. Large scale privatization, a restructuring of the banking sector and improvements in the business environment have underpinned economic progress.

FDI has been significant in bolstering the economy, averaging 6·7% of GDP in the period 2002–07 according to the World Bank. However, political uncertainty has hindered progress and reform, while high unemployment (at around 17% in 2010) is a concern. EU accession is a major objective and an application for membership was submitted in Dec. 2009.

Growth stalled with the global financial crisis and a Stand-By Arrangement with the IMF was signed in 2009. The economy began to recover in 2010 as exports rebounded. However, economic developments in the EU pose risks to the recovery. Increasing inflation and the high external deficit are other key concerns.

Currency

The unit of currency of Serbia is the *dinar* (RSD) of 100 *paras*. On 1 Jan. 2001 Yugoslavia adopted a managed float regime. The National Bank of Yugoslavia began setting the exchange rate of the dinar daily in the foreign exchange market on the previous day. In Kosovo both the dinar and the euro are legal tender. Inflation was 12·4% in 2008, decreased to 6·2% in 2010 but went up again to 11·1% in 2011.

Budget

(Excluding Kosovo and Metohija). Central government revenue totalled 1,049,548m. dinars in 2008 (914,361m. dinars in 2007) and expenditure 1,043,623m. dinars (867,589m. dinars in 2007).

VAT is 20% (reduced rate 8%).

Performance

The economy contracted by 3·5% in 2009 but grew by 1·0% in 2010 and 1·6% in 2011. Total GDP was US$45·8bn. in 2011.

Banking and Finance

The National Bank is the bank of issue responsible for the monetary policy, stability of the currency of Serbia, the dinar, control of the money supply and prescribing the method of maintaining internal and external liquidity. The dinar became fully convertible in May 2002. The present *Governor* of the National Bank of Serbia is Jorgovanka Tabaković.

Foreign debt amounted to US$32,222m. in 2010, representing 84·3% of GNI.

There is a stock exchange in Belgrade.

ENERGY AND NATURAL RESOURCES

Electricity

In 2007 installed capacity was an estimated 11·8m. kW. Production in 2007 was 36·52bn. kWh and consumption per capita 3,485 kWh.

Minerals

(Excluding Kosovo and Metohija, in 1,000 tonnes). 2008: lignite, 38,284; copper ore, 8,680 tonnes.

Agriculture

(Excluding Kosovo and Metohija). In 2007 the sown area was 3,095,006 ha. Yields in 2008 (in 1,000 tonnes): maize, 6,158; sugar beets, 2,300; wheat (2007), 1,864; potatoes, 844; plums, 607; sunflower seeds, 454; grapes, 373; soybeans, 351; barley (2007), 259; apples, 236. Livestock in 2009 (in 1,000): cattle, 1,002; pigs, 3,631; sheep, 1,504; poultry, 22,821.

Forestry

Forests covered 2·71m. ha. in 2010 (31% of the land area). Timber cut in 2007: 2,981,000 cu. metres.

Fisheries

Total catch, 2008, 3,197 tonnes (all from inland fishing).

INDUSTRY

Production, 2008 (excluding Kosovo and Metohija, in 1,000 tonnes): cement, 2,843; crude steel, 1,662; pig iron, 1,582. 2007: cotton fabrics, 19,557,000 sq. metres; woollen fabrics, 11,000 sq. metres; passenger cars, 9,400 units; trucks, 473 units.

Labour

In April 2010 there were 2,412,106 workers employed (without Kosovo and Metohija), including 549,816 in agriculture, forestry and water management; 405,485 in manufacturing; 346,038 in wholesale and retail trade and repair; 170,146 in health and social work; 148,943 in education; and 142,514 in transport, storage and communications. In April 2010 there were 1,582,455 employees and 641,712 self-employed persons. Average annual salary in 2009 (without Kosovo and Metohija) was 31,733 dinars. Unemployment in April 2010 (without Kosovo and Metohija) was running at 19·2%.

INTERNATIONAL TRADE

Imports and Exports

In 2007 imports (c.i.f.) totalled US$18,553·6m. (US$13,172·3m. in 2006); exports (f.o.b.), US$8,824·7m. (US$6,427·9m. in 2006). The leading import sources in 2007 were Russia (14·2%), Germany (11·8%), Italy (9·7%) and China (7·4%). Principal export markets in 2007 were Italy (12·4%), Bosnia and Herzegovina (11·8%), Montenegro (10·8%) and Germany (10·6%). The main imports are machinery and transport equipment, chemicals, petroleum, and iron and steel. Main exports are food and livestock, machinery and transport equipment, iron and steel, and chemicals.

COMMUNICATIONS

Roads

The length of roads in 2007 was 39,184 km, including 374 km of motorway and 5,133 km of main roads. In 2007 there were 1,476,600 passenger cars in use, 162,900 lorries and vans, 24,900 motorcycles and mopeds, and 8,900 buses and coaches. There were 962 deaths as a result of road accidents in 2007.

Rail

Railways are operated by Železnice Srbije; total length of network in 2005 (excluding Kosovo and Metohija) was 3,809 km. In 2009, 8·4m. passengers and 10·4m. tonnes of freight were carried (without Kosovo and Metohija). In Sept. 2010 the state-owned railway companies of Serbia, Croatia and Slovenia announced the creation of a joint venture called Cargo 10 to improve the management of freight trains along the route known as Corridor 10 that passes through all three countries.

Civil Aviation

The national airline (and the former national carrier of Yugoslavia) is Jat Airways. In Jan. 2010 it flew to 30 destinations in 23 countries. The main airport is Belgrade Nikola Tesla

Airport, which handled 2,386,402 passengers and 7,690 tonnes of cargo in 2009.

Telecommunications

There were 3,110,300 landline telephone subscriptions in 2010 (383·0 per 1,000 inhabitants) and 9,915,300 mobile phone subscriptions (1,220·8 per 1,000 inhabitants). In 2009, 36·7% of households had internet access. In March 2012 there were 3·2m. Facebook users.

SOCIAL INSTITUTIONS

Justice

In 2002 there was one Supreme Court, 30 District Courts and 138 Communal Courts with 2,180 judges and 17 Economic Courts of Law with 237 judges.

Education

In 2008–09 there were: 158,000 pupils and 11,000 teaching staff in pre-primary education; 282,000 pupils and 17,000 teaching staff in primary education; 604,000 pupils and 60,000 teaching staff in secondary education; and 236,000 students and 15,000 teaching staff in tertiary education.

RELIGION

Serbia has been traditionally Orthodox. Muslims are found in the south as a result of the Turkish occupation. The Serbian Orthodox Church with its seat in Belgrade has 27 bishoprics within the boundaries of former Yugoslavia and 12 abroad (five in the USA and Canada, five in Europe and two in Australia). The Serbian Orthodox Church numbers about 2,000 priests. Its *Patriarch* is Irinej (enthroned 23 Jan. 2010).

CULTURE

World Heritage Sites

There are four sites on the UNESCO World Heritage List: Stari Ras and Sopoćani (inscribed on the list in 1979); Studenica Monastery (1986); Medieval monuments in Kosovo (2004 and 2006); and Gamzigrad-Romuliana, the palace of Galerius (2007).

Press

In 2006 there were 11 daily newspapers. The two largest newspapers are *Blic* (readership of 650,000 in 2006) and *Večernje novosti* (636,000). In the 2010 *World Press Freedom Index*, compiled by Reporters Without Borders, Serbia was ranked 85th out of 178 countries.

Tourism

In 2011, 764,000 non-resident tourists stayed in holiday accommodation (up from 682,000 in 2010 and 645,000 in 2009). There were 280 hotels in 2011, with 16,034 rooms and 25,841 beds.

Festivals

The National Day is 15 Feb. The two biggest music festivals are the Exit Festival (July) in Novi Sad and Guča Trumpet Festival (Aug.). Jazz festivals are held annually in Belgrade (Oct.) and Novi Sad (Nov.) Belgrade Summer Festival (BELEF) is held throughout July and Aug. and features theatre, music and art. FEST (the international film festival in Belgrade) is in Feb.–March and the International Theatre Festival of Belgrade (BITEF) takes place in Sept.

DIPLOMATIC REPRESENTATIVES

Of Serbia in the United Kingdom (28 Belgrave Sq., London, SW1X 8QB)
Ambassador: Dejan Popović.

Of the United Kingdom in Serbia (Resavska 46, 11000 Belgrade)
Ambassador: Michael Davenport.

Of Serbia in the USA (2134 Kalorama Rd, NW, Washington, D.C., 20008)
Ambassador: Vladimir Petrović.

Of the USA in Serbia (Kneza Miloša 50, 11000 Belgrade)
Ambassador: Michael D. Kirby.

Of Serbia to the United Nations
Ambassador: Feodor Starčević.

Of Serbia to the European Union
Ambassador: Vacant.

FURTHER READING

Anzulovic, Branimir, *Heavenly Serbia: From Myth to Genocide.* 1999
Cox, John, *The History of Serbia.* 2002
Judah, Tim, *The Serbs: History, Myth and the Destruction of Yugoslavia.* 1997
Pavolwitch, Stevan K., *Serbia: The History of an Idea.* 2002
Stojanovic, Svetozar, *Serbia: The Democratic Revolution.* 2003
Thomas, Robert, *Serbia Under Milošević: Politics in the 1990s.* 1999
Vladisavljević, Nebojša, *Serbia's Antibureaucratic Revolution: Milošević, the Fall of Communism and Nationalist Mobilization.* 2008

National Statistical Office: Statistical Office of the Republic of Serbia, 5 Milana Rakića St., 11000 Belgrade.
Website: http://webrzs.stat.gov.rs/axd/index.php

Kosovo and Metohija

KEY HISTORICAL EVENTS

Kosovo has a large ethnic Albanian majority. Following Albanian-Serb conflicts, the Kosovo and Serbian parliaments adopted constitutional amendments in March 1989 surrendering much of Kosovo's autonomy to Serbia. Renewed Albanian rioting broke out in 1990. The Prime Minister and six other ministers resigned in April 1990 over ethnic conflicts. In July 1990, 114 of the 130 Albanian members of the National Assembly voted for full republican status for Kosovo but the Serbian National Assembly declared this vote invalid and unanimously voted to dissolve the Kosovo Assembly. Direct Serbian rule was imposed causing widespread violence. Western demands for negotiations in granting Kosovo some kind of special status were rejected. Ibrahim Rugova, the leader of the main Albanian party, the Democratic League of Kosovo (LDK), declared himself 'president' demanding talks on independence. In 1998 armed conflict between Yugoslavia and the Kosovo Liberation Army led 200,000 people, or a tenth of the population of the province, to flee the fighting. Further repression by Serbian forces led to the threat of NATO direct action. Air strikes against Yugoslavian military targets began on 24 March 1999. Retaliation against Albanian Kosovars led to a massive exodus of refugees. On 9 June after 78 days of air attacks NATO and Yugoslavia signed an accord on the Serb withdrawal from Kosovo, and on 11 June NATO's peacekeeping force, KFOR, entered Kosovo. In Nov. 2001 the Organization for Security and Co-operation in Europe mounted elections for a provincial assembly that were deemed fair and democratic.

The worst fighting between Serbs and Albanians since the end of the war claimed 19 lives in March 2004. The following March President Ramush Haradinaj resigned when he was indicted to face charges of war crimes at the UN tribunal in The Hague. In 2006 the UN sponsored talks on the future status of Kosovo and ethnic Serbian and Kosovan leaders met for the first time since 1999. In Oct. 2006 Serbia held a referendum that approved a new constitution keeping Kosovo as an integral part of the country. However, the Kosovan Albanian majority rejected the poll. In Feb. 2007 the UN announced plans for Kosovo's eventual independence that were immediately rejected by Serbia and heavily revised in July 2007 after Russian protests.

In Nov. 2007 Hashim Thaçi's Democratic Party of Kosovo won elections that were boycotted by the Serb minority. On 17 Feb. 2008 Kosovo made a unilateral declaration of independence though international recognition was patchy. Independence was recognized by the USA and EU member countries including Germany, France and the UK. However, Serbia rejected the declaration as did Russia, while Spain refused to support it and China expressed 'grave concern'. As such, Kosovo lacks recognition as a sovereign country by the UN and the EU although on 22 July 2010 the UN International Court of Justice gave an advisory ruling that the independence declaration was legal.

TERRITORY AND POPULATION

Area: 10,887 sq. km. The capital is Priština. The 1991 and 2002 censuses were not taken. According to the Statistical Office of Kosovo the population in Jan. 2008 was an estimated 2·2m., of whom 92% are Albanians; density, 202 per sq. km. About 5% of the population are Serbs, mostly in the north of the country. Population estimate of Priština, 2002, 564,800. Other major towns include Prizren, Peć and Kosovska Mitrovica.

SOCIAL STATISTICS

Statistics for 2007: live births, 33,312; deaths, 6,681; marriages, 16,824; divorces, 1,558.

CONSTITUTION AND GOVERNMENT

The constitution of Serbia defines the autonomous province of Kosovo and Metohija as an 'integral part' of the territory of Serbia with 'substantial autonomy'.

In April 2008 Kosovo's 120-member multi-ethnic parliamentary assembly (first convened on 10 Dec. 2001) approved a new constitution, which came into force on 15 June 2008. Its promulgation followed Kosovo's unilateral declaration of independence from Serbia in Feb. 2008 as the Republic of Kosovo. The constitution envisages a handover of executive power from the UN, which has been responsible for administration in the region since 1999, to the majority ethnic Albanian parliament, under the supervision of an EU team. It also specifies that Kosovo will 'have no territorial claim against, and shall seek no union with' any other state. However, the constitution was rejected by Serbia as it considers Kosovo to be part of its sovereign territory, while Russia claimed any EU involvement would be illegal as it had yet to be approved by the UN Security Council. In June 2008 an ethnic Serb assembly set up a rival administration in Mitrovica. The UN referred the declaration of independence to the International Court of Justice in Oct. 2008.

Kosovo came under interim international administration on 10 June 1999, in accordance with the terms of UN Security Council resolution 1244. The United Nations Interim Administration Mission in Kosovo (UNMIK) administered Kosovo following the arrival of KFOR (NATO-led peacekeeping force) and remained in place even after the Feb. 2008 declaration of independence. The unilateral declaration received a mixed response from the international community, with Serbia, Russia and Spain prominent among those nations who refused to recognize it. In Dec. 2008 EULEX, the European Union Rule of Law Mission in Kosovo, took over responsibility from the UN for policing, justice and customs services, with the agreement of Serbia.

There is a 120-member multi-ethnic parliamentary assembly, which first convened on 10 Dec. 2001. The new assembly brought together representatives of Kosovo's ethnic Albanian majority and its Serbian minority for the first time in more than a decade. Of the 120 members, 100 are directly elected with ten seats reserved for ethnic Serbs and ten for other ethnic minorities.

RECENT ELECTIONS

Parliamentary elections held on 12 Dec. 2010; turnout was 47·5%. The Democratic Party of Kosovo won 34 seats with 32·1% of the vote, the Democratic League of Kosovo 27 with 24·7%, Self-Determination 14 with 12·7%, the Alliance for New Kosovo 12 with 11·0%, the New Kosovo Coalition 8, Independent Liberal Party 8 and the United Serbia List 4. Ten other parties won three seats or fewer. The election was seriously hampered by a number of irregularities.

Atifete Jahjaga was elected president by parliament on 7 April 2011, receiving 80 out of a possible 120 votes.

CURRENT GOVERNMENT

President: Atifete Jahjaga; b. 1975 (ind.; since 7 April 2011).

Prime Minister: Hashim Thaçi; b. 1969 (Democratic Party of Kosovo; since 9 Jan. 2008).

Head of EULEX Kosovo: Bernd Borchardt (Germany; since 1 Feb. 2013).

INTERNATIONAL RELATIONS

As at March 2013 Kosovo was recognized as an independent state by 99 of the 193 UN members.

ECONOMY

Budget

Central government revenue in 2011 was €1,650·0m. (€1,417·3m. in 2010) and expenditure €1,612·3m. (€1,458·4m. in 2010).

Performance

Total GDP was US$6·5bn. in 2010. Real GDP growth was 3·9% in 2010 and 5·0% in 2011.

Banking and Finance

In Aug. 1999 the Deutsche Mark became legal tender alongside the Yugoslav dinar, and on 1 Jan. 2002 the euro became the official currency of Kosovo. The Serb dinar is also legal tender in Kosovo but is used only by ethnic Serbs.

ENERGY AND NATURAL RESOURCES

Electricity

Electricity production in 2004 was 3·48bn. kWh.

Minerals

Production (2010): lignite, 7,958,000 tonnes; limestone, 2,606,000 cu. metres.

Agriculture

The cultivated area in 2004 was 264,340 ha. Yields in 2004 (in 1,000 tonnes): wheat, 197; maize, 92; potatoes, 56; peppers, 40; plums, 16. Livestock (in 1,000): cattle, 241; dairy cows, 129; sheep, 124; pigs, 47; chickens, 1,617.

Forestry

Timber cut in Kosovo in 2007 was 192,000 cu. metres according to official data; however, significant illegal logging also takes place.

INDUSTRY

Production (2008): cement, 590,000 tonnes (estimate).

Labour

In 1997 there were 120,763 workers in the public sector, including 54,223 in industry, 10,471 in education and culture, 9,245 in trade, catering and tourism, 8,933 in transport and communications, 7,880 in communities and organizations and 1,526 in commercial services. In Oct. 1997 in the private sector there were 35,869 self-employed and employed, including 15,113 in trade, 5,023 in catering and tourism, 4,364 in arts and crafts and 2,006 in transport and communications. Average monthly salary in Dec. 1998 was 1,066 dinars.

COMMUNICATIONS

Roads

In 2007 there were 1,924 km of main and regional roads in Kosovo. Total vehicle registrations in April 2003 were 234,297.

Rail

Total length of railways in 2007 was 430 km, of which 97 km were freight only. In 2005 the state-owned Kosovo Railways was established to take over the running of railways from the UN Mission in Kosovo.

Civil Aviation

There is an international airport at Priština, which handled 990,952 passengers in 2007.

Telecommunications

In 2003 there were 101,059 main telephone lines and 315,000 mobile phones.

SOCIAL INSTITUTIONS

Justice

In 2004 there were five district courts and 23 municipal courts.

Education

In 2002–03 there were: 465 pre-schools and nurseries with 1,018 teachers and 20,365 pupils; 992 primary schools with 15,733 teachers and 299,934 pupils; and 128 secondary schools with 5,439 teachers and 89,387 pupils. In 2001–02, 21,216 students attended the University of Priština.

RELIGION

The population of Kosovo is predominantly Muslim.

CULTURE

World Heritage Sites

Kosovo and Metohija has one site on the UNESCO World Heritage List: medieval monuments in Kosovo (inscribed on the list in 2004 and 2006) that contain frescoes dating from the 13th through to the 17th century.

FURTHER READING

Judah, Tim, *Kosovo: War and Revenge.* 2000
King, Iain and Mason, Whit, *Peace at any Price: How the World Failed Kosovo.* 2006
Malcolm, N., *Kosovo: a Short History.* 2nd ed. 2002
Vickers, M., *Between Serb and Albanian: A History of Kosovo.* 1998

Vojvodina

KEY HISTORICAL EVENTS

After the Battle of Kosovo in 1389 Turkish attacks on the Balkans led to a mass migration of Serbians to Vojvodina. Turkish rule ended after their 1716–18 war with Austria and the Požarevac peace agreement. In exchange for acting as frontier protectors, the Austrians granted the people of Vojvodina religious autonomy. However, in 1848 a short-lived revolution led to a Serbian alliance with the Croats. Vojvodina was briefly declared an independent dukedom. After the First World War Vojvodina became part of the first Yugoslav state. In 1974 President Tito granted autonomy to Vojvodina, but this status was brought into question after Vojvodina's largely anti-Milošević provincial assembly resigned in 1988. In 1989 the Serbian government, led by Slobodan Milošević, stripped Vojvodina of most of its autonomous rights and secured Serbian control. After the fall of Milošević in 2000 there was a growing demand for autonomy, which was granted by statute in Dec. 2009.

TERRITORY AND POPULATION

Area: 21,506 sq. km. The capital is Novi Sad. Population of Vojvodina at the 2002 census, 2,031,992 (1,321,807 Serbs, 290,207 Hungarians); density, 94·5 per sq. km. Population of Novi Sad, 2002, 191,405. There are six official languages: Serbian, Hungarian, Slovak, Romanian, Croatian and Rusyn.

SOCIAL STATISTICS

In 2003 there were a total of 20,381 live births in Vojvodina, a rate of 9·9 per 1,000 inhabitants. There were 29,741 deaths (14·4 per 1,000) and 11,127 marriages (5·4 per 1,000). Rate of natural increase in 2003: –4·5 per 1,000.

CONSTITUTION AND GOVERNMENT

Vojvodina's autonomous status, rescinded by the Yugoslav government in 1990, was restored by statute in Dec. 2009. Novi Sad was defined as the province's chief administrative centre and Vojvodina is permitted to establish representative offices in Europe with the consent of the Serbian government.

RECENT ELECTIONS

Parliamentary elections were held on 6 and 20 May 2012. Choice for a Better Vojvodina won 58 seats (48·3% of the vote), ahead of Let's Get Vojvodina Moving with 22 seats and 18·3%, the coalition of the Socialist Party of Serbia, the Party of United Pensioners of Serbia, United Serbia and the Social Democratic Party of Serbia with 13 seats and 10·8%, the League of Social Democrats of Vojvodina with 10 and 8·3%, the Alliance of Vojvodina Hungarians with 7 and 5·8%, the Serbian Radical Party with 5 and 4·2%, the Democratic Party of Serbia with 4 and 3·3% and Vojvodina U-Turn with 1 and 0·8%.

CURRENT GOVERNMENT

President of the Assembly: István Pásztor; b. 1956 (in office since 22 June 2012).
    Chairman of the Executive Council: Bojan Pajtić; b. 1970 (in office since 30 Oct. 2004).

Government Website: http://www.vojvodina.gov.rs

ECONOMY

Budget

In 2003 total revenue was 66,538m. dinars; total expenditure was 59,234m. dinars.

ENERGY AND NATURAL RESOURCES

Electricity

Electricity production in 2004 was 526m. kWh.

Agriculture

The cultivated area in 2004 was an estimated 1,648,000 ha. Yield (in 1,000 tonnes): maize, 3,726; sugar beets, 2,689; wheat, 1,563; potatoes, 283. Livestock estimates (in 1,000): cattle, 212; sheep, 195; pigs, 1,190; poultry, 5,823.

Forestry

Timber cut in 2004: 699,000 cu. metres.

INDUSTRY

Production (2004): cement, 908,000 tonnes; fertilizers, 651,166 tonnes; crude petroleum, 640,000 tonnes; plastics, 179,000 tonnes.

Labour

In Oct. 2004 there were 748,809 persons employed, including 175,673 in manufacturing; 163,738 in agriculture, forestry and water works supply; 114,868 in wholesale and retail trade and repair; 42,170 in construction; 40,979 in health and social work; and 38,176 in transport, storage and communications. In Oct. 2004 there were 569,488 employees and 154,801 self-employed persons. Unemployment was 18·8% in Oct. 2004.

SOCIAL INSTITUTIONS

Education

In 2003–04 there were: 620 kindergartens and pre-schools with 46,696 pupils; 535 primary schools with 178,905 pupils; 124 secondary schools with 78,008 pupils; and 50 high and higher schools with 46,273 students.

SEYCHELLES

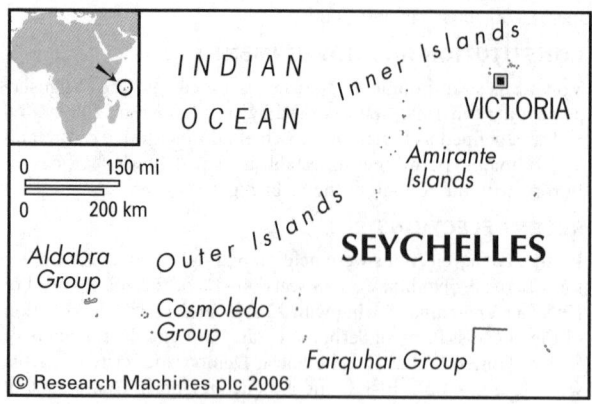

© Research Machines plc 2006

Republic of Seychelles

Capital: Victoria
Population, 2010: 91,000
GNI per capita, 2011: (PPP$) 16,729
HDI/world rank: 0·773/52
Internet domain extension: .sc

KEY HISTORICAL EVENTS

The Seychelles were colonized by the French in 1756 to establish spice plantations to compete with the Dutch monopoly. The islands were captured by the English in 1794. Subsequently, Britain offered to return Mauritius and its dependencies (which included the Seychelles) to France if that country would renounce all claims in India. France refused and the Seychelles were formally ceded to Britain as a dependency of Mauritius. In Nov. 1903 the Seychelles archipelago became a separate British Crown Colony. Internal self-government was achieved on 1 Oct. 1975 and independence as a republic within the British Commonwealth on 29 June 1976.

The first president, James Mancham, was deposed in a coup on 5 June 1977. Under the new constitution, the Seychelles People's Progressive Front became the sole legal party. There were several attempts to overthrow the regime but in 1979 and 1984 Albert René was the only candidate in the presidential elections. Under the new constitution approved in June 1993, President René was re-elected against two opponents. He stood down in 2004 and was succeeded by James Michel.

TERRITORY AND POPULATION

The Seychelles consist of 115 islands in the Indian Ocean, north of Madagascar, with a combined area of 455 sq. km (175 sq. miles) in two distinct groups and a 2010 census population of 90,945. The Granitic group of 40 islands cover 244 sq. km (94 sq. miles); the principal island is Mahé, with 160 sq. km (62 sq. miles) and 78,539 inhabitants (2010 census), the other inhabited islands of the group being Praslin, La Digue, Silhouette, Fregate, North and Denis, which together had 12,406 inhabitants in 2010.

The Outer or Coralline group comprises 75 islands spread over a wide area of ocean between the Mahé group and Madagascar, with a total land area of 211 sq. km (81 sq. miles). The main islands are the Amirante Isles (including Desroches, Poivre, Daros and Alphonse), Coetivy Island and Platte Island, all lying south of the Mahé group; the Farquhar, St Pierre and Providence Islands, north of Madagascar; and Aldabra, Astove, Assumption and the Cosmoledo Islands, about 1,000 km southwest of the Mahé

group. Aldabra (whose lagoon covers 142 sq. km), Farquhar and Desroches were transferred to the new British Indian Ocean Territory in 1965, but were returned by Britain to the Seychelles on the latter's independence in 1976.

Victoria, the chief town, had a census population of 26,450 in 2010. In 2011, 55·9% of the population were urban.

The official languages are Creole, English and French but 91% of the population speak Creole.

SOCIAL STATISTICS

2006 births, 1,467; deaths, 664. 2006 rates per 1,000 population, birth, 17·3; death, 7·8; infant mortality (2010), 12 per 1,000 births. Annual population growth rate, 2000–05, 0·4%. Life expectancy at birth in 2008 was estimated to be 68 years for males and 79 for females. Fertility rate, 2008, 1·9 births per woman.

CLIMATE

Though close to the equator, the climate is tropical. The hot, wet season is from Dec. to May, when conditions are humid, but southeast trades bring cooler conditions from June to Nov. Temperatures are high throughout the year. Victoria, Jan. 80°F (26·7°C), July 78°F (25·6°C). Annual rainfall 95" (2,287 mm).

CONSTITUTION AND GOVERNMENT

Under the 1979 constitution the Seychelles People's Progressive Front (SPPF) was the sole legal Party. There is a unicameral People's Assembly consisting of 34 seats, of which 25 are directly elected and nine are allocated on a proportional basis, and an executive *President* directly elected for a five-year term (with a maximum of three successive terms). A constitutional amendment of Dec. 1991 legalized other parties. A commission was elected in July 1992 to draft a new constitution. The electorate was some 50,000; turnout was 90%. The SPPF gained 14 seats on the commission, the Democratic Party, eight; the latter, however, eventually withdrew. At a referendum in Nov. 1992 the new draft constitution failed to obtain the necessary 60% approval votes. The commission was reconvened in Jan. 1993. At a further referendum on 18 June 1993 the constitution was approved by 73·6% of votes cast. The elections of 1993 were the first multiparty ones since 1974.

National Anthem

'Koste Seselwa' ('Come Together Seychellois'); words and tune by D. F. M. André and G. C. R. Payet.

RECENT ELECTIONS

In parliamentary elections held on 29 Sept.–1 Oct. 2011 the ruling Seychelles People's Progressive Front (SPPF) won all 31 available seats with 88·6% of the vote. The Popular Democratic received 10·9% of the vote but failed to win any seats. The main opposition party, the Seychelles National Party, boycotted the election. Turnout was 74·3%; 31·9% of the votes cast were spoiled.

In presidential elections held on 19–21 May 2011 incumbent James Michel (People's Party) won 55·5% of the vote against Wavel Ramkalawan (Seychelles National Party) with 41·4%, Philippe Boullé (ind.) with 1·7% and Ralph Volcere (New Democratic Party) 1·4%. Turnout was 85·3%.

CURRENT GOVERNMENT

On 14 April 2004 France-Albert René stepped down as president, a post he had held since 1977.

President: James Michel; b. 1944 (SPPF; took office on 14 April 2004 and then elected in July 2006 and re-elected in May 2011).

The President is *Minister of Defence, Legal Affairs, Information, Youth and Hydrocarbons.*

Vice-President, Minister of Public Administration, and Information Technology and Communications: Danny Faure.

In March 2013 the government comprised:

Minister of Education: Macsuzy Mondon. *Environment and Energy:* Rolph Payet. *Finance, Trade and Investment:* Pierre Laporte. *Foreign Affairs:* Jean-Paul Adam. *Health:* Mitcy Larue. *Home Affairs and Transport:* Joel Morgan. *Natural Resources and Industry:* Peter Sinon. *Labour and Human Resources Development:* Idith Alexander. *Land Use and Housing:* Christian Lionnet. *Social Affairs, Community Development and Sport:* Vincent Meriton. *Tourism and Culture:* Alain St Ange.

Government Website: http://www.egov.sc

CURRENT LEADERS

James Michel

Position
President

Introduction
Former Vice-President James Michel came to power in April 2004, handpicked by then president France-Albert René to succeed to the presidency on René's retirement after 27 years in power. Michel had been vice-president since 1996 and had previously held a variety of ministerial positions. He was elected to the post in July 2006 and re-elected in May 2011.

Early Life
James Alix Michel was born in the Seychelles on 18 Aug. 1944. He was a teacher before deciding to pursue a career in politics. His profile rose in the mid-1970s because of his involvement in the country's booming tourism industry.

In 1976, just before independence, he joined René's left-of-centre Seychelles People's United Party (SPUP)—renamed the Seychelles People's Progressive Front (SPPF) in 1978. He was a member of the SPUP's central committee when the party staged a bloodless coup in 1977, overthrowing the country's first president, James Mancham, and replacing him with René. There followed a 16-year one-party socialist dictatorship, during which time Michel held a series of important ruling party and ministerial positions. For several periods he was in charge of the highly-regulated Seychellois economy. On René's retirement in April 2004, Michel was sworn in as president.

Career in Office
Despite his allegiance to René, Michel was under pressure to speed up the country's democratization process, which had begun with multi-party elections in 1993. He also pledged to introduce more open political dialogue, particularly over matters concerning the Seychellois economy, and to develop the private sector. In Jan. 2005 Michel granted the Emirates Group the rights to operate non-stop flights three times a week between the Seychelles and Dubai in order to enhance the tourism industry and to increase trade for the business and cargo communities.

In March 2005 Michel detailed his foreign policy, underpinned by a desire to cement stronger regional ties in the Indian Ocean region—particularly in light of the Seychelles' exit in July 2004 from the Southern African Development Community, ostensibly because of high membership fees. He particularly focused on strengthening relations with Mauritius, working alongside the then Mauritian prime minister, Paul Bérenger, to strengthen the Indian Ocean Commission. Michel has favoured increased promotion of the Seychelles as a high-quality and safe tourist resort, and sought to make the country a leader in environmental issues.

In June 2005 Michel announced plans for a new national pension fund and a scheme to set-up a savings account of R1,000 for every Seychellois child, both of which came into effect in Jan. 2006.

After a close electoral contest in July 2006 he retained the presidency, defeating his Seychelles National Party rival. In the May 2007 parliamentary elections the SPPF retained its majority. Michel brought the poll forward after opposition MPs had boycotted parliamentary proceedings over moves to ban political parties (and also religious groups) from owning radio stations.

Confronted by a balance of payments and public debt crisis in 2008, Michel launched an economic reform programme with the help of the International Monetary Fund, which approved a two-year US$26m. support arrangement in Nov. As part of the programme, the government floated the currency and lifted foreign exchange controls. In Jan. 2009 he appealed for debt relief from international creditors as the Seychelles economy was hit by reduced tourist traffic and the effects of the global financial crisis. In Nov. that year the World Bank agreed a US$9m. loan to help restore economic stability.

In response to the expansion of Somali piracy in the Indian Ocean, the Seychelles government entered into agreements with European Union countries and the USA to enhance naval and air patrol and surveillance to deter attacks on international shipping.

In Aug. 2010 the Seychelles ratified the statute establishing the International Criminal Court.

Michel was re-elected for a further term in May 2011 and his party won all 31 seats in parliamentary polling in Sept.–Oct. following an opposition boycott.

DEFENCE

The Defence Force comprises all services. Personnel (2007) Army, 200; paramilitary national guard, 250; paramilitary coastguard, 200 including 80 marines.

Defence expenditure totalled US$8m. in 2008 (US$98 per capita), representing 0·9% of GDP.

Coastguard
The Seychelles Coast Guard superseded the former navy and air force in 1992. Based at Port Victoria it operates nine patrol and coastal combatants.

ECONOMY

Services accounted for 71·5% of GDP in 2006, industry 25·5% and agriculture 3·0%.

Overview
Since the early 1990s the government has favoured a free market economy. Tourism forms the backbone of the economy followed by the fisheries sector, where tuna fishing dominates. More than 70% of foreign exchange earnings derive from tourism, which employs 30% of the labour force. Aware of its over-reliance on imports, the government is making efforts to diversify the economy by encouraging agriculture and domestic (small-scale) industrial production. In recent years the country has attempted to develop an offshore sector as a third pillar of the economy and to position itself as a provider of business and financial services.

The economy suffered a balance of payments and debt crisis in 2008, although strong fiscal adjustment in 2009 and 2010 as well as international assistance have helped the economy recover. Tourism, communications and construction have been the main drivers of recent growth. The debt level, as well as piracy, climate change and increasing food and energy prices, challenge longer-term recovery.

Currency
The unit of currency is the *Seychelles rupee* (SCR) divided into 100 *cents*. In July 2005 foreign exchange reserves were US$44m. and total money supply was 1,220m. rupees. Following deflation in 2010 of 2·4% there was inflation in 2011 of 2·6%. 2008 and 2009 had seen inflation of 37·0% and 31·7% respectively.

Budget

Fiscal budget in 1m. rupees, for calendar years:

	2003	2004	2005	2006
Total revenues and grants	1,867·2	1,891·1	2,167·6	2,476·2
Total expenditures	1,597·2	1,788·8	1,815·6	2,301·8

VAT at 15% was introduced on 1 Jan. 2013.

Performance

There was a recession in both 2003 and 2004, with the economy contracting by 5·9% and 2·9% respectively. In 2005 the economy recovered to grow by 8·0%. Since then there has been growth of 8·9% in 2006 and 9·7% in 2007 but 2008 saw the economy shrink by 1·0%. There was then real GDP growth of 0·5% in 2009, 6·7% in 2010 and 5·1% in 2011. Total GDP was US$1·0bn. in 2011.

Banking and Finance

The Central Bank of Seychelles (established in 1983; *Governor*, Caroline Abel), which is the bank of issue, and the Development Bank of Seychelles provide long-term lending for development purposes. There are also six commercial banks, including two local banks (the Seychelles Savings Bank and the Seychelles International Mercantile Banking Co-operation or NOUVOBANQ), and four branches of foreign banks (Barclays Bank, Banque Française Commerciale, Habib Bank and Bank of Baroda).

Foreign debt totalled US$1,308m. in 2007.

ENERGY AND NATURAL RESOURCES

Environment

Carbon dioxide emissions from the consumption and flaring of fossil fuels were the equivalent of 12·4 tonnes per capita in 2008.

Electricity

Installed capacity on Mahé and Praslin combined was an estimated 95,000 kW in 2007. Production in 2007 was 271m. kWh and consumption per capita 3,071 kWh.

Water

There are two raw water reservoirs, the Rochon Dam and La Gogue Dam, which have a combined holding capacity of 1·05bn. litres.

Agriculture

Since the rise of the tourism industry in the 1970s there has been a general decline in the production of traditional cash crops, notably cinnamon bark, of which 158 tonnes were exported in 2002 (down from 289 tonnes in 1998). 261 tonnes of tea (green leaf) were produced in 2003. Other crops grown for local consumption include bananas, oranges, cassava, sweet potatoes, paw-paw, yams and vegetables. The staple food crop, rice, is imported from Asia. Livestock, 2003 estimates: 18,000 pigs, 5,000 goats, 1,000 cattle and 1m. chickens. In 2002 there were approximately 1,000 ha. of arable land and 6,000 ha. of permanent crop land.

Forestry

In 2010 forests covered 41,000 ha., or 88% of the total land area. The Ministry of Environment has a number of ongoing forestry projects that aim at preserving and upgrading the local system. There are also a number of terrestrial nature reserves including three national parks, four special reserves and an 'area of outstanding natural beauty'.

Fisheries

The fisheries sector is the Seychelles' largest foreign exchange earner. In 2006 it accounted for 52% of export revenue. Total catch in 2010 was 87,108 tonnes, exclusively from sea fishing. 2006 fish production (in tonnes) included: canned tuna, 40,222; fish landed, 4,050; crustaceans, 606. Fisheries exports in 2006 amounted to 1,096·8m. rupees, of which: canned tuna, 1,030·4m.; fish meal 25·1m.; frozen prawns, 23·7m.

INDUSTRY

Local industry is expanding, the major development in recent years being in tuna canning; in 2003 output totalled 36,436 tonnes, up from 7,500 tonnes in 1995. This is followed by brewing, with 7·1m. litres in 2003. Other main activities include production of cigarettes (50m. in 2003), dairy production, prawn production, paints and processing of cinnamon barks.

Labour

Some 41% of employed persons work in the services sector. In 2003, 7,195 people worked in trade, restaurants and hotels. In 2003, 17,425 were formally employed in the private sector, 11,973 in the public sector and 5,463 in the parastatal sector.

INTERNATIONAL TRADE

Imports and Exports

In 2005 imports (c.i.f.) totalled 3,712·2m. rupees (2,731·8m. rupees in 2004); exports (f.o.b.), 1,868·6m. rupees (1,599·9m. rupees in 2004). Domestic exports constitute around two-thirds of exports and re-exports a third. Principal imports: mineral fuel; food and live animals; machinery and transport equipment; manufactured goods; chemicals. Principal origins of imports, 2005: Saudi Arabia (23·0%), Spain (7·8%), Singapore (7·6%), France (6·5%). Principal exports: canned tuna; petroleum products; medicaments and medical appliances; fish meal (animal feed); frozen prawns. Main export markets (for domestic exports), 2005: UK (45·4%), France (23·1%), Italy (12·4%), Germany (10·2%).

COMMUNICATIONS

Roads

In 2006 there were 502 km of roads, of which 96·0% were surfaced. There were 6,800 private cars in 2006 (80 per 1,000 inhabitants), 2,600 commercial vehicles, 300 taxis and 215 buses.

Rail

There are no railways in the Seychelles.

Civil Aviation

Seychelles International airport is on Mahé. In 2003 Air Seychelles flew on domestic routes and to Comoros, Dubai, Frankfurt, Johannesburg, London, Malé, Mauritius, Mumbai, Munich, Paris, Réunion, Rome, Singapore and Zürich. In 2003 scheduled airline traffic of Seychelles-based carriers flew 12m. km, carrying 413,000 passengers (187,000 on international flights). In 2001 Seychelles International handled 598,133 passengers (330,726 on international flights) and 5,607 tonnes of freight.

Shipping

The main port is Victoria, which is also a tuna-fishing and fuel and services supply centre. In Jan. 2009 there were 12 ships of 300 GT or over registered, totalling 165,000 GT. Sea freight (2006) comprised: imports, 534,000 tonnes; exports, 4,604,000 TEUs (twenty foot equivalent units); transhipments (fish), 74,000 tonnes.

Telecommunications

There were 100,000 telephone subscribers in 2007, or 1,154·7 per 1,000 population. Mobile phone subscribers numbered 77,300 in 2007. There were 32,000 internet users (including 3,500 broadband subscribers) in 2007.

SOCIAL INSTITUTIONS

Justice

The death penalty was abolished for all crimes in 1993. The population in penal institutions in 2007 was 221 (270 per 100,000 of national population).

Education

Adult literacy was an estimated 91·8% in 2009 (91·4% among males and 92·3% among females). Education is free from five to

12 years in primary schools, and 13 to 17 in secondary schools. There are three private schools providing primary and secondary education and one dealing only with secondary learning. Education beyond 18 years of age is funded jointly by the government and parents. The University of Seychelles opened in Sept. 2009. In 2007 there were 8,864 pupils and 711 teaching staff in primary schools, 7,816 pupils and 588 teaching staff in secondary schools and (2003) 1,652 students and 193 teaching staff at polytechnic level.

Public expenditure on education came to 6·6% of GNI in 2006 or 12·6% of total government spending.

Health
In 2003 there were 107 doctors, 16 dentists and 422 nurses. In 2003 there were seven hospitals with 419 beds. The health service is free.

Welfare
Social security is provided for people of 63 years and over, for the disabled and for families needing financial assistance. There is also assistance via means testing for those medically unfit to work and for mothers who remain out of work for longer than their designated maternity leave. Orphanages are also subsidized by the government.

RELIGION
87% of the inhabitants are Roman Catholic, the remainder of the population being followers of other religions (mainly Anglicans, with some 7th Day Adventists, Bahais, Muslims, Hindus, Pentecostalists, Jehovah's Witnesses, Buddhists and followers of the Grace and Peace church).

CULTURE
World Heritage Sites
Entered on the UNESCO World Heritage List in 1982, the four coral islands of Aldabra Atoll protect a shallow lagoon. A heritage site since 1983, the Vallée de Mai Nature Reserve is a natural palm forest on the small island of Praslin.

Press
In 2006 there was one daily newspaper (circulation of 3,000), as well as three weekly papers.

Tourism
Tourism is the main foreign exchange earner. Visitor numbers were a record 208,034 in 2012, up from 194,753 in 2011.

Festivals
There are numerous religious festivals including Kavadi, an annual procession organized by the Hindu Association of Seychelles. Secular festivals include the annual Youth Festival, Jazz Festival, Creole Festival, Kite Festival and the Subios Festival, a celebration of the underwater world.

DIPLOMATIC REPRESENTATIVES
Of the Seychelles in the United Kingdom (4th Floor, 11 Grosvenor Cres., London, SW1X 7EE)
High Commissioner: Marie-Pierre Lloyd.

Of the United Kingdom in the Seychelles (3rd Floor, Oliaji Trade Centre, Francis Rachel St., PO Box 161, Victoria, Mahé)
High Commissioner: Lindsay Skoll.

Of the Seychelles in the USA and to the United Nations (800 2nd Ave., Suite 400C, New York, NY 10017)
Ambassador: Marie-Louise Potter.

Of the USA in the Seychelles
Ambassador: Shari Villarosa (resides in Port Louis, Mauritius).

Of the Seychelles to the European Union
Ambassador: Vivianne Fock Tave.

FURTHER READING
Scarr, D., *Seychelles Since 1970: History of a Slave and Post-Slavery Society.* 2000

National Statistical Office: Statistics and Database Administration Section (MISD), P. O. Box 206, Victoria, Mahé. *Seychelles in Figures*
Website: http://www.nsb.gov.sc

SIERRA LEONE

© Research Machines plc 2006

Republic of Sierra Leone

Capital: Freetown
Population projection, 2015: 6·51m.
GNI per capita, 2011: (PPP$) 737
HDI/world rank: 0·336/180
Internet domain extension: .sl

KEY HISTORICAL EVENTS

The ancestors of the Bulom, Nalou, Baga and Krim people are thought to have been the earliest settlers in coastal Sierra Leone. The Kissi and Gola lived inland to the east and the Limba inhabited the foothills of the Wara Wara mountains from at least the 10th century AD. Following the break-up of the Malian empire in the late 14th century much of Sierra Leone was settled by the Mande whose domination and interaction with the original inhabitants gave rise to new ethnic groups, including the Vai and Loko.

The Portuguese mariner Alvaro Fernandez sailed into the Rokel estuary near present-day Freetown in 1447 and located a source of fresh water. The inlet became a trading post for Portuguese, Dutch, French and British explorers on their way to and from India. Having established plantations in America and the Caribbean, European colonists sought labour from the Atlantic slave trade, much of it centred on Sierra Leone. In 1672 British traders from the Royal African Company established trading forts along the coast, although they repeatedly came under attack from European rivals.

When Britain outlawed slavery in 1772 the government established Granville Town (later Freetown) as a home for freed slaves. In 1791 the Sierra Leone Company was founded as a trading concession, with Sierra Leone becoming a British crown colony in 1808. By 1850 more than 40,000 freed slaves, many of them from Nova Scotia and Jamaica, had settled in Freetown and surrounding areas, intermarrying with natives and Europeans to forge the 'Krio' culture. British administrators and Krio traders led expeditions inland in the late 19th century and on 21 Aug. 1896 the hinterland was declared a British protectorate, governed by Frederic Cardew. The early years of the 20th century saw the development of the country's interior while Lebanese merchants increasingly dominated business and trade.

Milton Margai became the prime minister in 1960 and led the country to independence on 27 April 1961. In 1967 Dr Siaka Stevens' All People's Congress (APC) came to power via the ballot box, despite an attempted military coup. Sierra Leone became a republic on 19 April 1971 with Stevens as executive president. Following a referendum in June 1978 a new constitution was instituted under which the ruling APC became the sole legal party.

On 28 Nov. 1985 Joseph Saidu Momoh succeeded Stevens as president. He presided over an economic collapse and the rise of the Revolutionary United Front (RUF). Under the leadership of Foday Sankoh (and with alleged links to the Liberian president, Charles Taylor), the RUF began taking control of diamond mines in the east of the country. President Momoh was overthrown in a military coup led by Valentine Strasser on 29 April 1992 and a National Provisional Ruling Council (NPRC) was set up, although it was unable to prevent the country's slide into civil war. By 1995 the RUF controlled much of the countryside and the NPRC sought assistance from foreign mercenaries.

The NPRC agreed to presidential and parliamentary elections in April 1996 to establish a civilian government. Ahmad Tejan Kabbah became president but was ousted in May 1997 by the Armed Forces Revolutionary Council, led by Maj. Johnny Paul Koroma. In Feb. 1998 a Nigerian-led intervention force (ECOMOG) launched an offensive against the military junta. On 10 March President Kabbah returned from exile in Guinea promising a 'new beginning' but in Jan. 1999 the country again erupted into civil war.

In July 1999 the government reached an agreement with the rebel movement to bring the civil war to an end. Under the terms of the accord, the RUF was to gain four key government posts along with control of the country's mineral resources in return for surrendering its weapons. However, civil war resumed in early 2000 and, responding to a government appeal, British forces were sent to back up the UN peacekeeping force (UNAMSIL). Foday Sankoh was captured and handed over to UN forces in May 2000. In July 2001 the RUF formally recognized the civil government of President Kabbah. In Jan. 2002 Kabbah declared the war over. Re-elected in May 2002, he has been at the forefront of reconstruction efforts although corruption remains an obstacle to progress. The last UN peacekeeping troops left Sierra Leone in Dec. 2005.

TERRITORY AND POPULATION

Sierra Leone is bounded on the northwest, north and northeast by Guinea, on the southeast by Liberia and on the southwest by the Atlantic Ocean. The area is 71,740 sq. km (27,699 sq. miles). Population (census 2004), 4,976,871; density, 69·4 per sq. km. In 2011, 38·8% of the population were urban.

The UN gives a projected population for 2015 of 6·51m.

The capital is Freetown, with a 2004 census population of 772,873.

Sierra Leone is divided into three provinces and one area:

	Sq. km	Census 2004	Capital	Census 2004
Eastern Province	15,553	1,191,539	Kenema	128,402
Northern Province	35,936	1,745,553	Makeni	82,840
Southern Province	19,694	1,092,657	Bo	149,957
Western Area	557	947,122	Freetown	772,873

The provinces are divided into districts as follows: Bo, Bonthe, Moyamba, Pujehun (Southern Province); Kailahun, Kenema, Kono (Eastern Province); Bombali, Kambia, Koinaduga, Port Loko, Toukolili (Northern Province).

The principal peoples are the Mendes (26% of the total) in the south, the Temnes (25%) in the north and centre, the Konos, Fulanis, Bullomes, Korankos, Limbas and Kissis. English is the official language; a Creole (Krio) is spoken.

SOCIAL STATISTICS

2008 estimates: births, 224,000; deaths, 88,000. Estimated birth rate in 2008 was 40·3 per 1,000 population; estimated death rate, 15·8. Annual population growth rate, 2000–08, 3·4%. Expectation of life at birth in 2007 was 48·5 years for females and 46·0 years for males. The World Health Organization's *World Health Statistics 2009* ranked Sierra Leone in last place in a 'healthy life expectancy' list, with an expected 35 years of healthy life for babies born in 2007. Infant mortality was 114 per 1,000 live births in 2010 (the highest in the world). Fertility rate, 2008, 5·2 births per woman.

CLIMATE

A tropical climate, with marked wet and dry seasons and high temperatures throughout the year. The rainy season lasts from about April to Nov., when humidity can be very high. Thunderstorms are common from April to June and in Sept. and Oct. Rainfall is particularly heavy in Freetown because of the effect of neighbouring relief. Freetown, Jan. 80°F (26·7°C), July 78°F (25·6°C). Annual rainfall 135" (3,434 mm).

CONSTITUTION AND GOVERNMENT

In a referendum in Sept. 1991 some 60% of the 2·5m. electorate voted for the introduction of a new constitution instituting multi-party democracy. The constitution has been amended several times since. The president, who is both head of state and head of government, is elected by popular vote for not more than two terms of five years. There is a 124-seat *National Assembly* (112 members elected by popular vote and 12 filled by paramount chiefs).

There is a *Supreme Council of State (SCS)* and a *Council of State Secretaries.*

National Anthem

'High We Exalt Thee, Realm of the Free'; words by C. Nelson Fyle, tune by J. J. Akar.

RECENT ELECTIONS

Presidential and parliamentary elections were held on 17 Nov. 2012.

In the presidential elections, Ernest Bai Koroma of the All People's Congress (APC) received 58·7% of the vote, Julius Maada Bio of the Sierra Leone People's Party (SLPP) 37·4%, Charles Margai of the People's Movement for Democratic Change (PMDC) 1·3%, Joshua Albert Carew of the Citizens Democratic Party 1·0%, Eldred Collins of the Revolutionary United Front 0·6%, Gibrilla Kamara of the People's Democratic Party 0·4%, Kandeh Baba Conteh of the Peace and Liberation Party 0·3%, Mohamed Bangura of the United Democratic Movement 0·2% and James Obai Fullah of the United National People's Party 0·2%. As Ernest Bai Komora received the 55% required to win outright a potential second round run-off was not needed. Turnout was 87·3%.

In the parliamentary elections, the All People's Congress won 67 of 124 seats with 53·7% of the vote and the Sierra Leone People's Party 42 with 38·3%; 12 seats were allocated for elected chiefs and three remained vacant.

CURRENT GOVERNMENT

President: Ernest Bai Koroma; b. 1953 (APC; sworn in 17 Sept. 2007 and re-elected 17 Nov. 2012).

Vice-President: Samuel Sam-Sumana.

In March 2013 the government comprised:

Minister of Agriculture, Food Security and Forestry: Sam Sesay. *Defence:* Major (retd) Paulo Conteh. *Education, Science and Technology:* Minkailu Bah. *Energy:* Oluniyi Robbin-Coker. *Finance and Economic Development:* Kaifala Marah. *Foreign Affairs:* Samura Kamara. *Health and Sanitation:* Miatta Kargbo. *Information and Communications:* Alhaji Alpha Sahid Bakar Kanu. *Internal Affairs:* Joseph B. Dauda. *Justice and Attorney General:* Frank Kargbo. *Labour and Social Security:* Matthew Teambo. *Lands, Country Planning and Environment:* Musa Tarawali. *Local Government and Rural Development:* Diana Konomanyi. *Marine Resources and Fisheries:* Capt. Allieu Pat Sowe. *Mines and Mineral Resources:* Alhaji Minkailu Mansaray. *Political and Public Affairs:* Alhaji Kemoh Sesay. *Social Welfare, Gender and Children's Affairs:* Moijue Kaikai. *Sports:* Paul Kamara. *Tourism and Cultural Affairs:* Peter Bayuku Konte. *Trade and Industry:* Usman Boie-Kamara. *Transport and Aviation:* Vandi Chidi Minah. *Water Resources:* Momodu Maligie. *Works, Housing and Infrastructural Development:* Alimamy P. Koroma. *Youth Affairs:* Alimamy Kamara.

Office of the President: http://www.statehouse.gov.sl

CURRENT LEADERS

Ernest Bai Koroma

Position
President

Introduction
Ernest Bai Koroma became president in Sept. 2007 following victory in the second round of elections. He won re-election outright in the first round of voting in Nov. 2012. He had previously run unsuccessfully for the presidency in 2002.

Early Life
Koroma was born in 1953 in Bombali, northern Sierra Leone. Though the region is predominantly Muslim, Koroma is a Christian. After primary and secondary schooling, he graduated in 1976 from Fourah Bay College, part of the University of Sierra Leone, in Freetown. He then worked as a teacher before joining the National Insurance Company in 1978. In 1985 he moved to the Reliance Insurance Trust Corporation (Ritcorps), becoming managing director in 1988 and holding the post for 14 years.

A latecomer to politics, Koroma was chosen as the APC's presidential candidate in March 2002 but was beaten into second place by Ahmad Tejan Kabbah of the SLPP. Under Koroma's leadership the APC grew in popularity and won a landslide victory in the 2004 local government elections, winning almost all seats in the densely-populated Western Area (which includes Freetown). In June 2005 Koroma was briefly stripped of his party leadership after the Supreme Court found him guilty of illegally altering his party's constitution. However, he was unanimously re-elected leader in Sept. 2005. Having spent most of his life in Freetown, his support base was strongest in the north though he also made inroads in the south ahead of the 2007 presidential election. In a run-off in Sept. 2007 Koroma won against Solomon Berewa and was sworn in on 17 Sept.

Career in Office
Koroma took over an almost bankrupt country, at the time ranked last out of 177 in the Human Development Index world rankings. He had to maintain the peace process that followed ten years of civil war which ended in 2001. After his inauguration, Koroma visited neighbouring Guinea and Liberia, which subsequently granted citizenship to 2,600 Sierra Leonean refugees.

Key challenges facing Koroma's presidency included promoting economic development and rejuvenating energy and public services. He pledged a zero tolerance approach to corruption and in Nov. 2007 his government published a report detailing inadequacies in tax collection, health care and security services, as well as suspect loans.

Also in Nov. 2007 Koroma signed a commercial investment deal to double the country's rutile (titanium ore) production capacity and secured China's cancellation of US$22m. worth of debt. A large offshore oil discovery was reported in 2009 and in 2010 large iron ore extraction leases were granted to two British companies.

In April 2011 Sierra Leone marked 50 years of independence from Britain.

Koroma was re-elected president for a second term in Nov. 2012, with the APC securing a majority in parliamentary polling held at the same time.

DEFENCE

In 2008 military expenditure totalled US$14m. (US$3 per capita), representing 0·6% of GDP.

The UN peacekeeping force (UNAMSIL) left Sierra Leone in Dec. 2005 after monitoring the ceasefire for six years. It was replaced in Jan. 2006 by the United Nations Integrated Office (UNIOSIL), and this in turn was replaced by the United Nations Integrated Peacebuilding Office in Sierra Leone (UNIPSIL) in Aug. 2008. UNIPSIL was appointed to help support the continuing peace process; its current mandate runs until 31 March 2014, when it is to draw down its operations.

Army

Following the civil war, the Army has disbanded and a new National Army has been formed with a strength of 10,500.

Navy

Based in Freetown there is a small naval force of around 200 operating four patrol and coastal combatants.

ECONOMY

Agriculture accounted for 47·4% of GDP in 2006 (one of the highest percentages of any country), industry 25·5% and services 27·1%.

Overview

The economy has made substantial progress since the end of civil war in 2002. Real GDP grew by just over 7% on average between 2004 and 2007, aided by remittances and investments from citizens working abroad. Agriculture, mining, construction and the services sectors make up the bulk of the economy. External assistance and foreign aid have helped finance the current account deficit, while programmes such as the Poverty Reduction and Growth Facility (2001–04) have improved public finance management.

However, growth decelerated in 2008 and 2009 as a result of the global economic downturn. In June 2010, the IMF approved a three-year Extended Credit Facility to spur growth in the medium-term. Structural reform priorities include improving tax administration, broadening the tax base, developing deeper financial services and strengthening public financial management. Social indicators remain poor, with the economy ranked 180th of 187 countries in the Human Development Index. Poverty is widespread, particularly in rural areas where 79% are estimated to live below the poverty line.

Currency

The unit of currency is the *leone* (SLL) of 100 *cents*. Inflation was 17·8% in 2010 and 18·5% in 2011. Exchange controls were liberalized in 1993. Total money supply in July 2005 was 346,027m. leones and foreign exchange reserves were US$94m.

Budget

In 2007 total revenue was 1,179bn. leones (42·7% grants and 21·8% import duties) and total expenditure 1,222bn. leones (63·4% current expenditures).

Performance

GNP per capita was US$183 in 2000 compared to US$536 in 1984, although it has since risen back up to US$322 in 2007. There was positive growth in 2000 for the first time since 1994, with a rate of 3·8%, rising to 18·2% in 2001 and further to 27·4% in 2002. Real GDP growth was 3·2% in 2009, 5·3% in 2010 and 6·0% in 2011. Total GDP in 2011 was US$2·2bn. Sierra Leone is among the world's bottom five countries in income and life expectancy.

Banking and Finance

The bank of issue is the Bank of Sierra Leone which was established in 1964 (*Governor*, Sheku Sambadeen Sesay). There are four commercial banks (two foreign).

Foreign debt totalled US$778m. in 2010, representing 40·8% of GNI.

ENERGY AND NATURAL RESOURCES

Environment

Carbon dioxide emissions from the consumption and flaring of fossil fuels in 2008 were the equivalent of 0·2 tonnes per capita. An *Environmental Performance Index* compiled in 2008 ranked Sierra Leone 147th in the world out of 149 countries analysed, with 40·0%. The index examined various factors in six areas—air pollution, biodiversity and habitat, climate change, environmental health, productive natural resources and water resources.

Electricity

Installed capacity was an estimated 52,000 kW in 2007. Production in 2007 was around 60m. kWh; consumption per capita in 2007 was an estimated 11 kWh.

Minerals

The chief minerals mined are diamonds (estimated at 692,000 carats in 2005) and rutile (73,600 tonnes in 2006). There are also deposits of gold, iron ore and bauxite. The presence of rich diamond deposits partly explains the close interest of neighbouring countries in the politics of Sierra Leone.

Agriculture

In 2007 the agricultural population was an estimated 3·34m., of which approximately 1·26m. were economically active. Cattle production is important in the north. Production (2003 estimates, in 1,000 tonnes): cassava, 377; rice, 250; palm oil, 36; bananas and plantains, 30; sweet potatoes, 25; sugarcane, 24. In 2007 there were an estimated 900,000 ha. of arable land and 80,000 ha. of permanent crops.

Livestock (2003 estimates): cattle, 400,000; sheep, 375,000; goats, 220,000; pigs, 52,000; chickens, 8m.

Forestry

In 2010 forests covered 2·73m. ha., or 38% of the total land area. Timber production in 2007 was 5·60m. cu. metres.

Fisheries

In 2008, 203,582 tonnes of fish were caught, of which marine fish 93% and freshwater fish 7%.

INDUSTRY

There are palm oil and rice mills; sawn timber, joinery products and furniture are produced.

Labour

The economically active workforce was 1,935,000 in 2004 (51% males). In 2004 around two-thirds of the economically active population were engaged in agriculture, fisheries, forestry and hunting. 68,250 persons were registered unemployed in 2004.

INTERNATIONAL TRADE

Imports and Exports

Total trade for 2008: imports (c.i.f.), 1,590,338m. leones; exports (f.o.b.), 641,797m. leones. Main exports are bauxite, diamonds, gold, coffee and cocoa. A UN-mandated diamond export certification scheme is in force. The Security Council has commended Sierra Leone's government for its efforts in monitoring trade to prevent diamonds from becoming a future source of conflict. The main import suppliers in 2001 were the UK (25·3%), Netherlands (10·1%), USA (7·9%), Germany (6·3%). Principal export markets in 2001 were Belgium (40·6%), USA (9·1%), UK (8·5%), Germany (7·8%).

COMMUNICATIONS

Roads

There were 11,300 km of roads in 2007 (8% paved). Much of the damage to the road network as a result of the civil war has now been repaired. In 2007 there were 16,400 passenger cars in use and 14,100 vans and lorries. There were 71 deaths as a result of road accidents in 2007.

Civil Aviation

Freetown Airport (Lungi) is the international airport. The national carrier is Leone Airways, operated by Arik Air (a Nigerian airline) under a joint venture agreement. In 2003 scheduled airline traffic of Sierra Leone-based carriers flew 1m. km, carrying 14,000 passengers (all on international flights).

Shipping

The port of Freetown has one of the largest natural harbours in the world. Iron ore is exported through Pepel, and there is a small port at Bonthe. In Jan. 2009 there were 248 ships of 300 GT or over registered, totalling 547,000 GT.

Telecommunications

In 2008 Sierra Leone had 31,500 main (fixed) telephone lines. The country's telecommunications network was virtually destroyed during the civil war, but since then the sector has been one of Sierra Leone's main successes. In 2007 internet users numbered 13,000. There were 1,009,000 mobile phone subscribers in 2008, up from 67,000 in 2002.

SOCIAL INSTITUTIONS

Justice

The High Court has jurisdiction in civil and criminal matters. Subordinate courts are held by magistrates in the various districts. Native Courts, headed by court Chairmen, apply native law and custom under a criminal and civil jurisdiction. Appeals from the decisions of magistrates' courts are heard by the High Court. Appeals from the decisions of the High Court are heard by the Sierra Leone Court of Appeal. Appeal lies from the Sierra Leone Court of Appeal to the Supreme Court, which is the highest court.

The death penalty is in force, but has not been used since Oct. 1998 when 24 soldiers were executed for their part in the May 1997 coup.

Education

The adult literacy rate in 2008 was 40%. Primary education is partially free but not compulsory. In 2004–05 there were 4,295 primary schools with 1,286,074 pupils and 19,316 teachers, and 282 secondary schools with 198,827 pupils and 6,622 teachers. There were also 196 technical/vocational establishments with 18,686 pupils and 918 staff. As a result of the 2005 Universities Act there are now two universities. The University of Sierra Leone comprises Fourah Bay College, the College of Medicine and Allied Health Sciences, and the Institute of Public Administration and Management. Njala University, until 2005 a constituent college of the University of Sierra Leone, is now an autonomous institution.

In 2005 public expenditure on education came to 3·9% of GNI.

Health

In 2000 there were 145 general practitioners, 1,331 nurses and five dentists. In 2000 there were 64 hospitals with 692 beds.

RELIGION

There were 2·49m. Muslims in 2001 (just under half the population). Traditional animist beliefs persist; there is also a Christian minority.

CULTURE

Press

In 2006 there were ten paid-for dailies with an average circulation of 22,000, plus 31 non-dailies. *For di People*, the oldest independent newspaper, had the highest circulation in 2006 (5,000).

Tourism

Tourism is in the initial stages of development. In 2007 there were 32,000 non-resident tourist arrivals by air, bringing revenue of US$22m. (excluding passenger transport).

DIPLOMATIC REPRESENTATIVES

Of Sierra Leone in the United Kingdom (41 Eagle St., London, WC1R 4TL)
High Commissioner: Edward Mohamed Turay.

Of the United Kingdom in Sierra Leone (6 Spur Rd, Freetown)
High Commissioner: Peter West.

Of Sierra Leone in the USA (1701 19th St., NW, Washington, D.C., 20009)
Ambassador: Bockari Kortu Stevens.

Of the USA in Sierra Leone (Leicester, Freetown IVG 798-5000)
Ambassador: Michael S. Owen.

Of Sierra Leone to the United Nations
Ambassador: Shekou Touray.

Of Sierra Leone to the European Union
Ambassador: Christian Sheka Kargbo.

FURTHER READING

Abdullah, Ibrahim, (ed.) *Between Democracy and Terror: The Sierra Leone Civil War.* 2004
Conteh-Morgan, E. and Dixon-Fyle, M., *Sierra Leone at the End of the Twentieth Century: History, Politics, and Society.* 1999
Ferme, M., *The Underneath of Things: Violence, History, and the Everyday in Sierra Leone.* 2001

National Statistical Office: Statistics Sierra Leone, A. J. Momoh Street, Tower Hill, P.M.B. 595, Freetown.
Website: http://www.statistics.sl

SINGAPORE

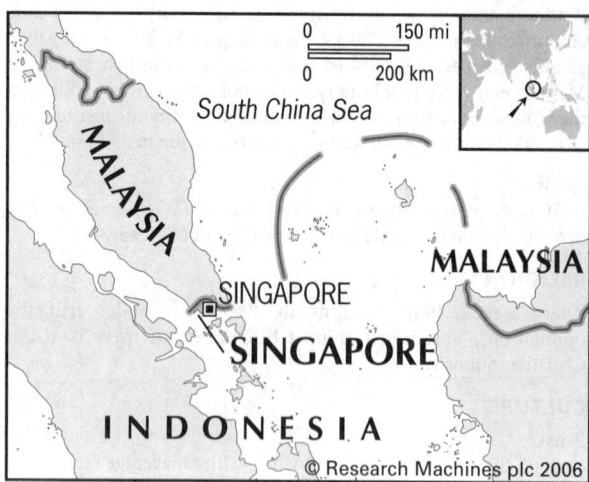

South China Sea

MALAYSIA

SINGAPORE

SINGAPORE

MALAYSIA

INDONESIA

© Research Machines plc 2006

Republik Singapura
(Republic of Singapore)
Population projection, 2015: 5·38m.
GNI per capita, 2011: (PPP$) 52,569
HDI/world rank: 0·866/26
Internet domain extension: .sg

KEY HISTORICAL EVENTS

The first known written account of the island of Singapore was by a Chinese explorer in the 3rd century AD. In a strategic location at the tip of the Malay peninsula, the island is likely to have been a port of call for sailors navigating the Melaka straits, although there is no evidence of settlement until the town of Temasek was described in the 13th century. Temasek was controlled by the Srivijaya Empire, centred on Palembang in Sumatra, during the 14th century, before falling to the Javanese Majapahit Empire. When Sultan Iksander Shah founded the Melaka Sultanate in the 1390s, he established a trading post on Singapore Island. After Portuguese forces sacked Melaka in 1511 the island came under the influence of the newly-created Johor Sultanate. In 1613 Singapore's main settlement was burnt down by Portuguese raiders and the island slipped into obscurity, with the ports of Melaka and Johor dominating the lucrative shipping routes that linked Europe and India with China and the East Indies.

In 1819 Sir Thomas Stamford Raffles, an administrator of the British East India Company based at the garrison of Bencoolen (Benkulu) in southwest Sumatra, established a trading settlement on Singapore Island to challenge Dutch supremacy of the region's ports. Raffles negotiated a deal with the ruling Sultan of Johor and left Col. William Farquhar in charge of the new settlement which was designated a free port. It grew rapidly, attracting Chinese, Malay, Bugis and Arab merchants who wanted to avoid the trade restrictions imposed at Dutch-controlled ports. In Aug. 1824 claims to Singapore by Britain and the East India Company were confirmed in treaties with the Dutch government and the Sultanate of Johor. Two years later Penang, Melaka and Singapore were combined as the Straits Settlements. With the opening of the Suez Canal in 1869 and the advent of ocean-going steamships, an era of prosperity began for Singapore. Growth was fuelled by the export of tin and rubber from the Malay peninsula, facilitated by the construction of a railway linking Singapore with Bangkok.

On 15 Feb. 1942, after capturing Malaya from the British in less than two months, Singapore was occupied by Japanese troops. Britain's return after Japan's surrender in 1945 was accompanied by growing nationalist sentiment. Britain formed the Federation of Malaya, administered by a high commissioner in Kuala Lumpur, and a separate Crown Colony of Singapore with a civil administration. Elections in both states in March 1948 were followed by outbreaks of anti-British violence, with the ethnic Chinese-dominated Malayan Communist Party demanding immediate independence and equality for all races. The British responded to 'the Emergency' by imposing hard-line restrictions on left-wing groups. Nevertheless, there was a gradual move towards self-government in Singapore in the 1950s and in June 1959, following the victory of the People's Action Party in the first legislative elections, Lee Kuan Yew became Singapore's first prime minister.

Singapore joined the Federation of Malaysia when it was formed in Sept. 1963 but tensions soon rose and in Aug. 1965 the Malaysian prime minister, Tunku Abdul Rahman, expelled Singapore from the Federation. It became an independent republic, which, under the authoritarian leadership of Lee Kuan Yew, saw rapid and sustained export-driven growth. Lee's resignation in Nov. 1990 saw Goh Chok Tong become prime minister. He served until Aug. 2004, when Lee Hsien Loong succeeded him.

TERRITORY AND POPULATION

The Republic of Singapore consists of Singapore Island and some 63 smaller islands. Singapore Island is situated off the southern extremity of the Malay peninsula, to which it is joined by a 1·1 km causeway carrying a road, railway and water pipeline across the Strait of Johor and by a 1·9 km bridge at Tuas, opened on 2 Jan. 1998. The Straits of Johor between the island and the mainland are 914 metres wide. The island is 682·3 sq. km in area, including the offshore islands.

Census of population (2000): Chinese residents 2,505,379 (76·8%), Malays 453,633 (13·9%), Indians 257,791 (7·9%) and others 46,406 (1·4%); resident population, 3,263,209. Total population in June 2010 was 5,076,700. The population is 100% urban. Population density, 7,441 per sq. km.

The UN gives a projected population for 2015 of 5·38m.

Malay, Chinese (Mandarin), Tamil and English are the official languages; Malay is the national language and English is the language of administration.

SOCIAL STATISTICS

2008 births, 39,826; deaths, 17,222. Birth rate per 1,000 population, 2008, 10·2; death rate, 4·4. Annual population growth rate, 2000–05, 1·5%; infant mortality, 2010, two per 1,000 live births (one of the lowest in the world); life expectancy, 2007, 77·8 years for males and 82·6 years for females. Fertility rate, 2008, 1·3 births per woman. In 2006 the most popular age range for marrying was 25–29 years for both males and females.

Source: Singapore Department of Statistics

CLIMATE

The climate is equatorial, with relatively uniform temperature, abundant rainfall and high humidity. Rain falls throughout the year but tends to be heaviest from Nov. to Jan. Average daily temperature is 26·8°C with a maximum daily average of 30·9°C and a minimum daily average of 23·9°C. Mean annual rainfall is 2,345 mm.

CONSTITUTION AND GOVERNMENT

Singapore is a republic with a parliamentary system of government. The organs of state—the executive, the legislature and the judiciary—are provided for by a written constitution. The Constitution is the supreme law of Singapore and any law enacted after the date of its commencement, which is inconsistent with its provisions, is void. The present constitution came into force on 3 June 1959 and was amended in 1965.

The Head of State is the *President*. The administration of the government is vested in the Cabinet headed by the *Prime Minister*. The Prime Minister and the other Cabinet Members are appointed by the President from among the Members of Parliament (MPs). The Cabinet is collectively responsible to Parliament.

Parliament is unicameral consisting of 87 elected members and three Non-Constituency MPs (NCMPs), elected by secret ballot from single-member and group representation constituencies, as well as nine Nominated Members of Parliament (NMPs) who are appointed for a term of two and a half years on the recommendation of a Special Select Committee of Parliament. With the customary exception of those serving criminal sentences, all citizens over 21 are eligible to vote. Voting in an election is compulsory. Group representation constituencies may return up to six Members of Parliament (four before 1996), one of whom must be from the Malay community, the Indian or other minority communities. To ensure representation of parties not in the government, provision is made for the appointment of three (or up to a maximum of six) NCMPs. The number of NCMPs is reduced by one for each opposition candidate returned. There is a common roll without communal electorates.

A Presidential Council to consider and report on minorities' rights was established in 1970. The particular function of this council is to draw attention to any Bill or to any subsidiary legislation which, in its opinion, discriminates against any racial or religious community.

Salaries for leading Singaporean politicians are the highest in the world. In 2012 the prime minister's salary was S\$2·2m. (US\$1·7m.) although this represented a 36% drop from 2011. Prime Minister Lee Hsien Loong's salary is more than four times that of President Barack Obama.

National Anthem

'Majulah Singapura' ('Onward Singapore'); words and tune by Zubir Said.

GOVERNMENT CHRONOLOGY

Prime Ministers since 1959. (PAP = People's Action Party)

1959–90	PAP	Lee Kuan Yew
1990–2004	PAP	Goh Chok Tong
2004–	PAP	Lee Hsien Loong

RECENT ELECTIONS

In parliamentary elections held on 7 May 2011 the ruling People's Action Party (PAP) won 81 of 87 seats (with 60·1% of votes cast), including five seats won automatically before the election because the opposition did not contest them. The Workers' Party took six seats (12·8%). Four other parties failed to gain any seats.

In presidential elections held on 27 Aug. 2011, Tony Tan Keng Yam won 35·2% of the vote, Tan Cheng Bock 34·8%, Tan Jee Say 25·0% and Tan Kin Lian 4·9%. Turnout was 94·6%.

CURRENT GOVERNMENT

President: Tony Tan Keng Yam; b. 1940 (sworn in 1 Sept. 2011).

In March 2013 the cabinet comprised:

Prime Minister: Lee Hsien Loong; b. 1952 (PAP; sworn in 12 Aug. 2004).

Deputy Prime Ministers: Teo Chee Hean (*Co-ordinating Minister for National Security and Minister for Home Affairs*); Tharman Shanmugaratnam (*Minister for Finance*).

Minister for Trade and Industry: Lim Hng Kiang. *Communications and Information:* Dr Yaacob Ibrahim (also *in Charge of Muslim Affairs*). *National Development:* Khaw Boon Wan. *Defence:* Ng Eng Hen. *Environment and Water Resources:* Vivian Balakrishnan. *Foreign Affairs and Law:* K. Shanmugam. *Health:* Gan Kim Yong. *Transport:* Lui Tuck Yew. *Education:* Heng Swee Keat. *Social and Family Development (acting):* Chan Chun Sing. *Manpower (acting):* Tan Chuan-Jin. *Culture, Community and Youth (acting):* Lawrence Wong. *Ministers in Prime Minister's Office:* Lim Swee Say; S. Iswaran (also *Second Minister for Home Affairs, and Trade and Industry*); Grace Fu Hai Yien (also *Second Minister for the Environment and Resources, and Foreign Affairs*).

Government Website: http://www.gov.sg

CURRENT LEADERS

Lee Hsien Loong

Position
Prime Minister

Introduction
When Lee Hsien Loong was sworn in as prime minister of Singapore on 12 Aug. 2004, it was only the second time the southeast Asian city-state had changed its leader since independence in the 1960s. His father, Lee Kuan Yew, was the country's charismatic leader for 31 years and oversaw a transformation from a third-world colony to a prosperous export-driven economy. Lee Hsien Loong, a former military strategist-turned-politician, pledged to sustain the vibrant economy while maintaining a cohesive society. In May 2011 his People's Action Party won parliamentary elections for the 11th successive time since independence.

Early Life
Lee Hsien Loong was born in Singapore on 10 Feb. 1952, the eldest son of a wealthy and well-connected Hakka-Chinese family. His father, Lee Kuan Yew, was Singapore's first prime minister, and led the former British colony as head of the People's Action Party (PAP) from its first period of self-governance in 1959, through independence in 1965, until 1990. Lee Hsien Loong attended state primary and secondary schools in Singapore, and was awarded a president's scholarship to study mathematics and computer science at Cambridge University in England. He graduated in 1974 with first class honours and returned to serve in the Singapore Armed Forces, rising through the ranks to become a Brig.-Gen. He gained a reputation for his analytical and problem-solving skills. He entered politics as an MP representing the PAP in the general election of Sept. 1984. The election was the first occasion since 1963 that two opposition parties—the Workers' Party and the Singapore Democratic Party—were able to win seats from the ruling PAP. Lee was elected to the Central Executive Committee of the PAP in 1986.

Lee Kuan Yew resigned in Nov. 1990 after 31 years in power and Goh Chok Tong became prime minister, appointing Lee Hsien Loong his deputy with responsibility for economic and civil service affairs. After three decades of authoritarian rule, Singapore underwent a cautious liberalization. Lee held his parliamentary seat in the 1991 and 1997 general elections, which were landslide victories for the PAP. In Jan. 1998 he was appointed chairman of the Monetary Authority of Singapore. When, in line with other southeast Asian economies, Singapore faced an economic downturn in 1998, the government responded by cutting wages, allowing its currency to adjust downward and positioning the country as a leading international financial centre. Despite continued economic pressures, the PAP won the 2001 general election by a large majority. Lee was re-elected, and was appointed minister of finance in Nov. 2001. He pursued a tax-cutting agenda and brought in pension reforms and policies to liberalize the financial sector.

Career in Office

On 12 Aug. 2004 Lee was sworn in as prime minister, handing the chairmanship of the Monetary Authority to Goh Chok Tong who became senior minister in the cabinet. Lee said his goal was to build a vibrant and competitive economy and he pledged to maintain the open, consultative style of the Goh Chok Tong era. He also signalled that social liberalization would continue, reflecting the demands of a highly-educated and increasingly less tractable population, as well as the government's realization that Singapore must move beyond manufacturing into 'knowledge-based' industries that depend more on individual creativity and entrepreneurship.

In April 2005 Lee announced his government's controversial decision to legalize gambling, paving the way for the building of two large casino resorts and perhaps hinting at a more permissive atmosphere within the country. Nevertheless, in 2007 parliament voted against a proposal to decriminalize sex between men, and international appeals for clemency failed to prevent the execution of two Nigerians for drug smuggling.

Singapore's usually fraught relations with neighbouring Malaysia improved in Jan. 2005 when the two countries settled a dispute over land reclamation work in their border waters.

In the first electoral test since his appointment in 2004, Lee's party won the May 2006 parliamentary elections overwhelmingly. In 2008 his government's budget included tax incentives to enhance business competition, but Singapore's export-led economy was one of the first in Asia to feel the impact of the global financial downturn. In Jan. 2009 the government announced a S$20·5bn. stimulus package to bolster the economy and by July the country was emerging from its deepest recession on record.

In May 2011 the PAP was re-elected with a large majority but with a lower share of the popular vote, which opposition parties portrayed as a significant shift in the political landscape.

DEFENCE

Compulsory military service in peacetime for all male citizens and permanent residents was introduced in 1967. The period of service is 24 months. Reserve liability continues to age 50 for officers, 40 for other ranks. In 2006 the SAF (Singapore Armed Forces) comprised 312,500 Operationally Ready National Servicemen and an estimated 72,500 regulars and Full-Time National Servicemen.

An agreement with the USA in Nov. 1990 provided for an increase in US use of naval and air force facilities.

Singapore is a member of the Five Powers Defence Arrangement, with Australia, New Zealand, Malaysia and the UK.

In 2008 defence expenditure totalled US$7,662m. (US$1,663 per capita—the highest of any Asian country), representing 4·2% of GDP.

Army

Strength (2006) 50,000 (including 35,000 conscripts) plus 300,000 reserves. In addition there is a Civil Defence Force totalling 81,800 including 3,200 conscripts, 23,000 Operationally Ready National Servicemen and more than 54,000 civil defence volunteers.

Navy

The Republic of Singapore Navy comprises four commands: Fleet, Coastal Command (COSCOM), Naval Logistics Command and Training Command. The fleet includes four diesel submarines. The Navy numbers an estimated 9,000 personnel (1,800 conscripts, 2,200 regulars and 5,000 active reservists). There are two naval bases: Tuas Naval Base and Changi Naval Base, the first phase of which completed in 2000 and replaces Brani Naval Base.

Air Force

The Republic of Singapore Air Force (RSAF) has fighter squadrons comprising the F16 Falcon and the F5S/F Tiger.

Personnel strength (2006) about 10,500 (3,000 conscripts). Equipment includes 111 combat-capable aircraft and eight attack helicopters.

ECONOMY

Services accounted for 72% of GDP in 2007 and industry 28%.

According to the anti-corruption organization *Transparency International*, Singapore ranked fifth in the world in a 2012 survey of the countries with the least corruption in business and government. It received 87 out of 100 in the annual index.

Overview

Singapore is credited with one of the least corrupt, most competitive and most open economies in the world. Manufacturing, particularly electronics, has been the main engine of growth. About half of Singapore's non-oil exports are electronic goods. Industry is dominated by foreign multinationals and a few large domestic enterprises with government links. However, the economy's openness leaves it vulnerable to external demand shocks.

The government is seeking to shift the economy away from manufacturing, where other Asian countries are seen as rising competitors, into knowledge-driven industries. Singapore has become an important offshore banking centre. Liberal rules on stem cell research have helped attract foreign scientists in a bid to make the country a leading biomedical centre.

Economic growth averaged 8% between 2004 and 2007 but fell to −1·0% in 2009 as a result of the global financial crisis. This represented the biggest output decline in twenty years. However, growth rebounded strongly in 2010, with electronics exports boosted by consumer demand in China and corporate IT investment in the USA.

Currency

The unit of currency is the *Singapore dollar* (SGD) of 100 *cents*. It is managed against a basket of currencies of Singapore's main trading partners. In Aug. 2009 foreign exchange reserves totalled US$174,606m. and total money supply was S$89,258m.

Inflation rates (based on IMF statistics):

2002	2003	2004	2005	2006	2007	2008	2009	2010	2011
−0·4%	0·5%	1·7%	0·5%	1·0%	2·1%	6·6%	0·6%	2·8%	5·2%

Budget

The fiscal year begins on 1 April. In 2007–08 budgetary central government revenue totalled S$51,007m. (S$41,577m. in 2006–07) and expenditure S$33,875m. (S$31,446m. in 2006–07).

Principal sources of revenue in 2007–08 were: taxes on income, profits and capital gains, S$14,938m.; taxes on goods and services S$12,193m.; taxes on property, S$2,582m. Main items of expenditure by economic type in 2007–08: use of goods and services, S$11,663m.; compensation of employees, S$8,575m.; social benefits, S$7,907m.

There is a Goods and Services Tax (GST) of 7%.

Performance

Real GDP growth rates (based on IMF statistics):

2003	2004	2005	2006	2007	2008	2009	2010	2011
4·6%	9·2%	7·4%	8·8%	8·9%	1·7%	−1·0%	14·8%	4·9%

Total GDP was US$239·7bn. in 2011. Singapore was ranked second in the Global Competitiveness Index in the World Economic Forum's *Global Competitiveness Report 2012–2013*. The index analyses 12 areas of competitiveness for over 100 countries including macroeconomy, higher education and training, institutions, innovation and infrastructure. In the 2012 *World*

Competitiveness Yearbook, compiled by the International Institute for Management Development, Singapore came fourth in the world ranking. This annual publication ranks and analyzes how a nation's environment creates and sustains the competitiveness of enterprises.

Banking and Finance

The Monetary Authority of Singapore (*Chairman*, Tharman Shanmugaratnam) performs the functions of a central bank, except the issuing of currency which is the responsibility of the Board of the Commissioners of Currency.

The Development Bank of Singapore and POSB (formerly the Post Office Savings Bank) were merged in 1998 to become the largest bank in southeast Asia and one of the leading banks in Asia. In 2003 the company was rebranded as DBS Bank although POSB still operates as a recognizable brand within the enterprise with a customer base of over 3·2m. in Singapore in 2009. Customer deposits at DBS Bank totalled S$183·9bn. in the second quarter of 2010 and total assets S$276·3bn.

In Oct. 2010 there were 120 commercial banks in Singapore, of which seven were local. There were 31 representative offices, 26 foreign banks with full licences, 49 with 'wholesale' licences and 38 with 'offshore' licences. The total assets/liabilities amounted to S$749,703·1m. in July 2010. Total deposits of non-bank customers in July 2010 amounted to S$408,837·5m. and loans and advances including bills financing totalled S$298,693·2m. There were 46 merchant banks in Oct. 2010.

Gross external debt amounted to US$1,113,298m. in June 2012.

The Singapore Exchange (SGX), a merger of the Stock Exchange of Singapore and the Singapore International Monetary Exchange, was officially launched on 1 Dec. 1999.

ENERGY AND NATURAL RESOURCES

Environment

Singapore's carbon dioxide emissions from the consumption and flaring of fossil fuels in 2008 were the equivalent of 34·6 tonnes per capita.

Electricity

Installed capacity was 10·9m. kW in 2007. Production (2007) 41,134m. kWh. Consumption per capita (2007) 8,964 kWh.

Oil and Gas

Replacing the Kallang Gasworks, the Senoko Gasworks started operations in Oct. 1996. It had a total gas production capacity of 1·6m. cu. metres per day. In Jan. 2001 a 640-km gas pipeline linking Indonesia's West Natuna field with Singapore came on stream. It is expected to provide Singapore with US$8bn. worth of natural gas over a 20-year period.

Water

Singapore uses an average of 1·25m. cu. metres of water per day. Singapore's water supply comes from local sources and sources in Johor, Malaysia. The total water supply system comprises 19 raw water reservoirs, nine treatment works, 15 storage or service reservoirs and 5,150 km of pipelines.

Agriculture

Only about 1·49% of the total area is used for farming. Local farms provide only about 35% of hen eggs, 1·6% of chickens and 2·4% of ducks. 18,928 tonnes of vegetables and fruits were produced for domestic consumption in 1999. In 2001 alone Singapore imported 44·1m. chickens, 7m. ducks, 722m. hen eggs, 210,077 tonnes of meat and meat products, 226,126 tonnes of fish and fish products, 352,919 tonnes of vegetables and 358,595 tonnes of fruits for local consumption.

Agro-technology parks house large-scale intensive farms to improve production of fresh food. As of the end of 2000, a total of 1,465 ha. of land in Murai, Sungei Tengah, Nee Soon, Loyang, Mandai and Lim Chu Kang had been developed into Agro-technology Parks. Through open tenders, auctions and direct allocations, 247 farms have been allocated 777 ha. of land for the production of livestock, eggs, milk, aquarium fish, food fish (fish for consumption), fruits, vegetables, orchids and ornamental and aquatic plants, as well as for the breeding of birds and dogs.

Forestry

In 2010 forests covered 2,000 ha., or 3% of the total land area.

Fisheries

The total catch in 2010 amounted to 1,732 tonnes. Imports of fishery commodities were valued at US$807·0m. in 2009.

INDUSTRY

The leading companies by market capitalization in Singapore in March 2012 were: Singapore Telecommunications (US$39·9bn.); Jardine Strategic, a diversified business group (US$34·2bn.); and DBS Group Holdings, a banking group (US$27·3bn.).

The largest industrial area is at Jurong. In 2007 there were an estimated 160,000 enterprises of which 99% were small and medium enterprises (SMEs). Only about 5% of SMEs are in the manufacturing sector but they generated 17% of SMEs' value-added in 2007.

Output, 2009 (in S$1m.), totalled 213,699·8: including electronic products and components, 67,226·1; refined petroleum products, 31,860·3; chemicals and chemical products, 25,873·3; machinery and equipment, 18,922·7; pharmaceutical products, 18,093·9; transport equipment, 16,431·7; fabricated metal products, 8,185·8; food, beverages and tobacco, 6,637·2.

According to the World Bank's *Doing Business 2012* Singapore is the easiest country in which to do business and the fourth easiest in which to start a business.

Labour

In June 2004 Singapore's labour force comprised 2,183,300 people, of whom 2,066,900 were employed. The principal areas of employment in June 2004 were manufacturing (356,700 people), wholesale and retail trade (319,700), business services (254,000), transport, storage and communications (212,500) and hotels and restaurants (129,300). The unemployment rate averaged 3·4% throughout 2004 (4·0% in 2003). The average worker put in 46·3 hours a week in 2004; average monthly earnings in 2004 were S$3,329.

Legislation regulates the principal terms and conditions of employment such as hours of work, sick leave and other fringe benefits. Young people of 14–16 years may work in industrial establishments, and children of 12–14 years may be employed in approved apprenticeship schemes. A trade dispute may be referred to the Industrial Arbitration Court. Singapore does not have a minimum wage.

The Ministry of Manpower operates an employment service and provides clients with disabilities with specialized on-the-job training. The Central Provident Fund was established in 1955 to make provision for employees in their old age. At the end of 2004 there were 3,018,000 members with S$111,874m. standing to their credit in the fund. The legal retirement age is 62.

Source: Singapore Department of Statistics

INTERNATIONAL TRADE

Foreign investment of up to 40% of the equity of domestic banks is permitted.

Imports and Exports

Total imports were S$333,191m. in 2005; and total exports S$382,532m. in 2005. Exports in 2005 were worth 244% of GDP.

Imports and exports (in S$1m.), by country, 2005:

	Imports (c.i.f.)	Exports (f.o.b.)
Australia	4,850	14,045
China	34,170	32,909
Germany	9,915	10,504

	Imports (c.i.f.)	Exports (f.o.b.)
Hong Kong	7,009	35,849
India	6,788	9,817
Indonesia	17,400	36,817
Japan	32,034	20,874
Korea, South	14,323	13,412
Malaysia	45,527	50,612
Saudi Arabia	14,894	708
Taiwan	19,720	14,938
Thailand	12,516	15,662
UK	6,554	10,525
USA	38,793	39,024

Main imports (2005, in S$1m.): machinery and equipment, 185,980; mineral fuels, 59,145; manufactured goods, 25,040; chemicals and chemical products, 20,744; food, 6,680; beverages and tobacco, 2,190; crude materials, 2,190; animal and vegetable oils, 479; miscellaneous manufactures, 26,526.

Main exports (2005, in S$1m.): machinery and transport equipment, 224,980; mineral fuels, 57,414; chemicals and chemical products, 43,611; manufactured goods, 17,498; food, 3,865; crude materials, 2,257; beverages and tobacco, 2,053; animal and vegetable oils, 422; miscellaneous manufactures, 26,049.

In May 2003 the USA and Singapore signed a free trade agreement removing tariffs on trade worth an estimated US$33bn. per annum.

Trade Fairs

Singapore ranked as the world's most important convention city in 2010 according to the Union des Associations Internationales (UAI). Singapore hosted 725 meetings recognized by UAI (6·5% of all meetings).

COMMUNICATIONS

Roads

In 2007 there were 3,297 km of public roads (100% asphalt-paved). Singapore has one of the densest road networks in the world. In 2007 there were 517,000 passenger cars, 14,500 buses and coaches, 151,000 vans and lorries, and 144,300 motorcycles and scooters.

Singapore was the top ranked nation for road infrastructure in the World Economic Forum's *Global Competitiveness Report 2009–2010*.

Rail

A 25·8-km main line runs through Singapore, connecting with the States of Malaysia and as far as Bangkok. Branch lines serve the port of Singapore and the industrial estates at Jurong. The total rail length of the Mass Rapid Transit (SMRT) metro is 93·2 km. The 20 km North-East Line (operated by SBS Transit), the world's first fully automated heavy metro, became operational in 2003. In late 1999 the Light Rapid Transit System (LRT) began operations, linking the Bukit Panjang Estate with Choa Chu Kang in the North West region.

Civil Aviation

As of Sept. 2010, Singapore Changi Airport was served by 96 airlines with more than 5,100 weekly flights to and from some 200 cities in 60 countries and territories worldwide. A total of 37,203,978 passengers and 1,633,791 tonnes of freight were handled in 2009. The national airline is Singapore Airlines, which carried 16,480,000 passengers in 2009–10. Its subsidiary, Silk Air, serves Asian destinations. Other Singapore-based carriers include Jetstar Asia and Tiger Airways.

In the World Economic Forum's *Global Competitiveness Report 2009–2010*, Singapore ranked first for quality of air transport infrastructure.

Shipping

Singapore has a large container port, the world's busiest in terms of containers handled in 2006 and second only to Shanghai in terms of shipping tonnage. The economy is dependent on shipping and entrepôt trade. A total of 146,265 vessels of 960m. gross tonnes (GT) entered Singapore during 2001. In 2001, 3,353 vessels with a total of 23·2m. GT were registered in Singapore. The Singapore merchant fleet ranked 7th among the principal merchant fleets of the world in 2001. Total cargo handled in 2006 was 448·5m. freight tons, and total container throughput in 2006 was 24,792,000 TEUs (twenty-foot equivalent units).

Singapore was ranked first in the World Economic Forum's *Global Competitiveness Report 2009–2010* for the quality of its port facilities.

Telecommunications

In 2008 there were 1,857,000 main (fixed) telephone lines. In the same year mobile phone subscribers numbered 6,376,000 (1,381·5 per 1,000 persons). In 1997 Singapore Telecom, one of the largest companies in Asia, lost its monopoly with the entry of a new mobile phone operator. In 2007 Singapore had three mobile phone operators—SingTel Mobile (owned by Singapore Telecom), M1 and StarHub Mobile. In 2008 there were 3·4m. internet users. There were 19·9 broadband subscribers per 100 inhabitants in June 2007. In March 2012 there were 2·6m. Facebook users. The Telecommunication Authority of Singapore (TAS) is the national regulator and promoter of the telecommunication and postal industries.

According to the World Economic Forum's *Global Information Technology Report 2010–11* Singapore is ranked second in the world in exploiting global information technology developments.

SOCIAL INSTITUTIONS

Justice

There is a Supreme Court in Singapore which consists of the High Court and the Court of Appeal. The Supreme Court is composed of a Chief Justice and 11 Judges. The High Court has unlimited original jurisdiction in both civil and criminal cases. The Court of Appeal is the final appellate court. It hears appeals from any judgement or order of the High Court in any civil matter. The Subordinate Courts consist of a total of 47 District and Magistrates' Courts, the Civil, the Family and Crime Registries, the Primary Dispute Resolution Centre, and the Small Claims Tribunal. The right of appeal to the UK Privy Council was abolished in 1994.

Penalties for drug trafficking and abuse are severe, including a mandatory death penalty. In 1994 there were 76 executions, although since then the average annual number has generally been declining. Amnesty International reported that there were four executions in 2011 but none in 2012.

The Technology Court was introduced in 1995 where documents were filed electronically. This process was implemented in Aug. 1998 in the Magistrates appeal and the Court of Appeal.

The average population in penal institutions in 2007 was 11,768 (267 per 100,000 of national population).

Education

The general literacy rate rose from 84% in 1980 to an estimated 94·7% in 2009 (male 97·5%; female 92·0%). Kindergartens are private and fee-paying. Compulsory primary state education starts at six years and culminates at 11 or 12 years with an examination which influences choice of secondary schooling. There are 17 autonomous and eight private fee-paying secondary schools. Tertiary education at 16 years is divided into three branches: junior colleges leading to university; four polytechnics; and ten technical institutes.

In 2007 there were 301,101 pupils with 14,743 teaching staff in primary schools, 232,100 pupils with 13,686 teaching staff in secondary schools and 183,627 students with 14,209 academic staff in tertiary education.

There are three universities: the National University of Singapore (established 1905) with 32,028 students in 2001–02, the Nanyang Technological University (established 1991) with 28,949 in 2007–08, and the Singapore Management University (established in 2000).

In 2009 public expenditure on education came to 3·1% of GDP and accounted for 11·6% of total government expenditure.

Health

There are 27 hospitals (five general hospitals, one community hospital, seven specialist hospitals/centres and 14 private), with 11,897 beds in 2001. In 2001 there were 5,747 doctors, 1,087 dentists, 17,398 registered nurses and midwives and 1,141 pharmacists.

The leading causes of death are cancer (4,238 deaths in 2000), heart disease (3,940) and pneumonia (1,794).

Welfare

The Central Provident Fund (CPF) was set up in 1955 to provide financial security for workers upon retirement or when they are no longer able to work. In 2001 there were 2,922,673 members with S$92,221m. standing to their credit in the Fund.

RELIGION

In 2001, 41·0% of the population were Buddhists and Taoists, 12·0% Muslims, 11·7% Christians and 3·2% Hindus; 0·5% belonged to other religions.

CULTURE

The National Arts Council (NAC) was established in 1991 to spearhead the development of the arts.

Press

In 2008 there were 11 daily newspapers, with a total daily circulation of 1,725,000 copies. The most popular paid-for daily is *The Straits Times*, with an average daily circulation of 389,000 in 2008.

Tourism

There were 10·3m. visitors in 2007. Most came from Indonesia, China, Australia, India, Malaysia, Japan, the UK and South Korea. The total tourism receipts for 2007 came to S$14·1bn. There were 98 gazetted hotels in 2007, providing 30,087 rooms.

Festivals

Every Jan. or Feb. the Lunar New Year is celebrated. Other Chinese festivals include Qing Ming (a time for the remembrance of ancestors), Yu Lan Jie (Feast of the Hungry Ghosts) and the Mid-Autumn Festival (Mooncake or Lantern festival).

Muslims in Singapore celebrate Hari Raya Puasa (to celebrate the end of a month-long fast) and Hari Raya Haji (a day of prayer and commemoration of the annual Mecca pilgrimage). There are also Muharram (a New Year celebration) and Maulud (Prophet Muhammad's birthday).

Hindus celebrate the Tamil New Year in mid-April. Thaipusam is a penitential Hindu festival popular with Tamils; and Diwali, the Festival of Lights, is celebrated by Hindus and Sikhs. Other festivals include Thimithi (a fire-walking ceremony) and Navarathiri (nine nights' prayer).

Buddhists observe Vesak Day, which commemorates the birth, enlightenment and Nirvana of the Buddha, and falls on the full moon day in May.

Christmas, Good Friday and Easter Sunday are also recognized.

DIPLOMATIC REPRESENTATIVES

Of Singapore in the United Kingdom (9 Wilton Cres., London, SW1X 8SP)
High Commissioner: Thambynathan Jasudasen.

Of the United Kingdom in Singapore (100 Tanglin Rd, Singapore 247919)
High Commissioner: Anthony Phillipson.

Of Singapore in the USA (3501 International Pl., NW, Washington, D.C., 20008)
Ambassador: Ashok Kumar Mirpuri.

Of the USA in Singapore (27 Napier Rd, Singapore 258508)
Ambassador: David I. Adelman.

Of Singapore to the United Nations
Ambassador: Albert Chua.

Of Singapore to the European Union
Ambassador: Ong Eng Chuan.

FURTHER READING

Department of Statistics. *Monthly Digest of Statistics.—Yearbook of Statistics.*
The Constitution of Singapore. 1992
Information Division, Ministry of Information and the Arts. *Singapore [year]: a Review of [the previous year].*
Ministry of Trade and Industry, *Economic Survey of Singapore.* (Quarterly and Annual)
Chee-Kiong, T., *The Making of Singapore Sociology: State and Society.* 2002
Chew, E. C. T., *A History of Singapore.* 1992
Huff, W. G., *Economic Growth of Singapore: Trade and Development in the Twentieth Century.* 1994
Myint, S., *The Principles of Singapore Law.* 4th ed. 2001
Tan, C. H., *Financial Markets and Institutions in Singapore.* 11th ed. 2005

National library: National Library of Singapore, 100 Victoria Street, Singapore 188064.
National Statistical Office: Department of Statistics, 100 High St. #05-01, The Treasury, Singapore 179434.
Website: http://www.singstat.gov.sg

SLOVAKIA

Slovenská Republika
(Slovak Republic)

Capital: Bratislava
Population projection, 2015: 5·51m.
GNI per capita, 2011: (PPP$) 19,998
HDI/world rank: 0·834/35
Internet domain extension: .sk

KEY HISTORICAL EVENTS

There is evidence of human habitation from 270,000 BC. In the Bronze Age the region was a centre for copper manufacture and was ruled by Carpathian, Celtic and Germanic tribes. The date of the Slavic arrival is contested but there is evidence of their presence from the sixth century under the Roman Empire. Waves of invasion and migration followed Roman withdrawal and control fell variously to the Avars, Franks and Magyars.

The first brief period of Slavic rule was the Samo Empire (623–658) followed by the Moravian Empire from 833–907. The region subsequently became part of the Kingdom of Hungary. In 1241 the region was invaded by Mongols and was stricken with famine but grew in prosperity through the medieval period. From the 16th–19th centuries Slovakia was at the centre of Hungary under the Habsburg dynasty. After 1867 it became part of the Austro-Hungarian Empire and underwent 'Magyarization', a repressive attempt to impose Magyar culture. In response, a nationalist movement at home and among immigrants in America gained momentum in the First World War.

On 28 Oct. 1918, after the dissolution of Austria-Hungary, the Czechoslovak State was founded. Two days later the Slovak National Council voted to unite with the Czechs. The Treaty of St Germain-en-Laye (1919) recognized the Czechoslovak Republic, consisting of the Czech lands (Bohemia, Moravia, part of Silesia) and Slovakia. The new state was numerically dominated by Czechs, giving rise to some tensions and calls for Slovakian independence. In 1939 negotiations between European powers resulted in Germany incorporating the Czech lands into the Reich as the 'Protectorate of Bohemia and Moravia'. Meanwhile the German-sponsored Slovak government declared independence. A government-in-exile, headed by Dr Edvard Beneš, was set up in London during the war. In 1944 Slovak resistance fighters began an uprising; liberation was completed by Soviet and US forces in

May 1945. Territories taken by Germans, Poles and Hungarians were restored to Czech sovereignty.

Elections in May 1946 returned a coalition government, under communist prime minister Klement Gottwald. On 20 Feb. 1948, 12 non-communist ministers resigned in protest at the infiltration of communists into the police. Gottwald formed a predominantly communist government and in May 1948, after the government won an 89% majority in rigged parliamentary elections, President Beneš resigned. During the next two decades the government banned other parties and followed Stalinist policies. On 14 May 1955 Czechoslovakia signed the Warsaw Pact, allying itself with the Soviet Union and other Eastern Bloc countries.

In 1968 pressure for liberalization culminated in the overthrow of the Stalinist leadership and under Alexander Dubček, new first secretary of the communist party, the 'Prague Spring' saw sweeping reforms including the abolition of censorship. Between May and Aug. 1968 the USSR put pressure on the government to abandon reforms. Finally, Warsaw Pact troops occupied Czechoslovakia on 21 Aug. The government was forced to reverse reforms and accept the stationing of Soviet troops. In April 1969, with Soviet support, Dubček was replaced by Gustáv Husák.

Demands for reform persisted and mass demonstrations began in Nov. 1989. When authorities used violence to break up a demonstration on 17 Nov., the communist leadership resigned. On 30 Nov. the federal assembly abolished the communists' sole right to govern. A new government was formed on 3 Dec. and another followed a week later as the protest movement grew. Gustáv Husák resigned as president and was replaced by Václav Havel by the unanimous vote of 323 members of the federal assembly on 29 Dec. This almost bloodless overthrow of communist rule became known as the 'Velvet Revolution'.

At the June 1992 elections the Movement for Democratic Slovakia, led by Vladimír Mečiar, campaigned for Slovak independence. On 17 July the Slovak National Council adopted a declaration of sovereignty by 113 to 24 votes, the 'Velvet Divorce'. Havel resigned as federal president on 20 July. A constitution ratified on 1 Sept. 1992 paved the way for independent Slovakia to come into being on 1 Jan. 1993. Economic property was divided between Slovakia and the Czech Republic, with government real estate remaining with the republic in which it was located. Other property was divided by special commissions in the proportion of two (Czech Republic) to one (Slovakia), on the basis of population size. Military equipment was also divided on the two-to-one principle and military personnel were invited to choose in which army to serve.

In the 1990s Slovakia resisted calls for economic reforms and closer ties with Western Europe. However, following the election of a coalition government under Mikuláš Dzurinda in Oct. 1998, the country implemented reforms and attracted foreign investment. It responded to criticism about its human rights record by improving conditions for its Romany and Hungarian minorities. Slovakia joined NATO in March 2004 and the European Union in May 2004. It elected a new coalition government under Prime Minister Robert Fico in June 2006. Fico was succeeded by Iveta Radičová in 2010 before reclaiming the premiership in March 2012.

TERRITORY AND POPULATION

Slovakia is bounded in the northwest by the Czech Republic, north by Poland, east by Ukraine, south by Hungary and southwest by Austria. Its area is 49,034 sq. km (18,932 sq. miles). Census population in 2011 was 5,397,036 (2,769,264 females and 2,627,772 males); density, 110·1 per sq. km.

The UN gives a projected population for 2015 of 5·51m.

In 2011, 54·9% of the population lived in urban areas. There are eight administrative regions *(Kraj)*, one of which is the capital, Bratislava. They have the same name as the main city of the region.

Region	Area in sq. km	2011 population
Banská Bystrica	9,455	660,563
Bratislava	2,053	602,436
Košice	6,753	791,723
Nitra	6,343	689,867
Prešov	8,993	814,527
Trenčín	4,501	594,328
Trnava	4,148	554,741
Žilina	6,788	688,851

The capital, Bratislava, had a population in 2011 of 411,228. The population of other principal towns (2011, in 1,000): Košice, 240; Prešov, 92; Žilina, 81; Banská Bystrica, 80; Nitra, 79; Trnava, 66; Martin, 57; Trenčín, 56.

The population is 80·7% Slovak, 8·5% Hungarian, 2·0% Roma, 0·6% Czech and 0·6% Ruthenian, with some Germans, Moravians, Poles and Ukrainians.

A law of Nov. 1995 makes Slovak the sole official language.

SOCIAL STATISTICS

Births, 2007, 54,424; deaths, 53,856; marriages, 27,437; divorces, 12,174. Rates (per 1,000 population), 2007: birth, 10·1; death, 10·0; marriage, 5·1; divorce, 2·3. Expectation of life, 2006, was 70·4 years for males and 78·2 for females. In 2006 the most popular age range for marrying was 25–29 for both males and females. Annual population growth rate, 1996–2006, 0·3%. Infant mortality, 2010 (per 1,000 live births), 7. Fertility rate, 2006, 1·2 births per woman (one of the lowest rates in the world).

CLIMATE

A humid continental climate, with warm summers and cold winters. Precipitation is generally greater in summer, with thunderstorms. Autumn, with dry, clear weather and spring, which is damp, are each of short duration. Bratislava, Jan. –0·7°C. June 19·1°C. Annual rainfall 649 mm.

CONSTITUTION AND GOVERNMENT

The constitution became effective on 1 Jan. 1993, creating a parliamentary democracy with universal suffrage from the age of 18. Parliament is the unicameral *National Council*. It has 150 members elected by proportional representation to serve four-year terms. The constitution was amended in Sept. 1998 to allow for the direct election of the *President*, who serves for a five-year term. The President may serve a maximum of two consecutive terms.

The Judicial Branch consists of a *Supreme Court*, whose judges are elected by the National Council, and a *Constitutional Court*, whose judges are appointed by the President from a group of nominees approved by the National Council.

Citizenship belongs to all citizens of the former federal Slovak Republic; other residents of five years standing may apply for citizenship. Slovakia grants dual citizenship.

National Anthem

'Nad Tatrou sa blýska' ('Storm over the Tatras'); words by J. Matúška, tune anonymous.

GOVERNMENT CHRONOLOGY

(DU = Democratic Union; HZD = Movement for Democracy; HZDS = Movement for a Democratic Slovakia; KDH = Christian Democratic Movement; SDK = Slovak Democratic Coalition; SDKÚ = Slovak Democratic and Christian Union; SDKÚ–DS = Slovak Democratic and Christian Union–Democratic Party; Smer–SD = Direction–Social Democracy; SOP = Party of Civic Understanding; n/p = non-partisan)

Presidents since 1993.

1993–98	n/p	Michal Kováč
1999–2004	SOP, n/p	Rudolf Schuster
2004–	HZD, n/p	Ivan Gašparovič

Prime Ministers since 1993.

1993–94	HZDS	Vladimír Mečiar
1994	DU	Jozef Moravčík
1994–98	HZDS	Vladimír Mečiar
1998–2006	KDH/SDK, SDKÚ	Mikuláš Dzurinda
2006–10	Smer–SD	Robert Fico
2010–12	SDKÚ–DS	Iveta Radičová
2012–	Smer–SD	Robert Fico

RECENT ELECTIONS

Elections to the National Council were held on 10 March 2012. Direction–Social Democracy (Smer–SD) won 83 seats with 44·4% of votes cast, ahead of the Christian Democratic Movement 8·8% (16), Ordinary People and Independent Personalities 8·6% (16), Most–Híd 6·9% (13), the Slovak Democratic and Christian Union–Democratic Party 6·1% (11) and Freedom and Solidarity 5·9% (11). Turnout was 59·1%.

In the first round of presidential elections on 21 March 2009, incumbent president Ivan Gašparovič of the Movement for Democracy/HZD won 46·7% of the vote against 38·1% for Iveta Radičová (Slovak Democratic and Christian Union–Democratic Party/SDKÚ–DS). There were five other candidates. Turnout was 43·6%. In the run-off held on 4 April Gašparovič won 55·5% against 44·5% for Radičová. Turnout in the second round was 51·7%.

European Parliament

Slovakia has 13 (14 in 2004) representatives. At the June 2009 elections turnout was 19·6% (17·0% in 2004)—the lowest in the EU. Smer won 5 seats with 32·0% of votes cast (political affiliation in European Parliament: Progressive Alliance of Socialists and Democrats); SDKÚ–DS won 2 seats with 17·0% (European People's Party); the Party of the Hungarian Coalition, 2 with 11·3% (European People's Party); the Christian Democratic Movement, 2 with 10·9% (European People's Party); the ĽS-HZDS, 1 with 9·0% (Alliance of Liberals and Democrats for Europe); the SNS, 1 with 5·6% (Europe of Freedom and Democracy).

CURRENT GOVERNMENT

President: Ivan Gašparovič; b. 1941 (ind.; sworn in 15 June 2004 and re-elected in April 2009).

The government was composed as follows in March 2013:

Prime Minister: Robert Fico; b. 1964 (Smer–SD; sworn in 4 April 2012, having previously held office from July 2006–July 2010).

Deputy Prime Minister and Minister of Finance: Peter Kažimír (Smer–SD). *Deputy Prime Minister and Minister of Foreign Affairs:* Miroslav Lajčák (ind.). *Deputy Prime Minister and Minister of Interior:* Robert Kaliňák (Smer–SD). *Deputy Prime Minister and Minister for Investments:* Ľubomír Vážny (Smer–SD).

Minister of Agriculture, Environment and Rural Development: Ľubomír Jahnátek (Smer–SD). *Culture:* Marek Maďarič (Smer–SD). *Defence:* Martin Glváč (Smer–SD). *Economy:* Tomáš Malatinský (ind.). *Education, Science, Research and Sport:* Dušan Čaplovič (Smer–SD). *Environment:* Peter Žiga (Smer–SD). *Government, Labour, Social Affairs and the Family:* Ján Richter (Smer–SD). *Health:* Zuzana Zvolenská (Smer–SD). *Justice:* Tomáš Borec (ind.). *Transport, Construction and Regional Development:* Ján Počiatek (Smer–SD).

The *Speaker* is Pavol Paška.

Government Website: http://www.government.gov.sk

CURRENT LEADERS

Ivan Gašparovič

Position
President

Introduction
Shortly before Slovakia became a member of the European Union on 1 May 2004, a respected lawyer, Ivan Gašparovič, was elected as the country's president. Instrumental in drawing up Slovakia's constitution prior to the dissolution of Czechoslovakia in 1993, he was also a close ally of the controversial nationalist former prime minister, Vladimír Mečiar, the man he beat in the second round of the presidential election. He was re-elected in April 2009.

Early Life
Ivan Gašparovič was born in Poltár, near Lučenec in southern Slovakia on 27 March 1941. His father, Vladimír Gašparovič, had migrated to the region from Rijeka, Croatia at the end of the First World War. The family moved to Bratislava, where Vladimír worked as a teacher in a secondary school. Having studied at the Law Faculty of the Komenský University in Bratislava from 1959–64, Ivan Gašparovič worked in the district prosecutor's office of Bratislava's Martin district (1965–66), and then became a prosecutor at the municipal prosecutor's office. In early 1968 he joined the Communist Party of Czechoslovakia and actively supported the reforms of Alexander Dubček, the party's Slovak first secretary. Under Dubček, in what became known as the Prague Spring, democratization went further than in any other Communist state—press censorship was reduced and Slovakia was granted political autonomy. However, opposition grew swiftly in the USSR and in other Warsaw Pact states which invaded Czechoslovakia on the night of 20 Aug. 1968. The following year Dubček was replaced by Gustáv Husák, who spearheaded a 'normalization' policy that turned Czechoslovakia into one of Central Europe's most repressive states.

Gašparovič left the Communist Party after the events of 1968 and began work as a teacher at the law faculty at the Komenský University. He remained there until 1990 when he became the vice chancellor in Feb. of that year, two months after the 'Velvet Revolution' had swept aside the Communists. Václav Havel, the playwright and former dissident who was elected federal president in Dec. 1989, nominated Gašparovič as prosecutor-general of Czechoslovakia. He moved to Prague and took up the post in July 1990, as the new government began to tackle the legacy of communism—a moribund economy, high unemployment and widespread social discontent. Under the 1968 constitution Czechoslovakia was a federal republic—each republic had a council and an assembly, but the federal government dealt with defence and foreign affairs. Arguments over the nature of the federation broke out and in 1991 Vladimír Mečiar formed the Movement for a Democratic Slovakia (HZDS). Gašparovič returned to Bratislava to teach at the Komenský University and joined the HZDS in 1992. Mečiar led the party to victory in the June 1992 elections, and Gašparovič became an HZDS member of the Slovak parliament. In late 1992 he was one of the authors of the constitution of Slovakia, which came into effect on 1 Jan. 1993 when the republic formally declared its independence.

Gašparovič was speaker of the Slovak parliament until Oct. 1998 and a close ally of Prime Minister Mečiar, whose controversial policies in the mid-1990s included stripping away the rights of the country's large Hungarian community and clamping down on the media. Slovakia became increasingly isolated from Western Europe until Mečiar's nationalist government was defeated in Sept. 1998 by an alliance of liberals, centrists, left-wingers and ethnic Hungarians. Mikuláš Dzurinda became prime minister and steered Slovakia through various reforms required for EU and NATO membership. From Oct. 1998–July 2002, when the HZDS was in opposition, Gašparovič

was a member of the parliamentary committee for the supervision of the SIS (the Slovak equivalent of the US Central Intelligence Agency).

In July 2002 Gašparovič and others left the HZDS after being struck off the list of candidates for the parliamentary elections in Sept. The HZDS went on to poll only 3·3% of the vote, not enough to win any seats. Gašparovič returned to Komenský University, but also established a new political party called the Movement for Democracy (HZD). In April 2004 he ran for president against Mečiar, who was attempting to make a comeback after losing the 2002 legislative elections. Although Mečiar won more votes in the first round, he failed to win a majority. In the second round, Gašparovič secured nearly 60% of the vote with the support of the eliminated candidates.

Career in Office
Ivan Gašparovič succeeded Rudolf Schuster as president of the Slovak Republic on 15 June 2004 and began a five-year term of office. Following elections to the National Council in June 2006, Gašparovič asked Robert Fico, the leader of the social democratic Direction Party, to form a new coalition government in place of Mikuláš Dzurinda's SDKÚ-led administration. Gašparovič secured a second term as president in April 2009, winning 55·5% of the vote in a second round run-off. Following further parliamentary elections in June 2010, the Direction Party remained the largest single party but Fico was unable to form a new coalition cabinet. Gašparovič instead turned to Iveta Radičová of the SDKÚ–DS, who took over as prime minister at the head of a four-party centre-right government in July.

In March 2010 Gašparovič vetoed a controversial nationalist law intended to instil patriotism by compulsory weekly national anthem-playing in state schools.

In Oct. 2011 the government was defeated in a parliamentary confidence vote but reappointed in a caretaker capacity. At elections in March 2012 the Direction–Social Democracy party won a majority and its leader, Robert Fico, became prime minister again.

Robert Fico

Position
Prime Minister

Introduction
Robert Fico became prime minister for a second time after his Direction–Social Democracy (Smer–SD) party achieved outright victory in legislative elections held in March 2012. He has pledged to lead Slovakia into closer fiscal co-operation with the European Union.

Early Life
Robert Fico was born in Topolčany, Czechoslovakia on 15 Sept. 1964. After studying law at Comenius University in Bratislava, he obtained a PhD in criminal law from the Slovak Academy of Sciences. On finishing military service in 1986 he worked for the law institute of the ministry of justice until 1995, holding the post of deputy director from 1992.

From 1994–2000 he represented the Slovak Republic at the European Court of Human Rights and the European Commission of Human Rights. He entered the Slovak parliament in 1992, serving on the parliamentary constitutional committee from 1992–2002, as chairman of the prison commission from 1995–2003 and on the parliamentary committee on human rights, national minorities and women's rights from 2002–06. From 1999 he was chair of Smer and has served as its parliamentary leader since 2002.

Fico took office as prime minister for the first time in July 2006, at the head of a coalition with the centre-left People's Party-Movement for a Democratic Slovakia and the right-wing Slovak National Party. As premier, he withdrew Slovakian troops from

Iraq and refused to deploy forces in southern Afghanistan. Scheduled visits to Libya and Venezuela in 2007 attracted criticism from opponents, and in Oct. 2006 the Alliance of European Socialists expelled Smer for going into coalition with the National Party.

On 1 Jan. 2009 Fico took Slovakia into the single European currency. Tensions with neighbouring Hungary rose in 2009 when Slovakia passed a law restricting the use of minority languages in government buildings. Relations worsened the following year when Fico fiercely criticized Hungarian legislation allowing dual citizenship, claiming it constituted a security threat to Slovakia.

Despite Smer winning the largest number of seats in the June 2010 election, it was unable to form a government. Fico went into opposition, against Prime Minister Iveta Radičová's centre-right coalition. In Oct. 2011, with the government divided over Slovakia's participation in the European Financial Stability Facility, Fico agreed to support the measure in return for early parliamentary elections. Initially expected to win, the ruling coalition was hit by allegations of ministerial corruption and in March 2012 Smer won 83 of 150 seats to claim an outright majority.

Career in Office
Fico was sworn in on 4 April 2012, promising to deliver on EU-agreed fiscal targets. He pledged to protect social benefits while reducing the budget deficit and reaffirmed his intention to raise tax rates for the wealthy. His main challenges include guiding the economy towards growth, meeting the expectations of unemployed and low-paid workers and building relations with Hungary.

DEFENCE

Since 1 Jan. 2006 Slovakia has had an all-volunteer professional army. In 2008 military expenditure totalled US$1,477m. (US$271 per capita), representing 1·6% of GDP.

Army

Personnel (2007), 7,324. In addition there is a national guard reserve force with an estimated strength of 20,000.

Air Force

There were 46 combat-capable aircraft in 2007 (Su-22, MiG-21 and MiG-29 fighters) and 16 attack helicopters. Personnel (2007), 4,280.

INTERNATIONAL RELATIONS

A referendum held on 16–17 May 2003 approved accession to the EU, with 92·5% of votes cast for membership and 7·5% against. Turnout was 52·2%. Slovakia became a member of NATO on 29 March 2004 and the EU on 1 May 2004.

Slovakia has had a long-standing dispute with Hungary over the Gabčíkovo-Nagymaros Project, involving the building of dam structures in both countries for the production of electric power, flood control and improvement of navigation on the Danube as agreed in a treaty signed in 1977 between Czechoslovakia and Hungary. In late 1998 Slovakia and Hungary signed a protocol easing tensions between the two nations and settling differences over the dam.

ECONOMY

Agriculture accounted for 3% of GDP in 2008, industry 38% and services 59%.

Overview

Slovakia has undergone almost constant restructuring since 1998 when policy shifted from state intervention towards pro-market reforms. Following economic transition the heavy industry and agriculture sectors shrank while the service sector increased its share of GDP to 59% in 2008.

From 1998–2008 the economy enjoyed rapid growth, driven by strong exports and foreign direct investment, particularly in the automotive industry. GDP grew at approximately 4% per annum and peaked in 2007 at 10·5% on the back of strong domestic demand and buoyant exports. Slovakia entered the ERM2 in 2005 and adopted the euro in Jan. 2009. However, the global financial crisis resulted in a 4·9% contraction in GDP. In early 2009 the government introduced a range of measures to counteract the effects of the downturn. The recovery in global demand for manufactured goods and the strong macroeconomic policy response saw robust growth return in the second half of 2009. Even so, the traditionally high unemployment rate rose beyond 14%.

Currency

On 1 Jan. 2009 the euro (EUR) replaced the *Slovak koruna* (SKK) as the legal currency of Slovakia at the irrevocable conversion rate of 30·126 koruny to one euro. Foreign exchange reserves in Sept. 2009 were US$54m. (US$17,493m. in Sept. 2008) and gold reserves 1·02m. troy oz. Inflation rates (based on OECD statistics):

2002	2003	2004	2005	2006	2007	2008	2009	2010	2011
3·5%	8·4%	7·5%	2·8%	4·3%	1·9%	3·9%	0·9%	0·7%	4·1%

Total money supply in Dec. 2008 was €545,969m.

Budget

In 2008 budgetary central government revenue totalled 344·24bn. koruny (341·23bn. koruny in 2007) and expenditure 401·94bn. koruny (357·98bn. koruny in 2007).

Slovakia's budget deficit in 2011 was 4·8% of GDP (2010, 7·7%; 2009, 8·0%). The required target set by the EU is a budget deficit of no more than 3%.

VAT, personal and company income tax, real estate taxes and inheritance taxes came into force in Jan. 1993. VAT is 20% (reduced rate, 10%).

Performance

Real GDP growth rates (based on OECD statistics):

2002	2003	2004	2005	2006	2007	2008	2009	2010	2011
4·6%	4·8%	5·1%	6·7%	8·3%	10·5%	5·8%	–4·9%	4·4%	3·2%

Slovakia's total GDP in 2011 was US$96·0bn.

Banking and Finance

The central bank and bank of issue is the Slovak National Bank, founded in 1993 (*Governor*, Jozef Makúch). It has an autonomous statute modelled on the German Bundesbank, with the duties of maintaining control over monetary policy and inflation, ensuring the stability of the currency, and supervising commercial banks. However, it is now proposed to amend the central bank law to allow the government to appoint half the members of the board and force the bank to increase its financing of the budget deficit.

In Oct. 1998 the Slovak National Bank abandoned its fixed exchange rate system, whereby the crown's value was fixed within a fluctuation band against a number of currencies, and chose to float the currency.

Decentralization of the banking system began in 1991, and private banks began to operate. The two largest Slovak banks were both privatized in 2001. The Austrian bank Erste Bank bought an 87·18% stake in Slovenská Sporiteľňa (Slovak Savings Bank) and the Italian bank IntesaBci bought a 94·47% stake in Všeobecná úverová banka (General Credit Bank). In 2006 Slovenská Sporiteľňa had assets of 298bn. koruny and Všeobecná úverová banka 241bn. koruny. In 2006 there were 24 commercial banks including three building savings banks.

Gross external debt amounted to US$65,592m. in June 2012.

Foreign direct investment in Slovakia in 2010 amounted to US$526m., down from US$4,687m. in 2008.

There is a stock exchange in Bratislava.

ENERGY AND NATURAL RESOURCES

In 2008, 8·4% of energy consumption came from renewables (wind power, solar power, hydro-electric power, tidal power, geothermal energy and biomass), compared to the European Union average of 10·3%. A target of 14% has been set by the EU for 2020.

Environment

Slovakia's carbon dioxide emissions from the consumption and flaring of fossil fuels in 2008 were the equivalent of 6·9 tonnes per capita.

Electricity

Installed capacity in 2007 was 7·3m. kW, of which 2·5m. kW is hydro-electric and 2·2m. kW nuclear. Production in 2007 was 28·06bn. kWh, with consumption per capita 5,517 kWh. There were four nuclear reactors in use in 2009. In 2007 about 55% of electricity was nuclear-generated, making Slovakia one of the most nuclear-dependent nations in the world.

Oil and Gas

In 2006 natural gas reserves were 14bn. cu. metres; oil reserves were 9m. bbls in 2007. Natural gas production in 2006 amounted to 123m. cu. metres. Slovakia relies on Russia for almost all of its oil and gas.

Minerals

In 2009, 2·57m. tonnes of lignite were produced. 6·61m. tonnes of limestone were extracted in 2006. There are also reserves of copper, lead, zinc, iron, dolomite, rock salt and others.

Agriculture

In 2006 there were 1·34m. ha. of arable land and 25,260 ha. of permanent crops. In 2006 agriculture employed 4·4% of the economically active population.

Production, 2006 (in 1,000 tonnes): sugar beets, 1,371; wheat, 1,343; maize, 838; barley, 642; potatoes, 263; rapeseed, 260; sunflower seeds, 229; grapes, 52; apples, 41, tomatoes, 36.

Livestock, 2006: pigs, 1·11m.; cattle, 508,000; sheep, 333,000; chickens, 13m. Livestock products (in 1,000 tonnes): meat (2006), 196; milk (2003), 1,166; cheese (2006), 72; eggs (2003), 68.

Forestry

The area under forests in 2010 was 1·93m. ha., or 40% of the total land area. In 2007 timber production was 8·87m. cu. metres.

Fisheries

In 2010 the total catch was 1,608 tonnes, exclusively freshwater fish.

INDUSTRY

The main industries in Slovakia are machine engineering, chemical products, electrical apparatus, textiles, clothing and footwear, metallurgy, food and beverages, paper, wood and woodworking. Output included (in 1m. tonnes): cement (2008), 6·4; crude steel (2007), 4·8; pig iron (2007), 4·0; distillate fuel oil (2007), 2·8; coke (2006), 1·9; petrol (2007), 1·6; residual fuel oil (2007), 0·5. Motor vehicle production (2008), 524,859 units. Slovakia has the highest per capita car production of any country.

Labour

Out of 2,351,400 people in employment in 2011, 568,000 were in manufacturing, 304,000 in wholesale and retail trade/repair of motor vehicles, motorcycles and personal and household goods, 243,700 in construction and 164,200 in education. The average monthly salary in 2011 was €855. In Jan. 2011 the monthly minimum wage was €317. Unemployment stood at 19·2% in 2001, but then fell to 16·2% in 2005 and still further to 9·6% in 2008. It rose again to 14·5% in 2010 and was 14·7% in Dec. 2012. Youth unemployment was 33·2% in 2011. Long-term unemployment is also very high, with 63·9% of jobless Slovakians in 2011 having been out of work for more than a year (although the rate has fallen from 73·1% in 2006). In 2011 part-time work accounted for around 4% of all employment in Slovakia.

INTERNATIONAL TRADE

Imports and Exports

In 2008 imports (c.i.f.) totalled US$74,034m. (US$62,102m. in 2007); exports (f.o.b.), US$70,982m. (US$57,766m. in 2007). Principal import sources in 2004 were: Germany, 23·8%; Czech Republic, 13·2%; Russia, 9·4%; Italy, 5·6%; Austria, 4·3%. The leading export markets in 2004 were: Germany, 28·6%; Czech Republic, 13·3%; Austria, 7·8%; Italy, 6·4%; Poland, 5·5%. In 2004 machinery and transport equipment accounted for 39·6% of Slovakia's imports and 45·9% of exports; chemicals, manufactured goods classified chiefly by material and miscellaneous manufactured articles 39·3% of imports and 41·4% of exports; mineral fuels, lubricants and related materials 12·5% of imports and 6·7% of exports.

COMMUNICATIONS

Roads

In 2006 there were 43,770 km of roads, including 328 km of motorways. In 2007 there were 1,468,600 passenger cars in use, 244,800 vans and lorries, 10,300 buses and coaches and 61,200 motorcycles and mopeds. In 2006 there were 7,988 road accidents resulting in 608 fatalities.

Rail

In 2011 the length of railway routes was 3,624 km. Most of the network is 1,435 mm gauge with short sections on three other gauges. In 2011, 47·5m. passengers were carried and 43·7m. tonnes of freight. There are tram/light rail networks in Bratislava, Košice and Trenčianske Teplice.

Civil Aviation

The main international airport is at Bratislava (M. R. Stefánik), which handled 2,218,545 passengers in 2008 (77% of all passengers using Slovakian airports) and 6,961 tonnes of freight. There are also some international flights from Košice (590,919 passengers in 2008). Slovak Airlines (formerly the Slovak flag carrier) ceased operations in Feb. 2007, as did Air Slovakia in March 2010. SkyEurope (central Europe's first low-cost airline), which operated domestic services and also flew to a number of destinations in Europe, ceased operations in Sept. 2009. The main airline now is Danube Wings, launched in 2008.

Shipping

In 2006 vessels registered by Slovak enterprises numbered 261. They carried 1·7m. tonnes of goods and 111,000 passengers using inland waterway transport.

Telecommunications

In 2008 there were 1,098,000 main (fixed) telephone lines. In 2007 mobile phone subscribers numbered 6,068,000 (112·5 per 100 persons). In 2000 Deutsche Telekom bought a 51% stake in the state-owned Slovakia Telecom. There were 3,556,500 internet users in 2008. In March 2012 there were 1·9m. Facebook users.

SOCIAL INSTITUTIONS

Justice

The post-Communist judicial system was established by a federal law of July 1991. This provided for a unified system of four types of court: civil, criminal, commercial and administrative. Commercial courts arbitrate in disputes arising from business activities. Administrative courts examine the legality of the decisions of state institutions when appealed by citizens. In addition, there are military courts which operate under the jurisdiction of the Ministry of Defence. There is a Supreme Court, and a hierarchy of courts under the Ministry of Justice at republic, region and district level. District courts are courts of first instance.

Cases are usually decided by senates comprising a judge and two associate judges, although occasionally by a single judge. (Associate judges are citizens in good standing over the age of 25 who are elected for four-year terms). Regional courts are courts of first instance in more serious cases and also courts of appeal for district courts. Cases are usually decided by a senate of two judges and three associate judges, although again occasionally by a single judge. The Supreme Court interprets law as a guide to other courts and functions also as a court of appeal. Decisions are made by senates of three judges. The judges of the Supreme Court are nominated by the President; other judges are appointed by the National Council.

The population in penal institutions in Dec. 2007 was 7,986 (148 per 100,000 of national population).

Education

In 2006–07 there were 2,928 pre-school institutions with 140,014 children and 13,149 teachers, 2,283 primary schools with 508,130 pupils and 33,736 teachers, 246 grammar schools with 99,931 students and 7,602 teachers, and 249 vocational schools with 80,339 pupils and 7,728 teachers. There were 210 secondary vocational apprentice training centres with 60,621 pupils and 3,223 teachers, and 376 special schools with 31,390 children and 4,299 teachers. There were 30 universities or university-type institutions with 125,213 students.

In 2006 spending on education was 15·1% of total state budget expenditure.

The adult literacy rate in 2003 was 99·6% (99·7% among males and 99·6% among females).

Health

In 2008 there were 18,121 physicians, 2,745 dentists, 33,778 nurses and 2,777 pharmacists. There were 46,742 beds in health establishments in total in 2008, of which 33,912 were in hospitals. In 2008 Slovakia spent 7·8% of its GDP on health.

Welfare

The age of retirement is 62 for men and is set to rise from 53–57 to 62 for women by 2015. Pensions rose by 6·95% on 1 Jan. 2009. The average monthly old-age pension at Feb. 2009 was €335·50 with approximately 925,700 beneficiaries. Maternity benefit is paid for a total of 28 weeks or 37 weeks for a single mother and for multiple births. State unemployment benefit is 50% of previous earnings during the first three months, thereafter 45% of previous earnings.

RELIGION

A federal Czechoslovakian law of July 1991 provides the basis for church-state relations and guarantees the religious and civic rights of citizens and churches. Churches must register to become legal entities but operate independently of the state. In 2011, 62·0% of the population were Roman Catholic, 5·9% members of the Evangelical Church of the Augsburg Confession, 5·8% Greek Catholic and 1·8% Calvinist. In Feb. 2013 there were two cardinals.

CULTURE

Košice is one of two European Capitals of Culture for 2013. The title attracts large European Union grants.

World Heritage Sites

There are seven UNESCO sites in Slovakia: Vlkolínec (inscribed on the list in 1993), a group of 45 traditional log houses; Banská Štiavnica (1993), a medieval mining town; Levoča, Spišský and the Associated Cultural Monuments (1993 and 2009)—13th century Spiš Castle is one of the largest castle complexes in central Europe; Bardejov Town Conservation Reserve (2000), a medieval fortified town; and Wooden Churches of the Slovak part of the Carpathian Mountain Area (2008).

Slovakia shares two UNESCO sites with other countries: the Caves of Aggtelek and Slovak Karst (1995, 2000 and 2008) with Hungary and the primeval beech forests of the Carpathians (2007) with Ukraine.

Press

Slovakia had ten daily newspapers in 2008 (nine paid-for and one free) with a combined average daily circulation of 508,000.

Tourism

In 2010, 1,327,000 non-resident tourists stayed in holiday accommodation (1,298,000 in 2009); there were 3,126 accommodation establishments in 2010 with 57,406 rooms and 147,492 beds.

Festivals

The Bratislava Rock Festival takes place in June and the Bratislava Music Festival and Interpodium is in Oct. The Myjava Folklore Festival is held each June, the Zvolen Castle Games in June–July, Theatrical Nitra is in Sept., and there is an annual Spring Music Festival in Košice.

DIPLOMATIC REPRESENTATIVES

Of Slovakia in the United Kingdom (25 Kensington Palace Gdns, London, W8 4QY)
Ambassador: Miroslav Wlachovský.

Of the United Kingdom in Slovakia (Panska 16, 81101 Bratislava)
Ambassador: Susannah Montgomery.

Of Slovakia in the USA (3523 International Court, NW, Washington, D.C., 20008)
Ambassador: Peter Kmec.

Of the USA in Slovakia (5 Hviezdoslavovo Sq., 81102 Bratislava)
Ambassador: Theodore Sedgwick.

Of Slovakia to the United Nations
Ambassador: František Ružička.

Of Slovakia to the European Union
Permanent Representative: Ivan Korčok.

FURTHER READING

Fisher, Sharon, *Political Change in Post-Communist Slovakia and Croatia: From Nationalist to Europeanist.* 2006
Kirschbaum, S. J., *A History of Slovakia: the Struggle for Survival.* 1995

National Statistical Office: Statistical Office of the Slovak Republic, Miletičova 3, 82467 Bratislava.
Website: http://portal.statistics.sk

SLOVENIA

0 25 mi
0 25 km

HUNGARY
AUSTRIA
ITALY
Jesenice
Maribor
Škofja Loka
LJUBLJANA
SLOVENIA
Gulf of Venice
Koper
CROATIA

© Research Machines plc 2006

Republika Slovenija
(Republic of Slovenia)

Capital: Ljubljana
Population projection, 2015: 2·05m.
GNI per capita, 2011: (PPP$) 24,914
HDI/world rank: 0·884/21
Internet domain extension: .si

KEY HISTORICAL EVENTS

The region was settled by Celts and Illyrians in pre-Roman times and fell to Rome in the 1st century AD. In the 6th century Slavic tribes arrived and part of the territory came under the Slavic Duchy of Karantania in the 7th century. In 745 Karantania became part of the Frankish Empire as an independent country with its own laws and language. After passing under the rule of Bavarian dukes and the Republic of Venice, it joined its neighbouring Slovene-inhabited areas to become part of the Habsburg dynasty in the 14th century.

The Slovenes retained a strong national identity, aided by the printing of the first Slovenian books in 1550 and the translation of the Protestant Bible into Slovenian in 1584. A 12-day peasant revolt in 1573 was bloodily suppressed. In the 19th century a nationalist movement developed, demanding Slovenian autonomy within the Habsburg monarchy. Administrative autonomy was granted in the province of Carinthia while Slovenes in other areas gained some cultural rights. With the demise of the Habsburg monarchy, Slovenia became part of Austria-Hungary. Following Germany's defeat in the First World War, the Slovenes joined their Slav neighbours in the Kingdom of the Serbs, Croats and Slovenes on 1 Dec. 1918. The country was renamed Yugoslavia in 1929. During the Second World War Slovenian territory was divided and annexed by the Axis powers. In 1945 Slovenia became a constituent republic of the Socialist Federal Republic of Yugoslavia under Josep Tito.

In the 1980s nationalism grew stronger throughout Yugoslavia, spearheaded by the Serbs' call for Serbian unity. In Oct. 1989 the Slovene Assembly passed a constitutional amendment giving it the right to secede from Yugoslavia and in a referendum on 23 Dec. 1990, 88·5% voted for independence. Slovenia declared independence on 25 June 1991 but, at EU-sponsored peace talks, suspended the claim for three months. Fighting between Slovenia

and federal troops ended with a federal withdrawal and on 8 Oct. 1991 Slovenia again declared its independence.

In 1992 Slovenia removed more than 18,000 non-Slovene residents from its records and cancelled their rights. Under pressure from the EU, which Slovenia joined in May 2004, partial restoration of citizenship was granted to about 12,000 people, though a 2004 referendum rejected proposals to restore full retrospective rights. Slovenia joined NATO in March 2004 and became the first former communist bloc country to adopt the euro in Jan. 2007.

TERRITORY AND POPULATION

Slovenia is bounded in the north by Austria, in the northeast by Hungary, in the southeast and south by Croatia and in the west by Italy. The length of coastline is 47 km. Its area is 20,273 sq. km. In Jan. 2011 the population at the register-based census was 2,050,189 (1,035,626 females); density per sq. km, 101·1. The capital is Ljubljana: 2011 census population, 272,220. Maribor (population of 95,171 in 2011) is the other major city. In 2011, 49·5% of the population lived in urban areas.

The UN gives a projected population for 2015 of 2·05m.

The official language is Slovene.

In April 2004 voters rejected plans to restore the civil rights of Slovenia's ethnic minorities, mainly nationals of other former Yugoslav republics, which were 'erased' in 1992.

SOCIAL STATISTICS

Statistics for calendar years:

	Live births	Deaths	Growth rate per 1,000	Marriages	Divorces
2004	17,961	18,523	−0·3	6,558	2,411
2005	18,157	18,825	−0·3	5,769	2,647
2006	18,932	18,180	0·4	6,368	2,334
2007	19,823	18,584	0·6	6,373	2,617
2008	21,817	18,308	1·7	6,703	2,246

Rates, 2008 (per 1,000 population): birth, 10·8; death, 9·1. Infant mortality, 2010: two per 1,000 live births (one of the lowest rates in the world). There were 529 suicides in 2006 (22·8 per 100,000 population).

In 2005 the most popular age range for marrying was 25–29 years for both males and females. Expectation of life, 2007, was 74·4 years for males and 81·7 for females. Annual population growth rate, 2000–05, 0·1%. Fertility rate, 2008, 1·4 births per woman.

CLIMATE

Summers are warm, winters are cold with frequent snow. Ljubljana, Jan. −4°C, July 22°C. Annual rainfall 1,383 mm.

CONSTITUTION AND GOVERNMENT

The constitution became effective on 23 Dec. 1991. Slovenia is a parliamentary democratic republic with an executive that consists of a directly-elected president and a prime minister, aided by a council of ministers. It has a bicameral parliament (*Skupščina Slovenije*), consisting of a 90-member *National Assembly* (*Državni Zbor*), 88 members elected for four-year terms by proportional representation with a 4% threshold and two members elected by ethnic minorities; and a 40-member, advisory *State Council* (*Državni Svet*), elected for five-year terms by interest groups and regions. It has veto powers over the National Assembly. Administratively the country is divided into 199 municipalities and 11 urban municipalities.

The Judicial branch consists of a *Supreme Court*, whose judges are elected by the National Assembly, and a *Constitutional Court*, whose judges are elected for nine-year terms by the National Assembly and nominated by the president.

National Anthem

'Zdravljica' ('A Toast'); words by Dr France Prešeren, tune by Stanko Premrl.

GOVERNMENT CHRONOLOGY

(LDS = Liberal Democracy of Slovenia; NSi = New Slovenia Christian People's Party; PS = Positive Slovenia; SD = Social Democrats; SDS = Slovenian Democratic Party; SKD = Slovenian Christian Democrats; SLS+SKD = Slovenian People's Party; n/p = non-partisan)

Presidents since 1990.

1990–2002	n/p	Milan Kučan
2002–07	LDS	Janez Drnovšek
2007–12	n/p	Danilo Türk
2012–	SD	Borut Pahor

Prime Ministers since 1990.

1990–92	SKD	Lojze Peterle
1992–2000	LDS	Janez Drnovšek
2000	SLS+SKD, NSi	Andrej Bajuk
2000–02	LDS	Janez Drnovšek
2002–04	LDS	Anton (Tone) Rop
2004–08	SDS	Janez Janša
2008–12	SD	Borut Pahor
2012–13	SDS	Janez Janša
2013–	PS	Alenka Bratušek

RECENT ELECTIONS

Presidential elections were held on 11 Nov. and 2 Dec. 2012. The turnout in the first round was 48·2% and in the second round 42·0%. Borut Pahor (Social Democrats) won 39·9% of the vote in the first round, Danilo Türk (ind.) 35·9% and Milan Zver (Slovenian Democratic Party) 24·2%. In the run-off held on 2 Dec. 2012 Borut Pahor received 67·4% of votes cast against 32·6% for Danilo Türk.

Elections were held for the National Assembly on 4 Dec. 2011; turnout was 65·6%. Positive Slovenia (PS) won 28 seats with 28·5% of votes cast; the Slovenian Democratic Party (SDS), 26 with 26·2%; the Social Democrats (SD), 10 with 10·5%; Gregor Virant's Civic List (LGV), 8 with 8·4%; the Democratic Party of Pensioners of Slovenia (DeSUS), 6 with 7·0%; the Slovenian People's Party (SLS), 6 with 6·8%; the New Slovenia–Christian People's Party (NSi), 4 with 4·9%.

European Parliament

Slovenia has eight representatives. At the June 2009 elections turnout was 28·3% (28·4% in 2004). The SDS won 2 seats with 26·9% of the vote (political affiliation in European Parliament: European People's Party); SD, 2 with 18·5% (Progressive Alliance of Socialists and Democrats); New Slovenia-Christian People's Party, 1 with 16·3% (European People's Party); LDS, 1 with 11·5% (Alliance of Liberals and Democrats for Europe); Zares, 1 with 9·8% (Alliance of Liberals and Democrats for Europe).

CURRENT GOVERNMENT

President: Borut Pahor; b. 1963 (Social Democrats; sworn in 22 Dec. 2012).

In March 2013 the government comprised:

Prime Minister: Alenka Bratušek; b. 1970 (PS; sworn in 20 March 2013).

Minister of Agriculture and Environment: Dejan Židan. *Culture:* Uroš Grilc. *Defence:* Roman Jakič. *Economic Development and Technology:* Stanko Stepišnik. *Education, Science and Sport:* Jernej Pikalo. *Finance:* Uroš Čufer. *Foreign Affairs:* Karl Erjavec. *Health:* Tomaž Gantar. *Infrastructure and Spatial Planning:* Igor Maher. *Interior and Public Administration:* Gregor Virant. *Justice:* Senko Pličanič. *Labour, Family, Social Affairs and Equal Opportunities:* Anja Kopač Mrak. *Minister without Portfolio Responsible for Slovenes Abroad:* Tina Komel.

President's Website: http://www.up-rs.si

CURRENT LEADERS

Borut Pahor

Position
President

Introduction
Borut Pahor of the Social Democrats was elected president in the second round of voting in Dec. 2012, defeating the incumbent Danilo Türk. He had previously served as prime minister at the head of a centre-left coalition from Nov. 2008 until Jan. 2012 when he was succeeded by Janez Janša of the Slovenian Democratic Party.

Early Life
Pahor was born in 1963 in Postojna, southeast of Ljubljana. He studied political science at the University of Ljubljana and, at the age of 26, became the youngest-ever member of the central committee of the Slovenian branch of the Communist Party, gaining a reputation as a reformist. In 1992, after the collapse of communism in eastern Europe and the fragmentation of Yugoslavia, he was elected a deputy in the National Assembly of the newly independent Slovenia.

A year later Pahor became deputy leader of the newly formed United List of Social Democrats (ZLSD). At the third ZLSD conference in March 1997 he was elected party leader on a centrist, 'third way' platform. At the 2000 elections, having led his party into a coalition with Janez Drnovšek's Liberal Democracy of Slovenia, Pahor won a third term in the National Assembly and became its president. In 2004 he was elected to the European Parliament, where he joined the Socialist Group. In the same year the centre-left coalition of which he was a member lost to the centre-right alliance led by the Slovenian Democratic Party. In 2005, on his initiative, the ZLSD was renamed the Social Democrats.

Parliamentary elections held on 21 Sept. 2008 saw the Social Democrats gain 29 of the 90 available seats to become the country's largest party. It formed a coalition government with Zares, Liberal Democracy of Slovenia and the Democratic Party of Pensioners of Slovenia. Pahor was sworn in as premier on 21 Nov. 2008.

Career in Office
Pahor promised to cut taxes, reform the national health and pension systems and increase investment incentives, but was forced to redraft the 2009 national budget in the wake of the global financial crisis, allowing for a significant increase in the budget deficit.

In Dec. 2008, owing to an unresolved border dispute, Slovenia obstructed Croatia's accession to the European Union and, in March 2009, was the last member of NATO to ratify Croatia's accession to the alliance. However, in Sept. 2009 Pahor said that the dispute would no longer prejudice Croatia's EU accession negotiations and in Nov. Slovenia lifted its embargo. In March 2010 he agreed to opposition demands for a referendum on allowing the border issue to go to international arbitration, and in June voters narrowly backed the plan.

Following a gas dispute between Russia and Ukraine in 2009, Pahor moved to diversify Slovenia's energy supply, so reducing dependence on Russia, and established the Strategic Energy Council to advise the government. He has also stated his commitment to the fight against climate change.

On 20 Sept. 2011, amid internal coalition disputes and an inability to implement economic reforms, parliament voted 51–36 against a motion of confidence. Pahor's government fell and assumed a caretaker role. Following inconclusive elections in Dec. 2011 he was succeeded as premier by former prime minister, Janez Janša.

Pahor nevertheless continued in politics and announced his candidacy for the presidential elections held in Nov. and Dec. 2012, in which he won a second round victory with 67% of the vote.

Alenka Bratušek

Position
Prime Minister

Introduction
Alenka Bratušek, leader of the Positive Slovenia Party (PS), became prime minister in March 2013 after the previous government fell to a no confidence vote. Heading a four-party coalition covering a wide spectrum of political opinion, Bratušek has pledged to balance austerity policies with measures to boost growth.

Early Life
Born on 31 March 1970 in Celje, Alenka Bratušek graduated from the faculty of natural sciences and technology at Ljubljana University in the early 1990s and earned a masters degree in management from the same institution in 2006. Entering public administration in 1995, she took a job at the ministry of the economy before moving to the ministry of finance in 1999. She became head of the ministry's budget directorate, responsible for administering the national budget, overseeing local government finances and drawing on European Union funding.

Bratušek stood unsuccessfully for parliament for the Zares party in 2008. Having joined the newly-formed PS in 2011, she was elected to parliament in the election of Dec. that year when the PS unexpectedly claimed the largest share of the vote. However, the party was unable to form a government and went into opposition to Janez Janša's centre-right administration. In this period, Bratušek chaired the commission for public finance control and served on the committee for finance and monetary policy, as well as on the committee for justice, public administration and local self-government. In line with PS policy she opposed plans for a government-subsidized 'bad bank' to deal with Slovenia's economic crisis, and advocated a relaxation of austerity together with policies to boost growth.

Following the publication of a report that alleged corruption against both Prime Minister Janša and PS leader Zoran Janković, Bratušek replaced Janković as acting party leader in Jan. 2013. When the government fell to a no confidence vote in Feb. 2013, Bratušek led all-party negotiations and emerged as leader of a broad coalition, uniting the centre-left PS and Social Democrats (SD) with the right-leaning Civic List (DL) and the Democratic Party of Pensioners of Slovenia (DeSUS).

Career in Office
Bratušek was voted in as prime minister on 20 March with the backing of 52 of 90 MPs. She took office with the country facing acute economic problems. She must attempt to address the immediate financial crisis while implementing policies to secure long-term stability.

DEFENCE

Compulsory military service for seven months ended in Sept. 2003. The army became fully professional in 2010 when the compulsory reserve was replaced by a new system of voluntary reserve service.

In 2008 military expenditure totalled US$834m. (US$415 per capita), representing 1·5% of GDP.

Army
Personnel (2007), 5,973 and an army reserve of 20,000. There is a paramilitary police force of 4,500 with 5,000 reserves.

Navy
There is an Army Maritime element numbering 47 personnel in 2007.

Air Force
The Army Air element numbers 530 with eight combat-capable helicopters.

INTERNATIONAL RELATIONS

Slovenia held a referendum on EU membership on 23 March 2003, in which 89·6% of votes cast were in favour of accession. It became a member of NATO on 29 March 2004 and the EU on 1 May 2004.

ECONOMY

Agriculture accounted for 2% of GDP in 2007, industry 34% and services 63%.

Overview
Slovenia was among the most developed of the ten countries that joined the EU in May 2004, benefiting from a well-educated, productive workforce, excellent infrastructure and a strategic location between Western and Eastern Europe. In Jan. 2007 it was the first of the ten to adopt the euro.

The economy enjoyed sound macroeconomic policy in the years prior to the global financial crisis, with tight monetary and fiscal policies contributing to small fiscal deficits and single-digit inflation. However, heavy dependence on exports (69% for EU markets in 2008) leaves the country vulnerable to wider economic trends. Despite its pre-2008 success, Slovenia was also criticized for having an unfriendly, inefficient business environment. The pace of privatization has been slower than in most other Central and Eastern European countries, with enterprises in key sectors remaining under state ownership. It has consequently attracted less FDI than other 2004 EU accession countries.

The economy contracted by nearly 8% in 2009, the first recession the country had experienced since independence. After growing by 1·2% in 2010, the economy recorded modest growth of just 0·6% in 2011. A government stimulus package in 2009 included subsidies for shorter working hours and R&D, as well as guarantee schemes in the financial sector. There followed a raft of austerity measures prompting popular protests that, along with a corruption scandal, led to the fall of Janez Janša's government in Feb. 2013.

Slovenia has one of the most generous welfare states in Europe and the European Commission warns of the heavy future cost of the pension system. Age-related expenditures are expected to reach 13% of GDP in the long term.

Currency
On 1 Jan. 2007 the euro (EUR) replaced the *tolar* (SLT) as the legal currency of Slovenia at the irrevocable conversion rate of 239·64 tolars to one euro. Inflation was 1·8% both in 2010 and 2011. Foreign exchange reserves were US$552m. and gold reserves 103,000 troy oz in Aug. 2010. Total money supply in July 2005 was 1,032·3bn. tolars.

Budget
In 2009 general government revenues totalled €15·29bn. and expenditures €17·35bn. In 2008 tax revenue accounted for 54·2% of revenue and social security contributions 33·7%; social protection accounted for 36·0% of expenditure and education 13·9%.

Slovenia's budget deficit in 2011 was 6·4% of GDP (2010, 6·0%; 2009, 6·1%). The required target set by the EU is a budget deficit of no more than 3%.

VAT is 20·0% (reduced rate, 8·5%).

Performance
The economy shrank by 7·8% in 2009 but grew by 1·2% in 2010 and 0·6% in 2011. Of all the central and eastern European countries that joined the European Union in May 2004, Slovenia has the highest per capita GDP at 85% of the EU average in 2010 (higher than in Portugal and only marginally lower than in Greece). Total GDP in 2011 was US$49·5bn.

Banking and Finance
The central bank and bank of issue, the Bank of Slovenia, was founded on 25 June 1991 upon independence. Its current *Governor* is Marko Kranjec, appointed on 3 July 2007 for a term of six years. In 2009 there were 20 commercial banks (five subsidiaries of foreign banks and one branch office of a foreign bank) and three savings banks. The largest bank is Nova Ljubljanska banka (NLB), which has a market share of around a third and had assets in 2009 of US$22·0bn. Other large banks are Nova Kreditna Banka Maribor (NKBM) and Abanka Vipa. In 2010 Slovenia received US$834m. of foreign direct investment, down from a high of US$1,947m. in 2008.

In June 2012 gross external debt amounted to US$52,416m.

Foreign debt amounted to €41,616m. at Aug. 2010. In 1997 Slovenia accepted 18% of the US$4,400m. commercial bank debt of the former Yugoslavia, thus gaining access to international capital markets.

There is a stock exchange in Ljubljana (LjSE).

ENERGY AND NATURAL RESOURCES
In 2008, 15·1% of energy consumption came from renewables (wind power, solar power, hydro-electric power, tidal power, geothermal energy and biomass), compared to the European Union average of 10·3%. A target of 25% has been set by the EU for 2020.

Environment
Slovenia's carbon dioxide emissions from the consumption and flaring of fossil fuels were the equivalent of 8·3 tonnes per capita in 2008. An *Environmental Performance Index* compiled in 2008 ranked Slovenia 15th in the world, with 86·3%. The index examined various factors in six areas—air pollution, biodiversity and habitat, climate change, environmental health, productive natural resources and water resources.

Electricity
Installed capacity was 3·0m. kW in 2007. There was one nuclear power station in operation. The total amount of electricity produced in 2007 was 15,043m. kWh (6,082m. kWh thermal, 5,695m. kWh nuclear and 3,266m. kWh hydro-electric). Consumption per capita in 2007 was 7,597 kWh.

Minerals
Brown coal production was 4,535,000 tonnes in 2007.

Agriculture
Only around 1·9% of the population worked in agriculture, forestry and fisheries in 2011. Output (in 1,000 tonnes) in 2011: maize, 349; wheat, 153; grapes, 121; apples, 105; potatoes, 96.

Livestock in 2011: cattle, 462,300; pigs, 347,310; sheep, 119,976; poultry, 4,006,718. Livestock products, 2010 (unless otherwise indicated): poultry meat, 60,700 tonnes; pork, bacon and ham, 44,100 tonnes; beef and veal, 35,800 tonnes; milk (2011), 584·1m. litres.

In 2011 there were 168,744 ha. of arable land and 26,867 ha. of permanent crops.

Forestry
In 2010 the area under forests was 1·25m. ha., or 62% of the total land area. Timber production in 2007 was 2·88m. cu. metres.

Fisheries
Total fish catch in 2010 was 932 tonnes. Freshwater farming produced 931 tonnes in 2009.

INDUSTRY
Industry contributed 19·8% of GDP in 2008. Traditional industries are metallurgy, furniture-making and textiles. The manufacture of electric goods and transport equipment is being developed.

Production in 2007 (in 1,000 tonnes): ready mixed concrete, 2,922; basic iron and steel and ferro alloys, 1,643; cement (2006), 1,269; paper and paperboard, 720; aluminium and aluminium products, 446; plastics in primary form, 291; passenger cars, 174,209 units.

Labour
Registered labour force was 934,658 in 2011. In 2003, 433,098 people worked in services, 308,059 in industry, and 36,092 in agriculture and forestry. In 2011 there were 110,692 registered unemployed. The unemployment rate in Dec. 2012 was 10·0% (compared to 8·2% in 2011 as a whole). In 2011 the average monthly gross wage per employee was €1,525.

INTERNATIONAL TRADE
Imports and Exports
Imports of goods (c.i.f.) in 2011 were worth US$31,253·8m. and exports of goods (f.o.b.) US$28,984·5m. Exports of goods accounted for 58·8% of GDP in 2011.

Major imports of goods in 2011 were petroleum and petroleum products (10·9%), road vehicles (9·6%), electrical machinery, apparatus and appliances (6·4%), and iron and steel (4·7%). Major exports of goods in 2011 were road vehicles (12·4%), electrical machinery, apparatus and appliances (10·4%), and medical and pharmaceutical products (9·2%).

Share of imports from principal markets in 2011: Germany, 18·7%; Italy, 17·8%; Austria, 11·6%; France, 4·7%; Hungary, 4·2%. Exports: Germany, 21·1%; Italy, 11·9%; Austria, 7·8%; France, 6·8%; Croatia, 6·7%. About 74% of trade is with fellow EU countries.

COMMUNICATIONS
Roads
In 2007 there were 38,708 km of road including 606 km of motorways. There were in 2007: 1,020,100 passenger cars; 2,300 buses and coaches; 81,500 vans and lorries; and 71,500 motorcycles and mopeds. 817m. passenger-km were travelled by road in 2007. There were 11,414 traffic accidents in 2007 in which 293 persons were killed. With 14·5 deaths per 100,000 population in 2007, Slovenia has among the highest death rates in road accidents of any industrialized country.

Rail
There were 1,228 km of 1,435 mm gauge in 2008, of which 503 km were electrified. In 2008, 16·7m. passengers and 17·3m. tonnes of freight were carried. In Sept. 2010 the state-owned railway companies of Slovenia, Croatia and Serbia announced the creation of a joint venture called Cargo 10 to improve the management of freight trains along the route known as Corridor 10 that passes through all three countries.

Civil Aviation
There is an international airport at Ljubljana (Brnik), which handled 1,433,855 passengers (all on international flights) and 14,333 tonnes of freight in 2009. The national carrier, Adria Airways, has flights to most major European cities and Tel Aviv.

In 2006 scheduled airline traffic of Slovenia-based carriers flew 15m. km, carrying 850,000 passengers.

Shipping

A total of 5,433 vessels arrived at or departed from Slovenia's ports in 2008 (4,447 cargo-carrying vessels and 986 passenger ships), including 4,474 at Koper. Goods traffic totalled 16·6m. tonnes in 2008 (Koper, 16·5m. tonnes).

Telecommunications

In 2007 Slovenia had 2,785,600 telephone subscribers (1,391·7 per 1,000 inhabitants), including 1,928,400 mobile phone subscribers. The leading telecommunications operator is the state-owned Telekom Slovenije. The number of internet users in 2007 was 1,060,800. In March 2012 there were 671,000 Facebook users.

SOCIAL INSTITUTIONS

Justice

There are 44 district courts, 11 regional courts, four higher courts, an administrative court and a supreme court. There are also four labour and social courts, and a higher labour and social court. The population in penal institutions in Sept. 2008 was 1,317 (65 per 100,000 of national population).

Education

Adult literacy rate in 2009 was an estimated 99·7%. In 2007 there were 95,173 pupils and 6,111 teaching staff in primary schools; and 165,467 pupils and 16,179 teaching staff in secondary schools. There were 115,944 students and 5,609 academic staff in tertiary education in 2007. There are three public universities—at Koper (University of Primorska), Ljubljana and Maribor—and one private university, the University of Nova Gorica.

In 2006 public expenditure on education came to 5·9% of GNI and 12·9% of total government spending.

Health

In 2008 there were 29 hospitals with 9,586 beds. In 2007 there were 4,981 doctors, 1,269 dentists, 15,897 nurses and 1,564 pharmacists. Slovenia spent 8·3% of its GDP on health In 2008.

Welfare

There were 551,258 people receiving pensions in 2008, of which 342,992 were old-age pensioners. The average retirement ages in 2008 were 61 years 11 months for men and 57 years 7 months for women. Plans to speed up the process of increasing the retirement age to 65 for both men and women, a move highlighted as key to long-term fiscal sustainability by the EU and the IMF, were put to the vote in a referendum in June 2011 but received only 28·0% backing. In 2006 spending on social protection accounted for 22·8% of GDP.

RELIGION

57·8% of the population were Roman Catholic according to the 2002 census. In Feb. 2013 there was one cardinal.

CULTURE

World Heritage Sites

Slovenia has three sites on the UNESCO World Heritage List: Škocjan Caves (inscribed on the list in 1986), consisting of limestone caves, passages and waterfalls more than 200 metres deep; the Prehistoric Pile dwellings around the Alps (2011), shared with Austria, France, Germany, Italy and Switzerland; and Heritage of Mercury: Almadén and Idrija (2012), shared with Spain.

Press

In 2008 there were eight daily newspapers with a combined circulation of 380,000 and 253 non-dailies. The most popular paid-for daily is *Slovenske novice*, with an average daily circulation in 2008 of 88,000.

Tourism

In 2010, 1,869,000 non-resident tourists stayed in holiday accommodation (1,824,000 in 2009) including: 412,000 from Italy; 202,000 from Austria; 194,000 from Germany; 103,000 from Croatia.

Festivals

Kurentovanje is a ten-day carnival to chase away the winter, which ends on Mardi Gras—the biggest celebrations occur in the town of Ptuj. Lent Festival in Maribor is a major performing arts festival held in late June to early July and similar summer festivals take place in Ljubljana and in the region of Primorska throughout July and Aug. Two of the biggest music festivals are the Druga Godba world music festival in Ljubljana in May–June and Rock Otocec at Novo Mesto in July.

DIPLOMATIC REPRESENTATIVES

Of Slovenia in the United Kingdom (10 Little College St., London, SW1P 3SH)
Ambassador: Iztok Jarc.

Of the United Kingdom in Slovenia (4th Floor, 3 Trg Republike, 1000 Ljubljana)
Ambassador: Andrew Page.

Of Slovenia in the USA (1525 New Hampshire Ave., NW, Washington, D.C., 20036)
Ambassador: Roman Kirn.

Of the USA in Slovenia (Presernova 31, 1000 Ljubljana)
Ambassador: Joseph Mussomeli.

Of Slovenia to the United Nations
Ambassador: Vacant.
Chargé d'affaires a.i: Matej Marn.

Of Slovenia to the European Union
Permanent Representative: Rado Genorio.

FURTHER READING

Benderly, J. and Kraft, E. (eds.) *Independent Slovenia: Origins, Movements, Prospects.* 1995
Cox, John K., *Slovenia.* 2005
Fink-Hafner, Danica and Robbins, John R. (eds.) *Making a New Nation: Formation of Slovenia.* 1997

National Statistical Office: National Statistical Office, Vožarski Pot 12, 1000 Ljubljana.
Website: http://www.stat.si

SOLOMON ISLANDS

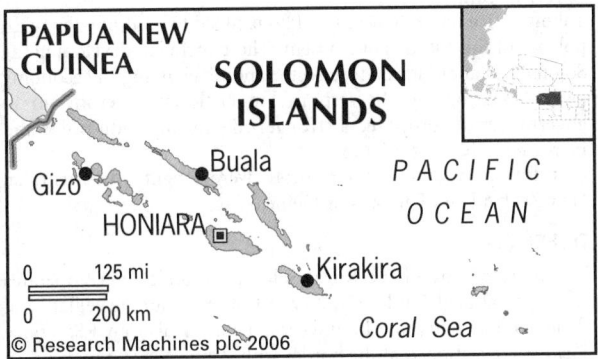

Capital: Honiara
Population projection, 2015: 610,000
GNI per capita, 2011: (PPP$) 1,782
HDI/world rank: 0·510/142
Internet domain extension: .sb

KEY HISTORICAL EVENTS

The Solomon Islands were discovered by Europeans in 1568 but 200 years passed before contact was made again. The southern Solomon Islands were placed under British protection in 1893; the eastern and southern outliers were added in 1898 and 1899. Santa Isabel and the other islands to the north were ceded by Germany in 1900. Full internal self-government was achieved on 2 Jan. 1976 and independence on 7 July 1978.

In 1997 Bartholomew Ulufa'alu, a Malaitan, was elected prime minister. The following year, fighting broke out between the Isatubu Freedom Movement, which claimed to be representative of the native peoples of the island of Guadalcanal, and the Malaitan Eagle Force. In 2000 the Malaita Eagles led a coup which deposed Ulufa'alu, who was held at gunpoint for two days. After the failure of a 2001 peace accord, conflict between the Malaita Eagles and the Isatabu Freedom Movement escalated in 2003 and an Australian-led peacekeeping force landed to restore order. After some initial success the force was scaled back towards the end of the year.

In 2006 Manasseh Sogavare became prime minister, having previously held the office following the 2000 coup. His appointment was marked by rioting in Honiara and the destruction of the city's Chinatown as tensions rose over the political and business influence of the local Chinese population.

TERRITORY AND POPULATION

The Solomon Islands lie within the area 5° to 12° 30' S. lat. and 155° 30' to 169° 45' E. long. The group includes the main islands of Guadalcanal, Malaita, New Georgia, San Cristobal (now Makira), Santa Isabel and Choiseul; the smaller Florida and Russell groups; the Shortland, Mono (or Treasury), Vella La Vella, Kolombangara, Ranongga, Gizo and Rendova Islands; to the east, Santa Cruz, Tikopia, the Reef and Duff groups; Rennell and Bellona in the south; Ontong Java or Lord Howe to the north; and many smaller islands. The land area is estimated at 28,370 sq. km (10,954 sq. miles). The larger islands are mountainous and forest clad, with flood-prone rivers of considerable energy potential. Guadalcanal has the largest land area and the greatest amount of flat coastal plain. Population at the census of Nov. 2009, 515,870 (251,415 females); density, 18·2 per sq. km. In 2011, 18·9% of the population lived in urban areas.

The UN gives a projected population for 2015 of 610,000.

The islands are administratively divided into nine provinces plus a Capital Territory. Area, population and capital:

Province	Sq. km	2009 census	Capital
Capital Territory	22	64,602	—
Central Islands	615	26,051	Tulagi
Choiseul	3,837	26,379	Taro
Guadalcanal	5,336	93,613	Honiara
Isabel	4,136	26,158	Buala
Makira and Ulawa	3,188	40,419	Kirakira
Malaita	4,225	137,596	Auki
Rennell and Bellona	671	3,041	Tigoa
Temotu	895	21,362	Lata (Santa Cruz)
Western	5,475	76,649	Gizo

The capital, Honiara, on Guadalcanal, is the largest urban area, with a population in 2009 of 64,602. 93% of the population are Melanesian; other ethnic groups include Polynesian, Micronesian, European and Chinese.

English is the official language, and is spoken by 1–2% of the population. In all 120 indigenous languages are spoken; Melanesian languages are spoken by 85% of the population.

SOCIAL STATISTICS

2008 estimates: births, 16,000; deaths, 3,000. Estimated birth rate in 2008 was 30·4 per 1,000 population; estimated death rate, 6·2. Life expectancy, 2007, 66·7 years for women and 64·9 for men. Annual population growth rate, 2000–08, 2·6%. Infant mortality, 2010, 23 per 1,000 live births; fertility rate, 2008, 3·9 births per woman.

CLIMATE

An equatorial climate with only small seasonal variations. Southeast winds cause cooler conditions from April to Nov., but northwest winds for the rest of the year bring higher temperatures and greater rainfall, with annual totals ranging between 80" (2,000 mm) and 120" (3,000 mm).

CONSTITUTION AND GOVERNMENT

The Solomon Islands are a constitutional monarchy with the British Sovereign (represented locally by a Governor-General, who must be a Solomon Island citizen) as Head of State. Legislative power is vested in the single-chamber *National Parliament* composed of 50 members, elected by universal adult suffrage for four years. Parliamentary democracy is based on a multi-party system. Executive authority is effectively held by the Cabinet, led by the Prime Minister.

The *Governor-General* is appointed for up to five years, on the advice of Parliament, and acts in almost all matters on the advice of the Cabinet. The Prime Minister is elected by and from members of Parliament. Other Ministers are appointed by the Governor-General on the Prime Minister's recommendation, from members of Parliament. The Cabinet is responsible to Parliament. Emphasis is laid on the devolution of power to provincial governments, and traditional chiefs and leaders have a special role within the arrangement.

National Anthem

'God save our Solomon Islands from shore to shore'; words and tune by P. Balekana.

RECENT ELECTIONS

National elections were held on 4 Aug. 2010. The Democratic Party won 14 seats, Our Party 4, SI Party for Rural Advancement 3, Direct Development Party 3, Reform and Democratic Party of the Solomon Islands 2, Independent Democratic Party 2, ind. 17 and five parties won one seat each. Gordon Darcy Lilo was elected prime minister on 16 Nov. 2011, defeating Milner Tozaka by 29 votes to 20 in the parliamentary vote.

Frank Kabui was elected governor-general by parliament on 15 June 2009. He defeated Edmund Andersen and the incumbent, Sir Nathaniel Waena.

CURRENT GOVERNMENT

Governor-General: Sir Frank Kabui (since 7 July 2009).

In March 2013 the government comprised:

Prime Minister: Gordon Darcy Lilo; b. 1965 (sworn in 16 Nov. 2011).

Deputy Prime Minister and Minister of the Interior: Manasseh Maelanga.

Minister of Agriculture and Livestock: David Tome. *Public Service:* Stanley Sofu. *National Planning and Aid Co-ordination:* Connelly Sandakabatu. *Finance and the Treasury:* Rick Hou. *Environment, Climate Change, Disaster Management and Meteorology:* Bradley Tovosia. *Police, National Security and Correctional Services:* Christopher Laore. *Justice and Legal Affairs:* Commins Aston Mewa. *Education and Human Resources Development:* Dickson Ha'Amori. *Health and Medical Services:* Charles Sigoto. *Foreign Affairs and External Trade:* Clay Forau. *Commerce, Industry, Labour and Immigration:* Elijah Doro Muala. *Culture and Tourism:* Samuel Manetoali. *Infrastructure Development:* Seth Gukuna. *Communication and Aviation:* Walter Folotalu. *Forestry:* Dickson Mua. *Fisheries and Marine Resources:* Alfred Ghiro. *Mines, Energy and Rural Electrification:* Moses Garu. *Provincial Government and Institutional Strengthening:* Silas Tausinga. *National Unity, Reconciliation and Peace:* Hypolite Taremae. *Women, Youth and Children's Affairs:* Peter Tom. *Lands, Housing and Survey:* Joseph Onika. *Rural Development and Indigenous Affairs:* Lional Alex.

Solomon Islands Parliament: http://www.parliament.gov.sb

CURRENT LEADERS

Gordon Darcy Lilo

Position
Prime Minister

Introduction
Gordon Darcy Lilo became prime minister in Nov. 2011 following the resignation of Danny Philip amid controversy over development funding. A finance minister in previous administrations, Lilo promised increased political transparency.

Early Life
Born on 28 Aug. 1965 in Ghatere on Kolombangara Island, Lilo studied economics at the University of Papua New Guinea. As a civil servant he was a permanent secretary at the ministries of environment and conservation and finance. He received a masters degree in development and administration from the Crawford School of Economics and Government at the Australian National University in 2001.

Having returned to the Solomon Islands, Lilo was elected to parliament as MP for Gizo/Kolombangara in Dec. 2001, serving as leader of the independent group in parliament from Dec. 2001 until April 2006. In May 2006 he was appointed finance minister by Prime Minister Sogavare. He then served as minister for environment and conservation under Prime Minister Sikua from Dec. 2007.

In Aug. 2010 Lilo was again appointed finance minister in the National Coalition for Rural Advancement (NCRA) government led by Danny Philip. He subsequently oversaw development projects across a range of sectors including agriculture, forestry, education and infrastructure. In Nov. that year a scandal erupted over the alleged misuse of Taiwanese-provided development funds

by Philip, forcing him to resign the premiership. In the ensuing parliamentary ballot Lilo was elected his successor by 29 votes to 20.

Career in Office
Taking office on 16 Nov. 2011, Lilo pledged to continue the main policies of the NCRA government. These included developing the Solomon Islands' agriculture sector and diversifying the economy away from logging. In response to criticism of cronyism in government, he proposed a stronger role for the audit office and increased public consultation.

In foreign affairs, closer relations were sought with Australia, New Zealand and Papua New Guinea.

DEFENCE

The marine wing of the Royal Solomon Islands Police operates three patrol boats and a number of fast crafts for surveillance of fisheries and maritime boundaries. There is also an RSI Police Field Force stationed at the border with Papua New Guinea.

In July 2003 an Australian-led peacekeeping force landed to restore stability after years of ethnic fighting and high-level corruption. The force included troops from the Fiji Islands, New Zealand, Papua New Guinea and Tonga.

ECONOMY

Agriculture accounted for 35·1% of GDP in 2006, industry 9·0% and services 55·9%.

Overview

The Solomon Islands is a small, open economy vulnerable to external shocks. Food and fuel account for about 40% of total imports. Timber is the main export, while gold mining operations were relaunched in March 2010 and began production a year later. Gold is projected to overtake logging as the main driver of growth in the medium to long term. Tourism accounts for less than 2% of GDP.

Following Australian intervention in 2003 to restore political stability after years of civil strife, economic performance proved robust with average annual growth of 7%. However, the global financial crisis led to a decline in commodity exports, a contraction in growth of nearly 5% in 2009 and a depletion of international reserves. An 18-month Standby Credit Facility arrangement with the IMF was approved in June 2010, helping to restore economic stability. Logging production surged in 2010 owing to strong demand from Asia, with growth rebounding to nearly 8%. In mid-2011 the trade balance shifted from deficit to surplus for the first time since 2004. China accounts for 50% of the country's exports.

Almost 25% of the population lives in poverty and recent growth has not yet resulted in poverty reduction. The precarious global economic outlook poses additional risks to recovery.

Currency

The *Solomon Island dollar* (SBD) of 100 *cents* was introduced in 1977. It was devalued by 20% in Dec. 1997 and 25% in March 2002. Inflation was 17·3% in 2008, falling to 0·9% in 2010 but increasing to 7·4% in 2011. In July 2005 foreign exchange reserves were US$83m. and total money supply was SI$447m.

Budget

In 2009 revenues totalled SI$1,704·8m. and expenditures were SI$1,701·6m. Tax revenue accounted for 76·7% of revenues in 2009; current expenditure accounted for 83·9% of expenditures.

Performance

The economy shrank by 4·7% in 2009 but there was growth of 7·8% in 2010 and 10·7% in 2011. Total GDP in 2011 was US$0·8bn.

Banking and Finance

The Central Bank of Solomon Islands is the bank of issue; its *Governor* is Denton Rarawa. There are three commercial banks and a development bank.

Total foreign debt in 2007 was US$178m.

ENERGY AND NATURAL RESOURCES

Environment

Carbon dioxide emissions from the consumption and flaring of fossil fuels in 2008 were the equivalent of 0·4 tonnes per capita.

Electricity

Installed capacity in 2007 was approximately 14,000 kW. Production in 2007 was 85m. kWh and consumption per capita 172 kWh.

Oil and Gas

The potential for oil, petroleum and gas production has yet to be tapped.

Minerals

In 1999 gold output from mining totalled 3,456 kg and silver output 2,138 kg. The only mine in the Solomon Islands closed in 2000 owing to the civil unrest. However, in March 2010 the mine was revived and production began again in March 2011. The value of gold exports in 1999 was SI$113·7m.

Agriculture

Land is held either as customary land (88% of holdings) or registered land. Customary land rights depend on clan membership or kinship. Only Solomon Islanders own customary land; only Islanders or government members may hold perpetual estates of registered land. Coconuts, cocoa, rice and other minor crops are grown. Production, 2003 estimates (in 1,000 tonnes): coconuts, 330; sweet potatoes, 83; taro, 38; palm oil, 34; yams, 28; copra, 30; palm kernels, 8. In 2007 there were an estimated 16,000 ha. of arable land and 60,000 ha. of permanent crops.

Livestock (2003 estimates): pigs, 68,000; cattle, 13,000.

Forestry

Forests covered 2·21m. ha. in 2010 (79% of the land area). Of the total area under forests 50% was primary forest in 2010, 49% other naturally regenerated forest and 1% planted forest. Timber production was 1·25m. cu. metres in 2007.

Fisheries

Solomon Islands' waters are among the richest in tuna. The total catch in 2010 was an estimated 35,200 tonnes.

INDUSTRY

Industries include palm oil manufacture (35,000 tonnes in 2002), processed fish production (13,700 tonnes in 2000), rice milling, fish canning, fish freezing, saw milling, food, tobacco and soft drinks. Other products include wood and rattan furniture, fibreglass articles, boats, clothing and spices.

Labour

The estimated economically active population in 2010 was 123,000 (69% males), up from 105,000 in 2005.

INTERNATIONAL TRADE

Imports and Exports

Imports 2007, US$285·0m.; exports, US$158·5m. Main imports, 2007: machinery and transport equipment, 28·4%; mineral fuels and lubricants, 24·9%; food and live animals, 17·6%. Main exports, 2007: timber, 66·6%; fish products, 13·0%; palm oil, 8·8%. Principal import suppliers (2007): Australia, 31·5%; Singapore, 27·0%; Japan, 8·2%. Principal export markets (2007): China, 46·5%; Thailand, 7·2%; South Korea, 6·1%.

Trade Fairs

An annual National Trade and Cultural Show is held in July to coincide with the anniversary of independence.

COMMUNICATIONS

Roads

In 2002 there was estimated to be a total of 1,360 km of roads, of which 34 km were paved. The unpaved roads included 800 km of private plantation roads.

Civil Aviation

A new terminal has been opened at Henderson International Airport in Honiara. The national carrier is Solomon Airlines. In 2006 scheduled airline traffic of Solomon Islands-based carriers flew 3m. km, carrying 101,000 passengers (32,000 on international flights).

Shipping

There are international ports at Honiara, Yandina in the Russell Islands and Noro in New Georgia, Western Province. In Jan. 2009 there were three ships of 300 GT or over registered, totalling 2,000 GT.

Telecommunications

Telecommunications are operated by Solomon Telekom, a joint venture between the government of Solomon Islands and Cable & Wireless (UK). Telecommunications between Honiara and provincial centres are facilitated by modern satellite communication systems. In 2008 there were 8,000 main (fixed) telephone lines; mobile phone subscribers numbered 30,000 (5·9 per 100 persons). In 2011 an estimated 6% of the population were internet users.

SOCIAL INSTITUTIONS

Justice

Civil and criminal jurisdiction is exercised by the High Court of the Solomon Islands, constituted 1975. A Solomon Islands Court of Appeal was established in 1982. Jurisdiction is based on the principles of English law (as applying on 1 Jan. 1981). Magistrates' courts can try civil cases on claims not exceeding SI$2,000, and criminal cases with penalties not exceeding 14 years' imprisonment. Certain crimes, such as burglary and arson, where the maximum sentence is for life, may also be tried by magistrates. There are also local courts, which decide matters concerning customary titles to land; decisions may be put to the Customary Land Appeal Court. There is no capital punishment.

The population in penal institutions in June 2007 was 211 (42 per 100,000 of national population).

Education

In 2005 there were 100,026 pupils at primary and 22,487 pupils at secondary level. The adult literacy rate in 1998 was 62·0%.

Training of teachers and trade and vocational training is carried out at the College of Higher Education. The University of the South Pacific Centre is at Honiara. Other rural training centres run by churches are also involved in vocational training.

In 2000–01 total expenditure on education came to 3·6% of GNP and in 1999–2000 accounted for 15·4% of total government spending.

Health

A free medical service is supplemented by the private sector. An international standard immunization programme is conducted in conjunction with the WHO for infants. Tuberculosis has been eradicated but malaria remains a problem. In 2003 there were 60 physicians and 653 nursing and midwifery personnel. There were 15 hospitals beds per 10,000 inhabitants in 2005.

RELIGION

92% of the population were Christians in 2001.

CULTURE

World Heritage Sites

The Solomon Islands has one site on the UNESCO World Heritage List: East Rennell (inscribed on the list in 1998), the largest raised coral atoll in the world.

Press

There are three main newspapers in circulation. *The Solomon Star* (circulation: 5,000) is daily and the *Solomon Express* and *The Island Sun* are weekly. The Government Information Service publishes a monthly issue of the *Solomon Nius* that exclusively disseminates news of government activities. Non-government organizations such as the Solomon Islands Development Trust (SIDT) also publish monthly papers on environmental issues.

Tourism

Tourism in the Solomon Islands is still in a development stage. The emphasis is on establishing major hotels in the capital and provincial centres, to be supplemented by satellite eco-tourism projects in the rural areas. The Solomon Islands Visitors Bureau is the statutory institution for domestic co-ordination and international marketing. In 2007 there were 13,748 foreign tourists.

Festivals

Festivities and parades in the capital and provincial centres normally mark the National Day of Independence. The highlight is the annual National Trade and Cultural Show.

DIPLOMATIC REPRESENTATIVES

Of the Solomon Islands in the United Kingdom
High Commissioner: Joseph Ma'ahanua (resides in Brussels).

Of the United Kingdom in the Solomon Islands (Telekom House, Mendana Ave., Honiara)
High Commissioner: Dominic Meiklejohn, OBE.

Of the Solomon Islands in the USA and to the United Nations (800 2nd Ave, Suite 400L, New York, NY 10017)
Ambassador: Collin Beck.

Of the USA in the Solomon Islands
Ambassador: Teddy B. Taylor (resides in Port Moresby, Papua New Guinea).

Of the Solomon Islands to the European Union
Ambassador: Joseph Ma'ahanua.

FURTHER READING

Bennett, J. A., *Wealth of the Solomons: A History of a Pacific Archipelago, 1800–1978.* 1987

Fraenkel, Jonathan, *Manipulation of Custom: From Uprising to Intervention in the Solomon Islands.* 2005

White, Geoffrey M., *Identity Through History: Living Stories in a Solomon Islands Society.* 2003

National Statistical Office: Solomon Islands National Statistical Office, PO Box G6, Department of Finance, Honiara.
Website: http://www.spc.int/prism/country/sb/stats

SOMALIA

Gulf of Aden

DJIBOUTI

Hargeisa

ETHIOPIA

SOMALIA

Baydhabo

INDIAN

OCEAN

Marka   MOGADISHU

KENYA

Kismayo   0   100 mi
         0   150 km

© Research Machines plc 2006

Jamhuuriyadda Federaalka Soomaaliya
(Federal Republic of Somalia)

Capital: Mogadishu
Population projection, 2015: 10·61m.
GNI per capita, 2010: US$110
Internet domain extension: .so

KEY HISTORICAL EVENTS

The origins of the Somali people can be traced back 2,000 years when they displaced an earlier Arabic people. They converted to Islam in the 10th century and were organized in loose Islamic states by the 19th century. The northern part of Somaliland was created a British protectorate in 1884. The southern part belonged to two local rulers who, in 1889, accepted Italian protection for their lands. The Italian invasion of Ethiopia in 1935 was launched from Somaliland and in 1936 Somaliland was incorporated with Eritrea and Ethiopia to become Italian East Africa. In 1940 Italian forces invaded British Somaliland but in 1941 the British, with South African and Indian troops, recaptured this territory as well as occupying Italian Somaliland. After the Second World War British Somaliland reverted to its colonial status and ex-Italian Somaliland became the UN Trust Territory of Somaliland, administered by Italy.

The independent Somali Republic came into being on 1 July 1960 as a result of the merger of the British Somaliland Protectorate, which first became independent on 26 June 1960, and the Italian Trusteeship Territory of Somaliland. On 21 Oct. 1969 Maj.-Gen. Mohammed Siyad Barre took power in a coup. Various insurgent forces combined to oppose the Barre regime in

a bloody civil war. Barre fled on 27 Jan. 1991 but interfactional fighting continued. In Aug. 1992 a new coalition government agreed a UN military presence to back up relief efforts to help the estimated 1·5–2m. victims of famine. On 11 Dec. 1992 the leaders of the two most prominent of the warring factions, Ali Mahdi Muhammad and Muhammad Farah Aidid, agreed to a peace plan under the aegis of the UN and a pact was signed on 15 Jan. 1993. At the end of March, the warring factions agreed to disarm and form a 74-member National Transitional Council. On 4 Nov. 1994 the UN Security Council unanimously decided to withdraw UN forces; the last of these left on 2 March 1995.

The principal insurgent group in the north of the country, the Somali National Movement, declared the secession of an independent '**Somaliland Republic**' on 17 May 1991. The Somalian government rejected the secession and Muhammad Aidid's forces launched a campaign to reoccupy the 'Republic' in Jan. 1996. Muhammad Farah Aidid was assassinated in July 1996 and succeeded by his son Hussein Aidid. In July 1998 leaders in the northeast of Somalia proclaimed an 'autonomous state' named **Puntland.**

Peace efforts in neighbouring Djibouti culminated in July 2000 in the establishment of a power-sharing agreement and a national constitution to see Somalia through a three-year transitional period. The election of members of parliament and a civilian government followed in Aug. 2000, and in Oct. the new government moved from Djibouti back to Somalia. In April 2002 '**Southwestern Somalia**' broke away from Mogadishu, thereby creating a third autonomous Somali state.

In June 2006 after months of fighting an Islamic militia, the Islamic Courts Union, took control of the capital, Mogadishu, and a large part of southern Somalia. In Dec. 2006, following a sustained assault by government forces backed by Ethiopian troops, the Islamist militias withdrew from Mogadishu. The southern city of Kismayo, the Islamists' last stronghold, was recaptured in Jan. 2007.

In June 2012 separatist leaders from Somaliland agreed to hold talks with the Somali government in London. In Aug. 2012 a Somali-elected parliament formally convened in Mogadishu for the first time in 20 years. It elected Hassan Sheikh Mohamud as president the following month, the first such vote undertaken on Somali territory since 1967. In Oct. 2012 a mix of government and African Union troops defeated the Al-Shabab militant Islamist group, ejecting them from Kismayo, the organization's last major urban stronghold.

TERRITORY AND POPULATION

Somalia is bounded north by the Gulf of Aden, east and south by the Indian ocean, and west by Kenya, Ethiopia and Djibouti. Total area 637,657 sq. km (246,201 sq. miles). A census has not been held since 1987, when the population was 7,114,431. The United Nations gave an estimated population for 2012 of 9·80m.; density, 15 per sq. km. Population counting is complicated owing to large numbers of nomads and refugee movements as a result of famine and clan warfare. Up to 1·5m. Somalis are displaced within the country and in 2010 there were 770,000 Somali refugees living abroad, mainly in Kenya and Yemen.

The UN gives a projected population for 2015 of 10·61m.

In 2011, 37·9% of the population were urban.

The country is administratively divided into 18 regions (with chief cities): Awdal (Baki), Bakol (Xuddur), Bay (Baydhabo), Benadir (Mogadishu), Bari (Bosaso), Galgudug (Duusa Marreeb), Gedo (Garbahaarrey), Hiran (Beledweyne), Jubbada Dexe (Jilib), Jubbada Hoose (Kismayo), Mudug (Gaalkacyo), Nogal (Garowe), Woqooyi Galbeed (Hargeisa), Sanaag (Ceerigabo), Shabeellaha

Dhexe (Jawhar), Shabeellaha Hoose (Marka), Sol (Las Anod), Togder (Burao). Somaliland comprises the regions of Awdal, Woqooyi Galbeed, Togder, Sanaag and Sol. Puntland consists of Bari, Nogal and northern Mudug. Southwestern Somalia consists of Bay, Bakol, Gedo, Jubbada Hoose and Shabeellaha Dhexe.

The capital is Mogadishu (2010 population estimate, 1,426,000). Other large towns are Baidoa, Bosaso, Gaalkacyo and Hargeisa.

The official language is Somali. Arabic, English and Italian are widely spoken.

SOCIAL STATISTICS

Births, 2008 estimate, 394,000; deaths, 140,000. Rates, 2008 estimate (per 1,000 population): birth, 44·1; death, 15·7. Infant mortality, 2010, 108 per 1,000 live births. Annual population growth rate, 2000–08, 2·4%. Life expectancy at birth, 2007, was 48·3 years for men and 51·2 years for women. Fertility rate, 2008, 6·4 births per woman.

CLIMATE

Much of the country is arid, although rainfall is more adequate towards the south. Temperatures are very high on the northern coasts. Mogadishu, Jan. 79°F (26·1°C), July 78°F (25·6°C). Annual rainfall 17" (429 mm). Berbera, Jan. 76°F (24·4°C), July 97°F (36·1°C). Annual rainfall 2" (51 mm).

CONSTITUTION AND GOVERNMENT

A new constitution was promulgated on 1 Aug. 2012 after over 20 years of non-functioning government. It replaced the constitution of 1979 that itself had lost authority after the ousting of President Siyad Barre in 1991. The 2012 constitution was adopted by the National Constitutional Assembly with 96% backing from the 645 community leaders present at the vote (from a total of 825).

The constitution includes a bill of rights enshrining the equality of all citizens regardless of clan or religion. Islam is the single recognized state religion, with Sharia law serving as the foundation of the legal system. The right to education up to the secondary level is guaranteed for all, while female circumcision and the deployment of children in armed conflict are proscribed. Provision is included for the establishment of a Truth and Reconciliation Commission and for the implementation of a federal system of government (though details of how power and resources are to be split remains to be decided).

There is a bicameral parliament with a proposed 275 seats in the lower house and a maximum of 54 seats in the senate. Parliament elects the *President*, who in turn appoints a *Prime Minister*.

Puntland. Puntland, in the northeast region of Somalia, declared itself an 'autonomous state' in July 1998 under the leadership of Abdullahi Yusuf. Since its creation, Puntland has been locked in dispute with Somaliland over control of the Sanaag and Sol areas. Abdirahman Farole became *President* in Jan. 2009.

Puntland covers 300,000 sq. km and had a population in 2000 of 2m. The capital is Garowe. Somali is the official language and the Somali shilling is the official currency. Puntland has not received international recognition.

Somaliland. An independent 'Somaliland Republic', based on the territory of the former British protectorate which ran from 1884 until Somali independence in 1960, was established on 17 May 1991 by the principal insurgent group in the north of the country, the Somali National Movement. The Somali government rejected the secession and an unsuccessful campaign to reoccupy Somaliland was launched in Jan. 1996. Somaliland is also engaged in a long-running dispute with Puntland over control of the Sanaag and Sol regions. The Republic has failed to secure international recognition although it has in effect seceded from Somalia.

Somaliland covers 137,600 sq. km. The capital is Hargeisa and there is a port at Berbera. There is a population of around 3·5m.

Somali is the official language and Arabic and English are also widely used.

There is a bicameral government with a house of representatives and one of elected elders. Ahmed Mohamed Silanyo became *President* in July 2010. The judiciary is independent. The official currency is the Somaliland shilling. The Bank of Somaliland, the central bank, was founded in 1994. The economy is reliant on livestock farming.

Southwestern Somalia. In April 2002 Southwestern Somalia broke away from Mogadishu and was declared an autonomous state by the Rahanwein Resistance Army. Hassan Muhammad Nur 'Shatigadud' was named president but fighting between Shatigadud and several of his deputies ensued, notably around the capital, Baydhabo.

National Anthem

'Qolobaa Calankeed' ('Every nation has its own flag'); words and tune by Abdullahi Qarshe.

RECENT ELECTIONS

Somalia's parliament elected Hassan Sheikh Mohamud president on 10 Sept. 2012. In the first round Sharif Sheikh Ahmed, who had been president from Jan. 2009 until Aug. 2012, won 64 votes, followed by Hassan Sheikh Mohamud with 60 votes. Prime Minister Abdiweli Mohamed Ali placed third with 32 votes. There were 19 other candidates. All candidates apart from Sharif Ahmed and Mohamud withdrew after the first round. A subsequent run-off was held in which Mohamud won 190 votes against 79 for Sharif Ahmed.

CURRENT GOVERNMENT

President: Hassan Sheikh Mohamud; b. 1955 (Peace and Development Party; sworn in 16 Sept. 2012).

In March 2013 the government comprised:

Prime Minister: Abdi Farah Shirdon; b. 1958 (ind.; sworn in 17 Oct. 2012).

Deputy Prime Minister and Minister of Foreign Affairs: Fowsiyo Yusuf Haji Adan.

Minister of Defence: Abdihakim Mohamoud Haji Fiqi. *Finance and Planning:* Mohamoud Hassan Suleiman. *Information and Telecommunication:* Abdullahi Elmoge Hersi. *Interior and National Security:* Abdikarim Hussein Guled. *Justice, Religious Affairs and Endowment:* Abdullahi Abyan Nur. *Natural Resources:* Abdirizak Omar Mohamed. *Public Works:* Muhyaddin Mohamed Kalmoi. *Social Development Services:* Maryan Kassim. *Trades and Industrialization:* Mohamud Ahmed Hassan.

Transitional Federal Government Website:
http://www.somaligov.net

CURRENT LEADERS

Hassan Sheikh Mohamud

Position
President

Introduction
Elected president by parliament in Sept. 2012, Hassan Sheikh Mohamud is a relative newcomer to frontline politics but is experienced in building civil institutions. He is a vocal critic of corruption and clan factionalism.

Early Life
Born on 29 Nov. 1955 in Jalalaqsi, Hassan Sheikh Mohamud graduated in technology from Somalia National University in Mogadishu in 1981. From 1981–84 he worked as a secondary school teacher and from 1984–86 as a lecturer in teacher training. In 1986 he moved to India, completing a masters degree in technical education at Bhopal University in 1988. Returning to

Somalia the same year, he worked with UNESCO on a project to expand technical and vocational training.

Following the onset of civil war, he became an education officer for UNICEF in 1993, focusing on rebuilding educational services in the central and southern regions. From 1995–97 he worked with civic organizations to establish communication between rival factions in Mogadishu. After the 1997 Cairo Agreement, he was key to negotiating the dismantling of the 'Green Line' that had divided Mogadishu since the outbreak of fighting in the early 1990s.

Mohamud co-founded the Somali Institute of Management and Administration Development in 1999, serving as its dean until 2010. He worked for the Center for Research and Development from 2001, where he oversaw the establishment of the Somalia Civil Society Forum, an umbrella group of activists and organizations engaged in rebuilding civil society. From 2007 he was consultant to a number of international and local NGOs, working on projects to rebuild different aspects of Somali society.

From 2007–09, in the period of Somalia's Transitional Federal Government (TFG), he was a consultant at the ministry of planning and international co-operation, overseeing aid management. In 2011 he co-founded the Peace and Development Party (PDP) and was elected its leader in April that year. In Sept. 2012 a presidential vote was held by parliament, with Mohamud achieving a surprise victory. He defeated the former president, Sharif Sheikh Ahmed, in a run-off, claiming 190 votes to 79.

Career in Office
Mohamud took office on 10 Sept. 2012, pledging to tackle corruption and combat the Islamic militia group, al-Shabab. He also promised to develop education and provide jobs for young people. In Jan. 2013 the USA recognized the Mogadishu government for the first time in over 20 years and the EU agreed to negotiate a 'New Deal' of support. Nonetheless, Mohamud faces major challenges, notably the continuing insurgency, clan rivalries, widespread poverty and piracy.

DEFENCE

With the breakdown of government following the 1991 revolution armed forces broke up into clan groupings, four of them in the north and six in the south.

Army
Following the 1991 revolution there are no national armed forces. In Northern Somalia the Somali National Movement controls an armed clan of 5–6,000 out of a total of 7,000 armed forces in the area. In the rest of the country several local groups control forces of which the Ali Mahdi Faction controls the largest, an armed clan of 10,000.

ECONOMY

Agriculture accounts for approximately 59% of GDP, industry 10% and services 31%.

Somalia was rated the joint most corrupt country in the world in a 2012 survey of 176 countries carried out by the anti-corruption organization *Transparency International*.

Overview
Somalia has been entangled in civil conflict with no effective central government since 1991. A large private sector, trading locally and with neighbouring Asian economies, has compensated for a weak public sector. Remittances, amounting to around US$1bn. per year, have helped stimulate private investment in a number of commercial ventures, as well as partially offsetting a decline in per capita income. Social conditions have deteriorated since the onset of civil war, with an estimated 43% of the population living in extreme poverty.

Somalia is one of three member countries to remain in protracted arrears to the IMF, accounting for 17% of total arrears to the Fund. Reviews of its overdue financial obligations have

been postponed since Oct. 1990, owing to the absence of a functioning government, continuous political and security problems, and the lack of official information on economic and financial developments. Economic growth is expected to remain minimal across most of the country until peace and stability return, although the formation of a new government in Oct. 2012 offers some hope of progress.

Currency
The unit of currency is the *Somali shilling* (SOS) of 100 *cents*.

Budget
A budget for 2012 of US$126m. was reportedly approved in Dec. 2011.

Performance
Real GDP growth was 3·5% in both 2002 and 2003. Total GDP in 2003 was US$1·5bn.

Banking and Finance
The bank of issue is the Central Bank of Somalia (*Governor*, Abdullahi Jama Ali). The separatist Somaliland Republic has its own functioning central bank in Hargeisa, the Bank of Somaliland (*Governor*, Abdi Dirir Abdib). Remittance companies (*hawala*) took the place of banks in the 1990s, channelling approximately US$800m. a year. Al-Barakaat, the largest *hawala*, was shut down in Nov. 2001. All national banks were bankrupted by 1990. The Universal Bank of Somalia, the first commercial bank in Mogadishu since 1990, opened with European backing in Jan. 2002.

In 2010 external debt totalled US$2,942m.

ENERGY AND NATURAL RESOURCES

Environment
Carbon dioxide emissions from the consumption and flaring of fossil fuels in 2008 were the equivalent of 0·1 tonnes per capita.

Electricity
In 2007 installed capacity was 65,000 kW. Production (2007): 326m. kWh.

Oil and Gas
Proven natural gas reserves were 5·7bn. cu. metres in 2007.

Minerals
There are deposits of chromium, coal, copper, gold, gypsum, lead, limestone, manganese, nickel, sepiolite, silver, titanium, tungsten, uranium and zinc.

Agriculture
Somalia is essentially a pastoral country, and about 80% of the inhabitants depend on livestock-rearing. Half the population is nomadic. In 2002 there were about 1·05m. ha. of arable land and 26,000 ha. of permanent crops. Around 200,000 ha. were irrigated in 2002. There were about 1,700 tractors in 2002. Output, 2003 estimates (in 1,000 tonnes): sugarcane, 200; maize, 164; sorghum, 121; cassava, 85; bananas, 35. Livestock (2003 estimates): 13·1m. sheep; 12·7m. goats; 7·0m. camels; 5·1m. cattle. Somalia has the greatest number of camels of any country in the world.

Forestry
In 2010 the area under forests was 6·75m. ha., or 11% of the total land area. In 2007, 11·57m. cu. metres of roundwood were cut. Wood and charcoal are the main energy sources. Frankincense and myrrh are produced.

Fisheries
An estimated 30,000 tonnes of fish were caught in 2010, almost entirely from marine waters.

INDUSTRY

A few small industries exist including sugar refining, food processing and textiles. Raw sugar production was an estimated 15,000 tonnes in 2005.

Labour

The estimated economically active population in 2010 was 3,627,000 (59% males), up from 3,267,000 in 2005.

INTERNATIONAL TRADE

Imports and Exports

Imports in 2003 were estimated at US$397m. and exports at US$95m. Main imports: manufactures, petroleum, foodstuffs. Main exports: livestock, hides and skins, bananas. Leading import suppliers, 2003: Djibouti, 32%; Kenya, 15%; Brazil, 11%; United Arab Emirates, 5%. Leading export markets, 2003: United Arab Emirates, 39%; Yemen, 24%; Oman, 11%; China, 6%.

COMMUNICATIONS

Roads

In 2002 there were an estimated 22,100 km of roads, of which 2,600 km were paved. Passenger cars numbered 12,700 in 2002, and there were 10,400 trucks and vans.

Civil Aviation

There are international airports at Mogadishu and Hargeisa. In 2010 there were flights to Aden, Djibouti, Dubai, Jeddah, Nairobi, Sharjah and Wajir in addition to internal services.

Shipping

The main ports are at Berbera, Bosaso, Kismayo, Marka and Mogadishu. In Jan. 2009 there were three ships of 300 GT or over registered, totalling 2,000 GT.

Piracy off the coast of Somalia was intensifying for several years with 111 attacks in the waters off Somalia recorded in 2008, 217 in 2009, 218 in 2010 and 237 in 2011. In 2012 there were only 75 recorded attacks, largely thanks to more patrolling of the waters off East Africa by international navies. There were 14 actual hijacks, down from 28 in 2011 and 49 in 2010.

Telecommunications

Somalia had 100,000 main telephone lines in 2008 (11·2 per 1,000 persons); mobile phone subscribers numbered 627,000 (70·2 per 1,000 persons). In 2008 there were 102,000 internet users.

SOCIAL INSTITUTIONS

Justice

There are 84 district courts, each with a civil and a criminal section. There are eight regional courts and two Courts of Appeal (at Mogadishu and Hargeisa), each with a general section and an assize section. The Supreme Court is in Mogadishu. The death penalty is in force; Amnesty International reported that there were at least six executions in 2012.

Education

The nomadic life of a large percentage of the population inhibits educational progress. Adult literacy was estimated at 38% in the period 2000–06 (50% among males and 26% among females). In 2004 there were an estimated 1,172 primary schools with over 285,500 children in attendance. In 2006, 23·5% of primary age children and 26·2% of secondary age children were attending school. The longest-established university was for many years the Somali National University in Mogadishu (founded 1954), but it was extensively damaged in the civil war and classes have been suspended indefinitely. A private university, Mogadishu University, was opened in 1997.

Health

In 2006 there were 300 physicians (3·5 per 100,000 population) and 965 nursing and midwifery personnel.

Somalia has among the highest percentages of undernourished people of any country, at 62% in the period 2005–07.

RELIGION

The population is almost entirely Sunni Muslims.

CULTURE

Press

The Somali press collapsed in 1991, with most of its facilities destroyed. Since 2000 several independent newspapers have emerged, including the daily *Wartire in Hargeisa* (Somaliland) and the weeklies *Yamayska* and *Bulsho* in Puntland. There were seven daily newspapers in 2008. Average daily circulation of newspapers in 2008 totalled 21,000.

Tourism

In 1998 there were 10,000 foreign tourists.

DIPLOMATIC REPRESENTATIVES

The Embassy of Somalia in the United Kingdom closed on 2 Jan. 1992.

Of the United Kingdom in Somalia
Ambassador: Matt Baugh (resides in Nairobi, Kenya).

The Embassy of Somalia in the USA closed on 8 May 1991. A liaison office opened in March 1994, and withdrew to Nairobi in Sept. 1994.

Of Somalia to the United Nations
Ambassador: Elmi Ahmed Duale.

Of Somalia to the European Union
Ambassador: Nur Hassan Hussein.

FURTHER READING

Lewis, I. M., *Blood and Bone: the Call of Kinship in Somali Society.* 1995.— *Understanding Somalia: a Guide to Culture, History and Social Institutions.* 2nd ed. 1995.—*A Modern History of the Somali: Nation and State in the Horn of Africa.* 2002

Omar, M. O., *The Road to Zero: Somalia's Self-Destruction.* 1995

Samatar, A. I. (ed.) *The Somali Challenge: from Catastrophe to Renewal?* 1994

Woodward, Peter, *The Horn of Africa: Politics and International Relations.* 2002

National Statistical Office: Central Statistical Department, State Planning Commission, Mogadishu.

SOUTH AFRICA

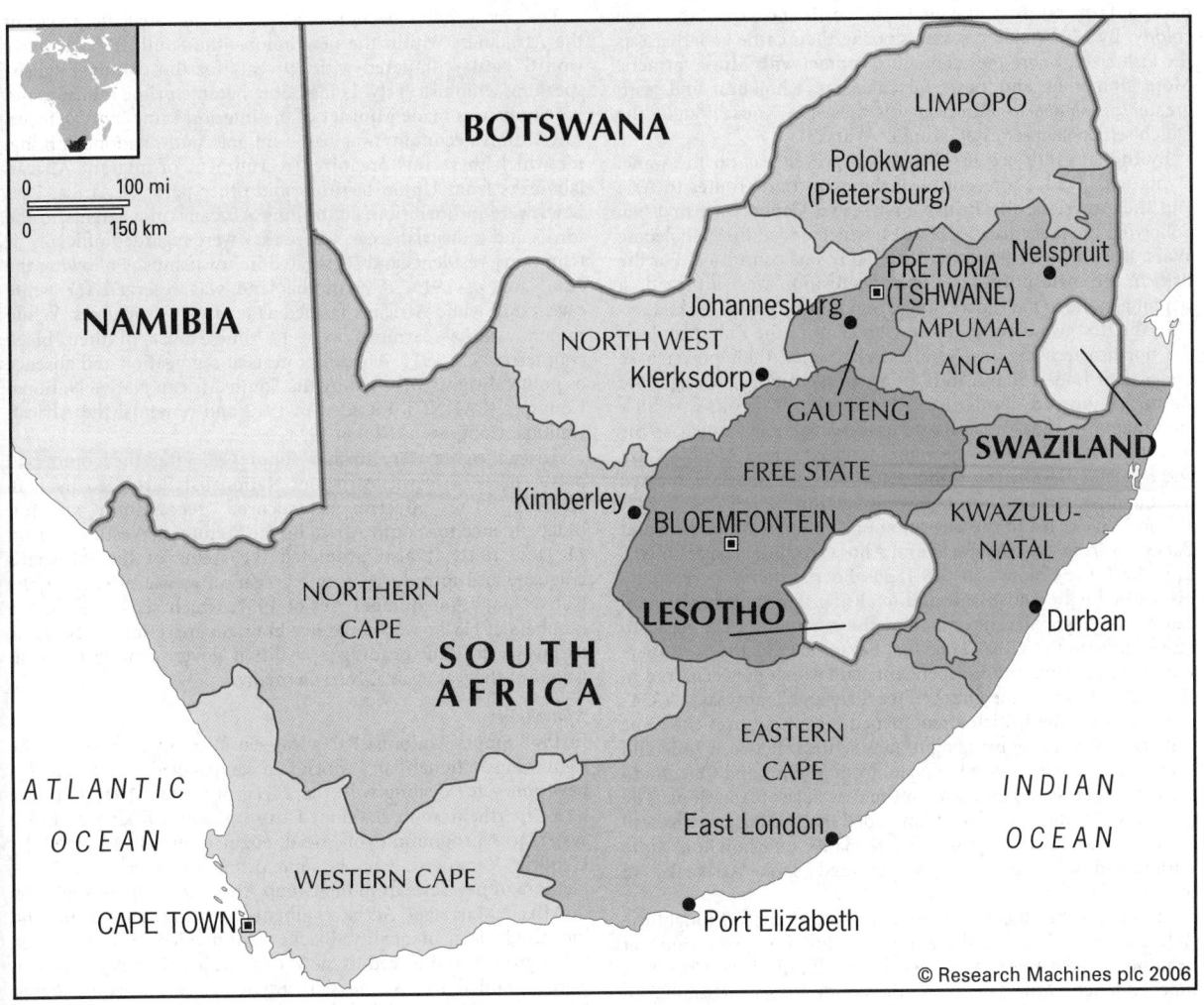

BOTSWANA

LIMPOPO

NAMIBIA

Polokwane ●
(Pietersburg)

Nelspruit ●

PRETORIA
□(TSHWANE)

Johannesburg ●

MPUMAL-
ANGA

NORTH WEST

Klerksdorp ●

GAUTENG

SWAZILAND

FREE STATE

Kimberley ●

BLOEMFONTEIN
□

KWAZULU-
NATAL

NORTHERN
CAPE

SOUTH
AFRICA

LESOTHO

Durban ●

ATLANTIC
OCEAN

EASTERN
CAPE

INDIAN
OCEAN

East London ●

WESTERN CAPE

CAPE TOWN ◎

Port Elizabeth ●

© Research Machines plc 2006

0   100 mi
0   150 km

Republic of South Africa

Capitals: Pretoria/Tshwane (Administrative), Cape Town
(Legislative), Bloemfontein (Judicial)
Seat of Parliament: Cape Town
Seats of Government: Cape Town, Pretoria
Population projection, 2015: 51·43m.
GNI per capita, 2011: (PPP$) 9,469
HDI/world rank: 0·619/123
Internet domain extension: .za

KEY HISTORICAL EVENTS

The San and the Khoikhoi were the indigenous peoples of
southern Africa. The San were nomadic hunter-gatherers who
had lived from the land at the edge of the Kalahari desert for
thousands of years. The Khoikhoi shared customs with the San
and spoke related languages but also herded cattle and lived in
more settled communities. The Khoikhoi settlements were most
numerous in the Orange River valley and around the Cape. From
the 4th century AD the eastern part of southern Africa was settled
by Bantu-speaking groups, moving south from the continent's

drier interior. They were mixed farmers: herding sheep and cattle,
hunting game, cultivating sorghum and making tools and
weapons from iron.

The hunting and herding communities of southern Africa came
into contact with the wider world at the end of the fifteenth
century. Portuguese mariners first rounded the Cape peninsula in
1487 and opened a trade route into the Indian Ocean. A century
later the route was used by Spanish, English, Dutch and French
seafarers. They landed occasionally on the Cape peninsula and
bartered sheep and cattle with Khoikhoi pastoralists in return for
iron and copper goods. In 1649 the Dutch East India Company,
the world's most powerful trading corporation, established a
trading post at the Cape. Three years later Jan van Riebeeck
arrived with orders to establish a fort at Table Bay and supply
passing ships with meat, fruit and vegetables. Within a decade
slaves were brought in to work on building and maintaining the
infrastructure, and settlers began to arrive from the Netherlands.
Relations between the Dutch and the Khoikhoi soon deteriorated:
quarrels over rights to graze cattle escalated into warfare as early
as 1659.

Over the next century the population of the Cape Colony
reached 10,000. It was a diverse community, where traders from

Europe and Asia converged and exchanged goods and news. Large farms, cultivating vines and grain, were established in the fertile valleys to the east of Cape Town. Devastated by smallpox in 1713, the Khoikhoi population was unable to prevent *trekboers* (Dutch pastoral farmers) from moving to the north and east of the Cape colony. By 1770 trekboers were grazing their cattle as far east as the Fish river, where they came into contact with Xhosa farmers. More numerous and powerful than the Khoikhoi, and with greater resistance to European diseases, the Xhosa fought the Dutch settlers in a series of 'Frontier Wars'.

By the late 18th century Dutch sea power was on the wane. Vying with France for control of the main trade routes to Asia and the Americas, the British first seized Cape Town in 1795. Following the peace treaties of 1814, which ended the Napoleonic Wars, British sovereignty over the colony was confirmed. For the British, the main purpose of their acquisition was to provide a stepping-stone to their increasingly important colonies in Asia.

In the first two decades of the 19th century the Zulu people of the northeastern region (Natal) strengthened their power-base under their leader, Shaka. In response to a prolonged drought the Zulus conquered lands from rival Nguni groups, which culminated in widespread havoc and destruction, known as the *Mfecane*. From the chaos new kingdoms emerged, notably Gaza and Swaziland, while the Sotho, under King Moshoeshoe, formed the mountain territory now known as Lesotho.

The *Mfecane* led to the migration of thousands of Basotho and Batswana from the High Veld and Xhosa from the coastal plains into the Cape Colony. In the 1830s Boer settlers, increasingly dissatisfied with British rule and, realising that the *Mfecane* had caused the depopulation of land to the north and east, began to move there. In the 'Great Trek' that began in 1836, the Afrikaners were seeking a free and independent state which they achieved in the establishment of the Orange Free State and Transvaal in 1854.

Meanwhile, the British strengthened their hold over the Cape Colony and Natal by bringing in new settlers. Between 1860 and 1866, 6,000 Indians arrived in Natal from Madras and Calcutta to work as indentured labourers on the new sugar plantations. The population of the Cape Colony included many Afrikaners as well as the 'coloured' community (descendants of Khoikhoi, white settlers and Malay slaves). Most Coloureds spoke Afrikaans, an offshoot of Dutch.

Britain annexed the Transvaal in 1877, and in 1879 fought the Zulus. Under King Ketshwayo the Zulus were victorious at Isandhlwana but were then defeated at Ulundi. Britain restored independence to the Transvaal (the South African Republic) in 1884 and annexed Zululand in 1887. Both the British and the Boers fought African resistance for many years, the last major rising being in Natal in 1906. However, the British and Boers continued to be rivals, especially after the discovery of diamonds at Kimberley in 1867 and of gold in the Transvaal in 1884. This led to an economic boom. Cecil Rhodes, owner of the De Beers company and for a time prime minister of the Cape, was the dominant entrepreneurial figure.

Boer War

In the 1890s the British, under Rhodes, sought control over the Transvaal goldfields. Despite being thwarted in their attempts to spark off rebellion amongst the Afrikaners of the South African Republic, the British continued to press for control. The Afrikaners, led by Paul Kruger, decided they would have to fight to keep their independence and declared war on Britain in late 1899. The contest appeared unequal, with the might of the British army against only 35,000 Boer soldiers. The Boers suffered a heavy defeat at Paardeberg in 1900, but then switched to guerrilla warfare. The British army, led by Gen. Kitchener, responded by setting up concentration camps and destroying crops and farmsteads. The 'scorched earth' policy was strongly criticized in Europe, but had the desired effect—in 1902 the Boer republics signed the Treaty of Vereeniging and came under British rule.

They were given self-government in 1907 and on 31 May 1910 the Cape Colony, Natal, the Transvaal and the Orange Free State combined to form the Union of South Africa, a self-governing dominion under the British Crown.

The first general election in 1910 demonstrated the power of the Afrikaners within the new union—the South African party won 67 seats compared with 39 seats for the mainly English-speaking Unionist Party. Louis Botha became prime minister and Jan Smuts was made Minister of the Interior, Mines and Defence. The Union's economy was based on gold and diamond mining, for which there was organized recruitment of migrant African labourers from Union territory and other parts of Africa. Pass Laws were in operation, controlling Africans' movements in the towns and industrial areas, where they were regarded officially as temporary residents and segregated in 'townships'. Following the Land Act of 1913, 87% of the land was reserved for white ownership while Africans farmed as tenants or squatters. White miners' annual earnings were 12 times those of their black counterparts in 1911. African protests at segregation and absence of political rights were led by the South African Native National Congress (SANNC), founded in 1912 and renamed the African National Congress (ANC) in 1923.

African rights were further suppressed after the coming to power in 1924 of the Afrikaner Nationalist Party, led by J. B. Hertzog. The government secured recognition of full independence for South Africa by the Statute of Westminster on 11 Dec. 1931. It also promoted the status of the Afrikaans language and introduced new segregation measures such as the Native Laws Amendment Act of 1937, which set limits on the numbers of blacks who could live in urban areas. Jan Smuts came to power in 1939 heading a coalition government broadly in favour of the war against Nazi Germany.

Apartheid

In 1948 Smuts' Unionist Party was sensationally defeated by the right-wing National Party which had campaigned for *apartheid*, a new policy for dealing with the 'racial problem'. After 1948 the term apartheid soon developed from a political slogan into a systematic programme of social engineering championed by Hendrik Verwoerd, who became prime minister in 1958. A plethora of new laws from the Group Areas Act to the Prohibition of Mixed Marriages Act strengthened existing segregation and increased racial inequality. Blacks were divided into one of ten tribal groups, and forced to move to so-called Homelands, which were intended to become self-sufficient, self-governing states. Chief Buthelezi was pivotal in the Inkatha movement which attempted, but ultimately failed, to unite Homeland leaders. The massacre by police of 69 protesters against the Pass Laws at Sharpeville on 21 March 1960 led to a major crisis from which, however, the government emerged even stronger. The ANC and the Pan African Congress were banned and the leaders, including Nelson Mandela, were jailed in 1964. After withdrawing from the British Commonwealth in 1961, South Africa became increasingly isolated. To the north, former European colonies were becoming independent, often socialist, republics.

On 16 June 1976 thousands of students demonstrated in Soweto, an African township outside Johannesburg, against mandatory schooling in Afrikaans. Many died when police broke up the demonstrations and rioting spread throughout the country. When P. W. Botha became prime minister in 1978, elements of the apartheid system were modified. Africans were allowed to form legal trade unions and the acts banning marriage and sexual relations between people of different races were repealed.

A new constitution, approved in a referendum of white voters on 2 Nov. 1983 and in force from 3 Sept. 1984, created a three-part parliament, with a House of Assembly for the Whites, a House of Representatives for the Coloureds and a House of Delegates for the Indians; Africans remained without representation. From late 1984 blacks in the cities and industrial areas staged large-scale protests. In June 1986 a state of emergency was imposed. Foreign condemnation led to the first economic sanctions against South Africa, imposed by a number of countries including the USA and Britain.

By 1989 a start had been made on dismantling apartheid and the government, led by F. W. de Klerk, announced its willingness to consider the extension of black South Africans' political rights. In Feb. 1990 a 30-year ban on the ANC was lifted and Nelson Mandela was released from prison on 11 Feb. 1990. In the Whites-only referendum on 17 March 1992, on the granting of constitutional equality to all races, 1,924,186 (68·7%) votes were in favour; 875,619 against.

On 22 Dec. 1993 parliament approved (by 237 votes to 45) a transitional constitution paving the way for a new multi-racial parliament which was elected on 29 April 1994. There was a decisive victory for the ANC and on 9 May 1994 Nelson Mandela was elected president. The new government included six ministers from the National Party and three from the Inkatha Freedom Party.

In 1997 the Truth and Reconciliation Commission, chaired by Archbishop Desmond Tutu, began hearings on human rights violations between 1960 and 1993. The commission promised amnesty to those who confessed their crimes under the apartheid system. Nelson Mandela, whose term as president cemented his reputation as a world statesman, retired in 1999. His deputy, Thabo Mbeki, elected president in a landslide vote, had already assumed many of Mandela's governing responsibilities. Mbeki wrestled with a developing economy, continuing inequality and a high crime rate. The nation remains in the grip of an AIDS epidemic which Mbeki was slow to acknowledge. He stood down from the presidency in Sept. 2008 after the ruling ANC called on him to resign.

TERRITORY AND POPULATION

South Africa is bounded in the north by Namibia, Botswana and Zimbabwe, northeast by Mozambique and Swaziland, east by the Indian Ocean, and south and west by the South Atlantic, with Lesotho forming an enclave. Area: 1,219,090 sq. km. This area includes the uninhabited Prince Edward Island (41 sq. km) and Marion Island (388 sq. km), lying 1,900 km southeast of Cape Town. The islands were handed over to South Africa in Dec. 1947 to prevent their falling into hostile hands. In 1994 Walvis Bay was ceded to Namibia, and Transkei, Bophuthatswana, Venda and Ciskei were reintegrated into South Africa.

At the census of 2011 the population was 51,770,560 (26,581,769 females), consisting of: Black African, 41,000,938 (79·2% of total population); Coloured, 4,615,401 (8·9%); White, 4,586,838 (8·9%); Indian/Asian, 1,286,930 (2·5%). The United Nations population estimate for 2011 was 50,460,000. The population has increased more than would have been expected since 2001 as huge numbers of migrants have entered South Africa from Zimbabwe, many of them undocumented.

The UN gives a projected population for 2015 of 51·43m.

62·2% of the population were urban in 2011. In 2000 cities with the largest populations were (estimate in 1,000): Johannesburg (Gauteng), 2,732; Cape Town (Western Cape), 2,715; Durban (KwaZulu-Natal), 2,370; Pretoria/Tshwane (Gauteng), 1,084; Port Elizabeth (Eastern Cape), 958.

There were 10,714 immigrants in 2004 (10,578 in 2003) and 16,165 emigrants in 2003 (10,890 in 2002).

Population by province, according to the 2011 census:

Province	Total (including unspecified)	Black African	White	Coloured	Indian/ Asian
Eastern Cape	6,562,053	5,660,230	310,450	541,850	27,929
Free State	2,745,590	2,405,533	239,026	83,844	10,398
Gauteng	12,272,263	9,493,684	1,913,884	423,594	356,574
KwaZulu-Natal	10,267,300	8,912,921	428,842	141,376	756,991
Limpopo	5,404,868	5,224,754	139,359	14,415	17,881

Province	Total (including unspecified)	Black African	White	Coloured	Indian/ Asian
Mpumalanga	4,039,939	3,662,219	303,595	36,611	27,917
Northern Cape	1,145,861	576,986	81,246	461,899	7,827
Limpopo	5,404,868	5,224,754	139,359	14,415	17,881
North-West	3,509,953	3,152,063	255,385	71,409	20,652
Western Cape	5,822,734	1,912,547	915,053	2,840,404	60,761

There are 11 official languages. Numbers of mother-tongue speakers at the 2011 census: isiZulu, 11,587,374 (22·7% of population); isiXhosa, 8,154,258 (16·0%); Afrikaans, 6,855,082 (13·5%); English, 4,892,623 (9·6%); Sepedi, 4,618,576 (9·1%); Setswana, 4,067,248 (8·0%); Sesotho, 3,849,563 (7·6%); Xitsonga, 2,277,148 (4·5%); SiSwati, 1,297,046 (2·5%); Tshivenda, 1,209,388 (2·4%); isiNdebele, 1,090,223 (2·1%). The use of any of these languages is a constitutional right 'wherever practicable'. Each province may adopt any of these as its official language. English is the sole language of command and instruction in the armed forces.

SOCIAL STATISTICS

Births: total number of registered live births in 2010 was 1,294,694 (down from a high of 1,677,415 in 2003).

Deaths: the number of deaths increased from 318,287 in 1997 to 591,213 in 2005, with AIDS as the factor underlying much of the increase. In 2009, 17·8% of all adults between 15 and 49 were infected with HIV. Estimated population growth rate, 2006–07, 1·0%. Fertility rate, 2009, 2·5 births per woman. Life expectancy at birth, 2007, was 49·8 years for males and 53·2 for females. Life expectancy, which fell in the late 1990s and in the early part of the 21st century, is now starting to rise again thanks to the development and improved availability of medical treatments for HIV.

Infant mortality, 2010, 41 per 1,000 live births.

From 1998, under a new bill, customary and traditional marriages are recognized in law as registered by the civil registration system. Same-sex marriage was legalized in Nov. 2006. Marriages in 2010 totalled 170,826 (186,522 in 2008). Western Cape had the highest marriage rate in 2005 (613·8 per 100,000), Northern Cape the second highest (515·7 per 100,000) followed by Gauteng (497·1 per 100,000). Limpopo had the lowest rate (186·9 per 100,000). Of the total marriages officially recorded in 2010, 52,817 (30·9%) were solemnized in religious ceremonies and 91,250 by civil rites. 26,759 were classed under 'unspecified'. In 2010 the most popular age range for marrying was 30–34 for males and 25–29 for females. Divorces granted in 2010 totalled 22,936 (37,098 in 1999). Gauteng had the highest divorce rate in 2005 (943·5 per 100,000 married couples); Western Cape (763·8). Limpopo had the lowest rate (132·1 per 100,000).

CLIMATE

There is abundant sunshine and relatively low rainfall. The southwest has a Mediterranean climate, with rain mainly in winter, but most of the country has a summer maximum, although quantities show a decrease from east to west. Pretoria, Jan. 73·4°F (23·0°C), July 53·6°F (12·0°C). Annual rainfall 26·5" (674 mm). Bloemfontein, Jan. 73·4°F (23·0°C), July 45·9°F (7·7°C). Annual rainfall 22" (559 mm). Cape Town, Jan. 69·6°F (20·9°C), July 54·0°F (12·2°C). Annual rainfall 20·3" (515 mm). Johannesburg, Jan. 68·2°F (20·1°C), July 50·7°F (10·4°C). Annual rainfall 28·1" (713 mm).

CONSTITUTION AND GOVERNMENT

An Interim *Constitution* came into effect on 27 April 1994 and was in force until 3 Feb. 1997. Under it, the National Assembly and Senate formed a Constitutional Assembly, which had the task of drafting a definitive constitution. This was signed into law in Dec. 1996 and took effect on 4 Feb. 1997. The 1996 constitution defines the powers of the President, Parliament (consisting of the

National Assembly and the National Council of Provinces—NCOP), the national executive, the judiciary, public administration, the security services and the relationship between the three spheres of government. It incorporates a Bill of Rights pertaining to, *inter alia*, education, housing, food and water supply, and security, in addition to political rights. All legislation must conform to the Constitution and the Bill of Rights. The Constitution was amended in 2001 to provide that Constitutional Court judges are appointed for a non-renewable 12-year term of office, or until they reach the age of 70 years, except where an Act of Parliament extends the term of office of a Constitutional Court judge. This Constitution Amendment Act also made the head of the Constitutional Court the Chief Justice. The head of the Supreme Court of Appeal is now the President of that Court.

A *Constitutional Court*, consisting of a president, a deputy president and nine other judges, was inaugurated in Feb. 1995. The Court's judges are appointed by the President of the Republic from a list provided by the Judicial Service Commission, after consulting the President of the Constitutional Court (now the Chief Justice) and the leaders of parties represented in the National Assembly.

Parliament is the legislative authority and has the power to make laws for the country in accordance with the Constitution. It consists of the National Assembly and the NCOP. Parliamentary sittings are open to the public.

The *National Assembly* consists of no fewer than 350 and no more than 400 members directly elected for five years, 200 from a national list and 200 from provincial lists in the following proportions: Eastern Cape, 28; Free State, 14; Gauteng, 44; KwaZulu-Natal, 42; Limpopo, 25; Mpumalanga, 11; Northern Cape, 4; North-West, 12; Western Cape, 20. In terms of the 1993 Constitution, which still regulated the 1999 elections, the nine provincial legislatures are elected at the same time and candidates may stand for both. If elected to both, they have to choose between sitting in the national or provincial assembly. In the former case, the runner-up is elected to the Provincial Assembly.

The *National Council of Provinces* (NCOP) consists of 54 permanent members and 36 special delegates and aims to represent provincial interests in the national sphere of government. Delegations from each province consist of ten representatives. Bills (except finance bills) may be introduced in either house but must be passed by both. A finance bill may only be introduced in the National Assembly. If a bill is rejected by one house it is referred back to both after consideration by a joint National Assembly-NCOP committee called the Mediation Committee. Bills relating to the provinces must be passed by the NCOP. By Aug. 2003 more than 780 pieces of legislation had been passed since 1994.

The Constitution mandates the establishment of *Traditional Leaders* by means of either provincial or national legislation. The National House of Traditional Leaders was established in April 1997. Each provincial House of Traditional Leaders nominated three members to be represented in the National House. The National House advises national government on the role of traditional leaders and on customary law. Following a six-year study by the Commission on Traditional Leadership Disputes and Claims, in July 2010 President Jacob Zuma announced that only seven of the 13 kingships would be officially recognized and six would be abolished. Five of the seven kingships and their kings were confirmed by the commission. The remaining two were awaiting the commission's recommendation on the rightful incumbent. In terms of the commission's findings, existing kings who were found not to qualify for the status of kingship were to be allocated a principal traditional leadership.

National Anthem

A combination of shortened forms of 'Die Stem van Suid-Afrika'/ 'The Call of South Africa' (words by C. J. Langenhoven; tune by M. L. de Villiers) and the ANC anthem 'Nkosi sikelel' iAfrika'/ 'God bless Africa' (words and tune by Enos Santonga).

GOVERNMENT CHRONOLOGY

Presidents from 1961. (ANC = African National Congress; NP = National Party; UP = United Party)

1961–67	NP	Charles Robberts Swart
1968–75	NP	Jacobus Johannes Fouché
1975–78	NP	Nicolaas Johannes Diederichs
1978–79	NP	Balthazar Johannes Vorster
1979–84	NP	Marais Viljoen
1984–89	NP	Pieter Willem Botha
1989–94	NP	Frederik Willem de Klerk
1994–99	ANC	Nelson Rolihlahla Mandela
1999–2008	ANC	Thabo Mvuyelwa Mbeki
2008–09	ANC	Kgalema Petrus Motlanthe
2009–	ANC	Jacob Gedleyihlekisa Zuma

Prime Ministers since 1939.

1939–48	military/UP	Jan Christiaan Smuts
1948–54	NP	Daniël François Malan
1954–58	NP	Johannes Gerhardus Strijdom
1958–66	NP	Hendrik Frensch Verwoerd
1966–78	NP	Balthazar Johannes Vorster
1978–84	NP	Pieter Willem Botha

RECENT ELECTIONS

Parliamentary elections were held on 22 April 2009. Turnout was 77·3%. The African National Congress (ANC) won 264 seats in Parliament's National Assembly with 65·9% of votes cast, Democratic Alliance (DA) 67 with 16·7%, Congress of the People (COPE) 30 with 7·4%, Inkatha Freedom Party (IFP) 18 with 4·6%, Independent Democrats (ID) 4 with 0·9%, United Democratic Movement (UDM) 4 with 0·9%, Freedom Front Plus (VF+) 4 with 0·8%, African Christian Democratic Party (ACDP) 3 with 0·8%, United Christian-Democratic Party (UCDP) 2 with 0·4%, Pan African Congress of Azania (PAC) 1 with 0·3%, Minority Front (MF) 1 with 0·3%, Azanian People's Organization (AZAPO) 1 with 0·2% and African People's Convention 1 with 0·2%.

CURRENT GOVERNMENT

President: Jacob Zuma; b. 1942 (ANC; sworn in 9 May 2009).

Deputy President: Kgalema Motlanthe.

In March 2013 the government comprised:

Minister of Agriculture, Fisheries and Forestry: Tina Joemat-Pettersson. *Arts and Culture:* Paul Mashatile. *Basic Education:* Angie Motshekga. *Communications:* Dina Pule. *Co-operative Governance and Traditional Affairs:* Richard Baloyi. *Correctional Services:* Subusiso Ndebele. *Defence and Military Veterans:* Nosiviwe Mapisa-Nqakula. *Economic Development:* Ebrahim Patel. *Energy:* Dipuo Peters. *Finance:* Pravin Gordhan. *Health:* Aaron Motsoaledi. *Higher Education and Training:* Blade Nzimande. *Home Affairs:* Naledi Pandor. *Human Settlements:* Tokyo Sexwale. *International Relations and Co-operation:* Maite Nkoana-Mashabane. *Justice and Constitutional Development:* Jeff Radebe. *Labour:* Mildred Oliphent. *Mining:* Susan Shabangu. *Policing:* Nathi Mthethwa. *Public Enterprises:* Malusi Gigaba. *Public Service and Administration:* Lindiwe Sisulu. *Public Works:* Thembelani Nxesi. *Rural Development and Land Affairs:* Gugile Nkwinti. *Science and Technology:* Derek Hanekom. *Social Development:* Bathabile Dlamini. *Sport and Recreation:* Fikile Mbalula. *State Security:* Siyabonga Cwele. *Tourism:* Marthinus van Schalkwyk. *Trade and Industry:* Rob Davies. *Transport:* Ben Martins. *Water and Environmental Affairs:* Edna Molewa. *Women, Youth, Children and People with Disabilities:* Lulu Xingwana. *Minister in the Presidency Responsible for the National Planning Commission:* Trevor Manuel. *Minister in the Presidency Responsible for Performance, Monitoring and Evaluation, and Administration:* Collins Chabane.

Government Website: http://www.gov.za

CURRENT LEADERS

Jacob Gedleyihlekisa Zuma

Position
President

Introduction
Jacob Zuma became president of the ANC on 18 Dec. 2007 after defeating incumbent Thabo Mbeki in a leadership election. He was elected national president after the ANC won the general election of 22 April 2009.

Early Life
Zuma was born in what is now the KwaZulu-Natal Province on 12 April 1942. He had only five years of formal schooling and spent his early life travelling between Zululand and Durban. Zuma's father died in the Second World War and by the time he was 15, Zuma was taking odd jobs to supplement his mother's income. Influenced by his family's trade unionist background, he became involved in politics at the age of 17.

Zuma joined the ANC in 1959. With the party banned by the apartheid government the following year, by 1962 he was an active member of Umkhonto we Sizwe (the military wing of the ANC), which was subsequently classified as a terrorist organization. In 1963 Zuma was arrested and convicted of conspiring to overthrow the government. He was jailed for ten years, serving part of the sentence on Robben Island alongside Nelson Mandela.

After his release Zuma helped re-establish the ANC as an underground movement in Natal. In 1975 he left South Africa, living in Swaziland and then Mozambique where he dealt with thousands of exiles fleeing South Africa after the Soweto uprising. Zuma was appointed to the ANC national executive committee in 1977 and served as deputy chief representative of the ANC in Mozambique until the signing of the Nkomati Accord between South Africa and Mozambique in 1984, when he became chief representative of the ANC. Forced to leave Mozambique in 1987, Zuma was appointed head of underground structures and chief of the intelligence department, serving on the ANC's political and military council.

When the ban on the ANC was lifted in 1990 Zuma returned to South Africa and was elected ANC chairperson for the Southern Natal region. In 1991 he was elected the party's deputy secretary general, and at the 1994 general election he agreed to Thabo Mbeki running unopposed for the deputy presidency. Zuma was appointed to the executive committee of economic affairs and tourism for the ANC in the KwaZulu-Natal provincial government. In 1997 he became the ANC's deputy president and two years later was chosen as South Africa's executive deputy president.

In 2005, following the conviction of Zuma's financial adviser Schabir Shaik on charges of corruption and fraud, Zuma was dismissed as deputy president by Mbeki.

Career in Office
In Dec. 2007 Zuma was elected ANC party president standing against Mbeki. His rhetoric found favour with many disadvantaged South Africans who felt marginalized by Mbeki's business-friendly policies. Zuma was thus clear favourite to become the next president of South Africa.

On 28 Dec. 2007 the directorate of special operations (or Scorpions) indicted Zuma to stand trial on charges of racketeering, money laundering, corruption and fraud. The allegations were linked to a US$5bn. arms procurement deal made by the government in 1999. The trial was scheduled to start on 4 Aug. 2008 but was delayed when the charges were declared unlawful because Zuma had not had a chance to make representations pre-indictment. The national directorate of public prosecutions announced an appeal.

Mbeki resigned as president of South Africa on 21 Sept. 2008 after losing the support of the ANC over claims that he had interfered in the case against Zuma. The charges against Zuma were dropped, although prosecutors were given leave to appeal the following month. In April 2009 the National Prosecuting Authority dismissed all charges. The ANC triumphed at the general election on 22 April 2009 and Zuma was elected president by parliament. He was sworn into office on 9 May 2009, succeeding Kgalema Motlanthe, who stepped aside after having replaced Mbeki.

Also in May 2009, the South African economy officially went into recession following a sharp downturn in the manufacturing and mining sectors. Unemployment accelerated and strikes and violent protests in July–Aug. were roundly condemned by Zuma. In a speech to the ANC in Jan. 2010 he warned that recovery from the economic crisis would be slow and that there would be a lag in job creation. A national strike by public sector workers over pay, which began in Aug. and paralysed hospitals, education and other services for three weeks before the action was suspended, undermined relations between the ANC and the Congress of South African Trade Unions.

In Oct. 2010 Zuma announced major changes to the composition of his cabinet, appointing a host of new ministers and deputy ministers. However, allegations of high-level corruption continued to blight his administration and in Oct. 2011 he sacked two senior ministers and suspended the chief of police. Local elections earlier in May had maintained the ANC's political dominance, but the opposition Democratic Alliance increased its vote share to about 24%. In Nov. the ANC suspended Julius Malema, its militant youth wing leader and a prominent Zuma critic, for five years for bringing the party into disrepute.

In Jan. 2012 Zuma addressed a rally in Bloemfontein in celebration of the ANC's 100th anniversary. In Aug. 2012, 34 platinum miners in the town of Marikana were killed and around 80 more were wounded by police during an industrial dispute over wages. Public condemnation of the treatment of the miners prompted Zuma to establish a judicial commission of inquiry in Oct.

In Dec. Zuma was re-elected leader of the ANC at the party's five-yearly conference.

DEFENCE

The South African National Defence Force (SANDF) comprises four services, namely the SA Army, the SA Air Force, the SA Navy and the SA Military Health Service (SAMHS). In 2007 the SANDF consisted of 62,334 active members (excluding 12,382 civilian employees). SAMHS personnel totalled 7,115 (including around 1,115 reservists) in 2007. South Africa ended conscription in 1994.

Defence expenditure totalled US$3,359m. in 2008 (equivalent to US$69 per capita), and represented 1·2% of GDP. Defence expenditure in 1985 represented 3·8% of GDP. In 2008 South Africa was responsible for 15% of Africa's total defence expenditure.

Army
Army personnel totalled 41,350 in 2007. Regular army reserves numbered 38,545 in 2007. The army territorial reserve was disbanded in 2009.

Navy
Navy personnel in 2007 totalled 5,801, with 861 reserves. The fleet is based at the naval bases at Simon's Town on the west coast and Durban on the east and includes three submarines and four corvettes.

Air Force
Strength (2007) 9,183, with 831 reserves. In 2007 the Air Force had 29 combat-capable aircraft (*Cheetah* Cs and *Cheetah* Ds) and 11 combat-capable helicopters.

ECONOMY

Agriculture accounted for 3·0% of GDP in 2009, industry 31·1% and services 65·8%.

Overview

South Africa is one of the few African countries to have reached the upper middle-income group. Its economy is the largest in the sub-Saharan region and heavily influences trade and investment flows on the continent. South Africa was admitted as a BRIC nation in Dec. 2010, joining the other leading developing economies of Brazil, Russia, India and China.

However, despite a rising black middle class, South Africa maintains high levels of income inequality. The country also has one of the highest HIV/AIDS infection rates, with 18% of those between the ages of 15 and 49 living with HIV in 2009. The unemployment rate was over 24% in 2010.

Until recently the economy was dominated by agriculture and precious metals. South Africa is the world's largest producer of platinum and was the world's largest producer of gold until 2006 (although it is now only the fourth largest). An advanced financial sector and the growing tourism sector have helped services to become the largest contributor to total output. Manufacturing principally involves metals and engineering, especially steel-related products.

In the early 1990s the removal of international sanctions following the end of apartheid and the adoption of structural reforms opened the economy to international competition, leading to productivity gains and greater penetration of international markets. In the decade from 1994 the economy grew at an average annual rate of 2·9%, with less volatility than the previous decade when average annual growth was 1%. Performance in the post-apartheid period was fostered by trade liberalization, increased private sector participation in the economy and sound macroeconomic management.

The economy grew by around 5% for three consecutive years up to 2007 on the back of strong household demand. However, there were increased inflationary pressures and a widening of the external current account deficit. In early 2008 an electricity crisis stemming from the state-owned supplier, Eskom, resulted in widespread blackouts and disruption to mining production.

Demand for South African exports fell sharply during the international financial crisis. An estimated 500,000 workers lost their jobs in 2009 as South Africa entered recession for the first time in 17 years. However, GDP growth began to recover in 2010. The economy's key challenges include reducing unemployment and income inequality.

Currency

The unit of currency is the *rand* (ZAR) of 100 *cents*. A single free-floating exchange rate replaced the former two-tier system on 13 March 1995. Inflation rates (based on IMF statistics):

2002	2003	2004	2005	2006	2007	2008	2009	2010	2011
9·2%	5·8%	1·4%	3·4%	4·7%	7·1%	11·5%	7·1%	4·3%	5·0%

Foreign exchange reserves were US$32,251m. in Sept. 2009 (US$4,171m. in 1998) and gold reserves 4·01m. troy oz. Total money supply was R419,506m. in Aug. 2009.

Budget

Consolidated national budget in R1bn.:

	2007–08	2008–09[1]	2009–10[1]	2010–11[1]
Revenue	580·4	650·0	720·1	789·0
Expenditure	560·1	631·5	704·1	768·5

[1]Budgeted figures.

Income tax is the Government's main source of income. As of 2001, South Africa's source-based income tax system was replaced with a residence-based system. With effect from the years of assessment commencing on or after 1 Jan. 2001, residents are (subject to certain exclusions) taxed on their worldwide income, irrespective of where their income is earned. Foreign taxes are credited against South African tax payable on foreign income. Foreign income and taxes are translated into the South African monetary unit, the Rand.

Value-added Tax (VAT) has remained at 14% since 1993. Corporate taxes were reduced to 28% in 2008. A tiered corporate tax was introduced in 2000 with taxes for small businesses reduced by half. Individual income tax rates range from 18% to 40%. A capital gains tax was introduced from 1 April 2001 and became effective on 1 Oct. 2001.

South Africa's fiscal year runs from 1 April to 31 March.

Performance

Real GDP growth rates (based on IMF statistics):

2002	2003	2004	2005	2006	2007	2008	2009	2010	2011
3·7%	2·9%	4·6%	5·3%	5·6%	5·5%	3·6%	−1·5%	2·9%	3·1%

In 2009 South Africa experienced its first recession in 17 years. Total GDP in 2011 was US$408·2bn.

Banking and Finance

The central bank and bank of issue is the South African Reserve Bank (SARB; established 1920), which functions independently. Its *Governor* is Gill Marcus. The Banks Act, 1990 governs the operations and prudential requirements of banks.

At the end of Dec. 2008, 33 banks (excluding two mutual banks) were registered with the Office of the Registrar of Banks. Furthermore, 43 foreign banks had authorized representative offices in South Africa. The combined assets of the banking institutions amounted to R3,170bn. (31 Dec. 2008), total liabilities to R2,989bn. and total equity to R181bn. Total assets of the four largest commercial banks (Standard Bank, FirstRand, Absa and Nedbank) came to R2,676bn.

Foreign debt totalled US$45,165m. in 2010, representing 12·7% of GNI.

The stock exchange, the JSE Securities Exchange, is based in Johannesburg. Foreign nationals have been eligible for membership since Nov. 1995.

ENERGY AND NATURAL RESOURCES

Environment

In 1998 the Committee for Environmental Co-ordination was established to harmonize the work of government departments on environmental issues, and to co-ordinate environmental implementation and national management plans at provincial level.

South Africa's carbon dioxide emissions from the consumption and flaring of fossil fuels in 2008 were the equivalent of 9·2 tonnes per capita (compared to the average for Africa as a whole of 1·1 tonnes per capita).

Electricity

South African households use over 25% of the country's energy. Coal supplies 75% of primary energy requirements, followed by oil (21%), nuclear (3%) and natural gas (1%). There is one nuclear power station (Koeberg) with two reactors, two gas turbine generators, two conventional hydroelectric plants and two pumped storage stations. Nuclear energy is being investigated as a future potential energy source and alternative to coal. Eskom, a public utility, generates 95% of the country's electricity (as well as two-thirds of the electricity for the African continent) and owns and operates the national transmission system. In 2011, 222,710 GWh of electricity were delivered throughout South Africa by Eskom.

Eskom electrified 149,914 homes in 2011, bringing the total number of homes connected since the start of the electrification programme in 1991 to 3,901,054. The Government aims to achieve universal access to electricity by 2014. Capacity shortages from late 2007 forced Eskom into emergency reductions to protect the power system from potential failure and a national electricity crisis was declared in Jan. 2008. According to a survey of Feb. 2007, 80·0% of households used electricity for lighting.

The energy sector contributes about 15% to GDP and employs about 250,000 people. Because of South Africa's large coal deposits, the country is one of the cheapest electricity suppliers in the world.

The first wind-energy farm in Africa was opened at Klipheuwel in the Western Cape in Feb. 2003.

Oil and Gas

South Africa has limited oil reserves and relies on coal for much of its oil production. It has a highly developed synthetic fuels industry. Sasol and PetroSA are the two major players in the synthetic fuel market. Synfuels meet approximately 40% of local demand. Natural gas production in 2007 amounted to 1·9bn. cu. metres.

PetroSA is responsible for exploration of both offshore natural gas and onshore coal-bed methane. The EM gas-field complex off Mossel Bay in the Western Cape started production in 2000, and was to ensure sufficient feedstock to PetroSA to maintain current liquid fuel production levels at 36,000 bbls of petroleum products a day until 2009 although 2008 saw a decline in production levels. Planning is under way for the construction of a new refinery near Port Elizabeth that is expected to come online in 2015. The oilfield, Sable, situated about 150 km south off the coast of Mossel Bay, is expected to produce 17% of South Africa's oil needs. Coming into operation in Aug. 2003, it was initially projected to produce 30,000 to 40,000 bbls of crude oil a day. PetroSA's gas-to-liquid plant supplies about 7% of South Africa's liquid fuel needs.

South Africa is one of the major oil refining nations in Africa with a crude refining capacity of 543,000 bbls per day.

Minerals

Total sales of primary minerals increased to R223·9bn. in 2007; the value of exports of primary minerals increased to R161·8bn. Mining and quarrying contributed 8·2% of GDP in 2006, up from 6·9% in 1996 although down from 14·4% in 1986.

In 2007 employment in the mining sector rose by 8·6% from 456,337 in 2006 to 495,474. Over 50 different minerals were produced in 2007 from 1,414 mines and quarries.

Mineral production (in tonnes), 2005: coal, 244·9m.; iron ore (metal content), 24·9m.; limestone and dolomite, 24·8m.; chromium, 7·5m.; manganese, 4·6m.; aluminium, 846,000; copper (metal content), 97,000; nickel (metal content), 42,000; zinc (metal content), 32,000; platinum-group metals, 303; gold, 295; silver, 88. Diamond production, 2005: 15,776,000 carats. In 2007 gold production fell by 7·2% from 272 tonnes in 2006 to 253 tonnes (its lowest level in over 80 years) although sales revenue rose by 1·6% to R38bn. South Africa is the world's leading producer of platinum (and has the largest reserves of both platinum and gold). In 2007 South Africa lost its status as the world's largest gold producer to China.

Agriculture

South Africa has a dual agricultural economy, comprising a well-developed commercial sector and a predominantly subsistence-orientated sector. Much of the land suitable for mechanized farming has unreliable rainfall. Of the total farming area, natural pasture occupies 81% (69·6m. ha.) and planted pasture 2% (2m. ha.). About 12% of South Africa's surface area can be used for crop production. High potential arable land comprises only 22% of the total arable land. Annual crops and orchards are cultivated on 9·9m. ha. of dry land and 1·3m. ha. under irrigation. In 2007 there were 39,982 commercial farming units with a gross farming income of R79·6m. Agriculture, forestry, hunting and fishing

contributed 2·9% of GDP in 2006, down from 4·2% in 1996. In 2007 there were 796,806 paid farm workers.

Production:

(*Field crops, 2006–07 unless otherwise indicated, in 1,000 tonnes*): sugarcane, 20,278; maize (2005–06), 6,947; wheat (2007), 1,913; lucerne hay, 1,232; sunflower seeds (2005–06), 541; soybeans (2005–06), 424; grain sorghum, 202; groundnuts, 66; seed cotton, 29; cotton seed, 18; tobacco, 14.

(*Horticulture, 2006–07, in 1,000 tonnes*): potatoes (2007), 1,917; oranges, 1,336; apples, 646; tomatoes, 453; grapefruit, 415; onions, 405; bananas, 357; pears, 337. Wine production totalled 1,012m. litres in 2006, ranking South Africa eighth in the world.

(*Animal products, 2007–08*): 5,812,000 sheep and goats slaughtered; 2,989,000 cattle and calves slaughtered; 2,579,000 pigs slaughtered; 717m. broilers slaughtered (2006); fresh milk (2007), 2,470m. litres; wool, 48·4m. kg.

Gross value of field crops in 2007 (R1,000), 25,287,668; horticulture, 25,683,799; animal products, 48,967,173.

Agricultural exports contribute about 4% of total imports and 5% of total exports. In 2007 the estimated value of agricultural imports was R19·5bn. and exports R21·9bn. Fruit and vegetables accounted for 59% of agricultural exports in 2007.

Forestry

South Africa has developed one of the largest man-made forestry resources in the world with plantations covering an area of 1·3m. ha. Production from these plantations in 2007 amounted to 20·3m. cu. metres, valued at almost R5·2bn., the most important products by value being pulpwood (R3·6bn.) and sawlogs (R1·4bn.). In terms of volume, 13·2m. cu. metres and 5·4m. cu. metres of the aforementioned products were produced respectively. Collectively, the forestry sector employs about 170,000 people. An equivalent of about 107,000 full-time staff are employed in the primary sector (growing and harvesting), while the balance is employed in the processing industries (sawmilling, pulp and paper, mining timber and poles, and board products). In 2007 the forestry and forest products industry contributed 0·9% to South Africa's GDP.

In 2007 the area under all types of forests was 9·2m. ha. Of this, commercial plantations covered 1·3m. ha. and indigenous forests 534,000 ha. or 0·4% of the country's surface. The balance was 'woodland' forests. The private sector owned or controlled 1,051,223 ha. (83%) of the plantation area of 1,266,194 ha. as well as 174 of the 178 primary processing plants in the country. The remaining 17% (214,973 ha.) were owned by the state.

The industry was a net exporter to the value of R2·84bn. in 2007, almost 99% of which was in the form of converted value-added products. The forest-products industry contributed 2·5% of total exports and 1·7% of total imports in 2007. In that year paper exports were the most important (R5·51bn. or 45% of the total), followed by pulp (R3·52bn. or 29% of the total), solid wood products (R2·84bn. or 23% of the total) and other products (R0·32bn. or 3% of the total). Woodchip exports, mainly to Japan, accounted for 68% (R1·93bn.) of the total solid wood products exports.

Fisheries

The commercial marine fishing industry is valued at approximately R2bn. annually and employs 27,000 people directly. In 2000 the commercial fishing fleet consisted of 4,477 vessels licensed by the department of environmental affairs and tourism.

The total number of fishing rights allocated stands at 2,200, 1,700 of which are small, medium and micro enterprises. The total catch in 2010 was 623,920 tonnes, over 99% of which came from marine fishing.

INDUSTRY

The leading companies by market capitalization in South Africa in March 2012 were: MTN Group, a mobile telecommunications company (US$33·2bn.); Sasol Ltd, a coal, oil and gas producer (US$31·1bn.); and Naspers, a multinational media company (US$23·1bn.).

Actual value of sales of the principal groups of industries (in R1m.) in 2005: coke, petroleum, chemical products, rubber and plastic, 228·0; basic metals, fabricated metal products, machinery and equipment, and office, accounting and computing machinery, 190·1; food products and beverages, 158·7; transport equipment, 149·7; wood and wood products, paper, publishing and printing, 79·8; textiles, clothing, leather and footwear, 38·6. Total actual value including other groups, R953·9m. In 2009 industry accounted for 31·1% of GDP, with manufacturing contributing 15·1%.

Labour

The Employment Equity Act, 1998 signalled the beginning of the final phase of transformation in the job market, which began with the implementation of the Labour Relations Act. It aims to avoid all discrimination in employment. The Basic Conditions of Employment Act, 1997 applies to all workers except for the South African National Defence Force (SANDF), the South African Secret Service (SASS) and the National Intelligence Agency (NIA). The new provisions include a reduction in the maximum hours of work from 46 to 45 hours per week (however, the Act allows for the progressive reduction of working hours to 40 per week).

The labour force in South Africa numbered 17·1m. in the fourth quarter of 2009, of which 4·2m. were unemployed. In the fourth quarter of 2009 the unemployment rate was 24·3%, up from 21·9% in the fourth quarter of 2008.

The Unemployment Insurance Fund (UIF) provides benefits to workers who become unemployed. All employees who work for more than 24 hours a month contribute to the Fund. In the year ending March 2009 there were 7·6m. contributors (7·3m. in 2008). In the same period the UIF paid benefits to 627,244 beneficiaries, a total amount of R3·8bn.

INTERNATIONAL TRADE

South Africa's four main trading partners in 2006 were Germany, USA, Japan and China. Germany is South Africa's number one trading partner in terms of total trade (the sum of exports and imports) recorded in 2006. Exports to Germany rose in nominal terms from R21·1bn. in 2005 to R24·4bn. in 2006. Imports from Germany increased in nominal terms from R49·2bn. in 2005 to R53·9bn. in 2006.

In 2005 Europe accounted for 38·9% (R116·9bn.) of South Africa's total exports and 40·3% (R140·4bn.) of total imports. Seven of South Africa's top ten trading partners are European countries. A trade, co-operation and development agreement was provisionally implemented on 1 Jan. 2000, under the terms of which South Africa will grant duty-free access to 86% of EU imports over a period of 12 years, while the EU will liberalize 95% of South Africa's imports over a ten-year period. The Agreement provides for ongoing EU financial assistance in grants and loans for development co-operation, which amounts to some R900m. per annum.

In 2010 the total value of South Africa's exports destined for Africa was R85·0bn. and imports R45·9bn. Within the Southern African Development Community (SADC), a smaller group of countries including South Africa, Botswana, Lesotho, Namibia and Swaziland have organized themselves into the Southern African Customs Union (SACU), sharing a common tariff regime without any internal barriers. Exports to SADC countries increased to R60·8bn. in 2010. Imports from within the region in 2010 totalled R23·7bn.

Japan is South Africa's largest trading partner in Asia. In 2006 total trade between the two countries amounted to R66·3bn.

Imports and Exports

Trade in US$1m.:

	2007	2008	2009	2010
Imports (f.o.b.)	79,872·6	87,593·1	63,766·1	80,139·3
Exports (f.o.b.)	64,026·6	73,965·5	53,863·9	71,484·3

Main imports (in US$1m.):

	2002	2003	2004
Petroleum and petroleum products	3,134·7	3,934·5	6,647·6
Road vehicles	1,662·5	2,441·5	3,818·7
Telecommunications, sound recording and reproducing equipment	1,613·5	1,722·7	2,614·7
Other transport equipment	737·7	1,486·6	2,391·6
Office machines and automatic data processing machines	1,080·4	1,616·0	2,329·7

Main exports (in US$1m.):

	2002	2003	2004
Non-ferrous metals (including platinum)	1,099·1	4,423·5	6,927·8
Iron and steel	2,411·4	3,877·4	5,642·0
Road vehicles	2,396·7	3,114·5	3,540·6
Coal, coke and briquettes	1,839·1	1,804·8	2,432·6
Non-metallic mineral manufactures (including diamonds)	1,750·1	2,010·3	2,326·5

In Oct. 1998 a transhipment facility for containers opened at Kidatu, southwest of Dar es Salaam, Tanzania, providing a link between the 1,067 mm gauge railways of the southern part of Africa and the 1,000 mm gauge lines of the north. With the opening up of new markets for South Africa elsewhere in the continent, it has helped to boost trade and facilitate the shipment of cargo to countries to the north.

COMMUNICATIONS

The public company Transnet Limited was established on 1 April 1990. It handles 176m. tonnes of rail freight per year, 2·8m. tonnes of road freight and 194m. tonnes of freight through the harbours, while 13·8m. litres are pumped through its petrol pipelines annually. For the financial year ended 31 March 2007 Transnet reported a profit of R7,404m. (R4,930m. for 2006).

The company, through South African Airways (SAA), flies 6·1m. domestic, regional and international passengers per year. In total, Transnet is worth R72bn. in fixed assets and has a workforce of some 80,000 employees.

Transnet Limited consists of nine main divisions, a number of subsidiaries and related businesses—Transnet Freight Rail, the National Ports Authority (NPA), South African Port Operations (SAPO), Petronet, Freightdynamics, Propnet, Metrorail, Transtel and Transwerk.

Roads

In 2006 the South African road network comprised some 754,600 km of roads and streets. There is a primary roads network of 9,600 km, with plans to extend it to 20,000 km. Toll roads, which are serviced by 32 mainline toll plazas, cover about 2,400 km. The network includes 1,437 km of dual-carriage freeway, 440 km of single-carriage freeway and 56,967 km of single-carriage main road with unlimited access. South Africa has the longest road network in Africa. As at 31 Dec. 2005 there were 7,971,187 registered motor vehicles. In 2009 a total of 13,768 people were killed in traffic accidents (14,920 in 2007 and 13,875 in 2008).

Rail

The Passenger Rail Agency of South Africa (PRASA) was formed in March 2009 as an umbrella organization to oversee the day-to-day running of rail services in South Africa. PRASA operates Metrorail, offering commuter rail services in urban areas and transporting 1·7m. passengers on weekdays to 478 stations over 2,400 km of track; and Shosholoza Meyl, providing regional and long-distance rail transport. South Africa's first high-speed rail service, Gautrain, was inaugurated in June 2010. It links Johannesburg's O. R. Tambo International Airport and the centre of the city with Pretoria.

Freight train services are provided by Transnet Freight Rail (formerly Spoornet), a public company owned solely by the South African government. Transnet transports 17% of the nation's freight annually and employs over 25,000 people.

Civil Aviation
Responsibility for civil aviation safety and security lies with the South African Civil Aviation Authority (SACAA). The Airports Company South Africa (ACSA) owns and operates South Africa's principal airports. The main international airports are: Johannesburg, Cape Town, Durban, Bloemfontein, Port Elizabeth, Pilanesberg, Lanseria and Upington. In April 2003 the Cabinet approved the status of the Kruger Mpumalanga Airport, near Nelspruit, as an international airport. ACSA also has a 35-year concession to operate Pilanesberg International Airport near Sun City in North-West Province.

The flag carrier South African Airways (SAA), along with Airlink, Comair, Interair and SA Express, operate scheduled international air services. Several other operators provide internal flights and cargo services.

In 2007–08 O. R. Tambo International Airport (formerly Johannesburg International) handled 19,457,498 passengers (11,009,841 on domestic flights), Cape Town handled 8,426,618 passengers (6,950,061 on domestic flights) and Durban handled 4,792,553 passengers (4,747,202 on domestic flights). O. R. Tambo Airport is also the busiest airport for freight, handling 280,095 tonnes of cargo in 2005.

Shipping
The South African Maritime Safety Authority (SAMSA) was established on 1 April 1998 as the authority responsible for ensuring the safety of life at sea and the prevention of sea pollution from ships. Approximately 98% of South Africa's exports are conveyed by sea.

The National Ports Authority supervises South Africa's major ports. The largest ports include the deep water ports of Richards Bay, with its multi-product dry bulk handling facilities, multi-purpose terminal and the world's largest bulk coal terminal, and Saldanha featuring a bulk ore terminal adjacent to a bulk oil jetty with extensive storage facilities. Durban, Cape Town and Port Elizabeth provide large container terminals for deep-sea and coastal container traffic. The Port of Durban handles 2.5m. containers per annum. East London, the only river port, has a multi-purpose terminal and dry dock facilities. Mossel Bay is a specialized port serving the south coast fishing industry and offshore gas fields. During 2008–09 the seven major ports handled a total of 184,628,480 tonnes of cargo (Richards Bay, 82,621,766 tonnes; Saldanha, 50,282,909 tonnes; Durban, 40,118,656 tonnes).

In Jan. 2009 there were four ships of 300 GT or over registered, totalling 32,000 GT.

Telecommunications
In 2008 there were 4.4m. main (fixed) telephone lines. In the same year mobile phone subscribers numbered 45.0m. (906.0 per 1,000 persons). The largest mobile phone networks are Vodacom and MTN.

Between 1997 and 2002 Telkom SA, for many years the national operator, concentrated on replacing analogue lines with digital technology, under the terms of its final exclusive licence; the transmission network is now almost wholly digital. Under the Telecommunications Acts of 1996 and 2001, South Africa has liberalized its telecommunications industry. Telkom lost its monopoly on the fixed-line market in 2006 with the launch of Neotel, the country's second national operator. Telkom was offered on the Johannesburg Securities Exchange and the New York Stock Exchange in March 2003, realizing R3.9bn. on the first day.

A new 14,000-km submarine cable, the West Africa Cable System, became operational in May 2012, allowing South Africa to greatly increase the capacity of its mobile phone and internet networks.

Internet users numbered 4,187,000 in 2008. In June 2012 there were 5.0m. Facebook users.

SOCIAL INSTITUTIONS

Justice
All law must be consistent with the Constitution and its Bill of Rights. Judgments of courts declaring legislation, executive action, or conduct to be invalid are binding on all organs of state and all persons. The common law of the Republic is based on Roman-Dutch law—that is the uncodified law of Holland as it was at the date of the cession of the Cape to the United Kingdom in 1806. South African law has, however, developed its own unique characteristics.

Judges hold office until they attain the age of 70 or, if they have not served for 15 years, until they have completed 15 years of service or have reached the age of 75, when they are discharged from active service. A judge discharged from active service must be ready to perform service for an aggregate of three months a year until the age of 75. The Chief Justice of South Africa, the Deputy Chief Justice, the President of the Supreme Court of Appeal and the Deputy President of the Supreme Court of Appeal are appointed by the President after consulting the Judicial Service Commission. In the case of the Chief Justice and Deputy Chief Justice, the President must also consult the leaders of parties represented in the National Assembly. The President on the advice of the Judicial Service Commission (JSC) appoints all other judges. No judge may be removed from office unless the JSC finds that the judge suffers from incapacity, is grossly incompetent or is guilty of gross misconduct, and the National Assembly calls for that judge to be removed by a resolution supported by at least two thirds of its members.

The higher courts include: 1) *The Constitutional Court* (CC), which consists of the Chief Justice of South Africa, the Deputy Chief Justice of South Africa and nine other judges. It is the highest court in all matters in which the interpretation of the Constitution or its application to any law, including the common law, is relevant; 2) *The Supreme Court of Appeal*, consisting of a President, a Deputy President and the number of judges of appeal determined by an Act of Parliament. It is the highest court of appeal in all other matters; 3) *The High Courts*, which may decide constitutional matters other than those which are within the exclusive jurisdiction of the Constitutional Court, and any other matter other than one assigned by Parliament to a court of a status similar to that of a High Court. Each High Court is presided over by a Judge President who may divide the area under his jurisdiction into circuit districts. In each such district there shall be held at least twice in every year and at such times and places determined by the Judge President, a court which shall be presided over by a judge of the High Court. Such a court is known as the circuit court for the district in question; 4) *The Land Claims Court*, established under the Restitution of Land Rights Act of 1994 deals with claims for restitution of rights in land to persons or communities dispossessed of such rights after 1913 as a result of past racially discriminatory laws or practices. It has jurisdiction throughout the Republic and the power to determine such claims and related matters such as compensation and rights of occupation; 5) *The Labour Court*, established under the Labour Relations Act, 1995 deals with labour disputes. It is a superior court that has authority, inherent powers and standing in relation to matters under its jurisdiction, equal to that the High Court has in relation to matters under its jurisdiction. Appeals from decisions of the Labour Court lie to the Labour Appeal Court which has authority in labour matters equivalent to that of the Supreme Court of Appeal in other matters.

The lower courts are called Magistrates' Courts. Magisterial districts have been grouped into 13 clusters headed by chief magistrates. From the magistrates court there is an appeal to the

High Court having jurisdiction in that area, and then to the Supreme Court of Appeal. In cases involving constitutional matters there is a further appeal to the Constitutional Court.

The death penalty was abolished in June 1995 and no executions have taken place since 1989. In 2008–09 there were 18,148 murders (down from a peak of 26,877 murders in 1995–96). Although steadily declining, South Africa's murder rate of 37 per 100,000 in 2008–09 is still one of the highest in the world. Budgeted spending on police, prisons and justice services for 2006–07 was R79·6bn. The population in penal institutions in Dec. 2006 was 189,748.

Education

The South African Schools Act, 1996 became effective on 1 Jan. 1997 and provides for: compulsory education for students between the ages of seven and 15 years of age, or students reaching the ninth grade, whichever occurs first. Pupils normally enrol for Grade 1 education at the beginning of the year in which they turn seven years of age although earlier entry at the age of six is allowed if the child meets specified criteria indicating that they have reached a stage of school readiness.

In 2008 the South African public education system accommodated 11·9m. school pupils, 799,387 university students (of which 140,330 at universities of technology) and 2·5m. further education and training college students. There were 25,875 primary, secondary, combined and intermediate schools with 400,953 educators.

In the 2010–11 financial year R165bn. was allocated to education. In 2007 public expenditure on education came to 5·4% of GDP and 17·4% of total government spending.

As a result of a major restructure of South African higher education during 2002–04, 36 universities and technikons (non-university higher education institutions) were reduced by means of mergers and incorporations to 17 universities and six universities of technology. The University of South Africa (UNISA) is the oldest and largest university in South Africa and one of the largest distance education institutions in the world. In 2009 UNISA had a total of 263,559 students and 9,344 staff (4,286 permanent). In 2005 the Tshwane University of Technology had 60,400 students, University of Pretoria 46,400, University of Johannesburg 45,500, University of Kwazulu-Natal 40,700, North West University 38,600 and the Cape Peninsula University of Technology 29,000.

The adult literacy rate in 2007 was an estimated 88% (89% for males and 87% for females).

Health

Some 40% of South Africans live in poverty and 75% of these live in rural areas with limited access to health services. There is an extensive network of public health clinics providing free public health services, plus mobile clinics run by the government to provide primary and preventive health care.

36,912 doctors were registered with the Health Profession Council of South Africa (HPCSA) in 2010. These include doctors working for the state, doctors in private practice and specialists. Doctors train at the medical schools of eight universities and the majority go on to practise privately. In 2010, 110,518 enrolled nurses and enrolled nursing auxiliaries were registered with the South African Nursing Council (SANC); there were 5,320 dental practitioners registered with the HPCSA. In 2010, 12,218 pharmacists were registered with the South African Pharmacy Council. Chris Hani Baragwanath Hospital, situated to the southwest of Johannesburg, with its 3,200 beds, is the largest hospital in the world.

In Oct. 1998 the first traditional hospital was opened in Mpumalanga—the Samuel Traditional Hospital. There are about 200,000 traditional healers in South Africa providing services to between 60% and 80% of their communities.

Approximately 5·5m. South Africans are HIV-infected, the highest number in the world (equivalent to nearly 12% of the population of South Africa and some 17% of global HIV/AIDS cases). Government expenditure on HIV and AIDS increased substantially from R30m. in 1994 to R3bn. in 2005–06. In Aug. 2003 the government announced plans to roll out the provision of anti-retrovirals (ARVs) in the public health sector which would see 1·4m. people on treatment. It also envisaged that there would be at least one service point in every local municipality across the country by 2008. Although the latter has been achieved, it is clear that the number of people receiving ARVs is well behind the initial target.

Welfare

In 2008–09 the department of social development was disbursing grants through its provincial offices to 13·0m. beneficiaries. Recipients are means-tested to determine their eligibility. 8·8m. people received the child support grant (CSG) of R240 per month. The age of children eligible for the CSG has been progressively increased to cover children up to and including the age of 17 years since 1 Jan. 2012. In 2008–09, 2·4m. received old-age grants of R1,010 per month. The age for male eligibility was lowered in phases over a two-year period from 65 to 60 in April 2010, the same as for women.

Other benefits paid are the disability, foster child, care dependency and war veterans' grants as well as institutional grants and grants in aid.

The total budget allocation for the payment of social assistance was R80·4bn. in 2009–10.

RELIGION

South Africa is a secular state and freedom of worship is guaranteed by the Constitution. Almost 80% of the population professes the Christian faith. Other major religious groups are Hindus, Muslims and Jews. A sizeable minority of the population subscribe to traditional African faiths. In 1992 the Anglican Church of Southern Africa voted by 79% of votes cast for the ordination of women. In Feb. 2013 there was one cardinal.

CULTURE

World Heritage Sites

UNESCO World Heritage sites under South African jurisdiction (with year entered on list) are: Greater St Lucia Wetland Park (1999), encompassing marine, wetland and savannah environments; Robben Island (1999), used since the 17th century as a prison, hospital and military base—it was the location for Nelson Mandela's incarceration; fossil hominid sites of Sterkfontein, Swartkrans, Kromdraai and environs (1999 and 2005), offering evidence of human evolution over 3·5m. years; uKhahlamba/Drakensberg Park (2000), including caves with 4,000-year old paintings; Mapungubwe Cultural Landscape (2003), a savannah landscape at the confluence of the Limpopo and Shashe rivers and the site of the largest kingdom in Africa in the 14th century; Cape Floral Region Protected Areas (2004); Vredefort Dome (2005), part of a meteorite impact structure; and Richtersveld cultural and botanical landscape (2007), covering 160,000 ha. of mountainous desert.

Press

The major press groups are Independent Newspapers (Pty) Ltd, Media24 Ltd, CTP/Caxton Publishers and Printers Ltd, and Johnnic Publishing Ltd. Other important media players include Primedia, Nail (New Africa Investments Limited) and Kagiso Media. Nail has unbundled into a commercial company (New Africa Capital) and a media company (New Africa Media).

In 2006 there were nine paid-for dailies, 13 paid-for Sunday newspapers and 71 paid-for non-daily newspapers. Newspapers with the highest circulations (Jan.–March 2008): *Sunday Times* (504,193); *Daily Sun* (499,436); *Rapport* (301,827); *Soccer-Laduma* (292,701); *Sunday World* (203,460); *Sunday Sun* (202,524); *City*

Press (201,790); *Sowetan* (145,173). *Beeld* is the largest Afrikaans daily (105,149) and *Isolezwe* the largest isiZulu daily (99,098).

Tourism

A record 9·59m. visitors travelled to the country in 2008, up from 9·09m. in 2007. Most visitors in 2008 came from Lesotho, Zimbabwe, Mozambique, Swaziland, Botswana, the UK, the USA and Germany. Gauteng and Western Cape were the most popular provinces visited in 2008 (with 32·3% and 26·9% of visitor nights respectively).

Festivals

Best-known arts festivals: the Klein Karoo Festival (Oudtshoorn, Western Cape), which has a strong Afrikaans component, is held in April; the Grahamstown Arts Festival in the Eastern Cape is held in June/July; the Mangaung African Cultural Festival (Macufe) is held in Sept. in Bloemfontein; and the Aardklop Arts Festival, in Potchefstroom in the North-West province, is held in Sept. The Encounters South African International Documentary Festival has been held since 1999.

DIPLOMATIC REPRESENTATIVES

Of South Africa in the United Kingdom (South Africa House, Trafalgar Square, London, WC2N 5DP)
High Commissioner: Dr Zola Sidney Themba Skweyiya.

Of the United Kingdom in South Africa (255 Hill St., Arcadia, Pretoria 0002)
High Commissioner: Dr Nicola Brewer.

Of South Africa in the USA (3051 Massachusetts Ave., NW, Washington, D.C., 20008)
Ambassador: Ebrahim Rasool.

Of the USA in South Africa (877 Pretorius St., Arcadia, Pretoria 0083)
Ambassador: Donald H. Gips.

Of South Africa to the United Nations
Ambassador: Baso Sangqu.

Of South Africa to the European Union
Ambassador: Sizo Mxolisi Nkosi.

FURTHER READING

Government Communication and Information System (GCIS), including extracts from the *South Africa Yearbook 2011/12*, compiled and published by GCIS.

Beinart, W., *Twentieth Century South Africa.* 1994
Butler, Anthony, *Contemporary South Africa.* 2nd ed. 2009
Davenport, T. R. H., *South Africa: a Modern History.* 5th ed. 2000
De Klerk, F. W., *The Last Trek—A New Beginning.* 1999
Ellis, Stephen, *External Mission: The ANC in Exile, 1960-1990.* 2012
Fine, B and Rustomjee, Z., *The Political Economy of South Africa.* 1996
Giliomee, Hermann, *The Afrikaners: Biography of a People.* 2003
Guelke, Adrian, *Rethinking the Rise of Apartheid.* 2004
Hough, M. and Du Plessis, A. (eds.) *Selected Documents and Commentaries on Negotiations and Constitutional Development in the RSA, 1989–1994.* 1994
Johnson, R. W. and Schlemmer, L. (eds.) *Launching Democracy in South Africa: the First Open Election, 1994.* 1996
Mandela, N., *Long Walk to Freedom: the Autobiography of Nelson Mandela.* 1994
Meredith, M., *South Africa's New Era: the 1994 Election.* 1994
Picard, Louis A., *The State of the State: Institutional Transformation, Capacity and Political Change in South Africa.* 2005
Sparks, Allister, *Beyond the Miracle: Inside the New South Africa.* 2006
Thompson, L., *A History of South Africa.* 3rd ed. 2001
The Truth and Reconciliation Commission of South Africa Report, 5 vols. 1999
Waldmeir, P., *Anatomy of a Miracle: the End of Apartheid and the Birth of the New South Africa.* 1997
Who's Who in South African Politics. Online only

National Statistical Office: Statistics South Africa, Private Bag X44, Pretoria 0001.
Website: http://www.statssa.gov.za

SOUTH AFRICAN PROVINCES

In 1994 the former provinces of the Cape of Good Hope, Natal, the Orange Free State and the Transvaal, together with the former 'homelands' or 'TBVC countries' of Transkei, Bophuthatswana, Venda and Ciskei, were replaced by nine new provinces. Transkei and Ciskei were integrated into Eastern Cape, Venda into Northern Province (now Limpopo), and Bophuthatswana into Free State, Mpumalanga and North-West.

The administrative powers of the provincial governments in relation to the central government are set out in the 1999 constitution after a revision of the original text, demanded by the Constitutional Court in 1996.

Eastern Cape

TERRITORY AND POPULATION

The area is 169,580 sq. km and the population at the 2011 census was 6,562,053 (3,472,353 female), the third largest population in South Africa. Of that number: Black African, 5,660,230 (86% of the population); Coloured, 541,850 (8%); White, 310,450 (5%); Indian/Asian, 27,929 (0·4%). In 2001, 38·8% of the population lived in urban areas. Density (2011), 39 per sq. km. Life expectancy at birth, 2007 estimate, was 48·4 years. At the 2011 census 78·8% spoke isiXhosa as their first language, 10·6% Afrikaans, 5·6% English and 2·5% Sesotho.

Eastern Cape comprises 77 administrative districts (including Umzimkulu district, an enclave within KwaZulu-Natal). The provincial capital is Bhisho.

SOCIAL STATISTICS

Registered live births in 2010 totalled 134,580; deaths in 2005 totalled 92,915. Total number of marriages officially recorded in 2010 was 22,329; divorces granted in 2005 numbered 1,928.

CONSTITUTION AND GOVERNMENT

There is a 63-seat provincial legislature.

RECENT ELECTIONS

At the provincial elections held on 22 April 2009, 44 seats were won by the African National Congress, nine by the Congress of the People, six by the Democratic Alliance, three by the United

Democratic Movement and one by the African Independent Congress.

CURRENT GOVERNMENT

In March 2013 the ANC Executive Council comprised:
   Premier: Noxolo Kiviet (took office on 6 May 2009).
   Minister of Economic Development and Environmental Affairs: Mcebisi Jonas. *Education and Training:* Mandla Makupula. *Finance and Provincial Planning:* Phumulo Masualle. *Health:* Sicelo Gqobana. *Human Settlements, Safety and Liaison:* Helen August-Sawls. *Local Government and Traditional Affairs:* Mlibo Qhoboshiane. *Public Works, Roads and Transport:* Thandiswa Marawu. *Rural Development and Agrarian Reform:* Zoleka Capa. *Social Development, Women, Children and People with Disabilities:* Pemmy Majodina. *Sport, Recreation, Arts and Culture:* Xoliswa Tom.
   Speaker: Fikile Xasa. *Director-General:* Mbulelo Sogoni.

Government Website: http://www.ecprov.gov.za

ENERGY AND NATURAL RESOURCES

Electricity

In 2010, 9,464 GWh of electricity were delivered by Eskom to Eastern Cape. According to a survey of Feb. 2007, 65·5% of households used electricity for lighting (the lowest percentage of any of the South African provinces).

Oil and Gas

There are plans to build a large oil refinery at Coega near Port Elizabeth.

Minerals

In 2006, 964 people were employed in mining.

Agriculture

There are around 4,000 commercial farms with an average area of 1,500 ha. Of this area only 7% is arable land with 45% not farmed at present owing to land ownership disputes in the former homelands (Transkei and Ciskei). Livestock accounts for 77% of commercial agricultural production; 18% comprises horticulture. Gross farming income for 2007 was R5·4bn. Gross farming income of field crops (R1,000), 369,086; horticulture, 1,396,208; animal and animal products, 3,616,267. The total number of paid farm workers in 2007 was 64,818.

Fisheries

There is a relatively small sea-fishing industry based on squid, sardines, hake, kingklip and crayfish. Aquaculture produces abalone for export to the Far East.

INDUSTRY

Manufacturing is based mainly in Port Elizabeth and East London with motor manufacturing as the prime industry. Wool, mohair and hides are an important area of the province's agro-industry.

Labour

In the fourth quarter of 2009 the labour force numbered 1,732,000, of whom 468,000 were unemployed (27·0%—the joint highest along with North-West). Eastern Cape is the centre of South Africa's motor manufacturing industry with the main production centres based at Port Elizabeth and East London.

COMMUNICATIONS

Roads

Total road network in 2008 was around 78,000 km. Under the government's S'hamba Sonke programme, R1bn. was allocated in April 2011 for road construction and maintenance. In Dec. 2009 Eastern Cape had 659,829 registered vehicles. There were 1,543 fatalities as a result of road traffic accidents in 2009.

Civil Aviation

The province has four airports: Port Elizabeth, East London, Umtata and Bulembu (Bhisho).

Shipping

There are three deep-water ports: Port Elizabeth, East London and Coega.

Telecommunications

According to a Statistics South Africa Community Survey of 2007, 61·2% of households had a mobile phone (21·5% at the 2001 census); landline telephone, 10·7% (15·7%); and computer, 7·5% (4·1%).

SOCIAL INSTITUTIONS

Education

In 2009 there were 2,076,400 children enrolled in schools and a total of 69,620 teaching staff. Following a nationwide restructure of higher education, the Province has four universities: Walter Sisulu University for Technology and Science (formed by the merger of University of Transkei and two technikons), Rhodes University, the University of Fort Hare and Nelson Mandela Metropolitan University (formed by the merger of Port Elizabeth Technikon and the Port Elizabeth campus of Vista University). At the 2001 census more than 22·8% of people aged 20 years and above had no schooling at all, while 6·3% had completed higher education.

Health

In 2009 there were 90 hospitals in the public sector with 14,456 beds. In the 2006–07 financial year a total of R233·9m. was allocated to the National School Nutrition Programme (formerly known as the Primary School Nutrition Programme) although actual spending only amounted to R158·5m.

CULTURE

Tourism

Overseas visitors to the province in 2008 totalled 403,000; domestic visitor trips, 5·4m.

Free State

TERRITORY AND POPULATION

The Free State lies in the centre of South Africa and is situated between the Vaal River in the north and the Orange River in the south. It borders on the Northern Cape, Eastern Cape, North-West, Mpumalanga, KwaZulu-Natal and Gauteng Province and shares a border with Lesotho. The area is 129,480 sq. km, 10·62% of South Africa's total surface area. The province is the third largest in South Africa but has the second smallest population and the second lowest population density. The population at the 2011 census was 2,745,590 (1,416,623 female). Of that number: Black African, 2,405,533 (87·6% of the population); White, 239,026 (8·7%); Coloured, 83,844 (3·1%); Indian/Asian, 10,398 (0·4%). In 2001, 75·8% of the population lived in urban areas. Density (2011), 21 per sq. km. Life expectancy at birth, 2007 estimate, was 46·5 years. At the 2011 census 64·2% (1,717,881) of the population spoke Sesotho as their first language, 12·7% (340,490) Afrikaans, 7·5% (201,145) isiXhosa, 5·2% (140,228) Setswana, 4·4% (118,126) isiZulu and 2·9% (78,782) English.

   Free State comprises 52 administrative districts. The provincial capital is Bloemfontein (meaning 'fountain of flowers'). Bloemfontein's indigenous name is Mangaung, which means 'place of the big cats'.

SOCIAL STATISTICS

Registered live births in 2010 totalled 58,581; deaths in 2005 totalled 50,210. The total number of marriages officially recorded in 2010 was 11,905; divorces granted in 2005 numbered 1,662.

CLIMATE

Temperatures are mild with averages ranging from 19·5°C in the west to 15°C in the east. Maximum temperatures in the west can reach 36°C in summer. Winter temperatures in the high-lying areas of the eastern Free State can drop as low as –15°C. The western and southern areas are semi-desert.

CONSTITUTION AND GOVERNMENT

There is a 30-seat provincial legislature. The Free State Executive Council, headed by the *Premier*, administers the province through ten Departments.

The Free State House of Traditional Leaders advises the Legislature on matters pertaining to traditional authorities and tribal matters.

RECENT ELECTIONS

At the provincial elections held on 22 April 2009 the African National Congress retained its majority and won 22 of the 30 seats; the Congress of the People four; the Democratic Alliance three; and Freedom Front Plus one.

CURRENT GOVERNMENT

In March 2013 the ANC Executive Council comprised:
Premier: Ace Magashule (took office on 6 May 2009).
Minister of Agriculture and Rural Development: Mamiki Qabathe. *Co-operative Governance, Traditional Affairs and Human Settlement:* Olly Mlamleli. *Economic Development, Tourism and Environmental Affairs:* Msebenzi Zwane. *Education:* Tate Makgoe. *Finance:* Elzabe Rockman. *Health:* Dr Benny Malakoane. *Police, Roads and Transport:* Butana Komphela. *Public Works:* Sisi Mabe. *Social Development:* Sisi Ntombela. *Sports, Arts, Culture and Recreation:* Dan Kgothule.
Speaker (acting): Ouma Tsopo. *Director-General (acting):* Kopung Ralikontsane.

Government Website: http://www.freestateonline.fs.gov.za

ENERGY AND NATURAL RESOURCES

Electricity

In 2010, 8,955 GWh of electricity were delivered by Eskom to the Free State. According to a survey of Feb. 2007, 86·6% of households used electricity for lighting.

Minerals

The province contributes about 16·5% of South Africa's total mineral output. Apart from rich gold and diamond deposits, the Free State is the source of numerous other minerals and is the founding home of South Africa's famous oil-from-coal industry centred on Sasolburg. Bentonite clays, gypsum, salt and phosphates are to be found while large concentrates of thorium-ilmenite-zircon also occur. In 2006, 46,067 people were employed in mining.

Agriculture

Good agricultural conditions allow for a wide variety of farming industries. Of the total 12·7m. ha., 90% (11·5m. ha.) is utilized as farmland. Of this, 63·9% is natural grazing; 2·1% is for nature conservation; and 1·1% is used for other purposes. Dryland cultivation is practised on 97% of the arable land, while the remaining 3% is under irrigation.

Free State has the highest number of commercial farming units of all the provinces: 7,515 in 2007 with a gross farming income of R11·9bn. Gross farming income of field crops in 2007 (R1,000), 4,226,749; horticulture, 984,203; animal and animal products, 6,718,152. Paid farm workers in the same year numbered 99,094.

The eastern region is the major producer of small grains; the northern region, maize and beef; and the southern region, mutton and wool. The province produces about 40% of total maize and 50% of total wheat output in South Africa.

INDUSTRY

Labour

In the fourth quarter of 2009 the labour force numbered 1,059,000, of whom 268,000 were unemployed (25·3%).

COMMUNICATIONS

Roads

Total road network in 2008 was around 111,000 km. Under the government's S'hamba Sonke programme, R447m. was allocated in April 2011 for road construction and maintenance. In Dec. 2009 Free State had 539,704 registered vehicles. There were 967 fatalities as a result of road traffic accidents in 2009.

Rail

Transnet Freight Rail is one of the biggest companies in the Free State with 4,217 employees. It transports most of the province's maize, wheat, gold ore, petroleum and fertilizer. The company's infrastructure consists of approximately 4,000 km of tracks, of which 1,300 km are electrified.

Telecommunications

According to a Statistics South Africa Community Survey of 2007, 68·3% of households had a mobile phone (24·8% at the 2001 census); landline telephone, 13·1% (20·4%); and computer, 11·1% (4·9%).

SOCIAL INSTITUTIONS

Education

In 2009 there were 651,785 pupils enrolled in schools and 23,741 teachers. Following a nationwide restructure of higher education, the Province has two universities: the University of the Free State (established in 1904) and the Central University of Technology, Free State. According to the 2001 census, 16·0% of those aged 20 and over had no schooling; 30·7% had some secondary education.

Health

In 2009 there were 32 public hospitals with 4,958 beds. There were 14 private hospitals in 2008 with 2,119 beds.

FURTHER READING

Free State: The Winning Province. 1997

Gauteng

TERRITORY AND POPULATION

Gauteng is the smallest province in South Africa, covering an area of 17,010 sq. km (approximately 1·4% of the total land surface of South Africa). The population at the 2011 census was 12,272,263 (6,082,388 female). Of that number: Black African, 9,493,684 (77%); White, 1,913,884 (16%); Coloured, 423,594 (3%); Indian/ Asian, 356,574 (3%). In 2001, 97·2% of the population lived in urban areas, making Gauteng the most urbanized South African province. Density (2011), 721 per sq. km. Estimated life expectancy at birth, 2007, was 52·2 years. At the 2011 census 19·8% spoke isiZulu as their first language, 13·3% English, 12·4% Afrikaans, 11·6% Sesotho, 10·6% Sepedi, 9·1% Setswana, 6·6% isiXhosa, 6·6% Xitsonga, 3·2% isiNdebele, 2·3% Tshivenda and 1·1% SiSwati.

The province of Gauteng, at first called Pretoria-Witwatersrand-Vereeniging (PWV), comprises 23 administrative districts. The provincial capital is Johannesburg. In the Sesotho language, Gauteng means 'Place of Gold'.

SOCIAL STATISTICS

Registered live births in 2010 totalled 395,097; deaths in 2005 totalled 107,528. The total number of marriages officially recorded in 2010 was 41,396; divorces granted in 2005 numbered 12,302.

CONSTITUTION AND GOVERNMENT

There is a 73-seat provincial legislature.

RECENT ELECTIONS

At the provincial elections held on 22 April 2009, 47 seats were won by the African National Congress, 16 by the Democratic Alliance, six by the Congress of the People and one each by the African Christian Democratic Party, the Independent Democrats, the Inkatha Freedom Party and Freedom Front Plus.

CURRENT GOVERNMENT

In March 2013 the ANC Executive Council comprised:
  Premier: Nomvula Mokonyane; b. 1963 (took office on 6 May 2009).
  Minister of Agriculture, Rural and Social Development: Nandi Mayathula-Khoza. *Community Safety:* Nonhlanhla Mazibuko. *Economic Development and Planning:* Nkosipendule Kolisile. *Education:* Barbara Creecy. *Finance:* Mandla Nkomfe. *Health:* Hope Papo. *Infrastructure Development:* Qedani Mahlangu. *Local Government and Housing:* Ntombi Mekgwe. *Roads and Public Transport:* Ismail Vadi. *Sport, Arts, Culture and Recreation:* Lebogang Maile.
  Speaker: Lindiwe Maseko. *Director-General:* Margaret-Ann Diedricks.

Gauteng Parliament: http://www.gpl.gov.za

ENERGY AND NATURAL RESOURCES

Electricity
In 2010, 61,247 GWh of electricity were delivered by Eskom to Gauteng. According to a survey of Feb. 2007, 83·5% of households used electricity for lighting.

Minerals
In 2006, 78,086 people were employed in mining.

Agriculture
There were 2,378 commercial farming units in 2007 with a gross farming income of R7·4bn. Gross farming income of field crops in 2007 (R1,000), 566,632; horticulture, 1,116,908; animal and animal products, 5,633,061. Paid farm workers in the same year numbered 34,936.

INDUSTRY

Labour
In the fourth quarter of 2009 the labour force numbered 5,046,000, of whom 1,297,000 were unemployed (25·7%). About 38,000 workers are employed by the motor manufacturing industry, which contributes an estimated 4·3% of the province's GDP. The aluminium industry is worth about US$20m.

COMMUNICATIONS

Roads
Total road network in 2008 was around 55,000 km. Under the government's S'hamba Sonke programme, R566m. was allocated in April 2011 for road construction and maintenance. In Dec. 2009 Gauteng had 3,680,158 registered vehicles. There were 2,485 fatalities as a result of road traffic accidents in 2009.

Civil Aviation
O. R. Tambo International Airport is the main airport in the province.

Telecommunications
According to a Statistics South Africa Community Survey of 2007, 80·3% of households had a mobile phone (44·7% at the 2001 census); landline telephone, 24·4% (31·8%); and computer, 24·2% (14·7%).

SOCIAL INSTITUTIONS

Education
In 2009 there were 1,903,838 children enrolled in schools with a total of 66,351 teaching staff. The Province has six universities: University of South Africa, University of Johannesburg (formed in 2005 by the merger of Rand Afrikaans University and Technikon Witwatersrand), University of Pretoria, University of Witwatersrand, Tshwane University of Technology and Vaal University of Technology. According to the 2001 census, 8·4% of those aged 20 and over had no schooling; 34·3% had some secondary education.

Health
In 2009 there were 31 public hospitals and 16,816 hospital beds. Private hospitals numbered 81 in 2008 (out of a total of 211 in South Africa as a whole) with 13,454 beds.

CULTURE

Tourism
Overseas visitors to the province in 2008 totalled 4·5m. (the highest of any of the South African provinces); domestic visitor trips, 4·8m.

KwaZulu-Natal

TERRITORY AND POPULATION

The area is 92,100 sq. km and the population at the 2011 census was 10,267,300 (5,388,625 female). Of that number: Black African, 8,912,921 (87% of the population); Indian/Asian, 756,991 (7%); White, 428,842 (4%); Coloured, 141,376 (1%). In 2001, 46·0% of the population lived in urban areas. Density (2011), 111 per sq. km. Life expectancy at birth, 2007 estimate, was 42·9 years. At the 2011 census 77·8% spoke isiZulu as their first language, 13·2% English, 3·4% isiXhosa and 1·6% Afrikaans.

  KwaZulu-Natal comprises 66 administrative districts. The provincial capital is Pietermaritzburg, chosen by referendum in 1995.

SOCIAL STATISTICS

Registered live births in 2010 totalled 258,286; deaths in 2005 totalled 138,206. The total number of marriages officially recorded in 2010 was 25,862; divorces granted in 2005 numbered 3,960.

CONSTITUTION AND GOVERNMENT

There is an 80-seat provincial legislature.

RECENT ELECTIONS

At the provincial elections held on 22 April 2009, 51 seats were won by the African National Congress, 18 by the Inkatha Freedom Party, seven by Democratic Alliance, two by the Minority Front, and one each by the African Christian Democratic Party and the Congress of the People.

CURRENT GOVERNMENT

In March 2013 the government comprised:
  Premier: Dr Zweli Mkhize; b. 1956 (ANC; took office on 6 May 2009).
  Minister of Agriculture, Environmental Affairs and Rural Development: Bonginkosi Meshack Radebe. *Arts, Culture, Sport*

and Recreation: Ntombikayise Sibhidla-Saphetha. *Co-operative Governance and Traditional Affairs:* Nomusa Dube. *Economic Development and Tourism:* Mike Mabuyakhulu. *Education:* Edward Senzo Mchunu. *Finance:* Catharina Magdalena Cronje. *Health:* Dr Sibongiseni Maxwell Dlomo. *Human Settlements and Public Works:* Ravi Pillay. *Social Development:* Weziwe Gcotyewa Thusi. *Transport, Community Safety and Liaison:* Thembinkosi Willies Mchunu.

Speaker: Neliswa Peggy Nkonyeni. *Director-General:* Nhlanhla Ngidi.

Government Website: http://www.kwazulunatal.gov.za

ENERGY AND NATURAL RESOURCES

Electricity
In 2010, 42,015 GWh of electricity were delivered by Eskom to KwaZulu-Natal. According to a survey of Feb. 2007, 71·5% of households used electricity for lighting.

Minerals
Coal is mined in the north of the province and titanium and zircon are produced at Richards Bay. In 2006, 9,198 people were employed in mining.

Agriculture
There were 3,560 commercial farming units in 2007 with a gross farming income of R10·1bn. Gross farming income of field crops in 2007 (R1,000), 2,867,839; horticulture, 1,086,975; animal and animal products, 5,794,379. Paid farm workers in the same year numbered 101,068. Sugarcane and maize are the principal crops.

INDUSTRY

Labour
In the fourth quarter of 2009 the labour force numbered 2,983,000, of whom 574,000 were unemployed (19·2%—the lowest unemployment rate of all the provinces).

COMMUNICATIONS

Roads
Total road network in 2008 was around 98,000 km. Under the government's S'hamba Sonke programme, R1·2bn. was allocated in April 2011 for road construction and maintenance. In Dec. 2009 KwaZulu-Natal had 1,308,090 registered vehicles. There were 2,854 fatalities as a result of road traffic accidents in 2009.

Civil Aviation
Durban International Airport is the main airport in the province.

Shipping
Durban harbour is the busiest in South Africa and one of the ten largest harbours in the world. Coal is exported from Richards Bay.

Telecommunications
According to a Statistics South Africa Community Survey of 2007, 71·9% of households had a mobile phone (28·2% at the 2001 census); landline telephone, 18·6% (23·8%); and computer, 11·7% (7·0%).

SOCIAL INSTITUTIONS

Education
Since 1995 education has been provided by a unified KwaZulu-Natal Education Department (KZNED). As at June 2011 there were 2·8m. children enrolled in schools with a total of 88,952 teaching staff. Following a nationwide restructure of higher education, the Province has four universities: University of KwaZulu-Natal, University of Zululand, Durban University of Technology and Mangosuthu University of Technology. According to the 2001 census, 21·9% of the population aged 20 and above had no schooling; 28·8% had some secondary education.

Health
In 2009 there were 75 public hospitals with 23,142 beds. Private hospitals numbered 32 in 2008 with 3,865 beds.

CULTURE

Tourism
Overseas visitors to the province in 2008 totalled 1·2m.; domestic visitor trips, 10·4m.

Limpopo

TERRITORY AND POPULATION

The area is 123,910 sq. km and the population at the 2011 census was 5,404,868 (2,880,732 female). Of that number: Black African, 5,224,754 (96·7% of the population); White, 139,359 (2·6%); Indian/Asian, 17,881 (0·3%); Coloured, 14,415 (0·3%). In 2001, 13·3% lived in urban areas. Density (2011), 44 per sq. km. Life expectancy at birth, 2007 estimate, was 55·8 years. At the 2011 census 52·9% spoke Sepedi as their first language, 17·0% Xitsonga, 16·7% Tshivenda, 2·6% Afrikaans and 2·0% isiNdebele.

Limpopo (Northern Province until March 2003) comprises 32 administrative districts. The provincial capital is Polokwane (Pietersburg).

SOCIAL STATISTICS

Registered live births in 2010 totalled 128,181; deaths in 2005 totalled 48,222. The total number of marriages officially recorded in 2010 was 9,699; divorces granted in 2005 numbered 996.

CONSTITUTION AND GOVERNMENT

There is a 49-seat provincial legislature.

RECENT ELECTIONS

At the provincial elections held on 22 April 2009, 43 seats were won by the African National Congress, four by the Congress of the People and two by the Democratic Alliance.

CURRENT GOVERNMENT

In March 2013 the ANC Executive Council comprised:

Premier: Cassel Mathale (sworn in 6 May 2009).

Minister of Agriculture: Jacob Marule. *Co-operative Governance, Human Settlement and Traditional Affairs:* Clifford Motsepe. *Economic Development, Environment and Tourism:* Pinky Kekana. *Education:* Dickson Masemola. *Health and Social Development:* Norman Mabasa. *Provincial Treasury:* David Masondo. *Public Works:* Thabitha Mohlala. *Roads and Public Transport:* Pitsi Moloto. *Safety, Security and Liaison:* Florence Dzhombere. *Sports, Arts and Culture:* Dipuo Letsatsi-Duba.

Speaker: Alfred Kgolane Phala. *Director-General:* Rachel Molepo-Modipa.

Government Website: http://www.limpopo.gov.za

ENERGY AND NATURAL RESOURCES

Electricity
In 2010, 12,212 GWh of electricity were delivered by Eskom to Limpopo. According to a survey of Feb. 2007, 81·0% of households used electricity for lighting.

Minerals
Mining is an important industry in the province with, in 2006, 63,347 people employed.

Agriculture
There were 2,657 commercial farming units in 2007 with a gross farming income of R5·5bn. Gross farming income of field crops in

2007 (R1,000), 497,679; horticulture, 2,904,969; animal and animal products, 2,027,780. Paid farm workers in the same year numbered 67,561.

INDUSTRY

Labour

In the fourth quarter of 2009 the labour force numbered 1,245,000, of whom 335,000 were unemployed (26·9%).

COMMUNICATIONS

Roads

Total road network in 2008 was around 66,000 km. Under the government's S'hamba Sonke programme, R934m. was allocated in April 2011 for road construction and maintenance. In Dec. 2009 Limpopo had 467,690 registered vehicles. There were 1,492 fatalities as a result of road traffic accidents in 2009.

Telecommunications

According to a Statistics South Africa Community Survey of 2007, 70·5% of households had a mobile phone (24·8% at the 2001 census); landline telephone, 4·6% (8·0%); and computer, 6·7% (2·4%).

SOCIAL INSTITUTIONS

Education

In 2009 there were 1,707,280 children enrolled in schools with a total of 58,563 teaching staff. Following a nationwide restructure of higher education, the Province has two universities: the University of Limpopo and the University of Venda. According to the 2001 census, 33·4% of those aged 20 and over had no schooling; 26·1% had some secondary education.

Health

In 2009 there were 41 public hospitals with 7,866 hospital beds.

Mpumalanga

TERRITORY AND POPULATION

The area is 78,490 sq. km and the population at the 2011 census was 4,039,939 (2,065,883 female). Of that number: Black African, 3,662,219 (90·7% of the population); White, 303,595 (7·5%); Coloured, 36,611 (0·9%); Indian/Asian, 27,917 (0·7%). In 2001, 41·3% lived in urban areas. Density (2011), 51 per sq. km. Life expectancy at birth, 2007 estimate, was 46·5 years. At the 2011 census 27·7% spoke SiSwati as their first language, 24·1% isiZulu, 10·4% Xitsonga, 10·1% isiNdebele, 9·3% Sepedi, 7·2% Afrikaans, 3·5% Sesotho, 3·1% English, 1·8% Setswana and 1·2% isiXhosa.

Mpumalanga comprises 28 administrative districts. The provincial capital is Nelspruit.

SOCIAL STATISTICS

Registered live births in 2010 totalled 90,637; deaths in 2005 totalled 45,234. The total number of marriages officially recorded in 2010 was 8,809; divorces granted in 2005 numbered 1,170.

CONSTITUTION AND GOVERNMENT

There is a 30-seat provincial legislature.

RECENT ELECTIONS

At the provincial elections held on 22 April 2009, 27 seats were won by the African National Congress, two by the Democratic Alliance and one by the Congress of the People.

CURRENT GOVERNMENT

In March 2013 the ANC government comprised:

Premier: David Mabuza (took office on 6 May 2009).

Minister of Agriculture, Rural Development and Land Administration: Violet Siwela. *Community Safety, Security and Liaison:* Vusi Shongwe. *Co-operative Governance and Traditional Affairs:* Simon Sikhosana. *Culture, Sports and Recreation:* Sbongile Manana. *Economic Development, Environment and Tourism:* Yvonne Pinky Phosa. *Education:* Regina Mhaule. *Finance:* Madala Masuku. *Health and Social Development:* Candith Mashego-Dlamini. *Human Settlement:* Andries Gamede. *Public Works, Roads and Transport:* Dikeledi Mahlangu.

Speaker: Sipho William Lubisi. *Director-General (acting):* Nonhlanhla Mkhize.

Government Website: http://www.mpumalanga.gov.za

ENERGY AND NATURAL RESOURCES

Electricity

In 2010, 34,453 GWh of electricity were delivered by Eskom to Mpumalanga. According to a survey of Feb. 2007, 81·7% of households used electricity for lighting.

Minerals

In 2006, 73,608 people were employed in mining. The province is rich in coal reserves and produces about 80% of the country's supplies.

Agriculture

There were 3,376 commercial farming units in 2007 with a gross farming income of R9·2bn. Gross farming income of field crops in 2007 (R1,000), 2,608,493; horticulture, 1,748,584; animal and animal products, 4,689,232. Paid farm workers in the same year numbered 79,346.

INDUSTRY

Labour

In the fourth quarter of 2009 the labour force numbered 1,200,000, of whom 319,000 were unemployed (26·6%).

COMMUNICATIONS

Roads

Total road network in 2008 was around 56,000 km. Under the government's S'hamba Sonke programme, R1bn. was allocated in April 2011 for road construction and maintenance. In Dec. 2009 Mpumalanga had 608,676 registered vehicles. There were 1,674 fatalities as a result of road traffic accidents in 2009.

Telecommunications

According to a Statistics South Africa Community Survey of 2007, 77·4% of households had a mobile phone (31·1% at the 2001 census); landline telephone, 9·0% (14·4%); and computer, 10·7% (4·3%).

SOCIAL INSTITUTIONS

Education

In 2009 there were 1,035,637 children enrolled in schools with a total of 35,221 teaching staff. According to the 2001 census 27·5% of those aged 20 years and over had no schooling; 26·6% had some secondary education.

Health

In 2005 there were 24 public hospitals, plus 198 clinics and 34 community health centres, with 4,173 beds; in 2006 there were ten private hospitals with 923 beds. In Oct. 1998 the first traditional hospital was opened in Mpumalanga—the Samuel Traditional Hospital.

CULTURE

Tourism

Overseas visitors to the province in 2008 totalled 1·3m.; domestic visitor trips, 2·2m.

Northern Cape

TERRITORY AND POPULATION

The area is 361,830 sq. km and the population at the 2011 census was 1,145,861 (580,889 female). Of that number: Black African, 576,986 (50% of the population); Coloured, 461,899 (40%); White, 81,246 (7%); Indian/Asian, 7,827 (0·7%). In 2001, 82·7% lived in urban areas. Density (2011), 3 per sq. km. Life expectancy at birth, 2007 estimate, was 56·8 years. At the 2011 census 53·8% spoke Afrikaans as their first language, 33·1% Setswana, 5·3% isiXhosa and 3·4% English.

Northern Cape comprises six administrative districts: Diamond Fields with Kimberley as the provincial and economic capital; Kalahari, which is the second richest and densely populated area in the province and includes the magisterial districts of Kuruman and Postmasburg; Hantam (North-West) with the towns of Calvinia, Sutherland, Williston, Fraserburg and Carnarvon; Benede-Orange with Upington as the agricultural, economic and cultural capital of the region; Bo-Karoo with De Aar as the capital of the area; and Namaqualand which is strong in mining.

SOCIAL STATISTICS

Registered live births in 2010 totalled 25,944; deaths in 2005 totalled 12,058. The total number of marriages officially recorded in 2010 was 4,552; divorces granted in 2005 numbered 520.

CONSTITUTION AND GOVERNMENT

There is a 30-seat provincial legislature.

RECENT ELECTIONS

At the provincial elections held on 22 April 2009, 19 seats were won by the African National Congress, five by the Congress of the People, four by the Democratic Alliance and two by the Independent Democrats.

CURRENT GOVERNMENT

In March 2013 the ANC Executive Council comprised:
Premier and Minister of Education: Grizelda Cjiekella (since 20 Feb. 2012).
Minister of Agriculture and Land Reform: Norman Shushu. *Co-operative Governance, Human Settlement and Traditional Affairs:* Mosimanegape Mmoiemang. *Environmental Affairs and Nature Conservation:* Sylvia Lucas. *Finance, Economic Affairs and Tourism:* John Block. *Health:* Mxolisi Sokatsha. *Roads and Public Works:* Dawid Rooi. *Social Development:* Alvin Botes. *Sport, Arts and Culture:* Pauline Williams. *Transport, Safety and Liaison:* Patrick Mabilo.
Speaker: Boeboe van Wyk. *Director-General:* Justice Bekebeke.

Government Website: http://www.northern-cape.gov.za

ENERGY AND NATURAL RESOURCES

Electricity
In 2010, 4,719 GWh of electricity were delivered by Eskom to Northern Cape. According to a survey of Feb. 2007, 87·3% of households used electricity for lighting.

Minerals
The province is well endowed with a variety of mineral deposits. Diamonds are found in shallow water at Port Nolloth, Hondeklipbaai and Lamberts Bay, and also mined inland along the entire coastal strip from the Orange river mouth in the north to Lamberts Bay in the south. Zircon is located along the west coast of Namaqualand. Limestone, asbestos and gypsum salt are also mined. Most of South Africa's reserves of zinc are found in the province in the Black Mountain, Broken Hill and Gamsberg. In 2006, 26,380 people were employed in mining.

Agriculture
Intensive irrigation takes place along the Orange River which supports vineyards and agribusiness. Stock farming predominates in the Bo-Karoo and Hantam areas. There were 5,226 commercial farming units in 2007 with a gross farming income of R4·8bn. Gross farming income of field crops in 2007 (R1,000), 1,148,288; horticulture, 1,243,491; animal and animal products, 2,371,143. Paid farm workers in the same year numbered 74,745.

INDUSTRY

Labour
In the fourth quarter of 2009 the labour force numbered 386,000, of whom 96,000 were unemployed (24·9%).

COMMUNICATIONS

Roads
Total road network in 2008 was around 111,000 km. Under the government's S'hamba Sonke programme, R308m. was allocated in April 2011 for road construction and maintenance. In Dec. 2009 Northern Cape had 214,226 registered vehicles. There were 337 fatalities as a result of road traffic accidents in 2009.

Rail
The main rail link is between Cape Town and Johannesburg, via Kimberley. Other main lines link the Northern Cape with Port Elizabeth via De Aar while another links Upington with Namibia.

Civil Aviation
Five airports are used for scheduled flights—Kimberley, Upington, Aggeneys, Springbok and Alexander Bay.

Telecommunications
According to a Statistics South Africa Community Survey of 2007, 61·8% of households had a mobile phone (24·5% at the 2001 census); landline telephone, 22·4% (27·4%); and computer, 13·2% (6·1%).

SOCIAL INSTITUTIONS

Education
In 2009 there were 267,709 children enrolled in schools with a total of 9,115 teaching staff. There is no university in the province but there are some technical colleges and a nursing college in Kimberley. According to the 2001 census 18·2% of those aged 20 years and over had no schooling; 29·9% had some secondary education.

Health
In 2009 there were 21 public hospitals with 1,958 beds. There were four private hospitals in 2008 with 335 beds. In 2010 there were 321 medical practitioners working in the public sector.

CULTURE

Tourism
Parks are a major tourism asset with the total area under protection being 1,080,200 ha. Provincial nature reserves occupy 50,240 ha. Hunting is a growing activity in the province.

North-West

TERRITORY AND POPULATION

The area is 116,320 sq. km and the population at the 2011 census was 3,509,953 (1,730,049 female). Of that number: Black African, 3,152,063 (89·8% of the total population); White, 255,385 (7·3%); Coloured, 71,409 (2·0%); Indian/Asian, 20,652 (0·6%). In 2001 the population was 41·8% urban. Density (2011), 30 per sq. km. Life expectancy at birth, 2007 estimate, was 49·8 years. At the 2011 census 63·4% spoke Setswana as their first language, 9·0% Afrikaans, 5·8% Sesotho, 5·5% isiXhosa, 3·7% Xitsonga, 3·5% English, 2·5% isiZulu, 2·4% Sepedi, 1·3% isiNdebele and 0·3% SiSwati.

North-West Province comprises 32 administrative districts. The provincial capital is Mmabatho.

SOCIAL STATISTICS

Registered live births in 2010 totalled 86,193; deaths in 2005 totalled 52,055. The total number of marriages officially recorded in 2010 was 13,193; divorces granted in 2005 numbered 1,874.

CONSTITUTION AND GOVERNMENT

There is a 33-seat provincial legislature.

RECENT ELECTIONS

At the provincial elections held on 22 April 2009 the African National Congress won 25 seats, the Congress of the People and the Democratic Alliance three each, and the United Christian Democratic Party two.

CURRENT GOVERNMENT

In March 2013 the ANC Executive Council comprised:
Premier: Thandi Modise; b. 1959 (took office on 19 Nov. 2010).
Minister of Agriculture and Rural Development: Desbo Mohono. *Economic Development, Environment and Tourism:* Motlalepula Rosho. *Education:* Louisa Moruakgomo Mabe. *Finance:* Paul Sebegoe. *Health:* Dr Magome Masike. *Human Settlement, Safety and Liaison:* Nono Maloi. *Local Government and Traditional Affairs:* China Dodovu. *Public Works, Roads and Transport:* Raymond Elisha. *Social Development, Women, Children and Persons with Disability:* Mosetsanagape Mokomela-Mothibi. *Sports, Arts and Culture:* Tebogo Modise.
Speaker: Supra Mahumapelo. *Director-General (acting):* Abram Tlaletsi.

Government Website: http://www.nwpg.gov.za

ENERGY AND NATURAL RESOURCES

Electricity

In 2010, 26,064 GWh of electricity were delivered by Eskom to North-West Province. According to a survey of Feb. 2007, 82·3% of households used electricity for lighting.

Minerals

In 2006, 157,565 people were employed in mining. Gold is mined at Klerksdorp, and diamonds at Lichtenburg, Koster, Christiana and Bloemhof.

Agriculture

There were 4,692 commercial farming units in 2007 with a gross farming income of R8·8bn. Gross farming income of field crops in 2007 (R1,000), 2,250,740; horticulture, 768,890; animal and animal products, 5,669,343. Paid farm workers in the same year numbered 85,749.

INDUSTRY

Labour

In the fourth quarter of 2009 the labour force numbered 1,076,000, of whom 290,000 were unemployed (27·0%—the joint highest along with Eastern Cape).

COMMUNICATIONS

Roads

Total road network in 2008 was around 73,000 km. Under the government's S'hamba Sonke programme, R501m. was allocated in April 2011 for road construction and maintenance. In Dec. 2009 North-West had 540,786 registered vehicles. There were 1,130 fatalities as a result of road traffic accidents in 2009.

Telecommunications

According to a Statistics South Africa Community Survey of 2007, 70·9% of households had a mobile phone (28·0% at the 2001 census); landline telephone, 18·5% (24·4%); and computer, 9·1% (4·3%).

SOCIAL INSTITUTIONS

Education

In 2009 there were 777,285 children enrolled in schools with a total of 26,697 teaching staff. The Province has one university, the North-West University, formed through a merger with Potchefstroom University for Christian Education in 2004. According to the 2001 census, 19·9% of those aged 20 and over had no schooling; 29·0% had some secondary education.

Health

In 2009 there were 20 public hospitals with 4,507 beds.

CULTURE

Tourism

Overseas visitors to the province in 2008 totalled 643,000; domestic visitor trips, 1·2m.

Western Cape

TERRITORY AND POPULATION

The area is 129,370 sq. km. Population, 2011 census, 5,822,734 (2,964,228 female). Of that number: Coloured, 2,840,404 (48·8%); Black African, 1,912,547 (32·8%); White, 915,053 (15·7%); Indian/Asian, 60,761 (1·0%). Density (2011), 45 per sq. km. Life expectancy at birth, 2007 estimate, was 61·2 years. At the 2011 census 49·7% spoke Afrikaans as their first language, 24·7% isiXhosa and 20·2% English. In 2001 the population was 90·4% urban.

There are 41 administrative districts. The capital is Cape Town.

SOCIAL STATISTICS

Registered live births in 2010 totalled 110,801; deaths in 2005 totalled 44,396. The total number of marriages officially recorded in 2010 was 26,855; divorces granted in 2005 numbered 5,381.

CONSTITUTION AND GOVERNMENT

There is a 42-seat provincial parliament.

RECENT ELECTIONS

At the provincial elections held on 22 April 2009, 22 seats were won by the Democratic Alliance (DA), 14 by the African National Congress, three by the Congress of the People, two by Independent Democrats and one by the African Christian Democratic Party.

CURRENT GOVERNMENT

In March 2013 the provincial cabinet comprised:
Premier: Helen Zille; b. 1951 (DA; took office on 6 May 2009).
Minister of Agriculture: Gerrit van Rensburg. *Community Safety:* Dan Plato. *Cultural Affairs and Sport:* Dr Ivan Meyer. *Education:* Donald Grant. *Finance, Economic Development and Tourism:* Alan Winde. *Health:* Theuns Botha. *Housing:* Bonginkosi Madikizela. *Local Government, Environmental Affairs and Development Planning:* Anton Bredell. *Social Development:* Albert Fritz. *Transport and Public Works:* Robin Carlisle.
Speaker: Thembekile Richard Majola. *Director-General:* Brent Gerber.

Government Website: http://www.capegateway.gov.za

ENERGY AND NATURAL RESOURCES

Electricity
In 2010, 23,122 GWh of electricity were delivered by Eskom to Western Cape. According to a survey of Feb. 2007, 94·0% of households used electricity for lighting (the highest percentage of any of the South African provinces).

Minerals
In 2006, 3,385 people were employed in mining.

Agriculture
There were 6,682 commercial farming units in 2007 with a gross farming income of R16·6bn. Gross farming income of field crops in 2007 (R1,000), 1,466,533; horticulture, 7,764,317; animal and animal products, 7,219,245. Paid farm workers in the same year numbered 189,489. The province is one of the world's finest grape-growing regions as well as producing other fruits such as apples, peaches and oranges. The Klein Karoo region is the centre of the ostrich-farming industry in South Africa with leatherware, feathers and meat exported worldwide.

INDUSTRY

Labour
In the fourth quarter of 2009 the labour force numbered 2,413,000, of whom 518,000 were unemployed (21·5%).

COMMUNICATIONS

Roads
Total road network in 2008 was around 92,000 km. Under the government's S'hamba Sonke programme, R411m. was allocated in April 2011 for road construction and maintenance. In Dec. 2009 Western Cape had 1,568,622 registered vehicles. There were 1,285 fatalities as a result of road traffic accidents in 2009.

Civil Aviation
Cape Town International Airport is the main airport in the province.

Telecommunications
According to a Statistics South Africa Community Survey of 2007, 74·5% of households had a mobile phone (41·4% at the 2001 census); landline telephone, 42·0% (50·5%); and computer, 30·1% (18·2%).

SOCIAL INSTITUTIONS

Education
In 2009 there were 980,694 children enrolled in schools with a total of 34,382 teaching staff. The Province has four universities: the University of the Western Cape, the University of Cape Town, Stellenbosch University and the Cape Peninsula University of Technology. The Western Cape has the highest adult-education level in South Africa with only 5·7% of the population aged 20 years and over with no schooling. According to the 2001 census, 36·5% had some secondary education—the highest rate in any of South Africa's provinces.

Health
In 2009 there were 58 public hospitals with 9,858 beds. Private hospitals numbered 34 in 2008 with 4,042 beds. Of the nine provinces Western Cape has the lowest prevalence of HIV.

CULTURE

Tourism
Overseas visitors to the province in 2008 totalled 1·6m.; domestic visitor trips, 4·1m.

SOUTH SUDAN

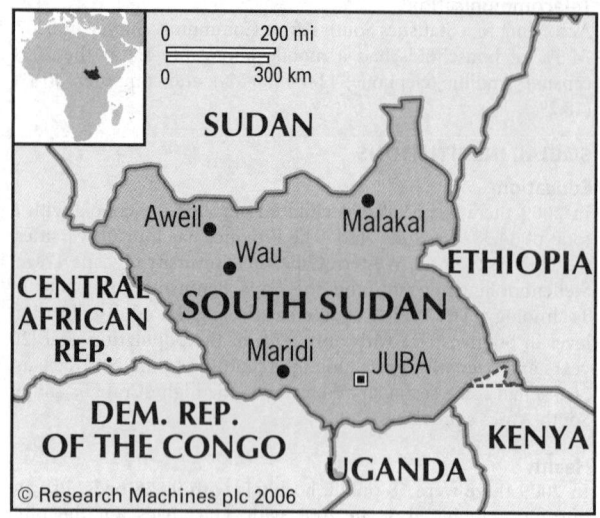

(Republic of South Sudan)

Capital: Juba
Population estimate, 2008: 8·3m.

KEY HISTORICAL EVENTS

South Sudan was administered separately from the north under Egyptian and British rule until 1946 when the British unified the regions. When independence as a single state was granted in 1956 and southern autonomy was compromised, tensions between the predominantly Muslim, Arabic north and the Christian and animist south escalated into civil war. An agreement between the government and the Southern Sudan Liberation Front signed in Addis Ababa in 1972 brought a brief respite from the fighting but in 1983 hostilities recommenced. They lasted 19 years and resulted in 1·9m. civilian deaths in the south of Sudan before a ceasefire was declared in Jan. 2002. A comprehensive peace agreement was signed between the Khartoum government and the southern rebels led by the Sudan People's Liberation Movement in Jan. 2005. They entered a unity government, despite ongoing arguments over disputed areas, and a southern autonomous government was formed. In a referendum on independence for the south on 9–15 Jan. 2011, 98·8% of votes cast were in favour. South Sudan became an independent country on 9 July 2011. On 14 July 2011 it became the 193rd member of the United Nations.

TERRITORY AND POPULATION

South Sudan is bounded in the north by Sudan, east by Ethiopia, southeast by Kenya, south by Uganda, southwest by the Democratic Republic of the Congo and west by the Central African Republic. Its area is 644,329 sq. km. In 2008 the census population was 8·26m. (disputed). More than half (51%) of the population is below the age of 18. 83% of the population is rural.

The country is composed of ten states (with capitals): Central Equatoria (Juba), East Equatoria (Kapoeta), Jungoli (Bor), Lakes (Rumbek), North Bahr Al Ghazal (Awil), Unity (Bantio), Upper Nile (Malakal), Warap (Warap), West Bahr Al Ghazal (Wau), West Equatoria (Yambio).

The capital, Juba, had a population of 230,000 in 2008. Other major cities are Wau (118,000), Malakal (115,000), Yei (111,000) and Yambio (106,000).

The official language is English.

CLIMATE

South Sudan's climate is tropical with wet and dry seasons. The winter is relatively cool and dry while the rainy season usually takes place from April to Dec. with most rain falling in the summer months. Juba, Jan. 81°F (27·3°C), July 76°F (24·5°C). Annual rainfall 38" (965 mm). Wau, Jan. 79°F (25·9°C), July 78°F (25·3°C). Annual rainfall 44" (1,118 mm).

CONSTITUTION AND GOVERNMENT

An interim constitution was ratified shortly before independence and came into force on 7 July 2011. Under the constitution the *President* is the Head of State, Head of Government and Commander-in-Chief of the armed forces and serves a four-year term. The National Legislature consists of two Houses: the *National Legislative Assembly*, comprising members of the former Southern Sudan Legislative Assembly and all South Sudanese who were members of the National Assembly of Sudan; and the *Council of States*, which consists of South Sudanese who had seats in the Council of States of Sudan plus 20 members appointed by the President. Members of both houses serve four-year terms.

National Anthem

'South Sudan Oyee'; words and tune by students and teachers of Juba University.

CURRENT GOVERNMENT

In March 2013 the government comprised:
  President: Salva Kiir Mayardit; b. 1951 (Sudan People's Liberation Movement; sworn in 9 July 2011).
  Vice President: Riek Machar.
  Minister of Cabinet Affairs: Deng Alor Kuol. *Defence and Veteran Affairs:* Gen. John Kong Nyuon. *Foreign Affairs and International Co-operation:* Nhial Deng Nhial. *National Security:* Gen. Oyay Deng Ajak. *Justice:* John Luk Jok. *Interior:* Gen. Alison Manani Magaya. *Parliamentary Affairs:* Michael Makuei Lueth. *Finance and Economic Planning:* Kosti Manibe Ngai. *Labour, Public Service and Human Resource Development:* Awut Deng Acuil. *Health:* Dr Michael Milly Hussein. *Information and Broadcasting:* Barnaba Marial Benjamin. *Agriculture and Forestry:* Betty Achan Ogwaro. *Roads and Bridges:* Gier Chuang Aloung. *Transport:* Agnes Poni Lokudu. *General Education:* Ustaz Joseph Ukel Abango. *Higher Education, Science and Technology:* Peter Adwok Nyaba. *Commerce, Industry and Investment:* Garang Diing Akuang. *Environment:* Alfred Lado Gore. *Housing and Physical Planning:* Jema Nunu Kumba. *Telecommunications and Postal Services:* Madut Biar Yel. *Petroleum and Mining:* Stephen Dhieu Dau. *Electricity and Dams:* David Deng Athorbei. *Gender, Child and Social Welfare:* Agnes Kwaje Lasuba. *Humanitarian Affairs and Disaster Management:* Joseph Lual Achuil. *Water Resources and Irrigation:* Paul Mayom Akec. *Wildlife Conservation and Tourism:* Gabriel Changson. *Animal Resources and Fisheries:* Martin Elia Lomuro. *Culture, Youth and Sports:* Cirino Hiteng Ofuho. *Minister in the President's Office:* Emmanuel Lowilla.

Government Website: http://www.goss-online.org

CURRENT LEADERS

Salva Kiir Mayardit

Position
President

Introduction
Salva Kiir Mayardit became president of the newly independent Republic of South Sudan in July 2011. A veteran of Sudan's long

civil war, he helped establish the dominant Marxist rebel movement in southern Sudan in the 1980s, leading its military wing.

Early Life

Salva Kiir Mayardit was born into the Dinka ethnic group in Bahr Al Ghazal state in southern Sudan in 1951. He became active in the Anya-Nya southern rebel movement in the civil war in the late 1960s. Following a settlement on regional autonomy in 1972, Kiir joined the Sudanese Army, graduating from its Military College at Omdurman.

The southern rebellion reignited in 1983 after President Nimeiry imposed Sharia law. Kiir joined John Garang in defecting from the Army and co-founded the Sudanese People's Liberation Movement (SPLM) that year. The Marxist-Leninist organization became the dominant southern rebel force.

Peace talks between the SPLM and Sudan's president, Omar al-Bashir, began in 2002. They culminated in a peace agreement of 2005, ending the war that had claimed 2m. lives and recognizing southern autonomy. Kiir became vice president of the Government of Southern Sudan (GOSS) in Juba. He became president following Garang's death in a helicopter accident in July 2005 but did not appoint a new vice-president. His ability to unite rival factions was underlined when he was re-elected president of the GOSS in April 2010 with 93% support. Nearly 99% of voters in the Jan. 2011 referendum then elected to separate from Sudan and, following approval from Khartoum, the Republic of South Sudan was formally established on 9 July that year, with Kiir as president.

Career in Office

In his inaugural address Kiir said the South Sudanese would forgive but not forget past injustices. He made conciliatory gestures to southern rebel groups previously opposed to the SPLM, and asked the United Nations for help to lift South Sudan from the 'abyss of poverty and deprivation'.

However, following deepening tensions with the Bashir regime in Sudan over transit fees (for South Sudan's only export route north to the Red Sea) and oil revenue-sharing, South Sudan shut down oil exports in early 2012, seriously undermining the economies of both countries. In April 2012 Sudan's air force raided its neighbour after South Sudanese troops occupied the disputed border region of Abyei and the Heglig oilfields. The UN threatened sanctions against both countries if they did not stop the escalating violence and return to negotiations. Kiir met President Bashir in Ethiopia in Sept. in an attempt to resolve the crisis, and although they concluded agreements on trade, security and a resumption of oil exports, they failed to resolve contested border issues.

DEFENCE

On independence the former rebel Sudan People's Liberation Army changed its name to the South Sudan Armed Forces.

ECONOMY

Overview

Oil exports began in 1999 and have been essential to the fortunes of what is now South Sudan, accounting for 98% of its budget. 75% of what was formerly Sudan's oil production is located in the South.

Despite a 2005 agreement to share oil revenues equally, there have been clashes between North and South over oil-rich border regions such as Abyei. When the 2005 agreement expired, tensions escalated over how to split future earnings. Other contentious issues include border definition and servicing the region's debts.

Despite oil wealth, South Sudan is among Africa's least developed economies, civil war having left the country impoverished. Infrastructure is poor, with few paved roads. Yet there is significant growth potential, with sizeable livestock, fishery and forestry resources, while fertile conditions make agriculture a promising non-oil sector.

Currency

The official unit of currency is the South Sudan pound (SSP) of 100 piastres, introduced on 18 July 2011. Inflation was 47·3% in 2011.

Performance

Total GDP in 2011 was US$19·2bn. Real GDP growth was 1·4% in 2011.

Banking and Finance

The Bank of South Sudan (BOSS) serves as the country's central bank. Its Governor is Kornelio Koriom Mayiek.

COMMUNICATIONS

Roads

Only a small proportion of South Sudan's road network is paved. It is estimated that the country has less than 100 km of paved road.

Rail

Total length of railway is 248 km, running from the Sudanese border to Wau.

Civil Aviation

There is an international airport at Juba with connections to Addis Ababa, Cairo, Entebbe, Khartoum and Nairobi. Other major airports include those at Malakal, Rumbek and Wau. Southern Sudan Airlines was established in 2005.

SOCIAL INSTITUTIONS

Justice

The Judiciary of Southern Sudan (JOSS) oversees the court systems of South Sudan. The highest court is the Supreme Court. The second tier of justice consists of three Courts of Appeal. Each state has a High Court, as well as county, town and city courts.

The death penalty is in force. There were at least five executions in 2012.

Education

Primary school enrolment in 2009 was 1,380,580. Secondary school enrolment (ages 14–17) was 44,027 in 2009. Higher education institutions include the University of Juba (founded in 1977), the University of Bahr el Ghazal (1991) in Wau and Upper Nile University (1991) in Malakal. The literacy rate for people aged 15 and above was 27% in 2009.

RELIGION

A large percentage of South Sudan's population are Christian— primarily Roman Catholic, Anglican and Presbyterian. There are also followers of African traditional animist religions as well as Muslims.

DIPLOMATIC REPRESENTATIVES

Of South Sudan in the United Kingdom (Hamilton House, Mabledon Place, London, WC1H 9BB)
Ambassador: Sabit Alley.

Of the United Kingdom in South Sudan (EU Compound, Thong Ping, Juba)
Ambassador: Ian Hughes.

Of South Sudan in the USA (1233 20th St., NW, Suite 602, Washington, D.C., 20036)
Ambassador: Akec Khoc Aciew Khoc.

Of the USA in South Sudan (Thong Ping, Juba)
Ambassador: Susan D. Page.

Of South Sudan to the United Nations
Ambassador: Francis Mading Deng.

FURTHER READING

National Statistical Office: National Bureau of Statistics, Near South Sudan High Court, May St., Juba.
Website: http://ssnbs.org

SPAIN

© Research Machines plc 2006

Reino de España
(Kingdom of Spain)

Capital: Madrid
Population projection, 2015: 47·53m.
GNI per capita, 2011: (PPP$) 26,508
HDI/world rank: 0·878/23
Internet domain extension: .es

KEY HISTORICAL EVENTS

A bridge between Europe and Africa, the Iberian peninsula has absorbed influences from both regions. The original inhabitants were Iberians, who spoke a non Indo-European language, and Celtic peoples, who were mainly to the north and west of the peninsula. From the 8th century BC the Phoenicians established trading colonies such as Gades (Cádiz), importing metalworking skills, music and literacy in the form of a semi-syllabic script. The Greeks established a trading settlement in Catalonia named Empirion (now Ampurias) around 575 BC, and there is evidence of other Greek and Phoenician settlements along the Mediterranean coast.

From 241 BC the Iberian peninsula came under the influence of Carthage in North Africa. The Carthaginians, led by Hamilcar Barca, landed at Cádiz and moved north and east. They eventually founded a new capital at Cartagena: the city grew rapidly and had a population of around 30,000 by 215 BC. A Roman presence began at this time, further north in Catalonia. The first legionnaires established their base at Tarragona, from where they waged war on the Carthaginians. Fighting between the two powers ebbed and flowed for years, until the Carthaginians were forced off the peninsula in 206 BC. Roman laws and customs were gradually adopted over the following six centuries, but there were frequent rebellions among the native peoples.

Roman rule was on the wane throughout Europe by AD 400, and Roman Hispania was no exception. In 409 Visigoths, Suevi and Vandals crossed the Pyrenees and began to establish themselves as the new rulers. By 470 most of the leading families were of Germanic origin. Toledo became the capital and seat of successive Visigothic monarchs until the early 700s. At this time, the Romans were defeated in North Africa by Muslim armies,

who began to turn their attention to the Iberian peninsula. Toledo fell to Arab and Berber forces and the death of King Roderic in 711 marked the end of Visigothic hegemony.

The Muslim conquerors brought a new language, religion and culture, which dominated large parts of the Iberian peninsula for the next five hundred years, though there were sizeable Jewish communities in the southern and eastern towns and there were some Christian principalities in the north. The Umayyad dynasty used Córdoba as the administrative centre of al-Andalus ('Land of the Vandals') until 1031. New trade links were established, connecting Córdoba with Egypt and Persia and most of the Islamic world. People and ideas flooded in and the great cities of Córdoba and Seville became beacons of modernity and creativity in fields ranging from architecture to botany, medicine, poetry and techniques for irrigation.

While al-Andalus prospered, the Christian principalities to the north in places such as Asturia, the Basque territories and northern Catalonia remained relatively poor and agrarian. However, from about 900 there was a gradual expansion southwards towards al-Andalus, described as the start of the *Reconquista*, or reconquest of Spain by the Christians. By 1000 there was considerable contact between the Christian principalities and France: Norman knights fought in Catalonia and French settlers arrived in towns along the pilgrimage route to Santiago de Compostela, bringing with them new ideas and skills.

From the 1120s the Muslim governors of al-Andalus found themselves under threat from both northern Christian rulers and native Andalusi. Alfonso VII of Leon-Castile eventually conquered Córdoba in 1146 and the strategically important Almería, on the Mediterranean coast, in 1147. Following these victories, the three most powerful Christian kingdoms of Aragon, Castile and Portugal pushed south and east and by 1300 the last remaining Islamic dominion was the amirate of Granada. Muslim inhabitants were expelled from many towns and cities, though in rural areas the Islamic faith and the Arabic language survived for centuries. While Córdoba declined, Barcelona blossomed: it emerged as a great economic success story on a par with Genoa and Venice.

Castile and Aragon were the dominant kingdoms by the early 1300s, but they were both characterized by infighting and rebellion. King Pedro the Cruel of Castile was challenged by a coalition of nobles led by his half brother Enrique de Trastámara. The English supported Pedro (Peter) and his heirs and the French backed Enrique (Henry) and his supporters. Enrique eventually prevailed, and was crowned Enrique II in 1369.

The Modern State

The Spanish monarchy was founded in 1469 following the marriage of Isabel (Isabella), princess of Castile and Fernando (Ferdinand), heir to the throne of Aragon. Under their joint reign, they laid the foundations for a unified Spain. In 1478 they established the notorious Spanish Inquisition, expelling and executing tens of thousands of Jews and other non-Christians. Four years later the last Islamic territory of Granada was besieged. It surrendered in 1492, the year in which Christopher Columbus reached the New World.

When Fernando and Isabel's son Juan died in 1497, the succession to the Spanish crowns passed to his sister, Juana *la loca* (the Mad). Juana married Philip (Felipe) the Handsome, heir through his father, Emperor Maximilian I, to the Habsburg domains in Germany and Flanders. When Fernando died in 1516, Juana and Felipe's son, Charles of Ghent, inherited Spain, its colonies in the New World, Naples and, following the death of Maximilian I in 1519, the Habsburg territories. Shortly afterwards he was elected Holy Roman Emperor, a title he held as Charles V (Carlos I of Spain). In the space of only a few years, Charles commanded one of the most extensive empires since Rome.

Columbus paved the way for the Spanish colonies in the New World but for 30 years after his discoveries attention focused solely on the Caribbean. It was only in 1521 that Hernando Cortés overthrew the Aztecs, with help from native Indian allies. After 1540 gold and silver began pouring into Spanish coffers from mines in Peru and Mexico. Sugar plantations were established in the Caribbean and the indigenous populations were gradually wiped out. Maintaining control of the new empire was a serious challenge and Charles V relied heavily on co-operation from Italians, Flemings and Germans. The rise of Spain as a military power began in the 1560s in the reign of Felipe II. He built a powerful navy and annexed Portugal in 1580. The new fleet patrolled the American supply routes, fended off attacks from the English and set up new colonies in the Philippines and at Buenos Aires.

Spain's Golden Age began to lose its lustre in the late 16th century, following a popular uprising in the Netherlands under William of Orange. The Dutch were beginning to establish their own colonies in Asia and started making inroads in Brazil. Spain was weakened by the cost of defending its empire against France and England and in 1640 the unity of the Iberian peninsula itself came under threat by rebellions in Catalonia and Portugal. The crowning of Carlos II, a disabled child, in 1665 symbolized Spain's growing vulnerability and isolation from the rest of Europe.

Carlos II died without issue in 1700 and left the throne to Philippe, duke of Anjou and grandson of King Louis XIV of France. Felipe (Philippe) V was the first in a line of five Bourbon monarchs, who reigned in Spain until 1833. Under Felipe V, and his successor Fernando VI, Spain restored some of its influence in Europe, particularly in Italy, where Naples and Sicily were recovered from Austria in 1734. Educational reforms led to a period of Enlightenment in the 1760s, with many universities replacing conservative Jesuit doctrines with modern physics, astronomy and political theory. The 1780s, when Spain was ruled by Carlos III, were a period of stability and prosperity. Catalonia became a centre of the early industrial revolution with its booming textile trade and Madrid saw a flowering of artistic expression encapsulated by the work of Goya.

When Louis XVI was guillotined in 1793, the new French Republic turned its attention to neighbouring Spain and Britain. In 1794–95 French forces invaded Catalonia and the Basque provinces. Following these defeats, King Carlos IV of Spain formed an alliance with France under its new emperor, Napoleon. Hostilities with Britain were resumed, though defeat for the Franco-Spanish naval forces at the Battle of Trafalgar further undermined links with the Spanish colonies and damaged the economy.

In 1808 the weak and unpopular Carlos IV abdicated and the Spanish crown passed to Fernando VII, though his right to the throne was ceded to Napoleon later that year. However, Napoleon misjudged the mood of the Spanish public who rioted in Madrid and began a five-year war of independence. In 1813 the French forces were finally expelled but ideas from revolutionary France were beginning to take root. The medieval Cortes (parliament) was revived and a Liberal reformist group secured Spain's first constitution. The following year, Fernando VII was restored to the Spanish throne, a move generally welcomed by a war-ravaged public. But Fernando's first act was to abolish the constitution, and his 20 year reign was characterized by a return to the old regime: the Inquisition was re-established, the Liberals were persecuted, free speech was repressed and Spain entered a severe economic recession.

End of Empire

Queen Isabel II inherited the Spanish throne as a child in 1833. During her reign there were various attempts by the Liberals and progressives to reinstate a constitution. Her support for

neo-catholic reactionaries in her governments of the 1860s fanned the flames of revolution. In 1868 Isabel was deposed and the Cortes approved a new constitution. Amadeo I of Savoy was chosen as monarch but he was unable to adapt to Spanish politics and abdicated in 1873. The Cortes immediately proclaimed a republic but in less than a year a coup restored the Bourbon monarchy, with Alfonso XII, the son of exiled Isabel II, as king.

The disastrous Spanish-American War of 1898 marked the end of the Spanish Empire. Spain was defeated by the USA in a series of one-sided naval battles, resulting in the loss of Cuba, Puerto Rico, Guam and the Philippines. Neutral in the First World War, Spain enjoyed a trade and industry boom. Barcelona's Hispano-Suiza factories produced aircraft engines for the French air force and luxury cars. However, prosperity did not filter through society and workers demonstrated against high food prices. In 1923 Gen. Miguel Primo de Rivera, marquis of Estella, led a coup, abolished the 1876 constitution and closed down the Cortes. Primo de Rivera was determined to clear out what he saw as corrupt, self-serving politicians. His alternative to the constitutional monarchy was the National Political Union but it attracted only opportunists and right-wing enthusiasts. There were improvements in the nation's infrastructure, but de Rivera's public works programmes were hit by financial difficulties in 1929, and the dictator resigned the following year.

1931 marked the beginning of a new genuinely democratic era for Spain. Municipal elections were held and won by a republican-socialist coalition. King Alfonso XIII went into exile and the Second Republic was declared. The 1936 elections saw the country split in two, with the Republican government and its supporters on one side (an uneasy alliance of communists, socialists and anarchists) and the Nationalists (the army, the Catholic church, monarchists and the fascist-style Falange Party) on the other.

Civil War

The assassination of the opposition leader José Calvo Sotelo by Republican police officers in July 1936 gave the army, led by Gen. Francisco Franco, an excuse to stage a coup. The failure of the military to overthrow the government led to a protracted civil war. The Nationalists received extensive military and financial support from fascist Germany and Italy, while the Republican government received support from the Soviet Union and, to a lesser degree, from the International Brigades, made up of foreign volunteers.

By 1939 the Nationalists, led by Franco, had prevailed. More than 350,000 Spaniards died in the fighting, but more bloodletting ensued. An estimated 100,000 Republicans were executed or died in prison after the civil war. Franco's cure for Spain's 'sick' economy was withdrawal from world markets and the establishment of a self-sufficient autarky which remained neutral in World War Two. But by the late 1940s inflation was rising steeply and Spain was losing ground to other European countries. Franco allowed a gradual liberalization of the economy, but despite this and the readmission of Spain to the UN in 1955 the economy remained in deep trouble. The desperate conditions endured by hundreds of thousands of workers led to nationwide strikes and growing opposition to the Franco regime among university students and intellectuals. In the 1960s the regime faced demands for independence in Catalonia and the Basque Country.

Franco died in 1975, having earlier named Juan Carlos, the grandson of Alfonso XIII, his successor. Under King Juan Carlos, Spain made the transition back to democracy. The first elections were held in 1977 and a new constitution was approved by referendum in 1978. In Feb. 1981 there was an attempted fascist coup, when for 18 hours the deputies of the lower house of parliament and the Cabinet were held hostage. The episode is now seen as the final, futile attempt to turn back the clock. The following year saw a spectacular victory for the socialist party which presided over a rapid expansion in the economy during its 14 years in power. In 1986 Spain joined the European Economic

Community, cementing the nation's status as a popular location for foreign investors and tourists, though its international reputation has suffered from the ongoing violent campaign waged by ETA, the separatist terrorist group attempting to secure an independent Basque homeland.

In 1996 Spaniards voted in a conservative party under the leadership of José María Aznar. In March 2000 he was re-elected with an absolute majority; his success was attributed to the buoyant state of the Spanish economy, which averaged in excess of 4% annual growth during Aznar's first term of office.

Madrid suffered Spain's worst terrorist attack on 11 March 2004 when four commuter trains were bombed, killing 191 and injuring over 1,800 people. The government initially blamed ETA but suspicion quickly moved to al-Qaeda and North African operatives. On 14 March the Socialists, led by José Luis Rodríguez Zapatero, defeated the People's Party in general elections. The last Spanish troops left Iraq in May 2004. In March 2006 ETA announced a permanent ceasefire, but the truce ended in Dec. when the group carried out a car bomb attack at Madrid's Barajas airport. ETA renounced the use of arms in Oct. 2011.

TERRITORY AND POPULATION

Spain is bounded in the north by the Bay of Biscay, France and Andorra, east and south by the Mediterranean and the Straits of Gibraltar, southwest by the Atlantic and west by Portugal and the Atlantic. Continental Spain has an area of 493,491 sq. km, and including the Balearic and Canary Islands and the towns of Ceuta and Melilla on the northern coast of Africa, 505,693 sq. km (195,249 sq. miles). Population (census, 2011), 46,815,916 (23,711,613 females). In 2011, 77·6% of the population lived in urban areas; population density in 2011 was 93 per sq. km. In 2011 foreigners resident in Spain numbered 5,252,473 (up from 3,730,610 in 2005), including 798,104 from Romania, 773,966 from Morocco, 316,756 from Ecuador, 312,098 from the UK and 250,087 from Colombia. Foreigners constituted 11·2% of the population in 2011 (8·5% in 2005 and 2·3% in 2000).

The UN gives a projected population for 2015 of 47·53m.

The growth of the population has been as follows:

Census year	Population	Rate of annual increase	Census year	Population	Rate of annual increase
1860	15,655,467	0·34	1960	30,903,137	1·05
1910	19,927,150	0·72	1970	33,823,918	0·95
1920	21,303,162	0·69	1981	37,746,260	1·05
1930	23,563,867	1·06	1991	38,872,268	0·30
1940	25,877,971	0·98	2001	40,847,371	0·51
1950	27,976,755	0·81	2011	46,815,916	1·46

Area and population of the autonomous communities (in italics) and provinces at the 2011 census:

Autonomous community/ Province	Area (sq. km)	Population	Per sq. km
Andalusia	*87,597*	*8,371,270*	*96*
Almería	8,774	688,736	78
Cádiz	7,436	1,244,732	167
Córdoba	13,771	802,575	58
Granada	12,647	922,100	73
Huelva	10,128	519,895	51
Jaén	13,496	667,484	49
Málaga	7,308	1,594,808	218
Seville (Sevilla)	14,036	1,930,941	138
Aragón	*47,720*	*1,344,509*	*28*
Huesca	15,636	225,962	14
Teruel	14,810	143,162	10
Zaragoza (Saragossa)	17,275	975,385	56
Asturias	*10,604*	*1,075,183*	*101*
Baleares	*4,992*	*1,100,503*	*220*
Basque Country	*7,230*	*2,185,393*	*302*
Álava	3,032	320,778	106

Autonomous community/ Province	Area (sq. km)	Population	Per sq. km
Guipúzcoa	1,980	708,425	358
Vizcaya	2,217	1,156,190	522
Canary Islands	*7,494*	*2,082,655*	*278*
Palmas de Gran Canaria, Las	4,066	1,087,225	267
Santa Cruz de Tenerife	3,381	995,429	294
Cantabria	*5,321*	*592,542*	*111*
Castilla-La Mancha	*79,462*	*2,106,331*	*27*
Albacete	14,926	401,580	27
Ciudad Real	19,813	526,628	27
Cuenca	17,141	215,165	13
Guadalajara	12,212	257,442	21
Toledo	15,370	705,516	46
Castilla y León	*94,227*	*2,540,188*	*27*
Ávila	8,050	171,647	21
Burgos	14,291	372,538	26
León	15,582	493,312	32
Palencia	8,053	170,513	21
Salamanca	12,350	350,018	28
Segovia	6,923	163,171	24
Soria	10,307	94,610	9
Valladolid	8,110	532,765	66
Zamora	10,561	191,613	18
Catalonia	*32,091*	*7,519,843*	*234*
Barcelona	7,728	5,522,565	715
Girona	5,910	751,806	127
Lleida	12,150	438,428	36
Tarragona	6,303	807,044	128
Extremadura	*41,635*	*1,104,499*	*27*
Badajoz	21,766	691,799	32
Cáceres	19,868	412,701	21
Galicia	*29,575*	*2,772,928*	*94*
Coruña, La	7,950	1,141,286	144
Lugo	9,857	348,067	35
Ourense	7,273	328,697	45
Pontevedra	4,495	954,877	212
Madrid	*8,028*	*6,421,874*	*800*
Murcia	*11,314*	*1,462,128*	*129*
Navarra	*10,390*	*640,129*	*62*
Rioja, La	*5,045*	*321,173*	*64*
Valencian Community	*23,254*	*5,009,931*	*215*
Alicante	5,817	1,852,166	318
Castellón	6,632	594,423	90
Valencia	10,806	2,563,342	237
Ceuta[1]	*19*	*83,517*	*4,176*
Melilla[1]	*13*	*81,323*	*6,777*
Total	*505,963*	*46,815,916*	*93*

[1]Ceuta and Melilla gained limited autonomous status in 1994.

The capitals of the autonomous communities are: *Andalusia:* Seville (Sevilla); *Aragón:* Zaragoza (Saragossa); *Asturias:* Oviedo; *Baleares:* Palma de Mallorca; *Basque Country:* Vitoria-Gasteiz; *Canary Islands:* dual capitals, Las Palmas de Gran Canaria and Santa Cruz de Tenerife; *Cantabria:* Santander; *Castilla-La Mancha:* Toledo; *Castilla y León:* Valladolid; *Catalonia:* Barcelona; *Extremadura:* Mérida; *Galicia:* Santiago de Compostela; *Madrid:* Madrid; *Murcia:* Murcia (but regional parliament in Cartagena); *Navarra:* Pamplona; *La Rioja:* Logroño; *Valencian Community:* Valencia.

The capitals of the provinces are the towns from which they take the name, except in the cases of Álava (capital, Vitoria-Gasteiz), Guipúzcoa (San Sebastián) and Vizcaya (Bilbao).

The islands that form the Balearics include Majorca (Mallorca), Minorca (Menorca), Ibiza and Formentera. Those which form the Canary Archipelago are divided into two provinces, under the name of their respective capitals: Santa Cruz de Tenerife and Las Palmas de Gran Canaria. The province of Santa Cruz de Tenerife is constituted by the islands of Tenerife, La Palma, Gomera and Hierro; that of Las Palmas by Gran Canaria, Lanzarote and Fuerteventura, with the small barren islands of Alegranza, Roque del Este, Roque del Oeste, Graciosa, Montaña Clara and Lobos.

Places under Spanish sovereignty in Africa (Alhucemas, Ceuta, Chafarinas, Melilla and Peñón de Vélez) constitute the two provinces of Ceuta and Melilla.

Populations of principal towns in 2011:

Town	Population	Town	Population
Albacete	171,999	Málaga	561,435
Alcalá de Henares	200,505	Marbella	135,124
Alcobendas	110,351	Mataró	123,367
Alcorcón	167,217	Móstoles	203,493
Algeciras	117,695	Murcia	437,667
Alicante	329,325	Ourense	107,314
Almería	189,680	Oviedo	225,005
Badajoz	151,214	Palma de Mallorca	402,044
Badalona	219,241	Palmas de Gran	
Barakaldo	100,064	Canaria, Las	381,271
Barcelona	1,611,013	Pamplona (Iruña)	195,943
Bilbao	351,356	Parla	122,045
Burgos	178,864	Reus	106,849
Cádiz	124,014	Sabadell	206,949
Cartagena	215,757	Salamanca	151,658
Castellón de la Plana	176,298	San Cristóbal de La	
Córdoba	328,326	Laguna	152,025
Coruña, La	245,053	San Sebastián	
Dos Hermanas	128,433	(Donostia)	185,512
Elche (Elx)	227,417	Santa Coloma de	
Fuenlabrada	196,986	Gramenet	119,391
Getafe	168,642	Santa Cruz de	
Gijón	276,969	Tenerife	204,476
Granada	241,003	Santander	178,095
Huelva	147,808	Seville	698,042
Jaén	116,469	Tarragona	133,223
Jerez de la Frontera	211,784	Telde	101,080
Leganés	185,758	Terrassa	214,406
León	131,411	Torrejón de Ardoz	123,213
L'Hospitalet de		Valencia	792,054
Llobregat	256,509	Valladolid	311,682
Lleida	137,283	Vigo	295,623
Logroño	152,698	Vitoria-Gasteiz	240,753
Madrid	3,198,645	Zaragoza	678,115

Languages

The Constitution states that 'Castilian is the Spanish official language of the State', but also that 'All other Spanish languages will also be official in the corresponding Autonomous Communities'. At the last linguistic census (2001) Catalan (an official EU language since 1990) was spoken in Catalonia by 74·5% of people and understood by 94·5%. It is also spoken in Baleares, Valencian Community (where it is frequently called Valencian), and in Aragón, a narrow strip close to the Catalonian and Valencian Community boundaries. Galician, a language very close to Portuguese, was understood in 1998 by 98·4% of people in Galicia and spoken by 89·2%; Basque by a significant and increasing minority in the Basque Country, and by a small minority in northwest Navarra. It is estimated that one-third of all Spaniards speaks one of the other three official languages as well as standard Castilian. In bilingual communities, both Castilian and the regional language are taught in schools and universities.

SOCIAL STATISTICS

Statistics for calendar years:

	Births	Deaths	Marriages	Divorces
2004	454,591	371,934	216,149	50,974
2005	466,371	387,355	208,146[1]	72,848
2006	482,957	371,478	203,453[1]	126,952
2007	492,527	385,361	201,579[1]	125,777
2008	518,967	385,954	193,064[1]	110,036

[1]Excluding same sex marriage, which was permitted from 3 July 2005.

Rate per 1,000 population, 2008: births, 11·7; deaths, 8·7; marriages, 4·3; divorces, 2·5. In 2005 the most popular age range for marrying was 25–29 for both males and females. Annual population growth rate, 2000–05, 1·5%. Suicide rate (per 100,000 population), 2005: 7·8. Expectation of life, 2007, was 77·5 years for males and 84·0 for females. Infant mortality, 2010, four per 1,000 live births; fertility rate, 2008, 1·4 births per woman. In 2009 Spain received 3,007 asylum applications, down from 7,662 in 2007.

A UNICEF report published in 2010 showed that 17·1% of children in Spain live in relative poverty (living in a household in which disposable income—when adjusted for family size and composition—is less than 50% of the national median income), compared to just 4·7% in Iceland.

In 2007–08, 19·6% of 15 to 34-year-olds used cannabis and 5·1% cocaine. A 2009 report found that use of cocaine in Spain was the highest in the European Union.

CLIMATE

Most of Spain has a form of Mediterranean climate with mild, moist winters and hot, dry summers, but the northern coastal region has a moist, equable climate, with rainfall well distributed throughout the year, mild winters and warm summers, and less sunshine than the rest of Spain. The south, in particular Andalusia, is dry and prone to drought.

Madrid, Jan. 41°F (5°C), July 77°F (25°C). Annual rainfall 16·8" (419 mm). Barcelona, Jan. 46°F (8°C), July 74°F (23·5°C). Annual rainfall 21" (525 mm). Cartagena, Jan. 51°F (10·5°C), July 75°F (24°C). Annual rainfall 14·9" (373 mm). La Coruña, Jan. 51°F (10·5°C), July 66°F (19°C). Annual rainfall 32" (800 mm). Seville, Jan. 51°F (10·5°C), July 85°F (29·5°C). Annual rainfall 19·5" (486 mm). Palma de Mallorca, Jan. 51°F (11°C), July 77°F (25°C). Annual rainfall 13·6" (347 mm). Santa Cruz de Tenerife, Jan. 64°F (17·9°C), July 76°F (24·4°C). Annual rainfall 7·72" (196 mm).

CONSTITUTION AND GOVERNMENT

Following the death of General Franco in 1975 and the transition to a democracy, the first democratic elections were held on 15 June 1977. A new constitution was approved by referendum on 6 Dec. 1978, and came into force 29 Dec. 1978. It has been amended twice since, in 1992 and 2011. It established a parliamentary monarchy.

The reigning king is **Juan Carlos I** (Juan Carlos de Borbón), born 5 Jan. 1938. The eldest son of Don Juan, Conde de Barcelona, Juan Carlos was given precedence over his father as pretender to the Spanish throne in an agreement in 1954 between Don Juan and General Franco. Don Juan, who resigned his claims to the throne in May 1977, died on 1 April 1993. King (then Prince) Juan Carlos married, in 1962, Princess Sophia of Greece, daughter of the late King Paul of the Hellenes and Queen Frederika. *Offspring:* Elena, born 20 Dec. 1963, married 18 March 1995 Jaime de Marichalar, divorced 25 Nov. 2009 (*offspring:* Felipe, b. 17 July 1998; Victoria, b. 9 Sept. 2000); Cristina, born 13 June 1965, married 4 Oct. 1997 Iñaki Urdangarín (*offspring:* Juan, b. 29 Sept. 1999; Pablo, b. 6 Dec. 2000; Miguel, b. 30 April 2002; Irene, b. 5 June 2005); Felipe, Prince of Asturias, heir to the throne, born 30 Jan. 1968, married 22 May 2004 Letizia Ortiz Rocasolano (*offspring:* Leonor, b. 8 Nov. 2005; Sofia, b. 29 April 2007).

The King receives an allowance, part of which is taxable, approved by parliament each year. For 2013 this is €7·9m. There is no formal court; the (private) *Diputación de la Grandeza* represents the interests of the aristocracy.

Legislative power is vested in the *Cortes Generales*, a bicameral parliament composed of the Congress of Deputies (lower house) and the Senate (upper house). The *Congress of Deputies* has not less than 300 nor more than 400 members (350 in the general election of 2011) elected in a proportional system under which electors choose between party lists of candidates in multi-member constituencies.

The *Senate* has 264 members of whom 208 are elected by a majority system: the 47 mainland provinces elect four senators each, regardless of population; the larger islands (Gran Canaria,

Majorca and Tenerife) elect three senators and each of the smaller islands or groups of islands (Ibiza-Formentera, Minorca, Fuerteventura, Gomera, Hierro, Lanzarote and La Palma) elect one senator. To these each self-governing community appoints one senator, and an additional senator for every million inhabitants in their respective territories. Currently 56 senators are appointed by the self-governing communities. Deputies and senators are elected by universal secret suffrage for four-year terms. The Prime Minister is elected by the Congress of Deputies.

The *Constitutional Court* is empowered to solve conflicts between the State and the Autonomous Communities; to determine if legislation passed by the Cortes is contrary to the Constitution; and to protect the constitutional rights of individuals violated by any authority. Its 12 members are appointed by the monarch. It has a nine-year term, with a third of the membership being renewed every three years.

National Anthem

'Marcha Real' ('Royal March'); no words, tune anonymous.

GOVERNMENT CHRONOLOGY

Heads of government since 1939. (PP = Popular Party; PSOE = Spanish Socialist Workers' Party; UCD = Central Democratic Union)

1939–73	military	Francisco Franco
1973	military	Luis Carrero
1973–76	civilian	Carlos Arias Navarro
1976	military	Fernando de Santiago
1976–81	UCD	Adolfo Suárez
1981–82	UCD	Leopoldo Calvo-Sotelo
1982–96	PSOE	Felipe González
1996–2004	PP	José María Aznar
2004–11	PSOE	José Luis Rodríguez Zapatero
2011–	PP	Mariano Rajoy Brey

RECENT ELECTIONS

A general election took place on 20 Nov. 2011. Turnout was 71·7%. In the *Congress of Deputies* the Popular Party (PP) won 186 seats with 44·6% of votes cast; the Spanish Socialist Workers' Party (PSOE), 110 with 28·7%; Convergence and Union (CiU; Catalan nationalists), 16 with 4·2%; the Communist-led United Left (IU), 11 with 6·9%; Amaiur, 7 with 1·4%; Union, Progress and Democracy (UPyD), 5 with 4·7%; the Basque Nationalist Party (PNV), 5 with 1·3%; the Catalan separatist Republican Left of Catalunya (ERC), 3 with 1·1%; Galician Nationalist Bloc (BNG), 2 with 0·8%; Canarian Coalition (CC–NC–PNC), 2 with 0·6%; Coalició Compromís, 1 with 0·5%; Asturian Forum (FAC), 1 with 0·4%; Geroa Bai (Basque Nationalist Party (PNV), Atarrabia Taldea, Zabaltzen), 1 with 0·2%. In the *Senate*, the PP won 136 seats; PSOE, 48; CiU, 9; Catalan Agreement of Progress (PSC–ICV–EUA), 7; Basque Nationalist Party, 4; Amaiur, 3; Canarian Coalition (CC–NC–PNC), 1.

European Parliament
Spain has 54 representatives. At the June 2009 elections turnout was 44·9% (45·1% in 2004). The PP won 23 seats with 42·2% of votes cast (political affiliation in European Parliament: European People's Party); the PSOE, 21 with 38·5% (Progressive Alliance of Socialists and Democrats); the Coalition for Europe (a coalition of CiU, PNV, CC, the Andalusian Party, the Majorcan Union and the Valencian Nationalist Bloc), 2 with 5·1% (Alliance of Liberals and Democrats for Europe); the Left (a coalition of the IU, Initiative for Catalonia-Greens, United and Alternative Left, and the Bloc for Asturias) 2 with 3·7% (one with Greens/European Free Alliance and one with European United Left/Nordic Green Left); UPD, 1 with 2·9% (non-attached); Europe of the Peoples-Greens (a coalition of Eusko Alkartasuna, the Greens, Aralar, the Galician Nationalist Bloc, the Republican Left of Catalonia and the Chunta Aragonesista), 1 with 2·5% (Greens/European Free Alliance).

CURRENT GOVERNMENT

In March 2013 the government comprised:
President of the Council and Prime Minister: Mariano Rajoy Brey; b. 1955 (PP; sworn in 21 Dec. 2011).
Vice-President and Minister for the Presidency: María Soraya Sáenz de Santamaría Antón.
Minister for Agriculture, Food and Environment: Miguel Arias Cañete. *Defence:* Pedro Morenés Eulate. *Economy and Competitiveness:* Luis de Guindos Jurado. *Education, Culture and Sport:* José Ignacio Wert Ortega. *Finance and Public Administration:* Cristóbal Montoro Romero. *Foreign Affairs and Co-operation:* José Manuel García Margallo. *Health, Social Policy and Equality:* Ana Mato Androver. *Industry, Energy and Tourism:* José Manuel Soria. *Interior:* Jorge Fernández Díaz. *Justice:* Alberto Ruíz Gallardón. *Labour and Social Security:* María Fátima Báñez García. *Public Works:* Ana Pastor Julián.

Government Website: http://www.la-moncloa.es

CURRENT LEADERS

Mariano Rajoy

Position
Prime Minister

Introduction
A veteran of the conservative Popular Party (PP), Mariano Rajoy became prime minister in 2011 during a period of economic crisis. He advocated sweeping economic reforms and has committed his government to accepting austere deficit reduction targets set by the European Union.

Early Life
Mariano Rajoy was born on 27 March 1955 in Santiago de Compostela, Galicia, into a political family. In 1978 he graduated in law from Santiago de Compostela University before working in the civil service as a property registrar from 1979–81. His political career began in 1981 when he won a seat in the newly inaugurated Galician parliament, representing the conservative People's Alliance (AP).

He served as regional minister of institutional relations from 1982–86 and as president of Pontevedra council from 1986–91. He was elected to Spain's national parliament in June 1986, serving briefly as a deputy before resigning to return to the Galician parliament in Nov. 1986, where he served as vice-president until Sept. 1987. In 1989 the AP merged with other parties to form the PP, with Rajoy appointed to its national executive committee.

He won re-election to parliament in 1993 and was minister of public administration in José María Aznar's government from 1996–99, before serving as minister of culture until 2000. After the PP retained power at the 2000 election, Rajoy was appointed deputy prime minister. From 2001–02 he served as minister of the interior and in 2004 fought the general election as party leader-designate. However, the PP unexpectedly lost to the Spanish Socialist Workers' Party (PSOE) in the aftermath of the train bombings in Madrid that the PP government had wrongly blamed on the Basque separatist group ETA.

Rajoy took over as party leader in Oct. 2004 and fought the 2008 election on a platform of liberal economic policies and social conservatism, including restrictions on immigration and opposition to further regional devolution. Although the DP gained seats, it narrowly failed to secure a majority.

As Spain struggled with recession and high unemployment, Rajoy adopted more moderate positions on social and cultural matters while calling for budget cuts. In the 2011 election campaign he committed his party to the ambitious deficit reduction targets set by the EU, balanced against promises to boost growth. In Nov. 2011 the PP won a decisive election victory, claiming 45% of the vote and 186 of 350 parliamentary seats.

Career in Office

Rajoy took office on 21 Dec. 2011 and announced spending cuts of at least €16·5bn. for 2012, along with increases in income and property taxes and tax breaks for companies hiring staff. However, given the scale of Spain's public debt, further austerity was expected if deficit targets were to be achieved. While Rajoy's parliamentary majority allowed him to pass his reforms, maintaining public support proved a major challenge against a background of continuing recession and rising unemployment in 2012, which prompted strikes and mass protests. In June that year his administration had to seek approval from its eurozone partners to access emergency financial assistance in order to bail out Spain's ailing banking sector, while a number of heavily-indebted regional governments applied for rescue funds from the central government.

Rajoy also faced separatist pressure from Catalonia, although the majority vote for pro-independence parties in regional elections in Nov. 2012 proved disparate and the government said that it would not recognize a referendum vote for secession. Also in Nov., the government rejected an offer from the Basque separatist group ETA to enter talks with Spain and France on a definitive end to its operations, stating that it would not negotiate with a terrorist organization.

DEFENCE

Conscription was abolished in 2001. The government had begun the phased abolition of conscription in 1996. In 2002 the armed forces became fully professional. However, a shortfall in recruitment in Spain meant that descendants of Spanish migrants, many of whom had never been to Europe, began to join. Since 1989 women have been accepted in all sections of the armed forces.

In 2008 defence expenditure totalled US$19,196m. (US$430 per capita). In 2007 defence spending represented 1·2% of GDP.

Army

A Rapid Reaction Force is formed from the Spanish Legion and the airborne and air-portable brigades. There is also an Army Aviation Brigade consisting of 153 helicopters (28 attack).

Strength (2004) 95,600. There were 265,000 army reservists in 2004.

Guardia Civil

The paramilitary *Guardia Civil* numbers 72,600.

Navy

The principal ships of the Navy is the light vertical/short take-off and landing aircraft carrier *Juan Carlos I*, commissioned in 2010. It carries AV-8B Harrier combat aircraft. There are also eight French-designed submarines and 16 frigates.

The Naval Air Service operates 17 combat aircraft and 37 armed helicopters. Personnel numbered 700 in 2004. There are 5,600 marines.

Main naval bases are at Ferrol, Rota, Cádiz, Cartagena, Palma de Mallorca, Mahón and Las Palmas de Gran Canaria (Canary Islands).

In 2004 personnel totalled 22,900 including the marines and naval air arm. There were 18,500 naval reservists in 2004.

Air Force

The Air Force is organized as an independent service, dating from 1939. It is administered through four operational commands. These are geographically oriented following a reorganization in 1991 and comprise Central Air Command, Strait Air Command, Eastern Air Command and Air Command of the Canaries.

There were 177 combat aircraft in 2004 including 91 EF/A-18s, 23 F-5Bs and 52 Mirage F-1s.

Strength (2004) 22,750. There were 45,000 air force reservists in 2004.

ECONOMY

Agriculture accounted for 3% of GDP in 2007, industry 30% and services 67%.

Spain's 'shadow' (black market) economy is estimated to constitute approximately 22% of the country's official GDP.

In 2011 Spain gave US$4·2bn. in international aid, equivalent to 0·29% of GNI (compared to the UN target of 0·7%).

Overview

The eurozone's fourth largest economy, Spain made significant economic strides in the 1980s when productivity growth was at its highest and performed strongly after the Europe-wide recession of the early 1990s. Structural reforms and sound macroeconomic policies encouraged growth and employment creation from the late 1990s. The annual budget deficit fell from 6%+ of GDP in the mid-1990s to near balance from 2001–07, while GDP growth was above 3% from 2003–07.

The services share of total GDP has grown at the expense of agriculture, forestry and fishing. Banking, retailing, tele-communications and tourism are the main components. Agriculture is concentrated on wine, olive oil, fruit and vegetables. Vehicle production for export is the leading manufacturing industry. Strong demand for tourist-related buildings, foreign demand for property and high levels of investment in infra-structure combined to make the construction sector, as a share of GDP, twice as large as in other big European economies.

After 15 years of strong growth, the global financial crisis burst the housing construction bubble, resulting in slower growth, cuts in household consumption and rising unemployment. At its height, construction accounted for 13% of Spanish employment but the collapse of the housing boom led to the loss of 900,000 jobs, largely among unskilled construction workers. The economy contracted by 3·7% in 2009 and the public finances recorded a deficit of over 11% of GDP. Unemployment stood at 22·9%. Real GDP further contracted in 2010 and 2012.

A large budget deficit and weak economic growth made Spain vulnerable to contagion from other highly-indebted European countries, prompting investor confidence to decline. In 2011 the cost of government borrowing soared. In July 2012 the yield on Spanish ten-year bonds climbed to a euro-era record of 7·75%, pushing Spain close to needing a full bailout. European lenders had already committed €100bn. to rescue Spain's banks. In July 2012 the president of the European Central Bank pledged to do whatever was necessary to save the euro, prompting ten-year bond yields to improve to 5·42% by Feb. 2013.

Government debt increased from 36·3% of GDP in 2007 to 89·6% in the first quarter of 2012. At the end of 2012 unemployment stood at 26·1%, the second highest rate in the EU (compared to an EU average of 10·7%). Youth unemployment stood at 55·6%, compared to an EU average of 23·4%. A large-scale labour reform package was introduced in Feb. 2012 and the implementation of financial reforms saw the contraction of the economy slow markedly between late 2012 and early 2013.

Currency

On 1 Jan. 1999 the euro (EUR) became the legal currency in Spain at the irrevocable conversion rate of 166·386 pesetas to one euro. The euro, which consists of 100 cents, has been in circulation since 1 Jan. 2002. On the introduction of the euro there was a 'dual circulation' period before the peseta ceased to be legal tender on 28 Feb. 2002. Euro banknotes in circulation on 1 Jan. 2002 had a total value of €68·6bn.

Foreign exchange reserves were US$12,657m. in Sept. 2009 (US$52,490m. in 1998) and gold reserves 9·05m. troy oz. Inflation rates (based on OECD statistics):

2002	2003	2004	2005	2006	2007	2008	2009	2010	2011
3·6%	3·1%	3·1%	3·4%	3·6%	2·8%	4·1%	−0·2%	2·0%	3·1%

Total money supply was €506,714m. in Aug. 2009.

Budget
In 2008 central government revenues totalled €272,978m. and expenditures €287,783m. Principal sources of revenue in 2008: social security contributions, €130,546m.; taxes on income, profits and capital gains, €76,877m.; taxes on goods and services, €39,501m. Main items of expenditure by economic type in 2008: social benefits, €133,819m.; grants, €81,065m.; compensation of employees, €25,764m.

Spain's budget deficit in 2011 was 8·5% of GDP (2010, 9·3%; 2009, 11·2%). The required target set by the EU is a budget deficit of no more than 3%.

VAT is 21% (since 1 Sept. 2012), with a rate of 10% on certain services (catering and hospitality) and 4% on basic foodstuffs.

Performance
Real GDP growth rates (based on OECD statistics):

2002	2003	2004	2005	2006	2007	2008	2009	2010	2011
2·7%	3·1%	3·3%	3·6%	4·1%	3·5%	0·9%	−3·7%	−0·3%	0·4%

The provisional real GDP growth rate in 2012 according to Spain's National Statistics Institute was −1·4%. Total GDP (2011): US$1,476·9bn.

Banking and Finance
The central bank is the Bank of Spain (*Governor*, Luis María Linde) which gained autonomy under an ordinance of 1994. Its Governor is appointed for a six-year term. The Banking Corporation of Spain, *Argentaria*, groups together the shares of all state-owned banks, and competes in the financial market with private banks. In 1993 the government sold 49·9% of the capital of Argentaria; the remainder in two flotations ending on 13 Feb. 1998.

Spanish banking is dominated by two main banks—Santander and BBVA (Banco Bilbao Vizcaya Argentaria). Santander had assets of €1,110·5bn. in Dec. 2009 and BBVA assets of €535·1bn.

Gross external debt amounted to US$2,264,614m. in June 2012.

There are stock exchanges in Madrid, Barcelona, Bilbao and Valencia.

ENERGY AND NATURAL RESOURCES
In 2008, 10·7% of energy consumption came from renewables (wind power, solar power, hydro-electric power, tidal power, geothermal energy and biomass), compared to the European Union average of 10·3%. A target of 20% has been set by the EU for 2020.

Environment
In 2008 Spain's carbon dioxide emissions from the consumption and flaring of fossil fuels were the equivalent of 8·9 tonnes per capita.

Electricity
Installed capacity was 88·9m. kW in 2007. The total electricity output in 2007 amounted to 303·3bn. kWh, of which 62·4% was thermal, 18·2% nuclear, 10·2% hydro-electric and 9·2% geothermal. In 2010 there were eight nuclear reactors in operation. Consumption per capita in 2007 was 6,631 kWh.

Oil and Gas
Spain is heavily dependent on imported oil; Mexico is its largest supplier. Crude oil production (2007), 142,000 tonnes.

The government sold its remaining stake in the oil, gas and chemicals group Repsol in 1997. Natural gas production (2007) totalled 98m. cu. metres. Ever increasing consumption means that Spain has to import large quantities of natural gas, primarily from Algeria.

Wind
Spain is one of the world's largest wind-power producers, with an installed capacity of 20,676 MW in 2010 (the fourth highest after China, the USA and Germany). Production of wind-generated electricity in 2010 totalled 42·7bn. kWh, a figure exceeded only by the USA.

Minerals
Coal production (2007), 7·87m. tonnes; other principal minerals (in 1,000 tonnes): limestone (2008 estimate), 270,000; dolomite (2008 estimate), 15,000; gypsum and anhydrite (2008 estimate), 15,000; marl (2008 estimate), 10,000; lignite (2007), 9,309; salt (2008), 4,141; feldspar (2008), 690; potash (2008), 435; aluminium (2008), 408; fluorspar (2008), 149; nickel (2008), 8. Gold production, 2008, 3,400 kg; silver production, 2008, 3,400 kg.

Agriculture
There were 1,287,000 farms in Spain in 2000. Agriculture employed about 5·9% of the workforce in 2002. It accounts for 15·8% of exports and 15·6% of imports.

There were 13·74m. ha. of arable land in 2002 and 4·98m. ha. of permanent crops. In 2002 there were 914,000 tractors, 52,000 harvester-threshers and 130,000 milking machines in use.

Principal crops	Area (in 1,000 ha.)			Yield (in 1,000 tonnes)		
	2000	2001	2002	2000	2001	2002
Barley	3,278	2,992	3,100	11,063	6,249	8,333
Sugar beets	125	107	115	7,930	6,755	8,040
Wheat	2,353	2,177	2,402	7,294	5,008	6,783
Maize	433	513	463	3,992	4,982	4,463
Potatoes	119	115	114	3,078	2,992	3,104
Oats	432	446	473	954	665	916
Rice	117	116	113	827	876	815
Sunflower seeds	839	858	754	919	871	757

Spain is ranked third among wine producers (behind France and Italy). Production of wine (2006), 38,137,000 hectolitres (13·5% of the world total); of grapes (2003), 6,480,000 tonnes.

The area planted with tomatoes in 2002 was 60,000 ha., yielding 3,878,000 tonnes; with onions, 23,000 ha., yielding 992,000 tonnes; peppers, 23,000 ha., yielding 980,000 tonnes.

Fruit production (2002, in tonnes): oranges, 2,867,000; tangerines, 1,952,000; peaches, 1,247,000; lemons, 920,000; apples, 653,000; pears, 603,000.

Production of olives, 2001–02, 6,983,000 tonnes; olive oil, 1,422,000 tonnes. Spain is the world's leading producer both of olives and olive oil.

Livestock (2003): cattle, 6·48m.; sheep, 23·81m.; goats, 3·05m.; pigs, 23·52m.; chickens, 128·0m.; asses and mules, 0·26m.; horses, 0·24m. Livestock products (2003, in 1,000 tonnes): pork, bacon and ham, 3,322; beef and veal, 700; mutton and lamb, 237; poultry meat, 1,042; milk, 6,917; cheese, 198; eggs, 686.

Forestry
In 2010 the area under forests was 18·17m. ha., or 36% of the total land area. In 2007 timber production was 14·53m. cu. metres.

Fisheries
Spain is one of the leading fishing nations in the EU; it is also the EU's leading importer of fishery commodities. Fishing vessels had a total tonnage of 461,071 GT in 2008, the highest in the EU and representing a quarter of the total tonnage; fleets have been gradually reduced from 18,385 boats in 1995 to 11,420 in 2008. Total catch in 2010 amounted to 968,662 tonnes (the highest since 2001), almost exclusively sea fish. Imports of fishery commodities were valued at US$7,101m. in 2008 and exports at US$3,465m.

INDUSTRY
The leading companies by market capitalization in Spain in March 2012 were: Telefónica SA (US$74·7bn.); Banco Santander (US$69·7bn.); and Inditex, a fashion distributor (US$59·6bn.).

In 2007 industry accounted for 30% of GDP, with manufacturing contributing 15%.

Industrial products, 2007 unless otherwise indicated (in tonnes): cement (2008), 43·1m.; distillate fuel oil, 23·9m.; crude steel (2006), 18·4m.; residual fuel oil, 9·3m.; petrol, 9·2m.; paper and paperboard, 6·7m.; pig iron (2005), 4·2m.; jet fuel, 2·6m.; lime (2005), 1·8m.; nitrogenous fertilizers, 836,270; cigarettes, 41·91bn. units.

Output of other products, 2007: passenger cars, 2,385,210 units; TV receivers, 3·01m. units; washing and drying machines; 2·48m. units; fridge-freezers, 1·15m. units. In 2007, 6,765·6m. litres of mineral water, 5,888·0m. litres of soft drinks and 3,350·2m. litres of beer were produced.

Labour

Out of 18,973,200 people in employment in 2005, 3,113,000 worked in manufacturing; 2,886,800 in wholesale and retail trade/repair of motor vehicles, motorcycles and personal and household goods; 2,357,200 in construction; 1,678,400 in real estate, renting and business activities; 1,291,100 in hotels and restaurants; and 1,196,700 in public administration and defence/compulsory social security. The monthly minimum wage for adults was €624·00 in 2009. The average working week in 2003 was 35·4 hours. The retirement age was traditionally 65 years but in July 2011 the government agreed to raise the age to 67. The change is being phased in gradually starting on 1 Jan. 2013 in a process that is set to continue through to 2027. In 2004 part-time work accounted for less than 9% of all employment in Spain—the lowest percentage in Western Europe.

Spain's unemployment rate reached a peak of nearly 25% in 1994 but then fell steadily, declining to 8·3% in 2007. However, as a result of the global economic crisis Spain's unemployment rate rose by more than seven percentage points in the space of 12 months from March 2008. In the period Jan.–March 2009 an average of 8,600 jobs were lost every day. In Dec. 2012 the rate stood at 26·1%, giving Spain the second highest unemployment rate in the EU (just below that of Greece). Youth unemployment is particularly high, rising to 47·8% in the third quarter of 2011.

Between 2005 and 2009 strikes cost Spain an average of 60 days per 1,000 employees a year. There were 1,001 industrial disputes in 2009 involving 653,483 workers.

INTERNATIONAL TRADE

Imports and Exports

Trade in US$1m.:

	2004	2005	2006	2007	2008
Imports c.i.f.	257,672	287,610	326,046	382,651	417,049
Exports f.o.b.	182,156	191,021	213,350	246,752	277,695

In 2004 chemicals, manufactured goods classified chiefly by material and miscellaneous manufactured articles accounted for 41·7% of imports and 43·4% of exports; machinery and transport equipment 37·7% of imports and 40·3% of exports; food, live animals, beverages and tobacco 8·4% of imports and 12·2% of exports; mineral fuels, lubricants and related materials 8·8% of imports and 1·1% of exports; inedible crude materials (except fuels), and animal and vegetable oil and fats 3·4% of imports and 3·0% of exports.

Leading import sources in 2004 were Germany (16·1%), France (15·2%), Italy (9·1%), United Kingdom (6·1%), Netherlands (4·1%); leading export markets in 2004 were: France (19·3%), Germany (11·7%), Portugal (9·4%), Italy (9·0%), United Kingdom (9·0%). In 2004 the EU accounted for 62·2% of Spain's imports and 70·0% of exports.

COMMUNICATIONS

Roads

In 2007 the total length of roads was 667,064; the network included 13,014 km of motorways, 12,832 km of highways/national roads and 140,165 km of secondary roads. In 2003 road transport totalled 397,117m. passenger-km; freight transport totalled 132,868m. tonne-km in 2003. Number of passenger cars in use (2007), 21,760,200; lorries and vans, 5,140,600; buses and coaches, 61,000; motorcycles and mopeds, 2,311,300. In 2007, 3,823 persons were killed in road accidents (5,604 in 1997).

Rail

The total length of the state railways in 2005 was 12,808 km, mostly broad (1,668-mm) gauge (6,942 km electrified). The state railway system was divided in two in 2005; Administrador de Infraestructuras Ferroviarias (ADIF) now manages the infrastructure and Renfe Operadora runs train operations. There is an ever-expanding high-speed standard-gauge (1,435-mm) network, totalling 2,665 km in 2010. Only China has a longer high-speed rail network. The first high-speed line, from Madrid to Seville, opened in 1992. It was extended northwards from Madrid initially to Lleida, with passenger services beginning in 2003, and further to Tarragona (2006) and Barcelona (2008). High-speed lines linking Madrid with Toledo and Valladolid were opened in 2005 and 2007 respectively, and linking Madrid with Albacete and Valencia in 2010. A high-speed link from Córdoba (on the Madrid to Seville line) to Málaga was also opened in 2007. Passenger-km travelled in 2005 came to 19·8bn. and freight tonne-km to 1·8bn. There are metros in Madrid (287 km), Valencia (152 km), Barcelona (105 km), Bilbao (38 km), Seville (18 km) and Palma de Mallorca (7 km).

In 2003 the construction of two 38 km-long rail tunnels under the Straits of Gibraltar was agreed with Morocco although there are ongoing talks as to the project's feasibility.

Civil Aviation

There are international airports at Madrid (Barajas), Barcelona (Prat del Llobregat), Alicante, Almería, Bilbao, Girona, Las Palmas de Gran Canaria, Ibiza, Lanzarote, Málaga, Palma de Mallorca, Santiago de Compostela, Seville, Tenerife (Los Rodeos and Reina Sofía), Valencia, Valladolid and Zaragoza. There are 43 airports open to civil traffic. A small airport in Seo de Urgel operates in Andorra. The former national carrier Iberia Airlines completed its privatization process in April 2001, when shares were listed for the first time on the stock exchange. In April 2010 it signed a deal with British Airways to merge and create a new company called International Airlines Group, which was founded in Jan. 2011. However, both carriers still operate under their own brands. Of other airlines, the largest are the low-cost carrier Vueling Airlines and Air Europa. Services are also provided by about 70 foreign airlines. In 2005 Iberia carried 27·4m. passengers (12·0m. on international flights); passenger-km totalled 49·0bn. Madrid was the busiest airport in 2001, handling 33,777,862 passengers (16,718,209 on domestic flights) and 294,692 tonnes of freight. Barcelona was the second busiest in 2001, with 20,545,680 (10,075,536 on domestic flights) and 76,966 tonnes of freight. Palma de Mallorca was the third busiest for passengers, with 19,122,832 (14,317,984 on international flights). Las Palmas was the third busiest for freight, with 40,615 tonnes in 2001.

Shipping

In Jan. 2009 there were 184 ships of 300 GT or over registered, totalling 2·39m. GT. Of the 184 vessels registered, 63 were passenger ships, 49 general cargo ships, 29 oil tankers, 20 container ships, 11 liquid gas tankers, nine bulk carriers and three chemical tankers. The Spanish-controlled fleet comprised 259 vessels of 1,000 GT or over in Jan. 2009, of which 128 were under the Spanish flag and 131 under foreign flags. The leading ports are Algeciras-La Linea (74,845,000 tonnes of cargo in 2008), Barcelona, Bilbao, Cartagena, Las Palmas de Gran Canaria, Santa Cruz de Tenerife, Tarragona and Valencia.

Telecommunications

In 2008 there were 20,200,000 main (fixed) telephone lines. In the same year mobile phone subscribers numbered 49,678,000

(1,116·7 per 1,000 persons). The government disposed of its remaining 21% stake in Telefónica in Feb. 1997, bringing 1·4m. shareholders into the company's equity base. A second operator, Retevisión, accounts for 3% of the domestic market, which was wholly deregulated in 1998. The mobile phone business was deregulated in 1995; the leading mobile network operators are Movistar, Vodafone, Orange and Yoigo.

There were 25·2m. internet users in 2008. The fixed broadband penetration rate stood at 23·4 subscribers per 100 inhabitants in Dec. 2010. In March 2012 there were 15·7m. Facebook users.

SOCIAL INSTITUTIONS

Justice

Justice is administered by Tribunals and Courts, which jointly form the Judicial Power. Judges and magistrates cannot be removed, suspended or transferred except as set forth by law. The constitution of 1978 established the *General Council of the Judicial Power*, consisting of a President and 20 magistrates, judges, attorneys and lawyers, governing the Judicial Power in full independence from the state's legislative and executive organs. Its members are appointed by the *Cortes Generales*. Its President is that of the Supreme Court (*Tribunal Supremo*), who is appointed by the monarch on the proposal of the General Council of the Judicial.

The Judicature is composed of the Supreme Court; 17 Higher Courts of Justice, one for each autonomous community; 52 Provincial High Courts; Courts of First Instance; Courts of Judicial Proceedings, not passing sentences; and Penal Courts, passing sentences.

The Supreme Court consists of a President, and various judges distributed among seven chambers: one for civil matters, three for administrative purposes, one for criminal trials, one for social matters and one for military cases. The Supreme Court has disciplinary faculties; is court of appeal in all criminal trials; for administrative purposes decides in first and second instance disputes arising between private individuals and the State; and in social matters makes final decisions.

A new penal code came into force in May 1996, replacing the code of 1848. It provides for a maximum of 30 years imprisonment in specified exceptional cases, with a normal maximum of 20 years. Sanctions with a rehabilitative intent include fines adjusted to means, community service and weekend imprisonment. The death penalty was abolished by the 1978 constitution. The prison population in Nov. 2008 was 73,687 (160 per 100,000 of national population). In 2007, 2,310,000 crimes were recorded by the police (2,267,000 in 2006). A jury system commenced operating in Nov. 1995 in criminal cases (first trials in May 1996). Juries consist of nine members.

A juvenile criminal law of 1995 lays emphasis on rehabilitation. It raised the age of responsibility from 12 to 14 years. Criminal conduct on the part of children under 14 is a matter for legal protection and custody. 14- and 15-year-olds are classified as 'minors'; 16- and 17-year-olds as 'young persons'; and the legal majority for criminal offences is set at 18 years. Persons up to the age of 21 may, at the courts' discretion, be dealt with as juveniles.

The *Audiencia Nacional* deals with terrorism, monetary offences and drug-trafficking where more than one province is involved. Its president is appointed by the General Council of the Judicial Power.

There is an Ombudsman (*Defensor del Pueblo*), who is elected for a five-year term (currently Soledad Becerril; b. 1944).

Education

In 1991 the General Regulation of the Educational System Act came into force. This Act gradually extends the school-leaving age to 16 years and determines the following levels of education: infants (3–5 years of age), primary (6–11), secondary (12–15) and baccalaureate or vocational and technical (16–17). Primary and secondary levels of education are now compulsory and free. Religious instruction is optional.

In Sept. 1997 a joint declaration with trade unions, parents' and schools' associations was signed in support of a new finance law guaranteeing that spending on education will reach 6% of GDP within five years, thus protecting it from changes in the political sphere. In 2006 public expenditure on education came to 4·4% of GNI and 11·1% of total government spending.

A new compulsory secondary education programme has replaced the Basic General Education programme which was in force since 1970. In addition, university entrance exams underwent reform in 1997, resulting in greater emphasis now being placed on the teaching of Humanities at secondary level.

In 2004–05 pre-primary education (under six years) was undertaken by 1,419,000 pupils; primary or basic education (6–14 years): 2,495,000 pupils. In 2008–09 there were 153,000 teaching staff in pre-primary and 213,000 teachers in primary schools. Secondary education (14–17 years), including high schools and technical schools, was conducted at 6,276 schools, with 3,054,000 pupils and 167,000 teachers in 2004–05.

In 2011 there were 76 universities: 50 public state universities and 26 private universities (including Catholic establishments). In 2004–05 there were 1,331,000 students at state universities; 132,000 at private universities.

The adult literacy rate is at least 99%.

Health

In 2009 there were 219,031 doctors, 63,337 pharmacists, 26,725 dentists and stomatologists, and 255,445 graduate nurses. Number of hospitals (2009), 770, with 146,310 beds. In 2008 Spain spent 9·0% of its GDP on health, with public spending accounting for 72·5% of total expenditure on health and private spending 27·5%.

Welfare

The social security budget was €82,425,871,000 in 2004, including €66·1bn. for pensions, €5·3bn. for temporary incapacity, €1·4bn. for health and €620m. for social services. The minimum pension in 2001 was the equivalent of just over €5,000 per year, made in 14 payments.

In 2003 the system of contributions to the social security and employment scheme was: for pensions, sickness, invalidity, maternity and children, a contribution of 28·3% of the basic wage (23·6% paid by the employer, 4·7% by the employee); for unemployment benefit, a contribution of 7·55% (6·0% paid by the employer, 1·55% by the employee). There are also minor contributions for a Fund of Guaranteed Salaries, working accidents and professional sicknesses, and for vocational training.

RELIGION

There is no official religion. Roman Catholicism is the religion of the majority. In Feb. 2013 there were ten cardinals. There are 11 metropolitan sees and 52 suffragan sees, the chief being Toledo, where the Primate resides. The archdioceses of Madrid-Alcalá and Barcelona depend directly from the Vatican. There are about 0·25m. other Christians, including several Protestant denominations, about 60,000 Jehovah's Witnesses and 29,000 Latter-day Saints (Mormons), and 0·45m. Muslims, including Spanish Muslims in Ceuta and Melilla. The first synagogue since the expulsion of the Jews in 1492 was opened in Madrid on 2 Oct. 1959. The number of people of Judaist faith is estimated at about 15,000.

CULTURE

World Heritage Sites

There are 44 UNESCO sites in Spain: the works of Antoni Gaudi in and around Barcelona (inscribed in 1984 and 2005), Burgos Cathedral (1984), Historic Centre of Córdoba (1984 and 1994), Alhambra, Generalife and Albayzin, Granada (1984 and 1994), Monastery and site of the Escurial, Madrid (1984), Altamira Cave

(1985 and 2008), Old Town of Segovia and its Aqueduct (1985), Monuments of Oviedo and the Kingdom of the Asturias (1985 and 1998), Santiago de Compostela (Old Town) (1985), Old Town of Ávila, with its Extra-Muros churches (1985 and 2007), Mudéjar Architecture of Aragón (1986 and 2001), Historic City of Toledo (1986), Garajonay National Park (1986), Old Town of Cáceres (1986), Cathedral, Alcazar and Archivo de Indias in Seville (1987), Old City of Salamanca (1988), Poblet Monastery (1991), Archaeological Ensemble of Mérida (1993), Royal Monastery of Santa María de Guadalupe (1993), Route of Santiago de Compostela (1993), Doñana National Park (1994 and 2005), Historic Walled Town of Cuenca (1996), La Lonja de la Seda de Valencia (1996), Las Médulas (1997), the Palau de la Música Catalana and the Hospital de Sant Pau, Barcelona (1997 and 2008), San Millán Yuso and Suso Monasteries (1997), University and Historic Precinct of Alcalá de Henares (1998), Rock-Art of the Mediterranean Basin on the Iberian Peninsula (1998), Ibiza, Biodiversity and Culture (1999), San Cristóbal de La Laguna (1999), the Archaeological Ensemble of Tárraco (2000), the Palmeral of Elche (2000), the Roman Walls of Lugo (2000), Catalan Romanesque Churches of the Vall de Boí (2000), Archaeological Site of Atapuerca (2000), Aranjuez Cultural Landscape (2001), Renaissance monumental ensembles of Úbeda and Baeza (2003), Vizcaya Bridge (2006), Teide National Park, Tenerife (2007), the Tower of Hercules, La Coruña (2009) and the Cultural Landscape of the Serra de Tramuntana (2011).

Spain shares the Pyrénées—Mount Perdu (1997 and 1999) with France, the Prehistoric Rock Art Sites in the Côa Valley and Siega Verde (1998) with Portugal and Heritage of Mercury: Almadén and Idrija (2012) with Slovenia.

Press

In 2008 there were 161 daily newspapers (140 paid-for and 21 free) with a total daily circulation of 8·21m. copies. The main paid-for titles are: *El País* (average daily circulation 435,000), *El Mundo* (336,000) and *As* (234,000), along with the dedicated sports paper, *Marca* (315,000). The leading free papers, notably *20 Minutos*, *Que!* and *ADN*, now have wider circulations than the paid-for dailies.

In 2009, 96,955 printed books were published.

Tourism

In 2010 Spain was behind only France, the USA and China in the number of foreign visitor arrivals, and behind only the USA for tourism receipts. In 2010, 52·7m. tourists visited Spain; receipts for 2010 amounted to US$52·5bn. In 2008 most tourists were from the UK (27·6%), followed by Germany (17·6%), France (14·2%), Italy (5·9%) and the Netherlands (4·3%). Of 268,552,000 overnight stays at hotels and inns in 2008, 49,633,000 were in the Balearics, 49,400,000 in the Canary Islands and 44,172,000 in Andalusia; overnight stays by visitors from abroad numbered 155,364,000 and by residents of Spain 113,118,000.

Festivals

Religious Festivals: Epiphany (6 Jan.), the Feast of the Assumption (15 Aug.), All Saints Day (1 Nov.) and Immaculate Conception (8 Dec.) are all public holidays. Cultural Festivals: Day of Andalusia (28 Feb.), the Feast of San José in Valencia (19 March) is the culmination of a 13-day festival; the Festival of the Sardine in Murcia is an end of Easter parade in which a huge papier mâché sardine is burned; Feria de Abril is a major festival in Seville at the end of April featuring flamenco dancing and bull-fighting; the San Fermines Festival, which takes place in mid-July, is most famous for the running of the bulls in the streets of Pamplona; La Tomatina, a battle of revellers armed with 50 tonnes of tomatoes, takes place on the last Wednesday in Aug. and is the highlight of the annual fiesta in Buñol, Valencia; National Day of Catalonia (11 Sept.); Spanish National Day (12 Oct.). The largest rock and pop music festivals are Festival Arte-Nativo Viña Rock in Villarrobledo (April), Primavera Sound (May) in Barcelona, Bilbao Live Festival (July) and Benicàssim International Festival (July).

DIPLOMATIC REPRESENTATIVES

Of Spain in the United Kingdom (39 Chesham Pl., London, SW1X 8SB)
Ambassador: Federico Trillo-Figueroa.

Of the United Kingdom in Spain (Torre Espacio, Paseo de la Castellana 259D, 28046 Madrid)
Ambassador: Giles Paxman, LVO.

Of Spain in the USA (2375 Pennsylvania Ave., NW, Washington, D.C., 20037)
Ambassador: Ramón Gil-Casares Satrústegui.

Of the USA in Spain (Serrano 75, 28006 Madrid)
Ambassador: Alan D. Solomont.

Of Spain to the United Nations
Ambassador: Fernando Arias González.

Of Spain to the European Union
Permanent Representative: Alfonso Dastis Quecedo.

FURTHER READING

Balfour, Sebastian, *The Politics of Contemporary Spain.* 2004
Barton, Simon, *A History of Spain.* 2nd ed. 2009
Carr, Raymond, (ed.) *Spain: A History.* 2000
Closa, Carlos and Heywood, Paul, *Spain and the European Union.* 2004
Conversi, D., *The Basques, The Catalans and Spain.* 1997
Gunther, Richard, *Democracy in Modern Spain.* 2004
Gunther, Richard and Montero, José Ramón, *The Politics of Spain.* 2009
Harrison, Joseph and Corkhill, David, *Spain: A Modern European Economy.* 2004
Heywood, P., *The Government and Politics of Spain.* 1995
Hooper, John, *The New Spaniards.* 2nd ed. revised. 2006
Payne, Stanley G., *Spain: A Unique History.* 2011
Péréz-Díaz, V. M., *The Return of Civil Society: the Emergence of Democratic Spain.* 1993
Phillips, William D., Jr and Phillips, Carla Rahn, *A Concise History of Spain.* 2010
Pierson, Peter, *The History of Spain.* 2008
Salvadó, Francisco J. Romero, *Twentieth-Century Spain: Politics and Society in Spain, 1898–1998.* 1999

National library: Biblioteca Nacional, Paseo de Recoletos, 20–22, 28071 Madrid.
National Statistical Office: Instituto Nacional de Estadística (INE), Paseo de la Castellana, 183, Madrid.
Website: http://www.ine.es

SRI LANKA

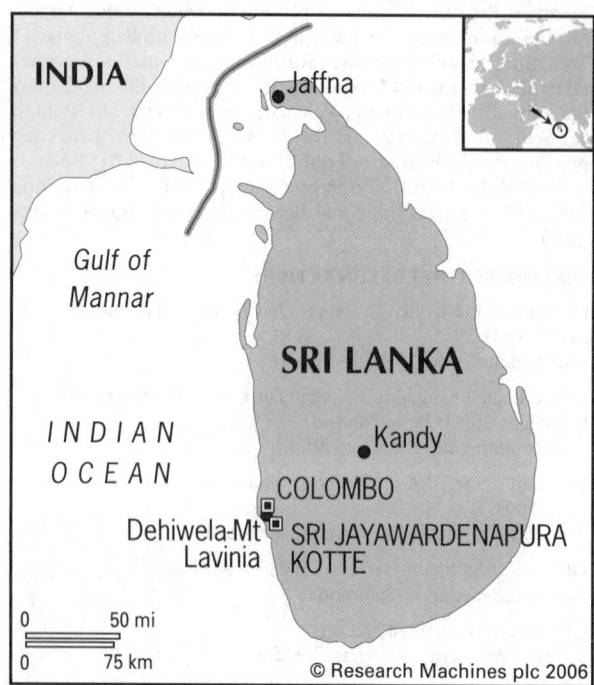

INDIA

Jaffna

Gulf of
Mannar

*INDIAN
OCEAN*

SRI LANKA

Kandy

COLOMBO
Dehiwela-Mt
Lavinia
SRI JAYAWARDENAPURA
KOTTE

0   50 mi
0   75 km

© Research Machines plc 2006

Sri Lanka Prajathanthrika Samajavadi Janarajaya
(Democratic Socialist Republic of Sri Lanka)

Capitals: Sri Jayawardenapura Kotte (Administrative and
Legislative), Colombo (Commercial)
Population projection, 2015: 21·71m.
GNI per capita, 2011: (PPP$) 4,943
HDI/world rank: 0·691/97
Internet domain extension: .lk

KEY HISTORICAL EVENTS

Archaeological evidence suggests Sri Lanka has been inhabited
since at least the Mesolithic era 34,000 years ago, possibly by
ancestors of the Vedda people, small numbers of whom live in the
central highlands. The island's recorded history begins in 483 BC,
when, according to the Sinhalese chronicle *Mahavamsa*, several
hundred men led by Vijaya, a prince from Bengal, reached the
island. Anuradhapura was founded in 377 BC, becoming the
principal settlement. Introduced in 250 BC by the Indian
Emperor, Ashoka, Buddhism was gradually adopted, becoming
central to the developing Sinhalese culture even while its influence
in India declined. Elaborate irrigation systems enabled rice
cultivation and brought prosperity to the northern plain around
Anuradhapura, which became a target for south Indian raiders.
South Indian Chola kings controlled Anuradhapura in the second
century BC until King Dutugemunu wrested control in 161 BC.

Invasions from south India continued while, at times, Sri
Lankan kings seized control of parts of southern India. A spate of
attacks in the 8th century AD prompted King Aggabodhi IV to
move Sri Lanka's seat of government to Polonnaruwa. A Chola
army under Rajarja I destroyed Anuradhapura in 993, although
Sinhalese control was re-established in 1070. Trade with southeast
Asia flourished in the 12th century under Parakramabahu I and
Sinhalese civilization reached its height under Nissanka Mala

(1187–96), the last Polonnaruwa king to rule the whole island.
During the 13th century a Tamil kingdom was established in the
northeast, centred on Jaffna. The seat of Sinhalese power
gradually shifted to the southwest, where coastal provinces thrived
on the spice trade with Arab merchants. King Alagakkonara
established a fort at Kotte in 1369.

The arrival of Portuguese mariners in 1505 heralded a new era
of European influence, initially centred on Kotte and the
southwest. Many coastal provinces, including Jaffna, came under
Portuguese rule from 1600, with missionaries promoting Roman
Catholicism. A military expedition failed to take the province of
Kandy in 1594 but Portuguese naval forces seized the Kandyan
ports of Trincomalee and Batticaola in the 1620s. Dutch seafarers
established a presence from the 1590s, gradually taking over
Portuguese-run forts and controlling Sri Lanka's maritime regions
from 1658. The trade in cinnamon was particularly lucrative. In
1670 the Dutch East Indian Company (VOC) declared a
monopoly over exports including elephants and pearls and
imports such as cotton. This prompted Kandyan attacks until a
truce was signed with King Rajasingha II in 1775.

As Dutch power in Asia waned, its possessions came under
attack from British and French forces. In 1796 Colombo,
Trincomalee, Jaffna and Kalpitya surrendered to British rule, with
control initially routed through Madras in India. In 1802 a
separate colony (Ceylon) under the British Crown was constituted
but an attempt to capture Kandy in 1803 failed. The Kandyan
Convention of 1815 annexed Kandy to British Ceylon while
recognizing most of the traditional rights of the chiefs.
Disillusionment culminated in a Kandyan rebellion in 1818 that
was harshly suppressed and the rights established by the
Convention abolished.

Ceylon was then unified under a single administration for the
first time in 400 years. Becoming aware of the island's economic
potential, the British established plantations, initially based on
coffee until an outbreak of *Hoemilia vastatrix* fungus in 1870
destroyed much of the crop. Large-scale cultivation of spices,
cocoa and rice followed but tea and rubber became the main cash
crops from the 1880s. When Kandyans refused plantation work,
thousands of indentured labourers were brought in from south
India.

As in other British colonies, the education system created an
English-speaking elite. Ethnic and caste rivalries continued,
notably between Kandyan villagers who were denied the welfare
benefits given to Indian Tamil labourers on the plantations.
Tensions between Sinhalese and Muslim merchants spilled into
violent riots in 1915. A movement for self-government gained
ground after the First World War through the Ceylon National
Congress, in which Sinhalese and Ceylon Tamil groups peacefully
negotiated with the British authorities.

Internal self-government under a directly elected State Council
was granted in 1931. Ceylon's economy was hit hard in the 1930s,
prompting demands for full independence. This followed on 4
Feb. 1948 with dominion status in the British Commonwealth.
D. S. Senanayake, leader of the United National Party (UNP),
became the first prime minister.

In 1956 Solomon Bandaranaike became prime minister at the
head of the People's United Front, advocating neutrality and the
promotion of Sinhalese national culture at home. Tamil demands
for official recognition of their language and a separate state
under a federal system caused riots in 1958, with considerable loss
of life. Bandaranaike was assassinated in Sept. 1959. His widow,
Sirimavo Bandaranaike, succeeded him the following year at the
head of an increasingly left-wing government. In May 1972
Ceylon became a republic and adopted the name Sri Lanka. In

July 1977 the UNP (dominant until 1956) returned to power and in 1978 a new constitution set up a presidential system. The problem of communal unrest remained unsolved and Tamil separatists were active. In 1983 Tamil United Liberation Front members of parliament were asked to renounce their objective for a separate Tamil state in the north and the east of the country. They refused and withdrew from parliament. Militant Tamils then began armed action that developed into civil war.

A state of emergency ended on 11 Jan. 1989 but violence continued. President Ranasinghe Premadasa was assassinated on 1 May 1993. A ceasefire was signed on 3 Jan. 1995 but fighting broke out again in April. The Tamil stronghold of Jaffna in the far north of the country was captured by government forces in Dec. 1995 and by mid-1997 was under government control. In April 2000 the Tamil Tigers captured a military garrison at Elephant Pass, the isthmus that links Jaffna to the rest of Sri Lanka, threatening a recapture of the Jaffna peninsula. A month-long ceasefire in Dec. 2001 led to peace negotiations. On 22 Feb. 2002 the government and Tamil Tiger leaders agreed to an internationally-monitored ceasefire, paving the way to the first full-scale peace talks for seven years. An estimated 61,000 people had died during the previous 19 years of conflict. In late 2002 the Tamil Tigers abandoned their ambitions for a separate state, settling instead for regional autonomy.

On 26 Dec. 2004 Sri Lanka, along with a number of other south Asian countries, was hit by a devastating tsunami. The death toll in Sri Lanka was put at 35,000.

A series of incidents in early 2006 brought the country back to the brink of civil war. Fighting intensified throughout the year, with the government claiming success in the east of the country. After failed peace talks in Geneva in Oct. 2006, the government withdrew from the 2002 ceasefire agreement in Jan. 2008.

In early 2009 Sri Lankan military forces launched a major offensive on the Tamil Tigers, during which the Tigers' leader, Velupillai Prabhakaran, was killed. On 19 May 2009 President Mahinda Rajapaksa formally declared an end to the civil war, stating the country had been liberated from terrorism. The UN estimated that around 7,000 Tamil civilians were killed in the 2009 military action while some 135,000 displaced people remained in camps in northern Sri Lanka in late 2009.

After a period of relative stability, in Aug. 2011 the government lifted a state of emergency in place since the assassination of the foreign minister, Lakshman Kadirgamar, in Aug. 2005.

TERRITORY AND POPULATION

Sri Lanka is an island in the Indian Ocean, south of the Indian peninsula from which it is separated by the Palk Strait. On 28 June 1974 the frontier between India and Sri Lanka in the Palk Strait was redefined, giving to Sri Lanka the island of Kachchativu.

Area (in sq. km) and population (2012 census):

District	Area	Population
Amparai	4,415	648,057
Anuradhapura	7,179	856,232
Badulla	2,861	811,758
Batticaloa	2,854	525,142
Colombo	669	2,309,809
Galle	1,652	1,058,771
Gampaha	1,387	2,294,641
Hambantota	2,609	596,617
Jaffna	1,025	583,378
Kalutara	1,598	1,217,260
Kandy	1,940	1,369,899
Kegalla	1,693	836,603
Kilinochchi	1,279	112,875
Kurunegala	4,816	1,610,299
Mannar	1,996	99,051
Matale	1,993	482,229
Matara	1,283	809,344
Moneragala	5,639	448,142
Mullaitivu	2,617	91,947

District	Area	Population
Nuwara Eliya	1,741	706,588
Polonnaruwa	3,293	403,335
Puttalam	3,072	759,776
Ratnapura	3,275	1,082,277
Trincomalee	2,727	378,182
Vavuniya	1,967	171,511
Total	65,610	20,263,723

Population (in 1,000) according to ethnic group and nationality in 2012 included: 15,173·8 Sinhalese, 2,270·9 Sri Lanka Tamils, 1,869·8 Sri Lanka Moors, 842·3 Indian Tamils, 40·2 Malays, 37·1 Burghers.

Of the population of 20,263,723 in 2012, 10,431,322 were females. Density, 309 per sq. km. In 2011, 14·3% of the population lived in urban areas.

The UN gives a projected population for 2015 of 21·71m.

Between the mid-1980s and the mid-1990s approximately 0·3m. Tamils left the country, one-third as refugees to India and two-thirds to seek political asylum in the West.

Colombo (the largest city) had an estimated 673,000 inhabitants in 2007. Other major towns and their populations (2007 estimates) are: Dehiwela-Mt Lavinia, 220,000; Moratuwa, 186,000; Jaffna, 152,000; Negombo, 150,000; Sri Jayawardenapura Kotte (now the administrative and legislative capital), 121,000; Kandy, 121,000; Kalmunai, 106,000; Galle, 95,000.

Sinhala and Tamil are the official languages; English is in use.

SOCIAL STATISTICS

Statistics for 2008: births, 373,575; deaths, 123,814. 2008 rates per 1,000 population: birth, 18·5; death, 6·1; infant mortality rate, 2010 (per 1,000 live births), 14. Life expectancy, 2007, 77·9 years for females and 70·3 for males. Annual population growth rate, 2000–05, 0·3%. Fertility rate, 2008, 2·3 births per woman. Sri Lanka has the third oldest population in Asia, after Japan and Singapore, thanks largely to relatively good health and a low fertility rate.

CLIMATE

Sri Lanka, which has an equatorial climate, is affected by the North-east Monsoon (Dec. to Feb.), the South-west Monsoon (May to July) and two inter-monsoons (March to April and Aug. to Nov.). Rainfall is heaviest in the southwest highlands while the northwest and southeast are relatively dry. Colombo, Jan. 79·9°F (26·6°C), July 81·7°F (27·6°C). Annual rainfall 95·4" (2,424 mm). Trincomalee, Jan. 78·8°F (26°C), July 86·2°F (30·1°C). Annual rainfall 62·2" (1,580 mm). Kandy, Jan. 73·9°F (23·3°C), July 76·1°F (24·5°C). Annual rainfall 72·4" (1,840 mm). Nuwara Eliya, Jan. 58·5°F (14·7°C), July 60·3°F (15·7°C). Annual rainfall 75" (1,905 mm).

On 26 Dec. 2004 an undersea earthquake centred off the Indonesian island of Sumatra caused a huge tsunami that flooded large areas along the southern and eastern coasts of Sri Lanka resulting in 35,000 deaths. In total there were more than 225,000 deaths in 14 countries.

CONSTITUTION AND GOVERNMENT

A new constitution for the Democratic Socialist Republic of Sri Lanka was promulgated on 7 Sept. 1978.

The executive *President* is directly elected for a six-year term. Under the terms of an amendment introduced in Sept. 2010, the previous bar on a president serving more than two terms was removed.

Parliament consists of one chamber, composed of 225 members (196 elected and 29 from the National List). Election is by proportional representation by universal suffrage at 18 years. The term of Parliament is six years. The Prime Minister and other Ministers, who must be members of Parliament, are appointed by the President.

National Anthem

'Sri Lanka Matha, Apa Sri Lanka' ('Mother Sri Lanka, thee Sri Lanka'); words and tune by A. Samarakone. There is a Tamil version, 'Sri Lanka thaaya, nam Sri Lanka'; words anonymous.

GOVERNMENT CHRONOLOGY

(UNP = United National Party; SLMP = Sri Lanka People's Party; SLFP = Sri Lanka Freedom Party; n/p = non-partisan)

Presidents since 1972.

1972–78	n/p	William Gopallawa
1978–89	UNP	Junius Richard Jayewardene
1989–93	UNP	Ranasinghe Premadasa
1993–94	UNP	Dingiri Banda Wijetunge
1994–2005	SLMP/SLFP	Chandrika Bandaranaike Kumaratunga
2005–	SLFP	Mahinda Rajapaksa

Prime Ministers since 1948.

1948–52	UNP	Don Stephen Senanayake
1952–53	UNP	Dudley Shelton Senanayake
1953–56	UNP	John Lionel Kotalawela
1956–59	SLFP	Solomon Ridgeway Dias Bandaranaike
1959–60	SLFP	Vijayananda Dahanayake
1960	UNP	Dudley Shelton Senanayake
1960–65	SLFP	Sirimavo Ratwatte Dias Bandaranaike
1965–70	UNP	Dudley Shelton Senanayake
1970–77	SLFP	Sirimavo Ratwatte Dias Bandaranaike
1977–78	UNP	Junius Richard Jayewardene
1978–89	UNP	Ranasinghe Premadasa
1989–93	UNP	Dingiri Banda Wijetunge
1993–94	UNP	Ranil Wickremasinghe
1994	SLMP/SLFP	Chandrika Bandaranaike Kumaratunga
1994–2000	SLFP	Sirimavo Ratwatte Dias Bandaranaike
2000–01	SLFP	Ratnasiri Wickremanayake
2001–04	UNP	Ranil Wickremesinghe
2004–05	SLFP	Mahinda Rajapaksa
2005–10	SLFP	Ratnasiri Wickremanayake
2010–	SLFP	Dissanayake Mudiyansalage Jayaratne

RECENT ELECTIONS

Presidential elections were held on 26 Jan. 2010. Incumbent President Mahinda Rajapaksa of the United People's Freedom Alliance (made up of several parties including the Sri Lanka Freedom Party) was re-elected with 57·9% of the vote, ahead of Sarath Fonseka, a former head of the army with 40·2%. There were 20 other candidates. Turnout was 74·5%.

At the parliamentary election of 8 and 20 April 2010 the United People's Freedom Alliance gained 144 seats with 60·3% of the vote; the United National Front (led by the Democratic People's Front) 60 with 29·3%; the Tamil National Alliance 14 with 2·9%; and the Democratic National Alliance 7 with 5·5%. Other parties received less than 1% of votes cast. Turnout was 61·3%.

CURRENT GOVERNMENT

In March 2013 the cabinet comprised:

President and Minister of Defence, Finance and Planning, Ports and Aviation, and Highways: Mahinda Rajapaksa; b. 1945 (Sri Lanka Freedom Party; sworn in 19 Nov. 2005 and re-elected in Jan. 2010).

Prime Minister and Minister of Buddha Sasana and Religious Affairs: D. M. Jayaratne; b. 1931 (Sri Lanka Freedom Party; sworn in 21 April 2010).

Senior Minister for Consumer Welfare: S. B. Navinne. *Food and Nutrition:* P. Dayaratne. *Good Governance and Infrastructure Facilities:* Ratnasiri Wickramanayake. *Human Resources:* D. E. W. Gunasekera. *International Monetary Co-operation:* Sarath Amunugama. *National Assets:* Piyasena Gamage. *Rural Affairs:* Athauda Seneviratne. *Scientific Affairs:* Prof. Tissa Vitharana. *Urban Affairs:* A. H. M. Fowzie.

Minister for Agrarian Services and Wildlife: S. M. Chandrasena. *Agriculture:* Mahinda Yapa Abeywardena. *Botanical Gardens and Public Recreation:* Jayaratne Herath. *Child Development and Women's Affairs:* Tissa Karaliyadde. *Civil Aviation:* Priyankara Jayaratna. *Coconut Development and State Plantations Development:* Jagath Pushpakumara. *Construction, Engineering Services, Housing and Common Amenities:* Wimal Weerawansa. *Co-operatives and Internal Trade:* Johnston Fernando. *Culture and Aesthetic Affairs:* T. B. Ekanayake. *Disaster Management:* Mahinda Amaraweera. *Economic Development:* Basil Rajapaksa. *Education:* Bandula Gunawardena. *Educational Service:* Duminda Dissanayake. *Environment and Renewable Energy:* Susil Premajayantha. *External Affairs:* Prof. G. L. Peiris. *Fisheries and Aquatic Resource Development:* Rajitha Senaratne. *Foreign Employment Promotion and Welfare:* Dilan Perera. *Health:* Maithripala Sirisena. *Higher Education:* S. B. Dissanayake. *Indigenous Medicine:* Salinda Dissanayake. *Industry and Commerce:* Rishad Bathiyutheen. *Investment Promotion:* Lakshman Yapa Abeywardena. *Irrigation and Water Resources Management:* Nimal Siripala de Silva. *Justice:* Rauff Hakeem. *Labour and Labour Relations:* Gamini Lokuge. *Land and Land Development:* Janaka Bandara Tennakoon. *Livestock and Rural Community Development:* Arumugam Thondaman. *Local Government and Provincial Councils:* A. L. M. Athaullah. *Mass Media and Information:* Keheliya Rambukwella. *National Heritage:* Jagath Balasuriya. *National Languages and Social Integration:* Vasudeva Nanayakkara. *Parliamentary Affairs:* Sumeda G. Jayasena. *Petroleum Industries:* Anura Priyadarshana Yapa. *Plantations:* Mahinda Samarasinghe. *Postal Services:* Jeevan Kumaranatunga. *Power and Energy:* Pavithra Wanniarachchi. *Private Transport Services:* C. B. Rathnayake. *Productivity Promotion:* Basir Segudawood. *Public Administration and Home Affairs:* W. D. J. Seneviratne. *Public Co-ordination and Public Affairs:* Mervin Silva. *Rehabilitation and Prison Reforms:* Chandrasiri Gajadeera. *Resettlement:* Gunaratne Weerakoon. *Small Export Crops Promotion:* Reginold Cooray. *Social Services:* Felix Perera. *Sports:* Mahindananada Aluthgamage. *State Assets and Enterprise Development:* Dayasritha Tissera. *State Management Reforms:* Navin Dissanayake. *Sugar Industry Development:* Lakshman Seneviratne. *Technology, Research and Atomic Energy:* Champika Ranawaka. *Telecommunication and Information Technology:* Ranjith Siyambalapitiya. *Traditional Industries and Small Enterprise Development:* Douglas Devananda. *Transport:* Kumara Welgama. *Water Supply and Drainage:* Dinesh Gunawardena. *Wildlife Conservation:* Gamini Vijith Vijayamuni Soysa. *Youth Affairs and Skills Development:* Dullas Alahaperuma.

Government Website: http://www.priu.gov.lk

CURRENT LEADERS

Mahinda Rajapaksa

Position
President

Introduction
Mahinda Rajapaksa succeeded Chandrika Kumaratunga as the executive president of Sri Lanka in Nov. 2005. A human rights lawyer and former prime minister, he rejected outright the demands of the Liberation Tigers of Tamil Eelam (LTTE) for an ethnic homeland and sought to crush the Tamil rebellion through military force. He achieved this objective by May 2009 and was re-elected president in Jan. 2010.

Early Life
Mahinda Rajapaksa was born on 18 Nov. 1945 in Weeraketiya in the southern district of Hambantota. He was educated at Richmond College, Galle, followed by Nalanda and Thurston

Colleges in Colombo. While studying law at Vidyodaya University he joined the centre-left Sri Lanka Freedom Party (SLFP) and in 1970 was elected as the party's parliamentary representative for Beliatta, Hambantota (a seat held by his father for the SLFP from 1948–65). Having graduated in 1974, Rajapaksa practised as a lawyer specializing in labour law and human rights and received plaudits for his work on behalf of the underprivileged.

Rajapaksa lost his parliamentary seat in the landslide defeat of the SLFP to the United National Party (UNP) in the general election of 1977. The UNP administration liberalized the economy and reduced unemployment but was unable to stem violence. The parliamentary elections in Feb. 1989 (in which Rajapaksa regained his seat) were preceded by terror campaigns by both the LTTE and the banned People's Liberation Front (JVP) in the south. Rajapaksa joined Mangala Samaraweera's 'Mother's Front', a group representing the mothers of those who 'disappeared' in the violence of 1988–89. He served on the central committee of the SLFP from the early 1990s and became an increasingly vocal critic of President Ranasinghe Premadasa's UNP government.

Following narrow victory for the SLFP (as part of the People's Alliance coalition) in the parliamentary elections of 1994, Rajapaksa was appointed minister for labour by President Chandrika Kumaratunga. His attempts to reform labour laws and introduce a workers' charter met with resistance. He was moved to the fisheries ministry, establishing a coast guard service and a university of oceanography. Following defeat for the People's Alliance in elections in Dec. 2001, Rajapaksa became leader of the parliamentary opposition. He forged alliances including, controversially, with the Sinhala-nationalist JVP to form the United People's Freedom Alliance (UPFA). The Alliance won the parliamentary elections that followed Kumaratunga's sacking of the UNP government in Feb. 2004. Kumaratunga then appointed Rajapaksa as prime minister and he was sworn in on 6 April 2004.

Career in Office
Without a commanding parliamentary majority, Rajapaksa's UPFA government struggled to implement its promises to halt privatization, increase wages and create new jobs. It was also criticized for its handling of the aftermath of the Indian Ocean tsunami in Dec. 2004, which killed 31,000 Sri Lankans and displaced nearly half a million.

Rajapaksa was chosen as the SLFP's presidential candidate for the election of Nov. 2005 and narrowly defeated the UNP's Ranil Wickremesinghe. He vowed a tougher approach to dealings with the LTTE, arguing that the 2002 ceasefire agreement had not brought peace, and appointed Ratnasiri Wickremanayake as prime minister. The security situation deteriorated seriously in 2006, and Rajapaksa reiterated his determination to defeat rebel violence as he revived draconian anti-terrorism legislation that had been suspended in 2002. Further escalation in 2007 of LTTE attacks and retaliatory offensives by state forces on LTTE positions in the north and east culminated in the government's formal abrogation of the 2002 ceasefire in Jan. 2008.

Rajapaksa intensified the military campaign against strategic LTTE positions through 2008, making significant territorial advances. In Jan. 2009 government troops captured Kilinochchi, the rebels' administrative headquarters, and also Elephant Pass linking the Jaffna peninsula with the mainland. Despite international concern over the safety of Tamil civilians trapped in the remaining LTTE-controlled enclave, government forces maintained their offensive (reportedly entering the last rebel-held town in Feb.) until Rajapaksa delivered a victory speech to parliament in May.

Meanwhile, the global financial crisis had a significant negative impact on the economy and in July 2009 the IMF approved a stand-by arrangement equivalent to US$2·6bn. to support recovery and help rebuild after the civil war.

A bitter breakdown in the relationship between Rajapaksa and his army chief Sarath Fonseka led the latter to resign and challenge the president in the elections in Jan. 2010. Capitalizing on his post-war popularity among the Sinhalese majority population, Rajapaksa was re-elected with almost 58% of the vote. The result was contested by Fonseka, but he was subsequently court-martialled, jailed and politically sidelined following his release in May 2012. In Sept. 2010 parliament endorsed a constitutional amendment allowing Rajapaksa to stand for an unlimited number of presidential terms.

In April 2011 Rajapaksa rejected as biased a United Nations report accusing both sides in the civil war of human rights abuses against civilians, although he did agree in Aug. to allow the expiry of long-standing and contentious state of emergency legislation. In March 2012 the UN Human Rights Council pressed the Sri Lankan government to investigate alleged serious violations committed during the final stages of the conflict.

Dissanayake Mudiyansalage Jayaratne

Position
Prime Minister

Introduction
On 21 April 2010 D. M. Jayaratne was sworn in as prime minister, a largely ceremonial position. He also serves as minister for Buddha Sasana and religious affairs. One of the country's longest serving politicians, he has headed various government ministries and is the senior member of the Sri Lanka Freedom Party (SLFP).

Early Life
D. M. Jayaratne was born on 7 June 1931 in the Central Province hill town of Gampola. The fifth child of nine, he was schooled at Doluwa Maha Vidyalaya (Gampola), Zahira College (Kandy) and Mahatma Gandhi College (Kandy). In 1951 he became a teacher at Doluwa Maha Vidyalaya and from 1960–62 he was postmaster of Gampola.

By 1950 he was politically active, working in the grassroots community centre networks where he rose through the ranks, acting as secretary in the Kandy council and then chair of the island-wide network. In 1951 he joined the newly-formed SLFP and in the 1970 general election he was elected to parliament as the representative for Gampola. In 1977 the SLFP suffered a landslide defeat and Jayaratne lost his seat. In 1989 he was elected MP for Kandy and appointed minister for agriculture, food and co-operatives.

Jayaratne was re-elected in 2000 and was reappointed to the agriculture, food and co-operatives portfolio. In 2004 he became minister of post and telecommunication and in 2007 took over at the ministry of plantation industries. He also served as chairman of the Asia-Pacific region of the Food and Agriculture Organization in 2001.

Career in Office
In April 2010 Jayaratne was appointed premier by President Mahinda Rajapaksa, heading up a seven-party United Popular Front coalition. With a coalition majority in parliament and backed by the popular president, Jayaratne began his tenure in a strong position. While his role is largely ceremonial, he is responsible for leading government business in parliament.

One of his key challenges was to oversee proposed constitutional reform, particularly the lifting of long-standing emergency powers and anti-terrorism laws. These, alongside reported human rights abuses and weak labour laws, have adversely affected the country's international trading relations, particularly with the European Union.

DEFENCE

Defence expenditure in 2008 totalled US$1,793m. (US$85 per capita), representing 4·2% of GDP. In 2008 Sri Lanka increased its military expenditure in real terms by 7·7%, the highest rate in South Asia.

Army

Strength (2007), 117,900 (including 39,900 recalled reservists). In addition there were 1,100 reserves. Paramilitary forces consist of the Ministry of Defence Police (61,600, including 1,000 women and a 3,000-strong anti-guerrilla force), the Home Guard (13,000) and the National Guard (some 15,000).

Navy

The main naval base is at Trincomalee. Personnel in 2007 numbered 15,000, including a reserve of about 2,400.

Air Force

Main Air Force bases are at Katunayake, Minneriya, Ratmalana, Vavuniya and China Bay, Trincomalee. Total strength (2007) 18,000 with 22 combat-capable aircraft and 13 attack helicopters. Main attack aircraft types included *Kfirs*, F-7s and MiG-27s.

ECONOMY

Agriculture accounted for 11·3% of GDP in 2006, industry 30·6% and services 58·0%.

The conflict with the minority separatists, the Tamil Tigers, which lasted 26 years, is estimated to have cost the country between 1–1·5% in growth per year.

Overview

Three decades of civil war disrupted inter-regional commerce, damaged infrastructure and weakened public finances. Nonetheless, the economy maintained growth, generating higher per capita income than neighbouring India.

In 2001 the economy contracted for the first time since independence as a result of the global slowdown, power shortages, and security and budgetary problems. With the start of the peace process in Feb. 2002, the government embarked on reforms to revive the economy, including exchange rate and trade liberalization, privatization and the introduction of VAT. In May 2009 the civil war ended, boosting economic prospects and providing renewed opportunity for reform.

The main sources of growth are domestic consumption, tourism and textile and garment exports. Tourism arrivals increased by 46·1% year-on-year in 2010. GDP growth averaged 6·5% from 2003–12 and the poverty rate has fallen, although the economy is strained by the costs of post-conflict relief and reconstruction. In July 2009 the IMF approved a US$2·6bn. loan to help the economy weather the effects of the financial crisis and to rebuild war-torn regions. Economic activity picked up in 2010, with real GDP growth reaching 8%.

At the end of 2011 public debt stood at 78·5% of GDP. Inflation was 9·5% in Aug. 2012, with the IMF expecting the rate to further increase in the coming years. About 60% of Sri Lankan exports go to Europe and the USA, making the economy vulnerable should there be a slow decline in demand from these regions.

Currency

The unit of currency is the *Sri Lankan rupee* (LKR) of 100 *cents*. Foreign exchange reserves were US$2,236m. and gold reserves 167,000 troy oz in June 2005. Inflation fell from 22·6% in 2008 to 3·5% in 2009 but increased to 6·2% in 2010 and 6·7% in 2011. Total money supply in March 2005 was Rs 208,095m.

Budget

Budgetary central government revenue and expenditure (in Rs 1bn.):

	2006	2007	2008
Revenue	506·9	595·0	686·5
Expenditure	621·5	717·4	847·4

Principal sources of revenue in 2007 were: taxes on goods and services, Rs 286·2bn.; taxes on income, profits and capital gains, Rs 107·2bn.; taxes on international trade and transactions, Rs 84·4bn. Main items of expenditure by economic type in

2007 were: compensation of employees, Rs 214·2bn.; interest, Rs 182·7bn.; social benefits, Rs 110·8bn.

VAT is 12%.

Performance

Real GDP growth was 3·5% in 2009, 7·8% in 2010 and 8·3% in 2011. Total GDP in 2011 was US$59·2bn.

Banking and Finance

The Central Bank of Sri Lanka is the bank of issue (*Governor*, Ajith Nivard Cabraal). There are 23 commercial banks with a total asset base of Rs 1,774bn. This includes two state-owned banks, the Bank of Ceylon and People's Bank, which account for about 40% of assets, and 12 foreign banks. There are 14 specialized banks including the National Savings Bank, which in total have an asset base of Rs 356bn. In the period 2005–10 Sri Lanka attracted US$2,989m. in foreign direct investment, including a record US$752m. in 2008.

Sri Lanka's foreign debt amounted to US$20,452m. in 2010, up from US$11,372m. in 2005 and US$9,087m. in 2000. However, as a proportion of GNI it declined from 56·7% in 2000 to 47·2% in 2005 and further to 41·8% in 2010.

There is a stock exchange in Colombo. In 2009 it was the best-performing stock market in the world, recording a gain of 125·2% on 2008.

ENERGY AND NATURAL RESOURCES

Environment

Carbon dioxide emissions from the consumption and flaring of fossil fuels in 2008 were the equivalent of 0·6 tonnes per capita.

Electricity

Installed capacity (2007), 2·4m. kW. Production, 2007, 9·8bn. kWh (40% hydro-electric). Consumption per capita in 2007 was 494 kWh.

Oil and Gas

There are plans for a huge oil refinery as part of the Hambantota port development.

Minerals

Gems are among the chief minerals mined and exported (particularly sapphires). Output of principal products: sapphires (2003), 773,547 carats; phosphate rock (2003), 41,357 tonnes; quartzite (2004), 18,139 tonnes; ilmenite (2004), 8,115 tonnes. Salt extraction is the oldest industry; the method is the solar evaporation of sea-water. Production, 2003, 78,713 tonnes.

Agriculture

There were approximately 970,000 ha. of arable land and 950,000 ha. of permanent crops in 2007. In 2006, 32·2% of the economically active population were engaged in agriculture. Main crops in 2003 (in 1,000 tonnes): rice, 3,071; coconuts, 1,850; sugarcane, 956; bananas and plantains, 610; tea, 303; cassava, 228; pumpkins and squash, 170. Sri Lanka ranks third in the world for tea production, behind India and China.

Livestock in 2003: 1,139,000 cattle; 635,000 buffaloes; 490,000 goats; 10m. chickens.

Forestry

The area under forests in 2010 was 1·86m. ha., or 29% of the land area. Of the total area under forests 9% was primary forest in 2010, 81% other naturally regenerated forest and 10% planted forest. In 2007, 6·13m. cu. metres of roundwood were cut.

Fisheries

Total catch in 2010 was 436,355 tonnes (of which 383,945 tonnes were from marine waters and 52,410 tonnes from inland waters).

INDUSTRY

The main industries are the processing of rubber, tea, coconuts and other agricultural commodities, tobacco, textiles, clothing and leather goods, chemicals, plastics, cement and petroleum refining. Industrial production rose by 8·1% in 2006.

Labour

The labour force in 2003 totalled 7,653,716 (67% males). In 2003 the economically active workforce numbered 7,012,755, of which 2,384,397 worked in agriculture, forestry and fishing, 1,156,682 in manufacturing and 867,131 in wholesale and retail trade, repair of motor vehicles and household goods. In 2003 the unemployment rate was 8·4%.

INTERNATIONAL TRADE

Imports and Exports

Trade in US$1m.:

	2005	2006	2007	2008	2009
Imports c.i.f.	8,307	9,773	11,386	13,629	9,432
Exports f.o.b.	6,160	6,760	7,661	8,177	7,122

Principal imports in 2009: petroleum and petroleum products, 18·5%; machinery and transport equipment, 16·6%; textile yarn, fabrics and finished articles, 15·2%; food and livestock, 14·2%. Principal exports in 2009: clothing and apparel, 45·9%; tea, 16·5%; diamonds and other precious and semi-precious stones, 5·1%; machinery and transport equipment, 4·3%;

In 2009 the leading import suppliers were India (18·0%), Singapore (11·7%), China (9·3%), Iran (9·0%) and Hong Kong (5·5%). The leading export markets in 2009 were the USA (22·3%), UK (14·4%), Italy (6·1%), Belgium (5·2%) and Germany (4·9%).

COMMUNICATIONS

Roads

In 2006 the road network totalled 91,907 km in length, including 11,716 km of national roads and 15,532 km of secondary roads. Number of motor vehicles, 2006, 2,269,575, comprising 338,608 passenger cars, 77,233 buses and coaches, 431,594 trucks and vans and 1,422,140 motorcycles and mopeds. There were 2,239 fatalities in road accidents in 2006.

Rail

In 2007 there were 1,463 km of railway (1,676 mm gauge). Passenger-km travelled in 2007 came to 4·77bn. and freight tonne-km to 135m.

Civil Aviation

There is an international airport at Colombo (Bandaranaike). The national carrier is SriLankan Airlines, which has been part-owned and managed by Emirates since 1998. Mihin Lanka, a low-cost airline fully owned and funded by the government, was launched in 2007. In 2006 SriLankan Airlines carried 2,900,068 passengers (all on international flights). Colombo handled 4,740,187 passengers and 169,038 tonnes of freight in 2006.

Shipping

In Jan. 2009 there were 36 ships of 300 GT or over registered, totalling 143,000 GT. Colombo is a modern container port; Galle and Trincomalee are natural harbours. The first of three phases of a new port at Hambantota was inaugurated in Nov. 2010. On completion it is set to be Sri Lanka's largest port.

Telecommunications

In 2008 there were 3,446,400 main (fixed) telephone lines; mobile phone subscribers numbered 11,082,500 in 2008 (55·2 per 100 persons). There were 1,163,500 internet users in 2008. In March 2012 there were 1·2m. Facebook users.

SOCIAL INSTITUTIONS

Justice

The systems of law which are valid are Roman-Dutch, English, Tesawalamai, Islamic and Kandyan.

Kandyan law applies in matters relating to inheritance, matrimonial rights and donations; Tesawalamai law applies in Jaffna as above and in sales of land. Islamic law is applied to all Muslims in respect of succession, donations, marriage, divorce and maintenance. These customary and religious laws have been modified by local enactments.

The courts of original jurisdiction are the High Court, Provincial Courts, District Courts, Magistrates' Courts and Primary Courts. District Courts have unlimited civil jurisdiction. The Magistrates' Courts exercise criminal jurisdiction. The Primary Courts exercise civil jurisdiction in petty disputes and criminal jurisdiction in respect of certain offences.

The Constitution of 1978 provided for the establishment of two superior courts, the Supreme Court and the Court of Appeal.

The Supreme Court is the highest and final superior court of record and exercises jurisdiction in respect of constitutional matters, jurisdiction for the protection of fundamental rights, final appellate jurisdiction in election petitions and jurisdiction in respect of any breach of the privileges of Parliament. The Court of Appeal has appellate jurisdiction to correct all errors in fact or law committed by any court, tribunal or institution.

The population in penal institutions in July 2007 was 25,537 (121 per 100,000 of national population). The death penalty, last used in 1976, was reactivated in Nov. 2004 after a 28-year moratorium.

Police

The strength of the police service in 2003 was 39,242.

Education

Education is free and is compulsory from age five to 14 years. The literacy rate in 2008 was 91%. Sri Lanka's rate compares very favourably with the rates of 65% in India and 54% in Pakistan.

In 2008 there were 9,335 schools with primary classes and 69,499 primary teachers for 1,626,285 pupils. The number of pupils in secondary classes in 2008 was 1,897,350, with 98,582 teachers. There were 401,666 students and 21,715 teachers in advanced level classes in 2008. There are 16 universities, including the Open University (distance) and the Buddhist and Pali University. In 2008, excluding at the two aforementioned universities, there were 66,675 undergraduates enrolled.

Health

In 2005 there were 609 hospitals and 413 central dispensaries. The hospitals had 60,237 beds. There were 9,290 physicians and 16,517 nurses in 2002; and 954 dentists, 907 pharmacists and 7,267 midwives in 2005. Total state budget expenditure on health, 2005, Rs 29,805m.

Welfare

To qualify for an old-age pension an individual must be above the age of 55 for men or 50 for women. However, a grant is payable at any age if the person is emigrating permanently. Old-age benefits are made up of a lump sum equal to total employee and employer contributions, plus interest.

The family allowances programme is being implemented in stages. Families earning below Rs 1,000 a month are entitled to Rs 100–Rs 1,000 a month benefit, depending on family size and income.

RELIGION

In 2012 the population was 70% Buddhist, 13% Hindu, 10% Muslim and 6% Roman Catholic. In Feb. 2013 there was one Roman Catholic cardinal.

CULTURE

World Heritage Sites

Sri Lanka has eight sites on the UNESCO World Heritage List: Sacred City of Anuradhapura (inscribed on the list in 1982); Ancient City of Polonnaruwa (1982); Ancient City of Sigiriya (1982); Sinharaja Forest Reserve (1988); Sacred City of Kandy (1988); Old Town of Galle and its Fortifications (1988); the Golden Temple of Dambulla (1991); and the Central Highlands (2010).

Press

In 2008 there were 18 paid-for daily newspapers with a combined circulation of 588,000. The papers with the highest circulation are *Lankadeepa* and *Divaina*, and the English-language *Daily News*.

Tourism

In 2010 there were a record 654,000 foreign tourists, bringing revenue of US$1,044m. The previous best year for tourist arrivals was 2004, the year of the Asian tsunami.

Festivals

Sri Lanka has an array of festivals principally centred on the Buddhist, Hindu, Muslim and Christian cultures. These include: Buddhist full moon celebrations (Poya Day), with the main festivals occurring in Jan., May, June, July/Aug. and Dec.; Diwali (the Hindu Festival of Lights, held in Oct./Nov.); Milad-un-Nabi (Dec.); and Christmas. New Year is also widely celebrated by both the Sinhalese and Tamil ethnic groups.

DIPLOMATIC REPRESENTATIVES

Of Sri Lanka in the United Kingdom (13 Hyde Park Gdns, London, W2 2LU)
High Commissioner: Dr Chrisantha Nicholas Anthony Nonis.

Of the United Kingdom in Sri Lanka (389 Bauddhaloka Mawatha, Colombo 7)
High Commissioner: John Rankin.

Of Sri Lanka in the USA (2148 Wyoming Ave., NW, Washington, D.C., 20008)
Ambassador: Jaliya Wickramasuriya.

Of the USA in Sri Lanka (210 Galle Rd, Kollupitiya, Colombo 3)
Ambassador: Michele J. Sison.

Of Sri Lanka to the United Nations
Ambassador: Palitha Kohona.

Of Sri Lanka to the European Union
Ambassador: Pakeer Mohideen Amza.

FURTHER READING

De Silva, C. R., *Sri Lanka: a History.* 1991
McGowan, W., *Only Man is Vile: the Tragedy of Sri Lanka.* 1992
Nira, Wickramasinghe, *Sri Lanka in the Modern Age: A History of Contested Identity.* 2005
Winslow, Deborah and Woost, Michael D., *Economy, Culture and Civil War in Sri Lanka.* 2004

National Statistical Office: Department of Census and Statistics, POB 563, Colombo 7.
Website: http://www.statistics.gov.lk

SUDAN

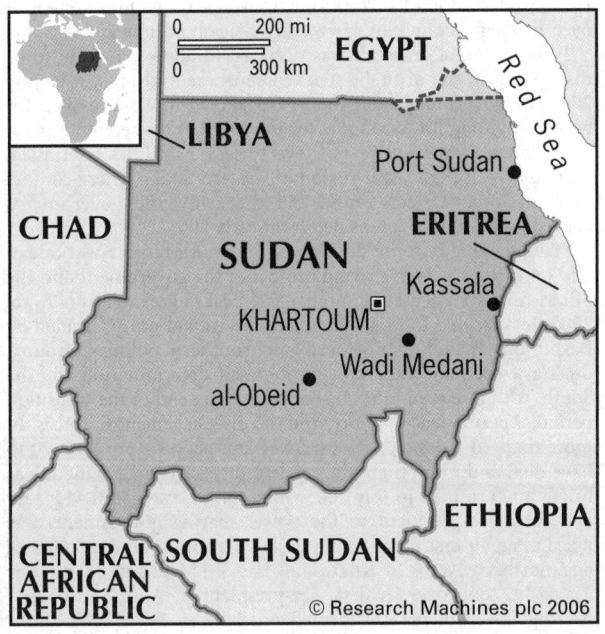

Jamhuryat es-Sudan
(The Republic of The Sudan)

Capital: Khartoum
Population projection, 2015: 49·07m.
GNI per capita, 2011: (PPP$) 1,894
HDI/world rank: 0·408/169
Internet domain extension: .sd

KEY HISTORICAL EVENTS

The earliest inhabitants of Sudan were Mesolithic hunter-gatherers, who lived and travelled in the region around Khartoum from as early as 30,000 BC. They had domesticated animals by 4000 BC. Cultural influences from Egypt rippled through to Nubia in north-eastern Sudan from around 3000 BC as Egypt's first dynasty moved south along the river Nile in search of construction materials and slaves. By 2000 BC it had reached as far south as the river Nile's fourth cataract, more than 700 km beyond Aswan. Egyptian-controlled Nubia was divided into Wawat in the north—centred on Aswan—and Kush in the south—based at Nepata (modern Marawi). When Egypt's power waned in the 11th century BC (the end of the New Kingdom) Kush, with its Egyptian and African influences, mineral resources and its position on trade routes linking the Nile to the Red Sea, became a powerful kingdom. At its height, under King Piantkhi in 750 BC, the whole of Egypt was brought under Kushite control. However, the invasion of Egypt by Assyrian forces in 671 BC forced a retreat to Nepata. From there, the kingdom of Kush continued to exert control over the middle Nile for much of the next millennium, developing a distinctive culture and language. By AD 200 Kush was in decline and was finally overthrown in 350 by the king of Aksum from the Ethiopian highlands.

Sudan was brought back into contact with the Mediterranean world in the 6th century by the arrival of Coptic Christian missionaries. They travelled south along the Nile and established churches in the three middle-Nile kingdoms that had superseded Kush: Nobatia in the north and Maqurrah and 'Alwah in the south, near modern Khartoum. Egypt was invaded by Arabs in 639 and came under Muslim rule. Raiding parties moved up the Nile and absorbed Nobatia. The king of Maqurrah engineered a truce at Dunqulah with an Arab military expedition, commanded by 'Abd Allah ibn Sa'd, preserving the kingdom for a further six centuries.

In 1250 Egypt came under the control of Mamluk sultans, supported by a caste of warrior slaves. They pushed south into Nubia, bringing chaos and devastation to Maqurrah and opening it up to waves of Arab immigrants, particularly the Juhaynah people. They intermarried with the Nubians and introduced Arab Muslim culture. Alwah, to the south, retained its Christian traditions until 1500 when, weakened by Bedouin raids, it collapsed under an Arab confederation led by Abd Allah Jamma. The Arabs themselves came under attack in the region around modern Khartoum from warriors of the Funj dynasty, a kingdom that had its origins in the Blue Nile's upper reaches. The Funj established their supremacy in the Al Jazirah region by 1607 and expanded northwards under Badi II Abu Daqn later in the 17th century. It was a relatively peaceful and stable period and the teaching of Islam flourished in schools and mosques along the Nile. The Funj themselves adopted Islam but retained a number of traditional African customs and beliefs.

Egyptian Ascendancy

In 1820 Muhammad Ali, viceroy of Egypt under the Ottoman Turks, sent an army southward to conquer Sudan. The Funj kingdom collapsed and within a year Ali's forces had taken control of the Nile valley from Nubia to the Ethiopian foothills. There was initial resistance but the appointment of Ali Kurshid Agha as governor general in 1826 led to the establishment of Khartoum as the administrative capital, improvements to agriculture and the development of the trade in slaves and ivory. Ismail Pasha became viceroy of Egypt in 1863 and announced a grand scheme to modernize and control the entire Nile river system from the Mediterranean to the Great Lakes of East Africa. Ismail needed the financial help of the European powers who, in return, demanded an end to the slave trade. The British had a particular interest in Egyptian affairs following the opening of the Suez Canal in 1869. Ismail commissioned the Englishmen Samuel Baker and, later, Charles Gordon to establish Egyptian control in southern Sudan and Central Africa and to crush slavery. The controllers of the slave trade were powerful and proved hard to defeat, especially away from the Nile. In addition, there was unease among Muslims over the crusading style of the English anti-slavers. In 1879, amid rising discontent, Ismail's financial backing collapsed and his grand project was abandoned. Ismail was exiled and Gordon resigned.

Rise of the Mahdi

The power vacuum was filled by Muhammad Ahmad in 1881; he declared himself the Mahdi ('divinely guided one') and led a movement that sought an end to Egyptian (Ottoman) influence and a return to the simplicity of early Islam. By 1882 the Mahdi had garnered the support of a least 30,000 armed followers (the Ansar). They captured the town of Al Abayyid, which led to a British order for the evacuation of Egyptians and foreigners from Khartoum, where the military commander was Charles Gordon. The campaign failed disastrously: Gordon was killed by Mahdists in early 1885. The Mahdi died in the same year, but his successor, the Khalifa Abdallahi, continued to build up the Mahdist state.

In the 1890s the European powers were vying for control of Africa and the British made plans for the control of the Nile valley and the reconquest of Sudan. In a series of attacks between 1896 and 1898, an Anglo-Egyptian force of 25,800 men under Herbert (later Lord) Kitchener destroyed the Mahdist state. Anglo-Egyptian agreements in 1899 established a joint (condominium) government of Sudan: in theory it was administered by a governor-general, appointed by Egypt with the consent of Great Britain. In practice the governor-general, Sir Reginald Wingate, controlled the condominium government from Khartoum. Sudanese resentment over colonial rule erupted in various Mahdist uprisings but with insufficient support to pose a threat to the government. In 1911 the Sudan Plantations syndicate launched a scheme to irrigate the Al Jazirah region and establish a large cotton plantation for Britain's textile industry. The cotton crop became the mainstay of the economy.

Sudanese nationalist sentiment grew in the early 1920s when 'Ali 'Abd al Latif, inspired by Egyptian nationalists, founded the White Flag League. When Governor-General Sir Lee Stack was assassinated in 1924 in Cairo (capital of newly-independent Egypt), the British ordered all Egyptian troops out of Sudan. British rule continued unchallenged until after the Second World War. The broad policy was to treat Sudan as two countries with the aim of integrating the southern provinces with their largely Christian and animist peoples with British East Africa. However, in 1948 a predominantly elective legislative assembly was convened for the whole territory. In the 1948 elections the Independence Front, which favoured the creation of an independent republic, gained a majority over the National Front, which aimed for union with Egypt.

Independence
Following the 1952 revolution in Egypt, Sudan became a parliamentary republic in 1956.

Sudan's democracy proved to be short-lived as the political parties became mired in internecine fighting. At the same time, a revolt raged in the south over Islamic domination. In 1958 Gen. Ibrahim Abboud led a military coup that ended the parliamentary system. In 1964, unable to improve Sudan's poor economic performance or to end the southern revolt, Abboud agreed to the re-establishment of civilian government. A coalition, headed by Muhammad Ahmad Mahjub, was weakened by factional disputes and little progress was made in solving the country's economic and social problems.

In 1969 Col. Muhammad Gaafar al-Nimeiry staged a successful coup. He banned all political parties and nationalized banks and numerous industries. The civil war was ended by an agreement between the government and the Southern Sudan Liberation Front signed in Addis Ababa in 1972. In the same year the Sudanese Socialist Union, the country's only political organization, elected a 'people's assembly' to draw up a new constitution, adopted in 1973. Nimeiry's regime was blamed for worsening economic conditions and came under Islamist attack for its support for Egypt's role in the Camp David Accords with Israel; in the late 1970s Nimeiry dismissed his cabinet and closed universities in an attempt to quell opposition.

Accelerating Violence
From the early 1980s political instability in southern Sudan worsened. Nimeiry responded by imposing Sharia law in 1983, inflaming a renewed civil war with the largely Christian and animist Sudan People's Liberation Movement (SPLM) led by John Garang. Having survived numerous earlier coup attempts, Nimeiry was overthrown in 1985. Following elections in 1986 a civilian government led by Sadiq al-Mahdi ruled until he was ousted three years later in a bloodless military coup. The new regime of Lieut.-Gen. Omar Ahmed al-Bashir strengthened ties with Libya, Iran, and Iraq, reinforced Islamic law, banned opposition parties and continued to pursue the war with the south against the backdrop of a stagnant economy and a catastrophic famine. Bashir officially became president in 1993 but significant political power was held by the National Islamic Front, a fundamentalist political organization led by Hassan al-Turabi, who became speaker of parliament. In 1999 Bashir declared a

state of emergency during a power struggle with Turabi, who was eventually toppled and parliament dissolved. Turabi subsequently formed his own opposition party, the Popular National Congress, although he and other party members were arrested in early 2001. He was released in Oct. 2003, only to be rearrested in March 2004 over an alleged coup plot. He was freed again in June 2005.

In Jan. 2002 a ceasefire was declared to allow relief aid to be distributed in the drought-stricken south-central region. In July 2002 the government and the SPLM agreed to a framework for peace providing for autonomy for the south and a referendum on independence after six years. Nevertheless, hostilities continued, particularly in the Darfur region of western Sudan where conflict between local African rebels and the army and government-backed Arab militias intensified from early 2003.

Having agreed on the division of oil wealth in post-war Sudan and on further power-sharing protocols, the government and the SPLM finally signed a comprehensive peace agreement on 9 Jan. 2005. It provided for a government of national unity (headed by Bashir but including northern and southern political groups) during a six-year transition period; self-determination for the South, with a referendum on secession at the end of the transition period; a permanent ceasefire; and the disengagement of forces. In more than 20 years of civil war, over 2m. people were thought to have died and more than 4m. made refugees. The agreement was briefly undermined in July 2005 when John Garang, SPLM leader and first vice-president in the power-sharing government, was killed in an air crash, provoking clashes between southern Sudanese and northern Arabs in Khartoum. His SPLM deputy, Salva Kiir Mayardit, took over as first vice-president in Aug. The power-sharing government was formed officially in Sept. 2005 and a devolved government of southern Sudan was established in Oct.

Although the comprehensive peace agreement focused mainly on the north-south civil war, some of its provisions for power-sharing and decentralization are applicable to Darfur. However, despite ongoing peace talks and UN and other international intervention to stop the violence, the conflict has continued. The government and Arab militias have been accused of systematic abuses of human rights, crimes that the UN Security Council referred to the International Criminal Court in March 2005. The targeting of civilians in the conflict has resulted in up to 300,000 deaths and the displacement of some 2·7m. people since 2003.

A referendum on full independence for the South took place on 9–15 Jan. 2011; 98·8% of votes were cast in favour of secession. South Sudan became an independent country on 9 July 2011.

TERRITORY AND POPULATION

Sudan is bounded in the north by Egypt, northeast by the Red Sea, east by Eritrea and Ethiopia, south by South Sudan, southwest by the Central African Republic, west by Chad and northwest by Libya. Its area is 1,881,000 sq. km. In 2008 the census population (provisional) was 39,154,490. In 2011, 40·8% of the population were urban.

The UN gives a projected population for 2015 of 49·07m.

The country is administratively divided into 17 states (with capitals): Al Qadarif (Al Qadarif), Blue Nile (Al Damazin), Central Darfur (Zalingei), East Darfur (Ed Daein), Gezira (Wadi Medani), Kassala (Kassala), Khartoum (Khartoum), North Darfur (Al Fashir), North Kordufan (Al Obeid), Northern (Dongola), Red Sea (Port Sudan), River Nile (Al Damar), Sennar (Singa), South Darfur (Nyala), South Kordufan (Kadugli), West Darfur (Geneina), White Nile (Rabak).

The capital, Khartoum, had a provisional census population of 1,410,858 in 2008. Other major cities, with 2008 provisional population, are Omdurman (1,849,659), Khartoum North (1,012,211), Nyala (492,984), Port Sudan (394,561), Al Obeid (345,126), Kassala (298,529), Wadi Medani (289,482) and Al Qadarif (269,395).

The country is mainly populated by Arab and Nubian peoples. Arabic and English are both official languages.

SOCIAL STATISTICS

2009 estimates: births, 1,402,000; deaths, 382,000. Rates, 2009 estimates (per 1,000 population): birth, 9; death, 33. Infant mortality, 2010 (per 1,000 live births), 66. Expectation of life in 2007 was 59·4 years for females and 56·3 for males. Annual population growth rate, 2000–05, 2·6%. Fertility rate, 2008, 4·2 births per woman.

CLIMATE

Lying wholly within the tropics, the country has a continental climate and only the Red Sea coast experiences maritime influences. Temperatures are generally high for most of the year, with May and June the hottest months. Winters are virtually cloudless and night temperatures are consequently cool. Summer is the rainy season inland, with amounts increasing from north to south, but the northern areas are virtually a desert region. On the Red Sea coast, most rain falls in winter. Khartoum, Jan. 64°F (18·0°C), July 89°F (31·7°C). Annual rainfall 6" (157 mm). Annual rainfall 39" (968 mm). Port Sudan, Jan. 74°F (23·3°C), July 94°F (34·4°C). Annual rainfall 4" (94 mm). Wadi Halfa, Jan. 50°F (10·0°C), July 90°F (32·2°C). Annual rainfall 0·1" (2·5 mm).

CONSTITUTION AND GOVERNMENT

The constitution was suspended after the 1989 coup and a 12-member Revolutionary Council then ruled. A 300-member Provisional National Assembly was appointed in Feb. 1992 as a transitional legislature pending elections. These were held in March 1996. On 26 May 1998 President Omar Hassan Ahmed al-Bashir approved a new constitution. Notably this lifted the ban on opposition political parties, although the government continued to monitor and control criticism until the constitution came legally into effect. The constitution was partially suspended in Dec. 1999.

In accordance with the peace deal agreed in Dec. 2004 to bring an end to the civil war and signed in Jan. 2005 there is a lower house, the 354-seat *National Assembly* (reduced from 450 following the independence of South Sudan), with members appointed by decree by the president, and an upper house, the *Council of States*, consisting of a maximum of 32 members (reduced from 52), of whom 30 are indirectly elected. The peace deal specified that 52% of National Assembly seats should go to the ruling National Congress Party, 28% to the former southern rebel Sudan People's Liberation Movement (the political wing of the Sudan People's Liberation Army), 14% to the northern opposition parties and 6% to their counterparts in the south. The agreement also allowed a referendum to be carried out in the south in 2011 to decide whether to become independent or maintain the unity of Sudan. This happened from 9–15 Jan. 2011, with 98·8% of votes cast in favour of independence. A new interim power-sharing constitution was adopted on 6 July 2005 giving the south some autonomy and allowing former rebels to take up seats in the country's government.

National Anthem

'Nahnu Jundullah, Jundu Al-Watlan' ('We are the Army of God and of Our Land'); words by A. M. Salih, tune by A. Murjan.

GOVERNMENT CHRONOLOGY

(DUP = Democratic Unionist Party; NCP = National Congress party; NUP = National Unionist Party; SSU = Sudan Socialist Union; Umma = 'Community of the Believers' Party; n/p = non-partisan)

Heads of State since 1956.

Commission of Sovereignty
1956–58　Abd al-Fattah Mohammad al-Mughrabi; Mohammad Uthman ad-Dardiri; Ahmad Mohammad Yasin; Ahmad Mohammad Salih; Siricio Iro Wani

Chairman of the Supreme Council of the Armed Forces
1958–64　military　　Ibrahim Abboud

Commission of Sovereignty (I)
1964–65　Abd al-Halim Mohammad; Tijani al-Mahi; Mubarak Shaddad; Ibrahim Yusuf Sulayman; Luigi Adwok Bong Gicomeho

Commission of Sovereignty (II)
1965　　Ismail al-Azhari; Abd Allah al-Fadil al-Mahdi, Luigi Adwok Bong Gicomeho; Abd al-Halim Mohammad; Khidr Hamad

Chairman of the Council of Sovereignty
1965–69　NUP　　Ismail al-Azhari

Chairmen of the Revolutionary Command Council
1969–71　military, SSU　Gaafur Muhammad al-Nimeiry
1971　　military, SSU　Abu Bakr an-Nur Uthman
1971　　military, SSU　Gaafur Muhammad al-Nimeiry

President
1971–85　military, SSU　Gaafur Muhammad al-Nimeiry

Chairman of the Transitional Military Council
1985–86　military　　Abd ar-Rahman Siwar ad-Dhahab

Chairman of the Council of Sovereignty
1986–89　DUP　　Ahmad Ali al-Mirghani

Chairman of the Revolutionary Command Council of National Salvation
1989–93　military　　Omar Hassan Ahmed al-Bashir

President
1993–　　military, NCP　Omar Hassan Ahmed al-Bashir

Heads of Government since 1952.

Chief Ministers
1952–53　Umma　　Abd ar-Rahman al-Mahdi
1954–56　NUP　　Ismail al-Azhari

Prime Ministers
1954–56　NUP　　　　Ismail al-Azhari
1956–58　Umma　　　Abd Allah Khalil
1958–64　military (de facto)　Ibrahim Abboud
1964–65　n/p　　　Sirr al-Khatim al-Khalifah
1965–66　Umma　　Muhammad Ahmad Mahgoub
1966–67　Umma　　Sadiq al-Mahdi
1967–69　Umma　　Muhammad Ahmad Mahgoub
1969　　n/p　　　Babiker Awadalla
1969–76　military, SSU　Gaafur Muhammad al-Nimeiry
1976–77　SSU　　Rashid Bakr
1977–85　military, SSU　Gaafur Muhammad al-Nimeiry
1985–86　n/p　　al-Jazuli Dafallah
1986–89　Umma　　Sadiq al-Mahdi

RECENT ELECTIONS

The first multi-party elections in 24 years were held from 11–15 April 2010 although they were boycotted by several of the main opposition parties. In presidential elections, incumbent Omar Hassan Ahmed al-Bashir was re-elected with 68·2% of votes cast. His nearest rival, Yasir Arman of the Sudan People's Liberation Movement (SPLM), gained 21·7% despite having withdrawn his candidature in late March owing to electoral irregularities and violence in Darfur. Abdullah Deng Nhial of the Popular Congress Party won 3·9% of the vote and Hatim Al-Sir of the Democratic Unionist Party 1·9%. There were eight other candidates who all gained less than 1% of the vote.

In concurrent National Assembly elections the National Congress Party of President Bashir won 323 of 450 seats, the SPLM 99, the Popular Congress Party 4, the Democratic Unionist Party 4 and the Federal Umma Party 3. Seven other parties received two seats or fewer with three seats going to independents. Four seats remained vacant.

CURRENT GOVERNMENT

President: Field Marshal Omar Hassan Ahmed al-Bashir; b. 1944 (NCP; appointed 1989, re-elected in March 1996, Dec. 2000 and April 2010).

First Vice-President: Ali Uthman Muhammad Taha. *Second Vice-President:* Al-Haj Adam Yousef.

In March 2013 the government comprised:

Minister of Agriculture and Irrigation: Abdul-Halim Ismail Al-Mutaafi. *Animal Resources and Fisheries:* Faisal Hassan Ibrahim. *Cabinet Affairs:* Ahmed Saad Omar Khadr. *Commerce:* Osman Omer Ali Al Shareef. *Culture and Information:* Ahmed Bilal Osman. *Defence:* Lieut.-Gen. Abdel Rahim Mohamed Hussein. *Environment, Forestry and Urban Development:* Hassan Abdel Qader Hilal. *Finance and National Economy:* Ali Mahmood Abdul-Rasool. *Foreign Affairs:* Ali Ahmed Karti. *Guidance and Endowments:* Al-Fateh Taj Essir Abdulla. *Health:* Bahar Idris Abu Gardah. *Higher Education and Scientific Research:* Khames Kajo Kundah. *Human Resources Development and Labour:* Ishraqa Sayed Mahmoud. *Industry:* Abdul Wahab Mohammed Osman. *Interior:* Ibrahim Mahmoud Hamid. *Justice:* Mohamed Bushara Dousa. *Minerals:* Kamal Abdul-Latif Abdul-Rahim. *Oil:* Awad Ahmed Al-Jaz. *Presidential Affairs:* Bakri Hassan Salih. *Public Education:* Su'ad Abdel Razik Mohamed Saeed. *Science and Communications:* Eissa Bushra Mohamed. *Tourism, Antiquities and Wildlife:* Mohamed Abdul Karim Al-Hud. *Transport, Roads and Bridges:* Ahmed Babiker Nahar. *Water Resources and Electricity:* Osama Abdalla Mohamed Al-Hassan. *Welfare and Social Security:* Amira Al-Fadil Mohamed Al-Fadil. *Youth and Sports:* Siddig Mohamed Tom Hammad.

Government Website: http://www.sudan.gov.sd

CURRENT LEADERS

Field Marshal Omar Hassan Ahmed al-Bashir

Position
President

Introduction
Omar al-Bashir is one of Africa's longest-serving presidents, having seized power in 1989. He has since been re-elected three times—in 1996, 2000 and 2010—although the polls were boycotted by the main opposition groups. His rule has been characterized by civil war and genocide, particularly in the western province of Darfur. He nevertheless signed a significant peace agreement in Jan. 2005 to end the long-running insurrection in South Sudan, which heralded the territory's drive to national independence in 2011.

Early Life
Bashir was born to a family of Sudanese peasants in 1944. He went to primary school in his home village before his family moved to Khartoum, where he completed secondary education. Having joined the Sudanese air force as a teenager, he soon made the grade as an officer and was sent to a military college in Egypt. He later served with a Sudanese unit that fought against Israel alongside Egyptian forces in the 1973 war. Promotions followed quickly and by the early 1980s he was a general. His political life began with the military coup of 1989 when, together with a group of middle-ranking officers, he overthrew the elected government of Sadiq al-Mahdi and installed a Revolutionary Command Council.

Career in Office
The National Islamic Front (later renamed the National Congress Party), led by Hassan al-Turabi, supported Bashir's new regime. Influenced by Turabi, and by the campaigns in South Sudan against animist and Christian secessionist rebels, Bashir began the Islamization of Sudan and introduced Sharia law. He moved Sudan into the radical Arab camp, inviting political and economic isolation and aggravating the distress caused by instability and civil war.

In 1993 Bashir declared himself president. Long-promised elections were held in March 1996, when Bashir was elected head of state in a poll that was regarded internationally as deeply flawed. For a time Sudan provided a haven for radical Islamic refugees who included al-Qaeda's Osama bin Laden. By 1998 Sudan was regarded as a pariah state by the USA. In that year, US missiles destroyed a pharmaceutical factory in Khartoum that was suspected, wrongly as it turned out, of producing chemical weapons.

In 1999 Bashir declared a state of emergency during a power struggle with Turabi, then the speaker of parliament, who had moved to reduce the president's powers. Turabi was toppled and parliament dissolved. In Dec. 2000 Bashir was re-elected as president, although the main opposition parties again boycotted the poll.

In the wake of the 11 Sept. 2001 attacks in the USA, Bashir made an effort to gain international acceptability. However, Sudan continued to be regarded by many in the West as a rogue state, crippled by poverty and conflict, and was also cited by the United Nations for gross human rights violations.

Meanwhile, the civil war in the South continued, by this time having claimed over 2m. lives and displaced more than 4m. refugees. In Jan. 2005, after three years of talks, Bashir's government and the Sudan People's Liberation Movement (SPLM) finally signed a comprehensive peace agreement. It provided for a government of national unity (headed by Bashir but also comprising members of the National Congress Party, the SPLM and other northern and southern political forces), self-determination for the South, a permanent ceasefire and the disengagement of forces. In July 2005 the agreement was briefly threatened when John Garang, the leader of the SPLM and first vice-president in the power-sharing government, was killed in a helicopter accident, provoking clashes between Sudanese in the south and Arab northerners. Garang was replaced as first vice-president by his deputy, Salva Kiir Mayardit, in Aug.

The formation of the national unity government was announced in Sept. 2005 and a devolved government in south Sudan was established in Oct. In Oct. 2007 the SPLM suspended its participation in the national unity administration, accusing the northern Sudanese of failing to honour the 2005 accord, but rejoined in Dec. In Jan. 2011 the results of a referendum in South Sudan on secession from the North showed an overwhelming vote in favour of self-determination. Bashir had earlier confirmed that he would accept the result, but the countdown to actual independence in July that year was marred by mutual recriminations between the North and South and violent clashes between their respective forces, particularly around the disputed oil-rich border area of Abyei. Post-independence, both sides agreed in Oct. to establish joint committees in an attempt to resolve outstanding flashpoints. However, in April 2012 Sudan's air force raided its neighbour after South Sudanese troops occupied the Abyei region and the Heglig oilfields. The United Nations threatened sanctions against both countries if they did not stop the escalating violence and return to negotiations. Bashir met the South Sudan president in Ethiopia in Sept. in an attempt to resolve the crisis, and although they concluded agreements on trade, security and a resumption of oil exports, they failed to resolve contested border issues.

In addition to the civil war in the South, Bashir's regime has overseen fierce fighting in the western province of Darfur between government-backed Arab militias and local black rebel forces. Since 2003 the conflict has escalated despite international attempts to stop the violence, the targeting of civilians having led to a refugee crisis and a humanitarian catastrophe. A UN report in March 2007 accused the government of orchestrating 'gross and systematic' human rights abuses in Darfur and complained that the response of the international community had been 'inadequate and ineffective'. After months of negotiations Bashir finally agreed on 31 Dec. 2007 to the deployment of a hybrid African Union-UN peacekeeping force (UNAMID, under UN control) in Darfur to replace an existing smaller African Union mission. The Darfur conflict also undermined Sudan's already fractious relations with neighbouring Chad until Jan. 2010, when both sides stated their readiness to normalize relations.

The International Criminal Court (ICC) has indicted Bashir for alleged genocide and war crimes in Darfur. Although the Sudanese government has rejected the accusation, violence across the region has continued to generate further casualties and internal displacement.

In April 2010 Bashir won another presidential term and the National Congress Party won a comfortable parliamentary majority in the country's first multi-party elections in 24 years, but the poll was criticized by international monitors.

In Oct. 2012 Sudan accused Israel of launching an attack on a munitions plant in Khartoum and complained to the UN Security Council.

DEFENCE

There is conscription for one to two years. Defence expenditure totalled US$524m. in 2006 (US$13 per capita), representing 1·5% of GDP. According to *Deadly Arsenals*, published by the Carnegie Endowment for International Peace, Sudan has both biological and chemical weapons research programmes.

In Aug. 2007 the Sudanese government agreed to a joint African Union/United Nations peacekeeping force being deployed in Darfur. The mission, known as UNAMID or UN-AU Mission in Darfur, began operations on 31 Dec. 2007 and had 20,852 uniformed personnel in Feb. 2013, making it the largest UN peacekeeping force in the world.

Army

Strength (2007) 105,000 (around 20,000 conscripts). There is a paramilitary People's Defence Force of 17,500 and an additional 102,500 reservists.

Navy

The navy operates in the Red Sea and also on the River Nile. The flotilla suffers from lack of maintenance and spares. Personnel in 2007 numbered 1,300. Major bases are at Port Sudan (HQ), Flamingo Bay and Khartoum.

Air Force

Personnel totalled (2007) 3,000, with 51 combat-capable aircraft, including A-5s (Chinese-built versions of MiG-19s), F-7s (Chinese-built versions of MiG-21s), MiG-23s and MiG-29s, and 23 attack helicopters.

ECONOMY

Agriculture accounted for 30·1% of GDP, industry 29·2% and services 40·8% in 2006.

Sudan featured among the ten most corrupt countries in the world in a 2012 survey of 176 countries carried out by the anti-corruption organization *Transparency International*.

Overview

In the 1990s the economy was near collapse. Greater stability came after 1997 with IMF-supported macroeconomic reforms. Until the turn of the century the economy was almost entirely agrarian. In 1999 a pipeline from the Muglad Basin to a Red Sea export terminal was opened and Sudan began exporting crude oil, recording its first trade surplus. Most oil is produced for export, with 65% going to China. 75% of oil production is now within the newly independent state of South Sudan.

When South Sudan became independent in July 2011, several key economic issues remain unresolved, including oil pricing. South Sudan needs pipelines in the north to distribute globally. Other key issues include external debt and water rights. With oil revenues set for a major decline, significant adjustments to a more diversified economy will be necessary to support Sudan's transition.

Oil production, infrastructure improvements, the development of export processing zones and increasing investment from Gulf

states have helped transform the economy. However, agriculture remains the leading sector, employing 80% of the population and accounting for a third of GDP.

Following the 2008 global financial crisis, GDP growth went into steep decline after a period of robust expansion. A high level of poverty and inequality remain, while Sudan's external position is vulnerable to fluctuations in oil prices and remittances from expatriates working in Gulf countries. The conflict in Darfur and tense relations with South Sudan are obstacles to economic development.

Currency

The unit of currency is the *Sudanese pound* (SDG) of 100 *piastres*, introduced in Jan. 2007 to replace the *Sudanese dinar* (SDD) at a rate of 1 Sudanese pound = 100 Sudanese dinars. The dinar ceased to be legal tender on 30 June 2007. Inflation, which has been in double figures every years since 2008, was 13·0% in 2010 and 18·3% in 2011. Foreign exchange reserves were US$2,107m. in July 2005 and total money supply was 705,505m. dinars.

Budget

In 2008 revenues totalled 26,424m. Sudanese pounds and expenditures 24,331m. Sudanese pounds.

VAT is 10%.

Performance

GDP growth was 4·6% in 2009 and 6·5% in 2010 but the economy contracted by 4·5% in 2011. Sudan's total GDP in 2011 was US$64·1bn.

Banking and Finance

The Bank of Sudan (*Governor*, Mohamed Khair Elzubair) opened in Feb. 1960 with an authorized capital of £S1·5m. as the central bank and bank of issue. Banks were nationalized in 1970 but in 1974 foreign banks were allowed to open branches. The application of Islamic law from 1 Jan. 1991 put an end to the charging of interest in official banking transactions, and seven banks are run on Islamic principles. Mergers of seven local banks in 1993 resulted in the formation of the Khartoum Bank, the Industrial Development Bank and the Savings Bank.

Sudan's external debt totalled US$21,846m. in 2010, up from US$17,474m. in 2005 and US$15,983m. in 2000. However, as a share of GNI it fell from 141·4% in 2000 to 67·1% in 2005 and further to 39·1% in 2010.

A stock exchange opened in Khartoum in 1995.

ENERGY AND NATURAL RESOURCES

Environment

Sudan's carbon dioxide emissions from the consumption and flaring of fossil fuels in 2008 were the equivalent of 0·3 tonnes per capita.

Electricity

Installed capacity was 1·1m. kW in 2007. Production in 2007 was 4·54bn. kWh, with consumption per capita 118 kWh.

Oil and Gas

In 2008 oil reserves totalled 6·7bn. bbls. In June 1998 Sudan began exploiting its reserves and on 31 Aug. 1999 it officially became an oil producing country; production in 2008 totalled 23·7m. tonnes. An oil refinery at Al-Jayli, with a capacity of 2·5m. tonnes, opened in 2000. Natural gas reserves in 2007 were 85bn. cu. metres.

Minerals

Mineral deposits include graphite, sulphur, chromium, iron, manganese, copper, zinc, fluorspar, natron, gypsum and anhydrite, magnesite, asbestos, talc, halite, kaolin, white mica, coal, diatomite (kieselguhr), limestone and dolomite, pumice, lead, wollastonite, black sands and vermiculite pyrites. Chromite and gold are mined. Production of salt, 2009: 35,793 tonnes; chromite, 2005: 21,654 tonnes; gold, 2005: 3,625 kg.

Agriculture

80% of the population depends on agriculture. Land tenure is based on customary rights; land is ultimately owned by the government. There were about 19·32m. ha. of arable land in 2007 and 225,000 ha. of permanent crops. 1·15m. ha. were irrigated in 2007. There were ten tractors and one harvester-thresher per 10,000 ha. in 2006.

Production (2003 estimates) in 1,000 tonnes: sugarcane, 5,500; sorghum, 5,188; groundnuts, 1,200; millet, 784; tomatoes, 700; wheat, 363; dates, 330; seed cotton, 315; cottonseed, 200.

Livestock (2003 estimates): sheep, 47·0m.; goats, 40·0m.; cattle, 38·3m.; camels, 3·2m.; chickens, 38m.

Livestock products (2003 estimates) in 1,000 tonnes: beef and veal, 325; lamb and mutton, 144; goat meat, 119; poultry meat, 30; cow's milk, 3,264; goat's milk, 1,295; sheep's milk, 463; cheese, 152; eggs, 47.

Forestry

Forests covered 69·95m. ha. in 2010, or 29% of the total land area. The annual loss of 589,000 ha. of forests between 1990 and 2000 was exceeded only in Brazil and Indonesia, although between 2000 and 2010 the pace of deforestation decreased significantly to 54,000 ha. annually. Of the total area under forests 20% was primary forest in 2010, 71% other naturally regenerated forest and 9% planted forest. In 2007, 20·28m. cu. metres of roundwood were cut.

Fisheries

In 2009 the total catch was 71,690 tonnes, of which approximately 92% were freshwater fish.

INDUSTRY

Production figures (2007 unless otherwise indicated, in 1,000 tonnes): distillate fuel oil, 1,873; petrol, 1,639; wheat and blended flour (2006), 1,200; raw sugar, 743; residual fuel oil, 677; cement, 326; jet fuel, 303; soap, 80. In 2000 an industrial complex assembling 12,000 vehicles a year opened.

Labour

The estimated total workforce in 2010 was 13,885,000 (70% males), up from 11,997,000 in 2005.

INTERNATIONAL TRADE

Imports and Exports

In 2009 imports (c.i.f.) amounted to US$8,589·9m. and exports (f.o.b.) to US$9,079·5m. The main imports are petroleum products, machinery and equipment, foodstuffs, manufactured goods, medicines and chemicals. Main exports are oil (which accounts for about 87% of export revenues), cotton, gum arabic, oil seeds, sorghum, livestock, sesame, gold and sugar. The main import sources in 2006 were China (18·8%), Saudi Arabia (8·9%), Japan (8·1%), India (6·7%) and United Arab Emirates (5·6%). Principal export markets in 2006 were China (78·9%), Japan (5·5%), United Arab Emirates (5·1%), Saudi Arabia (2·8%) and Egypt (1·3%).

COMMUNICATIONS

Roads

In 2002 there were estimated to be 11,900 km of roads, of which 4,320 km were paved. There were an estimated 768,000 passenger cars and 300,000 trucks and vans in 2007.

Rail

Total length in 2005 was 4,578 km. In 2008 the railways carried 100,000 passengers and 1·1m. tonnes of freight.

Civil Aviation

There is an international airport at Khartoum, which handled 2,178,097 passengers and 59,299 tonnes of freight in 2009. The national carrier is the government-owned Sudan Airways, which operates domestic and international services. In 2006 scheduled airline traffic of Sudan-based carriers flew 9m. km, carrying 563,000 passengers (365,000 on international flights).

Shipping

Supplementing the railways are regular steamer services of the Sudan Railways. Port Sudan is the major seaport; Suakin port opened in 1991. In Jan. 2009 there were five ships of 300 GT or over registered, totalling 23,000 GT.

Telecommunications

In 2008 there were 366,200 main (fixed) telephone lines; mobile phone subscribers numbered 11,991,500 in 2008 (29·0 per 100 persons). There were 4·2m. internet users in 2008.

SOCIAL INSTITUTIONS

Justice

The judiciary is a separate independent department of state, directly and solely responsible to the President of the Republic. The general administrative supervision and control of the judiciary is vested in the High Judicial Council.

Civil Justice is administered by the courts constituted under the Civil Justice Ordinance, namely the High Court of Justice—consisting of the Court of Appeal and Judges of the High Court, sitting as courts of original jurisdiction—and Province Courts—consisting of the Courts of Province and District Judges. The law administered is 'justice, equity and good conscience' in all cases where there is no special enactment. Procedure is governed by the Civil Justice Ordinance.

Justice for the Muslim population is administered by the Islamic law courts, which form the Sharia Divisions of the Court of Appeal, High Courts and Kadis Courts; President of the Sharia Division is the Grand Kadi. In Dec. 1990 the government announced that Sharia would be applied in the non-Muslim southern parts of the country as well.

Criminal Justice is administered by the courts constituted under the Code of Criminal Procedure, namely major courts, minor courts and magistrates' courts. Serious crimes are tried by major courts, which are composed of a President and two members and have the power to pass the death sentence. There were at least seven executions in 2011 and at least 19 in 2012. Major Courts are, as a rule, presided over by a Judge of the High Court appointed to a Provincial Circuit or a Province Judge. There is a right of appeal to the Chief Justice against any decision or order of a Major Court, and all its findings and sentences are subject to confirmation by him.

Lesser crimes are tried by Minor Courts consisting of three Magistrates and presided over by a Second Class Magistrate, and by Magistrates' Courts.

The population in penal institutions in mid-2009 was 19,144 (45 per 100,000 of national population).

Education

In 2007 there were 28,185 teaching staff for 490,808 pupils at pre-primary schools; 107,933 teaching staff for 3·96m. pupils at primary schools; and 79,122 secondary school teaching staff for 1·46m. pupils. In 2006 there were 29 public higher education institutions (including two Islamic universities, one university of science and technology, and an institute of advanced banking) and eight private higher education institutions. Adult literacy rate was 69% in 2008.

Health

In 2006 there were 11,083 physicians, 944 dentists, 33,354 nurses and midwives and 1,531 pharmacists. Hospital bed provision in 2006 was seven per 10,000 population.

RELIGION

Islam is the state religion. In 2001, 70% of the population were Sunni Muslims. There are also some Christians and traditional animists. In Feb. 2013 the Roman Catholic church had one cardinal.

CULTURE

World Heritage Sites

Sudan has two sites on the UNESCO World Heritage List: Gebel Barkal and the Sites of the Napatan Region (inscribed on the list in 2003), a collection of tombs, pyramids and palaces of the Second Kingdom of Kush (900 BC to AD 350); and the Archaeological Sites of the Island of Meroe (2011), once the heartland of the Kingdom of Kush that includes the royal city of the Kushite kings at Meroe.

Press

In 2008 there were 29 paid-for daily newspapers with a combined circulation of 90,000. Opposition newspapers are permitted although they are vetted by an official censor.

Tourism

In 2005 there were 246,000 foreign tourists, spending a total of US$89m.

DIPLOMATIC REPRESENTATIVES

Of Sudan in the United Kingdom (3 Cleveland Row, London, SW1A 1DD)
Ambassador: Abdullahi Hamad Ali Alazreg.

Of the United Kingdom in Sudan (off Sharia Al Baladia, Khartoum East)
Ambassador: Dr Peter Tibber.

Of Sudan in the USA (2210 Massachusetts Ave., NW, Washington, D.C., 20008)
Ambassador: Vacant.
Chargé d'Affaires a.i.: Emad Mirghani Altohamy.

Of the USA in Sudan (POB 699, Kilo 10, Soba, Khartoum)
Ambassador: Vacant.
Chargé d'Affaires a.i.: Joseph Stafford.

Of Sudan to the United Nations
Ambassador: Daffa-Alla Elhag Ali Osman.

Of Sudan to the European Union
Ambassador: Al-Tigani Salih Fedail.

FURTHER READING

Idris, Amir, *Conflict and Politics of Identity in Sudan.* 2006
Iyob, Ruth, *Sudan: The Elusive Quest for Peace.* 2006
Sidahmed, Alsir, *Sudan.* 2004
Woodward, Peter, *The Horn of Africa: Politics and International Relations.* 2002

National Statistical Office: Central Bureau of Statistics. PO Box 700, Khartoum.
Website: http://www.cbs.gov.sd

SURINAME

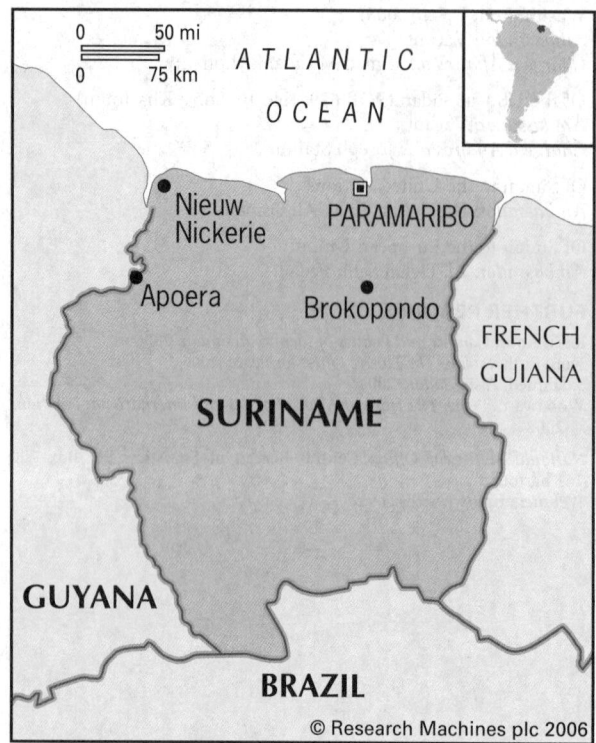

© Research Machines plc 2006

Republiek Suriname
(Republic of Suriname)

Capital: Paramaribo
Population projection, 2015: 548,000
GNI per capita, 2011: (PPP$) 7,538
HDI/world rank: 0·680/104
Internet domain extension: .sr

KEY HISTORICAL EVENTS

The first Europeans to reach the area were the Spanish in 1499 but it was the British who established a colony in 1650. At the peace of Breda (1667), Suriname was assigned to the Netherlands in exchange for the colony of New Netherland in North America. Suriname was twice in British possession during the Napoleonic Wars, in 1799–1802 and 1804–16, when it was returned to the Netherlands.

On 25 Nov. 1975 Suriname gained full independence. On 25 Feb. 1980 the government was ousted in a coup and a National Military Council (NMC) established. A further coup on 13 Aug. replaced several members of the NMC and the State President. Other attempted coups took place in 1981 and 1982, with the NMC retaining control. In Oct. 1987 a new constitution was approved by referendum. Following elections in Nov. Suriname returned to democracy in Jan. 1988 but on 24 Dec. 1990 a further military coup deposed the government. There was a peace agreement with rebel groups in Aug. 1992 and elections were held in May 1996.

TERRITORY AND POPULATION

Suriname is located on the northern coast of South America between 2–6° North latitude and 54–59° West longitude. It is bounded in the north by the Atlantic Ocean, east by French Guiana, west by Guyana, and south by Brazil. Area, 163,820 sq. km. Census population, 2004, 492,829; density, 3·0 per sq. km.

The UN gives a projected population for 2015 of 548,000.

The capital, Paramaribo, had (2004 census) 242,946 inhabitants.

Suriname is divided into ten districts. They are (with 2004 census population and chief town): Brokopondo, population 14,215 (Brokopondo); Commewijne, 24,649 (Nieuw Amsterdam); Coronie, 2,887 (Totness); Marowijne, 16,642 (Albina); Nickerie, 36,639 (Nieuw Nickerie); Para, 18,749 (Onverwacht); Paramaribo, 242,946—representing 49% of Suriname's total population (Paramaribo); Saramacca, 15,980 (Groningen); Sipaliwini, 34,136 (local authority in Paramaribo); Wanica, 85,986 (Lelydorp).

There is an ongoing unresolved dispute between Suriname and Guyana for the return of a triangle of uninhabited rainforest in Guyana between the New River and the Courantyne River, near the Brazilian border. In Sept. 2007 the UN settled a long-standing maritime boundary dispute between the two countries. The coastal area off both countries is believed to hold significant oil and gas deposits.

Major ethnic groups in percentages of the population in 2004: Indo-Pakistani, 26%; Creole, 18%; Javanese, 15%; Bushnegroes (Blacks), 15%; Amerindian, 4%. 69·8% of the population lived in urban areas in 2011.

The official language is Dutch. English is widely spoken next to Hindi, Javanese and Chinese as inter-group communication. A vernacular, called 'Sranan' or 'Surinamese', is used as a *lingua franca*. In 1976 it was decided that Spanish was to become the nation's principal working language.

SOCIAL STATISTICS

2007: births, 9,769; deaths, 3,374. Rates per 1,000 population: birth rate, 19·2; death rate, 6·6. Expectation of life, 2007, was 65·3 years for males and 72·5 for females. Annual population growth rate, 2000–05, 1·5%. Infant mortality, 2010, 27 per 1,000 live births; fertility rate, 2008, 2·4 births per woman. Abortion is illegal.

CLIMATE

The climate is equatorial, with uniformly high temperatures and rainfall. The temperature is an average of 27°C throughout the year; there are two rainy seasons (May–July and Nov.–Jan.) and two dry seasons (Aug.–Oct. and Feb.–April). Paramaribo, Jan. 21°C, July 32·4°C. Average rainfall 182·3 mm.

CONSTITUTION AND GOVERNMENT

The current constitution was ratified on 30 Sept. 1987. Parliament is a 51-member *National Assembly*. The head of state is the *President*, elected for a five-year term by a two-thirds majority by the National Assembly, or, failing that, by an electoral college, the United People's Assembly, enlarged by the inclusion of regional and local councillors, by a simple majority.

National Anthem

'God zij met ons Suriname' ('God be with our Suriname'); words by C. A. Hoekstra, tune by J. C. de Puy. There is a Sranan version, 'Opo kondreman oen opo'; words by H. de Ziel.

RECENT ELECTIONS

Parliamentary elections were held on 25 May 2010. The Mega Combination coalition (including the National Democratic Party) won 23 of the available 51 seats (40·2% of the vote), the New Front for Democracy and Development won 14 with 31·6%,

A-Combination took 7 (4·7%), the People's Alliance took 6 (13·0%) and the Party for Democracy and Development through Unity took 1 (5·1%).

On 19 July 2010 former dictator Dési Bouterse was elected president by the National Assembly, gaining 36 out of 50 votes against 13 for Chandrikapersad Santokhi with one abstention.

CURRENT GOVERNMENT

President: Dési Bouterse; b. 1945 (National Democratic Party; sworn in 12 Aug. 2010).

Vice-President: Robert Ameerali.

In March 2013 the government comprised:

Minister of Agriculture, Livestock and Fisheries: Hendrik Setrowidjojo. *Defence:* Lamuré Latour. *Education and Community Development:* Raymond Sapoen. *Finance:* Adeline Wijnerman. *Foreign Affairs:* Winston Lackin. *Health:* Celsius Waterberg. *Interior:* Soewarto Moestadja. *Justice and Police:* Martin Misiedjan. *Labour, Technological Development and Environment:* Ginmardo Kromosoeto. *Natural Resources:* Jim Hok. *Physical Planning, Land and Forestry Management:* Simon Martosatiman. *Public Works:* Ramon Abrahams. *Regional Development:* Linus Diko. *Social Affairs and Housing:* Alice Amafo. *Trade and Industry:* Michael Miskin. *Transport, Communication and Tourism:* Valisi Pinas. *Youth and Sports:* Paul Abena.

National Assembly Website (Dutch only): http://www.dna.sr

CURRENT LEADERS

Desiré Delano 'Dési' Bouterse

Position
President

Introduction
Former military dictator Dési Bouterse was sworn in as president in Aug. 2010 following his election by the National Assembly in July. He had earlier led the National Democratic Party to victory within the Mega Combination coalition at parliamentary elections in May that year. He succeeded Ronald Venetiaan.

Early Life
Bouterse was born on 13 Oct. 1945 in Domburg, Suriname. After attending the Middelbare Handelsschool, he moved to the Netherlands for military training.

On 25 Feb. 1980 a group of Surinamese army sergeants, including Bouterse, led a coup. By Aug. they had forced out President Johan Ferrier, who had overseen the nation's transition to independence from Dutch rule five years earlier. The military took over the government and declared a socialist republic, with Bouterse named chairman of the National Military Council. Over the next seven years he appointed puppet presidents while ruling as *de facto* leader.

Democratic elections in 1988 stripped Bouterse of much of his power, though a new constitution in 1987 confirmed his control of the military. In Dec. 1990 he dismissed the government, holding power until elections the following year. His position weakened as civil war took hold and a strong ethnic Maroon rebel movement developed under Ronnie Brunswijk, a former bodyguard of Bouterse.

Bouterse gradually entered into mainstream politics as leader of the National Democratic Party, which he had established in 1987. However, he was dogged by allegations of extra judicial murders following the 1980 coup. Convicted in absentia in the Netherlands for drug trafficking offences, he was sentenced in 1999 to 11 years imprisonment.

Campaigning on a populist platform, he secured the support of several small parties and formed the Mega Combination coalition to contest the general election of May 2010, winning 23 of a possible 51 seats. In the presidential election of July 2010, Bouterse secured 36 parliamentary votes, exceeding the two-thirds

majority needed to defeat the minister of justice, Chandrikapersad Santokhi.

Career in Office
Bouterse's election was widely criticized, with the suggestion that his run for the presidency was motivated by a desire to secure immunity from criminal prosecution. In April 2012 the National Assembly passed an amnesty law covering alleged violations under his earlier military rule. He has appealed to the nation's poor, blaming their plight on the austerity measures of previous governments.

DEFENCE

In 2008 defence expenditure totalled US$31m. (US$65 per capita), representing 1·3% of GDP.

Army
Total strength was 1,400 in 2007.

Navy
In 2007 personnel, based at Paramaribo, totalled around 240.

Air Force
Estimated personnel (2007): 200. There were four combat-capable aircraft.

ECONOMY

In 2006 agriculture contributed 5·2% of GDP, industry 35·7% and services 59·1%.

Overview
The economy is heavily dependent on the export of oil, gold and alumina, which jointly account for over 50% of GDP and about 85% of exports.

After a slowdown in 2008–09, growth picked up in 2010 thanks largely to the mineral sector. Rising food and fuel prices combined with civil service wage increases contributed to a rise in inflation from 1·3% at the end of 2009 to 10·3% at the end of 2010. Public debt increased from 18·0% of GDP in 2008 to 21·6% in 2010, while the fiscal account went from a surplus in 2007–08 to a deficit in 2009–10. In Jan. 2011 the government devalued the currency by 20%, raised fuel taxes by 40% and introduced other deficit-reduction measures.

The IMF projects a favourable economic outlook with strong investment in the mineral and energy sectors as well as high commodity prices. Containing rising inflation is a priority.

Currency
The unit of currency is the *Suriname dollar* (SRD) of 100 *cents*, introduced on 1 Jan. 2004 to replace the *Suriname guilder* (SRG) at a rate of one Suriname dollar = 1,000 Suriname guilders. Foreign exchange reserves totalled US$140m. and gold reserves were 29,000 troy oz in July 2005. Total money supply in June 2005 was 777m. Suriname dollars. The rate of inflation—which had been 98·7% in 1999—was 15·0% in 2008, falling to 0·0% in 2009. In 2010 the rate increased to 6·9% and in 2011 to 17·7%.

Budget
2007 revenue was 2,002m. Suriname dollars and expenditure 1,806·5m. Suriname dollars. Tax revenue accounted for 79·1% of revenues in 2007; current expenditure accounted for 87·5% of expenditures.

VAT is 10% for goods and 8% for services.

Performance
After two years of recession in 1999 and 2000 the economy has enjoyed steady growth. Real GDP growth was 4·1% in 2010 and 4·2% in 2011. In 2010 total GDP was US$4·4bn.

Banking and Finance
The Central Bank of Suriname (*Governor*, Gilmore Hoefdraad) is a bankers' bank and also the bank of issue. There are three

commercial banks; the Suriname People's Credit Bank operates under the auspices of the government. There is a post office savings bank, a mortgage bank, an investment bank, a long-term investments agency, a National Development Bank and an Agrarian Bank.

ENERGY AND NATURAL RESOURCES

Environment
Suriname's carbon dioxide emissions from the consumption and flaring of fossil fuels in 2008 were the equivalent of 4·5 tonnes per capita.

Electricity
Installed capacity in 2007 was 0·4m. kW. Production (2007) 1·62bn. kWh; consumption per capita in 2007 was 3,173 kWh.

Oil and Gas
Crude oil production (2007), 656,000 tonnes. Reserves in 2007 were 111m. bbls.

Minerals
Bauxite is the most important mineral. Suriname is the seventh largest bauxite producer in the world. Production (2007), 5,054,000 tonnes.

Agriculture
Agriculture is restricted to the alluvial coastal zone; in 2002 there were an estimated 57,000 ha. of arable land and 10,000 ha. of permanent crops. The staple food crop is rice: production, 163,000 tonnes in 2003. Other crops (2002 in 1,000 tonnes): sugarcane, 120; oranges, 11; plantains, 11; coconuts, 10; cassava, 4. Livestock in 2003 (estimates): cattle, 137,000; pigs, 24,000; sheep, 8,000; goats, 7,000; chickens, 4m.

Forestry
Forests covered 14·76m. ha. in 2010, or 95% of the land area. In terms of percentage coverage, Suriname was the world's most heavily forested country in 2010. Of the total area under forests 95% in 2010 was primary forest and 5% other naturally regenerated forest. Production of roundwood in 2007 was 214,000 cu. metres.

Fisheries
The catch in 2010 amounted to 34,402 tonnes, almost entirely from marine waters.

INDUSTRY
There is no longer any aluminium smelting, but there are food-processing and wood-using industries. Production: alumina (2007), 2,181,000 tonnes; residual fuel oil (2007), 360,000 tonnes; cement (2008 estimate), 65,000 tonnes; distillate fuel oil (2007), 41,000 tonnes; sawnwood (2007), 57,000 cu. metres.

Labour
Out of 156,705 people in employment in 2004, 27,995 were in public administration and defence; 25,012 in wholesale and retail trade; 14,031 in construction; 12,593 in agriculture, fishing, hunting and forestry; and 10,971 in manufacturing. In 2004 there were 16,425 unemployed persons, or 9·5% of the workforce.

INTERNATIONAL TRADE

Imports and Exports
Trade in US$1m.:

	2007	2008	2009	2010
Imports c.i.f.	1,044·3	1,304·3	1,390·1	1,397·5
Exports f.o.b.	1,359·1	1,743·4	1,401·8	2,025·6

Principal imports, 2004: nonelectrical machinery, 14·4%; food products, 11·9%; road vehicles, 9·5%. Principal exports, 2004: alumina, 40·8%; gold, 29·3%; crustaceans and molluscs, 3·6%.

In 2004 imports (in US$1m.) were mainly from the USA (165·2), Netherlands (147·8), Trinidad and Tobago (131·4), Japan (96·6) and the Netherlands Antilles (20·8); exports were mainly to Norway (173·0), USA (138·0), Canada (73·3), France (60·0) and the Netherlands (14·5).

COMMUNICATIONS

Roads
The road network covered 4,304 km in 2003, of which 26·2% were paved. In 2006 there were 81,778 passenger cars, 25,745 trucks and vans, 3,029 buses and coaches and 40,889 motorcycles and mopeds. There were 69 fatalities in road accidents in 2004.

Rail
There are two single-track railways.

Civil Aviation
There is an international airport at Paramaribo (Johan Adolf Pengel). The national carrier is Surinam Airways, which in 2003 had flights to Amsterdam, Belem, Cayenne, Curaçao, Georgetown, Haiti, Miami and Port of Spain. In 2003 scheduled airline traffic of Suriname-based carriers flew 5m. km, carrying 258,000 passengers (253,000 on international flights). In 2004 there were 149,589 passenger arrivals and 148,353 departures.

Shipping
In Jan. 2009 there were four ships of 300 GT or over registered, totalling 4,000 GT. In 2004 vessels totalling 1,518,000 NRT entered ports and vessels totalling 2,142,000 NRT cleared.

Telecommunications
Telephone subscribers numbered 401,500 in 2006, equivalent to 888·3 for every 1,000 persons. There were 320,000 mobile phone subscribers in 2006 and 38,000 internet users.

SOCIAL INSTITUTIONS

Justice
Members of the court of justice are nominated by the President. There are three cantonal courts. Suriname was one of ten countries to sign an agreement in Feb. 2001 establishing a Caribbean Court of Justice to replace the British Privy Council as the highest civil and criminal court. In the meantime the number of signatories has risen to 12. The court was inaugurated at Port-of-Spain, Trinidad on 16 April 2005 but it has not yet been accepted as Suriname's final court of appeal.

The estimated population in penal institutions in Dec. 2009 was 915 (175 per 100,000 of national population).

Education
Adult literacy was 94·6% in 2008 (95·5% among males and 93·8% among females). In 2003–04, 298 primary schools out of a total of 312 had 3,096 teachers and 62,086 pupils; 124 secondary schools had 41,904 pupils. In 2000–01 the university had 2,745 students. There is a teacher training college with (2000–01) 1,942 students.

Health
In 2004 there were 1,611 general hospital beds. In 2003 there were 295 physicians.

RELIGION
At the 2004 census there were 200,744 Christians of varying denominations, 98,240 Hindus and 66,307 Muslims.

CULTURE

World Heritage Sites
Suriname has two sites on the UNESCO World Heritage List: Central Suriname Nature Reserve (inscribed on the list in 2000); and the Historic Inner City of Paramaribo (2002).

Press
There were four daily newspapers in 2008 with a combined circulation of 55,000.

Tourism

In 2005 there were 160,000 non-resident tourist arrivals; tourist receipts totalled US$96m.

Festivals

The people of Suriname celebrate Chinese New Year (Jan.); Phagwa, a Hindu celebration (March–April); Id-Ul-Fitre, the sugar feast at the end of Ramadan (May); Avondvierdaagse, a carnival (during the Easter holidays); Suriflora, a celebration of plants and flowers (April–May); Keti koti, an Afro-Surinamese holiday to commemorate the abolition of slavery (1 July); Suri-pop, a popular music festival (July); Nationale Kunstbeurs, arts and crafts (Oct.–Nov.); Diwali, the Hindu ceremony of light (Nov.); Djaran Kepang, a Javanese dance held on feast days; Winti-prey, a ceremony for the Winti gods.

DIPLOMATIC REPRESENTATIVES

Of Suriname in the United Kingdom
Ambassador: Harvey Harold Naarendorp (resides in The Hague).
Honorary Consul: Dr Amwedhkar Jethu (89 Pier House, 31 Cheyne Walk, London, SW3 5HN).

Of the United Kingdom in Suriname
Ambassador: Andrew Ayre (resides in Georgetown, Guyana).

Of Suriname in the USA (4301 Connecticut Ave., NW, Washington, D.C., 20008)
Ambassador: Subhas-Chandra Mungra.

Of the USA in Suriname (Dr Sophie Redmondstraat 129, Paramaribo)
Ambassador: Jay N. Anania.

Of Suriname to the United Nations
Ambassador: Henry Leonard Mac-Donald.

Of Suriname to the European Union
Ambassador: Wilfred Eduard Christopher.

FURTHER READING

Dew, E. M., *Trouble in Suriname, 1975–1993.* 1995

National Statistical Office: Algemeen Bureau voor de Statistiek, POB 244, Paramaribo.
Website (limited English): http://www.statistics-suriname.org

Explore the world at
www.statesmansyearbook.com

SWAZILAND

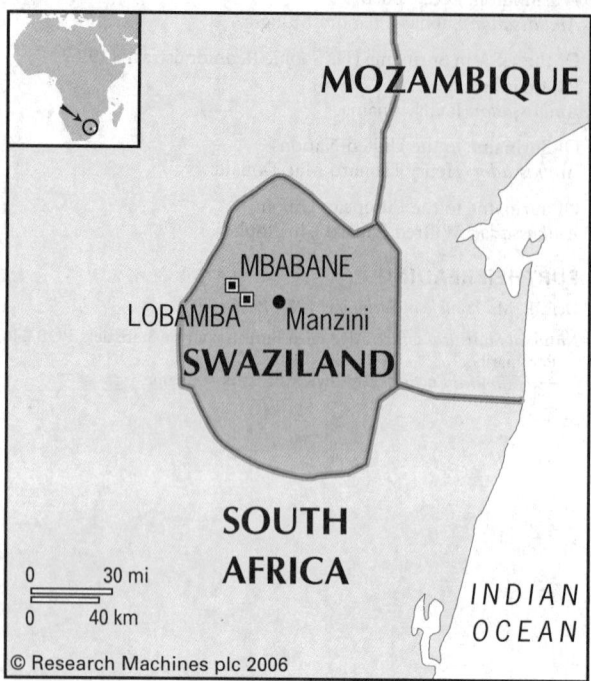

© Research Machines plc 2006

Umbuso weSwatini
(Kingdom of Swaziland)

Capitals: Mbabane (Administrative), Lobamba (Legislative)
Population projection, 2015: 1·27m.
GNI per capita, 2011: (PPP$) 4,484
HDI/world rank: 0·522/140
Internet domain extension: .sz

KEY HISTORICAL EVENTS

The Swazi migrated into the country to which they have given their name in the last half of the 18th century. The independence of the Swazis was guaranteed in the conventions of 1881 and 1884 between the British Government and the Government of the South African Republic. In 1894 the South African Republic was given powers of protection and administration. In 1902, after the conclusion of the Boer War, a special commissioner took charge, and under an order-in-council in 1903 the Governor of the Transvaal administered the territory. Swaziland became independent on 6 Sept. 1968. A state of emergency imposed in 1973 is still in force. On 25 April 1986 King Mswati III was installed as King of Swaziland.

TERRITORY AND POPULATION

Swaziland is bounded in the north, west and south by South Africa, and in the east by Mozambique. The area is 17,364 sq. km (6,704 sq. miles). Population (2007 census), 1,018,449 (537,021 females); density, 58·7 per sq. km.

The UN gives a projected population for 2015 of 1·27m.

In 2011, 21·3% of the population were urban. The country is divided into four regions: Hhohho, Lubombo, Manzini and Shiselweni.

Main urban areas: Mbabane, the administrative capital (60,281 inhabitants in 2007); Manzini; Big Bend; Mhlume; Nhlangano.

The population is 84% Swazi and 10% Zulu. The official languages are Swazi and English.

SOCIAL STATISTICS

2008 estimates: births, 35,000; deaths, 18,000. Estimated rates, 2008 (per 1,000 population): births, 29·9; deaths, 15·6. As a result of the impact of AIDS, expectation of life declined sharply. It was 59 years in 1990–95, but by 2000–05 was down to 45·9 years for females and 45·6 years for males. However, it has now started to rise again and in the period 2005–10 was 47·0 years for females and 47·6 years for males. In 2009, 25·9% of all adults between 15 and 49 were infected with HIV—the highest rate in any country. In 2010, 23% of Swazi children were orphans. In Sept. 2001 King Mswati III told the teenage girls of the country to stop having sex for five years as part of the country's drive to reduce the spread of HIV. Annual population growth rate, 2000–08, 1·0%. Infant mortality, 2010, 55 per 1,000 live births; fertility rate, 2008, 3·5 births per woman.

CLIMATE

A temperate climate with two seasons. Nov. to March is the wet season, when temperatures range from mild to hot, with frequent thunderstorms. The cool, dry season from May to Sept. is characterized by clear, bright sunny days. Mbabane, Jan. 68°F (20°C), July 54°F (12·2°C). Annual rainfall 56" (1,402 mm).

CONSTITUTION AND GOVERNMENT

The reigning King is **Mswati III** (b. 1968; crowned 25 April 1986), who succeeded his father, King Sobhuza II (reigned 1921–82). The King rules in conjunction with the Queen Mother (his mother, or a senior wife). Critics of the King or his mother run the risk of arrest.

A new constitution was signed into law on 26 July 2005 and came into force in Jan. 2006. There is a *House of Assembly* of 66 members, 55 of whom are elected each from one constituency (*inkhundla*) and ten appointed by the King plus the Speaker; and a *House of Senators* of 30 members, ten of whom are elected by the House of Assembly and 20 appointed by the King. Elections are held in two rounds, the second being a run-off between the five candidates who come first in each constituency.

There is also a traditional *Swazi National Council* headed by the King and Queen Mother at which all Swazi men are entitled to be heard.

National Anthem

'Nkulunkulu mnikati wetibusiso temaSwati' ('O Lord our God bestower of blessings upon the Swazi'); words by A. E. Simelane, tune by D. K. Rycroft.

RECENT ELECTIONS

At the elections of 19 Sept. 2008 only non-partisans were elected. Political parties are illegal and advocates of multi-party politics are considered to be troublemakers.

CURRENT GOVERNMENT

In March 2013 the cabinet comprised:
Prime Minister: Barnabas Sibusiso Dlamini; b. 1942 (sworn in 23 Oct. 2008).
Deputy Prime Minister: Themba Masuku.
Minister for Agriculture: Clement Dlamini. *Commerce, Industry and Trade:* Jabulile Mashwama. *Economic Planning and Development:* Prince Hlangusemphi. *Education and Training:* Wilson Future Ntshangase. *Finance:* Majozi Sithole. *Foreign Affairs and International Co-operation:* Mtiti Fakudze. *Health:*

Bennedict Xaba. *Home Affairs:* Prince Gcokoma. *Information, Communications and Technology:* Winnie Magagula. *Justice and Constitutional Affairs:* Chief Mgwagwa Gamedze. *Labour and Social Security:* Lutfo Dlamini. *Local Government and Housing:* Pastor Lindiwe Gwebu. *Natural Resources and Energy:* Princess Tsandzile. *Public Service:* Patrick Mamba. *Public Works and Transport:* Ntuthuko Dlamini. *Sports, Culture and Youth:* Hlobsile Ndlovu. *Tinkhundla Administration and Development:* Rogers Mamba. *Tourism and Environmental Affairs:* Macford Nsibandze.

Government Website: http://www.gov.sz

CURRENT LEADERS

Mswati III

Position
King

Introduction
Mswati came to the throne in 1986. Effectively an absolute monarchy, he has received domestic and international criticism for his suppression of political opposition and for economic mismanagement. Among the greatest challenges of Mswati's reign has been containing the rapid increase in cases of HIV and AIDS in Swaziland.

Early Life
Mswati was born on 19 April 1968 in Manzini to one of the wives of King Sobhuza II and given the name Makhosetive (King of All Nations). Sobhuza's death in 1982 left a power vacuum that led to several years of infighting between various queens regent, crown princes and members of Liqoqo (the traditional advisory body which wielded significant power over the crown). In Oct. 1985 Makhosetive's mother dismissed several leading Liqoqo figures and recalled her son from his schooling in England. Makhosetive was crowned as Mswati III in April 1986.

Career in Office
Among Mswati's first acts as King was to dissolve the Liqoqo. Popular discontent grew at the prohibition on opposition political parties and Mswati's increasingly autocratic rule, leading to the establishment of the illegal People's United Democratic Movement (Pudemo). In 1990 Mswati agreed to open dialogue on the nation's political future. The national assembly was directly elected for the first time in 1993 and Mswati announced plans for a new constitution the following year.

With little progress having been made by 1996, Mswati established a constitutional commission. Pudemo continued to co-ordinate opposition, criticizing the King for filling the commission with his conservative supporters and boycotting the national assembly elections of Oct. 1998. In 2000 Pudemo leader Mario Masuku demanded an end to the state of emergency called 27 years earlier and was arrested for sedition. He was imprisoned pending his trial which collapsed in 2002. On his release he stated his belief that government could only be reformed when the monarchy was 'wiped out'.

In 2001 the constitutional commission reported back, providing the framework for the writing of a new constitution but asserting that the majority of the population did not favour the formation of new parties. In Dec. 2002, amid declining relations with the judiciary, six court of appeal judges resigned in protest at the King's use of rule by decree. The judges claimed that Mswati's repeal of several court decisions was unconstitutional. The crisis gave renewed impetus to the opposition alliance who called for a series of mass strikes. A new draft constitution was presented to the King in 2003, introducing a bill of rights but maintaining the executive role of the monarchy and the ban on political parties. Having been adopted by parliament and signed by the King in 2005, the constitution came into force in Jan. 2006.

Swaziland has one of the highest AIDS rates in the world. In Oct. 2001 Mswati ordered that all virgins should abstain from sex for five years or face a fine. The following month Mswati made a gift of a cow in recompense for taking an 18 year-old bride, Zena Mahlangu. Mahlangu's mother accused aides of the King of kidnapping her daughter and undertook legal proceedings to secure her return. Although the case collapsed, it received international attention and highlighted the growing challenges to Mswati's autocratic style.

Opposition groups boycotted the Sept. 2008 elections and were critical of the lavish celebrations of Mswati's 40th birthday, coupled with the 40th anniversary of Swaziland's independence.

Pudemo leader Masuku, who had again been detained under anti-terror laws in Nov. 2008, was released in Sept. 2009 vowing to continue campaigning for democracy. In Sept. 2010 around 50 pro-democracy activists, including Masuku, were arrested by police as they prepared for a protest march through the capital, Mbabane.

In 2011, despite a rapidly deteriorating economic situation and declining public revenues, the government continued to resist popular and international demands for reforms. In 2012 the International Monetary Fund reiterated its warnings that Swaziland's budget problems were reaching crisis point.

DEFENCE

Army Air Wing
There are two Israeli-built Arava transports with weapon attachments for light attack duties.

ECONOMY
Agriculture accounted for 8·1% of GDP in 2006, industry 46·4% and services 45·5%.

At the core of Swazi society is Tibiyo Taka Ngwane. Created in 1968 by Royal Charter, Tibiyo is a national development fund that operates outside the government and falls directly under the King, who holds it in trust for the nation. Its money derives from its stake in virtually every sector of Swazi commerce and industry.

Overview
Despite strong ties with the South African economy— accounting for 90% of imports and 70% of exports—Swaziland's economic growth has lagged behind that of its neighbours. Real GDP growth has averaged about 2% per year since the late 1990s, lower than other members of the South African Customs Union (SACU). Contributing factors include the slow pace of economic reform, the loss of preferential trade agreements for exports of textiles and sugar, repeated droughts and declining competitiveness.

SACU revenues accounted for around 60% of total government income and 25% of GDP in 2008–09 but collapsed in 2010–11 when economic contraction in South Africa led to a sharp decline in SACU imports. The drop in SACU revenues and an unbudgeted wage increase granted to civil servants and politicians in 2010 caused a fiscal crisis, with the central bank reporting a fall in national reserves to US$666m. In 2011, the government failed to satisfy the criteria for IMF or African Development Bank loans. In Aug. 2011 South Africa made an emergency loan of R2·4bn. (US$350m.), in return demanding that the Swazi lilangeni be pegged to the rand.

About 75% of the population is employed in subsistence agriculture. According to the 2010 Human Development Report, 21% of the population live below the international poverty line. The country has the world's highest HIV prevalence rate and the highest death rate, while nearly a quarter of Swazi children are orphans. Low levels of economic growth, a poor business environment, high income inequality and persistent unemployment are major hurdles to development.

Currency

The unit of currency is the *lilangeni* (plural *emalangeni*) (SZL) of 100 *cents* but Swaziland remains in the Common Monetary Area and the South African rand is legal tender. Inflation was 4·5% in 2010 and 6·1% in 2011. In July 2005 foreign exchange reserves were US$291m. and total money supply was 1,229m. emalangeni.

Budget

The fiscal year begins on 1 April. Total revenue in 2008–09 came to 9,208m. emalangeni and total expenditure to 9,538m. emalangeni.

There is a sales tax of 14%.

Performance

Real GDP growth was 1·2% in 2009, 2·0% in 2010 and 0·3% in 2011. Total GDP in 2011 was US$4·0bn.

Banking and Finance

The central bank and bank of issue is the Central Bank of Swaziland (*Governor*, Martin Dlamini), established in 1974. In 2004 there were four banking institutions, three foreign (South African-owned) private banks, Swazibank (state-owned) and a housing bank. In 2003 there were 178 credit and saving unions.

Foreign debt totalled US$616m. in 2010, representing 17·2% of GNI.

In 1990 Swaziland Stock Brokers was established to trade in stocks and shares for institutional and private clients. A fully-fledged stock exchange, the Swaziland Stock Exchange, was inaugurated in 1999.

ENERGY AND NATURAL RESOURCES

Environment

Swaziland's carbon dioxide emissions from the consumption and flaring of fossil fuels were the equivalent of 0·8 tonnes per capita in 2008.

Electricity

Installed capacity was an estimated 0·1m. kW in 2007. Production was about 454m. kWh in 2007; total consumption was an estimated 1,389m. kWh. Swaziland imports about two-thirds of its electricity needs from South Africa.

Minerals

Output (in tonnes): coal (2007 estimate), 462,000; quarry stone (2008), 240,997 cu. metres. Swaziland's diamond mine closed down in 1996 (1996 production, 75,000 carats) and its asbestos mine in 2000 (2000 production, 12,690 tonnes).

The oldest known mine (iron ore) in the world, dating back to 41,000 BC, was located at the Lion Cavern Site on Ngwenya Mountain.

Agriculture

In 2007 the agricultural population was around 354,000, of which about 141,000 were economically active. There were approximately 178,000 ha. of arable land and 14,000 ha. of permanent cropland in 2007. Production of principal crops (2003 estimates, in 1,000 tonnes): sugarcane, 4,000; maize, 70; grapefruit and pomelos, 37; oranges, 36; pineapples, 32; potatoes, 6; seed cotton, 6; cottonseed, 4; groundnuts, 4.

Livestock (2003 estimates): cattle, 580,000; goats, 422,000; pigs, 30,000; sheep, 27,000; chickens, 3m.

Livestock products, 2003 estimates (in 1,000 tonnes): milk, 38; meat, 22.

Forestry

Forests covered 0·56m. ha. in 2010, or 33% of the land area. In 2007 timber production was 1·34m. cu. metres.

Fisheries

Estimated total catch, 2010, 70 tonnes, exclusively from inland waters.

INDUSTRY

Most industries are based on processing agricultural products and timber. Footwear and textiles are also manufactured, and some engineering products.

Labour

In 2002, 92,654 persons were in formal employment; 5,714 Swazis worked in gold mines in South Africa in 2002. Unemployment was about 40% in 2003.

INTERNATIONAL TRADE

Swaziland has a customs union with South Africa and receives a pro rata share of the dues collected.

Imports and Exports

Trade in US$1m.:

	2002	2003	2004	2005	2006
Imports f.o.b.	1,027·0	1,540·2	1,715·3	1,949·4	1,960·7
Exports f.o.b.	1,078·5	1,666·8	1,806·2	1,965·5	1,975·7

Main import products are motor vehicles, machinery, transport equipment, foodstuffs, petroleum products and chemicals; main export commodities are soft drink concentrates, sugar, wood pulp and cotton yarn. By far the most significant trading partner is South Africa. In 2005, 88·3% of imports came from South Africa; 74·6% of exports went to South Africa in 2005.

COMMUNICATIONS

Roads

The total length of roads in 2002 was 3,594 km, of which 1,465 km were main roads. There were 52,200 passenger cars in use in 2007 plus 41,800 lorries and vans and 8,100 buses and coaches. There were 235 fatalities in road accidents in 2007.

Rail

In 2005 the system comprised 301 km of route (1,067 mm gauge). There are north and south connections to South Africa's rail system, and a link in the northeast with Mozambique and the port of Maputo. In 2009, 4m. tonnes of freight were transported.

Civil Aviation

There is an international airport at Manzini (Matsapha). The national carrier is Swaziland Airlink, which had direct flights from Manzini to Johannesburg in 2012. The unrelated Airlink also operated on the same route in 2012.

Telecommunications

In 2008 there were 44,000 main (fixed) telephone lines; mobile phone subscribers numbered 531,600 in 2008 (45·5 per 100 persons). There were 80,000 internet users in 2008.

SOCIAL INSTITUTIONS

Justice

The constitutional courts practice Roman-Dutch law. The judiciary is headed by the Chief Justice. There is a High Court and various Magistrates and Courts. A Court of Appeal with a President and three Judges deals with appeals from the High Court. There are 16 courts of first instance. There are also traditional Swazi National Courts.

The population in penal institutions in June 2008 was 2,546 (231 per 100,000 of national population).

Education

In 2007 there were 232,572 primary school pupils with 7,169 teaching staff. The teacher/pupil ratio has decreased from 40/1 in the 1970s to 32/1. About half the children of secondary school age attend school. There are also private schools. In 2007 there were 83,049 pupils in secondary schools with 4,358 teaching staff. Many secondary and high schools teach agricultural activities.

There were 5,692 students in higher education in 2006 with 462 academic staff. The University of Swaziland (UNISWA), with its main campus at Kwaluseni, was founded in 1982 and has an enrolment of over 5,000 students.

Rural education centres offer formal education for children and adult education geared towards vocational training. The adult literacy rate in 2008 was 87%.

In 2006 public expenditure on education came to 7·9% of GNI and 24·4% of total government spending.

Health

In 2005 there were 400 health institutions, of which nine were hospitals and 19 were health centres. There were 184 physicians, 3,345 nurses, 20 dentists and 46 pharmacists in 2000.

RELIGION

In 2001 there were 480,000 African Christians, 160,000 Protestants and the remainder of the population followed other religions (including traditional beliefs).

CULTURE

Press

In 2008 there were two daily newspapers: *The Times of Swaziland* (English-language with a circulation of 22,000 in 2008), founded in 1897, and *The Swazi Observer* (English, 15,000).

Tourism

There were 870,000 non-resident tourists staying at hotels and similar establishments in 2007 (up from 461,000 in 2003), bringing revenue of US$32m.

Festivals

The annual Umhlanga (Reed Dance) takes place in Aug. or early Sept. in honour of the Queen Mother. Bush Fire is a festival of music and performing arts held in the Malkerns Valley south of Mbabane in May.

DIPLOMATIC REPRESENTATIVES

Of Swaziland in the United Kingdom (20 Buckingham Gate, London, SW1E 6LB)
High Commissioner: Dumsile T. Sukati.

Of the United Kingdom in Swaziland (High Commission in Mbabane closed in Aug. 2005)
High Commissioner: Dr Nicola Brewer (resides in Pretoria, South Africa).

Of Swaziland in the USA (1712 New Hampshire Ave., NW, Washington, D.C., 20009)
Ambassador: Abednego Ntshangase.

Of the USA in Swaziland (7th Floor, Central Bank Building, Mahlokohla St., Mbabane)
Ambassador: Makila James.

Of Swaziland to the United Nations
Ambassador: Zwelethu Mnisi.

Of Swaziland to the European Union
Ambassador: Joel Musa Nhleko.

FURTHER READING

Booth, Alan R., *Historical Dictionary of Swaziland.* 2000
Gillis, D. Hugh, *The Kingdom of Swaziland: Studies in Political History.* 1999
Matsebula, J. S. M., *A History of Swaziland.* 3rd ed. 1992

National Statistical Office: Central Statistical Office, POB 456, Mbabane.

SWEDEN

Norwegian Sea
Umeå
SWEDEN
FINLAND
NORWAY
Gulf of Bothnia
STOCKHOLM
ESTONIA
Gothenburg
LATVIA
DENMARK
Baltic Sea
Malmö
LITHUANIA
© Research Machines plc 2006

0   150 mi
0   200 km

**Konungariket Sverige
(Kingdom of Sweden)**

Capital: Stockholm
Population projection, 2015: 9·65m.
GNI per capita, 2011: (PPP$) 35,837
HDI/world rank: 0·904/10
Internet domain extension: .se

KEY HISTORICAL EVENTS

Sweden was covered by a thick ice cap until 14,000 years ago, when the ice began to retreat. The first human traces, in southern Sweden, date from 10,000 BC. Between 8000 and 6000 BC the country was populated by hunters and fishermen, using simple stone tools. Artefacts found in graves show that the Bronze Age was marked by a relatively advanced culture. From 500 BC to AD 800 agriculture became the basis for society and the economy. The Viking Age (800–1050) took expansion eastwards. Swedish Vikings reached into today's Russia, where they set up trading stations and principalities, such as Novgorod and Rurik. The Vikings also travelled to the Black and Caspian Seas and developed trading links with the Byzantine Empire and the Arabs.

In 830 the Frankish monk Ansgar introduced Christianity to Sweden but made few converts. In the 11th century, English missionaries had greater success and by the end of the century the country was fully Christianized. Olof Skötkonung, who proclaimed himself ruler of Sweden, supported the new religion.

By 1200 Sweden had become a united kingdom with largely the same borders as it has today, except that Skåne, Halland and Blekinge in the south of Sweden formed part of Denmark, and

Jämtland, Härjedalen and Bohuslän in the west belonged to Norway. There was a struggle for power between the Sverker and Erik families, who ruled alternately in 1160–1250. However, by the middle of the 13th century with the building of royal castles and introduction of provincial administration, the crown was able to assert the authority of the central government and impose laws valid for the whole kingdom. Among the most important figures of the 13th century was Birger Jarl, who promoted the newly founded city of Stockholm. At a time when Hanseatic merchants traded in Sweden, Stockholm contained a large German population while, on the southeast coast, Kalmar and Gotland were controlled by German immigrants. The Hanseatic trading posts to the east included Finland, which was brought into the Swedish kingdom. A new code of law, for the entire country, was introduced in 1350 by King Magnus Eriksson. In 1340 Valdemar Atterdag, King of Denmark, went to war with Sweden over the southern provinces of Skåne, Halland and Blekinge. He attacked Gotland in 1361 in one of the bloodiest battles in Nordic history, to secure a base for further assaults on Sweden and to overthrow the Hanseatic League.

In the 14th century trade increased and until the mid-16th century the Hanseatic League dominated Sweden's trade. In 1350 the Black Death decimated the population. Inheritance and marriage ties united the crowns of Denmark, Norway and Sweden in 1389, under the rule of Queen Margaret of Denmark. In 1397 the loose association known as the Union of Kalmar confirmed her five-year old nephew, Erik of Pomerania, as ruler of the three countries. The Swedish nationalist Sten Sture, who was made king of Sweden in 1470, defeated the Danes at the Battle of Brunkeberg in Stockholm. Following this success, Sten Sture promoted nationalistic sentiment by the public display of a great wood carving of St George slaying the dragon (now placed in Stockholm cathedral) and the setting up of the first Swedish university at Uppsala. The union period (1397–1521) was characterized by a struggle for power between the king, the nobility and the burghers and peasants. These conflicts culminated in the Stockholm Bloodbath in 1520, when eighty leading men in Sweden were executed on orders of the Danish king, Christian II. In 1521 Christian II was overthrown by Gustav Vasa, a Swedish nobleman, who was elected king in 1523.

Empire Building

Gustav Vasa (ruled 1523–60) laid the foundations of the Swedish national state. With the Reformation, Sweden was converted to Lutheranism and the church became a national institution. At the same time power was concentrated in the hands of the king and in 1544 a hereditary monarchy was established. By the 16th century Scandinavia was divided into two states, Sweden-Finland and Denmark-Norway. Since the dissolution of the union with Denmark and Norway, Swedish foreign policy focused on dominating the Baltic Sea, and this led to wars with Denmark from the 1560s.

In 1611 Gustavus Adolphus (Gustaf II Adolf) consolidated Sweden's position on the Russian side of the Baltic Sea. Sweden defeated Denmark in two wars (1643 and 1657) to take control of the previously Danish provinces of Skåne, Halland, Blekinge and Gotland and the Norwegian provinces of Bohuslän, Jämtland and Härjedalen. Gustavus Adolphus gave his support to the Protestant alliance in the Thirty Years War. This cost him his life but earned Sweden territory in the north of Germany. Finland and the present-day Baltic republics also belonged to Sweden making it a great power in northern Europe.

Following the death of King Karl XII in 1718 the Swedish Parliament (Riksdag) introduced a constitution that abolished royal absolutism. But royal authority was soon reasserted. After defeat in the Great Northern War (1700–21) against Denmark, Poland and Russia, Sweden lost most of its Baltic territories, including a part of Finland and all its north-German possessions except west Pomerania. During the Napoleonic Wars, Sweden lost Finland to Russia and withdrew from its remaining German provinces. In 1810 Napoleon's marshal Jean-Baptiste Bernadotte assumed power as Karl XIV Johan. He tried to win back Finland from Russia, but had to make do with a union of Sweden and Norway, confirmed by the treaty of Kiel in 1814.

The short war against Norway in 1814 was Sweden's last military adventure. Since then Sweden has favoured neutrality. For this reason, there was no early application to join the EEC (European Economic Community). But in line with a commitment to liberalize trade, Sweden was a founder member of EFTA (European Free Trade Agreement) in 1959. After the Cold War and the collapse of the Soviet Union, the policy of neutrality was seen by many as obsolete. Sweden became a member of the EU (European Union) in 1995.

Following constitutional reforms in 1974 the remaining powers of the king were reduced to purely ceremonial functions. Carl XVI Gustaf, who succeeded the throne in 1973, is the first Swedish king to be bound by the new constitution. In 1980 the order of succession was amended to allow for a female member of the royal family to inherit the crown. Consequently, Princess Victoria is the heir apparent rather than her younger brother Prince Carl Philip.

Modern Economy

After the Napoleonic Wars, Sweden suffered economic stagnation. The country was poor with 90% of the population living off the land. Out of a population of five million, over one million emigrated between 1866 and 1914, mostly to North America. Industry did not start to grow until the 1890s. However, it then developed rapidly and after the Second World War Sweden was transformed into one of the leading industrial nations in Europe. In the six years to 1951 the country's GNP rose by 20%. Economic success was partly thanks to the early utilization of hydro-electric power which supported the pulp and paper industries of the northern forests. Swedish inventors created the ball-bearing, the adjustable spanner, the primus stove and the cream separator, as well as the safety match and dynamite. By 1956 the economy was booming, poverty had almost disappeared and unemployment was at a minimum. Sweden was one of the richest countries in Europe.

Social democracy began as the political offshoot of the trade unions. The working class was supported by intellectuals, such as the scientist Hjalmar Branting, who was the first Scandinavian socialist prime minister. The first representative of social democracy entered the government in 1917. Universal suffrage was introduced for men in 1909 and for women in 1921. In the 1930s, when the Social Democrats had become the governing party, plans for the welfare society were laid. Reforms included restrictions on child and female labour, free elementary education and old-age pensions. A four-party coalition for the duration of the Second World War was succeeded by a Social Democrat government with Per Albin Hansson as prime minister. Following his death in 1946, Tage Erlander became prime minister and stayed in office until 1969. He was succeeded by Olof Palme who was prime minister between 1969 and 1976. Owing to the rise in oil prices in 1973, unemployment increased. From the mid-1970s, improvements in living standards slowed. Economic crisis drove the Social Democrats out of government and in 1976 a non-socialist coalition was formed under Centre Party chairman, Thorbjörn Fälldin. Conflicts over the expansion of nuclear power led to several government reshuffles. In 1982 the Social Democrats resumed office with Olof Palme as prime minister. The assassination of Palme in 1986 shook the country which had been spared political violence for nearly 200 years. Ingvar Carlsson

took over as head of government. In the 1990s industrial production fell and unemployment rose, which led to a high budget deficit and increased national debt. Popular dissatisfaction showed in the 1991 election, when a non-socialist coalition government was formed with Carl Bildt as prime minister. Launching a programme of deregulation and privatization, Bildt did much to prepare the economy for closer involvement with Europe. Capital gains taxes were reduced, as too were social benefits. But the government failed to reduce unemployment, the budget deficit or the national debt. Hence, the 1994 election put the Social Democrats back in power, with Ingvar Carlsson again as prime minister. In 1996 he stepped down to be replaced by Göran Persson. Persson led a series of minority governments. His third term, which began in 2002, was marked by the murder of his foreign minister, Anna Lindh, in 2003 and the rejection by referendum of Sweden's entry into the single European currency. In Sept. 2006 the Social Democrats emerged from elections as the single biggest party but Fredrik Reinfeldt of the Moderate Party was able to form a centre-right 'Alliance for Sweden' coalition to end 12 years of Social Democratic government. Despite economic problems, the country still boasts one of the highest standards of living and one of the most advanced welfare systems.

TERRITORY AND POPULATION

Sweden is bounded in the west and northwest by Norway, east by Finland and the Gulf of Bothnia, southeast by the Baltic Sea and southwest by the Kattegat. The area is 450,295 sq. km, including water (96,000 lakes) totalling 39,960 sq. km. At the last census, in 1990, the population was 8,587,353. Parliament decided in 1995 to change to a register-based method of calculating the population. The recorded population at 31 Dec. 2009 was 9,340,682; density 23 per sq. km. In 2011, 84·8% of the population lived in urban areas.

The UN gives a projected population for 2015 of 9·65m.

Area, population and population density of the counties (*län*):

	Land area (in sq. km)	Population (1990 census)	Population (31 Dec. 2009)	Density per sq. km (31 Dec. 2009)
Stockholm	6,519	1,640,389	2,019,182	310
Uppsala	8,208	268,503	331,898	40
Södermanland	6,103	255,546	269,053	44
Östergötland	10,605	402,849	427,106	40
Jönköping	10,495	308,294	336,044	32
Kronoberg	8,468	177,880	183,162	22
Kalmar	11,219	241,149	233,639	21
Gotland	3,151	57,132	57,221	18
Blekinge	2,947	150,615	152,591	52
Skåne	11,035	1,068,587	1,231,062	112
Halland	5,462	254,568	296,825	54
Västra Götaland	23,956	1,458,166	1,569,458	65
Värmland	17,591	283,148	273,257	16
Örebro	8,546	272,474	278,882	33
Västmanland	5,145	258,544	251,353	49
Dalarna	28,197	288,919	276,454	10
Gävleborg	18,200	289,346	276,220	15
Västernorrland	21,685	261,099	243,042	11
Jämtland	49,343	135,724	126,666	3
Västerbotten	55,190	251,846	258,548	5
Norrbotten	98,249	263,546	249,019	3

There are some 17,000 Sami (Lapps).

On 31 Dec. 2008 foreign-born persons in Sweden numbered 1,281,581. Of these, 269,681 were from Nordic countries; 459,139 from the rest of Europe; 90,733 from Africa; 28,750 from North America; 60,878 from South America; 361,333 from Asian countries; 6,437 from the former USSR; 3,957 from Oceania; and 673 country unknown. Of the total 175,113 were born in Finland. 13·8% of the population of Sweden in Dec. 2008 was foreign-born, the highest proportion in any of the Nordic countries.

Immigration: 2007, 99,485; 2008, 101,171; 2009, 102,280. Emigration: 2007, 45,418; 2008, 45,294; 2009, 39,240.

Population of the 50 largest communities, 31 Dec. 2009:

Stockholm	829,417	Luleå	73,950
Gothenburg (Göteborg)	507,330	Kungsbacka	73,938
Malmö	293,909	Skellefteå	71,770
Uppsala	194,751	Solna	66,909
Linköping	144,690	Järfälla	65,295
Västerås	135,936	Sollentuna	63,347
Örebro	134,006	Karlskrona	63,342
Norrköping	129,254	Täby	63,014
Helsingborg	128,359	Kalmar	62,388
Jönköping	126,331	Mölndal	60,381
Umeå	114,075	Östersund	59,136
Lund	109,147	Varberg	57,439
Borås	102,458	Gotland	57,221
Huddinge	95,798	Norrtälje	55,927
Eskilstuna	95,577	Falun	55,685
Sundsvall	95,533	Örnsköldsvik	55,128
Gävle	94,352	Trollhättan	54,873
Halmstad	91,087	Uddevalla	51,518
Nacka	88,085	Nyköping	51,209
Södertälje	85,270	Skövde	50,984
Karlstad	84,736	Hässleholm	50,036
Växjö	82,023	Borlänge	48,681
Botkyrka	81,195	Lidingö	43,445
Kristianstad	78,788	Tyresö	42,602
Haninge	76,237	Trelleborg	41,891

A 16-km long fixed link with Denmark was opened in July 2000 when the Öresund motorway and railway bridge between Malmö and Copenhagen was completed.

The *de facto* official language is Swedish.

SOCIAL STATISTICS

Statistics for calendar years:

	Total living births	To mothers single, divorced or widowed	Stillborn	Marriages	Divorces	Deaths exclusive of still-born
2004	100,928	55,991	318	43,088	20,106	90,532
2005	101,346	56,238	301	44,381	20,000	91,710
2006	105,913	58,820	319	45,551	20,295	91,177
2007	107,421	58,819	326	47,898	20,669	91,729
2008	109,301	59,832	396	50,332	21,377	91,449

Rates, 2008, per 1,000 population: births, 11·9; deaths, 9·9; marriages, 5·4; divorces, 2·3. Sweden has one of the highest rate of births outside marriage in Europe, at 55% in 2008. In 2008 the average age at first marriage was 35·1 years for males and 32·5 years for females. Expectation of life in 2007: males, 78·6 years; females, 83·0. Annual population growth rate, 2000–05, 0·4%. Infant mortality, 2010, two per 1,000 live births (one of the lowest rates in the world). Fertility rate, 2008, 1·9 births per woman. Sweden legalized same-sex marriage in May 2009. In 2008 Sweden received 24,353 asylum applications, equivalent to 2·6 per 1,000 inhabitants.

CLIMATE

The north has severe winters, with snow lying for 4–7 months. Summers are fine but cool, with long daylight hours. Further south, winters are less cold, summers are warm and rainfall well distributed throughout the year, although slightly higher in the summer. Stockholm, Jan. −2·8°C, July 17·2°C. Annual rainfall 385 mm.

CONSTITUTION AND GOVERNMENT

The reigning King is **Carl XVI Gustaf**, b. 30 April 1946, succeeded on the death of his grandfather Gustaf VI Adolf, 15 Sept. 1973, married 19 June 1976 to Silvia Renate Sommerlath, b. 23 Dec. 1943 (Queen of Sweden). *Daughter* and *Heir Apparent:* Crown Princess Victoria Ingrid Alice Désirée, Duchess of Västergötland, b. 14 July 1977, married 19 June 2010 Daniel Westling, b. 15 Sept. 1973 (*offspring:* Princess Estelle Silvia Ewa

Mary, b. 23 Feb. 2012); *son:* Prince Carl Philip Edmund Bertil, Duke of Värmland, b. 13 May 1979; *daughter:* Princess Madeleine Thérèse Amelie Josephine, Duchess of Hälsingland and Gästrikland, b. 10 June 1982. *Sisters of the King.* Princess Margaretha, b. 31 Oct. 1934, married 30 June 1964 to John Ambler, died 31 May 2008; Princess Birgitta (Princess of Sweden), b. 19 Jan. 1937, married 25 May 1961 (civil marriage) and 30 May 1961 (religious ceremony) to Johann Georg, Prince of Hohenzollern; Princess Désirée, b. 2 June 1938, married 5 June 1964 to Baron Niclas Silfverschiöld; Princess Christina, b. 3 Aug. 1943, married 15 June 1974 to Tord Magnuson. *Uncles of the King.* Count Sigvard Bernadotte of Wisborg, b. 7 June 1907, died 4 Feb. 2002; Count Carl Johan Bernadotte of Wisborg, b. 31 Oct. 1916, died 5 May 2012.

Under the 1975 constitution Sweden is a representative and parliamentary democracy. The King is Head of State, but does not participate in government. Parliament is the single-chamber *Riksdag* of 349 members elected for a period of four years in direct, general elections.

The manner of election to the *Riksdag* is proportional. The country is divided into 29 constituencies. In these constituencies 310 members are elected. The remaining 39 seats constitute a nationwide pool intended to give absolute proportionality to parties that receive at least 4% of the votes. A party receiving less than 4% of the votes in the country is, however, entitled to participate in the distribution of seats in a constituency if it has obtained at least 12% of the votes cast there.

A parliament, the *Sameting*, was instituted for the Sami (Lapps) in 1993.

National Anthem

'Du gamla, du fria' ('Thou ancient, thou free'); words by R. Dybeck; folk-tune.

GOVERNMENT CHRONOLOGY

Prime Ministers since 1936. (C = Centre Party; FP = Liberal Party; M = Moderate Party/New Moderates; SAP = Swedish Social Democratic Labour Party)

1936–46	SAP	Per Albin Hansson
1946–69	SAP	Tage Fritiof Erlander
1969–76	SAP	Sven Olof Joachim Palme
1976–78	C	Thorbjörn Fälldin
1978–79	FP	Ola Ullsten
1979–82	C	Thorbjörn Fälldin
1982–86	SAP	Sven Olof Joachim Palme
1986–91	SAP	Ingvar Gösta Carlsson
1991–94	M	Carl Bildt
1994–96	SAP	Ingvar Gösta Carlsson
1996–2006	SAP	Göran Persson
2006–	M	John Fredrik Reinfeldt

RECENT ELECTIONS

In parliamentary elections held on 19 Sept. 2010 the Swedish Social Democratic Labour Party (SAP) won 112 seats with 30·7% of votes cast (down from 130 with 35·2% in 2006), the New Moderates 107 with 30·1% (up from 97 with 26·1%), the Green Party 25 with 7·3% (19 with 5·2%), the Liberal Party 24 with 7·1% (28 with 7·5%), the Centre Party 23 with 6·6% (29 with 7·9%), the far-right Sweden Democrats 20 with 5·7%, the Christian Democratic Party 19 with 5·6% (24 with 6·6%) and the Left Party 19 with 5·6% (22 with 5·8%). Turnout was 84·6%. The election represented the worst result for the Social Democrats since 1914 and resulted in a hung parliament with the centre-right 'Alliance for Sweden' coalition of the New Moderates, the Centre Party, the Christian Democratic Party and the Liberal Party falling short of a majority by two seats.

European Parliament

Sweden has 20 (19 in 2004) representatives. At the June 2009 elections turnout was 45·5% (37·9% in 2004). The SAP won

5 seats with 24·4% of votes cast (political affiliation in European Parliament: Progressive Alliance of Socialists and Democrats); the New Moderates, 4 with 18·8% (European People's Party); the Liberal Party, 3 with 13·6% (Alliance of Liberals and Democrats for Europe); the Green Party, 2 with 11·0% (Greens/European Free Alliance); the Pirate Party, 1 with 7·1% (Greens/European Free Alliance); the Left Party, 1 with 5·7% (European United Left/ Nordic Green Left); the Centre Party, 1 with 5·5% (Alliance of Liberals and Democrats for Europe); the Christian Democratic Party, 1 with 4·7% (European People's Party).

CURRENT GOVERNMENT

Following parliamentary elections in Sept. 2010 a new minority centre-right coalition government was formed. In March 2013 the cabinet comprised:

Prime Minister: Fredrik Reinfeldt; b. 1965 (New Moderates; sworn in 6 Oct. 2006).

Deputy Prime Minister and Minister of Education: Jan Björklund.

Minister of Children and Elderly: Maria Larsson. *Culture and Sport:* Lena Adelsohn Liljeroth. *Defence:* Karin Enström. *Employment:* Hillevi Engström. *Enterprise and Regional Affairs:* Annie Lööf. *Environment:* Lena Ek. *European Union Affairs:* Birgitta Ohlsson. *Finance:* Anders Borg. *Financial Markets:* Peter Norman. *Foreign Affairs:* Carl Bildt. *Gender Equality:* Maria Arnholm. *Health and Social Affairs:* Göran Hägglund. *Information Technology and Energy:* Anna-Karin Hatt. *Infrastructure:* Catharina Elmsäter-Svärd. *Integration:* Erik Ullenhag. *International Development Co-operation:* Gunilla Carlsson. *Justice:* Beatrice Ask. *Migration and Asylum Policy:* Tobias Billström. *Public Administration and Housing:* Stefan Attefall. *Rural Affairs:* Eskil Erlandsson. *Social Security:* Ulf Kristersson. *Trade:* Ewa Björling.

The *Speaker* is Per Westerberg.

Government Website: http://www.sweden.gov.se

CURRENT LEADERS

Fredrik Reinfeldt

Position
Prime Minister

Introduction
Fredrik Reinfeldt led a centre-right alliance to victory in the legislative elections of Sept. 2006, ousting the Social Democrats (SAP) from over a decade in power. After taking over the leadership of the Moderate Party in 2003, Reinfeldt rebranded it as a centrist party advocating entrepreneurship and job creation coupled with reform of Sweden's cherished welfare system. He retained the premiership following the Sept. 2010 elections.

Early Life
John Fredrik Reinfeldt was born in Stockholm on 4 Aug. 1965. He joined the youth wing of the conservative Moderate Party in 1983. Having completed military service, in 1990 he graduated in business and economics from Stockholm University, where he was active in student politics. He embarked on a political career, becoming chairman of the Moderate Youth League and standing in the legislative elections of Sept. 1991. He won a seat in parliament as the Moderate Party emerged as the leading non-socialist party. The SAP remained the largest single party, but without an overall majority, and Moderate leader Carl Bildt became the first Conservative prime minister since 1930, heading a four-party coalition. The government attempted to tackle the economic crisis that gripped Sweden in 1992 by introducing market reforms, imposing spending cuts and privatizing publicly-owned enterprises. It also sought accession to the European Union.

In the Sept. 1994 elections the Moderates held on to the 80 seats they had won in 1991, but some of their centre-right coalition partners fared badly and the Social Democrats were returned to power. After Bo Lundgren succeeded Bildt as Moderate leader in 1999, Reinfeldt was promoted to chairman of the parliamentary justice committee in 2001–02.

The Moderate Party's poor performance in the 2002 elections was compounded by a scandal in 2003 in which some members were accused of racism. Lundgren was forced to resign as leader in Oct. 2003 and Reinfeldt was elected unanimously to succeed him. Reinfeldt rebranded the party as the New Moderates and shifted the focus to the centre ground. In the run-up to the Sept. 2006 legislative elections, he formed the Alliance for Sweden, aiming to unite a four-party centre-right coalition (New Moderates plus the Centre Party, Liberal Party and Christian Democrats). Presenting a joint manifesto, the alliance narrowly beat the SAP. The New Moderates took 26·1% of the vote, a record for the party, and Reinfeldt was nominated prime minister on 5 Oct. 2006.

Career in Office
On taking office, Reinfeldt initiated a programme of reforms aimed at strengthening incentives to work, reducing welfare dependency and streamlining the state's role in the economy. However, in the wake of the global financial crisis, Sweden slipped into recession in 2008, and in Dec. Reinfeldt proposed a stimulus package to boost the economy. The downturn nevertheless had a significant impact on Sweden's trade-oriented economy in 2009 as exports declined and job losses mounted.

Reinfeldt's own party has favoured joining the single European currency and, in principle, supports membership of NATO. In 2008 the Alliance, supported by the SAP, ratified the Lisbon Treaty (signed in Dec. 2007) on EU institutional and administrative reform. From July–Dec. 2009 Sweden held the rotating EU presidency and oversaw the treaty's implementation. The Swedish presidency also pressed the EU to take the lead in fighting climate change. In Feb. 2009 Reinfeldt's government announced its intention to lift a 30-year-old ban on building new nuclear energy capacity.

In the Sept. 2010 legislative elections the Alliance fell narrowly short of a parliamentary majority but Reinfeldt formed a new minority government the following month with no changes in key ministerial portfolios.

DEFENCE

The Supreme Commander is, under the government, in command of the three services. The Supreme Commander is assisted by the Swedish Armed Forces HQ. There is also a Swedish Armed Forces Logistics Organization.

Conscripted military service has been phased out and ended officially on 1 July 2010. 'Required military service' will only be enacted if the country feels threatened.

In 2008 military expenditure totalled US$6,659m. (US$736 per capita), representing 1·4% of GDP. Sweden's national security policy is currently undergoing a shift in emphasis. Beginning with the decommissioning of obsolete units and structures, the main thrust of policy is the creation of contingency forces adaptable to a variety of situations.

When Sweden joined the European Union in 1995 the government stressed that membership did not imply any change in the country's traditional policy of non-participation in military alliances, with the option of staying neutral in the event of war in its vicinity.

Sweden has modern air raid shelters with capacity for some 7m. people. Since this falls short of providing protection for the whole population, evacuation and relocation operations would be necessary in the event of war.

The Swedish Civil Contingencies Agency is responsible for civil protection and emergency preparedness and has a mandate spanning the entire spectrum of threats and risks, including wartime defence.

Army

The Army consists of one division HQ and divisional units and three army brigade command and control elements. Army strength, 2007, 10,200 (4,300 conscripts). The Army can mobilize a reserve of 225,000. Voluntary auxiliary organizations numbered 42,000.

Navy

The Navy has two naval warfare flotillas and one submarine flotilla.

The personnel of the Navy in 2007 totalled 7,900 (active manpower, including 2,000 conscripts, 1,300 coastal defence and 320 naval aviation). Reserve strength, 20,000. In addition there is a paramilitary coast guard numbering 600.

Air Force

The Air Force consists of eight air-base battalions, four fighter squadrons, four air transport squadrons, one fighter control and air surveillance battalion, one anti-submarine warfare squadron, one signal intelligence squadron, one airborne early warning squadron and a training school.

Strength (2007) 5,900 (1,500 conscripts), plus 17,000 reserves. There were 130 combat-capable aircraft in 2007 (JAS 39 *Gripens*).

In 1998 the helicopter units of the Swedish Army and Navy were merged with those of the Air Force to form a single helicopter wing, consisting of four squadrons and an independent unit, which since 2003 has fallen under the direct authority of the Air Force. Strength (2007), 1,050 (250 conscripts).

INTERNATIONAL RELATIONS

In a referendum held on 14 Sept. 2003 Swedish voters rejected their country's entry into the common European currency, 56·1% opposing membership of the euro against 41·8% voting in favour. Turnout was 81·2%.

ECONOMY

Services accounted for 70% of GDP in 2007, industry 29% and agriculture 2%.

According to the anti-corruption organization *Transparency International*, Sweden ranked fourth in the world in a 2012 survey of the countries with the least corruption in business and government. It received 88 out of 100 in the annual index.

Sweden gave US$5·6bn. in international aid in 2011, which at 1·02% of GNI made it the world's most generous developed country as a percentage of its gross national income.

Overview

Sweden combines an extensive welfare state with a market economy. Tax revenue as a percentage of GDP is among the highest in the world. Monetary policy has targeted a 2% inflation rate since 1993 and the IMF has praised the design and operation of the inflation-targeting framework implemented by the Riksbank. In Sept. 2003 a referendum rejected adoption of the euro.

After the 2001 global technology crash that damaged Ericsson, one of Sweden's largest companies, the economy experienced an upswing until 2008, with strong recovery in the telecommunications and automobile sectors fuelling export growth from 2003–07. Growing exports combined with rising production and low interest rates prompted a revival in business investment. Growth was further buoyed by gains in consumer wealth. However, the economy was hit hard by the global financial crisis, with exports significantly reduced and recession hitting in 2008. A rescue package for the banking sector was announced in early 2009 and Sweden enjoyed a stronger recovery than many other advanced countries. The employment rate was 65·7% in Oct. 2011. Effective employment is lower than that portrayed because of large numbers of employees benefiting from sick leave, social assistance, labour market programmes and mid-life sabbaticals. So far Sweden's strong growth has failed to make an impact on unemployment, as increased employment has been offset by new entrants into the labour market. The unemployment rate, at 8·4% in 2010, has remained above 6%. The youth unemployment rate is particularly high, at over 22% for those aged 15–24 in the period July–Sept. 2011.

26·2% of the labour force was employed by government in 2008, down 1·5% since 2000. General government expenditure as a percentage of GDP was the fourth highest in the OECD in 2009. Sweden had the EU's third highest level of expenditure on social protection per inhabitant in 2007.

A number of sectors, including electricity, telecommunications and parts of transport, have been deregulated. Sweden is facing pressure from the EU to privatize its state monopolies in the pharmaceutical, construction and alcoholic beverage sectors.

Reducing unemployment remains a priority. The turbulence in the euro is a risk to economic recovery. The IMF warns that Swedish house prices are over-inflated.

Currency

The unit of currency is the *krona* (SEK), of 100 *öre*. Inflation rates (based on OECD statistics):

2002	2003	2004	2005	2006	2007	2008	2009	2010	2011
2·2%	1·9%	0·4%	0·5%	1·4%	2·2%	3·4%	−0·5%	1·2%	3·0%

Foreign exchange reserves were US$38,704m. and gold reserves 4·04m. troy oz in Sept. 2009. Total money supply was 1,486·7bn. kr. in Aug. 2009.

Budget

Revenue of 901·3bn. kr. and expenditure of 766·1bn. kr. was estimated for the total budget (Current and Capital) for financial year 2008.

Revenue and expenditure for 2008 (1m. kr.):

Revenue	
Tax revenues	750,174
From government activities	52,964
From sale of property	76,519
Loans repaid	1,881
Computed revenues	8,700
Contributions, etc., from the EU	11,036
Total revenue	901,274
Expenditure	
The Swedish political system	10,966
Economy and fiscal administration	11,378
Tax administration and collection	9,446
Justice	32,693
Foreign policy administration and international co-operation	1,750
Total defence	43,030
International development assistance	27,453
Immigrants and refugees	6,134
Health care, medical care, social services	49,131
Financial security in the event of illness and disability	115,862
Financial security in old age	42,591
Financial security for families and children	66,393
The labour market	51,781
Working life	1,031
Study support	19,110
Education and university research	44,524
Culture, the media, religious organizations and leisure	10,117
Community planning, housing supply and construction	2,075
Regional balance and development	2,799
General environment and conservation	4,667
Energy	2,079
Communications	61,484
Agriculture and forestry, fisheries, etc.	16,508
Business sector	12,761
General grants to municipalities	64,770
Interest on central government debt, etc.	48,206
Contribution to the European Community	31,526
Other expenditure	−24,188
Total expenditure	766,076

Sweden registered a budget surplus in 2011 of 0·3% of GDP. It also had a surplus of 0·3% in 2010 but a deficit of 0·7% in 2009. The required target set by the EU is a budget deficit of no more than 3%.

VAT is 25% (reduced rates, 12% and 6%).

Performance

Real GDP growth rates (based on OECD statistics):

2002	2003	2004	2005	2006	2007	2008	2009	2010	2011
2·5%	2·5%	3·7%	3·2%	4·6%	3·4%	−0·8%	−5·0%	6·3%	3·9%

Sweden's total GDP in 2011 was US$539·7bn.

Sweden was ranked fourth in the Global Competitiveness Index in the World Economic Forum's *Global Competitiveness Report 2012–2013*. The index analyses 12 areas of competitiveness for over 100 countries including macroeconomy, higher education and training, institutions, innovation and infrastructure.

In 2008 the state debt amounted to 1,119bn. kr.

Banking and Finance

The central bank and bank of issue is the Sveriges Riksbank. The bank has 11 trustees, elected by parliament, and is managed by a directorate, including the governor, appointed by the trustees. The *Governor* is Stefan Ingves, appointed for a six-year term. In 2008 there were 118 banks (30 Swedish commercial banks, 33 foreign banks, 53 savings banks and two co-operative banks). Bank deposits from the public in 2008 amounted to 2,201bn. kr.; lending to the public in 2008 amounted to 3,033bn. kr. The largest banks are Nordea Bank AB (generally known as Nordea; previously MeritaNordbanken, formed in 1997 when Nordbanken of Sweden merged with Merita of Finland), Handelsbanken, Skandinavska Enskilda Banken (generally known as SEB) and Swedbank (formerly FöreningsSparbanken). In April 2000 MeritaNordbanken acquired Denmark's Unidanmark, thereby becoming the Nordic region's biggest bank in terms of assets. It became Nordea Bank AB in Dec. 2001. By 2009 approximately 71% of the Swedish population were using e-banking.

In June 2012 Sweden's gross external debt totalled US$1,006,665m.

There is a stock exchange in Stockholm.

ENERGY AND NATURAL RESOURCES

In 2007 Sweden obtained 31% of its energy from oil, down from 77% in 1970. It aims to end fossil fuel dependency completely by 2020. In 2008, 44·4% of energy consumption came from renewables (wind power, solar power, hydro-electric power, tidal power, geothermal energy and biomass), compared to the European Union average of 10·3%. Sweden has by some margin the highest proportion of its energy consumption coming from renewables of any EU member country. A target of 49% has been set by the EU for 2020.

Environment

Sweden's carbon dioxide emissions from the consumption and flaring of fossil fuels in 2008 were the equivalent of 6·2 tonnes per capita. An *Environmental Performance Index* compiled in 2008 ranked Sweden second in the world behind Switzerland, with 93·1%. The index examined various factors in six areas—air pollution, biodiversity and habitat, climate change, environmental health, productive natural resources and water resources.

Electricity

Sweden is rich in hydro-power resources. Installed capacity was 34,199 MW in 2007, of which 16,505 MW was in hydro-electric plants, 8,975 MW in nuclear plants and 7,890 MW in thermal plants. Electricity production in 2007 was 148,557m. kWh; consumption was 160,759 kWh. In 2007 consumption per capita was 16,478 kWh. A referendum of 1980 called for the phasing out of nuclear power by 2010. In Feb. 1997 the government began

denuclearization by designating one of the 12 reactors for decommissioning. However, in Feb. 2009 the government announced an end to the 30-year ban on nuclear plant construction. In 2010 there were ten nuclear reactors in operation.

Minerals

Sweden is a leading producer of iron ore with around 2% of the world's total output. It is the largest iron ore exporter in Europe. There are also deposits of copper, gold, lead, zinc and alum shale containing oil and uranium. Iron ore produced, 2005, 23·3m. tonnes; zinc (mine output, zinc content), 214,600 tonnes; copper (mine output, copper content), 97,800 tonnes.

Agriculture

In 2005 agricultural land totalled 3,216,839 ha. There were 2,703,333 ha. of arable land in 2005 and 513,505 ha. of natural pasture on agricultural holdings of more than 2 ha. Of the land used for arable farming in 2007, 2–5 ha. holdings covered a total area of 49,162 ha.; 5·1–10 ha. holdings covered 99,272 ha.; 10·1–20 ha., 194,006; 20·1–30 ha., 175,331; 30·1–50 ha., 322,764; 50·1–100 ha., 629,899 and holdings larger than 100 ha. covered 1,177,258 ha. There were 72,609 agricultural enterprises in 2007 compared to 150,014 in 1971 and 282,187 in 1951. Around 37% of the enterprises were between 5 and 20 ha. In 2005 Sweden set aside 222,268 ha. (7·3% of its agricultural land—one of the highest proportions in the world) for the growth of organic crops.

Agriculture accounts for 5·8% of exports and 7·9% of imports. The agricultural sector employs 3% of the workforce.

Chief crops	Area (1,000 ha.)			Production (1,000 tonnes)		
	2006	2007	2008	2006	2007	2008
Ley	1,055·1	1,081·1	1,114·3	2,425·0	2,720·8	2,326·1
Wheat	360·9	361·6	361·5	1,967·4	2,255·7	2,202·2
Sugar beet	44·2	40·7	36·8	2,189·0	2,137·7	1,974·9
Barley	315·1	326·7	405·8	1,110·6	1,439·0	1,671·6
Potatoes	28·2	28·4	26·9	777·8	789·0	853·2
Oats	206·1	207·9	227·6	624·4	889·8	820·0
Rye	23·5	24·7	27·6	115·4	137·6	168·8

Production (in 1,000 tonnes) in 2008: milk, 3,019; meat, 406; cheese, 114; butter, 39.

Livestock, 2008: pigs, 1,609,289; cattle, 1,558,381; sheep and lambs, 524,780; poultry, 7,194,759. There were 249,191 reindeer in Sami villages in 2007. Harvest of moose during open season 2008: 83,554.

Forestry

Forests form one of the country's greatest natural assets. The growing stock includes 42% spruce, 39% pine and 16% broad-leaved. In 2010 forests covered 28·20m. ha. (69% of the land area). Sweden's largest forest owner is Sveaskog, with holdings of 4·3m. ha. Since 2001 the company has been completely state-owned after taking over the part-privatized AssiDomän corporation. Public ownership (including the state) accounts for 19% of the forests, limited companies own 25% and the remaining 56% is in private hands. Of the 78·2m. cu. metres of wood felled in 2007, 40·1m. cu. metres were sawlogs, 31·7m. cu. metres pulpwood, 5·9m. cu. metres fuelwood and 0·5m. cu. metres other.

Fisheries

In 2008 the total saltwater catch was 218,955 tonnes, worth 968·4m. kr. In 2008 the saltwater fishing fleet comprised 1,464 vessels and there were 1,599 professional fishermen.

INDUSTRY

The leading companies by market capitalization in Sweden in March 2012 were: Hennes & Mauritz (H&M), a clothing company (US$52·7bn.); Nordea Bank AB (US$36·7bn.); and Ericsson, a technology hardware and equipment company (US$33·8bn.).

Manufacturing is mainly based on metals and forest resources. Chemicals (especially petrochemicals), building materials and decorative glass and china are also important.

Industry groups	Sales value of production (gross) in 1m. kr. 2007
Manufacturing industry	1,601,718
Food products, beverages and tobacco	123,671
Textiles and textile products, leather and leather products	11,965
Wood and wood products	90,112
Pulp, paper and paper products, publishers and printers	181,237
Coke, refined petroleum products and nuclear fuel	13,854
Chemicals, chemical products and man-made fibres	118,795
Other non-metallic mineral products	31,672
Basic metals	156,552
Fabricated metal products, machinery and equipment	787,711
Other manufacturing industries	46,549
Mines and quarries	28,166

Labour

In 2008 there were 4,898,000 persons in the labour force, of which 93·8% were employed. The main areas of employment were as follows: trade and communication (838,000); financial services and business activities (739,000); health and social work (721,000); manufacturing, mining, quarrying, electricity and water supply (689,000); education, research and development (537,000); personal services and cultural activities, and sanitation (397,000); construction (306,000); public administration (261,000); agriculture, forestry and fishing (101,000). The unemployment rate in Dec. 2012 was 7·8%, but with youth unemployment of 24·0% in the third quarter of 2012. In 2008, 69·6% of men and 63·8% of women were in employment. The average monthly salary in 2008 was 27,100 kr. (29,400 kr. for men and 24,700 kr. for women).

In 2008 a total of 106,801 working days were lost through strikes, compared to 1,971 in 2006.

INTERNATIONAL TRADE

Imports and Exports
Imports and exports (in 1m. kr.):

	2004	2005	2006	2007	2008
Imports	739,203	833,757	939,730	1,030,100	1,088,990
Exports	904,532	977,280	1,089,095	1,140,032	1,194,559

Breakdown by Standard International Trade Classification (SITC, revision 3) categories (value in 1bn. kr.):

	Imports		Exports	
	2006	2007	2006	2007
0. Food and live animals	59·7	64·6	31·4	33·4
1. Beverages and tobacco	7·6	8·1	5·8	6·5
2. Crude materials	31·4	37·3	63·9	69·7
3. Fuels and lubricants	117·5	113·6	64·4	63·0
4. Animal and vegetable oils	3·5	3·0	1·5	1·2
5. Chemicals	98·2	110·5	125·2	126·1
6. Manufactured goods	143·6	165·5	218·3	238·5
7. Machinery and transport equipment	366·4	398·8	481·2	502·2
8. Miscellaneous manufactured items	111·4	118·7	94·3	97·8
9. Other	0·5	1·1	3·2	3·1

Principal imports in 2007 (in 1bn. kr.): road vehicles, 111·7; petroleum and petroleum products, 98·9; electrical machines, apparatus and appliances, 65·5; telecommunications, sound recording and similar appliances, 54·6; iron and steel, 50·1. Principal exports in 2007 (in 1bn. kr.): road vehicles, 154·8; iron

and steel, 77·4; telecommunications, sound recording and similar appliances, 77·0; paper, paperboard and manufactures thereof, 73·2; medical and pharmaceutical preparations, 59·2. Machinery and transport equipment accounts for some 44% of Swedish exports. This includes the mobile phone sector, which is the largest product group in the Swedish export market. The telecommunications company Ericsson is now the leading export company, ahead of Volvo.

Imports and exports by countries (value in 1bn. kr.):

	Imports from		Exports to	
	2007	2008	2007	2008
Denmark	94·1	102·7	84·1	88·3
Finland	64·0	62·4	71·3	75·8
France	50·8	54·6	57·0	58·4
Germany	188·6	190·9	119·1	123·9
Netherlands	62·5	64·7	57·6	60·9
Norway	88·0	97·2	107·3	113·5
UK	74·8	68·8	81·6	87·5
USA	32·0	33·7	86·5	78·7

In 2008 fellow EU member countries accounted for 69·7% of imports and 60·0% of exports.

COMMUNICATIONS

Roads
In 2009 there were 215,597 km of roads open to the public of which 98,467 km were state-administered roads (main roads, 15,329 km; secondary roads, 83,138 km). There were also 1,855 km of motorway. 79% of all roads in 2005 were surfaced. Motor vehicles in 2008 included 4,279,000 passenger cars, 510,000 lorries, 13,000 buses and 489,000 motorcycles and mopeds. There were 1,015,997 Volvos, 434,757 Saabs, 343,060 Fords and 327,379 Volkswagens registered in 2006. Sweden has one of the lowest death rates in road accidents of any industrialized country, at 5·1 deaths per 100,000 people in 2007. 397 people were killed in traffic accidents in 2008.

Rail
Total length of railways at 31 Dec. 2008 was 11,022 km (7,866 km electrified). In 2008, 179m. passengers and 67m. tonnes of freight were carried. There is a metro in Stockholm (110 km), and tram/light rail networks in Stockholm (8 km), Gothenburg (118 km) and Norrköping (13 km).

Civil Aviation
The main international airports are at Stockholm (Arlanda), Gothenburg (Landvetter), Stockholm (Skavsta) and Malmö (Sturup). The principal carrier is Scandinavian Airlines System (SAS), which resulted from the 1950 merger of the three former Scandinavian airlines. SAS Sverige AB is the Swedish partner (SAS Denmark A/S and SAS Norge ASA being the other two). The Swedish government is the principal shareholder of SAS with a 21·4% share. The governments of Denmark and Norway each own 14·3%. Since 2001, 50% of SAS shares have been listed on the stock exchanges of Stockholm, Copenhagen and Oslo. SAS had a market capitalization in Oct. 2009 of 11,918m. kr. and an operating revenue in 2008 of 53,195m. kr.

Malmö Aviation, a Sweden-based carrier, operates some international as well as domestic flights.

In 2008 Stockholm (Arlanda) handled 18,136,165 passengers (13,281,466 on international flights) and 187,000 tonnes of freight. Gothenburg (Landvetter) was the second busiest airport, handling 4,303,722 passengers (3,158,822 on international flights) and 100,000 tonnes of freight. Malmö handled 1,882,428 passengers in 2006 (1,181,970 on domestic flights).

Shipping
The mercantile marine consisted on 31 Dec. 2008 of 1,036 vessels of 4·53m. GT. Cargo vessels entering Swedish ports in 2008 numbered 19,396 (125·74m. GT) while there were 75,343

passenger ferries (1,011·83m. GT). The number of cargo vessels leaving Swedish ports in 2008 totalled 19,389 (125·47m. GT) and the number of passenger ferries leaving was 75,636 (1,015·82m. GT).

The busiest port is Gothenburg. In 2007 a total of 42·33m. tonnes of goods were loaded and unloaded there (39·46m. tonnes unloaded from and loaded to foreign ports). Other major ports are Brofjorden, Trelleborg, Malmö and Luleå.

Telecommunications

In 2008 there were 5,323,000 main (fixed) telephone lines. In the same year mobile phone subscribers numbered 10,892,000 (1,183·3 per 1,000 persons). In June 2000 the state sold off a 30% stake in the Swedish telecommunications operator Telia. In Dec. 2002 Telia and the Finnish telecommunications operator Sonera merged to become TeliaSonera. The Swedish state owns 37·3% and the Finnish state 11·7%. Internet users numbered 8·1m. in 2008, or 87·8% of the total population (the second highest percentage in the world, after Iceland). In Dec. 2010 there were 82·9 wireless broadband subscribers per 100 inhabitants and 31·8 fixed broadband subscribers per 100. In March 2012 there were 4·5m. Facebook users.

According to the World Economic Forum's *Global Information Technology Report 2010–11* Sweden is ranked first in the world in exploiting global information technology developments.

SOCIAL INSTITUTIONS

Justice

Sweden has two parallel types of courts—general courts that deal with criminal and civil cases and general administrative courts that deal with cases related to public administration. The general courts have three instances: district courts, courts of appeal and the Supreme Court. There are 60 district courts, of which 23 also serve as real estate courts and four courts of appeal. The administrative courts also have three instances: 23 county administrative courts, four administrative courts of appeal and the Supreme Administrative Court. In addition, a number of special courts and tribunals have been established to hear specific kinds of cases and matters.

Every district court, court of appeal, county administrative court and administrative court of appeal has a number of lay judges. These take part in the adjudication of both specific concrete issues and matters of law; each has the right to vote.

Criminal cases are normally tried by one judge and three lay judges. Civil disputes are normally heard by a single judge or three judges. In the courts of appeal, criminal cases are determined by three judges and two lay judges. Civil cases are tried by three or four judges. In the settlement of family cases, lay judges take part in the proceedings in both the district court and in the court of appeal. Proceedings in the general administrative courts are in writing; i.e. the court determines the case on the basis of correspondences between the parties. Nevertheless, it is also possible to hold a hearing. The cases are determined by a single judge or one judge and three lay judges. In the administrative court of appeal, cases are normally heard by three judges or three judges and two lay judges.

Those who lack the means to take advantage of their rights are entitled to legal aid. Everyone suspected of a serious crime or taken into custody has the right to a public counsel (advocate). The title advocate can only be used by accredited members of the Swedish Bar Association. Qualifying as an advocate requires extensive theoretical and practical training. All advocates in Sweden are employed in the private sector.

The control over the way in which public authorities fulfil their commitments is exercised by the Parliamentary Ombudsmen and the Chancellor of Justice. In 2008–09 the Ombudsmen received 6,918 cases altogether, of which 68 were instituted on their own initiative. Sweden has no constitutional court. However, in each particular case the courts do have a certain right to ascertain

whether a statute meets the standards set out by superordinate provisions.

The population in penal institutions in April 2006 was 7,450 (82 per 100,000 of national population). There are 56 prisons spread throughout the country.

There were 209 incidents of murder, manslaughter and assault resulting in death in 2008 (121 in 1990 and 258 in 2007).

Education

In 2008–09 there were 906,189 pupils in 4,755 compulsory schools. In secondary education at the higher stage (the integrated upper secondary school) there were 396,336 pupils in Oct. 2008 (excluding pupils in the fourth year of the technical course regarded as third-level education). The folk high schools, 'people's colleges', had 27,508 pupils on courses of more than 15 weeks in the autumn of 2008.

In municipal adult education there were 170,318 students in 2007–08.

There are also special schools for pupils with visual and hearing impairments (516 pupils in 2008) and for those who are intellectually disabled (22,595 pupils).

In 2007–08 there were 384,733 students enrolled for undergraduate studies in integrated institutions for higher education. The number of students enrolled for postgraduate studies in 2008 was 16,922.

In 2007 public expenditure on education came to 6·6% of GDP (and accounted for 12·7% of total government expenditure). The adult literacy rate is at least 99%.

Health

In 2008 there were 29,100 doctors, 4,100 dentists, 71,400 nurses and midwives and 25,899 hospital beds. In 2008 Sweden spent 9·4% of its GDP on health.

In 2007, 14·5% of Swedes aged 15 and over smoked on a daily basis (males, 12%; females, 17%), the lowest percentage of any major industrialized country.

Welfare

Social insurance benefits are granted mainly according to uniform statutory principles. All persons resident in Sweden are covered, regardless of citizenship. All schemes are compulsory, except for unemployment insurance. Benefits are usually income-related. Most social security schemes are at present undergoing extensive discussion and changes.

Type of social insurance scheme	Payments 2008 (in 1m. kr.)
Old-age pension	219,376
Sickness insurance	96,748
Parental insurance	28,705
Child allowance	23,389
Attendance allowance	19,858
Survivor's pension	16,700
Unemployment insurance	13,592
Housing supplement	11,472
Work injury insurance	5,425

Under a Pension Reform Plan Sweden is one of the world's leaders in the shift to private pension systems. In the new system each worker's future pension will be based on the amount of money accumulated in two separate individual accounts. The bulk of retirement income will come from a notional account maintained by the government on behalf of the individual, but a significant portion of retirement income will come from a private individual account. There are two types of pension—the income pension and the premium pension. The income pension comes under a pay-as-you-go system, with the premium pension based on contributions invested in a fund chosen by the insured person.

There is also a guarantee pension for those aged 65 and resident in Sweden for the last three years but without an earnings-related pension. Its value in 2010 was 90,312 kr. for a single pensioner and 80,560 kr. for a married pensioner.

RELIGION

The Swedish Lutheran Church was disestablished in 2000. It is headed by Archbishop Anders Wejryd (b. 1948) and has its metropolitan see at Uppsala. In 2008 there were 13 bishoprics and 1,802 parishes. The clergy are chiefly supported from the parishes and the proceeds of the church lands. Around 70% of the population, equivalent to 6·6m. people, belong to the Church of Sweden. Other denominations, in 2010: Pentecostal Movement, 82,769 members; The Mission Covenant Church of Sweden, 60,445; InterAct, 32,138; Salvation Army, 5,159 soldiers; The Baptist Union of Sweden, 17,441; Swedish Alliance Mission, 13,687. There were also 96,950 Roman Catholics (under a Bishop resident at Stockholm). The Orthodox and Oriental churches number around 120,000 members.

Although there are no official statistics on the number of Muslims, their numbers were estimated at 450,000–500,000 in 2010. An estimated 20,000 Jews lived in Sweden in 2010.

CULTURE

Umeå will be one of two European Capitals of Culture for 2014. The title attracts large European Union grants.

World Heritage Sites

There are 15 sites under Swedish jurisdiction that appear on the UNESCO World Heritage List: the royal palace of Drottningholm (1991); the Viking settlements of Birka and Hovgården (1993); the Engelsberg ironworks (1993); the Bronze Age rock carvings in Tanum (1994); Skogskyrkogården cemetery (1994); the Hanseatic town of Visby (1995); the Lapponian area (home of the Sami people in the Arctic circle) (1996); the church town of Gammelstad in Luleå (1996); the naval port of Karlskrona (1998); the Kvarken Archipelago and High Coast (2000 and 2006), shared with Finland; the agricultural landscape of Southern Öland (2000); the Mining Area of the Great Copper Mountain in Falun (2001); the Varberg Radio Station (2004) at Grimeton in southern Sweden; the Struve Geodetic Arc (2005), a chain of survey triangulations spanning from Norway to the Black Sea that helped establish the exact shape and size of the earth and is shared with nine other countries; and the Decorated Farmhouses of Hälsingland (2012), an ensemble of timber buildings in northeastern Sweden, a fusion of local building and folk art traditions.

Press

In 2008 there were 168 daily newspapers with an average weekday net circulation of 3·7m. The leading papers in terms of circulation in 2008 were the free *Metro*, with an average daily circulation of 634,000 copies; the Social Democratic *Aftonbladet*, with an average daily circulation of 378,000; the independent *Dagens Nyheter*, with an average daily circulation of 340,000; and the liberal tabloid *Expressen*, with an average daily circulation of 304,000. In 2008 a total of 26,182 book titles were published.

Tourism

In 2008 Swedes stayed 20,042,193 nights in hotels in Sweden and 4,770,359 in holiday villages and youth hostels; and foreign visitors stayed 5,830,414 nights in hotels and 1,559,688 in holiday villages and youth hostels. There were 2,976 accommodation establishments in 2008 with 299,875 beds. Of 12·5m. trips abroad with an overnight stay undertaken by Swedes in 2006, 2·5m. were for business purposes and 10·0m. were leisure trips.

Festivals

Important traditional festivals include Lucia, held in Dec, and Walpurgis Night, a spring celebration held in April. The eight-day Malmö Festival includes live music and other cultural events and is held annually in Aug.

DIPLOMATIC REPRESENTATIVES

Of Sweden in the United Kingdom (11 Montagu Pl., London, W1H 2AL)
Ambassador: Nicola Clase.

Of the United Kingdom in Sweden (Skarpögatan 6–8, S-115 93 Stockholm)
Ambassador: Paul Johnston.

Of Sweden in the USA (1501 M St., NW, Suite 900, Washington, D.C., 20005-1702)
Ambassador: Jonas Hafström.

Of the USA in Sweden (Dag Hammarskjölds Väg 31, S-115 89 Stockholm)
Ambassador: Mark Brzezinski.

Of Sweden to the United Nations
Ambassador: Mårten Grunditz.

Of Sweden to the European Union
Permanent Representative: Dag Hartelius.

FURTHER READING

Statistics Sweden. *Statistik Årsbok/Statistical Yearbook of Sweden.*—*Historisk statistik för Sverige* (Historical Statistics of Sweden). 1955 ff.—*Allmän månadsstatistik* (Monthly Digest of Swedish Statistics).—*Statistiska meddelanden* (Statistical Reports). From 1963

Henrekson, M., *An Economic Analysis of Swedish Government Expenditure.* 1992

Nordstrom, Byron J., *The History of Sweden.* 2002

Petersson, O., *Swedish Government and Politics.* 1994

Schön, Lennart, *An Economic History of Sweden.* 2012

Sejersted, Francis, *The Age of Social Democracy: Norway and Sweden in the Twentieth Century.* 2011

Sveriges statskalender. Annual, from 1813

National library: Kungliga Biblioteket, PO Box 5039, SE–102 41 Stockholm.

National Statistical Office: Statistics Sweden, PO Box 24300, SE–104 51 Stockholm.

Website: http://www.scb.se

Swedish Institute Website: http://www.si.se

SWITZERLAND

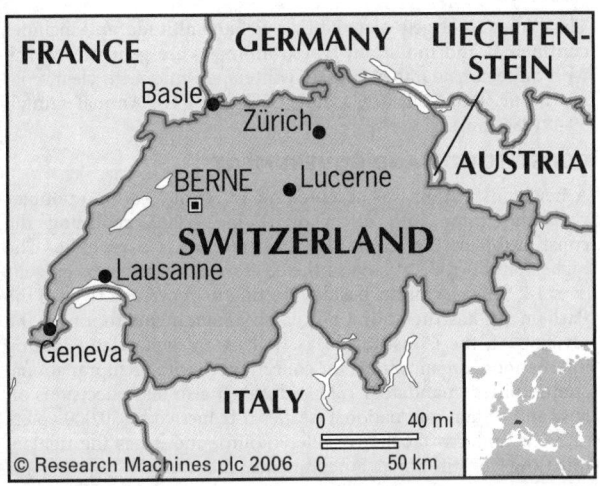

FRANCE · GERMANY · LIECHTEN-STEIN
Basle · Zürich · AUSTRIA
BERNE · Lucerne
SWITZERLAND
Lausanne
Geneva
ITALY
0   40 mi
0   50 km
© Research Machines plc 2006

Schweizerische Eidtgenossenschaft—
Confédération Suisse—
Confederazione Svizzera[1]
(Swiss Confederation)

Capital: Berne
Population projection, 2015: 7·81m.
GNI per capita, 2011: (PPP$) 39,924
HDI/world rank: 0·903/11
Internet domain extension: .ch

KEY HISTORICAL EVENTS

Neolithic settlements from around 3000 BC have been found. Celtic clans settled in fertile valleys in parts of present-day Switzerland from around 1500 BC, with the Raetians in the east and the Helvetti to the west. A Bronze Age Celtic civilization reached its height around 100 BC. An attempt by the Helvetti to spread west into Gaul was quashed by Julius Caesar in 58 BC. As the Roman Empire expanded northward and westward, Switzerland came under its domain, centred on Aventicum (Avenches). The Romans constructed a road network from the strategically important Alpine passes but attempts to conquer Germanic tribes to the north and east of the Rhine were thwarted in AD 9. Garrisons along the Rhine from Lake Constance to Basle were maintained until Roman forces withdrew in 401.

The Germanic Alemanni tribe became dominant in northern and central Switzerland as Rome's influence declined, while Latin-speaking Burgundians held sway in the Jura mountains. Celtic tribes were gradually subsumed over the following centuries. Frankish rulers established monasteries, enabling the spread of Christianity and feudalism throughout west-central Europe in the seventh and eighth centuries. Following the signing of the Treaty of Verdun in 840 western and southwest Switzerland came under the jurisdiction of the Burgundian king, Lothair I, and the north and east formed part of the domain of Louis the German. The Burgundian lands became part of the Holy Roman Empire in 1033, while various independent dukedoms emerged in the north and east, notably Swabia, Zahringen, Savoy and Kyberg. The Kyberg domains of central Switzerland passed to the Habsburgs in 1264. The expansion of this dynasty led to three mountain-based clans—the Uri, the Schwyz and the Unterwalden—forming a

defensive league. Their agreement, renewed in 1291, is considered the founding document of the Swiss nation. The league defeated the Habsburgs at Mortgarten in 1315 and by 1353 the confederation had added the cantons of Glarus and Zug and the city states of Lucerne, Zürich and Berne, forming the 'Old Federation' of eight states within the Holy Roman Empire.

Defeat at the hands of French forces at Marignano in 1515 led the Swiss confederation to form a 'perpetual alliance' with France and marked the start of a neutral stance. Relations between the cantons deteriorated in the 16th century and during the Reformation, when the city-states of Zürich, Berne, Basle and St Gallen adopted Protestantism while Catholicism was retained in the four forest cantons. The 1531 Treaty of Kappel ended the civil war and preserved Catholicism in the rural south, though religious tensions continued in the late 16th century with the rural cantons and city states linked only by neutrality in the Thirty Years War. The War's end in 1648 and the subsequent Peace of Westphalia saw Switzerland declared independent of the Holy Roman Empire.

Geneva, Basle, Berne and Zürich grew prosperous in the 18th century, becoming Enlightenment centres of intellectual and cultural achievement. Power remained with the oligarchs until the arrival of French Revolutionary troops in their offensive against Austrian and Russian forces in 1798. Napoleon's Act of Mediation in 1803 partially restored political power to the cantons but it was not until 1815 that the Congress of Vienna re-established Switzerland's independence, its perpetual neutrality guaranteed by Austria, France, Great Britain, Portugal, Prussia, Spain and Sweden.

In 1848 a new constitution was approved following disputes between Protestant and Catholic cantons. The 22 cantons were linked by a federal government (consisting of a bicameral parliament that elected a seven-member governing council) and a federal tribunal to rule on intra-cantonal disputes. This constitution was revised in 1874 to allow for national and local referenda on a range of issues.

Switzerland maintained its status of armed neutrality in the First World War but was unable to avoid mass unemployment, leading to a national strike in 1918 and demands for social security. The 1919 Treaty of Versailles reaffirmed Switzerland's neutrality and a year later it joined the League of Nations (based in Geneva). The Federal Council issued a declaration of neutrality at the start of the Second World War and much effort went in to shoring up defences and remaining self-sufficient while surrounded by the Axis powers. The Swiss government subsequently expressed regrets about the country's behaviour in the Second World War following a report by an independent panel of historians on relations with the Nazis.

In 1959 Switzerland became a founding member of the European Free Trade Association but has remained outside the European Union. It joined the UN by a narrow majority in a referendum in 2002. In 2005 Switzerland signed up to the Schengen accord.

TERRITORY AND POPULATION

Switzerland is bounded in the west and northwest by France, north by Germany, east by Austria and Liechtenstein and south by Italy. Area and population by canton (with date of establishment):

Canton	Area (sq. km) (1 Jan. 2008)	Census Population (1 Dec. 2000)	Population Estimate (1 Jan. 2010)
Uri (1291)	1,077	34,777	35,335
Schwyz (1291)	908	128,704	144,686
Obwalden (1291)	491	32,427	35,032

[1]The Latin 'Confoederatio Helvetica' is also in use.

Canton	Area (sq. km) (1 Jan. 2008)	Census Population (1 Dec. 2000)	Population Estimate (1 Jan. 2010)
Nidwalden (1291)	276	37,235	40,794
Lucerne (1332)	1,493	350,504	372,964
Zürich (1351)	1,729	1,247,906	1,351,297
Glarus (Glaris) (1352)	685	38,183	38,479
Zug (1352)	239	100,052	110,890
Fribourg (Freiburg) (1481)	1,671	241,706	273,159
Solothurn (Soleure) (1481)	790	244,341	252,748
Basel-Town (Bâle-V.) (1501)	37	188,079	187,898
Basel-Country (Bâle-C.) (1501)	518	259,374	272,815
Schaffhausen (Schaffhouse) (1501)	298	73,392	75,657
Appenzell-Outer Rhoden (1513)	243	53,504	53,043
Appenzell-Inner Rhoden (1513)	173	14,618	15,681
Berne (Bern) (1553)	5,959	957,197	974,235
St Gallen (St Gall) (1803)	2,026	452,837	474,676
Graubünden (Grisons) (1803)	7,105	187,058	191,861
Aargau (Argovie) (1803)	1,404	547,493	600,040
Thurgau (Thurgovie) (1803)	991	228,875	244,805
Ticino (Tessin) (1803)	2,812	396,846	335,720
Vaud (Waadt) (1803)	3,212	640,657	701,526
Valais (Wallis) (1815)	5,224	272,399	307,392
Neuchâtel (Neuenburg) (1815)	803	167,949	171,647
Geneva (Genève) (1815)	282	413,673	453,292
Jura (1979)	839	68,224	70,134
Total	41,285	7,228,010	7,785,806

In Dec. 2011 there were 4,032,400 females and 1,816,000 resident foreign nationals out of a total population of 7,954,700. In 2011 foreign nationals made up 22·8% of the population, one of the highest proportions in western Europe. In 2011, 73·7% of the population lived in urban areas. Population density in 2010 was 189 per sq. km.

The UN gives a projected population for 2015 of 7·81m.

German, French, Italian and Romansch (which is spoken mostly in Graubünden) are the official languages. German is spoken by the majority of inhabitants in 19 of the 26 cantons, French in Fribourg, Vaud, Valais, Neuchâtel, Jura and Geneva, and Italian in Ticino. At the 2000 census 63·7% of the population gave German as their mother tongue, 20·4% French, 6·5% Italian, 0·5% Romansch and 9·0% other languages.

At the end of 2011 the five largest cities were Zürich (377,000); Geneva (188,200); Basle (164,500); Lausanne (129,400); Berne (125,700). In 2011 the population figures of conurbations were: Zürich, 1,204,000; Geneva, 530,700; Basle, 500,600; Berne, 355,600; Lausanne, 342,200; other towns, 2008 (and their conurbations), Winterthur, 98,200 (135,000); St Gallen, 72,000 (148,500); Lucerne, 59,200 (205,400); Lugano, 54,400 (133,400); Biel, 50,000 (92,300).

SOCIAL STATISTICS

Statistics for calendar years:

	Live births	Marriages	Divorces	Deaths
2004	73,082	39,460	17,949	60,180
2005	72,903	40,139	21,332	61,124
2006	73,371	39,817	20,981	60,283
2007	74,494	40,330	19,882	61,089
2008	76,691	41,534	19,613	61,233

Rates (2008, per 1,000 population): birth, 10·0; death, 8·0; marriage, 5·4; divorce, 2·6. In 2005 the most popular age range for marrying was 30–34 for males and 25–29 for females. Expectation of life, 2008: males, 79·7 years; females, 84·4. In 2007 the suicide rate per 100,000 population was 15·1 (males, 21·9; females, 9·1). Annual population growth rate, 2000–05, 0·7%. Infant mortality, 2008, four per 1,000 live births; fertility rate, 2008, 1·5 births per woman. In 2008 Switzerland received 16,606 asylum applications, up from 10,844 in 2007. The World Health Organization's *World Health Statistics 2009* put citizens of Switzerland in equal second

place in a 'healthy life expectancy' list (level with San Marino and only behind Japan), with an expected 75 years of healthy life for babies born in 2007.

CLIMATE

The climate is largely dictated by relief and altitude, and includes continental and mountain types. Summers are generally warm, with quite considerable rainfall; winters are fine, with clear, cold air. Berne, Jan. 32°F (0°C), July, 65°F (18·5°C). Annual rainfall 39·4" (986 mm).

CONSTITUTION AND GOVERNMENT

A new constitution was accepted on 18 April 1999 in a popular vote and came into effect on 1 Jan. 2000, replacing the constitution dating from 1874. Switzerland is a republic. The highest authority is vested in the electorate, i.e. all Swiss citizens over 18. This electorate, besides electing its representatives to the Parliament, has the voting power on amendments to, or on the revision of, the Constitution as well as on Switzerland joining international organizations for collective security or supranational communities (mandatory referendum). It also takes decisions on laws and certain international treaties if requested by 50,000 voters or eight cantons (facultative referendum), and it has the right of initiating constitutional amendments, the support required for such demands being 100,000 voters (popular initiative). The Swiss vote in more referendums—three or four a year—than any other nation. A mandatory referendum and a constitutional amendment demanded by popular initiative require a double majority (a majority of the voters and a majority of the cantons voting in favour of the proposal) to be accepted while a facultative referendum is accepted if a majority of the voters vote in favour of the proposal. Between 1893 and March 2013, 183 initiatives were put to the vote but only 20 were adopted. The highest turnout for a popular initiative has been 80·5% and the lowest 32·1%.

The Federal government is responsible for legislating matters of foreign relations, defence (within the framework of its powers), professional education and technical universities, protection of the environment, water, public works, road traffic, nuclear energy, foreign trade, social security, residence and domicile of foreigners, civil law, banking and insurance, monetary policy and economic development. It is also responsible for formulating policy concerning statistics gathering, sport, forests, fishery and hunting, post and telecommunications, radio and television, private economic activity, competition policy, alcohol and gambling.

The legislative authority is vested in a parliament of two chambers: the Council of States (*Ständerat/Conseil des États*) and the National Council (*Nationalrat/Conseil National*). The Council of States is composed of 46 members, chosen and paid by the 23 cantons of the Confederation, two for each canton. The mode of their election and the term of membership depend on the canton. Three of the cantons are politically divided—Basle into Town and Country, Appenzell into Outer-Rhoden and Inner-Rhoden, and Unterwalden into Obwalden and Nidwalden. Each of these 'half-cantons' sends one member to the State Council. The Swiss parliament is a militia/semi-professional parliament.

The National Council has 200 members directly elected for four years, in proportion to the population of the cantons, with the proviso that each canton or half-canton is represented by at least one member. The members are paid from federal funds. The parliament sits for at least four ordinary three-week sessions annually. Extraordinary sessions can be held if necessary and if demanded by the Federal Council, 25% of the National Council or five cantons.

The 200 seats are distributed among the cantons according to population size:

Zürich	34	Aargau (Argovie)	15
Berne	26	St Gallen (St Gall)	12
Vaud (Waadt)	18	Geneva	11

Lucerne	10	Schwyz	4
Ticino (Tessin)	8	Zug	3
Basel-Country (Bâle-C.)	7	Jura	2
Fribourg (Freiburg)	7	Schaffhausen (Schaffhouse)	2
Solothurn (Soleure)	7	Appenzell Inner-Rhoden	1
Valais (Wallis)	7	Appenzell Outer-Rhoden	1
Thurgau (Thurgovie)	6	Glarus	1
Basel-Town (Bâle-V.)	5	Nidwalden	1
Graubünden (Grisons)	5	Obwalden	1
Neuchâtel (Neuenburg)	5	Uri	1

A general election takes place by ballot every four years. Every citizen of the republic who has entered on his 18th year is entitled to a vote, and any voter may be elected a deputy. Laws passed by both chambers may be submitted to direct popular vote, when 50,000 citizens or eight cantons demand it; the vote can be only 'Yes' or 'No'. This principle, called the *referendum*, is frequently acted on.

The chief executive authority is deputed to the *Bundesrat*, or Federal Council, consisting of seven members, elected for four years by the *United Federal Assembly*, i.e. joint sessions of both chambers, such as to represent both the different geographical regions and language communities. The members of this council must not hold any other office in the Confederation or cantons, nor engage in any calling or business. In the Federal Parliament legislation may be introduced either by a member, or by either chamber, or by the Federal Council (but not by the people). Every citizen who has a vote for the National Council is eligible to become a member of the executive.

The *President* of the Federal Council (called President of the Confederation) and the *Vice-President* are the first magistrates of the Confederation. Both are elected by the United Federal Assembly for one calendar year from among the Federal Councillors, and are not immediately re-eligible to the same offices. The Vice-President, however, may be, and usually is, elected to succeed the outgoing President.

The seven members of the Federal Council act as ministers, or chiefs of the seven administrative departments of the republic. The city of Berne is the seat of the Federal Council and the central administrative authorities.

National Anthem

'Trittst im Morgenrot daher'/'Sur nos monts quand le soleil'/ 'Quando il ciel' di porpora' ('When the morning skies grow red'); German words by Leonard Widmer, French by C. Chatelanat, Italian by C. Valsangiacomo, tune by Alberik Zwyssig.

GOVERNMENT CHRONOLOGY

Presidents since 1945. (BDP/PBD = Conservative Democratic Party of Switzerland; CVP/PDC = Christian Democratic People's Party; FDP/PLR = FDP.The Liberals; FDP/PRD = Free Democratic Party/Radical Democratic Party; SPS/PSS = Social Democratic Party of Switzerland; SVP/UDC = Swiss People's Party/Centre Democratic Union)

1945	SVP/UDC	Adolf Eduard von Steiger
1946	FDP/PRD	Karl Kobelt
1947	CVP/PDC	Philipp Etter
1948	CVP/PDC	Enrico Celio
1949	SPS/PSS	Ernst Nobs
1950	FDP/PRD	Max-Édouard Petitpierre
1951	SVP/UDC	Adolf Eduard von Steiger
1952	FDP/PRD	Karl Kobelt
1953	CVP/PDC	Philipp Etter
1954	FDP/PRD	Rodolphe Rubattel
1955	FDP/PRD	Max-Édouard Petitpierre
1956	SVP/UDC	Markus Feldmann
1957	FDP/PRD	Hans Streuli
1958	CVP/PDC	Thomas Emil Leo Holenstein
1959	FDP/PRD	Paul Chaudet
1960	FDP/PRD	Max-Édouard Petitpierre

1961	SVP/UDC	Friedrich Traugott Wahlen
1962	FDP/PRD	Paul Chaudet
1963	SPS/PSS	Willy Spühler
1964	CVP/PDC	Ludwig von Moos
1965	SPS/PSS	Hans-Peter Tschudi
1966	FDP/PRD	Hans Schaffner
1967	CVP/PDC	Roger Bonvin
1968	SPS/PSS	Willy Spühler
1969	CVP/PDC	Ludwig von Moos
1970	SPS/PSS	Hans-Peter Tschudi
1971	SVP/UDC	Rudolf Gnägi
1972	FDP/PRD	Nello Celio
1973	CVP/PDC	Roger Bonvin
1974	FDP/PRD	Ernst Brugger
1975	SPS/PSS	Pierre Graber
1976	SVP/UDC	Rudolf Gnägi
1977	CVP/PDC	Kurt Furgler
1978	SPS/PSS	Willi Ritschard
1979	CVP/PDC	Hans Hürlimann
1980	FDP/PRD	Georges-André Chevallaz
1981	CVP/PDC	Kurt Furgler
1982	FDP/PRD	Fritz Honegger
1983	SPS/PSS	Pierre Aubert
1984	SVP/UDC	Leon Schlumpf
1985	CVP/PDC	Kurt Furgler
1986	CVP/PDC	Alphons Egli
1987	SPS/PSS	Pierre Aubert
1988	SPS/PSS	Otto Stich
1989	FDP/PRD	Jean-Pascal Delamuraz
1990	CVP/PDC	Arnold Koller
1991	CVP/PDC	Flavio Cotti
1992	SPS/PSS	René Felber
1993	SVP/UDC	Adolf Ogi
1994	SPS/PSS	Otto Stich
1995	FDP/PRD	Kaspar Villiger
1996	FDP/PRD	Jean-Pascal Delamuraz
1997	CVP/PDC	Arnold Koller
1998	CVP/PDC	Flavio Cotti
1999	SPS/PSS	Ruth Dreifuss
2000	SVP/UDC	Adolf Ogi
2001	SPS/PSS	Moritz Leuenberger
2002	FDP/PRD	Kaspar Villiger
2003	FDP/PRD	Pascal Couchepin
2004	CVP/PDC	Joseph Deiss
2005	SVP/UDC	Samuel Schmid
2006	SPS/PSS	Moritz Leuenberger
2007	SPS/PSS	Micheline Calmy-Rey
2008	FDP/PRD	Pascal Couchepin
2009	FDP/PLR	Hans-Rudolf Merz
2010	CVP/PDC	Doris Leuthard
2011	SPS/PSS	Micheline Calmy-Rey
2012	BDP/PBD	Eveline Widmer-Schlumpf
2013	SVP/UDC	Ueli Maurer

RECENT ELECTIONS

In elections to the *National Council* on 23 Oct. 2011 the Swiss People's Party/Centre Democratic Union (SVP) took 26·6% of the vote (54 seats), the Social Democratic Party of Switzerland (SPS) 18·7% (46), FDP.The Liberals (FDP) 15·1% (30), the Christian Democratic People's Party (CVP) 12·3% (28), the Green Party (GPS) 8·4% (15), the Green Liberal Party (GLP) 5·4% (12), the Conservative Democratic Party of Switzerland (BDP) 5·4% (9), Evangelical People's Party (EVP) 2·0% (2) and the Ticino League (LdT) 0·8% (2). The Christian Social Party and the Geneva Citizens' Movement won one seat each. Turnout was 49·1%.

In the Council of States the CVP hold 13 seats, the FDP 11, the SPS 11, the SVP 5, the GLP 2, the GPS 2, the BDP 1 and ind. 1.

At an election held in the United Federal Assembly on 4 Dec. 2012 Ueli Maurer was elected president for 2013 with 148 votes out of 202; Didier Burkhalter was elected vice-president.

CURRENT GOVERNMENT

In March 2013 the Federal Council comprised:

President of the Confederation and Chief of the Department of Defence, Civil Protection and Sports: Ueli Maurer; b. 1950 (Swiss People's Party/SVP; sworn in 1 Jan. 2013).

Vice President and Chief of the Department of Foreign Affairs: Didier Burkhalter (FDP.The Liberals/FDP; sworn in 1 Jan. 2013).

Minister of Economic Affairs: Johann Schneider-Ammann (FDP). *Environment, Transport, Energy and Communications:* Doris Leuthard (CVP). *Finance:* Eveline Widmer-Schlumpf (BDP). *Home Affairs:* Alain Berset (SPS). *Justice and Police:* Simonetta Sommaruga (SPS).

Federal Authorities Website: http://www.admin.ch

CURRENT LEADERS

Ueli Maurer

Position
President

Introduction
Ueli Maurer joined the Swiss Federal Council on 1 Jan. 2009 and assumed the rotating one-year presidency on 1 Jan. 2013. A member of the Swiss People's Party (SVP), he also serves as defence minister on the seven-member council.

Early Life
Maurer was born on 1 Dec. 1950 in Wetzikon, in the canton of Zürich. He served a commercial apprenticeship and managed a farmers' co-operative from 1974 to 1994. He was elected to the council of his home town, Hinwil, in 1978 and five years later entered the cantonal parliament of Zürich. In 1991 he became president of the cantonal parliament and was elected to the National Council. He was manager of the Zürich Farmers' Association from 1994–2008, leaving to become president of the Swiss Vegetable Producers' Association (a role he relinquished on nomination to the Federal Council).

In 1996 Maurer was chosen as president of the SVP. Initially regarded as an acolyte of Christoph Blocher, the figurehead of the party's nationalist wing, Maurer proved himself an effective political operator. Under his sometimes controversial leadership—which included bullish stances on Europe, immigration and women's rights—the party became the country's most popular, doubling its share of the vote.

In Oct. 2007 Maurer resigned the SVP leadership after taking the party to its largest ever electoral victory. After narrowly failing to win a seat in the Council of States (the Federal Assembly's upper house), he took on the presidency of the SVP's Zürich division. Elected to the Federal Council in Dec. 2008 he assumed responsibility for defence, civil protection and sports.

In the parliamentary vote to decide the new president held on 5 Dec. 2012, Maurer received 148 of a possible 202 votes and received backing from all parliamentary groups bar the Green Party.

Career in Office
Maurer has pledged to use his presidency to focus on domestic issues, delegating international matters to his deputy, the foreign affairs minister, Didier Burkhalter.

DEFENCE

There are fortifications in all entrances to the Alps and on the important passes crossing the Alps and the Jura. Large-scale destruction of bridges, tunnels and defiles are prepared for an emergency.

Conscripts complete 18–21 weeks of basic training and then regular annual refresher training up to a set number of service days. In 2008 military expenditure totalled US$4,110m. (US$542 per capita), representing 0·8% of GDP.

Army

There are about 4,000 regular soldiers, but some 220,000 conscripts undergo training annually (18 or 21 weeks recruit training at 20; six or seven refresher courses of 19 days every year between 21 and 30). Proposals ('Army XXI') implemented in 2004 envisaged an Armed Forces based on the three areas of promoting peace, defence and general civil affairs support. Troop levels were cut to 220,000 (120,000 conscripts, 20,000 recruits, 80,000 reservists).

Since 2004 Switzerland has a Chief of the Armed Forces in the rank of a lieutenant-general. In peacetime the Army has no general; in time of war the Federal Assembly in joint session of both Houses appoints a general.

In 1999 for the first time a small Swiss contingent was deployed outside the country, in Kosovo.

Navy

There is no Navy in the Swiss Armed Forces but the Land Forces include a small Marine component with patrol boats.

Air Force

The Air Force has five air base commands. The fighter squadrons are equipped with Swiss-built F-5E Tiger IIs and F/A-18s. Personnel (2005), 19,000 on mobilization, with 85 combat aircraft.

INTERNATIONAL RELATIONS

In a referendum in 1986 the electorate voted against UN membership, but in a further referendum on 4 March 2002, 54·6% of votes cast were in favour of joining. Switzerland officially became a member at the UN's General Assembly in Sept. 2002. An official application for membership of the EU was made in May 1992, but in Dec. 1992 the electorate voted against joining the European Economic Area. At a referendum in March 2001, 76·7% of voters rejected membership talks with the EU, with just 23·3% in favour; turnout was 55·1%.

ECONOMY

Services accounted for 71·2% of GDP in 2006, industry 27·5% and agriculture 1·2%.

According to the anti-corruption organization *Transparency International*, Switzerland ranked sixth in the world in a 2012 survey of the countries with the least corruption in business and government. It received 86 out of 100 in the annual index.

Overview

Switzerland is a small economy with one of the highest living standards in the world. Owing to a lack of raw materials, prosperity is built on labour skills, a business-friendly environment and technological expertise. As well as banking and insurance, the economy is particularly focused on microtechnology, hi-tech industry, biotechnology and pharmaceuticals. Small- and medium-sized businesses dominate. According to a 2008 business census, more than 99% of enterprises had fewer than 250 full-time workers, employing about two-thirds of the total workforce.

The central government carries the responsibility for foreign policy, defence, pensions, postal services, telecommunications, railway services and currency. All other responsibilities are dealt with at the canton level, notably economic regulation, education, healthcare and the judiciary. This system has led to large income disparities across cantons, with tax revenue per capita differing by as much as a factor of two. In 2004 a referendum approved the New Financial Equalization System, which provided a clearer division between federal and local responsibilities.

In 2000 the National Bank introduced a monetary policy framework aimed at keeping inflation below 2%. Inflation has remained low for many years, contained primarily by strong retail competition and a flexible labour market. In 2004 the authorities launched a reform agenda to open sheltered sectors, reduce the role of the state, encourage external economic relations and improve the education system in a bid to boost competition and growth. Following these reforms growth performed above trend and became broad-based, benefiting from buoyant global markets and supported by strong investment and private consumption.

With Switzerland home to the world's third largest financial market, the financial sector accounts for over 11% of GDP. At the end of 2009, 6% of the total working population was directly employed by banks, insurance companies and other financial institutions. As a result, Switzerland suffered in the global financial crisis. Banks made large losses in 2008–09 and the country's largest bank, UBS, had to be rescued by the government in late 2008. The economy went into recession in 2009, although economic activity picked up in 2010 as the global economy began to recover and domestic demand grew. Reforms to strengthen the stability of the financial system have since been initiated. A countercyclical capital buffer, which requires banks to build up capital gradually as imbalances in the credit market develop, was implemented in July 2012. The eurozone debt crisis has driven up demand for the Swiss franc as a safe haven currency. Growth stood at 1·9% in 2011, down from 3·0% the previous year.

Pressures on the pension and health care systems are building in the face of a projected 16% increase in the old-age dependency ratio by 2035. A lack of consensus on the solution to long-term fiscal challenges has led to the rejection of proposals to raise the retirement age and increase the VAT rate to finance social security.

Currency

The unit of currency is the *Swiss franc* (CHF) of 100 *centimes* or *Rappen*. Foreign exchange reserves were US$78,864m. in Sept. 2009 and gold reserves were 33·44m. troy oz (77·79m. troy oz in 2000). Inflation rates (based on OECD statistics):

2002	2003	2004	2005	2006	2007	2008	2009	2010	2011
0·6%	0·6%	0·8%	1·2%	1·1%	0·7%	2·4%	−0·5%	0·7%	0·2%

Total money supply in July 2009 was 339,809m. Swiss francs.

Budget

Revenue and expenditure of the Confederation, in 1m. Swiss francs, for calendar years:

	2005	2006	2007	2008	2009
Revenue	52,985	58,506	58,739	64,243	68,082
Expenditure	52,607	53,096	62,178	64,189	58,704

VAT is 8%, with reduced rates of 3·8% and 2·5%.

Performance

Real GDP growth rates (based on OECD statistics):

2002	2003	2004	2005	2006	2007	2008	2009	2010	2011
0·2%	0·0%	2·4%	2·7%	3·8%	3·8%	2·2%	−1·9%	3·0%	1·9%

Total GDP was US$659·3bn. in 2011.

Switzerland was ranked first in the Global Competitiveness Index in the World Economic Forum's *Global Competitiveness Report 2012–2013*. The index analyses 12 areas of competitiveness for over 100 countries including macroeconomy, higher education and training, institutions, innovation and infrastructure. In the 2012 *World Competitiveness Yearbook*, compiled by the International Institute for Management Development, Switzerland came third in the world ranking. This annual publication ranks and analyzes how a nation's environment creates and sustains the competitiveness of enterprises.

Banking and Finance

The National Bank, with headquarters divided between Berne and Zürich, opened on 20 June 1907. It has the exclusive right to issue banknotes. The *Chairman* is Thomas Jordan.

On 31 Dec. 2004 there were 338 banks with total assets of 2,490,768m. Swiss francs. They included 24 cantonal banks, three big banks, 83 regional and saving banks, one Raiffeisen (consisting of around 420 member banks) and 277 other banks. The number of banks has come down from over 495 in 1990. In 2012 the largest banks in order of market capitalization were UBS (US$54·2bn.) and Crédit Suisse Groupe (US$36·2bn.). UBS ranks sixth in Europe by market capitalization and is Europe's ninth largest bank by assets, which totalled US$1,509,000m. in 2012. Banking, insurance and other finance activities is one of Switzerland's most successful industries, and contributed 10·3% of the country's GDP in 2011. Switzerland is the capital of the offshore private banking industry. It is reckoned that a third of the internationally invested private assets worldwide are managed by Swiss banks.

In June 2012 Switzerland's gross external debt totalled US$1,440,975m.

Money laundering was made a criminal offence in Aug. 1990. Complete secrecy about clients' accounts remains intact, but anonymity is lifted in cases of criminal offences such as money laundering, corruption and terrorism.

The stock exchange system has been reformed under federal legislation of 1990 on securities trading and capital market services. The four smaller exchanges have been closed and activity concentrated on the major exchanges of Zürich, Basle and Geneva, which harmonized their operations with the introduction of the Swiss Electronic Exchange (EBS) in Dec. 1995. Zürich is a major international insurance centre.

In Aug. 1998 Crédit Suisse and UBS AG agreed a deal to pay US$1·25bn. (£750m.) to Holocaust survivors over a three-year-period in an out-of-court settlement. The deal brought to an end the issue of money left in Holocaust victims' Swiss Bank accounts which were allowed to remain dormant after the war.

ENERGY AND NATURAL RESOURCES

Environment

In 2008 carbon dioxide emissions from the consumption and flaring of fossil fuels were the equivalent of 6·1 tonnes per capita. An *Environmental Performance Index* compiled in 2008 ranked Switzerland first in the world, with 95·5%. The index examined various factors in six areas—air pollution, biodiversity and habitat, climate change, environmental health, productive natural resources and water resources.

Switzerland is one of the world leaders in recycling. In 2005, 51% of all household waste was recycled, including 95% of glass and 90% of aluminium cans

Electricity

Installed capacity was 19·2m. kW in 2007. Production was 67·9bn. kWh in 2007. 54·1% of electricity produced in 2007 was hydro-electric, 41·1% nuclear and 4·8% from conventional thermal. In 1990, 54% of citizens voted for a ten-year moratorium on the construction of new nuclear plants. A referendum was held in 2003 on proposals to write a commitment to phase out nuclear power altogether into the constitution. However, the proposal was rejected. In May 2011 the government decided to abandon plans to build any further nuclear reactors in the wake of the Fukushima disaster in Japan. There are currently five nuclear reactors in use. Consumption per capita in 2007 was 8,726 kWh.

Minerals

In 2008, 4,832 people were employed in mining and quarrying. Production estimates in 2006 (in 1,000 tonnes): salt, 560; gypsum, 300; lime, 75.

Agriculture

The country is self-sufficient in milk. Agriculture is protected by subsidies and import controls. Farmers are guaranteed an income equal to industrial workers. In 2005 agriculture occupied 3·6% of the total workforce. There were 286,300 ha. of open arable land in 2005, 119,100 ha. of cultivated grassland and 625,100 ha. of natural grassland and pastures. In 2005 there were 12,900 ha. of vineyards. There were 63,300 farms in 2005 (41% in mountain or hill regions), of which 2,800 were under 1 ha., 19,900 over 20 ha. and 17,700 in part-time use. In 2005 there were 405,400 ha. of arable land and 22,900 ha. of permanent crops. Approximately 11·0% of all agricultural land is used for organic farming—one of the highest proportions in the world.

Area harvested, 2005 (in 1,000 ha.): cereals, 168; sugar beets, 18; potatoes, 13. Production, 2005 (in 1,000 tonnes): sugar beets, 1,409; wheat, 539; potatoes, 468; barley, 231; maize, 199; rapeseed, 56; carrots, 36. Fruit production (in 1,000 tonnes) in 2005 was: apples, 204; grapes, 127; pears, 65. Wine is produced in 25 of the cantons. In 2007 vineyards produced 104m. litres of wine.

Livestock, 2005 (in 1,000): pigs, 1,609; cattle, 1,554; sheep, 446; goats, 74; horses, 55; chickens, 8,117. Livestock products, 2005 (in 1,000 tonnes): meat, 450; milk, 3,934; cheese, 168.

Forestry

In 2010 the area under forests was 1·24m. ha., or 31% of the total land area. In 2007, 5·69m. cu. metres of roundwood were cut.

Fisheries

Total catch, 2010, 1,653 tonnes, exclusively freshwater fish.

INDUSTRY

The leading companies by market capitalization in Switzerland in March 2012 were: Nestlé SA, a world leader in food and beverages (US$207·4bn.); Novartis AG, a pharmaceuticals company (US$151·8bn.); and Roche AG, a health care company (US$151·5bn.).

The chief food producing industries, based on Swiss agriculture, are the manufacture of cheese, butter, sugar and meat. Among the other industries, the manufacture of textiles, clothing and footwear, chemicals and pharmaceutical products, the production of machinery (including electrical machinery and scientific and optical instruments) and watch and clock making are the most important. The leading industries in 2007 in terms of value added (in 1m. Swiss francs) were: construction, 26,516 (5·1% of GDP); chemical industry and oil processing, 20,394 (3·9%); medical and optical instruments and watches, 14,802 (2·8%); machinery and equipment, 13,723 (2·6%); electricity, gas, steam and distribution of water, 9,702 (1·9%).

Labour

In 2011 the total working population was 4,366,000, of whom 670,000 people were in manufacturing, 615,000 in trade and 566,000 in health. The unemployment rate in the second quarter of 2012 was 3·7%. In 2011, 85·4% of men and 73·3% of women between the ages of 15 and 64 were in employment. The percentage of men in employment is one of the highest among the major industrialized nations.

The foreign labour force was 1,014,000 in 2011 (410,000 women). Of these 187,000 were German, 158,000 Italian, 143,000 Portuguese and 60,000 French. In 2011 approximately 698,000 EU citizens worked in Switzerland.

INTERNATIONAL TRADE

Legislation of 1991 increased the possibilities of foreign ownership of domestic companies.

Imports and Exports

Imports and exports, excluding gold (bullion and coins) and silver (coins), were (in 1m. Swiss francs):

	2006	2007	2008	2009	2010
Imports	177,148	193,216	197,521	168,998	183,436
Exports	185,216	206,252	215,984	187,448	203,484

In 2004 the EU accounted for 81·4% of imports (112·9bn. Swiss francs) and 61·9% of exports (91·3bn. Swiss francs). Main import suppliers in 2004 (share of total trade): Germany, 32·8%; Italy, 11·3%; France, 9·9%; Netherlands, 5·0%; USA, 4·7%. Main export markets: Germany, 20·2%; USA, 10·4%; France, 8·7%; Italy, 8·3%; UK, 5·1%.

Main imports in 2004 (in 1m. Swiss francs): consumer goods, 55,318; raw materials and semi-manufactures, 35,680; equipment goods, 34,946.

Main exports in 2004 (in 1m. Swiss francs): chemicals, 49,445; machinery and electronics, 33,479; precision instruments, clocks and watches and jewellery, 24,195.

COMMUNICATIONS

Roads

In 2011 there were 71,452 km of roads, comprising 1,415 km of motorways, 18,411 km of highways and national roads and 51,638 km of secondary and local roads. Motor vehicles in 2011 (in 1,000): passenger cars, 4,163; motorcycles and mopeds, 834; vans and lorries, 349; buses and coaches, 16. Freight transported by road in 2010 totalled 17·1bn. tonne-km. Switzerland has one of the lowest death rates in road accidents of any industrialized country, at 4·1 deaths per 100,000 people in 2011. Road accidents injured 23,242 people in 2011 and killed 320 (down from 954 in 1990).

Switzerland was ranked fourth for its road infrastructure in the World Economic Forum's *Global Competitiveness Report 2009–2010.*

Rail

In 2007 the length of the general traffic railways was 5,021 km, of which the Swiss Federal Railways (SBB) 3,011 km. In 2009 Swiss railway companies carried 437m. passengers and 61m. tonnes of freight. In Oct. 2010 the world's longest rail tunnel was created— the 57-km long tunnel under the Gotthard mountain range in the Alps linking Erstfeld and Bodio. The tunnel is scheduled to open in 2017. There is a metro in Lausanne and a number of tram/light rail networks, notably in Basle, Berne, Geneva, Lausanne, Neuchâtel and Zürich. There are many other lines, the most important of which are the Berne–Lötschberg–Simplon (114 km from Berne to Brig) and Rhaetian (397 km) networks.

Switzerland was ranked first for rail infrastructure in the World Economic Forum's *Global Competitiveness Report 2009–2010.*

Civil Aviation

Switzerland owns seven airports with international scheduled and charter traffic: Basle (the binational Euroairport, which also serves Mulhouse in France), Berne (Belp), Geneva (Cointrin), Lugano (Agno), Sion, St Gallen (Altenrhein) and Zürich (Kloten). In 2010 these airports handled 39,009,046 passengers and 379,389 tonnes of freight and mail. Swissair, the former national carrier, faced collapse and grounded flights in Oct. 2001. In April 2002 a successor airline, Swiss International Air Lines (Swiss), took over as the national carrier. Services were also provided in 2003 by over 80 foreign airlines. Zürich is the busiest airport, handling 22,854,358 passengers in 2010 and 304,166 tonnes of freight. Geneva handled 11,748,972 passengers and 31,405 tonnes of freight in 2010. Together these two airports accounted for 89% of Swiss passenger traffic in 2010.

In the World Economic Forum's *Global Competitiveness Report 2009–2010* Switzerland ranked fifth for quality of air transport infrastructure.

Shipping
In 2010 there were 1,226 km of navigable waterways. 5·5m. tonnes of freight were transported on the Rhine and Swiss lakes in 2011. A merchant marine was created in 1941, the place of registry of its vessels being Basle. In 2007 it totalled 581,683 GRT.

Telecommunications
In 2008 there were 4·8m. main (fixed) telephone lines. In the same year mobile phone subscribers numbered 8·9m. (1,179·7 per 1,000 persons). There were 5·8m. internet users in 2008. The fixed broadband penetration rate in Dec. 2010 was 38·1 subscribers per 100 inhabitants. In March 2012 there were 2·7m. Facebook users.

SOCIAL INSTITUTIONS

Justice
The Federal Court, which sits at Lausanne, consists of 30 judges and 30 supplementary judges, elected by the Federal Assembly for six years and eligible for re-election; the President and Vice-President serve for two years and re-election is not practised. The Tribunal has original and final jurisdiction in suits between the Confederation and cantons; between different cantons; between the Confederation or cantons and corporations or individuals; between parties who refer their case to it; or in suits which the constitution or legislation of cantons places within its authority. It is a court of appeal against decisions of other federal authorities, and of cantonal authorities applying federal laws. The Tribunal comprises two courts of public law, two civil courts, a chamber of bankruptcy, a chamber of prosecution, a court of criminal appeal, a court of extraordinary appeal and a Federal Criminal Court.

A Federal Insurance Court sits in Lucerne, and comprises 11 judges and 11 supplementary judges elected for six years by the Federal Assembly.

A federal penal code replaced cantonal codes in 1942. It abolished capital punishment except for offences in wartime; this latter proviso was abolished in 1992.

The population in penal institutions in 2008 was 5,780 (75 per 100,000 population), of which 69·7% were non-nationals.

Education
Education is administered by the confederation, cantons and communes and is free and compulsory for nine years. Compulsory education consists of four years (Basel-Town and Vaud), five years (Aargau, Basel-Country, Neuchâtel and Ticino) or six years (other cantons) of primary education, and the balance in Stage I secondary education. This is followed by three to five years of Stage II secondary education in general or vocational schools. Tertiary education is at universities, universities of applied science, higher vocational schools and advanced vocational training institutes.

In 2009–10 there were 147,200 children in pre-primary schools. There were 769,314 pupils in compulsory education (436,111 at primary, 294,405 at lower secondary and 38,798 at special schools), 343,297 in Stage II secondary education and 250,073 students in higher education, including 126,940 students at universities and 69,676 at universities of applied sciences.

There are ten universities (date of foundation and students in 2010–11): Basle (1460, 12,367), Berne (1528, 14,442), Fribourg (1889, 9,651), Geneva (1559, 15,666), Lausanne (1537, 12,066), Lucerne (16th century, 2,450); Neuchâtel (1866, 4,215), St Gallen (1899, 6,996), Svizzera italiana (1996, 2,848), Zürich (1523, 26,134); and two institutions of equivalent status: Federal Institute of Technology Lausanne (1853, 8,009); Federal Institute of Technology Zürich (1854, 15,984). There are nine universities of applied sciences (date of foundation and enrolment figures for 2010–11): Berne (1997, 6,369); Central Switzerland (1997, 5,823); Eastern Switzerland (1997, 4,818); Kalaidos (2005, 1,664); Les Roches-Gruyère (2008, 117); Northwestern Switzerland (1997, 9,938); Southern Switzerland (1997, 3,661); Western Switzerland

(1997, 16,208); Zürich (2007, 16,727). 17% of university students in 2007 were foreign.

In 2006 public expenditure on education came to 5·1% of GNI and accounted for 16·3% of total government expenditure. The adult literacy rate is at least 99%.

Health
In 2011 there were 30,849 doctors, 4,123 dentists and 4,284 pharmacists. There were 300 hospitals with 38,373 beds in 2011. In 2007, 27·9% of the population were smokers. In 2011 Switzerland spent 8·1% of its GDP on health. Although active euthanasia is illegal in Switzerland, doctors may help patients die if they have given specific consent.

Welfare
The Federal Insurance Law against accident and illness, of 13 June 1911, entitled all citizens to insurance against illness; foreigners could also be admitted to the benefits. Major reform of the law was ratified in 1994 and came into effect in 1996, making it compulsory for all citizens. Subsidies are paid by the Confederation and the Cantons only for insured persons with low incomes. Also compulsory are the Old-Age and Survivors' Insurance (OASI, since 1948), Invalidity Insurance (II, since 1960) and Accident Insurance (1984/1996). Unemployment Insurance (1984) and Occupational benefit plans (Second Pillar, 1985) are compulsory for employees only.

The following amounts (in 1m. Swiss francs) were paid in social security benefits:

	2006	2007
Occupational pension plans	36,944	39,089
Old-age and survivors' insurance	31,559	33,171
Mandatory health insurance	19,095	19,897
Disability insurance	11,292	11,736
Accident insurance for employees	6,514	6,477
Family allowances	4,860	4,973
Unemployment insurance	5,080	4,311
Supplementary benefits (OASI/II)	3,080	3,246
Total (including other benefits)	127,496	132,364

RELIGION
There is liberty of conscience and of creed. At the 2000 census 41·8% of the population were Roman Catholic, 35·3% Protestant and 11·1% without religion. In 2000 the figures were estimated to be: Roman Catholics, 3,048,000; Protestants, 2,569,000; other, 1,671,000. In Feb. 2013 the Roman Catholic church had four cardinals with Swiss nationality.

CULTURE

World Heritage Sites
There are 11 sites in Switzerland that appear on the UNESCO World Heritage List. They are (with the year entered on list): the Abbey-Cathedral of St Gallen (1983), the 9th-century Benedictine convent of St John at Müstair (1983), the Old City of Berne (1983), the three castles and city walls of Bellinzona (2000), the Jungfrau-Aletsch-Bietschhorn mountain region (2001 and 2007), Monte San Giorgio (2003), the Lavaux vineyard terraces (2007), the Swiss Tectonic Arena Sardona (2008) and La Chaux-de-Fonds/Le Locle watchmaking town-planning (2009). The Rhaetian Railway in the Albula/Bernina Landscapes (2008) is shared with Italy and the Prehistoric Pile dwellings around the Alps (2011) is shared with Austria, France, Germany, Italy and Slovenia.

Press
There were 95 daily newspapers in 2008 (87 paid-for) and 101 paid-for non-daily papers; the combined circulation of paid-for papers was 2,650,000 in 2008. The average circulation of free dailies rose from 619,000 in 2004 to 1,886,000 in 2008.

Tourism

Tourism is an important industry. In 2008 there were 8·61m. non-resident tourists staying at hotels and similar establishments, bringing revenue of US$17,567m. Overnight stays by tourists in hotels and health establishments totalled 37,334,000 in 2008 (21,508,000 by foreigners). The main countries of origin of foreign tourists were Germany (6,313,000 overnight stays in 2008), the UK (2,282,000) and the USA (1,518,000). 11·15m. Swiss citizens travelled abroad in 2008.

Festivals

The Lucerne Festival is one of Europe's leading cultural events and since 2001 has been split into three festivals: Ostern during Lent, Sommer in Aug.–Sept. and Piano in Nov. The 2008 summer festival was attended by 108,200 people. The Montreux Jazz Festival is held annually in July. The 2007 festival attracted 220,000 people. The biggest rock and pop festivals are OpenAir St Gallen (June), Paléo Festival in Nyon (July), Gurtenfestival in Berne (July) and Rock Oz'Arènes in Avenches (Aug.).

DIPLOMATIC REPRESENTATIVES

Of Switzerland in the United Kingdom (16–18 Montagu Pl., London, W1H 2BQ)
Ambassador: Anton Thalmann.

Of the United Kingdom in Switzerland (Thunstrasse 50, 3005 Berne)
Ambassador: Sarah Gillett, CMG, MVO.

Of Switzerland in the USA (2900 Cathedral Ave., NW, Washington, D.C., 20008)
Ambassador: Manuel Sager.

Of the USA in Switzerland (Sulgeneckstrasse 19, 3007 Berne)
Ambassador: Donald S. Beyer, Jr.

Of Switzerland to the United Nations
Ambassador: Paul R. Seger.

Of Switzerland to the European Union
Ambassador: Roberto Balzaretti.

FURTHER READING

Office Fédéral de la Statistique. *Annuaire Statistique de la Suisse.*

Bewes, Diccon, *Swiss Watching: Inside Europe's Landlocked Island.* 2010

Butler, Michael, Pender, Malcolm and Charnley, Joy, *Making of Modern Switzerland, 1848–1998.* 2000

Church, Clive, *Politics and Government of Switzerland.* 2003

Kriesi, Hanspeter, Farago, Peter, Kohli, Martin and Zarin-Nejadan, Milad, *Contemporary Switzerland.* 2005

New, M., *Switzerland Unwrapped: Exposing the Myths.* 1997

National library: Bibliothèque Nationale Suisse, Hallwylstr. 15, 3003 Berne.

National Statistical Office: Office Fédéral de la Statistique, Espace de l'Europe 10, 2010 Neuchâtel.
SFSO Information Service email: *information@bfs.admin.ch*
Website: http://www.bfs.admin.ch

SYRIA

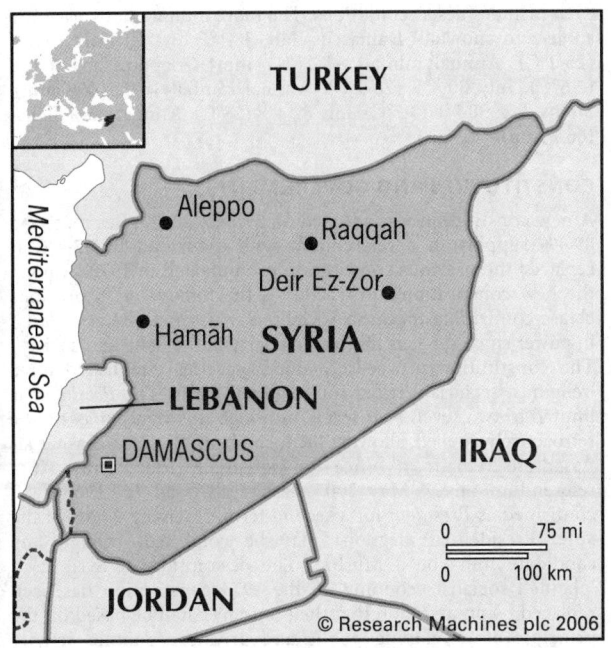

© Research Machines plc 2006

Jumhuriya al-Arabya as-Suriya
(Syrian Arab Republic)

Capital: Damascus
Population projection, 2015: 22·18m.
GNI per capita, 2011: (PPP$) 4,243
HDI/world rank: 0·632/119
Internet domain extension: .sy

KEY HISTORICAL EVENTS

Ancient Syria, a region encompassing modern Israel, Palestine, Lebanon and Jordan, was home to some of the world's earliest civilizations. From the city of Ebla, founded around 3000 BC, the Semitic empire developed. This was succeeded around 2260 BC by the Akkadian empire, then by the Amorites whose cities fell to the Hittites in the mid-2nd millennium BC. During the next 500 years Canaanites, Phoenicians, Aryans, Aramaeans and Hebrews settled different parts of the region. From the 9th–7th centuries BC the Assyrian empire dominated until, weakened by Cimmerian and Scythian immigration, it gave way to Babylonian rule.

The Babylonian empire, which saw the enslavement of the Jews, was defeated in 539 BC by the Persian King Cyrus. In the 4th century BC Alexander the Great overthrew the Persians and Syria came under Greek rule until the expansion of Rome in the early 2nd century BC. Syria became a Roman province in 64 BC. The Greek cities of the interior (the Decapolis) were rebuilt, including Damascus. Palmyra, a key city on the trade routes to the Euphrates, rose against Rome under Queen Zenobia but was defeated in AD 272.

Syria became an important frontier zone under Diocletian, who established lines of defence (*limes*) against eastern invaders. Syrian cities such as Edessa had contained the earliest Christian communities, and Antioch, where St Peter preached, grew in significance when Emperor Constantine moved the seat of the Empire east to Byzantium in 303. In 451 it was made a patriarchate.

Syria prospered under Byzantium until the Persian invasions of the 6th century. In the 620s Emperor Heraclius briefly regained the region for Byzantium, before ceding it to Muslim Arab forces at the Battle of the Yarmuk River in 636. Syria was at the heart, geographically and politically, of the Ummayad empires for the next 100 years. From 661 Damascus was capital of the Ummayad Caliphate, though its influence declined after 750 when the Abbasid Caliphate moved the capital to Baghdad.

From 969 the resurgent Byzantine Empire challenged for control of the region, twice conquering its principal cities. In 1085 Syria was conquered by the Seljid Turks. With the beginning of the Crusades Muslim-held strongholds were repeatedly attacked by Christian invaders. Antioch and Edessa fell to the crusaders in 1098 and Jerusalem in 1099. In the 12th century Muslim tribes won back much land under the successive leaderships of Zengi of Mosul, Nur-ad Din and Salah ad-Din (Saladin). Salah ad-Din founded the Ayyubid dynasty and in 1187 his forces recaptured Jerusalem from the Christians.

In 1260 a series of Mongol invasions began from the east, destroying cities and agriculture before being repelled by the Egyptian Mamluks. The Mamluks also drove the last crusaders from the Holy Land in 1302. The area remained under Mamluk rule until it fell to the Ottoman Turks in 1516. In the 1830s Egyptian forces invaded Syria as part of a wider war against the Ottomans. European powers became involved, brokering several agreements and finally forcing Egypt to withdraw in 1840.

During the First World War, in which Turkey joined the Central Powers, the French and British drew up the Sykes–Picot agreement, which planned the division of the Middle East into areas of French and British control. After the end of the war, the Ottoman Empire was dissolved and Lebanon, Palestine and Transjordan were made separate territories, reducing Syria to its modern borders.

In March 1920 Faisal ibn Husayn of Mecca became king of Syria but was deposed by the French, who were awarded the Syrian mandate by the League of Nations later that year. A widespread revolt against French rule was suppressed in 1925, with French troops bombarding Damascus. An abortive attempt to negotiate independence in 1936 was followed by more fruitful discussions in 1941, when British and Free French forces occupied the country. Syria held elections in 1943, electing the nationalist Shukri al-Kuwatli as president. Syrian independence was recognized on 1 Jan. 1944 and European forces withdrew in 1946.

A series of military coups from 1949–54 interrupted civilian government. The 1948 Arab–Israeli War, in which Syria fought for the Palestinians, ushered in decades of mutual hostility with neighbouring Israel. In 1956 the Suez Crisis prompted a period of martial law. In the same year Syria signed a pact with the Soviet Union which ensured supplies of military equipment in return for Communist influence. Popular enthusiasm for the Pan-Arabist movement led Syria to unite with Egypt in 1958, creating the United Arab Republic. However, Syria seceded in Sept. 1961 and a series of military coups ensued, culminating in the Ba'ath Party taking power in 1963. After the Syrian Ba'athists split from the Iraqi Ba'athists in 1966, long-term tensions arose between the two countries and led to Syrian support for Iran.

The 1967 war with Israel resulted in the loss of the Golan Heights. In 1970 Hafez al-Assad (of the minority Alawite sect) seized power. He was elected president in 1971 and embarked on a 'corrective' movement to end corruption. Domestic opposition was suppressed and the Sunni fundamentalist Muslim Brotherhood was destroyed along with the city of Hamah in 1982. Syrian forces invaded Lebanon in 1976 to prevent a Palestinian victory over the Maronite Christians, with whom Syria had political ties.

Assad joined the international coalition against the Iraqi occupation of Kuwait in 1991 and engaged in unsuccessful talks with Israel in the 1990s. His son, Bashar, took over on Assad's death in 2000. In 2003 Syria refused to back the US-led invasion of Iraq and relations with Washington were further strained over US accusations of Syrian support for terrorism. Though Syrian influence in Lebanon remained strong, it withdrew its last troops from the country in 2005, after involvement in the assassination of the former Lebanese prime minister, Rafiq al-Hariri. Following French-brokered talks, in Oct. 2008 Syria established diplomatic relations with Lebanon for the first time.

Beginning in March 2011 there were popular anti-government protests throughout the country, echoing similar demonstrations in the region and leading to the dismissal of the government. In April 2011 the state of emergency in place since 1963 was lifted. The USA and the EU imposed sanctions from May 2011 in response to the violent repression of protests that have resulted in more than 60,000 deaths according to the United Nations. Over 700,000 people have left the country in the wake of the crisis.

TERRITORY AND POPULATION

Syria is bounded by the Mediterranean and Lebanon in the west, by Israel and Jordan in the south, by Iraq in the east and by Turkey in the north. The frontier between Syria and Turkey was settled by the Franco-Turkish agreement of 22 June 1929. The area is 185,180 sq. km (71,498 sq. miles). The census of 2004 gave a population of 17,921,000; density, 97 per sq. km. Estimate, 2011, 21,377,000. In 2011, 56·2% of the population lived in urban areas.

The UN gives a projected population for 2015 of 22·18m.

Area and population (2011 estimate, in 1,000) of the 14 districts (*mohafaza*):

	Sq. km	Population
Aleppo (Halab)	18,500	4,868
Damascus City	105	1,754
Damascus District	18,032	2,836
Dará	3,730	1,027
Deir Ez-Zor	33,060	1,239
Hamah	8,883	1,628
Hasakah	23,334	1,512
Homs (Hims)	42,223	1,803
Idlib	6,097	1,501
Lattakia (Ladhiqiyah)	2,297	1,008
Qunaytirah	1,861	90
Raqqah	19,616	944
Suwaydá	5,550	370
Tartous	1,892	797

The capital is Damascus (Dimashq), with a 2004 population of 1,414,913. Other principal towns (population, 2004 in 1,000): Aleppo, 2,132·1; Homs, 652·6; Lattakia, 383·8; Hamah, 313·0; Raqqah, 220·5; Deir Ez-Zor, 211·9; Hasakah, 188·2; Al-Kamishli, 184·2.

Over 1m. Iraqi refugees entered Syria between 2003 and 2007, although many started to return in late 2007. There were an estimated 755,000 refugees in Syria in 2011, a figure exceeded only in Pakistan and Iran.

Arabic is the official language, spoken by 90% of the population, while 9% speak Kurdish (chiefly in Hasakah in the northeast of the country) and 1% other languages.

SOCIAL STATISTICS

2008 births, estimate, 594,000; deaths, 72,000. Rates, 2008 estimate (per 1,000 population): birth, 28·0; death, 3·4. Infant mortality, 2010 (per 1,000 live births), 14. Expectation of life, 2007, was 72·2 years for males and 76·0 for females. Annual population growth rate, 2000–08, 3·1%. Fertility rate, 2008, 3·2 births per woman.

CLIMATE

The climate is Mediterranean in type, with mild wet winters and dry, hot summers, though there are variations in temperatures and rainfall between the coastal regions and the interior, which even includes desert conditions. The more mountainous parts are subject to snowfall. Damascus, Jan. 38·1°F (3·4°C), July 77·4°F (25·2°C). Annual rainfall 8·8" (217 mm). Aleppo, Jan. 36·7°F (2·6°C), July 80·4°F (26·9°C). Annual rainfall 10·2" (258 mm). Homs, Jan. 38·7°F (3·7°C), July 82·4°F (28°C). Annual rainfall 3·4" (86·7 mm).

CONSTITUTION AND GOVERNMENT

A new constitution was adopted on 27 Feb. 2012, after receiving 89·4% support in a referendum with a turnout of 57·4%. It replaced the previous constitution promulgated in 1973. Among the new constitution's provisions is the removal of a previous clause confirming the Arab Socialist Renaissance (*Ba'ath*) Party, in power since 1963, as the 'leading party in the State and society'. The constitution proceeds to outlaw parties established on a 'religious, sectarian, tribal [or] regional' basis. The *President* is limited to two seven-year terms, although this clause may not be retroactively applied, allowing the incumbent, President Bashar al-Assad, to remain in office for potentially four terms. At a referendum on 27 May 2007 Bashar al-Assad (b. 1965) was confirmed as *President* for a second term, receiving 97·6% of the vote. Presidential elections may be contested, though any candidate must be a Muslim. The description of Syria as a 'planned socialist economy' in the 1973 constitution has been replaced by an assertion that the economy 'shall be based on the principle of developing public and private economic activity through economic and social plans'.

The amended constitution was widely perceived as an attempt by President al-Assad to appease the opposition movement that emerged during the 2011 Arab Spring and placate international opinion against him. Nonetheless, the referendum was boycotted by leading opposition groups and received little support from the international community. Legislative power is held by a 250-member People's Assembly (*Majlis al-Sha'ab*), renewed every four years in 15 multi-seat constituencies.

National Anthem

'Humata al Diyari al aykum salaam' ('Defenders of the Realm, on you be peace'); words by Khalil Mardam Bey, tune by M. S. and A. S. Flayfel.

GOVERNMENT CHRONOLOGY

Heads of State since 1943. (HS = People's Party; HSQ = Syrian National Party; KW = National Bloc)

President

1943–49	KW	Shukri al-Kuwatli

Chairmen of Supreme Military Council

1949	military	Husni al-Zaim
1949	military	Muhammad Sami Hilmi al-Hinnawi

President

1949–51	KW	Hashim Bay Khalid al-Atassi

Chairman of Supreme Military Council

1951	military	Adib ash-Shishakli

Presidents

1951–53	military	Fawzi Silu
1953–54	military	Adib ash-Shishakli
1954–55	KW	Hashim Bay Khalid al-Atassi
1955–58	HSQ	Shukri al-Kuwatli

United Arab Republic

1958–61

President
1961–63   HS   Nazim al-Qudsi

Chairmen of National Revolutionary Command Council
1963   military/Ba'ath   Lu'ayy al-Atassi
1963–64   military/Ba'ath   Muhammad Amin al-Hafez

Chairman of Presidential Council
1964–66   military/Ba'ath   Muhammad Amin al-Hafez

Heads of State
1966–70   Ba'ath   Nur ad-Din Mustafa al-Atassi
1970–71   Ba'ath   Ahmad al-Hasan al-Khatib
1971   military/Ba'ath   Abu Sulayman Hafez al-Assad

Presidents
1971–2000   Ba'ath   Abu Sulayman Hafez al-Assad
2000–   Ba'ath   Bashar al-Assad

RECENT ELECTIONS

Elections were held on 7 May 2012. The ruling National Progressive Front (led by the Ba'ath Party) won 168 of 250 seats, the Popular Front for Change and Liberation 5 and non-partisan candidates the remaining 77. Turnout was 51·3%. The opposition described the election as a farce.

CURRENT GOVERNMENT

Following the death of Lieut.-Gen. Hafez al-Assad on 10 June 2000, a presidential referendum was held on 10 July 2000. The former president's son Bashar al-Assad won 97·3% of the vote.

President: Bashar al-Assad; b. 1965 (Ba'ath; sworn in 17 July 2000).

Vice-Presidents: Farouk al-Shara; Najah al-Attar.

In March 2013 the government comprised:

Prime Minister: Wael al-Halki; b. 1964 (Ba'ath; sworn in 11 Aug. 2012).

Deputy Prime Ministers: Gen. Fahad Jassim al-Freij (also *Minister of Defence*); Walid Muallem (also *Minister of Foreign Affairs*); Omar Ghalawanji (also *Minister of Local Administration*); Qadri Jamil (also *Minister of Domestic Trade and Consumer Protection*).

Minister of the Interior: Mohammad al-Shaar. *Communications and Technology:* Imad Abdel-Ghani Sabouni. *Awqaf:* Mohammad Abdel-Sattar Sayyed. *Presidential Affairs:* Mansour Fadlallah Azzam. *Justice:* Najm Hamad al-Ahmad. *Finance:* Ismail Ismail. *Health:* Saad Abdul-Salam al-Nayef. *Tourism:* Hala Mohammad al-Nasser. *Electricity:* Imad Mohammad Dib Khamis. *Hydraulic Resources:* Bassam Hana. *Agriculture:* Ahmad al-Qadri. *Higher Education:* Mohammad Yahia Moalla. *Education:* Hazwan al-Wazz. *Economy and Foreign Trade:* Mohammad Mhabak. *Industry:* Adnan Abdo al-Sukhni. *Transportation:* Mahmud Said. *Housing and Urban Development:* Hussein Farzat. *Public Works:* Hussein Arnus. *Oil and Mineral Resources:* Sleiman Abbas. *Culture:* Lubana Mushaweh. *Social Affairs:* Kinda Shmat. *Labour:* Hassan Hijazi. *Information:* Omran al-Zohbi.

Syrian Parliament (Arabic only): http://www.parliament.gov.sy

CURRENT LEADERS

Bashar al-Assad

Position
President

Introduction
Bashar al-Assad was confirmed as president in a national referendum in July 2000 following the death of his father the previous month. He had not been groomed for a political career, pursuing instead a medical education in England. However, on the death of his elder brother Basil—their father's chosen successor—in an accident in 1994, Assad was recalled to Damascus. Thereafter he rose through the senior ranks of the armed forces, consolidating his influence and authority within his father's regime to achieve the first-ever father-to-son succession to the highest office in an Arab republic. From March 2011 his repressive rule was dogged by street protests, echoing those in several other Arab states. By early 2012 the discontent had descended into open civil war between the security forces and disparate opposition groups across the country. Despite mounting pressure on Assad from the international community—with the notable exceptions of Russia and China—to accept political reform or stand down, there seemed little prospect of a resolution to the conflict by early 2013, by which time an estimated 60,000 Syrians had died and many more had fled the country as refugees.

Early Life
Assad was born in Damascus on 11 Sept. 1965. After attending high school in the capital, he went to London, England, to study ophthalmology. Having returned to Syria upon the death of his brother, he became commander of the Syrian army's armoured division. Assad reportedly used this position to install his own supporters and remove ageing senior figures and potential rivals from the army and security services. He was appointed to the rank of colonel in 1999.

When President Hafez al-Assad died suddenly on 10 June 2000, Syria's political establishment was quick to demonstrate support for his son. The People's Assembly voted to change the constitution to lower the minimum age for a president from 40 to 34—Assad's age at that time. The Assembly and the dominant Ba'ath Party approved his nomination for the presidency (as the only candidate) and the party elected him as its secretary-general. He was also declared commander-in-chief of the armed forces, his military rank having been elevated to lieutenant-general. In a national referendum held on 10 July 2000, Assad was endorsed as president with 97·3% of the votes cast.

Career in Office
In his inaugural address to the People's Assembly, Assad spoke of the need for economic reform. He called for the restructuring of the state-dominated economy and improved competitiveness, the dismantling of bureaucracy and the ending of corruption. Private investment has since been encouraged. However, initial signs of political liberalization—a partial lifting of censorship, the release of some political prisoners, tolerance of criticism of the government and party, and the limited introduction of the Internet—faded in 2001 as dissidents were again arrested and detained. On the international stage, peace with Israel remained elusive, given Syria's stipulation that the Israelis give up the whole of the Golan Heights seized in the Six-Day War of 1967. Also, the renewed Palestinian *intifada* against Israeli occupation polarized the already volatile politics of the Middle East and made a Syrian-Israeli accord increasingly unlikely.

In 2002–03 Syria opposed US military threats against Iraq, fearing the consequences of another war in the Middle East, and claimed that the UN Security Council did not endorse an invasion. US-led forces nevertheless invaded Iraq in March 2003 and Saddam Hussein was toppled the following month. The USA subsequently threatened Assad with economic, diplomatic or other undefined sanctions, suggesting that Syria was harbouring members of Saddam Hussein's regime and had been involved in the development of chemical weapons.

In Jan. 2004 Assad visited Turkey, the first Syrian leader to do so, improving several decades of cool relations between the two countries. The following May the USA imposed economic sanctions on Syria for alleged support for terrorism and failure to stop militants entering Iraq. A UN Security Council resolution adopted in Sept. 2004 and the subsequent assassination of former Lebanese prime minister Rafiq al-Hariri in Beirut in Feb. 2005 (allegedly with Syrian involvement) increased the international pressure on Assad to remove Syria's forces from Lebanon completely. Although the withdrawal was completed in April that year, the UN continued to probe al-Hariri's murder, implicating senior Syrian officials and chiding Assad's government for its perceived lack of co-operation with UN investigators. Syria's frosty diplomatic relationship with the USA was aggravated by Assad's support for Hizbollah during the radical Lebanese militia's war against the Israeli military in July–Aug. 2006.

Following the restoration of Iraqi-Syrian diplomatic ties in Nov. 2006, President Jalal Talabani became the first Iraqi head of state to visit Damascus for 30 years in Jan. 2007. In March the European Union reopened a dialogue with the Syrian government, and in May the US Secretary of State met the Syrian foreign minister in Egypt for the first high-level bilateral contact in two years. Relations with Israel, however, deteriorated further after an Israeli air strike against an undefined site in northern Syria in Sept. (which US intelligence sources claimed in April 2008 to have been a covert nuclear reactor plant).

Despite the ruling National Progressive Front's overwhelming victory in parliamentary elections in April 2007 and Assad's endorsement as president for a further term in a national referendum in May, a crackdown on dissent remained in force. During that year several government critics and human rights campaigners were sentenced to terms of imprisonment, and in Jan. 2008 Syria's leading dissident, Riyad Seif, was detained by security services.

In Oct. 2008 Syria and Lebanon signed an accord establishing diplomatic relations for the first time in their turbulent post-independence history. After several postponements, the Damascus stock exchange was launched in March 2009, marking a further step in the liberalization of Syria's state-controlled economy.

From March 2011 Assad faced popular protests against his rule, echoing similar demonstrations in countries throughout North Africa and the Middle East. The initial release of several dozen political prisoners failed to quell discontent, which spread rapidly from Damascus to other major cities. In April 2011 he lifted the state of emergency in place for 48 years, but violent repression of the uprising by government forces continued through the rest of the year and into 2012 in defiance of international opinion. The USA and EU imposed sanctions, including asset freezing and travel bans, and the Arab League suspended Syria's membership. However, China and Russia vetoed two UN resolutions critical of Syria in Oct. 2011 and again in Feb. 2012.

At the end of 2011 Assad had agreed to allow observers from the Arab League into the country to monitor the security situation but, at the end of Jan. 2012 against the backdrop of intensifying violence, the League abandoned its observer mission. Assad continued to resist the pressure for political change through the rest of the year, maintaining that his opponents were terrorists and outlaws and did not reflect public opinion. However, his hold on the country became increasingly tenuous as civil war escalated, encompassing all Syria's major cities. In late 2012 anti-Assad groups united to form the Syrian National Coalition for Revolutionary and Opposition Forces, which promptly gained diplomatic recognition from the major Western nations, the Gulf states and also from Turkey (the Assad regime's relations with which had by then been further soured by a number of violent border incidents and the tide of Syrian refugees).

DEFENCE

Military service is compulsory for a period of 18 months. Defence expenditure in 2008 totalled US$1,941m. (US$91 per capita), representing 3·8% of GDP. Syria had 14,000 troops based in Lebanon in early 2005, but in March 2005 the two countries agreed that Syria would begin to redeploy the troops to the Bekaa Valley in the east of the country. They were subsequently all withdrawn from Lebanon.

In Sept. 2007 an Israeli air strike destroyed a suspected nuclear reactor in the Deir Ez-Zor region.

Army

Strength (2007) about 215,000 (including conscripts) with a further 280,000 available reservists. In addition there is a gendarmerie of 8,000 and a Workers Militia of approximately 100,000.

Navy

The Navy included two small frigates and 18 patrol and coastal combatants in 2007. A small naval aviation branch of the Air Force operates 13 attack helicopters. Personnel in 2007 numbered 7,600. The main base is at Tartous (where the Russian military also has a naval base) with additional bases located at Lattakia and Minet el-Baida.

Air Force

The Air Force, including Air Defence Command, had (2007) 30,000 personnel with an additional 10,000 reservists. Air Defence Command numbered 60,000 in 2007. There were 583 combat-capable aircraft (including MiG-21, MiG-23, MiG-25 and MiG-29 supersonic interceptors and Su-22 and Su-24 fighter-bombers), and 71 attack helicopters.

ECONOMY

In 2006 agriculture accounted for 18·3% of GDP, industry 32·2% and services 49·5%.

Overview

Syria experienced GDP growth above 3% every year from 2004 to 2010. With civil war from 2011 the economy contracted by 2–3%. Prior to 2011 oil accounted for 20% of government revenues and 35% of export receipts, with agriculture responsible for 20% of GDP. Since 2011 the EU has imposed sanctions on oil exports and the once prosperous tourism sector has been devastated. Inflation rose above 30% in 2012 and the economy is estimated to have contracted by 20%.

Foreign direct investment declined from 10% before the crisis to 2% in 2012 and the Syrian pound experienced a 51% decline against the US dollar. The Heritage Foundation has rated the country the fourth worst for economic freedom. Property rights are often unclear and corruption at the government and judicial levels rife. The authoritarianism and bureaucracy of the Assad regime has marginalized private enterprise.

Syria's economic future is dependent on ending the civil war. Structural reforms and higher educational standards are vital to recovery, as is the restoration of foreign investment. The Washington-based Institute for International Finance has predicted that foreign exchange reserves will run out by the end of 2013.

Currency

The monetary unit is the *Syrian pound* (SYP) of 100 *piastres*. Inflation was 15·2% in 2008, falling to 2·8% in 2009. It stood at 4·4% in 2010. Gold reserves were 833,000 troy oz in July 2004. Total money supply in March 2004 was £Syr.642,259m.

Budget

The fiscal year is the calendar year. In 2007 revenues were £Syr.458·6bn. and expenditures £Syr.520·5bn.

Performance

There was real GDP growth of 5·9% in 2009 and 3·4% in 2010; total GDP in 2010 was US$59·1bn.

Banking and Finance

The Central Bank is the bank of issue. Commercial banks were nationalized in 1963. The *Governor* of the Central Bank is Adib Mayaleh. In 2007 there were nine private banks.

In Aug. 2000 it was announced that private banks were to be established for the first time in nearly 40 years. Syria's first private banks since 1961 opened in 2004.

Foreign debt totalled US$4,729m. in 2010, equivalent to 8·2% of GNI.

A stock exchange opened in Damascus in March 2009 with six listed companies.

ENERGY AND NATURAL RESOURCES

Environment
Syria's carbon dioxide emissions from the consumption and flaring of fossil fuels in 2008 were the equivalent of 2·5 tonnes per capita.

Electricity
Installed capacity was 7·9m. kW in 2007. Production in 2007 was 38·64bn. kWh and consumption per capita 2,037 kWh.

Oil and Gas
Oil reserves in 2008 were 2·5bn. bbls; production, 19·8m. tonnes. Natural gas reserves (2008), 280bn. cu. metres; production, 5·5bn. cu. metres.

Minerals
Phosphate production, 2004, 2,883,000 tonnes; other minerals are gypsum (432,000 tonnes in 2004) and salt (141,000 tonnes in 2004). There are indications of lead, copper, antimony, nickel, chrome and other minerals widely distributed. Sodium chloride and bitumen deposits are being worked.

Agriculture
The arable area in 2001 was 4·64m. ha. and there were 815,000 ha. of permanent cropland. 1·27m. ha. were irrigated in 2001. In 2001 there were 100,347 tractors and 4,500 harvester-threshers in use. Production of principal crops, 2001 (in 1,000 tonnes): wheat, 4,745; sugar beets, 1,175; seed cotton, 1,010; olives, 866; tomatoes, 732; cottonseed, 656; potatoes, 480; oranges, 465. Livestock (2003 estimates, in 1,000): sheep, 13,500; goats, 1,000; cattle, 880; asses, 217; chickens, 30,000. Livestock products, 2003 (in 1,000 tonnes): milk, 1,768; meat, 368; eggs, 166; cheese, 91.

Forestry
In 2010 there were 0·49m. ha. of forest (3% of the land area). Timber production in 2007 was 65,000 cu. metres.

Fisheries
The total catch in 2010 was 6,635 tonnes (55% freshwater fish).

INDUSTRY
Production (in tonnes): cement (2009), 5,497,000; residual fuel oil (2007), 4,862,000; distillate fuel oil (2007), 3,824,000; petrol (2007), 1,278,000; fertilizers (2003), 304,000; vegetable oil (2001), 89,000; cotton yarn (2001), 83,000; refrigerators (2002), 113,000 units; washing machines (2002), 85,000 units; cigarettes (2001), 12·0bn. units; woollen carpets (2002), 2·2m. sq. metres.

Labour
In 2005 the labour force totalled 5,312,000. Unemployment was 11·5% in 2005.

INTERNATIONAL TRADE

Imports and Exports
Imports (c.i.f.) in 2008 totalled US$18,104·7m. (US$14,655·1m. in 2007) and exports (f.o.b.) US$14,380·0m. (US$11,545·7m. in 2007). Main imports, 2003, included: machinery and equipment, 18·3%; foodstuffs, 17·7%; chemicals and chemical products, 14·9%; base and fabricated metals, 14·6%. Main exports in 2003 included: crude petroleum, 62·5%; refined petroleum, 8·8%; live animals and meat, 4·2%; textiles, 4·2%. In 2003 imports came mainly from China (5·9%), Ukraine (5·8%), Turkey (5·7%) and USA (5·0%). Exports in 2003 went mainly to Italy (33·2%), France (14·4%), Turkey (7·5%) and Saudi Arabia (5·9%).

COMMUNICATIONS

Roads
In 2006 there were 40,032 km of roads, including 1,103 km of motorways, 5,971 km of main roads and 31,849 km of secondary roads; 95·8% of roads were paved. There were in 2007 a total of 446,100 passenger cars in use (22 per 1,000 inhabitants), 50,800 buses and coaches and 528,300 vans and lorries. In 2007 there were 13,465 road accidents involving injury resulting in 2,818 deaths.

Rail
In 2008 the Syrian Railways operated 1,801 km of 1,435 mm gauge; in 2005 the smaller Hedjaz-Syrian Railway operated 338 km of 1,050 mm gauge. Passenger-km travelled on the Syrian Railways in 2008 came to 1·1bn. and freight tonne-km to 2·4bn.; passenger-km travelled on the Hedjaz-Syrian Railway in 2005 came to 412,000 and freight tonne-km to 1·4m.

Civil Aviation
The main international airport is at Damascus, with some international traffic at Aleppo and Lattakia. The national carrier is the state-owned Syrian Arab Airlines. In 2003 scheduled airline traffic of Syrian-based carriers flew 9m. km, carrying 940,000 passengers (908,000 on international flights). Damascus handled an estimated 1,747,000 passengers in 2000 (1,660,000 on international flights) and 25,000 tonnes of freight.

Shipping
In Jan. 2009 there were 102 ships of 300 GT or over registered, totalling 314,000 GT. Vessels totalling 4,397,000 NRT entered ports in 2005 and vessels totalling 3,927,000 NRT cleared.

Telecommunications
In 2006 there were 3,243,000 main (fixed) telephone lines but 2,294,000 people were on the waiting list for a line. Mobile phone subscribers numbered 7,056,200 in 2008 (33·2 per 100 persons). There were an estimated 3·6m. internet users in 2008.

SOCIAL INSTITUTIONS

Justice
Syrian law is based on both Islamic and French jurisprudence. There are two courts of first instance in each district, one for civil and one for criminal cases. There is also a Summary Court in each sub-district, under Justices of the Peace. There is a Court of Appeal in the capital of each governorate, with a Court of Cassation in Damascus. The death penalty is still in force. Executions may be held in public.

The population in penal institutions in 2004 was 10,599 (57 per 100,000 of national population).

Education
In 2007 there were 145,781 pre-primary school children, 2,310,168 primary school pupils and 2,549,444 secondary school pupils. In 2001, 17 teacher colleges had 791 teachers and 8,204 students; 593 schools for technical education had 16,849 teachers and 139,551 students. Adult literacy in 2009 was an estimated 84·2% (male, 90·4%; female, 78·0%).

In 2005–06 there were five universities: Al-Baath University, the Syrian Virtual University, Tishreen University, the University of Aleppo and the University of Damascus. The establishment of private universities has been permitted since 2001.

In 2007 total expenditure on education came to 4·9% of GDP and accounted for 16·7% of total government expenditure.

Health
In 2001 there were 19,716 beds in 406 hospitals, and 1,046 health centres. There were 30,702 physicians, 16,169 dentists, 38,070 nurses and midwives and 16,579 pharmacists in 2008.

RELIGION

In 2001 there were an estimated 14·39m. Muslims (namely Sunni with some Shias and Ismailis). There are also Druzes and Alawites. Christians (920,000 in 2001) include Greek Orthodox, Greek Catholics, Armenian Orthodox, Syrian Orthodox, Armenian Catholics, Protestants, Maronites, Syrian Catholics, Latins, Nestorians and Assyrians. There are also Jews and Yezides.

CULTURE

World Heritage Sites

There are six UNESCO sites in Syria: the old city of Damascus, dating from the 3rd millennium BC and including the Umayyid Mosque (inscribed in 1979); the old city of Bosra, once the capital of the Roman province of Arabia and an important stopover on the ancient caravan routes (1980); Palmyra (Tadmur), a desert oasis northeast of Damascus, containing the ruins of a city that was one of the most prosperous centres of the ancient world (1980); the old city of Aleppo, located at the crossroads of various trade routes since the 2nd millennium BC (1986); the two castles of Crac des Chevaliers and Qal'at Salah El-Din (2006); and the Ancient Villages of Northern Syria (2011), which date from the 1st to the 7th centuries AD.

Press

In 2006 there were four national daily newspapers with a combined circulation of 130,000. In the 2010 *World Press Freedom Index* compiled by Reporters Without Borders, Syria ranked 173rd out of 178 countries.

Tourism

In 2005 there were 3,368,000 non-resident tourists; receipts totalled US$2·28bn.

Festivals

Islamic religious festivals are observed throughout the year. Among the most colourful is Milad al-Nabi, held in April to commemorate the birth of Muhammad. The Silk Road Festival takes place in Sept./Oct., when all major cities hold celebrations and Damascus in particular is transformed back to the city that once served as a meeting place for the caravans travelling the Silk Road. Independence Day is widely observed on 17 April.

DIPLOMATIC REPRESENTATIVES

Of Syria in the United Kingdom (8 Belgrave Sq., London, SW1X 8PH)
Temporarily closed.

Of the United Kingdom in Syria (Kotob Building, 11 Mohammad Kurd Ali St., Malki, Damascus POB 37)
All staff have been withdrawn.

Of Syria in the USA (2215 Wyoming Ave., NW, Washington, D.C., 20008)
Ambassador: Vacant.
Chargé d'Affaires a.i.: Mounir Koudmani.

Of the USA in Syria (embassy in Damascus closed in Feb. 2012).

Of Syria to the United Nations
Ambassador: Bashar Jaafari.

Of Syria to the European Union
Ambassador: Mohammad Ayman Jameel Soussan.

FURTHER READING

Choueiri, Y., *State and Society in Syria and Lebanon.* 1994
George, Alan, *Syria: Neither Bread nor Freedom.* 2003
Goodarzi, Jubin, *Syria and Iran: Diplomatic Alliance and Power Politics in the Middle East.* 2006
Guo, Luc, *Understanding Syria: History, Geography, Economy.* 2011
Hinnebusch, Raymond, *Syria: Revolution From Above.* 2002
Hitti, Philip K., *History of Syria Including Lebanon and Palestine.* 2002
Kienle, Eberhard, *Contemporary Syria: Liberalization Between Cold War and Peace.* 1997
Lesch, David W., *The New Lion of Damascus: Bashar Al Asad and Modern Syria.* 2005
Moubayed, Sami, *Steel and Silk: Men and Women Who Shaped Syria 1900–2000.* 2005
Perthes, Volker, *Syria Under Bashar Al-Asad: Modernisation and the Limits of Change.* 2005
Starr, Stephen, *Revolt in Syria: Eye-Witness to the Uprising.* 2012
Van Dam, Nikolaos, *The Struggle for Power in Syria: Politics and Society Under Asad and the Ba'th Party.* 2011

National Statistical Office: Central Bureau of Statistics, Nizar Kabbani St., Abu Romanneh, Damascus.
Website: http://www.cbssyr.org

**Explore the world at
www.statesmansyearbook.com**

TAJIKISTAN

© Research Machines plc 2006

Jumkhurii Tojikiston
(Republic of Tajikistan)

Capital: Dushanbe
Population projection, 2015: 7·40m.
GNI per capita, 2011: (PPP$) 1,937
HDI/world rank: 0·607/127
Internet domain extension: .tj

KEY HISTORICAL EVENTS

The Tajik Soviet Socialist Republic was formed from those regions of Bokhara and Turkestan where the population consisted mainly of Tajiks. It was admitted as a constituent republic of the Soviet Union on 5 Dec. 1929. In Aug. 1990 the Tajik Supreme Soviet adopted a declaration of republican sovereignty and in Sept. 1991 Tajikistan declared independence. In Dec. 1991 the republic became a member of the CIS. After demonstrations and fighting, the Communist government was replaced by a Revolutionary Coalition Council on 7 May 1992. Following further demonstrations, President Nabiyev was ousted on 7 Sept. Civil war broke out, and the government resigned on 10 Nov. On 30 Nov. it was announced that a CIS peacekeeping force would be sent to Tajikistan. A state of emergency was imposed in Jan. 1993. On 23 Dec. 1996 a ceasefire was signed. A further agreement on 8 March 1997 provided for the disarmament of the Islamic-led insurgents, the United Tajik Opposition, and their eventual integration into the regular armed forces. A peace agreement brokered by Iran and Russia was signed in Moscow on 27 June 1997 stipulating that the opposition should have 30% of ministerial posts in a Commission of National Reconciliation. President Rakhmon (formerly Rakhmonov), first elected in 1994, won a second term in 1999. The country's first multi-party parliamentary election was held in Feb. 2000, although it was criticized by observers for failing to meet democratic standards.

Ethnic conflict and terrorist attacks continue to plague Tajikistan, with Russia offering military support. Fighting in the Fergana Valley, involving the Islamist Movement of Uzbekistan, is a cause for concern for all Central Asian governments.

TERRITORY AND POPULATION

Tajikistan is bordered in the north and west by Uzbekistan and Kyrgyzstan, in the east by China and in the south by Afghanistan. Area, 143,100 sq. km (55,200 sq. miles). It includes two regions (Sughd and Khatlon), one autonomous region (Gorno-Badakhshan Autonomous Region), the city of Dushanbe and regions of republican subordination. 2010 provisional census population, 7,565,000; density, 53 per sq. km. 80% of the population in 2000 were Tajiks, 15% Uzbeks and 1% Russians.

The UN gives a projected population for 2015 of 7·40m.

In 2011 only 26·4% of the population lived in urban areas, making it the most rural of the former Soviet republics.

The capital is Dushanbe (2010 provisional population, 724,000). Other large towns are Khujand (formerly Leninabad), Kulyab (Kŭlob) and Kurgan-Tyube.

The official language is Tajik, written in Arabic script until 1930 and after 1992 (the Roman alphabet was used 1930–40; the Cyrillic, 1940–92).

SOCIAL STATISTICS

Estimates, 2008: births, 192,000; deaths, 44,000. Rates, 2008 estimate (per 1,000 population): births, 28·1; deaths, 6·4. Life expectancy, 2007, 63·7 years for men and 69·3 for women. Annual growth, 2000–08, 1·3%. Infant mortality, 2010, 52 per 1,000 live births; fertility rate, 2008, 3·4 births per woman.

In the Human Development Index, or HDI (measuring progress in countries in longevity, knowledge and standard of living), Tajikistan is ranked the lowest of any of the former Soviet republics at 127th in the 2011 rankings published in the annual Human Development Report.

CLIMATE

Considering its altitude, Tajikistan is a comparatively dry country. July to Sept. are particularly dry months. Winters are cold but spring comes earlier than farther north. Dushanbe, Jan. –10°C, July 25°C. Annual rainfall 375 mm.

CONSTITUTION AND GOVERNMENT

In Nov. 1994 a new constitution was approved by a 90% favourable vote by the electorate, which enhanced the President's powers. The head of state is the *President*, elected by universal suffrage. When the 1994 constitution took effect the term of office was five years. However, an amendment to the Constitution prior to the 1999 election extended the presidential term to seven years, although a president could only serve one term. A further referendum approved in June 2003 allowed President Rakhmonov (now Rakhmon) to serve two additional terms after the expiry of the one that he was serving at the time, in Nov. 2006, theoretically enabling him to remain in office until 2020. The Organization for Security and Co-operation in Europe and the USA expressed concerns at the result. Tajikistan has a bicameral legislature. The lower chamber is the 63-seat *Majlisi Namoyandagon* (*Assembly of Representatives*), with 41 members elected in single-seat constituencies and 22 by proportional representation for five-year terms. The upper chamber is the 34-seat *Majlisi Milliy* (*National Assembly*), with 25 members chosen for five-year terms by local deputies, eight appointed by the president and one seat reserved for the former president.

National Anthem

'Zinda bosh, ey Vatan, Tochikistoni ozodi man' ('Live long, O Nation, my free Tajikistan'); words by Gulnazar Keldi, tune by Suleiman Yudakov.

RECENT ELECTIONS

At presidential elections on 6 Nov. 2006 President Rakhmon was re-elected with 79·3% of votes cast. Opposition parties boycotted the poll.

In elections to the Assembly of Representatives held on 28 Feb. 2010 the People's Democratic Party of Tajikistan (PDPT) won 54 of 63 seats (71·0% of the vote), the Islamic Renaissance Party of Tajikistan (IRP) 2 (8·2%), the Communist Party (CP) 2 (7·0%), Agrarian Party 2 (5·1%) and the Party of Economic Reforms of Tajikistan 2 (5·1%). Turnout was 90·8%. The last seat went to the PDPT in a second round of voting held on 14 March bringing their total to 55. Elections to the National Assembly were held on 25 March 2010. 25 of the 34 seats were voted for by local majlisi deputies, eight were appointed by the president and the other member is the former president.

CURRENT GOVERNMENT

President: Emomalii Rakhmon; b. 1952 (PDPT; as Speaker elected by the former Supreme Soviet 19 Nov. 1992, re-elected 6 Nov. 1994, 6 Nov. 1999 and 6 Nov. 2006).

In March 2013 the government comprised:

Prime Minister: Akil Akilov; b. 1944 (PDPT; sworn in 20 Dec. 1999).

First Deputy Prime Minister: Matlubkhon Davlatov. *Deputy Prime Ministers:* Murodali Alimardon; Ruqiya Qurbonova.

Minister of Agriculture and Environmental Protection: Qosim Qosimov. *Culture:* Mirzoshorukh Asrori. *Defence:* Col.-Gen. Sherali Khairullaev. *Economic Development and Trade:* Sharif Rahimzoda. *Education:* Nuriddin Saidov. *Energy and Industry:* Sherali Gul. *Finance:* Safarali Najmuddinov. *Foreign Affairs:* Hamrokhon Zarifi. *Health:* Nusratullo Salimov. *Internal Affairs:* Ramazon Rahimov. *Justice:* Rustam Mengliyev. *Labour and Social Protection:* Mahmadamin Mahmadaminov. *Land Reclamation and Water Resources:* Said Yoqubzod. *Transport:* Nizom Hakimov.

Office of the President: http://www.prezident.tj

CURRENT LEADERS

Emomalii Rakhmon

Position
President

Introduction
Emomalii Rakhmon (formerly Rakhmonov), a former cotton-farm administrator, became Tajikistan's head of state in Nov. 1992, after the country's first post-Soviet leader, Rakhmon Nabiyev, was forced to resign. Rakhmon has survived civil war and an assassination attempt, but reforming institutions and raising living standards in one of the region's poorest countries has proved a hard challenge. In March 2007 he announced that he was dropping the Russian suffix (-ov) from his surname.

Early Life
Emomalii Sharipovich Rakhmon was born on 5 Oct. 1952 in the Danghara district of the Kulob province of the Tajik Soviet Socialist Republic (SSR). He studied electronics and from 1969 worked at a vegetable-oil extraction factory in Qurghonteppa. After three years in the Soviet navy, Rakhmon became an administrator at Lenin *Kholkov* (collective farm) in Danghara, constructing a power base as chairman of the farm's trade union committee.

He studied economics by correspondence and in 1982 graduated from the Tajik State University. He was elected people's deputy of the supreme council of the Tajik SSR in 1990. Tajikistan declared independence in Sept. 1991 but hopes of an economically viable state were undermined by civil war. On 19 Nov. 1992, following Nabiev's forced resignation and the annulment of the office of president, Rakhmon was elected chairman of the supreme council and head of state. On 6 Nov. 1994, following

inter-Tajik peace talks, he won presidential elections, claiming 58·3% of the vote.

Career in Office
Civil war continued through the early years of Rakhmon's presidency and by the time hostilities between the Islamist-led opposition and his Moscow-backed administration ended in June 1997, at least 50,000 people had died. In March 1998 he joined the centrist People's Democratic Party of Tajikistan (PDPT). He was re-elected president on 6 Nov. 1999 with 97% of the vote, and on 22 June 2003 won a referendum to allow him to run for two further seven-year terms. Rakhmon's grip on power was underlined in the general elections of Feb. 2005 and Feb. 2010 when his PDPT won landslide parliamentary victories. The opposition Islamic and communist parties alleged fraud in both polls and observers said the vote failed to meet international standards. Meanwhile, Rakhmon was again re-elected over-whelmingly to the presidency in Nov. 2006 as the opposition boycotted the vote.

Tajikistan suffered a particularly severe winter in 2007–08 and also faced an energy crisis. Many industries were forced to shut down, some rural areas had no electricity and even the capital faced food shortages. In April 2008 the International Monetary Fund ordered the Tajik authorities to repay IMF disbursements that had been obtained on the basis of false data and misreporting. Rakhmon subsequently removed some senior National Bank officials from their positions.

In April 2009 the government finalized an agreement allowing the USA to transport non-military equipment to Afghanistan across Tajik territory.

Tensions increased between Tajikistan and Uzbekistan through 2012 as the Tajik government accused its neighbour of disrupting gas and electricity deliveries, while the Uzbek government objected to Tajik plans for a large dam project that it claimed would restrict access to regional water resources.

In Oct. 2012 Tajikistan agreed to extend Russia's lease on a Soviet-era army base, scheduled to expire in 2014, for 30 more years.

DEFENCE

Conscription is compulsory for two years. In 2007 the active armed forces had a strength of 8,800. Paramilitary forces totalled 7,500 including 3,800 interior troops and 2,500 emergencies ministry troops. 5,500 Russian Army personnel were stationed in the country in 2007.

Defence expenditure in 2008 totalled US$80m. (US$11 per capita), representing 1·7% of GDP.

Army
Personnel strength (2007) 7,300.

Air Force
Air Force/Air Defence strength (2007) 1,500.

ECONOMY

In 2006 agriculture accounted for 24·8% of GDP, industry 27·4% and services 47·8%.

Overview
Though arable land accounts for only 10% of the total, Tajikistan is primarily an agrarian economy, with the sector employing two-thirds of the labour force and contributing 11% of export revenues. As a result of war and repeated changes of political leadership in the 1990s, the economy deteriorated faster than in other former Soviet satellites. In 1996 a reform programme was launched, supported by the IMF and the World Bank. Despite a slow start and the Russian financial crisis of 1998, the focus on reconstruction stimulated a market-driven economy.

Economic growth, which averaged 8·6% from 2000–08, depended on cotton and aluminium exports, as well as rising remittances from overseas. Sound macroeconomic management has enabled inflation to be controlled, having reached levels between 30–40% in the period 1998–2001. Other achievements include the halving of foreign debt, stabilization of the exchange rate and a reduction in poverty levels. Between 1999 and 2009, poverty rates fell from 83% to 47%. However, governance and institutions are weak and unemployment is widespread.

With the global financial crisis growth slowed to 3·9% in 2009, caused by a sharp reduction in remittances (down 31% from the 2008 peak level) and lower prices and demand for aluminium and cotton. Nonetheless, a pick-up in remittances initiated a recovery in 2010.

Currency
The unit of currency is the *somoni* (TJS) of 100 *dirams*, which replaced the Tajik rouble on 30 Oct. 2000 at 1 somoni = 1,000 Tajik roubles. The introduction of the new currency was intended to strengthen the national banking system. The IMF voiced their support for the new currency, which it believed would contribute to macroeconomic stability and expedite the transition to a market economy. Inflation in 1993 was 2,195%, declining to 418% in 1996 and still further to 7·2% in 2004, the reduction being helped by a US$22m. IMF loan in 1996 and maintenance of a tighter monetary regime. Although the rate had risen back up to 20·4% in 2008, inflation fell to 6·5% in 2010 but went up again to 12·4% in 2011. Total money supply was 241m. somoni in Dec. 2004. Gold reserves stood at 47,000 troy oz in July 2005.

Budget
Revenues in 2009 totalled 4,175m. somoni and expenditure 5,643m. somoni.
VAT is 20%.

Performance
Annual real GDP growth was negative for four consecutive years in the mid-1990s. Since then the economy has recovered—more recently there has been growth of 3·9% in 2009, 6·5% in 2010 and 7·4% in 2011. Total GDP in 2011 was US$6·5bn. Tajikistan is more reliant than any other country on remittances from abroad, which accounted for 31% of total GDP in 2010. In particular remittances from Russia contributed more than 21% of Tajikistan's GDP.

Banking and Finance
The central bank and bank of issue is the National Bank (*Chairman*, Abdujabbor Shirinov). In 1998 there were 27 commercial and private banks but the number had fallen to 14 by 2002 after a process of consolidation.

Foreign debt totalled US$2,955m. in 2010, representing 53·1% of GNI.

ENERGY AND NATURAL RESOURCES

Environment
In 2008 Tajikistan's carbon dioxide emissions from the consumption and flaring of fossil fuels were the equivalent of 0·9 tonnes per capita.

Electricity
Estimated installed capacity in 2007 was 4·4m. kW. Production was 17·5bn. kWh in 2007 and consumption per capita 2,612 kWh.

Oil and Gas
Natural gas output in 2007 was 14m. cu. metres, with reserves of 5·7bn. cu. metres in 2007.

Minerals
There are deposits of brown coal, lead, zinc, iron ore, antimony, mercury, gold, silver, tungsten and uranium. Lignite production, 2007, 15,000 tonnes. Aluminium production, 2005, 380,000 tonnes.

Agriculture
In 2007 there were approximately 710,000 ha. of arable land and 101,000 ha. of permanent crops. Cotton is the major cash crop, with various fruits, sugarcane, jute, silk, rice and millet also being grown.

Output of main agricultural products (in 1,000 tonnes) in 2003: wheat, 660; seed cotton, 537; potatoes, 473; cottonseed, 215; tomatoes, 170; onions, 153. Livestock, 2003: 1·59m. sheep; 1·14m. cattle; 842,000 goats; 2m. chickens. Livestock products, 2003 (in 1,000 tonnes): meat, 30; milk, 412.

Forestry
Forests covered 0·41m. ha. in 2010, or 3% of the land area. Timber production in 2007 was 90,000 cu. metres.

Fisheries
The estimated total catch in 2008 was 146 tonnes, exclusively from inland waters.

INDUSTRY
Major industries: aluminium, electro-chemical plants, textile machinery, carpet weaving, silk mills, refrigerators, hydro-electric power. Output in 2007 (unless otherwise specified): cement (2006), 281,500 tonnes; multi-nutrient fertilizer, 24,500 tonnes; cotton woven fabrics, 30·5m. sq. metres; woven carpets and other floor coverings, 941,700 sq. metres; woven silk fabrics, 40,900 sq. metres; footwear, 172,400 pairs.

Labour
The economically active force in 2005 totalled 2,154,000. The principal areas of activity were: agriculture, 1,424,000; education, 186,000; industry, 121,000. In 2005 the unemployment rate was 3·8%.

INTERNATIONAL TRADE
Imports and Exports
In 2006 imports were valued at US$1,954·6m. (US$1,430·9m. in 2005) and exports at US$1,511·8m. (US$1,108·1m. in 2005). Main imports: petroleum products, grain, manufactured consumer goods; main exports: cotton and aluminium. Principal import suppliers, 2000: Uzbekistan, 28·8%; Russia, 16·1%; Ukraine, 13·1%; Kazakhstan, 12·8%. Principal export markets in 2000: Russia, 37·4%; Netherlands, 25·7%; Uzbekistan, 14·1%; Switzerland, 10·4%.

COMMUNICATIONS
Roads
In 2000 there were 27,767 km of roads. There were 193,000 passenger cars in use in 2007, 59,000 lorries and vans, 10,700 motorcycles and mopeds, and 5,400 buses and coaches. In 2007 there were 464 fatalities as a result of road accidents.

Rail
Length of railways, 2008, 617 km. In 2008, 834,000 passengers were carried and 14·5m. tonnes of freight.

Civil Aviation
There are international airports at Dushanbe and Khujand. The national carrier is Tajik Air, which has flights to 12 international destinations as well as operating domestic services. In 2003 Tajik Air flew 10m. km, carrying 413,000 passengers (291,000 on international flights).

Telecommunications
In 2008 there were 286,900 main (fixed) telephone lines in Tajikistan and mobile phone subscribers numbered 3,673,500 (53·7 per 100 persons). There were 600,000 internet users in 2008.

SOCIAL INSTITUTIONS

Justice

In 2008, 11,658 crimes were reported (13,161 in 1998). The population in penal institutions on 1 Jan. 2008 was 7,350 (109 per 100,000 of national population). The death penalty is retained but has been subject to a moratorium since July 2004.

Education

The adult literacy rate in 2008 was over 99%. In 2007 there were 680,308 pupils and 31,482 teaching staff at primary schools; 1,012,275 pupils and 61,186 teaching staff at secondary schools; and 147,294 students and 7,761 academic staff at higher education institutions. There were 32 higher education institutions in total in 2007.

In 2007 public expenditure on education came to 3·5% of GNI and 18·2% of total government spending.

Health

In 2006 there were 61 hospital beds per 10,000 inhabitants. There were 13,267 physicians in 2006, 1,003 dentistry personnel and 33,165 nursing and midwifery personnel.

Welfare

At the beginning of 2009 there were 537,700 pensioners; average monthly pension was 87·36 somoni.

RELIGION

The Tajiks are predominantly Sunni Muslims (80%); Shia Muslims, 5%.

CULTURE

World Heritage Sites

Tajikistan has one site on the UNESCO World Heritage List: the Proto-urban site of Sarazm (inscribed on the list in 2010), an archaeological site displaying some of the early human settlements in Central Asia.

Press

Media freedom suffered during the civil war between 1992 and 1997 when around 60 journalists were killed and many others fled the country. *Imruz News*, the first daily newspaper since 1992, was launched in Aug. 2010.

Tourism

There were 450,000 foreign visitors in 2008.

DIPLOMATIC REPRESENTATIVES

Of Tajikistan in the United Kingdom (Grove House, 27 Hammersmith Grove, London, W6 7BA).
Ambassador: Erkin Kasymov.

Of the United Kingdom in Tajikistan (65 Mirzo Tursunzade St., Dushanbe)
Ambassador: Robin Ord-Smith.

Of Tajikistan in the USA (1005 New Hampshire Ave., NW, Washington, D.C., 20037)
Ambassador: Nuriddin Shamsov.

Of the USA in Tajikistan (109-A Ismoili Somoni Ave., Dushanbe)
Ambassador: Susan M. Elliott.

Of Tajikistan to the United Nations
Ambassador: Sirodjidin Aslov.

Of Tajikistan to the European Union
Ambassador: Rustamjon Soliev.

FURTHER READING

Abdullaev, K. and Akbarzadeh, S., *Historical Dictionary of Tajikistan.* 2002
Akiner, S., *Tajikistan: Disintegration or Reconciliation?* 2001
Djalili, M. R. (ed.) *Tajikistan: The Trials of Independence.* 1998
Jonson, Lena, *Tajikistan in the New Central Asia: Geopolitics, Great Power Rivalry and Radical Islam.* 2006

National Statistical Office: Statistical Agency under President of the Republic of Tajikistan, 17 Bokhtar St., Dushanbe.
Website: http://www.stat.tj

Gorno-Badakhshan Autonomous Region

Comprising the Pamir massif along the borders of Afghanistan and China, the province was set up on 2 Jan. 1925, initially as the Special Pamir Province. Area, 63,700 sq. km (24,590 sq. miles). The population in 2007 was 218,000 (mainly Tajiks with a Kirghiz minority). Capital, Khorog (2007: 29,000). The inhabitants are predominantly Ismaili Muslims.

Mining industries are developed (gold, rock-crystal, mica, coal, salt). Wheat, fruit and fodder crops are grown, and cattle and sheep are bred in the western parts. Total area under cultivation, 16,236 ha. In 2004 the region was 69% self-sufficient in food; humanitarian aid had comprised 85% of all food consumed in 1993.

The area is the most impoverished in Tajikistan, with 84% of the population falling below the poverty line in 2003, compared to a national average of 64%. Around 20% of the population of working age is employed abroad, mainly in the Russian Federation. Unemployment is approximately 70%.

The Khorog State University was founded in 1992. The private University of Central Asia has a campus located in Khorog (the other two being in Tekeli, Kazakhstan and Naryn, Kyrgyzstan). In 2004 there was one doctor per 476 inhabitants and 91 hospital beds per 10,000 inhabitants.

TANZANIA

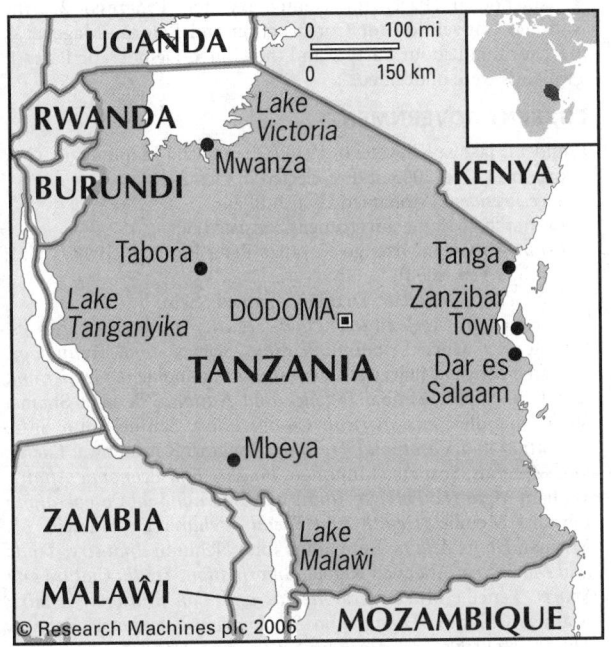

Germany become the dominant influence over most of mainland Tanzania, while the Sultan of Zanzibar retained control of a strip of coastal territories and Britain ruled Zanzibar as a protectorate.

German East Africa was conquered by the Allies in the First World War and subsequently divided between the Belgians, the Portuguese and the British. The country was administered as a League of Nations mandate until 1946, and then as a UN trusteeship territory until 9 Dec. 1961. Tanganyika achieved responsible government in Sept. 1960 and full self-government on 1 May 1961. On 9 Dec. 1961 Tanganyika became a sovereign independent member state of the Commonwealth of Nations. On 9 Dec. 1962 the country adopted a republican form of government (still within the British Commonwealth) and Dr Nyerere was elected as the first president.

Zanzibar gained internal self-government on 24 June 1963, followed by full independence on 9 Dec. 1963. On 12 Jan. 1964 the sultanate was overthrown by a revolt of the Afro-Shirazi Party leaders who established the People's Republic of Zanzibar. Also in Jan. 1964 there was an attempted coup against Nyerere who had to seek British military help. On 26 April 1964 Tanganyika, Zanzibar and Pemba combined to form the United Republic of Tanzania. The first multi-party elections were held in 1995.

TERRITORY AND POPULATION

Tanzania is bounded in the northeast by Kenya, north by Lake Victoria and Uganda, northwest by Rwanda and Burundi, west by Lake Tanganyika, southwest by Zambia and Malaŵi, and south by Mozambique. Total area 942,799 sq. km (364,881 sq. miles), including the offshore islands of Zanzibar (1,554 sq. km) and Pemba (906 sq. km) and inland water surfaces (59,050 sq. km). 2012 census population, 44,929,002, giving a density of 50·8 per sq. km.

The UN gives a projected population for 2015 of 52·31m.

In 2011, 26·9% of the population lived in urban areas. 0·5m. Hutu refugees were forcibly repatriated to Rwanda in Dec. 1996. In 2000 Tanzania hosted the highest number of refugees in Africa, with a population of over 680,000. By Nov. 2009 repatriation programmes had allowed the number of refugees to drop below 100,000 for the first time in 15 years.

The chief towns (2002 census populations) are Dar es Salaam, the chief port and former capital (2,339,910), Arusha (270,485), Mbeya (232,596) and Mwanza (209,806). Dodoma, the capital, had a population of 150,604 in 2002.

The United Republic is divided into 30 administrative regions of which 25 are in mainland Tanzania, three in Zanzibar and two in Pemba. Areas and 2002 populations of the regions:

Jamhuri ya Muungano wa Tanzania (United Republic of Tanzania)

Capital: Dodoma
Population projection, 2015: 52·31m.
GNI per capita, 2011: (PPP$) 1,328
HDI/world rank: 0·466/152
Internet domain extension: .tz

KEY HISTORICAL EVENTS

Archaeological evidence suggests that present-day Tanzania was inhabited by Khoisan-speaking hunter-gatherers from at least 10,000 BC. The Sandawe and Hadze of north-central Tanzania are descendents of these groups. Cushitic-speaking cattle herders migrated south from Ethiopia and Sudan from around 1000 BC. Beginning in the first millennium AD, Tanzania was settled by Bantu-speaking iron-working farmers, whose origins are considered to be in the borderlands of present-day Nigeria and Cameroon.

Seafarers from Arabia established a trading settlement on the coast at Kilwa around AD 800, and Persian merchants settled on the islands of Zanzibar and Pemba. Nilotic-speaking pastoralists (including the Maasai and Luo) moved south into Tanzania between AD 900 and AD 1700. Sultan Hassan bin Sulaiman I established control of Kilwa around 1270. Islam spread and a thriving Afro-Arab 'Swahili' culture took hold in coastal areas.

Portuguese explorers arrived off Kilwa in 1500, heralding two centuries of Portuguese control over various East African trading ports. Zanzibar came under Omani control in the 1650s and prospered as a centre of the slave trade, extending its influence over the coastal hinterland and into the mainland interior. Britain attempted to end the slave trade by signing the Treaty of Moresby with the Sultan of Zanzibar in 1822. During the 1880s the German imperialist, Dr Carl Peters, founded German East Africa by signing agreements with local rulers. A series of agreements between Britain and Germany and the Sultan of Zanzibar saw

Region	Sq. km	Population
Arusha	36,486	1,288,088
Dar es Salaam	1,393	2,487,288
Dodoma	41,311	1,692,025
Geita[1]	—	—
Iringa	56,864	1,490,892
Kagera	28,388	2,028,157
Katavi[1]	—	—
Kigoma	37,037	1,674,047
Kilimanjaro	13,309	1,376,702
Lindi	66,046	787,624
Manyara	45,820	1,037,605
Mara	19,566	1,363,397
Mbeya	60,350	2,063,328
Morogoro	70,799	1,753,362
Mtwara	16,707	1,124,481
Mwanza	19,592	2,929,644
Njombe[1]	—	—
Pwani (Coast)	32,407	885,017
Rukwa	68,635	1,136,354

Region	Sq. km	Population
Ruvuma	63,498	1,113,715
Shinyanga	50,781	2,796,630
Simiyu[1]	—	—
Singida	49,341	1,086,748
Tabora	76,151	1,710,465
Tanga	26,808	1,636,280
Zanzibar and Pemba	2,460	981,754
Pemba North	574	185,326
Pemba South	332	175,471
Zanzibar North	470	136,639
Zanzibar South	854	94,244
Zanzibar West	230	390,074

[1]Four new regions—Geita, Katavi, Njombe and Simiyu— were created in March 2012.

The official languages are Swahili (spoken as a mother tongue by only 8·8% of the population, but used as a *lingua franca* by 91%) and English.

SOCIAL STATISTICS

2008 estimates: births, 1,765,000; deaths, 482,000. Rates, 2008 estimates (per 1,000 population): births, 41·5; deaths, 11·4. Annual population growth rate, 2000–08, 2·7%. Life expectancy in 2007 was 54·2 years for men and 55·8 for women. 45% of the population was below 15 years old in 2008. Infant mortality, 2010, 50 per 1,000 live births; fertility rate, 2008, 5·6 births per woman.

CLIMATE

The climate is very varied and is controlled largely by altitude and distance from the sea. There are three climatic zones: the hot and humid coast, the drier central plateau with seasonal variations of temperature, and the semi-temperate mountains. Dodoma, Jan. 75°F (23·9°C), July 67°F (19·4°C). Annual rainfall 23" (572 mm). Dar es Salaam, Jan. 82°F (27·8°C), July 74°F (23·3°C). Annual rainfall 43" (1,064 mm).

CONSTITUTION AND GOVERNMENT

The current constitution dates from 25 April 1977 but underwent major revisions in Oct. 1984. The *President* is head of state, chairman of the party and commander-in-chief of the armed forces. The *Prime Minister* is also the leader of government business in the National Assembly.

The 357-member *Bunge (National Assembly)* is composed of 239 constituency representatives, 102 appointed women, ten Union presidential nominees (five of whom must be women), five representatives of the Zanzibar House of Representatives (two women), and one *ex officio* member (the Attorney General). In Dec. 1979 a separate constitution for Zanzibar was approved. Although at present under the same Constitution as Tanzania, Zanzibar has, in fact, been ruled by decree since 1964. The formation of a government of national unity was approved by 66·4% of voters in a referendum in July 2010. Following elections on 21 Oct. 2010 the new government was inaugurated in Nov. 2010.

National Anthem

'God Bless Africa/Mungu ibariki Afrika'; words collective, tune by M. E. Sontonga and V. E. Webster.

GOVERNMENT CHRONOLOGY

Presidents since 1964. (TANU = Tanganyika African National Union; CCM = Chama Cha Mapinduzi (Revolutionary State Party))

1964–85	TANU/CCM	Julius Kambarage Nyerere
1985–95	CCM	Ali Hassan Mwinyi
1995–2005	CCM	Benjamin William Mkapa
2005–	CCM	Jakaya Mrisho Kikwete

RECENT ELECTIONS

Presidential and parliamentary elections were held on 31 Oct. 2010. Incumbent Jakaya Kikwete of Chama Cha Mapinduzi (Revolutionary State Party) retained the presidency with 62·8% of votes cast against six other candidates. Turnout was 42·8%. In the parliamentary elections Chama Cha Mapinduzi gained 186 of 239 elected seats, the Civic United Front 23, Chama Cha Democracia na Maendeleo (Party for Democracy and Progress) 22, the National Convention for Construction and Reform–Mageuzi 4, the Tanzania Labour Party 1 and the United Democratic Party 1. Two seats were undecided.

CURRENT GOVERNMENT

President: Jakaya Kikwete; b. 1950 (Chama Cha Mapinduzi/CCM; sworn in 21 Dec. 2005 and re-elected in Oct. 2010).
Vice-President: Mohamed Gharib Bilal.
In March 2013 the government consisted of:
Prime Minister: Mizengo Kayanza Peter Pinda; b. 1948 (CCM; sworn in 9 Feb. 2008).
President of Zanzibar: Dr Ali Mohamed Shein.
Minister of Agriculture, Food Security and Co-operatives: Christopher Chiza. *Communication, Science and Technology:* Makame Mnyaa Mbarawa. *Community Development, Gender and Children:* Sophia Simba. *Defence and National Service:* Shamsi Vuai Nahodha. *East African Co-operation:* Samuel John Sitta. *Education and Vocational Training:* Shukuru Kawambwa. *Energy and Minerals:* Sospeter Muhongo. *Finance and Economic Affairs:* William Mgimwa. *Foreign Affairs and International Co-operation:* Bernard Membe. *Health and Social Welfare:* Dr Hussein Ali Mwinyi. *Home Affairs:* Emmanuel John Nchimbi. *Industry, Trade and Marketing:* Abdallah Kigoda. *Information, Youth, Culture and Sports:* Fenella Mukangara. *Justice and Constitutional Affairs:* Mathias Chikawe. *Labour and Employment:* Gaudensia Kabaka. *Lands, Housing and Human Settlements Development:* Anna Tibaijuka. *Livestock and Fisheries Development:* Mathayo David Mathayo. *Natural Resources and Tourism:* Khamis Kagasheki. *Transport:* Harrison Mwakyembe. *Water and Irrigation:* Jumanne Maghembe. *Works:* John Pombe Magufuli. *Minister without Portfolio:* Mark Mwandosya.

Government Website: http://www.tanzania.go.tz

CURRENT LEADERS

Jakaya Kikwete

Position
President

Introduction
Jakaya Kikwete became president of Tanzania on 14 Dec. 2005, winning an overwhelming majority in national elections that were generally considered free and fair. A Muslim from the coastal district of Bagamoyo, Kikwete was a military leader in the 1970s and 1980s and served as foreign minister for ten years from 1995. He was re-elected in Oct. 2010.

Early Life
Jakaya Mrisho Kikwete was born on 7 Oct. 1950 in Msoga, Bagamoyo District on the coast of Tanganyika. He attended schools in Msoga and Kibaha, before studying economics at the University of Dar es Salaam. In 1975, while at university, Kikwete joined the ruling Tanganyika African National Union, which later became the Chama Cha Mapinduzi (CCM, Revolutionary State Party). Following his graduation in 1978, Kikwete joined the Tanzania People's Defence Force (TPDF), where he served as a lieutenant from 1972–79 and subsequently as captain.

In 1984, having spent a year at the Monduli military officers college in Arusha, Kikwete became chief political instructor of the TPDF. In 1988 he was elected to represent Bagamoyo parliamentary constituency, a post he held for three consecutive terms. He was deputy minister of energy, water and minerals from 1988–90 before being promoted to minister and serving under President Ali Hassan Mwinyi for four years. Following constitutional reform that legalized opposition parties in 1992, Kikwete retired from the army.

In 1995, having served as finance minister for a year, Kikwete became one of 14 challengers for the CCM leadership. He lost to Benjamin Mkapa, who led the party to victory in national elections in Oct. 1995 amid widespread allegations of voting irregularities. Kikwete was appointed foreign minister, a post he held until 2005, winning praise for his mediation work in war-torn Burundi and the Democratic Republic of the Congo. His department was credited with advancing regional integration within the East African Community and in the Southern African Development Community. Kikwete won the right to lead his party into the 2005 national elections and emerged victorious from the poll on 14 Dec. He received 80% of the vote and replaced Mkapa as president. The CCM retained its overwhelming majority in parliament, with 206 out of 232 seats.

Career in Office
In his inauguration speech, Kikwete vowed to continue the free-market policies of Mkapa and prioritized the improvement of relations with the semi-autonomous islands of Zanzibar. He inherited a country in which poverty is widespread but whose economy had been growing at a rate of 6% a year. He was expected to maintain political stability as the country benefited from rising gold production and donor-supported investment. In Aug. 2006, in recognition of the government's economic reform efforts, the African Development Bank cancelled US$645m. of Tanzanian debt and also agreed a loan of US$74m. for poverty reduction programmes.

In Feb. 2008 Prime Minister Edward Lowassa and two other ministers resigned in the wake of a corruption scandal involving an energy deal with a US-based electricity company, in response to which Kikwete dissolved the cabinet and appointed Mizengo Kayanza Peter Pinda as the new premier. Kikwete served as chairman of the African Union for 2008.

In Oct. 2010 Kikwete and the CCM were returned to power in presidential and parliamentary elections. Following allegations of ministerial misuse of public funds, Kikwete sacked six members of his cabinet in May 2012.

In Aug. 2012 the government confirmed that Iranian oil tankers had been operating under the Tanzanian flag to circumvent international sanctions on Iranian oil trading. The USA, in response, threatened possible action against Tanzania unless the practice was stopped.

DEFENCE

Defence expenditure totalled US$184m. in 2008 (US$5 per capita), representing 0·9% of GDP.

Army
Strength (2007), 23,000. There is a paramilitary Police Field Force of 1,400.

Navy
Personnel in 2007 totalled about 1,000. The principal bases are at Dar es Salaam, Zanzibar and Mwanza.

Air Force
The Tanzanian People's Defence Force Air Wing was built up initially with the help of Canada, but combat equipment has been acquired from China. Air Defence Command personnel totalled 3,000 in 2007. Although there were a reported 19 combat-capable aircraft in 2007 virtually no air defence assets were serviceable.

INTERNATIONAL RELATIONS

In Nov. 1999 a treaty was signed between Tanzania, Kenya and Uganda to create a new East African Community as a means of developing East African trade, tourism and industry and laying the foundations for a future common market and political federation.

ECONOMY

Agriculture accounted for 28·8% of GDP in 2009, industry 24·3% and services 46·9%.

Overview
Real GDP grew by an average 6% per year from 2009–11, driven by mining and construction, the two fastest-growing sectors of the economy. Flourishing tourism and the implementation of economic reforms further boosted growth. Per capita income was US$1,500 in 2011. An estimated 40% of the state budget is financed by aid. Agriculture employs over 80% of the workforce and accounts for 85% of exports. In 2011, 7% of the government budget went to the agricultural sector.

From the late 1960s to the mid-1980s the economy stagnated under corrupt state control. From 1985 popular discontent encouraged the government of Ali Hassan Mwinyi to loosen control over prices and trade. Subsequent banking reforms and a stronger private sector have spurred investment.

Tanzania is one of the largest producers of gold in Africa, with exports growing by 30% in the year to June 2012. Gold exports in 2011 were 40·4 tonnes, up from 35·6 tonnes in 2010. The country is also a major producer of coffee. Tanzania's main export market is China (accounting for 14·2%), followed by India (9·9%) and Japan (7·7%).

Currency
The monetary unit is the *Tanzanian shilling* (TZS) of 100 *cents*. Foreign exchange reserves were US$3,533m. in Sept. 2009. Inflation, which had been 26·5% in 1995, was down to 4·1% in 2004, the lowest rate for more than 20 years. It has gone up since then and was 7·2% in 2010 and 12·7% in 2011. Total money supply in Aug. 2009 was Sh. 3,429·5bn.

Budget
The fiscal year ends 30 June. In 2006–07 revenues were Sh. 3,691·2bn. and expenditures Sh. 4,474·7bn. Tax revenue accounted for 68·5% of revenues in 2006–07 and current expenditure 70·1% of expenditures.

VAT is 18%.

Performance
Real GDP growth was 6·0% in 2009, 7·0% in 2010 and 6·4% in 2011. Total GDP in 2011 was US$23·9bn. (mainland Tanzania only).

Banking and Finance
The central bank is the Bank of Tanzania (*Governor*, Prof. Benno Ndulu).

On 6 Feb. 1967 all commercial banks with the exception of National Co-operative Banks were nationalized, and their interests vested in the National Bank of Commerce on the mainland and the Peoples' Bank in Zanzibar. However, in 1993 private-sector commercial banks were allowed to open. In 1997 the National Bank of Commerce (NBC) was split into a trade bank, a regional rural bank and a micro-finance bank. It was privatized in 2000, with the South African concern Absa Group Limited purchasing a 55% stake. The government retained 30% with the International Finance Corporation holding 15%. In 2010 it had 53 branches; total assets at 31 Dec. 2009, Sh. 1,294,606m. The NBC's market dominance has severely declined since 1997. In Dec. 2008 the Co-operatives Rural and Development Bank (CRDB) held 17% of the banking sector's total assets, making it the market leader (the National Microfinance Bank held 16% and the NBC 13%). In Oct. 2010 there were 28 commercial banks registered in Tanzania.

Foreign debt totalled US$8,664m. in 2010, representing 37·7% of GNI.

A stock exchange opened in Dar es Salaam in 1996.

ENERGY AND NATURAL RESOURCES

Environment
Tanzania's carbon dioxide emissions from the consumption and flaring of fossil fuels in 2008 were the equivalent of 0·1 tonnes per capita.

Electricity
Installed capacity was an estimated 1·0m. kW in 2007, but there are plans to increase this to 2·8m. kW by 2015. Only 14% of the population has access to electricity. Production in 2007 was 4·17bn. kWh, with consumption per capita 107 kWh.

Oil and Gas
A number of international companies are exploring for both gas and oil. In 2007 proven natural gas reserves were 6·5bn. cu. metres.

Minerals
Tanzania's mineral resources include gold, nickel, cobalt, silver and diamonds. International funds injected to improve Tanzania's economy have resulted in notable increases, particularly in gold production. The first commercial gold mine began operating in Mwanza in 1998. By 2005 the value of gold exports had reached US$642m., up from US$121m. in 2000. Gold production in 2005 totalled 52,236 kg. Large deposits of coal and tin exist but mining is on a small scale. Estimated diamond production in 2005 was 205,000 carats; exports totalled US$33·7m. in 2004.

Agriculture
About 80% of the workforce are engaged in agriculture, chiefly in subsistence farming. Agricultural produce contributes around 85% of exports. There were an estimated 4·0m. ha. of arable land in 2002 and 1·1m. ha. of permanent crops. Approximately 170,000 ha. were irrigated in 2002. There were about 7,600 tractors in 2002.

Production of main agricultural crops in 2002 (in 1,000 tonnes) was: cassava, 6,888; maize, 2,705; sugarcane, 1,600; sweet potatoes, 950; sorghum, 834; rice, 640; plantains, 602; coconuts, 370; millet, 300; dry beans, 270; potatoes, 240; seed cotton, 222. Zanzibar is a major producer of cloves.

Livestock (2003 estimates): 17·70m. cattle; 12·56m. goats; 3·52m. sheep; 30m. chickens.

Livestock products (2003 estimates, in 1,000 tonnes): beef and veal, 246; goat meat, 31; pork, bacon and ham, 13; lamb and mutton, 10; poultry meat, 45; goat's milk, 835; cow's milk, 100; eggs, 34; honey, 26.

Forestry
Forests covered 33·43m. ha. in 2010 (38% of the total land area). In 2007, 24·44m. cu. metres of roundwood were cut.

Fisheries
Catch (2008) 325,476 tonnes, of which 281,690 tonnes were from inland waters.

INDUSTRY
Industry is limited, and is mainly textiles, petroleum and chemical products, food processing, tobacco, brewing and paper manufacturing.

INTERNATIONAL TRADE

Imports and Exports
In 2010 imports (c.i.f.) amounted to US$8,012·9m. (US$6,530·8m. in 2009); exports (f.o.b.) US$4,050·5m. (US$2,982·4m. in 2009).

Principal imports, 2002: consumer goods, 31·0%; machinery and apparatus, 22·2%; transport equipment, 13·2%; crude and refined petroleum, 11·8%. Principal exports, 2002: minerals (notably gold), 42·4%; cashew nuts, 5·8%; tobacco, 5·6%; coffee, 4·0%; tea, 3·4%.

Main import suppliers, 2002: South Africa, 11·4%; Japan, 8·4%; India, 6·5%; Russia, 6·1%; UAE, 5·9%. Main export markets, 2002: UK, 18·5%; France, 17·4%; Japan, 11·0%; India, 7·3%; Netherlands, 6·2%.

COMMUNICATIONS

Roads
In 2008 there were 87,524 km of roads, including 10,042 km of highways or national roads. Passenger cars in use in 2007 numbered 80,900; there were also 369,900 lorries and vans, 23,100 buses and coaches, and 52,000 motorcycles and mopeds.

Rail
In 1977 the independent Tanzanian Railway Corporation was formed. The network totalled 2,707 km (metre-gauge) in 2005, excluding the joint Tanzania-Zambia (Tazara) railway's 961 km in Tanzania (1,067 mm gauge) operated by a separate administration. In 2008 the state railway carried 0·5m. passengers and 0·5m. tonnes of freight, and in 2005 the Tazara carried 0·9m. passengers and 0·6m. tonnes of freight.

In Oct. 1998 a transhipment facility for containers opened at Kidatu, southwest of Dar es Salaam, providing a link between the 1,067 mm gauge railways of the southern part of Africa and the 1,000 mm gauge lines of the north.

Civil Aviation
There are three international airports: Dar es Salaam, Zanzibar and Kilimanjaro (Moshi/Arusha). Air Tanzania, the national carrier, provides domestic services and in 2003 had flights to Abu Dhabi, Blantyre, Johannesburg, Lilongwe, Mombasa, Muscat and Nairobi. In 2003 scheduled airline traffic of Tanzanian-based carriers flew 4m. km, carrying 150,000 passengers (61,000 on international flights). Dar es Salaam is the busiest airport, handling 1,155,000 passengers in 2006 (659,000 on international flights), followed by Zanzibar with 486,000 (197,000 on international flights).

Shipping
In Jan. 2009 there were 20 ships of 300 GT or over registered, totalling 31,000 GT. The main seaports are Dar es Salaam, Mtwara, Tanga and Zanzibar. There are also ports on the lakes.

Telecommunications
In 2008 there were 123,800 main (fixed) telephone lines; mobile phone subscribers numbered 13,006,800 in 2008 (30·6 per 100 persons). There were 520,000 internet users in 2008. In June 2012 there were 518,000 Facebook users.

SOCIAL INSTITUTIONS

Justice
The Judiciary is independent in both judicial and administrative matters and is composed of a four-tier system of Courts: Primary Courts; District and Resident Magistrates' Courts; the High Court; and the Court of Appeal. The Chief Justice is head of the Court of Appeal and the Judiciary Department. The Court's main registry is at Dar es Salaam; its jurisdiction includes Zanzibar. The Principal Judge is head of the High Court, also headquartered at Dar es Salaam, which has resident judges at seven regional centres.

The population in penal institutions in Sept. 2006 was 43,911 (113 per 100,000 of national population).

Education
In 1999–2000 there were 11,409 primary schools with 103,731 teachers for 4·19m. pupils. At secondary level there were 247,579 pupils with 12,496 (1997) teachers in 826 schools, and at university level in 2000–01 there were 21,960 students with 2,192 academic staff. Primary school fees were abolished in Jan. 2002.

Technical and vocational education is provided at several secondary and technical schools, and at the Dar es Salaam Technical College. There are 53 teacher training colleges, including the college at Chang'ombe for secondary-school teachers.

There is one university, one university of agriculture and one open university. There are also nine other institutions of higher education.

Adult literacy rate in 2003 was 69·4% (male, 77·5%; female, 62·2%). In 1998–99 total expenditure on education came to 2·2% of GNP.

Health

In 2006 there were 1,807 doctors and dentists and 39,432 nurses and midwives. In 2005 there were 219 hospitals, 481 health centres and 4,679 dispensaries.

RELIGION

In 2001 there were 18·3m. Christians (including Roman Catholics, Anglicans and Lutherans) and 11·5m. Muslims. Muslims are concentrated in the coastal towns; Zanzibar is 99% Muslim. The remainder of the population follow traditional religions. In Feb. 2013 the Roman Catholic church had one cardinal.

CULTURE

World Heritage Sites

Tanzania has seven sites on the UNESCO World Heritage List: Ngorongoro Conservation Area (inscribed on the list in 1979); the Ruins of Kilwa Kisiwani and of Songo Mnara (1981); Serengeti National Park (1981); Selous Game Reserve (1982); Kilimanjaro National Park (1987); the Stone Town of Zanzibar (2000); and the Kondoa Rock-Art sites (2006).

Press

In 2006 there were 14 dailies with a combined circulation of 115,000.

Tourism

There were 15 national parks in Tanzania in 2008. In 2004 there were 583,000 foreign tourists (excluding same-day visitors), bringing revenue of US$610m. Tourism is the country's second largest foreign exchange earner after agriculture.

DIPLOMATIC REPRESENTATIVES

Of Tanzania in the United Kingdom (3 Stratford Pl., London, W1C 1AS)
High Commissioner: Peter Kallaghe.

Of the United Kingdom in Tanzania (Umoja House, Garden Ave., PO Box 9200 Dar es Salaam)
High Commissioner: Dianna Melrose.

Of Tanzania in the USA (2139 R. St., NW, Washington, D.C., 20008)
Ambassador: Mwandaidi Sinare Maajar.

Of the USA in Tanzania (686 Old Bagamoyo Rd, Msasani, PO Box 9123, Dar es Salaam)
Ambassador: Alfonso E. Lenhardt.

Of Tanzania to the United Nations
Ambassador: Tuvako Nathaniel Manongi.

Of Tanzania to the European Union
Ambassador: Simon Uforosia Ralph Mlay.

FURTHER READING

National Statistical Office: National Bureau of Statistics, Box 796, Dar es Salaam.
Website: http://www.tanzania.go.tz/statistics.html

The world in focus at
www.statesmansyearbook.com

THAILAND

LAOS
Chiang Mai
MYANMAR
THAILAND
Khon Kaen
Nakhon Ratchasima
BANGKOK
Samut Prakan
CAMBODIA
Gulf of Thailand
Nakhon Si Thammarat
Songkhla
Andaman Sea
0   100 miles
0   150 km
© Research Machines plc 2006
MALAYSIA

Prathet Thai
(Kingdom of Thailand)

Capital: Bangkok
Population projection, 2015: 70·88m.
GNI per capita, 2011: (PPP$) 7,694
HDI/world rank: 0·682/103
Internet domain extension: .th

KEY HISTORICAL EVENTS

Excavations at Ban Chiang on the Khorat plateau in northeast Thailand suggest rice farming was under way by as early as 2500 BC. From around 300 BC the Indianized Funan kingdom held sway across much of southeast Asia, including eastern and central Thailand. Artifacts discovered at the Funan capital of Ba Phnom, in modern Cambodia, point to trading links with China and India and as far as the Middle East and Rome. At its height in the 6th century AD, Funan control included part of the Malay peninsula.

Mon, Tai and Khmer peoples first entered northern and eastern Thailand from southern China in the 5th century AD. Taking advantage of Funan's decline after the 6th century, the Mon began

to establish independent kingdoms. Among them was Dvaravati, which in the 10th century was absorbed by the Indian-influenced Khmer empire, centred on Angkor. Mongol incursions into Yunnan in southern China in the mid-13th century forced a new wave of Tai migration that culminated in the Sukhothai kingdom of north-central Thailand. Under Ramkhamhaeng (1279–98), its influence stretched southward to the Malay peninsula. Trade with India and China flourished and the Siamese language developed in written form.

The Tai kingdom of Lan Na emerged as the dominant power in northern Thailand in the 14th century while further south the declining Sukhothai came under the influence of Rama Tibodi, prince of U Tong, who, around 1350, established a Buddhist dynasty centred on Ayutthaya. Rice cultivation and trade brought prosperity and power to Siam over the next four centuries, though tempered by frequent warfare with the Khmer empire and the Lao state of Chiang Mai. Trade with Europe, the Middle East, China and Japan expanded considerably in the 16th and 17th centuries, notably under King Narai (1656–88). Relations with France developed in the 1680s but soured amid attempts to convert Narai to Christianity: French troops were expelled in 1688 and most foreigners were barred from Siam over the next century.

Ayutthaya suffered a series of attacks from Burma in the mid-1700s and fell in 1767 although Thai forces, led by Gen. Phya Tak (King Taksin), re-established control with the help of Chinese merchants within a decade. His successor, Chao Phraya Chakkri (Rama I), established Bangkok as his capital and restored the Buddhist religion during his reign (1782–1809). Subsequent Chakkri kings resumed relations with the West but, through skilful diplomacy, managed to preserve Siam's independence. The nation remained an absolute monarchy until 24 June 1932 when a group of rebels calling themselves the People's Party precipitated a bloodless coup. After King Prajadhipok tried to dissolve the newly appointed general assembly the army moved against him to become the dominant political force, a position it has held ever since. In 1939 Field Marshal Pibul Songgram became premier and embarked on a pro-Japanese policy that brought Thailand into the Second World War on Japan's side.

After 1945 periods of military rule were interspersed with attempts at democratic, civilian government. Democratic government was reintroduced for a short time after 1963 and again from 1969–71, until a military coup was staged aimed at checking crime and the communist insurgence. A moderately democratic constitution was introduced in 1978. On 23 Feb. 1991 a military junta seized power in Thailand's 17th coup since 1932. Following the appointment of Gen. Suchinda Kraprayoon as prime minister on 17 April 1992 there were violent anti-government demonstrations. Gen. Suchinda resigned and in May the legislative assembly voted that future prime ministers should be elected by its members rather than appointed by the military. The 1995 election was fought against a background of political and financial corruption. After the 1996 election a new constitution was drafted allowing for the separation of the executive, legislative and judicial branches of government.

On 26 Dec. 2004 Thailand, along with a number of other south Asian countries, was hit by a devastating tsunami. The death toll in Thailand was put at 8,000. The government of Thaksin Shinawatra was overthrown in a bloodless military coup on 19 Sept. 2006. Elections held in Dec. 2007 were won by the People's Power Party (PPP), the successor to Thaksin's banned Thai Rak Thai. Samak Sundaravej became prime minister in Feb. 2008, marking a return to civilian rule. When a Constitutional Court ruling removed him from office in Sept. 2008 he was replaced by

Somchai Wongsawat. Wongsawat in turn had to step down in Dec. 2008 when the Constitutional Court disbanded the ruling PPP following accusations of electoral fraud. Opposition leader Abhisit Vejjajiva formed a parliamentary coalition and was elected to the premiership. The pro-Thaksin United Front for Democracy against Dictatorship ('red-shirts') held anti-government protests in April 2009 and again in March–May 2010. On both occasions the government imposed a state of emergency, the most recent of which lasted until Dec. 2010.

TERRITORY AND POPULATION

Thailand is bounded in the west by Myanmar, north and east by Laos and southeast by Cambodia. In the south it becomes a peninsula bounded in the west by the Indian Ocean, south by Malaysia and east by the Gulf of Thailand. The area is 513,120 sq. km (198,117 sq. miles).

At the 2010 census the population was 65,479,453; density, 127·6 per sq. km. In 2011, 34·4% of the population lived in urban areas.

The UN gives a projected population for 2015 of 70·88m.

Thailand is divided into six regions, 76 provinces and Bangkok, the capital. Population of Bangkok (2000 census figure), 6,355,144. Other towns (2000 census figures): Samut Prakan (378,741), Nonthaburi (291,555), Udon Thani (222,425), Nakhon Ratchasima (204,641), Hat Yai (187,920).

Thai is the official language, spoken by 53% of the population as their mother tongue. 27% speak Lao (mainly in the northeast), 12% Chinese (mainly in urban areas), 3·7% Malay (mainly in the south) and 2·7% Khmer (along the Cambodian border). There are more Lao speakers in Thailand than in Laos.

SOCIAL STATISTICS

2005–06 births, 705,639; deaths, 440,024; marriages (2005), 345,234; divorces (2005), 90,688. Rates (per 1,000 population, 2005–06): birth, 10·9; death, 6·8; marriage (2005), 5·2; divorce (2005), 1·4. Annual population growth rate, 2000–05, 1·0%. Expectation of life (2007): 65·4 years for men; 72·1 years for women. Infant mortality, 2010, 11 per 1,000 live births; fertility rate, 2008, 1·8 births per woman.

CLIMATE

The climate is tropical, with high temperatures and humidity. Over most of the country, three seasons may be recognized. The rainy season is June to Oct., the cool season from Nov. to Feb. and the hot season is March to May. Rainfall is generally heaviest in the south and lightest in the northeast. Bangkok, Jan. 78°F (25·6°C), July 83°F (28·3°C). Annual rainfall 56" (1,400 mm).

On 26 Dec. 2004 an undersea earthquake centred off the Indonesian island of Sumatra caused a huge tsunami that flooded coastal areas in western Thailand resulting in 8,000 deaths. In total there were more than 225,000 deaths in 14 countries.

CONSTITUTION AND GOVERNMENT

The reigning King is **Bhumibol Adulyadej**, born 5 Dec. 1927. King Bhumibol married on 28 April 1950 Princess Sirikit (born 12 Aug. 1932), and was crowned 5 May 1950 (making him currently the world's longest-reigning monarch). *Offspring*: Princess Ubol Ratana (born 5 April 1951, married Aug. 1972 Peter Ladd Jensen, divorced 1998); Crown Prince Vajiralongkorn (born 28 July 1952, married 3 Jan. 1977 Soamsawali Kitiyakra, divorced 1993; married 10 Feb. 2001 Srirasmi Akharaphongpreecha); Princess Maha Chakri Sirindhorn (born 2 April 1955); Princess Chulabhorn (born 4 July 1957, married 7 Jan. 1982 Virayudth Didyasarin, divorced 1984).

Following the coup of Sept. 2006 an interim constitution was introduced on 1 Oct. 2006 to replace the 1997 constitution. In the country's first ever referendum, held on 19 Aug. 2007, 56·7% of votes cast were in favour of a new draft constitution (the 18th constitution since independence in 1932) that paved the way for elections before the end of the year, set a limit of two four-year terms for the prime minister and made it easier to impeach the prime minister. Turnout was low.

There is a 150-seat *Senate* (76 senators elected in the 76 provinces and 74 appointed by a selection panel) and a 500-seat *House of Representatives* (375 members elected by popular vote from constituencies and 125 selected on a proportional basis from party lists).

National Anthem

'Prathet Thai ruam nua chat chua Thai' ('Thailand, cradle of Thais wherever they may be'); words by Luang Saranuprapan, tune by Phrachen Duriyang.

GOVERNMENT CHRONOLOGY

Heads of Government since 1944. (PCT = Thai Nation Party; PKS = Social Action Party; PKWM = New Aspiration Party; PP = Democrat Party; PPP = People's Power Party; PTP = Pheu Thai Party; SP = United Thai People's Party; ST = Free Thai Movement; TRT = Thai Rak Thai; n/p = non-partisan)

Prime Ministers

1944–45	military	Khuang Aphaiwong
1945	n/p	Tawee Boonyaket
1945–46	ST	Seni Pramoj
1946	military	Khuang Aphaiwong
1946	n/p	Pridi Phanomyong
1946–47	military	Thamrong Nawasawat
1948–57	military	Plaek Pibulsongkram
1957	n/p	Pote Sarasin
1958	military	Thanom Kittikachorn
1959–63	military	Sarit Thanarat
1963–73	military, SP	Thanom Kittikachorn
1973–75	n/p	Sanya Thammasak
1975	PP	Seni Pramoj
1975–76	PKS	Kukrit Pramoj
1976	PP	Seni Pramoj

Chairman of the National Administrative Reform Council

1976–80	military (*de facto* ruler)	Sangad Chaloryu

Prime Ministers

1976–77	n/p	Thanin Kraivichien
1977–80	military	Kriangsak Chomanan
1980–88	military	Prem Tinsulanonda
1988–91	PCT	Chatichai Choonhavan

Chairman of the National Peacekeeping Council

1991	military	Sunthorn Kongsompong

Prime Ministers

1991–92	n/p	Anand Panyarachun
1992	military	Suchinda Kraprayoon
1992	n/p	Anand Panyarachun
1992–95	PP	Chuan Leekpai
1995–96	PCT	Banharn Silpa-Archa
1996–97	PKWM	Chavalit Yongchaiyudh
1997–2001	PP	Chuan Leekpai
2001–06	TRT	Thaksin Shinawatra
2006	TRT	Chidchai Vanasatidya (acting for Thaksin Shinawatra)

Chairman of the Council for Democratic Reform under Constitutional Monarchy

2006	military	Sonthi Boonyaratkalin

Prime Ministers

2006–08	n/p	Surayud Chulanont
2008	PPP	Samak Sundaravej
2008	PPP	Somchai Wongsawat
2008	PPP	Chaovarat Chanweerakul (acting)

| 2008–11 | PP | Abhisit Vejjajiva |
| 2011– | PTP | Yingluck Shinawatra |

RECENT ELECTIONS

At the elections to the House of Representatives on 3 July 2011 the Pheu Thai Party (PTP) won a majority with 265 seats out of 500 and 53·0% of the vote, ahead of the ruling Democrat Party with 159 seats (31·8%), Bhumjaithai with 34 seats (6·8%), Chartthaipattana with 19 (3·8%), the Chart Pattana Puea Pandin Party with 7 (1·4%) and the Phalang Chon Party also with 7 and 1·4%. Five other parties won four seats or fewer. Turnout was 66%. In a special session of the House of Representatives held on 5 Aug. 2011 the Thai parliament elected Yingluck Shinawatra prime minister by 296 to 3 with 197 abstentions.

There were elections to the 150-seat Senate on 2 March 2008. 76 members were elected, one for each of the country's provinces. The other 74 seats were appointed by a selection committee headed by a military-installed chief. Among the latter, 18 were former MPs or relatives of top politicians.

CURRENT GOVERNMENT

In April 2013 the coalition government comprised:

Prime Minister: Yingluck Shinawatra; b. 1967 (PTP; in office since 8 Aug. 2011).

Deputy Prime Ministers: Yukol Limlamthong (also *Minister of Agriculture and Co-operatives*); Phongthep Thepkanjana (also *Minister of Education*); Kittirat Na Ranong (also *Minister of Finance*); Surapong Towijakchaikul (also *Minister of Foreign Affairs*); Plodprasob Suraswadi; Chalerm Ubumrung.

Ministers in the Prime Minister's Office: Niwatthamrong Boonsongpaisan; Varathep Ratanakorn; Sansanee Nakpong.

Minister of Commerce: Boonsong Teriyaphirom. *Culture:* Sontaya Kunplome. *Defence:* Sukumpol Suwanatat. *Energy:* Pongsak Raktapongpaisarn. *Industry:* Prasert Boonchaisuk. *Information and Communications Technology:* Anudith Nakornthap. *Interior:* Charupong Ruangsuwan. *Justice:* Pracha Promnok. *Labour:* Padermchai Sasomsap. *Natural Resources and Environment:* Preecha Rengsomboonsuk. *Public Health:* Pradit Sintavanarong. *Science and Technology:* Woravat Auapinyakul. *Social Development and Human Security:* Santi Prompat. *Tourism and Sports:* Somsak Purisrisak. *Transport:* Chadchart Sittipunt.

Government Website: http://www.thaigov.go.th

CURRENT LEADERS

Yingluck Shinawatra

Position
Prime Minister

Introduction
Yingluck Shinawatra became prime minister in Aug. 2011 after leading the Pheu Thai Party (PTP) to a landslide victory. The younger sister of former prime minister Thaksin Shinawatra, she is the country's first female premier.

Early Life
Yingluck was born in June 1967 in the northern province of Chiang Mai, the youngest of nine children. Her father, Lert Shinawatra, was a businessman and MP while her mother, Yindi Ramingwong, was the daughter of Princess Jantip Na Chiang Mai. In 1988 Yingluck graduated in public administration from Chiang Mai University and in 1991 completed a masters degree at Kentucky State University in the USA.

She joined Shinawatra Directories Co. in 1993, one of a number of businesses run by her brother, Thaksin. In 1994 she worked with Rainbow Media before returning to the family business as general productions manager and, from 1997, vice-president.

In 1999 Thaksin's businesses were brought together under Shin Corporation and in 2002 Yingluck became CEO of the group's flagship company and Thailand's biggest mobile phone enterprise, AIS (Advance Info Service). She resigned in 2006 when Shin Corporation was sold to Temasek Holdings and took over as executive president of the family-owned property development company, SC Asset. She also served on the committee of the Thaicom Foundation, which provides educational opportunities for underprivileged children.

Thaksin was elected prime minister in 2001 but was ousted in a military coup in Sept. 2006. His party was dissolved by the constitutional court in 2007 and he has subsequently spent much of his time in exile. However, his followers established the Pheu Thai Party (PTP) in 2008 and convinced Yingluck, a political newcomer, to run in the July 2011 elections. The PTP won a majority in parliament and elected Yingluck prime minister the following month.

Career in Office
Expected to continue the economic liberalization pursued by her brother, Yingluck surrounded herself with a team of long-established political operators. Among her key challenges was addressing the country's long-standing political instability, which had resulted in violent social clashes in April 2009 and April–May 2010. She pledged to support an Independent Truth and Reconciliation Commission, although work on reconciliation legislation and a proposed amnesty provoked opposition through 2012 from protesters fearful that it would pave the way for Thaksin's return to power. She also promised to eliminate poverty by 2020, cut corporate tax and raise the minimum wage. In addition, she backed proposals for free public wi-fi access and a computer for every school child.

In Jan. 2012 she appointed ten new ministers and in Nov. survived a parliamentary no confidence motion in the wake of a major anti-government demonstration in Bangkok.

DEFENCE

Conscription is for two years; if there are not enough volunteers a conscription lottery is held to fill the quota. In 2008 defence expenditure totalled US$4,294m. (US$65 per capita), representing 1·6% of GDP. Year-on-year expenditure rose in real terms by 25% in 2007 and 17% in 2008.

Army

Strength (2007) 190,000. In addition there were 45,000 National Security Volunteer Corps, around 20,000 *Thahan Phran* ('Hunter Soldiers', a volunteer irregular force), 41,000 Border Police and a 50,000 strong paramilitary provincial police force (including an estimated 500-strong special action force).

Navy

The Royal Thai Navy is, next to the Chinese, the most significant naval force in the South China Sea. The fleet includes a small Spanish-built vertical/short-take-off-and-land carrier *Chakri Naruebet*, which entered service in 1997 and operates nine ex-Spanish AV-8A Harrier aircraft (although their serviceability is in doubt) and helicopters, and ten frigates. Manpower was 70,600 (2007) including 25,849 conscripts and 1,940 naval aviation.

The main bases are at Bangkok, Sattahip, Songkla and Phang Nga, with the riverine forces based at Nakhon Phanom.

Air Force

The Royal Thai Air Force had a strength (2007) of 46,000 personnel and 165 combat-capable aircraft, including F-16s and F-5Es. The RTAF is made up of four air divisions.

ECONOMY

In 2006 agriculture accounted for 10·7% of GDP, industry 44·5% and services 44·8%.

Thailand's 'shadow' (black market) economy is estimated to constitute approximately 48% of the country's official GDP.

Overview

Thailand has transformed into a diverse, industrialized economy in the last 30 years. An export-oriented, labour-intensive manufacturing sector has developed thanks largely to foreign investment, which in 2011 reached US$5·8bn. Exports account for 71·3% of income, with the major trading partners being the USA (12·6%), Japan (11·9%) and China (9·7%).

The Asian financial crisis of the late 1990s saw GDP contract by 10·5% in 1998 as high inflation, rising unemployment and poverty took grip. The economy recovered quickly and from 1999–2007 grew at over 4% every year except for 2001, when it grew by 2·2% against the backdrop of a global slowdown.

A prudent macroeconomic policy, low public debt and low inflation helped Thailand recover from the Asian financial crisis, with average GDP growth from 2002 to 2006 of 5·7%. However, the global financial crisis combined with domestic political turmoil to restrict exports and undermine business confidence, resulting in one of the steepest contractions in Southeast Asia. Performance improved after 2009, with growth of nearly 8% in 2010, 0·1% in 2011 and over 6% in 2012. The low growth of 2011 reflected devastating floods that year, prompting the government to invest £40bn. in infrastructure reconstruction and improvements designed to limit the threat posed by future flooding.

The World Bank improved Thailand's economic status to one of upper middle income nation in July 2011 and the country is set to reach most of its UN Millennium Development targets. The poverty rate has declined to 8·1%, with the government set to raise the minimum wage in a bid to improve living standards.

Currency

The unit of currency is the *baht* (THB) of 100 *satang*. After being pegged to the US dollar, the baht was devalued and allowed to float on 2 July 1997. It was the devaluation of the baht that sparked the financial turmoil that spread throughout the world over the next year. Foreign exchange reserves were US$127,165m. in Sept. 2009 (US$28,434m. in 1998) and gold reserves 2·70m. troy oz. Total money supply in Aug. 2009 was 1,022·2bn. baht. Inflation rates (based on IMF statistics):

2002	2003	2004	2005	2006	2007	2008	2009	2010	2011
0·7%	1·8%	2·8%	4·5%	4·6%	2·2%	5·5%	−0·8%	3·3%	3·8%

Budget

The fiscal year runs from 1 Oct.–30 Sept. In 2006–07 budgetary central government revenue was 1,488·6bn. baht and expenditure 1,462·2bn. baht. Main sources of revenue in 2006–07: taxes on income, profits and capital gains, 614·9bn. baht; taxes on goods and services, 610·4bn. baht; taxes on international trade and transactions, 87·8bn. baht. Main items of expenditure by economic type in 2006–07: compensation of employees, 555·9bn. baht; grants, 358·4bn. baht; use of goods and services, 317·7bn. baht.

VAT is 7%.

Performance

Real GDP growth rates (based on IMF statistics):

2002	2003	2004	2005	2006	2007	2008	2009	2010	2011
5·3%	7·1%	6·3%	4·6%	5·1%	5·1%	2·6%	−2·3%	7·8%	0·1%

According to the National Economic and Social Development Board, the real GDP growth rate was 6·4% (provisional) in 2012. Thailand's total GDP in 2011 was US$345·7bn.

Banking and Finance

The Bank of Thailand (founded in 1942) is the central bank and bank of issue, an independent body although its capital is government-owned. Its assets and liabilities in 2002 were 2,853,897m. baht. Its *Governor* is Prasarn Trairatvorakul. In 2002 there were 30 commercial banks, 13 domestic banks and 21 foreign banks. In addition the Thai government controlled four banks in 2002: the Bank of Agriculture and Agricultural Co-operatives, the Government Housing Bank, the Government Savings Bank and the Export-Import Bank of Thailand. Total assets of commercial banks, 2002, 6,900,947m. baht. Deposits, 2001, 5,109,973m. baht.

Thailand received US$9·6bn. worth of foreign direct investment in 2011 (US$9·7bn. in 2010).

External debt amounted to US$71,263m. in 2010, up from US$46,3424m. in 2005 although down from US$79,720m. in 2000. As a share of GNI foreign debt declined from 66·0% in 2000 to 27·6% in 2005 and further to 23·4% in 2010.

There is a stock exchange (SET) in Bangkok.

ENERGY AND NATURAL RESOURCES

Environment

Thailand's carbon dioxide emissions from the consumption and flaring of fossil fuels were the equivalent of 3·9 tonnes per capita in 2008.

Electricity

Installed capacity, 2007, was 34·3m. kW. Output, 2007, 143·38bn. kWh, with consumption per capita 2,256 kWh.

Oil and Gas

Proven crude petroleum reserves in 2008 were 0·5bn. bbls. Production of oil (2008), 13·4m. tonnes. Thailand and Vietnam settled an offshore dispute in 1997 that stretched back to 1973. Demarcation allowed for petroleum exploration in the Gulf of Thailand, with each side required to give the other some revenue if an underground reservoir is discovered that straddles the border.

Production of natural gas (2008), 28·9bn. cu. metres. Reserves, 2008, 300bn. cu. metres. In April 1998 Thailand and Malaysia agreed to share equally the natural gas jointly produced in an offshore area (the Malaysian-Thailand Joint Development Area) that both countries claim as their own territory.

Minerals

The mineral resources include antimony, cassiterite (tin ore), copper, diatomite, dolomite, gold, gypsum, kaolin, lignite, limestone, manganese, marl, potash, rubies, sapphires, silica sand, silver and zinc. Production, 2005 unless otherwise indicated (in tonnes): limestone (2008), 142·12m.; lignite, 21·43m.; gypsum, 7·11m.; salt, 1·17m.; feldspar, 1·15m.; kaolin, 746,000; iron ore (metal content), 116,000 (estimate); zinc ore (metal content), 31,000 (estimate).

Agriculture

In 2002 there were an estimated 15·9m. ha. of arable land and 3·5m. ha. of permanent cropland. About 4·96m. ha. were irrigated in 2002. The chief produce is rice, a staple of the national diet. Output of the major crops in 2002 was (in 1,000 tonnes): sugarcane, 74,258; rice, 26,057; cassava, 16,868; maize, 4,230; natural rubber, 2,456; bananas, 1,800; pineapples, 1,739; mangoes, 1,750; coconuts, 1,418. Thailand is the world's leading producer of natural rubber. Livestock, 2003: pigs, 7,059,000; cattle, 5,048,000; buffaloes, 1,613,000; chickens, 177m.; ducks, 20m.

Forestry

Forests covered 18·97m. ha. in 2010, or 37% of the land area. Of the total area under forests 35% was primary forest in 2010, 44% other naturally regenerated forest and 21% planted forest. Teak and other hardwoods grow in the deciduous forests of the north; elsewhere tropical evergreen forests are found, with the timber yang the main crop (a source of yang oil). In 2007, 28·32m. cu. metres of roundwood were cut.

Fisheries

In 2010 the total catch came to 1,827,199 tonnes with marine fishing accounting for 89% of all fish caught. Thailand is the third largest exporter of fishery commodities in the world (after China and Norway), with exports in 2009 totalling US$6·24bn. Imports of fishery commodities in 2009 totalled US$1·98bn. Thailand is also the third largest producer of fishmeal (after Peru and Chile), with 381,200 tonnes in 2009. Aquaculture production in 2010 was the sixth highest in the world, at 1,286,122 tonnes.

INDUSTRY

The leading companies by market capitalization in Thailand in March 2012 were: PTT, an oil and gas company (US$32·8bn.); Siam Commercial Bank (US$23·5bn.); and PTTEP, a petroleum exploration and production company (US$18·8bn.).

Production (2002 unless otherwise indicated): 31·65m. tonnes of cement (2008), 18·38m. tonnes of distillate fuel oil (2007), 7·33m. tonnes of residual fuel oil (2007), 6·31m. tonnes of petrol (2007), 5·95m. tonnes of sugar, 5·21m. tonnes of crude steel (2008), 768,098 tonnes of synthetic fibre, 519,006 tonnes of galvanized iron sheets, 208,000 tonnes of tin plate (2001), 1,636·0m. litres of soft drinks (2001), 1,238·0m. litres of beer (2001), 30·8bn. cigarettes, 169,304 automobiles and 415,593 commercial vehicles, and 6,096,000 televisions.

Labour

In the period Sept.–Dec. 2003 the total labour force was 35·5m.; 14·2m. persons were employed in agriculture, hunting and forestry, 5·3m. in manufacturing, 5·2m. in wholesale and retail trade and 2·1m. in hotels and restaurants. The unemployment rate was 2·4% in June 2002. There is no nationwide minimum wage but a minimum wage is set at different levels at the provincial level. It varied between 159 baht and 221 baht per day in July 2011.

INTERNATIONAL TRADE

Imports and Exports

In 2010 imports (c.i.f.) totalled US$182,393·4m. (US$133,769·6m. in 2009); exports (f.o.b.), US$195,311·5m. (US$152,497·2m. in 2009). Main imports in 2003: machinery and transport equipment, 43·5%; chemicals and related products, 11·2%; petroleum and petroleum products, 10·6%; iron and steel, 5·4%. Exports: machinery and transport equipment, 43·8%; chemicals and related products, 6·5%; fish and seafood, 4·9%; clothing and apparel, 4·5%. Main import sources in 2003: Japan (24·1%), USA (9·5%), China (8·0%), Malaysia (6·0%), South Korea (3·8%). Principal export destinations (2003): USA (17·0%), Japan (14·2%), Singapore (7·3%), China (7·1%), Hong Kong (5·4%).

COMMUNICATIONS

Roads

In 2006 there were 180,053 km of roads, of which 450 km were motorways. Vehicles in use in 2006 included: 3·80m. passenger cars, 4·99m. lorries and vans and 15·67m. motorcycles and mopeds.

Rail

The State Railway totalled 4,071 km in 2005. Passenger-km travelled in 2008 came to 8·0bn.; freight tonne-km transported in 2007 totalled 3·2bn. A metro ('Skytrain'), or elevated transit system, was opened in Bangkok in 1999. A second (underground) mass transit system in Bangkok, the Bangkok Subway, was opened in 2004.

Civil Aviation

There are international airports at Bangkok (Suvarnabhumi), Chiang Mai, Phuket and Hat Yai. The national carrier, Thai Airways International, is 51·03% state-owned. In 2005 scheduled airline traffic of Thai-based carriers flew 213·8m. km, carrying 21,507,900 passengers. Bangkok handled 28,808,422 passengers in 2001 (21,395,311 on international flights) and 840,033 tonnes of freight. Phuket is the second busiest airport for passenger traffic, with 3,557,319 passengers in 2001 (2,225,031 on domestic flights), and Chiang Mai the second busiest for freight, with 23,786 tonnes in 2001. Suvarnabhumi, Bangkok's newest international airport, opened in Sept. 2006.

Shipping

In Jan. 2009 there were 612 ships of 300 GT or over registered, totalling 2,738,000 GT. Of the 612 vessels registered, 209 were oil tankers, 205 general cargo ships, 72 liquid gas tankers, 52 bulk carriers, 26 chemical tankers, 26 passenger ships and 22 container ships. The busiest ports are Laem Chabang and Bangkok.

Telecommunications

In 2008 there were 7·0m. main (fixed) telephone lines. In the same year mobile phone subscribers numbered 62·0m. (920·1 per 1,000 persons). There were 16·1m. internet users in 2008. In March 2012 there were 14·2m. Facebook users.

SOCIAL INSTITUTIONS

Justice

The judicial power is exercised in the name of the King, by (a) courts of first instance, (b) the court of appeal (Uthorn) and (c) the Supreme Court (Dika). The King appoints, transfers and dismisses judges, who are independent in conducting trials and giving judgment in accordance with the law.

Courts of first instance are subdivided into 20 magistrates' courts (Kwaeng) with limited civil and minor criminal jurisdiction; 85 provincial courts (Changwad) with unlimited civil and criminal jurisdiction; the criminal and civil courts with exclusive jurisdiction in Bangkok; the central juvenile courts for persons under 18 years of age in Bangkok.

The court of appeal exercises appellate jurisdiction in civil and criminal cases from all courts of first instance. From it appeals lie to Dika Court on any point of law and, in certain cases, on questions of fact.

The Supreme Court is the supreme tribunal of the land. Besides its normal appellate jurisdiction in civil and criminal matters, it has semi-original jurisdiction over general election petitions. The decisions of Dika Court are final. Every person has the right to present a petition to the government who will deal with all matters of grievance.

The death penalty is still in force; there were two confirmed executions in 2009 (the first since 2003), but none since. The population in penal institutions in March 2011 was 224,292 (328 per 100,000 of national population).

Education

Education is compulsory for children for nine years and is free in local municipal schools. In 2007 there were 5,703,756 primary school pupils with 321,930 teaching staff. There were 4,789,339 secondary school pupils in 2007 with 227,929 teaching staff. There were 1,005,481 students in vocational education in 2005. In higher education there were 2,503,572 students in 2007 with 66,431 academic staff. In 2005 there were 78 public and 61 private institutions of higher education, 146 industry and community colleges, 110 technical colleges, 54 polytechnic colleges and 44 agricultural and technology colleges.

The adult literacy rate in 2005 was 94%.

In 2007 public expenditure on education came to 4·0% of GNI and 20·9% of total government spending.

Health

In 2002 there were 3,658 hospitals, with a provision of 69 beds per 10,000 population. In 2000 there were 18,025 doctors, 4,141 dentists, 6,384 pharmacists, 70,978 nurses and (in 1997) 2,677 midwives. Thailand is considered to be the most successful country in the world in preventing the spread of HIV/AIDS. The number of annual new HIV cases has fallen to under 20,000 from a high of 140,000 in the mid 1990s.

RELIGION

At the 2000 census 94·6% of the population were Buddhists and 4·6% Muslims. In Feb. 2013 the Roman Catholic church had one cardinal.

CULTURE

World Heritage Sites

There are five UNESCO sites in Thailand. They are: the Thung Yai-Huai Kha Khaeng wildlife sanctuaries (inscribed in 1991); the palace, temples, Buddhas, etc. of the historic town of Sukhothai (1991); the 15th–18th century historic town of Ayutthaya (1991); the Bronze Age Ba Chiang archaeological site (1992); and the mountainous Dong Phayayen-Khao Yai forest complex (2005).

Press

In 2008 there were 46 daily newspapers (45 paid-for and one free), with a combined circulation of 7·4m. The newspapers with the highest circulation figures are *Thai Rath*, *Daily News* and *Kom Chad Luek*.

Tourism

In 2010 there were 15,936,000 tourist arrivals, up from 14,150,000 in 2009. Tourist numbers have doubled since 1998. The leading nationalities of tourists in 2010 were Malaysia (2,059,000), China (1,122,000), Japan (994,000) and the United Kingdom (811,000).

Festivals

Songkran, celebrated on 13 April each year, is the traditional Thai New Year festival, although 1 Jan. was made the official New Year in 1940. Other notable festivals include: the colourful Bosang Umbrella Fair in Jan. and a Flower Carnival in Feb., both in Chiang Mai; a Candle Festival held in Ubon Ratchathani on Khao Phansa Day in July (the day after the full moon of the eighth lunar month, marking the start of Buddhist Lent); Buffalo Races, held during Oct. at Chonburi; and an annual Elephant Round-up in the third week of Nov. at Surin.

DIPLOMATIC REPRESENTATIVES

Of Thailand in the United Kingdom (29–30 Queen's Gate, London, SW7 5JB)
Ambassador: Pasan Teparak.

Of the United Kingdom in Thailand (14 Wireless Rd, Bangkok 10330)
Ambassador: Mark Kent.

Of Thailand in the USA (1024 Wisconsin Ave., NW, Washington, D.C., 20007)
Ambassador: Chaiyong Satjipanon.

Of the USA in Thailand (120 Wireless Rd, Bangkok 10330)
Ambassador: Kristie Kenney.

Of Thailand to the United Nations
Ambassador: Norachit Sinhaseni.

Of Thailand to the European Union
Ambassador: Apichart Chinwanno.

FURTHER READING

National Statistical Office. *Thailand Statistical Yearbook.*

Krongkaew, M. (ed.) *Thailand's Industrialization and its Consequences.* 1995

National Statistical Office: National Statistical Office, The Government Complex, Building B, Chang Watthana Rd, Laksi, Bangkok 10210.
Website: http://web.nso.go.th

TIMOR-LESTE

© Research Machines plc 2006

República Democrática de Timor-Leste
(Democratic Republic of East Timor)

Capital: Dili
Population projection, 2015: 1·30m.
GNI per capita, 2011: (PPP$) 3,005
HDI/world rank: 0·495/147
Internet domain extension: .tl

KEY HISTORICAL EVENTS

Portugal abandoned its former colony, with its largely Roman Catholic population, in 1975, when it was occupied by Indonesia and claimed as the province of Timor Timur. The UN did not recognize Indonesian sovereignty over the territory. An independence movement, the Revolutionary Front for an Independent East Timor (FRETILIN), maintained a guerrilla resistance to the Indonesian government which resulted in large-scale casualties and alleged atrocities. On 24 July 1998 Indonesia announced a withdrawal of troops from Timor-Leste and an amnesty for some political prisoners, although no indication was given of how many of the estimated 12,000 troops and police would pull out. On 5 Aug. 1998 Indonesia and Portugal reached agreement on the outlines of an autonomy plan which would give the Timorese the right to self-government except in foreign affairs and defence.

In a referendum on the future of Timor-Leste held on 30 Aug. 1999 the electorate was some 450,000 and turnout was nearly 99%. 78·5% of voters opted for independence, but pro-Indonesian militia gangs wreaked havoc both before and after the referendum. The militias accused the UN of rigging the poll. There was widespread violence in and around Dili, the provincial capital, with heavy loss of life, and thousands of people were forced to take to the hills after intimidation. Timor-Leste's first democratic election took place on 30 Aug. 2001 in a ballot run by the UN, with FRETILIN winning 57% of the vote and 55 of the 88 seats in the new constituent assembly. Timor-Leste became an independent country on 20 May 2002.

TERRITORY AND POPULATION

Timor-Leste (East Timor) has a total land area of 14,954 sq. km (5,774 sq. miles), consisting of the mainland (13,987 sq. km), the enclave of Oecussi-Ambeno in West Timor (817 sq. km), and the islands of Ataúro to the north (140 sq. km) and Jaco to the east (10 sq. km). The mainland area incorporates the eastern half of the island of Timor. Oecussi-Ambeno lies westwards, separated from the main portion of Timor-Leste by a distance of some 100

km. The island is bound to the south by the Timor Sea and lies approximately 500 km from the Australian coast.

Population at the census of July 2010, 1,066,409 (544,198 males); density, 71 per sq. km. The largest city is Dili, Timor-Leste's capital. In 2010 its population was 192,652. In 2011, 28·6% of the population were urban.

The UN gives a projected population for 2015 of 1·30m.

The ethnic East Timorese form the majority of the population. Non-East Timorese, comprising Portuguese and West Timorese as well as persons from Sumatra, Java, Sulawesi and other parts of Indonesia, are estimated to constitute approximately 20% of the total population.

During Indonesian occupation the official language was Bahasa Indonesia. Timor-Leste's constitution designates Portuguese and Tetum (the region's *lingua franca*) as the official languages, and English and Bahasa Indonesia as working languages.

SOCIAL STATISTICS

2008 estimates: births, 44,000; deaths, 9,500. Rates, 2008 estimates (per 1,000 population): births, 40·0; deaths, 8·7. Annual population growth rate in 2000–08, 3·7%. Fertility rate, 2008, 6·5 children per woman. In 2007 life expectancy at birth was 59·8 years for males and 61·5 years for females.

From having the world's highest rate of infant mortality in the early 1980s, Timor-Leste's infant mortality rate dropped to 46 per 1,000 live births in 2010, although the figure varies widely between urban and rural areas.

CLIMATE

In the north there is an average annual temperature of over 24°C (75°F), weak precipitation—below 1,500 mm (59") annually—and a dry period lasting five months. The mountainous zone, between the northern and southern parts of the island, has high precipitation—above 1,500 mm (59")—and a dry period of four months. The southern zone has precipitation reaching 2,000 mm (79") and is permanently humid. The monsoon season extends from Nov. to May.

CONSTITUTION AND GOVERNMENT

The constitution promulgated in 2002 created a unicameral system with a *National Parliament* with a minimum requirement of 52 directly-elected seats and a maximum of 65. For the first term after independence the parliament had 88 members but this was reduced after the June 2007 legislative elections.

The *President* is directly elected for a period of five years and may not serve more than two terms.

National Anthem

'Pátria, Pátria, Timor-Leste, nossa Nação' ('Fatherland, fatherland, East Timor our Nation'); words by F. Borja da Costa, tune by A. Araujo.

RECENT ELECTIONS

Presidential elections were held on 17 March and 16 April 2012. In the first round Francisco Guterres of the ruling FRETILIN party received 28·8% of votes cast against Taur Matan Ruak (ind.) with 25·7%, José Ramos-Horta (ind.) with 17·5% and Fernando de Araújo of the Democratic Party with 17·3%. Eight other candidates received less than 4% each. In the second round Ruak was elected president with 61·2% of the vote against 38·8% for Guterres. Turnout was 78·2% in the first round and 73·1% in the second.

Elections to the 65-member National Parliament took place on 7 July 2012. The Congresso Nacional de Reconstrução de

Timor-Leste (CNRT; National Congress for Timorese Reconstruction) won 36·7% of votes cast and 30 seats, ahead of the Frente Revolucionária de Timor-Leste Independente (FRETILIN; Revolutionary Front for an Independent East Timor) with 29·9% of the votes and 25 seats, the Partido Democrático (PD; Democratic Party) 10·3% and 8 seats and the Frente de Reconstrução Nacional de Timor-Leste–Mudança (Front for National Reconstruction of Timor-Leste–Change) 3·1% and 2 seats. 17 other parties failed to win a seat. Turnout was 74·8%.

CURRENT GOVERNMENT

President: Taur Matan Ruak; b. 1956 (ind.; since 20 May 2012).

In March 2013 the government was comprised as follows:

Prime Minister: Xanana Gusmão; b. 1946 (National Congress for Timorese Reconstruction; since 8 Aug. 2007).

Deputy Prime Minister and Co-ordinator of Social Affairs: Fernando La Sama de Araújo.

Minister for Agriculture and Fisheries: Mariano Assanami Sabino. *Commerce, Industry and Environment:* António da Conceição. *Defence and Security:* Cirilo José Cristóvão. *Education:* Bendito dos Santos Freitas. *Finance:* Emília Pires. *Foreign Affairs and Co-operation:* José Luís Guterres. *Health:* Sérgio Gama Lobo. *Justice:* Dionísio Babo Soares. *Petroleum and Mineral Resources:* Alfredo Pires. *Public Works:* Gastão Francisco de Sousa. *Social Solidarity:* Isabel Guterres. *State Administration:* Jorge da Conceição Teme. *Tourism:* Francisco Kalbuadi Lay. *Transport and Communications:* Pedro Lay. *Minister of State:* Hermenegildo Pereira.

Office of the Prime Minister and Government Website:
http://www.pm.gov.tp

CURRENT LEADERS

Taur Matan Ruak

Position
President

Introduction
Taur Matan Ruak became president in May 2012, taking the largely ceremonial role as UN peacekeepers stationed in the country since 1999 prepared to withdraw.

Early Life
Born José Maria de Vasconcelhos on 10 Oct. 1956 in Baucau District, the president is popularly known by his *nom de guerre*, Taur Matan Ruak, meaning 'two sharp eyes' in the local Tetun language.

During the Indonesian invasion in Dec. 1975, Ruak joined FALINTIL (Forças Armadas da Libertação Nacional de Timor-Leste)—the military wing of FRETILIN (Frente Revolucionária de Timor-Leste Independente). On 13 March 1979 he was captured by Indonesian troops but escaped after 23 days to re-join FALINTIL. Rising through the ranks, he became chief-of-staff in 1992, commander in 1998 and commander-in-chief in 2000.

With the dissolution of FALINTIL, Ruak became a brigadier general in the newly established national armed force (F-FDTL) on 1 Feb. 2001. In 2006 he faced criticism over the military's transfer of weapons to civilians during unrest that year, with a UN commission of inquiry recommending his prosecution. However, a Timorese inquiry cleared him of all charges and Ruak was promoted to major general in 2009.

He resigned from the F-FDTL on 1 Sept. 2011 and announced his presidential candidacy a month later. Running as an independent, he was backed by Prime Minister Xanana Gusmão and the Conselho Nacional de Reconstrução de Timor (CNRT) party. After coming second with 26% in the first round of voting, Ruak won the second round with 61% of votes cast.

Career in Office
Ruak was sworn in on 20 May 2012, the tenth anniversary of Timor-Leste's independence. He reappointed Gusmão as prime minister after the July 2012 parliamentary elections. Despite having no executive powers, Ruak will work closely with the premier and his government on issues including the diversification of the economy, national defence strategy and forging closer ties with regional partners, especially Indonesia and Australia.

Xanana Gusmão

Position
Prime Minister

Introduction
Independent Timor-Leste's first president, Xanana Gusmão, having led the independence movement for over two decades, came to power in a landslide victory at elections in April 2002. He stood down in May 2007 but was subsequently appointed prime minister in Aug. that year.

Early Life
Xanana Gusmão was born José Alexandre Gusmão on 20 June 1946 in Laleia, Manatuto. After studying at a Jesuit seminary in Soibada and then at Dare, he became a civil servant.

He joined FRETILIN in 1974, becoming its leader in 1978. In 1981 he was elected commander-in-chief of its military wing and worked to integrate the various groups fighting for independence.

In Nov. 1992 he was captured by the Indonesian army and sentenced to life imprisonment on charges of subversion but remained the figurehead of the independence movement. Following an appeal from then UN Secretary-General Kofi Annan, Gusmão was released after the referendum of Sept. 1999 in which an overwhelming majority of Timorese voted for independence.

Timor-Leste gained independence on 20 May 2002 and Gusmão was inaugurated as president, having won a landslide victory in elections the previous month.

Career in Office
Gusmão appealed for reconciliation and an end of violence against those who opposed independence. The authority of the state and its institutions, however, remained fragile. In April 2006, 600 striking soldiers who had been sacked by Prime Minister Mari Alkatiri demonstrated in Dili. The protests turned into wider factional violence across the country and the government called in foreign troops led by Australia in May to restore law and order. At the same time, relations between president and prime minister broke down. In June Alkatiri stood down and was replaced by José Ramos-Horta. A UN peacekeeping mission was then set up in Aug. 2006.

Gusmão did not contest presidential elections held in April–May 2007, announcing his intention to run instead for prime minister as the leader of a new National Congress for Timorese Reconstruction (CNRT), having become disillusioned with FRETILIN. His close ally José Ramos-Horta succeeded him as president. Despite FRETILIN winning the largest number of seats in parliamentary elections in June 2007, Gusmão formed a coalition government and was sworn in as prime minister on 8 Aug. 2007.

Like then President Ramos-Horta, Gusmão was also targeted by rebel soldiers in Feb. 2008 in a separate attack, but was unhurt. He described the incident as a coup attempt and imposed a state of emergency.

In Oct. 2009 Gusmão's government survived an opposition vote of confidence in parliament over its controversial release of a pro-Indonesian militia leader accused of war crimes against Timorese citizens in 1999.

Acknowledging public concern over allegations of corruption against senior officials, in Feb. 2010 the government appointed the country's first anti-corruption commissioner.

In parliamentary elections in July 2012, Gusmão's CNRT became the largest party with 30 seats, ahead of FRETILIN with 25.

DEFENCE

The Timor-Leste Defence Force comprises an army and a small naval element. In 2005 there was a 1,250-strong army, but nearly half of the personnel were dismissed in early 2006 for going on strike.

In 2006 the UN initiated the 'Integrated Mission in East Timor' (UNMIT) in a bid to bring stability to the country. Totalling 2,745 personnel in mid-2012, it was withdrawn at the end of Dec. 2012. The Australian-led International Stabilisation Force, also deployed in 2006, left in March 2013.

ECONOMY

Overview

Timor-Leste is one of Asia's poorest countries. Following a referendum in 1999 in which the population voted over-whelmingly in favour of independence from Indonesia, militia groups led a campaign of violence that ravaged the infrastructure. Since independence in May 2002, authorities have relied on the UN and other development agencies to help rebuild the economy.

Oil resources in the Timor Sea have transformed the prospects of the country. Expansionary fiscal policy driven by a boom in oil revenues since 2005 has resulted in strong growth, although this was interrupted in 2006 with an outbreak of violence leading to military and humanitarian assistance from neighbouring economies and the UN. The global economic crisis has affected the economy largely through lower oil and gas revenues. Maintaining fiscal discipline, ensuring good use of public monies and upholding security will be critical in sustaining growth.

Currency

The official currency is the US dollar. The Australian dollar and the Indonesian rupiah, both previously used, no longer serve as legal tender. Inflation was 6·8% in 2010, increasing to 13·5% in 2011.

Budget

Revenues in 2005–06 were US$485m. (oil and gas revenue, 93·1%) and expenditures US$93m. (current expenditure, 71·3%).

Performance

Total GDP in 2011 was US$1·1bn. Real GDP growth was 12·8% in 2009, 9·5% in 2010 and 10·6% in 2011.

ENERGY AND NATURAL RESOURCES

Environment

Timor-Leste's carbon dioxide emissions from the consumption and flaring of fossil fuels were the equivalent of 0·3 tonnes per capita in 2008.

Electricity

Electricity produced in 2007 totalled an estimated 135m. kWh.

Oil and Gas

Although current production is small, the Timor Gap, an area of offshore territory between Timor-Leste and Australia, is one of the richest oilfields in the world outside the Middle East. Potential revenue from the area is estimated at US$11bn. The area is split into three zones with a central 'zone of occupation' (occupying 61,000 sq. km). Royalties on oil discovered within the central zone were split equally between Indonesia and Australia following the Timor Gap Treaty which came into force on 9 Feb. 1991. Questions over Timor-Leste's rights to oil revenue from the area

have arisen following the 1999 independence referendum. There are also extensive offshore gas fields, but the region generally remains underexplored.

Minerals

Gold, iron sands, copper and chromium are present.

Agriculture

Although the presence of sandalwood was one of the principal reasons behind Portuguese colonization, its production has declined in recent years. In 2007 there were an estimated 170,000 ha. of arable land and 68,000 ha. of permanent crops. Principal crops are maize, rice, cassava, sweet potatoes, coconuts and coffee.

Forestry

Forests covered 0·74m. ha. in 2010, representing 50% of the total land area.

Fisheries

The total fish catch in 2008 was 3,125 tonnes.

INDUSTRY

Labour

In 2010 unemployment was officially 9·8% of the labour force between 15 and 64.

INTERNATIONAL TRADE

Imports and Exports

Imports totalled US$268·5m. in 2008 and exports US$49·2m. Major imports are mineral fuels, motor vehicles and cereals. Coffee and cattle are important exports. Leading import suppliers are Indonesia, Singapore, Australia and Vietnam. Leading export destinations are Germany, the USA, Indonesia and Singapore.

COMMUNICATIONS

Civil Aviation

There is an international airport at Dili (Presidente Nicolau Lobato International Airport).

Telecommunications

In 2010 there were 2,907 landline telephone subscriptions (2,334 in 2005) and 350,891 mobile phone subscriptions (33,072 in 2005).

SOCIAL INSTITUTIONS

Justice

Timor-Leste's judiciary currently conforms to a UN-drafted legal system based on Indonesian law. This is scheduled to be replaced by a civil and penal judiciary system which will be based on Portuguese law. The Supreme Court of Justice is the highest court of law. There are four District Courts and a Court of Appeal.

Health

Plans for a medical system include 64 community health centres, 88 health posts, 117 mobile clinics and 21 doctors.

RELIGION

Over 90% of Timor-Leste's population are Roman Catholic, with Protestants, Muslims, Hindus and Buddhists accounting for the remainder.

CULTURE

Press

In 2007 there were three daily newspapers: *Suara Timor Lorosae*, *Timor Post* and *Jornal Nacional Diario*. There were also three non-dailies in 2007.

Tourism

In 2009, 44,131 non-resident tourists—excluding same-day visitors—arrived by air (up from 35,999 in 2008).

DIPLOMATIC REPRESENTATIVES

Of the United Kingdom in Timor-Leste (embassy in Dili closed in Oct. 2006)
Ambassador: Mark Canning, CMG (resides in Jakarta, Indonesia).

Of Timor-Leste in the USA (4201 Connecticut Ave., NW, Suite 504, Washington, D.C., 20008)
Ambassador: Constancio C. Pinto.

Of the USA in Timor-Leste (Avenida de Portugal, Pantai Kelapa, Dili)
Ambassador: Judith Fergin.

Of Timor-Leste to the United Nations
Ambassador: Sofia Mesqíta Borges.

Of Timor-Leste to the European Union
Ambassador: Nelson Santos.

FURTHER READING

Dunn, James, *East Timor: A Rough Passage to Independence.* 2003
Hainsworth, Paul and McCloskey, Stephen, (eds.) *The East Timor Question: The Struggle for Independence from Indonesia.* 2000
Kohen, Arnold S., *From the Place of the Dead: Bishop Belo and the Struggle for East Timor.* 2000
Kingsbury, Damien and Leach, Michael, (eds.) *East Timor: Beyond Independence.* 2007
Nevins, Joseph, *A Not-So-Distant Horror: Mass Violence in East Timor.* 2005
Robinson, Geoffrey, *"If You Leave Us Here, We Will Die": How Genocide Was Stopped in East Timor.* 2009
Tanter, Richard, Ball, Desmond and Van Klinken, Gerry, (eds.) *Masters of Terror: Indonesia's Military and Violence in East Timor.* 2006

National Statistical Office: Direcção Nacional de Estatística, Rua de Caicoli, P. O. Box 10, Dili.
Website: http://www.dne.mof.gov.tl

TOGO

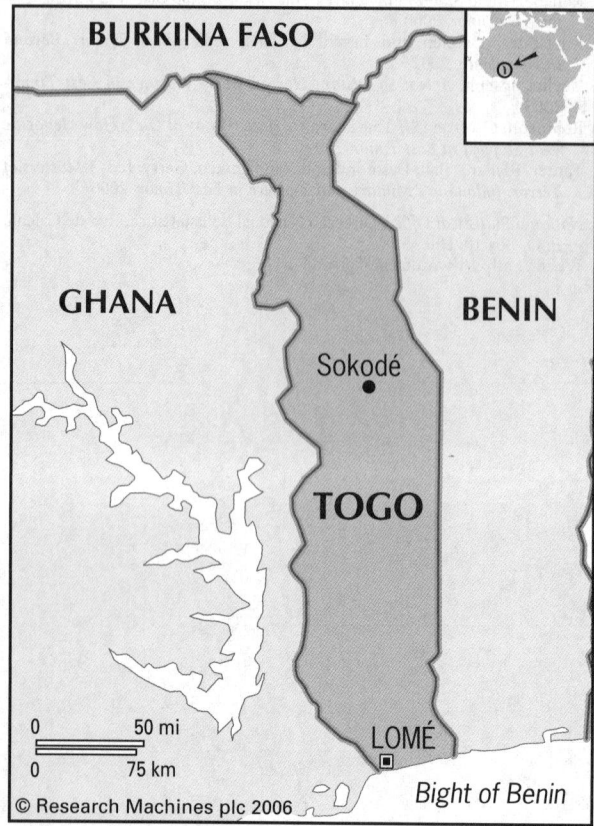

République Togolaise
(Togolese Republic)

Capital: Lomé
Population projection, 2015: 6·67m.
GNI per capita, 2011: (PPP$) 798
HDI/world rank: 0·435/162
Internet domain extension: .tg

KEY HISTORICAL EVENTS

Eleventh century records relate Togo's settlement by a succession of tribes. The Kwa people from the Volta region were joined by the Ewe people from the Niger Valley, while later migrations included the Ana, Mina and Guin people from Ghana and the Ivory Coast, and the Nawdba from the east, who settled in northern Togo.

During 1471–72 Portuguese explorers and traders began trading along the coast in gold, silver and pepper. In the 16th century Britain, Denmark, Holland, France, Germany, Sweden and Portugal all shipped slaves from the region in co-operation with tribal leaders. The 18th century saw conflicts with the Akwamu Confederacy and the Ashanti Kingdom in the west, and the Kingdom of Dahomey in the east, as tribal rulers tried to consolidate their territory. Meanwhile, Britain, France and Germany became the dominant European powers in the region.

Following the abolition of slavery in the 19th century, British, French and German traders built up a flourishing palm oil export trade from Anecho, Agoue and Porto Seguro. As a result several Togolese families of partly Brazilian or Portuguese origin came to prominence, some retaining considerable influence to the present day.

Despite competition from Britain and France, Germany signed a treaty with chief Mlapa III on 5 July 1884 in the village of Togo (modern Togoville) to establish colonial rule on the coast. German control was then extended inland and, following agreement with Britain and France, the borders of German Togoland were defined. Using slave labour, the German administration developed the agricultural sector, focusing on cacao, coffee and cotton.

In 1914 the Allies overran German Togoland and in 1922 it was partitioned into a western British Mandated Territory and an eastern French Mandated Territory under the League of Nations. In 1946, following the Second World War, British Togoland and French Togoland became Trust territories under the United Nations. On 9 May 1956 a UN-sponsored referendum in British Togoland resulted in a majority vote for union with the neighbouring colony of Gold Coast. The territory soon merged with Gold Coast to become Ghana, despite objections from the southern Togolese, most of whom had voted for union with French Togo. In 1956 French Togo was granted partial self-government to become the Republic of Togo. It gained full independence on 27 April 1960.

Sylvanus Olympio was elected president in 1958. In 1961 a new constitution established a seven-year-term and broad executive powers for the presidency, alongside an elected national assembly. Olympio won the 1961 election and his party, Unité Togolaise (UT), took all 51 national assembly seats. In 1962 Olympio disbanded the main opposition parties, alleging they had plotted against his government. On 13 Jan. 1963 Olympio was assassinated by soldiers and the army voted in Nicolas Grunitsky as head of state. A military coup ousted Grunitsky in Jan. 1967 and on 14 April 1967 Gen. (then Col.) Etienne Gnassingbé Eyadéma took the presidency.

Eyadéma developed the country's investment banking sector and its phosphates exports, prompting economic growth until recession in the 1980s caused a price collapse. Eyadéma ruled as head of Togo's only officially recognized political party, the Rassemblement du Peuple Togolais (RPT), and survived several attempted coups. In Aug. 1991 a national conference of Togo's leading political movements led to a reduction in Eyadéma's executive powers. A transitional administration, the High Council of the Republic, was installed, led by Kokou Koffigoh.

In Nov. 1991 the new government banned the RPT and the army, which backed Eyadéma, attempted a coup. Eyadéma negotiated the return of some of his powers and won an election in Aug. 1992, though its validity was disputed and the main opposition parties boycotted it. In 1994 a coalition of the Comité d'Action pour le Renouveau and the Union Togolaise pour la Démocratie won a majority in the national assembly elections. The RPT returned to power in the election of June 1998 and Eyadéma retained the presidency throughout the entire period until his death in Feb. 2005.

On Eyadéma's death the military installed his son, Faure Gnassingbé, as president and the following day the constitution was amended to legalize his succession. Under domestic and international pressure he stepped down on 25 Feb. 2005 and parliament speaker Abbas Bonfoh became interim president. Faure Gnassingbé won the presidential election held in April 2005 but the opposition alleged vote rigging and 400–500 people died in riots. Gnassingbé was subsequently confirmed as president.

TERRITORY AND POPULATION

Togo is bounded in the west by Ghana, north by Burkina Faso, east by Benin and south by the Gulf of Guinea. The area is 56,600 sq. km. 2010 census population, 6,191,155 (3,182,060 females); density, 109 per sq. km.

The UN gives a projected population for 2015 of 6·67m.

In 2011, 44·1% of the population lived in urban areas. In 2005, 43% were below the age of 15. The capital is Lomé (2010 census population, 837,437), other towns being Sokodé (95,070), Kara (94,878), Kpalimé (75,084), Atakpamé (69,261), Dapaong (58,071) and Tsévié (54,474).

Area, 2010 census population and chief town of the five regions:

Region	Area in sq. km	Population	Chief town
Centrale	13,317	617,871	Sokodé
De La Kara	11,738	769,940	Kara
Des Plateaux	16,975	1,375,165	Atakpamé
Des Savanes	8,470	828,224	Dapaong
Maritime	6,100	2,599,955	Lomé

There are 37 ethnic groups. The south is largely populated by Ewe-speaking peoples (forming 23% of the population), Watyi (10%) and other related groups, while the north is mainly inhabited by Hamitic groups speaking Kabre (14%), Tem (6%) and Gurma (3%). The official language is French but Ewe and Kabre are also taught in schools.

SOCIAL STATISTICS

2008 estimates: births, 212,000; deaths, 53,000. Estimated rates, 2008 (per 1,000 population): births, 32·9; deaths, 8·2. Expectation of life (2007) was 60·4 years for males and 63·9 for females. Annual population growth rate, 2000–05, 2·8%. Infant mortality, 2010, 66 per 1,000 live births; fertility rate, 2008, 4·3 births per woman.

CLIMATE

The tropical climate produces wet seasons from March to July and from Oct. to Nov. in the south. The north has one wet season, from April to July. The heaviest rainfall occurs in the mountains of the west, southwest and centre. Lomé, Jan. 81°F (27·2°C), July 76°F (24·4°C). Annual rainfall 35" (875 mm).

CONSTITUTION AND GOVERNMENT

A referendum on 27 Sept. 1992 approved a new constitution by 98·11% of votes cast. Under this the *President* and the *National Assembly* were directly elected for five-year terms. Initially the president was allowed to be re-elected only once. However, on 30 Dec. 2002 parliament approved an amendment to the constitution lifting the restriction on the number of times that the president may be re-elected. The National Assembly has 81 seats and is elected for a five-year term.

National Anthem

'Terre de nos aïeux' ('Land of our forefathers').

RECENT ELECTIONS

In presidential elections held on 4 March 2010 Faure Gnassingbé was re-elected with 60·9% of the vote, ahead of Jean-Pierre Fabre of the Union of Forces for Change (Union des Forces de Changement) with 33·9% and Yawovi Agboyibo of the Action Committee for Renewal (Comité d'Action pour la Renouveau) with 3·0%. There were four other candidates. Turnout was 65·7%.

At the parliamentary elections on 14 Oct. 2007 the ruling RPT (reconfigured as the Union for the Republic/UNIR in April 2012) won 50 of 81 seats with 32·7% of votes cast, the Union of Forces for Change 27 with 30·8% and the Action Committee for Renewal 4 with 6·8%. Turnout was 94·8%.

CURRENT GOVERNMENT

President: Faure Gnassingbé; b. 1966 (UNIR—formerly RPT; sworn in 4 May 2005 and re-elected 4 March 2010).

In March 2013 the government comprised:

Prime Minister: Kwesi Ahoomey-Zunu; b. 1958 (Patriotic Pan-African Convergence; sworn in 23 July 2012).

Minister of State, Minister of Foreign Affairs and Co-operation: Elliot Ohin. *Minister of State, Minister of Primary and Secondary Education, and Literacy:* Solitoki Magnim Esso.

Minister of Advancement of Women: Patricia Dagban-Zonvide. *Agriculture, Animal Husbandry and Fisheries:* Ouro Koura Agadazi. *Arts and Culture:* Fiatuwo Kwadjo Sessenou. *Basic Development, Handicrafts and Youth Employment:* Victoire Sidéméo Tomegah Togbé. *Civil Service and Administrative Reform:* Kokou Dzifa Adjeoda. *Commerce and Promotion of the Private Sector:* Bernadette Léguezim-Balouki. *Communications:* Djimon Orhé. *Economy and Finance:* Adji Ayassor. *Environment and Forest Resources:* Dédé Ahoéfa Ekoué. *Health:* Charles Condji Agba. *Higher Education and Research:* Octave Nicoué Broohm. *Human Rights, Promotion of Democracy and Civic Education:* Rita Boris De Souza. *Industry, Free Zones and Technological Innovation:* François Agbéviadé Galley. *Justice and Keeper of the Seals:* Tchitchao Tchalim. *Labour, Employment and Social Security:* Yacoubou Hamadou. *Post and Telecommunications:* Sina Lawson. *Public Works:* Ninsao Gnofam. *Security and Civil Protection:* Col. Yark Damehane. *Social Action and National Solidarity:* Afi Ntifa Amenyo. *Sport and Leisure:* Bakalawa Fofana. *Technical Education and Professional Training:* Amadou Bouraïma Diabacté. *Territorial Administration, Decentralization and Local Collectivities:* Gilbert Bawara. *Tourism:* Padumhékou Christophe Tchao. *Transport, and Mines and Energy (acting):* Dammipi Noupokou. *Urban Affairs and Housing:* Komlan Nunyabu. *Water, Sanitation and Village Hydraulics:* Bissoune Nabagou. *Minister at the Presidency in Charge of Planning, Development and Land Management:* Kokou Sémondji.

Government Website (French only):
http://www.republicoftogo.com

CURRENT LEADERS

Faure Gnassingbé

Position
President

Introduction
Faure Gnassingbé was installed as president after the death of his father, one of Africa's longest-serving leaders. The appointment led to violent protests and international condemnation. Forced to step down, he contested a presidential election and emerged victorious in April 2005 with just over 60% of the vote. Faure then pursued a policy of political reconciliation, paving the way for peaceful parliamentary elections in Oct. 2007. He was re-elected to the presidency in March 2010.

Early Life
Faure Essozimna Gnassingbé was born in Afagnan, Togo on 6 June 1966. He is the son of Gnassingbé Eyadéma, a general who led a coup in 1963, declared himself president in 1967 and remained head of state until his death in Feb. 2005. Faure attended school in the capital, Lomé, followed by the Sorbonne University in Paris and the George Washington University in the USA. Returning to an unstable and economically crippled Togo in the mid-1990s, Faure began work as a civil servant.

He entered politics in 2002, when he was elected the representative of the ruling Togolese People's Assembly (RPT; since reconfigured as the UNIR) for Blitta constituency in central Togo, and in June 2003 he took office as minister of public works, mines and telecommunications.

On 6 Feb. 2005, the day after his father's death, Faure was proclaimed as his successor by the Togolese army, sparking widespread protest. Amid mounting international criticism and threats of sanctions, Faure announced that a presidential election would be held on 24 April 2005 and stepped down from office. Official results gave Faure over 60% of the vote at the subsequent elections and, despite violence erupting in Lomé amid claims of electoral fraud, he was sworn in as president on 4 May 2005.

Career in Office

Faure promised to push for economic growth, to reform institutions and to improve Togo's image abroad. Although most of the key government posts went to members of the ruling party, Edem Kodjo, the leader of a moderate opposition party, was named prime minister in June 2005. In Sept. 2006 Faure appointed another veteran opposition leader, Yawovi Madji Agboyibo of the Action Committee for Renewal, as prime minister with the task of forming a unity government. In Oct. 2007 the ruling RPT won parliamentary elections in which opposition parties took part for the first time in almost two decades and which were declared by international observers to be free and fair. Faure then appointed Komlan Mally prime minister and a new government was formed in Dec. When Mally resigned in Sept. 2008, Faure appointed Gilbert Houngbo, an independent, as the new premier. In April 2009 the government claimed that there had been a foiled coup plot against Faure involving his half-brother and former defence minister Kpatcha Gnassingbé. Faure was returned to office in presidential elections on 4 March 2010 that were deemed relatively fair by international observers. In May veteran opposition leader Gilchrist Olympio agreed to join a government of national unity, ending years of hostility towards the then RPT but causing a deep split within his Union of Forces for Change which had alleged widespread fraud in the election.

Parliamentary elections scheduled for Oct. 2012 were postponed after violent protests in June against electoral law amendments that demonstrators claimed favour the ruling UNIR. In July Faure appointed Kwesi Ahoomey-Zunu as prime minister following the resignation of Gilbert Houngbo.

DEFENCE

There is selective conscription that lasts for two years. Defence expenditure totalled US$56m. in 2008 (US$10 per capita), representing 1·9% of GDP.

Army

Strength (2007) around 8,100, with a further 750 in a paramilitary gendarmerie.

Navy

In 2007 the Naval wing of the armed forces numbered about 200 and was based at Lomé.

Air Force

The Air Force—established with French assistance—numbered (2007) 250, with 16 combat aircraft although their serviceability is in doubt.

ECONOMY

Agriculture contributed 33% of GDP in 2009, industry 16% and services 51%.

Overview

Togo's economic performance is among the weakest in Sub-Saharan Africa, with poverty increasing significantly since the early 1980s. Social indicators are low, with the economy ranked 162nd out of 187 countries on the Human Development Index. In the World Bank's Doing Business 2010 report, Togo ranked 160th of 183 countries. The main economic activities are mining, agriculture and re-exporting. The primary sector, port activities

and public investment have been the drivers of growth in recent years.

After civil and economic turmoil in the early 1990s, a structural redevelopment programme launched in 1994 supported growth. Following a downturn in 2005, the government acted to restore fiscal discipline and strengthen governance. The surge in global food and fuel prices in 2008, as well as heavy flooding, hurt the economy. Further flooding in large parts of the country between Sept. and Oct. 2010 caused more damage to infrastructure and crops.

In Dec. 2010 Togo completed the Heavily Indebted Poor Countries (HIPC) process, leading to a cancellation of around 82% of its external debt. As a result, total government debt fell from 67% of GDP to about 30%.

Currency

The unit of currency is the *franc CFA* (XOF) with a parity of 655·957 francs CFA to one euro. Foreign exchange reserves were US$270m. in June 2005 and total money supply was 177,138m. francs CFA. Inflation was 8·7% in 2008, falling to 3·2% in 2010. It stood at 3·6% in 2011.

Budget

In 2008 revenues were 249·9bn. francs CFA and expenditures 253·3bn. francs CFA. Tax revenue accounted for 84·5% of revenues in 2008; current expenditure accounted for 80·2% of expenditures.

VAT is 18%.

Performance

Real GDP growth was 3·5% in 2009, 4·0% in 2010 and 4·9% in 2011. Total GDP in 2011 was US$3·6bn.

Banking and Finance

The bank of issue is the Central Bank of West African States (BCEAO). The *Governor* is Tiémoko Meyliet Koné. In 2003 there were six commercial banks, three development banks, a savings bank and a credit institution. Foreign debt totalled US$1,728m. in 2010, representing 61·1% of GNI.

ENERGY AND NATURAL RESOURCES

Environment

Togo's carbon dioxide emissions from the consumption and flaring of fossil fuels in 2008 were the equivalent of 0·5 tonnes per capita.

Electricity

Installed capacity in 2007 was an estimated 84,000 kW. In 2007 production totalled 196m. kWh. Additional electricity is imported, mainly from Ghana. Consumption per capita in 2007 was 130 kWh.

Minerals

Output of phosphate rock in 2007 was 750,000 tonnes. Other minerals are limestone, iron ore and marble.

Agriculture

Agriculture supports about 80% of the population. Most food production comes from individual holdings under 3 ha. Inland, the country is hilly; dry plains alternate with arable land. There were an estimated 2·46m. ha. of arable land in 2007 and 0·17m. ha. of permanent crops. There are considerable plantations of oil and cocoa palms, coffee, cacao, kola, cassava and cotton. Production, 2003 (in 1,000 tonnes): cassava, 724; yams, 569; maize, 516; sorghum, 177; seed cotton, 159; cotton lint, 76; cottonseed, 75; rice, 68; millet, 50; dry beans, 44; groundnuts, 37.

Livestock (2003 estimates, in 1,000): sheep, 1,800; goats, 1,470; pigs, 310; cattle, 279; chickens, 8,000.

Forestry

Forests covered 0·29m. ha. in 2010, or 5% of the land area. Teak plantations covered 8,600 ha. In 2007, 6·04m. cu. metres of roundwood were cut.

Fisheries

The catch in 2010 totalled an estimated 27,500 tonnes (82% from marine waters).

INDUSTRY

Industry is small-scale. Cement and textiles are produced and food processed. In 2005 industry accounted for 24% of GDP, with manufacturing contributing 10%.

Labour

In 2010 the estimated labour force was 3,059,000 (56% males), up from 2,182,000 in 2000. In Aug. 2008 the statutory monthly minimum wage was raised to 28,000 francs CFA.

INTERNATIONAL TRADE

A free trade zone was established in 1990.

Imports and Exports

In 2008 imports (c.i.f.) amounted to US$1,540m. (US$1,483m. in 2007); exports (f.o.b.) US$790m. (US$705m. in 2007). The main import suppliers in 2005 were France (17·6%), China (13·2%), Côte d'Ivoire (6·5%), Italy (4·5%) and Spain (4·2%). Principal export destinations in 2005 were Ghana (20·3%), Burkina Faso (18·4%), Benin (11·6%), Mali (7·4%) and India (5·9%). Leading imports are food, refined petroleum, and chemicals and chemical products; main exports are cement, phosphates and cotton.

COMMUNICATIONS

Roads

There were 11,652 km of roads in 2007, including 3,067 km of highways or national roads. In 2007 there were 10,600 passenger cars in use, 2,200 lorries and vans and 34,200 motorcycles and mopeds.

Rail

There are four metre-gauge railways connecting Lomé, with Aného (continuing to Cotonou in Benin), Kpalimé, Tabligbo and (via Atakpamé) Blitta; total length in 2005, 532 km. In 2005 the railways carried 1·1m. tonnes of freight. There has been no passenger rail service since 1996.

Civil Aviation

In 2003 Trans African Airlines flew from Tokoin airport, near Lomé, to Abidjan, Bamako, Brazzaville, Cotonou, Dakar and Pointe-Noire. There were also international flights with other airlines to Addis Ababa, Brussels, Douala, Kinshasa, Lagos, Libreville, Ouagadougou and Paris. In 2001 Tokoin handled 151,000 passengers (all on international flights) and 5,100 tonnes of freight. In 2003 scheduled airline traffic of Togo-based carriers flew 1m. km, carrying 46,000 passengers (all on international flights).

Shipping

In Jan. 2009 there were 18 ships of 300 GT or over registered, totalling 33,000 GT.

Telecommunications

In 2008 there were 140,900 main (fixed) telephone lines in Togo; mobile phone subscribers numbered 1,549,500 in 2008 (24·0 per 100 persons). There were 350,000 internet users in 2008.

SOCIAL INSTITUTIONS

Justice

The Supreme Court and two Appeal Courts are in Lomé, one for criminal cases and one for civil and commercial cases. Each receives appeal from a series of local tribunals.

The death penalty was abolished in June 2009. The population in penal institutions in Oct. 2010 was 4,116 (59 per 100,000 of national population).

Education

The adult literacy rate in 2008 was 65%. In 2007 there were 1,021,617 pupils and 26,103 teaching staff in primary schools, and 408,964 pupils in secondary schools with 11,518 teaching staff. In 2006 there were 32,502 students in higher education and 455 academic staff. In 2007 about 77% of children of primary school age were attending school. The University of Benin at Lomé (founded in 1970) is the leading institution of tertiary education.

In 2007 public expenditure on education came to 3·8% of GNI and accounted for 17·2% of total government expenditure.

Health

In 2005 hospital bed provision was nine per 10,000 population. In 2004 there were 225 physicians, 19 dentistry personnel and 1,937 nursing and midwifery personnel. Government expenditure on health in 2005 came to 6·9% of total government spending.

RELIGION

In 2001, 38% of the population followed traditional animist religions; 35% were Christian and 19% Muslim.

CULTURE

World Heritage Sites

There is one UNESCO site in Togo: Koutammakou, the land of the Batammariba (inscribed on the list in 2004).

Press

There is one government-controlled daily newspaper, *Togo-Presse* (circulation of 5,000 in 2006).

Tourism

In 2007 there were 86,000 non-resident tourists staying at hotels and similar establishments; spending by tourists totalled US$23m. in 2006.

DIPLOMATIC REPRESENTATIVES

Of Togo in the United Kingdom
Ambassador: Calixte Madjoulba (resides in Paris).

Of the United Kingdom in Togo
Ambassador: Peter Jones (resides in Accra, Ghana).

Of Togo in the USA (2208 Massachusetts Ave., NW, Washington, D.C., 20008)
Ambassador: Edawe Limbaye Kadangha Bariki.

Of the USA in Togo (4332 Boulevard Gnassingbé Eyadema, BP 852, Cité-OUA, Lomé)
Ambassador: Robert E. Whitehead.

Of Togo to the United Nations
Ambassador: Kodjo Menan.

Of Togo to the European Union
Ambassador: Félix Kodjo Sagbo.

FURTHER READING

National Statistical Office: Direction Générale de la Statistique et de la Comptabilité Nationale, B. P. 118, Lomé.
Website (French only): http://www.stat-togo.org

TONGA

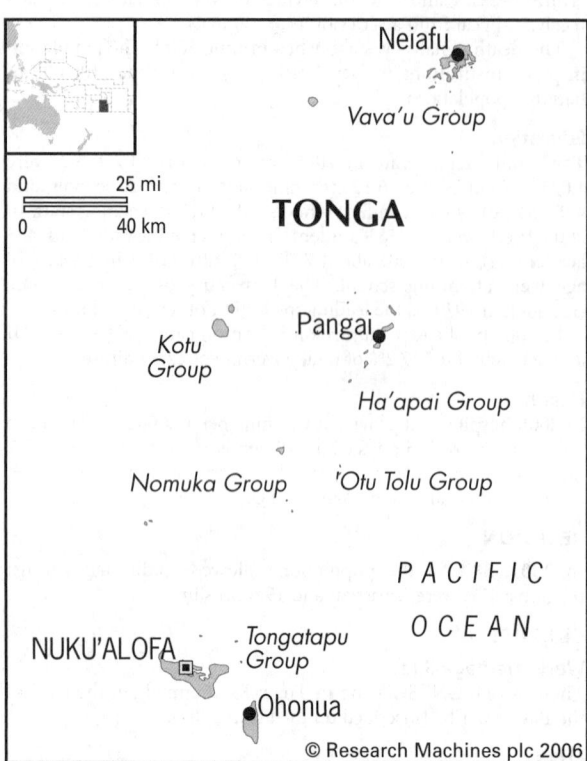

Pule'anga Fakatu'i 'o Tonga
(Kingdom of Tonga)

Capital: Nuku'alofa
Population projection, 2015: 106,000
GNI per capita, 2011: (PPP$) 4,186
HDI/world rank: 0·704/90
Internet domain extension: .to

KEY HISTORICAL EVENTS

The Tongatapu group of islands in the south western Pacific Ocean were discovered by Tasman in 1643. The Kingdom of Tonga attained unity under Taufa'ahau Tupou (George I) who became ruler of his native Ha'apai in 1820, of Vava'u in 1833 and of Tongatapu in 1845. By 1860 the kingdom had converted to Christianity. In 1862 the king granted freedom from arbitrary rule of minor chiefs and extended land rights. These institutional changes, together with the establishment of a parliament of chiefs, paved the way towards a democratic constitution. By the Anglo-German Agreement of 14 Nov. 1899, the Tonga Islands became a British protectorate. The protectorate was dissolved on 4 June 1970 when Tonga, the only ancient kingdom surviving from the pre-European period in Polynesia, achieved independence within the Commonwealth.

TERRITORY AND POPULATION

The Kingdom consists of some 169 islands and islets with a total area, including 30 sq. km of inland waters plus uninhabited islands, of 748 sq. km (289 sq. miles), and lies between 15° and 23° 30' S. lat and 173° and 177° W. long, its western boundary being the eastern boundary of the Fiji Islands. The islands are split

up into the following groups (reading from north to south): the Niuas, Vava'u, Ha'apai, Tongatapu and 'Eua. The three main groups, both from historical and administrative significance, are Tongatapu in the south, Ha'apai in the centre and Vava'u in the north. Census population (2011, provisional) 103,036; density, 138 per sq. km. In 2011, 23·5% of the population lived in urban areas.

The UN gives a projected population for 2015 of 106,000.

The capital is Nuku'alofa on Tongatapu; population (2011 provisional), 35,778 (urban population, 24,158).

There are five divisions comprising 23 districts:

Division	Sq. km	Census 2011 (provisional)	Capital
Niuas	72	1,281	Hihifo
Vava'u	119	14,938	Neiafu
Ha'apai	110	6,650	Pangai
Tongatapu	261	75,158	Nuku'alofa
'Eua	87	5,011	Ohonua

Both Tongan and English are recognized as official languages.

SOCIAL STATISTICS

Births, 2008 estimates, 2,900; deaths, 600; marriages (Tongatapu only), 892; divorces, 95. Expectation of life, 2007: males, 69·0 years; females, 74·6. Annual population growth rate, 2000–08, 0·6%. Infant mortality, 2010, 13 per 1,000 live births. Fertility rate, 2008, 4·0 births per woman.

CLIMATE

Generally a healthy climate, although Jan. to March hot and humid, with temperatures of 90°F (32·2°C). Rainfall amounts are comparatively high, being greatest from Dec. to March. Nuku'alofa, Jan. 25·8°C, July 21·3°C. Annual rainfall 1,643 mm. Vava'u, Jan. 27·3°C, July 23·4°C. Annual rainfall 2,034 mm.

CONSTITUTION AND GOVERNMENT

The reigning King is **Tupou VI ('Aho'eitu 'Unuaki'otonga Tuku'aho Tupou VI)**, born 12 July 1959, succeeded on 18 March 2012 on the death of his brother, George Tupou V.

The current Constitution is based on the one granted in 1875. It was last amended in 2003 to increase the constitutional powers of the King and restrict media freedom. There is a Privy Council, Cabinet, Legislative Assembly and Judiciary. The 28-member *Legislative Assembly* is composed of 17 elected representatives of the people, nine nobles elected by their peers and two *ex officio* members. Prior to the Nov. 2010 election there were 32 members, of which only nine were elected representatives of the people (plus the nine nobles elected by their peers and also 14 appointed ministers).

National Anthem

'E 'Otua, Mafimafi, ko ho mau 'eiki Koe' ('Oh Almighty God above, thou art our Lord and sure defence'); words by Prince Uelingtoni Ngu Tupoumalohi, tune by K. G. Schmitt.

RECENT ELECTIONS

Elections were held on 25 Nov. 2010 for the 17 elected seats. The Democratic Party of the Friendly Islands won 12 seats and independent candidates five.

CURRENT GOVERNMENT

In March 2013 the government comprised:

Prime Minister and Minister of Foreign Affairs and Trade, Defence, and Information and Communications: Lord Tu'ivakano (in office since 22 Dec. 2010).

Deputy Prime Minister and Minister for Infrastructure: Samiu Kuita Vaipulu.

Minister for Agriculture, Food, Forests and Fisheries: Sangster Saulala. *Commerce, Tourism and Labour:* Viliami Uasike Latu. *Education and Training:* 'Ana Maui Taufe'ulungaki. *Finance and National Planning:* Lisiate 'Aloveita 'Akolo. *Health:* Lord Tu'i'afitu. *Internal Affairs:* Lord Vaea. *Justice:* Clive Edwards. *Lands, Environment, Climate Change and Natural Resources:* Lord Ma'afu. *Police, Prisons and Fire Services, and Revenue Services:* Siosifa Tu'utafaiva. *Public Enterprises:* Sosefo Fe'aomoeata Vakata.

Government Website: http://www.mic.gov.to

CURRENT LEADERS

Lord Tu'ivakano

Position
Prime Minister

Introduction
Lord Tu'ivakano became prime minister in Dec. 2010 after winning a majority in the legislative assembly. His victory came as a blow to pro-democracy hopes that recent reforms would reduce the influence of the traditional power brokers.

Early Life
Lord Tu'ivakano was born Siale 'Ataongo Kaho on 15 Jan. 1952 in Niutoua, in the Western District of Tongatapu. He was educated in New Zealand at Three Kings School, Auckland, and at Wesley College, Paerata. In 1974 he graduated with a diploma in teaching from Ardmore Teachers' College, Auckland.

On his return to Tonga in 1975, Tu'ivakano joined the staff at of Tonga High School before moving to the physical education department of the ministry of education. In 1980 he was appointed head of the youth, sports and culture division, where he developed a physical health curriculum for primary schools.

On his father's death in Jan. 1986 he became the 17th Lord Tu'ivakano and inherited four estates in Tongatapu. In 1991 he graduated in political science from Flinders University in South Australia. The following year he returned to the ministry of education as senior education officer for youth, sport and culture but resigned in 1996 when he entered parliament as a Nobles' Representative for the island of Tongatapu.

From 2002–04 Tu'ivakano served as speaker of the Legislative Assembly and in March 2005 joined the cabinet as minister for works. Following a reshuffle in May 2006 he was given the portfolio of training, employment, youth and sport. In the Nov. 2010 elections he was re-elected as representative for Tongatapu and, following constitutional reforms, became the first prime minister elected by parliament, winning 14 of the 26 votes cast in a secret ballot. He was sworn into office on 22 Dec. 2010.

Career in Office
Tu'ivakano's election was regarded as a victory for traditionalists over the pro-democracy movement. He was expected to put the revitalization of the agriculture and tourism sectors at the heart of his premiership and to spur manufacturing growth. However, in Oct. 2012 he narrowly survived a parliamentary no confidence motion after the opposition accused his government of mis-appropriating funds.

DEFENCE

Army
The Tonga Defence Services number around 450 troops.

Navy
A coastal naval force operating several small patrol boats is based at Touliki, in Nuku'alofa.

Air Force
An Air Force was created in 1996. There is also a small naval aviation unit.

ECONOMY

In 2010 agriculture accounted for 19% of GDP, industry 20% and services 61%.

Overview
Tonga is heavily reliant on remittances, averaging around 32% of GDP since the early 2000s. The main source of remittances is the USA, followed by New Zealand and Australia. Agriculture accounts for 19% of GDP. Tourism is the second largest source of foreign exchange and a key sector of growth, accounting for 8% of GDP.

The global downturn saw growth fall to 0·9% in 2009 as a result of a sharp fall in remittances, as well as the increased cost of food and fuel imports. Domestic economic activity rebounded with growth of 1·5% in 2011, driven by donor-funded construction and infrastructure projects as well as increased tourism. Tonga suffers from chronic trade deficits, with a fiscal deficit of 6% of GDP.

The economy has underperformed relative to other Pacific Island countries and is vulnerable to external shocks since it relies heavily on imports, particularly of food and fuel. High labour costs, limited export diversification and lack of long-term investment to improve productivity and efficiency impede efforts towards sustainable growth.

Currency
The unit of currency is the *pa'anga* (TOP) of 100 *seniti*. There was inflation of 3·9% in 2010 and 5·3% in 2011. In July 2005 foreign exchange reserves were US$49m. and total money supply was T$65m.

Budget
Revenues were T$200·3m. in 2008–09, with expenditures T$185·3m.

There is a consumption tax of 15%.

Performance
The economy grew by 0·9% in 2009, 1·6% in 2010 and 1·5% in 2011. Total GDP in 2011 was US$0·4bn. Tonga ranks among the countries most reliant on remittances from abroad, which accounted for 28% of total GDP in 2009.

Banking and Finance
The National Reserve Bank of Tonga (*Governor*, Siosi Cocker Mafi) was established in 1989 as a bank of issue and to manage foreign reserves. The Bank of Tonga and the Tonga Development Bank are both situated in Nuku'alofa with branches in the main islands. Other commercial banks in Nuku'alofa are ANZ Banking Group Ltd, the MBF Bank Ltd, the National Reserve Bank of Tonga and the Westpac Banking Corp. Foreign debt in 2007 amounted to US$91m.

ENERGY AND NATURAL RESOURCES

Environment
Tonga's carbon dioxide emissions from the consumption and flaring of fossil fuels in 2008 were the equivalent of 1·7 tonnes per capita.

Electricity
Production (2007 estimate) 43m. kWh. Installed capacity (2007 estimate) 12,000 kW.

Agriculture
In 2007 there were approximately 15,000 ha. of arable land and 12,000 ha. of permanent crops. Production (2003 estimates, in 1,000 tonnes): coconuts, 58; pumpkins and squash, 20; cassava, 9; sweet potatoes, 6; taro, 4; yams, 4; plantains, 3.

Livestock (2003 estimates): pigs, 81,000; goats, 12,000; cattle, 11,000; horses, 11,000.

Forestry
Forests covered 9,000 ha. in 2010, or 13% of the land area.

Fisheries
In 2008 the estimated total catch was 2,141 tonnes.

INDUSTRY
The main industries produce food and beverages, paper, chemicals, metals and textiles.

INTERNATIONAL TRADE
Imports and Exports
In 2010 imports (c.i.f.) were valued at US$158·8m. and exports (f.o.b.) at US$8·3m. Main imports are food and live animals, basic manufactures, machinery and transport equipment, and mineral fuels and lubricants; main exports are coconut oil, vanilla beans, root crops, desiccated coconut and watermelons. The leading import suppliers in 2006 were New Zealand (33·1%), Fiji Islands (28·2%), Australia (12·5%) and the USA (9·8%); principal export markets were Japan (40·6%), USA (25·0%), Australia (14·6%) and South Korea (8·3%).

COMMUNICATIONS
Roads
In 2002 there were 680 km of roads (184 km paved). Vehicles in use in 2000 numbered approximately 8,400 passenger cars, 8,700 trucks and vans, and (1996) 40 buses and coaches.

Civil Aviation
There is an international airport at Nuku'alofa on Tongatapu. The national carrier was the state-owned Royal Tongan Airlines, but it ceased operations in May 2004 owing to financial difficulties. In 2009 Nuku'alofa (Fua'Amotu International) handled 222,612 passengers and 1,417 tonnes of freight.

Shipping
In Jan. 2009 there were 29 ships of 300 GT or over registered, totalling 62,000 GT. The main port is Nuku'alofa.

Telecommunications
The operation of the National Telecommunication Network and Services is the responsibility of the Tonga Telecommunication Commission (TCC). In 2008 there were 25,500 main (fixed) telephone lines; mobile phone subscribers numbered 50,500 in 2008 (48·7 per 100 persons). There were 8,400 internet users in 2008. Ucall mobile GSM digital has been in operation in Tonga since Dec. 2001.

SOCIAL INSTITUTIONS
Justice
The judiciary is presided over by the Chief Justice. The enforcement of justice is the responsibility of the Attorney-General and the Minister of Police. In 1994 the UK ceased appointing Tongan judges and subsidizing their salaries.

The population in penal institutions in mid-2007 was 86 (74 per 100,000 of national population).

Education
In 2002 there were a total of 17,105 pupils with 773 teachers in primary schools and 14,567 pupils with 1,012 teachers in secondary schools. There is an extension centre of the University of the South Pacific at Nuku'alofa, a teacher training college and three technical institutes.

Adult literacy in 2006 was 99·0%. In 2004 public expenditure on education came to 5·0% of GDP.

Health
In 2004 there were four hospitals and 29 hospital beds per 10,000 inhabitants. There were 41 physicians in 2004, 29 dentists, 316 nurses and 20 health officers.

RELIGION
In 2001 there were 44,000 adherents of the Free Wesleyan Church and 16,000 Roman Catholics, with the remainder of the population being followers of other religions (notably Latter-day Saints).

CULTURE
Press
There are no daily newspapers. There were three paid-for non-daily newspapers in 2008: the *Tonga Chronicle* (a government-owned weekly), the *Times of Tonga* and *Matangi Tonga*.

Tourism
There were 46,040 tourist arrivals by air in 2007. Receipts in 2007 (excluding passenger transport) totalled US$16m.

DIPLOMATIC REPRESENTATIVES
Of Tonga in the United Kingdom (36 Molyneux St., London, W1H 5BQ)
High Commissioner: Dr Sione Ngongo Kioa.

Of the United Kingdom in Tonga (High Commission in Nuku'alofa closed in March 2006)
Acting High Commissioner: Martin Fidler (resides in Suva, Fiji Islands).

Of Tonga in the USA and to the United Nations (250 E. 51st St., New York, NY 10022)
Ambassador: Sonatane Tu'a Taumoepeau Tupou.

Of the USA in Tonga
Ambassador: Frankie A. Reed (resides in Suva, Fiji Islands).

Of Tonga to the European Union
Ambassador: Sione Ngongo Kioa.

FURTHER READING
Campbell, I. C., *Island Kingdom: Tonga, Ancient and Modern.* 1994
Wood-Ellem, E., *Queen Salote of Tonga, The Story of an Era 1900–1965.* 2000

National Statistical Office: Tonga Statistics Department, P. O. Box 149, Nuku'alofa.
Website: http://www.spc.int/prism/Country/to/stats

TRINIDAD AND TOBAGO

Caribbean Sea

Tobago
Plymouth

TRINIDAD AND TOBAGO

0 — 20 mi
0 — 25 km

Trinidad

Arima

PORT OF SPAIN

San Fernando

Pierreville

Guayaguayare

ATLANTIC OCEAN

VENEZUELA

© Research Machines plc 2006

Republic of Trinidad and Tobago

Capital: Port-of-Spain
Population projection, 2015: 1·36m.
GNI per capita, 2010: US$15,641
HDI/world rank: 0·760/62
Internet domain extension: .tt

KEY HISTORICAL EVENTS

When Columbus visited Trinidad in 1498 the island was inhabited by Arawak Indians. Tobago was occupied by the Caribs. Trinidad remained a neglected Spanish possession for almost 300 years until it was surrendered to a British naval expedition in 1797. The British first attempted to settle Tobago in 1721 but the French captured the island in 1781 and transformed it into a sugar-producing colony. In 1802 the British acquired Tobago and in 1899 it was administratively combined with Trinidad. When slavery was abolished in the late 1830s, the British subsidized immigration from India to replace plantation labourers. Sugar and cocoa declined towards the end of the 19th century. Oil and asphalt became the main sources of income. On 31 Aug. 1962 Trinidad and Tobago became an independent member of the Commonwealth. A Republican Constitution was adopted on 1 Aug. 1976.

The government shut down the sugar industry after the 2007 harvest. The industry underpinned the economy for centuries but had become unsustainable after the withdrawal of EU subsidies. In Aug. 2011 Prime Minister Kamla Persad-Bissessar declared a state of emergency in a bid to confront the increasingly violent threat posed by the illegal drugs and weapons trade. Lasting until Dec. 2011, it was the first state of emergency since 1990, when the Jamaat al Muslimeen group unsuccessfully attempted a coup.

TERRITORY AND POPULATION

The island of Trinidad is situated in the Caribbean Sea, about 12 km off the northeast coast of Venezuela; several islets, the largest being Chacachacare, Huevos, Monos and Gaspar Grande, lie in the Gulf of Paria which separates Trinidad from Venezuela. The smaller island of Tobago lies 30·7 km further to the northeast.

Altogether, the islands cover 5,128 sq. km (1,980 sq. miles), of which Trinidad (including the islets) has 4,828 sq. km (1,864 sq. miles) and Tobago 300 sq. km (116 sq. miles). In 2000 the census population was 1,262,366 (Trinidad, 1,208,282; Tobago, 54,084); density, 246 per sq. km.

The UN gives a projected population for 2015 of 1·36m.

In 2011, 14·2% of the population lived in urban areas. Capital, Port-of-Spain (2000 census, 49,031); other important towns, San Fernando (55,419), Arima (32,278) and Point Fortin (19,056). The main towns on Tobago are Scarborough and Plymouth. Those of African descent are (2000) 39·2% of the population; East Indians, 38·6%; mixed races, 16·3%; European, Chinese and others, 5·9%.

The official language is English.

SOCIAL STATISTICS

Births, 2006, 18,090; deaths (2005), 9,885. 2006 birth rate (per 1,000 population), 13·9; 2005 death rate, 7·6. Expectation of life, 2007, was 65·6 years for males and 72·8 for females. Annual population growth rate, 1998–2008, 0·4%. Infant mortality, 2010, 24 per 1,000 live births; fertility rate, 2008, 1·6 births per woman.

CLIMATE

A tropical climate cooled by the northeast trade winds. The dry season runs from Jan. to June, with a wet season for the rest of the year. Temperatures are uniformly high the year round. Port-of-Spain, Jan. 76·3°F (24·6°C), July 79·2°F (26·2°C). Annual rainfall 1,870 mm.

CONSTITUTION AND GOVERNMENT

The 1976 constitution provides for a bicameral legislature of a *Senate* and a *House of Representatives*, who elect the *President*, who is head of state. The *Senate* consists of 31 members, 16 being appointed by the President on the advice of the *Prime Minister*, six on the advice of the Leader of the Opposition and nine at the discretion of the President.

The *House of Representatives* consists of 41 (39 for Trinidad and two for Tobago) elected members and a Speaker elected from within or outside the House.

Executive power is vested in the Prime Minister, who is appointed by the President, and the Cabinet.

National Anthem

'Forged from the love of liberty'; words and music by P. Castagne.

GOVERNMENT CHRONOLOGY

Presidents since 1976.

1976–87	Ellis Emmanuel Innocent Clarke
1987–97	Noor Mohammed Hassanali
1997–2003	Arthur Napoleon Raymond Robinson
2003–13	George Maxwell Richards
2013–	Anthony Thomas Aquinas Carmona

Prime Ministers since independence. (PNM = People's National Movement; NAR = National Alliance for Reconstruction; UNC = United National Congress)

1962–81	PNM	Eric Eustace Williams
1981–86	PNM	George Michael Chambers
1986–91	NAR	Arthur Napoleon Raymond Robinson
1991–95	PNM	Patrick Augustus Mervyn Manning
1995–2001	UNC	Basdeo Panday
2001–10	PNM	Patrick Augustus Mervyn Manning
2010–	UNC	Kamla Persad-Bissessar

RECENT ELECTIONS

Indirect presidential elections were held on 15 Feb. 2013. Anthony Carmona stood unopposed and was duly elected.

In parliamentary elections held on 24 May 2010 the People's Partnership coalition (comprising the United National Congress, the Congress of the People, the Tobago Organization of the People, the National Joint Action Committee and the Movement for Social Justice) won 29 of 41 seats with 59·8% of votes cast, against 12 seats for the ruling People's National Movement (PNM) with 39·5%. Turnout was 69·5%.

CURRENT GOVERNMENT

President: Anthony Carmona; b. 1953 (ind.; sworn in 18 March 2013).

In April 2013 the cabinet comprised:

Prime Minister: Kamla Persad-Bissessar; b. 1952 (UNC; sworn in 26 May 2010).

Minister of Arts and Multiculturalism: Lincoln Douglas. *Communication:* Jamal Mohammed. *Community Development:* Winston Peters. *Education:* Dr Tim Gopeesingh. *Energy and Energy Affairs:* Kevin Ramnarine. *Environment and Water Resources:* Ganga Singh. *Finance and the Economy:* Larry Howai. *Food Production:* Devant Maharaj. *Foreign Affairs:* Winston Dookeran. *Gender, Youth Affairs and Child Development:* Marlene Coudray. *Health:* Dr Fuad Khan. *Housing:* Roodal Moonilal. *Justice:* Christlyn Moore. *Labour, and Small and Micro Enterprise Development:* Errol McLeod. *Legal Affairs:* Prakash Ramadhar. *Local Government, and Works and Infrastructure:* Surujrattan Rambachan. *National Diversity and Social Integration:* Clifton de Coteau. *National Security:* Emmanuel George. *People and Social Development:* Glenn Ramadharsingh. *Planning:* Bhoendradatt Tewarie. *Public Administration:* Carolyn Seepersad-Bachan. *Public Utilities:* Nizam Baksh. *Science and Technology:* Rupert Griffith. *Sports:* Anil Roberts. *Tertiary Education:* Fazal Karim. *Tobago Development:* Delmon Baker. *Tourism:* Stephen Cadiz. *Trade, Industry and Investment:* Vasant Bharath. *Transport:* Chandresh Sharma. *Attorney General:* Anand Ramlogan.

Government Website: http://www.gov.tt

CURRENT LEADERS

Anthony Carmona

Position
President

Introduction
Anthony Carmona became president in March 2013. Previously a high court judge, he served on the International Criminal Tribunal for the former Yugoslavia (ICTY) and was scheduled to become a judge at the International Criminal Court (ICC). He is regarded as a political independent.

Early Life
Anthony Carmona was born on 7 March 1953 in Fyzabad and educated at Presentation College, San Fernando. After studying English and political science at the University of the West Indies (UWI) in Jamaica, he studied law, first at the UWI in Barbados, then at the Sir Hugh Wooding Law School in Trinidad and Tobago, graduating in 1983. In this period he also taught in schools and at the UWI.

Carmona served as deputy director of public prosecutions from 1995 until 1999, when he was appointed acting director. He was legal adviser to President A. N. R. Robinson on matters of criminal law relating to the establishment of the ICC, and was a member of the Trinidad and Tobago legislative review committee. He also advised the ministry of foreign affairs and the ministry of the attorney general on international treaties and conventions.

From 2001–04 he served on the ICTY, dealing with the appeals of those convicted of war crimes, crimes against humanity and genocide. From 2004–13 he was a high court judge in Trinidad and Tobago. Elected one of six ICC judges in 2012, he was scheduled to take up his post in The Hague in March 2013.

However, on 4 Feb. 2013 Prime Minister Persad-Bissessar nominated him for the presidency. Carmona was elected to the post in a vote by the coalition government on 15 Feb. 2013, resigning from the ICC in order to take up the role.

Career in Office
Carmona took office on 18 March 2013. He has pledged to champion a more open style of government.

Kamla Persad-Bissessar

Position
Prime Minister

Introduction
Kamla Persad-Bissessar was sworn in as the country's first female prime minister in May 2010.

Early Life
Persad-Bissessar was born on 22 April 1952 in Siparia, southern Trinidad. She studied at the University of the West Indies, the Hugh Wooding Law School in Trinidad and Tobago, Norwood Technical College in England and the Arthur Lok Jack Graduate School of Business, Trinidad.

While in England, Persad-Bissessar was a social worker with the Church of England Children's Society of London. She then worked as a teacher in Jamaica. She later taught at the St Augustine campus of the University of the West Indies in Trinidad. She also lectured at the Jamaica College of Insurance. After six years in education, Persad-Bissessar became a full-time attorney-at-law.

From 1987–91 she served as an alderman for St Patrick County Council. She joined the senate in 1994 representing the United National Congress (UNC). From 1995 she was the MP for Siparia. She twice served as attorney general, in 1995 and in 2001. In Dec. 2000 she became minister of education. The UNC returned to the opposition benches in 2002, with Persad-Bissessar elected leader of the parliamentary opposition in April 2006.

Career in Office
On 24 Jan. 2010 Persad-Bissessar was elected party leader of the UNC. After a landslide victory for the People's Partnership coalition (of which the UNC is a part) at a snap general election on 24 May 2010, she replaced Patrick Manning as prime minister, promising increased transparency in government.

In March 2012 she defeated a no confidence motion in parliament brought by the opposition leader, who accused her government of not properly managing the economic, political and social issues facing the country.

DEFENCE

The Trinidad and Tobago Defence Force consists of the Trinidad and Tobago Regiment, the Coast Guard, the Air Guard and the Defence Force Reserves. Personnel in 2007 totalled around 2,700.

In 2008 defence expenditure totalled US$143m. (US$116 per capita), representing 0·6% of GDP.

Army

The Trinidad and Tobago Regiment (the Army) is part of the Trinidad and Tobago Defence Force. It has approximately 2,000 personnel organized into a Regiment Headquarters and four battalions.

Navy

In 2007 there was a Coast Guard of about 700.

Air Force

The Air Guard, formerly part of the Coast Guard, had 50 personnel and five aircraft in 2007.

ECONOMY

Industry accounted for 59% of GDP in 2007 and services 41%.

Overview

Trinidad and Tobago has one of the highest per capita incomes in the Latin American and Caribbean region. The country is a major financial centre in the Caribbean and a leader in regional economic integration, playing a key role in the CARICOM Single Market Economy since its establishment in 2006.

The islands are rich in natural resources, with oil and gas accounting for 44% of GDP, 83% of merchandise exports and 58% of government revenue. However, the country is vulnerable to international market fluctuations. Additionally, few have benefited from the wealth created by oil and gas, which employs only 3% of the labour force. Tourism constituted 11% of GDP and 15% of total employment in 2009. An estimated 17% of the population lives in poverty.

The global economic crisis caused a sharp decline in the prices of major exports, including crude oil, resulting in a lengthy recession. The largest privately-held conglomerate in the country, the CL Financial Group, received a government bailout in 2009. The debt-to-GDP ratio has increased sharply since 2008, with public sector debt standing at 32·4% of GDP in 2011.

Currency

The unit of currency is the *Trinidad and Tobago dollar* (TTD) of 100 *cents*. Inflation was 10·5% in 2010 and 5·1% in 2011. In April 1994 the TT dollar was floated and managed by the Central Bank at TT$6·06 to US$1·00. Foreign exchange reserves in July 2005 were US$3,918m. and gold reserves 61,000 troy oz. Total money supply in May 2005 was TT$9,438m.

Budget

The fiscal year for the budget is 1 Oct. to 30 Sept. In 2009–10 central government revenue was TT$41,983m. and expenditure was TT$45,805m. Energy accounted for 52·5% of revenues; current expenditure accounted for 83·7% of expenditures.

VAT is 15%.

Performance

After the economy contracted by 3·3% in 2009 there was zero growth in 2010. In 2011 there was negative growth again, of –1·5%. Total GDP in 2011 was US$22·5bn.

Banking and Finance

The Central Bank of Trinidad and Tobago began operations in 1964 (*Governor*, Jwala Rambarran). Its net reserves were US$1,281·1m. in Aug. 2000. There are seven commercial banks. Government savings banks are established in 69 offices, with a head office in Port-of-Spain. The stock exchange in Port-of-Spain participates in the regional Caribbean exchange.

ENERGY AND NATURAL RESOURCES

Environment

Carbon dioxide emissions from the consumption and flaring of fossil fuels were the equivalent of 41·0 tonnes per capita in 2008.

Electricity

In 2007 the estimated installed capacity was 1·49m. kW, electricity production was 7·66bn. kWh and consumption per capita 5,748 kWh.

Oil and Gas

Oil production is one of Trinidad's leading industries. Commercial production began in 1908; production of oil in 2008 was 6·9m. tonnes. Reserves in 2008 totalled 0·8bn. bbls. Crude oil is also imported for refining.

In 2008 production of natural gas was 39·3bn. cu. metres; proven reserves of natural gas were 480bn. cu. metres. A major discovery of approximately 50bn. cu. metres was made by BP in 2000, followed by a further discovery of approximately 30bn. cu. metres in 2002.

Agriculture

In 2007 the agricultural population was an estimated 94,000, of which some 49,000 were economically active. Production of main crops (2003 estimates, in 1,000 tonnes): sugarcane, 873; coconuts, 16; bananas, 7; pumpkins and squash, 6; oranges, 5; cucumbers and gherkins, 4; pineapples, 4; plantains, 4. There were around 25,000 ha. of arable land and 22,000 ha. of permanent cropland in 2007. Livestock (2003 estimates): pigs, 76,000; goats, 23,000; cattle, 29,000; chickens, 28m. Livestock products, 2003: meat, 60,000 tonnes (including poultry, 57,000 tonnes); milk, 9,000 tonnes.

Forestry

Forests covered 0·23m. ha. in 2010, or 44% of the land area. Timber production for 2007 was 99,000 cu. metres.

Fisheries

The catch in 2010 totalled 13,931 tonnes.

INDUSTRY

Industrial production includes (in tonnes): ammonia and urea (1998), 3,946,700; iron and steel (2007), 3,440,400; residual fuel oil (2007), 3,027,000; methanol (1999), 2,149,800; distillate fuel oil (2007), 1,736,000; petrol (2007), 1,273,000; cement (2006), 883,000; sugar (2002), 104,000; rum (1998), 3,916,000 proof gallons; beer (2000), 62·5m. litres; cigarettes (2000), 2,050,000 units. Trinidad and Tobago ranks among the world's largest producers of ammonia and methanol.

Labour

The working population in the first quarter of 2003 was 588,300. The number of unemployed was 65,000. 77,300 people worked in construction (including electricity and water); 55,500 in manufacturing (including other mining and quarrying); 38,600 in transport storage and communication; 37,800 in agriculture; 17,500 in petroleum and gas; other services, 295,300. Total employment: 523,300. The unemployment rate in the fourth quarter of 2005 was a record low 6·7%.

INTERNATIONAL TRADE

External debt was US$2,652m. in 2005.

Imports and Exports

In 2010 imports (c.i.f.) totalled US$6,479·6m. and exports (f.o.b.) US$10,981·7m. In 2005 crude petroleum accounted for 19% of imports, and refined petroleum 29% of exports. Trinidad and Tobago is the world's leading exporter of ammonia and methanol. The principal import sources in 2005 were the USA (29·2%), Brazil (13·5%), Venezuela (6·0%) and Colombia (5·6%). The main export markets in 2005 were the USA (58·6%), Jamaica (7·5%), France (4·4%) and Barbados (4·3%).

COMMUNICATIONS

Roads

In 2002 there were about 8,320 km of roads, of which 51·1% were paved. There were 468,255 vehicles in use in 2007.

Civil Aviation

There is an international airport at Port-of-Spain (Piarco) and in Tobago (A. N. R. Robinson International Airport). In 2001 Piarco handled 1,725,111 passengers (1,317,811 on international flights) and 29,673 tonnes of freight. The national carrier is Caribbean Airlines, which has flights to 11 international destinations as well as operating domestic services. In 2003 scheduled airline traffic of Trinidad and Tobago-based carriers flew 31m. km, carrying 1,084,000 passengers (972,000 on international flights).

Shipping
In Jan. 2009 there were 11 ships of 300 GT or over registered, totalling 29,000 GT. The largest port is Port-of-Spain. The other main harbour is Point Lisas. There is a deep-water harbour at Scarborough (Tobago). A ferry service links Port-of-Spain with Scarborough.

Telecommunications
International and domestic communications are provided by Telecommunications Services of Trinidad and Tobago (TSTT) by means of a satellite earth station and various high-quality radio circuits. The marine radio service is also maintained by TSTT. In 2008 there were 307,000 main (fixed) telephone lines; mobile phone subscribers numbered 1,505,000 in 2008 (112·9 per 100 persons). There were 227,000 internet users in 2008. In Dec. 2011 there were 441,000 Facebook users.

SOCIAL INSTITUTIONS
Justice
The High Court consists of the Chief Justice and 11 puisne judges. In criminal cases a judge of the High Court sits with a jury of 12 in cases of treason and murder, and with nine jurors in other cases. The Court of Appeal consists of the Chief Justice and seven Justices of Appeal. In hearing appeals, the Court is comprised of three judges sitting together except when the appeal is from a Summary Court or from a decision of a High Court judge in chambers. In such cases two judges would comprise the Court. There is a limited right of appeal from it to the British Privy Council. There are three High Courts and 12 magistrates' courts. There is an *Ombudsman*. Trinidad and Tobago was one of ten countries to sign an agreement in Feb. 2001 establishing a Caribbean Court of Justice (CCJ) to replace the British Privy Council as the highest civil and criminal court. In the meantime the number of signatories has risen to 12. The court was inaugurated at Port-of-Spain on 16 April 2005. However, despite its location Trinidad and Tobago has yet to accept the CCJ as its final court of appeal.

The death penalty is authorized. There were ten executions in 1999, although none since. There were 550 homicides in 2008 (41 per 100,000 persons), up from just 98 in 1998.

The population in penal institutions at 31 Dec. 2007 was 3,510 (270 per 100,000 of national population).

Education
In 2007 there were 130,242 pupils enrolled in primary schools with 8,171 teaching staff and 98,490 pupils in secondary schools with 7,041 teaching staff. In 2005 there were 16,920 students in higher education and 1,800 academic staff. The University of the West Indies campus in St Augustine (2006–07) had 13,629 students and 572 academic staff. 1,358 of the students were from other countries.

Adult literacy was estimated at 98·7% in 2009.

In 2001 public spending on education came to 4·2% of GDP and accounted for 13·4% of total government expenditure.

Health
In 1999 there were 1,171 physicians, 189 dentists, 500 pharmacists and 71 hospitals and nursing homes with 4,384 beds. There were 1,936 nurses and midwives and 1,486 nursing assistants in government institutions.

RELIGION
In 2001, 29·9% of the population were Roman Catholics (under the Archbishop of Port-of-Spain), 24·2% Hindus, 19·2%

Protestants, 11·2% Anglicans (under the Bishop of Trinidad and Tobago) and 6·0% Muslims.

CULTURE
Press
There were four daily newspapers in 2006 (*Trinidad Express*, *Trinidad Guardian*, *Newsday* and *The Wire*), with a total circulation of 160,000. There were nine paid-for non-dailies in 2006.

Tourism
There were 457,387 tourist arrivals in 2006, plus a record 85,859 cruise ship visitors. Receipts from tourism in 2007 totalled US$621m. (US$517m. in 2006).

Festivals
Religious festivals: the Feast of La Divina Pastora, or Sipari Mai, a Catholic and Hindu celebration of the Holy Mother Mary; Saint Peter's Day Celebration, the Patron Saint of Fishermen; Hosein, or Hosay, a Shia Muslim festival; Phagwah, a Hindu spring festival; Santa Rosa, a Caribbean Amerindian festival; Eid-al-Fitr, the Muslim festival at the end of Ramadan; Diwali, the Hindu festival of light; Christmas. Cultural festivals: Carnival (on 3 and 4 March in 2014); Spiritual Baptist Shouter Liberation Day, a recognition of the Baptist religion; Indian Arrival Day, commemorating the arrival of the first East Indian labourers; Sugar and Energy Festival; Pan Ramajay, a music festival of all types; Emancipation, a recognition of the period of slavery; Tobago Heritage Festival, celebrating Tobago's traditions and customs; Parang Festival, traditional folk music of Christmas; Pan Jazz Festival; Music Festival, predominantly classical music but Indian and Calypso are included.

DIPLOMATIC REPRESENTATIVES
Of Trinidad and Tobago in the United Kingdom (42 Belgrave Sq., London, SW1X 8NT)
High Commissioner: Garvin Nicholas.

Of the United Kingdom in Trinidad and Tobago (19 St Clair Ave., Port-of-Spain)
High Commissioner: Arthur Snell.

Of Trinidad and Tobago in the USA (1708 Massachusetts Ave., NW, Washington, D.C., 20036)
Ambassador: Neil Parsan.

Of the USA in Trinidad and Tobago (15 Queen's Park West, Port-of-Spain)
Ambassador: Vacant.
Chargé d'Affaires a.i.: Thomas Smitham.

Of Trinidad and Tobago to the United Nations
Ambassador: Rodney Charles.

Of Trinidad and Tobago to the European Union
Ambassador: Margaret Allison King-Rousseau.

FURTHER READING
Meighoo, Kirk, *Politics in a Half-Made Society: Trinidad and Tobago, 1925–2001.* 2003
Williams, E., *History of the People of Trinidad and Tobago.* 1993

Central library: The Central Library of Trinidad and Tobago, Queen's Park East, Port-of-Spain.
National Statistical Office: Central Statistical Office, 80 Independence Square, Port-of-Spain.
Website: http://cso.gov.tt

TUNISIA

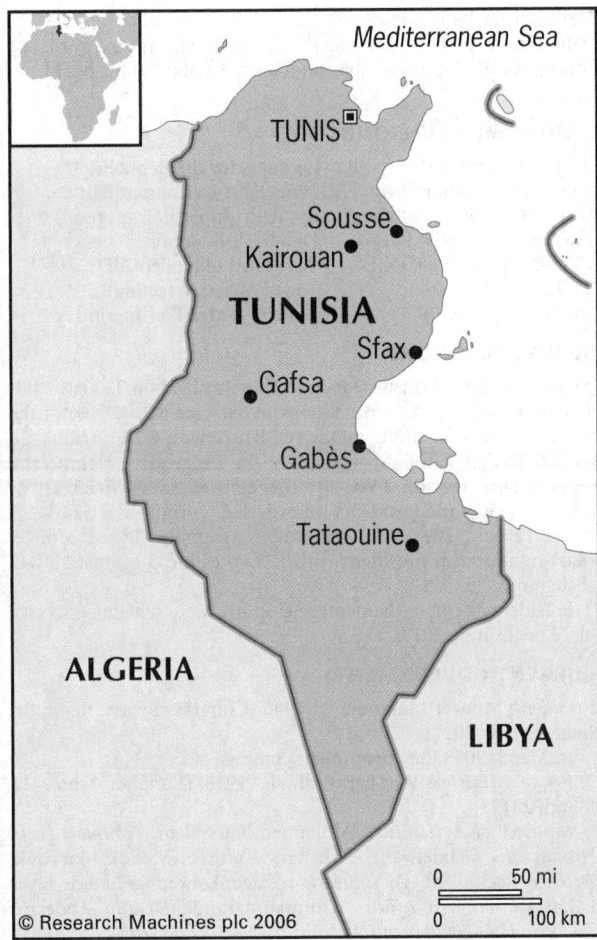

Mediterranean Sea

TUNIS

Sousse
Kairouan
TUNISIA
Sfax
Gafsa
Gabès
Tataouine

ALGERIA

LIBYA

| 0 | 50 mi |
| 0 | 100 km |

© Research Machines plc 2006

Jumhuriya at-Tunisiya
(Republic of Tunisia)

Capital: Tunis
Population projection, 2015: 11·03m.
GNI per capita, 2011: (PPP$) 7,281
HDI/world rank: 0·698/94
Internet domain extension: .tn

KEY HISTORICAL EVENTS

Tunisia's earliest inhabitants included the semi-nomadic Berbers, whose descendants still live in North Africa's Atlas Mountains. Phoenician merchants established trading settlements throughout the central and western Mediterranean from the 10th century BC and founded the port of Carthage in 814 BC. By the 5th century Carthage had become the most powerful city in the western Mediterranean with an empire extending from present-day Morocco to Egypt and controlling Sardinia, the Balearic Islands, Malta and much of Sicily. A rival to the Roman Empire, the city was eventually destroyed in the Third Punic War. From 146 BC Tunisia was absorbed into the Roman Empire and its people sold into slavery.

Emperor Augustus supervised the reconstruction of Carthage in the 2nd century AD, although frequent revolts broke out as the Roman Empire waned. In 429 Tunisia was captured by the Vandals who ruled until ousted by the Byzantines in 534. Uqba ibn Nafa led an Arab Muslim army into Tunisia in 670 and founded the city of Kairouan. Most of the indigenous Berber population converted to Islam and the region, known as Ifriqiya, became part of the powerful caliphate centred on Baghdad where it reached its apotheosis under Harun al Rashid (786–809). Tunisia subsequently came under the control of the local Aghlabid dynasty and then under the Fatimids from 909. Bedouin Arabs from the Nile Valley (the Banu Hilal) entered the region around 1000 and from the 12th century Tunisia came under the orbit of the al-Muwahhid empire. The Berber general Abd al-Wahid ibn Abi Hafs wrested control in 1207, heralding three centuries of rule by the Hafsid dynasties. The port of Tunis was made capital and the term 'Tunisia' gradually replaced 'Ifriqiya'.

Ottoman troops advanced from the east in the early 16th century, forcing the Hafsids into an alliance with the Spanish Habsburgs in 1534. However, the Ottomans seized Tunis in 1574. Direct control from İstanbul was superseded by rule through local governors known as beys, such as the Muradid beys who made powerful alliances in the Tunisian hinterland and held sway from the 1640s to 1705. Al-Husayn ibn Ali (ruled 1705–40) founded the Husaynid dynasty and oversaw a period of prosperity through trade and piracy.

As the Ottoman Empire declined Tunisia became increasingly autonomous but attempts to establish a republic in the mid-19th century failed owing to a weak economy and political unrest. In 1869 Tunisia declared itself bankrupt and an international commission from France, Great Britain and Italy took over the country's finances. In 1881 Tunisia was invaded by France which alleged that Tunisian troops had been threatening the French colony of Algeria. Tunisia became a French protectorate despite Italian opposition.

Nationalist sentiment increased after the First World War and the Destour (Constitutional) Party was set up in 1920. Its more radical successor, the Neo-Destour Party, was established by Habib Bourguiba in 1934. Seen as a threat to colonial rule, Bourguiba was imprisoned in 1938. After the Second World War France offered increased autonomy. Nationalists, dismayed at the slow pace of reform, staged mass strikes and demonstrations in 1950. Violent protest increased until France granted internal self-government in 1955, with full independence on 20 March 1956. A constitutional assembly was established and Bourguiba became prime minister. The monarchy was abolished and a republic established the following year, with Bourguiba as president.

Taking a socialist secular line, Bourguiba promoted social and economic development. In 1975 the constitution was amended so that Bourguiba could be president-for-life. The late 1970s saw hardship and growing discontent. A general strike in Jan. 1978 erupted into riots and violence flared again in 1980 and 1984. Islamist groups became increasingly influential and Bourguiba was overthrown in a bloodless coup in 1987. His successor, Zine El Abidine Ben Ali, introduced democratic reforms, eased restrictive press laws and opened negotiations with Islamic groups, although he refused to recognize the prominent al-Nahda party. He was re-elected president in 1994, 1999, 2004 and 2009. However, widespread anti-government protests in Dec. 2010 and Jan. 2011 led to Ben Ali going into exile. A national unity government took power in his place. A state of emergency declared in Jan. 2011 remains in place.

TERRITORY AND POPULATION

Tunisia is bounded in the north and east by the Mediterranean Sea, west by Algeria and south by Libya. The area is 163,610 sq. km, including 9,080 sq. km of inland waters. In 2004 the census population was 9,910,872; density, 64 per sq. km. Estimate, 2011, 10,673,800. In 2011, 67·7% of the population were urban.

The UN gives a projected population for 2015 of 11·03m.

The areas and populations (2011 estimate) of the 24 governorates:

	Land area in sq. km	Population
Aryanah (Ariana)	498	510,500
Bajah (Béja)	3,558	307,300
Banzart (Bizerta)	3,685	551,500
Bin Arus (Bin Arous)	761	588,700
Jundubah (Jendouba)	3,102	426,000
Kaf (Le Kef)	4,965	258,100
Madaniyin (Médénine)	8,588	460,000
Mahdiyah (Mahdia)	2,966	400,400
Manubah (Manouba)	1,060	375,300
Munastir (Monastir)	1,019	525,500
Nabul (Nabeul)	2,788	762,600
Qabis (Gabès)	7,175	366,100
Qafsah (Gafsa)	8,990	341,600
Qasrayn (Kassérine)	8,066	437,200
Qayrawan (Kairouan)	6,712	564,900
Qibili (Kebili)	22,084	152,200
Safaqis (Sfax)	7,545	944,500
Sidi Bu Zayd (Sidi Bouzid)	6,994	415,900
Silyanah (Siliana)	4,631	235,300
Susah (Sousse)	2,621	622,100
Tatawin (Tataouine)	38,889	148,000
Tawzar (Tozeur)	4,719	104,800
Tunis	346	1,002,900
Zaghwan (Zaghouan)	2,768	172,300

Tunis, the capital, had 728,500 inhabitants at the 2004 census. Other main cities (2004 census in 1,000): Sfax, 265·1; Ariana, 240·7; Sousse, 173·0; Ettadhamen, 118·5; Kairouan, a holy city of the Muslims, 117·9; Gabès, 116·3; Bizerta, 114·4.

The official language is Arabic but French is the main language in the media, commercial enterprise and government departments. Berber-speaking people form less than 1% of the population.

SOCIAL STATISTICS

2008 estimates: births, 183,000; deaths, 61,000; marriages (2005), 74,000. Rates (2008 estimates): birth, 18 per 1,000 population; death, 6. Annual population growth rate, 2000–05, 1·0%. In 2005 the most popular age range for marrying was 30–34 for males and 25–29 for females. Expectation of life, 2007, was 71·8 years for males and 76·0 for females. Infant mortality, 2010, 14 per 1,000 live births; fertility rate, 2008, 1·8 births per woman.

CLIMATE

The climate ranges from warm temperate in the north, where winters are mild and wet and the summers hot and dry, to desert in the south. Tunis, Jan. 48°F (8·9°C), July 78°F (25·6°C). Annual rainfall 16" (400 mm). Bizerta, Jan. 52°F (11·1°C), July 77°F (25°C). Annual rainfall 25" (622 mm). Sfax, Jan. 52°F (11·1°C), July 78°F (25·6°C). Annual rainfall 8" (196 mm).

CONSTITUTION AND GOVERNMENT

The Constitution was promulgated on 1 June 1959 and reformed in 1988. The office of President-for-life was abolished and Presidential elections were to be held every five years. On 26 May 2002 a referendum was held that abolished the three-term limit on the presidency with 99% of votes cast in favour of doing so and of raising the age limit for incumbent presidents from 70 to 75 years. The results were viewed with scepticism by human rights groups and opposition figures who saw the referendum as an attempt by then President Zine El Abidine Ben Ali to retain power. Ben Ali was scheduled to retire in 2004 after his third presidential term but remained the president until the revolution of Jan. 2011.

Following the revolution, elections were held in Oct. 2011 for a new 217-seat Constituent Assembly, with a mandate to draft and approve a new constitution. Approval requires two-thirds support from the Assembly. In the event that a draft constitution is not approved after two readings, it will be submitted to a popular referendum.

National Anthem

'Humata al Hima' ('Defenders of the Homeland'); words by Mustapha al Rafi and Abdoul Kacem Chabbi, tune by M. A. Wahab.

GOVERNMENT CHRONOLOGY

Presidents since 1957. (CPR = Congress for the Republic; ND = Neo-Destour Party; PSD = Socialist Destourian Party; RCD = Constitutional Democratic Rally; n/p = non-partisan)

1957–87	ND, PSD	Habib Ali Bourguiba
1987–2011	PSD, RCD	Zine El Abidine Ben Ali
2011	n/p	Foued Mebazaa (acting)
2011–	CPR	Moncef Marzouki (interim)

RECENT ELECTIONS

An election for a Constituent Assembly was held on 23 Oct. 2011. The moderate Islamist Ennahda won 89 seats with 37·0% of the vote; Congress for the Republic (CPR), 29 with 8·7%; Aridha, 26 with 6·7%; Ettakatol, 20 with 7·0%; the Progressive Democratic Party (PDP), 16 with 3·9%. 14 other parties gained five seats or fewer and eight independents were elected. Turnout was 52·0%.

On 12 Dec. 2011 the Constituent Assembly elected Moncef Marzouki interim president with 155 votes for, 3 against and 42 abstentions.

Presidential and parliamentary elections were scheduled to take place on 23 June 2013.

CURRENT GOVERNMENT

President: Moncef Marzouki; b. 1945 (Congress for the Republic; since 13 Dec. 2011).

In March 2013 the government comprised:

Prime Minister: Ali Larayedh; b. 1955 (Ennahda; since 13 March 2013).

Minister of Agriculture: Mohamed Ben Salem. *Commerce and Handicrafts:* Abdelwaheb Maatar. *Culture:* Mehdi Mabrouk. *Defence:* Rachid Sabagh. *Education:* Salem Labyadh. *Finance:* Elyes Fakhfakh. *Foreign Affairs:* Othman Jarandi. *Health:* Abdelatif Mekki. *Higher Education and Scientific Research:* Moncef Ben Salem. *Human Rights and Transitional Justice:* Samir Dilou. *Industry:* Mehdi Jomaa. *Information Technology and Communication:* Mongi Marzouk. *Interior:* Lotfi Ben Jeddou. *Development and International Co-operation:* Lamine Dohgri. *Justice:* Nadhir Ben Ammou. *Public Works and Environment:* Mohamed Salmane. *Religious Affairs:* Noureddine El Khadmi. *Social Affairs:* Khalil Zaouia. *State Property and Land Affairs:* Slim Ben Hamidane. *Tourism:* Jamel Gamra. *Transport:* Abdelkarim Harouni. *Vocational Training and Employment:* Naoufel Jammeli. *Women and Family:* Sihem Bardi. *Youth and Sports:* Tarek Dhiab.

Deputy Ministers in the Prime Minster's Office: Ridha Saïdi (for *Economic Affairs*); Noureddine Bhiri (for *Political Affairs*); Abderahmane Ladgham (for *Transparency and the Fight Against Corruption*).

Government Website: http://www.ministeres.tn

CURRENT LEADERS

Moncef Marzouki

Position
President

Introduction
Moncef Marzouki was elected interim president in Dec. 2011, a year on from the start of the revolution that ignited the Arab Spring. A French-trained physician, human rights activist and

leader of a secular, left-wing political party, he was a long-standing critic of the autocratic Ben Ali regime and spent many years in exile in France.

Early Life
Mohamed Moncef ben-Bedoui Marzouki was born in Grombalia in the northeast of the French Protectorate of Tunisia on 7 July 1945, the son of a magistrate. He attended Sadiki College in Tunis and then the Lycee Régnault in Tangier after his family moved to Morocco in 1956 amid the upheaval of Tunisia's independence.

In 1973 he graduated in medicine from Strasbourg University in France, specializing in neurology and practising in Alsace. Returning to Tunisia in 1979, he settled in Sousse and became a professor of community medicine at the city's university and a member of the Tunisian League for Human Rights (LTDH). Elected leader of the LTDH in 1989, he was an increasingly vocal critic of President Zine El Abidine Ben Ali, particularly in the wake of the government's violent crackdown on the Islamist Ennahda movement in 1991.

Marzouki failed in an attempt to contest the presidential election of March 1994, in which Ben Ali was the only official candidate. Marzouki was subsequently jailed and only released after an international outcry and the intervention of Nelson Mandela. In the 1990s he campaigned on behalf of Amnesty International and the Cairo-based Arab Organization for Human Rights.

In July 2001 he formed the Congress for the Republic (CPR), a centre-left, secular political party that was declared illegal. Having been forced out of the University of Sousse, Marzouki left Tunisia in 2002 for France, where he worked as a doctor in the suburbs of Paris. He published books and articles on community medicine, human rights and politics, cautioning against the 'Arab malaise' of societies caught 'between dictatorship and religious fundamentalism'.

Marzouki returned to Tunisia on 18 Jan. 2011 during a popular uprising fired by high unemployment, rising prices, corruption and lack of civil freedoms, which culminated in the overthrow of Ben Ali. The CPR came second behind the moderate Islamist Ennahda party in elections for the constituent assembly on 23 Oct. 2011. After the parties reached a coalition deal, Marzouki was elected interim president. He took office on 13 Dec. 2011.

Career in Office
Marzouki's mandate was to oversee the drafting of a new constitution and preparations for fresh elections. He was expected to direct foreign policy and to serve as commander in chief of the armed forces, although key decisions would require consultation with the prime minister and leader of Ennahda, Hamadi Jebali.

Religious tensions sparked protests and some violence in 2012. Extreme Islamists staged a number of demonstrations demanding the incorporation of Sharia law into the new constitution, and in June 2012 the government imposed a curfew in some areas in response to riots. In Aug. there were also protests against constitutional provisions that some feared could limit sexual equality and women's rights.

In June 2012 the exiled former president, Ben Ali, received a life sentence *in absentia* for the deaths of demonstrators in the 2011 revolution.

DEFENCE
Selective conscription is for one year. Defence expenditure in 2008 totalled US$534m. (US$51 per capita), representing 1·3% of GDP.

Army
Strength (2007) 27,000 (22,000 conscripts). There is also a National Guard numbering 12,000.

Navy
In 2007 naval personnel totalled around 4,800. Forces are based at Bizerta, Sfax and Kelibia.

Air Force
The Air Force operated 27 combat-capable aircraft in 2007, including 12 F-5E/F Tiger II fighters and 12 Aero L-59s. Personnel (2007) 4,000.

ECONOMY
In 2006 agriculture accounted for 11·1% of GDP, industry 27·8% and services 61·1%.

Overview
Tunisia's economic record compares favourably with other developing countries, particularly those on the African continent. After a balance of payments crisis in the mid-1980s, steps were taken to improve macroeconomic policy, foster the non-public sector and liberalize prices and controls. In the 1990s the economy grew steadily at an average annual rate of 5%. Inflation and annual budget deficits have fallen and poverty has been reduced.

Tunisia's trade is oriented towards the EU, with France, Italy, Germany and Spain the country's most important trading partners in descending order. However, Tunisia's textile exporters lost EU market share to Asia as a result of the end of the Multi-Fibre Agreement quota system in Jan. 2005. The economy is diversified relative to many of its non-European neighbours, with significant mining, tourism, energy, manufacturing and agricultural sectors.

A swift policy response left the economy largely unaffected by the global financial crisis. Growth was 3·1% in 2009, with exports and domestic demand rebounding. Following the Jasmine Revolution of early 2011 and the neighbouring Libyan crisis, the short-term economic outlook weakened, with tourism and foreign direct investment suffering.

Unemployment remains persistently high, particularly among young people, with the unemployment rate for university graduates at 25%. In May 2011 the G8 group of countries promised US$20bn. in loans and grants to Tunisia and Egypt over a three-year period. In 2011 the interim government announced a US$1·5bn. stimulus package, plus an emergency economic and social development plan. This includes measures to tackle security problems and unemployment, boost private sector growth and regional development and improve the lot of the poorest families.

Currency
The unit of currency is the *Tunisian dinar* (TND) of 1,000 *millimes*. The currency was made convertible on 6 Jan. 1993. Foreign exchange reserves were US$4,069m. and gold reserves 218,000 troy oz in July 2005. Inflation was 4·4% in 2010 and 3·5% in 2011. Total money supply was 8,339m. dinars in June 2005.

Budget
The fiscal year is the calendar year. Budgetary central government revenue totalled 13,266m. dinars in 2008 and expenditure 11,544m. dinars. Taxes accounted for 85·4% of total revenues in 2008. Principal sources of revenue in 2008: taxes on goods and services, 5,061m. dinars; taxes on income, profits and capital gains, 4,561m. dinars; taxes on international trade and transactions, 965m. dinars. Main items of expenditure by economic type in 2008: compensation of employees, 5,164m. dinars; subsidies, 2,713m. dinars; interest, 1,143m. dinars.

VAT is 18% (reduced rates, 12% and 6%).

Performance
Real GDP growth was 3·1% in both 2009 and 2010 but the economy contracted by 1·8% in 2011. Tunisia's total GDP in 2011 was US$45·9bn.

Banking and Finance
The Central Bank of Tunisia (*Governor*, Chedly Ayari) is the bank of issue. In 2003 there were 12 commercial banks, six development banks, two merchant banks and five 'offshore' banks.

In 2010 external debt totalled US$21,584m., equivalent to 51·1% of GNI.

There is a small stock exchange (51 companies trading in 2007).

ENERGY AND NATURAL RESOURCES

Environment
Tunisia's carbon dioxide emissions from the consumption and flaring of fossil fuels in 2008 were the equivalent of 2·1 tonnes per capita.

Electricity
Installed capacity was 3·2m. kW in 2007. Production in 2007 was 14·06bn. kWh; consumption per capita was 1,375 kWh.

Oil and Gas
Oil production (2008) was 4·2m. tonnes with 0·6bn. bbls in proven reserves. Natural gas production (2007), 2·3bn. cu. metres; proven reserves were 65bn. cu metres in 2007.

Minerals
Mineral production (in 1,000 tonnes) in 2004: phosphate rock, 8,051; salt, 1,117; iron ore, 256; zinc ore (concentrated), 53; lead, 5.

Agriculture
There are five agricultural regions: the *north*, mountainous with large fertile valleys; the *northeast*, with the peninsula of Cap Bon, suited for the cultivation of oranges, lemons and tangerines; the *Sahel*, where olive trees abound; the *centre*, a region of high tablelands and pastures; and the *desert* of the south, where dates are grown.

In 2007 the economically active agricultural population was an estimated 787,000. Large estates predominate; smallholdings are tending to fragment, partly owing to inheritance laws. In 2006 356,000 ha. were irrigated. There were 2·76m. ha. of arable land in 2007 and 2·17m. ha. of permanent crops. There were 140 tractors and ten harvester-threshers per 10,000 ha. of arable land in 2006. The main crops are cereals, citrus fruits, tomatoes, melons, olives, dates, grapes and olive oil. Production, 2003 estimates (in 1,000 tonnes): wheat, 1,150; tomatoes, 800; melons and watermelons, 520; olives, 500; barley, 345; potatoes, 345; chillies and green peppers, 206; onions, 120; dates, 115; grapes, 115; oranges, 106; apples, 100; peaches and nectarines, 82; olive oil, 73; grapefruit and pomelos, 72; pears, 68.

Livestock, 2003 estimates (in 1,000): sheep, 6,850; goats, 1,400; cattle, 760; camels, 231; asses, 230; mules, 81; horses, 57. Livestock products, 2003 estimates (in 1,000 tonnes): meat, 250; milk, 989; eggs, 83.

Forestry
In 2010 there were 1·01m. ha. of forests (6% of the land area). Timber production in 2007 was 2·38m. cu. metres.

Fisheries
In 2010 the catch amounted to 97,743 tonnes, almost exclusively from marine waters.

INDUSTRY
Production (in 1,000 tonnes): cement (2007), 7,379; phosphoric acid (2008), 2,113; residual fuel oil (2007), 646; distillate fuel oil (2007), 556; lime (2008), 369; crude steel (2007), 61. There were 5,837 industrial enterprises of ten employees or more in 2010 employing a total of 498,929 people. Industry accounted for 28·1% of GDP in 2006, with manufacturing contributing 16·9%.

Labour
The economically active population totalled 3,593,200 in 2007. Out of 3,085,100 in employment, 48·6% were engaged in commerce and services, 32·1% in industry, mining, energy and construction, and 19·3% in agriculture and fisheries. Unemployment was 14·1% in 2007.

INTERNATIONAL TRADE
In Feb. 1989 Tunisia signed a treaty of economic co-operation with the other countries of Maghreb: Algeria, Libya, Mauritania and Morocco. Tunisia was the first country to sign a partnership agreement with the European Union, becoming fully integrated in its free trade zone in Jan. 2008.

Imports and Exports
Trade in US$1m.:

	2004	2005	2006	2007	2008
Imports c.i.f.	12,818	13,177	14,865	18,980	24,612
Exports f.o.b.	9,685	10,494	11,513	15,029	19,319

Main imports in 2004: textiles, 18·8%; electrical machinery, 11·9%; petroleum, 9·6%; motor vehicles, 7·3%; iron and steel, 6·0%. Main exports in 2004: textiles, 37·2%; electrical machinery, 13·9%; petroleum, 9·5%; olive oil, 5·9%; leather products, 5·2%.

The main import suppliers in 2004 were France (24·9%), Italy (18·9%), Germany (8·4%) and Spain (5·3%). Main export markets in 2004 were France (33·1%), Italy (25·3%), Germany (9·2%) and Spain (6·1%).

COMMUNICATIONS

Roads
The road network covered 19,371 km in 2008, including 359 km of motorways and 4,738 km of national roads. In 2007 there were 746,700 passenger cars, 300,500 lorries and vans, 10,100 buses and coaches, and 5,300 motorcycles and mopeds. There were 10,681 road accidents in 2007 resulting in 1,497 fatalities.

Rail
In 2007 there were 2,165 km of railways on metre and 1,435 mm gauge track. Passenger-km travelled in 2007 came to 1,487m. and freight tonne-km to 2,197m. There is a tramway in Tunis (32 km).

Civil Aviation
The national carrier, Tunisair, is 74·5% state-owned. Scheduled airline traffic of Tunisian-based carriers flew 43·2m. km and carried 2,098,300 passengers in 2005. There are six international airports. In 2001 Monastir (Habib Bourguiba) handled 3,894,000 passengers (3,885,000 on international flights) and 800 tonnes of freight. Tunis-Carthage handled 3,315,000 (3,061,000 on international flights) and 21,800 tonnes of freight. Djerba handled 2,161,000 passengers (1,945,000 on international flights) and 700 tonnes of freight.

Shipping
There are ports at Tunis, its outer port Tunis-Goulette, Sfax, Sousse and Bizerta, all of which are directly accessible to ocean-going vessels. The ports of La Skhirra and Gabès are used for the shipping of Algerian and Tunisian oil. In Jan. 2009 there were 17 ships of 300 GT or over registered, totalling 125,000 GT.

Telecommunications
Tunisia had 1,273,300 fixed line telephone subscribers (126·5 per 1,000 persons) in 2007 and 7,842,600 mobile phone subscribers. There were 1,722,190 internet users in 2007. In June 2012 there were 3·1m. Facebook users.

SOCIAL INSTITUTIONS

Justice
There are 51 magistrates' courts, 13 courts of first instance, three courts of appeal (in Tunis, Sfax and Sousse) and the High Court in Tunis.

A Personal Status Code was promulgated on 13 Aug. 1956 and applied to Tunisians from 1 Jan. 1957. This raised the status of women, made divorce subject to a court decision, abolished polygamy and decreed a minimum marriage age.

The population in penal institutions in Jan. 2011 was approximately 31,000 (297 per 100,000 of national population).

Education

The adult literacy rate in 2008 was 78%. All education is free from primary schools to university. Attendance at school is compulsory between the ages of six and 16. In 2007 there were 1,068,822 pupils in primary schools with 58,879 teaching staff and 1,268,219 pupils in secondary schools with 79,735 teaching staff.

In 2007 there were 13 public universities including the Virtual University of Tunis. There are a number of other public and private higher education institutions. There were 326,185 students in higher education in 2007 with 18,117 academic staff.

In 2005 public expenditure on education came to 7·2% of GDP and accounted for 20·8% of total government expenditure.

Health

There were 172 hospitals and 2,079 basic health centres in 2007 with a capacity of 17,998 beds. In 2006 there were 9,653 doctors, 1,858 dentists, 30,812 paramedical staff and 2,255 pharmacists.

RELIGION

The constitution recognizes Islam as the state religion. In 2001 there were 9·72m. Sunni Muslims. The remainder of the population follow other religions, including Roman Catholicism.

CULTURE

World Heritage Sites

Tunisia has eight sites on the UNESCO World Heritage List: the Amphitheatre of El Jem (inscribed on the list in 1979); the Site of Carthage (1979); the Medina of Tunis (1979); Ichkeul National Park (1980); the Punic Town of Kerkuane and its Necropolis (1985 and 1986); the Medina of Sousse (1988); Kairouan (1988); and Dougga/Thugga (1997).

Press

In 2009 there were nine paid-for daily newspapers (four in Arabic and five in French). In the 2010 *World Press Freedom Index* compiled by Reporters Without Borders, Tunisia ranked 164th out of 178 countries.

Tourism

In 2005 there were 6,378,000 foreign tourists, spending US$2·78bn.

DIPLOMATIC REPRESENTATIVES

Of Tunisia in the United Kingdom (29 Prince's Gate, London, SW7 1QG)
Ambassador: Hatem Atallah.

Of the United Kingdom in Tunisia (Rue du Lac Windermere, Les Berges du Lac, 1053, Tunis)
Ambassador: Christopher O'Connor.

Of Tunisia in the USA (1515 Massachusetts Ave., NW, Washington, D.C., 20005)
Ambassador: Mohamed Salah Tekaya.

Of the USA in Tunisia (Les Berges du Lac, 1053 Tunis)
Ambassador: Jacob Walles.

Of Tunisia to the United Nations
Ambassador: Mohamed Khaled Khiari.

Of Tunisia to the European Union
Ambassador: Mohamed Ridha Farhat.

FURTHER READING

Hassan, Fareed M. A., *Tunisia: Understanding Successful Socioeconomic Development.* 2005
Murphy, Emma C., *Economic and Political Change in Tunisia: From Bourguiba to Ben Ali.* 2003

National Statistical Office: Institut National de la Statistique, 70 Rue Ech-cham, BP 265 CEDEX, Tunis.
Website: http://www.ins.nat.tn

TURKEY

Türkiye Cumhuriyeti
(Republic of Turkey)

Capital: Ankara
Population projection, 2015: 77·00m.
GNI per capita, 2011: (PPP$) 12,246
HDI/world rank: 0·699/92
Internet domain extension: .tr

KEY HISTORICAL EVENTS

There is evidence of human habitation in Anatolia (Asia Minor) from around 7500 BC. Catal Huyuk (on the Konya Plain) flourished between 6500 and 5800 BC to become one of the world's largest and most important Neolithic sites. Between 1800 and 1200 BC much of Anatolia came under Hittite rule, initially centred on Cappadocia. The artistic work of the Hittites shows a high level of culture with Babylonian and Assyrian influence. Greek colonies were established around the Anatolian coast from around 700 BC including Byzantium, which was founded by Greeks from Megara in 667 BC. Anatolia was conquered by Persians in the 6th century BC.

Alexander the Great defeated the Persians around 330 BC. After his death there was a long civil war between the Seleucids and the Ptolemies, while the kingdoms of Galatia, Armenia, Pergamum, Cappadocia, Bithynia and Pontus all established footholds in the region. Rome gained dominance around the 2nd century BC and brought stability and prosperity. Turkey was home to some of the earliest centres of Christianity, such as Antioch (modern Antalya) and Ephesus. In AD 324 the Emperor Constantine began the construction of a new capital at Byzantium. Constantinople became the centre of the Byzantine (Eastern Roman) Empire, which peaked under Justinian in the mid-6th century.

Muslim Arab forces attacked Constantinople in the 670s and besieged the city again in 716 but were repelled, thwarting the expansion of the Umayyad Caliphate ruled by Umar II. Constantine V (741–75) led Christian Byzantine forces eastward to recover lands in Anatolia. The Seljuk Turks, whose origins were in central Asia, established dominance over much of Anatolia during the 11th century, led by Alp Arslan. They came under threat during the Crusades and were overrun by the Mongol hordes from 1243. The Ottoman principality was one of a number of small Turkish states that emerged in Anatolia amid the retreat

of the Mongols and the waning of the Seljuk and Byzantine empires. Osman I led the early phase of Ottoman expansion, conquering Byzantine towns in northwest Anatolia in the early 14th century. Sultan Mehmed II seized Constantinople in 1453 and went on to establish Ottoman dominance in the Balkans and the Aegean. The empire expanded to its fullest extent under Suleiman the Magnificent (1494–1566), taking in North Africa, the Levant, Persia, Anatolia, the Balkans and the Caucasus.

From the early 17th century the Ottoman empire fell into a long decline, its power weakening rapidly in the 19th century. The Kingdom of Greece broke away from Ottoman rule in 1832, with Serbs, Romanians, Armenians, Albanians, Bulgarians and Arabs demanding independence soon afterwards. Attempts by Turkey to redefine itself were further hindered in the 20th century by the First World War, during which it sided with Germany. In fighting with Greece over disputed territory from 1920–22, the Turkish National Movement was led by Mustafa Kemal (Atatürk: 'Father of the Turks'), who wanted a republic based on a modern secular society. Turkey became a republic on 29 Oct. 1923, with Ankara its capital. Islam ceased to be the official state religion in 1928 and women were given the same rights to employment and education as men, although they were not able to vote in national elections until 1934. İsmet İnönü became president following Kemal's death in 1938 and steered a neutral course through the Second World War. The 1950s were marked by a policy of firm alignment with the West, and Turkey joined NATO in 1952.

On 27 May 1960 the Turkish army overthrew the government and party activities were suspended. A new constitution was approved in a referendum held on 9 July 1961. On 12 Sept. 1980 the Turkish armed forces again drove the government from office. A new constitution was enforced after a national referendum on 7 Nov. 1982. In the face of mounting Islamization of government policy, the Supreme National Security Council reaffirmed its commitment to the secular state. On 6 March 1997 Prime Minister Necmettin Erbakan, leader of the pro-Islamist Welfare Party, promised to combat Muslim fundamentalism but in June he was forced to resign by a campaign led by the army.

There are ongoing quarrels with Greece over the division of Cyprus, oil rights under the Aegean and ownership of uninhabited islands close to the Turkish coast. In Feb. 2000 the Kurdish Workers' Party (PKK) formally abandoned its 15-year rebellion and adopted the democratic programme urged by its imprisoned leader, Abdullah Öçalan. Despite being a long-term ally of the USA, the Turkish parliament voted against allowing US troops to attack Iraq from its southeastern border in 2003, with the incumbent Justice and Development Party (AKP) government harbouring concerns about the possibility of an independent Kurdish state arising from a divided Iraq.

An associate member of the EU since 1964, Turkey is in the process of accession pending the completion of negotiations. However, significant hurdles remain, including the status of Northern Cyprus, the issue of human rights in Turkey and lukewarm support for its accession in some EU states.

In Aug. 2007 Abdullah Gül became president, following several months of political wrangling over concerns that his Islamist background might compromise Turkey's constitutional secularism.

TERRITORY AND POPULATION

Turkey is bounded in the west by the Aegean Sea and Greece, north by Bulgaria and the Black Sea, east by Georgia, Armenia and Iran, and south by Iraq, Syria and the Mediterranean. The area (including lakes) is 783,562 sq. km (302,535 sq. miles). The last traditional census was in 2000. In 2007 an address-based population registration system was established to replace ten-yearly censuses. The population at the census of 31 Dec. 2009 using this method was 72,561,312, giving a density of 94·3 per sq. km. In 2011, 70·1% of the population lived in urban areas.

The UN gives a projected population for 2015 of 77·00m.

Turkish is the official language. Kurdish and Arabic are also spoken.

Some 14m. Kurds live in Turkey. In Feb. 1991 limited use of the Kurdish language was sanctioned, and in Aug. 2002 parliament legalized Kurdish radio and television broadcasts.

Land area and population of the 81 provinces at the 2009 census:

	Area in sq. km	Population		Area in sq. km	Population
Adana	13,915	2,062,226	Kahraman-		
Adıyaman	7,033	588,475	maraş	14,346	1,037,491
Afyon-			Karabük	4,109	218,564
karahisar	14,314	701,326	Karaman	8,845	231,872
Ağrı	11,470	537,665	Kars	10,127	306,536
Aksaray	7,570	376,907	Kastamonu	13,153	359,823
Amasya	5,690	324,268	Kayseri	17,043	1,205,872
Ankara	24,521	4,650,802	Kilis	1,428	122,104
Antalya	20,723	1,919,729	Kırıkkale	4,534	280,834
Ardahan	4,842	108,169	Kırklareli	6,278	333,179
Artvin	7,367	165,580	Kırşehir	6,352	223,102
Aydın	7,851	979,155	Kocaeli	3,612	1,522,408
Balıkesir	14,299	1,140,085	Konya	38,873	1,992,675
Bartın	2,080	188,449	Kütahya	11,977	571,804
Batman	4,659	497,998	Malatya	11,776	736,884
Bayburt	3,739	74,710	Manisa	13,096	1,331,957
Bilecik	4,302	202,061	Mardin	8,806	737,852
Bingöl	8,253	255,745	Mersin	15,485	1,640,888
Bitlis	7,021	328,489	Muğla	12,851	802,381
Bolu	8,320	271,545	Muş	8,059	404,484
Burdur	6,840	251,550	Nevşehir	5,379	284,025
Bursa	10,422	2,550,645	Niğde	7,352	339,921
Çanakkale	9,933	477,735	Ordu	5,952	723,507
Çankırı	7,490	185,019	Osmaniye	3,124	471,804
Çorum	12,792	540,704	Rize	3,922	319,569
Denizli	11,692	926,362	Sakarya	4,838	861,570
Diyarbakır	15,058	1,515,011	Samsun	9,083	1,250,076
Düzce	2,567	335,156	Şanlıurfa	18,765	1,613,737
Edirne	6,074	395,463	Siirt	5,473	303,622
Elazığ	8,455	550,667	Sinop	5,792	201,134
Erzincan	11,619	213,288	Şırnak	7,152	430,424
Erzurum	25,323	774,207	Sivas	28,549	633,347
Eskişehir	13,842	755,427	Tekirdağ	6,313	783,310
Gaziantep	6,819	1,653,670	Tokat	9,958	624,439
Giresun	6,832	421,860	Trabzon	4,664	765,127
Gümüşhane	6,437	130,976	Tunceli	7,432	83,061
Hakkâri	7,179	256,761	Uşak	5,341	335,860
Hatay	5,828	1,448,418	Van	19,299	1,022,310
Iğdır	3,588	183,486	Yalova	847	202,531
Isparta	8,276	420,796	Yozgat	14,072	487,365
İstanbul	5,196	12,915,158	Zonguldak	3,304	619,812
İzmir	12,012	3,868,308			

Population of cities of over 250,000 inhabitants in 2009:

İstanbul	12,611,910	Gebze	505,240
Ankara	4,097,051	Denizli	488,768
İzmir	2,797,573	Samsun	482,873
Bursa	1,713,646	Sakarya	402,310
Adana	1,556,238	Malatya	388,590
Gaziantep	1,278,676	Kahramanmaraş	384,953
Konya	1,003,373	Erzurum	368,146
Antalya	885,486	Van	360,810
Kayseri	882,563	Elazığ	323,420
Mersin	842,230	Batman	313,355
Diyarbakır	834,854	Sivas	300,795
Eskişehir	617,215	İzmit	293,339
Urfa	517,077	Manisa	291,374

SOCIAL STATISTICS

Births, 2009, 1,241,617; deaths, 367,971. 2009 birth rate per 1,000 population, 17·3; death rate, 5·1. 2009 marriages, 591,472 (rate of 8·2 per 1,000 population); divorces, 114,162 (rate of 1·6 per 1,000 population). Population growth rate, 2009, 1·3%. Expectation of life, 2007, was 69·4 years for males and 74·2 for females. Infant mortality, 2010, 12 per 1,000 live births, declining significantly from 66 per 1,000 live births in 1990. Fertility rate, 2008, 2·1 births per woman. In 2006 the most popular age for marrying was 25–29 for males and 20–24 for females.

CLIMATE

Coastal regions have a Mediterranean climate, with mild, moist winters and hot, dry summers. The interior plateau has more extreme conditions, with low and irregular rainfall, cold and snowy winters, and hot, almost rainless summers. Ankara, Jan. 32·5°F (0·3°C), July 73°F (23°C). Annual rainfall 14·7" (367 mm). İstanbul, Jan. 41°F (5°C), July 73°F (23°C). Annual rainfall 28·9" (723 mm). İzmir, Jan. 46°F (8°C), July 81°F (27°C). Annual rainfall 28" (700 mm).

CONSTITUTION AND GOVERNMENT

On 7 Nov. 1982 a new constitution was adopted, which has subsequently undergone several revisions. Following a referendum on 21 Oct. 2007, it was amended so that the *President* will be directly elected by the people, rather than by *Parliament*, as is currently the case. Furthermore, the President will be able to serve for up to two five-year terms, rather than being limited to a single seven-year term. This reform will come into force at the next presidential election (scheduled for Aug. 2014), and does not apply to the incumbent President. Further amendments were introduced after acceptance in a referendum on 12 Sept. 2010. Under their terms, military officers accused of crimes against the state may be tried in civilian courts. Legal protection previously granted to participants in the 1980 coup was removed. Government workers are granted the right to collective bargaining and restrictions on striking were loosened. The *Constitutional Court* was expanded, with the president and parliament having a greater say in judicial appointments.

The Presidency is not an executive position; the President may not be linked to a political party but can veto laws and official appointments. There is a 550-member Turkish Grand National Assembly, elected by universal suffrage (at 18 years and over) for four-year terms by proportional representation.

National Anthem

'Korkma! Sönmez bu şafaklarda yüzen al sancak' ('Be not afraid! Our flag will never fade'); words by Mehmed Akif Ersoy, tune by Zeki Üngör.

GOVERNMENT CHRONOLOGY

(AKP = Justice and Development Party; ANAP = Motherland Party; AP = Justice Party; CGP = Republican Reliance Party; CHP = Republican People's Party; DP = Democrat Party; DSP = Democratic Left Party; DYP = True Path Party; RP = Welfare Party; n/p = non-partisan)

Heads of State since 1938.

Presidents of the Republic

1938–50	CHP	İsmet İnönü
1950–60	DP	Mahmut Celal Bayar

Chairman of the Committee of National Unity (MBK) and Head of State

1950–61	military	Cemal Gürsel

Presidents of the Republic

1961–66	n/p (ex-military)	Cemal Gürsel
1966–73	n/p (ex-military)	Cevdet Sunay
1973–80	n/p (ex-military)	Fahri Korutürk

Chairman of the National Security Council (MGK) and Head of State

1980–82	military	Kenan Evren

Presidents of the Republic

1982–89	n/p (ex-military)	Kenan Evren
1989–93	ANAP	Turgut Özal
1993–2000	DYP	Süleyman Demirel
2000–07	n/p	Ahmet Necdet Sezer
2007–	AKP	Abdullah Gül

Prime Ministers since 1921.

1921–22	military	Mustafa Fevzı Çakmak
1922–23	n/p	Hüseyin Rauf Bey
1923–23	CHP	Ali Fehti Okyar
1923–24	CHP	Mustafa İsmet İnönü
1924–25	CHP	Ali Fethi Okyar
1925–37	CHP	Mustafa İsmet İnönü
1937–39	CHP	Mahmut Celal Bayar
1939–42	CHP	Refık İbrahım Saydam
1942–46	CHP	Mehmet Şükrü Saraçoğlu
1946–47	CHP	Mehmet Recep Peker
1947–49	CHP	Hasan Saka
1949–50	CHP	Mehmet Şemsettin Günaltay
1950–60	DP	Adnan Menderes
1960–61	military	Cemal Gürsel
1961–65	CHP	Mustafa İsmet İnönü
1965	n/p	Suat Hayri Ürgüplü
1965–71	AP	Süleyman Demirel
1971–72	n/p	İsmaıl Nıhat Erım
1972–73	CGP	Ferit Melen
1973–74	n/p	Mehmet Naim Talu
1974–74	CHP	Mustafa Bülent Ecevıt
1975–77	AP	Süleyman Demirel
1977–77	CHP	Mustafa Bülent Ecevıt
1977–78	AP	Süleyman Demirel
1978–79	CHP	Mustafa Bülent Ecevıt
1979–80	AP	Süleyman Demirel
1980–83	n/p	Saim Bülent Ulusu
1983–89	ANAP	Turgut Özal
1989–91	ANAP	Yıldırım Akbulut
1991	ANAP	Ahmet Mesut Yılmaz
1991–93	DYP	Süleyman Demirel
1993–96	DYP	Tansu Çiller
1996	ANAP	Ahmet Mesut Yılmaz
1996–97	RP	Necmettin Erbakan
1997–99	ANAP	Ahmet Mesut Yılmaz
1999–2002	DSP	Mustafa Bülent Ecevıt
2002–03	AKP	Abdullah Gül
2003–	AKP	Recep Tayyip Erdoğan

RECENT ELECTIONS

Parliamentary elections were held on 12 June 2011. The ruling Justice and Development Party (AKP)—former Islamists—won 326 of the 550 seats with 49·9% of votes cast, against 135 seats and 25·9% for the Republican People's Party (CHP) and 53 seats and 13·0% for the Nationalist Movement Party. The remaining seats went to independents. There were 12 parties that failed to secure the 10% of votes needed to gain parliamentary representation. Turnout was 86·7%.

In the presidential elections held on 27 April and 6 May 2007, the AKP nominee Abdullah Gül failed to gain the backing of enough members of the assembly. Amid concerns that Gül's Islamist background would compromise the country's secular status, the presidential vote was postponed until after early parliamentary elections that were brought forward to July. In the second presidential elections of 2007 the first round was held on 20 Aug. Abdullah Gül received 347 votes, Sabahattin Çakmakoğlu (Nationalist Movement Party) 70 and Tayfun İçli (Democratic Left Party) 13, with 23 votes blank and one invalid. As 367 votes (a two-thirds majority) were required to be elected in either the first or second round a second round was held on 24 Aug., with Gül receiving 337 votes, Çakmakoğlu 71 and İçli 14; 24 votes were

blank. A third round was therefore held on 28 Aug. at which only a simple majority was required. Gül received 339 votes, Çakmakoğlu 70 and İçli 13.

CURRENT GOVERNMENT

President: Abdullah Gül; b. 1950 (sworn in 28 Aug. 2007).

In March 2013 the government comprised:

Prime Minister: Recep Tayyip Erdoğan; b. 1954 (AKP; sworn in 14 March 2003).

Deputy Prime Ministers: Bülent Arınç; Beşir Atalay; Ali Babacan; Bekir Bozdağ.

Minister of Culture and Tourism: Ömer Çelik. *Customs and Trade:* Hayati Yazıcı. *Defence:* İsmet Yılmaz. *Development:* Cevdet Yılmaz. *Economy:* Zafer Çağlayan. *Energy and Natural Resources:* Taner Yıldız. *Environment and Urbanization:* Erdoğan Bayraktar. *EU Affairs and Chief Negotiator:* Egemen Bağış. *Family and Social Policy:* Fatma Şahin. *Finance:* Mehmet Şimşek. *Food, Agriculture and Animal Husbandry:* Mehmet Mehdi Eker. *Foreign Affairs:* Ahmet Davutoğlu. *Forestry and Water Resources:* Veysel Eroğlu. *Health:* Mehmet Müezzinoğlu. *Interior:* Muammer Güler. *Justice:* Sadullah Ergin. *Labour and Social Security:* Faruk Çelik. *National Education:* Nabi Avcı. *Science, Industry and Technology:* Nihat Ergün. *Transport, Maritime Affairs and Communications:* Binali Yıldırım. *Youth and Sports:* Suat Kılıç.

Office of the President: http://www.tccb.gov.tr

CURRENT LEADERS

Abdullah Gül

Position
President

Introduction
Abdullah Gül was elected president on 28 Aug. 2007. His background in Islamist politics and membership of political parties banned under the country's secular constitution stoked widespread concern when he was nominated as a presidential candidate. A former prime minister and close ally of the incumbent premier, Recep Tayyip Erdoğan, Gül has taken a moderate line since 2001, advocating a pro-Western agenda and eventual EU membership.

Early Life
Abdullah Gül was born on 29 Oct. 1950 in Kayseri, central Turkey. He graduated in economics from İstanbul University in 1971 and began an academic career there. From 1980–83 he taught economics at the Sakarya School of Engineering and Architecture. As a devout Muslim, and having received a PhD in 1983, he joined the Islamic Development Bank (in Jeddah, Saudi Arabia) as an economist, a position he held for eight years.

Returning to Turkey in 1991, Gül entered politics. Campaigning for the Islamist Welfare Party, he was elected representative for Kayseri. He rose through the party ranks to become state minister and speaker for the government of Necmettin Erbakan in 1996. He was initially critical of Turkey's overtures towards the West and opposed EU membership. His ambitions were curtailed in 1997 by a military-backed campaign to oust the government. The following year a ban was imposed on the Welfare Party which was said to threaten the secular constitution. Gül, along with around 100 Welfare Party members, joined the Virtue Party, contesting its leadership in 2000. This party was banned in June 2001.

In Aug. 2001 Gül and other ex-party members joined Recep Tayyip Erdoğan's newly-formed Justice and Development Party (AKP), which presented itself as pro-Western and democratic. In the 2002 parliamentary elections Erdoğan led a high profile campaign, but was barred from standing because he had a criminal conviction for reading an Islamic poem at a political rally. The AKP found popularity with voters dissatisfied with the ruling government and won an outright victory to replace the three-party coalition. Two weeks after the election the party nominated Gül for the premiership.

Career in Office

As prime minister, Gül wanted to prove that Turkey could operate as both a Muslim and democratic state. The AKP campaigned on a pro-Western agenda and was committed to steering Turkey towards EU membership. Gül announced plans to reform the laws on the freedom of expression and human rights that had been partly responsible for Turkey's omission from the EU expansion plans in 2002. He supported further privatization and sought to achieve a modernized and efficient administration, reducing his cabinet from 36 to 26 places. In Dec. 2002 President Ahmet Necdet Sezer agreed to constitutional changes that would allow Erdoğan to stand for a parliamentary seat and thus become eligible for the premiership.

In the run-up to the US-led invasion of Iraq in March 2003, parliament refused to allow the USA to deploy troops on Turkish territory, though permission was given to use its airspace. Erdoğan returned to parliament in a by-election the same month and was appointed prime minister. Gül became foreign minister, working to achieve an EU accession date, although this was thwarted by the continued impasse over the status of Cyprus.

Prime Minister Erdoğan announced in April 2007 that Gül would be the AKP candidate in the 2007 presidential election. This sparked Turkey's most serious political crisis in a decade, with mass protests in the big cities in support of secularism. The military also warned that it would defend secularism. The AKP was forced to call early elections for 22 July, which it won decisively. Gül was re-nominated as the AKP candidate and on 28 Aug. he was elected president in the third round of voting. The chief of the general staff absented himself from the swearing-in ceremony.

In his inauguration speech, Gül sought to dispel secularist fears of an AKP Islamist agenda. However, parliament's vote to remove the ban on women wearing headscarves at universities in Feb. 2008 was seized on by secularists as evidence that Gül was attempting to introduce Islamic rule. In June the Constitutional Court rejected the move in a ruling that was viewed as a setback for the AKP government.

In July 2009 Gül approved controversial government legislation allowing civilian courts to prosecute military personnel for offences against the state. Then in Aug. 2011, after Turkey's top military leadership had resigned en masse in a dispute over promotions, he moved quickly to appoint replacements and assert civilian control over the powerful and traditionally secular armed forces.

Recep Tayyip Erdoğan

Position
Prime Minister

Introduction
Recep Tayyip Erdoğan became prime minister in March 2003. He led the Justice and Development Party (AKP) to victory at the general elections of Nov. 2002 but, because of a previous criminal conviction, was banned from standing for a parliamentary seat thus making him ineligible for the premiership. A constitutional amendment allowed him to stand for election in early 2003 and he subsequently replaced his party deputy, Abdullah Gül, as prime minister. For many years a prominent Islamist spokesman, Erdoğan has remoulded himself as a pro-European moderate conservative, although he continues to cause unease among many of Turkey's secularists. He identified Turkey's admission to the European Union as his government's top priority and introduced reforms that paved the way for the opening of membership talks from Oct. 2005. However, the EU negotiations have since been hampered by Turkey's continuing refusal to recognize the government of Greek Cyprus. Erdoğan was returned as premier in 2007 as the AKP won parliamentary elections with almost 47% of the vote, and again in 2011 as the party won a third consecutive victory at the polls with an increased vote share.

Early Life
Erdoğan was born in 1954 in Rize and his family later moved to İstanbul. He attended a Koranic college before graduating in economics in 1981 from Marmara University in İstanbul where he met Necmettin Erbakan, who would become Turkey's first Islamist premier. From the mid-1980s Erdoğan became active in the pro-Islamist Welfare Party. In 1994 he was made mayor of İstanbul and was noted for running an effective administration free of corruption.

The Welfare Party was outlawed in 1998 for contravening Turkey's secularist constitution. In the same year Erdoğan was imprisoned for inciting racial hatred when he read a pro-Islamist poem at a political rally. He served four months of a ten-month sentence. Following the banning of the Virtue Party (the successor party to Welfare) in June 2001, Erdoğan established the AKP, espousing pro-Western and democratic policies, and the party won an outright victory at the general elections of Nov. 2002. Erdoğan remained the AKP's figurehead while Abdullah Gül, his deputy and a former foreign minister, was named prime minister.

Following constitutional changes in Dec. 2002, Erdoğan was able to successfully contest a by-election in Feb. 2003. Gül stood down to be replaced by Erdoğan the following month.

Career in Office
Despite the AKP's Islamic roots, Erdoğan believes that Turkey can operate as both a Muslim and democratic state within Europe. He has voiced his commitment to democratization—including liberalizing laws on freedom of expression and human rights that were partly responsible for Turkey's omission from the EU expansion plans advanced in 2002—and confirmed his support for further privatization.

Erdoğan's early tenure was dominated by the US-led invasion of neighbouring Iraq from March 2003. Mirroring Turkish popular opinion, the government refused to allow the deployment of US ground troops on its territory, endangering aid and loans from the USA and IMF until the Turkish parliament agreed to the use of its airspace by the US air force. Turkey's deployment of troops in Kurdish-held northern Iraq to block any attempts to establish a Kurdish separatist state meanwhile caused international concern.

Turkey's wish for early entry into the EU was undermined by the failure in 2003 of the leaders of the Greek and Turkish sectors of Cyprus to agree on UN proposals for the island's reunification. A revised UN reunification plan was put to both sides in twin referenda in April 2004, which was endorsed by Turkish Cypriots but rejected by Greek Cypriots. Because both sides had to approve the proposals, the island remained divided as it joined the EU the following month. To fulfil the political criteria for EU membership, Erdoğan had pushed through parliament a series of reform packages in 2003 to bring Turkey into line with EU legislation. In 2004 a protocol abolishing the death penalty was signed and penal reforms introduced tougher measures to prevent torture and violence against women. Once Erdoğan's government had introduced the necessary legislative and constitutional reforms, and made a deal accepting Cyprus as an EU member, the European Council agreed in Dec. 2004 to open accession negotiations with Turkey which began in Oct. 2005. However, in late 2006 the EU partially suspended them because of the Turkish government's failure to open its ports and airports to Cypriot traffic. Erdoğan emphasized that the accession process would be maintained but there has been little substantive progress as he continues to press the EU to take steps to end the Turkish Cypriot community's economic isolation.

In early 2007 Erdoğan decided not to stand for election to the state presidency in view of strong secular opposition, and instead put forward foreign minister Abdullah Gül as the AKP candidate. Gül was equally unpalatable to secular and particularly military opinion, leading to a political stand-off in the National Assembly. However, Erdoğan consequently called an early general election for July, which returned the AKP to power, and in Aug. the new parliament endorsed Gül as president.

Secular concerns over the AKP's Islamist intentions resurfaced in Feb. 2008 when parliament voted to remove the ban on women wearing headscarves at universities. The Constitutional Court overturned the proposed law in June that year and then only narrowly ruled in July against a petition by state prosecutors to ban the AKP for alleged anti-secular activities, which could have led to a political crisis and Erdoğan's disqualification from politics. Tensions between the AKP and the secular opposition have since continued. In Sept. 2010 Erdoğan's government won a referendum on constitutional reforms increasing parliamentary control over the army and judiciary, and in Dec. the trial began of 196 active and retired military officers charged with a longstanding alleged plot to destabilize the country and overthrow the AKP administration. In Jan. 2012 former military chief of staff Gen. İlker Başbuğ was arrested and accused of planning a coup. His trial began in March, while in Sept. three other generals were jailed for 20 years.

Erdoğan's harder line towards Kurdish separatist insurgents based in northern Iraq had led to cross-border Turkish air and artillery assaults in 2007–08. In a new initiative to end the conflict his government put forward measures in parliament in Dec. 2009 to extend Kurdish linguistic and cultural rights and to limit the military presence in the mainly Kurdish southeast of the country. In July 2010 the PKK indicated its willingness to consider a truce in return for greater political and cultural rights, but the offer met with no official response. There was instead an escalation of Turkish military and penal measures against the PKK and pro-Kurdish activists in 2011 and 2012.

In foreign affairs, Erdoğan clashed publicly in 2009 with the president of Israel over Israeli military action in the Palestinian Gaza Strip, and relations were damaged further in 2010 as nine Turkish activists were killed in an Israeli commando raid on a flotilla of ships carrying aid and supplies to Gaza. In Sept. 2011 the Turkish government suspended all defence links after expelling the Israeli ambassador. Turkey's diplomatic ties with the USA and France have also come under strain over US congressional, and French parliamentary, recognition of the mass killing of Armenians in 1915 during Ottoman rule as genocide. Despite Turkey's rejection of Armenia's claims about the episode, the two governments agreed in Oct. 2009 on a framework to normalize relations, subject to parliamentary ratification by both sides, although this has yet to be realized. In 2011, while critical of NATO's military action against the Gaddafi regime in Libya, Erdoğan was supportive of the 'Arab Spring' protest movements that erupted against established regimes across much of the Middle East and North Africa.

The escalation of the civil war in Syria in 2012 led to a serious deterioration in Turkey's relations with the neighbouring regime of President Assad. In June a Turkish reconnaissance aircraft was shot down by a Syrian missile, prompting threats of retaliation from Erdoğan. In Oct. a series of exchanges of artillery fire across the border followed a Syrian mortar attack that killed five Turkish civilians. In Nov. the Turkish government formally requested the deployment of NATO Patriot missiles on its territory as a defensive deterrent to Syrian attack. Turkey was meanwhile faced with a surge of Syrian refugees from the civil conflict.

DEFENCE

The President of the Republic is C.-in-C. of the armed forces. The *National Security Council*, chaired by the Prime Minister and comprising military leaders and the ministers of defence and the economy, also functions as a *de facto* constitutional watchdog. Reforms passed in July 2003 in preparation for EU membership aimed to reduce the influence of the military in the political system. In Oct. 2003 the Turkish parliament voted to send 10,000 troops to Iraq, which would have made it the third largest force in the country after the USA and the UK, but the Iraqi Governing Council rejected the plan.

Conscription is 15 months for privates, 12 months for reserve officers and six months for privates who have completed a university degree.

In 2008 defence expenditure totalled US$13,531m., with spending per capita US$179. The 2008 expenditure represented 1·9% of GDP.

Army

Strength (2007) 660,700 (including 325,000 conscripts and 258,700 reservists). There is also a paramilitary gendarmerie-cum-national guard of 150,000. In addition around 36,000 Turkish troops are stationed in Northern Cyprus.

Navy

The fleet includes 13 diesel submarines and 24 frigates. The main naval base is at Gölcük in the Gulf of İzmit. Other bases are located at Aksaz-Karaağaç, Antalya, Bartın, Çanakkale, Erdek, Eregli, Foça, İskenderun, İstanbul, İzmir and Mersin. There are three naval shipyards: Gölcük, İzmir and Taşkızak.

The naval air component operates 11 combat-capable helicopters. There is a 3,100-strong Marine Regiment. The Coast Guard numbers 3,250 (including 1,400 conscripts).

Personnel in 2007 totalled 48,600 (34,500 conscripts) including marines and coast guard.

Air Force

The Air Force is organized as two tactical air forces, with headquarters at Eskişehir and Diyarbakır. There were 435 combat-capable aircraft in operation in 2007 including F-5A/Bs, F-4E Phantoms and F-16C/Ds.

Personnel strength (2007), 60,000.

INTERNATIONAL RELATIONS

Relations between Turkey and Iraq have long been strained over activity in the borderlands of northern Iraq by Kurdish separatist movements including the PKK. In Dec. 2007 Turkey launched air strikes on Iraqi territory against the PKK and in Feb. 2008 Turkish troops made a week-long incursion into northern Iraq to fight rebels.

Turkey has applied to join the European Union. At the EU's Helsinki Summit in Dec. 1999 Turkey was awarded candidate status. Talks on membership began in Oct. 2005 but Turkey is unlikely to join the EU in the foreseeable future.

ECONOMY

Agriculture accounted for 9·5% of GDP in 2006, industry 28·7% and services 61·8%.

Overview

In recent decades strong growth has been interrupted by sharp recessions in 1994, 1999, 2001 and 2009. In 2000 Turkey committed to a programme of wide-ranging structural reforms, strong fiscal adjustment and a pre-announced exchange rate crawl. However, the financial and currency crisis in 2001 caused the collapse of the three-year exchange rate-based stabilization programme and brought the country to the brink of debt default.

A strengthened programme was introduced in May 2001 with additional IMF support. Key structural reforms emphasized public sector standards, liberalizing markets and building a strong banking sector. Tight fiscal policies, IMF-inspired reforms and central bank independence brought Turkey improved macroeconomic health. From 2003–07 annual GDP growth averaged nearly 7% and public debt fell from 74% to 39% of GDP.

The subsequent global economic crisis saw oil and food import prices surge while unemployment reached 16% in 2009, higher than during the 2001 crisis. GDP growth dropped to 0·7% in 2008, with domestic political tensions adding to the country's vulnerability. The economy contracted in 2009, with export earnings and private investment falling by more than 30% in the early part of the year. However, growth recovered in 2010 and unemployment fell back to its pre-crisis level of around 10%. Real GDP growth stood above 8% in 2010 and 2011, but slowed in 2012. The current account deficit and inflation rate declined, while net exports increased.

Agriculture accounts for roughly 25% of total employment. The largest industrial sector is textiles and clothing, while the auto, autoparts and electronics industries have grown strongly. In 2005 the Baku-Tblisi-Ceyhan oil pipeline opened, bringing up to 1m. bbls per day from the Caspian to the Ceyhan Marine Terminal in Turkey. The country also has one of the most successful tourism sectors in the region. In 1996 a customs union was established with the EU and in 2005 Turkey began the EU accession process, providing further incentive to maintain macroeconomic and structural reforms. In the long term, the IMF emphasizes the need to develop competitiveness and improve the business climate, as well as diversifying energy sources.

Currency
The unit of currency is the Turkish *lira* (TRY) of 100 *kuruş*. It was introduced on 1 Jan. 2005 as the new Turkish lira—officially abbreviated as YTL—replacing the Turkish lira (TRL) at 1 new Turkish lira = 1m. Turkish lira. On 1 Jan. 2009 the 'new' was removed and its official name is again just 'Turkish lira'. Gold reserves were 3·73m. troy oz in Sept. 2009 and foreign exchange reserves US$69,387m. Inflation rates (based on OECD statistics):

2002	2003	2004	2005	2006	2007	2008	2009	2010	2011
45·0%	21·6%	8·6%	8·2%	9·6%	8·8%	10·4%	6·3%	8·6%	6·5%

Total money supply in July 2009 was YTL88,267m.

Budget
The fiscal year is the calendar year. Budgetary central government revenue totalled YTL218,858m. in 2007 (YTL189,578m. in 2006) and expenditure YTL206,695m. (YTL167,990m. in 2006). Tax revenues were YTL157,913m. in 2007.

VAT is 18%, with reduced rates of 8% and 1%.

Performance
Real GDP growth rates (based on OECD statistics):

2002	2003	2004	2005	2006	2007	2008	2009	2010	2011
6·2%	5·3%	9·4%	8·4%	6·9%	4·7%	0·7%	−4·8%	9·2%	8·5%

Total GDP was US$775·0n. in 2011. GDP per capita nearly trebled in the space of eight years, rising from US$3,492 in 2002 to US$10,440 in 2008.

Banking and Finance
The Central Bank (Merkez Bankası; *Governor*, Erdem Başçı) is the bank of issue. In 2003 there were 36 commercial banks (three state-owned, two under the Deposit Insurance Fund, 18 private, 13 foreign), and 14 development and investment banks. The Central Bank's assets were US$51·66bn. in 2003. The assets and liabilities of deposit money banks were US$25·8bn. Turkey's two state-owned banks, Ziraat Bankası (the Agricultural Bank, with a public mission to lend to farmers) and Halk Bankası (with a public mission to lend to small and medium sized enterprises), together accounted for 27% of total assets in the Turkish banking sector in Dec. 2002. Ziraat Bankası is Turkey's largest bank, with 18·1% of total assets as of March 2003.

In 2010 external debt totalled US$293,872m., equivalent to 40·4% of GNI.

Foreign direct investment in 2011 was US$15,876m., up from US$9,038m. in 2010. In Dec. 2000 the IMF gave Turkey an emergency loan of US$7·5bn. as the country experienced a financial crisis after ten banks were placed in receivership. The economic crisis continued as the lira was floated on the international market and lost 30% of its value against the US dollar in the space of 12 hours in Feb. 2001. Within a week the lira had been devalued by approximately 40%. In April 2001 Turkey secured a further US$10bn. loan from the IMF and the World Bank. This was followed in Feb. 2002 with a three-year US$16bn. loan from the IMF, taking total loans paid or pledged to US$31bn.

There is a stock exchange in İstanbul (ISE).

ENERGY AND NATURAL RESOURCES
Environment
In 2008 Turkey's carbon dioxide emissions from the consumption and flaring of fossil fuels were the equivalent of 3·6 tonnes per capita. Turkey's greenhouse gas emissions rose by 96·0% between 1990 and 2008.

Electricity
In 2007 installed capacity was 40·84m. kW (13·39m. kW hydro-electric); production in 2007 was 191·6bn. kWh and consumption per capita 2,572 kWh. There are plans to build two nuclear plants to meet the country's growing energy needs.

Oil and Gas
Crude oil production (2007) was 2,134,000 tonnes. Reserves in 2005 were 296m. bbls. In 2007, 23,446,000 tonnes of crude petroleum were imported. Natural gas output was 891m. cu. metres in 2007.

Accords for the construction of an oil pipeline from Azerbaijan through Georgia to the Mediterranean port of Ceyhan in southern Turkey (the BTC pipeline) were signed in Nov. 1999. Work on the pipeline began in Sept. 2002 and it was officially opened in May 2005.

A gas pipeline from Baku in Azerbaijan (the South Caucasus pipeline) through Georgia to Erzurum was commissioned in June 2006.

Minerals
Turkey is rich in minerals, and is a major producer of chrome.

Production of principal minerals (in 1,000 tonnes, in 2007 unless otherwise indicated) was: lignite, 72,902; iron, 4,849; copper (gross weight), 4,806; boron, 4,407; coal, 2,462; salt, 2,366; magnesite (2006), 2,088; chrome, 1,679.

Agriculture
In 2002 there were 6,745,000 households engaged in farming, of which 148,190 were engaged purely in animal farming. Agriculture accounts for 45% of the workforce but only 11·5% of GDP. In 2002 Turkey had 25·94m. ha. of arable land and 2·59m. ha. of permanent crops. 5·2m. ha. were irrigated in 2002. Vineyards, orchards and olive groves occupied 2,776,000 ha. in 2005.

Production (2005, in 1,000 tonnes) of principal crops: wheat, 21,500; sugar beets, 15,181; tomatoes, 10,050; barley, 9,500; melons and watermelons, 5,795; maize, 4,200; potatoes, 4,090; grapes, 3,850; apples, 2,570; dry onions, 2,070; peppers, 1,829; cucumbers, 1,745; cottonseed (2004), 1,426; oranges, 1,202; sunflower seeds, 975; cotton lint (2004), 936; aubergines, 930; apricots, 860; olives, 800; cabbage, 675; chick-peas, 600; lemons, 600; hazelnuts, 530. Turkey is the largest producer of apricots and hazelnuts.

Livestock, 2002 (in 1,000): sheep, 26,972; cattle, 10,548; goats, 7,022; asses, 462; horses, 271; buffaloes, 138; mules, 97; chickens, 218,000. Livestock products, 2002 (in 1,000 tonnes): milk, 8,409; meat, 1,376; eggs, 543; cheese, 113; honey, 75.

Forestry

There were 11·33m. ha. of forests in 2010, or 15% of the total land area. Timber production was 17·66m. cu. metres in 2007.

Fisheries

The catch in 2010 totalled 485,939 tonnes (445,680 tonnes from marine waters). Aquaculture production in 2010 came to 167,721 tonnes.

INDUSTRY

The leading companies by market capitalization in March 2012 were: Garanti Bankası (US$16·6bn.); and Akbank (US$15·7bn.).

Production in 2005 (in 1,000 tonnes unless otherwise stated): cement, 41,100; crude steel, 20,965; iron and steel bars, 11,854; distillate fuel oil (2007), 7,016; residual fuel oil (2007), 6,399; petrol (2007), 4,098; coke (2007), 3,335; sugar, 1,928; nitrogenous fertilizers, 1,525; paper and paperboard, 1,005; cotton yarn, 459; olive oil, 301; polyethylene, 274; pig iron, 178; cotton woven fabrics, 609m. metres; woollen woven fabrics, 32m. metres; carpets, 58,034,681 sq. metres; TV sets, 20,790,123 units; refrigerators, 5,098,866 units; cars, 635,137 units; lorries, 39,324 assembled units; tractors, 38,800 units; cigarettes, 104,170 tonnes.

Labour

Out of 22,047,000 people in employment in 2005 (16,346,000 men), 6,493,000 were engaged in agriculture, hunting, forestry and fisheries, 4,083,000 in manufacturing, 3,610,000 in wholesale and retail trade/repair of motor vehicles, motorcycles and personal and household goods and 1,246,000 in public administration and defence/compulsory social security. The unemployment rate in 2011 was 8·8%. The gross monthly minimum wage was YTL666 in Jan. 2009.

INTERNATIONAL TRADE

A customs union with the EU came into force on 1 Jan. 1996.

Imports and Exports

Imports (c.i.f.) in 2010 totalled US$185,541m. (US$140,869m. in 2009) and exports (f.o.b.) US$113,980m. (US$102,139m. in 2009). Chief imports (2005) in US$1m.: machinery and transport equipment, 37,809; manufactured goods, 19,990; chemicals and related products, 16,167; petroleum and petroleum products, 12,413; crude materials excluding fuels, 7,661. Chief exports: machinery and transport equipment, 21,509; apparel and clothing accessories, 11,833; textile yarn, fabrics and finished articles, 7,076; food and live animals, 6,512; iron and steel, 5,827.

The main import suppliers in 2006 (in US$1m.) were: Russia, 17,645; Germany, 14,653; China, 9,601; Italy, 8,597; France, 7,212; USA, 6,221. Main export markets, 2006: Germany, 9,684; UK, 6,813; Italy, 6,753; USA, 5,061; France, 4,604; Spain, 3,721. The EU accounted for 40·7% of imports and 54·3% of exports in 2006.

COMMUNICATIONS

Roads

In 2006 there were 427,099 km of roads, including 1,987 km of motorway. In 2007 road vehicles in use included 6,472,200 passenger cars, 2,619,700 lorries and vans, 561,700 buses and coaches and 2,003,500 motorcycles and mopeds. There were 5,002 fatalities from road accidents in 2007.

Rail

Total length of railway lines in 2005 was 8,697 km (1,435 mm gauge), of which 2,336 km were electrified. An undersea rail tunnel to link European and Asian Turkey is currently under construction; it is expected to open in 2015. Passenger-km travelled in 2005 came to 5·04bn. and freight tonne-km to 9·15bn. There are metro systems operating in Adana, Ankara, Bursa, İstanbul and İzmir.

Civil Aviation

There are international airports at İstanbul (Atatürk and Sabiha Gökçen), Dalaman (Muğla), Ankara (Esenboga), İzmir (Adnan Menderes), Adana and Antalya. The national carrier is Turkish Airlines, which is 49·1% state-owned. In 2006 it flew 207·2m. km and carried 16,946,000 passengers (8,041,000 on international flights). In 2009 İstanbul's Atatürk Airport handled 29,854,119 passengers (18,396,050 on international flights) and 381,174 tonnes of freight. Antalya was the second busiest airport for passenger traffic, with 18,403,617 passengers (15,210,554 on international flights) and İstanbul's Sabiha Gökçen Airport third with 6,640,230 passengers (4,510,895 on domestic flights).

Shipping

In Jan. 2009 there were 912 ships of 300 GT or over registered, totalling 5·05m. GT. Of the 912 vessels registered, 469 were general cargo ships, 148 passenger ships, 123 oil tankers, 100 bulk carriers, 37 container ships, 27 chemical tankers and eight liquid gas tankers. The Turkish-controlled fleet comprised 1,156 vessels of 1,000 GT or over in Jan. 2009, of which 520 were under the Turkish flag and 636 under foreign flags. In 2007 Turkish ports handled 288·1m. tonnes of cargo (114·6m. tonnes loaded and 173·5m. tonnes unloaded), more than double the 140·2m. tonnes handled in 2003.

Telecommunications

In 2008 there were 17,502,000 main (fixed) telephone lines. In the same year mobile phone subscribers numbered 65,824,000 (890·5 per 1,000 persons). In Nov. 2005 the government sold a 55% stake in Türk Telecom to a consortium led by Saudi Arabia's Oger Telecom and Telecom Italia. The government's stake fell to 30% in May 2008 through a public offering. There were 25·4m. internet users in 2008. In March 2012 there were 31·0m. Facebook users.

SOCIAL INSTITUTIONS

Justice

The unified legal system consists of: (1) justices of the peace (single judges with limited but summary penal and civil jurisdiction); (2) courts of first instance (single judges, dealing with cases outside the jurisdiction of (3) and (4)); (3) central criminal courts (a president and two judges, dealing with cases where the crime is punishable by imprisonment over five years); (4) commercial courts (three judges); (5) state security courts, to prosecute offences against the integrity of the state (a president and two judges).

The civil and military High Courts of Appeal sit at Ankara. The Council of State is the highest administrative tribunal; it consists of five chambers. Its 31 judges are nominated from among high-ranking personalities in politics, economy, law, the army, etc. The Military Administrative Court deals with the judicial control of administrative acts and deeds concerning military personnel. The Court of Jurisdictional Disputes is empowered to resolve disputes between civil, administrative and military courts. The Supreme Council of Judges and Public Prosecutors appoints judges and prosecutors to the profession and has disciplinary powers.

The Civil Code and the Code of Obligations have been adapted from the corresponding Swiss codes. The Penal Code is largely based upon the Italian Penal Code, and the Code of Civil Procedure closely resembles that of the Canton of Neuchâtel. The Commercial Code is based on the German.

The population in penal institutions in Nov. 2008 was 101,100 (142 per 100,000 of national population).

The death penalty, not used since 1984, was abolished in peacetime in Aug. 2002. The government signed a European Convention protocol abolishing the death penalty entirely in Jan. 2004.

Education

Adult literacy in 2009 was 90·8% (male, 96·4%; female, 85·3%). Compulsory primary and secondary education is free of charge in state schools between the ages of six and 18. In Aug. 2002 parliament legalized education in Kurdish. In 2003–04 there were 583 religious schools with 84,898 pupils.

Statistics for 2005–06	Number	Teachers	Students
Pre-school institutions	18,539	20,910	550,146
Primary schools	34,990	389,859	10,673,935
High schools	3,406	102,581	2,075,617
Vocational and technical high schools	4,029	82,736	1,182,637

In 2007–08 there were 53 state universities, including two higher institutes of technology, and 24 private universities. In 2002–03 a total of 1,894,000 students enrolled at 1,379 establishments of higher education (including the universities); teaching staff numbered 74,134. In 2001, 41,867 students were studying abroad.

In 2006 public expenditure on education came to 2·9% of GDP.

Health

In 2006 there were 116,014 physicians, 23,798 dentists, 217,685 nurses and midwives and 24,740 pharmacists. There were 1,347 hospitals with 179,649 beds in 2009.

Welfare

In 2000, 1,349,151 beneficiaries received TRL2,273,278,239m. from the Government Employees Retirement Fund. Of these, 820,167 persons were retired and 376,131 were widows, widowers or orphans of retired persons. There were 3,339,327 beneficiaries from the Social Insurance Institution in 2000.

RELIGION

Islam ceased to be the official religion in 1928. The Constitution guarantees freedom of religion but forbids its political exploitation or any impairment of the secular character of the republic.

In 2001 there were 64·36m. Muslims, two-thirds Sunni and one-third Shia (Alevis). The Greek Orthodox, Gregorian Armenian, Armenian Apostolic and Roman Catholic Churches are represented in İstanbul, and there are small Uniate, Protestant and Jewish communities.

CULTURE

World Heritage Sites

UNESCO World Heritage sites under Turkish jurisdiction (with year entered on list) are: Historic Areas of İstanbul (1985), including the ancient Hippodrome of Constantine, the 6th-century Hagia Sophia and the 16th-century Suleymaniye Mosque; Göreme National Park and the Rock Sites of Cappadocia (1985); Great Mosque and Hospital of Divriği (1985), founded in the early 13th century; Hattusha (1986), the former capital of the Hittite Empire; Nemrut Dağ (1987), including the 1st century BC mausoleum of Antiochus I; Xanthos-Letoon (1988), the capital of Lycia; Hierapolis-Pamukkale (1988), including mineral forests, petrified waterfalls and the ruins of ancient baths, temples and other Greek monuments; City of Safranbolu (1994), a caravan station from the 13th century; Archaeological Site of Troy (1998); the Selimiye Mosque and its Social Complex (2011) in Edirne, the former capital of the Ottoman Empire; and the Neolithic Site of Çatalhöyük (2012), vast archaeological remains of Neolithic occupation dating from 7400–6200 BC, a unique illustration of the evolution of social organization and cultural practices as humans adapted to a sedentary life.

Press

In 2006 there were 81 daily newspapers with a combined average daily circulation of 5·1m. The best-selling newspapers are *Posta* and *Zaman*, with average daily circulations of 635,000 and 565,000 respectively. In the 2010 *World Press Freedom Index* compiled by Reporters Without Borders, Turkey ranked 138th out of 178 countries. In March 2011, 57 journalists were in prison—more than in any other country.

Tourism

In 2009, 25·5m. international tourists visited Turkey, making it the seventh most popular tourist destination; receipts from tourism in 2009 totalled US$21·3bn.

Festivals

Republic Day is on 29 Oct. The most important Islamic festivals are Şeker Bayramı ('sugar festival', Eid al-Fitr) at the end of Ramadan and Kurban Bayramı (Eid al-Adha). Their dates vary depending on the Islamic calendar. İstanbul hosts a variety of festivals including International İstanbul Film Festival (April), the One Love music festival (June), the International İstanbul Music Festival (June), International İstanbul Jazz Festival (July) and the Rock'n Coke music festival (July). Aspendos International Opera and Ballet Festival takes place in Antalya in June–July.

DIPLOMATIC REPRESENTATIVES

Of Turkey in the United Kingdom (43 Belgrave Sq., London, SW1X 8PA)
Ambassador: Ahmet Ünal Çeviköz.

Of the United Kingdom in Turkey (Sehit Ersan Caddesi 46/A, Cankaya, Ankara)
Ambassador: David Reddaway, CMG, MBE.

Of Turkey in the USA (2525 Massachusetts Ave., NW, Washington, D.C., 20008)
Ambassador: Namik Tan.

Of the USA in Turkey (110 Atatürk Blvd, Ankara)
Ambassador: Francis J. Ricciardone.

Of Turkey to the United Nations
Ambassador: Yaşar Halit Çevik.

Of Turkey to the European Union
Ambassador: İzzet Selim Yenel.

FURTHER READING

State Institute of Statistics. *Türkiye İstatistik Yilliği/Statistical Yearbook of Turkey.—Diş Ticaret İstatistikleri/Foreign Trade Statistics* (Annual).— *Aylik İstatistik Bülten* (Monthly).

Abramowitz, Morton, (ed.) *Turkey's Transformation and American Policy.* 2000

Howe, Marvin, *Turkey Today: A Nation Divided over Islam's Revival.* 2000
İnalcık, H., Faroqhi, S., McGowan, B., Quataert, D. and Pamuk, Ş., *An Economic and Social History of the Ottoman Empire.* 1994
Jenkins, Gareth, *Political Islam in Turkey: Running West, Heading East.* 2008
Joseph, Joseph S., *Turkey and the European Union: Internal Dynamics and External Challenges.* 2006
Kalaycioğlu, Ersin, *Turkish Dynamics: Bridge Across Troubled Lands.* 2006. —*Turkish Democracy Today.* 2006
LaGro, Esra and Jørgensen, Knud Erik, *Turkey and the European Union: Prospects for a Difficult Encounter.* 2007
Stone, Norman, *Turkey: a Short History.* 2011

National Statistical Office: Turkstat, Necatibey Caddesi no. 114, 06100 Ankara.
Website: http://www.tuik.gov.tr

TURKMENISTAN

Türkmenistan

Capital: Ashgabat
Population projection, 2015: 5·36m.
GNI per capita, 2011: (PPP$) 7,306
HDI/world rank: 0·686/102
Internet domain extension: .tm

KEY HISTORICAL EVENTS

Until 1917 Russian Central Asia was divided politically into the Khanate of Khiva, the Emirate of Bokhara and the Governor-Generalship of Turkestan. The Khan of Khiva was deposed in Feb. 1920 and a People's Soviet Republic was set up. In Aug. 1920 the Amir of Bokhara suffered the same fate. The former Governor-Generalship of Turkestan was constituted an Autonomous Soviet Socialist Republic within the RSFSR on 11 April 1921. In the autumn of 1924 the Soviets of the Turkestan, Bokhara and Khiva Republics decided to redistribute their territories on a nationality basis. The redistribution was completed in May 1925 when the new states of Uzbekistan, Turkmenistan and Tadzhikistan were accepted into the USSR as Union Republics. Following the break-up of the Soviet Union, Turkmenistan declared independence in Oct. 1991. Saparmurad Niyazov was elected president and founded the Democratic Party of Turkmenistan, the country's only legal party. Also prime minister and supreme commander of the armed forces, parliament proclaimed Niyazov head of state for life in Dec. 1999. He held the official title of 'Turkmenbashi', leader of all Turkmen. In July 2000 Niyazov introduced a law requiring all officials to speak Turkmen. He died of a heart attack in Dec. 2006. Gurbanguly Berdymukhammedov succeeded him.

TERRITORY AND POPULATION

Turkmenistan is bounded in the north by Kazakhstan, in the north and northeast by Uzbekistan, in the southeast by Afghanistan, in the southwest by Iran and in the west by the Caspian Sea. Area, 448,100 sq. km (186,400 sq. miles). The 1995 census population was 4,483,251; density 10·0 per sq. km. Estimate, 2010, 5·04m.; density, 11 per sq. km. In 1999, 85% of the population were Turkmen, 7% Russian, 5% Uzbek and 3% other. Since then the Russian population has declined dramatically as the rights of Russians living in Turkmenistan deteriorated considerably. A dual-citizenship treaty between Turkmenistan and

Russia has been rescinded. In 2011, 50·0% of the population lived in rural areas.

The UN gives a projected population for 2015 of 5·36m.

There are five administrative regions (*velayaty*): Ahal, Balkan, Dashoguz, Lebap and Mary, comprising 42 rural districts, 15 towns and 74 urban settlements. The capital is Ashgabat (formerly Ashkhabad; 2004 estimated population, 827,500); other large towns are Turkmenabat (formerly Chardzhou), Mary (Merv), Balkanabad (Nebit-Dag) and Dashoguz.

The official language is Turkmen, spoken by 77% of the population; Uzbek is spoken by 9% and Russian by 7%.

SOCIAL STATISTICS

2008 estimates: births, 111,000; deaths, 39,000. Estimated rates, 2008 (per 1,000 population): births, 21·9; deaths, 7·7. Annual population growth rate, 2000–08, 1·4%. Life expectancy, 2007: 60·6 years for males and 68·8 for females. Infant mortality, 2010, 47 per 1,000 live births; fertility rate, 2008, 2·5 births per woman.

CLIMATE

The summers are warm to hot but the humidity is relatively low. The winters are cold but generally dry and sunny over most of the country. Ashgabat, Jan. –1°C, July 25°C. Annual rainfall 375 mm.

CONSTITUTION AND GOVERNMENT

A new constitution was adopted on 26 Sept. 2008. It provided for a head of state who is elected by popular vote for a five-year term and abolished the 2,500-member *Khalk Maslakhaty* (People's Council), formerly the highest representative body. The *Majlis* (Assembly), which now serves as the sole legislative body, was increased from 65 to 125 members. The constitution also allows for a multiparty system.

At a referendum on 16 Jan. 1994, 99·99% of votes cast were in favour of prolonging President Niyazov's term of office to 2002. In 1999 the *Khalk Maslakhaty* declared him president for life.

National Anthem

'Turkmenbasyn guran beyik binasy' ('The country which Turkmenbashi has built'); composed by Veli Muhatov.

RECENT ELECTIONS

In the presidential election of 12 Feb. 2012 incumbent Gurbanguly Berdymukhammedov was re-elected with 97·0% of the vote (89·2% in 2007), ahead of Yarmukhammet Orazgulyev with 1·2%. Turnout was 96·3%.

Majlis elections were held on 14 Dec. 2008. Most of the candidates were from the Democratic Party (DP; former Communists). 123 of 125 seats were filled on 14 Dec. 2008 and results in the remaining two constituencies were decided by run-offs held on 28 Dec. 2008 and 8 Feb. 2009. The official turnout figure of 93·9% was disputed by human rights groups.

CURRENT GOVERNMENT

President and Prime Minister: Gurbanguly Berdymukhammedov; b. 1957 (DP; in office since 21 Dec. 2006—acting until 14 Feb. 2007).

In March 2013 the government comprised:

Deputy Prime Ministers: Shamukhammet Durdylyev; Annamuhammet Gochiyev; Baymyrat Hojamuhammedov; Yagshygeldi Kakayev; Rashid Meredov (also *Minister of Foreign Affairs*); Hojamuhammed Muhammedov; Byagul Nurmuradova; Sapardurdy Toylyev; Annageldy Yazmyradov; Akmyrat Yegeleyev.

Minister of Agriculture: Rejep Bazarov. *Communications:* Bayramgeldi Ovezov. *Construction:* Dzhumageldi Bayramov.

Culture: Guncha Mammedova. *Defence:* Begench Gundogdiyev. *Economic Policy and Development:* Babamyrat Taganov. *Education:* Goulshat Mammedova. *Energy:* Myrat Artikov. *Environmental Protection:* Babageldi Annabayramov. *Finance:* Dovletgeldy Sadykov. *Health and Pharmaceutical Industry:* Gurbanmammet Elyasov. *Industry:* Babaniyaz Italmazov. *Internal Affairs:* Isgender Mulikov. *Justice:* Begench Charyev. *Labour and Social Welfare:* Bekmyrat Shamyradov. *Public Services:* Satlyk Satlykov. *National Security:* Yaylim Berdiyev. *Oil and Gas, and Mineral Resources:* Muhammetnur Khalylov. *Railways:* Bayram Annameredov. *Road Transport:* Mele Gurbandurdyev. *Textile Industry:* Saparmyrat Batyrov. *Trade and Foreign Economic Relations:* Bayar Abayev. *Water Resources and Irrigation:* Seyitmyrat Taganov.

Chairman, Supreme Council (Majlis): Akja Nurberdiyeva.

Government Website: http://www.turkmenistan.gov.tm

CURRENT LEADERS

Gurbanguly Berdymukhammedov

Position
President

Introduction
Gurbanguly Berdymukhammedov came to power in 2006 following the death of President Saparmurad Niyazov, known as Turkmenbashi. Berdymukhammedov had served as minister of health in Niyazov's government since 1997, implementing the closure of rural hospitals. Though not initially regarded as a frontrunner for the presidency, he became acting president when Niyazov's constitutional successor, Ovezgeldi Atayev, was charged with criminal offences. Having won the Feb. 2007 presidential election, Berdymukhammedov began to strengthen ties with the outside world and introduced some constitutional reforms, although critics questioned their democratic validity. He was re-elected in Feb. 2012.

Early Life
Berdymukhammedov was born in Babaarap village in the region of Ashgabat in 1957. He graduated in dentistry from the Turkmen state medical institute in 1979 and later completed a PhD in medical sciences in Moscow. In 1995 he was appointed head of the dentistry centre of the ministry of health and became associate professor and dean of the dentistry faculty of the state medical institute. He was Niyazov's personal dentist and in 1997 entered political life as the minister for health and the medical industry.

In April 2001 Berdymukhammedov was appointed deputy chairman of the council of ministers. Under Niyazov's autocratic regime the health service suffered acute financial problems. In April 2004 Niyazov announced that since state healthcare workers were not being paid, Berdymukhammedov would also forfeit his pay for three months. In the same year Berdymukhammedov implemented drastic cuts, closing all rural hospitals and sacking 15,000 healthcare workers, replacing them with untrained army conscripts.

When Niyazov died of a heart attack on 21 Dec. 2006, the Turkmen constitution provided for Ovezgeldi Atayev, chairman of the *Khalk Maslakhaty* (People's Council), to become acting president. However, Berdymukhammedov assumed the role and announced that Atayev was the subject of criminal investigation. On 26 Dec. an extraordinary session of the *Khalk Maslakhaty* amended the constitution to allow the acting president to stand in presidential elections. It also blocked the candidacy of leading opposition figures based abroad.

Berdymukhammedov was said to have the backing of Akmurad Rejepo, head of the presidential security service, and although five other candidates stood in the elections, his victory was widely predicted. During his campaign Berdymukhammedov promised to reform the agricultural sector and to improve living conditions.

He pledged that the government would continue to provide free natural gas, electricity, salt and water, and that salaries and pensions would be increased regularly.

Career in Office
On 11 Feb. 2007 Berdymukhammedov was elected president with nearly 90% of the vote. While making no decisive break with Niyazov's isolationist policies, there were early indications of a more outward-looking stance. Representatives of many foreign governments attended his inauguration and he subsequently received delegations from China, Russia and the USA. He confirmed that Turkmenistan would honour its gas contracts with Russia and moved towards stronger relations with neighbouring Afghanistan by writing off US$4m. of debt. The government also announced that it was reinforcing commercial ties with Iran.

Berdymukhammedov extended the period of formal schooling, reintroduced foreign languages to the curriculum and lifted restrictions on travel within Turkmenistan. He raised the prospect of reopening rural hospitals and promised to widen Internet access (available to 1% of the population when he came to power). However, these reforms would be dependent on an economic upturn. A new constitution was adopted in Sept. 2008 that increased the number of parliamentary seats to 125 and increased the influence of the *Majlis*. Nevertheless, Berdymukhammedov's hold on power was reinforced as parliamentary elections in Dec. that year returned an overwhelming number of candidates from his dominant Democratic Party.

The cult of personality around the presidency has lessened under Berdymukhammedov, but the government retains tight control over political organizations. It also continues to attract criticism for human rights and a lack of transparency in the electoral system, under which the president was returned for a second term in Feb. 2012.

The government has sought to diversify its gas markets and break free of Russian dominance of the export routes. An export pipeline to China was inaugurated in Dec. 2009 and in Jan. 2010 a second pipeline to Iran opened.

DEFENCE

Conscription is compulsory for two years. Defence expenditure in 2008 totalled US$84m. (US$17 per capita), representing 0·7% of GDP.

Army
In 2007 the Army was 18,500-strong.

Navy
A Navy/Coast Guard is in the process of being formed and is expected to be completed in 2015. In 2007 it numbered 500 and operated from a minor base at Turkmenbashi with six patrol and coastal combatants. The Caspian Sea Flotilla is operating as a joint Russian, Kazakhstani and Turkmenistani flotilla under Russian command. It is based at Astrakhan.

Air Force
The Air Force, with 3,000 personnel (including Air Defence), had 94 combat-capable aircraft in 2007 including Su-17s and MiG-29s.

ECONOMY

In 2008 agriculture accounted for 12·3% of GDP, industry 53·7% and services 34·0%.

Turkmenistan featured among the ten most corrupt countries in the world in a 2012 survey of 176 countries carried out by the anti-corruption organization *Transparency International*.

Overview
Turkmenistan is home to the fifth largest oil and gas reserves in the world. Hydrocarbons comprise 80% of exports and are the chief factor in the country's recent economic upturn. Other major

exports include textiles and raw cotton. Despite a privatization programme launched in 1994, most businesses remain state-run and foreign investment levels are low.

When Turkmenistan gained independence from the Soviet Union in 1991 the economy suffered from the break in traditional economic ties, poor harvests and mismanagement of state-run industries. The situation worsened in 1997 when natural gas exports were halted as a result of non-payment by Commonwealth of Independent States (CIS) countries but improved the following year with the resumption of gas exports to Ukraine and Russia.

Government sources report growth of 17% annually since 1999 but research by international organizations suggest a lower figure. Nevertheless, growth in the past has been impressive, driven by state investment (around 30% of GDP) in oil refineries, food processing, transportation and textiles.

Turkmenistan reached agreement with Russia in 2003 and China in 2005 to supply natural gas, guaranteeing income in the medium term. However, gas exports are hindered by reliance on Russian pipelines while government expenditures have focused on infrastructure and national prestige projects that bring little revenue.

Currency

The unit of currency is the *new manat* (TMT) of 100 *tenge*, introduced on 1 Jan. 2009 to replace the *manat* (TMM) at a rate of 1 TMT = 5,000 TMM. There was inflation of 14·5% in 2008, followed by deflation of 2·7% in 2009. In 2010 there was inflation again, of 4·4%, rising to 5·3% in 2011.

Budget

Revenues were 10·1bn. manat in 2008 and expenditures 5·4bn. manat.

VAT is 15%.

Performance

Total GDP in 2011 was US$28·1bn. Annual real GDP growth averaged –10·6% between 1994 and 1997. However, a spectacular revival led to average annual economic growth of 15·9% between 1999 and 2006. Real GDP growth was 14·7% in 2008, 6·1% in 2009, 9·2% in 2010 and 14·7% again in 2011. The rapid growth of recent years is largely down to large-scale gas exports to Russia.

Banking and Finance

There are two types of bank in Turkmenistan—state commercial banks and joint stock open-end commercial banks. The central bank is the State Central Bank of Turkmenistan (*Chairman*, Tuvakmammed Japarov). A government-led restructuring of the banking sector in 1999 saw the total number of banks reduced from 67 to 12 by 2002. In 2010 foreign debt totalled US$422m., equivalent to just 2·1% of GNI.

ENERGY AND NATURAL RESOURCES

Environment

Carbon dioxide emissions from the consumption and flaring of fossil fuels in 2008 were the equivalent of 11·8 tonnes per capita.

Electricity

Installed capacity in 2007 was an estimated 3·9m. kW. Production was 14·88bn. kWh in 2007, with consumption per capita 2,703 kWh.

Oil and Gas

Turkmenistan possesses the world's fourth largest reserves of natural gas and substantial oil resources, but disputes with Russia have held up development. The 1,833-km Turkmenistan–China gas pipeline, taking natural gas to Xinjiang in China via Kazakhstan and Uzbekistan, was inaugurated in Dec. 2009. Oil production in 2008 was 10·2m. tonnes.

In 2008 natural gas reserves were estimated at 7,940bn. cu. metres and oil reserves at 0·6bn. bbls. In 2008 natural gas production was 66·1bn. cu. metres.

Minerals

There are reserves of coal, sulphur, magnesium, potassium, lead, barite, viterite, bromine, iodine and salt.

Agriculture

Cotton and wheat account for two-thirds of agricultural production. Barley, maize, corn, rice, wool, silk and fruit are also produced. Production of main crops (2003, in 1,000 tonnes): wheat, 2,534; seed cotton, 714; cottonseed, 480; sugar beets, 255; watermelons, 230; tomatoes, 150; cotton lint, 140; grapes, 130; rice, 110. There were approximately 1·85m. ha. of arable land in 2007 and 63,000 ha. of permanent crops.

Livestock, 2003 estimates: sheep, 6·0m.; cattle, 860,000; goats, 370,000; pigs, 45,000; chickens, 5m.

Forestry

There were 4·13m. ha. of forests (9% of the land area) in 2010.

Fisheries

There are fisheries in the Caspian Sea. The estimated total catch in 2008 was 15,000 tonnes, exclusively freshwater fish.

INDUSTRY

Main industries: oil refining, gas extraction, chemicals, manufacture of machinery, fertilizers, textiles and clothing. Output, 2007 (in tonnes): distillate fuel oil, 2,908,000; residual fuel oil, 2,020,000; petrol, 1,464,000; cement, 941,000; cotton woven fabrics (2001), 61·0m. sq. metres; footwear (2001), 375,000 pairs.

Labour

The estimated labour force in 2010 totalled 2,509,000 (53% males), up from 1,826,000 in 2000.

INTERNATIONAL TRADE

Imports and Exports

Imports, 2003, US$2,512m.; exports, US$2,632m. Main imports: light manufactured goods, processed food, metalwork, machinery and parts. Main exports: gas, oil and cotton. The main import suppliers in 1998 were Ukraine (16·1%), Turkey (13·1%), Russia (11·6%), Germany (6·9%) and the USA (6·4%). The leading export markets were Iran (24·1%), Turkey (18·3%), Azerbaijan (6·9%), UK (4·9%) and Russia (4·7%).

COMMUNICATIONS

Roads

Length of roads in 2002, 58,592 km (of which 81·2% were paved). In 2006 there were 650 fatalities as a result of road accidents.

Rail

Length of railways in 2008, 3,095 km of 1,520 mm gauge. A rail link to Iran was opened in 1996. In 2008, 6·2m. passengers and 25·4m. tonnes of freight were carried.

Civil Aviation

Turkmenistan Airlines, founded in 1992, is the flag carrier. In 2005 scheduled airline traffic of Turkmenistan-based carriers flew 9·5m. km, carrying 1,899,800 passengers.

Shipping

In Jan. 2009 there were ten ships of 300 GT or over registered, totalling 23,000 GT. The main port is Turkmenbashi, on the Caspian Sea.

Telecommunications

Telephone subscribers numbered 805,500 in 2007 (162·2 per 1,000 population), including 347,600 mobile phone subscribers. There were 70,000 internet users in 2007. The internet was banned under the former president, Saparmurad Niyazov, and has only been available since early 2007.

SOCIAL INSTITUTIONS

Justice

The population in penal institutions in 2006 was 10,953 (224 per 100,000 of national population). The death penalty was abolished in 1999 (there were over 100 executions in 1996).

Education

Free compulsory education from age seven was extended from nine years' duration to ten in 2007. The duration of higher education programmes is generally five years (longer in the case of medicine and some other disciplines). In 2003 there were 1,705 general secondary schools, 116 professional schools and 15 secondary vocational training schools. About 1,018,600 students were enrolled in general secondary education. In the same year there were approximately 65,100 teachers. In 2003, 129,000 children (24% of those eligible) were enrolled in 944 pre-schools.

In 2009 adult literacy was estimated at 99·6%.

Health

In 2006 there were 43 hospital beds per 10,000 inhabitants. There were 12,210 physicians, 703 dentistry personnel and 23,026 nursing and midwifery personnel in 2006.

Welfare

A Social Security Code was adopted in 2007. In 2010 there were 274,171 pensioners; in Jan. 2010 the minimum pension was 121 new manat a month and the maximum pension 532 new manat.

RELIGION

Around 87% of the population in 2001 were Muslims (mostly Sunni).

CULTURE

World Heritage Sites

Turkmenistan has three sites on the UNESCO World Heritage List: the State Historical and Cultural Park 'Ancient Merv' (inscribed on the list in 1999), the oldest and best-preserved Central Asian Silk Route city, dominated by Seljuk architecture; Kunya-Urgench (2005), the ancient capital of the Khorezem region; and the Parthian Fortresses of Nisa (2007), the site of one of the earliest and most important cities of the Parthian Empire.

Press

In 2008 there were two daily newspapers with a combined average circulation of 56,000. Approval is required from the president's office before publication. In the 2010 *World Press Freedom Index* compiled by Reporters Without Borders, Turkmenistan ranked 176th out of 178 countries.

Tourism

In 2005 there were 12,000 non-resident tourists.

Calendar

In Aug. 2002 President Saparmurad Niyazov renamed the days of the week and the months, for example with Jan. becoming 'Turkmenbashi' after the president's official name, meaning 'head of all the Turkmen'. April was renamed in honour of the president's mother. Tuesday was renamed 'Young Day' and Saturday 'Spiritual Day'. However, in April 2008 President Gurbanguly Berdymukhammedov reversed his predecessor's decisions.

DIPLOMATIC REPRESENTATIVES

Of Turkmenistan in the United Kingdom (131 Holland Park Avenue, London, W11 4UT)
Ambassador: Yazmurad N. Seryaev.

Of the United Kingdom in Turkmenistan (301–308 Office Building, Ak Atin Plaza Hotel, Ashgabat)
Ambassador: Keith Allan.

Of Turkmenistan in the USA (2207 Massachusetts Ave., NW, Washington, D.C., 20008)
Ambassador: Meret Bairamovich Orazov.

Of the USA in Turkmenistan (9 Puskin St., Ashgabat)
Ambassador: Robert Patterson.

Of Turkmenistan to the United Nations
Ambassador: Aksoltan T. Ataeva.

Of Turkmenistan to the European Union
Ambassador: Karadjan Mommadov.

FURTHER READING

Abazov, Rafis, *Historical Dictionary of Turkmenistan.* 2005

National Statistical Office: State Statistical Committee of Turkmenistan, 72 Magtymgyly Ave., Ashgabat 744000.
Website (Turkmen and Russian only): http://www.stat.gov.tm

TUVALU

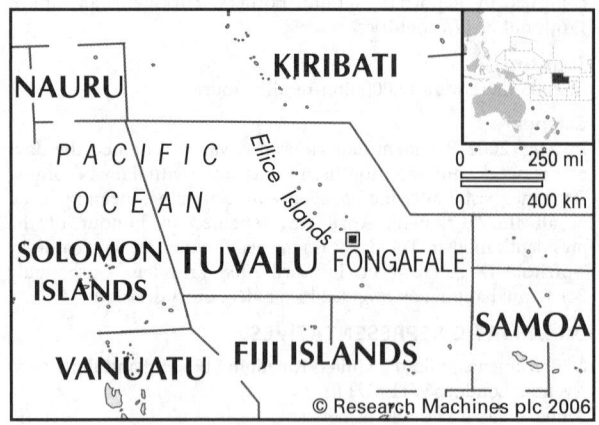

Capital: Fongafale
Population, 2002: 10,000
GNI per capita, 2010: US$4,607
Internet domain extension: .tv

KEY HISTORICAL EVENTS

Formerly known as the Ellice Islands, Tuvalu is a group of nine islands in the western central Pacific. Joining the British controlled Gilbert Islands Protectorate in 1916, they became the Gilbert and Ellice Islands colony.

After the Japanese occupied the Gilbert Islands in 1942, US forces occupied the Ellice Islands. A referendum held in 1974 produced a large majority in favour of separation from the Gilbert Islands. In 1975 the Ellice Islands were renamed Tuvalu. Independence was achieved on 1 Oct. 1978. Early in 1979 the USA signed a treaty of friendship with Tuvalu and relinquished its claim to the four southern islands in return for the right to veto any other nation's request to use any of Tuvalu's islands for military purposes. In 2011 a state of emergency was declared on two occasions; the first in Jan. as a result of protests and the second in Oct. owing to a shortage of clean drinking water.

TERRITORY AND POPULATION

Tuvalu lies between 5° 30' and 11° S. lat. and 176° and 180° E. long. and comprises Nanumea, Nanumaga, Niutao, Nui, Vaitupu, Nukufetau, Funafuti (administrative centre; 2002 estimated population, 4,492), Nukulaelae and Niulakita. Population (census 2002) 9,561, excluding an estimated 1,500 who were working abroad, mainly in Nauru and Kiribati. Area approximately 26 sq. km (10 sq. miles). Density, 2002, 373 per sq. km.

In 2011, 50·9% of the population lived in urban areas. The population is of a Polynesian race.

The official languages are Tuvaluan and English.

SOCIAL STATISTICS

2005 births (est.), 230; deaths (est.), 60. Infant mortality, 2010, 27 per 1,000 live births. Expectation of life, 2008: males, 64 years; females, 63. Annual population growth rate, 1998–2008, 0·5%; fertility rate, 2008, 3·2 births per woman.

CLIMATE

A pleasant but monotonous climate with temperatures averaging 86°F (30°C), though trade winds from the east moderate conditions for much of the year. Rainfall ranges from 120" (3,000 mm) to over 160" (4,000 mm). Funafuti, Jan. 84°F (28·9°C), July 81°F (27·2°C). Annual rainfall 160" (4,003 mm). Although the

islands are north of the recognized hurricane belt they were badly hit by hurricanes in the 1990s, raising fears for the long-term future of Tuvalu as the sea level continues to rise.

CONSTITUTION AND GOVERNMENT

The Head of State is the British sovereign, represented by an appointed Governor-General. The Constitution provides for a Prime Minister and the cabinet ministers to be elected from among the 15 members of the *Fale I Fono* (*Parliament*).

National Anthem

'Tuvalu mote Atua' ('Tuvalu for the Almighty'); words and tune by A. Manoa.

RECENT ELECTIONS

Elections were held on 16 Sept. 2010. Only non-partisans were elected as there are no political parties. On 21 Dec. 2010, after only three months in power, Maatia Toafa and his government were ousted following a motion of no confidence. Willy Telavi, minister of home affairs, was elected the new prime minister by parliament on 24 Dec., defeating Enele Sopoaga by eight votes to seven.

CURRENT GOVERNMENT

Governor-General: Sir Iakoba Taeia Italeli (sworn in 16 April 2010).

In March 2013 the cabinet comprised:

Prime Minister and Minister of Natural Resources: Willy Telavi (took office on 24 Dec. 2010).

Deputy Prime Minister and Minister of Transport, Communication and Public Utilities: Kausea Natano.

Minister of Education, Youth and Sports: Falesa Pitoi. *Finance and Economic Development:* Lotoala Metia. *Foreign Affairs, Environment, Trade, Labour and Tourism:* Apisai Ielemia. *Health:* Taom Tanukale. *Home Affairs:* Pelenike Isaia.

Speaker: Kamuta Latasi.

CURRENT LEADERS

Willy Telavi

Position
Prime Minister

Introduction
Tuvalu's long-standing police commissioner, Willy Telavi, was first elected to parliament in 2006 and served as minister for home affairs. He was elected prime minister in Dec. 2010 following a vote of no confidence in the previous incumbent, Maatia Toafa.

Early Life
Willy Telavi was born on Nanumea, the north-westernmost of Tuvalu's nine islands. He completed a diploma in legal studies at the University of the South Pacific in 1999 and received a masters degree in international management from the University of the Northern Territory (now Charles Darwin University), Australia, in 2000.

Between 1993 and 2009 he served as Tuvalu's police commissioner and in 2006 he was elected to the 15-member parliament as one of two representatives of Nanumea. He was appointed minister of home affairs in the government of Apisai Ielemia, a position he retained following the general election of Sept. 2010 when Maatia Toafa became prime minister.

Toafa's premiership soon came under pressure amid widespread concerns that the government could no longer cover the full costs of medical treatment for patients needing to go abroad. When a motion of no confidence in Toafa was tabled in

Dec. 2010 Telavi, along with two other MPs, withdrew his support from the prime minister and joined the opposition, enabling the motion to be carried and parliament to be dissolved. Willy Telavi was elected premier on 24 Dec. 2010.

Career in Office
Telavi's administration came under pressure in early Jan. 2011 when protesters from the island of Nukufetau peacefully demanded the resignation of Lotoala Metia, the finance and economic development minister, after he allegedly snubbed a meeting with island elders. On 13 Jan., following a protest on the main island of Funafuti that was supported by five members of the opposition and appeared to be linked to a campaign for another change in government, Telavi declared a state of emergency—the first in the nation's history—and imposed a 14-day ban on public gatherings. Metia was placed under police guard and the country's naval patrol boat was deployed to guard the coastline near Telavi's home and those of other government ministers.

In Aug. 2012 Telavi agreed to stop letting ships owned by Iran operate under its national flag, having been accused by the USA of breaking international sanctions on Iranian oil trading.

ECONOMY

Finance, real estate, public administration, defence and services accounted for 50·5% of GDP in 2008; agriculture and fishing, 18·2%; trade, hotels and restaurants, 13·9%; transport and communications, 12·2%.

Overview
With few natural resources, the economy is dependent on foreign aid, the sale of tuna fishing licences and interest from the Tuvalu Trust Fund, established in 1987 by Australia, New Zealand and the UK. Revenues are also derived from the sale of the '.tv' internet suffix to a US-based company.

Government spending on administration, construction and social development has helped offset a decline in remittances. However, the volatility of fishing licence income and overruns in the government's scholarship and medical insurance programmes are a threat to financial stability.

Currency
The unit of currency is the Australian *dollar* although Tuvaluan coins up to $A1 are in local circulation.

Budget
In 2008 revenues totalled $A45·4m. and expenditures $A42·9m.

Performance
The economy contracted by 1·7% in 2009 and 2·9% in 2010 but there was growth of 1·1% in 2011. In 2011 total GDP was US$36m.

Banking and Finance
The Tuvalu National Bank was established at Funafuti in 1980, and is a joint venture between the Tuvalu government and Westpac International. There is also a development bank.

ENERGY AND NATURAL RESOURCES

Electricity
Installed capacity was an estimated 2,000 kW in 2007; production was 4m. kWh.

Agriculture
Coconut palms are the main crop. Production of coconuts (2009 estimate), 1,700 tonnes. Fruit and vegetables are grown for local consumption. Livestock, 2009 estimate: pigs, 13,600; chickens, 45,000.

Fisheries
Total catch, 2009, 5,097 tonnes. A seamount was discovered in Tuvaluan waters in 1991 and is a good location for deep-sea fish. The sale of fishing licences to American, Japanese, New Zealand, South Korean and Taiwanese fleets provides a significant source of income, in some years contributing up to 50% of total government revenue.

INDUSTRY
Small amounts of copra, handicrafts and garments are produced.

INTERNATIONAL TRADE

Imports and Exports
Main sources of income are copra, stamps, handicrafts and remittances from Tuvaluans abroad. 2007 imports, $A18·5m.; 2007 exports, $A0·1m. Leading import suppliers are Australia, China and Japan. Significant export destinations are Australia, India and Indonesia.

COMMUNICATIONS

Roads
In 2002 there were 20 km of roads.

Civil Aviation
In 2010 Air Pacific operated two flights a week from Funafuti International to Suva in the Fiji Islands.

Shipping
Funafuti is the only port and a deep-water wharf was opened in 1980. In Jan. 2009 there were 96 ships of 300 GT or over registered, totalling 1·03m. GT. Of the 96 vessels registered, 40 were oil tankers, 36 general cargo ships, nine chemical tankers, five bulk carriers, four passenger ships and two container ships. Tuvalu is a 'flag of convenience' country.

Telecommunications
In 2008 there were approximately 1,500 main telephone lines in operation. There were some 2,000 mobile phone subscribers and 4,200 internet users in 2008.

SOCIAL INSTITUTIONS

Justice
There is a High Court presided over by the Chief Justice of the Fiji Islands. A Court of Appeal is constituted if required. There are also eight Island Courts with limited jurisdiction.

Education
There were 1,798 pupils at nine primary schools in 2001, and 558 pupils at Motufoua Secondary School in 2001. The Fetuvalu High School reopened in 2002 with Form 3 only. Education is free and compulsory from the ages of six to 15. There is a Maritime Training School at Funafuti, and the University of the South Pacific, based in the Fiji Islands, has an extension centre at Funafuti.

In 1999–2000 total expenditure on education came to 16·8% of total government expenditure.

Health
In 2002 there was one central hospital situated at Funafuti and clinics on each of the other eight islands; there were seven doctors and 34 nurses.

RELIGION
The majority of the population are Christians, mainly Protestant, but with small groups of Roman Catholics, Seventh Day Adventists, Jehovah's Witnesses and Bahais. There are some Muslims and Latter-day Saints (Mormons).

CULTURE

Press
The Government Broadcasting and Information Division produces *Tuvalu Echoes*, a fortnightly publication, and *Te Lama*, a monthly religious publication.

Tourism
There were 1,130 visitor arrivals in 2007.

DIPLOMATIC REPRESENTATIVES

Of Tuvalu in the United Kingdom (Tuvalu House, 230 Worple Rd, London, SW20 8RH)
Honorary Consul: Dr Iftikhar A. Ayaz.

Of the United Kingdom in Tuvalu
Acting High Commissioner: Martin Fidler (resides in Suva, Fiji Islands).

Of Tuvalu in the USA and to the United Nations (800 Second Ave., Suite 400D, New York, NY, 10017)
Ambassador: Aunese Makoi Simati.

Of the USA in Tuvalu
Ambassador: Frankie A. Reed (resides in Suva, Fiji Islands).

Of Tuvalu to the European Union
Ambassador: Tine Leuelu.

FURTHER READING

Bennetts, P. and Wheeler, T., *Time and Tide: The Islands of Tuvalu.* 2001

National Statistical Office: Ministry of Finance, Economic Planning and Industries, Private Bag, Vaiaku, Funafuti.
Website: http://www.spc.int/prism/country/tv/stats

The world in focus at
www.statesmansyearbook.com

UGANDA

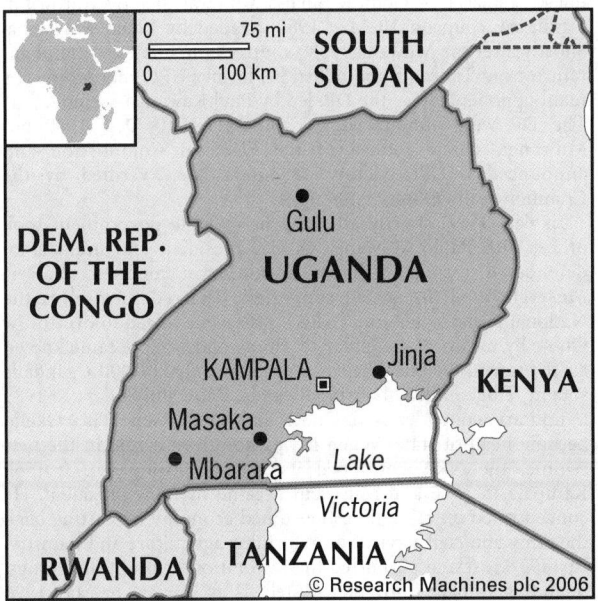

Jamhuri ya Uganda
(Republic of Uganda)

Capital: Kampala
Population projection, 2015: 39·11m.
GNI per capita, 2011: (PPP$) 1,124
HDI/world rank: 0·446/161
Internet domain extension: .ug

KEY HISTORICAL EVENTS

Bantu-speaking mixed farmers first migrated into southwest Uganda from the west around 500 BC. There is evidence that they smelted iron for tools and weapons. In the following centuries Nilotic-speaking pastoralists entered northern Uganda from the upper Nile valley (now southern Sudan). By AD 1300 several kingdoms (the Chwezi states) had been established in southern Uganda. In 1500 Nilotic-speaking Luo people invaded the Chwezi states and established the kingdoms of Buganda, Bunyoro and Ankole. At this time, northern Uganda became home to the Alur and Acholi ethnic groups. During the 17th century Bunyoro was southern Uganda's most powerful state, controlling an area that stretched into present-day Rwanda and Tanzania. From about 1700 the kingdom of Buganda expanded (largely at the expense of Bunyoro), and a century later it dominated a large territory bordering Lake Victoria from the Victoria Nile to the Kagera River. The *kabaka* (king) maintained a large court and a powerful army and traded in cattle, ivory and slaves.

Arab traders from Zanzibar on Africa's east coast reached Lake Victoria by 1844. A prominent trader, Ahmad bin Ibrahim, introduced the kabaka to foreign trade; imported cloth and firearms were exchanged for ivory and slaves. Ibrahim also introduced Islam to the region. In 1862 John Speke, a British explorer who was attempting to find the source of the Nile, became the first European to visit Buganda, by then a highly developed state supported by an army of more than 150,000 and a navy.

He met with Kabaka Mutesa I; as did Henry Stanley, who reached Buganda in 1875. Mutesa, fearful of attacks from Egypt, agreed to Stanley's proposal to allow Christian missionaries to enter his realm. Members of the British Protestant Church Missionary Society arrived in 1877 and were followed two years later by representatives of the French Roman Catholic White Fathers. Both were successful in attracting converts but by the 1880s they were in fierce competition. Trade with the Indian Ocean ports continued, bringing with it greater Islamic influence.

Mutesa was succeeded by Mwanga in 1884. He was wary of the new foreign ideologies and attempted to halt their spread but was deposed by Christian and Muslim converts in 1888. He was later reinstated but with considerably reduced power and influence. In 1889 Mwanga was visited by Carl Peters, a German doctor, and the kabaka subsequently signed a treaty of friendship with Germany. Britain was concerned by the growth of German influence and the potential threat to its position on the Nile. In 1890 the two European powers signed a treaty giving Britain rights to what was to become Uganda and giving Germany control over land to the southeast (now Tanzania). Frederick Lugard, acting as an agent of the Imperial British East Africa Company (IBEA), arrived in Buganda with a detachment of troops and in 1892 he backed Protestant converts in an attack on the French Catholic mission.

British Rule

In 1894 Britain made Uganda a protectorate. Allying with the Protestant Baganda chiefs, the British set about conquering the rest of the country, assisted by Nubian mercenary troops, formerly in the service of the khedive of Egypt. The British deposed Mwanga and replaced him with his infant son Daudi Chwa. Bunyoro had been spared the religious civil wars of Buganda and was firmly united by its king, Kabarega. Following five years of conflict, the British occupied Bunyoro and conquered Acholi and the northern region. Other African chiefdoms, such as Ankole in the southwest, signed treaties with the British, as did the chiefdoms of Busoga. In 1900 an agreement was signed between the British administration under Sir Harry Johnston and Buganda, giving the kingdom considerable autonomy and transforming it into a constitutional monarchy controlled largely by Protestant chiefs. Half of Bunyoro's conquered territory was also awarded to Buganda, including the historic heartland of the kingdom containing several royal tombs. Buganda doubled in size from ten to 20 counties (*sazas*), but the 'lost counties' of Bunyoro remained a grievance.

Economic Development

In 1901 a railway from Mombasa on the Indian Ocean reached Kisumu, on Lake Victoria, connected by boat with Uganda. The railway was later extended to Kampala. The railway had cost far more than was anticipated and the British, anxious for a return on their investment, turned to cotton to provide raw materials for British mills. Buganda, with its strategic location on the north shore of Lake Victoria, reaped the benefits of cotton growing; it soon became the major export crop and made the Buganda kingdom relatively prosperous. Coffee and sugar production accelerated in the 1920s. The country attracted few permanent European settlers and the cash crops were mostly produced by African smallholders, rather than the plantation system used in other colonies. Many South Asians were encouraged to settle in Uganda, where they played a leading role in the country's commerce. In 1921 a legislative council for the protectorate was established (although its first African member was admitted only in 1945).

The colonial government regulated the buying and processing of cash crops, setting prices and reserving the role of intermediary for Asians, who were thought to be more efficient. The British and Asians repelled African attempts to break into cotton ginning, leading to resentment among the Baganda. In addition, on the

Asian-owned sugar plantations established in the 1920s, labour for sugarcane and other cash crops was increasingly provided by migrants from the fringes of Uganda and beyond. In 1949 discontented Baganda rioted and burned down the houses of pro-government chiefs in Kampala. The rioters had three demands: the right to bypass government price controls on the sales of cotton, the removal of the Asian monopoly over cotton ginning, and the right to have their own representatives in local government. They were also critical of the young kabaka, Frederick Walugembe Mutesa II. The British governor, Sir John Hall, regarded the riots as the work of communist-inspired agitators such as the Uganda African Farmers Union (UAFU), founded by I. K. Musazi in 1947. The UAFU was banned and none of the requested reforms were implemented. Musazi's Uganda National Congress replaced the UAFU in 1952 but remained a discussion group rather than an organized political party.

Meanwhile, the British began to prepare for an independent Uganda. Britain's post-war withdrawal from India, nationalism in West Africa and a more liberal philosophy all had an effect. Sir Andrew Cohen was installed as governor in 1952 and pursued economic and political reforms: removing obstacles to African cotton ginning, encouraging co-operatives, establishing the Uganda Development Corporation and reorganizing the Legislative Council to include Africans elected from districts across the country for the first time. There was also talk of a future federation of east African territories (Kenya, Uganda and Tanganyika). However, there was resistance among the Baganda, who feared the erosion of their power-base. Mutesa II refused to co-operate with Cohen's plan for an integrated Buganda. Cohen deported him to exile in London, setting off a storm of protest. Two year later Mutesa II was reinstated, officially as a constitutional monarch, but in reality having considerable political clout. In 1960 a political organizer from Lango, Milton Obote, formed a new party, the Uganda People's Congress (UPC), as a coalition of all those who opposed Buganda dominance (apart from the Catholic-dominated Democratic Party (DP)).

Independence

On 9 Oct. 1962 Uganda became independent, with Obote as prime minister and the kabaka as head of state. Buganda was given considerable autonomy. In 1963 Uganda became a republic and Mutesa II was elected president. The first years of independence were dominated by a struggle between the central government and Buganda. In 1966 Obote introduced a new constitution that ended Buganda's autonomy and restored the 'lost counties' to the Bunyoro. Obote then captured the kabaka's palace at Mengo and forced the kabaka to flee the country. In 1967 a new constitution was introduced giving the central government—especially the president—greater power and dividing Buganda into four districts. The traditional kingships were also abolished (in the case of Buganda retroactive to 1966). The 1960s saw a steady build-up of military power in Uganda, under Major Gen. Idi Amin Dada.

In Jan. 1971 Obote was deposed in a coup by Idi Amin. Amin was faced with opposition within the army by officers and troops loyal to Obote but by the end of 1971 he was in firm control. In 1972 he ordered Asians who were not citizens of Uganda to leave the country and within three months all 60,000 had left, most of them for Britain. Their expulsion hit the Ugandan economy hard. Amin's rule became increasingly dictatorial and brutal; it is estimated that over 300,000 Ugandans were killed during the 1970s. His corrupt administration led to divisions in the military and a number of coup attempts. Israel conducted a successful raid on the Entebbe airport in 1976 to rescue passengers on a plane hijacked by Palestinian terrorists. Amin's expulsion of Israeli technicians won him the support of Arab nations such as Libya.

In 1976 Amin declared himself president for life and two years later he invaded Tanzania in an attempt to annex the Kagera region. The following year Tanzania launched a successful counter-invasion, unifying anti-Amin forces under the Uganda National Liberation Front (UNLF). Amin's forces were driven out and he fled to exile in Saudi Arabia. Tanzania left an occupation force in Uganda. Yusufu Lule was installed as president but was quickly replaced by Godfrey Binaisa, who was then overthrown in a military coup on 10 May 1980, headed by Paulo Muwanga. Shortly after the Muwanga 1980 coup, Obote made a triumphant return from Tanzania and rallied his former UPC supporters. His main opponents were the DP, led by Paul Kawanga Ssemogerere. The DP were announced as winners on 10 Dec. 1980 but Muwanga seized control of the Electoral Commission and announced a UPC victory 18 hours later, verified by the Commonwealth Observer Group.

In Feb. 1981, shortly after the new Obote government took office, with Paulo Muwanga as vice-president and minister of defence, a former Military Commission member, Yoweri Museveni, and his armed supporters declared themselves the National Resistance Army (NRA). Museveni vowed to overthrow Obote by means of a popular rebellion, and what became known as 'the war in the bush' began. Approximately 200,000 Ugandans sought refuge in neighbouring Rwanda, Zaïre and Sudan. In 1985 a military coup deposed Obote and Lieut.-Gen. Tito Okello became head of state. When it was not given a role in the new regime, the NRA continued its guerrilla campaign. It took Kampala in 1986 and Museveni became the new president. He concentrated on rebuilding the ruined economy by cutting back the army and civil service and reforming agriculture and industry. In 1993 Museveni permitted the restoration of traditional kings, including Ronald Muwenda Mutebi II as kabaka. In May 1996 Museveni was returned to office in the country's first direct presidential elections. A new parliament, chosen in elections in June, was dominated by Museveni supporters.

Museveni was re-elected in March 2001, following a period of relative stability and economic growth. However, his popularity was diminished by discontent with Uganda's intervention in the Democratic Republic of the Congo's (formerly Zaïre) civil war and signs of corruption in the government. Uganda's forces were largely withdrawn from the Democratic Republic of the Congo by the end of 2002. In 2005 the International Court in The Hague ordered Uganda to pay compensation to the DRC for its activities there between 1998 and 2003.

In Feb. 2006 Museveni won a new presidential term. In July 2006 peace talks commenced between the government and the Lord's Resistance Army (LRA), a fanatically religious group led by Joseph Kony that has terrorized northern Uganda for many years. A ceasefire was declared in Aug. 2006 but subsequent talks have been marred by disputes and walkouts.

TERRITORY AND POPULATION

Uganda is bounded in the north by South Sudan, in the east by Kenya, in the south by Tanzania and Rwanda, and the west by the Democratic Republic of the Congo. Total area 241,550 sq. km, including 41,740 sq. km of inland waters.

The 2002 census population was 24,442,084 (11,929,803 males, 12,512,281 females). Estimate, 2011, 32,939,800; density, 164 per sq. km. The largest city is Kampala, the capital (population estimate of 1,659,600 in 2011). Other major towns are Gulu, Lira, Jinja, Mbale, Mbarara, Masaka and Entebbe. In 2011, 13·5% of the population lived in urban areas.

The UN gives a projected population for 2015 of 39·11m.

The country is administratively divided into one city and 111 districts, which are grouped in four geographical regions (which do not have administrative status). Area and population of the regions in 2002 (census) and 2011 (estimate):

Region	Area in sq. km	2002 population in 1,000	2011 population in 1,000
Central Region	61,510	6,575·4	8,465·5
Eastern Region	39,953	6,204·9	8,623·7

Region	Area in sq. km	2002 population in 1,000	2011 population in 1,000
Northern Region	84,658	5,363·7	7,620·6
Western Region	54,917	6,298·1	8,230·0

The official languages are English and (since 2005) Kiswahili. About 70% of the population speak Bantu languages; Nilotic languages are spoken in the north and east.

SOCIAL STATISTICS

2008 estimates: births, 1,461,000; deaths, 401,000. Rates, 2008 estimates (per 1,000 population): births, 46·2; deaths, 12·7. Uganda has one of the youngest populations of any country, with 76% of the population under the age of 30 and 48% under 15. Uganda's life expectancy at birth in 2007 was 51·4 years for males and 52·4 years for females. Life expectancy declined dramatically until the late 1990s, largely owing to the huge number of people in the country with HIV. However, for both males and females expectation of life is now starting to rise again. Annual population growth rate, 2000–08, 3·2%. Infant mortality, 2010, 63 per 1,000 live births; fertility rate, 2008, 6·3 births per woman.

CLIMATE

Although in equatorial latitudes, the climate is more tropical because of its elevation, and is characterized by two distinct rainy seasons, March–May and Sept.–Nov. In comparison, June–Aug. and Dec.–Feb. are relatively dry. Temperatures vary little over the year. Kampala, Jan. 74°F (23·3°C), July 70°F (21·1°C). Annual rainfall 46·5" (1,180 mm). Entebbe, Jan. 72°F (22·2°C), July 69°F (20·6°C). Annual rainfall 63·9" (1,624 mm).

CONSTITUTION AND GOVERNMENT

The *President* is head of state and head of government, and is elected for a five-year term by adult suffrage. In Aug. 2005 Parliament amended the constitution to allow an incumbent to hold office for more than two terms, thus enabling President Museveni to serve another term in office. Having lapsed in 1966, the kabakaship was revived as a ceremonial office in 1993. Ronald Muwenda Mutebi (b. 13 April 1955) was crowned Mutebi II, 36th Kabaka, on 31 July 1993.

Until 1994 the national legislature was the 278-member National Resistance Council, but this was replaced by a 284-member *Constituent Assembly* in March 1994. A new constitution was adopted on 8 Oct. 1995 and the Constituent Assembly dissolved. Uganda's parliament is now the 386-member *National Assembly* (238 members elected by popular vote, 137 indirectly elected from special interest groups—including women and the army—and 11 *ex officio* members). A referendum on the return of multi-party democracy was held on 29 June 2000, but 88% of voters supported President Museveni's 'no-party' Movement system of government. Turnout was 51%. In Feb. 2003 President Museveni pledged to lift the ban on political parties. In a referendum held on 28 July 2005, 92·4% of voters backed the restoration of a multi-party political system, although the opposition called for a boycott.

National Anthem

'Oh, Uganda, may God uphold thee'; words and tune by G. W. Kakoma.

RECENT ELECTIONS

Presidential elections were held on 18 Feb. 2011. President Museveni was re-elected by 68·4% of votes cast, with his main rival, Kizza Besigye, receiving 26·0% of the vote. There were six other candidates. Turnout was 59·3%.

Parliamentary elections were held on the same day. The National Resistance Movement won 279 seats, the Forum for Democratic Change 34, the Democratic Party 11, Uganda People's Congress 9 and the Conservative Party and the Justice Forum both gained one seat each. 37 seats went to independents. Turnout was 59·3%.

CURRENT GOVERNMENT

President: Yoweri K. Museveni; b. 1944 (sworn in 27 Jan. 1986; re-elected in 1996, 2001, 2006 and 2011).

In March 2013 the government comprised:

Vice-President: Edward Kiwanuka Ssekandi; b. 1943 (sworn in 24 May 2011).

Prime Minister: Amama Mbabazi; b. 1949 (sworn in 24 May 2011).

First Deputy Prime Minister and Minister of East African Affairs: Eriya Kategaya. *Second Deputy Prime Minister and Minister of Public Service:* Henry Muganwa Kajura. *Third Deputy Prime Minister and Deputy Leader of Government Business:* Moses Ali.

Minister of Agriculture, Animal Industry and Fisheries: Tress Buchanayande. *Communication and Information Communication Technology:* Ruhakana Rugunda. *Defence:* Crispus Kiyonga. *Education and Sports:* Jessica Arupo. *Energy and Minerals:* Irene Muloni. *Finance:* Maria Kiwanuka. *Foreign Affairs:* Sam Kutesa. *Gender, Labour and Social Development:* Tarsis Kabwegyere. *Health:* Christine Androa. *Information and National Guidance:* Mary Karooro Okurut. *Internal Affairs:* Hilary Onek. *Justice and Constitutional Affairs:* Kahinda Otafiire. *Karamoja Affairs:* Janet Museveni. *Lands, Housing and Urban Development:* Daud Migereko. *Local Government:* Adolf Mwesigye. *Presidency:* Frank Tumwebaze. *Relief and Disaster Preparedness:* Dr Steven Mallinga. *Security:* Wilson Muruli Mukasa. *Tourism, Wildlife and Antiquities:* Maria Mutagamba. *Trade and Industry:* Amelia Kyambadde. *Transport and Works:* James Abraham Byandaala. *Water and Environment:* Ephraim Kamuntu. *Minister without Portfolio:* Richard Todwong. *Attorney General:* Peter Nyombi. *Chief Whip:* Justine Kasule Lumumba. *Minister of General Duties in the Office of the Prime Minister:* John Nasasira.

Speaker of Parliament: Rebecca Kadaga.

Ugandan Parliament: http://www.parliament.go.ug

CURRENT LEADERS

Yoweri Museveni

Position
President

Introduction
Yoweri Museveni became president of Uganda in 1986 and has been largely credited with transforming the country's economy after the years of misrule by Idi Amin Dada and Milton Obote. He won the first direct presidential elections in 1996 and was re-elected in 2001. In July 2005 a national referendum approved the lifting of restrictions on multi-party politics (in force since Museveni came to power) and the National Assembly abolished a constitutional limit on presidential terms. He won further terms in Feb. 2006 and Feb. 2011. His government has been in faltering talks with the Lord's Resistance Army (LRA) since 2006 in an attempt to end the long-standing conflict with the rebel group.

Early Life
Yoweri Kaguta Museveni was born in 1944 in Ankole, western Uganda, where he attended Mbarara High School and Ntare School. He studied economics and political science at the University of Dar es Salaam, Tanzania, graduating in 1970. While at university Museveni was politically active and became the chairman of a leftist student group linked to African liberation movements. In 1971 Idi Amin Dada came to power in Uganda and Museveni went back to Tanzania. He was a founder of the Front for National Salvation, one of the rebel groups that overthrew Amin in 1979. Museveni held various ministerial posts before running for president in 1980. Defeated by Milton Obote, he formed the National Resistance Army, which took power on 26

Jan. 1986 when Museveni declared himself president and minister of defence. His movement was supported by Col. Gaddafi of Libya.

Career in Office

For much of his presidency, Museveni has been favoured by Western nations and foreign aid donors for opening up the Ugandan economy and reducing poverty. Primary school education has increased markedly and, thanks to anti-AIDS campaigns, he has succeeded in reducing HIV levels. However, Museveni's image has been tarnished internationally by Ugandan military interference in neighbouring Democratic Republic of the Congo and by human rights concerns.

After coming to power, Museveni claimed that political parties divided poor countries like Uganda into ethnic, religious and tribal groups. His preferred system therefore had individuals competing for political office on individual merit. However, after 2001 calls for a return to multi-party democracy in Uganda became more persistent, and this was approved in a national referendum in July 2005, although the turnout was low. Museveni's government nevertheless supported the restoration. Parliament meanwhile voted to lift the constitutional limit on the office of president. Museveni stood for re-election in Feb. 2006 and won a further term with almost 60% of the vote.

In Oct. 2005 Kizza Besigye—Museveni's main opposition rival in the 2001 presidential poll which had been tainted by violence—returned to Uganda from exile in South Africa to contest the 2006 elections. His subsequent arrest for treason provoked violent street protests before his release on bail in Jan. 2006. Meanwhile, concern over alleged human rights abuses by Museveni's government led the UK and other European countries to suspend direct development aid in Dec. 2005. Britain later agreed to resume aid in Nov. 2007 at the biennial Commonwealth Heads of Government conference hosted by Museveni in Kampala.

In July 2006 the government opened peace talks with the rebel LRA, which has conducted an insurgency in northern Uganda since the 1980s involving atrocities against the local population. A truce was signed in Aug., but subsequent political progress was limited amid reported divisions within the LRA itself. In early 2008 the LRA announced that its leader, Joseph Kony, was ready to sign a peace agreement. Optimism proved unfounded, however, leading to a new offensive in Dec. against the rebels by troops from Uganda, Sudan and the Democratic Republic of the Congo. In Jan. 2009 the LRA called for a new ceasefire and in March Ugandan troops ended their offensive. Museveni's government claimed that the campaign had seriously hampered LRA capabilities and also rescued many kidnap victims. The LRA has nevertheless remained a destabilizing force in the region.

Following the announcement in Jan. 2009 of a significant oil discovery in Uganda by a British exploration company, Museveni said that the income from the oil, if well invested, could transform the country's economy and development. Subsequent allegations of bribery in the award of oil contracts led parliament to suspend new oil deals in Oct. 2011 and Museveni to publicly reject claims that government ministers were engaged in corrupt practices. At the same time, however, Sam Kutesa resigned as foreign minister ahead of an investigation (although he was reinstated in Aug. 2012).

In Sept. 2009 there were riots in Kampala by supporters of Ronald Mutebi, the king of Buganda (one of five ancient kingdoms that had been restored by Museveni) and leader of Uganda's largest ethnic group. Tensions between Museveni's government and Buganda grew when parliament passed new land legislation that the king had opposed because it eroded his ownership rights.

In Dec. 2009 the National Resistance Movement endorsed Museveni as its candidate for the presidential election in Feb. 2011. He secured a further term, taking 68% of the vote, but Kizza Besigye, his main challenger again, rejected the result amid claims of vote-rigging. In April and May Besigye organized opposition protests in Kampala, partly in reaction to rising food and fuel prices, but they were met with a heavy-handed response by security forces and Besigye himself was attacked and injured.

In late 2012 there were further allegations from the United Nations of Ugandan support for rebel forces opposed to the Kabila regime in the Democratic Republic of the Congo. In Nov. Museveni's government responded by declaring that Ugandan troops would be withdrawn from UN-sponsored peacekeeping missions.

DEFENCE

Defence expenditure in 2008 totalled US$277m. (US$9 per capita), representing 1·7% of GDP.

Army

The Uganda People's Defence Forces had a strength estimated at 45,000 in 2007. There is a Border Defence Unit about 600-strong and local defence units estimated at 10,000.

Navy

There is a Marine unit of the police (about 400-strong in 2007).

Air Force

In 2007 the Army's aviation wing operated 14 combat-capable aircraft (although the serviceability of some was in doubt) and one attack helicopter. There was also a police air wing around 800-strong.

INTERNATIONAL RELATIONS

In Nov. 1999 Uganda, Tanzania and Kenya created a new East African Community to develop East African trade, tourism and industry and to lay the foundations for a future common market and political federation.

ECONOMY

In 2006 agriculture accounted for 31·1% of GDP, industry 18·1% and services 50·7%.

Overview

After the economic turmoil under Idi Amin in the 1970s, Uganda has become a relatively stable and prosperous economy. GDP growth rose from an average of 7% per year in the 1990s to over 8% in the seven years up to 2007–08. Growth remained robust throughout the global financial crisis. The main export commodities are coffee, and fish and fish products. Recent trends have shown a shift away from agriculture towards industry and services.

A Poverty Eradication Action Plan (PEAP) was launched in 1997, aimed at reducing the number of people in poverty to less than 10% of the population by 2017. While the long-term growth rate falls short of this target, poverty fell from 44% of the population in 1997 to 31·5% in 2006. In April 2010 President Museveni unveiled the successor to PEAP, a five-year National Development Plan to turn Uganda into a middle-income economy. Strategies include improvements in the transport infrastructure, increased employment opportunities and nurturing the private sector, while reducing the proportion below the poverty line to 25%.

Currency

The monetary unit is the *Uganda shilling* (UGX) notionally divided into 100 *cents*. In 1987 the currency was devalued by 77% and a new 'heavy' shilling was introduced worth 100 old shillings. Inflation was 13·1% in 2009, falling to 4·0% in 2010. It increased to 18·0% in 2011, the highest rate since 1993. Foreign exchange reserves in June 2005 were US$1,326m. Total money supply in July 2005 was Shs 1,513·0bn.

Budget

The financial year runs from 1 July–30 June. Revenues for 2006–07 were Shs 3,574bn. and expenditures Shs 4,032bn. Tax revenue accounted for 63·3% of revenues in 2006–07; current expenditure accounted for 60·6% of expenditures.

VAT is 18%.

Performance

Real GDP growth was 7·0% in 2009, 6·1% in 2010 and 5·1% in 2011. In recent times Uganda has consistently been among Africa's best performers. In spite of growth rates that averaged 6·4% over ten years to 1998, per capita income is only just around the level of 1971, when Gen. Idi Amin came to power. Uganda's total GDP in 2011 was US$16·8bn.

Banking and Finance

The Bank of Uganda (*Governor*, Emmanuel Tumusiime Mutebile) was established in 1966 and is the central bank and bank of issue. In addition there are five foreign, six commercial and two development banks. There is also the state-owned Uganda Development Bank, which is scheduled for eventual privatization.

External debt totalled US$2,994m. in 2010, representing 17·9% of GNI.

In 2010 foreign direct investment totalled US$848m.

ENERGY AND NATURAL RESOURCES

Environment

Uganda's carbon dioxide emissions from the consumption and flaring of fossil fuels were the equivalent of 0·1 tonnes per capita in 2008.

Electricity

Installed capacity in 2007 was approximately 0·4m. kW, about 75% of which was provided by the Owen Falls Extension Project (a hydro-electric scheme). Production (2007) 2·0bn. kWh. Per capita consumption (2007) 67 kWh. Only 10% of the population has access to electricity.

Oil and Gas

Oil was first discovered in Uganda in 2006. There was a further major find in 2009, which may represent the largest onshore discovery in sub-Saharan Africa. Reserves have been estimated at 2·5bn. bbls. Production was expected to begin in 2012 but it is now not forecast to start until 2017.

Minerals

In Nov. 1997 extraction started on the first of an estimated US$400m. worth of cobalt from pyrites. Tungsten and tin concentrates are also mined. There are also significant quantities of clay and gypsum.

Agriculture

80% of the workforce is involved with agriculture. In 2007 the agricultural area included an estimated 5·5m. ha. of arable land and about 2·2m. ha. of permanent crops. Agriculture is one of the priority areas for increased production, with many projects funded both locally and externally. It contributes 90% of exports. Production (2003 estimates) in 1,000 tonnes: cassava, 13,500; plantains, 10,000; sweet potatoes, 2,600; sugarcane, 1,600; maize, 1,200; bananas, 615; millet, 584; potatoes, 546; dry beans, 535; sorghum, 395; coffee, 186. Coffee is the mainstay of the economy, accounting for more than 50% of the annual commodity export revenue. Uganda is the world's leading producer of plantains.

Livestock (2003 estimates): goats, 6·9m.; cattle, 6·1m.; pigs, 1·7m.; sheep, 1·2m.; chickens, 33m. Livestock products, 2003 estimates (in 1,000 tonnes): milk, 700; meat, 293.

Forestry

In 2010 the area under forests was 2·99m. ha., or 15% of the total land area. Exploitable forests consist almost entirely of hardwoods. Timber production in 2007 totalled 41·08m. cu. metres. Uganda has great potential for timber-processing for export, manufacture of high-quality furniture and wood products, and various packaging materials.

Fisheries

In 2010 fish landings totalled 413,805 tonnes, entirely from inland waters. Fish farming (especially catfish and tilapia) is a fast-growing industry, with production having increased from 5,539 tonnes in 2004 to 76,654 tonnes in 2009. Exports of fishery commodities, valued at US$109m. in 2009, are now one of Uganda's leading foreign currency earners.

INDUSTRY

Production (in 1,000 tonnes): cement (2001), 416; sugar (2002), 160; soap (2002), 92; beer (2002), 98·9m. litres. In 2001 industry accounted for 20·9% of GDP, with manufacturing contributing 9·8%. Industrial production grew by 5·4% in 2001.

Labour

The labour force in 2002–03 totalled 9,772,600 (47% males). In 2002–03 the unemployment rate was 3·2%.

INTERNATIONAL TRADE

Imports and Exports

In 2010 imports (c.i.f.) amounted to US$4,664·3m. (US$4,247·4m. in 2009); exports (f.o.b.) US$1,618·6m. (US$1,567·6m. in 2009). Coffee, fish and fish products are the principal exports. Major imports are machinery and transport equipment, and food, beverages and tobacco products. The main import suppliers in 2006 were Kenya (15·7%), UAE (12·7%) and India (8·2%). In 2006 the main export markets were UAE (19·4%), Sudan (9·5%) and Kenya (9·1%).

COMMUNICATIONS

Roads

In 2003 there were 70,746 km of roads, of which 23% were paved. There were 81,300 passenger cars in use in 2007, 79,300 lorries and vans, 40,500 buses and coaches, and 176,500 motorcycles and mopeds. In 2007 there were 17,428 road accidents resulting in 2,779 deaths.

In 2007 the government established the Uganda Road Fund, a body responsible for financing road maintenance. The estimated budget in 2007–08 totalled US$111·35m.

Rail

In 2005 the Uganda Railways network totalled 1,241 km (metre gauge). In 1996 passenger services were suspended and have not been reinstated in the meantime. Freight tonne-km in 2003 came to 213m.

Civil Aviation

There is an international airport at Entebbe, 40 km from Kampala. Air Uganda, formed in 2007, is the national airline. In 2003 scheduled airline traffic of Uganda-based carriers flew 2m. km, carrying 40,000 passengers (all on international flights). In 2001 Entebbe handled 370,063 passengers (343,722 on international flights) and 37,195 tonnes of freight.

Telecommunications

In 2008 there were 168,500 main (fixed) telephone lines; mobile phone subscribers numbered 8,554,900 in 2008 (27·0 per 100 persons). There were 2·5m. internet users in 2008. In June 2012 there were 415,000 Facebook users.

SOCIAL INSTITUTIONS

Justice

The Supreme Court of Uganda, presided over by the Chief Justice, is the highest court. There is a Court of Appeal and a High Court below that. Subordinate courts, presided over by Chief Magistrates and Magistrates of the first, second and third grade, are established in all areas: jurisdiction varies with the grade of Magistrate. Chief and first-grade Magistrates are professionally qualified; second- and third-grade Magistrates are trained to diploma level at the Law Development Centre, Kampala. Chief Magistrates exercise supervision over and hear appeals from second- and third-grade courts, and village courts.

The population in penal institutions in April 2007 was 26,273 (88 per 100,000 of national population). The death penalty is still in force. There were two executions in 2006 but none since.

Education

In 2007 there were 7,537,971 pupils and 132,325 teaching staff at primary schools. In 1995, 93·9% of primary schools were government-aided and 6·1% private. There were 1,000,580 students and 54,267 teaching staff at secondary schools in 2007. In 1995 there were 13,174 students in 94 primary teacher training colleges; 13,360 students in 24 technical institutes and colleges; 22,703 students in ten national teachers' colleges; 1,628 students in five colleges of commerce; 504 students in the Uganda Polytechnic, Kyambogo; 800 students in the National College of Business Studies, Nakawa. In 1995–96 there was one university and one university of science and technology in the public sector, and one Christian, one Roman Catholic and one Islamic university in the private sector. In 2008–09 there were 124,000 students in tertiary education and 4,000 academic staff. The adult literacy rate was 71·4% in 2006 (81·4% among males and 62·1% among females).

School attendance has trebled since Yoweri Museveni became president in 1986. In 1997 free primary education was introduced, initially for four children in every family but from 2003 for all children. In 2008–09 public expenditure on education came to 3·2% of GDP and 15·0% of total government spending.

Health

In 2001 there were 946 health centres (189 private) and 104 hospitals (49 private). In 2002 there were 1,175 physicians, 75 dentists, 1,350 nurses and 850 midwives. Uganda has been one of the most successful African countries in the fight against AIDS. A climate of free debate, with President Museveni recognizing the threat as early as 1986 and making every government department take the problem seriously, resulted in HIV prevalence among adults declining from approximately 30% in 1992 to 11% in 2000.

RELIGION

In 2001 there were 10·05m. Roman Catholics, 9·45m. Anglicans and 1·25m. Muslims. In Feb. 2013 there was one Roman Catholic cardinal. Traditional beliefs are also widespread.

CULTURE

World Heritage Sites

Uganda has three sites on the UNESCO World Heritage List: Bwindi Impenetrable National Park (inscribed on the list in 1994); Rwenzori Mountains National Park (1994); and the Tombs of the Buganda Kings at Kasubi (2001).

Press

There were five daily newspapers in 2008 with a combined average daily circulation of 110,000.

Tourism

In 2005 there were 468,000 non-resident tourists; spending by tourists totalled US$357m.

Festivals

The main festivals are for Islamic holidays (March and June), Martyrs' Day (3 June), Heroes' Day (9 June) and Independence Day (9 Oct.).

DIPLOMATIC REPRESENTATIVES

Of Uganda in the United Kingdom (Uganda House, 58/59 Trafalgar Square, London, WC2N 5DX)
High Commissioner: Joan Kakima Nyakatuura Rwabyomere.

Of the United Kingdom in Uganda (4 Windsor Loop, PO Box 7070, Kampala)
High Commissioner: Alison Blackburne.

Of Uganda in the USA (5911 16th St., NW, Washington, D.C., 20011)
Ambassador: Perezi Kamunanwire.

Of the USA in Uganda (1577 Ggaba Rd, Kampala)
Ambassador: Scott DeLisi.

Of Uganda to the United Nations
Ambassador: Vacant.
Chargé d'Affaires a.i.: Adonia Ayebare.

Of Uganda to the European Union
Ambassador: Stephen T. Kapimpina Katenta-Apuli.

FURTHER READING

Museveni, Y., *What is Africa's Problem?* 1993.—*The Mustard Seed.* 1997
Mutibwa, P., *Uganda since Independence: a Story of Unfulfilled Hopes.* 1992
Ofcansky, Thomas P., *Uganda: Tarnished Pearl of Africa.* 1999

National Statistical Office: Uganda Bureau of Statistics, P. O. Box 7186, Kampala.
Website: http://www.ubos.org

UKRAINE

Ukraina

Capital: Kyiv (formerly Kiev)
Population projection, 2015: 44·22m.
GNI per capita, 2011: (PPP$) 6,175
HDI/world rank: 0·729/76
Internet domain extension: .ua

KEY HISTORICAL EVENTS

Kyiv (formerly Kiev) was the centre of the Rus principality in the 11th and 12th centuries and is still known as the Mother of Russian cities. The western Ukraine principality of Galicia was annexed by Poland in the 14th century. At about the same time, Kyiv and the Ukrainian principality of Volhynia were conquered by Lithuania before being absorbed by Poland. Poland, however, could not subjugate the Ukrainian cossacks, who allied themselves with Russia. Ukraine, except for Galicia (part of the Austrian Empire, 1772–1919), was incorporated into the Russian Empire after the second partition of Poland in 1793.

In 1917, following the Bolshevik revolution, the Ukrainians in Russia established an independent republic. Austrian Ukraine proclaimed itself a republic in 1918 and was federated with its Russian counterpart. The Allies ignored Ukrainian claims to Galicia, however, and in 1918 awarded that area to Poland. From 1922 to 1932, drastic efforts were made by the USSR to suppress Ukrainian nationalism. Ukraine suffered from the forced collectivization of agriculture and the expropriation of foodstuffs; the result was the famine of 1932–33 when more than 7m. people died. Following the Soviet seizure of eastern Poland in Sept. 1939, Polish Galicia was incorporated into the Ukrainian SSR. When the Germans invaded Ukraine in 1941 hopes that an autonomous or independent Ukrainian republic would be set up under German protection were disappointed. Ukraine was retaken by the USSR in 1944. The Crimean region was joined to Ukraine in 1954.

On 5 Dec. 1991 the Supreme Soviet declared Ukraine's independence. Ukraine was one of the founder members of the Commonwealth of Independent States in Dec. 1991. After independence Crimea, which was part of Russia until 1954, became a source of contention between Moscow and Kyiv. The Russian Supreme Soviet laid claim to the Crimean port city of Sevastopol, the home port of the 350-ship Black Sea Fleet, despite an agreement to divide the fleet. There was also conflict between Ukraine and Russia over possession and transfer of nuclear weapons, delivery of Russian fuel to Ukraine and military and political integration within the CIS. Leonid Kuchma was elected president in 1994 and re-elected in 1999. Support for him fell after public demonstrations against maladministration including the accusation that he was responsible for the murder of a radical journalist. Conflicts between the presidential administration and government led to the sacking of reform-minded prime minister Viktor Yushchenko in April 2001, who was replaced by Kuchma loyalist Anatolii Kinakh at the end of May.

The Pope's historic visit to Ukraine in June 2001 was accompanied by disturbances, particularly in the capital. Presidential elections in Oct. and Nov. 2004 were won by Kuchma's chosen successor, Viktor Yanukovych, who defeated Viktor Yushchenko in the second round run-off. But observers claimed the election failed to meet democratic standards and in Kyiv widespread protests came to be known as the 'Orange Revolution'. After the poll was declared invalid Yushchenko was elected president in a repeat of the run-off. However, infighting between the leaders of the revolution led to growing popular discontent. In Feb. 2010 Yanukovych was elected president, defeating Yuliya Tymoshenko, a figurehead of the 2004 protests.

TERRITORY AND POPULATION

Ukraine is bounded in the east by the Russian Federation, north by Belarus, west by Poland, Slovakia, Hungary, Romania and Moldova, and south by the Black Sea and Sea of Azov. Area, 603,628 sq. km (233,062 sq. miles). In 2001 the census population was 48,457,102, of whom 26,015,758 were female; Jan. 2012 estimate, 45,633,637, giving a density of 76 per sq. km. In 2001, 78% of the population were Ukrainians, 17% Russians and 5% others—Belarusians, Moldovans, Hungarians, Bulgarians, Poles and Crimean Tatars (most of the Tatars were forcibly transported to Central Asia in 1944 for anti-Soviet activities during the Second World War). Ukraine's population is projected to drop to 41·62m. by 2025, a reduction of 8% in 15 years. In 2011, 69·1% of the population lived in urban areas.

The UN gives a projected population for 2015 of 44·22m.

Ukraine is divided into 24 provinces, two municipalities (Kyiv and Simferopol) and the Autonomous Republic of Crimea. Area and estimated population in Jan. 2012:

	Area (sq. km)	Population
Cherkaska	20,900	1,277,303
Chernihivska	31,865	1,088,509
Chernivetska	8,097	905,264
Crimea	26,081	1,963,008
Dnipropetrovska	31,974	3,320,299
Donetska	26,517	4,403,178
Ivano-Frankivska	13,928	1,380,128
Kharkivska	31,415	2,742,180
Khersonska	28,461	1,083,367
Khmelnitska	20,645	1,320,171
Kirovohradska	24,588	1,002,420
Kyiv	839	2,814,258
Kyivska	28,131	1,719,558
Luhanska	26,684	2,272,676
Lvivska	21,833	2,540,938
Mykolaïvska	24,598	1,178,223
Odeska	33,310	2,388,297
Poltavska	28,748	1,477,195
Rivnenska	20,047	1,154,256

	Area (sq. km)	Population
Sevastopol	864	381,234
Sumska	23,834	1,152,333
Ternopilska	13,823	1,080,431
Vinnytska	26,513	1,634,187
Volynska	20,144	1,038,598
Zakarpatska	12,777	1,250,759
Zaporizhska	27,180	1,791,668
Zhytomyrska	29,832	1,273,199

The capital is Kyiv (estimated population 2,814,258 in Jan. 2012). Other towns with Jan. 2012 estimated populations over 0·2m. are:

	Population		Population
Kharkiv	1,441,362	Chernihiv	296,723
Odesa	1,008,162	Cherkasy	286,163
Dnipropetrovsk	999,577	Zhytomyr	271,895
Donetsk	955,041	Sumy	269,663
Zaporizhzhya	772,627	Khmelnitsky	263,703
Lviv	729,842	Horlivka	258,879
Kryvy Rih	660,203	Chernivtsi	255,929
Mykolaïv	497,032	Rivne	250,174
Mariupol	464,457	Dniprodzerzhynsk	242,646
Luhansk	427,187	Kirovohrad	234,919
Vinnytsya	370,814	Kremenchuk	226,434
Makiïvka	356,118	Ivano-Frankivsk	224,660
Sevastopol	340,559	Ternopil	217,300
Simferopol	335,582	Lutsk	213,063
Kherson	300,666	Bila Tserkva	210,551
Poltava	297,589		

The 1996 constitution made Ukrainian the sole official language. Russian (the language of 33% of the population), Romanian, Polish and Hungarian are also spoken. Additionally, the 1996 constitution abolished dual citizenship, previously available if there was a treaty with the other country (there was no such treaty with Russia). Anyone resident in Ukraine since 1991 may be naturalized.

SOCIAL STATISTICS

2009 births, 512,525; deaths, 706,739; marriages, 318,198; divorces, 145,439. Rates (per 1,000 population), 2009: births, 11·1; deaths, 15·3. Annual population growth rate, 2000–05, –0·9%. Life expectancy, 2007: males, 62·7 years, females, 73·8. In 2006 the most popular age range for marrying was 20–24 for both males and females. Infant mortality, 2010, 11 per 1,000 live births; fertility rate, 2008, 1·3 births per woman (one of the lowest rates of any country).

CLIMATE

Temperate continental with a subtropical Mediterranean climate prevalent on the southern portions of the Crimean Peninsula. The average monthly temperature in winter ranges from 17·6°F to 35·6°F (–8°C to 2°C), while summer temperatures average 62·6°F to 77°F (17°C to 25°C). The Black Sea coast is subject to freezing, and no Ukrainian port is permanently ice-free. Precipitation generally decreases from north to south; it is greatest in the Carpathians where it exceeds more than 58·5" (1,500 mm) per year, and least in the coastal lowlands of the Black Sea where it averages less than 11·7" (300 mm) per year.

CONSTITUTION AND GOVERNMENT

In a referendum on 1 Dec. 1991, 90·3% of votes cast were in favour of independence. Turnout was 83·7%.

A new constitution was adopted on 28 June 1996. It defines Ukraine as a sovereign, democratic, unitary state governed by the rule of law and guaranteeing civil rights. The head of state is the *President*, elected directly by the people for a five-year term. An amendment to the constitution that came into effect on 1 Jan. 2006 gives increased powers to parliament, including the right to appoint and dismiss the prime minister. However, after

parliament dismissed the prime minister and the cabinet on 10 Jan. 2006 President Yushchenko stated that only the new parliament that was to be elected in March 2006 would have such powers.

Parliament is the 450-member unicameral *Verkhovna Rada* (*Supreme Council*), elected for four-year terms. Prior to the March 2006 election half of the members were chosen from party lists by proportional vote and half from individual constituencies, but in accordance with a constitutional amendment for the 2006 election all 450 members were chosen from party lists. The same method applied for the elections in 2007 but for the 2012 election it was decided to revert to the pre-2006 mixed voting system.

There is an 18-member *Constitutional Court*, six members being appointed by the President, six by parliament and six by a panel of judges. Constitutional amendments may be initiated at the President's request to parliament, or by at least one-third of parliamentary deputies. The Communist Party was officially banned in the country in 1991, but was renamed the Socialist Party of Ukraine. Hard-line Communists protested against the ban, which was rescinded by the Supreme Council in May 1993.

National Anthem

'Shche ne vmerla, Ukraïny i slava, i volya' ('Ukraine's freedom and glory has not yet perished'); words by P. Chubynsky, tune by M. Verbytsky.

GOVERNMENT CHRONOLOGY

Presidents since 1991.

1991–94	Leonid Makarovich Kravchuk
1994–2005	Leonid Danylovich Kuchma
2005–10	Viktor Andriyovich Yushchenko
2010–	Viktor Fedorovych Yanukovych

Prime Ministers since 1990.

1990–92	Vitold Pavlovich Fokin
1992	Valentyn Kostyantynovich Symonenko
1992–93	Leonid Danylovich Kuchma
1994–95	Vitaliy Anriyovich Masol
1995–96	Yevhen Kyrylovich Marchuk
1996–97	Pavlo Ivanovich Lazarenko
1997–99	Valeriy Pavlovich Pustovoytenko
1999–2001	Viktor Andriyovich Yushchenko
2001–02	Anatolii Kyrylovich Kinakh
2002–05	Viktor Fedorovych Yanukovych
2005	Yuliya Volodymyrivna Tymoshenko
2005–06	Yuriy Ivanovich Yekhanurov
2006–07	Viktor Fedorovych Yanukovych
2007–10	Yuliya Volodymyrivna Tymoshenko
2010–	Mykola Yanovych Azarov

RECENT ELECTIONS

Presidential elections were held in two rounds on 17 Jan. and 7 Feb. 2010. In the first round Viktor Yanukovych won 35·3% of the vote against 25·1% for prime minister Yuliya Tymoshenko, 13·1% for Sergei Tigipko, 7·0% for Arseniy Yatsenyuk and 5·5% for incumbent President Viktor Yushchenko. There were 13 other candidates. Turnout was 66·8%. Yanukovych won the second round with 49·0% against 45·5% for Tymoshenko. However, the results were suspended after Tymoshenko accused Yanukovych of electoral fraud prompting an inquiry. She withdrew her appeal on 20 Feb. but did not attend Yanukovych's inauguration on 25 Feb. and refused to recognize his position. On 3 March, Tymoshenko's government lost a parliamentary vote of confidence and she was ousted as prime minister.

In parliamentary elections held on 28 Oct. 2012 the Party of Regions of President Viktor Yanukovych won 187 seats with 30·0% of the vote under the proportional party-list system, the All-Ukrainian Union 'Fatherland' of Yuliya Tymoshenko 102 seats (with 25·5%), the Ukrainian Democratic Alliance for Reform of WBC boxing world heavyweight champion Vitali Klitschko 40

(14·0%), the far-right Svoboda 38 (10·4%) and the Communist Party of Ukraine 32 (13·2%). Four other parties gained three seats or fewer and 44 went to independents. Turnout was 58·0%.

CURRENT GOVERNMENT

President: Viktor Yanukovych; b. 1950 (ind.; sworn in 25 Feb. 2010).

In March 2013 the government comprised:

Prime Minister: Mykola Azarov; b. 1947 (Party of Regions; sworn in 11 March 2010).

First Deputy Prime Minister: Serhiy Arbuzov. *Deputy Prime Ministers:* Yuriy Boyko; Kostyantyn Hryschenko; Oleksandr Vilkul.

Minister of Agrarian Policy and Food: Mykola Prysiazhniuk. *Culture:* Leonid Novohatko. *Defence:* Pavlo Lebedev. *Economic Development and Trade:* Ihor Prasolov. *Education, Science, Youth and Sport:* Dmytro Tabachnyk. *Energy and Coal:* Eduard Stavytsky. *Environment and Natural Resources:* Oleh Proskuryakov. *Finance:* Yuriy Kolobov. *Foreign Affairs:* Leonid Kozhara. *Health:* Raisa Bohatyriova. *Industrial Policy:* Mykhaylo Korolenko. *Infrastructure:* Volodymyr Kozak. *Interior:* Vitaliy Zakharchenko. *Justice:* Oleksandr Lavrynovych. *Regional Development, Construction, Housing and Communal Services:* Hennady Temnyk. *Revenues and Duties:* Oleksander Klymenko. *Social Policy:* Natalia Korolevska. *Head of the State Emergency Services:* Mykhaylo Bolotsky.

Government Website: http://www.kmu.gov.ua

CURRENT LEADERS

Viktor Yanukovych

Position
President

Introduction
Having served as prime minister under President Kuchma from 2002–05, Viktor Yanukovych was briefly declared president after the bitterly contested 2004 election. When the result was annulled he lost the re-run and resigned as prime minister. In March 2006 his Party of Regions won the largest number of seats in parliament and in Aug. 2006 he became prime minister in a coalition government. His power base is in eastern Ukraine, where he has strong links with industrialists, and he favours close ties between Ukraine and Russia. On 18 Dec. 2007 Yanukovych was formally dismissed by parliament. In 2010 he fought a bitterly contested presidential election against Yuliya Tymoshenko, narrowly defeating her in the second round of voting. Tymoshenko contested the results, claiming electoral misconduct, but Yanukovych was sworn in as president on 25 Feb.

Early Life
Viktor Yanukovych was born on 9 July 1950 in Yenakiyeve, Donetsk Oblast, in Russian-speaking eastern Ukraine. His mother, an ethnic Ukrainian nurse, died when he was two, and his father, an ethnic Belarusian train driver, died when he was in his teens, leaving him in the care of his grandmother. He served a prison sentence in 1967 for robbery and another in 1970 for bodily injury, although he claims to have been later cleared of both crimes. In 1972 he began working in the Donetsk coal industry and completed his education, graduating in mechanical engineering from Donetsk Polytechnic Institute, in 1980. He joined the Communist Party and rose rapidly as a manager in Donetsk regional transport.

He entered politics in Aug. 1996 as deputy head of Donetsk Oblast administration and was appointed head in May 1997. From May 1999–May 2001 he was head of the Donetsk Oblast regional council and became closely associated with a group of business and political figures known as the 'Clan of Donetsk', led by the coal and steel oligarch Rinat Akhmetov. His lobbying for them brought him strong political and financial support but also

fuelled rumours of links to organized crime. In Nov. 2002 he was appointed prime minister by President Leonid Kuchma.

Yanukovych oversaw the continuing liberalization of the economy, cutting higher rates of income tax and encouraging land privatization. He often favoured the interests of Ukrainian industrialists over international investors, helping to power the domestic economy but leading to allegations of corruption. He maintained strong links with Russia and spoke against Ukraine joining the European Union and NATO. In 2003 his government signed an agreement to take Ukraine into a free trade zone and customs alliance with Russia, Belarus and Kazakhstan, although negotiations subsequently stalled. Kuchma chose not to fight the 2004 presidential election and Yanukovych stood as his successor, openly supported by Russia's President Putin.

Yanukovych lost the first round to Viktor Yushchenko of the pro-west, liberal Our Ukraine party but won the second round on 21 Nov. 2004. The result was challenged and Yushchenko's supporters staged huge street protests dubbed the 'Orange Revolution'. The Supreme Court annulled the result and a re-run of the second round saw Yushchenko triumph. Following a parliamentary vote of no confidence in his government, Yanukovych resigned as prime minister in Jan. 2005.

Yanukovych spent 2005 building on his grassroots support in eastern Ukraine. He profited from disillusionment with Yushchenko's government, which was divided and indecisive. Yanukovych's Party of Regions emerged as the largest party at the March 2006 parliamentary elections with 186 seats out of 450, ahead of the Tymoshenko Bloc and Our Ukraine. However, with no overall parliamentary majority, months of wrangling over a new government ensued before Yushchenko was forced to nominate Yanukovych as prime minister in Aug. 2006. Yanukovych agreed a coalition deal with Yushchenko's party, but the working relationship was uneasy from the outset. In Dec. 2007, following further elections in Sept., he was replaced by Tymoshenko as prime minister of a new coalition between her supporters and Yushchenko's Our Ukraine.

Career in Office
Yanukovych regained the presidency in Feb. 2010 after comprehensively defeating Yushchenko in the first round of elections and then prime minister Tymoshenko in the second, winning by a 3·5% margin. Tymoshenko accused him of vote-rigging and mounted a legal challenge against the results. However, she subsequently dropped her action stating that she would not receive a fair hearing. She and her party nevertheless refused to recognize Yanukovych's election and boycotted the inauguration ceremony. In March Yanukovych appointed a longstanding ally, Mykola Azarov, to succeed Tymoshenko as prime minister. The following month parliament agreed to extend the lease of the Russian Black Sea naval base in Crimea for 25 years in exchange for accords on cheaper gas supplies and in June voted to abandon Ukraine's NATO membership aspirations.

In Oct. 2010 Yanukovych forced through constitutional changes overturning limits on presidential power that had been introduced in 2004. At the end of the year Tymoshenko and former interior minister Yuriy Lutsenko were charged with abuse of state funds while in office. Both rejected the accusation as a politically motivated manoeuvre by the Yanukovych government, but in Oct. 2011 Tymoshenko was found guilty and jailed for seven years. Her conviction prompted international condemnation, particularly from the European Union whose high representative for foreign policy warned of 'profound implications' for EU–Ukraine relations, including for Ukraine's hope of concluding a trade and association agreement.

In Oct. 2012 the Party of Regions retained power after a decisive win in parliamentary elections, although international monitors criticized the conduct of the poll. In Dec. Yanukovych appointed a reconstituted but largely unchanged government under Prime Minister Azarov.

Mykola Azarov

Position
Prime Minister

Introduction
Mykola Azarov was appointed prime minister on 11 March 2010. A close ally of President Viktor Yanukovych, his predecessor as leader of the Party of Regions, Azarov is regarded as a technocrat. Ethnically Russian, he is a veteran of the country's turbulent political system, having previously served as deputy prime minister, foreign minister and, on two occasions, acting prime minister.

Early Life
Mykola Azarov was born on 17 Dec. 1947 in the town of Kaluga, in what is now Russia. He studied geology at Moscow State University before working at a coal mine in the Russian city of Tula. In 1984 he moved to Donetsk to serve first as deputy director and later as director of the Ukrainian State Geological Institute. He became a member of Ukraine's parliament in 1994 and two years later was appointed head of the state tax authority.

During his six years there, Azarov had a reputation for authoritarianism and controversy. It was alleged that he promoted electoral fraud and the intimidation of journalists to secure the re-election of President Leonid Kuchma in 1999. Azarov staunchly denied the accusations, citing a plot against him by rivals.

In 2002 the European Choice parliamentary group, of which he was chairman, nominated Azarov for prime minister. He declined and stood aside for Yanukovych, who assumed both the leadership of the Party of Regions and the premiership. Yanukovych appointed Azarov as deputy prime minister and finance minister in his first cabinet, with Azarov serving until 2005 and then again in 2006–07. His most notable reform was to bring the variable tax rate on personal income down to a uniform flat rate.

During the political upheaval that accompanied the Orange Revolution over the winter of 2004–05, Azarov twice served briefly as acting prime minister. After masterminding Yanukovych's victorious presidential election campaign in 2010, he assumed the leadership of the Party of Regions and was confirmed as prime minister by parliament on 11 March 2010.

Career in Office
Azarov declared that his primary task was to restore the struggling economy. Admitting that 'the coffers are empty', he pledged to push through a programme of budgetary restraints to encourage the IMF to resume funding suspended in 2009. However, in Nov. 2011 the IMF postponed loan instalment negotiations because of a lack of progress by the Azarov government on pension reform and the removal of domestic gas subsidies.

Criticized by opponents for his poor grasp of the Ukrainian language, Azarov promised that government affairs would be conducted in Ukrainian.

In foreign affairs he has sought to improve strained relations with Moscow.

Following the Party of Regions' victory in parliamentary elections in Oct. 2012, President Yanukovych nominated Azarov in Dec. for a new term as prime minister of a reconstituted government.

DEFENCE

The 1996 constitution bans the stationing of foreign troops on Ukrainian soil, but permits Russia to retain naval bases. Ukraine hosts Russia's Black Sea Fleet at Sevastopol in Crimea under a lease that is scheduled to end in 2042 (having been extended by 25 years in April 2010). Conscription is for 12 months. Although the former government announced its intention to end conscription and move towards a professional military, this process is now unlikely to begin until at least 2015. On 31 May 1997 the presidents of Ukraine and Russia signed a Treaty of Friendship and Co-operation which provided inter alia for the division of the former Soviet Black Sea Fleet and shore installations. There were around 1m. armed forces reserves in 2007.

Military expenditure in 2008 totalled US$1,804m. (US$39 per capita), representing 1·0% of GDP.

Army
In 2007 ground forces numbered 70,753. Equipment included 2,984 main battle tanks (T-55s, T-64s, T-72s, T-80s and T-84s) and 139 attack helicopters.

In addition there were around 39,900 Ministry of Internal Affairs troops, 45,000 Border Guards and some 9,500 civil defence troops.

Navy
In 2007 the Navy numbered 13,932, including some 2,500 Naval Aviation and 3,000 naval infantry. The main base is located at Sevastopol. The operational forces include one frigate.

The aviation forces of the former Soviet Black Sea Fleet under Ukrainian command operate ten combat-capable aircraft and 77 helicopters.

Air Force
There are three air commands—West, South and Central—plus a Task Force 'Crimea'.

Equipment includes 211 combat-capable aircraft (MiG-29s, Su-24s, Su-25s and Su-27s).

Personnel, 2007, 45,240.

ECONOMY

In 2006 agriculture accounted for 8·6% of GDP, industry 34·1% and services 57·3%.

Overview
Following independence in 1991, Ukraine's transition to a market economy was tumultuous. The economy was in recession for much of the 1990s, losing 60% of GDP between 1991 and 1999. Prices stabilized after the introduction of a new currency, the hryvnia, in 1996 and structural reforms helped bring the economy under control in 2000. Despite a sharp slowdown in 2005 following the political upheaval of the 2004 presidential elections, economic growth averaged 7·0% between 2001 and 2008 (peaking at 12·1% in 2004), the second highest rate in Europe after that of Latvia.

While the economy has proved resilient in the face of domestic political uncertainties and rising energy prices, global financial turmoil sparked a withdrawal of investment and a collapse in steel prices. Amid forecasts of an imminent recession, the IMF approved a US$16·4bn. loan in Oct. 2008. Nonetheless, in 2009 GDP shrank by nearly 15%, reflecting the collapse of steel output and declines in the construction and retail sectors. Beginning in Oct. 2008 the hryvnia lost 40% of its value against the US dollar. Economic growth resumed in 2010 following moderate improvements in external demand but significant challenges remain. In Aug. 2010 the IMF approved a US$15·1bn. loan to put the country on the path to fiscal sustainability, restructure the gas sector and to stabilize the banking system.

Currency
The unit of currency is the *hryvnia* (UAH) of 100 *kopiykas*, which replaced karbovanets on 2 Sept. 1996 at 100,000 karbovanets = 1 hryvnia. 2000 saw the introduction of a floating exchange rate for the hryvnia.

Inflation rates (based on IMF statistics):

2003	2004	2005	2006	2007	2008	2009	2010	2011
5·2%	9·0%	13·5%	9·1%	12·8%	25·2%	15·9%	9·4%	8·0%

Inflation had been 4,735% in 1993. Foreign exchange reserves in Sept. 2009 were US$25,189m., gold reserves were 864,000 troy oz and total money supply was 221,530m. hryvnias.

Budget
In 2008 revenues were 229,598m. hryvnias and expenditures 241,490m. hryvnias. Tax revenue accounted for 73·8% of revenues in 2008; public services accounted for 33·4% of expenditures in 2008, social protection 22·0%, and education and health 12·4%.

VAT is 20%.

Performance
Real GDP growth rates (based on IMF statistics):

2003	2004	2005	2006	2007	2008	2009	2010	2011
9·6%	12·1%	2·7%	7·3%	7·9%	2·3%	–14·8%	4·1%	5·2%

Between 1994 and 1998 average annual real GDP growth was –10·0%, and it was still negative in 1999, at –0·2%. In 2000 the economy expanded considerably and this resurgence continued in the following years with per capita income more than tripling in the period 2000–06. Ukraine's economy did, however, experience a major downturn again in the wake of the global financial crisis that started in 2008. Ukraine's total GDP in 2011 was US$165·2bn.

Banking and Finance
A National Bank was founded in March 1991. It operates under government control, its Governor being appointed by the President with the approval of parliament. The *Governor* is Igor Sorkin. There were 198 banks in all as at 1 Jan. 2012, with assets totalling 1,054,280m. hryvnias. The largest banks are PrivatBank, Ukreximbank and Oschadbank.

In 2010 external debt totalled US$116,808m., equivalent to 85·9% of GNI.

There is a stock exchange in Kyiv.

ENERGY AND NATURAL RESOURCES
Environment
Carbon dioxide emissions from the consumption and flaring of fossil fuels in 2008 were the equivalent of 7·6 tonnes per capita. Ukraine's greenhouse gas emissions fell by 53·9% between 1990 and 2008, mainly owing to the decline of polluting industries from the Soviet era.

Electricity
Installed capacity was an estimated 53·9m. kW in 2007. In 2007 production was 196·25bn. kWh; consumption per capita in 2007 was 4,049 kWh. A Soviet programme to greatly expand nuclear power-generating capacity in the country was abandoned in the wake of the 1986 accident at Chernobyl. Chernobyl was closed down on 15 Dec. 2000. In 2007 there were 15 nuclear reactors in use supplying 47·5% of output.

Oil and Gas
In 2007 output of crude petroleum was 3·3m. tonnes; in 2008 production of natural gas was 18·7bn. cu. metres, with 920bn. cu. metres of proven natural gas reserves.

Minerals
Ukraine's industrial economy, accounting for more than a quarter of total employment, is based largely on the republic's vast mineral resources. The Donetsk Basin contains huge reserves of coal, and the nearby iron-ore reserves of Kryvy Rih are equally rich. Among Ukraine's other mineral resources are manganese, bauxite, nickel, titanium and salt. Production in 2004 (in 1m. tonnes): iron ore, 65·5m.; coal, 59·2m.; manganese, 2·4m.; salt, 2·3m.

Agriculture
Ukraine has extremely fertile black-earth soils in the central and southern portions, totalling nearly two-thirds of the territory. In 2002 there were 32·54m. ha. of arable land and 0·91m. ha. of permanent crops. Output (in 1,000 tonnes) in 2002: wheat, 20,556; potatoes, 16,620; sugar beets, 14,452; barley, 10,364; maize,

4,180; sunflower seeds, 3,270; rye, 1,509; cabbage, 1,229; tomatoes, 1,038. Livestock, 2003: 9,203,000 pigs, 9,108,000 cattle, 1,034,000 goats, 950,000 sheep, 148m. chickens, 20m. ducks. Livestock products, 2003 (in 1,000 tonnes): milk, 13,660; meat, 1,649; eggs, 656.

Forestry
The area under forests in Ukraine in 2010 was 9·71m. ha. (17% of the total land area). In 2007, 16·88m. cu. metres of timber were produced.

Fisheries
In 2010 the catch totalled 186,021 tonnes, of which 181,381 tonnes were from sea fishing. The total catch in 1988 had been 1,048,157 tonnes.

INDUSTRY
In 2007 industry accounted for 28·2% of GDP, with manufacturing contributing 20·5%. Industrial production grew by 10·2% in 2007. Output, 2007 (in tonnes unless otherwise stated): pig iron, 35·6m.; crude steel, 29·0m.; rolled ferrous metals, 24·5m.; cement, 15·0m.; petrol, 4·2m.; distillate fuel oil, 4·1m.; residual fuel oil, 3·4m.; mineral fertilizer, 2·8m.; bread and bakery products, 2·0m.; sugar, 1·9m.; sulphuric acid, 1·6m.; fabrics, 114m. sq. metres; footwear, 22·5m. pairs; refrigerators, 824,000 units; television sets, 507,000 units; passenger cars, 380,000 units; washing machines, 173,000 units; cigarettes, 129bn. units.

Labour
In 2011 a total of 20,324,000 persons aged 15 to 70 were in employment. The principal areas of activity were (in 1,000): wholesale and retail trade, and restaurants and hotels, 4,865; agriculture, hunting, forestry and fishing, 3,394; manufacturing, 3,353. In 2011 there were 1,733,000 unemployed and the level of unemployment was 7·9%.

INTERNATIONAL TRADE
Imports and Exports
In 2008 imports (c.i.f.) amounted to US$85,534m. (US$60,618m. in 2007); exports (f.o.b.) US$67,003m. (US$49,296m. in 2007). Main import suppliers in 2004: Russia, 40·7%; Germany, 9·4%; Turkmenistan, 6·7%; Poland, 3·3%; Italy, 2·8%. Main exports markets in 2004: Russia, 18·0%; Germany, 5·8%; Turkey, 5·7%; Italy, 5·0%; USA, 4·6%. Main imports, 2004: crude petroleum, 16·7%; machinery, 16·3%; chemicals and chemical products, 12·6%; natural gas, 12·4%. Main exports: ferrous and nonferrous metals, 39·9%; food and raw materials, 10·6%; chemicals and chemical products, 9·9%; machinery, 9·3%.

COMMUNICATIONS
Roads
In 2007 there were 169,422 km of roads, including 20,497 km of national roads. There were 5,939,600 passenger cars in use in 2007 and 714,300 motorcycles and mopeds. There were 63,554 road accidents involving injury in 2007 (9,574 fatalities).

Rail
Total length was 22,302 km in 2009. Passenger-km travelled in 2009 came to 48·3bn. and freight tonne-km to 196·2bn. There are metros in Kyiv, Kharkiv, Kryvy Rih and Dnipropetrovsk.

Civil Aviation
The main international airport is Kyiv (Boryspil), and there are international flights from seven other airports. There are two major Ukrainian carriers. Ukraine International Airlines operated international flights in 2005 to Amsterdam, Barcelona, Berlin, Brussels, Dubai, Düsseldorf, Helsinki, Kuwait, Lisbon, London, Madrid, Paris, Rome, Vienna and Zürich. Aerosvit had international flights in 2005 to Ashgabat, Athens, Baku, Bangkok, Beijing, Belgrade, Birmingham, Budapest, Cairo, Delhi, Dubai,

Hamburg, İstanbul, Larnaca, Moscow, New York, Prague, St Petersburg, Sofia, Stockholm, Tel Aviv, Toronto and Warsaw.

In 2001 Kyiv handled 1,517,130 passengers (1,458,524 on international flights) and 11,875 tonnes of freight. Simferopol was the second busiest airport for passenger traffic, with 325,323 passengers (248,574 on international flights), and Odesa the second busiest for freight, with 3,588 tonnes.

Shipping

In 2007, 2m. passengers and 15m. tonnes of freight were carried by inland waterways. In Jan. 2009 there were 224 ships of 300 GT or over registered, totalling 770,000 GT. The main seaports are Illichivsk, Izmail, Mariupol, Mykolaïv, Odesa and Yuzhny. Odesa is the leading port, in 2006 handling 28,010,000 tonnes of freight (23,674,000 tonnes loaded and 4,336,000 tonnes discharged).

Telecommunications

In 2008 there were 13,177,000 main (fixed) telephone lines. In the same year mobile phone subscribers numbered 55,695,000 (1,210·9 per 1,000 persons). Ukraine had 6·4m. internet subscribers in 2007, of which 0·8m. were broadband subscribers. In March 2012 there were 1·7m. Facebook users.

SOCIAL INSTITUTIONS

Justice

A new civil code was voted into law in June 1997. Justice is administered by the Constitutional Court of Ukraine and by courts of general jurisdiction. The Supreme Court of Ukraine is the highest judicial organ of general jurisdiction. The death penalty was abolished in 1999. Over the period 1991–95, 642 death sentences were awarded and 442 carried out; there were 169 executions in 1996. In March 1997 death penalties were still being awarded but not carried out. 553,994 crimes were reported in 2000.

The population in penal institutions in Jan. 2008 was 149,690 (323 per 100,000 of national population).

Education

In 2003–04 the number of pupils in 21,900 primary and secondary schools was 5·9m.; 339 further education establishments had 1,843,800 students, and 670 technical colleges had 592,900 students; 977,000 children were attending pre-school institutions.

In 2005–06 there were 16 universities including an international university of information systems, management and business.

Adult literacy rate in 2008 was over 99%.

In 2006 public expenditure on education came to 6·2% of GDP and 19·3% of total government spending.

Health

In 2007 there were 223,294 physicians, 25,450 dentists, 340,986 nurses, 23,645 midwives and 21,745 pharmacists. There were 439,549 beds in 2,843 hospitals in 2007.

Welfare

There were 10·6m. old-age pensioners as at 1 Jan. 2012 and 3·2m. other pensioners. The total included 881,636 Chernobyl victims. In 2002 social insurance and pension security programmes totalled 27·4bn. hryvnias, representing 12·4% of GDP.

RELIGION

The majority faith is the Orthodox Church, which is split into three factions. The largest is the Ukrainian Orthodox Church, Moscow Patriarchate (the former exarchate of the Russian Orthodox Church), headed by Metropolitan Volodymyr (Sabodan), Metropolitan of Kyiv and All Ukraine, which recognizes Kirill I (Vladimir Mikhailovich Gundyayev) as Patriarch of Moscow and All Russia and insists that all Ukrainian churches should be under Moscow's jurisdiction. There were 9·5m. adherents in 2001. The second largest is the Ukrainian

Orthodox Church, Kyivan Patriarchate, headed by Metropolitan Filaret (Denysenko), Patriarch of Kyiv and All Rus-Ukraine, which was created in June 1992. It had 4·8m. adherents in 2001. Metropolitan Filaret was excommunicated by the Ukrainian Orthodox Church, Moscow Patriarchate in Feb. 1997. The third faction is the Ukrainian Autocephalous Orthodox Church, headed by Metropolitan Mefodiy (Kudryakov) of Ternopil, which favours the unification of the three bodies. Only the Ukrainian Orthodox Church, Moscow Patriarchate is in communion with world Orthodoxy.

The hierarchy of the Roman Catholic Church (*Primate*, Cardinal Mieczysław Mokrzycki, Archbishop Metropolitan of Lviv) was restored by the Pope's confirmation of ten bishops in Jan. 1991. In Feb. 2013 there were two cardinals. The Ukrainian Greek Catholic Church (*Head*, Bishop Sviatoslav Shevchuk, Major Archbishop, Metropolitan of Kyiv-Halych) is a Church of the Byzantine rite, which is in full communion with the Roman Church. Catholicism is strong in the western half of the country.

CULTURE

World Heritage Sites

Ukraine has five sites on the UNESCO World Heritage List: Kyiv—Saint Sophia Cathedral and Related Monastic Buildings, Kyiv—Pechersk Lavra (inscribed on the list in 1990 and 2005); Lviv—Ensemble of the Historic Centre (1998 and 2008); the Struve Geodetic Arc (2005), a chain of survey triangulations spanning from Norway to the Black Sea that helped establish the exact shape and size of the earth, which is shared with nine other countries; the primeval beech forests of the Carpathians (2007), shared with Slovakia; and the Residence of Bukovinian and Dalmatian Metropolitans (2011), a property built from 1864–82.

Press

In 2006 there were 41 daily newspapers with an average combined circulation of 3,511,000. The newspapers with the highest circulation figures are *Fakty i Kommentarii* and *Komsomolskaja Pravda v Ukraine*.

Tourism

There were 23,122,000 non-resident tourists in 2007; total receipts were US$5,317m.

Festivals

The arts festival Day of Kyiv occurs on the last weekend of May. The main music festivals are KaZantip Republic at Popovka in Crimea (Aug.), Koktebel Jazz Festival (Sept.) and Jazz Carnival in Odesa (Sept.). The Molodist International Film Festival takes place in Kyiv in Oct.

DIPLOMATIC REPRESENTATIVES

Of Ukraine in the United Kingdom (60 Holland Park, London, W11 3SJ)
Ambassador: Volodymyr Khandogiy.

Of the United Kingdom in Ukraine (9 Desyatinna St., 01025 Kyiv)
Ambassador: Simon Smith.

Of Ukraine in the USA (3350 M St., NW, Washington, D.C., 20007)
Ambassador: Olexander Motsyk.

Of the USA in Ukraine (4 A. I. Sikorsky St., 04112 Kyiv)
Ambassador: John F. Tefft.

Of Ukraine to the United Nations
Ambassador: Yuriy Serheyev.

Of Ukraine to the European Union
Ambassador: Kostiantyn Yelisieiev.

FURTHER READING

Encyclopedia of Ukraine, 5 vols. 1984–93

Aslund, Anders, *Revolution in Orange: The Origins of Ukraine's Democratic Breakthrough.* 2006

D'Anieri, Paul, *Economic Interdependence in Ukrainian–Russian Relations.* 2000.—*Understanding Ukrainian Politics: Power, Politics and Institutional Design.* 2006

Kuzio, Taras, Kravchuk, Robert and D'Anieri, Paul, *State and Institution Building in Ukraine.* 2000

Magocsi, P. R., *A History of Ukraine.* 1997

Motyl, A. J., *Dilemmas of Independence: Ukraine after Totalitarianism.* 1993

Nahaylo, B., *Ukrainian Resurgence.* 2nd ed. 2000

Reid, A., *Borderland: A Journey Through the History of Ukraine.* 1997

Wilson, Andrew, *The Ukrainians: Unexpected Nation.* 2000.—*Ukraine's Orange Revolution.* 2006

National Statistical Office: State Committee of Statistics of Ukraine, 3 Shota Rustavely St., Kyiv 01023.

Website: http://www.ukrstat.gov.ua

Crimea

The Crimea is a peninsula extending southwards into the Black Sea with an area of 26,100 sq. km. Population (Oct. 2005), 1,990,000 (ethnic groups, 2001 census: Russians, 58·3%; Ukrainians, 24·3%; Tatars, 12·0%). The capital is Simferopol (2001 census, 363,000).

It was occupied by Tatars in 1239, conquered by Ottoman Turks in 1475 and retaken by Russia in 1783. In 1921 after the Communist revolution it became an autonomous republic, but was transformed into a province (*oblast*) of the Russian Federation in 1945, after the deportation of the Tatar population in 1944 for alleged collaboration with the German invaders in the Second World War. 46% of the total Tatar population perished during the deportation. Crimea was transferred to Ukraine in 1954 and became an autonomous republic in 1991. About half the surviving Tatar population of 0·5m. had returned from exile by 2000. The Tatar population is disproportionately disadvantaged within Crimea, with an unemployment rate of over 60% in 2000.

At elections held in two rounds on 16 and 30 Jan. 1994 Yuri Meshkov was elected *President*. The post of president was abolished by Ukraine in March 1995 after calls for a referendum on Crimean independence. Parliamentary elections were held on 26 March 2006. The Bloc For Yanukovych! (Party of Regions and the Russian Bloc) obtained 44 seats, Soiuz 10, Kunitsyn's Electoral Bloc 10, the Communist Party of Ukraine 9, the People's Movement of Ukraine 8, the Yuliya Tymoshenko Electoral Bloc 8, the People's Opposition Bloc of Natalia Vitrenko 7 and the Opposition Bloc Ne Tak 4. The *Prime Minister* is Anatoliy Mohylyov and the *Chairman of Parliament* Volodymyr Konstantynov.

On 2 Nov. 1995 parliament adopted a new constitution which defines the Crimea as 'an autonomous republic forming an integral part of Ukraine'. The status of 'autonomous republic' was confirmed by the 1996 Ukrainian Constitution, which provides for Crimea to have its own constitution as approved by its parliament. The Prime Minister is appointed by the Crimean parliament with the approval of the Ukrainian parliament.

The Tatar National Kurultay (Parliament) elects an executive board (*Mejlis*). The *Chairman* is Mustafa Jemilev. A power-sharing agreement of 12 May 2005 guarantees the Crimean Tatars two ministry portfolios and the post of deputy prime minister in the Crimean local government.

In 2004 there were 624 pre-school institutions with 38,000 pupils and 638 primary and secondary schools with 241,000 students and 20,396 teachers. There were 68,000 students studying at 35 institutes of higher education in 2004.

UNITED ARAB EMIRATES

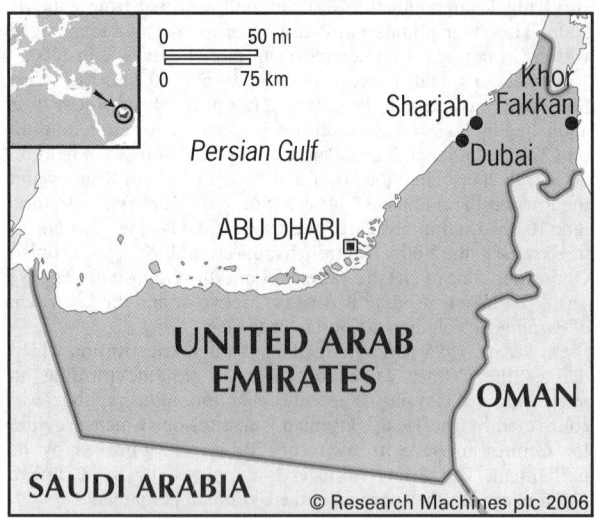

Imarat al-Arabiya al-Muttahida

Capital: Abu Dhabi
Population projection, 2015: 8·37m.
GNI per capita, 2011: (PPP$) 59,993
HDI/world rank: 0·846/30
Internet domain extension: .ae

KEY HISTORICAL EVENTS

Archaeological evidence indicates that in the 3rd millennium BC a culture known as Umm al-Nar developed in modern-day Abu Dhabi, its influence spreading inland and along the coast of Oman to the south. There was trade with both the Mesopotamian civilization and the Indus culture, particularly the export of copper (then the most valuable natural resource) from the Hajar mountains. Later settlements, with Hellenistic features and dating from between the 3rd century BC and the 3rd century AD, have been discovered at Meleiha, near the Sharjah coast, and at Al-Dur in the emirate of Umm al Qaiwain. There are indications that the coastal areas of the United Arab Emirates (UAE) and Oman came under Sassanian (Persian) influence from the 4th century AD until the early 7th century when the Islamic era began. After the death of the Prophet Muhammad, tribes in the Dibba region along the eastern coast rebelled before Islamic forces won a decisive battle in AD 632.

In the Middle Ages much of the region was part of the Persian Kingdom of Hormuz (from 1300), which controlled the approach to the Gulf and most of the trade. European intervention in the Gulf began in the early 16th century when the Portuguese established a commercial monopoly, building a number of forts including Julfar (in modern-day Ras al-Khaimah), a major medieval trading centre. Portuguese ascendancy was later challenged by the Dutch and then by the British, who exercised their naval power in the Gulf in the 18th century to protect trade with India.

By that time two major tribal confederations had grown powerful along the coast of the lower Gulf. The largest tribal grouping, the Bani Yas, was established on the coast by the late 16th century. A large Bani Yas settlement was founded in Abu Dhabi in the 1760s, following the discovery of water on Abu Dhabi island, and in 1793 it became the seat of government of the al-Nahyan branch of the tribal confederation. There is further evidence of elements of the Bani Yas population extending from Qatar in the west to Dubai in the east and inland to the Liwa oasis belt. The Qawasim, a branch of the Huwalah tribe, were a maritime people (largely operating from Ras al-Khaimah) who emerged as an important group once the Omani empire of the late 17th and early 18th centuries was destroyed at the end of the Omani-Persian war in 1720. They established control over other strands of the Huwalah between Sharjah and Musandam and with a large fleet, posed a serious challenge to British shipping.

Piracy was rife until the early 19th century when the British sent several naval expeditions to the Gulf ports to suppress the raiders. In 1820 Britain signed the General Treaty of Peace against piracy and the slave trade with the principal Arab sheikhdoms. From this and later agreements the area became known as the Trucial Coast from the 1850s (the term Trucial referring to the fact that the component sheikhdoms were bound by the truces concluded with Britain). Britain assumed responsibility for the defence of the territory under the maritime treaty of 1853 and, under the exclusive treaties of 1892, for external relations of each of the Trucial sheikhdoms. The sheikhdoms were otherwise autonomous and followed the traditional form of Arab monarchy, with each ruler having virtually absolute power over his subjects.

The collapse of the pearl market, which was believed to have originated in the Gulf in the late Stone Age, and the world economic depression of the early 20th century undermined the economies of the Trucial States. Their subsequent transformation began with the discovery of oil off the coast of Abu Dhabi. Sheikh Shakhbut bin Sultan al-Nahyan, the ruler of Abu Dhabi from 1928–66, granted the first of several oil exploration concessions to foreign companies in 1939. However, the Second World War delayed exploration. The first commercial discovery was made in the late 1950s and the first exports began in 1962.

Sheikh Shakhbut, who was seen as an obstacle to the development of the oil industry, was deposed in 1966 in favour of his younger brother Sheikh Zayed bin Sultan al-Nahyan. The president of the UAE for more than 30 years until his death in 2004, he was re-elected by the rulers of the other emirates at five-year intervals. In the late 1960s oil was discovered in Dubai and in Sharjah, and then in Ras al-Khaimah in the 1980s.

In Jan. 1968 Britain announced the withdrawal of its military forces from the area by 1971. In March the Trucial States joined Bahrain and Qatar (which were also under British protection) in what was named the Federation of Arab Emirates. It was intended that the Federation should become fully independent but the interests of Bahrain and Qatar proved to be incompatible with those of the other sheikhdoms and both seceded from the Federation in 1971 to become separate independent entities. Six of the Trucial States (Abu Dhabi, Dubai, Sharjah, Umm al Qaiwain, Ajman and Fujairah) had agreed a federal constitution for achieving independence as the United Arab Emirates. The British accordingly terminated its special treaty relationship, and the UAE became independent on 2 Dec. 1971. The remaining sheikhdom, Ras al-Khaimah, joined the UAE in Feb. 1972. At independence Sheikh Zayed of Abu Dhabi took office as the first President of the loose federation. Sheikh Rashid bin Said al-Maktoum, the ruler of Dubai for over 30 years from 1958, became Vice-President. The al-Maktoum family, like the al-Nahyan rulers of Abu Dhabi, are a dynastic line of the Bani Yas tribe. The UAE instituted the Federal National Council, a 40-member consultative body appointed by the seven rulers. After the oil price increases of 1973–74—the UAE having given support to the Arab cause in the 1973 war with Israel—the economy and wealth of the new federation developed rapidly.

An attempted coup took place in Sharjah in 1987. Sheikh Sultan Bin Mohammad al-Qassimi abdicated in favour of his brother after admitting mismanagement of the emirate's economy but was restored to power by the Supreme Council of Rulers. In 1991 the UAE became involved in a major international financial scandal when the Bank of Credit and Commerce International (BCCI), in which the Abu Dhabi ruling family held a controlling 77% interest, collapsed. Abu Dhabi sued BCCI for damages in 1993 and several executives were convicted of fraud, given prison sentences and ordered to pay compensation.

Sheikh Rashid bin Said al-Maktoum died in 1990 and was succeeded by his son Sheikh Maktoum bin Rashid al-Maktoum as ruler of Dubai and UAE vice-president. In June 1996 the Federal National Council approved a permanent constitution replacing the provisional document that had been renewed every five years since 1971. At the same time, Abu Dhabi City was designated as the UAE's permanent capital.

In foreign relations the UAE has adopted a largely pro-Western and anti-Iranian stance. Following the fall of the Shah of Iran in 1979 and the outbreak of the Iran–Iraq war in 1980, the stability of the entire region was threatened. Partly in response to Iranian threats to close the Strait of Hormuz to shipping carrying oil exports from Gulf countries, the UAE joined with five other states to form the Gulf Co-operation Council (GCC) in 1981 to work towards political and economic integration. There has been further tension with Iran relating to territorial claims over the island of Abu Musa and the Greater and Lesser Tunb islands, strategically located in the Strait of Hormuz. Iran claimed sovereignty over the islands in the early 1990s and rejected a proposal by the GCC for the claim to be resolved by the International Court of Justice. In 1996 Iran opened an airport on Abu Musa and a power station on Greater Tunb, further damaging relations. Following UAE criticism of a rapprochement between Saudi Arabia and Iran, the GCC reiterated its support for the Emirates in the dispute in 2001. Iran rejected what it claimed was a biased decision. An agreement on the demarcation of the border between the UAE and Oman was ratified in 2003.

UAE forces joined the US-led international coalition against Iraq after the invasion of Kuwait in 1990. Following the Sept. 2001 attacks in the USA, the UAE's banking sector was subject to international scrutiny, and the government consequently ordered financial institutions to freeze the assets of 62 organizations and individuals suspected of funding terrorist movements. The USA stationed forces in the UAE during the invasion of Iraq in March 2003.

TERRITORY AND POPULATION

The Emirates are bounded in the north by the Persian Gulf, northeast by Oman, east by the Gulf of Oman and Oman, south and west by Saudi Arabia. Their area is approximately 83,600 sq. km (32,300 sq. miles), excluding over 100 offshore islands. The total population at the 2005 census was 4,106,427 (68·3% male); density, 49 per sq. km. Estimate, 1 July 2008, 4,765,000. The United Nations population estimate for 2008 was 6,207,000. About one-tenth are nomads. In 2011, 84·4% of the population lived in urban areas. Approximately 80% of the population are foreigners, the highest percentage of any country.

The UN gives a projected population for 2015 of 8·37m. The population of the United Arab Emirates has doubled since the early 2000s.

Populations of the seven Emirates, 2005 census (in 1,000): Abu Dhabi, 1,399; Ajman, 207; Dubai, 1,321; Fujairah, 126; Ras al-Khaimah, 210; Sharjah, 794; Umm al Qaiwain, 49. The chief cities are Dubai (2005 census population of 1,305,060), Abu Dhabi, the federal capital (959,692 in 2005), Sharjah and Al Ain. In addition to being the most populous Emirate, Abu Dhabi is also the wealthiest, ahead of Dubai.

The official language is Arabic; English is widely spoken.

SOCIAL STATISTICS

2008 births, 68,779; deaths, 9,775. 2008 birth rate (per 1,000 population), 14·4; death rate, 1·6; infant mortality rate (per 1,000 live births), 6 (2010). Life expectancy, 2007, 76·6 years for men and 78·7 years for women. Annual population growth rate, 1998–2008, 4·4%; fertility rate, 2008, 1·9 births per woman. The UAE has had one of the largest reductions in its fertility rate of any country in the world over the past quarter of a century, having had a rate of 4·4 births per woman in 1990.

CLIMATE

The country experiences desert conditions, with rainfall both limited and erratic. The period May to Sept. is generally rainless. Abu Dhabi, Jan. 65°F (18·3°C), July 95°F (35·0°C). Annual rainfall 3·5" (89 mm). Dubai, Jan. 66°F (18·9°C), July 94°F (34·4°C). Annual rainfall 3·7" (94 mm).

CONSTITUTION AND GOVERNMENT

The Emirates is a federation, headed by a *Supreme Council of Rulers* which is composed of the seven rulers which elects from among its members a *President* and *Vice-President* for five-year terms, and appoints a *Council of Ministers*. The Council of Ministers drafts legislation and a federal budget; its proposals are submitted to a *Federal National Council* of 40 appointed members which may propose amendments but has no executive power. It was announced in Dec. 2005 that 20 of the 40 members would in future be elected through councils for each of the seven Emirates. There is a *National Consultative Council* made up of citizens.

The current constitution came into force on 2 Dec. 1971 and was made permanent in 1996.

National Anthem

'Ishy Biladi' (Long live my Homeland); words by Abdullah Al Hassan, tune by Mohamed Abdel Wahab.

GOVERNMENT CHRONOLOGY

Presidents since 1971.
1971–2004   Sheikh Zayed bin Sultan al-Nahyan
2004–   Sheikh Khalifa bin Zayed al-Nahyan

RECENT ELECTIONS

A parliamentary election was held on 24 Sept. 2011, with 20 seats in the Federal National Council elected by the population and the other 20 chosen by the rulers of the Emirates. One woman was among the 20 candidates elected. There are no political parties. Turnout was 27·8%.

CURRENT GOVERNMENT

President: HH Sheikh Khalifa bin Zayed al-Nahyan, Ruler of Abu Dhabi (b. 1948; appointed 3 Nov. 2004).

Members of the Supreme Council of Rulers:
President: HH Sheikh Khalifa bin Zayed al-Nahyan.

Vice-President and Prime Minister: HH Sheikh Muhammad bin Rashid al-Maktoum, Ruler of Dubai (b. 1949).

HH Dr Sheikh Sultan bin Mohammed al-Qassimi, Ruler of Sharjah.

HH Sheikh Saud ibn Saqr al-Qasimi, Ruler of Ras al-Khaimah.

HH Sheikh Hamad bin Mohammed al-Sharqi, Ruler of Fujairah.

HH Sheikh Humaid bin Rashid al-Nuaimi, Ruler of Ajman.

HH Sheikh Saud bin Rashid al-Mualla, Ruler of Umm al Qaiwain.

In March 2013 the cabinet comprised:
Prime Minister and Minister of Defence: HH Sheikh Muhammad bin Rashid al-Maktoum; b. 1949 (sworn in 5 Jan. 2006).

Deputy Prime Ministers: HH Maj.-Gen. Sheikh Saif bin Zayed al-Nahyan (also *Minister of the Interior*); HH Sheikh Mansour bin Zayed al-Nahyan (also *Minister of Presidential Affairs*).

Minister of Finance and Industry: HH Sheikh Hamdan bin Rashid al-Maktoum. *Economy:* Sultan bin Said al-Mansouri. *Culture, Youth and Community Development:* Sheikh Nahyan bin Mubarak al-Nahyan. *Development and International Co-operation:* Sheikha Lubna bin Khalid al-Qasimi. *Education:* Humaid Mohammed Obeid al-Qattami. *Higher Education and Scientific Research:* Sheikh Hamdan bin Mubarak al-Nahyan. *Energy:* Suhail Mohammad al-Mazroui. *Labour:* Saqr Ghobash Said Ghobash. *Justice:* Hadef bin Juaan al-Dhaheri. *Foreign Affairs:* Sheikh Abdullah bin Zayed al-Nahyan. *Health:* Abdul Rahman Mohammad al-Owais. *Public Works:* Abdullah Bel Haif al-Nuaimi. *Social Affairs:* Mohammed Khalfan al-Roumi. *Environment and Water:* Rashid Ahmed bin Fahad. *Cabinet Affairs:* Mohammed al-Gergawi.

Government Website: http://www.government.ae

CURRENT LEADERS

Sheikh Khalifa bin Zayed al-Nahyan

Position
President

Introduction
Sheikh Khalifa bin Zayed al-Nahyan was appointed president on 3 Nov. 2004, following the death of his father, Sheikh Zayed, who helped establish the country in 1971 and presided over it for 33 years. Sheikh Khalifa has continued his father's policies of co-operation with neighbouring Arab countries and with the USA, and reducing the UAE's economy's dependence on oil and gas extraction.

Early Life
Sheikh Khalifa bin Zayed al-Nahyan, the eldest son of his father, was born in 1948 in Al-'Ayn, Abu Dhabi. In 1966, following his father's promotion to the post of Ruler of Abu Dhabi, he was appointed as Ruler's Representative in the Emirate's eastern province. Three years later he was nominated as Crown Prince and head of Abu Dhabi's new department of defence.

Following Sheikh Zayed's initiative in 1971 to bring together the rulers of the Trucial States to form the United Arab Emirates (and his assumption of the federation presidency), Sheikh Khalifa assumed two posts—deputy prime minister in the UAE's federal cabinet and chairman of the Abu Dhabi Executive Council. He oversaw the implementation of a massive infrastructure development programme in Abu Dhabi, funded by revenue from the Emirate's vast oil reserves. In May 1976, following the unification of the armed forces of the seven Emirates, Sheikh Khalifa was nominated as deputy supreme commander of the UAE Armed Forces. He went on to establish numerous military training institutions and was responsible for the procurement of equipment and weapons.

In 1981 Sheikh Khalifa established the Abu Dhabi Department of Social Services and Commercial Buildings to offer low-interest loans for house building. The scheme was extended in 1991 and 2000, in line with the rapid increase in the country's population. From the late 1980s, he was Chairman of the Supreme Petroleum Council, and helped develop the UAE's petrochemicals and industrial complex at Ruwais in an attempt to diversify the economy. He also served as chairman of the Abu Dhabi Fund for Development (responsible for the country's overseas aid programme), chairman of the Abu Dhabi Investment Authority and head of the Environmental Research and Wildlife Development Agency. In late 1991 and early 1992 Sheikh Khalifa and his father were mired in the scandal surrounding the collapse of the BCCI bank, 77% of which was owned by the Abu Dhabi government. This followed a fraud in which its founder, Agha Hassan Abedi, and other officers stole billions of dollars. Sheikh Khalifa's scheme to compensate creditors resulted in a payout of US$1·8bn.

When Sheikh Zayed's health began to decline in the late 1990s, Sheikh Khalifa became the public face of the UAE, along with Sheikh Maktoum, Ruler of Dubai. On 3 Nov. 2004, the day after the death of Sheikh Zayed, Sheikh Khalifa was appointed president of the UAE.

Career in Office
Sheikh Khalifa said his key objective as president was to continue on the path laid down by his father. A strong supporter of the six-member Gulf Co-operation Council, he has sought to promote solidarity between Arab states, to support the Palestinian people and to help restore stability in Iraq. In Dec. 2005 he announced plans for the UAE's first elections, in which half of the members of the consultative Federal National Council would be elected by a limited number of citizens. The non-party elections were held in Dec. 2006, and the government said that they marked the start of a wider process to extend political participation.

In May 2009 the UAE withdrew from plans for Gulf monetary union, undermining prospects for further economic integration in the region. This move followed the decision to locate the headquarters of the Gulf Co-operation Council Monetary Council in Saudi Arabia. Meanwhile, the UAE and particularly Dubai were adversely affected by the global financial crisis, as banks were overexposed to a serious reversal in the property market. In Feb. 2009, and again in Dec., the Abu Dhabi government intervened to support the banking sector, to prevent a debt default by Dubai and to reassure the financial markets.

In 2011 the UAE was not disturbed significantly by the wave of political protest that erupted across much of the Arab world, although the government expanded the voter franchise for elections to the Federal National Council in Sept. that year. There was, however, increasing nervousness in 2012 over the continuing regional pressures for political liberalization and the rise of Islamist movements, prompting the arrest of a number of activists accused of plotting to destabilize the country.

In 2011 the UAE joined the NATO-led military operation that helped dislodge the authoritarian Gaddafi regime in Libya.

DEFENCE

In 2008 defence expenditure totalled US$13,733m. (US$2,972 per capita), representing 5·1% of GDP. The expenditure per capita in 2008 was the highest in the world.

Army
The strength was (2007) 44,000 (including Dubai independent forces).

Navy
The combined naval flotilla of the Emirates includes two frigates and two corvettes. Personnel in 2007 numbered around 2,500. The main base is at Abu Dhabi, with minor bases in the other Emirates.

Air Force
Personnel (2007) 4,500, with 184 combat-capable aircraft (including F-16s, Mirage 2000s and *Hawks*), and some 40 attack helicopters.

ECONOMY

Crude petroleum and natural gas accounted for 36·8% of GDP in 2008; trade and hotels, 17·4%; finance and real estate, 15·1%; construction, 7·4%.

Overview
The UAE has a strong record of market-oriented economic reform. It is the fourth largest oil exporter in the world and ranks fifth globally in proven reserves of natural gas and oil. Diversification has stimulated manufacturing, tourism, media, shipping, and financial and commercial services. 80% of the population are

expatriates. Unemployment is high among nationals, particularly the young and those living in the Northern Emirates.

From 2003–08 growth was consistently strong, buoyed by high oil prices and the rapid expansion of construction, financial services and trade services. The economy ran significant current account surpluses (including US$36bn. in 2006), primarily from oil although non-oil exports in manufacturing, agriculture and services also contributed.

However, the global financial crisis caused a slowdown in real estate and construction, weakened tourism, trade and financial services, and lowered oil prices. Dubai World, one of three government-backed conglomerates, was forced to restructure its debt, resulting in higher sovereign debt. Non-performing loans doubled while property values fell. Strong tourism and trade, large-scale public investment in Abu Dhabi, and higher oil prices boosted GDP in 2010–11 as a recovery gained strength. Nonetheless, excess supply of property in Dubai, high dependency on the hydrocarbon sector, and regional turmoil are risks to long-term prospects.

Currency
The unit of currency is the *dirham* (AED) of 100 *fils*. Gold reserves are negligible. In March 2009 foreign exchange reserves were US$34,040m. and total money supply was DH 211,314m. Inflation rates (based on IMF statistics):

2002	2003	2004	2005	2006	2007	2008	2009	2010	2011
2·9%	3·1%	5·0%	6·2%	9·3%	11·1%	12·3%	1·6%	0·9%	0·9%

In 2001 the six Gulf Arab states—the United Arab Emirates, along with Bahrain, Kuwait, Oman, Qatar and Saudi Arabia—signed an agreement to establish a single currency by 2010. In June 2009 it was agreed to postpone the implementation of the new currency, the *khaleeji*, until 2013. It is now likely to be even later. Both Oman and the United Arab Emirates have now withdrawn from the scheme, in 2007 and 2009 respectively.

Budget
The fiscal year is the calendar year. Revenue in 2008 totalled DH 292,600m. and expenditure DH 289,000m. Revenue is principally derived from oil-concession payments. Defence, education, and public order and safety are the main items of expenditure.

Performance
Real GDP growth rates (based on IMF statistics):

2002	2003	2004	2005	2006	2007	2008	2009	2010	2011
−0·7%	16·4%	10·1%	8·6%	8·8%	6·5%	5·3%	−4·8%	1·3%	5·2%

In 2011 total GDP was US$360·2bn.

Banking and Finance
The UAE Central Bank was established in 1980 (*Governor*, Sultan bin Nasser Al-Suwaidi). The largest banks are Emirates NBD, with assets of US$76·9bn. in Dec. 2008, the National Bank of Abu Dhabi (assets of US$44·6bn.) and the Abu Dhabi Commercial Bank (US$40·2bn.). Foreign banks are restricted to eight branches each.

There are stock exchanges in Abu Dhabi and Dubai.

ENERGY AND NATURAL RESOURCES
Environment
In 2008 carbon dioxide emissions from the consumption and flaring of fossil fuels were the equivalent of 43·1 tonnes per capita, among the highest in the world.

Electricity
Installed capacity was 18·5m. kW in 2007. Production in 2007 was 76·11bn. kWh, with consumption per capita (2007) 17,376 kWh. Construction of the first of four nuclear plants began in July 2012. The United Arab Emirates is the first country to begin construction of its first nuclear power plant since construction was started on China's first plant in 1985. The first plant is scheduled to become operational in 2017. All four plants are projected to be completed by 2020.

Oil and Gas
Oil and gas provided about 33·7% of GDP in 2002. Oil production, 2008, 139·5m. tonnes. The UAE had reserves in 2008 amounting to 97·8bn. bbls. Oil production in Abu Dhabi is 85% of the UAE's total.

Abu Dhabi has reserves of natural gas, nationalized in 1976. There is a gas liquefaction plant on Das Island. Proven natural gas reserves (2008) were 6,430bn. cu. metres. Natural gas production, 2008, 50·2bn. cu. metres.

Minerals
Sulphur, gypsum, chromite and lime are mined.

Agriculture
The fertile Buraimi Oasis, known as Al Ain, is largely in Abu Dhabi territory. A lack of water and good soil means few natural opportunities for agriculture but there is a programme of fostering agriculture by desalination, dam-building and tree-planting; strawberries, flowers and dates are now cultivated for export. In 2002 there were 75,000 ha. of arable land and 191,000 ha. of permanent cropland. Output, 2002 (in 1,000 tonnes): dates, 758; tomatoes, 231; cabbage, 30; cucumbers and gherkins, 26; pumpkins and squash, 21.

Livestock products, 2003 estimates (in 1,000 tonnes): meat, 74; milk, 56; eggs, 18. Livestock (2002): goats, 1·4m.; sheep, 554,000; camels, 246,000; cattle, 107,000; chickens, 12m.

Forestry
In 2010, 0·32m. ha. were under forests (4% of the total land area).

Fisheries
Total catch, 2010, 79,610 tonnes (exclusively marine fish).

INDUSTRY
The largest company in the United Arab Emirates by market capitalization in March 2012 was Emirates Telecom (US$19·2bn.).

In 2005 industry accounted for 57% of GDP, with manufacturing contributing 13%. Products include aluminium, cable, cement, chemicals, fertilizers (Abu Dhabi), rolled steel and plastics (Dubai, Sharjah), and tools and clothing (Dubai). The diamond business is becoming increasingly important in Dubai.

Labour
Males constituted 85% of the economically active labour force in 2005 (one of the highest percentages of any country in the world). Foreign workers make up over 90% of the workforce in the private sector. A total of 2,660,000 persons were in employment in 2005, with the leading areas of activity as follows: community, social and personal services, 587,900; construction, 502,700; wholesale and retail trade, restaurants and hotels, 460,200; manufacturing, 292,600. In 2005 the unemployment rate was 1·9%.

INTERNATIONAL TRADE
There are free trade zones at Jebel Ali (administered by Dubai), Sharjah and Fujairah. Foreign companies may set up wholly owned subsidiaries. In 2010 there were over 6,400 companies in the Jebel Ali zone.

The United Arab Emirates, along with Bahrain, Kuwait, Oman, Qatar and Saudi Arabia entered into a customs union in Jan. 2003.

Imports and Exports
Imports in 2001 totalled DH 120·6bn.; exports DH 176·9bn. Principal imports: machinery and transport equipment, food and textiles. Crude petroleum and natural gas are the main exports.

Main import suppliers, 2001: Japan (10·2%), USA (9·6%), UK (8·8%), China (8·6%). Main export markets: Japan (36·4%), India (7·5%), South Korea (7·1%), Singapore (6·3%).

COMMUNICATIONS

Roads

In 2008 there were 4,080 km of roads. There were 1,279,100 passenger cars (293 per 1,000 inhabitants), 48,200 buses and coaches and 39,400 lorries and vans in 2007.

Rail

Etihad Rail, a rail network linking the seven Emirates, is under construction and is expected to become operational by 2018. The first of three phases is scheduled for completion during 2013.

A metro system opened in Dubai in Sept. 2009.

Civil Aviation

There are international airports at Abu Dhabi, Al Ain, Dubai, Fujairah, Ras al-Khaimah and Sharjah. Dubai is the busiest airport, handling 40,901,752 passengers and 1,927,520 tonnes of freight in 2009 (up from 37,441,440 passengers and 1,824,992 tonnes of freight in 2008). In the meantime passenger numbers have grown still further and Dubai is now the 13th busiest airport in the world overall and the fourth busiest for international passenger traffic. As recently as 2006 it did not even rank among the 30 busiest airports in the world. Dubai set up its own airline, Emirates, in 1985. It now operates internationally, and in 2006–07 flew to 89 destinations worldwide and carried 17,544,140 passengers. Etihad Airways, the national airline of the United Arab Emirates, began operations in Nov. 2003. Air Arabia is a low-cost airline based in Sharjah, flying to 35 destinations (mainly in Asia).

In the World Economic Forum's *Global Competitiveness Report 2009–2010* the UAE ranked third for quality of air transport infrastructure.

Shipping

There are 15 commercial seaports, of which five major ports are on the Persian Gulf (Zayed in Abu Dhabi, Rashid and Jebel Ali in Dubai, Khalid in Sharjah, and Saqr in Ras al-Khaimah) and two on the Gulf of Oman: Fujairah and Khor Fakkan. Rashid and Fujairah are important container terminals. In Jan. 2009 there were 124 ships of 300 GT or over registered, totalling 913,000 GT.

Telecommunications

In 2008 there were 1,508,000 main (fixed) telephone lines. In the same year active mobile phone subscribers numbered 9·4m. (2,086·5 per 1,000 persons—the highest rate of any country). There were 2·9m. internet users in 2008. In March 2012 there were 2·9m. Facebook users.

SOCIAL INSTITUTIONS

Justice

The basic principles of the law are Islamic. Legislation seeks to promote the harmonious functioning of society's multi-national components while protecting the interests of the indigenous population. Each Emirate has its own penal code. A federal code takes precedence and ensures compatibility. There are federal courts with appellate powers, which function under federal laws. Emirates have the option to merge their courts with the federal judiciary.

The death penalty is in force; it was used in Feb. 2011 for the first time since 2008. There was also an execution in Dec. 2012.

Education

In 2007 there were 100,269 pre-primary pupils with 4,823 teaching staff, 284,034 primary pupils with 16,523 teaching staff and 310,999 secondary pupils with 24,152 teaching staff. In 2005–06 there were 14,984 students at the United Arab Emirates University and 2,925 students at the five colleges of Zayed University. The largest higher education institution is the system of the Higher Colleges of Technology, with 16,894 students in 2005–06. It has 17 campuses throughout the UAE. The adult literacy rate in 2005 was 90·0%. In 2009 public expenditure on education came to 1·2% of GDP and 23·4% of total government spending.

Health

In 2003 there were 38 government hospitals with 5,722 beds. In 2003 there were 27 private hospitals, 131 government health centres and 1,281 private clinics. There were 5,825 physicians in 2001 and 954 dentists, 12,045 nurses and 1,086 pharmacists.

RELIGION

Most inhabitants are Sunni Muslims, with a small Shia minority.

CULTURE

World Heritage Sites

The Cultural Sites of Al Ain (Hafit, Hili, Bidaa Bint Saud and Oases Areas), a property that includes circular stone tombs, wells, towers, palaces and administrative buildings dating back to 2500 BC, was inscribed on the UNESCO World Heritage List in 2011.

Press

In 2006 there were 14 daily newspapers with a combined circulation of 901,000.

Tourism

In 2005, 7,126,000 tourists stayed in hotels and similar accommodation; spending by tourists in 2005 totalled US$3,218m.

DIPLOMATIC REPRESENTATIVES

Of the UAE in the United Kingdom (30 Prince's Gate, London, SW7 1PT)
Ambassador: Abdul Rahman Ghanim Al Mutaiwee.

Of the United Kingdom in the UAE (POB 248, Abu Dhabi)
Ambassador: Dominic Jermey, OBE.

Of the UAE in the USA (3522 International Court, NW, Washington, D.C., 20008)
Ambassador: Yousef Al Otaiba.

Of the USA in the UAE (POB 4009, Abu Dhabi)
Ambassador: Michael H. Corbin.

Of the UAE to the United Nations
Ambassador: Ahmad Abdul Rahman Al Jarman.

Of the UAE to the European Union
Ambassador: Sulayman Hamid Al Mazroui.

FURTHER READING

Davidson, Christopher M., *The United Arab Emirates: A Study in Survival.* 2005
Vine, P. and Al Abed, I., *United Arab Emirates: A New Perspective.* 2001

National Statistical Office: Ministry of Economy, P. O. Box 901–904, Abu Dhabi.
Website: http://www.economy.ae/English/EconomicAndStatisticReports/Pages/default.aspx

UNITED KINGDOM OF GREAT BRITAIN AND NORTHERN IRELAND

© Research Machines plc 2006

Capital: London
Population projection, 2015: 63·94m.
GNI per capita, 2011: (PPP$) 33,296
HDI/world rank: 0·863/28
Internet domain extension: .uk

KEY HISTORICAL EVENTS

Remains of Stone Age settlements of hunters and fishermen suggest that the first inhabitants crossed from the low countries of Continental Europe on one or more wide causeways. By the time their successors had turned to subsistence farming, the land links to the continent had disappeared under the sea. These offshore islands created at the ending of the Ice Age shared, with nearside Europe, a slowly evolving agricultural economy using bronze and iron tools. The Ancient Britons were Celts, whose ancestors had migrated from the valleys of the Rhine, the Rhône and the Danube. Having asserted their command of northern Italy and France (Gaul), the Celts established a bridgehead to Ireland and thence to Britain. By 600 BC they were the undisputed dominant force of Western Europe and were to remain so until challenged by the Romans.

The Romans were dominant from AD 78. From the 3rd century they were increasingly harried by tribes of Celts from Scotland and Ireland and by Angles and Saxons from northern Germany. Celtic tradition, presided over by druids (religious leaders) and bards (storytellers), survived most successfully in Ireland and Wales where Roman influence was barely visible. Scotland resisted the Roman legions; Hadrian's Wall was built as a northern frontier between the Tyne and Solway Firth in the early 2nd century AD. Roman authority was challenged, notably by Boudicca, queen of the Iceni tribe of East Anglia. The rebellion and the brutal repression that followed led to a long period of peaceful settlement, during which the Romans established a road network linking new towns such as Londinium (London) and Eboracum (York). But by the 5th century Roman Britain had disintegrated into a collection of warring kingdoms. The English and Welsh economies thrived on the export of silver, lead, gold, iron and other minerals. With the spread of Christianity, chiefly by Irish missionaries, came the beginnings of an education and legal system.

After the withdrawal of the Roman legions in the early 5th century, the Romano-British were pushed back to higher land in the west by waves of invading Saxons, Angles and Jutes. Danish invasions in 865 established the Danelaw in northern England. Alfred the Great of Wessex resisted Danish expansion, strengthening Anglo-Saxon unity.

Norman Conquest

William, duke of Normandy, led the Norman Conquest and was crowned king in 1066. When William died in 1087 he left Normandy to his eldest son Robert, thus separating it from England. The French dialect known as Anglo-Norman was spoken by the ruling class in England for two centuries after the Conquest. The Norman heritage was preserved also in the overlap between French and English feudal lords. Henry II, the founder of the Plantagenet dynasty, was feudatory lord of half of France. But most of the French possessions were lost by Henry's son John. Thereafter, the Norman baronage came to regard themselves as English. The ambitions of Edward III began and those of Henry V renewed the Hundred Years War (1338–1453) with France, which ended with the loss of all the remaining French possessions except Calais.

The dynastic struggle between the rival houses of York and Lancaster was concluded by the invasion of Henry (VII) Tudor in 1485. His son, Henry VIII, asserted royal authority over the church and rejected papal authority. Tudor power reached its zenith with Elizabeth I, under whom Protestantism became firmly established in England. The Spanish Armada—an attempt by Catholic Spain to return England to the papal fold—was repelled in 1588.

The accession of James VI of Scotland to the English throne in 1603 brought the two countries into dynastic union. A struggle for supremacy between Crown and Parliament culminated in the Civil War, which started in 1642. Charles I was executed by Parliament in 1649, beginning the rule of Protector Oliver Cromwell. The Stuart monarchy was restored in 1660, on terms which conceded financial authority and thus decision-making power to Parliament. The attempt of James II, a Catholic, to restore the royal prerogative led to the intervention of William of Orange. James fled the country and the crown was taken by William (III) and his wife Mary as queen regnant. The accession of William involved England in a protracted war against France.

The parliaments of England and Scotland were united in 1707 under Queen Anne, the first British monarch. With the accession of the Hanoverian George I in 1714, the system of Parliamentary party government took hold. By the mid-18th century London had taken over from Amsterdam as the leading financial centre. With easy access to capital, entrepreneurs were able to invest in new, improved methods of production. With the harnessing of steam power made possible by the engineering genius of Thomas Newcomen and James Watt, economic enterprise shifted away from the southeast to the north of England, Scotland and South Wales where there were large reserves of coal. The demand for raw materials and the pursuit of markets for finished goods opened up trade throughout the civilized world and extended British influence.

American Colonies
Britain's first successful colonies in North America were established in the reign of James I of England (1603–25) and, soon afterwards, Bermuda, St Kitts, Barbados and Nevis were colonized. By the mid-17th century, Britain controlled the American east coast and had strong bases in India and in the West Indies, where the sugar economy was dependent on slave labour imported from West Africa. Critical to imperial expansion was Britain's rivalry with France. Britain emerged much strengthened from the War of the League of Augsburg (1689–97) and the War of Spanish Succession (1702–13) while French ambitions in Europe and beyond were severely curtailed. But it was the Seven Years' War (1756–63), in which France and Prussia were the chief contenders, that deprived France of her remaining territorial claims in North America and India and confirmed Britain as the world's leading maritime power.

Relations between Parliament and Crown went through an unsettled period in the reign of George III, who was blamed for the loss of the American colonies. The War of Independence ended with Britain's recognition of American right to self-government in 1783. In 1793 revolutionary France declared war and was not finally defeated until 1815. The demands of war further stimulated the new, steam-powered industries. Despite Britain finding itself the pre-eminent world power, after 1815 there was frequent unrest as an increasingly urban and industrial society found its interests poorly represented by a parliament composed chiefly of landowners. The Reform Act of 1832 extended representation in Parliament and further acts (1867, 1884, 1918 and 1928) led gradually to universal adult suffrage.

Ireland was brought under direct rule from Westminster in 1801, creating the United Kingdom of Great Britain and Ireland. The accession of Victoria in 1837 was the beginning of an era of unprecedented material progress. Early industrial development produced great national wealth but its distribution was uneven and the condition of the poor improved slowly. Whereas early Victorian reforms were responses to obvious distress, governments after 1868 were more inclined towards preventive state action.

The Victorian empire included India, Canada, Australasia and vast territories in Africa and Eastern Asia. There was war with Russia in the Crimea (1854–56); most wars, however, were fought to conquer or pacify colonies. After 1870 the Suez Canal enabled Britain to control the empire more efficiently; Britain became a 40% shareholder in 1875 and the controlling power in Egypt in 1882. The most serious imperial wars were the Boer Wars of 1881 and 1899–1902 against the Dutch settlers in South Africa. After a less than glorious victory, Britain negotiated a Union of South Africa, by which South Africa enjoyed the same autonomy agreed for Canada (1867), Australia (1901) and later New Zealand (1907). The 'dominion status' of these countries was clarified by the Statute of Westminster (1931).

With the spread of trade unionism and the emergence of the Labour Party, the gap between right- and left-wing politics widened after 1900. Labour had to wait until 1924 to form its first government but the Liberal landslide of 1906 carried forward the programme of social reform. David Lloyd George's People's Budget led to the abolition of the House of Lords' right to override the House of Commons while a contributory insurance scheme to cover basic health care, a modest benefit for the unemployed, free school meals and non-contributory old age pensions were all introduced.

On 3 Aug. 1914 Germany invaded Belgium. Britain was obliged by treaty to retaliate by declaring war. Four years of bloody trench warfare ensued in northern France and Belgium, with American intervention in 1917 helping to break the stalemate. The United Kingdom alone lost 715,000 soldiers and another 200,000 from the empire. Rebellions broke out in Ireland, born of the failure of successive attempts to agree a formula for Irish Home Rule. The issue was complicated by factional disagreement in southern Ireland and the wish of northern Ireland (Ulster) to remain in the United Kingdom. In 1920, after four years' conflict, the Government of Ireland Act partitioned the country. The northern six counties remained British, a parliament was created and a Unionist government took office. The southern 26 counties moved by stages to complete independence as the Irish Free State in 1922.

Second World War
A post-war boom was followed by a lengthy recession and heavy unemployment, exacerbated by the reluctance of politicians to adopt Keynesian economics. Germany revived as a military power in the 1930s, unchecked by reluctant neighbours after the punitive Treaty of Versailles. British Prime Minister Neville Chamberlain agreed to the German acquisition of parts of Czechoslovakia at the Munich Agreement in 1938. His policy of appeasement was much criticized, though it is arguable that Britain was in no position to go to war in 1938. Germany invaded Poland on 1 Sept. 1939. Britain, bound once more by treaty, declared war. In May 1940 Chamberlain was replaced as prime minister by Winston Churchill, who formed a national unity government. Although British military casualties were less than in the 1914–18 war, the civilian population was hit much worse during the Second World War; over 90,000 died, many as a result of German bombing in the Battle of Britain in 1940.

The war ended with German and Japanese defeat in 1945, by which stage the United Kingdom was virtually bankrupt. In 1939 the country had had assets of around £3,000m. By the end of the war, it owed about the same amount. A pre-war balance of payments deficit averaging £43m. a year had jumped to £750m. It was a time of great social upheaval. In the 1945 election a Labour government under Clement Attlee was returned with a large majority and a socialist programme, which emphasized wealth distribution above wealth creation, was implemented. It undertook to establish a free National Health Service, an ambitious housing programme and the state control of major industries. Subsequent governments modified but generally accepted the changes.

With Britain bankrupted, the United States stepped into the breach as the now undisputed free world leader. Fearing a European breakdown and a Communist takeover, the Marshall Plan was implemented by the USA, providing massive investment to rebuild Europe. An essential condition of the Marshall Plan was a joint effort of the participating nations to put their economies in order. But when continental leaders made the first tentative moves towards European unity, the UK was unwilling to be closely involved. With the independence of India (and Pakistan), the centrepiece of the British Empire, in 1947, decolonization took root, reaching its climax in the 1960s. Rather than to Europe, Britain now looked instead to a Commonwealth of freely associated states, recognizing the British monarch as symbolic Commonwealth head (some states chose to retain the monarch as head of state), and to the 'special relationship' with the United States.

In March 1957 France, Germany, Italy, Belgium, the Netherlands and Luxembourg signed the Treaty of Rome, which laid down terms for the European Economic Community. Two years later seven of the European countries outside the Common Market—Austria, Denmark, Norway, Portugal, Sweden, Switzerland and the UK—formed the European Free Trade Association. When the United Kingdom moved to join the EEC in 1962, five of the six members of the Community were willing to support the application but France vetoed it. A second application, in 1967, also failed but admission was achieved in 1973 under the Conservative government of Edward Heath. Membership of the Community was endorsed by referendum in 1975.

On the wider international scene, the limits of independent military action were made clear by the Suez crisis of 1956 when the UK, in collusion with France and Israel, used force to stop President Nasser of Egypt nationalizing the Suez Canal. Assumed American support was not forthcoming and the enterprise collapsed when the UK was left alone to cope with a potentially disastrous run on sterling. In the 1960s and 1970s the UK began to come to terms with advanced technology. Old-established industries such as textiles, shipbuilding, iron and steel and coal mining, the leaders of the first industrial revolution, gave way to manufacturing that relied on the microchip. Service industries, particularly in the financial sector, occupied an increasing share of the economy and trade restrictions were dismantled throughout the world.

In 1979 a Conservative government led by Margaret Thatcher came to power, committed to a free market economy. State industry was returned to private enterprise, the trade unions (blamed for the crippling 1978–79 Winter of Discontent) lost much of their power to direct government policy, and high earners were to benefit from lower taxation. A period of readjustment climaxed with a coal miners' strike during 1984–85 that turned into a trial of strength between the government and organized labour. The Labour Party and allied unions, themselves in the process of modernization, distanced themselves from the socialist rhetoric of the miners' leaders and the strike collapsed.

Despite rising living standards, there was concern about the quality of essential services such as education and health and disillusionment with a Conservative administration unable to construct a coherent European policy. In 1997 a Labour government, led by Tony Blair, was returned with a large Commons majority. Like Thatcher, he believed in the free market. In addition he introduced reforms in the system of government including the abolition of voting rights of hereditary peers in the House of Lords and the setting up of directly elected assemblies for Scotland, Wales and Northern Ireland. Blair showed greater enthusiasm for involvement in Europe while the war in Iraq went some way to rementing the 'special relationship' with the USA. Fears of terrorist reprisals for British involvement in the Iraq war were realised on 7 July 2005 when bombs planted on three London underground trains and a bus killed 52 people. Gordon Brown, the chancellor during Blair's premiership, succeeded him as prime minister in June 2007. He was replaced by Conservative leader David Cameron, who took office in May 2010 in a coalition with the Liberal Democrats to end 13 years of Labour rule.

TERRITORY AND POPULATION

Area (in sq. km) and population at the census taken on 27 March 2011:

Divisions	Area	Population
England	130,432	53,012,456
Wales	20,780	3,063,456
Scotland	78,808	5,295,000[1]
Northern Ireland	14,130	1,810,863
	244,150	63,182,000[1]

[1]Figure rounded to nearest thousand.

Population of the United Kingdom (present on census night) at the four previous decennial censuses:

Divisions	1971	1981	1991	2001
England[1]	46,018,371	46,226,100[2]	46,382,050	49,138,831
Wales	2,731,204	2,790,500[2]	2,811,865	2,903,085
Scotland	5,228,963	5,130,700	4,998,567	5,062,011
Northern Ireland	1,536,065	1,532,196[3]	1,577,836	1,685,267
United Kingdom	55,514,603	55,679,496[2]	55,770,318	58,789,194

[1]Areas now included in Wales formed the English county of Monmouthshire until 1974.
[2]The final counts for England and Wales are believed to be over-stated as a result of an error in processing. The preliminary counts presented here rounded to the nearest hundred are thought to be more accurate.
[3]There was a high level of non-enumeration in Northern Ireland during the 1981 census mainly as a result of protests in Catholic areas about the Republican hunger strikes.

The land area of the United Kingdom in 2011 was 242,509 sq. km; density, 261 per sq. km. 79·8% of the population lived in urban areas in 2011. London had a 2011 population of 8,174,000.

The UN gives a projected population for 2015 of 63·94m.

Population of the United Kingdom by sex at census day 2011:

Divisions	Males	Females
England	26,069,148	26,943,308
Wales	1,504,228	1,559,228
Scotland[1]	2,567,000	2,728,000
Northern Ireland	887,323	923,540
United Kingdom	31,028,000[1]	32,154,000[1]

[1]Figures rounded to nearest thousand.

Households in the United Kingdom at the 2011 census: England, 22,063,000; Wales, 1,303,000; Scotland (2011 estimate), 2,368,000; Northern Ireland, 703,000.

The age distribution in the United Kingdom at census day in 2011 (provisional) was as follows (in 1,000):

Age-group	England and Wales	Scotland	Northern Ireland	United Kingdom
Under 5	3,497	293	124	3,914
5 and under 10	3,136	270	111	3,517
10 „ 15	3,259	292	119	3,670
15 „ 20	3,539	331	126	3,997
20 „ 25	3,807	364	126	4,297
25 „ 35	7,521	668	244	8,433
35 „ 45	7,831	735	254	8,820
45 „ 55	7,702	787	249	8,738
55 „ 65	6,561	667	194	7,421
65 „ 70	2,674	261	82	3,017
70 „ 75	2,179	221	63	2,463
75 „ 85	3,116	300	87	3,502
85 and upwards	1,255	108	31	1,394

In 2011, 17·6% of the population of the UK were under the age of 15, 66·0% between 15 and 64 and 16·4% aged 65 and over. In 1911 only 5·3% of the population had been 65 and over.

England and Wales. The census population (present on census night) of England and Wales 1801 to 2011:

Date of enumeration	Population	Pop. per sq. mile[1]	Date of enumeration	Population	Pop. per sq. mile[1]
1801	8,892,536	152	1911	36,070,492	618
1811	10,164,256	174	1921	37,886,699	649
1821	12,000,236	206	1931	39,952,377	685
1831	13,896,797	238	1951	43,757,888	750
1841	15,914,148	273	1961	46,104,548	791
1851	17,927,609	307	1971	48,749,575	323
1861	20,066,224	344	1981	49,016,600	325
1871	22,712,266	389	1991	49,193,915	330
1881	25,974,439	445	2001	52,041,916	345
1891	29,002,525	497	2011	56,075,909	371
1901	32,527,843	558			

[1]Per sq. km from 1971.

The birthplaces of the population of England and Wales at census day 2011 were: England, 44,882,858; Wales, 2,732,624; Scotland, 733,218; Northern Ireland, 214,988; Ireland, 407,357; other European Union countries, 2,035,619 (including: Poland, 579,121; Germany, 273,564; Italy, 134,619; France, 129,804); elsewhere, 5,062,034 (including: India, 694,148; Pakistan, 482,137; Bangladesh, 211,500; Nigeria, 191,183).

Ethnic Groups. The 1991 census was the first to include a question on ethnic status.

Percentage figures from the 2011 census relating to ethnicity in England and Wales:

	England and Wales (%)	England (%)	Wales (%)
White			
British	80·6	79·8	93·2
Irish	1·0	1·0	0·5
Other	0·6	0·6	0·3
Mixed			
White and Black Caribbean	0·8	0·8	0·4
White and Black African	0·3	0·3	0·1
White and Asian	0·6	0·6	0·3
Other Mixed	0·1	0·1	0·1
Asian or Asian British			
Indian or British Indian	2·5	2·6	0·6
Pakistani or British Pakistani	2·0	2·1	0·4
Bangladeshi or British Bangladeshi	0·8	0·8	0·3
Chinese	0·7	0·7	0·5
Other Asian	0·1	0·2	0·0
Arab	0·4	0·4	0·3
Black or Black British			
African	1·9	2·0	0·4
Caribbean	1·1	1·1	0·1
Other Black	0·1	0·1	0·0
Other ethnic groups	0·1	0·1	0·0

In Scotland about 2% of the population in 2001 were from a minority (non-White) ethnic group, compared with 1·3% in 1991. Pakistanis formed the largest such group, constituting 0·3%.

The following table shows the distribution of the urban and rural population in the United Kingdom since 1960:

	Population in thousands		Percentage	
	Urban areas	Rural areas	Urban	Rural
1960	41,218	11,327	78·4	21·6
1970	42,912	12,734	77·1	22·9
1980	44,187	12,116	78·5	21·5
1990	44,708	12,507	78·1	21·9
2000	46,305	12,569	78·7	21·3
2010	49,323	12,712	79·5	20·5

Urban and rural areas were re-defined for the 1981 and 1991 censuses on a land use basis. In Scotland 'localities' correspond to urban areas. The 1981 census gave the usually resident population of England and Wales as 48,521,596, of which 43,599,431 were in urban areas; and of Scotland as 5,035,315, of which 4,486,140 were in localities.

British Citizenship. Under the British Nationality Act 1981 there are three main forms of citizenship: citizenship for persons closely connected with the UK; British Dependent Territories citizenship; British Overseas citizenship. British citizenship is acquired automatically at birth by a child born in the UK if his or her mother or father is a British citizen or is settled in the UK. A child born abroad to a British citizen is a British citizen by descent. British citizenship may be acquired by registration for stateless persons, and for children not automatically acquiring such citizenship or born abroad to parents who are citizens by descent; and, for other adults, by naturalization. Requirements for the latter include five years' residence (three years for applicants married to a British citizen). The Hong Kong (British Nationality) Order 1986 created the status of British National (Overseas) for citizens connected with Hong Kong before 1997, and the British

Nationality (Hong Kong) Act 1990 made provision for up to 50,000 selected persons to register as British citizens.

Emigration and Immigration. Immigration is mainly governed by the Immigration Act 1970 and Immigration Rules made under it. British and Commonwealth citizens with the right of abode before 1983 are not subject to immigration control, nor are citizens of European Economic Area countries. Other persons seeking to work or settle in the UK must obtain a visa or entry clearance.

Total international migration estimates for recent years are as follows.

Inflows (in 1,000):

	Total	British	Non-British
2007	574	74	500
2008	590	85	505
2009	567	96	471
2010	591	93	498

Outflows (in 1,000):

	Total	British	Non-British
2007	341	171	169
2008	427	173	255
2009	368	140	228
2010	339	136	203

The number of immigrants into the UK in 2006, at 596,000, was the highest on record for a calendar year. The number of emigrants from the UK in 2008, at 427,000, was the highest on record for a calendar year. The number of emigrants in 2010 was the lowest since 2001. Net migration—the difference between immigration and emigration—was 252,000 in 2010 (the highest on record for a calendar year), up from 198,000 in 2009. Emigration from the UK last exceeded immigration into the UK in 1993.

In 2010 there were 241,192 grants of settlement in the UK (194,781 in 2009 and 148,936 in 2008), including from Asia, 121,812; and Africa, 64,875. Main individual countries were: India, 37,436; Pakistan, 21,382; China, 14,616; Nigeria, 10,031; Philippines, 9,934; Zimbabwe, 9,848. In 2008, 6·6% of the UK's population were foreign citizens, compared to the EU-wide average of 6·2%.

Asylum. In 2010 there were 17,916 applications for asylum, down from 24,487 in 2009 and the high of 84,132 in 2002. The main countries of origin in 2010 were Iran, Afghanistan, Zimbabwe, Pakistan, Sri Lanka and China. Applications, including dependants, were 22,644 in 2010. While respecting its obligations to political refugees under the UN Convention and Protocol relating to the status of Refugees, the government has powers under the Asylum and Immigration Act 1996 to weed out applicants seeking entry for non-political reasons and to designate certain countries as not giving risk of persecution.

Coleman, D. and Salt, J., *The British Population: Patterns, Trends and Processes.* 1992

See also ENGLAND, SCOTLAND, WALES and NORTHERN IRELAND: Territory and Population.

Language. Although there is no legally defined official language in the United Kingdom, English is the *de facto* official language.

SOCIAL STATISTICS

UK statistics, 2010: births, 807,271; deaths, 561,666; marriages (provisional), 277,736; divorces (provisional), 132,338. UK rates (per 1,000 population), 2010: birth, 13·0; death, 9·0; marriage (provisional), 4·5; divorce (provisional), 2·1. The number of births in the UK in 2010 was the highest since 1972; the number of deaths in 2009 was the lowest since the early 1930s. Provisional statistics, 2011: births, 807,776; deaths, 552,232. In 1976, for the only time in the 20th century, deaths in the UK (680,800) exceeded births (675,500). In 2007 cancer caused 159,000 deaths (28% of all deaths in the UK, making it the biggest killer, ahead of coronary heart disease, at 91,000 (16%) and respiratory diseases, at 78,000 (14%)). UK life expectancy, 2008–10: males, 78·2 years; females, 82·3. The World Health Organization's *World Health Statistics 2009* put the UK in joint 21st place in a 'healthy life expectancy' list, with an expected 72 years of healthy life for babies born in 2007. Annual population growth rate, 2001–10, 0·6%. In 2007, 16·0% of the total population was over 65, up from 11·7% in 1960. In 2009 there were 5,675 suicides (4,304 of whom were men), giving a suicide rate of nine per 100,000 population. Infant mortality, 2008, 4·7 per 1,000 live births. Fertility rate, 2010, 2·0 births per woman. Of the 794,383 live births in the UK in 2008, 45·4% were to unmarried women, up from 6% in 1961 and 20% in 1986. In 1999 for the first time there were more births to women in the 30–34 age group in the UK than in the 25–29 bracket. 63% of dependent children lived in married couple families in the UK in 2010 and 23% in single-parent families.

Great Britain statistics, 2009: births, 765,924; deaths, 545,204; marriages, 259,967; divorces, 124,343; abortions, 208,748. In 2010 the average household in Great Britain consisted of 2·4 people, down from 3·1 in 1961. A UNICEF report published in 2010 showed that 12·1% of children in the UK live in relative poverty (living in a household in which disposable income—when adjusted for family size and composition—is less than 50% of the national median income), compared to just 4·7% in Iceland.

England and Wales statistics (in 1,000), 2009 (and 2008): births, 706 (709); deaths, 491 (509); marriages, 232 (236); divorces, 114 (122). In 2003 there was a rise in the number of births for the first time since 1996, a trend which has generally continued since. Provisional totals for 2010 suggest that the annual number of births in England and Wales is now at its highest level since the early 1970s. The under 18 conception rate for 2010 was the lowest since 1969, at 35·5 conceptions per 1,000 females aged 15–17. By 2010 the number of centenarians had reached an estimated 11,600 in England and Wales. The average age of first marriage in England and Wales in 2009 was 32 years for men and 30 years for women, up from 25 years for men and 23 years for women in 1971. 67% of marriages in England and Wales in 2009 were civil and 33% religious, compared to 51% religious and 49% civil in 1991. Same-sex civil partnerships were legalized in Dec. 2005.

Britain has one of the highest rates of drug usage in Europe. Figures released in 2005 showed that 38% of schoolchildren in England and Wales aged 15 and 16 have used cannabis. In 2007–08, 15·4% of 16 to 30-year-olds had used cannabis and 5·0% cocaine. A 2009 report found that use of ecstasy and amphetamines in England and Wales was the highest in the European Union and cocaine use the second highest after Spain. There were 2,878 recorded drug-related deaths in England and Wales in 2009, down slightly from 2,928 in 2008 and from a peak of 3,110 in 1999, although up from 2,570 in 2006.

See also NORTHERN IRELAND: Social Statistics.

CLIMATE

The climate is cool temperate oceanic, with mild conditions and rainfall evenly distributed over the year, though the weather is very changeable because of cyclonic influences. In general, temperatures are higher in the west and lower in the east in winter and rather the reverse in summer. Rainfall amounts are greatest in the west, where most of the high ground occurs.

London, Jan. 39°F (3·9°C), July 64°F (17·8°C). Annual rainfall 25" (635 mm). Aberdeen, Jan. 38°F (3·3°C), July 57°F (13·9°C). Annual rainfall 32" (813 mm). Belfast, Jan. 40°F (4·5°C), July 59°F (15·0°C). Annual rainfall 37·4" (950 mm). Birmingham, Jan. 38°F (3·3°C), July 61°F (16·1°C). Annual rainfall 30" (749 mm). Cardiff, Jan. 40°F (4·4°C), July 61°F (16·1°C). Annual rainfall 42·6" (1,065 mm). Edinburgh, Jan. 38°F (3·3°C), July 58°F (14·5°C). Annual rainfall 27" (686 mm). Glasgow, Jan. 39°F (3·9°C), July 59°F (15·0°C). Annual rainfall 38" (965 mm). Manchester, Jan. 39°F (3·9°C), July 61°F (16·1°C). Annual rainfall 34·5" (876 mm).

CONSTITUTION AND GOVERNMENT

The reigning Queen, Head of the Commonwealth, is **Elizabeth II** Alexandra Mary, b. 21 April 1926, daughter of King George VI and Queen Elizabeth; married on 20 Nov. 1947 Lieut. Philip Mountbatten (formerly Prince Philip of Greece), created Duke of Edinburgh, Earl of Merioneth and Baron Greenwich on the same day and created Prince Philip, Duke of Edinburgh, 22 Feb. 1957; succeeded to the crown on the death of her father, on 6 Feb. 1952.

Offspring. Prince Charles Philip Arthur George, Prince of Wales (Heir Apparent), b. 14 Nov. 1948; married Lady Diana Frances Spencer on 29 July 1981; after divorce, 28 Aug. 1996, Diana, Princess of Wales (died in Paris in a road accident on 31 Aug. 1997); married Camilla Parker Bowles on 9 April 2005. *Offspring of first marriage:* William Arthur Philip Louis, b. 21 June 1982; married Catherine 'Kate' Middleton on 29 April 2011; Henry Charles Albert David 'Harry', b. 15 Sept. 1984. Princess Anne Elizabeth Alice Louise, the Princess Royal, b. 15 Aug. 1950; married Mark Anthony Peter Phillips on 14 Nov. 1973; divorced, 1992; married Cdr Timothy Laurence on 12 Dec. 1992. *Offspring of first marriage:* Peter Mark Andrew, b. 15 Nov. 1977; married Autumn Patricia Kelly on 17 May 2008 (*offspring:* Savannah Phillips, b. 29 Dec. 2010; Isla Elizabeth Phillips, b. 29 March 2012); Zara Anne Elizabeth, b. 15 May 1981; married Michael James 'Mike' Tindall on 30 July 2011. Prince Andrew Albert Christian Edward, created Duke of York, 23 July 1986, b. 19 Feb. 1960; married Sarah Margaret Ferguson on 23 July 1986; after divorce, 30 May 1996, Sarah, Duchess of York. *Offspring:* Princess Beatrice Mary, b. 8 Aug. 1988; Princess Eugenie Victoria Helena, b. 23 March 1990. Prince Edward Antony Richard Louis, created Earl of Wessex and Viscount Severn, 19 June 1999, b. 10 March 1964; married Sophie Rhys-Jones, Countess of Wessex, on 19 June 1999. *Offspring:* Louise Alice Elizabeth Mary, Lady Louise Windsor, b. 8 Nov. 2003; James Alexander Philip Theo, Viscount Severn, b. 17 Dec. 2007.

Sister of the Queen. Princess Margaret Rose, Countess of Snowdon, b. 21 Aug. 1930; married Antony Armstrong-Jones (created Earl of Snowdon, 3 Oct. 1961) on 6 May 1960; divorced, 1978; died 9 Feb. 2002. *Offspring:* David Albert Charles (Viscount Linley), b. 3 Nov. 1961, married Serena Alleyne Stanhope on 8 Oct. 1993. *Offspring:* Charles Patrick Inigo Armstrong-Jones, b. 1 July 1999; Margarita Elizabeth Rose Alleyne Armstrong-Jones, b. 14 May 2002. Lady Sarah Frances Elizabeth Chatto, b. 1 May 1964; married Daniel Chatto on 14 July 1994. *Offspring:* Samuel David Benedict Chatto, b. 28 July 1996; Arthur Robert Nathaniel Chatto, b. 5 Feb. 1999.

The Queen's legal title rests on the statute of 12 and 13 Will. III, ch. 3, by which the succession to the Crown of Great Britain and Ireland was settled on the Princess Sophia of Hanover and the 'heirs of her body being Protestants'. By proclamation of 17 July 1917 the royal family became known as the House and Family of Windsor. On 8 Feb. 1960 the Queen issued a declaration varying her confirmatory declaration of 9 April 1952 to the effect that while the Queen and her children should continue to be known as the House of Windsor, her descendants, other than descendants entitled to the style of Royal Highness and the title of Prince or Princess, and female descendants who marry and their descendants should bear the name of Mountbatten-Windsor.

Lineage to the throne. 1) Prince of Wales. 2) Prince William of Wales. 3) Prince Henry of Wales. 4) Duke of York. 5) Princess Beatrice of York. 6) Princess Eugenie of York.

By letters patent of 30 Nov. 1917 the titles of Royal Highness and Prince or Princess are restricted to the Sovereign's children, the children of the Sovereign's sons and the eldest living son of the eldest son of the Prince of Wales.

Traditionally provision has been made for the support of the royal household, after the surrender of hereditary revenues, by the settlement of the Civil List soon after the beginning of each reign.

The Civil List Act of 1 Jan. 1972 provided for a decennial, and the Civil List (Increase of Financial Provision) Order 1975 for an annual review of the List, but in July 1990 it was again fixed for one decade. The Civil List of 2012 provided for an annuity of £7,900,000 to the Queen annually; and £359,000 to Prince Philip. These amounts were the same as for the periods 2001–10 and 1991–2000. The income of the Prince of Wales derives from the Duchy of Cornwall. The Civil List was exempted from taxation in 1910. The Queen has paid income tax on her private income since April 1993. In Oct. 2010 Chancellor George Osborne announced that from 2013 the Civil List would be abolished. It has been replaced by a Sovereign Support Grant combining the Civil List and three Grants-in-Aid into one payment, with the Queen receiving 15% of the profits from the Crown Estate but two years in arrears. The Sovereign Grant is set at £36·1m. for 2013–14.

The supreme legislative power is vested in Parliament, which consists of the Crown, the House of Lords and the House of Commons, and evolved into something resembling its present form in the mid-14th century. A Bill which is passed by both Houses and receives Royal Assent becomes an Act of Parliament and part of statute law.

Parliament is summoned, and a General Election is called, by the sovereign on the advice of the Prime Minister. Under the Fixed Term Parliaments Act 2011 a Parliament lasts five years, normally divided into annual sessions. A session is ended by prorogation, and most Public Bills which have not been passed by both Houses then lapse, unless they are subject to a carry over motion. A Parliament ends by dissolution at the end of five years, if a motion of no confidence is passed and no alternative government is found or if a motion for an early general election is agreed by at least two-thirds of the House of Commons.

Under the Parliament Act 1911 all Money Bills (so certified by the Speaker of the House of Commons), if not passed by the Lords without amendment, may become law without their concurrence within one month of introduction in the Lords. Under the Parliament Acts 1911 and 1949 Public Bills introduced in the House of Commons, other than Money Bills or a Bill extending the maximum duration of Parliament, if passed by the Commons in two successive sessions and rejected each time by the Lords, may become law without being passed by the Lords provided that one year has elapsed between Commons second reading in the first session and passing of the bill by the Commons in the second session, and that the Bill reaches the Lords at least one month before the end of the second session. The Parliament Acts have been used four times since 1949: in 1991 for the War Crimes Act, in 1999 for the European Parliamentary Elections Act, in 2000 for the Sexual Offences (Amendment) Act and for the Hunting Act in 2004.

Peerages are created by the sovereign, on the advice of the Prime Minister, with no limits on their number. The following are the main categories of membership (composition at 6 Sept. 2011, excluding 24 members who were on leave of absence, 13 disqualified as senior members of the judiciary, one disqualified as an MEP and one suspended):

Party	Excepted Life Peers	Hereditary Peers[1]	Bishops	Total
Conservative	170	48	...	218
Labour	236	4	...	240
Liberal Democrat	87	4	...	91
Crossbench	152	32	...	184
Archbishops and Bishops	...	...	24	24
Other	29	2	...	31
Total	674	90	24	788

[1]Made up of 74 peers elected by parties and groups, 15 peers elected by the whole House and one royal office-holder (the Earl Marshal). The second royal office-holder (the Lord Great Chamberlain) and one peer elected by the Crossbench peers were on leave of absence as at 6 Sept. 2011.

Composition by type:

Archbishops and bishops	24
Life Peers under the Appellate Jurisdiction Act 1876	23 (1 woman)
Life Peers under the Life Peerages Act 1958	688 (178 women)
Peers under the House of Lords Act 1999	92 (2 women)
Total	827

The House of Commons consists of Members (of both sexes) representing constituencies determined by the Boundary Commissions. Persons under 18 years of age (since July 2006—formerly 21 years), Clergy of the Church of England and of the Scottish Episcopal Church, Ministers of the Church of Scotland, Roman Catholic clergymen, civil servants, members of the regular armed forces, policemen, most judicial officers and other office-holders named in the House of Commons (Disqualification) Act are disqualified from sitting in the House of Commons. In general hereditary peers are no longer disqualified from membership of the Commons, following the passage of the House of Lords Act 1999. However, the 92 hereditary peers who still sit in the Lords remain disqualified, as do the few hereditary peers who were given life peerages after the passing of the 1999 Act. Life peers are disqualified from membership of the House of Commons.

The Representation of the People Act 1948 abolished the business premises and university franchises, thus ending the practice of plural voting previously afforded certain graduates and business-owners. Only registered persons can vote at Parliamentary elections. No person may vote in more than one constituency at a general election. All persons may apply to vote by post if they are unable to vote in person, or if they fulfil certain legal requirements they may also be entitled to vote by proxy. Elections are held on the first-past-the-post system, in which the candidate who receives the most votes is elected.

All persons over 18 years old and not subject to any legal incapacity to vote and who are either British subjects or citizens of Ireland are entitled to be included in the register of electors for the constituency containing the address at which they were residing on the qualifying date for the register, and are entitled to vote at elections held during the period for which the register remains in force.

Members of the armed forces, Crown servants employed abroad and the wives accompanying their husbands, are entitled, if otherwise qualified, to be registered as 'service voters' provided they make a 'service declaration'. To be effective for a particular register, the declaration must be made on or before the qualifying date for that register. In certain circumstances, British subjects living abroad may also vote.

The Parliamentary Constituencies Act 1986, as amended by the Boundary Commissions Act 1992, provided for the setting up of Boundary Commissions for England, Wales, Scotland and Northern Ireland. The Commissions' last reports were made in Feb. 2007, and future reports will be due at intervals of not less than eight and not more than 12 years; and may be submitted from time to time with respect to the area comprised in any particular constituency or constituencies where some change appears necessary. Any changes giving effect to reports of the Commissions are to be made by Orders in Council laid before Parliament for approval by resolution of each House. The Parliamentary electorate of the United Kingdom and Northern Ireland in the register in Dec. 2009 numbered 45,420,808 (38,129,082 in England, 2,261,269 in Wales, 3,869,700 in Scotland and 1,160,757 in Northern Ireland).

At the UK general election held on 6 May 2010, 649 out of 650 members were returned, 532 from England, 59 from Scotland, 40 from Wales and 18 from Northern Ireland. Every constituency returns a single member. Voting was postponed until 27 May in Thirsk and Malton owing to the death of a candidate shortly before the election.

On 5 May 2011 a referendum was held on proposals to replace the 'first past the post' electoral system (in which each constituency returns the candidate polling the largest number of votes) with an 'alternative vote' system (AV). The promise of a referendum on the issue was a key condition for the Liberal Democrats joining the coalition government following the 2010 general election. Under the AV system, voters rank candidates in order of preference, with the votes of the candidate receiving least support redistributed until one candidate has over 50% of votes. The Liberal Democrats and the Labour Party campaigned in support of moving to AV while the Conservatives opposed it. Turnout was 42·0%, with 67·9% of votes against the change and 32·1% in favour.

One of the main aspects of the former Labour government's programme of constitutional reform was Scottish and Welsh devolution. In the referendum on Scottish devolution on 11 Sept. 1997, 1,775,045 votes (74·3%) were cast in favour of a Scottish parliament and 614,400 against (25·7%). The turnout was 60·4%, so around 44·8% of the total electorate voted in favour. For the second question, on the Parliament's tax-raising powers, 1,512,889 votes were cast in favour (63·5%) and 870,263 against (36·5%). This represented 38·4% of the total electorate.

On 18 Sept. 1997 in Wales there were 559,419 votes cast in favour of a Welsh assembly (50·3%) and 552,698 against (49·7%). The turnout was 51·3%.

For MPs' salaries *see below*. Members of the House of Lords are unsalaried but may recover expenses incurred in attending sittings of the House by claiming a flat rate attendance allowance of £150 or £300. Additionally, Members of the House who are disabled may recover the extra cost of attending the House incurred by reason of their disablement. In connection with attendance at the House and parliamentary duties within the UK, Lords may also recover the cost of travelling to and from home.

The executive government is vested nominally in the Crown, but practically in a committee of Ministers, called the Cabinet, which is dependent on the support of a majority in the House of Commons. The head of the Cabinet is the *Prime Minister*, a position first constitutionally recognized in 1905. The Prime Minister's colleagues in the Cabinet are appointed on his recommendation.

Salaries. Members of Parliament received an annual salary of £65,738 in 2012–13 (unchanged since 2010–11). Ministers who are MPs also receive a ministerial salary. Salary entitlements for 2012–13 (including parliamentary salaries where applicable): Prime Minister, £142,500 (£76,762 ministerial salary); Cabinet Ministers, £134,565 (in the House of Lords, £101,038); Ministers of State, £98,740 (in the Lords £78,891); Parliamentary Under-Secretaries, £89,435 (in the Lords, £68,710); Government Chief Whip, £134,565 (in the Lords, £78,891); Leader of the Opposition, £128,836 (in the Lords, £68,710); Speaker, £141,504; Attorney General, £161,510; Solicitor General, £124,986; Advocate General for Scotland, £91,775. Following a scandal over MPs' abuse of expenses, the Parliamentary Standards Act 2009 was passed in July 2009 creating the Independent Parliamentary Standards Authority (IPSA). Reforms to the expenses system were introduced by the IPSA in May 2010.

The Privy Council. Before the development of the Cabinet System, the Privy Council was the chief source of executive power, but now its functions are largely formal. It advises the monarch to approve Orders in Council and on the issue of royal proclamations, and has some independent powers such as the supervision of the registration of the medical profession. It consists of all Cabinet members, the Archbishops of Canterbury and York, the Speaker of the House of Commons and senior British and Commonwealth statesmen. There are a number of advisory Privy Council committees. The Judicial Committee is the final court of appeal from courts of the UK dependencies, the Channel Islands and the Isle of Man, and some Commonwealth countries.

Freedom of Information Act. The Freedom of Information Act 2000 was implemented gradually between Nov. 2002 and Jan. 2005 when the General Right of Access to all information became law. Not to be confused with the Data Protection Act of 1998, the FOIA allows individuals to gain access to information held by public authorities in England, Wales and Northern Ireland. A separate Act applies in Scotland. Some information is exempted from release, for example security-related documents. An independent Commissioner for Information oversees the process.

Local Government

Administration is carried out by four types of bodies: (i) local branches of some central ministries, such as the Departments of Health and Social Security; (ii) local sub-managements of nationalized industries; (iii) specialist authorities such as the National Rivers Authority; and (iv) the system of local government described below. The phrase 'local government' has come to mean that part of the local administration conducted by elected councils. There are separate systems for England, Wales, Scotland and Northern Ireland.

The Local Government Act 1992 provided for the establishment of new unitary councils (authorities) in England, responsible for all services in their areas, though the two-tier structure of district and county councils remained for much of the country. In 1996 all of Wales and Scotland was given unitary local government systems. In April 2009 a further nine single-tier unitary authorities were created in a bid to simplify the system.

Local authorities have statutory powers and claims on public funds. Relations with central government are maintained through the Department for Communities and Local Government in England, and through the Welsh and Scottish Executives. In England the Home Office is concerned with some local government functions. (These are performed by departments within the Welsh and Scottish Offices.) Ministers have powers of intervention to protect individuals' rights and safeguard public health, and the government has the power to cap (i.e. limit) local authority budgets.

The chair of the council (known as the Mayor in boroughs and cities) is traditionally one of the councillors elected by the rest. However, the Mayor of London has been directly elected since 1999 and following the Local Government Act 2000, 15 councils in England have also introduced direct mayoral elections. Mayors of cities may have the title of Lord Mayor conferred on them. 51 towns in England, seven in Scotland, six in Wales and five in Northern Ireland have the status of city. Brighton and Hove, Wolverhampton and Inverness were awarded city status in 2000. In 2002 Preston, Newport, Stirling, Lisburn and Newry were given city status to mark Queen Elizabeth II's golden jubilee; and in 2012 Chelmsford, Perth and St Asaph were granted city status to mark her diamond jubilee. This status is granted by the personal command of the monarch and confers no special privileges or powers. In Scotland, the chair of city councils is deemed Lord Provost, and is elsewhere known as Convenor or Provost. In Wales, the chair is called Chairman in counties and Mayor in county boroughs. Any parish or community council can by simple resolution adopt the style 'town council' and the status of town for the parish or community.

Functions. Legislation in the 1980s initiated a trend for local authorities to provide services by, or in collaboration with, commercial or voluntary bodies rather than provide them directly. Savings are encouraged by compulsory competitive tendering. In England, county councils are responsible for strategic planning, transport planning, non-trunk roads and regulation of traffic, personal social services, consumer protection, disposal of waste, the fire and library services and, partially, for education. District councils are responsible for environmental health, housing, local planning applications (in the first instance) and refuse collection. Unitary authorities combine the functions of both levels.

Finance. Revenue is derived from the Council Tax, which supports about one-fifth of current expenditure, the remainder being funded by central government grants and by the redistribution of revenue from the national non-domestic rate (property tax). Capital expenditure is financed by borrowing within government-set limits and sales of real estate.

Elections. Elections. England: The 36 metropolitan districts are divided into wards, each represented by three councillors. One-third of the councillors are elected each year for three years out of four. All metropolitan districts had an election on 3 May 2012. The 201 district councils and the 56 English unitary authorities are divided into wards. Each chooses either to follow the metropolitan district system, or to have all seats contested once every four years, or to elect by halves every two years. 92 district councils had an election on 3 May 2012. The 27 county councils have one councillor for each electoral division, elected every four years, with elections next scheduled for 2017.

In London there are 33 councils (including the City of London), the whole of which are elected every four years. London borough elections took place on 6 May 2010. The Greater London Authority has a 25-member Assembly, elected using AMS (Additional Member System), and a directly elected mayor, elected by the SV (Supplementary Vote) system. For the election of London Assembly members London is divided into 14 constituencies. Each constituency elects one member, in addition to which there are 11 'London Member' seats. Boris Johnson (Con.) was re-elected mayor on 3 May 2012.

Wales: The 22 unitary authorities are split between single and multi-member wards, elected every four years. Elections for 21 authorities were held on 3 May 2012.

Scotland: The 32 unitary authorities generally hold elections every four years. The last elections, held on 3 May 2012, were delayed by one year to avoid a clash with elections to the Scottish parliament in May 2011. The next elections are scheduled for 2017 and thereafter local elections will revert to a four-year cycle.

Resident citizens of the UK, Ireland, a Commonwealth country or an EU country may vote and stand for election at age 18.

Election Results. English elections for 35 councils (of which 27 non-metropolitan county councils, seven unitary authorities and the Isles of Scilly) took place on 2 May 2013. Excluding the Isles of Scilly—where all the councillors are independents—the Conservatives secured control over 18 councils (a net loss of 10), Labour 3 (a net gain of 2) and no overall control in the remaining 13. The Conservatives lost 335 seats (for a total of 1,116), Labour gained 291 (538), the Liberal Democrats lost 124 (352), the UK Independence Party gained 139 (147) and others gained 29 (209).

English elections for 36 metropolitan boroughs, 18 unitary authorities and 74 districts on 3 May 2012 resulted in Labour control of 61 councils, Conservative 42, the Liberal Democrats 6 and no overall control in 19. Labour gained 534 seats (1,188 overall), the Conservatives lost 328 (total 785), the Liberal Democrats lost 190 (total 289) and others lost 52 (total 152).

English elections for 36 metropolitan boroughs, 49 unitary authorities and 194 districts on 5 May 2011 resulted in Conservative control of 157 councils, Labour 57, the Liberal Democrats 10 and others 55. The Conservatives gained 81 seats (4,820 overall), Labour gained 800 (total 2,392), the Liberal Democrats lost 695 (total 1,056) and others lost 199 (total 761).

English elections for 36 metropolitan boroughs, 32 London boroughs, 20 unitary authorities and 76 shire district councils were held on 6 May 2010. The Conservatives emerged with control of 65 councils (a net loss of 8), Labour 37 (net gain 15), Liberal Democrats 13 (net loss 4), with no overall control in 45.

Elections for 21 Welsh unitary authorities held on 3 May 2012 resulted in Labour control of ten councils, independent control of two and no overall control in nine councils. An election for Anglesey's unitary authority on 2 May 2013 resulted in a continuation of no party having overall control.

Elections for all 32 unitary authorities in Scotland held on 3 May 2012 resulted in Labour control of 4 councils, Scottish National Party control of 2 and independents 3, with no overall control in the remaining 23. The Scottish National Party won 424 seats; Labour, 394; the Conservatives, 115; the Liberal Democrats, 71; the Greens, 14; the Scottish Socialist Party, 1; and others, 201.

Elections for London's Mayor and a 25-member London Assembly took place on 3 May 2012. Boris Johnson (Con.) won with 51·53% of the vote (including second preferences). He gained 1,054,811 votes (971,931 as first votes) against 992,273 votes (889,918 as first votes) for former mayor Ken Livingstone (Lab.).

National Anthem

'God Save the Queen' (King) (words and tune anonymous; earliest known printed source, 1744).

GOVERNMENT CHRONOLOGY

Governments and Prime Ministers since the Second World War (Con = Conservative Party; Lab = Labour Party):

1945–51	Lab	Clement Attlee
1951–55	Con	Winston Churchill
1955–57	Con	Sir Anthony Eden
1957–63	Con	Harold Macmillan
1963–64	Con	Sir Alec Douglas-Home
1964–70	Lab	Harold Wilson
1970–74	Con	Edward Heath
1974–76	Lab	Harold Wilson
1976–79	Lab	James Callaghan
1979–90	Con	Margaret Thatcher
1990–97	Con	John Major
1997–2007	Lab	Tony Blair
2007–10	Lab	Gordon Brown
2010–	Con	David Cameron

RECENT ELECTIONS

At the general election of 6 May 2010, 29,653,638 votes were cast. The Conservative Party won 306 seats with 36·1% of votes cast (197 with 32·3% in 2005); the Labour Party 258 with 29·0% (356 seats with 35·2%); the Liberal Democrats 57 with 23·0% (62 with 22·1%); Green 1 (others 3). Regional parties (Scotland): the Scottish National Party won 6 seats (6 in 2005); (Wales): Plaid Cymru 3 (3); (Northern Ireland): the Democratic Unionist Party 8 (9); Sinn Féin 5 (5); the Social and Democratic Labour Party 3 (3); the Alliance Party 1 (0). Voting was postponed in one constituency (Thirsk and Malton) following the death of a candidate but a subsequent by-election held on 27 May saw the Conservatives win the seat, bringing their total to 307. The Conservatives gained 100 seats and lost 3; Labour gained 3 seats and lost 94; the Liberal Democrats gained 8 seats and lost 13. Turnout was 65·1% (61·3% in 2005).

European Parliament

The United Kingdom has 73 (78 in 2004) representatives. At the June 2009 elections turnout was 34·7% (38·5% in 2004). The Conservative Party won 25 seats with 27·0% of votes cast (political affiliation in European Parliament: 24 with European Conservatives and Reformists and one non-attached); UK Independence Party, 13 with 16·1% (Europe of Freedom and Democracy); the Labour Party, 13 with 15·3% (Progressive Alliance of Socialists and Democrats); the Liberal Democrats, 11 with 13·4% (Alliance of Liberals and Democrats for Europe); the Green Party, 2 with 8·4% (Greens/European Free Alliance); the British National Party, 2 with 6·0% (non-attached); the Scottish National Party, 2 with 2·1% (Greens/European Free Alliance); Plaid Cymru, 1 with 0·8% (Greens/European Free Alliance). Voting for these parties was on a proportional system. Voting in Northern Ireland was by the transferable vote system: Sinn Féin (European United Left/Nordic Green Left), the Democratic Unionist Party (non-attached) and the Ulster Unionist Party (European Conservatives and Reformists) gained 1 seat each.

CURRENT GOVERNMENT

In March 2013 the cabinet comprising the Conservative Party (Con.) and the Liberal Democrats (LD) consisted of the following:

(a) 23 MEMBERS OF THE CABINET

Prime Minister and First Lord of the Treasury: David Cameron (Con.), b. 1966.

Deputy Prime Minister: Nick Clegg (LD), b. 1967.

Secretary of State for Foreign and Commonwealth Affairs: William Hague (Con.), b. 1961.

Chancellor of the Exchequer: George Osborne (Con.), b. 1971.

Chief Secretary to the Treasury: Danny Alexander (LD), b. 1972.

Secretary of State for the Home Department: Theresa May (Con.), 1956.

Secretary of State for Defence: Philip Hammond (Con.), b. 1955.

Secretary of State for Business, Innovation and Skills: Vince Cable (LD), b. 1943.

Secretary of State for Work and Pensions: Iain Duncan Smith (Con.), b. 1954.

Secretary of State for Justice and Lord Chancellor: Chris Grayling (Con.), b. 1962.

Secretary of State for Education: Michael Gove (Con.), b. 1967.

Secretary of State for Communities and Local Government: Eric Pickles (Con.), b. 1952.

Secretary of State for Health: Jeremy Hunt (Con.), b. 1966.

Secretary of State for Environment, Food and Rural Affairs: Owen Paterson (Con.), b. 1956.

Secretary of State for International Development: Justine Greening (Con.), b. 1969.

Secretary of State for Scotland: Michael Moore (LD), b. 1965.

Secretary of State for Energy and Climate Change: Edward Davey (LD), b. 1965.

Secretary of State for Transport: Patrick McLoughlin (Con.), b. 1957.

Secretary of State for Culture, Media and Sport: Maria Miller (Con.), b. 1964 (*also Minister for Women and Equalities*).

Secretary of State for Northern Ireland: Theresa Villiers (Con.), b. 1968.

Secretary of State for Wales: David Jones (Con.), b. 1952.

Leader of the House of Lords and Chancellor of the Duchy of Lancaster: Lord Hill of Oareford (Con.), b. 1960.

(Non-cabinet members but attend cabinet meetings): Ken Clarke (Con.), b. 1940, *Minister without Portfolio;* Andrew Lansley (Con.), b. 1956, *Leader of the House of Commons and Lord Privy Seal;* Sir George Young (Con.), b. 1941, *Chief Whip;* Francis Maude (Con.), b. 1953, *Minister for the Cabinet Office and Paymaster General;* Oliver Letwin (Con.), b. 1956, *Minister for Government Policy;* David Laws (LD), b. 1965, *Minister of State for Schools and the Cabinet Office;* Grant Shapps (Con.), b. 1968, *Minister without Portfolio and Chairman of the Conservative Party;* Baroness Warsi (Con.), b. 1971, *Senior Minister of State;* David Willetts (Con.), b. 1956, *Minister of State for Universities and Science.*

(b) LAW OFFICERS

Attorney General: Dominic Grieve, QC (Con.), b. 1956 (*invited to attend Cabinet when required*).

Solicitor General: Oliver Heald (Con.), b. 1954.

Advocate General for Scotland: Jim Wallace, Baron Wallace of Tankerness, QC (LD), b. 1954.

(c) MINISTERS OF STATE (BY DEPARTMENT)

Department for Business, Innovation and Skills: David Willetts (Con.), b. 1956, *Minister for Universities and Science;* Michael Fallon (Con.), b. 1952, *Minister for Business and Enterprise;* Lord Green of Hurstpierpoint (Con.), b. 1948 (*also Foreign and Commonwealth Office*), *Minister for Trade and Investment.*

Cabinet Office: David Laws (LD), b. 1965 (*also Department for Education*).

Department for Communities and Local Government: Baroness Warsi (Con.), b. 1971 (*also Foreign and Commonwealth Office*),

Minister for Faith and Communities; Mark Prisk (Con.), b. 1962, *Minister for Housing.*

Department for Culture, Media and Sport: Hugh Robertson (Con.), b. 1962, *Minister for Sport.*

Ministry of Defence: Andrew Robathan (Con.), b. 1951, *Minister for the Armed Forces;* Mark Francois (Con.), b. 1965, *Minister for Defence Personnel, Welfare and Veterans.*

Department for Education: David Laws (LD), b. 1965 (*also Cabinet Office*), *Minister for Schools.*

Department for Energy and Climate Change: Gregory Barker (Con.), b. 1966, *Minister for Climate Change;* John Hayes (Con.), b. 1958, *Minister for Energy.*

Department for Environment, Food and Rural Affairs: David Heath (LD), b. 1954, *Minister for Agriculture and Food.*

Foreign and Commonwealth Office: Baroness Warsi (Con.), b. 1971 (*also Department for Communities and Local Government*); David Lidington (Con.), b. 1956; Hugo Swire (Con.), b. 1959; Lord Green of Hurstpierpoint (Con.), b. 1948 (*also Department for Business, Innovation and Skills*), *Minister for Trade and Investment.*

Department of Health: Norman Lamb (LD), b. 1957, *Minister for Care Services.*

Home Office: Mark Harper (Con.), b. 1970, *Minister for Immigration;* Damian Green (Con.), b. 1956 (*also Ministry of Justice*), *Minister for Policing and Criminal Justice;* Jeremy Browne (LD), b. 1970, *Minister for Crime Prevention.*

Department for International Development: Alan Duncan (Con.), b. 1957.

Ministry of Justice: Lord McNally (LD), b. 1943; Damian Green (Con.), b. 1956 (*also Home Office*), *Minister for Policing and Criminal Justice.*

Northern Ireland Office: Mike Penning (Con.), b. 1957.

Department for Transport: Simon Burns (Con.), b. 1952.

Department for Work and Pensions: Mark Hoban (Con.), b. 1964, *Minister for Employment;* Steve Webb (LD), b. 1965, *Minister for Pensions.*

(d) PARLIAMENTARY SECRETARIES AND UNDER-SECRETARIES (BY DEPARTMENT)

Department for Business, Innovation and Skills: Jo Swinson (LD), b. 1980 (*also Department of Culture, Media and Sport*), *Minister for Employment Relations and Consumer Affairs;* Matthew Hancock (Con.), b. 1978 (*also Department for Education*), *Minister for Skills;* Lord Younger (Con.), b. 1955.

Cabinet Office: Nick Hurd (Con.), b. 1962; Chloe Smith (Con.), b. 1982.

Department for Communities and Local Government: Nicholas Boles (Con.), b. 1965, *Minister for Planning;* Don Foster (LD), b. 1947; Brandon Lewis (Con.), b. 1971; Baroness Hanham (Con.), b. 1939.

Department for Culture, Media and Sport: Ed Vaizey (Con.), b. 1968, *Minister for Culture, Communications and Creative Industries;* Helen Grant (Con.), b. 1961 (*also Ministry of Justice*); Jo Swinson (LD), b. 1980 (*also Department for Business, Innovation and Skills*).

Ministry of Defence: Andrew Murrison (Con.), b. 1961, *Minister for International Security Strategy;* Philip Dunne (Con.), b. 1958, *Minister for Defence Equipment, Support and Technology;* Lord Astor of Hever (Con.), b. 1946.

Department for Education: Matthew Hancock (Con.), b. 1978 (*also Department for Business, Innovation and Skills*); John Nash (Con.), b. 1949; Edward Timpson (Con.), b. 1973; Elizabeth Truss (Con.), b. 1975.

Department of Energy and Climate Change: Baroness Verma (Con.), b. 1959.

Department for Environment, Food and Rural Affairs: Richard Benyon (Con.), b. 1960; Lord de Mauley (Con.), b. 1957.

Foreign and Commonwealth Office: Mark Simmonds (Con.), b. 1964; Alistair Burt (Con.), b. 1955.

Department of Health: Anna Soubry (Con.), b. 1956; Daniel Poulter (Con.), b. 1978; Earl Howe (Con.), b. 1951.

Home Office: James Brokenshire (Con.), b. 1968; Lord Taylor of Holbeach (Con.), b. 1943.

Department for International Development: Lynne Featherstone (LD), b. 1951.

Ministry of Justice: Helen Grant (Con.), b. 1961 (*also Department for Culture, Media and Sport*), *Minister for Victims and the Courts;* Jeremy Wright (Con.), b. 1972, *Minister for Prisons and Rehabilitation.*

Scotland Office: David Mundell (Con.), b. 1962.

Department for Transport: Norman Baker (LD), b. 1957; Stephen Hammond (Con.), b. 1962.

HM Treasury: Greg Clark (Con.), b. 1967, *Financial Secretary;* David Gauke (Con.), b. 1971, *Exchequer Secretary;* Sajid Javid (Con.), b. 1969, *Economic Secretary;* Lord Sassoon (Con.), b. 1955, *Commercial Secretary.*

Wales Office: Stephen Crabb (Con.), b. 1973; Baroness Randerson (LD), b. 1948.

Department for Work and Pensions: Lord Freud (Con.), b. 1950, *Minister for Welfare Reform;* Esther McVey (Con.), b. 1967, *Minister for Disabled People.*

Leader of the House of Commons: Tom Brake (LD), b. 1962.

(e) OPPOSITION FRONT BENCH
Leader of the Opposition: Ed Miliband, b. 1969.

Shadow Leader of the House of Lords: Baroness Royall of Blaisdon, b. 1955.

The *Speaker* of the House of Commons is John Bercow (Con.), elected on 22 June 2009.

Government Website: http://www.direct.gov.uk

CURRENT LEADERS

David Cameron

Position
Prime Minister

Introduction
David Cameron became prime minister of a coalition government with the Liberal Democrats on 11 May 2010, five days after the general election that returned the Conservatives to power following 13 years of Labour rule. On taking office he became the UK's youngest prime minister in nearly 200 years. After winning the Conservative leadership in 2005, Cameron aligned his party with the political centre. He has embraced broadly the same foreign policy as the previous government, aligning Britain with the USA and supporting British military involvement in Iraq and Afghanistan. However, he has adopted a more sceptical stance on further integration with the European Union, pledging to negotiate a new relationship with the EU and to put the outcome to a national referendum after the next general election scheduled for 2015. Regarding economic policy, he promised to reduce the large deficit in the public finances, while protecting key public services, but has had to contend with the recurring threat of recession and minimal growth in GDP through the first half of his premiership. Meanwhile, his working relationship in government with his Liberal Democrat partners has at times proved fractious.

Early Life
David William Duncan Cameron was born on 9 Oct. 1966 in London, the son of a stockbroker and a baronet's daughter. He grew up in Peasemore, Berkshire and was educated at Heatherdown Preparatory School and Eton College. From 1985–88 he studied politics, philosophy and economics at Brasenose College, Oxford, after which he worked for the research department of the ruling Conservative party. He rose to lead the department's political section and in 1992 headed the economic section of the Conservatives' general election campaign team. Following their

victory, he was appointed special adviser. He was working for the chancellor of the exchequer, Norman Lamont, when the government was forced to withdraw sterling from the European Exchange Rate Mechanism on 'Black Wednesday', 16 Sept. 1992.

In May 1993 he was appointed adviser to the Home Office where, under home secretary Michael Howard, he helped introduce criminal justice legislation and promoted initiatives for private companies to build and run prisons. Cameron left politics in 1994 to work in corporate affairs at media company Carlton Communications, where for seven years he worked closely with chairman Michael Green, handling public relations and communications. At the 1997 general election he unsuccessfully contested Stafford for the Conservatives. During this campaign he broke with official Conservative party policy to oppose joining the single European currency. In the June 2001 election, he was elected MP for Witney in Oxfordshire, increasing the Conservative share of the vote. In the subsequent Conservative leadership contest he initially supported modernizer Michael Portillo, before finally voting for Iain Duncan Smith, who was seen as an outsider candidate and a eurosceptic.

Cameron was appointed to the Home Affairs Select Committee, where he argued for a review of legislation on illegal drugs. He supported British participation in the 2003 Iraq war, in line with Conservative party policy, despite expressing doubts in the press. In June 2003 he was appointed shadow minister in the Privy Council office and in Nov. that year became deputy chairman of the Conservative party under newly elected leader Michael Howard. Cameron was named spokesman on local government finance in March 2004 and in June 2004, following poor local election results for the Conservatives, he was given responsibility for party policy co-ordination.

Cameron played a key role in developing the party's 2005 general election manifesto, which promised to cut taxes, increase numbers of police and prisons, and oppose the introduction of a European constitution. In May 2005 the Conservatives gained votes but failed to prevent a third successive Labour victory. Cameron was subsequently appointed shadow education secretary and when Howard resigned in Sept. 2005, he entered the contest for the party leadership. He positioned himself as a modernizer, able to appeal to a wide range of voters, while also winning the support of eurosceptics with a pledge to withdraw Conservative Members of the European Parliament (MEPs) from the European People's Party (EPP—the EU's main centre-right grouping) on the grounds that it was too federalist. After initially trailing his main rival, right-winger David Davis, he won support with his speech at the party conference in Oct. 2005. In the final ballot on 6 Dec. 2005 he defeated Davis, taking 67% of the vote.

In his acceptance speech he promised to address the under-representation of women among Conservative MPs, to reformulate the party's approach to the inner cities and to end 'Punch and Judy politics' by supporting government initiatives where they were in line with Conservative thinking. In July 2006 Cameron fulfilled his campaign promise and withdrew Conservative MEPs from the EPP, forming a new alliance with Czech and Polish MEPs, a move criticized by some as taking the party too far to the right. He supported the 2006 Education Bill that diluted local council control over schools. Identifying himself as a liberal conservative on social matters, he spoke in support of civil partnerships for gay couples at the 2006 party conference. During the same year he introduced the 'A-list' initiative, which required constituency parties to choose their parliamentary candidate from a centrally approved list. Intended to improve representation of women and ethnic minority candidates, this met resistance at local level.

In March 2007 Cameron launched a review into the quality of childhood in Britain and promised to promote marriage through taxation reforms. He repeatedly depicted Britain as a society in moral and economic decline, an interpretation summed up in his phrase 'broken Britain', and declared his support for economic liberalism combined with the promotion of voluntary and charity work.

During 2007 he said that a Conservative government would aim for lower taxes, with priority given to reducing inheritance tax liability. These proposals were widely credited with prompting Labour government reforms to inheritance tax thresholds in Oct. 2007. Cameron also announced his intention to diversify provision of public services, in particular in education and the health service, proposing that funds could be diverted from state agencies to private enterprise or the voluntary sector. Though widely welcomed by Conservative voters, some were disappointed by his refusal to consider creating more grammar schools.

In foreign policy Cameron continued to support British troop deployment in Iraq and Afghanistan. On domestic security issues, in June 2008 he opposed increasing the maximum period of detention without charge from 28 to 42 days, arguing that it threatened civil liberties. He advocated a new border protection service that would include armed police officers. In Feb. 2009 he pledged to repeal the Human Rights Act and replace it with a British bill of rights, in response to public anxieties that the Act could be used to protect criminals. Cameron advocated increased prioritization of environmental policies, including proposals to expand nuclear energy, and he opposed plans to build a third runway at Heathrow airport.

During the international financial crisis in the autumn of 2008, he supported the Labour government's bank rescue package, though he strongly criticized its record of financial management, particularly the high level of the public finance deficit. The subsequent financial downturn forced him to reassess his party's policies. In 2009 he announced that tax cuts would be contingent on economic improvement and confirmed that this meant postponing further inheritance tax reforms. In Dec. 2009 ratification of the Lisbon Treaty on the European constitution obviated his previous pledge to hold a referendum on the issue. Instead he promised to introduce a sovereignty bill that would prevent any further transfer of power to the EU without a referendum. Cameron fought the 2010 general election on a platform of reduced government spending (with the exception of the NHS and overseas aid), fewer business regulations and increased choice in health and education services.

Career in Office
Cameron became prime minister on 11 May 2010. Although the Conservatives failed to secure an absolute majority in the election of 6 May they did win 306 seats compared to 258 for Labour and succeeded in forming a coalition (the UK's first since the Second World War) with the Liberal Democrats, who had won 57 seats.

Cameron's key challenges were to address the faltering economy and the growing gap between rich and poor. Following an initial austerity budget in June, the coalition government announced in Oct. a raft of deep, and unpopular, public spending cuts in a comprehensive spending review aimed at reducing the UK's large budget deficit. Then, in Feb. 2011, details of proposed legislation to radically overhaul the state welfare system were unveiled.

Security has remained a pressing issue as the UK continues to face the threat of terrorism, both imported and home-grown. Cameron is maintaining Britain's military commitment to, and offensive engagement in, Afghanistan until 2014, when the anticipated withdrawal of NATO combat troops from the country should take place. He also concluded a defence and security accord with France in Nov. 2010, which provides for co-operation in testing nuclear warheads, shared aircraft-carrier capability and a joint expeditionary force. In 2011 Britain played a prominent military role in UN-sanctioned NATO intervention in Libya to protect civilians against the Gaddafi regime which was overthrown in Oct. Then in early 2013 the UK government provided logistical support for French military intervention in

Mali to counter Islamist extremists controlling much of the north of the country.

In June 2010 a long-running independent inquiry into 'Bloody Sunday', in which unarmed Catholic civil rights demonstrators in Northern Ireland were shot dead by British troops in 1972, concluded that the killings were unjustified, prompting an apology by Cameron as prime minister on behalf of the British state.

In May 2011 voters in a national referendum rejected by a large margin a proposal to replace the first-past-the-post electoral system for the House of Commons with the alternative vote (AV) system. Cameron had strongly opposed the plan, placing considerable strain on his governing partnership with the Liberal Democrats, for whom such a change had been a central policy objective. There was further political controversy in Oct. that year as the secretary of state for defence, Liam Fox, resigned from the cabinet over a conflict of interest involving the lobbying links of a close friend.

In Aug. 2011 the worst rioting and looting in decades erupted in London and other English major cities and was only contained by police after several days of disorder and mass arrests. The violence prompted Cameron to recall Parliament from its recess to debate appropriate responses.

The effects of the debt crisis in several eurozone countries further aggravated the UK's parlous financial situation in 2011 and 2012, and highlighted the political divisions within the Conservative Party towards the EU. In Oct. 2011 Cameron suffered a major revolt by Conservative MPs who, in defiance of government policy, supported a motion calling for a referendum on continued membership. The motion was only defeated with the support of Liberal Democrat and opposition Labour members. In Dec. 2011, aware of Conservative hostility to further European integration, Cameron vetoed an EU-wide treaty change proposing greater fiscal union on the grounds that it would undermine London's position as a leading international financial hub. Although popular with his own parliamentary eurosceptics, the move put Cameron's relationship with his pro-EU Liberal Democrat coalition partners under pressure. This was exacerbated in Jan. 2013 when Cameron, acting after rising support for the anti-EU UK Independence Party in several parliamentary by-elections in 2012, promised a renegotiation of Britain's relationship with the EU and, more controversially, a referendum on continued membership after 2015. The following month Cameron claimed credit for spearheading successful inter-governmental negotiations to reduce the EU's seven-year budget for the first time in the Union's history.

Meanwhile, the UK economy continued to struggle, registering minimal growth, falling demand and stubborn long-term unemployment, particularly among young people. In Jan. 2012 official figures indicated net public sector debt (excluding bank bail-outs) had risen above £1trn. for the first time and in Dec. the Chancellor of the Exchequer warned of extended austerity until 2018. Discontent with the government's austerity measures was also aggravated by public concerns over Cameron's championing of controversial reforms to the National Health Service and deep cuts in the level of social welfare provision.

Cameron's director of communications, Andy Coulson, had resigned in Jan. 2011 amid allegations that he was aware of illegal telephone hacking by the national Sunday newspaper of which he was editor before his political appointment in 2007. The subsequent escalation of the controversy, which saw some commentators cast doubt on Cameron's personal judgment, led to an inquiry under Lord Justice Leveson that published its report in Nov. 2012 into the culture, practices and ethics of the news-paper industry. Cameron's rejection of Leveson's central recommendation of statutory press regulation again highlighted differences with his Liberal Democrat coalition partners. Tensions were aggravated further by the Conservatives' failure to fully embrace Liberal Democrat aspirations to reform the House of

Lords and by the Liberal Democrat refusal to back Conservative efforts to update parliamentary constituency boundaries.

In addition to the coalition friction, Cameron courted further dissent within his own party in Jan. 2013 as government legislation to approve same-sex marriage failed to attract the support of over half of Conservative MPs and only passed with Liberal Democrat and opposition backing.

In Oct. 2012 Cameron and Alex Salmond, the First Minister of Scotland, agreed to stage a referendum in 2014 on Scottish independence from the rest of the United Kingdom, although Cameron has stressed his opposition to a breakaway.

DEFENCE

The Defence Council was established on 1 April 1964 under the chairmanship of the Secretary of State for Defence, who is responsible to the Sovereign and Parliament for the defence of the realm. Vested in the Defence Council are the functions of commanding and administering the Armed Forces. The Secretary of State heads the Department of Defence.

Defence policy decision-making is a collective governmental responsibility. Important matters of policy are considered by the full Cabinet or, more frequently, by the Defence and Overseas Policy Committee under the chairmanship of the Prime Minister.

Total full-time trained strength in 2008 numbered 174,000, untrained regulars 18,400 and (2005) reserve personnel 235,600. In 2007 UK armed forces abroad included 21,400 personnel based in Germany, 7,400 in Afghanistan (in most cases serving as part of ISAF), 6,400 in Iraq and 3,000 in Cyprus. Deaths in the UK regular armed forces totalled 129 in 2012, down from 187 in 2010. The last British troops left Iraq in July 2009. In April 2012 there were approximately 9,500 British troops in Afghanistan. British troop deaths in Iraq between 2003 and July 2009 totalled 179; deaths in Afghanistan between 2001 and April 2013 totalled 444.

In Nov. 2010 the UK and France signed a Defence and Security Cooperation Treaty providing for the creation of a rapid reaction force, with troops from both nations called up as required after a joint political decision. Aircraft carriers may be jointly used under certain circumstances and there will be joint training exercises, pooling of resources for the maintenance and logistics of the A400M transport aircraft, and joint work on several other projects. A separate Joint Radiographic/Hydrodynamics Facilities treaty will see the sharing of testing facilities at atomic weapons establishments in the UK (Aldermaston) and France (Valduc), and the development of a new joint hydrodynamic facility.

The ban on homosexuals serving in the armed forces, which had been upheld by a House of Commons vote in May 1996, was suspended in Sept. 1999 after the European Court of Human Rights ruled that the current ban was unlawful.

Defence Budget. In accordance with the Spending Review 2010 the planned defence budget for 2013–14 is £34·1bn. (£24·9bn. resource budget); 2014–15, £33·5bn. (£24·7bn. resource budget). Defence spending in 2008 represented 2·3% of GDP (the lowest level since the 1930s), down from 5·2% in 1985. Per capita defence expenditure in 2008 totalled £638 (US$1,070).

Nuclear Weapons. Having carried out its first test in 1952, there have been 45 tests in all (the last in 1991). The nuclear arsenal consisted of about 160 Trident submarine-launched ballistic missile warheads in Jan. 2012 according to the Stockholm International Peace Research Institute. In addition there were some 65 non-deployed weapons in the nuclear stockpile.

Arms Trade. The UK is a net exporter of arms and in 2010 was the world's fifth largest supplier of major conventional weapons after the USA, Russia, Germany and China (with sales worth US$1·1bn., or 4·6% of the world total).

In 2010 BAE Systems was the UK's largest arms-producing company and the second largest in the OECD. It accounted for US$32·9bn. of arms sales. Rolls Royce, Babcock International Group and Cobham are other significant UK arms-producing companies.

The UK was the 14th largest recipient of major conventional weapons in the world during the period 2007–11, spending US$2,800m. over the five-year period.

Army

The Chief of the General Staff (CGS) is the 4-star commander and professional head of the Army and is responsible for implementing departmental decisions within it. He provides advice to ministers on managerial and operational issues and sets the strategic direction for the Army. CGS is also accountable for the efficiency and fighting effectiveness of the Army and the delivery of military capability.

CGS commands the Army through a single Army Staff that works out of the Army Headquarters in Andover, Hants. Subordinate to CGS are three 3-star commanders: Commander Land Forces (CLF), the Adjutant General (AG) and Commander Force Development and Training (FDT). In line with the 2010 Defence Reform Review, the four-star post of Commander-in-Chief Land Forces ceased to exist in Nov. 2010. On operations command rests with the Ministry of Defence (MOD) via the tri-service Permanent Joint Headquarters (PJHQ) at Northwood, north London.

The British Army's military tasks fall into four core categories: defending the United Kingdom and its Overseas Territories; supporting the civil emergency organizations in time of crisis; defending the UK's interests by projecting power strategically and through expeditionary intervention; and providing security for stabilization operations.

As at 1 Jan. 2012 the established strength of the Regular Army was 100,630. In line with the 2010 Defence Security and Strategic Review, however, the Regular Army strength will reduce to a size of around 82,000 by 2020. This reduction will be offset in part by an enhancement to the Territorial Army (TA), which will become a progressively larger and more integrated element of a trained Army of around 112,000 by 2020. Women serve throughout the Army although they do not currently serve in the infantry or armoured roles.

Successful transition in Afghanistan remains the top priority for the Army. There were approximately 9,500 British servicemen and women deployed as part of the International Security Assistance Force (ISAF) as at 1 Jan. 2012. Current planning sees the Army withdrawn from combat operations in Afghanistan by the end of 2014.

The British Army is currently equipped with a wide range of equipment to meet the present-day threats. This includes the tracked range of Challenger 2 main battle tanks, Warrior Infantry Armoured Fighting vehicles, AS90 self-propelled artillery and the Combat Vehicle Reconnaissance (Tracked) series. The wheeled fleet includes Mastiff, Ridgeback, Husky and Jackal, and since 2012 Foxhound, the army's latest agile vehicle, which offers unprecedented levels of blast protection. Additionally the British Army has a fleet of AH-64D Apache and Lynx helicopters.

Chandler, David G. and Beckett, Ian, (eds.) *The Oxford History of the British Army.* 2003

Marquess of Anglesey, *A History of the British Cavalry 1816–1919.* 8 vols. 2007

Navy

Control of the Royal Navy is vested in the Defence Council and is exercised through the Admiralty Board, chaired by the Secretary of State for Defence.

The First Sea Lord and Chief of Naval Staff is the professional head of the Royal Navy and is responsible to the Secretary of State for Defence for the fighting effectiveness, efficiency and morale of the Naval Service. Subordinate to the First Sea Lord, the Fleet Commander, based at Navy Command Headquarters (NCHQ) in Portsmouth, is responsible for operations, while the Second Sea

Lord is responsible for naval personnel. Main naval bases are at Devonport, Portsmouth and Faslane.

The roles of the Royal Navy are war-fighting, maritime security and international engagement. The Navy are also the custodians of the UK's Continuous-At-Sea-Deterrent (CASD).

The strength of the fleet's major units in the respective years:

	2007	2008	2009	2010	2011	2012
Strategic Submarines	4	4	4	4	4	4
Nuclear Submarines	9	9	8	7	6	5
Aircraft Carriers	2	2	2	2	1	1
Destroyers	8	7	6	6	6	6
Frigates	17	17	17	17	13	13
Landing Platform Docks	2	2	2	2	2	2
Landing Platform Helicopters	1	1	1	1	1	1

The five Trafalgar class nuclear submarines in active service are scheduled to be replaced by seven Astute class nuclear submarines by 2024. In Dec. 2012, two of the new Astute class submarines (HMS *Astute* and HMS *Ambush*) were undergoing sea trials and a further three were under construction.

The Continuous-At-Sea-Deterrent is borne by four Vanguard class Trident submarines—*Vanguard, Victorious, Vigilant* and *Vengeance*. They are each capable of deploying 16 US-built Trident II D5 missiles. The principal surface ship is the Light vertical/short take-off and landing Aircraft Carrier of the *Invincible* class, HMS *Illustrious,* commissioned in 1982. However, after the retirement of the Harrier jump jets in Dec. 2010, it no longer launches any fixed-wing fighter aircraft. Its sister ship, *Ark Royal*, was decommissioned in Jan. 2011 following the government's Strategic Defence and Security Review in 2010. A third ship in the class, *Invincible*, was decommissioned in 2005. *Illustrious* is set to be replaced by two large *Queen Elizabeth* class aircraft carriers. The first, HMS *Queen Elizabeth*, will launch for sea trials in 2015–16, joining the Fleet later in the decade. Her sister ship, HMS *Prince of Wales*, is scheduled to enter service by 2020. A Helicopter Carrier, HMS *Ocean*, specifically designed for amphibious operations, entered service in 1998 and was joined by two amphibious Landing Platform Docks (LPD), HMS *Albion* and HMS *Bulwark*, in 2003 and 2004 respectively.

The current class of Type 23 Frigates are set to be replaced by the Type 26 Global Combat Ship in the early 2020s. Along with six Type 45 Destroyers, these ships will form the backbone of the future surface fleet.

The Fleet Air Arm is currently going through a period of transition. The Sea King Surveillance and Control helicopter continues to provide crucial capability both at sea and ashore whilst its counterpart, the Sea King Mk5, will continue to provide the Royal Navy's contribution to the UK's Search and Rescue effort until 2016. The Navy has taken delivery of its first Wildcat helicopters (which replace the Lynx Mk3 and 8 helicopters), the Merlin Mk2 (replacing the Merlin Mk1) and, alongside the RAF will, in the near future, operate a Carrier Strike capability from the new aircraft carriers with a fleet of F-35 Lightning IIs (otherwise known as the Joint Strike Fighter), which are scheduled to come into service in the second half of the decade.

The Maritime Reserve (MR) comprises the Royal Naval Reserve (RNR) and the Royal Marines Reserve (RMR), which are volunteer forces that currently number around 2,600 but are expected to grow to 4,150. The MR provides trained personnel to supplement regular forces.

The Royal Marines Command, 6,880-strong in Dec. 2012, provides a commando brigade comprising three commando groups. The Special Boat Squadron and specialist defence units complete their operational strength.

The total number of trained naval service personnel was 32,000 in Dec. 2012 (down from 45,600 in April 1996).

Air Force

The Royal Air Force was formed on 1 April 1918 through the merger of the Royal Flying Corps and the Royal Naval Air Service. It consists of one Command (AIR), which is divided into three Groups.

Number 1 Group, with bases such as Coningsby (Lincs), Leuchars (Fife), Lossiemouth (Moray) and Marham (Norfolk), is home to combat aircraft, support helicopters, and Intelligence, Surveillance, Target Acquisition and Reconnaissance platforms.

The Typhoon has replaced the Tornado F3 in the air defence role and continues to develop a potent air-to-ground capability, with the new Paveway IV bomb as well as Stormshadow missiles for long-range attack. The Harrier was retired in Dec. 2010 following the government's Strategic Defence and Security Review; its successor, the F-35 Lightning II (otherwise known as the Joint Strike Fighter or Joint Combat Aircraft), is set to come into service from 2020. Attack versions of the Tornado have been upgraded and fitted with RAPTOR (Reconnaissance Airborne Pod for Tornado). Remotely Piloted Air Systems (RPAS) continue to be widely employed in support of operations in Afghanistan.

There are two main pillars within Number 2 Group: Air Transport (AT)/Air-to-Air Refuelling (AAR) and Force Protection. AT/AAR provides rapid strategic and tactical reach, including the delivery of the airbridge to the Middle East and South Atlantic. Force protection comprises the RAF Regiment and RAF Police. The Operational Support Squadrons of the Royal Auxiliary Air Force are also included in No. 2 Group, as are Regiment Auxiliaries and the Mountain Rescue Service.

Mainstays of the AT and AAR force are the Hercules, VC10, TriStar and C-17s based at Brize Norton (Oxon). The A330 Voyager entered service in April 2012 and will gradually supplant the VC10 and TriStar tankers as part of the Future Strategic Tanker Aircraft programme. The A400M Atlas is scheduled to enter operational service in late 2014 to replace the Hercules fleet.

The UK Air Surveillance and Control Systems (ASACS) Force retains a fixed, static structure and continues to hold responsibility for the security of UK airspace; in platform terms this mission is delivered through Quick Reaction Alert Typhoons operating in the air-to-air role. In addition the Force has a deployable radar capability provided through No. 1 Air Control Centre, which has provided valuable support to ongoing operations in Afghanistan. Another key element is the space surveillance and missile warning radar at Fylingdales (N. Yorks).

Number 22 (Training) Group recruits and provides trained specialist personnel, and is also responsible for the Air Cadet Organisation and the University Air Squadrons. No. 22 Group also has responsibility both for the Royal Air Force Aerobatic Team (the Red Arrows) and the Royal Air Force Battle of Britain Memorial Flight.

In terms of personnel, on 1 Oct. 2011 Royal Air Force strength stood at 41,580 (9,390 officers and 32,190 other ranks).

ECONOMY

In 2009 services accounted for 78% of GDP, industry 21% and agriculture 1%.

According to the anti-corruption organization *Transparency International*, in 2012 the United Kingdom ranked equal 17th in the world in a survey of the countries with the least corruption in business and government. It received 74 out of 100 in the annual index.

In 2011 the UK gave US$13·8bn. in international aid, representing 0·56% of its GNI. In actual terms this made the UK the fourth most generous country in the world, but as a percentage of GNI only the ninth most generous.

Overview

The UK ranks among the world's largest economies. From 1993–2004 it enjoyed sustained non-inflationary growth, the longest period of expansion in 30 years. However, the economy was severely dented by the global economic crisis.

Despite almost closing the GDP per capita gap with other European countries, the disparity with the most successful OECD countries (Canada, USA and Australia) remains, partly because of weaker productivity levels in the UK. In addition, low levels of gross fixed investment (approximately 17% of GDP) have constrained productivity growth. Private consumption accounted for 63% of GDP in 2012.

The service sector is responsible for more than three-quarters of GDP. The financial services sector represents 20% of GDP thanks to the strength of the City of London and the growth in business services catering to an international market. The strong growth performance over the decade to 2007 was partly attributable to the high share of value-added produced in high growth sectors, especially knowledge intensive services.

In 2009 manufacturing accounted for 11% of national output. The UK is the world's sixth largest producer of manufactured products. Despite the fall in manufacturing's share of overall output (in 1970 the sector accounted for over 30% of GDP), in absolute terms, output of manufacturing increased decade by decade until its peak in 2007. Pharmaceuticals, electronics and the automotive industry are important contributors to the UK's manufacturing base, while the country has a 15% global market share in aerospace.

From 1979–97 the Conservative government introduced reforms making the UK one of Europe's freest economies. The subsequent Labour government continued with privatization, deregulation and competition reforms. Low product market regulation, low barriers to foreign investment and labour market flexibility prompted higher levels of foreign direct investment than in most other EU countries. The UK has exceptionally high numbers of non-EU businesses when compared with its European neighbours.

Monetary policy has been conducted by the Bank of England since 1997. From 1997–2004 average RPIX inflation was stable at around the 2·5% target, despite housing appreciation. Rising commodity and services prices resulted in inflation above target in March 2007, prompting the first open explanatory letter from the governor to the chancellor since the Bank assumed responsibility for monetary policy. Consumer price index (CPI) inflation continued to exceed the 2% target by more than one percentage point, resulting in further open letters in May, Aug. and Nov. 2008. CPI inflation remained one percentage point or more above the 2% target for over two years until April 2012, following a VAT rise to 17·5% in Jan. 2010 and a further increase to 20% in Jan. 2011.

British bank Northern Rock was taken into 'temporary' public ownership in Feb. 2008 as a result of credit market problems. In Oct. 2008 continued instability in the financial markets prompted a government injection of £37bn. into three of the UK's biggest banks—Royal Bank of Scotland, Lloyds TSB and Halifax Bank of Scotland. The Bank of England reduced interest rates six times from Oct. 2008 to a record low of 0·5% in March 2009 in an attempt to stave off a deep recession, a rate still in place in March 2013. By Feb. 2010 the Bank of England had injected a total of £200bn. into the economy as part of an asset-purchase programme labelled 'quantitative easing'.

The economy fell into recession following contractions in the second and third quarters of 2008 and continued to shrink up to Sept. 2009. In the fourth quarter of 2009 GDP expanded by 0·5% but the economy contracted by 4·0% over the full year. Recovery in 2011 was impeded by high commodity prices and the effects of the eurozone debt crisis—the value of shipments to eurozone countries, the destination of two-fifths of British exports, fell significantly. The economy relapsed into recession in the first half of 2012, before the London Olympics provided a boon in the third quarter. A contraction in the following quarter sparked fears of a triple-dip recession, although in fact this was narrowly avoided.

Unemployment rose to 8·4% in Dec. 2011, its highest level in over 15 years, before dipping in 2012. In Feb. 2012 the Bank of England extended its quantitative easing programme by £50bn and another £50bn. was injected in July 2012, taking the total to £375bn.

The Code for Fiscal Stability, introduced in 1998, stipulated that the government may borrow only to invest and not to support current spending. The 'sustainable investment' or 'golden' rule looked to maintain the public sector net debt below 40% of GDP over the economic cycle. In April 2009 the then Chancellor, Alistair Darling, confirmed in his Budget report that borrowing would be taken to record levels to help the government manage the escalating economic crisis. By Sept. 2009 overall government debt stood at 57·5% of GDP, its highest level since 1974. In June 2010 Chancellor George Osborne outlined plans for an emergency Budget to tackle the budget deficit, which reached £155bn. in 2009–10. In Oct. 2010 the Chancellor revealed details of a spending review, which would see nearly 500,000 public sector jobs lost by 2014–15, an increase in the retirement age from 65 to 66 by 2020 and £7bn. in additional welfare budget cuts.

Osborne subsequently focused policy on managing the large national deficit despite weak growth, with the debt-to-GDP ratio expected to peak at 86% in 2016–17. Since 2012 he has announced a reduction in the top rate of income tax from 50% to 45%, increases to the personal tax-free allowance and several freezes to fuel duty. Welfare reforms, including plans to change to a 'universal credit' system that merges several benefits into a single monthly payment, are scheduled to take effect nationally in Oct. 2013.

Spending on health care and education are around the OECD average but public services are overstretched. As an increasing number of companies close final salary pension schemes, the government has encouraged individuals to take more responsibility for their retirement. In March 2011 the government accepted Lord Hutton's review of public sector pension contributions. Hutton proposed linking pensions to career earnings rather than final salaries to make them more 'affordable'.

For further developments *see* www.statesmansyearbook.com.

Currency

The unit of currency is the *pound sterling* (£; GBP) of 100 *pence* (p.). Before decimalization on 15 Feb. 1971 £1 = 20 shillings (s) of 12 pence (d). A gold standard was adopted in 1816, the sovereign, a £1, or twenty-shilling gold coin, weighing 7·98805 grams. It is eleven-twelfths pure gold and one-twelfth alloy. Currency notes for £1 and 10s. were first issued by the Treasury in 1914, replacing the circulation of sovereigns. The issue of £1 and 10s. notes was taken over by the Bank of England in 1928. 10s. notes ceased to be legal tender in 1970 and £1 notes (in England and Wales) in 1988. Sterling was a member of the exchange rate mechanism of the European Monetary System from 8 Oct. 1990 until 16 Sept. 1992 ('Black Wednesday').

Inflation. Consumer Price Index (CPI) inflation rates (based on OECD statistics):

2002	2003	2004	2005	2006	2007	2008	2009	2010	2011
1·3%	1·4%	1·3%	2·0%	2·3%	2·3%	3·6%	2·2%	3·3%	4·5%

Coinage. Estimated number of coins in circulation at 31 March 2010: 28,441m.; £2, 345m.; £1, 1,474m.; 50p, 845m.; 20p, 2,473m.; 10p, 1,651m.; 5p, 3,774m.; 2p, 6,664m.; 1p, 11,215m. Total value: £3,681m.

Banknotes. The Bank of England issues notes in denominations of £5, £10, £20 and £50 up to the amount of the fiduciary issue. Under the provisions of the Currency Act 1983 the amount of the fiduciary issue is limited, but can be altered by direction of HM Treasury on the advice of the Bank of England.

All current series Bank of England notes are legal tender in England and Wales. Some banks in Scotland (Bank of Scotland, Clydesdale Bank and the Royal Bank of Scotland) and Northern Ireland (Bank of Ireland, First Trust Bank, Northern Bank and Ulster Bank) have note-issuing powers.

The total amount of Bank of England notes in circulation at 28 Feb. 2010 was £50,220m.

Foreign exchange reserves were US$39,736m. and gold reserves 9·98m. troy oz in Sept. 2009 (22·98m. troy oz in April 1999, before the Treasury's announcement of its intention to sell nearly 60% of UK gold reserves over the medium term).

Budget

The fiscal year runs from 6 April to 5 April. The March 2013 Budget estimated public sector net borrowing for 2012–13 at £86bn. (against £92bn. predicted in the March 2012 Budget), with a forecast of £108bn. in 2013–14, then £97bn. in 2014–15, £87bn. in 2015–16 and £61bn. in 2016–17. Public sector net debt as a proportion of GDP is put at 75·9% for 2012–13, rising to 85·6% by 2016–17 before falling to 84·8% in 2017–18. As a share of GDP, this will see borrowing fall from 5·6% in 2012–13 to 2·2% by 2017–18.

The United Kingdom's budget deficit in 2011–12 was 6·0% of GDP and is forecast to be within the required target set by the EU of no more than 3% by 2016–17.

Current spending for 2012–13 is set at £657·2bn., increasing to £713·0bn. by 2017–18. Net investment is to increase from –£6bn. in 2012–13 to £26bn. by 2017–18.

The independent Office for Budget Responsibility forecasts economic growth of 0·6% in 2013 (a decrease on the Dec. 2012 forecast of 1·2%), increasing to 1·8% in 2014 and 2·3% in 2015. Among the Budget's provisions was a rise in the personal income tax allowance for under-65s from £8,105 for 2012–13 to £10,000 from April 2014 (a year earlier than planned). In the same time frame the income threshold (including personal allowance) qualifying for higher rate (40%) income tax falls from £42,475 to £41,865.

Plans were announced to make £11·5bn. of additional government cuts in the 2015–16 spending review, up from £10bn. The budgets of most government departments are expected to fall by 1% in each of 2013–14 and 2014–15, although spending on schools and the NHS is protected. A 1% cap on public sector pay was extended to 2016. Tax relief was announced for investors in social enterprises and also in shale gas exploration.

Corporation tax is scheduled to fall from 21% to 20% in 2015. A new employment allowance will cover the first £2,000 of the national insurance bill for every company in the country. Measures to address tax avoidance and evasion—including agreements with Guernsey, Jersey and the Isle of Man—are expected to recover up to £3bn. in unpaid taxes. From 2015 there is to be tax relief on childcare costs up to £6,000 per child. In 2016, a year earlier than planned, the flat-rate pension of £144 will be introduced. The cap on social care costs will protect savings of more than £72,000 from 2017.

An additional £15bn. was earmarked for spending on large infrastructure projects, beginning with a £3bn. injection in 2015–16. Guarantees to support £130bn. of new mortgage lending over three years are to come into effect in 2014. In addition, there are to be interest-free government loans to cover up to 20% of the buyer's cost on new-build properties. A scheduled 3p rise in beer duty was replaced by a 1p cut and a rise in fuel duty planned for Sept. 2013 was also scrapped.

The *Financial Times* commented on 'Osborne's shrewd politics but dismal economics': 'Normalising diminished expectations has become the job description of George Osborne, the chancellor of the exchequer. The government, he insists, has the best strategy to deal with the toxic legacy it inherited and dangerous economic environment it confronts. Given this, he claims, he has done what he can by delivering a "Budget for people who aspire to work hard

and get on"... The politics look shrewd... The economics are vastly more problematic, which in the long run, may also determine the politics. The chancellor cannot disfigure the brutal fact that outcomes for economic activity and public finances are slipping still further from the expectations with which he launched the government's programme in the emergency Budget of June 2010.' (*Financial Times*, 20 March 2013).

Current Budget (in £1bn.)	2011–12 Outturn	2012–13 Forecast	2013–14 Forecast
Current Receipts	572·6	586·8	612·4
Current Expenditure	643·8	657·2	672·9

Surplus on Current Budget (in £1bn.)	2011–12 Outturn	2012–13 Forecast	2013–14 Forecast
	–92	–93	–84

Current Receipts (in £1bn.)	2011–12 Outturn	2012–13 Forecast	2013–14 Forecast
National Accounts Taxes	549·5	553·7	574·3
Current Receipts	572·6	586·8	612·4

Departmental Expenditure Limits (Resource Budget, in £1bn.)	2012–13 Estimate	2013–14 Planned	2014–15 Planned
Education	51·4	53·1	53·8
Health (NHS)	102·9	106·9	109·8
Transport	4·4	4·8	4·4
CLG Communities	1·4	2·0	1·3
CLG Local Government	24·0	23·9	21·7
Business, Innovation and Skills	15·4	14·9	13·8
Home Office	7·9	8·0	7·4
Justice	8·1	7·2	6·8
Law Officers' Departments	0·6	0·6	0·5
Defence	27·1	26·5	24·5
Foreign and Commonwealth Office	2·0	1·8	1·1
International Development	6·1	8·8	8·3
Energy and Climate Change	1·2	1·4	1·1
Environment, Food and Rural Affairs	1·9	1·9	1·7
Culture, Media and Sport	1·9	1·2	1·1
Work and Pensions	7·1	7·6	7·4
Scotland	25·0	25·3	25·3
Wales	13·3	13·5	13·5
Northern Ireland	9·5	9·5	9·5
Chancellor's Departments	3·3	3·7	3·5
Cabinet Office	2·1	2·1	2·3

VAT, introduced on 1 April 1973, was raised from 17·5% to 20·0% in Jan. 2011. The reduced rate is 5·0%.

Rates of Income Tax for 2013–14:

Income between	%[1]
£9,440–£41,450 (basic rate)	20
Between £41,450–£150,000 (higher rate)	40
Over £150,000	45

[1]The rate of tax applicable to savings income is 10% for income up to £2,790, then 20% up to the basic rate limit, 40% to the higher rate limit and thereafter 45%. The rates applicable to dividends are 10% for income up to the basic rate limit, 32·5% to the higher rate limit and 37·5% above that.

Performance

In 2011 total GDP was US$2,445·4bn. (£1,579·7bn.), the seventh highest in the world.

Real GDP growth rates (based on OECD statistics):

2002	2003	2004	2005	2006	2007	2008	2009	2010	2011
2·4%	3·8%	2·9%	2·8%	2·6%	3·6%	–1·0%	–4·0%	1·8%	1·0%

The real GDP growth rate in 2012 according to the Office of National Statistics was 0·3%. With the economy contracting in both the second and third quarters of 2008 the UK went into recession for the first time since 1991. There were five consecutive quarters of negative growth before the economy expanded by 0·4% in the third quarter of 2009. It shrank again in the fourth quarter of 2010, by 0·4%. After a slight recovery there was a further recession in late 2011 and early 2012.

The Office for Budget Responsibility forecasts economic growth of 0·6% for 2013 and 1·8% in 2014.

The UK was ranked eighth in the Global Competitiveness Index in the World Economic Forum's *Global Competitiveness Report 2012–2013*. The index analyses 12 areas of competitiveness for over 100 countries including macroeconomy, higher education and training, institutions, innovation and infrastructure.

Banking and Finance

The Bank of England is the government's banker and the 'banker's bank'. It has the sole right of note issue in England and Wales. It was founded by Royal Charter in 1694 and nationalized in 1946. The capital stock has, since 1 March 1946, been held by HM Treasury. The *Governor* (appointed for five-year terms) is Mervyn King (b. 1948; took office 2003, reappointed in Jan. 2008). He was set to stand down in June 2013, to be succeeded by Mark Carney.

The statutory Bank Return is published weekly. End-Dec. figures are as follows (in £1):

	Notes in circulation	Reserve balances	Other liabilities
2008	46,885,821,070	48,628,063,155	142,976,600,357
2009	52,886,230,380	144,025,592,134	40,801,813,208
2010	54,806,680,990	138,349,127,533	53,750,398,819

Major British Banking Groups' statistics at end Aug. 2011: total deposits (sterling and currency), £3,820bn.; sterling market loans, £392bn.; market loans (sterling and currency), £1,010bn.; advances (sterling and currency), £2,330bn.; sterling investments, £437bn.

In Dec. 2012 Britain's largest bank both by market capitalization and assets was HSBC with US$195,340m. and US$2,693bn. respectively.

Between 2000 and 2007 the number of adults banking online increased by 505%. In 2007, 35% of the population were using e-banking.

In May 1997 the power to set base interest rates was transferred from the Treasury to the Bank of England. The government continues to set the inflation target but the Bank has responsibility for setting interest rates to meet the target. Base rates are now set by a nine-member Monetary Policy Committee at the Bank; members include the Governor. Membership of the Court (the governing body) was widened. The 1998 Act provides for Court to consist of the Governor, two Deputy Governors and 16 Directors. The Act also established the MPC as a Committee of the Bank and sets a framework for its operations. Responsibility for supervising banks was transferred from the Bank to the Financial Services Authority (FSA). The bank rate was lowered from 1·0% to 0·5% on 5 March 2009 (the lowest since the Bank of England was founded in 1694).

National Savings Bank. Statistics for 2006–07 and 2007–08:

	Ordinary accounts		Investment accounts		Premium bonds	
	2006–07	2007–08	2006–07	2007–08	2006–07	2007–08
	in £1,000	in £1,000	in £1,000	in £1,000	in £1,000	in £1,000
Amounts—						
Received	760	1,598	555,032	541,871	8,422,391	6,637,597
Interest credited	3,571	5,665	202,644	223,766	—	—
Paid	(20,069)	(12,965)	(1,181,818)	(974,945)	(5,346,953)	(6,341,256)
Due to depositors at 31 March	312,994	307,292	5,643,572	5,434,264	35,249,099	36,923,456

The London Stock Exchange (called International Stock Exchange until May 1991) originated over 300 years ago, although a regulated stock exchange did not come into existence until 1801. In July 1991 the 91 shareholders voted unanimously for a new memorandum and articles of association which devolves power to a wider range of participants in the securities industry, and replaces the Stock Exchange Council with a 14-member board. The Financial Times Stock Exchange 100 (FTSE 100) ended 2012 at 5,897·8, up from 5,572·3 at the end of 2011 (rising 5·8% during the year).

Gross external debt totalled US$9,944,667m. in June 2012 (compared to US$9,964,658m. in June 2011).

The UK received US$53·95bn. worth of foreign direct investment in 2011, up from US$50·60bn. in 2009 but down from a high of US$196·39bn. in 2007.

Conaghan, Dan, *The Bank: Inside the Bank of England.* 2012
Roberts, R. and Kynaston, D. (eds.) *The Bank of England: Money, Power and Influence, 1694–1994.* 1995

ENERGY AND NATURAL RESOURCES

In 2008 just 2·2% of energy consumption came from renewables (wind power, solar power, hydro-electric power, tidal power, geothermal energy and biomass), compared to the European Union average of 10·3%. A target of 15% has been set by the EU for 2020.

Environment

The UK's carbon dioxide emissions from the consumption and flaring of fossil fuels in 2008 were the equivalent of 9·4 tonnes per capita. The UK's total emission of greenhouse gases is estimated to have fallen from 809m. tonnes in 1990 to 707m. tonnes by 2007. An *Environmental Performance Index* compiled in 2008 ranked the UK 14th in the world, with 86·3%. The index examined various factors in six areas—air pollution, biodiversity and habitat, climate change, environmental health, productive natural resources and water resources.

Electricity

The Electricity Act of 1989 implemented the restructuring and privatization of the electricity industry. In 1999 the domestic electricity market (peak load below 100 kW) was opened to competition.

Generators. Under the 1989 Act, National Power and Powergen took over the fossil fuel and hydro-electric power stations previously owned by the Central Electricity Generating Board, and were privatized in 1991. A succession of takeovers and mergers saw the rapid diversification of the UK generation market, with 30 major power producers operating in 2007 compared to seven in 1990. There were a total of 18 nuclear reactors in use in the UK at ten nuclear power stations in July 2011. However, one of these, Oldbury, owned by BNFL Magnox, ceased generation in Feb. 2012. Wylfa, also owned by BNFL Magnox, is set to close in 2014. The remaining eight stations are operated by British Energy, which was acquired by EDF Energy in Jan. 2009. EDF Energy plans to build several new nuclear power stations in the UK. A variety of companies are planning investment in new generation capacity in order to meet rising demand and replace ageing power stations.

Transmission. Since 2005 the electricity systems of England, Wales and Scotland have been integrated and operate under the British Electricity Trading and Transmission Arrangements (BETTA). National Grid became the operator of the UK transmission networks under the new arrangements.

Distribution and Supply. The 12 Area Boards were replaced under the 1989 Act by regional electricity companies (RECs), which were privatized in 1990; 14 public electricity suppliers then came into operation which, under the terms of the Utilities Act of 2000, needed separate licences for their supply businesses and distribution networks. Subsequent market liberalization has seen a sharp increase in the number of electricity suppliers in operation, among the largest of which are British Gas, EDF Energy, E.ON, RWE npower, Scottish & Southern Energy and ScottishPower.

See also SCOTLAND.

Electricity Associations. The Electricity Association, formerly the trade association of the UK electricity companies, was replaced in Oct. 2003 by three industry bodies. The Energy Networks Association (ENA) represents the transmission and distribution companies for both gas and electricity. The Association of Electricity Producers (AEP) represented companies that generate electricity using coal, gas and nuclear power as well as renewable sources such as wind, biomass and water. The Energy Retail Association (ERA) represented Britain's domestic electricity and gas suppliers in the internal market. AEP and ERA were merged with the UK Business Council for Sustainable Energy (UKBCSE) in April 2012 to create Energy UK. A trade association for both the electricity and the gas sector, it represents a wide range of interests. It includes small, medium and large companies working in electricity generation, energy networks and gas and electricity supply, as well as a number of businesses that provide equipment and services to the industry.

Regulation. The Office of Electricity Regulation ('Offer') was set up under the 1989 Act to protect consumer interests following privatization. In 1999 it was merged with the Office of Gas Supply ('Ofgas') to form the Office of Gas and Electricity Markets ('Ofgem'), reflecting the opening up of all markets for electricity and gas supply to full competition from May that year, with many suppliers now offering both gas and electricity to customers.

Statistics. The electricity industry contributed about 1·05% of the UK's Gross Domestic Product in 2007. The installed capacity of all UK power stations in 2007 was 82,951 MW. In 2007 the fuel generation mix was: gas 41·9%, coal 34·8%, nuclear 16·1%, hydro and renewables 5·3%, oil 1·2% and other 0·7%. Final consumption totalled 341,945 GWh, of which domestic users took 33·6%, industrial users 34·4% and commercial and other users the remaining 32·0%. Consumption per capita in 2007 was 6,582 kWh.

Electrica Services. *Electricity Industry Review.* Annual

Surrey, J. (ed.) *The British Electricity Experience: Privatization—the Record, the Issues, the Lessons.* 1996

Oil and Gas

Production in 1,000 tonnes, in 2010: throughput of crude and process oils, 73,200; refinery use, 4,478. Refinery output: gas/diesel oil, 24,837; motor spirit, 19,918; fuel oil, 6,912; aviation turbine fuel, 5,781; butane, propane and other petroleum gases, 2,764; burning oil, 2,570; naphtha, 1,596; bitumen, 1,276. Total output of refined products, 68,394. Total indigenous oil production (2011), 52·0m. tonnes. The UK had proven oil reserves of 2·8bn. bbls in 2011. The UK became a net importer of oil in 2005, having been a net exporter since 1980.

In 2007 total income from sales of oil and gas produced was £30·1bn. (oil, £20·7bn.). The UK ranks 15th among the world's largest gas producers and 19th among the largest oil producers.

The first significant offshore gas discovery was made in 1965 in the North Sea, followed in 1969 by the first commercial oil offshore. Offshore production of gas began in 1967 and oil in 1975.

Oil and gas have played an important part in providing the UK's energy needs. In 2007, either through direct use or as a source of energy to produce electricity, oil and gas accounted for some 74% (41% gas and 33% oil) of total UK energy consumption, with UK-based production supplying some 93% of all the oil and gas consumed. However, oil production peaked in 1999 as did natural gas production in 2000, and annual production of both has gradually been declining in the years since then.

Oil products also provide important contributions to other industries, such as feedstocks for the petrochemical industry and lubricants for various uses. While the importance of oil as a source of energy for electrical generation and use by industry and commercial operations has declined with the increasing use of gas, oil still makes up around 14% of total industrial uses of energy. Its prime importance is in the transport sector, where it provides 99% of the total energy used.

The reform of the old nationalized gas industry began with the Gas Act of 1986, which paved the way for the privatization later that year of the British Gas Corporation, and established the Director General of Gas Supply (DGSS) as the independent regulator. This had a limited effect on competition, as British Gas retained a monopoly on tariff (domestic) supply. Competition progressively developed in the industrial and commercial (non-tariff) market.

The Gas Act 1995 amended the 1986 Act to prepare the way for full competition, including the domestic market. It created three separate licences—for Public Gas Transporters who operate pipelines, for Shippers (wholesalers) who contract for gas to be transported through the pipelines, and for Suppliers (retailers) who then market gas to consumers. It also placed the DGSS under a statutory duty to secure effective competition.

The domestic market was progressively opened to full competition from 1996 until May 1998.

In 1997 British Gas took a commercial decision to de-merge its trading business. Centrica plc (a new company) was formed to handle the gas sales, gas trading, services and retail businesses of BG, together with the gas production businesses of the North and South Morecambe Field. The remaining parts of the business, including transportation and storage and the international downstream activities, were contained in BG plc. As a result of subsequent changes, National Grid Gas plc now owns and operates the UK's national gas transmission system.

The Department of Energy and Climate Change (DECC) was created on 3 Oct. 2008 to oversee the UK oil and gas industry. The regulator for Britain's gas and electricity industries is *Ofgem* (Office of the Gas and Electricity Markets), created in 1999 through the merger of *Ofgas* (Office of Gas Supply) and *Offer* (Office of Electricity Regulation). Its role is to protect and advance the interests of consumers by promoting competition where possible.

A second European Directive was published in June 2003 with rules for the internal market in natural gas. Member states were allowed one year to execute its provisions; the UK implemented the directive in July 2004.

The UK became a net importer of gas in 2004. To help ensure a secure gas supply as the UK becomes more dependent on imported gas, several gas infrastructure projects have been developed. These supply the UK with gas from a number of sources, including Norway and the Netherlands. There is already a pipeline (the Interconnector) linking the UK and European gas grids in Belgium. This link to Continental Europe opened in Oct. 1998 and has an export capacity of 20·0bn. cu. metres a year and an import capacity of 25·5bn. cu. metres a year.

Proven natural gas reserves in 2009 were 290bn. cu. metres. Production was 45·2bn. cu. metres in 2011. The UK's natural gas output has declined every year since 2000, when it totalled 108·4bn. cu. metres. In 2007, 33·0% of the UK's total gas supply was used by domestic users and 33·4% by electricity generators.

Wind

In 2007 there were 164 wind farms and 1,941 turbines with a capacity of 2,388·4 MW for electricity generation.

Minerals

Legislation to privatize the coal industry was introduced in 1994 and established the Coal Authority to take over certain activities from British Coal Corporation. The Coal Authority is the owner of almost all the UK's coal reserves; it licenses private coal-mining and disposes of property not required for operational purposes. The Coal Authority also deals with the historic legacy of coal mining including handling subsidence claims in former mining areas, treating mine water discharges and dealing with surface hazards. In 2010 there were seven former British Coal collieries, eleven additional deep mines and 37 surface mines, employing some 6,081 mineworkers.

Total production from deep mines was 8·1m. tonnes in 2008 (204·7m. tonnes in 1958 and 83·8m. tonnes in 1988). Output from opencast sites in 2008 was 9·5m. tonnes (15·0m. tonnes in 1958 and 20·3m. tonnes in 1988). In 2008 inland coal consumption was 58·2m. tonnes (113·3m. tonnes in 1988).

Output of non-fuel minerals in Great Britain, 2009 (in 1,000 tonnes): sand and gravel, 65,301; limestone, 58,092; igneous rock, 38,860; sandstone, 8,542; salt, 5,855; clay and shale, 5,310; chalk, 4,047; industrial sand, 3,755; dolomite, 3,164; china clay, 988.

Steel and Metals
Steel production in recent years (in 1m. tonnes):

2006	13·9
2007	14·4
2008	13·5
2009	10·1
2010	9·7

Deliveries of finished steel products from UK mills in 2010 were worth £8·5bn. in product sales and comprised 4·7m. tonnes to the UK domestic market and 4·8m. tonnes for export. About 71% of UK steel exports went to other EU countries. UK steel imports in 2010 were about 6·0m. tonnes, with 71% coming from other EU countries. The UK steel industry's main markets are construction, engineering, automotive and metal goods. Tata Steel Europe (formerly Corus) is the UK's largest steel producer and in 2010 made around 80% of UK crude steel.

Agriculture
Land use in 2008: agriculture, 73%; forests, 12%; other, 15%. In 2011 agricultural land in the UK totalled (in 1,000 ha.) 18,263, comprising agricultural holdings, 17,064; and common rough grazing, 1,199. Land use of agricultural holdings in 2011 (in 1,000 ha.): permanent grassland (including rough grazing), 9,858; crops, 4,673; temporary grass under five years old; 1,278; uncropped arable land, 156; other, 1,100. Area sown to crops in 2011 (in 1,000 ha.): arable crops, 4,497 (of which wheat, 1,968; barley, 968; oilseed rape, 705); horticultural crops (including fruit), 176.

In 2005 there were 5·73m. ha. of arable land and 47,000 ha. of permanent crops.

In 2010 the area of fully organic farmland in the UK was 667,600 ha. Including land in conversion, 3·9% of the agricultural land was managed organically in 2010. Organic food sales for the UK in 2009 totalled £1·8bn., up from £1·1bn. in 2003–04.

Farmers receiving financial support under the EU's Common Agricultural Policy are obliged to 'set-aside' land in order to control production. In 2007 such set-aside totalled 440,000 ha.

There were 820 tractors and 77 harvester-threshers per 1,000 ha. of arable land in 2006.

The number of people working on agricultural holdings was, in June 2010, 466,000 of whom 295,000 were farmers, partners, directors or spouses; 11,000 salaried managers; and 160,000 other workers including 56,000 seasonal, casual or gang workers. Of the 160,000 regular workers, 39,000 were part-time. There were some 315,900 farm holdings in 2006. Average size of holdings, 56·3 ha.

Total farm incomes dropped from £5·3bn. to £1·7bn. between 1995 and 2000, before rising to £2·3bn. in 2006. Food and live animals accounted for 3·4% of exports and 7·5% of imports in 2008, down from 4·5% of exports and 8·6% of imports in 1991.

Area given over to principal crops in the UK:

	Wheat	Sugar beets	Potatoes	Barley	Oilseed rape	Oats
			Area (1,000 ha.)			
2006	1,836	130	140	881	568	121
2007	1,830	125	140	898	674	129
2008	2,080	120	144	1,032	598	135
2009	1,775	114	144	1,143	570	129
2010	1,939	118	138	921	642	124

Production of principal crops in the UK:

	Wheat	Sugar beets	Potatoes	Barley	Oilseed rape	Oats
			Total production (1,000 tonnes)			
2006	14,755	7,400	5,727	5,239	1,890	728
2007	13,221	6,733	5,564	5,079	2,108	712
2008	17,227	7,641	6,145	6,144	1,973	784
2009	14,076	8,457	6,396	6,668	1,912	744
2010	14,878	6,484	6,045	5,252	2,230	685

Horticultural crops. 2010 output (in 1,000 tonnes): carrots, 748; onions, 364; cabbage, 248; apples, 228; peas, 162; lettuce, 134; cauliflowers, 109; turnips and swedes, 99.

Livestock in the UK as at June in each year (in 1,000):

	2004	2005	2006	2007	2008
Sheep	35,817	35,416	34,722	33,946	33,131
Cattle	10,588	10,440	10,324	10,304	10,107
(dairy)	(2,129)	(2,063)	(2,066)	(1,954)	(1,909)
(beef)	(1,736)	(1,762)	(1,733)	(1,698)	(1,670)
Pigs	5,159	4,862	4,933	4,834	4,714
Poultry	181,759	173,909	173,081	167,667	166,200

Livestock products in 2009, provisional (1,000 tonnes): beef and veal, 856; pork, bacon and ham, 706; lamb and mutton, 314; poultry meat, 1,459; cheese, 365; hens' eggs, 8,964m. (units). Milk production in 2009 (provisional) totalled 13,014m. litres.

In March 1996 the government acknowledged the possibility that bovine spongiform encephalopathy (BSE) might be transmitted to humans as a form of Creutzfeldt-Jakob disease via the food chain. Confirmed cases of BSE in cattle in the UK: 1988, 2,188; 1989, 7,166; 1990, 14,294; 1991, 25,202; 1992, 37,056; 1993, 34,829; 1994, 24,290; 1995, 14,475; 1996, 8,090; 1997, 4,336; 1998, 3,198; 1999, 2,283; 2000, 1,430; 2001, 1,187; 2002, 1,137; 2003, 611; 2004, 343; 2005, 225; 2006, 114; 2007, 67; 2008, 37; 2009, 12; 2010, 11; 2011, 7. Confirmed deaths attributed to nvCJD (new variant Creutzfeldt-Jakob Disease, the form of the disease thought to be linked to BSE): 1995, 3; 1996, 10; 1997, 10; 1998, 18; 1999, 15; 2000, 28; 2001, 20; 2002, 17; 2003, 18; 2004, 9; 2005, 5; 2006, 5; 2007, 5; 2008, 2; 2009, 3; 2010, 3; 2011, 5.

British beef was widely banned overseas and in March 1996 the European Commission introduced a ban on the export of bovine animals, semen and embryos, beef and beef products and mammalian meat and bonemeal from the UK. The government introduced a number of preventive measures including bans on sales of older meat and the use of meat in animal feed and fertilizer, and compensation schemes. Following inspections, the European Commission allowed for the export of deboned beef and beef products beginning 1 Aug. 1999. In March 2006 the ban was also lifted on exporting live animals born after 1 Aug. 1996 and exporting beef and beef products made from cattle slaughtered after 15 June 2005.

In Feb. 2001 the UK was hit by a major foot-and-mouth disease epidemic for the first time since 1967–68, with 2,030 confirmed cases and 4,050,000 animals being slaughtered during the months which followed. In the 1967–68 epidemic there had been 2,364 cases with approximately 434,000 animals slaughtered.

Forestry
In March 2011 the area of woodland in the United Kingdom was 3,078,000 ha., of which the Forestry Commission/Forest Service owned or managed 870,000 ha. In the year to March 2011, 8,200 ha. of new woodland was created (800 ha., Forestry Commission/ Forest Service; 7,300 ha., private woodlands) and 14,000 ha. restocked after harvesting. UK production of roundwood in 2010 was 9·7m. cu. metres underbark.

In 2010 imports of wood products (wood, panels, pulp and paper) were equivalent to 41·6m. cu. metres underbark; exports 5·5m. cu. metres underbark.

Forestry Commission (*Website:* http://www.forestry.gov.uk). *Forestry Facts and Figures.* Annual

Fisheries

Quantity (in 1,000 tonnes) and value (in £1m.) of all fish landings into the UK and UK vessels' landings abroad:

Quantity	2007	2008	2009	2010	2011
Wet fish	466·2	437·8	447·0	454·7	441·9
Shell fish	147·7	150·4	137·3	153·5	157·7
	613·9	588·2	584·3	608·2	599·6
Value					
Wet fish	365·2	370·4	437·2	453·4	537·4
Shell fish	281·1	265·2	242·3	266·5	290·8
	646·3	635·6	679·5	719·9	828·2

In Dec. 2010 the fishing fleet comprised 5,800 registered vessels excluding Channel Islands and the Isle of Man. Major fishing ports: (England) Plymouth, Brixham, Newlyn; (Scotland) Peterhead, Lerwick, Fraserburgh, Scrabster; (Northern Ireland) Ardglass. Peterhead is the UK's leading port, with 106,600 tonnes of fish landed by UK vessels in 2011 (with a value of £132·5m.).

In the period 2005–07 the average person in the UK consumed 20·5 kg of fish and fishery products a year, compared to the European Union average of 23·0 kg.

INDUSTRY

The UK's largest company by market capitalization on 22 March 2013 was HSBC at £129,354m. (US$195,367m.); Vodafone was the second largest at £89,756m. (US$135,561m.); and BP the third largest at £87,722m. (US$132,489m.).

In 2008 there were 162,785 manufacturing firms, of which 600 employed 500 or over persons, and 94,570 employed four or fewer. Manufacturing contributed 11% of GDP in 2009.

Chemicals and chemical products. Manufacturers' sales, (in £1m.) in 2007: primary plastics and other plastic products (excluding plastic packing goods), 15,815; pharmaceutical preparations and basic pharmaceutical products, 11,757; organic basic chemicals, 7,101; paints, etc., 2,822; rubber products (2006), 2,105; perfumes and toilet products, 1,888; soap, polish and detergents, 1,753; inorganic basic chemicals, 1,211; dyes, 1,047; fertilizers, etc., 945.

Construction. Total value (in £1m.) of constructional work in Great Britain in 2009 was 106,692, including new work, 55,537 (of which housing, 14,590). Cement production, 2007, 11,892,000 tonnes; building brick production, 2008, 1,932m. units.

Electrical Goods. Manufacturers' sales (in £1m.) for 2007: electric motors, generators and transformers, 2,756; electricity distribution and control apparatus, 2,682; electronic valves and tubes and other electronic components, 2,304; television and radio receivers, sound or video recording, 2,092; electric domestic appliances, 1,741; radio and electronic capital goods, 1,730.

Engineering, machinery and instruments. Manufacturers' sales (in £1m.) for 2007: motor vehicles, 25,543; aircraft and spacecraft, 10,667; parts and accessories for motor vehicles and engines, 9,428; appliances for measuring, checking and testing, 5,731; non-domestic cooling and ventilation equipment, 3,275; lifting and handling equipment, 3,261; medical and surgical equipment and orthopaedic appliances, 2,771. Car production, 2008, 1,446,619 units (down 5·7% from 2007).

Foodstuffs, etc. Manufacturers' sales (in £1m.) for 2007: operation of dairies, 6,110; bread, fresh pastry goods and cakes, 4,596; meat production and preservation, 4,320; cocoa, chocolate and sugar confectionery, 3,609; beer, 3,578; mineral water and soft drinks, 3,273; biscuits, rusks, preserved pastry goods and cakes (2006), 3,089; grain mill products, 2,911; prepared feeds for farm animals, 2,685; fruit and vegetable processing and preservation, 2,637; poultry production and preservation, 2,432; distilled alcoholic beverages (2004), 2,216; fish and fish products processing and preservation, 1,805; tobacco products, 1,626. Alcoholic beverage

production, 2008: beer, 4,946·9m. litres (5,955·2m. litres in 1991); wine, 1,348·3m. litres (658·3m. litres in 1991); spirits, 694·9m. litres (447·6m. litres in 1991).

Metals. Manufacturers' sales (in £1m.) for 2007: metal structures and parts of structures, 7,328; general mechanical engineering, 3,732; aluminium production, 2,339; forging, pressing, stamping and roll forming of metal, 2,135; steel tubes, 1,690; treatment and coating of metals, 1,335; builders' carpentry and joinery of metal, 1,259.

Textiles and clothing. Manufacturers' sales (in £1m.) in 2007: women's outerwear and underwear (2006), 990; carpets and rugs, 773; household textiles, 686; soft furnishings, 646; textile weaving, 578; finishing of textiles, 468; preparation and spinning of textile fibres, 386; men's outerwear and underwear, 230.

Wood products, furniture, paper and printing. Manufacturers' sales (in £1m.) in 2007: furniture of whatever construction, 7,318; journals and periodicals, 7,304; wood products except furniture, 6,774; newspapers, 4,120; publishing of books, 3,458; cartons, boxes and cases, 3,050; paper and paperboard, 2,759.

Labour

In 2009 the UK's total economically active population (i.e. all persons in employment plus the claimant unemployed) was (in 1,000) 31,374 (13,452 females), of whom 28,979 (12,542 females) were in employment, including 24,937 (12,280 females) as employees and 3,850 (1,103 females) as self-employed. In 1999 only 27,167,000 people had been in employment, representing an increase of 1,812,000 in ten years. However, the recession of 2008–09 resulted in the number of employees in 2009 being the lowest since 2005. UK employees by form of employment in 2009 (in 1,000): real estate renting and business activities, 4,504; wholesale and retail trade, repair of motor vehicles, motorcycles and household goods, 4,435; health and social work, 3,527; retail trade except motor vehicles/motorcycles and repair of household goods, 2,808; manufacturing industry, 2,642; education, 2,460; hotels and restaurants, 1,787; transport, storage and communications, 1,530; public administration and defence, compulsory social security, 1,465; construction, 1,277; wholesale trade and commission trade except motor vehicles, 1,102; financial intermediation, 1,002; agriculture, hunting, forestry and fishing, 255. Between June 2005 and June 2009 employment in service industries increased by 117,000 (despite the recession) while employment in manufacturing declined by 460,000 over the same period.

Registered unemployed in UK (in 1,000; figures seasonally adjusted): 2004, 1,424 (4·8%); 2005, 1,465 (4·9%); 2006, 1,669 (5·4%); 2007, 1,653 (5·3%); 2008, 1,776 (5·7%); 2009, 2,395 (7·6%). Of the 2,395,000 unemployed people in 2009, 1,465,000 were men and 930,000 women. The number of jobless people was 2,516,000 in the period Nov. 2012–Jan. 2013 (up slightly from 2,510,000 in the three months from Aug.–Oct. 2012 but down from 2,652,000 in the period Nov. 2011–Jan. 2012). The number of unemployed people on benefits (the 'claimant count') was 1·54m. in Jan. 2013— giving a rate of 4·7%, down from 4·9% in Jan. 2012. The unemployment rate on the International Labour Organization (ILO) definition, which includes all those who are looking for work whether or not claiming unemployment benefits, was 7·8% in the period Oct.–Dec. 2012 (down from 8·4% in the fourth quarter of 2011). In the period Nov. 2012–Jan. 2013, 993,000 young people in the UK aged 16–24 were unemployed (a rate of 21·2%), down from 1·04m. in Nov. 2011–Jan. 2012 (the highest level since 1986–87). Long-term unemployment rose from 22·6% of the labour force between 16 and 64 having been out of work for more than a year in the period Oct.–Dec. 2008 to 32·2% in the period Oct.–Dec. 2011.

There were 4,794,000 private sector businesses in the UK at the start of 2012 (up from 3,559,000 in 2002), of which 4,788,000 were small and medium-sized enterprises (fewer than 250 employees) and 4,580,000 had fewer than ten employees. 2·6m. businesses were registered at Companies House in March 2012.

Workers (in 1,000) involved in industrial stoppages (and working days lost): 2004, 293 (0·90m.); 2005, 93 (0·16m.); 2006, 713 (0·76m.); 2007, 745 (1·04m.); 2008, 511 (0·76m.). In 1975, 6m. working days had been lost through stoppages. Between 2001 and 2007 strikes cost Britain an average of 28 working days per 1,000 employees a year.

The Wages Councils set up in 1909 to establish minimum rates of pay (in 1992 of 2·5m. workers) were abolished in 1993. The former Labour government that came to power in May 1997 was committed to the introduction of a National Minimum Wage and established a Low Pay Commission to advise on its implementation. It is currently £6·19 an hour for adults, £4·98 for 18–20 year olds and £3·68 for 16–17 year olds above school leaving age. In April 2010 the median gross salary for full-time employees was £25,900 (£28,100 for men and £22,500 for women). Median hourly pay for full-time employees excluding overtime in April 2010 in the UK was £11·09 (£12·35 for males and £9·90 for females). Median weekly earnings in April 2010 were highest in London, at £642·30, and lowest in Northern Ireland, at £440·80.

Britons in full-time employment worked an average of 37·3 hours a week in 2009, compared to the EU average of 38·7 hours. In 2008, 3·28m. Britons (2·55m. men and 730,000 women) worked more than an average of 48 hours a week.

INTERNATIONAL TRADE

Imports and Exports

Imports of goods in 2011 totalled £399,330m. and exports £298,987m. In 2010 the UK's goods imports from other EU member countries totalled £186,048m. and from non-EU member countries £178,128m., compared to £162,676m. and £148,302m. respectively in 2009. Goods exports to other EU member countries in 2010 totalled £142,208m. and to non-EU member countries £123,506m., compared to £124,700m. and £103,426m. respectively in 2009.

In 2010 other EU members accounted for 52·1% of the UK's foreign trade in goods (51·1% of imports and 53·5% of exports). The USA accounted for 10·3% of foreign trade and the rest of the world 37·6%. Germany was the UK's biggest trading partner overall in 2010, followed by the USA, the Netherlands, France and China.

Figures for trade in goods by country and groups of countries (in £1m.):

EU-27 countries	Imports from 2009	Imports from 2010	Exports to 2009	Exports to 2010
EU total	162,676	186,048	124,700	142,208
Austria	2,280	2,633	1,290	1,463
Belgium	15,005	17,081	10,882	13,377
Bulgaria	183	231	197	249
Cyprus	123	106	620	562
Czech Republic	3,336	3,988	1,441	1,827
Denmark	3,837	4,117	2,477	2,757
Estonia	127	166	140	193
Finland	2,113	2,164	1,323	1,498
France	20,439	21,572	17,204	19,228
Germany	39,984	46,415	24,270	27,913
Greece	558	701	1,621	1,381
Hungary	2,540	3,251	850	1,082
Ireland	12,430	12,873	15,936	16,932
Italy	12,314	13,940	8,348	8,853
Latvia	339	440	109	168
Lithuania	370	551	172	225
Luxembourg	616	937	193	246
Malta	106	170	401	393
Netherlands	21,945	26,449	18,205	21,250
Poland	4,669	6,106	2,798	3,804
Portugal	1,423	1,749	1,543	1,837
Romania	788	1,241	686	781
Slovakia	1,604	1,620	378	466
Slovenia	251	357	175	220
Spain	9,570	10,381	9,224	9,940
Sweden	5,726	6,809	4,217	5,563

Other foreign countries	Imports from 2009	Imports from 2010	Exports to 2009	Exports to 2010
Europe—				
Croatia	80	98	223	173
Iceland	489	433	128	132
Norway	16,212	21,053	2,839	3,114
Russia	4,649	5,258	2,402	3,595
Switzerland	5,319	7,429	3,979	5,214
Turkey	4,665	5,351	2,369	3,248
Ukraine	145	283	589	474
Other in Europe	372	384	714	1,073
Africa—				
Egypt	687	659	1,006	1,198
Morocco	336	343	311	565
South Africa	3,854	4,423	2,254	2,890
Other in Africa	4,853	6,357	4,684	5,600
Asia—				
China	24,627	30,637	5,401	7,611
Hong Kong	7,757	8,172	3,735	4,456
India	4,618	5,812	2,949	4,068
Indonesia	1,240	1,388	367	461
Iran	211	206	405	307
Israel	1,099	1,566	1,145	1,387
Japan	6,670	8,107	3,570	4,334
Korea, South	2,862	2,563	2,167	2,344
Malaysia	1,665	1,837	1,052	1,272
Pakistan	702	809	478	461
Philippines	396	525	268	284
Saudi Arabia	610	766	2,363	2,488
Singapore	3,589	4,146	2,955	3,447
Taiwan	2,268	3,118	795	1,116
Thailand	2,321	2,683	914	1,137
Other in Asia	6,790	9,589	7,851	9,118
Oceania—				
Australia	2,225	2,315	2,960	3,362
New Zealand	819	844	350	414
Other in Oceania	168	107	91	82
Americas—				
Argentina	669	666	255	357
Brazil	2,566	3,111	1,788	2,217
Canada	4,464	5,770	3,329	4,127
Chile	600	584	514	630
Colombia	583	678	173	230
Mexico	757	1,032	751	952
USA	24,222	26,992	33,951	37,925
Venezuela	428	421	300	273
Other in America	1,715	1,613	1,051	1,370
Total, foreign countries	310,978	364,176	228,126	265,714

In 2010 finished manufactured goods accounted for 50·5% of the UK's imports and 46·7% of exports; semi-manufactured goods 24·3% of imports and 30·2% of exports; oil 9·0% of imports and 11·8% of exports; oil 9·9% of imports; and food, beverages and tobacco 9·2% of imports and 6·1% of exports.

The UK's trade deficit in goods in 2010 totalled £98,462m., up from £33,030m. in 2000 and £82,852m. in 2009. The last trade surplus in goods was in 1982. The trade surplus in services in 2010 was £58,778m., giving a trade deficit in goods and services combined of £39,684m. The last trade surplus in goods and services combined was in 1997. The UK is the second biggest exporter of services (after the USA) but only the eleventh biggest exporter of goods.

COMMUNICATIONS

Roads

Responsibility for the construction and maintenance of trunk roads belongs to central government. Roads not classified as trunk roads are the responsibility of county or unitary councils.

In 2009 there were 394,428 km of public roads in Great Britain, classified as: motorways, 3,560 km; trunk roads, 8,596 km; other major roads, 38,173 km; minor roads, 344,099 km.

In 2008 journeys by car, vans and taxis totalled 679bn. passenger km (less than 60bn. in the early 1950s). Even in the early 1950s passenger km in cars, vans and taxis exceeded the annual total at the end of the 20th century by rail. Licensed motor vehicles in 2008 included 27,021,000 passenger cars, 1,160,000 mopeds, scooters and motorcycles, 111,000 public transport vehicles and 3,303,000 other private and light goods vehicles. In 2010, 75% of households had regular use of a car/van with 33% of households having use of two or more cars/vans. New vehicle registrations in 2008, 2,672,200. Driving tests, 2008–09 (in 1,000): applications, 1,796; tests held, 1,717; tests passed, 777; pass rate, 45% (49% among males and 42% among females). The driving test was extended in July 1996 to include a written examination.

Road casualties in Great Britain in 2011, 203,950 including 1,901 killed (up slightly from 2010, which saw the lowest total of road deaths since records began in 1926). Britain has one of the lowest death rates in road accidents of any industrialized country, at 4·2 deaths per 100,000 people in 2008.

Inter- and intra-urban bus and coach journeys average 50bn. passenger-km annually. Passenger journeys by local bus services, 2007–08, 5,164m. *For London buses see* Transport for London *under* RAIL, *below*.

Rail

In 1994 the nationalized railway network was restructured to allow for privatization. Ownership of the track, stations and infrastructure was vested in a government-owned company, Railtrack, which was privatized in May 1996.

Passenger operations were reorganized into 25 train-operating companies, which were transferred to the private sector by Feb. 1997. By March 1997 all freight operations were also privatized. On 3 Oct. 2002 a new private sector not-for-dividend company limited by guarantee, Network Rail, took over from Railtrack plc as network owner and operator. The train-operating companies pay Network Rail for access to the rail network, and lease the rolling stock from three private-sector companies.

The rail network comprises 15,754 route km (around a third electrified). Annual passenger-km were 51·1bn. in 2009–10. There were 1·26bn. passenger journeys in 2009–10 on franchised operated services (1·27bn. in 2008–09). The amount of freight moved declined gradually over many years to 13·0bn. tonne-km in 1994–95 but has since risen and totalled 19·1bn. tonne-km in 2009–10. In 2009 a total of 15 people (excluding suicides and trespassers) were fatally injured on the railways and ten people on level crossings (compared to 2,222 deaths in road accidents in 2009).

Eurotunnel PLC holds a concession from the government to operate the Channel Tunnel (49·4 km), through which vehicle-carrying and Eurostar passenger trains are run in conjunction with French and Belgian railways. Since Nov. 2007 a new dedicated high-speed line connects the Channel Tunnel to London St Pancras. Domestic trains began using the line on regular services in Dec. 2009.

Transport for London (TfL) is accountable to the Mayor of London and is responsible for implementing his Transport Strategy as well as planning and delivering a range of transport facilities. TfL's remit covers London Underground (since July 2003), London Buses, the Docklands Light Railway and Croydon Tramlink. It is also responsible for London River Services, Victoria Coach Station and London's Transport Museum, and provides transport for users with reduced mobility via Dial-a-Ride. As well as running the central London congestion charging scheme, TfL manages a 580 km network of London's main roads, all 4,600 traffic lights and the private hire trade. It also provides grants to London Boroughs to fund local transport improvements.

Every weekday in Greater London, 5·4m. journeys are made on London's buses, 3m. on the underground, 7m. on foot, 0·3m. by bicycle, 0·2m. by taxi, 160,000 on the Docklands Light Railway and 60,000 on Croydon Tramlink.

The privately franchised Docklands Light Railway is operated in east inner London.

There are metros in Glasgow and Newcastle, and light rail systems in Birmingham/Wolverhampton, Blackpool, Manchester, Nottingham and Sheffield.

Civil Aviation

All UK airports handled a total of 219·6m. passengers in 2011 (211·2m. in 2010). London area airports (Heathrow, Gatwick, London City, Luton, Southend and Stansted) handled 127·4m. passengers in 2010.

Busiest airports in 2010:

	Passengers	International		Freight (tonnes)
Heathrow	65,881,660	60,904,418	Heathrow	1,472,988
Gatwick	31,375,290	27,845,885	East Midlands	
Stansted	18,573,592	16,838,531	International	273,669
Manchester	17,759,173	15,425,106	Stansted	202,238
Luton	8,738,712	7,796,385	Manchester	115,922
			Gatwick	104,032

Heathrow was the world's fourth busiest airport for passenger traffic in 2010 and Europe's busiest. More international passengers use Heathrow than any other airport in the world.

Following the Civil Aviation Act 1971, the Civil Aviation Authority (CAA) was established as an independent public body responsible for the economic and safety regulation of British civil aviation. A CAA wholly owned subsidiary, National Air Traffic Services, operates air traffic control. Highlands and Islands Airports Ltd is owned by the Scottish Ministers and operates 11 airports.

There were 20,161 civil aircraft registered in the UK at 3 Oct. 2011.

British Airways is the largest UK airline in terms of numbers of aircraft and distance flown, with a total of 233 aircraft in service at 31 Dec. 2009. It operates long- and short-haul international services, as well as an extensive domestic network. British Airways also has franchise agreements with two other operators: Comair and Sun-Air of Scandinavia. In April 2010 it signed a deal with the Spanish airline Iberia to merge and create a new company called International Airlines Group, which was formed in Jan. 2011. However, both carriers still operate under their own brands. Other major airlines in 2009 (with numbers of aircraft): BMI Group (70); easyJet (169); Flybe (71); Thomas Cook Airlines (44); Thomson Airways (74); Virgin Atlantic (38). BMI Group has since ceased operations and was merged into British Airways in Oct. 2012. According to CAA airline statistics, in 2008 easyJet overtook British Airways in terms of passengers carried to become the largest airline as measured by passenger numbers—in 2010 easyJet carried 42,400,581 passengers (37,664,647 on international flights) and flew 376·7m. km while British Airways carried 29,733,241 passengers in 2010 (25,999,007 on international flights), although it flew 580·9m. km. In April 2003 British Airways announced that Concorde, the world's first supersonic jet which began commercial service in 1976, would be permanently grounded from Oct. 2003. In recent years low-cost airlines such as Ryanair and easyJet have become increasingly popular. Serving mostly domestic and European destinations, they recovered quickly from the slump of the airline business following the attacks on New York and Washington on 11 Sept. 2001.

The most frequently flown route into and out of the UK in 2010 was Heathrow–New York John F. Kennedy and vice-versa (2,517,896 passengers), followed by Heathrow–Dublin and vice-versa (1,493,613) and Heathrow–Hong Kong Chek Lap Kok and vice-versa (1,386,779).

Shipping

The UK-owned merchant fleet (trading vessels over 100 GT) in Dec. 2008 totalled 754 ships of 21·3m. DWT and 18·8m. GT.

The UK-owned and registered fleet totalled 400 ships of 6·6m. DWT. The average age (DWT) of the UK-owned trading vessels was 16·8 years.

Total gross international revenue in 2008 was £13,176m. The net direct contribution to the UK balance of payments was £5,293m.; there were import savings of £1,764m., giving a total contribution of £7,057m.

The ports handling the most domestic and international passengers on short sea routes are (with 1m. passengers handled in 2008, excluding inter-island traffic): Dover (13·8), Portsmouth (2·1), Holyhead (2·0), Belfast (1·3), Stranraer (1·1). Domestic sea passengers totalled 3·7m. in 2008. The principal ports in terms of freight are (with 1m. tonnes of cargo handled in 2008): Grimsby and Immingham (65·3), London (53·0), Tees and Hartlepool (45·4), Southampton (41·0), Forth, including Grangemouth, Leith and Rosyth (39·1). Total traffic in 2008 was 562·2m. tonnes.

Inland Waterways

There are approximately 3,500 miles (5,630 km) of navigable canals and river navigations in Great Britain. In July 2012 a new waterways charity, the Canal & River Trust (CRT), took over the management of the network of waterways in England and Wales from British Waterways. In Scotland the 137 miles (220 km) of inland waterways remain under the control of British Waterways (operating as Scottish Canals), which is a stand-alone public body of the Scottish government.

River navigations and canals managed by other authorities include the Thames, Great Ouse and Nene, Norfolk Broads and Manchester Ship Canal.

The Association of Inland Navigation Authorities (AINA) represents some 30 navigation authorities providing an almost complete UK coverage.

Telecommunications

In 2009 there were around 120 operators offering fixed telecommunication services and five mobile networks. BT (then British Telecom) was established in 1981 to take over the management of telecommunications from the Post Office. In 1984 it was privatized as British Telecommunications plc, changing its trading name from British Telecom to BT in 1991.

By 1998 all of the BT system was served by digital exchanges. In 2010 there were 33,409,000 fixed telephone subscribers (equivalent to 538·5 per 1,000 population). 84% of UK households had a fixed telephone in 2012, a fall from a peak of 95% in the late 1990s as increasingly UK consumers use email, speaking on a mobile phone and in particular text messaging. In June 2009, 71% of fixed lines were residential and 29% business. There were 67,000 public payphones in 2012 (146,000 in 2002). BT had the largest share of the landline phone market at the end of 2011 with 36%, followed by Virgin Media with 12%. The other main providers are Sky and TalkTalk.

In 2010 there were 81,115,000 mobile telephone subscribers in the UK (1,307·6 per 1,000 persons), up from 43,452,000 in 2000 and 1,114,000 in 1990. The leading operators are O2, including Tesco Mobile (with an estimated 27·1% share of the market); T-Mobile, including Virgin Mobile (21·2%); Orange (20·6%); Vodafone (20·2%); and 3 (6·2%). 3 was launched by Hutchison on 3 March 2003 and is the UK's first mainland third generation mobile network.

Telecommunications services are regulated by the Office of Communications ('*Ofcom*') in the interests of consumers. According to its *Communications Market Report 2012*, 58% of adults in a survey stated that they use text messaging to communicate with friends and family at least once a day, 49% face-to-face, 47% voice call on a mobile phone, 32% social networking, 30% email, 29% voice call on a landline and 26% instant messaging. Among those aged 16–24, 90% stated that they used text messaging, 63% face-to-face, 67% voice call on a mobile

phone, 73% social networking, 43% email, 15% voice call on a landline and 62% instant messaging.

In March 2012 there were 30·5m. Facebook users (48% of the total population of the UK).

Internet

At 30 June 2012 there were 52·7m. internet users in the UK (the third highest total in Europe after Russia and Germany), just over 83% of the total population. According to a report published in March 2006, 48% of children aged 8–11 and 65% of children aged 12–15 used the internet at home. In 2006, 67% of households had a home computer. In 2012, 80% of households in Great Britain had internet access, up from 57% in 2006; 93% of households with internet access were using fixed broadband, up from 69% in 2006. In 2011, 45% of internet users used a mobile phone to connect to the internet, up from 23% in 2009. Internet commerce, or e-commerce, amounted to £50bn. in 2011, the second highest in the world after the USA. In Dec. 2010 there were 36·9 wireless broadband subscribers per 100 inhabitants and 31·9 fixed broadband subscribers per 100.

SOCIAL INSTITUTIONS

Justice

England and Wales. The legal system of England and Wales, divided into civil and criminal courts, has at the head of the superior courts, as the ultimate court of appeal, the Supreme Court of the United Kingdom, which hears appeals for all civil law cases in the UK and for all criminal cases in England, Wales and Northern Ireland (Scotland's highest court for criminal cases is the High Court of Justiciary). The Supreme Court was created as a result of the Constitutional Reform Act 2005 and came into being on 1 Oct. 2009, replacing the Appellate Committee of the House of Lords. In order that civil cases may go from the Court of Appeal (or Court of Session in Scotland) to the Supreme Court, it is necessary to obtain the leave of either the respective lower court, although in certain limited cases an appeal may lie direct to the Supreme Court from the decisions of the High Courts. Appeals may be brought to the Supreme Court provided that the lower Court is satisfied that a point of law 'of general public importance' is involved, and that it is in the public interest that a further appeal should be brought. As a judicial body, the Supreme Court consists of 12 Justices drawn from the different jurisdictions of the United Kingdom, and is led by the President of the Supreme Court, the Rt Hon. the Lord Phillips of Worth Matravers. The final court of appeal for certain of the Commonwealth countries, the UK overseas territories and the British Crown dependencies is the Judicial Committee of the Privy Council which includes Justices of the Supreme Court, other Lords of Appeal and Privy Councillors who hold or have held high judicial office in the UK or Privy Councillors who are or have been Chief Justices or Judges of certain Superior Courts of Commonwealth countries.

Civil Law. The main courts of original civil jurisdiction are the High Court and county courts.

The High Court has exclusive jurisdiction to deal with specialist classes of case, e.g. judicial review. It has concurrent jurisdiction with county courts in cases involving contract and tort although it will only hear those cases where the issues are complex or important. The High Court also has appellate jurisdiction to hear appeals from lower tribunals.

The judges of the High Court are attached to one of its three divisions: Chancery, Queen's Bench and Family; each with its separate field of jurisdiction. The Heads of the three divisions are the Lord Chief Justice (Queen's Bench), the Vice-Chancellor (Chancery) and the President of the Family Division. In addition there are 107 High Court judges (100 men and seven women). For the hearing of cases at first instance, High Court judges sit singly. Appellate jurisdiction is usually exercised by Divisional Courts consisting of two (sometimes three) judges, though in certain circumstances a judge sitting alone may hear the appeal. High Court business is dealt with in the Royal Courts of Justice and by over 130 District Registries outside London.

County courts can deal with all contract and tort cases, and recovery of land actions, regardless of value. They have upper financial limits to deal with specialist classes of business such as equity and Admiralty cases. Certain county courts have been designated to deal with family, bankruptcy, patents and discrimination cases.

There are about 220 county courts located throughout the country, each with its own district. A case may be heard by a circuit judge or by a district judge. Defended claims are allocated to one of three tracks—the small claims track, the fast track and the multi-track. The small claims track provides a simple and informal procedure for resolving disputes, mainly in claims for debt, where the value of the claim is no more than £5,000. Parties should be able to do this without the need for a solicitor. Other claims valued between £5,000 and £15,000 will generally be allocated to the fast track, and higher valued claims which could not be dealt with justly in the fast track may be allocated to the multi-track.

Specialist courts include the Patents Court, which deals only with matters concerning patents, registered designs and appeals against the decision of the Comptroller General of Patents. Cases suitable to be heard by a county court are dealt with at Central London County Court.

The Court of Appeal (Civil Division) hears appeals in civil actions from the High Court and county courts, and tribunals. Its President is the Master of the Rolls, aided by up to 38 Lords Justices of Appeal (as at July 2008) sitting in six or seven divisions of two or three judges each.

Civil proceedings are instituted by the aggrieved person, but as they are a private matter, they are frequently settled by the parties through their lawyers before the matter comes to trial. In very limited classes of dispute (e.g. libel and slander), a party may request a jury to sit to decide questions of fact and the award of damages.

Criminal Law. At the base of the system of criminal courts in England and Wales are the magistrates' courts which deal with over 95% of criminal cases. In general, in exercising their summary jurisdiction, they have power to pass a sentence of up to six months' imprisonment (or 12 months for consecutive sentences) and to impose a fine of up to £5,000 on any one offence. They also deal with the preliminary hearing of cases triable at the Crown Court. In addition to dealing summarily with over 2·0m. cases, which include thefts, assaults, drug abuse, etc., they also have a limited civil and family jurisdiction.

Magistrates' courts normally sit with a bench of three lay justices. Although unpaid they are entitled to loss of earnings and travel and subsistence allowance. They undergo training after appointment and they are advised by a professional legal adviser. Full-time District Judges (magistrates' courts), formerly known as stipendiary magistrates, also deal with cases in magistrates' courts. Generally they possess the same powers as the lay bench, but they sit alone. On 1 April 2010 the total strength of the lay magistracy was 29,270 including 14,540 women. Justices of the Peace are appointed on behalf of the Queen by the Lord Chancellor.

Justices are selected and trained specially to sit in Youth and Family Proceedings Courts. Youth Courts deal with cases involving children and young persons up to the age of 18 charged with criminal offences (other than homicide and other grave offences). These courts normally sit with three justices, including at least one man and one woman, and are accommodated separately from other courts.

Family Proceedings Courts deal with matrimonial applications and Children Act matters, including care, residence and contact and adoption. These courts normally sit with three justices including at least one man and one woman.

Above the magistrates' courts is the Crown Court. This was set up by the Courts Act 1971 to replace quarter sessions and assizes. Unlike quarter sessions and assizes, which were individual courts, the Crown Court is a single court which is capable of sitting anywhere in England and Wales. It has power to deal with all trials on indictment and has inherited the jurisdiction of quarter sessions to hear appeals, proceedings on committal of persons from the magistrates' courts for sentence, and certain original proceedings on civil matters under individual statutes.

The jurisdiction of the Crown Court is exercisable by a High Court judge, a Circuit judge or a Recorder (a part-time judge) sitting alone, or, in specified circumstances, with Justices of the Peace. The Lord Chief Justice has given directions as to the types of case to be allocated to High Court judges (the more serious cases) and to Circuit judges or Recorders respectively.

Appeals from magistrates' courts go either to a Divisional Court of the Queen's Bench Division of the High Court (when a point of law alone is involved) or to the Crown Court where there is a complete rehearing on appeals against conviction and/or sentence. Appeals from the Crown Court in cases tried on indictment lie to the Court of Appeal (Criminal Division). Appeals on questions of law go by right, and appeals on other matters by leave. The Lord Chief Justice or a Lord Justice of Appeal sits with judges of the High Court to constitute this court. Thereafter, appeals in England and Wales can be made to the Supreme Court.

There remains as a last resort the invocation of the royal prerogative of mercy exercised on the advice of the Home Secretary. In 1965 the death penalty was abolished for murder and in 1998 abolished for all crimes.

All contested criminal trials, except those which come before the magistrates' courts, are tried by a judge and a jury consisting of 12 members. The jury decides whether the accused is guilty or not. The judge is responsible for summing up on the facts and directing the jury on the relevant law. He or she sentences offenders who have been convicted by the jury (or who have pleaded guilty). If, after at least two hours and ten minutes of deliberation, a jury is unable to reach a unanimous verdict it may, on the judge's direction—and provided that in a full jury of 12 at least ten of its members are agreed—bring in a majority verdict. The failure of a jury to agree on a unanimous verdict or to bring in a majority verdict may involve the retrial of the case before a new jury.

The Employment Appeal Tribunal. The Employment Appeal Tribunal, which is a superior Court of Record with the like powers, rights, privileges and authority of the High Court, was set up in 1976 to hear appeals on questions of law against decisions of employment tribunals and of the Certification Officer. The appeals are heard by a judge sitting alone or with two members (in exceptional cases four) appointed for their special knowledge or experience of industrial relations either on the employer or the trade union side, with always an equal number on each side. The great bulk of their work is concerned with the problems which can arise between employees and their employers.

Military Courts. Under the Armed Forces Act 2006, criminal offences and disciplinary offences alleged against service personnel subject to service law (or civilians overseas who are subject to service discipline) may be tried in the Court Martial. Lower-level offences by service personnel may be dealt with at a summary hearing by their commanding officer, subject to appeal to the Summary Appeal Court.

The Personnel of the Law. All judicial officers are independent of Parliament and the Executive. They are appointed by the Crown on the advice of the Prime Minister or the Lord Chancellor, or directly by the Lord Chancellor himself, and hold office until retiring age. Under the Judicial Pensions and Retirement Act 1993 judges normally retire by age 70 years.

The legal profession is divided; barristers, who advise on legal problems and can conduct cases before all courts, usually act for the public only through solicitors, who deal directly with the legal business brought to them by the public and have rights to present cases before certain courts. The distinction between the two branches of the profession has been weakened since the passing of the Courts and Legal Services Act 1990, which has enabled solicitors to obtain the right to appear as advocates before all courts. Long-standing members of both professions are eligible for appointment to most judicial offices.

For all judicial appointments up to and including the level of Circuit Judge, it is necessary to apply in writing to be considered for appointment. Vacancies are advertised. A panel consisting of a judge, an official and a lay member decide whom to invite for interview and also interview the shortlisted applicants. They make recommendations to the Lord Chancellor, who retains the right of final recommendation to the Sovereign or appointment, as appropriate.

Legal Services. The system of legal aid in England and Wales was established after the Second World War under the Legal Aid and Advice Act 1949. The Legal Aid Board was then set up under the Legal Aid Act 1988, and took over the administration of legal aid from the Law Society in 1989. The Legal Services Commission (LSC) was set up (as a non-departmental public body) under the Access to Justice Act 1999 and replaced the Legal Aid Board on 1 April 2000.

In 2010–11 the Commission funded over 2·7m. acts of assistance overall at a cost of £2·1bn. on the provision of legal aid services and spent £99·2m. on administration costs. The Community Legal Service spend of £985·4m. funded 1·25m. acts of assistance. The Criminal Defence Service spend of £1,129·8m. funded 1·47m. acts of assistance.

The LSC was abolished as a result of the Legal Aid, Sentencing and Punishment of Offenders Act (LASPO) 2012 and replaced from 1 April 2013 with the Legal Aid Agency, which is an executive agency of the Ministry of Justice. A new office of the Director of Legal Casework was created to ensure independence of decision-making. The Director will take decisions on the funding of individual cases.

The Legal Aid Agency includes contracted solicitors and advice agencies that provide civil and family legal advice and representation. It manages Civil Legal Advice, which includes telephone and internet-based services, and contracts with quality assured providers to deliver face-to-face civil legal aid services across a range of categories such as debt and housing.

The Agency also provides legal advice and representation to people being investigated or charged with a criminal offence. It manages the duty solicitor schemes for police stations and magistrates' courts so that those who need advice and representation can see a solicitor, and funds services in the higher courts. The Public Defender Service provides criminal defence services directly to the public.

See also SCOTLAND.

CIVIL JUDICIAL STATISTICS
ENGLAND AND WALES

	Number of cases 2007
Appellate Courts	
Judicial Committee of the Privy Council	71
House of Lords	51
Court of Appeal	1,114
High Court of Justice (appeals and special cases from inferior courts)	6,690
Courts of First Instance (excluding Magistrates' Courts and Tribunals)	
High Court of Justice:	
Chancery Division	45,541
Queen's Bench Division	18,505

CIVIL JUDICIAL STATISTICS
ENGLAND AND WALES

County courts: Matrimonial suits	137,465[1]
County courts: Non-family work	2,014,962

[1]Includes dissolutions of civil partnerships.

CRIMINAL STATISTICS
ENGLAND AND WALES

	Total number of offenders[1] (in 1,000)		Indictable offences[1] (in 1,000)	
	2007	2008	2007	2008
Aged 10 and over				
Proceeded against in magistrates' courts	1,733	1,640	405	397
Found guilty at all courts	1,416	1,363	313	317
Cautioned	363	327	205	180
Aged 10 and under 18				
Proceeded against in magistrates' courts	126	111	67	59
Found guilty at all courts	98	88	51	46
Cautioned[2]	127	98	75	58

[1]On the principal offence basis. [2]From 1 June 2000 the Crime and Disorder Act 1998 came into force nationally and removed the use of cautions for persons under 18 and replaced them with reprimands and final warnings.

British Crime Survey (BCS) interviews in 2010–11 estimate that there were approximately 9·6m. crimes against adults aged 16 or over living in private households in England and Wales, up slightly from 9·5m. in 2009–10. Property crime accounts for 77% of BCS crime and violence 23%. The 2009–10 total was the lowest since the BCS was introduced in 1981. There were about 4·2m. crimes recorded by the police in 2010–11, down slightly from 4·3m. in 2009–10.

In June 2011 the prison population in England and Wales was 85,374 (85,002 in June 2010). The annual average prison population rose 30% between 2001 and 2011, from 66,301 to 85,951. During this time the annual average female prison population rose by 12%, from 3,740 to 4,188; the annual average male population rose by 31%, from 62,560 to 81,763. These figures do not include prisoners held in police cells.

See also SCOTLAND *and* NORTHERN IRELAND.

Police

In England and Wales there are 43 police forces, each maintained by a police authority typically comprising nine local councillors, three magistrates and five independent members. London is policed by the Metropolitan Police Service (responsible to the 23-member Metropolitan Police Authority, 11 of whom are members of the Greater London Assembly, plus the Mayor) and the City of London Police (whose police authority is the City of London Corporation). A similar tripartite arrangement (Cabinet Secretary, Chief Constable and Joint Police Board/Police Authority) exists for the accountability of the police service in Scotland.

Figures show that the total strength of the police service in England and Wales at 31 March 2008 was 140,230 (including 32,931 women). Police officers are supported by police staff and at the end of March 2008 there were 77,350 police staff (approximately 47,000 female), including 1,903 designated officers (investigation officers, detention officers and escort officers). In addition there were 15,805 police community support officers on 31 March 2008 (including approximately 6,800 women). There were 14,547 special constables in March 2008 (including 3,828 women). The Police Service in England and Wales has benefited from a significant increase in resources over a sustained period. On a like-for-like basis government grant for the police will have increased by more than 60% or over £3·7bn. between 1997–98 and 2010–11.

Education

Adult Literacy and Numeracy. The government published the *Skills for Life Strategy* in 2001 in response to the recommendations in the 1999 Moser report (*A Fresh Start. Improving Literacy and Numeracy*). The strategy covers adults aged 16 and above at skills levels of pre-entry up to and including Level 2. The results of the 2003 *Skills for Life Needs and Impact Survey* showed that in England 5·2m. adults aged 16–65 have literacy levels below Level 1 (equivalent to the level expected of an average 11-year-old) and 15m. have numeracy skills below Level 1. From April 2001 to May 2006, 1,416,000 learners achieved at least one qualification in literacy, numeracy or language. In terms of participation in literacy-, language- or numeracy-learning a total of 4·5m. learners took up 9·7m. learning opportunities between April 2001 and May 2006.

The Publicly Maintained System of Education. Compulsory schooling begins at the age of five (four in Northern Ireland). The minimum leaving age for all pupils was 16 but as a result of the Education and Skills Act of Nov. 2008 this will be raised in stages to 18. From 2015 all young people will remain in education or training to 18. No tuition fees are payable in any publicly maintained school (but parents can choose to pay for their children to attend independent schools run by individuals, companies or charitable institutions). The post-school or tertiary stage, which is voluntary, includes universities, further education establishments and other higher education establishments, as well as adult education and youth services. Financial assistance (grants and loans) is generally available to students in higher education and to some students on other courses in further education.

National Curriculum. The National Curriculum was introduced in 1988 and has undergone a number of revisions—the latest taking place in 2008. It determines the content of what will be taught, sets attainment targets for learning and determines how performance will be assessed and reported.

The National Curriculum comprises the core subjects of English, mathematics and science; and foundation subjects of information communication technology, design and technology, history, geography, modern foreign languages, art and design, music, physical education and citizenship.

At Key Stage 4 (ages 14–16) schools must provide access for each pupil to a minimum of one course in the arts (art and design, music, dance, drama and media arts); one course in the humanities (history and geography); at least one modern foreign language; and design and technology. However, since Sept. 2004 these subject areas are no longer compulsory for key stage 4 pupils.

In addition, pupils must be taught religious education during all four key stages, although parents have the right to withdraw their children from this provision. Careers and sex education are compulsory at Key Stages 3 and 4, and work-related learning is compulsory at Key Stage 4. The subject of personal, social and health education is not statutory but should be taught across all four key stages. Every school must also provide a form of daily collective worship, but with the right to withdraw.

In Sept. 2008 the Early Years Foundation Stage (EYFS) replaced the Foundation Stage curriculum which was a distinct phase of education for children aged three to the end of the reception year of primary school. The EYFS is a national play-based framework for supporting the learning, development and safety of children from birth to the age of five. The EYFS does not form part of the National Curriculum. All state and independent schools and 'registered' early years providers are required to meet the learning and development requirements of the EYFS.

Early Learning. All three- and four-year-olds are entitled to 15 hours per week of free early learning for 38 weeks per year until they reach compulsory school age. Free early-learning places can be delivered by state nursery schools; nursery classes in primary schools and reception classes; and private, voluntary and independent providers and registered childminders who are part of a quality assured network. There are around 38,900 sites currently involved in delivering free early-learning. Local authorities have a duty to ensure that there are sufficient free early learning places for all three- and four-year-olds.

Primary Schools. These provide compulsory education for pupils from the age of five up to the age of 11 (12 in Scotland). Most public sector primary schools take boys and girls in mixed classes. Some pre-compulsory age pupils attend nursery classes within primary schools, however, and in England some middle schools cater for pupils at either side of the secondary education transition age. In 2008–09 there were 21,568 public sector mainstream primary schools in the United Kingdom, with an average of 21 pupils per teacher.

Middle Schools. A number of local authorities operate a middle school system. These provide for pupils from the age of eight, nine or ten up to the age of 12, 13 or 14, and are deemed either primary or secondary according to the age range of the pupils.

Secondary Schools. In 2008–09 there were 4,183 state-funded secondary schools in the United Kingdom providing for pupils from the age of 11 upwards. Some local authorities have retained a selective admissions policy at age 11 (mainly for entry to grammar schools), and some 233 state-funded secondary schools in the United Kingdom operate a selective admissions policy. There are 169 secondary modern schools in England providing a general education up to the minimum school leaving age of 16, although some pupils stay on beyond that age. In state-funded secondary schools in the United Kingdom there are an average 15 pupils per teacher.

Almost all local authorities operate a system of comprehensive schools to which pupils are admitted without reference to ability or aptitude. There are 3,247 such schools in the United Kingdom with over 3·9m. pupils. With the development of comprehensive education, various patterns of secondary schools have come into operation. Principally these are: 1) All-through schools with pupils aged 11 to 18 or 11 to 16; pupils over 16 being able to transfer to an 11 to 18 school or a sixth form college providing for pupils aged 16 to 19. (Since 1 April 1993, sixth form colleges have been part of the further education sector—there were 95 sixth form colleges in 2007–08). 2) Local authorities operating a three-tier system involving middle schools where transfer to secondary school is at ages 12, 13 or 14. These correspond to 12 to 18, 13 to 18 and 14 to 18 comprehensive schools respectively. 3) In areas where there are no middle schools a two-tier system of junior and senior comprehensive schools for pupils aged 11 to 18, with optional transfer to these schools at age 13 or 14.

Specialist Schools. The Specialist Schools Programme began with specialist Technology Colleges in 1994; by Oct. 2008 over 2,900 had specialized in a diverse range of subject areas. Specialist Schools have access to additional funding and support that allows them to focus on a particular part of the curriculum both in their own school and with other schools whilst continuing to cover the full National Curriculum.

Academies. Academies (only in England) are independent state schools open to all abilities. They are usually in disadvantaged areas and are established by sponsors from business, faith, or voluntary groups. In July 2006 the endowment model of sponsorship was announced. Sponsors now establish an endowment fund with the Academy Trust using the revenue generated from the endowment to support the objectives of the Academy. The Department for Education provides the capital and running costs. In Sept. 2008 there were 130 Academies.

City Technology Colleges. CTCs are independent all-ability secondary schools established in partnership between government and business sponsors under the Education Reform Act 1988. They teach the full National Curriculum but give special emphasis to technology, science and mathematics. The government meets all recurrent costs. Although there were originally 15 CTCs, there are now only three remaining as 12 have been converted to Academies.

Music and Dance Scheme (formerly the Music and Ballet Scheme). The 'Aided Pupil Scheme' for boys and girls with outstanding talent in music or dance (principally ballet) helps parents with the fees and boarding costs at eight specialist private schools in England. Since 2004 the scheme has been developed to include 21 new centres for advanced training and a national grants scheme for out-of-school-hours training.

Special Education. It is estimated that, nationally, 20% of the school population will have special educational needs at some time during their school career. For some 2·7% of pupils the local authority will need to make a statutory assessment of special educational needs under the Education Act 1996 and draw up a legal document, the statement, which sets out the extra provision a child needs. (In Scotland pupils are assessed for a Coordinated Support Plan.)

Maintained schools must use their best endeavours to make provision for such pupils. The Special Educational Needs Code of Practice, a revised version of which came into force on 1 Jan. 2002, gives practical guidance.

Further Education (Non-University). The English Further Education (FE) system provides a wide range of education and training opportunities for individuals and employers. Learning opportunities are provided, from age 14 upwards, at all levels from basic skills to higher education. The FE system's primary purpose is to help people gain the skills they need to improve their employability.

Following the abolition of the former Learning and Skills Council, from April 2010 responsibility for funding full- and part-time education and training provision for people aged 16 to 19 fell to local education authorities and the Young People's Learning Agency (YPLA). The Skills Funding Agency, an agency of the Department for Business, Innovation and Skills, funds and regulates adult FE and skills training in England. Under the Education Act 2011 the YPLA, which was sponsored by the Department for Education and supported the delivery of training and education to 16- to 19-year-olds in England, was to close on 31 March 2012. On 1 April 2012 responsibilities of the YPLA were transferred to the Education Funding Agency (EFA), which is an executive agency of the Department for Education. The EFA funds the education system for 3- to 19-year-olds and manages the school and sixth form college estate. It is responsible for the allocation and distribution of approximately £50 bn. revenue and capital funding each year, as well as the delivery of capital investment in schools.

Further education is the largest sector providing educational opportunities for the over 16s. There are 373 FE colleges in England, with around 941,000 full-time and 1,786,000 part-time students in 2007–08. The SFA operates with an average annual budget of £4·6bn. Until its closure in March 2012, the YPLA provided direct financial support to 750,000 young people to stay in learning and realise their potential.

Youth Work. The priority age group for agencies and services providing youth work is 13- to 19-year-olds, but the target age group may extend to 11- to 25-year-olds. Provision is usually in the form of positive activities delivered in youth clubs and centres, or through 'detached' or outreach work aimed at young people at risk from alcohol or drug misuse, or of drifting into crime. There is an increasing emphasis on youth workers working with disaffected, and socially excluded, young people and providing services at times and in places where young people want them. Youth work can be delivered by local authority youth services, the voluntary and community sector, and other specialist youth agencies.

Independent/State School Partnerships Grant Scheme. The aim is to promote collaborative working between the independent and state school sectors to raise standards in education. A total of 24

projects are receiving more than £4m. of funding over the period 2008–11. In excess of 330 projects received funding totalling £10m. in the first nine years of the scheme (1998–2007).

Higher Education (HE) Student Support. All students are expected to make a contribution towards their tuition fees. In Dec. 2010 the coalition government agreed to increase the maximum tuition fee rate to £9,000 for 2012–13 for students beginning their studies in Sept. 2012 or later. No eligible student (new or existing) has to pay their fees either before or during their course, as a Tuition Fee Loan is available. Students in the fifth or later years of medical or dental courses and on NHS-funded courses in professions allied to medicine may be eligible for a means-tested grant from the NHS Bursary system. Postgraduate trainee teachers may be eligible for tax-free bursaries; these range from £6,000 to £9,000 depending on the teaching subject. Non-repayable Maintenance Grants of up to £3,250 are also available, targeted at students from lower income backgrounds.

Universities and colleges wishing to charge maximum fees must also make bursaries available to students who are in receipt of the full Maintenance Grant.

A Maintenance Loan is available to help with students' living costs. All students are entitled to 72% of the maximum loan, with the balance subject to income assessment. The maximum loans available for students who started their courses in 2010–11 or 2011–12 are: £6,928 (living away from home and studying in London); £4,950 (living away from home and studying outside London); and £3,838 (living at home). For students who started their courses in 2012–13 or are starting in 2013–14 the maximum loans available are: £7,675 (living away from home and studying in London); £5,500 (living away from home and studying outside London); and £4,375 (living at home). Neither the Tuition Fee Loan nor the Maintenance Loan has to be repaid until the student has left university or college and is earning over £15,000 a year (if the course started before Sept. 2012) or over £21,000 (Sept. 2012 or later).

Applications for student finance are made through Local Authorities or, under a pilot scheme operating in some areas, direct to the Student Loans Company. The Student Loans Company manages loan accounts and the payment of the student finance package.

Postgraduate studentships and research grants are available from the Arts and Humanities Research Council (AHRC) or similar councils. There are six other grant-awarding Research Councils which each report to the Department of Business, Innovation and Skills. They offer awards to students studying within the broad spectrum of economics, engineering, astronomy and medical, biological and physical sciences.

In 2010–11 the AHRC was scheduled to receive approximately £112m. from the government to support research and post-graduate study in the arts and humanities, from languages and law, archaeology and English literature to design and creative and performing arts. In any one year the AHRC makes approximately 700 research awards and around 1,350 post-graduate awards. Awards are made after a rigorous peer review process, to ensure that only applications of the highest quality are funded.

Professional and Career Development Loans (CDLs). These loans are specifically designed to help individuals acquire and improve vocational skills, and are aimed at those who would otherwise not have reasonable or adequate access to the funds. Loans of between £300 and £10,000 can be applied for to support up to two years of education or learning (plus up to one year's practical work experience where it forms part of the course).

Teachers. Qualified teacher status (QTS) is obtained through either an undergraduate or postgraduate course of initial teacher training. Training courses are either college-based or employment-based, the latter allowing trainees to teach as unqualified teachers and also study for QTS at the same time.

As well as meeting common professional standards, trainees must also pass computerized skills tests in information and communication technology, numeracy and literacy in order to be awarded QTS. Newly qualified teachers are then required to complete an induction programme during their first year of teaching.

Those who are recognized as qualified teachers in Scotland or Northern Ireland are also entitled to apply to the General Teaching Council for England or the General Teaching Council for Wales for QTS without undertaking further training. Nationals of European Economic Area (EEA) countries who are recognized as schoolteachers in an EEA member state can apply to the GTCE for QTS without undertaking further training.

Teachers who qualified as teachers in countries outside of the EEA are allowed to teach for four years in state maintained and non-maintained special schools in England. This is subject to satisfying Home Office rules on employing overseas workers. They are only allowed to teach beyond four years if they have been awarded QTS which can be obtained through undertaking an employment-based training programme.

In 2007–08, 31,300 trainees began college-based initial teacher training courses and a further 7,120 trainees were expected to enter employment-based initial teacher training.

In 2007–08, 477,100 full-time equivalent teachers were employed in maintained nursery, primary and secondary schools in the United Kingdom.

Finance. Total education expenditure by central and local government in the United Kingdom for 2008–09 was £79·9bn. representing 6·1% of GDP, compared with £44·4bn. and 4·6% of GDP in 2000–01.

Independent Schools. Independent schools which belong to an association affiliated to the Independent Schools Council (accounting for 80% of pupils) are subject to an inspection regime agreed between the government and the ISC. The Schools Inspection Service inspects schools affiliated to the Focus Learning Trust, and the Bridge Schools Inspectorate inspects schools affiliated to the Christian Schools Trust and the Association of Muslim Schools, also under arrangements agreed with the government. Non-association schools are inspected by Ofsted on a regular cycle.

The earliest of the independent schools were founded by medieval churches. Many were founded as 'grammar' (classical) schools in the 16th century, receiving charters from the reigning sovereign. Reformed mainly in the middle of the 19th century, among the best-known are Eton College, founded in 1440 by Henry VI; Winchester College (1394), founded by William of Wykeham, Bishop of Winchester; Harrow School, founded in 1560 as a grammar school by John Lyon, a yeoman; and Charterhouse (1611). Among the earliest foundations are King's School, Canterbury, founded 600; King's School, Rochester (604) and St Peter's, York (627).

Higher Education. In 2008–09 there were almost 2·4m. students in the UK at 166 higher education institutions, of which 116 were universities. The higher education student population was 57·1% female in 2007–08. 37% of the UK population between the ages of 25 and 34 have attained a tertiary qualification (OECD average, 34%), compared to only 25% of those aged between 55 and 64.

Total funding for higher education institutions in the UK was around £23·4bn. in 2007–08. Of this £8·5bn. came from funding council grants; £6·3bn. from tuition fees and education grants and contracts; £3·7bn. from research grants and contracts; £0·5bn. from endowment and investment income; and £4·4bn. from other income. Higher education institutions are funded by four UK bodies, one each for England, Scotland, Wales and Northern Ireland. Their roles include: allocating funds for teaching and research; promoting high-quality education and research; advising government on the needs of higher education; informing students about the quality of higher education available; and ensuring the proper use of public funds.

The *Open University* (OU) received its Royal Charter on 23 April 1969 and is an independent, self-governing institution, awarding its own degrees at undergraduate and postgraduate level. It is financed by students' fees and by the government through the Higher Education Funding Council for England for its students in England, Northern Ireland and other EU countries, the Scottish Funding Council and the Higher Education Funding Council for Wales.

Study materials provided may include specially written text-books, online teaching materials, audio CDs, DVDs and computer software. The OUAnywhere initiative allows students to access learning materials through their smartphones and tablets. Support is usually provided by tutors or study advisers who are available to students through group tutorials and seminars, online or by telephone.

No formal qualifications are required for entry to the majority of undergraduate courses. 69% of undergraduates have no previous higher education qualifications on entry. The University launched with 25,000 students and in 2011–12 there were 246,626 students (including 37,121 on OU validated programmes). 342 undergraduate modules are offered by the OU along with 141 postgraduate modules. There are 13 national and regional centres in addition to 350 study centres across the UK and more than 6,340 associate lecturers. 80 of the FTSE 100 companies have sponsored staff to study with the OU. In 2010–11 the OU attracted more than £17·5m. in external research income.

The only university independent of the state system is the *University of Buckingham*, which opened in 1976 and received a Royal Charter in 1983. It offers two-year honours degrees, the academic year commencing in Jan., July or Sept., and consisting of four 10-week terms. There are a wide variety of degree courses available, the most popular being in business, humanities, law and science. In Jan. 2011 there were over 1,000 students.

All universities charge fees, but financial help is available to students from several sources, and the majority of students receive some form of financial assistance.

See also ENGLAND, SCOTLAND *and* NORTHERN IRELAND.

British Council

The British Council is the UK's international organization for educational opportunities and cultural relations. The British Council works in over 100 countries in the arts, education and English-teaching to build opportunity and trust through the exchange of knowledge and ideas between people. This work contributes to the security and prosperity both of the UK and the countries where it works. In 2010–11 it engaged face to face with 30m. people and reached 578m. It has 6,800 staff worldwide. Its total turnover in 2010–11 was £693m., of which £190m. was a grant-in-aid from the Foreign and Commonwealth Office. The remainder was generated through trading activities such as English-language teaching.

Chair: Vernon Ellis.

Director-General: Martin Davidson, CMG.

Headquarters: 10 Spring Gdns, London, SW1A 2BN.

Website: http://www.britishcouncil.org

Health

The National Health Service (NHS) in England and Wales started on 5 July 1948. There is a separate Act for Scotland.

The NHS is a charge on the national income in the same way, for example, as the armed forces. Every person normally resident in the UK is entitled to use any part of the service, and no insurance qualification is necessary.

Since its inception, the NHS has been funded from general taxation and National Insurance (NI) contributions, and the present government has maintained the original principle that the NHS should be a service provided to all those who need it, regardless of their ability to pay or where they live. In 2008–09 the NHS in England was funded 17·5% by NI contributions, 71·9% by general taxation with the remainder coming from charges and receipts, including land sales and proceeds from income generation schemes. Health authorities may raise funds from voluntary sources; hospitals may take private, paying patients.

Health is the second largest government spending sector after social protection. In 2008–09 estimated NHS expenditure on health was £110·5bn. When combined with private expenditure on health care, the estimated overall percentage of UK GDP spent on health was 8·9%.

Organization. The Health and Social Care Act 2012 provided for a major restructuring of the NHS. On 1 April 2013 the ten Strategic Health Authorities (SHAs), responsible for implementing Department of Health policy at regional level and overseeing the operations of NHS trusts at the local level since 2002, were abolished. Primary Care Trusts—introduced in 2000 to control local health care and hold to account provider organizations for delivery of services that they commissioned—also ceased to exist.

Under the terms of the 2012 Act, responsibility for commissioning NHS care was given over to more than 200 clinical commissioning groups (CCGs), each made up of GPs and other clinicians. CCGs are directly responsible for a combined budget of some £60bn. Every GP practice is required to be a member of a CCG. Each CCG has a governing body comprising at least one registered nurse and a secondary care specialist alongside GPs.

CCGs are supported by the NHS Commissioning Board, which authorizes clinical commissioning groups, allocates resources and commissions certain services itself, such as primary care. The Commissioning Board operates 27 area teams.

The NHS Trust Development Authority oversees the performance and governance of NHS Trusts, including clinical quality and managing their progress towards NHS Foundation Trusts (which have greater managerial and financial independence than NHS Trusts). The Quality Care Commission, meanwhile, guarantees the safety and quality of services, and Monitor regulates economic efficiency. Separate bodies have overall responsibility for professional education, training and public health programmes.

Services. The NHS broadly consists of hospital and specialist services, general medical, dental and ophthalmic services, pharmaceutical services, community health services and school health services. In general these services are free of charge; the main exceptions are prescriptions, spectacles, dental and optical examination, dentures and dental treatment, amenity beds in hospitals, and some community services, for which contributory charges are made with certain exemptions.

Private. In recent years increasing numbers of people have turned to private medical insurance. This covers the costs of private medical treatment (PMI) for curable short-term medical conditions. PMI includes the costs of surgery, specialists, nursing and accommodation at a private hospital or in a private ward of an NHS hospital. Approximately 13% of the UK population have private medical insurance. The leading companies are BUPA Healthcare, AXA PPP Healthcare and PruHealth.

In 2009, 21% of the population of the UK aged 16 and over smoked. In 1974 the percentage had been 45%, with 51% of males and 41% of females smoking. Over the years the difference between the percentage of men and of women who smoke has been declining—in 2009 the rates were 22% for men and 20% for women. The overall percentage of the UK population who are smokers is similar to the average for the EU as a whole, but among men the percentage of smokers in the UK is lower than in the EU as a whole whereas among women it is higher. Alcohol consumption has increased in recent years. Whereas in 1961 the average Briton consumed the equivalent of 4·5 litres of pure alcohol a year, by 2006 this figure had risen to 8·9 litres (although down from a peak of 9·4 litres in 2004).

There were an estimated 86,500 people living with HIV in the UK in 2009 (0·1% of the population).

See also NORTHERN IRELAND.

Personal Social Services

Under the Local Authority Social Services Act, 1970, and in Scotland the Social Work (Scotland) Act, 1968, the welfare and social work services provided by local authorities were made the responsibility of a new local authority department—the Social Services Department in England and Wales, and Social Work Departments in Scotland headed by a Director of Social Work, responsibility in Scotland passing in 1975 to the local authorities. The social services thus administered include: the fostering, care and adoption of children, welfare services and social workers for people with learning difficulties and the mentally ill, the disabled and the aged, and accommodation for those needing residential care services. Legislation of 1996 permits local authorities to make cash payments as an alternative to community care. In Scotland the Social Work Departments' functions also include the supervision of persons on probation, of adult offenders and of persons released from penal institutions or subject to fine supervision orders.

Expenditure is reviewed by the Social Services Inspectorate and the Audit Commission (in Scotland by the Social Work Services Inspectorate and the Accounts Commission).

Welfare

The National Insurance Act 1965 now operates under the Social Security Contributions and Benefits Act 1992 and the Social Security Administration Act 1992.

Since 1975 Class 1 contributions have been related to the employee's earnings and are collected with PAYE income tax. Class 2 and Class 3 contributions remain flat-rate, but, in addition to Class 2 contributions, those who are self-employed may be liable to pay Class 4 contributions, which for the year 2013–14 are at the rate of 9% on profits or gains between £7,755 and £41,450 (with a further 2% contribution on any profit exceeding the upper limit), which are assessable for income tax under Schedule D. The non-employed and others whose contribution record is not sufficient to give entitlement to benefits are able to pay a Class 3 contribution of £13·55 per week in 2013–14 voluntarily, to qualify for a limited range of benefits. Class 2 weekly contributions for 2013–14 for men and women are £2·70. Class 1A contributions are paid by employers who provide employees with a car and fuel for their private use.

The Social Security Pensions Act 1975 introduced earnings-related retirement, invalidity and widows' pensions. Members of occupational pension schemes may be contracted out of the earnings-related part of the state scheme relating to retirement and widows' benefits. Employee's national insurance contribution liability depends on whether he/she is in contracted-out or not contracted-out employment.

Full-rate contributions for non-contracted-out employment in 2013–14:

Weekly Earnings (in £1)	Employee pays	Employer pays
Below 109 (Lower Earnings Limit)	Nil	Nil
109–148 (Secondary Threshold)	Nil	Nil
148–149 (Secondary Threshold to Primary Threshold)	Nil	13·8%
149–797 (Primary Threshold to Upper Earnings Limit)	12%	13·8%
Over 797 (Upper Earnings Limit)	2%	13·8%

For contracted-out employment, the contracted-out rebate for primary contributions (employee's contribution) is 1·4% of earnings between the lower earnings limit and the upper earnings limit for all forms of contracting-out; the contracted-out rebate for secondary contributions (employer's contributions) is 3·4% of earnings between the lower earnings limit and the upper earnings limit.

Contributions together with interest on investments form the income of the *National Insurance Fund* from which benefits are paid. 29,390,000 persons (13,420,000 women) paid contributions in 2008–09, including 25,210,000 employees at standard rate.

Receipts, 2009–10 (in £1m.), 129,320, including: contributions, 73,817; investment income, 1,684; transfers from Great Britain, 395; compensation for Statutory Sick Pay/Statutory Maternity Pay, 237. Disbursements (in £1m.), 79,793, including: retirement pensions, 66,442; Incapacity, 6,177; Personal Pensions, 2,623; administration, 1,401; Jobseeker's Allowance (Contributory), 1,106; Bereavement Benefits, 647; redundancy payments, 531; transfers to Northern Ireland, 395; maternity, 343.

Statutory Sick Pay (SSP). Employers are responsible for paying SSP to their employees who are absent from work through illness or injury for up to 28 weeks in any three-year period. All employees aged between 16 and 65 (60 for women) with earnings above the Lower Earnings Limit are covered by the scheme whenever they are sick for four or more days consecutively. The weekly rate is £86·70. For most employees SSP completely replaces their entitlement to state incapacity benefit which is not payable as long as any employer's responsibility for SSP remains.

Contributory benefits. Qualification for these depends upon fulfilment of the appropriate contribution conditions, except that persons who are incapable of work as the result of an industrial accident may receive incapacity benefit followed by invalidity benefit without having to satisfy the contributions conditions.

Jobseeker's Allowance. Unemployed persons claiming the allowance must sign a 'Jobseeker's Agreement' setting out a plan of action to find work. The allowance is not payable to persons who left their job voluntarily or through misconduct. Claimants with sufficient National Insurance contributions are entitled to the allowance for six months regardless of their means; otherwise, recipients qualify through a means test and the allowance is fixed according to family circumstances, at a rate corresponding to Income Support for an indefinite period. In May 2010 there were 1,354,620 people receiving the Jobseeker's Allowance (973,100 males), down from 1,443,000 a year earlier. Payments start at £56·80 per week.

Incapacity benefit. Entitlement begins when entitlement to SSP (if any) ends. There are three rates: a lower rate for the first 28 weeks; a higher rate between the 29th and 52nd week; and a long-term rate from the 53rd week of incapacity. It also comprises certain age additions and increases for adult and child dependants. A more objective medical test of incapacity for work was introduced for incapacity benefit as well as for other social security benefits paid on the basis of incapacity for work. This test applies after 28 weeks' incapacity for work and assesses ability to perform a range of work-related activities rather than the ability to perform a specific job. Benefit is taxable after 28 weeks. Some 1,892,990 claims were being made in May 2010. In Oct. 2008 Incapacity Benefit was replaced by Employment and Support Allowance for all new claims.

Statutory Maternity Pay. Pregnant working women may be eligible to receive Statutory Maternity Pay directly from their employer for a maximum of 39 weeks if average gross earnings are £109 a week or more (2013–14). There are two rates: a higher rate (90% of average earnings for the first six weeks), and a lower rate of £136·78 or 90% of earnings (whichever is less) for up to 33 weeks. For women who do not qualify for Statutory Maternity Pay, including self-employed women, there is a Maternity Allowance.

A payment of £500 from the Social Fund (Sure Start Maternity Grant) may be available if the mother or her partner are receiving Income Support, income-based Jobseeker's Allowance, Pension Credit, Child Tax Credit (at a rate higher than the family element) or Working Tax Credit (in cases of disability). It is also available if a parent adopts a baby, is granted a parental order on a surrogate birth or is granted a residence order on a child (subject to certain conditions). It is restricted to the first child in a family.

Statutory Paternity Pay. Since 6 April 2003 working fathers have had the right to two weeks paid paternity leave providing average earnings are £109 a week or more (2013–14). This will be paid at the same rate as the lower rate of Statutory Maternity Pay (£136·78 a week or 90% of average weekly earnings if this is less than £136·78).

Statutory Adoption Pay. Paid adoption leave is for up to 39 weeks at the same rate as Statutory Paternity Pay. It is available to employed people adopting a child on their own, or for one member of a couple adopting together. Parents adopting from overseas are also eligible, although conditions may differ.

Bereavement Benefits. Available to both men and women, Bereavement Benefits were introduced from 9 April 2001 to replace the former Widows' Benefit scheme. There are three main types of Bereavement Benefits available to men and women widowed on or after 9 April 2001: bereavement payment, widowed parent's allowance and bereavement allowance. *Bereavement Payment* is a single tax-free lump sum of £2,000 payable immediately on bereavement. A widower/widow may be able to get this benefit if their late spouse has paid enough National Insurance Contributions (NIC) and was under 60 at death; or was not getting a Category A State Retirement Pension at death. *Widowed Parent's Allowance* is a weekly benefit payable when the widower/widow is receiving Child Benefit. The amount of Widowed Parent's Allowance is based on the late spouse's NIC record. He/she may also get benefit for the eldest dependent child and further higher benefit for each subsequent child; also an additional pension based on their late spouse's earnings. If the late spouse was a member of a contracted-out occupational scheme or a personal pension scheme, that scheme is responsible for paying the whole or part of the additional pensions. Widowed Parent's Allowance is taxable. *Bereavement Allowance* is a weekly benefit payable to widows and widowers without dependent children and is payable between age 45 and State Pension age. The amount of Bereavement Allowance payable to a widower/widow between 45 and 54 is related to their age at the date of entitlement. Their weekly rate is reduced by 7% for each year they are aged under 55 so that they get 93% rate at age 54, falling to 30% at age 45. Those aged 55 or over at the date of entitlement will get the full rate of Bereavement Allowance. The amount of Bereavement Allowance is based on the late spouse's NIC record and is payable for a maximum of 52 weeks from the date of bereavement. A widower/widow cannot get a Bereavement Allowance at the same time as a Widowed Parent's Allowance. Women widowed before 9 April 2001 continue to receive their Widows' Benefit entitlement on the arrangements that existed before that date so long as they continue to satisfy the qualifying conditions. There were 50,750 recipients of Widows' Benefits and 63,310 recipients of Bereavement Benefits in May 2010.

Retirement Pension. The state retirement ('old-age') pension scheme has two components: a basic pension and an additional state pension (state second pension or S2P). The amount of the first is subject to National Insurance contributions made; the amount of the second is earnings-related although it is scheduled to become a flat-rate top-up payment by 2030. Payments are made automatically when the basic pension is claimed. Qualifying National Insurance contributions are made for employed persons earning over a lower threshold (£5,668 in 2013–14), anyone looking after children under the age of 12 and in receipt of Child Benefit, carers working more than 20 hours a week claiming Carer's Credit and foster carers claiming Carer's Credit, and persons receiving certain other benefits as a result of illness or disability.

Pensions are payable to men at 65 years of age. Until April 2010 they were payable to women at 60 years of age but since then the state pension age for women has been gradually rising, with the age differential in the process of being phased out. Under the 2011 Pensions Act the equalization of the state pension age will take place by Nov. 2018 (instead of April 2020) and the state pension age for both men and women will be increased to 66 by Oct. 2020 and eventually to 68 by 2044–46. There are standard rates for single persons and for married couples. Proportionately reduced pensions are payable where contribution records are deficient.

Employees may contract out of the additional state pension if they join an employers contracted-out occupational pension scheme or have an appropriate stakeholder or personal pension scheme. The self-employed, unemployed, those in full-time training and anyone earning below a lower threshold (£5,668 in 2013–14) do not qualify.

Self- and non-employed persons may contribute voluntarily for retirement pension.

In Jan. 2013 the government announced plans to introduce a simplified state pension scheme from April 2017. A weekly flat-rate pension of £144 (adjusted for inflation) is scheduled to replace the existing basic state pension and state second pension. Means-tested pension credits will also be ended. Claimants must have made 35 years of National Insurance contributions to receive the new pension, up from the previously required 30 years.

Persons who defer claiming their pension during the five years following retirement age are paid an increased amount, as do men and women who had paid graduated contributions. 12,277,360 persons were receiving National Insurance retirement pensions in May 2009 (7,650,400 women and 4,626,960 men). The full basic state pension in 2013–14 is £110·15 per week for a single person and £176·15 per week for a married couple. Since 1 Oct. 1989 the pension for which a person has qualified may be paid in full whether a person continues in work or not irrespective of the amount of earnings. Although for males the official retirement age is 65, in 2010 the average actual retirement age among males was 64·6 years (up from 63·8 years in 2004). For women the average actual retirement age rose from 61·2 years to 62·3 years between 2004 and 2010.

At the age of 80 a small age addition is payable. In addition non-contributory pensions are now payable, subject to residence conditions, to persons aged 80 and over who do not qualify for a retirement pension or qualify for one at a low rate.

Pensioners whose pension is insufficient to live on may qualify for Income Support.

Non-Contributory Benefits

Universal Credit. Universal Credit is being introduced nationally in Oct. 2013 but has been trialled since April 2013 on a limited basis. It replaces Child Tax Credit, Housing Benefit, Income Support, Income-related Employment Support and Allowance, Income-related Jobseeker's Allowance and Working Tax Credit.

Child Benefit. Child Benefit is a tax-free cash allowance normally paid to the mother. The weekly rates are highest for the eldest qualifying child (£20·30 weekly in 2013–14) and less for each other child (£13·40 weekly in 2013–14). Child Benefit is payable for children under 16, for 16- and 17-year-olds registered for work or training, and for those under 20 receiving full-time non-advanced education. Some 7,841,675 families received benefit in Aug. 2010. From Jan. 2013 families with at least one partner earning £50,000 or more are not able to claim the total amount of child benefit.

Child Support Agency. The Child Support Agency (CSA) is responsible for calculating, collecting and enforcing child maintenance payments. The non-resident parent pays 15% of their net income if they have one child, 20% for two and 30% for three or more children. The agency currently deals with around 1·5m. child support cases. In Dec. 2006 the Child Maintenance White Paper announced the establishment of a new and radically different organization called the Child Maintenance and Enforcement Commission. It assumed responsibility for the Child Support Agency in Nov. 2008, and was to introduce a tougher enforcement regime that encouraged parents to take greater responsibility for the financial support of their children. However,

it was abolished in July 2012, with the CSA becoming the delivery arm of the Child Maintenance Group within the Department for Work and Pensions.

Working Tax Credit. This tackles poor work incentives and persistent poverty among working people. For families with children, credit is available for those with low incomes. It also extends support to low-income working people without children aged 25 or over working 30 hours or more a week. The Working Tax Credit is not just restricted to those with children; the amount of the award varies considerably depending on the prevailing circumstances. Both single persons and couples may be eligible.

Child Tax Credit. The Child Tax Credit aims at creating a single system of support for families with children, payable irrespective of the work status of the adults in the household. This means that the Child Tax Credit forms a stable and secure income bridge as families move off welfare and into work. It also provides a common framework of assessment, so that all families are part of the same inclusive system. The Child Tax Credit provides a family element of up to £545 per year and a child element of up to £2,720 per child per year in addition to Child Benefit. The amount paid varies depending on the number of children and the gross annual joint income.

Guardian's Allowance. A person responsible for an orphan child may be entitled to a Guardian's Allowance in addition to Child Benefit. Normally, both the child's parents must be dead but when they never married or were divorced, or one is missing or serving a long sentence of imprisonment, the allowance may be paid on the death of one parent only. The weekly rate of Guardian's Allowance is £15·90 in 2013–14.

Attendance Allowance. This is a tax-free Social Security benefit for disabled people over 65 who need help with personal care. The rates are increased for the terminally ill. There were 1,614,270 recipients in May 2010.

Carers' Allowance. This is a taxable benefit paid to those who care for a disabled person for at least 35 hours per week. The carer must be at least 16, not in full-time education of 21 hours or more per week, and not earn more than £100 per week after certain deductions have been made—such as income tax. This is a weekly rate (£59·75 in 2013–14), with increases for dependants. In May 2010 there were 536,900 recipients.

Disability Living Allowance. This is a non-taxable benefit available to people disabled before the age of 65, who need help with getting around or with personal care for at least three months. The mobility component has two weekly rates, the care component has three. There were 3,157,310 recipients in May 2010. From 8 April 2013 Disability Living Allowance is being gradually phased out and replaced by Personal Independence Payment.

Industrial Injuries Disablement and Death Benefits. The scheme provides a system of insurance against 'personal injury by accident arising out of and in the course of employment' and against certain prescribed diseases and injuries owing to the nature of the employment. There are no contribution conditions for the payment of benefit. There were 260,690 recipients in March 2009. Two types of benefit are provided:

—*Disablement Benefit.* This is payable where, as the result of an industrial accident or prescribed disease, there is a loss of physical or mental faculty. The loss of faculty will be assessed as a percentage by comparison with a person of the same age and sex whose condition is normal. If the assessment is between 14–100% benefit will be paid as weekly pension. The rates vary from 20% disabled to 100% disablement. Assessments of less than 14% do not normally attract basic benefit except for certain progressive chest diseases. Pensions for persons under 18 are at a reduced rate. When injury benefit was abolished for industrial accidents occurring and prescribed diseases commencing on or after 6 April

1983, a common start date was introduced for the payment of Disablement Benefit 90 days (excluding Sundays) after the date of the relevant accident or onset of the disease.

—Death Benefit. This is payable to the widow, widower or children of a person who died before 11 April 1988 as the result of an industrial accident or a prescribed disease. For deaths which occurred on or after 11 April 1988, standard Bereavement Benefits apply (and, until 2001, Widows' Benefits). Allowances may be paid to people who are suffering from pneumoconiosis or byssinosis or certain other slowly developing diseases due to employment before 5 July 1948. They must not at any time have been entitled to benefit under the Industrial Injuries provision of the Social Security Act, or compensation under Workmen's Compensation Acts, or have received damages through the courts.

War Pensions. Pensions are payable for disablement or death as a result of service in the armed forces. Similar schemes exist for other groups such as merchant seamen injured as a result of war or for civilians injured by enemy action in the Second World War. The amount depends on the degree of disablement. There were 180,400 recipients in March 2010.

Housing Benefit. The Housing Benefit scheme assists persons who need help to pay their rent, using general assessment rules and benefit levels similar to those for the income support scheme. The scheme sets a limit of £16,000 on the amount of capital a person may have and still remain entitled. Restrictions on the granting of benefit to persons under 25 were introduced in 1995. In May 2010 there were 4,751,530 beneficiaries. Since April 2013 claimants in social housing have their Housing Benefit reduced by 14% if they have one spare bedroom and by 25% if they have two or more spare bedrooms.

Income Support. Income Support is a non-contributory benefit for people aged 16 or over, not working 16 hours or more a week or with a partner not working more than 24 hours or more per week, and not required to be available for employment. These include single parents, long-term sick or disabled persons, and those caring for them who qualify for Invalid Care Allowance. Income Support is not payable if the claimant (or claimant and partner together) has capital assets that total more than £16,000. These include savings, investments or property other than their home. Savings/capital assets worth under £6,000 are ignored. Savings between £6,000 and £16,000 are treated as if each £250 or part of £250 brings in an income of £1 per week. Income Support claimants whose partners are of pensionable age may have up to £12,000 and still be entitled to Income Support. Claimants in residential care and nursing homes are allowed to have up to £16,000 and still be entitled to Income Support. From 6 Oct. 2003 a new Pension Credit replaced the Minimum Income Guarantee (Income Support for people aged 60 and over). In Oct. 2008 Income Support paid on the grounds of incapacity was replaced by Employment and Support Allowance for all new claims. In May 2010 there were 2,734,170 Pension Credit claimants, 1,842,490 Minimum Income Guarantee claimants and 9,780 Income Support claimants. The average weekly award was £84·94 in May 2009.

Council Tax Benefit. Council Tax Benefit was abolished with effect from 1 April 2013 and has been replaced by a system of localized support (Council Tax Support).

The Social Fund. This comprises: *Sure Start Maternity Grant* (a payment of up to £500 for the first baby expected, born or adopted, payable to persons receiving Income Support, Income-based Jobseeker's Allowance, Child Tax Credit, Working Tax Credit or Pension Credit); *Funeral Payments* (a payment of fees levied by the burial authorities and crematoria, plus up to £700 for other funeral expenses, to persons receiving Income Support, Income-based Jobseeker's Allowance, Housing Benefit, Child Tax Credit, Working Tax Credit or Pension Credit); *Cold Weather Payments* (a payment of £25 for any consecutive seven days when the temperature is below freezing to persons receiving income

support who are pensioners, disabled or have a child under five); *Winter Fuel Payments* (a payment of £250 to every household with a person aged 60 or over providing they do not live permanently in a hospital, residential care or nursing home, or £400 if the household has someone aged 80 years old or over). The Discretionary Social Fund comprises: *Community Care Grants* (payments to help persons receiving income support to move into the community or avoid institutional care); *Budgeting Loans* (interest-free loans to persons receiving income support for expenses difficult to budget for); *Crisis Loans* (interest-free loans to anyone without resources in an emergency where there is no other means of preventing serious risk to health or safety). Savings over £500 (£1,000 for persons aged 60 or over) are taken into account before payments are made.

Fraser, Derek, *The Evolution of the British Welfare State: A History of Social Policy since the Industrial Revolution.* 2009

Harris, Bernard, *The Origins of the British Welfare State: Social Welfare in England and Wales, 1800–1945.* 2004

Hill, M., *The Welfare State in Britain: a Political History since 1945.* 1993

Lowe, Rodney, *The Welfare State in Britain since 1945.* 2004

Timmins, N., *The Five Giants: a Biography of the Welfare State.* 1995

RELIGION

The Anglican Communion originated from the Church of England and parallels in its fellowship of autonomous churches the evolution of British influence beyond the seas from colonies to dominions and independent nations. The Archbishop of Canterbury presides as *primus inter pares* at the decennial meetings of the bishops of the Anglican Communion at the Lambeth Conference and at the biennial meetings of the Primates and the Anglican Consultative Council. The 2008 Conference was held in Canterbury and was attended by 670 bishops. Average attendance at Sunday worship in 2004 numbered 1·0m., compared to 3·5m. in 1950. There were 1,543,000 Anglicans on the electoral roll in 2005.

The Anglican Communion (Anglican Episcopal family) consists of an estimated 80m. Christians who are members of 44 different Churches (including six Extra-Provincial Dioceses). These are: The Anglican Church in Aotearoa, New Zealand and Polynesia; The Anglican Church of Australia; The Church of Bangladesh; The Anglican Church of Bermuda (Extra-Provincial to the Archbishop of Canterbury); The Episcopal Anglican Church of Brazil; The Anglican Church of Burundi; The Anglican Church of Canada; The Church of the Province of Central Africa; The Anglican Church of the Central America Region; The Church of Ceylon (Sri Lanka) (Extra-Provincial to the Archbishop of Canterbury); The Province of the Anglican Church of the Congo; The Episcopal Church of Cuba (under a Metropolitan Council); The Church of England; The Parish of the Falkland Islands (Extra-Provincial to the Archbishop of Canterbury); Hong Kong Sheng Kung Hui; The Church of the Province of the Indian Ocean; The Church of Ireland; The Nippon Sei Ko Kai; The Episcopal Church in Jerusalem and the Middle East; The Anglican Church of Kenya; The Anglican Church of Korea; The Lusitanian Church (Portugal) (Extra-Provincial to the Archbishop of Canterbury); The Church of the Province of Melanesia; The Anglican Church of Mexico; The Church of the Province of Myanmar (Burma); The Church of Nigeria (Anglican Communion); The Church of North India; The Church of Pakistan; The Anglican Church of Papua New Guinea; The Episcopal Church in the Philippines; The Episcopal Church of Rwanda; The Scottish Episcopal Church; The Church of the Province of South East Asia; The Church of South India; The Anglican Church of Southern Africa; The Anglican Church of the Southern Cone of America; The Reformed Episcopal Church of Spain (Extra-Provincial to the Archbishop of Canterbury); The Episcopal Church of the Sudan; The Anglican Church of Tanzania; The Church of the Province of Uganda; The Episcopal Church in the United States of America; The Church in Wales; The Church of the Province of West Africa; and The Church in the Province of the West Indies. New provinces are also currently in formation. Churches in Communion include the Mar Thoma Syrian Church, the Philippine Independent Church, and some Lutheran and Old Catholic Churches in Europe. The Church in China is known as a 'post-denominational' Church whose formation included Anglicans in the Holy Catholic Church in China.

England and Wales. The established Church of England, which baptizes about 13% of infants born in England (i.e. excluding Wales but including the Isle of Man and the Channel Islands), is Anglican. Civil disabilities on account of religion do not attach to any class of British subject. Under the Welsh Church Acts, 1914 and 1919, the Church in Wales and Monmouthshire was disestablished as from 1 April 1920, and Wales was formed into a separate Province.

The Queen is, under God, the supreme governor of the Church of England, with the right, regulated by statute, to nominate to the vacant archbishoprics and bishoprics. The Queen, on the advice of the First Lord of the Treasury, also appoints to such deaneries, prebendaries and canonries as are in the gift of the Crown, while a large number of livings and also some canonries are in the gift of the Lord Chancellor.

There are two archbishops (at the head of the two Provinces of Canterbury and York), and 44 diocesan sees including the diocese in Europe, which is part of the Province of Canterbury. Justin Welby was enthroned as *Archbishop of Canterbury* on 21 March 2013. Each archbishop also has his own particular diocese, wherein he exercises episcopal, as in his Province he exercises metropolitan, jurisdiction. In Dec. 2008 there were 42 serving bishops, 64 suffragan and assistant bishops, 42 deans and provosts of cathedrals and 110 archdeacons. The *General Synod*, which replaced the Church Assembly in 1970 in England, consists of a House of Bishops, a House of Clergy and a House of Laity, and has power to frame legislation regarding Church matters. The first two Houses consist of the members of the Convocations of Canterbury and York, each of which consists of the diocesan bishops and elected representatives of the suffragan bishops, five for Canterbury province and three for York (forming an Upper House); deans and archdeacons, and a certain number of proctors elected as the representatives of the priests and deacons in each diocese, together with, in the case of Canterbury Convocation, four representatives of the Universities of Oxford, Cambridge, London and the Southern Universities, and in the case of York two representatives of the Universities of Durham and Newcastle and the other Northern Universities, and three archdeacons to the Armed Forces, the Chaplain General of Prisons and two representatives of the Religious Communities (forming the Lower House). The House of Laity is elected by the lay members of the Deanery Synods but also includes two representatives of the Religious Communities. The Houses of Clergy and Laity also include a small number of *ex officio* members. Every Measure passed by the General Synod must be submitted to the Ecclesiastical Committee, consisting of 15 members of the House of Lords nominated by the Lord Chancellor and 15 members of the House of Commons nominated by the Speaker. This committee reports on each Measure to Parliament, and the Measure receives the Royal Assent and becomes law if each House of Parliament resolves that the Measure be presented to the Queen.

Parochial affairs are managed by annual parochial church meetings and parochial church councils. In 2008 there were 12,702 ecclesiastical parishes, inclusive of the Isle of Man and the Channel Islands. These parishes do not, in many cases, coincide with civil parishes. Although most parishes have their own churches, not every parish nowadays can have its own incumbent or priest. About 3,290 non-stipendiary clergy hold a bishop's licence to officiate at services.

In 2008 there were 4,431 incumbents excluding dignitaries, 1,913 other clergy of incumbent status and 1,301 assistant curates working in the parishes.

Women have been admitted to Holy Orders (but not the Episcopate) as deacons since 1987 and as priests since 1994. At 31 Dec. 2009 there were 1,649 full-time stipendiary women clergy, 1,548 of whom were in the parochial ministry. In the years following the vote in 1992 in favour of the ordination of women, 441 clergymen resigned because they disagreed with the decision. 11 clergymen subsequently re-entered the Church of England ministry and of the 441 who resigned, 260 are estimated to have joined the Roman Catholic Church and 30 the Orthodox Church. In Nov. 2010 five bishops (two retired) resigned in opposition to plans to admit women to the Episcopate.

Private persons possess the right of presentation to over 2,000 benefices; the patronage of the others belongs mainly to the Queen, the bishops and cathedrals, the Lord Chancellor, the colleges of the Universities of Oxford and Cambridge and other patronage trusts. More than 750 benefices include patronage trusts among their patrons. In addition to the dignitaries and parochial clergy already identified there were, in 2008, 143 cathedral and 342 full-time non-parochial clergy working within the diocesan framework, giving a total of 8,346 full-time stipendiary clergy working within the diocesan framework as at Dec. 2008. In addition there were 311 part-time stipendiary clergy. Although these figures account for the majority of active clergy in England, there are many others serving in institutions and elsewhere who cannot be quantified with any certainty. They include 1,609 chaplains in hospitals, the forces, prisons, schools and colleges, and those in mission agencies and religious communities.

Of the 40,405 buildings registered for the solemnization of marriages at 30 June 2007 (statistics from the Office of National Statistics), 16,418 belonged to the Church of England and the Church in Wales, and 23,987 to other religious denominations (Methodist, 6,362; Roman Catholic, 3,310; Baptist, 3,075; United Reformed, 1,587; Congregationalist, 1,245; Calvinistic Methodist, 1,057; Jehovah's Witnesses, 830; Brethren, 741; Salvation Army, 722; Unitarians, 161; other Christian, 4,291; Sikhs, 161; Muslims, 164; other non-Christian, 281). Of the 235,794 marriages celebrated in 2008 (331,150 in 1990), 57,057 were in the Established Church and the Church in Wales (115,328 in 1990), 21,444 in other denominations (43,837 in 1990) and 157,296 were civil marriages in Register Offices (156,875 in 1990).

The Roman Catholic population in England and Wales (the number of adherents) was 4,084,351 in 2009; 898,852 regularly attended Mass in England and Wales in 2009. There are 22 dioceses in five provinces and one Bishopric of the Forces (also covers Scotland), and the Personal Ordinariate of Our Lady of Walsingham. Vincent Nichols was installed as *Archbishop of Westminster* on 21 May 2009. In Feb. 2013 there were two Roman Catholic cardinals, one of whom is in Scotland. In England and Wales there are five archbishops, 17 other diocesan bishops and nine auxiliary or assistant bishops. In 2009 there were 4,645 priests in active ministry, 2,524 parish churches and 1,036 convents.

Membership of other denominations in the UK in 2005 (and 1975): Presbyterians, 876,970 (1·65m.); Methodists, 303,973 (0·61m.); Baptists, 199,171 (0·27m.); other Protestants, 298,744; independent churches, 171,993; Orthodox, 271,158 (0·2m.); Pentecostals (including Afro-Caribbean churches), 288,183; Latter-day Saints (Mormons), 175,000; Jehovah's Witnesses, 128,333; Spiritualists, 30,000; Muslims, 893,700 active members (0·4m.); Sikhs, 184,000 active members (0·12m.); Hindus, 305,000 active members (0·1m.); Jews (2001), 267,373 (0·11m.).

In 2001 for the first time the census asked an optional question about religion. In England and Wales 37·3m. people described themselves as Christian. In England, 3·1% of the population stated their religion as Muslim, 1·1% Hindu, 0·7% Sikh, 0·5% Jewish and 0·3% Buddhist. In Wales, 0·7% of the population stated their religion as Muslim, 0·2% Buddhist, 0·2% Hindu, 0·1% Jewish and

0·1% Sikh. In England and Wales 7·7m. people said they had no religion (14·6% in England and 18·5% in Wales). Just over 4m. people chose not to answer the religion question.

In Scotland the 2001 census asked two questions on religion—religion of upbringing and current religion. For religion of upbringing, the largest groups were Church of Scotland (47%), no religion (18%) and Roman Catholic (17%). The equivalent percentages for current religion were 42%, 28% and 16%.

Across all denominations, adult church attendance in Great Britain was 6·5% of the population in 2005.

The Salvation Army is an international Christian church and charity, working in 124 countries, and offers unconditional friendship, support and practical help to people of all ages, backgrounds and needs. In the UK and Ireland (as at 2011) this work includes more than 800 community churches and social centres with more than 5,200 employees, some 1,200 ministers of religion and close to 50,000 members.

There is a 400-member Board of Deputies of British Jews.

In 2010 there were approximately 1·89m. to St Paul's Cathedral, London, 1·39m. to Westminster Abbey, London, 1·03m. visits to Canterbury Cathedral and 512,000 visits to York Minster.

See also SCOTLAND *and* NORTHERN IRELAND.

Bradley, I., *Marching to the Promised Land: Has the Church a Future?* 1992
Chapman, Mark, *Anglicanism: A Very Short Introduction.* 2006
Davie, Martin, *A Guide to the Church of England.* 2008
De La Noy, M., *The Church of England: a Portrait.* 1993

CULTURE

World Heritage Sites

Sites under UK jurisdiction which appear on UNESCO's World Heritage List are (with year entered on list): Giant's Causeway and Causeway Coast (1986), rock formations on the Antrim Plateau in Northern Ireland; Durham Castle and Cathedral (1986 and 2008), the largest example of a Norman cathedral; Ironbridge Gorge (1986), built in the 18th century and considered the emblem of the industrial revolution; Studley Royal Park, including the Ruins of Fountains Abbey (1986), developed from the 18th century on the site of a former Cistercian abbey in Yorkshire; Stonehenge, Avebury and Associated Sites (1986 and 2008), among the world's most famous pre-historic monoliths; Castles and Town Walls of King Edward in Gwynedd (1986), a testament to the early period of English colonization in the late 13th century; St Kilda (1986, 2004 and 2005), a volcanic archipelago on the coast of the Hebrides; Blenheim Palace (1987), seat of the Dukes of Marlborough near Oxford and birthplace of Sir Winston Churchill; City of Bath (1987), with remains from its time as a Roman spa town, and home to many examples of neo-classical Georgian architecture; Westminster Palace, Westminster Abbey and Saint Margaret's Church (1987 and 2008)—the palace is the medieval seat of parliament rebuilt in the 19th century, the abbey the site of all coronations since the 11th century and Saint Margaret's is a small medieval gothic church; Henderson Island (1988), a South Pacific atoll; Tower of London (1988), a Norman fortress built to guard London; Canterbury Cathedral, St Augustine's Abbey and St Martin's Church (1988), the spiritual seat of the Church of England; Old and New Towns of Edinburgh (1995), the Scottish capital; Gough and Inaccessible Islands (1995 and 2004), two of the least disturbed islands and marine eco-systems in the South Atlantic; Maritime Greenwich (1997), including Britain's first Palladian building, designed by Inigo Jones, Christopher Wren's Royal Naval College and the Royal Observatory; Heart of Neolithic Orkney (1999), comprising several important neolithic monuments; Historic Town of St George's and Related Fortifications, Bermuda (2000), an example of early English New World colonialism; Blaenavon Industrial Landscape (2000), a symbol of South Wales' role as a coal and iron provider in the 19th century; Dorset and East Devon Coast (2001), which demonstrate rock formations and fossil remains from the Mesozoic Era; Derwent Valley Mills (2001), 18th-century cotton mills at the forefront of the Industrial Revolution; New Lanark (2001), Robert Owen's model industrial community and cotton mills of the early 19th century; Saltaire (2001), a mid-19th century planned industrial community for the textile industry; Royal Botanical Gardens, Kew (2003), containing important botanical collections in a historic landscape; Liverpool—Maritime Mercantile City (2004); the Cornwall and West Devon mining landscape (2006); and the Pontcysyllte Aqueduct and Canal near Wrexham (2009), a feat of civil engineering by Thomas Telford during the Industrial Revolution.

The UK also shares the Frontiers of the Roman Empire sites (1987, 2005 and 2008) with Germany, containing the border line of the Roman Empire at its greatest extent in the 2nd century AD (specifically Hadrian's Wall).

Press

In Feb. 2013 there were 11 national dailies with a combined average daily circulation of 7,984,376 and ten national Sunday newspapers (7,537,144). In Jan. 2013 there were also 109 morning, evening and Sunday regional newspapers and 945 weeklies (420 of these for free distribution). In 2010 there were 3,212 consumer magazines and 4,765 business magazines. The most widely read daily is the tabloid *The Sun*, with an average daily circulation of 2,281,990 in Feb. 2013. The most widely read Sunday paper, until its demise in July 2011, was the tabloid *News of the World*, which had an average circulation of 2,789,560 in Jan. 2011. *The Sun (Sunday)* is now the UK's most widely read Sunday paper, with an average daily circulation of 1,912,643 in Feb. 2013.

In Jan. 1991 the Press Complaints Commission replaced the former Press Council. It has 15 members and a chair (Lord Hunt), including five editors. It is funded by the newspaper industry.

In 2011 a total of 149,800 book titles were published in the UK (151,969 in 2010).

Tourism

In 2010 UK residents made 119·4m. trips within the UK, passing 373·3m. nights in accommodation and spending £20,835m. Of these, 78·7m. were holidaymakers. Visits from foreign tourists to the UK totalled 29·8m. in 2010 (down from a record 32·8m. in 2007). Spending was £16·9bn. in 2010. In 2010 the UK ranked sixth for international tourism arrivals behind the USA, Spain, France, Italy and China. The main countries of origin for foreign visitors in 2010 were: France (3·6m.), Germany (3·0m.), USA (2·7m.), Ireland (2·6m.) and Spain (1·8m.).

In 2010 there were 2·6m. people working in tourism-related industries.

UK residents made 55·6m. trips abroad in 2010 (39·6m. in 1994 but 69·6m. in 2006). Spain is the most popular destination for Britons travelling abroad for leisure (10·4m. visits), followed by France (9·1m.), the USA (3·2m.) and Ireland (3·0m.).

Festivals

Among the most famous music festivals are the Promenade Concerts or 'Proms', which take place at the Royal Albert Hall in London every year from July to Sept; the Glyndebourne season in Sussex (May to Aug.); the Aldeburgh Festival in Suffolk (June); the Glastonbury Festival in Somerset (June); and the Buxton Festival in Derbyshire (July). The annual London Film Festival takes place in Oct. Literary festivals include the Oxford Literary Festival in March, the Hay Festival at Hay-on-Wye in Powys (late May/early June) and the Cheltenham Festival of Literature in Gloucestershire (Oct.). The Edinburgh Festival and the Fringe Festival both take place in Aug./early Sept. and are major international festivals of culture. The Brighton Festival in May is England's largest arts festival. The multicultural Notting Hill Carnival in London takes place at the end of Aug. Other major events in the annual calendar are the New Year's Day Parade in London, the Crufts Dog Show at the Birmingham National Exhibition Centre (March), the Ideal Home Exhibition in London (March–April), the London Marathon (April), the Chelsea Flower Show (May), Royal Ascot (horse racing, in June), Wimbledon (tennis, in June–July), Henley Royal Regatta (July), Cowes (yachting, in Aug.) and the Lord Mayor's Show in London (Nov.).

DIPLOMATIC REPRESENTATIVES

Of the USA in Great Britain (24/31 Grosvenor Sq., London, W1A 1AE)
Ambassador: Louis B. Susman.

Of Great Britain in the USA (3100 Massachusetts Ave., NW, Washington, D.C., 20008)
Ambassador: Sir Peter Westmacott, KCMG, LVO.

Of Great Britain to the United Nations
Ambassador: Sir Mark Lyall Grant.

Of Great Britain to the European Union
Permanent Representative: Sir Jon Cunliffe, CB.

FURTHER READING

The Office for National Statistics publishes data online at http://www.ons.gov.uk. The Stationery Office (TSO), formerly HMSO, offers publications covering legislation, official reports, and government and parliamentary papers.

Bache, Ian and Jordan, Andrew, *The Europeanization of British Politics.* 2006
Beech, Matt and Lee, Simon, (eds.) *Ten Years of New Labour.* 2008
Black, Jeremy, *A History of the British Isles.* 2nd ed. 2003
Bogdanor, Vernon, *Devolution in the United Kingdom.* 1999.—*The New British Constitution.* 2009
Butler, David and Butler, Gareth, *British Political Facts.* 10th ed. 2010
Cairncross, A., *The British Economy Since 1945: Economic Policy and Performance, 1945–1995.* 2nd ed. 1995
Casey, Terence, (ed.) *The Blair Legacy: Politics, Policy, Governance, and Foreign Affairs.* 2009
Clark, Alistair, *Political Parties in the UK.* 2012
Davies, Norman, *The Isles: A History.* 1999

Denver, David, Carman, Christopher and Johns, Robert, (eds.) *Elections and Voters in Britain.* 3rd ed. 2012
Floud, Roderick and Johnson, Paul, (eds.) *The Cambridge Economic History of Modern Britain.* 3 vols. 2004
Gascoigne, B. (ed.) *Encyclopedia of Britain.* 1994
Geddes, Andrew, *Britain and the European Union.* 2013
Hannay, David, *Britain's Quest for a Role: A Diplomatic Memoir from Europe to the UN.* 2012
Harrison, B., *The Transformation of British Politics, 1860–1995.* 1996
Heffernan, Richard, Cowley, Philip and Hay, Colin, (eds.) *Developments in British Politics 9.* 2011
Kavanagh, Dennis and Cowley, Philip, *The British General Election of 2010.* 2010
Kellner, Peter, *Democracy: 1,000 Years in Pursuit of British Liberty.* 2009
Leese, Peter, *Britain Since 1945.* 2006
Leventhal, F. M. (ed.) *20th-Century Britain: an Encyclopedia.* 1995
Marquand, David, *Britain Since 1918: The Strange Career of British Democracy.* 2008
McCormick, John, *Contemporary Britain.* 3rd ed. 2012
Moran, Michael, *Politics and Governance in the UK.* 2nd ed. 2011
Oakland, John, *British Civilization: an Introduction.* 7th ed. 2010
Oxford History of the British Empire. 2 vols. 1999
Palmer, A. and Palmer, V., *The Chronology of British History.* 1995
Penguin History of Britain. 9 vols. 1996
Robbins, Keith, *A Bibliography of British History 1914–1989.* 1996
Wall, Stephen, *A Stranger in Europe: Britain and the EU from Thatcher to Blair.* 2008.—*The Official History of Britain and the European Community, Volume II: From Rejection to Referendum, 1963–75.* 2012
Waller, R. and Criddle, B., *The Almanac of British Politics.* 8th ed. 2007
Wilson, David and Game, Chris, *Local Government in the United Kingdom.* 5th ed. 2011

Other more specialized titles are listed under TERRITORY AND POPULATION; ARMY; BANKING AND FINANCE; ELECTRICITY; WELFARE; *and* RELIGION, *above. See also* Further Reading *in* SCOTLAND, WALES *and* NORTHERN IRELAND.

National Statistical Office: UK Statistics Authority, Statistics House, Tredegar Park, Newport, Gwent, NP10 8XG. *National Statistician:* Jil Matheson.
Website: http://www.statistics.gov.uk

ENGLAND

KEY HISTORICAL EVENTS

Emperor Claudius' invasion in AD 43 established Roman rule in southern England. After the failed rebellions in AD 60 of Queen Boudicca of the Iceni and the suppression of Wales by AD 78, there was a long period of peaceful settlement, during which the Romans established new towns such as Londinium (London) and Eboracum (York). After the withdrawal of the Roman legions in the early 5th century, Pictish and Saxon raiders harassed the British towns. Defensive Saxon settlements were at first encouraged by the authorities but their rebellion soon threatened the Roman way of life. The Romano-British were pushed back to higher land in the west by waves of invading Saxons, Angles and Jutes. After a period of Mercian supremacy under Offa in the 8th century, the West Saxons (Wessex) dominated southern England. Danish invasions in 865 established the Danelaw in northern England. Alfred the Great of Wessex and his son Edward resisted Danish expansion, strengthening Anglo-Saxon unity under Alfred's successors—Athelstan became the first king of all England in 927.

Danish rule over England was reasserted by Sweyn in 994 and his son, Canute. The Anglo-Saxon restoration was short-lived; William, duke of Normandy led the Norman Conquest in 1066, defeating Harold II at the Battle of Hastings. When William died in 1087, he left Normandy to his eldest son Robert, thus separating it from England. Henry II, the founder of the Plantagenet dynasty, was feudatory lord of half of France but Henry's son John lost most of the French possessions. The barons forced John to sign the Magna Carta in 1215, later interpreted as the source of English civil liberties. Thereafter, the Norman baronage came to regard themselves as English.

The Hundred Years War (1338–1453) with France ended with the loss of all remaining French possessions except Calais. In 1387 and in later outbreaks, the Black Death reduced the population by over a third. A dynastic struggle between the rival houses of York and Lancaster was concluded by the invasion of Henry Tudor in 1485. His son, Henry VIII, asserted royal authority over the church, breaking with Rome. Tudor power reached its zenith with Elizabeth I. Philip II's Spanish Armada, destroyed in 1588, was sent to turn back the Protestant tide in England and to counter English ambitions in the New World.

The accession of James VI of Scotland to the English throne in 1603 brought the two countries into personal union. Charles I's defeat in the Civil War resulted in a republican Commonwealth but the Stuart monarchy was restored in 1660. England and Scotland were united in 1707 under Anne, queen of Great Britain.

TERRITORY AND POPULATION

At the census taken on 27 March 2011 the area of England was 130,432 sq. km (of which land 130,278 sq. km) and the population 53,012,456, giving a density of 407 per sq. km. England covers 53·4% of the total area of the United Kingdom. Households at the 2011 census: 22,063,368.

Population (present on census night) at the four previous decennial censuses:

1971	1981	1991	2001
46,018,371[1]	46,226,100[2]	46,382,050	49,138,831

[1]Area now included in Wales formed the English county of Monmouthshire until 1974. [2]The final count is believed to be over-stated as a result of an error in processing. The preliminary counts presented here rounded to the nearest hundred are thought to be more accurate.

Population at census day 2011:

Males	Females	Total
26,069,148	26,943,308	53,012,456

For further statistical information, see under Territory and Population, UNITED KINGDOM.

The population on census day in 2011 in the nine English Government Office regions (created in 1994) was as follows: East, 5,846,965; East Midlands, 4,533,222; London, 8,173,941; North East, 2,596,886; North West, 7,052,177; South East, 8,634,750; South West, 5,288,935; West Midlands, 5,601,847; Yorkshire and the Humber, 5,283,733.

Following the local government reorganization in the mid-1990s, there is a mixed pattern to local government in England. Apart from Greater London, England is divided into 27 counties with two tiers of administration; a county council and district councils. There are six metropolitan county areas containing 36 single-tier metropolitan districts.

In addition, there are 56 single-tier unitary authorities which, with the exception of the Isle of Wight, were formerly district councils in the shire counties of England. The Isle of Wight is a unitary county council. The Isles of Scilly have a unitary council but are considered a district of 'Cornwall and the Isles of Scilly'.

As a consequence of the establishment of the 56 unitary authorities, a number of county areas were abolished. These were Avon, Cleveland and Humberside. Berkshire County Council was also abolished but the county itself is retained for ceremonial purposes. Greater London comprises 32 boroughs and the City of London.

Area in sq. km of English counties and unitary authorities, and population at census day 2011:

	Land area (sq. km)	Population		Land area (sq. km)	Population
Metropolitan counties			Hampshire		
Greater			(Hants)	3,679	1,317,788
Manchester	1,276	2,682,528	Hertfordshire		
Merseyside	645	1,381,189	(Herts)	1,643	1,116,062
South Yorkshire	1,552	1,343,601	Kent	3,544	1,463,740
Tyne and Wear	540	1,104,825	Lancashire		
West Midlands	902	2,736,460	(Lancs)	2,903	1,171,339
West Yorkshire	2,029	2,226,058	Leicestershire		
			(Leics)	2,083	650,489
Non-metropolitan counties			Lincolnshire		
Buckinghamshire			(Lincs)	5,921	713,653
(Bucks)	1,565	505,283	Norfolk	5,371	857,888
Cambridgeshire			Northamptonshire		
(Camb)	3,046	621,210	(Northants)	2,364	691,952
Cumbria	6,767	499,858	North Yorkshire		
Derbyshire	2,547	769,686	(N. Yorks)	8,038	598,376
Devon	6,564	746,399	Nottinghamshire		
Dorset	2,542	412,905	(Notts)	2,085	785,802
East Sussex	1,709	526,671	Oxfordshire		
Essex	3,464	1,393,587	(Oxon)	2,605	653,798
Gloucestershire			Somerset (Som)	3,451	529,972
(Gloucs)	2,653	596,984			

	Land area (sq. km)	Population		Land area (sq. km)	Population
Staffordshire			Middlesbrough	54	138,412
(Staffs)	2,620	848,489	Milton Keynes	309	248,821
Suffolk	3,800	728,163	North East		
Surrey	1,663	1,132,390	Lincolnshire	192	159,616
Warwickshire	1,975	545,474	North		
West Sussex	1,990	806,892	Lincolnshire	846	167,446
Worcestershire	1,741	566,169	North Somerset	374	202,566
Unitary Authorities			Northumber-		
Bath and North			land	5,013	316,028
East Somerset	346	176,016	Nottingham	75	305,680
Bedford	476	157,479	Peterborough	343	183,631
Blackburn with			Plymouth	80	256,384
Darwen	137	147,489	Poole	65	147,645
Blackpool	35	142,065	Portsmouth	40	205,056
Bournemouth	46	183,491	Reading	40	155,698
Bracknell Forest	109	113,205	Redcar and		
Brighton and			Cleveland	245	135,177
Hove	83	273,369	Rutland	382	37,369
Bristol, City of	110	428,234	Shropshire	3,197	306,129
Central			Slough	33	140,205
Bedfordshire	716	254,381	South		
Cheshire East	1,166	370,127	Gloucestershire	497	262,767
Cheshire West			Southampton	50	236,882
& Chester	917	329,608	Southend-on-		
Cornwall	3,546	532,273	Sea	42	173,658
Darlington	197	105,564	Stockton-on-		
Derby	78	248,752	Tees	204	191,610
Durham	2,226	513,242	Stoke-on-Trent	93	249,008
East Riding of			Swindon	230	209,156
Yorkshire	2,408	334,179	Telford and		
Halton	79	125,746	Wrekin	290	166,641
Hartlepool	94	92,028	Thurrock	163	157,705
Herefordshire,			Torbay	63	130,959
County of	2,180	183,477	Warrington	181	202,228
Isle of Wight	380	138,265	West Berkshire	704	153,822
Isles of Scilly	16	2,203	Wiltshire		
Kingston upon			(Wilts)	3,255	470,981
Hull, City of	71	256,406	Windsor and		
Leicester	73	329,839	Maidenhead	197	144,560
Luton	43	203,201	Wokingham	179	154,380
Medway	192	263,925	York	272	198,051

Source: Office of National Statistics

In 2011 London had a population of 8,173,941. Populations of next largest cities in 2011 were: Birmingham, 1,073,045; Leeds, 751,485; Sheffield, 552,698; Bradford, 522,452; Manchester, 503,127; Liverpool, 466,415; Bristol, 428,234.

Greater London Boroughs. Total area 1,572 sq. km. Population at census day 2011: 8,173,941 (inner London, 3,231,901). Population by borough (census day 2011):

Barking and		Islington[1]	206,125
Dagenham	185,911	Kensington and	
Barnet	356,386	Chelsea[1]	158,649
Bexley	231,997	Kingston	
Brent	311,215	upon Thames	160,060
Bromley	309,392	Lambeth[1]	303,086
Camden[1]	220,338	Lewisham[1]	275,885
Croydon	363,378	Merton	199,693
Ealing	338,449	Newham[1]	307,984
Enfield	312,466	Redbridge	278,970
Greenwich	254,557	Richmond upon	
Hackney[1]	246,270	Thames	186,990
Hammersmith		Southwark[1]	288,283
and Fulham[1]	182,493	Sutton	190,146
Haringey[1]	254,926	Tower Hamlets[1]	254,096
Harrow	239,056	Waltham Forest	258,249
Havering	237,232	Wandsworth[1]	306,995
Hillingdon	273,936	Westminster,	
Hounslow	253,957	City of[1]	219,396

[1]Inner London borough.

Source: Office of National Statistics

The City of London (677 acres) is administered by its Corporation which retains some independent powers. Population at census day 2011: 7,375.

CLIMATE

For more detailed information, see under Climate, UNITED KINGDOM.

London, Jan. 39°F (3·9°C), July 64°F (17·8°C). Annual rainfall 25" (635 mm). Birmingham, Jan. 38°F (3·3°C), July 61°F (16·1°C). Annual rainfall 30" (749 mm). Manchester, Jan. 39°F (3·9°C), July 61°F (16·1°C). Annual rainfall 34·5" (876 mm).

CONSTITUTION AND GOVERNMENT

The Parliamentary electorate of England in the register in Dec. 2009 numbered 38,129,082.

RECENT ELECTIONS

At the UK general election held in May 2010, 532 members were returned from England. Voting in Thirsk and Malton was postponed owing to the death of a candidate. A by-election was subsequently held on 27 May 2010.

See also Constitution and Government, Recent Elections *and* Current Government *in* UNITED KINGDOM.

DEFENCE

For information on defence, see UNITED KINGDOM.

ECONOMY

For information on the economy, see UNITED KINGDOM.

ENERGY AND NATURAL RESOURCES

For information on energy and natural resources, see UNITED KINGDOM.

Environment

35·5% of household waste was recycled in 2007–08.

Water

The Water Act of Sept. 1989 privatized the nine water and sewerage authorities in England: Anglian; North West (now United Utilities Water plc); Northumbrian; Severn Trent; South West; Southern; Thames; Wessex; Yorkshire. There are also 12 water only companies in England and Wales. The Act also inaugurated the National Rivers Authority, with environmental and resource management responsibilities, and the 'regulator' *Office of Water Services (Ofwat)*, charged with protecting consumer interests.

INDUSTRY

Labour

The ILO unemployment rate in the period Oct.–Dec. 2011 was 8·4%, the same as the UK as a whole. Unemployment was lowest in the southwest (6·1%) and highest in the northeast (11·2%).

INTERNATIONAL TRADE

For information on international trade, see UNITED KINGDOM.

COMMUNICATIONS

For information on communications, see UNITED KINGDOM.

Shipping

In 2009 English ports handled 340·8m. tonnes of traffic, down from 400·6m. tonnes in 2006.

SOCIAL INSTITUTIONS

Education

For details on the nature and types of school, see under Education, UNITED KINGDOM.

In 2006–07 education expenditure by central and local government in England was just under £58bn.

In Jan. 2008 there were 445 public sector and two direct grant nursery schools in England with provision for children under five. In 2008 there were 37,380 pupils under five attending public sector nursery schools. Some of these children were attending part-time.

In Jan. 2008 there were 4,087,790 pupils at 17,205 primary schools in England. Nearly all primary schools take both boys and girls. Almost 15% of primary schools had 100 full-time pupils or fewer.

In Jan. 2008 there were 320 middle schools deemed either primary or secondary according to the age range of the school concerned.

In Jan. 2008 there were 3,383 state-funded secondary schools in England (including City Technology Colleges and Academies). Some local authorities continue to operate a selective admissions policy at age 11, as do 164 state-funded secondary schools in England. There were 172 secondary modern schools in the 2006, providing a general education up to the minimum school leaving age of 16, although some pupils stay on beyond that age.

Almost all local education authorities operate a system of comprehensive schools to which pupils are admitted without reference to ability or aptitude. In Jan. 2008 there were 2,704 such schools in England with almost 2·8m. pupils. With the development of comprehensive education, various patterns of secondary schools have come into operation. Principally these are: 1) All-through schools with pupils aged 11 to 18 or 11 to 16; pupils over 16 being able to transfer to an 11 to 18 school or a sixth form college providing for pupils aged 16 to 19—there are currently 98 sixth form colleges in England; 2) Local authorities operating a three-tier system involving middle schools where transfer to secondary school is at ages 12, 13 or 14. These correspond to 12 to 18, 13 to 18 and 14 to 18 comprehensive schools respectively; or 3) In areas where there are no middle schools a two-tier system of junior and senior comprehensive schools for pupils aged 11 to 18 with optional transfer to these schools at age 13 or 14.

Under the Education Act 1996 children have special educational needs if they have a learning difficulty which calls for special educational provision to be made for them. In some cases the local authority will need to make a statutory assessment of special educational needs under the Education Act 1996, which may ultimately lead to a 'statement'. In England the total number of pupils with statements in 2008 was 223,600. In 2008 there were 993 maintained special schools and 72 non-maintained special schools.

Outside the state system of education there were 2,327 independent schools (excluding direct grant nursery schools) in England in Jan. 2008, ranging from large prestigious schools to small local ones. Some provide boarding facilities but the majority include non-resident day pupils. There are about 582,330 pupils in these schools, which represent about 7% of the total pupil population in England.

Further Education (Non-University). In 2006–07, 4·2m. students were enrolled at FE colleges in England and around 6m. in the sector as a whole. Total funding for the FE sector in 2006–07 was £4·8bn.

Higher Education. As of Aug. 2008 there were 132 higher education institutions in England, of which 91 were universities. The main funding body is the Higher Education Funding Council for England (HEFCE), which distributes public money for teaching and research to universities and colleges. It works in partnership with the higher education sector and advises government on higher education policy. In 2008–09 HEFCE distributed a total of £7·48bn., including £4·63bn. for teaching and £1·46bn. for research.

a) *Universities*

Name (Location)	No. of students (2007–08)	No. of academic staff (2005–06)
Anglia Ruskin University (Chelmsford)	19,005	1,150
Aston University (Birmingham)	9,570	470
University of Bath	12,965	1,420
Bath Spa University	7,470	330
University of Bedfordshire	14,195	370
University of Birmingham	28,240	2,465
Birmingham City University	23,765	1,750
University of Bolton	8,590	270
Bournemouth University (Poole)	17,875	740
University of Bradford	12,375	870
University of Brighton	21,220	1,320
University of Bristol	21,740	2,320
Brunel University (Uxbridge)	14,265	1,185
Buckinghamshire New University	9,380	570
University of Cambridge[1]	22,745	3,945
Canterbury Christ Church University	15,545	535
University of Central Lancashire (Preston)	31,245	1,170
University of Chester	13,515	490
University of Chichester	4,805	345
City University (London)	21,410	1,660
Coventry University	20,505	1,090
University for the Creative Arts (Canterbury, Epsom, Farnham, Maidstone and Rochester)	7,755	240
University of Cumbria (Carlisle)	12,045	530[2]
De Montfort University (Leicester)	21,215	1,430
University of Derby	22,865	1,005
University of Durham	16,275	1,205
University of East Anglia (Norwich)	15,695	1,475
University of East London	19,430	650
Edge Hill University (Ormskirk)	20,140	735
University of Essex (Colchester)	11,510	980
University of Exeter	14,705	1,045
University of Gloucestershire (Cheltenham)	8,920	625
University of Greenwich (London)	24,505	1,010
University of Hertfordshire (Hatfield)	23,005	1,515
University of Huddersfield	20,430	1,305
University of Hull	21,005	950
Imperial College London	13,845	3,175
Keele University (Newcastle-under-Lyme)	11,415	705
University of Kent (Canterbury)	17,805	1,240
Kingston University (Kingston upon Thames)	24,135	1,510
University of Lancaster	13,720	1,395
University of Leeds	32,250	2,675
Leeds Metropolitan University	41,245	1,495
University of Leicester	15,355	1,260
University of Lincoln	16,115	670
University of Liverpool	19,380	1,950
Liverpool Hope University	7,110	285
Liverpool John Moores University	24,445	1,235
University of London[1]	123,080	16,965[3]
London Metropolitan University	28,525	1,050
London South Bank University	23,225	800
Loughborough University	17,650	1,530
University of Manchester	37,360	3,960
Manchester Metropolitan University	33,155	2,015
Middlesex University (London)	21,625	845
University of Newcastle upon Tyne	19,050	2,295
University of Northampton	11,585	480
University of Northumbria at Newcastle	30,470	1,080
University of Nottingham	31,830	2,745
Nottingham Trent University	23,845	1,460
Open University[4]	181,695	7,645
University of Oxford[1]	23,985	4,190
Oxford Brookes University	18,385	1,240
University of Plymouth	29,375	1,110
University of Portsmouth	19,805	1,240
University of Reading	14,470	1,210
Roehampton University	8,235	595
University of Salford	19,180	1,465
University of Sheffield	24,560	2,595
Sheffield Hallam University	31,090	1,750
University of Southampton	23,765	2,300
Southampton Solent University	18,170	645
Staffordshire University (Stoke-on-Trent)	15,735	745
University of Sunderland	17,820	880
University of Surrey (Guildford)	15,070	1,100
University of Sussex (Brighton)	12,450	1,455
Teesside University (Middlesbrough)	26,210	655
University of Warwick (Coventry)	28,445	1,815
University of the West of England, Bristol	31,700	1.630
University of West London	18,135	1,170
University of Westminster (London)	23,225	1,650
University of Winchester	5,235	360
University of Wolverhampton	21,305	1,020
University of Worcester	7,765	335
University of York	13,185	1,305
York St John University	6,205	235

[1]See listing of colleges below. [2]Excluding staff of the former Carlisle and Penrith campuses of the University of Central Lancashire. [3]Excluding Imperial College London, which became independent of the University of London in July 2007. [4]Entirely distance learning—see page 1289.

b) *University of Cambridge; University of London; University of Oxford*

University of Cambridge Colleges:
Christ's College; Churchill College; Clare College; Clare Hall[1]; Corpus Christi College; Darwin College[1]; Downing College; Emmanuel College; Fitzwilliam College; Girton College; Gonville and Caius College; Homerton College; Hughes Hall; Jesus College; King's College; Lucy Cavendish; Magdalene College; New Hall; Newnham College; Pembroke College; Peterhouse; Queen's College; Robinson College; St Catharine's College; St Edmund's College; St John's College; Selwyn College; Sidney Sussex College; Trinity College; Trinity Hall; Wolfson College.

[1]Postgraduate only.

University of London Colleges (total no. of students 2007–08/no. of full-time equivalent academic staff 2007–08):
Birkbeck (17,225/554); Central School of Speech and Drama (880/55[1]); Courtauld Institute of Art (440/35); Goldsmiths College (7,605/388); Heythrop College (745/39); Institute of Cancer Research (300/511[2]); Institute of Education (7,385/359); King's College London (21,110/2,827); London Business School (1,555/125); London School of Economics and Political Science (9,105/890); London School of Hygiene and Tropical Medicine (1,100/266); Queen Mary, University of London (13,610/670); Royal Academy of Music (730/79); Royal Holloway, University of London (8,385/1,264[1]); Royal Veterinary College (1,795/129[1]); St George's Hospital Medical School (4,160/448[2]); School of Oriental and African Studies (4,730/415); School of Pharmacy (1,230/49); University College London (20,990/4,078[3]). In 2005–06 the University of London had 36,000 external programme students.

[1]Total number of staff. [2]2005–06. [3]2009–10.

University of Oxford Colleges:
All Souls College; Balliol College; Brasenose College; Christ Church; Corpus Christi College; Exeter College; Green Templeton College[1]; Harris Manchester College; Hertford College; Jesus College; Keble College; Kellogg College[1]; Lady Margaret Hall; Linacre College[1]; Lincoln College; Magdalen College; Mansfield College; Merton College; New College; Nuffield College[1]; Oriel College; Pembroke College; The Queen's College; St Anne's College; St Antony's College[1]; St Catherine's College; St Cross College[1]; St Edmund Hall; St Hilda's College; St Hugh's College; St John's College; St Peter's College; Somerville College; Trinity College; University College; Wadham College; Wolfson College[1]; Worcester College. *Permanent Private Halls:* Blackfriars; Campion Hall; Greyfriars; Regent's Park College; St Benet's Hall; St Stephen's House; Wycliffe Hall.

[1]Postgraduate only.

c) *Colleges of Art, Dance, Drama and Music*
2011–12: Arts University College at Bournemouth; Conservatoire for Dance and Drama (London)[1]; Guildhall School of Music and Drama (London); Leeds College of Music; Liverpool Institute for Performing Arts; University of the Arts, London; Norwich University College of the Arts; Ravensbourne (Bromley); Rose Bruford College of Theatre & Performance (Sidcup); Royal College of Art (London); Royal College of Music (London); Royal Northern College of Music (Manchester); Trinity Laban Conservatoire of Music and Dance (London).

[1]Affiliate schools: Bristol Old Vic Theatre School; Central School of Ballet; The Circus Space; London Academy of Music and Dramatic Art; London Contemporary Dance School; Northern School of Contemporary Dance; Rambert School of Ballet and Contemporary Dance; The Royal Academy of Dramatic Art.

d) *Other Institutions*
2011–12: University College Birmingham; Bishop Grosseteste College Lincoln; Cranfield University[1]; University College Falmouth Incorporating Dartington College of Arts; Harper Adams University College (Newport); Leeds Trinity University College; University of London (Institutes and Activities); Newman University College (Birmingham); University College Plymouth St Mark and St John; Royal Agricultural College (Cirencester); St Mary's University College, Twickenham; University Campus Suffolk; Writtle College (Chelmsford).

[1]Postgraduate only.

Health

As at 30 Sept. 2008 there were 34,010 general medical practitioners in England excluding registrars and retainers (an increase of 1·9% since Sept. 2007), with an average of 1,586 patients per doctor. There were 20,887 NHS dentists in England as at 31 Dec. 2006. In England in 2008 there were 408,160 qualified nursing and midwifery staff, including GP practice nurses. As at 30 Sept. 2008 there were 34,910 consultants in England (24,401 in 2000) and 3,203 GP registrars. In 2008–09 provision of beds in England was 31 per 10,000 population. There were 159,386 hospital beds in England in 2008–09 (282,918 in 1988–89).

In 2007, 24% of men and 24% of women were obese (having a body mass index over 30), up from 17% and 20% in 1997 respectively. In 2006, 23% of men and 21% of women aged 16 and over in England were smokers. Among 11- to 15-year-olds 10% of girls but only 7% of boys smoked in 2004.

Personal Social Services staff numbered 256,100 in England at 30 Sept. 2008. The total expenditure (2007–08) for PSS was £18,500m.

CULTURE

Tourism
The leading free admission attraction in 2009 was the British Museum, with 5·6m. visits. The leading tourist attractions charging admission in 2009 were: the Tower of London, with 2·4m. visits; St Paul's Cathedral, with 1·8m.; Westminster Abbey, with an estimated 1·4m., Flamingo Land Theme Park and Zoo in Malton, N. Yorks, with an estimated 1·4m.; and Windermere Lake Cruises in Bowness, Cumbria, with 1·3m.

FURTHER READING

See Further Reading *in* UNITED KINGDOM.

SCOTLAND

KEY HISTORICAL EVENTS

Earliest evidence of human settlement in Scotland dates from the Middle Stone Age. Hunters and fishermen on the west coast were succeeded by farming communities as far north as Shetland. The Romans, who were active in the 1st century AD, built Hadrian's Wall between the Tyne and Solway Firth as their northern frontier. At this time, the Picts formed two kingdoms north of the Firth of Clyde. From the 6th century, the Celtic Scots from Dalriada, in the north of Ireland, fought with Angles and Britons for control of southern Scotland.

In 843 Kenneth MacAlpine united the Scots and the Picts to found the kingdom of Scotland. A legal and administrative uniformity was established by David I (reigned 1124–53). William the Lion abandoned claims to Northumbria in 1209 but began the alliance with France. In 1286 Edward I of England asserted his claim as overlord of Scotland and appointed his son to succeed to the crown. Resistance to English rule was led by William Wallace and later by Robert Bruce, who defeated the English at Bannockburn in 1314. His grandson, Robert II, became the first Stewart (Stuart) king in 1371.

Royal minorities undermined the authority of the crown in the 15th century until the accession of James IV in 1488. Relations with England improved after his marriage to Margaret Tudor in 1503 but when Henry VIII invaded France, James attacked England and was killed at the Battle of Flodden in 1513. The young James V was assailed by conflicting pressures from pro-French and pro-English factions but having secured his personal rule, he entered into two successive French marriages. His daughter, Mary Queen of Scots, married the French Dauphin in 1558. Protestant opposition to French influence was bolstered by Elizabeth I of England, who sent troops. Mary was in France when the Scottish parliament renounced papal authority, bolstering the reformist movement, led by John Knox. Returning to Scotland after her husband's death in 1561, Mary was forced to take refuge in England. Her son, James VI, survived the animosity between his own and his mother's followers to make an alliance with England. Deemed a threat because of her claim to the English throne, Mary was executed on Elizabeth's orders in 1587.

Elizabeth died without issue in 1603 and was succeeded by James. Although he styled himself 'king of Great Britain', England and Scotland remained independent. Charles I alienated much of the Scottish nobility and was defeated in the Bishops' Wars by the Covenanters, who rejected English interference in the Scottish church. Scottish armies fought for both sides in the English Civil War, which led to the execution of Charles I in 1649. However, the Scots soon united to accept Charles II as their king. Having established dominance in England, Cromwell moved against Scotland forcing Charles II into exile. His restoration in 1660 was welcomed in both kingdoms. His successor, James VII (James II of England), was less astute in managing religious and political differences. The collapse of his regime in 1688 and the arrival of William of Orange confirmed the Protestant ascendancy in Scotland and England.

The union of parliament in 1707 brought Scotland more directly under English authority. However, Scotland retained its own legal and ecclesiastical systems. The remaining supporters of James VII, the Jacobites, led two abortive risings on behalf of James' son and grandson (the old and young Pretenders) but were defeated decisively at Culloden in 1746.

TERRITORY AND POPULATION

The total area of Scotland is 78,808 sq. km (2011), including its islands—186 in number—and inland water. Scotland covers 32·3% of the total area of the United Kingdom.

Population (including military in the barracks and seamen on board vessels in the harbours) at the dates of each census:

Date of enumeration	Population	Pop. per sq. mile[1]
1801	1,608,420	53
1811	1,805,864	60
1821	2,091,521	70
1831	2,364,386	79
1841	2,620,184	88
1851	2,888,742	97
1861	3,062,294	100
1871	3,360,018	113
1881	3,735,573	125
1891	4,025,647	135
1901	4,472,103	150
1911	4,760,904	160
1921	4,882,497	164
1931	4,842,980	163
1951	5,096,415	171
1961	5,179,344	174
1971	5,228,963	67
1981	5,130,735	66
1991	4,998,567	60
2001	5,062,011	65
2011	5,295,000[2]	67

[1]Per sq. km from 1971. [2]Figure rounded to nearest thousand.

Population at census day 2011:

Males	Females	Total
2,567,000	2,728,000	5,295,000

In 2001, 58,652 people aged three and over spoke Gaelic (65,978 in 1991). Households in 2011 (estimate): 2,368,000.

The age distribution in Scotland at census day in 2011 was as follows (in 1,000):

Age-group		
Under 5		293
5 and under 10		270
10 „ 15		292
15 „ 20		331
20 „ 25		364
25 „ 35		668
35 „ 45		735
45 „ 55		787
55 „ 65		667
65 „ 70		261
70 „ 75		221
75 „ 85		300
85 and upwards		108

Land area and population (27 March 2011) by administrative area:

Council Area	Area (sq. km)	Population
Aberdeen City	186	223,000
Aberdeenshire	6,313	253,000
Angus	2,182	116,000
Argyll and Bute	6,908	88,000
Clackmannanshire	159	51,000
Dumfries and Galloway	6,426	151,000
Dundee City	60	147,000
East Ayrshire	1,262	123,000
East Dunbartonshire	175	105,000
East Lothian	679	100,000
East Renfrewshire	174	91,000
Edinburgh, City of	263	477,000
Eilean Siar[1]	3,056	28,000
Falkirk	297	156,000
Fife	1,325	365,000
Glasgow City	174	593,000
Highland	25,684	232,000
Inverclyde	160	81,000

Council Area	Area (sq. km)	Population
Midlothian	354	83,000
Moray	2,238	93,000
North Ayrshire	885	138,000
North Lanarkshire	470	338,000
Orkney Islands	990	21,000
Perth and Kinross	5,285	147,000
Renfrewshire	262	175,000
Scottish Borders	4,732	114,000
Shetland Islands	1,467	23,000
South Ayrshire	1,222	113,000
South Lanarkshire	1,772	314,000
Stirling	2,187	90,000
West Dunbartonshire	159	91,000
West Lothian	428	175,000
Total	77,933	5,295,000

[1]Formerly Western Isles.

Glasgow is Scotland's largest city, with a population of 593,000 in 2011, followed by Edinburgh, the capital (477,000), and Aberdeen, with 223,000.

The birthplaces of the 2001 census day population in Scotland were: Scotland, 4,410,400; England, 408,948; Northern Ireland, 33,528; Ireland 21,774; Wales, 16,623; other European Union countries, 44,432; elsewhere, 126,306.

SOCIAL STATISTICS

	Estimated resident population at 30 June[1]	Total births	Live births outside marriage	Deaths	Marriages	Divorces, annulments and dissolutions
2004	5,078,400	53,957	25,202	56,187	32,154	11,234
2005	5,094,800	54,386	25,617	55,747	30,881	10,875
2006	5,116,900	55,690	26,584	55,093	29,898	13,012
2007	5,144,200	57,781	28,377	55,986	29,866	12,781
2008	5,168,500	60,041	30,055	55,700	28,903	11,461
2009	5,194,000	59,046	29,710	53,856	27,524	10,395

[1]Includes merchant navy at home and forces stationed in Scotland.

Birth rate, 2009, per 1,000 population, 11·4; death rate, 10·4; marriage, 5·3; infant mortality per 1,000 live births, 4·0; sex ratio, 1,044 male births to 1,000 female. Average age of marriage in 2009: males, 36·9, females, 34·4. Expectation of life, 2008–10: males, 75·8 years, females, 80·3.

CLIMATE

For more detailed information, see under Climate, UNITED KINGDOM.

Aberdeen, Jan. 38°F (3·3°C), July 57°F (13·9°C). Annual rainfall 32" (813 mm). Edinburgh, Jan. 38°F (3·3°C), July 58°F (14·5°C). Annual rainfall 27" (686 mm). Glasgow, Jan. 39°F (3·9°C), July 59°F (15°C). Annual rainfall 38" (965 mm).

CONSTITUTION AND GOVERNMENT

In a referendum on devolution on 11 Sept. 1997, Scotland's voters opted for devolved government, calling for the reinstatement of a separate parliament in Scotland, the first since union with England in 1707. 1,775,045 votes (74·3%) were cast in favour of a Scottish parliament and 614,400 against (25·7%). On a turnout of 60·4%, around 44·8% of the total electorate voted in favour. For the second question, on the Parliament's tax-raising powers, 1,512,889 votes were cast in favour (63·5%) and 870,263 against (36·5%). This represented 38·4% of the total electorate. The Scottish government plans to hold a referendum on independence from the United Kingdom in the autumn of 2014. The Parliamentary electorate of Scotland in the register in Dec. 2009 numbered 3,869,700.

The Scottish Parliament is made up of 129 members and managed a budget of £31·3bn. in 2008–09. The parliament may pass laws and has limited tax raising powers; it is also responsible for devolved issues, including health, education, police and fire services; however, 'reserved issues' (foreign policy, constitutional matters, and many domestic areas including social security, trade and industry, and employment legislation) remain the responsibility of the British Parliament in Westminster.

RECENT ELECTIONS

At the UK general election held in May 2010, 59 members were returned from Scotland. Labour won 41 seats; the Liberal Democrats, 11; the Scottish National Party, 6; Conservative, 1. At the June 2009 European Parliament elections Labour won 2 seats, the Scottish National Party 2, the Conservatives 1 and Liberal Democrats 1.

In elections to the Scottish Parliament on 5 May 2011, the Scottish National Party (SNP) won 69 seats (16 by regional list), against 37 (22 by regional list) for Labour, 15 (12 by regional list) for the Conservatives, 5 (3 by regional list) for the Liberal Democrats and 2 (both by regional list) for the Greens. One independent was also elected (by regional list). Of the 129 seats, 73 were won on a first-past-the-post basis and 56 through proportional representation (regional list). The election delivered the first majority government since the opening of the Scottish Parliament in 1999. The result means there is sufficient support in the Scottish Parliament for a referendum on Scottish independence, which is scheduled for 18 Sept. 2014.

See also Constitution and Government, Recent Elections *and* Current Government *in* UNITED KINGDOM.

CURRENT GOVERNMENT

First Minister: Alex Salmond; b. 1954 (Scottish National Party).
  Presiding Officer: Tricia Marwick.

Scottish Executive: http://www.scotland.gov.uk

DEFENCE

For information on defence, *see* UNITED KINGDOM.

ECONOMY

Overview
After legislative powers were devolved to a reconstituted Scottish Parliament in 1999, the economy achieved a degree of financial autonomy, including the power to set the cost of university education and to vary the level of income tax.

Traditionally, heavy industry—including mining, shipbuilding and engineering—has dominated the economy. Industrial decline during the second half of the twentieth century was countered by the discovery of oil in the North Sea in the mid-1960s and a shift towards a service-oriented economy. Tourism and financial services are key contributors, with Edinburgh one of the leading financial centres in Europe according to the Global Financial Centres Index 2010. Other primary exports include textiles, whisky and shortbread.

In 2008 the near collapse of the Royal Bank of Scotland and HBOS (the holding company of the Bank of Scotland) threatened a UK financial sector meltdown until both institutions were bailed out by the government. Nonetheless, figures compiled by the Government Expenditure and Revenue Scotland (GERS) for 2009–10 illustrated the relative health of the economy compared to the rest of the UK in the wake of the global financial crisis. Scotland's budget deficit was estimated to be £9bn. or 6·8% of GDP, compared to a UK-wide deficit of £107·3bn. or 7·6% of GDP, while the Scottish economy was able to generate 9·4% of UK tax revenues with 8·4% of the total population. The Scottish National Party, the largest party in the Scottish Parliament, has proposed bringing forward a referendum on full independence during its five-year term, with a view to securing overall control of its revenue base.

Currency
The Bank of Scotland, Clydesdale Bank and the Royal Bank of Scotland have note-issuing powers.

Budget
Government expenditure in Scotland came to £56·5bn. in 2008–09 (including social protection £18·6bn., health £10·2bn. and education £7·6bn.). Revenues (excluding North Sea revenue) totalled £43·5bn. (including income tax £10·7bn., national insurance contributions £8·0bn. and VAT £7·5bn.).

Performance
The real GDP growth rate in 2010 was 0·9%.

ENERGY AND NATURAL RESOURCES

Environment
35% of household waste was recycled in 2008–09.

Electricity
The Electricity Act 1989 led to the privatization of the industry in Scotland and the creation of three new companies: ScottishPower, Scottish Hydro-Electric and Scottish Nuclear. After a series of acquisitions and mergers, in Dec. 1998 Scottish Hydro-Electric became part of the newly-formed Scottish and Southern Energy, a vertically integrated company that covers generation, transmission, distribution and supply in northern Scotland. ScottishPower runs distribution in central and southern Scotland and in 2007 became a subsidiary of the Spanish company Iberdrola. Scottish Nuclear, responsible for operating the two Scottish nuclear power stations, merged with British Energy in 1996, which in turn became a subsidiary of EDF in Jan. 2009.

Water
Water supply is the responsibility of the Regional and Island local authorities. Seven river purification boards are responsible for environmental management.

Agriculture
In 2010 total agricultural area was 5,643,054 ha., of which 3,191,593 ha. were used for rough grazing and 1,932,274 ha. for crops and grass.

Selected crop production, 2010 (in 1,000 tonnes): barley, 1,665; potatoes, 1,472; wheat, 918; oats, 135.

Livestock, 2010 (in 1,000): sheep, 6,753; cattle, 1,826; pigs, 409; poultry, 14,593.

Forestry
Total forest area in March 2011 was 1,390,000 ha., of which 481,000 ha. was owned or managed by the Forestry Commission.

Fisheries
The major fishing ports in terms of value of fish landed are Peterhead, Lerwick, Fraserburgh and Aberdeen. In 2009 there were 2,174 fishing vessels that landed 378,000 tonnes of fish worth £443m.

INDUSTRY

Labour
In 2007 the economically active population numbered 2,670,000 (1,262,000 females), of whom 135,000 (60,000 females) were unemployed. The ILO unemployment rate in the period Oct.–Dec. 2011 was 8·6% (compared to 8·4% in the UK as a whole), up from 8·0% in the fourth quarter of 2010. In Sept. 2009, 37·1% of employee jobs were in public and other services, 22·0% in retail, wholesale and hotels, 18·7% in finance and business, and 8·7% in manufacturing.

COMMUNICATIONS

Roads
Responsibility for the construction and maintenance of trunk roads belongs to the Scottish Office. Roads not classified as trunk roads are the responsibility of county or unitary councils. As at 1 April 2009 there were 55,420 km of public roads, of which 559 km were motorways. There were 2·38m. licensed private and light goods vehicles.

Rail

Total railway length in 2007–08 was 2,745 km. In 2007–08 a total of 84·8m. passengers travelled by rail and 11·0m. tonnes of freight were carried. There is a metro in Glasgow.

Civil Aviation

There are major airports at Aberdeen, Edinburgh, Glasgow and Prestwick. In 2008 Edinburgh was the seventh busiest for passenger traffic in the UK, with 8,992,178 passengers (5,281,038 on domestic flights) and Glasgow was the eighth busiest with 8,135,260 (4,192,121). In 2008, 24,348,159 passengers and 45,554 tonnes of freight were carried by Scottish airports.

Shipping

The principal Scottish port is Forth (including Grangemouth, Leith and Rosyth), which handled 39·1m. tonnes of cargo in 2008.

SOCIAL INSTITUTIONS

Justice

The High Court of Justiciary is the supreme criminal court in Scotland and has jurisdiction in all cases of crime committed in any part of Scotland, unless expressly excluded by statute. It consists of the Lord Justice General, the Lord Justice Clerk and 30 other Judges, who are the same Judges who preside in the Court of Session, the Scottish Supreme Civil Court. One Judge is seconded to the Scottish Law Commission. The Court is presided over by the Lord Justice General, whom failing, by the Lord Justice Clerk, and exercises an appellate jurisdiction as well as being a court of first instance. The home of the High Court is Edinburgh, but the court visits other towns and cities in Scotland on circuit and indeed the busiest High Court sitting is in Glasgow. The court sits in Edinburgh both as a Court of Appeal (the *quorum* being two judges if the appeal is against sentence or other disposals, and three in all other cases) and on circuit as a court of first instance. Although the decisions of the High Court are not subject to review by the Supreme Court of the United Kingdom, with the Scotland Act 1998 coming into force on 20 May 1999, there is a limited right of appeal against the termination of a devolution issue to the Supreme Court of the United Kingdom. One Judge sitting with a Jury of 15 persons can, and usually does, try cases, but two or more Judges (with a Jury) may do so in important or complex cases. The court has a privative jurisdiction over cases of treason, murder, rape, breach of duty by Magistrates and certain statutory offences under the Official Secrets Act 1911 and the Geneva Conventions Act 1957. It also tries the most serious crimes against person or property and those cases in which a sentence greater than imprisonment for three years is likely to be imposed.

The Sheriff Court has an inherent universal criminal jurisdiction (as well as an extensive civil one), limited in general to crimes and offences committed within a sheriffdom (a specifically defined region), which has, however, been curtailed by statute or practice under which the High Court of Justiciary has exclusive jurisdiction in relation to the crimes mentioned above. The Sheriff Court is presided over by a Sheriff Principal or a Sheriff, who when trying cases on indictment sits with a Jury of 15 people. His powers of awarding punishment involving imprisonment are restricted to a maximum of three years, but he may under certain statutory powers remit the prisoner to the High Court for sentence if this is felt to be insufficient. The Sheriff also exercises a wide summary criminal jurisdiction and when doing so sits without a Jury; and he has concurrent jurisdiction with every other court within his Sheriff Court district in regard to all offences competent for trial in summary courts. The great majority of offences which come before courts are of a more minor nature and as such are disposed of in the Sheriff Summary Courts or in the District Courts (*see* below). Where a case is to be tried on indictment either in the High Court of Justiciary or in the Sheriff Court, the Judge may, before the trial, hold a preliminary or first diet to decide questions of a preliminary nature, whether relating to the competency or relevancy of proceedings or otherwise. Any decision at a preliminary diet (other than a decision to adjourn the first or preliminary diet or discharge trial diet) can be the subject of an appeal to the High Court of Justiciary prior to the trial. The High Court also has the exclusive power to provide a remedy for all extraordinary occurrences in the course of criminal business where there is no other mode of appeal available. This is known as the Nobile Officium powers of the High Court and all petitions to the High Court as the Nobile Officium must be heard before at least three judges.

In cases to be tried on indictment in the Sheriff Court a first diet is mandatory before the trial diet to decide questions of a preliminary nature and to identify cases which are unlikely to go to trial on the date programmed. Likewise in summary proceedings, an intermediate diet is again mandatory before trial. In High Court cases such matters may be dealt with at a preliminary diet.

District Courts have jurisdiction in more minor offences occurring within a district which before recent local government reorganization corresponded to district council boundaries. These courts are presided over by Lay Magistrates, known as Justices, who have limited powers for fine and imprisonment. In Glasgow District there are also Stipendiary Magistrates, who are legally qualified, and who have the same sentencing powers as Sheriffs.

The Court of Session, presided over by the Lord President (the Lord Justice General in criminal cases), is divided into an inner-house comprising two divisions of five judges each with a mainly appellate function, and an outer-house comprising 22 single Judges sitting individually at first instance; it exercises the highest civil jurisdiction in Scotland, with the Supreme Court of the United Kingdom as the final Court of Appeal.

CIVIL JUDICIAL STATISTICS

	2007[1]
House of Lords (Appeals from Court of Session)	10
Court of Session—	
General Department	3,689
Petition Department	3,196
Sheriff Courts—Ordinary Cause	59,459
Sheriff Courts—Summary Cause	34,253
Small Claims	24,997

[1]Provisional. The Scottish Court Service, the main provider of the data concerned, is working with the Justice Analytical Services to improve the accuracy and level of detail of the civil judicial statistics it collects.

CRIMINAL STATISTICS

	2006–07	2007–08	2008–09
Crimes and offences[1] recorded by the police:			
All crimes	419,257	385,509	377,433
Non-sexual crimes of violence	14,099	12,874	12,612
All offences	607,406	572,068	560,291
Persons with a charge proved in court:			
All crimes and offences[1]	134,371	133,583	125,430
All crimes	47,028	46,969	45,079
Persons aged 8–15[2]	130	159	124
Average prison population	7,183	7,376	7,835

[1]Contraventions of Scottish criminal law are divided for statistical purposes into crimes and offences. The term 'crime' is generally used for more serious criminal acts; and 'offence' for less serious ones, although the term 'offence' may also be used in relation to serious breaches of criminal law. The distinction is made only for working purposes and the 'seriousness' is generally related to the maximum sentence that can be imposed. [2]Except for serious offences which qualify for solemn proceedings, children aged 8–15 are not proceeded against in Scottish courts. Children within this age group that commit crime are generally referred to the reporter of the children's panel or are given a police warning.

Police

In Scotland, the unitary councils have the role of police authorities. Establishment levels were abolished in Scotland on 1 April 1996. The actual strength at 30 Sept. 2010 was 17,371 officers. There were 1,567 special constables. Total police funding in Scotland for 2010–11 was £1,412m.

Education

In Sept. 2007 there were 2,729 publicly funded (local authority, grant-aided and self-governing) primary, secondary and special schools. All teachers employed in these schools are required to be qualified.

Pre-school Education. In Jan. 2007 there were 2,823 pre-school centres that were in partnership with their local authority and 106,060 pupils enrolled in these centres.

Primary Education. In Sept. 2007 there were 2,168 publicly funded primary schools with 375,946 pupils and 23,508 full-time equivalent teachers.

Secondary Education. In Sept. 2007 there were 378 publicly funded secondary schools with 309,560 pupils and 26,365 full-time equivalent teachers. All but 21 schools provided a full range of Scottish Certificate of Education courses and non-certificate courses. Pupils who start their secondary education in schools which do not cater for a full range of courses may be transferred at the end of their second or fourth year to schools where a full range of courses is provided.

Independent schools. There were 157 independent schools in Sept. 2007, with a total of 30,981 pupils and 3,396 full-time equivalent teachers. A small number of the Scottish independent schools are of the 'public school' type, but they are not known as 'public schools' since in Scotland this term is used to denote education authority (i.e. state) schools.

Special Education. In Sept. 2007 there were 183 publicly funded special schools with 6,709 pupils.

Further Education. Under the Further and Higher Education (Scotland) Act 1992 funding of the Further Education colleges was transferred to central government in 1993. Scotland's FE colleges are funded by the Scottish Funding Council (SFC), which replaced the Scottish Further Education Funding Council and the Scottish Higher Education Funding Council in Oct. 2005.

There are 43 incorporated FE colleges as well as the FE colleges in Orkney and Shetland, which are run by the local education authorities, and two privately managed colleges, Sabhal Mor Ostaig and Newbattle Abbey College. The colleges offer training in a wide range of vocational areas and co-operate with the Scottish Qualifications Authority, the Enterprise, Energy and Lifelong Learning Directorate and the Education Directorate of the Scottish Executive in the development of new courses. The qualifications offered by colleges aim to improve the skills of the nation's workforce and increase the country's competitiveness.

In 2005–06 there were 359,530 students and a total of 446,619 enrolments on courses at Scotland's 43 further education institutions; the full-time equivalent staff number in the colleges was 12,338.

Full-time students resident in Scotland (and EU students) undertaking non-advanced (further education) courses are mainly supported through discretionary further education bursaries which are administered locally by further education colleges within National Policy Guidelines issued by the SFC. The Colleges have delegated discretionary powers for some aspects of the bursary support award.

In May 2000 the Scottish Executive announced the abolition of tuition fees for all eligible Scottish (and EU) full-time further education students from autumn 2000. The Executive also made a commitment to take steps to align, from autumn 2001, the levels of support available on a weekly basis for FE students with those that will apply for HE students and to begin to align the systems of assessment of parental/family contributions.

Higher Education. In Scotland in 2006 there were 21 institutions of higher education funded by the Scottish Higher Education Funding Council (now incorporated into the SFC), with the exception of the Scottish Agricultural College which is funded by the Scottish

Executive Rural Affairs Department. Included in this total is the Open University. The Scottish Higher Education Funding Council (SHEFC) took over the responsibility for funding the Open University in Scotland at the start of the 2000–01 academic session. University education in Scotland has a long history. Four universities—St Andrews, Glasgow, Aberdeen and Edinburgh, known collectively as the 'ancient Scottish universities'—were founded in the 15th and 16th centuries. Four further universities—Strathclyde, Heriot-Watt, Stirling and Dundee—were formally established as independent universities between 1964 and 1967, and four others—Napier, Paisley (now the University of the West of Scotland), Robert Gordon and Glasgow Caledonian—were granted the title of university in 1992, with a fifth, the University of Abertay, Dundee, being added in 1994.

Of the remaining higher education institutions, which all offer courses at degree level (although not themselves universities), four were formerly Central Institutions: Glasgow School of Art; Queen Margaret University College (Edinburgh); Royal Conservatoire of Scotland (Glasgow); and SRUC (Scotland's Rural College) (Perth).

Two additional higher education institutions were established in 2001. UHI Millennium Institute was designated as a higher education institution on 1 April when it took over from the local colleges of further education and other non-SHEFC funded institutions responsibility for all HE provision and all students on courses of HE in the Academic Partner institutions. Bell College of Technology became a higher education institution on 1 Aug. when its transfer from the further to the higher education sector was completed. On 1 Aug. 2007 Bell College merged with the University of Paisley. The merged institution operated under the name University of Paisley until the Privy Council approved the new name—University of the West of Scotland—on 30 Nov. 2007.

Further education colleges may also provide higher education courses.

University and HE student and staff figures:

Name (and Location)	Students (2005–06)	Staff (2006–07)
Aberdeen Univ.	15,225	1,435
Abertay Dundee Univ.	4,100	215
Bell College (Hamilton)[1]	4,860	190
Dundee Univ.	19,230	1,395
Edinburgh College of Art[2]	1,680	140
Edinburgh Univ.[2]	24,370	2,890
Glasgow School of Art	1,535	155
Glasgow Caledonian Univ.	17,180	895
Glasgow Univ.	23,145	2,580
Heriot-Watt Univ. (Edinburgh)	24,205	665
Napier Univ. (Edinburgh)	15,785	865
Paisley Univ.[1]	13,400	410
Queen Margaret University College (Edinburgh)	5,400	200
Robert Gordon Univ. (Aberdeen)	12,530	735
Royal Conservatoire of Scotland (Glasgow)[3]	680	60
St Andrews Univ.	8,365	890
SRUC (Perth)[4]	750	195
Stirling Univ.	8,875	990
Strathclyde Univ. (Glasgow)	25,700	1,440
Univ. of the Highlands and Islands (Inverness)[5]	7,205	10

[1]Merged in Aug. 2007 to form the University of the West of Scotland. [2]Merged in Aug. 2011. [3]Known as Royal Scottish Academy of Music and Drama until Sept. 2011. [4]Created in Oct. 2012 through the merger of the Scottish Agricultural College and three further education colleges. [5]Known as UHI Millennium Institute until Feb. 2011 when it was granted university status.

All the higher education institutions are independent and self-governing. In addition to funding through the higher education funding councils, they receive tuition fees from the Students Awards Agency for Scotland for students domiciled in Scotland, and through local education authorities for students domiciled in England and Wales. Institutions which carry out research may also receive funding through the five Research Councils administered by the Office of Science and Technology.

Health

As at 30 Sept. 2009 there were 4,941 general medical practitioners in Scotland. At 31 March 2009 there were 2,507 general dental practitioners in Scotland.

In 2009, 27% of men and 26% of women aged 16–64 were obese (having a body mass index of 30 or over), up from 16% and 17% in 1995 respectively.

Scottish Social Work Services staff numbered around 53,100 in Oct. 2009. Scotland's expenditure on social work (2008–09) was £2,684m.

Welfare

In Feb. 2010 there were 1,003,760 retirement pensioners, 213,460 beneficiaries of incapacity benefit, 340,540 recipients of disability living allowance, 185,990 claimants of income support, 278,480 of pension credit, 147,030 recipients of attendance allowance and 141,840 claimants of Jobseeker's Allowance. A total of 468,960 households were receiving housing benefit in July 2010 and 562,680 council tax benefit. There were 517,300 families with child tax credit or working tax credit awards, or with children and receiving out-of-work benefits in Aug. 2008.

RELIGION

The Church of Scotland, which was reformed in 1560, subsequently developed a presbyterian system of church government which was established in 1690 and has continued to the present day.

The supreme court is the General Assembly, which now consists of some 750 voting members, ministers and elders in equal numbers, together with some members of the diaconate, all commissioned by presbyteries. It meets annually in May, under the presidency of a Moderator appointed by the Assembly. The Queen is normally represented by a Lord High Commissioner, but has occasionally attended in person. The royal presence in a special throne gallery in the hall (but outside the Assembly) symbolizes the independence from state control of what is nevertheless recognized as the national Church in Scotland.

There are also 43 presbyteries in Scotland, providing governance and supervision at regional level, together with the presbyteries of England, Europe and Jerusalem. At the base of this conciliar structure of Church courts are the Kirk Sessions, of which there were 1,454 on 31 Dec. 2009. The total communicant membership of the Church on 31 Dec. 2009 was 464,355.

The Scottish Episcopal Church is a province of the Anglican Communion and is one of the historic Scottish churches. It consists of seven dioceses. As at 31 Dec. 2009 it had 320 churches, 524 clergy and 37,047 members, of whom 25,776 were communicants.

There are in Scotland some small outstanding Presbyterian bodies and also Baptists, Congregationalists, Methodists and Unitarians.

The Roman Catholic Church had in Scotland (as at 1 Oct. 2010) one cardinal archbishop, one archbishop, six bishops, three bishops emeriti, 59 permanent deacons, 715 clergy, 449 parishes and 643,075 adherents.

The proportion of marriages in Scotland according to the rites of the various Churches in 2009 was: Church of Scotland, 22·3%; Roman Catholic, 6·5%; Humanist Society of Scotland, 5·6%; others 13·9%; civic; 51·7%.

CULTURE

World Heritage Sites

Scotland has four sites on UNESCO's World Heritage List (with year entered on list): St Kilda (1986, 2004 and 2005), a volcanic archipelago on the coast of the Hebrides; Old and New Towns of Edinburgh (1995), the Scottish capital; the Heart of Neolithic Orkney (1999), comprising several important neolithic monuments; and New Lanark (2001), Robert Owen's model industrial community and cotton mills of the early 19th century.

Press

Average daily circulation in Jan. 2011 for the daily *Scotsman* was 43,362 and the *Daily Record* 306,872; and for *Scotland on Sunday* 56,256 and the *Sunday Mail* 366,325.

Tourism

There were 2·59m. overseas visitors to Scotland in 2009, spending £1·4bn. Overall tourism receipts totalled £4·1bn. The leading free admission attraction was Kelvingrove Art Gallery and Museum in Glasgow, with 1·4m. visitors in 2009. The leading tourist attraction charging admission in 2009 was Edinburgh Castle, with 1·2m. visits. In 2007, 13% of the workforce was employed in the tourism industry.

Festivals

St Andrew's Day on 30 Nov. is Scotland's official national day. However, Burns Night on 25 Jan. to commemorate the life and works of Scots poet Robert Burns has become the day on which Scottish culture is celebrated throughout the UK. The Edinburgh Festival and the Fringe Festival both take place in Aug./early Sept. and are major international festivals of culture. New Year's Eve is known as Hogmanay and the largest celebrations occur in Edinburgh.

FURTHER READING

Scottish Executive. Scottish Economic Report. Twice yearly—*Scottish Abstract of Statistics.* Annual

Brown, A., *et al.*, *Politics and Society in Scotland.* 1996
Devine, T. M. and Finlay, R. J. (eds.) *Scotland in the 20th Century.* 1996
Harvie, C., *Scotland and Nationalism: Scottish Society and Politics, 1707 to the Present.* 4th ed. 2004
Keay, J. and J., *Collins Encyclopedia of Scotland: The Story of a Nation.* 2000
Macleod, J., *Highlanders: A History of the Gaels.* 1997
Magnusson, M., *Scotland: The Story of a Nation.* 2000
McCaffrey, J. F., *Scotland in the Nineteenth Century.* 1998
McGarvey, Neil and Cairney, Paul, (eds.) *Scottish Politics: An Introduction.* 2008
Mitchell, James, *Governing Scotland.* 2003

Statistical office: Room 1N.04, St Andrew's House, Regent Road, Edinburgh, EH1 3DG.
Website: http://www.scotland.gov.uk/Topics/Statistics

WALES

KEY HISTORICAL EVENTS

After the Roman evacuation, Wales divided into tribal kingdoms. Cunedda Wledig, a prince from southern Scotland, founded a dynasty in the northwest region of Gwynedd—to become the focus for Welsh unity—while the Irish exerted an influence in the kingdom of Dyfed. Offa's Dyke, a defensive earthwork, was the dividing line between England and Wales. In the late 9th century the kings of southern Wales swore fealty to Alfred of Wessex, a relationship assumed by the English crown. Gruffydd ap Llywelyn of Gwynedd briefly united Wales from 1055–63. His death was followed by Norman expansion into southern Wales, where the Marcher lordships were created.

With the accession of Llywelyn the Great (1194–1240), the house of Gwynedd overcame rival claims from Powys and Deheubarth to forge a stable political state under English suzerainty. His grandson, Llywelyn ap Gruffydd (1246–82), was recognized as prince of Wales by Henry III but Llywelyn intrigued against Edward I, who reduced Gwynedd's hegemony. Wales was annexed and subdued by a network of castles. Edward's infant son, born at Caernarfon, was made Prince of Wales.

Loyalty to Henry VIII, who was of Welsh descent, was rewarded with political influence. The 'Act of Union' in 1536 made English law general, admitted Welsh representatives to Parliament and established the Council of Wales and the Marches.

TERRITORY AND POPULATION

At the census taken on 27 March 2011 the population was 3,063,456. The area of Wales is 20,780 sq. km. (of which land 20,735 sq. km). Population density, 2011 census: 148 per sq. km. Wales covers 8·5% of the total area of the United Kingdom.

Population at census day 2011:

Males	Females	Total
1,504,228	1,559,228	3,063,456

Population (present on census night) at the four previous decennial censuses:

1971	1981	1991	2001
2,731,204[1]	2,790,500[2]	2,811,865	2,903,805

[1]Areas now recognized as Monmouthshire and small sections of various other counties formed the county of Monmouthshire in England until 1974. [2]The final count is believed to be over-stated as a result of an error in processing. The preliminary counts presented here rounded to the nearest hundred are thought to be more accurate.

Cardiff, the capital and largest city, had a population in 2011 of 346,090; Swansea, the second largest city, had a population of 239,023.

In 2011, 630,062 people aged three and over were able to speak, read or write Welsh. Households at the 2011 census: 1,302,676.

For further statistical information, see under Territory and Population, UNITED KINGDOM.

Wales is divided into 22 unitary authorities (cities and counties, counties and county boroughs).

Designations, areas and populations of the unitary authority areas at census day 2011:

Unitary Authority	Designation	Land area (sq. km)	Population
Blaenau Gwent	County Borough	109	69,814
Bridgend	County Borough	251	139,178
Caerphilly	County Borough	277	178,806
Cardiff	City and County	140	346,090
Carmarthenshire	County	2,370	183,777
Ceredigion	County	1,785	75,922
Conwy	County Borough	1,126	115,228
Denbighshire	County	837	93,734
Flintshire	County	438	152,506
Gwynedd	County	2,535	121,874
Isle of Anglesey	County	711	69,751
Merthyr Tydfil	County Borough	111	58,802
Monmouthshire	County	849	91,323
Neath Port Talbot	County Borough	441	139,812
Newport	County Borough	191	145,736
Pembrokeshire	County	1,619	122,439
Powys	County	5,180	132,976
Rhondda Cynon Taff	County Borough	424	234,410
Swansea	City and County	380	239,023
The Vale of Glamorgan	County Borough	331	126,336
Torfaen	County Borough	126	91,075
Wrexham	County Borough	504	134,844

SOCIAL STATISTICS

2009: births, 34,937 (11·6 per 1,000 population); deaths, 31,006 (10·3 per 1,000 population); infant deaths, 166 (4·8 per 1,000 live births); marriages, 12,553.

CLIMATE

For more detailed information, see under Climate, UNITED KINGDOM.

Cardiff, Jan. 40°F (4·4°C), July 61°F (16·1°C). Annual rainfall 42·6" (1,065 mm).

CONSTITUTION AND GOVERNMENT

One of the main aspects of the former British Labour government's programme of constitutional reform was devolution. On 18 Sept. 1997 in the referendum there were 559,419 votes cast in favour of a Welsh assembly (50·3%) and 552,698 against (49·7%). The turnout was 51·3%.

The Parliamentary electorate of Wales in the register in Dec. 2009 numbered 2,261,269.

RECENT ELECTIONS

At the UK general election in May 2010, 40 members were returned from Wales. Labour won 26 seats (29 in 2005), Conservatives 8 seats (3), Liberal Democrats 3 (4), Plaid Cymru 3 seats (3).

At the 2009 European Parliamentary elections the Conservatives, Labour, Plaid Cymru and UKIP won one seat each.

In the elections to the Welsh Assembly on 5 May 2011, Labour won 30 seats (2 by regional list), followed by the Conservatives with 14 (8 by regional list), Plaid Cymru with 11 (6 by regional list) and the Liberal Democrats with 5 (4 by regional list). Of the 60 seats, 40 seats were won on a first-past-the-post basis and 20 through proportional representation (regional list).

See also Constitution and Government, Recent Elections *and* Current Government *in* UNITED KINGDOM.

CURRENT GOVERNMENT

First Secretary: Carwyn Jones; b. 1967 (Labour).
Presiding Officer: Rosemary Butler.

National Assembly for Wales: http://wales.gov.uk

DEFENCE

For information on defence, see UNITED KINGDOM.

ECONOMY

For information on the economy, see UNITED KINGDOM.

Overview

Wales has enjoyed an increasing degree of economic autonomy since the Welsh Assembly was created in 1998. In 2006 extended powers were delegated from Westminster to the Assembly. A referendum held in March 2011 approved law-making powers similar to those of the Scottish Parliament.

With rich mineral deposits, development in the 19th and early 20th centuries centred on coal, slate and steel exports. Steel also supported a substantial manufacturing industry. The post-Second World War era marked a shift away from traditional heavy industry towards light industry and financial services. Tourism is a vibrant part of the economy with 45 beaches and marinas receiving Blue Flag status (awarded by the Foundation for Environmental Education). Tourism, financial services and light industry remain the keystones to the devolved government's development plans.

Gross Value Added for the economy in 2009 was £44·5bn. or 74·3% of the UK average, the lowest among the devolved countries and English regions. Unemployment levels in Wales are slightly higher than the UK average. The small population and lack of large metropolitan centres have been cited as factors in uneven development and the failure to attract high value-added employment in areas such as finance.

ENERGY AND NATURAL RESOURCES

For information on energy and natural resources, see UNITED KINGDOM.

Environment

41% of municipal waste was recycled in April–June 2009.

Water

The Water Act of Sept. 1989 privatized Welsh Water (Dŵr Cymru Cyfyngedig), along with the nine water authorities in England.

Agriculture

In 2009 there were 39,024 agricultural holdings. Of these, 15,287 were under 5 ha., 8,209 were between 5 and 20 ha., 6,491 were between 20 and 50 ha. and 9,037 were over 50 ha. The average size of a holding in Wales in 2009 was 38 ha.

The area of tillage in 2009 was 83,584 ha. (82,788 ha. for crops and 796 ha. bare fallow). Major crops, 2009–10 (1,000 tonnes): wheat, 140; barley, 134; potatoes (2010), 57; oats, 23.

Livestock, 2009: sheep and lambs, 8,237,737; cattle and calves, 1,129,968; pigs, 22,303; poultry, 7,250,598.

Forestry

In March 2011 there were 114,000 ha. of Forestry Commission woodland and 190,000 ha. of non-Forestry Commission woodland.

Fisheries

The major fishing port is Milford Haven. In 2009, in all ports in Wales, 15,904 tonnes of fish and shellfish worth £25,693,649 were landed. There were 481 fishing vessels registered in Wales in 2009.

INDUSTRY

Main industrial production (gross value added), 2007 (£1m.): basic metals and fabricated metal products, 1,381; food products, beverages and tobacco, 1,046; transport equipment, 1,017; electrical and optical equipment, 1,014; chemicals, chemical products and man-made fibres, 856.

Labour

In the year ending 31 March 2012 there were 1,305,500 people in employment, of whom 178,900 were self-employed. The number of claimant count unemployed in March 2012 was 79,800. The largest sectors in terms of jobs in 2011 were: wholesale and retail trade, 212,000; human health and social work activities, 209,000; manufacturing, 144,000; education, 134,000. As a proportion of total employment, just over half (51%) of the workforce in 2011 were in these four industry sectors. The ILO unemployment rate in the period Oct.–Dec. 2011 was 9·0% (compared to 8·4% in the UK as a whole), up from 8·4% in the fourth quarter of 2010. In 2006, 62,000 working days were lost as a result of industrial disputes.

INTERNATIONAL TRADE

For information on international trade, see UNITED KINGDOM.

COMMUNICATIONS

Roads

Responsibility for the construction and maintenance of trunk roads belongs to the Welsh Assembly Government. Roads not classified as trunk roads are the responsibility of county or unitary councils. In 2009 there were 133 km of motorway, 1,577 km of trunk roads and 2,742 km of principal roads. 1,742,400 vehicles were licensed in 2008, including 1,493,200 private and light goods vehicles. In 2008 there were 7,783 reported accidents which led to 11,185 casualties, including 142 deaths.

Civil Aviation

Cardiff Airport handled 1,631,236 passengers in 2009 (1,247,167 on international flights) and 178 tonnes of freight.

Shipping

The principal ports are (with 1m. tonnes of cargo handled in 2009): Milford Haven (39·3) and Port Talbot (5·2).

SOCIAL INSTITUTIONS

Justice

As at 31 March 2007 police strength amounted to 7,627. During the financial year 2007–08 there were 243,623 notable offences, including 49,376 violent and 2,574 sexual offences. The clear-up rate was 31·0%. 14,959 people were found guilty of indictable offences in Magistrates' Courts in 2005 and 3,068 in Crown Courts.

Education

In April 2006 ACCAC (the qualification, curriculum and assessment authority), Dysg (the Welsh operation of the Learning and Skills Development Agency), ELWa (the National Council for Education and Training Wales) and the Wales Youth Agency merged with the Welsh Assembly Government's Department for Training and Education to form a new department, the Department for Children, Education, Lifelong Learning and Skills (DCELLS) in the Welsh Assembly Government.

There were 25 maintained nursery schools in Jan. 2010, and 67,093 pupils under five years provided for in nursery schools and in nursery or infants classes in primary schools.

In Jan. 2010 there were 257,445 pupils at 1,462 primary schools. Of these, 476 primary schools use Welsh as the sole or main medium of instruction. Such schools are to be found in all parts of Wales but are mainly concentrated in the predominantly Welsh-speaking areas of west and northwest Wales. Generally, children transfer from primary to secondary schools at 11 years of age.

In Jan. 2010 there were 223 secondary schools. All maintained secondary schools are classified as comprehensive; there are no middle schools in Wales. In 2009–10, 58 of the secondary schools were classed as Welsh-speaking as defined in section 354(b) of the Education Act 1996.

Under the Education Act 1996, children have special educational needs if they have a learning difficulty that calls for special educational provision to be made for them. In a minority of cases the local education authority will need to make a statutory assessment of special educational needs under the Education Act 1996, which may ultimately lead to a 'statement of Special Educational Needs'. The total number of pupils with statements in Jan. 2010 was 14,327. Since April 2002 Special Educational Needs (SEN) guidance for Wales has been set out in the SEN Code of Practice for Wales.

In Jan. 2010, 8,943 full-time pupils and 279 part-time pupils attended 64 independent schools.

Post-16 Learning. The responsibilities of DCELLS (*see* above) include the funding, planning and promotion of children's services, education, learning and skills in Wales. In 2010–11 DCELLS funding for post-16 education and training Wales (excluding higher education provision) providers totalled £577m. This is split between FE institutions, school sixth forms, work-based learning providers, HE institutions and Local Authority (LA) community learning providers. In 2008–09, 45,040 full-time students and 160,250 students studying part-time in the further education sector (excluding work-based learning) were supported at 24 further education institutions and at several higher education institutions. The proportion of 16- to 18-year-olds with no qualifications fell from 12% in 2001 to 8% in 2009. The proportion of working age adults with no qualifications fell from 21% in 2001 to 14% in 2009; and 50% of working age adults had the equivalent of a National Qualification Framework Level 3 or above in 2009, compared to 40% in 2001.

Higher Education. In 2008–09 there were 11 higher education institutions in Wales with a total income of £1·17bn., of which 37·6% came from funding council grants. There were 134,565 students in the higher education sector in 2008–09, including those registered with the Open University in Wales, of which 80,990 were full-time and 53,580 part-time students, excluding those enrolled on higher education provision at further education colleges.

Higher Education Institutes (HEIs)	Full-time/sandwich HE students at HEIs (2008–09)	No. of academic staff (2008–09)
Aberystwyth University	7,275	1,005
Bangor University	8,340	745
Cardiff Metropolitan University[1]	8,435	585
Cardiff University	20,045	2,785
Glyndŵr University (Wrexham)	2,990	330
Swansea Metropolitan University	3,690	310
Swansea University	10,760	1,100
Trinity College Carmarthen[2]	1,240	110
University of Glamorgan (Pontypridd)	12,545	1,425
University of Wales, Lampeter[2]	1,765	135
University of Wales, Newport	3,900	445

[1]Known as University of Wales Institute, Cardiff until Nov. 2011.
[2]In July 2010 Trinity College Carmarthen and the University of Wales, Lampeter merged to form the University of Wales Trinity Saint David.

Health

As at 30 Sept. 2008 there were 1,940 general medical practitioners in Wales, with an average of 1,605 patients per doctor. In 2005–06 there were 28,152 whole-time equivalent nursing, midwifery and health visiting staff. At 31 March 2007 there were 1,141 general dental practitioners in Wales. The average daily number of hospital beds available in 2009–10 was 12,900, of which 10,500 were occupied. 510,900 in-patient cases were reported, with stays lasting an average 7·5 days. At 31 March 2011, 354,766 people were waiting to start treatment of whom 20,744 had waited for more than 26 weeks.

The 2008 Welsh Health Survey found that 21% of adults were obese (having a body mass index over 30).

Personal Social Services (PSS) staff numbered 27,287 in Wales at 31 March 2009. The total expenditure (2007–08) for PSS was £1,303m. in Wales.

Welfare

In Feb. 2008, 609,000 people received state retirement pensions; 284,000 people received some form of income support including 162,000 who received pension credit; and in Aug. 2007, 361,000 families received child benefit.

RELIGION

Under the Welsh Church Acts, 1914 and 1919, the Church in Wales and Monmouthshire was disestablished as from 1 April 1920, and Wales was formed into a separate Province.

CULTURE

World Heritage Sites

Wales has three sites on UNESCO's World Heritage List (with year entered on list): Castles and Town Walls of King Edward in Gwynedd (1986), a testament to the early period of English colonization in the late 13th century; Blaenavon Industrial Landscape (2000), a symbol of South Wales' role as a coal and iron provider in the 19th century; and the Pontcysyllte Aqueduct and Canal near Wrexham (2009), a feat of civil engineering by Thomas Telford during the Industrial Revolution.

Tourism

In 2009 there were some 9m. domestic trips (from elsewhere in the UK) into Wales. Visitors stayed 32·9m. nights and spent £1·4bn. The leading free admission attraction in 2009 was the Wales Millennium Centre in Cardiff with 1·1m. visits. The leading paid admission attraction was the LC, a leisure complex and waterpark in Swansea, with 597,000 visitors in 2009.

Festivals

Every year there are local and national *eisteddfods* (festivals for musical competitions, etc.). The National Eisteddfod of Wales takes place every Aug., traditionally alternating between north and south Wales. In 2014 it will be held at Llanelli.

FURTHER READING

National Assembly. Digest of Welsh Statistics. National Statistics. Great Britain (annual)

Davies, J., *History of Wales.* 1993
History of Wales. vols. 3, 4 (1415–1780). 2nd ed. 1993
Jenkins, G. H., *The Foundations of Modern Wales 1642–1780.* 1988.—*The Welsh Language and its Social Domains 1801–1911: A Social History of the Welsh Language.* 2000
Jones, G. E., *Modern Wales: a Concise History.* 2nd ed. 1994
Morgan, K. and Mungham, G., *Redesigning Democracy. The Making of the Welsh Assembly.* 2000

Statistical office: Statistical Directorate, Welsh Assembly Government, Cathays Park, Cardiff CF10 3NQ.
Website: http://www.statswales.wales.gov.uk

NORTHERN IRELAND

KEY HISTORICAL EVENTS

The Government of Ireland Act 1920 granted Northern Ireland its own bicameral parliament (Stormont). The rejection of home rule by the rest of Ireland (which pursued independence) forced a separation along primarily religious lines, with a large Catholic minority in the six northern counties. Between 1921–72 Stormont had full responsibility for local affairs except for taxation and customs; Northern Ireland was on the whole neglected by Westminster, allowing the virtual exclusion of Catholics from political office. The (predominantly Protestant) Unionist government ignored demands from London and the Catholic community to end communal discrimination.

In the late 1960s a Civil Rights campaign and reactions to it escalated into serious rioting and sectarian violence involving the Irish Republican Army (IRA, a terrorist organization aiming to unify Northern Ireland with the Republic of Ireland) and loyalist paramilitary organizations, such as the Ulster Defence Association. The British Army was deployed to protect civilians and was at first welcomed by the Catholic community. However, British soldiers shot dead 13 Catholic civil rights protesters in (London)Derry on 30 Jan. 1972—'Bloody Sunday'—prompting the Republic of Ireland's foreign minister to demand United Nations intervention. 467 people died in 1972, on account of 'the Troubles', and nearly 1,800 between 1971–77. The Northern Ireland government resigned and direct rule from Westminster was imposed.

Attempts have been made by successive governments to find a means of restoring greater power to Northern Ireland's political representatives on a widely acceptable basis, including a Constitutional Convention (1975–76), a Constitutional Conference (1979–80) and 78-member Northern Ireland Assembly elected by proportional representation in 1982. This was dissolved in 1986, partly in response to Unionist reaction to the Anglo-Irish Agreement signed on 15 Nov. 1985, which established an

Intergovernmental Conference of British and Irish ministers to monitor issues of concern to the nationalist community. The Provisional IRA bombing of a Remembrance Day service in Enniskillen in 1987 killed 11. Universally condemned, it galvanized the anti-violence campaign.

On 15 Dec. 1993 the British and Irish prime ministers, John Major and Albert Reynolds, issued a joint declaration as a basis for all-party talks to achieve a political settlement. They invited Sinn Féin, the political wing of the IRA, to join the talks in an All-Ireland Forum after the cessation of terrorist violence. The IRA announced 'a complete cessation of military operations' on 31 Aug. 1994. On 13 Oct. 1994 the anti-IRA Combined Loyalist Military Command also announced a ceasefire 'dependent upon the continued cessation of all nationalist republican violence'.

Elections were held on 30 May 1996 to constitute a 110-member forum to take part in talks with the British and Irish governments. The Ulster Unionist Party won 30 seats, the Democratic Unionist Party 24 seats, the Social Democratic and Labour Party 21 seats and Sinn Féin 17 seats. Opening plenary talks, excluding Sinn Féin, began under the chairmanship of US Senator George Mitchell on 12 June 1996. A marathon negotiating struggle on 9–10 April 1998 led to agreement on a framework for sharing power designed to satisfy Protestant demands for a reaffirmation of their national identity as British, Catholic desires for a closer relationship with the Republic of Ireland and Britain's wish to return to Northern Ireland the powers London assumed in 1972.

Under the Good Friday Agreement, there was to be a democratically elected legislature in Belfast, a ministerial council giving the governments of Northern Ireland and Ireland joint responsibilities in areas like tourism, transportation and the environment, and a consultative council meeting twice a year to bring together ministers from the British and Irish parliaments, and the three assemblies being created in Northern Ireland and in Scotland and Wales. The Irish government eliminated from its constitution its territorial claim on Northern Ireland.

In the referendum on 22 May 1998, 71·1% of votes in Northern Ireland were cast in favour of the Good Friday peace agreement and 94·4% in the Republic of Ireland. As a consequence, in June, Northern Ireland's 1·2m. voters elected the first power-sharing administration since the collapse of the Sunningdale Agreement in 1974.

On 15 Aug. 1998 a 200 kg bomb exploded in the centre of Omagh. The dissident republican group the 'Real IRA' claimed responsibility. 29 people died and over 200 were injured, making it the single bloodiest incident of the Troubles—about 3,500 deaths had been recorded since the Troubles began by the end of 2001.

In Nov. 1999 the Mitchell talks finally produced an agreement between the Ulster Unionists and Sinn Féin, paving the way for devolved government. The new Northern Ireland Assembly met on 29 Nov. 1999 and on 2 Dec. legislative powers were fully devolved from London to Belfast. However, on 11 Feb. 2000 the Assembly was suspended following a breakdown in negotiations on the decommissioning of IRA weapons. Direct rule from London was restored. Devolved government resumed on 30 May after the IRA agreed to open their arms dumps to independent inspection. First Minister David Trimble resigned on 30 June 2001 to pressure republicans over decommissioning but on 22 Oct. Sinn Féin president Gerry Adams announced that he had recommended a 'ground-breaking' step on the arms issue. The IRA made a start on decommissioning arms, ammunition and explosives. David Trimble was re-elected first minister on 6 Nov. 2001.

On 15 Oct. 2002 the Assembly executive was again suspended over allegations of IRA spying at the Northern Ireland Office, although all charges were dropped in Jan. 2006. Direct rule from London was reimposed and on 30 Oct. the IRA cut off its links with the weapons decommissioning body. The Ulster Volunteer Force followed suit on 17 Jan. 2003. Elections for the Northern Ireland Assembly took place on 26 Nov. 2003. The theft of £26·5m. from the Northern Bank in Belfast in Dec. 2004 suggested closer than acknowledged associations between Sinn Féin and the IRA. Controversy surrounding the raid put the peace process on hold. In July 2005 the IRA formally announced an end to its armed campaign. In Sept. 2005 it claimed to have destroyed its arsenal of weapons.

The Stormont assembly met in May 2006 for the first time in more than three years, charged with restoring devolution by 24 Nov. On that date a transitional assembly was established and in Jan. 2007 Sinn Féin voted to support policing policies in Northern Ireland, a key requirement for a workable power-sharing agreement. The transitional assembly was dissolved at the end of that month ahead of assembly elections in March 2007, in which the DUP and Sinn Féin made strong gains. Both parties came under pressure to compromise and on 26 March reached an historic agreement to share power from 8 May with a devolved Northern Ireland government replacing direct rule from London.

TERRITORY AND POPULATION

Population of Northern Ireland at census day in 2011 and areas of districts in 2011 were as follows:

District	Population	Area in ha. (including inland water)
Antrim	53,428	57,686
Ards	78,078	37,799
Armagh	59,340	67,060
Ballymena	64,044	63,202
Ballymoney	31,224	41,820
Banbridge	48,339	45,263
Belfast	280,962	10,998
Carrickfergus	39,114	8,181
Castlereagh	67,242	8,514
Coleraine	59,067	48,352
Cookstown	37,013	62,244
Craigavon	93,023	37,842
Derry (Londonderry)	107,877	37,986
Down	69,731	64,825
Dungannon	57,852	78,360
Fermanagh	61,805	187,126
Larne	32,180	33,579
Limavady	33,536	58,562
Lisburn	120,165	44,684
Magherafelt	45,038	57,280
Moyle	17,050	49,419
Newry and Mourne	99,480	90,070
Newtownabbey	85,139	15,056
North Down	78,937	8,150
Omagh	51,356	113,045
Strabane	39,843	85,868
Northern Ireland	*1,810,863*	*1,412,972*

Northern Ireland's area of 14,130 sq. km represents 5·8% of the total area of the United Kingdom. Chief town (census, 2011): Belfast, 280,962.

Population at census day 2011:

Males	Females	Total
887,323	923,540	1,810,863

SOCIAL STATISTICS

In 2010 there were 25,315 births, 14,457 deaths, 8,156 marriages and 2,600 divorces.

CLIMATE

For more detailed information, see under Climate, UNITED KINGDOM.

Belfast, Jan. 40°F (4·5°C), July 59°F (15·0°C). Annual rainfall 37·4" (950 mm).

CONSTITUTION AND GOVERNMENT

Under the Northern Ireland Act 1998 power that was previously exercised by the NI Departments was devolved to the Northern Ireland Assembly and its Executive Committee of Ministers. In April 2010 a new justice department came into existence to take responsibility for policing and judicial matters, which had previously been under the jurisdiction of the Secretary of State. Constitutional and national security issues along with firearms and explosives licensing and legislation remain the domain of the Secretary of State, who is also involved in public inquiries and some policy-making decisions.

The Parliamentary electorate of Northern Ireland in the register in Dec. 2009 numbered 1,160,757.

Secretary of State for Northern Ireland. Theresa Villiers.

RECENT ELECTIONS

At the general election of 6 May 2010, 18 members were returned from Northern Ireland. The Democratic Unionist Party won 8 seats (9 in 2005); Sinn Féin 5 (5); the Social and Democratic Labour Party 3 (3); the Alliance Party 1 (0); others 1 (1).

In the Northern Ireland Assembly elections on 5 May 2011 the Democratic Unionist Party won 38 of the 108 seats (30·0% of first preference votes), Sinn Féin 29 (26·9%), the Ulster Unionist Party 16 (13·2%), the Social Democratic and Labour Party 14 (14·2%), Alliance Party of Northern Ireland 8 (7·7%), the Traditional Unionist Voice 1 (2·5%) and Green Party 1 (0·9%). One other candidate was elected. Turnout was 54·5%.

At the June 2009 European Parliament elections, voting was by the single transferable vote system: Sinn Féin (26·0%), the Democratic Unionist Party (18·2%) and the Ulster Conservatives and Unionists–New Force (17·1%) gained 1 seat each. Turnout was 42·8%.

CURRENT GOVERNMENT

Peter Robinson (Democratic Unionist Party) was sworn in as first minister of the Northern Ireland Assembly on 5 June 2008. Martin McGuinness (Sinn Féin) was initially sworn in as deputy first minister on 8 May 2007 and reappointed on 5 June 2008. They are joint leaders of the administration. On 11 Jan. 2010 Peter Robinson temporarily stood down as first minister after he was accused of political misconduct concerning illegal financial dealings carried out by his wife. Enterprise minister Arlene Foster (also Democratic Unionist Party) became acting first minister for a six-week period. The Assembly was restored on 8 May 2007 for the first time since its suspension on 14 Oct. 2002.

Northern Ireland Executive: http://www.northernireland.gov.uk

ECONOMY

Overview

Northern Ireland is the smallest of the four economies that make up the United Kingdom. Traditionally, it has been led by manufacturing, heavy industry and agriculture, but there has been a shift towards a more service-based economy since the 1980s. Northern Ireland is noted for its strong dependence on the public sector (accounting for 30% of all jobs), its low wages and low labour productivity levels relative to the rest of the UK, reflecting a legacy of 30 years of domestic conflict, its demographic structure and its peripheral location.

The economy has benefited from the peace process that culminated in the 1998 Good Friday Agreement, although implementation of the agreement continued to cause difficulties over the following decade. Gross Value Added per capita is lower than other parts of the United Kingdom at £15,800 in 2009— slightly higher than Wales but significantly below England and Scotland, owing to the economy's low productivity levels and high

rates of economic inactivity. Tourism currently plays an important role and offers potential for further expansion.

Northern Ireland has close economic links with neighbouring Ireland, which accounted for 28·4% of manufacturing exports along with a quarter of construction output in 2007 prior to the global recession. Two of the country's four main banks are Irish-owned. This has led to significant knock-on effects from Ireland's severe recession, with unemployment up to 8·0% in Dec. 2010 compared to 4·3% in Dec. 2007 and levels of economic inactivity remaining high.

Currency

Banknotes are issued by Bank of Ireland, First Trust Bank, Northern Bank and Ulster Bank.

Banking and Finance

The Department of Finance and Personnel is responsible for control of the expenditure of Northern Ireland departments, involving liaison with HM Treasury, the European Commission and the Northern Ireland Office on financial matters, economic and social research and analysis; to review and develop rating policy and legislation; procurement for the Northern Ireland public sector; formulation of policy for central personnel management, and legal services, including law reform. The Department's Agencies are: Land Registers NI; Northern Ireland Statistics and Research Agency; and Land and Property Services Agency.

Public income of Northern Ireland (in £1,000 sterling):

	2003–04	2004–05	2005–06
Block grant	8,505,000	8,950,000	9,030,000
Regional and district rates	650,852	717,010	784,684
Interest on loans made from the Consolidated Fund	126,322	123,690	123,183
Miscellaneous receipts EU	314,947	111,660	217,523
Other	96,700	382,305	380,671
Total public income	9,693,821	10,284,665	10,536,061

ENERGY AND NATURAL RESOURCES

Environment

29% of municipal waste was recycled in 2007–08.

Electricity

There are three power stations with an installed capacity of some 2,100 MW.

In addition, electricity is also supplied through a 500 MW interconnector linking the Northern Ireland Electricity (NIE) and Scottish Power networks and a number of interconnectors linking the NIE network with the Electricity Supply Board (ESB) network in the Republic of Ireland.

Oil and Gas

In Sept. 2001 the Northern Ireland executive approved grant support for the development of the gas network outside the Greater Belfast area, to the North/North West region and for the construction of a South/North pipeline. The North West gas pipeline was completed in late 2004 and supplies gas for the Combined Cycle Turbine power station at Coolkeeragh outside Londonderry, which opened in June 2005. The South/North pipeline was commissioned in Oct. 2006, connecting the grids of Northern Ireland and the Republic of Ireland for the first time. The Northern Ireland Authority for Energy Regulation has granted a license to supply gas to the major towns along the route of both these.

Minerals

Output of minerals (in 1,000 tonnes), 2010: sandstone, 2,768; basalt and igneous rock (other than granite), 5,438; limestone, 3,689; sand and gravel, 2,178; other minerals (rock salt, fireclay, dolomite, granite and chalk), 2,087. There are extensive lignite deposits but they have not yet been developed. The United Kingdom's only operational gold mine, located near Omagh, opened in 2007; production in 2011 was 202 kg. Silver production in 2011 was 531 kg.

Agriculture

Provisional gross output in 2011:

	Quantity	Value (£1m.)
Cattle and calves	463,288	375·2
Sheep and lambs	693,457	55·0
Pigs	1,020,600	104·0
Poultry (1,000 tonnes)	258·1	239·0
Eggs (m. dozen)	82·3	55·4
Milk (1m. litres)	1,966·0	542·9
Other livestock products	—	13·0
Cereals (1,000 tonnes)	239·8	44·1
Potatoes (1,000 tonnes)	196·3	23·4
Fruit (1,000 tonnes)	48·1	7·4
Vegetables (1,000 tonnes)	37·4	17·4
Mushrooms (1,000 tonnes)	15·3	18·5
Other crops	—	12·1
Flowers, ornamentals and nursery stock	—	10·9
Capital formation	—	97·2
Contract work	—	72·0
Other items	—	17·6
Gross output	—	1,705·1

Area (in 1,000 ha.) on farms:

	2009	2010	2011
Cereals	39	38	38
Potatoes	5	5	5
Horticulture	3	3	3
Other crops	10	9	8
Grass	791	780	777
Rough grazing	142	141	141
Other land	18	18	19
Total area	1,008	994	991

Livestock (in 1,000 heads) on farms at June census:

	2009	2010	2011
Dairy cows	285	281	283
Beef cows	257	258	270
Other cattle	1,057	1,065	1,038
Ewes	892	876	895
Sows	38	39	38
Laying hens	2,316	2,099	2,430
Broilers	11,418	11,915	14,069

INDUSTRY

Labour

The main sources of employment statistics are the Census of Employment, conducted every two years, and the Quarterly Employment Survey. In June 2007 there were 716,760 employees, of whom 350,010 were males. Employment in services amounted to 567,050 (79% of all employees in employment) and in manufacturing and construction to 132,480 (18%). There were 88,410 people working in manufacturing and 44,070 in construction. The ILO unemployment rate in the period Oct.–Dec. 2011 was 7·2% (compared to 8·4% in the UK as a whole), down from 8·0% in the fourth quarter of 2010.

COMMUNICATIONS

Roads

In June 2009 the total length of public roads was 25,167 km, graded for administrative purposes as follows: motorway, 133 km (including 19 km slip roads); 'A' roads dual carriageway, 168 km; 'A' roads single carriageway, 2,101 km; 'B' roads, 2,883 km; 'C' roads, 4,705 km; unclassified, 15,195 km.

The Northern Ireland Transport Holding Company (NITHC) oversees the provision of public transport services in Northern Ireland. Its subsidiary companies, Ulsterbus, Citybus and Northern Ireland Railways, are responsible for the delivery of most bus and rail services under the brand name of Translink.

At 31 March 2007 there were 2,379 professional hauliers and 6,609 vehicles licensed to engage in road haulage.

The number of motor vehicles licensed at 31 Dec. 2006 was 958,677, including private light goods, 800,969; heavy goods vehicles, 24,806; motorcycles, scooters and mopeds, 27,083.

Rail

Northern Ireland Railways, a subsidiary of the Northern Ireland Transport Holding Company, provides rail services within Northern Ireland and cross-border services to Dublin, jointly with Irish Rail (Iarnród Éireann). The number of track km operated is 340. In 2008–09 railways carried 10·2m. passengers, generating passenger receipts of £29·0m.

Civil Aviation

There are scheduled air services to three airports in Northern Ireland: Belfast International, George Best Belfast City and City of Derry. Scheduled services are provided by easyJet, British Airways, Aer Lingus, Flybe, Jet2.com, United Airlines, Ryanair, Thomas Cook, Thomson and Citywing. In 2009 the airports collectively handled approximately 7·5m. passengers. Belfast International, the busiest airport, is Northern Ireland's main charter airport. In 2005 it launched a direct service to New York with Continental Airlines (now United Airlines). Belfast International handled 4·5m. passengers in 2009.

In 2009 the airports collectively handled approximately 7·5m. passengers. Belfast International, the busiest airport, is Northern Ireland's main charter airport. In 2005 it launched a direct service to New York with Continental Airlines (now United Airlines). Belfast International handled 4·5m. passengers in 2009.

Belfast City Airport offers services to 18 regional airports in Great Britain including services to London Heathrow. The City of Derry Airport provides services from the northwest of Ireland to four United Kingdom destinations including London Stansted. There are two other licensed airfields at St Angelo and Newtownards.

Shipping

There are five commercial ports in Northern Ireland. Belfast is the largest port, competing with Larne for the majority of the passenger and Roll-on Roll-off services that operate to and from Northern Ireland. Passenger services are currently available to Liverpool, Stranraer, Cairnryan and Troon. In addition, Belfast, Londonderry and Warrenpoint ports offer bulk cargo services mostly for British and European markets. They also occasionally service other international destinations direct.

Total tonnage of goods through the principal ports in Northern Ireland in 2009 was 20·8m. tonnes. Belfast handled 12·1m. tonnes of cargo in 2009.

SOCIAL INSTITUTIONS

Justice

The Lord Chief Justice of Northern Ireland is President of the Courts of Northern Ireland and Head of the Judiciary and as such is responsible for assigning the judiciary to the courts. The Lord Chief Justice is also Chairman of the Judicial Appointments Commission which is responsible for selecting and recommending to the Lord Chancellor candidates for judicial appointment in Northern Ireland. The court structure in Northern Ireland has three tiers: the Court of Judicature of Northern Ireland (comprising the Court of Appeal, the High Court and the Crown Court), the County Courts and the Magistrates' Courts. There are 21 Petty Sessions districts which when grouped together for administration purposes form seven County Court Divisions and four Crown Court Circuits.

The County Court has general civil jurisdiction subject to an upper monetary limit. Appeals from the Magistrates' Courts lie to the County Court, or to the Court of Appeal on a point of law, while appeals from the County Court lie to the High Court or, on a point of law, to the Court of Appeal.

Police

Following legislation introduced in the House of Commons in May 2000, the name of the Royal Ulster Constabulary was changed to the Police Service of Northern Ireland (PSNI). The Police Authority for Northern Ireland has been replaced by the Northern Ireland Policing Board. The Police Service continues to undergo significant changes arising from the recommendations of the Commission into the future of policing in Northern Ireland published in 1999. In 2010 the PSNI comprised 7,212 regular officers including those student officers undergoing training, 305 full-time reserve officers and 677 part-time reserves. The proportion of Catholic regular officers, which was around 19·1% by Jan. 2006, had increased to 29·6% by Dec. 2010.

The population in penal institutions in Nov. 2008 was 1,562 (88 per 100,000 population).

Education

Public education, other than university education, is presently administered by the Department of Education, the Department of Employment and Learning, and locally by five Education and Library Boards. The Department of Education is concerned with the range of education from nursery education through to secondary, youth services and for the development of community relations within and between schools. The Department of Employment and Learning is responsible for higher education, further education, student support, postgraduate awards, and the funding of teacher training.

Integrated Schools. The Department of Education has a statutory duty to encourage and facilitate the development of integrated education. It does not seek to impose integration but responds to parental demand for new integrated schools where this does not involve unreasonable public expenditure. The emphasis for future development of the integrated sector has increasingly been on the transformation of existing schools to integrated status. In Dec. 2005 there were 56 grant-aided integrated schools, with a total enrolment of 17,134 pupils, over 5% of all pupils.

Irish Medium Education. Following a commitment in the Belfast Agreement, the 1998 Education Order placed a statutory duty on the Department to encourage and facilitate the development of Irish-medium education. It also provided for the funding of an Irish-medium promotional body, and funding of Irish-medium schools on the same basis as integrated schools. In Dec. 2005 there were 19 Irish-medium primary schools, one post-primary and twelve units, two of which are post-primary catering for 2,935 pupils.

Pre-school Education is provided in nursery schools or nursery classes in primary schools, reception classes and in funded places in voluntary and private settings. There were 100 nursery schools in 2005–06 with 6,175 pupils, and 8,049 nursery pupils in primary schools. A further 754 reception pupils were enrolled in primary schools. In addition there were 5,952 children in funded places in voluntary and private pre-school education centres in 2004–05.

Primary Education is from four to 11 years. In 2005–06 there were 886 primary schools with 158,665 pupils. There were also 17 preparatory departments of grammar schools with 2,478 pupils. In 2004–05 there were 8,353 FTE primary school teachers and 149 FTE preparatory department teachers.

Secondary Education is compulsory from 11 to 16 years. In 2005–06 there were 69 grammar schools with 62,419 pupils and 161 secondary schools with 89,421 pupils. In 2004–05 there were 6,527 FTE secondary school teachers and 4,171 FTE grammar school teachers.

Further Education. There are 16 institutions of further education. In 2003–04 there were 1,703 full-time and 3,068 part-time teachers, approximately 31,000 full-time enrolments and approximately 110,000 part-time assembled enrolments. There were about 72,000 students on non-vocational (mostly evening) courses.

Special Education. The Education and Library Boards provide for children with special educational needs up to the age of 19. This provision may be made in ordinary classes in primary or secondary schools or in special units attached to those schools, or in special schools. In 2005–06 there were 48 special schools with 4,895 pupils. This includes three hospital schools.

Universities. There are two universities: the Queen's University of Belfast (founded in 1849 as a college of the Queen's University of Ireland and reconstituted as a separate university in 1908), which had 23,675 students, 1,368 full-time and 81 part-time academic staff in 2003–04; and the University of Ulster, formed on 1 Oct. 1984, which has campuses in Belfast, Coleraine, Jordanstown and Londonderry. In 2003–04 it had 26,202 students, 1,112 full-time and 360 part-time academic staff.

Full-Time Initial Teacher Education takes place at both universities and at two university colleges of education—Stranmillis and St Mary's—the latter mainly for the primary school sector, in respect of which four-year (Hons) BEd courses are available. The training of teachers for secondary schools is provided, in the main, in the education departments of the two universities, but four-year (Hons) BEd courses are also available in the colleges for intending secondary teachers of religious education, business studies and craft, design and technology. There were a total of 2,015 students (1,642 women) in training at the two university colleges and the two universities during 2003–04.

Health

The Department of Health, Social Services and Public Safety has three main business responsibilities: Health and Personal Social Services (HPSS), which includes policy and legislation for hospitals, family practitioner services, and community health and personal social services; Public Health, which covers policy, legislation and administrative action to promote and protect the health and well-being of the population; and Public Safety, which covers policy and legislation for fire, rescue and ambulance services.

The four Health and Social Services Boards commission health and personal social services for their resident populations from providers including HSS Trusts, and voluntary and private sector bodies. The 19 Health and Social Services Trusts, established under the Health and Personal Social Services (NI) Order 1991, are managerially independent but accountable to the Minister. They are the main providers of health and personal social services as commissioned by the HSS Boards and are responsible for the management of staff and services of hospitals and other health and personal social services establishments. Seven provide hospital services only, five provide community and personal social services only and six provide both. In 2007 there were 1,127 doctors (principals) with an average of 1,626 patients each.

The Northern Ireland Health and Social Welfare Survey 2005/06 found that around 24% of adults were obese (having a body mass index over 30).

Welfare

The Social Security Agency's remit is now part of the Department for Social Development, and social security schemes are similar to those in Great Britain.

National Insurance. During the year ended 31 March 2009 the receipts of the National Insurance Fund at £2,293m. exceeded payments by £314m. Total benefit expenditure was £1,879m. Jobseeker's Allowance contributions amounted to £19·5m. State Pensions amounted to £1,463·4m. and Bereavement Benefits to

£24·9m. Incapacity Benefits totalled £336·7m. Maternity Allowance of £9·2m. was paid and employers were reimbursed £59·0m. in respect of Statutory Sick, Maternity, Adoption and Paternity Pay. £50·6m. was given to personal pension plan providers.

Child Benefit. In 2008–09 a total of £362m. was paid. 195,000 families benefited from Child and Working Tax Credits.

RELIGION

According to the 2001 census there were: Roman Catholics, 678,462; Presbyterians, 348,742; Church of Ireland, 257,788; Methodists, 59,173; other Christian, 102,221; other religions and philosophies, 5,028. There were also 233,853 persons with no religion or religion was not stated.

CULTURE

World Heritage Sites
Northern Ireland has one site on UNESCO's World Heritage List: the Giant's Causeway and Causeway Coast (inscribed on the list in 1986), rock formations on the Antrim Plateau.

Tourism
There were 2·11m. visits to Northern Ireland in 2007, contributing £376m. to the economy. Domestic holiday makers contributed a further £134m. Nine Areas of Outstanding Natural Beauty and 47 National Nature Reserves have been declared, and there are many country and regional parks. The leading paid admission attraction

in 2009 was the Giant's Causeway Visitor Centre with 715,000 visits. The leading free admission attraction was Oxford Island National Nature Reserve at Craigavon with 325,000 visits in 2009.

FURTHER READING

Adshead, Maura and Tonge, Jonathan, *Politics in Ireland: Convergence and Divergence in a Two-Polity Island.* 2009
Aughey, A. and Morrow, D. (eds.) *Northern Ireland Politics.* 1996
Bardon, Jonathan, *A History of Ulster.* 1992
Bloomfield, D., *Peacemaking Strategies in Northern Ireland.* 1998
Bourke, Richard, *Peace in Ireland: The War of Ideas.* 2003
Bow, P. and Gillespie, G., *Northern Ireland: a Chronology of the Troubles, 1968–1993.* 1993
Dixon, Paul, *Northern Ireland: The Politics of War and Peace.* 2nd ed. 2008
Fay, Marie-Thérèse, Morrisey, Mike and Smyth, Marie, *Northern Ireland's Troubles.* 1999
Hennessey, T., *A History of Northern Ireland 1920–96.* 1998
Loughlin, James, *The Ulster Question Since 1945.* 1998
McGarry, J. and O'Leary, B. (eds.) *Explaining Northern Ireland: Broken Images.* 1995
Neumann, Peter R., *Britain's Long War: British Strategy in the Northern Ireland Conflict, 1969–98.* 2003
Patterson, Henry, *Ireland Since 1939: The Persistence of Conflict.* 2006
Rose, Peter, *How the Troubles Came to Northern Ireland.* 1999
Ruane, J. and Todd, J., *The Dynamics of Conflict in Northern Ireland: Power, Conflict and Emancipation.* 1997
Tonge, Jonathan, *The New Northern Irish Politics?* 2004

Statistical office: Northern Ireland Statistics and Research Agency (NISRA), McAuley House, 2–14 Castle St., Belfast BT1 1SA.
Website: http://www.nisra.gov.uk

ISLE OF MAN

KEY HISTORICAL EVENTS

The Isle of Man was first inhabited 10,000 years ago. Part of Norway in the 9th century, in 1266 it was ceded to Scotland but came under English control in 1333.

The Isle of Man has been a British Crown dependency since 1765, with the British government responsible for its defence and foreign policy. Otherwise it has extensive right of self-government.

A special relationship exists between the Isle of Man and the European Union providing for free trade, and adoption by the Isle of Man of the EU's external trade policies with third countries. The island remains free to levy its own taxes.

TERRITORY AND POPULATION

Area, 572 sq. km (221 sq. miles); resident population census in March 2011, 84,497, giving a density of 147·7 per sq. km. In 2006 an estimated 76% of the population lived in urban areas. The principal towns are (March 2011 population) Douglas (27,935), Onchan (adjoining Douglas; 9,283), Ramsey (7,809) and Peel (5,092). The island is divided into six sheadings—Ayre, Garff, Glenfaba, Michael, Middle and Rushen. Garff is further subdivided into two parishes and the others each have three parishes. Just over half the population at the 2006 census was born outside of the island.

SOCIAL STATISTICS

2011: births, 938; deaths, 816; marriages (2004), 399. Annual growth rate, 2001–06, 1·0%.

CLIMATE

Lying in the Irish Sea, the island's climate is temperate and lacking in extremes. Thunderstorms, snow and frost are infrequent, although the island tends to be windy. July and Aug.

are the warmest months with an average daily maximum temperature of around 17·6°C (63°F).

CONSTITUTION AND GOVERNMENT

As a result of Revestment in 1765, the Isle of Man became a dependency of the British Crown. The UK government is responsible for the external relations of the island, including its defence and international affairs, and the island makes a financial contribution to the cost of these services. The Isle of Man has a special relationship with the European Union. It neither contributes funds to, nor receives money from, the EU. The Isle of Man is not represented in either the UK or European Parliaments.

The island is administered in accordance with its own laws by the High Court of *Tynwald*, consisting of the President of Tynwald, the *Legislative Council* and the *House of Keys*. The Legislative Council is composed of the Lord Bishop of Sodor and Man, eight members selected by the House of Keys and the Attorney General, who has no vote. The House of Keys is an assembly of 24 members chosen by adult suffrage. The minimum age for voting was lowered to 16 in 2006. The President of Tynwald is chosen by the Legislative Council and the House of Keys, sitting together as Tynwald. An open-air Tynwald ceremony is held in early July each year at St Johns. Until 1990 the Lieut.-Governor, appointed by the UK government, presided over Tynwald.

A Council of Ministers was instituted in 1990, replacing the Executive Council which had acted as an advisory body to the Lieut.-Governor. The Council of Ministers consists of the Chief Minister (elected for a five-year term) and the ministers of the nine major departments, being the Treasury; Agriculture, Fisheries and Forestry; Education; Health and Social Security; Home Affairs; Local Government and the Environment; Tourism and Leisure; Trade and Industry; and Transport.

RECENT ELECTIONS

Elections to the House of Keys were held on 29 Sept. 2011. Independents took 21 of the 24 seats and the Liberal Vannin Party 3. Turnout was 54%.

CURRENT GOVERNMENT

Lieut.-Governor: Adam Wood.
   President: Clare Christian (elected July 2011).
   In March 2013 the *Chief Minister* was Allan Bell. *Treasury Minister:* Eddie Teare.

Website: http://www.gov.im

ECONOMY

Currency

The Isle of Man government issues its own notes and coins on a par with £ sterling. Various commemorative coins have been minted. Inflation was 3·0% in 2006.

Budget

The Isle of Man is statutorily required to budget for a surplus of revenue over expenditure. Revenue is raised from income tax, taxes on expenditure, health and social security contributions, and fees and charges for services.

The standard rate of tax is 10% for personal income, and there is a higher rate of 18%. Banking and land property businesses are liable at 10% on their first £100m. of taxable income and 18% on the balance.

There is a Customs and Excise Agreement with the UK, and rates of tax on expenditure are the same as those in the UK with very few exceptions. In addition, there is a reciprocal agreement on social security with the UK, and the rates of most health and social security (National Insurance) contributions are the same as in the UK.

In 2007–08 the Isle of Man government budgeted for expenditure of £809m. and revenue of £845m.

Performance

In 2007 GDP was £4,076m. Real GDP growth in 2005–06 was 5·7%. Just over 80% of national income is generated from services with the finance sector being the single largest contributor (36%).

Banking and Finance

The banking sector is regulated by the Financial Supervision Commission which is responsible for the licensing and supervision of banks, deposit-takers and financial intermediaries giving financial advice, and receiving client monies for investment and management. A compensation fund to protect investors was set up in 1991 under the Commission.

In Dec. 2006 the deposit base was £43bn., and there were 42 licensed banks, 65 investment businesses and three building societies with Isle of Man licences.

The insurance industry is regulated by the Insurance and Pensions Authority. In June 2006 there were 178 insurance companies.

ENERGY AND NATURAL RESOURCES

Electricity

The Manx Electricity Authority generates most of the island's electricity by oil-fired power stations although there is a small hydro-electric plant. A cable link with the UK power grid came into operation in Nov. 2000. In 2006, 379m. kWh were sold.

Oil and Gas

All oil and gas needs are met from imports, with gas being imported via a link to the Scotland–Eire gas pipeline. The island's gas suppliers and distributors are in the private sector.

Minerals

Although lead and tin mining industries were major employers in the past, they have long since shut down and the only mining activity in the island is now for aggregates. The Lady Isabella, built in 1854 to drain the mines above Laxey, is one of the largest waterwheels in Europe.

Agriculture

The area farmed is about 104,000 acres, being 74% of a total land area of around 141,500 acres. 65,000 acres are grassland with a further 27,000 acres for rough grazing. There are approximately 137,000 sheep, 30,000 cattle, 20,000 poultry and 600 pigs on the island's 658 farms. Agriculture now contributes less than 2% of the island's GDP.

Forestry

The Department of Agriculture, Fisheries and Forestry has a forestry estate of some 6,800 acres. Commercial forestry is directed towards softwood production. The Manx National Glens and other amenity areas are maintained for public use by the Department, which owns some 18,000 acres of the island's hills and uplands open for public use.

Fisheries

The Isle of Man is noted for the Manx kipper, a gutted smoked herring. Scallops and the related queen scallops (queenies) are the economic mainstay of the Manx fishing fleet. In 2006 the total catch was about 2,000 tonnes.

INDUSTRY

Labour

The economically active population in 2006 was 41,793, of whom 6,381 were self-employed and 1,010 were unemployed. Employment by sector: finance, 23%; professional services, 20%; distributive services, 11%; construction, 8%; manufacturing, 5%.

At the end of 2006 there were 577 persons on the unemployment register, giving an unemployment rate of 1·4%.

INTERNATIONAL TRADE

The Isle of Man forms part of the customs union of the European Union, although the island is not part of the EU itself. The relationship with the EU provides for free trade and the adoption of the EU's external trade policies and tariffs with non-EU countries.

Imports and Exports

The Isle of Man is in customs and excise union with the United Kingdom, which is also its main trading partner.

COMMUNICATIONS

Roads

There are 800 km of good roads. At the end of March 2010 there were 81,985 licensed vehicles, with 72,621 of these being private cars. Omnibus services operate to all parts of the island. The TT (Tourist Trophy) motorcycle races take place annually on the 60·75-km Mountain Circuit.

Rail

Several novel transport systems operate on the island during the summer season from May to Sept. Horse-drawn trams run along Douglas promenade, and the Manx Electric Railway links Douglas, Laxey, Ramsey and Snaefell Mountain (621 metres) in the north. The Isle of Man Steam Railway also operates between Douglas and Port Erin in the south.

Civil Aviation

Ronaldsway Airport in the south handles scheduled services linking the island with Belfast, Birmingham, Blackpool, Bristol, Brussels, Dublin, East Midlands, Edinburgh, Glasgow, Jersey, Leeds, Liverpool, London, Manchester, Prestwick and Southampton. Air taxi services also operate.

Shipping

Car ferries run between Douglas and the UK and the Irish Republic. In 2006 there were 316 merchant vessels on the island's shipping register.

Telecommunications

Manx Telecom Limited, a wholly owned subsidiary of O2, holds the telecommunications licence issued by the Communications Commission for the Isle of Man.

SOCIAL INSTITUTIONS

Justice

The First Deemster is the head of the Isle of Man's judiciary. The Isle of Man Constabulary numbered 236 all ranks in 2007.

The average size of the prison population during 2006 was 72·8, equivalent to 91 per 100,000 of national population. A further four persons are serving their sentences in the United Kingdom.

Education

Education is compulsory between the ages of five and 16. In 2006 there were 6,610 pupils in the 35 infant and junior schools and 5,667 pupils in the five secondary schools operated by the Department of Education. The Department also runs a college of further education and a special school. Government expenditure on education was budgeted to be £100m. in 2007–08. The island has a private primary school, a private secondary school and an international business school.

Health

The island has had its own National Health Service since 1948, providing medical, dental and ophthalmic services. In 2007–08 government expenditure on the NHS was budgeted to be £129m. There are two hospitals, one of which opened in 2003. In 2006 there were 120 full-time equivalent physicians, 41 full-time and seven part-time general practitioners, 42 full-time and two part-time dentists, and 24 pharmacies.

Welfare

Numbers receiving certain benefits at July 2007: Retirement Pension, 16,713; Child Benefit, 9,781; Sick and Disablement Benefits, 5,518; Income Support, 3,008; Jobseekers' Allowance, 163. Total government expenditure on the social security system in 2007–08 was budgeted to be £201m.

RELIGION

The island has a rich heritage of Christian associations, and the Diocese of Sodor and Man, one of the oldest in the British Isles, has existed since 476.

CULTURE

Press

In 2007 there were three weekly newspapers and one quarterly newspaper. There are also various magazines concentrating on Manx issues.

Tourism

During the late 19th century through to the middle of the 20th century, tourism was one of the island's main sources of income and employment. Tourism now contributes around 5% of the island's GDP. There were 219,000 visitors during 2006.

FURTHER READING

Additional information is available from: Economic Affairs Division, Illiam Dhone House, 2 Circular Rd, Douglas, Isle of Man, IM1 1PQ. *Email:* economics@gov.im
Isle of Man Digest of Economic and Social Statistics. Annual

Belchem, J. (ed.) *A New History of the Isle of Man, Volume V—The Modern Period 1830–1999.* 2000
Kermode, D. G., *Offshore Island Politics: The Constitutional and Political Development of the Isle of Man in the Twentieth Century.* 2001
Moore, A. W., *A History of the Isle of Man.* 1900; reprinted 1992

Manx National Heritage publishes a series of booklets including *Early Maps of the Isle of Man, The Art of the Manx Crosses, The Ancient & Historic Monuments of the Isle of Man, Pre-historic Sites of the Isle of Man.*

CHANNEL ISLANDS

KEY HISTORICAL EVENTS

The Channel Islands consist of Jersey, Guernsey and the following dependencies of Guernsey: Alderney, Brechou, Great Sark, Little Sark, Herm, Jethou and Lihou. They were an integral part of the Duchy of Normandy at the time of the Norman Conquest of England in 1066. Since then they have belonged to the British Crown and are not part of the UK. The islands have created their own self-government, with the British government at Westminster being responsible for defence and foreign policy. The Lieut.-Governors of Jersey and Guernsey, appointed by the Crown, are the personal representatives of the Sovereign as well as being the commanders of the armed forces. The legislature of Jersey is 'The States of Jersey', and that of Guernsey is 'The States of Deliberation'.

Left undefended from 1940 to 1945 the islands were the only part of Britain to fall to Germany.

TERRITORY AND POPULATION

The Channel Islands cover a total of 194 sq. km (75 sq. miles), and in 2008 had a population of approximately 150,000.

The official languages are French and English, but English is now the main language.

CLIMATE

The climate is mild, with an average temperature for the year of 11·5°C. Average yearly rainfall totals: Jersey, 862·9 mm; Guernsey, 858·9 mm. The wettest months are in the winter. Highest temperatures recorded: Jersey (St Helier), 36·0°C; Guernsey (airport), 33·7°C. Maximum temperatures usually occur in July and Aug. (daily maximum 20·8°C in Jersey, slightly lower in Guernsey). Lowest temperatures recorded: Jersey, –10·3°C; Guernsey, –7·4°C. Jan. and Feb. are the coldest months (mean temperature approximately 6°C).

CONSTITUTION AND GOVERNMENT

The Lieut.-Governors and Cs.-in-C. of Jersey and Guernsey are the personal representatives of the Sovereign, the Commanders of the Armed Forces of the Crown, and the channel of communication between the Crown and the insular governments. They are appointed by the Crown and have a voice but no vote in the islands' legislatures. The Secretaries to the Lieut.-Governors are their staff officers.

ECONOMY

Performance

Total GDP in 2007 was US$11·5bn.

ENERGY AND NATURAL RESOURCES

Fisheries

Total catch in 2010 was 3,373 tonnes, exclusively from sea fishing.

EXTERNAL ECONOMIC RELATIONS

The Channel Islands are not members of the European Union but under a special relationship accept a number of European laws, including certain EU customs regulations. Trade with the UK is classed as domestic.

COMMUNICATIONS

Civil Aviation

Scheduled air services are maintained by Aer Lingus, Air Berlin, Aurigny Air Services, Blue Islands, British Airways, Citywing, easyJet, Flybe, Jet2.com and Lufthansa.

Shipping

Passenger and cargo services between Jersey, Guernsey, England (Poole, Portsmouth and Weymouth) and France (St Malo) are maintained by Condor Ferries. Local companies run between Guernsey, Alderney and England, and between Guernsey and Sark.

SOCIAL INSTITUTIONS

Justice

Justice is administered by the Royal Courts of Jersey and Guernsey, each of which consists of the Bailiff and 12 Jurats (magistrates), the latter being elected by an electoral college. There is an appeal from the Royal Courts to the Courts of Appeal of Jersey and of Guernsey. A final appeal lies to the Privy Council in certain cases. A stipendiary magistrate in each, Jersey and Guernsey, deals with minor civil and criminal cases.

RELIGION

Jersey and Guernsey each constitutes a deanery under the jurisdiction of the Bishop of Winchester. The rectories (12 in Jersey; 10 in Guernsey) are in the gift of the Crown. The Roman Catholic and various Nonconformist Churches are represented.

FURTHER READING

Lemprière, R., *History of the Channel Islands.* Rev. ed. 1980
Turner, Barry, *Outpost of Occupation: How the Channel Islands Survived Nazi Rule, 1940–45.* 2010

Jersey

TERRITORY AND POPULATION

The area is 118·2 sq. km (45·6 sq. miles). Resident population (2011 census), 97,857 (49,561 females); density, 828 per sq. km. The chief town is St Helier on the south coast. It had a population of 33,522 in 2011. The official language is English (French until 1960). The island has its own language, known as Jersey French, or Jérriaise. French, Portuguese and Polish are also spoken.

SOCIAL STATISTICS

In 2009 there were 1,010 births and 758 deaths. In 2009 there were 541 marriages. Life expectancy, 2006–09: males, 72 years; females, 79 years.

CONSTITUTION AND GOVERNMENT

The island parliament is the *States of Jersey.* The States comprises the Bailiff, the Lieut.-Governor, the Dean of Jersey, the Attorney-General and the Solicitor-General, and 51 members elected by universal suffrage: ten Senators (four elected for a three-year term only), the Constables of the 12 parishes (every third year) and 29 Deputies (every third year). They all have the right to speak in the Assembly, but only the 51 elected members have the right to vote; the Bailiff has a casting vote. Except in specific instances,

enactments passed by the States require the sanction of The Queen-in-Council. The Lieut.-Governor has the power of veto on certain forms of legislation.

A new post of Chief Minister was inaugurated in 2005. The Chief Minister, who is elected by the States, presides over a nine-member Council of Ministers responsible for government policy.

RECENT ELECTIONS

On 14 Nov. 2011 parliament elected Ian Gorst Chief Minister by 27 votes to 24.

CURRENT GOVERNMENT

Lieut.-Governor and C.-in-C. of Jersey: Gen. Sir John McColl, KCB, CBE, DSO.
    Secretary and Aide-de-Camp to the Lieut.-Governor: Lieut. Col. C. Woodrow, OBE, MC, QGM, LVO.
    Bailiff of Jersey and President of the States: Michael Birt.
    Chief Minister: Ian Gorst.

Government Website: http://www.gov.je

ECONOMY

Currency

The States issue banknotes in denominations of £50, £20, £10, £5 and £1. Coinage from 1p to 50p is struck in the same denominations as the UK. There were £75·6m. worth of States of Jersey banknotes and £6·7m. worth of coinage in circulation in 2007. Inflation in Sept. 2007 was 3·9%.

Budget

2009 net general revenue income, £674m.; expenditure, £565m. Income from taxation was forecast to be £507m.
    Parochial rates are payable by owners and occupiers.

Performance

In 2009 total GDP was £3·7bn.

Banking and Finance

In 2007 there were 48 banks; combined deposits were £219·5bn. There were 4,050 registered companies in 2007. Jersey is a leading offshore financial centre.
    A 0% rate of company tax was introduced on 1 Jan. 2009.

ENERGY AND NATURAL RESOURCES

Agriculture

2008 total agricultural exports, £29,397,944. Jersey Royal New Potatoes account for 79% of Jersey's agricultural exports to the UK. 55·2% of the island's land area was farmed commercially in 2008. In 2008 there were 28 dairy farmers, 5,092 cows and heifers in milk, 703 sheep and 615 pigs.

Fisheries

There were 160 fishing vessels in 2010. The total catch of shellfish in 2009 was 1,181 tonnes and of wet fish 72 tonnes. Aquaculture production (oysters, mussels and scallops) totalled 1,007 tonnes in 2009. The value of the fishing industry at first sale in 2010 was £7m.

INDUSTRY

Principal activities: light industry, mainly electrical goods, textiles and clothing.

Labour

In Dec. 2011, 53,790 persons were economically active; 1,540 persons were registered unemployed. Financial and legal services was the largest employment sector, followed by wholesale and retail trades and the public sector.

EXTERNAL ECONOMIC RELATIONS

Imports and Exports

Since 1980 the Customs have ceased recording imports and exports. Principal imports: machinery and transport equipment, manufactured goods, food, mineral fuels, and chemicals. Principal exports: machinery and transport equipment, food, and manufactured goods.

COMMUNICATIONS

Roads

In 2009 there were 113,605 registered motor vehicles. There were 1·57 cars/vans per private household in 2008.

Civil Aviation

Jersey airport is situated at St Peter. It covers approximately 375 acres. In 2011 the airport handled 1,474,373 passengers.

Shipping

All vessels arriving in Jersey from outside Jersey waters report at St Helier or Gorey on first arrival. There is a harbour of minor importance at St Aubin. There were 730,000 passenger arrivals and departures in 2009; 395,000 tonnes of freight (exports and imports) were shipped through St Helier harbour.

Telecommunications

Postal, and overseas telephone and telegraph services, are maintained by the Postal Administration of Jersey. The local telephone service is maintained by the Insular Authority. In 2009 main telephone lines numbered 73,900. There were 83,900 mobile phone subscribers in 2004.

SOCIAL INSTITUTIONS

Justice

Justice is administered by the Royal Court, consisting of the Bailiff and 12 Jurats (magistrates). There is a final appeal in certain cases to the Sovereign in Council. There is also a Court of Appeal, consisting of the Bailiff and two judges. Minor civil and criminal cases are dealt with by a stipendiary magistrate.

In 2009 there were 4,529 crimes recorded. Customs and Excise was responsible for 108 drug seizures with a street value of £2·6m. in 2009 (up from 92 seizures and a street value of £1·3m. in 2008). In 2009 the daily average prison population was 184.

Education

In 2007 there were six States secondary schools (two fee-paying), one high school and four special needs secondary schools. There were 24 States primary schools (two fee-paying). 4,196 pupils attended secondary schools and 5,675 attended primary schools, and there were 288 pupils with special needs. There were five private primary schools with 1,330 pupils and three private secondary schools with 1,078 pupils. Public expenditure on education amounted to 2·8% of GNI in 2006.

Health

Gross revenue expenditure on health and social services in 2011 was £188,687,500. In 2010 there was one general hospital with about 245 beds. In 2011 there were 95 doctors (general practitioners).

Welfare

A contributory Health Insurance Scheme is administered by the Social Security Department. In 2009 expenditure on benefits paid out from the Social Security Fund was £172m., and income from contributions, supplementation, interest and rent was £217m. Over 25,000 people received Old Age Pension and 1,012 claimed Maternity Allowance.

CULTURE

Tourism

In 2011 there were 689,700 visitors to the island, spending £242m.

FURTHER READING

Balleine, G. R., *A History of the Island of Jersey*. Rev. ed. 1981

States of Jersey Library: Halkett Place, St Helier.

Statistical Office: Statistics Unit, P. O. Box 140, Cyril Le Marquand House, The Parade, St Helier, Jersey, JE4 8QT.
Website: http://www.gov.je/Government/JerseyWorld/StatisticsUnit

---

Guernsey

TERRITORY AND POPULATION

The area is 63·4 sq. km. Population (March 2011) 62,915. The main town is St Peter Port (2001 population of 16,488).

English is the most widely spoken language. A Norman-French dialect (called Guernsey French or Guernesiais) is spoken by a small number of (mainly older) people. It is now being reintroduced into some school curriculums.

SOCIAL STATISTICS

Births during 2011 were 605; deaths, 503.

CONSTITUTION AND GOVERNMENT

The States of Deliberation, the Parliament of Guernsey, is composed of the following members: the Bailiff, who is President *ex officio*; H.M. Procureur and H.M. Comptroller (Law Officers of the Crown), who have a voice but no vote; 45 People's Deputies elected by popular franchise; ten Douzaine Representatives elected by their Parochial Douzaines; two representatives of the States of Alderney. Since May 2004 there has been a slimmed-down States of Deliberation, and an executive form of government has been introduced. For the first time a Chief Minister has been appointed. There are also ministers, a deputy chief minister, members of departmental committees, chairmen and members of committees.

The States of Election, an electoral college, elects the Jurats (magistrates). It is composed of the following members: the Bailiff (President *ex officio*); the 12 Jurats or 'Jurés-Justiciers'; H.M. Procureur and H.M. Comptroller; the 45 People's Deputies and 34 representatives from the 10 Parochial Douzaines.

Since Jan. 1949 all legislative powers and functions (with minor exceptions) formerly exercised by the Royal Court have been vested in the States of Deliberation. Projets de Loi (Bills) require the sanction of The Queen-in-Council.

RECENT ELECTIONS

Elections for People's Deputies were held on 18 April 2012.

CURRENT GOVERNMENT

Lieut.-Governor and C.-in-C. of Guernsey and its Dependencies: Air Marshal Peter Walker CB, CBE.

Secretary and Aide-de-Camp to the Lieut.-Governor: Colonel R. H. Graham, MBE.

Bailiff of Guernsey and President of the States: Richard Collas.

Chief Minister: Peter Harwood.

Government Website: http://www.gov.gg

ECONOMY

Budget

Year ended 31 Dec. 2007: revenue £365,000,000; expenditure, £294,000,000. The standard rate of income tax is 20p in the pound. States and parochial rates are very moderate. No super-tax or death duties are levied.

Banking and Finance

In March 2008 there were 346 employers in the finance and legal sector, employing a total of 7,893 people. Financial services accounts for approximately 35% of Guernsey's GDP. Guernsey is a leading offshore financial centre.

The general rate of income tax payable by Guernsey companies, formerly 20%, has been 0% since 1 Jan. 2008.

EXTERNAL ECONOMIC RELATIONS

Imports and Exports

In 2007, 125,538,000 litres of oil were imported. Horticulture exports (2006) in £1m.: plants, 34·95; postal flowers, 6·62; food, 3·26; cut flowers, 2·08.

COMMUNICATIONS

Civil Aviation

The airport is situated at La Villiaze. There were direct flights in 2012 to Alderney, Birmingham, Bristol, Dinard, Dublin, Düsseldorf, East Midlands, Exeter, Geneva, Hannover, Isle of Man, Jersey, London (Gatwick and Stansted), Manchester, Rotterdam, Southampton and Zürich. In 2008 passenger movements totalled 918,978.

Shipping

The principal port is St Peter Port. There is also a harbour at St Sampson's (mainly for commercial shipping). In 2008 sea passenger movements totalled 333,865. There were 277 fishing vessels registered in 2008 and more than 5,000 other craft.

Telecommunications

There were 45,100 main telephone lines in 2006, or 898 per 1,000 population. Mobile phone subscribers numbered 43,800 in 2004. Guernsey Telecom was sold to Cable and Wireless in May 2002 and now trades as C & W Guernsey.

SOCIAL INSTITUTIONS

Justice

The total number of criminal offences reported to the police in 2009 was 2,954, up from 2,648 in 2008 but down from 3,465 in 2005. In 2009 the average prison population was 81 (87 in 2008).

Education

There are two public schools, one grammar school, a number of modern secondary and primary schools, and a College of Further Education. The total number of schoolchildren in Jan. 2008 was 9,033. Facilities are available for the study of art, domestic science and many other subjects of a technical nature.

Health

Guernsey is not covered by the UK National Health Service. In 2012 acute services were based at the Princess Elizabeth Hospital, elderly care services at King Edward VII Hospital and mental health services at Castel Hospital. There are plans to build a new mental health facility, potentially to open in 2015.

CULTURE

Press

The *Guernsey Evening Press* is published daily except Sundays.

Tourism

There were 316,000 visitors to Guernsey in 2006 (332,000 in 2005). In 2006 there were 220,000 holidaymakers (218,000 in 2005), 54,000 people visiting friends and relatives (70,000 in 2005) and 36,000 business visitors (41,000 in 2005).

FURTHER READING

Marr, L. J., *A History of Guernsey.* 1982

Statistical office: Policy and Research Unit, Sir Charles Frossard House, La Charroterie, St. Peter Port, GY1 1FH.
Website: http://www.gov.gg/pru

Alderney

GENERAL DETAILS

Population (2001 estimate), 2,400. The main town is St Anne's. The island has an airport.

The Constitution of the island (reformed 1987) provides for its own popularly elected President and States (10 members), and its own Court. Elections were held for the five members of the States in Dec. 2008. Alderney levies its taxes at Guernsey rates and passes the revenue to Guernsey, which charges for the services it provides.

President of the States: Stuart Trought.
Chief Executive: Roy Burke.
Greffier: Sarah Kelly.

FURTHER READING

Coysh, V., *Alderney.* 1974

Sark

GENERAL DETAILS

2001 population estimate, 580. In order to comply with European human rights legislation, the constitution was amended in Jan. 2008 to make the Chief Pleas (parliament) democratically electable. Previously 40 out of 52 seats were reserved for landowners. Elections took place in Dec. 2008 for a fully-elected 28-seat chamber that held its first session on 21 Jan. 2009. In addition, the powers of the Seigneur (who is head of the island) have been restricted. These changes are a decisive move away from the previously feudal system. Sark has no income tax. Motor vehicles, except tractors, are not allowed.

Seigneur: J. M. Beaumont.
Seneschal: R. J. Guille.

FURTHER READING

Hathaway, S., *Dame of Sark: An Autobiography.* 1961

UNITED KINGDOM OVERSEAS TERRITORIES

There are 14 British Overseas Territories: Anguilla; Bermuda; British Antarctic Territory; British Indian Ocean Territory; British Virgin Islands; Cayman Islands; Falkland Islands; Gibraltar; Montserrat; Pitcairn Islands; St Helena, Ascension and Tristan da Cunha; South Georgia and the South Sandwich Islands; the Sovereign Base Areas of Akrotiri and Dhekelia in Cyprus; and the Turks and Caicos Islands. Three (British Antarctic Territory, British Indian Ocean Territory and South Georgia and the South Sandwich Islands) have no resident populations and are administered by a commissioner instead of a governor.

Gibraltar is a peninsula bordering the south coast of Spain; the Sovereign Base Areas are in Cyprus and the remainder are islands in the Caribbean, Pacific, Indian Ocean and South Atlantic. Gibraltar and the Falkland Islands are the subjects of territorial claims by Spain and Argentina respectively.

The Overseas Territories are constitutionally not part of the United Kingdom. They have separate constitutions, and most of them have elected governments with varying degrees of responsibilities for domestic matters. The Governor, who is appointed by, and represents, HM the Queen, retains responsibility for external affairs, internal security, defence, and in most cases the public service.

At the launch of the White Paper 'Partnership for Progress and Prosperity', in March 1999, the UK Foreign Secretary at the time, Robin Cook, outlined four underlying principles for the relationship between Britain and the Overseas Territories: self-determination for the Territories; mutual obligations and responsibilities; freedom for the Territories to run their own affairs to the greatest degree possible; and Britain's firm commitment to help the territories develop economically and to assist them in emergencies. He also offered British citizenship, with the right of abode in the UK, to those citizens of the Overseas Territories who did not already enjoy it. The Overseas Territories Consultative Council was established in 1999. The Council, which meets annually, is a forum for discussion of key policy issues between British government ministers and heads of territory governments. On 21 May 2002 the citizenship provisions of the British Overseas Territories Act came into force. It granted British citizenship to the citizens of all Britain's Overseas Territories (except those who derived their British nationality by virtue only of a connection with the Sovereign Base Areas of Akrotiri and Dhekelia in Cyprus).

Anguilla

KEY HISTORICAL EVENTS

Anguilla was probably given its name by the Spaniards or the French because of its eel-like shape. It was inhabited by Arawaks for several centuries before the arrival of Europeans. Anguilla was colonized in 1650 by English settlers from neighbouring St Kitts. In 1688 the island was attacked by a party of Irishmen who then settled. Anguilla was subsequently administered as part of the Leeward Islands and from 1825 became even more closely associated with St Kitts. In 1875 a petition sent to London requesting separate status and direct rule from Britain was rejected. Again, in 1958, the islanders petitioned the Governor requesting a dissolution of the political and administrative association with St Kitts, but this too failed. From 1958 to 1962 Anguilla was part of the Federation of the West Indies.

Opposition to rule from St Kitts erupted on 30 May 1967 when St Kitts policemen were evicted from the island and Anguilla refused to recognize the authority of the State government any longer. During 1968–69 the British government maintained a 'Senior British Official' to advise the local Anguilla Council and devise a solution to the problem. In March 1969, following the ejection from the island of a high-ranking British civil servant, British security forces occupied Anguilla. A Commissioner was installed, and in 1969 Anguilla became de facto a separate dependency of Britain, a situation rendered de jure on 19 Dec. 1980 under the Anguilla Act 1980 when Anguilla formally separated from the territory of St Kitts-Nevis-Anguilla. A new constitution came into effect in 1982 providing for a large measure of autonomy under the Crown.

TERRITORY AND POPULATION

Anguilla is the most northerly of the Leeward Islands, some 112 km (70 miles) to the northwest of St Kitts and 8 km (5 miles) to the north of St Martin/Sint Maarten. The territory also comprises the island of Sombrero and several other off-shore islets or cays. The total area of the territory is about 155 sq. km (60 sq. miles). May 2011 census population (provisional) was 13,452; density of 86·8 per sq. km. Average annual population increase between 2001 and 2011 was 1·5%. People of African descent make up 90% of the population, mixed origins 5% and white 4%. The capital is The Valley.

The official language is English.

SOCIAL STATISTICS

Births, 2001, 183; deaths, 66. In 2001 life expectancy at birth for females was 78·0 years and for males 77·9 years. Households numbered 3,788 in 2001.

CLIMATE

Tropical oceanic climate with rain throughout the year, particularly between May and Dec. Tropical storms and hurricanes may occur between July and Nov. Generally summers are hotter than winters although there is little variation in temperatures.

CONSTITUTION AND GOVERNMENT

A set of amendments to the constitution came into effect in 1990, providing for a Deputy Governor, a Parliamentary Secretary and an Opposition Leader. The House of Assembly consists of a Speaker, Deputy Speaker, seven directly elected members for five-year terms, two nominated members and two ex officio members: the Deputy Governor and the Attorney-General. The Governor discharges his executive powers on the advice of an Executive Council comprising a Chief Minister, three Ministers and two ex officio members: the Deputy Governor, Attorney-General and the Secretary to the Executive Council.

RECENT ELECTIONS

In parliamentary elections held on 15 Feb. 2010 the Anguilla United Movement won four of seven seats, the United Front (Anguilla National Alliance and Anguilla Democratic Party) two and the Anguilla Progressive Party one. Turnout was 82·1%.

CURRENT GOVERNMENT

Governor: Alistair Harrison (took office on 21 April 2009).

Chief Minister: Hubert Hughes; b. 1933 (Anguilla United Movement; sworn in 16 Feb. 2010).

Government Website: http://www.gov.ai

ECONOMY

Currency

The East Caribbean dollar (see ANTIGUA AND BARBUDA: Currency).

Budget

In 2008 central government revenue was EC$171·0m. and expenditure EC$180·9m. The main sources of revenue are custom duties, tourism and bank licence fees. There is little taxation.

Performance

Real GDP growth was an impressive 18·3% in 2006 and 21·0% in 2007.

Banking and Finance

The East Caribbean Central Bank based in St Kitts and Nevis functions as a central bank. The Governor is Sir Dwight Venner. There is a small offshore banking sector. In 2012 there were four commercial banks: Caribbean Commercial Bank (Anguilla), National Bank of Anguilla, CIBC FirstCaribbean and Scotiabank Anguilla.

ENERGY AND NATURAL RESOURCES

Electricity

Production (2007) 80m. kWh.

Agriculture

Because of low rainfall, agriculture potential is limited. About 1,200 ha. are cultivable. Main crops are pigeon peas, maize and sweet potatoes. Livestock consists of sheep, goats, pigs and poultry. The island relies on imports for food.

Forestry

In 2010 the area under forests in was 6,000 ha., or 60% of the total land area.

Fisheries

Fishing is a thriving industry (mainly lobster). The total catch in 2010 was 701 tonnes.

INDUSTRY

Labour

The unemployment rate was 7·8% in July 2002.

EXTERNAL ECONOMIC RELATIONS

Imports and Exports

Imports (c.i.f.) in 2008 (and 2007) totalled US$271·7m. (US$247·9m.); exports (f.o.b.) in 2008 (and 2007) totalled US$11·5m. (US$9·2m.).

COMMUNICATIONS

Roads

There are about 63 km of main roads and 112 km of secondary roads. In 2004 there were 6,681 vehicles in use, including 4,193 passenger cars and 219 vans and lorries.

Civil Aviation

Wallblake is the airport for The Valley. Anguilla is linked to neighbouring islands by services operated by American Airlines, Coastal Air Transport, LIAT and WINAIR.

Shipping

The main seaports are Sandy Ground and Blowing Point, the latter serving passenger and cargo traffic to and from St Martin.

Telecommunications

There is a modern internal telephone service with (2008) 5,800 main lines in operation; and internet services. In 2008 there were 13,100 mobile phone subscribers.

SOCIAL INSTITUTIONS

Justice

Justice is based on UK common law as exercised by the Eastern Caribbean Supreme Court on St Lucia. Final appeal lies to the UK Privy Council.

Education

Adult literacy was 80% in 1995. Education is free and compulsory between the ages of five and 17 years. There are six government primary schools with (1996) 1,540 pupils and one comprehensive school with (1996) 1,060 pupils. Higher education is provided at regional universities and similar institutions.

In 2008 expenditure on education came to 10·7% of total government expenditure.

Health

In 2003 there was one hospital with a total of 36 beds; there were also four health centres and a government dental clinic. There were nine government-employed and five private doctors, two dentists and 32 nurses in 2003.

RELIGION

There were in 2001 Anglicans (29%), Methodists (24%), plus Seventh Day Adventists, Pentecostalists, Church of God, Baptists and Roman Catholics as significant minorities.

CULTURE

Press

In 2006 there were two weeklies, *The Anguillan* and *The Light.*

Tourism

Tourism accounts for 50% of GDP. In 2007 there were 78,000 non-resident tourists, with revenue (excluding passenger transport) totalling US$119m.

FURTHER READING

Petty, C. L., *A Handbook History of Anguilla.* 1991

Statistical office: Anguilla Statistics Department, PO Box 60, The Valley, Anguilla.
Website: http://www.gov.ai/statistics

Bermuda

KEY HISTORICAL EVENTS

The islands were discovered by Juan Bermúdez, probably in 1503, but were uninhabited until British colonists were wrecked there in 1609. A plantation company was formed; in 1684 the Crown took over the government. A referendum in Aug. 1995 rejected independence from the UK.

TERRITORY AND POPULATION

Bermuda consists of a group of 138 islands and islets (about 20 inhabited), situated in the western Atlantic (32° 18' N. lat., 64° 46' W. long.); the nearest point of the mainland, 940 km distant, is Cape Hatteras (North Carolina). The area is 53·3 sq. km (20·6 sq. miles). In June 1995 the USA surrendered its lease on land used since 1941 for naval and air force bases. The 2010 census population was 64,237; density, 1,205 per sq. km. Capital, Hamilton; population (2010), 1,010. Population of St George's (2010), 1,743.

Ethnic composition, 2010: Black, 54%; White and others, 46%.
The official language is English.

SOCIAL STATISTICS

In 2006 there were 798 live births, 876 marriages and 461 deaths. Average annual growth rate, 2000–05, 0·2%. Life expectancy at birth, 2009: 79·4 years.

CLIMATE

A pleasantly warm and humid climate, with up to 60" (1,500 mm) of rain spread evenly throughout the year. Hamilton, Jan. 63°F (17·2°C), July 79°F (26·1°C). Annual rainfall 58" (1,463 mm).

CONSTITUTION AND GOVERNMENT

Under the 1968 constitution the *Governor*, appointed by the Crown, is normally bound to accept the advice of the Cabinet in matters other than external affairs, defence, internal security and the police, for which he retains special responsibility. The legislature consists of a Senate of 11 members, five appointed by the Governor on the recommendation of the Premier, three by the Governor on the recommendation of the Opposition Leader and three by the Governor in his own discretion. The members of the *House of Assembly* are elected, one from each of 36 constituencies (as of 2003) by universal suffrage.

At a referendum on 17 Aug. 1995, 16,369 votes were cast against the option of independence, and 5,714 were in favour. The electorate was 38,000; turnout was 58%.

RECENT ELECTIONS

A general election was held on 17 Dec. 2012. Turnout was 71·1%. The One Bermuda Alliance/OBA—created in 2011 through the merger of the two main opposition parties, the United Bermuda Party and the Bermuda Democratic Alliance—won 19 of the 36 seats in parliament with 51·7% of votes cast. The ruling Progressive Labour Party (PLP) won 17 seats with 46·1%. The PLP is largely representative of the black population, while the OBA membership is mostly white.

CURRENT GOVERNMENT

Governor: George Fergusson; b. 1955 (took office on 23 May 2012).
  Premier: Craig Cannonier; b. 1963 (took office on 18 Dec. 2012).

Government Website: http://www.gov.bm

DEFENCE

The Bermuda Regiment numbers 600 personnel, mostly part-time. There are 29 professional staff. Bermuda, unlike the rest of the UK, retains conscription.

ECONOMY

Bermuda is the world's third largest insurance market after London and New York. Reserves of insurance companies total BD$39bn. Bermuda has one of the highest per capita incomes of any country thanks to the ever-increasing role of the offshore banking sector.

Currency

The unit of currency is the *Bermuda dollar* (BMD) of 100 *cents* at parity with the US dollar. Inflation was 3·1% in both 2005 and 2006.

Budget

The fiscal year ends on 31 March. Total revenue in 2007–08 was BD$1,008·9m. and total expenditure BD$1,064·6m. Chief sources of revenue (in BD$1m.) in 2007–08: payroll tax, 341·3; customs duty, 230·1; international company tax, 72·8. Chief items of expenditure (in BD$1m.) in 2007–08: salaries and wages, 442·4; other goods and services, 251·2; grants and contributions, 215·2.

Performance

Total GDP was US$5·8bn. in 2010. Real GDP growth was 5·4% in 2006.

Banking and Finance

Bermuda is a leading offshore financial centre with tax exemption facilities. In 2008 there were 15,631 international companies registered in Bermuda. There are four commercial banks, with total assets of BD$24,277m. in 2008. HSBC bought the Bank of Bermuda in 2003 for US$1·3bn. The Bermuda Monetary Authority (*Chairman,* Alan Cossar) acts as a central bank. There is a stock exchange, the BSX.

ENERGY AND NATURAL RESOURCES

Environment

Bermuda's carbon dioxide emissions from the consumption and flaring of fossil fuels in 2008 were the equivalent of 10·2 tonnes per capita.

Electricity

Installed capacity was 0·2m. kW in 2007. Production in 2007 was 643m. kWh, with consumption per capita 9,892 kWh in 2007.

Minerals

Bermuda is rich in limestone.

Agriculture

The chief products are fresh vegetables, bananas and citrus fruit. The value of agricultural products in 2008 was BD$8,587,000. In 2008 there were 968 skilled agricultural and fishery workers in employment. Livestock, 2008 estimates: 1,000 horses, 800 pigs, 650 cattle, 350 goats, 50,000 chickens.

Forestry

Approximately 20% of land is woodland.

Fisheries

In 2008 there were 204 licensed fishing vessels and 313 registered fishermen. The total catch in 2010 was 382 tonnes. Fishing is centred on reef-dwelling species such as groupers and lobsters.

INDUSTRY

Bermuda's leading industry is tourism, with annual revenue in excess of US$350m.

Labour

The number of people in employment in 2008 totalled 40,213 including 4,869 in hotels and restaurants, 4,776 in international business activity and 4,770 in wholesale, retail trade and repairs.

EXTERNAL ECONOMIC RELATIONS

Foreign firms conducting business overseas only are not subject to a 60% Bermuda ownership requirement. In 2009 some 400 international companies had a physical presence in Bermuda.

Imports and Exports

The visible adverse balance of trade is more than compensated for by invisible exports, including tourism and off-shore insurance business.
  Merchandise imports totalled BD$964m. in 2005 and exports BD$25m. In 2005 the USA accounted for 74·5% of imports, Canada 4·4% and the UK 4·2%.
  Principal imports are food, beverages and tobacco, machinery, chemicals, clothing, fuels and transport equipment. The bulk of exports comprise sales of fuel to aircraft and ships, and re-exports of pharmaceuticals.

COMMUNICATIONS

Roads

There are 225 km of public highway and 222 km of private roads. In 2008 there were a total of 51,215 vehicles including: 22,793 private cars; 760 buses, taxis and limousines; 4,018 trucks and tank wagons; and 15,192 motorcycles. There are heavy fines for breaking the speed limit of 35 km/h (22 mph). Bermuda limits cars to one per household and bans hire vehicles.

Civil Aviation

L. F. Wade International Airport (formerly known as Bermuda International Airport) is 19 km from Hamilton. It handled 805,802 passengers and 5,662 tonnes of freight in 2009. Air Canada, American Airlines, British Airways, Continental Airlines, Delta Air Lines, jetBlue, United Airlines, US Airways and WestJet serve Bermuda with regular scheduled services.

Shipping

There are three ports: Hamilton, St George's and Dockyard. There is an open shipping registry. In Jan. 2009 there were 142 ships of 300 GT or over registered, totalling 9·31m. GT. Of the 142 vessels registered, 33 were passenger ships, 33 liquid gas tankers, 23 bulk carriers, 22 container ships, 22 oil tankers and nine general cargo ships.

Telecommunications

In 2008 there were 57,600 main (fixed) telephone lines; mobile phone subscribers numbered 79,000 in 2008 (122·1 per 100 persons). There were 51,000 internet users in 2008. In 2010 Bermuda had the highest fixed broadband penetration rate in the world, at 36·7 subscribers per 100 inhabitants.

SOCIAL INSTITUTIONS

Justice

There are four magistrates' courts, three Supreme Courts and a Court of Appeal. The police had a strength of 468 men and women in 2009.

Bermuda is the only country in the world where McDonald's restaurants are banned by law.

Education

Education is compulsory between the ages of five and 16, and government assistance is given by the payment of grants and, where necessary, school fees. In 2008 there were 5,785 pupils in government schools and 3,669 in private schools. There were 969 full-time students attending the Bermuda College in 2006.

The adult literacy rate is at least 98%. In 2010 public expenditure on education came to 2·6% of GDP and 13·4% of total government spending.

Health

There is one general hospital and one psychiatric hospital. In 2008 there were 127 doctors, 31 dentists, 30 pharmacists and 522 professional nurses.

RELIGION

Many religions are represented, but the larger number of worshippers are attracted to the Anglican, Methodist, Roman Catholic, Seventh Day Adventist, African Methodist Episcopal and Baptist faiths.

CULTURE

World Heritage Sites

The Historic Town of St George's and Related Fortifications, an example of early English New World colonialism, were inscribed on UNESCO's World Heritage List in 2000.

Press

In 2010 there was one daily newspaper (*The Royal Gazette*) with a circulation of about 16,000 and two weeklies.

Tourism

In 2011 there were 655,236 visitor arrivals in 2011 (down slightly on the peak figure of 663,767 in 2007), including a record 415,711 cruise passengers; there were 236,038 air arrivals and the rest were yacht passengers. 73% of the air passengers visiting Bermuda in 2011 were from the USA, 12% from Canada and 9% from the UK.

FURTHER READING

Government Department of Statistics. *Bermuda Facts and Figures.* Annual. Ministry of Finance. *Economic Review.* Annual.

Zuill, W. S., *The Story of Bermuda and Her People.* 3rd ed. 1999

National library: The Bermuda National Library, Hamilton.
Statistical office: Government Department of Statistics, Hamilton.

British Antarctic Territory

KEY HISTORICAL EVENTS

The British Antarctic Territory was established on 3 March 1962, as a consequence of the entry into force of the Antarctic Treaty, to separate those areas of the then Falkland Islands Dependencies which lay within the Treaty area from those which did not (i.e. South Georgia and the South Sandwich Islands).

TERRITORY AND POPULATION

The territory encompasses the lands and islands within the area south of 60°S latitude lying between 20°W and 80°W longitude (approximately due south of the Falkland Islands and the Dependencies). It covers an area of some 1,700,000 sq. km, and its principal components are the South Orkney and South Shetland Islands, the Antarctic Peninsula (Palmer Land and Graham Land), the Filchner and Ronne Ice Shelves and Coats Land.

There is no indigenous or permanently resident population. There is, however, an itinerant population of scientists and logistics staff of about 300, manning a number of research stations.

CURRENT GOVERNMENT

Commissioner: Peter Hayes (non-resident).
  Administrator: Henry Burgess.

British Indian Ocean Territory

KEY HISTORICAL EVENTS

This territory was established to meet UK and US defence requirements by an Order in Council on 8 Nov. 1965, consisting then of the Chagos Archipelago (formerly administered from Mauritius) and the islands of Aldabra, Desroches and Farquhar (all formerly administered from Seychelles). The latter islands became part of Seychelles when that country achieved independence on 29 June 1976. In Nov. 2000 the High Court ruled that the 2,000 Ilois people (native to the archipelago) deported between 1967 and 1973 to accommodate a US military base on Diego Garcia had been removed unlawfully. Chagos islanders subsequently lost a UK High Court case for compensation and the right to return in 2003. However, further High Court rulings in 2006 and 2007 went against the UK government. Nonetheless, return by the Ilois to the archipelago (excepting Diego Garcia) is unlikely until all appeal processes have been exhausted.

TERRITORY AND POPULATION

The group, with a total land area of 60 sq. km (23 sq. miles), comprises five coral atolls (Diego Garcia, Peros Banhos, Salomon, Eagle and Egmont), of which the largest and southernmost, Diego Garcia, covers 44 sq. km (17 sq. miles) and lies 725 km (450 miles) south of the Maldives. A US Navy support facility has been established on Diego Garcia. There is no permanent population.

CURRENT GOVERNMENT

Commissioner: Peter Hayes (non-resident).
  Administrator: John McManus.

British Virgin Islands

KEY HISTORICAL EVENTS

Discovered by Columbus on his second voyage in 1493, British Virgin Islands were first settled by the Dutch in 1648 and taken over in 1666 by a group of English planters. The islands were annexed to the British Crown in 1672. Constitutional government was granted in 1773, but was later surrendered in 1867. A Legislative Council formed in that year was abolished in 1902. In 1950 a partly nominated and partly elected Legislative Council was restored. A ministerial system of government was introduced in 1967.

TERRITORY AND POPULATION

The Islands form the eastern extremity of the Greater Antilles and number 60, of which 16 are inhabited. The largest, with census populations (2001), are Tortola, 19,282; Virgin Gorda, 3,203; Anegada, 250; and Jost Van Dyke, 244. Other islands had a total population (2001 census) of 182 (including marine population).

Total area 151 sq. km (58 sq. miles); total population (2001 census), 23,161. In 2000, 61·1% of the population were urban. The capital, Road Town, on the southeast of Tortola, is a port of entry; population (estimate 2000), 7,974.

The official language is English. Spanish and Creole are also spoken.

SOCIAL STATISTICS

Birth rate, 2001, was 15·4 per 1,000 population; death rate, 4·9 per 1,000. Life expectancy in 2001 was an estimated 75·5 years. Annual growth rate, 1·96% in 2000.

CLIMATE

A pleasant healthy sub-tropical climate with summer temperatures lowered by sea breezes and cool nights. Road Town (1999), Jan. 21°C, July 27°C; rainfall (1998), 1471 mm.

CONSTITUTION AND GOVERNMENT

The constitution became effective on 15 June 2007. It granted the British Virgin Islands greater autonomy and self-determination. There is a *Premier* (formerly *Chief Minister*) and a *House of Assembly* (formerly *Legislative Council*). The Premier is appointed by the *Governor*. The House of Assembly consists of 13 members elected for a four-year term (five directly elected members from constituencies and four members from 'at large' seats covering the territory as a whole), a Speaker and the Attorney General *ex officio*. The Cabinet consists of the Premier, four other Ministers and the Attorney General *ex officio*.

RECENT ELECTIONS

In parliamentary elections on 7 Nov. 2011 the National Democratic Party (NDP) won nine of the 13 available seats, ahead of the Virgin Islands Party (VIP) with four. The remaining two seats are held by a Speaker and the Attorney General. Turnout was 68·8%.

CURRENT GOVERNMENT

Governor: Boyd McCleary.
   Premier: Orlando Smith (NDP; sworn in 9 Nov. 2011).

Government Website: http://www.bvi.org.uk

ECONOMY

The economy is based on tourism and international financial services.

Currency
The official unit of currency is the US dollar.

Budget
In 2000 revenue was US$183·1m. and expenditure US$134·6m. Outstanding debt, in 2000, US$37·1m.

Performance
Real GDP growth was 8·7% in 2001 following growth of 4·4% in 2000. Total GDP was US$943m. in 2004; GDP per capita amounted to US$43,366.

Banking and Finance
In 2003 there were 13 banks. As of Sept. 2001 total deposits were US$1,143·8m. Financial services have surpassed the performance of the tourism industry to become the largest contributor to the GDP. As of 30 June 2001, 448,767 International Business Companies were registered in the British Virgin Islands.

ENERGY AND NATURAL RESOURCES

Environment
Carbon dioxide emissions from the consumption and flaring of fossil fuels in 2008 were the equivalent of 4·9 tonnes per capita.

Electricity
Production, 2007 estimate, 48m. kWh. In 2007 installed capacity was 10,000 kW.

Agriculture
Agricultural production is limited, with the chief products being livestock (including poultry), fish, fruit and vegetables.
   Livestock (2009 estimates): cattle, 2,400; goats, 10,000; pigs, 1,500; sheep, 6,100.

Forestry
The area under forests in 2010 was 4,000 ha., or 24% of the total land area.

Fisheries
The total catch was an estimated 1,200 tonnes in 2010.

INDUSTRY

The construction industry is a significant employer. There are ice-making plants, cottage industries producing tourist items and a rum distillery.

Labour
In 2005 there were 16,232 employed persons (51% males). 5,142 persons were employed in government services, 2,573 in hotels and restaurants and 1,624 in wholesale and retail trade.

EXTERNAL ECONOMIC RELATIONS

Imports and Exports
In 2000 imports were US$237·6m. and exports US$26·6m.

COMMUNICATIONS

Roads
In 2000 there were 362·09 km of paved roads and 10,631 registered vehicles.

Civil Aviation
Beef Island Airport, about 16 km from Road Town, is capable of receiving 80-seat short-take-off-and-landing jet aircraft. Several airlines serve the British Virgin Islands, notably LIAT. There are scheduled flights to Puerto Rico and a number of islands in the Eastern Caribbean.

Shipping
There are two deep-water harbours, Port Purcell and Road Town, and four other harbours.

Telecommunications
In 2008 there were 18,900 main telephone lines and 23,000 mobile phone subscribers (100·9 per 100 persons).

SOCIAL INSTITUTIONS

Justice
Law is based on UK common law. There are courts of first instance. The appeal court is in the UK.

Education
In 2004 adult literacy was 98·2%. Education is free and compulsory from 5 years to 16. There are several government schools as well as private schools. Total number of pupils in primary schools (2010), 3,201; total number of secondary level pupils (2010), 2,029. There is a community college, H. Lavity Stoutt Community College, located at the eastern end of Tortola. In 1986 a branch of the Hull University (England) School of Education was established.

Health
As of 31 Dec. 2000 there were 19 doctors, 74 nurses, 44 public hospital beds and one private hospital with ten beds. Expenditure, 2000 (estimate) was US$7·6m.

RELIGION

There are Anglican, Methodist, Seventh-Day Adventist, Roman Catholic, Baptist, Pentecostal and other Christian churches in the Territory. There are also Jehovah's Witness and Hindu congregations.

CULTURE

Press

In 2008 there were three weekly newspapers and five other non-dailies.

Tourism

Tourism is one of the mainstays of the economy, along with financial services. There were 356,271 overnight visitors and 473,987 cruise passenger arrivals in 2006.

FURTHER READING

Statistical Office: The Development Planning Unit, Central Administration Complex, Road Town, Tortola.
Website: http://www.dpu.gov.vg

Cayman Islands

KEY HISTORICAL EVENTS

The islands were discovered by Columbus on 10 May 1503 and (with Jamaica) were recognized as British possessions by the Treaty of Madrid in 1670. Grand Cayman was settled in 1734 and the other islands in 1833. They were administered by Jamaica from 1863, but remained under British sovereignty when Jamaica became independent on 6 Aug. 1962.

TERRITORY AND POPULATION

The Islands consist of Grand Cayman, Cayman Brac and Little Cayman. They are located in the Caribbean Sea, about 305 km (190 miles) northwest of Jamaica; area, 259 sq. km (100 sq. miles). 2010 census population, 55,456, giving a density of 214 per sq. km. The official language is English. The chief town is George Town with a population of 28,089 (2010).

The areas and populations of the islands are:

	Sq. km	1999	2010
Grand Cayman	197	37,473	53,160
Cayman Brac	36	1,822	2,296
Little Cayman	26	115	

SOCIAL STATISTICS

2005: births, 699; deaths, 170. 2005: resident marriages, 810. Population growth rate, 2000–05, 3·7%.

CLIMATE

The climate is tropical maritime, with a cool season from Nov. to March. The average yearly temperature is 27°C, and rainfall averages 57" (1,400 mm) a year at George Town. Hurricanes may be experienced between July and Nov.

CONSTITUTION AND GOVERNMENT

The Cayman Islands are a self-governing overseas territory of the United Kingdom. A new draft constitution granting the Islands more political autonomy was approved at a national referendum on 20 May 2009, with 63·1% of votes cast in favour. It took effect on 6 Nov. 2009, creating the office of premier and enlarging the Legislative Assembly to 18 members. The premier is limited to serving two consecutive four-year terms of office.

RECENT ELECTIONS

At the Legislative Assembly elections on 20 May 2009 the United Democratic Party won 9 of the 15 available seats; the People's Progressive Movement, 5; ind., 1. Turnout was 80·6%.

CURRENT GOVERNMENT

Governor: Duncan Taylor.
     Premier: Juliana O'Connor-Connolly.
Government Website: http://www.gov.ky

ECONOMY

Currency

The unit of currency is the *Cayman Island dollar* (KYD), usually written as CI$, of 100 *cents*.

Budget

In 2007 revenues totalled CI$513·0m. and expenditures CI$552·0m.

Performance

Real GDP growth was negative in 2008, with the economy contracting by 0·7%; there was then a much more severe decline in 2009 when it shrank by 7·0%.

Banking and Finance

Financial services, the Islands' chief industry, are monitored by the Cayman Islands Monetary Authority (*Chairman,* George McCarthy). At June 2008, 416 banks and trust companies held licenses that permit the holders to offer services to the public, 19 domestically. Most of the world's leading banks have branches or subsidiaries in the Cayman Islands. In 2007, 87,109 companies, almost all offshore, were registered as well as 10,037 mutual funds and 800 insurance companies. Net domestic assets of Cayman-registered banks totalled CI$2,409·0m.

ENERGY AND NATURAL RESOURCES

Environment

Carbon dioxide emissions from the consumption and flaring of fossil fuels in 2008 were the equivalent of 10·3 tonnes per capita.

Electricity

Installed capacity on Grand Cayman was 152·6 MW in 2009. Production in 2009 was 609m. kWh.

Agriculture

Mangoes, bananas, citrus fruits, yams, cassava, breadfruit, tomatoes, honey, beef, pork and goat meat are produced for local consumption.

Forestry

The area under forests in 2010 was 13,000 ha., or 50% of the total land area.

Fisheries

In 2010 the total catch was estimated at 125 tonnes.

INDUSTRY

Labour

Unemployment rate: 6·0% of workforce in 2009 (4·0% in 2008).

EXTERNAL ECONOMIC RELATIONS

Imports and Exports

2005: imports, US$1·19bn.; exports, US$1·56m.

COMMUNICATIONS

Roads

In 2007 there were about 304 miles of paved roads on the Cayman Islands and 34,031 licensed motor vehicles.

Civil Aviation

George Town (Owen Roberts) on Grand Cayman and Cayman Brac (Gerrard Smith) have international airports. George Town handled 909,500 passengers and 3,650 tonnes of freight in 2007. Cayman Airways provides a regular inter-island service and also flies to Chicago, Miami, New York, Tampa, Washington D.C., Cuba and Jamaica. Eight additional international airlines provide services to London, Toronto, the Bahamas, Honduras, Jamaica and the USA.

Shipping

Motor vessels ply regularly between the Cayman Islands, Cuba, Jamaica and Florida. In 2007, 329,133 tonnes of cargo were offloaded at George Town.

Telecommunications

At the end of 2007 there were 130,622 fixed and mobile telephone lines.

SOCIAL INSTITUTIONS

Justice

There is a Grand Court, sitting six times a year for criminal sessions at George Town under a Chief Justice and two puisne judges. There are three Magistrates presiding over the Summary Court.

The population in penal institutions in Aug. 2008 was 207 (equivalent to 380 per 100,000 population).

Education

In 2007 there were 17 government schools with 4,637 pupils and 2,692 students were enrolled in ten private schools. There are two government facilities for special educational needs: a school for children and a training centre for adults. Four institutions—a private four-year college, a private medical and veterinary college, the government community college and a law school—provide tertiary education.

Health

The government's health services complex in George Town includes a 101-bed hospital, a dental clinic and an eye clinic. On Grand Cayman there are four district health centres. There is a hospital on Cayman Brac (18 beds) and a health centre on Little Cayman. In 2007 there were 46 doctors in government service (including five on Cayman Brac) and 44 in private practice.

RELIGION

The residents are primarily Christian (85%) and over 12 denominations meet regularly; Church of God, Presbyterian/United, Roman Catholic, Baptist and Seventh-Day Adventists are the largest. Other religions, including Ba'hai, Buddhism, Hinduism, Islam and Judaism, have representation in the community.

CULTURE

Press

In 2008 there were two daily newspapers, the *Caymanian Compass* and the *Caymanian Net News*, with a combined average daily circulation of 18,000.

Tourism

Tourism is the chief industry after financial services, and in 2007 there were 2,197 rooms in hotels and 287 rooms in apartments, guest houses and cottages. In 2007 there were 291,503 tourist arrivals by air and 1,715,666 cruise passenger arrivals. Tourism receipts in 2007 totalled CI$399·1m.

FURTHER READING

Cayman Islands' Compendium of Statistics 2011. 2012

Statistical Office: The Information Centre, Economics & Statistics Office, Government Administration Building, Grand Cayman, KY1-9000.
Website: http://www.eso.ky

Falkland Islands

KEY HISTORICAL EVENTS

France established a settlement in 1764 and Britain a second settlement in 1765. In 1770 Spain bought out the French and drove off the British. This action on the part of Spain brought that country and Britain to the verge of war. The Spanish restored the settlement to the British in 1771, but the settlement was withdrawn in 1774. In 1806 Spanish rule was overthrown in Argentina, and the Argentine claimed to succeed Spain in the French and British settlements in 1820. The British objected and reclaimed their settlement in 1832 as a Crown Colony.

On 2 April 1982 Argentine forces occupied the Falkland Islands. On 3 April the UN Security Council called, by 10 votes to one, for Argentina's withdrawal. After a military campaign, but without a formal declaration of war, the UK regained possession on 14–15 June when Argentina surrendered. In April 1990 Argentina's Congress declared the Falkland and other British-held South Atlantic islands part of the new Argentine province of Tierra del Fuego though the threat of hostilities has been lifted.

TERRITORY AND POPULATION

The Territory comprises numerous islands situated in the South Atlantic Ocean about 480 miles northeast of Cape Horn covering 12,200 sq. km. The main East Falkland Island, 6,760 sq. km; the West Falkland, 5,410 sq. km, including the adjacent small islands. The population at the census of April 2012 was 2,841. The only town is Stanley, in East Falkland, with a 2012 population of 2,121. 59% of the residents consider their national identity to be 'Falkland Islander' and 29% consider themselves to be British. The population is nearly all of British descent. In 2000, 78·8% lived in urban areas. There is a British garrison of about 500 servicemen, stationed in East Falkland.

The official language is English.

SOCIAL STATISTICS

In 2008 there were 29 births and 22 deaths on the islands.

CLIMATE

A cool temperate climate, much affected by strong winds, particularly in spring. Stanley, Jan. 49°F (9·4°C), July 35°F (1·7°C). Annual rainfall 24" (625 mm).

CONSTITUTION AND GOVERNMENT

A new constitution came into force in Jan. 2009, updating the previous constitution of 1985 and the 1997 amendments. Provisions on the right of self-determination contained in international covenants were affirmed in the main body of the constitution. It also provided for a Public Accounts Committee and a Complaints Commissioner to allow for greater transparency in financial matters. The human rights chapter was updated to bring it in line with international agreements.

Domestic policy is the responsibility of the Executive Council although the Governor has the right to override the council and decide policy 'in the interests of good governance' or in matters relating to external affairs, defence, internal security (including police), the administration of justice, audit and public service management.

There is a *Legislative Council* consisting of eight members (five from Stanley and three from Camp, elected every four years) and two *ex officio* members, the Chief Executive and Financial Secretary. Only elected members have a vote.

British citizenship was withdrawn by the British Nationality Act 1981, but restored after the Argentine invasion of 1982.

In a referendum on the Territory's political status held on 10–11 March 2013, 99·8% of votes cast were in favour of remaining a British territory.

RECENT ELECTIONS

Elections to the Legislative Assembly were held on 5 Nov. 2009. Only non-partisans were elected.

CURRENT GOVERNMENT

Governor: Nigel Haywood, CVO.
   Chief Executive: Keith Padgett.

Government Website: http://www.falklands.gov.fk

DEFENCE

Since 1982 the Islands have been defended by a 2,000-strong garrison of British servicemen. In addition there is a local volunteer defence force.

ECONOMY

The economy of the Islands grew considerably in the second half of the 1980s as a result of the expansion of the fishing industry. In 2007 the GDP was estimated at £104m., up from £5m. in 1980.

Currency

The unit of currency is the *Falkland Islands pound* (FKP) of 100 *pence*, at parity with £1 sterling.

Budget

Revenue and expenditure (in £ sterling) for fiscal year ending 30 June 2000 was: revenue, 52·3m.; expenditure, 40·4m.

Banking and Finance

There is not a central bank—the only bank is Standard Chartered Bank, which has one branch in Stanley.

ENERGY AND NATURAL RESOURCES

Electricity

Electricity production in 2007 totalled about 17m. kWh. Installed capacity in 2007 was estimated at 9,000 kW.

Oil and Gas

In 1996 the Falkland Islands government awarded production licences to Shell, Amerada Hess, Desire Petroleum and International Petroleum Corporation (Sodra), allowing them to begin oil exploration. The licensed areas were situated 150 km north of the Islands over the North Falkland Basin. Six exploration wells were drilled in 1998 but the project was abandoned as crude oil prices fell to around US$10 a barrel.

Interest revived in 2008 when oil prices rose above US$100 a barrel, as the areas around the Falkland Islands are believed to have reserves of oil estimated at about 60bn. bbls. In 2010 five companies held licences for oil exploration in the region—Desire, Falkland Oil & Gas, Rockhopper, Borders & Southern and Argos Resources. With the potential for significant oil riches, Argentina has appealed to the United Nations to bring the UK into renewed talks over the islands.

Agriculture

The economy was formerly based solely on agriculture, principally sheep farming. Following a programme of sub-division, much of the land is divided into family-size units. There were 85 farms in 2009–10, averaging 13,053 ha. and 5,630 sheep. Wool is the principal product; 1,587 tonnes were produced in 2009–10.

Livestock: in May 2010 there were 478,525 sheep, 4,738 cattle and 519 horses. With a population of 2,841 in 2012 the Falkland Islands have the equivalent of approximately 170 sheep per resident.

Fisheries

The total catch in 2010 was 99,560 tonnes, up from 27,190 tonnes in 1995.

INDUSTRY

Labour

In 2001 there were 2,025 people employed full-time, including 358 in construction and 326 in agriculture, hunting and fishing. The growth of the fishing industry has ensured practically zero unemployment.

EXTERNAL ECONOMIC RELATIONS

Around 85% of trade is with the UK, the rest with Latin America, mainly Chile. In 2007 imports totalled £35m.; exports, £134m. Main import commodities are refined petroleum oils and surveying, hydrographic, oceanographic, hydrological, meteorological or geophysical instruments and appliances. Significant export commodities are molluscs (which constitute around three-quarters of exports), frozen fish and wool.

COMMUNICATIONS

Roads

There are over 50 km of surfaced roads and another 400 km of unsurfaced road. This includes the 80 km between Stanley and Mount Pleasant Airport. Other settlements outside Stanley are linked by tracks. There were 1,754 motor vehicles in 2006 (1,264 vehicles in Stanley).

Civil Aviation

Air communication is currently via Ascension Island. An airport, completed in 1986, is sited at Mount Pleasant on East Falkland. RAF Tristar aircraft operate a twice-weekly service between the Falklands and the UK. A Chilean airline, LAN Airlines, runs a weekly service to Puerto Montt, Punta Arenas, Rio Gallegos and Santiago.

Shipping

A charter vessel calls four or five times a year to and from the UK. Vessels of the Royal Fleet Auxiliary run regularly to South Georgia. Sea links with Chile and Uruguay began in 1989.

Telecommunications

Number of telephone main lines in 2008 was 2,000. International direct dialling is available, as are international fax links. In 2008 there were 2,800 internet users.

SOCIAL INSTITUTIONS

Justice

There is a Supreme Court, and a Court of Appeal sits in the UK; appeals may go from that court to the judicial committee of the Privy Council. The senior resident judicial officer is the Senior Magistrate. There is an Attorney General and a Senior Crown Counsel.

Education

Education is compulsory between the ages of five and 16 years. In Stanley in 2002 there were 30 pre-school pupils, 190 primary pupils (18 teachers) and 160 pupils in the 11–16 age range (18 teachers). In rural areas students attend small settlement schools or are visited by one of seven travelling teachers. Lessons may also be carried out over the radio or telephone.

Health

The Government Medical Department is responsible for all medical services to civilians. Primary and secondary health care facilities are based at the King Edward VII Memorial Hospital, the only hospital on the islands. It has 28 beds. It is staffed by five doctors, six sisters (including four midwives), eight staff nurses, a health visitor, counsellor, physiotherapist, social worker and an auxiliary nursing staff. The Royal Army Medical Corps staff the surgical facilities. There are two dentists on the island.

Welfare

In 1998 total amount spent on old age pension payments was £504,075. Total amount spent on family allowance payments was £336,365.

CULTURE

Press

In 2006 there was one weekly newspaper, the *Penguin News*.

Tourism

In the 2008–09 season there were a record 62,488 cruise ship visitors. There are a variety of lodges around the Islands. Stanley has one major hotel and some guest houses.

FURTHER READING

Gough, B., *The Falkland Islands/Malvinas: the Contest for Empire in the South Atlantic.* 1992

Gibraltar

KEY HISTORICAL EVENTS

The Rock of Gibraltar was settled by Moors in 711. In 1462 it was taken by the Spaniards, from Granada. It was captured by Admiral Sir George Rooke on 24 July 1704, and ceded to Great Britain by the Treaty of Utrecht, 1713. The cession was confirmed by the treaties of Paris (1763) and Versailles (1783). In 1830 Gibraltar became a British crown colony.

On 10 Sept. 1967 a UN resolution on the decolonization of Gibraltar led to a referendum to ascertain whether the people of Gibraltar wished to retain their link with the UK. Out of an electorate of 12,762, an overwhelming majority voted to retain the British connection.

The border was closed by Spain in 1969, opened to pedestrians in 1982 and fully opened in 1985. In 1973 Gibraltar joined the European Community as a dependent territory of the United Kingdom. In 2001 talks were held between Britain and Spain over the colony's sovereignty. In a joint statement, the British and Spanish foreign ministers said they would work towards a comprehensive agreement by the summer of 2002. Gibraltar's government held an unofficial referendum on sharing sovereignty with Spain on 7 Nov. 2002 in which 98·97% of votes cast were against joint sovereignty. While Britain sees the principle of shared sovereignty as the definitive solution, Spain maintains its historic claim to outright control. An agreement reached in Sept. 2006 eased border controls and opened up Gibraltar to flights from Spain and other European countries.

TERRITORY AND POPULATION

Gibraltar is situated in latitude 36°07′ N and longitude 05°21′ W. Area, 6·5 sq. km (2½ sq. miles) including port and harbour. Total population (2009), 29,431 (of whom 23,907 were British Gibraltarian, 3,129 Other British and 2,395 Non-British); density, 4,528 per sq. km. The population is mostly of Genoese, Portuguese, Maltese and Spanish descent.

The official language is English; Spanish is also spoken.

SOCIAL STATISTICS

Statistics (2009): births, 417; deaths, 234; marriages, 966. Rates per 1,000 population, 2006: birth, 12·6; death, 8·0.

CLIMATE

The climate is warm temperate, with westerly winds in winter bringing rain. Summers are pleasantly warm and rainfall is low. Mean maximum temperatures: Jan. 16°C, July 28°C. Annual rainfall 722 mm.

CONSTITUTION AND GOVERNMENT

A new constitution was approved in a referendum on 30 Nov. 2006 and came into effect on 2 Jan. 2007, giving Gibraltar full internal self-government. The *Gibraltar House of Assembly* (itself an amalgamation of the former Legislative and City Councils) was renamed the *Gibraltar Parliament*. The legislature of Gibraltar consists of the Queen as head of state and the Gibraltar Parliament. The Governor remains Commander-in-Chief but only retains direct constitutional responsibility for matters relating to defence, external affairs and some aspects of internal security. There is a Council of Ministers presided over by the Chief Minister who is appointed by the Governor. The constitution also abolished the former Gibraltar Council.

The Gibraltar Parliament consists of a Speaker appointed by the parliament and at least 17 elected members. A Mayor of Gibraltar is elected by the members of parliament (excluding the Speaker).

Gibraltarians have full UK citizenship.

RECENT ELECTIONS

At the elections of 8 Dec. 2011 turnout was 81·4%. The opposition alliance of the Gibraltar Socialist Labour Party and the Liberal Party won 48·9% of the vote, gaining seven and three seats respectively. The Gibraltar Social Democratic Party took 46·8% of the vote with seven seats and the Progressive Democratic Party received 4·4% but did not win any seats.

CURRENT GOVERNMENT

Governor and C.-in-C: Vice Adm. Sir Adrian Johns, KCB, CBE, ADC (sworn in 26 Oct. 2009).

Chief Minister: Fabian Picardo; b. 1972 (Gibraltar Socialist Labour Party; elected in Dec. 2011).

Deputy Chief Minister: Joseph Garcia. *Minister for Education, Financial Services, Gaming, Telecommunications and Justice:* Gilbert Licudi. *Enterprise, Training and Employment:* Joe Bossano. *Equality and Social Services:* Samantha Sacramento. *Health and Environment:* John Cortes. *Housing and the Elderly:* Charles Bruzon. *Sports, Culture, Heritage and Youth:* Steven Linares. *Tourism, Public Transport and the Port:* Neil Costa. *Traffic, Health and Safety, and Technical Services:* Paul Balban.

The *Speaker* is Adolfo Canepa.

Government Website: http://www.gibraltar.gov.gi

DEFENCE

The Ministry of Defence presence consists of a tri-service garrison numbering approximately 900 uniformed personnel. Supporting the garrison are approximately 1,100 locally-employed civilian personnel. The garrison supports a NATO Headquarters.

ECONOMY

Overview

The economy is primarily dependent on service industries and port facilities, with income derived from tourism, transhipment and, perhaps most importantly in terms of growth, the provision of financial services.

Currency

The legal tender currency is UK sterling. Also legal tender are Government of Gibraltar Currency notes and coins. The *Gibraltar pound* (GIP) of 100 *pence* is at parity with the UK £1 sterling. The total of Government of Gibraltar notes in circulation at 31 March 2009 was £23·4m. The annual rate of inflation was 3·4% in Jan. 2010.

Budget

Departmental revenue credited to the consolidated fund for the year ending 31 March 2009 totalled £244m. whilst expenditure amounted to £201m. The main sources of consolidated fund revenues were income tax (£109m.), import duties (£47m.), company taxes (£26m.) and general rates and salt water charges (£14m.). Main items of consolidation fund expenditure: health (£24m.), education (£24m.), social affairs (£22m.) and utilities (£14m.).

Performance
In 2006–07 Gibraltar's GDP was £660m., equivalent to £22,900 per head.

Banking and Finance
At June 2010 there were 19 authorized credit institutions. The majority of these are either subsidiaries or branches of major UK or other European Economic Area (EEA) credit institutions. 12 of these credit institutions are incorporated in Gibraltar and are licensed by the Financial Services Commission, one of which is only authorized as an e-money institution. There are five branches of EEA-authorized credit institutions operating in Gibraltar and two non-EEA branches, one from the Channel Islands and one from Switzerland. The banking sector provides services to both local and non-resident customers.

ENERGY AND NATURAL RESOURCES

Environment
Gibraltar's carbon dioxide emissions from the consumption and flaring of fossil fuels in 2008 were the equivalent of 161·6 tonnes per capita, the highest in the world.

Electricity
Production in 2009 amounted to 174·01m. kWh.

Oil and Gas
Gibraltar is dependent on imported petroleum for its energy supplies.

Agriculture
Gibraltar lacks agricultural land and natural resources; the territory is dependent on imports of foodstuffs and fuels.

INDUSTRY
The industrial sector (including manufacturing, construction and power) employed around 16% of the working population in 2009.

Labour
The total number of employee jobs at Oct. 2009 was 20,450. Principal areas of employment (Oct. 2009): community, social and personal services, 4,496; trade, restaurants and hotels, 4,076; construction, 2,557; manufacturing, 473; electricity and water, 299; other, 8,848. (Figures cover only non-agricultural activities, excluding mining and quarrying). Under 5% of the labour force was unemployed in 2009.

EXTERNAL ECONOMIC RELATIONS
Gibraltar has a special status within the EU which exempts it from the latter's fiscal policy.

Imports and Exports
Imports in 2009 totalled £479·3m. and exports £169·5m. (excluding petroleum products).

Britain provided 25% of imports in 2009 and is the largest source. Other major trade partners include Spain, the Netherlands and Germany. Foodstuffs accounted for 5% of total imports in 2009 (excluding petroleum products). Mineral fuels comprised about 71% of the value of total imports in 2008. Exports are mainly re-exports of petroleum and petroleum products supplied to shipping, and include manufactured goods, wines, spirits, malt and tobacco. Gibraltar depends largely on tourism, offshore banking and other financial sector activity, online gaming, the entrepôt trade and the provision of supplies to visiting ships. Exports of domestic produce are negligible. In 2009 Gibraltar recorded a visible trade deficit of £350·5m.

COMMUNICATIONS

Roads
There are 56 km of roads including 6·8 km of pedestrian way. In 2006 there were 14,637 private vehicles, 7,024 motorcycles, 1,391 goods vehicles, 112 taxis and 99 buses.

Civil Aviation
There is an international airport, Gibraltar North Front. Scheduled flights were operated in 2009 by easyJet to London (Gatwick), by British Airways to London (Gatwick and Heathrow), by Monarch Airlines to London (Luton) and Manchester and by Andalus Airlines to Madrid and Barcelona. In 2009, 183,703 passengers arrived by air and 186,049 departed; 10·7 tonnes of freight were loaded and 134 tonnes were unloaded (figures exclude military freight).

Shipping
The Strait of Gibraltar is a principal ocean route between the Mediterranean and Black Sea areas and the rest of the world. Approximately 110,000 vessels pass through the Straits per year. Tax concessions are available to ship-owners who register their ships at Gibraltar. A total of 10,042 merchant ships of 253·1m. GRT entered port during 2009, including 7,818 deep-sea ships of 208·4m. GRT. In 2009, 349 calls were made by yachts of 372,925 GRT. 238 cruise liners called during 2009 involving 348,508 passengers.

Telecommunications
Gibtelecom, jointly owned by Telekom Slovenije and the Government of Gibraltar, is the main provider of fixed, wireless and internet services to residential and business customers.

As at 1 Jan. 2010 the Group's fixed exchange lines stood at 23,293. At the end of 2006 there were 8,000 internet accounts in Gibraltar.

SOCIAL INSTITUTIONS

Justice
The judicial system is based on the English system. There is a Court of Appeal, a Supreme Court, presided over by the Chief Justice, a Court of First Instance and a Magistrates' Court.

The population in penal institutions in Sept. 2008 was 43 (equivalent to 154 per 100,000 population).

Education
Free compulsory education is provided between the ages of four and 15 years. The medium of instruction is English. The comprehensive system was introduced in Sept. 1972 and all schools currently follow a locally adapted version of the National Curriculum for England and Wales. In the 2009–10 academic year there were 11 primary and two secondary schools. Primary schools are divided into first schools for children aged 4–8 years and middle schools for children aged 8–12 years. All primary schools are mixed though secondary schools are single-sex.

Vocational education and training is available at the Gibraltar College (a post-15 institution), the Construction Training Centre and the Gibdock Training Centre; the former two are managed by the Gibraltar government and the latter by Cammell Laird, a shipbuilding company (and part-funded by the Government). In Sept. 2009 there were 2,926 pupils at government primary schools, 326 at private primary schools and 156 at the Services school. 1,140 pupils were enrolled at the boys' comprehensive school and 1,040 at the girls' comprehensive. There were 338 students in the Gibraltar College. Government expenditure on education in the year ended 31 March 2010 was £25·2m.

Health
The Gibraltar Health Authority is the organization responsible for providing health care in Gibraltar. The Authority operates a Group Practice Medical Scheme, which is a contributory scheme and enables registered persons to access free medical treatment.

The St Bernard's Hospital, opened in 2005, has a total of 201 beds. A military hospital, the Princess Royal Medical Centre, opened in 2009, and replaced the old Royal Naval Hospital. Total expenditure on medical and health services for the financial year 2009–10 was £74·1m.

Welfare

The social security system consists of: the Social Security (Employment Injuries Insurance) Scheme which only applies to employed persons; the Social Security (Short-Term Benefits) Scheme which provides for payments of maternity grants. maternity allowance, death grants and unemployment benefit; and the Social Security (Open Long-Term Benefits) Scheme which provides for pensions and widows' allowances.

RELIGION

According to the 2001 census 78·1% of the population were Roman Catholic, 7·0% Church of England, 4·0% Muslim, 2·1% Jewish and 1·8% Hindu. In 2004 there were seven Roman Catholic and three Anglican churches (including one Catholic and one Anglican cathedral), one Presbyterian and one Methodist church, four synagogues and two mosques.

CULTURE

Press

In 2009 there were two daily and two weekly newspapers.

Tourism

In 2009 just over 10·3m. tourists visited Gibraltar (including day-visitors) bringing in revenue of £257·6m. There are approximately 900 hotel beds in Gibraltar. Tourism accounts for more than a quarter of GDP.

FURTHER READING

Gibraltar Year Book. Annual

Morris, D. S. and Haigh, R. H., *Britain, Spain and Gibraltar, 1940–90: the Eternal Triangle.* 1992

Statistical Office: Statistics Office, 99 Harbours Walk, The New Harbours, Gibraltar.
Website: http://www.gibraltar.gov.gi/statistics

Montserrat

KEY HISTORICAL EVENTS

Montserrat was discovered by Columbus in 1493 and colonized by Britain in 1632, who brought Irish settlers to the island. Montserrat formed part of the federal colony of the Leeward Islands from 1871 until 1958, when it became a separate colony following the dissolution of the Federation.

On 18 July 1995 the Soufriere Hills volcano erupted for the first time in recorded history, which led to over half the inhabitants being evacuated to the north of the island, and the relocation of the chief town, Plymouth. Another major eruption on 25 June 1997 caused a number of deaths and led to further evacuation.

TERRITORY AND POPULATION

Montserrat is situated in the Caribbean Sea, 43 km southwest of Antigua. The area is 102·3 sq. km (39·5 sq. miles). Census population, 2011, 4,922. What was previously the capital, Plymouth, is now deserted as a result of the continuing activity of the Soufriere Hills volcano. The safe area is in the north of the island, where Brades, the *de facto* capital, is located.

The official language is English.

CLIMATE

A tropical climate with an average annual rainfall of 60" (1,500 mm) the wettest months being Sept.–Dec., with a hurricane season June–Nov. Plymouth, Jan. 76°F (24·4°C), July 81°F (27·2°C).

CONSTITUTION AND GOVERNMENT

Montserrat is a British Overseas Territory. A new constitution was promulgated in 2011, replacing one enacted in 1989. The head of state is Queen Elizabeth II, represented by a *Governor.* The 2011 constitution provides for greater control of international relations by the government of Montserrat, the strengthening of fundamental rights and freedoms, and the establishment of a national advisory council along with commissions to deal with complaints, integrity, elections and clemency. The *Legislative Assembly* consists of nine elected members and two *ex officio* members (namely the Attorney-General and the Financial Secretary); it sits for five year terms. The cabinet includes a minimum of the *Premier* (formerly the Chief Minister), three other ministers and two *ex officio* members (the Attorney-General and the Financial Secretary).

RECENT ELECTIONS

In elections to the Legislative Council on 8 Sept. 2009 the Movement for Change and Prosperity won six of nine seats with ind. winning the other three.

CURRENT GOVERNMENT

Governor: Adrian Davis (since 8 April 2011).
Premier: Reuben Meade (since 27 Sept. 2011, having been Chief Minister from 10 Sept. 2009–26 Sept. 2011).

ECONOMY

Currency

Montserrat's currency is the *East Caribbean dollar* (*see* ANTIGUA AND BARBUDA: Currency).

Budget

In 2008 the estimated current revenue was EC$40·2m. and estimated current expenditure EC$99·2m.

Performance

In 2006 the economy shrank by 3·8% but there was a slight recovery in 2007 with growth of 2·8%.

Banking and Finance

The East Caribbean Central Bank based in St Kitts and Nevis functions as a central bank. The *Governor* is Sir Dwight Venner. In 2010 there were two commercial banks and 11 offshore banks. Responsibility for overseeing offshore banking rests with the Governor.

ENERGY AND NATURAL RESOURCES

Environment

Carbon dioxide emissions from the consumption and flaring of fossil fuels in 2008 were the equivalent of 18·2 tonnes per capita.

Electricity

Production (2007 estimate) 23m. kWh. Installed capacity (2007 estimate): 10,000 kW.

Agriculture

The volcanic eruptions in 1997 dramatically reduced the area of land under cultivation. Agriculture, now concentrated in the north, is showing signs of recovering. In 2002 there were 2,000 ha. of arable and permanent crop land. The main products have traditionally been potatoes, tomatoes, onions, mangoes and limes. Meat production began in 1994 and the island soon became self-sufficient in chicken, mutton and beef.

Livestock (2002); cattle, 10,000; pigs, 1,000; sheep, 5,000; goats, 7,000.

Forestry
The area under forests in 2010 was 3,000 ha., or 24% of the total land area.

Fisheries
The total catch in 2010 was 24 tonnes.

INDUSTRY
Manufacturing has in recent years contributed about 6% to GDP and accounted for 10% of employment, but has been responsible for about 80% of exports. It has been limited to rice milling and the production of light consumer goods such as electronic components, light fittings, plastic bags and leather goods. The volcanic activity has put a halt to the milling of rice in the exclusion zone and curtailed the production of light consumer goods.

EXTERNAL ECONOMIC RELATIONS
Imports and Exports
Imports in 2005 totalled US$26·2m.; exports, US$1·8m. The USA is the main trading partner.

COMMUNICATIONS
Roads
The percentage of usable roads has declined considerably since the volcanic eruptions of 1995 and 1997. Since then, improving and maintaining Montserrat's road network to a high standard has been one of the island's main objectives. A number of road work projects are under way. In Dec. 2008 there were 2,471 registered vehicles.

Civil Aviation
At the W. H. Bramble airport LIAT used to provide services to Antigua with onward connections to the rest of the eastern Caribbean, but it was closed in June 1997 as volcanic activity increased. A new airport opened in Feb. 2005.

Shipping
Plymouth is the port of entry, but alternative anchorage was provided at Old Bay Road during the volcanic crisis.

Telecommunications
In 2008 there were 2,800 main (fixed) telephone lines; mobile phone subscribers numbered 3,000 in 2008 (50·8 per 100 persons). There were 1,200 internet users in 2008.

SOCIAL INSTITUTIONS
Justice
Law is based on UK common law as exercised by the Eastern Caribbean Supreme Court. Final appeal lies to the UK Privy Council. Law is administered by the West Indies Associated States Court, a Court of Summary Jurisdiction and Magistrate's Courts.

Education
In 1996–97 there were 11 primary schools (only four open), a comprehensive secondary school with three campuses, and a technical college. Schools are run by the government, the churches and the private sector. There is a medical school, the American University of the Caribbean.

In 2000–01 total expenditure on education came to 7·9% of total government spending.

Health
In 2011 Montserrat had one 30-bed hospital, five doctors, and 36 nurses and midwives.

RELIGION
In 2001, 24·4% of the population were Anglican, 19·0% Methodist, 15·9% Pentecostal, 13·0% Roman Catholic and 11·8% Seventh Day Adventist.

CULTURE
Press
In 2006 there were two weekly newspapers.

Tourism
Tourism at one time contributed about 30% of GDP. There were 36,077 visitors including 11,636 cruise ship arrivals in 1994. However, after the volcanic eruptions the tourist industry declined dramatically; over half the island is closed. There were 7,991 visitors in 2006.

FURTHER READING
Fergus, H. A., *Montserrat: History of a Caribbean Colony*. 1994

Pitcairn Island

KEY HISTORICAL EVENTS
Pitcairn was discovered by Carteret in 1767, but remained uninhabited until 1790, when it was occupied by nine mutineers of HMS *Bounty*, with 12 women and six men from Tahiti. Nothing was known of their existence until the island was visited in 1808.

In 2004 six men, comprising some 12% of the island's resident population and including several prominent public figures one of whom was the mayor, were convicted of child sex offences. The scandal divided the island, with defendants arguing that English law did not have jurisdiction on Pitcairn and that underage sex was a long-standing tradition. Sentences ranged up to six-and-a-half years in prison. Several of the convicted served part of their sentences in home detention to enable them to carry on essential island work.

TERRITORY AND POPULATION
Pitcairn Island (4·6 sq. km; 1·75 sq. miles) is situated in the Pacific Ocean, nearly equidistant from New Zealand and Panama (25° 04' S. lat., 130° 06' W. long.). Adamstown is the only settlement. The population in 2008 was 66. The uninhabited islands of Henderson (31 sq. km), Ducie (3·9 sq. km) and Oeno (5·2 sq. km) were annexed in 1902. English is the official language but Pitkern is also spoken.

CLIMATE
An equable climate, with average annual rainfall of 80" (2,000 mm) spread evenly throughout the year. Mean monthly temperatures range from 75°F (24°C) in Jan. to 66°F (19°C) in July.

CONSTITUTION AND GOVERNMENT
The Local Government Ordinance of 1964 constitutes a *Council* of ten members; four councillors and the chairman of the internal committee are elected annually, one is nominated by the Council and two—including the island secretary—are appointed by the *Governor*. It is presided over by the island mayor. There is also a commissioner liaising between the governor and the council. No political parties exist. The Island Magistrate, who is elected triennially, presides over the Council; other members hold office for only one year. Liaison between Governor and Council is through a Commissioner in the Auckland, New Zealand, office of the British Consulate-General.

CURRENT GOVERNMENT

Governor: Vicki Treadell.
  Mayor: Mike Warren.

Government Website: http://www.government.pn

ECONOMY

Currency
New Zealand currency is used.

Budget
For the financial year 2003–04 revenue was NZ$742,000 and expenditure NZ$1,347,000. There is no taxation. Instead all able-bodied men and women are required to perform public work.

ENERGY AND NATURAL RESOURCES

Forestry
The area under forests in 2010 was 4,000 ha., or 83% of the total land area.

Fisheries
The catch in 2010 was approximately three tonnes.

COMMUNICATIONS

Roads
There were 6·4 km of roads in 2011 (all paved). Four-wheeled quad bikes are the main means of transport.

SOCIAL INSTITUTIONS

Justice
The Island Court consists of the Island Magistrate and two assessors.

Education
In 2004 there was one teacher and nine pupils.

CULTURE

World Heritage Sites
Henderson Island was inscribed on the UNESCO World Heritage List in 1988.

FURTHER READING

Murray, S., *Pitcairn Island: the First 200 Years.* 1992

St Helena, Ascension and Tristan da Cunha

KEY HISTORICAL EVENTS

The island of St Helena was uninhabited when discovered by the Portuguese in 1502. It was administered by the East India Company from 1659 and became a British colony in 1834. Napoleon died there in exile in 1821. In 2009 a new constitution saw the territory renamed from St Helena and Dependencies to St Helena, Ascension and Tristan da Cunha.

TERRITORY AND POPULATION

St Helena, of volcanic origin, is 3,100 km from the west coast of Africa. Area, 122 sq. km (47 sq. miles), with a cultivable area of 243 ha. The resident population of St Helena at the 2008 census was 3,981 (excluding the then dependencies of Ascension and Tristan da Cunha) of which St Helena itself had a population of 3,867 with 114 on board the RMS *St Helena* and on yachts in the harbour. In 2010, 60·5% of the population were rural. The capital and port is Jamestown, population (2008) 640.

The official language is English.

Ascension is a small island of volcanic origin, of 88 sq. km (34 sq. miles), 700 miles northwest of St Helena. There are 120 ha. providing fresh meat, vegetables and fruit. The population in 2008 was 1,122.

The island is the resort of sea turtles, rabbits, the sooty tern or 'wideawake', and feral donkeys.

A cable station connects the island with St Helena, Sierra Leone, St Vincent, Rio de Janeiro and Buenos Aires. There is an airstrip (Miracle Mile) near the settlement of Georgetown; the Royal Air Force maintains an air link with the Falkland Islands.

Tristan da Cunha is the largest of a small group of islands in the South Atlantic, lying 2,124 km (1,320 miles) southwest of St Helena, of which they became dependencies on 12 Jan. 1938. Tristan da Cunha has an area of 98 sq. km and a population (2008) of 284, all living in the settlement of Edinburgh. Inaccessible Island (10 sq. km) lies 20 miles west, and the three Nightingale Islands (2 sq. km) lie 20 miles south of Tristan da Cunha; they are uninhabited. Gough Island (90 sq. km) is 220 miles south of Tristan and has a meteorological station.

Tristan consists of a volcano rising to a height of 2,060 metres, with a circumference at its base of 34 km. The volcano, believed to be extinct, erupted unexpectedly early in Oct. 1961. The whole population was evacuated without loss and settled temporarily in the UK; in 1963 they returned to Tristan. Potatoes remain the chief crop. Cattle, sheep and pigs are now reared, and fish are plentiful.

The original inhabitants were shipwrecked sailors and soldiers who remained behind when the garrison from St Helena was withdrawn in 1817.

At the end of April 1942 Tristan da Cunha was commissioned as HMS *Atlantic Isle*, and became an important meteorological and radio station. In Jan. 1949 a South African company commenced crawfishing operations. An Administrator was appointed at the end of 1948 and a body of basic law brought into operation. The Island Council, which was set up in 1932, consists of a Chief Islander, three nominated and eight elected members, under the chairmanship of the Administrator.

SOCIAL STATISTICS

2008 figures for St Helena: births, 36; deaths, 44. Expectation of life at birth (2007): females, 77·3 years; males, 70·8 years. Annual average growth rate of resident population, 1998–2008, –2·4%.

CLIMATE

A mild climate, with little variation. Temperatures range from 75–85°F (24–29°C) in summer to 65–75°F (18–24°C) in winter. Rainfall varies between 13" (325 mm) and 37" (925 mm) according to altitude and situation.

CONSTITUTION AND GOVERNMENT

A new constitution came into force on 1 Sept. 2009, replacing its 20-year old predecessor. Under the terms of the new constitution, the territory changed its official name from St Helena and Dependencies to St Helena, Ascension and Tristan da Cunha. It included a bill of rights, allowing citizens to appeal to local courts on human rights issues rather than address the European Court of Human Rights in Strasbourg as they had done previously. Constraints were placed on the powers of the Governor (who no longer holds the title of 'Commander-in-Chief') and the independence of the judiciary and public service were constitutionally enshrined.

The territory's *Legislative Council* consists of twelve elected members, three non-voting *ex officio* members (the Chief Secretary, the Financial Secretary and the Attorney General) and the Speaker and Deputy Speaker. The Governor is advised by an *Executive Council* comprising three non-voting *ex officio* members (the Chief Secretary, the Financial Secretary and the Attorney General) and five elected members from the Legislative Council.

The Governor must, except in certain prescribed circumstances, act in accordance with the advice of the Executive Council.

RECENT ELECTIONS

The last Legislative Council elections were on 4 Nov. 2009. Only non-partisans were elected.

CURRENT GOVERNMENT

Governor: Mark Capes.
   Administrator of Ascension: Colin Wells.
   Administrator of Tristan da Cunha: Sean Burns.

Government Website: http://www.sainthelena.gov.sh

ENERGY AND NATURAL RESOURCES

Environment

Carbon dioxide emissions from the consumption and flaring of fossil fuels in 2008 were the equivalent of 1·9 tonnes per capita.

Electricity

Production in 2007 totalled 8m. kWh. Installed capacity in 2007 was 4,000 kW.

Agriculture

In 2007 there were about 4,000 ha. of arable land.

Fisheries

The total catch in 2010 was 864 tonnes.

INDUSTRY

Labour

In 2000 there were 270 registered unemployed persons.

COMMUNICATIONS

Roads

There were (2003) 94 km of all-weather motor roads. There were 1,931 vehicles in 2002.

Shipping

There is a service from Portland (UK) twice a year, and links with South Africa and neighbouring islands.

Telecommunications

In 2006 there were 2,200 main telephone lines in operation. There were 800 internet users in 2008.

SOCIAL INSTITUTIONS

Justice

Police force, 32; cases are dealt with by a police magistrate.

Education

In 2002–03 there were eight schools with, in 1999–2000, 87 teachers and 860 pupils. The Prince Andrew School (opened in 1989) offers vocational courses leading to British qualifications.

Health

There were four doctors, one dentist and one hospital in 2001.

RELIGION

There are ten Anglican churches, four Baptist chapels, three Salvation Army halls, one Seventh Day Adventist church and one Roman Catholic church.

CULTURE

World Heritage Sites

St Helena, Ascension and Tristan da Cunha has one site on UNESCO's World Heritage List: Gough and Inaccessible Islands (inscribed on the list in 1995 and 2004), two of the least disturbed islands and marine eco-systems in the South Atlantic.

FURTHER READING

Statistical office: Development and Economic Planning Department, 3 Main Street, Jamestown, St Helena Island, South Atlantic Ocean, STHL 1ZZ.
Website: http://www.sainthelena.gov.sh/pages/statistics.html

South Georgia and the South Sandwich Islands

KEY HISTORICAL EVENTS

The first landing and exploration was undertaken by Capt. James Cook, who formally took possession in the name of George III on 17 Jan. 1775. British sealers arrived in 1788 and American sealers in 1791. Sealing reached its peak in 1800. A German team was the first to carry out scientific studies there in 1882–83. Whaling began in 1904 and ceased in 1966, and the civil administration was withdrawn. Argentine forces invaded South Georgia on 3 April 1982. A British naval task force recovered the Island on 25 April 1982.

TERRITORY AND POPULATION

South Georgia lies 1,300 km southeast of the Falkland Islands and has an area of 3,760 sq. km. The South Sandwich Islands are 760 km southeast of South Georgia and have an area of 340 sq. km. In 1993 crown sovereignty and jurisdiction were extended from 19 km (12 miles) to 322 km (200 miles) around the islands. There is no permanent population. The British Antarctic Survey operate a fisheries science facility at King Edward Point and a biological station on Bird Island. The South Sandwich Islands are uninhabited.

CLIMATE

The climate is wet and cold, with strong winds and little seasonal variation. 15°C is occasionally reached on a windless day. Temperatures below –15°C at sea level are unusual.

CONSTITUTION AND GOVERNMENT

Under the new constitution which came into force on 3 Oct. 1985 the Territories ceased to be dependencies of the Falkland Islands. The Government of South Georgia and the South Sandwich Islands (GSGSSI) administers the islands. The local administration is the responsibility of the Government Officer based at King Edward Point. Executive power is vested in a Commissioner, who is also the Governor of the Falkland Islands. On matters relating to defence, the Commissioner consults the officer commanding Her Majesty's British Forces in the South Atlantic. The Commissioner, whenever practicable, consults the Executive Council of the Falkland Islands on the exercise of functions that in his opinion might affect the Falkland Islands. There is no Legislative Council. Laws are made by the Commissioner (Nigel Haywood, CVO, resident in the Falkland Islands).

ECONOMY

Budget

The total projected revenue of the Territories (2006) was £4,037,700, of which 85% from fishing licenses, 8% landing fees, 3% philatelic sales and 2% harbour dues. Expenditure (projected), £4,265,900, includes 50% fisheries research and protection, 12% King Edward Point running costs, 11% observer fees and 6% Grytviken remediation and maintenance projects.

COMMUNICATIONS

The bases at King Edward Point and Bird Island have modern satellite communication systems. King Edward Point is regularly visited by the GSGSSI Fishery Patrol Vessel. Other visiting vessels include cruise ships, BAS research ships, warships and auxiliaries, fishing vessels and yachts.

SOCIAL INSTITUTIONS

Justice

There is a Supreme Court for the Territories and a Court of Appeal in the United Kingdom. Appeals may go from that court to the Judicial Committee of the Privy Council. The British Antarctic Survey base commander at King Edward Point is usually appointed a magistrate.

CULTURE

Tourism
In the region of 4,000 tourists visit the island annually.

FURTHER READING

Headland, R.K., *The Island of South Georgia*. 1984

Sovereign Base Areas of Akrotiri and Dhekelia in Cyprus

KEY HISTORICAL EVENTS

The Sovereign Base Areas (SBAs) are those parts of the island of Cyprus that stayed under British jurisdiction and remained British sovereign territory when the 1960 Treaty of Establishment created the independent Republic of Cyprus. The Akrotiri facility formed a strategic part of the West's nuclear capacity during the Cold War. The SBAs were used for the deployment of troops in the Gulf War in 1991. Military intelligence is now the key role of the SBAs. The construction of massive antennae at the RAF communications base at Akrotiri sparked violent riots in 2001 and 2002, led by a Greek Cypriot MP. In Feb. 2003 the British Government offered to surrender approximately half the area of the SBAs as an incentive for a settlement between the Greek and Turkish administrations in Cyprus.

TERRITORY AND POPULATION

The Sovereign Base Areas (SBAs), with a total land area of 254 sq. km (98 sq. miles), comprise the Western SBA (123 sq. km), including Episkopi Garrison and RAF Akrotiri (opened 1956), and the Eastern SBA (131 sq. km), including Dhekelia Garrison. The SBAs cover 3% of the land area of the island of Cyprus. There are approximately 3,000 military personnel and approximately 5,000 civilians. The British Government has declared that it will not develop the SBAs other than for military purposes. Citizens and residents of the Republic of Cyprus are guaranteed freedom of access and communications to and through the SBAs.

The SBAs are administered as military bases reporting to the Ministry of Defence in London. The Administrator is the Commander, British Forces Cyprus. The joint force headquarters are at Episkopi. Greek and English are spoken.

CURRENT GOVERNMENT

Commander British Forces Cyprus: Air Vice Marshal Graham Stacey (since 4 Nov. 2010).

The Turks and Caicos Islands

KEY HISTORICAL EVENTS

After a long period of rival French and Spanish claims the islands were secured to the British Crown in 1766, and became a separate colony in 1973 after association with the colonies of the Bahamas and Jamaica. In 2009 the British government imposed direct rule after allegations of widespread corruption among the islands' ruling class.

TERRITORY AND POPULATION

The Islands are situated between 21° and 22°N. lat. and 71° and 72°W. long, about 80 km east of the Bahamas, of which they are geographically an extension. There are over 40 islands, covering an estimated area of 500 sq. km (193 sq. miles). Only seven are inhabited: Grand Caicos, the largest, is 48 km long by 3 to 5 km broad; Grand Turk, the capital and main political and administrative centre, is 11 km long by 2 km broad. Population (2012 census, provisional) 31,458; Grand Turk, 4,831; Middle Caicos, 168; North Caicos, 1,312; Parrot Cay, 131; Providenciales, 23,769; Salt Cay, 108; South Caicos, 1,139. 54·8% of the population were rural in 2000.

The official language is English.

SOCIAL STATISTICS

2004: births, 300; deaths, 218. Population growth rate, 2004, 9·4%.

CLIMATE

An equable and healthy climate as a result of regular trade winds, though hurricanes are sometimes experienced. Grand Turk, Jan. 76°F (24·4°C), July 83°F (28·3°C). Annual rainfall 21".

CONSTITUTION AND GOVERNMENT

A new constitution entered force on 9 Aug. 2006. It granted the Turks and Caicos Islands further self-government, while at the same time incorporating provisions enabling the Government to fulfil their responsibilities for the territory. It made provision for the establishment of an Advisory National Security Council, and changed the title of Chief Minister and Deputy Chief Minister to Premier and Deputy Premier. Following the resignation of premier Michael Misick in March 2009 on corruption charges, the UK government dissolved the cabinet and the House of Assembly in Aug. 2009 and placed the island under direct rule. Despite an initial commitment to hold House of Assembly elections by July 2011 at the latest, direct rule continued to be in place until elections were held in Nov. 2012. The Cabinet (formerly Executive Council) comprises the Governor, the Premier, no more than six other Ministers, the Deputy Governor and the Attorney General. The House of Assembly (formerly Legislative Council) consists of a Speaker, one official member (the Attorney General), 15 elected members and four appointed members.

RECENT ELECTIONS

At elections held on 9 Nov. 2012 for the 15 elective seats in the House of Assembly, the Progressive National Party won 8 seats and the People's Democratic Movement 7. The new government subsequently resumed local administration following three years of direct rule by Britain.

CURRENT GOVERNMENT

Governor: Damian Roderic Todd; b. 1959 (since 12 Sept. 2011).
  Premier: Rufus Ewing (since 13 Nov. 2012).

Government Website: http://www.gov.tc

ECONOMY

Overview
The economy is based on free-market private sector-led development. The focus is on the service sector, with tourism and finance still the dominant industries.

Currency
The US dollar is the official currency. Inflation was 2·9% in 2007 (3·7% in 2006).

Budget
In 2004–05 current revenues were US$118m. and current expenditures US$122m.

Performance
Real GDP growth was 12·7% in 2005 (11·6% in 2004).

Banking and Finance
There were six commercial banks in 2004. Offshore finance is a major industry.

ENERGY AND NATURAL RESOURCES

Environment
Carbon dioxide emissions from the consumption and flaring of fossil fuels in 2008 were the equivalent of 0·7 tonnes per capita.

Electricity
Electrical services are provided to all of the inhabited islands. Total electricity production for 2007 was 182m. kWh. Installed capacity in 2007 was an estimated 50,000 kW. Consumption per capita in 2007 was 5,221m. kWh.

Agriculture
Farming is done on a small scale mainly for subsistence.

Forestry
Forests covered 34,000 ha. in 2010, or 80% of the land area.

Fisheries
In 2010 the total catch was 5,446 tonnes.

INDUSTRY

Labour
In 2001, out of a total population of 13,436 aged 15 or over, 10,181 were working, 1,094 unemployed and 2,161 economically inactive.

EXTERNAL ECONOMIC RELATIONS

Imports and Exports
Imports (c.i.f.) in 2009 totalled US$375·4m.; exports (f.o.b.) in 2009 were US$20·8m. The USA accounted for over 99% of foreign trade in 2009. The main export are seafood products, and machinery and transport equipment. Major imports are food and beverages, manufactured goods, and machinery and transport equipment.

COMMUNICATIONS

Civil Aviation
The international airports are on Grand Turk and Providenciales. International services are provided by Air Canada, Air Jamaica, American Airlines, Bahamasair, British Airways, Delta Airlines, TCI Skyking, Tropical Airways d'Haiti and US Airways.

An internal air service provides regular daily flights between the inhabited islands.

Shipping
The main ports are at Grand Turk, Cockburn Harbour and Providenciales. There is a service to Miami.

Telecommunications
There are internal and international cable, telephone, telegraph and fax services.

SOCIAL INSTITUTIONS

Justice
Laws are a mixture of Statute and Common Law. There is a Magistrates Court and a Supreme Court. Appeals lie from the Supreme Court to the Court of Appeal which sits in Nassau, Bahamas. There is a further appeal in certain cases to the Privy Council in London.

Education
The adult literacy rate is 98%. Education is free between the ages of five and 14 in the ten government primary schools; there are also four private primary schools. Total school enrolment in 2004–05 was 1,931 in government primary schools and 1,282 in government secondary schools. There were 1,670 pupils enrolled in the private schools.

In 2005 public expenditure on education came to 11·8% of total government spending.

Health
In 2004 there were 17 doctors, two dentists and 43 hospital beds.

RELIGION
There are Anglican, Catholic, Methodist, Baptist and Evangelist groups.

CULTURE

Press
In 2006 there were two weekly newspapers.

Tourism
There were 248,300 non-resident tourists in 2006 (of which 221,400 were from elsewhere in the Americas), up from 176,100 in 2005.

FURTHER READING
Statistical Office: Department of Economic Planning and Statistics, Ministry of Finance, South Base, Grand Turk.
Website: http://www.depstc.org

UNITED STATES OF AMERICA

CT	CONNECTICUT	NJ	NEW JERSEY
DE	DELAWARE	PV	PENNSYLVANIA
MA	MASSACHUSETTS	RI	RHODE ISLAND
MD	MARYLAND	VT	VERMONT
NH	NEW HAMPSHIRE	WV	WEST VIRGINIA

Capital: Washington, D.C.
Population projection, 2015: 323·89m.
GNI per capita, 2011: (PPP$) 43,017
HDI/world rank: 0·910/4
Internet domain extension: .us

KEY HISTORICAL EVENTS

The earliest inhabitants of the north American continent can be traced back to Palaeolithic times. The Pueblo culture in modern-day Colorado and New Mexico flourished from the 11th to the 14th century AD. In the 12th century permanent settlements appeared in the east where cultivation and fishing supported major fortified towns. The first Europeans to make their presence felt were the Spanish, who based themselves in Florida before venturing north and west. Santa Fe in New Mexico was founded in 1610. But by the mid-17th century there was competition centred on Quebec from the French who colonized the banks of the St Lawrence River.

Elizabethan adventurers were eager to exploit the New World but it was not until 1607 that an English colony was established.

This was at Jamestown in what is now southern Virginia. After a perilous start when disease and malnutrition carried off most of the settlers, Virginia's population grew rapidly to meet the European demand for tobacco. Maryland, originally a refuge for persecuted Catholics, also thrived on the tobacco trade. To make up for the shortage of labour, slaves were imported from Africa.

In 1620 a hundred pilgrims landed at Plymouth Rock to found a Puritan enclave, which became the colony of Massachusetts. Other settlements soon followed, accommodating a broad range of Christian radicals fleeing persecution. Not all were tolerant of beliefs that differed from their own. Pennsylvania, the colony named after the Quaker William Penn, was exceptional in offering freedom of worship to 'all persons who confess and acknowledge the one almighty and eternal God'. In 1664 the British took control of neighbouring Dutch colonies. New Amsterdam became New York. Almost all of the eastern seaboard was now claimed by British settlers who were also venturing inland.

Their main European rivals were the French who claimed a vast area around and to the southwest of the Great Lakes. With American Indian tribes allied to both sides, there was heavy fighting in 1744 and 1748. But within a decade British forces had captured most of the French strongholds. After the Treaty of Paris in 1763, Britain commanded the whole of North America east of the Mississippi while Spain, having surrendered Florida, gained Louisiana from France. For a brief period colonization was restricted to the area east of the Appalachians, the rest of the territory being reserved for Indian tribes. This soon became a point of issue between the settlers who were intent on expansion and the government in London, which wanted a settled, self-supporting community benefiting British trade. Having disposed of the French threat, the colonists felt confident enough to defy orders that ignored their interests. In particular, they objected to the Navigation Acts which required goods to be carried in British vessels and to various taxes imposed without consultation. 'No taxation without representation' became a rallying cry for disaffected colonists. The centre of opposition was Boston, scene of the infamous 'tea party' when, in 1773, militants destroyed a cargo of East India tea. In 1775 the arrest of rebel ringleaders served only to provoke the 13 colonies to co-operate in further acts of rebellion, including the setting up of a *de facto* government which appointed George Washington commander of American forces.

Independence
The War of Independence was by no means a clear-cut affair. British forces, never more than 50,000 strong, were supported by a powerful body of colonists who remained loyal to the Crown. The war lasted for seven years from 1776 with both sides often getting close to a conclusive victory. The decisive moment came at last with the surrender of Gen. Burgoyne and his 8,000 troops in upper New York state in Oct. 1777, a defeat that persuaded a cautious France to enter the war. Under the peace terms secured in 1783 Britain kept Canada leaving the new United States with territory stretching from the Atlantic to the Mississippi. A constitution based on democratic principles buttressed by inalienable rights including the ownership of property came into force in 1789. It allowed for a federal government headed by a president and executive, a legislature with a House of Representatives and a Senate, and a judiciary with ultimate authority on constitutional matters exercised by a Supreme Court. The first president was George Washington, who was elected in 1789. In 1800 Washington, D.C. was declared the national capital.

Hostilities with Britain resumed in 1812 amidst accusations that Britain was using the excuse of the Napoleonic wars to harass American shipping and to encourage Indian resistance to expansion into the Midwest. Most of the fighting took place on the Canadian border where an attempted invasion was decisively repulsed. But Louisiana, having reverted to French rule and subsequently sold to the USA, was secured for the Union. With the exception of Louisiana, other American territories that had once been part of the Spanish empire fell to Mexico. But not for long. In 1836 Texas broke away from Mexico, surviving as an independent republic until 1845 when it was annexed by the USA. This provoked war with Mexico which ended in 1848 with the USA taking over what are now the states of California, Arizona, Colorado, Utah, Nevada and New Mexico. Any temptation there might have been for European involvement in the struggle was removed by the Monroe Doctrine, a declaration by President Monroe that interference from the Old World in matters concerning the western hemisphere would not be tolerated. It was a measure of the growing military and economic self confidence of the USA that such a warning, delivered in 1823, was taken seriously.

The westward expansion began soon after independence but accelerated with the destruction of Indian power and the removal of the native population to designated reservations. In 1846 a long-running dispute with Britain confirming US title to Oregon acted as a spur to migration as did the Californian gold rush of 1848. By the 1850s the railway network was bringing people and economic prosperity to the mid-west. Population quadrupled between 1815 to 1860, from 8m. to almost 31m. In 1862 the Homestead Act allocated 160 acres to anyone who was ready to farm it. By 1890 the west was won.

Civil War
The transition from a rural society to an industrial power of world importance created tensions, not least between the slave-owning southern states and the rest of the Union which favoured the abolition of slavery. Economic as well as humanitarian factors were in play since the North resented the advantage cheap labour gave to the South. The opposing view held that the South, by now the world's largest cotton producer, depended on slavery for its commercial survival. Mutual antagonism came to a head with the secession of the southern states from the Union in 1860–61 and their formation as a Confederacy. Despite sporadic outbreaks of violence, civil war was not in prospect until Confederate troops fired on the US flag at Fort Sumter. President Abraham Lincoln ordered a blockade of the South. The recruitment of rival armies followed within weeks. The war turned out to be much bloodier than anyone had expected. More American lives were lost in the Civil War than in the two world wars combined. The military balance was maintained until 1863 when the North secured a crushing victory at the Battle of Gettysburg. However, the war continued until April 1865 when Robert E. Lee surrendered to Ulysses S. Grant at Appomattox Courthouse in Virginia. A few days later Lincoln was assassinated, a loss that the southern states had subsequent cause to regret. Contrary to Lincoln's hopes, a generous settlement was now out of the question. Instead of a gradual transition to a new society, the South was rushed into a social revolution. This in turn led to terrorist violence and acts of vengeance against freed slaves. From this carnage emerged the notorious Ku Klux Klan as the standard bearer of lynch law. While the 13th amendment prohibited slavery, political freedom was denied to the black community by state-imposed literacy tests and discriminatory property taxes.

That America had interests beyond its own borders was made evident by the Spanish war of 1898 which resulted in the USA becoming the dominant power in the Caribbean. However, the effort to take over in the Philippines came up against Filipino resistance and led to a heavy death toll.

By 1900 the USA rivalled Britain and Germany as the world's dominant power. With vast natural resources and a manufacturing capacity that secured 11% of world trade, it was clear that Europe was soon to lose its grip on world affairs. Ironically, though, it was Europe as the chief supplier of labour that gave the USA the impetus it needed to fulfil its promise. Between 1881 and 1920, 23m. immigrants entered the USA, the largest population movement ever recorded.

Given the heterogeneous background of the American population in the early 20th century it is scarcely surprising that popular opinion was against involvement in the First World War. But events, including German U-boat harassment of American shipping, soon proved that isolationism was not an option. It was not until 1917 that America joined the hostilities but the resurgence of energy created by the arrival of the American Expeditionary Force was critical to the Allied breakthrough.

Post-war America, relatively unscathed by the European conflict, was unquestionably the most powerful nation and as such was able to dictate terms at the Versailles peace conference. But President Wilson's 'fourteen points' which set out a plan for collective security policed by a League of Nations failed to win support in the one country that was critical to its success. The Treaty was rejected by the Senate in 1920 and America retreated once again into isolationism.

A resumption of economic growth was accompanied by a struggle to impose a common set of values, chiefly white and Protestant, on a diverse population. To outsiders the most extraordinary experiment in social engineering was Prohibition, a federal imposed attempt to outlaw all alcoholic drinks. Whatever gain there was to the health of the nation, the chief beneficiaries were the bosses of organized crime.

New Deal

Dreams of everlasting prosperity were shattered by the 1929 Stock Market Crash. A succession of bank failures was followed by widespread bankruptcies and mass unemployment which sent the economy into a further downward spin. The beginning of the end to the agony came with the election to the presidency of Franklin D. Roosevelt, who pushed through Congress a series of radical measures known collectively as the New Deal, aimed at revitalizing the nation. The abandonment of the gold standard, cheap loans to restart factories and farms and huge investment in public works proved to be the key to recovery. Nonetheless, unemployment remained high until production was boosted by the demands of another world war. In 1935 Roosevelt's social security act provided the bare bones of an American welfare state.

Roosevelt was well aware of the dangers to the USA if the fascist dictators were allowed to triumph, but as in 1914, there was formidable opposition to direct involvement. Roosevelt compromised by supplying Britain with much needed armaments on favourable terms. But it was events in Asia rather than in Europe that eventually persuaded America of the need for direct action. Opposition to Japanese expansion into China and southeast Asia, including an oil embargo and a freezing of Japanese assets in the USA, brought a savage retaliation at Pearl Harbor, when much of the US fleet was destroyed. America declared war on Japan while Germany declared war on America. The US military effort focused initially on the Pacific but after 1942 American forces were also committed to the campaign in north Africa and Europe. With the D-Day landings in June 1944, US troops led the attack on Germany and in May 1945, within a month of Roosevelt's death, Germany surrendered. By then Vice President Harry Truman had been confirmed as Roosevelt's successor and forced a Japanese surrender by sacrificing Hiroshima and Nagasaki to the atomic bomb.

This time, in the aftermath of war, the USA needed no encouragement to assume the leadership of the free world. The threat of a Soviet takeover in Europe was countered by the formation of NATO in 1949 and the provision of dollar aid under the 1947 Marshall Plan to kick-start European economic recovery. In addition, the Truman doctrine provided a $400m. aid package for the Turkish and Greek governments. The risk of a return to isolationism receded still further when China fell to communism. In 1950 American troops went to the aid of South Korea when it was invaded by the communist North. Though technically under the aegis of the UN, the campaign was an almost entirely American affair led by General Douglas MacArthur. When Chinese forces became involved, MacArthur spoke openly of extending the war to the Chinese mainland, a threat countered strongly by President Truman who forced MacArthur's resignation to establish undisputed political control over the military.

A ceasefire was negotiated after Dwight Eisenhower was elected president in 1953. The USA took the lead in setting up the South East Asia Treaty Organization on the same lines as NATO. Eisenhower had a decisive influence on the Suez crisis in 1956 when he refused to support the invasion of Egypt by British, French and Israeli forces. Domestically he made little headway against a powerful Democratic opposition in both Houses of Congress. His attempts to thaw the Cold War also met with frustration. He handed over the Republican presidential candidacy to his vice-president Richard Nixon, who lost the 1960 election by a slim margin to John F. Kennedy.

Civil Rights

In the early 1960s civil rights were high on the political agenda. The thuggish tactics of Senator Joe McCarthy and the House Committee on Un-American Activities during the previous decade brought into focus basic democratic freedoms guaranteed by the Constitution while growing protests against racial discrimination led to legislation to enforce equality of opportunity in education and employment. The civil rights movement peaked in the early 1960s when the imposition of federal law in the South led to acts of violence against liberal protesters. Foremost among the campaigners for racial equality was Martin Luther King, Jr. who was awarded the Nobel Peace Prize in 1964 and who was assassinated four years later.

Social tensions were exacerbated by the Cold War confrontation. In his first year of office Kennedy was embarrassed by the failed Bay of Pigs invasion when anti-Castro Cubans, trained and supported by the CIA, attempted to overthrow the country's communist regime. The building of the Berlin Wall in Aug. 1961 symbolized a hardening of the Cold War. In 1962 Kennedy had to confront the prospect of the Soviets placing missiles in Cuba. The prospect of world war was only too real until an agreement between the two nations allowed for a withdrawal of the missiles on the condition of a US promise not to invade Cuba. The incident prompted a thawing in East–West relations and in 1963 the USA, UK and USSR signed the Limited Test Ban Treaty which, for the first time, put a brake on the spread of nuclear weapons. Less hopeful was the acceleration of the conflict in Vietnam. Kennedy increased the American military presence in South Vietnam from 700 at the beginning of his term in office to 15,000 to counter the threat of communist domination by the North. Domestically, Kennedy's government pledged $1·2bn. for social and housing programmes.

Kennedy was assassinated in Dallas in Nov. 1963 and his vice president, Lyndon Johnson, was inaugurated as his successor. Johnson oversaw the implementation of civil rights legislation initiated by the Kennedy administration, epitomized by the Voting Rights Act, and also introduced Medicare (health insurance for the elderly). By 1966 over 350,000 American troops were in Vietnam and by the following year almost 80,000 Americans had been killed or wounded. The public turned against involvement in southeast Asia, not least because increased military expenditure led to a delay in domestic reforms.

Watergate

Johnson decided not to contest the 1968 presidential election. His likely successor for the Democrat nomination was John Kennedy's brother, Bobby, but he was assassinated in June of that year. The Republican nominee, Richard Nixon, won the presidency. With falling support for US involvement in Vietnam, he reduced the number of troops stationed there from 550,000 in 1969 to 30,000 three years later but authorized military operations in North Vietnam, Laos and Cambodia in the hope of forcing North Vietnam to the negotiating table. Elsewhere, he signed the Strategic Arms Limitation Treaty (SALT) with Moscow in 1972 and relaxed trade restrictions against China.

Nixon was re-elected as president in 1973 and shortly afterwards agreed a ceasefire with North Vietnam. However, his second term of office was cut short by the Watergate scandal. The charges against him centred on White House-released taped transcripts of discussions in which Nixon authorized a cover-up of a break-in at the Democratic party headquarters in the Watergate complex, Washington, D.C. in 1972. Threatened with Congressional impeachment, Nixon announced his resignation in Aug. 1974.

Gerald Ford, who replaced Nixon, granted his predecessor a controversial 'full, free and absolute pardon'. Ford lost the 1976 presidential election to Democrat Jimmy Carter. Perceived as a Washington outsider, Carter's often strained relations with Congress and the Senate obstructed his domestic agenda. The economy suffered and by 1980 inflation and unemployment were both running high. Internationally, he secured the neutrality of the Panama Canal and brokered the influential Camp David talks between Egypt and Israel. He also re-established diplomatic ties with China. Henry Kissinger, who was secretary of state for part of Nixon's and all of Carter's years in office, oversaw America's withdrawal from southeast Asia, winning the Nobel Peace Prize jointly with his North Vietnamese counterpart Le Duc Tho in 1974. Attempts at further improving US–Soviet relations were scuppered when the signing of the Strategic Arms Limitation Treaty (SALT II) was postponed because of the Soviet invasion of

Afghanistan in 1979. The incursion also led to a US boycott of the 1980 Moscow Olympics. Radical Iranian students stormed the US embassy in Tehran in late 1979 and seized over 50 US hostages. After a year of negotiations, a secret US military rescue mission failed and contributed to Republican Ronald Reagan's landslide victory at the 1980 presidential polls.

Reagan's economic policies, known as 'Reaganomics', redefined American society in the 1980s. In his first year of office he introduced a 25% tax cut for individuals and corporations. He slashed welfare but increased military expenditure. In terms of governmental structure, he was intent on delegating many federal programmes to state and local levels. A recession in 1982 prompted tax increases and set the pattern of boom and bust that characterized his tenure. In 1986 he reduced the number of tax rates, abolishing tax altogether for many low-income earners. In Oct. 1987 the stock market collapsed, losing a third of its value over two months, and by the end of his presidency, Reagan had seen the national debt more than triple to $2·5trn.

End to the Cold War

Relations between the USA and USSR deteriorated in the early 1980s. The shooting down of a South Korean airliner carrying American citizens in 1983 led to a further deployment of US missiles in Western Europe while US proposals for the Strategic Defense Initiative (known as 'Star Wars') added to tensions. In 1983 the USA invaded Grenada, scene of a coup, in a bid to curb Soviet–Cuban influence in the Caribbean. However, relations between the two superpowers improved in the mid-eighties after successful negotiations on nuclear arms limitations. Reagan met Soviet leader Mikhail Gorbachev in 1985 and in 1987 the two leaders signed a treaty in Washington, D.C. agreeing to destroy a range of intermediate-range nuclear weapons.

Reagan's foreign policy elsewhere was unstinting in its protection of US interests. In 1986 he bombed Tripoli after Libya was accused of involvement in the bombing of a nightclub in West Berlin which killed two American servicemen. The following year he became embroiled in the Iran-Contra affair. The CIA was found to have sold arms to Iran to fund anti-communist guerrillas in Nicaragua. Reagan and his deputy, George Bush, were cleared of direct involvement but Reagan was censured for allowing the affair to develop.

Bush took over the presidency in 1989 and continued an active foreign policy. At the end of 1989 he authorized the invasion of Panama to remove Gen. Manuel Antonio Noriega from power. The collapse of the Soviet empire in 1990 extended US economic aid to Eastern Europe and Bush signed a non-aggression pact with Soviet leader Mikhail Gorbachev which effectively ended the Cold War. In 1990–91 Bush led a coalition of European and Arab states to counter the Iraqi invasion of Kuwait. Around 500,000 US troops were stationed in the Persian Gulf and when trade embargoes and diplomacy failed to persuade Iraq to withdraw, Bush authorized a military offensive in Jan. 1991. By the end of Feb. Kuwaiti independence had been restored. Domestically, Bush was badly damaged when he was forced to raise taxes despite his election promise of 'no new taxes'.

The Democrats regained control of the White House with the election of Bill Clinton in 1992. Clinton combined economic recovery at home with an active foreign policy which underlined America's role as the only superpower. He secured the passage of the North American Free Trade Agreement, which created a free-trade zone between the United States, Canada and Mexico, cut the United States budget deficit by 50% in his first term and in his second term authorized America's first tax cut since 1981. Unemployment reached its lowest levels since the late-1960s and in 1998 there was a federal budget surplus for the first time in almost 30 years. His social legislation included anti-crime provisions, the Family and Medical Leave Act, a welfare reform bill and an increase in the minimum wage. He also appointed Madeleine Albright as the first-ever female secretary of state.

On the international scene Clinton brokered talks between the Palestinian leader Yasser Arafat and Israeli Prime Minister Yitzhak Rabin which resulted in limited Palestinian self-rule. He sent peacekeeping troops both to Bosnia and Herzegovina and to Haiti. In 1995 he was instrumental in securing the Dayton accords that offered peace between Yugoslavia, Croatia and Bosnia and Herzegovina. Relations with Vietnam were normalized and diplomatic and trade links with China much improved. He was also an important figure in the formulation of the 1998 Good Friday agreement which sought to reach a peace settlement in Northern Ireland.

Clinton retained a hard-line stance against Iraq, sending forces against Saddam Hussein in 1994, 1996 and 1998. Sudan and Afghanistan were attacked in 1998 having been linked with the al-Qaeda terrorist network held responsible for the bombing of US embassies in Tanzania and Kenya. In 1999 there was a Clinton-led NATO campaign of air strikes against Yugoslavia when the country's leaders refused to end a campaign of violence against ethnic Albanians in Kosovo. Yugoslav president Slobodan Milošević was forced to withdraw his troops and allow an international peacekeeping force in Kosovo.

However, Clinton's second term of office was dominated by scandal. He reached an out-of-court agreement with Paula Jones, a state government employee who had accused him of sexual harassment. Clinton and his wife, Hillary, were also accused of criminal wrongdoing over a land deal in Arkansas, known as Whitewater, though they were both eventually cleared of the charges. Most damagingly, Clinton had an affair with Monica Lewinsky, a White House intern. Having denied the sexual nature of the affair under oath, Clinton was impeached for perjury and obstruction of justice although the senate trial ended when neither motion gained a simple majority.

In 2000 Clinton's vice president, Al Gore, lost the presidential election to George W. Bush, son of the earlier President George Bush. Gore was defeated despite winning the popular vote. In 2001 Bush's first budget included a $1·25trn. tax cut. He attracted international criticism in his early months in office for refusing to ratify the Kyoto Agreement on global warming and climate change and for his bid to replace the 1972 Anti-Ballistic Missile Treaty with a new accord allowing for a missile defence system in the United States. Following talks with Russian President Vladimir Putin in June 2001 the two leaders signed an anti-nuclear deal to reduce their respective strategic nuclear warheads by two-thirds over the next ten years.

War on Terrorism

On 11 Sept. 2001 the heart of New York City was devastated after hijackers flew two jet airliners into the World Trade Center. A plane also crashed into the Pentagon, in Washington, D.C., and a fourth hijacked plane crashed near the town of Shanksville, Pennsylvania. The death toll, initially put at 6,700, was eventually lowered to 2,753, with 67 countries reporting dead or missing citizens. Osama bin Laden, the Saudi dissident leader of the al-Qaeda terrorist network and believed to be living in Afghanistan at the invitation of the ruling Taliban, immediately became the chief suspect and military action against Afghanistan followed, with air strikes beginning on 7 Oct. 2001. Despite the UN establishing a fragile multi-party government in Afghanistan, the USA continues to carry out special missions against Taliban and al-Qaeda targets.

In early 2002 Bush declared North Korea, Iran and Iraq 'an axis of evil' and by Sept. 2002 was pressing the UN to act against Iraq. Bush's foreign policy was played out against a background of domestic recession. On 20 March 2003 US forces, supported by the UK, launched attacks on Iraq, and initiated a war aimed at 'liberating Iraq'. On 9 April 2003 American forces took control of central Baghdad, effectively bringing an end to Saddam Hussein's rule. In Nov. 2004 Bush won a second term as president, which was dominated by continuing military engagement in Iraq and Afghanistan and by the collapse in 2007 of the US sub-prime mortgage sector. He was succeeded by the Democrat Barack Obama, who won the presidential election of 2008. Obama took office with the global financial system in turmoil. In May 2011 he ordered the military strike that led to the death of Osama bin Laden in Pakistan. He won a second term with victory in the 2012 presidential election.

TERRITORY AND POPULATION

The United States is bounded in the north by Canada, east by the North Atlantic, south by the Gulf of Mexico and Mexico, and west by the North Pacific Ocean. The area of the 50 states of the USA plus the District of Columbia is 3,796,742 sq. miles (9,833,517 sq. km), of which 3,531,905 sq. miles (9,147,593 sq. km) are land and 264,837 sq. miles (685,924 sq. km) are water (comprising Great Lakes, inland and coastal water).

Population at each census from 1790 to 2010 (including Alaska and Hawaii from 1960). Figures do not include Puerto Rico, Guam, American Samoa or other Pacific islands, or the US population abroad. Residents of Indian reservations not included before 1890.

	White	Black	Other races	Total
1790	3,172,464	757,208	—	3,929,672
1800	4,306,446	1,002,037	—	5,308,483
1810	5,862,073	1,377,808	—	7,239,881
1820	7,866,797	1,771,562	—	9,638,359
1830	10,537,378	2,328,642	—	12,866,020
1840	14,195,805	2,873,648	—	17,069,453
1850	19,553,068	3,638,808	—	23,191,876
1860	26,922,537	4,441,830	78,954	31,443,321
1870	34,337,292	5,392,172	88,985	39,818,449
1880	43,402,970	6,580,793	172,020	50,155,783
1890	55,101,258	7,488,676	357,780	62,947,714
1900	66,868,508	8,834,395	509,265	76,212,168
1910	81,812,405	9,828,667	587,459	92,228,531
1920	94,903,540	10,463,607	654,421	106,021,568
1930	110,395,753	11,891,842	915,065	123,202,660
1940	118,357,831	12,865,914	941,384	132,165,129
1950	135,149,629	15,044,937	1,131,232	151,325,798
1960	158,831,732	18,871,831	1,619,612	179,323,175
1970	177,748,975	22,580,289	2,882,662	203,211,926
1980	188,371,622	26,495,025	11,679,158	226,545,805
1990	199,686,070	29,986,060	19,037,743	248,709,873
2000[1]	211,460,626	34,658,190	35,303,090	281,421,906
2010[1]	223,553,265	38,929,319	46,262,954	308,745,538

[1]'White' refers to the White-alone population, 'Black' to the Black-alone population and 'Other races' to the population in all the remaining race groups, including those reporting two or more races.

The mid-year population estimate for 2012 was 313,914,040.

The 2010 census population of 308,745,538 represented an increase of 9·7% since 2000 (the smallest percentage increase between ten-yearly US censuses since the Second World War). Minorities accounted for 92% of the growth. There were 156,964,212 females at the 2010 census, or 50·8% of the total population.

The UN gives a projected population for 2015 of 323·89m.

2010 density, 33·8 per sq. km (87·4 per sq. mile). Urban population (persons living in places with at least 2,500 inhabitants) at the 2010 census was 249,253,271 (80·7%); rural, 59,492,267. In 2000 it was 79·0%; in 1990, 75·2%; in 1980, 73·7%; in 1970, 73·6%.

Sex distribution by race of the population at the 2000 census:

	Males	Females
White	103,773,194	107,687,432
Black or African American	16,465,185	18,193,005
American Indian and Alaska Native	1,233,982	1,241,974
Asian	4,948,741	5,294,257
Native Hawaiian and Other Pacific Islander	202,629	196,206
Other Race	8,009,214	7,349,859
Two or More Races	3,420,618	3,405,610
Total	138,053,563	143,368,343

Alongside these racial groups, and applicable to all of them, a category of 'Hispanic origin' comprised 50,477,594 persons in 2010 (including 31,798,258 of Mexican ancestry), up 15,171,776 from 35,305,818 in 2000. Hispanics are now the largest ethnic minority in the USA.

Among ten-year age groups the 45–54 age group contained most people according to the 2010 census, with a total of 45,006,716 (14·6% of the population). At the 2000 census the 35–44 age group had most people.

At the 2010 census there were 116,716,292 households, up from 105,480,101 in 2000. There were 31·7m. single-person households in 2010 representing 27% of all households, up from 9% in 1950.

At the 2000 census there were 50,454 people aged 100 or over, compared to 37,306 in 1990. Of the 50,454 centenarians in 2000, 40,397 were female, and of the 37,306 in 1990, 29,405 were female.

There is no official language, though English has been provided with official status in 30 states. The 2000 census showed that 47·0m. persons five years and over spoke a language other than English in the home, including Spanish or Spanish Creole by 28·1m.; French or French Creole by 2·1m.; Chinese by 2·0m.; German by 1·4m.; Tagalog by 1·2m.; Italian by 1·0m.; Vietnamese by 1·0m. 82% of the population has English as its mother tongue.

The following table includes population statistics, the year in which each of the original 13 states (Connecticut, Delaware, Georgia, Maryland, Massachusetts, New Hampshire, New Jersey, New York, North Carolina, Pennsylvania, Rhode Island, South Carolina, Virginia) ratified the constitution, and the year when each of the other states was admitted into the Union. Two-letter abbreviated postal codes for use in addresses are shown in brackets.

The USA is divided into four geographic regions comprised of nine divisions. These are, with their 2010 census populations: Northeast (comprised of the New England and Middle Atlantic divisions), 53,317,240; Midwest (East North Central, West North Central), 66,927,001; South (South Atlantic, East South Central, West South Central), 114,555,744; West (Mountain, Pacific), 71,945,553.

Geographic divisions and states		Land area: sq. miles, 2010	Census population, 1 April 2010	Pop. per sq. mile, 2010
United States		3,531,905	308,745,538	87·4
New England		62,689	14,444,865	230·4
Connecticut (1788)	(CT)	4,842	3,574,097	738·1
Maine (1820)	(ME)	30,843	1,328,361	43·1
Massachusetts (1788)	(MA)	7,800	6,547,629	839·4
New Hampshire (1788)	(NH)	8,953	1,316,470	147·0
Rhode Island (1790)	(RI)	1,034	1,052,567	1,018·1
Vermont (1791)	(VT)	9,217	625,741	67·9
Middle Atlantic		99,223	40,872,375	411·9
New Jersey (1787)	(NJ)	7,354	8,791,894	1,195·5
New York (1788)	(NY)	47,126	19,378,102	411·2
Pennsylvania (1787)	(PA)	44,743	12,702,379	283·9
East North Central		242,903	46,421,564	191·1
Illinois (1818)	(IL)	55,519	12,830,632	231·1
Indiana (1816)	(IN)	35,826	6,483,802	181·0
Michigan (1837)	(MI)	56,539	9,883,640	174·8
Ohio (1803)	(OH)	40,861	11,536,504	282·3
Wisconsin (1848)	(WI)	54,158	5,686,986	105·0
West North Central		507,621	20,505,437	40·4
Iowa (1846)	(IA)	55,857	3,046,355	54·5
Kansas (1861)	(KS)	81,759	2,853,118	34·9
Minnesota (1858)	(MN)	79,627	5,303,925	66·6
Missouri (1821)	(MO)	68,742	5,988,927	87·1

Geographic divisions and states		Land area: sq. miles, 2010	Census population, 1 April 2010	Pop. per sq. mile, 2010
Nebraska (1867)	(NE)	76,824	1,826,341	23·8
North Dakota (1889)	(ND)	69,001	672,591	9·7
South Dakota (1889)	(SD)	75,811	814,180	10·7
South Atlantic		265,062	59,777,037	225·5
Delaware (1787)	(DE)	1,949	897,934	460·8
Dist. of Columbia (1791)[1]	(DC)	61	601,723	9,856·2
Florida (1845)	(FL)	53,625	18,801,310	350·6
Georgia (1788)	(GA)	57,513	9,687,653	168·4
Maryland (1788)	(MD)	9,707	5,773,552	594·8
North Carolina (1789)	(NC)	48,618	9,535,483	196·1
South Carolina (1788)	(SC)	30,061	4,625,364	153·9
Virginia (1788)	(VA)	39,490	8,001,024	202·6
West Virginia (1863)	(WV)	24,038	1,852,994	77·1
East South Central		178,289	18,432,505	103·4
Alabama (1819)	(AL)	50,645	4,779,736	94·4
Kentucky (1792)	(KY)	39,486	4,339,367	109·9
Mississippi (1817)	(MS)	46,923	2,967,297	63·2
Tennessee (1796)	(TN)	41,235	6,346,105	153·9
West South Central		425,066	36,346,202	85·5
Arkansas (1836)	(AR)	52,035	2,915,918	56·0
Louisiana (1812)	(LA)	43,204	4,533,372	104·9
Oklahoma (1907)	(OK)	68,595	3,751,351	54·7
Texas (1845)	(TX)	261,232	25,145,561	96·3
Mountain		855,767	22,065,451	25·8
Arizona (1912)	(AZ)	113,594	6,392,017	56·3
Colorado (1876)	(CO)	103,642	5,029,196	48·5
Idaho (1890)	(ID)	82,643	1,567,582	19·0
Montana (1889)	(MT)	145,546	989,415	6·8
Nevada (1864)	(NV)	109,781	2,700,551	24·6
New Mexico (1912)	(NM)	121,298	2,059,179	17·0
Utah (1896)	(UT)	82,170	2,763,885	33·6
Wyoming (1890)	(WY)	97,093	563,626	5·8
Pacific		895,287	49,880,102	55·7
Alaska (1959)	(AK)	570,641	710,231	1·2
California (1850)	(CA)	155,779	37,253,956	239·1
Hawaii (1960)	(HI)	6,423	1,360,301	211·8
Oregon (1859)	(OR)	95,988	3,831,074	39·9
Washington (1889)	(WA)	66,456	6,724,540	101·2

Geographic divisions and states	Land area: sq. miles, 2010	Census population, 1 April 2010	Pop. per sq. mile, 2010
Outlying Territories, total	4,032	4,100,954	1,017·1
American Samoa	76	55,519	730·5
Guam	210	159,358	758·8
Johnston Atoll	1	0[2]	0
Midway Islands	2	0[2]	0
Northern Marianas	182	53,883	296·1
Puerto Rico	3,424	3,725,789	1,088·1
Virgin Islands	134	106,405	794·1
Wake Island	3	0[2]	0

[1]District of Columbia selected as site of national government in 1791.
[2]No permanent residents. There may be occasional visits from military and scientific personnel.

The 2000 census showed 31,107,889 foreign-born persons. The ten countries contributing the largest numbers who were foreign-born were: Mexico, 9,177,487; Philippines, 1,369,070; India, 1,022,552; China, 988,857; Vietnam, 988,174; Cuba, 872,716; Korea, 864,125; Canada, 820,771; El Salvador, 817,336; Germany, 706,704. In fiscal year 2000 a total of 841,002 persons obtained legal permanent resident status (1,535,872 in fiscal year 1990). In 2007 the estimate for the foreign-born population was 38·1m., representing 12·6% of the population.

Population of cities with over 100,000 inhabitants at the censuses of 2000 and 2010:

Cities	Census 2000	Census 2010	Cities	Census 2000	Census 2010
New York, NY	8,008,278	8,175,133	Wichita, KS	344,284	382,368
Los Angeles, CA	3,694,820	3,792,621	Arlington, TX	332,969	365,438
Chicago, IL	2,896,016	2,695,598	Bakersfield, CA	247,057	347,483
Houston, TX	1,953,631	2,099,451	New Orleans, LA	484,674	343,829
Philadelphia, PA	1,517,550	1,526,006	Urban Honolulu, HI[1, 2]	—	337,256
Phoenix, AZ	1,321,045	1,445,632	Anaheim, CA	328,014	336,265
San Antonio, TX	1,144,646	1,327,407	Tampa, FL	303,447	335,709
San Diego, CA	1,223,400	1,307,402	Aurora, CO	276,393	325,078
Dallas, TX	1,188,580	1,197,816	Santa Ana, CA	337,977	324,528
San Jose, CA	894,943	945,942	St Louis, MO	348,189	319,294
Indianapolis, IN	791,926	829,718	Pittsburgh, PA	334,563	305,704
Jacksonville, FL	735,617	821,784	Corpus Christi, TX	277,454	305,215
San Francisco, CA	776,733	805,235	Riverside, CA	255,166	303,871
Austin, TX	656,562	790,390	Cincinnati, OH	331,285	296,943
Columbus, OH	711,470	787,033	Lexington-Fayette, KY	260,512	295,803
Fort Worth, TX	534,694	741,206	Anchorage, AK[3]	260,283	291,826
Louisville/Jefferson County, KY	256,231	741,096	Stockton, CA	243,771	291,707
Charlotte, NC	540,828	731,424	Toledo, OH	313,619	287,208
Detroit, MI	951,270	713,777	St Paul, MN	287,151	285,068
El Paso, TX	563,662	649,121	Newark, NJ	273,546	277,140
Memphis, TN	650,100	646,889	Greensboro, NC	223,891	269,666
Nashville-Davidson, TN	569,891	626,681	Buffalo, NY	292,648	261,310
Baltimore, MD	651,154	620,961	Plano, TX	222,030	259,841
Boston, MA	589,141	617,594	Lincoln, NE	225,581	258,379
Seattle, WA	563,374	608,660	Henderson, NV	175,381	257,729
Washington, DC	572,059	601,723	Fort Wayne, IN	205,727	253,691
Denver, CO	554,636	600,158	Jersey City, NJ	240,055	247,597
Milwaukee, WI	596,974	594,833	St Petersburg, FL	248,232	244,769
Portland, OR	529,121	583,776	Chula Vista, CA	173,556	243,916
Las Vegas, NV	478,434	583,756	Norfolk, VA	234,403	242,803
Oklahoma City, OK	506,132	579,999	Orlando, FL	185,951	238,300
Albuquerque, NM	448,607	545,852	Chandler, AZ	176,581	236,123
Tucson, AZ	486,699	520,116	Laredo, TX	176,576	236,091
Fresno, CA	427,652	494,665	Madison, WI	208,054	233,209
Sacramento, CA	407,018	466,488	Winston-Salem, NC	185,776	229,617
Long Beach, CA	461,522	462,257	Lubbock, TX	199,564	229,573
Kansas City, MO	441,545	459,787	Baton Rouge, LA	227,818	229,493
Mesa, AZ	396,375	439,041	Durham, NC	187,035	228,330
Virginia Beach, VA	425,257	437,994	Garland, TX	215,768	226,876
Atlanta, GA	416,474	420,003	Glendale, AZ	218,812	226,721
Colorado Springs, CO	360,890	416,427	Reno, NV	180,480	225,221
Omaha, NE	390,007	408,958	Hialeah, FL	226,419	224,669
Raleigh, NC	276,093	403,892	Paradise, NV[1]	186,070	223,167
Miami, FL	362,470	399,457	Chesapeake, VA	199,184	222,209
Cleveland, OH	478,403	396,815	Scottsdale, AZ	202,705	217,385
Tulsa, OK	393,049	391,906	North Las Vegas, NV	115,488	216,961
Oakland, CA	399,484	390,724	Irving, TX	191,615	216,290
Minneapolis, MN	382,618	382,578	Fremont, CA	203,413	214,089
			Irvine, CA	143,072	212,375
			Birmingham, AL	242,820	212,237
			Rochester, NY	219,773	210,565
			San Bernardino, CA	185,401	209,924

Cities	Census 2000	Census 2010
Spokane, WA	195,629	208,916
Gilbert, AZ[4]	109,697	208,453
Arlington, VA[1]	189,453	207,627
Montgomery, AL	201,568	205,764
Boise City, ID	185,787	205,671
Richmond, VA	197,790	204,214
Des Moines, IA	198,682	203,433
Modesto, CA	188,856	201,165
Fayetteville, NC	121,015	200,564
Augusta-Richmond County, GA	199,775	200,549
Shreveport, LA	200,145	199,311
Akron, OH	217,074	199,110
Tacoma, WA	193,556	198,397
Aurora, IL	142,990	197,899
Oxnard, CA	170,358	197,899
Fontana, CA	128,929	196,069
Yonkers, NY	196,086	195,976
Mobile, AL	198,915	195,111
Little Rock, AR	183,133	193,524
Moreno Valley, CA	142,381	193,365
Glendale, CA	194,973	191,719
Amarillo, TX	173,627	190,695
Huntington Beach, CA	189,594	189,992
Columbus, GA	186,291	189,885
Sunrise Manor, NV[1]	156,120	189,372
Grand Rapids, MI	197,800	188,040
Salt Lake City, UT	181,743	186,440
Tallahassee, FL	150,624	181,376
Worcester, MA	172,648	181,045
Newport News, VA	180,150	180,719
Huntsville, AL	158,216	180,105
Knoxville, TN	173,890	178,874
Spring Valley, NV[1]	117,390	178,395
Providence, RI	173,618	178,042
Santa Clarita, CA	151,088	176,320
Grand Prairie, TX	127,427	175,396
Brownsville, TX	139,722	175,023
Jackson, MS	184,256	173,514
Overland Park, KS	149,080	173,372
Garden Grove, CA	165,196	170,883
Santa Rosa, CA	147,595	167,815
Chattanooga, TN	155,554	167,674
Oceanside, CA	161,029	167,086
Fort Lauderdale, FL	152,397	165,521
Rancho Cucamonga, CA	127,743	165,269
Port St Lucie, FL	88,769	164,603
Ontario, CA	158,007	163,924
Vancouver, WA	143,560	161,791
Tempe, AZ	158,625	161,719
Springfield, MO	151,580	159,498
Lancaster, CA	118,718	156,633
Eugene, OR	137,893	156,185
Pembroke Pines, FL	137,427	154,750
Salem, OR	136,924	154,637
Cape Coral, FL	102,286	154,305
Peoria, AZ	108,364	154,065

Cities	Census 2000	Census 2010
Sioux Falls, SD	123,975	153,888
Springfield, MA	152,082	153,060
Elk Grove, CA[2]	—	153,015
Rockford, IL	150,115	152,871
Palmdale, CA	116,670	152,750
Corona, CA	124,966	152,374
Salinas, CA	151,060	150,441
Pomona, CA	149,473	149,058
Pasadena, TX	141,674	149,043
Joliet, IL	106,221	147,433
Paterson, NJ	149,222	146,199
Kansas City, KS	146,866	145,786
Torrance, CA	137,946	145,438
Syracuse, NY	147,306	145,170
Bridgeport, CT	139,529	144,229
Hayward, CA	140,030	144,186
Fort Collins, CO	118,652	143,986
Escondido, CA	133,559	143,911
Lakewood, CO	144,126	142,980
Naperville, IL	128,358	141,853
Dayton, OH	166,179	141,527
Hollywood, FL	139,357	140,768
Sunnyvale, CA	131,760	140,081
Alexandria, VA	128,283	139,966
Mesquite, TX	124,523	139,824
Metairie, LA[1]	146,136	138,481
Hampton, VA	146,437	137,436
Pasadena, CA	133,936	137,122
Orange, CA	128,821	136,416
Savannah, GA	131,510	136,286
Cary, NC[4]	94,536	135,234
Fullerton, CA	126,003	135,161
Warren, MI	138,247	134,056
Clarksville, TN	103,445	132,929
McKinney, TX	54,369	131,117
McAllen, TX	106,414	129,877
New Haven, CT	123,626	129,779
Sterling Heights, MI	124,471	129,699
West Valley City, UT	108,896	129,480
Columbia, SC	116,278	129,272
Killeen, TX	86,911	127,921
Topeka, KS	122,377	127,473
Thousand Oaks, CA	117,005	126,683
East Los Angeles, CA[1]	124,283	126,496
Cedar Rapids, IA	120,758	126,326
Olathe, KS	92,962	125,872
Elizabeth, NJ	120,568	124,969
Waco, TX	113,726	124,805
Hartford, CT	121,578	124,775
Visalia, CA	91,565	124,442
Gainesville, FL	95,447	124,354
Simi Valley, CA	111,351	124,237
Stamford, CT	117,083	122,643
Bellevue, WA	109,569	122,363
Concord, CA	121,780	122,067
Miramar, FL	72,739	122,041
Coral Springs, FL	117,549	121,096
Lafayette, LA	110,257	120,263
Charleston, SC	96,650	120,083
Carrollton, TX	109,576	119,097
Roseville, CA	79,921	118,788
Thornton, CO	82,384	118,772
Beaumont, TX	113,866	118,296
Allentown, PA	106,632	118,032
Surprise, AZ	30,848	117,517
Evansville, IN	121,582	117,429
Abilene, TX	115,930	117,063
Frisco, TX	33,714	116,989

Cities	Census 2000	Census 2010
Independence, MO	113,288	116,830
Athens-Clarke County, GA	101,489	116,714
Santa Clara, CA	102,361	116,468
Springfield, IL	111,454	116,250
Vallejo, CA	116,760	115,942
Victorville, CA	64,029	115,903
Peoria, IL	112,936	115,007
Lansing, MI	119,128	114,297
Ann Arbor, MI	114,024	113,934
El Monte, CA	115,965	113,475
Denton, TX	80,537	113,383
Berkeley, CA	102,743	112,580
Provo, UT	105,166	112,488
Downey, CA	107,323	111,772
Midland, TX	94,996	111,147
Norman, OK	95,694	110,925
Waterbury, CT	107,271	110,366
Costa Mesa, CA	108,724	109,960
Inglewood, CA	112,580	109,673
Manchester, NH	107,006	109,565
Murfreesboro, TN	68,816	108,755
Columbia, MO	84,531	108,500
Enterprise, NV[1]	14,676	108,481
Elgin, IL	94,487	108,188
Clearwater, FL	108,787	107,685
Miami Gardens, FL[2]	—	107,167
Rochester, MN	85,806	106,769
Pueblo, CO	102,121	106,595
Lowell, MA	105,167	106,519
Wilmington, NC	75,838	106,476

Cities	Census 2000	Census 2010
Arvada, CO	102,153	106,433
San Buenaventura (Ventura), CA	100,916	106,433
Westminster, CO	100,940	106,114
West Covina, CA	105,080	106,098
Gresham, OR	90,205	105,594
Fargo, ND	90,599	105,549
Norwalk, CA	103,298	105,549
Carlsbad, CA	78,247	105,328
Fairfield, CA	96,178	105,321
Cambridge, MA	101,355	105,162
Wichita Falls, TX	104,197	104,553
High Point, NC	85,839	104,371
Billings, MT	89,847	104,170
Green Bay, WI	102,313	104,057
West Jordan, UT	68,336	103,712
Richmond, CA	99,216	103,701
Brandon, FL[1]	77,895	103,483
Murrieta, CA	44,282	103,466
Burbank, CA	100,316	103,340
Palm Bay, FL	79,413	103,190
Everett, WA	91,488	103,019
Flint, MI	124,943	102,434
Antioch, CA	90,532	102,372
Erie, PA	103,717	101,786
South Bend, IN	107,789	101,168
Daly City, CA	103,621	101,123
Centennial, CO[2]	—	100,377

[1]Designated by the United States Census Bureau as a CDP (census designated place) and not incorporated as a city. [2]Incorporated since the 2000 census. [3]Officially designated as a Municipality. [4]Officially designated as a Town.

Immigration and naturalization. The Immigration and Nationality Act, as amended, provides for the numerical limitation of most immigration. The Immigration Act of 1990 established major revisions in the numerical limits and preference system regulating legal immigration. The numerical limits are imposed on visas issued and not admissions. The maximum number of visas allowed to be issued under the preference categories in fiscal year 2008 was 388,704: 226,000 for family-sponsored immigrants and 162,704 for employment-based immigrants. Within the overall limitations the per-country limit for independent countries is set to 7% of the total family and employment limits, while dependent areas are limited to 2% of the total. Immigrants not subject to any numerical limitation are spouses, children, and parents of US citizens who are 21 years of age or older; certain former US citizens; ministers of religion; certain long-term US government employees; refugees and asylum-seekers adjusting to immigrant status; and certain other groups of immigrants.

Immigrant aliens admitted to the USA for permanent residence, by country or region of birth, for fiscal years:

Country or region of birth	Immigrants admitted			
	2005	2006	2007	2008
All countries	1,122,257	1,266,129	1,052,415	1,107,126
Europe	176,516	164,244	120,821	119,138
Germany	9,264	8,436	7,582	7,091
Poland	15,351	17,051	10,355	8,354
Russia	18,055	13,159	9,426	11,695
Ukraine	22,745	17,140	11,001	10,813
UK	19,800	17,207	14,545	14,348

Country or region of birth	2005	Immigrants admitted 2006	2007	2008
Other Europe	91,301	91,251	67,912	66,837
Asia	400,098	422,284	383,508	383,608
Bangladesh	11,487	14,644	12,074	11,753
China (mainland)	69,933	87,307	76,655	80,271
India	84,680	61,369	65,353	63,352
Iran	13,887	13,947	10,460	13,852
Japan	8,768	8,265	6,748	6,821
Korea (North and South)	26,562	24,386	22,405	26,666
Pakistan	14,926	17,418	13,492	19,719
Philippines	60,746	74,606	72,596	54,030
Taiwan	9,196	8,086	8,990	9,073
Thailand	5,505	11,749	8,751	6,637
Uzbekistan	2,887	4,015	4,665	6,375
Vietnam	32,784	30,691	28,691	31,497
Other Asia	58,737	65,801	52,628	53,562
North and Central America	345,561	414,075	339,355	393,253
Canada	21,878	18,207	15,495	15,109
Cuba	36,261	45,614	29,104	49,500
Dominican Republic	27,503	38,068	28,024	31,879
El Salvador	21,359	31,782	21,127	19,659
Guatemala	16,818	24,133	17,908	16,182
Haiti	14,524	22,226	30,405	26,007
Honduras	7,012	8,177	7,646	6,540
Jamaica	18,345	24,976	19,375	18,477
Mexico	161,445	173,749	148,640	189,989
Other North America	183	333	171	216
Other Central America	20,233	26,810	21,460	19,695
South America	103,135	137,986	106,525	98,555
Brazil	16,662	17,903	14,295	12,195
Colombia	25,566	43,144	33,187	30,213
Ecuador	11,608	17,489	12,248	11,663
Guyana	9,317	9,552	5,726	6,823
Peru	15,676	21,718	17,699	15,184
Venezuela	10,645	11,341	10,692	10,514
Other South America	13,661	16,839	12,678	11,963
Africa	85,098	117,422	94,711	105,915
Egypt	7,905	10,500	9,267	8,712
Ethiopia	10,571	16,152	12,786	12,917
Ghana	6,491	9,367	7,610	8,195
Kenya	5,347	8,779	7,030	6,998
Liberia	4,880	6,887	4,102	7,193
Nigeria	10,597	13,459	12,448	12,475
Somalia	5,829	9,462	6,251	10,745
Other Africa	33,478	42,816	35,217	38,680
Other countries	11,849	10,118	7,495	6,657

The total number of immigrants admitted from 1820 to 2008 was 74,225,904; this included 7,476,092 from Mexico, 7,275,320 from Germany and 5,455,888 from Italy.

The number of immigrants admitted for legal permanent residence in the United States in fiscal year 2009 was 1,130,818. Included in this total were 463,042 aliens previously living abroad who obtained immigrant visas through the US Department of State and became legal permanent residents upon entry into the United States. The remaining 667,776 legal immigrants, including former undocumented immigrants, refugees and asylees, had adjusted status through the Citizenship and Immigration Services (USCIS). The USA has by far the largest annual net gain in migrants of any country. It also receives the most asylum seekers on an annual basis.

A total of 743,715 persons were naturalized in fiscal year 2009 (including 111,630 persons born in Mexico).

The refugee admissions ceiling for fiscal year 2010 was fixed at 80,000.

SOCIAL STATISTICS

Figures only include Alaska and Hawaii from 1960 onwards.

	Live births	Deaths	Marriages	Divorces	Deaths under 1 year
1900	—	343,217	709,000	56,000	—
1910	2,777,000	696,856	948,000	83,000	—
1920	2,950,000	1,118,070	1,274,476	170,505	170,911
1930	2,618,000	1,327,240	1,126,856	195,961	143,201
1940	2,559,000	1,417,269	1,595,879	264,000	110,984
1950	3,632,000	1,452,454	1,667,231	385,144	103,825
1960	4,257,850	1,711,982	1,523,000	393,000	110,873
1970	3,731,386	1,921,031	2,158,802	708,000	74,667
1980	3,612,258	1,989,841	2,390,252	1,189,000	45,526
1990	4,158,212	2,148,463	2,443,489	1,182,000	38,351
2000	4,058,814	2,403,351	2,315,000	—	28,035
2001	4,025,933	2,416,425	2,326,000	—	27,568
2002	4,021,726	2,443,387	2,290,000	—	28,034
2003	4,089,950	2,448,288	2,245,000	—	28,025
2004	4,112,052	2,397,615	2,279,000	—	27,936
2005	4,138,349	2,448,017	2,249,000	—	28,440
2006	4,265,555	2,426,264	2,193,000[1]	—	28,527
2007	4,316,233	2,432,712	2,197,000	—	29,138
2008	4,247,694	2,471,984	2,157,000	—	28,059
2009	4,130,665	2,437,163	2,080,000	—	26,412
2010	3,999,386	2,468,435	2,096,000	—	24,548[2]

[1]Excluding Louisiana. [2]Provisional.

Rates (per 1,000 population):

	Birth	Death	Marriage
2000	14·4	8·5	8·2
2001	14·1	8·5	8·2
2002	13·9	8·5	8·0
2003	14·1	8·4	7·7
2004	14·0	8·2	7·8
2005	14·0	8·3	7·6
2006	14·2	8·1	7·5[1]
2007	14·3	8·0	7·3
2008	14·0	8·1	7·1
2009	13·5	7·9	6·8
2010	13·0	8·0	6·8

[1]Excluding Louisiana.

Although divorce figures are not available since 1997 as not all states maintain complete statistics, it is estimated that the rate fell from 4·3 per 1,000 population in 1997 to 3·6 per 1,000 in 2010. Rate of natural increase per 1,000 population: 6·1 in 2006; 6·3 in 2007. Population growth rate, 2009, 0·9%.

Even though the marriage rate shows a gradual decline, it remains much higher than in most other industrial countries. The most popular age range for marrying is 25–29 for males and 20–24 for females. In 2008, 14% of black men married a white women (up from 5% in 1980) and 6% of black women married a white man (up from 1% in 1980). The number of births to unmarried women in 2009 was 1,693,658 (41% of all births), compared to 666,000 in 1980 and 1,726,566 in 2008. The rate of births to teenagers was 34·3 per 1,000 women in 2010. The number of babies born to women aged 15–19 was 367,752 in 2010, down from 409,802 in 2009 and the fewest reported in more than 60 years.

Between 1970 and 2008 the annual number of births rose by 13·8%. The number of births within marriage declined by 19·8% between 1970 and 2005 whereas the number outside of marriage rose by 283·0%. Whereas in 1970 as many as 83·4% of children lived with both biological parents, by 2004 only 59·9% were living with married biological parents and 2·5% with unmarried biological parents. In 2009, 27·3% of children lived in one-parent households—the highest proportion in the industrialized world.

Infant mortality rates, per 1,000 live births: 29·2 in 1950; 12·9 in 1980; 6·4 in 2009. Fertility rate, 2010, 1·9 births per woman (the lowest since the late 1980s).

There were a reported 1,212,400 abortions in 2008, down from a peak in 1990 of 1,608,600.

Expectation of life, 1970: males, 67·1 years; females, 74·7 years. 2009: males, 76·0 years; females, 80·9 years.

Numbers of deaths by principal causes, 2009 (and as a percentage of all deaths): heart disease, 599,413 (24·6%); cancer, 567,628 (23·3%); chronic lower respiratory disease, 137,353 (5·6%); stroke, 128,842 (5·3%); accidents, 118,021 (4·8%); Alzheimer's disease, 79,003 (3·2%); diabetes, 68,705 (2·8%); pneumonia and influenza, 53,692 (2·2%); kidney diseases, 48,935 (2·0%); suicide, 36,909 (1·5%).

The number of Americans living in poverty in 2009 was 43·6m. or 14·3% of the total population, down from 15·1% in 1993 but up 1·1% on 2008.

A UNICEF report published in 2010 showed that 23·1% of children in the USA live in relative poverty (living in a household in which disposable income—when adjusted for family size and composition—is less than 50% of the national median income), compared to just 4·7% in Iceland.

CLIMATE

For temperature and rainfall figures, *see* entries on individual states as indicated by regions, below, of mainland USA.

Pacific Coast. The climate varies with latitude, distance from the sea and the effect of relief, ranging from polar conditions in North Alaska through cool to warm temperate climates further south. The extreme south is temperate desert. Rainfall everywhere is moderate. *See* Alaska, California, Oregon, Washington.

Mountain States. Very varied, with relief exerting the main control; very cold in the north in winter, with considerable snowfall. In the south, much higher temperatures and aridity produce desert conditions. Rainfall everywhere is very variable as a result of rain-shadow influences. *See* Arizona, Colorado, Idaho, Montana, Nevada, New Mexico, Utah, Wyoming.

High Plains. A continental climate with a large annual range of temperature and moderate rainfall, mainly in summer, although unreliable. Dust storms are common in summer and blizzards in winter. *See* Nebraska, North Dakota, South Dakota.

Central Plains. A temperate continental climate, with hot summers and cold winters, except in the extreme south. Rainfall is plentiful and comes at all seasons, but there is a summer maximum in western parts. *See* Mississippi, Missouri, Oklahoma, Texas.

Mid-West. Continental, with hot summers and cold winters. Rainfall is moderate, with a summer maximum in most parts. *See* Indiana, Iowa, Kansas.

Great Lakes. Continental, resembling that of the Central Plains, with hot summers but very cold winters because of the freezing of the lakes. Rainfall is moderate with a slight summer maximum. *See* Illinois, Michigan, Minnesota, Ohio, Wisconsin.

Appalachian Mountains. The north is cool temperate with cold winters, the south warm temperate with milder winters. Precipitation is heavy, increasing to the south but evenly distributed over the year. *See* Kentucky, Pennsylvania, Tennessee, West Virginia.

Gulf Coast. Conditions vary from warm temperate to sub-tropical, with plentiful rainfall, decreasing towards the west but evenly distributed over the year. *See* Alabama, Arkansas, Florida, Louisiana.

Atlantic Coast. Temperate maritime climate but with great differences in temperature according to latitude. Rainfall is ample at all seasons; snowfall in the north can be heavy. *See* Delaware, District of Columbia, Georgia, Maryland, New Jersey, New York State, North Carolina, South Carolina, Virginia.

New England. Cool temperate, with severe winters and warm summers. Precipitation is well distributed with a slight winter maximum. Snowfall is heavy in winter. *See* Connecticut, Maine, Massachusetts, New Hampshire, Rhode Island, Vermont. *See* also Hawaii and Outlying Territories.

CONSTITUTION AND GOVERNMENT

The form of government of the USA is based on the constitution adopted on 17 Sept. 1787 and effective from 4 March 1789.

By the constitution the government of the nation is composed of three co-ordinate branches, the executive, the legislative and the judicial.

The Federal government has authority in matters of general taxation, treaties and other dealings with foreign countries, foreign and inter-state commerce, bankruptcy, postal service, coinage, weights and measures, patents and copyright, the armed forces (including, to a certain extent, the militia), and crimes against the USA; it has sole legislative authority over the District of Columbia and the possessions of the USA.

The 5th article of the constitution provides that Congress may, on a two-thirds vote of both houses, propose amendments to the constitution, or, on the application of the legislatures of two-thirds of all the states, call a convention for proposing amendments, which in either case shall be valid as part of the constitution when ratified by the legislatures of three-fourths of the several states, or by conventions in three-fourths thereof, whichever mode of ratification may be proposed by Congress. Ten amendments (called collectively 'the Bill of Rights') to the constitution were added 15 Dec. 1791; two in 1795 and 1804; a 13th amendment, 6 Dec. 1865, abolishing slavery; a 14th in 1868, including the important 'due process' clause; a 15th, 3 Feb. 1870, establishing equal voting rights for white and black; a 16th, 3 Feb. 1913, authorizing the income tax; a 17th, 8 April 1913, providing for popular election of Senators; an 18th, 16 Jan. 1919, prohibiting alcoholic liquors; a 19th, 18 Aug. 1920, establishing woman suffrage; a 20th, 23 Jan. 1933, advancing the date of the President's and Vice-President's inauguration and abolishing the 'lameduck' sessions of Congress; a 21st, 5 Dec. 1933, repealing the 18th amendment; a 22nd, 27 Feb. 1951, limiting a President's tenure of office to two terms, or two full terms in the case of a Vice-President who has succeeded to the office of President and has served two years or less of another President's term, or one full term in the case of a Vice-President who has succeeded to the office of President and has served more than two years of another President's term; a 23rd, 30 March 1961, granting citizens of the District of Columbia the right to vote in national elections; a 24th, 4 Feb. 1964, banning the use of the poll-tax in federal elections; a 25th, 10 Feb. 1967, dealing with Presidential disability and succession; a 26th, 22 June 1970, establishing the right of citizens who are 18 years of age and older to vote; a 27th, 7 May 1992, providing that no law varying the compensation of Senators or Representatives shall take effect until an election has taken place.

National motto. 'In God we trust'; formally adopted by Congress 30 July 1956.

Presidency

The executive power is vested in a president, who holds office for four years, and is elected, together with a vice-president chosen for the same term, by electors from each state, equal to the whole number of Senators and Representatives to which the state may be entitled in the Congress. The President must be a natural-born citizen, resident in the country for 14 years, and at least 35 years old.

The presidential election is held every fourth (leap) year on the Tuesday after the first Monday in Nov. Technically, this is an election of presidential electors, not of a president directly; the electors thus chosen meet and give their votes (for the candidate to whom they are pledged, in some states by law, but in most states by custom and prudent politics) at their respective state capitals on the first Monday after the second Wednesday in Dec. next following their election; and the votes of the electors of all the states are opened and counted in the presence of both Houses of Congress on the sixth day of Jan. The total electorate vote is one for each Senator and Representative. Electors may not be a member of Congress or hold federal office. If no candidate secures the minimum 270 college votes needed for outright victory, the

12th Amendment to the Constitution applies, and the House of Representatives chooses a president from among the first three finishers in the electoral college. (This last happened in 1824).

If the successful candidate for President dies before taking office the Vice-President-elect becomes President; if no candidate has a majority or if the successful candidate fails to qualify, then, by the 20th amendment, the Vice-President acts as President until a president qualifies. The duties of the Presidency, in absence of the President and Vice-President by reason of death, resignation, removal, inability or failure to qualify, devolve upon the Speaker of the House under legislation enacted on 18 July 1947. In case of absence of a Speaker for like reason, the presidential duties devolve upon the President *pro tem.* of the Senate and successively upon those members of the cabinet in order of precedence, who have the constitutional qualifications for President.

The presidential term, by the 20th amendment to the constitution, begins at noon on 20 Jan. of the inaugural year. This amendment also installs the newly elected Congress in office on 3 Jan. instead of—as formerly—in the following Dec. The President's salary is $400,000 per year (taxable), with an additional $50,000 to assist in defraying expenses resulting from official duties. Also he may spend up to $100,000 non-taxable for travel and $19,000 for official entertainment. In 1999 the presidential salary was increased for the president taking office in Jan. 2001, having remained at $200,000 a year since 1969. The office of Vice-President carries a salary of $230,700 and $20,000 allowance for expenses, all taxable. The Vice-President is *ex officio* President of the Senate, and in the case of 'the removal of the President, or of his death, resignation, or inability to discharge the powers and duties of his office', he becomes the President for the remainder of the term.

Cabinet. The administrative business of the nation has been traditionally vested in several executive departments, the heads of which, unofficially and *ex officio*, formed the President's cabinet. Beginning with the Interstate Commerce Commission in 1887, however, an increasing amount of executive business has been entrusted to some 60 so-called independent agencies, such as the Housing and Home Finance Agency, Tariff Commission, etc.

All heads of departments and of the 60 or more administrative agencies are appointed by the President, but must be confirmed by the Senate.

Congress. The legislative power is vested by the Constitution in a Congress, consisting of a Senate and House of Representatives.

Electorate. By amendments of the constitution, disqualification of voters on the ground of race, colour or sex is forbidden. The electorate consists of all citizens over 18 years of age. Literacy tests have been banned since 1970. In 1972 durational residency requirements were held to violate the constitution. In 1973 US citizens abroad were enfranchised.

With limitations imposed by the constitution, it is the states which determine voter eligibility. In general states exclude from voting: persons who have not established residency in the jurisdiction in which they wish to vote; persons who have been convicted of felonies whose civil rights have not been restored; persons declared mentally incompetent by a court.

Illiterate voters are entitled to receive assistance in marking their ballots. Minority-language voters in jurisdictions with statutorily prescribed minority concentrations are entitled to have elections conducted in the minority language as well as English. Disabled voters are entitled to accessible polling places. Voters absent on election days or unable to go to the polls are generally entitled under state law to vote by absentee ballot.

The Constitution guarantees citizens that their votes will be of equal value under the 'one person, one vote' rule.

Senate. The Senate consists of two members from each state (but not from the District of Columbia), chosen by popular vote for six years, approximately one-third retiring or seeking re-election

every two years. Senators must be no less than 30 years of age; must have been citizens of the USA for nine years, and be residents in the states for which they are chosen. The Senate has complete freedom to initiate legislation, except revenue bills (which must originate in the House of Representatives); it may, however, amend or reject any legislation originating in the lower house. The Senate is also entrusted with the power of giving or withholding its 'advice and consent' to the ratification of all treaties initiated by the President with foreign powers, a two-thirds majority of Senators present being required for approval. (However, it has no control over 'international executive agreements' made by the President with foreign governments; such 'agreements' cover a wide range and are more numerous than formal treaties.)

The Senate has 21 Standing Committees to which all bills are referred for study, revision or rejection. The House of Representatives has 20 such committees. In both Houses each Standing Committee has a chairman and a majority representing the majority party of the whole House; each has numerous sub-committees. The jurisdictions of these Committees correspond largely to those of the appropriate executive departments and agencies. Both Houses also have a few select or special Committees with limited duration.

House of Representatives. The House of Representatives consists of 435 members elected every second year. The number of each state's Representatives is determined by the decennial census, in the absence of specific Congressional legislation affecting the basis. The number of Representatives for each state in the 113th congress, which began in Jan. 2013 (based on the 2010 census), is given below:

Alabama	7	Louisiana	6	Ohio	16
Alaska	1	Maine	2	Oklahoma	5
Arizona	9	Maryland	8	Oregon	5
Arkansas	4	Massachusetts	9	Pennsylvania	18
California	53	Michigan	14	Rhode Island	2
Colorado	7	Minnesota	8	South Carolina	7
Connecticut	5	Mississippi	4	South Dakota	1
Delaware	1	Missouri	8	Tennessee	9
Florida	27	Montana	1	Texas	36
Georgia	14	Nebraska	3	Utah	4
Hawaii	2	Nevada	4	Vermont	1
Idaho	2	New Hampshire	2	Virginia	11
Illinois	18	New Jersey	12	Washington	10
Indiana	9	New Mexico	3	West Virginia	3
Iowa	4	New York	27	Wisconsin	8
Kansas	4	North Carolina	13	Wyoming	1
Kentucky	6	North Dakota	1		

The constitution requires congressional districts within each state to be substantially equal in population. Final decisions on congressional district boundaries are taken by the state legislatures and governors. By custom the Representative lives in the district from which he is elected. Representatives must be not less than 25 years of age, citizens of the USA for seven years and residents in the state from which they are chosen.

In addition, five delegates (one each from the District of Columbia, American Samoa, Guam, the US Virgin Islands and Puerto Rico) are also members of Congress. They have a voice but no vote, except in committees. The delegate from Puerto Rico is the resident commissioner. Puerto Ricans vote at primaries, but not at national elections. Each of the two Houses of Congress is sole 'judge of the elections, returns and qualifications of its own members'; and each of the Houses may, with the concurrence of two-thirds, expel a member. The period usually termed 'a Congress' in legislative language continues for two years, terminating at noon on 3 Jan.

The salary of a Senator is $174,000 per annum, with tax-free expense allowance and allowances for travelling expenses and for clerical hire. The salary of the Speaker of the House of

Representatives is $223,500 per annum, with a taxable allowance. The salary of a Member of the House is $174,000 ($193,400 for the Majority Leader and Minority Leader).

No Senator or Representative can, during the time for which he is elected, be appointed to any *civil* office under authority of the USA which shall have been created or the emoluments of which shall have been increased during such time; and no person holding *any* office under the USA can be a member of either House during his continuance in office. No religious text may be required as a qualification to any office or public trust under the USA or in any state.

American Indians. By an Act passed on 2 June 1924 full citizenship was granted to all American Indians born in the USA, though those remaining in tribal units were still under special federal jurisdiction. The Indian Reorganization Act of 1934 gave the tribal Indians, at their own option, substantial opportunities of self-government and the establishment of self-controlled corporate enterprises empowered to borrow money and buy land, machinery and equipment; these corporations are controlled by democratically elected tribal councils. Recently a trend towards releasing American Indians from federal supervision has resulted in legislation terminating supervision over specific tribes. In 1988 the federal government recognized that it had a special relationship with, and a trust responsibility for, federally recognized American Indian entities in continental USA and tribal entities in Alaska. In 2010 the Bureau of Indian Affairs listed 564 'Indian Entities Recognized and Eligible to Receive Services'. American Indian lands covered 112,637 sq. miles in 2000. Indian lands are held free of taxes. Total American Indian and Alaska Native population at the 2010 census was 2,932,248, of which California (362,801), Oklahoma (321,687), Arizona (296,529) and New Mexico (193,222) accounted for more than 40%.

The **District of Columbia**, ceded by the State of Maryland for the purposes of government in 1791, is the seat of the US government. It includes the city of Washington, and embraces a land area of 61 sq. miles. The Reorganization Plan No. 3 of 1967 instituted a Mayor Council form of government with appointed officers. In 1973 an elected Mayor and elected councillors were introduced; in 1974 they received power to legislate in local matters. Congress retains power to enact legislation and to veto or supersede the Council's acts. Since 1961 citizens have had the right to vote in national elections. On 23 Aug. 1978 the Senate approved a constitutional amendment giving the District full voting representation in Congress. This has still to be ratified.

The Commonwealth of the Northern Mariana Islands, the Commonwealth of the Puerto Rico, American Samoa, Guam and the Virgin Islands each have a local legislature, whose acts may be modified or annulled by Congress, though in practice this has seldom been done. Puerto Rico, since its attainment of commonwealth status on 25 July 1952, enjoys practically complete self-government, including the election of its governor and other officials. The conduct of foreign relations, however, is still a federal function and federal bureaus and agencies still operate in the island.

General supervision of territorial administration is exercised by the Office of Territories in the Department of Interior.

Local Government

The Union comprises 13 original states, seven states which were admitted without having been previously organized as territories, and 30 states which had been territories—50 states in all. Each state has its own constitution (which the USA guarantees shall be republican in form), deriving its authority, not from Congress, but from the people of the state. Admission of states into the Union has been granted by special Acts of Congress, either (1) in the form of 'enabling Acts' providing for the drafting and ratification of a state constitution by the people, in which case the territory

becomes a state as soon as the conditions are fulfilled, or (2) accepting a constitution already framed, and at once granting admission.

Each state is provided with a legislature of two Houses (except Nebraska, which since 1937 has had a single-chamber legislature), a governor and other executive officials, and a judicial system. Both Houses of the legislature are elective, but the senators (having larger electoral districts usually covering two or three counties compared with the single county or, in some states, the town, which sends one representative to the Lower House) are less numerous than the representatives, while in 38 states their terms are four years; in 12 states the term is two years. Of the four-year senates, Illinois, Montana and New Jersey provide for two four-year terms and one two-year term in each decade. Terms of the lower houses are usually shorter; in 45 states, two years. The trend is towards annual sessions of state legislatures; most meet annually now whereas in 1939 only four did.

The Governor is elected by direct vote of the people over the whole state for a term of office ranging in the various states from two to four years, and with a salary ranging from $70,000 (Maine) to $183,255 (Pennsylvania). His duty is to see to the faithful administration of the law, and he has command of the military forces of the state. He may recommend measures but does not present bills to the legislature. In some states he presents estimates. In all but one of the states (North Carolina) the Governor has a veto upon legislation, which may, however, be overridden by the two Houses, in some states by a simple majority, in others by a three-fifths or two-thirds majority. In some states the Governor, on his death or resignation, is succeeded by a Lieut.-Governor who was elected at the same time and has been presiding over the state Senate. In several states the Speaker of the Lower House succeeds the Governor.

National Anthem

The Star-spangled Banner, 'Oh say, can you see by the dawn's early light'; words by F. S. Key, 1814, tune by J. S. Smith; formally adopted by Congress 3 March 1931.

GOVERNMENT CHRONOLOGY

PRESIDENTS OF THE USA

Name	Party[1]	From state	Term of service	Born	Died
George Washington	(F.)	Virginia	1789–97	1732	1799
John Adams	(F.)	Massachusetts	1797–1801	1735	1826
Thomas Jefferson	(D.)	Virginia	1801–09	1743	1826
James Madison	(D.)	Virginia	1809–17	1751	1836
James Monroe	(D.)	Virginia	1817–25	1759	1831
John Quincy Adams	(n.p.)	Massachusetts	1825–29	1767	1848
Andrew Jackson	(D.)	Tennessee	1829–37	1767	1845
Martin Van Buren	(D.)	New York	1837–41	1782	1862
William H. Harrison	(W.)	Ohio	Mar.–Apr. 1841	1773	1841
John Tyler	(W.)	Virginia	1841–45	1790	1862
James K. Polk	(D.)	Tennessee	1845–49	1795	1849
Zachary Taylor	(W.)	Louisiana	1849–July 1850	1784	1850
Millard Fillmore	(W.)	New York	1850–53	1800	1874
Franklin Pierce	(D.)	New Hampshire	1853–57	1804	1869
James Buchanan	(D.)	Pennsylvania	1857–61	1791	1868
Abraham Lincoln	(R.)	Illinois	1861–Apr. 1865	1809	1865
Andrew Johnson	(D.)	Tennessee	1865–69	1808	1875
Ulysses S. Grant	(R.)	Illinois	1869–77	1822	1885
Rutherford B. Hayes	(R.)	Ohio	1877–81	1822	1893
James A. Garfield	(R.)	Ohio	Mar.–Sept. 1881	1831	1881
Chester A. Arthur	(R.)	New York	1881–85	1830	1886
Grover Cleveland	(D.)	New York	1885–89	1837	1908
Benjamin Harrison	(R.)	Indiana	1889–93	1833	1901
Grover Cleveland	(D.)	New York	1893–97	1837	1908
William McKinley	(R.)	Ohio	1897–Sept. 1901	1843	1901
Theodore Roosevelt	(R.)	New York	1901–09	1858	1919
William H. Taft	(R.)	Ohio	1909–13	1857	1930
Woodrow Wilson	(D.)	New Jersey	1913–21	1856	1924

Name	Party[1]	From state	Term of service	Born	Died
Warren Gamaliel Harding	(R.)	Ohio	1921–Aug. 1923	1865	1923
Calvin Coolidge	(R.)	Massachusetts	1923–29	1872	1933
Herbert C. Hoover	(R.)	California	1929–33	1874	1964
Franklin D. Roosevelt	(D.)	New York	1933–Apr. 1945	1882	1945
Harry S. Truman	(D.)	Missouri	1945–53	1884	1972
Dwight D. Eisenhower	(R.)	New York	1953–61	1890	1969
John F. Kennedy	(D.)	Massachusetts	1961–Nov. 1963	1917	1963
Lyndon B. Johnson	(D.)	Texas	1963–69	1908	1973
Richard M. Nixon	(R.)	California	1969–74	1913	1994
Gerald R. Ford	(R.)	Michigan	1974–77	1913	2006
James Earl Carter	(D.)	Georgia	1977–81	1924	—
Ronald W. Reagan	(R.)	California	1981–89	1911	2004
George H. Bush	(R.)	Texas	1989–93	1924	—
Bill (William J.) Clinton	(D.)	Arkansas	1993–2001	1946	—
George W. Bush	(R.)	Texas	2001–09	1946	—
Barack H. Obama	(D.)	Illinois	2009–	1961	—

[1]F. = Federalist; D. = Democrat; n.p. = no party; W. = Whig; R. = Republican.

VICE-PRESIDENTS OF THE USA

Name	Party[1]	From state	Term of service	Born	Died
John Adams	(F.)	Massachusetts	1789–97	1735	1826
Thomas Jefferson	(R.)	Virginia	1797–1801	1743	1826
Aaron Burr	(R.)	New York	1801–05	1756	1836
George Clinton	(R.)	New York	1805–12[2]	1739	1812
Elbridge Gerry	(R.)	Massachusetts	1813–14[2]	1744	1814
Daniel D. Tompkins	(R.)	New York	1817–25	1774	1825
John C. Calhoun	(NR./D.)	South Carolina	1825–32[2]	1782	1850
Martin Van Buren	(D.)	New York	1833–37	1782	1862
Richard M. Johnson	(D.)	Kentucky	1837–41	1780	1850
John Tyler	(D.)	Virginia	Mar.–Apr. 1841[2]	1790	1862
George M. Dallas	(D.)	Pennsylvania	1845–49	1792	1864
Millard Fillmore	(W.)	New York	1849–50[2]	1800	1874
William R. King	(D.)	Alabama	Mar.–Apr. 1853[2]	1786	1853
John C. Breckinridge	(D.)	Kentucky	1857–61	1821	1875
Hannibal Hamlin	(R.)	Maine	1861–65	1809	1891
Andrew Johnson	(D.)	Tennessee	Mar.–Apr. 1865[2]	1808	1875
Schuyler Colfax	(R.)	Indiana	1869–73	1823	1885
Henry Wilson	(R.)	Massachusetts	1873–75[2]	1812	1875
William A. Wheeler	(R.)	New York	1877–81	1819	1887
Chester A. Arthur	(R.)	New York	Mar.–Sept. 1881[2]	1830	1886
Thomas A. Hendricks	(D.)	Indiana	Mar.–Nov. 1885[2]	1819	1885
Levi P. Morton	(R.)	New York	1889–93	1824	1920
Adlai Stevenson	(D.)	Illinois	1893–97	1835	1914
Garret A. Hobart	(R.)	New Jersey	1897–99[2]	1844	1899
Theodore Roosevelt	(R.)	New York	Mar.–Sept. 1901[2]	1858	1919
Charles W. Fairbanks	(R.)	Indiana	1905–09	1855	1920
James S. Sherman	(R.)	New York	1909–12[2]	1855	1912
Thomas R. Marshall	(D.)	Indiana	1913–21	1854	1925
Calvin Coolidge	(R.)	Massachusetts	1921–Aug. 1923[2]	1872	1933
Charles G. Dawes	(R.)	Illinois	1925–29	1865	1951
Charles Curtis	(R.)	Kansas	1929–33	1860	1935
John N. Garner	(D.)	Texas	1933–41	1868	1967
Henry A. Wallace	(D.)	Iowa	1941–45	1888	1965
Harry S. Truman	(D.)	Missouri	1945–Apr. 1945[2]	1884	1972
Alben W. Barkley	(D.)	Kentucky	1949–53	1877	1956
Richard M. Nixon	(R.)	California	1953–61	1913	1994
Lyndon B. Johnson	(D.)	Texas	1961–Nov. 1963[2]	1908	1973
Hubert H. Humphrey	(D.)	Minnesota	1965–69	1911	1978
Spiro T. Agnew	(R.)	Maryland	1969–73	1918	1996

Name	Party[1]	From state	Term of service	Born	Died
Gerald R. Ford	(R.)	Michigan	1973–74	1913	2006
Nelson Rockefeller	(R.)	New York	1974–77	1908	1979
Walter Mondale	(D.)	Minnesota	1977–81	1928	—
George H. Bush	(R.)	Texas	1981–89	1924	—
Danforth Quayle	(R.)	Indiana	1989–93	1947	—
Albert Gore	(D.)	Tennessee	1993–2001	1948	—
Richard B. Cheney	(R.)	Wyoming	2001–09	1941	—
Joseph R. Biden	(D.)	Delaware	2009–	1942	—

[1]F. = Federalist; R. = Republican; NR. = National Republican; D. = Democrat; W. = Whig. [2]Position vacant thereafter until commencement of the next presidential term.

RECENT ELECTIONS

At the presidential election on 6 Nov. 2012 Barack Obama (Democrat) was re-elected president with 332 electoral college votes against 206 for Mitt Romney (Republican).

a) Won by Obama

State	Electoral college votes	State	Electoral college votes
California	55	Nevada	6
Colorado	9	New Hampshire	4
Connecticut	7	New Jersey	14
Delaware	3	New Mexico	5
D.C.	3	New York	29
Florida	29	Ohio	18
Hawaii	4	Oregon	7
Illinois	20	Pennsylvania	20
Iowa	6	Rhode Island	4
Maine	4	Vermont	3
Maryland	10	Virginia	13
Massachusetts	11	Washington	12
Michigan	16	Wisconsin	10
Minnesota	10		

b) Won by Romney

State	Electoral college votes	State	Electoral college votes
Alabama	9	Montana	3
Alaska	3	Nebraska	5
Arizona	11	North Carolina[1]	15
Arkansas	6	North Dakota	3
Georgia	16	Oklahoma	7
Idaho	4	South Carolina	9
Indiana[1]	11	South Dakota	3
Kansas	6	Tennessee	11
Kentucky	8	Texas	38
Louisiana	8	Utah	5
Mississippi	6	West Virginia	5
Missouri	10	Wyoming	3

[1]Won by Obama in 2008.

Following the elections of 6 Nov. 2012 the 113th Congress (2013–15) is constituted as follows: Senate—53 Democrats, 45 Republicans and 2 ind. who are expected to caucus with the Democrats (51 Democrats, 47 Republicans and 2 ind. for the 112th Congress at the time of the 2012 elections); House of Representatives—233 Republicans and 200 Democrats and two vacant (240 Republicans, 190 Democrats and five vacant for the 112th Congress at the time of the 2012 elections).

The Speaker of the House of Representatives is John Boehner (R.). The Majority Leader of the Senate is Harry Reid (D.).

CURRENT GOVERNMENT

President of the United States: Barack Obama, of Hawaii; b. 1961. Majored in Political Science at Columbia (1983); MA at Harvard Law (1991); Fellow of University of Chicago Law School teaching constitutional law (1992–2004); Senator for the 13th District of Illinois (1996–2004); US Senator (2004–08).

Vice President: Joe Biden, b. Pennsylvania, 1942. First elected to the Senate in 1972; Chair of the Senate Judiciary Committee (1987–95); Adjunct Professor at Widener University School of Law since 1991.

In April 2013 the cabinet consisted of the following:

1. *Secretary of State* (created 1789). John Kerry; b. 1943.
2. *Secretary of the Treasury* (1789). Jacob J. Lew; b. 1955.
3. *Secretary of Defense* (1947). Charles T. Hagel; b. 1946.
4. *Attorney General* (Department of Justice, 1870). Eric Holder; b. 1951.
5. *Secretary of the Interior* (1849). Sally Jewell; b. 1955.
6. *Secretary of Agriculture* (1889). Tom Vilsack; b. 1950.
7. *Acting Secretary of Commerce* (1903). Rebecca M. Blank; b. 1955.
8. *Acting Secretary of Labor* (1913). Seth D. Harris; b. 1962.
9. *Secretary of Health and Human Services* (1953). Kathleen Sebelius; b. 1948.
10. *Secretary of Housing and Urban Development* (1966). Shaun L. S. Donovan; b. 1966.
11. *Secretary of Transportation* (1967). Ray LaHood; b. 1945.
12. *Acting Secretary of Energy* (1977). Daniel Poneman; b. 1956.
13. *Secretary of Education* (1979). Arne Duncan; b. 1964.
14. *Secretary of Veterans' Affairs* (1989). Eric K. Shinseki; b. 1942.
15. *Secretary of Homeland Security* (2002). Janet Napolitano; b. 1957.

Each of the above cabinet officers receives an annual salary of $199,700 and holds office during the pleasure of the President.

The following also have cabinet status:

Chair of the Council of Economic Advisers: Alan B. Krueger; Environmental Protection Agency Administrator (acting): Bob Perciasepe; Director of the Office of Management and Budget: Sylvia Burwell; US Trade Representative (acting): Demetrios Marantis; US Ambassador to the United Nations: Susan Rice; White House Chief of Staff: Denis McDonough; Administrator of the Small Business Administration: Karen Mills.

Office of the President: http://www.whitehouse.gov

CURRENT LEADERS

Barack Obama

Position
President

Introduction
Barack Obama became the 44th president of the USA in Jan. 2009 and the first African American to hold the office. Having secured the Democratic candidacy, he contested the presidential election in Nov. 2008 against a background of deepening economic crisis. Viewed as on the centre-liberal wing of his party, Obama's principal electoral pledges included the introduction of a national health insurance plan and a scaling down of the US troop presence in Iraq. He also sought to counter the economic crisis with stimulus measures, while his Republican opponent advocated a curb on spending. On foreign policy, Obama confirmed the US military commitment to Afghanistan as a central plank in the campaign against terrorism, while seeking to promote compliance with international law and the adoption of democratic values across the world political stage. He was re-elected for a second and final term on 6 Nov. 2012 and was inaugurated in Jan. 2013.

Early Life
Barack Hussein Obama was born on 4 Aug. 1961 in Honolulu, Hawaii, to a Kenyan father and white American mother. His parents divorced and, following his mother's remarriage in 1967, the family moved to Indonesia, where Obama was educated until the age of ten. He attended Punahou School in Honolulu and Occidental College, Los Angeles, before graduating from Columbia University, New York, in 1983 with a BA in political science. From 1983–85 he worked at Business International

Corporation and at the New York Public Interest Research Group, then moved to Chicago to become director of the church-based Developing Communities Project (DCP). From 1985–88 he led the DCP, expanding its staff and budget and establishing new projects. He attended Harvard Law School from 1988–91 and was elected president of the *Harvard Law Review* in 1990.

Following his graduation Obama took up a fellowship with the University of Chicago Law School, where he taught constitutional law from 1992–2004. In 1992 he directed 'Illinois Project Vote!', a campaign to register African Americans to vote, and from 1992–2002 he served on the boards of various community organizations and foundations, including the Joyce Foundation and Public Allies. In 1993 he joined law firm Davis, Miner, Barnhill & Galland, practising first as an associate then as a counsel. In 1995 he published a memoir, *Dreams from My Father: A Story of Race and Inheritance*.

In 1996 Obama was elected senator for the 13th District of Illinois, subsequently winning re-election in 1998 and 2002. As senator he supported health care and welfare reforms, sponsored a law to increase tax credits for low paid workers and promoted tighter regulation of the mortgage industry. In 2003, in co-operation with Republican senators, he led legislation to monitor police procedures in the state of Illinois, requiring police to profile the ethnicity of motorists they stopped and making it compulsory to videotape interrogations of homicide suspects.

After an unsuccessful run for the House of Representatives in 2000, Obama mounted a campaign for the 2004 US Senate elections. He attracted national attention at that year's National Democratic Convention when he gave a keynote speech, 'The Audacity of Hope', in which he spoke of the shared aspirations and efforts of American citizens and set out government's obligations towards them. In Nov. 2004 Obama was elected to the US Senate with 70% of the vote, the largest winning margin in Illinois state history. In office he supported legislation to reduce carbon emissions, voted for robust border controls and immigration reform, and campaigned for controls on political financing, in particular gifts and funding provided by lobbyists.

Having opposed military action against Iraq in 2003, he continued to criticize the conduct of the war. As a member of the Senate's foreign relations committee, he explored ways of reducing the threat from conventional weapons and, with Republican Senator Richard G. Lugar, co-authored a law extending US co-operation in identifying and disposing of stockpiled weapons. He supported successive bills calling for international intervention in Sudan and in 2006 voted for a no-fly zone over Darfur. Obama also served on the health, education, labour and pensions committees, the committee on veterans' affairs and the committee on homeland security and governmental affairs. In these areas he supported moves to expand early years schooling, to increase financial help for low-income high school and college students and to provide funding for veterans to attend college.

In Feb. 2007 Obama announced his candidacy for the Democratic presidential nomination. Campaigning on the themes of change and unity, he promised to address the key issues of Iraq, health care and the USA's dependence on oil. Obama fought a vigorous contest with main rival Hillary Clinton throughout 2007 and early 2008, gaining praise for his oratory while defending himself against charges of inexperience. In May 2007 he pledged a national health insurance plan open to all. Criticized by environmentalists for supporting liquefied coal, he subsequently modified his position. On the Iraq War, he argued for the phased redeployment of US forces and the withdrawal of combat troops, as proposed in his Iraq War De-Escalation Act of 2007.

By June 2008 Obama had secured the support of a majority of Democratic Party delegates and was confirmed at the Democratic National Convention of Aug. 2008. He selected Joe Biden, the long-serving senator of Delaware, as his running mate.

Some commentators claimed that during the campaign Obama softened his line on troop withdrawal from Iraq and on gun control. His decision to use private donations for his presidential campaign reversed an earlier pledge to work within federal public funding limits. He accused his opponents of 'gaming this broken system' when turning his back on $84m. that would have been available to him and, in doing so, becoming the first presidential candidate in over three decades to bypass the federal system in favour of raising unlimited private finance. During the early weeks of campaigning, polls showed a close contest between Obama and Republican candidate John McCain, with Obama being seen as inexperienced in foreign affairs. However, the failure of key US financial institutions in late 2008 focused attention on the economy, prompting a spike in Obama's support. He called for regulatory reforms and a bipartisan approach to tackling the crisis and voted in favour of President Bush's $700bn. package to buy up mortgage-related securities. Obama won the election on 4 Nov. 2008 with 53% of the vote to McCain's 46% and by 365 electoral college votes to 173.

In the transition between his election and inauguration, Obama appointed Rahm Emanuel as chief of staff and former election rival, Hillary Clinton, as secretary of state. Both appointments were seen as an indication that he would tap the experience of long-serving politicians and officials and retain much of the previous administration's foreign policy. He also gathered an economic team and began preparing a stimulus plan to aid economic recovery through investment.

Career in Office

Obama was inaugurated on 20 Jan. 2009. Because of a minor misreading of the oath of office on the part of Chief Justice John Roberts, which caused Obama to make a similar error, he took the oath for a second time on 21 Jan. 2009. Among his first presidential aims was the fulfilment of an election pledge that the administration would run down and eventually close the detention facility for terrorist suspects at the US naval base in Guantanamo Bay. However, this aspiration proved elusive throughout his first term of office. Other early measures included tightening restrictions on lobbyists joining the administration, introducing stricter curbs on fuel emissions, enacting equal pay legislation and expanding children's health care.

Obama sought bipartisan support for a $825bn. stimulus package, which aimed to boost economic recovery through sustained investment programmes. However, most Republicans opposed the package, arguing for less direct government spending and tax reductions. After Republicans forced substantial amendments, the American Recovery and Reinvestment Act was passed on 13 Feb. 2009, relying almost exclusively on Democrat support. Worth a slightly reduced $787bn., it detailed plans for unprecedented levels of investment in education, health care, infrastructure, the environment, employment and tax reduction.

On foreign policy, Obama's early months saw a move away from the hawkish tone of the previous administration. In Feb. 2009 he announced that most US troops would be withdrawn from Iraq by 31 Aug. 2010, with residual forces leaving by the end of 2011. He also signalled a change of approach on Afghanistan, indicating that, although troop numbers would initially be increased in an echo of the 'surge' tactics employed in Iraq, he was reviewing strategy and did not believe the region could be stabilized by military means alone. Obama also changed the tone of the USA's dealings with the Middle East, sending envoys to Syria in March 2009 and expressing a willingness to talk to Iran's leaders, subject to their compliance with UN directives on nuclear development. He was similarly cautious on the Israeli–Palestinian conflict, reaffirming the USA's commitment to pursuing a two-state solution while maintaining the previous administration's stance of refusing to talk to the militant Hamas leadership in Gaza. He gave moderate encouragement to the idea of closer dialogue between the USA and China and opened up the possibility

of negotiations between the USA and Russia on cutting nuclear stockpiles and on curbing the development of new weapons.

However, one year on from his inauguration Obama had yet to secure any major policy objectives and his personal approval rating among voters had slipped markedly according to opinion polls. Despite the significant injections of borrowed money to stimulate demand and boost the economy, job creation proved slow and unemployment rose to 10% in 2009. Economic weakness in turn further undermined the country's fiscal position.

Obama's radical health care reform plan to extend insurance cover to all Americans proved particularly contentious and remained the subject of fierce public and congressional division. The House of Representatives and the Senate each passed their own bills on health care reform in late 2009. In Jan. 2010 the loss of the Democrats' critical 60–40 majority in the Senate upset plans to reconcile the two bills. Nonetheless, the health care reform was passed in March 2010, by 56–43 votes in the Senate and by 220–207 in the House of Representatives, and it was hoped that under its terms coverage would extend to a further 32m. Americans. However, the legislation continued to be widely challenged as unconstitutional before its validity was upheld by a majority vote of the Supreme Court in June 2012.

In Sept. 2009 Obama eased friction with Russia as he announced the abandonment of a missile defence deployment in the Czech Republic and Poland. In March 2010 the two countries agreed a treaty to replace the START Treaty on nuclear arms reduction. Obama was awarded the Nobel Peace Prize in Oct. 2009 for his efforts to create 'a new international atmosphere'. Relations with China, however, worsened over trade and currency policy, US weapons sales to Taiwan and, in Feb. 2010, the president's meeting with the Dalai Lama of Tibet.

In April 2010 an explosion in the Gulf of Mexico led to the biggest oil spill in US history, causing huge damage to local communities and the economy. The political fallout, together with the rise in support for the conservative libertarian Tea Party movement, hit Obama's ratings. In mid-term congressional elections in Nov., the Republicans made sweeping gains to regain control of the House of Representatives, heralding a legislative gridlock with the Democratic majority in the Senate.

Obama's pressure on the Israeli government to stop settlement building on Palestinian land in the Middle East has been largely resisted by Prime Minister Netanyahu, undermining any further US-brokered talks between the Israeli and Palestinian authorities. Although critical of Israeli policy, the US government did, however, oppose the UN General Assembly's recognition in Nov. 2012 of Palestine's enhanced status as a non-member observer state. Meanwhile, US moves to curb the nuclear ambitions of Iran and North Korea have also stalled, despite several rounds of multilateral negotiations and the imposition of punitive sanctions on both countries. After some progress in improving relations with Syria, with a US ambassador appointed in Feb. 2010 after a five-year gap, President Assad's violent reaction to domestic dissent starting in 2011 has since ruptured US–Syrian ties. In March 2011 Obama committed the USA to join the NATO military intervention in Libya to protect civilians against the Gaddafi regime, stating that Americans could not brush aside 'responsibilities to our fellow human beings'.

In Afghanistan Obama initially intensified the war, announcing further troop deployments during 2009 to fight the Taliban insurgency. There has since been a gradual scaling down of foreign involvement in the country, and in May 2012 NATO endorsed a planned withdrawal of its combat forces (including US troops) by the end of 2014.

In May 2011 US special forces killed Osama bin Laden, having traced him to a compound in Abbottabad in northwestern Pakistan. Bin Laden, the leader of the al-Qaeda movement responsible for the 11 Sept. attacks on New York and Washington, D.C. in 2001, had been in hiding for almost ten years. Obama commented that 'his demise should be welcomed by all who believe in peace and human dignity'. The operation strained already uneasy relations between the USA and Pakistan.

In April 2011 Obama called for a \$4trn. reduction over 12 years in the federal budget deficit through spending cuts and increased taxes on higher earners. In May that year the government reached its federal debt ceiling (by law) of \$14·3trn., heralding several months of political wrangling before a last-minute cross-party debt-reduction agreement was reached in Aug. to raise the ceiling and stave off the risk of a default. Nevertheless, concerns over US public finances in the long term prompted the Standard & Poor's rating agency to downgrade the USA's triple-A rating for the first time. Obama dismissed the downgrade, claiming: 'We've always been and always will be a triple-A country.' Entering an election year in 2012, Obama announced defence plans in Jan. aimed at reconciling budget savings with more focused strategic priorities (including a reduced troop presence in Europe) while maintaining US military superiority. The economy meanwhile registered growth of 1·8% in 2011 and 2·2% in 2012.

In the presidential election on 6 Nov. 2012 Obama defeated the Republican candidate, Mitt Romney, winning by 332 electoral votes to Romney's 206. In his inaugural speech in Jan. 2013 Obama mapped out a programme of economic and social change, including a pledge to sustain recovery and plans for a ban on assault weapons prompted by the murder in Dec. 2012 of 20 children and six adults by a gunman in Newtown, Connecticut.

Following his re-election Obama was embroiled in renewed congressional discord over US debt and federal expenditure. However, despite Republican obstruction, swingeing spending cuts and tax rises from the start of 2013 were narrowly averted when the House of Representatives and the Senate approved a compromise deal allowing for limited tax increases on higher incomes.

DEFENCE

The President is C.-in-C. of the Army, Navy and Air Force.

The National Security Act of 1947 provides for the unification of the Army, Navy and Air Forces under a single Secretary of Defense with cabinet rank. The President is also advised by a National Security Council and the Office of Civil and Defense Mobilization.

Defence expenditure in 2010 totalled \$698,281m. (\$2,156 per capita). Defence spending in 2009 represented 4·7% of GDP (down from 37·8% in 1944, 14·2% in 1953, 9·4% in 1968 and 6·2% of GDP in 1986 although up from the post-war low of 3·0% in 1999). The USA spent more on defence in 2008 than the next 15 biggest spenders combined. US expenditure was 41·5% of the world total, although its population is less than 5% of the world total. In 1997 the Quadrennial Defense Review (QDR) was implemented—a plan to transform US defence strategy and military forces.

Conscription was first introduced during the American Civil War in 1862 and operated during all subsequent major periods of conflict including World Wars I and II, the Korean War and the Vietnam War. A limited draft was also employed in peacetime in the Cold War and early 1960s. The final draft ended in 1973 when the USA converted to an all-volunteer military. Although conscription is not currently in force the Military Selective Service Act requires all males between the ages of 18 and 26 to register for compulsory military service should the need arise.

Active duty military personnel in Sept. 2011 numbered 1,468,364, of which 214,098 were women. In Dec. 2010 US armed forces abroad numbered 291,700, including 103,700 in and around Afghanistan and 85,600 in and around Iraq. In Dec. 2011 the last US troops left Iraq although 4,000 remained in neighbouring Kuwait. Active duty military deaths in the US armed forces totalled 1,483 in 2010, down from 1,515 in 2009. In June 2011 President Obama announced that 10,000 troops would be withdrawn from Afghanistan by the end of 2011 and a further 23,000 by the end of Sept. 2012. US troop deaths in Iraq between 2003 and April 2012 totalled 4,486; deaths in Afghanistan between 2001 and April 2013 totalled 2,200.

The USA is the world's largest supplier of major conventional weapons, with sales in 2011 worth \$10·0bn., or 33·3% of the world total. In 2010 Lockheed Martin and Boeing were the two largest arms producing companies in the USA, accounting for \$35·7bn. and \$32·9bn. worth of sales respectively.

The USA's last nuclear test was in 1993. In accordance with START I—the treaty signed by the US and USSR in 1991 to reduce strategic offensive nuclear capability—the number of strategic nuclear warheads (intercontinental ballistic missiles, submarine-launched ballistic missiles and bombers) in Jan. 2012 was approximately 1,950. There were also 200 non-strategic warheads in Jan. 2012, making a total of around 2,150 deployed warheads. There are a further 2,750 warheads held in reserve and another 3,100 scheduled to be dismantled. In May 2010 the Obama administration announced that the USA had a total of 5,113 active nuclear warheads, down from a peak of 31,225 in 1967. Strategic nuclear delivery vehicles were made up as follows:

Intercontinental ballistic missiles: 500 Minuteman III.
Submarine-launched ballistic missiles: 288 Trident II.
Bombers: 91 B-52H; 20 B-2.

START I expired in Dec. 2009. In July 2007 the Bush administration decided not to extend the treaty beyond the original expiry date. In May 2001 President Bush called for the development of an anti-missile shield to move beyond the constraints of the Anti-Ballistic Missile Treaty. In Dec. 2001 he announced that the USA was unilaterally abandoning the Treaty. On 24 May 2002 the USA and Russia signed an arms control treaty (the Strategic Offensive Reductions Treaty or Moscow Treaty) to reduce the number of US and Russian warheads, from between 6,000 and 7,000 each to between 1,700 and 2,200 each. The treaty was in force from June 2003 until Feb. 2011. It was then superseded by the Measures to Further Reduction and Limitation of Strategic Offensive Arms (New START), signed by the US and Russian presidents in April 2010 and ratified by the US Senate in Dec. 2010 and the Russian Federal Assembly in Jan. 2011. It laid out terms for the number of warheads on each side to be limited to 1,550 by 2017, a 30% drop on previous levels.

Estimates of the number of firearms in the country are around 310m., equivalent to 99 firearms for every 100 people, making USA the world's most heavily armed country.

Army

Secretary of the Army: John McHugh.

The Secretary of the Army is the head of the Department of the Army. Subject to the authority of the President as C.-in-C. and of the Secretary of Defense, he is responsible for all affairs of the Department.

The Army consists of the Active Army, the Army National Guard of the US, the Army Reserve and civilian workforce; and all persons appointed to or enlisted into the Army without component; and all persons serving under call or conscription, including members of the National Guard of the States, etc., when in the service of the US. The active duty strength of the Army was 565,463 (79,694 women) in Sept. 2011.

The Army budget for fiscal year 2012 was \$201,387m., with a request for fiscal year 2013 of \$184,640m.

The US Army Forces Command (FORSCOM), with headquarters at Fort Bragg, North Carolina, is the largest United States Army Command and consists of more than 750,000 Active Army, US Army Reserve and Army National Guard soldiers. The other Army Commands are US Army Training and Doctrine Command (TRADOC) and US Army Materiel Command (AMC). The Army is presently undergoing a period of transformation. The Army Service Component Commands are: United States Army Central (USARCENT)/Third Army; United States Army Europe (USAREUR)/Seventh Army; United States Army North (USARNORTH)/Fifth Army; United States Army South (USARSO)/Sixth Army; United States Army Pacific (USARPAC); United States Army Africa (USARAF); United States Army

Special Operations Command (USASOC); Military Surface Deployment and Distribution Command (SDDC). The US Army Space and Missile Defense Command (SMDC), created in 1997 and with its headquarters at Redstone Arsenal, Alabama, is a specialized major command within the United States Army.

The First United States Army now serves as a mobilization, readiness and training command; the Second United States Army, which was reactivated in 2010, supports US Cyber Command.

Approximately 32% of the Active Army is deployed outside the continental USA. Several divisions, which are located in the USA, keep equipment in Germany and can be flown there in 48–72 hours. Headquarters of US Seventh and Eighth Armies are in Europe and Korea respectively.

Combat vehicles of the US Army are the tank, armoured personnel carrier, infantry fighting vehicle, and the armoured command vehicle. The Army's main battle tank is the M1A2 Abrams; its standard infantry fighting vehicle is the M2A3 Bradley.

The Army has nearly 4,900 aircraft, all but about 300 of them helicopters, including AH-1 Cobra and AH-64 Apache attack helicopters.

Over 95% of recruits enlisting in the Army have a high-school education and over 50% of the Army is married. Women serve in both combat support and combat service support units.

The National Guard is a reserve military component with both a state and a federal role. Enlistment is voluntary. The members are recruited by each state, but are equipped and paid by the federal government (except when performing state missions). As the organized militia of the several states, the District of Columbia, Puerto Rico and the Territories of the Virgin Islands and Guam, the Guard may be called into service for local emergencies by the chief executives in those jurisdictions; and may be called into federal service by the President to thwart invasion or rebellion or to enforce federal law. In its role as a reserve component of the Army, the Guard is subject to the order of the President in the event of national emergency. In 2006 it numbered 458,030 (Army, 351,350; Air Force, 106,680).

The Army Reserve is designed to supply qualified and experienced units and individuals in an emergency. Members of units are assigned to the Ready Reserve, which is subject to call by the President in case of national emergency without declaration of war by Congress. The Standby Reserve and the Retired Reserve may be called only after declaration of war or national emergency by Congress. In 2006 the Army Reserve numbered 324,100.

Navy
Secretary of the Navy: Ray Mabus.

The Navy's Operating Forces include the Atlantic Fleet, divided between the 2nd fleet (home waters) and 6th fleet (Mediterranean) and the Pacific Fleet, similarly divided between the 3rd fleet (home waters), the 7th fleet (West Pacific) and the 5th fleet (Indian Ocean), which was formally activated in 1995 and maintained by units from both Pacific and Atlantic.

The Navy budget for fiscal year 2012 was $172,510m., with a request for fiscal year 2013 of $170,132m.

The active duty strength of the Navy was 325,123 (53,385 women) in Sept. 2011.

The active ship force levels of the Navy in the year indicated:

Category	2000	2005	2010	2011
Aircraft Carriers	12	12	11	11
Cruisers	27	23	22	22
Destroyers	54	46	59	61
Frigates	35	30	29	26
Littoral Combat Ships	—	—	2	2
Surface Warships	128	111	123	122
Attack Submarines	56	54	53	53
Ballistic Missile-Carrying Submarines (SSBNs)	18	14	14	14

Category	2000	2005	2010	2011
Guided-Missile Submarines (SSGNs)	—	4	4	4
Mine Warfare	18	17	14	14
Amphibious	41	37	33	31
Auxiliary	57	45	47	47
Total Active	318	282	288	285

Ships in the Naval Reserve Force, Military Sealift Command and Naval Fleet Auxiliary Force are included in this table.

Submarine Forces. A principal part of the US naval task is to deploy the seaborne strategic deterrent from nuclear-powered ballistic missile-carrying submarines (SSBN), of which there were 14 in 2012, all of the Ohio class. The listed total of 57 tactical submarines in 2012 comprised four Ohio class guided-missile submarines, plus 41 attack submarines of the Los Angeles class, three of the Seawolf class and nine of the Virginia class.

Surface Combatant Forces. The surface combatant forces are comprised of modern cruisers, destroyers and frigates. These ships provide multi-mission capabilities to achieve maritime dominance in the crowded and complex littoral warfare environment.

As of 2012 the cruiser force consisted of 22 Ticonderoga class ships. There were 62 active guided-missile Arleigh Burke Aegis class destroyers and 22 active Oliver Hazard Perry class guided-missile frigates.

Aircraft carriers. There were ten nuclear-powered Nimitz class carriers in 2012, the first of which, USS *Nimitz*, was commissioned on 3 May 1975 and the tenth and last of which, USS *George H. W. Bush*, was commissioned on 10 Jan. 2009. USS *Enterprise*, completed in 1961, was the prototype nuclear-powered carrier and was in service until Dec. 2012 when it was retired, temporarily reducing the number of aircraft carriers to ten. Each of the Nimitz class carriers can carry between 85 and 90 fixed-wing aircraft and helicopters. Strike fighter planes are primarily F/A-18 Hornets and Super Hornets.

Naval Aviation. The principal function of the naval aviation organization (strength in 2009 of 98,588) is to train and provide combat ready aviation forces. The main carrier-borne combat aircraft in the 2009 inventory were 753 F/A-18 Hornet dual-purpose fighter/attack aircraft out of a total of 900 combat-capable aircraft.

The Marine Corps
While administratively part of the Department of the Navy, the Corps ranks as a separate armed service, with the Commandant serving in his own right as a member of the Joint Chiefs of Staff, and responsible directly to the Secretary of the Navy. Its strength was 204,261 in 2009.

The role of the Marine Corps is to provide specially trained and equipped amphibious expeditionary forces. The Corps includes an autonomous aviation element numbering 34,700 in 2009.

The US Coast Guard
The Coast Guard operates under the Department of Homeland Security in time of peace and as part of the Navy in time of war or when directed by the President. The act of establishment stated the Coast Guard 'shall be a military service and branch of the armed forces of the United States at all times'.

The Coast Guard is the country's oldest continuous sea-going service and its missions include maintenance of aids to navigation, icebreaking, environmental response (oil spills), maritime law enforcement, marine licensing, port security, search and rescue and waterways management.

The workforce in 2009 was made up of 43,598 military personnel augmented by 7,659 civilians. On an average Coast Guard day, the service saves 14 lives, boards 193 ships and boats, seizes $12·9m. worth of illegal drugs, conducts 74 search and rescue cases, processes 375 seaman's documents, investigates 24 marine casualty accidents, inspects 18 commercial fishing vessels, assists 98 people in distress, services 135 aids to navigation and interdicts 17 illegal immigrants.

Air Force

Secretary of the Air Force: Michael B. Donley.

The Department of the Air Force was activated within the Department of Defense on 18 Sept. 1947, under the terms of the National Security Act of 1947.

The USAF has the mission to defend the USA through control and exploitation of air, space and cyberspace. For operational purposes the service is divided into nine major commands, 35 field operating agencies and four direct-reporting units. In addition there are two reserve components: the Air Force Reserve and the Air National Guard.

Major commands accomplish designated phases of USAF worldwide activities. They also organize, administer, equip and train their subordinate elements. In descending order, elements of major commands include numbered air forces, wings, groups, squadrons and flights. The basic unit for generating and employing combat capability is the wing, considered to be the Air Force's prime war-fighting instrument. The bulk of the combat forces are grouped under the Air Combat Command, which controls strategic bombing, tactical strike, air defence and reconnaissance assets in the USA.

Air Force bombers include the B-1B Lancer, the B-2A Spirit and the B-52H Stratofortress, which has been the primary manned strategic bomber for 50 years. In the fighter category are the F-22A Raptor, F-15 Eagle and Strike Eagle and the F-16 Fighting Falcon.

The Air Force budget for fiscal year 2012 was $161,666m., with a request for fiscal year 2013 of $154,337m.

The active duty strength of the Air Force was 333,370 (63,552 women) in Sept. 2011—19·1% of personnel were women, compared to 16·4% in the Navy and 13·6% in the Army. Since 1991 women have been authorized to fly combat aircraft, but not until 1993 were they allowed to fly fighters.

ECONOMY

Services accounted for 77% of GDP in 2007, industry 22% and agriculture 1%.

Per capita personal income in 2010 was $39,945, up from $19,477 in 1990 although down from $40,947 in 2008.

According to the anti-corruption organization *Transparency International*, the United States ranked 19th in the world in a 2012 survey of the countries with the least corruption in business and government. It received 73 out of 100 in the annual index.

In 2011 the USA gave $30·9bn. in international aid, the highest figure of any country. In terms of a percentage of GNI, however, USA was one of the least generous major industrialized countries, giving just 0·20% (compared to more than 0·6% in the early 1960s).

Overview

The USA is the world's leading economy, although in 2007 the 27-member European Union replaced it as the world's largest in terms of total GDP. The USA's volume of trade is the largest, but the value of the external sector as a percentage of GDP is relatively low. It is self-sufficient in most raw materials, with the notable exception of oil. Core industries include motor vehicles, steel, aerospace, chemicals, telecommunications, electronics and computers. Since 1992 the economy has grown at higher average rates than the OECD and G7 in most years, and per capita GDP is higher than in other G7 countries.

In 1995 labour productivity in advanced European countries had reached US levels but since then US labour and total factor productivity growth has outpaced that of Europe and Japan. The principal sectors in which the USA outperforms its rivals are retail, wholesale and finance. It has also been able to extract greater efficiency gains from IT-related investments. US firms enjoy greater flexibility than their counterparts in Western Europe

and Japan in laying off workers and in introducing labour-saving equipment. However, income inequality is higher than in other advanced economies and the gap between skilled and unskilled labour incomes has been growing over the last three decades. Most income gains have accrued to the richest 20% of Americans.

The attacks of 11 Sept. 2001 posed serious challenges to the dollar payment system and the stock market. Business fixed investment relative to GDP plummeted in 2001–02 after reaching record highs in 2000–01. The collapse of energy giant Enron in Dec. 2002 was followed by other accounting scandals. With investor confidence low, the dollar fell to near parity with the euro for the first time since the introduction of the euro in Jan. 2002. In the third quarter of 2003 the economy recorded an annualized growth rate of 6·9%, fuelled by tax cuts, low interest rates and increased consumer and business spending. Interest rates began rising from historic lows in 2003 with the federal funds rate reaching 5·25% by the end of 2006. Growth remained strong through 2006 despite the impact of Hurricane Katrina and high oil prices. The domestic economy was buoyed by strong consumer demand, which in turn increased personal debt.

The sub-prime mortgage crisis of 2007 damaged the wider economy. A sharp rise in defaults and foreclosures as a result of the rapid decline in house market prices began in 2006, spreading panic through the banking sector. A stimulus package enacted in early 2008 (consisting of targeted tax rebates and investment incentives) together with interest rate cuts amounting to 3·75% by Oct. 2008 were aimed at raising confidence in the financial markets. However, nervousness increased with tougher lending criteria and declining asset prices. In Sept. 2008 Lehman Brothers became the first major bank to collapse since the start of the credit crisis. The US government announced rescue packages for mortgage giants Fannie Mae and Freddie Mac—between them responsible for half of the outstanding mortgages in the economy—along with AIG, the country's biggest insurance company.

In Oct. 2008 the House of Representatives passed a $700bn. government plan aimed at purchasing bad debts of failing institutions. However, continued market turmoil resulted in a further $250bn. investment plan that gave the government stakes in a range of banks. Interest rates reached their lowest recorded level at between 0–0·25% in Dec. 2008. In Feb. 2009 Congress approved President Obama's $787bn. economic stimulus package, comprising tax breaks and money for social programmes, a move that substantially increased the fiscal deficit.

The government reported growth of 1·4% in the third quarter of 2009, the first quarterly increase since the second quarter of 2008, indicating that the recession had ended. Recovery was underpinned by strong government stimulus, a rebound in world trade (with increasing demand from large emerging-market economies) and housing market stabilization. The economy nonetheless contracted by 3·1% in 2009 while real GDP growth in the 2000s, at an annual average of 1·7%, was the lowest in a full decade since the 1930s. Subsequent recovery has been slow by historical standards owing to a combination of external events such as rising oil prices and Europe's debt crisis, and domestic drivers such as stuttering consumer spending and a sluggish housing market. However, GDP grew by 2·4% in 2010 and positive employment and retail sales figures in early 2012 saw the stock market jump to its highest level since May 2008, although employment figures remain well below pre-crisis levels.

Further fiscal pressures have emerged with the retirement of the baby-boom generation and the rise in life expectancy, increasing the strain on entitlement programmes. In March 2010 President Obama signed into law a health care reform bill that was expected to extend health insurance to an additional 32m. Americans, impose new taxes on the wealthy and outlaw restrictive insurance practices. According to the Congressional Budget Office, the health care bill was expected to cut the federal deficit by $138bn. over ten years.

Federal fiscal deficits have increased as a result of the growth slowdown and stimulus packages. In Feb. 2011 Obama unveiled his 2012 Budget, which aimed to cut $1·1trn. from the US deficit over the next decade. However, political wrangling over raising the debt ceiling caused disruption on the financial markets as fears mounted that the economy would default. In Aug. 2011 Congress approved a bipartisan compromise deal that raised the debt ceiling by $2·4trn. from $14·3trn and paved the way for savings of at least $2·1trn. over ten years. The credit-rating agency Standard & Poor's downgraded the USA's AAA credit-rating to AA+ days after the deal had been reached, causing further volatility in the markets.

A combination of scheduled tax increases and automatic spending cuts—known as the 'fiscal cliff'—threatened to push the economy into recession in 2013. Failure to reach agreement on the budget by 31 Dec. 2012 would have triggered automatic tax increases of over $500bn. and spending cuts of over $100bn. from domestic and military programmes, amounting to a 4–5% cut in output. A deal was reached on the eve of the deadline, based on a package of measures including tax rises for families earning over $450,000 and individuals above $400,000 a year, a 2% increase in employee payroll taxes, a one-year extension for unemployment benefits and a five-year extension to tax cuts from 2009, including child credit and credit for attending university. Nonetheless, a first instalment of budget cuts, known as the 'sequester' and worth $85bn., was not averted and came into force on 1 March 2013.

Currency

The unit of currency is the *dollar* (USD) of 100 *cents*. Notes are issued by the 12 Federal Reserve Banks, which are denoted by a branch letter (A = Boston, MA; B = New York, NY; C = Philadelphia, PA; D = Cleveland, OH; E = Richmond, VA; F = Atlanta, GA; G = Chicago, IL; H = St Louis, MO; I = Minneapolis, MN; J = Kansas City, MO; K = Dallas, TX; L = San Francisco, CA).

Inflation rates (based on OECD statistics):

2002	2003	2004	2005	2006	2007	2008	2009	2010	2011
1·6%	2·3%	2·7%	3·4%	3·2%	2·9%	3·8%	−0·3%	1·6%	3·1%

The inflation rate in 2012 according to the Bureau of Labor Statistics was 2·1%. Foreign exchange reserves in Sept. 2009 were $51,840m. and gold reserves were 261·50m. troy oz. The USA has the most gold reserves of any country, and more than the combined reserves of the next two (Germany and Italy). Total money supply in June 2009 was $1,497·4bn.

Budget

The budget covers virtually all the programmes of federal government, including those financed through trust funds, such as for social security, Medicare and highway construction. Receipts of the government include all income from its sovereign or compulsory powers; income from business-type or market-orientated activities of the government is offset against outlays. The fiscal year ends on 30 Sept. (before 1977 on 30 June). Budget receipts and outlays, including off-budget receipts and outlays (in $1m.):

Fiscal year ending in	Receipts	Outlays	Surplus (+) or deficit (−)
1950	39,443	42,562	−3,119
1960	92,492	92,191	+301
1970	192,807	195,649	−2,842
1980	517,112	590,941	−73,830
1990	1,031,958	1,252,994	−221,036
2000	2,025,191	1,788,950	+236,241
2001	1,991,082	1,862,846	+128,236
2002	1,853,136	2,010,894	−157,758
2003	1,782,314	2,159,899	−377,585
2004	1,880,114	2,292,841	−412,727
2005	2,153,611	2,471,957	−318,346
2006	2,406,869	2,655,050	−248,181
2007	2,567,985	2,728,686	−160,701
2008	2,523,991	2,982,544	−458,553
2009	2,104,989	3,517,677	−1,412,688
2010	2,162,706	3,457,079	−1,294,373

Fiscal year ending in	Receipts	Outlays	Surplus (+) or deficit (−)
2011	2,303,466	3,603,059	−1,299,593
2012	2,450,164	3,537,127	−1,086,963
2013[1]	2,712,045	3,684,947	−972,902
2014[1]	3,033,618	3,777,807	−744,189

[1]Estimates.

President Obama released the first budget of his second term in April 2013. The budget of $3·77trn. proposed cutting the deficit from $973bn. to $774bn. in 2014 and by $1·8trn. over ten years. Additional revenue would be raised from establishing the 'Buffet Rule' (which would involve taxing households making more than $1m. a year at a minimum rate of 30%) and through a combination of cuts from defence, Medicare, agricultural subsidies and retiree programmes. The budget outlines plans for spending $50bn. on 'Fix-It-First' projects aimed at improving nationwide infrastructure and innovation programmes, as well as a pre-school expansion initiative offset by an increased tobacco tax. A key change is switching to the chained Consumer Price Index, an alternative formula to measuring inflation that would cut spending long term on government benefit programmes by $130bn. over ten years.

Budget and off-budget receipts, by source, for fiscal years (in $1m.):

Source	2012	2013[1]	2014[1]
Individual income taxes	1,132,206	1,234,012	1,383,172
Corporation income taxes	242,289	287,716	332,819
Social insurance and retirement receipts	845,314	951,093	1,030,655
Excise taxes	79,061	85,346	104,949
Other	151,294	153,878	182,023
Total	2,450,164	2,712,045	3,033,618

[1]Estimates.

Budget and off-budget outlays, by function, for fiscal years (in $1m.):

Function	2012	2013[1]	2014[1]
National defence	677,856	660,037	626,755
Education, training, employment and social service	90,823	84,555	129,041
Health	346,742	371,664	442,697
Medicare	471,793	510,544	530,893
Income security	541,344	563,994	541,816
Social security	773,290	818,402	865,635
Veterans' benefits and services	124,595	139,568	148,222
Energy	14,857	15,101	12,680
Natural resources and environment	41,628	37,660	40,156
Commerce and housing credit	40,823	17,730	−30,126
Transportation	93,019	94,486	103,839
Community and regional development	25,132	37,998	34,663
Net interest	220,408	222,750	222,887
International affairs	47,189	56,929	55,883
General science, space and technology	29,060	30,734	30,157
Agriculture	17,791	27,049	23,454
Administration of justice	56,277	60,580	58,737
General government	28,036	30,469	28,949
Allowances	—	−177	4,083
Undistributed offsetting receipts	−103,536	−95,126	−92,614
Total	3,537,127	3,684,947	3,777,807

[1]Estimates.

Budget and off-budget outlays, by agency, for fiscal years (in $1m.):

Agency	2012	2013[1]	2014[1]
Legislative Branch	4,440	5,037	4,922
Judicial Branch	7,227	7,567	7,679
Agriculture	139,717	156,045	143,642

Agency	2012	2013[1]	2014[1]
Commerce	10,273	9,894	9,480
Defence—Military	650,867	633,287	597,553
Education	57,249	48,084	75,394
Energy	32,484	28,888	31,322
Health and Human Services	848,056	907,699	967,196
Homeland Security	47,422	60,814	53,436
Housing and Urban Development	49,600	60,489	49,267
Interior	12,891	10,448	13,497
Justice	31,159	35,346	34,475
Labor	104,588	95,212	86,766
State	26,947	31,022	31,048
Transportation	75,149	80,130	88,247
Treasury	464,714	490,727	507,684
Veterans' Affairs	124,124	139,237	147,733
Corps of Engineers	7,777	5,470	7,078
Defence—Civil	77,313	59,678	61,184
Environmental Protection Agency	12,796	9,178	8,637
Executive Office of the President	405	390	412
General Services Administration	1,753	855	361
International Assistance Programmes	20,009	24,518	25,333
National Aeronautics and Space Administration	17,190	17,797	17,936
National Science Foundation	7,255	8,429	7,479
Office of Personnel Management	79,457	89,780	96,922
Small Business Administration	2,936	1,040	1,133
Social Security Administration	821,144	872,629	922,550
Other independent agencies	32,867	47,130	18,041
Allowances	—	-184	4,069
Undistributed offsetting receipts	-230,682	-251,689	-242,669
Total	3,537,127	3,684,947	3,777,807

[1]Estimates.

National Debt. Federal debt held by the public (in $1m.), and per capita debt (in $1) on 30 June to 1976 and on 30 Sept. since then:

	Public debt	Per capita
1920	24,299	229
1930	16,185	132
1940	42,772	324
1950	219,023	1,447
1960	236,840	1,321
1970	283,198	1,394
1980	711,923	3,143
1990	2,411,558	9,696
2000	3,409,804	12,085
2005	4,592,212	15,540
2006	4,828,972	16,184
2007	5,035,129	16,715
2008	5,803,050	19,083
2009	7,544,707	24,594
2010	9,018,882	29,157
2011	10,128,187	32,505
2012	11,281,124	35,937

National Income

The Bureau of Economic Analysis of the Department of Commerce prepares detailed estimates on the national income and product. In Dec. 2003 the Bureau revised these accounts back to 1929. The principal tables are published monthly in *Survey of Current Business;* the complete set of national income and product tables are published in the *Survey* normally each Aug., showing data for recent years.

Gross Domestic Product

			(in $1,000m.)		
	2007	2008	2009	2010	2011
Gross Domestic Product	14,028·7	14,291·5	13,973·7	14,498·9	15,075·7
Personal consumption expenditures	9,772·3	10,035·5	9,845·9	10,215·7	10,729·0
Goods	3,363·9	3,381·7	3,194·4	3,364·9	3,624·8
Durable goods	1,188·4	1,108·9	1,029·6	1,079·4	1,146·4
Nondurable goods	2,175·5	2,272·8	2,164·8	2,285·5	2,478·4
Services	6,408·3	6,653·8	6,651·5	6,850·9	7,104·2
Gross private domestic investment	2,295·2	2,087·6	1,549·3	1,737·3	1,854·9
Fixed investment	2,266·1	2,128·7	1,703·5	1,679·0	1,818·3
Nonresidential	1,637·5	1,656·3	1,349·3	1,338·4	1,479·6
Structures	524·9	586·3	451·1	376·3	404·8
Equipment and software	1,112·6	1,070·0	898·2	962·1	1,074·7
Residential	628·7	472·4	354·1	340·6	338·7
Change in private inventories	29·1	-41·1	-154·2	58·4	36·6
Net exports of goods and services	-713·1	-709·7	-388·7	-511·6	-568·1
Exports	1,661·7	1,846·8	1,587·4	1,844·4	2,094·2
Goods	1,162·0	1,297·5	1,064·7	1,278·5	1,474·5
Services	499·7	549·3	522·7	565·9	619·7
Imports	2,374·8	2,556·5	1,976·2	2,356·1	2,662·3
Goods	2,000·7	2,146·3	1,587·5	1,947·0	2,229·2
Services	374·0	410·1	388·7	409·1	433·0
Government consumption expenditures and gross investment	2,674·2	2,878·1	2,967·2	3,057·5	3,059·8
Federal	976·3	1,080·1	1,143·6	1,223·1	1,222·1
National defence	662·3	737·8	776·0	817·7	820·8
Nondefence	314·0	342·3	367·6	405·3	401·3
State and local	1,697·9	1,798·0	1,823·6	1,834·4	1,837·7

Relation of Gross Domestic Product, Gross National Product, Net National Product, National Income and Personal Income

			(in $1,000m.)		
	2007	2008	2009	2010	2011
Gross domestic product	14,028·7	14,291·5	13,973·7	14,498·9	15,075·7
Plus: Income receipts from the rest of the world	871·0	856·1	642·4	716·5	783·7
Less: Income payments to the rest of the world	747·7	686·9	498·9	507·2	531·8
Equals: Gross national product	14,151·9	14,460·7	14,117·2	14,708·2	15,327·5
Less: Consumption of fixed capital	1,767·5	1,854·1	1,866·3	1,873·4	1,936·8
Private	1,476·2	1,542·9	1,542·8	1,539·9	1,587·4

Relation of Gross Domestic Product, Gross National Product,
Net National Product, National Income and Personal Income

(in $1,000m.)

	2007	2008	2009	2010	2011
Domestic business	1,190·7	1,248·3	1,250·0	1,245·0	1,285·7
Capital consumption allowances	1,087·2	1,325·2	1,282·9	1,316·0	1,509·5
Less: Capital consumption adjustment	–103·6	76·9	32·9	71·0	223·8
Households and institutions	285·5	294·6	292·7	294·9	301·7
Government	291·3	311·2	323·5	333·5	349·4
General government	243·2	259·6	270·4	278·2	291·0
Government enterprises	48·1	51·5	53·1	55·3	58·4
Equals: Net national product	12,384·4	12,606·6	12,250·9	12,834·8	13,390·8
Less: Statistical discrepancy	–12·0	–2·4	118·3	23·3	31·9
Equals: National income	12,396·4	12,609·1	12,132·6	12,811·4	13,358·9
Less: Corporate profits with inventory valuation and capital					
consumption adjustments	1,510·6	1,248·4	1,342·3	1,702·4	1,827·0
Taxes on production and imports less subsidies	972·6	985·7	963·5	998·0	1,036·2
Contributions for government social insurance (domestic)	959·5	987·3	963·1	983·3	919·3
Net interest and miscellaneous payment on assets	731·6	870·1	640·5	567·9	527·4
Business current transfer payments (net)	103·3	123·0	133·4	140·0	132·6
Current surplus of government enterprises	–11·8	–16·0	–15·6	–19·5	–26·5
Wage accruals less disbursements	–6·3	–5·0	5·0	0·0	0·0
Plus: Personal income receipts on assets	2,057·0	2,165·4	1,626·5	1,598·3	1,685·1
Personal current transfer receipts	1,718·5	1,879·2	2,140·1	2,284·3	2,319·2
Equals: Personal income	11,912·3	12,460·2	11,867·0	12,321·9	12,947·3
Addenda:					
Gross domestic income	14,040·7	14,294·0	13,855·4	14,475·6	15,043·8
Gross national income	14,163·9	14,463·1	13,998·9	14,684·9	15,295·7
Gross national factor income	13,099·8	13,370·4	12,917·6	13,566·4	14,153·4
Net domestic product	12,261·2	12,437·5	12,107·4	12,625·5	13,138·9
Net domestic income	12,273·2	12,439·9	11,989·1	12,602·1	13,107·0
Net national factor income	11,332·3	11,516·3	11,051·3	11,692·9	12,216·6
Net domestic purchases	12,974·2	13,147·2	12,496·1	13,137·1	13,707·0

National Income by Type of Income

(in $1,000m.)

	2007	2008	2009	2010	2011
National income	12,396·4	12,609·1	12,132·6	12,811·4	13,358·9
Compensation of employees	7,855·9	8,068·3	7,799·4	7,970·0	8,295·2
Wage and salary accruals	6,415·5	6,545·9	6,275·3	6,404·6	6,661·3
Government	1,089·0	1,144·1	1,175·2	1,191·3	1,195·3
Other	5,326·4	5,401·8	5,100·1	5,213·3	5,466·0
Supplements to wages and salaries	1,440·4	1,522·5	1,524·0	1,565·4	1,633·9
Employer contributions for employee pension and insurance					
funds	980·5	1,052·4	1,067·2	1,097·3	1,139·0
Employer contributions for government social insurance	459·9	470·1	456·9	468·1	494·9
Proprietors' income with inventory valuation and capital					
consumption adjustments	1,090·4	1,097·9	979·4	1,103·4	1,157·3
Farm	37·8	51·8	39·9	44·3	54·6
Nonfarm	1,052·6	1,046·1	939·5	1,059·1	1,102·8
Rental income of persons with capital consumption adjustment	143·7	231·6	289·7	349·2	409·7
Corporate profits with inventory valuation and capital					
consumption adjustments	1,510·6	1,248·4	1,342·3	1,702·4	1,827·0
Taxes on corporate income	445·5	309·0	269·4	373·3	379·0
Profits after tax with inventory valuation and capital					
consumption adjustments	1,065·2	939·4	1,073·0	1,329·1	1,447·9
Net dividends	794·5	786·9	554·1	600·9	697·2
Undistributed profits with inventory valuation and capital					
consumption adjustments	270·7	152·5	518·8	728·2	750·7
Net interest and miscellaneous payments	731·6	870·1	640·5	567·9	527·4
Taxes on production and imports	1,027·2	1,038·6	1,023·2	1,055·0	1,097·9
Less: Subsidies	54·6	52·9	59·7	57·0	61·6
Business current transfer payments (net)	103·3	123·0	133·4	140·0	132·6
To persons (net)	30·5	36·8	39·6	47·4	44·9
To government (net)	66·8	79·0	96·0	94·5	94·8
To the rest of the world (net)	6·0	7·2	–2·3	–1·9	–7·1
Current surplus of government enterprises	–11·8	–16·0	–15·6	–19·5	–26·5
Addenda for corporate cash flow:					
Net cash flow with inventory valuation adjustment	1,244·1	1,245·2	1,632·8	1,774·9	1,850·7
Undistributed profits with inventory valuation and capital					
consumption adjustments	270·7	152·5	518·8	728·2	750·7
Consumption of fixed capital	973·4	1,028·5	1,030·4	1,026·5	1,061·9
Less: Capital transfers paid (net)	0·0	–64·2	–83·6	–20·2	–38·1

National Income by Type of Income

(in $1,000m.)

	2007	2008	2009	2010	2011
Addenda:					
Proprietors' income with inventory valuation and capital consumption adjustments	1,090·4	1,097·9	979·4	1,103·4	1,157·3
Farm	37·8	51·8	39·9	44·3	54·6
Proprietors' income with inventory valuation adjustment	43·9	58·3	46·2	50·1	60·5
Capital consumption adjustment	−6·1	−6·5	−6·3	−5·8	−5·9
Nonfarm	1,052·6	1,046·1	939·5	1,059·1	1,102·8
Proprietors' income (without inventory valuation and capital consumption adjustments)	959·9	885·7	782·9	898·1	902·6
Inventory valuation adjustment	−7·0	−6·3	1·0	−5·8	−8·9
Capital consumption adjustment	99·6	166·7	155·6	166·8	209·0
Rental income of persons with capital consumption adjustment	143·7	231·6	289·7	349·2	409·7
Rental income of persons (without capital consumption adjustment)	160·3	247·8	304·8	363·9	424·4
Capital consumption adjustment	−16·6	−16·2	−15·1	−14·7	−14·7
Corporate profits with inventory valuation and capital consumption adjustments	1,510·6	1,248·4	1,342·3	1,702·4	1,827·0
Corporate profits with inventory valuation adjustment	1,691·1	1,315·5	1,443·6	1,777·7	1,791·6
Profits before tax (without inventory valuation and capital consumption adjustments)	1,738·4	1,359·9	1,440·5	1,816·3	1,854·1
Taxes on corporate income	445·5	309·0	269·4	373·3	379·0
Profits after tax (without inventory valuation and capital consumption adjustments)	1,292·9	1,050·9	1,171·1	1,443·0	1,475·1
Net dividends	794·5	786·9	554·1	600·9	697·2
Undistributed profits (without inventory valuation and capital consumption adjustments)	498·4	264·0	617·0	842·1	777·9
Inventory valuation adjustment	−47·2	−44·5	3·2	−38·7	−62·6
Capital consumption adjustment	−180·5	−67·1	−101·3	−75·2	35·4

Real Gross Domestic Product

(in 1,000m. chained [2005] dollars[1])

	2007	2008	2009	2010	2011
Gross domestic product	13,206·4	13,161·9	12,757·9	13,063·0	13,299·1
Personal consumption expenditures	9,262·9	9,211·7	9,032·6	9,196·2	9,428·8
Goods	3,273·5	3,192·9	3,098·2	3,209·1	3,331·0
Durable goods	1,232·4	1,171·8	1,109·1	1,178·3	1,262·6
Nondurable goods	2,042·9	2,019·1	1,982·8	2,029·3	2,075·2
Services	5,990·2	6,017·0	5,930·6	5,987·6	6,101·5
Gross private domestic investment	2,159·5	1,939·8	1,458·1	1,658·0	1,744·0
Fixed investment	2,130·6	1,978·6	1,602·2	1,598·7	1,704·5
Nonresidential	1,550·0	1,537·6	1,259·8	1,268·5	1,378·2
Structures	438·2	466·4	368·1	310·6	319·2
Equipment and software	1,106·8	1,059·4	885·2	963·9	1,070·0
Residential	584·2	444·4	344·8	332·2	327·6
Change in private inventories	27·7	−36·3	−139·0	50·9	31·0
Net exports of goods and services	−648·8	−494·8	−355·2	−419·7	−408·0
Exports	1,554·4	1,649·3	1,498·7	1,665·5	1,776·9
Goods	1,088·1	1,157·0	1,018·6	1,164·1	1,247·6
Services	466·3	492·3	479·6	501·9	529·8
Imports	2,203·2	2,144·0	1,853·8	2,085·2	2,184·9
Goods	1,856·1	1,784·8	1,506·4	1,730·3	1,820·0
Services	347·1	359·8	347·8	356·6	366·6
Government consumption expenditures and gross investment	2,434·2	2,497·4	2,589·4	2,605·8	2,523·9
Federal	906·1	971·1	1,030·6	1,076·8	1,047·0
National defence	611·8	657·7	696·9	717·6	699·1
Nondefence	294·2	313·3	333·7	359·2	347·9
State and local	1,528·1	1,528·1	1,561·8	1,534·1	1,482·0
Residual	−1·3	16·3	39·9	20·2	−7·8

[1]In 1996 the chain-weighted method of estimating GDP replaced that of constant base-year prices. In chain-weighting the weights used to value different sectors of the economy are continually updated to reflect changes in relative prices.

Performance

Total GDP in 2011 was $14,991·3bn. (more than twice that of China, the second largest economy), representing 21% of the world's total GDP. Real GDP growth rates (based on OECD statistics):

2002	2003	2004	2005	2006	2007	2008	2009	2010	2011
1·8%	2·5%	3·5%	3·1%	2·7%	1·9%	−0·3%	−3·1%	2·4%	1·8%

The real GDP growth rate in 2012 according to the Bureau of Economic Analysis was 2·2%. The USA's economy shrank by 3·7% in the third quarter of 2008, by 8·9% in the fourth quarter (the steepest quarterly fall since 1958) and by 5·3% in the first quarter of 2009. There was also negative growth in the second quarter of 2009, of 0·3%. The recession ended with growth of 1·4% in the third quarter of 2009.

The USA was ranked seventh in the Global Competitiveness Index in the World Economic Forum's *Global Competitiveness Report 2012–2013*, down from fifth in the 2011–2012 report. The index analyses 12 areas of competitiveness for over 100 countries including macroeconomy, higher education and training, institutions, innovation and infrastructure. In the 2012 *World Competitiveness Yearbook*, compiled by the International Institute for Management Development, the USA came second in the world ranking. This annual publication ranks and analyzes how a nation's environment creates and sustains the competitiveness of enterprises.

Banking and Finance

The Federal Reserve System, established under The Federal Reserve Act of 1913, comprises the Board of seven Governors, the 12 regional Federal Reserve Banks with their 25 branches, and the Federal Open Market Committee. The seven members of the Board of Governors are appointed by the President with the consent of the Senate. Each Governor is appointed to a full term of 14 years or an unexpired portion of a term, one term expiring every two years. The Board exercises broad supervisory authority over the operations of the 12 Federal Reserve Banks, including approval of their budgets and of the appointments of their presidents and first vice presidents; it designates three of the nine directors of each Reserve Bank including the Chairman and Deputy Chairman. The Chairman of the Federal Reserve Board is appointed by the President for four-year terms. The *Chairman* is Ben Bernanke. The Board has supervisory and regulatory responsibilities over banks that are members of the Federal Reserve System, bank holding companies, bank mergers, Edge Act and agreement corporations, foreign activities of member banks, international banking facilities in the USA, and activities of the US branches and agencies of foreign banks. Legislation of 1991 requires foreign banks to prove that they are subject to comprehensive consolidated supervision by a regulator at home, and have the Board's approval to establish branches, agencies and representative offices. The Board also assures the smooth functioning and continued development of the nation's vast payments system. Another area of the Board's responsibilities involves the implementation by regulation of major federal laws governing consumer credit.

From 1968 the Congress passed a number of consumer financial protection acts, the first of which was the Truth in Lending Act, for which it has directed the Board to write implementing regulations and assume partial enforcement responsibility. Others include the Equal Credit Opportunity Act, Home Mortgage Disclosure Act, Consumer Leasing Act, Fair Credit Billing Act, Truth in Savings Act and Electronic Fund Transfer Act. To manage these responsibilities the Board has established a Division of Consumer and Community Affairs. To assist it, the Board consults with a Consumer Advisory Council, established by the Congress in 1976 as a statutory part of the Federal Reserve System.

Another statutory body, the Federal Advisory Council, consists of 12 members (one from each district); it meets in Washington four times a year to advise the Board of Governors on economic and banking developments. Following the passage of the Monetary Control Act of 1980, the Board of Governors established the Thrift Institutions Advisory Council to provide information and views on the special needs and problems of thrift institutions. The group is comprised of representatives of mutual savings banks, savings and loan associations, and credit unions.

All depository institutions (commercial and savings banks, savings and loan associations, credit unions, US agencies and branches of foreign banks, and Edge Act and agreement corporations) must meet reserve requirements set by the Federal Reserve and hold the reserves in the form of vault cash or deposits at Federal Reserve Banks.

Banks which participate in the federal deposit insurance fund have their deposits insured against loss up to $100,000 for each account. The fund is administered by the Federal Deposit Insurance Corporation established in 1933; it obtains resources through annual assessments on participating banks. All members of the Federal Reserve System are required to insure their deposits through the Corporation, and non-member banks may apply and qualify for insurance.

The Federal Deposit Insurance Corporation Improvement Act of 1992 originated with bank reform initiatives. It imposed new capital rules on banks, new reporting requirements and a code of 'safety and soundness' standards. The main aim of the Act is to reduce risk through rigorous enforcement of capital requirements. Regulators are required to take action where banks fail to observe these standards.

In April 2013 the largest banks in the USA in terms of market value were: Wells Fargo ($197·2bn.); J. P. Morgan Chase ($186·6bn.); and Citigroup ($136·3bn.).

The key stock exchanges are the New York Stock Exchange (NYSE Euronext) and the Nasdaq Stock Exchange (NASDAQ). There are several other stock exchanges, in Philadelphia, Boston, San Francisco (Pacific Stock Exchange) and Chicago, although trading is very limited in them.

Gross external debt totalled $15,415,265m. in June 2012 (compared to $15,199,591m. in June 2011).

The USA received $226·94bn. worth of foreign direct investment in 2011, up from $197·91bn. in 2010 although down from a record $314·01bn. in 2000. By the end of 2011 the total stock of foreign direct investment was $3,509·4bn.

By Aug. 2010, 45·1% of internet users in the USA were using e-banking.

ENERGY AND NATURAL RESOURCES

In 2010 fossil fuels accounted for 78·0% of primary energy production, nuclear power 11·2% and renewables (including biomass, hydropower, wind, geothermal and solar) 10·8% (up from 8·6% in 2000). Provisional figures suggested that generation of renewables surpassed nuclear power in 2011.

Environment

The USA's carbon dioxide emissions from the consumption and flaring of fossil fuels in 2008 accounted for 19·2% of the world total (the second highest after China) and were equivalent to 19·2 tonnes per capita (down from 20·4 tonnes per capita in 2004). The population of the USA is only 4·6% of the world total. An *Environmental Performance Index* compiled in 2008 ranked the USA 39th in the world, with 81·0%. The index examined various factors in six areas—air pollution, biodiversity and habitat, climate change, environmental health, productive natural resources and water resources.

In March 2001 then President Bush rejected the 1997 Kyoto Protocol, which aimed to combat the rise in the earth's temperature through the reduction of industrialized nations' carbon dioxide emissions from the consumption and flaring of fossil fuels by an average 5·2% below 1990 levels by 2012. In Feb. 2002 he unveiled an alternative climate-change plan to the Kyoto Protocol, calling for voluntary measures to reduce the rate of increase of US carbon dioxide emissions from the consumption and flaring of fossil fuels.

In 2009 President Obama endorsed an American Clean Energy and Security bill that laid out plans for an emissions trading scheme echoing the EU model. It was passed by the House of Representatives in 2009 but did not become law owing to inaction in the Senate. Obama had previously called for a reduction in US greenhouse gas emissions of 80% on 1990 levels by 2050.

The USA recycled 33·2% of its municipal solid waste in 2008.

Electricity

Net summer capacity in 2009 was 1,027·6m. kW. Fossil fuel accounts for approximately 69% of electricity generation. In 2009, 20% of electricity was produced by nuclear reactors. (The last one to begin commercial operation was in 1996.) The USA has more nuclear reactors in use than any other country in the world. In 2009 the USA had a nuclear generating capacity of 101,004 MW, with 104 nuclear reactors. Two reactors are under construction in

Georgia, which are the first new ones to be built in the USA since the 1970s. Electricity generation in 2011 was 4,100,656m. kWh, the second highest in the world behind China (which overtook the USA as the largest producer that year). Consumption per capita in 2007 was 14,522 kWh.

Oil and Gas

Crude oil production (2011), 2,060m. bbls. Up until 2008 production had been gradually declining since the mid-1980s, when annual production was 3,275m. bbls. In 2008 production was at its lowest level since 1946, at 1,830m. bbls. However, it has since risen sharply, largely thanks to the development of hydraulic fracturing (or 'fracking') and horizontal drilling. Only Russia and Saudi Arabia produce more crude oil, but it has been suggested that the USA may overtake both of them to be the leading producer by 2020. Proven reserves were 30·9bn. bbls in 2011. Output (2007) was valued at $122·96bn. Offshore oil accounts for about 33% of total production, but this is likely to increase in the future. Crude oil imports began to exceed production in 1993 and in 2008 totalled 3,581m. bbls, with Canada supplying nearly a fifth of US oil imports. In Oct. 2002 the USA took its first delivery of Russian oil for its Strategic Petroleum Reserve as a consequence of an energy dialogue declared by then Presidents George W. Bush and Vladimir Putin at their summit in May 2002. The USA is by far the largest single consumer of oil (888·5m. tonnes in 2008). In 2011 the USA exported more petroleum products than it imported on an annual basis for the first time since 1949.

The USA is by some distance the greatest single consumer of natural gas and also the largest producer (having overtaken Russia in 2010). Natural gas production, 2011, was a record 23·00trn. cu. ft. Shale gas production was 4·87trn. cu. ft in 2010 (22% of total US natural gas production), up from 0·4trn. cu. ft in 2000. Proven natural gas reserves in 2008 totalled 245trn. cu. ft.

Wind

The USA is the largest producer of wind power, with 70·8bn. kWh in 2009 (up from 55·4bn. kWh in 2008 and 14·1bn. kWh in 2004). In 2010 total installed capacity amounted to 40,180 MW, second only behind China.

Ethanol

The USA is the largest producer of ethanol (from maize). Production totalled 10,600m. gallons in 2009 (9,000m. gallons in 2008 and 6,500m. gallons in 2007).

Water

The total area within the 50 states of the USA plus the District of Columbia covered by water is 264,837 sq. miles. Americans' average annual water usage is nearly 67,000 cu. ft per person—more than twice the average for an industrialized nation.

Non-Fuel Minerals

The USA is wholly dependent upon imports for columbium, bauxite, mica sheet, manganese, strontium and graphite, and imports over 80% of its requirements of industrial diamonds, fluorspar, platinum, tantalum, tungsten, chromium and tin.

Total value of non-fuel minerals produced in 2011 (provisional) was $74bn. ($33bn. in 1990). Details of some of the main minerals produced are given in the following tables.

Production of metals:

	Unit	Quantity 2007
Copper	1,000 tonnes	1,170
Gold	tonnes	238
Iron ore	1m. tonnes	51
Lead	1,000 tonnes	434
Silver	tonnes	1,260
Zinc	1,000 tonnes	769

In 2008 the value of metal mine production was $27·4bn.

Precious metals are mined mainly in California and Utah (gold); and Nevada, Arizona and Idaho (silver).

Production of non-metals:

	Unit	Quantity 2007
Barite	1,000 tonnes	455
Boron	1,000 tonnes	1,150[1]
Bromine	1,000 tonnes	243[2]
Cement	1m. short tons	94
Clays	1,000 tonnes	36,800
Diatomite	1,000 tonnes	687
Feldspar (including aplite)	1,000 tonnes	730
Garnet (industrial)	1,000 tonnes	61
Gypsum	1m. tonnes	18
Lime	1m. tonnes	20
Phosphate rock	1m. tonnes	30
Pumice	1,000 tonnes	1,270
Salt	1m. tonnes	45
Sand and gravel	1m. tonnes	1,260
Stone (crushed/broken)	1m. tonnes	1,600

[1]2005. [2]2006.

In 2008 the value of non-metal mineral production was $43·9bn.

Aluminium production for 2007, 2·55m. tonnes; uranium production for 2009, 1,453 tonnes. The USA is the world's leading producer of salt.

Coal

Proven recoverable coal reserves were 237,295m. short tons in 2008, more than a quarter of the world total. Output in 2008 (in 1m. short tons): 1,171·8 including bituminous coal, 555·3; sub-bituminous coal, 539·1; lignite, 75·7; anthracite, 1·7. 2008 output from surface workings, 814·7m. short tons; underground mines, 357·1m. short tons. Value of total output, 2008, $38·18bn. The USA is the world's second largest coal producer, after China.

Agriculture

Agriculture in the USA is characterized by its ability to adapt to widely varying conditions, and still produce an abundance and variety of agricultural products. From colonial times to about 1920 the major increases in farm production were brought about by adding to the number of farms and the amount of land under cultivation. During this period nearly 320m. acres of virgin forest were converted to crop land or pasture, and extensive areas of grasslands were ploughed. Improvident use of soil and water resources was evident in many areas.

During the next 20 years the number of farms reached a plateau of about 6·5m., and the acreage planted to crops held relatively stable around 330m. acres. The major source of increase in farm output arose from the substitution of power-driven machines for horses and mules. Greater emphasis was placed on development and improvement of land, and the need for conservation of basic agricultural resources was recognized. A successful conservation programme, highly co-ordinated and on a national scale—to prevent further erosion, to restore the native fertility of damaged land and to adjust land uses to production capabilities and needs—has been in operation since early in the 1930s.

Since the Second World War the uptrend in farm output has been greatly accelerated by increased production per acre and per farm animal. These increases are associated with a higher degree of mechanization; greater use of lime and fertilizer; improved varieties, including hybrid maize and grain sorghums; more effective control of insects and disease; improved strains of livestock and poultry; and wider use of good husbandry practices, such as nutritionally balanced feeds, use of superior sites and better housing. During this period land included in farms decreased slowly, crop land harvested declined somewhat more rapidly, but the number of farms declined sharply.

All land in farms totalled less than 500m. acres in 1870, rose to a peak of over 1,200m. acres in the 1950s and declined to 920m. acres in 2008, even with the addition of the new States of Alaska and Hawaii in 1960. The number of farms declined from 6·35m. in 1940 to 2·20m. in 2008, as the average size of farms doubled. The average size of farms in 2008 was 418 acres, but ranged from a few acres to many thousand acres. In 2007 the total value of land and buildings was $1,744,295m. The average value of land and buildings per acre in 2007 was $1,892.

At the 2000 census 59,063,597 persons (21·0% of the population) were rural, of whom 2,987,531 (just over 1% of the total population) lived on farms. In 2007 there were 1,906,335 farms managed by families or individuals (86·5% of all farms); 1,522,033 farms (69·0% of all farms) were managed by full owners (farmers who own all the land they operate). Hired farmworkers numbered 2,636,509 in 2007. There were 4·4m. tractors in 2007 and an estimated 411,000 harvester-threshers. In 2007 there were an estimated 170·43m. ha. of arable land and 2·73m. ha. of permanent crops. 22·9m. ha. were irrigated in 2007.

Cash receipts from farm marketings and government payments (in $1bn.):

	Crops	Livestock and livestock products	Total
2007	149·9	138·6	288·5
2008	176·8	141·5	318·3
2009	163·7	119·8	283·4

Net farm income was $62·2bn. in 2009 ($86·6bn. in 2008).

The harvest area and production of the principal crops for 2007 and 2008 were:

	2007 Harvested 1m. acres	2007 Pro-duction 1m.	2007 Yield per acre	2008 Harvested 1m. acres	2008 Pro-duction 1m.	2008 Yield per acre
Corn for grain (bu.)	86·5	13,038	151	78·6	12,101	154
Soybeans (bu.)	64·1	2,677	41·7	74·6	2,959	39·6
Wheat (bu.)	51·0	2,051	40·2	55·7	2,500	44·9
Cotton (bales)[1]	10·5	19·2	879	7·7	13·0	810
Potatoes (cwt.)	1·1	445	396	1·0	413	395
Hay (sh. tons)	61·0	147	2·41	60·1	146	2·43

[1]Yield in lb.

The USA is the world's leading producer of maize, soybeans, sorghum and tree nuts and the second largest producer of tomatoes and apples.

Fruit. Utilized production, in 1,000 tons:

	2006	2007	2008
Apples	4,865	4,523	4,970
Grapefruit	1,232	1,627	1,569
Grapes	6,366	7,036	7,421
Oranges	9,021	7,625	10,167
Peaches	987	1,116	1,100
Strawberries	1,202	1,223	1,266

The farm value of the above crops in 2008 was: apples, $2,599m.; grapefruit, $265m.; grapes, $3,342m.; oranges, $2,135m.; peaches, $539m.; strawberries, $1,885m.

In 2008 there were 2,643,221 acres of organic crops. Certified organic producers numbered 12,941 in 2008. Organic food sales for the USA in 2005 totalled $13·8bn. (the highest in the world).

Dairy produce. In 2008 production of milk was 189,992m. lb; cheese, 9,935m. lb; butter, 1,644m. lb; ice cream, 943m. gallons; non-fat dry milk, 1,519m. lb; yoghurt, 3,599m. lb. The USA is the world's largest producer of both cheese and milk.

Livestock. In 2008 there were 9,009m. broilers and 273m. turkeys. Eggs produced, 2007, 90·2bn.

Value of production (in $1m.) was:

	2006	2007	2008
Cattle and calves	35,491	35,973	34,859
Hogs and pigs	12,714	13,468	14,435
Broilers	17,739	21,514	23,118
Turkeys	3,468	3,954	4,477
Eggs	4,460	6,719	8,225

Livestock numbered, in 2008 (1m.): cattle and calves (including dairy cows), 96·0; hogs and pigs, 68·2; sheep and lambs, 6·0. Approximate value of livestock (in $1bn.), 2008: cattle, 95·1; hogs and pigs, 5·0; sheep and lambs (in $1m.), 823.

Forestry

Forests covered a total area of 751·25m. acres (304·02m. ha.) in 2010, or 33% of the land area. Of the total area under forests 25% was primary forest in 2010, 67% other naturally regenerated forest and 8% planted forest. The national forests had an area of 147,181,000 acres in 2007. In 2007 there were 514m. acres of timberland (99m. acres national forest, 59m. acres state, county or municipality owned, 356m. acres private). Timber production was 15,680m. cu. ft in 2007. The USA is the world's largest producer of roundwood (12·4% of the world total in 2007). It is also the highest consumer of roundwood; timber consumption in 2007 totalled 15·41bn. cu. ft.

In 2008 there were 704 designated wilderness areas throughout the USA, covering a total of 107·4m. acres (43·5m. ha.). More than half of the areas are in Alaska (53·5%), followed by California (13·3%), Arizona, Washington and Idaho.

Fisheries

In 2007 the domestic catch was 9,231·8m. lb, valued at $4,089·0m. (including 1,021·1m. lb of shellfish valued at $2,013·0m.). Main species landed in terms of value ($1m.): crabs, 454·0; sea scallops, 385·0; salmon, 381·1; shrimp, 370·0; American lobster, 349·1. Disposition of the domestic catch in 2007 (1m. lb): fresh or frozen, 7,387; tinned, 501; cured, 121; reduced to meal or oil, 1,223. The USA's imports of fishery commodities in 2009 ($13·86bn.) were the largest of any country.

In the period 2005–07 the average American citizen consumed 53·4 lb (24·2 kg) of fish and fishery products a year, compared to an average 37·0 lb (16·8 kg) for the world as a whole.

Tennessee Valley Authority

Established by Act of Congress, 1933, the TVA is a multiple-purpose federal agency which carries out its duties in an area embracing some 41,000 sq. miles in the seven Tennessee River Valley states: Tennessee, Kentucky, Mississippi, Alabama, North Carolina, Georgia and Virginia. In addition, 76 counties outside the Valley are served by TVA power distributors. It is the largest public power company in the USA. Its three directors are appointed by the President, with the consent of the Senate; headquarters are in Knoxville (TN).

INDUSTRY

The largest companies in the USA by market capitalization in March 2013 were: Apple Inc ($411·2bn.), a computer company; the Exxon Mobil Corporation ($399·5bn.), the world's largest integrated oil company; and Google Inc ($275·2bn.), an internet-related products and services company. They were also the three largest companies by market value in the world in March 2013. According to a survey published by the New York-based Interbrand in Oct. 2012, Coca-Cola is the world's most valuable brand, worth $77·84bn.

The following table presents industry statistics of manufactures as reported at various censuses from 1909 to 1980 and from the Annual Survey of Manufactures for years in which no census was taken.

The Annual Surveys of Manufactures carry forward the key measures of manufacturing activity which are covered in detail by the Census of Manufactures. The large plants in the surveys account for approximately two-thirds of the total employment in operating manufacturing establishments in the USA.

	Production workers (average for year)	Production workers' wages total ($1,000)	Value added by manufacture ($1,000)
1909	3,261,736	3,205,213	8,160,075
1919	9,464,916	9,664,009	23,841,624
1929	8,369,705	10,884,919	30,591,435
1933	5,787,611	4,940,146	14,007,540
1939	7,808,205	8,997,515	24,487,304
1950	11,778,803	34,600,025	89,749,765
1960	12,209,514	55,555,452	163,998,531
1970	13,528,000	91,609,000	300,227,600
1980	13,900,100	198,164,000	773,831,300
1990	12,232,700	275,208,400	1,346,970,100
2000	11,943,646	363,380,819	1,973,622,421
2005	9,235,635	337,980,878	2,210,349,247
2006	9,175,328	344,192,729	2,285,928,967
2007	9,385,789	353,051,004	2,390,643,058
2008	8,872,902	343,373,530	2,274,366,727

The total number of employees in the manufacturing industry in 2009 was approximately 11,051,000; there were 331,355 manufacturing establishments in 2007. Manufacturing employment has declined every year since 1998. Much of the decline reflects the recession that began in 2001 and the relatively weak recovery in demand that followed. In 2008 manufacturing contributed 11·5% of GDP, down from 14·5% of GDP in 2000. The leading industries in 2009 in terms of value added by manufacture (in $1m.) were: chemicals, 328,871; food, 258,615; transportation equipment, 229,642; computer and electronic products, 193,242; fabricated metal products, 146,876. In 2008 a total of 8,673,000 motor vehicles were made in the USA, the lowest annual total since the 1980s.

In 2008 principal commodities produced (by value of shipments, in $1m.) were: chemicals, 689,000; petroleum and coal products, 674,000; transportation equipment, 619,000; food, 611,000; computer and electronic products, 380,000.

Net profits (2008) for manufacturing corporations were $418bn. before tax ($294bn. after tax). Hourly earnings of production workers in Dec. 2008 were $18·06 in manufacturing and $22·52 in construction.

The USA is the second largest beer producer after China, with 6,151m. gallons in 2007; and second after China for cigarette production, with 484bn. units in 2006.

Iron and Steel. Output of the iron and steel industries (in 1m. net tons of 2,000 lb) in recent years was:

	Pig iron	Raw steel	Steel by method of production[1] Electric	Basic oxygen
2005	37·2	94·9	52·2	42·7
2006	37·9	98·2	42·1	56·1
2007	36·3	98·1	41·0	57·1
2008	33·7	91·9	52·8	39·1
2009	19·0	59·4	36·7	22·7

[1]The sum of these two items should equal the total in the preceding column; any difference is due to rounding.

In 2008 iron and steel mills and ferroalloy manufacturing employed 109,308 persons (an average 87,400 production workers). The total payroll in 2008 amounted to $7,668·0m.

Labour

The Bureau of Labor Statistics estimated that in 2009 the civilian labour force was 154,142,000 (65·4% of those 16 years and over), of whom 139,877,000 were employed and 14,265,000 (9·3%) were unemployed. The unemployment rate was 7·8% in Dec. 2012,

down from 8·5% in Dec. 2011, 9·3% in Dec. 2010 and 9·9% in Dec. 2009. Payroll employment fell by 820,000 in Jan. 2009, the largest monthly fall since 1945. In 2009 as a whole 4·8m. jobs were lost—the highest number since records began in 1939. Long-term unemployment rose from 10·0% of the labour force between 16 and 64 having been out of work for more than a year in 2007 to 29·0% in 2010. In Jan. 2010 for the first time there were more women (64·2m.) than men (63·4m.) on US payrolls. Employment by industry in 2009:

Industry Group	Male	Female	Total	Percentage distribution
Employed (1,000 persons):	73,669	66,207	139,877	100·0
Agriculture, forestry, fisheries, and hunting	1,607	496	2,103	1·5
Mining	613	94	707	0·5
Construction	8,782	919	9,702	6·9
Manufacturing: Durable goods	6,724	2,204	8,927	6·4
Manufacturing: Non-durable	3,409	1,865	5,275	3·8
Wholesale and retail trade	10,766	8,918	19,684	14·1
Transportation and utilities	5,583	1,662	7,245	5·2
Information	1,879	1,360	3,239	2·3
Financial activities	4,426	5,196	9,622	6·9
Professional and business services	8,744	6,264	15,008	10·7
Education and health services	7,895	23,925	31,819	22·7
Leisure and hospitality	6,173	6,562	12,736	9·1
Other services	3,322	3,613	6,935	5·0
Public administration	3,746	3,129	6,875	4·9

A total of 15 strikes and lockouts of 1,000 workers or more occurred in 2008, involving 72,000 workers and 1,954,000 idle days.

On 24 July 2007 the federal hourly minimum wage was raised from $5·15 to $5·85 an hour. It had been $5·15 for nearly ten years, having previously been increased from $4·75 an hour on 1 Sept. 1997. It was raised again on 24 July 2008, to $6·55 an hour, and on 24 July 2009, to $7·25. Americans worked an average 1,703 hours per person in 2008. Median weekly earnings were $739 in 2009 ($819 among men and $657 among women).

Labour relations are legally regulated by the National Labor Relations Act, amended by the Labor–Management Relations (Taft–Hartley) Act, 1947 as amended by the Labor–Management Reporting and Disclosure Act, 1959, again amended in 1974, and the Railway Labor Act of 1926, as amended in 1934 and 1936.

INTERNATIONAL TRADE

The North American Free Trade Agreement (NAFTA) between the USA, Canada and Mexico was signed on 7 Oct. 1992 and came into effect on 1 Jan. 1994. The Central America-Dominican Republic-United States Free Trade Agreement (CAFTA-DR) between the USA, Costa Rica, the Dominican Republic, El Salvador, Guatemala, Honduras and Nicaragua entered into force for the USA on 1 March 2006. The UK has had 'most-favoured-nation' status since 1815.

Imports and Exports

Total value of imports and exports of goods (in $1bn.):

	Imports	Exports	Trade balance
2006	1,853·9	1,026·0	−828·0
2007	1,957·0	1,148·2	−808·8
2008	2,103·6	1,287·4	−816·2
2009	1,559·6	1,056·0	−503·6
2010	1,913·2	1,278·3	−634·9
2011	2,207·4	1,480·6	−726·7

In 2009 the USA's merchandise trade deficit fell to its lowest level since 2002. Its deficit with China went up from $83·1bn. in 2001 to $295·5bn. in 2011, although despite generally rising during those ten years it actually fell by $41·2bn. in 2009 at the height of the recession. Its largest surplus is with the Netherlands

($19·4bn. in 2011). The USA last recorded a trade surplus in goods in 1975. The USA is both the world's leading importer and the leading trading nation, although only the third largest exporter of goods after China and Germany (but it is the largest exporter of goods and services combined). In 2009 its trade in goods accounted for 12·7% of the world's imports and 8·5% of exports.

Principal imports and exports (in $1m.), 2010:

	Imports	Exports
Agricultural commodities		
Animal feeds	1,349	8,996
Cereal flour	4,519	3,037
Corn	300	10,181
Cotton, raw and linters	8	5,896
Meat and preparations	5,071	13,216
Soybeans	220	18,589
Vegetables and fruits	20,915	15,712
Wheat	563	6,769
Manufactured goods		
Alcoholic beverages, distilled	5,608	1,126
Aluminium	10,815	5,171
Artwork/antiques	6,268	3,034
Basketware, etc.	13,316	8,360
Chemicals – cosmetics	9,564	12,488
Chemicals – dyeing	3,105	7,407
Chemicals – fertilizers	6,647	3,731
Chemicals – inorganic	13,833	11,806
Chemicals – medicinal	65,170	41,960
Chemicals – organic	45,792	37,494
Chemicals – plastics	17,825	42,019
Chemicals – misc.	11,379	24,136
Clothing	78,518	3,197
Computer and telecommunications equipment, etc.	113,476	22,238
Copper	7,821	3,496
Cork, wood, lumber	4,479	4,732
Crude fertilizers	2,257	2,367
Electrical machinery	119,634	77,019
Fish and preparations	14,576	4,223
Footwear	20,902	728
Furniture and bedding	31,124	4,821
Gem diamonds	18,599	2,862
General industrial machinery	60,426	51,793
Glass	2,588	3,380
Gold, non-monetary	12,491	17,458
Iron and steel mill products	24,440	15,720
Jewellery	10,085	4,848
Lighting, plumbing	7,397	2,509
Metal manufactures, misc.	25,913	17,491
Metal ores; scrap	7,293	28,366
Metalworking machinery	5,565	5,330
Nickel	2,976	1,130
Optical goods	5,507	3,278
Paper and paperboard	15,285	14,920
Photographic equipment	2,048	3,345
Plastic articles, misc.	16,042	9,710
Platinum	4,146	1,370
Power generating machinery	42,465	33,013
Printed materials	4,585	5,879
Pulp and waste paper	3,887	8,640
Records/magnetic media	5,296	4,424
Rubber articles, misc.	3,310	2,049
Rubber tyres and tubes	10,673	4,159
Scientific instruments	37,795	44,276
Ships/boats	1,588	2,498
Specialized industrial machinery	30,912	46,754
Televisions, etc.	137,305	21,511
Textile yarn, fabric	22,120	11,384
Toys/games/sporting goods	30,360	4,245
Travel goods	8,012	463
Vehicles	178,946	88,119
Watches/clocks/parts	3,747	379
Wood manufactures	6,920	2,053
Mineral fuel		
Coal	2,018	10,100
Crude oil	260,105	1,368

	Imports	Exports
Liquefied propane/butane	2,541	2,448
Mineral fuel, misc.	3,410	7,434
Natural gas	17,402	4,921
Petroleum preparations	67,409	53,528

Imports and exports by selected countries for the calendar years 2009 and 2010 (in $1m.):

	General imports		Exports incl. re-exports	
Country	2009	2010	2009	2010
Australia	8,012	8,583	19,599	21,798
Belgium	13,826	15,552	21,608	25,456
Brazil	20,070	23,958	26,095	35,425
Canada	226,248	277,647	204,658	249,105
China	296,374	364,944	69,497	91,881
Colombia	11,323	15,659	9,451	12,069
France	34,236	38,355	26,493	26,969
Germany	71,498	82,429	43,306	48,161
Hong Kong	3,571	4,296	21,051	26,570
India	21,166	29,533	16,441	19,250
Indonesia	12,939	16,478	5,107	6,946
Ireland	28,101	33,848	7,465	7,276
Israel	18,744	20,982	9,559	11,294
Italy	26,430	28,505	12,268	14,219
Japan	95,804	120,545	51,134	60,486
South Korea	39,216	48,875	28,612	38,846
Malaysia	23,283	25,900	10,403	14,080
Mexico	176,654	229,908	128,892	163,473
Netherlands	16,098	19,055	32,242	34,939
Nigeria	19,128	30,516	3,687	4,068
Russia	18,200	25,691	5,332	6,006
Saudi Arabia	22,053	31,413	10,792	11,556
Singapore	15,705	17,427	22,232	29,017
Spain	7,857	8,553	8,717	10,178
Switzerland	16,053	19,136	17,504	20,687
Taiwan	28,362	35,846	18,486	26,043
Thailand	19,082	22,693	6,918	8,977
United Kingdom	47,480	49,775	45,704	48,414
Venezuela	28,059	32,707	9,315	10,649
Vietnam	12,288	14,868	3,097	3,709

COMMUNICATIONS

Roads

On 31 Dec. 2007 the total public road mileage was 4,032,126 miles (urban, 1,044,368; rural, 2,987,758). Of the urban roads, 14% were state controlled and 86% under local control. 21% of rural roads were controlled by the states, 75% of rural roads were under local control and the remainder were federal park and forest roads. State highway funds were $130,306m. in 2007.

Motor vehicles registered in 2007: 247,265,000, of which 135,933,000 automobiles, 110,497,000 trucks and 834,000 buses. There were 205,742,000 licensed drivers in 2007 and 7,093,000 motorcycle registrations. The average distance travelled by a passenger car in the year 2007 was 12,300 miles. There were 33,883 fatalities in road accidents in 2009 and 32,885 in 2010 (the lowest total since 1949).

Rail

Freight service is provided by nine major independent railroad companies and several hundred smaller operators. The largest companies are the Union Pacific Railroad and the BNSF Railway (formerly the Burlington Northern and Santa Fe Railway). Long-distance passenger trains are run by the National Railroad Passenger Corporation (Amtrak), which is federally assisted. Amtrak was set up in 1971 to maintain a basic network of long-distance passenger trains, and is responsible for almost all non-commuter services. In 2009 the operational Amtrak rail system measured 21,178 miles. Outside the major conurbations, there are almost no regular passenger services other than those of Amtrak, which carried 27·2m. passengers in fiscal year 2009. Passenger revenue for Amtrak in fiscal year 2009 was $1,814m.; revenue passenger miles (fiscal year 2008), 6,160m.

Civil Aviation

The busiest airport in 2011 was Atlanta (Hartsfield-Jackson), which handled 92,389,023 passengers. The second busiest was Chicago (O'Hare) with 66,659,709 passengers, followed by Los Angeles International, with 61,862,052 passenger enplanements. As well as being the three busiest airports in the USA for passenger traffic in 2011, they are also three of the six busiest in the world. The six busiest in the world in 2011 were Atlanta, Beijing Capital International, London Heathrow, Chicago O'Hare, Tokyo Haneda and Los Angeles. New York (John F. Kennedy) was the busiest airport in the USA for international passengers in 2010, with 22,702,882, ahead of Miami International with 16,207,353.

There were 24 airports with more than 10m. enplanements in 2007. These were, in descending order: Atlanta (Hartsfield-Jackson); Chicago (O'Hare); Dallas/Fort Worth; Los Angeles; Denver; Las Vegas (McCarran); Phoenix; Houston (George Bush Intercontinental); Detroit (Metropolitan-Wayne County); Minneapolis/St Paul; New York (John F. Kennedy); Newark International; Orlando International; Charlotte; Philadelphia; Seattle; San Francisco; Miami; Boston; New York (LaGuardia); Salt Lake City; Fort Lauderdale-Hollywood International; Washington, D.C.; Baltimore.

The leading airports in 2010 on the basis of aircraft departures completed were Atlanta Hartsfield-Jackson (465,000); Chicago O'Hare (426,600); Dallas/Fort Worth (315,100).

In 2011 Delta Air Lines carried the most scheduled passengers of any airline in the world with 163,838,000, ahead of United Airlines, with 141,799,000, and the low-cost carrier Southwest Airlines, with 135,274,000. Delta Airlines carried the most international passengers of any US carrier in 2010, with 21,029,000 (ranking it eighth in the world for international passengers carried). Delta Air Lines filed for bankruptcy in Sept. 2005, but emerged from bankruptcy protection in April 2007. AMR, the parent company for American Airlines, filed for bankruptcy in Nov. 2011. Delta Air Lines bought Northwest Airlines in Oct. 2008, in the process creating the world's largest airline. However, this was superseded following the merger of Continental Airlines and United Airlines in Oct. 2010. The two airlines are continuing to operate separately under the name United Continental Holdings, Inc.

In 2008 US flag carriers in scheduled service enplaned 741·4m. revenue passengers.

Shipping

At the end of 2007 the cargo-carrying US-owned fleet comprised 40,250 vessels, of which 39,695 were US-flag vessels. There were 38,936 US-flag tugs and barges for domestic coastwise, Great Lakes and inland waterway trade, 523 US-flag offshore supply vessels (which service offshore oil exploration and production) and 236 US-flag ocean and Great Lakes self-propelled vessels (10,000 DWT or greater) for US coastwise and international trade (of which 55 tankers, 76 containerships, 37 roll-on/roll-off carriers, 61 dry bulk carriers and seven general cargo carriers). Of the ocean-going fleet, 146 have coastwise trading privileges (Jones Act), which means that they were built or reconstructed in the USA or foreign-built but seized for violation of US law and registered under the US flag.

The busiest port is South Louisiana, which handled 212,581,000 tons of cargo in 2009. Other major ports are Houston (211,341,000 tons in 2009), New York-New Jersey (144,690,000 tons in 2009), Long Beach, Corpus Christi, New Orleans and Beaumont. South Louisiana handles the most domestic cargo but Houston the most foreign trade cargo.

Telecommunications

Regional private companies formed from the American Telephone and Telegraph Co. after its dissolution in 1995 ('Baby Bells') operate the telephone and electronic transmission services system at the national and local levels. Telegram services are still available through iTelegram and Globegram. In 2009 there were 141·0m. main telephone lines in operation (448·1 per 1,000 inhabitants), down from 182·93m. in 2003. There were 285·6m. cellphone subscribers in 2009 (907·8 per 1,000 persons), up from 160·6m. in 2003. In addition the proportion of households that only have cellphones has rapidly increased, to such an extent that by Dec. 2008 the number of cellphone-only households had surpassed landline-only households, with 20% of households being cellphone-only. The leading cellphone operators are Verizon Wireless (with more than 115m. subscribers), AT&T Mobility, Sprint Nextel and T-Mobile. Internet users numbered 245·4m. in 2009, or 78·0% of the population. Internet commerce, or e-commerce, amounted to $194bn. in 2011, the highest in the world. In Dec. 2010 there were 53·5 wireless broadband subscribers per 100 inhabitants and only 27·7 fixed broadband subscribers per 100. In Dec. 2011 there were 157·4m. Facebook users (about three times as many as any other country and 50% the total population of the USA).

SOCIAL INSTITUTIONS

Justice

Legal controversies may be decided in two systems of courts: the federal courts, with jurisdiction confined to certain matters enumerated in Article III of the Constitution, and the state courts, with jurisdiction in all other proceedings. The federal courts have jurisdiction exclusive of the state courts in criminal prosecutions for the violation of federal statutes, in civil cases involving the government, in bankruptcy cases and in admiralty proceedings, and have jurisdiction concurrent with the state courts over suits between parties from different states, and certain suits involving questions of federal law.

The highest court is the Supreme Court of the US, which reviews cases from the lower federal courts and certain cases originating in state courts involving questions of federal law. It is the final arbiter of all questions involving federal statutes and the Constitution; and it has the power to invalidate any federal or state law or executive action which it finds repugnant to the Constitution. This court, consisting of nine justices appointed by the President who receive salaries of $213,900 a year (the Chief Justice, $223,500), meets from Oct. until June every year. For the term beginning Oct. 2006 it disposed of 8,923 cases, deciding 78 on their merits. In the remainder of cases it either summarily affirms lower court decisions or declines to review. A few suits, usually brought by state governments, originate in the Supreme Court, but issues of fact are mostly referred to a master.

The US courts of appeals number 13 (in 11 circuits composed of three or more states and one circuit for the District of Columbia and one Court of Appeals for the Federal Circuit); the 179 circuit judges receive salaries of $184,500 a year. Any party to a suit in a lower federal court usually has a right of appeal to one of these courts. In addition, there are direct appeals to these courts from many federal administrative agencies. In the year ending 30 Sept. 2007, 58,410 appeals were filed in the courts of appeals, in addition to 1,406 in the US Court of Appeals for the Federal Circuit.

The trial courts in the federal system are the US district courts, of which there are 94 in the 50 states, one in the District of Columbia and one each in the Commonwealth of Puerto Rico and the Territories of the Virgin Islands, Guam and the Northern Marianas. Each state has at least one US district court, and three states have four apiece. Each district court has from one to 28 judgeships. There are 674 US district judges ($174,000 a year), who received 257,507 civil cases in 2006–07.

In addition to these courts of general jurisdiction, there are special federal courts of limited jurisdiction. The US Court of Federal Claims (16 judges at $174,000 a year) decides claims for money damages against the federal government in a wide variety of matters; the Court of International Trade (13 judges at

$174,000) determines controversies concerning the classification and valuation of imported merchandise.

The judges of all these courts are appointed by the President with the approval of the Senate; to assure their independence, they hold office during good behaviour and cannot have their salaries reduced. This does not apply to judges in the Territories, who hold their offices for a term of ten years or to judges of the US Court of Federal Claims. The judges may retire with full pay at the age of 70 years if they have served a period of ten years, or at 65 if they have 15 years of service, but they are subject to call for such judicial duties as they are willing to undertake.

In 2006–07, of the 257,507 civil cases filed in the district courts, 159,916 arose under various federal statutes (such as labour, social security, tax, patent, securities, antitrust and civil rights laws); 61,359 involved personal injury or property damage claims; 33,939 dealt with contracts; and 5,180 were actions concerning real property. In the year ending Sept. 2009, 1,402,816 cases were filed in the US Bankruptcy Courts (up 34·5% from 1,042,993 in fiscal year 2008 and the highest since fiscal year 2005), 1,344,095 of which involved individuals or non-businesses.

In 2000 the number of lawyers in the USA passed the 1m. mark, the equivalent to 363 per 100,000 people, reaching 1·2m. by 2008.

There were 68,413 criminal cases filed for the year ending Sept. 2007 of which 17,046 involved drugs. Among the 79,904 offenders convicted in 2006 in the district courts, 27,361 persons were charged with alleged infractions of drug laws, 17,017 with immigration offences, 11,303 with property offences, 8,831 with weapon offences, 6,045 with public order offences and 2,452 with violent offences. All other people convicted were charged with miscellaneous general offences.

Persons convicted of federal crimes may be fined, released on probation under the supervision of the probation officers of the federal courts, confined in prison, or confined in prison with a period of supervised release to follow, also under the supervision of probation officers of the federal courts. Federal prisoners are confined in 87 institutions incorporating various security levels that are operated by the Bureau of Prisons. On 30 June 2008 the total number of prisoners under the jurisdiction of Federal or State adult correctional authorities was 1,525,428. A record 2,310,984 inmates were held in Federal or State prisons or local jails in June 2008, giving a rate of 762 per 100,000 population (the highest of any country). Although the USA has less than 5% of the world's population it has around 24% of the world's prisoners.

The state courts have jurisdiction over all civil and criminal cases arising under state laws, but decisions of the state courts of last resort as to the validity of treaties or of laws of the USA, or on other questions arising under the Constitution, are subject to review by the Supreme Court of the US. The state court systems are generally similar to the federal system, to the extent that they generally have a number of trial courts and intermediate appellate courts, and a single court of last resort. The highest court in each state is usually called the Supreme Court or Court of Appeals with a Chief Justice and Associate Justices, usually elected but sometimes appointed by the Governor with the advice and consent of the State Senate or other advisory body; they usually hold office for a term of years, but in some instances for life or during good behaviour. The lowest tribunals are usually those of Justices of the Peace; many towns and cities have municipal and police courts, with power to commit for trial in criminal matters and to determine misdemeanours for violation of the municipal ordinances.

There were no executions from 1968 to 1976. The US Supreme Court had held the death penalty, as applied in general criminal statutes, to contravene the eighth and fourteenth amendments of the US constitution, as a cruel and unusual punishment when used so irregularly and rarely as to destroy its deterrent value. The death penalty was reinstated by the Supreme Court in 1976, but has not been authorized in Alaska, the District of Columbia,

Hawaii, Iowa, Kansas, Maine, Massachusetts, Michigan, Minnesota, New Jersey, New York, North Dakota, Rhode Island, Vermont, West Virginia and Wisconsin. At 1 April 2012 there were 3,170 (including 61 women) prisoners under sentence of death. 77 people were sentenced to death in 2012. In 2012 there were 43 executions (a fall from a peak of 98 in 1999, though there were only 11 in 1988). From 1977–2012 there were 1,320 executions of which 492 were in Texas and 109 in Virginia. The death penalty for offenders under the age of 18 was abolished in March 2005. For the first time since 1963, there were two executions under federal jurisdiction in 2001. In Sept. 2003 the federal Court of Appeals in San Francisco overturned over 100 death sentences in Arizona, Idaho and Montana on the grounds that judges, not juries, had passed sentence, contravening a Supreme Court ruling of 2002.

There were 14,612 murders in 2011, the lowest total since 1968. The murder rate in 2011 was 4·7 per 100,000 persons, the lowest since 1963 (4·6 per 100,000) and down from 10·2 per 100,000 in 1980. 67·7% of all murders in 2011 were carried out with firearms.

Education

The adult literacy rate is at least 99%.

Elementary and secondary education is mainly a state responsibility. Each state and the District of Columbia has a system of free public schools, established by law, with courses covering 12 years plus kindergarten. There are three structural patterns in common use; the K8-4 plan, meaning kindergarten plus eight elementary grades followed by four high school grades; the K6-3-3 plan, or kindergarten plus six elementary grades followed by a three-year junior high school and a three-year senior high school; and the K5-3-4 plan, kindergarten plus five elementary grades followed by a three-year middle school and a four-year high school. All plans lead to high-school graduation, usually at age 17 or 18. Vocational education is an integral part of secondary education. Most states also have two-year colleges in which education is provided at a nominal cost. Each state has delegated a large degree of control of the educational programme to local school districts (numbering 13,809 in school year 2008–09), each with a board of education (usually three to nine members) selected locally and serving mostly without pay. The school policies of the local school districts must be in accord with the laws and the regulations of their state Departments of Education. While regulations differ from one jurisdiction to another, in general it may be said that school attendance is compulsory from age seven to 16.

'Charter schools' are legal entities outside the school boards administration. They retain the basics of public school education, but may offer unconventional curricula and hours of attendance. Founders may be parents, teachers, public bodies or commercial firms. Organization and conditions depend upon individual states' legislation. The first charter schools were set up in Minnesota in 1991. By 2008, 4,694 charter schools were operating in 40 states and Washington, D.C.

Since 1940 data have been tabulated using a definition of 'functionally illiterate', comprising those who had completed fewer than five years of elementary schooling; for persons 25 years of age or over this percentage was 1·4 in March 2009 (for the Black population as a whole it was 1·1%); it was 0·1% for Whites, 0·3% for Blacks and 2·6% for Hispanics in the age group 25 to 29. It was reported in March 2010 that 87·1% of all persons 25 years old and over had completed four years of high school or more, and that 29·5% had completed a bachelor's degree or more. In the age group 25 to 29, 88·8% had completed four years of high school or more, and 31·7% had completed a bachelor's degree or more. However, according to a study conducted in 2009 about a third of American fourth graders (aged 9–10) are unable to read at a basic level.

In the fall of 2009, 20·4m. students (12·7m. full-time and 11·7m. women) were enrolled in 4,495 colleges and universities; 3·2m.

were first-time students. It is projected that in 2019 the student population will number 23·4m.

In 2007–08 expenditure for public elementary and secondary education totalled $596·6bn., comprising $506·8bn. for current operating expenses, $65·8bn. for capital outlay, $15·7bn. for interest on school debt and $8·3bn. for other expenses. The current expenditure per pupil in fall 2007 enrolment was $10,297.

In 2009–10 total expenditure on education came to 7·9% of GDP, of which 3·3% (about two-thirds private funding and a third public) was on tertiary education. Spending as a proportion of GDP on tertiary education (3·1% in 2007) is the highest in the world, although spending on non-tertiary education is only around the average for an industrialized country.

Estimated total expenditures for private elementary and secondary schools in 2009–10 were about $48bn. In 2009–10 college and university spending totalled about $461bn., of which about $289bn. was spent by institutions under public control. In 2009–10 the federal government contributed about 12% of total revenue for public institutions; state governments, 42%; student tuition and fees, 16%; and local sources, 30%. Federal support for vocational education in fiscal year 2010 amounted to about $1·9bn.

Summary of statistics of regular schools (public and private), teachers and pupils for 2008–09 (compiled by the US National Center for Education Statistics):

Schools by level	Number of schools	Teachers (in 1,000)	Enrolment (in 1,000)
Elementary and secondary schools:			
Public	98,706	3,222	49,266
Private	33,740[1]	455	5,969
Higher education:			
Public	1,676	885	13,972
Private	2,733	505	5,131
Total	136,855	5,066	74,338

[1]Data from 2007–08.

In the fall of 2008 there were 15·3 pupils per teacher in public schools in the USA and 13·1 pupils per teacher in private schools.

Most of the private elementary and secondary schools are affiliated with religious denominations. In 2007–08 there were 7,510 Catholic schools with 2,308,000 pupils and 147,000 teachers, and 15,400 schools of other religious affiliations with 2,283,000 pupils and 183,000 teachers.

During the school year 2007–08 high-school graduates numbered 3,314,000 (of whom 3m. were from public schools). Institutions of higher education conferred 1,601,000 bachelor's degrees during the year 2008–09; 787,000 associate's degrees; 657,000 master's degrees; 68,000 doctorates; and 92,000 first professional degrees. In the fiscal year 2010 the US Department of Education provided $32bn. in grants, loans, work-study programmes and other financial assistance to post-secondary students. An additional $109bn. was provided to students through private lending institutions supported by federal programmes.

During the academic year 2008–09, 671,616 foreign students were enrolled in American colleges and universities. The countries with the largest numbers of students in American colleges were: India, 103,260; China, 98,235; South Korea, 75,065; Canada, 29,697; Japan, 29,264; Taiwan, 28,065.

In 2008, 52,328 US students were enrolled in degree programmes in colleges and universities outside of the USA. The country attracting the most students from the USA was the United Kingdom, with 13,895. In addition to these students, well over 250,000 US college students attend short programmes in other countries every year.

School enrolment, Oct. 2009, embraced 94·1% of the children who were 5 and 6 years old; 98·2% of the children aged 7–13 years; 96·3% of those aged 14–17; 68·9% of those aged 18–19; and 38·7% of those aged 20–24.

The US National Center for Education Statistics estimates the total enrolment in the fall of 2010 at all of the country's elementary, secondary and higher educational institutions (public and private) at 75·9m. (68·7m. in the fall of 2000).

The number of teachers in public and private elementary and secondary schools in 2010 increased slightly to about 3,633,000. The estimated average annual salary of public school teachers was $55,350 in 2009–10.

Health

Admission to the practice of medicine (for both doctors of medicine and doctors of osteopathic medicine) is controlled in each state by examining boards directly representing the profession and acting with authority conferred by state law. Although there are a number of variations, the usual time now required to complete training is eight years beyond the secondary school with up to three or more years of additional graduate training. Certification as a specialist may require between three and five more years of graduate training plus experience in practice. In Dec. 2007 the estimated number of physicians (MD and DO—in all forms of practice) in the USA, Puerto Rico and outlying US areas was 941,300 (615,400 in 1990 and 467,700 in 1980).

Dental employment in 2008 numbered around 141,900.

Number of hospitals listed by the American Hospital Association in 2007 was 5,708, with 945,000 beds (equivalent to 3·1 beds per 1,000 population). Of the total, 213 hospitals with 46,000 beds were operated by the federal government; 1,111 with 131,000 beds by state and local government; 2,913 with 554,000 beds by non-profit organizations (including church groups); 873 with 116,000 beds were investor-owned. The categories of non-federal hospitals were (2007): 4,897 short-term general and special hospitals with 801,000 beds; 135 non-federal long-term general and special hospitals with 17,000 beds; 444 psychiatric hospitals with 79,000 beds; one tuberculosis hospital with fewer than 500 beds.

Patient admissions to community hospitals (2007) was 35,346,000; average daily census was 533,300. There were 603·3m. outpatient visits.

Personal health care costs in 2007 totalled $1,878,275m., distributed as follows: hospital care, $696,539m.; physicians and clinical services, $478,768m.; prescription drugs, $227,453m.; nursing-home care, $131,338m.; dental services, $95,171m.; home health care, $59,031m.; medical durables, $24,450m.; other personal health care, $165,525m. Total national health expenditure in 2007 amounted to $2,241·2bn. In 2008 the USA spent 16·0% of its GDP on health—nearly 5% more than any other leading industrialized nation. Public spending on health amounted to 46·5% of total health spending in 2008 (the lowest percentage of any major industrialized nation). In March 2010 President Obama secured the passage of a health care reform package that was expected to increase insurance coverage to a further 32m. citizens. The US Census Bureau had estimated in 2008 that 46·3m. people in America were uninsured.

In 2007, 19·8% of Americans (21·3% of males and 18·4% of females) were smokers, down from a peak of over 40% in 1964. In 2005–06, 34·3% of the adult population were considered obese (having a body mass index over 30), compared to 14·6% in the early 1970s.

Welfare

Social welfare legislation was chiefly the province of the various states until the adoption of the Social Security Act of 14 Aug. 1935. This as amended provides for a federal system of old-age, survivors and disability insurance; health insurance for the aged and disabled; supplemental security income for the aged, blind and disabled; federal state unemployment insurance; and federal grants to states for public assistance (medical assistance for the aged and aid to families with dependent children generally and for maternal and child health and child welfare services).

Legislation of Aug. 1996 began the transfer of aid administration back to the states, restricted the provision of aid to a maximum period of five years, and abolished benefits to immigrants (both legal and illegal) for the first five years of their residence in the USA. The Social Security Administration (formerly part of the Department of Health and Human Services but an independent agency since March 1995) has responsibility for a number of programmes covering retirement, disability, Medicare, Supplemental Security Income and survivors. The Administration for Children and Families (ACF), an agency of the Department of Health and Human Services, is responsible for federal programmes which promote the economic and social wellbeing of families, children, individuals and communities. ACF has federal responsibility for the following programmes: Temporary Assistance for Needy Families; low income energy assistance; Head Start; child care; child protective services; and a community services block grant. The ACF also has federal responsibility for social service programmes for children, youth, native Americans and persons with developmental disabilities.

The Administration on Aging (AoA), an agency in the US Department of Health and Human Services, is one of the nation's largest providers of home- and community-based care for older persons and their caregivers. Created in 1965 with the passage of the Older Americans Act (OAA), AoA is part of a federal, state, tribal and local partnership called the National Network on Aging. It serves about 9m. older persons and their caregivers, and consists of 56 State Units on Aging, 655 Area Agencies on Aging, 236 Tribal and Native organizations, two organizations that serve Native Hawaiians, 29,000 service providers and thousands of volunteers. These organizations provide assistance and services to older individuals and their families in urban, suburban, and rural areas throughout the USA.

The Centers for Medicare and Medicaid Services (formerly the Health Care Financing Administration), an agency of the Health and Human Services Department, has federal responsibility for health insurance for the aged and disabled. Unemployment insurance is the responsibility of the Department of Labor.

In 2007 an average of 1,673,000 families (3,896,000 recipients) were receiving payments under Temporary Assistance for Needy Families. Total payments under Temporary Assistance for Needy Families were $26,922m. in 2007. The role of Child Support Enforcement is to ensure that children are supported by their parents. Money collected is for children who live with only one parent because of divorce, separation or birth outside marriage. In 2007, $24,855m. was collected on behalf of these children.

The Social Security Act provides for protection against the cost of medical care through Medicare, a two-part programme of health insurance for people age 65 and over, people of any age with permanent kidney failure, and for certain disabled people under age 65 who receive Social Security disability benefits. In 2008 payments totalling $230,240m. were made under the hospital portion of Medicare. During the same period, $224,835m. was paid under the voluntary medical insurance portion of Medicare. Medicare enrolment in July 2008 totalled 45·2m.

In 2009 about 52m. beneficiaries were on the rolls. Full retirement benefits are now payable at age 66, with reduced benefits available as early as age 62. Pensions may also be deferred up to age 70. The age for full retirement benefits is gradually increasing until it reaches 67 in 2027. Claimants must have at least 40 credits of insurance coverage. This equates to ten years of work. In 2008 the average actual retirement age for both males and females was 63·6 years. The maximum monthly payment for claimants retiring at the full retirement age in 2009 was $2,323. The minimum social security benefit was eliminated in Jan. 1982 for all workers becoming eligible for retirement or disability insurance benefits after Dec. 1981. A means-tested supplemental income benefit is available to over-65s and disabled and blind individuals with limited income and limited resources.

Medicaid is a jointly-funded, Federal-State health insurance programme for certain low-income and vulnerable people. It covered 57·8m. individuals in 2006 including children, the aged, blind, and/or disabled, and people who are eligible to receive federally-assisted income maintenance payments.

In Dec. 2007, 7·36m. persons were receiving Supplementary Security Income payments. 1,205,000 old-age persons received $5,301m. in benefits; 72,000 blind people received $419m.; and 6,083,000 disabled people received $35,485m. Payments, including supplemental amounts from various states, totalled $41,205m. in 2007.

In 2008 the food stamp programme helped 28,408,000 persons at a cost of $34,611m.; and 31·0m. persons received help from the national school lunch programme at a cost of $8,258m.

RELIGION

The leading religious bodies according to the *2010 U.S. Religion Census: Religious Congregations & Membership Study* and based on the number of congregations are as follows:

Religious bodies	Tradition	Congregations	Adherents
Southern Baptist Convention	Evangelical Protestant	50,816	19,896,279
United Methodist Church	Mainline Protestant	33,323	9,860,653
Catholic Church[1]	Catholic	20,589	58,963,835
Church of Jesus Christ of Latter-day Saints	Other	13,601	6,144,582
Churches of Christ	Evangelical Protestant	12,584	1,584,162
Assemblies of God	Evangelical Protestant	12,258	2,944,887
Presbyterian Church (USA)	Mainline Protestant	10,487	2,451,980
Evangelical Lutheran Church in America	Mainline Protestant	9,846	4,181,219
Episcopal Church	Mainline Protestant	6,794	1,951,907
Church of God (Cleveland, Tennessee)	Evangelical Protestant	6,100	1,109,992

[1]In Feb. 2013 there were 19 cardinals.

In 2010 (unless otherwise stated) there were 2,600,000 Muslims (estimate), 2,257,000 Jews (estimate), 1,162,686 Jehovah's Witnesses (2009), 992,000 Buddhists (estimate) and 641,186 Hindus. Other major religious bodies are: African Methodist Episcopal Church; American Baptist Churches in the USA; Christian Churches and Churches of Christ; Lutheran Church–Missouri Synod; National Baptist Convention, USA, Inc.; Pentecostal Assemblies of the World, Inc.; Seventh-day Adventist Church; United Church of Christ.

CULTURE

World Heritage Sites

There are 21 sites under American jurisdiction that appear on the UNESCO World Heritage List. They are (with year entered on list): Mesa Verde National Park, Colorado (1978); Yellowstone National Park, Wyoming/Idaho/Montana (1978); Everglades National Park, Florida (1979); Grand Canyon National Park, Arizona (1979); Independence Hall, Pennsylvania (1979); Redwood National and State Parks, California (1980); Mammoth Cave National Park, Kentucky (1981); Olympic National Park, Washington State (1981); Cahokia Mounds State Historic Site, Illinois (1982); Great Smoky Mountains National Park, North Carolina/Tennessee (1983); San Juan National Historic Site and La Fortaleza, Puerto Rico (1983); the Statue of Liberty, New York (1984); Yosemite National Park, California (1984); Monticello and the University of Virginia, Charlottesville, Virginia (1987); Chaco Culture National Historic Park, New Mexico (1987); Hawaii

Volcanoes National Park, including Mauna Loa, Hawaii (1987); Pueblo de Taos, New Mexico (1992); Carlsbad Caverns National Park, New Mexico (1995); Papahanaumokuakea, Hawaii (2010).

Two UNESCO World Heritage sites fall under joint US and Canadian jurisdiction: Kluane/Wrangell-St Elias/Glacier Bay/Tatshenshini-Alsek (1979, 1992 and 1994), parks in Alaska, the Yukon Territory and British Columbia; Waterton Glacier International Peace Park (1995), in Montana and Alberta.

Press

In 2008 there were 1,408 daily newspapers with a combined daily circulation of 48·6m., the fourth highest in the world behind India, China and Japan. There were 872 morning papers and 546 evening papers, plus 902 Sunday papers (circulation, 49·1m.). Unlike China and India, where circulation is rising, in the USA it has fallen since 1985, when daily circulation was 62·8m. The most widely read newspapers are *USA Today* (average daily circulation in the period April–Sept. 2008 of 2·3m.), followed by the *Wall Street Journal* (2·0m.) and the *New York Times* (1·0m.). According to research carried out by the Pew Research Centre, in 2008 for the first time more Americans obtained national and international news from the internet than from newspapers.

Total book sales in 2008 reached a record high of 561,580 (up from 266,322 in 2003), largely as a result of print on demand sales and reprints that totalled 271,851 (up from 21,936 in 2006). Of the record 289,729 traditional print books published in 2008, 53,058 were fiction, 29,825 children's books, 24,737 sociology and economics and 18,296 religion. US publishers' net sales revenue rose from $26·5bn. in 2008 to $27·1bn. in 2009 and $27·9bn. in 2010.

Tourism

In 2011 the USA received 62,711,000 foreign visitors (59,796,000 in 2010), of whom 21,337,000 were from Canada and 13,491,000 from Mexico. 19% of all tourists were from Europe. Only France received more tourists than the USA in 2011.

In 2010 visitors to the USA spent $103·5bn., giving the USA by far the highest annual revenue from tourists of any country (Spain, which received the second most, had $52·5bn.). Expenditure by US travellers in foreign countries for 2010 was $75·5bn., second only to spending by German travellers in foreign countries.

Festivals

Independence Day is celebrated on 4 July and Thanksgiving on the fourth Thursday of Nov. There are major opera festivals at Cooperstown (Glimmerglass), New York State (July–Aug.); Santa Fe, New Mexico (June–Aug.); and Seattle, Washington State (Aug.). Among the many famous film festivals are the Sundance Film Festival in Jan. and the New York Film Festival in late Sept./early Oct. Leading rock and pop festivals include SXSW in Austin, Texas (March), Coachella in Palm Springs, California (April), Bonnaroo in Manchester, Tennessee (June), Summerfest in Milwaukee, Wisconsin (June–July) and Lollapalooza in Chicago, Illinois (Aug.).

DIPLOMATIC REPRESENTATIVES

Of the USA in the United Kingdom (24 Grosvenor Sq., London, W1A 1AE)
Ambassador: Louis B. Susman.

Of the United Kingdom in the USA (3100 Massachusetts Ave., NW, Washington, D.C., 20008)
Ambassador: Sir Peter Westmacott, KCMG, LVO.

Of the United States to the United Nations
Ambassador: Susan Rice.

Of the United States to the European Union
Ambassador: William E. Kennard.

FURTHER READING

OFFICIAL STATISTICAL INFORMATION

The Office of Management and Budget, Washington, D.C., 20503 is part of the Executive Office of the President; it is responsible for co-ordinating all the statistical work of the different Federal government agencies. The Office does not collect or publish data itself. The main statistical agencies are as follows:

(1) Data User Services Division, Bureau of the Census, Department of Commerce, Washington, D.C., 20233. Responsible for decennial censuses of population and housing, quinquennial census of agriculture, manufactures and business; current statistics on population and the labour force, manufacturing activity and commodity production, trade and services, foreign trade, state and local government finances and operations. (*Statistical Abstract of the United States*, annual, and others).

(2) Bureau of Labor Statistics, Department of Labor, 441 G Street NW, Washington, D.C., 20212. (*Monthly Labor Review* and others).

(3) Information Division, Economic Research Service, Department of Agriculture, Washington, D.C., 20250. (*Agricultural Statistics*, annual, and others).

(4) National Center for Health Statistics, Department of Health and Human Services, 3700 East-West Highway, Hyattsville, MD 20782. (*Vital Statistics of the United States*, monthly and annual, and others).

(5) Bureau of Mines Office of Technical Information, Department of the Interior, Washington, D.C., 20241. (*Minerals Yearbook*, annual, and others).

(6) Office of Energy Information Services, Energy Information Administration, Department of Energy, Washington, D.C., 20461.

(7) Statistical Publications, Department of Commerce, Room 5062 Main Commerce, 14th St and Constitution Avenue NW, Washington, D.C., 20230; the Department's Bureau of Economic Analysis and its Office of Industry and Trade Information are the main collectors of data.

(8) Center for Education Statistics, Department of Education, 555 New Jersey Avenue NW, Washington, D.C., 20208.

(9) Public Correspondence Division, Office of the Assistant Secretary of Defense (Public Affairs P.C.), The Pentagon, Washington, D.C., 20301-1400.

(10) Bureau of Justice Statistics, Department of Justice, 633 Indiana Avenue NW, Washington, D.C., 20531.

(11) Public Inquiry, APA 200, Federal Aviation Administration, Department of Transportation, 800 Independence Avenue SW, Washington, D.C., 20591.

(12) Office of Public Affairs, Federal Highway Administration, Department of Transportation, 400 7th St. SW, Washington, D.C., 20590.

(13) Statistics Division, Internal Revenue Service, Department of the Treasury, 1201 E St. NW, Washington, D.C., 20224.

Statistics on the economy are also published by the Division of Research and Statistics, Federal Reserve Board, Washington, D.C., 20551; the Congressional Joint Committee on the Economy, Capitol; the Office of the Secretary, Department of the Treasury, 1500 Pennsylvania Avenue NW, Washington, D.C., 20220.

OTHER OFFICIAL PUBLICATIONS

Economic Report of the President. Annual. Bureau of the Census. *Statistical Abstract of the United States.* Annual. *Historical Statistics of the United States, Colonial Times to 1970.*

United States Government Manual. Annual.

The official publications of the USA are issued by the US Government Printing Office and are distributed by the Superintendent of Documents, who issued in 1940 a cumulative *Catalog of the Public Documents of the Congress and of All Departments of the Government of the United States.* This *Catalog* is kept up to date by *Monthly Catalog of United States Government Publications* with annual index and supplemented by *Price Lists.* Each *Price List* is devoted to a special subject or type of material.

Treaties and other International Acts of the United States of America (Edited by Hunter Miller), 8 vols. 1929–48. This edition stops in 1863. It may be supplemented by *Treaties, Conventions, International Acts, Protocols and Agreements Between the US and Other Powers, 1776–1937* (Edited by William M. Malloy and others). 4 vols. 1909–38. A new Treaty Series, *US Treaties and Other International Agreements,* was started in 1950.

Writings on American History. Washington, annual from 1902 (except 1904–5 and 1941–47).

OTHER PUBLICATIONS

The Cambridge Economic History of the United States. vol. 1. 1996; vol. 2. 2000; vol. 3. 2000

Bacevich, Andrew J., *American Empire: The Realities and Consequences of US Diplomacy.* 2002

Brands, H. W., *American Dreams: The United States Since 1945.* 2010

Brogan, H., *The Longman History of the United States of America.* 2nd ed. 1999

Daalder, Ivo H. and Lindsay, James M., *America Unbound: the Bush Revolution in Foreign Policy*. 2003

Duncan, Russell and Goddard, Joe, *Contemporary America*. 3rd ed. 2009

Foner, E. and Garraty, J. A. (eds.) *The Reader's Companion to American History*. 1992

Grunwald, Michael, *The New New Deal: The Hidden Story of Change in the Obama Era*. 2012

Haass, Richard, *The Reluctant Sheriff: The United States After the Cold War*. 1998

Heilemann, John and Halperin, Mark, *Game Change: Obama and the Clintons, McCain and Palin, and the Race of a Lifetime*. 2010

Jenkins, Philip, *A History of the United States*. 4th ed. 2012

Jennings, F., *The Creation of America*. 2000

Jentleson, B. W. and Paterson, T. G. (eds.) *Encyclopedia of US Foreign Relations*. 4 vols. 1997

Kuklick, Bruce, *A Political History of the USA*. 2009

Little, Douglas, *American Orientalism: The United States and the Middle East since 1945*. 2002

Lord, C. L. and E. H., *Historical Atlas of the US*. Rev. ed. 1969

Merriam, L. A. and Oberly, J. (eds.) *United States History: an Annotated Bibliography*. 1995

Morison, S. E. with Commager, H. S., *The Growth of the American Republic*. 2 vols. 5th ed. 1962–63

Norton, M. B., *A People and a Nation: the History of the United States*. 9th ed. 2 vols. 2011

Peele, Gillian, Bailey, Christopher, J., Cain, Bruce and Peters, B. Guy, (eds.) *Developments in American Politics 6*. 2010

Pfucha, F. P., *Handbook for Research in American History: a Guide to Bibliographies and Other Reference Works*. 2nd ed. 1994

Prestowitz, Clyde, *Rogue Nation: American Unilateralism and the Failure of Good Intentions*. 2003

Remnick, David, *The Bridge: the Life and Rise of Barack Obama*. 2010

Who's Who in America. Annual

National library: The Library of Congress, Independence Ave. SE, Washington, D.C., 20540. *Librarian*: James H. Billington.

National statistical office: Bureau of the Census, Washington, D.C., 20233. *Acting Director*: Thomas L. Mesenbourg.

Website: http://www.census.gov

STATES AND TERRITORIES

GENERAL DETAILS

Against the names of the Governors, Lieut.-Governors and the Secretaries of State, (D.) stands for Democrat and (R.) for Republican.

See also Local Government on page 1345.

FURTHER READING

Official publications of the various states and insular possessions are listed in the *Monthly Check-List of State Publications*, issued by the Library of Congress since 1910.

The Book of the States. Biennial. 1953 ff.

State Government Finances. Annual. 1966 ff.

Bureau of the Census. *State and Metropolitan Area Data Book*. Irregular.— *County and City Data Book*. Irregular.

Hill, K. Q., *Democracy in the 50 States*. 1995

Alabama

KEY HISTORICAL EVENTS

The early European explorers were Spanish, but the first permanent European settlement was French, as part of French Louisiana after 1699. During the 17th and 18th centuries the British, Spanish and French all fought for control of the territory; it passed to Britain in 1763 and thence to the USA in 1783, except for a Spanish enclave on Mobile Bay, which lasted until 1813. Alabama was organized as a Territory in 1817 and was admitted to the Union as a state on 14 Dec. 1819.

The economy was then based on cotton, grown in white-owned plantations by black slave labour imported since 1719. Alabama seceded from the Union at the beginning of the Civil War (1861) and joined the Confederate States of America; its capital Montgomery became the Confederate capital. After the defeat of the Confederacy the state was readmitted to the Union in 1878. Attempts made during the reconstruction period to find a role for the newly freed black slaves—who made up about 50% of the population—largely failed, and when whites regained political control in the 1870s a strict policy of segregation came into force. At the same time Birmingham began to develop as an important centre of iron- and steel-making. Most of the state was still rural. In 1915 a boll-weevil epidemic attacked the cotton and forced diversification into other farm produce. More industries developed from the power schemes of the Tennessee Valley Authority in the 1930s. The black population remained mainly rural, poor and without political power, until the 1960s when confrontations on the issue of civil rights produced reforms.

TERRITORY AND POPULATION

Alabama is bounded in the north by Tennessee, east by Georgia, south by Florida and the Gulf of Mexico and west by Mississippi. Land area, 50,645 sq. miles (131,171 sq. km). Census population, 1 April 2010, was 4,779,736, an increase of 7·5% since 2000. July 2012 estimate, 4,822,023.

Population in five census years was:

	White	Black	American Indian	Asiatic	Total	Per sq. mile
1930	1,700,844	944,834	465	105	2,646,248	51·3
			All others			
1980	2,872,621	996,335	24,932		3,893,888	74·9
1990	2,975,797	1,020,705	44,085		4,040,587	79·6
2000	3,162,808	1,155,930	128,362		4,447,100	87·6
2010	3,275,394	1,251,311	253,031		4,779,736	94·4

Of the total population in 2010, 2,459,548 were female and 3,647,277 were 18 years old or older. In 2010 the Hispanic population was 185,602, up from 75,830 in 2000 (an increase of 144·8%).

The large cities (2010 census) were (with metropolitan areas in brackets): Birmingham, 212,237 (Birmingham–Hoover metropolitan area, 1,128,047); Montgomery (the capital), 205,764 (374,536); Mobile, 195,111 (412,992); Huntsville, 180,105 (417,593); Tuscaloosa, 90,468 (219,461).

SOCIAL STATISTICS

Births, 2010, 60,050 (12·6 per 1,000 population); deaths, 2010 (provisional), 48,022 (10·0). Infant deaths, 2009, 8·3 per 1,000 live births. 2009 (provisional): marriages, 37,300 (8·3 per 1,000 population); divorces and annulments, 20,200 (4·4).

CLIMATE

Birmingham, Jan. 46°F (7·8°C), July 80°F (26·7°C). Annual rainfall 54" (1,372 mm). Mobile, Jan. 52°F (11·1°C), July 82°F (27·8°C).

Annual rainfall 62" (1,575 mm). Montgomery, Jan. 49°F (9·4°C), July 81°F (27·2°C). Annual rainfall 52" (1,321 mm). The growing season ranges from 190 days (north) to 270 days (south). Alabama belongs to the Gulf Coast climate zone (*see* UNITED STATES: Climate).

CONSTITUTION AND GOVERNMENT

The current constitution dates from 1901; it has had 835 amendments (as at June 2010). The legislature consists of a Senate of 35 members and a House of Representatives of 105 members, all elected for four years. The Governor and Lieut.-Governor are elected for four years.

For the 113th Congress, which convened in Jan. 2013, Alabama sends seven members to the House of Representatives. It is represented in the Senate by Richard Shelby (D. 1987–94; R. 1994–2017) and Jeff Sessions (R. 1997–2015).

Applicants for registration must take an oath of allegiance to the United States and fill out an application showing evidence that they meet State voter registration requirements.

Montgomery is the capital.

RECENT ELECTIONS

In the 2012 presidential election Mitt Romney took Alabama with 60·7% of the vote (John McCain won it in 2008).

CURRENT GOVERNMENT

Governor: Robert Bentley (R.), 2011–15 (salary: $120,935·80, but opted not to receive a salary).

Lieut.-Governor: Kay Ivey (R.), 2011–15 (varies considerably depending on expenses and whether or not the Legislature is in session).

Secretary of State: Beth Chapman (R.), 2011–15 ($85,247·76).

Government Website: http://www.alabama.gov

ECONOMY

Per capita personal income (2010) was $33,516.

Budget

In 2011 total state revenue was $26,305m. Total expenditure was $28,061m. (education, $10,938m.; public welfare, $5,962m.; hospitals, $1,948m.; highways, $1,616m.; health, $609m.) Outstanding debt in 2011, $9,067m.

Performance

Gross Domestic Product by state was $150,330m. in 2011 (provisional), ranking Alabama 26th in the United States. In 2011 state real GDP growth was –0·8% (provisional).

ENERGY AND NATURAL RESOURCES

Electricity

In 2009 production was 143·3bn. kWh, of which 55·6bn. kWh was from coal.

Oil and Gas

In 2008 Alabama produced 7·5m. bbls of crude petroleum; marketed production of natural gas totalled 258bn. cu. ft.

Water

The total area covered by water is 1,775 sq. miles.

Minerals

Principal minerals, 2005–06 (in net 1,000 tons): limestone, 51,127; coal, 19,270; sand and gravel, 12,925. Value of non-fuel mineral production in 2009 was $1,020m.

Agriculture

The number of farms in 2009 was 48,500, covering 9·0m. acres; the average farm had 186 acres and was valued at $2,100 per acre (Jan. 2010).

Cash receipts from farm marketings, 2009: crops, $883m.; livestock and poultry products, $3,335m.; total, $4,218m. The net farm income in 2009 was $1,019m. Principal sources: broilers, cattle and calves, eggs, hogs, dairy products, greenhouse and nursery products, peanuts, soybeans, cotton and vegetables. In 2009 broilers accounted for the largest percentage of cash receipts from farm marketings; cattle and calves were second; and eggs third.

Forestry

Alabama had 22·69m. acres of forested land in 2007 of which 746,000 acres were national forest. Harvest volumes in 2004, 334·15m. cu. ft pine saw timber, 86·81m. cu. ft hardwood saw timber, 839·11m. cu. ft pulp wood and 8·83m. cu. ft poles. Total harvest, 2004, was 1,268·90m. cu. ft. Georgia is the only state with a larger annual harvest. The estimated delivered timber value of forest products in 2004 was $1·6bn.

Fisheries

Alabama had aquaculture sales worth $102·8m. in 2005, ranking it third behind Mississippi and Arkansas.

INDUSTRY

In 2007 the state's 4,928 manufacturing establishments had 272,000 employees, earning $11,352m. Total value added by manufacturing in 2009 was $36,184m. Alabama is both an industrial and service-oriented state. The chief industries are lumber and wood products, food and kindred products, textiles and apparel, non-electrical machinery, transportation equipment and primary metals.

Labour

In 2010, 1,869,000 were employed in non-agricultural sectors, of whom 386,800 were in government; 360,100 in trade, transportation and utilities; 236,100 in manufacturing; 214,400 in education and health services; 208,300 in professional and business services. In Dec. 2010 the unemployment rate was 9·1%.

COMMUNICATIONS

Roads

Total road length in 2007 was 97,323 miles, comprising 75,299 miles of rural road and 22,024 miles of urban road. Registered motor vehicles numbered 4,677,771.

Rail

In 2009 there were 3,271 miles of freight railroad (excluding trackage rights). There were 24 freight railroads operating in 2009. Rail traffic originating in Alabama in 2009 totalled 30·0m. tons and rail traffic terminating in the state came to 45·3m. tons.

Civil Aviation

In 2011 there were five primary airports (commercial service airports with more than 10,000 passenger boardings annually) with a combined total of 2,566,909 enplanements, up from 2,562,567 in 2010.

Shipping

There are 1,600 miles of navigable inland water and 50 miles of Gulf Coast. The only deep-water port is Mobile, with a large ocean-going trade; total tonnage (2009), 52·2m. tons. The Alabama State Docks also operates a system of ten inland docks; there are several privately run inland docks.

SOCIAL INSTITUTIONS

Justice

In Dec. 2008 the prison population totalled 30,508. Following the reinstatement of the death penalty by the US Supreme Court in 1976 death sentences have been awarded since 1983. There were six executions in 2011 but none in 2012.

In 41 counties the sale of alcoholic beverages is permitted, and in 26 counties it is prohibited; but it is permitted in eight cities within those 26 counties. Draught beverages are permitted in 22 counties.

Education

In 2007–08 there were 1,605 public elementary and secondary schools with 50,420 teachers and 744,865 students enrolled in grades K–12. Average public school teacher salary in 2008–09 was $48,906. Spending per student in fiscal year 2008 was $9,197.

As of fall 2005 there were 66 degree-granting institutions (39 public and 27 private) with 256,389 enrolled students. Enrolment for four-year courses at Auburn University totalled 23,333, the University of Alabama at Tuscaloosa 21,793, University of Alabama at Birmingham 16,572 and Troy University 14,957.

Health

In 2009 there were 108 community hospitals with 15,300 beds. A total of 666,000 patients were admitted during the year. In 2009 there were 10,265 active physicians (22 per 10,000 population).

Welfare

Medicare enrolment in July 2010 totalled 836,560. In fiscal year 2009 a total of 876,741 people in Alabama received Medicaid. In Dec. 2008 there were 952,511 Old-Age, Survivors, and Disability Insurance (OASDI) beneficiaries. A total of 43,347 people were receiving payments under Temporary Assistance for Needy Families (TANF) in Dec. 2008.

RELIGION

The main religious traditions in 2010 were: Evangelical Protestants, with 2,009,448 members; Mainline Protestants, 388,252; Black Protestants, 346,519; Catholics, 200,657; Latter-day Saints (Mormons), 36,820.

FURTHER READING

Alabama Official and Statistical Register. Quadrennial
Alabama County Data Book. Annual
Directory of Health Care Facilities.

Alaska

KEY HISTORICAL EVENTS

Discovered in 1741 by Vitus Bering, Alaska's first settlement, on Kodiak Island, was in 1784. The area known as Russian America with its capital (1806) at Sitka was ruled by a Russo-American fur company and vaguely claimed as a Russian colony. Alaska was purchased by the United States from Russia under the treaty of 30 March 1867 for $7·2m. Settlement was boosted by gold workers in the 1880s. In 1884 Alaska became a 'district' governed by the code of the state of Oregon. By Act of Congress approved 24 Aug. 1912 Alaska became an incorporated Territory; its first legislature in 1913 granted votes to women, seven years in advance of the Constitutional Amendment.

During the Second World War the Federal government acquired large areas for defence purposes and for the construction of the strategic Alaska Highway. In the 1950s oil was found. Alaska became the 49th state of the Union on 3 Jan. 1959. In the 1970s new oilfields were discovered and the Trans-Alaska pipeline was opened in 1977. The state obtained most of its income from petroleum by 1985.

Questions of land-use predominate; there are large areas with valuable mineral resources, other large areas held for the native peoples and some still held by the Federal government. The population increased by over 400% between 1940 and 1980.

TERRITORY AND POPULATION

Alaska is bounded north by the Beaufort Sea, west and south by the Pacific and east by Canada. The total area is 665,384 sq. miles (1,723,337 sq. km), making it the largest state of the USA; 570,641 sq. miles (1,477,953 sq. km) are land and 94,743 sq. miles (245,383 sq. km) are water. It is also the least densely populated state. The federal government owned 225,848,164 acres in 2010 (the largest area in any state), equivalent to 61·8% of the state area (the third highest percentage after Nevada and Utah). Census population, 1 April 2010, was 710,231, an increase of 13·3% since 2000. July 2012 estimate, 731,449.

Population in five census years was:

	White	Black	All others	Total	Per sq. mile
1950	92,808	—	35,835	128,643	0·23
1980	309,728	13,643	78,480	401,851	1·00
1990	415,492	22,451	112,100	550,043	1·00
2000	434,534	21,787	170,611	626,932	1·10
2010	473,576	23,263	213,392	710,231	1·24

Of the total population in 2010, 369,628 were male and 522,853 were 18 years old or older. Alaska's Hispanic population was 39,249 in 2010, up from 25,852 in 2000. As of 2010, 14·8% of Alaska's population was identified as Alaska Native or American Indian.

The largest county equivalent and city is in the borough of Anchorage, which had a 2010 census population of 291,826. Census populations of the other 14 county equivalents, 2010: Fairbanks North Star, 97,581; Matanuska-Susitna, 88,995; Kenai Peninsula, 55,400; Juneau (the capital), 31,275; Bethel, 17,013; Kodiak Island, 13,592; Ketchikan Gateway, 13,477; Valdez-Cordova, 9,636; Nome, 9,492; North Slope, 9,430; Sitka, 8,881; Northwest Arctic, 7,523; Wade Hampton, 7,459; Southeast Fairbanks, 7,029. Largest incorporated places in 2010 were: Anchorage, 291,826; Fairbanks, 31,535; Juneau, 31,275; Sitka, 8,881; Ketchikan, 8,050; Wasilla, 7,831; Kenai, 7,100; Kodiak, 6,130; Bethel, 6,080; Palmer, 5,937.

SOCIAL STATISTICS

Births, 2010, 11,471 (16·2 per 1,000 population); deaths, 2010 (provisional), 3,727 (5·2—the lowest rate in any US state). Infant mortality, 2009, 6·8 per 1,000 live births. 2009 (provisional): marriages, 5,500 (7·8 per 1,000 population); divorces and annulments, 3,300 (4·4).

CLIMATE

Anchorage, Jan. 12°F (−11·1°C), July 57°F (13·9°C). Annual rainfall 15" (371 mm). Fairbanks, Jan. −11°F (−23·9°C), July 60°F (15·6°C). Annual rainfall 12" (300 mm). Sitka, Jan. 33°F (0·6°C), July 55°F (12·8°C). Annual rainfall 87" (2,175 mm). Alaska belongs to the Pacific Coast climate zone (*see* UNITED STATES: Climate).

CONSTITUTION AND GOVERNMENT

The state has the right to select 103·55m. acres of vacant and unappropriated public lands in order to establish 'a tax basis'; it can open these lands to prospectors for minerals, and the state is to derive the principal advantage in all gains resulting from the discovery of minerals. In addition, certain federally administered lands reserved for conservation of fisheries and wild life have been transferred to the state. Special provision is made for federal control of land for defence in areas of high strategic importance.

The constitution of Alaska was adopted by public vote, 24 April 1956. The state legislature consists of a Senate of 20 members (elected for four years) and a House of Representatives of 40 members (elected for two years).

For the 113th Congress, which convened in Jan. 2013, Alaska sends one member to the House of Representatives. It is represented in the Senate by Lisa Murkowski (R. 2002–17) and

Mark Begich (D. 2009–15). The franchise may be exercised by all citizens over 18.

The capital is Juneau.

RECENT ELECTIONS

In the 2012 presidential election Mitt Romney took Alaska with 55·3% of the vote (John McCain won it in 2008).

CURRENT GOVERNMENT

Governor: Sean R. Parnell (R.), Dec. 2010–Dec. 2014 (salary: $145,000).

Lieut.-Governor: Mead Treadwell (R.), Dec. 2010–Dec. 2014 ($115,000).

Government Website: http://www.alaska.gov

ECONOMY

Per capita personal income (2010) was $44,205.

Budget

In 2011 total state revenue was $14,921m. Total expenditure was $11,320m. (education, $2,475m.; public welfare, $1,918m.; highways, $1,412m.; government administration, $586m.; natural resources, $391m.) Outstanding debt in 2011, $6,418m.

Performance

2011 Gross Domestic Product by state was $44,702m. (provisional), ranking Alaska 44th in the United States. In 2011 state real GDP growth was 2·5% (provisional).

ENERGY AND NATURAL RESOURCES

Oil and Gas

Alaska ranks fourth among the leading oil producers in the USA, although production is steadily declining and is now less than a third of what it was in the late 1980s. Commercial production of crude petroleum began in 1959 and by 1961 had become the most important mineral by value. Production in 2008 totalled 250m. bbls (value in 2007, $16,788m.). Proven reserves in 2006 were 3,879m. bbls. Oil comes mainly from Prudhoe Bay, the Kuparuk River field and several Cook Inlet fields. Oil from the Prudhoe Bay Arctic field is now carried by the Trans-Alaska pipeline to Prince William Sound on the south coast, where there is a tanker terminal at Valdez.

Natural gas marketed production, 2008, 398bn. cu. ft, with (2006) reserves of 10,245bn. cu. ft.

Water

The total area covered by water is 94,743 sq. miles (the most of any state), of which 19,304 sq. miles are inland.

Minerals

Estimated value of production, 2003, in $1,000: zinc, 486,916; gold, 191,986; industrial minerals (including sand, gravel and building stone), 100,000; silver, 90,773; lead, 70,094; coal, 37,975; peat, 175. Total 2003 value, $980·3m. Value of non-fuel mineral production in 2009 was $2,620m.

Agriculture

In some parts of the state the climate during the brief spring and summer (about 100 days in major areas and 152 days in the southeastern coastal area) is suitable for agricultural operations, thanks to the long hours of sunlight, but Alaska is a food-importing area. In 2007 there were 686 farms covering a total of 882,000 acres. The average farm had 1,285 acres in 2007 and was valued at $391 per acre.

Farm income, 2006: crops, $25m.; livestock and products, $39m. The net farm income in 2006 was $20m. Principal sources: greenhouse products, hay, dairy products and potatoes.

In 2007 there were 14,823 cattle and calves, 522 sheep and lambs, 757 hogs and pigs, and 3,600 laying hens.

Forestry

Of the 126·87m. forested acres of Alaska (the highest acreage of any state), 10·46m. acres are national forest land. The interior forest covers 115m. acres; more than 13m. acres are considered commercial forest, of which 3·4m. acres are in designated parks or wilderness and unavailable for harvest. The coastal rain forests provide the bulk of commercial timber volume; of their 13·6m. acres, 7·6m. acres support commercial stands, of which 1·9m. acres are in parks or wilderness and unavailable for harvest. In 2006 timber removals were 66m. cu. ft (59m. cu. ft softwoods and 7m. cu. ft hardwoods).

In 2008 there were 704 designated wilderness areas throughout the USA, covering a total of 107·4m. acres (43·5m. ha.). More than 53% of the system is in Alaska (57·4m. acres or 23·2m. ha.).

Fisheries

In 2009 commercial fishing landed 4,064m. lb of fish and shellfish at a value of $1·3bn. Alaska is by the far the most important state for fishery landings, accounting for 50·6% of the total by weight in 2009. The most important species are salmon, crab, herring, halibut and pollock.

INDUSTRY

In 2007 the state's 544 manufacturing establishments had 13,000 employees, earning $490m. Total value added by manufacturing in 2009 was $1,895m. The largest manufacturing sectors are wood processing, seafood products and printing and publishing.

Labour

Total non-agricultural employment, 2010: 324,400. Employees by branch, 2010: government, 85,200; trade, transportation and utilities, 62,800; education and health services, 41,700; leisure and hospitality, 31,500; professional and business services, 26,200. The unemployment rate in Dec. 2010 was 7·9%.

COMMUNICATIONS

Roads

Alaska's highway and road system, 2007, totalled 14,438 miles comprising 2,357 miles of urban road and 12,081 miles of rural road. Registered motor vehicles numbered 680,141.

The Alaska Highway extends 1,523 miles from Dawson Creek, British Columbia, to Fairbanks, Alaska. It was built by the US Army in 1942, at a cost of $138m. The greater portion of it, because it lies in Canada, is maintained by Canada.

Rail

There is a railroad from Skagway to the town of Whitehorse, the White Pass and Yukon route, in the Canadian Yukon region (this service operates seasonally, although only the section between Skagway and Carcross is in service). The government-owned Alaska Railroad runs from Seward to Fairbanks. This is a freight service with only occasional passenger use. In 2009 there were 506 miles of freight railroad.

Civil Aviation

Alaska's largest international airports are Anchorage and Fairbanks. There were 30 primary airports in 2011 (commercial service airports with more than 10,000 passenger boardings annually), more than in any other state. General aviation aircraft in the state per 1,000 population is about ten times the US average. Anchorage handled 2,543,105 tonnes of freight in 2011 (including transit freight), ranking it second in the USA behind Memphis for cargo handled (and fourth in the world). In 2008 Alaska Airlines carried 16·8m. passengers and flew to 59 destinations. There were 4,560,483 passenger enplanements statewide in 2007.

Shipping

Regular shipping services to and from the USA are furnished by two steamship and several barge lines operating out of Seattle and other Pacific coast ports. A Canadian company also furnishes a regular service from Vancouver, BC. Anchorage is the main port.

A 1,435 nautical-mile ferry system for motor cars and passengers (the 'Alaska Marine Highway') operates from Bellingham, Washington and Prince Rupert (British Columbia) to Juneau, Haines (for access to the Alaska Highway) and Skagway. A second system extends throughout the south-central region of Alaska linking the Cook Inlet area with Kodiak Island and Prince William Sound.

SOCIAL INSTITUTIONS

Justice
The death penalty was abolished in Alaska in 1957. In Dec. 2008 the jail and prison population totalled 5,014.

Education
In 2007–08 there were 501 elementary and secondary schools with 131,029 pupils and 7,613 teachers; total expenditure on public elementary and secondary education in 2006–07 was $1,940m. Average teacher salary in 2008–09 was $58,916.

There are seven degree-granting institutions (five public and two private). The University of Alaska (founded in 1922) had 16,412 students at fall 2005 and comprised eight teaching units at the main campus in Anchorage.

Health
In 2009 there were 22 community hospitals with 1,500 beds. A total of 57,000 patients were admitted during the year. In 2009 there were 1,574 active physicians (23 per 10,000 population).

Welfare
Medicare enrolment in July 2010 totalled 54,300. In fiscal year 2009 a total of 118,694 people in Alaska received Medicaid. In Dec. 2008 there were 71,145 Old-Age, Survivors, and Disability Insurance (OASDI) beneficiaries. A total of 7,546 people were receiving payments under Temporary Assistance for Needy Families (TANF) in Dec. 2008.

RELIGION

The main religious traditions in 2010 were: Evangelical Protestants, with 100,960 members; Catholics, 50,866; Mainline Protestants, 32,550; Latter-day Saints (Mormons), 32,671; Orthodox Christians, 13,480.

FURTHER READING

Statistical Information: Department of Commerce and Economic Development, Economic Analysis Section, POB 110804, Juneau 99811. Publishes *The Alaska Economy Performance Report*. *Annual Financial Report.*

Naske, C.-M. and Slotnick, H. E., *Alaska: a History of the 49th State.* 2nd ed. 1995

State library: POB 110571, Juneau, Alaska 99811-0571.

Arizona

KEY HISTORICAL EVENTS

Spaniards looking for sources of gold or silver entered Arizona in the 16th century finding there American natives, including Tohono O'odham, Navajo, Hopi and Apache. The first Spanish Catholic mission was founded in the early 1690s by Father Eusebio Kino. Settlements were made in 1752 and a Spanish army headquarters was set up at Tucson in 1776. The area was governed by Mexico after the collapse of Spanish colonial power. Mexico ceded it to the USA in the Treaty of Guadalupe Hidalgo after the Mexican-American war (1848). Arizona was then part of

New Mexico; the Gadsen Purchase (of land south of the Gila River) was added to it in 1853. The whole was organized as the Arizona Territory on 24 Feb. 1863.

Miners and ranchers began settling in the 1850s. Conflicts between Indian and immigrant populations intensified when troops were withdrawn to serve in the Civil War. The Navajo surrendered in 1865, but the Apache continued to fight, under Geronimo and other leaders, until 1886. Arizona was admitted to the Union as the 48th state in 1912.

Large areas of the state have been retained as Indian reservations and as parks to protect the exceptional desert and mountain landscape. In recent years this landscape and the Indian traditions have been used to attract tourists.

TERRITORY AND POPULATION

Arizona is bounded north by Utah, east by New Mexico, south by Mexico, west by California and Nevada. Land area, 113,594 sq. miles (294,207 sq. km). Of the total area in 2009, 28% was Indian Reservation, 18% was in individual or corporate ownership, 17% was held by the US Bureau of Land Management, 15% by the US Forest Service, 13% by the State and 10% by others. Census population, 1 April 2010, was 6,392,017, an increase of 24·6% since 2000. The rate of Arizona's population increase during the 2000s was the second fastest in the USA (behind Nevada), at 24·6%. July 2012 estimate, 6,553,255.

Population in five census years:

	White	Black	American Indian/Alaska Native	Chinese	Japanese	Total	Per sq. mile
1910	171,468	2,009	29,201	1,305	371	204,354	1·8
			All others				
1980	2,260,288	74,159	162,854	383,768		2,718,215	23·9
1990	2,963,186	110,524	203,527	387,991		3,665,228	32·3
2000	3,873,611	158,873	255,879	842,269		5,130,632	45·2
2010	4,667,121	259,008	296,529	1,169,359		6,392,017	56·3

Of the total population in 2010, 3,216,194 were female and 4,763,003 were 18 years old or older. Arizona's Hispanic population was 1,895,149 in 2010 (29·6%) up from 1,295,617 in 2000 (an increase of 46·3%).

The large cities (2010 census) were: Phoenix (the capital) 1,445,632; Tucson, 520,116; Mesa, 439,041; Chandler, 236,123; Glendale, 226,721; Scottsdale, 217,385; Gilbert, 208,453; Tempe, 161,719; Peoria, 154,065; Surprise, 117,517. The Phoenix–Mesa–Glendale metropolitan area had a 2010 census population of 4,192,887.

SOCIAL STATISTICS

Births, 2010, 87,477 (13·7 per 1,000 population); deaths, 2010 (provisional), 46,765 (7·3). Infant mortality, 2009, 6·0 per 1,000 live births. 2009 (provisional) marriages, 35,300 (5·4); divorces and annulments, 23,100 (3·5).

CLIMATE

Phoenix, Jan. 53·6°F (12°C), July 93·5°F (34°C). Annual rainfall 7·66" (194 mm). Yuma, Jan. 56·5°F (13·6°C), July 93·7°F (34·3°C). Annual rainfall 3·17" (80 mm). Flagstaff, Jan. 28·7°F (−1·8°C), July 66·3°F (19·1°C). Annual rainfall 22·8" (579 mm). Arizona belongs to the Mountain States climate zone (*see* UNITED STATES: Climate).

CONSTITUTION AND GOVERNMENT

The state constitution (1911, with 146 amendments) placed the government under direct control of the people through the initiative, referendum and the recall provisions. The state Senate consists of 30 members, and the House of Representatives consists of 60, all elected for two years.

For the 113th Congress, which convened in Jan. 2013, Arizona sends nine members to the House of Representatives. It is represented in the Senate by John McCain (R. 1987–2017) and Jeff Flake (R. 2013–19).

The state capital is Phoenix. The state is divided into 15 counties.

RECENT ELECTIONS

In the 2012 presidential election Mitt Romney took Arizona with 54·2% of the vote (John McCain won it in 2008).

CURRENT GOVERNMENT

Governor: Janice K. Brewer (R.), 2011–15 (salary: $95,000).
 Secretary of State: Ken Bennett (R.), 2011–15 ($70,000).

Government Website: http://az.gov

ECONOMY

Per capita personal income (2010) was $34,553.

Budget

In 2011 total state revenue was $36,611m. Total expenditure was $32,875m. (public welfare, $9,511m.; education, $9,122m.; highways, $2,038m.; health, $1,822m.; correction, $906m.) Outstanding debt in 2011, $14,163m.

Performance

Gross Domestic Product by state was $227,098m. in 2011 (provisional), ranking Arizona 20th in the United States. In 2011 state real GDP growth was 1·5% (provisional).

ENERGY AND NATURAL RESOURCES

Primary energy sources are coal (35·2%), nuclear (24·5%), gas (19·5%) and hydroelectric (18·9%).

Electricity

In 2009 production was 112·0bn. kWh, of which 39·7bn. kWh was from coal.

Oil and Gas

In 2008 oil production totalled 52,000 bbls; marketed production of natural gas totalled 523m. cu. ft.

Water

The total area covered by water is 396 sq. miles.

Minerals

The mining industry historically has been and continues to be a significant part of the economy. By value the most important mineral produced is copper. Production in 2007 was 731,000 tonnes. Most of the state's silver and gold are recovered from copper ore. Other minerals include sand and gravel, molybdenum, coal and gemstones. Value of non-fuel mineral production in 2009 was $5,180m.

Agriculture

Arizona, despite its dry climate, is well suited for agriculture along the water-courses and where irrigation is practised on a large scale from great reservoirs constructed by the USA as well as by the state government and private interests. Irrigated area in 2007 was 876,158 acres. The wide pasture lands are favourable for the rearing of cattle and sheep, but numbers are either stationary or declining compared with 1920.

In 2007 Arizona contained 15,637 farms and ranches and the total farm and pastoral area was 26·1m. acres; in 2007 there were 1,205,425 acres of crop land. In 2007 the average farm was 1,670 acres (the fifth largest average size in the USA) and was valued at $748 per acre. Farming is highly commercialized and mechanized and concentrated largely on cotton picked by machines.

Area under cotton in 2009: upland cotton, 145,000 acres (440,000 bales harvested); American Pima cotton, 1,700 acres (4,000 bales harvested).

In 2009 the cash receipts from crops were $1,766m., and from livestock and products $1,178m. The net farm income in 2009 was $203m. Most important cereals are wheat, corn and barley; most important crops include lettuce, cotton, citrus fruit, broccoli, spinach, cauliflower, melons, onions, potatoes and carrots. In Dec. 2009 there were 930,000 cattle, 167,000 hogs, 160,000 sheep and 43,000 goats.

Forestry

The state had a forested area of 18,671,000 acres in 2007, of which 7,663,000 acres were national forest.

INDUSTRY

In 2007 the state's 5,074 manufacturing establishments had 172,000 employees, earning $8,774m. Total value added by manufacturing in 2009 was $23,938m.

Labour

In 2010 total non-agricultural employment was 2,377,000. Employees by branch, 2010 (in 1,000): trade, transportation and utilities, 468; government, 417; education and health services, 344; professional and business services, 339; leisure and hospitality, 253. The unemployment rate in Dec. 2010 was 9·6%.

COMMUNICATIONS

Roads

In 2007 there were 60,593 miles of roads comprising 22,918 miles of urban road and 37,675 miles of rural road. There were 4,372,035 registered vehicles.

Rail

In 2009 there were 1,679 miles of freight railroad (excluding trackage rights). Rail traffic originating in Arizona in 2009 totalled 2·8m. tons and rail traffic terminating in the state came to 24·7m. tons. A light rail system opened in Phoenix in 2008.

Civil Aviation

In 2011 Arizona had nine primary airports (commercial service airports with more than 10,000 passenger boardings annually) with a combined total of 22,735,131 enplanements, up from 21,834,569 in 2010. The busiest airport, Phoenix Sky Harbor International, had 19,750,306 enplanements in 2011 (ranking it ninth in the USA).

SOCIAL INSTITUTIONS

Justice

A 'right-to-work' amendment to the constitution, adopted 5 Nov. 1946, makes illegal any concessions to trade-union demands for a 'closed shop'.

In Dec. 2008 the prison population totalled 39,589. Chain gangs were reintroduced into prisons in 1995. The death penalty is authorized. There were six executions in 2012 (four in 2011).

Education

School attendance is compulsory between the ages of six and 16. There were 2,135 public elementary and secondary schools in 2007–08 of which 457 were charter schools. Public school teachers numbered 54,032 in fall 2007; K-12 enrolment totalled 1,087,447 students. There were 64,910 students enrolled at 360 private schools with 4,220 teachers in fall 2007. In 2006–07 the total expenditure on public elementary and secondary education was $9,539m. The state maintains three universities: the University of Arizona (Tucson) with 36,337 students in 2006–07; Northern Arizona University (Flagstaff) with 20,562 in 2006–07; Arizona State University (four campuses) with 63,278 in 2005–06.

Health

In 2009 the state had 72 community hospitals with 13,500 beds. A total of 705,000 patients were admitted during the year. In 2009 there were 14,051 active physicians (21 per 10,000 population).

Welfare

Medicare enrolment in July 2010 totalled 919,910. In fiscal year 2009 a total of 1,588,257 people in Arizona received Medicaid. Old-age assistance (maximum depending on the programme) is given to needy citizens 65 years of age or older through the federal supplemental security income (SSI) programme. In Dec. 2008 SSI payments went to 13,807 aged, and 89,436 disabled and blind (average of $487·57 each). In Dec. 2008 there were 986,539 Old-Age, Survivors, and Disability Insurance (OASDI) beneficiaries. A total of 82,639 people were receiving payments under Temporary Assistance for Needy Families (TANF) in Dec. 2008.

RELIGION

The principal religious traditions in 2010 were: Catholics, with 930,702 members; Evangelical Protestants, 762,376; Latter-day Saints (Mormons), 394,844; Mainline Protestants, 171,792; Hindus, 32,887.

CULTURE

Tourism

In 2011 overseas visitors to Arizona—excluding those from Canada and Mexico—numbered 864,000, up from 765,000 in 2010 and 563,000 in 2006.

FURTHER READING

Statistical information: College of Business and Public Administration, Univ. of Arizona, Tucson 85721. Publishes *Arizona Statistical Abstract.*

Arizona Commission of Indian Affairs. *2007–2008 Tribal-State Resource Directory.* 2007

Arizona Department of Health Services, Center for Health Statistics. *Arizona Health Status and Vital Statistics, 2010.* Online only

Arizona Historical Society. *1999/2000 Official Directory, Arizona Historical Museums and Related Support Organizations.* 1999

August, Jack L., *Vision in the Desert: Carl Hayden and the Hydropolitics in the American Southwest.* 1999

Leavengood, Betty, *Lives Shaped by Landscape: Grand Canyon Women.* 1999

Office of the Secretary of State. *Arizona Blue Book, 2011–12.* 2011

Shillingberg, William B., *Tombstone, A. T.: A History of Early Mining, Milling and Mayhem.* 1999

Arizona State Library, Archives and Public Records (ASLAPR): 1700 West Washington, Suite 200, Phoenix. *Website:* http://www.lib.az.us

---

Arkansas

KEY HISTORICAL EVENTS

In the 16th and 17th centuries French and Spanish explorers encountered tribes of Chaddo, Osage and Quapaw. The first European settlement was French, at Arkansas Post in 1686, and the area became part of French Louisiana. The USA bought Arkansas from France as part of the Louisiana Purchase in 1803; it was organized as a Territory in 1819 and entered the Union on 15 June 1836 as the 25th state.

The eastern plains by the Mississippi were settled by white plantation-owners who grew cotton with black slave labour. The rest of the state attracted a scattered population of small farmers. The plantations were the centre of political power. Arkansas seceded from the Union in 1861 and joined the Confederate States of America. At that time the slave population was about 25% of the total.

In 1868 the state was readmitted to the Union. Attempts to integrate the black population into state life achieved little, and a policy of segregation was rigidly adhered to until the 1950s. In 1957 federal authorities ordered that high school segregation must end. The state governor called on the state militia to prevent desegregation; there was rioting, and federal troops entered Little Rock, the capital, to restore order. It was another ten years before school segregation finally ended.

The main industrial development followed the discovery of large reserves of bauxite.

TERRITORY AND POPULATION

Arkansas is bounded north by Missouri, east by Tennessee and Mississippi, south by Louisiana, southwest by Texas and west by Oklahoma. Land area, 52,035 sq. miles (134,771 sq. km). Census population, 1 April 2010, was 2,915,918, an increase of 9·1% since 2000. July 2012 estimate, 2,949,131.

Population in five census years was:

	White	Black	American Indian	Asiatic	Total	Per sq. mile
1910	1,131,026	442,891	460	472	1,574,449	30·0
			All others			
1980	1,890,332	373,768	22,335		2,286,435	43·9
1990	1,944,744	373,912	32,069		2,350,725	45·1
2000	2,138,598	418,950	115,852		2,673,400	51·3
2010	2,245,229	449,895	220,794		2,915,918	56·0

Of the total population in 2010, 1,484,281 were female and 2,204,443 were 18 years old or older. In 2010 the Hispanic population of Arkansas was 186,050, up from 86,866 in 2000, an increase of 114·2%.

Little Rock (capital) had a population of 193,524 in 2010; Fort Smith, 86,209; Fayetteville, 73,580; Springdale, 69,797; Jonesboro, 67,263; North Little Rock, 62,304; Conway, 58,908; Rogers, 55,964. The population of the largest metropolitan statistical areas in 2010 was: Little Rock–North Little Rock–Conway, 699,757; Fayetteville–Springdale–Rogers, 463,204; Fort Smith, 298,592; Texarkana, 136,027; Jonesboro, 121,026; Pine Bluff, 100,258.

SOCIAL STATISTICS

Births, 2010, were 38,540 (13·2 per 1,000 population); deaths, 2010 (provisional), 28,913 (9·9). Infant mortality, 2009, 7·7 per 1,000 live births. 2009 (provisional): marriages, 31,600 (10·7 per 1,000 population); divorces and annulments, 16,300 (5·7).

CLIMATE

Little Rock, Jan. 39·9°F, July 84°F. Annual rainfall 52·4". Arkansas belongs to the Gulf Coast climate zone (*see* UNITED STATES: Climate).

CONSTITUTION AND GOVERNMENT

The General Assembly consists of a Senate of 35 members elected for four years, partially renewed every two years, and a House of Representatives of 100 members elected for two years. The sessions are biennial and usually limited to 60 days. The Governor and Lieut.-Governor are elected for four years.

For the 113th Congress, which convened in Jan. 2013, Arkansas sends four members to the House of Representatives. It is represented in the Senate by Mark Pryor (D. 2003–15) and John Boozman (R. 2011–17).

The state is divided into 75 counties; the capital is Little Rock.

RECENT ELECTIONS

In the 2012 presidential election Mitt Romney took Arkansas with 60·6% of the vote (John McCain won it in 2008).

CURRENT GOVERNMENT

Governor: Mike Beebe (D.), 2011–15 (salary: $86,890).
  Lieut.-Governor: Mark Darr (R.), 2011–15 ($41,896).
  Secretary of State: Mark Martin (R.), 2011–15 ($54,305).

Government Website: http://www.ar.gov

ECONOMY

Per capita personal income (2010) was $32,678.

Budget

In 2011 total revenue was $22,808m. Total expenditure was $18,862m. (education, $7,508m.; public welfare, $4,494m.; highways, $1,081m.; hospitals, $857m.; government administration, $576m.) Outstanding debt in 2011, $3,749m.

Performance

2011 Gross Domestic Product by state was $91,496m. (provisional), ranking Arkansas 34th in the United States. In 2011 state real GDP growth was 0·3% (provisional).

ENERGY AND NATURAL RESOURCES

Oil and Gas

2008 production of crude oil was 6·1m. bbls; natural gas, 447bn. cu. ft.

Water

The total area covered by water is 1,143 sq. miles.

Minerals

Employment in mining, quarrying, and oil and gas extraction totalled 6,364 in March 2007. Crushed stone was the leading mineral commodity produced, in terms of value, followed by bromine. Value of domestic non-fuel mineral production in 2009 was $636m.

Agriculture

In 2007, 49,346 farms had a total area of 13·9m. acres; average farm was 281 acres and was valued at $2,343 per acre. 7·37m. acres were harvested cropland. Arkansas ranked first in the acreage and production of rice in 2007 (48·4% of US total production), second in the production of broilers (1,172m. birds) and third in turkeys (29·2m. birds).

Farm income, 2006: crops, $2,397m.; livestock and products, $3,767m. The net farm income in 2006 was $1,951m.

Forestry

In 2007 the state had a forested area of 18,830,000 acres, of which 2,546,000 acres were national forest.

Fisheries

Arkansas had aquaculture sales worth $110·5m. in 2005, ranking it second behind Mississippi.

INDUSTRY

In 2007 the state's 3,088 manufacturing establishments had 185,000 employees, earning $6,518m. Total value added by manufacturing in 2009 was $19,208m.

Labour

Total non-agricultural employment, 2010: 1,613,200. Employees by branch, 2010 (in 1,000): trade, transportation and utilities, 234; government, 218; education and health services, 166; manufacturing, 160; professional and business services, 118. The unemployment rate in Dec. 2010 was 7·9%.

COMMUNICATIONS

Roads

Total road mileage (2007), 99,558 miles—urban, 11,966; rural, 87,592. There were 2,010,301 registered motor vehicles.

Rail

In 2009 there were 2,780 miles of freight railroad (excluding trackage rights). Rail traffic originating in Arkansas in 2009 totalled 14·4m. tons and rail traffic terminating in the state came to 25·1m. tons.

Civil Aviation

There were four primary airports—commercial service airports with more than 10,000 passenger boardings annually—in 2011 with a combined total of 1,715,357 enplanements, down from 1,757,190 in 2010.

Shipping

There are about 1,000 miles of navigable rivers, including the Mississippi, Arkansas, Red, White and Ouachita Rivers. The Arkansas River/Kerr-McClellan Channel flows diagonally eastward across the state and gives access to the sea via the Mississippi River.

SOCIAL INSTITUTIONS

Justice

In Dec. 2008 there were 14,716 federal and state prisoners. There was one execution in 2004 and one in 2005, but the death penalty was suspended in 2006.

Education

In 2007–08 there were 1,121 public elementary and secondary schools with 479,016 enrolled pupils and 33,882 teachers. Total expenditure on public elementary and secondary education in 2006–07 was $4,644m.; average teacher salary was $47,472 in 2008–09. Spending per student in fiscal year 2008 was $8,677.

In 2008–09 there were 33 public and 17 private degree-granting institutions. Total student enrolment in 2005 was 140,700 (127,949 public and 12,751 private). Arkansas State University (founded in 1909) includes the Jonesboro, Beebe, Mountain Home, Newport, Heber Springs, Marked Tree and Searcy campuses; total student enrolment was 16,937 in fall 2005.

Health

In 2009 there were 86 community hospitals with 9,600 beds. A total of 380,000 patients were admitted during the year. In 2009 there were 5,902 active physicians (20 per 10,000 population).

Welfare

Medicare enrolment in July 2010 totalled 526,568. In fiscal year 2009 a total of 825,121 people in Arkansas received Medicaid. In Dec. 2008 there were 602,017 Old-Age, Survivors, and Disability Insurance (OASDI) beneficiaries. A total of 19,791 people were receiving payments under Temporary Assistance for Needy Families (TANF) in Dec. 2008.

RELIGION

The principal religious traditions in 2010 were: Evangelical Protestants, with 1,136,611 members; Mainline Protestants, 205,332; Catholics, 122,662; Black Protestants, 107,202; Latter-day Saints (Mormons), 29,645.

FURTHER READING

Statistical information: Arkansas Institute for Economic Advancement, Univ. of Arkansas at Little Rock, Little Rock 72204. Publishes *Arkansas State and County Economic Data.*
Agricultural Statistics for Arkansas. Annual
Current Employment Developments. Monthly
Statistical Summary for the Public Schools of Arkansas. Annual

---

California

KEY HISTORICAL EVENTS

There were many small Indian tribes, but no central power, when the area was discovered in 1542 by the Spanish navigator Juan Cabrillo. The Spaniards did not begin to establish missions until the 18th century, when the Franciscan friar Junipero Serra settled at San Diego in 1769. The missions became farming and ranching villages with large Indian populations. When the Spanish empire collapsed in 1821, the area was governed from newly independent Mexico.

The first wagon-train of American settlers arrived from Missouri in 1841. In 1846, during the war between Mexico and the USA, Americans in California proclaimed it to be part of the

USA. The territory was ceded by Mexico on 2 Feb. 1848 and became the 31st state of the Union on 9 Sept. 1850.

Gold was discovered in 1848–49 and there was an immediate influx of population. The state remained isolated, however, until the development of railways in the 1860s. From then on the population doubled on average every 20 years. The sunny climate attracted fruit-growers, market-gardeners and wine producers. In the early 20th century the bright lights and cheap labour attracted film-makers to Hollywood, Los Angeles.

Southern California remained mainly agricultural with an Indian or Spanish-speaking labour force until after the Second World War. Now more than 90% of the population is urban, with the manufacturing emphasis on hi-technology equipment, much of it for the aerospace, computer and office equipment industries.

TERRITORY AND POPULATION

Land area, 155,779 sq. miles (403,466 sq. km). California is the third largest US state behind Alaska and Texas. The federal government owned 47,797,533 acres in 2010, equivalent to 47·7% of the state area. Census population, 1 April 2010, was 37,253,956, an increase of 10·0% since 2000. The growth rate reflects continued high though somewhat reduced natural increase (excess of births over deaths) as well as substantial net immigration. California's population is 12·1% of the total US population. July 2012 estimate, 38,041,430.

Population in five census years was:

	White	Black	Japanese	Chinese	Total (incl. all others)	Per sq. mile
1910	2,259,672	21,645	41,356	36,248	2,377,549	15·2
1960	14,455,230	883,861	157,317	95,600	15,717,204	100·8

	White	Black	Asian/other	Hispanic	Total	Per sq. mile
1990	20,524,327	2,208,801	7,026,893	7,687,938	29,760,021	190·8
2000	20,170,059	2,263,882	11,437,707	10,966,556	33,871,648	217·2
2010	21,453,934	2,299,072	13,500,950	14,013,719	37,253,956	239·1

Of the total population in 2010, 18,736,126 were female and 27,958,916 were 18 years old or older.

In addition to having the highest population of any state in the USA, California has the largest Hispanic population of any state in terms of numbers and the third largest in terms of percentage of population. In 2010 there were 14,013,719 Hispanics living in California (37·6% of the overall population), up from 10,966,556 in 2000, the largest numeric rise of any state over the same period. By 2020 Hispanics are projected to form a majority.

The 50 largest cities (2012 population estimates) are:

City	Population	City	Population
Los Angeles	3,825,297	Rancho Cucamonga	169,498
San Diego	1,321,315	Oceanside	169,319
San Jose	971,372	Santa Rosa	168,841
San Francisco	812,538	Ontario	166,134
Fresno	505,009	Lancaster	157,826
Sacramento (capital)	470,956	Elk Grove	155,937
Long Beach	464,662	Corona	154,520
Oakland	395,341	Palmdale	153,708
Bakersfield	354,480	Salinas	152,401
Anaheim	343,793	Pomona	149,950
Santa Ana	327,731	Hayward	147,113
Riverside	308,511	Torrance	146,115
Stockton	295,707	Escondido	146,064
Chula Vista	249,382	Sunnyvale	142,896
Irvine	223,729	Pasadena	139,222
Fremont	217,700	Orange	138,010
San Bernardino	211,674	Fullerton	137,481
Modesto	203,085	Thousand Oaks	128,031
Oxnard	200,390	Visalia	126,864
Fontana	199,898	Simi Valley	125,317
Moreno Valley	196,495	Concord	123,206
Glendale	192,654	Roseville	122,060
Huntington Beach	192,524	Victorville	119,059
Santa Clarita	177,445	Santa Clara	118,813
Garden Grove	172,648	Vallejo	115,928

Largest metropolitan areas (2010 census): Los Angeles–Long Beach–Santa Ana, 12,828,837; San Francisco–Oakland–Fremont, 4,335,391; Riverside–San Bernardino–Ontario, 4,224,851; San Diego–Carlsbad–San Marcos, 3,095,313; Sacramento–Arden-Arcade–Roseville, 2,149,127; San Jose–Sunnyvale–Santa Clara, 1,836,911; Fresno, 930,450.

SOCIAL STATISTICS

Births, 2010, 509,979 (13·7 per 1,000 population); deaths, 2010, 233,143 (6·0). Marriages, 2011, 218,484. Infant deaths, 2010, 5.1 per 1,000 live births. California was the second state after Massachusetts to allow same-sex marriage when it became legal in June 2008, but the ruling was nullified when 52·5% of votes cast in a referendum held on 4 Nov. 2008 were in favour of a ban. On 4 Aug. 2010 a federal court declared the ban unconstitutional, a decision which was upheld by the United States Court of Appeals for the Ninth Circuit on 7 Feb. 2012.

CLIMATE

Los Angeles, Jan. 58°F (14·4°C), July 74°F (23·3°C). Annual rainfall 15" (381 mm). Sacramento, Jan. 45°F (7·2°C), July 76°F (24·4°C). Annual rainfall 18" (457 mm). San Diego, Jan. 57°F (13·9°C), July 71°F (21·7°C). Annual rainfall 10" (259 mm). San Francisco, Jan. 51°F (10·6°C), July 59°F (15°C). Annual rainfall 20" (508 mm). Death Valley, Jan. 52°F (11°C), July 100°F (38°C). Annual rainfall 1·6" (40 mm). California belongs to the Pacific Coast climate zone (see UNITED STATES: Climate).

CONSTITUTION AND GOVERNMENT

The present constitution became effective from 4 July 1879; it has had numerous amendments since 1962. The Senate is composed of 40 members elected for four years—half being elected every two years—and the Assembly, of 80 members, elected for two years. Two-year regular sessions convene in Dec. of each even numbered year. The Governor and Lieut.-Governor are elected for four years.

For the 113th Congress, which convened in Jan. 2013, California sends 53 members to the House of Representatives. It is represented in the Senate by Dianne Feinstein (D. 1993–2019) and Barbara Boxer (D. 1993–2017).

The capital is Sacramento. The state is divided into 58 counties.

RECENT ELECTIONS

In the 2012 presidential election Barack Obama took California with 60·2% of the vote, having also won it in 2008.

CURRENT GOVERNMENT

Governor: Jerry Brown (D.), 2011–15 (salary: $165,288).
   Lieut.-Governor: Gavin Newsom (D.), 2011–15 ($123,965).
   Secretary of State: Debra Bowen (D.), 2011–15 ($123,965).
   Attorney General: Kamala D. Harris (D.), 2011–15 ($143,571).

Government Website: http://www.ca.gov

ECONOMY

Per capita personal income (2011) was $43,647.

California's economy, one of the largest and most diverse in the world, has major components in high-technology, trade, entertainment, finance, agriculture, manufacturing, government, tourism, construction and services. California's economy is very dependent on international trade.

Farm production in California has risen despite declines in acreage. Farm-related sales have more than quadrupled over the past three decades. Largest production categories are fruits and nuts, livestock and poultry, vegetables and melons.

Budget

For the year ending 30 June 2011, total state revenues were $122·5bn. Total expenditures were $131·0bn. (education, $50·1bn.;

health and human services, $41·9bn.; corrections and rehabilitation, $9·7bn.) Debt outstanding (2012) $84·4bn.

Performance

California's economy, the largest among the 50 states and one of the largest in the world, has major components in high technology, trade, entertainment, agriculture, manufacturing, tourism, construction and services. California is home to leading innovators and entrepreneurs and to firms like Apple, Cisco and Intel in technology; eBay, Facebook, Google and Yahoo in pioneering the use of the internet; Amgen and Genentech in biotech; and DreamWorks and Pixar in combining technology and entertainment. California experienced a severe economic recession that began at the end of 2007, from which the state is still slowly recovering. The principal cause of the recession was a financial crisis instigated by risky financial activity that led to the bursting of the housing bubble.

If California were a country in its own right it would be the world's 12th largest economy. 2011 Gross Domestic Product by state (provisional) was $1,735,360m., the highest in the United States and representing more than 13% of the USA's total GDP. In 2011 state real GDP growth was 2·0% (provisional). Taxable sales in 2011 totalled $517,299m.

Banking and Finance

In 2007 there were 10,461 establishments of depository institutions including 7,386 commercial banks, 1,580 credit unions and 1,486 savings institutions.

In 2007 savings and loan associations had deposits of $46,507m. Total mortgage loans in 2007 were $54,118m. On 31 Dec. 2007 all insured commercial banks had demand deposits of $23,030m. and time and savings deposits of $221,565m. Total loans reached $270,619m., of which real-estate loans were $190,656m. Credit unions had assets totalling $120,607m. and total loans outstanding were $86,562m.

ENERGY AND NATURAL RESOURCES

Electricity

In 2011 production was 200,414 m. kWh of which 90,751m. kWh was from natural gas, 36,666m. kWh from nuclear energy and 42,727m. kWh from conventional hydroelectric energy. California leads the nation in electricity generation from non-hydroelectric renewable energy sources. California generates electricity using wind, geothermal, solar, fuel wood and municipal solid waste/landfill gas resources. The world's largest solar power facility operates in California's Mojave Desert. Consumption is growing at a rate of 2% annually.

Oil and Gas

California is the nation's fourth largest oil producing state. Total onshore and offshore production fell to an average of 539,200 bbls per day in 2011. California ranks tenth out of US states for the production of natural gas. Net natural gas production in 2011 was 243bn. cu. ft.

Water

The total area covered by water is approximately 7,916 sq. miles. Water quality is judged to be good along 83% of the 960 miles of assessed coastal shoreline.

Minerals

Gold production was 181,959 troy oz in 2010 (compared to less than 20,000 troy oz in 2007). The value of gold production increased to $239·7m. from $138·5m. in 2009. Clays, crushed stone, dimension stone, feldspar, fuller's earth, gemstones, gypsum, iron ore (used in cement manufacture), kaolin clay, lime, magnesium compounds, perlite, pumice, pumicite, salt, soda ash and zeolites are also produced.

In 2010 California ranked sixth among the states in non-fuel mineral production, accounting for approximately 4·2% of the US total. California was the leading state in the production of diatomite and natural sodium sulphate, and was the only producer of boron and rare earth minerals. The only metals produced in California were gold and silver. The market value of non-fuel minerals produced was $2·9bn.; the mining industry employed around 26,800 persons in 2011 (compared to 48,000 in the early 1980s).

Agriculture

California is the most diversified agricultural economy in the world, producing more than 400 agricultural commodities. It is by far the largest agricultural producer and exporter in the United States. The state grows nearly half of the nation's total of fruits, nuts and vegetables. Many of these commodities are specialty crops and almost solely produced in California. In 2011 there were 81,500 farms and ranches. The net farm income in 2011 was $16,276m. (the largest of any state). In 2011 cash income from marketings reached $43·5bn. (11·6% of the US total). Fruit and nut cash receipts, at $15·32bn. in 2011, were 11% above the previous year and comprised 32% of the total. Vegetable receipts were up 2% from the previous year at $7·24bn., comprising 15% of total sales. Livestock and poultry receipts rose 26% from 2010 at $12·36bn. and comprised 26% of total sales. California's leading commodity in cash receipts is milk and cream with $7·68bn. in 2011.

Production of cotton lint in 2011 was 321,800 short tons; other field and seed crops included (in 1m. short tons): hay and alfalfa, 8; rice, 2; sugar beets, 1; wheat, 1. Principal fruit, nut and vegetable crops in 2011 (in 1,000 short tons): tomatoes, 12,562; grapes, 6,612; lettuce, 3,247; oranges, 2,500; strawberries, 1,023; almonds, 1,015; broccoli, 1,012; onions, 973; carrots, 945; lemons, 820; celery, 773; peaches, 773.

In 2011 there were 1·7m. dairy cows; 5·2m. all cattle and calves; 570,000 sheep and lambs; and 105,000 hogs and pigs.

Forestry

In 2011 California had 32·94m. acres of forested land—of which 20,802,641 acres were national forest. There are about 16·6m. acres of productive forest land, from which about 2,900m. bd ft are harvested annually. Total value of timber harvest, 2011, $272m. Lumber production, 2011, 1,288m. bd ft.

Fisheries

The catch in 2011 was 408m. lb; leading species in landings were sardine, squid, anchovy, mackerel, crab, urchin, sole, whiting, sablefish and tuna.

INDUSTRY

Manufacturers in California accounted for 11·7% of the total output in the state in 2011; total output from manufacturing was $229·9bn. Employment in the sector was 8·8% of overall non-farm employment in 2011. There were 40,557 manufacturing establishments in California in 2009.

Labour

In 2011 the civilian labour force was 18·4m., of whom 16·2m. were employed. A total of 126,200 jobs were added in 2011, mainly in professional and business services, educational and health services, trade, transportation and utilities, and leisure and hospitality. The unemployment rate was 10·1% in Oct. 2012.

INTERNATIONAL TRADE

Imports and Exports

Estimated foreign trade through Californian ports totalled $559bn. in 2011. Exports of made-in-California goods totalled $159·4bn., an increase of 11·3% from 2010. Of the total $46·1bn. was computers and electronics (accounting for 29% of all the state's exports in 2011), $15·0bn. transportation equipment, $14·8bn. non-electrical machinery, $12·5bn. chemicals and $10·6bn. agricultural products.

Total agricultural exports for 2010 were a record $14·7bn. California's top markets are Canada, the European Union, Japan, China/Hong Kong, Mexico, South Korea, India, the United Arab Emirates, Taiwan and Australia.

COMMUNICATIONS

Roads

In 2010 California had 172,139 miles of public roads (including 15,160 of state highways) and 22·0m. registered automobiles. In 2007 there were a total of 33,935,386 registered motor vehicles. Motor vehicle collision fatalities in 2010 were 2,520. Average commute time was 26·9 minutes.

Rail

In addition to Amtrak's long-distance trains, local and medium-distance passenger trains run in the San Francisco Bay area sponsored by the California Department of Transportation, and a network of commuter trains around Los Angeles opened in 1992. There are metro and light rail systems in San Francisco and Los Angeles, and light rail lines in Sacramento, San Diego and San Jose.

Civil Aviation

In 2011 there were a total of 968 public and private airports, heliports, stolports and seaplane bases.

A total of 61,862,052 passengers (16,731,324 international; 45,130,728 domestic) embarked/disembarked at Los Angeles airport in 2011. It handled approximately 1,853,658 tonnes of freight. At San Francisco airport, in 2011, 40,800,352 passengers (9,013,021 international; 31,787,331 domestic) embarked/disembarked, and 421,096 tonnes of freight were handled. There were a total of 51,311,722 passenger enplanements at Los Angeles and San Francisco in 2011.

Shipping

The chief ports are Long Beach and Los Angeles. In 2010 Long Beach was the fourth busiest port in the USA, handling 68·4m. tons of cargo, and Los Angeles the ninth busiest, with 56·6m. tons.

SOCIAL INSTITUTIONS

Justice

A 'three strikes law', making 25-years-to-life sentences mandatory for third felony offences, was adopted in 1994 after an initiative (i.e. referendum) was 72% in favour. However, the state's Supreme Court ruled in June 1996 that judges may disregard previous convictions in awarding sentences. In Nov. 2012 voters approved Proposition 36, which modifies the three strikes law, to impose a life sentence only when the third felony conviction is 'serious or violent' and authorizes resentencing for offenders currently serving life sentences if their third strike conviction was not serious or violent. In Jan. 2009 there were 33 adult prisons. As of Dec. 2012 there were 8,905 adult inmates serving 'three strikes' sentences, of which 2,842 were eligible for resentencing under Proposition 36. The death penalty is authorized following its reinstatement by the US Supreme Court in 1976. Since 1978 the state has executed 14 condemned inmates. The first execution since 2002 was carried out in Jan. 2005. There was one execution in total in 2006. A moratorium on death sentences was ordered in Dec. 2006. The California Department of Corrections and Rehabilitation (CDCR) has not indicated when it expects to successfully overcome legal challenges related to executions although a new execution chamber has been completed at San Quentin State Prison. As of Dec. 2012, CDCR housed 725 condemned inmates.

Education

Full-time attendance at school is compulsory for children from six to 18 years of age for a minimum of 175 days per annum. In fall 2011 there were 6·7m. pupils enrolled in both public and private elementary and secondary schools. Total state expenditure on public education, 2010–11, was $50·1bn. Average teacher salary in 2010–11 was $67,871 (the fourth highest in the USA).

Community colleges had 2,423,958 students in fall 2012.

California has two publicly-supported higher education systems: the University of California (1868) and the California State University and Colleges. In fall 2010 the University of California, with campuses for resident instruction and research at Berkeley, Los Angeles (UCLA), San Francisco and seven other centres, had 234,464 students. California State University and Colleges with campuses at Sacramento, Long Beach, Los Angeles, San Francisco and 18 other cities had 412,372 students. In addition to the 32 publicly-supported institutions for higher education there are 117 private colleges and universities that had a total estimated enrolment of 433,832 in the fall of 2008.

Health

In 2008 there were 6,906 state licensed facilities; capacity, 260,036 beds. On 30 June 2008 state hospitals for the mentally disabled had 6,080 patients. In 2009 there were 100,131 active physicians (27 per 10,000 population).

Welfare

Medicare beneficiaries in 2010 totalled 4,684,845. In fiscal year 2009 a total of 11,518,565 people in California received Medicaid. On 1 Jan. 1974 the federal government (Social Security Administration) assumed responsibility for the Supplemental Security Income/State Supplemental Program, which replaced the State Old-Age Security. The SSI/SSP provides financial assistance for needy aged (65 years or older), blind or disabled persons. An individual recipient may own assets up to $2,000; a couple up to $3,000, subject to specific exclusions. In Jan. 2010 there were 1·23m. SSI recipients. In fiscal year 2011–12 the caseload of the welfare programme CalWORKS averaged 575,988, a decrease of 21·3% from fiscal year 1997–98 (the year CalWORKS was implemented).

RELIGION

The principal religious traditions in 2010 were: Catholics, with 10,234,036 members; Evangelical Protestants, 3,502,250; Mainline Protestants, 880,981; Latter-day Saints (Mormons), 770,437; Buddhists, 326,940.

CULTURE

Tourism

The travel and tourism industry provides 2·5% of the state's $2·0trn. economy. Visitors in 2011 spent $95·3bn. generating $6·3bn. in state and local tax revenues. Tourist spending has increased at an average annual rate of 2·9% since 2003. California was the state most visited by overseas travellers in 2011, when international tourists spent $19·1bn. (approximately 20% of all tourist spending).

FURTHER READING

California Government and Politics. Hoeber, T. R., *et al.*, (eds.) Annual
California Statistical Abstract. Annual; online only

Bean, W. and Rawls, J. J., *California: an Interpretive History.* 10th ed. 2011
Gerston, L. N. and Christensen, T., *California Politics and Government: a Practical Approach.* 12th ed. 2013

State library: The California State Library, Library-Courts Bldg, Sacramento 95814.

Colorado

KEY HISTORICAL EVENTS

Spanish explorers claimed the area for Spain in 1706; it was then the territory of the Arapaho, Cheyenne, Ute and other Plains and

Great Basin Indians. Eastern Colorado, the hot, dry plains, passed to France in 1802 and then to the USA as part of the Louisiana Purchase in 1803. The rest remained Spanish, becoming Mexican when Spanish power in the Americas ended. In 1848, after war between Mexico and the USA, Mexican Colorado was ceded to the USA. A gold rush in 1859 brought a great influx of population, and in 1861 Colorado was organized as a Territory. The Territory officially supported the Union in the Civil War of 1861–65, but its settlers were divided and served on both sides.

Colorado became a state in 1876. Mining and ranching were the mainstays of the economy. In the 1920s the first large projects were undertaken to exploit the Colorado River. The Colorado River Compact was agreed in 1922, and the Boulder Dam (now Hoover Dam) was authorized in 1928. Since then irrigated agriculture has overtaken mining as an industry and is as important as ranching. In 1945 the Colorado-Big Thompson project diverted water by tunnel beneath the Rocky Mountains to irrigate 700,000 acres (284,000 ha.) of northern Colorado. Now more than 80% of the population is urban, with the majority engaged in telecommunications, aerospace and computer technology.

TERRITORY AND POPULATION

Colorado is bounded north by Wyoming, northeast by Nebraska, east by Kansas, southeast by Oklahoma, south by New Mexico and west by Utah. Land area, 103,642 sq. miles (268,431 sq. km).

Census population, 1 April 2010, was 5,029,196, an increase of 16·9% since 2000. July 2012 estimate, 5,187,582.

Population in five census years was:

	White	Black	American Indian	Asiatic	Total	Per sq. mile
1910	783,415	11,453	1,482	2,674	799,024	7·7
			All others			
1980	2,571,498	101,703	216,763		2,889,964	27·9
1990	2,905,474	133,146	255,774		3,294,394	31·8
2000	3,560,005	165,063	576,193		4,301,261	41·5
2010	4,089,202	201,737	738,257		5,029,196	48·5

Of the total population in 2010, 2,520,662 were male and 3,803,587 were 18 years old or older. The Hispanic population in 2010 was 1,038,687, up from 735,601 in 2000 (an increase of 41·2%). Large cities, with 2008 populations: Denver City (the capital), 566,974; Colorado Springs, 372,437; Aurora, 305,582; Lakewood, 140,024; Fort Collins, 129,467; Westminster, 105,753; Arvada, 104,830; Pueblo, 103,730.

Main metropolitan areas (2008): Denver–Aurora, 2,506,626; Colorado Springs, 617,714; Boulder, 293,161; Fort Collins–Loveland, 292,825; Greeley, 249,715; Pueblo, 156,737; Grand Junction, 143,171.

SOCIAL STATISTICS

Births, 2010, were 66,355 (13·2 per 1,000 population); deaths, 2010 (provisional), 31,465 (6·3). Infant mortality, 2009, 6·3 per 1,000 live births. 2009 (provisional): marriages, 37,400 (6·8 per 1,000 population); divorces and annulments, 21,200 (4·2).

CLIMATE

Denver, Jan. 31°F (–0·6°C), July 73°F (22·8°C). Annual rainfall 14" (358 mm). Pueblo, Jan. 30°F (–1·1°C), July 83°F (28·3°C). Annual rainfall 12" (312 mm). Colorado belongs to the Mountain States climate zone (see UNITED STATES: Climate).

CONSTITUTION AND GOVERNMENT

The constitution adopted in 1876 is still in effect with (2006) 153 amendments. The General Assembly consists of a Senate of 35 members elected for four years, one-half retiring every two years, and of a House of Representatives of 65 members elected for two years. Sessions are annual, beginning 1951. Qualified as electors

are all citizens, male and female (except convicted, incarcerated criminals), 18 years of age, who have resided in the state and the precinct for 32 days immediately preceding the election. There is a seven-member State Supreme Court.

For the 113th Congress, which convened in Jan. 2013, Colorado sends seven members to the House of Representatives. It is represented in the Senate by Michael Bennet (D. 2009–17) and Mark Udall (D. 2009–15).

The capital is Denver. There are 64 counties.

RECENT ELECTIONS

In the 2012 presidential election Barack Obama took Colorado with 51·2% of the vote, having also won it in 2008.

CURRENT GOVERNMENT

Governor: John Hickenlooper (D.), 2011–15 (salary: $90,000).
  Lieut.-Governor: Joe Garcia (D.), 2011–15 ($68,500).
  Secretary of State: Scott Gessler (R.), 2011–15 ($68,500).

Government Website: http://www.colorado.gov

ECONOMY

Per capita personal income (2010) was $42,226.

Budget

In 2011 total revenue was $30,248m. and total expenditure $29,169m. Major areas of expenditure were: education, $9,250m.; public welfare, $5,663m.; highways, $1,436m.; health, $1,232m.; correction, $1,024m. Debt outstanding, in 2011, was $16,335m.

Performance

2011 Gross Domestic Product by state was $234,308m. (provisional), ranking Colorado 18th in the United States. In 2011 state real GDP growth was 1·9% (provisional).

ENERGY AND NATURAL RESOURCES

Oil and Gas

In 2008 Colorado produced 1,436bn. cu. ft of natural gas and 24m. bbls of crude oil. It ranked fifth in the USA for daily gas production, and eleventh in crude oil production. Total production value of all hydrocarbons was $9·8bn.

Water

The Rocky Mountains of Colorado form the headwaters for four major American rivers: the Colorado, Rio Grande, Arkansas and Platte. The total area covered by water is 452 sq. miles.

Minerals

Coal (2009): 28·3m. short tons were produced. In 2008 there were 29,900 people employed in mining, including 7,882 in extracting oil and natural gas. Value of domestic non-fuel mineral production in 2009 was $1,420m.

Agriculture

In 2009 farms and ranches numbered 36,200, with a total of 31·3m. acres of agricultural land. 5,781,000 acres were harvested crop land; average farm, 865 acres. Average value of farmland and buildings per acre in Jan. 2010 was $1,080. Cash receipts from farm marketings, 2009: from crops, $2,230m.; from livestock and products, $3,323m. The net farm income in 2009 was $745m.

Production of principal crops in 2009: corn for grain, 151·5m. bu.; wheat, 100·6m. bu.; barley, 10·4m. bu.; sorghum for grain, 6·8m. bu.; millet, 5·3m. bu.; hay, 4,778,000 tons; corn for silage, 1,998,000 tons; sugar beets, 945,000 tons; potatoes, 23·6m. cwt.

In Dec. 2009 the number of farm animals was: 2,600,000 cattle and calves, 710,000 swine and 375,000 sheep. Wool production in 2007 totalled 2,916,000 lb.

Forestry

The state had a forested area of 22,612,000 acres in 2007, of which 11,259,000 acres were national forest.

INDUSTRY

In 2007 the state's 5,288 manufacturing establishments had 138,000 employees, earning $6,790m. Total value added by manufacturing in 2009 was $20,717m.

Labour

Total non-agricultural employment, 2010: 2,220,000. Employees by branch, 2010 (in 1,000): trade, transportation and utilities, 397; government, 393; professional and business services, 329; education and health services, 265; leisure and hospitality, 263. The unemployment rate in Dec. 2010 was 8·9%.

INTERNATIONAL TRADE

Imports and Exports

In 2008 Colorado exported $7·7bn. in goods. The largest trading partners were Canada, China (including Hong Kong), Mexico, Malaysia and Japan. Largest export categories are industrial machinery (including computers), electrical machinery, and optic, photographic and medical/surgical instruments.

Trade Fairs

The National Western Stock Show and Rodeo is the largest event of its kind in the USA, drawing over 600,000 visitors.

COMMUNICATIONS

Roads

In 2007 there were 88,163 miles of road, of which 19,229 miles were urban roads and 68,934 miles rural roads. There were 1,707,139 motor vehicle registrations.

Rail

In 2009 there were 2,684 miles of freight railroad (excluding trackage rights). There were 14 freight railroads operating in 2009. Rail traffic originating in Colorado in 2009 totalled 27·3m. tons and rail traffic terminating in the state came to 27·5m. tons. A light rail system opened in Denver in 1994.

Civil Aviation

In 2007 there were 74 airports open to the public; 14 with commercial service, 60 public non-commercial (general aviation) and 14 private non-commercial. Denver International Airport, the largest airport in the state and 11th busiest in the world, handled 52,699,298 passengers in 2011.

SOCIAL INSTITUTIONS

Justice

In Dec. 2008 there were 23,274 federal and state prisoners. The death penalty is authorized but has not been used since 1997.

Education

In 2007–08 there were 1,757 public elementary and secondary schools with 801,867 pupils and 47,761 teachers. In 2008–09 teachers' salaries averaged $48,707.

Enrolments in four-year state universities and colleges in fall 2005 were: University of Colorado at Boulder, 31,589 students; University of Colorado at Denver and Health Sciences Center, 19,766; University of Colorado at Colorado Springs, 9,333; Colorado State University (Fort Collins), 27,780; University of Northern Colorado (Greeley), 13,622; Colorado School of Mines (Golden), 4,318; Metropolitan State College of Denver, 21,010; Colorado State University-Pueblo (was University of Southern Colorado), 5,870; Mesa State College (Grand Junction), 6,062; Fort Lewis College (Durango), 3,946; Adams State College (Alamosa), 9,157; Western State College of Colorado (Gunnison), 2,253.

Total enrolment in private degree-granting universities and colleges in fall 2007 was 82,653.

Health

In 2009 there were 81 community hospitals with 10,400 beds. A total of 445,000 patients were admitted during the year. In 2009 there were 13,047 active physicians (26 per 10,000 population).

Welfare

Medicare enrolment in July 2010 totalled 620,678. In fiscal year 2009 a total of 677,712 people in Colorado received Medicaid. In Dec. 2008 there were 635,816 Old-Age, Survivors, and Disability Insurance (OASDI) beneficiaries. A total of 20,640 people were receiving payments under Temporary Assistance for Needy Families (TANF) in Dec. 2008.

RELIGION

The principal religious traditions in 2010 were: Catholics, with 811,630 members; Evangelical Protestants, 601,009; Mainline Protestants, 246,706; Latter-day Saints (Mormons), 145,033; Buddhists, 23,920.

CULTURE

Tourism

In 2011 overseas visitors to Colorado—excluding those from Canada and Mexico—numbered 446,000, up from 343,000 in 2010.

FURTHER READING

Statistical information: Business Research Division, Univ. of Colorado, Boulder 80309. Publishes *Statistical Abstract of Colorado.*

Griffiths, M. and Rubright, L., *Colorado: a Geography.* 1983

State library: Colorado State Library, 201 E. Colfax, Rm. 314, Denver 80203.

Connecticut

KEY HISTORICAL EVENTS

Formerly territory of Algonquian-speaking Indians, Connecticut was first colonized by Europeans during the 1630s, when English Puritans moved there from Massachusetts Bay. Settlements were founded in the Connecticut River Valley at Hartford, Saybrook, Wethersfield and Windsor in 1635. They formed an organized commonwealth in 1637. A further settlement was made at New Haven in 1638 and was united to the commonwealth under a royal charter in 1662. The charter confirmed the commonwealth constitution, drawn up by mutual agreement in 1639 and called the Fundamental Orders of Connecticut.

The area was agricultural and its population of largely English descent until the early 19th century. After the War of Independence, Connecticut was one of the original 13 states of the Union. Its state constitution came into force in 1818 and lasted with amendments until 1965 when a new one was adopted.

In the early 1800s a textile industry thrived on water power. By 1850 the state had more employment in industry than in agriculture, and immigration from Europe (and especially from southern and eastern Europe) grew rapidly throughout the 19th century. Some immigrants worked in whaling and iron-mining, but most sought industrial employment. Settlement was spread over a large number of small towns, with no single dominant culture.

Yale University was founded at New Haven in 1701. The US Coastguard Academy was founded in 1876 at New London, a former whaling port.

TERRITORY AND POPULATION

Connecticut is bounded in the north by Massachusetts, east by Rhode Island, south by the Atlantic and west by New York. Land area, 4,842 sq. miles (12,542 sq. km).

Census population, 1 April 2010, was 3,574,097, an increase of 4·9% since 2000. July 2012 estimate, 3,590,347.

Population in five census years was:

	White	Black	American Indian	Asian	Total	Per sq. mile
1910	1,098,897	15,174	152	533	1,114,756	231·3
1980	2,799,420	217,433	4,533	18,970	3,107,576	634·3

	White	Black	American Indian/Alaska Native	Asian	Others	Total	Per sq. mile
1990	2,859,353	274,269	6,654	50,078	96,762	3,287,116	678·6
2000	2,780,355	309,843	9,639	82,313	148,567	3,405,565	702·9
2010	2,772,410	362,296	11,256	135,565	292,570	3,574,097	738·1

Of the total population in 2010, there were 479,087 persons of Hispanic origin, up from 320,323 in 2000 (an increase of 49·6%). Of the total population in 2010, 1,834,483 were female and 2,757,082 were 18 years old or older. There were 183 residents in five American Indian Reservations in 2000.

The chief cities and towns are (2010 census populations):

Bridgeport	144,229	Danbury	80,893
New Haven	129,779	New Britain	73,206
Hartford (capital)	124,775	West Hartford	63,268
Stamford	122,643	Meriden	60,868
Waterbury	110,366	Bristol	60,477
Norwalk	85,603	West Haven	55,564

SOCIAL STATISTICS

Births, 2010, 37,708 (10·6 per 1,000 population); deaths, 2010 (provisional), 28,719 (8·0). Infant mortality rate, 2009, 5·5 per 1,000 live births. 2009 (provisional): marriages, 19,800 (5·9 per 1,000 population); divorces and annulments, 10,800 (3·1). Connecticut became the third state after Massachusetts and California to allow same-sex marriage when it became legal in Oct. 2008 (although it was subsequently banned in California).

CLIMATE

New Haven: Jan. 25°F (−3·8°C), July 74°F (23·4°C). Annual rainfall 45" (1,143 mm). Connecticut belongs to the New England climate zone (see UNITED STATES: Climate).

CONSTITUTION AND GOVERNMENT

The 1818 constitution was revised in 1955. On 30 Dec. 1965 a new constitution went into effect, having been framed by a constitutional convention in the summer of 1965 and approved by the voters in Dec. 1965.

The General Assembly consists of a Senate of 36 members and a House of Representatives of 151 members. Members of each House are elected for the term of two years. Legislative sessions are annual.

For the 113th Congress, which convened in Jan. 2013, Connecticut sends five members to the House of Representatives. It is represented in the Senate by Richard Blumenthal (D. 2011–17) and Christopher Murphy (D. 2013–19).

There are eight counties. The state capital is Hartford.

RECENT ELECTIONS

In the 2012 presidential election Barack Obama took Connecticut with 58·4% of the vote, having also won it in 2008.

CURRENT GOVERNMENT

Governor: Dannel P. Malloy (D.), 2011–15 (salary: $150,000).
   Lieut.-Governor: Nancy Wyman (D.), 2011–15 ($110,000).
   Secretary of State: Denise Merrill (D.), 2011–15 ($110,000).

Government Website: http://www.ct.gov

ECONOMY

Per capita personal income (2010) was $54,877, the second highest in the country (after the District of Columbia).

Budget

In 2011 total state revenue was $28,928m. Total expenditure was $28,094m. (education, $6,748m.; public welfare, $6,362m.; hospitals, $1,476m.; government administration, $1,178m.; highways, $1,013m.) Outstanding debt in 2011, $30,524m.

Performance

Gross Domestic Product by state in 2011 was $201,386m. (provisional), ranking Connecticut 24th in the United States. In 2011 state real GDP growth was 2·0% (provisional).

ENERGY AND NATURAL RESOURCES

Water

The total area covered by water is 701 sq. miles.

Minerals

The state has some mineral resources: crushed stone, sand, gravel, clay, dimension stone, feldspar and quartz. Total non-fuel mineral production in 2009 was valued at more than $160m.

Agriculture

In 2007 the state had 4,916 farms with a total area of 405,616 acres; the average farm size was 83 acres, valued at $12,667 per acre in 2007. Farm income 2006: crops $372m., and livestock and products $151m. The net farm income in 2006 was $183m. Principal crops are greenhouse and nursery products, grains, hay, tobacco, vegetables, maize, melons, fruit, nuts and berries.

In 2007 there were 50,213 all cattle (value $81·7m.), 5,767 sheep and 3,645 swine.

Forestry

Total forested area was 1,859,000 acres in 2008.

INDUSTRY

In 2007 the state's 4,294 manufacturing establishments had 191,000 employees, earning $10,345m. Total value added by manufacturing in 2009 was $27,829m.

Labour

Total non-agricultural employment, 2010, 1,608,000. Employees by branch, 2010 (in 1,000): education and health services, 307; trade, transportation and utilities, 289; government, 245; professional and business services, 190; manufacturing, 166. The average annual wage per employee in 2009 was $57,771—the highest of any state. The unemployment rate in Dec. 2010 was 9·0%.

COMMUNICATIONS

Roads

The total length of highways in 2007 was 21,295 miles comprising 15,108 miles of urban road and 6,187 miles of rural road. Motor vehicles registered in 2007 numbered 3,047,330.

Rail

In 2010 there were 629 miles of railroad route.

Civil Aviation

In 2011 there were two primary airports (commercial service airports with more than 10,000 passenger boardings annually). The main airport is Bradley International at Windsor Locks, which had 2,772,315 enplanements in 2011.

SOCIAL INSTITUTIONS

Justice

In Dec. 2008 the jail and prison population totalled 20,661. In April 2012 Connecticut became the 17th US state to abolish the death penalty.

Education

Instruction is free for all children and young people between the ages of four and 21 years, and compulsory for all children between the ages of five and 18 years. In 2007–08 there were 1,117 public schools. In fall 2007 there were 570,626 public school pupils and

39,304 teachers. In 2006–07 total expenditure on public elementary and secondary education was $9,301m. Private schools numbered 420 in fall 2007. Average teacher salary was $63,976 in 2008–09. In fiscal year 2008 spending per pupil was $14,610.

In fall 2009 there were 124,185 students enrolled in the state's public colleges and universities (five state universities, one external degree college, 12 community colleges and a US Coast Guard Academy). The University of Connecticut (founded in 1881), with a main campus at Storrs and six smaller campuses, had 29,517 students in fall 2009. The Connecticut State University System (founded in 1983 but with its oldest institution established in 1849) comprises four universities and had 36,503 students in fall 2009. Enrolment in independent institutions of higher education in 2009 totalled 69,028 students. In 2005 Yale University, New Haven (founded in 1701) had 11,483 students; Wesleyan University, Middletown (1831), 3,205 students; Trinity College, Hartford (1823), 2,470 students; Connecticut College, New London (1915), 1,898 students; and the University of Hartford (1877), 7,260 students. There were 20 independent (four-year course) colleges and three independent (two-year course) colleges as well as two seminaries and the International College of Hospitality Management.

Health
In 2009 there were 35 community hospitals with 7,900 beds. A total of 408,000 patients were admitted during the year. In 2009 there were 13,370 active physicians (38 per 10,000 population).

Welfare
Medicare enrolment in July 2010 totalled 562,355. In fiscal year 2009 a total of 558,201 people in Connecticut received Medicaid. In Dec. 2008 there were 599,533 Old-Age, Survivors, and Disability Insurance (OASDI) beneficiaries. A total of 32,959 people were receiving payments under Temporary Assistance for Needy Families (TANF) in Dec. 2008.

RELIGION
The principal religious traditions in the state in 2010 were: Catholics, with 1,254,340 members; Mainline Protestants, 281,174; Evangelical Protestants, 157,336; Jews, 47,639; Black Protestants, 22,969.

CULTURE
Tourism
Overseas visitors to Connecticut—excluding those from Canada and Mexico—numbered 307,000 in 2011, up from 290,000 in 2010.

FURTHER READING
State Register and Manual. Annual

Halliburton, W. J., *The People of Connecticut.* 1985

State library: Connecticut State Library, 231 Capitol Avenue, Hartford (CT) 06105.

State Book Store: Dept. of Environmental Protection, 79 Elm St., Hartford (CT) 06106.

Business Incentives: Connecticut Economic Resource Center, 805 Brook St., Rocky Hill (CT) 06067.

Connecticut Tourism: Dept. of Economic and Community Development, 865 Brook St., Rocky Hill (CT) 06067.

Delaware

KEY HISTORICAL EVENTS
Delaware was the territory of Algonquian-speaking Indians who were displaced by European settlement in the 17th century. The first settlers were Swedes who came in 1638 to build Fort Christina (now Wilmington), and colonize what they called New Sweden. In 1655 their colony was taken by the Dutch, who were based in New Amsterdam. In 1664 the British took the whole New Amsterdam colony, including Delaware, and called it New York.

In 1682 Delaware was granted to William Penn, who wanted access to the coast for his Pennsylvania colony. Union of the two colonies was unpopular, and Delaware gained its own government in 1704, although it continued to share a royal governor with Pennsylvania until the War of Independence. Delaware then became one of the 13 original states of the Union and the first to ratify the federal constitution (on 7 Dec. 1787).

The population was of Swedish, Finnish, British and Irish extraction. The land was low-lying and fertile, and the use of slave labour was legal. There was a significant number of black slaves, but Delaware was a border state during the Civil War (1861–65) and did not leave the Union.

19th-century immigrants were mostly European Jews, Poles, Germans and Italians. The north became industrial and densely populated, more so after the Second World War with the rise of the petrochemical industry. Industry in general profited from the opening of the Chesapeake and Delaware Canal in 1829; it was converted to a toll-free deep channel for ocean-going ships in 1919.

TERRITORY AND POPULATION
Delaware is bounded in the north by Pennsylvania, northeast by New Jersey, east by Delaware Bay, south and west by Maryland. Land area 1,949 sq. miles (5,047 sq. km). Census population, 1 April 2010, was 897,934, an increase of 14·6% since 2000. July 2012 estimate, 917,092. Population in five census years was:

	White	Black	American Indian	Asiatic	Total	Per sq. mile
1910	171,102	31,181	5	34	202,322	103·0
			All others			
1980	488,002	96,157	10,179		594,338	290·8
1990	535,094	112,460	18,614		666,168	325·9
2000	584,773	150,666	48,161		783,600	401·0
2010	618,617	191,814	87,503		897,934	460·8

Of the total population in 2010, 462,995 were female and 692,169 were 18 years old or older. The Hispanic population in 2010 was 73,221, up from 37,277 in 2000 (an increase of 96·4%).

The 2010 census figures show Wilmington with a population of 70,851; Dover (the capital), 36,047; Newark, 31,454; Bear, 19,371; Middletown, 18,871.

SOCIAL STATISTICS
Births, 2010, 11,364 (12·7 per 1,000 population); deaths, 2010 (provisional), 7,706 (8·6). 2009 infant mortality, 7·9 per 1,000 live births. 2009 (provisional): marriages, 5,100 (5·4 per 1,000 population); divorces and annulments, 3,400 (3·6). The abortion rate, at 40 for every 1,000 women aged 15 to 44 in 2008, is the highest in the USA.

CLIMATE
Wilmington, Jan. 31°F (−0·6°C), July 76°F (24·4°C). Annual rainfall 43" (1,076 mm). Delaware belongs to the Atlantic Coast climate zone (*see* UNITED STATES: Climate).

CONSTITUTION AND GOVERNMENT
The present constitution (the fourth) dates from 1897, and has had 51 amendments; it was not ratified by the electorate but promulgated by the Constitutional Convention. The General Assembly consists of a Senate of 21 members elected for four years and a House of Representatives of 41 members elected for two years.

For the 113th Congress, which convened in Jan. 2013, Delaware sends one member to the House of Representatives. It is represented in the Senate by Thomas Carper (D. 2001–19) and Christopher Coons (D. 2011–15).

The state capital is Dover. Delaware is divided into three counties.

RECENT ELECTIONS

In the 2012 presidential election Barack Obama took Delaware with 58·6% of the vote, having also won it in 2008.

CURRENT GOVERNMENT

Governor: Jack Markell (D.), 2013–17 (salary: $171,000).
  Lieut.-Governor: Matt Denn (D.), 2013–17 ($78,553).
  Secretary of State: Jeffrey W. Bullock (D.), since Jan. 2009 ($127,590).

Government Website: http://www.delaware.gov

ECONOMY

Per capita personal income (2010) was $39,664.

Budget

In 2011 total revenue was $9,106m. Total expenditure was $7,936m. (education, $2,544m.; public welfare, $1,762m.; highways, $461m.; government administration, $440m.; health, $395m.) Debt outstanding in 2011, $5,808m.

Performance

2011 Gross Domestic Product by state was $57,293m. (provisional), ranking Delaware 39th in the United States. In 2011 state real GDP growth was 1·6% (provisional).

Banking and Finance

Delaware National Bank has branches statewide. Also based in Delaware, MBNA is the world's largest independent credit card issuer, with managed loans of $97·5bn.

ENERGY AND NATURAL RESOURCES

Electricity

In 2009 production was 4·8bn. kWh, of which 2·8bn. kWh was from coal.

Water

The total area covered by water is 540 sq. miles.

Minerals

The mineral resources of Delaware are not extensive, consisting chiefly of clay products, stone, sand and gravel and magnesium compounds. Total non-fuel mineral production in 2009 was valued at more than $25m.

Agriculture

There were 510,000 acres in 2,546 farms in 2007. The average farm was 200 acres and was valued (land and buildings) at $10,347 per acre in 2007. Farm income 2007: crops $209m., and livestock and products $795m. The net farm income in 2006 was $388m. The major product is broilers, with a value of $734·9m. in 2007.

The chief crops are corn for feed, greenhouse products and soybeans.

Forestry

Total forested area was 383,000 acres in 2007.

INDUSTRY

In 2007 the state's 673 manufacturing establishments had 35,000 employees, earning $1,760m. Total value added by manufacturing in 2009 was $6,257m. Main manufactures are chemicals, transport equipment and food.

Labour

Total non-agricultural employment, 2010, 413,000. Employees by branch, 2010 (in 1,000): trade, transportation and utilities, 74; education and health services, 65; government, 64; professional and business services, 55; financial activities, 43. The unemployment rate in Dec. 2010 was 8·5%.

COMMUNICATIONS

Roads

In 2007 there were 6,242 miles of roads comprising 2,963 miles of urban road and 3,279 miles of rural road. In 2007 total vehicles registered numbered 851,223.

Rail

In 2009 there were 227 miles of freight railroad (excluding trackage rights). Rail traffic originating in Delaware in 2009 totalled 0·5m. tons and rail traffic terminating in the state came to 4·2m. tons. An important component of Delaware's freight infrastructure is the rail access to the Port of Wilmington.

Civil Aviation

Delaware is the only state not to have a primary airport (a commercial service airport with more than 10,000 passenger boardings annually). It had four general aviation airports in 2011.

SOCIAL INSTITUTIONS

Justice

In Dec. 2008 the jail and prison population totalled 7,075. The death penalty is authorized. There was one execution in 2011 and one in 2012, although prior to these the death penalty had not been implemented since 2005.

Education

The state has free public schools and compulsory school attendance to age 16. In 2007–08 the 235 elementary and secondary public schools had 122,574 enrolled pupils and 8,198 classroom teachers. Another 32,520 children were enrolled in private schools in fall 2007. Total expenditure for public elementary and secondary education in 2006–07 was $1,740m. and average teacher salary in 2008–09 was $55,994. Expenditure per pupil was $12,153 in fiscal year 2008.

The state supports the University of Delaware at Newark (founded in 1834) which had 1,077 full-time faculty members and 20,982 students in 2005; Delaware State University, Dover (1891), with 170 full-time faculty members in 2004 and 3,722 students in 2005; and the campuses of Delaware Technical and Community College at Newark with 7,473 students in 2005, Dover with 2,569 students and Georgetown with 3,936 students.

Health

In 2009 there were seven community hospitals with 2,200 beds. A total of 102,000 patients were admitted during the year. In 2009 there were 2,177 active physicians (25 per 10,000 population).

Welfare

Medicare enrolment in July 2010 totalled 148,009. In fiscal year 2009 a total of 208,904 people in Delaware received Medicaid. In Dec. 2008 there were 161,314 Old-Age, Survivors, and Disability Insurance (OASDI) beneficiaries. A total of 12,831 people were receiving payments under Temporary Assistance for Needy Families (TANF) in Dec. 2008.

RELIGION

The main religious traditions in the state in 2010 were: Catholics, with 182,532 members; Mainline Protestants, 86,858; Evangelical Protestants, 64,625; Black Protestants, 13,371; Hindus, 7,805.

FURTHER READING

Statistical information: Delaware Economic Development Office, Dover, DE 19901. Publishes *Delaware Statistical Overview.*
State Manual, Containing Official List of Officers, Commissions and County Officers. Annual

Smeal, L., *Delaware Historical and Biographical Index.* 1984

District of Columbia

KEY HISTORICAL EVENTS

The District of Columbia, organized in 1790, is the seat of the government of the USA, for which the land was ceded by the states of Maryland and Virginia to the USA as a site for the national capital. It was established under Acts of Congress in 1790 and 1791. Congress first met in it in 1800 and federal authority over it became vested in 1801. In 1846 the land ceded by Virginia (about 33 sq. miles) was given back.

TERRITORY AND POPULATION

The District forms an enclave on the Potomac River, where the river forms the southwest boundary of Maryland. The land area of the District of Columbia is 61 sq. miles (158 sq. km).

Census population, 1 April 2010, was 601,723, an increase of 5·2% since 2000. July 2012 estimate, 632,323. The population was 100% urban in 2010. Metropolitan area of Washington, D.C.–Arlington–Alexandria (2010), 5,582,170. The Hispanic population in 2010 was 54,749, up from 44,953 in 2000 (an increase of 21·8%). Of the total population in 2010, 317,501 were female and 500,908 were 18 years old or older.

Population in five census years was:

	White	Black	American Indian	Chinese and Japanese	Total	Per sq. mile
1910	236,128	94,446	68	427	331,069	5,517·8
			All others			
1980	171,768	448,906	17,659		638,333	10,464·4
1990	179,667	339,604	87,629		606,900	9,949·2
2000	176,101	343,312	52,646		572,059	9,378·0
2010	231,471	305,125	65,127		601,723	9,856·5

Blacks constituted 50·7% of the population at the 2010 census—the highest proportion in the USA.

SOCIAL STATISTICS

Births, 2010, 9,165 (15·2 per 1,000 population); deaths, 2010 (provisional), 4,672 (7·8). Infant mortality rate, 2009, 9·9 per 1,000 live births. 2009 (provisional): marriages, 1,900 (4·7 per 1,000 population); divorces and annulments, 1,300 (2·6). Same-sex marriage became legal in March 2010.

CLIMATE

Washington, Jan. 34°F (1·1°C), July 77°F (25°C). Annual rainfall 43" (1,064 mm). The District of Columbia belongs to the Atlantic Coast climate zone (*see* UNITED STATES: Climate).

CONSTITUTION AND GOVERNMENT

Local government, from 1 July 1878 until Aug. 1967, was that of a municipal corporation administered by a board of three commissioners, of whom two were appointed from civil life by the President, and confirmed by the Senate, for a term of three years each. The other commissioner was detailed by the President from the Engineer Corps of the Army. The Commission form of government was abolished in 1967 and a new Mayor Council instituted with officers appointed by the President with the advice and consent of the Senate. On 24 Dec. 1973 the appointed officers were replaced by an elected Mayor and councillors, with full legislative powers in local matters as from 1974. Congress retains the right to legislate, to veto or supersede the Council's acts. The 23rd amendment to the federal constitution (1961) conferred the right to vote in national elections. The District has one delegate and one shadow delegate to the House of Representatives and two shadow senators. The Congressman may participate but not vote on the House floor.

RECENT ELECTIONS

In the 2012 presidential election Barack Obama took the District of Columbia with 91·4% of the vote, having also won it in 2008.

CURRENT GOVERNMENT

Mayor: Vincent Gray (D.), 2011–15 (salary: $200,000).

Secretary of the District: Cynthia Brock-Smith (D.), appointed Dec. 2010 ($140,000).

Government Website: http://www.dc.gov

ECONOMY

Per capita personal income (2010) was $70,044, the highest in the country.

Budget

The District's revenues are derived from a tax on real and personal property, sales taxes, taxes on corporations and companies, licences for conducting various businesses and from federal payments. The District of Columbia has no bonded debt not covered by its accumulated sinking fund.

Performance

Gross Domestic Product by state in 2011 was $91,643m. (provisional). In 2011 state real GDP growth was 1·9% (provisional).

ENERGY AND NATURAL RESOURCES

Water

The total area covered by water is 7 sq. miles.

INDUSTRY

In 2007 there were 137 manufacturing establishments with 2,000 employees, earning $81m. Total value added by manufacturing in 2009 was $151m. The main industries are communications, finance, government service, insurance, real estate, services, transport, utilities, and wholesale and retail trade.

Labour

Total non-agricultural employment, 2010, 711,000. Employees by branch, 2010 (in 1,000): government, 246; professional and business services, 149; education and health services, 108; leisure and hospitality, 59; trade, transportation and utilities, 27. In Dec. 2010 the unemployment rate was 9·6%.

COMMUNICATIONS

Roads

In 2007 there were 1,505 miles of roads. There were 217,521 registered vehicles in 2007.

Rail

There is a metro in Washington extending to 130 km, and two commuter rail networks.

Civil Aviation

The District is served by three general airports; across the Potomac River in Arlington County, Va., is Ronald Reagan Washington National Airport; in Dulles, Va., is Washington Dulles International Airport; and in Maryland is Baltimore/Washington International Thurgood Marshall Airport.

SOCIAL INSTITUTIONS

Justice

The death penalty was declared unconstitutional in the District of Columbia on 14 Nov. 1973.

The District's Court system is the Judicial Branch of the District of Columbia. It is the only completely unified court system in the United States, possibly because of the District's unique city-state jurisdiction. Until the District of Columbia Court Reform and Criminal Procedure Act of 1970, the judicial system was almost entirely in the hands of Federal government. Since that time, the system has been similar in most respects to the autonomous systems of the states.

Education

In 2007–08 there were 78,422 pupils enrolled at 244 elementary and secondary public schools with 6,347 teachers. Expenditure per pupil in fiscal year 2008 was $16,353.

Higher education is given through the Consortium of Universities of the Metropolitan Washington Area, which consists of six universities and three colleges: Georgetown University, founded in 1795 by the Jesuit Order; George Washington University, non-sectarian founded in 1821; Howard University, founded in 1867; Catholic University of America, founded in 1887; American University (Methodist), founded in 1893; University of District of Columbia, founded 1976; Gallaudet College, founded 1864; Trinity College in Washington, D.C. (women's college), founded 1897. There were 16 degree-granting institutions altogether in 2008–09.

Health

In 2009 there were ten community hospitals with 3,500 beds. A total of 138,000 patients were admitted during the year. In 2009 there were 4,900 active physicians (82 per 10,000 population).

Welfare

Medicare enrolment in July 2010 totalled 77,372. In fiscal year 2009 a total of 175,167 people in the District of Columbia received Medicaid. In Dec. 2008 there were 71,468 Old-Age, Survivors, and Disability Insurance (OASDI) beneficiaries. A total of 12,510 people were receiving payments under Temporary Assistance for Needy Families (TANF) in Dec. 2008.

RELIGION

The main religious traditions in 2010 were: Mainline Protestants, with 82,388 members; Catholics, 75,948; Evangelical Protestants, 75,306; Black Protestants, 50,602; Jews, 17,664.

FURTHER READING

Statistical Information: The Metropolitan Washington Board of Trade publications.
Reports of the Commissioners of the District of Columbia. Annual.

Bowling, K. R., *The Creation of Washington D.C.: the Idea and the Location of the American Capital.* 1991

Florida

KEY HISTORICAL EVENTS

Of the French and Spanish settlements in Florida in the 16th century, the Spanish, at St Augustine from 1565, survived. Florida was claimed by Spain until 1763 when it passed to Britain. Although regained by Spain in 1783, the British used it as a base for attacks on American forces during the war of 1812. Gen. Andrew Jackson captured Pensacola for the USA in 1818. In 1819 a treaty was signed which ceded Florida to the USA with effect from 1821 and it became a Territory of the USA in 1822.

Florida had been the home of the Apalachee and Timucua Indians. After 1770 groups of Creek Indians began to arrive as refugees from the European-Indian wars. These 'Seminoles' or runaways attracted other refugees including slaves, the recapture of whom was the motive for the first Seminole War of 1817–18. A second war followed in 1835–42, when the Seminoles retreated to the Everglades swamps. After a third war in 1855–58 most Seminoles were forced or persuaded to move to reserves in Oklahoma.

Florida became a state in 1845. About half of the population were black slaves. At the outbreak of Civil War in 1861 the state seceded from the Union.

During the 20th century Florida continued to grow fruit and vegetables, but real-estate development (often for retirement) and the growth of tourism and the aerospace industry set it apart from other ex-plantation states.

TERRITORY AND POPULATION

Florida is a peninsula bounded in the west by the Gulf of Mexico, south by the Straits of Florida, east by the Atlantic, north by Georgia and northwest by Alabama. Land area, 53,625 sq. miles (138,887 sq. km). Census population, 1 April 2010, was 18,801,310, an increase of 17·6% since 2000. July 2012 estimate, 19,317,568.

Population in five federal census years was:

	White	Black	All others	Total	Per sq. mile
1950	2,166,051	603,101	2,153	2,771,305	51·1
1980	8,319,448	1,342,478	84,398	9,746,324	180·1
1990	10,749,285	1,759,534	429,107	12,937,926	238·9
2000	12,465,029	2,335,505	1,181,844	15,982,378	296·4
2010	14,109,162	2,999,862	1,692,286	18,801,310	350·6

Of the total population in 2010, 9,611,955 were female and 14,799,219 were 18 years old or older. The Hispanic population in 2010 was 4,223,806, up from 2,682,715 in 2000 (a rise of 57·4% and the third largest numeric increase of any state in the USA).

The largest cities in the state, 2010 census, are: Jacksonville, 821,784; Miami, 399,457; Tampa, 335,709; St Petersburg, 244,769; Orlando, 238,300; Hialeah, 224,669; Tallahassee (the capital), 181,376; Fort Lauderdale, 165,521; Port St Lucie, 164,603; Pembroke Pines, 154,750; Cape Coral, 154,305; Hollywood, 140,768; Gainesville, 124,354; Miramar, 122,041; Coral Springs, 121,096; Clearwater, 107,685; Miami Gardens, 107,167; Brandon, 103,483; Palm Bay, 103,190; West Palm Beach, 99,919; Pompano Beach, 99,845; Spring Hill, 98,621; Lakeland, 97,422; Davie, 91,992; Miami Beach, 87,779; Lehigh Acres, 86,784; Deltona, 85,182; Plantation, 84,955; Sunrise, 84,439; Boca Raton, 84,392.

Population of the largest metropolitan areas (2010): Miami–Fort Lauderdale–Pompano Beach, 5,564,635; Tampa–St Petersburg–Clearwater, 2,783,243; Orlando–Kissimmee–Sanford, 2,134,411; Jacksonville, 1,345,596.

SOCIAL STATISTICS

Births, 2010, 214,590 (11·4 per 1,000 population); deaths, 2010 (provisional), 173,763 (9·2). Infant mortality, 2009, 6·9 per 1,000 live births. 2009 (provisional): marriages, 141,200 (7·5 per 1,000 population); divorces and annulments, 79,900 (4·2).

CLIMATE

Jacksonville, Jan. 55°F (12·8°C), July 81°F (27·2°C). Annual rainfall 54" (1,353 mm). Key West, Jan. 70°F (21·1°C), July 83°F (28·3°C). Annual rainfall 39" (968 mm). Miami, Jan. 67°F (19·4°C), July 82°F (27·8°C). Annual rainfall 60" (1,516 mm). Tampa, Jan. 61°F (16·1°C), July 81°F (27·2°C). Annual rainfall 51" (1,285 mm). Florida belongs to the Gulf Coast climate zone (*see* UNITED STATES: Climate).

CONSTITUTION AND GOVERNMENT

The 1968 Legislature revised the constitution of 1885. The state legislature comprises the Senate and House of Representatives. The Senate has 40 members elected for four years. Half of the membership is elected every two years. The House has 120 members, all of whom are elected every two years during elections held in even-numbered years. Sessions of the legislature are held annually, and are limited to 60 days. Senate and House districts are based on population, with each senator and member representing approximately the same number of residents. The Senate and House are reapportioned every ten years when the federal census is released. In addition to the Governor and Lieut.-Governor (who are elected for four years), the constitution provides for a cabinet composed of an attorney general, a chief financial officer and a commissioner of agriculture.

For the 113th Congress, which convened in Jan. 2013, Florida sends 27 members to the House of Representatives. It is represented in the Senate by Bill Nelson (D. 2001–19) and Marco Rubio (R. 2011–17).

The state capital is Tallahassee. The state is divided into 67 counties.

RECENT ELECTIONS

In the 2012 presidential election Barack Obama took Florida with 50·0% of the vote, having also won it in 2008.

CURRENT GOVERNMENT

Governor: Rick Scott (R.), 2011–15 (salary: $0·12).

Lieut.-Governor: Jennifer Carroll (R.), 2011–15 ($124,851).

Secretary of State: Kenneth W. Detzner (R.), appointed Jan. 2012 ($140,000·04).

Government Website: http://www.myflorida.com

ECONOMY

Per capita personal income (2010) was $38,222.

Budget

In 2011 total state revenue was $106,166m. Total expenditure was $84,633m. (including: education, $24,883m.; public welfare, $22,303m.; highways, $5,449m.; health, $3,750m.; government administration, $2,758m.) Outstanding debt in 2011, $43,472m.

Performance

2011 Gross Domestic Product by state was $661,091m. (provisional), ranking Florida 4th in the United States. In 2011 state real GDP growth was 0·5% (provisional).

Banking and Finance

In 2002 there were 301 financial institutions in Florida insured by the US Federal Deposit Insurance Corporation, with assets worth $99,900m. They had 4,626 offices with total deposits of $242,800m.

ENERGY AND NATURAL RESOURCES

Electricity

In 2009 production was 218·0bn. kWh, of which 118·3bn. kWh was from natural gas.

Oil and Gas

2008 production of crude oil was 2m. bbls.

Water

The total area covered by water is 12,133 sq. miles (the third largest of any state after Alaska and Michigan), of which 5,027 sq. miles are inland.

Minerals

The chief mineral is phosphate rock, of which marketable production in 2002 was 27m. tonnes. This was approximately 75% of US and 25% of the world supply of phosphate in 2002. Other important non-fuel minerals include crushed stone, cement, and sand and gravel. Total non-fuel mineral production for 2009 was valued at $4,250m.

Agriculture

In 2007 there were 9·23m. acres of farmland; 47,463 farms with an average of 195 acres per farm. The total value of land and buildings was $52,054m. in 2007; average value (2007) of land and buildings per acre, $5,639.

Farm income from crops and livestock (2006) was $6,974m., of which crops provided $5,669m. and livestock $1,305m. Major crop contributors are greenhouse products, oranges, sugarcane, tomatoes, grapefruit, peppers, other winter vegetables and indoor and landscaping plants. The net farm income in 2006 was $2,340m. In 2007 the state had 1·71m. cattle, including 119,900 dairy cows, and 19,900 hogs and pigs, plus 11·79m. laying hens.

Forestry

In 2007 Florida had 16·15m. acres of forested land (11·43m. acres privately-owned), including 1·07m. acres of national forests.

Fisheries

Florida has extensive fisheries with shrimp the highest value fish commodity. Other important catches are spiny lobster, snapper, crabs, hard clams, swordfish and tuna. Commercial catch (2009) totalled 92·8m. lb of fish with a value of $156·0m.

INDUSTRY

In 2007 the state's 14,324 manufacturing establishments had 355,000 employees, earning $15,227m. Total value added by manufacturing in 2009 was $43,793m. Main industries include: printing and publishing, machinery and computer equipment, apparel and finished products, fabricated metal products, and lumber and wood products.

Labour

Total non-agricultural employment, 2010, 7,175,000. Employees by branch, 2010 (in 1,000): trade, transportation and utilities, 1,455; government, 1,115; education and health services, 1,079; professional and business services, 1,036; leisure and hospitality, 918. In Dec. 2010 the unemployment rate was 12·0%.

INTERNATIONAL TRADE

Imports and Exports

Export sales of merchandise in 2000 totalled $24·2bn. Florida exported to 213 foreign markets in 2000: Canada was the biggest (10·3% of exports), followed by Brazil (8·4%) and Mexico (8·1%). Other important markets include Japan, Venezuela, Dominican Republic, UK, Colombia, Germany, Argentina and China. The leading export category is computers and electronic products (accounting for 33% of total exports in 2000). Other manufactured exports include transportation equipment, machinery, chemicals, electrical equipment, appliances and parts, and miscellaneous manufactures. The state also exports significant quantities of farm products and other non-manufactured commodities. Total agricultural exports were worth $1·2bn. in 2001.

COMMUNICATIONS

Roads

The state (2007) had 121,526 miles of highways, roads and streets (81,270 miles being urban roads and 40,256 rural). In 2007 there were 16,473,908 vehicle registrations and 3,214 traffic accident fatalities.

Rail

In 2009 there were 2,875 miles of freight railroad (excluding trackage rights). There were 14 freight railroads operating in 2009. Rail traffic originating in Florida in 2009 totalled 37·7m. tons and rail traffic terminating in the state came to 60·5m. tons. There is a metro of 22 miles, a peoplemover and a commuter rail route in Miami.

Civil Aviation

In 2011 Florida had 19 primary airports (commercial service airports with more than 10,000 passenger boardings annually), seven of which had more than 1m. passenger enplanements. There were 69,319,400 passenger enplanements at the primary airports in 2011 (up from 66,716,652 in 2010). The busiest airports in 2011 were Miami International (18,342,158 enplanements), Orlando International (17,250,415) and Fort Lauderdale/Hollywood International (11,332,466).

Shipping

There are 14 deepwater ports: those on the Gulf coast handle mainly domestic trade and those on the Atlantic coast primarily international trade and cruise ship traffic. The major ports are Tampa (which handled 34·9m. short tons of cargo in 2009), Port Everglades (20·1m. short tons in 2009), Jacksonville, Miami and Port Manatee.

SOCIAL INSTITUTIONS

Justice

The state resumed the use of the death penalty in 1979. There have been 74 executions since 1977, including two in 2011 and three in 2012. In Dec. 2008 there were 102,388 federal and state prisoners, up 4·2% from 98,219 in Dec. 2007. Chain gangs were introduced in 1995.

Education

Attendance at school is compulsory between six and 16. In 2007–08 there were 3,935 public elementary and secondary schools with 2,666,811 enrolled pupils and 168,737 teachers. According to the National Center for Education Statistics, Florida's public elementary and high schools have the highest average enrolment in the country: in the 2006–07 school year there were 654 pupils per elementary school (compared to a national average of 446), 933 pupils per middle school (ranking Florida second behind Nevada—national average 593) and 1,717 pupils per high school (nearly twice the national average of 876). Total expenditure on public elementary and secondary education in 2006–07 was $30,108m. and the average teacher salary was $48,126 in 2008–09. Spending per pupil in fiscal year 2008 was $9,084.

In 2008–09 there were 188 degree-granting institutions (40 public and 148 private); there were 913,793 students at public degree-granting institutions in fall 2007. There are 11 state universities with a total of 258,874 students in 2002: the University of Florida at Gainesville (founded 1853) with 46,850 students; the Florida State University at Tallahassee (founded in 1857) with 36,651; the University of South Florida at Tampa (founded 1960) with 37,764; Florida A. & M. (Agricultural and Mechanical) University at Tallahassee (founded 1887) with 12,467; Florida Atlantic University (founded 1964) at Boca Raton with 23,996; the University of West Florida at Pensacola with 9,206; the University of Central Florida at Orlando with 38,795; the University of North Florida at Jacksonville with 13,460; Florida International University at Miami with 33,799; Florida Gulf Coast University (founded 1997) at Fort Myers with 5,236; and New College of Florida (founded 2001) at Sarasota with 650. There are 28 private colleges and universities belonging to the Independent Colleges and Universities of Florida (ICUF) association. Their enrolments vary from fewer than 100 to more than 22,000 students.

Health

In 2009 there were 210 community hospitals with 53,300 beds. A total of 2,453,000 patients were admitted during the year. There were 46,645 active physicians in 2009 (25 per 10,000 population).

Welfare

Medicare enrolment in July 2010 totalled 3,334,266. In fiscal year 2009 a total of 3,261,031 people in Florida received Medicaid. In Dec. 2008 there were 3,547,492 Old-Age, Survivors, and Disability Insurance (OASDI) beneficiaries. A total of 94,824 people were receiving payments under Temporary Assistance for Needy Families (TANF) in Dec. 2008.

RELIGION

The principal religious traditions in the state in 2010 were: Evangelical Protestants, with 3,049,524 members; Catholics, 2,515,243; Mainline Protestants, 870,959; Black Protestants, 333,390; Muslims, 165,000 (estimate).

CULTURE

Tourism

Overseas visitors to Florida—excluding those from Canada and Mexico—numbered 5,688,000 in 2011 (down from 5,826,000 in 2010 although up from 4,117,000 in 2006). There were 3,102,000 visits by Canadians in 2010. Florida ranks among the most visited states and Miami and Orlando among the most visited cities.

FURTHER READING

Statistical information: Bureau of Economic and Business Research, Univ. of Florida, Gainesville 32611. Publishes *Florida Statistical Abstract.*

Benton, J. E. (ed.) *Government and Politics in Florida.* 3rd ed. 2008
Morris, A., *The Florida Handbook.* Biennial

State library: 500 S Bronough Street, Tallahassee 32399.

Georgia

KEY HISTORICAL EVENTS

Originally the territory of Creek and Cherokee tribes, Georgia was first settled by Europeans in the 18th century. James Oglethorpe founded Savannah in 1733, intending it as a colony offering a new start to debtors, convicts and the poor. Settlement was slow until 1783, when growth began in the cotton-growing areas west of Augusta. The Indian population was cleared off the rich cotton land and moved beyond the Mississippi. Georgia became one of the original 13 states of the Union.

A plantation economy developed rapidly, using slave labour. In 1861 Georgia seceded from the Union and became an important source of supplies for the Confederate cause, although some northern areas never accepted secession and continued in sympathy with the Union during the Civil War. At the beginning of the war 56% of the population were white, descendants of British, Austrian and New England immigrants; the remaining 44% were black slaves.

The city of Atlanta, which grew as a railway junction, was destroyed during the war but revived to become the centre of southern reconstruction in the post-war period. It was confirmed as the state capital in 1877. Successive movements for black freedom in social, economic and political life have developed in the city, notably the Southern Christian Leadership Conference, led by Martin Luther King, who was assassinated in 1968.

TERRITORY AND POPULATION

Georgia is bounded north by Tennessee and North Carolina, northeast by South Carolina, east by the Atlantic, south by Florida and west by Alabama. Land area, 57,513 sq. miles (148,959 sq. km). Census population, 1 April 2010, was 9,687,653, an increase of 18·3% since 2000. July 2012 estimate, 9,919,945.

Population in five census years was:

	White	Black	American Indian	Asiatic	Total	Per sq. mile
1910	1,431,802	1,176,987	95	237	2,609,121	44·4
			All others			
1980	3,948,007	1,465,457	50,801		5,464,265	92·7
1990	4,600,148	1,746,565	131,503		6,478,216	110·0
2000	5,327,281	2,349,542	509,630		8,186,453	141·4
2010	5,787,440	2,950,435	949,778		9,687,653	168·4

Of the total population in 2010, 4,958,482 were female and 7,196,101 were 18 years old or older. The estimated Hispanic population was 853,689 in 2010, up from 435,277 in 2000 (an increase of 96·1%).

The largest cities are: Atlanta (capital), with a population (2010 census) of 420,003; Augusta-Richmond County, 200,549; Columbus, 189,885; Savannah, 136,286; Athens-Clarke County, 116,714. The Atlanta–Sandy Springs–Marietta metropolitan area had a 2010 census population of 5,268,860; Augusta-Richmond County, 556,877; Savannah, 347,611.

SOCIAL STATISTICS

Births, 2010, 133,947 (13·8 per 1,000 population); deaths, 2010 (provisional), 71,323 (7·4). Infant mortality, 2009, 7·4 per 1,000 live births. Marriages, 2009 (provisional), 63,600 (6·5 per 1,000 population).

CLIMATE

Atlanta, Jan. 43°F (6·1°C), July 78°F (25·6°C). Annual rainfall 49" (1,234 mm). Georgia belongs to the Atlantic Coast climate zone (see UNITED STATES: Climate).

CONSTITUTION AND GOVERNMENT

A new constitution was ratified in the general election of 2 Nov. 1976, proclaimed on 22 Dec. 1976 and became effective on 1 Jan. 1977. The General Assembly consists of a Senate of 56 members and a House of Representatives of 180 members, both elected for two years. Legislative sessions are annual, beginning the 2nd Monday in Jan. and lasting for 40 days.

Georgia was the first state to extend the franchise to all citizens 18 years old and above.

For the 113th Congress, which convened in Jan. 2013, Georgia sends 14 members to the House of Representatives. It is represented in the Senate by Saxby Chambliss (R. 2003–15) and Johnny Isakson (R. 2005–17).

The state capital is Atlanta. Georgia is divided into 159 counties.

RECENT ELECTIONS

In the 2012 presidential election Mitt Romney took Georgia with 53·4% of the vote (John McCain won it in 2008).

CURRENT GOVERNMENT

Governor: Nathan Deal (R.), 2011–15 (salary: $139,339·44).
    Lieut.-Governor: Casey Cagle (R.), 2011–15 ($91,609·44).
    Secretary of State: Brian Kemp (R.), 2011–15 ($130,690·80).

Government Website: http://www.georgia.gov

ECONOMY

Per capita personal income (2010) was $34,800.

Budget

In 2011 total state revenue was $52,295m. Total expenditure was $44,750m. (education, $17,429m.; public welfare, $10,367m.; highways, $1,641m.; correction, $1,462m.; health, $1,159m.) Outstanding debt in 2011, $13,403m.

Performance

Gross Domestic Product by state was $365,809m. in 2011 (provisional), ranking Georgia 11th in the United States. In 2011 state real GDP growth was 1·7% (provisional).

ENERGY AND NATURAL RESOURCES

Electricity

In 2009 production was 128·7bn. kWh, of which 69·5bn. kWh was from coal.

Water

The total area covered by water is 1,912 sq. miles.

Minerals

Georgia is the leading producer of kaolin. The state ranks first in production of crushed and dimensional granite, and second in production of fuller's earth and marble (crushed and dimensional). Total value of non-fuel mineral production for 2009 was $1,410m.

Agriculture

In 2007, 47,846 farms covered 10·15m. acres; the average farm was of 212 acres. In 2007 the average value of farmland and buildings was $3,117 per acre. The major product is broilers; in 2007, 1·40bn. were produced (the most in any state) with a value of $3·19bn. For 2007 cotton output was 1·6m. bales (of 480 lb). Other major crops include tobacco, corn, wheat, soybeans, peanuts and pecans. Cash receipts from farm marketings, 2006:

crops, $2,240m.; livestock and products, $3,765m.; total, $6,005m. The net farm income in 2006 was $2,388m.

In 2007 farm animals included 1·12m. cattle and 236,500 swine; there were also 19·27m. laying hens.

Forestry

The forested area in 2007 was 24·78m. acres with 736,000 acres of national forest. Annual timber removals are the highest of any state, at 1,341m. cu. ft in 2006 (1,044m. cu. ft softwoods and 297m. cu. ft hardwoods).

INDUSTRY

In 2007 the state's 8,699 manufacturing establishments had 411,000 employees, earning $16,128m. Total value added by manufacturing in 2009 was $53,437m.

Labour

Total non-agricultural employment, 2010, 3,826,000. Employees by branch, 2010 (in 1,000): trade, transportation and utilities, 808; government, 678; professional and business services, 519; education and health services, 486; leisure and hospitality, 374. Georgia's unemployment rate in Dec. 2010 was 10·4%.

INTERNATIONAL TRADE

Imports and Exports

In 2010 exports from Georgia totalled $29·0bn., up from $23·7bn. in 2009.

COMMUNICATIONS

Roads

In 2007 there were 118,779 miles of roads comprising 37,683 miles of urban road and 81,096 miles of rural road. There were 8,512,511 motor vehicles registered.

Rail

In 2011 there were 4,971 miles of freight railroad, including 3,538 miles of Class I railroads. There is a 48-mile heavy-rail subway system serving Metropolitan Atlanta.

Civil Aviation

In 2011 there were seven primary airports (commercial service airports with more than 10,000 passenger boardings annually) with a combined total of 45,649,069 enplanements, up from 44,347,182 in 2010. Hartsfield–Jackson Atlanta International Airport handled a record 44,414,121 passenger enplanements in 2011—the highest number of any airport in the world.

Shipping

There are deepwater ports at Savannah, the principal port, and Brunswick.

SOCIAL INSTITUTIONS

Justice

In Dec. 2008 there were 52,719 federal and state prisoners. The death penalty is authorized for capital offences. There were four executions in 2011 but none in 2012.

Under a Local Option Act, the sale of alcoholic beverages is prohibited in some counties.

Education

Since 1945 education has been compulsory; tuition is free until the age of 18 and school attendance is compulsory for pupils between the ages of six and 16 years. In 2007–08 there were 2,452 public elementary and secondary schools with 1·65m. pupils and 116,857 teachers; total expenditure on public elementary and secondary education (2006–07) was $17,301m. Teachers' salaries averaged $53,270 in 2008–09. In 2006–07 Georgia's public elementary schools had the nation's second highest average enrolment at 624 students.

The University of Georgia (Athens) was founded in 1785 and was the first chartered State University in the USA (34,885 students in fall 2009). Other institutions of higher learning include Georgia Institute of Technology, Atlanta (20,291); Emory University, Atlanta (12,930); Georgia State University, Atlanta (30,427); Georgia Southern University, Statesboro (19,086). The Atlanta University Center, devoted primarily to Black education, includes co-educational Clark Atlanta University (3,873); Morehouse College (2,689), a liberal arts college for men; Interdenominational Theological Center (421), a co-educational school; and Spelman College (2,229), the first liberal arts college for Black women in the USA. Wesleyan College (685), near Macon, is the oldest chartered women's college in the world.

Health

In 2009 there were 152 community hospitals with 25,400 beds. A total of 957,000 patients were admitted during the year. There were 21,269 active physicians in 2009 (22 per 10,000 population).

Welfare

Medicare enrolment in July 2010 totalled 1,224,221. In fiscal year 2009 a total of 1,804,917 people in Georgia received Medicaid. In Dec. 2008 there were 1,347,932 Old-Age, Survivors, and Disability Insurance (OASDI) beneficiaries. A total of 39,222 people were receiving payments under Temporary Assistance for Needy Families (TANF) in Dec. 2008.

RELIGION

The principal religious traditions in the state in 2010 were: Evangelical Protestants, with 2,853,360 members; Mainline Protestants, 855,259; Catholics, 596,384; Black Protestants, 381,421; Latter-day Saints (Mormons), 78,177.

CULTURE

Tourism

Overseas visitors to Georgia in 2011—excluding those from Canada and Mexico—numbered 669,000, down from 817,000 in 2010 although up from 520,000 in 2006.

FURTHER READING

Statistical information: Selig Center for Economic Growth, Univ. of Georgia, Athens 30602. Publishes *Georgia Statistical Abstract.*

State Law Library: Judicial Building, Capital Sq., Atlanta.

Hawaii

KEY HISTORICAL EVENTS

The islands of Hawaii were settled by Polynesian immigrants, probably from the Marquesas Islands, about AD 400. A second major immigration, from Tahiti, occurred around 800–900. In the late 18th century all the islands were united into one kingdom by Kamehameha I. Western exploration began in 1778, and Christian missions were established after 1820. Europeans called Hawaii the Sandwich Islands. The USA, Britain and France all claimed an interest. Kamehameha III placed Hawaii under US protection in 1851. US sugar-growing companies became dominant and in 1887 the USA obtained a naval base at Pearl Harbor. A struggle developed between forces for and against annexation by the USA. In 1893 the monarchy was overthrown. The republican government agreed to be annexed to the USA in 1898, and Hawaii became a US Territory in 1900.

The islands and the naval base were of great strategic importance during the Second World War, when the Japanese attack on Pearl Harbor brought the USA into the war.

Hawaii became the 50th state of the Union in 1959. The 19th-century plantation economy encouraged the immigration of workers, especially from China and Japan. Hawaiian laws,

religions and culture were gradually adapted to the needs of the immigrant community.

TERRITORY AND POPULATION

The Hawaiian Islands lie in the North Pacific Ocean, between 18° 54' and 28° 15' N. lat. and 154° 40' and 178° 25' W. long., about 2,090 nautical miles southwest of San Francisco. There are 137 named islands and islets in the group, of which seven major and five minor islands are inhabited. Land area, 6,423 sq. miles (16,635 sq. km). Census population, 1 April 2010, was 1,360,301, an increase of 12·3% since 2000. July 2012 estimate, 1,392,313. Of the total population in 2010, 681,243 were male and 1,056,483 were 18 years old or older.

The principal islands are Hawaii, 4,028 sq. miles, population 2010, 185,079; Maui, 727 sq. miles, population 144,444; Oahu, 600 sq. miles, population 953,207; Kauai, 552 sq. miles, population 66,921; Molokai, 260 sq. miles, population 7,345; Lanai, 141 sq. miles, population 3,135; Niihau, 70 sq. miles, population 170; Kahoolawe, 45 sq. miles (uninhabited). The capital Honolulu—on the island of Oahu—had a population in 2010 of 387,170.

Figures for main racial groups, 2010, were: 336,599 White; 197,497 Filipino; 185,502 Japanese; 80,337 Native Hawaiian; 54,955 Chinese; 24,203 Korean; 21,424 Black or African American; 18,287 Samoan; 9,779 Vietnamese.

SOCIAL STATISTICS

Births, 2010, 18,988 (14·0 per 1,000 population); deaths, 2010 (provisional), 9,617 (7·1). Infant deaths, 2009, were at a rate of 6·1 per 1,000 live births. Marriages, 2009 (provisional), 22,200 (17·9 per 1,000 population). Inter-marriage between the races is common. In 2007, of the 9,401 resident couples married, 55·1% were inter-racial. 65·6% were non-resident marriages.

CLIMATE

All the islands have a tropical climate, with an abrupt change in conditions between windward and leeward sides, most marked in rainfall. Temperatures vary little. Average temperatures in Honolulu: Jan. 73·0°F, July 80·8°F. Average annual rainfall in Honolulu: 18·29".

CONSTITUTION AND GOVERNMENT

Hawaii was officially admitted into the United States on 21 Aug. 1959. However, the constitution of the State of Hawaii was created by the 1950 Constitutional Convention, ratified by the voters of the Territory on 7 Nov. 1950, and amended on 27 June 1959. The Legislature consists of a Senate of 25 members elected for four years and a House of Representatives of 51 members elected for two years. There have been two constitutional conventions since 1950, in 1968 and 1978. In addition to amendments proposed by these conventions the Legislature is able to propose amendments to voters during the general election. This has resulted in numerous amendments.

For the 113th Congress, which convened in Jan. 2013, Hawaii sends two members to the House of Representatives. It is represented in the Senate by Brian Schatz (D. 2012–17) and Mazie Hirono (D. 2013–19).

The state capital is Honolulu. There are five counties.

RECENT ELECTIONS

In the 2012 presidential election Barack Obama took Hawaii with 70·6% of the vote, having also won it in 2008.

CURRENT GOVERNMENT

Governor: Neil Abercrombie (D.), Dec. 2010–Dec. 2014 (salary: $117,312).

Lieut.-Governor: Shan S. Tsutsui (D.), Dec. 2010–Dec. 2014 ($114,420).

Government Website: http://www.ehawaii.gov

ECONOMY

Per capita personal income (2010) was $41,661.

Budget
Revenue is derived mainly from taxation of sales and gross receipts, real property, corporate and personal income, and inheritance taxes, licences, public land sales and leases.

In 2011 total state revenue was $12,932m. Total expenditure was $11,476m. (education, $3,345m.; public welfare, $2,087m.; hospitals, $667m.; health, $620m.; government administration, $417m.) Outstanding debt in 2011, $7,913m.

Performance
2011 Gross Domestic Product by state was $57,977m. (provisional), ranking Hawaii 38th in the United States. In 2011 state real GDP growth was –0·2% (provisional).

ENERGY AND NATURAL RESOURCES

Electricity
In 2009 production was 11,011m. kWh, of which 8,289m. kWh was from petroleum.

Oil and Gas
In 2011, $139·5m. was generated by gas sales.

Water
The total area covered by water is 4,509 sq. miles.

Minerals
Production in 2008: crushed stone, 7·5m. tonnes; construction sand and gravel, 1·4m. tonnes. Total value of non-fuel mineral production in 2009 was $116m.

Agriculture
Farming is highly commercialized and highly mechanized. In 2007 there were 7,521 farms covering an area of 1·12m. acres; average number of acres per farm, 149, valued at $7,688 per acre.

Greenhouse products, pineapples, sugarcane and macadamia nuts are the staple crops. Farm income, 2006, from crop sales was $467m., and from livestock $88m. The net farm income in 2006 was $106m.

Forestry
Hawaii had 1·75m. acres of forested land in 2007 (1·16m. acres private and 0·59m. acres public). Timberland area totalled 0·70m. acres in 2007.

Fisheries
In 2009 the commercial fish catch was 26·91m. lb with a value of $71·2m.

INDUSTRY

In 2007 the state's 984 manufacturing establishments had 14,000 employees, earning $511m. Total value added by manufacturing in 2009 was $1,703m.

Labour
Total non-agricultural employment amounted to 587,000 in 2010. Employees by branch, 2010 (in 1,000): government, 125; trade, transportation and utilities, 109; leisure and hospitality, 100; education and health services, 76; professional and business services, 71. The unemployment rate in Dec. 2010 was 6·3%.

COMMUNICATIONS

Roads
In 2007 there were 4,341 miles of roads comprising 2,300 miles of urban road and 2,041 miles of rural road. There were 993,117 registered motor vehicles.

Civil Aviation
Hawaii had seven primary airports in 2011 (commercial service airports with more than 10,000 passenger boardings annually), with a combined total of 14,603,529 enplanements. The busiest airport, Honolulu International, had 8,689,699 enplanements in 2011 (down from 8,740,077 in 2010). Hawaiian Airlines carried a record 8,666,319 passengers in 2011, up from 8,424,288 in 2010.

Shipping
Several lines of steamers connect the islands with the mainland USA, Canada, Australia, the Philippines, China and Japan. In 2002, 1,270 overseas and 2,663 inter-island vessels entered the port of Honolulu carrying a total of 130,792 overseas and 19,952 inter-island passengers as well as 6,425,288 tonnes of overseas and 1,796,910 tonnes inter-island cargo.

SOCIAL INSTITUTIONS

Justice
The death penalty was abolished in Hawaii in 1948. In Dec. 2008 the jail and prison population totalled 5,955.

Education
Education is free and compulsory between the ages of six and 18. The language in the schools is English. In 2007–08 there were 287 public schools with 179,897 pupils and 11,397 teachers. In 2007 there were 140 private schools with 37,300 pupils and 2,880 teachers. In 2006–07, $2,300m. was spent on public elementary and secondary education; average teacher salary was $55,733 in 2008–09. In fall 2007 the number of students enrolled in degree-granting institutions was 66,601.

Health
In 2009 there were 25 community hospitals with 3,000 beds. A total of 112,000 patients were admitted during the year. There were 4,800 active physicians in 2009 (37 per 10,000 population).

Welfare
Medicare enrolment in July 2010 totalled 205,161. In fiscal year 2007 a total of 223,687 people in Hawaii received Medicaid. In Dec. 2008 there were 212,890 Old-Age, Survivors, and Disability Insurance (OASDI) beneficiaries. A total of 14,331 people were receiving payments under Temporary Assistance for Needy Families (TANF) in Dec. 2008.

RELIGION

The principal religious traditions in the state in 2010 were: Catholics, with 249,619 members; Evangelical Protestants, 130,265; Latter-day Saints (Mormons), 69,872; Buddhists, 65,759; Mainline Protestants, 40,833.

CULTURE

Tourism
Tourism is outstanding in Hawaii's economy. There were 6·4m. tourist arrivals by air in 2009. Tourist expenditure (air visitors only) contributed $9,794·3m. to the state's economy in 2009.

FURTHER READING

Statistical information: Hawaii State Department of Business, POB 2359, Honolulu 96804. Publishes *The State of Hawaii Data Book.*
Atlas of Hawaii. 3rd ed. 1998

Oliver, Anthony M., *Hawaii Facts and Reference Book: Recent Historical Facts and Events in the Fiftieth State.* 1995

Idaho

KEY HISTORICAL EVENTS

Kutenai, Kalispel, Nez Percé and other tribes lived on the Pacific watershed of the northern Rocky Mountains. European exploration began in 1805, and after 1809 there were trading posts and small settlements, with fur-trapping as the primary activity. The area was disputed between Britain and the USA until 1846 when British claims were dropped. In 1860 the discovery of gold and silver brought a rush of immigrant prospectors. An area including present-day Montana was created a Territory in March 1863. Montana was separated from it in 1864. Population growth was stimulated by refugees from the Confederate states after the Civil War and by settlements of Mormons from Utah.

Fur-trapping and mining gave way to arable farming. Idaho became a state in 1890, with its capital at Boise. The population of the Territory capital, Idaho City, a gold-mining boom town in the 1860s (about 40,000 at its height), was the largest in the Pacific Northwest. By 1869 the population was down to 1,000.

In the 20th century the Indian population shrank to 1%. The Mormon community has grown to include much of southeastern Idaho and more than half the church-going population of the state.

The Snake River of southern Idaho has supported hydro-electricity and irrigation. Food processing, minerals and timber are important. So too are the high technology companies in Idaho's metropolitan areas. Much of the state, however, remains sparsely populated and rural.

TERRITORY AND POPULATION

Idaho is within the Rocky Mountains and bounded north by Canada, east by Montana and Wyoming, south by Nevada and Utah, west by Oregon and Washington. Land area, 82,643 sq. miles (214,045 sq. km). In 2010 the federal government owned 61·7% of the state area. Census population, 1 April 2010, was 1,567,582, an increase of 21·1% since 2000. July 2012 estimate, 1,595,728.

Population in five census years was:

	White	Black	American Indian/Alaska Native	Asian	Total (including others)	Per sq. mile
1910	319,221	651	3,488	2,234	325,594	3·9
1980	901,641	2,716	10,521	5,948	943,935	11·3
1990	950,451	3,370	13,780	9,365	1,006,749	12·2
2000	1,177,304	5,456	17,645	13,197	1,293,953	15·6
2010	1,396,487	9,810	21,441	19,069	1,567,582	19·0

Of the total population in 2010, 785,324 were male and 1,138,510 were 18 years old or older. In 2010 Idaho's Hispanic population was 175,901, up from 101,690 in 2000 (an increase of 73·0%).

The largest cities are: Boise City (the capital), with a 2010 population of 205,671; Nampa, 81,557; Meridian, 75,092; Idaho Falls, 56,813; Pocatello, 54,255; Caldwell, 46,237; Coeur d'Alene, 44,137; Twin Falls, 44,125; Lewiston, 31,894.

SOCIAL STATISTICS

Births, 2010, 23,198 (14·8 per 1,000 population); deaths, 2010 (provisional), 11,430 (7·3). Infant mortality rate, 2009, 5·4 per 1,000 live births. 2009 (provisional): marriages, 13,900 (8·9 per 1,000 population); divorces and annulments, 7,700 (5·5).

CLIMATE

Boise City, Jan. 29°F (−1·7°C), July 74°F (23·3°C). Annual rainfall 12" (303 mm). Idaho belongs to the Mountain States climate zone (*see* UNITED STATES: Climate).

CONSTITUTION AND GOVERNMENT

The constitution adopted in 1890 is still in force; it has had 135 amendments as of 2011. The Legislature consists of a Senate of 35 members and a House of Representatives of 70 members, all the legislators being elected for two years. It meets annually.

For the 113th Congress, which convened in Jan. 2013, Idaho sends two members to the House of Representatives. It is represented in the Senate by Michael Crapo (R. 1999–2017) and James Risch (R. 2009–15).

The state is divided into 44 counties. The capital is Boise City.

RECENT ELECTIONS

In the 2012 presidential election Mitt Romney took Idaho with 64·5% of the vote (John McCain won it in 2008).

CURRENT GOVERNMENT

Governor: C. L. 'Butch' Otter (R.), 2011–15 (salary: $117,000).
  Lieut.-Governor: Brad Little (R.), 2011–15 ($35,100).
  Secretary of State: Ben Ysursa (R.), 2011–15 ($99,450).

Government Website: http://www.idaho.gov

ECONOMY

Per capita personal income (2010) was $31,986.

Budget

In 2011 total state revenue was $10,776m. Total expenditure was $8,733m. (education, $2,701m.; public welfare, $2,175m.; highways, $836m.; government administration, $273m.; correction, $219m.) Outstanding debt in 2011, $3,928m.

Performance

Gross Domestic Product by state in 2011 was $51,463m. (provisional), ranking Idaho 42nd in the United States. In 2011 state real GDP growth was 0·6% (provisional).

ENERGY AND NATURAL RESOURCES

Electricity

In 2009 production was 13,100m. kWh, of which 10,434m. kWh was from conventional hydro-electric energy.

Water

The total area covered by water is 926 sq. miles.

Minerals

Principal non-fuel minerals are processed phosphate rock, silver, gold, molybdenum and sand and gravel. Value of non-fuel mineral output for 2009 was $935m.

Agriculture

Agriculture is the second largest industry, despite a great part of the state being naturally arid. Extensive irrigation works have been carried out, bringing an estimated 3·5m. acres under irrigation, and there are over 50 soil conservation districts.

In 2009 there were 25,500 farms with a total area of 11·4m. acres; average value per acre (Jan. 2010), $2,100. In 2009 the average farm was 447 acres.

Value of crop production in 2009, $2,664m.; and livestock and products, $2,549m. The most important crops are potatoes and wheat. Other crops are sugar beets, hay, barley, field peas and beans, onions and apples. The net farm income in 2009 was $927m. In Dec. 2009 there were 2·1m. cattle and calves, 220,000 sheep and 36,000 hogs and pigs. There were 30m. food-sized trout produced on fish farms in 2009. The dairy industry is the fastest-growing sector in Idaho agriculture.

Forestry

In 2007 there was a total of 21·43m. acres of forest, of which 16·38m. acres were national forest (the highest acreage of any state).

Fisheries

73% of the commercial trout processed in the USA was produced in Idaho in 2009. Idaho ranked first in state trout production by producing fish to a value of $35·6m.

INDUSTRY

In 2007 Idaho's 1,942 manufacturing establishments had 65,000 employees, with a payroll of $2,829m. Total value added by manufacturing in 2009 was $8,117m.

Labour

Total non-agricultural employment, 2010, 603,000. Employees by branch, 2010 (in 1,000): trade, transportation and utilities, 121; government, 119; education and health services, 84; professional and business services, 73; leisure and hospitality, 58. The unemployment rate in Dec. 2010 was 9·7%.

COMMUNICATIONS

Roads

In 2007 there were 48,416 miles of public roads (42,810 miles rural, 5,606 urban). There were 1,281,899 registered motor vehicles in 2007.

Rail

The state had (2009) 1,630 miles of railroads (including one Amtrak route).

Civil Aviation

There were 68 municipally-owned airports in 2009. There were 1,904,446 passenger enplanements statewide in 2008.

Shipping

Water transport is provided from the Pacific to the port of Lewiston, by way of the Columbia and Snake rivers, a distance of 464 miles.

SOCIAL INSTITUTIONS

Justice

The death penalty may be imposed for first degree murder or aggravated kidnapping, but the judge must consider mitigating circumstances before imposing a sentence of death. Since 1976 there have only been three executions, in 1994, 2011 and 2012. In Dec. 2008 there were 7,290 prisoners in federal and state prisons.

Education

In 2007–08 there were 727 public schools with 272,229 pupils and 15,013 teachers. Average salary (2008–09) of teachers was $45,439. Total expenditure on public elementary and secondary education in 2006–07 was $2,149m.

The University of Idaho, founded at Moscow in 1889, had 928 faculty in fall 2009, and a total enrolment of 11,957. Boise State University had 611 faculty in fall 2009 and a total enrolment of 18,936. Idaho State University had 602 full-time faculty in fall 2009 and a total enrolment of 13,493. Total enrolment in degree-granting institutions in fall 2007 was 78,846 (60,526 in public institutions).

Health

In 2009 there were 41 community hospitals with 3,400 beds. A total of 130,000 patients were admitted during the year. There were 2,649 active physicians in 2009 (17 per 10,000 population).

Welfare

Medicare enrolment in July 2010 totalled 227,873. In fiscal year 2009 a total of 252,785 people in Idaho received Medicaid. Old-age, Survivors, and Disability Insurance (OASDI) is granted to persons if they paid sufficiently into the system or meet other qualifications; in Dec. 2008 there were 247,847 beneficiaries. A total of 2,396 people were receiving payments under Temporary Assistance for Needy Families (TANF) in Dec. 2008.

RELIGION

In 2010 the chief religious traditions were: Latter-day Saints (Mormons), with 410,289 members; Evangelical Protestants, 201,546; Catholics, 123,400; Mainline Protestants, 57,056.

FURTHER READING

Statistical information: Idaho Commerce and Labor, 700 West State St., Boise 83720. Publishes *County Profiles of Idaho, Community Profiles of Idaho* and *Profile of Rural Idaho* on the Internet.

Schwantes, C. A., *In Mountain Shadows: a History of Idaho.* 1996

Website: http://labor.idaho.gov

Illinois

KEY HISTORICAL EVENTS

Home to Algonquian-speaking tribes, Illinois was explored first by the French in 1673. France claimed the area until 1763 when, after the French and Indian War, it was ceded to Britain along with all the French land east of the Mississippi. In 1783 Britain recognized US claims to Illinois, which became part of the North West Territory of the USA in 1787, and of Indiana Territory in 1800. Illinois became a Territory in its own right in 1809, and a state in 1818.

Immigration increased greatly with the opening in 1825 of the Erie Canal from New York, along which farmer settlers could move west and their produce back east for sale. Chicago was incorporated as a city in 1837 and quickly became the transport, trading and distribution centre of the mid west. Industrial growth brought a further wave of immigration in the 1840s, mainly of European refugees. This movement continued with varying force until the 1920s, when it was largely replaced by immigration of black work-seekers from the southern states.

In the 20th century the population was urbanized and heavy industry was established along a network of rail and waterway routes. Chicago recovered from a destructive fire in 1871 to become the hub of this network and at one time the second largest American city.

TERRITORY AND POPULATION

Illinois is bounded north by Wisconsin, northeast by Lake Michigan, east by Indiana, southeast by the Ohio River (forming the boundary with Kentucky), and west by the Mississippi River (forming the boundary with Missouri and Iowa). Land area: 55,519 sq. miles (143,793 sq. km). Census population, 1 April 2010, was 12,830,632, an increase of 3·3% since 2000. July 2012 estimate, 12,875,255.

Population in five census years was:

	White	Black	American Indian	All others	Total	Per sq. mile
1910	5,526,962	109,049	188	2,392	5,638,591	100·6
			All others			
1980	9,233,327	1,675,398	517,793		11,426,518	203·0

	White	Black	American Indian/ Alaska Native	Asian/ Native Hawaiian/ Pacific Islander	Other	Total	Per sq. mile
1990	8,957,923	1,690,855	24,077	284,944	472,803	11,430,602	205·6
2000	9,125,471	1,876,875	31,006	428,213	957,728	12,419,293	223·4
2010	9,177,877	1,866,414	43,963	590,984	1,151,394	12,830,632	231·1

Of the total population in 2010, 6,538,356 were female and 9,701,453 were 18 years old or older. In 2010 the Hispanic population was 2,027,578 (1,530,262 in 2000).

The most populous cities (2010) are: Chicago, 2,695,598; Aurora, 197,899; Rockford, 152,871; Joliet, 147,433; Naperville, 141,853; Springfield (the capital), 116,250; Peoria, 115,007; Elgin, 108,188; Waukegan, 89,078.

Largest metropolitan area populations, 2010 census: Chicago–Joliet–Naperville, 9,461,105; Peoria, 379,186; Rockford, 349,431; Champaign–Urbana, 231,891; Springfield, 210,170.

SOCIAL STATISTICS

Births, 2010, 165,200 (12·9 per 1,000 population); deaths, 2010 (provisional), 99,838 (7·8). Infant mortality rate, 2009, 6·9 per 1,000 live births. 2009 (provisional): marriages, 72,700 (5·6 per 1,000 population); divorces and annulments, 32,700 (2·5).

CLIMATE

Chicago, Jan. 25·3°F (−3·7°C), July 75·4°F (24·1°C). Annual rainfall 38·0". Illinois belongs to the Great Lakes climate zone (*see* UNITED STATES: Climate).

CONSTITUTION AND GOVERNMENT

The present constitution became effective on 1 July 1971. The General Assembly consists of a House of Representatives of 118 members elected for two years, and a Senate of 59 members who are divided into three groups; in one, they are elected for terms of four years, four years, and two years; in the next, for terms of four years, two years, and four years; and in the last, for terms of two years, four years, and four years. Sessions are annual. The state is divided into legislative districts, in each of which one senator is chosen; each district is divided into two representative districts, in each of which one representative is chosen.

For the 113th Congress, which convened in Jan. 2013, Illinois sends 18 members to the House of Representatives. It is represented in the Senate by Richard Durbin (D. 1997–2015) and Mark Kirk (R. 2011–17).

The capital is Springfield.

RECENT ELECTIONS

In the 2012 presidential election Barack Obama took Illinois with 57·3% of the vote, having also won it in 2008.

CURRENT GOVERNMENT

Governor: Patrick Quinn (D.), 2011–15 (salary: $177,412).
  Lieut.-Governor: Sheila Simon (D.), 2011–15 ($135,669).
  Secretary of State: Jesse White (D.), 2011–15 ($156,541).

Government Website: http://www.illinois.gov

ECONOMY

Per capita personal income (2010) was $42,057.

Budget

In 2011 total state revenues amounted to $79,512m. Total expenditure was $74,655m. (public welfare, $19,508m.; education, $17,131m.; highways, $5,110m.; health, $2,234m.; correction, $1,514m.) Debt outstanding, in 2011, $64,801m.

Performance

Gross Domestic Product by state in 2011 was $582,094m. (provisional), ranking Illinois 5th in the United States. In 2011 state real GDP growth was 1·3% (provisional).

ENERGY AND NATURAL RESOURCES

Electricity

In 2009 production was 193·9bn. kWh, of which 95·5bn. kWh was from nuclear energy.

Oil and Gas

In 2008 Illinois produced 9·4m. bbls of crude petroleum.

Water

The total area covered by water is 2,395 sq. miles.

Minerals

The chief mineral product is coal. In 2009 there were 22 operative mines; output was 33·7m. short tons. Mineral production also includes sand, gravel and limestone. Value of non-fuel mineral production in 2009 was $929m.

Agriculture

In 2007 there were 76,860 farms in Illinois that contained 26·78m. acres of land. The average farm had 348 acres in 2007 and was valued at $3,792 per acre. In 2007 cash receipts from farm marketings in Illinois totalled $11·68bn. The net farm income in 2007 was $3,244m. In 2008 Illinois was the second largest producer among US states of corn and soybeans, producing 2·13bn. bu. and 428m. bu. respectively. Cash receipts for corn totalled $9·3bn. in 2007; for soybeans, $3·7bn. In 2007 there were 4·30m. hogs and pigs, 1·23m. cattle including 429,100 beef cows and 99,700 dairy cows, and 52,400 sheep and lambs.

Forestry

In 2007 there was a total of 4·53m. acres of forest (3·73m. acres private), of which 290,000 acres were national forest. Timberland area totalled 4·36m. acres in 2007.

INDUSTRY

Important industries include financial services, manufacturing, retail and transportation. In 2007 the state's 15,704 manufacturing establishments had 664,000 employees, earning $31,716m. Total value added by manufacturing in 2009 was $97,756m.

Labour

Total non-agricultural employment, 2010, 5,611,000. Employees by branch, 2010 (in 1,000): trade, transportation and utilities, 1,125; government, 857; education and health services, 833; professional and business services, 799; manufacturing, 559. In Dec. 2010 the unemployment rate was 9·2%.

INTERNATIONAL TRADE

Imports and Exports

In 2010 exports from Illinois totalled $50·1bn., up from $41·6bn. in 2009 although down from $53·7bn. in 2008.

COMMUNICATIONS

Roads

In 2007 there were 139,157 miles of roads comprising 40,952 miles of urban road and 98,205 miles of rural road. There were 9,757,004 registered motor vehicles in 2007.

Rail

Union Station, Chicago is the home of Amtrak's national hub. Amtrak trains provide service to cities in Illinois to many destinations in the USA. Illinois is also served by a metro (CTA) system, and by seven groups of commuter railroads controlled by METRA, which has many stations and serves several Illinois counties. In 2009 there were 7,313 miles of freight railroad (excluding trackage rights), including 5,864 miles of Class I railroads. Only Texas among US states has a larger rail network. There were 42 freight railroads operating in 2009. Rail traffic originating in Illinois in 2009 totalled 99·8m. tons (the second highest after Wyoming) and rail traffic terminating in the state came to 149·0m. tons (the second highest after Texas). There is also a metro system in Chicago (108 miles).

Civil Aviation

In 2011 Illinois had nine primary airports (commercial service airports with more than 10,000 passenger boardings annually) with a combined total of 42,242,807 enplanements, up from 41,944,295 in 2010. The busiest airport, Chicago O'Hare International, had 31,892,301 enplanements in 2011 (ranking it second in the USA behind Hartsfield–Jackson Atlanta International). Chicago Midway International had 9,134,576 enplanements in 2011.

Shipping

The port of Chicago handled 19,228,125 tons of cargo in 2009.

SOCIAL INSTITUTIONS

Justice

There were 45,474 federal and state prisoners in Dec. 2008.

Executions began in 1990 following the US Supreme Court's reinstatement of capital punishment in 1976, with the most recent execution being in March 1999. However, on 31 Jan. 2000 the death penalty was suspended and it was formally abolished on 9 March 2011.

A Civil Rights Act (1941), as amended, bans all forms of discrimination by places of public accommodation, including inns, restaurants, retail stores, railroads, aeroplanes, buses, etc., against persons on account of 'race, religion, colour, national ancestry or physical or mental handicap'; another section similarly mentions 'race or colour'.

The Fair Employment Practices Act of 1961, as amended, prohibits discrimination in employment based on race, colour, sex, religion, national origin or ancestry, by employers, employment agencies, labour organizations and others. These principles are embodied in the 1971 constitution.

The Illinois Human Rights Act (1979) prevents unlawful discrimination in employment, real property transactions, access to financial credit and public accommodations, by authorizing the creation of a Department of Human Rights to enforce, and a Human Rights Commission to adjudicate, allegations of unlawful discrimination.

Education

Education is free and compulsory for children between seven and 17 years of age. In 2007–08 there were 4,399 public schools (elementary, junior high, secondary, special education and others) and 1,920 private schools. There were 2,112,805 students enrolled in public schools in fall 2007 with 136,571 teachers; 312,270 students were enrolled in private schools with 20,750 teachers. In 2008–09 the average teacher salary was $62,787. Total expenditure on public elementary and secondary education in 2006–07 was $23,157m.; spending per pupil amounted to $10,353 in fiscal year 2008. In fall 2007 degree-granting institutions had a total enrolment of 837,018. There were 180 degree-granting institutions (60 public, 84 not-for-profit independent and 36 for-profit independent) in 2008–09.

Major colleges and universities (fall 2005):

Founded	Name	Place	Control	Enrolment
1851	Northwestern University	Evanston	Independent	18,065
1857	Illinois State University	Normal	Public	20,653
1867	University of Illinois	Urbana/		
		Champaign	Public	41,938
		Springfield		
		(1969)		4,517
		Chicago		
		(1946)		24,812
1867	Chicago State University	Chicago	Public	7,131
1869	Southern Illinois			
	University	Carbondale	Public	21,441
		Edwardsville		
		(1957)		13,460
1890	Loyola University of	Chicago	Roman	
	Chicago		Catholic	14,764
1891	University of Chicago	Chicago	Independent	14,150
1895	Eastern Illinois			
	University	Charleston	Public	12,129
1895	Northern Illinois			
	University	DeKalb	Public	25,208
1897	Bradley University	Peoria	Independent	6,154
1899	Western Illinois			
	University	Macomb	Public	13,404
1940	Illinois Institute of			
	Technology	Chicago	Independent	6,472
1945	Roosevelt University	Chicago	Independent	7,234
1961	Northeastern Illinois			
	University	Chicago	Public	12,227
1969	Governors State	University		
	University	Park	Public	5,405

Health

In 2009 there were 189 community hospitals, with 33,900 beds. A total of 1,558,000 patients were admitted during the year.

There were 36,528 active physicians in 2009 (28 per 10,000 population).

Welfare

Medicare enrolment in July 2010 totalled 1,823,045. In fiscal year 2009 a total of 2,626,372 people in Illinois received Medicaid. In Dec. 2008 there were 1,948,578 Old-Age, Survivors, and Disability Insurance (OASDI) beneficiaries. A total of 52,083 people were receiving payments under Temporary Assistance for Needy Families (TANF) in Dec. 2008.

RELIGION

In 2010 the chief religious traditions were: Catholics, with 3,648,907 members; Evangelical Protestants, 1,649,402; Mainline Protestants, 933,690; Muslims, 359,000 (estimate); Black Protestants, 220,435.

CULTURE

Tourism

In 2011 overseas visitors to Illinois—excluding those from Canada and Mexico—numbered 1,255,000, up from 1,186,000 in 2010 and 1,083,000 in 2006.

FURTHER READING

Statistical information: Department of Commerce and Community Affairs, 620 Adams St., Springfield 62701. Publishes *Illinois State and Regional Economic Data Book.* Bureau of Economic and Business Research, Univ. of Illinois, 1206 South 6th St., Champaign 61820. Publishes *Illinois Statistical Abstract.*

Blue Book of the State of Illinois. Edited by Secretary of State. Biennial

Miller, D. L., *City of the Century: The Epic of Chicago and the Making of America.* 1996

The Illinois State Library: Springfield, IL 62756.

Indiana

KEY HISTORICAL EVENTS

The area was inhabited by Algonquian-speaking tribes when the first European explorers (French) laid claim to it in the 17th century. They established fortified trading posts but there was little settlement. In 1763 the area passed to Britain, with other French-claimed territory east of the Mississippi. In 1783 Indiana became part of the North West Territory of the USA; it became a separate territory in 1800 and a state in 1816. Until 1811 there was continuing conflict with the Indian inhabitants, who were then defeated at Tippecanoe.

Early farming settlement was by families of British and German descent, including Amish and Mennonite communities. Later industrial development offered an incentive for more immigration from Europe, and, subsequently, from the southern states. In 1906 the town of Gary was laid out by the United States Steel Corporation and named after its chairman, Elbert H. Gary. The industry benefited from navigable water to supplies of iron ore and of coal. Indiana Port on Lake Michigan was a thriving trade centre, especially after the opening of the St Lawrence Seaway in 1959. The Ohio River also carried freight.

Indianapolis was built after 1821 and became the state capital in 1825. Natural gas was discovered in the neighbourhood in the late 19th century. This stimulated the growth of a motor industry, celebrated by the Indianapolis 500 race, held annually since 1911.

TERRITORY AND POPULATION

Indiana is bounded west by Illinois, north by Michigan and Lake Michigan, east by Ohio and south by Kentucky across the Ohio

River. Land area, 35,826 sq. miles (92,789 sq. km). Census population, 1 April 2010, was 6,483,802, an increase of 6·6% since 2000. July 2012 estimate, 6,537,334.

Population in five census years was:

	White	Black	American Indian/Alaska Native	Asian	Other	Total	Per sq. mile
1930	3,125,778	111,982	285	458	—	3,238,503	89·4
1980	5,004,394	414,785	7,836	20,557	42,652	5,490,224	152·8
1990	5,020,700	432,092	12,720	37,617	41,030	5,544,159	154·6
2000	5,320,022	510,034	15,815	61,131	173,483	6,080,485	169·5
2010	5,467,906	591,397	18,462	102,474	303,563	6,483,802	181·0

Of the total population in 2010, 3,294,065 were female and 4,875,504 were 18 years old or older. Indiana's Hispanic population was 389,707 in 2010, an 81·7% increase on the 2000 total of 214,536.

The largest cities with census population, 2010, are: Indianapolis (capital), 820,445; Fort Wayne, 253,691; Evansville, 117,429; South Bend, 101,168; Hammond, 80,830; Bloomington, 80,405; Gary, 80,294; Carmel, 79,191; Fishers, 76,794; Muncie, 70,085.

SOCIAL STATISTICS

Births, 2010, 83,940 (12·9 per 1,000 population); deaths, 2010 (provisional), 56,739 (8·8). Infant mortality rate, 2009, 7·8 per 1,000 live births. 2009 (provisional): marriages, 52,900 (7·9 per 1,000 population).

CLIMATE

Indianapolis, Jan. 29°F (−1·7°C), July 76°F (24·4°C). Annual rainfall 41" (1,034 mm). Indiana belongs to the Mid-West climate zone (see UNITED STATES: Climate).

CONSTITUTION AND GOVERNMENT

The present constitution (the second) dates from 1851. The General Assembly consists of a Senate of 50 members elected for four years, and a House of Representatives of 100 members elected for two years. It meets annually.

For the 113th Congress, which convened in Jan. 2013, Indiana sends nine members to the House of Representatives. It is represented in the Senate by Daniel Coats (R. 2011–17) and Joe Donnelly (D. 2013–19).

The state capital is Indianapolis. The state is divided into 92 counties and 1,008 townships.

RECENT ELECTIONS

In the 2012 presidential election Mitt Romney took Indiana with 54·3% of the vote (Barack Obama won it in 2008).

CURRENT GOVERNMENT

Governor: Mike Pence (R.), 2013–17 (salary: $111,687·94).
    Lieut.-Governor: Sue Ellspermann (R.), 2013–17 ($85,880·60).
    Secretary of State: Connie Lawson (R.), 2011–14 ($74,580·74).

Government Website: http://www.in.gov

ECONOMY

Per capita personal income (2010) was $34,042.

Budget

In 2011 total state revenue was $38,895m. Total expenditure was $35,262m. (including: education, $14,055m.; public welfare, $8,397m.; highways, $2,680m.; correction, $661m.; health, $566m.) Outstanding debt in 2011, $22,144m.

Performance

In 2011 Gross Domestic Product by state was $240,933m. (provisional), ranking Indiana 17th in the United States. In 2011 state real GDP growth was 1·1% (provisional).

ENERGY AND NATURAL RESOURCES

Electricity

In 2009 production was 116·7bn. kWh, of which 108·3bn. kWh was from coal.

Oil and Gas

2008 production of crude oil was 1·9m. bbls; natural gas, 5bn. cu. ft.

Water

The total area covered by water is 593 sq. miles.

Minerals

The state produced 44,100,000 tonnes of crushed stone and 206,000 tonnes of dimension stone in 2009. Production of coal (2010) was 35·3m. short tons. Value of domestic non-fuel mineral production in 2009 was $806m.

Agriculture

Indiana is largely agricultural, about 75% of its total area being in farms. In 2007, 60,938 farms had 14·77m. acres (average, 242 acres). The average value of land and buildings per acre was $3,583 in 2007.

Farm income 2007: crops, $5,211m.; livestock and products, $2,565m.; total, $7,776m. The net farm income in 2007 was $2,315m. The four most important products were corn, soybeans, hogs and dairy products. Cash receipts for corn totalled $4·3bn. in 2007; for soybeans, $2·2bn. The livestock in 2007 included 875,000 cattle (166,000 dairy cows), 49,000 sheep and lambs, 3·67m. hogs and pigs, and 24·24m. laying hens.

Forestry

In 2007 there were 4·66m. acres of forest including 189,000 acres of national forest.

INDUSTRY

In 2007 Indiana's 9,015 manufacturing establishments had 537,000 employees, earning $24,475m. Total value added by manufacturing in 2009 was $80,665m. The steel industry is the largest in the country.

Labour

Total non-agricultural employment, 2010, 2,793,000. Employees by branch, 2010 (in 1,000): trade, transportation and utilities, 541; manufacturing, 446; government, 438; education and health services, 425; professional and business services, 275. The unemployment rate in Dec. 2010 was 9·5%.

COMMUNICATIONS

Roads

In 2007 there were 95,469 miles of public roads (73,320 miles rural). There were 4,955,539 registered motor vehicles.

Rail

In 2011 there were 3,884 miles of mainline railroad of which 2,596 miles were Class I.

Civil Aviation

Of airports in 2011, 113 were for public use and 631 were for private use. There were 7,006,250 passenger enplanements statewide in 2010. The busiest airport, Indianapolis International, had 3,728,698 enplanements in 2010.

SOCIAL INSTITUTIONS

Justice

Following the US Supreme Court's reinstatement of the death penalty in 1976, death sentences have been given since 1980.

There was one execution in 2009 but none in 2010 or 2011. In Dec. 2008, 28,322 prisoners were under the jurisdiction of state and federal correctional authorities.

The Civil Rights Act of 1885 forbids places of public accommodation to bar any persons on grounds not applicable to all citizens alike; no citizen may be disqualified for jury service 'on account of race or colour'. An Act of 1947 makes it an offence to spread religious or racial hatred.

A 1961 Act provided 'all of its citizens equal opportunity for education, employment and access to public conveniences and accommodations' and created a Civil Rights Commission.

Education

School attendance is compulsory from seven to 18 years. In fall 2007 there were 1,970 public schools with 1,046,766 pupils and 62,334 teachers. The average expenditure per pupil was $8,867 in fiscal year 2008. Teachers' salaries averaged $49,198 (2008–09). Total expenditure for public elementary and secondary education, 2006–07, $10,869m.

Some leading institutions for higher education were (2009):

Founded	Institution	Control	Students
1801	Vincennes University	State	13,947
1824	Indiana University, Bloomington	State	42,347
1832	Wabash College, Crawfordsville	Independent	883
1837	De Pauw University, Greencastle	Methodist	2,396
1842	University of Notre Dame	R.C.	11,816
1847	Earlham College, Richmond	Quaker	1,266
1850	Butler University, Indianapolis	Independent	4,505
1859	Valparaiso University, Valparaiso	Evangelical Lutheran Church	4,065
1870	Indiana State University, Terre Haute	State	10,534
1874	Purdue University, Lafayette	State	41,052
1898	Ball State University, Muncie	State	21,401
1902	University of Indianapolis, Indianapolis	Methodist	5,055
1963	Ivy Tech Community College, Indianapolis	State	110,359
1969	Indiana University-Purdue University, Indianapolis	State	30,383
1985	University of Southern Indiana, Evansville	State	10,516

Health

In 2009 there were 123 community hospitals with 17,300 beds. A total of 713,000 patients were admitted during the year. In 2009 there were 13,938 active physicians (22 per 10,000 population).

Welfare

Medicare enrolment in July 2010 totalled 996,053. In fiscal year 2009 a total of 1,109,310 people in Indiana received Medicaid. In Dec. 2008 there were 1,121,662 Old-Age, Survivors, and Disability Insurance (OASDI) beneficiaries. A total of 105,995 people were receiving payments under Temporary Assistance for Needy Families (TANF) in Dec. 2008.

RELIGION

In 2010 the chief religious traditions were: Evangelical Protestants, with 1,238,154 members; Catholics, 747,706; Mainline Protestants, 689,902; Black Protestants, 94,705; Latter-day Saints (Mormons), 42,608.

FURTHER READING

Statistical information: Indiana Business Research Center, Indiana Univ., Indianapolis 46202. Publishes *Indiana Factbook.*

Gray, R. D. (ed.) *Indiana History: a Book of Readings.* 1994
Martin, J. B., *Indiana: an Interpretation.* 1992

State library: Indiana State Library, 140 North Senate, Indianapolis 46204.

Iowa

KEY HISTORICAL EVENTS

Originally the territory of the Iowa Indians, the area was explored by the Frenchmen Marquette and Joliet in 1673. French trading posts were set up, but there were few other settlements. In 1803 the French sold their claim to Iowa to the USA as part of the Louisiana Purchase. The land was still occupied by Indians but, in the 1830s, the tribes sold their land to the US government and migrated to reservations. Iowa became a US Territory in 1838 and a state in 1846.

The state was settled by immigrants drawn mainly from neighbouring states to the east. Later there was more immigration from Protestant states of northern Europe. The land was extremely fertile and most immigrants came to farm. Not all the Indian population had accepted the cession and there were some violent confrontations, notably the murder of settlers at Spirit Lake in 1857. The capital, Des Moines, was founded in 1843 as a fort to protect Indian rights. It expanded rapidly along with coal mining after 1910.

TERRITORY AND POPULATION

Iowa is bounded east by the Mississippi River (forming the boundary with Wisconsin and Illinois), south by Missouri, west by the Missouri River (forming the boundary with Nebraska), northwest by the Big Sioux River (forming the boundary with South Dakota) and north by Minnesota. Land area, 55,857 sq. miles (144,669 sq. km). Census population, 1 April 2010, was 3,046,355, an increase of 4·1% since 2000. July 2012 estimate, 3,074,186.

Population in five census years was:

	White	Black	American Indian	Asiatic	Total	Per sq. mile
1870	1,188,207	5,762	48	3	1,194,020	21·5
			All others			
1980	2,839,225	41,700	32,882		2,913,808	51·7
1990	2,683,090	48,090	45,575		2,776,755	49·7
2000	2,748,640	61,853	115,831		2,926,324	52·4
2010	2,781,561	89,148	175,646		3,046,355	54·5

Of the total population in 2010, 1,538,036 were female and 2,318,362 were 18 years old or older. In 2010 the Hispanic population was 151,544, up from 82,473 in 2000 (an increase of 83·7%).

The largest cities in the state, with their population in 2010, are: Des Moines (capital), 203,433; Cedar Rapids, 126,326; Davenport, 99,685; Sioux City, 82,684; Waterloo, 68,406; Iowa City, 67,862; Council Bluffs, 62,230; Ames, 58,965; Dubuque, 57,637; West Des Moines, 56,609; Ankeny, 45,582; Urbandale, 39,463; Cedar Falls, 39,260; Marion, 34,768; Bettendorf, 33,217.

SOCIAL STATISTICS

Births, 2010, 38,719 (12·7 per 1,000 population); deaths, 2010 (provisional), 27,745 (9·1). Infant mortality, 2009, 4·6 per 1,000 live births. 2009 (provisional): marriages, 21,200 (7·0 per 1,000 population); divorces and annulments, 7,300 (2·4). Same-sex marriage became legal in April 2009.

CLIMATE

Cedar Rapids, Jan. 17·6°F, July 74·2°F. Annual rainfall 34". Des Moines, Jan. 19·4°F, July 76·6°F. Annual rainfall 33". Iowa belongs to the Mid-West climate zone (*see* UNITED STATES: Climate).

CONSTITUTION AND GOVERNMENT

The constitution of 1857 still exists; it has had 48 amendments as of Aug. 2011. The General Assembly comprises a Senate of 50 and a House of Representatives of 100 members, meeting annually for an unlimited session. Senators are elected for four years, half retiring every second year: Representatives for two years. The Governor and Lieut.-Governor are elected for four years.

For the 113th Congress, which convened in Jan. 2013, Iowa sends four members to the House of Representatives. It is represented in the Senate by Chuck Grassley (R. 1981–2017) and Tom Harkin (D. 1985–2015).

Iowa is divided into 99 counties; the capital is Des Moines.

RECENT ELECTIONS

In the 2012 presidential election Barack Obama took Iowa with 52·1% of the vote, having also won it in 2008.

CURRENT GOVERNMENT

Governor: Terry Branstad (R.), 2011–15 (salary: $130,000).
  Lieut.-Governor: Kim Reynolds (R.), 2011–15 ($103,212).
  Secretary of State: Matt Schultz (R.), 2011–15 ($103,212).

Government Website: http://www.iowa.gov

ECONOMY

Per capita personal income (2010) was $38,084.

Budget

In 2011 total state revenue amounted to $24,089m. Total state expenditure was $19,937m. (education, $6,270m.; public welfare, $4,901m.; highways, $1,591m.; hospitals, $1,090m.; government administration, $551m.) Outstanding debt in 2011, $7,574m.

Performance

Gross Domestic Product by state was $128,597m. in 2011 (provisional), ranking Iowa 30th in the United States. In 2011 state real GDP growth was 1·9% (provisional).

ENERGY AND NATURAL RESOURCES

Water

The total area covered by water is 416 sq. miles.

Minerals

Production in 2008: crushed stone, 37·8m. tonnes; sand and gravel, 15·6m. tonnes. The value of domestic non-fuel mineral products in 2009 was $590m.

Agriculture

Iowa is the wealthiest of the agricultural states, partly because nearly the whole area (92%) is arable and included in farms. The total farm area, 2007, is 30·7m. acres. The average farm in 2007 was 331 acres. The average value of buildings and land per acre was, in 2007, $3,388. The number of farms has declined since 1960, from 174,000 to 92,856 in 2007.

Farm income 2007: crops, $10,180m.; livestock and products, $8,857m.; total, $19,037m. (the third highest total, behind California and Texas). The net farm income in 2007 was $5,334m. In 2007 production of corn was 2,377m. bu.[1], value $10,197m.; and soybeans, 449m. bu.[1], value $4,712m. In 2007 livestock included: swine, 19·3m.[1]; dairy cows, 215,000; all cattle, 3·98m.; sheep and lambs, 209,000; laying hens, 53·79m.[1] Wool production in 2007 totalled 1·25m. lb.

[1]More than any other state.

Forestry

Total forested area was 2·88m. acres in 2007.

INDUSTRY

In 2007 Iowa's 3,802 manufacturing establishments had 223,000 employees, earning $9,526m. Total value added by manufacturing in 2009 was $35,798m.

Labour

Total non-agricultural employment, 2010, 1,469,000. Employees by branch, 2010 (in 1,000): trade, transport and utilities, 300; government, 254; education and health services, 214; manufacturing, 200; leisure and hospitality, 130. Iowa had an unemployment rate of 6·1% in Dec. 2010.

COMMUNICATIONS

Roads

In 2007 there were 114,193 miles of streets and highways, of which 102,905 miles were rural and 11,288 urban. There were 3,360,196 motor vehicle registrations.

Rail

In 2009 there were 3,925 miles of freight railroad (excluding trackage rights). There were 16 freight railroads operating in 2009. Rail traffic originating in Iowa in 2009 totalled 37·5m. tons and rail traffic terminating in the state came to 40·3m. tons.

Civil Aviation

In 2011 Iowa had seven primary airports (commercial service airports with more than 10,000 passenger boardings annually) with a combined total of 1,473,744 enplanements, up from 1,465,337 in 2010. The busiest airport, Des Moines International, had 932,828 enplanements in 2011.

SOCIAL INSTITUTIONS

Justice

The death penalty was abolished in Iowa in 1965. There were 8,766 federal and state prisoners in Dec. 2008.

Education

School attendance is compulsory for 24 consecutive weeks annually during school age (6–16). In 2010–11 there were 1,396 public primary and secondary schools with 473,493 pupils in attendance and 33,916 teachers. There were 198 private schools with 33,804 pupils and 2,410 teachers in 2010–11. Average teacher's salary in 2010–11 was $49,794. In the 2009–10 school year the state spent an average of $8,603 on each elementary and secondary school student.

Leading institutions for higher education enrolment figures (fall 2010) were:

Founded	Institution	Control	Professors	Full-time students
1843	Clarke College, Dubuque	Independent	128	1,255
1846	Grinnell College, Grinnell	Independent	156	1,655
1847	University of Iowa, Iowa City	State	2,156	30,825
1851	Coe College, Cedar Rapids	Independent	80	1,343
1852	Wartburg College, Waverly	Evangelical Lutheran	109	1,775
1853	Cornell College, Mount Vernon	Independent	119	1,191
1854	Upper Iowa University, Fayette	Independent	83	6,765
1858	Iowa State University, Ames	State	1,766	28,682
1859	Luther College, Decorah	Evangelical Lutheran	249	2,481
1876	Univ. of Northern Iowa, Cedar Falls	State	824	13,201
1881	Drake University, Des Moines	Independent	362	5,616
1882	St Ambrose University, Davenport	Roman Catholic	356	3,663
1891	Buena Vista University, Storm Lake	Presbyterian	97	2,706
1894	Morningside College, Sioux City	Methodist	77	1,991

Health

In 2009 the state had 118 community hospitals with 10,300 beds. A total of 355,000 patients were admitted during the year. In 2009 there were 5,696 active physicians in the state (19 per 10,000 population).

Welfare

Medicare enrolment in July 2010 totalled 511,942. In fiscal year 2009 a total of 481,599 people in Iowa received Medicaid. In Dec. 2008 there were 563,610 Old-Age, Survivors, and Disability Insurance (OASDI) beneficiaries. In 2008 Temporary Assistance to Needy Families (TANF) was received by on average 39,071 recipients monthly.

RELIGION

The chief religious traditions in Iowa in 2010 were: Mainline Protestants, with 666,637 members; Catholics, 503,080; Evangelical Protestants, 402,376; Latter-day Saints (Mormons), 32,283; Black Protestants, 17,902.

FURTHER READING

Annual Survey of Manufactures.
Government Finance.
Official Register. Secretary of State. Biennial
State Government Website: http://www.iowa.gov

State Library of Iowa: Des Moines 50319.

Kansas

KEY HISTORICAL EVENTS

The area was explored from Mexico in the 16th century, when Spanish travellers encountered Kansas, Wichita, Osage and Pawnee tribes. The French claimed Kansas in 1682, establishing a valuable fur trade with local tribes in the 18th century. In 1803 the area passed to the USA as part of the Louisiana Purchase and became a base for pioneering trails further west. After 1830 it was 'Indian Territory' and a number of tribes displaced from eastern states were settled there. In 1854 the Kansas Territory was created and opened for white settlement. The early settlers were farmers from Europe or New England, but the Territory's position also brought it into contact with southern culture. Slavery was prohibited by the Missouri Compromise of 1820 but the 1854 Kansas-Nebraska Act affirmed the principle of 'popular sovereignty' to settle the issue, which was then fought out by opposing factions throughout 'Bleeding Kansas'.

Kansas entered the Union (as a non-slave state) in 1861, minus the territory that is now in Colorado.

The economy was based on cattle-ranching and railways. Herds were driven to the railheads and shipped from vast stockyards, or slaughtered and processed in railhead meat-packing plants. Wheat and sorghum also became important once the plains could be ploughed on a large scale.

TERRITORY AND POPULATION

Kansas is bounded north by Nebraska, east by Missouri, with the Missouri River as boundary in the northeast, south by Oklahoma and west by Colorado. Land area, 81,759 sq. miles (211,754 sq. km). Census population, 1 April 2010 was 2,853,118, an increase of 6·1% since 2000. July 2012 estimate, 2,885,905.

Population in five federal census years was:

	White	Black	American Indian	Asiatic	Total	Per sq. mile
1870	346,377	17,108	914	—	364,399	4·5
			All others			
1980	2,168,221	126,127	69,888		2,364,236	28·8
1990	2,231,986	143,076	102,512		2,477,574	30·3
2000	2,313,944	154,198	220,276		2,688,418	32·9
2010	2,391,044	167,864	294,210		2,853,118	34·9

Of the total population in 2010, 1,437,710 were female and 2,126,179 were 18 years old or older. In 2010 the Hispanic population was 300,042, up from 188,252 in 2000 (an increase of 59·4%).

Cities, with 2010 census population: Wichita, 382,368; Overland Park, 173,372; Kansas City, 145,786; Topeka (capital), 127,473; Olathe, 125,872; Lawrence, 87,643.

SOCIAL STATISTICS

Births, 2010, 40,469 (14·2 per 1,000 population); deaths, 2010 (provisional), 24,502 (8·6). Infant mortality, 2009, 7·0 per 1,000 live births. 2009 (provisional): marriages, 18,500 (6·5 per 1,000 population); divorces and annulments, 10,300 (3·7 per 1,000 population).

CLIMATE

Dodge City, Jan. 29°F (−1·7°C), July 78°F (25·6°C). Annual rainfall 21" (518 mm). Kansas City, Jan. 30°F (−1·1°C), July 79°F (26·1°C). Annual rainfall 38" (947 mm). Topeka, Jan. 28°F (−2·2°C), July 78°F (25·6°C). Annual rainfall 35" (875 mm). Wichita, Jan. 31°F (−0·6°C), July 81°F (27·2°C). Annual rainfall 31" (777 mm). Kansas belongs to the Mid-West climate zone (*see* UNITED STATES: Climate).

CONSTITUTION AND GOVERNMENT

The year 1861 saw the adoption of the present constitution; it has had 183 amendments. The Legislature includes a Senate of 40 members, elected for four years, and a House of Representatives of 125 members, elected for two years. Sessions are annual.

For the 113th Congress, which convened in Jan. 2013, Kansas sends four members to the House of Representatives. It is represented in the Senate by Pat Roberts (R. 1997–2015) and Jerry Moran (R. 2011–17).

The capital is Topeka. The state is divided into 105 counties.

RECENT ELECTIONS

In the 2012 presidential election Mitt Romney took Kansas with 60·0% of the vote (John McCain won it in 2008).

CURRENT GOVERNMENT

Governor: Sam Brownback (R.), 2011–15 (salary: $110,707).
 Lieut.-Governor: Jeff Colyer (R.), 2011–15 ($31,313).
 Secretary of State: Kris Kobach (R.), 2011–15 ($86,003).

Government Website: http://www.kansas.gov

ECONOMY

Per capita personal income (2010) was $39,005.

Budget

In 2011 total state revenue was $18,613m. Total expenditure was $16,687m. (including: education, $5,966m.; public welfare, $3,531m.; highways, $1,239m.; hospitals, $1,228m.; government administration, $456m.) Outstanding debt in 2011, $6,893m.

Performance

Gross Domestic Product by state in 2011 was $113,367m. (provisional), ranking Kansas 31st in the United States. In 2011 state real GDP growth was 0·5% (provisional).

ENERGY AND NATURAL RESOURCES

Oil and Gas

In 2008 Kansas produced 39·6m. bbls of crude petroleum; marketed production of natural gas totalled 374bn. cu. ft.

Water

The total area covered by water is 520 sq. miles.

Minerals

Important fuel minerals are coal, petroleum and natural gas. Principal non-fuel minerals are cement, salt and crushed stone. Total value of non-fuel mineral output in 2009 was $953m.

Agriculture

Kansas is pre-eminently agricultural, but sometimes suffers from lack of rainfall in the west. In 2007 there were 65,531 farms with a total acreage of 46·35m. Average number of acres per farm was 707. Average value of farmland and buildings per acre, in 2007, was $911. Farm income 2007: from crops, $4,518m.; and from livestock and products, $7,212m. Chief crops: wheat, corn and soybeans. The net farm income in 2007 was $2,156m. Wheat production was 283·8m. bu. in 2007. Kansas was the USA's second largest wheat producer in 2007, after North Dakota, although it had been the leading producer in 2006. There is an extensive livestock industry, including, in 2007, 6·67m. cattle (only Texas had more), 84,000 sheep, and 1·89m. hogs and pigs.

Forestry

The state had a forested area of 2·11m. acres in 2007.

INDUSTRY

In 2007 the state's 3,170 manufacturing establishments had 178,000 employees, earning $7,983m. Total value added by manufacturing in 2009 was $23,617m.

Labour

Total non-agricultural employment, 2010, 1,323,000. Employees by branch, 2010 (in 1,000): government, 262; trade, transportation and utilities, 251; education and health services, 180; manufacturing, 160; professional and business services, 142. In Dec. 2010 the state unemployment rate was 6·8%.

COMMUNICATIONS

Roads

In 2007 there were 140,271 miles of roads (127,612 miles rural). There were 2,429,064 registered motor vehicles.

Rail

There were 4,721 miles of railroad in 2009.

Civil Aviation

There were four primary airports in 2011 (commercial service airports with more than 10,000 passenger boardings annually) with a combined total of 822,434 enplanements, up from 815,385 in 2010.

SOCIAL INSTITUTIONS

Justice

In Dec. 2008 there were 8,539 federal and state prisoners. The death penalty was declared unconstitutional in Kansas in 2004. The last execution was in 1965.

Education

In 2007–08 there were 1,422 public elementary and secondary schools. In fall 2007, 468,295 pupils were enrolled in public schools with 35,359 teachers. Total expenditure on public elementary and secondary education in 2006–07 was $4,868m.; average teacher salary was $46,987 in 2008–09. Spending per pupil in fiscal year 2008 was $9,883.

The Kansas Board of Regents governs six state universities: Kansas State University, Manhattan (founded in 1863); University of Kansas, Lawrence (1864); Emporia State University, Emporia; Pittsburg State University, Pittsburg; Fort Hays State University, Hays; and Wichita State University, Wichita. It also supervises and co-ordinates 19 community colleges, five technical colleges, six technical schools and a municipal university.

Health

In 2009 there were 133 community hospitals with 10,100 beds. A total of 316,000 patients were admitted during the year. In 2009 there were 6,436 active physicians in the state (23 per 10,000 population).

Welfare

Medicare enrolment in July 2010 totalled 428,472. In fiscal year 2009 a total of 354,752 people in Kansas received Medicaid. In Dec. 2008 there were 464,699 Old-Age, Survivors, and Disability Insurance (OASDI) beneficiaries. A total of 30,874 people were receiving payments under Temporary Assistance for Needy Families (TANF) in Dec. 2008.

RELIGION

In 2010 the chief religious traditions were: Evangelical Protestants, with 516,818 members; Catholics, 426,611; Mainline Protestants, 386,980; Black Protestants, 41,666; Latter-day Saints (Mormons), 40,251.

FURTHER READING

Statistical information: Institute for Public Policy and Business Research, Univ. of Kansas, 607 Blake Hall, Lawrence 66045. Publishes *Kansas Statistical Abstract.*
Annual Economic Report of the Governor.

State library: Kansas State Library, Topeka.

Kentucky

KEY HISTORICAL EVENTS

Lying west of the Appalachians and south of the Ohio River, the area was the meeting place and battleground for the eastern Iroquois and the southern Cherokees. Northern Shawnees were also present. The first successful white settlement took place in 1769 when Daniel Boone reached the Bluegrass plains from the eastern, trans-Appalachian, colonies. After 1783 immigration from the east was rapid, settlers travelling by river or crossing the mountains by the Cumberland Gap. The area was originally attached to Virginia but became a separate state in 1792.

Large plantations dependent on slave labour were established, as were small farms worked by white owners. The state became divided on the issue of slavery, although plantation interests (mainly producing tobacco) dominated state government. In the event the state did not secede in 1861, and the majority of citizens supported the Union. Public opinion was more favourable to the south in the hard times of the reconstruction period.

The eastern mountains became an important coal-mining area, tobacco-growing continued and the Bluegrass plains produced livestock, including especially fine thoroughbred horses.

TERRITORY AND POPULATION

Kentucky is bounded in the north by the Ohio River (forming the boundary with Illinois, Indiana and Ohio), northeast by the Big Sandy River (forming the boundary with West Virginia), east by Virginia, south by Tennessee and west by the Mississippi River (forming the boundary with Missouri). Land area, 39,486 sq. miles (102,269 sq. km). Census population, 1 April 2010, was 4,339,367, an increase of 7·4% since 2000. July 2012 estimate, 4,380,415.

Population in five census years was:

	White	Black	All others	Total	Per sq. mile
1930	2,388,364	226,040	185	2,614,589	65·1
1980	3,379,006	259,477	22,294	3,660,777	92·3
1990	3,391,832	262,907	30,557	3,685,296	92·8
2000	3,640,889	295,994	104,886	4,041,769	101·7
2010	3,809,537	337,520	192,310	4,339,367	109·9

Of the total population in 2010, 2,204,415 were female and 3,315,996 were 18 years old or older. Kentucky's Hispanic population was 132,836, up 121·6% on the 2000 census figure of 59,939.

The principal cities with census population in 2010 are: Louisville, 597,337; Lexington-Fayette, 295,803; Bowling Green, 58,067; Owensboro, 57,265; Covington, 40,640; Hopkinsville, 31,577; Richmond, 31,364; Florence, 29,951; Georgetown, 29,098; Henderson, 28,757; Elizabethtown, 28,531; Nicholasville, 28,015; Jeffersontown, 26,595; Frankfort (capital), 25,527.

SOCIAL STATISTICS

Births, 2010, 55,784 (12·9 per 1,000 population); deaths, 2010 (provisional), 41,980 (9·7). Infant mortality, 2009, 6·9 per 1,000 live births. 2009 (provisional): marriages, 33,400 (7·6 per 1,000 population); divorces and annulments, 19,900 (4·6).

CLIMATE

Kentucky is in the Appalachian Mountains climatic zone (see UNITED STATES: Climate). It has a temperate climate. Temperatures are moderate during both winter and summer, precipitation is ample without a pronounced dry season, and winter snowfall amounts are variable. Mean annual temperatures range from 52°F in the northeast to 58°F in the southwest. Annual rainfall averages at about 45". Snowfall ranges from 5 to 10" in the southwest of the state, to 25" in the northeast, and 40" at higher altitudes in the southeast.

CONSTITUTION AND GOVERNMENT

The constitution dates from 1891; there had been three preceding it. The 1891 constitution was promulgated by convention and provides that amendments be submitted to the electorate for ratification. The General Assembly consists of a Senate of 38 members elected for four years, one half retiring every two years, and a House of Representatives of 100 members elected for two years. It has annual sessions. All citizens of 18 or over are qualified as electors.

For the 113th Congress, which convened in Jan. 2013, Kentucky sends six members to the House of Representatives. It is represented in the Senate by Mitch McConnell (R. 1985–2015) and Rand Paul (R. 2011–17).

The capital is Frankfort. The state is divided into 120 counties.

RECENT ELECTIONS

In the 2012 presidential election Mitt Romney took Kentucky with 60·5% of the vote (John McCain won it in 2008).

CURRENT GOVERNMENT

Governor: Steve Beshear (D.), Dec. 2011–Dec. 2015 (salary: $133,644·00).

Lieut.-Governor: Jerry Abramson (D.), Dec. 2011–Dec. 2015 ($133,615·80).

Secretary of State: Alison Lundergan Grimes (D.), since Jan. 2012 ($115,593·60).

Government Website: http://kentucky.gov

ECONOMY

Per capita personal income (2010) was $32,376.

Budget

In 2011 total state revenue was $31,056m. Total expenditure was $29,370m. (including: education, $9,415m.; public welfare, $7,334m.; highways, $1,936m.; hospitals, $1,135m.; government administration, $880m.) Debt outstanding in 2011, $14,522m.

Performance

Gross Domestic Product by state in 2011 was $141,266m. (provisional), ranking Kentucky 28th in the United States. In 2011 state real GDP growth was 0·5% (provisional).

ENERGY AND NATURAL RESOURCES

Electricity

In 2009 production was 90,630m. kWh, of which 84,038m. kWh was from coal.

Oil and Gas

Production of crude oil in 2008 was 2·6m. bbls; natural gas, 114bn. cu. ft.

Water

The total area covered by water is 921 sq. miles.

Minerals

The principal mineral is coal: 107·3m. short tons were mined in 2009, value $6·3bn. In 2008, 51·0m. tonnes of crushed stone were mined, value $411m.; 7·6m. tonnes of sand and gravel, value $41·6m.; 0·4m. tonnes of clay, value $8·2m. Other minerals include fluorspar, ball clay, gemstones, dolomite, cement and lime. Total value of non-fuel mineral production for 2009 was $668m.

Agriculture

In 2007, 85,260 farms covered an area of 13·99m. acres. The average farm was 164 acres. In 2007 the average value of farmland and buildings per acre was $2,682.

Farm income, 2006: from crops, $1,299m.; and from livestock, $2,708m. The net farm income in 2006 was $1,742m. The chief crop is tobacco: production, in 2007, 196·3m. lb. Kentucky is the USA's second largest tobacco producer, after North Carolina. Other principal crops include corn (172m. bu. in 2007), soybeans, hay and wheat.

Stock-raising is important in Kentucky, which has long been famous for its horses. There were 175,500 horses in the state in 2007, a number exceeded only in Texas. The livestock in 2007 included 2·40m. all cattle and calves, 90,000 dairy cows, 37,000 sheep, 348,000 swine and 4·58m. laying hens.

Forestry

In 2007 Kentucky had 11·97m. acres forested land, of which 744,000 acres were national forest.

INDUSTRY

In 2007 Kentucky's 4,165 manufacturing establishments had 247,000 employees, earning $10,773m. The value added by manufacture in 2009 was $31,994m.

Labour

Total non-agricultural employment, 2010, 1,770,000. Employees by branch, 2010 (in 1,000): trade, transportation and utilities, 359; government, 331; education and health services, 250; manufacturing, 209; professional and business services, 180. The unemployment rate in Dec. 2010 was 10·3%.

COMMUNICATIONS

Roads

In 2007 there were 78,587 miles of roads comprising 12,479 miles of urban road and 66,108 miles of rural road. There were 3,546,620 registered motor vehicles.

Rail

In 2009 there were 2,558 miles of freight railroad (excluding trackage rights). There were 13 freight railroads operating in 2009. Rail traffic originating in Kentucky in 2009 totalled 76·8m. tons and rail traffic terminating in the state came to 41·0m. tons.

Civil Aviation

In 2011 Kentucky had five primary airports—commercial service airports with more than 10,000 passenger boardings annually—with a combined total of 5,642,399 enplanements, down from 6,133,451 in 2010. The busiest airports are Cincinnati/Northern Kentucky International (which had 3,422,466 enplanements in 2011) and Louisville International (1,650,707 enplanements in

2011). Louisville handled 2,188,422 tonnes of freight in 2011, ranking it third in the USA for cargo handled behind Memphis and Anchorage.

Shipping
There is barge traffic on the 1,100 miles of navigable rivers. There are six public river ports, over 30 contract terminal facilities and 150 private terminal operations. Kentucky's waterways have access to the junction of the upper and lower Mississippi, Ohio and Tennessee-Tombigbee navigation corridors.

SOCIAL INSTITUTIONS

Justice
There are 12 adult prisons within the Department of Corrections Adult Institutions and three privately run adult institutions. In Dec. 2008 there were 21,706 prison inmates. The death penalty is authorized for murder and kidnapping. As of Dec. 2011 there were 35 persons (including one female) under sentence of death. There was one execution in 2008 but none since.

Education
Attendance at school between the ages of six and 16 years (inclusive) is compulsory, the normal term being 175 days. In fall 2007 there were 666,225 pupils and 43,536 teachers in public elementary and secondary schools. Public school classroom teachers' salaries (2008–09) averaged $49,539. The average total expenditure per pupil in fiscal year 2008 was $8,740.

There were also 5,640 teachers working in 400 private elementary and secondary schools with 76,140 students in fall 2007.

In 2008–09 the state had 24 public, 27 not-for-profit independent and 22 for-profit independent degree-granting institutions. There were 211,234 students at public degree-granting institutions in fall 2007 and 46,979 students at private degree-granting institutions. In 2005–06 there were 23 community colleges with a total enrolment of 84,669 students. The largest of the institutions of higher learning are (fall 2009): University of Kentucky, with 26,295 students; University of Louisville, 21,016; Western Kentucky University, 20,712; Eastern Kentucky University, 16,268; Northern Kentucky University, 15,378; Murray State University, 10,071; Morehead State University, 8,822; Kentucky State University, 2,834. Five of the several privately-endowed colleges of standing are Berea College, Berea; Centre College, Danville; Transylvania University, Lexington; Georgetown College, Georgetown; and Bellarmine College, Louisville.

Health
In 2009 there were 104 community hospitals with 14,100 beds. A total of 597,000 patients were admitted during the year. In 2009 there were 10,076 active physicians in the state (23 per 10,000 population).

Welfare
Medicare enrolment in July 2010 totalled 752,863. In fiscal year 2009 a total of 941,990 people in Kentucky received Medicaid. In Dec. 2008 there were 844,573 Old-Age, Survivors, and Disability Insurance (OASDI) beneficiaries. A total of 60,627 people were receiving payments under Temporary Assistance for Needy Families (TANF) in Dec. 2008.

RELIGION
The principal religious traditions in the state in 2010 were: Evangelical Protestants, with 1,448,947 members; Catholics, 359,783; Mainline Protestants, 305,955; Black Protestants, 64,958; Latter-day Saints (Mormons), 32,559.

CULTURE
The Kentucky Center for the Arts hosts productions by the Kentucky Opera Association, the Louisville Ballet, the Louisville Orchestra and Broadway touring productions.

FURTHER READING
Kentucky Deskbook of Economic Statistics, Lackey, Brent, (ed.) Kentucky Cabinet for Economic Development, Frankfort

Ulack, R. (ed.) *Atlas of Kentucky*. 1998

Louisiana

KEY HISTORICAL EVENTS
Originally the territory of Choctaw and Caddo tribes, the area was claimed for France in 1682. In 1718 the French founded New Orleans which became the centre of a crown colony in 1731. France ceded the area west of the Mississippi (most of the present state) to Spain in 1762 and the eastern area, north of New Orleans, to Britain in 1763. The British section passed to the USA in 1783 but France bought back the rest from Spain in 1800, including New Orleans and the mouth of the Mississippi. The USA, fearing exclusion from a strategically important and commercially promising shipping area, persuaded France to sell Louisiana in 1803. The present states of Missouri, Arkansas, Iowa, North Dakota, South Dakota, Nebraska and Oklahoma were included in the purchase.

The area became the Territory of New Orleans in 1804 and was admitted to the Union as a state in 1812. The economy initially depended on cotton and sugarcane plantations. The population was of French, Spanish and black descent, with a growing number of American settlers. Plantation interests succeeded in achieving secession in 1861 but New Orleans was occupied by the Union in 1862. Planter influence was reasserted in the late 19th century, imposing rigid segregation and denying black rights.

The state has become mainly urban industrial, with the Mississippi ports growing rapidly. There is petroleum and natural gas, and a strong tourist industry based on the French culture and Caribbean atmosphere of New Orleans.

Louisiana, and New Orleans in particular, suffered widespread damage and loss of life after Hurricane Katrina struck the Gulf Coast on 31 Aug. 2005.

TERRITORY AND POPULATION
Louisiana is bounded north by Arkansas, east by Mississippi, south by the Gulf of Mexico and west by Texas. Land area, 43,204 sq. miles (111,898 sq. km). Census population, 1 April 2010, was 4,533,372, an increase of 1·4% since 2000. July 2012 estimate, 4,601,893.

Population in five census years was:

	White	Black	American Indian	Asiatic	Total	Per sq. mile
1930	1,322,712	776,326	1,536	1,019	2,101,593	46·5
			All others			
1980	2,911,243	1,237,263	55,466		4,205,900	93·5
1990	2,839,138	1,299,281	81,554		4,219,973	96·9
2000	2,856,161	1,451,944	160,871		4,468,976	102·6
2010	2,836,192	1,452,396	244,784		4,533,372	104·9

Of the total population in 2010, 2,314,080 were female and 3,415,357 were 18 years old or older. The Hispanic population was 192,560 in 2010, an increase of 78·7% on the 2000 census figure of 107,738.

The largest cities with their 2010 census population are: New Orleans, 343,829 (484,764 in 2000); Baton Rouge (the capital),

229,493; Shreveport, 199,311; Metairie, 138,481; Lafayette, 120,623; Lake Charles, 71,993; Kenner, 66,702; Bossier City, 61,315; Monroe, 48,815. In Jan. 2006 the population of New Orleans was estimated at 144,000 in the wake of Hurricane Katrina, making Baton Rouge temporarily the most populous city in Louisiana.

SOCIAL STATISTICS

Births, 2010, 62,379 (13·8 per 1,000 population); deaths, 2010 (provisional), 40,671 (9·0). Infant deaths, 2009, 8·7 per 1,000 live births. Marriages, 2009 (provisional), 28,700 (7·1 per 1,000 population).

CLIMATE

New Orleans, Jan. 54°F (12·2°C), July 83°F (28·3°C). Annual rainfall 58" (1,458 mm). Louisiana belongs to the Gulf Coast climate zone (*see* UNITED STATES: Climate).

CONSTITUTION AND GOVERNMENT

The present constitution dates from 1974. The Legislature consists of a Senate of 39 members and a House of Representatives of 105 members, both chosen for four years. Sessions are annual; a fiscal session is held in even years.

For the 113th Congress, which convened in Jan. 2013, Louisiana sends six members to the House of Representatives. It is represented in the Senate by Mary Landrieu (D. 1997–2015) and David Vitter (R. 2005–17).

Louisiana is divided into 64 parishes (corresponding to the counties of other states). The capital is Baton Rouge.

RECENT ELECTIONS

In the 2012 presidential election Mitt Romney took Louisiana with 57·8% of the vote (John McCain won it in 2008).

CURRENT GOVERNMENT

Governor: Bobby Jindal (R.), 2012–16 (salary: $130,000).
  Lieut.-Governor: Jay Dardenne (R.), 2012–16 ($115,000).
  Secretary of State: Tom Schedler (R.), 2012–16 ($115,000).

Government Website: http://www.louisiana.gov

ECONOMY

Per capita personal income (2010) was $37,021.

Budget

In 2011 total revenue was $33,975m. Total expenditure was $33,396m. (including: education, $8,904m.; public welfare, $6,426m.; hospitals, $2,205m.; highways, $2,185m.; government administration, $982m.) Debt outstanding, in 2011, $18,447m.

Performance

Gross Domestic Product by state in 2011 was $205,877m. (provisional), ranking Louisiana 23rd in the United States. In 2011 state real GDP growth was 0·5% (provisional).

ENERGY AND NATURAL RESOURCES

Electricity

In 2009 production was 91·0bn. kWh, of which 44·0bn. kWh was from natural gas.

Oil and Gas

Louisiana ranks fourth among states of the USA for oil production and sixth for production of natural gas. Production in 2008 of crude oil was 73m. bbls; marketed production of natural gas totalled 1,377bn. cu. ft.

Water

The total area covered by water is 9,174 sq. miles, of which 4,562 sq. miles are inland.

Minerals

Principal non-fuel minerals are salt, sand, gravel and lime. Total non-fuel mineral production in 2009 was more than $460m.

Agriculture

The state is divided into two parts, the uplands and the alluvial and swamp regions of the coast. A delta occupies about one-third of the total area. Manufacturing is the leading industry, but agriculture is important. The number of farms in 2007 was 30,106 covering 8·11m. acres; the average farm had 269 acres. Average value of farmland per acre, in 2007, was $2,058.

Farm income, 2006: from crops, $1,322m.; and from livestock, $864m. The net farm income in 2006 was $766m. Principal crops, 2007 production, were: soybeans, 24·72m. bu.; sugarcane, 14·09m. tons; rice, 23·12m. cwt; corn, 114·67m. bu.; cotton, 699,000 bales; sorghum, 22·40m. bu.

Forestry

In 2007 the state had 14·22m. acres of forested land, of which 695,000 acres were national forest. Private, non-industrial landowners own 62% of the state's forestland, forest products industries own 29% and the general public owns 9%. Production 2011: sawtimber, 828,637,892 bd ft; cordwood, 5,844,089 standard cords. The economic impact of forestry and forest products industries in Louisiana was $3·1bn. in 2010.

Fisheries

In 2009 Louisiana's commercial fisheries catch for all species totalled 1,147·4m. lb, valued at $271·7m. Louisiana's commercial fishery landings in terms of weight are second only behind those of Alaska. In 2005 Louisiana had 873 fish farms, more than any other state.

INDUSTRY

Louisiana's leading manufacturing activity is the production of chemicals, followed, in order of importance, by the processing of petroleum and coal products, the production of transportation equipment and production of paper products. In 2007 the state's 3,442 manufacturing establishments had 148,000 employees, earning $7,565m. Total value added by manufacturing in 2009 was $41,820m.

Labour

Total non-agricultural employment, 2010, 1,884,000. Employees by branch, 2010 (in 1,000): government, 366; trade, transportation and utilities, 364; education and health services, 271; leisure and hospitality, 194; professional and business services, 193. The unemployment rate was 7·7% in Dec. 2010.

COMMUNICATIONS

Roads

In 2007 there were 61,008 miles of road (44,731 miles rural). Registered motor vehicles numbered 3,926,741.

Rail

In 2009 there were 2,830 miles of freight railroad (excluding trackage rights). There were 17 freight railroads operating in 2009. Rail traffic originating in Louisiana in 2009 totalled 25·2m. tons and rail traffic terminating in the state came to 30·0m. tons. There is a tramway in New Orleans.

Civil Aviation

In 2011 there were seven primary airports—commercial service airports with more than 10,000 passenger boardings annually—with a combined total of 5,496,614 enplanements, up from 5,274,620 in 2010. By far the busiest airport, Louis Armstrong New Orleans International, had 4,255,411 enplanements in 2011.

Shipping

The port of South Louisiana is the busiest in the country and the second busiest after Houston for foreign trade cargo. In 2009

South Louisiana handled 212·6m. tons of cargo (109·5m. tons of domestic cargo and 103·1m. tons of foreign trade cargo). Other major ports are New Orleans, Lake Charles, Baton Rouge and Plaquemines, which ranked as the USA's 6th, 12th, 14th and 15th busiest ports respectively in 2009. The Mississippi and other waterways provide 7,500 miles of navigable water.

SOCIAL INSTITUTIONS

Justice
In Dec. 2008 there were 38,381 federal and state prisoners. There was one execution in 2010 but none in 2011.

Education
School attendance is compulsory between the ages of seven and 18. In 2007–08 there were 1,470 public schools with 681,038 pupils and 48,610 teachers. There were 390 private schools in fall 2007 with 137,460 pupils and 9,080 teachers. Teachers' average salary in 2008–09 was \$49,284. In 2008–09 the state had 51 public and 34 non-public degree-granting institutions. There were 193,316 students at public degree-granting institutions in fall 2007 and 31,438 students at private degree-granting institutions.

Enrolment, 2003–04, in the University of Louisiana System was 83,303 (Lafayette, 16,208; Southeastern, 15,662; Louisiana Tech., 11,960; Northwestern, 10,505; Monroe, 8,592; McNeese, 8,447; Nicholls, 7,260; Grambling, 4,669); Louisiana State University, 62,841 (with campuses at Alexandria, Baton Rouge, Eunice, New Orleans and Shreveport); Southern University System, 15,044. Major private institutions: Tulane University, 9,920; Loyola University, 5,900; Xavier University, 3,994; Dillard University, 1,953.

Health
In 2009 there were 128 community hospitals with 15,900 beds. A total of 639,000 patients were admitted during the year. In 2009 there were 11,974 active physicians in the state (27 per 10,000 population).

Welfare
Medicare enrolment in July 2010 totalled 680,326. In fiscal year 2009 a total of 1,184,335 people in Louisiana received Medicaid. In Dec. 2008 there were 748,171 Old-Age, Survivors, and Disability Insurance (OASDI) beneficiaries. A total of 25,427 people were receiving payments under Temporary Assistance for Needy Families (TANF) in Dec. 2010.

RELIGION

The principal religious traditions in the state in 2010 were: Catholics, with 1,200,900 members; Evangelical Protestants, 1,064,486; Black Protestants, 217,176; Mainline Protestants, 202,751; Latter-day Saints (Mormons), 29,107.

FURTHER READING

Louisiana State Census Data Center. Online only

Calhoun, Milburn and McGovern, Bernie, (eds.) *Louisiana Almanac 2012 Edition.* 2012

Wall, Bennett H., *et al.*, (eds.) *Louisiana: a History, Fifth Edition.* 2008

Wilds, J., *et al.*, (eds.) *Louisiana Yesterday and Today: a Historical Guide to the State.* 1996

State library: The State Library of Louisiana, 701 North 4th St., Baton Rouge.

Maine

KEY HISTORICAL EVENTS

Originally occupied by Algonquian-speaking tribes, the Territory was disputed between groups of British settlers, and between the British and French, throughout the 17th and most of the 18th centuries. After 1652 Maine was governed as part of Massachusetts, and French claims finally failed in 1763. Most of the early settlers were English and Protestant Irish, with many Quebec French.

The Massachusetts settlers gained control when the first colonist, Sir Ferdinando Gorges, supported the losing royalist side in the English civil war. During the English-American war of 1812, Maine residents claimed that the Massachusetts government did not protect them against British raids. Maine was separated from Massachusetts and entered the Union as a state in 1820.

Maine is a mountainous state and even the coastline is rugged, but the coastal belt is where most settlement has developed. In the 19th century there were manufacturing towns making use of cheap water-power and the rocky shore supported a shell-fish industry. The latter still flourishes, together with intensive horticulture, producing potatoes and fruit. The other main economic activity is forestry for timber, pulp and paper.

The capital is Augusta, a river trading post which was fortified against Indian attacks in 1754, incorporated as a town in 1797 and chosen as capital in 1832.

TERRITORY AND POPULATION

Maine is bounded west, north and east by Canada, southeast by the Atlantic, south and southwest by New Hampshire. Land area, 30,843 sq. miles (79,883 sq. km). Census population, 1 April 2010, was 1,328,361, an increase of 4·2% since 2000. July 2012 estimate, 1,329,192.

Population for five census years was:

	White	Black	American Indian	Asiatic	Total	Per sq. mile
1910	739,995	1,363	992	121	742,371	24·8
			All others			
1980	1,109,850	3,128	12,049		1,125,027	36·3
1990	1,208,360	5,138	14,430		1,227,928	39·8
2000	1,236,014	6,760	32,149		1,274,923	41·3
2010	1,264,971	15,707	47,683		1,328,361	43·1

Of the total population in 2010, 678,305 were female and 1,053,828 were 18 years old or older. In 2010 the Hispanic population was 16,935, an increase of 80·9% on the 2000 census figure of 9,360. Only North Dakota and Vermont have fewer persons of Hispanic origin in the USA.

The largest city in the state is Portland, with a census population of 66,194 in 2010. Other cities (with population in 2010) are: Lewiston, 36,592; Bangor, 33,039; South Portland, 25,002; Auburn, 23,055; Biddeford, 21,277; Augusta (capital), 19,136; Saco, 18,482; Westbrook, 17,494.

SOCIAL STATISTICS

Births, 2010, 12,970 (9·8 per 1,000 population—the joint lowest rate in any US state); deaths, 2010 (provisional), 12,755 (9·6). Infant mortality rate, 2009, 5·6 per 1,000 live births. 2009 (provisional): marriages, 9,400 (7·2 per 1,000 population); divorces and annulments, 5,300 (4·1). Same-sex marriage was legalized in Nov. 2012 when it was approved in a referendum on the same day as the presidential election.

CLIMATE

Average maximum temperatures range from 56·3°F in Waterville to 48·3°F in Caribou, but record high (since c. 1950) is 103°F. Average minimum ranges from 36·9°F in Rockland to 28·3°F in Greenville, but record low (also in Greenville) is –42°F. Average annual rainfall ranges from 48·85" in Machias to 36·09" in Houlton. Average annual snowfall ranges from 118·7" in Greenville to 59·7" in Rockland. Maine belongs to the New England climate zone (*see* UNITED STATES: Climate).

CONSTITUTION AND GOVERNMENT

The constitution of 1820 is still in force, but it has been amended 171 times. In 1951, 1967, 1973, 1983, 1993 and 2003 the Legislature approved recodifications of the constitution as arranged by the Chief Justice under special authority.

The Legislature consists of the Senate with 35 members and the House of Representatives with 151 members, both Houses being elected simultaneously for two years. Sessions are annual.

For the 113th Congress, which convened in Jan. 2013, Maine sends two members to the House of Representatives. It is represented in the Senate by Susan Collins (R. 1997–2015) and Angus King (Independent, 2013–19).

The capital is Augusta. The state is divided into 16 counties.

RECENT ELECTIONS

In the 2012 presidential election Barack Obama took Maine with 56·0% of the vote, having also won it in 2008.

CURRENT GOVERNMENT

Governor: Paul R. LePage (R.), 2011–15 (salary: $70,000).
   Senate President: Justin Alford (D.), 2013–15 ($35,269).
   Secretary of State: Matthew Dunlap (D.), 2013–15 ($69,264).

Government Website: http://www.maine.gov

ECONOMY

Per capita personal income (2010) was $36,717.

Budget

In 2011 total state revenue was $10,611m. Total expenditure was $9,099m. (public welfare, $2,905m.; education, $2,121m.; highways, $647m.; health, $446m.; government administration, $286m.) Outstanding debt in 2011, $5,904m.

Performance

Gross Domestic Product by state was $44,821m. in 2011 (provisional), ranking Maine 43rd in the United States. In 2011 state real GDP growth was –0·4% (provisional).

ENERGY AND NATURAL RESOURCES

Water

The total area covered by water is 4,537 sq. miles.

Minerals

Minerals include sand and gravel, stone, lead, clay, copper, peat, silver and zinc. Total value of non-fuel mineral production for 2009 was $125m.

Agriculture

In 2007, 8,136 farms occupied 1·35m. acres; the average farm was 166 acres. Average value of farmland and buildings per acre in 2007 was $2,203. Farm income, 2006: from crops, $303m.; and from livestock and products, $289m. The net farm income in 2006 was $217m. Principal commodities are potatoes, dairy products, blueberries and chicken eggs.

Forestry

There were 17·67m. acres of forested land in 2007, of which 53,000 acres were national forests. Commercial forest includes pine, spruce and fir. Wood products industries are of great economic importance.

Fisheries

In 2009 the commercial catch was 184·6m. lb, valued at $285·9m.

INDUSTRY

In 2007 the state's 1,825 manufacturing establishments had 59,000 employees, earning $2,524m. Total value added by manufacturing in 2009 was $8,080m.

Labour

Total non-agricultural employment, 2010, 593,000. Employees by branch, 2010 (in 1,000): education and health services, 119; trade, transportation and utilities, 117; government, 103; leisure and hospitality, 60; professional and business services, 56. The unemployment rate in Dec. 2010 was 7·5%.

COMMUNICATIONS

Roads

In 2007 there were 22,792 miles of road (19,805 miles rural). There were 1,079,843 registered motor vehicles.

Rail

In 2009 there were 1,151 miles of freight railroad (excluding trackage rights). Rail traffic originating in Maine in 2009 totalled 2·0m. tons and rail traffic terminating in the state came to 2·3m. tons.

Civil Aviation

There are international airports at Portland and Bangor. There were 1,058,062 passenger enplanements statewide in 2007.

SOCIAL INSTITUTIONS

Justice

In Dec. 2008 there were 2,195 federal and state prisoners. Capital punishment was abolished in 1887.

Education

Education is free for pupils from five to 21 years of age, and compulsory from seven to 17. In fall 2007 there were 196,245 pupils and 16,558 teachers in 670 public elementary and secondary schools. Expenditure on public education in 2006–07 was $2,421m.

In 2008–09 there were 30 degree-granting institutions (15 public); there were 67,173 students enrolled in fall 2007 (48,357 in public institutions). The University of Maine System, created by Maine's state legislature in 1965, consists of seven universities: the University of Maine (founded in 1865); the University of Maine at Augusta, at Farmington, at Fort Kent (1878), at Machias (1909), at Presque Isle (1903); and the University of Southern Maine (1878, campuses at Portland, Gorham and Lewiston-Auburn).

There are several independent universities, including: Bowdoin College, founded in 1794 at Brunswick; Bates College at Lewiston; Colby College at Waterville; Husson College at Bangor; Westbrook College at Westbrook; Unity College at Unity; and the University of New England (formerly St Francis College) at Biddeford.

Health

In 2009 there were 37 community hospitals with 3,600 beds. A total of 150,000 patients were admitted during the year. In 2009 there were 3,663 active physicians in the state (28 per 10,000 population).

Welfare

Medicare enrolment in July 2010 totalled 262,772. In fiscal year 2009 a total of 314,809 people in Maine received Medicaid. In Dec. 2008 there were 286,123 Old-Age, Survivors, and Disability Insurance (OASDI) beneficiaries. A total of 24,470 people were receiving payments under Temporary Assistance for Needy Families (TANF) in Dec. 2008.

RELIGION

The principal religious traditions in 2010 were: Catholics, with 190,106 members; Mainline Protestants, 93,580; Evangelical Protestants, 59,052; Latter-day Saints (Mormons), 11,704.

FURTHER READING

Statistical information: Maine Department of Economic and Community Development, State House Station 59, Augusta 04333. Publishes *Maine: a Statistical Summary.*

Palmer, K. T., *et al., Maine Politics and Government.* 1993

Maryland

KEY HISTORICAL EVENTS

The first European visitors found Algonquian-speaking tribes, often under attack by Iroquois from further north. The first white settlement was made by the Calvert family, British Roman Catholics, in 1634. The settlers received some legislative rights in 1638. In 1649 their assembly passed the Act of Toleration, granting freedom of worship to all Christians. A peace treaty was signed with the Iroquois in 1652, after which it was possible for farming settlements to expand north and west. The capital (formerly at St Mary's City) was moved to Annapolis in 1694. Baltimore, which became the state's main city, was founded in 1729.

The first industry was tobacco-growing, which was based on slave-worked plantations. There were also many immigrant British small farmers, tradesmen and indentured servants.

At the close of the War of Independence, the treaty of Paris was ratified in Annapolis. Maryland became a state of the Union in 1788. In 1791 the state ceded land for the new federal capital, Washington, and its economy has depended on the capital's proximity ever since. Baltimore also grew as a port and industrial city, attracting European immigration in the 19th century. Although in sympathy with the south, Maryland remained in the Union in the Civil War albeit under the imposition of martial law.

TERRITORY AND POPULATION

Maryland is bounded north by Pennsylvania, east by Delaware and the Atlantic, south by Virginia and West Virginia, with the Potomac River forming most of the boundary, and west by West Virginia. Chesapeake Bay almost cuts off the eastern end of the state from the rest. Land area, 9,707 sq. miles (25,142 sq. km). Census population, 1 April 2010, was 5,773,552, an increase of 9·0% since 2000. July 2012 estimate, 5,884,563.

Population for five federal censuses was:

	White	Black	American Indian	Asiatic	Total	Per sq. mile
1920	1,204,737	244,479	32	400	1,449,661	145·8
1960	2,573,919	518,410	1,538	5,700	3,100,689	314·0
			All others			
1990	3,393,964	1,189,899	197,605		4,781,468	489·2
2000	3,391,308	1,477,411	427,767		5,296,486	541·9
2010	3,359,284	1,700,298	713,970		5,773,552	594·8

Of the total population in 2010, 2,981,790 were female and 4,420,588 were 18 years old or older. In 2010 Maryland's Hispanic population was 470,632, up from 227,916 in 2000 (an increase of 106·5%).

The largest city in the state (containing 10·8% of the population) is Baltimore, with 620,961 (2010 census). Baltimore–Towson, metropolitan area, 2,710,4894 (2010). Other main population centres (2010 census) are Columbia (99,615); Germantown (86,395); Silver Spring (71,452); Waldorf (67,752); Glen Burnie (67,639); Ellicott City (65,834); Frederick (65,239); Dundalk (63,597); Rockville (61,209); Bethesda (60,858). Annapolis (the capital) had a population of 38,394 in 2010.

SOCIAL STATISTICS

Births, 2010, 73,801 (12·8 per 1,000 population); deaths, 2010 (provisional), 43,324 (7·5). Infant mortality, 2009, 7·3 per 1,000 live births. 2009 (provisional): marriages, 32,400 (5·8 per 1,000 population); divorces and annulments, 15,200 (2·8). Same-sex marriage was legalized in Nov. 2012 when it was approved in a referendum on the same day as the presidential election.

CLIMATE

Baltimore, Jan. 36°F (2·2°C), July 79°F (26·1°C). Annual rainfall 42" (1,066 mm). Maryland belongs to the Atlantic Coast climate zone (see UNITED STATES: Climate).

CONSTITUTION AND GOVERNMENT

The present constitution dates from 1867; it has had 224 amendments as of Aug. 2011. Amendments are proposed and considered annually by the General Assembly and must be ratified by the electorate. The General Assembly consists of a Senate of 47, and a House of Delegates of 141 members, both elected for four years, as are the Governor and Lieut.-Governor. Voters are citizens who have the usual residential qualifications.

For the 113th Congress, which convened in Jan. 2013, Maryland sends eight members to the House of Representatives. It is represented in the Senate by Barbara Mikulski (D. 1987–2017) and Benjamin Cardin (D. 2007–19).

The state capital is Annapolis. The state is divided into 23 counties and Baltimore City.

RECENT ELECTIONS

In the 2012 presidential election Barack Obama took Maryland with 61·7% of the vote, having also won it in 2008.

CURRENT GOVERNMENT

Governor: Martin O'Malley (D.), 2011–15 (salary: $150,000).

Lieut.-Governor: Anthony Brown (D.), 2011–15 ($125,000).

Secretary of State: John P. McDonough (D.), appointed July 2008 ($87,500).

Government Website: http://www.maryland.gov

ECONOMY

Per capita personal income (2010) was $49,070. Maryland had the highest average household income in 2010, at $68,854.

Budget

In 2011 total state revenue was $41,717m. Total expenditure was $37,673m. (education, $11,213m.; public welfare, $9,274m.; highways, $2,196m.; health, $1,849m.; correction, $1,384m.) Outstanding debt in 2011, $25,250m.

Performance

Gross Domestic Product by state in 2011 was $264,373m. (provisional), ranking Maryland 15th in the United States. In 2011 state real GDP growth was 0·9% (provisional).

ENERGY AND NATURAL RESOURCES

Electricity

In 2009 production was 43,775m. kWh, of which 24,162m. kWh was from coal.

Water

The total area covered by water is approximately 2,699 sq. miles.

Minerals

Value of non-fuel mineral production in 2009 was more than $300m. The leading mineral commodities by weight are crushed stone (24·8m. tonnes in 2008) and sand and gravel (12·0m. tonnes in 2008). Stone is the leading mineral commodity by value followed by Portland cement, coal, and sand and gravel. In 2007 output of crushed stone was valued at $282m. and Portland cement at an estimated $265m. Coal output was 2·31m. short tons in 2009.

Agriculture

In 2007 there were 12,834 farms with an area of 2·05m. acres. The average number of acres per farm was 160. The average value per acre in 2007 was $7,034.

Farm animals, 2007 were: dairy cows, 57,200; all cattle, 190,500; swine (2001), 52,000; sheep and lambs, 22,100; and laying hens,

2·66m. Farm income cash receipts, 2006: crops, $726m.; livestock and products, $872m.; total, $1,598m. The net farm income in 2006 was $595m. Broilers (2007 value, $732·3m.), greenhouse and nursery products ($396·1m. in 2007) and dairy products ($207·6m. in 2007) are the leading agricultural commodities.

Forestry
Total forested area was 2·57m. acres in 2007.

Fisheries
In 2009, 55·8m. lb of seafood was landed at a dockside value of $67·3m. In 2011 there were 70 processing plants employing 1,217 people.

INDUSTRY
In 2007 the state's 3,680 manufacturing establishments had 128,000 employees, earning $6,454m. Total value added by manufacturing in 2009 was $20,848m.

Labour
Total non-agricultural employment, 2010, 2,513,000. Employees by branch, 2010 (in 1,000): government, 501; trade, transportation and utilities, 438; education and health services, 400; professional and business services, 386; leisure and hospitality, 229. The unemployment rate in Dec. 2010 was 7·4%.

COMMUNICATIONS

Roads
In 2007 there were 31,300 miles of road comprising 17,276 miles of urban road and 14,024 miles of rural road. There were 4,510,464 registered vehicles in 2007.

Rail
Maryland is served by CSX Transportation, Norfolk Southern Railroad as well as by six short-line railroads. Metro lines also serve Maryland in suburban Washington, D.C. Amtrak provides passenger service linking Baltimore and BWI Airport to major cities on the Atlantic Coast. MARC commuter rail serves the Baltimore–Washington metropolitan area.

Civil Aviation
In 2011 there were two primary airports (commercial service airports with more than 10,000 passenger boardings annually). The main airport, Baltimore/Washington International Thurgood Marshall, had 11,067,319 enplanements in 2011.

Shipping
In 2009 Baltimore handled 30·1m. tons of cargo. It is located about 200 miles further inland than any other Atlantic seaport.

SOCIAL INSTITUTIONS

Justice
Prisons in Dec. 2008 held 23,324 inmates. Maryland's prison system has conducted a work-release programme for selected prisoners since 1963. All institutions have academic and vocational training programmes. There was one execution in 2004, the first since 1998, and one in 2005, but the death penalty was suspended in 2006.

Education
Education is compulsory from five to 16 years of age. In 2007–08 there were 1,453 public schools with 845,700 pupils and 59,320 teachers. Average teacher salary in 2008–09 was $60,844. Total expenditure on public elementary and secondary education in 2006–07 was $11,547m. Per pupil spending in fiscal year 2008 was $13,235.

In 2008–09 there were 57 degree-granting institutions (29 public and 28 private). The largest is the University System of Maryland (created in 1988), with 128,425 students (fall 2005), consisting of 11 campuses, two major research institutions, two regional higher education centres and a system office. The USM

colleges and universities are: Bowie State University; Coppin State College; Frostburg State University; Salisbury University; Towson University; the University of Baltimore; and the five campuses of the University of Maryland (Baltimore, Baltimore County, College Park, Eastern Shore and University College).

Health
In 2009 there were 49 community hospitals with 11,900 beds. A total of 715,000 patients were admitted during the year. In 2009 there were 24,118 active physicians in the state (42 per 10,000 population).

Welfare
Medicare enrolment in July 2010 totalled 778,844. In fiscal year 2009 a total of 845,519 people in Maryland received Medicaid. In Dec. 2008 there were 802,066 Old-Age, Survivors, and Disability Insurance (OASDI) beneficiaries. A total of 53,866 people were receiving payments under Temporary Assistance for Needy Families (TANF) in Dec. 2008.

RELIGION
Maryland was the first US state to give religious freedom to all who came within its borders. The principal religious traditions in 2010 were Catholics (837,338 members), Evangelical Protestants (693,990 members), Mainline Protestants (500,112 members), Black Protestants (157,854 members) and Jews (83,284 members).

CULTURE
Cultural venues include: Frostburg Performing Arts Center, Strathmore Hall Arts Center and Center Stage. Performing arts institutions include the Baltimore Opera Company, Peabody Music Conservatory and Arena Players.

Tourism
In 2011 there were 335,000 overseas visitors to Maryland, excluding those from Canada and Mexico.

FURTHER READING
Statistical Information: Maryland Department of Economic and Employment Development, 217 East Redwood St., Baltimore 21202.

DiLisio, J. E., *Maryland.* 1982
Rollo, V. F., *Maryland's Constitution and Government.* 1982

State library: Maryland State Library, Annapolis.

Massachusetts

KEY HISTORICAL EVENTS
The first European settlement was at Plymouth, where the *Mayflower* landed its English religious separatists in 1620. In 1626–30 more colonists arrived, the main body being English Puritans who founded a Puritan commonwealth. This commonwealth, of about 1,000 colonists led by John Winthrop, became the Massachusetts Bay Colony and was founded under a company charter. Following disagreement between the English government and the colony, the charter was withdrawn in 1684, but in 1691 a new charter united a number of settlements under the name of Massachusetts Bay. The colony's government was rigidly theocratic.

Shipbuilding, iron-working and manufacturing were more important than farming, the land being poor. The colony was Protestant and of English descent until the War of Independence. The former colony adopted its present constitution in 1780. In the struggle which ended in the separation of the American colonies from the mother country, Massachusetts took the foremost part, and on 6 Feb. 1788 became the 6th state to ratify the US constitution. The state acquired its present boundaries (having previously included Maine) in 1820.

During the 19th century, industrialization and immigration from Europe increased while Catholic Irish and Italian immigrants began to change the population's character. The main inland industry was textile manufacture, the main coastal occupation, whaling; both have now gone. Boston has remained the most important city of New England, attracting a large black population since 1950.

TERRITORY AND POPULATION

Massachusetts is bounded north by Vermont and New Hampshire, east by the Atlantic, south by Connecticut and Rhode Island and west by New York. Land area, 7,800 sq. miles (20,202 sq. km). Census population, 1 April 2010, was 6,547,629, an increase of 3·1% since 2000. July 2012 estimate, 6,646,144.

Population at five federal census years was:

	White	Black	Other	Total	Per sq. mile
1950	4,611,503	73,171	5,840	4,690,514	598·4
1980	5,362,836	221,279	152,922	5,737,037	732·0
1990	5,405,374	300,130	310,921	6,016,425	767·6
2000	5,367,286	343,454	638,357	6,349,097	809·8
2010	5,265,236	434,398	847,995	6,547,629	839·4

Of the total population in 2010, 3,381,001 were female and 5,128,706 were 18 years old or older. In 2010 the Hispanic population was 627,654, up from 428,729 in 2000 (an increase of 46·4%).

Population of the largest cities at the 2010 census: Boston (the capital), 617,594; Worcester, 181,045; Springfield, 153,060; Lowell, 106,519; Cambridge, 105,162; New Bedford, 95,072; Brockton, 93,810; Quincy, 92,271; Lynn, 90,329; Fall River, 88,857; Newton, 85,146. The Boston–Cambridge–Quincy metropolitan area had a 2010 census population of 4,552,402; Worcester, 798,552; Springfield, 692,942.

SOCIAL STATISTICS

Births, 2010, 72,865 (11·1 per 1,000 population); deaths, 2010 (provisional), 52,595 (8·0). Infant mortality, 2009, 5·1 per 1,000 live births. 2009 (provisional): marriages, 36,700 (5·5 per 1,000 population); divorces and annulments, 12,700 (2·2). Massachusetts was the first state to allow same-sex marriage.

CLIMATE

Boston, Jan. 28°F (–2·2°C), July 71°F (21·7°C). Annual rainfall 41" (1,036 mm). Massachusetts belongs to the New England climate zone (*see* UNITED STATES: Climate).

CONSTITUTION AND GOVERNMENT

The constitution dates from 1780 and has had 120 amendments as of 2011. The legislative body, styled the General Court of the Commonwealth of Massachusetts, meets annually, and consists of the Senate with 40 members and the House of Representatives of 160 members, both elected for two years.

For the 113th Congress, which convened in Jan. 2013, Massachusetts sends nine members to the House of Representatives. It is represented in the Senate by Elizabeth Warren (D. 2013–19) and William Cowan (D. interim senator following John Kerry's resignation to become secretary of state).

The capital is Boston. The state has 14 counties.

RECENT ELECTIONS

In the 2012 presidential election Barack Obama took Massachusetts with 60·7% of the vote, having also won it in 2008.

CURRENT GOVERNMENT

Governor: Deval Patrick (D.), 2011–15 (salary: $137,315).
Lieut.-Governor: Timothy P. Murray (D.), 2011–15 ($122,058).

Secretary of the Commonwealth: William F. Galvin (D.), 2011–15 ($130,916).

Government Website: http://www.mass.gov

ECONOMY

Per capita personal income (2010) was $51,302, the third highest in the country.

Budget

In 2011 total state revenue was $56,637m. Total expenditure was $52,551m. (public welfare, $14,716m.; education, $12,334m.; highways, $1,908m.; government administration, $1,648m.; health, $1,112m.) Outstanding debt in 2011, $74,316m.

Performance

Gross State Product by state in 2011 was $348,577m. (provisional), ranking Massachusetts 12th in the United States. In 2011 state real GDP growth was 2·2% (provisional).

ENERGY AND NATURAL RESOURCES

Water

The total area covered by water is 2,754 sq. miles.

Minerals

Total domestic non-fuel mineral output in 2009 was valued at more than $210m., most of which came from sand, gravel, crushed stone and lime.

Agriculture

In 2007 there were 7,691 farms with an average area of 67 acres and a total area of 517,879 acres. Average value per acre in 2007 was $12,313. Farm income 2006: from crops, $344m.; and from livestock and products, $89m. Principal commodities are greenhouse products, cranberries, dairy products and sweetcorn. The net farm income in 2006 was $115m.

Forestry

About 62% of the state is forest. In 2007 state forests covered about 603,000 acres, with total forest land covering 3·17m. acres. Commercially important hardwoods are sugar maple, northern red oak and white ash; softwoods are white pine and hemlock.

Fisheries

In 2009 commercial fishing produced 356·0m. lb of fish with a value of $400·2m.

INDUSTRY

In 2007 the state's 7,737 manufacturing establishments had 289,000 employees, earning $15,712m. Total value added by manufacturing in 2009 was $41,297m.

Labour

Total non-agricultural employment, 2010, 3,186,000. Employees by branch, 2010 (in 1,000): education and health services, 664; trade, transportation and utilities, 544; professional and business services, 461; government, 438; leisure and hospitality, 306. The state unemployment rate was 8·3% in Dec. 2010.

COMMUNICATIONS

Roads

In 2007 there were 36,008 miles of public road (28,041 miles urban, 7,967 rural). There were 5,366,708 registered motor vehicles.

Rail

In 2009 there were 952 miles of freight railroad (excluding trackage rights). There were 11 freight railroads operating in 2009. Rail traffic originating in Massachusetts in 2009 totalled 1·9m. tons and rail traffic terminating in the state came to 6·2m. tons. There are metro, light rail, tramway and commuter networks in and around Boston.

Civil Aviation

In 2011 there were seven primary airports (commercial service airports with more than 10,000 passenger boardings annually) with a combined total of 14,532,785 enplanements, up from 13,963,433 in 2010. By far the busiest airport is General Edward Lawrence Logan International, Boston's airport, with 14,180,730 enplanements in 2011.

Shipping

The state has three deep-water harbours, the busiest of which is Boston (with 20·5m. tons of cargo in 2009—13·5m. tons of foreign trade cargo and 7·0m. tons of domestic trade cargo). Other ports are Fall River and New Bedford.

SOCIAL INSTITUTIONS

Justice

There were 11,408 federal and state prisoners in Dec. 2008. The death penalty was abolished in 1984.

Education

School attendance is compulsory for ages six to 16. In 2007–08 there were 1,878 public elementary and secondary schools with 962,958 pupils and 70,719 teachers; total expenditure on public schools in 2006–07 was $13,409m. Teachers' salaries in 2008–09 averaged $62,769.

Some leading higher education institutions are:

Year opened	Name and location of universities and colleges	Students (fall 2009)
1636	Harvard University, Cambridge	27,651
1839	Framingham State College	5,989
1839	Westfield State College	5,675
1840	Bridgewater State College	10,774
1852	Tufts University, Medford[1]	10,252
1854	Salem State College	10,125
1861	Mass. Institute of Technology, Cambridge	10,384
1863	University of Massachusetts, Amherst	27,016
1863	Boston College (RC), Chestnut Hill	15,036
1865	Worcester Polytechnic Institute, Worcester	4,961
1869	Boston University, Boston	31,960
1874	Worcester State College	5,473
1894	Fitchburg State College	7,043
1894	University of Massachusetts, Lowell	13,602
1895	University of Massachusetts, Dartmouth	9,302
1898	Northeastern University, Boston[2]	27,537
1899	Simmons College, Boston[3]	5,003
1905	Wentworth Institute of Technology	3,808
1906	Suffolk University	9,148
1917	Bentley University[4]	5,628
1919	Western New England College	3,710
1919	Babson College	3,445
1947	Merrimack College	2,090
1948	Brandeis University, Waltham	5,598
1964	University of Massachusetts, Boston	14,912

[1]Includes Jackson College for women. [2]Includes Forsyth Dental Center School. [3]For women only. [4]Name change from Bentley College, effective Oct. 2008.

Health

In 2009 there were 78 community hospitals with 15,500 beds. A total of 820,000 patients were admitted during the year. In 2009 there were 31,252 active physicians in the state (47 per 10,000 population).

Welfare

Medicare enrolment in July 2010 totalled 1,052,461. In fiscal year 2008 a total of 1,230,063 people in Massachusetts received Medicaid. In Dec. 2008 there were 1,094,012 Old-Age, Survivors, and Disability Insurance (OASDI) beneficiaries. A total of 92,961 people were receiving payments under Temporary Assistance for Needy Families (TANF) in Dec. 2008.

RELIGION

The principal religious traditions in 2010 were Catholics (2,940,199 members), Mainline Protestants (308,286 members), Evangelical Protestants (224,726 members), Jews (80,502 members) and Orthodox Christians (61,544 members).

CULTURE

Tourism

Overseas visitors to Massachusetts (excluding those from Canada and Mexico) numbered 1,422,000 in 2011, up from 1,292,000 in 2010 and 1,105,000 in 2006.

FURTHER READING

Levitan, D. with Mariner, E. C., *Your Massachusetts Government*. 1984

Michigan

KEY HISTORICAL EVENTS

The French were the first European settlers, establishing a fur trade with the local Algonquian Indians in the late 17th century. They founded Sault Ste Marie in 1668 and Detroit in 1701. In 1763 Michigan passed to Britain, along with other French territory east of the Mississippi, and from Britain it passed to the USA in 1783. Britain, however, kept a force at Detroit until 1796, and recaptured Detroit in 1812. Regular American settlement did not begin until later. The Territory of Michigan (1805) had its boundaries extended after 1818 and 1834. It was admitted to the Union as a state (with its present boundaries) in 1837.

During the 19th century there was rapid industrial growth, especially in mining and metalworking. The largest groups of immigrants were British, German, Irish and Dutch. Other groups came from Scandinavia, Poland and Italy. Many settled as miners, farmers and industrial workers. The motor industry became dominant, especially in Detroit. Lake Michigan ports shipped bulk cargo of iron ore and grain.

Detroit was the capital until 1847, when that function passed to Lansing. Detroit remained, however, an important centre of flour-milling and shipping and, after the First World War, of the motor industry.

TERRITORY AND POPULATION

Michigan is divided into two by Lake Michigan. The northern part is bounded south by the lake and by Wisconsin, west and north by Lake Superior, east by the North Channel of Lake Huron; between the two latter lakes the Canadian border runs through straits at Sault Ste Marie. The southern part is bounded in the west and north by Lake Michigan, east by Lake Huron, Ontario and Lake Erie, south by Ohio and Indiana. Total area is 96,714 sq. miles (250,487 sq. km) of which 56,539 sq. miles (146,435 sq. km) are land and 40,175 sq. miles (104,052 sq. km) water. Census population, 1 April 2010, was 9,883,640, a fall of 0·6% since 2000. Michigan was the only state whose population fell between 2000 and 2010. July 2012 estimate, 9,883,360.

Population of five federal census years was:

	White	Black	American Indian	Asiatic	Total	Per sq. mile
1910	2,785,247	17,115	7,519	292	2,810,173	48·9
			All others			
1980	7,872,241	1,199,023	190,814		9,262,078	162·6
1990	7,756,086	1,291,706	247,505		9,295,297	160·0
2000	7,966,053	1,412,742	559,649		9,938,444	175·0
2010	7,803,120	1,400,362	680,158		9,883,640	174·8

Of the total population in 2010, 5,035,526 were female and 7,539,572 were 18 years old or more. In 2010 the Hispanic population was 436,358, up from 323,877 in 2000 (an increase of 34·7%).

Populations of the chief cities in 2010 were: Detroit, 713,777; Grand Rapids, 188,040; Warren, 134,056; Sterling Heights, 129,699; Lansing (the capital), 114,297; Ann Arbor, 113,934; Flint, 102,434; Dearborn, 98,153. The Detroit–Warren–Livonia metropolitan area had a 2010 census population of 4,296,250.

SOCIAL STATISTICS

Births, 2010, 114,531 (11·6 per 1,000 population); deaths, 2010 (provisional), 85,561 (8·7). Infant mortality, 2009, 7·5 per 1,000 live births. 2009 (provisional): marriages, 53,100 (5·4 per 1,000 population); divorces and annulments, 32,500 (3·3).

CLIMATE

Detroit, Jan. 23·5°F (–5·0°C), July 72°F (22·5°C). Annual rainfall 32" (810 mm). Grand Rapids, Jan. 22°F (–5·5°C), July 71·5°F (22·0°C). Annual rainfall 34" (860 mm). Lansing, Jan. 22°F (–5·5°C), July 70·5°F (21·5°C). Annual rainfall 29" (740 mm). Michigan belongs to the Great Lakes climate zone (*see* UNITED STATES: Climate).

CONSTITUTION AND GOVERNMENT

The present constitution became effective on 1 Jan. 1964. The Senate consists of 38 members, elected for four years, and the House of Representatives of 110 members, elected for two years. Sessions are biennial.

For the 113th Congress, which convened in Jan. 2013, Michigan sends 14 members to the House of Representatives. It is represented in the Senate by Carl Levin (D. 1979–2015) and Debbie Stabenow (D. 2001–19).

The capital is Lansing. The state is organized in 83 counties.

RECENT ELECTIONS

In the 2012 presidential election Barack Obama took Michigan with 54·3% of the vote, having also won it in 2008.

CURRENT GOVERNMENT

Governor: Rick Snyder (R.), 2011–15 (salary: $159,300).
 Lieut.-Governor: Brian Calley (R.), 2011–15 ($111,510).
 Secretary of State: Ruth Johnson (R.), 2011–15 ($112,410).

Government Website: http://www.michigan.gov

ECONOMY

Per capita personal income (2010) was $34,691.

Budget

In 2011 total state revenue was $64,440m. Total expenditure was $63,109m. (education, $23,146m.; public welfare, $14,927m.; hospitals, $2,522m.; highways, $2,464m.; correction, $1,663m.) Outstanding debt in 2011, $30,975m.

Performance

Gross Domestic Product by state in 2011 was $337,427m. (provisional), ranking Michigan 13th in the United States. In 2011 state real GDP growth was 2·3% (provisional).

ENERGY AND NATURAL RESOURCES

Electricity

In 2009 production was 101·2bn. kWh, of which 66·8bn. kWh was from coal.

Oil and Gas

Natural gas production in 2008 was 272bn. cu. ft; production of crude oil was 6·2m. bbls.

Water

The total area covered by water is 40,175 sq. miles (the second largest area covered by water after Alaska).

Minerals

Domestic non-fuel mineral output in 2009 was valued at $1,760m. according to the US Geological Survey. Output was mainly iron ore, cement, crushed stone, sand and gravel.

Agriculture

The state, formerly agricultural, is now chiefly industrial. It contained 55,000 farms in 2008 with a total area of 10·0m. acres; the average farm was 182 acres. The farm real estate average value per acre in 2009 was $3,750. Principal crops are corn, soybeans, wheat, sugar beets, dry beans, potatoes and hay. Principal fruit crops include apples, blueberries, cherries (tart and sweet), grapes, peaches and strawberries. In 2008 there were 353,000 dairy cows, 92,000 beef cows and 1·02m. pigs. Output in 2008 included 110m. lb of blueberries, 22·2m. pots of geraniums and 1·7m. cwt of black beans. Farm income in 2008: total $6,607m.; crops, $4,078m.; livestock and products, $2,529m. The net farm income in 2008 was $2·03bn.

Forestry

Forests covered 19·3m. acres in 2009, with 2·6m. acres of national forest (2008). In 2009 about 18·7m. acres was timberland acreage. Three-quarters of the timber volume was hardwoods, principally hard and soft maples, aspen, oak and birch. Christmas trees are another important forest crop. Net annual growth of growing stock and saw timber was 697m. cu. ft and 2·8bn. bd ft respectively in 2008.

Fisheries

In 2009 recreational fishing licences were purchased by 1,023,433 residents and 285,416 non-residents. Recreational fishing revenue (2006) was approximately $1·67bn.

INDUSTRY

Manufacturing is important; among principal products are motor vehicles and trucks, machinery, fabricated metals, primary metals, cement, chemicals, furniture, paper, foodstuffs, rubber, plastics and pharmaceuticals. In 2007 Michigan's 13,675 manufacturing establishments had 582,000 employees, earning $29,910m. Total value added by manufacturing in 2009 was $71,019m.

Labour

Total non-agricultural labour force in 2010 was 3,861,400. Employees by branch, 2010 (in 1,000): trade, transportation and utilities, 709; government, 636; education and health services, 617; professional and business services, 514; manufacturing, 474. The unemployment rate in Dec. 2010 was 11·1%.

INTERNATIONAL TRADE

Imports and Exports

In 2010 exports from Michigan totalled $44·8bn., up from $32·7bn. in 2009 although down from $45·1bn. in 2008.

COMMUNICATIONS

Roads

In 2007 there were 121,593 miles of road (85,837 miles of rural road and 35,756 miles of urban road). Vehicle registrations in 2007 numbered 8,191,748.

Rail

In 2009 there were 3,600 miles of railroad in Michigan and a 3-mile light rail peoplemover in Detroit.

Civil Aviation

There are international airports at Detroit, Flint, Grand Rapids, Kalamazoo, Port Huron, Saginaw and Sault Ste Marie. There were 20,064,544 passenger enplanements statewide in 2007.

Shipping

There are over 90 commercial and recreational ports spanning the state's 3,200 miles of shoreline. In 2006, 40 of these ports served commercial cargoes. Stone, sand, salt, iron ore, coal and cement accounted for 96% of approximately 76m. tonnes of traffic in 2009.

SOCIAL INSTITUTIONS

Justice

A Civil Rights Commission was established, and its powers and duties were implemented by legislation in the extra session of 1963. Statutory enactments guaranteeing civil rights in specific areas date from 1885. The legislature has a unique one-person grand jury system. The Michigan Supreme Court consists of seven non-partisan elected justices. In Dec. 2008 there were 48,738 prisoners in state or federal correctional institutions. Capital punishment was officially abolished in 1964 but there has never been an execution in Michigan.

Education

Education is compulsory for children from six to 16 years of age. In 2007–08 there were 1,692,739 pupils and 96,204 teachers in 4,096 public schools. Total expenditure on public elementary and secondary education in 2006–07 was $19,932m.; average teacher salary was $57,327 in 2008–09. Spending per pupil in fiscal year 2008 was $10,075.

In 2008–09 there were 106 degree-granting institutions (45 public and 61 private); there were 643,279 students in total in fall 2007 (519,449 at public institutions).

Universities and students (fall 2009):

Founded	Name	Students
1817	University of Michigan, Ann Arbor	41,674
(1956	University of Michigan, Flint	7,773)
(1959	University of Michigan, Dearborn	8,379)
1849	Eastern Michigan University	22,893
1855	Michigan State University	47,071
1868	Wayne State University	31,786
1884	Ferris State University	13,865
1885	Michigan Technological University	7,136
1892	Central Michigan University	27,247
1899	Northern Michigan University	9,428
1903	Western Michigan University	24,576
1946	Lake Superior State University	2,588
1957	Oakland University	18,918
1960	Grand Valley State University	24,408
1963	Saginaw Valley State University	10,498

Health

In 2009 there were 158 community hospitals with 25,900 beds. A total of 1,220,000 patients were admitted during the year. In 2009 there were 25,697 active physicians in the state (26 per 10,000 population).

Welfare

Medicare enrolment in July 2010 totalled 1,637,687. In fiscal year 2009 a total of 1,890,198 people in Michigan received Medicaid. In Dec. 2008 there were 1,840,547 Old-Age, Survivors, and Disability Insurance (OASDI) beneficiaries. A total of 158,943 people were receiving payments under Temporary Assistance for Needy Families (TANF) in Dec. 2008.

RELIGION

The principal religious traditions in the state in 2010 were: Catholics, with 1,717,296 members; Evangelical Protestants, 1,277,144; Mainline Protestants, 653,898; Black Protestants, 214,114; Muslims, 120,000 (estimate).

FURTHER READING

Michigan Manual. Biennial
Michigan Economic Development Corporation. *Economic Profiler.* Online only

Browne, W. P. and Verburg, K., *Michigan Politics and Government: Facing Change in a Complex State.* 1995
Dunbar, W. F. and May, G. S., *Michigan: A History of the Wolverine State.* 3rd ed. 1995

State Library Services: Library of Michigan, Lansing 48909.

Minnesota

KEY HISTORICAL EVENTS

Minnesota remained an Indian territory until the middle of the 19th century, the main groups being Chippewa and Sioux. In the 17th century there had been some French exploration, but no permanent settlement. After passing under the nominal control of France, Britain and Spain, the area became part of the Louisiana Purchase and was sold to the USA in 1803.

Fort Snelling was founded in 1819. Early settlers came from other states, especially New England, to exploit the great forests. Lumbering gave way to homesteading, and the American settlers were joined by Germans, Scandinavians and Poles. Agriculture, mining and forest industries became the mainstays of the economy. Minneapolis, founded as a village in 1856, grew first as a lumber centre, processing the logs floated down the Minnesota River, and then as a centre of flour-milling and grain marketing. St Paul, its twin city across the river, became Territorial capital in 1849 and state capital in 1858. St Paul also stands at the head of navigation on the Mississippi which rises in Minnesota.

The Territory (1849) included parts of North and South Dakota, but at its admission to the Union in 1858, the state of Minnesota had its present boundaries.

TERRITORY AND POPULATION

Minnesota is bounded north by Canada, east by Lake Superior and Wisconsin, with the Mississippi River forming the boundary in the southeast, south by Iowa, west by South and North Dakota, with the Red River forming the boundary in the northwest. Land area, 79,627 sq. miles (206,232 sq. km). Census population, 1 April 2010, was 5,303,925, an increase of 7·8% since 2000. July 2012 estimate, 5,379,139.

Population in five census years was:

	White	Black	American Indian	Asiatic	Total	Per sq. mile
1910	2,059,227	7,084	9,053	344	2,075,708	25·7
			All others			
1980	3,935,770	53,344	86,856		4,075,970	51·4
1990	4,130,395	94,944	149,760		4,375,099	55·0
2000	4,400,282	171,731	347,466		4,919,479	61·8
2010	4,524,062	274,412	505,451		5,303,925	66·6

Of the total population in 2010, 2,671,793 were female and 4,019,862 were 18 years old or older. In 2010 the Hispanic population was 250,258, up from 143,382 in 2000 (an increase of 74·5%).

The largest cities (with 2010 census population) are Minneapolis (382,578), St Paul (the capital; 285,068), Rochester (106,769), Duluth (86,265) and Bloomington (82,893). The Minneapolis–St Paul–Bloomington metropolitan area had a 2010 census population of 3,279,833.

SOCIAL STATISTICS

Births, 2010, 68,610 (12·9 per 1,000 population); deaths, 2010 (provisional), 38,971 (7·3). Infant mortality, 2009, 4·6 per 1,000 live births. 2009 (provisional): marriages, 28,400 (5·3 per 1,000 population).

CLIMATE

Duluth, Jan. 8°F (−13·3°C), July 63°F (17·2°C). Annual rainfall 29" (719 mm). Minneapolis-St. Paul, Jan. 12°F (−11·1°C), July 71°F (21·7°C). Annual rainfall 26" (656 mm). Minnesota belongs to the Great Lakes climate zone (*see* UNITED STATES: Climate).

CONSTITUTION AND GOVERNMENT

The original constitution dated from 1857; it was extensively amended and given a new structure in 1974. The Legislature consists of a Senate of 67 members, elected for four years, and a House of Representatives of 134 members, elected for two years. It meets for 120 days within each two years.

For the 113th Congress, which convened in Jan. 2013, Minnesota sends eight members to the House of Representatives. It is represented in the Senate by Amy Klobuchar (D. 2007–19) and Al Franken (D. 2009–15).

The capital is St Paul. There are 87 counties.

RECENT ELECTIONS

In the 2012 presidential election Barack Obama took Minnesota with 52·8% of the vote, having also won it in 2008.

CURRENT GOVERNMENT

Governor: Mark Dayton (Democratic-Farmer-Labor), 2011–15 (salary: $120,303).

Lieut.-Governor: Yvonne Prettner Solon (Democratic-Farmer-Labor), 2011–15 ($78,196).

Secretary of State: Mark Ritchie (Democratic-Farmer-Labor), 2011–15 ($90,227).

Government Website: http://www.state.mn.us

ECONOMY

Per capita personal income (2010) was $42,847.

Budget

In 2011 total state revenue was $45,684m. Total expenditure was $38,488m. (education, $12,406m.; public welfare, $10,872m.; highways, $2,565m.; government administration, $905m.; natural resources, $691m.) Outstanding debt in 2011, $12,897m.

Performance

In 2011 Gross Domestic Product by state was $244,912m. (provisional), ranking Minnesota 16th in the United States. In 2011 state real GDP growth was 1·2% (provisional).

ENERGY AND NATURAL RESOURCES

Water

The total area covered by water is 7,309 sq. miles, of which 4,763 sq. miles are inland.

Minerals

The iron ore and taconite industry is important in the USA. Production of usable iron ore in 2007 was 38·8m. tons, value $2,320m. Other important minerals are sand and gravel, crushed and dimension stone, clays and peat. Total value of non-fuel mineral production in 2009 was $2,050m

Agriculture

In 2007 there were 80,992 farms with a total area of 26·92m. acres; the average farm was of 332 acres. Average value of land and buildings per acre, 2007, $2,569. Farm income, 2006: from crops, $5,128m.; and from livestock and products, $4,642m. The net farm income in 2006 was $2,494m. Important products: corn, soybeans, sugar beets, spring wheat, processing sweetcorn, oats, dry milk, cheese, mink, turkeys, wild rice, butter, eggs, flaxseed, dairy cows, barley, swine, cattle for market, honey, potatoes, rye, chickens, sunflower seed and dry edible beans. In 2007 there were 2·40m. cattle (460,000 dairy cows), 7·65m. hogs and pigs and 144,600 sheep and lambs.

Forestry

In 2007 Minnesota had 16,391,000 acres of forested land, including 2,459,000 acres of national forest.

INDUSTRY

In 2007 the state's 7,951 manufacturing establishments had 341,000 employees, earning $15,999m. Total value added by manufacturing in 2009 was $46,048m.

Labour

Total non-agricultural employment, 2010, 2,637,000. Employees by branch, 2010 (in 1,000): trade, transportation and utilities, 490; education and health services, 458; government, 417; professional and business services, 313; manufacturing, 292. In Dec. 2010 the unemployment rate was 6·9%.

COMMUNICATIONS

Roads

In 2007 there were 137,693 miles of public roads (119,310 miles rural). There were 4,755,753 registered motor vehicles in 2007.

Rail

In 2009 there were 4,528 miles of freight railroad (excluding trackage rights). There were 20 freight railroads operating in 2009. Rail traffic originating in Minnesota in 2009 totalled 67·4m. tons and rail traffic terminating in the state came to 60·2m. tons.

Civil Aviation

In 2011 Minnesota had seven primary airports (commercial service airports with more than 10,000 passenger boardings annually) with a combined total of 16,221,481 enplanements, up from 15,846,286 in 2010. By far the busiest airport is Minneapolis-St Paul International (MSP), with 15,895,653 enplanements in 2011.

SOCIAL INSTITUTIONS

Justice

In Dec. 2008 there were 9,406 federal and state prisoners. Capital punishment was abolished in 1911.

Education

In fall 2007 there were 837,578 students and 52,975 teachers in public elementary and secondary schools. In 2007–08 there were 2,679 public schools including 169 charter schools. There were 101,740 students enrolled in 580 private schools with 7,180 teachers in fall 2007.

The Minnesota State Colleges and Universities System (created in 1995) is composed of 32 public colleges and universities (25 two-year colleges and seven state universities). In 2007 enrolled students at public degree-granting institutions numbered 250,397. The seven state universities are: St Cloud State University, with 18,123 students in fall 2009; Minnesota, Mankato, 14,955; Winona, 8,657; Minnesota, Moorhead, 7,510; Metropolitan State University (in Minneapolis and St Paul), 7,354; Bemidji State University, 5,175; Southwest Minnesota State University (in Marshall), 6,740. Minnesota State University's Akita campus in Japan closed in 2003.

The University of Minnesota (founded in 1851) has four campuses at Crookston, Duluth, Morris and Twin Cities.

Health

In 2009 there were 132 community hospitals with 15,600 beds. A total of 624,000 patients were admitted during the year. In 2009 there were 15,620 active physicians in the state (30 per 10,000 population).

Welfare

Medicare enrolment in July 2010 totalled 779,609. In fiscal year 2009 a total of 801,908 people in Minnesota received Medicaid. In Dec. 2008 there were 831,763 Old-Age, Survivors, and Disability Insurance (OASDI) beneficiaries. A total of 45,300 people were receiving payments under Temporary Assistance for Needy Families (TANF) in Dec. 2008.

RELIGION

The principal religious traditions in the state in 2010 were: Catholics, with 1,150,367 members; Mainline Protestants, 974,156; Evangelical Protestants, 744,910; Latter-day Saints (Mormons), 31,569; Jews, 23,940.

FURTHER READING

Statistical Information: Department of Trade and Economic Development, 500 Metro Square, St Paul 55101. Publishes *Compare Minnesota: an Economic and Statistical Factbook.—Economic Report to the Governor.*
Legislative Manual. Biennial
Minnesota Agriculture Statistics. Annual

Mississippi

KEY HISTORICAL EVENTS

Mississippi was one of the territories claimed by France and ceded to Britain in 1763. The indigenous people were Choctaw and Natchez. French settlers at first traded amicably, but in the course of three wars (1716, 1723 and 1729) the French allied with the Choctaw to drive the Natchez out. The Natchez massacred the settlers of Fort Rosalie, which the French had founded in 1716 and which was later renamed Natchez.

In 1783 the area became part of the USA except for Natchez which was under Spanish control until 1798. The United States then made it the capital of the Territory of Mississippi. The boundaries of the Territory were extended in 1804 and again in 1812. In 1817 it was divided into two territories, with the western part becoming the state of Mississippi. (The eastern part became the state of Alabama in 1819.) The city of Jackson was laid out in 1822 as the new state capital.

A cotton plantation economy developed, based on black slave labour and by 1860 the majority of the population was black. Mississippi joined the Confederacy during the Civil War. After defeat and reconstruction there was a return to rigid segregation and denial of black rights. This situation lasted until the 1960s. There was a black majority until the Second World War, when out-migration began to change the pattern. By 1990 about 35% of the population was black, and manufacture (especially clothing and textiles) had become the largest single employer of labour.

Mississippi suffered widespread damage and loss of life after Hurricane Katrina struck the Gulf Coast on 31 Aug. 2005.

TERRITORY AND POPULATION

Mississippi is bounded in the north by Tennessee, east by Alabama, south by the Gulf of Mexico and Louisiana, and west by the Mississippi River forming the boundary with Louisiana and Arkansas. Land area, 46,923 sq. miles (121,531 sq. km). Census population, 1 April 2010, was 2,967,297, an increase of 4·3% since 2000. July 2012 estimate, 2,984,926.

Population of five federal census years was:

	White	Black	American Indian	Asiatic	Total	Per sq. mile
1910	786,111	1,009,487	1,253	263	1,797,114	38·8
			All others			
1980	1,615,190	887,206	18,242		2,520,638	53·0
1990	1,633,461	915,057	24,698		2,573,216	54·8
2000	1,746,099	1,033,809	64,750		2,844,658	60·6
2010	1,754,684	1,098,385	114,228		2,967,297	63·2

Of the total population in 2010, 1,526,057 were female and 2,211,742 were 18 years old or older. In 2010 Mississippi's Hispanic population was 81,481, up from 39,569 in 2000 (an increase of 105·9%).

The largest city (2010 census) is Jackson (the capital), 173,514. Others (2010 census) are: Gulfport, 67,793; Southaven, 48,982; Hattiesburg, 45,989; Biloxi, 44,054; Meridian, 41,148; Tupelo, 34,546; Greenville, 34,400; Olive Branch, 33,484; Horn Lake, 26,066; Clinton, 25,216.

SOCIAL STATISTICS

2010: births, 40,036 (13·5 per 1,000 population); deaths, 28,964 (9·8 per 1,000 population). Infant mortality, 2010, 9·7 per 1,000 live births. 2010: marriages, 14,621; divorces, 12,703.

Mississippi has the highest proportion of people living in poverty of any state, at 21·8% in 2010.

CLIMATE

Jackson, Jan. 45°F (7·2°C), July 81°F (27·2°C). Annual rainfall 56" (1,422 mm). Vicksburg, Jan. 47°F (8·3°C), July 82°F (28·0°C). Annual rainfall 58" (1,473 mm). Mississippi belongs to the Central Plains climate zone (*see* UNITED STATES: Climate).

CONSTITUTION AND GOVERNMENT

The present constitution was adopted in 1890 without ratification by the electorate; there were 123 amendments by 2009.

The Legislature consists of a Senate (52 members) and a House of Representatives (122 members), both elected for four years. Electors are all citizens who have resided in the state, in the county and in the election district for 30 days prior to the election and have been registered according to law.

For the 113th Congress, which convened in Jan. 2013, Mississippi sends four members to the House of Representatives. It is represented in the Senate by Thad Cochran (R. 1977–2015) and Roger Wicker (R. 2007–19).

The capital is Jackson; there are 82 counties.

RECENT ELECTIONS

In the 2012 presidential election Mitt Romney took Mississippi with 55·5% of the vote (John McCain won it in 2008).

CURRENT GOVERNMENT

Governor: Phil Bryant (R.), 2012–16 (salary: $122,160).
   Lieut.-Governor: Tate Reeves (R.), 2012–16 ($61,714).
   Secretary of State: Delbert Hosemann (R.), 2012–16 ($90,000).

Government Website: http://www.ms.gov

ECONOMY

Per capita personal income (2010) was $31,046, the lowest in the country. Mississippi also had the lowest average household income in 2010, at $36,851.

Budget

In 2011 total state revenue was $23,606m. Total expenditure was $20,157m. (education, $5,519m.; public welfare, $5,437m.; highways, $1,378m.; hospitals, $1,073m.; health, $434m.) Outstanding debt in 2011, $6,768m.

Performance

Gross Domestic Product by state in 2011 was $84,272m. (provisional), ranking Mississippi 35th in the United States. In 2011 state real GDP growth was –0·8% (provisional).

ENERGY AND NATURAL RESOURCES

Oil and Gas

Petroleum and natural gas account for about 90% (by value) of mineral production. Output of petroleum, 2010, was 21m. bbls and of natural gas 86bn. cu. ft. There are three oil refineries.

Water

The total area covered by water is 1,509 sq. miles.

Minerals
The value of domestic non-fuel mineral production in 2010 was $183m.

Agriculture
Agriculture is the leading industry of the state because of the semi-tropical climate and a rich productive soil. In 2010 farms numbered 42,400 with an area of 11·2m. acres. Average size of farm was 263 acres. This compares with an average farm size of 176 acres in 1967. Average value of farm land and farm buildings per acre in 2011 was $2,120.

Cash income from all crops and livestock in 2010 was $4,890m. Cash income from crops was $1,914m., and from livestock and products $2,975m. The net farm income in 2010 was $1,407m. The chief product is soybeans, cash income (2010) $856m. from 1,980,000 acres producing 76,230,000 bu. Cotton, rice, corn, hay, wheat, oats, sorghum, peanuts, pecans, sweet potatoes, peaches, blueberries, other vegetables, nursery and forest products continue to contribute.

On 1 Jan. 2010 there were 900,000 head of cattle and calves on Mississippi farms. In Dec. 2010 dairy cows totalled 17,000; beef cows, 495,000; hogs and pigs, 385,000. Of cash income from livestock and products, 2010, $169m. was credited to cattle and calves. Cash income from poultry and eggs, 2010, totalled $2·5bn.; swine, $96m.; dairy products, $74m.

Forestry
In 2010 income from forestry amounted to $1·04bn. Output (2010): pine logs 897m. bd ft; hardwood lumber, 319m. bd ft; pulpwood, 5·08m. cords. There were 19·8m. acres of forest in 2009, with 1·2m. acres of national forest area.

Fisheries
Commercial catch, in 2010, totalled 111m. lb of fish with a value of $21·9m. Mississippi has the largest aquaculture industry of any state; value in 2010 was $290m. (of which catfish sales accounted for $198m.).

INDUSTRY
In 2009 the 2,545 manufacturing establishments had average monthly employment of 141,340 workers, earning $5,588,932,859. The average annual wage was $39,542. Total value added by manufacturing in 2009 was $22,810m.

Labour
In 2010 total non-agricultural employment was 1,084,900. Employees by branch, 2010 (in 1,000): government, 249; services, 168; wholesale and retail trade, 167; manufacturing, 137. The unemployment rate in Dec. 2010 was 10·2%.

COMMUNICATIONS
Roads
The state as of 1 July 2011 maintained 14,609 miles of highways, of which 14,606 miles were paved. In fiscal year 2011, 1·8m. passenger vehicles and pick-ups were registered.

Rail
In 2011 the state had 2,542 main-line and short-line miles of railroad.

Civil Aviation
There were 80 public airports in 2010, 73 of them general aviation airports. There were 1,231,879 passenger enplanements statewide in 2010.

SOCIAL INSTITUTIONS
Justice
The death penalty is authorized; there were six executions in 2012 (two in 2011). As of 1 Jan. 2011 the state prison system had 24,718 inmates.

Education
Attendance at school is compulsory as laid down in the Education Reform Act of 1982. The public elementary and secondary schools in 2009–10 had 447,806 pupils and 33,210 classroom teachers. In 2009–10 teachers' average salary was $42,308. The expenditure per pupil in average daily attendance, 2009–10, was $8,930.

There are 18 universities and senior colleges, of which eight are state-supported. In fall 2010 the University of Mississippi, Oxford had 1,846 faculty and 19,954 students; Mississippi State University, Starkville, 1,289 faculty and 19,725 students; Mississippi University for Women, Columbus, 174 faculty and 2,730 students; University of Southern Mississippi, Hattiesburg, 885 faculty and 17,254 students; Jackson State University, Jackson, 504 faculty and 9,615 students; Delta State University, Cleveland, 254 faculty and 4,416 students; Alcorn State University, Lorman, 226 faculty and 3,682 students; Mississippi Valley State University, Itta Bena, 154 faculty and 2,840 students. State support for the universities (2010–11) was $400,842,200.

Community and junior colleges had (2009–10) 80,550 full-time equivalent students and 2,534 full-time instructors. The state appropriation for junior colleges, 2009–10, was $168,422,707.

Health
In 2010 the state had 106 acute general hospitals (11,484 beds) listed by the State Department of Health; 17 hospitals with facilities for the care of the mentally ill had 777 licensed beds. In addition, 12 rehabilitation hospitals had 327 beds. In 2009 there were 5,281 active physicians in the state (18 per 10,000 population).

Welfare
Medicare enrolment in 2011 totalled 637,781. The Division of Medicaid paid (fiscal year 2011) $3·62bn. for medical services, including $617·5m. for hospital services, $727m. for skilled nursing home care and $302m. for drugs. There were 66,990 persons eligible for Aged Medicaid benefits as of 30 June 2011 and 165,523 persons eligible for Disabled Medicaid benefits. In the fiscal year 2010–11, 11,609 families with 17,896 dependent children received $1,617,869 in the Temporary Assistance to Needy Families programme. The average monthly payment was $139·85 per family or $66·26 per recipient.

RELIGION
The principal religious traditions in 2010 were: Evangelical Protestants, with 1,168,450 members; Mainline Protestants, 244,121; Black Protestants, 182,556; Catholics, 112,488; Latter-day Saints (Mormons), 22,308.

CULTURE
Tourism
Total receipts in 2010 amounted to $5·97bn.; an estimated 12m. overnight tourists visited the state.

FURTHER READING
Secretary of State. *Mississippi Official and Statistical Register* (quadrennial).— *Blue Book* http://www.sos.state.ms.us/ed_pubs/BlueBook
Mississippi Library Commission: 3881 Eastwood Drive, Jackson, MS 39211.

Missouri

KEY HISTORICAL EVENTS
Territory of several Indian groups, including the Missouri, the area was not settled by European immigrants until the 18th century. The French founded Ste Geneviève in 1735, partly as a lead-mining community. St Louis was founded as a fur-trading

base in 1764. The area was nominally under Spanish rule from 1770 until 1800 when it passed back to France. In 1803 the USA bought it as part of the Louisiana Purchase.

St Louis was made the capital of the whole Louisiana Territory in 1805, and of a new Missouri Territory in 1812. In that year American immigration increased markedly. The Territory became a state in 1821. Bitter disputes between slave-owning and anti-slavery factions led to the former obtaining statehood without the prohibition of slavery required of all other new states north of latitude 36° 30'. This was achieved by the Missouri Compromise of 1820. The Compromise was repealed in 1854 and declared unconstitutional in 1857. During the Civil War the state held to the Union side, although St Louis was placed under martial law.

With the development of steamboat traffic on the Missouri and Mississippi rivers, and the expansion of railways, the state became the transport hub of all western movement. Lead and other mining remained important, as did livestock farming. European settlers came from Germany, Britain and Ireland.

TERRITORY AND POPULATION

Missouri is bounded north by Iowa, east by the Mississippi River forming the boundary with Illinois and Kentucky, south by Arkansas, southeast by Tennessee, southwest by Oklahoma, west by Kansas and Nebraska, with the Missouri River forming the boundary in the northwest. Land area, 68,742 sq. miles (178,040 sq. km).

Census population, 1 April 2010, was 5,988,927, an increase of 7·0% since 2000. July 2012 estimate, 6,021,988.

Population of five federal census years was:

	White	Black	American Indian	Asiatic	Total	Per sq. mile
1930	3,403,876	223,840	578	1,073	3,629,367	52·4
			All others			
1980	4,345,521	514,276	56,889		4,916,686	71·3
1990	4,486,228	548,208	82,637		5,117,073	74·3
2000	4,748,083	629,391	217,737		5,595,211	81·2
2010	4,958,770	693,391	336,766		5,988,927	87·1

Of the total population in 2010, 3,055,450 were female and 4,563,491 were 18 years old or older. In 2010 Missouri's Hispanic population was 212,470, up from 118,592 in 2000 (an increase of 79·2%).

The principal cities at the 2010 census were:

Kansas City	459,787	O'Fallon	79,329
St Louis	319,294	St Joseph	76,780
Springfield	159,498	St Charles	65,794
Independence	116,830	Blue Springs	52,575
Columbia	108,500	St Peters	52,575
Lee's Summit	91,364	Florissant	52,158

The capital, Jefferson City, had a population in 2010 of 43,079. Largest metropolitan areas, 2010: St Louis, 2,812,896; Kansas City, 2,035,334; Springfield, 436,712.

SOCIAL STATISTICS

Births, 2010, 76,759 (12·8 per 1,000 population); deaths, 2010 (provisional), 55,276 (9·2). Infant mortality, 2009, 7·2 per 1,000 live births. 2009 (provisional): marriages, 39,800 (6·5 per 1,000 population); divorces and annulments, 23,300 (3·7).

CLIMATE

Kansas City, Jan. 30°F (−1·1°C), July 79°F (26·1°C). Annual rainfall 38" (947 mm). St Louis, Jan. 32°F (0°C), July 79°F (26·1°C). Annual rainfall 40" (1,004 mm). Missouri belongs to the Central Plains climate zone (*see* UNITED STATES: Climate).

CONSTITUTION AND GOVERNMENT

A new constitution, the fourth, was adopted on 27 Feb. 1945; it has had 108 amendments in the meantime. The General Assembly consists of a Senate of 34 members elected for four years (half for re-election every two years), and a House of Representatives of 163 members elected for two years. The Governor and Lieut.-Governor are elected for four years.

For the 113th Congress, which convened in Jan. 2013, Missouri sends eight members to the House of Representatives. It is represented in the Senate by Claire McCaskill (D. 2007–19) and Roy Blunt (R. 2011–17).

Jefferson City is the state capital. The state is divided into 114 counties and the city of St Louis.

RECENT ELECTIONS

In the 2012 presidential election Mitt Romney took Missouri with 53·9% of the vote (John McCain won it in 2008).

CURRENT GOVERNMENT

Governor: Jay Nixon (D.), 2013–17 (salary: $133,821).
   Lieut.-Governor: Peter Kinder (R.), 2013–17 ($86,484).
   Secretary of State: Jason Kander (D.), 2013–17 ($107,746).

Government Website: http://www.mo.gov

ECONOMY

Per capita personal income (2010) was $36,965.

Budget

In 2011 total state revenue was $38,607m. Total expenditure was $30,647m. (education, $8,855m.; public welfare, $7,587m.; highways, $2,033m.; hospitals, $1,493m.; health, $1,402m.) Outstanding debt in 2011, $20,682m.

Performance

In 2011 Gross Domestic Product by state was $216,099m. (provisional), ranking Missouri 22nd in the United States. In 2011 state real GDP growth was 0% (provisional).

ENERGY AND NATURAL RESOURCES

Water

The total area covered by water is 965 sq. miles.

Minerals

The three leading mineral commodities are lead, Portland cement and crushed stone. Value of domestic non-fuel mineral production was $1,810m. in 2009.

Agriculture

In 2007 there were 107,825 farms in Missouri producing crops and livestock on 29·03m. acres; the average farm had 269 acres and was valued at $2,179 per acre. Production of principal crops, 2007: corn, 439·4m. bu.; soybeans, 165·9m. bu.; wheat, 36·3m. bu.; sorghum, 9·9m. bu.; rice, 12·27m. cwt; cotton, 723,043 bales (of 480 lb). Farm income 2006: crops, $2,628m.; livestock and products, $2,994m.; total, $5,621m. The net farm income in 2006 was $1,697m.

Forestry

The state had a forested area of 15,078,000 acres in 2007, of which 1,493,000 acres were national forest.

INDUSTRY

In 2007 the state's 6,886 manufacturing establishments had 295,000 employees, earning $12,997m. Total value added by manufacturing in 2009 was $41,363m.

Labour

Total non-agricultural employment, 2010, 2,647,000. Employees by branch, 2010 (in 1,000): trade, transportation and utilities, 510; government, 451; education and health services, 406; professional and business services, 319; leisure and hospitality, 271. The unemployment rate was 9·6% in Dec. 2010.

INTERNATIONAL TRADE

Imports and Exports
In 2010 exports from Missouri totalled $12,926m., up from $9,522m. in 2009. The leading destinations for exports in 2010 were Canada (value of $3,996m.), followed by Mexico, China (excluding Hong Kong) and South Korea.

COMMUNICATIONS

Roads
In 2007 there were 129,122 miles of road (106,412 miles rural) and 4,916,993 registered motor vehicles.

Rail
In 2009 there were 4,050 miles of freight railroad (excluding trackage rights). There were 17 freight railroads operating in 2009. Rail traffic originating in Missouri in 2009 totalled 12·5m. tons and rail traffic terminating in the state came to 72·8m. tons. There is a light rail line in St Louis.

Civil Aviation
There were five primary airports—commercial service airports with more than 10,000 passenger boardings annually—in 2011 with a combined total of 11,587,550 enplanements, up from 11,414,476 in 2010. The busiest airports are Kansas City International and Lambert-St Louis International.

SOCIAL INSTITUTIONS

Justice
In Dec. 2008 there were 30,186 federal and state prisoners. The death penalty was reinstated in 1978. Executions were suspended between June 2006 and June 2007. There was one execution in 2011 but none in 2012. The Missouri Law Enforcement Assistance Council was created in 1969 for law reform. With reorganization of state government in 1974 the duties of the Council were delegated to the Department of Public Safety. The Department of Corrections was organized as a separate department of State by an Act of the Legislature in 1981.

Education
School attendance is compulsory for children from seven to 16 years. In 2007–08 there were 2,417 public schools (kindergarten through grade 12) with 917,188 pupils and 68,430 teachers. Total expenditure on public elementary and secondary education in 2006–07 was $9,346m.; teacher salaries averaged $44,712 in 2008–09. Spending per pupil in fiscal year 2008 was $9,532.

Institutions for higher education include the University of Missouri, founded in 1839, with campuses at Columbia (with 27,930 enrolled students in fall 2005), Kansas City (14,310), Rolla (5,600 students) and St Louis (15,548). Washington University at St Louis, an independent college founded in 1857, had 13,383 students in fall 2005; St Louis University, an independent Roman Catholic college founded in 1818, had 14,966 students. There were 384,366 students enrolled in degree-granting institutions in fall 2007.

Health
In 2009 there were 125 community hospitals with 19,100 beds. A total of 825,000 patients were admitted during the year. In 2009 there were 14,789 active physicians in the state (25 per 10,000 population).

Welfare
Medicare enrolment in July 2010 totalled 995,011. In fiscal year 2008 a total of 1,054,099 people in Missouri received Medicaid. In Dec. 2008 there were 1,106,923 Old-Age, Survivors, and Disability Insurance (OASDI) beneficiaries. A total of 84,962 people were receiving payments under Temporary Assistance for Needy Families (TANF) in Dec. 2008.

RELIGION
The principal religious traditions in 2010 were: Evangelical Protestants, with 1,518,847 members; Catholics, 724,315; Mainline Protestants, 462,246; Black Protestants, 93,900; Latter-day Saints (Mormons), 92,248.

FURTHER READING
Statistical information: Business and Public Administration Research Center, Univ. of Missouri, Columbia 65211. Publishes *Statistical Abstract for Missouri.*
Missouri Area Labor Trends. Monthly
Missouri Farm Facts. Annual
Report of the Public Schools of Missouri. Annual

Montana

KEY HISTORICAL EVENTS
Originally the territory of many Indian hunters including the Sioux, Cheyenne and Chippewa, Montana was not settled by American colonists until the 19th century. The area passed to the USA with the Louisiana Purchase of 1803, but the area west of the Rockies was disputed with Britain until 1846. Trappers and fur-traders were the first immigrants, and the fortified trading post at Fort Benton (1846) became the first permanent settlement. Colonization increased when gold was found in 1862. Montana was created a separate Territory (out of Idaho and Dakota Territories) in 1864. In 1866 large-scale grazing of sheep and cattle provoked violent confrontation with the indigenous people whose hunting lands were invaded. Indian wars led to the defeat of federal forces at Little Bighorn in 1876 and at Big Hole Basin in 1877, but by 1880 the Indians had been moved to reservations. Montana became a state in 1889.

Helena, the capital, was founded as a mining town in the 1860s. In the early 20th century there were many European immigrants who settled as farmers or as copper-miners, especially at Butte.

TERRITORY AND POPULATION
Montana is bounded north by Canada, east by North and South Dakota, south by Wyoming and west by Idaho and the Bitterroot Range of the Rocky Mountains. Land area, 145,546 sq. miles (376,962 sq. km). In 2000 American Indian lands covered 13,358 sq. miles (13,094 sq. miles in reservations and 264 sq. miles in off-reservation trust land). Census population, 1 April 2010, was 989,415, an increase of 9·7% since 2000. July 2012 estimate, 1,005,141.

Population in five census years was:

	White	Black	American Indian/ Alaska Native	Asian	Total (including others)	Per sq. mile
1910	360,580	1,834	10,745	2,870	376,053	2·6
1980	740,148	1,786	37,270	2,503	786,690	5·3
1990	741,111	2,381	47,679	4,259	799,065	5·4
2000	817,229	2,692	56,068	5,161	902,195	6·2
2010	884,961	4,027	62,555	6,253	989,415	6·8

Of the total population in 2010, 496,667 were male and 765,852 were 18 years old or older. Median age, 39·8 years. Households, 409,607. In 2010 Montana's Hispanic population was 28,565, up from 18,081 in 2000 (an increase of 58·0%).

The largest cities, 2010, are Billings, 104,170; Missoula, 66,788; Great Falls, 58,505; Bozeman, 37,280; Butte-Silver Bow, 33,525; Helena (capital), 28,190; Kalispell, 19,927.

SOCIAL STATISTICS

Births, 2010, 12,060 (12·2 per 1,000 population); deaths, 2010 (provisional), 8,827 (8·9). Infant mortality rate, 2009, 5·9 per 1,000 live births. 2009 (provisional): marriages, 7,100 (7·4 per 1,000 population); divorces and annulments, 3,900 (4·1).

CLIMATE

Helena, Jan. 18°F (−7·8°C), July 69°F (20·6°C). Annual rainfall 13" (325 mm). Montana belongs to the Mountain States climate zone (*see* UNITED STATES: Climate).

CONSTITUTION AND GOVERNMENT

A new constitution came into force on 1 July 1973. The Senate consists of 50 senators, elected for four years, one half at each biennial election. The 100 members of the House of Representatives are elected for two years.

For the 113th Congress, which convened in Jan. 2013, Montana sends one member to the House of Representatives. It is represented in the Senate by Max Baucus (D. 1978–2015) and Jon Tester (D. 2007–19).

The capital is Helena. The state is divided into 56 counties.

RECENT ELECTIONS

In the 2012 presidential election Mitt Romney took Montana with 55·3% of the vote (John McCain won it in 2008).

CURRENT GOVERNMENT

Governor: Steve Bullock (D.), 2013–17 (salary: $108,167).
   Lieut.-Governor: John Walsh (D.), 2013–17 ($86,362).
   Secretary of State: Linda McCulloch (D.), 2013–17 ($86,108).

Government Website: http://www.mt.gov

ECONOMY

Per capita personal income (2010) was $35,068.

Budget

In 2011 total state revenue was $7,951m. Total expenditure was $7,105m. (education, $1,841m.; public welfare, $1,390m.; highways, $709m.; government administration, $420m.; natural resources, $260m.) Outstanding debt in 2011, $4,267m.

Performance

Gross Domestic Product by state in 2011 was $31,983m. (provisional), ranking Montana 48th in the United States. In 2011 state real GDP growth was 0% (provisional).

ENERGY AND NATURAL RESOURCES

Oil and Gas

Montana has vast technically recoverable oil reserves in the northeast of the state in an area known as the Bakken Formation. Marketed natural gas production in 2008 was 113bn. cu. ft; production of crude oil was 31·5m. bbls.

Water

The total area covered by water is 1,494 sq. miles.

Minerals

The total value of non-fuel mineral production for 2009 was $982m. Principal minerals include copper, gold, platinum-group metals, molybdenum and silver. Production of coal (2009) was 39·5m. short tons.

Agriculture

In 2007 there were 29,524 farms and ranches with an area of 61·39m. acres. Large-scale farming predominates; in 2007 the average size per farm was 2,079 acres. The average value per acre in 2007 was $775. Area harvested, 2007, 9,163,867 acres, including 5·1m. acres of wheat.

The chief crops are wheat, hay, barley, oats, sugar beets, potatoes, corn, dry beans and cherries. Wheat production in 2007

totalled 147·5m. bu., a figure exceeded only by North Dakota and Kansas. Farm income, 2006: from crops, $1,070m.; and from livestock and products, $1,279m. In 2007 there were 2·59m. cattle and calves, 182,000 hogs and pigs and 272,000 sheep and lambs. The net farm income in 2006 was $257m.

Forestry

In 2007 there were 25·01m. acres of forested land with 15·00m. acres in 11 national forests.

INDUSTRY

In 2007 the state's 1,324 manufacturing establishments had 20,000 employees, earning $808m. Total value added by manufacturing in 2009 was $2,248m.

Labour

Total non-agricultural employment, 2010, 428,000. Employees by branch, 2010 (in 1,000): government, 91; trade, transportation and utilities, 87; education and health services, 64; leisure and hospitality, 56; professional and business services, 39. In Dec. 2010 the unemployment rate was 7·4%.

INTERNATIONAL TRADE

Imports and Exports

In 2010 exports from Montana totalled $1,389m., up from $1,053m. in 2009 although down from $1,395m. in 2008. The leading destinations for exports in 2010 were Canada (value of $506m.), followed by South Korea, China (excluding Hong Kong) and Japan.

COMMUNICATIONS

Roads

In 2007 there were a total of 73,203 miles of road comprising 2,983 miles of urban road and 70,220 miles of rural road. There were 948,528 registered motor vehicles.

Rail

In 2009 there were 3,173 miles of freight railroad (excluding trackage rights). Rail traffic originating in Montana in 2009 totalled 44·2m. tons and rail traffic terminating in the state came to 5·9m. tons.

Civil Aviation

There were seven primary airports—commercial service airports with more than 10,000 passenger boardings annually—in 2011 with a combined total of 1,570,469 enplanements, up from 1,496,777 in 2010.

SOCIAL INSTITUTIONS

Justice

In Dec. 2008 there were 3,607 prison inmates. The death penalty is authorized; there was one execution in 2006, the first since 1998, but none since.

Education

In 2007–08 the 831 public elementary and secondary schools had 142,823 pupils and 10,519 teachers. Total expenditure on public school education in 2006–07 was $1,447m.; average teacher salary was $44,426 in 2008–09. Spending per pupil in fiscal year 2008 was $9,786.

In fall 2007 there were 47,371 students enrolled at 23 degree-granting institutions (18 public). The Montana State University System (created in 1994) consists of the Montana State University at Bozeman (2005 enrolment: 12,143 students), founded 1893; Montana State University-Billings (3,832); Montana State University-Northern at Havre (1,347); and Montana State University-Great Falls (1,875). The University of Montana System comprises the University of Montana at Missoula, founded in 1893 (2005 enrolment: 13,569); Montana Tech at Butte (1,813); and the University of Montana-Western at Dillon (1,159). The private University of Great Falls (founded in 1932) had 778 students in fall 2005.

Health

In 2009 there were 48 community hospitals with 3,800 beds. A total of 101,000 patients were admitted during the year. In 2009 there were 2,138 active physicians in the state (22 per 10,000 population).

Welfare

Medicare enrolment in July 2010 totalled 168,471. In fiscal year 2009 a total of 112,584 people in Montana received Medicaid. In Dec. 2008 there were 180,802 Old-Age, Survivors, and Disability Insurance (OASDI) beneficiaries. A total of 8,766 people were receiving payments under Temporary Assistance for Needy Families (TANF) in Dec. 2008.

RELIGION

The principal religious traditions in 2010 were: Catholics, with 127,612 members; Evangelical Protestants, 121,064; Mainline Protestants, 76,869; Latter-day Saints (Mormons), 47,380.

FURTHER READING

Statistical information. Census and Economic Information Center, Montana Department of Commerce, 1425 9th Ave., Helena 59620.

Nebraska

KEY HISTORICAL EVENTS

The Nebraska region was first reached by Europeans from Mexico under the Spanish general Coronado in 1541. It was ceded by France to Spain in 1763, returned to France in 1801, and sold by Napoleon to the USA as part of the Louisiana Purchase in 1803. During the 1840s the Platte River valley was the trail for thousands of pioneers' wagons heading for Oregon and California. The need to serve and protect the trail led to the creation of Nebraska as a Territory in 1854. In 1862 the Homestead Act opened the area for settlement, but colonization was slow until the Union Pacific Railroad was completed in 1869. Omaha, developed as the starting point of the Union Pacific, became one of the largest railway towns in the country.

Nebraska became a state in 1867, with approximately its present boundaries except that it later received small areas from the Dakotas. Many early settlers were from Europe, brought in by railway-company schemes, but from the late 1880s eastern Nebraska suffered catastrophic drought. Crop and stock farming recovered but crop growing was only established in the west by means of irrigation.

TERRITORY AND POPULATION

Nebraska is bounded in the north by South Dakota, with the Missouri River forming the boundary in the northeast and the boundary with Iowa and Missouri to the east, south by Kansas, southwest by Colorado and west by Wyoming. Land area, 76,824 sq. miles (198,974 sq. km). Census population, 1 April 2010, was 1,826,341, an increase of 6·7% since 2000. July 2012 estimate, 1,855,525.

Population in five census years was:

	White	Black	American Indian	Asiatic	Total	Per sq. mile
1910	1,180,293	7,689	3,502	730	1,192,214	15·5
			All others			
1980	1,490,381	48,390	31,054		1,569,825	20·5
1990	1,480,558	57,404	40,423		1,578,385	20·5
2000	1,533,261	68,541	109,461		1,711,263	22·3
2010	1,572,838	82,885	170,618		1,826,341	23·8

Of the total population in 2010, 920,045 were female and 1,367,120 were 18 years old or older. In 2010 the estimated Hispanic population of Nebraska was 167,405, up from 94,425 in 2000 (a rise of 77·3%). The largest cities in the state are: Omaha, with a census population, 2010, of 408,958; Lincoln (the capital), 258,379; Bellevue, 50,137; Grand Island, 48,520; Kearney, 30,787; Fremont, 26,397; Hastings, 24,907; North Platte, 24,733; Norfolk, 24,210.

In 2006 the Bureau of Indian Affairs administered 64,932 acres, of which 21,742 acres were allotted to tribal control.

SOCIAL STATISTICS

Births, 2010, 25,918 (14·2 per 1,000 population); deaths, 2010 (provisional), 15,171 (8·3). Infant mortality rate, 2009, 5·4 per 1,000 live births. 2009 (provisional): marriages, 12,500 (6·7 per 1,000 population); divorces and annulments, 5,400 (3·4).

CLIMATE

Omaha, Jan. 22°F (−5·6°C), July 77°F (25°C). Annual rainfall 29" (721 mm). Nebraska belongs to the High Plains climate zone (*see* UNITED STATES: Climate).

CONSTITUTION AND GOVERNMENT

The present constitution was adopted in 1875; it had been amended 194 times by 2011. By an amendment of 1934 Nebraska has a single-chambered legislature (elected for four years) of 49 members elected on a non-party ballot and classed as senators—the only state in the USA to have one. It meets annually.

For the 113th Congress, which convened in Jan. 2013, Nebraska sends three members to the House of Representatives. It is represented in the Senate by Mike Johanns (R. 2009–15) and Deb Fischer (R. 2013–19).

The capital is Lincoln. The state has 93 counties.

RECENT ELECTIONS

In the 2012 presidential election Mitt Romney took Nebraska with 60·5% of the vote (John McCain won it in 2008, although Barack Obama did win in one of the five congressional districts).

CURRENT GOVERNMENT

Governor: David Heineman (R.), 2011–15 (salary: $105,000).
 Lieut.-Governor: Lavon Heidemann (ind.), 2013–15 ($75,000).
 Secretary of State: John Gale (R.), 2011–15 ($85,000).

Government Website: http://www.nebraska.gov

ECONOMY

Per capita personal income (2010) was $39,674.

Budget

In 2011 total state revenue was $11,523m. Total expenditure was $9,356m. (education, $3,330m.; public welfare, $2,091m.; highways, $602m.; health, $456m.; hospitals, $261m.) Outstanding debt in 2011, $2,346m.

Performance

Gross Domestic Product by state was $79,889m. in 2011 (provisional), ranking Nebraska 36th in the United States. In 2011 state real GDP growth was 0·1% (provisional).

ENERGY AND NATURAL RESOURCES

Oil and Gas

Natural gas production in 2008 was 3bn. cu. ft; production of crude oil was 2·4m. bbls.

Water

The total area covered by water is 524 sq. miles.

Minerals

Output of non-fuel minerals, 2008 (in 1,000 tonnes): sand and gravel for construction, 13,700; stone, 7,960; clays, 109 (estimate).

Other minerals include limestone, potash, pumice, slate and shale. Total value of non-fuel mineral output in 2009 was $248m.

Agriculture

Nebraska is one of the most important agricultural states. In 2007 it contained 47,712 farms, with a total area of 45·48m. acres. The average farm was 953 acres and was valued in 2007 at $1,159 per acre. In 2007 the total acreage harvested was 18·17m. acres.

In 2006 net farm income was $2,297m. Farm income 2006: from crops, $4,359m.; and from livestock and products, $7,683m. Principal commodities are cattle, corn, soybeans and hogs. Livestock, 2007: cattle, 6·58m.; hogs and pigs, 3·27m.; sheep and lambs, 76,400; laying hens, 10·49m.

Forestry

The state had a forested area of 1,245,000 acres in 2007, of which 48,000 acres were national forest.

INDUSTRY

In 2007 the state's 1,984 manufacturing establishments had 100,000 employees, earning $3,789m. Total value added by manufacturing in 2009 was $15,820m.

Labour

Total non-agricultural employment, 2010, 939,000. Employees by branch in 2010 (in 1,000): trade, transportation and utilities, 196; government, 169; education and health services, 136; professional and business services, 101; manufacturing, 92. In Dec. 2010 the unemployment rate was 4·3%.

INTERNATIONAL TRADE

Imports and Exports

In 2010 exports from Nebraska totalled $5,820m., up from $4,873m. in 2009. The leading destinations for exports in 2010 were Canada (value of $1,607m.), followed by Mexico, Japan and China (excluding Hong Kong).

COMMUNICATIONS

Roads

In 2007 there were 93,398 miles of road (87,176 miles rural). Registered motor vehicles in 2007 numbered 1,739,072.

Rail

In 2009 there were 3,215 miles of freight railroad (excluding trackage rights). There were ten freight railroads operating in 2009. Rail traffic originating in Nebraska in 2009 totalled 30·0m. tons and rail traffic terminating in the state came to 20·4m. tons.

Civil Aviation

There were five primary airports—commercial service airports with more than 10,000 passenger boardings annually—in 2011 with a combined total of 2,251,850 enplanements, down from 2,296,210 in 2010. The busiest airport, Eppley Airfield (Omaha's airport), had 2,047,055 enplanements in 2011.

SOCIAL INSTITUTIONS

Justice

A 'Civil Rights Act' revised in 1969 provides that all people are entitled to a full and equal enjoyment of public facilities. In Dec. 2008 there were 4,520 prison inmates. The last execution was in 1997.

Education

School attendance is compulsory for children from six to 18 years of age. There were 1,143 public elementary and secondary schools in 2007–08. In fall 2007, 291,244 pupils were enrolled in public schools with 21,930 teachers and 40,320 pupils in private schools with 2,820 teachers. Total expenditure on public elementary and

secondary education in 2006–07 was $3,243m. Spending per pupil in fiscal year 2008 was $10,565. Total enrolment in public degree-granting institutions in fall 2007 was 96,680; there were also 30,698 students in private degree-granting institutions.

Founded	Institution	Students (fall 2005)
1867	Peru State College	1,959
	University of Nebraska (State)	38,621
1869	Lincoln	19,513
1902	Medical Center	2,735
1905	Kearney	5,542
1908	Omaha	10,831
1965	College of Technical Agriculture, Curtis	262
1872	Doane College, Crete (United Church of Christ)	2,394
1878	Creighton University, Omaha (Roman Catholic)	6,791
1882	Hastings College (Presbyterian)	1,189
1883	Midland Lutheran College, Fremont (Lutheran Church of America)	926
	Nebraska Methodist College of Nursing and Allied Health, Omaha (Private)	565
1884	Dana College, Blair (American Lutheran)	673
1887	Nebraska Wesleyan University (Private)	2,016
1888	Clarkson College, Omaha (Private)	711
1890	York College[1] (Private)	450
1891	Union College, Lincoln (Seventh Day Adventist)	930
1894	Concordia University Nebraska, Seward (Lutheran)	1,330
1910	Wayne State College	3,322
1911	Chadron State College	2,472
1923	College of St Mary (Roman Catholic)	955
1943	Grace University, Omaha (Mennonite)	440
1945	Nebraska Christian College (Church of Christ)	143
1966	Bellevue University (Private)	5,929
1971	Nebraska Community Colleges (Local government)	39,851
	Central Area	6,564
	Metropolitan Area	13,237
	Mid Plains Area	2,607
	Northeast Area	5,101
	Southeast Area	10,059
	Western Area	2,283
1972	Nebraska Indian Community College	107

[1]Two-year college.

Health

In 2009 there were 87 community hospitals with 7,400 beds. A total of 210,000 patients were admitted during the year. In 2009 there were 4,511 active physicians in the state (25 per 10,000 population).

Welfare

Medicare enrolment in July 2010 totalled 276,341. In fiscal year 2009 a total of 255,629 people in Nebraska received Medicaid. In Dec. 2008 there were 297,811 Old-Age, Survivors, and Disability Insurance (OASDI) beneficiaries. A total of 17,909 people were receiving payments under Temporary Assistance for Needy Families (TANF) in Dec. 2008.

RELIGION

The principal religious traditions in 2010 were: Catholics, with 372,838 members; Mainline Protestants, 297,522; Evangelical Protestants, 288,965; Latter-day Saints (Mormons), 25,611; Black Protestants, 13,106.

FURTHER READING

Statistical information: Department of Economic Development, Box 94666, Lincoln 68509.

Nebraska Blue Book. Biennial

Olson, J. C., *History of Nebraska*. 3rd ed. 1997

State library: Nebraska State Library, PO Box 98931, State Capitol Bldg, Lincoln.

Nevada

KEY HISTORICAL EVENTS

The area was part of Spanish America until 1821 when it became part of the newly independent state of Mexico. Following a war between Mexico and the USA, Nevada was ceded to the USA as part of California in 1848. Settlement began in 1849 and the area was separated from California and joined with Utah Territory in 1850. In 1859 a rich deposit of silver was found in the Comstock Lode. Virginia City was founded as a mining town and immigration increased rapidly. Nevada Territory was formed in 1861. During the Civil War the Federal government, allegedly in order to obtain the wealth of silver for the Union cause, agreed to admit Nevada to the Union as the 36th state. This was in 1864. Areas of Arizona and Utah Territories were added in 1866–67.

The mining boom lasted until 1882, by which time cattle ranching in the valleys, where the climate is less arid, had become equally important. Carson City, the capital, developed in association with the nearby mining industry. The largest cities, Las Vegas and Reno, grew in the 20th century with the building of the Hoover dam, the introduction of legal gambling and of easy divorce.

After 1950 much of the desert area was adopted by the Federal government for weapons testing and other military purposes.

TERRITORY AND POPULATION

Nevada is bounded north by Oregon and Idaho, east by Utah, southeast by Arizona, with the Colorado River forming most of the boundary, south and west by California. Land area, 109,781 sq. miles (284,332 sq. km). In 2010 the federal government owned 56,961,778 acres, equivalent to 81·1% of the state area (the highest percentage in any state).

Census population, 1 April 2010, was 2,700,551, an increase of 35·1% since 2000. Nevada had the fastest-growing population of any state between the censuses of 2000 and 2010, continuing a trend stretching back to 1950. July 2012 estimate, 2,758,931.

Population in five census years was:

	White	Black	American Indian/Alaska Native	Asian	Total (including others)	Per sq. mile
1910	74,276	513	5,240	1,846	81,875	0·7
1980	700,360	50,999	13,308	35,841	800,508	7·2
1990	1,012,695	78,771	19,637	90,730	1,201,833	10·9
2000	1,501,886	135,477	26,420	334,474	1,998,257	18·2
2010	1,786,688	218,626	32,062	195,436	2,700,551	24·6

Of the total population in 2010, 1,363,616 were male and 2,035,543 were 18 years old or older. In 2010 the Hispanic population was 716,501, up from 393,970 in 2000 (an increase of 81·9%).

The largest cities in 2010 were: Las Vegas, 583,756; Henderson, 257,729; Reno, 225,221; North Las Vegas, 216,961; Sparks, 90,264; Carson City (the capital), 55,274.

SOCIAL STATISTICS

Births, 2010, were 35,934 (13·3 per 1,000 population); deaths, 2010 (provisional), 19,623 (7·3). Infant mortality rate, 2009, 5·9 per 1,000 live births. 2009 (provisional): marriages, 108,200 (40·9 per 1,000 population); divorces and annulments, 17,700 (6·7). Nevada's marriage rate in 2009 was more than double that of any other state and six times the national average, but less than half that of 20 years earlier.

CLIMATE

Las Vegas, Jan. 57°F (14°C), July 104°F (40°C). Annual rainfall 4·13" (105 mm). Reno, Jan. 45°F (7°C), July 91°F (33°C). Annual rainfall 7·53" (191 mm). Nevada belongs to the Mountain States climate zone (see UNITED STATES: Climate).

CONSTITUTION AND GOVERNMENT

The constitution adopted in 1864 is still in force, with 149 amendments as of 2011. The Legislature meets biennially (and in special sessions) and consists of a Senate of 21 members elected for four years, with half their number elected every two years, and an Assembly of 42 members elected for two years. The Governor may be elected for two consecutive four-year terms.

For the 113th Congress, which convened in Jan. 2013, Nevada sends four members to the House of Representatives. It is represented in the Senate by Harry Reid (D. 1987–2017) and Dean Heller (R. 2011–19).

The state capital is Carson City. There are 16 counties, 18 incorporated cities and 49 unincorporated communities and one city-county (the Capitol District of Carson City).

RECENT ELECTIONS

In the 2012 presidential election Barack Obama took Nevada with 52·3% of the vote, having also won it in 2008.

CURRENT GOVERNMENT

Governor: Brian Sandoval (R.), 2011–15 (salary: $135,054).
   Lieut.-Governor: Brian Krolicki (R.), 2011–15 ($57,470).
   Secretary of State: Ross Miller (D.), 2011–15 ($92,910).

Government Website: http://www.nv.gov

ECONOMY

Per capita personal income (2010) was $36,919.

Budget

In 2011 total state revenue was $17,597m. Total expenditure was $13,203m. (including: education, $4,148m.; public welfare, $2,128m.; highways, $773m.; government administration, $283m.; correction, $276m.) Outstanding debt in 2011, $4,201m.

Performance

Gross Domestic Product by state in 2011 was $112,503m. (provisional), ranking Nevada 32nd in the United States. In 2011 state real GDP growth was 1·2% (provisional).

ENERGY AND NATURAL RESOURCES

Electricity

In 2009 production was 37·7bn. kWh, of which 25·9bn. kWh was from natural gas.

Water

The total area covered by water is 791 sq. miles.

Minerals

Nevada has led the nation in gold production since 1981, producing 76% of gold in 2008. It is ranked second in silver production, accounting for 19% of the nation's silver in 2008. In 2008 Nevada produced 178,000 kg of gold and 235,000 kg of silver. Nevada also produces other minerals such as aggregates, clays, copper, diatomite, dolomite, geothermal energy, gypsum, lapidary, lime and limestone. The total value of Nevada's non-fuel mineral production in 2009 was $6,020m.

Agriculture

In 2007 there were 3,131 farms. Farms averaged 1,873 acres; farms and ranches totalled 5·87m. acres. Average value per acre in 2007 was $613.

In 2006 farm income from crops totalled $166m., and from livestock and products $280m. The four most important commodities were cattle, hay, dairy products and onions. The net farm income in 2006 was $85m.

In 2007 there were 442,000 cattle and 69,000 sheep and lambs.

Forestry

Nevada had, in 2007, 11·09m. acres of forested land with 3·36m. acres of national forest.

INDUSTRY

The main industry is the service industry, especially tourism and legalized gambling. Gaming industry total revenue for 2011 was $10,701m. In June 2012 there were 2,859 active licenses in force, including 443 for non-restricted casinos (those with more than 15 slot machines and/or gaming tables).

In 2007 Nevada's 2,035 manufacturing establishments had 52,000 employees, earning $2,291m. Total value added by manufacturing in 2009 was $7,664m.

Labour

Total non-agricultural employment in 2010 was 1,116,000. Employees by branch in 2010 (in 1,000): leisure and hospitality, 309; trade, transportation and utilities, 209; government, 155; professional and business services, 136; education and health services, 100. The unemployment rate in Dec. 2010 was 14·9%, the highest of any US state.

INTERNATIONAL TRADE

Imports and Exports

In 2010 exports from Nevada totalled $5,912m., up from $5,672m. in 2009 although down from $6,121m. in 2008. The leading destinations for exports in 2010 were Switzerland (value of $2,425m.), followed by Canada, China (excluding Hong Kong) and Mexico.

COMMUNICATIONS

Roads

In 2007 there were 33,872 miles of road, of which 26,794 miles were rural roads and 7,078 urban. Vehicle registrations in 2007 numbered 1,424,322.

Rail

In 2009 there were 1,192 miles of freight railroad (excluding trackage rights). Rail traffic originating in Nevada in 2009 totalled 1·9m. tons and rail traffic terminating in the state came to 7·7m. tons. Las Vegas has a 4-mile monorail metro system.

Civil Aviation

In 2011 Nevada had four primary airports (commercial service airports with more than 10,000 passenger boardings annually) with a combined total of 21,907,927 enplanements, up from 21,046,012 in 2010. By far the busiest airport is McCarran International (the airport for Las Vegas), with 19,872,617 enplanements in 2011.

SOCIAL INSTITUTIONS

Justice

Capital punishment was reintroduced in 1978, and executions began in 1979. There was one execution in 2006 but none since. In Dec. 2008 there were 12,743 prison inmates in state or federal correctional institutions.

Education

School attendance is compulsory for children from seven to 18 years of age. In 2007–08 there were 610 public elementary and secondary schools with 429,362 pupils and 23,423 teachers. Expenditure per pupil in fiscal year 2008 was $8,187. Teachers' salaries in public schools in 2008–09 averaged $50,067. Pupils in the state's 160 private elementary and secondary schools in fall 2007 numbered 29,820 with 1,550 teachers.

The University of Nevada System comprises the University of Nevada, Las Vegas and Reno, the Nevada State College (at Henderson), the Community College of Southern Nevada (at Las Vegas and Henderson), Great Basin College (at Elko), Truckee Meadows Community College (at Reno) and Western Nevada Community College (at Carson City, Minden and Fallon). In fall 2007 there were 116,276 students in degree-granting institutions.

Health

In 2009 there were 35 community hospitals with 5,100 beds. A total of 246,000 patients were admitted during the year. In 2009 there were 4,967 active physicians in the state (19 per 10,000 population).

Welfare

Medicare enrolment in July 2010 totalled 353,830. In fiscal year 2009 a total of 280,864 people in Nevada received Medicaid. In Dec. 2008 there were 374,289 Old-Age, Survivors, and Disability Insurance (OASDI) beneficiaries. A total of 20,021 people were receiving payments under Temporary Assistance for Needy Families (TANF) in Dec. 2008.

RELIGION

The principal religious traditions in the state in 2010 were: Catholics, with 451,070 members; Evangelical Protestants, 213,188; Latter-day Saints (Mormons), 175,466; Mainline Protestants, 41,558; Buddhists, 14,727.

CULTURE

Tourism

In 2011 overseas visitors to Nevada—excluding those from Canada and Mexico—numbered 2,872,000, up from 2,504,000 in 2010 and 1,690,000 in 2006. Nevada ranks among the most visited states and Las Vegas among the most visited cities.

FURTHER READING

Statistical information: Budget and Planning Division, Department of Administration, Capitol Complex, Carson City, Nevada 89710. Publishes *Nevada Statistical Abstract* (Biennial).

Bowers, Michael W., *The Stagebrush State: Nevada's History, Government, and Politics.* 1996
Hulse, J. W., *The Nevada Adventure: a History.* 6th ed. 1990.—*The Silver State: Nevada's Heritage Reinterpreted.* 1998

State Government Website: http://www.nv.gov
Nevada State Library: Nevada State Library and Archives, Carson City.

New Hampshire

KEY HISTORICAL EVENTS

The area was part of a grant by the English crown to John Mason and fellow-colonists and was first settled in 1623. In 1629 an area between the Merrimack and Piscataqua rivers was called New Hampshire. More settlements followed, and in 1641 they were taken under the jurisdiction of the governor of Massachusetts. New Hampshire became a separate colony in 1679.

After the War of Independence New Hampshire was one of the 13 original states of the Union, ratifying the US constitution in 1788. The state constitution, which dates from 1776, was almost totally rewritten in 1784 and amended again in 1792.

The settlers were Protestants from Britain and Northern Ireland. They developed manufacturing industries, especially shoe-making, textiles and clothing, to which large numbers of French Canadians were attracted after the Civil War.

Portsmouth, originally a fishing settlement, was the colonial capital and is the only seaport. In 1808 the state capital was moved to Concord (having had no permanent home since 1775); Concord produced the Concord Coach which was widely used on the stagecoach routes of the West until 1900.

TERRITORY AND POPULATION

New Hampshire is bounded in the north by Canada, east by Maine and the Atlantic, south by Massachusetts and west by Vermont. Land area, 8,953 sq. miles (23,187 sq. km). Census population, 1 April 2010, was 1,316,470, an increase of 6·5% since 2000. July 2012 estimate, 1,320,718.

Population at five federal censuses was:

	White	Black	American Indian	Asiatic	Total	Per sq. mile
1910	429,906	564	34	68	430,572	47·7
			All others			
1980	910,099	3,990	6,521		920,610	101·9
1990	1,087,433	7,198	14,621		1,109,252	123·7
2000	1,186,851	9,035	39,900		1,235,786	137·8
2010	1,236,050	15,035	65,385		1,316,470	147·0

Of the total population in 2010, 667,076 were female and 1,029,236 were 18 years old or older. In 2010 the Hispanic population was 36,704, up from 20,489 in 2000 (an increase of 79·1%). The largest city in the state is Manchester, with a 2010 census population of 109,565. The capital is Concord, 42,695. Other main cities and towns (with 2010 populations) are: Nashua, 86,494; Dover, 29,987; Rochester, 29,752; Keene, 23,409; Derry, 22,015; Portsmouth, 20,779; Laconia, 15,951.

SOCIAL STATISTICS

Births, 2010, 12,874 (9·8 per 1,000 population—the joint lowest rate in any US state); deaths, 2010 (provisional), 10,201 (7·7). Infant mortality rate, 2009, 4·9 per 1,000 live births. 2009 (provisional): marriages, 8,500 (6·4 per 1,000 population); divorces and annulments, 4,900 (3·7). Same-sex marriage became legal in Jan. 2010.

CLIMATE

New Hampshire is in the New England climate zone (*see* UNITED STATES: Climate). Manchester, Jan. 22°F (−5·6°C), July 70°F (21·1°C). Annual rainfall 40" (1,003 mm).

CONSTITUTION AND GOVERNMENT

While the present constitution dates from 1784, it was extensively revised in 1792 when the state joined the Union. Since 1775 there have been 16 state conventions with 49 amendments adopted to change the constitution.

The Legislature (called the General Court) consists of a Senate of 24 members, elected for two years, and a House of Representatives, of 400 members, elected for two years. It meets annually. The Governor and five administrative officers called 'Councillors' are also elected for two years.

For the 113th Congress, which convened in Jan. 2013, New Hampshire sends two members to the House of Representatives. It is represented in the Senate by Jeanne Shaheen (D. 2009–15) and Kelly Ayotte (R. 2011–17).

The capital is Concord. The state is divided into ten counties.

RECENT ELECTIONS

In the 2012 presidential election Barack Obama took New Hampshire with 52·2% of the vote, having also won it in 2008.

CURRENT GOVERNMENT

Governor: Maggie Hassan (D.), 2013–15 (salary: $110,418).

Senate President: Peter Bragdon (R.), 2012–14 ($250 per term).

Secretary of State: William M. Gardner (D.), first elected by legislature in 1976 ($104,364).

Government Website: http://www.nh.gov

ECONOMY

Per capita personal income (2010) was $43,586.

Budget

New Hampshire has no general sales tax or state income tax but does have local property taxes. Other government revenues come from rooms and meals tax, business profits tax, motor vehicle licences, fuel taxes, fishing and hunting licences, state-controlled sales of alcoholic beverages, and cigarette and tobacco taxes.

In 2011 total state revenue was $8,521m. Total expenditure was $7,638m. (including: education, $2,016m.; public welfare, $1,945m.; highways, $553m.; government administration, $256m.; correction, $113m.) Outstanding debt in 2011, $8,450m.

Performance

Gross Domestic Product by state in 2011 was $56,572m. (provisional), ranking New Hampshire 40th in the United States. In 2011 state real GDP growth was 1·5% (provisional).

ENERGY AND NATURAL RESOURCES

Water

The total area covered by water is 397 sq. miles.

Minerals

Minerals are little worked; they consist mainly of sand and gravel, stone, and clay for building and highway construction. Value of domestic non-fuel mineral production in 2009 was $108m.

Agriculture

In 2007 there were 4,166 farms covering 472,000 acres; average farm was 113 acres. Average value per acre in 2007, $4,929. Farm income 2006: from crops, $98m.; from livestock and products, $64m. The net farm income in 2006 was $43m.

The chief field crops are hay and vegetables; the chief fruit crop is apples. Livestock, 2007: cattle, 37,000; sheep, 8,000; hogs and pigs, 3,000; laying hens, 211,000.

Forestry

In 2007 the state had a forested area of 4,850,000 acres, of which 719,000 acres were national forest.

Fisheries

2009 commercial fishing landings amounted to 13·9m. lb worth $17·7m.

INDUSTRY

Principal manufactures: electrical and electronic goods, machinery and metal products. In 2007 the state's 2,104 manufacturing establishments had 82,000 employees, earning $4,196m. Total value added by manufacturing in 2009 was $8,944m.

Labour

Total non-agricultural employment, 2010, 623,000. Employees by branch, 2010 (in 1,000): trade, transportation and utilities, 132; education and health services, 110; government, 97; manufacturing, 66; professional and business services, 64. In Dec. 2010 the unemployment rate was 5·6%.

INTERNATIONAL TRADE

Imports and Exports

In 2010 exports from New Hampshire totalled $4,367m., up from $3,061m. in 2009. The leading destinations for exports in 2010 were Mexico (value of $1,049m.), followed by Canada, China (excluding Hong Kong) and Germany.

COMMUNICATIONS

Roads

In 2007 there were 15,839 miles of road (11,017 miles rural). There were 1,184,842 registered motor vehicles.

Rail

In 2009 there were 415 miles of freight railroad (excluding trackage rights). There were nine freight railroads operating in 2009.

Civil Aviation

New Hampshire's only primary airport—a commercial service airport with more than 10,000 passenger boardings annually—is Manchester. In 2011 it had 1,342,308 enplanements.

SOCIAL INSTITUTIONS

Justice

There were 2,904 prison inmates in Dec. 2008. The death penalty was abolished in May 2000—the last execution had been in 1939.

Education

School attendance is compulsory for children from six to 18 years of age (since 1 July 2009—previously school attendance had only been compulsory to 16). Employed illiterate minors between 16 and 21 years of age must attend evening or special classes, if provided by the district.

In 2007–08, 488 public elementary and secondary schools had 200,772 pupils and 15,484 teachers. Public school teachers' salaries in 2008–09 averaged $48,934. An average of $11,951 was spent on education per pupil in fiscal year 2008.

Of the 4-year colleges, the University of New Hampshire (founded in 1866) had 13,349 students in 2005; Dartmouth College (1769), 5,529; Keene State College (1909), 4,463; Plymouth State University (1871), 4,453; Southern New Hampshire University (1932, was New Hampshire College), 3,744. There were 28 degree-granting institutions in 2008–09; total enrolment, fall 2007, was 70,724.

Health

In 2009 there were 28 community hospitals with 2,900 beds. A total of 123,000 patients were admitted during the year. In 2009 there were 3,828 active physicians in the state (29 per 10,000 population).

Welfare

Medicare enrolment in July 2010 totalled 216,551. In fiscal year 2009 a total of 141,380 people in New Hampshire received Medicaid. In Dec. 2008 there were 237,498 Old-Age, Survivors, and Disability Insurance (OASDI) beneficiaries. A total of 11,818 people were receiving payments under Temporary Assistance for Needy Families (TANF) in Dec. 2008.

RELIGION

The principal religious traditions in the state in 2010 were: Catholics, with 311,028 members; Mainline Protestants, 81,111; Evangelical Protestants, 47,128; Latter-day Saints (Mormons), 8,231; Orthodox Christians, 4,926.

FURTHER READING

Delorme, D. (ed.) *New Hampshire Atlas and Gazetteer*. 1983

New Jersey

KEY HISTORICAL EVENTS

Originally the territory of Delaware Indians, the area was settled by immigrant colonists in the early 17th century, when Dutch and Swedish traders established fortified posts on the Hudson and Delaware Rivers. The Dutch gave way to the English in 1664. In 1676 the English divided the area; the eastern portion was assigned to Sir George Carteret and the western granted to Quaker settlers. This lasted until 1702 when New Jersey was united as a colony of the Crown and placed under the jurisdiction of the governor of New York. It became a separate colony in 1738.

During the War of Independence crucial battles were fought at Trenton, Princeton and Monmouth. New Jersey became the 3rd state of the Union in 1787. Trenton, the state capital since 1790, began as a Quaker settlement and became an iron-working town. Industrial development grew rapidly, there and elsewhere in the state, after the opening of canals and railways in the 1830s. Princeton, also a Quaker settlement, became an important post on the New York road; the college of New Jersey (Princeton University) was transferred there from Newark in 1756.

The need for supplies in the Civil War stimulated industry and New Jersey became a manufacturing state. The growth of New York and Philadelphia, however, encouraged commuting to employment in both centres. By 1980 about 60% of the state's population lived within 30 miles of New York.

TERRITORY AND POPULATION

New Jersey is bounded north by New York, east by the Atlantic with Long Island and New York City to the northeast, south by Delaware Bay and west by Pennsylvania. Land area, 7,354 sq. miles (19,047 sq. km). Census population, 1 April 2010, was 8,791,894, an increase of 4·5% since 2000. July 2012 estimate, 8,864,590.

Population at five federal censuses was:

	White	Black	Asian	Others	Total	Per sq. mile
1910	2,445,894	89,760	1,345	168	2,537,167	337·7
1980	6,127,467	925,066	103,848	208,442	7,364,823	986·2
1990	6,130,465	1,036,825	272,521	290,377	7,730,188	1,042·0
2000	6,104,705	1,141,821	483,605	684,219	8,414,350	1,134·4
2010	6,029,248	1,204,826	725,726	832,094	8,791,894	1,195·5

Of the total population in 2010, 4,512,294 were female and 6,726,680 were 18 years old or older. In 2010 the Hispanic population was 1,555,144, up from 1,117,191 in 2000 (an increase of 39·2%).

Census populations of the largest cities and towns in 2010 were:

Newark	277,140	Bayonne	63,024
Jersey City	247,597	North Bergen	
Paterson	146,199	Township	60,773
Elizabeth	124,969	Vineland	60,724
Edison	99,967	Union Township	56,642
Toms River	88,791	New Brunswick	55,181
Trenton (capital)	84,913	Wayne	54,717
Clifton	84,136	Irvington	53,926
Camden	77,344	Lakewood	53,805
Brick Township	75,072	Parsippany-Troy Hills	
Passaic	69,781	Township	53,238
Union City	66,455	Perth Amboy	50,814
East Orange	64,270	Hoboken	50,005

SOCIAL STATISTICS

Births, 2010, 106,922 (12·2 per 1,000 population); deaths, 2010 (provisional), 69,499 (7·9). Infant mortality, 2009, 5·1 per 1,000 live births. 2009 (provisional): marriages, 46,300 (5·0 per 1,000 population); divorces and annulments, 24,000 (2·8).

CLIMATE

Jersey City, Jan. 31°F (−0·6°C), July 75°F (23·9°C). Annual rainfall 41" (1,025 mm). Trenton, Jan. 32°F (0°C), July 76°F (24·4°C). Annual rainfall 40" (1,003 mm). New Jersey belongs to the Atlantic Coast climate zone (*see* UNITED STATES: Climate).

CONSTITUTION AND GOVERNMENT

The present constitution, ratified by the registered voters on 4 Nov. 1947, has been amended 45 times. There is a 40-member Senate and an 80-member General Assembly. Assembly members serve two years, senators four years, except those elected at the election following each census, who serve for two years. Sessions are held throughout the year.

For the 113th Congress, which convened in Jan. 2013, New Jersey sends 12 members to the House of Representatives. It is represented in the Senate by Frank Lautenberg (D. 1982–2001, 2003–15) and Robert Menendez (D. 2007–19).

The capital is Trenton. The state is divided into 21 counties, which are subdivided into 566 municipalities—cities, towns, boroughs, villages and townships.

RECENT ELECTIONS

In the 2012 presidential election Barack Obama took New Jersey with 58·0% of the vote, having also won it in 2008.

CURRENT GOVERNMENT

Governor: Chris Christie (R.), 2010–14 (salary: $175,000).

Lieut.-Governor and Secretary of State: Kim Guadagno (R.), 2010–14 ($141,000).

Government Website: http://www.nj.gov

ECONOMY

Per capita income (2010) was $51,167, the fourth highest in the country.

Budget

In 2011 total state revenue was $70,798m. Total expenditure was $67,114m. (including: education, $15,710m.; public welfare, $14,214m.; highways, $3,179m.; hospitals, $2,119m.; government administration, $1,690m.) Outstanding debt in 2011, $64,005m.

Performance

Gross Domestic Product by state in 2011 was $426,765m. (provisional), ranking New Jersey 7th in the United States. In 2011 state real GDP growth was –0·5% (provisional).

ENERGY AND NATURAL RESOURCES

Water

The total area covered by water is 1,368 sq. miles.

Minerals

In 2008 the chief minerals were stone (17·9m. tons, value $155m.) and sand and gravel (15·1m. tons, value $191m.); others are clays, peat and gemstones. New Jersey is a leading producer of greensand marl, magnesium compounds and peat. Total value of domestic non-fuel mineral products for 2009 was more than $270m.

Agriculture

Livestock raising, market-gardening, fruit-growing, horticulture and forestry are pursued. In 2007 there were 10,327 farms covering a total of 733,000 acres with an average farm size of 71 acres. Average value per acre in 2007 was $15,346—making it the second most valuable land per acre in the USA, after Rhode Island.

Cash receipts from farm marketings, 2006: crops, $763m.; livestock and products, $161m. The net farm income in 2006 was $305m. Principal commodities are greenhouse products, horses/mules, blueberries and peaches.

Livestock, 2007: 10,000 dairy cows, 38,000 all cattle, 14,800 sheep and lambs, 8,600 swine and 1·56m. laying hens.

Forestry

Total forested area was 2,132,000 acres in 2007.

Fisheries

2009 commercial fishing landings amounted to 161·6m. lb worth $149·0m.

INDUSTRY

In 2007 the state's 9,250 manufacturing establishments had 311,000 employees, earning $16,399m. Total value added by manufacturing in 2009 was $45,386m.

Labour

Total non-agricultural employment, 2010, 3,855,000. Employees by branch, 2010 (in 1,000): trade, transportation and utilities, 808; government, 643; education and health services, 606; professional and business services, 582; leisure and hospitality, 335. The unemployment rate in Dec. 2010 was 9·1%.

INTERNATIONAL TRADE

Imports and Exports

In 2010 exports from New Jersey totalled $32,154m., up from $27,244m. in 2009 although down from $35,643m. in 2008. The leading destinations for exports in 2010 were Canada (value of $6,254m.), followed by the United Kingdom, Mexico and Japan.

COMMUNICATIONS

Roads

In 2007 there were 38,752 miles of road, of which 31,455 miles were urban. There were 6,247,130 registered vehicles in 2007.

Rail

In 2009 there were 983 miles of freight railroad (excluding trackage rights). There were 18 freight railroads operating in 2009. Rail traffic originating in New Jersey in 2009 totalled 9·4m. tons and rail traffic terminating in the state came to 20·7m. tons.

There is a metro link to New York (22 km), a light rail line (7 km) and extensive commuter railroads around Newark.

Civil Aviation

In 2011 there were two primary airports (commercial service airports with more than 10,000 passenger boardings annually). The main airport is Newark Liberty International, with 16,814,092 enplanements in 2011 (up from 16,571,754 in 2010). Atlantic City International had 668,930 enplanements in 2011 (down from 669,470 in 2010).

SOCIAL INSTITUTIONS

Justice

In Dec. 2008 there were 25,953 prison inmates. The death penalty was abolished on 17 Dec. 2007, after its initial suspension in Jan. 2006. The death penalty was last used in 1963.

Education

Elementary instruction is compulsory for all from six to 16 years of age and free to all from five to 20 years of age. 128 school districts with high concentrations of disadvantaged children must offer free pre-school education to three- and four-year olds. In 2007–08 there were 2,591 public elementary and secondary schools with 1,382,348 pupils and 111,500 teachers; total expenditure on public schools in 2006–07 was $24,912m. Spending per pupil in fiscal year 2008 came to $17,620 (the highest in the USA). Teachers' salaries averaged $63,018 in 2008–09.

There are 31 public universities and colleges (including 19 community colleges) in New Jersey. In fall 2009 public institutions had 348,934 students (177,173 in community colleges). Enrolment in fall 2009: Rutgers, the State University (founded as Queen's College in 1766), had 54,648 students at campuses in Camden, Newark and New Brunswick; College of New Jersey (1855; formerly Trenton State College), 6,980; Kean University, at Union City (1855), 15,051; Montclair State University (1908), 18,171; Rowan University, at Glassboro (1923), 11,006; William Paterson University, at Wayne (1855), 10,820.

There are 32 independent institutions, of which 14 are senior colleges and universities with a public mission, two independent two-year religious colleges, ten rabbinical schools and theological seminaries, and six proprietary institutions with degree-granting authority. Independent institutions had 83,233 students in fall 2009: Fairleigh Dickinson University, at Teaneck (1941), had 8,804 students; Princeton University (founded in 1746) 7,592; Seton Hall University, at South Orange (1856), 9,616.

Health

In 2009 there were 74 community hospitals with 21,100 beds. A total of 1,095,000 patients were admitted during the year. In 2009 there were 27,433 active physicians in the state (32 per 10,000 population).

Welfare

Medicare enrolment in July 2010 totalled 1,315,062. In fiscal year 2009 a total of 1,150,571 people in New Jersey received Medicaid. In Dec. 2008 there were 1,407,621 Old-Age, Survivors, and Disability Insurance (OASDI) beneficiaries. A total of 79,134 people were receiving payments under Temporary Assistance for Needy Families (TANF) in Dec. 2008.

RELIGION

The principal religious traditions in the state in 2010 were: Catholics, with 3,235,290 members; Mainline Protestants, 502,797; Evangelical Protestants, 380,347; Jews, 216,706; Muslims, 161,000 (estimate).

CULTURE

Tourism

Overseas visitors to New Jersey (excluding those from Canada and Mexico) numbered 976,000 in 2011, up from 975,000 in 2010 and 845,000 in 2006.

FURTHER READING

Statistical information: New Jersey State Data Center, Department of Labor, CN 388, Trenton 08625. Publishes *New Jersey Statistical Factbook.*
Legislative District Data Book. Annual
Manual of the Legislature of New Jersey. Annual

Cunningham, J. T., *New Jersey: America's Main Road.* Rev. ed. 1976

State library: 185 W. State Street, Trenton, CN 520, NJ 08625.

New Mexico

KEY HISTORICAL EVENTS

The first European settlement was established in 1598. Until 1771 New Mexico was the Spanish 'Kingdom of New Mexico'. In 1771 it was annexed to the northern province of New Spain. When New Spain won its independence in 1821, it took the name of Republic of Mexico and established New Mexico as its northernmost department. Ceded to the USA in 1848 after war between the USA and Mexico, the area was recognized as a Territory in 1850, by which time its population was Spanish and Indian. There were frequent conflicts between new settlers and raiding parties of Navajo and Apaches. The Indian war lasted from 1861–66, and from 1864–68 about 8,000 Navajo were imprisoned at Bosque Redondo.

The boundaries were altered several times when land was taken into Texas, Utah, Colorado and lastly (1863) Arizona. New Mexico became a state in 1912.

Settlement proceeded by means of irrigated crop-growing and Mexican-style ranching. During the Second World War the desert areas were used as testing zones for atomic weapons. Mineral related industries developed after the discovery of uranium and petroleum.

TERRITORY AND POPULATION

New Mexico is bounded north by Colorado, northeast by Oklahoma, east by Texas, south by Texas and Mexico and west by Arizona. Land area, 121,298 sq. miles (314,161 sq. km).

Census population, 1 April 2010, was 2,059,179, an increase of 13·2% since 2000. July 2012 estimate, 2,085,538. Of the total population in 2010, 1,041,758 were female and 1,540,507 were 18 years old or older.

The population in five census years was:

	White	Black	American Indian/ Alaska Native	Asian/ Native Hawaiian/ Pacific Islander	Other	Total	Per sq. mile
1910	304,594	1,628	20,573	506	—	327,301	2·7
1980	977,587	24,020	106,119	6,825	188,343	1,302,894	10·7
1990	1,146,028	30,210	134,355	14,124	190,352	1,515,069	12·5
2000	1,214,253	34,343	173,483	20,758	376,209	1,819,046	15·0
2010	1,407,876	42,550	193,222	30,018	385,513	2,059,179	17·0

Before 1930 New Mexico was largely a Spanish-speaking state, but after 1945 an influx of population from other states considerably reduced the percentage of persons of Spanish origin or descent. However, in recent years the percentage of the Hispanic population has begun to rise again. In 2010 the Hispanic population was 953,403, up from 765,386 in 2000 (an increase of 24·6%). At 46·3%, New Mexico has the largest percentage of persons of Hispanic origin of any state in the USA.

The largest cities are Albuquerque, with a 2010 census population of 545,852; Las Cruces, 97,618; Rio Rancho, 87,521; Santa Fe (the capital), 67,947; Roswell, 48,366.

SOCIAL STATISTICS

Births, 2010, 27,850 (13·5 per 1,000 population); deaths, 2010 (provisional), 15,931 (7·7). Infant mortality, 2009, 5·3 per 1,000 live births. 2009 (provisional): marriages, 10,200 (5·1 per 1,000 population); divorces and annulments, 8,000 (4·0).

CLIMATE

Santa Fe, Jan. 26·4°F (−3·1°C), July 68·4°F (20°C). Annual rainfall 15·2" (386 mm). New Mexico belongs to the Mountain States climate zone (*see* UNITED STATES: Climate).

CONSTITUTION AND GOVERNMENT

The constitution of 1912 is still in force with 161 amendments as of Nov. 2010. The state Legislature, which meets annually, consists of 42 members of the Senate, elected for four years, and 70 members of the House of Representatives, elected for two years.

For the 113th Congress, which convened in Jan. 2013, New Mexico sends three members to the House of Representatives. It is represented in the Senate by Tom Udall (D. 2009–15) and Martin Heinrich (D. 2013–19).

The state capital is Santa Fe. The state is divided into 33 counties.

RECENT ELECTIONS

In the 2012 presidential election Barack Obama took New Mexico with 52·9% of the vote, having also won it in 2008.

CURRENT GOVERNMENT

Governor: Susana Martinez (R.), 2011–15 (salary: $110,000).
    Lieut.-Governor: John Sanchez (R.), 2011–15 ($85,000).
    Secretary of State: Dianna J. Duran (R.), 2011–15 ($85,000).

Government Website: http://www.newmexico.gov

ECONOMY

Per capita personal income (2010) was $33,368.

Budget

In 2011 total state revenue was $19,867m. Total expenditure was $17,865m. (including: education, $5,392m.; public welfare, $4,329m.; health, $921m.; highways, $806m.; government administration, $637m.) Outstanding debt in 2011, $8,119m.

Performance

Gross Domestic Product by state in 2011 was $70,497m. (provisional), ranking New Mexico 37th in the United States. In 2011 state real GDP growth was 0·2% (provisional).

ENERGY AND NATURAL RESOURCES

Oil and Gas

In 2008 New Mexico produced 59·4m. bbls of crude petroleum; marketed production of natural gas totalled 1,446bn. cu. ft. New Mexico ranks fourth in the USA for natural gas production.

Water

The total area covered by water is 292 sq. miles.

Minerals

New Mexico is one of the largest energy producing states in the USA. Production in 2001: potash, 1,086,410 short tons; copper (2008), 104,000 short tons; coal (2009), 25,124,000 short tons. New Mexico is the country's leading potash producer and ranked third for copper production in 2008. The value of coal output in 2001 was $584·9m.; total non-fuel mineral output had a value of $888m. in 2009.

Agriculture

New Mexico produces grains, vegetables, hay, livestock, milk, cotton and pecans. In 2007 there were 20,930 farms covering 43·24m. acres; average farm size 2,066 acres. In 2007 average value of farmland and buildings per acre was $337.

2006 cash receipts from crops, $602m.; and from livestock products, $1,861m. The net farm income in 2006 was $423m. Principal commodities are dairy products, cattle, hay and pecans. Farm animals in 2007 included 326,000 dairy cows, 1·53m. all cattle, 127,000 sheep and lambs, and 2,000 swine.

Forestry

The state had a forested area of 16,682,000 acres in 2007, of which 8,092,000 acres were national forest.

INDUSTRY

In 2007 the state's 1,574 manufacturing establishments had 35,000 employees, earning $1,560m. Total value added by manufacturing in 2009 was $5,445m.

Labour

Total non-agricultural employment, 2010, 802,000. Employees by branch, 2010 (in 1,000): government, 199; trade, transportation and utilities, 133; education and health services, 120; professional and business services, 98; leisure and hospitality, 84. The unemployment rate in Dec. 2010 was 8·6%.

INTERNATIONAL TRADE

Imports and Exports

In 2010 exports from New Mexico totalled $1,541m., up from $1,270m. in 2009 although down from $2,783m. in 2008. The leading destinations for exports in 2010 were Mexico (value of $429m.), followed by Canada, Germany and China (excluding Hong Kong).

COMMUNICATIONS

Roads

In 2007 there were 68,339 miles of road (60,351 miles rural). There were 1,599,333 registered motor vehicles.

Rail

In 2009 there were 1,835 miles of freight railroad (excluding trackage rights). Rail traffic originating in New Mexico in 2009 totalled 12·1m. tons and rail traffic terminating in the state came to 3·1m. tons.

Civil Aviation

In 2011 there were four primary airports—commercial service airports with more than 10,000 passenger boardings annually—with a combined total of 2,865,348 enplanements, down from 2,924,466 in 2010. By far the busiest airport, Albuquerque International Sunport, had 2,768,435 enplanements in 2011.

SOCIAL INSTITUTIONS

Justice

In Dec. 2008 there were 6,402 prison inmates in state or federal correctional institutions. The death penalty was abolished with effect from 1 July 2009 for crimes committed after that date, although two prisoners remained on death row. It was most recently used in 2001 (one execution) for the first time since 1960.

Since 1949 the denial of employment by reason of race, colour, religion, national origin or ancestry has been forbidden. A law of 1955 prohibits discrimination in public places because of race or colour. An 'equal rights' amendment was added to the constitution in 1972.

Education

Elementary education is free, and compulsory between five and 18 years. In fall 2007 the 89 school districts had an enrolment of 329,040 students in 851 public elementary and secondary schools and 22,300 teachers; average teacher salary in 2008–09 was $47,341. Total expenditure on public school education in 2006–07 was $3,240m. Spending per pupil in fiscal year 2008 was $9,291.

In fall 2007 there were 124,773 students attending public degree-granting institutions and 9,602 at private institutions. 64,137 students attended community colleges in 2005–06. The state-supported four-year institutes of higher education are (fall 2009 enrolment):

	Students
University of New Mexico, Albuquerque	27,241
New Mexico State University, Las Cruces	18,526
Eastern New Mexico University, Portales	4,679
New Mexico Highlands University, Las Vegas	3,739
Western New Mexico University, Silver City	3,370
New Mexico Institute of Mining and Technology, Socorro	1,761

Health

In 2009 there were 37 community hospitals with 3,900 beds. A total of 183,000 patients were admitted during the year. In 2009 there were 4,877 active physicians in the state (24 per 10,000 population).

Welfare

Medicare enrolment in July 2010 totalled 311,065. In fiscal year 2009 a total of 561,758 people in New Mexico received Medicaid. In Dec. 2008 there were 335,471 Old-Age, Survivors, and Disability Insurance (OASDI) beneficiaries. A total of 39,813 people were receiving payments under Temporary Assistance for Needy Families (TANF) in Dec. 2008.

RELIGION

The principal religious traditions in 2010 were: Catholics, with 584,941 members; Evangelical Protestants, 277,326; Mainline Protestants, 71,118; Latter-day Saints (Mormons), 68,192; Buddhists, 8,180.

FURTHER READING

Bureau of Business and Economic Research, Univ. of New Mexico—*Census in New Mexico* (Continuing series. Vols. 1–5, 1992–).—*Economic Census: New Mexico* (Continuing series. Vols. 1–3).—*New Mexico Business*. Monthly; annual review in Jan.–Feb. issue.

Etulain, R., *Contemporary New Mexico, 1940–1990*. 1994

New York State

KEY HISTORICAL EVENTS

The first European immigrants came in the 17th century, when there were two powerful Indian groups in rivalry: the Iroquois confederacy (Mohawk, Oneida, Onondaga, Cayuga and Seneca) and the Algonquian-speaking Mohegan and Munsee. The Dutch made settlements at Fort Orange (now Albany) in 1624 and at New Amsterdam in 1625, trading with the Indians for furs. In the 1660s there was conflict between the Dutch and the British in the Caribbean; as part of the concluding treaty the British, in 1664, received Dutch possessions in the Americas, including New Amsterdam, which they renamed New York.

In 1763 the Treaty of Paris ended war between the British and the French in North America (in which the Iroquois had allied themselves with the British). Settlers of British descent in New England then felt confident enough to expand westward. The climate of northern New York being severe, most settled in the Hudson river valley. After the War of Independence New York became the 11th state of the Union (1778), having first declared itself independent of Britain in 1777.

The economy depended on manufacturing, shipping and other means of distribution and trade. During the 19th century New York became the most important city in the USA. Its industries, especially clothing, attracted thousands of European immigrants. Industrial development spread along the Hudson-Mohawk valley, which was made the route of the Erie Canal (1825) linking New York with Buffalo on Lake Erie and thus with the developing farmlands of the middle west.

On 11 Sept. 2001 two hijacked commercial airliners were flown into the World Trade Center in central New York. The complex was destroyed in the attack and 2,749 people died.

TERRITORY AND POPULATION

New York is bounded west and north by Canada with Lake Erie, Lake Ontario and the St Lawrence River forming the boundary; east by Vermont, Massachusetts and Connecticut, southeast by the Atlantic, south by New Jersey and Pennsylvania. Land area, 47,126 sq. miles (122,057 sq. km). Census population, 1 April 2010, was 19,378,102, an increase of 2·1% since 2000. July 2012 estimate, 19,570,261.

Population in five census years was:

	White	Black	American Indian	Asiatic	Total	Per sq. mile
1910	8,966,845	134,191	6,046	6,532	9,113,614	191·2
			All others			
1980	13,961,106	2,401,842	1,194,340		17,557,288	367·0
1990	13,385,255	2,859,055	1,746,145		17,990,455	381·0
2000	12,893,689	3,014,385	3,068,383		18,976,457	401·9
2010	12,740,974	3,073,800	3,563,328		19,378,102	411·2

Of the total population in 2010, 10,000,955 were female and 15,053,173 were 18 years old or older. In 2010 the Hispanic population was 3,416,922, up from 2,867,583 in 2000 (an increase of 19·2%).

The population of New York City, by boroughs, census of 1 April 2010 was: Bronx, 1,385,108; Brooklyn, 2,504,700; Manhattan, 1,585,873; Queens, 2,230,722; Staten Island, 468,730; total, 8,175,133.

Population of other large cities and incorporated places at the 2010 census was:

Buffalo	261,310	Cheektowaga	75,178
Rochester	210,565	Mount Vernon	67,292
Yonkers	195,976	Schenectady	66,135
Syracuse	145,170	Utica	62,235
Albany (capital)	97,856	Brentwood	60,664
New Rochelle	77,062	Tonawanda	58,144

White Plains	56,853	West Babylon	43,213
Hempstead	53,891	Freeport	42,860
Levittown	51,881	Hicksville	41,547
Irondequoit	51,692	Coram	39,113
Niagara Falls	50,193	East Meadow	38,132
Troy	50,129	Valley Stream	37,511
Binghamton	47,376	Brighton	36,609
West Seneca	44,711	Commack	36,124

The New York–Northern New Jersey–Long Island metropolitan area had, in 2010, a population of 18,897,109. Other large urbanized areas, census 2010: Buffalo–Niagara Falls, 1,135,509; Rochester, 1,054,323; Albany–Schenectady–Troy, 870,716.

SOCIAL STATISTICS

Births, 2010, 244,375 (12·6 per 1,000 population); deaths, 2010 (provisional), 146,413 (7·6). Infant mortality rate, 2009, 5·3 per 1,000 live births. 2009 (provisional): marriages, 120,100 (6·4 per 1,000 population); divorces and annulments, 46,100 (2·6). Same-sex marriage became legal in July 2011.

CLIMATE

Albany, Jan. 24°F (–4·4°C), July 73°F (22·8°C). Annual rainfall 34" (855 mm). Buffalo, Jan. 24°F (–4·4°C), July 70°F (21·1°C). Annual rainfall 36" (905 mm). New York, Jan. 30°F (–1·1°C), July 74°F (23·3°C). Annual rainfall 43" (1,087 mm). New York belongs to the Atlantic Coast climate zone (see UNITED STATES: Climate).

CONSTITUTION AND GOVERNMENT

New York State has had five constitutions, adopted in 1777, 1821, 1846, 1894 and 1938. The constitution produced by the 1938 convention (which was substantially a modification of the 1894 one), forms the fundamental law of the state (as modified by subsequent amendments). A proposed new constitution in 1967 was rejected by the electorate. In 1997 voters rejected a proposal to hold a new constitutional convention.

The Legislature comprises the Senate, with 62 members, and the Assembly, with 150. All members are elected in even-numbered years for two-year terms. The Legislature meets every year, typically for several days a week from Jan.–June and, if recalled by leaders of the Legislature, at other times during the year. The Governor can also call the Legislature into extraordinary session. The state capital is Albany. For local government the state is divided into 62 counties, five of which constitute the city of New York.

Each of the state's 62 cities is incorporated by charter, under special legislation. The government of New York City is vested in the mayor (Michael Bloomberg), elected for four years, and a city council, whose president and members are elected for four years. The council has a President and 51 members, each elected from a district wholly within the city. The mayor appoints all the heads of departments, except the comptroller (the chief financial officer), who is elected. Each of the five city boroughs (Manhattan, Bronx, Brooklyn, Queens and Staten Island) has a president, elected for four years. Each borough is also a county, although Manhattan borough, as a county, is called New York, Brooklyn is called Kings, and Staten Island is called Richmond.

For the 113th Congress, which convened in Jan. 2013, New York State sends 27 members to the House of Representatives. It is represented in the Senate by Charles Schumer (D. 1999–2017) and Kirsten Gillibrand (D. 2009–19), who succeeded Hillary Clinton (D. 2001–09) following the latter's appointment as Secretary of State. Gillibrand won a special election in Nov. 2010 to take on the remainder of Clinton's term, which was set to end in Jan. 2013.

RECENT ELECTIONS

In the 2012 presidential election Barack Obama took New York State with 62·6% of the vote, having also won it in 2008.

CURRENT GOVERNMENT

Governor: Andrew Cuomo (D.), 2011–15 (salary: $179,000).

Lieut.-Governor: Robert Duffy (D.), 2011–15 ($151,500).

Secretary of State: Cesar A. Perales (D.), since June 2011 ($120,800).

Government Website: http://www.ny.gov

ECONOMY

Per capita personal income (2010) was $48,450.

Budget

In 2011 total state revenue was $205,546m. Total expenditure was $184,009m. (including: public welfare, $51,132m.; education, $45,219m.; health, $8,828m.; hospitals, $6,114m.; government administration, $5,699m.) Outstanding debt in 2011 was $134,929m.

Performance

Gross Domestic Product by state was $1,016,350m. in 2011 (provisional), ranking New York third after California and Texas. In 2011 state real GDP growth was 1·1% (provisional).

Banking and Finance

In 2002 there were 211 financial institutions in New York State insured by the US Federal Deposit Insurance Corporation, with assets worth $1,620bn. They had 4,526 offices with total deposits of $516bn.

ENERGY AND NATURAL RESOURCES

Electricity

In 2009 production was 133·0bn. kWh, of which 43·5bn. kWh was from nuclear energy.

Oil and Gas

Production of crude oil in 2008 was 0·4m. bbls; natural gas, 50bn. cu. ft.

Water

The total area covered by water is 7,429 sq. miles.

Minerals

Principal minerals are: sand and gravel, salt, titanium concentrate, talc, abrasive garnet, wollastonite and emery. Quarry products include trap rock, slate, marble, limestone and sandstone. Value of domestic non-fuel mineral output in 2009 was $1,370m.

Agriculture

New York has large agricultural interests. In 2007 it had 36,352 farms, with a total area of 7·17m. acres; average farm was 197 acres. Average value per acre in 2007 was $2,275.

Farm income, 2006: from crops, $1,527m.; and from livestock, $1,982m. The net farm income in 2006 was $869m. Dairying is an important type of farming. Field crops comprise maize, winter wheat, oats and hay. New York ranks second in the USA in the production of apples and maple syrup. Other products are grapes, tart cherries, peaches, pears, plums, strawberries, raspberries, cabbage, onions, potatoes and maple sugar. Farm animals, 2007, included 1,443,000 all cattle, 626,000 dairy cows, 85,700 hogs and pigs, 63,000 sheep and lambs, and 3·95m. laying hens.

Forestry

Total forested area was 18,669,000 acres in 2007, of which 11,000 acres were national forest. There were state parks and recreation areas covering 330,000 acres in 2011.

INDUSTRY

Leading industries are clothing, non-electrical machinery, printing and publishing, electrical equipment, instruments, food and allied products and fabricated metals. In 2007 the state's 18,629 manufacturing establishments had 534,000 employees, earning $24,268m. Total value added by manufacturing in 2009 was $80,668m.

Labour

Total non-agricultural employment, 2010, 8,553,300. Employees by branch, 2010 (in 1,000): education and health services, 1,704; government, 1,510; trade, transportation and utilities, 1,457; professional and business services, 1,100; leisure and hospitality, 733. In Dec. 2010 the unemployment rate was 8·2%.

INTERNATIONAL TRADE

Imports and Exports

In 2010 exports from New York State totalled $69,696m., up from $58,743m. in 2009 although down from $81,386m. in 2008. The leading destinations for exports in 2010 were Canada (value of $14,693m.), followed by the United Kingdom, Hong Kong and Switzerland.

COMMUNICATIONS

Roads

In 2007 there were 113,740 miles of road (65,874 miles rural). The New York State Thruway extends 559 miles from New York City to Buffalo. The Northway, a 176-mile toll-free highway, is a connecting road from the Thruway at Albany to the Canadian border at Champlain, Quebec.

There were 11,494,513 motor vehicle registrations in 2007 and 1,333 traffic accident fatalities.

Rail

In 2009 there were 3,494 miles of freight railroad (excluding trackage rights). There were 37 freight railroads operating in 2009. Rail traffic originating in New York State in 2009 totalled 6·6m. tons and rail traffic terminating in the state came to 20·5m. tons. New York City has NYCTA and PATH metro systems, and commuter railroads run by Metro-North, New Jersey Transit and Long Island Rail Road. Buffalo has a 7-mile metro line.

Civil Aviation

There were 14 primary airports—commercial service airports with more than 10,000 passenger boardings annually—in New York State in 2011 with a combined total of 44,211,872 enplanements, up from 43,617,917 in 2010. Five of the primary airports had more than 1m. passenger enplanements in 2011: New York City's John F. Kennedy International (with 23,664,832 enplanements, ranking it sixth in the USA), New York City's La Guardia (11,989,227), Buffalo Niagara International (2,582,597), Albany International (1,216,626) and Greater Rochester International (1,190,967).

Shipping

The canals of the state, combined in 1918 in what is called the Improved Canal System, have a length of 524 miles, of which the Erie or Barge canal has 340 miles.

SOCIAL INSTITUTIONS

Justice

The State Human Rights Law was approved on 12 March 1945, effective on 1 July 1945. The State Division of Human Rights is charged with the responsibility of enforcing this law. The division may request and utilize the services of all governmental departments and agencies; adopt and promulgate suitable rules and regulations; test, investigate and pass judgment upon complaints alleging discrimination in employment, in places of public accommodation, resort or amusement, education, and in housing, land and commercial space; hold hearings, subpoena witnesses and require the production for examination of papers relating to matters under investigation; grant compensatory damages and require repayment of profits in certain housing cases among other provisions; apply for court injunctions to prevent frustration of orders of the Commissioner.

In Dec. 2008 there were 60,347 federal and state prisoners, down 3·6% from 62,620 in Dec. 2007—the largest fall in any state over the same period.

The death penalty was declared unconstitutional in New York in 2004; the last person was removed from death row in 2007. The last execution was in 1963.

Education

Education is compulsory between the ages of six and 16. In fall 2007 the 4,631 public elementary and secondary schools had 2,765,435 pupils and 211,854 teachers. There were 518,850 pupils at 2,130 private schools.

The state's educational system, including public and private schools and secondary institutions, universities, colleges, libraries, museums, etc., constitutes (by legislative act) the 'University of the State of New York', which is governed by a Board of Regents consisting of 15 members appointed by the Legislature. Within the framework of this 'University' was established in 1948 a 'State University' (SUNY), which controls 64 colleges and educational centres, 30 of which are locally operated community colleges. The 'State University' is governed by a board of 16 Trustees, appointed by the Governor with the consent and advice of the Senate.

In fall 2009 there were 269 degree-granting colleges and universities in New York State; enrolled students numbered 1,248,908.

Student enrolment (fall 2009) in degree-granting institutions in the state included:

Founded	Name and place	Students
1754	Columbia University, New York City	24,230
1795	Union College, Schenectady and Albany	2,194
1824	Rensselaer Polytechnic Institute, Troy	6,901
1829	Rochester Institute of Technology, Rochester	15,445
1831	New York University, New York City	43,404
1836	Alfred University, Alfred	2,319
1841	Manhattanville College, New York City	2,993
1846	Colgate University, Hamilton	2,837
1846	Fordham University, New York City	14,544
1847	The City University of New York (CUNY), New York City	259,515
1848	University of Rochester, Rochester	9,506
1854	Polytechnic Institute of New York University, New York City	4,514
1856	St Lawrence University, Canton	2,401
1859	Cooper Union for the Advancement of Science and Art, NYC	995
1861	Vassar College, Poughkeepsie	2,453
1863	Manhattan College, New York City	3,461
1865	Cornell University, Ithaca	20,633
1870	Syracuse University, Syracuse	19,638
1870	St John's University, New York City	20,352
1892	Ithaca College, Ithaca	6,894
1906	Pace University, New York City and Westchester	12,706
1926	Long Island University	21,682
1929	Marist College, Poughkeepsie	6,179
1935	Hofstra University, Hempstead	12,068
1948	State University of New York (SUNY)	464,981

Health

In 2009 there were 189 community hospitals with 60,400 beds. A total of 2,534,000 patients were admitted during the year. In 2009 there were 77,042 active physicians in the state (39 per 10,000 population).

Welfare

Medicare enrolment in July 2010 totalled 2,965,774. In fiscal year 2009 a total of 4,984,578 people in New York State received Medicaid. In Dec. 2008 there were 3,143,642 Old-Age, Survivors, and Disability Insurance (OASDI) beneficiaries. A total of 257,205 people were receiving payments under Temporary Assistance for Needy Families (TANF) in Dec. 2008.

RELIGION

The principal religious traditions in 2010 were: Catholics, with 6,287,618 members; Mainline Protestants, 1,027,403; Evangelical Protestants, 871,326; Jews, 784,106; Muslims, 393,000 (estimate).

CULTURE

Tourism

Overseas visitors to New York State—excluding those from Canada and Mexico—numbered 9,508,000 in 2011 (the most of any state), up from 8,647,000 in 2010 and 6,614,000 in 2006. There were 3,446,000 visits by Canadians in 2010.

FURTHER READING

Statistical information: Nelson Rockefeller Institute of Government, 411 State St., Albany 12203. Publishes *New York State Statistical Yearbook.*
New York Red Book. Biennial.
Legislative Manual. Biennial.
The Modern New York State Legislature: Redressing the Balance. 1991
State library: The New York State Library, Albany 12230.

North Carolina

KEY HISTORICAL EVENTS

The early inhabitants were Cherokees. European settlement was attempted in 1585–87, following an exploratory visit by Sir Walter Raleigh, but this failed. Settlers from Virginia came to the shores of Albemarle Sound after 1650 and in 1663 Charles II chartered a private colony of Carolina. In 1691 the north was put under a deputy governor who ruled from Charleston in the south. The colony was formally separated into North and South Carolina in 1712. In 1729 control was taken from the private proprietors and vested in the Crown, whereupon settlement grew, and the boundary between north and south was finally fixed (1735).

After the War of Independence, North Carolina became one of the original 13 states of the Union. The city of Raleigh was laid out as the new capital. Having been a plantation colony North Carolina continued to develop as a plantation state, growing tobacco with black slave labour. It was also an important source of gold before the western gold-rushes of 1848.

In 1861 at the outset of the Civil War, North Carolina seceded from the Union, but General Sherman occupied the capital unopposed. A military governor was admitted in 1862, and civilian government restored with readmission to the Union in 1868.

TERRITORY AND POPULATION

North Carolina is bounded north by Virginia, east by the Atlantic, south by South Carolina, southwest by Georgia and west by Tennessee. Land area, 48,618 sq. miles (125,920 sq. km). Census population, 1 April 2010, was 9,535,483, an increase of 18·5% since 2000. July 2012 estimate, 9,752,073.

Population in five census years was:

	White	Black	American Indian	Asiatic	Total	Per sq. mile
1910	1,500,511	697,843	7,851	82	2,206,287	45·3
			All others			
1980	4,453,010	1,316,050	105,369		5,874,429	111·5
1990	5,008,491	1,456,323	163,823		6,628,637	136·1
2000	5,804,656	1,737,545	507,112		8,049,313	165·2
2010	6,528,950	2,048,628	957,905		9,535,483	196·1

Of the total population in 2010, 4,889,991 were female and 7,253,848 were 18 years old or older. In 2010 North Carolina's Hispanic population was 800,120, up from 378,963 in 2000. This represented a rise of 111·1%.

The principal cities (with census population in 2010) are: Charlotte, 731,424; Raleigh (the capital), 403,892; Greensboro, 269,666; Winston-Salem, 229,617; Durham, 228,330; Fayetteville, 200,564; Cary, 135,234; Wilmington, 106,476; High Point, 104,371.

SOCIAL STATISTICS

Births, 2010, 122,350 (12·8 per 1,000 population); deaths, 2010 (provisional), 78,761 (8·3). Infant mortality rate, 2009, 7·9 per 1,000 live births. 2009 (provisional): marriages, 65,800 (6·7 per 1,000 population); divorces and annulments, 36,700 (3·8).

CLIMATE

Climate varies sharply with altitude; the warmest area is in the southeast near Southport and Wilmington; the coldest is Mount Mitchell (6,684 ft). Raleigh, Jan. 42°F (5·6°C), July 79°F (26·1°C). Annual rainfall 46" (1,158 mm). North Carolina belongs to the Atlantic Coast climate zone (see UNITED STATES: Climate).

CONSTITUTION AND GOVERNMENT

The present constitution dates from 1971 (previous constitutions, 1776 and 1868); it has had 29 amendments as of 2010. The General Assembly consists of a Senate of 50 members and a House of Representatives of 120 members; all are elected by districts for two years. It meets in odd-numbered years in Jan.

The Governor and Lieut.-Governor are elected for four years; they can be elected to only one additional consecutive term. There are also 19 executive departments—eight have elected heads (for four-year terms) and ten have heads appointed by the Governor; the other department is the North Carolina Community College System, under a president.

For the 113th Congress, which convened in Jan. 2013, North Carolina sends 13 members to the House of Representatives. It is represented in the Senate by Richard Burr (R. 2005–17) and Kay Hagan (D. 2009–15).

The capital is Raleigh. There are 100 counties.

RECENT ELECTIONS

In the 2012 presidential election Mitt Romney took North Carolina with 50·6% of the vote (Barack Obama won it in 2008).

CURRENT GOVERNMENT

Governor: Pat McCrory (R.), 2013–17 (salary: $141,265).
　Lieut.-Governor: Dan Forest (R.), 2013–17 ($124,676).
　Secretary of State: Elaine Marshall (D.), 2013–17 ($124,676).

Government Website: http://www.nc.gov

ECONOMY

Per capita personal income (2010) was $34,977.

Budget

In 2011 total state revenue was $63,199m. Total expenditure was $53,089m. (education, $19,311m.; public welfare, $11,619m.; highways, $3,433m.; hospitals, $1,622m.; health, $1,567m.) Outstanding debt in 2011, $18,556m.

Performance

Gross Domestic Product by state in 2011 was $385,092m. (provisional), ranking North Carolina 9th in the United States. In 2011 state real GDP growth was 1·8% (provisional).

ENERGY AND NATURAL RESOURCES

Electricity

In 2009 production was 118·4bn. kWh, of which 65·1bn. kWh was from coal.

Water

The total area covered by water is 5,201 sq. miles, of which 4,052 sq. miles are inland.

Minerals

Principal minerals are stone, sand and gravel, phosphate rock, feldspar, lithium minerals, olivine, kaolin and talc. North Carolina is a leading producer of bricks, making more than 1bn. bricks a year. Value of domestic non-fuel mineral production in 2009 was $846m.

Agriculture

In 2007 there were 52,913 farms covering 8·47m. acres; average size of farms was 160 acres and average value per acre in 2007 was $4,096.

Farm income, 2006: from crops, $2,925m.; and from livestock and products, $5,274m. The net farm income in 2006 was $3,702m. Principal commodities are broilers, hogs, greenhouse products and tobacco. North Carolina is the USA's largest tobacco producer (366·0m. lb in 2007, representing 47% of overall production).

Livestock, 2007: cattle, 820,000; hogs and pigs, 10·13m.; laying hens, 12·75m.

Forestry

Forests covered 18·45m. acres in 2007, with 1·17m. acres of national forest. In 2006 timber removals were 1,075m. cu. ft (636m. cu. ft softwoods and 439m. cu. ft hardwoods). Main products are hardwood veneer and hardwood plywood, furniture woods, pulp, paper and lumber.

Fisheries

Commercial fish catch, 2009, had a value of approximately $77·2m. and produced 69·0m. lb. The catch is mainly of blue crab, menhaden, Atlantic croaker, flounder, shark, sea trout, mullet, blue fish and shrimp.

INDUSTRY

The leading industries by employment are textiles, clothing, furniture, electrical machinery and equipment, non-electrical machinery and food processing. In 2007 the state's 10,150 manufacturing establishments had 506,000 employees, earning $19,590m. Total value added by manufacturing in 2009 was $84,451m.

Labour

Total non-agricultural employment, 2010, 3,862,000. Employees by branch, 2010 (in 1,000): trade, transport and utilities, 711; government, 704; education and health services, 539; professional and business services, 481; manufacturing, 431. The unemployment rate in Dec. 2010 was 9·8%.

INTERNATIONAL TRADE

Imports and Exports

In 2010 exports from North Carolina totalled $24,905m., up from $21,793m. in 2009 although down from $25,091m. in 2008. The leading destinations for exports in 2010 were Canada (value of $5,448m.), followed by China (excluding Hong Kong), Mexico and Japan.

COMMUNICATIONS

Roads

In 2007 there were 104,412 miles of road (71,306 miles rural). There were 6,317,148 registered motor vehicles.

Rail

In 2009 there were 3,230 miles of freight railroad (excluding trackage rights). There were 22 freight railroads operating in 2009. Rail traffic originating in North Carolina in 2009 totalled 9·6m. tons and rail traffic terminating in the state came to 52·3m. tons.

Civil Aviation

In 2011 North Carolina had nine primary airports (commercial service airports with more than 10,000 passenger boardings annually) with a combined total of 25,751,825 enplanements, up from 25,331,266 in 2010. By far the busiest airport is Charlotte/Douglas International, with 19,022,535 enplanements in 2011.

Shipping

There are two ocean ports, Wilmington and Morehead City.

SOCIAL INSTITUTIONS

Justice

There were five executions in 2005 and four in 2006, but the death penalty was suspended in Jan. 2007. There were 166 death row inmates as at 1 Jan. 2012. In Dec. 2008 there were 39,482 federal and state prisoners.

Education

School attendance is compulsory between seven and 16. In fall 2007 there were 1,489,492 pupils and 106,562 teachers at 2,516 public schools; there were 121,660 pupils at 660 private schools. Total expenditure on public schools was $12,829m. in 2006–07; teachers' salaries in 2008–09 averaged $48,603.

The 16 senior universities are all part of the University of North Carolina system (with 222,322 enrolled students in fall 2009). The largest institution is the North Carolina State University (founded 1887), at Raleigh, with 33,819 students. The University of North Carolina at Chapel Hill (founded in 1789; the first state university to open in America in 1795) had 28,916 students in fall 2009; East Carolina University (founded in 1907), at Greenville, had 27,654. There were 58 private degree-granting institutions in 2008–09; enrolment totalled 91,584 in fall 2007.

Health

In 2009 there were 115 community hospitals with 22,800 beds. A total of 1,034,000 patients were admitted during the year. In 2009 there were 24,072 active physicians in the state (26 per 10,000 population).

Welfare

Medicare enrolment in July 2010 totalled 1,476,516. In fiscal year 2009 a total of 1,782,185 people in North Carolina received Medicaid. In Dec. 2008 there were 1,631,266 Old-Age, Survivors, and Disability Insurance (OASDI) beneficiaries. A total of 49,256 people were receiving payments under Temporary Assistance for Needy Families (TANF) in Dec. 2008.

RELIGION

The principal religious traditions in 2010 were: Evangelical Protestants, with 2,585,530 members; Mainline Protestants, 1,130,241; Catholics, 392,912; Black Protestants, 248,257; Latter-day Saints (Mormons), 77,627.

CULTURE

Tourism

In 2011 overseas visitors to North Carolina—excluding those from Canada and Mexico—numbered 335,000, down from 343,000 in 2010.

FURTHER READING

Statistical information: Office of State Planning, 116 West Jones St., Raleigh 27603. Publishes *Statistical Abstract of North Carolina Counties. North Carolina Manual.* Biennial

Fleer, J. D., *North Carolina: Government and Population.* 1995

North Dakota

KEY HISTORICAL EVENTS

The original inhabitants were Plains Indians. French explorers and traders were active in the 18th century, often operating from French possessions in Canada. France claimed the area until 1803, when it passed to the USA as part of the Louisiana Purchase, except for the northeastern part which was held by the British until 1818.

Trading with the Indians, mainly for furs, continued until the 1860s, with American traders succeeding the French. In 1861 the Dakota Territory (North and South) was established. In 1862 the Homestead Act was passed (allowing 160 acres of public land free to any family who had worked and lived on it for five years) and this greatly stimulated settlement. Farming settlers came to the wheat lands in great numbers, many of them from Canada, Norway and Germany.

Bismarck, the capital, began as a crossing-point on the Missouri and was fortified in 1872 to protect workers building the Northern Pacific Railway. There followed a gold-rush nearby and the town became a service centre for prospectors. In 1889 North and South Dakota were admitted to the Union as separate states with Bismarck as the Northern capital. The largest city, Fargo, was also a railway town, named after William George Fargo, the express-company founder.

The population grew rapidly until 1890 and steadily until 1930 by which time it was about one-third European in origin. Between 1930 and 1970 there was a steady population drain, increasing whenever farming was affected by the extremes of the continental climate.

TERRITORY AND POPULATION

North Dakota is bounded north by Canada, east by the Red River (forming a boundary with Minnesota), south by South Dakota and west by Montana. Land area, 69,001 sq. miles (178,711 sq. km). In 2000 American Indian lands covered 3,542 sq. miles (3,467 sq. miles in reservations and 75 sq. miles in off-reservation trust land). Census population, 1 April 2010, was 672,591, an increase of 4·7% since 2000. July 2012 estimate, 699,628.

Population at five census years was:

	White	Black	American Indian	Asiatic	Total	Per sq. mile
1910	569,855	617	6,486	98	577,056	8·2
			All others			
1980	625,557	2,568	24,692		652,717	9·5
1990	604,142	3,524	31,134		638,800	9·3
2000	593,182	3,916	45,102		642,200	9·3
2010	605,449	7,960	59,182		672,591	9·7

Of the total population in 2010, 339,864 were male and 522,720 were 18 years old or older. Only Vermont has fewer persons of Hispanic origin than North Dakota. In 2010 the Hispanic population was 13,467, up from 7,786 in 2000 (an increase of 73·0%).

The largest cities are Fargo with a population, census 2010, of 105,549; Bismarck (capital), 61,272; Grand Forks, 52,838; and Minot, 40,888.

SOCIAL STATISTICS

Births, 2010, 9,104 (13·5 per 1,000 population); deaths, 2010 (provisional), 5,946 (8·8). Infant mortality rate, 2009, 6·1 per 1,000 live births. 2009 (provisional): marriages, 4,300 (6·6 per 1,000 population); divorces and annulments, 1,600 (2·9).

CLIMATE

Bismarck, Jan. 8°F (−13·3°C), July 71°F (21·1°C). Annual rainfall 16" (402 mm). Fargo, Jan. 6°F (−14·4°C), July 71°F (21·1°C). Annual rainfall 20" (503 mm). North Dakota belongs to the High Plains climate zone (*see* UNITED STATES: Climate).

CONSTITUTION AND GOVERNMENT

The present constitution dates from 1889; it has had 154 amendments as of 2011. The Legislative Assembly consists of a Senate of 47 members elected for four years, and a House of

Representatives of 94 members elected for four years. The Governor and Lieut.-Governor are elected for four years.

For the 113th Congress, which convened in Jan. 2013, North Dakota sends one member to the House of Representatives. It is represented in the Senate by John Hoeven (R. 2011–17) and Heidi Heitkamp (D. 2013–19).

The capital is Bismarck. The state has 53 organized counties.

RECENT ELECTIONS

In the 2012 presidential election Mitt Romney took North Dakota with 58·7% of the vote (John McCain won it in 2008).

CURRENT GOVERNMENT

Governor: Jack Dalrymple (R.), Dec. 2012–Dec. 2016 (salary: $117,001).

Lieut.-Governor: Drew Wrigley (R.), Dec. 2012–Dec. 2016 ($90,829).

Secretary of State: Alvin A. Jaeger (D.), Dec. 2010–Dec. 2014 ($93,071).

Government Website: http://www.nd.gov

ECONOMY

Per capita personal income (2010) was $42,764.

Budget

In 2011 total state revenue was $7,806m. Total expenditure was $5,516m. (education, $1,781m.; public welfare, $913m.; highways, $717m.; natural resources, $262m.; government administration, $160m.) Outstanding debt in 2011, $2,061m.

Performance

Gross Domestic Product by state in 2011 was $34,262m. (provisional), ranking North Dakota 47th in the United States. In 2011 state real GDP growth was 7·6% (provisional), the highest rate of any state.

ENERGY AND NATURAL RESOURCES

Oil and Gas

Oil was discovered in 1951. North Dakota is now the USA's second largest oil-producing state (accounting for 11% of output). Crude petroleum production has increased fivefold since 2003 and in 2011 was 152m. bbls; marketed production of natural gas in 2008 totalled 61bn. cu. ft.

Water

The total area covered by water is 1,698 sq. miles.

Minerals

Production of lignite (2009) was 29·9m. short tons. Total value of domestic non-fuel mineral production in 2009 was more than $50m.

Agriculture

In 2007 there were 31,970 farms (61,963 in 1954) in an area of 39·67m. acres and with an average farm acreage of 1,241. In 2007 the average value of farmland and buildings per acre was $771.

Farm income, 2006: from crops, $3,088m.; and from livestock, $892m. The net farm income in 2006 was $606m. Production, 2007, included: wheat, 293·5m. bu. (the most in any state); corn, 275·3m. bu.; soybeans, 106·6m. bu.; barley, 75·4m. bu. (the most in any state); honey, 31m. lb (the most in any state).

The state has also an active livestock industry, chiefly cattle raising. Livestock, 2007: cattle, 1·81m.; hogs and pigs, 181,700; sheep, 89,000; laying hens, 109,300.

Forestry

Forest area, 2007, was 724,000 acres, of which 72,000 acres were national forest.

INDUSTRY

Although the state is still mainly agricultural it is diversifying into high tech and information technology industries. In 2007 the state's 767 manufacturing establishments had 26,000 employees, earning $991m. Total value added by manufacturing in 2009 was $3,548m.

Labour

Total non-agricultural employment, 2010, 376,000. Employees by branch, 2010 (in 1,000): government, 80; trade, transportation and utilities, 80; education and health services, 55; leisure and hospitality, 34; professional and business services, 28. The unemployment rate in Dec. 2010 was just 3·8%, the lowest of any US state.

INTERNATIONAL TRADE

Imports and Exports

In 2010 exports from North Dakota totalled $2,536m., up from $2,193m. in 2009 although down from $2,772m. in 2008. The leading destinations for exports in 2010 were Canada (value of $1,570m.), followed by Mexico, Australia and Belgium.

COMMUNICATIONS

Roads

In 2007 there were 86,842 miles of road (84,945 miles rural). There were 710,537 registered motor vehicles.

Rail

In 2009 there were 3,413 miles of freight railroad (excluding trackage rights). Rail traffic originating in North Dakota in 2009 totalled 31·1m. tons and rail traffic terminating in the state came to 13·8m. tons.

Civil Aviation

In 2011 there were six primary airports—commercial service airports with more than 10,000 passenger boardings annually—with a combined total of 861,139 enplanements, up from 798,204 in 2010.

SOCIAL INSTITUTIONS

Justice

In Dec. 2008 there were 1,452 federal and state prisoners. The Missouri River Correctional Center is a minimum custody institution. The death penalty was abolished in 1973.

Education

School attendance is compulsory between the ages of seven and 16. In 2007–08 there were 528 public schools. In fall 2007 there were 95,059 pupils enrolled in public schools with 8,068 teachers. According to the National Center for Education Statistics, North Dakota has the smallest average high school enrolment in the country with 204 pupils in the 2006–07 school year. State expenditure per pupil in elementary and secondary schools in fiscal year 2008 was $9,324. Average teacher salary was $41,534 in 2008–09.

The University of North Dakota in Grand Forks, founded in 1883, had 12,954 students in fall 2005; North Dakota State University in Fargo, 12,099 students. There were 22 degree-granting institutions in 2008–09 (14 public and eight private); total enrolment in fall 2007 was 49,945.

Health

In 2009 there were 41 community hospitals with 3,400 beds. A total of 93,000 patients were admitted during the year. In 2009 there were 1,617 active physicians in the state (25 per 10,000 population).

Welfare

Medicare enrolment in July 2010 totalled 108,118. In fiscal year 2009 a total of 76,866 people in North Dakota received Medicaid.

In Dec. 2008 there were 117,130 Old-Age, Survivors, and Disability Insurance (OASDI) beneficiaries. A total of 5,855 people were receiving payments under Temporary Assistance for Needy Families (TANF) in Dec. 2008.

RELIGION

The main religious traditions in 2010 were: Mainline Protestants, with 196,839 members; Catholics, 167,349; Evangelical Protestants, 78,607; Latter-day Saints (Mormons), 7,206.

FURTHER READING

Statistical information: Bureau of Business and Economic Research, Univ. of North Dakota, Grand Forks 58202. Publishes *Statistical Abstract of North Dakota.*
North Dakota Blue Book.

Ohio

KEY HISTORICAL EVENTS

The land was inhabited by Delaware, Miami, Shawnee and Wyandot Indians. It was explored by French and British traders in the 18th century and confirmed as part of British North America in 1763. After the War of Independence it became part of the Northwest Territory of the new United States. Former independence fighters came in from New England in 1788 to make the first permanent white settlement at Marietta, at the confluence of the Ohio and Muskingum rivers. In 1803 Ohio was separated from the rest of the Territory and admitted to the Union as the 17th state.

In the early 19th century there was steady immigration from Europe, mainly of Germans, Swiss, Irish and Welsh. Industrial growth began with the processing of farm, forest and mining products; it increased rapidly with the need to supply the Union armies in the Civil War of 1861–65.

As the industrial cities grew, so immigration began again, with many whites from eastern Europe and the Balkans and blacks from the southern states looking for work in Ohio.

Cleveland, which developed rapidly as a Lake Erie port after the opening of commercial waterways to the interior and the Atlantic coast (1825, 1830 and 1855), became an iron-and-steel town during the Civil War.

TERRITORY AND POPULATION

Ohio is bounded north by Michigan and Lake Erie, east by Pennsylvania, southeast and south by the Ohio River (forming a boundary with West Virginia and Kentucky) and west by Indiana. Land area, 40,861 sq. miles (105,829 sq. km). Census population, 1 April 2010, was 11,536,504, an increase of 1·6% since 2000. July 2012 estimate, 11,544,225.

Population at five census years was:

	White	Black	American Indian	Asiatic	Total	Per sq. mile
1910	4,654,897	111,452	127	645	4,767,121	117·0
			All others			
1980	9,597,458	1,076,748	123,424		10,797,630	263·2
1990	9,521,756	1,154,826	170,533		10,847,115	264·5
2000	9,645,453	1,301,307	406,380		11,353,140	277·3
2010	9,539,437	1,407,681	589,386		11,536,504	282·3

Of the total population in 2010, 5,904,348 were female and 8,805,753 were 18 years old or older. In 2010 the Hispanic population was 354,674, up from 217,123 in 2000 (an increase of 63·4%).

Census population of chief cities on 1 April 2010 was:

Columbus		Lorain	64,097	Newark	47,573
(capital)	787,033	Hamilton	62,477	Mentor	47,159
Cleveland	396,815	Springfield	60,608	Cleveland	
Cincinnati	296,943	Kettering	56,163	Heights	46,121
Toledo	287,208	Elyria	54,533	Beavercreek	45,193
Akron	199,110	Lakewood	52,131	Strongsville	44,750
Dayton	141,527	Cuyahoga Falls	49,652	Fairfield	42,510
Parma	81,601	Euclid	48,920	Dublin	41,751
Canton	73,007	Middletown	48,694	Warren	41,557
Youngstown	66,982	Mansfield	47,821	Findlay	41,202

Largest metropolitan areas, 2010 census: Cincinnati–Middleton, 2,130,151; Cleveland–Elyria–Mentor, 2,077,240; Columbus (the capital), 1,836,536; Dayton, 841,502; Akron, 703,200; Toledo, 651,429; Youngstown–Warren–Boardman, 565,773; Canton–Massillon, 404,422.

SOCIAL STATISTICS

Births, 2010, 139,128 (12·1 per 1,000 population); deaths, 2010 (provisional), 108,710 (9·4). Infant mortality, 2009, 7·7 per 1,000 live births. 2009 (provisional): marriages, 64,800 (5·8 per 1,000 population); divorces and annulments, 36,900 (3·3).

CLIMATE

Cincinnati, Jan. 30·6°F, July 76·8°F, annual rainfall 39·6"; Cleveland, Jan. 25·7°F, July 71·9°F, annual rainfall 38·7"; Columbus, Jan. 28·3°F, July 74·7°F, annual rainfall 40·0". Ohio belongs to the Great Lakes climate zone (*see* UNITED STATES: Climate).

CONSTITUTION AND GOVERNMENT

The question of a general revision of the constitution drafted by an elected convention is submitted to the people every 20 years. The constitution dates from 1851, since when there have been 161 amendments adopted to change the constitution.

The Senate consists of 33 members and the House of Representatives of 99 members. The Senate is elected for four years, half every two years; the House is elected for two years; the Governor, Lieut.-Governor and Secretary of State for four years. Qualified as electors are (with necessary exceptions) all citizens 18 years of age who have the usual residential qualifications.

For the 113th Congress, which convened in Jan. 2013, Ohio sends 16 members to the House of Representatives. It is represented in the Senate by Sherrod Brown (D. 2007–19) and Rob Portman (R. 2011–17).

The capital (since 1816) is Columbus. Ohio is divided into 88 counties.

RECENT ELECTIONS

In the 2012 presidential election Barack Obama took Ohio with 50·1% of the vote, having also won it in 2008.

CURRENT GOVERNMENT

Governor: John Kasich (R.), 2011–15 (salary: $148,865·60).
  Lieut.-Governor: Mary Taylor (R.), 2011–15 ($150,405).
  Secretary of State: John Husted (R.), 2011–15 ($109,985·92).

Government Website: http://ohio.gov

ECONOMY

Per capita personal income (2010) was $36,180.

Budget

In 2011 total state revenue was $98,560m. Total expenditure was $79,153m. (education, $22,408m.; public welfare, $18,422m.; highways, $3,494m.; hospitals, $2,660m.; health, $2,482m.) Outstanding debt in 2011, $30,926m.

Performance
In 2011 Gross Domestic Product by state was $418,881m. (provisional), ranking Ohio 8th in the United States. In 2011 state real GDP growth was 1·1% (provisional).

ENERGY AND NATURAL RESOURCES
Electricity
In 2009 production was 136·1bn. kWh, of which 113·7bn. kWh was from coal.

Oil and Gas
In 2008, 5·72m. bbls of crude oil and 84,858m. cu. ft of gas were produced. In 2007 the value of oil and gas production was $1·04bn.

Water
Lake Erie supplies northern Ohio with its water. The total area covered by water is 3,965 sq. miles.

Minerals
Ohio has extensive mineral resources, of which coal is the most important by value: production (2011), 27,929,089 short tons. Coal production in 2011 was valued at $1,194,931,060. Production of other minerals totalled 88,556,399 short tons, of which limestone, dolomite, sand and gravel accounted for 91%. The remainder was comprised of various amounts of salt, sandstone, clay, shale, gypsum and peat. The combined value of all non-fuel industrial minerals in 2011 was $886,417,044.

Agriculture
Ohio is extensively devoted to agriculture. In 2006, 76,200 farms covered 14·3m. acres; average farm value per acre, $3,480. The average size of a farm in 2006 was 188 acres.

Farm income 2006: from crops, $3,448m.; from livestock, $2,031m.; total, $5,480m. The net farm income in 2006 was $1,614m. Production (2006): corn for grain (470·6m. bu.), soybeans (217·1m. bu.), wheat (65·3m. bu.), oats (4·13m. bu.). In 2006 there were 1·56m. pigs, 1·28m. cattle and 141,000 sheep.

Forestry
Forest area, 2007, 7,894,000 acres. State forest lands area, 2007, 191,142 acres. In 2008 there were 74 state parks covering 323,215 acres.

INDUSTRY
In 2007, 16,237 manufacturing establishments employed 760,000 persons, earning $35,485m. Total value added by manufacturing in 2009 was $98,409m. The largest industries were manufacturing of transport equipment, fabricated metal products and machinery.

Labour
Total non-agricultural employment, 2010, 5,031,000. Employees by branch, 2010 (in 1,000): trade, transportation and utilities, 947; education and health services, 843; government, 782; professional and business services, 623; manufacturing, 620. In Dec. 2010 the unemployment rate was 9·5%.

INTERNATIONAL TRADE
Imports and Exports
In 2010 exports from Ohio totalled $41,494m., up from $34,104m. in 2009 although down from $45,628m. in 2008. The leading destinations for exports in 2010 were Canada (value of $17,221m.), followed by Mexico, China (excluding Hong Kong) and France. Ohio was the 8th largest exporting state in the USA in 2010.

COMMUNICATIONS
Roads
In 2007 there were 125,160 miles of road; there were 10,848,476 registered motor vehicles in the same year.

Rail
In 2009 there were 5,286 miles of freight railroad (excluding trackage rights). There were 36 freight railroads operating in 2009. Rail traffic originating in Ohio in 2009 totalled 50·3m. tons and rail traffic terminating in the state came to 66·4m. tons. Cleveland has a 19-mile metro system.

Civil Aviation
In 2011 there were six primary airports—commercial service airports with more than 10,000 passenger boardings annually—with a combined total of 9,715,163 enplanements, down from 9,873,248 in 2010. The busiest airports, Cleveland-Hopkins International and Port Columbus International, had 4,401,033 and 3,134,379 enplanements respectively in 2011.

Shipping
Ohio has more than 700 miles of navigable waterways, with Lake Erie having a 265-mile shoreline. There are nine deep-draft ports in the state. The busiest port is Cincinnati, which handled 11·8m. tons of cargo in 2009.

SOCIAL INSTITUTIONS
Justice
In June 2006 there were 46,839 inmates (43,560 males) in the 32 adult correctional institutions. There were 187 death-row inmates (185 male) in Nov. 2007: 97 were African Americans; 82 Caucasians (two females); four Hispanics; two Native Americans; and two Arab Americans. There were five executions in 2011 and three in 2012.

Education
School attendance during full term is compulsory for children from six to 18 years of age. In 2005–06 public schools had 1,842,943 enrolled pupils. Teachers' salaries (2005–06) averaged $50,654. Estimated public expenditure on elementary and secondary schools for 2006 was $8,349m., 39·0% of the total state budget. In 2003–04 total expenditure for the co-ordination of higher education in Ohio (controlled by the Board of Regents) was $5·54bn.

Public colleges and universities had a total enrolment (2006–07) of 457,322 students. Independent colleges and universities enrolled 132,139 students. Average annual charge (for undergraduates in 2006–07): $8,553 (state); $19,111 (private) (2005–06).

Main campuses, fall 2006:

Founded	Institutions	Enrolments
1804	Ohio University, Athens (State)	20,408
1809	Miami University, Oxford (State)	15,726
1819	University of Cincinnati (State)	28,327
1826	Case Western Reserve University, Cleveland	9,592
1850	University of Dayton (R.C.)	10,502
1870	University of Akron (State)	21,882
1870	Ohio State University, Columbus (State)	51,818
1872	University of Toledo (State)	19,374
1908	Youngstown University (State)	13,183
1910	Bowling Green State University (State)	19,108
1910	Kent State University (State)	22,317
1964	Cleveland State University (State)	15,471
1964	Wright State University, Dayton (State)	16,093
1986	Shawnee State University, Portsmouth (State)	3,889

Health
In 2009 the state had 183 registered community hospitals with 33,900 beds. A total of 1,531,000 patients were admitted during the year. In 2006 state mental retardation facilities had ten developmental centres serving 1,597 residents. In 2009 there were 31,315 active physicians (27 per 10,000 population).

Welfare
Public assistance is administered through the Ohio Works First programme (OWF). In 2006–07 OWF-Combined assistance

groups had 169,218 recipients and money payments totalled $308,786,133. OWF-Regular assistance groups had 156,124 recipients with $292,020,311 paid out in 2006–07. OWF-Unemployed had 13,094 recipients and money payments were $16,765,822 in 2006–07. Disability Assistance had 13,991 recipients in 2006–07; and food stamps, 1,069,561 recipients.

In 2006–07 Disability Assistance totalled $22,847,689; food stamps in 2006–07 totalled $1,287,406,968; and foster care totalled $175,540,449. Optional State Supplement is paid to aged, blind or disabled adults. Free social services available to those eligible by income or circumstances. Medicare enrolment in July 2010 totalled 1,880,459. In fiscal year 2009 a total of 2,238,140 people in Ohio received Medicaid. In Dec. 2010 there were 2,124,650 Old-Age, Survivors, and Disability Insurance (OASDI) beneficiaries. A total of 238,143 people were receiving payments under Temporary Assistance for Needy Families (TANF) in Dec. 2010.

RELIGION

The principal religious traditions in Ohio in 2010 were: Catholics, with 1,992,567 members; Evangelical Protestants, 1,491,845; Mainline Protestants, 1,154,461; Black Protestants, 170,388; Jews, 64,479.

CULTURE

Tourism

Overseas visitors to Ohio—excluding those from Canada and Mexico—numbered 279,000 in 2011, down from 316,000 in 2010 and 390,000 in 2006.

FURTHER READING

Official Roster: Federal, State, County Officers and Department Information. Biennial

Shkurti, W. J. and Bartle, J. (eds.) *Benchmark Ohio.* 1991

Oklahoma

KEY HISTORICAL EVENTS

Francisco Coronado led a Spanish expedition in 1541, claiming the land for Spain. There were several Indian groups but no strong political unit. In 1714 Juchereau de Saint Denis made the first French contact. During the 18th century French fur-traders were active. A French and Spanish struggle for control was resolved by the French withdrawal in 1763. France returned briefly in 1800–03, and the territory then passed to the USA as part of the Louisiana Purchase.

In 1828 the Federal government set aside the area of the present state as Indian Territory (a reservation and sanctuary for Indian tribes who had been driven off their lands elsewhere by white settlement). About 70 tribes came, among whom were Creeks, Choctaws and Cherokees from the southeastern states, and Plains Indians.

In 1889 the government took back about 2·5m. acres of the Territory and opened it to white settlement. About 10,000 homesteaders gathered at the site of Oklahoma City on the Santa Fe Railway in the rush to stake their land claims. The settlers' area, and others subsequently opened to settlement, were organized as the Oklahoma Territory in 1890. In 1907 the Oklahoma and Indian Territories were combined and admitted to the Union as a state. Indian reservations were established within the state.

The economy first depended on ranching and farming, with packing stations on the railways. A mining industry grew in the 1870s attracting foreign immigration, mainly from Europe. In 1901 oil was found near Tulsa.

TERRITORY AND POPULATION

Oklahoma is bounded north by Kansas, northeast by Missouri, east by Arkansas, south by Texas (the Red River forming part of the boundary) and, at the western extremity of the 'panhandle', by New Mexico and Colorado. Land area, 68,595 sq. miles (177,660 sq. km). Census population, 1 April 2010, was 3,751,351, an increase of 8·7% since 2000. July 2012 estimate, 3,814,820.

The population at five federal censuses was:

	White	Black	American Indian/Alaska Native	Other	Total	Per sq. mile
1930	2,130,778	172,198	92,725	339	2,396,040	34·6
1980	2,597,783	204,658	169,292	53,557	3,025,486	43·2
1990	2,583,512	233,801	252,420	119,723	3,189,456	44·5
2000	2,628,434	260,968	273,230	288,022	3,450,654	50·3
2010	2,706,845	277,644	321,687	445,175	3,751,351	54·7

Of the total population in 2010, 1,894,374 were female and 2,821,685 were 18 years old or older. Oklahoma is home to 39 recognized Indian tribes. In 2010 Oklahoma's Hispanic population was 332,007, up from 179,304 in 2000 (an increase of 85·2%).

The most important cities with population, 2010, are Oklahoma City (capital), 579,999; Tulsa, 391,906; Norman, 110,925; Broken Arrow, 98,850; Lawton, 96,867; Edmond, 81,405; Moore, 55,081; Midwest City, 54,371; Enid, 49,379; Stillwater, 45,688; Muskogee, 39,223; Bartlesville, 35,750.

SOCIAL STATISTICS

Births, 2010, 53,238 (14·2 per 1,000 population); deaths, 2010 (provisional), 36,544 (9·7). Infant mortality rate, 2009, 7·9 per 1,000 live births. 2009 (provisional): marriages, 23,500 (6·9 per 1,000 population); divorces and annulments, 16,900 (4·9).

CLIMATE

Oklahoma City, Jan. 34°F (1°C), July 81°F (27°C). Annual rainfall 31·9" (8,113 mm). Tulsa, Jan. 34°F (1°C), July 82°F (28°C). Annual rainfall 33·2" (8,438 mm). Oklahoma belongs to the Central Plains climate zone (*see* UNITED STATES: Climate). Oklahoma's average temperature in July 2011 was 88·9°F (31·6°C), the highest for any state in US history.

CONSTITUTION AND GOVERNMENT

The constitution, dating from 1907, provides for amendment by initiative petition and legislative referendum; it has had 200 amendments (as of Sept. 2005).

The Legislature consists of a Senate of 48 members, who are elected for four years, and a House of Representatives elected for two years and consisting of 101 members. The Governor and Lieut.-Governor are elected for four-year terms; the Governor can only be elected for two terms in succession. Electors are (with necessary exceptions) all citizens 18 years or older, with the usual qualifications.

For the 113th Congress, which convened in Jan. 2013, Oklahoma sends five members to the House of Representatives. It is represented in the Senate by James Inhofe (R. 1994–2015) and Tom Coburn (R. 2005–17).

The capital is Oklahoma City. The state has 77 counties.

RECENT ELECTIONS

In the 2012 presidential election Mitt Romney took Oklahoma with 66·8% of the vote (John McCain won it in 2008).

CURRENT GOVERNMENT

Governor: Mary Fallin (R.), 2011–15 (salary: $147,000).
  Lieut.-Governor: Todd Lamb (R.), 2011–15 ($114,713).
  Secretary of State: Glenn Coffee (R.), 2011–15 ($90,000).

Government Website: http://www.ok.gov

ECONOMY

Per capita personal income (2010) was $35,396.

Budget

In 2011 total state revenue was $26,225m. Total expenditure was $22,378m. (education, $7,491m.; public welfare, $5,457m.; highways, $2,049m.; health, $804m.; government administration, $590m.) Outstanding debt in 2011, $10,255m.

Performance

Gross Domestic Product by state in 2011 was $134,146m. (provisional), ranking Oklahoma 29th in the United States. In 2011 state real GDP growth was 1·0% (provisional).

ENERGY AND NATURAL RESOURCES

Oil and Gas

In 2008 Oklahoma produced 64·1m. bbls of crude petroleum; marketed production of natural gas totalled 1,913bn. cu. ft. Oklahoma ranks third in the USA for natural gas production behind Texas and Wyoming. In 2005 there were 9,407 persons employed in crude petroleum and natural gas extraction.

Water

The total area covered by water is 1,304 sq. miles.

Minerals

Coal production (2009), 956,000 short tons. Principal minerals are: crushed stone, cement, sand and gravel, iodine, glass sand, gypsum. Other minerals are helium, clay and sand, zinc, lead, granite, tripoli, bentonite, lime and volcanic ash. Total value of domestic non-fuel minerals produced in 2009 was $675m.

Agriculture

In 2007 the state had 86,565 farms and ranches with a total area of 35.09m. acres; average size was 405 acres and average value per acre was $1,157. Area harvested, 2007, 7,650,080 acres. Livestock, 2007: cattle, 5·39m.; sheep and lambs, 76,200; hogs and pigs, 2·40m.; laying hens, 3·32m.

Farm income 2006: from crops, $974m.; from livestock and products, $4,120m. The net farm income in 2006 was $877m. The major cash grain is wheat (167m. bu. in 2008 with a value of $1,082m.). Other crops include barley, oats, rye, grain, corn, soybeans, grain sorghum, cotton, peanuts and peaches. Value of cattle and calves produced, 2008, $1,954m.

The Oklahoma Conservation Commission works with 91 conservation districts, universities, and state and federal government agencies. The early work of the conservation districts, beginning in 1937, was limited to flood and erosion control: since 1970, they also include urban areas.

Irrigated production has increased in the Oklahoma 'panhandle'. The Ogalala aquifer is the primary source of irrigation water there and in western Oklahoma, a finite source because of its isolation from major sources of recharge. Declining groundwater levels necessitate the most effective irrigation practices.

Forestry

There were 7,665,000 acres of forested land in 2007, with 245,000 acres of national forest. The forest products industry is concentrated in the 118 eastern counties. There are three forest regions: Ozark (oak, hickory); Ouachita highlands (pine, oak); Cross-Timbers (post oak, black jack oak). Southern pine is the chief commercial species, at almost 80% of saw-timber harvested annually. Replanting is essential.

INDUSTRY

In 2007 Oklahoma's 3,964 manufacturing establishments had 142,000 employees, earning $5,971m. Total value added by manufacturing in 2009 was $22,886m.

Labour

Total non-agricultural employment in 2010 was 1,526,000. Employees by branch, 2010 (in 1,000): government, 340; trade, transportation and utilities, 277; education and health services, 204; professional and business services, 169; leisure and hospitality, 138. Oklahoma's unemployment rate was 6·8% in Dec. 2010.

INTERNATIONAL TRADE

Imports and Exports

In 2010 exports from Oklahoma totalled $5,353m., up from $4,415m. in 2009. The leading destinations for exports in 2010 were Canada (value of $1,867m.), followed by Mexico, Japan and China (excluding Hong Kong).

COMMUNICATIONS

Roads

In 2007 there were 112,922 miles of road comprising 15,633 miles of urban road and 97,289 miles of rural road. There were 3,224,653 registered motor vehicles.

Rail

In 2009 there were 3,275 miles of freight railroad (excluding trackage rights). There were 19 freight railroads operating in 2009. Rail traffic originating in Oklahoma in 2009 totalled 16·9m. tons and rail traffic terminating in the state came to 33·2m. tons.

Civil Aviation

There were three primary airports—commercial service airports with more than 10,000 passenger boardings annually—in 2011 with a combined total of 3,148,469 enplanements, down from 3,161,942 in 2010. The main airport is Will Rogers World, at Oklahoma City, which had 1,738,438 enplanements in 2011.

Shipping

The McClellan-Kerr Arkansas Navigation System provides access from east central Oklahoma to New Orleans through the Verdigris, Arkansas and Mississippi rivers. In 2004, 12·9m. tons were shipped inbound and outbound on the Oklahoma Segment. Commodities shipped are mainly chemical fertilizer, farm produce, petroleum products, iron and steel, coal, sand and gravel.

SOCIAL INSTITUTIONS

Justice

There were 25,864 federal and state prisoners in Dec. 2008. In 2006 there were 17 state correctional facilities, 15 community work centres and seven community correctional centres. The death penalty was suspended in 1966 and reimposed in 1976. There were six executions in 2012 (two in 2011). Oklahoma's total of 103 executions between 1977 and 2012 is the third highest in the USA behind Texas and Virginia.

Education

Public elementary and secondary schools numbered 1,798 in 2007–08. In fall 2007 there were 642,065 pupils and 46,735 teachers at public schools. The average teacher salary per annum was $45,702 in 2008–09. In 2006–07 total expenditure on public elementary and secondary education was $5,230m. There were 59 degree-granting establishments (29 public) in 2008–09; enrolment totalled 206,382 in fall 2007.

Institutions of higher education include:

Founded	Name	Place	2004 enrolment
1890	University of Oklahoma	Norman	31,529
1890	Oklahoma State University	Stillwater	27,419
1890	University of Central Oklahoma	Edmond	18,107
1894	The University of Tulsa	Tulsa	4,629
1897	Langston University	Langston	3,827
1897	Northeastern State University	Tahlequah	11,217
1897	Northwestern Oklahoma State University	Alva	2,731

Founded	Name	Place	2004 enrolment
1897	Southwestern Oklahoma State University	Weatherford	6,352
1908	Cameron University	Lawton	7,917
1909	East Central University	Ada	5,606
1909	Oklahoma Panhandle State University	Goodwell	1,425
1909	Southeastern Oklahoma State University	Durant	5,150
1909	Rogers State College	Claremore	4,896
1950	Oklahoma Christian University of Science and Arts	Oklahoma City	1,725
1969	Rose State College	Midwest City	13,804
1970	Tulsa Community College	Tulsa	26,838
1972	Oklahoma City Community College	Oklahoma City	19,700

Health

In 2009 there were 116 community hospitals with 11,300 beds. A total of 442,000 patients were admitted during the year. In 2009 there were 6,467 active physicians (18 per 10,000 population).

Welfare

Medicare enrolment in July 2010 totalled 597,228. In fiscal year 2009 a total of 808,808 people in Oklahoma received Medicaid. In Dec. 2008 there were 669,673 Old-Age, Survivors, and Disability Insurance (OASDI) beneficiaries. A total of 18,948 people were receiving payments under Temporary Assistance for Needy Families (TANF) in Dec. 2008.

RELIGION

The principal religious traditions in the state in 2010 were: Evangelical Protestants, with 1,531,381 members; Mainline Protestants, 383,786; Catholics, 178,430; Black Protestants, 51,621; Latter-day Saints (Mormons), 46,693.

FURTHER READING

Center for Economic and Management Research, Univ. of Oklahoma, 307 West Brooks St., Norman 73019. *Statistical Abstract of Oklahoma.*
Oklahoma Department of Libraries. *Oklahoma Almanac.* Biennial

Goins, Charles Robert and Goble, Danney, *Historical Atlas of Oklahoma.* 4th ed. 2006

State library: Oklahoma Department of Libraries, 200 Northeast 18th Street, Oklahoma City 73105.

Oregon

KEY HISTORICAL EVENTS

The area was divided between many Indian tribes including the Chinook, Tillamook, Cayuse and Modoc. In the 18th century English and Spanish visitors tried to establish claims, based on explorations of the 16th century. The USA also laid claim by right of discovery when an expedition entered the mouth of the Columbia River in 1792.

Oregon was disputed between Britain and the USA. An American fur trading settlement established at Astoria in 1811 was taken by the British in 1812. The Hudson Bay Company was the most active force in Oregon until the 1830s when American pioneers began to migrate westwards along the Oregon Trail. The dispute between Britain and the USA was resolved in 1846 with the boundary fixed at 49°N. lat. Oregon was organized as a Territory in 1848 but with wider boundaries; it became a state with its present boundaries in 1859.

Early settlers were mainly American. They came to farm in the Willamette Valley and to exploit the western forests. Portland developed as a port for ocean-going traffic, although it was 100 miles inland at the confluence of the Willamette and Columbia rivers. Industries followed when the railways came and the rivers were exploited for hydro-electricity. The capital of the Territory from 1851 was Salem, a mission for Indians on the Willamette river; it was confirmed as state capital in 1864. Salem became the processing centre for the farming and market-gardening of the Willamette Valley.

TERRITORY AND POPULATION

Oregon is bounded in the north by Washington, with the Columbia River forming most of the boundary, east by Idaho, with the Snake River forming most of the boundary, south by Nevada and California and west by the Pacific. Land area, 95,988 sq. miles (248,608 sq. km). In 2010 the federal government owned 53·0% of the state area. Census population, 1 April 2010, was 3,831,074, an increase of 12·0% since 2000. July 2012 estimate, 3,899,353.

Population at five federal censuses was:

	White	Black	American Indian	Asiatic	Total	Per sq. mile
1930	938,598	2,234	4,776	8,179	953,786	9·9
1980	2,490,610	37,060	27,314	34,775	2,633,105	27·3

	White	Black	American Indian/ Alaska Native	All Others	Total	Per sq. mile
1990	2,636,787	46,178	38,496	120,860	2,842,321	29·6
2000	2,961,623	55,662	45,211	358,903	3,421,399	35·6
2010	3,204,614	69,206	53,203	504,051	3,831,074	39·9

Of the total population in 2010, 1,935,072 were female and 2,964,621 were 18 years old or older. In 2010 the Hispanic population was 450,062, up from 275,314 in 2000 (an increase of 63·5%).

In 2000 American Indian lands covered 1,377 sq. miles (1,342 sq. miles in reservations and 35 sq. miles in off-reservation trust land).

The largest cities (2010 census figures) are: Portland, 583,776; Eugene, 156,185; Salem (the capital), 154,637; Gresham, 105,594; Hillsboro, 91,611; Beaverton, 89,803; Bend, 76,639; Medford, 74,907; Springfield, 59,403; Corvallis, 54,462; Albany, 50,158. Primary statistical (metropolitan) areas: Portland–Vancouver–Hillsboro, 2,226,009; Salem, 390,738; Eugene-Springfield, 351,715.

SOCIAL STATISTICS

Births, 2010, 45,540 (11·9 per 1,000 population); deaths, 2010 (provisional), 31,886 (8·3). Infant mortality rate, 2009, 4·8 per 1,000 live births. 2009 (provisional): marriages, 23,500 (6·6 per 1,000 population); divorces and annulments, 13,300 (3·9).

CLIMATE

Jan. 32°F (0°C), July 66°F (19°C). Annual rainfall 28" (710 mm). Oregon belongs to the Pacific coast climate zone (*see* UNITED STATES: Climate).

CONSTITUTION AND GOVERNMENT

The present constitution dates from 1859; some 250 items in it have been amended. The Legislative Assembly consists of a Senate of 30 members, elected for four years (half their number retiring every two years), and a House of 60 representatives, elected for two years. The Governor is elected for four years. The constitution reserves to the voters the rights of initiative and referendum and recall.

For the 113th Congress, which convened in Jan. 2013, Oregon sends five members to the House of Representatives. It is represented in the Senate by Ron Wyden (D. 1996–2017) and Jeff Merkley (D. 2009–15).

The capital is Salem. There are 36 counties in the state.

RECENT ELECTIONS

In the 2012 presidential election Barack Obama took Oregon with 54·5% of the vote, having also won it in 2008.

CURRENT GOVERNMENT

Governor: John Kitzhaber (D.), 2011–15 (salary: $93,600).
   Secretary of State: Kate Brown (D.), 2009–13 ($72,000).

Government Website: http://www.oregon.gov

ECONOMY

Per capita personal income (2010) was $36,427.

Budget

In 2011 total state revenue was $34,991m. Total expenditure was $27,335m. (including: education, $7,102m.; public welfare, $6,027m.; highways, $1,672m.; hospitals, $1,629m.; government administration, $786m.) Outstanding debt in 2011, $14,069m.

Performance

Gross Domestic Product by state was $186,228m. in 2011 (provisional), ranking Oregon 25th in the United States. In 2011 state real GDP growth was 4·7% (provisional), the second highest rate of any state after North Dakota.

ENERGY AND NATURAL RESOURCES

Water

The total area covered by water is 2,391 sq. miles.

Minerals

Mineral resources include gold, silver, lead, mercury, chromite, sand and gravel, stone, clays, lime, silica, diatomite, expansible shale, scoria, pumice and uranium. There is geothermal potential. The total value of non-fuel mineral production in 2009 was $314m.

Agriculture

Oregon, which has an area of 61,557,184 acres, is divided by the Cascade Range into two distinct climate zones. West of the Cascade Range there is a good rainfall and almost every variety of crop common to the temperate zone is grown; east of the Range stock-raising and wheat-growing are the principal industries and irrigation is needed for row crops and fruits.

There were, in 2008, 38,600 farms with an acreage of 16·4m. and an average farm size of 425 acres; most are family-owned corporate farms. Average value per acre (2007), $1,890.

Farm income in 2006: from crops, $2,961m.; from livestock and products, $1,030m. The net farm income in 2006 was $876m. Principal crops in 2008: greenhouse and nursery products (estimate, $880·1m.), hay ($613·3m.), grass seed (estimate, $510·3m.), wheat ($340·2m.), potatoes ($211·0m.), Christmas trees (estimate, $122·8m.), onions ($97·5m.), pears ($92·5m.).

Livestock, 2009: cattle and calves, 1·24m. (including 115,000 dairy cows); sheep and lambs, 220,000; goats, 38,000; hogs, 20,000 (2008).

Forestry

Oregon had 30,169,000 acres of forest in 2007 (the second largest area in the USA after that of California), with 14,012,000 acres of national forest. In 2006 ownership was as follows (acres): US Forestry Service, 12·5m.; US Bureau of Land Management, 2·4m.; other federal, 338,000; State of Oregon, 969,000; tribal, 622,000; local government, 287,000; industrial private, 5·3m.; non-industrial private, 5·0m. Oregon's commercial forest lands provided an estimated 2007 harvest of 3·8bn. bd ft of logs, as well as the benefits of recreation, water, grazing, wildlife and fish. Trees vary from the coastal forest of hemlock and spruce to the state's primary species, Douglas-fir, throughout much of western Oregon. In eastern Oregon, ponderosa pine, lodgepole pine and true firs are found. Here, forestry is often combined with livestock grazing to provide an economic operation. Along the Cascade summit and in the mountains of northeast Oregon, alpine species are found.

Total covered payroll in the forestry and logging industry in 2008 was $515·2m.

Fisheries

Commercial fish and shellfish landings in 2008 was 260·3m. lb and amounted to a value of $126·5m. The most important are: ground fish, crab, shrimp, tuna, whiting and salmon.

INDUSTRY

Forest products manufacturing is Oregon's leading industry, followed by high technology. In 2007 the state's 5,717 manufacturing establishments had 184,000 employees, earning $8,139m. Total value added by manufacturing in 2009 was $29,679m.

Labour

Total non-agricultural employment, 2010, 1,600,000. Employees by branch, 2010 (in 1,000): trade, transportation and utilities, 308; government, 300; education and health services, 228; professional and business services, 181; manufacturing, 164. The unemployment rate was 10·6% in Dec. 2010.

INTERNATIONAL TRADE

Imports and Exports

In 2010 exports from Oregon totalled $17,671m., up from $14,907m. in 2009 although down from $19,352m. in 2008. The leading destinations for exports in 2010 were China (excluding Hong Kong) (value of $4,047m.), followed by Malaysia, Canada and Japan.

COMMUNICATIONS

Roads

In 2007 there were 59,758 miles of road (46,975 miles rural). There were 3,088,313 registered vehicles in 2007.

Rail

In 2008 there were 2,388 total miles of active track and 22 federally franchised freight railroads. There is a light rail network in Portland.

Civil Aviation

There were six primary airports in 2011 (commercial service airports with more than 10,000 passenger boardings annually) with a combined total of 7,772,049 enplanements, up from 7,533,325 in 2010. The busiest airport is Portland International, with 6,808,486 enplanements in 2011.

Shipping

Portland is a major seaport for large ocean-going vessels and is 101 miles inland from the mouth of the Columbia River. In 2009 Portland handled 21,001,697 tons of cargo; Coos Bay, the second busiest port, handled 1,328,340 tons of cargo in 2009. Portland was the 31st busiest US port overall in 2009 but the 13th busiest for exports.

SOCIAL INSTITUTIONS

Justice

There are 14 correctional institutions in Oregon. In Dec. 2008 there were 14,167 federal and state prisoners. The sterilization law, originally passed in 1917, was amended in 1967 and abolished in 1993. Some categories of euthanasia were legalized in Dec. 1994.

The death penalty is authorized but there have been no executions since 2001.

Education

School attendance is compulsory from seven to 18 years of age if the twelfth year of school has not been completed; those between the ages of 16 and 18 years, if legally employed, may attend part-time or evening schools. Others may be excused under certain circumstances. In fall 2011 the 1,355 public elementary and secondary schools had 561,698 students and 30,157 teachers; average salary for teachers (2010–11), $56,203. The State's share of the K-12 education budget was $5·75bn. for the two years ending 30 June 2011. For the same two years the Federal Government is estimated to have provided $1·5bn. for Oregon K-12 programmes.

Leading state-supported institutions of higher education (fall 2012) included:

	Students
Portland State University, Portland	28,731
Oregon State University, Corvallis	26,393
University of Oregon, Eugene	24,593
Southern Oregon State College, Ashland	6,481
Western Oregon State College, Monmouth	6,187
Eastern Oregon State College, La Grande	4,208
Oregon Institute of Technology, Klamath Falls	4,001
Oregon Health and Science University, Portland	3,929

There were 26 public and 34 private degree-granting institutions in 2011–12; enrolment in all institutions totalled 202,928 in fall 2007. Largest of the privately endowed universities are University of Portland, with 3,887 students in fall 2012; Lewis and Clark College, Portland, with 3,712 students; George Fox College, Newberg, 3,519 students; Willamette University, Salem, 2,800 students; Linfield College, McMinnville, 2,664 students; Marylhurst College, 1,797 students; Reed College, Portland, 1,474 students.

Health
In 2009 there were 58 community hospitals with 6,500 beds. A total of 324,000 patients were admitted during the year. In 2009 there were 10,753 active physicians (28 per 10,000 population).

Welfare
Medicare enrolment in July 2010 totalled 615,798. In fiscal year 2009 a total of 564,307 people in Oregon received Medicaid. In Dec. 2008 there were 659,719 Old-Age, Survivors, and Disability Insurance (OASDI) beneficiaries. A total of 46,376 people were receiving payments under Temporary Assistance for Needy Families (TANF) in Dec. 2008.

RELIGION
The principal religious traditions in the state in 2010 were: Evangelical Protestants, with 447,009 members; Catholics, 399,440; Latter-day Saints (Mormons), 150,097; Mainline Protestants, 140,248; Buddhists, 14,785.

FURTHER READING
Oregon Blue Book. Biennial

Friedman, R., *The Other Side of Oregon.* 1993
McArthur, L. A., *Oregon Geographic Names.* 7th ed. 2003
Orr, E. L., *et al., Geology of Oregon.* 1992

State library: The Oregon State Library, 250 Winter St. NE, Salem 97301–3950.

Pennsylvania

KEY HISTORICAL EVENTS
Pennsylvania was occupied by four powerful tribes in the 17th century: Delaware, Susquehannock, Shawnee and Iroquois. The first white settlers were Swedish, arriving in 1643. The British became dominant in 1664 and in 1681 William Penn, an English Quaker, was given a charter to colonize the area as a sanctuary for his fellow Quakers. Penn's ideal was peaceful co-operation with the Indians and religious toleration within the colony. Several religious groups were attracted to Pennsylvania, including Protestant sects from Germany and France. In the 18th century, co-operation with the Indians failed as the settlers extended their territory.

The Declaration of Independence was signed in Philadelphia while Pennsylvania became one of the original 13 states of the Union. In 1812 the state capital was moved to its current location in Harrisburg, originally a trading post and ferry point on the Susquehanna River in the south-central part of the state. The Mason-Dixon line, the state's southern boundary, was the dividing line between free and slave states in the build-up to the Civil War. Gettysburg and other crucial battles were fought in the state. Industrial growth was rapid after the war. Pittsburgh, founded as a British fort in 1761 during war with the French, had become an iron-making town by 1800 and grew rapidly when canal and railway links opened in the 1830s. The American Federation of Labor was founded in Pittsburgh in 1881, by which time the city was of national importance producing coal, iron, steel and glass.

At the beginning of the 20th century, industry attracted immigration from Italy and eastern Europe. In farming areas the early sect communities survive, notably Amish and Mennonites. (The Pennsylvania 'Dutch' are of German extraction.)

TERRITORY AND POPULATION
Pennsylvania is bounded north by New York, east by New Jersey, south by Delaware and Maryland, southwest by West Virginia, west by Ohio and northwest by Lake Erie. Land area, 44,743 sq. miles (115,883 sq. km). Census population, 1 April 2010, was 12,702,379, an increase of 3·4% since 2000. July 2012 estimate, 12,763,536.

Population at five census years was:

	White	Black	American Indian	All others	Total	Per sq. mile
1910	7,467,713	193,919	1,503	1,976	7,665,111	171·0
			All others			
1980	10,652,320	1,046,810	164,765		11,863,895	264·7
1990	10,520,201	1,089,795	271,647		11,881,643	265·1
2000	10,484,203	1,224,612	572,239		12,281,054	274·0
2010	10,406,288	1,377,689	918,402		12,702,379	283·9

Of the total population in 2010, 6,512,016 were female and 9,910,224 were 18 years old or older. In 2010 Pennsylvania's Hispanic population was 719,660, up from 394,088 in 2000 (a rise of 82·6%).

The population of the largest cities and townships, 2010 census, was:

Philadelphia	1,526,006	Scranton	76,089
Pittsburgh	305,704	Bethlehem	74,982
Allentown	118,032	Lancaster	59,322
Erie	101,786	Levittown	52,983
Reading	88,082		

The Philadelphia–Camden–Wilmington metropolitan area had a 2010 census population of 5,965,343. The capital, Harrisburg, had a population in 2010 of 49,528.

SOCIAL STATISTICS
Births, 2010, 143,321 (11·3 per 1,000 population); deaths, 2010 (provisional), 124,599 (9·8). Infant mortality, 2009, 7·2 per 1,000 live births. 2009 (provisional): marriages, 64,200 (5·3 per 1,000 population); divorces and annulments, 28,800 (2·7).

CLIMATE
Philadelphia, Jan. 32°F (0°C), July 77°F (25°C). Annual rainfall 40" (1,006 mm). Pittsburgh, Jan. 31°F (−0·6°C), July 74°F (23·3°C).

Annual rainfall 37" (914 mm). Pennsylvania belongs to the Appalachian Mountains climate zone (*see* UNITED STATES: Climate).

CONSTITUTION AND GOVERNMENT

The present constitution dates from 1968. The General Assembly consists of a Senate of 50 members chosen for four years, one-half being elected biennially, and a House of Representatives of 203 members chosen for two years. The Governor and Lieut.-Governor are elected for four years. Every citizen 18 years of age, with the usual residential qualifications, may vote. Registered voters in Nov. 2004, 8,366,663.

For the 113th Congress, which convened in Jan. 2013, Pennsylvania sends 18 members to the House of Representatives. It is represented in the Senate by Robert Casey, Jr (D. 2007–19) and Patrick Toomey (R. 2011–17).

The state capital is Harrisburg. The state is organized in counties (numbering 67), cities, boroughs, townships and school districts.

RECENT ELECTIONS

In the 2012 presidential election Barack Obama took Pennsylvania with 52·0% of the vote, having also won it in 2008.

CURRENT GOVERNMENT

Governor: Tom Corbett (R.), 2011–15 (salary: $183,255).
    Lieut.-Governor: Jim Cawley (R.), 2011–15 ($153,907).
    Secretary of the Commonwealth: Carol Aichele (R.), appointed Jan. 2011 ($131,922).

Government Website: http://www.pa.gov

ECONOMY

Per capita personal income (2010) was $40,599.

Budget

In 2011 total state revenue was $91,705m. Total expenditure was $90,792m. (including: public welfare, $23,707m.; education, $22,849m.; highways, $7,790m.; hospitals, $3,531m.; government administration, $2,957m.) Outstanding debt in 2011, $45,267m.

Performance

Gross Domestic Product by state in 2011 was $500,443m. (provisional), ranking Pennsylvania 6th in the United States. In 2011 state real GDP growth was 1·2% (provisional).

ENERGY AND NATURAL RESOURCES

Electricity

In 2009 production was 219·5bn. kWh, of which 105·5bn. kWh was from coal.

Oil and Gas

In 2008 Pennsylvania produced 4m. bbls of crude petroleum; marketed production of natural gas totalled 198bn. cu. ft.

Water

The total area covered by water is 1,312 sq. miles.

Minerals

Pennsylvania is almost the sole producer of anthracite coal. Production, 2007: crushed stone, 122m. tons; construction sand and gravel, 20m. tons. Bituminous coal production in 2009 totalled 56,248,000 tons; anthracite coal production in 2009, 1,731,000 tons. Non-fuel mineral production was worth more than $1,620m. in 2009.

Agriculture

Agriculture, market-gardening, fruit-growing, horticulture and forestry are pursued within the state. In 2007 there were 63,163 farms with a total farm area of 7·81m. acres. Average number of acres per farm in 2007 was 124 and the average value per acre was $4,775. Cash receipts, 2007: from crops, $1,929m.; and from livestock and products, $3,831m. The net farm income in 2007 was $2,155m.

In 2007–08 Pennsylvania ranked first in the USA for the production of mushrooms (496·7m. lb, value in 2007 $486·7m.). Cash receipts for other leading commodities in 2007: dairy products, $2,219·2m.; cattle and calves, $462·3m.; chicken eggs, $389·1m.; broilers, $381·0m.; corn, $380·5m.

In 2007 there were on farms: 1·61m. cattle and calves, 96,900 sheep and lambs, 1·17m. hogs and pigs and 21·98m. laying hens.

Forestry

In 2007 the total forested area was 16,577,000 acres, of which 497,000 acres were national forest. In 2009 state forest land totalled 2,145,804 acres; state park land, 274,105 acres; state game land, 1,400,000 acres.

INDUSTRY

In 2007 the state's 15,406 manufacturing establishments had 651,000 employees, earning $29,433m. Total value added by manufacturing in 2009 was $92,781m.

Labour

Total non-agricultural employment, 2010, 5,616,000. Employees by branch, 2010 (in 1,000): education and health services, 1,136; trade, transportation and utilities, 1,080; government, 757; professional and business services, 685; manufacturing, 561. The unemployment rate in Dec. 2010 was 8·5%.

INTERNATIONAL TRADE

Imports and Exports

In 2010 exports from Pennsylvania totalled $34,928m., up from $28,381m. in 2009. The leading destinations for exports in 2010 were Canada (value of $10,287m.), followed by China (excluding Hong Kong), Mexico and Japan.

COMMUNICATIONS

Roads

In 2007 highways and roads in the state (federal, local and state combined) totalled 121,581 miles (76,451 miles rural). Registered motor vehicles numbered 9,937,941.

Rail

In 2009 there were 4,973 miles of freight railroad (excluding trackage rights), including 2,326 miles of Class I railroads. There were 55 freight railroads operating in 2009 (the most in any state). Rail traffic originating in Pennsylvania in 2009 totalled 48·9m. tons and rail traffic terminating in the state came to 56·3m. tons. There are metro, light rail and tramway networks in Philadelphia and Pittsburgh, and commuter networks around Philadelphia.

Civil Aviation

In 2011 there were nine primary airports (commercial service airports with more than 10,000 passenger boardings annually) with a combined total of 20,584,069 enplanements, up from 20,540,939 in 2010. By far the busiest airport is Philadelphia International, with 14,883,180 enplanements in 2011.

Shipping

The major ports are Pittsburgh (which handled 32·9m. tons of cargo in 2009), Philadelphia (31·8m. tons in 2009), Marcus Hook, Penn Manor and Chester.

SOCIAL INSTITUTIONS

Justice

The death penalty is authorized. The last execution was in 1999, but there were 211 death row inmates as at 1 Jan. 2012. There were 50,147 prisoners in state correctional institutions in Dec.

2008, up 9·1% from 45,969 in Dec. 2007—the largest rise in any state over the same period.

Education

School attendance is compulsory for children eight to 17 years of age. In 2007–08 there were 3,246 public elementary and secondary schools with 1,801,971 pupils and 135,234 teachers; total expenditure on public elementary and secondary education in 2006–07 was $24,460m. Teachers' salaries averaged $56,906 in 2008–09.

Leading senior academic institutions include:

Founded	Institutions	Faculty[1] (fall 2008)	Students (fall 2009)
1740	University of Pennsylvania (non-sect.)	6,288	24,599
1787	University of Pittsburgh (all campuses)	7,428	35,394
1832	Lafayette College, Easton (Presbyterian)	247	2,406
1833	Haverford College	158	1,190
1842	Villanova University (R.C.)	1,245	10,375
1846	Bucknell University (Baptist)	425	3,737
1851	St Joseph's University, Philadelphia (R.C.)	604	8,337
1852	California University of Pennsylvania	390	9,017
1855	Pennsylvania State University (all campuses)	5,944[2]	87,309[3]
1855	Millersville University of Pennsylvania	562	8,427
1863	La Salle University, Philadelphia (R.C.)	516	6,470
1864	Swarthmore College	232	1,525
1866	Lehigh University, Bethlehem (non-sect.)	1,160	6,996
1871	West Chester University of Pennsylvania	1,009	14,211
1875	Indiana University of Pennsylvania	1,094	14,638
1878	Duquesne University, Pittsburgh (R.C.)	1,414	10,294
1884	Temple University, Philadelphia	4,282	36,507
1885	Bryn Mawr College	271	1,771
1888	University of Scranton (R.C.)	573	5,811
1891	Drexel University, Philadelphia	2,943	22,493
1900	Carnegie-Mellon University, Pittsburgh	3,350	11,197

[1]Instructional faculty only. [2]Fall 2009. [3]Fall 2010.

Health

In 2009 there were 194 community hospitals with 39,200 beds. A total of 1,842,000 patients were admitted during the year. In 2009 there were 38,676 active physicians (31 per 10,000 population).

Welfare

Medicare enrolment in July 2010 totalled 2,258,365. In fiscal year 2008 a total of 2,134,331 people in Pennsylvania received Medicaid. In Dec. 2008 there were 2,481,695 Old-Age, Survivors, and Disability Insurance (OASDI) beneficiaries. A total of 112,592 people were receiving payments under Temporary Assistance for Needy Families (TANF) in Dec. 2008.

RELIGION

The principal religious traditions in 2010 were Catholics (3,503,028 members), Mainline Protestants (1,773,491 members), Evangelical Protestants (1,078,477 members), Black Protestants (114,337 members) and Jews (102,312 members).

CULTURE

Tourism

In 2011 overseas visitors to Pennsylvania—excluding those from Canada and Mexico—numbered 920,000, down from 923,000 in 2010 but up from 672,000 in 2006.

FURTHER READING

Statistical information: Pennsylvania State Data Center, 777 West Harrisburg Pike, Middletown 17057. Publishes *Pennsylvania Statistical Abstract.*

Downey, D. B. and Bremer, F. (eds.) *Guide to the History of Pennsylvania.* 1994

Rhode Island

KEY HISTORICAL EVENTS

The earliest white settlement was founded by Roger Williams, an English Puritan who was expelled from Massachusetts because of his dissident religious views and his insistence on the land-rights of the Indians. At Providence he bought land from the Narragansetts and founded a colony there in 1636. A charter was granted in 1663. Religious toleration attracted Jewish and nonconformist settlers; later there was French Canadian settlement.

Shipping and fishing developed strongly, especially at Newport and Providence. These two cities were twin capitals until 1900, when the capital was fixed at Providence.

Significant actions took place in Rhode Island during the War of Independence. In 1790 the state accepted the federal constitution and was admitted to the Union.

Early farming development was most successful in dairying and poultry. Early industrialization from the 1790s was mainly in textiles. Thriving on abundant water power, the industry began to decline after the First World War. British, Irish, Polish, Italian and Portuguese workers settled in the state, working in the mills or in the shipbuilding, shipping, fishing and naval ports. The growth of the cities led to the abolition of the property qualification for the franchise in 1888.

TERRITORY AND POPULATION

Rhode Island is bounded north and east by Massachusetts, south by the Atlantic and west by Connecticut. Land area, 1,034 sq. miles (2,678 sq. km). Census population, 1 April 2010, was 1,052,567, an increase of 0·4% since 2000. July 2012 estimate, 1,050,292. Population of five census years was:

	White	Black	American Indian/ AlaskaNative	Asian	Total	Per sq. mile
1910	532,492	9,529	284	305	542,610	508·5
			All others			
1980	896,692	27,584	22,878		947,154	903·0
					Total (including others)	
1990	917,375	38,861	4,071	18,325	1,003,164	960·3
2000	891,191	46,908	5,121	24,232	1,048,319	1,003·2
2010	856,869	60,189	6,058	30,457	1,052,567	1,018·1

Of the total population in 2010, 544,167 were female and 828,611 were 18 years old or older. In 2010 the Hispanic population was 130,655, up from 90,820 in 2000 (an increase of 43·9%).

The chief cities and their population (census, 2010) are Providence (the capital), 178,042; Warwick, 82,672; Cranston, 80,387; Pawtucket, 71,148; East Providence, 47,037.

SOCIAL STATISTICS

Births, 2010, 11,177 (10·6 per 1,000 population); deaths, 2010 (provisional), 9,581 (9·1). Infant mortality rate, 2009, 6·2 per 1,000 live births. 2009 (provisional): marriages, 6,500 (5·9 per 1,000 population); divorces and annulments, 3,300 (3·0).

CLIMATE

Providence, Jan. 28°F (–2·2°C), July 72°F (22·2°C). Annual rainfall 43" (1,079 mm). Rhode Island belongs to the New England climate zone (*see* UNITED STATES: Climate).

CONSTITUTION AND GOVERNMENT

The present constitution dates from 1843; it has had 62 amendments. The General Assembly consists of a Senate of 38 members and a House of Representatives of 75 members, both elected for two years. The Governor and Lieut.-Governor are now elected for four years. Every citizen, 18 years of age, who has resided in the state for 30 days, and is duly registered, is qualified to vote.

For the 113th Congress, which convened in Jan. 2013, Rhode Island sends two members to the House of Representatives. It is represented in the Senate by Jack Reed (D. 1997–2015) and Sheldon Whitehouse (D. 2007–19).

The capital is Providence. The state has five counties but no county governments. There are 39 municipalities, each having its own form of local government.

RECENT ELECTIONS

In the 2012 presidential election Barack Obama took Rhode Island with 62·7% of the vote, having also won it in 2008.

CURRENT GOVERNMENT

Governor: Lincoln Chafee (ind.), 2011–15 (salary: $129,210).
  Lieut.-Governor: Elizabeth H. Roberts (D.), 2011–15 ($108,808).
  Secretary of State: A. Ralph Mollis (D.), 2011–15 ($108,808).

Government Website: http://www.ri.gov

ECONOMY

Per capita personal income (2010) was $42,095.

Budget

In 2011 total state revenue was $9,372m. Total expenditure was $8,271m. (including: public welfare, $2,403m.; education, $1,815m.; government administration, $321m.; highways, $255m.; correction, $182m.) Outstanding debt in 2011, $9,174m.

Performance

Gross Domestic Product in 2011 was $43,663m. (provisional), ranking Rhode Island 45th in the United States. In 2011 state real GDP growth was 0·8% (provisional).

ENERGY AND NATURAL RESOURCES

Water

The total area covered by water is 511 sq. miles.

Minerals

The small non-fuel mineral output—mostly stone, sand and gravel—was valued at more than $40m. in 2009.

Agriculture

In 2007 there were 1,219 farms with an area of 68,000 acres. The average size of a farm was 56 acres. In 2007 the average value of land and buildings per acre was $16,828—making it the most valuable of any state in the USA. Farm income 2006: from crops, $56m.; livestock and products, $10m. The net farm income in 2006 was $26m. Principal commodities are greenhouse products, dairy products, sweetcorn and aquaculture.

Forestry

Total forested area was 356,000 acres in 2007.

Fisheries

In 2009 the commercial catch was 84·5m. lb valued at $61·7m.

INDUSTRY

Manufacturing is the chief source of income and the largest employer. Principal industries are jewellery and silverware, electrical machinery, electronics, plastics, metal products, instruments, chemicals and boat building. In 2007 the state's 1,831 manufacturing establishments had 54,000 employees, earning $2,375m. Total value added by manufacturing in 2009 was $4,632m.

Labour

In 2007 total non-agricultural employment was 459,000. Employees by branch, 2010 (in 1,000): education and health services, 102; trade, transportation and utilities, 73; government, 62; professional and business services, 53; leisure and hospitality, 50; manufacturing, 40. The unemployment rate in Dec. 2010 was 11·5%.

INTERNATIONAL TRADE

Imports and Exports

In 2010 exports from Rhode Island totalled $1,949m., up from $1,496m. in 2009 although down from $1,974m. in 2008. The leading destinations for exports in 2010 were Canada (value of $591m.), followed by Mexico, Germany and Turkey.

COMMUNICATIONS

Roads

In 2007 there were 6,510 miles of roads (5,240 miles urban). There were 796,683 registered motor vehicles.

Rail

Amtrak's New York-Boston route runs through the state, serving Providence.

Civil Aviation

In 2011 there were two primary airports (commercial service airports with more than 10,000 passenger boardings annually). The main airport is Theodore Francis Green Memorial State at Warwick, near Providence, which had 1,920,699 enplanements in 2011.

Shipping

The leading port is Providence, which handled 6·9m. tons of cargo in 2009.

SOCIAL INSTITUTIONS

Justice

In Dec. 2008 the jail and prison population totalled 4,045. The last execution was in 1845. Capital punishment was abolished in 1852, but reinstated in 1872 for murder committed by a life prisoner. However, it was never carried out and the death penalty was completely abolished in 1984.

Education

In 2007–08 there were 328 public elementary and secondary schools. In fall 2007, 147,629 pupils attended public schools with 11,271 teachers. In fall 2007, 28,260 pupils were enrolled in 230 private schools which had 2,530 teachers. Total expenditure on public elementary and secondary education in 2006–07 was $2,169m. Spending per pupil in fiscal year 2008 was $14,459.

There were 13 degree-granting institutions (three public and ten private) in 2008–09. The state maintained Rhode Island College, Providence, with 8,871 students in fall 2005; the University of Rhode Island, Kingston, with 15,095; and the Community College of Rhode Island, Warwick, with 16,042. Among the private institutions, Brown University at Providence, founded in 1764, had 8,261 students; Providence College at Providence, founded in 1917, by the Dominican Order of Preachers, had 5,457 students; Bryant University at Smithfield had 3,642 students; Rhode Island School of Design in Providence had 2,258 students; and Johnson and Wales University in Providence had 10,171 students.

Health

In 2009 there were 11 community hospitals with 2,500 beds. A total of 127,000 patients were admitted during the year. In 2009 there were 4,020 active physicians (38 per 10,000 population).

Welfare

Medicare enrolment in July 2010 totalled 181,026. In fiscal year 2009 a total of 203,223 people in Rhode Island received Medicaid.

In Dec. 2008 there were 196,161 Old-Age, Survivors, and Disability Insurance (OASDI) beneficiaries. A total of 18,488 people were receiving payments under Temporary Assistance for Needy Families (TANF) in Dec. 2008.

RELIGION

The principal religious traditions in the state in 2010 were: Catholics, with 466,598 members; Mainline Protestants, 57,103; Evangelical Protestants, 26,242; Jews, 8,845; Orthodox Christians, 7,625.

FURTHER READING

Statistical information: Rhode Island Economic Development Corporation, 1 West Exchange Street, Providence, RI 02903. Publishes *Rhode Island Basic Economic Statistics.*
Rhode Island Manual.

Wright, M. I. and Sullivan, R. J., *Rhode Island Atlas.* 1983

State library: Rhode Island State Library, State House, Providence 02908.

South Carolina

KEY HISTORICAL EVENTS

Originally the territory of Yamasee Indians, the area attracted French and Spanish explorers in the 16th century. There were attempts at settlement on the coast, none of which lasted. Charles I of England made a land grant in 1629 but the first permanent white settlement began at Charles Town in 1670, moving to Charleston in 1680. This was a proprietorial colony including North Carolina until 1712; both passed to the Crown in 1729.

The coastlands developed as plantations worked by slave labour. In the hills there were small farming settlements and many trading posts, dealing with Indian suppliers.

After active campaigns during the War of Independence, South Carolina became one of the original states of the Union in 1788.

In 1793 the cotton gin was invented, enabling the speedy mechanical separation of seed and fibre. This made it possible to grow huge areas of cotton and meet the rapidly-growing needs of new textile industries. Plantation farming spread widely and South Carolina became hostile to the anti-slavery campaign which was strong in northern states. The state first attempted to secede from the Union in 1847, but was not supported by other southern states until 1860, when secession led to civil war.

At that time the population was about 703,000, of whom 413,000 were black. During the reconstruction periods there was some political power for black citizens but control was back in white hands by 1876. The constitution was amended in 1895 to disenfranchise most black voters and they remained with hardly any voice in government until the Civil Rights movement of the 1960s. Columbia became the capital in 1786.

TERRITORY AND POPULATION

South Carolina is bounded in the north by North Carolina, east and southeast by the Atlantic, southwest and west by Georgia. Land area, 30,061 sq. miles (77,857 sq. km). Census population, 1 April 2010, was 4,625,364, an increase of 15·3% since 2000. July 2012 estimate, 4,723,723.

The population in five census years was:

	White	Black	American Indian	Asiatic	Total	Per sq. mile
1910	679,161	835,843	331	65	1,515,400	49·7
			All others			
1980	2,150,507	948,623	22,703		3,121,833	100·3
1990	2,406,974	1,039,884	39,845		3,486,703	115·8
2000	2,695,560	1,185,216	131,236		4,012,012	133·2
2010	3,060,000	1,290,684	274,680		4,625,364	153·9

Of the total population in 2010, 2,375,263 were female and 3,544,890 were 18 years old or older. In 2010 the Hispanic population of South Carolina was 235,682, up from 95,076 in 2000 (an increase of 147·9%, the largest percentage rise in any US state).

Populations of large towns in 2010: Columbia (capital), 129,272; Charleston, 120,083; North Charleston, 97,471; Mount Pleasant, 67,843; Rock Hill, 66,154; Greenville, 58,409.

SOCIAL STATISTICS

Births, 2010, 58,342 (12·6 per 1,000 population); deaths, 2010 (provisional), 41,604 (9·0). Infant deaths, 2009, 7·1 per 1,000 live births. 2009 (provisional): marriages, 29,200 (7·4 per 1,000 population); divorces and annulments, 12,200 (3·0).

CLIMATE

Columbia, Jan. 44·7°F (7°C), Aug. 80·2°F (26·9°C). Annual rainfall 49·12" (1,247·6 mm). South Carolina belongs to the Atlantic Coast climate zone (*see* UNITED STATES: Climate).

CONSTITUTION AND GOVERNMENT

The present constitution dates from 1895, when it went into force without ratification by the electorate. The General Assembly consists of a Senate of 46 members, elected for four years, and a House of Representatives of 124 members, elected for two years. It meets annually. The Governor and Lieut.-Governor are elected for four years.

For the 113th Congress, which convened in Jan. 2013, South Carolina sends seven members to the House of Representatives. It is represented in the Senate by Lindsey Graham (R. 2003–15) and Tim Scott (R. 2013–17). Scott was appointed to fill the seat formerly held by Jim DeMint until a special election scheduled for 4 Nov. 2014. DeMint's term was to end in 2017.

The capital is Columbia. There are 46 counties.

RECENT ELECTIONS

In the 2012 presidential election Mitt Romney took South Carolina with 54·6% of the vote (John McCain won it in 2008).

CURRENT GOVERNMENT

Governor: Nikki Haley (R.), 2011–15 (salary: $106,078).
   Lieut.-Governor: Ken Ard (R.), 2011–15 ($46,545).
   Secretary of State: Mark Hammond (R.), 2011–15 ($92,007).

Government Website: http://www.sc.gov

ECONOMY

Per capita personal income (2010) was $32,460.

Budget

In 2011 total state revenue was $31,733m. Total expenditure was $29,352m. (including: education, $8,081m.; public welfare, $6,831m.; hospitals, $1,478m.; highways, $1,171m.; health, $952m.) Outstanding debt in 2011, $15,341m.

Performance

Gross Domestic Product by state was $143,278m. in 2011 (provisional), ranking South Carolina 27th in the United States. In 2011 state real GDP growth was 1·2% (provisional).

ENERGY AND NATURAL RESOURCES

Electricity

In 2009 production was 100·1bn. kWh, of which 52·1bn. kWh was from nuclear energy.

Water

The total area covered by water is 1,960 sq. miles.

Minerals

Gold is found, though non-metallic minerals are of chief importance: value of non-fuel mineral output in 2009 was more than \$440m., chiefly from cement (Portland), stone and gold. Production of kaolin, vermiculite and scrap mica is also important.

Agriculture
In 2007 there were 25,867 farms covering a farm area of 4·89m. acres. The average farm was of 189 acres. The average value of farmland and buildings per acre was \$2,858 in 2007.

Farm income, 2007: from crops, \$785m.; from livestock and products, \$1,242m. The net farm income in 2007 was \$535m. Cash receipts for leading commodities in 2007: broilers, \$666·0m.; greenhouse products, \$272·9m.; turkeys, \$198·5m.; cattle and calves, \$133·4m.; corn, \$105·8m.

Livestock on farms, 2007: 401,000 all cattle, 293,800 hogs and pigs, 8,000 sheep and lambs, and 4·71m. laying hens.

Forestry
The forest industry is important; total forest land (2007), 12·75m. acres. National forests amounted to 641,000 acres.

INDUSTRY
In 2007 the state's 4,335 manufacturing establishments had 242,000 employees, earning \$10,061m. Total value added by manufacturing in 2009 was \$31,477m.

Labour
Total non-agricultural employment, 2010, 1,805,000. Employees by branch, 2010 (in 1,000): government, 345; trade, transportation and utilities, 344; professional and business services, 214; education and health services, 213; manufacturing, 207. The unemployment rate in Dec. 2010 was 10·9%.

INTERNATIONAL TRADE
Imports and Exports
In 2010 exports from South Carolina totalled \$20,329m., up from \$16,488m. in 2009. The leading destinations for exports in 2010 were Canada (value of \$3,187m.), followed by Germany, China (excluding Hong Kong) and Mexico.

COMMUNICATIONS
Roads
In 2007 there were 66,248 miles of road comprising 16,422 miles of urban road and 49,826 miles of rural road. There were 3,521,026 registered motor vehicles.

Rail
In 2009 there were 2,292 miles of freight railroad (excluding trackage rights). There were 14 freight railroads operating in 2009. Rail traffic originating in South Carolina totalled 9·8m. tons in 2009 and rail traffic terminating in the state came to 33·7m. tons.

Civil Aviation
There were six primary airports—commercial service airports with more than 10,000 passenger boardings annually—in 2011 with a combined total of 3,593,332 enplanements, up from 3,082,283 in 2010. The main airport is Charleston International, which had 1,247,459 enplanements in 2011.

Shipping
The state has three deep-water ports.

SOCIAL INSTITUTIONS
Justice
In Dec. 2008 there were 24,326 federal and state prisoners. The death penalty is authorized. There was one execution in 2011 but none in 2012.

Education
In fall 2007 there were 712,317 pupils and 47,382 teachers in 1,195 public schools. Total expenditure on public elementary and secondary education in 2006–07 was \$7,677m.; average teaching

salary was \$47,704 in 2008–09. Spending per pupil in fiscal year 2008 was \$9,060. In fall 2007 there were 410 private schools with total enrolment of 71,430 pupils and 5,550 teachers.

For higher education the state operates the University of South Carolina (USC), founded at Columbia in 1801, with (fall 2005) 27,065 enrolled students; USC Aiken, with 3,303 students; USC Spartanburg, with 4,484 students; Clemson University, founded in 1889, with 17,165 students; Citadel Military College at Charleston with 3,386 students; Winthrop University, Rock Hill, with 6,480 students; Medical University of South Carolina, at Charleston, with 2,499 students; South Carolina State University, at Orangeburg, with 4,446 students; Francis Marion University, at Florence, with 4,008 students; the College of Charleston with 11,332 students; and Lander University, at Greenwood, with 2,703 students.

Health
In 2009 there were 70 community hospitals with 12,500 beds. A total of 528,000 patients were admitted during the year. In 2009 there were 10,403 active physicians (23 per 10,000 population).

Welfare
Medicare enrolment in July 2010 totalled 766,324. In fiscal year 2009 a total of 906,277 people in South Carolina received Medicaid. In Dec. 2008 there were 850,368 Old-Age, Survivors, and Disability Insurance (OASDI) beneficiaries. A total of 39,081 people were receiving payments under Temporary Assistance for Needy Families (TANF) in Dec. 2008.

RELIGION
The principal religious traditions in 2010 were: Evangelical Protestants, with 1,410,988 members; Mainline Protestants, 482,103; Black Protestants, 256,178; Catholics, 181,743; Latter-day Saints (Mormons), 37,863.

FURTHER READING
Statistical information: Budget and Control Board, R. C. Dennis Bldg, Columbia 29201. Publishes *South Carolina Statistical Abstract.* *South Carolina Legislative Manual.* Annual

Edgar, W. B., *South Carolina in the Modern Age.* 1992
Graham, C. B. and Moore, W. V., *South Carolina Politics and Government.* 1995

State library: South Carolina State Library, Columbia.

South Dakota

KEY HISTORICAL EVENTS
The area was part of the hunting grounds of nomadic Dakota (Sioux) Indians. French explorers visited the site of Fort Pierre in 1742–43 and claimed the area for France. In 1763 the claim, together with French claims to all land west of the Mississippi, passed to Spain. Spain held the Dakotas until defeated by France in the Napoleonic Wars when France regained the area and sold it to the USA as part of the Louisiana Purchase in 1803.

Fur-traders were active but there was no settlement until Fort Randall was founded on the Missouri river in 1856. In 1861 North and South Dakota were united as the Dakota Territory. The Homestead Act of 1862 stimulated settlement, mainly in the southeast until there was a gold-rush in the Black Hills of the west in 1875–76. Colonization was by farming communities in the east with miners and ranchers in the west. Livestock farming predominated, attracting European settlers from Scandinavia, Germany and Russia.

In 1889 the North and South were separated and admitted to the Union as states. Pierre, founded as a railhead in 1880, was confirmed as the capital in 1904. It faces Fort Pierre, the former centre of the fur trade, across the Missouri river. During the 20th century there have been schemes to exploit the Missouri for power and irrigation.

TERRITORY AND POPULATION

South Dakota is bounded in the north by North Dakota, east by Minnesota, southeast by the Big Sioux River (forming the boundary with Iowa), south by Nebraska (with the Missouri River forming part of the boundary) and west by Wyoming and Montana. Land area, 75,811 sq. miles (196,350 sq. km). In 2000 American Indian lands covered 14,966 sq. miles (14,061 sq. miles in reservations and 905 sq. miles in off-reservation trust land).

Census population, 1 April 2010, was 814,180, an increase of 7·9% since 2000. July 2012 estimate, 833,354.

Population in five federal censuses was:

	White	Black	American Indian	Asiatic	Total	Per sq. mile
1910	563,771	817	19,137	163	583,888	7·6
			All others			
1980	638,955	2,144	49,079		690,178	9·0
			American Indian/Alaska Native	Asian/other		
1990	637,515	3,258	50,575	4,656	696,004	9·2
2000	669,404	4,685	62,283	18,472	754,844	9·9
2010	699,392	10,207	71,817	32,764	814,180	10·7

Of the total population in 2010, 407,381 were male and 611,383 were 18 years old or older. In 2010 the Hispanic population was 22,119, up from 10,903 in 2000 (an increase of 102·9%).

Population of the chief cities (census of 2010) was: Sioux Falls, 153,888; Rapid City, 67,956; Aberdeen, 26,091; Brookings, 22,056; Watertown, 21,482; Mitchell, 15,254; Pierre (the capital), 13,646.

SOCIAL STATISTICS

Births, 2010, 11,811 (14·5 per 1,000 population); deaths, 2010 (provisional), 7,100 (8·7). Infant deaths, 2009, 6·7 per 1,000 live births. 2009 (provisional): marriages, 5,900 (7·2 per 1,000 population); divorces and annulments, 2,600 (3·3).

CLIMATE

Rapid City, Jan. 25°F (–3·9°C), July 73°F (22·8°C). Annual rainfall 19" (474 mm). Sioux Falls, Jan. 14°F (–10°C), July 73°F (22·8°C). Annual rainfall 25" (625 mm). South Dakota belongs to the High Plains climate zone (see UNITED STATES: Climate).

CONSTITUTION AND GOVERNMENT

Voters are all citizens 18 years of age or older. The people reserve the right of the initiative and referendum. The Senate has 35 members, and the House of Representatives 70 members, all elected for two years; the Governor and Lieut.-Governor are elected for four years.

For the 113th Congress, which convened in Jan. 2013, South Dakota sends one member to the House of Representatives. It is represented in the Senate by Tim Johnson (D. 1997–2015) and John Thune (R. 2005–17).

The capital is Pierre. The state is divided into 66 organized counties.

RECENT ELECTIONS

In the 2012 presidential election Mitt Romney took South Dakota with 57·9% of the vote (John McCain won it in 2008).

CURRENT GOVERNMENT

Governor: Dennis Daugaard (R.), 2011–15 (salary: $100,972).
Lieut.-Governor: Matt Michels (R.), 2011–15 ($61,800).
Secretary of State: Jason Gant (R.), 2011–15 ($80,713).

Government Website: http://www.sd.gov

ECONOMY

Per capita personal income (2010) was $39,593.

Budget

In 2011 total state revenue was $6,017m. Total expenditure was $4,498m. (education, $1,300m.; public welfare, $966m.; highways, $607m.; government administration, $171m.; natural resources, $165m.) Outstanding debt in 2011, $3,545m.

Performance

Gross Domestic Product by state in 2011 was $34,443m. (provisional), ranking South Dakota 46th in the United States. In 2011 state real GDP growth was 0·8% (provisional).

ENERGY AND NATURAL RESOURCES

Oil and Gas

In 2008 South Dakota produced 1·7m. bbls of crude petroleum; marketed production of natural gas totalled 2bn. cu. ft.

Water

The total area covered by water is 1,305 sq. miles.

Minerals

Gold is one of the leading mineral commodities, with 1,900 kg produced in 2008. Gross value rose from $29m. in 2003 to $53m. in 2008. In 2008 sand and gravel was the major non-metallic industrial mineral commodity with 12·3m. tonnes produced. Other major minerals were: crushed stone (5·4m. tonnes); limestone (2·8m.); sandstone and quartzite (2·1m.); clays (155,000). Value of non-fuel mineral production in 2009 was $230m.

Agriculture

In 2007 there were 31,169 farms with an acreage of 43·67m. and an average farm size of 1,401 acres. Average value of farmland and buildings per acre in 2007 was $896. Farm income, 2006: from crops, $2,065m.; from livestock and products, $2,652m. The net farm income in 2006 was $742m.

Principal commodities, with 2007 production figures, are: corn, 542m. bu.; wheat, 144m. bu.; soybeans, 136m. bu.; cattle, 1,472m. lb. The farm livestock in 2007 included 3·69m. cattle; 335,900 sheep and lambs; and 1·49m. hogs and pigs. In 2007, 13·4m. lb of honey were produced, a total exceeded only by North Dakota and California.

Forestry

South Dakota had 1,682,000 acres of forested land in 2007, of which 1,039,000 acres were national forest.

INDUSTRY

In 2007 the state's 1,052 manufacturing establishments had 41,000 employees, earning $1,539m. Total value added by manufacturing in 2009 was $4,933m.

Labour

Total non-agricultural employment, 2010, 403,000. Employees by branch, 2010 (in 1,000): trade, transportation and utilities, 81; government, 79; education and health services, 64; leisure and hospitality, 43; manufacturing, 37. The average annual wage per employee in 2009 was $33,352—the lowest of any state. The state unemployment rate in Dec. 2010 was 4·7%.

INTERNATIONAL TRADE

Imports and Exports

In 2010 exports from South Dakota totalled $1,259m., up from $1,011m. in 2009 although down from $1,654m. in 2008. The leading destinations for exports in 2010 were Canada (value of $416m.), followed by Mexico, Germany and Japan.

COMMUNICATIONS

Roads

In 2007 there were 83,744 miles of road comprising 2,878 miles of urban road and 80,866 miles of rural road. There were 864,838 registered vehicles.

Rail

In 2009 there were 1,741 miles of freight railroad (excluding trackage rights). There were ten freight railroads operating in 2009. Rail traffic originating in South Dakota totalled 16·5m. tons in 2009 and rail traffic terminating in the state came to 4·4m. tons.

Civil Aviation

There were four primary airports in 2011 (commercial service airports with more than 10,000 passenger boardings annually) with a combined total of 716,885 enplanements, up from 674,840 in 2010.

SOCIAL INSTITUTIONS

Justice

In Dec. 2008 there were 3,342 adults in state prisons. The death penalty is authorized and was used in 2007 for the first time in 60 years. After a five-year gap there were two more executions in 2012.

Education

School attendance is compulsory between the ages of six and 18 (since 1 July 2009—previously school attendance had only been compulsory to 16). In fall 2007 there were 121,606 pupils at 730 public schools with 9,416 teachers. In fall 2007 the 80 private schools had 12,280 pupils and 930 teachers. Public school teacher salaries in 2008–09 averaged $38,017. Total expenditure on public elementary and secondary education in 2006–07 was $1,106m. Spending per pupil in fiscal year 2008 was $8,535.

Higher (public) education in fall 2005: the School of Mines at Rapid City, established 1885, had 2,313 students; South Dakota State University at Brookings, 10,938; the University of South Dakota, founded at Vermillion in 1882, 8,641; Northern State University, Aberdeen, 2,631; Black Hills State University at Spearfish, 3,919; and Dakota State University at Madison, 2,319. There were 12 private degree-granting institutions in 2008–09; enrolment in fall 2007 totalled 10,830 students.

Health

In 2009 there were 53 community hospitals with 4,100 beds. A total of 102,000 patients were admitted during the year. In 2009 there were 1,818 active physicians (22 per 10,000 population).

Welfare

Medicare enrolment in July 2010 totalled 135,142. In fiscal year 2009 a total of 141,131 people in South Dakota received Medicaid. In Dec. 2008 there were 146,991 Old-Age, Survivors, and Disability Insurance (OASDI) beneficiaries. A total of 6,129 people were receiving payments under Temporary Assistance for Needy Families (TANF) in Dec. 2008.

RELIGION

The principal religious traditions in 2010 were: Mainline Protestants, with 196,001 members; Catholics, 148,883; Evangelical Protestants, 118,142; Latter-day Saints (Mormons), 10,001.

FURTHER READING

Statistical information: State Data Center, Univ. of South Dakota, Vermillion 57069.
Governor's Budget Report. South Dakota Bureau of Finance and Management. Annual
South Dakota Historical Collections. 1902–82
South Dakota Legislative Manual. Biennial

Berg, F. M., *South Dakota: Land of Shining Gold.* 1982

State library: South Dakota State Library, 800 Governor's Drive, Pierre, S.D. 57501–2294.

Tennessee

KEY HISTORICAL EVENTS

Bordered on the west by the Mississippi, Tennessee was part of an area inhabited by Cherokee. French, Spanish and British explorers navigated the Mississippi to trade with the Cherokee in the late 16th and 17th centuries. French claims were abandoned in 1763. Colonists from the British colonies of Virginia and Carolina then began to cross the Appalachians westwards, but there was no organized Territory until after the War of Independence. In 1784 there was a short-lived, independent state called Franklin. In 1790 the South West Territory (including Tennessee) was formed and Tennessee entered the Union as a state in 1796.

The state was active in the war against Britain in 1812. After the American victory, colonization increased and pressure for land mounted. The Cherokee were forcibly removed during the 1830s and taken to Oklahoma, a journey on which many died.

Tennessee was a slave state and seceded from the Union in 1861, although eastern Tennessee was against secession. There were important battles at Shiloh, Chattanooga, Stone River and Nashville. In 1866 Tennessee was readmitted to the Union.

Nashville, the capital since 1843, Memphis, Knoxville, and Chattanooga all developed as river towns, Memphis becoming an important cotton and timber port. Growth was greatly accelerated by the creation of the Tennessee Valley Authority in the 1930s, producing power for industry. With an expanding economy, by 1970, the southern pattern of emigration and population loss had been reversed.

TERRITORY AND POPULATION

Tennessee is bounded north by Kentucky and Virginia, east by North Carolina, south by Georgia, Alabama and Mississippi and west by the Mississippi River (forming the boundary with Arkansas and Missouri). Land area, 41,235 sq. miles (106,798 sq. km). Census population, 1 April 2010, was 6,346,105, an increase of 11·5% since 2000. July 2012 estimate, 6,456,243.

Population in five census years was:

	White	Black	American Indian	Asiatic	Total	Per sq. mile
1910	1,711,432	473,088	216	53	2,184,789	52·4
			All others			
1980	3,835,452	725,942	29,726		4,591,120	111·6
1990	4,048,068	778,035	51,082		4,877,185	115·7
2000	4,563,310	932,809	193,164		5,689,283	138·0
2010	4,921,948	1,057,315	366,842		6,346,105	153·9

Of the total population in 2010, 3,252,601 were female and 4,850,104 were 18 years old or older. In 2010 the Hispanic population of Tennessee was 290,059, up from 123,838 in 2000 (an increase of 134·2%).

The cities, with population (2010) are Memphis, 646,889; Nashville (capital), 601,222; Knoxville, 178,874; Chattanooga, 167,674; Clarksville, 132,929; Murfreesboro, 108,755; Jackson, 65,211; Johnson City, 63,152; Franklin, 62,487; Bartlett, 54,613; Hendersonville, 51,372. Largest Metropolitan Statistical Areas, with 2010 populations: Nashville-Davidson–Murfreesboro–Franklin, 1,589,934; Memphis, 1,316,100; Knoxville, 698,030; Johnson City, 198,716; Chattanooga, 528,143; Clarksville, 273,949.

SOCIAL STATISTICS

Births, 2010, 79,495 (12·5 per 1,000 population); deaths, 2010 (provisional), 59,574 (9·4). Infant mortality, 2009, 8·0 per 1,000 live births. 2009 (provisional): marriages, 55,200 (8·4 per 1,000 population); divorces and annulments, 25,800 (3·9).

CLIMATE

Memphis, Jan. 41°F (5°C), July 82°F (27·8°C). Annual rainfall 49" (1,221 mm). Nashville, Jan. 39°F (3·9°C), July 79°F (26·1°C). Annual rainfall 48" (1,196 mm). Tennessee belongs to the Appalachian Mountains climate zone (see UNITED STATES: Climate).

CONSTITUTION AND GOVERNMENT

The state has operated under three constitutions, the last of which was adopted in 1870 and has been since amended 35 times (first in 1953). Voters at an election may authorize the calling of a convention limited to altering or abolishing one or more specified sections of the constitution. The General Assembly consists of a Senate of 33 members and a House of Representatives of 99 members, senators elected for four years and representatives for two years. Qualified as electors are all citizens (usual residential and age (18) qualifications).

For the 113th Congress, which convened in Jan. 2013, Tennessee sends nine members to the House of Representatives. It is represented in the Senate by Lamar Alexander (R. 2003–15) and Bob Corker (R. 2007–19).

The capital is Nashville. The state is divided into 95 counties.

RECENT ELECTIONS

In the 2012 presidential election Mitt Romney took Tennessee with 59·5% of the vote (John McCain won it in 2008).

CURRENT GOVERNMENT

Governor: Bill Haslam (R.), 2011–15 (salary: $178,356).
Lieut.-Governor: Ronald L. Ramsey (R.), 2013–15 ($57,027).
Secretary of State: Tre Hargett (R.), 2013–17 ($187,452).

Government Website: http://www.tennessee.gov

ECONOMY

Per capita personal income (2010) was $34,955.

Budget

In 2011 total state revenue was $34,681m. Total expenditure was $30,841m. (including: public welfare, $10,747m.; education, $9,177m.; highways, $1,705m.; health, $1,119m.; correction, $796m.) Outstanding debt in 2011, $5,899m.

Performance

Gross Domestic Product by state in 2011 was $233,997m. (provisional), ranking Tennessee 19th in the United States. In 2011 state real GDP growth was 1·9% (provisional).

ENERGY AND NATURAL RESOURCES

Water

The total area covered by water is 909 sq. miles.

Minerals

Domestic non-fuel mineral production was worth $675m. in 2009.

Agriculture

In 2007, 79,280 farms covered 10·97m. acres. The average farm was of 138 acres. In 2007 the average value of farmland and buildings per acre was $3,378.

Farm income 2006: from crops, $1,373m.; from livestock, $1,192m. The net farm income in 2006 was $722m. Main crops were cotton and greenhouse products.

In 2007 there were on farms: 61,000 dairy cows, 2·12m. all cattle, 138,200 hogs and pigs, and 29,800 sheep and lambs.

Forestry

Forests occupied 14·48m. acres in 2007. The forest industry and industries dependent on it employ about 0·04m. workers. Wood products are valued at over $500m. per year. National forest system land (2007) 741,000 acres.

INDUSTRY

The manufacturing industries include iron and steel working, but the most important products are chemicals, including synthetic fibres and allied products, electrical equipment and food. In 2007 the state's 6,752 manufacturing establishments had 369,000 employees, earning $15,166m. Total value added by manufacturing in 2009 was $48,282m.

Labour

In 2010 total non-agricultural employment was 2,613,000. Employees by branch, 2010 (in 1,000): trade, transportation and utilities, 555; government, 432; education and health services, 373; professional and business services, 305; manufacturing, 298. The unemployment rate in Dec. 2010 was 9·4%.

INTERNATIONAL TRADE

Imports and Exports

In 2010 exports from Tennessee totalled $25,943m., up from $20,484m. in 2009. The leading destinations for exports in 2010 were Canada (value of $7,210m.), followed by Mexico, China (excluding Hong Kong) and Japan.

COMMUNICATIONS

Roads

In 2007 there were 91,058 miles of roads (69,345 miles rural). There were 5,339,946 registered motor vehicles.

Rail

In 2009 there were 2,635 miles of freight railroad (excluding trackage rights). There were 25 freight railroads operating in 2009. Rail traffic originating in Tennessee totalled 14·4m. tons in 2009 and rail traffic terminating in the state came to 24·1m. tons. There is a tramway in Memphis.

Civil Aviation

There were five primary airports—commercial service airports with more than 10,000 passenger boardings annually—in 2011 with a combined total of 10,383,482 enplanements, down from 10,663,323 in 2010. The main airports are Nashville International (which had 4,673,047 enplanements in 2011) and Memphis International (with 4,344,213 enplanements in 2011). Memphis International handled 3,916,410 tonnes of freight in 2011, ranking it second in the world behind Hong Kong (which surpassed Memphis as the world's busiest cargo airport in 2010).

SOCIAL INSTITUTIONS

Justice

There were two executions in 2009 but none in 2010, 2011 or 2012. In Dec. 2008 there were 27,228 prison inmates.

Education

School attendance has been compulsory since 1925 and the employment of children under 16 years of age in workshops, factories or mines is illegal.

In fall 2007 there were 1,718 public schools with 964,259 pupils and 64,659 teachers. Total expenditure on public elementary and secondary education was $7,756m. in 2006–07 and average teacher salary $46,278 in 2008–09. Spending per pupil in fiscal year 2008 was $7,820.

Tennessee has 22 public colleges and universities (nine four-year and 13 two-year institutions). In fall 2005 the universities included the University of Tennessee, Knoxville (founded 1794), with 28,512 students; the University of Memphis (1912) with 20,465; Vanderbilt University, Nashville (1873) with 11,479; Tennessee State University (1912) with 8,880; the University of

Tennessee at Chattanooga (1886) with 8,656; and Fisk University (1866) with 920.

Health

In 2009 there were 137 community hospitals with 21,000 beds. A total of 859,000 patients were admitted during the year. In 2009 there were 16,754 active physicians (27 per 10,000 population).

Welfare

Medicare enrolment in July 2010 totalled 1,047,285. In fiscal year 2009 a total of 1,479,028 people in Tennessee received Medicaid. In Dec. 2008 there were 1,168,699 Old-Age, Survivors, and Disability Insurance (OASDI) beneficiaries. A total of 147,865 people were receiving payments under Temporary Assistance for Needy Families (TANF) in Dec. 2008.

RELIGION

The principal religious traditions in 2010 were: Evangelical Protestants, with 2,384,381 members; Mainline Protestants, 533,145; Black Protestants, 273,889; Catholics, 223,045; Latter-day Saints (Mormons), 47,391.

FURTHER READING

Statistical information: Center for Business and Economic Research, Univ. of Tennessee, Knoxville 37996. Publishes *Tennessee Statistical Abstract* *Tennessee Blue Book.*

Dykeman, W., *Tennessee.* Rev. ed. 1984

State library: State Library and Archives, 403 7th Avenue North, Nashville.

Texas

KEY HISTORICAL EVENTS

A number of Indian tribes occupied the area before French and Spanish explorers arrived in the 16th century. In 1685 La Salle established a colony at Fort St Louis, but Texas was confirmed as Spanish in 1713. Spanish missions increased during the 18th century with San Antonio (1718) as their headquarters.

In 1820 a Virginian colonist, Moses Austin, obtained permission to begin a settlement in Texas. In 1821 the Spanish empire in the Americas came to an end and Texas, together with Coahuila, formed a state of the newly independent Mexico. The Mexicans agreed to the Austin venture and settlers of British and American descent came in.

Discontented with Mexican government, the settlers declared independence in 1836. Warfare, including the siege of the Alamo fort, ended with the foundation of the independent Republic of Texas which lasted until 1845. During this period the Texas Rangers were organized as a police force and border patrol. Texas was annexed to the Union in Dec. 1845, as the Federal government feared its vulnerability to Mexican occupation. This led to war between Mexico and the USA from 1845 to 1848. In 1861 Texas left the Union and joined the southern states in the Civil War, being readmitted in 1869. Ranching and cotton-growing were the main activities before the discovery of oil in 1901.

TERRITORY AND POPULATION

Texas is bounded north by Oklahoma, northeast by Arkansas, east by Louisiana, southeast by the Gulf of Mexico, south by Mexico and west by New Mexico. Land area, 261,232 sq. miles (676,587 sq. km). Texas is the second largest US state behind Alaska. Census population, 1 April 2010, was 25,145,561, an increase of 20·6% since 2000. July 2012 estimate, 26,059,203.

Population for five census years was:

	White	Black	American Indian	Asian	Total	Per sq. mile
1910	3,204,848	690,049	702	943	3,896,542	14·8
			All others			
1980	11,197,663	1,710,250	1,320,470		14,228,383	54·2
			American Indian/ Alaska Native	All Others		
1990	12,774,762	2,021,632	65,877	2,124,239	16,986,510	64·9
2000	14,799,505	2,404,566	118,362	3,529,387	20,851,820	79·7
2010	17,701,552	2,979,598	170,972	4,293,439	25,145,561	96·3

Of the total population in 2010, 12,673,281 were female and 18,279,737 were 18 years old or older. In 2010 the Hispanic population was 9,460,921, up from 6,669,666 in 2000 (an increase of 41·8%). The numerical increase was the second largest in the Hispanic population of any state in the USA, after California. Only New Mexico has a greater percentage of Hispanics in the state population.

The largest cities, with census population in 2010, are:

Houston	2,099,451	Brownsville	175,023
San Antonio	1,327,407	Pasadena	149,043
Dallas	1,197,816	Mesquite	139,824
Austin (capital)	790,390	McKinney	131,117
Fort Worth	741,206	McAllen	129,877
El Paso	649,121	Killeen	127,921
Arlington	365,438	Waco	124,805
Corpus Christi	305,215	Carrollton	119,097
Plano	259,841	Beaumont	118,296
Laredo	236,091	Abilene	117,063
Lubbock	229,573	Frisco	116,989
Garland	226,876	Denton	113,383
Irving	216,290	Midland	111,147
Amarillo	190,695	Wichita Falls	104,553
Grand Prairie	175,396		

The population of the largest Metropolitan Statistical Areas in 2010 was: Dallas–Fort Worth–Arlington, 6,371,773; Houston–Sugarland–Baytown, 5,946,800; San Antonio–New Braunfels, 2,142,508; Austin–Round Rock–San Marcos, 1,716,289.

SOCIAL STATISTICS

Births, 2010, 386,118 (15·4 per 1,000 population); deaths, 2010 (provisional), 166,525 (6·6). Infant mortality, 2009, 6·0 per 1,000 live births. 2009 (provisional): marriages, 179,800 (7·1 per 1,000 population); divorces and annulments, 76,900 (3·3).

CLIMATE

Dallas, Jan. 45°F (7·2°C), July 84°F (28·9°C). Annual rainfall 38" (945 mm). El Paso, Jan. 44°F (6·7°C), July 81°F (27·2°C). Annual rainfall 9" (221 mm). Galveston, Jan. 54°F (12·2°C), July 84°F (28·9°C). Annual rainfall 46" (1,159 mm). Houston, Jan. 52°F (11·1°C), July 83°F (28·3°C). Annual rainfall 48" (1,200 mm). Texas suffered its worst one-year drought on record in 2011, when the average rainfall for the state was 14·9" (378 mm). Texas belongs to the Central Plains climate zone (*see* UNITED STATES: Climate).

CONSTITUTION AND GOVERNMENT

The present constitution dates from 1876; it has been amended 467 times as of Aug. 2011. The state legislature consists of the Senate and House of Representatives. The Senate has 31 members elected for four-year terms. Half of the membership is elected every two years. The House has 150 members, elected for two-year terms during polling held in even-numbered years. The legislature meets in regular session for about five months every other year. The session begins in Jan. of odd-numbered years and lasts no more than 140 days (although special sessions can be called by the Governor). The Governor and Lieut.-Governor are elected for four years.

For the 113th Congress, which convened in Jan. 2013, Texas sends 36 members to the House of Representatives. It is represented in the Senate by John Cornyn (R. 2002–15) and Ted Cruz (R. 2013–19).

The capital is Austin. The state has 254 counties.

RECENT ELECTIONS

In the 2012 presidential election Mitt Romney took Texas with 57·2% of the vote (John McCain won it in 2008).

CURRENT GOVERNMENT

Governor: Rick Perry (R.), 2011–15 (salary: $150,000).

Lieut.-Governor: David Dewhurst (R.), 2011–15 ($7,200 plus legislature session salary).

Secretary of State: John T. Steen, Jr. (R.), appointed Nov. 2012 ($125,880).

Government Website: http://www.tx.gov

ECONOMY

Per capita personal income (2010) was $37,706.

Budget

In 2011 total state revenue was $134,345m. Total expenditure was $125,940m. (education, $48,809m.; public welfare, $31,269m.; highways, $6,558m.; hospitals, $4,196m.; correction, $3,765m.) Outstanding debt in 2011, $38,530m.

Performance

In 2011 Gross Domestic Product by state was $1,149,908m. (provisional), ranking Texas second after California. If Texas were a country in its own right it would be the world's 15th largest economy. In 2011 state real GDP growth was 3·3% (provisional).

Banking and Finance

In 2002 there were 751 financial institutions in Texas insured by the US Federal Deposit Insurance Corporation, with assets worth $216,900m. They had 4,980 offices with total deposits of $256,600m.

As at Dec. 2003 there were 351 state-chartered banks operating in Texas, with total assets of $121,970m. The largest banks were International Bank of Commerce, Laredo (with assets of $5,294·2m.), Texas State Bank, McAllen ($4,215·6m.), Sterling Bank, Houston ($3,110·1m.), Prosperity Bank, El Campo ($2,395·0m.) and PlainsCapital Bank, Lubbock ($2,061·8m.).

ENERGY AND NATURAL RESOURCES

Electricity

In 2009 production was 396·6bn. kWh, of which 189·0bn. kWh was from natural gas.

Oil and Gas

Texas is the leading producer in the USA of both oil and natural gas. In 2007 it produced 21% of the country's oil and 22% of its natural gas. Production, 2008: crude petroleum, 398m. bbls (value in 2007, $27,108m.); marketed production of natural gas, 6,921bn. cu. ft (value in 2007, $42,734m.). Natural gasoline, butane and propane gases are also produced.

Water

The total area covered by water is 7,365 sq. miles, of which 5,616 sq. miles are inland. Only Alaska has a larger area covered by inland water.

Minerals

Minerals include helium, crude gypsum, granite and sandstone, salt and cement. Production of coal (2009) was 35·1m. short tons. Total value of domestic non-fuel mineral products in 2009 was $2,650m.

Agriculture

Texas is one of the most important agricultural states. In 2007 it had 247,437 farms covering 130·40m. acres (up from 228,926 farms covering 129·88m. acres in 2002); average farm was of 527 acres. Both the number of farms and the total area covered are the highest in the USA. In 2007 land and buildings were valued at $1,270 per acre. Large-scale commercial farms, highly mechanized, dominate in Texas; farms of 1,000 acres or more in number far exceed that of any other state, but small-scale farming persists. Soil erosion is a serious problem in some parts.

Production: corn, barley, beans, cotton, hay, oats, peanuts, rye, sorghum, soybeans, sunflowers, wheat, oranges, grapefruit, peaches, sweet potatoes. Farm income, 2006: from crops $5,703m.; from livestock and products, $10,324m. The net farm income in 2006 was $4,866m.

The state has an important livestock industry, leading in the number of all cattle (13·71m.) and sheep (945,000); it also had 0·44m. dairy cows, 1·16m. hogs and pigs, and 19·12m. laying hens in 2007. Wool production in 2007 totalled 4·60m. lb, the most of any state.

Forestry

There were 17,273,000 acres of forested land in 2007, with 682,000 acres of national forest.

INDUSTRY

In 2007 the state's 21,115 manufacturing establishments had 894,000 employees, earning $42,836m. Total value added by manufacturing in 2009 was $174,881m.

Labour

Total non-agricultural employment, 2010, 10,342,000. Employees by branch, 2010 (in 1,000): trade, transportation and utilities, 2,050; government, 1,860; education and health services, 1,388; professional and business services, 1,273; leisure and hospitality, 1,006. The unemployment rate in Dec. 2010 was 8·3%.

INTERNATIONAL TRADE

Imports and Exports

In 2010 exports from Texas totalled $207·0bn., the most of any state and representing 16% of all US exports. The state's major export markets in 2010 were Mexico (with a value of $72·6m.), Canada ($18·8m.), China ($10·3m.) and Brazil ($7·2m.).

COMMUNICATIONS

Roads

In 2007 there were 305,855 miles of road comprising 84,194 miles of urban road and 221,661 miles of rural road. There were 18,072,148 registered motor vehicles and 3,363 traffic accident fatalities.

Rail

In 2009 there were 10,405 miles of freight railroad (excluding trackage rights), including 8,375 miles of Class I railroads. Texas has more rail miles than any other state. There were 45 freight railroads operating in 2009. Rail traffic originating in Texas in 2009 totalled 82·4m. tons and rail traffic terminating in the state came to 186·6m. tons (the most of any state). There are light rail systems in Austin, Dallas and Houston.

Civil Aviation

There were 24 primary airports—commercial service airports with more than 10,000 passenger boardings annually—in Texas in 2011 with a combined total of 68,524,498 enplanements, up from 67,664,379 in 2010. Dallas/Fort Worth International had 27,518,358 enplanements in 2011 (ranking it fourth in the USA) and George Bush Intercontinental/Houston had 19,306,660. A further five primary airports had more than 1m. passenger enplanements in 2011: William P. Hobby (Houston's second airport), Austin-Bergstrom International, San Antonio International, Dallas Love Field and El Paso International.

Shipping

The port of Houston, connected by the Houston Ship Channel (50 miles long) with the Gulf of Mexico, is the second busiest in the country after South Louisiana and the busiest for foreign trade cargo. In 2009 it handled 211·3m. tons of cargo (148·0m. tons of foreign trade cargo and 63·4m. tons of domestic cargo). Other major ports are Corpus Christi, Beaumont, Texas City and Port Arthur, which ranked as the USA's 5th, 7th, 11th and 19th busiest ports respectively in 2009. There were 1,021 miles of inland waterways (271 miles of deep-draft channels and 750 miles of shallow-draft channels) in 2007.

SOCIAL INSTITUTIONS

Justice

In Dec. 2008 there were 172,506 prison inmates. Between 1977 and 2012 Texas was responsible for 492 of the USA's 1,320 executions (more than four times as many as any other state), although it was not until 1982 that Texas reintroduced the death penalty. In 2012, 15 people were executed (13 in 2011). There were 312 death row inmates as at 1 Jan. 2012. In 2000, 40 people had been executed, the highest number in a year in any state since the authorities began keeping records in 1930.

Education

School attendance is compulsory from six to 18 years of age.

In the 2007–08 school year there were 8,758 public elementary and secondary schools with 4,674,832 enrolled pupils; there were 321,929 teachers. Total expenditure on public schools in 2006–07 was $45,189m.

In 2010 there were 146 higher education institutions (38 public universities, 39 independent colleges and universities, 50 public community college districts, four campuses of the Texas State Technical College System, three public Lamar state colleges, nine public health-related institutions, one independent medical school and two independent junior colleges). Enrolled students in fall 2008 totalled about 1·4m.

Public universities and student enrolment, fall 2009:

Institutions	Students
University of Texas System	202,240
Texas A&M University System	114,691
Texas State University System	76,290[1]
University of Houston System	61,059
University of North Texas System	35,003
Texas Technical University System	30,049
Stephen F. Austin State University, Nacogdoches	12,845
Texas Southern University, Houston	9,394
Midwestern State University, Wichita Falls	6,341

[1]Fall 2010 (provisional).

Independent colleges and universities with the largest student enrolments, fall 2009:

Institutions	Students
Baylor University, Waco	14,614
Southern Methodist University, Dallas	10,891
Texas Christian University, Fort Worth	8,853
Wayland Baptist University, Plainview	5,886
Rice University, Houston	5,576
Dallas Baptist University, Dallas	5,400
Abilene Christian University, Abilene	4,813
University of St Thomas, Houston	3,132

Health

In 2009 there were 428 community hospitals with 62,100 beds. A total of 2,621,000 patients were admitted during the year. In 2009 there were 53,546 active physicians (22 per 10,000 population).

Welfare

Medicare enrolment in July 2010 totalled 2,976,720. In fiscal year 2009 a total of 4,282,564 people in Texas received Medicaid. In Dec. 2008 there were 3,192,227 Old-Age, Survivors, and Disability Insurance (OASDI) beneficiaries. A total of 112,929 people were receiving payments under Temporary Assistance for Needy Families (TANF) in Dec. 2008.

RELIGION

In 2010 the leading religious traditions were: Evangelical Protestants, with 6,456,168 members; Catholics, 4,673,500; Mainline Protestants, 1,553,959; Muslims, 422,000 (estimate); Black Protestants, 345,998.

CULTURE

Tourism

In 2011 overseas visitors to Texas—excluding those from Canada and Mexico—numbered 1,283,000, up from 1,028,000 in 2010 and 975,000 in 2006.

FURTHER READING

Texas Almanac. Biennial

Kingston, M., *Texas Almanac's Political History of Texas.* 1992
Kraemer, R., Newell, C. and Prindle, D., *Essentials of Texas Politics.* 10th ed. 2007

Legislative Reference Library: Box 12488, Capitol Station, Austin, Texas 78711-2488.

Utah

KEY HISTORICAL EVENTS

Spanish Franciscan missionaries explored the area in 1776, finding Shoshoni Indians. Spain laid claim to Utah and designated it part of Spanish Mexico. As such it passed into the hands of the Mexican Republic when Mexico rebelled against Spain and gained independence in 1821.

In 1848, at the conclusion of war between the USA and Mexico, the USA received Utah along with other southwestern territory. Settlers had already arrived in 1847 when the Mormons (the Church of Jesus Christ of Latter-day Saints) arrived, having been driven on by hostility in Ohio, Missouri and Illinois. Led by Brigham Young, they entered the Great Salt Valley and colonized it. In 1849 they applied for statehood but were refused. In 1850 Utah and Nevada were joined as one Territory. The Mormon community continued to ask for statehood but this was only granted in 1896, after they had renounced polygamy and disbanded their People's Party.

Mining, especially of copper, and livestock farming were the basis of the economy. Settlement had to adapt to desert conditions and the main centres of population were in the narrow belt between the Wasatch Mountains and the Great Salt Lake. Salt Lake City, the capital, was founded in 1847 and laid out according to Joseph Smith's plan for the city of Zion. It was the centre of the Mormons' provisional 'State of Deseret' and Territorial capital from 1856 until 1896, except briefly in 1858 when federal forces occupied it during conflict between territorial and Union governments.

TERRITORY AND POPULATION

Utah is bounded north by Idaho and Wyoming, east by Colorado, south by Arizona and west by Nevada. Land area, 82,170 sq. miles (212,818 sq. km). In 2000 American Indian lands covered 8,937 sq. miles (8,910 sq. miles in reservations and 27 sq. miles in off-reservation trust land). The federal government owned 35,033,603 acres in 2010, equivalent to 66·5% of the state area (the second highest percentage after Nevada).

Census population, 1 April 2010, was 2,763,885, an increase of 23·8% since 2000. July 2012 estimate, 2,855,287.

Population at five federal censuses was:

	White	Black	American Indian/Alaska Native	Asian	Total (including others)	Per sq. mile
1910	366,583	1,144	3,123	2,501	373,851	4·5
1980	1,382,550	9,225	19,256	15,076	1,461,037	17·7
1990	1,615,845	11,576	24,283	25,696	1,722,850	21·0
2000	1,992,975	17,657	29,684	37,108	2,233,169	27·2
2010	2,379,560	29,287	32,927	55,285	2,763,885	33·6

Of the total population in 2010, 1,388,317 were male and 1,892,858 were 18 years old or older. In 2010 the Hispanic population was 358,340, up from 201,559 in 2000 (an increase of 77·8%).

The largest cities are Salt Lake City (the capital), with a population (census, 2010) of 186,440; West Valley City, 129,480; Provo, 112,488; West Jordan, 103,712; Orem, 88,328; Sandy, 87,461.

SOCIAL STATISTICS

Births, 2010, 52,258 (18·9 per 1,000 population—the highest rate in any US state); deaths, 2010 (provisional), 14,776 (5·3). Infant mortality rate, 2009, 5·3 per 1,000 live births. 2009 (provisional): marriages, 23,900 (8·2 per 1,000 population); divorces and annulments, 10,700 (3·6). Fertility rate, 2008, 2·6 births per woman (the highest of any American state).

CLIMATE

Salt Lake City, Jan. 29°F (-1·7°C), July 77°F (25°C). Annual rainfall 16" (401 mm). Utah belongs to the Mountain States climate region (*see* UNITED STATES: Climate).

CONSTITUTION AND GOVERNMENT

Utah adopted its present constitution in 1896; it has had numerous amendments since then. The Legislature consists of a Senate (in part renewed every two years) of 29 members, elected for four years, and of a House of Representatives of 75 members elected for two years. It sits annually in Jan. The Governor is elected for four years. The constitution provides for the initiative and referendum.

For the 113th Congress, which convened in Jan. 2013, Utah sends four members to the House of Representatives. It is represented in the Senate by Orrin Hatch (R. 1977–2019) and Mike Lee (R. 2011–17).

The capital is Salt Lake City. There are 29 counties in the state.

RECENT ELECTIONS

In the 2012 presidential election Mitt Romney took Utah with 72·8% of the vote (John McCain won it in 2008).

CURRENT GOVERNMENT

Governor: Gary R. Herbert (R.), 2013–17 (salary: $109,900).
Lieut.-Governor: Gregory S. Bell (R.), 2013–17 ($104,405).

Government Website: http://www.utah.gov

ECONOMY

Per capita personal income (2010) was $32,473.

Budget

In 2011 total state revenue was $16,985m. Total expenditure was $16,683m. (including: education, $6,516m.; public welfare, $2,812m.; highways, $1,514m.; hospitals, $956m.; government administration, $800m.) Outstanding debt in 2011, $7,206m.

Performance

Gross Domestic Product by state in 2011 was $108,329m. (provisional), ranking Utah 33rd in the United States. In 2011 state real GDP growth was 2·0% (provisional).

ENERGY AND NATURAL RESOURCES

Oil and Gas

In 2008 Utah produced 22m. bbls of crude petroleum; marketed production of natural gas totalled 434bn. cu. ft.

Water

The total area covered by water is 2,727 sq. miles.

Minerals

The principal minerals are: copper, gold, magnesium, petroleum, lead, silver and zinc. The state also has natural gas, clays, tungsten, molybdenum, uranium and phosphate rock. Production of coal (2009) was 21·7m. short tons. The value of domestic non-fuel mineral production in 2009 was $3,910m.

Agriculture

In 2007 Utah had 16,700 farms covering 11·09m. acres. Average number of acres per farm was 664 and the average value per acre was $1,249. Farm income, 2006: from crops, $313m.; and from livestock and products, $931m. The net farm income in 2006 was $264m. The principal crops are: barley, wheat (spring and winter), oats, potatoes, hay (alfalfa, sweet clover and lespedeza) and maize. Livestock, 2007: cattle, 843,000; hogs and pigs, 760,000; sheep and lambs, 278,000; laying hens, 3·59m.

Forestry

Forest area, 2007, was 17,962,000 acres and included 6,259,000 acres of national forest.

INDUSTRY

Leading manufactures by value added are primary metals, ordinances and transport, food, fabricated metals and machinery, and petroleum products. In 2007 Utah's 3,368 manufacturing establishments had 123,000 employees, earning $5,508m. Total value added by manufacturing in 2009 was $18,962m.

Labour

Total non-agricultural employment, 2010, 1,181,000. Employees by branch, 2010 (in 1,000): trade, transportation and utilities, 229; government, 216; education and health services, 155; professional and business services, 153; manufacturing, 111. The unemployment rate in Dec. 2010 was 7·5%.

INTERNATIONAL TRADE

Imports and Exports

In 2010 exports from Utah totalled $13,809m., up from $10,337m. in 2009. The leading destinations for exports in 2010 were the United Kingdom (value of $4,408m.), followed by Canada, India and Hong Kong.

COMMUNICATIONS

Roads

In 2007 there were 44,221 miles of road (32,672 miles rural). There were 2,320,171 registered motor vehicles.

Rail

In 2009 there were 1,358 miles of freight railroad (excluding trackage rights). Rail traffic originating in Utah totalled 16·6m. tons in 2009 and rail traffic terminating in the state came to 12·3m. tons. A light rail system opened in Salt Lake City in 1999.

Civil Aviation

There is an international airport at Salt Lake City. There were 10,673,984 passenger enplanements statewide in 2007.

SOCIAL INSTITUTIONS

Justice

In Dec. 2008 there were 6,546 prison inmates. The death penalty is authorized; there was one execution in 2010 (the first since 1999) but none in 2011 or 2012.

Education

School attendance is compulsory for children from six to 18 years of age. There are 40 school districts. Teachers' salaries, 2008–09, averaged $42,335. In 2007 there were 576,244 pupils and 24,336 teachers in 1,010 public elementary and secondary schools. In 2006–07 total expenditure on public elementary and secondary education was $3,807m. Spending per pupil in fiscal year 2008 was $5,978.

In fall 2007 there were 203,679 students enrolled in degree-granting institutions. There were 14 public and 24 private degree-granting institutions in 2008–09. Among the public institutions, the University of Utah (founded in 1850) in Salt Lake City had 30,558 students in fall 2005; Utah State University (1890) in Logan, 14,458; Weber State University in Ogden, 18,142; Southern Utah University in Cedar City, 6,859; College of Eastern Utah in Price, 2,178; Snow College in Ephraim, 3,333; Dixie State College in St George, 8,945; Utah Valley State College in Orem, 24,180; and Salt Lake Community College in Salt Lake City, 24,111. The Mormon Church maintains the private Brigham Young University at Provo (1875) with 30,067 students in fall 2005.

Health

In 2009 there were 44 community hospitals with 5,000 beds. A total of 226,000 patients were admitted during the year. In 2009 there were 5,903 active physicians (21 per 10,000 population).

Welfare

Medicare enrolment in July 2010 totalled 280,838. In fiscal year 2007 a total of 242,650 people in Utah received Medicaid. In Dec. 2008 there were 299,088 Old-Age, Survivors, and Disability Insurance (OASDI) beneficiaries. A total of 13,821 people were receiving payments under Temporary Assistance for Needy Families (TANF) in Dec. 2008.

RELIGION

In 2010 the leading religious traditions were: Latter-day Saints (Mormons), with 1,911,047 members; Catholics, 160,125; Evangelical Protestants, 63,040; Mainline Protestants, 23,546; Buddhists, 8,602.

CULTURE

Tourism

Overseas visitors to Utah (excluding those from Canada and Mexico) numbered 502,000 in 2011, up from 475,000 in 2010.

FURTHER READING

Statistical information: Bureau of Economic and Business Research, Univ. of Utah, 401 Kendall D. Garff Bldg., Salt Lake City 84112. Publishes *Statistical Abstract of Utah.*

Utah Foundation. *Statistical Review of Government in Utah.* 1991

Vermont

KEY HISTORICAL EVENTS

The original Indian hunting grounds of the Green Mountains and lakes was explored by the Frenchman, Samuel de Champlain, in 1609. He reached Lake Champlain on the northwest border. The first attempt at permanent settlement was also French, on Isle la Motte in 1666. In 1763 the British gained the area by the Treaty of Paris. The Treaty, which also brought peace with the Indian allies of the French, opened the way for settlement, but in a mountain area transport was slow and difficult. Montpelier, the state capital from 1805, was chartered as a township site in 1781 to command the main pass through the Green Mountains.

In the War of Independence Vermont declared itself an independent state to avoid being taken over by New Hampshire and New York. In 1791 it became the 14th state of the Union.

Most early settlers were New Englanders of British and Protestant descent. After 1812 a granite-quarrying industry grew around the town of Barre, attracting immigrant workers from Italy and Scandinavia. French Canadians settled in Winooski. Textile and engineering industries developed in the 19th century attracted more European workers.

Vermont saw the only Civil War action north of Pennsylvania when a Confederate raiding party attacked from Canada in 1864.

During the 20th century the textile and engineering industries have declined but paper and lumber industries flourish. Settlement is still mainly rural or in small towns.

TERRITORY AND POPULATION

Vermont is bounded in the north by Canada, east by New Hampshire, south by Massachusetts and west by New York. Land area, 9,217 sq. miles (23,871 sq. km). Census population, 1 April 2010, was 625,741, an increase of 2·8% since 2000. July 2012 estimate, 626,011.

Population at five census years was:

	White	Black	American Indian/ Alaska Native	Asian	Total (including others)	Per sq. mile
1910	354,298	1,621	26	11	355,956	39·0
1980	506,736	1,135	984	1,355	511,456	55·1
1990	555,088	1,951	1,696	3,215[1]	562,758	60·8
2000	589,208	3,063	2,420	5,358[1]	608,827	65·8
2010	596,292	6,277	2,207	8,107[1]	625,741	67·9

[1]Includes Native Hawaiian and other Pacific Islander.

Of the total population in 2010, 317,535 were female and 496,508 were 18 years old or older. In 2000, 414,480 (66·2%) were rural, the highest rural population percentage of any state in the USA. In 2010 the Hispanic population was 9,208, the lowest total of any state. However, this figure represents a rise of 67·3% compared to the 2000 census figure of 5,504. The largest cities are Burlington, with a population (2010) of 42,417; South Burlington, 17,904; Rutland City, 16,495. The capital, Montpelier, had a population in 2010 of 7,855.

SOCIAL STATISTICS

Births, 2010, 6,223 (9·9 per 1,000 population); deaths, 2010 (provisional), 5,380 (8·6). Infant deaths, 2009, 6·2 per 1,000 live births. 2009 (provisional): marriages, 4,700 (8·7 per 1,000 population); divorces and annulments, 2,100 (3·5). Same-sex marriage became legal in Sept. 2009.

CLIMATE

Burlington, Jan. 17°F (−8·3°C), July 70°F (21·1°C). Annual rainfall 33" (820 mm). Vermont belongs to the New England climate zone (*see* UNITED STATES: Climate).

CONSTITUTION AND GOVERNMENT

The constitution was adopted in 1793 and has since been amended. Amendments are proposed by two-thirds vote of the Senate every four years, and must be accepted by two sessions of the legislature; they are then submitted to popular vote. The state Legislature, consisting of a Senate of 30 members and a House of Representatives of 150 members (both elected for two years), meets in Jan. every year. The Governor and Lieut.-Governor are elected for two years. Electors are all citizens who possess certain residential qualifications and have taken the freeman's oath set forth in the constitution.

For the 113th Congress, which convened in Jan. 2013, Vermont sends one member to the House of Representatives. It is represented in the Senate by Patrick Leahy (Democrat, 1975–2017) and Bernard Sanders (Independent Democrat, 2007–19).

The capital is Montpelier. There are 14 counties and 251 cities, towns and other administrative divisions.

RECENT ELECTIONS

In the 2012 presidential election Barack Obama took Vermont with 67·0% of the vote, having also won it in 2008.

CURRENT GOVERNMENT

Governor: Peter Shumlin (D.), 2013–15 (salary: $150,067).
  Lieut.-Governor: Phil Scott (R.), 2013–15 ($63,701).
  Secretary of State: Jim Condos (D.), 2013–15 ($95,156).

Government Website: http://vermont.gov

ECONOMY

Per capita personal income (2010) was $40,098.

Budget

In 2011 total state revenue was $6,506m. Total expenditure was $5,854m. (education, $2,323m.; public welfare, $1,463m.; highways, $419m.; health, $191m.; government administration, $138m.) Outstanding debt in 2011, $3,485m.

Performance

Gross Domestic Product by state was $22,968m. in 2011 (provisional), ranking Vermont 50th in the United States. In 2011 state real GDP growth was 0·5% (provisional).

Banking and Finance

In 2010 there were 14 banking institutions domiciled in Vermont, and seven out-of-state banks operating.

ENERGY AND NATURAL RESOURCES

Water

The total area covered by water is 400 sq. miles.

Minerals

Stone, chiefly granite, marble and slate, is the leading mineral produced in Vermont, contributing about 60% of the total value of mineral products. Other products include asbestos, talc, sand and gravel. Value of domestic non-fuel mineral products in 2009 was more than $120m.

Agriculture

Agriculture is the most important industry. In 2007 the state had 6,984 farms covering 1·23m. acres; the average farm was of 177 acres and the average value per acre of land and buildings was $2,903. In 2007 farm income from crops totalled $93m.; from livestock and products, $581m. The net farm income in 2007 was $263m. Principal commodities are dairy products (cash receipts of $517·9m. in 2007), cattle, greenhouse products and hay. In 2007 Vermont had 265,000 cattle and calves and 2,700 hogs and pigs.

Forestry

The state is 78% forest, with 17% in public ownership. In 2007 Vermont had 4,618,000 acres of forested land with 337,000 acres of national forest. State-owned forests, parks, fish and game areas (2007), 475,655 acres. In 2006 timber removals were 44m. cu. ft (27m. cu. ft hardwoods and 17m. cu. ft softwoods).

INDUSTRY

In 2007 the state's 1,108 manufacturing establishments had 36,000 employees, earning $1,650m. Total value added by manufacturing in 2009 was $3,844m.

Labour

Total non-agricultural employment, 2010, 298,000. Employees by branch, 2010 (in 1,000): education and health services, 59; trade, transportation and utilities, 56; government, 55; leisure and hospitality, 32; manufacturing, 31. The unemployment rate in Dec. 2010 was 5·8%.

INTERNATIONAL TRADE

Imports and Exports

In 2010 exports from Vermont totalled $4,277m., up from $3,219m. in 2009. The leading destinations for exports in 2010 were Canada (value of $2,029m.), followed by China (excluding Hong Kong), Malaysia and Hong Kong.

COMMUNICATIONS

Roads

In 2007 there were 14,400 miles of road comprising 1,421 miles of urban road and 12,979 miles of rural road. Motor vehicle registrations totalled 564,967.

Rail

In 2009 there were 590 miles of freight railroad (excluding trackage rights). Rail traffic originating in Vermont totalled 0·4m. tons in 2009 with rail traffic terminating in the state totalling 1·2m. tons.

Civil Aviation

Vermont's only primary airport—a commercial service airport with more than 10,000 passenger boardings annually—is Burlington International. In 2011 it had 636,019 enplanements.

SOCIAL INSTITUTIONS

Justice

In Dec. 2008 the jail and prison population totalled 2,116. The death penalty was officially abolished in 1987 but effectively in 1964.

Education

School attendance during the full school term is compulsory for children from six to 16 years of age, unless they have completed the 10th grade or undergo approved home instruction. In fall 2007 public elementary and secondary schools had 94,038 pupils and 8,749 teachers. Average teacher's salary in 2008–09 was $47,697. Total expenditure on public elementary and secondary education in 2006–07 totalled $1,394m.

There were 25 degree-granting institutions in 2008–09 (six public); enrolment in fall 2007 totalled 42,191. In fall 2009 the University of Vermont (1791), in Burlington, had 13,391 students; Norwich University (1834, founded as the American Literary, Scientific and Military Academy in 1819), had 3,378; St Michael's College (1904), 2,466.

Health

In 2009 there were 14 community hospitals with 1,300 beds. A total of 51,000 patients were admitted during the year. In 2009 there were 2,313 active physicians (37 per 10,000 population).

Welfare

Medicare enrolment in July 2010 totalled 110,666. In fiscal year 2009 a total of 170,619 people in Vermont received Medicaid. In Dec. 2008 there were 120,249 Old-Age, Survivors, and Disability Insurance (OASDI) beneficiaries. A total of 6,445 people were receiving payments under Temporary Assistance for Needy Families (TANF) in Dec. 2008.

RELIGION

The leading religious traditions in the state in 2010 were: Catholics, with 128,293 members; Mainline Protestants, 48,029; Evangelical Protestants, 22,630; Latter-day Saints (Mormons), 4,384.

FURTHER READING

Statistical information: Office of Policy Research and Coordination, Montpelier 05602

Legislative Directory. Biennial

Vermont Annual Financial Report. Annual

Vermont Atlas and Gazetteer. 12th ed. 2007

Vermont Year-Book, formerly *Walton's Register.* Annual

State library: Vermont Dept. of Libraries, 109 State St., Montpelier.

Virginia

KEY HISTORICAL EVENTS

In 1607 a British colony was founded at Jamestown, on a peninsula in the James River, to grow tobacco. The area was marshy and unhealthy but the colony survived and in 1619 introduced a form of representative government. The tobacco plantations expanded and African slaves were imported. Jamestown was later abandoned but tobacco-growing continued and spread through the eastern part of the territory.

In 1624 control of the colony passed from the Virginia Company of London to the Crown. Growth was rapid during the 17th and 18th centuries. The movement for American independence was strong in Virginia; George Washington and Thomas Jefferson were both Virginians, and crucial battles of the War of Independence were fought there.

When the Union was formed, Virginia became one of the original states, but with reservations because of its attachment to slave-owning. In 1831 there was a slave rebellion. The tobacco plantations began to decline, and plantation owners turned to the breeding of slaves. While the eastern plantation lands seceded from the Union in 1861, the small farmers and miners of the western hills refused to secede and remained in the Union as West Virginia.

Richmond, the capital, became the capital of the Confederacy. Much of the Civil War took place in Virginia, with considerable damage to the economy. After the war the position of the black population was little improved. Blacks remained without political or civil rights until the 1960s.

TERRITORY AND POPULATION

Virginia is bounded northwest by West Virginia, northeast by Maryland and the District of Columbia, east by the Atlantic, south by North Carolina and Tennessee and west by Kentucky. Land area, 39,490 sq. miles (102,279 sq. km). Census population, 1 April 2010, was 8,001,024, an increase of 13·0% since 2000. July 2012 estimate, 8,185,867.

Population for five federal census years was:

	White	Black	American Indian/ Alaska Native	Asian/Other	Total	Per sq. mile
1910	1,389,809	671,096	539	168	2,061,612	51·2
			All others			
1980	4,230,000	1,008,311	108,517		5,346,818	134·7
1990	4,791,739	1,162,994	15,282	217,343	6,187,358	155·9
2000	5,120,110	1,390,293	21,172	546,940	7,078,515	178·8
2010	5,486,852	1,551,399	29,225	933,548	8,001,024	202·6

Of the total population in 2010, 4,075,041 were female and 6,147,347 were 18 years old or older. In 2010 the Hispanic population was 631,825, up from 329,540 in 2000 (an increase of 91·7%).

The population (2010 census population) of the principal cities was: Virginia Beach, 437,994; Norfolk, 242,803; Chesapeake, 222,209; Arlington CDP, 207,627; Richmond (the capital), 204,214; Newport News, 180,719; Alexandria, 139,966; Hampton, 137,436.

SOCIAL STATISTICS

Births, 2010, 103,002 (12·9 per 1,000 population); deaths, 2010 (provisional), 59,031 (7·4). Infant mortality, 2009, 7·2 per 1,000 live births. 2009 (provisional): marriages, 54,100 (7·0 per 1,000 population); divorces and annulments, 28,500 (3·7).

CLIMATE

Average temperatures in Jan. are 41°F (5°C) in the Tidewater coastal area and 32°F (0°C) in the Blue Ridge mountains; July averages, 78°F (25·5°C)and 68°F (20°C) respectively. Precipitation averages 36" (914 mm) in the Shenandoah valley and 44" (1,118 mm) in the south. Snowfall is 5–10" (125–250 mm) in the Tidewater and 25–30" (625–750 mm) in the western mountains. Norfolk, Jan. 41°F (5°C), July 79°F (26°C). Annual rainfall 46" (1,145 mm). Virginia belongs to the Atlantic Coast climate zone (*see* UNITED STATES: Climate).

CONSTITUTION AND GOVERNMENT

The present constitution became effective in 1971. The General Assembly consists of a Senate of 40 members, elected for four years, and a House of Delegates of 100 members, elected for two years. It sits annually in Jan. The Governor and Lieut.-Governor are elected for four years.

For the 113th Congress, which convened in Jan. 2013, Virginia sends 11 members to the House of Representatives. It is represented in the Senate by Mark Warner (D. 2009–15) and Tim Kaine (D. 2013–19).

The state capital is Richmond; the state contains 95 counties and 40 independent cities.

RECENT ELECTIONS

In the 2012 presidential election Barack Obama took Virginia with 50·8% of the vote, having also won it in 2008.

CURRENT GOVERNMENT

Governor: Bob McDonnell (R.), 2010–14 (salary: $175,000).

Lieut.-Governor: William T. Bolling (R.), 2010–14 ($36,321).

Secretary of the Commonwealth: Janet Vestal Kelly (R.), since Jan. 2010 ($152,793).

Government Website: http://www.virginia.gov

ECONOMY

Per capita personal income (2010) was $44,246.

Budget

In 2011 total state revenue was $50,782m. Total expenditure was $5,549m. (education, $14,372m.; public welfare, $9,348m.; highways, $3,328m.; health, $3,254m.; correction, $1,679m.) Outstanding debt in 2011, $26,479m.

Performance

Gross Domestic Product by state in 2011 was $375,747m. (provisional), ranking Virginia 10th in the United States. In 2011 state real GDP growth was 0·3% (provisional).

ENERGY AND NATURAL RESOURCES

Oil and Gas

In 2008 marketed production of natural gas totalled 128bn. cu. ft.

Water

The total area covered by water is 3,285 sq. miles.

Minerals

Coal is the most important mineral, with output (2009) of 21·2m. short tons. Lead and zinc ores, stone, sand and gravel, lime and titanium ore are also produced. Total domestic non-fuel mineral output was valued at $955m. in 2009.

Agriculture

In 2007 there were 47,383 farms with an area of 8·10m. acres; the average farm had 171 acres, and the average value per acre was $4,213. Farm income, 2006: from crops, $834m.; and from livestock and products, $1,855m. The net farm income in 2006 was $678m. The leading commodities are broilers (cash receipts of $559·4m. in 2007), cattle, dairy products and turkeys. Livestock, 2007: cattle and calves, 1·57m.; dairy cows, 99,000; sheep and lambs, 78,000; hogs and pigs, 371,200; laying hens, 3·21m.

Forestry

Forests covered 15,766,000 acres in 2007, including 1,692,000 acres of national forest.

Fisheries

Commercial catch (2009) totalled 426·3m. lb of fish, worth $152·7m.

INDUSTRY

The manufacture of cigars and cigarettes, of rayon and allied products, and the building of ships lead in value of products. In 2007 the state's 5,777 manufacturing establishments had 277,000 employees, earning $12,170m. Total value added by manufacturing in 2009 was $48,658m.

Labour

Total non-agricultural employment, 2010, 3,627,000. Employees by branch, 2010 (in 1,000): government, 703; professional and business services, 648; trade, transportation and utilities, 620; education and health services, 456; leisure and hospitality, 338. The unemployment rate in Dec. 2010 was 6·6%.

INTERNATIONAL TRADE

Imports and Exports

In 2010 exports from Virginia totalled $17,163m., up from $15,052m. in 2009 although down from $18,942m. in 2008. The leading destinations for exports in 2010 were Canada (value of $2,975m.), followed by China (excluding Hong Kong), Singapore and the United Kingdom.

COMMUNICATIONS

Roads

In 2007 there were 72,662 miles of roads (50,350 miles rural). There were 6,613,781 registered motor vehicles.

Rail

In 2009 there were 3,212 miles of freight railroad (excluding trackage rights). Rail traffic originating in Virginia totalled 34·6m. tons in 2009 and rail traffic terminating in the state came to 63·4m. tons. A light rail system opened in Norfolk in 2011.

Civil Aviation

There are international airports at Norfolk, Dulles, Richmond, Arlington and Newport News. There were 25,607,060 passenger enplanements statewide in 2007.

SOCIAL INSTITUTIONS

Justice

In Dec. 2008 there were 38,276 prison inmates. The death penalty is authorized. Between 1977 and 2012 there were 109 executions in Virginia, after Texas the most of any state. There was one execution in 2011 but none in 2012.

Education

Elementary and secondary instruction is free, and for ages 5–18 attendance is compulsory.

There are 134 school districts. In fall 2007 there were 1,230,857 pupils in 2,027 elementary and secondary schools, with 71,861 teachers. Average annual salary in 2008–09 for public elementary and secondary teachers was $48,554. Total expenditure on education, 2006–07, was $14,234m.

In 2008–09 there were 114 degree-granting education institutions (39 public and 75 private). The leading institutions include:

Founded	Name and place of college	Students 2006–07
1693	College of William and Mary, Williamsburg (State)	7,709
1749	Washington and Lee University, Lexington	2,149
1776	Hampden-Sydney College, Hampden-Sydney (Presbyterian)	1,108
1819	University of Virginia, Charlottesville (State)	24,068
1832	Randolph-Macon College, Ashland (Methodist)	1,146
1832	University of Richmond, Richmond (Baptist)	4,496
1838	Virginia Commonwealth University, Richmond	30,189
1839	Virginia Military Institute Lexington (State)	1,397
1865	Virginia Union University, Richmond	1,599
1868	Hampton University	6,152
1872	Virginia Polytechnic Institute and State University, Blacksburg	28,470
1882	Virginia State University, Petersburg	4,872
1908	James Madison University, Harrisonburg	17,393
1910	Radford University (State)	9,220
1930	Old Dominion University, Norfolk	16,490
1935	Norfolk State University (State)	6,238
1957	George Mason University (State), Fairfax	29,889

Health

In 2009 there were 90 community hospitals with 17,500 beds. A total of 793,000 patients were admitted during the year. In 2009 there were 21,931 active physicians (28 per 10,000 population).

Welfare

Medicare enrolment in July 2010 totalled 1,127,246. In fiscal year 2009 a total of 916,577 people in Virginia received Medicaid. In Dec. 2008 there were 1,207,101 Old-Age, Survivors, and Disability Insurance (OASDI) beneficiaries. A total of 69,563 people were receiving payments under Temporary Assistance for Needy Families (TANF) in Dec. 2008.

RELIGION

The principal religious traditions in the state in 2010 were: Evangelical Protestants, with 1,531,731 members; Mainline Protestants, 870,842; Catholics, 674,555; Muslims, 213,000 (estimate); Black Protestants, 111,028.

CULTURE

Tourism

In 2011 overseas visitors to Virginia—excluding those from Canada and Mexico—numbered 362,000, down from 369,000 in 2010.

FURTHER READING

Statistical information: Cooper Center for Public Service, Univ. of Virginia, 918 Emmet St. N., Suite 300, Charlottesville 22903-4832. Publishes *Virginia Statistical Abstract.—Population Estimates of Virginia Cities and Counties.*

Rubin, L. D. Jr., *Virginia: a Bicentennial History.* 1977

Salmon, E. J. and Campbell Jr., E. D. C., *The Hornbook of Virginia History: A Ready-Reference Guide the Old Dominion's People, Places, and Past.* 1994

State library: Library of Virginia, Richmond 23219.

Washington State

KEY HISTORICAL EVENTS

The strongest Indian tribes in the 18th century were Chinook, Nez Percé, Salish and Yakima. The area was designated by

European colonizers as part of the Oregon Country. Between 1775 and 1800 it was claimed by Spain, Britain and the USA; the dispute between the two latter nations was not settled until 1846.

The first small white settlements were Indian missions and fur-trading posts. In the 1840s American settlers began to push westwards along the Oregon Trail, making a settlement with Britain a matter of urgency. When this was achieved the whole area was organized as the Oregon Territory in 1848. Washington was made a separate Territory in 1853.

Apart from trapping and fishing, the chief industry was supplying timber for the new settlements of California. After 1870 the westward extension of railways encouraged settlement. Statehood was granted in 1889. Settlers were mostly Americans from neighbouring states to the east and Canadians. Scandinavian immigrants followed. Seattle, laid out in 1853 as a saw-milling town, was named after the Indian chief who had ceded the land and befriended the settlers. It grew as a port during the Alaskan and Yukon gold-rushes of the 1890s. The economy thrived on exploiting the Columbia River for hydro-electric power.

TERRITORY AND POPULATION

Washington is bounded north by Canada, east by Idaho, south by Oregon with the Columbia River forming most of the boundary, and west by the Pacific. Land area, 66,456 sq. miles (172,119 sq. km). Lands owned by the federal government, 2003, were 13,246,559 acres or 31·0% of the total area. Census population, 1 April 2010, was 6,724,540, an increase of 14·1% since 2000. July 2012 estimate, 6,897,012.

Population in five federal census years was:

	White	Black	American Indian/Alaska Native	Asian/ Other	Total	Per sq. mile
1910	1,109,111	6,058	10,997	15,824	1,141,990	17·1
1980	3,779,170	105,574	60,804	186,608	4,132,156	62·1
1990	4,308,937	149,801	81,483	326,471	4,866,692	73·1
2000	4,821,823	190,267	93,301	788,730	5,894,121	88·6
2010	5,196,362	240,042	103,869	1,184,267	6,724,540	101·2

Of the total population in 2010, 3,374,833 were female and 5,143,186 were 18 years old or older. In 2010 the Hispanic population was 755,790, up from 441,509 in 2000 (a rise of 71·2%).

There were 26 Indian reservations in 2000. American Indian lands covered 5,963 sq. miles (5,157 sq. miles in reservations).

Leading cities are Seattle, with a population in 2010 of 608,660; Spokane, 208,916; Tacoma, 198,397; Vancouver, 161,791; Bellevue, 122,363. Others: Everett, 103,019; Kent, 92,411; Yakima, 91,067; Renton, 90,927; Spokane Valley, 89,755; Federal Way, 89,306; Bellingham, 80,885; Kennewick, 73,917; Auburn, 70,180; Marysville, 60,020; Pasco, 59,781; Lakewood, 58,163; Redmond, 54,144; Shoreline, 53,007. The Seattle–Tacoma–Bremerton metropolitan area had a 2010 census population of 3,439,809. The capital, Olympia, had a population in 2010 of 46,478.

SOCIAL STATISTICS

Births, 2010, 86,539 (12·9 per 1,000 population); deaths, 2010 (provisional), 48,145 (7·2). Infant mortality rate, 2009, 4·9 per 1,000 live births. 2009 (provisional): marriages, 40,400 (6·0 per 1,000 population); divorces and annulments, 26,300 (3·9). Same-sex marriage was legalized in Nov. 2012 when it was approved in a referendum on the same day as the presidential election.

CLIMATE

Seattle, Jan. 40°F (4·4°C), July 63°F (17·2°C). Annual rainfall 34" (848 mm). Spokane, Jan. 27°F (−2·8°C), July 70°F (21·1°C).

Annual rainfall 14" (350 mm). Washington belongs to the Pacific Coast climate zone (see UNITED STATES: Climate).

CONSTITUTION AND GOVERNMENT

The constitution, adopted in 1889, has had 104 amendments as of Aug. 2011. The Legislature consists of a Senate of 49 members elected for four years, half their number retiring every two years, and a House of Representatives of 98 members, elected for two years. The Governor and Lieut.-Governor are elected for four years.

For the 113th Congress, which convened in Jan. 2013, Washington sends ten members to the House of Representatives. It is represented in the Senate by Patty Murray (D. 1993–2017) and Maria Cantwell (D. 2001–19).

The capital is Olympia. The state contains 39 counties.

RECENT ELECTIONS

In the 2012 presidential election Barack Obama took Washington State with 55·8% of the vote, having also won it in 2008.

CURRENT GOVERNMENT

Governor: Jay Inslee (D.), 2013–17 (salary: $166,891).
    Lieut.-Governor: Brad Owen (D.), 2013–17 ($93,948).
    Secretary of State: Kim Wyman (R.), 2013–17 ($116,950).

Government Website: http://access.wa.gov

ECONOMY

Per capita personal income (2010) was $42,570.

Budget

In 2011 total state revenue was $50,420m. Total expenditure was $46,000m. (education, $14,886m.; public welfare, $8,669m.; highways, $3,030m.; hospitals, $2,051m.; health, $1,797m.) Outstanding debt in 2011, $28,154m.

Performance

In 2011 Gross Domestic Product by state was $310,906m. (provisional), ranking Washington 14th in the United States. In 2011 state real GDP growth was 2·0% (provisional).

ENERGY AND NATURAL RESOURCES

Electricity

In 2009 production was 104·5bn. kWh, of which 72·9bn. kWh was from conventional hydro-electric energy.

Water

The total area covered by water is 4,842 sq. miles.

Minerals

Mining and quarrying are not as important as forestry, agriculture or manufacturing. Total value of non-fuel mineral production in 2009 was $650m.

Agriculture

Agriculture is constantly growing in value as a result of more intensive and diversified farming, and because of the 1m.-acre Columbia Basin Irrigation Project.

In 2007 there were 39,284 farms with an acreage of 15·32m.; the average farm was 381 acres. Average value of farmland and buildings per acre in 2007 was $1,992. Apples, milk, wheat, cattle and calves and potatoes are the top five commodities. Washington is the USA's largest producer of apples, cherries and pears. In 2007 livestock included 1·09m. all cattle (274,000 beef cows and 243,100 dairy cows), 53,200 sheep and lambs, 28,500 hogs and pigs, and 5·79m. laying hens.

Farm income, 2006: from crops, $4,524m.; from livestock and products, $1,615m. The net farm income in 2006 was $958m.

Forestry

Forests covered 22·28m. acres in 2007, of which 8·2m. acres were national forest. In 2006 timber harvested totalled 3,484m. bd ft. Production of wood and bark residues, 2004, was 5,956,000 tons.

Fisheries

Salmon and shellfish are important; total commercial catch, 2009, was 163·9m. lb and was worth $227·8m.

INDUSTRY

Principal manufactures are aircraft, pulp and paper, lumber and plywood, aluminium, processed fruit and vegetables. In 2007 the state's 7,650 manufacturing establishments had 270,000 employees, earning $13,275m. Total value added by manufacturing in 2009 was $42,628m.

Labour

In 2010 total non-agricultural employment was 2,777,000. Employees by branch, 2010 (in 1,000): government, 547; trade, transportation and utilities, 517; education and health services, 375; professional and business services, 326; leisure and hospitality, 266. The unemployment rate in Dec. 2010 was 9·3%.

INTERNATIONAL TRADE

Imports and Exports

In 2010 exports from Washington totalled $53,353m., up from $51,851m. in 2009 although down from $54,498m. in 2008. The leading destinations for exports in 2010 were China (excluding Hong Kong) (value of $10,303m.), followed by Canada, Japan and South Korea. Washington was the fifth largest exporting state in the USA in 2010.

COMMUNICATIONS

Roads

In 2007 there were 83,431 miles of road comprising 22,551 miles of urban road and 60,880 miles of rural road. There were 5,757,943 registered motor vehicles.

Rail

In 2009 there were 3,169 miles of freight railroad (excluding trackage rights). There were 22 freight railroads operating in 2009. Rail traffic originating in Washington totalled 18·5m. tons in 2009 and rail traffic terminating in the state came to 59·1m. tons. A light rail system opened in Seattle in 2009.

Civil Aviation

There are international airports at Seattle/Tacoma, Spokane and Boeing Field. There were 17,892,317 passenger enplanements statewide in 2007.

SOCIAL INSTITUTIONS

Justice

In Dec. 2008 there were 17,926 prison inmates. There was one execution in 2010, the first since 2001, but none in 2011 or 2012.

Education

Education is given free to all children between the ages of five and 21 years, and is compulsory for children from eight to 18 years of age. In fall 2007 there were 1,030,247 pupils in public elementary and secondary schools with 53,960 teachers; teachers' salaries in 2008–09 averaged $51,970. In fall 2007 there were 104,070 pupils in 730 private schools.

The University of Washington, founded 1861, at Seattle, had (in fall 2009) 45,943 students; and Washington State University at Pullman, founded 1890, for science and agriculture, had 26,101 students. Eastern Washington University had 11,300; Central Washington University, 11,357; Evergreen State College, 4,891; Western Washington University, 14,575. In 2005–06 there were 35 community colleges with a total of 190,423 students.

Health

In 2009 there were 87 community hospitals with 11,300 beds. A total of 589,000 patients were admitted during the year. In 2009 there were 18,090 active physicians (27 per 10,000 population).

Welfare

Medicare enrolment in July 2010 totalled 965,207. In fiscal year 2009 a total of 1,176,868 people in Washington received Medicaid. In Dec. 2008 there were 1,008,804 Old-Age, Survivors, and Disability Insurance (OASDI) beneficiaries. A total of 132,476 people were receiving payments under Temporary Assistance for Needy Families (TANF) in Dec. 2008.

RELIGION

The principal religious traditions in the state in 2010 were: Evangelical Protestants, with 820,643 members; Catholics, 784,332; Mainline Protestants, 308,292; Latter-day Saints (Mormons), 271,414; Buddhists, 49,065.

CULTURE

Tourism

In 2011 overseas visitors to Washington—excluding those from Canada and Mexico—numbered 502,000, up from 501,000 in 2010 and 390,000 in 2006. There were 2,331,000 visits by Canadians in 2010.

FURTHER READING

Statistical information: State Office of Financial Management, POB 43113, Olympia 98504-3113. Publishes *Washington State Data Book*

Dodds, G. B., *American North-West: a History of Oregon and Washington.* 1986

West Virginia

KEY HISTORICAL EVENTS

In 1861 the slave-owning state of Virginia seceded from the Union. The 40 western counties, mostly hilly country and settled by miners and small farmers who were not slave-owners, were declared a new state. On 20 June 1863 West Virginia became the 35th state of the Union.

The capital, Charleston, was an 18th-century fortified post on the early westward migration routes across the Appalachians. In 1795 local brine wells were tapped and the city grew as a salt town. Coal, oil, natural gas and a variety of salt brines were all found in due course. Huntington, the next largest town, served the same industrial area as a railway terminus and port on the Ohio river. Wheeling, the original state capital, located on the major transportation routes of the Ohio River, Baltimore and Ohio Railroad and the National Road, was a well established, cosmopolitan city when it hosted the statehood meetings in 1861.

Three-quarters of the state is forest. Settlement has been concentrated in the mineral-bearing Kanawha valley, along the Ohio river and in the industrial Monongahela valley of the north. More than half of the population is still classified as rural. However, the majority commute to industrial employment.

TERRITORY AND POPULATION

West Virginia is bounded in the north by Pennsylvania and Maryland, east and south by Virginia, southwest by the Big Sandy River (forming the boundary with Kentucky) and west by the Ohio River (forming the boundary with Ohio). Land area, 24,038 sq. miles (62,259 sq. km). Census population, 1 April 2010, was 1,852,994, an increase of 2·5% since 2000. July 2012 estimate, 1,855,413.

Population in five federal census years was:

	White	Black	American Indian/Alaska Native	Asian/Other	Total	Per sq. mile
1910	1,156,817	64,173	36	93	1,221,119	50·8
1980	1,874,751	65,051	1,610	5,194	1,949,644	80·3
1990	1,725,523	56,295	2,458	7,459	1,793,477	74·0
2000	1,718,777	57,232	3,606	9,834	1,808,344	75·1
2010	1,739,988	63,124	3,787	46,095	1,852,994	77·1

Of the total population in 2010, 939,408 were female and 1,465,576 were 18 years old or older. In 2010 the Hispanic population was 22,268, up from 12,279 in 2000 (an increase of 81·4%). Even with this increase West Virginia has the smallest Hispanic percentage in its population (at 1·2%) of any state.

The 2010 census population of the principal cities was: Charleston (the capital), 51,400; Huntington, 49,130. Others: Parkersburg, 31,492; Morgantown, 29,660; Wheeling, 28,486; Weirton, 19,746; Fairmont, 18,704; Beckley, 17,614; Martinsburg, 17,227; Clarksburg, 16,578.

SOCIAL STATISTICS

Births, 2010, 20,470 (11·0 per 1,000 population); deaths, 2010 (provisional), 21,274 (11·5—the highest rate in any US state). West Virginia was the only state in the USA where deaths exceeded births in 2010. Infant mortality, 2009, 7·8 per 1,000 live births. 2009 (provisional): marriages, 12,400 (6·9 per 1,000 population); divorces and annulments, 9,200 (5·2).

CLIMATE

Charleston, Jan. 34°F (1·1°C), July 76°F (24·4°C). Annual rainfall 40″ (1,010 mm). West Virginia belongs to the Appalachian Mountains climate zone (see UNITED STATES: Climate).

CONSTITUTION AND GOVERNMENT

The present constitution was adopted in 1872; it has had 17 amendments. The Legislature consists of the Senate of 34 members elected for a term of four years, one-half being elected biennially, and the House of Delegates of 100 members, elected biennially. The Governor is elected for four years and may serve one successive term.

For the 113th Congress, which convened in Jan. 2013, West Virginia sends three members to the House of Representatives. It is represented in the Senate by Jay Rockefeller IV (D. 1985–2015) and Joe Manchin III (D. 2011–19).

The state capital is Charleston. There are 55 counties.

RECENT ELECTIONS

In the 2012 presidential election Mitt Romney took West Virginia with 62·3% of the vote (John McCain won it in 2008).

CURRENT GOVERNMENT

Governor: Earl Ray Tomblin (D.), 2013–17 (salary: $150,000).

Senate President: Jeffrey V. Kessler (D.), 2013–15 (base salary: $20,000).

Secretary of State: Natalie Tennant (D.), 2013–17 ($95,000).

Government Website: http://www.wv.gov

ECONOMY

Per capita personal income (2010) was $31,999.

Budget

Total revenues in 2011 were $15,329m. Total expenditures were $13,000m. (education, $4,162m.; public welfare, $3,288m.; highways, $1,224m.; government administration, $418m.; health, $318m.) Outstanding debt in 2011, $7,406m.

Performance

Gross Domestic Product by state in 2011 was $55,765m. (provisional), ranking West Virginia 41st in the United States. In 2011 state real GDP growth was 4·5% (provisional).

ENERGY AND NATURAL RESOURCES

Electricity

In 2009 total production was 70,783m. kWh., of which 96% was from coal (the highest percentage of any state).

Oil and Gas

In 2008 West Virginia produced 2m. bbls of crude petroleum; marketed production of natural gas totalled 246bn. cu. ft.

Water

The total area covered by water is 192 sq. miles.

Minerals

West Virginia is the nation's second largest coal-producing state after Wyoming; output was 137·0m. short tons in 2009. Coal-related employment (21,665 jobs in 2009) is higher than in any other state. Salt, sand and gravel, sandstone and limestone are also produced. The total value of non-fuel mineral production in 2009 was $215m.

Agriculture

In 2007 the state had 23,618 farms with an area of 3·70m. acres; average size of farm was 157 acres, valued at $2,385 per acre. Livestock farming predominates. Principal commodities are broilers, cattle, turkeys and dairy products.

Cash income, 2006: from crops was $80m.; from livestock and products, $370m. The net farm income in 2006 was $50m. The most important crops are hay, corn and apples.

Livestock on farms, 2007, included 411,000 cattle, of which 204,000 were beef cows and 12,000 dairy cows; sheep, 38,000; hogs and pigs, 9,000; laying hens, 1·22m.

Forestry

Forests covered 12,007,000 acres in 2007, with 1,073,000 acres of national forest. 78·5% of the state is woodland.

INDUSTRY

In 2007 the state's 1,413 manufacturing establishments had 60,000 employees, earning $2,646m. Total value added by manufacturing in 2009 was $9,307m.

Labour

Total non-agricultural employment, 2010, 746,000. Employees by branch, 2010 (in 1,000): government, 153; trade, transportation and utilities, 135; education and health services, 121; leisure and hospitality, 72; professional and business services, 61. The state unemployment rate in Dec. 2010 was 9·7%.

INTERNATIONAL TRADE

Imports and Exports

In 2010 exports from West Virginia totalled $6,449m., up from $4,826m. in 2009. The leading destinations for exports in 2010 were Canada (value of $1,474m.), followed by Japan, Brazil and the Netherlands.

COMMUNICATIONS

Roads

In 2007 there were 38,274 miles of road (32,995 miles rural). There were 1,413,467 registered motor vehicles.

Rail

In 2009 there were 2,231 miles of freight railroad (excluding trackage rights). Rail traffic originating in West Virginia totalled 99·7m. tons in 2009 (the third highest after Wyoming and Illinois) with rail traffic terminating in the state totalling 18·2m. tons.

Civil Aviation

There were five primary airports—commercial service airports with more than 10,000 passenger boardings annually—in 2011 with a combined total of 435,193 enplanements, up from 414,317 in 2010.

Shipping

There are some 420·5 miles of navigable rivers.

SOCIAL INSTITUTIONS

Justice

The state court system consists of a Supreme Court, 31 circuit courts, and magistrate courts in each county. The Supreme Court of Appeals, exercising original and appellate jurisdiction, has five members elected by the people for 12-year terms. Each circuit court has from one to seven judges (as determined by the Legislature on the basis of population and case-load) chosen by the voters within each circuit for eight-year terms.

In Dec. 2008, 6,059 prisoners were under the jurisdiction of state and federal correctional authorities. Capital punishment was abolished in 1965. The last execution was in 1959.

Education

School attendance is compulsory for all between the ages of six and 16. In fall 2007, 762 public elementary and secondary schools had 282,535 pupils and 20,306 teachers. Total expenditure on public elementary and secondary education in 2006–07 was $2,845m.; average teacher salary was $44,625 in 2008–09. Spending per pupil in fiscal year 2008 was $10,059.

In 2008–09 there were 23 public and 21 private degree-granting institutions; leading public institutions of higher education include:

Founded		Students (fall 2005)
1837	Marshall University, Huntington	13,988[1]
1837	West Liberty State College, West Liberty	2,248
1867	Fairmont State University, Fairmont	4,740
1868	West Virginia University, Morgantown	26,051
1872	Concord College, Athens	2,826
1872	Glenville State College, Glenville	1,392
1872	Shepherd University, Shepherdstown	3,901
1891	West Virginia State University, Institute	3,491
1895	West Virginia Univ. Inst. of Technology, Montgomery	1,551
1895	Bluefield State College, Bluefield	1,708
1901	Potomac State College of West Virginia Univ., Keyser	1,279
1961	West Virginia Univ. at Parkersburg, Parkersburg	3,772
1976	School of Osteopathic Medicine, Lewisburg	394

[1]Includes Marshall Univ. Graduate College, South Charleston, founded in 1972.

Health

In 2009 there were 56 community hospitals with 7,400 beds. A total of 280,000 patients were admitted during the year. In 2009 there were 4,295 active physicians (24 per 10,000 population).

Welfare

The Department of Health Human Resources, originating in the 1930s as the Department of Public Assistance, is both state and federally financed. Medicare enrolment in July 2010 totalled 377,811. In fiscal year 2009 a total of 386,427 people in West Virginia received Medicaid. In Dec. 2008 there were 429,613 Old-Age, Survivors, and Disability Insurance (OASDI) beneficiaries. A total of 20,743 people were receiving payments under Temporary Assistance for Needy Families (TANF) in Dec. 2008.

RELIGION

The principal religious traditions in the state in 2010 were: Mainline Protestants, with 274,766 members; Evangelical Protestants, 249,756; Catholics, 95,849; Latter-day Saints (Mormons), 17,555; Black Protestants, 11,505.

FURTHER READING

West Virginia Blue Book. Annual, since 1916

Rice, O. K., *West Virginia: A History.* 2nd ed. 1994

State library: Archives and History, Division of Culture and History, Charleston.

Wisconsin

KEY HISTORICAL EVENTS

The French were the first European explorers of the territory; Jean Nicolet landed at Green Bay in 1634, a mission was founded in 1671 and a permanent settlement at Green Bay followed. In 1763 French claims were surrendered to Britain. In 1783 Britain ceded the area to the USA which designated the Northwest Territory, of which Wisconsin was part. In 1836 a separate Territory of Wisconsin included the present Iowa, Minnesota and parts of the Dakotas.

In 1836 James Duane Doty founded Madison and, even before it was inhabited, successfully pressed its claim to be the capital of the Territory. In 1848 Wisconsin became a state, with its present boundaries.

The city of Milwaukee was founded on Lake Michigan when Indian tribes gave up their claims to the land in 1831–33. It grew rapidly as a port and industrial town, attracting German settlers in the 1840s and Poles and Italians 50 years later. The Lake Michigan shore was developed as an industrial area; the rest of the south proved suitable for dairy farming; the north, mainly forests and lakes, has remained sparsely settled except for tourist bases.

There are 11 Indian reservations where more than 15,500 of Wisconsin's 47,000 Indians live. Since the Second World War there has been black immigration from the southern states to the industrial lake-shore cities.

TERRITORY AND POPULATION

Wisconsin is bounded north by Lake Superior and the Upper Peninsula of Michigan, east by Lake Michigan, south by Illinois, and west by Iowa and Minnesota, with the Mississippi River forming most of the boundary. Land area, 54,158 sq. miles (140,268 sq. km). Census population, 1 April 2010, was 5,686,986, an increase of 6·0% since 2000. July 2012 estimate, 5,726,398.

Population in five census years was:

	White	Black	All others	Total	Per sq. mile
1910	2,320,555	2,900	10,405	2,333,860	42·2
1980	4,443,035	182,592	80,015	4,705,642	86·4
1990	4,512,523	244,539	134,707	4,891,769	90·1
2000	4,769,857	304,460	289,358	5,363,675	98·8
2010	4,902,067	359,148	425,771	5,686,986	105·0

Of the total population in 2010, 2,864,586 were female and 4,347,494 were 18 years old or older. In 2010 Wisconsin's Hispanic population was 336,056, up from 192,921 in 2000 (an increase of 74·2%).

Population of the large cities, 2010 census, was as follows:

Milwaukee	594,833	West Allis	60,411
Madison (capital)	233,209	La Crosse	51,320
Green Bay	104,057	Sheboygan	49,288
Kenosha	99,218	Wauwatosa	46,396
Racine	78,860	Fond du Lac	43,021
Appleton	72,623	New Berlin	39,584
Waukesha	70,718	Wausau	39,106
Oshkosh	66,083	Brookfield	37,920
Eau Claire	65,883	Beloit	36,966
Janesville	63,575	Greenfield	36,720

Population of largest metropolitan areas, 2010 census: Milwaukee–Waukesha–West Allis, 1,555,908; Madison, 568,593; Green Bay, 306,241.

SOCIAL STATISTICS

Births, 2010, 68,487 (12·0 per 1,000 population); deaths, 2010 (provisional), 47,308 (8·3). Infant deaths, 2009, 6·1 per 1,000 live births. 2009 (provisional): marriages, 30,300 (5·3 per 1,000 population); divorces and annulments, 17,300 (3·0).

CLIMATE

Milwaukee, Jan. 19°F (−7·2°C), July 70°F (21·1°C). Annual rainfall 29" (727 mm). Wisconsin belongs to the Great Lakes climate zone (*see* UNITED STATES: Climate).

CONSTITUTION AND GOVERNMENT

The constitution, which dates from 1848, has 141 amendments. The legislative power is vested in a Senate of 33 members elected for four years, one-half elected alternately, and an Assembly of 99 members all elected simultaneously for two years. The Governor and Lieut.-Governor are elected for four years.

For the 113th Congress, which convened in Jan. 2013, Wisconsin sends eight members to the House of Representatives. It is represented in the Senate by Ronald Johnson (R. 2011–17) and Tammy Baldwin (D. 2013–19).

The capital is Madison. The state has 72 counties.

RECENT ELECTIONS

In the 2012 presidential election Barack Obama took Wisconsin with 52·8% of the vote, having also won it in 2008.

CURRENT GOVERNMENT

Governor: Scott Walker (R.), 2011–15 (salary: $144,423).
  Lieut.-Governor: Rebecca Kleefisch (R.), 2011–15 ($76,261).
  Secretary of State: Douglas LaFollette (D.), 2011–15 ($68,556).

Government Website: http://www.wisconsin.gov

ECONOMY

Per capita personal income (2010) was $38,177.

Budget

Total state revenues in 2011 were $44,807m.; total expenditure, $39,350m. (including: education, $11,345m.; public welfare, $8,969m.; highways, $2,432m.; hospitals, $1,188m.; correction, $1,161m.) Outstanding debt in 2011, $22,879m.

Performance

Gross Domestic Product by state in 2011 was $221,741m. (provisional), ranking Wisconsin 21st in the United States. In 2011 state real GDP growth was 1·1% (provisional).

ENERGY AND NATURAL RESOURCES

Electricity

In 2009 production was 59,959m. kWh, of which 37,280m. kWh was from coal.

Water

The total area covered by water is 11,339 sq. miles.

Minerals

Construction sand and gravel, crushed stone, industrial or specialty sand and lime are the chief mineral products. Mineral production in 2009 was valued at $546m.

Agriculture

In 2007 there were 78,463 farms with a total acreage of 15·19m. acres and an average size of 194 acres, compared with 142,000 farms with a total acreage of 22·4m. acres and an average of 158 acres in 1959. In 2007 the average value per acre was $3,225. Farm income, 2006: from crops $2,135m.; from livestock and products, $4,656m. The net farm income in 2006 was $1,091m.

Dairy farming is important, with 1·25m. dairy cows in 2007. Production of cheese was 2·5bn. lb in 2008 and accounted for over 25% of the USA's total. Cash receipts from dairy products were $4·57bn. in 2008. Other important commodities are corn, cattle, soybeans, potatoes, cranberries and wheat. Wisconsin is the leading cranberry producer, with 224,000 tons in 2008, accounting for 57% of the USA's total.

Forestry

Wisconsin had (2007) 16,275,000 acres of forested land, with 1,407,000 acres of national forest. In 2006 timber removals were 454m. cu. ft (349m. cu. ft hardwoods and 105m. cu. ft softwoods).

INDUSTRY

Wisconsin has much heavy industry, particularly in the Milwaukee area. Three-fifths of manufacturing employees work on durable goods. Industrial machinery is the major industrial group (17% of all manufacturing employment) followed by fabricated metals, food and kindred products, printing and publishing, paper and allied products, electrical equipment and transportation equipment. In 2007 the state's 9,659 manufacturing establishments had 488,000 employees, earning $21,850m. Total value added by manufacturing in 2009 was $58,900m.

Labour

Total non-agricultural employment, 2010, 2,735,000. Employees by branch, 2010 (in 1,000): trade, transportation and utilities, 508; manufacturing, 431; government, 421; education and health services, 418; professional and business services, 268. Average annual pay per worker (2009) was $39,131. The state unemployment rate in Dec. 2010 was 7·5%.

INTERNATIONAL TRADE

Imports and Exports

In 2010 exports from Wisconsin totalled $19,790m., up from $16,725m. in 2009 although down from $20,570m. in 2008. The leading destinations for exports in 2010 were Canada (value of $6,053m.), followed by Mexico, China (excluding Hong Kong) and Germany.

COMMUNICATIONS

Roads

In 2007 the state had 114,705 miles of road of which 92,557 miles were rural roads. There were 5,017,895 registered motor vehicles.

Rail

In 2009 there were 3,510 miles of freight railroad (excluding trackage rights). There were ten freight railroads operating in 2009. Rail traffic originating in Wisconsin totalled 12·9m. tons in 2009 with rail traffic terminating in the state totalling 56·6m. tons.

Civil Aviation

There were eight primary airports—commercial service airports with more than 10,000 passenger boardings annually—in 2011 with a combined total of 6,292,628 enplanements, down from 6,459,046 in 2010. By far the busiest airport is General Mitchell International (Milwaukee's airport), with 4,671,976 enplanements in 2011.

Shipping

Lake Superior and Lake Michigan ports handled 47·8m. tons of freight in 2002; 87% of it at Superior, one of the world's biggest grain ports, and much of the rest at Milwaukee and Green Bay.

SOCIAL INSTITUTIONS

Justice

In Dec. 2008 there were 23,380 prison inmates. The death penalty was abolished in 1853.

Education

All children between the ages of six and 18 are required to attend school full-time to the end of the school term in which they become 18 years of age. In fall 2007, 2,268 public elementary and secondary schools had 874,633 pupils; there were 58,914 teachers. Private schools enrolled 138,290 students. Public pre-schools enrolled 37,773 children. Average public school teacher salary in 2008–09 was $50,424. Total expenditure for public elementary and secondary education was $10,373m. in 2006–07. Spending per pupil in fiscal year 2008 was $10,791.

There are three non-profit higher educational sectors in Wisconsin: the public colleges and universities, the technical colleges, and the private colleges and universities.

The University of Wisconsin, established in 1848, was joined by law in 1971 with the Wisconsin State Universities System to become the University of Wisconsin System with 13 degree granting campuses, 13 two-year campuses in the Center System and the state-wide University Extension. The system had, in 2009–10, 6,898 faculty members. In fall 2009, 178,909 students enrolled (11,216 at Eau Claire, 6,638 at Green Bay, 10,009 at La Crosse, 41,654 at Madison, 30,418 at Milwaukee, 13,192 at Oshkosh, 5,303 at Parkside, 7,803 at Platteville, 6,728 at River Falls, 9,209 at Stevens Point, 9,017 at Stout, 2,794 at Superior, 11,139 at Whitewater and 13,789 at the Center System freshman-sophomore centres).

The Wisconsin Technical College System has 16 technical college districts, which award two-year associate degrees, one- and two-year technical diplomas and short-term technical diplomas. Approximately 400,000 students enrol in the colleges each year.

UW-Extension enrolled 187,809 students in its continuing education programmes in year ending June 2009. Independent institutions of higher education include: Marquette University (Jesuit), in Milwaukee (with 11,689 students in fall 2009); Cardinal Strich University (Franciscan), with campuses in Milwaukee, Madison and Edina, Minnesota (6,276); Concordia University Wisconsin (Lutheran), in Mequon (7,178); and Lawrence University, Appleton (1,483). The state's educational and broadcasting service is licensed through the UW Board of Regents.

Health

In 2009 there were 126 community hospitals with 13,600 beds. A total of 609,000 patients were admitted during the year. In 2009 there were 14,816 active physicians (26 per 10,000 population).

Welfare

Medicare enrolment in July 2010 totalled 902,894. In fiscal year 2008 a total of 1,531,520 people in Wisconsin received Medicaid. In Dec. 2010 there were 1,061,501 Old-Age, Survivors, and Disability Insurance (OASDI) beneficiaries. A total of 59,364 people were receiving payments under Temporary Assistance for Needy Families (TANF) in Dec. 2010.

RELIGION

In 2010 the main religious traditions were: Catholics, with 1,425,523 members; Evangelical Protestants, 806,028; Mainline Protestants, 684,035; Black Protestants, 44,203; Latter-day Saints (Mormons), 25,738.

FURTHER READING

Wisconsin Blue Book. Biennial
State Historical Society of Wisconsin: *The History of Wisconsin.* Vol. IV [J. Buenker]. 1999

State Information Agency: Legislative Reference Bureau, One East Main St., Suite 200, Madison, WI 53703-2037.
Website: http://www.legis.state.wi.us

Wyoming

KEY HISTORICAL EVENTS

The territory was inhabited by Plains Indians (Arapahoes, Sioux and Cheyenne) in the early 19th century. There was some trading with white Americans, but very little white settlement. In the 1840s the great western migration routes, the Oregon and the Overland Trails, ran through the territory with Wyoming offering mountain passes accessible to wagons. Once migration became a steady flow forts were built to protect the route from Indian attack.

In 1867 coal was discovered. In 1868 Wyoming was organized as a separate Territory and in 1869 the Sioux and Arapaho were confined to reservations. At the same time the route of the Union Pacific Railway brought railway towns to southern Wyoming. Settlement of the north was delayed until after the final defeat of hostile Indians in 1876.

The economy was based on ranching. Cheyenne, made Territorial capital in 1869, also functioned as a railway town moving cattle. Casper, on the site of a fort on the Pony Express route, was also a railway town on the Chicago and North Western. Laramie started as a Union Pacific construction workers' shanty town in 1868. In 1890 oil was discovered at Casper, and Wyoming became a state the same year. Subsequently, mineral extraction became the leading industry, as natural gas, uranium, bentonite and trona were exploited as well as oil and coal.

TERRITORY AND POPULATION

Wyoming is bounded north by Montana, east by South Dakota and Nebraska, south by Colorado, southwest by Utah and west by Idaho. Land area, 97,093 sq. miles (251,470 sq. km). The Yellowstone National Park occupies about 2·22m. acres; the Grand Teton National Park has 307,000 acres. The federal government in 2003 owned 49,268 sq. miles (50·6% of the total area of the state). The Federal Bureau of Land Management administers 17·4m. acres.

Census population, 1 April 2010, was 563,626, an increase of 14·1% since 2000. Wyoming has the smallest population of any of the states of the USA, but in the year 1 July 2008–30 June 2009 the state's population showed the largest percentage growth of any in the USA. July 2012 estimate, 576,412.

Population in five census years was:

	White	Black	American Indian	Asiatic	Total	Per sq. mile
1910	140,318	2,235	1,486	1,926	145,965	1·5
			All others			
1980	446,488	3,364	19,705		469,557	4·8

	White	Black	American Indian/ Alaska Native	Asian/ Native Hawaiian/ Pacific Islander	Other	Total	Per sq. mile
1990	427,061	3,606	9,479	2,806	10,636	453,588	4·7
2000	454,670	3,722	11,133	3,073	21,184	493,782	5·1
2010	511,279	4,748	13,336	4,853	29,410	563,626	5·8

Of the total population in 2010, 287,437 were male and 428,224 were 18 years old or older. At the 2010 census the Hispanic population of Wyoming was 50,231, up from 31,669 in 2000 (an increase of 58·6%).

The largest towns (with 2010 census population) are Cheyenne (the capital), 59,466; Casper, 55,316; Laramie, 30,816; Gillette, 29,087; Rock Springs, 23,036; Sheridan, 17,444.

SOCIAL STATISTICS

Births, 2010, 7,556 (13·4 per 1,000 population); deaths, 2010 (provisional), 4,438 (7·9). Infant mortality rate, 2009, 6·0 per 1,000 live births. 2009 (provisional): marriages, 4,700 (8·2 per 1,000 population); divorces and annulments, 2,800 (5·2). The abortion rate, at less than 1 for every 1,000 women aged 15 to 44 in 2008, is the lowest of any US state.

CLIMATE

Cheyenne, Jan. 25°F (−3·9°C), July 66°F (18·9°C). Annual rainfall 15" (376 mm). Yellowstone Park, Jan. 18°F (−7·8°C), July 61°F (16·1°C). Annual rainfall 18" (444 mm). Wyoming belongs to the Mountain States climate region (see UNITED STATES: Climate).

CONSTITUTION AND GOVERNMENT

The constitution, drafted in 1890, has since had 76 amendments. The Legislature consists of a Senate of 30 members elected for staggered four-year terms, and a House of Representatives of 60 members elected for two years. It sits annually in Jan. or Feb. The Governor is elected for four years.

For the 113th Congress, which convened in Jan. 2013, Wyoming sends one member to the House of Representatives. It is represented in the Senate by Michael Enzi (R. 1997–2015) and John Barrasso (R. 2007–19).

The capital is Cheyenne. The state contains 23 counties.

RECENT ELECTIONS

In the 2012 presidential election Mitt Romney took Wyoming with 69·3% of the vote (John McCain won it in 2008).

CURRENT GOVERNMENT

Governor: Matt Mead (R.), 2011–15 (salary: $105,000).
  Secretary of State: Max Maxfield (R.), 2011–15 ($92,000).

Government Website: http://wyoming.gov

ECONOMY

Per capita personal income (2010) was $44,861.

Budget

In 2011 total state revenue was $7,494m. Total expenditure was $5,674m. (education, $1,679m.; public welfare, $740m.; highways, $537m.; natural resources, $354m.; health, $282m.) Outstanding debt in 2011, $1,364m.

Performance

Gross Domestic Product by state was $31,542m. in 2011 (provisional), ranking Wyoming 49th in the United States. In 2011 state real GDP growth was −1·2% (provisional).

ENERGY AND NATURAL RESOURCES

Oil and Gas

Wyoming produces significant quantities of oil and natural gas. In 2008 the output of oil was 53m. bbls; marketed production of natural gas totalled 2,275bn. cu. ft (the second highest total after Texas).

Water

The total area covered by water is 720 sq. miles.

Minerals

In 2010 the output of coal was 442·1m. short tons; trona, 16·5m. short tons; uranium (2010 estimate), 1·8m. lb. Wyoming is the USA's leading coal and uranium producer, accounting for 40% of the country's coal output in 2010. It also has 40% of the country's recoverable coal reserves. Wyoming's recoverable coal reserves total more than 42bn. tons. Total value of non-fuel mineral production in 2010 was $1,860m.

Agriculture

Wyoming is semi-arid, and agriculture is carried on by irrigation and dry farming. In 2007 there were 11,069 farms and ranches;

total farm area was 34·4m. acres; average size of farm in 2007 was 2,726 acres (the largest of any state). In 2007 the average value of farmland was $513 per acre.

Total value, 2006, of crops produced, $162m.; of livestock and products, $859m. The net farm income in 2006 was $65m. Principal commodities are cattle, hogs, hay, sheep, sugar beets, barley, wheat and corn. Animals on farms in 2007 included 1·31m. cattle, 413,000 sheep and lambs, and 107,000 hogs and pigs.

Forestry

The state had a forested area of 11,445,000 acres in 2007, of which 6,028,000 acres were national forest.

INDUSTRY

In 2007 the state's 596 manufacturing establishments had 12,000 employees, earning $574m. Total value added by manufacturing in 2009 was $3,744m. In 2009 there were 852 mining establishments. A large portion of the manufacturing in the state is based on natural resources, mainly oil and farm products.

Labour

Total non-agricultural employment, 2010, 283,000. Employees by branch, 2010 (in 1,000): government, 73; trade, transportation and utilities, 52; leisure and hospitality, 33; education and health services, 26; mining, 25.

INTERNATIONAL TRADE

Imports and Exports

In 2010 exports from Wyoming totalled $983m., up from $926m. in 2009 although down from $1,081m. in 2008. The leading destinations for exports in 2010 were Canada (value of $238m.), followed by Brazil, Mexico and Indonesia.

COMMUNICATIONS

Roads

In 2007 there were 2,642 miles of urban roads and 25,410 miles of rural roads. There were 652,102 motor vehicle registrations in 2007.

Rail

In 2007, 1,857 miles of railroad were operated.

Civil Aviation

In 2011 there were eight primary airports (commercial service airports with more than 10,000 passenger boardings annually) with a combined total of 496,642 enplanements, down from 498,358 in 2010.

SOCIAL INSTITUTIONS

Justice

In Dec. 2008 there were 2,084 prison inmates. Capital punishment is authorized but has been used only once, in 1992, since the US Supreme Court reinstated the death penalty in 1976.

Education

In fall 2007, 368 public elementary and secondary schools had 86,422 pupils and 6,915 teachers. In fall 2007 pupils in private elementary and secondary schools numbered 2,930. The average expenditure per pupil for fiscal year 2008 was $13,856. Total expenditure on public elementary and secondary education in 2006–07 was $1,383m.

The University of Wyoming, founded at Laramie in 1887 had, in the academic year 2004–05, 13,207 students. There were seven community colleges in 2005–06 with 19,485 students.

Health

In 2009 the state had 24 community hospitals with 2,000 beds. A total of 52,000 patients were admitted during the year.

In 2009 there were 1,020 active physicians (19 per 10,000 population).

Welfare

In fiscal year 2009 a total of 72,277 people in Wyoming received Medicaid, with total payments of $552m. Medicare enrolment in July 2010 totalled 79,548. In Dec. 2008 there were 85,755 Old-Age, Survivors, and Disability Insurance (OASDI) beneficiaries. A total of 555 people were receiving payments under Temporary Assistance for Needy Families (TANF) in Dec. 2008.

RELIGION

In 2010 the main religious traditions were: Latter-day Saints (Mormons), with 63,316 members; Catholics, 61,222; Evangelical Protestants, 59,247; Mainline Protestants, 36,539.

FURTHER READING

Equality State Almanac 2010. Wyoming Department of Administration and Information. Division of Economic Analysis. Cheyenne, WY 82002
Wyoming Official Directory. Secretary of State. Annual

Treadway, T., *Wyoming*. 1982

Statistics Website: http://eadiv.state.wy.us

OUTLYING TERRITORIES

The outlying territories of the USA comprise the two Commonwealths of the Northern Mariana Islands and Puerto Rico, the incorporated territory of Palmyra Atoll, a number of unincorporated territories (including American Samoa, Guam and the US Virgin Islands) in the Pacific Ocean and one unincorporated territory in the Caribbean Sea.

Commonwealth of the Northern Mariana Islands

KEY HISTORICAL EVENTS

In 1889 Spain ceded Guam (largest and southernmost of the Marianas Islands) to the USA and sold the rest to Germany. Occupied by Japan in 1914, the islands were administered by Japan under a League of Nations mandate until occupied by US forces in Aug. 1944. In 1947 they became part of the US-administered Trust Territory of the Pacific Islands. On 17 June 1975 the electorate voted for a Commonwealth in association with the USA; this was approved by the US government in April 1976 and came into force on 1 Jan. 1978. In Nov. 1986 the islanders were granted US citizenship. The UN terminated the Trusteeship status on 22 Dec. 1990.

TERRITORY AND POPULATION

The Northern Marianas form a single chain of 16 mountainous islands extending north of Guam for about 560 km, with a total area of 5,117 sq. km (1,976 sq. miles) of which 472 sq. km (182 sq. miles) are dry land, and with a population (2010 census) of 53,883.

The areas and populations of the islands are as follows:

Island(s)	Sq. km	2000 Census	2010 Census
Northern Group[1]	160	6	0
Saipan	119	62,392	48,220
Tinian (with Aguijan)	108[2]	3,540	3,136
Rota	85	3,283	2,527

[1]Pagan, Agrihan, Alamagan and nine uninhabited islands.
[2]Including uninhabited Aguijan.

In 2003, 23% spoke Chinese, 22% Chamorro and 24% Filipino languages. English remains an official language along with Carolinian and Chamorro. The largest town is Chalan Kanoa on Saipan.

SOCIAL STATISTICS

Births, 2010, 1,072 (22·0 per 1,000 population); deaths, 2009, 203 (3·7). Infant mortality, 2002, 6 per 1,000 live births.

CONSTITUTION AND GOVERNMENT

The Constitution was approved by a referendum on 6 March 1977 and came into force on 9 Jan. 1978. The legislature comprises a nine-member *Senate*, with three Senators elected from each of the main three islands for a term of four years, and an 20-member *House of Representatives*, elected for a term of two years.

The Commonwealth is administered by a Governor and Lieut.-Governor. The present term of the Governor is five years, but future terms will be four years, with elections taking place in even-numbered years as from 2014.

As from Jan. 2009 the Commonwealth sends one delegate to the US House of Representatives. The Congressman may participate but not vote on the House floor.

RECENT ELECTIONS

At the elections of 7 Nov. 2009 the Republican Party won 9 seats in the House of Representatives and the Covenant Party 7. Four independents were elected.

In the gubernatorial elections, also on 7 Nov. 2009, Heinz S. Hofschneider received 36·3% of votes cast, incumbent Benigno R. Fitial 36·2%, Juan Guerrero 19·3% and Ramon Guerrero 8·0%. In a run-off on 23 Nov. between Fitial and Hofschneider, Fitial retained power after winning 51·4% of the vote to Hofschneider's 48·6%.

CURRENT GOVERNMENT

Governor: Eloy S. Inos (Covenant Party), 2013–15 (salary: $70,000).

Lieut.-Governor: Jude U. Hofschneider (R.), 2013–15 ($60,000).

Legislature Website: http://www.cnmileg.gov.mp

ENERGY AND NATURAL RESOURCES

Water

The total area covered by water is 1,793 sq. miles, of which 6 sq. miles are inland.

Fisheries

In 2010 total catch was 536,000 lb (243 tonnes), entirely from marine waters.

INDUSTRY

Labour

In 2000 the labour force totalled 44,471 people (46% male) of which 42,753 were employed. Unemployment was 3·8% in 2000.

INTERNATIONAL TRADE

Imports and Exports

In 1997 imports totalled $836·2m.; in 1999 exports totalled $1,049·0m. Most imports came from other US Pacific territories, Hong Kong and Japan.

COMMUNICATIONS

Roads

There are about 381 km of roads.

Civil Aviation

In 2011 there were three primary airports (commercial service airports with more than 10,000 passenger boardings annually). The main airport, Francisco C. Ada/Saipan International, had 382,386 enplanements in 2011 (433,557 in 2010). The airports at Rota and Tinian have much smaller passenger numbers.

Telecommunications

There were 24,700 main telephone lines in 2008, equivalent to 289·3 per 1,000 inhabitants, and 20,500 mobile phone subscribers in 2004.

SOCIAL INSTITUTIONS

Education

In fall 2007 there were 462 pupils enrolled in nursery school and pre-school, 689 in kindergarten, 6,989 in elementary school (grades 1–8), 3,159 in high school (grades 9–12) and 901 in higher education.

Health

In 1999 there were 31 doctors, three dentists, 123 nursing personnel, four pharmacists and 14 midwives. In 2001 there was one hospital with 86 beds.

RELIGION

The population is predominantly Roman Catholic.

CULTURE

Tourism

In 2007 there were 385,000 tourist arrivals by air.

Commonwealth of Puerto Rico

KEY HISTORICAL EVENTS

A Spanish dependency since the 16th century, Puerto Rico was ceded to the USA in 1898 after the Spanish defeat in the Spanish-American war. In 1917 US citizenship was conferred and in 1932 there was a name change from Porto Rico to Puerto Rico. In 1952 Puerto Rico was proclaimed a commonwealth with a representative government and a directly elected governor.

TERRITORY AND POPULATION

Puerto Rico is the easternmost of the Greater Antilles and lies between the Dominican Republic and the US Virgin Islands. The total area is 13,791 sq. km (5,325 sq. miles), of which 8,868 sq. km (3,424 sq. miles) are dry land; the population, according to the census of 1 April 2010, was 3,725,789, a fall of 2·2% from 2000. July 2012 estimate, 3,667,084. The urban population was 3,595,521 in 2000, representing 94·4% (73·3% in 1995) of the total population. Population density was 1,088 per sq. mile in 2010. Of the total population in 2010, 1,940,618 were female. The UN gives a projected population for 2015 of 3·74m.

Chief towns, 2010 census, are: San Juan (the capital), 395,326; Bayamón, 208,116; Carolina, 176,762; Ponce, 166,327; Caguas, 142,893.

The Puerto Rican island of Vieques, 10 miles to the east, has an area of 51·7 sq. miles and 9,301 (2010) inhabitants. The island of Culebra, between Puerto Rico and St Thomas, has an area of 10 sq. miles and 1,818 (2010) inhabitants. Both islands have good harbours.

Spanish and English are the joint official languages.

SOCIAL STATISTICS

Births, 2010, 42,153 (11·3 per 1,000 population); deaths, 2010 (provisional), 29,133 (7·8). Marriages, 2006, 23,185; infant mortality rate, 2009, 7·9 per 1,000 live births. Population growth rate, 2007–08, 0·3%. In 2006 the most popular age range for marrying was 20–24 for both males and females. Fertility rate, 2008, 1·6 births per woman.

CLIMATE

Warm, sunny winters with hot summers. The north coast experiences more rainfall than the south coast and generally does not have a dry season as rainfall is evenly spread throughout the year. San Juan, Jan. 25°C, July 28°C. Annual rainfall 1,246 mm.

CONSTITUTION AND GOVERNMENT

Puerto Rico is a self-governing commonwealth (*Estado Libre Asociado*) in association with the United States. The chief of state is the President of the United States of America. The head of government is an elected Governor. There are two legislative chambers: the 51-member House of Representatives and the 27-member Senate. Both houses meet annually in Jan. The executive power is exercised by the Governor, elected every four years, who leads a cabinet of 15 ministers.

A new constitution was drafted by a Puerto Rican Constituent Assembly and approved by the electorate at a referendum on 3 March 1952. It was then submitted to Congress, which struck out Section 20 of Article 11 covering the 'right to work' and the 'right to an adequate standard of living'; the remainder was passed and proclaimed by the Governor on 25 July 1952.

Puerto Rico broadly has authority over its internal affairs, but the USA controls areas generally regulated by the federal government. Puerto Ricans are US citizens and possess most of the rights and obligations of citizens from the 50 states, such as paying Social Security and receiving federal welfare. Differences include Puerto Rico's local taxation system and partial exemption from Internal Revenue Code, its lack of voting representation in either house of the US Congress (they have one non-voting representative) and the ineligibility of Puerto Ricans to vote in presidential elections. Puerto Rican men are subject to subscription in the US Armed Forces, but the commonwealth sends independent teams to the Olympics.

The US Constitution remains the supreme legal document. Puerto Rico's authority over internal affairs is subject to the agreement of the US Congress and may, in theory, be rescinded at any time. Its status has long been the subject of debate in Washington and San Juan, centred on the options of remaining as a Commonwealth or adopting statehood, a compact of Free Association or independence.

At a plebiscite on 14 Nov. 1993 on Puerto Rico's future status, 48·6% of votes cast were for Commonwealth (status quo), 46·3% for Statehood (51st State of the USA) and 4·4% for full independence. In a further plebiscite in Dec. 1998, some 52·2% of voters backed the opposition's call for no change, while 46·5% supported statehood. Independence was supported by 2·5%, while free association received 0·3%. A Presidential Task Force on Puerto Rico's Status was established by President Clinton in 2000 and continued to report during the tenures of George W. Bush and Barack Obama. President Obama has reiterated his commitment to its status being decided by the will of the people of Puerto Rico as expressed in a plebiscite or referendum.

A non-binding referendum of Puerto Rico's status was held on 6 Nov. 2012. Voters were asked two questions: firstly, whether they wished to continue with the current territorial status; and, secondly, which status they favoured of statehood, independence or sovereign nation in free association with the USA. Turnout approached 80%, with 54·0% voting no to the first question. Of those, 73·0% cast a vote on the second question, with 61·1% opting for statehood, 33·3% for sovereign free association and 5·5% for independence. Nonetheless, before Puerto Rico can become the 51st state of the USA, legislation would be subject to approval by the Congress in Washington before being signed off by the President.

RECENT ELECTIONS

At the gubernatorial election on 6 Nov. 2012 Alejandro García Padilla (Popular Democratic Party/PPD) won with 47·8% of the vote, ahead of incumbent Luis Guillermo Fortuño (New Progressive Party/PNP) with 47·1% and Juan Dalmau Ramírez (Puerto Rican Independence Party/PIP) with 2·5%. A number of other candidates each received less than 1% of the vote.

In elections to the Chamber of Representatives on 6 Nov. 2012 the Popular Democratic Party (PPD) won 28 of the 51 seats and the New Progressive Party (PNP) 23 seats. In the Senate elections of the same day PPD took 18 seats, PNP 8 and PIP 1.

CURRENT GOVERNMENT

Governor: Alejandro García Padilla (PPD), 2013–17 (salary: $70,000).

Secretary of State: David Bernier (PPD), since Jan. 2013 ($90,000).

Government Website (Spanish only): http://www.gobierno.pr

ECONOMY

Budget

Revenues in 2005 totalled $12,444·0m. and expenditures were $25,205·0m. Tax revenues accounted for 60·8% of revenue. Main items of expenditure were social development (52%), economic development (20%), debt service (13%) and protection and security (7%).

Per capita personal income (2005) was $12,502.

Performance

Real GDP growth was 0·3% in 2005. Total GDP in 2010 was $96,261m.

ENERGY AND NATURAL RESOURCES

Environment

Puerto Rico's carbon dioxide emissions from the consumption and flaring of fossil fuels in 2008 were the equivalent of 8·0 tonnes per capita.

Electricity

Installed capacity was an estimated 5·4m. kW in 2007. Production in 2007 was 23·7bn. kWh. Consumption per capita in 2007 was 6,014 kWh.

Water

The total area covered by water is 1,901 sq. miles, of which 76 sq. miles are inland.

Agriculture

Gross agricultural income in 2009 was $794m. In 2009, 1·6% of the economically active population was employed in agriculture. Production estimates in 2008 (in 1,000 tonnes): plantains, 80; bananas, 54; oranges, 20; tomatoes, 19; pineapples, 17; pumpkins and squash, 14; mangoes and guavas, 14. Livestock (2008 estimates): cattle, 380,000; pigs, 50,000; horses, 6,600; sheep, 6,300;

chickens, 13m. Livestock products, 2008 estimates (in 1,000 tonnes): poultry, 50; pork, 11; beef, 10; milk, 350; eggs, 12.

Forestry

In 2010 the area under forests was 0·55m. ha., or 62% of the total land area.

Fisheries

The total catch in 2010 was 4,184,000 lb (1,898 tonnes), exclusively from sea fishing.

INDUSTRY

Manufacturing contributed $33,132m. to total GDP in 2005. There is some production of cement (1·58m. tonnes in 2004).

Labour

There were 1,238,000 people in employment in 2005, including 274,000 people in public administration, 261,000 in wholesale and retail trade and 138,000 in manufacturing. The unemployment rate was 15·7% in Dec. 2010.

INTERNATIONAL TRADE

Imports and Exports

In 2005 imports amounted to $38,905·2m., of which $19,133·7m. came from the USA; exports were valued at $56,543·2m., of which $46,703·0m. went to the USA.

Main imports in 2005 were (in $1m.): chemicals, 17,086·8; petroleum and coal products, 2,924·5; computers and electronic products, 2,833·8. Main exports in 2005 were (in $1m.): chemical products, 38,618·9; computer and electronic products, 7,452·1; medical equipment and supplies, 4,993·8.

Puerto Rico is not permitted to levy taxes on imports.

COMMUNICATIONS

Roads

In 2007 there were 16,397 miles of roads and 2,531,199 registered motor vehicles.

Rail

There are 96 km of railroad, although no passenger service. There is a 17·2-km urban train system in use.

Civil Aviation

There were six primary airports in 2011 (commercial service airports with more than 10,000 passenger boardings annually) with a combined total of 4,413,509 enplanements, down from 4,679,049 in 2010. By far the busiest airport is San Juan's Luis Muñoz Marin International, which had 3,983,130 enplanements in 2011.

Shipping

The leading ports are San Juan (which handled 11·3m. tons of cargo in 2009) and Ponce (2·2m. tons).

Telecommunications

In 2011 there were 826,100 landline telephone subscriptions (equivalent to 220·6 per 1,000 inhabitants) and 3,108,400 mobile phone subscriptions (or 829·9 per 1,000 inhabitants). In 2011, 48·0% of the population were internet users.

SOCIAL INSTITUTIONS

Justice

The Judicial power of the commonwealth is vested in a unified judicial system with regard to jurisdiction, operation and administration. It consists of the Supreme Court, the Court of Appeals and the Court of First Instance, which jointly constitute the General Court of Justice. The Supreme Court is the court of last instance, and is composed of nine members (a chief justice and eight associate justices) named by the Governor with the consent of the Senate. Judgments on federal laws are subject to review by the US Supreme Court.

The Court of First Instance is a court of original general jurisdiction consisting of a Superior Court (253 judges) and municipal courts (85 judges). There are 13 judicial regions in the First Instance Court. Final judgments made by the Court of First Instance must be appealed at the Court of Appeals. The United States Court of Appeals for the First Circuit, based in Boston, Massachusetts, sits in San Juan for two weeks each year.

The population in penal institutions in Dec. 2008 was 12,130 (303 per 100,000 population).

Education

Education was made compulsory in 1899. The percentage of literacy in 2002 was 94·1% of those 15 years of age or older. Total enrolment in public day schools, 2004–05, was 575,993. All private schools had a total enrolment of 138,560 pupils in 2004–05. Puerto Rico has 20 specialized schools in music, arts, theatre, science and mathematics, and communication in radio and television.

In 2004–05 the University of Puerto Rico had eleven units with a total of 66,389 students. Other institutions of higher education in the public sector had a total of 4,655 students in 2004–05: San Juan Technology College, 936 students; Corporation of Music of Puerto Rico, 298 students; School of Plastic Arts, 461 students; Technology Institute of Puerto Rico of the Department of Education, 2,960 students. Private sector higher education institutions in 2004–05 (with enrolment): the American University, 3,403 students; Caribbean University, 5,078; Columbia College, 1,086; Electronic Data Processing College of Puerto Rico, 2,017; Huertas Junior College, 1,459; Commercial Institute of Puerto Rico Junior College, 1,300; National University College, 4,460; Pontifical Catholic University of Puerto Rico, 10,124; Fundación Ana G. Méndez, 34,217; Sacred Heart University, 5,206; Inter-American University of Puerto Rico, 43,937. Other private universities and colleges had 24,274 students.

Health

There were 71 hospitals in 2009, with a hospital bed provision of 32 per 10,000 population. In the same year there were 12,698 non-federal physicians.

RELIGION

In 2001 about 65% of the population were Roman Catholic. In Feb. 2012 there was one cardinal.

CULTURE

Tourism

There were 4,379,200 non-resident visitors in 2010 (4,415,300 in 2009 and 5,213,100 in 2008) with spending from such visitors totalling $3,210·7m.

FURTHER READING

Statistical Information: The Program of Economic Research and Social Planning of the Puerto Rico Planning Board publishes: *(a) Economic Report to the Governor* (annual); *(b) External Trade Statistics* (annual); *(c) Reports on national income and balance of payments; and other socioeconomic statistics* (since 1940). There is a weekly economic summary (in Spanish) at http://www.jp.gobierno.pr

Office of Economic Studies and Analysis: Government Development Bank for Puerto Rico, PO Box 42001, San Juan 00940-2001.

Website: http://www.gdb-pur.com
Commonwealth Library: Univ. of Puerto Rico Library, Rio Piedras.

American Samoa

KEY HISTORICAL EVENTS

The first recorded visit by Europeans was in 1722. On 14 July 1889 a treaty between the USA, Germany and Great Britain proclaimed the Samoan Islands neutral territory, under a four-power government consisting of the three treaty powers and the local native government. By the Tripartite Treaty of 7 Nov. 1899, ratified 19 Feb. 1900, Great Britain and Germany renounced, in favour of the USA, all rights over the islands of the Samoan group east of 171° long. west of Greenwich. The islands to the west of that meridian, now the independent state of Samoa, were assigned to Germany. The islands of Tutuila and Aunu'u were ceded to the USA by their High Chiefs on 17 April 1900, and the islands of the Manu'a group on 16 July 1904. Congress accepted the islands under a Joint Resolution approved 20 Feb. 1929. Swain's Island, 210 miles north of the Samoan Islands, was annexed in 1925 and is administered as an integral part of American Samoa.

TERRITORY AND POPULATION

The islands (Tutuila, Aunu'u, Ta'u, Olosega, Ofu and Rose) are approximately 650 miles east-northeast of the Fiji Islands. The total area is 1,505 sq. km (581 sq. miles), of which 198 sq. km (76 sq. miles) are dry land; population (2010 census), 55,519, nearly all Polynesians or part-Polynesians. Population density was 278 per sq. km in 2010.

In 2000, 88·8% of the population lived in urban areas. The capital is Pago Pago, which had a population of 3,656 in 2010. The island's three Districts are Eastern (population, 2010, 23,030), Western (31,329) and Manu'a (1,143). There is also Swain's Island, with an area of 1·9 sq. miles and 17 inhabitants (2010), which lies 210 miles to the northwest. Rose Island (uninhabited) is 0·4 sq. mile in area.

The official languages are Samoan and English.

SOCIAL STATISTICS

Births, 2010, 1,234 (22·2 per 1,000 population); deaths, 2010 (provisional), 224 (4·0). Infant mortality, 2006, 11·8 per 1,000 live births.

CLIMATE

A tropical maritime climate with a small annual range of temperature and plentiful rainfall. Pago Pago, Jan. 83°F (28·3°C), July 80°F (26·7°C). Annual rainfall 194" (4,850 mm).

CONSTITUTION AND GOVERNMENT

American Samoa is constitutionally an unorganized, unincorporated territory of the USA administered under the Department of the Interior. Its indigenous inhabitants are US nationals and are classified locally as citizens of American Samoa with certain privileges under local laws not granted to non-indigenous persons. Polynesian customs (not inconsistent with US laws) are respected.

Fagatogo is the seat of the government.

The islands are organized in 15 counties grouped in three districts; these counties and districts correspond to the traditional political units. On 25 Feb. 1948 a bicameral legislature was established, at the request of the Samoans, to have advisory legislative functions. With the adoption of the constitution of 22 April 1960, and the revised Constitution of 1967, the legislature was vested with limited law-making authority. The lower house, or House of Representatives, is composed of 20 members elected by universal adult suffrage and one non-voting member for Swain's Island. The upper house, or Senate, is comprised of 18 members elected, in the traditional Samoan manner, in meetings of the chiefs. The Governor and Lieut.-Governor have been popularly elected since 1978. American Samoa also sends one delegate to the US House of Representatives. The Congressman may participate but not vote on the House floor.

RECENT ELECTIONS

At elections to the House of Representatives on 6 Nov. 2012, only non-partisans were elected.

At gubernatorial elections on the same day Lolo Matalasi Moliga received 33·5% of votes cast, Faoa Aitofele Sunia 33·1%, Afoa Moega Lutu 19·3%, Salu Hunkin-Finau 6·8%, Save Liuato Tuitele 5·8% and Tim Jones 1·4%. In the run-off between the two leading candidates held on 20 Nov. 2012, Moliga took 52·9% of the vote against 47·1% for Sunia.

CURRENT GOVERNMENT

Governor: Lolo Matalasi Moliga (ind.), 2013–17 (salary: $85,000).
  Lieut.-Governor: Lemanu Peleti Mauga (ind.), 2013–17 ($75,000).

Government Website: http://www.government.as

ECONOMY

Overview

The Economic Development and Planning Office promotes economic expansion and outside investment.

Budget

The chief sources of revenue are annual federal grants from the USA, local revenues from taxes, duties, receipts from commercial operations (enterprise and special revenue funds), utilities, rents and leases, and liquor sales. In 2004–05 revenues were $182·0m. and expenditures $192·5m.

ENERGY AND NATURAL RESOURCES

Environment

American Samoa's carbon dioxide emissions from the consumption and flaring of fossil fuels in 2008 were the equivalent of 9·2 tonnes per capita.

Electricity

Installed capacity was estimated at 60,000 kW in 2007. Production in 2007 was about 196m. kWh. Per capita consumption in 2007 was an estimated 2,874 kWh. All the Manu'a islands have electricity.

Water

The total area covered by water is 505 sq. miles, of which 8 sq. miles are inland.

Agriculture

Of the 48,640 acres of land area, 12,000 acres are used for tropical crops; most commercial farms are in the Tafuna plains and west Tutuila. Principal crops are coconuts, taro, bread-fruit, yams and bananas.
  Livestock (2008): hogs and pigs, 16,904; chickens, 35,709.

Forestry

Forests covered a total area of 18,000 ha. in 2010, or 89% of the land area.

Fisheries

Total catch in 2010 was 11,599,000 lb (5,261 tonnes).

INDUSTRY

Fish canning is important, employing the second largest number of people (after government). Attempts are being made to provide a variety of light industries. Tuna fishing and local inshore fishing are both expanding.

Labour

In 2000 the civilian labour force numbered 17,627, of whom 16,718 were employed. The unemployment rate in 2000 was 5·2%.

INTERNATIONAL TRADE

Imports and Exports

Imports in 2006 totalled $579m. and exports $439m.
  Chief imports are fish for canning, building materials, fuel oil, food, jewellery, machines and parts, alcoholic beverages and cigarettes. Chief exports are canned tuna, watches, pet foods and handicrafts.

COMMUNICATIONS

Roads

There are about 150 km of paved roads and 200 km of unpaved roads in all. Motor vehicles registered, 2006, 9,215 (including 7,758 private vehicles).

Civil Aviation

American Samoa has one primary airport—a commercial service airport with more than 10,000 passenger boardings annually. Pago Pago International had 45,486 enplanements in 2011.

Shipping

The harbour at Pago Pago, which nearly bisects the island of Tutuila, is the only good harbour for large vessels in American Samoa.

Telecommunications

There were an estimated 10,400 main telephone lines in operation in 2010 and 2,300 mobile phone subscribers in 2004.

SOCIAL INSTITUTIONS

Justice

Judicial power is vested firstly in a High Court. The trial division has original jurisdiction of all criminal and civil cases. The probate division has jurisdiction of estates, guardianships, trusts and other matters. The land and title division decides cases relating to disputes involving communal land and Matai title court rules on questions and controversy over family titles. The appellate division hears appeals from trial, land and title, and probate divisions as well as having original jurisdiction in selected matters. The appellate court is the court of last resort. Two American judges sit with five Samoan judges permanently. In addition there are temporary judges or assessors who sit occasionally on cases involving Samoan customs. There is also a District Court with limited jurisdiction and there are 69 village courts.
  The population in penal institutions in Dec. 2007 was 236 (equivalent to 410 per 100,000 population).

Education

Education is compulsory between the ages of six and 18. In 2006–07 there were 16,427 pupils in public elementary and secondary schools, of which 16,191 were in regular schools, 187 in vocational education and 49 in special education. There were 1,767 students in higher education in fall 2007.

Welfare

In Dec. 2001 there were 5,320 beneficiaries including 1,470 survivors, 1,370 retired workers and 1,240 disabled workers. Total payments came to $2m. with average monthly benefits of $442.

RELIGION

In 2001 about 41% of the population belonged to the Congregational Church and 19% were Roman Catholics. Methodists and Latter-day Saints (Mormons) are also represented.

CULTURE

Tourism

In 2008 there were 31,277 visitor arrivals, including 7,084 tourists.

Guam

Guahan

KEY HISTORICAL EVENTS

Magellan is said to have discovered the island in 1521; it was ceded by Spain to the USA by the Treaty of Paris (10 Dec. 1898).

The island was captured by the Japanese on 10 Dec. 1941 and retaken by American forces following the Battle of Guam (21 July–8 Aug. 1944). Guam is of strategic importance; substantial numbers of naval and air force personnel occupy about one-third of the usable land.

TERRITORY AND POPULATION

Guam is the largest and most southern island of the Marianas Archipelago, in 13° 26' N. lat., 144° 45' E. long. Total area, 210 sq. miles (543 sq. km). Hagåtña (previously Agaña), the seat of government, is about eight miles from the anchorage in Apra Harbor. The census in 2010 showed a population of 159,358; density, 290 per sq. km. In 2010 an estimated 93·2% of the population lived in urban areas. The UN gives a projected population for 2015 of 191,000. The Malay strain is predominant. Chamorro, the native language, and English are the official languages.

SOCIAL STATISTICS

Births, 2010, 3,416 (21·4 per 1,000 population); deaths, 2010 (provisional), 857 (5·4). Infant mortality rate, 2009, 10·5 per 1,000 live births. Life expectancy at birth, 2005–10, was 73·3 years for males and 77·9 years for females. Fertility rate, 2008, 2·7 births per woman.

CLIMATE

Tropical maritime, with little difference in temperatures over the year. Rainfall is copious at all seasons, but is greatest from July to Oct. Hagåtña, Jan. 81°F (27·2°C), July 81°F (27·2°C). Annual rainfall 93" (2,325 mm).

CONSTITUTION AND GOVERNMENT

Guam's constitutional status is that of an 'unincorporated territory' of the USA. In Aug. 1950 the President transferred the administration of the island from the Navy Department to the Interior Department. The transfer conferred full citizenship on the Guamanians, who had previously been 'nationals' of the USA. There was a referendum on status on 30 Jan. 1982. 38% of eligible voters voted; 48·5% of those favoured Commonwealth status.

The Governor and Lieut.-Governor are elected for four-year terms. The legislature is a 15-member elected Senate; its powers are similar to those of an American state legislature. Guam sends one non-voting delegate (Madeleine Bordallo, D., 2013–15) to the US House of Representatives.

RECENT ELECTIONS

At the election of 2 Nov. 2010 for the Guam Legislature the Republicans won ten seats and the Democrats won five. In gubernatorial elections, also held on 2 Nov. 2010, initial results indicated that Eddie Calvo (Republican) had won against Carl Gutierrez (Democrat). The Guam Election Commission ordered a recount owing to the slim majority but it confirmed Calvo as the victor with 50·6% of the vote against his opponent's 49·4%.

CURRENT GOVERNMENT

Governor: Eddie Calvo (R.), 2011–15 (salary: $90,000).
  Lieut.-Governor: Ray Tenorio (R.), 2011–15 ($85,000).

Government Website: http://www.guam.gov

ECONOMY

Budget
Total revenue (2007) $514·4m.; expenditure $394·8m.

ENERGY AND NATURAL RESOURCES

Environment
Guam's carbon dioxide emissions from the consumption and flaring of fossil fuels in 2008 were the equivalent of 9·9 tonnes per capita.

Electricity
Installed capacity was an estimated 0·6m. kW in 2007. Production was 1,879m. kWh in 2007. Consumption per capita in 2007 was 10,835 kWh.

Water
The total area covered by water is 361 sq. miles, of which 8 sq. miles are inland.

Agriculture
The major products of the island are sweet potatoes, cucumbers, watermelons and beans. In 2002 there were approximately 5,000 acres of arable land and 9,000 acres of permanent cropland. Production (2002 estimates, in 1,000 tonnes): coconuts, 52; copra, 2; watermelons, 2. Livestock (2002) included 1,000 goats and 5,000 pigs. There is an agricultural experimental station at Inarajan.

Forestry
There were 26,000 ha. of forest in 2010, or 47% of the total land area.

Fisheries
In 2010 total catch was 642,000 lb (291 tonnes), exclusively from sea fishing.

INDUSTRY

Guam Economic Development Authority controls three industrial estates: Cabras Island (32 acres); Calvo estate at Tamuning (26 acres); Harmon estate (16 acres). Industries include textile manufacture, cement and petroleum distribution, warehousing, printing, plastics and ship-repair. Other main sources of income are construction and tourism.

Labour
In 2000 there were 105,014 persons of employable age, of whom 68,894 were in the workforce (64,452 civilian). 7,399 were unemployed.

INTERNATIONAL TRADE

Guam is the only American territory which has complete 'free trade'; excise duties are levied only upon imports of tobacco, liquid fuel and liquor.

Imports and Exports
In 2002 imports were valued at $389m. and exports at $37m. Main export destinations in 2001 were Japan, 50·0%; Palau, 9·4%; Micronesia, 9·1%.

COMMUNICATIONS

Roads
There are 674 km of all-weather roads. In 2002 there were 59,700 passenger cars and 20,100 commercial vehicles in use.

Civil Aviation
Guam has one primary airport (a commercial service airport with more than 10,000 passenger boardings annually). Guam International, which is at Tamuning, had 1,369,586 enplanements in 2011.

Shipping
There is a port at Apra Harbor.

Telecommunications
Telephone subscribers numbered 145,300 in 2003 (887·3 per 1,000 inhabitants). Mobile phone subscribers numbered 98,000 in 2004. There were 506·4 internet users per 1,000 inhabitants in 2009.

SOCIAL INSTITUTIONS

Justice
The Organic Act established a District Court with jurisdiction in matters arising under both federal and territorial law; the judge is

appointed by the President subject to Senate approval. There is also a Supreme Court and a Superior Court; all judges are locally appointed except the Federal District judge. Misdemeanours are under the jurisdiction of the police court. The Spanish law was superseded in 1933 by five civil codes based upon California law.

The population in penal institutions in July 2008 was 559 (318 per 100,000 population).

Education

Education is compulsory from five to 16. Bilingual teaching programmes integrate the Chamorro language and culture into public school courses. Total primary and secondary school enrolment in fall 2008 was 39,875, including 30,329 in the public system and 7,053 in private schools; there were 34 public schools and 27 private schools. The University of Guam is in Mangilao and Guam Community College is in Barrigada.

Health

There is a hospital, eight nutrition centres, a school health programme and an extensive immunization programme. Emphasis is on disease prevention, health education and nutrition.

Welfare

In 2000, $38·9m. was paid in public assistance to individuals, including $19·1m. temporary aid to needy families, $11·5m. Medicaid and $1·5m. old-age assistance.

RELIGION

About 75% of the Guamanians are Roman Catholics; the other 25% are Baptists, Episcopalians, Bahais, Lutherans, Latter-day Saints (Mormons), Presbyterians, Jehovah's Witnesses and members of the Church of Christ and Seventh Day Adventists.

CULTURE

Tourism

In 2011, 1,227,000 overseas visitors—excluding those from Canada and Mexico—visited Guam (down from 1,318,000 in 2010 although up from 1,170,000 in 2006).

FURTHER READING

Report (Annual) of the Governor of Guam to the US Department of Interior Guam Annual Economic Review.

Rogers, R. F., *Destiny's Landfall: a History of Guam.* 1995
Wuerch, W. L. and Ballendorf, D. A., *Historical Dictionary of Guam and Micronesia.* 1995

Statistical Office: P. O. Box 2950, Hagåtña, Guam 96932.
Website: http://bsp.guam.gov

Virgin Islands of the United States

KEY HISTORICAL EVENTS

The Virgin Islands of the United States, formerly known as the Danish West Indies, were named and claimed for Spain by Columbus in 1493. They were later settled by Dutch and English planters, invaded by France in the mid-17th century and abandoned by the French *c.* 1700, by which time Danish influence had been established. St Croix was held by the Knights of Malta between two periods of French rule.

The Virgin Islands were purchased from Denmark by the United States for $25m. on 31 March 1917. Their value was wholly strategic, inasmuch as they commanded the Anegada Passage from the Atlantic Ocean to the Caribbean Sea and the approach to the Panama Canal. Although the inhabitants were made US citizens in 1927, the islands are constitutionally an 'unincorporated territory'.

TERRITORY AND POPULATION

The Virgin Islands group, lying about 40 miles due east of Puerto Rico, comprises the islands of St Thomas (31 sq. miles), St Croix (83 sq. miles), St John (20 sq. miles) and 65 small islets or cays, mostly uninhabited. The total area is 1,898 sq. km (733 sq. miles), of which 348 sq. km (134 sq. miles) are dry land.

The population according to the 2010 census was 106,405; density 794 per sq. mile. 92·6% of the population were urban in 2000.

Population (2010 census) of St Thomas, 51,634; St Croix, 50,601; St John, 4,170. In 2000, 69·8% of the population were native born.

The UN gives a projected population for 2015 of 108,000.

The capital and only city, Charlotte Amalie, on St Thomas, had a population (2010 census) of 10,354. There are two towns on St Croix with 2010 census populations of: Christiansted, 2,433; Frederiksted, 859. The official language is English. Spanish is also spoken.

SOCIAL STATISTICS

Births, 2010, 1,600 (15·1 per 1,000 population); deaths, 2009, 675 (6·3). Infant mortality, 2005–07, 11·5 per 1,000 live births.

CLIMATE

Average temperatures vary from 77°F to 82°F throughout the year; humidity is low. Average annual rainfall, about 45". The islands lie in the hurricane belt; tropical storms with heavy rainfall can occur in late summer.

CONSTITUTION AND GOVERNMENT

The Organic Act of 22 July 1954 gives the US Department of the Interior full jurisdiction; some limited legislative powers are given to a single-chambered legislature, composed of 15 senators elected for two years representing the two legislative districts of St Croix and St Thomas-St John.

The Governor is elected by the residents. Since 1954 there have been four attempts to redraft the Constitution, to provide for greater autonomy. Each has been rejected by the electorate. The latest was defeated in a referendum in Nov. 1981, with 50% of the electorate participating.

For administration, there are 14 executive departments, 13 of which are under commissioners and the other, the Department of Justice, under an Attorney-General. The US Department of the Interior appoints a Federal Comptroller of government revenue and expenditure.

The franchise is vested in residents who are citizens of the United States, 18 years of age or over. They do not participate in the US presidential election but they have a non-voting representative in Congress.

The capital is Charlotte Amalie, on St Thomas Island.

RECENT ELECTIONS

In elections for governor held on 2 Nov. 2010 John deJongh, Jr (Democrat) won 56·3% of the votes, against 43·6% for Kenneth Mapp (ind.). In Senate elections held on 6 Nov. 2012 the Democratic Party of the Virgin Islands won 10 out of 15 seats.

CURRENT GOVERNMENT

Governor: John P. deJongh, Jr (D.), 2011–15 (salary: $138,000).
Lieut.-Governor: Gregory R. Francis (D.), 2011–15 ($115,000).

US Virgin Islands Government: http://ltg.gov.vi

ECONOMY

Currency

United States currency became legal tender on 1 July 1934.

Budget

Under the 1954 Organic Act finances are provided partly from local revenues—customs, federal income tax, real and personal property tax, trade tax, excise tax, pilotage fees, etc.—and partly from Federal Matching Funds, being the excise taxes collected by the federal government on such Virgin Islands products transported to the mainland as are liable.

Per capita income, 2000, $13,139.

Revenues in 2004 totalled $557·9m.; expenditures were $592·0m.

ENERGY AND NATURAL RESOURCES

Environment

Carbon dioxide emissions from the consumption and flaring of fossil fuels in 2008 were the equivalent of 126·5 tonnes per capita, the second highest in the world.

Electricity

The Virgin Islands Water and Power Authority provides electric power from generating plants on St Croix and St Thomas; St John is served by power cable and emergency generator. Production in 2007 was about 1·07bn. kWh. Per capita consumption in 2007 was an estimated 9,743 kWh. Installed capacity in 2007 was an estimated 323,000 kW.

Water

The total area covered by water is 599 sq. miles, of which 18 sq. miles are inland.

Agriculture

Land for fruit, vegetables and animal feed is available on St Croix, and there are tax incentives for development.

Livestock (2002): cattle, 8,000; goats, 4,000; pigs, 3,000; sheep, 3,000.

Forestry

Forests covered 20,000 ha. in 2010, or 58% of the land area.

Fisheries

The total catch in 2010 was 1,867,000 lb (847 tonnes).

INDUSTRY

The main occupations on St Thomas are tourism and government service; on St Croix manufacturing is more important. Manufactures include rum (the most valuable product), watches, pharmaceuticals and fragrances. Industries in order of revenue: tourism, refining oil, watch assembly, rum distilling, construction.

Labour

In 2000 the total labour force was 51,042, of whom 7,351 were employed in the arts, entertainment, recreation, accommodation and food services, 6,742 were employed in educational, health and social services, 6,476 in retail trade, 4,931 in public administration and 4,900 in construction. In 2000 there were 4,368 registered unemployed persons, or 5·6% of the workforce.

INTERNATIONAL TRADE

Imports and Exports

Imports, 2005, totalled $10,243m. and exports $10,476m. The main import is crude petroleum, while the principal exports are petroleum products.

COMMUNICATIONS

Roads

In 2008 the Virgin Islands had an estimated 1,260 km of roads. There were 74,500 vehicles registered in 2010.

Civil Aviation

In 2011 there were two primary airports (commercial service airports with more than 10,000 passenger boardings annually). The main airport is Cyril E. King at Charlotte Amalie on St Thomas, which had 596,832 enplanements in 2011. The smaller Henry E. Rohlsen airport at Christiansted on St Croix had 184,331 enplanements in 2011.

Shipping

The whole territory has free port status. There is an hourly boat service between St Thomas and St John and a 75-minute catamaran service between St Croix and St Thomas two to three times a day.

Telecommunications

Telephone subscribers numbered an estimated 75,800 in 2010 (686·7 per 1,000 population). In 2005 there were an estimated 80,300 mobile phone subscribers. Internet users numbered 30,000 in 2007.

SOCIAL INSTITUTIONS

Justice

The population in penal institutions in Dec. 2007 was 555 (512 per 100,000 population).

Education

In fall 2007 there were 10,770 in elementary school, 5,133 in high school and 2,384 students in higher education; there were 1,518 school teachers. In fall 2005 the University of the Virgin Islands (St Thomas and St Croix campuses, with an ecological research station on St John) had 2,392 students (1,263 full-time and 1,129 part-time) and 107 full-time instructional staff; 77% of the students were female.

Health

In 2004 there were 208 active physicians and 45 licensed dentists. The Roy Lester Schneider Hospital on St Thomas had 169 beds in 2008. The Governor Juan F. Luis Hospital, Christiansted, serves St Croix, with 188 beds in 2008.

Welfare

In 2001 federal direct payments for individuals totalled $233·4m., including: retirement insurance, $72·0m.; housing assistance, $53·4m.; survivors insurance, $20·2m.; disability insurance, $18·0m.; food stamps, $17·6m.

RELIGION

At the 2000 census 42% of the population were Baptists, 34% were Roman Catholics and 17% were Episcopalians.

There are places of worship of the Protestant, Roman Catholic and Jewish faiths in St Thomas and St Croix, and Protestant and Roman Catholic churches in St John.

CULTURE

Tourism

Tourism accounts for some 70% of GDP. In 2006 there were 671,362 staying visitors and 1,903,533 cruise passenger arrivals, down from 697,033 and 1,912,548 respectively in 2005. Receipts from tourism amounted to $1,493m. in 2005.

Other Unincorporated Territories

Baker Island

A small Pacific island 2,600 km southwest of Hawaii. Administered under the US Department of the Interior. Area 0·6 sq. mile; population (2000), nil. The islands are considered part of the United States Pacific Island Wildlife Refuges.

Howland Island

A small Pacific island 2,600 km southwest of Hawaii. Administered under the US Department of the Interior. Area 0·7 sq. mile; population (2000), nil. The islands are considered part of the United States Pacific Island Wildlife Refuges.

Jarvis Island

A small Pacific island 2,100 km south of Hawaii. Administered under the US Department of the Interior. Area 1·7 sq. miles; population (2000), nil. The islands are considered part of the United States Pacific Island Wildlife Refuges.

Johnston Atoll

Two small Pacific islands 1,100 km southwest of Hawaii, administered by the US Air Force. Area, 1·0 sq. mile; population (2000) numbered 315. The islands are considered part of the United States Pacific Island Wildlife Refuges.

Kingman Reef

Small Pacific reef 1,500 km southwest of Hawaii, administered by the US Navy. Area one tenth of a sq. mile; population (2000), nil. The islands are considered part of the United States Pacific Island Wildlife Refuges.

Midway Islands

Two small Pacific islands at the western end of the Hawaiian chain, administered by the US Navy. Area, 2·4 sq. miles; population (2000), nil. The islands are considered part of the United States Pacific Island Wildlife Refuges.

Navassa Island

Small Caribbean island 48 km west of Haiti, administered by US Coast Guards. Area 1·4 sq. miles; population (2000), nil.

Wake Island

Three small Pacific islands 3,700 km west of Hawaii, administered by the US Air Force. Area, 2·5 sq. miles; population (2000) numbered 1.

Incorporated Territories

Palmyra Atoll

Small atoll 1,500 km southwest of Hawaii, administered by the US Department of the Interior. It is part federally-owned and part privately-owned. Area 5 sq. miles; population (2000), nil.

URUGUAY

República Oriental del Uruguay
(Oriental Republic of Uruguay)

Capital: Montevideo
Population projection, 2015: 3·43m.
GNI per capita, 2011: (PPP$) 13,242
HDI/world rank: 0·783/48
Internet domain extension: .uy

KEY HISTORICAL EVENTS

From around 4000 BC Uruguay was populated principally by Charrúa and Guaraní Indians. The Charrúa migrated seasonally between coastal and inland areas, while the Guaraní settled in the eastern forests and in the north. Smaller groups also settled the region and lands were fought over vigorously. In 1516 the first Europeans to enter the territory, Spanish navigator Juan Díaz de Solis and his party, were killed. Other Spanish and Portuguese expeditions followed and in the late 16th century the Spanish laid claim to the Río de la Plata. In 1603 Spanish governor Hernando Arias de Saavedra is said to have shipped cattle and horses from the Paraguay region into Río de la Plata, and Spanish, Portuguese and English settlers began livestock farming. The native peoples resisted the European colonizers but were killed in large numbers in warfare and by European disease.

In 1680 the Portuguese established the settlement of Colonia do Sacramento on the Río de la Plata. In 1726 the Spanish founded San Felipe de Montevideo as a fortified port and settled its hinterland. For the next 50 years the two powers fought over the coastal region and in 1776 the Spanish established the viceroyalty of the Río de la Plata, with Buenos Aires as its capital. While the Napoleonic wars were weakening Spain, an autonomous junta was established in Montevideo and in 1810 *criollos* (ethnic Spanish

born in South America) seized power from the Spanish in Buenos Aires. A federalist movement originating in the inland region of the Banda Oriental challenged both Montevideo and Buenos Aires. In the ensuing conflict Portuguese Brazil annexed the Banda Oriental from 1820–25, until Uruguayan federalists led a successful uprising and regained control. Banda Oriental declared its independence and its incorporation into the United Provinces of Río de la Plata on 25 Aug. 1825. Fighting continued until 1828 when negotiations began with Brazil and Argentina. On 18 July 1830 the constitution of the Oriental Republic of Uruguay was approved.

Conflict between the two main political parties, the *colorados* and *blancos*, dominated the next 90 years, factionalizing the nation and leading to the economically damaging War of the Triple Alliance (1864–70), fought between Paraguay and the unified forces of Uruguay, Argentina and Brazil. In 1905 the first democratic elections took place, returning the *colorados* to government under President José Batlle y Ordóñez. In the early 20th century ranching brought prosperity and an influx of immigrants, and a welfare state was developed. In 1919 a new constitution aimed to protect against dictatorship by providing for a plural executive known as a *colegiado*. However, economic depression led to instability and in 1933 presidential government and quadrennial elections were introduced.

Collective leadership returned from 1951–56 but a series of strikes and riots in the 1960s led to the extension of army influence. The military took repressive measures and the presidency was restored in 1967. Marxist urban guerrillas, the Tupamaro, fought a violent campaign against the regime but were finally defeated in 1972. The military took over the government in 1973. A period of harsh repression followed during which thousands were arrested and human rights abuses were rife. Economic difficulties in the 1980s weakened the military government and the country returned to civilian rule in 1985. Economic growth in the 1990s was followed by a downturn from 1999–2002. In 2005 Uruguay elected its first left-wing leader, Tabaré Vázquez of the Progressive Encounter-Broad Front-New Majority coalition.

TERRITORY AND POPULATION

Uruguay is bounded on the northeast by Brazil, on the southeast by the Atlantic, on the south by the Río de la Plata and on the west by Argentina. The area is 176,215 sq. km (68,037 sq. miles), including 1,199 sq. km (463 sq. miles) of inland waters. The following table shows the area and the population of the 19 departments at census 2011:

Departments	Sq. km	Census 2011	Capital
Artigas	11,928	73,378	Artigas
Canelones	4,536	520,187	Canelones
Cerro-Largo	13,648	84,698	Melo
Colonia	6,106	123,203	Colonia
Durazno	11,643	57,088	Durazno
Flores	5,144	25,050	Trinidad
Florida	10,417	67,048	Florida
Lavalleja	10,016	58,815	Minas
Maldonado	4,793	164,300	Maldonado
Montevideo	530	1,319,108	Montevideo
Paysandú	13,922	113,124	Paysandú
Río Negro	9,282	54,765	Fray Bentos
Rivera	9,370	103,493	Rivera
Rocha	10,551	68,088	Rocha
Salto	14,163	124,878	Salto
San José	4,992	108,309	San José
Soriano	9,008	82,595	Mercedes
Tacuarembó	15,438	90,053	Tacuarembó
Treinta y Tres	9,529	48,134	Treinta y Tres

1469

The total population at the 2011 census was 3,286,314; density, 18·8 per sq. km.

The UN gives a projected population for 2015 of 3·43m.

In 2011 Montevideo (the capital) accounted for 39·7% of the total population. It had a population in 2011 of 1,304,687. Other major cities are Salto (population of 104,011 in 2011) and Ciudad de la Costa (95,176 in 2011). 92·6% of the population lived in urban areas in 2011.

13% of the population are over 65; 24% are under 15; 63% are between 15 and 64.

The official language is Spanish.

SOCIAL STATISTICS

2009: births, 47,152; deaths, 32,179. Rates (per 1,000 population), 2009: birth, 14·1; death, 9·6. Annual population growth rate, 2000–05, 0·0%. Infant mortality, 2010 (per 1,000 live births), 9. Life expectancy in 2007 was 72·6 years among males and 79·8 years among females. Fertility rate, 2008, 2·1 births per woman.

CLIMATE

A warm temperate climate, with mild winters and warm summers. The wettest months are March to June, but there is really no dry season. Montevideo, Jan. 72°F (22·2°C), July 50°F (10°C). Annual rainfall 38" (950 mm).

CONSTITUTION AND GOVERNMENT

The Constitution was adopted on 27 Nov. 1966 and became effective in Feb. 1967; it has been amended in 1989, 1994, 1996 and 2004.

Congress consists of a *Senate* of 31 members and a *Chamber of Deputies* of 99 members, both elected by proportional representation for five-year terms although in the case of the Senate only 30 members are elected with one seat reserved for the Vice-President. The electoral system provides that the successful presidential candidate be a member of the party which gains a parliamentary majority. Electors vote for deputies on a first-past-the-post system, and simultaneously vote for a presidential candidate of the same party. The winners of the second vote are credited with the number of votes obtained by their party in the parliamentary elections. Referendums may be called at the instigation of 10,000 signatories. Voting is compulsory.

National Anthem

'Orientales, la patria o la tumba' ('Easterners, the fatherland or the tomb'); words by F. Acuña de Figueroa, tune by F. J. Deballi.

GOVERNMENT CHRONOLOGY

Heads of State since 1943. (FA = Broad Front; PC = Colorado Party; PN = National Party (Blancos); PS = Socialist Party of Uruguay)

Presidents of the Republic

1943–47	PC	Juan José Amézaga Landaraso
1947	PC	Tomás Berreta Gandolfo
1947–51	PC	Luis Conrado Batlle Berres
1951–52	PC	Andrés Martínez Trueba

Chairman of the 1st National Council of Government

1952–55	PC	Andrés Martínez Trueba

Chairmen of the 2nd National Council of Government

1955–56	PC	Luis Conrado Batlle Berres
1956–57	PC	Alberto Fermín Zubiría Urtiague
1957–58	PC	Arturo Lezama Bagez
1958–59	PC	Carlos Lorenzo Fischer Brusoni

Chairmen of the 3rd National Council of Government

1959–60	PN	Martín Recaredo Echegoyen Machicote
1960–61	PN	Benito Nardone Cetrulo
1961–62	PN	Eduardo Víctor Haedo
1962–63	PN	Faustino Harrison Usoz

Chairmen of the 4th National Council of Government

1963–64	PN	Daniel Fernández Crespo
1964–65	PN	Luis Giannattasio Finocchietti
1965–66	PN	Washington Beltrán Mullin
1966–67	PN	Alberto Heber Usher

Presidents of the Republic

1967	PC	Óscar Diego Gestido Pose
1967–72	PC	Jorge Pacheco Areco
1972–76	PC[1]	Juan María Bordaberry Arocena
1976–81	PN[2]	Aparicio Méndez Manfredini
1981–85	military	Gregorio Conrado Álvarez Armellino
1985–90	PC	Julio María Sanguinetti Coirolo
1990–95	PN	Luis Alberto Lacalle de Herrera
1995–2000	PC	Julio María Sanguinetti Coirolo
2000–05	PC	Jorge Luis Batlle Ibáñez
2005–10	PS, FA	Tabaré Ramón Vázquez Rosas
2010–	FA	José Alberto Mujica

[1]Civilian president under military rule 1973–76.
[2]Civilian president under military rule.

RECENT ELECTIONS

Elections for the General Assembly were held on 25 Oct. 2009. In elections to the Chamber of Deputies, the Broad Front (FA) won 50 seats, the National Party (PN) 30, the Colorado Party (PC) 17 and the Independent Party 2. In the Senate election FA won 16 seats, PN 9 and PC 5.

In the presidential election, also held on 25 Oct. 2009, José Alberto Mujica Cordano (FA) received 48·0% of the vote, Luis Alberto Lacalle de Herrera (PN) 29·1%, Pedro Bordaberry Herrán (Colorado Party) 17·0% and Pablo Mieres (Independent Party) 2·5%. Three other candidates received less than 1% of the vote each. Mujica won the 29 Nov. run-off with 54·8% of the vote to Lacalle's 45·2%. Turnout was 89·9% in the first round and 89·1% in the run-off.

CURRENT GOVERNMENT

President: José Alberto Mujica; b. 1935 (FA; sworn in 1 March 2010).

Vice-President: Danilo Astori.

In March 2013 the government comprised:

Minister of Defence: Eleuterio Fernández Huidobro. *Economy and Finance:* Fernando Lorenzo. *Education and Culture:* Ricardo Ehrlich. *Foreign Relations:* Luis Almagro. *Housing, Land Management and Environment:* Graciela Muslera. *Industry, Energy and Mining:* Roberto Kreimerman. *Interior:* Eduardo Bonomi. *Livestock, Agriculture and Fisheries:* Tabaré Aguerre. *Public Health:* Jorge Venegas. *Social Development:* Daniel Olesker. *Tourism and Sports:* Héctor Lescano. *Transport and Public Works:* Enrique Pintado. *Work and Social Security:* Eduardo Brenta.

Presidency Website (Spanish only): http://www.presidencia.gub.uy

CURRENT LEADERS

José Mujica

Position
President

Introduction
José Mujica took office on 1 March 2010 after winning a presidential run-off in Nov. 2009. A former leftist guerrilla, Mujica received 55% of the vote after campaigning on a platform of continued economic growth and policies to tackle crime and poverty.

Early Life
Mujica was born in Montevideo on 20 May 1935. He was a member of the centrist Partido Nacional (National Party) in his youth before joining the newly-formed Movimiento de Liberación

Nacional (popularly known as the Tupamaros) in the 1960s, an armed guerrilla movement inspired by the Cuban revolution.

In 1971 he was convicted by a military tribunal under the government of Jorge Pacheco Areco of killing a police officer. He escaped from Punta Carretas prison but was re-arrested in 1972. Following the 1973 military coup Mujica was transferred to a military prison where he was subjected to torture and solitary confinement. When the military dictatorship ended in 1985 he was freed under a general amnesty covering political crimes since 1962.

On his release and the restoration of democracy, Mujica steered the Tupamaros away from its guerrilla past and remodelled it into the Movimiento de Participación Popular (Movement of Popular Participation), a legitimate political party that later joined the left-wing Frente Amplio (Broad Front) coalition. He was elected to the Chamber of Deputies in 1994 and in 1999 won a seat in the Senate, gaining re-election five years later.

On 1 March 2005 he resigned from the Senate when he was appointed minister of livestock, agriculture and fisheries by then president, Tabaré Vázquez. During his tenure Mujica intervened to keep down the price of beef, a staple of the Uruguayan people, winning popular acclaim. In 2008 he returned to the Senate after losing his cabinet post in a reshuffle.

On 28 June 2009 Mujica became Frente Amplio's presidential candidate after winning the coalition's primary election. He pledged to maintain the policies of outgoing President Vázquez, whose term of office had seen prolonged economic growth and strong social interventions by the government. In the first round of elections held in Oct. 2009 Mujica received 48% of the vote and on 30 Nov. 2009 he was declared winner of a run-off against Luis Alberto Lacalle of the National Party, with 55% of the vote.

Career in Office
Keen to maintain his predecessor's legacy, Mujica left the vice-president, Danilo Astori, in charge of economic policy. Mujica was expected to improve Uruguay's often strained relations with its neighbours, and he expressed his support for MERCOSUR, the regional economic bloc. His other priorities included the improvement of educational standards and the maintenance of energy supplies.

Mujica opposed a parliamentary bid to annul legislation giving officers immunity from prosecution for crimes committed during the years of military rule and, despite support in the Senate, the bill was rejected by the Chamber of Deputies in May 2011. The legislation was subsequently revoked in a further congressional vote in Oct. that year.

In Oct. 2012 Mujica signed into law a controversial bill legalizing abortion for all women during the first 12 weeks of pregnancy, as Uruguay became only the second Latin American country after Cuba to decriminalize terminations.

DEFENCE

Defence expenditure totalled US$260m. in 2008 (US$75 per capita), representing 0·8% of GDP.

Army
The Army consists of volunteers who enlist for one to two years service. There are four military regions with divisional headquarters. Strength (2007), 17,000. In addition there are government paramilitary forces numbering 920.

Navy
The Navy includes two frigates. A 280-strong naval aviation service operates one combat-capable aircraft. Personnel in 2007 totalled 5,000 including 450 naval infantry and 1,600 coast guards. The main base is at Montevideo.

Air Force
Organized with US aid, the Air Force had (2007) 3,000 personnel and 19 combat-capable aircraft.

ECONOMY

In 2006 agriculture contributed 9·2% of GDP, industry 32·4% and services 58·4%.

According to the anti-corruption organization *Transparency International*, Uruguay ranked equal 20th in the world in a 2012 survey of the countries with the least corruption in business and government. It received 72 out of 100 in the annual index.

Overview
Along with a favourable climate for agriculture and tourism, the economy has substantial hydropower potential. Economic links with its larger neighbours, Brazil and Argentina, have proved a mixed blessing but the creation of the Southern Common Market (MERCOSUR) in the 1990s cemented trade relations and increased exports to those markets.

From 1990–98 the economy grew 3·9% per annum, above the country's long-term trend of 2% from 1960–2000. In 1999 the economy shrank by 2·8% after the devaluation of the Brazilian *real* and a serious drought. In 2002 the weak banking environment and strong ties with Argentina exposed Uruguay to Argentina's economic collapse. This led to a run on Uruguay's banks, the forced flotation of the *peso* in June 2002 and a deep recession from 2002–03. The events of 1999–2003 highlighted macroeconomic and structural weaknesses, particularly the need for banking reform and reduced dependence on MERCOSUR partners.

However, unlike Argentina, Uruguay did not default on its debt and the country has relatively effective public institutions. Since the crisis, farming, tourism and finance—the traditional pillars of the economy—have performed well. Annual growth was 6·7% from 2004–08, and poverty declined from 33% to 21·7% between 2002 and 2008. There were structural reforms to competition and bankruptcy laws, as well as fiscal reform. In Jan. 2007 Uruguay signed a framework agreement on trade with the USA, a first step towards a free trade agreement. The global economic downturn slowed Uruguay's growth, especially in export oriented sectors, but positive growth rates were maintained.

Currency
The unit of currency is the *Uruguayan peso* (UYU), of 100 *centésimos*, which replaced the *nuevo peso* in March 1993 at 1 Uruguayan peso = 1,000 nuevos pesos. In June 2002 Uruguay allowed the peso to float freely. Inflation, which had been over 100% in 1990, was 6·7% in 2010 and 8·1% in 2011. In July 2005 total money supply was 22,794m. pesos, foreign exchange reserves were US$2,578m. and gold reserves were 8,000 troy oz (1·8m. troy oz in Nov. 1999).

Budget
In 2008 budgetary central government revenues totalled 123,171m. pesos (of which taxes 116,046m. pesos) and expenditures 120,462m. pesos. Main items of expenditure by economic type in 2008: compensation of employees, 36,025m. pesos; grants, 31,565m. pesos; use of goods and services, 18,583m. pesos.

Standard rate of VAT is 22% (reduced rate, 10%).

Performance
Uruguay depends heavily on its two large neighbours, Brazil and Argentina. Uruguay suffered four successive years of negative growth between 1999 and 2002, culminating in the economy shrinking by 7·1% in 2002, as the general downturn in the world economy was exacerbated by Argentina's crisis. However, the economy has since recovered to achieve GDP growth rates of 2·4% in 2009, 8·9% in 2010 and 5·7% in 2011. Total GDP in 2011 was US$46·7bn.

Banking and Finance
The Central Bank (*President*, Mario Bergara) was inaugurated on 16 May 1967. It is the bank of issue and supreme regulatory

authority. In 2003 there were three other state banks, three principal commercial banks, ten foreign banks and two major credit co-operatives. Savings banks deposits were 1,993,029m. pesos in 1995.

In 2010 foreign debt totalled US$11,347m., equivalent to 29·0% of GNI.

There is a stock exchange in Montevideo.

ENERGY AND NATURAL RESOURCES

Environment
Uruguay's carbon dioxide emissions from the consumption and flaring of fossil fuels in 2008 were the equivalent of 2·2 tonnes per capita.

Electricity
Installed capacity was 2·3m. kW in 2007. Production in 2007 was 9·42bn. kWh; consumption per capita in 2007 was 2,760 kWh.

Agriculture
Rising investment has helped agriculture, which has given a major boost to the country's economy. Some 41m. acres are devoted to farming, of which 90% to livestock and 10% to crops. Some large *estancias* have been divided up into family farms; the average farm is about 250 acres. In 2007 there were approximately 1·35m. ha. of arable land and 33,000 ha. of permanent crops. 169,000 ha. were irrigated in 2006.

Main crops (in 1,000 tonnes), 2003: rice, 1,250; wheat, 326; barley, 324; sunflower seeds, 234; soybeans, 183; maize, 178; sugarcane, 165; potatoes, 151; oranges, 122; grapes, 103; tangerines and mandarins, 75; apples, 72; wine, 72; sweet potatoes, 65; sorghum, 60; lemons and limes, 40; oats, 40. The country has some 6m. fruit trees, principally peaches, oranges, tangerines and pears.

Livestock, 2003: cattle, 11·69m.; sheep, 9·78m.; horses, 380,000; pigs, 240,000; chickens, 13m.

Livestock products, 2003 (in 1,000 tonnes): beef and veal, 424; lamb and mutton, 42; pork, bacon and ham, 17; poultry meat, 54; milk 1,495; greasy wool, 43; eggs, 42.

Forestry
In 2010 the area under forests was 1·74m. ha. (mainly eucalyptus and pine), representing 10% of the total land area. In 2007, 7·17m. cu. metres of roundwood were cut.

Fisheries
The total catch in 2010 was 73,105 tonnes, almost entirely marine fish.

INDUSTRY
In 2001 industry accounted for 26·6% of GDP, with manufacturing contributing 16·6%. Industries include meat packing, oil refining, cement manufacture, foodstuffs, beverages, leather and textile manufacture, chemicals, light engineering and transport equipment. Output in 1,000 tonnes (2007 unless otherwise indicated): distillate fuel oil, 637; cement (2008), 620; residual fuel oil, 368; petrol, 352; meat-packing (1991), 1,132,000 head; 9·6bn. cigarettes (2001).

Labour
In 1996 the retirement age was raised from 55 to 60 for women; it remains 60 for men. The labour force in 2005 totalled 1,269,300 (54% males). In 2001, 22·4% of the urban workforce was engaged in wholesale and retail trade/repair of motor vehicles, motorcycles and personal and household goods/hotels and restaurants; 15·5% in manufacturing/electricity, gas and water supply; 9·2% in private households with employed persons; and 9·1% in financial intermediation and real estate, renting and business activities. In 2001 the unemployment rate in urban areas was 15·3%.

INTERNATIONAL TRADE

Imports and Exports
Trade in US$1m.:

	2004	2005	2006	2007	2008
Imports c.i.f.	3,114	3,879	4,757	5,667	8,943
Exports f.o.b.	2,931	3,405	3,953	4,485	6,421

Main imports in 2004: petroleum, 24·1%; chemicals and chemical products, 16·0%; machinery and appliances, 14·1%; food, beverages and tobacco, 8·5%; plastic products, 6·2%. Main exports in 2004: beef, 20·6%; hides and leather goods, 9·5%; textiles, 8·1%; dairy products, 7·0%; rice, 6·2%.

In 2004 the main import suppliers were Argentina (22·2%), Brazil (21·7%), Russia (11·0%), USA (7·1%) and China (5·5%). Leading export destinations in 2004 were the USA (19·8%), Brazil (16·5%), Argentina (7·6%), Germany (5·2%) and Mexico (4·0%).

COMMUNICATIONS

Roads
In 2004 there were 77,732 km of roads, including 3,877 km of national roads and 4,819 km of regional roads. Passenger cars in 2007 numbered 553,200 (151 per 1,000 inhabitants in 2005). There were 150 fatalities as a result of road accidents in 2005.

Rail
The total railway system open for traffic in 2005 was 1,508 km of 1,435 mm gauge. Passenger services, which had been abandoned in 1988, were resumed on a limited basis in 1993. In 2007 the railways carried 600,000 passengers and 1·4m. tonnes of freight.

Civil Aviation
There is an international airport at Montevideo (Carrasco). The national carrier is Pluna. In 2003 it operated domestic services and maintained routes to Asunción, Buenos Aires, Madrid, Porto Alegre, Rio de Janeiro, Santiago and São Paulo. There were 60 airports in 1996, 45 with paved runways and 15 with unpaved runways. In 1999 Montevideo handled an estimated 1,423,000 passengers (1,115,000 on international flights) and 25,500 tonnes of freight. In 2003 scheduled airline traffic of Uruguay-based carriers flew 8m. km, carrying 464,000 passengers (all on international flights).

Shipping
In Jan. 2009 there were 25 ships of 300 GT or over registered, totalling 53,000 GT. In 2004 vessels totalling 5,067,000 NRT entered ports and vessels totalling 22,262,000 NRT cleared.

Telecommunications
The telephone system in Montevideo is controlled by the State; small companies operate in the interior. Uruguay had 3,969,500 telephone subscribers in 2007 (1,188·6 for every 1,000 persons). There were 3,004,300 mobile phone subscribers in 2007. Internet users numbered 968,000 in 2007 (including 165,000 broadband subscribers). In March 2012 there were 1·5m. Facebook users.

SOCIAL INSTITUTIONS

Justice
The Supreme Court is elected by Congress; it appoints all other judges. There are six courts of appeal, each with three judges. There are civil and criminal courts. Montevideo has ten courts of first instance, Paysandú and Salto have two each and the other departments have one each. There are approximately 300 lower courts.

The population in penal institutions in Aug. 2006 was 6,947 (193 per 100,000 of national population).

Education

Adult literacy in 2006 was 98·3% (male, 97·6%; female, 98·6%). The female literacy rate is the second highest in South America, behind Chile. Primary education is obligatory; both primary and secondary education are free. In 2007 there were 122,089 pupils in pre-primary schools with 5,220 teaching staff, 359,439 primary school pupils with 23,175 teaching staff and at secondary level there were 294,852 pupils and 21,369 teaching staff.

There is one state university, one independent Roman Catholic university and one private institute of technology. In 2007 there were 158,841 students and 15,789 academic staff in tertiary education.

In 2006 public expenditure on education came to 3·0% of GNI and represented 11·6% of total government expenditure.

Health

In 2003 there were 13,071 physicians, 4,154 dentists, 3,118 nurses, 1,323 pharmacists and 572 midwives. There were 107 hospitals with 6,661 beds.

Welfare

The welfare state dates from the beginning of the 1900s. In 2002 there were 0·5m. recipients of pensions and benefits. A private pension scheme inaugurated in 1996 had 315,000 members at 31 Dec. 1996. State spending on social security has been capped at 15% of GDP.

RELIGION

State and Church are separate, and there is complete religious liberty. In 2001 there were 2·6m. Roman Catholics and 710,000 persons with other beliefs.

CULTURE

World Heritage Sites

Uruguay has one site on the UNESCO World Heritage List: the Historic Quarter of the City of Colonia del Sacramento (inscribed on the list in 1995), founded in 1680 by the Portuguese.

Press

In 2008 there were 34 paid-for dailies with an average circulation of 145,000. The newspaper with the highest circulation is *El País*, which sold a daily average of 46,000 copies in 2008.

Tourism

There were 1·81m. non-resident tourists in 2005, mainly from Argentina. Receipts totalled US$690m.

DIPLOMATIC REPRESENTATIVES

Of Uruguay in the United Kingdom (4th Floor, 150 Brompton Rd., London SW3 1XH)
Ambassador: Julio Moreira Móran.

Of the United Kingdom in Uruguay (Calle Marco Bruto 1073, 11300 Montevideo)
Ambassador: Ben Lyster-Binns.

Of Uruguay in the USA (1913 I St., NW, Washington, D.C., 20006)
Ambassador: Juan Carlos Pita Alvariza.

Of the USA in Uruguay (Lauro Muller 1776, Montevideo)
Ambassador: Julissa Reynoso.

Of Uruguay to the United Nations
Ambassador: José Luis Cancela Gómez.

Of Uruguay to the European Union
Ambassador: Walter Cancela.

FURTHER READING

González, L. E., *Political Structures and Democracy in Uruguay.* 1992
Sosnowski, S. (ed.) *Repression, Exile and Democracy: Uruguayan Culture.* 1993

National library: Biblioteca Nacional del Uruguay, 18 de julio de 1790, Montevideo.
National Statistical Office: Instituto Nacional de Estadística (INE), Rio Negro 1520, Montevideo.
Website (Spanish only): http://www.ine.gub.uy

**Explore the world at
www.statesmansyearbook.com**

UZBEKISTAN

© Research Machines plc 2006

Uzbekiston Respublikasy
(Republic of Uzbekistan)

Capital: Tashkent
Population projection, 2015: 29·06m.
GNI per capita, 2011: (PPP$) 2,967
HDI/world rank: 0·641/115
Internet domain extension: .uz

KEY HISTORICAL EVENTS

Evidence of human settlement from at least 2200 BC is believed to be that of the Oxus civilization which extended across central Asia from Turkmenistan to Tajikistan. The region came under the influence of the first Persian Empire, centred on Persepolis, from around 550 BC when it was known as Sogdiana. Alexander the Great conquered Sogdiana and the ancient Greek kingdom of Bactria in 327 BC, marrying Roxane, daughter of a Sogdian chieftain.

Turkic nomads entered the area from the 5th century AD and control subsequently passed to Arabs, who introduced Islam to Transoxiana in the 8th century. The Persian Samanid dynasty, centred on the cities of Bukhara, Samarkand and Heart, held sway from around AD 875 for over a century, before falling to the Khara-Khanid Khanate.

Much of present-day Uzbekistan came under the control of Seljuk Turks from the 11th century. Led by Alp Arslan, they went on to conquer Georgia, Armenia, Syria and most of Anatolia. Khwarazm in northwest Uzbekistan gained independence from the Seljuks in the late 12th century, expanding westward as far as the Caspian Sea. Genghis Khan's Mongol Hordes overran Central Asia from 1221 and retained power until the rise of Timur, who made his native Samarkand the capital of an empire that by 1405 stretched from western India to the Black Sea. The subsequent Timurid dynasty was ruled by Shahrukh and, from 1447, by Ulugh-beg. Located on the trade routes between Europe, Persia, India and China, the oasis cities of Samarkand, Bukhara and Tashkent became prosperous centres of culture and learning.

The Uzbeks were Turkic-speaking tribes who moved into the region from the steppes to the north of the Aral Sea from the early 16th century. They established separate principalities, notably the Emirate of Bukhara and the Khanates of Khiva and Kokand. The fall of the Timurid Empire presaged the gradual decline of the Central Asian trade routes, partly as a result of the opening of shipping lines between Western Europe and India and East Asia.

During the 1850s the Russian empire began to expand into central Asia. In 1865 Russian forces seized the city of Tashkent and a year later the Khanate of Kokand was dissolved and incorporated into the Governor-Generalship of Turkestan. The Khanate of Khiva and the Emirate of Bukhara became protectorates. Russian colonists, settling throughout Central Asia, developed the region's infrastructure while exploiting the abundant minerals and promoting the growth of cash crops such as cotton. There were periodic revolts against Russian rule, notably the Andijan Uprising of 1898. The chaos surrounding the Bolshevik Revolution of 1917 precipitated the growth of an Uzbek guerrilla army, 'the Basmachi', although their efforts to establish a democratic republic were unsuccessful. In 1924 the Khorezm and Bukhara People's Republics were incorporated into the Uzbekistan Soviet Socialist Republic (SSR), which also included Tajikistan until it became a separate SSR in 1929. The Soviet period brought collectivization and rapid industrialization, particularly around the capital, Tashkent, leading to an influx of Russian migrant workers. Uzbekistan SSR became the primary cotton-producing region of the Soviet Union, although the large-scale diversion of rivers for irrigation had disastrous consequences for the Aral Sea, which lost two-thirds of its volume.

On 20 June 1990 the Supreme Soviet adopted a declaration of sovereignty, although the Communist Party, led by Islam Karimov, remained the only official political party. Following the collapse of the Soviet Union, Uzbekistan was declared an independent republic on 31 Aug. 1991. In Dec. 1991 Uzbekistan became a member of the Commonwealth of Independent States and Karimov was elected president by popular vote. He subsequently cracked down on political opponents, notably the Birlik party and the Islamic Renaissance Party, citing the need for 'stability'.

Bomb blasts in Tashkent, one of which almost killed the president in Feb. 1999, were blamed on extremists from the Islamist Movement of Uzbekistan (IMU), which operates largely in the Ferghana valley and aims to create a pan-Central Asian Islamic state. The IMU leader, Juma Namangoniy, was reportedly killed in Aug. 2002. Nearly 50 people died in a series of bombings and shootings in March 2004, allegedly perpetrated by Islamic militants. Suicide bombers targeted the US and Israeli embassies in Tashkent in July 2004.

On 13 May 2005 several hundred demonstrators were killed in Andijan after troops fired into a crowd protesting against the imprisonment of local businessmen. The USA joined calls for an international inquiry into the shootings. In Nov. 2005 the USA closed its military air base at Karshi-Khanabad which had been used for operations in Afghanistan.

TERRITORY AND POPULATION

Uzbekistan is bordered in the north by Kazakhstan, in the east by Kyrgyzstan and Tajikistan, in the south by Afghanistan and in the west by Turkmenistan. Area, 447,400 sq. km (172,700 sq. miles), including 22,000 sq. km (8,500 sq. miles) of inland water. A census has not been held since 1989, when the population was 19,810,077. A 'mini-census' based on 10% of the population was conducted in April 2011 but there are no future plans for a full census. Estimate, 2011, 28·54m.; density, 67 per sq. km. The vast majority of the population are Uzbeks, with small Tajik, Kazakh, Tatar and Russian minorities. In 2008, 63·3% of the population lived in rural areas.

The UN gives a projected population for 2015 of 29·06m.

The areas and populations of the 12 regions, the Karakalpak Autonomous Republic (Karakalpakstan) and the city of Tashkent are as follows (Uzbek spellings in brackets):

Region	Area (in sq. km)	Population (2008 estimate)	Capital	Population (2006 estimate)
Andizhan (Andijon)	4,200	2,477,900	Andizhan	356,800
Bukhara (Bukhoro)	39,400	1,576,800	Bukhara	241,300
Dzhizak (Jizzakh)	20,500	1,090,900	Dzhizak	139,200
Ferghana (Farghona)	7,100	2,997,400	Ferghana	188,100
Khorezm (Khorazm)	6,300	1,517,600	Urgench (Urganch)	140,700
Kashkadar (Qashqadaryo)	28,400	2,537,600	Karshi (Qarshi)	217,400
Karakalpak Autonomous Republic (Qoraqalpoghiston)	164,900	1,612,300	Nukus (Nuqus)	259,700
Namangan	7,900	2,196,200	Namangan	415,000
Navoi (Nawoiy)	110,800	834,100	Nawoiy	121,200
Samarkand (Samarqand)	16,400	3,032,000	Samarkand	364,200
Syr-Darya (Sirdaryo)	5,100	698,100	Gulistan (Guliston)	55,600[1]
Surkhan-Darya (Surkhondaryo)	20,800	2,012,600	Termez (Termiz)	122,700
Tashkent (Toshkent)	15,600	4,730,200	Tashkent	2,140,600

[1]2005.

Regions are further subdivided into 227 districts and cities.

The capital is Tashkent (2006 population estimate, 2,140,600); other large towns are Namangan, Samarkand, Andizhan, Nukus, Bukhara, Karshi, Kokand, Ferghana, Margilan, Chirchik and Urgench.

The Roman alphabet (in use 1929–40) was reintroduced in 1994. Arabic script was in use prior to 1929, and Cyrillic from 1940–94.

The official language is Uzbek. Russian and Tajik are also spoken.

SOCIAL STATISTICS

2009 births, 649,700; deaths, 130,700; marriages, 227,600; divorces, 17,200. Rates, 2009: birth (per 1,000 population), 23·3; death, 4·7; marriage, 10·0; divorce, 0·6. Life expectancy, 2007, 64·5 years for men and 70·9 for women. Annual population growth rate, 1998–2008, 1·2%. Infant mortality, 2010, 44 per 1,000 live births; fertility rate, 2008, 2·3 births per woman.

CLIMATE

The summers are warm to hot but the heat is made more bearable by the low humidity. The winters are cold but generally dry and sunny. Tashkent, Jan. –1°C, July 25°C. Annual rainfall 14·8" (375 mm).

CONSTITUTION AND GOVERNMENT

A new constitution was adopted on 8 Dec. 1992 stating that Uzbekistan is a pluralist democracy. The constitution restricts the president to standing for two five-year terms. In Jan. 2002 a referendum was held at which 91% of the electorate voted in favour of extending the presidential term from five to seven years. Voters were also in favour of changing from a single-chamber legislature to a bicameral parliament. Based on the constitution President Karimov's term of office that started in Jan. 2000 ended in Jan. 2007, but according to election law a vote must be held in Dec. of the year in which the president's term expires. Pro-Karimov legislators maintained that he was eligible to stand again in the Dec. 2007 elections as he had only served one seven-year term despite having been president since 1990.

Uzbekistan switched to a bicameral legislature in Jan. 2005 with the establishment of the 100-member *Senate* (with 16 members appointed by the president and 84 elected from the ranks of regional, district and city legislative councils). The lower house is the 150-member *Oliy Majlis* (Supreme Assembly). 135 seats are elected by popular vote for five-year terms and 15 are reserved for the Ecological Movement.

National Anthem

'Serquyosh, hur o'lkam, elga baxt najot' ('Stand tall, my free country, good fortune and salvation to you'); words by Abdulla Aripov, tune by Mutal Burhanov.

GOVERNMENT CHRONOLOGY

President since 1990. (n/p = non-partisan)
1990–   n/p   Islam Abduganiyevich Karimov

RECENT ELECTIONS

Presidential elections were held on 23 Dec. 2007. Incumbent Islam Karimov was elected against three opponents with 90·8% of the vote, although the elections were condemned by the OSCE. Turnout was 90·6%.

In parliamentary elections held in two rounds on 27 Dec. 2009 and 10 Jan. 2010, the Liberal-Democratic Party won 53 of the 150 seats, followed by the People's Democratic Party with 32, the National Revival Democratic Party with 31 and the Justice Social Democratic Party with 19. A further 15 seats were reserved for the Ecological Movement of Uzbekistan. All parties taking part in the election were loyal to President Islam Karimov—opposition parties were barred from participating.

CURRENT GOVERNMENT

President: Islam Karimov; b. 1938 (sworn in 24 March 1990).

In March 2013 the government comprised:

Prime Minister: Shavkat Mirziyayev; b. 1957 (People's Democratic Party; in office since 11 Dec. 2003).

First Deputy Prime Minister: Rustam Azimov (also *Minister of Finance*).

Deputy Prime Ministers: Elmira Basitkhanova; Gulomjon Ibragimov; Adkham Ikramov; Ulugbek Rozukulov; Batir Zakirov.

Minister of Agriculture and Water Resources: Zafar Ruziev. *Culture and Sports:* Minhojiddin Hojimatov. *Defence:* Kobil Berdiev. *Economy:* Galina Saidova. *Emergency Situations:* Tursinkhon Khudayberganov. *Foreign Affairs:* Abdulaziz Kamilov. *Foreign Economic Relations, Investments and Trade:* Elyor Ganiyev. *Health:* Anvar Alimov. *Higher and Secondary Specialized Education:* Bakhodir Khodiev. *Internal Affairs:* Bakhodir Matlyubov. *Justice:* Nigmatilla Yuldoshev. *Labour and Social Protection:* Aktam Haitov. *Public Education:* Ulugbek Inoyatov.

Chairman, Oliy Majlis: Dilorom Tashmuhamedova.

Office of the President: http://www.gov.uz

CURRENT LEADERS

Islam Abduganiyevich Karimov

Position
President

Introduction
Islam Karimov, a former Soviet official, has been president since Uzbekistan declared independence in 1990. His regime has been characterized by the suppression of domestic political and religious opposition, and his electoral victories have been questioned for their irregularities. Karimov has sought to build ties with the West, and won US favour for co-operation in the war against terrorism in the aftermath of the 11 Sept. 2001 attacks. However, reports of torture and other human rights violations, culminating in an alleged massacre of Uzbek civilians in Andizhan in May 2005, have provoked increasing international criticism of his regime. In Dec. 2007 he retained the presidency, claiming a landslide victory in an election widely condemned as undemocratic and constitutionally illegal.

Early Life

Karimov was born on 30 Jan. 1938 in Samarkand. He qualified as a mechanical engineer at the Central Asian Polytechnical Institute and graduated in economics from the Tashkent Institute of National Economy. He then worked in Tashkent at farm machinery and aircraft plants. In 1966 he moved to the state planning committee of Uzbekistan, attaining the rank of vice-chairman.

In 1983 Karimov was appointed finance minister for Uzbekistan and three years later became deputy head of government, as well as chairman of the state planning committee. In 1989 he was named head of the Uzbek Communist Party. The following year Uzbekistan claimed sovereignty from the USSR and Karimov was chosen as president. Following the attempted coup against Mikhail Gorbachev in Moscow in 1991, Karimov declared full independence.

Career in Office

Against little organized opposition, Karimov dominated presidential elections and led Uzbekistan into the Commonwealth of Independent States. In 1992 he continued his campaign against domestic opposition, banning two leading parties—Birlik (Unity) and Erk (Freedom)—and imprisoning many members. In 1994 he agreed an economic integration treaty with Russia and signed a co-operation pact with Kazakhstan and Kyrgyzstan which was developed into a single economic community in 1996. In 1995 Karimov won a further five years in office by plebiscite.

In 1999 Tashkent was the scene of several car bombings which Karimov blamed on the Islamic Movement of Uzbekistan (IMU). Government and IMU forces clashed several times, the culmination of growing tensions between the two sides since the mid-1990s. In the same year Karimov withdrew Uzbekistan from the CIS agreement on collective security, increasing the nation's isolation among Central Asian nations predominantly loyal to Moscow.

Karimov was re-elected to the presidency in 2000 with over 90% of the vote, although the electoral process was severely criticized by the international community, notably the opposition candidate's assertion that he himself would vote for Karimov. Following the 11 Sept. attacks in New York and Washington, Karimov permitted the USA to use Uzbek air bases for the war in Afghanistan. In the same year he signed up to the Shanghai Co-operation Society (with China, Russia, Kazakhstan, Kyrgyzstan and Tajikistan), established to promote regional economic co-operation and fight religious and ethnic militancy.

In Jan. 2002 Karimov secured a constitutional change, accepted by referendum, extending the presidential term from five to seven years. His assistance in the US campaign in Afghanistan was meanwhile rewarded by US$160m. worth of aid from Washington. Also in 2002 a long-running border feud with Kazakhstan was settled.

In March 2004 a series of shootings and explosions in the Tashkent and Bukhara regions left dozens of people dead. Further bombings near the US and Israeli embassies and in the Prosecutor General's Office in Tashkent occurred in July. Karimov's government blamed Islamic militants. Then in May 2005 several hundred civilians, protesting against the trial of local businessmen accused of Islamic extremism, were reportedly killed by security forces in Andizhan. Unrest also spread to the towns of Paktabad and Kara Suu before troops reasserted government control. At the end of July, in a punitive response to international criticism of the massacre, Karimov gave the USA six months to close its military airbase in Uzbekistan. His relations with the European Union also became increasingly strained in the wake of the events in Andizhan and the subsequent convictions of those accused of instigating unrest. The EU imposed an arms embargo and a visa ban, and these sanctions were extended in 2006. Uzbekistan meanwhile sought to strengthen its military and economic co-operation with Russia.

In Dec. 2007 Karimov's re-election as president failed to meet democratic standards, according to international observers, and was condemned by opposition activists as a sham.

In late 2008 and early 2009 there was some improvement in relations with the West as Karimov allowed the USA and NATO to transport non-lethal supplies through Uzbekistan to their armed forces in Afghanistan and the EU eased sanctions despite continuing concerns about human rights. The EU's decision followed the release of some political prisoners and the abolition of the death penalty.

In mid-2009 two terrorist incidents in Andizhan and Khanabad were attributed to increasing Islamist militancy.

In Sept. 2012 Karimov warned of potential confrontations with neighbouring Tajikistan and Kyrgyzstan over their plans for large dam projects that he claimed would give the two countries unfair control of regional water resources.

DEFENCE

Conscription is for 12 months. Defence expenditure in 2006 totalled US$85m. (US$3 per capita), representing 0·5% of GDP.

Army

Personnel, 2007, 50,000. There are, in addition, paramilitary forces totalling up to 20,000.

Air Force

Personnel, 2007, 17,000. There were 135 combat-capable aircraft in operation (including Su-17s, Su-24s, Su-25s, Su-27s and MiG-29s) and 29 attack helicopters.

ECONOMY

Agriculture accounted for 26·1% of GDP in 2006, industry 27·4% and services 46·5%.

Uzbekistan featured among the ten most corrupt countries in the world in a 2011 survey of 183 countries carried out by the anti-corruption organization *Transparency International*.

Overview

After the collapse of the Soviet Union, Uzbekistan embarked on a cautious economic transition. There is heavy dependence on state controls and planning, foreign exchange and trade restrictions and public investments aimed at achieving import-substitution industrialization and self-sufficiency in food and energy.

Heavily populated river valleys are intensely cultivated and irrigated. Agriculture employs a third of the labour force. The land is rich in natural resources and primary commodities (cotton, gold, copper, energy resources and precious stones) account for the majority of exports. Uzbekistan is the world's second largest cotton exporter behind the USA. In recent years strong cotton and gold prices, with growing exports of natural gas and manufactured products, have strengthened the economy and its external financial position. However, the business climate is unfavourable to private enterprise.

The country receives significant remittances from citizens who work abroad, primarily in Russia and Kazakhstan. Uzbekistan has increasingly looked towards Russia, China and other central Asian countries to build economic links. Neighbouring countries are keen to develop Uzbekistan's oil and gas industry.

Economic growth increased from 4% in the period 1996–2003 to over 9% in 2007–08. However, the global downturn led to weaker demand for exports and there was a 30% decline in remittances in 2009.

Currency

A coupon for a new unit of currency, the *soum* (UZS), was introduced alongside the rouble on 15 Nov. 1993. This was replaced by the *soum* proper at 1 soum = 1,000 coupons on 1 July 1994. In 1994 inflation was 1,568%, but it reached more stable levels of 9·4% in 2010 and 12·8% in 2011.

Budget

In 2006 general government consolidated revenue amounted to 6,406bn. soums and expenditure to 6,331bn. soums.

Performance

Real GDP growth was 8·1% in 2009, 8·5% in 2010 and 8·3% in 2011; total GDP in 2011 was US$45·4bn.

Banking and Finance

The Central Bank is the bank of issue (*Chairman*, Faizulla Mullajanov). In 2001 there were 38 commercial banks, of which 16 were privately owned. Foreign debt amounted to US$7,404m., equivalent to 19·0% of GNI.

ENERGY AND NATURAL RESOURCES

Environment

Irrigation of arid areas has caused the drying up of the Aral Sea. Uzbekistan's carbon dioxide emissions from the consumption and flaring of fossil fuels in 2008 were the equivalent of 4·6 tonnes per capita.

Electricity

Installed capacity was an estimated 11·7m. kW in 2007. Production was 48·9bn. kWh in 2007 and consumption per capita 1,785 kWh.

Oil and Gas

Oil production was 4·8m. tonnes in 2008; natural gas output was 62·2bn. cu. metres. In 2008 there were proven oil reserves of 0·6bn. bbls and natural gas reserves of 1,580bn. cu. metres.

Minerals

Lignite production in 2007 was 3·28m. tonnes. In 2005, 84 tonnes of gold and an estimated 60 tonnes of silver were produced. There are also large reserves of uranium, copper, lead, zinc and tungsten; all uranium mined (2,629 tonnes in 2005) is exported.

Agriculture

Farming is intensive and based on irrigation. In 2002 there were 4·48m. ha. of arable land and 0·34m. ha. of permanent cropland. Approximately 4·28m. ha. were irrigated in 2002.

Output of main agricultural products (2003, in 1,000 tonnes): wheat, 5,331; seed cotton, 2,856; cottonseed, 1,720; tomatoes, 1,100; cotton lint, 946; cabbage, 900; potatoes, 760; grapes, 510; apples, 503; watermelons, 460; rice, 311; cucumbers and gherkins, 300. Livestock, 2003 estimates: 8·2m. sheep; 5·4m. cattle; 820,000 goats; 14m. chickens. Livestock products, 2003 estimates (in 1,000 tonnes): meat, 508; milk, 3,790; eggs, 83.

Forestry

In 2010 the area under forests was 3·28m. ha., accounting for 8% of the total land area. In 2007, 31,000 cu. metres of timber were produced.

Fisheries

The total catch in 2009 was 6,051 tonnes, exclusively freshwater fish.

INDUSTRY

Industrial production grew by 3·4% in 2002. Major industries include fertilizers, agricultural and textile machinery, aircraft, metallurgy and chemicals. Output (in tonnes): cement (2009 estimate), 6,700,000; distillate fuel oil (2007), 1,391,000; residual fuel oil (2007), 1,342,000; petrol (2007), 1,284,000; sulphuric acid (2000), 823,000; mineral fertilizer (2000), 800,000; cotton woven fabrics (2000), 360m. sq. metres; 1,000 tractors (2000); 26,000 TV sets (2000).

Labour

In 1999 a total of 8,885,000 persons were in employment, including: 3,421,000 engaged in agriculture, hunting, forestry and fishing; 1,968,000 in community, social and personal services;

1,142,000 in manufacturing, mining and quarrying, electricity, gas and water; and 734,000 in wholesale and retail trade, restaurants and hotels. In 2000 the unemployment rate was 0·6%. Average monthly salary in 1999 was 8,823 soums. A minimum wage of 6,530 soums a month was imposed on 1 Aug. 2004.

INTERNATIONAL TRADE

Imports and Exports

In 2008 imports were valued at US$7,504m. and exports at US$11,573m. Principal imports, 2002, were machinery and equipment, 41·4%; chemicals and chemical products, 15·1%; foodstuffs, 12·5%; ferro and non-ferro metals, 8·0%.

The main import sources in 2002 were Russia (20·5%), South Korea (17·4%), Germany (8·9%) and Kazakhstan (7·5%). Principal export markets in 2002 were Russia (17·3%), Ukraine (10·2%), Italy (8·3%) and Tajikistan (7·8%).

COMMUNICATIONS

Roads

Length of roads, 2000, was 86,496 km (87·3% paved).

Rail

The total length of railway in 2005 was 3,986 km of 1,520 mm gauge (619 km electrified). In 2008, 14·3m. passengers and 78·3m. tonnes of freight were carried. There is a metro in Tashkent.

Civil Aviation

The main international airport is in Tashkent (Vostochny). Andizhan, Namangan and Samarkand also have airports. The national carrier is the state-owned Uzbekistan Airways, which in 2003 operated domestic services and flew to Almaty, Amritsar, Ashgabat, Athens, Baku, Bangkok, Beijing, Birmingham, Bishkek, Chelyabinsk, Delhi, Dhaka, Ekaterinburg, Frankfurt, İstanbul, Kazan, Khabarovsk, Krasnodar, Krasnoyarsk, Kuala Lumpur, Kyiv, London, Mineralnye Vody, Moscow, New York, Novosibirsk, Omsk, Osaka, Paris, Rome, Rostov, St Petersburg, Samara, Seoul, Sharjah, Simferopol, Tel Aviv, Tokyo, Tyumen and Ufa. In 2003 scheduled airline traffic of Uzbekistan-based carriers flew 40m. km, carrying 1,466,000 passengers (1,048,000 on international flights). In 2001 Tashkent handled 1,387,000 passengers (884,000 on international flights) and 36,200 tonnes of freight.

Telecommunications

In 2008 there were 1,849,600 main (fixed) telephone lines; mobile phone subscribers numbered 12,733,700 in 2008 (46·8 per 100 persons). There were 2,469,000 internet users in 2008. In March 2012 there were 129,000 Facebook users.

SOCIAL INSTITUTIONS

Justice

In 1994, 73,561 crimes were reported, including 1,219 murders and attempted murders. The death penalty was abolished in Jan. 2008. The population in penal institutions in 2009 was approximately 42,000 (153 per 100,000 of national population).

Education

In 2007 there were 562,000 pre-primary pupils with 60,642 teaching staff, 2·16m. primary pupils with 118,676 teaching staff and 4·60m. secondary pupils with 352,001 teaching staff. There were 288,550 students and 23,354 academic staff in tertiary education in 2007. There are universities and medical schools in Tashkent and Samarkand. Adult literacy rate in 2009 was estimated at 99·3% (99·6% among males and 99·1% among females).

Health

In 2005 there were 52 hospital beds per 10,000 population and 70,564 physicians, 5,194 dentistry personnel and 290,162 nursing and midwifery personnel.

Welfare

There is a statutory pension system consisting of two pillars: a public pay-as-you-go defined-benefit pension scheme and (since 2005) a mandatory public funded defined-contribution scheme. The larger pay-as-you-go pillar provides income to approximately 2·8m. people.

RELIGION

The Uzbeks are predominantly Sunni Muslims.

CULTURE

World Heritage Sites

Uzbekistan has four sites on the UNESCO World Heritage List: Itchan Kala (inscribed on the list in 1990); the Historic Centre of Bukhara (1993); the Historic Centre of Shakhrisyabz (2000); and Samarkand—Crossroads of Cultures (2001).

Press

In 2006 there were four daily newspapers with a combined circulation of 30,000.

Tourism

There were 903,000 non-resident tourists in 2007. Tourism expenditure (excluding passenger transport) totalled US$51m.

DIPLOMATIC REPRESENTATIVES

Of Uzbekistan in the United Kingdom (41 Holland Park, London, W11 3RP)
Ambassador: Otabek Akbarov.

Of the United Kingdom in Uzbekistan (Ul. Gulyamova 67, 700000 Tashkent)
Ambassador: George Edgar.

Of Uzbekistan in the USA (1746 Massachusetts Ave., NW, Washington, D.C., 20036)
Ambassador: Ilkhom Nematov.

Of the USA in Uzbekistan (3 Moyqorghon St., 5th Block, Yunusobod District, 100093 Tashkent)
Ambassador: George A. Krol.

Of Uzbekistan to the United Nations
Ambassador: Murad Askarov.

Of Uzbekistan to the European Union
Ambassador: Bakhtiyar Gulyamov.

FURTHER READING

Bohr, A. (ed.) *Uzbekistan: Politics and Foreign Policy.* 1998
Kalter, J. and Pavaloi, M., *Uzbekistan: Heir to the Silk Road.* 1997
Melvin, N. J., *Uzbekistan: Transition to Authoritarianism on the Silk Road.* 2000
Yalcin, Resul, *The Rebirth of Uzbekistan: Politics, Economy and Society in the Post-Soviet Era.* 2002

National Statistical Office: State Committee of the Republic of Uzbekistan on Statistics, Mustakillik Avenue 63, Tashkent 100077.
Website: http://www.stat.uz

Karakalpak Autonomous Republic (Karakalpakstan)

Area, 166,600 sq. km (64,320 sq. miles); population (2008 estimate), 1,612,300. Capital, Nukus (2006 estimate, 259,700). The Karakalpaks (Qoraqalpoghs) came under Russian rule in the second half of the 19th century. On 11 May 1925 the territory was constituted within the then Kazakh Autonomous Republic (of the Russian Federation) as an Autonomous Region. On 20 March 1932 it became an Autonomous Republic within the Russian Federation, and on 5 Dec. 1936 it became part of the Uzbek SSR. The main ethnic groups are Qoraqalpoghs, Uzbeks and Kazakhs.

Its manufactures are in the field of light industry—bricks, leather goods, furniture, canning and wine. The principal crops are rice and cotton.

The shrinking of the Aral Sea has had a detrimental effect on agriculture and public health in the region. In 2002 the incidence of poverty was 36·4%, compared to 9·2% in Tashkent, the Uzbek capital. Estimates place unemployment at around the 70% mark.

In 2004–05 there were 6,800 students at Karakalpak State University. There is a branch of the Uzbek Academy of Sciences.

Total spending on health care in 2002 was US$10m., or US$6·50 per capita.

VANUATU

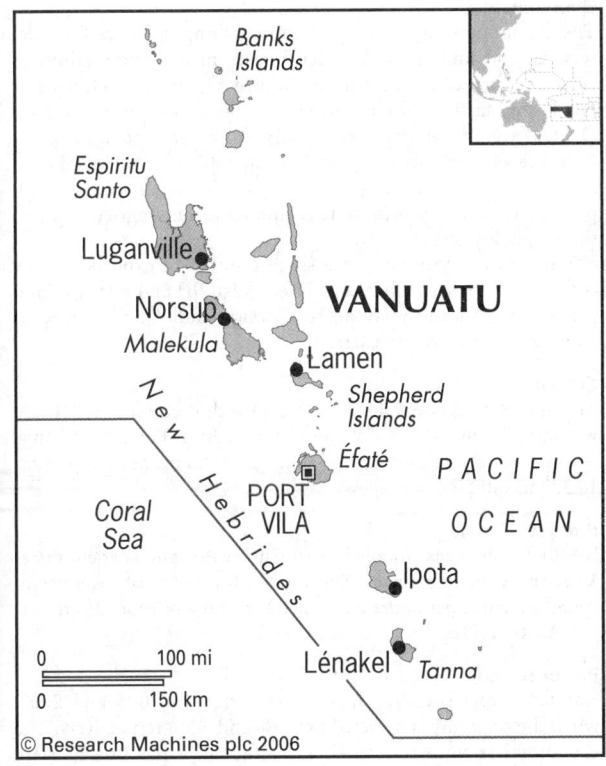

© Research Machines plc 2006

Ripablik blong Vanuatu
(Republic of Vanuatu)

Capital: Port Vila
Population projection, 2015: 270,000
GNI per capita, 2011: (PPP$) 3,950
HDI/world rank: 0·617/125
Internet domain extension: .vu

KEY HISTORICAL EVENTS

Vanuatu occupies the group of islands formerly known as the New Hebrides, in the southwestern Pacific Ocean. Capt. Bligh and his companions, cast adrift by the *Bounty* mutineers, sailed through part of the island group in 1789. Sandalwood merchants and European missionaries came to the islands in the mid-19th century and were then followed by cotton planters—mostly French and British—in 1868. In response to Australian calls to annexe the islands, Britain and France agreed on joint supervision. Joint sovereignty was held over the indigenous Melanesian people but each nation retained responsibility for its own nationals according to a protocol of 1914. The island group escaped Japanese invasion during the Second World War and became an Allied base. On 30 July 1980 New Hebrides became an independent nation under the name of Vanuatu, meaning 'Our Land Forever'.

TERRITORY AND POPULATION

Vanuatu comprises 83 islands (65 of which are inhabited), which lie roughly 800 km west of the Fiji Islands and 400 km northeast of New Caledonia. The estimated land area is 12,190 sq. km (4,706 sq. miles). The larger islands of the group are: (Espiritu

Santo, Malekula, Epi, Pentecost, Aoba, Maewo, Paama, Ambrym, Efate, Erromanga, Tanna and Aneityum. They also claim Matthew and Hunter islands. Population at the 2009 census, 234,023 giving a density of 19·2 per sq. km.

The UN gives a projected population for 2015 of 270,000.

In 2009, 75·6% of the population lived in rural areas. Port Vila (the capital) had a 2009 census population of 44,040 and Luganville 13,167.

39% of the population is under 15 years of age, 55% between the ages of 15 and 59 and 6% 60 or over.

The national language is Bislama (spoken by 57% of the population): English and French are also official languages; about 30,000 speak French.

SOCIAL STATISTICS

2008 estimates: births, 7,100; deaths, 1,200. Rates, 2008 estimates (per 1,000 population): births, 30·2; deaths, 5·0. Annual population growth rate, 2000–08, 2·6%. Life expectancy, 2007, was 68·1 years for males and 72·0 years for females. Infant mortality, 2010, 12 per 1,000 live births; fertility rate, 2008, 4·0 births per woman.

CLIMATE

The climate is tropical, but moderated by oceanic influences and by trade winds from May to Oct. High humidity occasionally occurs and cyclones are possible. Rainfall ranges from 90" (2,250 mm) in the south to 155" (3,875 mm) in the north. Vila, Jan. 80°F (26·7°C), July 72°F (22·2°C). Annual rainfall 84" (2,103 mm).

CONSTITUTION AND GOVERNMENT

Legislative power resides in a 52-member unicameral Parliament elected for a term of four years. The *President* is elected for a five-year term by an electoral college comprising Parliament and the presidents of the 11 regional councils. Executive power is vested in a Council of Ministers, responsible to Parliament, and appointed and led by a Prime Minister who is elected from and by Parliament.

There is also a *Council of Chiefs,* comprising traditional tribal leaders, to advise on matters of custom.

National Anthem

'Yumi, yumi, yumi i glat blong talem se, yumi, yumi, yumi i man blong Vanuatu' ('We, we, we are glad to tell, we, we, we are the people of Vanuatu'); words and tune by F. Vincent Ayssav.

RECENT ELECTIONS

Iolu Abil was elected president on 2 Sept. 2009 by an electoral college after a series of votes in the third round of voting, receiving 41 votes against 16 for outgoing president Kalkot Mataskelekele.

Parliamentary elections were held on 30 Oct. 2012. The Party of Our Land (Vanua'aka Pati/VP) won 8 of 52 seats, the People's Progress Party (Parti Progressiste Populaire/PPP) 6, the Union of Moderate Parties (Union des Partis Modérés/UMP) 5, the Ground and Justice Party (Graon mo Jastis Pati/GJP) 4, the National United Party/NUP 4, the Green Confederation Party/GCP 3, Iauko Group 3, Nagriamel 3 and Reunification of Movement for Change 3. A number of smaller parties each took one or two seats. Turnout was 63·2%.

CURRENT GOVERNMENT

President: Iolu Abil; b. 1942 (ind.; since 2 Sept. 2009).

In March 2013 the government comprised:

Prime Minister: Moana Carcasses Kalosil; b. 1963 (GCP; since 23 March 2013).

Deputy Prime Minister, and Minister of Foreign Affairs and External Trade: Edward Natapei.

Minister for Agriculture, Livestock, Forestry, Fisheries and Biosecurity: David Tosul. *Education:* Bob Loughman. *Finance:* Willie Jimmy Tapangararua. *Health:* Serge Vohor. *Infrastructure and Public Utilities:* Esmon Sae. *Internal Affairs:* Patrick Crowby. *Justice and Social Welfare:* Maki Simelum. *Lands, Geology, Mines, Energy and Water Resources:* Ralph Regenvanu. *Planning and Climate Change Adaptation:* Thomas Laken. *Tourism and Commerce:* Marcellino Pipite. *Youth Development and Sports:* Tony Wright.

Speaker: George Wells.

Government Website: http://www.governmentofvanuatu.gov.vu

CURRENT LEADERS

Moana Carcasses Kalosil

Position
Prime Minister

Introduction
Moana Kalosil became prime minister in March 2013. A naturalized citizen, he is the first non-native-born head of government in the country's history.

Early Life
Kalosil was born on 27 Jan. 1963 in Taravoa, Tahiti, to a French father and a Tahitian mother. He joined the Green Confederation Party in 2000 and successfully stood as its candidate for the parliamentary seat in the capital, Port Vila, in 2002. From 2003–04 he served as foreign minister in the cabinet of Edward Natapei.

In July 2004 Kalosil became finance minister in Serge Vohor's government, staying in the post after Ham Lini became premier. Losing his cabinet seat when Lini resigned in 2005, Kalosil led the opposition at the 2008 general election and retained his Port Vila seat.

With Natapei forming the government, Kalosil was the opposition whip until joining the government in 2009 as minister of internal affairs and labour (despite having been threatened with suspension from parliament earlier in the year amid allegations of abetting a prison breakout, although the charges were subsequently dropped).

When Natapei left office in 2010 after a vote of no confidence, Kalosil joined the administration of Sato Kilman as minister of finance and economic management, positions he kept until Kilman was himself ousted by a no confidence vote in 2011.

Kalosil retained his seat at the 2012 election and sat as a backbencher in the reinstated Kilman government. In March 2013 he and eight other MPs defected to the opposition, forcing Kilman's resignation. Three days later, Kalosil was voted premier by parliamentary ballot, winning support from 34 of 52 MPs.

Career in Office
Only the world's second Green prime minister, Kalosil began his tenure by establishing a ministry for planning and climate change. He is expected to pursue closer ties with foreign partners including France, Australia and Vanuatu's Melanesian neighbours. He is a strong advocate of self-determination for the people of West Papua—a province of Indonesia in the island of New Guinea—and is expected to stand against a decision made by the previous administration to grant Indonesia observer status at the intergovernmental Melanesian Spearhead Group.

DEFENCE

Vanuatu does not have an army but there is a Vanuatu Police Force and a paramilitary Vanuatu Mobile Force.

ECONOMY

Agriculture accounted for 14·7% of GDP in 2006, industry 8·6% and services 76·7%.

Overview
The economy is based on agriculture, fishing, offshore financial services and tourism. Efforts have been made to boost tourism since the turn of the century by improving air connections and cruise ship facilities. Remittances from seasonal workers in New Zealand are the country's second largest foreign exchange earner. GDP grew at over 5% per year between 2005 and 2008, and even 3·5% in 2009 at the height of the global financial crisis, with particularly strong growth in telecommunications, agriculture and the hospitality sector.

Vanuatu is dependent on a small number of exports, mainly copra, beef, cocoa, timber and coffee. Australia and New Zealand provide significant amounts of foreign aid. The country is vulnerable to natural disasters.

Currency
The unit of currency is the *vatu* (VUV) with no minor unit. There was inflation in 2010 of 2·8% and 0·9% in 2011. Foreign exchange reserves in July 2005 were US$60m. and total money supply was 14,373m. vatu.

Budget
In 2008 revenues totalled 16,997m. vatu and expenditures 15,121m. vatu. Tax revenue accounted for 69·5% of revenues in 2008; current expenditure accounted for 77·6% of expenditures.

VAT is 12·5%.

Performance
Vanuatu experienced a two-year recession in 2001 and 2002, when the economy contracted by 3·5% and 4·4% respectively. The economy has since recovered, and although it has slowed there were still GDP growth rates of 3·5% in 2009, 1·5% in 2010 and 2·5% in 2011. Total GDP in 2011 was US$0·8bn.

Banking and Finance
The Reserve Bank blong Vanuatu (*Governor,* Odo Tevi) is the central bank and bank of issue. There is also the state-owned National Bank, plus three other banks—ANZ Bank Ltd, Westpac Banking Corp. and BRED. Total assets and liabilities of commercial banks at 31 Dec. 2009 were 80,527m. vatu. Foreign debt in 2007 amounted to US$94m.

ENERGY AND NATURAL RESOURCES

Environment
Vanuatu's carbon dioxide emissions from the consumption and flaring of fossil fuels in 2008 were the equivalent of 0·5 tonnes per capita.

Electricity
Electrical capacity in 2007 was an estimated 12,000 kW. Production in 2007 was about 46m. kWh and consumption per capita an estimated 204 kWh.

Agriculture
About 65% of the labour force are employed in agriculture. In 2007 there were approximately 20,000 ha. of arable land and 85,000 ha. of permanent crops. The main commercial crops are copra, coconuts, cocoa and coffee. Production (2003 estimates, in 1,000 tonnes): coconuts, 206; copra, 23; bananas, 14; groundnuts, 2. 80% of the population are engaged in subsistence agriculture; yams, taro, cassava, sweet potatoes and bananas are grown for local consumption.

Livestock (2003 estimates): cattle, 130,000; pigs, 62,000; goats, 12,000; horses, 3,000.

Forestry
There were 0·44m. ha. of forest in 2010 (36% of the land area). In 2007, 119,000 cu. metres of roundwood were cut.

Fisheries
Fish landings in 2010 totalled 97,807 tonnes. The principal catch is tuna.

INDUSTRY
Principal industries include copra processing, meat canning and fish freezing, a saw-mill, soft drinks factories and a print works.

In 2006 industry accounted for 8·6% of GDP, with manufacturing contributing 3·2%.

INTERNATIONAL TRADE
Imports and Exports
In 2007 imports (f.o.b.) amounted to US$201·7m. (US$159·1m. in 2006); exports (f.o.b.) US$29·9m. (US$36·7m. in 2006). In 2007 the main import suppliers were Australia, New Zealand, Singapore, the Fiji Islands and China. The main export destinations were the Philippines, New Caledonia, the Fiji Islands, Japan and Singapore.

The main exports are copra, beef, timber, cocoa and coffee.

COMMUNICATIONS
Roads
In 2002 there were 1,070 km of roads, about 260 km paved, mostly on Efate Island and Espiritu Santo. There were estimated to be 8,200 passenger cars and 2,100 commercial vehicles in use in 2002.

Civil Aviation
There is an international airport at Bauerfield Port Vila. In 2003 the state-owned Air Vanuatu flew to Auckland, Brisbane, Honiara, Nadi, Nouméa and Sydney. Domestic services were provided by Vanair, a subsidiary of Air Vanuatu. In 2003 scheduled airline traffic of Vanuatu-based carriers flew 3m. km, carrying 83,000 passengers (all on international flights).

Shipping
In Jan. 2009 there were 57 ships of 300 GT or over registered, totalling 1·41m. GT. Vanuatu is a 'flag of convenience' country. The chief ports are Port Vila and Santo.

Telecommunications
In 2008 there were 10,400 main (fixed) telephone lines; mobile phone subscribers numbered 36,000 in 2008 (15·4 per 100 persons). There were 17,000 internet users in 2008.

SOCIAL INSTITUTIONS
Justice
The legal system is based on British common law and French civil law. There is a Supreme Court with a Chief Justice and a Magistrates Court.

The death penalty was abolished in 1980. The population in penal institutions in June 2007 was 117 (53 per 100,000 of national population).

Education
In 2006 there were 1,316 pupils with 111 teaching staff in pre-primary schools. There were 38,000 pupils in primary schools in 2007 with 2,000 teaching staff, and 18,000 pupils in secondary schools in 2009. There is a campus of the University of the South Pacific at Port Vila and a Vanuatu Institute of Technology. The adult literacy rate in 2009 was an estimated 82·0%. In 2009 public expenditure on education came to 5·0% of GDP.

Health
There were 30 physicians and 360 nursing and midwifery personnel in 2004; and 41 hospital beds per 10,000 population in 2005.

RELIGION
About two-thirds of the population are Christians, but animist beliefs are still prevalent.

CULTURE
World Heritage Sites
Vanuatu has one site on the UNESCO World Heritage List: Chief Roi Mata's Domain (inscribed on the list in 2008), three sites on the islands of Efate, Lelepa and Artok that contain the chief's residence and burial chamber.

Press
In 2008 there was one daily newspaper (the *Vanuatu Daily Post*) with a circulation of 3,000.

Tourism
In 2011 there were a record 248,898 non-resident visitor arrivals (154,938 by cruise ship and 93,960 by air), up from 237,648 in 2010 and 225,452 in 2009.

DIPLOMATIC REPRESENTATIVES
Of Vanuatu in the United Kingdom
High Commissioner: Roy Mickey Joy (resides in Brussels).

Of the United Kingdom in Vanuatu (High Commission in Port Vila closed in Oct. 2005)
Acting High Commissioner: Martin Fidler (resides in Suva, Fiji Islands).

Of Vanuatu in the USA
Ambassador: Vacant.

Of the USA in Vanuatu
Ambassador: Teddy B. Taylor (resides in Port Moresby, Papua New Guinea).

Of Vanuatu to the United Nations
Ambassador: Donald Kalpokas.

Of Vanuatu to the European Union
Ambassador: Roy Mickey Joy.

FURTHER READING
Miles, W. F. S., *Bridging Mental Boundaries in a Postcolonial Microcosm: Identity and Development in Vanuatu*. 1998

National Statistical Office: Vanuatu Statistics Office, Private Mail Bag 019, Port Vila.
Website: http://www.vnso.gov.vu

VATICAN CITY STATE

Stato della Città del Vaticano

Population estimate, 2011: 800

Internet domain extension: .va

KEY HISTORICAL EVENTS

The history of the Vatican as a papal residence in Rome dates from the 5th century, following the construction of St Peter's Basilica by Emperor Constantine I. For many centuries the Popes bore temporal sway over much of the Italian peninsula. In 1860, following prolonged civil unrest, Victor Emmanuel's army seized the Papal States, leaving only Rome and surrounding coastal regions under papal control. When Rome was captured in 1871 and declared the capital of the Kingdom of Italy, papal temporal power was brought to an end. On 11 Feb. 1929 a treaty between the Italian Government and the Vatican recognized the sovereignty of the Holy See in the city of the Vatican.

TERRITORY AND POPULATION

The area of the Vatican City is 44 ha. or 0·44 sq. km (108·7 acres or 0·17 sq. miles), making it the smallest independent country in the world. It includes the Piazza di San Pietro (St Peter's Square), which is to remain normally open to the public and subject to the powers of the Italian police. It has its own railway station (for freight only), postal facilities, coins and radio. Twelve buildings in and outside Rome enjoy extra-territorial rights, including the Basilicas of St John Lateran, St Mary Major and St Paul without the Walls, the Pope's summer villa at Castel Gandolfo and a further Vatican radio station on Italian soil. *Radio Vaticana* broadcasts an extensive service in 40 languages from the transmitters in Vatican City and in Italy. The Holy See and the Vatican are not synonymous—the Holy See, referring to the primacy of the Pope, is located in Vatican City. The *de facto* official language is Latin.

Vatican City had about 800 inhabitants in 2011.

CONSTITUTION AND GOVERNMENT

Vatican City State is governed by a Commission appointed by the Pope. The reason for its existence is to provide an extra-territorial, independent base for the Holy See, the government of the Roman Catholic Church. The Pope exercises sovereignty and has absolute legislative, executive and judicial powers. The judicial power is delegated to a tribunal in the first instance, to the Sacred Roman Rota in appeal and to the Supreme Tribunal of the Signature in final appeal.

A new Fundamental Law was promulgated by Pope John Paul II on 26 Nov. 2000 and became effective on 22 Feb. 2001; this replaced the first Fundamental Law of 1929. The Pope is elected by the College of Cardinals, meeting in secret conclave. The election is by scrutiny and requires a two-thirds majority.

National Anthem

'Inno e Marcia Pontificale' ('Hymn and Pontifical March'); words by Raffaello Lavagna, tune by Charles-François Gounod.

GOVERNMENT CHRONOLOGY

Popes since 1939.

1939–58	Pius XII (Eugenio Maria Pacelli)	Italian
1958–63	John XXIII (Angelo Giuseppe Roncalli)	Italian
1963–78	Paul VI (Giovanni Battista Montini)	Italian
1978	John Paul I (Albino Luciani)	Italian
1978–2005	John Paul II (Karol Józef Wojtyła)	Polish
2005–13	Benedict XVI (Joseph Alois Ratzinger)	German
2013–	Francis (Jorge Mario Bergoglio)	Argentinian

CURRENT GOVERNMENT

Supreme Pontiff: Francis (Jorge Mario Bergoglio), born in Buenos Aires, Argentina, 17 Dec. 1936. Archbishop of Buenos Aires 1998–2013, created Cardinal in 2001, elected Pope 13 March 2013, inaugurated 19 March 2013. Pope Francis is the first Latin American to be elected Pope, the first Jesuit and the first non-European Pope since 741.

Secretary of State: Tarcisio Cardinal Bertone.

Secretary for Relations with Other States: Archbishop Dominique Mamberti.

Office of the Sovereign of the Vatican City: http://www.vatican.va

CURRENT LEADERS

Pope Francis

Introduction

Jorge Bergoglio became Pope Francis after being elected head of the Roman Catholic Church on 13 March 2013 following the resignation of Benedict XVI. He is the first pope from Latin America and the first Jesuit to hold the post. Seen as an outsider untarnished by the internal power struggles within the Curia and free from the taint of association with the Church's sex abuse scandals, he is known for his austerity. Bergoglio chose his papal name in honour of St Francis of Assisi, famous for his pledge of poverty and humility.

Early Life

Jorge Mario Bergoglio was born in the Flores district of Buenos Aires on 17 Dec. 1936 to Italian immigrant parents. Having completed a high school diploma in chemistry, in 1955 he enrolled at the Inmaculada Concepción seminary in Buenos Aires. In March 1958 he joined the Society of Jesus (Jesuits) as a novice. In 1960 he graduated in philosophy from the Colegio Máximo San José.

On 13 Dec. 1969 Bergoglio was ordained as a priest. He completed his tertianship, a required period of strict discipline undertaken by Jesuits, at Alcalá de Henares in Spain and took his final vows on 22 April 1973. He served as the Jesuit Provincial Superior for Argentina from 1973–79 and as rector of the philosophical and theological faculty at the Jesuit University in San Miguel from 1980–86.

Bergoglio was appointed titular Bishop of Auca and Auxiliary of Buenos Aires in May 1992. In Feb. 1998 he became Archbishop of Buenos Aires. During his tenure, Bergoglio became known for his work with the poor and his tense relationship with the governments of Néstor Kirchner and Cristina Fernández de Kirchner, which championed liberal social measures including gay marriage. On 21 Feb. 2001 Bergoglio was proclaimed Cardinal by John Paul II and at the 2005 conclave was regarded as a frontrunner for the papacy that eventually fell to Joseph Ratzinger. After Ratzinger resigned on 28 Feb. 2013, Bergoglio was elected his successor in the fifth ballot of the conclave.

Career in Office
Bergoglio's papacy coincides with troubled times for the Church. He faces the challenge of dwindling congregations and the fallout from sex abuse and corruption scandals involving Catholic priests. He must also contend with the question of the role of women within the Church and elucidate the Vatican's position on other social equality debates including same-sex marriage.

ECONOMY
Overview
The economy is supported by donations from Roman Catholics across the world, the sale of postage stamps and tourist souvenirs, admission fees to museums and publication sales. The global economic turmoil reduced donations from US$79·8m. in 2007 to US$75·8m. in 2008 and US$67·7m. in 2010. However, increased ticket sales at the Vatican Museums helped make up much of this shortfall.

The Vatican Bank, officially called the Institute for the Works of Religion (IOR), was established in 1942 and is used by Vatican agencies, church organizations, bishops and religious orders around the world. It offers currency exchange services and interest-bearing accounts, and has an investment portfolio. The bank's surplus is used for religious or charitable purposes.

In Sept. 2010 Italian treasury police investigated the then-president of the IOR for money-laundering. In Dec. 2010 the Pope instituted a new agency to monitor all Vatican financial operations to ensure they meet international norms.

Currency
Since 1 Jan. 2002 the Vatican City has been using the euro (EUR). Italy has agreed that the Vatican City may mint a small part of the total Italian euro coin contingent with their own motifs.

Budget
Revenues in 2006 were €227·8m. and expenditures €225·4m.

Performance
Real GDP growth was 1·7% in 2001.

COMMUNICATIONS
Civil Aviation
The Vatican launched a charter airline, Mistral Air, in Aug. 2007 to fly pilgrims to holy sites across the world.

SOCIAL INSTITUTIONS
Justice
In 2006 the Vatican City's legal system hosted 341 civil cases and 486 criminal cases. Most of the offences are committed by outsiders, principally at St Peter's Basilica and the museums.

ROMAN CATHOLIC CHURCH
As the Vicar of Christ and the Successor of St Peter, the Pope is held to be by divine right the centre of all Catholic unity and exercises universal governance over the Church. He is also the sovereign ruler of Vatican City State. He has for advisers the Sacred College of Cardinals, consisting in March 2013 of 206 cardinals from 66 countries (two created by Pope Paul VI, 121 created by Pope John Paul II and 83 created by Pope Benedict

XVI), of whom 113 are cardinal electors—those under the age of 80 who may enter into conclave to elect a new Pope. Cardinals, addressed by the title of 'Eminence', are appointed by the Pope from senior ecclesiastics who are either the bishops of important Sees or the heads of departments at the Roman Curia. In addition to the College of Cardinals, there is a Synod of Bishops, created by Pope Paul VI and formally instituted on 15 Sept. 1965. This consists of the Patriarchs and certain Metropolitans of the Catholic Church of Oriental Rite, of elected representatives of the national episcopal conferences and religious orders of the world, of the cardinals in charge of the Roman Congregations and of other persons nominated by the Pope. The Synod meets in both general (global) and special (regional) assemblies.

The central administration of the Roman Catholic Church is carried out by permanent organisms called Congregations, Council, Commissions and Offices. The Congregations are composed of cardinals and diocesan bishops (both appointed for five-year periods), with Consultors and Officials. There are nine Congregations, viz.: Doctrine, Oriental Churches, Bishops, the Sacraments and Divine Worship, Clergy, Religious, Catholic Education, Evangelization of the Peoples and Causes of the Saints. Pontifical Councils have replaced some of the previously designated Secretariats and Prefectures and now represent the Laity, Christian Unity, the Family, Justice and Peace, Cor Unum, Migrants, Health Care Workers, Interpretation of Legislative Texts, Inter-Religious Dialogue, Culture, Preserving the Patrimony of Art and History, and a Commission for Latin America.

CULTURE
World Heritage Sites
The Holy See has two sites on the UNESCO World Heritage List: Vatican City (inscribed on the list in 1984)—the centre of the Roman Catholic Church, it contains some of the greatest pieces of European art and architecture, including St Peter's Basilica.

The Historic Centre of Rome, the properties of the Holy See in that city enjoying extraterritorial rights (inscribed on the list in 1980 and 1990), is shared with Italy.

Press
In 2008 there was one daily evening paper, *L'Osservatore Romano*.

DIPLOMATIC REPRESENTATIVES
In its diplomatic relations with foreign countries the Holy See is represented by the Secretariat of State and the Second Section (Relations with States) of the Council for Public Affairs of the Church. It maintains permanent observers to the UN.

Of the Holy See in the United Kingdom (54 Parkside, London, SW19 5NE)
Apostolic Nuncio: Archbishop Antonio Mennini.

Of the United Kingdom at the Holy See (Osborne House, Via XX Settembre 80A, 00187 Rome)
Ambassador: Nigel Baker, OBE, MVO.

Of the Holy See in the USA (3339 Massachusetts Ave., NW, Washington, D.C., 20008)
Apostolic Nuncio: Carlo Maria Viganò.

Of the USA at the Holy See (Villa Domiziana, Via Delle Terme Deciane 26, 00153 Rome)
Ambassador: Vacant.
Chargé d'Affaires a.i.: Mario Mesquita.

Of the Holy See to the European Union
Apostolic Nuncio: Archbishop Alain Paul Lebeaupin.

FURTHER READING
Reese, T., *Inside the Vatican.* 1997

Permanent Observer Mission to the UN: http://www.holyseemission.org

VENEZUELA

República Bolivariana de Venezuela
(Bolivarian Republic of Venezuela)

Capital: Caracas
Population projection, 2015: 31·23m.
GNI per capita, 2011: (PPP$) 10,656
HDI/world rank: 0·735/73
Internet domain extension: .ve

KEY HISTORICAL EVENTS

Present-day Venezuela was inhabited by hunter-gatherers from at least 3000 BC. The Arawaks and Carib lived mainly in the north and around the Orinoco river system. Christopher Columbus landed at Macuro with three Spanish ships on 5 Aug. 1498. A year later the area was explored by Alonso de Ojeda and Amerigo Vespucci. They named it Venezuela (Little Venice) after the indigenous villages built on stilts over water. Spanish settlements were established on the Caribbean coast from the early 16th century and ruled from Santo Domingo (Dominican Republic). Santiago de León de Caracas, founded in 1567, became the seat of the government of the province of Venezuela in 1578. Cocoa plantations developed slowly and much of the forested interior remained unexplored.

In 1717 Venezuela came under the Spanish Viceroyalty of New Granada, centred on Bogotá. In May 1795 rumblings of discontent over Spanish rule erupted into a revolt, led by José Leonardo Chirino who was inspired by the French Revolution. Francisco de Miranda and Simón Bolívar led further rebellions from 1805 and, following Napoleon's defeat of Spain, independence was declared in 1811. Spanish control was restored but the colonial power's defeat in the battle of Carabobo in June 1821 ensured that Venezuela became part of the federal republic of Greater Colombia. Gen. José Páez opposed the Colombian leadership and led revolts, culminating in a congress in 1830 that created a constitution for a new Republic of Venezuela and elected Páez the first president.

Páez was succeeded by José Tadeo Monagas and then by his brother, José Gregorio Monagas, who governed Venezuela until he was overthrown in 1858. Gen. Antonio Guzmán Blanco came to power in the April Revolution of 1870 and dominated politics for the next 18 years. He improved education, modernized the country's infrastructure and forced the separation of church and state. Blanco's autocratic rule was followed by the dictatorships of Joaquín Crespo (1892–98) and Cipriano Castro (1899–1908). Castro's corrupt and inefficient administration led to civil strife and, in 1902, to a major international incident known as the Venezuela Claims when Great Britain, Germany and Italy dispatched a joint naval force to Venezuela to seek redress for unpaid loans.

The brutal dictatorship of Juan Vicente Gómez began in Dec. 1908 when he seized power from the ailing Castro. Gómez's tyrannical regime, which endured until his death in 1935 did, however, develop the nation's oil industry. The election of Isaías Medina Angarita as president on 28 April 1941 led to the re-establishment of political parties, including *Acción Democrática* (AD), founded in Sept. 1941. In 1945 a three-day revolt against the government of Gen. Isaías Medina led to constitutional and economic reforms. In 1947 the writer Rómulo Gallegos was elected president but it was to be a short-lived democratic spell as the government was overthrown in a military coup in Nov. 1948 and the subsequent dictatorship of Marcos Jiménez lasted until 1958.

In 1961 a new constitution provided for a presidential election every five years, a national congress, and state and municipal legislative assemblies. Twenty political parties participated in the 1983 elections, with the economy in crisis and corruption linked to drug trafficking widespread. In Feb. 1992 there were two abortive coups and a state of emergency was declared. In Dec. 1993 Dr Rafael Caldera Rodríguez's election as president reflected disenchantment with the established political parties. He took office in the early stages of a banking crisis that cost 15% of GDP to resolve. Fiscal tightening backed by the IMF brought rapid recovery. Hugo Chávez Frías, who became president in Feb. 1999, continued with economic reforms and amended the constitution to increase presidential powers. The country was given a new name—the Bolivarian Republic of Venezuela. In Dec. 1999 the north coast of Venezuela was hit by devastating floods and mudslides which resulted in 30,000 deaths.

President Chávez was deposed and arrested on 12 April 2002 in a coup following a general strike but was back in the presidential palace within 48 hours. Opposition pressure intensified in 2002 and 2003 with more protests, a prolonged general strike and an attempt to petition for a referendum on the president's rule. Chávez faced a referendum on 15 Aug. 2004 and emerged victorious. His 'socialist revolution' and opposition to US policies has won him support but his long term impact has still to be assessed. 14 years in power, President Chávez died on 5 March 2013 from a heart attack and after a long battle with cancer. His death triggered a presidential election on 14 April 2013, narrowly won by Acting President Nicolás Maduro.

TERRITORY AND POPULATION

Venezuela is bounded to the north by the Caribbean with a 2,813 km coastline, east by the Atlantic and Guyana, south by Brazil, and southwest and west by Colombia. The area is 916,445 sq. km (353,839 sq. miles) including 72 islands in the Caribbean. Population at the 2011 census was 27,227,930 (13,678,178 females and 13,549,752 males); density, 29·7 per sq. km. Venezuela has the highest percentage of urban population in South America, with 93·4% living in urban areas in 2010.

The UN gives a projected population for 2015 of 31·23m.

The official language is Spanish. English is taught as a mandatory second language in high schools.

Area, population and capitals of the 23 states (*estados*), one federal dependency (*dependencias federales*) and one capital district (*distrito capital*):

State	Area (sq. km)	2011 census population	Capital	Density; inhabitants per sq. km
Distrito Capital	433	1,943,901	Caracas	4,489·4
Amazonas	180,145	146,480	Puerto Ayacucho	0·8
Anzoátegui	43,300	1,469,747	Barcelona	33·9
Apure	76,500	459,025	San Fernando	6·0
Aragua	7,014	1,630,308	Maracay	235·6
Barinas	35,200	816,264	Barinas	23·2
Bolívar	238,000	1,410,964	Ciudad Bolívar	5·9
Carabobo	4,650	2,245,744	Valencia	514·0
Cojedes	14,800	323,165	San Carlos	21·8
Delta Amacuro	40,200	167,676	Tucupita	4·2
Falcón	24,800	902,847	Coro	36·4
Guárico	64,986	747,739	San Juan de los Morros	11·5
Lara	19,800	1,774,867	Barquisimeto	89·6
Mérida	11,300	828,592	Mérida	73·3
Miranda	7,950	2,675,165	Los Teques	336·5
Monagas	28,900	905,443	Maturín	31·3
Nueva Esparta	1,150	491,610	La Asunción	427·5
Portuguesa	15,200	876,496	Guanare	57·7
Sucre	11,800	896,291	Cumaná	76·0
Táchira	11,100	1,168,908	San Cristóbal	105·3
Trujillo	7,400	686,367	Trujillo	92·8
Vargas	1,497	352,920	La Guaira	235·8
Yaracuy	7,100	600,852	San Felipe	84·6
Zulia	63,100	3,704,404	Maracaibo	73·7
Dependencias Federales	120	2,155		18·0

37·3% of all Venezuelans are under 15 years of age, 58·7% are between the ages of 15 and 64, and 4·0% are over the age of 65.

Caracas, Venezuela's largest city, is the political, financial, commercial, communications and cultural centre of the country. Caracas had a 2009 population estimate of 2,097,000. Maracaibo, the nation's second largest city (estimated 2009 population of 1,892,000), is located near Venezuela's most important petroleum fields and richest agricultural areas. Other major cities are Valencia, Barquisimeto and Ciudad Guayana.

SOCIAL STATISTICS

2008 births, 581,480; deaths, 124,062. 2008 birth rate per 1,000 population, 20·8; death rate, 4·4. Annual population growth rate, 2008–10, 1·6%. Life expectancy, 2007, was 70·7 years for males and 76·7 years for females. Infant mortality, 2010, 16 per 1,000 live births; fertility rate, 2008, 2·5 births per woman. In 2006 the most popular age for marrying was 25–29 for men and 20–24 for women.

CLIMATE

The climate ranges from warm temperate to tropical. Temperatures vary little throughout the year and rainfall is plentiful. The dry season is from Dec. to April. The hottest months are July and Aug. Caracas, Jan. 65°F (18·3°C), July 69°F (20·6°C). Annual rainfall 32″ (833 mm). Ciudad Bolívar, Jan. 79°F (26·1°C), July 81°F (27·2°C). Annual rainfall 41″ (1,016 mm). Maracaibo, Jan. 81°F (27·2°C), July 85°F (29·4°C). Annual rainfall 23″ (577 mm).

CONSTITUTION AND GOVERNMENT

The present constitution was approved in a referendum held on 15 Dec. 1999. Venezuela is a federal republic, comprising 34 federal dependencies, 23 states and one federal district. Executive power is vested in the *President*. The ministers, who together constitute the Council of Ministers, are appointed by the President and head various executive departments. There are 17 ministries and seven officials who also have the rank of Minister of State.

92% of votes cast in a referendum (the first in Venezuela's history) on 25 April 1999 were in favour of the plan to rewrite the constitution proposed by then President Hugo Chávez. As a result, on 25 July the public was to elect a constitutional assembly to write a new constitution, which was subsequently to be voted on in a national referendum. In Aug. 1999 the constitutional assembly declared a national state of emergency. It subsequently suspended the Supreme Court, turned the elected Congress into little more than a sub-committee, stripping it of all its powers, and assumed many of the responsibilities of government. In Dec. 1999 Chávez's plan to redraft the constitution was approved by over 70% of voters in a referendum. Consequently presidents were able to serve two consecutive six-year-terms instead of terms of five years which could not be consecutive, the senate was abolished and greater powers were given to the state and the armed forces. Chávez effectively took over both the executive and the judiciary. The constitution provides for procedures by which the president may reject bills passed by Congress, as well as provisions by which Congress may override such presidential veto acts. In Aug. 2007 Chávez presented a set of constitutional reforms, including an end to presidential term limits. The proposals were rejected in a national referendum held on 2 Dec. 2007, with 49% of votes cast in favour of the amendments to the constitution and 51% against. However, a referendum on 15 Feb. 2009 to abolish presidential term limits (and those of various other elected officials including National Assembly deputies) was approved with 54% of votes cast in favour and 46% against.

Since the senate was dissolved under the constitution adopted in Dec. 1999 Venezuela has become a unicameral legislature, the 165-seat *National Assembly*, with members being elected for five-year terms.

National Anthem

'Gloria al bravo pueblo' ('Glory to the brave people'); words by Vicente Salias, tune by Juan Landaeta.

GOVERNMENT CHRONOLOGY

Heads of State since 1941. (AD = Democratic Action; CD = Democratic Convergence; COPEI = Social Christian Party; MVR = Movement for the Fifth Republic; PSUV = United Socialist Party of Venezuela; n/p = non-partisan)

President of the Republic
1941–45   military   Isaías Medina Angarita

Revolutionary Junta of Government
1945–48   Rómulo E. Betancourt Bello (AD) (chair); Luis Beltrán Prieto Figueroa (AD); Carlos Román Delgado Chalbaud (military); Raúl Leoni Otero (AD); Gonzalo Barrios Bustillos (AD); Mario Ricardo Vargas Cárdenas (military); Edmundo Fernández (n/p)

President of the Republic
1948   AD   Rómulo Ángel Gallegos Freire

Junta of Government
1948–52   Carlos Román Delgado Chalbaud (military); Germán Suárez Flamerich (n/p); Marcos Evangelista Pérez Jiménez (military); Luis Felipe Llovera Páez (military)

President of the Republic
1952–58   military   Marcos Evangelista Pérez Jiménez

Junta of Government (I)

1958	Wolfgang Enrique Larrazábal Ugueto (military) (chair); Pedro José Quevedo (military); Roberto Casanova (military); Carlos Luis Araque (military); Abel Romero Villate (military); Eugenio Mendoza Goiticoa (n/p); Blas Lamberti Cano (n/p); Arturo Sosa Fernández (n/p); Edgar Sanabria Arcia (n/p)

Junta of Government (II)

1958–59	Edgard Sanabria Arcia (n/p) (chair); Arturo Sosa Fernández (n/p); Miguel J. Rodríguez Olivares (military); Carlos Luis Araque (military); Pedro José Quevedo (military)

Presidents of the Republic

1959–64	AD	Rómulo Ernesto Betancourt Bello
1964–69	AD	Raúl Leoni Otero
1969–74	COPEI	Rafael Caldera Rodríguez
1974–79	AD	Carlos Andrés Pérez Rodríguez
1979–84	COPEI	Luis Antonio Herrera Campins
1984–89	AD	Jaime Lusinchi
1989–93	AD	Carlos Andrés Pérez Rodríguez
1994–99	CD	Rafael Caldera Rodríguez
1999–2013	MVR, PSUV	Hugo Rafael Chávez Frías
2013–	PSUV	Nicolás Maduro

RECENT ELECTIONS

A special presidential election was held on 14 April 2013 following the death of President Hugo Chávez Frías on 5 March. Nicolás Maduro (Great Patriotic Pole/GPP, led by the United Socialist Party of Venezuela/PSUV) was elected president with 50·7% of the vote, ahead of Henrique Capriles Radonski (Democratic Unity Roundtable/MUD) with 49·1%. There were five other candidates. Turnout was 78·7%. Capriles refused to accept the result and demanded a recount.

In elections to the Congress, held on 26 Sept. 2010, 165 seats were contested. The Partido Socialista Unido de Venezuela (United Socialist Party of Venezuela, PSUV) of then President Chávez won 96 seats, the opposition Mesa de la Unidad Democrática (Coalition for Democratic Unity, MUD) 64, Patria para Todos (Fatherland for All) 2 with the remaining three seats reserved for indigenous peoples. Turnout was 66·5%.

CURRENT GOVERNMENT

President: Nicolás Maduro; b. 1962 (PSUV; since 5 March 2013—acting until 19 April 2013).

Executive Vice-President: Jorge Arreaza.

In April 2013 the government comprised:

Minister of Agriculture and Lands: Iván Gil. *Air and Water Transportation:* Herbert García Plaza. *Commerce:* Alejandro Fleming. *Communes and Social Protection:* Reinaldo Iturriza. *Communications and Information:* Ernesto Villegas. *Culture:* Fidel Barbarito. *Defence:* Adm. Diego Molero. *Education:* Maryann Hanson. *Electricity:* Jesse Chacón. *Environment:* Dante Rivas. *Finance:* Nelson Merentes. *Food Affairs:* Félix Osorio Guzmán. *Foreign Affairs:* Elías Jaua. *Health:* Isabel Iturria. *Higher Education:* Pedro Calzadilla. *Housing and Habitat:* Ricardo Molina. *Indigenous Peoples:* Aloha Núñez. *Industry:* Ricardo Menéndez. *Interior and Justice:* Miguel Rodríguez Torres. *Labour and Social Security:* María Cristina Iglesias. *Land Transport:* Juan García Toussaintt. *Oil and Mining:* Rafael Ramírez. *Penitentiary Services:* Iris Varela. *Planning and Development:* Jorge Giordani. *Science and Technology:* Manuel Fernández. *Sports:* Alejandra Benítez. *Tourism:* Andrés Izarra. *Women and Gender Equality:* Andreina Tarazón. *Youth:* Héctor Rodríguez. *Presidential Secretariat:* Carmen Meléndez.

Office of the President (Spanish only): http://www.presidencia.gob.ve

CURRENT LEADERS

Nicolás Maduro

Position
President

Introduction
Nicolás Maduro took office as interim president on 5 March 2013 following the death of Hugo Chávez. He was confirmed in the role when he won the presidential election held on 14 April 2013. A long-time ally of Chávez, he served as his minister of foreign affairs for seven years.

Early Life
Nicolás Maduro was born on 23 Nov. 1962 in Caracas. The son of a prominent union leader, he was schooled in the capital city before becoming a bus driver. As a public transport workers' trade unionist, Maduro met Chávez in 1992 after the latter was imprisoned for his involvement in an attempted coup. Maduro campaigned for his release, meeting his future wife, Chávez's defence attorney Cilia Flores, in the process.

After Chávez was released in 1994, Maduro helped him establish the Movement for a Fifth Republic, the Bolivarian party that launched Chávez to the presidency in 1998. In the same year Maduro was elected to the Chamber of Deputies, the lower house of the old bicameral parliament. The following year a unicameral parliament was introduced, with Maduro appointed its Speaker in 2005. In Aug. 2006 he was named foreign minister, a position he held until March 2013. His tenure was characterized by the 'oil diplomacy' that saw Venezuela provide subsidized oil to neighbouring South American countries. He also attracted international attention by stating his support for Muammar Gaddafi's regime in Libya and promoted the creation of a union of emerging economies to counter US influence.

When Chávez won a third presidential term in Oct. 2012, he selected Maduro as his vice-president. In Dec. 2012 Chávez announced he was again suffering from cancer and named Maduro as his preferred successor. Chávez died in March 2013 and Maduro succeeded him on an interim basis. In the presidential election of 14 April 2013 Maduro took 50·7% of the vote, defeating Henrique Capriles, the candidate for a coalition of opposition groups.

Career in Office
In the aftermath of Chávez's death, Maduro publicly venerated his predecessor, even suggesting he may have played a divine role in the election in March 2013 of the first South American pope. Maduro is expected to continue Chávez's work and must address a soaring crime rate while attempting to rebalance an economy over-reliant on oil and subject to inflation of up to 30%.

Rapprochement with the USA is considered unlikely given that, shortly before Chávez's death, Maduro expelled two junior US diplomats and suggested that Chávez's cancer (which was first diagnosed in June 2011) was the result of an attack by 'historic enemies' of the country.

DEFENCE

There is selective conscription for a minimum of 12 months. Defence expenditure totalled US$3,328m. in 2008 (US$126 per capita), representing 1·0% of GDP.

Army

The Army has six divisions (four infantry, one armoured and one cavalry), one combat engineer corps, one aviation command and one logistics command. Equipment includes 81 main battle tanks. Strength (2007) around 63,000. There were an additional 8,000 reserves.

A 23,000-strong volunteer National Guard is responsible for internal security.

Navy

Strength (2007) around 17,500 (3,200 conscripts). The combatant fleet comprises two submarines and six frigates. Naval Aviation, 500 strong, operates ten combat-capable aircraft. Main bases are at Caracas, Puerto Cabello and Punto Fijo.

Air Force

The Air Force was 11,500 strong in 2007 and had 94 combat-capable aircraft. Main aircraft types include CF-5s, Mirage 50s, F-16A/Bs and Su-30s.

INTERNATIONAL RELATIONS

In March 2008 Venezuela sent large numbers of troops to the border with Colombia. The action was in sympathy with Ecuador, whose border was violated when Colombian troops crossed it to attack forces of the rebel FARC movement. Within seven days of the mobilization, a diplomatic solution saw the troops stand down.

ECONOMY

In 2010 industry accounted for 52% of GDP, services 42% and agriculture 6%.

Overview

Venezuela has the fifth largest economy in South America but since the turn of the century has been among its worst performers. The country has the largest mineral and oil reserves in the region, with oil providing 30% of GDP, 90% of exports and 50% of fiscal revenue. Overly-reliant on the international oil price, growth rates fluctuate wildly. In 2004 the economy grew by 18·3% but contracted by 3·2% in 2009 and 1·5% in 2010 as a result of the global crisis. Growth resumed in 2011 at 4·2%.

In 2009, 28·5% of the population lived in poverty, down from 43·7% in 2005. Long-time President Hugo Chávez portrayed himself as a hero of the poor, introducing programmes of free health care, subsidized food and land reform. His tenure saw large-scale nationalization in sectors including oil, banking and roads. The development of the private sector has been constrained as a result of corruption and the threat of government expropriation. In 2010 the government's reputation was dealt a serious blow when thousands of tons of food destined for the poorest in the country was discovered to have rotted. The US Heritage Foundation ranked the economy 174th out of 177 nations in its 2013 index of economic freedom.

Currency

The unit of currency is the *bolívar fuerte* (VEF) of 100 *céntimos*. It was introduced on 1 Jan. 2008, replacing the *bolívar* (VEB) at a rate of one bolívar fuerte = 1,000 bolívares. In Aug. 2009 foreign exchange reserves were US$18,559m., gold reserves were 11·46m. troy oz and total money supply was 186,679m. Bs.F. Exchange controls were abolished in April 1996. The bolívar was devalued by 12·6% in 1998, and in Feb. 2002 it was floated, ending a regime that permitted the bolívar to trade only within a fixed band. However, in Feb. 2003 it was pegged to the dollar at Bs 2·15 = US$1 (and after the introduction of the *bolívar fuerte* in Jan. 2008 officially at 2·15 Bs.F = US$1). In Jan. 2010 the bolívar was devalued by 17% to 2·6 Bs.F = US$1 and a second rate, known as the 'oil bolívar' for its aim of boosting revenue from oil exports, was introduced at 4·3 Bs.F = US$1. A further devaluation of the currency on 1 Jan. 2011 eliminated the rate of 2·6 Bs.F to the dollar and created a single official exchange rate of 4·3 Bs.F to the dollar. The currency was again devalued on 8 Feb. 2013, by 31·%. As a result the official exchange rate became 6·3 Bs.F to the dollar. Inflation rates (based on IMF statistics):

2002	2003	2004	2005	2006	2007	2008	2009	2010	2011
22·4%	31·1%	21·7%	16·0%	13·7%	18·7%	30·4%	27·1%	28·2%	26·1%

Inflation has not been in single figures since 1983.

Budget

The fiscal year is the calendar year. In 2006 revenue totalled Bs 117,326bn. and expenditure Bs 117,255bn. Petroleum income accounted for 52·9% of revenues in 2006; current expenditure accounted for 75·0% of expenditures. VAT is 12% (reduced rate, 11%).

Performance

Real GDP growth rates (based on IMF statistics):

2002	2003	2004	2005	2006	2007	2008	2009	2010	2011
−8·9%	−7·8%	18·3%	10·3%	9·9%	8·8%	5·3%	−3·2%	−1·5%	4·2%

Total GDP in 2011 was US$316·5bn.

Banking and Finance

A law of Dec. 1992 provided for greater autonomy for the Central Bank. Its *President*, currently Nelson Merentes, is appointed by the President for five-year terms. Since 1993 foreign banks have been allowed a controlling interest in domestic banks. In 2003 there were 24 commercial banks and three foreign banks.

Venezuela's foreign debt totalled US$55,572m. in 2010, representing 14·3% of GNI.

The total stock of FDI at the end of 2010 was US$38·02bn.

There is a stock exchange in Caracas. In 2011 it was the best-performing stock market in the world, recording a gain of 80·8% on 2010.

ENERGY AND NATURAL RESOURCES

Environment

Carbon dioxide emissions from the consumption and flaring of fossil fuels in 2008 were the equivalent of 7·0 tonnes per capita.

Electricity

Installed capacity in 2007 was 22·5m. kW; production was 114·85bn. kWh in 2007 and consumption per capita 4,219 kWh. Venezuela imports electricity to meet increasing domestic demand.

Oil and Gas

Proven reserves of oil were 296·5bn. bbls in 2011, the most of any country and 18% of the world total. However, the reserves in Saudi Arabia—which ranks second behind Venezuela—are much more accessible. Venezuela has significant reserves of heavy crude oil, which is more expensive and difficult to extract than conventional crude oil. The oil sector was nationalized in 1976. Private and foreign investment were permitted after 1992, before then President Chávez instigated a 'renationalization' following strikes in Dec. 2002–Feb. 2003. In Feb. 2007 Chávez announced that the last foreign-controlled oil production sites would be brought under government control. Oil production in 2008 was 131·6m. tonnes. Oil provides about 50% of Venezuela's revenues. Natural gas production in 2008 was 31·5bn. cu. metres. Natural gas reserves in 2008 were 4,840bn. cu. metres, the largest in Latin America.

Minerals

Output (in 1,000 tonnes) in 2004: iron ore, 19,196; limestone (2002), 13,434; coal, 6,748; bauxite, 5,842; alumina (2003), 1,882; gold, 9,690 kg. Estimated diamond production in 2005 was 115,000 carats.

Agriculture

Coffee, cocoa, sugarcane, maize, rice, wheat, tobacco, cotton, beans and sisal are grown. 50% of farmers are engaged in subsistence agriculture. There were approximately 2·44m. ha. of arable land in 2002 and 0·81m. ha. of permanent crops. About 575,000 ha. were irrigated in 2002.

Production in 2003 in 1,000 tonnes: sugarcane, 6,825; maize, 1,505; plantains, 760; rice, 701; bananas, 639; sorghum, 600;

cassava, 490; pineapples, 384; oranges, 316; melons and water-melons, 308; potatoes, 298; onions, 237.

Livestock (2003): cattle, 16·07m.; pigs, 2·92m.; goats, 2·70m.; sheep, 820,000; horses, 500,000; chickens, 110m.

Forestry
In 2010 the area under forests was 46·28m. ha., or 52% of the total land area. Timber production in 2007 was 6·06m. cu. metres.

Fisheries
In 2008 the total catch was 296,266 tonnes (mostly from marine waters). Imports of fishery commodities were valued at US$290m. in 2008.

INDUSTRY
Production (2007 unless otherwise indicated, in tonnes): petrol, 16·4m.; residual fuel oil, 14·6m.; distillate fuel oil, 14·3m.; cement (2007 estimate), 11m.; crude steel (2006), 4·9m.; raw sugar (2005), 690,000.

Labour
Out of 9,698,900 people in employment in 2002, 2,932,700 were in community, social and personal services, 2,585,300 in whole-sale and retail trade, restaurants and hotels, 1,150,300 in manufacturing and 949,000 in agriculture, hunting, fishing and forestry. In Sept. 2005, 11·5% of the workforce was unemployed, down from 14·5% a year earlier.

In late 2002 and early 2003 a two-month long general strike intended to oust then President Chávez ended in failure, instead crippling an already depressed economy.

INTERNATIONAL TRADE
Imports and Exports
Trade in US$1m.:

	2008	2009	2010
Imports c.i.f.	47,450	38,677	32,343
Exports f.o.b.	83,478	56,583	66,963

Exports of oil in 2001 were valued at US$19bn., the third highest export revenues after Saudi Arabia and Iran. Oil revenues account for 93% of all export revenues.

The main import sources in 2000 were the USA (37·8%), Colombia (7·4%), Brazil (5·0%) and Italy (4·4%). The main markets for exports in 2000 were the USA (59·6%), Netherlands Antilles (5·6%), Brazil (3·6%) and Colombia (2·8%).

COMMUNICATIONS
Roads
In 2002 there were 96,155 km of roads, of which 33·6% were paved. There were 2,952,100 passenger cars in use in 2007 (107 per 1,000 inhabitants) plus 84,000 lorries and vans. There were 6,218 fatalities as a result of road accidents in 2006.

Rail
The railway network comprises 742 km of 1,435 gauge track. In 2004, 50·6m. tonnes of freight were carried. In 2006 Venezuela's first inter-city passenger service in nearly 70 years was opened with the inauguration of a line from Caracas to Cúa. Several other new lines are planned or currently under construction.

There are metros in Caracas, Los Teques, Maracaibo and Valencia.

Civil Aviation
The main international airport is at Caracas (Simon Bolívar), with some international flights from Maracaibo. The national carrier is Conviasa, founded in 2004 as the successor to Viasa, which had ceased operations in 1997. In 2005 scheduled airline traffic of Venezuela-based carriers flew 31·0m. km, carrying 3,240,200 passengers.

Shipping
In Jan. 2009 there were 82 ships of 300 GT or over registered, totalling 784,000 GT. La Guaira, Maracaibo, Puerto Cabello, Puerto Ordaz and Guanta are the chief ports. The principal navigable rivers are the Orinoco and its tributaries the Apure and Arauca.

Telecommunications
In 2008 there were 6,304,000 main (fixed) telephone lines. There were 23,820,000 mobile phone subscribers in 2007 (861·3 per 1,000 persons). CANTV, the national telephone company, lost its 50-year monopoly on fixed-line telephony in 2000. In 2007 the number of internet users was 5,720,000. In June 2012 there were 9·7m. Facebook users.

SOCIAL INSTITUTIONS
Justice
A new penal code was implemented on 1 July 1999. The new, US-style system features public trials, verbal arguments, prosecutors, citizen juries and the presumption of innocence, instead of an inquisitorial system inherited from Spain which included secretive trials and long exchanges of written arguments.

In Aug. 1999 the new constitutional assembly declared a judicial emergency, granting itself sweeping new powers to dismiss judges and overhaul the court system. The assembly excluded the Supreme Court and the national Judicial Council from a commission charged with reorganizing the judiciary. Then President Chávez declared the assembly the supreme power in Venezuela.

The court system is plagued by chronic corruption and a huge case backlog. Only about 40% of the country's prisoners in 1999 had actually been convicted. In Oct. 1999 over 100 judges accused of corruption were suspended. The population in penal institutions in April 2008 was approximately 22,000 (79 per 100,000 population).

Venezuela's murder rate, at 52 per 100,000 population in 2008, is among the highest in the world.

Education
In 2007 there were 3,521,139 primary school pupils (184,409 teaching staff in 2005) and 2,174,619 secondary school pupils (187,737 teaching staff in 2005).

The leading institute of higher education is the Central University of Venezuela (Universidad Central de Venezuela), founded in 1721 in Caracas. The Bolivarian University of Venezuela (Universidad Bolivariana de Venezuela) was founded in 2003 by then President Chávez in order to give lower-income students the opportunity to study free of charge regardless of academic qualifications or prior education. More than 75% of students are from poor backgrounds. There were 1,381,126 students in tertiary education in 2006 with 108,594 academic staff.

Adult literacy was 95% in 2007.

Public expenditure on education came to 3·7% of GNI in 2007.

Health
In 2002 there were 567 hospitals with 19 beds per 10,000 inhabitants. There were 48,000 physicians and 13,680 dentists in 2001; and 46,305 nurses and 8,751 pharmacists in 1997.

Welfare
The official retirement age is 60 years (men) or 55 years (women). However, the pensionable age is lower for those in arduous or unhealthy employment. The old-age pension is the minimum urban wage (967·50 Bs.F a month), plus 30% of the reference salary (20% of covered earnings in the last five years or 10% in the last ten years—whichever is higher) and an increment of 1% of earnings for every 50-week period of contributions beyond 750 weeks. The minimum pension is equal to the minimum urban wage.

Unemployment benefit is 60% of the insured person's average monthly earnings during the previous 12 months. The benefit is paid for up to five months.

RELIGION

In 2001 there were 22·05m. Roman Catholics. There are four archbishops, one at Caracas, who is Primate of Venezuela, two at Mérida and one at Ciudad Bolívar. There are 19 bishops. There was one cardinal in Feb. 2013. The remainder of the population follow other religions, notably Protestantism.

CULTURE

World Heritage Sites

Venezuela has three sites on the UNESCO World Heritage List: Coro and its Port (inscribed on the list in 1993); Canaima National Park (1994); and La Ciudad Universitaria de Caracas (2000).

Press

In 2008 there were 108 daily newspapers (106 paid-for and two free) with a circulation of 2·53m.

Tourism

In 2009 there were 615,000 non-resident tourists (excluding same-day visitors), down from 771,000 in 2007 and 745,000 in 2008. Of the 615,000 tourists in 2009, 340,000 were from elsewhere in the Americas and 241,000 were from Europe.

Festivals

Among the country's most celebrated festivals is the Procession of the Holy Shepherdess (Jan.), which has run annually since 1856. Its centrepiece is a grand procession taking a statue of the divine shepherdess from Santa Rosa to the city of Barquisimeto, before the return trip is made at Easter. Celebrations to mark the initial declaration of independence from Spain in 1810 fall on 19 April, which is also designated as Day of the Indian. The Festival Internacional de Teatro in Caracas takes place in April and the Festival Internacional de Música El Hatillo is in Oct. Independence Day is observed on 5 July while the birthday of Simón Bolívar is celebrated on 24 July.

DIPLOMATIC REPRESENTATIVES

Of Venezuela in the United Kingdom (1 Cromwell Rd, London, SW7 2HW)
Ambassador: Samuel Moncada.

Of the United Kingdom in Venezuela (Torre La Castellana, Piso 11, Avenida Principal de La Castellana, Caracas 1061)
Ambassador: Catherine Nettleton.

Of Venezuela in the USA (1099 30th St., NW, Washington, D.C., 20007)
Ambassador: Vacant.
Chargé d'Affaires a.i.: Dr Angelo Rivero Santos.

Of the USA in Venezuela (Calle Suapure, con calle F. Colinas de Valle Arriba, Caracas)
Ambassador: Vacant.
Chargé d'Affaires a.i.: James Derham.

Of Venezuela to the United Nations
Ambassador: Jorge Valero Briceño.

Of Venezuela to the European Union
Ambassador: Alejandro Antonio Fleming Cabrera.

FURTHER READING

Dirección General de Estadística, Ministerio de Fomento, Boletín Mensual de Estadística.—Anuario Estadístico de Venezuela. Annual

Canache, D., *Venezuela: Public Opinion and Protest in a Fragile Democracy.* 2002
Carroll, Rory, *Comandante: Hugo Chávez's Venezuela.* 2013
Frederick, Julia C. and Tarver, H. Micheal, *The History of Venezuela.* 2006
Rudolph, D. K. and Rudolph, G. A., *Historical Dictionary of Venezuela.* 2nd ed. 1995
Wilpert, Greg, *Changing Venezuela by Taking Power: the History and Policies of the Chavez Government.* 2006

National Statistical Office: Instituto Nacional de Estadística, Avenida Boyacá Edificio Fundación La Salle, Piso 4, Maripérez, Caracas.
Website (Spanish only): http://www.ine.gov.ve

VIETNAM

CHINA

HANOI □

Hai Phong

LAOS

Gulf of Tonkin

THAILAND

Da Nang

VIETNAM

CAMBODIA

Nha Trang

Ho Chi Minh City

Gulf of Thailand

Can Tho

South China Sea

0   125 mi
0   200 km

© Research Machines plc 2006

Công Hòa Xã Hôi Chu Nghĩa Viêt Nam
(Socialist Republic of Vietnam)

Capital: Hanoi
Population projection, 2015: 92·44m.
GNI per capita, 2011: (PPP$) 2,805
HDI/world rank: 0·593/128
Internet domain extension: .vn

KEY HISTORICAL EVENTS

Archaeological evidence suggests Neolithic settlements in northern Vietnam existed from around 5000 BC. There were small rice-dependent villages in the Red River Delta from around 2000 BC and by AD 1500 there was bronze working at Dong Dau, near present-day Hanoi.

Much of present-day northern Vietnam was closely linked with China from the third century BC, initially as part of Nam Viet, regionally-administered from Canton, and from 111 BC under Han dynasty control. Chinese rule of Vietnam's north continued under the Tsin (265–420), Sui (581–618) and T'ang (618–907) dynasties. There followed several short-lived kingdoms until

relative stability was restored by the Ly dynasty (1010–1225), notably under Emperor Ly Thanh Tong (1127–38) who adopted the name Dai Viet. The region's political structure remained Chinese-influenced but independent, although it came under attack from Sino-Mongol forces in 1285 and succumbed to Ming Dynasty control for 20 years from 1407.

From the late 1420s the Le dynasty expanded southward. In 1471 Le Thanh Tong seized control of much of the Kingdom of Champa (now Vietnam's central area), although complete defeat of the Cham took several centuries. By the 1500s the Le dynasty was in decline and was overthrown in 1527. The powerful Mac, Trinh and Nguyen families then fought for control. By 1590 the Trinh had established dominance in the north, centred on Thang Long (Hanoi)—the kingdom of Tonkin. The Nguyen ruled the southern provinces from Hue, a kingdom that became known as Cochin-China.

Both kingdoms benefited from growing trade with China and Japan and, to a lesser extent, in the 17th century with Portuguese, Dutch and English merchants. Catholicism, introduced from Europe, took root across much of Vietnam while the Nguyen pushed south into the Mekong Delta, which was home to thousands of Chinese refugees fleeing after the collapse of the Ming dynasty.

In 1749 Cambodia ceded the remainder of the lower Mekong delta to Nguyen control. In the 1770s a rebellion led by three brothers from the village of Tay Son, in Quy Nhan province, toppled the Nguyen and the Trinh, although the Nguyen wrested back control in 1802. At the end of the 18th century, France helped establish Emperor Gia-Long as ruler of a unified Vietnam. French forces captured Saigon in 1859 and took control of Cochin-China. Tonkin and Annam became French protectorates in 1884 and, with Cambodia already annexed (1863), a union of Indochina was formed in 1887, with Laos added six years later.

A Marxist/Leninist-inspired nationalist movement grew during the 1920s. In 1930 Nguyen Ai Quoc (Ho Chi Minh) established the Indochinese Communist Party, which became the Viet Minh movement. In 1940 Vietnam was occupied by Japan but after Japanese defeat in 1945, the Viet Minh established a republic with its capital in Hanoi. On 6 March 1946 France recognized the Democratic Republic of Vietnam as a 'Free State within the Indo-Chinese Federation'. In Nov. 1946 French forces attacked the Viet Minh in Haiphong, sparking a war of resistance until a peace agreement was reached in July 1954.

By the Paris Agreement of Dec. 1954 France transferred sovereignty to Vietnam, which was divided along the 17th parallel into Communist North Vietnam and non-Communist South. From 1959 the North promoted insurgency in the south, and in 1963 the Viet Cong (Communist guerrillas operating in the south) defeated several units of the South Vietnamese Army, prompting the overthrow and execution of President Ngo Dinh Diem. The Gulf of Tonkin incident, in which two US destroyers were allegedly attacked by North Vietnam, prompted retaliation by Washington. Aerial bombing raids began in March 1965 and the first combat troops soon followed, peaking at 540,000 in 1968. Opposition to the war in America mounted from 1967.

On 27 Jan. 1973 an agreement was signed in Paris ending the war although hostilities continued between North and South until the latter's defeat in 1975. Between 150,000 and 200,000 South Vietnamese fled the country. North and South Vietnam unified as the Socialist Republic of Vietnam on 2 July 1976. Vietnam invaded Cambodia in Dec. 1978, prompting a bloody but indecisive war with Chinese forces in Feb.–March 1979. In 1986 Vietnam began a shift towards a multi-sectoral market economy under state regulation. In 1995 Vietnam normalized relations with the USA and in July that year joined ASEAN and signed trade agreements with the EU.

TERRITORY AND POPULATION

Vietnam is bounded in the west by Cambodia and Laos, north by China and east and south by the South China Sea. It has a total area of 331,212 sq. km and is divided into eight regions, 58 provinces and five municipalities (Can Tho, Da Nang, Hai Phong, Hanoi and Thanh Pho Ho Chi Minh). The areas and 2009 census populations were as follows:

Region/Province	Area (sq. km)	Census population 2009	Capital
Dac Lac	13,139	1,733,624	Buon Me Thuot
Dac Nong	6,517	489,392	Gia Nghia
Gia Lai	15,537	1,274,412	Play Cu
Kon Tum	9,691	430,133	Kon Tum
Lam Dong	9,776	1,187,574	Da Lat
Central Highlands	54,660	5,115,135	
An Giang	3,537	2,142,709	Long Xuyen
Bac Lieu	2,584	856,518	Bac Lieu
Ben Tre	2,360	1,255,946	Ben Tre
Ca Mau	5,332	1,206,938	Ca Mau
Can Tho	1,402	1,188,435	Can Tho
Dong Thap	3,376	1,666,467	Sa Dec
Hau Giang	1,601	757,300	Vi Thanh
Kien Giang	6,348	1,688,248	Rach Gia
Long An	4,494	1,436,066	Tan An
Soc Trang	3,312	1,292,853	Soc Trang
Tien Giang	2,484	1,672,271	My Tho
Tra Vinh	2,295	1,003,012	Tra Vinh
Vinh Long	1,479	1,024,707	Vinh Long
Mekong River Delta	40,605	17,191,470	
Ha Tinh	6,027	1,227,038	Ha Tinh
Nghe An	16,499	2,912,041	Vinh
Quang Binh	8,065	844,893	Dong Hoi
Quang Tri	4,760	598,324	Dong Ha
Thanh Hoa	11,136	3,400,595	Thanh Hoa
Thua Thien-Hue	5,065	1,087,420	Hue
North Central Coast	51,552	10,070,311	
Bac Can	4,868	293,826	Bac Can
Bac Giang	3,827	1,554,131	Bac Giang
Cao Bang	6,725	507,183	Cao Bang
Ha Giang	7,946	724,537	Ha Giang
Lang Son	8,331	732,515	Lang Son
Lao Cai	6,384	614,595	Lao Cai
Phu Tho	3,528	1,316,389	Phu Tho
Quang Ninh	6,099	1,144,988	Ha Long
Thai Nguyen	3,547	1,123,116	Thai Nguyen
Tuyen Quang	5,870	724,821	Tuyen Quang
Yen Bai	6,900	740,397	Yen Bai
North East	64,025	9,476,498	
Dien Bien	9,563	490,306	Dien Bien Phu
Hoa Binh	4,684	785,217	Hoa Binh
Lai Chau	9,112	370,502	Lai Chau
Son La	14,174	1,076,055	Son La
North West	37,534	2,722,080	
Bac Ninh	823	1,024,472	Bac Ninh
Ha Nam	860	784,045	Phu Ly
Hai Duong	1,653	1,705,059	Hai Duong
Hai Phong	1,521	1,837,173	Hai Phong
Hanoi[1]	3,120	6,451,909	Hanoi
Hung Yen	924	1,127,903	Hung Yen
Nam Dinh	1,651	1,828,111	Nam Dinh
Ninh Binh	1,392	898,999	Ninh Binh
Thai Binh	1,547	1,781,842	Thai Binh
Vinh Phuc	1,373	999,786	Vinh Yen
Red River Delta	14,863	18,439,299	
Binh Dinh	6,040	1,486,465	Quy Nhon
Da Nang	1,257	887,435	Da Nang
Khanh Hoa	5,218	1,157,604	Nha Trang
Phu Yen	5,061	862,231	Tuy Hoa
Quang Nam	10,438	1,422,319	Tam Ky
Quang Ngai	5,153	1,216,773	Quang Ngai
South Central Coast	33,166	7,032,827	

Region/Province	Area (sq. km)	Census population 2009	Capital
Ba Ria (Vung Tau)	1,990	996,682	Vung Tau
Binh Duong	2,696	1,481,550	Thu Dau Mot
Binh Phuoc	6,884	873,598	Dong Xoai
Binh Thuan	7,837	1,167,023	Phan Thiet
Dong Nai	5,904	2,486,154	Bien Hoa
Ninh Thuan	3,363	564,993	Phan Rang
Tay Ninh	4,036	1,066,513	Tay Ninh
Thanh Pho Ho Chi Minh	2,099	7,162,864	Ho Chi Minh City
South East[2]	34,808	15,799,377	

[1]Includes former province of Ha Tay. [2]Formerly North East South.

At the 2009 census the population was 85,846,997; density, 259 per sq. km. 31·0% of the population live in urban areas (2011).

The UN gives a projected population for 2015 of 92.44m.

Major cities (with 2009 populations): Ho Chi Minh City (5,880,615), Hanoi (2,316,772), Da Nang (770,911), Hai Phong (769,739), Can Tho (731,545).

86% of the population are Vietnamese (Kinh). There are also 53 minority groups thinly spread in the extensive mountainous regions. The largest minorities are: Tay, Thai, Muong, Khmer, Mong and Nung.

The official language is Vietnamese. Chinese, French and Khmer are also spoken.

SOCIAL STATISTICS

2008 estimates: births, 1,494,000; deaths, 469,000. Estimated birth rate in 2008 was 17·2 per 1,000 population; estimated death rate, 5·4. Life expectancy, 2007, was 72·3 years for males and 76·1 years for females. Annual population growth rate, 2000–08, 1·3%. Infant mortality, 2010, 19 per 1,000 live births; fertility rate, 2008, 2·1 births per woman. Vietnam has had one of the largest reductions in its fertility rate of any country in the world in recent years, having had a rate of 5·8 births per woman in 1975. Sanctions are imposed on couples with more than two children. The rate at which Vietnam has reduced poverty, from 58% of the population in 1993 to 20% in 2004, is among the most dramatic of any country in the world.

CLIMATE

The humid monsoon climate gives tropical conditions in the south, with a rainy season from May to Oct., and sub-tropical conditions in the north, though real winter conditions can affect the north when polar air blows south over Asia. In general, there is little variation in temperatures over the year. Hanoi, Jan. 62°F (16·7°C), July 84°F (28·9°C). Annual rainfall 72" (1,830 mm).

CONSTITUTION AND GOVERNMENT

The National Assembly unanimously approved a new constitution on 15 April 1992. Under this the Communist Party retains a monopoly of power and the responsibility for guiding the state according to the tenets of Marxism-Leninism and Ho Chi Minh, but with certain curbs on its administrative functions. The powers of the National Assembly are increased. The 500-member *National Assembly* is elected for five-year terms. Candidates may be proposed by the Communist Party or the Fatherland Front (which groups various social organizations), or they may propose themselves as individual Independents. The Assembly convenes three times a year and appoints a prime minister and cabinet. It elects the *President*, the head of state. The latter heads a *State Council* which issues decrees when the National Assembly is not in session.

The ultimate source of political power is the Communist Party of Vietnam, founded in 1930; it had 3·6m. members in 2011.

National Anthem

'Doàn quân Việt Nam di chung lòng cúu quôc' ('Soldiers of Vietnam, we are advancing'); words and tune by Van Cao.

GOVERNMENT CHRONOLOGY

General Secretaries of the Communist Party since 1976.

1976–86	Le Duan
1986	Truong Chinh
1986–91	Nguyen Van Linh
1991–97	Do Muoi
1997–2001	Le Kha Phieu
2001–11	Nong Duc Manh
2011–	Nguyen Phu Trong

Heads of State since 1976.

Presidents

1976–80	Ton Duc Thang
1980–81	Nguyen Huu Tho

Chairmen of the State Council

1981–87	Truong Chinh
1987–92	Vo Chi Cong

Presidents

1992–97	Le Duc Anh
1997–2006	Tran Duc Luong
2006–11	Nguyen Minh Triet
2011–	Truong Tan Sang

Prime Ministers since 1976.

1976–87	Pham Van Dong
1987–88	Pham Hung
1988	Vo Van Kiet
1988–91	Do Muoi
1991–97	Vo Van Kiet
1997–2006	Phan Van Khai
2006–	Nguyen Tan Dung

RECENT ELECTIONS

In parliamentary elections held on 22 May 2011 Communist Party members won 458 of 500 seats, with 42 seats going to non-party candidates. All but four of the elected candidates were members of the Vietnamese Fatherland Front, a political coalition organization led by the Communist Party. Turnout was more than 99%.

Truong Tan Sang was elected president by the National Assembly on 25 July 2011, receiving 487 of 496 votes.

CURRENT GOVERNMENT

President (titular head of state): Truong Tan Sang; b. 1949 (in office since 25 July 2011).

Vice-President: Nguyen Thi Doan.

Full members of the Politburo of the Communist Party of Vietnam: Nguyen Phu Trong (b. 1944; *Secretary General*); Le Hong Anh; Nguyen Tan Dung; Truong Tan Sang; Phung Quang Thanh; Le Thanh Hai; Nguyen Sinh Hung; Pham Quang Nghi; To Huy Rua; Tran Dai Quang; Tong Thi Phong; Ngo Van Du; Dinh The Huynh; Nguyen Xuan Phuc.

In March 2013 the government comprised:

Prime Minister: Nguyen Tan Dung; b. 1949 (in office since 27 June 2006).

Deputy Prime Ministers: Nguyen Xuan Phuc; Hoang Trung Hai; Nguyen Thien Nhan; Vu Van Ninh.

Minister of National Defence: Phung Quang Thanh. *Public Security:* Tran Dai Quang. *Foreign Affairs:* Pham Binh Minh. *Home Affairs:* Nguyen Thai Binh. *Justice:* Ha Hung Cuong. *Planning and Investment:* Bui Quang Vinh. *Finance:* Vuong Dinh Hue. *Industry and Trade:* Vu Huy Hoang. *Agriculture and Rural Development:* Cao Duc Phat. *Transport:* Dinh La Thang. *Construction:* Trinh Dinh Dung. *Natural Resources and Environment:* Nguyen Minh Quang. *Information and Communications:* Nguyen Bac Son. *Labour, War Invalids and Social Affairs:* Pham Thi Hai Chuyen. *Culture, Sports and Tourism:* Hoang Tuan Anh. *Science and Technology:* Nguyen Quan. *Education and Training:* Pham Vu Luan. *Health:* Nguyen Thi Kim Tien.

Vietnamese parliament: http://www.na.gov.vn

CURRENT LEADERS

Nguyen Phu Trong

Position

Secretary General of the Communist Party of Vietnam

Introduction

Nguyen Phu Trong was elected secretary general of the Communist Party of Vietnam in Jan. 2011 at the Party's 11th national congress, replacing Nong Duc Manh. Chairman of the National Assembly since 2006, Trong was previously the party secretary for Hanoi.

Early Life

Trong was born in Hanoi on 14 April 1944. He attended the Nguyen Gia Thieu school in Hanoi district and from 1963–67 studied linguistics at the Hanoi General University. After graduating, Trong became a civil servant and in 1968 was inducted into the Communist Party of Vietnam. From 1968–73 he worked on the *Communist Review*, the party's official journal, and was a prominent representative of the party's youth contingent. He then undertook post-graduate studies at Ho Chi Minh National Academy of Politics and Public Administration before returning as editor of the *Communist Review* in 1976. In 1983 he graduated with a history PhD from the Soviet Academy of Social Sciences. He returned to the *Communist Review* and held senior posts, joining the editorial board in 1991 and serving as editor-in-chief until 1996.

In 1994 Trong became a member of the Communist Party central committee and from 1996–98 was deputy secretary of its Hanoi section. In 1997 he joined the politburo, focusing on political research and theory. He was appointed head of 'ideological-cultural and scientific-educational affairs' and from 2001–06 served as chairman of the 'theoretical council'.

From 2000–06 he was party secretary in Hanoi and from 2002–06 a deputy in the National Assembly. In 2006 he was elected secretary of the Assembly's party organization, as well as chairman of the Assembly itself, and became a member of the council for defence and security.

Career in Office

At 67 years of age, Trong was past the mandatory retirement age when he became secretary general and is considered a conservative hard-liner. In 2011 he sought closer ties with China (signing an agreement in Oct. aimed at resolving disputed control over parts of the South China Sea) and held talks with Laotian delegates aimed at reinforcing relations. In seeking to build consensus, he liaises between the prime minister, Nguyen Tan Dung, and the state president, Truong Tan Sang (who was elected in July 2011).

Trong adopted a tough stance towards the media and moved against critical journalism, including banning the use of unnamed sources. However, against a background of continuing scandals in state-owned enterprises, he acknowledged in Oct. 2012 that the government had failed to curb corruption in its top ranks.

Nguyen Tan Dung

Position

Prime Minister

Introduction

Nguyen Tan Dung was appointed prime minister in June 2006. Once seen as part of the new generation aspiring to leadership of Vietnam's single-party government, he is a proponent of economic reform and liberalization. He was reappointed for a second term in Jan. 2011.

Early Life
Nguyen Tan Dung was born in Ca Mau in the south of Vietnam on 17 Nov. 1949. While serving in the army he joined the Communist Party in 1967, and then enrolled in the elite Nguyen Ai Quoc Party School in 1981 to study political theory.

Dung advanced rapidly within the party, serving on influential committees and supporting Vietnam's 'doi moi' programme of economic reform to move the country towards a market economy. In Jan. 1995 he became deputy minister for home affairs and in May the following year he became the youngest person ever to be appointed to the politburo. Tipped as a future party leader, he was appointed deputy prime minister and also director of the central committee's economic commission in charge of the party's finances in 1997. The following year he took over as governor of the state bank.

From 1998–2006 Dung was groomed for leadership by the then prime minister Phan Van Khai, a fellow modernizer. Appointed to a range of key party posts during that time, he oversaw the continuing liberalization of Vietnam's economy. Dung expanded foreign trade relations and prepared Vietnam's accession to the World Trade Organization (WTO). He was also given responsibility for tackling domestic corruption and organized crime. When Khai resigned the premiership, Dung was confirmed as prime minister by the National Assembly on 27 June 2006.

Career in Office
Dung reiterated his intention to proceed with economic and social reform. His first action was to replace several government figures who had been implicated in corruption. He sought to strengthen commercial links internationally, notably with the European Union. In Nov. 2006 Vietnam was approved for membership of the WTO and also hosted the annual Asia-Pacific Economic Co-operation meeting.

However, the economy remained in a malaise. In Nov. 2009 the government devalued the Vietnamese currency (for the third time since June 2008) by about 5% against the US dollar, at the same time increasing interest rates in a bid to dampen rising inflation. Vietnam also experienced a sharp decline in exports and in 2009 and in Dec. that year the World Bank approved a loan for the first time to Vietnam worth US$500m. By 2011 the country was facing further increases in inflation, a weakening currency (devalued twice in 2010) and uncertainty over the country's financing that resulted in a credit rating downgrade in 2010.

During an official visit to Moscow in Dec. 2009, Dung announced multi-billion dollar contracts to buy submarines, fighter jets and other military hardware from Russia, as well as agreements on co-operation relating to oil and gas, mining and financial services.

Dung has been closely associated with the communist regime's strategy of building up large state-run conglomerates and he faced criticism in the National Assembly in 2010 over the financial collapse of Vinashin, the giant state-owned shipbuilder. Dung conceded that his administration was partly responsible for the inadequate supervision of the company's management. Despite speculation that his opponents would use the failure as a way of removing him from office, in Jan. 2011 the Communist Party congress gave him a second term as prime minister. However, against a background of continuing scandals in state-owned enterprises, the Communist Party secretary general's acknowledgment in Oct. 2012 that the government had failed to address corruption in its top ranks was widely viewed as an indictment of Dung's premiership.

DEFENCE
Conscription is for 18 months (army) or three years (air force and navy). For specialists it is also three years.

In 2008 defence expenditure totalled US$2,907m. (US$33 per capita), representing 3·2% of GDP.

Army
There are nine military regions (including the capital). Strength (2007) was estimated to be 412,000. Paramilitary Local Defence forces number around 5m. and include the Peoples' Self-Defence Force (urban) and the People's Militia (rural). There is also a paramilitary Border Defence Corps numbering some 40,000.

Navy
The fleet includes two diesel submarines (although their serviceability is in doubt), five frigates and six corvettes. Vietnam has ordered six *Kilo* class submarines from Russia. The first one was scheduled to be handed over in Aug. 2013, with the last set to be delivered in 2016. In 2007 personnel was estimated at 13,000 plus an additional Naval Infantry force of about 27,000.

Air Force
In 2007 the People's Air Force had 30,000 personnel, with 219 combat-capable aircraft (Su-22s, Su-27s, Su-30s and MiG-21s) and 26 attack helicopters.

ECONOMY
Agriculture accounted for 20·4% of GDP in 2006, industry 41·6% and services 38·1%.

Overview
Over the last decade, the economy has been one of the best performing in Asia. Real GDP grew on average by 7·5% per year between 2000 and 2008, while poverty fell from over 50% in the mid-1990s to 16% in 2006. The oil sector plays an important role, with exports of crude oil accounting for 20% of total exports. Other key exports include rice, garments and footwear.

Vietnam joined the WTO in Jan. 2007 and signed bilateral trade agreements with the USA in 2000 and the EU in 2004. Economic activity remained strong during the global financial crisis, thanks to a sizable stimulus package (5% of GDP) and loose monetary policy. Though real GDP growth fell to 5·3% in 2009, its slowest pace since 2000, it remained among the strongest in Asia. Growth rebounded to 6·8% in 2010. Increased macroeconomic risks stemming from the stimulus led to interest rate rises, while in Feb. 2011 the government announced a stabilization package including measures to reduce credit growth and the budget deficit.

Currency
The unit of currency is the *dong* (VND). The dong was devalued by 5·4% in Nov. 2009, 3·4% in Feb. 2010, 2·1% in Aug. 2010 and 8·5% in Feb. 2011. Foreign exchange reserves were US$8,268m. in May 2005 and total money supply was 192,281bn. dong. Inflation was 23·1% in 2008, falling to 6·7% in 2009 before rising to 9·2% in 2010 and further still to 18·7% in 2011.

Budget
In 2008 revenues were 323trn. dong and expenditures 364trn. dong. Tax revenue accounted for 89·0% of revenues in 2008, non-tax revenue 9·9% and grants 1·1%; current expenditure accounted for 72·6% of expenditures and capital expenditure 27·4%.

VAT is 10% (reduced rate, 5%).

Performance
Real GDP growth rates have been consistently strong since 1994, and averaged 7·7% between 2001 and 2007. This robust performance continued into 2008, 2009, 2010 and 2011 with growth rates of 6·3%, 5·3%, 6·8% and 5·9% respectively. GDP per head, which was US$98 in 1990, had risen to US$815 by 2007. Vietnam's total GDP in 2011 was US$123·6bn.

Banking and Finance
The central bank and bank of issue is the State Bank of Vietnam (founded in 1951; *Governor*, Nguyen Van Binh). There were 42

banks in 2011. The leading commercial banks are Vietnam Bank for Agriculture and Rural Development (usually referred to as Agribank), Vietinbank, Bank for Investment and Development of Vietnam (BIDV) and Joint Stock Commercial Bank for Foreign Trade of Vietnam (Vietcombank).

External debt totalled US$35,139m. in 2010, representing 36·5% of GNI. Foreign direct investment in Vietnam was US$7·6bn. in 2009, up from US$2·0bn. in 2005 although down from the high of US$9·6bn. in 2008.

There are stock exchanges in Ho Chi Minh City, which opened in July 2000, and Hanoi, which opened in March 2005.

ENERGY AND NATURAL RESOURCES

Environment
Vietnam's carbon dioxide emissions from the consumption and flaring of fossil fuels were the equivalent of 1·1 tonnes per capita in 2008.

Electricity
Total installed capacity of power generation in 2007 was an estimated 12·5m. kW. In 2007, 69·48bn. kWh of electricity were produced (39·60bn. kWh thermal and 29·88bn. kWh hydro-electric); consumption per capita was 816 kWh. The proportion of households with electricity has doubled in the past 20 years, to 94% by 2007.

Oil and Gas
Oil reserves in 2008 totalled 4·7bn. bbls. In Aug. 2001 an offshore oil mine containing more than 400m. bbls of petroleum was discovered. Oil production in 2008, 15·4m. tonnes. Natural gas reserves in 2008 were 560bn. cu. metres; production was 7·9bn. cu. metres.

Minerals
Vietnam is endowed with an abundance of mineral resources such as coal (3·5bn. tonnes), bauxite (3bn. tonnes), apatite (1bn. tonnes), iron ore (700m. tonnes), chromate (10m. tonnes), copper (600,000 tonnes) and tin (70,000 tonnes); coal production was 34·1m. tonnes in 2005. There are also deposits of manganese, titanium, a little gold and marble. 2005 output (in 1,000 tonnes): sand and gravel, 146,400; lime, 1,718; salt, 925.

Agriculture
Agriculture employs 70% of the workforce. Ownership of land is vested in the state, but since 1992 farmers may inherit and sell plots allocated on 20-year leases. There were an estimated 6·35m. ha. of arable land in 2007 and 3·08m. ha. of permanent crops.

Production in 1,000 tonnes in 2003: rice, 34,519; sugarcane, 16,525; cassava, 5,228; maize, 2,934; sweet potatoes, 1,592; bananas, 1,221; coconuts, 920; coffee, 771; cabbage, 606; oranges, 500. Vietnam is the second largest coffee producer in the world after Brazil, and the second largest exporter of rice behind Thailand.

Livestock, 2003: pigs, 24·89m.; cattle, 4·39m.; buffaloes, 2·84m.; goats, 780,000; chickens, 185m.; ducks, 69m.

Livestock products (2003): meat, 2,487,000 tonnes; eggs, 234,000 tonnes; milk, 158,000 tonnes.

There were 257 tractors and 365 harvester-threshers per 10,000 ha. of arable land in 2006.

Forestry
In 2010 forests covered 13·80m. ha., or 44% of the land area. An export ban on logs has been in place since 1992. Timber production was 34·17m. cu. metres in 2007, nearly all of it for fuel.

Fisheries
Total catch, 2010, 2,420,800 tonnes (92% from sea fishing). Vietnam's aquaculture production is the third largest in the world behind those of China and India, at 2,462,000 tonnes in 2008.

INDUSTRY
Estimated total industrial output in 2002 was 260,202·0bn. dong. In 2002 estimated production (in 1,000 tonnes) was: processed sea produce, 288,701; cement, 19,482; steel, 2,429; fertilizers, 1,176; sugar, 1,074; paper (2003), 800; detergents, 381; beer (2003), 1,049·8m. litres; clothes, 47·6m. items.

Labour
In 2004 a total of 41·6m. persons were in employment. Agriculture and forestry accounted for 23·0m. people; industry, 5·3m.; trade, 4·8m.; construction, 1·9m.; culture, health and education, 1·7m. A liberal Enterprise Law was adopted in 2000, leading to the creation of over 50,000 new private businesses and more than 1·5m. new jobs during the next three years. Official statistics put unemployment at 7·4% of the workforce in early 2000.

INTERNATIONAL TRADE
In Feb. 1994 the USA lifted the trade embargo it had imposed in 1975, and in Nov. 2001 a trade agreement with the USA was ratified. The agreement allows Vietnam's exports access to the US market on the same terms as those enjoyed by most other countries. The 1992 constitution regulates joint ventures with western firms; full repatriation of profits and non-nationalization of investments are guaranteed.

Imports and Exports
Trade is conducted through the state import-export agencies. Imports (c.i.f.) in 2008, US$80,714m. (US$36,761m. in 2005); exports (f.o.b.), US$62,685m. (US$32,447m. in 2005). In 2002 the main imports (by value) were: machinery (19%), petroleum (10%), textiles and garments (9%) and steel (7%). Other significant imports were plastics, vehicles, fertilizers and chemicals. Main exports: crude oil (20%), textiles and garments (16%), sea produce (12%), footwear (11%) and rice (4%). Other significant exports were electronics, coffee, latex, cashew nuts, pepper and coal.

The main import suppliers in 2000 were Singapore (15·1%), Japan (14·0%), South Korea (11·9%) and China (10·9%). Principal export markets in 2000 were Japan (18·6%), Australia (9·7%), Germany (7·7%) and China (6·6%).

COMMUNICATIONS

Roads
There were 160,089 km of roads in 2007, of which 47·6% were paved. In 2007 there were 1,146,300 passenger cars in use and around 21·78m. motorcycles and mopeds. There were 13,200 fatalities in road accidents in 2007.

Rail
There were 2,402 km of railways in 2005, mostly metre gauge. Rail links with China were reopened in Feb. 1996. In 2008, 11·3m. passengers and 8·4m. tonnes of freight were carried.

Civil Aviation
There are international airports at Hanoi (Noi Bai) and Ho Chi Minh City (Tan Son Nhat) and 13 domestic airports. The national carrier is Vietnam Airlines, which provides domestic services and in 2003 had international flights to Bangkok, Beijing, Dubai, Guangzhou, Hong Kong, Kaohsiung, Kuala Lumpur, Kunming, Manila, Melbourne, Moscow, Osaka, Paris, Phnom Penh, Seoul, Siem Reap, Singapore, Sydney, Taipei, Tokyo and Vientiane. In 2005 scheduled airline traffic of Vietnam-based carriers flew 43·7m. km, carrying 3,762,200 passengers. The busiest airport is Ho Chi Minh City, which in 2001 handled 4,306,143 passengers and 96,560 tonnes of freight. Hanoi handled 2,207,052 passengers and 40,668 tonnes of freight in 2001.

Shipping

In Jan. 2009 there were 918 ships of 300 GT or over registered (including 769 general cargo ships and 76 oil tankers), totalling 2,683,000 GT. The major ports are Hai Phong, Ho Chi Minh City and Da Nang. There are regular services to Hong Kong, Singapore, Thailand, Cambodia and Japan. There are some 19,500 km of navigable waterways.

Telecommunications

Vietnam Posts and Telecommunications and the military operate telephone systems with the assistance of foreign companies. In 2008 there were 29·6m. main (fixed) telephone lines; mobile phone subscribers numbered 70m. in 2008 (80·4 per 100 persons). There were 20·8m. internet users in 2008. In March 2012 there were 3·2m. Facebook users.

SOCIAL INSTITUTIONS

Justice

A new penal code came into force on 1 Jan. 1986 'to complete the work of the 1980 constitution'. Penalties (including death) are prescribed for opposition to the people's power and for economic crimes. The judicial system comprises the Supreme People's Court, provincial courts and district courts. The president of the Supreme Court is responsible to the National Assembly, as is the Procurator-General, who heads the Supreme People's Office of Supervision and Control.

The death penalty is still in force. There were five confirmed executions in 2011 although none in 2012.

The population in penal institutions in mid-2010 (sentenced prisoners only) was 108,557 (122 per 100,000 of national population).

Education

Adult literacy rate in 2009 was estimated at 92·8% (95·2% among males and 90·5% among females). Primary education consists of a ten-year course divided into three levels of four, three and three years respectively. In 2007 there were 7,041,312 pupils and 344,547 teaching staff in primary schools, and 9,845,407 pupils and 451,165 teaching staff at secondary schools. About 75% of children of secondary school age were in school by 2003, up from around a third in 1990. In 1995–96 there were seven universities, two open (distance) universities and nine specialized universities (agriculture, three; economics, two; technology, three; water resources, one). In 2007 there were 1,587,609 students in higher education with 53,518 academic staff.

Health

In 2001 there were 42,327 physicians, 44,539 nurses, 14,662 midwives and 5,977 pharmacists. There were 842 hospitals in 2003 with a provision of 24 beds per 10,000 population.

RELIGION

Taoism is the traditional religion but Buddhism is widespread. At the census of 2009 the principal denominations were: Buddhists, 6,802,318; Catholics, 5,677,086; Hoa Hao (a tradition based on Buddhism), 1,433,252; Cao Dai (a synthesis of Christianity, Buddhism and Confucianism), 807,915; Protestants, 734,168; no religion, 70,193,377. In Feb. 2013 there was one cardinal. There are 26 dioceses, including three archdioceses.

CULTURE

World Heritage Sites

Vietnam has seven sites on the UNESCO World Heritage List: the Complex of Hue Monuments (inscribed on the list in 1993); Ha Long Bay (1994 and 2000); Hoi An Ancient Town (1999); My Son Sanctuary (1999); Phong Nha-Ke Bang National Park (2003); the Central Sector of the Imperial Citadel of Thang Long, Hanoi (2010); and the Citadel of the Ho Dynasty (2011).

Press

In 2008 there were 55 paid-for daily newspapers with a combined circulation of 2·8m. The Communist Party controls all print media but some criticism of government policy is allowed.

Tourism

There were a record 6,014,000 international visitors in 2011 (up from 5,050,000 in 2010). Tourist numbers have doubled since 2003. The main nationalities of tourists in 2010 were China (905,000), South Korea (496,000) and Japan (442,000).

DIPLOMATIC REPRESENTATIVES

Of Vietnam in the United Kingdom (12–14 Victoria Rd, London, W8 5RD)
Ambassador: Vu Quang Minh.

Of the United Kingdom in Vietnam (Central Building, 31 Hai Ba Trung, Hanoi)
Ambassador: Dr Antony Stokes, LVO.

Of Vietnam in the USA (1233 20th St., NW, Suite 400, Washington, D.C., 20036)
Ambassador: Nguyen Quoc Cuong.

Of the USA in Vietnam (7 Lang Ha, Ba Dinh District, Hanoi)
Ambassador: David B. Shear.

Of Vietnam to the United Nations
Ambassador: Le Hoai Trung.

Of Vietnam to the European Union
Ambassador: Pham Sanh Chau.

FURTHER READING

Trade and Tourism Information Centre with the General Statistical Office. *Economy and Trade of Vietnam* [various 5-year periods]

Gilbert, Marc Jason, (ed.) *Why the North Won the Vietnam War.* 2002
Harvie, C. and Tran Van Hoa V., *Reforms and Economic Growth.* 1997
Karnow, S., *Vietnam: a History.* 2nd ed. 1992
Morgan, Ted, *Valley of Death: The Tragedy at Dien Bien Phu That Led America into the Vietnam War.* 2010
Morley, J. W. and Nishihara M., *Vietnam Joins the World.* 1997

National Statistical Office: General Statistical Office, 6B Hoang Dieu, Ba Dinh District, Hanoi.
Website: http://www.gso.gov.vn

YEMEN

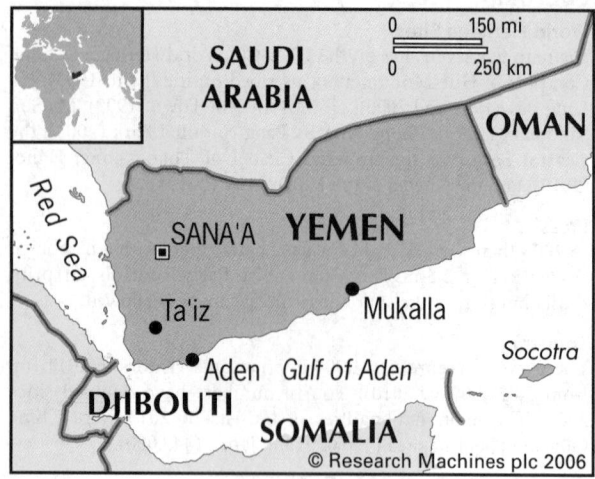

**Jamhuriya al Yamaniya
(Republic of Yemen)**

Capitals: Sana'a (Legislative and Administrative),
Aden (Commercial)
Population projection, 2015: 27·98m.
GNI per capita, 2011: (PPP$) 2,213
HDI/world rank: 0·462/154
Internet domain extension: .ye

KEY HISTORICAL EVENTS

One of the earliest recorded pre-Islamic Arab civilizations was the Sabaean culture, which flourished in what is now Yemen and southwestern Saudi Arabia during the 1st millennium BC. The wealth of the kingdom of Saba (or Sheba) was based on the incense and spice trade and on agriculture. Beginning in about 115 BC, the Himyarites gradually absorbed Saba and Hadhramaut (to the east) to claim control of all of the southwest Arabian peninsula by the 4th century AD. Himyarite dominance came to an end in the 6th century as Abyssinian (Ethiopian) forces invaded in AD 525. Abyssinian rule was overthrown in 575 by Persian military intervention, and Persian control then endured until the advent of Islam in 628.

Yemen became a province of the Muslim caliphate. Thereafter its fortunes reflected the fluctuating power of the imams (kings and spiritual leaders) of the Zaidi sect—who built the theocratic political structure of Yemen that endured from the 9th century until 1962—and of rival dynasties and conquerors. These included the Fatimid caliphs of Egypt, who occupied most of Yemen from about 1000 until 1175 and, subsequently, the Ayyubids, who ruled until about 1250. The Rasulids, who had served as governors of Yemen under the Ayyubid dynasty, then exercised a measure of sovereignty. However, central authority had fragmented by the time the Ottoman Turks, fearful of Christian Portuguese influence in southern Arabia, intervened in Yemen in the first half of the 16th century. They held nominal but tenuous sovereignty until the end of the First World War (when Yemen became independent), and conflict with the Zaidi imams was frequent.

The southern Yemeni port of Aden was a coveted commercial location on the trading routes to the wider East from ancient times. By the end of the 18th century its strategic importance had increased as Britain sought to contain the French threat to colonial communications with British India following Napoleon's conquest of Egypt. With the coming of the steamship, Britain's need for a military and refuelling base in the region became more pressing. In 1839 the British captured Aden, attaching it administratively to India. In the 1850s the Perim, Kamaran and Kuria Muria islands were made part of Aden which became a free port. Britain also purchased areas on the mainland from local rulers and entered into protectionist agreements with them. The opening of the Suez Canal in 1869 further enhanced both Aden's strategic significance and British colonial consolidation.

In the early 1870s, tribal and religious warfare led the Ottoman Turks to reassert authority over northern Yemen, an occupation that lasted until the armistice of 1918. After the Ottoman evacuation, Imam Yahya, who had supported the Turks during the First World War despite leading an earlier revolt against them in 1911, sought to expand Yemeni territory. However, in 1934, after brief hostilities with Saudi Arabia (by then under the rule of Ibn Saud) and skirmishes with British forces from Aden, a trilateral treaty to fix Yemen's boundaries was agreed. The treaty heralded a period of generally peaceful coexistence, which lasted for the rest of Imam Yahya's reign. Aden, meanwhile, was formally made a British crown colony in 1937 and the surrounding region became known as the Aden protectorate.

Opposition to the theocratic and despotic rule of Imam Yahya led to his assassination in Feb. 1948. His son, Crown Prince Ahmad, succeeded him and put down the insurgents. Ahmad's reign was marked by further repression, renewed friction over the British presence in southern Yemen and growing pressure in the 1950s to support the Arab nationalist objectives of the new Nasser regime in Egypt. Following Ahmad's death his son, Muhammad al-Badr, was deposed a week after his accession in Sept. 1962 by revolutionary forces aided by Egypt. The rebels took control of the capital, Sana'a, and proclaimed the Yemen Arab Republic (YAR). Saudi Arabia and Jordan supported al-Badr's royalist forces against the new republic and conflict continued periodically until 1970. Despite its instability, the YAR regime retained power and secured international recognition.

In 1963 the Aden colony was merged into the Federation of South Arabia. A power struggle between rival nationalist groups after the British withdrawal in 1967 led to an independent Marxist state with Aden as the capital. This was renamed the People's Democratic Republic of Yemen (PDRY) in 1970. Mistrust and frequent border clashes between the YAR and the PDRY characterized the next decade, despite an accord in 1972 to merge the two entities.

A coup in the YAR brought Lieut.-Col. Ibrahim al-Hamadi to power in 1974. However, he was assassinated in Oct. 1977, as was his successor, Lieut.-Col. Ahmad al-Ghashmi, in June 1978. The following month Lieut.-Col. Ali Abdullah Saleh, the commander of the Ta'iz military area, was elected president by the Constitutional Assembly. Saleh's first months in power were turbulent. In early 1979 sporadic fighting erupted into full-scale war between North and South Yemen. Arab League mediation brought the hostilities to an end and both sides acknowledged the need to effect permanent Yemeni union. Saleh was meanwhile re-elected as YAR President by the Constituent Assembly in May 1983 and again in July 1988 by a new Consultative Council.

In the PDRY the chairman of the Presidential Council, Salem Rubayyi Ali, was overthrown in June 1978. Prime Minister Ali Nasser Muhammad briefly assumed the chairmanship before Abdul Fattah Ismail, a hard-line orthodox Marxist, was elected as head of state by the new Presidium of the People's Supreme Assembly in Dec. 1978. Ismail resigned unexpectedly in April 1980 and went to the Soviet Union, to be replaced by Ali Nasser

Muhammad. While maintaining South Yemen's close relations with the Soviet Union, Muhammad favoured reconciliation with moderate Arab states. In particular, he viewed improved relations with Saudi Arabia as necessary to further the proposed merger with the YAR. In Feb. 1985 Muhammad resigned as prime minister but remained head of state and secretary-general of the dominant Yemeni Socialist Party (YSP). The rest of that year was marked by the re-emergence of political rivalries, aggravated by the return of Ismail from the Soviet Union, which triggered a brief civil war in Jan. 1986. Ismail was killed and Muhammad fled into exile, after which a new government was formed.

In May 1988 the YAR and PDRY governments reached an agreement on renewing unification discussions and demilitarizing their borders. Their respective leaders, Ali Abdullah Saleh and Ali Salem Albidh, agreed in late 1989 a draft unity constitution (originally drawn up in 1981) and on 22 May 1990 the Republic of Yemen was declared. A five-member Presidential Council assumed power and Saleh was appointed as president for a transitional period. By 1993 relations between the North and South had again deteriorated, Vice-President Albidh having withdrawn to Aden to demand political reforms. Sporadic military clashes escalated into full civil war in May 1994 between disaffected southern forces and Yemen's northern-based government. Southern officials announced their secession from Yemen on 21 May 1994, but northern forces quickly prevailed and Aden was captured on 7 July 1994. The former vice-president and other southern leaders went into exile.

Confirmed in office by parliament in Oct. 1994 for a five-year term, President Saleh's political position was further consolidated when he was re-elected by popular vote in Sept. 1999 (though the turnout was low, particularly in the south). An extension of his term of office from five to seven years was approved in a referendum in Feb. 2001. In April 2003 Saleh's ruling General People's Congress retained power in parliamentary elections with over two-thirds of the seats in the 301-member Assembly of Representatives. However, the main opposition parties claimed that polling was tainted by ballot rigging and intimidation.

In foreign affairs Yemen suffered diplomatic isolation and economic sanctions in the early 1990s for its equivocal response to Iraq's invasion of Kuwait. In 1995 Yemen clashed with Eritrea over control of the Hanish Islands in the Red Sea. Following arbitration by an international panel, Yemen assumed control of the main islands in 1998.

Longstanding tensions with neighbouring Saudi Arabia were eased when a border dispute was settled in June 2000. Attacks on Western targets in Yemeni territory—most notably the suicide bombing of the US naval vessel *USS Cole* in Oct. 2000 and a bomb attack on a French supertanker, the *Limburg*, off the Yemeni coast in Oct. 2002—fuelled concerns that Yemen might be a haven for Islamic extremists. Nevertheless, President Saleh pledged full support for the USA's global campaign against terrorism in the wake of the events of 11 Sept. 2001.

Attacks linked to al-Qaeda continued throughout the decade. In Oct. 2010 parcel bombs intended for US-bound cargo planes were suspected of originating in the country. In Jan. 2011 the US government expressed its 'urgent concern' at Yemen's extremist links. Popular anti-government protests starting the same month led to President Saleh relinquishing power in Nov. 2011. He was succeeded by Abdo Rabu Mansour al-Hadi, who won the presidential election held in Feb. 2012 unopposed.

TERRITORY AND POPULATION

Yemen is bounded in the north by Saudi Arabia, east by Oman, south by the Gulf of Aden and west by the Red Sea. The territory includes 112 islands including Kamaran (181 sq. km) and Perim (300 sq. km) in the Red Sea and Socotra (3,500 sq. km) in the Gulf of Aden. The islands of Greater and Lesser Hanish are claimed by both Yemen and Eritrea. On 15 Dec. 1995 Eritrean troops occupied them, and Yemen retaliated with aerial

bombardments. A ceasefire was agreed at presidential level on 17 Dec. On 20 Dec. the UN resolved to send a good offices mission to the area. In an agreement of 21 May 1996 brokered by France, Yemen and Eritrea renounced the use of force to settle the dispute and agreed to submit it to arbitration. Following a ruling issued by the Permanent Court of Arbitration in the Hague, Yemen assumed control of the main islands in 1998. The area is 555,000 sq. km excluding the desert Empty Quarter (Rub Al-Khahi).

A dispute with Saudi Arabia broke out in Dec. 1994 over some 1,500–2,000 km of undemarcated desert boundary. A memorandum of understanding signed on 26 Feb. 1995 reaffirmed the border agreement reached at Taif in 1934, and on 12 June 2000 a 'final and permanent' border treaty between the two countries was signed. An agreement of June 1995 completed the demarcation of the border with Oman.

Census population, 2004: 19,685,161; density, 35 persons per sq. km. In 2011, 32·4% of the population lived in urban areas.

The UN gives a projected population for 2015 of 27·98m.

In 2004 there were 20 governorates plus the capital city, Sana'a:

	2004 census population		2004 census population
Abyan	433,819	Hajjah	1,479,568
Aden	589,419	Ibb	2,131,861
Al-Baidha	577,369	Lahj	722,694
Al-Dhale	470,564	Mareb	238,522
Al-Hodeidah	2,157,552	Raymah	394,448
Al-Jawf	443,797	Sa'adah	695,033
Al-Mahrah	88,594	Sana'a (city)	1,747,834
Al-Mahwit	494,557	Sana'a	919,215
Amran	877,786	Shabwa	470,440
Dhamar	1,330,108	Ta'iz	2,393,425
Hadhramaut	1,028,556		

The population of the capital, Sana'a, was 1,707,531 in 2004. The commercial capital is the port of Aden, with a population of (2004) 588,938. Other important towns are Ta'iz, the port of Hodeida, Mukalla, Ibb and Abyan. Sana'a is currently the fastest-growing city in the world, with a population increase of 832·6% in the period 1975–2000 and a projected increase of 128·2% between 2000–15, by when it is expected to have 3·03m. inhabitants.

The official language is Arabic.

SOCIAL STATISTICS

2009 estimates: births, 886,000; deaths, 163,000. Rates, 2009 estimates (per 1,000 population): birth, 38; death, 7. Yemen has one of the youngest populations of any country, with 75% of the population under the age of 30 and 44% under 15. Life expectancy, 2007, was 60·9 years for males and 64·1 years for females. Infant mortality, 2010, 57 per 1,000 live births. Annual population growth rate, 1998–2008, 2·9%; fertility rate, 2008, 5·2 births per woman.

CLIMATE

A desert climate, modified by relief. Sana'a, Jan. 57°F (13·9°C), July 71°F (21·7°C). Aden, Jan. 75°F (24°C), July 90°F (32°C). Annual rainfall 20" (508 mm) in the north, but very low in coastal areas: 1·8" (46 mm).

CONSTITUTION AND GOVERNMENT

Parliament consists of a 301-member *Assembly of Representatives* (*Majlis al-Nuwaab*), elected for a six-year term in single-seat constituencies and, since 2001, a 111-member *Shura Council* (*Majlis al-Shura*), appointed by the president.

The constitution was adopted in May 1991 but was drastically amended in 1994 following the civil war. On 28 Sept. 1994 the Assembly of Representatives unanimously adopted the amended constitution founded on Islamic law. It abolished the former five-member Presidential Council and installed a *President* elected by parliament for a five-year term, subsequently amended to a seven-year term through a referendum held on 20 Feb. 2001. As a result of the same referendum the term for MPs was extended from four to six years. Following the popular protests in 2011 that unseated President Saleh, his successor, Abdo Rabu Mansour al-Hadi, is expected to oversee the drafting of a new constitution.

National Anthem

'Raddidi Ayyatuha ad Dunya nashidi' ('Repeat, O World, my song'); words by A. Noman, tune by Ayub Tarish.

RECENT ELECTIONS

In presidential elections held on 21 Feb. 2012 the sole candidate, Vice President Abdo Rabu Mansour al-Hadi, won 99·8% of the vote. Turnout was 65%.

Parliamentary elections were held on 27 April 2003, in which the General People's Congress (MSA) gained 238 seats (58·0% of the vote), Yemeni Congregation for Reform (Islah) 46 seats (22·6%), Yemeni Socialist Party 8 seats (3·8%), Nasserite Unionist People's Organization (TWSN) 3 seats (1·9%), the Arab Socialist Rebirth Party (Baath) 2 seats (0·7%) and ind. 4 seats. Turnout was 76·0%.

CURRENT GOVERNMENT

President: Abdo Rabu Mansour al-Hadi; b. 1945 (MSA; in office since 25 Feb. 2012).

In March 2013 the government comprised:

Prime Minister: Mohammed Basindawa; b. 1935 (ind.; in office since 7 Dec. 2011).

Minister of Finance: Sakhr Ahmed Abbas. *Defence:* Mohammed Nasser Ahmed. *Foreign Affairs:* Abubakr Al-Qirbi. *Interior:* Abdelqader Qahtan. *Oil and Mineral Resources:* Ahmed Abdullah Daris. *Legal Affairs:* Mohammed Ahmed al-Makhlafi. *Justice:* Morshed Ali al-Arshani. *Higher Education and Scientific Research:* Hisham Sharaf Abdullah. *Labour and Social Affairs:* Amat Al-Razaq Ali Hamad. *Telecommunications and Information Technology:* Ahmed Obeif bin Dagher. *Fisheries:* Awad Mohammed Saqtari. *Transport:* Waed Abdullah Batheeb. *Information:* Ali Ahmed al-Amrani. *Human Rights:* Houriya Mashhour Ahmed. *Youth and Sports:* Moammar al-Iryani. *Electricity:* Saleh Hasan Sami. *Agriculture and Irrigation:* Farid Ahmed al-Moujawar. *Trade and Industry:* Saad Eddine Ali Salem bin Taleb. *Culture:* Abdullah Manthouq. *Technical and Vocational Training:* Abdel Hafeth Thabet Noman. *Planning and International Co-operation:* Mohammed Said al-Saadi. *Public Health and Population:* Ahmed Qasem al-Ansi. *Education:* Abdelrazzaq Yahya al-Ashwal. *Religious Endowments and Guidance:* Hamoud Mohammad Abad. *Water and Environment:* Abdo Razaz Saleh Khaled. *Civil Service and Social Security:* Nabil Abdo Shamsan. *Tourism:* Qasim Salam. *Local Government Affairs:* Ali Mohammed al-Yazidi. *Public Works:* Omar Al-Qurshumi. *Expatriate Affairs:* Mujahid al-Qahali.

Minister of State for Parliamentary Affairs: Rashad Ahmed Rasas. *Cabinet Affairs:* Jawhara Hammoud Thabet. *Ministers of State without Portfolio:* Shaef Aziz Sagheer; Hasan Ahmed Sharaf Eddine.

Office of the President: http://www.presidentsaleh.gov.ye

CURRENT LEADERS

Abdo Rabu Mansour al-Hadi

Position
President

Introduction
Abdo Rabu Mansour al-Hadi became president in Feb. 2012. His appointment ended 33 years of rule by Ali Abdullah Saleh, who was toppled following months of mass protests inspired by the

Arab Spring. A career military officer from southern Yemen, Hadi is considered a moderate and was granted two years to manage a transition towards stability and unity.

Early Life
Hadi was born on 1 May 1945 in Thukian, Abyan, in southern Yemen. He attended Aden Protectorate Army School, a private establishment for the sons of army officers in the former British colony. Graduating in 1964, he won a scholarship to the UK's Royal Military Academy at Sandhurst.

Following South Yemen's independence in Nov. 1967, Hadi served as leader of an armoured division at the Al-Anad military base. He undertook further training in Cairo in 1970, before being posted to the Al-Dhale region and serving on the mediation committee that successfully resolved a conflict with North Yemen in 1972. For much of the 1970s he managed South Yemen's military training establishment and spent time in Moscow studying military leadership. Promoted to deputy chief of staff in 1983, Hadi was responsible for military liaison with the then Soviet Union.

Following Yemen's unification in May 1990, Hadi was a leading figure in the republic's new army. During the 1994 civil war he supported Ali Abdullah Saleh (the president and former leader of North Yemen) and was rewarded for his loyalty when he was appointed vice president and minister of defence—the most senior position in the government held by a southerner. He was a key negotiator on issues including the Yemeni–Saudi border dispute, the controversial maritime frontier with Eritrea and the ongoing conflict between government forces and Houthi rebels in control of the northern Sa'adah province.

Saleh stood down in Nov. 2011 after more than ten months of popular unrest interspersed with violent government crackdowns. In accordance with a UN Security Council resolution, he agreed to transfer power to Hadi for an interim period ahead of presidential elections. On 21 Feb. 2012 Hadi, the sole candidate, was elected president, with turnout reported as 65%.

Career in Office
Hadi was sworn in to office on 25 Feb. 2012. A popular figure within Yemen, he also received support from many neighbouring states. However, he faced a trying tenure in view of continued threats from al-Qaeda, the Houthi rebellion, a secessionist movement in the south, a weak economy and crumbling infrastructure.

DEFENCE

Conscription is for two years. Defence expenditure in 2008 totalled US$1,492m. (US$67 per capita), representing 6·4% of GDP.

Estimates of the number of small arms in the country are around 12m., equivalent to 61 firearms for every 100 people, making Yemen second only behind the USA as the world's most heavily armed country.

Army

Strength (2007), 60,000 (including conscripts). There are paramilitary tribal levies numbering at least 20,000 and a Ministry of Security force of 50,000.

Navy

Navy forces are based at Aden and Al-Hodeidah, with other facilities at Mukalla, Perim and Socotra. Personnel in 2007 numbered 1,700.

Air Force

The unified Air Forces of the former Arab Republic and People's Democratic Republic are now under one command, although this unity was broken by the attempted secession of the south in

1994 which resulted in heavy fighting between the air forces of Sana'a and Aden. Personnel (2007), 3,000. There were 79 combat-capable aircraft in 2007 including Su-17/20/22s, MiG-21s and MiG-29s.

ECONOMY

Crude petroleum and natural gas accounted for 28% of GDP in 2007; trade, restaurants and hotels 16%; transport and communication 12%; and finance and real estate 11%.

Overview

Yemen is the poorest country in the Middle East and its outlook is unpromising. It has limited resources, including declining oil reserves, little arable land and scarce water supplies. Oil dominates the economy, accounting for 70% of government revenue and more than 90% of export earnings. With one of the highest population growth rates in the world, over 40% of the population lives in poverty, mostly in rural areas. Its problems were aggravated by a 60% increase in food prices in 2007–08. The economy has also been hampered by long-running political and social unrest, including a civil war from 1990–94.

Despite a relatively insulated financial sector, Yemen suffered during the global economic crisis with declining FDI and remittances, sharp decreases in oil revenues and reduced public expenditure. Since Jan. 2011 mass civilian protests, echoing similar unrest throughout the region, have threatened to send the economy into freefall. Foreign donors are expected to put limits on aid amid ongoing security threats and the volatile political situation. Coupled with soaring prices and a failing currency, the medium-term outlook is gloomy.

Currency

The unit of currency is the *riyal* (YER) of 100 *fils*. During the transitional period to north-south unification the northern *riyal* of 100 *fils* and the southern *dinar* of 1,000 *fils* co-existed. There were three foreign exchange rates operating: an internal clearing rate, an official rate and a commercial rate. In 1996 the official rate was abolished. Total money supply in June 2005 was 372,105m. riyals, gold reserves totalled 50,000 troy oz and foreign exchange reserves were US$5,253m. Inflation was 11·2% in 2010, increasing to 19·5% in 2011.

Budget

The fiscal year is the calendar year. General government revenues in 2009 (provisional) were 1,276bn. riyals and expenditures 1,476bn. riyals. Oil revenue accounted for 58·3% of total revenues in 2009. The main items of expenditure in 2009 were wages and salaries (35·8%) and subsidies (26·9%).

Performance

Real GDP growth was 3·9% in 2009 and 7·7% in 2010 but the economy contracted by 10·5% in 2011. Total GDP in 2011 was US$33·8bn.

Banking and Finance

The *Governor* of the Central Bank of Yemen is Mohamed Awad Bin Humam. Total reserves of the Central Bank were 81,089m. riyals in 2002 and there were 446,287m. riyals in deposits. In 2009 foreign debt amounted to US$6,370m., which represented 25·6% of GNI.

ENERGY AND NATURAL RESOURCES

Environment

Yemen's carbon dioxide emissions from the consumption and flaring of fossil fuels were the equivalent of 0·9 tonnes per capita in 2008. An *Environmental Performance Index* compiled in 2008 ranked Yemen 141st in the world out of 149 countries analysed, with 49·7%. The index examined various factors in six areas—air pollution, biodiversity and habitat, climate change,

environmental health, productive natural resources and water resources.

Electricity

Installed capacity was an estimated 1·3m. kW in 2007. Production in 2007 was 6·03bn. kWh; consumption per capita was 280 kWh.

Oil and Gas

In 2008 there were oil reserves of 2·7bn. bbls; mostly near the former north–south border. Oil production (2008): 14·4m. tonnes. Natural gas reserves in 2008 were 490bn. cu. metres.

Minerals

In 2008, 65,000 tonnes of salt were produced. In 2008, 100,000 tonnes of gypsum were extracted and 4·95m. cu. metres of quarried stone (including marble).

Agriculture

In 2007 there were around 1·38m. ha. of arable land and 250,000 ha. of permanent cropland. 772,000 ha. were irrigated in 2006. In the south, agriculture is largely of a subsistence nature, sorghum, sesame and millet being the chief crops, and wheat and barley being widely grown at the higher elevations. Cash crops include cotton. Fruit is plentiful in the north. Estimated production (2003, in 1,000 tonnes): tomatoes, 273; sorghum, 260; alfalfa, 242; potatoes, 213; oranges, 191; grapes, 169. Livestock in 2003 (estimates): goats, 7·3m.; sheep, 6·5m.; cattle, 1·4m.; chickens, 35m. Livestock products, 2003 estimates (in 1,000 tonnes): milk, 237; meat, 215.

Forestry

There were 0·55m. ha. of forest in 2010 (1% of the total land area). Timber production in 2007 was 395,000 cu. metres.

Fisheries

Fishing is a major industry. Total catch in 2008 was 127,132 tonnes, exclusively marine fish.

INDUSTRY

Output (2007 unless otherwise indicated, in 1,000 tonnes): cement (2009), 2,118; petrol, 1,127; distillate fuel oil, 991; residual fuel oil, 676; wheat flour (2003), 667; jet fuel, 486; kerosene, 97. In 2001 industry accounted for 49·2% of GDP, with manufacturing contributing 6·7%.

Labour

Of 3,621,700 persons in employment in 2002, 1,927,700 were engaged in agriculture, hunting and forestry; 394,200 in wholesale and retail trade/repair of motor vehicles, motorcycles and personal and household goods; 358,000 in public administration and defence/compulsory social security; and 238,200 in construction. Unemployment was 18% in 2004.

INTERNATIONAL TRADE

Imports and Exports

Trade in US$1m.:

	2005	2006	2007	2008	2009
Imports c.i.f.	5,399·9	6,080·9	8,510·7	10,546·2	9,184·8
Exports f.o.b.	5,607·7	6,654·1	6,298·9	7,583·8	6,259·0

Main import suppliers, 2009: United Arab Emirates, 9·9%; China, 9·3%; USA, 6·4%; Japan, 5·6%. Main export markets, 2009: China, 25·2%; India, 20·1%; Thailand, 18·4%; Singapore, 6·9%.

Oil and fish are major exports, the largest imports being food and live animals, and transport and machinery equipment. Oil and oil products account for nearly 90% of exports. A large transhipment and entrepôt trade is centred on Aden, which was made a free trade zone in 1991.

COMMUNICATIONS

Roads
There were 71,300 km of roads in 2005 (8·7% paved). In 2007 there were 777,700 vehicles in use.

Civil Aviation
There are international airports at Sana'a and Aden. In 2001 Sana'a handled 881,000 passengers (707,000 on international flights) and 15,900 tonnes of freight. The national carrier is Yemenia, which operates internal services and in 2003 had international flights to Abu Dhabi, Addis Ababa, Amman, Asmara, Bahrain, Beirut, Cairo, Damascus, Dar es Salaam, Djibouti, Doha, Dubai, Frankfurt, Jakarta, Jeddah, Khartoum, Kuala Lumpur, London, Marseille, Milan, Moroni, Mumbai, Paris, Riyadh and Rome. In 2003 Yemenia flew 18m. km, carrying 844,000 passengers (622,000 on international flights).

Shipping
In Jan. 2009 there were seven ships of 300 GT or over registered, totalling 17,000 GT. There are ports at Aden, Mokha, Al-Hodeidah, Mukalla and Nashtoon.

Telecommunications
In 2008 there were 1·1m. main (fixed) telephone lines in Yemen; mobile phone subscribers numbered 3·7m. in 2008 (16·1 per 100 persons). There were 370,000 internet users in 2008. In March 2012 there were 437,000 Facebook users.

SOCIAL INSTITUTIONS

Justice
A civil code based on Islamic law was introduced in 1992. Amnesty International reported that there were at least 28 executions in 2012 (at least 41 in 2011).

Education
In 2005 there were 17,993 children in pre-primary schools, 3·21m. pupils in primary schools and 1·46m. pupils in secondary schools. In 2006 there were 209,386 students in higher education. The adult literacy rate in 2009 was estimated at 62·4% (79·9% among males but only 44·7% among females, one of the biggest differences in literacy rates between the sexes of any country).

In 2008 total expenditure on education came to 5·2% of GDP and accounted for 16·0% of total government expenditure.

Health
In 2003 there were 68 hospitals with 8,871 beds. There were 4,078 physicians, 222 dentists, 8,342 nurses and 750 pharmacists in 2001; and 385 midwives in 1994.

RELIGION
In 2001 there were some 18·05m. Muslims (mostly Sunnis) and approximately 20,000 followers of other religions.

CULTURE

World Heritage Sites
There are four sites under Yemeni jurisdiction that appear in the UNESCO World Heritage List. They are (with year entered on the list): the old walled city of Shibam (1982); the old city of Sana'a (1986); the historic town of Zabid (1993); and the Socotra Archipelago (2008).

Press
In 2006 there were three daily newspapers with a combined average daily circulation of 40,000.

Tourism
There were 336,000 foreign tourists in 2005, bringing revenue of US$262m.

DIPLOMATIC REPRESENTATIVES
Of Yemen in the United Kingdom (57 Cromwell Rd, London, SW7 2ED)
Ambassador: Abdulla Ali Al-Radhi.

Of the United Kingdom in Yemen (POB 1287, 938 Thayer Himiyar St., East Ring Rd, Sana'a)
Ambassador: Nicholas Hopton.

Of Yemen in the USA (2319 Wyoming Ave., NW, Washington, D.C., 20008)
Ambassador: Vacant.
Chargé d'Affaires a.i.: Adel Ali Ahmed Alsunaini.

Of the USA in Yemen (Sa'awan St., Himyar Zone, Sana'a)
Ambassador: Gerald M. Feierstein.

Of Yemen to the United Nations
Ambassador: Jamal Abdullah al-Sallal.

Of Yemen to the European Union
Ambassador: Abdulwahab Mohammed Alshawkani.

FURTHER READING
Central Statistical Organization. *Statistical Year Book*

Al-Rasheed, Madawi and Vitalis, Robert, (eds.) *Counter-Narratives: History, Contemporary Society, and Politics in Saudi Arabia and Yemen.* 2004

Bruck, Gabriele vom, *Islam, Memory and Morality in Yemen: Ruling Families in Transition.* 2005

Clark, Victoria, *Yemen: Dancing on the Heads of Snakes.* 2010

Dresch, Paul, *A History of Modern Yemen.* 2001

Leach, Hugh, *Seen in the Yemen: Travelling with Freya Stark and Others.* 2011

Mackintosh-Smith, T., *Yemen—Travels in Dictionary Land.* 1997

Manea, Elham, *Regional Politics in the Gulf: Saudi Arabia, Oman and Yemen.* 2005

National Statistical Office: Central Statistical Organization, Ministry of Planning and International Co-operation.

Website: http://www.cso-yemen.org

ZAMBIA

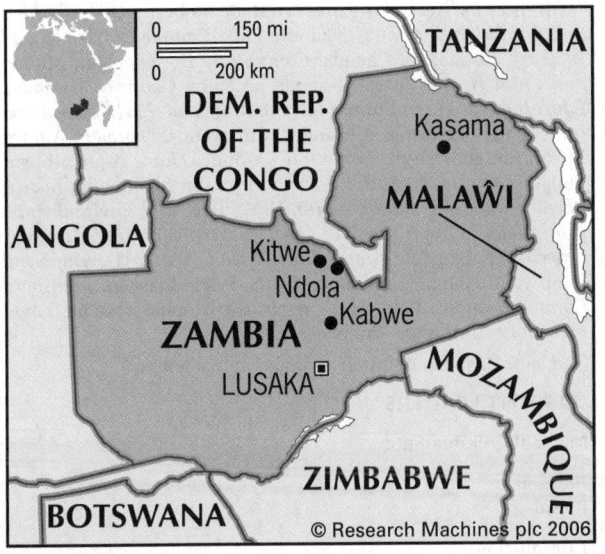

© Research Machines plc 2006

Republic of Zambia

Capital: Lusaka
Population projection, 2015: 15·24m.
GNI per capita, 2011: (PPP$) 1,254
HDI/world rank: 0·430/164
Internet domain extension: .zm

KEY HISTORICAL EVENTS

The earliest known inhabitants were nomadic bushmen. From the 4th century AD, Bantu tribes farmed the region and established villages. Copper was mined for weapons and tools and from the 11th century, trade developed with neighbouring regions in copper and textiles. From 1500–1900 the region was divided into four tribal kingdoms: the Kazembe-Lunda in the north, the Bemba in the northeast, the Chewa in the east and the Barotse, later known as the Lozi, in the west. An inland region, it was not penetrated by non-Africans until the late 18th century, when Portuguese traders arrived near Lake Mweru. The Scottish explorer David Livingstone followed in the mid-19th century, and in the 1880s the British colonialist and mining magnate, Cecil Rhodes, arrived.

As part of his expansion north from the Transvaal, Rhodes persuaded the British government to secure Bechuanaland in 1885, and in 1888 he obtained mining rights in the territory of the Ndebele, which reached to the Zambezi river. In 1890 the chief of the Lozi, Lewanika, granted Rhodes mining rights in Lozi territory (known as Barotseland) in return for British protection. In 1891 the British government granted Rhodes' British South Africa Company a charter to administer the area from the Zambezi river to Lake Tanganyika.

From 1900 the territory was administered as two separate protectorates, Northwestern and Northeastern Rhodesia. Lead, zinc and copper were discovered, attracting more European prospectors. In 1911 the two protectorates were merged to become Northern Rhodesia and in 1924 the British Crown took over direct administration, while the British South Africa Company retained mineral rights. The territory was governed through a legislative council whose members were elected by the European population. The detection of rich copper seams—known as the Copper Belt—led to a huge increase in mining

activity. The European population grow tenfold to 40,000 over the next three decades. Meanwhile, the African majority (98% of the country) was increasingly voicing its opposition to European rule. In 1948 some African members were added to the legislative council. In the same year the country's first African political party, the Northern Rhodesian Congress, was established by members of welfare societies from mining and rural communities. Connected to the African National Congress (ANC), in 1951 it renamed itself the Northern Rhodesian African National Congress.

In 1953 the British government imposed a federation consisting of Northern Rhodesia, its southern self-governing neighbour Rhodesia and Nyasaland. The Federation was welcomed by most Europeans in the territories as a means of consolidating their economic and political control but was opposed by most Africans for the same reason. Political dissent grew, with each region pressing its own claims for independence. In 1958 Kenneth Kaunda led a breakaway group from the Northern Rhodesian ANC to found the more radical Zambian African National Congress. After he was jailed for leading a civil disobedience campaign, the United National Independence Party (UNIP) was formed and elected Kaunda its leader on his release in Jan. 1960. Membership grew rapidly, convincing the British to accept it as a negotiating partner. At a London conference in Dec. 1960, Kaunda and other UNIP members secured a timetable for independence.

Elections in 1962 resulted in a UNIP–ANC coalition. The Federation was dissolved on 31 Dec. 1963 and in general elections held that year, with universal adult suffrage, UNIP won a clear victory. On 24 Oct. 1964 the country declared independence as the Republic of Zambia, named after the Zambezi river. The government took over mineral rights and oversaw a period of prosperity until copper prices collapsed in the 1970s. Further difficulties were caused by UN-imposed sanctions on Rhodesia—a key trading partner for Zambia—and rising oil prices, so that by the late 1970s Zambia was greatly indebted to the IMF.

In response to domestic dissent, Kaunda brought in a series of centralizing and authoritarian measures, including a new constitution making Zambia a one-party state. Living standards fell as copper production, Zambia's biggest foreign exchange earner, almost halved. Food riots and widespread unrest resulted in the 1991 electoral victory of the Movement for Multi-Party Democracy (MMD). However, allegations of corruption dogged the MMD government throughout successive terms, while the economy remained weak. Privatization of the copper mines in 2000 and other market-orientated reforms in the early 2000s reduced inflation but food security remained a serious issue. Failed harvests in 2001 resulted in the country receiving food aid.

TERRITORY AND POPULATION

Zambia is bounded by the Democratic Republic of the Congo in the north, Tanzania in the northeast, Malawi in the east, Mozambique in the southeast, Zimbabwe and Namibia in the south, and by Angola in the west. The area is 752,612 sq. km (290,584 sq. miles). Population (2010 census), 13,092,666; population density, 17·4 per sq. km. In 2011, 35·9% of the population were urban.

The UN gives a projected population for 2015 of 15·24m.

The republic is divided into ten provinces. Area, population at the 2010 census and chief towns:

Province	Area (in sq. km)	Population	Chief Town
Central	94,394	1,307,111	Kabwe
Copperbelt	31,328	1,972,317	Ndola
Eastern	51,620	1,592,661	Chipata
Luapula	50,567	991,927	Mansa

Province	Area (in sq. km)	Population	Chief Town
Lusaka	21,896	2,191,225	Lusaka
Muchinga[1]	87,410	711,657	Chinsali
Northern	77,900	1,105,824	Kasama
North-Western	125,826	727,044	Solwezi
Southern	85,283	1,589,926	Livingstone
Western	126,386	902,974	Mongu

[1]Muchinga Province, consisting of districts from Northern and Eastern provinces, was established in 2011.

The capital is Lusaka, which had a census population in 2010 of 1,460,566. Other major towns (with 2010 census population in 1,000) are: Kitwe, 548; Ndola, 495; Kabwe, 215; Chingola, 178; Mufulira, 141; Livingstone 134; Luanshya, 132.

The population consists of over 70 Bantu-speaking ethnic groups, with the main groups being the Bemba (18%), Tonga (10%), Nyanja (8%) and Lozi (6%). The official language is English.

SOCIAL STATISTICS

Estimates, 2008: births, 541,000; deaths, 218,000. Estimated birth rate in 2008 was 42·9 per 1,000 population; estimated death rate, 17·3. Zambia's life expectancy at birth in 2007 was 44·0 years for males and 45·0 for females. Life expectancy was declining for many years, largely owing to the huge number of people in the country with HIV, although it has now begun to rise again slowly. In 2009, 13·5% of all adults between 15 and 49 were infected with HIV. Annual population growth rate, 2000–08, 2·3%. Infant mortality, 2010, 69 per 1,000 live births; fertility rate, 2008, 5·8 births per woman.

CLIMATE

The climate is tropical, but has three seasons. The cool, dry one is from May to Aug., a hot dry one follows until Nov., when the wet season commences. Frosts may occur in some areas in the cool season. Lusaka, Jan. 70°F (21·1°C), July 61°F (16·1°C). Annual rainfall 33" (836 mm). Livingstone, Jan. 75°F (23·9°C), July 61°F (16·1°C). Annual rainfall 27" (673 mm). Ndola, Jan. 70°F (21·1°C), July 59°F (15°C). Annual rainfall 52" (1,293 mm).

CONSTITUTION AND GOVERNMENT

Zambia has a unicameral legislature. In 2009 the number of seats in the *National Assembly* was increased from 159 to 280 with effect from the 2011 elections, with 240 members elected for a five-year term in single-member constituencies, 30 members elected by proportional representation and ten members appointed by the president. Candidates for election as president must have both parents born in Zambia (this excludes ex-president Kaunda). The constitution was adopted on 24 Aug. 1991 and was amended in 1996, shortly before the parliamentary and presidential elections. The amendment restricts the president from serving more than two terms of office.

National Anthem

'Lumbanyeni Zambia' ('Stand and Sing of Zambia'); words collective, tune by M. E. Sontonga.

RECENT ELECTIONS

Presidential elections took place on 20 Sept. 2011. Michael Sata (Patriotic Front/PF) won 42·0% of the vote, Rupiah Banda (Movement for Multi-Party Democracy) 35·4% and Hakainde Hichilema (United Party for National Development) 18·2%.

In the parliamentary elections held on the same day the Patriotic Front gained 61 of the 150 elected seats in the National Assembly; the Movement for Multi-Party Democracy took 55, the United Party for National Development 29, the Alliance for Democracy and Development 1 and the Forum for Democracy and Development also 1. Three seats went to independents and two were vacant. Turnout was 54·0%.

CURRENT GOVERNMENT

President: Michael Sata; b. 1937 (PF; since 23 Sept. 2011).

In March 2013 the government comprised:

Vice-President: Guy Scott.

Minister for Agriculture and Livestock: Robert Sichinga. *Chiefs and Traditional Affairs:* Nkandu Luo. *Commerce, Trade and Industry:* Emmanuel Chenda. *Community Development, Mother and Child Health:* Joseph Katema. *Defence:* Geoffrey Mwamba. *Education, Science, Vocational Training and Early Education:* John Phiri. *Finance:* Alexander Chikwanda. *Foreign Affairs:* Effron Lungu. *Health:* Dr Joseph Kasonde. *Home Affairs:* Edgar Lungu. *Labour and Social Security:* Fackson Shamenda. *Justice:* Wynter Kabimba. *Lands, Natural Resources and Environmental Protection:* Wilber Simusa. *Local Government and Housing:* Emerine Kabanshi. *Mines, Energy and Water Development:* Yamfwa Mukanga. *Tourism and Arts:* Sylvia Masebo. *Transport, Communications, Works and Supply:* Christopher Yaluma. *Youth and Sport:* Chishimba Kambwili.

Zambian Parliament: http://www.parliament.gov.zm

CURRENT LEADERS

Michael Chilufya Sata

Position
President

Introduction
Michael Sata has been head of state since 23 Sept. 2011. He appointed Guy Scott as his vice-president a week after taking office.

Early Life
Born in 1937, Sata was raised in Mpika in the Northern Province of what was then Northern Rhodesia. Before entering politics he worked as a policeman, railwayman and trade unionist. After Zambia won independence in 1964 he became a member of the ruling United National Independence Party until joining the Movement for Multi-Party Democracy (MMD) in 1991. In the 1990s he variously served as minister for local government, health and labour, as well as minister without portfolio.

When Levy Mwanawasa was chosen as the MMD's 2001 presidential candidate, Sata left to form the Patriotic Front. However, he made little impact at the polls and in parliamentary elections his new party won only a single seat. He stood for the presidency in 2006, finishing second behind Mwanawasa who secured a second term. Sata was arrested for making a false declaration of his assets when registering for the campaign but the charge was dropped. In 2007 he was deported on arrival in neighbouring Malawi on allegations that he was plotting a coup. Later the same year his passport was withdrawn by the Zambian authorities, who accused him of bypassing regulations.

After an apparent reconciliation with President Mwanawasa in May 2008, Sata suffered a heart attack. When Mwanawasa died in Aug. that year, Sata was banned from attending the funeral. On 30 Aug. 2008 he was unanimously chosen as the Patriotic Front's presidential candidate but lost to Rupiah Banda of the MMD.

At the presidential poll on 23 Sept. 2011, Sata gained 42% of the vote against Banda's 35% after a campaign marred by violence. Sata was sworn in to office in Lusaka within hours of the results being declared. His victory ended two decades of MMD rule.

Career in Office
Having previously expressed his admiration for Robert Mugabe, Sata (known as 'King Cobra' for his fiery rhetoric) was accused of authoritarianism. He in turn accused the electoral commission of plotting to rig the election with pre-marked ballots.

Sata campaigned on promises to help the nation's poor and fight corruption. He pledged to reinstate a 25% windfall tax on mining revenues abolished by the MMD in 2009. Nonetheless, the kwacha dropped to a 12-month low against the US dollar following Sata's victory. He has encouraged foreign investment while demanding improved working conditions, having criticized the employment practices of companies from China, a heavy investor in Zambia's mining industries.

DEFENCE

In 2008 defence expenditure totalled US$262m. (US$22 per capita), representing 1·8% of GDP.

Army
Strength (2007) 13,500. There are also two paramilitary police units totalling 1,400.

Air Force
In 2007 the Air Force had 29 combat-capable aircraft including F-6 (Chinese-built MiG-19s) and MiG-21s. Serviceability of most types is reported to be low. Personnel (2007) 1,600.

ECONOMY

In 2006 agriculture accounted for 20·9% of GDP, industry 32·9% and services 46·2%.

Overview
Zambia is among the world's poorest economies despite rich natural resources. Copper accounts for the majority of export earnings. The principal drivers of growth in the past decade have been strong macroeconomic performance and expansion of mining, construction and communications.

From 2001–09 GDP grew on average 5·4% per year but this has yet to translate into significant poverty reduction, with 59% of the population living below the poverty line. Zambia's aim of attaining middle-income status by 2030 requires average annual growth of 6–7%. Policy reforms, including a more-efficient pre-budget consultation process and increases in electricity tariffs, are expected to boost growth, but the economy remains heavily dependent on foreign aid and is vulnerable to world copper and oil price changes.

Currency
The unit of currency is the *kwacha* (ZMK) of 100 *ngwee*. Foreign exchange reserves were US$477m. in July 2005. In Dec. 1992 the official and free market exchange rates were merged and the kwacha devalued 29%. Inflation, which was 183·3% in 1993, was down to 9·0% in 2006. The rate rose slightly in the following years but fell again to 8·5% in 2010 and stood at 8·7% in 2011. Total money supply in June 2005 was 2,140·3bn. kwacha.

Budget
The fiscal year is the calendar year. Revenues in 2007 totalled 10,095bn. kwacha and expenditures 12,034bn. kwacha. Tax revenues accounted for 77·3% of total revenue in 2007; education accounted for 16·9% of total expenditure, economic affairs 14·1%, and housing and community amenities 12·2%.

VAT is 16%.

Performance
Real GDP growth was 6·4% in 2009, 7·6% in 2010 and 6·6% in 2011. Total GDP in 2011 was US$19·2bn.

Banking and Finance
The central bank is the Bank of Zambia (*Governor*, Michael Gondwe). In 2003 there were five commercial banks, six foreign banks and four development banks. The Bank of Zambia monitors and supervises the operations of financial institutions. Banks and building societies are governed by the Banking and Financial Services Act 1994. Foreign debt totalled US$3,689m. in 2010, representing 25·8% of GNI.

There is a stock exchange in Lusaka.

ENERGY AND NATURAL RESOURCES

Environment
Zambia's carbon dioxide emissions from the consumption and flaring of fossil fuels in 2008 were the equivalent of 0·3 tonnes per capita.

Electricity
Installed capacity in 2007 was 1·8m. kW. Production in 2007 was 9·85bn. kWh, almost exclusively hydro-electric; consumption per capita was 788 kWh.

Oil and Gas
Oil and gas were both discovered in the northwest of Zambia in 2006. Exploration licences have been issued both to local and to foreign companies.

Minerals
Minerals produced (in 1,000 tonnes): copper (2009), 697; cobalt (2009), 1·5; silver (2009 estimate), 6,000 kg; gold (2009 estimate), 3,100 kg. Zambia is well-endowed with gemstones, especially emeralds, amethysts, aquamarine, tourmaline and garnets. Zambia Consolidated Copper Mines, privatized in 2000, is the country's largest employer. In 2007, 276,000 tonnes of coal were produced.

Agriculture
70% of the population is dependent on agriculture. There were an estimated 5·26m. ha. of arable land in 2007 and 29,000 ha. of permanent crops. Principal agricultural products (2003 estimates, in 1,000 tonnes): sugarcane, 1,800; maize, 1,161; wheat, 100; seed cotton, 62; sweet potatoes, 53; groundnuts, 42.

Livestock (2003 estimates): cattle, 2·6m.; goats, 1·3m.; pigs, 340,000; sheep, 150,000; chickens, 30m.

Forestry
Forests covered 49·47m. ha. in 2010, or 67% of the total land area. Timber production in 2007 was 10·03m. cu. metres, most of it for fuel.

Fisheries
Total catch, 2010, 76,396 tonnes (exclusively from inland waters).

INDUSTRY

In 2006 industry accounted for 32·9% of GDP, with manufacturing contributing 10·8%. Industrial production grew by 9·1% in 2006. Zambia's economy is totally dependent upon its mining sector. Other industries include construction, foodstuffs, beverages, chemicals and textiles.

Labour
The labour force totalled 3,165,200 in 2000 (59% males). 71·6% of the economically active population in 2000 were engaged in agriculture, 7·5% in community services and 6·8% in trade.

INTERNATIONAL TRADE

Imports and Exports
In 2007 imports were valued at US$3,971·1m. and exports at US$4,618·6m.

Imports declined every year from 1997 to 1999 before increasing by 12% in 2000, and exports declined every year from 1997 to 2000, However, between 2003 and 2007 there was a 150% increase in imports and exports increased by 370%. In 2007 copper provided 71% of all exports (by value). Since 1990 non-copper exports have increased in value from US$50m. to US$1·1bn.

The main import sources in 2007 were South Africa (47·4%), United Arab Emirates (6·4%), China (5·9%) and India (4·1%). Principal export markets were Switzerland (41·8%), South Africa (12·0%), Thailand (5·9%) and the Democratic Republic of the Congo (5·3%).

COMMUNICATIONS

Roads
There were, in 2001, 91,440 km of roads, including 4,222 km of highway. 131,100 passenger cars were in use in 2007 and there were 75,500 trucks and vans.

Rail
In 2005 there were 1,271 km of the state-owned Zambia Railways (ZR) and 891 km of the Tanzania-Zambia (Tazara) Railway, both on 1,067 mm gauge. A 27-km stretch of railway linking Chipata in the east of the country with Mchinji in Malawi was opened in Aug. 2010. This links with the existing railway to Nacala, one of Mozambique's leading ports.

Civil Aviation
The former flag carrier, Zambian Airways, operated internal flights and in 2007 flew to Dar es Salaam, Harare and Johannesburg as well as operating domestic services, but ceased flying in Jan. 2009. Lusaka is the principal international airport. In 2001 Lusaka International handled 410,000 passengers (359,000 on international flights) and 24,800 tonnes of freight. In 2003 scheduled airline traffic of Zambian-based carriers flew 2m. km, carrying 45,000 passengers (17,000 on international flights).

Telecommunications
In 2008 there were 90,600 main (fixed) telephone lines; mobile phone subscribers numbered 3,539,000 in 2008 (28·0 per 100 persons). Telecel (2) Ltd. has been licensed to run a mobile telecommunications service in addition to the Zambia Telecommunications Company (ZAMTEL) since 1996. Internet services are provided by Zambia Communications Systems (ZAMNET), a private company of ZAMTEL. There were some 700,000 internet users in 2008. In June 2012 there were 236,000 Facebook users.

SOCIAL INSTITUTIONS

Justice
The Judiciary consists of the Supreme Court, the High Court and four classes of magistrates' courts; all have civil and criminal jurisdiction.

The Supreme Court hears and determines appeals from the High Court. Its seat is at Lusaka. The High Court exercises the powers vested in the High Court in England, subject to the High Court ordinance of Zambia. Its sessions are held where occasion requires, mostly at Lusaka and Ndola. All criminal cases tried by subordinate courts are subject to revision by the High Court.

The death penalty is authorized, the last execution having taken place in 1997. The population in penal institutions in mid-2009 was 15,544 (120 per 100,000 of national population).

Education
Schooling is for nine years. In April 2002 then President Mwanawasa announced the reintroduction of universal free primary education, abolished under former President Chiluba. In 2007 there were 2,790,312 pupils in primary schools with 56,557 teaching staff and 607,296 pupils in secondary schools with 14,246 teaching staff.

There are three universities and a National Institute of Public Administration. The University of Zambia, at Lusaka, was founded in 1965 and had 10,122 students in 2007. Copperbelt University, at Kitwe, founded in 1987, had 4,273 students in 2007. Mulungushi University, near Kabwe, was founded in 2008.

The adult literacy rate in 2009 was estimated at 70·9% (80·6% among males and 61·3% among females).

In 2008 public expenditure on education came to 1·3% of GDP.

Health
In 2004 there were 1,264 physicians, 19,014 nurses, 2,996 midwives, 491 dentists and 1,039 pharmacists. There were 88 public hospitals in 2004, with a total of 22,800 beds (207 per 100,000 population).

RELIGION
In 1993 the president declared Zambia to be a Christian nation, but freedom of worship is a constitutional right. In 2001 there were 3·89m. Christians. In Feb. 2013 there was one Roman Catholic cardinal. Traditional beliefs are also widespread.

CULTURE

World Heritage Sites
Zambia shares one site with Zimbabwe on the UNESCO World Heritage List: the Victoria Falls/Mosi-oa-Tunya (inscribed on the list in 1989), waterfalls on the Zambezi River.

Press
In 2008 there were three paid-for daily papers, *The Post*, the *Times of Zambia* and the *Zambia Daily Mail*. *The Post* is privately owned and the *Times of Zambia* and the *Zambia Daily Mail* state-owned.

Tourism
In 2005, 668,862 foreign tourists visited Zambia. Most visitors in 2005 were from Zimbabwe (148,436), followed by South Africa (110,272), Tanzania (65,881) and the UK (44,369). Spending by tourists in 2005 totalled US$98m.

Festivals
The N'cwala ceremony is held in Feb. by the Ngoni people to commemorate their arrival in Zambia in 1835 and the first produce of the year. The Kuomboka, in Feb. or March, is the canoe procession of the Lozi chief and his family from the palace at Lealui down the Zambezi to Limulunga for the rainy season. The National Fishing Competition is held at Lake Tanganyika in March. Likumbi Lya Mize, held at Mize in July, celebrates the Luvale tribe's cultural heritage. The Livingstone Cultural and Arts Festival, held annually in Sept. since 1994, brings together many of Zambia's tribes and their traditional rulers. Independence Day is celebrated on 24 Oct.

DIPLOMATIC REPRESENTATIVES
Of Zambia in the United Kingdom (2 Palace Gate, London, W8 5NG)
High Commissioner: Prof. Royson Mukwena.

Of the United Kingdom in Zambia (5210 Independence Ave., 15101 Ridgeway, Lusaka)
High Commissioner: James Thornton.

Of Zambia in the USA (2419 Massachusetts Ave., NW, Washington, D.C., 20008)
Ambassador: Palan Mulonda.

Of the USA in Zambia (Eastern end of Kabulonga Rd, Ibex Hill, PO Box 31617, Lusaka)
Ambassador: Mark C. Storella.

Of Zambia to the United Nations
Ambassador: Patricia Mwaba Kasese-Bota.

Of Zambia to the European Union
Ambassador: Irene Mumba Kamanga.

FURTHER READING

Chiluba, F., *Democracy: the Challenge of Change.* 1995
Sardanis, Andrew, *Africa: Another Side of the Coin: Northern Rhodesia's Final Years and Zambia's Nationhood.* 2003
Simon, David J., Pletcher, James R. and Siegel, Brian V., *Historical Dictionary of Zambia.* 2008

Central Statistical Office. *Monthly Digest of Statistics.*
National Statistical Office: Central Statistical Office, PO Box 31908, Lusaka.
Website: http://www.zamstats.gov.zm

ZIMBABWE

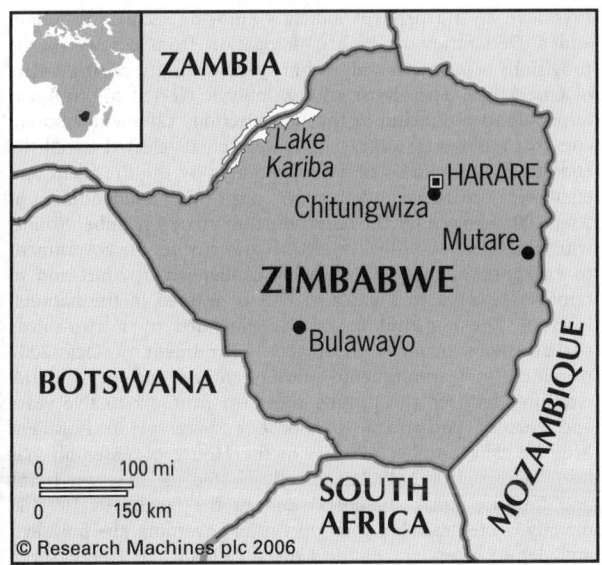

Republic of Zimbabwe

Capital: Harare
Population projection, 2015: 14·00m.
GNI per capita, 2011: (PPP$) 376
HDI/world rank: 0·376/173
Internet domain extension: .zw

KEY HISTORICAL EVENTS

Archaeological evidence shows human settlement dating back several thousand years. The Khoisan people were early inhabitants of the region, followed from around AD 500 by the Bantu-speaking Gokomere. Trading civilizations flourished from the 9th century, culminating in the Mwene Mutapa Empire (Empire of Great Zimbabwe) from the 15th century. Its stronghold was a fortified stone town known as Great Zimbabwe ('houses of stone'), which was founded around 1000 and had a population of up to 18,000 at its peak. During the 16th and 17th centuries the area came under partial control by the Portuguese until the Shona people defeated them in 1693 and established the Rozwi Empire. This fell to the migrating Ndebele (Matabele) in 1834.

King Lobengula of Matabeleland signed the Rudd Concession on 13 Oct. 1888, giving the British mining rights. The British South Africa Company, under Cecil Rhodes, colonized the territory from 1890. European immigration increased after white settlers won the first Matabele war in 1893–94. Following unsuccessful Ndebele and Shona uprisings in 1896 and 1897, indigenous people were increasingly displaced from their land. In 1911 the region was divided into Southern and Northern Rhodesia (*see* ZAMBIA: Key Historical Events) and in 1923 became a self-governing colony. In 1953 Southern and Northern Rhodesia joined with Nyasaland to form the Federation of Rhodesia and Nyasaland, a move opposed by African nationalists who feared it would entrench white minority rule. The South Rhodesian African National Congress was founded in 1957 and campaigned for self-rule by the African majority until it was banned in 1960. After unrest mounted in Nyasaland and neighbouring states, the federation was dissolved on 31 Dec. 1963

and Southern Rhodesia reverted to self-governing colony status within the British Commonwealth under the name of Rhodesia.

On 11 Nov. 1965 Rhodesia's white-dominated government issued a unilateral declaration of independence (UDI) to forestall any British attempt to grant black majority rule. The British governor dismissed the prime minister, Ian Smith, and his cabinet, and the British government reasserted its formal responsibility for Rhodesia although the Smith cabinet was effectively left to run the country. Britain imposed economic sanctions, as did the UN in 1968. On 2 March 1970 the Smith government declared Rhodesia a republic and adopted a new constitution. From 1966–79 the Zimbabwe African People's Union (ZAPU) and its off-shoot, the Zimbabwe African National Union (ZANU)—both of which were banned—led a sporadic guerrilla campaign. On 3 March 1978 Smith signed a constitutional agreement with the internationally-backed nationalist leaders. A draft constitution was published in Jan. 1979 and accepted by the white electorate in a referendum. Following the Commonwealth Conference in Lusaka in Aug. 1979, elections in March 1980 resulted in victory for ZANU. Southern Rhodesia became the Republic of Zimbabwe on 18 April 1980.

Political conflict between ZANU and ZAPU supporters in Matabeleland led to allegations of atrocities by government forces in 1983–84. Land reform—the redistribution of good farming land from the minority white population to the African population—became a pressing issue. President Mugabe's government bought 3·5m. ha. of land from white farmers, with the UK funding the £44m. bill. However, only 70,000 families benefited amid allegations of government profiteering. In early 2000 a policy of land occupation began, with black settlers taking over white-owned farms by force. President Mugabe ignored international pressure to restore the rule of law and violence escalated. During campaigning for the 2002 presidential election, which failed to meet international democratic standards, the opposition leader, Morgan Tsvangirai, was charged with treason. Zimbabwe was suspended from the Commonwealth in March 2002 over the conduct of the election and when the suspension was extended in 2003 it pulled out altogether.

In the early 2000s agricultural production fell drastically, the result of drought and of the continuing land seizures. In Sept. 2005 President Mugabe nationalized all land and ended the right of landowners to challenge government expropriations in the courts. In May 2005 the government began the mass demolition of urban slums, claiming it would improve law and order. Around 700,000 people were left homeless. The conduct of parliamentary and senate elections in 2005, which resulted in victory for President Mugabe's party, was criticized by the international community and in March 2007 evidence that Morgan Tsvangirai had been tortured by police caused an international outcry. In 2006 and 2007 the population experienced mass hunger, with UNICEF estimating that 4m. needed aid. The country's AIDS pandemic continued to worsen in 2007, with 24% of children under 17 orphaned by the disease.

In 2008 elections held against a backdrop of hyperinflation, Mugabe's ZANU-PF lost their parliamentary majority—to the Movement for Democratic Change (MDC)—for the first time since independence. Mugabe retained the presidency when Morgan Tsvangirai, leader of the MDC, pulled out of a run-off, claiming his supporters had suffered intimidation at the hands of Mugabe's supporters. An international outcry forced Mugabe into talks with Tsvangirai that resulted in a power-sharing agreement in Sept. 2008. The agreement faltered over the allocation of ministries until the composition of the new cabinet was finalized in Feb. 2009. By the end of 2008 millions of civilians relied on food relief, while a cholera outbreak had claimed over 2,000 lives by Jan. 2009, with 1,500 new cases a day being reported.

TERRITORY AND POPULATION

Zimbabwe is bounded in the north by Zambia, east by Mozambique, south by South Africa and west by Botswana and the Caprivi Strip of Namibia. The area is 390,757 sq. km (150,871 sq. miles). The 2002 census population was 11,631,657 (5,997,477 female). Provisional population at the 2012 census, 12,973,808; density, 33·2 per sq. km. In 2011, 38·8% of the population were urban. Although the population is still rising, it has done so at a lower rate than had been previously expected, partly owing to the large number of AIDS-related deaths and partly as a result of over 3m. Zimbabweans having left the country as President Mugabe's grip on power strengthened, with the vast majority choosing to live in South Africa. However, since 2010 the growth rate has accelerated again.

The UN gives a projected population for 2015 of 14·00m.

There are eight provinces and two cities, Harare and Bulawayo, with provincial status. Area and population (2012 census provisional):

	Area (sq. km)	Population
Bulawayo	479	655,675
Harare	872	2,098,199
Manicaland	36,459	1,755,000
Mashonaland Central	28,347	1,139,940
Mashonaland East	32,230	1,337,059
Mashonaland West	57,441	1,449,938
Masvingo	56,566	1,486,604
Matabeleland North	75,025	743,871
Matebeleland South	54,172	685,046
Midlands	49,166	1,662,476

Harare, the capital, had a provisional population in 2012 of 1,468,767. Other main cities (with 2012 provisional census populations) were Bulawayo (655,675), Chitungwiza (354,472), Mutare (188,243) and Epworth (161,840l). The population is approximately 98% African, 1% mixed and Asian and there are around 70,000 whites. The main ethno-linguistic groups are the Shona (71%), Ndebele (16%), Ndau (3%) and Nyanja (3%). Other smaller ones include Kalanga, Manyika, Tonga and Lozi.

The official language is English.

SOCIAL STATISTICS

2008 estimates: births, 373,000; deaths, 199,000. Rates (2008 estimates, per 1,000 population); birth, 29·9; death, 16·0. Annual population growth rate, 2000–08, 0·0%. Zimbabwe's expectation of life at birth in 2007 was 43·6 years for females and 42·6 for males, down from an average of 54 years in 1993. The sharp decline is largely attributed to the huge number of people in the country with HIV. In 2009, 14·3% of all adults between 15 and 49 were infected with HIV. Zimbabwe had both the lowest overall life expectancy of any country in 2007 and the lowest for males, although Afghanistan's life expectancy for females was lower. Infant mortality, 2010, 51 per 1,000 live births; fertility rate, 2008, 3·4 births per woman.

CLIMATE

Though situated in the tropics, conditions are remarkably temperate throughout the year because of altitude, and an inland position keeps humidity low. The warmest weather occurs in the three months before the main rainy season, which starts in Nov. and lasts until March. The cool season is from mid-May to mid-Aug. and, though days are mild and sunny, nights are chilly. Harare, Jan. 69°F (20·6°C), July 57°F (13·9°C). Annual rainfall 33" (828 mm). Bulawayo, Jan. 71°F (21·7°C), July 57°F (13·9°C). Annual rainfall 24" (594 mm). Victoria Falls, Jan. 78°F (25·6°C), July 61°F (16·1°C). Annual rainfall 28" (710 mm).

CONSTITUTION AND GOVERNMENT

The 1979 Constitution, with 18 amendments, provides for a single-chamber 150-member Parliament (*House of Assembly*), universal suffrage for citizens over the age of 18, an executive *President*, an independent judiciary enjoying security of tenure and a Declaration of Rights, derogation from certain of the provisions being permitted, within specified limits, during a state of emergency. The House of Assembly is elected for five-year terms: up to and including the 2005 elections 120 members were elected by universal suffrage, ten were chiefs elected by all the country's tribal chiefs, 12 were appointed by the President and eight were provincial governors. A constitutional amendment of Aug. 2005 allowed for the reintroduction of a 66-member *Senate*, which had been abolished in 1987. It also enables the government to expropriate land without being challenged in court and to remove the right to a passport if it is deemed in the national interest. The constitution can be amended by a two-thirds parliamentary majority. A further amendment of Oct. 2007 provided for simultaneous parliamentary and presidential elections, reduced the presidential term from six to five years and enabled parliament to choose a successor to President Mugabe. The number of seats in the House of Assembly has been increased to 214 (210 directly elected for five-year terms and four *ex officio* members) and in the Senate to 100 (60 directly elected members, ten provincial governors, the president and deputy president of the Chiefs' Council, 16 chiefs, seven additional members and five senators appointed by the President).

In a referendum on 16 March 2013 on the adoption of a new constitution, 92·9% of votes cast were in favour. Supported by both ZANU-PF and the MDC, the constitution provides for a strengthened Bill of Rights and limits the president to two five-year terms (not to be implemented retrospectively). To become law, the draft constitution required two-thirds support in both the House of Assembly and the Senate, plus presidential approval. This process began in May 2013.

National Anthem

'Kalibusiswe Ilizwe leZimbabwe' ('Blessed be the Land of Zimbabwe'); words by Dr Solomon M. Mutswairo; tune by Fred Changundega.

GOVERNMENT CHRONOLOGY

Presidents since 1980. (ZANU = Zimbabwe African National Union; ZANU-PF = Zimbabwe African National Union-Patriotic Front)

| 1980–87 | ZANU | Canaan Sodindo Banana |
| 1987– | ZANU-PF | Robert Gabriel Mugabe |

Prime Ministers since 1980. (MDC-T = Movement for Democratic Change-Tsvangirai; ZANU = Zimbabwe African National Union)

| 1980–87 | ZANU | Robert Gabriel Mugabe |
| 2009– | MDC-T | Morgan Richard Tsvangirai |

RECENT ELECTIONS

Parliamentary and presidential elections took place on 29 March 2008. Of the 207 seats contested in the House of Assembly the Movement for Democratic Change-Tsvangirai (MDC-T) won 99 seats, against 97 for the Zimbabwe African National Union-Patriotic Front (ZANU-PF). This was the first time that ZANU-PF, the party of incumbent president Robert Mugabe, had lost its legislative majority since the country gained independence in 1980. Of the remaining seats, ten were claimed by the Movement for Democratic Change-Mutambara (MDC-M), a splinter faction of the MDC, and one by an independent candidate. In the Senate elections, the 60 available seats were divided equally between the

ZANU-PF (30 seats) and the combined opposition of the MDC-T (24) and the MDC-M (6).

The results of the presidential election were contested. On 13 April the Zimbabwe Electoral Commission announced a full recount of both parliamentary and presidential votes in 23 constituencies. Following the recount it was confirmed that Morgan Tsvangirai, leader of the MDC-T, had won the highest proportion of the votes, but had not gained the absolute majority necessary to avoid a second round run-off against incumbent president Robert Mugabe. On 22 June Tsvangirai withdrew from the second round citing a campaign of violence against the MDC's supporters. Mugabe won the second round on 27 June with 85·5% of votes cast, against 9·3% for Tsvangirai (as his name was still on the ballot). Under international pressure, Mugabe entered into talks with Tsvangirai, resulting in a power-sharing deal in Sept. 2008 that saw Mugabe remain president while Tsvangirai eventually became prime minister in Feb. 2009.

CURRENT GOVERNMENT

Executive President: Robert G. Mugabe; b. 1924 (ZANU-PF; sworn in 30 Dec. 1987, having previously been prime minister from 1980 to 1987; re-elected in April 1990, March 1996, March 2002 and June 2008).

Vice-President: Joyce Mujuru (ZANU-PF).

In March 2013 a coalition government was formed that in Oct. 2012 comprised:

Prime Minister: Morgan Tsvangirai; b. 1952 (MDC-T; sworn in 11 Feb. 2009).

Deputy Prime Ministers: Arthur Mutambara (MDC-M); Thokozani Khuphe (MDC-T).

Minister for Agriculture, Mechanization and Irrigation Development: Joseph Mtakwese Made (ZANU-PF). *Constitutional and Parliamentary Affairs:* Eric Matinenga (MDC-T). *Defence:* Emmerson Mnangagwa (ZANU-PF). *Economic Planning and Investment Promotion:* Tapiwa Mashakada (MDC-T). *Education, Sport, Art and Culture:* David Coltart (MDC-M). *Energy and Power Development:* Elton Mangoma (MDC-T). *Environment and Natural Management:* Francis Nhema (ZANU-PF). *Finance:* Tendai Biti (MDC-T). *Foreign Affairs:* Simbarashe Mumbengegwi (ZANU-PF). *Health and Child Welfare:* Dr Henry Madzorera (MDC-T). *Higher and Tertiary Education:* Vacant. *Home Affairs:* Theresa Makone (MDC-T); Kembo Mohadi (ZANU-PF). *Industry and Commerce:* Welshman Ncube (MDC-M). *Information Communication Technology:* Nelson Chamisa (MDC-T). *Justice and Legal Affairs:* Patrick Chinamasa (ZANU-PF). *Labour and Social Welfare:* Paurina Mpariwa (MDC-T). *Lands and Land Resettlement:* Herbert Murerwa (ZANU-PF). *Local Government, and Urban and Rural Development:* Ignatius Chombo (ZANU-PF). *Media, Information and Publicity:* Webster Shamu (ZANU-PF). *Mines and Mining Development:* Obert Mpofu (ZANU-PF). *National Housing and Social Amenities:* Giles Mutsekwa (MDC-T). *Public Service:* Lucia Matibenga (MDC-T). *Public Works:* Joel Gabuza Gabuza (MDC-T). *Regional Integration and International Co-operation:* Priscilla Misihairabwi-Mushonga (MDC-M). *Science and Technology Development:* Henry Dzinotyiwei (MDC-T). *Small and Medium Enterprises, and Co-operative Development:* Sithembiso Nyoni (ZANU-PF). *State Enterprise and Parastatals:* Gorden Moyo (MDC-T). *Tourism and Hospitality Industry:* Walter Mzembi (ZANU-PF). *Transport and Infrastructural Development:* Nicholas Goche (ZANU-PF). *Water Resources Development and Management:* Samuel Nkomo (MDC-T). *Women's Affairs, Gender and Community Development:* Olivia Muchena (ZANU-PF). *Youth Development, Indigenization and Empowerment:* Savior Kasukuwere (ZANU-PF).

Speaker: Lovemore Moyo (MDC-T).

Government Website: http://www.zim.gov.zw

CURRENT LEADERS

Robert Mugabe

Position
President

Introduction
Robert Mugabe came to power as newly-independent Zimbabwe's (formerly Rhodesia) first prime minister in 1980, becoming president in 1987. Although initially hailed as a democratic reformer, his economic mismanagement of the country, violent electoral campaigns and controversial programme of land seizures have tarnished his image at home and abroad. He has defended his land reform programme as the conclusion of the process of decolonization, but his policies have been widely perceived as short-term political expediency for the maintenance of personal power. In the wake of the disputed presidential elections in 2008, Mugabe conceded to power-sharing with the opposition Movement for Democratic Change (MDC) after protracted negotiations, but he has shown no inclination to relax his autocratic grip on the country.

Early Life
The son of a carpenter, Robert Gabriel Mugabe was born 21 Feb. 1924 at Kutama mission, northwest of Harare. After an early education at a Roman Catholic mission school, he studied at the University College of Fort Hare, South Africa, marking the beginning of an academic career boasting seven university degrees, three of which he completed during imprisonment. He worked as a primary school teacher in Ghana from 1956–60 when he returned to Rhodesia and joined Joshua Nkomo's Zimbabwe African People's Union (ZAPU). In 1963 he became a founding member of the breakaway Zimbabwe African National Union (ZANU) with Rev. Ndabaningi Sithole. A year later Mugabe was arrested for subversion and imprisoned, without trial, for ten years. Despite imprisonment, he remained politically active and was able to orchestrate, in 1974, a coup against Sithole to become party leader. In 1975, freed from prison, Mugabe joined Nkomo as joint leader of the Patriotic Front of Zimbabwe which waged a guerrilla war against Ian Smith's white Rhodesian Front government. In 1980 independence was achieved and parliamentary elections took place in which Mugabe, at the head of ZANU, won a landslide victory to become prime minister.

Career in Office
In office Mugabe appeared set to usher in a bright new era for the country. Having built a coalition government with ZAPU, he adopted a conciliatory stance towards the white, landowning minority. He introduced higher wages, credit programmes and food subsidies for poor farmers, a better infrastructure and equal land rights for women. Reform in the education system saw primary school enrolment trebled and secondary school enrolment increased five-fold during the first ten years of his rule (with Zimbabwe laying claim to the highest literacy rate of any African nation).

Troubles began in 1982 when ethnic turmoil between the Shona majority (represented by ZANU) and the Ndebele minority (represented by ZAPU) broke out after Mugabe dismissed Nkomo and ZAPU from government. The ensuing violence prompted much of the white population to emigrate, in turn creating an economic downturn. Centred in Matabeleland, the conflict drew international attention after the discovery of mass graves and alleged atrocities.

In 1987 Mugabe won the presidential elections and set about bringing ZAPU back into government. A unity agreement was signed and Nkomo became senior minister in a newly-formed Zimbabwe African National Union-Patriotic Front (ZANU-PF) government. Mugabe was again re-elected in 1990 (in polling marred by violence) and 1996, but throughout the 1990s he adopted a series of unpopular policies. His military support for President Kabila's beleaguered government in the Democratic Republic of the Congo led to strikes within his own country. An announcement of pay increases for himself and his party officials in 1998 prompted rioting, coming as it did amidst a growing economic crisis. Plans to raise food and fuel prices and to introduce a tax to support war veterans from the 1970s were blocked by trades unions and further diminished his popularity.

In Feb. 2000 Mugabe lost a referendum in which he sought to increase his presidential powers. Blaming the white minority for the defeat, he then targeted the issue of land ownership. A programme of violent land seizure followed, with black settlers taking over white-owned farms. A court order to halt the seizures was ignored and Mugabe subsequently replaced high court judges with political allies. In June 2000 parliamentary elections were held. Mugabe won the elections, but only by a narrow margin and after a campaign of intimidation which led to more than 30 deaths.

Mugabe was re-elected president in March 2002. Final results gave him 56·2% of the vote against 42·0% for opposition rival, Morgan Tsvangirai, the leader of the MDC. However, the elections failed to meet international democratic standards. They were preceded by violence against opposition supporters, the passing of a law limiting press freedom, the withdrawal of the European Union monitoring team and the arrest of Mugabe's main political rival on charges of treason. As a result, Zimbabwe was suspended from the Commonwealth and a range of targeted sanctions from the UK, the USA and the EU were placed on Mugabe and his cabinet. In March 2003 the USA froze Zimbabwean assets and forbade US citizens from undertaking economic dealings with Mugabe and his government colleagues.

The state of political uncertainty and violence following the 2000 referendum damaged investor confidence, causing export prices to decline and unemployment and food shortages to rise. Coupled with this, severe drought in early 2002 raised the threat of mass starvation. In April 2002 Mugabe declared a state of disaster, allowing him the temporary use of 'extraordinary measures' to cope with the situation. Although little was done in practical terms to relieve the threat of famine, Mugabe pushed ahead with the land redistribution programme. In June 2002 he ordered almost 3,000 white farmers to leave their land within 45 days, or face imprisonment. In Sept. 2002 new legislation was passed allowing farmers only a week's notice after receiving an eviction order. In March 2003 Amnesty International reported that up to 500 people had been arrested following a general strike, with members of the MDC especially targeted. In the same month the Commonwealth extended Zimbabwe's suspension until at least Dec. 2003. In June 2003 police detained Tsvangirai, who had called for mass popular protests against Mugabe's government, and in Jan. 2004 he went on trial for treason. Although he was acquitted in Oct. 2004 of charges relating to an assassination plot against Mugabe, he still faced a separate treason charge. Meanwhile, relations with the international community worsened. In Dec. 2003 the Commonwealth (despite South African disapproval) again extended Zimbabwe's suspension, prompting Mugabe's withdrawal from the organization.

At the parliamentary elections in March 2005, ZANU-PF took 78 of 150 seats. The MDC claimed that there had been widespread vote rigging and intimidation. Then, in May, Mugabe's government launched a demolition of urban slum dwellings and illegal settlements, including business premises, around the country without compensation. The policy drew international condemnation as an estimated 700,000 people (according to the UN) lost their homes, or source of livelihood, or both. In Aug. 2005 parliament approved amendments to the constitution, including the reintroduction of the Senate, which had been abolished in 1990. Other changes provided for the government to confiscate passports of those deemed to pose a threat to national security and to strengthen control over land redistribution with

no right of appeal. Also in Aug., the authorities dropped the remaining treason charge against Tsvangirai. ZANU-PF won the Nov. 2005 elections to the new Senate, securing an overwhelming majority of 66 seats amid low voter turnout and opposition calls for a boycott.

Repression of the MDC and wider opposition intensified from 2006, particularly in March 2007 when Tsvangirai was beaten by security forces and hospitalized after his arrest at a political rally. A meeting of regional leaders subsequently invited South Africa's president, Thabo Mbeki, to mediate in Zimbabwe's political and economic crisis. In Dec. 2007 Mugabe's presence at the EU–Africa summit in Lisbon provoked criticism of his regime's abuse of human rights.

Mugabe was endorsed as the ZANU-PF candidate for the March 2008 presidential (and parliamentary) elections. After a relatively peaceful campaign the MDC made a strong showing and ZANU-PF lost its legislative majority. Tsvangirai also claimed outright victory in the presidential race but Mugabe challenged the results. Despite international pressure, the electoral commission delayed the publication of results and a second round run-off was scheduled. There followed an orchestrated campaign of brutality against supporters of the opposition, which led to the withdrawal of Tsvangirai from the race in June. The international community was united in its condemnation of Mugabe's actions. The electoral crisis took place against a backdrop of economic meltdown.

Under international pressure, talks between Mugabe and Tsvangirai were brokered in Aug. 2008, which resulted in a deal the following month that saw Mugabe remain as president while Tsvangirai was to become executive prime minister. However, implementation of the agreement then stalled for several months over the allocation of cabinet posts between ZANU-PF and the MDC. The political deadlock was further exacerbated by the collapse of the economy and of basic services, which contributed to a serious outbreak of the disease cholera in Nov. 2008. In Jan. 2009, after months of acrimony and pressure from neighbouring states, Mugabe agreed to put power-sharing into effect and on 11 Feb. Tsvangirai was sworn in as prime minister. However, the failure of Mugabe's military and security service chiefs to attend the inauguration ceremony and the controversial detention of an MDC ministerial nominee did not represent an auspicious start for the new unity government.

In Sept. 2009 the EU sent its first high-level delegation to Zimbabwe for several years but refused to lift targeted sanctions. Donors were also cautious about releasing aid money to the government, fearing that it could be misused. In Oct. Mugabe called for improved relations with the West but added that the lifting of sanctions, which he blamed for ruining the country's economy, was an essential prerequisite. Mugabe's supporters had meanwhile continued to harass MDC activists, leading in Oct. to a stand-off between Mugabe and Tsvangirai in which the latter led an MDC boycott of cabinet meetings. The boycott was called off, however, after the intervention of the Southern African Development Community, which insisted that all 'outstanding issues' in the power-sharing pact be finally settled.

In Dec. 2009 Mugabe was re-elected as ZANU-PF leader for a further five years at a party congress in Harare. From 2010 he became increasingly outspoken about the shortcomings of the power-sharing agreement. He was again endorsed at the ZANU-PF congress in Dec. 2010 as the party's candidate for presidential elections scheduled to be held in the course of 2013. In Feb. 2011, and again in Feb. 2012, the EU eased some sanctions on Zimbabwe but expressed concern at ZANU-PF's continuing failure to honour the unity government pact and at the reported rise in political violence and intimidation directed against the MDC in the run-up to the 2013 elections.

Morgan Tsvangirai

Position
Prime Minister

Introduction
Morgan Tsvangirai is leader of the Movement for Democratic Change (MDC), the only significant opposition to President Robert Mugabe's regime since independence was established in 1980. Tsvangirai became an increasingly credible alternative to Mugabe as dissatisfaction with the ever more authoritarian regime and its vast mishandling of the national economy grew. This electoral viability made Tsvangirai and his supporters a target for violence and intimidation, culminating in the nationwide bloodshed that followed the disputed 2008 presidential elections. Subsequently, however, the implementation of a power-sharing agreement brokered with Mugabe under international pressure led to Tsvangirai's inauguration as executive prime minister of a unity government in Feb. 2009. Despite some progress in Zimbabwe's economic fortunes, political instability has nevertheless persisted and Tsvangirai's ability to carry out his executive functions has been impeded.

Early Life
Tsvangirai was born in 1952 in Buhera, then part of Southern Rhodesia. He left school at 16 and took a job in a textile factory and then at a nickel mine in Bindura, eventually becoming general foreman and branch chairman of the Associated Mineworkers' Union (AMU). A strong supporter of Mugabe's ZANU-PF, Tsvangirai was brought into the first post-independence government. However, Tsvangirai's lack of direct involvement in the guerrilla war that led to the end of minority white rule in 1980 has been used against him by Mugabe's supporters.

In 1985 Tsvangirai left to become vice president of the AMU, having been voted onto the national executive two years earlier. In 1987 he was named secretary-general of the Southern Africa Miners' Federation. After a period studying in the UK, Tsvangirai was named general secretary of the Zimbabwe Congress of Trade Unions (ZCTU) in 1988. Six years later he became secretary general of the Southern African Trade Union co-ordinating council.

Under his leadership the ZCTU began to dissent from the ZANU-PF line. In the late-1980s and early-1990s Tsvangirai and Mugabe repeatedly clashed, notably over a programme of structural reform adopted by Mugabe. Though in line with IMF demands, Tsvangirai condemned the programme as an attack on workers' rights. He led several mass protests against the proposals, which forced Mugabe to back down. Tsvangirai was later jailed on unproven charges of being a South African spy.

Over the course of the 1990s the ZCTU became the focus of opposition to Mugabe's government. In Sept. 1999 Tsvangirai formed the MDC to formally challenge the dominance of ZANU-PF. In Feb. 2000 the MDC co-ordinated the defeat in a referendum of a government-championed constitutional amendment that would have extended Mugabe's personal power. It represented Mugabe's first significant defeat in a public vote since taking office. In Sept. 2000 Tsvangirai was charged with treason for comments made against Mugabe but the charges were later judged to be unconstitutional.

At the parliamentary elections of June 2000 the MDC won 57 of 150 seats, only five behind ZANU-PF. Mugabe's pre-election campaign received widespread international condemnation as white-owned farms were illegally seized and opposition supporters intimidated. Tsvangirai was defeated at the presidential elections of March 2002, winning 42·1% against Mugabe's 56·1%. However, observers claimed the elections failed to meet international standards and it was widely believed that Tsvangirai would have triumphed in a free vote.

Shortly before the election a video tape was exhibited allegedly showing Tsvangirai discussing an assassination attempt against Mugabe. Tsvangirai was charged with treason, carrying a possible death sentence, and his trial began in Feb. 2003. In June 2003 he called for mass action against Mugabe via a general strike and anti-government rallies. The MDC was targeted by government forces in the preceding weeks and Tsvangirai was arrested and subjected to new treason charges. He was acquitted of all charges in 2004.

Mugabe's regime met with international condemnation in March 2007 when Tsvangirai suffered serious head injuries whilst in police custody. More than 50 opposition leaders were arrested during a prayer meeting although no-one was charged. In June Tsvangirai toured Western Europe with rival opposition leader Arthur Mutambara, calling on European politicians to support their struggle for democracy.

In the general election of 29 March 2008 the MDC gained a majority in the House of Assembly, removing ZANU-PF from overall parliamentary control for the first time since Zimbabwe gained independence. Tsvangirai claimed the largest number of votes in the presidential election of the same date, winning 47·9% of the ballot against 43·2% for President Mugabe (according to the Zimbabwe Electoral Commission). However, the margin of victory was disputed, with the MDC claiming that Tsvangirai had gained the absolute majority required to avoid a second round run-off against the incumbent. These claims were rejected by the government and a run-off was scheduled for 27 June. There followed a sustained campaign of intimidation and violence against supporters of Tsvangirai. The MDC reported the death of 85 and the displacement of 200,000 of its followers over the following weeks. Five days before the ballot, having been forced to take refuge in the Dutch embassy, Tsvangirai withdrew from the race, professing that he could no longer force the Zimbabwean people to 'suffer this torture'. Over the following days he appealed to the international community, contending that 'the words of indignation from global leaders [must] be backed by the moral rectitude of military force'. Nevertheless, he subsequently entered into talks with Mugabe in Aug. 2008, which resulted in a deal the following month that provided for Mugabe to remain as president with Tsvangirai taking the premiership.

Career in Office
Implementation of the power-sharing deal was hampered by disagreements over the allocation of ministerial portfolios in a unity government until the end of Jan. 2009 when Tsvangirai accepted an arrangement giving the Movement for Democratic Change-Tsvangirai/MDC-T and the Movement for Democratic Change-Mutambara/MDC-M 16 and four posts respectively in a 35-member cabinet. On 11 Feb. Tsvangirai was sworn in as prime minister. However, there remained deep-seated domestic and international reservations about the viability of the new government, given the animosity between prime minister and president, Mugabe's autocratic track record and the conspicuous absence of the pro-Mugabe leaders of the security forces from Tsvangirai's inauguration ceremony.

With the finance portfolio under MDC control, the use of foreign currency was authorized, effectively replacing the worthless Zimbabwean dollar by the US dollar and South African rand and helping to bring about an end to years of hyperinflation. During a tour of the USA and Europe in June 2009, Tsvangirai successfully lobbied Western donors and the IMF to restore aid for essential services, although they refused to lift sanctions or release more substantive aid until the new administration had undertaken political and other reforms.

However, progress towards a new constitution was hindered by further political wrangling and in Oct. 2009 Tsvangirai temporarily boycotted meetings of the cabinet in protest at Mugabe's failure to honour agreements. The MDC meanwhile continued to face violence and intimidation from ZANU-PF supporters.

In Jan. 2010 Tsvangirai again urged the easing of targeted international sanctions in recognition of the unity government's economic record. The EU made some concessions in 2011 and 2012 but continued to express concern at political intimidation in Zimbabwe. Meanwhile, relations between Tsvangirai and Mugabe deteriorated further as the latter sought to sideline the prime minister and maintain control over executive decision-making. In Oct. 2012, against a background of rising violence directed against the MDC and its supporters in the approach to elections expected before the end of 2013, Tsvangirai once more threatened to withdraw from the unity government.

On 6 March 2009 Tsvangirai survived a car crash in which his wife, Susan, died.

DEFENCE

In 2006 military expenditure totalled US$156m. (US$13 per capita), representing 2·8% of GDP.

Army
Strength in 2007 was estimated at 25,000. There were a further 21,800 paramilitary police including a police support unit of 2,300.

Air Force
The Air Force (ZAF) had a strength in 2007 of about 4,000 personnel. The headquarters of the ZAF and the main ZAF stations are in Harare; the second main base is at Gweru, with many secondary airfields throughout the country. There were 45 combat-capable aircraft (including *Hunters* and F-7s (MiG-21)) in 2007 although the serviceability of some aircraft was in doubt, and six attack helicopters.

ECONOMY

Agriculture accounted for 19% of GDP in 2007, industry 24% and services 57%.

Since Robert Mugabe came to power in 1980 the economy has collapsed with the country experiencing sustained periods of roaring inflation and heavy unemployment. Shortages of food and other necessities culminated in the authorities making an international appeal for food in July 2001. In Feb. 2000 the country ran out of gasoline because it could not pay the import bills.

Zimbabwe's 'shadow' (black market) economy is estimated to constitute approximately 63% of the country's official GDP, one of the highest percentages of any country in the world.

Overview
At independence Zimbabwe was the second largest economy in Africa but the macroeconomic situation has deteriorated sharply since 1998. Between 1999 and 2008 output declined by more than 40%. According to the IMF the economic crisis was attributable to loose fiscal and monetary policies, an overvalued fixed exchange rate, excessive administrative controls and regulations, and chronic shortages of goods and foreign exchange. The effect of these policies has been magnified by collapsing health and education systems, the HIV pandemic, the fast track land reform programme and recurring droughts.

In Feb. 2009 the government launched a Short-Term Emergency Recovery Program focused on stabilizing the economy. In May 2010 the Medium Term Plan 2010–15 was announced, with the goals of restoring the economy to its 1990 level and developing the information technology and mining industries.

In 2007–08 Zimbabwe experienced hyperinflation as a result of excessive money supply. An extremely large public sector, insufficient external financing, the collapse of agricultural and mining exports and an unsustainable external debt burden fuelled this rapid monetary expansion. Most Western donors reduced or withdrew operations, with financial support confined to humanitarian aid. While China remains a major donor of both finance and food, it made some aid cuts in 2007 under pressure from Western governments. The government subsequently countered hyperinflation by eliminating the use of the Zimbabwe dollar in 2009, removing price controls and adopting a multi-currency regime. The economy grew for the first time in a decade in 2009 while FDI also increased.

Nonetheless, unemployment and poverty rates remain high. Despite boasting several major attractions, tourism has declined since 2000 as a result of internal crises, with numerous airlines suspending flights to the country. Zimbabwe is heavily dependent on aid to meet basic needs, with the IMF allocating US$328m. to the country in Feb. 2010. Political uncertainty remains a major impediment to growth.

Currency
The use of the Zimbabwean dollar as an official currency was effectively abandoned on 12 April 2009. Currencies such as the South African rand, the Botswana pula, the pound sterling and the US dollar are used instead. Until 12 April 2009 the unit of currency was the *Zimbabwe fourth dollar* (ZWL), introduced on 2 Feb. 2009, with 12 zeros being removed to make 1trn. dollars (ZWR) equal to one new dollar. The *Zimbabwe third dollar* (ZWR) had replaced the *Zimbabwe second dollar* (ZWD) on 1 Aug. 2008, with a conversion rate of 1 revalued dollar = 10bn. old dollars (ZWD). The currency was devalued 17% in Jan. 1994 and made fully convertible. Its value dropped by 65% in 1998. It was devalued again in Aug. 2000 by 24%, in May 2005 by 45% and in July 2005 by 94%. It was further devalued by 60% in July 2006 and the following day the *Zimbabwean new (second) dollar* became the new currency. Inflation, which was 18·9% in 1997, rose to 133·2% in 2002 and 6,723·7% in 2007. The last official annual rate calculated in Zimbabwe dollars was 231,150,888·9% in July 2008, although the Central Statistical Office released monthly inflation rates calculated in US dollars for Oct. and Nov. 2009 of 0·8% and −0·1% respectively. Total money supply was Z$425,445·0bn. in Dec. 2007, up from Z$0·4bn. in Dec. 2002 and Z$44·7bn. in Dec. 2005.

Budget
Revenues in 2009 totalled US$933·6m. and expenditures US$920·9m. Tax revenues accounted for 94·5% of total revenue in 2009; current expenditures accounted for 87·3% of total expenditure. VAT was reduced from 17·5% to 15% in Jan. 2006.

Performance
Real GDP growth was negative every year from 2002 through to 2008. The economy contracted by 9·6% in 2002, 16·9% in 2003, 6·0% in 2004, 5·5% in 2005, 3·5% in 2006, 3·8% in 2007 and 18·3% in 2008. However, there was then growth of 6·3% in 2009, 9·6% in 2010 and 9·4% in 2011. Total GDP in 2011 was US$9·7bn.

Banking and Finance
The Reserve Bank of Zimbabwe is the central bank (established 1965; *Governor*, Dr Gideon Gono). It acts as banker to the government and to the commercial banks, is the note-issuing authority and co-ordinates the application of the government's monetary policy. The Zimbabwe Development Bank, established in 1983 as a development finance institution, is 30·6% government-owned. In 2003 there were seven commercial and four merchant banks. In 1997 there were five registered finance houses, three of which are subsidiaries of commercial banks.

Zimbabwe's foreign debt totalled US$5,016m. in 2010, representing 71·8% of GNI.

In Aug. 2003 Zimbabwe's banks ran out of banknotes as inflation reached 360%.

There is a stock exchange in Harare.

ENERGY AND NATURAL RESOURCES

Environment

Carbon dioxide emissions from the consumption and flaring of fossil fuels were the equivalent of 1·0 tonnes per capita in 2008.

Electricity

Installed capacity was 2·0m. kW in 2007. Production in 2007 was 9·18bn. kWh. Consumption per capita in 2007 was 887 kWh.

Minerals

The mining sector accounts for about 4% of GDP. 2006 production: coal, 2·11m. tonnes; chromite (gross weight), 700,000 tonnes; asbestos, 97,000 tonnes; nickel, 8,825 tonnes; gold, 11·4 tonnes. Diamond production in 2005 totalled 251,152 carats. A huge diamond field was discovered near Mutare in 2006 that the government believes could produce up to a fifth of global diamond output.

Agriculture

Agriculture is the largest employer, providing jobs for 25% of the workforce. In 2002 there were an estimated 3·22m. ha. of arable land and 0·13m. ha. of permanent crops. Approximately 117,000 ha. were irrigated in 2002. There were about 24,000 tractors in 2002 and 800 harvester-threshers.

A constitutional amendment providing for the compulsory purchase of land for peasant resettlement came into force in March 1992. A provision to seize white-owned farmland for peasant resettlement was part of the government's new draft constitution that was rejected in the referendum of Feb. 2000. Various deadlines were given for white farmers to abandon their property during Aug. and Sept. 2002. The government claims that 300,000 landless black Zimbabweans have been resettled on seized land.

The staple food crop is maize, but 2003 production (at 803,000 tonnes) was less than a half of the 2000 figure. Tobacco is the most important cash crop, although production fell from 237,000 tonnes in 2000 to 68,000 tonnes in 2004. Production of other leading crops, 2002, in 1,000 tonnes: sugarcane, 4,100; seed cotton, 200; cassava, 175; wheat, 150; cottonseed, 127; groundnuts, 110; oranges, 93; bananas, 85; soybeans, 83; sorghum, 80; cotton lint, 72.

Livestock (2003 estimates): cattle, 5·75m.; goats, 2·97m.; sheep, 610,000; pigs, 605,000; chickens, 22m. Livestock products (2003 estimates, in 1,000 tonnes): milk, 280; meat, 206.

Forestry

In 2010 forests covered 15·62m. ha., or 40% of the total land area. Timber production in 2007 was 9·23m. cu. metres.

Fisheries

The catch in 2010 was approximately 10,500 tonnes (all from inland waters).

INDUSTRY

Metal products account for over 20% of industrial output. Important agro-industries include food processing, textiles, furniture and other wood products.

Labour

The labour force in 1996 totalled 5,281,000 (56% males). Unemployment in March 2007 was around 80%.

INTERNATIONAL TRADE

Imports and Exports

Imports (c.i.f.) in 2010 were US$9,051·5m. (US$3,526·8m. in 2009); exports (f.o.b.), US$3,199·2m. (US$2,268·9m. in 2009).

Main imports in 2004 (in US$1m.): machinery and transport equipment, 581·7; manufactured goods, 370·3; food and livestock, 282·2; petroleum and petroleum products, 278·1; chemicals and related products, 270·6. Main exports in 2004: tobacco and tobacco manufactures, 443·3; textile fibres, 243·4; metalliferous ore and scrap metal, 212·9; iron and steel 211·5.

Main import suppliers, 2005: Zambia, 40·9%; South Africa, 15·0%; Mozambique, 9·9%; Botswana, 5·0%; Kuwait, 4·0%. Main export destinations, 2005: South Africa, 41·5%; USA, 6·9%; Switzerland, 6·4%; Zambia, 5·6%; UK, 5·3%.

COMMUNICATIONS

Roads

The road network covers some 97,000 km but much of it is in poor condition. Number of vehicles in use, 2007: passenger cars, 1,214,100; lorries and vans, 186,800; buses and coaches, 15,600; motorcycles and mopeds, 109,000. There were 1,037 road accident fatalities in 2006.

Rail

In 2005 the National Railways of Zimbabwe had 2,759 km (1,067 mm gauge) of route ways (483 km electrified). In 2005 the railways carried 3m. passengers and 6·1m. tonnes of freight (including the Beitbridge-Bulawayo Railway).

Civil Aviation

There are three international airports: Harare (the main airport), Bulawayo and Victoria Falls. Air Zimbabwe, the state-owned national carrier, ceased operations in Feb. 2012 but resumed flying on a limited basis in May. After the government took over Air Zimbabwe's debts it began flying on international routes again in Nov. 2012. In 2003 scheduled airline traffic of Zimbabwe-based carriers flew 6m. km, carrying 201,000 passengers (102,000 on international flights). In 1999 Harare handled an estimated 1,276,000 passengers (995,000 on international flights).

Shipping

Zimbabwe's outlets to the sea are Maputo and Beira in Mozambique, Dar es Salaam, Tanzania and the South African ports.

Telecommunications

In 2008 there were 348,000 main (fixed) telephone lines; mobile phone subscribers numbered 1,654,700 in 2008 (13·3 per 100 persons). There were 1,421,000 internet users in 2008.

SOCIAL INSTITUTIONS

Justice

The general common law of Zimbabwe is the Roman Dutch law as it applied in the Colony of the Cape of Good Hope on 10 June 1891, as subsequently modified by statute. Provision is made by statute for the application of African customary law by all courts in appropriate cases.

The death penalty is authorized. In 2003 there were four executions.

The Supreme Court consists of the Chief Justice and at least two Supreme Court judges. It is the final court of appeal. It exercises appellate jurisdiction in appeals from the High Court and other courts and tribunals; its only original jurisdiction is that conferred on it by the Constitution to enforce the protective provisions of the Declaration of Rights. The Court's permanent seat is in Harare but it also sits regularly in Bulawayo.

The High Court is also headed by the Chief Justice, supported by the Judge President and an appropriate number of High Court judges. It has full original jurisdiction, in both Civil and Criminal cases, over all persons and all matters in Zimbabwe. The Judge President is in charge of the Court, subject to the directions of the Chief Justice. The Court has permanent seats in both Harare and Bulawayo and sittings are held three times a year in three other principal towns.

Regional courts, established in Harare and Bulawayo but also holding sittings in other centres, exercise a solely criminal jurisdiction which is intermediate between that of the High Court and the Magistrates' courts. Magistrates' courts, established in

20 centres throughout the country, and staffed by full-time professional magistrates, exercise both civil and criminal jurisdiction.

Primary courts consist of village courts and community courts. Village courts are presided over by officers selected for the purpose from the local population, sitting with two assessors. They deal with specific classes of civil cases and have jurisdiction only where African customary law is applicable. Community courts are presided over by officers in full-time public service, who may also be assisted by assessors. They have jurisdiction in all civil cases determinable by African customary law and also deal with appeals from village courts. They also have limited criminal jurisdiction in respect of petty offences.

The population in penal institutions in June 2007 was 17,967 (136 per 100,000 of national population).

Education
Education is compulsory. 'Manageable' school fees were introduced in 1991; primary education had hitherto been free to all. All instruction is given in English. In 2006 there were 2,445,520 pupils at primary schools (64,001 teaching staff) and 831,488 pupils at secondary schools (33,964 teaching staff in 2003). In May 2004 the 45 private schools were closed down for increasing fees without state approval although they have for the most part since reopened. Private school fees are out of reach of the majority of families. Tens of thousands of teachers have left Zimbabwe and thousands more have left the profession. In 2009 the adult literacy rate was an estimated 91·9%. Both the overall rate and the rate for males are the highest in Africa.

There are 12 universities, the oldest and largest of which is the University of Zimbabwe, founded in 1952. Although historically well-regarded, the university experienced a series of problems ranging from staff shortages to electricity supply issues that resulted in its closure for much of 2008 and 2009.

In 2003 there were 55,689 students in tertiary education.

Health
There were 30 hospital beds per 10,000 inhabitants in 2006. All mission health institutions get 100% government grants-in-aid for recurrent expenditure. In 2004 there were 2,086 physicians, 310 dentistry personnel and 9,357 nursing and midwifery personnel. It is estimated that one in three adults is HIV infected.

Welfare
It is a statutory responsibility of the government in many areas to provide: processing and administration of war pensions and old age pensions; protection of children; administration of remand, probation and correctional institutions; registration and super-vision of welfare organizations.

RELIGION
In 2001, 4·58m. persons were African Christians, 1·40m. Protestants, 1·09m. Roman Catholics and 870,000 followers of other religions. There were also 3·43m. followers of animist beliefs in 2001.

CULTURE
World Heritage Sites
Zimbabwe has five sites on the UNESCO World Heritage List: Mana Pools National Park, Sapi and Chewore Safari Areas (inscribed on the list in 1984); the Great Zimbabwe National Monument (1986); the Khambi Ruins National Monument (1986); and Matobo Hills (2003).

Zimbabwe shares with Zambia the Victoria Falls/Mosi-oa-Tunya (1989), waterfalls on the Zambezi River.

Press
In 2007 there were two daily newspapers, both controlled by the government, with a combined circulation of 115,000. In Jan. 2002 parliament passed an Access to Information Bill restricting press freedom, making it an offence to report from Zimbabwe unless registered by a state-appointed commission. In Sept. 2003 the independent *Daily News* was shut down for contraventions of the new press law. Zimbabwe's High Court ordered the government to allow its reopening but the order was ignored.

Tourism
There were 1,559,000 foreign tourists in 2005 (down from 1,854,000 in 2004); spending by tourists totalled US$99m.

Festivals
Of particular importance are the Harare International Festival of the Arts (April) and the Bulawayo Music Festival (May). The Zimbabwe International Film Festival is held in Harare in Aug./Sept.

DIPLOMATIC REPRESENTATIVES
Of Zimbabwe in the United Kingdom (Zimbabwe House, 429 Strand, London, WC2R 0JR)
High Commissioner: Gabriel Mharadze Machinga.

Of the United Kingdom in Zimbabwe (3 Norfolk Rd, Mount Pleasant, Harare, PO Box 4490)
High Commissioner: Deborah Bronnert.

Of Zimbabwe in the USA (1608 New Hampshire Ave., NW, Washington, D.C., 20009)
Ambassador: Dr Machivenyika Mapuranga.

Of the USA in Zimbabwe (172 Herbert Chitepo Ave., Harare)
Ambassador: Bruce Wharton.

Of Zimbabwe to the United Nations
Ambassador: Chitsaka Chipaziwa.

Of Zimbabwe to the European Union
Ambassador: Mary Margaret Muchada.

FURTHER READING
Central Statistical Office. *Monthly Digest of Statistics.*

Hatchard, J., *Individual Freedoms and State Security in the African Context: the Case of Zimbabwe.* 1993
Hill, Geoff, *What Happens After Mugabe? Can Zimbabwe Rise From the Ashes?* 2005
Meredith, Martin, *Mugabe: Power and Plunder in Zimbabwe.* 2002
Skålnes, T., *The Politics of Economic Reform in Zimbabwe: Continuity and Change in Development.* 1995
Weiss, R., *Zimbabwe and the New Elite.* 1994

National Statistical Office: Central Statistical Office, POB 8063, Causeway, Harare.
Website: http://www.zimstat.co.zw

KEY WORLD FACTS

- World population in 2013 — 7,130 million (3,596 million males and 3,534 million females)

- World population under 30 in 2013 — 3,655 million

- World population over 60 in 2013 — 840 million

- World population over 100 in 2013 — 372,000

- Number of births worldwide every day — 372,000

- Number of deaths worldwide every day — 159,000

- World economic growth rate in 2012 — 2·3% (3·8% in 2011)

- Number of illiterate adults — 775 million

- Number of unemployed people — 197 million

- Average world life expectancy — 71·9 years for females; 67·4 years for males

- Annual world population increase — 77·88 million people

- Number of people living outside country of birth — 214 million, or more than 3% of the world's population

- Fertility rate — 2·4 births per woman

- Urban population — 52·1% of total population

- World trade in 2011 — US$36·7 billion

- Annual world defence expenditure — US$1,738 billion

- Number of cigarettes smoked — 5,600 billion a year

- Number of internet users — 2·4 billion

- Number of Facebook users — 836 million

- Number of mobile phone users — 6·0 billion

- Number of motor vehicles on the road — 1 billion

- Number of people who cross international borders every day — 2 million

- Number of people living in extreme poverty — 1·5 billion

- Number of people living in urban slums — 1 billion

- Number of undernourished people — 870 million

- Number of overweight adults — 1 billion

- Number of obese adults — 475 million

- Number of people dying of starvation — 24,000 every day

- Number of people lacking clean water — 783 million

- Number of people lacking basic sanitation — 2·5 billion

- Number of people worldwide exposed to indoor air pollution that exceeds WHO guidelines — 1 billion

- Annual carbon dioxide emissions — 34·7 billion tonnes

CHRONOLOGY

April 2012–March 2013

Week beginning 1 April 2012

In San Marino, Maurizio Rattini and Italo Righi were sworn in as captains regent.

Fernando Herrero resigned as Costa Rica's finance minister. Two weeks later, Edgar Ayales was appointed to the post.

Following the resignation of Pál Schmitt, the president of the National Assembly, László Kövér, became Hungary's acting president. János Áder was elected president a month later by 262 votes to 40.

In a cabinet reshuffle in New Zealand, Chris Tremain was appointed internal affairs minister.

Jules Baillet became finance minister in a cabinet reshuffle in Niger.

In Senegal, Macky Sall was sworn in as president. Abdoul Mbaye was appointed prime minister and the new government included Alioune Badara Cissé as foreign minister, Augustin Tine as defence minister, Mbaye Ndiaye as interior minister and Amadou Kane as finance minister.

With the resignation of Boris Tadić, speaker of parliament Slavica Đukić Dejanović became Serbia's acting president.

In Slovakia, Robert Fico was appointed prime minister. His government included Miroslav Lajčák as foreign minister, Martin Glváč as defence minister, Robert Kaliňák as interior minister and Peter Kažimír as finance minister.

Following the death of Bingu wa Mutharika, Malaŵi's vice president Joyce Banda took over as president. A week later Khumbo Hastings Kachali was appointed vice president. The new cabinet included Ephraim Mganda Chiume as foreign minister, Ken Kandodo as defence minister and Uladi Mussa as home affairs minister.

Week beginning 8 April 2012

In a cabinet reshuffle in Benin, president Yayi Boni took charge of the defence portfolio and Jonas Gbian was appointed finance minister.

In Haiti, the Senate endorsed the nomination of Laurent Lamothe as prime minister by 19 votes to three. Three weeks later, the Chamber of Deputies also approved Lamothe as prime minister with 62 votes to three against.

In North Korea, Kim Jong-gak was appointed defence minister. Kim Jong-un was elected First Secretary of the Workers' Party of Korea and was also named First Chairman of the National Defence Commission two days later.

In parliamentary elections in South Korea, the Saenuri Party won 152 out of 300 seats with 42·8% of votes cast; the Democratic United Party 127 with 36·5%; the Unified Progressive Party 13 with 10·3%; and the Liberty Forward Party 5 with 3·2%. Three seats went to independents.

Following the coup a month earlier in Mali, Dioncounda Traoré was sworn in as president. Cheick Modibo Diarra was subsequently appointed acting prime minister. The new government, announced a week later, included Sadio Lamine Sow as foreign minister, Col. Yamoussa Camara as defence minister, Col. Moussa Sinko Coulibaly as territorial administration (interior) minister and Tiéna Coulibaly as finance minister.

Guinea-Bissau's interim president Raimundo Pereira was deposed in a military coup and the presidential run-off was aborted. The following week, Manuel Serifo Nhamadjo was named transitional president but he refused the nomination.

In Lithuania, Artūras Melianas was appointed interior minister.

Week beginning 15 April 2012

In the presidential run-off in Timor Leste, Taur Matan Ruak was elected president with 61·2% of the vote against 38·8% for Francisco Guterres.

In a cabinet reshuffle in the Gambia, Mambury Njie became foreign minister, Lamin Kaba Bajo interior minister and Abdou Kolley finance minister.

Jim Yong Kim was named president of the World Bank, to take office on 1 July.

Augustin Matata Ponyo was appointed prime minister of the Democratic Republic of the Congo. His government was named ten days later and included Raymond Tshibanda as foreign minister and Alexandre Lubal Tamu as defence minister.

In Sweden, Karin Enström was appointed defence minister.

In a cabinet reshuffle in Tonga, Lisiate 'Aloveita 'Akolo was named finance minister.

Week beginning 22 April 2012

In the first round of French presidential elections, François Hollande gained the largest number of votes (28·63% of those cast) against nine opponents. His nearest rivals were the incumbent president Nicolas Sarkozy, who came second with 27·18% of votes cast, and Marine Le Pen, with 17·90%.

In a cabinet reshuffle in Ecuador, Miguel Carvajal was appointed defence minister.

In the Netherlands, prime minister Mark Rutte's cabinet resigned but stayed on as caretaker government until elections in September.

With the resignation of Awn Khasawneh as Jordan's prime minister, King Abdullah II designated Fayez Tarawneh to form a government. In the new cabinet sworn in the following week, Tarawneh also assumed the defence portfolio. Ghabib Al Zu'bi was appointed interior minister and Suleiman Al Hafez finance minister.

In the Czech Republic, prime minister Petr Nečas' government won a confidence vote in parliament (105–93).

In Romania, Mihai-Răzvan Ungureanu's government was defeated in a no-confidence vote that was supported by 235 of the 460 members of parliament. President Traian Băsescu designated Victor Ponta as prime minister. A week later parliament approved Ponta's government, which included Andrei Marga as foreign minister, Gen. Corneliu Dobrițoiu as defence minister, Ioan Rus as interior minister and Florin Georgescu as finance minister.

Week beginning 29 April 2012

In a cabinet reshuffle in Tanzania, Shamsi Vuai Nahodha was named defence minister, William Mgimwa finance minister and Emmanuel John Nchimbi home affairs minister.

In the second round of parliamentary elections in Iran, with 65 of the 290 seats contested, President Mahmoud Ahmadinejad's opponents won 41 seats and his supporters 13. 11 seats went to independents.

Week beginning 6 May 2012

In parliamentary elections in Armenia, the Republican Party of Armenia won 69 of 131 seats with 44·1% of votes cast; Prosperous Armenia 36 with 30·2%; Armenian National Congress 7 with 7·1%; Armenian Revolutionary Federation 6 with 5·7%; and Rule of Law 6 with 5·5%. Tigran Sargsyan was reappointed prime minister four weeks later.

In the presidential run-off in France, François Hollande was elected president with 51·6% of votes cast against 48·4% for Nicolas Sarkozy. Hollande was sworn in nine days later and named Jean-Marc Ayrault as prime minister. The government, sworn in the following day, included Laurent Fabius as foreign minister, Jean-Yves Le Drian as defence minister, Manuel Valls as interior minister and Pierre Moscovici as finance minister.

In parliamentary elections in Greece, New Democracy won 108 of 300 seats with 18·9% of votes cast; Coalition of the Radical Left 52 with 16·8%; Pasok (Panhellenic Socialist Movement) 41 with 13·2%; Independent Greeks, 33 with 10·6%; Communist Party 26 with 8·5%; Golden Dawn 21 with 7·0%; and Democratic Left 19 with 6·1%. The leaders of New Democracy, Coalition of the Radical Left and Pasok were each unable to agree on terms for setting up a government. The president's efforts to form a national unity government also failed and new elections were scheduled for June. Panagiotis Pikrammenos was sworn in as interim prime minister, leading a government that included Petros Molyviatis as foreign minister, Frangoulis Frangos as defence minister, Antonios Manitakis as interior minister and Georgios Zanias as finance minister.

In presidential elections in Serbia, Boris Tadić gained 25·3% of the vote, followed by Tomislav Nikolić with 25·0%, Ivica Dačić with 14·2% and Vojislav Koštunica with 7·4%. There were eight other candidates. In the run-off held two weeks later, Nikolić won with 49·5% of the vote against 47·3% for Tadić. Nikolić took office as president at the end of May.

In parliamentary elections in The Bahamas, the Progressive Liberal Party won 48·6% of votes cast and 29 out of 38 seats against the ruling Free National Movement with 42·1% and 9 seats. The following day, Perry Christie was sworn in as prime minister and finance minister. Bernard Nottage was later sworn in as national security minister with Frederick Mitchell as foreign minister.

Vladimir Putin was inaugurated as Russia's president. He nominated Dmitry Medvedev as prime minister. The new government, named three weeks later, included Vladimir Kolokoltsev as interior minister. Other key portfolios remained unchanged.

In the Democratic Republic of the Congo, prime minister Augustin Matata Ponyo's government won an investiture vote in parliament (324–53).

In parliamentary elections in Algeria, the Front de Libération Nationale/National Liberation Front won 208 out of 462 seats; Rassemblement National Démocratique/National Rally for Democracy took 68 seats; the Green Algeria Alliance, 49; Front des Forces Socialistes/Front of Socialist Forces, 27; and Parti des Travailleurs/Workers' Party, 24. Independents gained 18 seats and the remainder went to minor parties.

János Áder took office as Hungary's president.

In Guinea-Bissau, Manuel Serifo Nhamadjo accepted nomination as transitional president for one year. Five days later, he named Rui Duarte Barros as prime minister. The new government, appointed the following week, included Faustino Fudut Imbali as foreign minister, Col. Celestino Carvalho as defence minister,

Antonio Suca Intchama as interior minister and Abubacar Demba Dahaba as finance minister.

Week beginning 13 May 2012

In Estonia, Urmas Reinsalu was sworn in as defence minister after the resignation of Mart Laar.

Following the resignation of Peru's interior minister Daniel Lozada and defence minister Alberto Otárola, Wilver Calle was named interior minister and José Urquizo defence minister.

In Grenada, prime minister Tillman Thomas and his government defeated a no-confidence motion by 8 votes to 5. Two days later foreign minister Karl Hood resigned and the prime minister took over the portfolio.

In Slovakia, prime minister Robert Fico's government won a confidence vote in parliament winning 82 votes.

In Haiti, Laurent Lamothe's new cabinet took office. Jean Rodolphe Joazile became defence minister and Marie-Carmelle Jean Marie finance minister.

Week beginning 20 May 2012

In presidential elections in the Dominican Republic, Danilo Medina Sánchez of the ruling Dominican Liberation Party won 51·2% of the votes and Hipólito Mejía of the Dominican Revolutionary Party 47·0%.

Taur Matan Ruak was sworn in as Timor-Leste's new president.

Following the resignation of Ignacio Milam Tang's government, Vincenté Ehaté Tomi was named Equatorial Guinea's new prime minister. Agapito Mba Mokuy was appointed foreign minister and Marcelino Owono Edu finance minister.

In the first round of presidential elections in Egypt, Mohamed Morsy (Freedom and Justice Party) came first with 24·8% of the vote, followed by Ahmed Shafik (ind.) with 23·7%, Hamdeen Sabahi (Dignity Party) 20·7%, Abdel Moneim Aboul Fotouh (ind.) 17·5% and Amr Moussa (ind.) 11·1%. There were eight other candidates.

In parliamentary elections in Lesotho, the Democratic Congress won 48 of 120 seats, All Basotho Convention 30 and Lesotho Congress for Democracy 26. Prime minister Pakalitha Mosisili resigned four days later.

Week beginning 27 May 2012

In Barbados, Elliot Belgrave took office as governor-general.

Tigran Sargsyan was reappointed prime minister of Armenia with no changes in the main cabinet posts.

In the Central African Republic, finance minister Sylvain Ndoutingaï was replaced by Albert Besse.

Syed Naveed Qamar was appointed defence minister of Pakistan.

Week beginning 3 June 2012

In a cabinet reshuffle in Japan, Satoshi Morimoto became defence minister.

In Pakistan, the Supreme Court suspended interior minister Rehman Malik's Senate membership. He was reappointed as adviser on interior affairs the following day. Two weeks later, the Supreme Court disqualified prime minister Yousaf Raza Gilani from office. President Asif Ali Zardari nominated Makhdoom Shahabuddin as prime minister but replaced him with Raja Pervez after a judge ordered Shahabuddin's arrest. Ashraf was elected in parliament with 211 votes. The interior portfolio was left vacant in the new cabinet. Other key portfolios remained unchanged.

In Malta, prime minister Lawrence Gonzi won a confidence vote in parliament by 35 votes to 34.

In Syria, Riyad Hijab was sworn in as prime minister. Key portfolios in the new cabinet, sworn in three weeks later, remained unchanged.

Tom Motsoahae Thabane was sworn in as Lesotho's prime minister, also taking the defence portfolio. Leketekete Ketso was appointed finance minister and Joang Molapo home affairs minister.

Week beginning 10 June 2012

In Kenya, interior minister George Saitoti was killed in a plane crash.

Nauru's president Sprent Dabwido appointed a new cabinet that included Kieren Keke as foreign minister and Roland Kun as finance minister.

Albania's parliament elected Bujar Nishani president in a fourth round after previous votes had failed to result in the required three-fifths majority.

In a cabinet reshuffle in South Africa, Nosiviwe Mapisa-Nqakula was appointed defence minister.

Prince Ahmed bin Abdulaziz Al-Saud was appointed Saudi Arabia's new interior minister after the death of Crown Prince Nayef.

Rubén Candia Amarilla was sworn in as Paraguay's new interior minister following the resignation of Carlos Filizzola. In an impeachment trial held six days later, the Senate voted to remove president Fernando Lugo by 39 votes to 4. Vice president Federico Franco was sworn in as president. The new cabinet included Carmelo Caballero as interior minister, José Félix Fernández Estigarribia as foreign minister and María Liz Arnold as defence minister.

Week beginning 17 June 2012

After two rounds of parliamentary elections in France, the Socialist Party along with its allies won with a total of 331 seats, providing the new government with an absolute parliamentary majority. The Socialist Party (PS) won 280 of the 577 available seats; the Union for a Popular Movement (UMP), 194; Miscellaneous Left (DVG), 22; Europe Ecology–the Greens (EELV), 17; Miscellaneous Right (DVD), 15; Radical Party of the Left (PRG), 12; the New Centre (NC), 12; the Left Front (FDG), 10; the Radical Party (PRV), 6; and others, 9. After resigning (a traditional procedure following French legislative elections) Jean-Marc Ayrault was reappointed prime minister with no changes in the key cabinet posts.

In a presidential run-off in Egypt, Mohamed Morsy won 51·7% of the vote and Ahmed Shafik 48·3%. The following week prime minister Kamal Ganzouri and his cabinet resigned. Morsy was sworn in as president five days later.

In the second parliamentary elections in Greece in six weeks, New Democracy won 129 seats with 29·7% of the vote; Syriza (Coalition of the Radical Left), 71 (26·9%); Pasok (Panhellenic Socialist Movement), 33 (12·3%); Independent Greeks, 20 (7·5%); Golden Dawn, 18 (6·9%); Democratic Left, 17 (6·3%); and the Communist Party, 12 (4·5%). Antonis Samaras was sworn in as prime minister three days later. His cabinet included Dimitris Avramopoulos as foreign minister, Panos Panagiotopoulos as defence minister, Evripidis Stylianidis as interior minister and Vassilis Rapanos as finance minister. Owing to ill health Rapanos failed to take office and was replaced by Yannis Stournaras.

In the USA, commerce secretary John Bryson resigned. Rebecca M. Blank, who had previously taken over when Bryson took a medical leave of absence, continued as acting secretary.

In Nigeria, defence minister Bello Mohammed was dismissed.

In a cabinet reshuffle in Trinidad and Tobago, Winston Dookeran was appointed foreign minister, Larry Howai finance minister and Jack Warner national security minister.

Week beginning 24 June 2012

Following the resignation of Pranab Mukherjee, India's prime minister Manmohan Singh took over the finance portfolio.

In Serbia, president Tomislav Nikolić asked Ivica Dačić to form a government.

In presidential elections in Iceland incumbent Ólafur Ragnar Grímsson won 52·8% of the vote, Thóra Arnórsdóttir 32·2% and Ari Trausti Guðmundsson 8·6%.

Week beginning 1 July 2012

In China, Leung Chun-ying took office as chief executive of Hong Kong.

In Kuwait, following the resignation of the government, the Amir reappointed Sheikh Jaber Mubarak al-Hamad al-Sabah as prime minister. The new cabinet was announced two weeks later with Nayef al-Hajraf as finance minister. Other key portfolios remained unchanged.

In presidential elections in Mexico, Enrique Peña Nieto of the Institutional Revolutionary Party (PRI) won 39·1% of the vote, Andrés Manuel López Obrador of the Party of the Democratic Revolution (PRD) 32·4% and Josefina Vázquez Mota of the National Action Party (PAN) 26·0%. Turnout was 63·1%.

In parliamentary elections in Senegal, the United in Hope coalition won 119 of 150 seats and the Senegalese Democratic Party 12. Turnout was 36·8%.

Jim Yong Kim took office as the president of the World Bank.

Following the resignation of Montenegro's foreign minister Milan Rocen, Nebojša Kaluđerović was appointed to the post.

In a cabinet reshuffle in Albania, Edmond Panariti was appointed foreign minister and Flamur Noka interior minister.

In Georgia, parliament confirmed the new cabinet of prime minister Vano Merabishvili, including Dimitri Shashkin as defence minister and Bacho Akhalaia as interior minister. Other key portfolios remained unchanged.

Yannis Stournaras was sworn in as finance minister of Greece.

In Libya's parliamentary elections, the National Forces Alliance won 39 of 80 seats allocated to registered parties and the Justice and Construction Party 17; 120 of the 200 members were elected as independents.

In parliamentary elections in Timor-Leste, the National Congress for Timorese Reconstruction won 36·7% of the vote and 30 of 65 seats, the Revolutionary Front for an Independent East Timor 29·9% and 25 seats, the Democratic Party 10·3% and 8 seats, and the Front for National Reconstruction of Timor-Leste–Change 3·1% and 2 seats. Turnout was 74·8%.

Week beginning 8 July 2012

In Italy, Vittorio Grilli replaced Mario Monti as economy and finance minister.

In Togo, the government of Gilbert Houngbo resigned. Kwesi Ahoomey-Zunu took office as prime minister eight days later. The cabinet subsequently named included Col. Yark Damehane as security and civil protection minister. Other key portfolios remained unchanged.

Week beginning 15 July 2012

Francisco Álvarez de Soto became acting foreign minister of Panama with Roberto Henríquez becoming minister of the presidency.

Parliamentary elections were held in Papua New Guinea between 23 June and 17 July 2012. Prime minister Peter O'Neill's People's National Congress won 27 of 111 seats; the Triumph Heritage Empowerment Party 12; PNG Party 8; the National Alliance Party 7; the United Resources Party 7; the People's Party 6; the People's Progress Party 6; and the Social Democratic Party 3. A number of smaller parties each took one or two seats, with 16 going to independents.

In Syria, defence minister Gen. Dawoud Rajha was killed in a bomb attack. Gen. Fahad Jassim al-Freij was sworn in as his successor.

Pranab Mukherjee was sworn in as the president of India following his victory in the indirect presidential elections. Mukherjee gained 713,763 votes while his opponent won 315,987.

Rajkeswur Purryag was elected president of Mauritius and sworn in the following day.

Week beginning 22 July 2012

In Peru, following the resignation of the government of Óscar Valdés, Juan Jiménez was sworn in as prime minister. His cabinet included Pedro Cateriano Bellido as defence minister and Wilfredo Pedraza as interior minister.

Bujar Nishani took office as the president of Albania.

In Egypt, Hisham Qandil was named prime minister by president Mohamed Morsy.

In Ghana, president John Atta Mills died. He was succeeded by vice president John Dramani Mahama. Kwesi Bekoe Amissah-Arthur was named vice president a week later.

In Pakistan, Rehman Malik was sworn in as interior minister.

The Serbian parliament voted in a government headed by prime minister Ivica Dačić. Dačić continued as interior minister while Ivan Mrkić was appointed foreign minister, Aleksandar Vučić defence minister and Mlađan Dinkić finance minister.

In Tunisia, finance minister Houcine Dimassi resigned. The secretary of state for finance, Slim Besbes, took over the portfolio.

Week beginning 29 July 2012

In a cabinet reshuffle in India, Palaniappan Chidambaram was appointed finance minister, while Sushil Kumar Shinde succeeded him as home affairs minister.

In Honduras, following the resignation of finance minister Héctor Guillén, Wilfredo Cerrato was appointed his successor.

In Egypt, Hisham Qandil's government was sworn in. Moumtaz Saïd continued as finance minister while Mohammed Kamel Amr was appointed foreign minister, Field Marshal Mohamed Hussein Tantawi defence minister and Ahmed Gamal Eddin interior minister. Ten days after later Mahmoud Mekki was appointed vice president and Abdel Fattah al-Sisi became defence minister after Tantawi was ordered to retire.

Week beginning 5 August 2012

Kwesi Bekoe Amissah-Arthur was sworn in as vice president of Ghana.

In a cabinet reshuffle in Haiti, Pierre Richard Casimir was appointed foreign minister and Ronsard Saint-Cyr interior minister.

Mircea Dușa became the Romanian interior minister following the resignation of Ioan Rus. Titus Corlățean was named foreign minister.

In Syria, prime minister Riyad Hijab defected and fled to Jordan. The Syrian regime released a statement saying Hijab had been dismissed but no official explanation was offered. Wael al-Halki was sworn in as his successor.

In Timor-Leste, a new cabinet took office with José Luís Guterres as foreign minister and Cirilo José Cristóvão as defence and security minister; Emília Pires remained finance minister.

Libya's newly formed national assembly elected former opposition leader Mohamed Magariaf as the country's interim president. Magariaf won 113 votes in the 200-member General National Congress, with his opponent Ali Zeidan receiving 85 votes.

Norov Altankhuyag took office as prime minister of Mongolia. A week later numerous cabinet members were approved, including Luvsanvandan Bold as foreign minister, Dashdemberel Bat-Erdene as defence minister and Chultem Ulaan as finance minister.

Week beginning 12 August 2012

In a cabinet reshuffle in Panama, Rómulo Roux was appointed foreign minister.

In Myanmar, following the resignation of vice president Tin Aung Myint Oo, Nyan Tun was sworn in as his successor.

Sam Kutesa was reinstated as Uganda's foreign minister after stepping down ten months earlier following allegations of corruption.

Danilo Medina Sánchez was sworn in as president of the Dominican Republic. Carlos Morales Troncoso continued as foreign minister and José Ramón Fadul as interior minister; Adm. Sigfrido Pared Pérez became defence minister and Simón Lizardo finance minister.

Week beginning 19 August 2012

In Belarus, Vladimir Makei was appointed foreign minister following the dismissal of Syarhey Martynau.

A national unity government formed in Mali included Tieman Coulibaly as foreign minister. Other key portfolios remained unchanged.

Russia joined the World Trade Organization.

Mauricio Cárdenas was named finance minister of Colombia.

In a minor cabinet reshuffle in the Gambia, Mamadou Tangara became foreign minister.

Week beginning 26 August 2012

In a cabinet reshuffle in Myanmar, Lieut.-Gen. Wai Lwin became defence minister and U Win Shein finance minister.

Romanian president Traian Băsescu resumed office, following an attempted parliamentary impeachment.

In parliamentary elections in Angola, the Popular Movement for the Liberation of Angola (MPLA) gained 175 seats in the National Assembly with 71·8% of votes cast, National Union for the Total Independence of Angola (UNITA) 32 with 18·7%, Broad Convergence for the Salvation of Angola–Electoral Coalition (CASA–CE) 8 with 6·0%, Social Renewal Party 3 with 1·7% and the National Front for the Liberation of Angola 2 with 1·1%. Turnout was 62·8%.

In the Philippines, following the death of interior secretary Jesse Robredo, Manuel 'Mar' Roxas took over the portfolio.

In Kyrgyzstan, following the resignation of the government, the parliament approved the new government with Zhantoro Satybaldiyev as prime minister, Erlan Abdyldayev as foreign minister and Olga Lavrova as finance minister; Taalaibek Omuraliyev remained defence minister and Zarylbek Rysaliyev interior minister.

Week beginning 2 September 2012

Abdelmalek Sellal was appointed prime minister of Algeria by president Abdelaziz Bouteflika. The government was named the following day with no changes in key portfolios.

Week beginning 9 September 2012

Hassan Sheikh Mohamud was elected president of Somalia. He was sworn in a week later.

Mustafa Abu Shagur was elected prime minister of Libya, but was dismissed the following month after failing to form a government.

In the Netherlands' parliamentary elections, the People's Party for Freedom and Democracy won 41 of 150 seats with 26·6% of the vote, the Labour Party 38 with 24·8%, the Party for Freedom 15 with 10·1%, the Socialist Party 15 with 9·7%, the Christian Democratic Appeal 13 with 8·5%, the Democrats 66 12 with 7·9%, the Christian Union 5 with 3·2%, Green Left 4 with 2·3%, the Reformed Political Party 3 with 2·1%, the Party for the Animals 2 with 1·9% and 50+ 2 with 1·9%. Turnout was 74·3%.

In a cabinet reshuffle in Afghanistan, Bismallah Mohammadi became defence minister and Mujtaba Patang interior minister.

In a major cabinet reshuffle in Bangladesh, Mahiuddin Khan Alamgir was appointed interior minister. Other key portfolios remained unchanged, but there were a number of changes among more minor posts.

Week beginning 16 September 2012

In Dominica, following the resignation of president Nicholas Liverpool, Eliud Williams was sworn in as his successor.

Following the resignation of Georgia's interior minister Bacho Akhlaia, Eka Zguladze took over the portfolio.

In Ethiopia, the acting prime minister Hailemariam Desalegn was sworn in as prime minister.

In a cabinet reshuffle in Norway, Espen Barth Eide was appointed foreign minister while Anne-Grete Strøm-Erichsen took over the defence portfolio.

Week beginning 23 September 2012

In parliamentary elections in Belarus, independent candidates won 104 of 109 seats, the Communist Party 3, the Agrarian Party 1 and the Republican Party of Labour and Justice 1. All the elected deputies were supporters of President Lukashenka as the two main opposition parties boycotted the election. Turnout was 74·2%.

In Kazakhstan, Serik Akhmetov was appointed prime minister following the resignation of Karim Masimov. Subsequently a cabinet was announced including Krymbek Kusherbayev as deputy prime minister and Yerlan Idrisov as foreign minister.

In a cabinet reshuffle in the Republic of the Congo, Charles Richard Mondjo was appointed defence minister and Raymond Zéphirin Mboulou interior minister.

In Thailand, deputy prime minister and interior minister Yongyuth Wichaidit resigned following allegations of corruption.

Week beginning 30 September 2012

Angola's new government was sworn in including Manuel Vicente as vice president and Ângelo de Barros Veiga Tavares as new interior minister. Other key portfolios remained unchanged.

In parliamentary elections in Georgia, the opposition Georgian Dream coalition won 83 of the 150 seats, with the ruling United National Movement taking 67 seats. Turnout was 59·8%.

In a cabinet reshuffle in Japan, Koriki Jojima became finance minister, Shinji Tarutoko interior minister and Keishu Takana justice minister.

Denise Bronzetti and Teodoro Lonfernini took office as captains regent of San Marino.

In a cabinet reshuffle in Guinea, Kerfalla Yansané was appointed economy and finance minister and Louncény Fall foreign minister.

Abdi Farah Shirdon was appointed prime minister of Somalia by president Hassan Sheikh Mohamud.

Week beginning 7 October 2012

In presidential elections in Venezuela, incumbent president Hugo Chávez was re-elected. He subsequently appointed Nicolás Maduro as vice president and Nestor Reverol as interior minister. Some weeks later Diego Molero was named defence minister.

In Mozambique, Alberto Vaquina was appointed prime minister following the dismissal of Aires Ali.

Abdullah Ensour was appointed prime minister of Jordan by King Abdullah II. The government was named the following day with Ensour also serving as defence minister; Awad Khlewait became interior minister, Suleiman Hafez remained finance minister and Nasser Judeh foreign minister.

The European Union was awarded the Nobel Peace Prize for 'promoting peace, democracy and human rights over six decades'.

Week beginning 14 October 2012

In the two rounds of Lithuania's parliamentary elections, the Social Democratic Party of Lithuania won 38 of the 141 seats, Homeland Union-Lithuanian Christian Democrats 33 and Labour Party 29. Turnout was 52·9%.

In Montenegro's parliamentary elections, the ruling Coalition for European Montenegro won 39 of 81 seats with 45·6% of the vote, the Democratic Front 20 with 22·8%, the Socialist People's Party 9 with 11·1% and Positive Montenegro 7 with 8·2%. Turnout was 72·8%.

Week beginning 21 October 2012

In Bosnia and Herzegovina, defence minister Muhamed Ibrahimović was dismissed.

In Georgia, Bidzina Ivanichvili's government was sworn in. Irakli Alasania was appointed deputy prime minister and defence minister, Nodar Khaduri finance minister, Maya Panjikidze foreign minister and Thea Tsulukiani justice minister.

In a major government reshuffle in India, Salman Khurshid was named foreign minister. Other key portfolios remained unchanged.

Week beginning 28 October 2012

In Ukraine's parliamentary elections, the Party of Regions won 187 seats with 30·0% of the vote, Yuliya Tymoshenko's Fatherland 102 with 25·5%, Vitali Klitschko's Ukrainian Democratic Alliance for Reform 40 with 14·0%, Svoboda 38 with 10·4% and the Communist Party of Ukraine 32 with 13·2%. 51 seats were divided between various small parties and independent candidates. Turnout was 58·0%.

In a government reshuffle in Thailand, Phongthep Thepkanjana was appointed deputy prime minister, Surapong Towijakchaikul deputy prime minister and foreign minister and Charupong Ruangsuwan interior minister. Other key posts remained unchanged.

In a major cabinet reshuffle in Senegal, Mankeur Ndiaye was appointed foreign minister and Pathé Seck interior minister. Other key portfolios remained unchanged.

In Vanuatu's parliamentary elections, the Party of Our Land won 8 of 52 seats, the People's Progressive Party 6, the Union of

Moderate Parties 5, the Ground and Justice Party 4, the National United Party 4 and the Iauko Group 3. Turnout was 63·2%.

In Libya, the government of Ali Zeidan took office. Senior ministers included Mohammed Mahmoud al-Bargati as defence minister, Alkilani al-Jazi as finance minister and Salah Bashir Margani as justice minister.

In a cabinet reshuffle in the Gambia, Susan Waffa Ogoo was appointed foreign minister.

Week beginning 4 November 2012

In a cabinet reshuffle in Chile, Rodrigo Hinzpeter was appointed defence minister and Andrés Chadwick interior minister.

In the Netherlands, Mark Rutte's new government was sworn in. Frans Timmermans became foreign minister, Jeanine Hennis-Plasschaert defence minister, Jeroen Dijsselbloem finance minister and Ronald Plasterk interior minister.

Prince Muhammad bin Nayef was appointed interior minister of Saudi Arabia by King Abdullah.

In the second round of presidential elections in Palau, Tommy Remengesau was elected president with 58·0% of the vote. He defeated incumbent leader Johnson Toribiong.

In Russia, following the dismissal of Anatoly Serdyukov, Sergey Shoigu was named defence minister.

Barack Obama was re-elected in the US presidential elections with 50·9% of the vote and 332 electoral college votes; the Republican candidate Mitt Romney received 47·3% of the vote and 206 electoral votes.

CIA director David Petraeus resigned.

In Bosnia and Herzegovina, Nebojša Radmanović took over the eight-month rotating presidency chairmanship.

Week beginning 11 November 2012

In parliamentary elections in San Marino, the San Marino Common Good coalition won 35 of 60 seats with 50·7% of the vote; the Agreement for the Country coalition won 12 with 22·3%; the Active Citizenship coalition 9 with 16·1%; and the Network Civic Movement 4 with 6·3%. Turnout was 63·8%. A new cabinet was announced two weeks later with Pasquale Valentini as foreign minister, Giancarlo Venturini as interior minister and Claudio Felici as finance minister.

Xi Jinping was elected General Secretary of the Chinese Communist Party and consequently new president. Li Keqiang, Zhang Dejiang, Yu Zhengsheng, Liu Yunshan, Wang Qishan and Zhang Gaoli were also chosen as members of the country's principal decision making body, the Standing Committee of the Politburo.

In Somalia, Abdi Farah Shirdon's government was sworn in. Fowsiyo Yusuf Haji Adan became foreign minister, Abdihakim Mohamoud Haji defence minister, Mohamoud Hassan Suleiman finance minister and Abdikarim Hussein Guled interior minister.

Presidential and parliamentary elections were held in Sierra Leone. The incumbent president Ernest Bai Koroma was re-elected with 58·7% of the vote. In parliamentary elections, the All People's Congress won 67 of 124 seats with 53·7% of the vote and the Sierra Leone People's Party 42 with 38·3%; 12 seats were allocated for elected chiefs.

Week beginning 18 November 2012

In Poland, the vice prime minister and minister of economy Waldemar Pawlak resigned. Two weeks later Janusz Piechocinski took over both posts.

Sato Kilman was re-elected prime minister of Vanuatu. He named Charlot Salwai finance minister and Toara Daniel Kalo internal affairs minister; Alfred Carlot remained foreign minister.

In Côte d'Ivoire, following the dismissal of Jeannot Ahoussou-Kouadio's government, Daniel Kablan Duncan was appointed prime minister. The cabinet subsequently named included Charles Koffi Diby as foreign minister, Hamed Bakayoko as interior minister and Paul Koffi Koffi as defence minister.

In a government reshuffle in Bosnia and Herzegovina, Zekerijah Osmić became defence minister.

Algirdas Butkevičius took office as the prime minister of Lithuania. The new cabinet was announced two weeks later with Linas Antanas Linkevičius as foreign minister, Rimantas Šadžius as finance minister, Dailis Alfonsas Barakauskas as interior minister and Juozas Bernatonis as justice minister.

Week beginning 25 November 2012

Following the resignation of Ecuador's defence minister Miguel Carvajal, María Fernanda Espinosa was sworn in as his successor.

Francis Zammit Dimech was named foreign minister of Malta.

In a cabinet reshuffle in Ethiopia, Tedros Adhanom was appointed foreign minister.

In North Korea, Kim Kyok-sik replaced Kim Jong-gak as defence minister.

Palestine was granted 'non-member observer' status in the United Nations.

Enrique Peña Nieto took office as president of Mexico. His cabinet included José Antonio Meade as foreign minister, Miguel Ángel Osorio Chong as interior minister and Gen. Salvador Cienfuegos Zepeda as defence minister.

Week beginning 2 December 2012

In parliamentary elections in Burkina Faso, the Congress for Democracy and Progress won 70 of 127 seats; the Alliance for Democracy and Federation–African Democratic Rally 19; and the Union for Progress and Change 19. Turnout was 76·0%.

In Montenegro, Milo Đukanović's government was sworn in. Igor Lukšić became deputy prime minister and foreign minister, Radoje Žugić finance minister and Raško Konjević interior minister. Milica Pejanović continued as defence minister.

In the second round of presidential elections in Slovenia, Borut Pahor was elected president with 67·4% of the vote. He defeated incumbent Danilo Türk. Pahor was sworn in three weeks later.

In a major government reshuffle in Namibia, Hage Geingob was appointed prime minister, Netumbo Nandi-Ndaitwah foreign minister and Nahas Angula defence minister. Untoni Nujoma was named justice minister and Pendukeni Iivula-Iithana home affairs minister.

In the presidential elections in Ghana, John Dramani Mahama of the National Democratic Congress (NDC) ensured victory in the first round with 50·7% of the vote. He defeated Nana Akufo-Addo of the New Patriotic Party (NPP), who received 47·7%. Turnout was 79·4%. In parliamentary elections held simultaneously, the NDC won 148 of 275 seats and the NPP 123. Turnout was 80·0%.

Week beginning 9 December 2012

In parliamentary elections in Romania, the Social Liberal Union (USL) alliance won 273 of 412 seats with 58·6% of the vote in the Chamber of Deputies; the Right Romania Alliance won 56 with 16·5%; and the People's Party–Dan Diaconescu 47 with 14·0%. Turnout was 41·7%. Two weeks later Victor Ponta's cabinet was sworn in with Daniel Chițoiu as deputy prime minister and finance minister and Radu Stroe as interior minister. Mircea Duşa became defence minister while Titus Corlăţean continued as foreign minister.

In São Tomé e Príncipe, following the dismissal of Patrice Trovoada's cabinet, Gabriel Arcanjo da Costa was appointed prime minister. Two days later a new government took office including Natália Pedro da Costa Umbelina Neto as foreign minister, Óscar Aguiar Sacramento e Sousa as interior and defence minister and Hélio Silva Vaz d'Almeida as finance minister.

In Estonia, following the resignation of justice minister Kristen Michal, Hanno Pevkur took over the portfolio.

In Mali, Diango Cissoko was appointed prime minister following the forced resignation of Cheick Modibo Diarra by the military. Subsequently a new cabinet was announced with no changes in key portfolios.

In a cabinet reshuffle in Czech Republic, Karolina Peake was named defence minister. She was dismissed a week later and prime minister Petr Nečas provisionally took over the portfolio.

In Israel, Binyamin Netanyahu became the acting foreign minister following the resignation of Avigdor Lieberman.

Week beginning 16 December 2012

In parliamentary elections in Japan, the Liberal Democratic Party (LDP) won 294 of 480 seats with 43·0% of the single-seat constituency vote; the ruling Democratic Party (DPJ) 57 with 22·8%; and the Restoration Party (JRP) 54 with 11·6%. Turnout was 59·3%, the lowest since the Second World War. Two weeks later Shinzo Abe was elected prime minister for a second time. He formed a cabinet with Fumio Kishida as foreign minister, Itsunori Onodera as defence minister, Yoshitaka Shindo as internal affairs minister and Taro Aso as finance minister.

In presidential elections in South Korea, Park Geun-hye was elected president with 51·6% of the vote. Turnout was 75·8%.

Elyes Fakhfakh was appointed finance minister of Tunisia.

Italian Prime Minister Mario Monti resigned following the passage of the 2013 budget.

Week beginning 23 December 2012

In Ukraine, Mykola Azarov was reappointed as prime minister following his resignation a few weeks earlier. A new cabinet was appointed by president Viktor Yanukovych including Leonid Kozhara as foreign minister and Pavlo Lebedev as defence minister. Other key portfolios remained unchanged.

President Mohamed Morsy ratified the new constitution of Egypt. The constitution had been approved in the previous weeks by the Egyptian people in a two-stage referendum.

Week beginning 30 December 2012

Ueli Maurer took office as president of Switzerland.

The US Senate passed a last-minute deal averting the so-called fiscal cliff, a combination of dramatic spending cuts and tax increases mandated to come into effect in January 2013. The agreement involved delaying federal budget cuts for two months and returning to Bill Clinton-era tax rates for top earners of 39·6% for individuals earning US$400,000 and above and households earning US$450,000 and above.

In Burkina Faso, Luc Adolphe Tiao's government was sworn in. Dramane Yaméogo became justice minister; other key portfolios remained unchanged.

In a cabinet reshuffle in Sierra Leone, Joseph B. Dauda was named internal affairs minister, Samura Kamara foreign minister and Kaifala Marah finance minister.

Week beginning 6 January 2013

In a cabinet reshuffle in Egypt, Gen. Mohamed Ibrahim became interior minister and El-Morsi Hegazy finance minister.

In Guatemala, foreign minister Harold Caballeros resigned. A week later Fernando Carrera took over the portfolio.

Mohamed Abdulaziz was appointed acting foreign minister of Libya by prime minister Ali Zeidan.

Week beginning 13 January 2013

Elías Jaua was named foreign minister of Venezuela.

Tommy Remengesau was sworn in as the president of Palau.

Nicolas Tiangaye took office as the prime minister of the Central African Republic. A new cabinet was announced two weeks later with Michel Djotodia as defence minister, Parfait-Anicet M'bay as foreign affairs minister and Enoch Derant Lakoué as economy minister.

Over 800 people were held hostage at a gas facility near In Amenas in Algeria. Following a siege lasting several days, Algerian special forces launched an attack on the site in order to free the hostages; the assault ending the crisis resulted in the death of around 80 people.

Week beginning 20 January 2013

Barack Obama was sworn in for his second term as the president of the United States.

In Chad, Djimrangar Dadnadji was appointed prime minister following the resignation of Emmanuel Nadingar. Subsequently a cabinet was announced including Atteib Habib Doutoum as finance minister and Abdoulaye Sabre Fadoul as justice minister.

In a cabinet reshuffle in Haiti, David Bazile was appointed interior minister.

In parliamentary elections in Israel, Likud Yisrael Beiteinu (an alliance of Likud and Yisrael Beiteinu) won 31 of 120 seats with 23·3% of the votes cast; the Yesh Atid won 19 with 14·3%; the Labour Party won 15 with 11·4%; the Jewish Home won 12 with 9·1%; and Shas won 11 with 8·8%. Turnout was 67·8%.

Muammer Güler was appointed interior minister in a cabinet reshuffle in Turkey.

In the second round of the Czech Republic's first ever direct presidential elections, Miloš Zeman was elected president with 54·8% of the vote to Karel Schwarzenberg's 45·2%. There had been nine candidates in the first round two weeks earlier. Turnout was 61·3% in the first round and 59·1% in the second.

Week beginning 27 January 2013

President Morsy declared a 30-day state of emergency and curfew in three Suez Canal provinces hit hardest by a wave of violent protests sparked by the second anniversary of the Egyptian revolution.

Queen Beatrix of the Netherlands announced her abdication as of 30 April with her eldest son, Prins Willem-Alexander, becoming the country's first king since 1890.

In Slovenia, following the resignation of ministers of finance and justice, prime minister Janez Janša temporarily took over the finance portfolio and Zvonko Černač became the acting justice minister.

John Kerry succeeded Hillary Clinton as the United States Secretary of State.

A new cabinet took office in Ghana with Hanna Tetteh as foreign minister and Seth Terkper as finance minister.

Week beginning 3 February 2013

In parliamentary elections in Liechtenstein, the Progressive Citizens' Party (FBP) won 10 of 25 seats with 40·0% of the vote; the Patriotic Union (VU) won 8 with 33·5%; the Independents won 4 with 15·3%; and the Free List (FL) won 3 with 11·1%. Turnout was 79·8%.

Reckya Madougou became minister of justice in a cabinet reshuffle in Benin.

In a cabinet reshuffle in St Kitts and Nevis, Dr Earl Asim Martin was appointed deputy prime minister and Patrice Nisbett foreign minister.

Ismail Ismail was named finance minister in a cabinet reshuffle in Syria. Other key portfolios remained unchanged.

Week beginning 10 February 2013

In parliamentary elections in Monaco, Horizon Monaco won 20 of 24 seats with 50·3% of the vote; the Monegasque Union won 3 with 39·0%; and Renaissance won one seat with 10·7%. Turnout was 74·5%.

Pope Benedict XVI announced that he would resign at the end of the month, citing health reasons.

Anthony Carmona was elected president of Trinidad and Tobago.

Week beginning 17 February 2013

In presidential elections in Ecuador, Rafael Vicente Correa Delgado was re-elected with 57·2% of the vote against 22·7% for Guillermo Lasso. There were six other candidates. In parliamentary elections, Alianza PAIS won 100 of 137 seats with 52·3% of the vote and Creating Opportunities 11 with 11·4%. Six other parties also won seats.

In presidential elections in Armenia, Serzh Sargsyan was re-elected with 58·6% of the vote. He defeated Raffi Hovannisian. Turnout was 60·1%. Protests over alleged electoral fraud followed.

Jiang Yi-huah took office as prime minister of Taiwan. His cabinet included Mao Chi-Kuo as deputy premier and Chang Chia-juch as economic affairs minister. Lin Yung-lo remained foreign minister.

In a cabinet reshuffle in Chad, Ali Mahamat Zene Ali Fadel became interior minister and Issa Ali Taher economy minister.

In parliamentary elections in Grenada, New National Party won all 15 seats taking 60·0% of the vote; the ruling National Democratic Congress failed to obtain any seats despite taking 39·3%. Turnout was 85·0%. Keith Mitchell was sworn in as prime minister the following day.

In Pakistan, following the resignation of finance minister Abdul Hafeez Shaikh, Saleem Mandviwalla took over his portfolio.

In Tunisia, the government of prime minister Hamadi Jebali resigned after failing to reshape the government as a response to a political crisis. The crisis was sparked by the assassination of opposition leader Chokri Belaid earlier in the month.

In Bulgaria, the government of prime minister Boyko Borisov resigned following nationwide street protests against austerity measures and high electricity prices.

In parliamentary elections in Barbados, the ruling Democratic Labour Party won 16 of 30 seats with 51·3% of the vote and the Barbados Labour Party won 14 with 48·3%. A week later prime minister Freundel Stuart announced his cabinet with no change in key portfolios.

In parliamentary elections in Djibouti, the Union for a Presidential Majority won 43 of 65 seats with 61·5% of the vote; the Union for National Salvation won 21 seats with 35·6%; the Centre of Unified Democrats took one seat with 3·0%. Turnout was 69·2%. The opposition rejected the result and claimed that the vote was rigged.

Abdylda Suranchiyev was appointed interior minister of Kyrgyzstan.

Week beginning 24 February 2013

In the second round of presidential elections in Cyprus, Nicos Anastasiadies was elected with 57·5% of the vote against 42·5% for Stavros Malas. Anastasiadies and his all male cabinet were sworn in a week later.

In parliamentary elections in Italy, Pier Luigi Bersani's centre-left coalition won 345 of 630 seats in the Chamber of Deputies (including the Democratic Party with 297 seats), former prime minister Silvio Berlusconi's centre-right coalition 125 (including People of Freedom, 98), Beppe Grillo's Five Star Movement 109 and Prime Minister Mario Monti's coalition 47. In the Senate, Bersani's coalition won 123 of 315 seats (including the Democratic Party 111), Berlusconi's coalition 117 (including People of Freedom 98), the Five Star Movement 54 and Monti 19.

Park Geun-hye was sworn in as the president of South Korea.

In a cabinet reshuffle in Poland, Jan-Vincent Rostowski became deputy prime minister and finance minister. Bartlomiej Sienkiewicz was named interior minister.

Claver Gatete was named finance minister of Rwanda.

In Panama, following the resignation of Rómulo Roux, Fernando Núñez Fábrega took over as foreign minister.

In the USA, Charles 'Chuck' Hagel and Jacob 'Jack' Lew were sworn in as defence secretary and treasury secretary respectively.

In Zambia, Effron Lungu was appointed foreign minister following the dismissal of Given Lubinda.

Week beginning 3 March 2013

The government of Grenada was sworn in with the finance, national security and home affairs portfolios assigned to prime minister Keith Mitchell. Elvin Nimrod was named deputy premier and Nickolas Steele foreign affairs minister.

In presidential elections in Kenya, Uhuru Kenyatta was elected with 50·1% of the vote against 43·3% for Raila Odinga. There were six other candidates. In parliamentary elections Uhuru Kenyatta's Jubilee alliance won 167 seats (of which The National Alliance won 89); Raila Odinga's Coalition for Reforms and Democracy alliance, 141 (of which the Orange Democratic Movement won 96); and Musalia Mudavadi's Amani coalition, 24 (of which the United Democratic Forum won 12). Five other parties also won seats.

In Belgium, following the resignation of finance minister Steven Vanackere, Koen Geens took over his portfolio.

Only non-partisans were elected in parliamentary elections in Micronesia.

Mihály Varga became the finance minister of Hungary.

Miloš Zeman was sworn in as the president of the Czech Republic.

John Brennan was sworn in as the director of the CIA.

Following the death of Hugo Chávez, Venezuela's vice president Nicolás Maduro took over as president. Jorge Arreaza was named vice president.

In parliamentary elections in Malta, the opposition Labour Party (MLP) won 39 of 69 seats with 54·8% of votes cast and the Nationalist Party (NP) won 30 with 43·3%.

Week beginning 10 March 2013

Gerald Klug was appointed defence minister of Austria.

A caretaker government in Bulgaria took office with Marin Raykov as prime minister and minister of foreign affairs, Petya Purvanova as interior minister, Todor Tagarev as defence minister and Kalin Hristov as finance minister.

The papal conclave elected Cardinal Jorge Mario Bergoglio to the papacy. He became Pope Francis and was inaugurated a week later as the 266th Supreme Pontiff.

In Tunisia, Ali Larayedh was sworn in as caretaker prime minister. His cabinet took office the following day including Rachid Sabagh as defence minister and Lotfi Ben Jeddou as interior minister.

Abdul Hamid was appointed acting president of Bangladesh as incumbent Zillur Rahman underwent medical treatment abroad. Rahman died a week later in Singapore after a long illness.

Xi Jinping was elected president of China by the National People's Congress. Li Yuanchao became vice president and Li Keqiang premier.

Khil Raj Regmi took office as interim prime minister of Nepal. He subsequently named his cabinet with Shankar Koirala as finance minister; Madhav Prasad Ghimire took over the foreign affairs and interior portfolios.

Week beginning 17 March 2013

In South Korea, president Park Geun-hye appointed Yun Byung-se as foreign minister, Hwang Kyo-ahn as justice minister and Bahk Jae-wan as finance minister. The cabinet list was finalized a week later when Kim Kwan-jin was appointed defence minister.

Binyamin Netanyahu was sworn in for his third term as the prime minister of Israel. Netanyahu also took on the foreign affairs portfolio. Moshe Ya'alon became defence minister, Gideon Sa'ar interior minister and Yair Lapid finance minister.

Vlastimil Picek was appointed defence minister of Czech Republic.

Alenka Bratušek was sworn in as the prime minister of Slovenia. Her cabinet included Karl Erjavec as foreign minister, Roman Jakič as defence minister, Uroš Čufer as finance minister and Gregor Virant as interior minister.

In Lebanon, the government of Najib Mikati resigned but stayed on in a caretaker capacity.

Following the resignation of Sato Kilman, Moana Carcasses Kalosil became the new prime minister of Vanuatu. His cabinet included Edward Natapei as foreign minister, Willie Jimmy Tapangararua as finance minister and Patrick Crowby as interior minister.

Week beginning 24 March 2013

Mir Hazar Khan Khoso was sworn in as caretaker prime minister of Pakistan. A week later Malik Habib was appointed interior minister.

In Italy, foreign minister Giulio Terzi resigned.

François Bozizé, the president of the Central African Republic, was forced to flee the country as the Séléka rebel coalition under the leadership of Michel Djotodia took control of the capital. Nicolas Tiangaye was reappointed prime minister and a transition government was subsequently named with Djotodia as defence minister, Charles Armel Doubane as foreign minister and Georges Bozanga as finance minister.

Adrian Hasler took office as the prime minister and finance minister of Liechtenstein. His cabinet included Thomas Zwiefelhofer as interior and justice minister with Aurelia Frick remaining foreign minister.

Week beginning 31 March 2013

Abdoulkader Kamil Mohamed was sworn in as the prime minister of Djibouti. Hassan Darar Houffaneh was appointed defence minister and Hassan Omar Mohamed interior minister. Mahamoud Ali Youssouf remained foreign minister and Ilyas Moussa Dawaleh finance minister.

Pak Pong-ju was appointed prime minister of North Korea.

Antonella Mularoni and Denis Amici were inaugurated as captains regent of San Marino.

In Cyprus, following the resignation of finance minister Michalis Sarris, Charis Georgiadis took over his portfolio.

Aldo Bumçi was appointed foreign minister in a cabinet reshuffle in Albania.

STATESMAN'S YEARBOOK SELECT SOURCES

The Statesman's Yearbook references the following sources to maintain accuracy and currency of information contained in our database:

- 2011 Global Go To Think Tank Rankings (http://www.gotothinktank.com/wp-content/uploads/2012/09/2011-Global-Go-To-Think-Tanks-Report_FINAL-VERSION.pdf)
- United Nations Statistical Yearbook (http://unstats.un.org/unsd/syb)
- Euromonitor International Marketing Data and Statistics (http://www.euromonitor.com/international-marketing-data-and-statistics-2012/book)
- United Nations Development Programme Human Development Report (http://hdr.undp.org/en)
- United Nations World Population Prospects (http://esa.un.org/unpd/wpp/index.htm)
- United Nations World Urbanization Prospects (http://esa.un.org/unup)
- United Nations Demographic Yearbook (http://unstats.un.org/unsd/demographic/products/dyb/dyb2.htm)
- Stockholm International Peace Research Institute Yearbook (http://www.sipri.org/yearbook)
- International Institute of Strategic Studies Military Balance (http://www.iiss.org/publications/military-balance)
- International Monetary Fund World Economic Outlook (http://www.imf.org/external/pubs/ft/weo/2012/01/index.htm)
- Selected International Monetary Fund Reports
- Selected European Union Reports
- Eurostat (http://epp.eurostat.ec.europa.eu/portal/page/portal/eurostat/home)
- Selected World Bank Reports
- Selected World Trade Organization Reports
- International Monetary Fund Government Finance Statistics Yearbook (http://www.imfbookstore.org/ProdDetails.asp?ID=GYIEA2011001&PG=1&Type=BL)
- World Investment Report (http://unctad.org/en/Pages/Publications/WorldInvestmentReports(1991-2009).aspx)
- United Nations Energy Statistics Yearbook (http://unstats.un.org/unsd/energy/yearbook/default.htm)
- BP Statistical Review of World Energy (http://www.bp.com/sectionbodycopy.do?categoryId=7500&contentId=7068481)
- Food and Agricultural Organization Forest Products Yearbook (http://www.fao.org/forestry/62283/en)
- Food and Agricultural Organization Fishery Statistics — Capture Production (http://www.fao.org/fishery/statistics/global-capture-production/en)
- United Nations Industrial Commodity Statistics Yearbook (http://unstats.un.org/unsd/industry/icsy_intro.asp)
- International Labour Organization Statistics (http://laborsta.ilo.org)
- United Nations International Trade Statistics Yearbook (http://comtrade.un.org/pb/first.aspx)
- International Road Federation World Road Statistics (http://www.irfnet.org/statistics.php)
- International Railway Statistics (http://www.uic.org/etf/publication/publication-resultat.php?domaine=3)
- Railway Directory (http://www.railwaydirectory.net)
- Selected Airports Council International Reports (http://www.aci.aero/Data-Centre)
- Shipping Statistics Yearbook (http://www.infoline.isl.org)
- International Telecommunication Union Yearbook of Statistics (http://www.itu.int/ITU-D/ict/publications/yb/index.html)
- UNESCO Institute for Statistics (http://stats.uis.unesco.org/unesco/TableViewer/document.aspx?ReportId=143&IF_Language=eng)
- Religious Trends No 7
- World Health Statistics (http://www.who.int/gho/publications/world_health_statistics/en/index.html)
- World Association of Newspapers World Press Trends (http://www.wan-ifra.org/microsites/world-press-trends)
- National and regional statistical offices and yearbooks, government departments and international organizations throughout the world
- Selected print and online national and international news media

ABBREVIATIONS

ACP	African Caribbean Pacific	ind.	independent(s)
Adm.	Admiral	ICT	information and communication technology
Adv.	Advocate		
a.i.	ad interim	ISO	International Organization for Standardization (domain names)
b.	born		
bbls	barrels	K	kindergarten
bd	board	kg	kilogramme(s)
bn.	billion (one thousand million)	kl	kilolitre(s)
Brig.	Brigadier	km	kilometre(s)
bu.	bushel	kW	kilowatt
		kWh	kilowatt hours
Capt.	Captain		
Cdr	Commander	lat.	latitude
CFA	Communauté Financière Africaine	lb	pound(s) (weight)
CFP	Comptoirs Français du Pacifique	Lieut.	Lieutenant
CGT	compensated gross tonnes	long.	longitude
c.i.f.	cost, insurance, freight		
C.-in-C.	Commander-in-Chief	m.	million
CIS	Commonwealth of Independent States	Maj.	Major
cm	centimetre(s)	MW	megawatt
Co.	County	MWh	megawatt hours
Cres.	Crescent		
cu.	cubic	NA	not available
CUP	Cambridge University Press	n.e.c.	not elsewhere classified
cwt	hundredweight	NRT	net registered tones
		NT	net tonnes
D.	Democratic Party	NTSC	National Television System Committee (525 lines 60 fields)
DWT	dead weight tonnes		
EEA	European Economic Area	OUP	Oxford University Press
EEZ	Exclusive Economic Zone	oz	ounce(s)
EMS	European Monetary System		
EMU	European Monetary Union	PAYE	Pay-As-You-Earn
ERM	Exchange Rate Mechanism	PPP	Purchasing Power Parity
est.	estimate		
		R.	Republican Party
f.o.b.	free on board	Rd	Road
FDI	foreign direct investment	retd	retired
ft	foot/feet	Rt Hon.	Right Honourable
FTE	full-time equivalent		
		SADC	Southern African Development Community
G8 Group	Canada, France, Germany, Italy, Japan, UK, USA, Russia	SDR	Special Drawing Rights
GDP	gross domestic product	sq.	square
Gdns	Gardens	St.	Street
Gen.	General	SSI	Supplemental Security Income
GNI	gross national income		
GNP	gross national product	TAFE	technical and further education
GRT	gross registered tonnes	TEU	twenty-foot equivalent units
GT	gross tonnes	trn.	trillion (one million million)
GW	gigawatt	TV	television
GWh	gigawatt hours		
		Univ.	University
ha.	hectare(s)	VAT	value-added tax
HDI	Human Development Index	v.f.d.	value for duty

CURRENT LEADERS INDEX

Explore the world at
www.statesmansyearbook.com

PLACE AND INTERNATIONAL ORGANIZATIONS INDEX

Italicized page numbers refer to extended entries

International Historical Statistics

The *International Historical Statistics* is a peerless collection of statistical data from around the world, covering a wide range of socio-economic topics. The collection not only includes data on the Americas and Europe, but also hard-to-find data on Africa, Asia and Oceania.

What's exciting about *International Historical Statistics*?

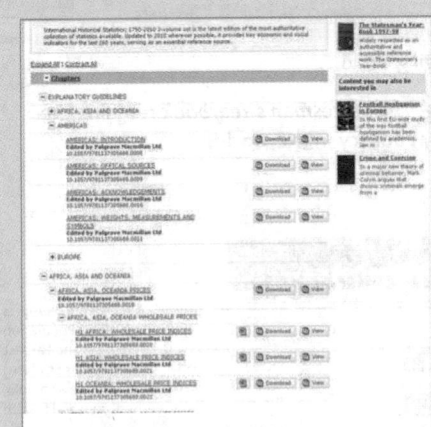

- 490 tables, 211 countries and territories
- Tables updated to include 2010 data, where available
- Excel format compatible with econometrics tools, with footnotes included on each table file (ePDF and Excel)
- Sources and method of collation are consistent, making it possible to compare tables from across volumes
- Difficult-to-source data on countries whose borders have changed (e.g. Czechoslovakia and India)
- Unique historical data for Africa, Asia and Oceania, ranging from 1750 to 2010. This data is not available anywhere else in this breadth and depth
- Editable Excel files with the ability to remove sections and merge tables from the other volumes and regions
- All the hard work of pulling data from various sources into one place is already done! All sources have been verified and are listed

University Student Numbers in Europe

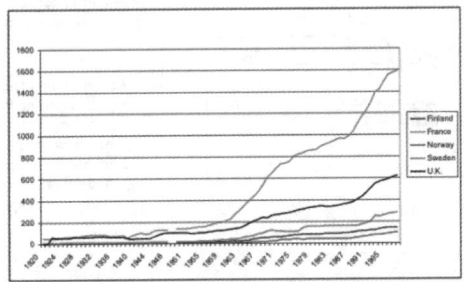

Coffee Output in South America

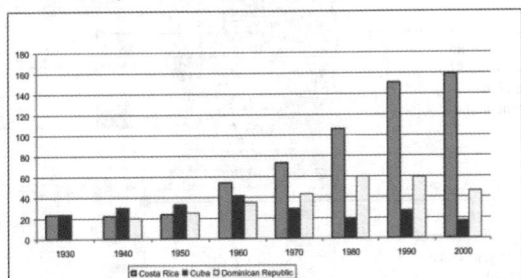

↳ *Turn over to see explanation from The Statesman's Yearbook on why these changes have taken place*

THE STATESMAN'S YEARBOOK
ONLINE, IN PRINT AND INDISPENSABLE

The Statesman's Yearbook features extensive updated profiles on every country in the world. Each profile includes a concise geographic, historical, economic, political, social and cultural overview of the country together with essential statistics.

It includes:

- Detailed maps of all 194 countries
- Biographical profiles of all current leaders
- Economic overviews for every country
- Over 1,700 links to key national websites
- Our complete archive from 1864 – a unique insight into a changing world

At a time when opinion and propaganda can be overwhelming, *The Statesman's Yearbook* remains the first point of reference for reliable, concise information on any country in the world.

Features on our website include:

Chronologies including a week-by-week summary of world events and a special feature on the Olympic Games

Top 50 think tanks

Focus: an extract from one of our scholarly titles with content analysis

City profile: series of quarterly, in-depth city profiles

The Comparison Tool allows you to compare a range of data fields including summary statistics, economy, key historical events and much more – for as many countries as you like

Fact sheets on topics including World Population Development and Largest Urban Agglomerations, Migration, News Media, Crime and Health

Spotlight: where our editor Dr Barry Turner focuses on a pivotal event in history and its relevance today

Editorial commentary on topics ranging from the current state of print and online media to lessons from history and the fallibility of experts

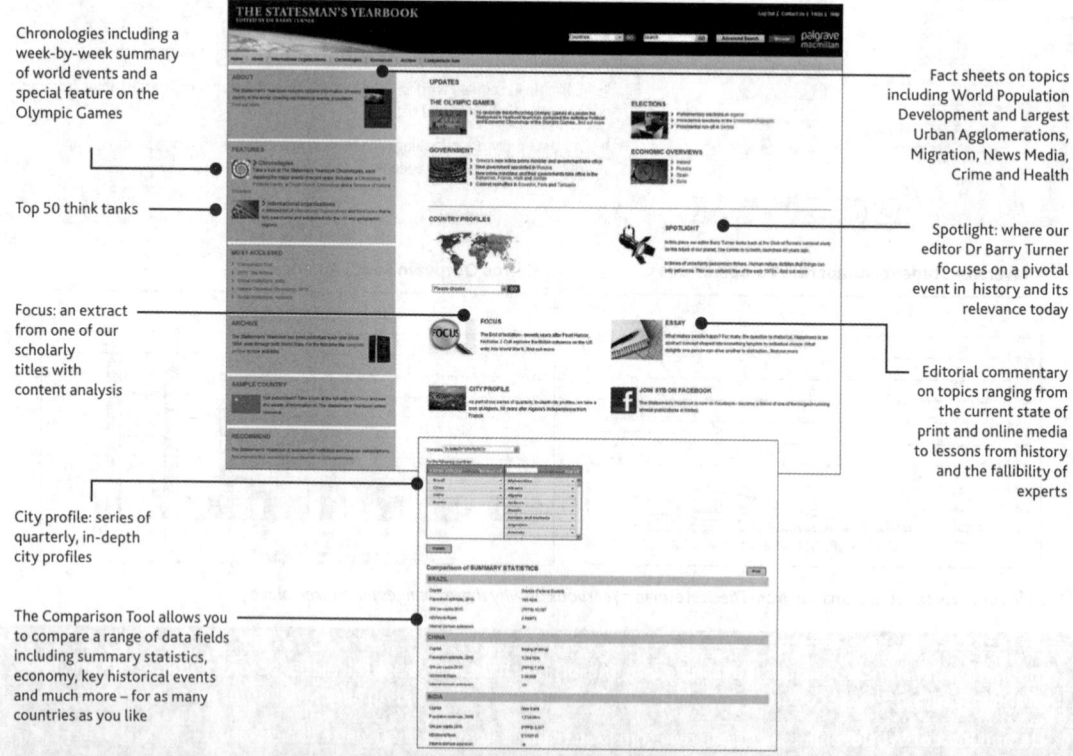

Find out more at www.statesmansyearbook.com

What you can do with this resource...

- Research a country or region in detail – consistent headings, sub-headings and research methods provide easy international cross-referencing
- Explore the impact of Key Historical Events
- Navigate Chronologies of Political Events, the Credit Crunch and a Timeline of Natural Disasters
- Discover how international organizations and the top 50 think tanks transcend borders – includes detailed information on the organizations together with links to their websites all in one place
- Use the archive to chart the history of a country, sometimes from its inception through to its end – for instance, track the demise of the Confederate States of America or learn how Abyssinia became Ethiopia
- Trace the shifting political and geographical landscape of the world through the contemporary narrative of successive editors

The facts behind the figures – *The Statesman's Yearbook* and *International Historical Statistics* – perfect partners!

Line Chart of University Student Numbers in Europe

On the previous page the chart shows how the number of students in Universities in selected European countries has grown since the 1920s. It demonstrates that France saw a dramatic growth in the number of students attending universities from the 1960s. In 1968 reforms* divided several older institutions into many more new ones, a restructuring that prompted an explosion in the number of HE institutions.

* The Statesman's Yearbook; France, Social Institutions, Education

Bar Chart of Coffee Output in South America

On the previous page the chart shows how Cuba's exports dropped off in 1960 when the US trade embargo* began, allowing Costa Rica to flourish. Perhaps this, allied to the growth in coffee shop numbers in recent decades, explains the sharp rise in output.

* The Statesman's Yearbook; Cuba, Key Historical Events

THE STATESMAN'S YEARBOOK ARCHIVE

Access to the 1864–2004 Archive is also available as a one-time purchase through Palgrave Connect

The Statesman's Yearbook Archive is an easily accessible compilation of interesting and useful facts and figures through the generations, offering a way to survey changes at the global and individual country level over 140 years.

The Archive presents a unique way to trace the shifting landscape of culture, politics and economics across the globe. The prefaces offer a snapshot of the changing world, and delving further into the Archive reveals how countries have evolved, merged or dissolved.

Perpetual access to The Statesman's Yearbook Archive is also available via Palgrave Connect

Printed in Hong Kong

REVIEWS OF PREVIOUS EDITIONS

'Why search on the internet from thousands of sources when it's all here in one?' – *The Diplomat*

'Few resources - yes, even online ones - package country information like this book does. And it delivers a highly objective international perspective … difficult to find anywhere else.' – *www.goodreads.com*

'A miracle of compression: the key facts about the entire world crisply collated in [its] tightly edited pages. It is an essential desktop guide for anyone who needs to think, write or talk about the nature and future of our unimaginably odd and constantly surprising planet.' – Godfrey Smith, *The Sunday Times*

'All you need to know about the population of various states and countries, officials, exports, constitutions, governments, diplomatic representatives, religion, finance and basic histories.' – *The New York Times*

'Originally designed for statesmen but now used by anyone needing information on the politics, cultures, and economies of the world, this yearbook is one of the longest-running annual publications in history.' – *Library Journal*

'A must for any enquiring mind wishing to be well-informed on world facts and events.' – *www.amazon.co.uk*